The Lignan Handbook

The Lignan Handbook

Edited by Norman G. Lewis, Laurence B. Davin, V. Ranjit N. Munasinghe, and Andrew D. Roberts

CRC Press
Taylor & Francis Group
Boca Raton London New York

CRC Press is an imprint of the
Taylor & Francis Group, an **informa** business

First edition published 2022
by CRC Press
6000 Broken Sound Parkway NW, Suite 300, Boca Raton, FL 33487-2742

and by CRC Press
2 Park Square, Milton Park, Abingdon, Oxon, OX14 4RN

CRC Press is an imprint of Taylor & Francis Group, LLC

Library of Congress Cataloging-in-Publication Data
Names: Lewis, Norman G., 1949- editor. | Laurence, Davin B., editor.
Title: The lignan handbook / edited by Norman G. Lewis, Davin B. Laurence.
Description: First edition. | Boca Raton : CRC Press, 2022. | Includes bibliographical references and index.
Identifiers: LCCN 2021040032 (print) | LCCN 2021040033 (ebook) | ISBN 9781498752190 (hardback) | ISBN 9781032197005 (paperback) | ISBN 9781315152523 (ebook)
Subjects: LCSH: Lignans.
Classification: LCC QD415.2 .L54 2022 (print) | LCC QD415.2 (ebook) | DDC 547/.613--dc23/eng/20211027
LC record available at https://lccn.loc.gov/2021040032
LC ebook record available at https://lccn.loc.gov/2021040033

ISBN: 9781498752190 (hbk)
ISBN: 9781032197005 (pbk)
ISBN: 9781315152523 (ebk)

DOI: 10.1201/9781315152523

Typeset in Times
by CRC Press/Taylor & Francis Group

John Buckingham
(21 March 1943–28 August 2015)

Our *Lignan Handbook* volume is dedicated to both the memory and the professional dedication of the late John Buckingham, internationally renowned for his expertise and infectious enthusiasm in compilation of natural products. John very sadly and unexpectedly lost his life during the planning and execution phases of our volume. His untimely demise in the Austrian mountains near Carinthia, on 28 August 2015, resulted from a horrific car crash during an Alvis vintage car club expedition in which he was participating. The car he was driving went off the road and plunged into a ravine, where he was found dead on emergency service arrival. A requiem mass at Hochosterwitz Castle Church, arranged by the Castle's Count, was held that same evening for the Alvis club members he was with. The Church has a beautiful and peaceful picturesque view overlooking its local surroundings.

Fiona Macdonald conveyed to us the news of this tragic accident on 30 August 2015. John is laid to rest in Kensal Green Cemetery in London, United Kingdom.

***The Lignan Handbook* Concept:** We were approached by both Fiona and John to consider putting together a dictionary handbook devoted to lignans. We agreed to the then daunting task. The photograph below is of John Buckingham and me at our one and only face-to-face meeting to discuss the handbook concept and planning.

Taylor & Francis (specifically Chapman & Hall/CRC) have long been the leaders in compiling natural products series that are of immeasurable value to both experts and enthusiasts alike in numerous fields of endeavour. *The Lignan Handbook* was the next in line!

As one example of John's infectious enthusiasm and untiring commitment, part of the success of the *Dictionary of Natural Products* (Chapman & Hall, 1993) with John as Executive Editor, was the august International Advisory Board that the team had assembled. This included Carl Djerassi, Stanford University (Lewis's academic grandfather), Joe D. Connolly, University of Glasgow (and a wonderful friend), Ralph A. Raphael, University of Cambridge (and Lewis's previous department chairman), and Koji Nakanishi, Columbia University (another wonderful friend and colleague). Additional subject editors included, among others, Jeffrey B. Harborne, Reading University, and Executive Editor, *Phytochemistry* (who Lewis had the pleasure and honour to work with on the Phytochemistry Editorial Board), Eddie Haslam (University of Sheffield), and Robert (Bob) A. Hill, University of Glasgow – all wonderful friends and colleagues.

In short, John brought the true leaders in their fields to the table – another example of his infectious enthusiasm and foresight.

A glimpse of John's life: John won a scholarship at age 11 to the highly prestigious Haberdasher Aske's Boy's School in Elstree, Hertfordshire, England, a public school that received its Royal Charter in 1690. There he developed a passion for chemistry, subsequently gaining First Class Honours in Chemistry at the University of Southampton followed by his Ph.D. from the University of Sussex. From there he went to the University of Zurich, and later became a lecturer in chemistry at the University of London.

Norman Lewis (left) and John Buckingham (right), Caledonian Club, London (30 August 2014), photo courtesy of Robert A. Hill.

His passion for organic compounds and natural products, however, led to him join (in editorial duties) Chapman and Hall (later CRC Press, and then Taylor and Francis). The various massive contributions he made include *Atlas of Stereochemistry* (Chapman & Hall 1974); *Dictionary of Natural Products* (Chapman and Hall 1993); *Dictionary of Organic Compounds, Sixth Edition* (Chapman & Hall 1995); *Chasing the Molecule* (Sutton Publishing 2004); *Bitter Nemesis: The Intimate History of Strychnine* (CRC Press 2007); and *Dictionary of Alkaloids, Second Edition* (CRC Press 2010).

Of these, just one example will perhaps suffice as to the massive impact of his contributions. As Fiona Macdonald eloquently summarized in a eulogy at his memorial service:

> "No other publication has ever achieved the breadth and depth of the *Dictionary of Natural Products* and today it is widely regarded as a benchmark in natural product information. John was very proud of its success, and it truly is his life's work and his legacy".

What a wonderful tribute to a job well done!

His experience as a young man at Haberdasher Aske's Boy's School (and later as an Old Haberdasher) placed him in good stead in later life. In addition to John's successes above, the Old Haberdashers alumni include numerous eminent politicians, scientists, academicians, musicians, senior members of the military, artists, businessmen, sports personalities and broadcasters, and so on. He was in good company from the very onset. [One of his contemporaries was Richard Wright, the vocalist, songwriter, and keyboardist in Pink Floyd.]

In memoriam: In life and in passing, John Buckingham had a most enthralling passage through life and a most productive odyssey. Our single true regret is that he is no longer with us to celebrate the completion of our handbook. Our thoughts also go out to his then partner, Julie Daurensan, who suffered extensive injuries from the car crash and continues in her rehabilitation.

We dedicate this volume to the memory of John Buckingham.

With our greatest respects,

Laurence B. Davin and Norman G. Lewis

Contents

Foreword

It is considered that there are more than 200,000 structurally defined specialized metabolites from the approximately 390,000 terrestrial plants estimated to occur worldwide. Many of these natural products as purified molecules have diverse uses, and one of the largest groups that occur are phenolic substances, constituted by about 10,000 structurally characterized representatives. A subset of the plant phenols are the lignans, which are mainly phenylpropanoid (C_6C_3) dimers, and are found very widely in terrestrial vascular plants, with several thousand of these compounds known, and approximately 200 new examples being reported in the scientific literature annually. These compounds exhibit a broad spectrum of biological activities, inclusive of antibacterial, antifungal, antidiabetic, anti-inflammatory, antioxidant, antiprotozoal, antitumor, antiviral, and neuroprotective effects. Lignans are of great interest to food scientists, since they occur in many plant foods in the human diet, and several beneficial effects on health have been ascribed to them. In terms of drug development, the aryltetralin lignan, podophyllotoxin, has been modified as a lead compound by chemical synthesis to produce three currently approved anticancer agents, namely, etoposide, etoposide phosphate (Etopophos®), and teniposide, which are used as chemotherapeutic agents against a range of cancer types.

In this highly insightful and timely *Lignan Handbook,* the two editors, Norman G. Lewis and Laurence B. Davin, have provided an in-depth examination of plant lignans from many different perspectives. This book is based on decades of their active research collaboration leading to detailed scientific investigations of various aspects of phenylpropanoid metabolism. An overall goal has been to investigate the biosynthetic origin of both lignans and structural cell wall polymers such as lignins and other more complex phenols, including proanthocyanidins (condensed tannins). Some 25 years ago, Professor Lewis and Dr. Davin collaborated on a landmark investigation that identified the first "dirigent" protein (DP), which was purified from the plant *Forsythia intermedia*. Under the conditions utilized, this 78-kilodalton protein was found to affect the relevant DP-engendered coupling stereoselectively, and to direct (+)-pinoresinol biosynthesis from intermediates derived from *E*-coniferyl alcohol (*Science* **1997**, *275*, 262–366).

While many authors continue to produce scientific updates on plant lignans that appear periodically in a range of specialist journals, most of these tend to have a somewhat narrow focus. In contrast, in the present volume, a more fundamental and deeper approach to our current knowledge of lignans is provided. The book is divided into a total of eleven chapters, with all of these being co-authored by the book editors. However, additional authors have contributed to Chapters 4, 7, 10, and 11. Each chapter is provided with lucid illustrations and a detailed bibliography. In addition, there is a dictionary section included at the end of the volume for readers to access specific information on compounds of interest.

Chapter 1 provides a historical introduction to the lignans and their related dimeric phenolic compound group, the "neolignans", and describes the original early structural work performed in the mid-20th century. A central consideration in this chapter is the trivial nomenclature of these phenolic substances, with it being recommended that the term "lignan" should be expanded to define all C_6C_3 coupled products, inclusive of dimers and higher oligomers. In Chapter 2, various monomeric plant phenolic substances that can be regarded as potential monomeric components of lignans are mentioned. Chapter 3 provides an overview of the occurrence of lignans and of phenylpropanoid "monomers" in algae, bacteria, fungi, marine animals, and higher and lower plants. Dirigent proteins involved in the biosynthesis of lignans and other vascular plant phenols are described in Chapter 4. In turn, Chapter 5 surveys, in detail, current knowledge of the biosynthesis of 8–8′-lignans, and mentions experimental enzymatic work performed on specific plants, in addition to well-known individual compounds like hinokinin, justicidin B, lariciresinol, pinoresinol, podophyllotoxin, and yatein. Next, in Chapter 6, information known so far on the biosynthesis of dimeric compounds with different types of linkages [8–2′, 8–3′, 8–5′, and 8–*O*–4′] is presented. Chapter 7 presents a discussion on allylphenol and propenyl phenol biosynthesis, inclusive of the synthase enzymes involved, and comments on the lack of oxygenation often observed at C9/C9′ in lignan derivatives produced. Chapter 8 offers a specific treatment of 8–8′ norlignans, in addition to 8–7′ norlignans such as agatharesinol and hinikoresinol. A consideration of "mixed biochemical pathway" lignan products, as typified by alkaloid lignans, flavolignans, and stilbenolignans, is presented in Chapter 9. It is pointed out in Chapter 10 that there are many lignan skeletal types found in vascular plants with only minimal information on their formation being available, and a number of such oligomeric lignans are described. In Chapter 11, the final major topic of this volume is the chirality of lignans, inclusive of information on how to separate such enantiomers and methods for their analysis using circular dichroism.

This thoughtful and well-organized handbook can be recommended for use by a wide range of readers, inclusive of agricultural, forestry, and food scientists, biochemists, organic and medicinal chemists, phytochemists, and those specifically interested in biosynthesis and drug discovery. It may be anticipated that the *Lignan Handbook* will be used as a major reference text on the specific topics covered for many years to come.

Prof. A. Douglas Kinghorn, Ph.D., D.Sc., D.Sc. (h.c.)

Professor and Jack L. Beal Chair
Distinguished Scholar Awardee
The Ohio State University

Preface

HANDBOOK BEGINNINGS AND ORGANIZATION

The starting point began with a meeting with John Buckingham, Fiona Macdonald (publisher) and Robert A. Hill in London 2014 to discuss the concept of, and contemplated need for, a Lignan Handbook. Thus began our consideration and development of the needed scope and content, which we earnestly hope the readers at different knowledge levels can and will appreciate — including for ready reference purposes. Completion of this very comprehensive volume has been an enjoyable endeavor, other than for the saddest low point in our collective journey, when John Buckingham unexpectedly lost his life in an automobile accident in the mountains of Austria in August 2015. The Handbook is dedicated to cherishing his memory.

Our collective endeavors eventually led to two distinct but complementary parts to the Handbook:

The first part is a series of 11 Chapters, each co-authored by both of us (Laurence B. Davin and Norman G. Lewis) with additional authors depending upon the topic area. First, in Chapter 1, we address current lignan/neolignan/nor-lignan nomenclature inconsistencies, and the urgent need for a simplified, consistent, and more readily understandable framework for same. The historical classifications of lignans (i.e. 8–8′ linked) and neo-lignans (all coupling modes other than 8–8′ linked, e.g. 8–5′, 8–*O*–4′) as being distinct metabolic classes is both inexact and inconsistent. They are better systematically defined as all being lignans – i.e. through designating their inter-unit linkage types, rather than being one type or the other. In many cases, plant natural products can have both lignan and neo-lignan sub-units in their structures, i.e. they are neither lignans nor neo-lignans in the current, somewhat archaic definitions. Chapter 1 also gives a brief description of contributions of the early pioneers in the field, and whose contributions were truly inspirational in the development of the overall field. They (Holger G. E. Erdtman, Robert D. Haworth, and Otto R. Gottlieb) must never be forgotten!

We have also explored the range of organisms (aquatic/vascular/non-vascular plants, as well as fungi, bacteria and animals) that harbor or potentially harbor lignans and/or possible monomeric phenylpropanoid (C_6C_3) precursors (Chapters 2 and 3).

We paid particular attention (Chapter 4) to our discovery of dirigent proteins (from the Latin *dirigere*, to guide or align) and their function as entry points to distinct lignan skeletal types. Since our discovery, distinct dirigent protein multi-gene family members have been found to biochemically function as entry points to a wide range of distinct plant phenol metabolic classes, including not only lignans, but also lignins, aromatic diterpenoids, pterocarpans and isoflavenes thus far. Other plant phenol metabolic classes with dirigent protein multi-gene family members as potential entry points are discussed, together with what is known thus far about the diverse lignan-related biochemical pathways that include "mixed" metabolic pathways (Chapter 9).

Given the enormous stereochemical and structural diversity of the lignans in Nature, various chapters systematically lay out our current knowledge of distinct biosynthetic pathway roadmaps to each of these (oligo)lignan types (e.g. 8–8′, 8–5′, 8–*O*–4′) that lead to the diverse metabolic sub-families encountered (Chapters 5, 6, and 10). This was particularly, in our opinion, greatly needed to help those unfamiliar with these classes of plant natural products – further complicated, as many of their chemical structures have been given trivial names that give little insight into either their biosynthetic origins or their structural relationships with each other. Moreover, we thank Professor Simo Sarkanen, University of Minnesota, for the "Roadmap" suggestion.

Chapter 7 has as its focus how C_9 oxygenation is lost during metabolism to afford the allylphenol and propenylphenols, and their dimeric counterparts etc. This was unresolved until 2005-2010, with various speculations as to how this biochemically occurred that preceded the discoveries made. Most of those earlier speculations have now been put to rest.

Chapter 8 addresses another peculiarity, in the sense of nor-lignan formation, which generally lack a carbon atom at C_9. The biochemical pathways to the 8–8′ and 8–7′ linked lignans are the main focus of this chapter, with the latter being the best understood biochemically at present. Various other lignan modifications, resulting in the loss of a carbon atom anywhere in the molecule, are often also described as nor-lignans, although such modifications may result from other biochemical processes.

A final contribution (Chapter 11) in this first part of the Handbook addresses the importance of chirality, and the danger of it being a "lost art" in this and related fields. Chirality, of course, cannot be ignored as we live and breathe in a chiral world. Furthermore, as stereochemical differences often have a profound impact on bioactivity and/or other biological properties, we give our reasoning as to the ongoing necessity for considering and understanding the implications and relevance of chirality throughout the Handbook.

The second part of the Handbook, done in close conjunction with our CRC Press co-editors V. Ranjit N. Munasinghe and Andrew D. Roberts, is an exhaustive compilation of the extensive structural diversity of lignans/neolignans/nor-lignans that have been reported thus far, both as natural and synthetic compounds, as well as their potential biochemical pathway precursors or similar. Each Handbook entry for a specific compound in the second part of the volume provides ready references, particularly as to their detailed chemical characterization and properties.

THE LIGNAN AND LIGNIN ENIGMA

Our interests in lignans, together with related phytochemical metabolic classes including polymeric lignins, initially came from our complementary but distinct perspectives and experience. For one of us (Norman Lewis), the interest in both was spawned during evenings in the library of the Atlantic Regional Laboratory of the National Research Council, Halifax, Nova Scotia, Canada in 1980. There the focus was on what research areas were being envisaged to pursue, and why. Coincidentally, in the library was a portrait of Dr. Arthur Neish from the 1960s, a previous Director.

During Fall 1980, Lewis moved to the Pulp and Paper Research Institute of Canada (PAPRICAN) in Pointe Claire, Quebec, Canada, eventually having offices and research laboratories both there and in the Pulp and Paper Building of McGill University, Montreal, Quebec, Canada. There he further developed interests in both lignin biopolymer assembly and lignin utilization, as well as on woody plant phytochemical "extractives" that included lignans.

One major difference between the metabolically related lignin biopolymers and (generally dimeric) lignans was that, while many were derived from the same potential precursors (monolignols, such as *p*-coumaryl, coniferyl, and sinapyl alcohols), lignans were frequently optically active whereas lignins were apparently not. (During that time, Lewis became aware of the early research of Arthur C. Neish on lignin formation and constitution that had been conducted mainly at the Prairie Regional Laboratory in Saskatoon, Sakatchewan, Canada). This chirality question had long been a conundrum in the lignin/lignan research area dating back to the 1930s.

Laurence Davin's entry into the lignan/lignin area began during her sojourn at the Prairie Reginal Laboratory (now the Plant Biotechnology Institute, PBI) where she then studied glucosinolate biochemical pathways with the late Ted Underhill. Joining Virginia Tech to study lignan formation and subsequently Washington State University, her protein biochemistry expertise helped design the pivotal biochemical experiments that led to Davin and Lewis' joint discovery of dirigent proteins. Thus began their work on monolignol, allyl/propenyl phenol, lignan, lignin and other biochemical pathways, such as to pterocarpans and isoflavenes, either alone or in collaboration with other researchers.

We hope the readers enjoy the Handbook!

The Editors

Laurence B. Davin' and Norman G. Lewis' Scientific Research Emphases: Dr. Lewis' and Dr. Davin's scientific research emphases are largely devoted to discovering, modifying and exploiting plant biochemical pathways, including mainly to plant phenolics/aromatics, such as lignans, many of which are medicinal compounds, specialty and commodity chemicals, as well as potential bioenergy bioproducts. They have also conducted various comprehensive studies on microgravity adaptation of plants on both the Space Shuttle and International Space Station, ISS, with the latter ongoing to the present day.

Research activities have largely been supported by the U.S. DOE, NASA, NIH, NSF, USDA/USDA NIFA, the National Sciences and Engineering Research Council, NSERC (Canada), National Research Council (Canada), as well as from Lewis B. and Dorothy Cullman, Thomas G. and Anita Hargrove and Arthur M. and Katie Eisig-Tode Foundations. Many researchers from the laboratory hold academic positions worldwide including in the U.S., Canada, Brazil, China, France, Japan, Korea, New Zealand, Thailand, and the United Kingdom.

Drs. Davin and Lewis co-founded Obiou Consulting Group LLC, where they provide expert guidance (including broad expertise in plant sciences, plant chemistry/phytochemistry, plant biochemistry/molecular biology, and integrated 'omics' [metabolomics/proteomics/transcriptomics]). Dr. Lewis also co-founded the plant biotechnology company, Ealasid, Inc.

Norman G. Lewis currently holds the positions of Regents Professor and Arthur M. and Katie Eisig-Tode Distinguished Professor, Institute of Biological Chemistry (IBC), Washington State University. He was elected as Corresponding Fellow of the Royal Society of Edinburgh (Corr. FRSE, Scotland's National Academy of Science and Letters); Fellow of the American Society for Plant Biologists (ASPB); Fellow of the American Association for the Advancement of Science (AAAS); Life Member of the Phytochemical Society for North America (PSNA); Phytochemical Pioneer (PSNA); and Fellow, International Academy of Wood Science (IAWS). He recently held a Fulbright Distinguished Professor Fellowship (Science without Borders) in Brazil and has had many leadership positions in various learned societies, such as President of the American Society of Gravitational and Space Biology (ASGSB) and President, PSNA. Over the last thirty plus years, he has served or currently serves on numerous scientific advisory committees, grant panels and editorial boards worldwide. In addition to the recent Lignan Handbook editorial responsibilities, he has served as Regional Editor/Editorial Board Member of Phytochemistry for a period of over 25 years until 2016, as Monitoring Editor (Plant Physiology), and as Executive Editor of Advances in Plant Biochemistry and Molecular Biology. Currently he serves on several other editorial boards, including Frontiers in Plant Science and Recent Patents on Biotechnology, as well as on various federal/international grant review panels, and scientific advisory boards worldwide. Professor Lewis has served as Panel Manager and/or Panel Member for the U.S. Department of Energy (DOE), National Aeronautics and Space Administration (NASA), National Institutes of Health (NIH), National Science Foundation (NSF) and United States Department of Agriculture (USDA), as well as on various government competitive grant panels worldwide. He has responsibilities as Adjunct Research Scientist at the National Center for Genome Resources (NCGR) and the New Mexico Consortium (NMC) on competitive federal NSF and DOE grants and is a consultant for various organizations. Professor Lewis served as IBC Director for nearly 25 years, with three of its small IBC faculty number honored as U.S. National Academy of Science Members.

Prof. Lewis completed his BSc (Honors, Chemistry, 1973) at the University of Strathclyde, his PhD (Chemistry, 1977) at the University of British Columbia, as well as post-doctoral studies at Cambridge University (Chemistry, 1978-80) in the late Professor Sir Alan R. Battersby's laboratory, and at the National Research Council (NRC) Atlantic Laboratory (1980). Prior to WSU, he held various research and technical positions in Imperial Chemical Industries (ICI), National Research Council of Canada (1980), the Pulp and Paper Research Institute of Canada (1980–1995), and Virginia Tech (1985–1990). At UBC, as a PhD student, he served as President of the Graduate Student Association.

Laurence B. Davin. Since 1990, Dr. Davin has been a Senior Research Scientist Faculty in the Institute of Biological Chemistry, Washington State University, Pullman, WA, USA, to which she brings her extensive plant biochemistry expertise. Dr. Davin's research activities include support as co-Principal Investigator from funding through U.S. DOE, NASA, NIH, NSF, USDA/USDA NIFA, as well as from Lewis B. and Dorothy Cullman, Thomas G. and Anita Hargrove and Arthur M. and Katie Eisig-Tode Foundations.

In addition to the recent Lignan Handbook editorial responsibilitics, Dr. Davin has also carried out extensive editorial duties with Phytochemistry through 2018, and as a member of the PSNA Editorial Board for Recent Advances in Phytochemistry. She has also played an integral part of the team efforts for both the organization and smooth running of various PSNA and American Society of Gravitational and Space Biology (ASGSB) Annual Meeting conferences galvanized with her tireless and infectious enthusiasm.

Dr. Davin received her BSc and PhD degrees in Plant Biochemistry/Physiology from l' Université Paul Sabatier, Toulouse. She then was awarded a post-doctoral appointment at the National Research Council (Saskatoon) of Canada's Plant Biotechnology Institute in the late Ted Underhill's Laboratory, where she focused on glucosinolate biogenesis. This was followed by a Research Associate appointment at Virginia Tech, prior to joining the WSU faculty ranks.

V. Ranjit N. Munasinghe was formerly a Senior Lecturer in organic chemistry at the University of Colombo, Sri Lanka, Research Officer and visiting lecturer at Birkbeck College, University of London and a Fellow at the chemistry department of Imperial College, University of London. He was also a senior research Scientific Officer at National Institute for Medical Research (NIMR, MRC), London.

With a BSc (Chemistry special) from University of Colombo, Sri Lanka, a PhD from Birkbeck College (1978), and a DIC from Imperial College, University of London (1994), he has about 40 years of research and teaching experience in organic chemistry. Specialising in carbohydrate chemistry, his main areas of research included photochemical synthesis, deoxy sugars, branched chain sugars, C-glycosides, trisaccharides and fluorescent conjugates of sialic acids.

He has been involved with Chapman &Hall chemical data bases since 1983 and has compiled (with Prof. P. M. Collins) *Carbohydrates* (1987) and *Dictionary of Carbohydrates* (1st Ed. 1998, 2nd Ed. 2006). He was also a consultant and compiler for *Dictionary of Organic Compounds*, *Dictionary of Natural Products* and *Dictionary of Food Compounds* (2nd Ed, 2013) and joint editor (with J. Buckingham) for *Dictionary of Flavonoids* (2015).

Andrew D. Roberts has been a freelance contributing editor for Chapman & Hall/CRC databases for 40 years. With a BSc (Hons) Chemistry from Nottingham Trent University he has also been a consultant for the Royal Society of Chemistry and related organisations.

Under the guiding editorship of John Buckingham, he contributed to the *Dictionary of Organic Compounds* (5th Ed 1982), and related publications, including *Dictionary of Antibiotics and Related Substances* (1988), *Dictionary of Food Compounds* (2013), and *Dictionary of Flavonoids* (2015).

As a major contributing editor to the *Dictionary of Natural Products*, Andrew specialises in evaluation of new natural products and their nomenclature. This culminated in the literature and revision of the nomenclature of lignans for this volume.

1 Redefining Lignan/Neolignan Nomenclature and Diversity

A Historical Perspective

Laurence B. Davin and Norman G. Lewis

1.1 INTRODUCTION

The *Lignan Handbook* begins with a brief discussion of the pioneering advances leading to lignan and (so-called) neolignan chemical structure discoveries, these being originally categorized as phenylpropanoid (C_6C_3) pathway dimers. Later, this led to different chemical nomenclature definitions being separately proposed for each 'class' of natural products. Each definition, formulated many decades ago, was — in hindsight — too restrictive.

This arose largely for two reasons. The first reason came from the consideration of a relatively small number of dimeric phenylpropanoids initially obtained and identified from different plant species. The second complicating factor was that the corresponding biosynthetic pathways involving dirigent proteins were then unknown, and were not to be discovered until much later.

The introduction thus brings into sharp focus why, with the advantage of hindsight, they were even given different natural product nomenclature definitions at all. This matter is worthy of mention, as both chemical nomenclature definitions continue to be frequently (and often ambiguously and confusingly) used in the scientific literature. First, the descriptor "neo-lignan" or "neolignan" is difficult, if not impossible, to continue to justify in terms of best usage of chemical nomenclature as explained in the following. Second, the term *lignan* was originally defined very narrowly for dimeric C_6C_3 phenylpropanoids with 8–8′ (β–β′) inter-unit linkages (Haworth 1937). Yet, this original lignan definition by Haworth can now conveniently be both expanded and systematically deployed to encompass all structural types of different inter-unit linkages, i.e. whether lignans or so-called neolignans, in a facile, straightforward manner, e.g. as 8–5′, 8–1′, etc.

Our redefinition of the term lignan/neolignan nomenclature to simply be lignan here is not a criticism of the previous excellent researchers in the field, who proposed these distinct nomenclature definitions, or of their highly valuable scientific contributions. Our redefinitions serve to remind that as a scientific field evolves, and more information and insights (e.g. chemical structures and biochemical pathways) come to light, this can massively alter our thinking. When this occurs, such as here, there comes a time when logical revision and streamlining can and must become necessary. Indeed, this critical juncture is where we have been for some time in the lignan/neolignan fields (Lewis et al. 1998, Lewis and Davin 1999, Vassão et al. 2010).

The rationale for revision of phenylpropanoid (C_6C_3) lignan and neolignan nomenclature definitions is best approached here, however, in terms of initially considering the key contributions, analyses, and recommendations made *at the time* by the pioneering scientists involved, and which spanned a period encompassing now more than a century.

1.2 PIONEERS

The seminal scientific contributions resulting in the lignan and neolignan nomenclature definitions that emerged, as well as how their biosynthetic formation ***might be achieved*** largely through some form of oxidative phenolic coupling, mainly came from the insights gained from research in the academic laboratories of Professors Robert D. Haworth, Holger G. H. Erdtman, and Otto R. Gottlieb. The brief summaries and background information given in the following may give some insights as to what led to their individual research interests and their contributions in this important field.

Professor Robert Downs Haworth (born 15 March 1898, passed away on 21 May 1990, Figure 1.1A) (Jones 1991). His main body of natural product research work was carried out at the University of Sheffield, England, where he contributed extensively to the chemistry of lignans and hydrolysable tannins. He received much of his early training in the laboratory of Professor William Henry Perkin, Jr. (17 June 1860–17 September 1929), the son of the legendary Professor Sir William Henry Perkin (1838–1907). The latter, at the age of 18, made an outstanding serendipitous synthetic achievement in producing aniline purple (mauveine) in the 1850s, which ultimately spawned the massive artificial dye and pharmaceutical industries (Chemical Heritage Foundation 2015, Editors of Encyclopædia Britannica 2016, Banerjee 2017).

All three (W. H. Perkin, W. H. Perkin Jr., and R. D. Haworth) were highly deserving recipients of the prestigious Royal Society of London's Davy Medal, a recognition long reserved for the world's most outstanding chemists and given annually since 1877 (Royal Society 2017). Indeed, recipients of the Davy Medal from the early 1900s to the 1960s were often subsequently Nobel Prize winners, such as Adolf von

FIGURE 1.1 Pioneers. **A.** Professor Robert Downs Haworth. **B.** Professor Holger Gustaf Henrik Erdtman. **C.** Professor Otto Richard Gottlieb. Photograph from Otto R. Gottlieb digital archive (Dr. Maria Renata Borin). Picture (A) reproduced from Biographical Memoirs of Fellows of the Royal Society: Jones E. (1991). Robert Downs Haworth. 15 March 1898-19 May 1990. *Biogr. mem.* Fellows R. Soc. 37:264-276. Copyright (1991) Royal Society of Chemistry. Picture (B) reproduced from Phytochemistry: Norin T. (1990). Holger Erdtman 1902–1989. *Phytochemistry* 29:1713-1714. Copyright (1990) Elsevier.

Baeyer, Emil Fischer, Ferdinand Moissan, Svante Arrhenius, Pierre and Marie Curie, Otto Wallach, Paul Sabatier, Hans Fischer, Richard Willstätter, Sir Robert Robinson, Norman Haworth, Arthur Harden, Linus Pauling, Lord Alexander R. Todd, Robert Burns Woodward, and Sir Derek H. R. Barton. Exalted company indeed!

Interestingly, R. D. Haworth originally planned to do his PhD studies with Sir Robert Robinson, who was then at the University of St. Andrews. However, this did not occur as Robinson moved to another institution (Jones 1991). Sir Robert also received his early training in the William H. Perkin, Jr. laboratory.

As an aside, in the early 1970s, while at Strathclyde University, one of the coauthors (Norman G. Lewis) had an undergraduate research thesis mentor, Professor Peter Ludwig Pauson (1925–2013). Scientifically, Professor Pauson perhaps became best known for the discovery of ferrocene (Kealy and Pauson 1951), which was later shown to be the first "sandwich" π-π complex (Wilkinson et al. 1952). Born in Bamberg, Bavaria, in 1925, Peter Pauson and his family fled in early 1939 from Nazi Germany to the United Kingdom.

Unbeknownst then to Lewis in the early 1970s, Professor Pauson had completed his PhD thesis under the supervision of Professor Haworth in 1949 at Sheffield University, where he elucidated the structure of the brilliant red natural product, the disubstituted benzotropolone, purpurogallin (Haworth et al. 1948).

Lewis's undergraduate academic "grandfather" was thus the scientist who coined the term *lignan*, the subject of this *Handbook*. On finding out this fact many, many years later, it served to remind that the world can be a very small place!

Professor Holger Gustaf Henrik Erdtman (born 25 April 1902) passed away on 13 December 1989 (Figure 1.1B). His main body of research work was carried out at the Swedish Forest Products Research Laboratory and at the Royal Institute of Technology (KTH), Sweden. He completed his BSc equivalent degree (studying chemistry, botany, and zoology) in 1926, and his licentiate (PhD equivalent) degree at the University of Stockholm in 1931 (Norin 1990). His PhD degree equivalent was on the chemistry of humic acids and on products derived from oxidative phenolic coupling, with the latter topic becoming a lifelong interest. While his thesis advisor was Professor Hans von Euler at Stockholm, his thesis research also included studies outside of Sweden from 1929 to 1931, i.e. as a Ramsay Fellow with Sir Robert Robinson, as well as subsequently working with Professors Fridrich Fichter in Basel and Ernst Spath in Vienna in 1931. Interestingly, his British advisor Sir Robert Robinson, the 1930 Davy Medal recipient, received the 1947 Nobel Prize in Chemistry "for his investigations on plant products of biological importance, particularly the alkaloids".

Later in his career, Professor Erdtman served as chairman of the 1973 Nobel Committee for Chemistry (Zydowsky 2015), with that year's Nobel Prize in Chemistry being shared by Ernst Otto Fischer and Geoffrey Wilkinson for their respective work on establishing the nature of "sandwich" compounds, i.e. of the ferrocene Peter Pauson had discovered serendipitously.

Professor Otto Richard Gottlieb (born 31 August 1920) passed away on 19 June 2011 (Figure 1.1C). His main body of outstanding research work was carried out at the University of São Paulo, Brazil, and was largely devoted to the phytochemistry of Brazilian plant species and their chemotaxonomic relationships.

Otto Gottlieb was born in Brno, in the then newly created Czechoslovakia. As a young man in his late teens in 1939, together with his mother, he fled from Nazi oppression in Europe and settled in Brazil. There, he subsequently received his undergraduate degree in industrial chemistry from the Universidade do Brasil, Rio de Janeiro, in 1945. After subsequent extensive industrial (in his father's essential oil company) and academic pursuits, he was awarded the degree of Doctorate and "Livre-Docente" by the Universidade Federal Rural do Rio de Janeiro in 1966 (Bolzani and Cragg 2004).

In addition to being an outstanding phytochemist, Professor Gottlieb rigorously trained (essentially single-handedly) many of the next (current) generation of organic chemists/phytochemists in Brazil. In 1999, the Brazilian scientific community enthusiastically campaigned for his nomination for consideration for the Nobel Prize in Chemistry (Righetti 2011). This campaign was warmly supported and endorsed by others, including Nobel prizewinner Roald Hoffman, also an escapee from the Nazi holocaust; he hid in Poland with his mother from early 1943 until June 1944 when they were freed by the Red Army (Nobelprize.org 2014).

Independently, Erdtman, Gottlieb, and Haworth made substantial, complementary, and important pioneering scientific contributions. Based on the research studies underway in their respective laboratories, and elsewhere, Professors Haworth and Gottlieb coined the terms *lignan(e)* in 1937 (Haworth 1937) and *neolignan* in 1972 (Gottlieb 1972), respectively. The nomenclature definitions fashioned for both, however, were based on the knowledge at the time, i.e. from trends and insights deduced from the then available chemical structure determinations from within the plant kingdom.

Synergistically, a major contribution by Erdtman in 1933 was to hypothesize that polymeric lignin formation could be achieved by a dehydrogenative polymerization process (Erdtman 1933a, b). This built on earlier speculations by Ferdinand Tiemann (1875) and Peter Klason (1908) that lignin was produced from the C_6C_3 phenylpropanoid, coniferyl alcohol (**1**), which was later coined to be a monolignol.

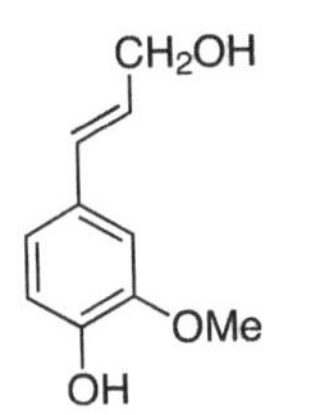

Coniferyl alcohol (**1**)

Erdtman based this thinking of a putative lignin polymerization mechanism on his results from oxidative phenolic coupling experiments *in vitro* with both eugenol (**2**), which produced the 5-5′ linked dehydrodieugenol (**3**) (Erdtman 1933a), and isoeugenol (**4**), which afforded the racemic 8–5′ linked (±)-dehydrodiisoeugenols (**5a/b**) (Erdtman 1933a, b) (Figure 1.2). This was an intriguing concept at the time, as other scientists had considered that lignins were produced from polymeric carbohydrates (as artifacts) (Phillips 1934) or were from terpenes (see Haworth 1937). Overall, his oxidative phenolic coupling studies ultimately gave important conceptual insights

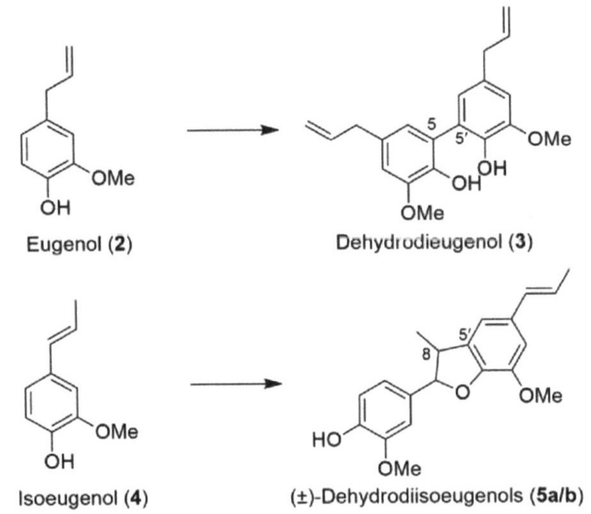

FIGURE 1.2 Oxidative (i.e. free radical) phenolic coupling products from eugenol (**2**) and isoeugenol (**4**).

into biosynthesis of not only lignins, but also other free radical coupling processes leading to lignans, lichen depsidones, and some alkaloids (e.g. morphine) (Norin 1990).

Having made this important hypothesis of (resonance stabilized) phenoxy radical coupling, Erdtman was nevertheless acutely aware that many lignans were optically active (Erdtman 1939). However, all efforts to demonstrate (stereoselective and regiospecific) control of phenoxy radical coupling *in vitro* failed as described previously. Instead, only racemic products with different inter-unit linkages were obtained, which resulted from non-regiospecific coupling.

It was to be almost 60 years later when stereoselective and regiospecific control over inter-molecular phenoxy radical coupling *in vitro* was finally demonstrated. This occurred with our discovery of dirigent (Latin: *dirigere,* to guide or to align) protein control of stereoselective phenoxy radical coupling, with the monolignol coniferyl alcohol (**1**) substrate giving (+)-pinoresinol (**6a**) as the first example (Davin et al. 1997), rather than a mixture of different racemic coupling products (e.g. 8–*O*–4′, 8–5′ and 8–8′ linked).

Each of the important scientific contributions and interpretations by Erdtman, Gottlieb, and Haworth must be considered, however, based on when each of their contributions was made, and also the corresponding technologies then available. With that perspective, each of these pioneers richly deserve our greatest gratitude and appreciation for the key advances they made in the phytochemical studies that they pursued, and for the new and significant knowledge generated.

1.3 THE NARROW 1937 DEFINITION OF LIGNANS: PHENYLPROPANOID DIMERS WITH 8–8′ (β–β′) INTER-UNIT LINKAGES

By 1937, a relatively small group of phytochemical compounds had been chemically identified, some of whose existence, but not their chemical structures, had been known for a very long time. In each case, their structures were ultimately deduced to result from two phenylpropanoid (C_6C_3) moieties of some type being linked together through their respective 8–8′ (or β–β′) positions. This class of structurally diverse dimeric natural products was thus given the nomenclature of lignanes (later shortened to lignans) by Haworth (1937).

The discoveries of various members of this apparently structurally related natural product class had emerged from the multiple investigations undertaken worldwide of plant natural products present in the wood, rhizomes, seeds, oils, exuded resins, and spent pulping liquors (from the pulp and paper industry) of commercially or medicinally important species. These ***initial*** definitive lignan structural determinations, which began shortly before the onset of the 20th century and spanned a period of nearly 50 years, came from analyses (including chemical degradation) of the main natural products present in extracts and resins from these species.

1.3.1 Lignans: Associated Shortfalls of Ad Hoc Trivial Names and IUPAC Designated Names

From near the end of the 19th century and up to 1937, various members of this group of 8–8′ (or β–β′) linked lignans were isolated and structurally identified, with each given an ad hoc trivial name. In every case shown in the following (i.e., lignans **6–17**), their names were trivially partly derived from the plant species from which they were (first) isolated. It is noteworthy, though, that in this small group of identified 8–8′ linked lignans, functionalization at the carbon 9 and 9′ positions also varied quite extensively in terms of being either oxygenated or non-oxygenated (i.e., with CH_2OX, C=O, CHOH, or CH_3 groups).

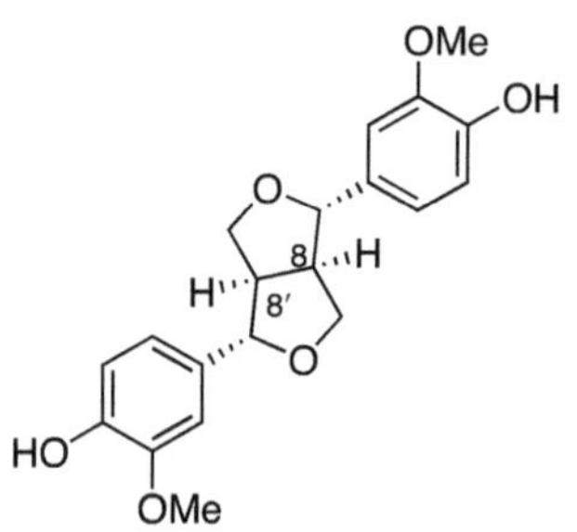

(+)-Pinoresinol (**6a**)
Forsythia intermedia

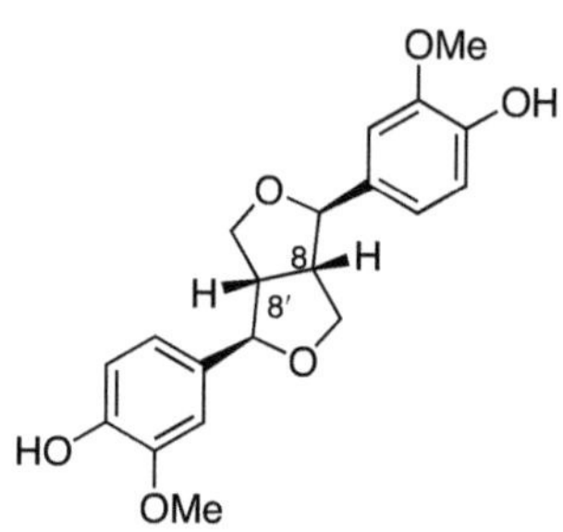

(−)-Pinoresinol (**6b**)
Pinus nigra, Picea abies.
Known as (−)-eudesmin
when first isolated from
kinos of *Eucalyptus* species

The 8–8′ (or β–β′) linked lignans identified by then included the following examples, the description of which also includes a brief explanation as to why these plant species were of scientific and/or commercial/medicinal interest:

Guaiaretic acid (**7**) was isolated and identified from heartwood of *Guaiacum officinale* (Schroeter et al. 1918, Haworth and Richardson 1935b), an evergreen tree species long used for both its wood and resin components; its absolute configuration was determined much later in 1956 (Schrecker and Hartwell). *G. officinale* was trivially known in commerce as "lignum vitae", being valued for its resin and its pharmaceutical uses since the early 1500s (Flückiger and Hanbury 1874). Its dense wood (bulk density of 1.33 g/cm³) (Longwood 1962) also found use as bearings for lubrication in hydroelectric dams (Unwin 1899), as its heartwood can support high compression stresses even when submerged.

α-Guaiaconic acid (**8**), also isolated from its resin (Richter 1906) and subsequently identified (Kratochvil et al. 1971, Stevenson and Williams 1976), is widely used for testing for the presence of blood in stool samples. This is the so-called guaiac fecal occult blood test (gFOBT) (Simon 1985), where it, in the presence of H_2O_2 and any blood (heme) in the stool, undergoes oxidation to give a diagnostic, blue colored (fully conjugated), bis-methylene quinone reaction product.

(−)-Matairesinol (**9**) was first isolated and identified as a constituent of the heartwood of the broad-leafed New Zealand evergreen, the matai (*Podocarpus spicatus*) (Easterfield and Bee 1910, Briggs et al. 1935, Haworth and Richardson 1935a), with its absolute configuration determined in 1950 (Haworth and Wilson). Matai has long been highly valued for the durability and straight grain of its (heart)wood.

(−)-Arctigenin (**10**) was isolated and identified from great burdock (*Arctium lappa*) (Shinoda and Kawagoye 1929). The position of the free hydroxyl group in **10** was established through cyclization studies (Haworth and Kelly 1936, Omaki 1937), and its absolute configuration was reported by Ozawa (1952). Burdock root, rich in dietary fibers, can grow up to 1-meter long and it is a popular foodstuff (*gobo*) in Japan. Its dried roots are also used in traditional medicine to treat sore throat and various skin problems (Chan et al. 2011), and is used as a diuretic, depurative, and digestive stimulant (dos Santos et al. 2008). Its seed hook system (for seed dispersal via "hitchhiking" on passing animals) served as the inspiration for studies that led to Velcro (Swiners 2013).

(−)-Hinokinin (**11**) (Yoshiki and Ishiguro 1933, Briggs 1935, Keimatsu and Ishiguro 1935, Mameli 1935) was obtained and identified from the resin of Japanese (hinoki) cypress (*Cupressus obtusa*), a species now classified as *Chamaecyparis obtusa*. It reportedly has antichagasic properties (Marcotullio et al. 2014). Its chemical synthesis was also achieved, albeit in racemic form (Keimatsu et al. 1935, Haworth 1937). On the other hand, (−)-hinokinin (**11**) found in *C. obtusa* is optically active ($[\alpha]_D^{22} = -90.51$) (Yoshiki and Ishiguro 1933). As for this and many other optically active naturally occurring lignans, it could not be ascertained at the time whether it or others were optically pure or in enantiomeric excess (e.e.).

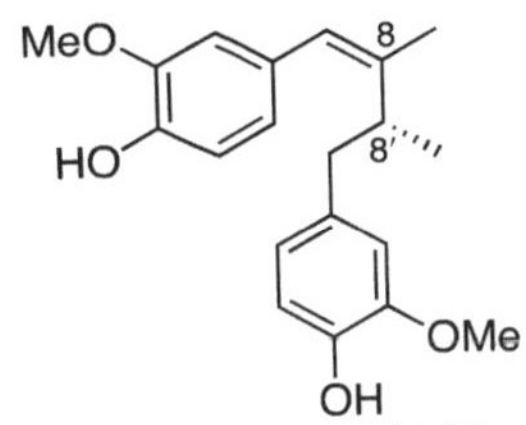
(–)-Guaiaretic acid (**7**)
Guaiacum officinale

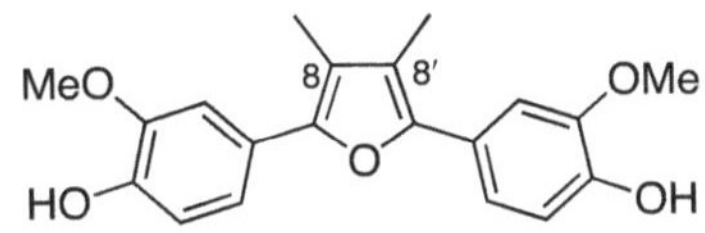
α-Guaiaconic acid (**8**)
Guaiacum officinale

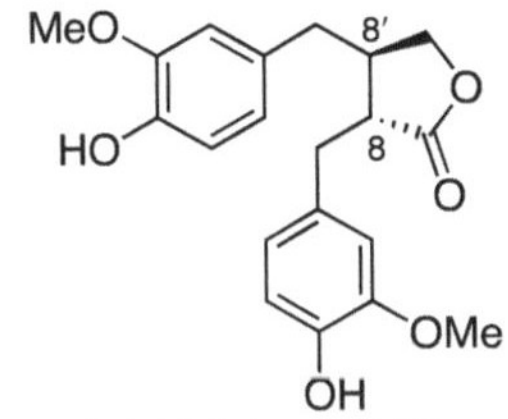
(–)-Matairesinol (**9**)
Podocarpus spicatus

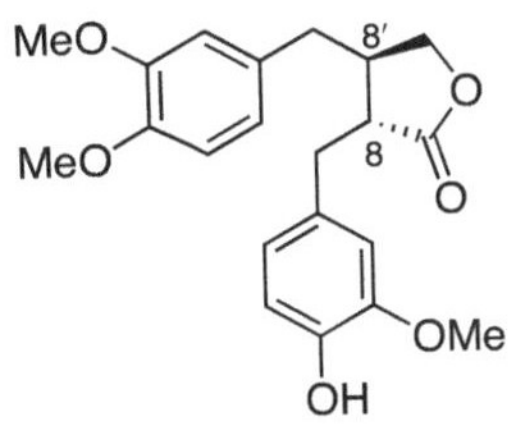
(–)-Arctigenin (**10**)
Arctium lappa

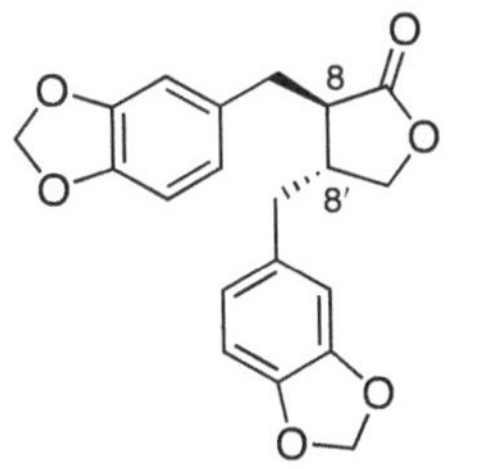
(–)-Hinokinin (**11**)
Chamaecyparis obtusa

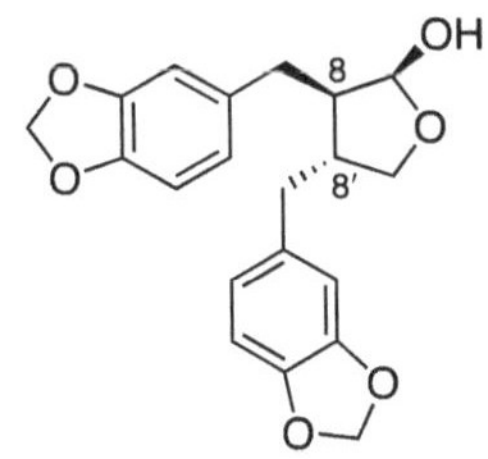
(–)-Cubebin (**12**)
Piper cubeba

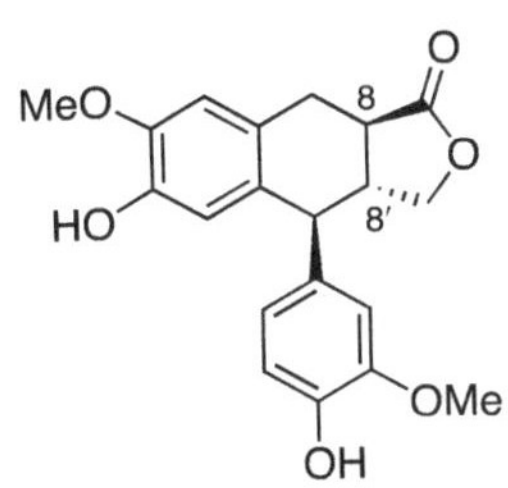
(–)-α-Conidendrin (**13**)
Picea abies
Tsuga sieboldii

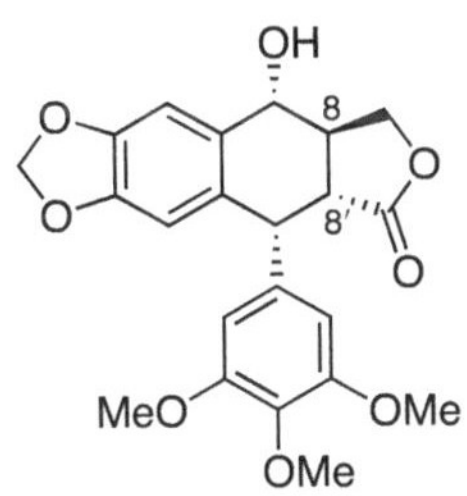
(–)-Podophyllotoxin (**14**)
Podophyllum hexandrum
and *Podophyllum peltatum*

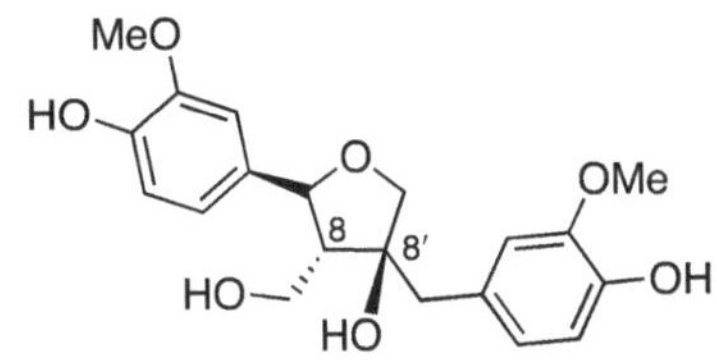
(–)-Olivil (**15**)
Olea europa

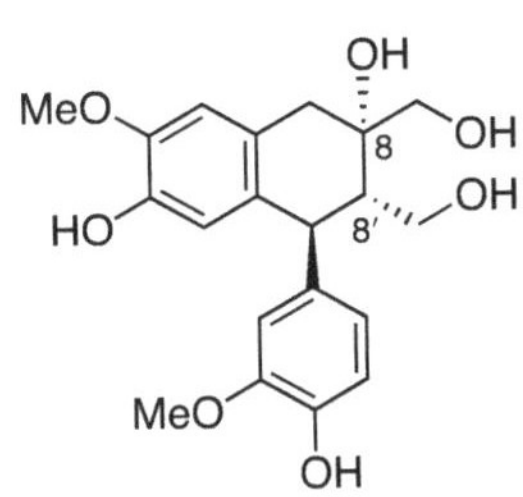
(+)-Cycloolivil (**16**)
Olea europa

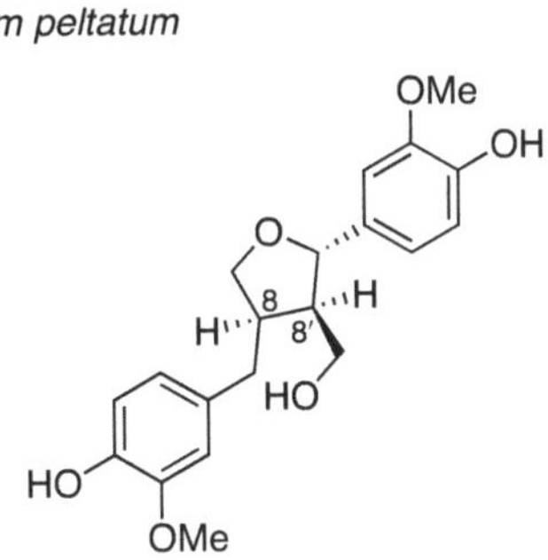
(+)-Lariciresinol (**17**)
Larix decidua

C. obtusa heartwood has also long been highly valued because of its decay resistance. For example, it was used for construction of the Horyu-ji temple (a 5-story, ~1300-year-old building) in Ikaruga (Nara Prefecture, Japan), which still stands today (Smith and Snow 2008).

(–)-Cubebin (**12**), a crystalline substance, was discovered by Capitaine and Soubeiran (1839) following its isolation from seeds of *Piper cubeba,* a plant named Cubeb pepper or Java pepper. It was identified by Mameli (1907), and the orientation of the OH group on the lactone ring at C9 was determined later (Chen et al. 1987). This species has long been used in medicinal and culinary (seed) applications (see Pereira et al. 2016), as well as in cigarettes as a treatment for asthma (Peinemann 1896).

(–)-Conidendrin (**13**) was given its name by Holger Erdtman (1935) following its structure determination. It was first isolated in 1892 from spent sulphite liquor (SSL) generated during commercial pulp and paper production from a softwood (conifer); it was initially named sulphite liquors lactone and was isolated from Norway spruce (*Picea abies*) (Lindsey and Tollens 1892).

(–)-Podophyllotoxin (**14**), obtained from resins of Indian and American podophyllin (resin obtained from *Podophyllum hexandrum* and *Podophyllum peltatum,* respectively), was identified in 1933 (Robertson and Waters 1933). Today, this natural product serves as the scaffold for some of Nature's blockbuster anticancer drugs, etoposide, teniposide, and Etopophos (see Canel et al. 2000).

(–)-Olivil (**15**) and *iso*-olivil (also called (+)-cycloolivil) (**16**) were isolated and identified from olive tree (*Olea europaea*) resin (Vanzetti 1929, Vanzetti and Dreyfuss 1934). Olive trees are cultivated for their fruits (oil) and their wood, which is very hard and highly valued.

(–)-Eudesmin (**6b**), now known as (–)-pinoresinol (**6b**), was first obtained in 1895 from kinos of some *Eucalyptus* species, e.g. *E. hemiphloia;* its structure was deduced by R. Robinson and H. G. Smith (1915).

(+)-Pinoresinol (**6a**), isolated by M. Bamberger (1894) from callus resin *(Überwallungsharz)* from Austrian pine or black pine *(Pinus nigra)* and Norway spruce (*Picea abies*), was later found to be the opposite antipode of (–)-eudesmin (**6b**) (Erdtman 1934, 1935).

M. Bamberger and A. Landsiedl also obtained lariciresinol (**17**) from larch (*Larix decidua*) callus resin in 1897, whose structure was correctly deduced later by W. Kelly and R. D. Haworth (see Haworth 1937).

These few early examples of lignans provide some historical context as to how the now thousands of known lignans received their chemical compound names, i.e. these largely being based on ad hoc adaptations from either the trivial or Latin name of the plant species, a practice that continues to the present time. Thus, names given to lignans (as for many other metabolic classes, e.g. terpenes, alkaloids) are certainly not intuitive.

Nor were their given names systematically linked to their metabolic pathways, as most of the latter pathways were unknown at the time of naming. This approach to lignan naming, coupled with the subsequent and often clumsy, cumbersome International Union of Pure and Applied Chemistry (IUPAC) rules-derived names for lignans, is a serious impediment for those wishing to rapidly assimilate advances in both this important field (see the next section) and for other metabolic classes.

1.4 FURTHER INCONGRUENCIES IN LIGNAN NOMENCLATURE: OTHER C_6C_3 INTER-UNIT LINKAGE TYPES

The early definition of lignans above, as being solely 8–8′ linked, did not take into account structures of other dimeric C_6C_3 natural products that had also been reported. This included compounds, such as the 5–5′ and 8–5′ linked dimers, dehydrodieugenol (**3**) and dehydrodiisoeugenol (**5**), obtained from eugenol (**2**) and isoeugenol (**4**), respectively, *in vitro* (Erdtman 1933a, b). Like the 8-8′ linked lignans, guaiaretic acid (**7**) and α-guaiaconic acid (**8**) discussed previously, both **3** and **5** lacked oxygenation at $C_9C_{9'}$.

Furthermore, it was established by the late 1950s that incubation of the monolignol coniferyl alcohol (**1**) with laccase/O_2 or peroxidase/H_2O_2 *in vitro* gave a mixture of racemic 8–5′, 8–*O*–4′ and 8–8′ dimers (or dilignols) **18a/b**, **19a/b** and **6a/b** (Figure 1.3) in a one-pot reaction (Higuchi 1957, Freudenberg et al. 1958, Higuchi and Ito 1958, Freudenberg 1959).

These compounds, as embodied in **6a/b**, **18a/b**, and **19a/b** (as well as in other lignin model compounds, generally dimers), were thus classified according to the type of inter-unit linkage (i.e. as 8–5′, 8–*O*–4′ or 8–8′ linked) present in the polymeric lignins. However, in contrast to the original narrow lignan definition of being solely 8–8′-linked, these different inter-unit linkage types were not considered to be separate natural product classes in lignin nomenclature (Sarkanen and Ludwig 1971).

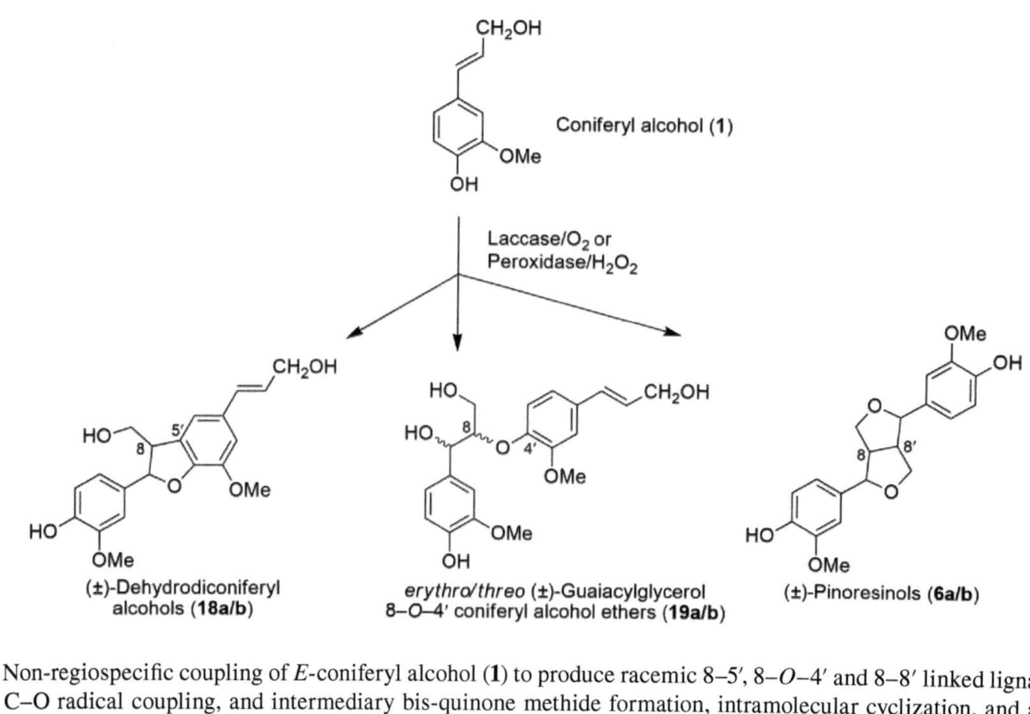

FIGURE 1.3 Non-regiospecific coupling of *E*-coniferyl alcohol (**1**) to produce racemic 8–5′, 8–*O*–4′ and 8–8′ linked lignans via 1e oxidation, C–C and C–O radical coupling, and intermediary bis-quinone methide formation, intramolecular cyclization, and aromatization to afford **18a/b** and **6a/b** or quenching through nucleophilic attack (OH^-) at C7 to produce **19a/b**.

In hindsight, the 1937 lignan nomenclature should have included other linkage types, such as those with 8–5′ and 5–5′ bonds. Had this been done, all C_6C_3 dimeric natural products would have been classified as lignans, albeit differing in terms of inter-unit linkage types. It was not until much later, however, when the biosynthetic pathways to monomers, such as coniferyl alcohol (**1**) and eugenol (**2**)/isoeugenol (**4**), and to various lignans, had been determined, that the basis for redefining all of these natural products as lignans became self-evident.

1.5 NEOLIGNANS

By the 1970s, the field of phytochemistry had rapidly blossomed with many, many additional plant species being phytochemically investigated. This was not only for study of their C_6C_3 monomeric and dimeric constituents and other natural products, but also for the purpose of deducing possible chemotaxonomic relationships between plant families.

A major contributor was the Gottlieb laboratory, which up to then (and later) had unfettered access to Brazilian flora that was essentially inaccessible outside of Brazil. During these pioneering investigations, it became increasingly clear that numerous plant families (particularly in the Lauraceae and Piperaceae (Gottlieb 1974)), as well as various other species worldwide, accumulated many different C_6C_3 dimers with linkages other than 8–8′. Moreover, because of a frequent lack of oxygenation at $C_9C_{9'}$, Gottlieb proposed that these C_6C_3 dimers were quite distinct from lignans with $C_9C_{9'}$ hydroxyl groups, i.e. such as pinoresinol (**6**) or the 8–5′ linked dehydrodiconiferyl alcohol (**18**) (Gottlieb 1974).

Consequently, Professor Gottlieb advanced the following hypothesis:

"I propose to call this group of metabolites neolignans, and do not believe that it keeps more than a mere structural relationship with the *bis*-arylpropanoids with oxygenated (γ)-carbons, as I propose to redefine the lignans". He also suggested that allyl/propenyl phenylpropanoid dimers (e.g. licarin A and B (**5b** and **20b**) (Aiba et al. 1973, Nascimento et al. 2000) lacked any close biosynthetic relationship to, for example, dehydrodiconiferyl alcohol (**18**) (Gottlieb 1974). A new form of neolignan nomenclature was thus proposed, in spite of the fact that guaiaretic acid (**7**) and α-guaiaconic acid (**8**) had long before been reported as lignans, and both lacked oxygenated functionalities at $C_9C_{9'}$.

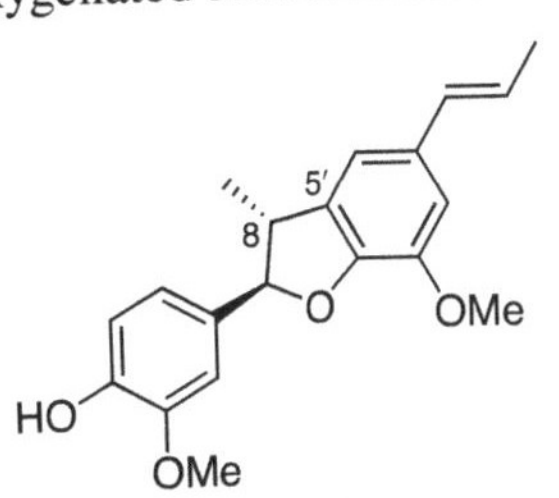

(–)-Licarin A also known as (–)-dehydrodiisoeugenol (**5b**)
Licaria aritu

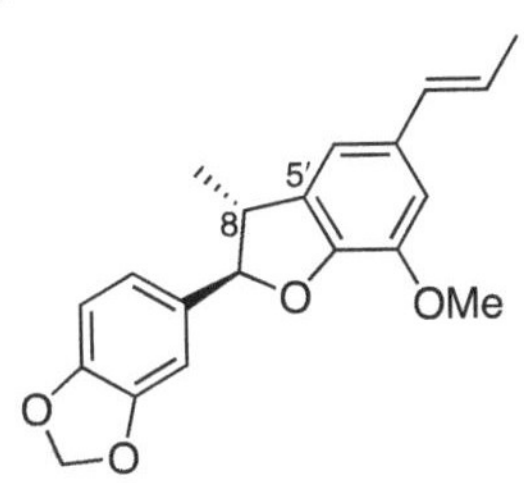

(–)-Licarin B (**20b**)
Licaria aritu

Later, yet another term, *oxyneolignan*, was introduced to define ether-linked structures (Moss 2000) such as with isomagnolol (**21**) (El-Feraly et al. 1983). Why this simple dimer **21** (with an aryl-*O*-aryl linkage) would be defined as an oxyneolignan is unclear. Again, the term lignan can be straightforwardly employed with stipulation of its inter-unit linkage (i.e. 3-*O*-4′).

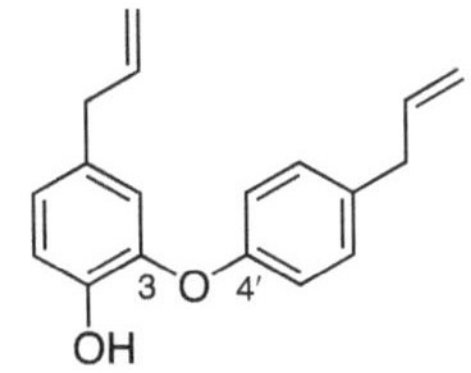

Isomagnolol (**21**)
Magnolia officinale

In addition, the usage of the prefix *neo-*, in chemical terms, is another complication for two reasons. First, it has been used chemically to describe entities, such as neopentane, to allow for ready distinction between a long-chain hydrocarbon and one containing a branched *t*-butyl group (such as with *n*-pentane and neopentane) (Mooney 2016). However, the prefix *neo-* was recently no longer recommended by IUPAC for hydrocarbon nomenclature in 2014. Second, the prefix neo- is derived from the Greek *neos*, being conventionally used for meaning new, such as in neonatal, neoplasm, neoplasia, etc. None of these descriptors appears to be apt for use in the term *neolignans* as envisaged above.

1.6 REDEFINING ALL C_6C_3 COUPLED PRODUCTS (DIMERS AND HIGHER OLIGOMERS) AS LIGNANS

It is now well established that structurally diverse lignans (and "neolignans") encompass an enormous array of distinct skeleta (Lewis and Davin 1999, Vassão et al. 2010). For examples, Figure 1.4 serves to illustrate just some of the structural (inter-unit linkage) diversity encountered through distinct C_6C_3 coupling modes in plant natural products **22–29** (Kupchan et al. 1978, Braz Filho et al. 1981, Lopes et al. 1984, Lopes et al. 1986, Chen et al. 1989, Carroll and Taylor 1991, Sehlapelo et al. 1993, Zschocke et al. 2000, Cui et al. 2014). For examples, (–)-megaphone (**22**) from *Aniba megaphylla* (Kupchan et al. 1978) has an 8–1′-linkage, whereas (–)-machilin D (**25**) from *Saururus chinensis* (Cui et al. 2014) is 8–*O*–4′-linked. Both of these linkage types are also present in lignins that are derived from precursors oxygenated at C_9/$C_{9'}$. Of these, the 8–*O*–4′ linkage predominates in lignins. In an analogous manner, the benzodioxane linkage type present in eusiderin B (**26**) has also been reported in modified lignins (Moinuddin et al. 2010). That is, while there is a broad range of inter-molecular coupling products with distinct inter-unit linkages such as in dimers **5** and **21–29**, several of these

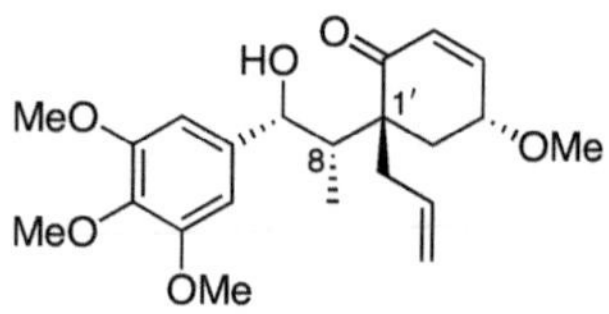

(–)-Megaphone (**22**)
Aniba megaphylla

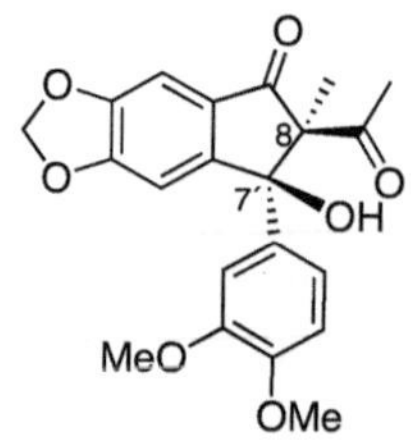

23
Virola sebifera

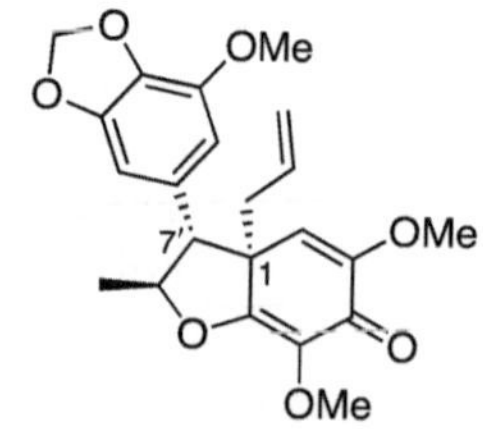

Chrysophyllon 1A
(**24**)
Licaria chrysophylla

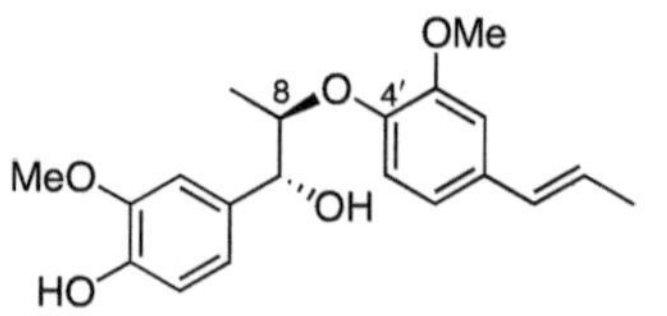

(–)-Machilin D (**25**)
Saururus chinensis

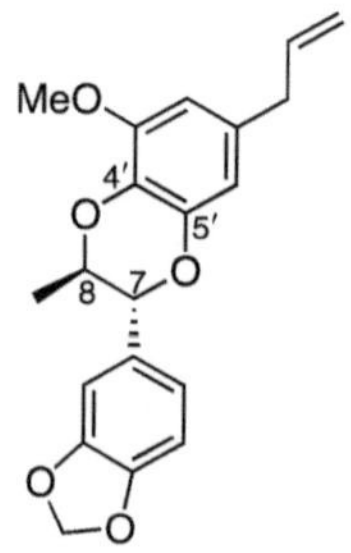

Eusiderin B (**26**)
Licaria rigida

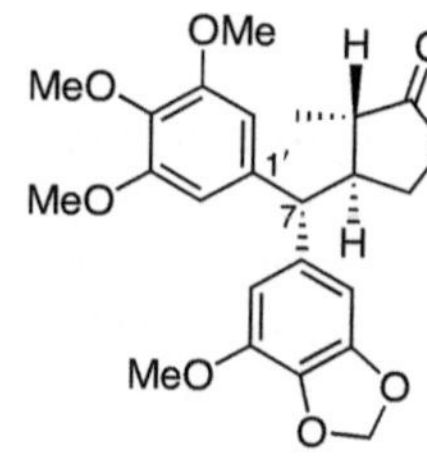

(+)-Peperomin B (**27**)
Peromia japonica

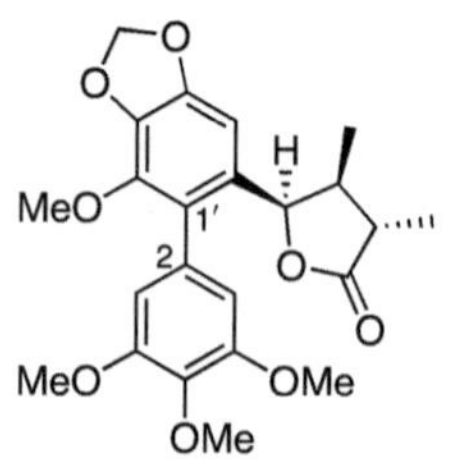

(–)-Eupomatilone 3 (**28**)
Eupomatia bennettii

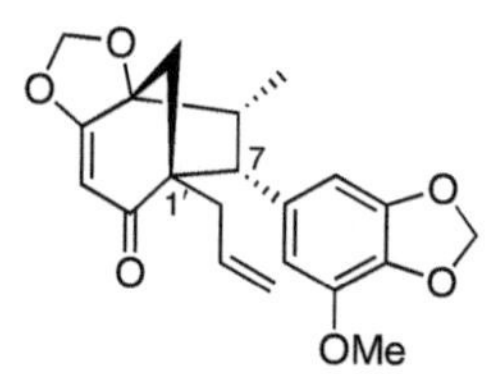

(+)-Ocobullenone (**29**)
Ocotea bullata

FIGURE 1.4 Examples of structural diversity in coupling modes (inter-unit linkage types) found in lignan skeleta.

linkage types are also present in C_6C_3 coupling products having $C_9/C_{9'}$ oxygenated functionalities.

All of these dimeric metabolites are thus best described as lignans rather than neolignans, by simple reference to the type of inter-unit linkage (or linkages) present. To put the current nomenclature definition incongruency into even sharper focus, the following two additional examples further demonstrate what we consider to be glaring inconsistencies.

The two selected examples chosen are (–)-saucerneol D (**30**), a C_6C_3 trimer from *Saururus chinensis* (Hwang et al. 2003), and the bi-lignan (+)-lappaol F (**31**), a C_6C_3 tetramer from *Arctium lappa* (Ichihara et al. 1978, Ichihara et al. 1979). Both contain conventional "lignan" and "neolignan" inter-unit linkages, and yet can also have – or not have – oxygenated $C_9/C_{9'}$ inter-unit linkages. Thus, under the current nomenclature usage, they are neither one nor the other!

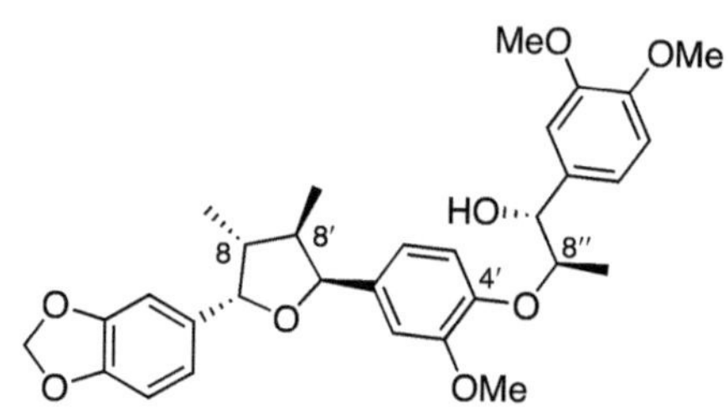

(–)-Saucerneol D (**30**)
Saururus chinensis

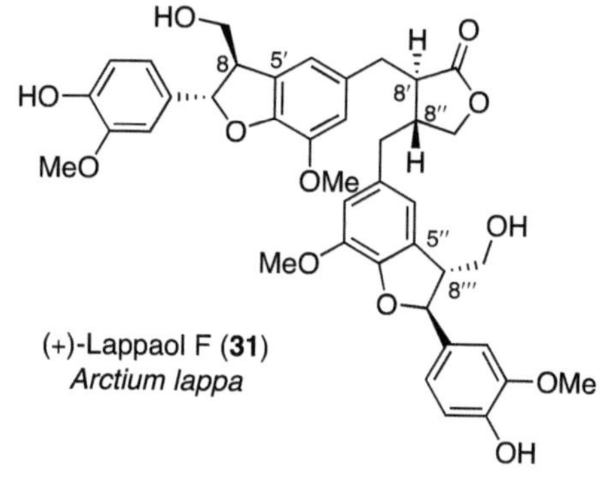

(+)-Lappaol F (**31**)
Arctium lappa

That is, the C_6C_3 oxidative phenolic coupling products now known can range from being C_6C_3 dimers to higher oligomers, linked through either C–C and/or C–O inter-unit bonds. Accordingly, the usage of various terms, such as lignans, neolignans, and oxyneolignans, results in unnecessary nomenclature inconsistencies and complications, i.e. including whether oxygenated or not at $C_9/C_{9'}$. In sum, we continue to strongly recommend that the term *neo-* in neolignans be generally discontinued, as done in this *Handbook*.

1.7 SYSTEMATIZING LIGNAN NAMES AND CHEMICAL STRUCTURES

This lignan introductory chapter (including neolignans) gives a brief glimpse into just a few of the structural variations in lignans that are to be found in this compound class in the *Handbook*, as well as some inklings as to how they received their ad hoc derived chemical names.

Additionally, the IUPAC rule-derived systematic naming of lignans becomes essentially unworkable when considering larger size lignan/neolignan oligomers possessing different substructure types in their overall molecules. Indeed, even the IUPAC carbon numbering system can change priority on a dimer when an additional substituent (such as an additional methoxyl group) is added to one of the aromatic rings.

Accordingly, we have made a serious, concerted attempt, in the following chapters, to help enable the reader to quickly assimilate lignan structural diversity in a systematic metabolic context, at least in terms of our current knowledge of same.

ACKNOWLEDGEMENTS

The authors thank the Chemical Sciences, Geosciences and Biosciences Division, DOE Office of Basic Energy Sciences (DE-FG-0397ER20259) as well as the Arthur and Katie Eisig-Tode Endowment.

REFERENCES

Aiba, C. J., R. G. Campos Corrêa, and O. R. Gottlieb. 1973. Natural occurrence of Erdtman's dehydrodiisoeugenol. *Phytochemistry* 12:1163–1164.

Bamberger, M. 1894. Zur Kenntniss der Überwallungsharze. *Monatsh. Chem.* 15:505–518.

Bamberger, M., A. Landsiedl. 1897. Zur Kenntniss der Überwallungsharze. *Monatsh. Chem.* 18:481–509.

Banerjee, J. 2017. Sir William Henry Perkin and the Coal-Tar Colours. The Victorian Web. http://www.victorianweb.org/science/perkin.html.

Bolzani, V. d. S., and G. M. Cragg. 2004. A tribute to Professor Otto Richard Gottlieb. *ARKIVOC* vi:1–4.

Braz Filho, R., M. G. De Carvalho, O. R. Gottlieb, J. G. S. Maia, and M. L. Da Silva. 1981. Neolignans from *Licaria rigida*. *Phytochemistry* 20:2049–2050.

Briggs, L. H. 1935. Hinokinin as the enantiomorph of cubebinolide. *J. Am. Chem. Soc.* 57:1383–1384.

Briggs, L. H., D. A. Peak, and J. L. D. Woolloxall. 1935. The constitution of matairesinol. *J. Proc. R. Soc. N. S. W.* 69:61–67.

Canel, C., R. M. Moraes, F. E. Dayan, and D. Ferreira. 2000. Podophyllotoxin. *Phytochemistry* 54:115–120.

Capitaine, H., and E. Soubeiran. 1839. Note sur le cubébin. *J. Pharm.* 25:355–358.

Carroll, A. R., and W. C. Taylor. 1991. Constituents of *Eupomatia* species. XIV. The structures of eupomatilone-1, -2, -3, -4, -5, -6, and -7 isolated from *Eupomatia bennettii*. *Aust. J. Chem.* 44:1705–1714.

Chan, Y.-S., L.-N. Cheng, J.-H. Wu, et al. 2011. A review of the pharmacological effects of *Arctium lappa* (burdock). *Inflammopharmacology* 19:245–254.

Chemical Heritage Foundation. 2015. William Henry Perkin. https://www.chemheritage.org/historical-profile/william-henry-perkin (accessed January 10, 2017).

Chen, C.-M., F.-Y. Jan, M.-T. Chen, and T.-J. Lee. 1989. Peperomins A, B, and C, novel secolignans from *Peperomia japonica*. *Heterocycles* 29:411–414.

Chen, W. M., R. Mayer, and G. Rücker. 1987. Orientierung der Hydroxyl-Gruppe des (–)-Cubebins und seines Epimeren. *Arch. Pharm.* 320:374–376.

Cui, H., B. Xu, T. Wu, et al. 2014. Potential antiviral lignans from the roots of *Saururus chinensis* with activity against Epstein-Barr virus lytic replication. *J. Nat. Prod.* 77:100–110.

Davin, L. B., H.-B. Wang, A. L. Crowell, et al. 1997. Stereoselective bimolecular phenoxy radical coupling by an auxiliary (dirigent) protein without an active center. *Science* 275:362–366.

dos Santos, A. C., C. H. Baggio, C. S. Freitas, et al. 2008. Gastroprotective activity of the chloroform extract of the roots from *Arctium lappa* L. *J. Pharm. Pharmacol.* 60:795–801.

Easterfield, T. H., and J. Bee. 1910. The resin acids of the Coniferae. Part II. Matairesinol. *J. Chem. Soc.*:1028–1032.

Editors of Encyclopædia Britannica. 2016. Sir William Henry Perkin. In *Encyclopædia Britannica*. https://www.britannica.com/biography/William-Henry-Perkin (accessed January 10, 2017).

El-Feraly, F. S., S. F. Cheatham, and R. L. Breedlove. 1983. Antimicrobial neolignans of *Sassafras randaiense* roots. *J. Nat. Prod.* 46:493–498.

Erdtman, H. 1933a. Dehydrierungen in der Coniferylreihe. I. Dehydrodi-eugenol und Dehydrodiisoeugenol. *Biochem. Z.* 258:172–180.

Erdtman, H. 1933b. Dehydrierungen in der Coniferylreihe. II. Dehydrodi-isoeugenol. *Justus Liebigs Ann. Chem.* 503:283–294.

Erdtman, H. 1934. Konstitution der Harzphenole und deren biogenetische Zusammenhänge. I. Über Pinoresinol. *Sven. Kem. Tidskr.* 46:229–233.

Erdtman, H. 1935. Die Konstitution der Harzphenole und ihre biogenetischen Zusammenhänge. II. Über Pinoresinol und seine Beziehungen zu Eudesmin. *Justus Liebigs Ann. Chem.* 516:162–176.

Erdtman, H. 1939. Översikt över naturprodukter av diaryl-butantyp. *Svensk Papp. Tidn.* 42:115–122.

Flückiger, F. A., and D. Hanbury. 1874. *Pharmacographia. A History of the Principal Drugs of Vegetable Origin, Met with in Great Britain and British India*. London: MacMillan and Co.

Freudenberg, K., J. M. Harkin, M. Reichert, and T. Fukuzumi. 1958. Die an der Verholzung beteiligten Enzyme. Die Dehydrierung des Sinapinalkohols. *Chem. Ber.* 91:581–590.

Freudenberg, K. 1959. Biosynthesis and constitution of lignin. *Nature* 183:1152–1155.

Gottlieb, O. R. 1972. Chemosystematics of the Lauraceae. *Phytochemistry* 11:1537–1570.

Gottlieb, O. R. 1974. Lignans and neolignans. *Rev. Latinoamer. Quim.* 5:1–10.

Haworth, R. D., and T. Richardson. 1935a. The constituents of natural phenolic resins. Part I. Matairesinol. *J. Chem. Soc.*:633–636.

Haworth, R. D., and T. Richardson. 1935b. The constituents of *Guaiacum* resin. Part III. Synthesis of *dl*-guaiaretic acid diethyl ether. *J. Chem. Soc.*:120–122.

Haworth, R. D., and W. Kelly. 1936. The constituents of natural phenolic resins. Part VII. Arctigenin. *J. Chem. Soc.*:998–1003.

Haworth, R. D. 1937. Natural resins. *Ann. Rep. Prog. Chem.* 33:266–279.

Haworth, R. D., B. P. Moore, and P. L. Pauson. 1948. Purpurogallin. Part I. *J. Chem. Soc.*:1045–1051.

Haworth, R. D., and L. Wilson. 1950. The constituents of natural phenolic resins. Part XXII. Reduction of some lactonic lignans with lithium aluminium hydride. *J. Chem. Soc.*:71–72.

Higuchi, T. 1957. Biochemical studies of lignin formation. *Physiol. Plant.* 10:356–372.

Higuchi, T., and Y. Ito. 1958. Dehydrogenation products of coniferyl alcohol formed by the action of mushroom phenol oxidase, *Rhus* laccase, and radish peroxidase. *J. Biochem.* 45:575–579.

Hwang, B. Y., J.-H. Lee, J. B. Nam, Y.-S. Hong, and J. J. Lee. 2003. Lignans from *Saururus chinensis* inhibiting the transcription factor NF-KB. *Phytochemistry* 64:765–771.

Ichihara, A., S. Kanai, Y. Nakamura, and S. Sakamura. 1978. Structures of lappaol F and H, dilignans from *Arctium lappa* L. *Tetrahedron Lett.* 19:3035–3038.

Ichihara, A., Y. Nakamura, H. Kawagishi, and S. Sakamura. 1979. Stereochemistry of lappaol A and F. *Tetrahedron Lett.* 20:3735–3736.

Jones, E. 1991. Robert Downs Haworth. 15 March 1898–19 May 1990. *Biographical Memoirs of Fellows of the Royal Society* 37:264–276.

Kealy, T. J., and P. L. Pauson. 1951. A new type of organo-iron compound. *Nature* 168:1039–1040.

Keimatsu, S., and T. Ishiguro. 1935. Constituents of hinokiol. II. A comparison of hinokinin and cubebinolide. *Yakugaku Zasshi* 55:96–99.

Keimatsu, S., T. Ishiguro, and Y. Nakamura. 1935. Über die Synthese des α,β-Dipiperonyl-butyrolactons. *Yakugaku Zasshi* 55:775–780.

Klason, P. 1908. Bidrag till närmare kännedom om granvedens kemiska sammansättning. *Arkiv för Kemi, Mineralogi och Geologi* 3:1–20.

Kratochvil, J. F., R. H. Burris, M. K. Seikel, and J. M. Harkin. 1971. Isolation and characterization of α-guaiaconic acid and the nature of guaiacum blue. *Phytochemistry* 10:2529–2531.

Kupchan, S. M., K. L. Stevens, E. A. Rohlfing, et al. 1978. New cytotoxic neolignans from *Aniba megaphylla* Mez. *J. Org. Chem.* 43:586–590.

Lewis, N. G., L. B. Davin, and S. Sarkanen. 1998. Lignin and lignan biosynthesis: Distinctions and reconciliations. In *Lignin and lignan biosynthesis* (N. G. Lewis and S. Sarkanen eds.), Vol. 697, ACS Symposium Series, Washington, DC. pp. 1–27.

Lewis, N. G., and L. B. Davin. 1999. Lignans: Biosynthesis and function. In *Comprehensive Natural Products Chemistry. Vol.1: Polyketides and Other Secondary Metabolites, Including Fatty Acids and Their Derivatives* (Sir D. H. R. Barton, K. Nakanishi and O. Meth-Cohn eds.). pp. 639–712.

Lindsey, J. B., and B. Tollens. 1892. Über Holz-Sulfitflüssigkeit und Lignin. *Justus Liebigs Ann. Chem.* 267:341–366.

Longwood, F. R. 1962. *Present and Potential Commercial Timbers of the Caribbean with Special Reference to the West Indies, the Guianas, and British Honduras. Agriculture Handbook No. 207.* Washington, DC: U.S. Department of Agriculture.

Lopes, L. M. X., M. Yoshida, and O. R. Gottleib. 1984. Aryltetralone and arylindanone neolignans from *Virola sebifera. Phytochemistry* 23:2021–2024.

Lopes, M. N., M. S. Da Silva, J. M. Barbosa F., et al. 1986. Unusual benzofuranoid neolignans from *Licaria chrysophylla. Phytochemistry* 25:2609–2612.

Mameli, E. 1907. Sulla cubebina. Nota I. *Gazzetta* 37:483–506.

Mameli, E. 1935. Cubebin. VIII. Identity of cubebinolide with hinokinin. *Gazz. Chim. Ital.* 65:886–888.

Marcotullio, M., C, A. Pelosi, and M. Curini. 2014. Hinokinin, an emerging bioactive lignan. *Molecules* 19:14862–14878.

Moinuddin, S. G. A., M. Jourdes, D. D. Laskar, et al. 2010. Insights into lignin primary structure and deconstruction from *Arabidopsis thaliana* COMT (caffeic acid *O*-methyl transferase) mutant *Atomt1. Org. Biomol. Chem.* 8:3928–3946.

Mooney, J. 2016. Nomenclature of alkanes. LibreTexts, Chemistry. http://chem.libretexts.org/Core/Organic_Chemistry/Alkanes/Nomenclature_of_Alkanes (accessed January 10, 2017).

Moss, G. P. 2000. Nomenclature of lignans and neolignans (IUPAC recommendations 2000). *Pure Appl. Chem.* 72:1493–1523.

Nascimento, I. R., L. M. X. Lopes, L. B. Davin, and N. G. Lewis. 2000. Stereoselective synthesis of 8,9-licarinediols. *Tetrahedron* 56:9181–9193.

Nobelprize.org. 2014. Roald Hoffmann – Biographical. Nobel Media AB 2014. http://www.nobelprize.org/nobel_prizes/chemistry/laureates/1981/hoffmann-bio.html (accessed October 2, 2017).

Norin, T. 1990. Holger Erdtman 1902–1989. *Phytochemistry* 29:1713–1714.

Omaki, T. 1937. Arctiin, a constituent of the seeds of *Arctium lappa*, L. III. The position of the hydroxyl group of arctigenin. *Yakugaku Zasshi* 57:22–25.

Ozawa, T. 1952. Studies on the component of the seed of *Arctium lappa* L. VII. Synthesis of *l*-arctigenin. *Yakugaku Zasshi* 72:551–553.

Peinemann, K. 1896. Beiträge zur pharmazeutischen und chemischen Kenntnis der Cubeben und der als Verfälschung derselben beobachteten Piperaceenfrüchte. *Arch. Pharm.* 234:204–224.

Pereira, C. C. S. S., F. F. Perazzo, G. H. B. Souza, F. L. A. Fonseca, and P. C. P. Rosa. 2016. Cubebin and semisynthetic dibenzyl butyrolactone derivatives: Biological activities. *Afr. J. Pharm. Pharmacol.* 10:526–532.

Phillips, M. 1934. The chemistry of lignin. *Chem. Rev.* 14:103–170.

Richter, P. 1906. Zur Kenntnis des Guajakharzes. *Arch. Pharm.* 244:90–119.

Righetti, S. 2011. Químico que chegou mais perto do Nobel no país morre aos 90 *Folha de S. Paulo*, June 21, 2011, Ciência. http://www1.folha.uol.com.br/ciencia/2011/06/932761-quimico-que-chegou-mais-perto-do-nobel-no-pais-morre-aos-90.shtml (accessed January 10, 2017).

Robertson, A., and R. B. Waters. 1933. Podophyllotoxin. *J. Chem. Soc.*:83–86.

Robinson, R., and H. G. Smith. 1915. Eudesmin and its derivatives. Part 1. *J. Proc. Soc. N. S. W.* 48:449–463.

Royal Society. 2017. Davy Medal. https://royalsociety.org/grants-schemes-awards/awards/davy-medal/ (accessed January 10, 2017).

Sarkanen, K. V., and C. H. Ludwig. 1971. *Lignins: Occurrence, Formation, Structure and Reactions.* New York, NY: Wiley Interscience.

Schrecker, A. W., and J. L. Hartwell. 1956. Components of podophyllin. XX. The absolute configuration of podophyllotoxin and related lignans. *J. Org. Chem.* 21:381–382.

Schroeter, G., L. Lichtenstadt, and D. Irineu. 1918. Über die Konstitution der Guajacharz-Subztanzen (I). *Chem. Ber.* 51:1587–1613.

Sehlapelo, B. M., S. E. Drewes, and P. Sandor. 1993. Ocobullenone: A bicyclo[3.2.1]octanoid neolignan from *Ocotea bullata. Phytochemistry* 32:1352–1353.

Shinoda, J., and M. Kawagoye. 1929. Über die Bestandteile des *Arctium lappa* L. *J. Pharm. Soc. Japan* 49:565–575.

Simon, J. B. 1985. Occult blood screening for colorectal carcinoma. A critical review. *Gastroenterology* 88:820–837.

Smith, I., and M. A. Snow. 2008. Timber: An ancient construction material with a bright future. *Forestry Chronicle* 84:504–510.

Stevenson, R., and J. R. Williams. 1976. Synthesis of α-guaiaconic acid (furoguaiacin). *Org. Prep. Proced. Int.* 8:179–183.

Swiners, J.-L. 2013. 1941. George de Mestral, ingénieur-électricien, a l'idée du Velcro. http://www.serendipite-strategique.com/exemples/velcro.html (accessed January 10, 2017).

Tiemann, F. 1875. Über eine synthetische Bildungsweise des Vanillins und einige durch Reduction des letzteren darstellbare Verbindungen. (Hydrovanilloïn und Vanillylakohol). *Ber. Deutsch. Chem. Ges.* 8:1123–1127.

Unwin, W. C. 1899. *The Elements of Machine Design. Part I. General Principles, Fastenings, and Transmissive Machinery.* 7th ed. London: Longmans, Green, and Co.

Vanzetti, B. L. 1929. Sul costituente principale della gomma (resina) d'olivo: Olivile e derivati. *Gazzetta* 59:373–378.

Vanzetti, B. L., and P. Dreyfuss. 1934. Sulla costituzione dell'olivile. Prodotti di ossidazione dell'isoolivile. *Gazzetta* 64:381–399.

Vassão, D. G., K.-W. Kim, L. B. Davin, and N. G. Lewis. 2010. Lignans (neolignans) and allyl/propenyl phenols: Biogenesis, structural biology, and biological/human health considerations. In *Comprehensive Natural Products II. Chemistry and Biology. Vol. 1: Natural Products Structural Diversity-I. Secondary Metabolites: Organization and Biosynthesis* (L. Mander and H.-W. Liu eds.), Elsevier, Oxford, UK. pp. 815–928.

Wilkinson, G., M. Rosenblum, M. C. Whiting, and R. B. Woodward. 1952. The structure of iron biscyclopentadienyl. *J. Am. Chem. Soc.* 74:2125–2126.

Yoshiki, Y., and T. Ishiguro. 1933. Über die krystallisierten Bestandteile des Hinokiöls. *Yakugaku Zasshi* 53:73–151.

Zschocke, S., J. van Staden, K. Paulus, et al. 2000. Stereostructure and anti-inflammatory activity of three diastereomers of ocobullenone from *Ocotea bullata. Phytochemistry* 54:591–595.

Zydowsky, T. M. 2015. Of sandwiches and Nobel prizes: Robert Burns Woodward. In *Culture of Chemistry* (B. Hargittai and I. Hargittai eds.), Springer, Boston, MA. pp.271–276.

2 The Occurrence and Diversity of Potential "Lignan Building Blocks" in Nature

Laurence B. Davin and Norman G. Lewis

2.1 SCOPE AND PURPOSE

Overall, the *Handbook* describes the occurrence in Nature of the various types of lignans (ranging from the simplest phenylpropanoid (C_6C_3) dimers to higher oligomers), their enormous structural diversity (including those of "mixed" biochemical pathway origin, such as flavonolignans, alkaloid lignans, etc.), and what is known thus far about their biosynthetic pathways, as well as other possible means of formation (discussed below).

Additionally, for needed context on the proposed nomenclature herein, we have redefined and expanded lignan definition from being solely 8-8′ linked phenylpropanoids, to embracing all interunit linkages, such as 8–5′, 8–*O*–4′, etc., regardless of the modifications and functionalization on either their respective aromatic rings or C3 side-chains (Davin and Lewis 2022a, Moinuddin et al. 2022).

In this contribution, we address the diversity of the corresponding "monomeric" species from which various lignan skeleta (e.g. 8–8′, 8–5′, 8–*O*–4′, etc., inter-unit linkages) could *potentially* be formed *in planta* and other organisms (see the following). This is necessary because, as rationalized in this contribution, there are many such "monomeric" natural products that could be considered as "***potential*** building blocks" to lignans, particularly to those of "mixed" biochemical pathway origin.

It must be emphasized that while the main biochemical pathways *in planta* to lignans (C_6C_3 dimers and higher oligomers) result from an established involvement of specific proteins and enzymes stipulating the outcome of coupling (Vassão et al. 2010), there are other considerations that cannot be ignored.

For example, some of the reports of lignans and their higher oligomers in Nature may be a consequence of ***artifact formation***. In this regard, the literature has an enormous number of reports of natural plant products, some of which include artifacts. In the latter case, their suspected formation may have resulted from prolonged plant storage or very lengthy multiple-step isolation protocols, often employing acidic or basic conditions, or exposure to light. That is, the formation of artifacts could be attained through many years of storage of harvested plant material at ambient temperature prior to analysis, or through exposure to light, acidic, or basic conditions during extraction and isolation protocols.

Such occurrences are, in particular, encountered when the natural products from organisms from which they were originally derived contain reactive group entities, such as free phenolic groups, aldehydic, or carboxylic acid functionalities, and so forth. Some of the reported lignans from plants here in the *Lignan Handbook* here, for example, cannot exclude the possibility that some may have been formed as artifacts during such manipulations. Moreover, some previously assigned chemical structures were incorrect and have been corrected (see Yang et al. 2015, Moinuddin et al. 2022).

This matter is raised here, given that the *Lignan Handbook* comprehensively and purposefully encompasses the broadest possible range of reported chemical structures of not only lignans, but also potential lignan "monomers" (i.e. those with free phenolic groups and provisionally able to undergo intermolecular coupling reactions).

For the sake of completeness, the *Handbook* also includes related derivatives, such as those monomers with "masked" or "protected" phenolic groups due to glycosylation, which could also be potential lignan monomers via deglycosylation. Related compounds, albeit obtained synthetically, are included in the *Handbook* compilation.

2.2 LIGNAN "BUILDING BLOCK" MONOMERS PROPER

Monomers that are biochemically produced via shikimate/chorismate and phenylpropanoid pathways, and perhaps metabolic offshoots thereof (Laskar et al. 2010, Vassão et al. 2010), are generally considered as lignan "building block" monomers proper. The vast majority of these (if not all) are derived from phenylalanine (Phe, **1**) and/or tyrosine (Tyr, **2**), which in turn originate from arogenate (**3**). In this regard, Figures 2.1–2.3, respectively, depict the biosynthetic pathways to Phe (**1**) and Tyr (**2**) from chorismate (**4**), as well as those of general phenylpropanoid metabolism to the monolignols, and those in downstream allylphenol/propenylphenol formation, respectively.

All of these pathways have been very well described and reviewed in the literature (Lewis et al. 1999, Laskar et al. 2010,

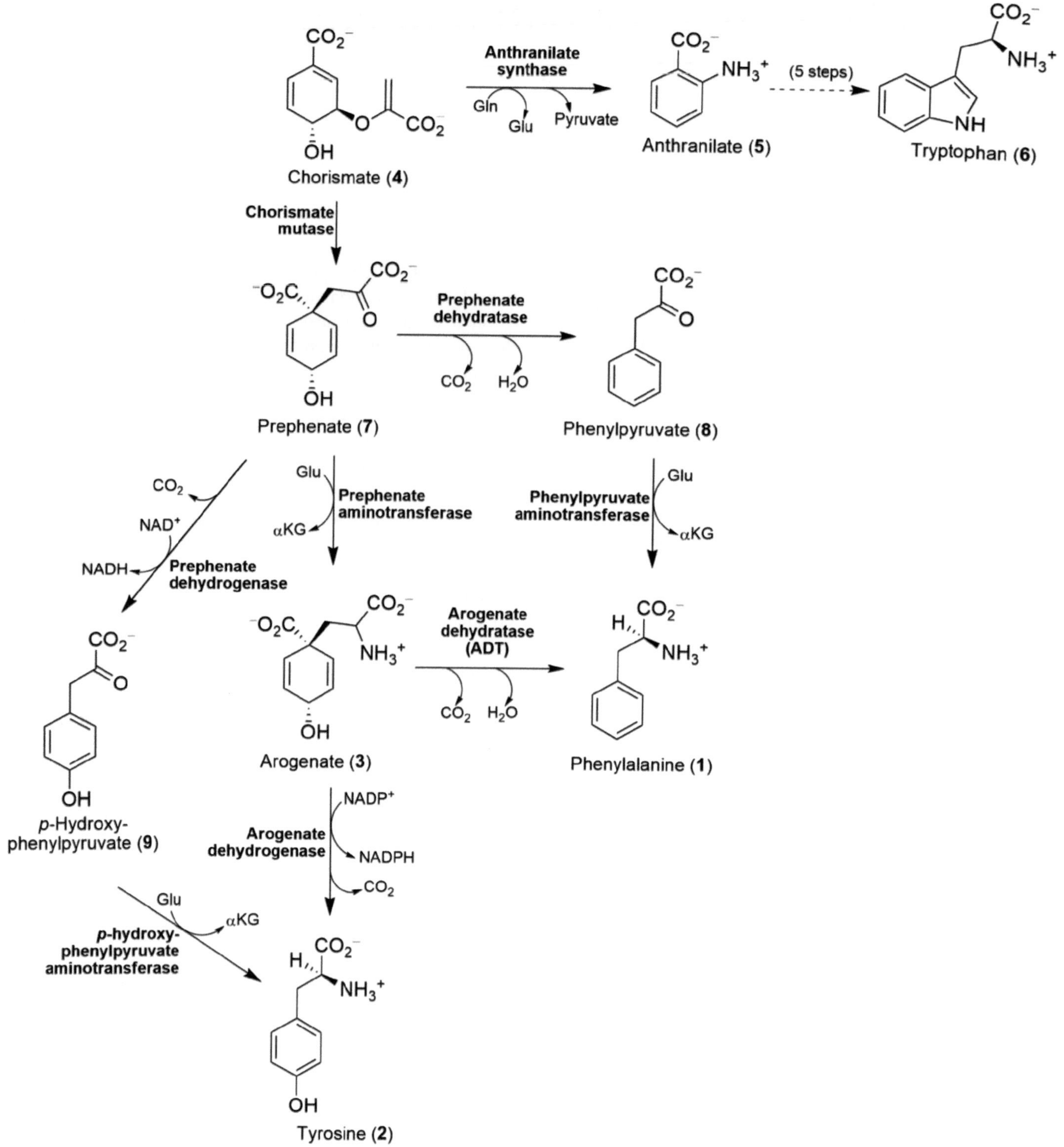

FIGURE 2.1 Proposed biosynthetic pathways to Phe (**1**), Tyr (**2**), and Trp (**6**) in plants and microorganisms. In plants, the dominant pathway is via arogenate (**3**).

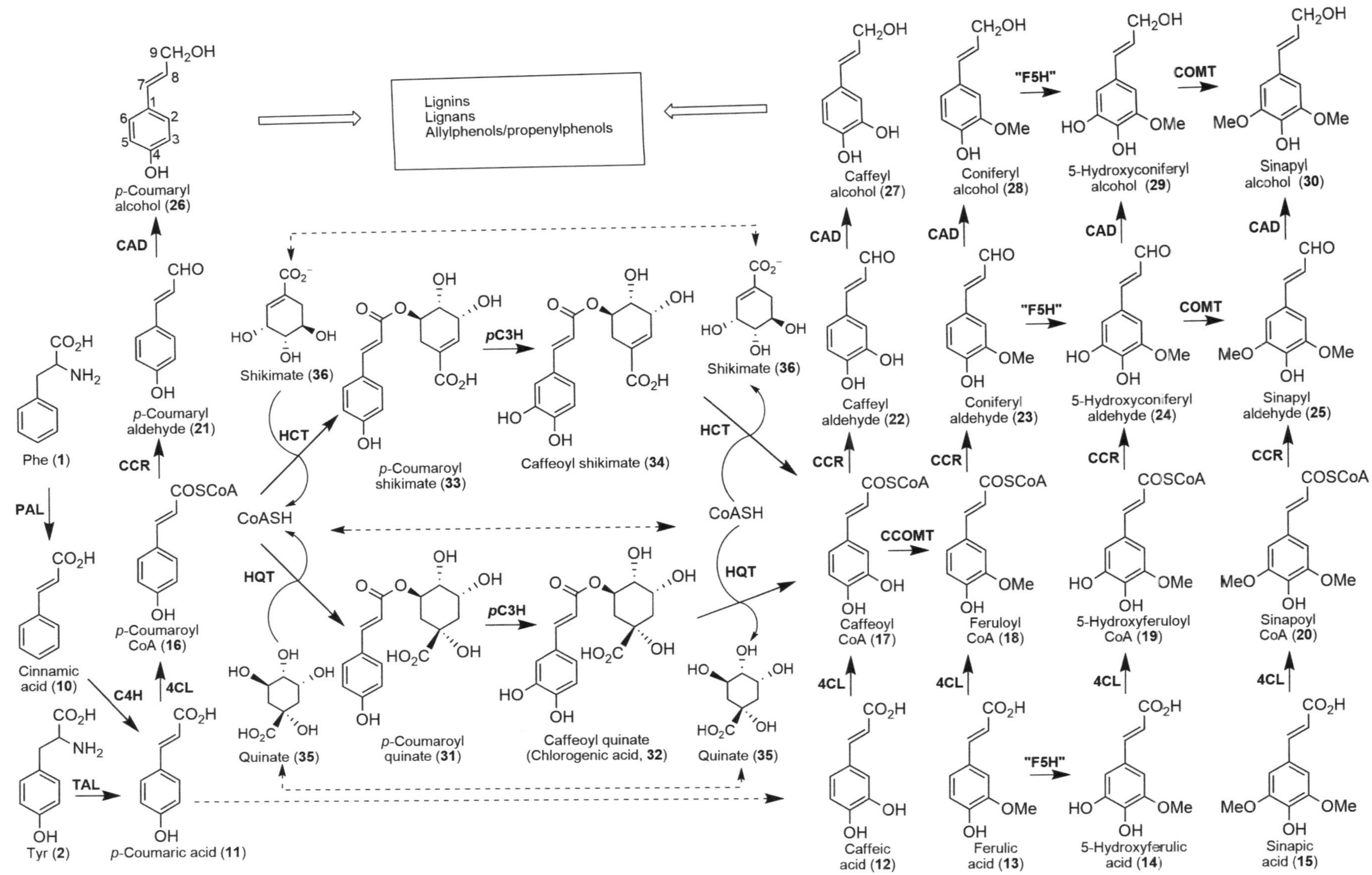

FIGURE 2.2 Current simplified view of the phenylpropanoid pathway to the monolignols **26–30**. 4CL, hydroxycinnamoyl CoA ligases; *p*C3H, *p*-coumarate 3-hydroxylase; C4H, cinnamate hydroxylase; CAD, cinnamyl alcohol dehydrogenase; CCOMT, hydroxycinnamoyl CoA *O*-methyltransferase; CCR, cinnamoyl-CoA oxidoreductase; COMT, caffeic acid *O*-methyltransferase; F5H, ferulate 5-hydroxylase; HCT, hydroxycinnamoyl-CoA shikimate hydroxycinnamoyltransferase; HQT, hydroxycinnamoyl-CoA quinate hydroxycinnamoyltransferase; PAL, phenylalanine ammonia lyase; TAL, tyrosine ammonia lyase.

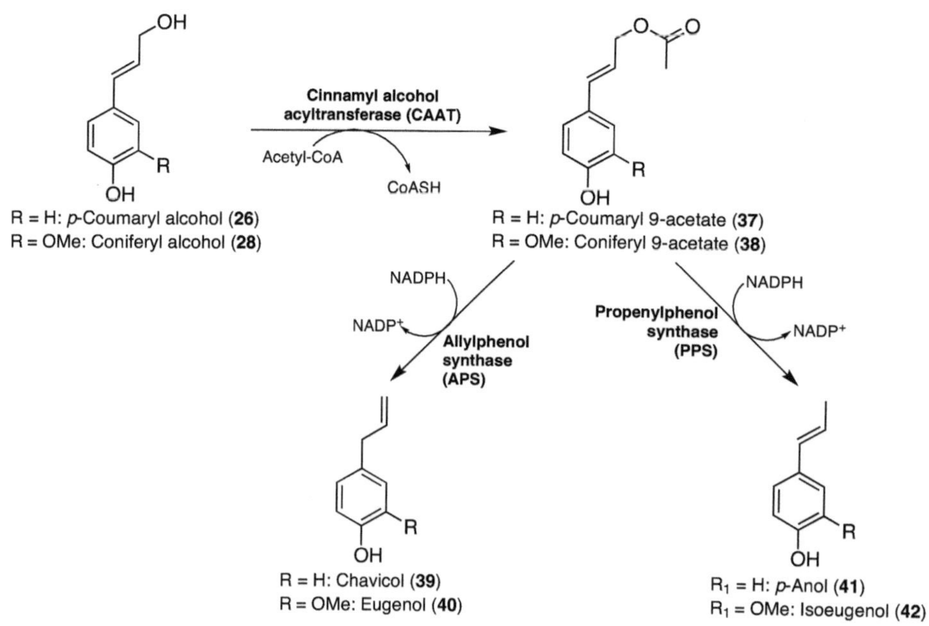

FIGURE 2.3 Biosynthetic pathway to allylphenol/propenylphenol (**39–42**). CAAT, cinnamyl alcohol acyltransferase; APS, allylphenol synthase; PPS, propenylphenol synthase.

Petersen et al. 2010). The interested reader is encouraged to read key papers that both summarize and explain the metabolites, enzymes, genes, transcription factors, and networks involved. Taken together, these pathways basically depict the overall biochemical steps to the hydroxycinnamyl alcohols (monolignols), *p*-coumaryl (**26**), caffeyl (**27**), coniferyl (**28**), 5-hydroxyconiferyl (**29**), and sinapyl (**30**) alcohols, from Phe (**1**) (and to a lesser extent from Tyr (**2**) in grasses), as well as to the allylphenols/propenylphenols, such as chavicol (**39**), eugenol (**40**), *p*-anol (**41**), and isoeugenol (**42**).

To this point, however, only monolignols have been demonstrated to act as substrates in dirigent protein (DP) engendered phenoxy radical stereoselective coupling leading to optically active lignans (Davin et al. 1997, Pickel et al. 2010, Vassão et al. 2010, Kim et al. 2012, Kim et al. 2015, Seneviratne et al. 2015, Corbin et al. 2018, Davin et al. 2022, Davin and Lewis 2022b). On the other hand, there are numerous dimeric and higher oligomeric lignan species reported in the literature, whose structures could be provisionally rationalized, for example, as being formed either through oxidative coupling of hydroxycinnamic acids and hydroxycinnamyl aldehydes, or through post-coupling transformations of monolignol-derived dimeric products (Vassão et al. 2010). To date, however, the body of scientific evidence suggests such modifications *appear* to be derived from post-monolignol coupling.

Nevertheless, given the paucity in our current understanding of how all of the different lignan skeletal types reported from plants are biosynthesized, essentially all of the pathway intermediates/metabolites in Figures 2.2 and 2.3 can also ***potentially*** be considered as core "building blocks" to the diverse lignan structures found in Nature, except, for example, the aromatic amino acid Phe (**1**) or phenylethyl amine (**43**, Figure 2.4). These are excluded as *bona fide* C_6C_3 lignan monomeric precursors, because they are unable to directly undergo phenoxy radical coupling (lacking a free phenol group functionality). Other unlikely lignan monomeric precursors proper are the coenzyme A (CoA) esters **16–20**, as such moieties are generally efficiently metabolized to produce the corresponding aldehydes **21–25** (Figure 2.2).

2.3 ADDITIONAL POTENTIAL LIGNAN-FORMING MONOMERS

Formation of lignans could also ***potentially*** occur from other C_6C_3 (phenylpropanoid) skeletal types, including those resulting through mixed biochemical pathway "monomer" formation, where the term *monomer* could be a further derivatized C_6C_3 moiety. This matter is raised because the diversity of "monomeric" species found in Nature that could potentially undergo coupling reactions to produce various lignan skeleta through stereoselective or non-stereoselective coupling (including as artifacts) is very extensive. Moreover, these "monomers" are not exclusive to plants, as other organisms, such as some sponges, fungi, bacteria, and animals, can also biosynthesize potential lignan monomeric precursors depending on the organism. These examples, albeit generally rare, are included here for completeness (see below and *Handbook* entries), which include their structures, their reported occurrence in various plant species, their physical properties, and references to various biological properties.

As indicated in Figures 2.4–2.10, there are many derivatives (and potential derivatives thereof) of such metabolites in Nature in different plant species, and other organisms. Note, though, that many of these potential lignan-forming monomers are frequently present in near-trace-amount quantities (i.e. <0.001%). Nevertheless, the description of "***potential***

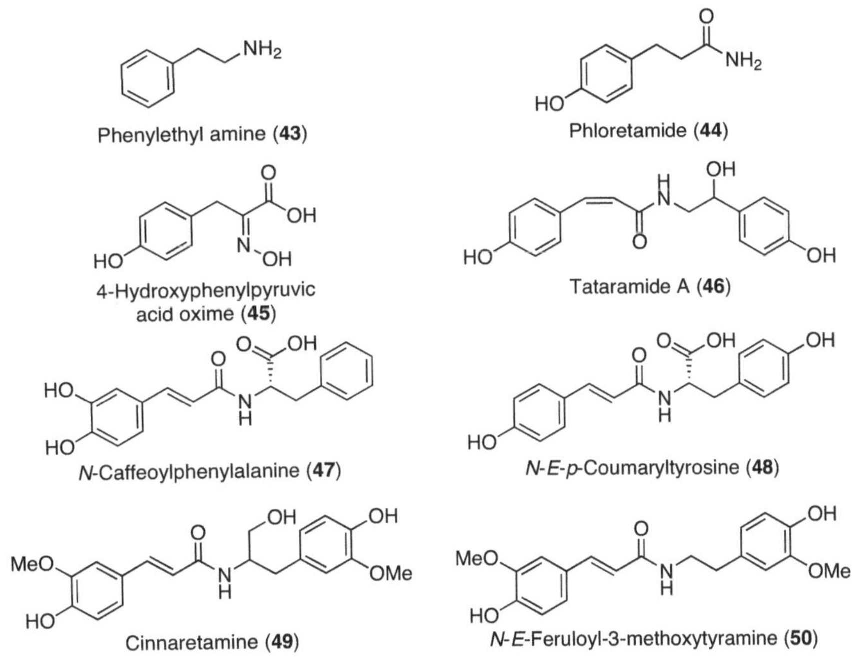

FIGURE 2.4 Examples of *N*-containing phenylpropanoids or C_6C_3 derivatives thereof most of which could generate various alkaloid lignan skeleta through derivatization and/or homo- or hetero-coupling with C_6C_3 (derived) entities.

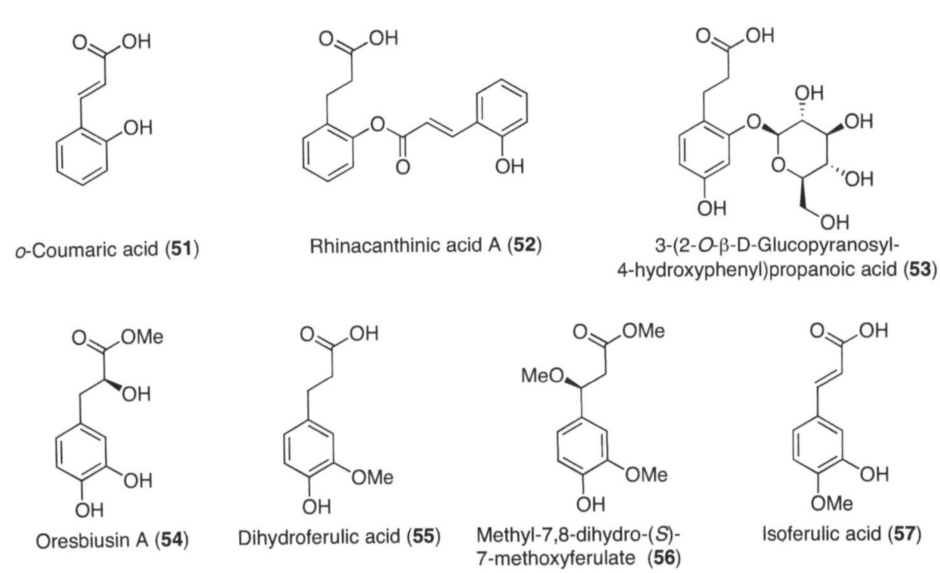

FIGURE 2.5 Examples of hydroxycinnamic acid (derivatives) that could generate various lignan skeleta through homo- or hetero-coupling with C_6C_3 (derived) entities. Note: numbering systems are per the published work.

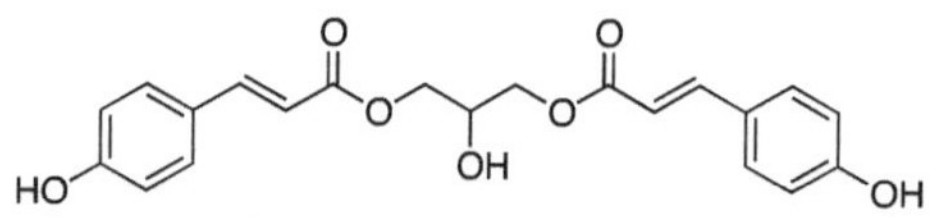
1,3-*O*-di-*E*-*p*-coumaroylglycerol (**58**)

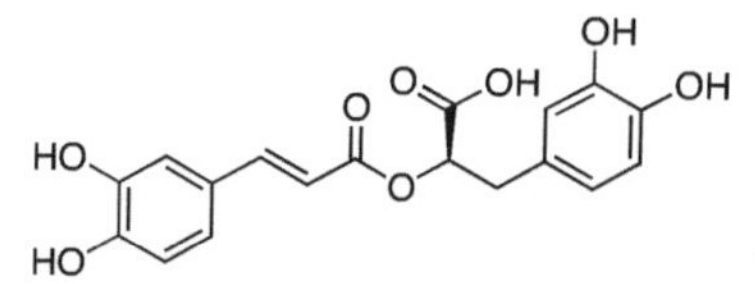
Rosmarinic acid (**59**)

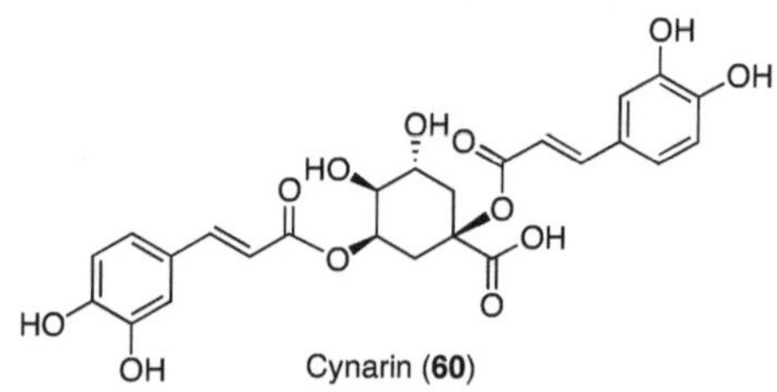
Cynarin (**60**)

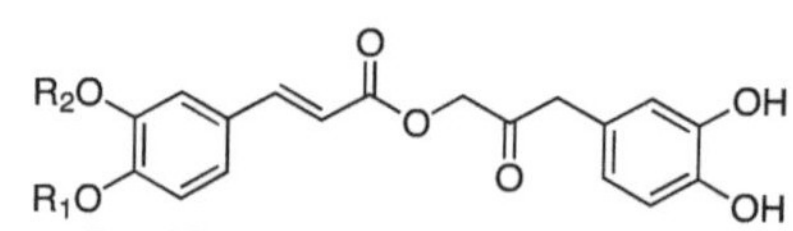
R_1 = Me, R_2 = H: Cimiracemate A (**61**)
R_1 = H, R_2 = Me: Cimiracemate B (**62**)

FIGURE 2.6 Examples of hydroxycinnamic acid esters that could generate various lignan skeleta through homo- or hetero-coupling with C_6C_3 (derived) entities.

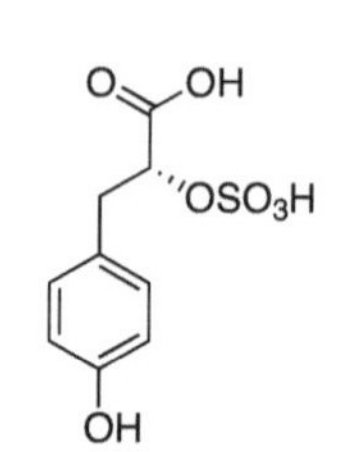
Tichocarpol A (**63**)

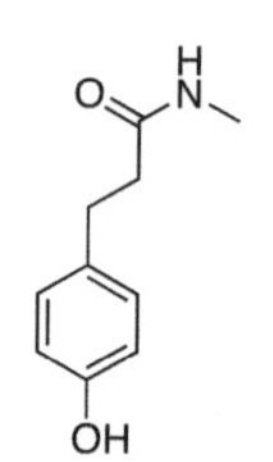
N-Methylphloretamide (**64**)

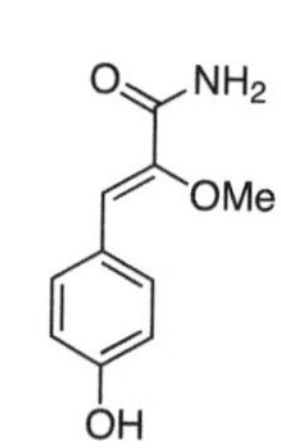
Botryllamide L (**65**)

FIGURE 2.7 Examples of C_6C_3 monomeric natural products potentially able to undergo homo- or hetero-coupling to produce (mixed biochemical pathway) lignan skeleta.

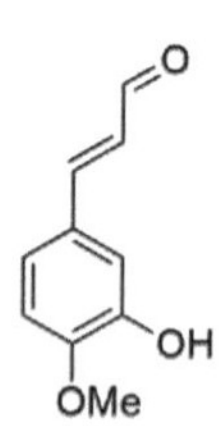
Isoconiferaldehyde (**66**)

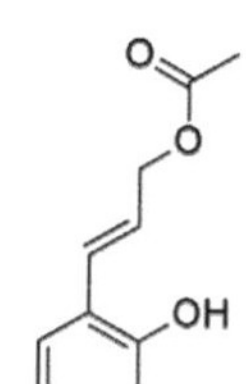
Cryptamygin A (**67**)

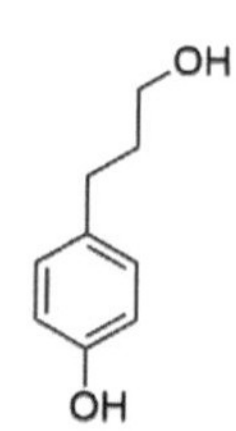
Dihydro-*p*-coumaryl alcohol (**68**)

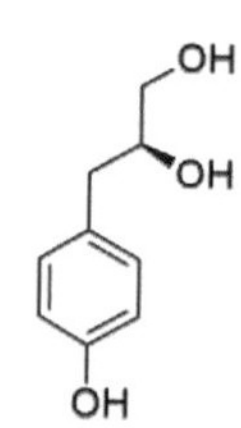
3-(*p*-Hydroxyphenyl)-1,2-propanediol (**69**)

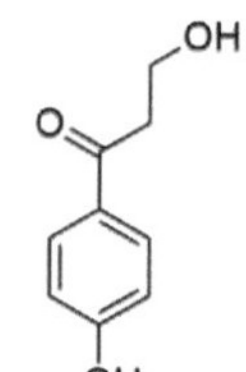
3,4′-Dihydroxy-propiophenone (**70**)

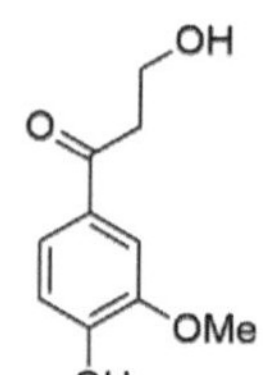
3′,4-Dihydroxy-3-methoxy-propiophenone (**71**)

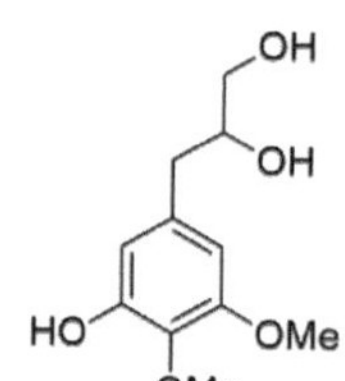
Todatriol (**72**)

FIGURE 2.8 Examples of phenolic monomers with potential to form lignan skeleta through homo- or hetero-coupling. Note: numbering systems are per the published work.

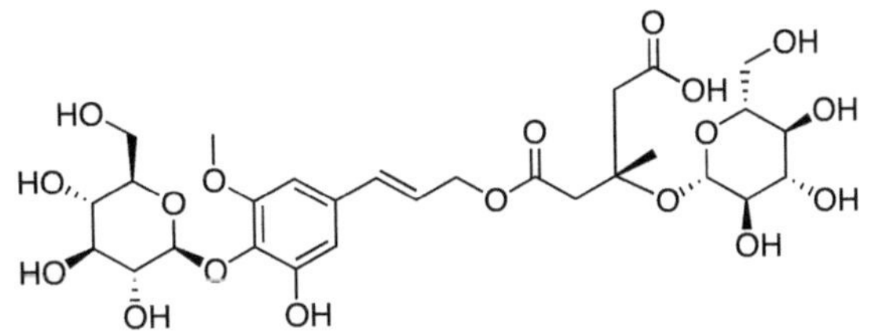

Barbatoside A (**73**)

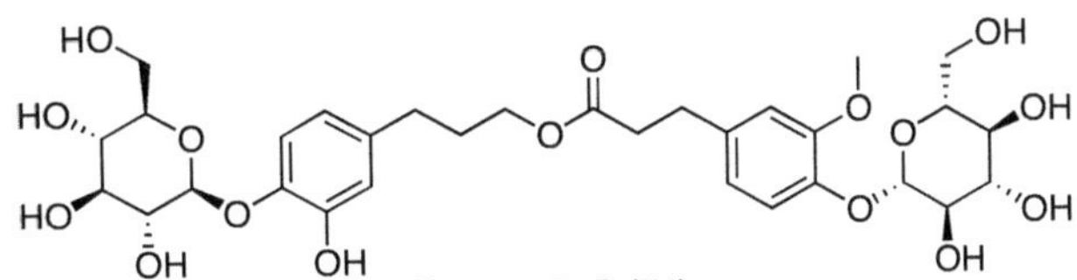

Strangusin B (**74**)

FIGURE 2.9 Examples of potential lignan "monomers" if undergoing homo- or hetero-coupling to afford mixed biochemical pathway lignan skeleta.

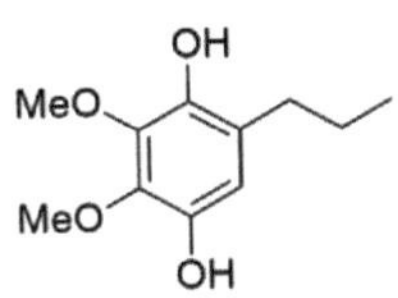

2,3-Dimethoxy-5-propyl-1,4-benzenediol (**75**)

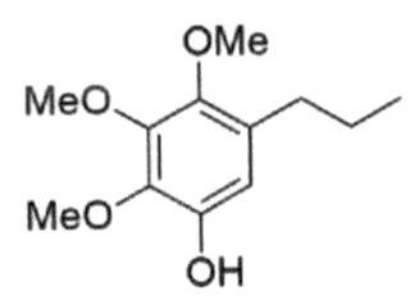

Bruguierol D (**76**)

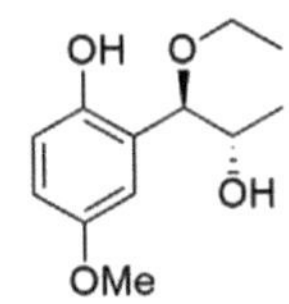

Thellungianin A (**77**)

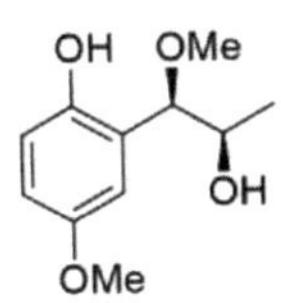

Thellungianin E (**78**)

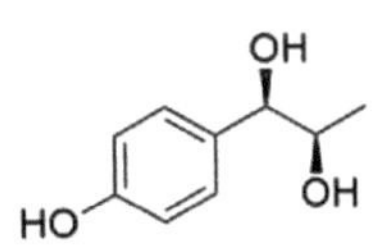

threo 4-Hydroxyphenyl propan-7,8-diol (**79**)

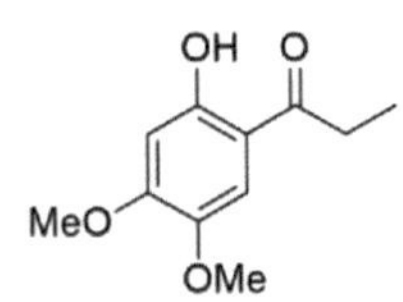

1-(2-Hydroxy-4,5-dimethoxyphenyl)-1-propanone (**80**)

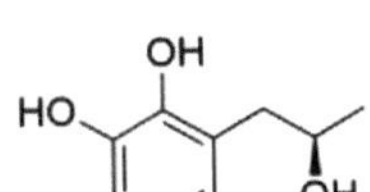

(2*R*) 3-(2-Hydroxypropyl)-1,2-benzenediol (**81**)

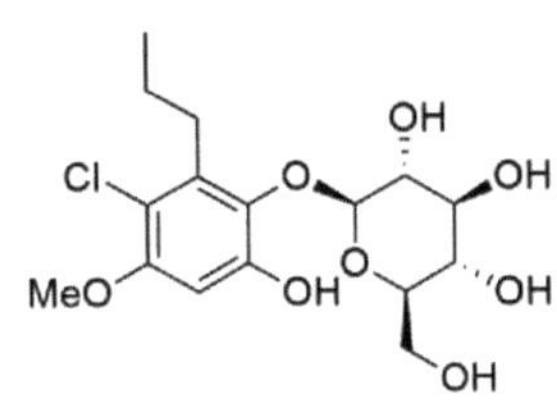

Globosumoside A (**82**)

FIGURE 2.10 Examples of allylphenol-/propenylphenol-derived metabolites with potential to form lignan skeleta through homo- or hetero-coupling. Note: numbering systems are per the published work.

lignan-forming monomeric precursors" is purposefully broadened here. These include not just the established or putative lignan building blocks proper in Section 2.2, but also other potential mixed biochemical pathway monomers and their biochemical pathway precursors that have been reported in Nature. These are also reported as *Handbook* entries, as some may potentially be true "building blocks" to lignans, given the broad structural diversity of lignan structures already described or being discovered. Some, however, may also be artifacts of isolation as indicated previously, but are included and considered as to how their formation might occur.

The diversity of C_6C_3-derived "monomeric" phenol or phenol-derived natural products that can be formed, through seemingly additional biochemical pathway involvement, include many that are potentially involved in coupling with other C_6C_3 moieties to generate lignan skeleta. If such moieties underwent additional homo- or hetero-coupling reactions with C_6C_3-derived metabolites to produce lignan skeleta (including as lignan (2 × C_6C_3) substructures in the overall natural product), for example, these would be classified as lignans or norlignans from mixed biochemical pathway origins. As described below, potential C_6C_3 precursors to "mixed" biochemical pathway lignans span not only kingdoms of vascular plants, but also fungi, algae, bacteria, and animals.

A potential lignan building block monomer might also even have Phe (**1**) or phenylethyl amine (**43**) covalently linked to the C_6C_3 phenolic compound. Alternatively, in the latter case, such a C_6C_3 phenolic compound might be linked to some other covalently linked entity that is not necessarily phenylpropanoid or allylphenol/propenylphenol derived. Such moieties, if undergoing either homo-coupling through their C_6C_3-derived substructures or hetero-coupling with another C_6C_3 phenylpropanoid (for some examples, see Figure 2.2), could produce molecules with lignan skeleta derived from mixed biochemical pathway origins.

Examples such as these could potentially include the following, with or without covalent attachment to other C_6C_3 moieties prior to coupling. These include a number of N-containing monomeric natural products: the amide phloretamide (**44**) from *Malus domestica* root (pressure) exudate (Rybicka 1984), 4-hydroxyphenylpyruvic acid oxime (**45**) from the marine sponge *Hymeniacidon sanguinea* (Cimino et al. 1975), tataramide A (**46**) from the roots of the traditional Chinese medicinal plant *Acorus tatarinowii* (Wang et al. 1997), *N*-caffeoylphenylalanine (**47**) from fronds of the lady fern *(Athyrium filix-femina)* (Adam 1995), *N-E-p*-coumaroyltyrosine (**48**) from cacao beans *(Theobroma cacao)* (Sanbongi et al. 1998), cinnaretamine (**49**) from stems of *Cinnamomum reticulatum* Hay (Chen and Yeh 2011), and *N-E*-feruloyl 3-methoxytyrosine (**50**) in seed balls of sugar beet (*Beta vulgaris* L. var. *saccharifera*) (Chiji et al. 1984) and the stems of *Cinnamomum burmannii* (Hong et al. 2011) (Figure 2.4). Clearly, each of these could produce alkaloid lignan (like) skeleta (see Moinuddin et al. 2022 for further discussion).

The same consideration can also potentially be extended to many other natural products. For illustration purposes, some examples include *o*-coumaric acid (**51**), which occurs in many plant species [e.g. sweet clover (*Melilotus officinalis* (Kosuge and Conn 1959), *Betula davurica* (Ibrahim and Towers 1960) and Persian lime (*Citrus latifolia*) (Ledesma-Escobar et al. 2018)]; rhinacanthinic acid A (**52**) from stems of the Thai medicinal plant *Rhinacanthus nasutus* (Thongchuai et al. 2015); 3-(2-*O*-β-D-glucopyranosyl-4-hydroxyphenyl) propanoic acid (**53**) from the leaves of *Peucedanum japonicum* (Hisamoto et al. 2004); oresbiusin A (**54**) from *Isodon oresbius* (Huang et al. 1996); dihydroferulic acid (**55**) from the orchid *Bulbophyllum vaginatum* (Leong et al. 1999); methyl-7,8-dihydro-(*S*)-7-methoxyferulate (**56**) from the roots of *Ligularia dentata* (Baba et al. 2007), and isoferulic acid (**57**) from the rhizomes of *Cimicifuga dahurica* (Liu et al. 1999) (Figure 2.5). All of these C_6C_3 monomers, if undergoing homo- or hetero-coupling, would produce lignan skeleta (including of mixed biochemical pathway origin).

Other examples of potential lignan-forming "monomeric precursors" of mixed biochemical pathway origin include 1,3-*O*-di-*E*-*p*-coumaroylglycerol (**58**) from the aerial parts of *Greigia sphacelata* (Flagg et al. 2000) and the endophytic fungus *Epichloe typhina* (Koshino et al. 1988); rosmarinic acid (**59**) from *Rosmarinus officinalis* (Scarpati and Oriente 1958) and *Coleus blumei* (Razzaque and Ellis 1977); cynarin (**60**) from artichoke (*Cynara scolymus*) (Panizzi and Scarpati 1954, 1965, Horman et al. 1984), and cimiracemates A (**61**) and B (**62**) from rhizomes of black cohosh (*Cimicifuga racemosa*) (Chen et al. 2002) (Figure 2.6).

Additional examples are in algae, bacteria, and sea animals (sponges and sea squirts). These include tichocarpol A (**63**) from the red alga (*Tichocarpus crinitus*) (Ishii et al. 2004); *N*-methylphloretamide (**64**) isolated from the culture of the bacterium *Micromonospora* sp. P1068 (Gutierrez-Lugo et al. 2005); 4-hydroxyphenylpyruvic acid (**9**) and its *p*-sulfoxy derivative from the red alga *Ceratodictyon spongiosum*, and its sponge symbiont *Haliclona cymaeformis* (Bugni et al. 2002), as well as numerous other plant species; and botryllamide L (**65**) from the sea squirt, *Aplidium altarium* (Yin et al. 2010) (Figure 2.7).

Comparable potential monomeric lignan precursor considerations can be given to the *p*-hydroxycinnamaldehydes **21–25**, as well as related natural products such as isoconiferyl aldehyde (**66**) from the stems of *Euonymus acanthocarpus* (Zhu et al. 2012), *Jatropha curcas* seeds (Yao et al. 2012), and *Tamarix nilotica* roots (Barakat et al. 1987) (Figure 2.8).

Other monolignols or derivatives thereof as potential lignan monomeric precursors are also encountered. These can include, but are not limited to, metabolites such as cryptamygin A (**67**) from the bark of *Cryptocarya amygdalina* (Chan et al. 2002); dihydro-*p*-coumaryl alcohol (**68**) from various plant species including gymnosperms, e.g. *Pinus sylvestris* (bark) (Oksanen 1966) and *Taxus baccata* (needles) (Das et al. 1993), as well as from the angiosperm *Zanthoxylum cuspidatum* (bark) (Ishii et al. 1976); 3-(*p*-hydroxyphenyl)-1,2-propanediol (**69**) from the bird's nest fungus *Nidula niveo-tomentosa* (Ayer and Singer 1980) and *Abies delavayi* (Yang et al. 2014); 3,4′ dihydroxypropiophenone (**70**) from *Betula platyphylla* var. *latifolia* leaves (Lee et al. 1992), as well as *Carissa edulis* roots (Achenbach et al. 1983) and *Pinus armandii* bark (Fang et al. 1991); 3′,4-dihydroxy-3-methoxypropiophenone, also called ω-hydroxypropioguaiacone (**71**), from the aerial parts of *Mikania rimachii* and *M. microptera* (Diaz et al. 1992) and stems of *Bauhinia manca* (Achenbach et al. 1988), and todatriol (**72**) in *Todaroa aurea* (González et al. 1988) (Figure 2.8), as well as barbatoside A (**73**) from *Campanula barbata* (Cuendet et al. 2001) and strangusin B (**74**) from *Serratula strangulata* (Dai et al. 2002) (Figure 2.9). As noted before, different lignan skeleta could potentially be obtained through homo- or hetero-coupling, either biochemically or even chemically or photochemically, as artifacts.

In a very similar manner, numerous lignan skeleta can also be obtained from allylphenols/propenylphenols, such as chavicol (**39**), eugenol (**40**), *p*-anol (**41**), and isoeugenol (**42**) (Figure 2.3). However, there also many derivatives or analogs of these simple allylphenols/propenylphenols (Figure 2.10). Examples include 2,3-dimethoxy-5-propyl-1,4-benzenediol (**75**) from the pyrenomycete (fungus) *Camarops microspora* (Volc et al. 1977); bruguierol D (**76**) from the mangrove tree, *Bruguiera*

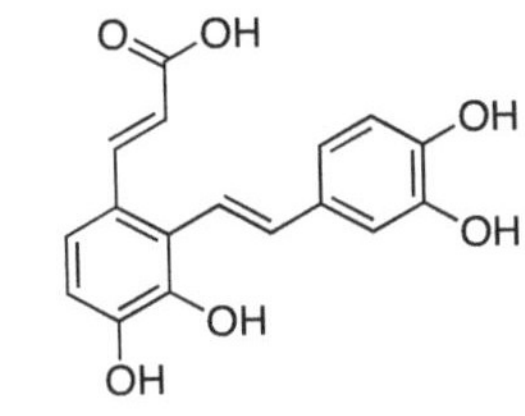

Salvianolic acid F (**83**)

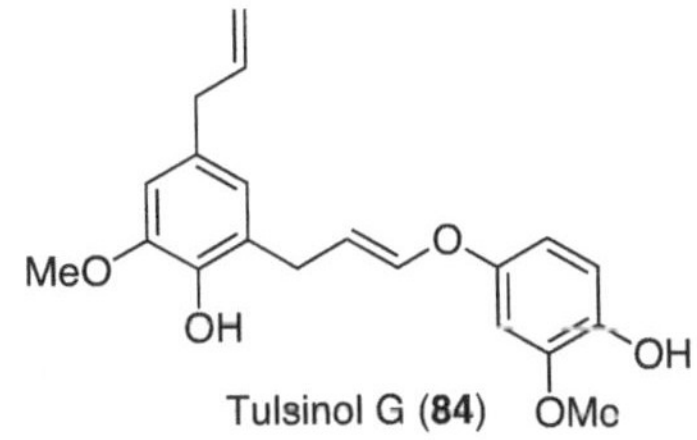

Tulsinol G (**84**)

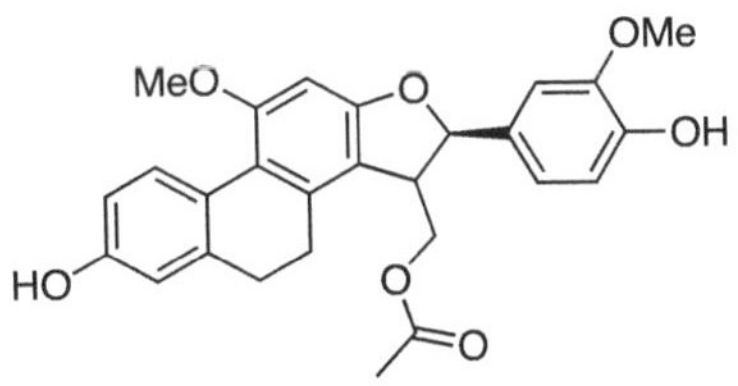

(–)-Shanciol G (**85**)

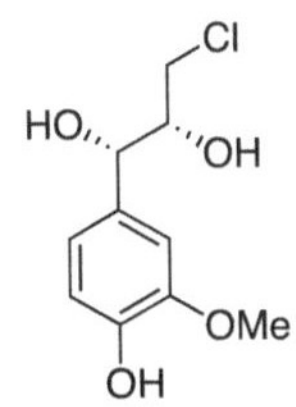

threo-3-Chloro-1-(4-hydroxy-3-methoxyphenyl)-1,2-propanediol (**86**)

FIGURE 2.11 Examples of various phenol-derived metabolites with potential to form mixed pathway lignan skeleta through homo- or hetero-coupling. Note: numbering systems are per the published work.

gymnorrhiza (Han et al. 2007); thellungianin A (**77**) and E (**78**) from *Pimpinella thellungiana* roots (Qiao et al. 1997, Li et al. 1998); *threo* 4-hydroxyphenylpropan-7,8-diol (**79**) from *Narvalina domingensis* (Balboul et al. 1996); 1-(2-hydroxy-4,5-dimethoxyphenyl)-1-propanone (**80**) from *Trigonostemon howii* stems (Ma et al. 2017); (2*R*)-3-(2-hydroxypropyl)-1,2-benzenediol (**81**) from the endophytic fungus *Lecythophora* sp. isolated from *Alyxia reinwardtii* (Sugijanto et al. 2009), and globosumoside A (**82**) from the fungus *Chaetomium elatum* (Chen et al. 2012). All of these structural moieties could potentially serve as lignan precursors.

The diversity of C_6C_3-derived phenols that could potentially undergo homo- or hetero-coupling to generate lignan skeleta of mixed biochemical pathway origin can further vary extensively. Some examples include salvianolic acid F (**83**) from roots of *Salvia miltiorrhiza* (Lu and Foo 2002) and stems of *Tournefortia sarmentosa* (Lin et al. 2002), tulsinol G (**84**) from the Nepalese medicinal plant, tulsi (*Ocimum sanctum*) (Suzuki et al. 2009), (–)-shanciol G (**85**) from tubers of *Pleione bulbocodioides* (Liu et al. 2008), and the chlorinated *threo*-3-chloro-1-(4-hydroxy-3-methoxyphenyl)-1,2-propanediol (**86**) in berries of *Pimenta dioica* (Kikuzaki et al. 1999) (Figure 2.11).

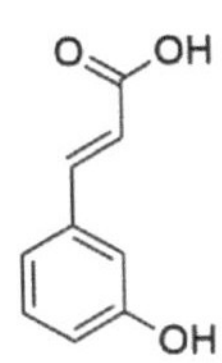

E-*m*-Coumaric acid (**87**)

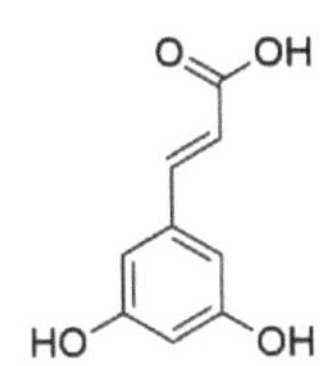

3,5-Dihydroxycinnamic acid (**88**)

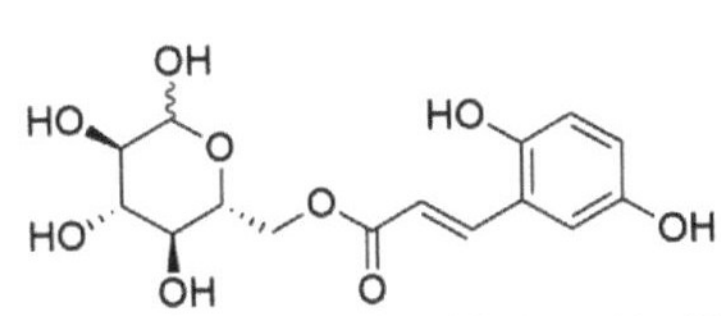

2,5-Dihydroxycinnamic acid glucoside (**89**)

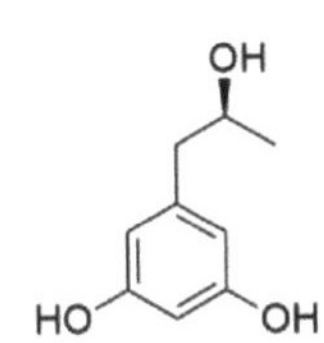

Orcinotriol (**90**)

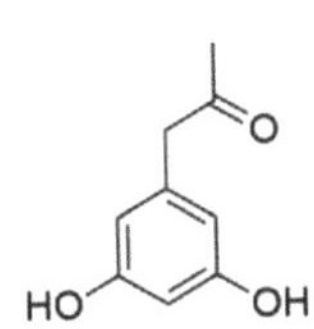

α-Acetylorcinol (**91**)

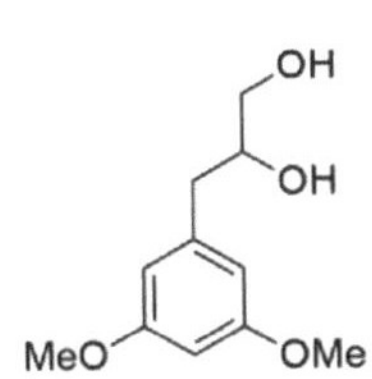

Todadiol (**92**)

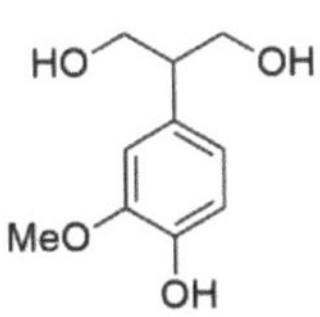

Junipediol A (**93**)

FIGURE 2.12 Examples of various monomeric phenol-derived metabolites with unusual hydroxylation patterns on their aromatic rings, but with potential to form mixed pathway lignan skeleta through homo- or hetero-coupling.

Other potential monomeric lignan precursors include reported natural products with unusual hydroxylation patterns on their aromatic rings (relative to those shown in Figure 2.2). This includes metabolites such as *E*-*m*-coumaric acid (**87**) in roots/leaves of *Citrus* sp. (Feldman and Hanks 1965); 3,5-dihydroxycinnamic acid (**88**) from the fruits of *Acanthopanax sessiliflorus* (In et al. 2012); 2,5-dihydroxycinnamic acid glucoside (**89**) from leaves of *Grevillea robusta* (Yamashita-Higuchi et al. 2012); orcinotriol (**90**) from the marine yeast *Aureobasidium pullulans* (Shigemori et al. 1998); α-acetylorcinol (**91**) from the fungus *Cochliobolus lunata* (Nukina and Marumo 1977); todadiol (**92**) from *T. aurea* (González et al. 1988), as well as the group of metabolites whose C3 side-chain seems to have undergone rearrangement, such as reportedly with junipediol A (**93**) from *Juniperus phœnicea* leaves (Comte et al. 1997) (Figure 2.12).

In sum, there are thus many, many possibilities to potentially produce different lignan skeleta, via mixed chemical pathways, at least conceptually, even if only as a lignan substructure in the overall product being formed. If formed, these are envisaged to occur via (in general) phenoxy radical coupling processes, either constitutively, inducibly, or even as artifacts of isolation protocols. Perhaps significantly, however, evidence for the presence of lignans proper in bacteria, yeast, fungi, algae, and animals is sparse to non-existent, suggesting strongly that the biochemical machinery to further process these reported monomeric precursors into lignans is absent in such life forms.

ACKNOWLEDGEMENTS

The authors thank the Chemical Sciences, Geosciences and Biosciences Division, DOE Office of Basic Energy Sciences (DE-FG-0397ER20259), as well as the Arthur and Katie Eisig-Tode Endowment.

REFERENCES

Achenbach, H., R. Waibel, and I. Addae-Mensah. 1983. Lignans and other constituents from *Carissa edulis*. *Phytochemistry* 22:749–753.

Achenbach, H., M. Stöcker, and M. A. Constenla. 1988. Flavonoid and other constituents of *Bauhinia manca*. *Phytochemistry* 27:1835–1841.

Adam, K.-P. 1995. Caffeic acid derivatives in fronds of the lady fern (*Athyrium filix-femina*). *Phytochemistry* 40:1577–1578.

Ayer, W. A., and P. P. Singer. 1980. Phenolic metabolites of the bird's nest fungus *Nidula niveo-tomentosa*. *Phytochemistry* 19:2717–2721.

Baba, H., Y. Yaoita, and M. Kikuchi. 2007. Constituents of the roots of *Ligularia dentata* Hara. *J. Nat. Med.* 61:472–473.

Balboul, B. A. A. A., A. A. Ahmed, H. Otsuka, and A. De Adams. 1996. 4-Hydroxyphenylpropan-7,8-diols and derivatives from *Narvalina domingensis*. *Phytochemistry* 42:1191–1193.

Barakat, H. H., M. A. M. Nawwar, J. Buddrus, and M. Linscheid. 1987. Niloticol, a phenolic glyceride and two phenolic aldehydes from the roots of *Tamarix nilotica*. *Phytochemistry* 26:1837–1838.

Bugni, T. S., G. P. Concepcíon, G. C. Mangalindan, et al. 2002. *p*-Sulfooxyphenylpyruvic acid from the red macro alga *Ceratodictyon spongiosum* and its sponge symbiont *Haliclona cymaeformis*. *Phytochemistry* 60:361–363.

Chan, Y.-Y., C.-H. Wu, S.-J. Wu, and T.-S. Wu. 2002. The constituents and synthesis of cryptamygin-A from the stem bark of *Cryptocarya amygdalina*. *J. Chin. Chem. Soc.* 49:263–268.

Chen, C.-Y., and H.-C. Yeh. 2011. A new amide from the stems of *Cinnamomum reticulatum* Hay. *Nat. Prod. Res.* 25:26–30.

Chen, G.-D., Y.-J. Li, H. Gao, et al. 2012. New azaphilones and chlorinated phenolic glycosides from *Chaetomium elatum* with caspase-3 inhibitory activity. *Planta Med.* 78:1683–1689.

Chen, S.-N., D. S. Fabricant, Z.-Z. Lu, et al. 2002. Cimiracemates A-D, phenylpropanoid esters from the rhizomes of *Cimicifuga racemosa*. *Phytochemistry* 61:409–413.

Chiji, H., T. Giga, M. Izawa, and S. Kiriyama. 1984. Two phenolic amides in the seed balls of sugar beet (*Beta vulgaris* L. var. *saccharifera* Alefeld). *Agric. Biol. Chem.* 48:1653–1654.

Cimino, G., S. De Stefano, and L. Minale. 1975. Occurrence of 4-hydroxyphenylpyruvic acid oxime in the marine sponge *Hymeniacidon sanguinea*. *Experientia* 31:756–757.

Comte, G., D. P. Allais, A. J. Chulia, J. Vercauteren, and N. Pinaud. 1997. Three phenylpropanoids from *Juniperus phœnicea*. *Phytochemistry* 44:1169–1173.

Corbin, C., S. Drouet, L. Markulin, et al. 2018. A genome-wide analysis of the flax (*Linum usitatissimum* L.) dirigent protein family: from gene identification and evolution to differential regulation. *Plant Mol. Biol.* 97:73–101.

Cuendet, M., O. Potterat, and K. Hostettmann. 2001. Flavonoids and phenylpropanoid derivatives from *Campanula barbata*. *Phytochemistry* 56:631–636.

Dai, J.-Q., Y.-P. Shi, L. Yang, and Y. Li. 2002. Two new components from *Serratula strangulata* Iljin. *Pharmazie* 57:340–342.

Das, B., M. Takhi, K. V. N. S. Srinivas, and J. S. Yadav. 1993. Phenolics from needles of Himalayan *Taxus baccata*. *Phytochemistry* 33:1489–1491.

Davin, L. B., H.-W. Wang, A. L. Crowell, et al. 1997. Stereoselective bimolecular phenoxy radical coupling by an auxiliary (dirigent) protein without an active center. *Science* 275:362–366.

Davin, L. B., J. R. Cort, C. A. Smith, et al. 2022. Dirigent protein roadmap to lignans and other vascular plant phenol classes. In *The Lignan Handbook* (N. G. Lewis and L. B. Davin eds.), Taylor & Francis, Boca Raton, London. pp 55–77.

Davin, L. B., and N. G. Lewis. 2022a. Redefining lignan/neolignan nomenclature and diversity: A historical perspective. In *The Lignan Handbook* (N. G. Lewis and L. B. Davin eds.), Taylor & Francis, Boca Raton, London. pp 1–11.

Davin, L. B., and N. G. Lewis. 2022b. 8–8′- and 8–7′-Norlignan biosynthesis roadmap. In *The Lignan Handbook* (N. G. Lewis and L. B. Davin eds.), Taylor & Francis, Boca Raton, London. pp 147–152

Díaz, J. G., V. L. Goedken, and W. Herz. 1992. Sesquiterpene lactones and other constituents of *Mikania rimachii* and *Mikania microptera*. *Phytochemistry* 31:597–608.

Fang, J.-M., W.-Y. Tsai, and Y.-S. Cheng. 1991. The constituents of the bark of armand pine. *J. Chin. Chem. Soc.* 38:61–64.

Feldman, A. W., and R. W. Hanks. 1965. Phenolic compounds in roots and leaves of four *Citrus* cultivars. *Nature* 207:985–986.

Flagg, M. L., G. A. Wächter, A. L. Davis, G. Montenegro, and B. N. Timmermann. 2000. Two novel flavanones from *Greigia sphacelata*. *J. Nat. Prod.* 63:1689–1691.

González, A. G., J. Bermejo Barrera, J. G. Díaz, L. Arancibia López, and P. P. De Paz. 1988. Distribution of secondary metabolites in two subspecies of *Todaroa aurea*. *Biochem. Syst. Ecol.* 16:641–645.

Gutierrez-Lugo, M.-T., G. M. Woldemichael, M. P. Singh, et al. 2005. Isolation of three new naturally occurring compounds from the culture of *Micromonospora* sp. P1068. *Nat. Prod. Res.* 19:645–652.

Han, L., X. S. Huang, I. Sattler, et al. 2007. Two new constituents from mangrove *Bruguiera gymnorrhiza*. *J. Asian Nat. Prod. Res.* 9:327–331.

Hisamoto, M., H. Kikuzaki, and N. Nakatani. 2004. Constituents of the leaves of *Peucedanum japonicum* Thunb. and their biological activity. *J. Agric. Food Chem.* 52:445–450.

Hong, Z.-L., J.-C. Huang, S.-Y. Kuo, and C.-Y. Chen. 2011. Amides from the stems of *Cinnamomum burmannii*. *Nat. Prod. Commun.* 6:1297–1298.

Horman, I., R. Badoud, and W. Ammann. 1984. Food-related applications of one- and two-dimensional high-resolution proton-nuclear magnetic resonance: structure and conformation of cynarin. *J. Agric. Food Chem.* 32:538–540.

Huang, H., H.-D. Sun, M.-S. Wang, and S.-X. Zhao. 1996. Phenolic compounds of *Isodon oresbius*. *J. Nat. Prod.* 59:1079–1080.

Ibrahim, R. K., and G. H. N. Towers. 1960. The identification, by chromatography, of plant phenolic acids. *Arch. Biochem. Biophys.* 87:125–128.

In, S.-J., D.-Y. Lee, K.-H. Seo, et al. 2012. Anti-oxidant activity of phenolic compounds isolated from the fruits of *Acanthopanax sessiliflorus* Seeman. *J. Appl. Biol. Chem.* 55:217–220.

Ishii, H., T. Ishikawa, S.-T. Lu, and I.-S. Chen. 1976. Studies on the chemical constituents of Rutaceous plants. XXXI. The chemical constituents of *Xanthoxylum cuspidatum* Champ. (*Fagara cuspidata* Engl.). *Yakugaku Zasshi* 96:1458–1467.

Ishii, T., T. Okino, M. Suzuki, and Y. Machiguchi. 2004. Tichocarpols A and B, two novel phenylpropanoids with feeding-deterrent activity from the red alga *Tichocarpus crinitus*. *J. Nat. Prod.* 67:1764–1766.

Kikuzaki, H., S. Hara, Y. Kawai, and N. Nakatani. 1999. Antioxidative phenylpropanoids from berries of *Pimenta dioica*. *Phytochemistry* 52:1307–1312.

Kim, K.-W., S. G. A. Moinuddin, K. M. Atwell, et al. 2012. Opposite stereoselectivities of dirigent proteins in *Arabidopsis* and *Schizandra* species. *J. Biol. Chem.* 287:33957–33972.

Kim, K.-W., C. A. Smith, M. D. Daily, et al. 2015. Trimeric structure of (+)-pinoresinol-forming dirigent protein at 1.95 Å resolution with three isolated active sites. *J. Biol. Chem.* 290:1308–1318.

Koshino, H., S. Terada, T. Yoshihara, et al. 1988. Three phenolic acid derivatives from stromata of *Epichloe typhina* on *Phleum pratense*. *Phytochemistry* 27:1333–1338.

Kosuge, T., and E. E. Conn. 1959. The metabolism of aromatic compounds in higher plants. I. Coumarin and *o*-coumaric acid. *J. Biol. Chem.* 234:2133–2137.

Laskar, D. D., O. R. A. Corea, A. M. Patten, et al. 2010. Vascular plant lignification: Biochemical and structural biology considerations of upstream aromatic amino acid and monolignol pathway biosynthesis. In *Comprehensive Natural Products II Chemistry and Biology* (L. Mander and H.-W. Lui eds.), Vol. 6. *Carbohydrates, Nucleosides and Nucleic Acids*, Elsevier, Oxford, UK. pp 541–604.

Ledesma-Escobar, C. A., F. Priego-Capote, V. J. Robles Olvera, and M. D. Luque de Castro. 2018. Targeted analysis of the concentration changes of phenolic compounds in Persian lime (*Citrus latifolia*) during fruit growth. *J. Agric. Food Chem.* 66:1813–1820.

Lee, M.-W., T. Tanaka, G.-I. Nonaka, and D.-R. Hahn. 1992. Phenolic compounds of the leaves of *Betula platyphylla* var. *latifolia*. *Arch. Pharm. Res.* 15:211–214.

Leong, Y.-W., L. J. Harrison, and A. D. Powell. 1999. Phenanthrene and other aromatic constituents of *Bulbophyllum vaginatum*. *Phytochemistry* 50:1237–1241.

Lewis, N. G., L. B. Davin, and S. Sarkanen. 1999. The nature and function of lignins. In *Comprehensive Natural Products Chemistry* (Sir D. H. R. Barton, K. Nakanishi and O. Meth-Cohn eds.), Vol. 3. *Carbohydrates and Their Derivatives Including Tannins, Cellulose and Related Lignins*, Elsevier, Oxford. pp 617–745.

Li, F.-X., C.-D. Wang, C.-F. Mi, H.-L. Shi, and B.-L. Qiao. 1998. Chemical constituents of the root of *Pimpinella thellungiana*. *Acta Bot. Sin.* 40:88–90.

Lin, Y.-L., Y.-Y. Chang, Y.-H. Kuo, and M.-S. Shiao. 2002. Anti-lipid-peroxidative principles from *Tournefortia sarmentosa*. *J. Nat. Prod.* 65:745–747.

Liu, I.-M., T.-C. Chi, F.-L. Hsu, C.-F. Chen, and J.-T. Cheng. 1999. Isoferulic acid as active principle from the rhizoma of *Cimicifuga dahurica* to lower plasma glucose in diabetic rats. *Planta Med.* 65:712–714.

Liu, X.-Q., Y.-Q. Guo, W.-Y. Gao, T.-J. Zhang, and L.-L. Yan. 2008. Two new phenanthrofurans from *Pleione bulbocodioides*. *J. Asian Nat. Prod. Res.* 10:453–457.

Lu, Y., and Y. L. Foo. 2002. Polyphenolics of *Salvia*—a review. *Phytochemistry* 59:117–140.

Ma, J., X. Yang, P. Wang, et al. 2017. Phytochemicals with NO inhibitory effects and interactions with iNOS protein from *Trigonostemon howii*. *Bioorg. Chem.* 75:71–77.

Moinuddin, S. G. A., L. B. Davin, and N. G. Lewis. 2022. Oligomeric lignan diversity *in planta*: A biochemical roadmap for the future. In *The Lignan Handbook* (N. G. Lewis and L. B. Davin eds.), Taylor & Francis, Boca Raton, London. pp 167–190

Nukina, M., and S. Marumo. 1977. α-Acetylorcinol from *Cochliobolus lunata*. *Agric. Biol. Chem.* 41:717.

Oksanen, H. 1966. Occurrence of dihydro-*p*-coumaryl alcohol in the bark of pine (*Pinus sylvestris* L.). *Suom. Kemistil. B* 39:56–57.

Panizzi, L., and M. L. Scarpati. 1954. Constitution of cynarine, the active principle of the artichoke. *Nature* 174:1062.

Panizzi, L., and M. L. Scarpati. 1965. 1,4- and 1,5-Dicaffeoylquinic acids. *Gazz. Chim. Ital.* 95:71–82.

Petersen, M., J. Hans, and U. Matern. 2010. Biosynthesis of phenylpropanoids and related compounds. *Annu. Plant Rev.* 40:182–257.

Pickel, B., M.-A. Constantin, J. Pfannstiel, et al. 2010. An enantiocomplementary dirigent protein for the enantioselective laccase-catalyzed oxidative coupling of phenols. *Angew. Chem. Int. Ed.* 49:202–204.

Qiao, B. L., C. D. Wang, C. F. Mi, et al. 1997. Separation and identification of thellungianin A and B from the root of *Pimpinella thellungiana* Wolff. *Acta Pharm. Sin.* 32:56–58.

Razzaque, A., and B. E. Ellis. 1977. Rosmarinic acid production in *Coleus* cell cultures. *Planta* 137:287–291.

Rybicka, H. 1984. 3-[*p*-Hydroxyphenyl]-propionic acid amide (phloretamide) in root exudate of *Malus*. *Biochem. Physiol. Pflanz.* 179:303–309.

Sanbongi, C., N. Osakabe, M. Natsume, et al. 1998. Antioxidative polyphenols isolated from *Theobroma cacao*. *J. Agric. Food Chem.* 46:454–457.

Scarpati, M. L., and G. Oriente. 1958. Isolation and constitution of rosmarinic acid, from *Rosmarinus officinalis*. *Ricerca Sci.* 28:2329–2333.

Seneviratne, H. K., D. S. Dalisay, K.-W. Kim, et al. 2015. Non-host disease resistance response in pea (*Pisum sativum*) pods: biochemical function of DRR206 and phytoalexin pathway localization. *Phytochemistry* 113:140–148.

Shigemori, H., M. Tenma, K. Shimazaki, and J. i. Kobayashi. 1998. Three new metabolites from the marine yeast *Aureobasidium pullulans*. *J. Nat. Prod.* 61:696–698.

Sugijanto, N. E., A. Diesel, R. Ebel, G. Indrayanto, and N. C. Zaini. 2009. Chemical constituents of the endophytic fungus *Lecythophora* sp. isolated from *Alyxia reinwardtii*. *Nat. Prod. Commun.* 4:1485–1488.

Suzuki, A., O. Shirota, K. Mori, et al. 2009. Leishmanicidal active constituents from Nepalese medicinal plant Tulsi (*Ocimum sanctum* L.). *Chem. Pharm. Bull.* 57:245–251.

Thongchuai, B., Y. Tragoolpua, P. Sangthong, and K. Trisuwan. 2015. Antiviral carboxylic acids and naphthoquinones from the stems of *Rhinacanthus nasutus*. *Tetrahedron Lett.* 56:5161–5163.

Vassão, D. G., K.-W. Kim, L. B. Davin, and N. G. Lewis. 2010. Lignans (neolignans) and allyl/propenyl phenols: biogenesis, structural biology, and biological/human health considerations. In *Comprehensive Natural Products II Chemistry and Biology* (L. Mander and H.-W. Liu eds.), Vol. 1: *Structural Diversity I*, Elsevier, Oxford, UK. pp. 815–928.

Volc, J., P. Sedmera, K. Roy, V. Sasek, and J. Vokoun. 1977. Two antibiotic benzoquinone-hydroquinone pairs from the pyrenomycete *Camarops microspora* (Karst.) Shear. *Collect. Czech. Chem. Commun.* 42:2957–2961.

Wang, M. F., A. N. Lao, and H. C. Wang. 1997. Two new amides from the roots of *Acorus tatarinowii* Schott. *Chin. Chem. Lett.* 8:35–36.

Yamashita-Higuchi, Y., S. Sugimoto, K. Matsunami, and H. Otsuka. 2012. Reinvestigation of structures of robustasides B and C, and isolation of (*E*)-2,5-dihydroxycinnamic acid esters of arbutin and glucose from the leaves of *Grevillea robusta*. *Chem. Pharm. Bull.* 60:1347–1350.

Yang, X.-W., S.-M. Li, Y.-L. Li, et al. 2014. Chemical constituents of *Abies delavayi*. *Phytochemistry* 105:164–170.

Yang, Y.-N., H. Zhu, Z. Chen, et al. 2015. NMR spectroscopic method for the assignment of 3,5-dioxygenated aromatic rings in natural products. *J. Nat. Prod.* 78:705–711.

Yao, L., C. Han, G. Chen, et al. 2012. A new asymmetric diamide from the seed cake of *Jatropha curcas* L. *Fitoterapia* 83:1318–1321.

Yin, S., C. Cullinane, A. R. Carroll, R. J. Quinn, and R. A. Davis. 2010. Botryllamides K and L, new tyrosine derivatives from the Australian ascidian *Aplidium altarium*. *Tetrahedron Lett.* 51:3403–3405.

Zhu, J. X., J. Ren, J. J. Qin, et al. 2012. Phenylpropanoids and lignanoids from *Euonymus acanthocarpus*. *Arch. Pharmacal Res.* 35:1739–1747.

3 Monomeric Phenylpropanoid (C_6C_3) and Lignan (C_6C_3 Dimeric) Natural Products in Plants, Bacteria, Fungi, and Animals

Structural Diversity, Chemotaxonomy, and Roadmap

Laurence B. Davin and Norman G. Lewis

3.1 C_6C_3 MONOMER NUMBERING SYSTEMS

Phenylpropanoid-derived natural products and their derivatives are (near) ubiquitous in vascular plants, whereas their occurrence is much more limited in other life forms. They have a core C_6C_3 skeleton, generally consisting of an intact aromatic (C_6) ring with a C_3 side-chain. There are two main numbering systems commonly used for C_6C_3 monomeric natural products, although note that the comprehensive entries in this *Handbook* (Lewis and Davin 2022) also include other systems for numbering (based on the International Union of Pure and Applied Chemistry [IUPAC] system) as well as these.

The most common phenylpropanoid monomer numbering system employed is C1–C9, as primarily used here. This is exemplified with *p-E*-coumaric acid (**1**) (Figure 3.1). Alternatively, and to a much lesser extent, a C1-C6 numbering system is sometimes used for the aromatic ring, with the side-chain being depicted in terms of α, β, and γ carbons, respectively. The numbering system for the side-chain is, however, different from that for the aromatic amino acids, phenylalanine (Phe, **2**) and tyrosine (Tyr, **3**), from which they are derived. There, C1-C3 is utilized for numbering the C_3 side-chain, or α and β for the two carbons adjacent to the carboxylic acid functionality as shown.

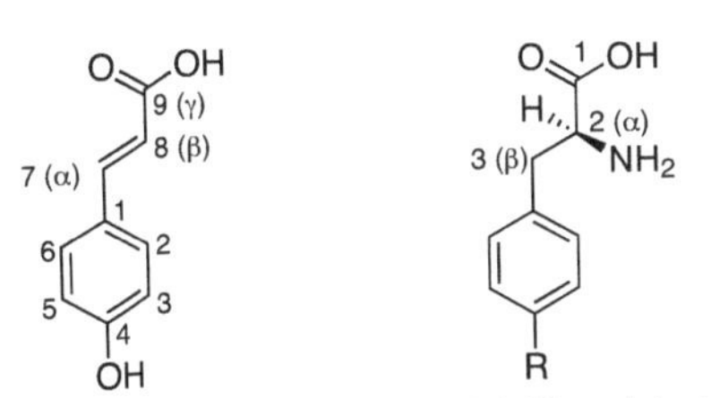

p-Coumaric acid (**1**)

R = H: L-Phenylalanine (**2**)
R = OH: L-Tyrosine (**3**)

FIGURE 3.1 Two of the numbering systems frequently used for monomeric phenylpropanoids [e.g. *p*-coumaric acid (**1**)] and aromatic amino acids Phe (2-amino-3-phenyl propanoic acid) (**2**) and Tyr (**3**).

3.2 C_6C_3 MONOMERS: COMMON VARIATIONS AND RARE C_3 SIDE-CHAIN MIGRATIONS AND REARRANGEMENTS

In terms of chemotaxonomic occurrence, the C_6C_3 natural products have a rather broad, but not uniform, pattern of distribution throughout the plant kingdom, and are found more sporadically in other kingdoms (bacteria, fungi, animals). In this contribution, the occurrence of phenylpropanoid (C_6C_3) monomers and dimers (lignans, or "mixed" biochemical pathway products, such as alkaloid lignans) are described from a class-of-organism perspective, (i.e. bacteria, fungi, plants, animals). The trivial compound names are, however, largely derived from the names (common or Latin) of the organism from which they were first isolated.

The C_6C_3 natural products are highly diverse structurally, in terms of aromatic ring substitution patterns (including mainly differing in, for example, hydroxylation, methoxylation, methylenedioxy group, and prenylation patterns) and in C_3 side-chain functionalization. The latter includes C-9 functional group variability, e.g. CH_2OH, CO_2H, CHO, or CH_3. Other variations can, for instance, be the presence or absence of a double bond at C7–C8 or at C8–C9.

To illustrate this, examples of such natural product diversity are the monomeric allylphenols with a C8–C9 terminal double bond [e.g. chavicol (**4**) and eugenol (**5**)], while propenylphenols have a C7–C8 double bond conjugated with the aromatic ring. The latter include *p*-anol (**6**) and isoeugenol (**7**), the monolignols, *p*-coumaryl (**8**), caffeyl (**9**), coniferyl (**10**), 5-hydroxyconiferyl (**11**) and sinapyl (**12**) alcohols, the corresponding hydroxycinnamic acids (**1, 13–16**) and hydroxycinnamyl aldehydes (**17–21**). The C_3 side-chain double bond can, for instance, also be reduced, e.g. as in dihydroconiferyl alcohol (**22**) or functionalized at C7 and C8, such as with ketone

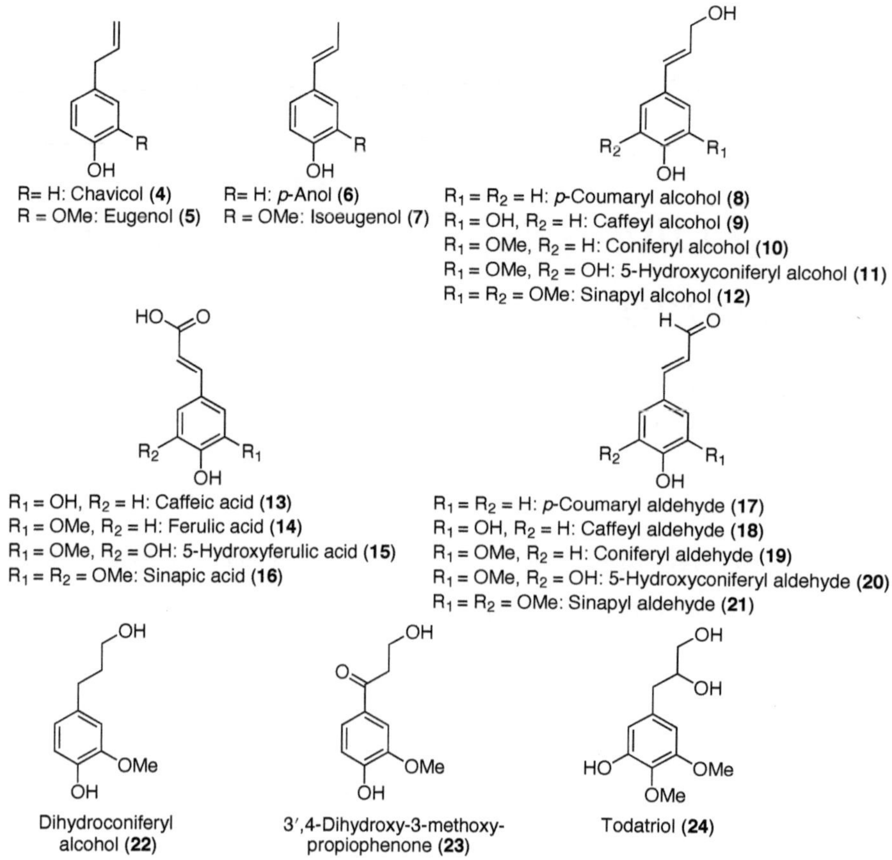

FIGURE 3.2 Examples of C_6C_3 monomer natural product diversity in the plant kingdom.

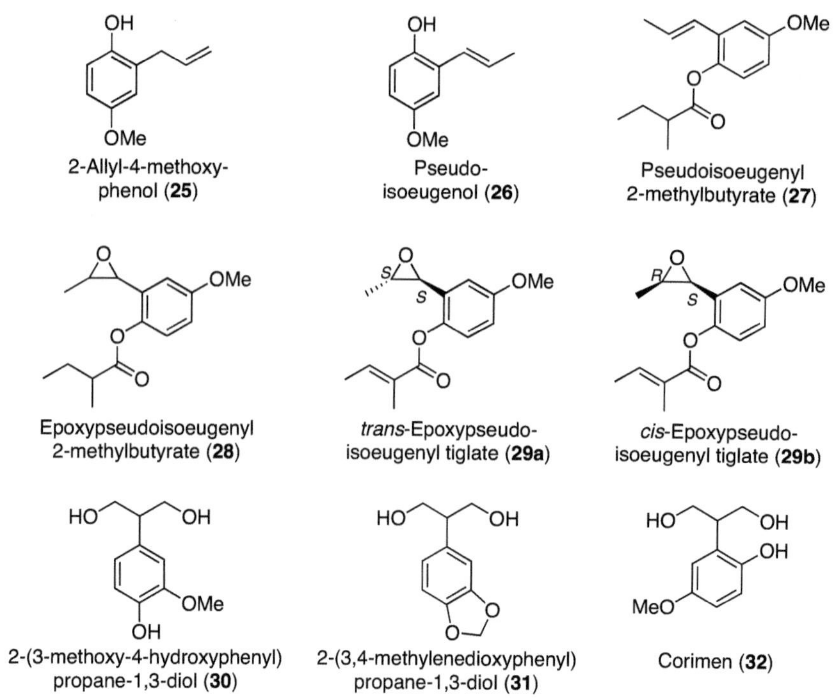

FIGURE 3.3 Additional examples of C_6C_3 monomer natural product diversity in the plant kingdom.

or hydroxyl groups as in 3′,4-dihydroxy-3-methoxypropiophenone (**23**) and todatriol (**24**), respectively (Figure 3.2).

While generally the C_6C_3 carbon skeleton is intactly maintained (Vassão et al. 2006), there are exceptions. These can include the migrated C_3 side-chain in the pseudoisoeugenols (e.g. **25–29**), which reportedly result from an NIH-shift (migration) of the C_3 side-chain (Martin and Reichling 1992) with concomitant aromatic ring hydroxylation (Figure 3.3), as well as those resulting from re-arrangement of the C_3 side-chain linkage to the aromatic ring. The latter is illustrated by 2-(3-methoxy-4-hydroxyphenyl) propane-1,3-diol (**30**) (Pérez et al. 1995), 2-(3,4-methylenedioxyphenyl)propane-1,3-diol (**31**) (Chattopadhyay et al. 1999), and corimen (**32**) (Mahjoub et al. 2012). In particular, corimen (**32**) may result from both a C3 migration and a C3 bond rearrangement *in vivo*.

Moreover, other than transient coenzyme A (CoA) intermediates in their biosynthetic pathways, there are generally no additional hetero-atoms, i.e. other than oxygen (O). Some apparently rare C_6C_3 exceptions can occur, such as with S, N, and Cl functionalization, as described in the following.

3.3 CHEMOTAXONOMICAL C_6C_3 DIVERSITY IN BACTERIA, FUNGI, ALGAE, AND ANIMALS

This section summarizes the chemotaxonomic range of C_6C_3 phenylpropanoids (monomers, lignans, i.e. with the latter herein being considered specifically as those formed by coupling of two C_6C_3 monomers), together with those reportedly present in bacteria, fungi, algae, and animals. Here, the term lignan refers to all C–C and C–O inter-unit linkage types that can be formed *in vivo* (see Vassão et al. 2010, Davin and Lewis 2022). With a few exceptions in this chapter, higher lignan oligomers (trimers with 3 C_6C_3 monomers, tetramers, etc.) are discussed in Chapter 10 on biosynthesis and occurrence/road map for the future (Moinuddin et al. 2022).

3.3.1 Bacteria

There are only very few documented examples of what appears to be a very limited extent of phenylpropanoid metabolism in bacteria, with two exceptions. The first is that of the now very routine transformation of organisms, such as *Escherichia coli*, to express various shikimate-chorismate and phenylpropanoid pathway enzymes and their enzymatic products (see for examples Kim et al. 2004, 2014, Hwang et al. 2020). The other exception is that of the so-called "mammalian lignans" produced by intestinal bacteria (discussed in Section 3.8), which are formed through digestion of dietary phenylpropanoids including lignans (Bambagiotti-Alberti et al. 1994, Heinonen et al. 2001).

As regards C_6C_3-derived natural products, some bacteria are known to contain phenylalanine ammonia lyase (PAL) and tyrosine ammonia lyase (TAL) genes and, in some cases, a few other genes/enzymes associated with a limited extent of phenylpropanoid pathway metabolism (Emiliani et al. 2009). For example, the sediment-derived bacterium *Streptomyces maritimus* reportedly produces cinnamic acid (**33**) and benzoyl CoA (**34**) (Figure 3.4) as intermediates to the polyketide antibiotic enterocin (Moore et al. 2002). Additionally, PAL/TAL activity in the actinobacterium *Saccharothrix espanaensis* produces the antibiotic saccharomicins A and B via caffeic acid (**13**) (Berner et al. 2006), and TAL is involved in forming the chromophore of the photoactive yellow protein photoreceptor in *Rhodobacter capsulatus* (Kyndt et al. 2002).

Other apparently quite rare examples of metabolites with a C_6C_3-like skeleton or C_6C_3-derived skeleton from bacteria are those of *N*-methylphloretamide (**35**), vanillic acid (**36**), and the isoflavone daidzein (**37**) from *Micromonospora* sp. P1068 (Gutierrez-Lugo et al. 2005). That report apparently provides evidence of early steps in phenylpropanoid pathway metabolism in this bacterium, at least to the isoflavone branch. It is not currently known, though, whether *N*-methylphloretamide (**35**) is formed directly from shikimate-chorismate pathway metabolism via *p*-hydroxyphenylpyruvic acid (**38**), or is derived from *p*-coumaric acid (**1**) through C7–C8 double bond reduction, amide formation, and *N*-methylation. Another unusual report is that of the presence of lignans [i.e. justicidin B (**39**)] in a sample of an Egyptian marine actinomycete *Nocardia* sp. ALAA 2000, isolated from the red alga *Laurenica spectabilis* (El-Gendy et al. 2008).

Importantly, both of these reports (Gutierrez-Lugo et al. 2005, El-Gendy et al. 2008) are so intriguing and unusual that the genome sequencing of *Micromonospora* sp. P1068 and *Nocardia* sp. ALAA 2000 is needed to further support these observations, i.e. in order to provide additional insights and further confirmation as to the extent of phenylpropanoid metabolism evolution in bacteria at the gene and enzyme levels. Until this is done, and the pathways are proven to exist in these two bacteria, it cannot yet be considered definitive that these organisms produce either isoflavones or lignans. Their presence could also be due to detritus from other plant species, for example.

Summary: There is currently no unequivocal and comprehensive evidence at the metabolite, gene, and enzyme level to support the concept that bacteria can produce downstream monomeric C_6C_3 metabolites, such as monolignols (**8–12**), allylphenols/propenylphenols (**4–7**), and lignans (regardless of inter-unit linkage type). Some bacteria, however, may have at best a very limited extent of phenylpropanoid pathway metabolism.

3.3.2 Fungi

Some fungi harbor homologues of genes (and the corresponding enzymes) associated with phenylpropanoid metabolism, with PAL in particular being reported in yeast, e.g. *Rhodotorula glutinis* (Hodgins 1971, Fritz et al. 1976) and *Sporobolomyces roseus* (Moore et al. 1968); ascomycetes, e.g. *Aspergillus* and *Neurospora* species (Sikora and Marzluf 1982, Seshime et al. 2005); the ectomycorrhizal fungus, *Amanita muscaria* (Nehls et al. 1999); and basidiomycete mushrooms,

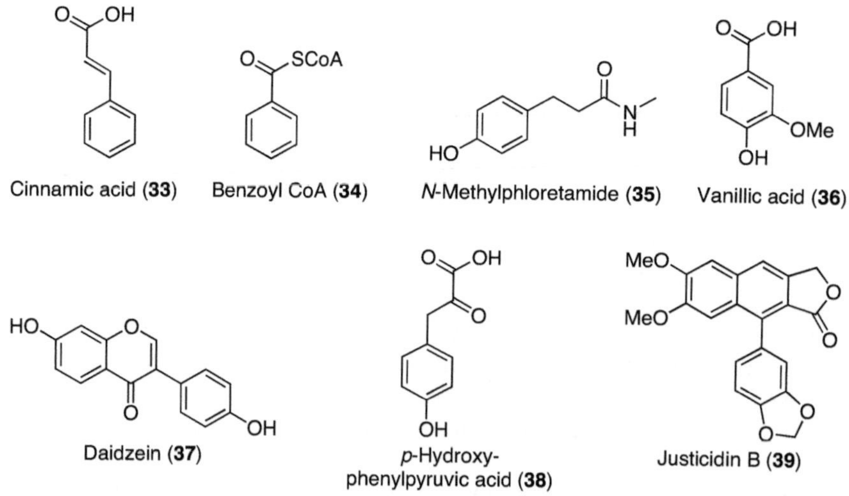

FIGURE 3.4 Various C_6C_3 and C_6C_3-derived monomers, a lignan and an isoflavone reported as present in different bacteria.

e.g., *Tricholoma matsutake* (Tasaki and Miyakawa 2015). Some fungal natural products that are reportedly derived from phenylpropanoid metabolism include styrylpyrones squarrosidine (**40**) and pinillidine (**41**) (Figure 3.5) from *Pholiota squarrosa* and *Phellinus pini*, respectively (Wangun and Hertweck 2007); brasiliamides A (**42**) and B (**43**) from the endophyte *Penicillium brasilianum* present in *Melia azedarach* root bark (Fill et al. 2009), and the putative prenylated phenylpropanoids marianin A (**44**) and marianin B (**45**) from *Mariannaea camptospora* (Fukuda et al. 2011); the latter is apparently a prenylated C_6C_2 rather than a C_6C_3 metabolite. However, whether these are biosynthesized via the general phenylpropanoid pathway has not been reported.

Other reports of C_6C_3 phenylpropanoids in fungi include 1,3-*O*-di-*E*-*p*-coumaroyl glycerol (**46**) from *Epichloe typhina* (Koshino et. 1988); 3-(*p*-hydroxyphenyl)-1,2-propanediol (**47**) from *Nidula niveo-tomentosa* (Ayer and Singer 1980); 2,3-dimethoxy-5-propyl 1,4 benzenediol (**48**) from *Camarops microspora* (Volc et al. 1977); (2*R*) 3-(2-hydroxypropyl)-1,2 benezenediol (**49**) from *Lecythophora* sp. (Sugijanto et al. 2009); the chlorinated globosumoside A (**50**) from *Chaetomium elatum* (Chen et al. 2012); orcinotriol (**51**) from the marine yeast *Aureobasidium pullulans* (Shigemori et al. 1998) and α-acetylorcinol (**52**) from *Cochliobolus lunata* (Nukina and Marumo 1977) (Figure 3.6). Globosumoside A (**50**) is quite unusual in having a chlorine in its chemical structure, particularly given that it is a root-colonizing fungus.

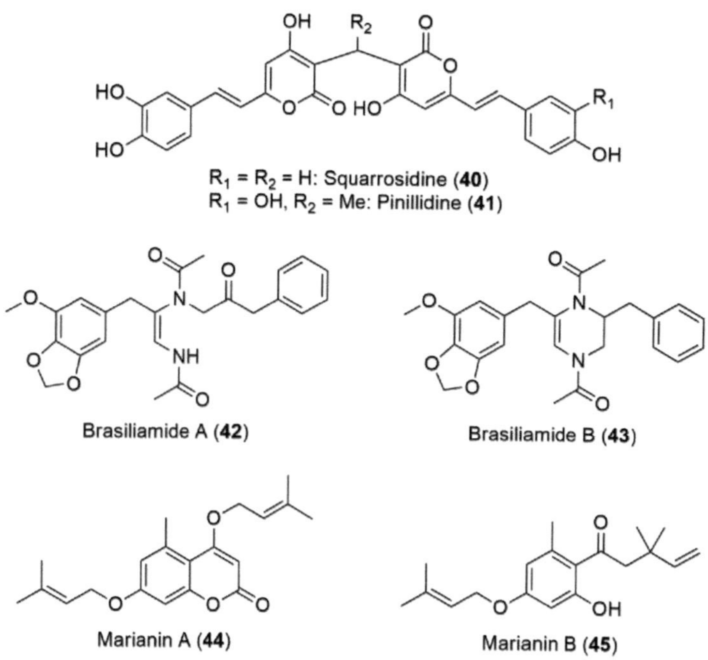

FIGURE 3.5 Examples of C_6C_3-derived phenylpropanoid natural products reported present in various fungi.

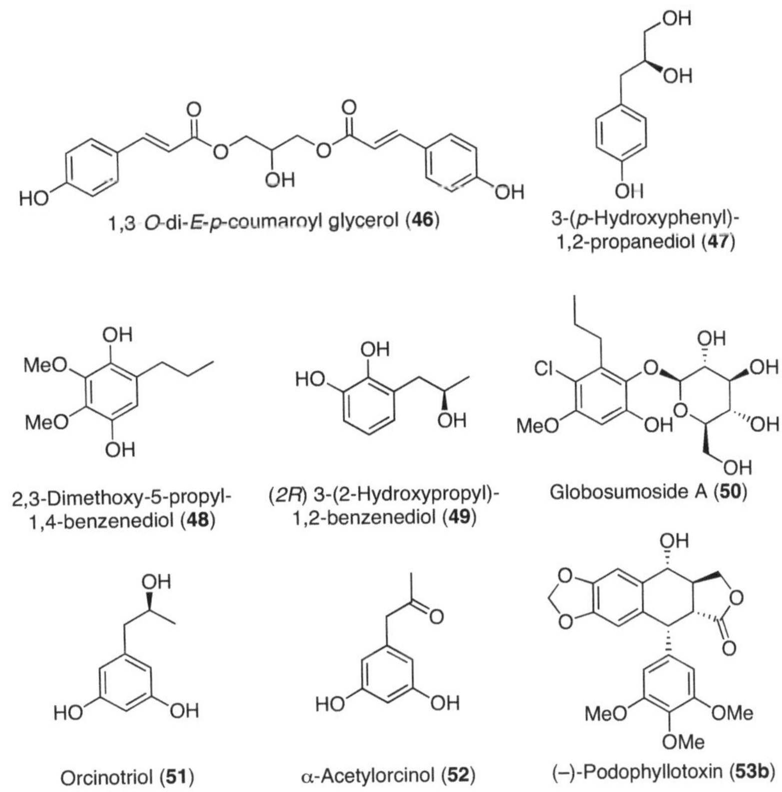

FIGURE 3.6 Further examples of C_6C_3-derived phenylpropanoids (including one lignan) reported present in various fungi. Note: Numbering systems are per published work, "b" is the (–)-enantiomer.

In terms of potential biosynthetic pathways to produce 1,3-*O*-di-*E*-*p*-coumaroyl glycerol (**46**), its structure suggests the presence of either PAL and cinnamate-4-hydroxylase (C4H) or TAL in this organism, whereas 3-(*p*-hydroxyphenyl)-1,2-propanediol (**47**) could be derived directly from *p*-hydroxyphenylpyruvic acid (**38**) or alternatively from further metabolism of *p*-coumaric acid (**1**).

Whether the other fungal natural products (**49–52**) are phenylpropanoids proper or more likely acetate/polyketide pathway derived remains to be determined; the latter is a distinct possibility given the reported aromatic ring hydroxylation patterns at C3 and C5. Nevertheless, conceptually the homo-coupling of natural products (**46–52**) or hetero-coupling with other C_6C_3 phenolics could lead to lignan skeleta (i.e. possessing C_6C_3-C_6C_3 substructures). However, to our knowledge, there are no reports to indicate that this occurs.

Some endophytic fungi are claimed to be able to produce the lignan podophyllotoxin (**53b**), e.g. *Phialocephala fortinii* and *Trametes hirsuta* isolated from the rhizomes of *Podophyllum peltatum* (Eyberger et al. 2006) and *P. hexandrum* (Puri et al. 2006), respectively, as well as *Fusarium solani* from the roots of *P. hexandrum* (Nadeem et al. 2012) and *F. oxysporum* from *Juniperus recurva* (Kour et al. 2008).

Such reports are, however, controversial (Heinig et al. 2013). They must be viewed as preliminary until there is unequivocal evidence that the biochemical pathways *in toto* truly exist in these organisms, e.g. through genome sequencing and assays of the putative enzymes involved. Until that is done, such reports, while provisionally interesting, must be considered inconclusive.

Summary: There appears to be very a limited extent of phenylpropanoid metabolism in fungi. In general, fungal phenylpropanoid pathway elaboration appears insufficient to result in monolignol (**8–12**), allylphenol/propenylphenol (**4–7**), or lignan formation.

3.3.3 Algae

There is currently no unequivocal evidence for substantial elaboration of the phenylpropanoid pathway in algae, i.e. in terms of demonstration of either genes or proteins/enzymes leading to the monolignols, allylphenols/propenylphenols, lignans (Lewis et al. 1999, Vassão et al. 2006, 2010), or polymeric lignins. Indeed, to date, there are only a few algae where PAL or TAL activities have been reported (Bouarab et al. 2004). Moreover, an *in silico* analysis of sequenced algae (as accessed at National Center for Biotechnology Information [NCBI]), only revealed fragments of putative phenylpropanoid pathway genes of low homology/identity, suggesting that complete phenylpropanoid pathways to the monolignols **8–12** and products thereof were absent (S. J. Kim, unpublished results).

Curiously, some reports have claimed the presence of polymeric lignins in algae in vanishingly small amounts, i.e. in

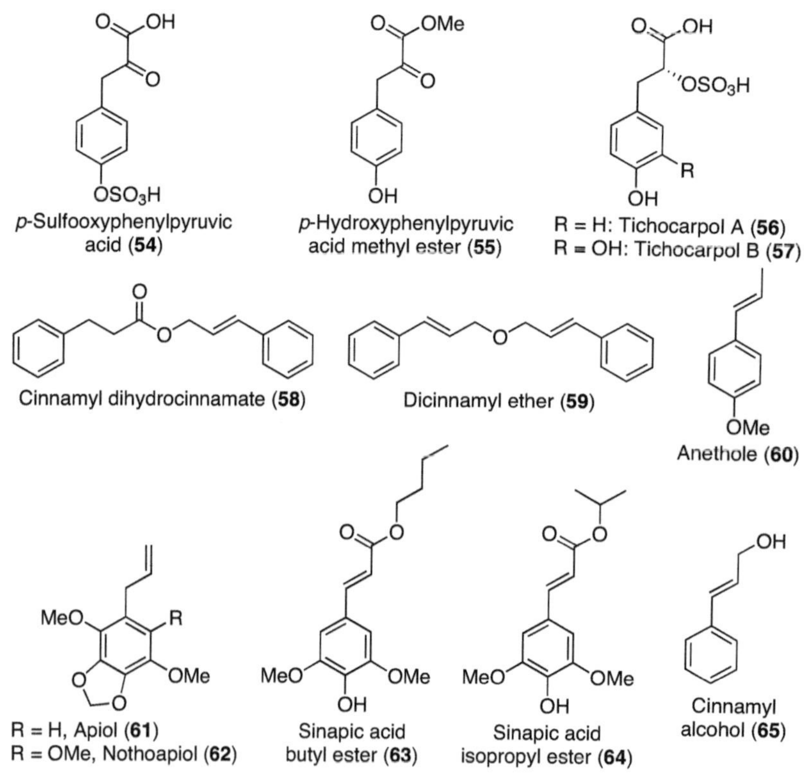

FIGURE 3.7 C_6C_3-derived phenylpropanoids reported present in red, green, and brown algae.

the brown algae *Fucus vesiculosus* (Reznikov et al. 1978) and *Cystoseira barbata* (Dovgan and Medvedeva 1983, Dovgan et al. 1983), in the coralline (calcified) red alga *Calliarthron cheilosporioides* (Martone et al. 2009), and in the freshwater green alga *Coleochaete* (Delwiche et al. 1989). However, this "algal lignin" detection may much more likely result from, for example, contaminants from terrestrial plants in the waters in which the algae were growing (Lewis et al. 1999, Laskar et al. 2010). At the very minimum, such claims of lignins in algae need to show that the intact phenylpropanoid biochemical pathways to the monolignols/lignins are present and functional (i.e. the corresponding genes and enzymes).

Yet, in spite of a lack of evidence of intact phenylpropanoid biochemical pathways in algae to the monolignols and allylphenols/propenylphenols, there have been a number of somewhat sporadic and rather preliminary reports of monomeric C_6C_3 or C_6C_3-like natural products present in green, red, and brown algae.

These include *p*-hydroxyphenylpyruvic acid (**38**) and its corresponding sulphonic acid (**54**) derivative from the red alga *Ceratodictyon spongiosum* (Bugni et al. 2002), its methyl ester (**55**) from the green alga *Caulerpa taxifolia* (Mancini et al. 1998), and the presumed sulphonated analogs tichocarpol A (**56**) and tichocarpol B (**57**) from the red alga (*Tichocarpus crinitus*) (Ishii et al. 2004) (Figure 3.7). As noted earlier, such natural products are most likely all derived from the shikimate-chorismate pathway product *p*-hydroxyphenylpyruvic acid (**38**), rather than being phenylpropanoid pathway proper derived. As noted, three of these [*p*-sulfooxyphenylpyruvic acid (**54**), tichocarpol A (**56**) and tichocarpol B (**57**)] have the unusual feature of an S atom in their molecules.

Other C_6C_3 natural products reported in trace amounts (less than 0.001%) in algae include cinnamyl dihydrocinnamate (**58**) and dicinnamyl ether (**59**) from *Caulerpa racemosa* (Anjaneyulu et al. 1992); anethole (**60**) from various green [*C. racemosa* and *Codium tomentosum* (Awad 2002b)] and red (*Jania rubens*) (Awad 2002a) algae; eugenol (**5**) and isoeugenol (**7**) from the green and red algae, *Enteromorpha compressa* and *Hypnea musciformis* (Moussa 2003), respectively, as well as apiol (**61**), nothoapiol (**62**), and the sinapic acid butyl and isopropyl esters (**63** and **64**) from the brown alga *Spatoglossum variabile* (Atta-ur-Rahman et al. 1999, Hayat et al. 2002).

The presence of cinnamyl dihydrocinnamate (**58**) and dicinnamyl ether (**59**) from *C. racemosa* (Anjaneyulu et al. 1992) would provisionally suggest a functional biochemical pathway to cinnamic acid (**33**) and perhaps to cinnamyl alcohol (**65**). However, this also needs to be established at the gene/enzyme level. Whether the other reports of allyl/propenyl phenols (**5**, **7**, **60–62**) are truly produced by these organisms requires more proof. For example, the tiny amounts of apiol (**61**) and nothoapiol (**62**) isolated from the brown alga *S. variabile* (Atta-ur-Rahman et al. 1999) growing on shoreline sublittoral rock near Karachi, Pakistan, could possibly have resulted from contaminating detritus. Furthermore, most of these reported algal C_6C_3 phenylpropanoids could also undergo homo-coupling or hetero-coupling with other

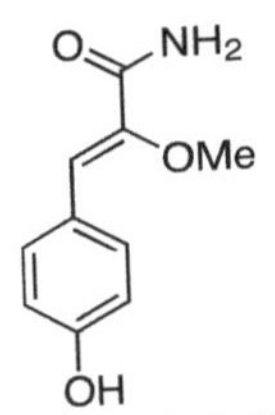

Botryllamide L (**66**)

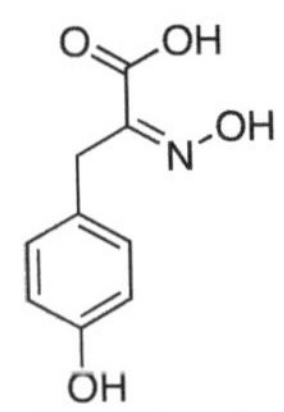

p-Hydroxyphenylpyruvic acid oxime (**67**)

FIGURE 3.8 C_6C_3 or C_6C_3-like natural products reported present in animals.

C_6C_3-derived moieties to produce various lignan skeleta. However, no lignans in algae have been reported to our knowledge.

Moreover, additional evidence in support of phenylpropanoid biochemical pathway elaboration to the allylphenol/propenylphenol metabolites in *C. racemosa*, *C. tomentosum*, *E. compressa*, *H. musciformis*, *J. rubens*, and *S. variabile* is required at the biochemical (gene and enzyme) level, as well as under controlled laboratory conditions. Such a proof would be of considerable scientific and evolutionary significance.

Summary: As regards lignans, to our knowledge, there are none reported from algae. Furthermore, it seems unlikely that any of the reported monomeric C_6C_3 algal natural products noted previously serve as precursors to lignans in these organisms.

3.3.4 Marine Animals

Other than the "mammalian lignans" derived from foodstuffs that are ingested and digested (Bambagiotti-Alberti et al. 1994, Heinonen et al. 2001) (see Section 3.8), there are only sporadic reports of C_6C_3 or C_6C_3-like natural products produced by marine animals. One is botryllamide L (**66**) from the sea squirt, *Aplidium altarium* (Yin et al. 2010), and the other is *p*-hydroxyphenylpyruvic acid oxime (**67**) from the marine sponge *Hymeniacidon sanguinea* (Cimino et al. 1975) (Figure 3.8). Based on their structures, both are likely produced directly from *p*-hydroxyphenylpyruvic acid (**38**), rather than being phenylpropanoid pathway products proper.

Summary: There is no evidence in support of animals producing monolignols, allylphenols/propenylphenols, or lignans *de novo*.

3.4 CHEMOTAXONOMICAL C_6C_3 DIVERSITY IN NON-VASCULAR PLANTS: BRYOPHYTES

The non-vascular bryophytes (liverworts, hornworts, and mosses) and later the vascular pteridophytes (lycophytes, horsetails, and ferns, Section 3.5) are considered to be some of the earliest plant divisions adapting to land. In this section, we address the phenylpropanoids (monomers and lignans) found in these different land plants, and the evolutionary elaboration of phenylpropanoid pathway metabolism from a chemotaxonomical perspective.

Bryophytes consist of three divisions, the liverworts, the hornworts, and the mosses.

3.4.1 Liverworts

Liverworts (Marchantiophyta) are considered to be the oldest division of the non-vascular land plants (bryophytes), with an estimated 4,500–5,000 extant species worldwide (Forrest et al. 2006, Crandall-Stotler et al. 2008) and whose fossil record dates back more than 475 million years (Wellman et al. 2003). Monomeric allylphenols/propenylphenols have been reported from liverworts including eugenol (**5**) and methyleugenol (**68**) in *Anthelia julacea*, *Conocephalum conicum*, *Marchesinia brachiata*, and *M. mackaii* (Asakawa et al. 1979, Asakawa et al. 1980, Gradstein et al. 1981) (Figure 3.9); α and β-asarones (**69**, **70**), as well as (*E*/*Z*)-methylisoeugenols (**72**/**73**), and methyleugenol (**68**) in *M. mackaii* (Figueiredo et al. 2002); and γ-asarone (**71**) in an unidentified Jamaican liverwort (Connolly 1985). Of the lignans reported from liverworts, only one thus far is apparently allylphenol/propenylphenol

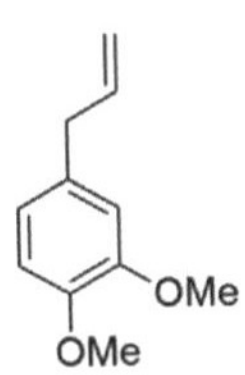

Methyleugenol (**68**)

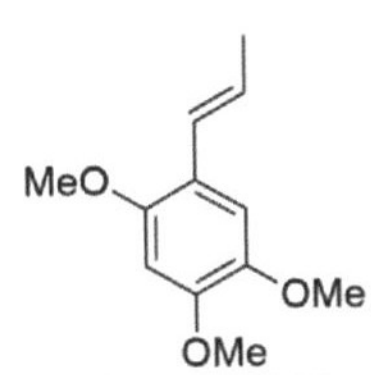

α-Asarone (**69**)

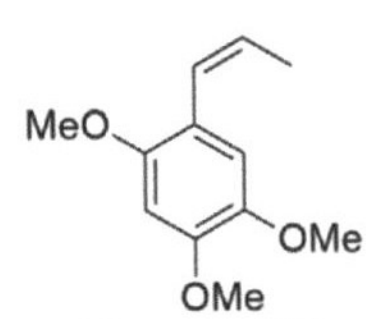

β-Asarone (**70**)

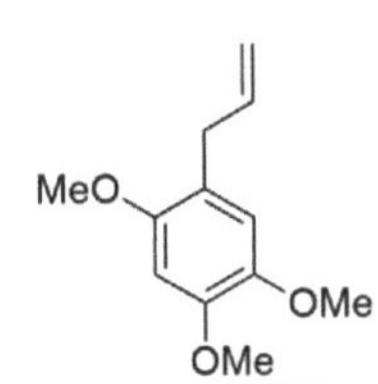

γ-Asarone (**71**)

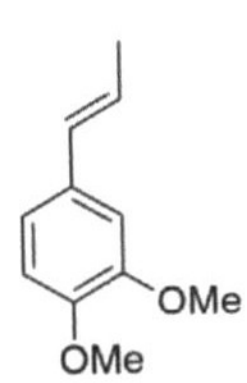

(*E*)-Methylisoeugenol (**72**)

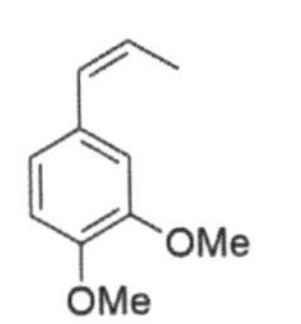

(*Z*)-Methylisoeugenol (**73**)

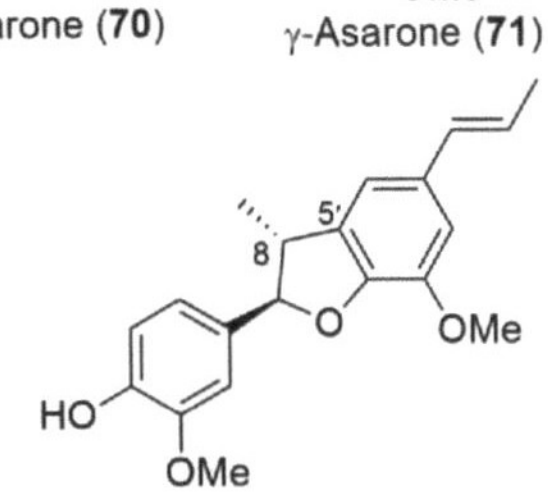

(–)-Licarin A (**74b**)

FIGURE 3.9. Liverwort C_6C_3 monomers and the lignan (–)-licarin A (**74b**), where "b" is the (–)-enantiomer.

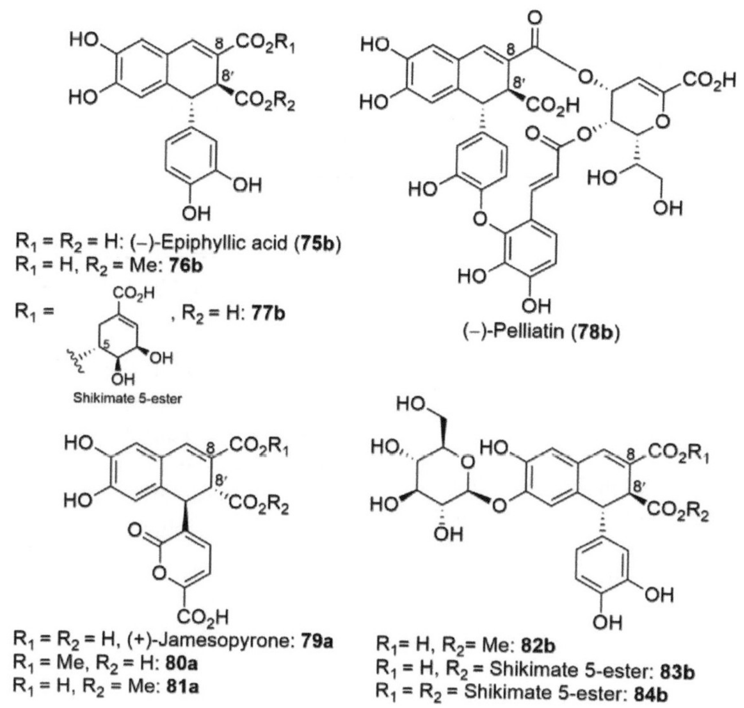

FIGURE 3.10 Liverwort aryldihydronaphthalene lignans **75–84**, where "a" refers to the (+)-enantiomers, and "b" to the (–)-antipodes.

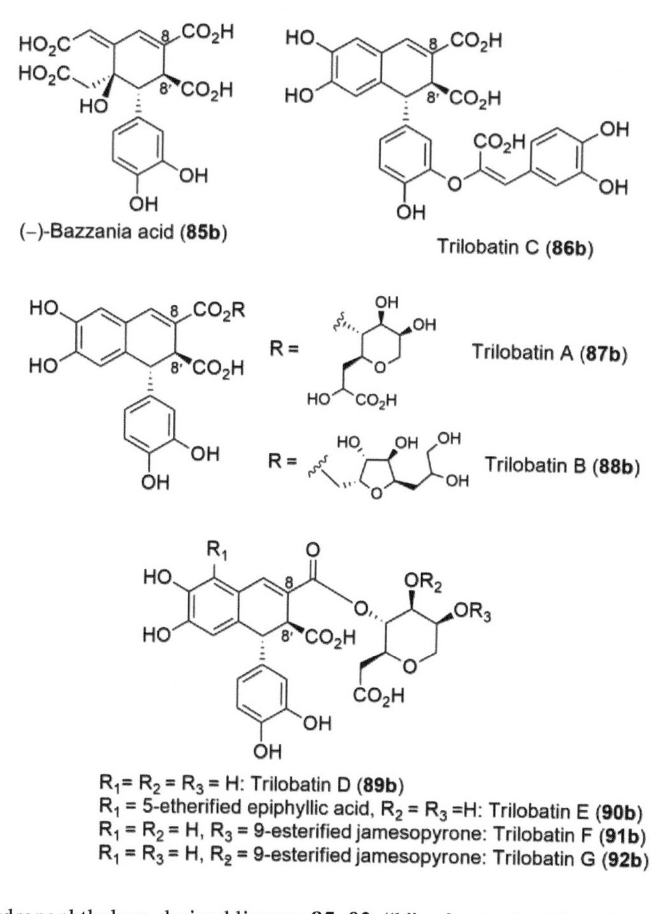

FIGURE 3.11 Liverwort aryldihydronaphthalene-derived lignans **85–92**; "b" refers to the (–)-antipodes.

(**5/7**) derived, namely the 8-5′-linked (–)-licarin A (**74b**) found in *Jackiella javanica* (Nagashima et al. 1990).

Their biochemical pathways overall suggest formation from the monolignol coniferyl alcohol (**10**) via eugenol (**5**) or isoeugenol (**7**) (Koeduka et al. 2006, Vassão et al. 2006, Vassão et al. 2007, Vassão et al. 2010). However, the α/β/γ-asarones (**69–71**) have additional methoxylation occurring at C-6, that provisionally may have occurred post-allylphenol/propenylphenol formation. Nevertheless, these C9 deoxygenated monomers (**5, 68–73**) and the phenylcoumaran lignan (–)-licarin A (**74b**) in this early basal land plant group suggests that phenylpropanoid pathway elaboration was an early feature in land plant evolution and colonization.

Liverworts mainly accumulate numerous 8–8′-linked lignans, such as aryldihydronaphthalene lignans (**75–97**) (Figures 3.10–3.12), arylnaphthalene lignans (e.g. **98**), and arylnaphthalene norlignans (**99–104**), where the carbon lost in the norlignans is that of C-9 (Figure 3.13). Provisionally, their chemical structures might suggest a biosynthetic pathway directly from caffeic acid (**13**), or if monolignol derived, then by post-coupling modifications, including demethylation, i.e. to account for their catechol and dicarboxylic acid groups in the final product.

Liverwort 8–8′-linked aryldihydronaphthalene lignans include (–)-epiphyllic acid (**75b**), its methyl/shikimate esters (**76b/77b**), and (–)-pelliatin (**78b**), respectively, from *Pellia epiphylla* (Cullmann et al. 1993, Cullmann et al. 1996). (–)-Epiphyllic acid (**75b**) is additionally found in *Lepidozia incurvata*, *Chiloscyphus polyanthos* (Cullmann et al. 1999), and *Lophocolea heterophylla* (Tazaki et al. 1999). *Jamesoniella autumnalis* accumulates (+)-jamesopyrone (**79a**) and its methyl derivatives (**80a/81a**) (Tazaki et al. 1995, Tazaki et al. 1999), whereas *Lepicolea ochroleuca* has the lignans **82b–84b** (Cullmann and Becker 1999). *Bazzania trilobata* also contains (–)-epiphyllic acid (**75b**), (+)-jamesopyrone (**79a**), as well as (–)-bazzania acid (**85b**) and trilobatins A–K (**86b–96b**), together with the corresponding 5′–5″-dimer **97b** of (–)-epiphyllic acid (**75b**) (Martini et al. 1998, Scher et al. 2003).

The arylnaphthalene lignan **98** was also reported from *J. autumnalis* (Tazaki et al. 1995), whereas the arylnaphthalene norlignans **99–104** were from *L. incurvata* (i.e. **99**, **100**), *C. polyanthos* (i.e. **99**, **101**), *Jungermannia exsertifolia* spp. *cordifolia* (i.e. **99**, **102**) (Cullmann et al. 1999), *J. autumnalis* (i.e. **99**, **103**, **104**) (Tazaki et al. 1995), and *B. trilobata* (i.e. **99**) (Martini et al. 1998), respectively. In addition to aryldihydronaphthalene and arylnaphthalene lignans, however, other interesting structural modifications occur. These include

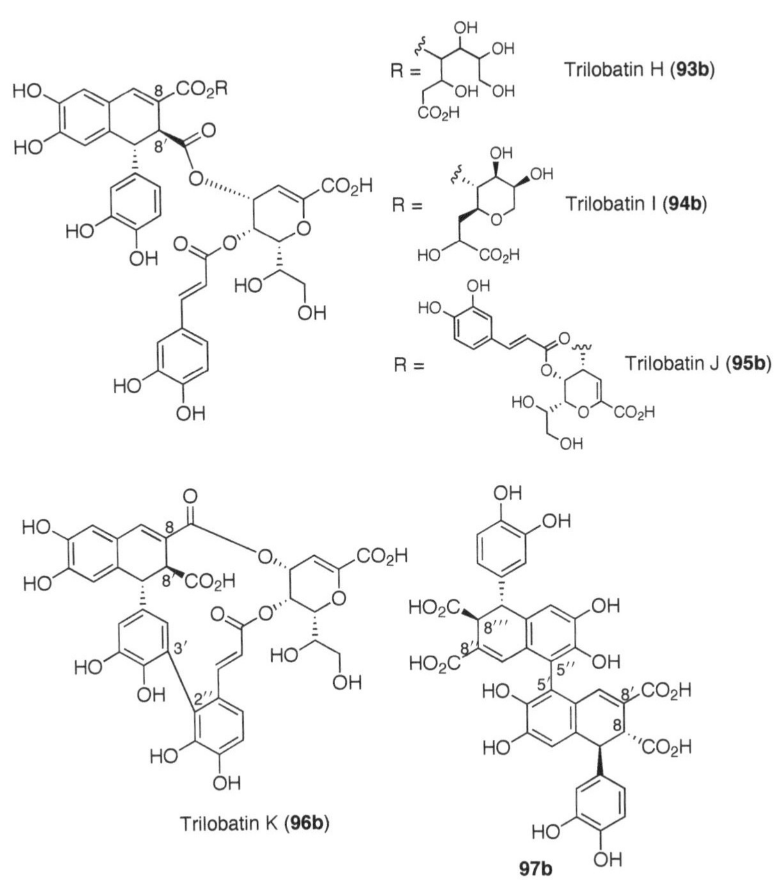

FIGURE 3.12 Liverwort aryldihydronaphthalene lignans **93–97**, where "b" refers to the (–)-enantiomeric forms.

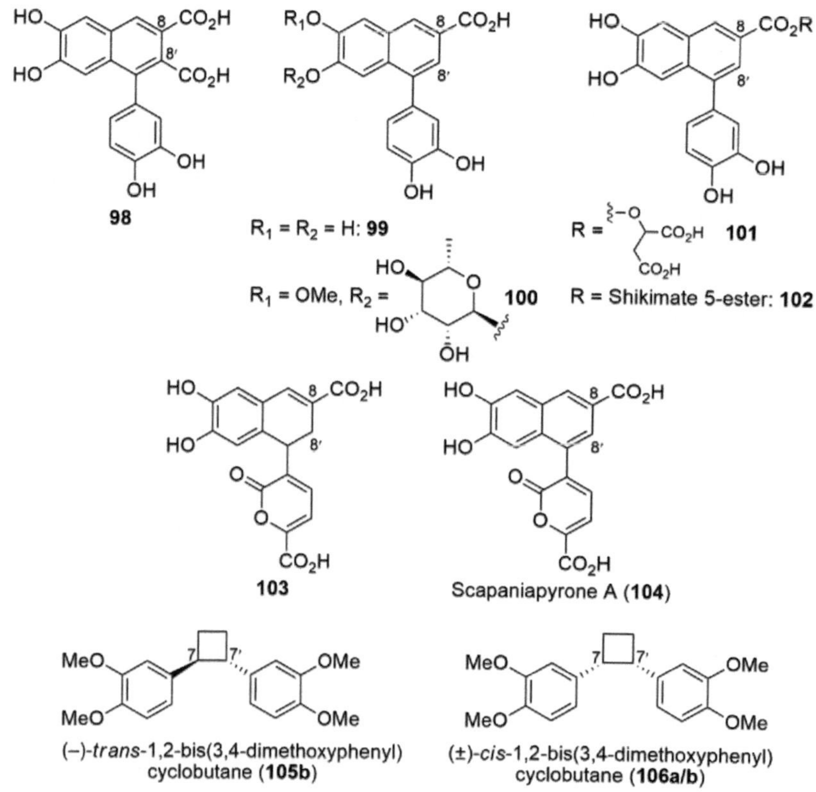

FIGURE 3.13. Arylnaphthalene and cyclobutane-derived norlignans **98–106** isolated from liverworts, where "a" refers to the (+)-enantiomers, and "b" to the (–)-antipodes.

fission of the pendant aryl group resulting in lactone formation (e.g. in **79a–81a**, **103**, and **104**) or of one of the dihydronaphthalene rings as in **85b**.

Using chiral column chromatography, it was shown that enantiomerically pure (–)-epiphyllic acid (**75b**) accumulates in *Calypogeia azurea*, *L. heterophylla* and *Aneura pinguis*, whereas the (+)-enantiomer **75a** is found in *Haplomitrium mnioides*, *J. autumnalis*, and *Marsupella emarginata*. In other liverwort species, both enantiomers of epiphyllic acid (**75**) have been reported, with one being in enantiomeric excess (e.e.) over the other, i.e. the (–)-enantiomer **75b** was in greater than 90% e.e. in *Bazzania yoshinagana* and *Heteroscyphus planus*, and the (+)-enantiomer **75a** was in greater than 80% e.e. in *Diplophyllum taxifolium* (Tazaki 2002).

Two rare cyclobutane norlignans have also been reported present in the liverwort, *Conocephalum japonica*, one being the optically active (–)-7*R*, 7′*R* (**105b**) and the other the (±)-7*R*, 7′*S* racemate (**106a/b**) (Liu et al. 2011). Formation of (+)-**105a** and (±)-**106a/b** are provisionally considered to result from loss of C9/C9′ functionalities. However, while perhaps less likely, they could be formed via (2 + 2) cycloaddition of styrene derivatives.

Summary: Phytochemical analyses of liverworts has established the presence of numerous 8–8′-linked lignans or derivatives thereof. These reports, together with that of allylphenols/propenylphenols and (–)-licarin A (**74b**), as well as the cyclobutane norlignans (+)-**105a** and (±)-**106a/b**, reveal extensive elaboration of phenylpropanoid metabolism in the liverworts.

3.4.2 Hornworts

Hornworts (Anthocerotophyta) are estimated to have ~300 extant species (Duff et al. 2007), with their oldest fossil records dating from the late Cretaceous period (~70–90 million years ago [MYA]) (Němejc and Pacltová 1974). A hypothesis that hornworts are evolutionarily more basal than liverworts has been both supported (Nickrent et al. 2000) and rejected (Qiu et al. 2006).

While the phytochemistry of hornworts even today remains still largely unexplored, there are no reports to our knowledge of any monomeric allylphenols/propenylphenols being isolated. Two hornwort lignans have been reported, however, namely the 8–7′-linked (+)-megacerotonic acid (**107a**) isolated from *Megaceros flagellaris*, *Dendroceros japonicus*, *Notothylas temperata*, and *Phaeoceros laevis*, and the 8–7′-linked, 9′–*O*-esterified (+)-anthocerotonic acid (**108a**), a cyclobutane lignan type from *Anthoceros punctatus* (Takeda et al. 1990a, Takeda et al. 1990b) (Figure 3.14).

Summary: Hornworts produce the unusual (rare) 8–7′-linked lignan structural types. From a chemical structure perspective, both lignans **107** and **108** appear provisionally to be derived from stereoselective *p*-coumaric (**1**) and caffeic (**13**) acid coupling, or alternatively, from two *p*-coumaric (**1**)

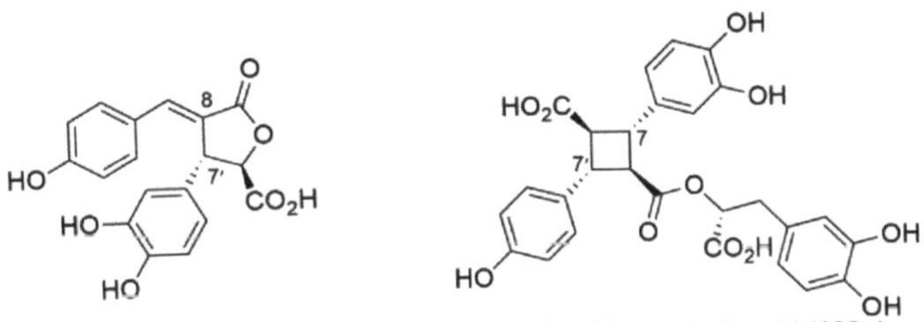

FIGURE 3.14 Lignans isolated from the hornworts *Megaceros flagellaris* (**107a**) and *Anthoceros punctatus* (**108a**), where "a" refers to the (+)-enantiomer.

acid moieties undergoing stereoselective coupling with subsequent hydroxylation of one of the aromatic rings. Another interesting structural feature is the presence of the ester-linked 2-hydroxy-3-(3,4-dihydroxyphenyl)propanoic acid.

As before for the majority of the 8–8′-linked lignans reported in liverworts, these metabolites have highly oxygenated C9 and C9′ end-group moieties and catechol groups (lacking *O*-methylation).

3.4.3 Mosses

Mosses may have emerged up to ~360 MYA (early Carboniferous) (Thomas 1972) and are the largest group of the bryophytes, totaling ~10,000 species (Troitsky et al. 2007). They have not been reported to accumulate lignans (or allyl/propenyl phenols), thus differing from the other two bryophyte divisions. The absence of these natural product classes underscores the chemotaxonomical differences in elaboration of various branches of phenylpropanoid metabolism during land colonization by early non-vascular plants.

Interestingly, the moss *Physcomitrella patens* had its genome sequenced about a decade ago (Rensing et al. 2008). Yet, in contrast to the chemotaxonomy findings discussed previously (i.e. absence of lignans, allyl/propenylphenols), it appeared to have genes of relatively high similarity (i.e. from 67%–86%) to most (if not all) *Arabidopsis thaliana* phenylpropanoid pathway genes (Labeeuw et al. 2015), i.e. from PAL to cinnamyl alcohol dehydrogenase (CAD). However, to our knowledge, their only enzyme activities and functions established to date are for PAL (Jendresen et al. 2015) and 4-coumarate:CoA ligase (4CL) (Silber et al. 2008). However, the presence of caffeic acid (**13**) in *P. patens* (Erxleben et al. 2012) and of the 4-*O*-β-glucoside of the *cis*-form of *p*-coumaric acid (**1**) in *Sphagnum fallax* (Rasmussen et al. 1996) suggests, at the minimum, partial elaboration of the phenylpropanoid pathway.

Summary: There is currently no evidence that mosses produce monolignols, allyl/propenylphenols, or lignans.

3.5 CHEMOTAXONOMICAL C_6C_3 DIVERSITY IN VASCULAR PLANTS: PTERIDOPHYTES

These consist of three divisions: the lycophytes, horsetails, and ferns. Lycophytes emerged more than 400 MYA during the early Devonian period (Gifford and Foster 1989). Present-day lycophytes consist of only three families, Lycopodiaceae, Isoetaceae, and Selaginellaceae (Ambrose 2013). Horsetails apparently emerged during the Late Devonian and Carboniferous periods (Schneider 2013). The fossil record also indicates that the earliest fern-like plants evolved during the Carboniferous period, some 345–280 MYA (Schneider 2013). There are ~13,000 pteridophyte species considered to be in existence currently (Plant List 2013).

3.5.1 Lycophytes

Selaginella sp. have long been known to produce sinapyl alcohol (**12**)-enriched lignin biopolymers (White and Towers 1967, Jin et al. 2005). This was an interesting finding, since before 1967 it was considered that biochemical pathways to sinapyl alcohol (**12**) derived moieties evolved in angiosperms (see Weng et al. 2008, for historical account).

Selaginella sp. also produce lignans. These are considered to be derived from stereo-selective monolignol coupling and subsequent metabolism (Vassão et al. 2010). Species accumulating lignans include *S. doederleinii* (Lin et al. 1994), *S. involven* (Long et al. 2014), *S. moellendorffii* (Zheng et al. 2008b, Feng et al. 2011, Wu and Wang 2011), and *S. tamariscina* (Zheng et al. 2004, Dat et al. 2017).

Their reported lignans are presumably monolignol derived, with 8–8′-linked furofuran and dibenzylbutyrolactone, 8–5′ phenylcoumaran and 8–*O*–4′ linked lignans. *S. doederleinii* (Lin et al. 1994) reportedly contains (+)-matairesinol (**109a**), (+)-nortrachelogenin (**110a** = wikstromol) and the glucoside of (–)-nortrachelogenin (**111b**), (–)-nortracheloside (**112b**), as well as the presumed sinapyl alcohol (**12**)-derived (–)-*epi*-syringaresinol [**113b** = (–)-lirioresinol A], and (–)-syringaresinol [**114b** = (–)-lirioresinol B] (Figure 3.15).

In addition, *S. involven* accumulates the 8–8′ dibenzylbutyrolactone lignan (**115,** Figure 3.15) (Long et al. 2014). *S. moellendorffii* also has several lignans and norlignans reported. These include the 8–5′-linked (–)-moellenoside A (**116b**) (Zheng et al. 2008b) and (–)-moellenoside B (**117b**), the 8–8′ linked lariciresinol (**118**) (Feng et al. 2011), secoisolariciresinol (**119**) and its corresponding olefin derivative **120**, as well as the (+)-8-*O*-4′ linked sinapyl/coniferyl alcohol-derived lignans (**121-124**) (Wu and Wang 2011) (Figure 3.16). (Of these, **116b** and **117b** are reported as norlignans.)

Additionally, *S. tamariscina* accumulates the 8–5′-linked norlignans tamariscinoside C (**125**) (Zheng et al. 2004), moellenoside B (**117b**), **126**, and the related tamariscinosides D-F (**127–129**), as well as the 8–8′-linked lignan liriodendrin (**130**) (Dat et al. 2017) (Figure 3.17).

The 8–5′-linked lignans (–)-lycocernuaside A (**131b**) and (–)-dihydrodehydrodiconiferyl alcohol

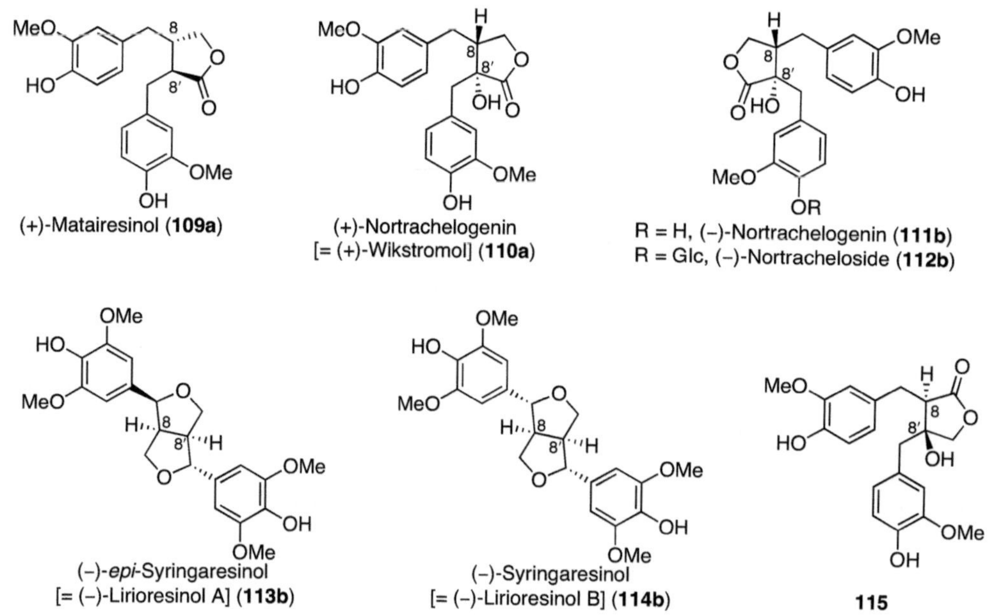

FIGURE 3.15 8–8′ Linked lignans isolated from *Selaginella doederleinii* (**109–114**) and *S. involven* (**115**), where "a" and "b" signify the (+)- and (−)-forms.

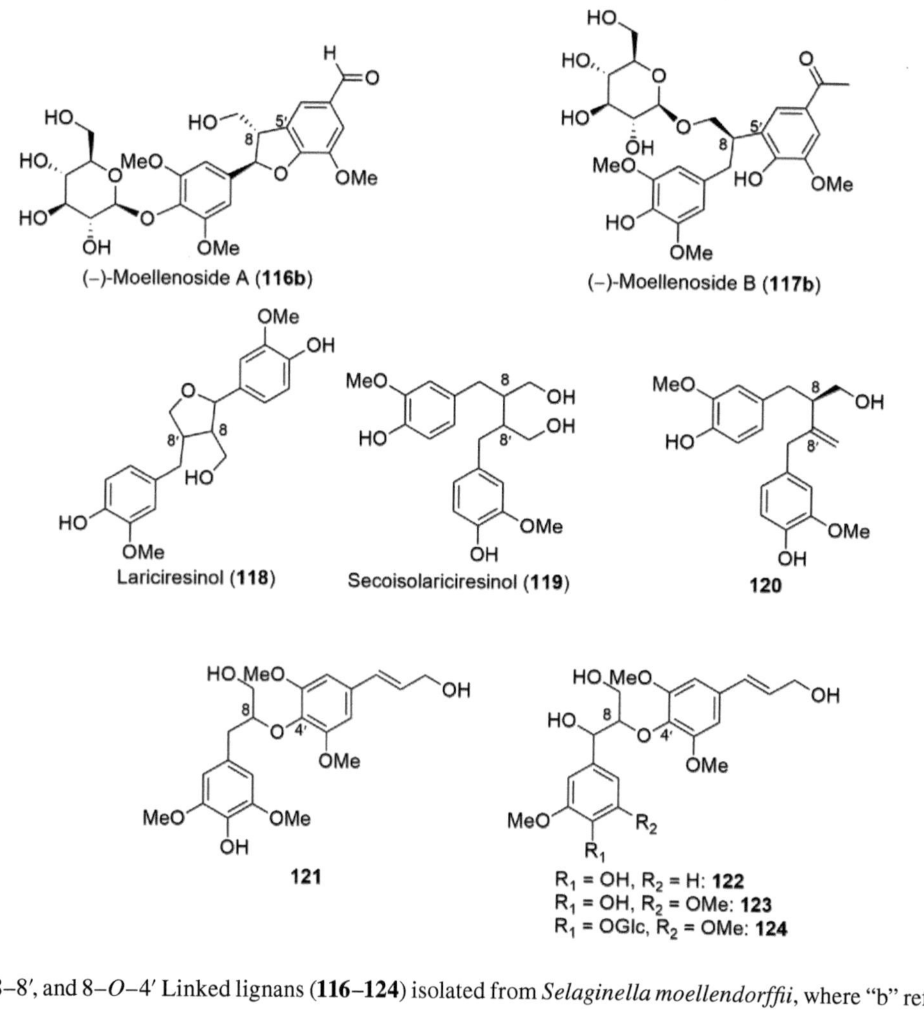

FIGURE 3.16. 8–5′, 8–8′, and 8–*O*–4′ Linked lignans (**116–124**) isolated from *Selaginella moellendorffii*, where "b" refers to the (−)-enantiomeric forms.

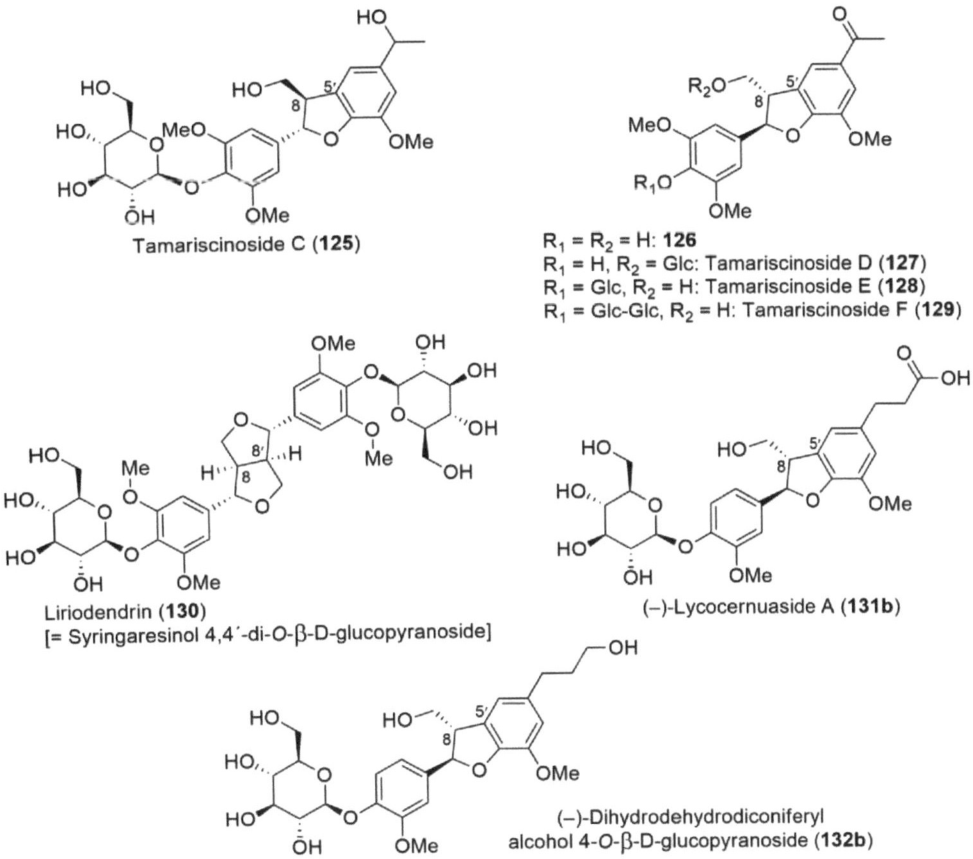

FIGURE 3.17. 8–5′ and 8–8′ Linked norlignans and lignans isolated from *Selaginella tamariscina* (**125–130**) and *Lycopodiella cernua* (**131** and **132**).

4-*O*-β-D-glucopyranoside (**132b**) have also been isolated from the lycopod *Lycopodiella cernua* (Hung et al. 2015).

Summary: Phenylpropanoid metabolism in the lycophytes enabled formation of 8–8′, 8–5′ and 8–*O*–4′ linked lignans and some norlignans, with the biochemical pathway to sinapyl alcohol (**12**) also sporadically being in place.

3.5.2 Horsetails

One horsetail sp. (*Equisetum arvense*, Equisetopsida) reportedly had the C_6C_3 monomer anethole (**60**) in *trace* amounts (Radulović et al. 2006). However, as mentioned earlier for algae, this needs to be fully biochemically established, i.e. as to whether this is a true natural product formed by this species *in vivo* or a minor contaminant in obtained samples. There is, however, one report of lignans in *Equisetum debile*, i.e. (+)-lariciresinol-9-*O*-β-D-glucopyranoside (**133a**, Figure 3.18), (+)-isolariciresinol-9-*O*-β-D-glucopyranoside (**134a**), guaiacylglycerol-8-*O*-4′-coniferyl ether (**135**) (no data provided), and its 9-*O*-glucoside named debilignanoside (**136**) (Tan et al. 2011). However, their combined total amounts by dry weight accounted for only ~0.001%.

Summary: The occurrence of C_6C_3 monomeric and dimeric (lignan) phenylpropanoids in horsetails is of a very preliminary nature, given that there are, to our knowledge, only two reports. More studies (at the genomic and enzymatic level) are needed to confirm and extend the observations that horsetails may be able to biosynthesize monolignols, allyl-/propenylphenols, and lignans.

3.5.3 Ferns

Lignans have been described from several fern species. These include Blechnaceae species (Wada et al. 1992, Davin et al. 2003), *Pteris vittata* (Pteridaceae) (Satake et al. 1978), and *Didymochlaena truncatula* (Dryopteridaceae) (Cao et al. 2006). Structurally, they can be classified into three quite distinct groups (i.e. whether containing carboxylic acid, alcohol, or allylphenol/propenylphenol moieties in their C3 side-chains).

Blechnaceae ferns mainly have the putatively hydroxycinnamic acid-derived lignans, the 8–2′-linked (–)-blechnic acid (**137b**) (and its C7 and C8 epimers, **138b** and **139b**, respectively), and its shikimate derivative (–)-brainic acid (**140b**) from *Blechnum spicant* (Figure 3.19) (Wada et al. 1992, Davin et al. 2003). However, the related Blechnaceae fern, *Brainea insignis,* also accumulates the 8–5′-linked icariside E5 (**141**) and icariside E3 (**142**), both of which provisionally appear to be coniferyl alcohol (**10**) derived, as well as (–)-7-*epi*-blechnic acid (**138b**) (Wang et al. 2010). Those from *Pteris vittata* are also presumed to be coniferyl alcohol (**10**) derived (Lewis and Davin 1999), i.e. the 8–5′ and 8–8′ linked 9-*O*-glucosides of

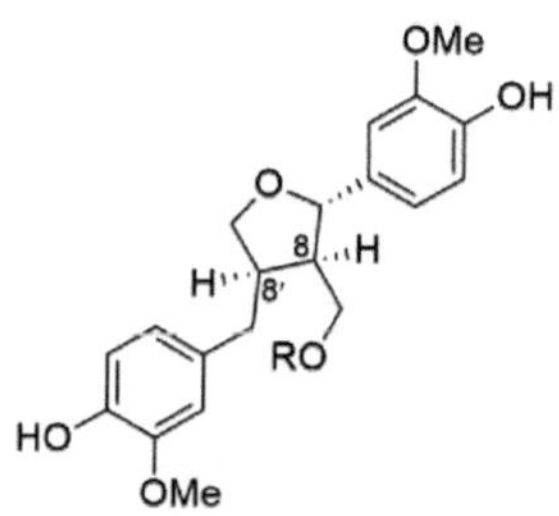

R = H: (+)-Lariciresinol (**118a**)
R = Glc: (+)-Lariciresinol 9-*O*-β-D-glucopyranoside (**133a**)

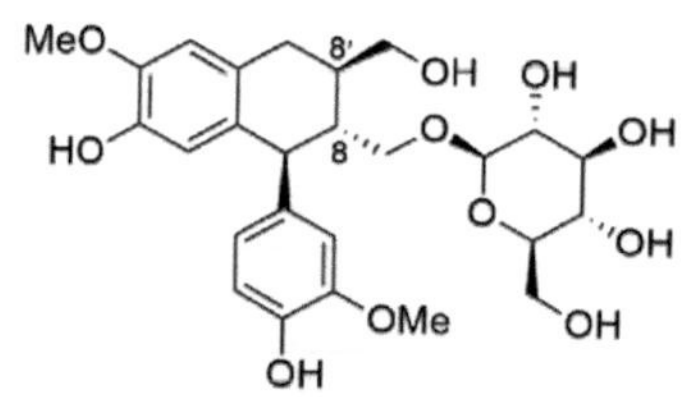

(+)-Isolariciresinol 9-*O*-β-D-glucopyranoside (**134a**)

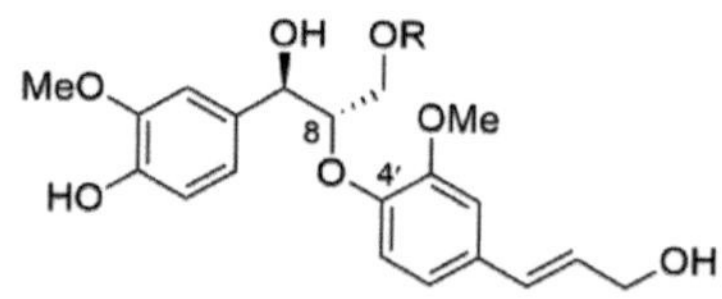

R = H: Guaiacylglycerol-8-*O*-4′-coniferyl alcohol ether (**135**)
R = Glc: Debilignanoside (**136**)

FIGURE 3.18. 8–8′, and 8–*O*–4′ Linked lignans reportedly isolated from the horsetail *Equisetum debile,* where "a" refers to the (+)-enantiomers.

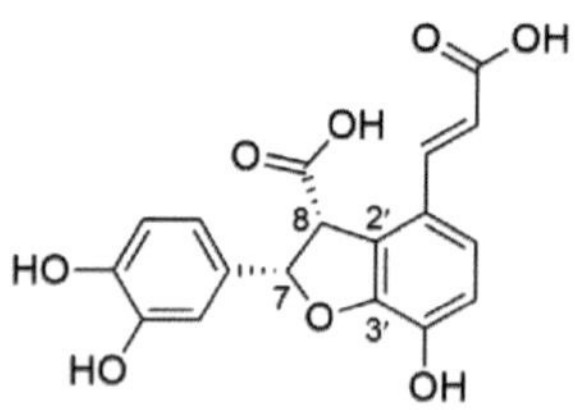

(–)-Blechnic acid (**137b**)

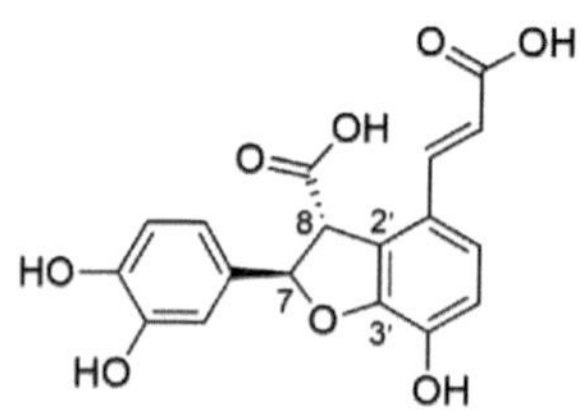

(–)-7-*epi*-Blechnic acid (**138b**)

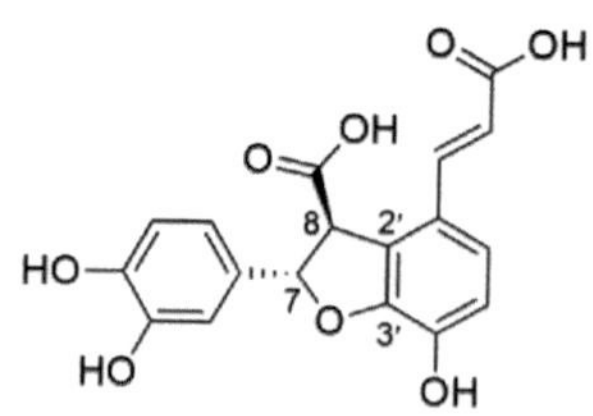

(–)-8-*epi*-Blechnic acid (**139b**)

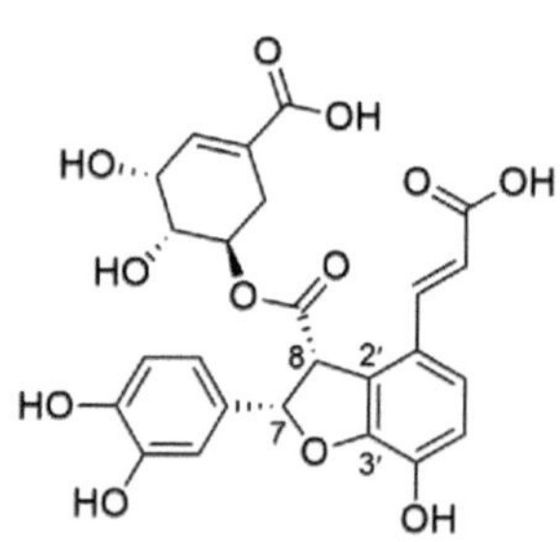

(–)-Brainic acid (**140b**)

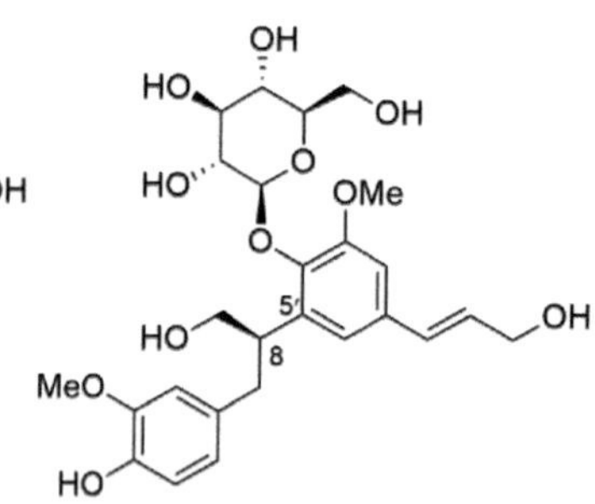

Icariside E5 (**141**)

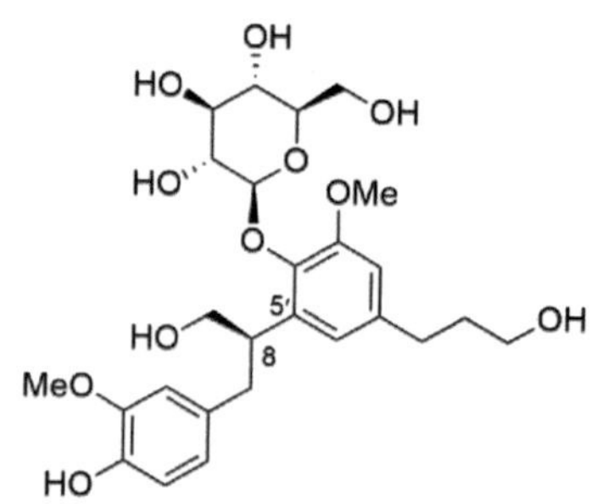

Icariside E3 (**142**)

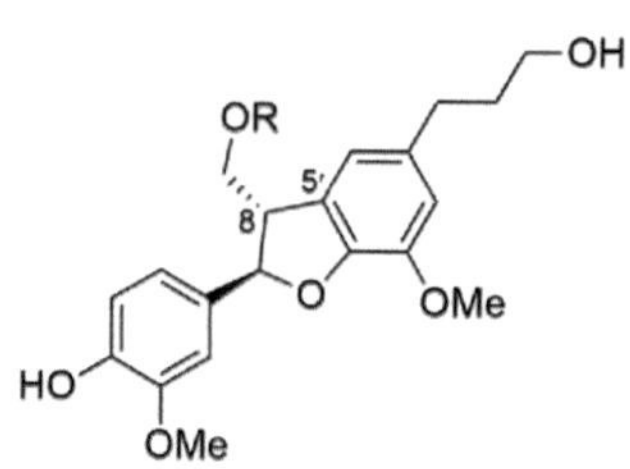

R = H: (–)-Dihydrodehydrodiconiferyl alcohol (**143b**)
R = Glc: (–)-Dihydrodehydrodiconiferyl alcohol 9-*O*-β-D-glucoside (**144b**)

FIGURE 3.19. Lignans isolated from the ferns *Blechnum spicant*, *Brainea insignis*, and *Pteris vittata,* where "b" refers to the (–)-enantiomers.

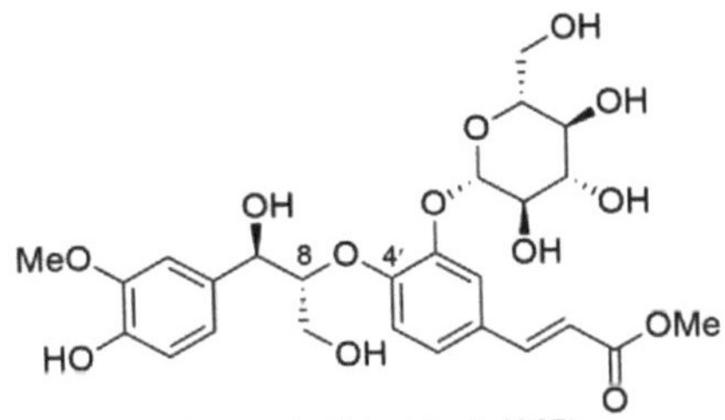

Matteustruthioside A (**145**)

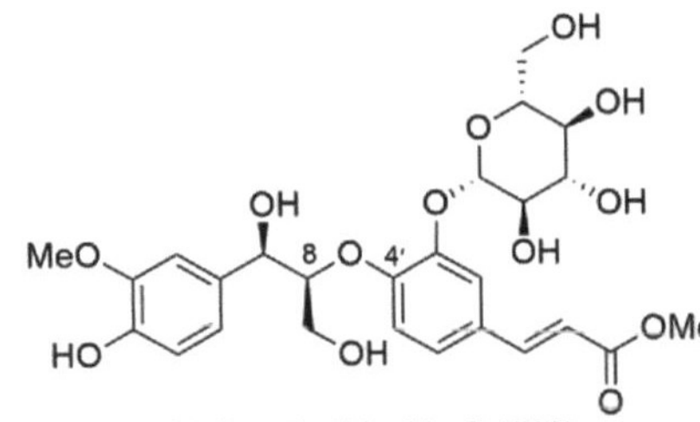

Matteustruthioside B (**146**)

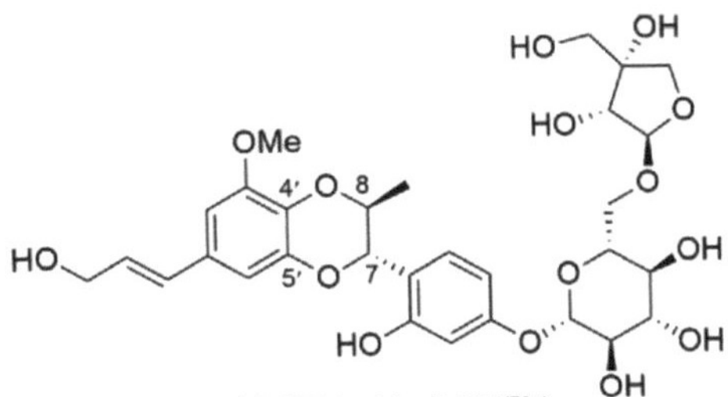

(–)-Multifidoside A (**147b**)

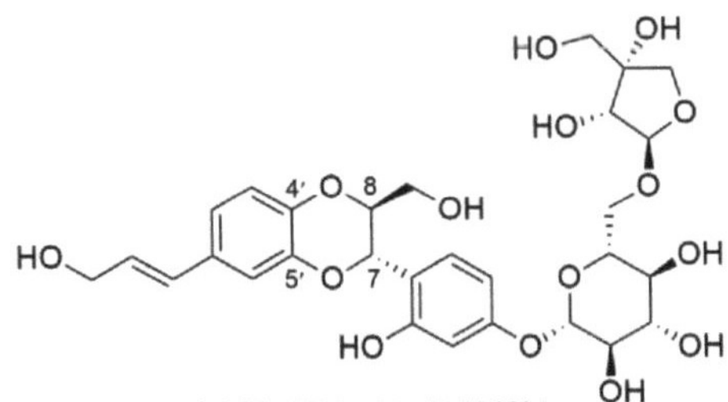

(–)-Multifidoside B (**148b**)

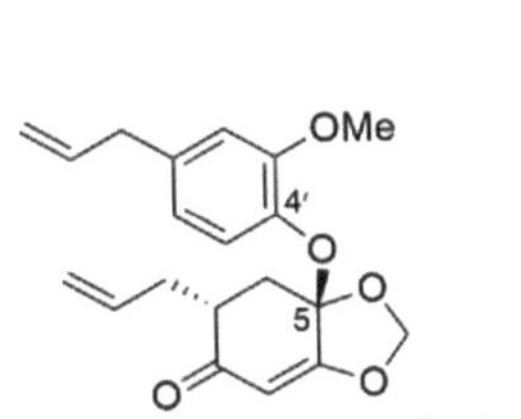

(+)-Didymochlaenone A (**149a**)

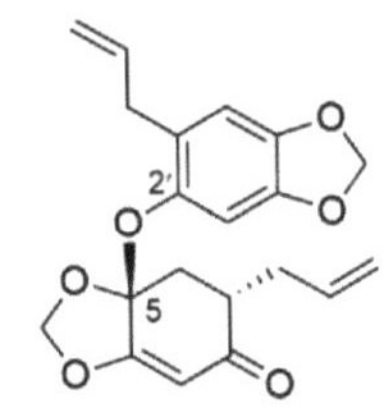

(–)-Didymochlaenone B (**150b**)

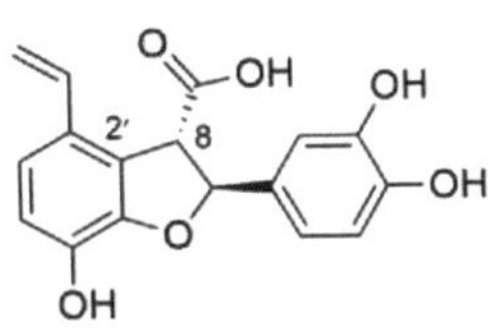

(+)-Penangianol A (**151a**)

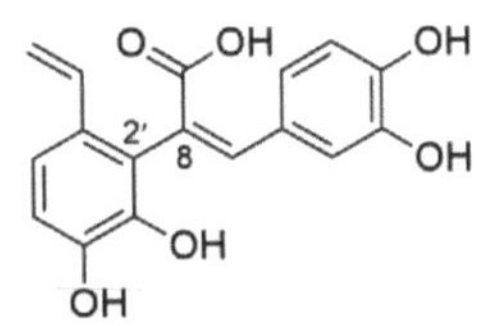

Penangianol B (**152**)

FIGURE 3.20. Lignan and norlignan skeletal types reported from various fern species, where "a" refers to the (+)-enantiomers, and "b" to the (–)-antipodes.

(–)-dihydrodehydrodiconiferyl alcohol (**143b**) and (+)-lariciresinol (**118a**) (Satake et al. 1978), respectively. Icariside E3 (**142**), **143** and **144** are noteworthy in having a (saturated) propanol side-chain.

Two lignans have been reported for the fern *Matteuccia struthiopteris*; both are 8–*O*–4′-linked and are named matteustruthiosides A (**145**) and B (**146**) (Zhu et al. 2016) (Figure 3.20). These perhaps result from coupling of coniferyl alcohol (**10**) and the methyl ester of caffeic acid (**13**), with subsequent glycosylation of the 3′-OH group, as one possibility.

Pteris multifida contains a third type of lignan of the 1,4 benzodioxin type (Zheng et al. 2008a). This type includes (–)-multifidoside A (**147b**) and B (**148b**). Among other possibilities, these could result from coupling of *p*-coumaryl alcohol (**8**)/*p*-anol (**6**) with subsequent hydroxylations of their aromatic rings and methylation of the 3′-OH group. Alternatively, they could result from coupling of *p*-coumaryl alcohol (**8**)/*p*-anol (**6**) with caffeyl alcohol (**9**) and/or 5-hydroxyconiferyl alcohol (**11**), respectively, with subsequent oxidative modifications. Additionally, *Didymochlaena truncatula* contains (–)-nortrachelogenin (**111b**), as well as the 5–*O*–4′- and 5–*O*–2′-ether linked, partially dearomatized, lignans (+)-didymochlaenone A (**149a**) and (–)-didymochlaenone B (**150b**) (Cao et al. 2006). In the latter two examples, both are optically active, have allylphenol-like C_3 side-chains, methylenedioxy groups, and additional oxygenation at their C-5 positions.

Norlignans have also been reported in ferns, e.g. the 8–2′-linked (+)-penangianol A (**151a**) and penangianol B (**152**) from *Abacopteris penangiana* (Jiao et al. 2015). The assumption made here, however, is that there is loss of the C9 functionality. While perhaps less likely, it could result from coupling of a styrene derivative with a C_6C_3 monomer.

Summary: The phenylpropanoid pathway is further elaborated in the ferns, based on phytochemical analysis of various species. This includes formation of 8–8′, 8–2′, 8–5′, 8–*O*–4′, 1,4 benzodioxin, 5–*O*–4′ and 5–*O*–2′ lignan skeleta, as well as 8–2′ norlignan skeletal types. These phytochemical analyses, in addition to the different skeletal types produced, also strongly indicate that the biochemical pathway to allylphenols/propenylphenols (lignans) appeared in some fern species.

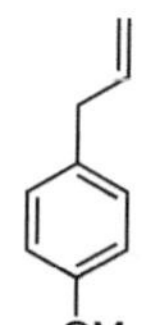

Estragole (**153**)
= methylchavicol
= 4-allylanisole)

FIGURE 3.21. Estragole (**153**) form *Cycas revoluta*.

3.6 CHEMOTAXONOMICAL C_6C_3 DIVERSITY IN VASCULAR PLANTS: GYMNOSPERMS

The earliest conifers emerged during the Carboniferous period, ~308 MYA, although it was not until the Late Triassic period 228–200 MYA when the first conifer wood similar to that observed in extant species appeared (Savidge 2007, Savidge 2008). Currently, there are considered to be ~1,000 conifer species (Plant List 2013).

Chemotaxonomically, the gymnosperms, and in particular the conifers, further expanded lignan/norlignan structural diversity, i.e. in terms of different linkage modes, oxygenation patterns of both aromatic groups, and C_3 side-chain modifications.

3.6.1 Cycads

Only one species (out of ~90 cycad species [Plant List 2013]) has C_6C_3-derived monomers reported, i.e. anethole (**60**), methyleugenol (**68**), and estragole (**153** = methylchavicol = 4-allylanisole) in volatiles emitted from *Cycas revoluta* male and female cones (Azuma and Kono 2006) (Figure 3.21). To our knowledge, no lignans or norlignans have been reported from cycads.

Summary: Based on current phytochemical studies, the cycads may not have fully elaborated phenylpropanoid biochemical pathways to the lignans/norlignans, although *C. revoluta* apparently has the pathway operative to allyl/propenyl phenols.

3.6.2 Conifers

Evolution of the different gymnosperm plant groups was accompanied by a further rich increase in the structural variety of their lignans and norlignans (Castro et al. 1996). Entries in the *Handbook* herein describe many gymnosperm lignans with selected examples described below. Frequently, their lignans/norlignans have guaiacyl-like aromatic rings, i.e. as in coniferyl alcohol (**10**).

3.6.2.1 8–8′-Linked Lignans

These are very abundant in conifers, and particularly in their heartwood (Gang et al. 1998) and knots (Holmbom et al. 2003, and references therein). They are also found in bark, roots, needles, and cones. They are typically found as (a) dibenzylfurofurans, (b) dibenzyltetrahydrofurans, (c) dibenzylbutyrolactones, (d) dibenzylbutanes, (e) arylnaphthalenes, and (f) aryltetrahydronaphthalenes.

They can be exemplified by (a) dibenzylfurofurans, e.g. (–)-sesamin (**154b**) in *Juniperus thurifera* (San Feliciano et al. 1989); (b) dibenzyltetrahydrofurans, e.g. shonanin (**155b**) in *Calocedrus formosana* (Fang et al. 1989) and *Pinus taeda* cell cultures (Nose et al. 1995); (c) dibenzylbutyrolactones, e.g. (–)-matairesinol (**109b**) in *Podocarpus spicatus* (Easterfield and Bee 1910); (d) dibenzylbutanes, e.g. (–)-secoisolariciresinol (**119b**) in *P. spicatus* (Briggs et al. 1959b, a); (e) arylnaphthalenes, e.g. junaphtoic acid (**156**) in *Juniperus sabina* (San Feliciano et al. 1991); and (f) aryltetrahydronaphthalenes, e.g. (–)-plicatic acid (**157b**) in *Thuja plicata* (Gardner et al. 1959, Gardner et al. 1960, Gardner et al. 1966) and (–)-α-conidendrin (**158b**) in *Tsuga heterophylla*] (Krahmer et al. 1970, Kawamura et al. 1996), respectively (Figure 3.22).

In many highly valued tree heartwoods, lignans and less frequently norlignans, are abundantly deposited, e.g. (–)-plicatic acid (**157b**) in *T. plicata* and its derivatives can account for up to 20% of dry weight (Johansson et al. 2000), and are in part responsible for the heartwood's durability and texture. Similarly, other abundant lignans include (–)-7-hydroxymatairesinol (**159b**, Figure 3.23) and (–)-α-conidendrin (**158b**) in *T. heterophylla* (Krahmer et al. 1970). In terms of molecular size, the largest documented higher oligomer estimates are those derived from (–)-plicatic acid (**157b**) with MW ~10,000 Da in *T. plicata* (Johansson et al. 2000).

As illustrated previously, many 8–8′ lignans have considerable C3 side-chain functional group variations, including a few having C9 and/or C9′- deoxygenated end groups, e.g. junaphtoic acid (**156**) in *J. sabina.* While these could be allylphenol/propenylphenol derived, this needs to be established biochemically as other possibilities exist.

3.6.2.2 8–5′- and 8–*O*–4′-Linked Lignans

Gymnosperm lignans with other inter-unit linkages and other modifications, such as C_3 side-chain reductions, are also common, i.e. 8–5′ and 8–*O*–4′ linked. Examples include (±)-dehydrodiconiferyl alcohols (**160a/b**) and guaiacylglycerol 8–*O*–4′ coniferyl alcohol ether (**135**) in *Pinus taeda* cell cultures (Nose et al. 1995); (–)-dihydrodehydrodiconiferyl alcohol (**143b**) in *Juniperus chinensis* (Fang et al. 1992) and (+)-cedrusin (**161a**) in *Cedrus deodara* (Agrawal et al. 1980) and *T. heterophylla* (Kawamura et al. 1996); as well as **162** in *J. chinensis* (Fang et al. 1992).

The 8–5′ linked per-acetylated lignan **163** is also found in *C. japonica* (Su et al. 1995), as well as monomers such as dihydromonolignols, dihydroconiferyl alcohol (**22**), and its *O*-glucoside (**164**) in *Picea glauca* knots (Kraus and Spiteller 1997) (Figure 3.23).

3.6.2.3 Norlignans

Most gymnosperm norlignans (Castro et al. 1996, Suzuki and Umezawa 2007), formed via presumed C9 loss, are 8–7′-linked dibenzylpentanes ($C_6C_5C_6$). These are reported present in the

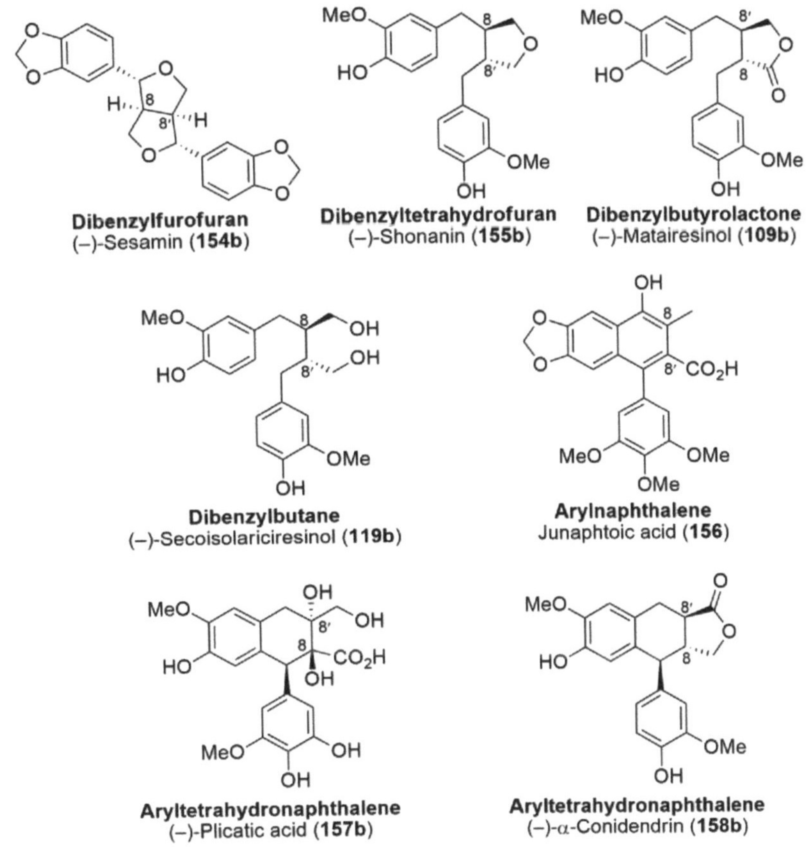

FIGURE 3.22. 8–8′ Linked lignan structural types present in some gymnosperms (conifers); "b" refers the (–)-antipode.

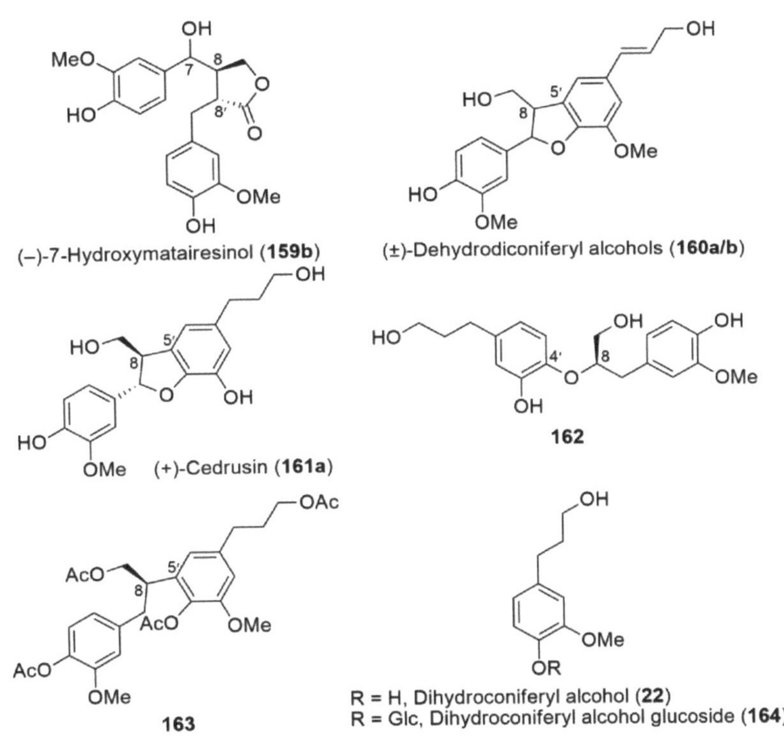

FIGURE 3.23. 8–5′ and 8–*O*–4′ Linked gymnosperm lignans, where "a" refers to the (+)-enantiomers, and "b" to the (–)-enantiomers.

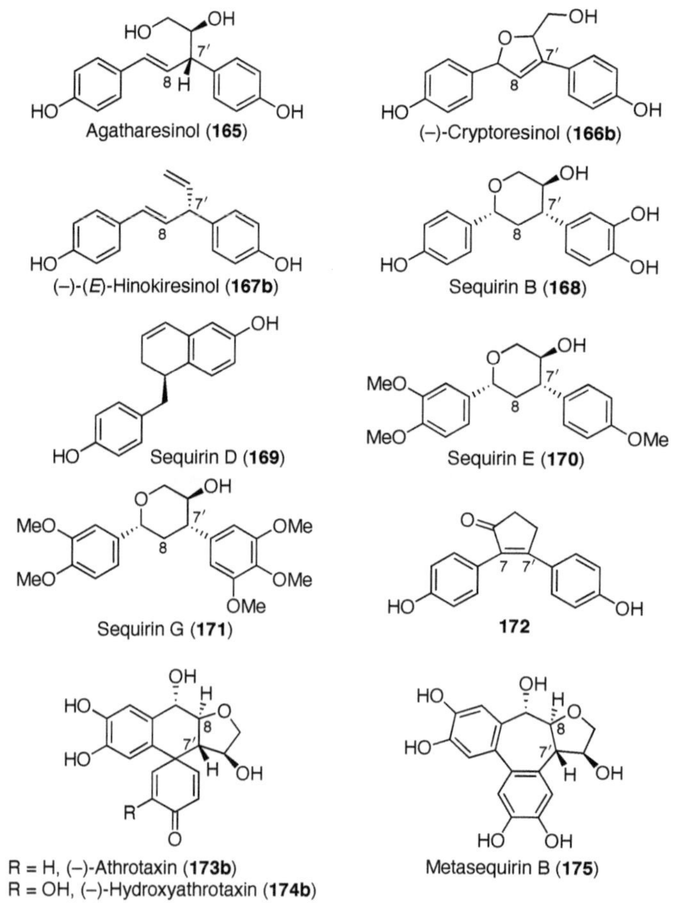

FIGURE 3.24 Gymnosperm norlignans, where "b" refers to the (−)-enantiomeric forms.

Cupressaceae, Araucariaceae, and Taxodiaceae, but not in the Pinaceae, Podocarpaceae, and other conifers (Castro et al. 1996, Suzuki and Umezawa 2007). Examples are agatharesinol (**165**) in *Agathis australis* (Araucariaceae) (Enzell and Thomas 1965, Enzell et al. 1967) and *Cryptomeria japonica* (Takahashi and Ogiyama 1986), as well as in *Sequoia sempervirens* and *Sequoiadendron gigantea* (Henley-Smith and Whiting 1976); (−)-cryptoresinol (**166b**) (Takahashi et al. 1988) and (−)-(*E*)-hinokiresinol (**167b**) in *C. japonica* (Takahashi and Ogiyama 1986, Suzuki et al. 2004); sequirin B (**168**) and D (**169**) in *S. sempervirens* (Hatam and Whiting 1967, 1969, Begley et al. 1973), and the permethyl ether analogs, sequirin E (**170**) and sequirin G (**171**) from *S. gigantea* (Henley-Smith and Whiting 1976), as well as the 7–7′-linked norlignan **172** in *Araucaria angustifolia* (Ohashi et al. 1992) (Figure 3.24).

Other interesting structural variations include the rare 8–7′-linked spiro-norlignan (−)-athrotaxin (**173b**) from *Athrotaxis selaginoides* (Daniels et al. 1972) and *Metasequoia glyptostroboides* (Enoki et al. 1977b), its hydroxylated derivative (−)-hydroxyathrotaxin (**174b**), and the cycloheptadiene lignan metasequirin B (**175**) from *M. glyptostroboides* (Enoki et al. 1977a) (Figure 3.24).

Lastly, the aromatic rings can vary greatly from being mono-oxygenated, as in **165**, **166**, **167**, and **169** to having catechol, dimethoxyaryl, and trimethoxyaryl moieties, e.g. as illustrated by sequirin B (**168**), sequirin E (**170**), and sequirin G (**171**).

Summary: Phytochemical analysis of gymnosperm lignans and norlignans has established the further elaboration of phenylpropanoid metabolism, including formation of additional inter-unit linkage types.

3.6.3 Ginkgo

Ginkgo biloba, the only living species in the Ginkgophyta, has fossil records dating back 270 million years. In terms of lignans, initially (+)-sesamin (**154a**) was found to accumulate in both heartwood (Kariyone et al. 1958) and leaves (Bedir et al. 2002). Since then, various other 8–8′ and 8–5′ linked lignans have been found in *G. biloba* seeds (Ngan et al. 2014). These include (+)-8-hydroxypinoresinol

4,4′-di-*O*-β-D-glucopyranoside (**176a**), cedrusin (**161**) and dihydrodehydrodiconiferyl alcohol 4-*O*-β-D-glucopyranoside (**132**), ceplignan 4-*O*-β-D-glucoside (**177**), together with the C_6C_3 monomers, ferulic acid 4-*O*-β-D-glucoside (**178**), coniferin (**179**), isoconiferin (**180**), and (2*R*)-3-(4-hydroxy-3-methoxyphenyl)propan-1,2-diol (**181**) (Figure 3.25).

In addition, *G. biloba* leaves also contain numerous truxinate (cyclobutane) lignan derivatives, such as the biginkgosides A-I (**182–190**) (Figures 3.26 and 3.27) (Ma et al. 2016). The latter are particularly interesting as related cyclobutane and truxinate compounds were also reported in the liverworts (see Section 3.4.1) and monocotyledons (see Section 3.7.2), respectively.

3.6.4 Gnetophytes

The oldest gnetophytes are reported from the Upper Permian (270 MYA) (Wang 2004). There are three extant genera, *Gnetum* (41 species), *Welwitschia* (1 species), and *Ephedra* (70 species) (Plant List 2013).

Although there are few reports of Gnetales lignans, most are 8–8′-linked. The most interesting observation is that some of these, i.e. (+)-syringaresinol (**114a**) in *Gnetum gnemon* roots (Iliya et al. 2003) and (±)-syringaresinols (**114a/b**) in *Ephedra alata* (Nawwar et al. 1985), have aromatic ring methoxylation patterns as found in sinapyl alcohol (**12**). As mentioned earlier, the latter – and derivatives thereof – are widespread in the angiosperms but rare in gymnosperms. Others, such as (+)-lariciresinol (**118a**), (+)-isolariciresinol (**192a**), and their 9-acetate derivatives **191a** and **193a** in *E. viridis* (Pullela et al. 2005), have aromatic ring methoxylation substitutions as for coniferyl alcohol (**10**) (Figure 3.28).

The 8–3′-linked spermine diamides, ephedradines A-C (**194a–196a,** Figure 3.28) in mao-kon, a crude drug from *Ephedra* roots (Tamada et al. 1979a, Tamada et al. 1979b, Konno et al. 1980) are rare examples of N-containing lignans (i.e. a sort of pseudo-lignan alkaloid hybrid). Allylphenols/propenylphenols have not been reported to date in the latter two gymnosperm families.

Summary: The emergence of the gymnosperms gave mainly the 8–8′, 8–5′ and 8–*O*–4′ linked lignans with considerable variety in types of lignans (e.g. dibenzyltetrahydrofuran, arylnaphthalene, etc.) as well as in both oxidative and reductive modifications. Of these, a few species are able to produce sinapyl alcohol (**12**)-derived lignans, which are most commonly encountered in the angiosperms. The gymnosperms also contain several norlignans with additional C7′-C8 linkages including those affording spiro-norlignans.

3.7 CHEMOTAXONOMICAL C_6C_3 DIVERSITY IN VASCULAR PLANTS: ANGIOSPERMS

Angiosperms are thought to have evolved from basal seed plants approximately 130 MYA, although whether they were

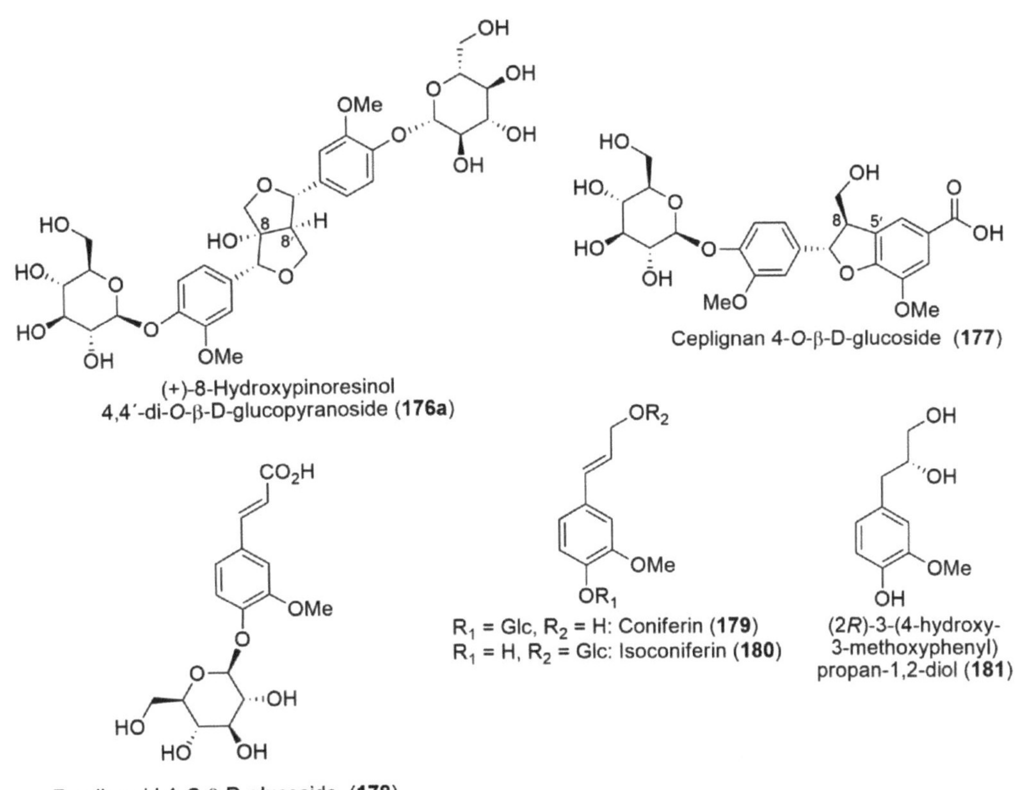

FIGURE 3.25. 8–*O*–4′ and 8–5′ Linked lignans (**176–177**) and C_6C_3 monomers (**178–181**) in *Ginkgo biloba*, where "a" refers to the (+)-enantiomers.

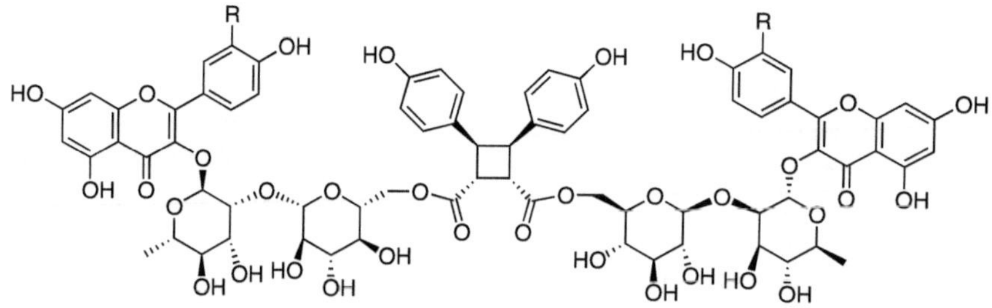

R = H, Biginkgoside A (**182**)
R = OH, Biginkgoside B (**183**)

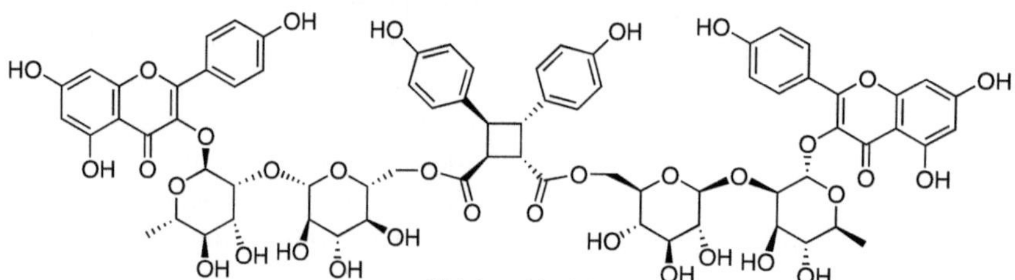

Biginkgoside C (**184**)

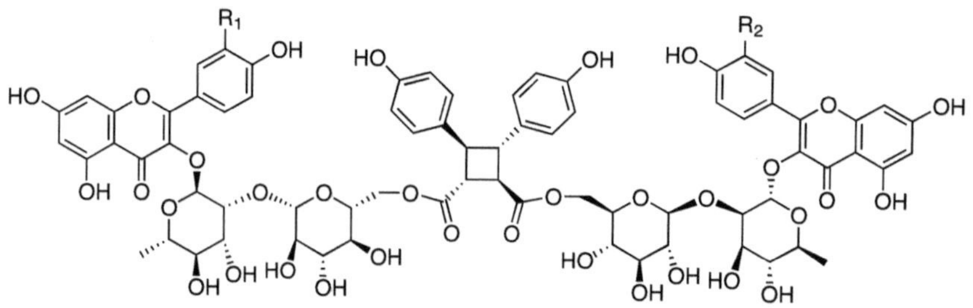

$R_1 = R_2 =$ H, Biginkgoside D (**185**)
$R_1 = R_2 =$ OH, Biginkgoside E (**186**)
$R_1 =$ H, $R_2 =$ OH, Biginkgoside F (**187**)

FIGURE 3.26. Biginkgosides, truxinate (cyclobutane) lignan derivatives (**182–187**) in *Ginkgo biloba*.

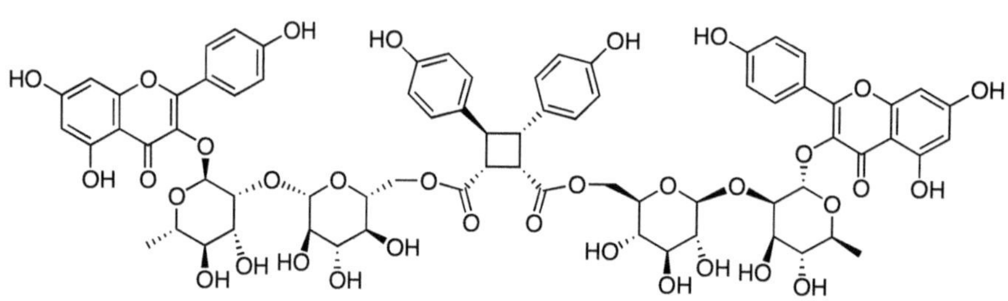

Biginkgoside G (**188**)

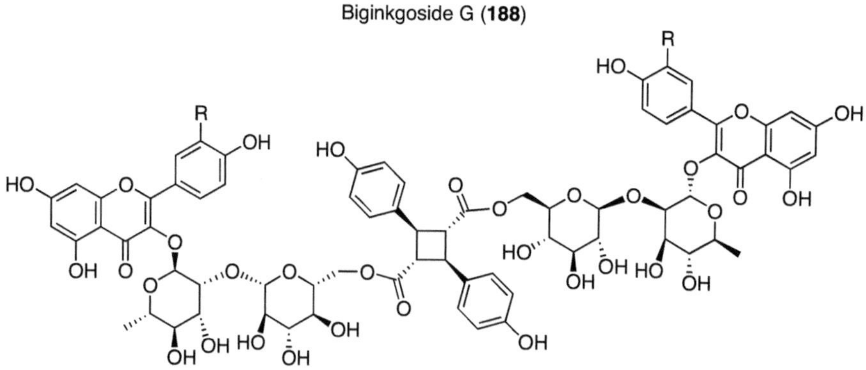

R = H, Biginkgoside H (**189**)
R = OH, Biginkgoside I (**190**)

FIGURE 3.27. Biginkgosides, truxinate (cyclobutane) lignan derivatives (**188–190**) in *Ginkgo biloba*.

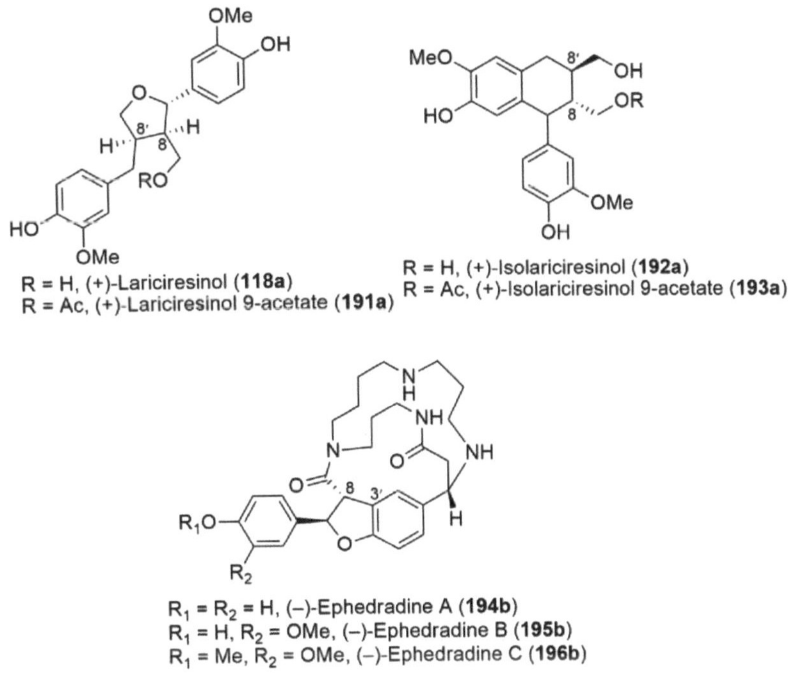

FIGURE 3.28. Gnetales lignans (**118, 191–196**), where "a" refers to the (+)-enantiomers, and "b" to the (–)-antipodes.

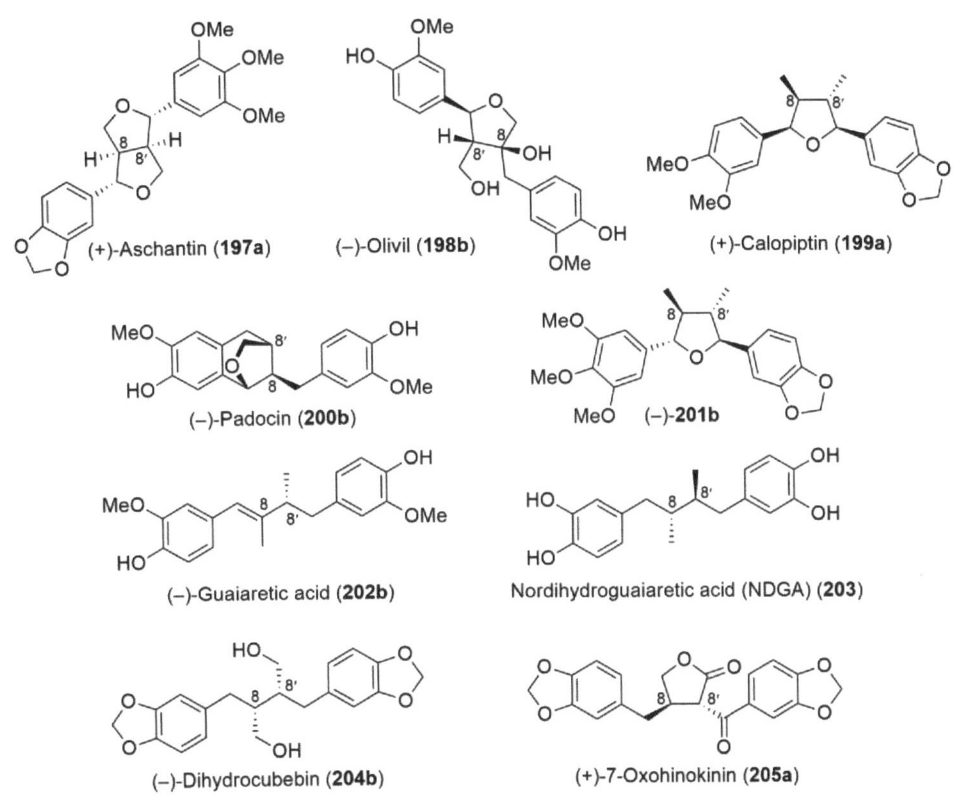

FIGURE 3.29. Some angiosperm lignans of dibenzylfurofuran, dibenzylfuran, dibenzylbutane, dibenzylbutyrolactone, and aryltetrahydronaphthalene classes, where "a" refers to the (+)-enantiomers, and "b" to the (–)-antipodes.

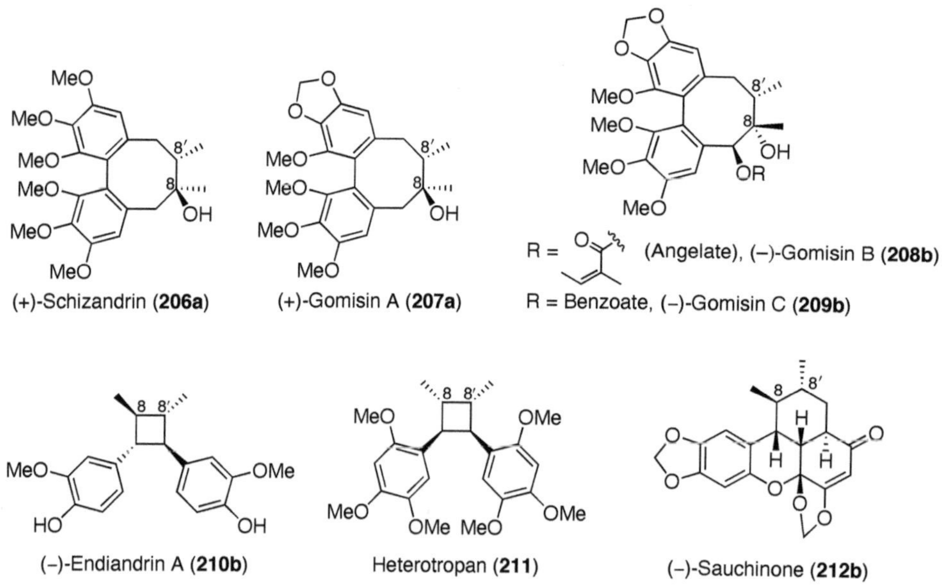

FIGURE 3.30. Examples of dibenzocyclooctadiene (**206–209**), cyclobutane (**210** and **211**), and polycyclic (**212**) angiosperm lignans, where "a" refers to the (+)-enantiomers, and "b" to the (–)-antipodes.

initially herbaceous or woody is uncertain. Earliest angiosperm wood fossils date from the Lower Cretaceous Period, ~145–100 MYA (Takahashi and Suzuki 2003, Feild et al. 2004, Soltis and Soltis 2004). Today, there is estimated to be approximately 252,000 species of angiosperms (Plant List 2013). Their emergence resulted in an enormous "explosion" of lignan/norlignan structural variations. Of these, most are found in woody and non-woody dicotyledons, with only a few reported in the monocotyledons.

3.7.1 Dicotyledons

The 8–8′ inter-unit linkage is overall probably the most frequent in reported dicotyledon lignans. They can exist in (a) dibenzylfurofuran, (b) dibenzyltetrahydrofuran, (c) dibenzylbutane, (d) dibenzylbutyrolactone, (e) aryltetrahydronaphthalene, (f) dibenzocyclooctadiene, (g) cyclobutane, as well as (h) rarer structural variant forms. Examples include the following:

(a) Dibenzylfurofuran lignans, such as (+)-sesamin (**154a**) from *Sesamum indicum* (Villavecchia and Fabris 1896, Böeseken and Cohen 1928), and (+)-aschantin (**197a**, Figure 3.29) from *Piper guineense* (Hänsel and Zander 1961)
(b) Dibenzylfurano lignans, such as (–)-olivil (**198b**) from *Olea europaea* (Vanzetti 1929, Vanzetti and Dreyfuss 1934, Ayres and Mhasalkar 1965), (+)-calopiptin (**199a**) from *Piptocalyx moorei* (Riggs and Stevens 1962), (–)-padocin (**200b**) from *Haplophyllum cappadocicum* (Gözler et al. 1994), and (–)-**201b** from *Virola surinamensis* (Lopes et al. 1996)
(c) Dibenzylbutane lignans, such as (–)-guaiaretic acid (**202b**) from *Guaiacum officinale* (Doebner and Lücker 1896, Schroeter et al. 1918, Haworth et al. 1934), nordihydroguaiaretic acid (**203**) from the creosote bush (*Larrea tridentata*) (Konno et al. 1990, Moinuddin et al. 2003), and (–)-dihydrocubebin (**204b**) from *Piper* sp. (Parmar et al. 1997)
(d) Dibenzylbutyrolactone lignans, such as (–)-matairesinol (**109b**) in *Forsythia intermedia*, (Nishibe et al. 1988) and (+)-7-oxohinokinin (**205a**) from *Calocedrus formosana* (Fang et al. 1990) and *Chamaecyparis obtusa* (Takaku et al. 2001)
(e) Aryltetrahydronaphthalene lignans, such as (–)-podophyllotoxin (**53b**) from *Podophyllum* spp. (Hartwell and Schrecker 1951)
(f) Dibenzocyclooctadiene lignans, such as (+)-schizandrin (**206a**) and gomisins A-C (**207a, 208b** and **209b**) from *Schizandra chinensis* (Figure 3.30) (Ikeya et al. 1979)
(g) Cyclobutane lignans, such as (–)-endiandrin A (**210b**) in *Endiandra anthropophagorum* (Davis et al. 2007), and heterotropan (**211**) from *Heterotropa takaoi* (Yamamura et al. 1978)
(h) Polycyclic lignans, such as (–)-sauchinone (**212b**) from *Saururus chinensis* (Wang et al. 1996, Sung and Kim 2000)

Many other inter-unit coupling modes also emerged in the angiosperms, and these and others are recorded in the entries in this *Handbook*. Some examples of this enormous diversity of structure and inter-unit linkage type include the 1–2′-linked (–)-eupomatilone 6 (**213b**) from *Eupomatia bennettii* (Carroll and Taylor 1991) (Figure 3.31); 7–1′ linked lignans chrysophyllon 1A (**214**) from *Licaria chrysophylla*

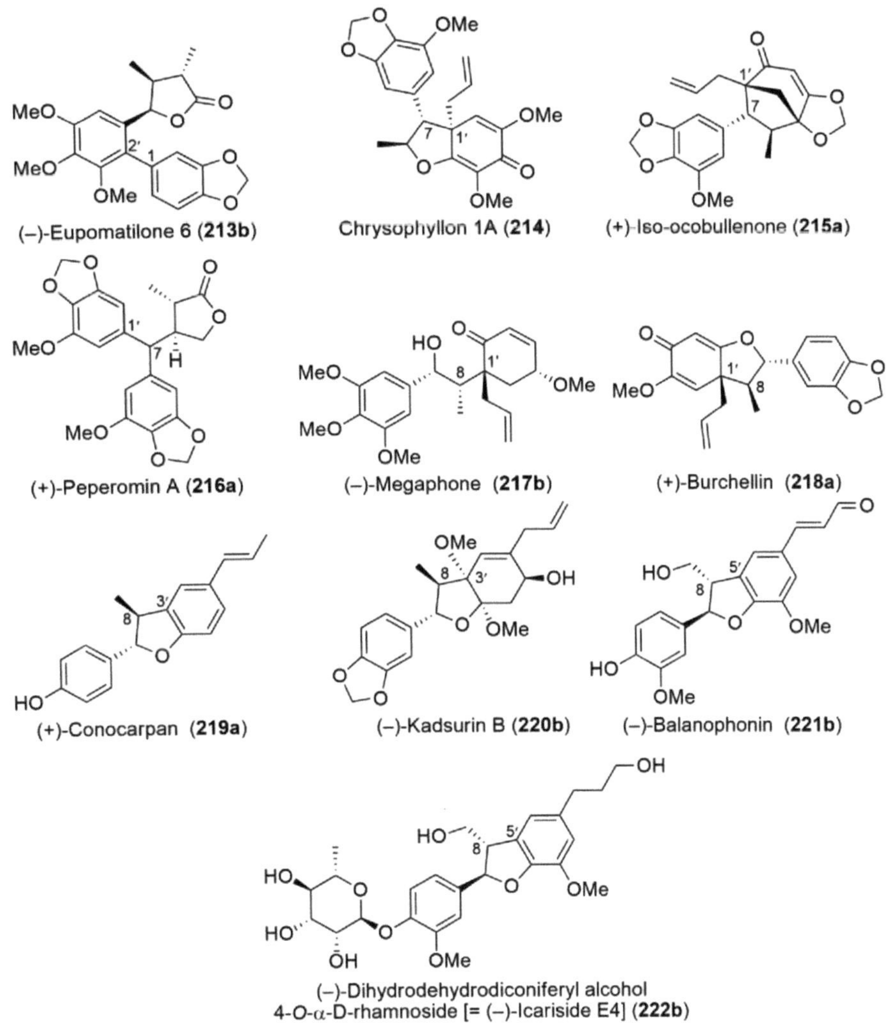

FIGURE 3.31 Various angiosperm lignans of different inter-unit linkage types, where "a" refers to the (+)-enantiomers, and "b" to the (–)-antipodes.

(Lopes et al. 1986), (+)-iso-ocobullenone (**215a**) from *Ocotea bullata* (Drewes et al. 1995) and (+)-peperomin A (**216a**) from *Peperomia japonica* (Chen et al. 1989); 8–1′ linked (–)-megaphone (**217b**) from *Aniba megaphylla* (Kupchan et al. 1978) and (+)-burchellin (**218a**) from *Piper wightii* (Tyagi et al. 1994); 8–3′ (+)-conocarpan (**219a**) and (–)-kadsurin B (**220b**) from *Piper regnellii* (Sartorelli et al. 2001) and *Piper futokadsura* (Chang et al. 1985), respectively; 8–5′ linked (–)-balanophonin (**221b**) from *Balanophora japonica* (Haruna et al. 1982, Yuen et al. 1998) and (–)-dihydrodehydrodiconiferyl alcohol 4-*O*-α-D-rhamnoside [= (–)-icariside E4] (**222b**) from *Epimedium diphyllum* (Miyase et al. 1989) and *Nymphaea odorata* (Zhang et al. 2003); 8–*O*–4′ linked lignans **223b** from *Caryodaphnopsis baviensis* (Anh et al. 1997) (Figure 3.32); (–)-nymphaeoside A (**224b**) from *N. odorata* (Zhang et al. 2003) and (+)-fargesiphenol A (**225a**) from *Magnolia fargesii* (Gao et al. 2012), as well as the benzodioxocin eusiderin B (**226**) from *Licaria rigida* (Braz Filho et al. 1981); 8–2′/3′ linked isofutoquinol A (**227**) from *Piper futokadzura* (Matsui and Munakata 1976) and presumed "rearrangement" products such as kadsurenin I (**228**) from *Piper kadzura* (Ma et al. 1992).

8–8′- or 8–3′-Linked norlignans have also been reported, i.e. such as the dibenzyltetrahydrofuran cestrumoside (**229**) in *Cestrum diurnum* (Mohamed et al. 2007), the benzofuran machicendiol (**230**) in *Machilus glaucescens* (Talapatra et al. 1976), and the arylnaphthalene vitrofolal A (**231**) in *Vitex rotundifolia* (Kawazoe et al. 1999) (Figure 3.33). Note, however, that many lignans purportedly lacking hydroxylated functionalities at C4/C4′ supposedly had instead 3,5-dihydroxylated functionalities. Many structures having this biochemically unexpected 3,5-dihydroxylated substitution pattern have since had their structures revised (Yang et al. 2015). That latter study demonstrated instead the expected patterns of aromatic hydroxylation at C4/C4′, and that the various reports of 3,5-dihydroxylation were incorrect. Whether cestrumoside (**229**) also falls into this category of needing structural revision needs to be established.

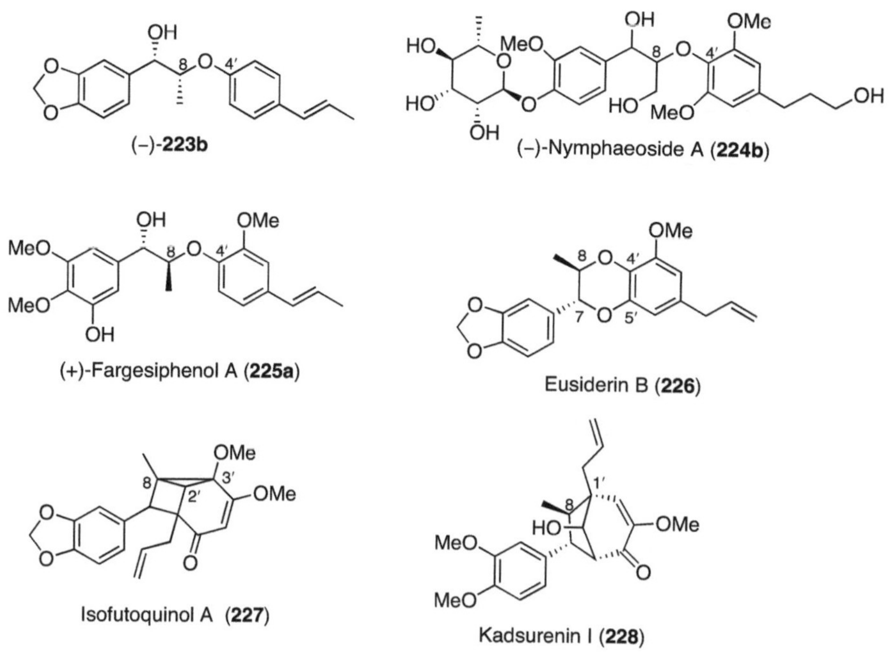

FIGURE 3.32. Various angiosperm lignans illustrating different inter-unit linkage types and structural diversity, where "a" refers to the (+)-enantiomers, and "b" to the (–)-antipodes.

These are just a few examples of the enormous structural diversity of lignans and, to a much lesser extent, of norlignans in the dicotyledons.

3.7.2 Monocotyledons

Few monocotyledons are known to accumulate lignans (Lewis and Davin 1999). Those reported include 5–5′-linked diferulic acid (**232**), which is only released by alkali treatment of monocot cell walls (Hartley and Jones 1976, Markwalder and Neukom 1976) and aryl cyclobutanes, i.e. dihydroxytruxillic/truxinic acids **233**/**234** in cereals and grasses (Hartley and Ford 1989, Ford and Hartley 1990), as well as the 8–8′-linked aryltetrahydronaphthalene lyoniresinol (**235**) and the dilignans phyllostadimers A (**236**) and B (**237**) (Suga et al. 2003) in bamboo (*Phyllostachis edulis*) stems (Figure 3.34). While all three bamboo lignans contain syringyl-like aromatic moieties, the latter two contain 7′–*O*–9/8–*O*–9′ linkages.

Norlignans: Angiosperms apparently have relatively few norlignans reported, these being most often found in the Asparagales.

They include 8–7′-linked (+)-(*Z*)-hinokiresinol (= nyasol, **167a**) in *Asparagus* spp. (Oketch-Rabah et al. 1997, Suzuki et al. 2007), and its (–)-form (**167b**) in *Anemarrhena asphodeloides*

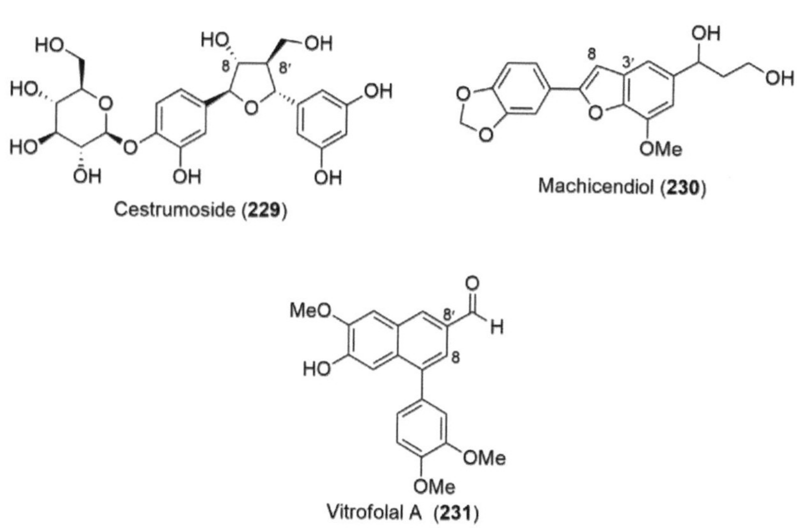

FIGURE 3.33. Examples of dicotyledon norlignans **229–231**.

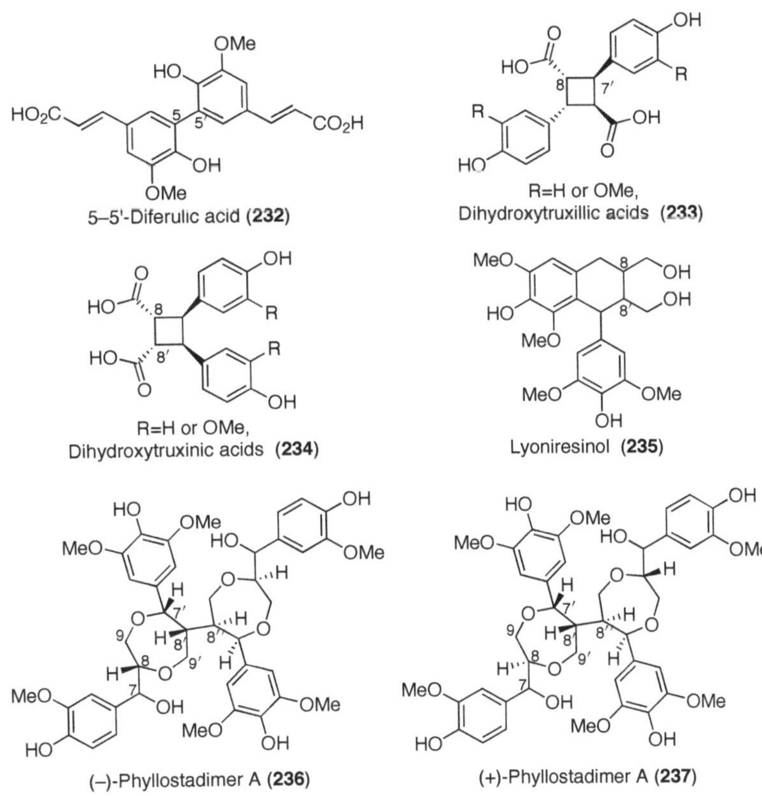

FIGURE 3.34. Examples of monocotyledon lignans (**232–235**) and dilignans (**236** and **237**).

(Nikaido et al. 1981), as well as glycosides **238–240** in *Hypoxis* spp. (Marini-Bettolo et al. 1985, Messana et al. 1989, Betto et al. 1992, Nicoletti et al. 1992) (Figure 3.35). Rare acetylenic norlignans have also been reported. These are asparenydiol (**241**), asparenyol (**242**), and asparenyn (**243**) present in *Asparagus officinalis* (Terada et al. 1995), which have an unusual oxygen insertion to afford aryl ethers; rooperol (**244**) and its glycoside hypoxoside (**245**) in *Hypoxis* spp. (Marini Bettolo et al. 1982, Drewes et al. 1984, Nair and Kanfer 2006), as well as nyasicoside (**246**) in *Hypoxis* spp. (Galeffi et al. 1987) and *Curculigo*

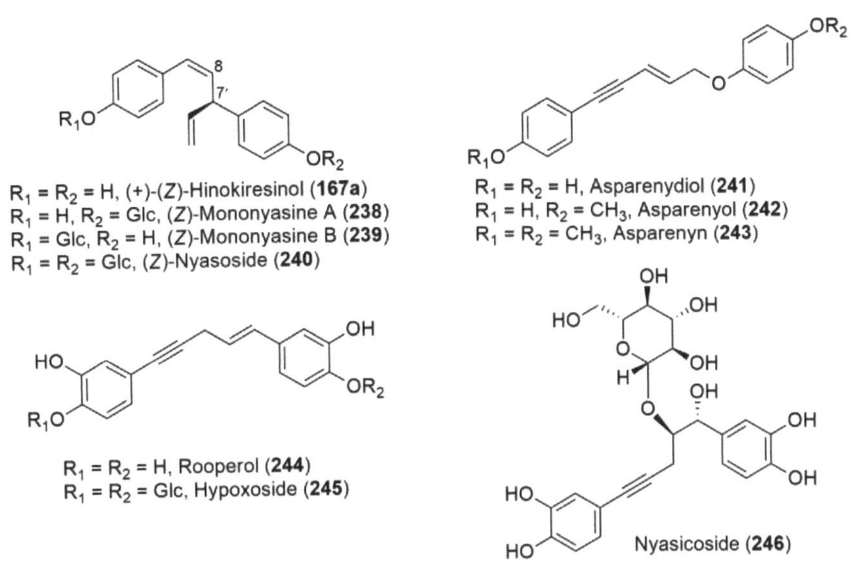

FIGURE 3.35. Examples of norlignans **167a**, **238–240**, and putative norlignans (**241–246**) in monocotyledons; "a" refers to the (+)-enantiomer.

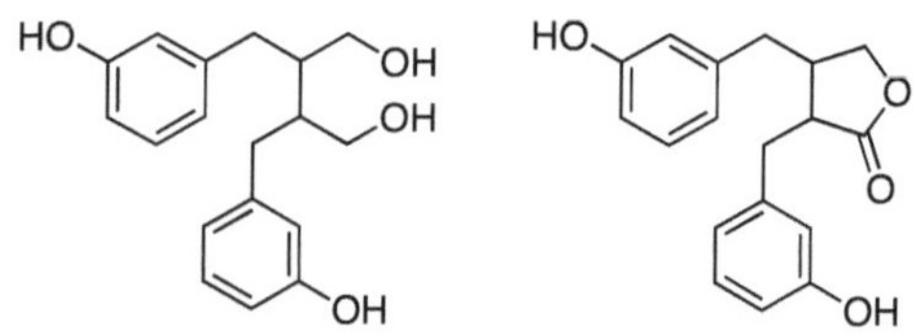

Enterodiol (**247**) Enterolactone (**248**)

FIGURE 3.36. Mammalian lignans.

capitulata (Chang et al. 1999). However, as discussed in Chapter 8 of the *Handbook*, whether these are true norlignans or are diarylpentanoids remains to be established.

Summary: The emergence of the angiosperms (dicotyledons and monocotyledons) was accompanied by an explosion in lignan and norlignan skeletal types and structural variations.

3.8 MAMMALIAN LIGNANS

Dietary ingestion of various plant foodstuffs, such as flax seed and cereals, results in intestinal bacteria that modifies the ingested lignans (and presumably lignins), which are then excreted in urine. The two most common "mammalian lignans" are enterodiol (**247**) and enterolactone (**248**) (Figure 3.36) (see Vassão et al. 2010). Individuals on antibiotics, which kill the intestinal bacteria, are unable to catabolize ingested lignans/lignins and enterodiol (**247**) and enterolactone (**248**) formation does not occur.

Summary: Mammalian lignans are produced by the action of intestinal bacteria on digested lignans and related phenylpropanoids. Their precursor structures are not produced by animals.

ACKNOWLEDGEMENTS

The authors thank the Chemical Sciences, Geosciences and Biosciences Division, DOE Office of Basic Energy Sciences (DE-FG-0397ER20259), as well as the Arthur and Katie Eisig-Tode Endowment.

REFERENCES

Agrawal, P. K., S. K. Agarwal, and R. P. Rastogi. 1980. A new neolignan and other phenolic constituents from *Cedrus deodara*. *Phytochemistry* 19:1260–1261.

Ambrose, B. A. 2013. The morphology and development of lycophytes. *Annual Plant Reviews* 45:91–114.

Anh, N. H., H. Ripperger, A. Porzel, T. V. Sung, and G. Adam. 1997. Neolignans from *Caryodaphnopsis baviensis*. *Phytochemistry* 46:569–571.

Anjaneyulu, A. S. R., C. V. S. Prakash, K. V. S. Raju, and U. V. Mallavadhani. 1992. Isolation of new aromatic derivatives from a marine algal species *Caulerpa racemosa*. *J. Nat. Prod.* 55:496–499.

Asakawa, Y., N. Tokunaga, M. Toyota, T. Takemoto, and C. Suire. 1979. Chemosystematics of bryophytes. 1. The distribution of terpenoids in bryophytes. *J. Hattori Bot. Lab.* 45:395–407.

Asakawa, Y., N. Tokunaga, T. Takemoto, et al. 1980. Chemosystematics of bryophytes. IV. The distribution of terpenoids and aromatic compounds in Hepaticae and Anthocerotae. *J. Hattori Bot. Lab.* 47:153–164.

Atta-ur-Rahman, M. I. Choudhary, S. Hayat, et al. 1999. Spatozoate and varninasterol from the brown alga *Spatoglossum variabile*. *Phytochemistry* 52:495–499.

Awad, N. E. 2002a. Biologically active constituents from the red alga *Jania rubens* (L) Lamx. *Bull. Fac. Pharm. Cairo Univ.* 40:169–174.

Awad, N. E. 2002b. Volatile constituents, carbohydrate compositions and topical anti-inflammatory of the two green algae *Caulerpa racemosa* (O. Dargent) and *Codium tomentosum* (Stackhouse). *Bull. Fac. Pharm. Cairo Univ.* 40:233–243.

Ayer, W. A., and P. P. Singer. 1980. Phenolic metabolites of the bird's nest fungus *Nidula niveo-tomentosa*. *Phytochemistry* 19:2717–2721.

Ayres, D. C., and S. E. Mhasalkar. 1965. Lignans. V. (–)-Olivil and (+)-cycloolivil. *J. Chem. Soc.*:3586–3589.

Azuma, H., and M. Kono. 2006. Estragole (4-allylanisole) is the primary compound in volatiles emitted from the male and female cones of *Cycas revoluta*. *J. Plant Res.* 119:671–676.

Bambagiotti-Alberti, M., S. A. Coron, C. Ghiara, G. Moneti, and A. Raffaelli. 1994. Investigation of mammalian lignan precursors in flax seed: First evidence of secoisolariciresinol diglucoside in two isomeric forms by liquid chromatography/mass spectrometry. *Rapid Commun. Mass Spectrom.* 8:929–932.

Bedir, E., I. I. Tatli, R. A. Khan, et al. 2002. Biologically active secondary metabolites from *Ginkgo biloba*. *J. Agric. Food Chem.* 50:3150–3155.

Begley, M. J., R. V. Davies, P. Henley-Smith, and D. A. Whiting. 1973. Constitution of sequirin-D (*Sequoia sempervirens*), a novel dihydronaphthalene norlignan. *J. Chem. Soc., Chem. Commun.*:649–650.

Berner, M., D. Krug, C. Bihlmaier, et al. 2006. Genes and enzymes involved in caffeic acid biosynthesis in the actinomycete *Saccharothrix espanaensis*. *J. Bacteriol.* 188:2666–2673.

Betto, P., R. Gabriele, and C. Galeffi. 1992. Determination of the norlignan glucosides of Hypoxidaceae by high-performance liquid chromatography. *J. Chromatogr.* 594:131–135.

Böeseken, J., and W. D. Cohen. 1928. Das Sesamin. *Biochem. Z.* 201:454–463.

Bouarab, K., F. Adas, E. Gaquerel, et al. 2004. The innate immunity of a marine red alga involves oxylipins from both the eicosanoid and octadecanoid pathways. *Plant Physiol.* 135:1838–1848.

Braz Filho, R., M. G. De Carvalho, O. R. Gottlieb, J. G. S. Maia, and M. L. Da Silva. 1981. Neolignans from *Licaria rigida*. *Phytochemistry* 20:2049–2050.

Briggs, L. H., R. C. Cambie, and J. L. Hoare. 1959a. Secoisolariciresinol, a new lignan from *Podocarpus spicatus*. *Tetrahedron Lett.* 1:14–15.

Briggs, L. H., R. C. Cambie, and J. L. Hoare. 1959b. Chemistry of the Podocarpaceae III: A new lignan, seco-isolariciresinol and further constituents of the heartwood of *Podocarpus spicatus*. *Tetrahedron* 7:262–269.

Bugni, T. S., G. P. Concepcíon, G. C. Mangalindan, et al. 2002. *p*-Sulfooxyphenylpyruvic acid from the red macro alga *Ceratodictyon spongiosum* and its sponge symbiont *Haliclona cymaeformis*. *Phytochemistry* 60:361–363.

Cao, S., M. M. Radwan, A. Norris, et al. 2006. Cytotoxic and other compounds from *Didymochlaena truncatula* from the Madagascar rain forest. *J. Nat. Prod.* 69:284–286.

Carroll, A. R., and W. C. Taylor. 1991. Constituents of *Eupomatia* species. XII. Isolation of constituents of the tubers and aerial parts of *Eupomatia bennettii* and determination of the structures of new alkaloids from the aerial parts of *E. bennettii* and minor alkaloids of *E. laurina*. *Aust. J. Chem.* 44:1615–1626.

Castro, M. A., M. Gordaliza, J. M. Miguel Del Corral, and A. San Feliciano. 1996. The distribution of lignanoids in the order coniferae. *Phytochemistry* 41:995–1011.

Chang, M. N., G.-Q. Han, B. H. Arison, et al. 1985. Neolignans from *Piper futokadsura*. *Phytochemistry* 24:2079–2082.

Chang, W.-L., C.-H. Chen, and S.-S. Lee. 1999. Three novel constituents from *Curculigo capitulata* and revision of C-2 stereochemistry in nyasicoside. *J. Nat. Prod.* 62:734–739.

Chattopadhyay, S. K., M. Kulshrestha, G. C. Saha, V. Tripathi, and R. P. Sharma. 1999. Studies on the Himalayan yew *Taxus wallichiana*. Part VI. Isolation of nontaxoid constituents. *Indian J. Chem., Sect B* 38:246–247.

Chen, C.-M., F.-Y. Jan, M.-T. Chen, and T.-J. Lee. 1989. Peperomins A, B and C, novel secolignans from *Peperomia japonica*. *Heterocycles* 29:411–414.

Chen, G.-D., Y.-J. Li, H. Gao, et al. 2012. New azaphilones and chlorinated phenolic glycosides from *Chaetomium elatum* with caspase-3 inhibitory activity. *Planta Med.* 78:1683–1689.

Cimino, G., S. De Stefano, and L. Minale. 1975. Occurrence of 4-hydroxyphenylpyruvic acid oxime in the marine sponge *Hymeniacidon sanguinea*. *Experientia* 31:756–757.

Connolly, J. D. 1985. Recent advances in the chemistry of the Hepaticae. In *Natural Products Chemistry* (R. I. Zalewski and J. J. Skolik eds.), Elsevier Science Publishers, Amsterdam. pp. 3–14.

Crandall-Stotler, B., R. E. Stotler, and D. G. Long. 2008. Morphology and classification of the Marchantiophyta. In *Bryophyte Biology* (B. Goffinet and A. J. Shaw eds.), 2nd ed., Cambridge University Press, Cambridge. pp. 1–54.

Cullmann, F., K.-P. Adam, and H. Becker. 1993. Bisbibenzyls and lignans from *Pellia epiphylla*. *Phytochemistry* 34:831–834.

Cullmann, F., K.-P. Adam, J. Zapp, and H. Becker. 1996. Pelliatin, a macrocyclic lignan derivative from *Pellia epiphylla*. *Phytochemistry* 41:611–615.

Cullmann, F., and H. Becker. 1999. Lignans from the liverwort *Lepicolea ochroleuca*. *Phytochemistry* 52:1651–1656.

Cullmann, F., A. Schmidt, F. Schuld, M. L. Trennheuser, and H. Becker. 1999. Lignans from the liverworts *Lepidozia incurvata, Chiloscyphus polyanthos* and *Jungermannia exsertifolia* ssp. *cordifolia*. *Phytochemistry* 52:1647–1650.

Daniels, P., H. Erdtman, K. Nishimura, et al. 1972. Athrotaxin, a C_{17}-phenolic constituent from *Athrotaxis selaginoides* *J. Chem. Soc., Chem. Commun.*:246–247.

Dat, L. D., B. T. Zhao, N. D. Hung, et al. 2017. Lignan derivatives from *Selaginella tamariscina* and their nitric oxide inhibitory effects in LPS-stimulated RAW 264.7 cells. *Bioorg. Med. Chem. Lett.* 27:524–529.

Davin, L. B., C.-Z. Wang, G. L. Helms, and N. G. Lewis. 2003. [^{13}C]-Specific labeling of 8-2′ linked (–)-*cis*-blechnic, (–)-*trans*-blechnic and (–)-brainic acids in the fern *Blechnum spicant*. *Phytochemistry* 62:501–511.

Davin, L. B., and N. G. Lewis. 2022. Redefining lignan/neolignan nomenclature and diversity: A historical perspective. In *The Lignan Handbook* (N. G. Lewis and L. B. Davin eds.), Taylor & Francis, Boca Raton, London. pp 1–11.

Davis, R. A., A. R. Carroll, S. Duffy, et al. 2007. Endiandrin A, a potent glucocorticoid receptor binder isolated from the Australian plant *Endiandra anthropophagorum*. *J. Nat. Prod.* 70:1118–1121.

Delwiche, C. F., L. E. Graham, and N. Thomson. 1989. Lignin-like compounds and sporopollenin in *Coleochaete*, an algal model for land plant ancestry. *Science* 245:399–401.

Doebner, O., and E. Lücker. 1896. *Guaiacum* resin. *Archiv. Pharm.* 234:590–610.

Dovgan, I. V., and E. I. Medvedeva. 1983. Change in the structural elements of the lignin of the brown alga *Cystoseira barbata* at different ages. *Chem. Nat. Compd.* 19:81–84.

Dovgan, I. V., E. I. Medvedeva, and E. N. Yanishevskaya. 1983. Cleavage of the lignins of the alga *Cystoseira barbata* by thioacetic acid. *Chem. Nat. Compd.* 19:84–87.

Drewes, S. E., A. J. Hall, R. A. Learmonth, and U. J. Upfold. 1984. Isolation of hypoxoside from *Hypoxis rooperi* and synthesis of (*E*)-1,5-*bis*(3′,4′-dimethoxyphenyl)pent-4-ene-1-yne. *Phytochemistry* 23:1313–1316.

Drewes, S. E., M. M. Horn, B. M. Sehlapelo, et al. 1995. *Iso*-ocobullenone and a neolignan ketone from *Ocotea bullata* bark. *Phytochemistry* 38:1505–1508.

Duff, R. J., J. C. Villarreal, D. C. Cargill, and K. S. Renzaglia. 2007. Progress and challenges toward developing a phylogeny and classification of the hornworts. *Bryologist* 110:214–243.

Easterfield, T. H., and J. Bee. 1910. The resin acids of the coniferae. Part II. Matairesinol. *J. Chem. Soc. Trans.* 97:1028–1032.

El-Gendy, M. M. A., U. W. Hawas, and M. Jaspars. 2008. Novel bioactive metabolites from a marine derived bacterium *Nocardia* sp. ALAA 2000. *J. Antibiot.* 61:379–386.

Emiliani, G., M. Fondi, R. Fani, and S. Gribaldo. 2009. A horizontal gene transfer at the origin of phenylpropanoid metabolism: A key adaptation of plants to land. *Biol. Direct* 4:7.

Enoki, A., S. Takahama, and K. Kitao. 1977a. The extractive of Metasekoia, *Metasequoia glyptostroboides* Hu et Cheng. II. The isolation of hydroxyathrotaxin, metasequirin-B and hydroxymetasequirin-A. *Mokuzai Gakkaishi* 23:587–593.

Enoki, A., S. Takahama, and K. Kitao. 1977b. The extractives of Metasekoia, *Metasequoia glyptostroboides* Hu et Cheng. I. The isolation of metasequirin-A, athrotaxin and agatharesinol from the heartwood. *Mokuzai Gakkaishi* 23:579–586.

Enzell, C. R., and B. R. Thomas. 1965. Chemistry of the order Araucariales. II. The wood resin of *Agathis australis*. *Acta Chem. Scand.* 19:913–919.

Enzell, C. R., Y. Hirose, and B. R. Thomas. 1967. The chemistry of the order Araucariales. VI. Absolute configurations of agatharesinol, hinokiresinol and sugiresinol. *Tetrahedron Lett.* 8:793–798.

Erxleben, A., A. Gessler, M. Vervliet-Scheebaum, and R. Reski. 2012. Metabolite profiling of the moss *Physcomitrella patens* reveals evolutionary conservation of osmoprotective substances. *Plant Cell Rep.* 31:427–436.

Eyberger, A. L., R. Dondapati, and J. R. Porter. 2006. Endophyte fungal isolates from *Podophyllum peltatum* produce podophyllotoxin. *J. Nat. Prod.* 69:1121–1124.

Fang, J.-M., K.-C. Hsu, and Y.-S. Cheng. 1989. Lignans from leaves of *Calocedrus formosana*. *Phytochemistry* 28:3553–3555.

Fang, J.-M., M.-Y. Liu, and Y.-S. Cheng. 1990. Lignans from wood of *Calocedrus formosana*. *Phytochemistry* 29:3048–3049.

Fang, J.-M., C.-K. Lee, and Y.-S. Cheng. 1992. Lignans from leaves of *Juniperus chinensis*. *Phytochemistry* 31:3659–3661.

Feild, T. S., N. C. Arens, J. A. Doyle, T. E. Dawson, and M. J. Donoghue. 2004. Dark and disturbed: A new image of early angiosperm ecology. *Paleobiology* 30:82–107.

Feng, W.-s., K.-k. Li, and X.-k. Zheng. 2011. A new norlignan lignanoside from *Selaginella moellendorffii* Hieron. *Acta Pharm. Sin. B* 1:36–39.

Figueiredo, A. C., M. Sim-Sim, J. G. Barroso, et al. 2002. Composition of the essential oil from the liverwort *Marchesinia mackaii* (Hook.) S. F. Gray grown in Portugal. *J. Essent. Oil Res.* 14:439–442.

Fill, T. P., R. M. Geris dos Santos, A. Barisson, E. Rodrigues-Filho, and A. Q. L. Souza. 2009. Co-production of bisphenylpropanoid amides and meroterpenes by an endophytic *Penicillium brasilianum* found in the root bark of *Melia azedarach*. *Z. Naturforsch., C: Biosci.* 64:355–360.

Ford, C. W., and R. D. Hartley. 1990. Cyclodimers of *p*-coumaric and ferulic acids in the cell walls of tropical grasses. *J. Sci. Food Agric.* 50:29–43.

Forrest, L. L., C. E. Davis, D. G. Long, et al. 2006. Unraveling the evolutionary history of the liverworts (Marchantiophyta): Multiple taxa, genomes and analyses. *Bryologist* 109:303–334.

Fritz, R. R., D. S. Hodgins, and C. W. Abell. 1976. Phenylalanine ammonia-lyase. Induction and purification from yeast and clearance in mammals. *J. Biol. Chem.* 251:4646–4650.

Fukuda, T., Y. Sudoh, Y. Tsuchiya, et al. 2011. Marianins A and B, prenylated phenylpropanoids from *Mariannaea camptospora*. *J. Nat. Prod.* 74:1327–1330.

Galeffi, C., G. Multari, J. D. Msonthi, M. Nicoletti, and G. B. Marini-Bettolo. 1987. Research on African medicinal plants. XIII. Nyasicoside, a new glucoside of *Hypoxis nyasica* Bak. *Tetrahedron* 43:3519–3522.

Gang, D. R., M. Fujita, L. B. Davin, and N. G. Lewis. 1998. The "abnormal lignins": Mapping heartwood formation through the lignan biosynthetic pathway. In *Lignin and Lignan Biosynthesis* (N. G. Lewis and S. Sarkanen eds.), Vol. 697, ACS Symposium Series, Washington, DC. pp. 389–421.

Gao, X., Y. Shen, L. Yang, et al. 2012. 8-*O*-4′-Neolignans from flower buds of *Magnolia fargesii* and their biological activities. *J. Braz. Chem. Soc.* 23:1274–1279.

Gardner, J. A. F., G. M. Barton, and H. MacLean. 1959. The polyoxyphenols of western red cedar (*Thuja plicata* Donn.) I. Isolation and preliminary characterization of plicatic acid. *Can. J. Chem.* 37:1703–1709.

Gardner, J. A. F., B. F. MacDonald, and H. MacLean. 1960. The polyoxyphenols of western red cedar (*Thuja plicata* Donn) II. Degradation studies on plicatic acid, a possible lignan acid. *Can. J. Chem.* 38:2387–2394.

Gardner, J. A. F., E. P. Swan, S. A. Sutherland, and H. MacLean. 1966. The polyoxyphenols of western red cedar (*Thuja plicata* Donn) III. Structure of plicatic acid. *Can. J. Chem.* 44:52–58.

Gifford, E. M., and A. S. Foster. 1989. *Morphology and Evolution of Vascular Plants*. 3rd ed. New York: W. H. Freeman and Company.

Gözler, B., B. Kivçak, G. Arar, T. Gözler, and M. Hesse. 1994. (–)-Padocin: A novel epoxylignan from *Haplophyllum cappadocicum*. *Heterocycles* 39:243–250.

Gradstein, S. R., R. Matsuda, and Y. Asakawa. 1981. Studies on Colombian cryptogams. XIII. Oil bodies and terpenoids in Lejeuneaceae and other selected Hepaticae. *J. Hattori Bot. Lab.* 50:231–248.

Gutierrez-Lugo, M.-T., G. M. Woldemichael, M. P. Singh, et al. 2005. Isolation of three new naturally occurring compounds from the culture of *Micromonospora* sp. P1068. *Nat. Prod. Res.* 19:645–652.

Hänsel, R., and D. Zander. 1961. Lignans of Congo cubebs. *Arch. Pharm. Ber. Dtsch. Pharm. Ges.* 294:699–713.

Hartley, R. D., and E. C. Jones. 1976. Diferulic acid as a component of cell walls of *Lolium multiflorum*. *Phytochemistry* 15:1157–1160.

Hartley, R. D., and C. W. Ford. 1989. Phenolic constituents of plant cell walls and wall biodegradability. In *Plant Cell Wall Polymers. Biogenesis and Biodegradation* (N. G. Lewis and M. G. Paice eds.), Vol. 399, ACS Symposium Series, Washington, DC. pp 137–145.

Hartwell, J. L., and A. W. Schrecker. 1951. Components of podophyllin. V. The constitution of podophyllotoxin. *J. Am. Chem. Soc.* 73:2909–2916.

Haruna, M., T. Koube, K. Ito, and H. Murata. 1982. Balanophonin, a new neo-lignan from *Balanophora japonica* Makino. *Chem. Pharm. Bull.* 30:1525–1527.

Hatam, N. A. R., and D. A. Whiting. 1967. Constituents of Californian redwood. The structures of sequirin B and C. *Tetrahedron Lett.* 8:781–785.

Hatam, N. A. R., and D. A. Whiting. 1969. The constituents of Californian redwood: The constitution, absolute stereochemistry, and chemistry of sequirin-B and sequirin-C. *J. Chem. Soc., C*:1921–1932.

Haworth, R. D., C. R. Mavin, and G. Sheldrick. 1934. The constituents of *Guaiacum* resin. Part II. Synthesis of *dl*-guaiaretic acid dimethyl ether. *J. Chem. Soc.*:1423–1429.

Hayat, S., Atta ur Rahman, M. I. Choudhary, K. M. Khan, and A. Abbaskhan. 2002. Two new cinnamic acid esters from marine brown alga *Spatoglossum variabile*. *Chem. Pharm. Bull.* 50:1297–1299.

Heinig, U., S. Scholz, and S. Jennewein. 2013. Getting to the bottom of Taxol biosynthesis by fungi. *Fungal Diversity* 60:161–170.

Heinonen, S., T. Nurmi, K. Liukkonen, et al. 2001. *In vitro* metabolism of plant lignans: New precursors of mammalian lignans enterolactone and enterodiol. *J. Agric. Food Chem.* 49:3178–3186.

Henley-Smith, P., and D. A. Whiting. 1976. New norlignans of *Sequoiadendron gigantea;* phytochemical comparison with *Sequoia sempervirens*. *Phytochemistry* 15:1285–1287.

Hodgins, D. S. 1971. Yeast phenylalanine ammonia-lyase: Purification, properties, and the identification of catalytically essential dehydroalanine. *J. Biol. Chem.* 246:2977–2985.

Holmbom, B., C. Eckerman, P. Eklund, et al. 2003. Knots in trees: A new rich source of lignans. *Phytochem. Rev.* 2:331–340.

Hung, T. M., J. S. Lee, N. N. Chuong, et al. 2015. Kinetics and molecular docking studies of cholinesterase inhibitors derived from water layer of *Lycopodiella cernua* (L.) Pic. Serm. (II). *Chem.-Biol. Interact.* 240:74–82.

Hwang, J. K., S. G. A. Moinuddin, L. B. Davin, and N. G. Lewis. 2020. Pinoresinol-lariciresinol reductase: Substrate versatility, enantiospecificity, and kinetic properties. *Chirality* 32:770–789.

Ikeya, Y., H. Taguchi, I. Yosioka, and H. Kobayashi. 1979. The constituents of *Schizandra chinensis* Baill. I. Isolation and structure determination of five new lignans, gomisin A, B, C, F and G, and the absolute structure of schizandrin. *Chem. Pharm. Bull.* 27:1383–1394.

Iliya, I., Z. Ali, T. Tanaka, et al. 2003. Stilbene derivatives from *Gnetum gnemon* Linn. *Phytochemistry* 62:601–606.

Ishii, T., T. Okino, M. Suzuki, and Y. Machiguchi. 2004. Tichocarpols A and B, two novel phenylpropanoids with feeding-deterrent activity from the red alga *Tichocarpus crinitus*. *J. Nat. Prod.* 67:1764–1766.

Jendresen, C. B., S. G. Stahlhut, M. Li, et al. 2015. Highly active and specific tyrosine ammonia-lyases from diverse origins enable enhanced production of aromatic compounds in bacteria and *Saccharomyces cerevisiae*. *Appl. Environ. Microbiol.* 81:4458–4476.

Jiao, Y., J. Fang, S. Tang, and H. Duan. 2015. Penangianol A and B: Two new norlignans from rhizomes of *Abacopteris penangiana*. *Chem. Nat. Compd.* 51:232–235.

Jin, Z., Y. Matsumoto, T. Tange, et al. 2005. Proof of the presence of guaiacyl-syringyl lignin in *Selaginella tamariscina*. *J. Wood Sci.* 51:424–426.

Johansson, C. I., J. N. Saddler, and R. P. Beatson. 2000. Characterization of the polyphenolics related to the colour of western red cedar (*Thuja plicata* Donn) heartwood. *Holzforschung* 54:246–254.

Kariyone, T., H. Kimura, and I. Nakamura. 1958. Studies on the components of *Ginkgo biloba* L. Isolation of bilobanone. *Yakugaku Zasshi* 78:1152–1155.

Kawamura, F., H. Ohashi, S. Kawai, F. Teratani, and Y. Kai. 1996. Photodiscoloration of western hemlock (*Tsuga heterophylla*) sapwood. II. Structures of constituents causing photodiscoloration. *Mokuzai Gakkaishi* 42:301–307.

Kawazoe, K., A. Yutani, and Y. Takaishi. 1999. Arylnaphthalene norlignans from *Vitex rotundifolia*. *Phytochemistry* 52:1657–1659.

Kim, S.-J., M.-R. Kim, D. L. Bedgar, et al. 2004. Functional reclassification of the putative cinnamyl alcohol dehydrogenase multigene family in *Arabidopsis*. *Proc. Natl. Acad. Sci. USA* 101:1455-1460.

Kim, S.-J., D. G. Vassão, S. G. A. Moinuddin, et al. 2014. Allyl/propenyl phenol synthases from the creosote bush and engineering production of specialty/commodity chemicals, eugenol/isoeugenol, in *Escherichia coli*. *Arch. Biochem. Biophys.* 541:37–46.

Koeduka, T., E. Fridman, D. R. Gang, et al. 2006. Eugenol and isoeugenol, characteristic aromatic constituents of spices, are biosynthesized via reduction of a coniferyl alcohol ester. *Proc. Natl. Acad. Sci. USA* 103:10128–10133.

Konno, C., M. Tamada, K. Endo, and H. Hikino. 1980. Structure of ephedradine C, a hypotensive principle of *Ephedra* roots. *Heterocycles* 14:295–298.

Konno, C., Z.-Z. Lu, H.-Z. Xue, et al. 1990. Furanoid lignans from *Larrea tridentata*. *J. Nat. Prod.* 53:396–406.

Koshino, K., S.-I. Terada, T. Yoshihara, et al. 1988. Three phenolic acid derivatives from stroma of *Epichloe typhina* on *Phleum pratense*. *Phytochemistry* 27:1333–1338.

Kour, A., A. S. Shawl, S. Rehman, et al. 2008. Isolation and identification of an endophytic strain of *Fusarium oxysporum* producing podophyllotoxin from *Juniperus recurva*. *World J. Microbiol. Biotechnol.* 24:1115–1121.

Krahmer, R. L., R. W. Hemingway, and W. E. Hillis. 1970. The cellular distribution of lignans in *Tsuga heterophylla* wood. *Wood Sci. Technol.* 4:122–139.

Kraus, C., and G. Spiteller. 1997. Comparison of phenolic compounds from galls and shoots of *Picea glauca*. *Phytochemistry* 44:59–67.

Kupchan, S. M., K. L. Stevens, E. A. Rohlfing, et al. 1978. New cytotoxic neolignans from *Aniba megaphylla* Mez. *J. Org. Chem.* 43:586–590.

Kyndt, J. A., T. E. Meyer, M. A. Cusanovich, and J. J. Van Beeumen. 2002. Characterization of a bacterial tyrosine ammonia lyase, a biosynthetic enzyme for the photoactive yellow protein. *FEBS Lett.* 512:240–244.

Labeeuw, L., P. T. Martone, Y. Boucher, and R. J. Case. 2015. Ancient origin of the biosynthesis of lignin precursors. *Biol. Direct* 10:23.

Laskar, D. D., O. R. A. Corea, A. M. Patten, et al. 2010. Vascular plant lignification: Biochemical/structural biology considerations of upstream aromatic amino acid and monolignol pathway. In *Comprehensive Natural Products II Chemistry and Biology* (L. Mander and H. W. Lui eds.), Vol. 6. *Carbohydrates, Nucleosides and Nucleic Acids, Elsevier*, Oxford, UK. pp. 541–604.

Lewis, N. G., and L. B. Davin. 1999. Lignans: Biosynthesis and function. In *Comprehensive Natural Products Chemistry* (Sir D. H. R. Barton, K. Nakanishi and O. Meth-Cohn eds.), Vol. 1. *Polyketides and Other Secondary Metabolites, Including Fatty Acids and Their Derivatives*. pp. 639–712.

Lewis, N. G., L. B. Davin, and S. Sarkanen. 1999. The nature and function of lignins. In *Comprehensive Natural Products Chemistry* (Sir D. H. R. Barton, K. Nakanishi and O. Meth-Cohn eds.), Vol. 3. *Carbohydrates and Their Derivatives Including Tannins, Cellulose and Related Lignins*, Elsevier, Oxford. pp. 617–745.

Lewis, N. G., and L. B. Davin. 2022. *The Lignan Handbook*. Boca Raton, London: Taylor & Francis.

Lin, R. C., A.-L. Skaltsounis, E. Seguin, F. Tillequin, and M. Koch. 1994. Phenolic constituents of *Selaginella doederleinii*. *Planta Med.* 60:168–170.

Liu, N., D.-X. Guo, Y.-Y. Wang, et al. 2011. Aromatic compounds from the liverwort *Conocephalum japonicum*. *Nat. Prod. Commun.* 6:49–52.

Long, H.-P., F.-S. Li, K.-P. Xu, et al. 2014. Bioactive compounds from *Selaginella involven* Spring that protect PC-12 cells. *Chin. Chem. Lett.* 25:805–808.

Lopes, M. N., M. S. Da Silva, J. M. Barbosa Filho, et al. 1986. Unusual benzofuranoid neolignans from *Licaria chrysophylla*. *Phytochemistry* 25:2609–2612.

Lopes, N. P., E. E. de Almeida Blumenthal, A. J. Cavalheiro, M. J. Kato, and M. Yoshida. 1996. Lignans, γ-lactones and propiophenones of *Virola surinamensis*. *Phytochemistry* 43:1089–1092.

Ma, G.-L., J. Xiong, G.-X. Yang, et al. 2016. Biginkgosides A-I, unexpected minor dimeric flavonol diglycosidic truxinate and truxillate esters from *Ginkgo biloba* leaves and their antineuroinflammatory and neuroprotective activities. *J. Nat. Prod.* 79:1354–1364.

Ma, Y., G. Q. Han, T. Y. Jiang, and X. B. Yang. 1992. Neolignans from *Piper kadsura*. II. *Chin. Chem. Lett.* 3:635–636.

Mahjoub, M. A., S. Ammar, K. Majouli, and Z. Mighri. 2012. Two new alkaloids and a new polyphenolic compound from *Cotula coronopifolia*. *Chem. Nat. Compd.* 47:955–958.

Mancini, I., G. Guella, A. Defant, et al. 1998. Polar metabolites of the tropical green seaweed *Caulerpa taxifolia* which is spreading in the Mediterranean Sea. Glycoglycerolipids and stable enols (α-keto esters). *Helv. Chim. Acta* 81:1681–1691.

Marini-Bettolo, G. B., M. Nicoletti, I. Messana, et al. 1985. Research on African medicinal plants, X. Glucosides of *Hypoxis nyasica* Bak. The structure of nyasoside, a new glucoside biologically related to hypoxoside. *Tetrahedron* 41:665–670.

Marini Bettolo, G. B., M. Patamia, M. Nicoletti, C. Galeffi, and I. Messana. 1982. Research on African medicinal plants. II. Hypoxoside, a new glycoside of uncommon structure from *Hypoxis obtusa* Busch. *Tetrahedron* 38:1683–1687.

Markwalder, H. U., and H. Neukom. 1976. Diferulic acid as a possible crosslink in hemicelluloses from wheat germ. *Phytochemistry* 15:836–837.

Martin, R., and J. Reichling. 1992. NIH-shift during biosynthesis of epoxy-pseudoisoeugenol(2-methylbutyrate) in tissue cultures of *Pimpinella anisum*. *Phytochemistry* 31:511–514.

Martini, U., J. Zapp, and H. Becker. 1998. Lignans from the liverwort *Bazzania trilobata*. *Phytochemistry* 49:1139–1146.

Martone, P. T., J. M. Estevez, F. Lu, et al. 2009. Discovery of lignin in seaweed reveals convergent evolution of cell-wall architecture. *Curr. Biol.* 19:169–175.

Matsui, K., and K. Munakata. 1976. Four new neolignans from *Piper futokadzura*. *Tetrahedron Lett.* 17:4371–4374.

Messana, I., J. D. Msonthi, Y. De Vicente, G. Multari, and C. Galeffi. 1989. Research on African medicinal plants. Part 16. Mononyasine A and mononyasine B: Two glucosides from *Hypoxis nyasica*. *Phytochemistry* 28:2807–2809.

Miyase, T., A. Ueno, N. Takizawa, H. Kobayashi, and H. Oguchi. 1989. Ionone and lignan glycosides from *Epimedium diphyllum*. *Phytochemistry* 28:3483–3485.

Mohamed, K. M., M. A. Fouad, K. Matsunami, M. S. Kamel, and H. Otsuka. 2007. A new norlignan glycoside from *Cestrum diurnum* L. *ARKIVOC*:63–70.

Moinuddin, S. G. A., L. B. Davin, and N. G. Lewis. 2022. Oligomeric lignan diversity *in planta*: A biochemical roadmap for the future. In *The Lignan Handbook* (N. G. Lewis and L. B. Davin eds.), Taylor & Francis, Boca Raton, London. pp 167–190.

Moinuddin, S. G. A., S. Hishiyama, M.-H. Cho, L. B. Davin, and N. G. Lewis. 2003. Synthesis and chiral HPLC analysis of the dibenzyltetrahydrofuran lignans, larreatricins, 8′-*epi*-larreatricins, 3,3′-didemethoxyverrucosins and *meso*-3,3′-didemethoxynectandrin B in the creosote bush (*Larrea tridentata*): Evidence for regiospecific control of coupling. *Org. Biomol. Chem.* 1:2307–2313.

Moore, B. S., C. Hertweck, J. N. Hopke, et al. 2002. Plant-like biosynthetic pathways in bacteria: From benzoic acid to chalcone. *J. Nat. Prod.* 65:1956–1962.

Moore, K., P. V. Subba Rao, and G. H. N. Towers. 1968. Degradation of phenylalanine and tyrosine by *Sporobolomyces roseus*. *Biochem. J.* 106:507–514.

Moussa, M. Y. 2003. Volatile constituents of the red alga: *Hypnea musciformis* (Wulfen) Lamouroux, green alga: *Enteromorpha compressa* (L.) Nees and their antimicrobial activities. *Bull. Fac. Pharm. Cairo Univ.* 41:139–144.

Nadeem, M., M. Ram, P. Alam, et al. 2012. *Fusarium solani*, P1, a new endophytic podophyllotoxin-producing fungus from roots of *Podophyllum hexandrum*. *Afr. J. Microbiol. Res.* 6:2493–2499.

Nagashima, F., M. Toyota, and Y. Asakawa. 1990. Terpenoids from some Japanese liverworts. *Phytochemistry* 29:2169–2174.

Nair, V. D. P., and I. Kanfer. 2006. High-performance liquid chromatographic method for the quantitative determination of hypoxoside in African potato (*Hypoxis hemerocallidea*) and in commercial products containing the plant material and/or its extracts. *J. Agric. Food Chem.* 54:2816–2821.

Nawwar, M. A. M., H. H. Barakat, J. Buddrus, and M. Linscheid. 1985. Alkaloidal, lignan and phenolic constituents of *Ephedra alata*. *Phytochemistry* 24:878–879.

Nehls, U., M. Ecke, and R. Hampp. 1999. Sugar- and nitrogen-dependent regulation of an *Amanita muscaria* phenylalanine ammonium lyase gene. *J. Bacteriol.* 181:1931–1933.

Němejc, F., and B. Pacltová. 1974. Hepaticae in the Senonian of South Bohemia. *Palaeobotanist* 21:23–26.

Ngan, N. T. T., T. H. Quang, S. B. Song, and Y. H. Kim. 2014. Anti-inflammatory and PPAR transactivational effects of components from *Ginkgo biloba* seeds. *Bull. Korean Chem. Soc.* 35:2843–2846.

Nickrent, D. L., C. L. Parkinson, J. D. Palmer, and R. J. Duff. 2000. Multigene phylogeny of land plants with special reference to bryophytes and the earliest land plants. *Mol. Biol. Evol.* 17:1885–1895.

Nicoletti, M., C. Galeffi, I. Messana, and G. B. Marini-Bettolo. 1992. Hypoxidaceae. Medicinal uses and the norlignan constituents. *J. Ethnopharmacol.* 36:95–101.

Nikaido, T., T. Ohmoto, H. Noguchi, et al. 1981. Inhibitors of cyclic AMP phosphodiesterase in medicinal plants. *Planta Med.* 43:18–23.

Nishibe, S., A. Sakushima, S. Kitagawa, et al. 1988. Phenolic compounds from *Forsythia* leaves (III). On the comparison of constituents between hybrid and parents. *Shoyakugaku Zasshi* 42:324–328.

Nose, M., M. A. Bernards, M. Furlan, et al. 1995. Towards the specification of consecutive steps in macromolecular lignin assembly. *Phytochemistry* 39:71–79.

Nukina, M., and S. Marumo. 1977. α-Acetylorcinol from *Cochliobolus lunata*. *Agric. Biol. Chem.* 41:717.

Ohashi, H., S. Kawai, Y. Sakurai, and M. Yasue. 1992. Norlignan from the knot resin of *Araucaria angustifolia*. *Phytochemistry* 31:1371–1373.

Oketch-Rabah, H. A., S. F. Dossaji, S. Brøgger Christensen, et al. 1997. Antiprotozoal compounds from *Asparagus africanus*. *J. Nat. Prod.* 60:1017–1022.

Parmar, V. S., S. C. Jain, K. S. Bisht, et al. 1997. Phytochemistry of the genus *Piper*. *Phytochemistry* 46:597–673.

Pérez, C., L. N. Almonacid, J. M. Trujillo, et al. 1995. A new irregular phenylpropanoid from *Apollonias barbujana*. *Planta Med.* 61:535–536.

Plant List, The. 2013. Version 1.1. Published on the Internet, http://www.theplantlist.org/ (accessed June 2018).

Pullela, S. V., S. Takamatsu, S. I. Khan, and I. A. Khan. 2005. Isolation of lignans and biological activity studies of *Ephedra viridis*. *Planta Med.* 71:789–791.

Puri, S. C., A. Nazir, R. Chawla, et al. 2006. The endophytic fungus *Trametes hirsuta* as a novel alternative source of podophyllotoxin and related aryl tetralin lignans. *J. Biotechnol.* 122:494–510.

Qiu, Y.-L., L. Li, B. Wang, et al. 2006. The deepest divergences in land plants inferred from phylogenomic evidence. *Proc. Natl. Acad. Sci. USA* 103:15511–15516.

Radulović, N., G. Stojanović, and R. Palić. 2006. Composition and antimicrobial activity of *Equisetum arvense* L. essential oil. *Phytother. Res.* 20:85–88.

Rasmussen, S., C. Wolff, and H. Rudolph. 1996. 4′-*O*-β-D-Glucosyl-*cis*-*p*-coumaric acid. A natural constituent of *Sphagnum fallax* cultivated in bioreactors. *Phytochemistry* 42:81–87.

Rensing, S. A., D. Lang, A. D. Zimmer, et al. 2008. The *Physcomitrella* genome reveals evolutionary insights into the conquest of land by plants. *Science* 319:64–69.

Reznikov, V. M., M. F. Mikhaseva, and M. A. Zil'bergleit. 1978. The lignin of the alga *Fucus vesiculosus*. *Chem. Nat. Compd.* 14:554–556.

Riggs, N. V., and J. D. Stevens. 1962. Some constituents of *Piptocalyx moorei* Oliv. *Aust. J. Chem.* 15:305–313.

San Feliciano, A., M. Medarde, J. L. Lopez, et al. 1989. Lignans from *Juniperus thurifera*. *Phytochemistry* 28:2863–2866.

San Feliciano, A., J. M. M. Del Corral, M. Gordaliza, and A. Castro. 1991. Acidic and phenolic lignans from *Juniperus sabina*. *Phytochemistry* 30:3483–3485.

Sartorelli, P., P. J. C. Benevides, R. M. Ellensohn, et al. 2001. Enantioselective conversion of *p*-hydroxypropenylbenzene to (+)-conocarpan in *Piper regnellii*. *Plant Sci.* 161:1083–1088.

Satake, T., T. Murakami, Y. Saiki, and C.-M. Chen. 1978. Chemische Untersuchungen der Inhaltsstoffe von *Pteris vittata* L. *Chem. Pharm. Bull.* 26:1619–1622.

Savidge, R. A. 2007. Wood anatomy of Late Triassic trees in Petrified Forest National Park, Arizona, USA, in relation to *Araucarioxylon arizonicum*, Knowlton, 1889. *Bulletin of Geosciences* 82:301–328.

Savidge, R. A. 2008. Learning from the past: The origin of wood. *The Forestry Chronicle* 84:498–503.

Scher, J. M., J. Zapp, and H. Becker. 2003. Lignan derivatives from the liverwort *Bazzania trilobata*. *Phytochemistry* 62:769–777.

Schneider, H. 2013. Evolutionary morphology of ferns (Monilophytes). *Annual Plant Reviews* 45:115–140.

Schroeter, G., L. Lichtenstadt, and D. Irineu. 1918. Über die Konstitution der Guajacharz-Subztanzen. *Chem. Ber.* 51:1587–1613.

Seshime, Y., P. R. Juvvadi, I. Fujii, and K. Kitamoto. 2005. Genomic evidences for the existence of a phenylpropanoid metabolic pathway in *Aspergillus oryzae*. *Biochem. Biophys. Res. Commun.* 337:747–751.

Shigemori, H., M. Tenma, K. Shimazaki, and J. i. Kobayashi. 1998. Three new metabolites from the marine yeast *Aureobasidium pullulans*. *J. Nat. Prod.* 61:696–698.

Sikora, L. A., and G. A. Marzluf. 1982. Regulation of L-phenylalanine ammonia-lyase by L-phenylalanine and nitrogen in *Neurospora crassa*. *J. Bacteriol.* 150:1287–1291.

Silber, M. V., H. Meimberg, and J. Ebel. 2008. Identification of a 4-coumarate:CoA ligase gene family in the moss, *Physcomitrella patens*. *Phytochemistry* 69:2449–2456.

Soltis, P. S., and D. E. Soltis. 2004. The origin and diversification of angiosperms. *Am. J. Bot.* 91:1614–1626.

Su, W.-C., J.-M. Fang, and Y.-S. Cheng. 1995. Flavonoids and lignans from leaves of *Cryptomeria japonica*. *Phytochemistry* 40:563–566.

Suga, A., Y. Takaishi, S. Goto, et al. 2003. Two lignan dimers from bamboo stems (*Phyllostachys edulis*). *Phytochemistry* 64:991–996.

Sugijanto, N. E., A. Diesel, R. Ebel, G. Indrayanto, and N. C. Zaini. 2009. Chemical constituents of the endophytic fungus *Lecythophora* sp. isolated from *Alyxia reinwardtii*. *Nat. Prod. Commun.* 4:1485–1488.

Sung, S. H., and Y. C. Kim. 2000. Hepatoprotective diastereomeric lignans from *Saururus chinensis* herbs. *J. Nat. Prod.* 63:1019–1021.

Suzuki, S., M. Yamamura, M. Shimada, and T. Umezawa. 2004. A heartwood norlignan, (*E*)-hinokiresinol, is formed from 4-coumaryl 4-coumarate by a *Cryptomeria japonica* enzyme preparation. *J. Chem. Soc., Chem. Commun.*:2838–2839.

Suzuki, S., and T. Umezawa. 2007. Biosynthesis of lignans and norlignans. *J. Wood Sci.* 53:273–284.

Suzuki, S., M. Yamamura, T. Hattori, T. Nakatsubo, and T. Umezawa. 2007. The subunit composition of hinokiresinol synthase controls geometrical selectivity in norlignan formation. *Proc. Natl. Acad. Sci. USA* 104:21008–21013.

Takahashi, K., and K. Ogiyama. 1986. Phenols of discolored sugi (*Cryptomeria japonica* D. Don) sapwood. IV. Norlignans of Ayasugi (cultivar) in the Kyushu region. *Mokuzai Gakkaishi* 32:457–461.

Takahashi, K., and M. Suzuki. 2003. Dicotyledonous fossil wood flora and early evolution of wood characters in the Cretaceous of Hokkaido, Japan. *IAWA J.* 24:269–309.

Takahashi, K., M. Yasue, and K. Ogiyama. 1988. A norlignan, cryptoresinol, from the heartwood of *Cryptomeria japonica*. *Phytochemistry* 27:1550–1552.

Takaku, N., D.-H. Choi, K. Mikame, et al. 2001. Lignans of *Chamaecyparis obtusa*. *J. Wood Sci.* 47:476–482.

Takeda, R., J. Hasegawa, and K. Shinozaki. 1990a. Phenolic compounds from Anthocerotae. In *Bryophytes. Their Chemistry and Chemical Taxonomy* (H. D. Zinsmeister and R. Mues eds.), Clarendon Press, Oxford, UK. pp. 201–207.

Takeda, R., J. Hasegawa, and M. Shinozaki. 1990b. The first isolation of lignans, megacerotonic acid and anthocerotonic acid, from non-vascular plants, Anthocerotae (hornworts). *Tetrahedron Lett.* 31:4159–4162.

Talapatra, B., T. Ray, and S. K. Talapatra. 1976. Structure and partial synthesis of machicendiol, a new norlignan from *Machilus glaucescens* (Lauraceae). *Indian J. Chem., Sect B* 14:613–614.

Tamada, M., K. Endo, and H. Hikino. 1979a. Structure of ephedradine B, a hypotensive principle of *Ephedra* roots. *Heterocycles* 12:783–786.

Tamada, M., K. Endo, H. Hikino, and C. Kabuto. 1979b. Structure of ephedradine A, a hypotensive principle of *Ephedra* roots. *Tetrahedron Lett.* 20:873–876.

Tan, J.-M., Y.-H. Qiu, X.-Q. Tan, C.-H. Tan, and K. Xiao. 2011. Chemical constituents of *Equisetum debile*. *J. Asian Nat. Prod. Res.* 13:811–816.

Tasaki, Y., and H. Miyakawa. 2015. Structure and expression of two phenylalanine ammonia-lyase genes of the basidiomycete mushroom *Tricholoma matsutake*. *Mycoscience* 56:503–511.

Tazaki, H., K.-P. Adam, and H. Becker. 1995. Five lignan derivatives from *in vitro* cultures of the liverwort *Jamesoniella autumnalis*. *Phytochemistry* 40:1671–1675.

Tazaki, H., T. Hayashida, F. Ishikawa, et al. 1999. Lignan biosynthesis in liverworts *Jamesoniella autumnalis* and *Lophocolea heterophylla*. *Tetrahedron Lett.* 40:101–104.

Tazaki, H. 2002. Lignans in bryophytes: Their structures and biosynthesis. *Curr. Top. Phytochem.* 5:141–151.

Terada, K., C. Honda, K. Suwa, et al. 1995. Acetylenic compounds isolated from cultured cells of *Asparagus officinalis*. *Chem. Pharm. Bull.* 43:564–566.

Thomas, B. A. 1972. A probable moss from the Lower Carboniferous of the Forest of Dean, Gloucestershire. *Ann. Bot.* 36:155–161.

Troitsky, A. V., M. S. Ignatov, V. K. Bobrova, and I. A. Milyutina. 2007. Contribution of genosystematics to current concepts of phylogeny and classification of bryophytes. *Biochemistry (Moscow)* 72:1368–1376.

Tyagi, O. D., J. Wengel, A. K. Prasad, et al. 1994. Benzofuranoid neolignans from *Piper wightii*. *Acta Chem. Scand.* 48:1007-1011.

Vanzetti, B. L. 1929. Sul costituente principale della gomma (resina) d'olivo: olivile e derivati. *Gazz. Chim. Ital.* 59:373–378.

Vanzetti, B. L., and P. Dreyfuss. 1934. Sulla costituzione dell'olivile. Prodotti di ossidazione dell'isoolivile. *Gazz. Chim. Ital.* 64:381–399.

Vassão, D. G., D. R. Gang, T. Koeduka, et al. 2006. Chavicol formation in sweet basil (*Ocimum basilicum*): Cleavage of an esterified C9 hydroxyl group with NAD(P)H-dependent reduction. *Org. Biomol. Chem.* 4:2733–2744.

Vassão, D. G., S.-J. Kim, J. K. Milhollan, et al. 2007. A pinoresinol-lariciresinol reductase homologue from the creosote bush (*Larrea tridentata*) catalyzes the efficient *in vitro* conversion of *p*-coumaryl/coniferyl alcohol esters into the allylphenols chavicol/eugenol, but not the propenylphenols *p*-anol/isoeugenol. *Arch. Biochem. Biophys.* 465:209–218.

Vassão, D. G., K.-W. Kim, L. B. Davin, and N. G. Lewis. 2010. Lignans (neolignans) and allyl/propenyl phenols: Biogenesis, structural biology, and biological/human health considerations. In *Comprehensive Natural Products II Chemistry and Biology* (L. N. Mander and H.-W. Liu eds.), Vol. 1. *Structural Diversity I,* Elsevier, Oxford, UK. pp. 815–928.

Villavecchia, V., and G. Fabris. 1896. Substances contained in sesame oil, and their relation to the characteristic colour reactions of the oil. *Ann. Lab. chim. centr. delle Gabelle* 3:13–26.

Volc, J., P. Sedmera, K. Roy, V. Šašek, and J. Vokoun. 1977. Two antibiotic benzoquinone-hydroquinone pairs from the pyrenomycete *Camarops microspora* (Karst.) Shear. *Collect. Czech. Chem. Commun.* 42:2957–2961.

Wada, H., T. Kido, N. Tanaka, et al. 1992. Chemical and chemotaxonomical studies of ferns. LXXXI. Characteristic lignans for Blechnaceous ferns. *Chem. Pharm. Bull.* 40:2099–2101.

Wang, E.-C., M.-H. Shih, M.-C. Liu, M.-T. Chen, and G.-H. Lee. 1996. Studies on constituents of *Saururus chinensis. Heterocycles* 43:969–976.

Wang, K., M.-M. Li, X.-Q. Chen, et al. 2010. Phenolic constituents from *Brainea insignis. Chem. Pharm. Bull.* 58:868–871.

Wang, Z.-Q. 2004. A new Permian gnetalean cone as fossil evidence for supporting current molecular phylogeny. *Ann. Bot.* 94:281–288.

Wangun, H. V. K., and C. Hertweck. 2007. Squarrosidine and pinillidine: 3,3′-fused bis(styrylpyrones) from *Pholiota squarrosa* and *Phellinus pini. Eur. J. Org. Chem.*:3292–3295.

Wellman, C. H., P. L. Osterloff, and U. Mohiuddin. 2003. Fragments of the earliest land plants. *Nature* 425:282–285.

Weng, J.-K., X. Li, J. Stout, and C. Chapple. 2008. Independent origins of syringyl lignin in vascular plants. *Proc. Natl. Acad. Sci. USA* 105:7887–7892.

White, E., and G. H. N. Towers. 1967. Comparative biochemistry of the lycopods. *Phytochemistry* 6:663–667.

Wu, B., and J. Wang. 2011. Phenolic compounds from *Selaginella moellendorfii. Chem. Biodiversity* 8:1735–1747.

Yamamura, S., M. Niwa, M. Nonoyama, and Y. Terada. 1978. Three novel neolignans from *Heterotropa takaoi* M. *Tetrahedron Lett.* 19:4891–4894.

Yang, Y.-N., H. Zhu, Z. Chen, et al. 2015. NMR spectroscopic method for the assignment of 3,5-dioxygenated aromatic rings in natural products. *J. Nat. Prod.* 78:705–711.

Yin, S., C. Cullinane, A. R. Carroll, R. J. Quinn, and R. A. Davis. 2010. Botryllamides K and L, new tyrosine derivatives from the Australian ascidian *Aplidium altarium. Tetrahedron Lett.* 51:3403–3405.

Yuen, M. S. M., F. Xue, T. C. W. Mak, and H. N. C. Wong. 1998. On the absolute structure of optically active neolignans containing a dihydrobenzo[*b*]furan skeleton. *Tetrahedron* 54:12429–12444.

Zhang, Z., H. N. ElSohly, X.-C. Li, et al. 2003. Phenolic compounds from *Nymphaea odorata. J. Nat. Prod.* 66:548–550.

Zheng, X.-d., H.-b. Hu, and H.-s. Hu. 2008a. Two new neolignan glycosides from *Pteris multifida* Poir. *Indian J. Chem., Sect B* 47:773–777.

Zheng, X.-k., S.-p. Shi, Y.-f. Bi, et al. 2004. Isolation and identification of a new lignanoside from *Selaginella tamariscina* (Beauv.) Spring. *Acta Pharm. Sin.* 39:719–721.

Zheng, X. K., K. K. Li, Y. Z. Wang, and W. S. Feng. 2008b. A new dihydrobenzofuran lignanoside from *Selaginella moellendorffii* Hieron. *Chin. Chem. Lett.* 19:79–81.

Zhu, L.-J., F. Yan, J.-P. Chen, et al. 2016. 8-*O*-4′ Neolignan glycosides from the aerial parts of *Matteuccia struthiopteris. Chin. Chem. Lett.* 27:63–65.

4 Dirigent Protein Roadmap to Lignans and Other Vascular Plant Phenol Classes

Laurence B. Davin, John R. Cort, Clyde A. Smith, Qingyan Meng, Syed G. A. Moinuddin, Michael A. Costa, Diana L. Bedgar, and Norman G. Lewis

4.1 INTRODUCTION

Gateway entry points to distinct vascular plant phenol metabolic classes (e.g. to lignans [neo-lignans], biopolymeric lignins and suberins [aromatic component], as well as bioactive dimeric terpenoids, dimeric stilbenes, and others) were crucial to successful colonization of plants on land from their aquatic forerunners. Yet how, for example, stereoselective or regiospecific control over phenoxy-radical coupling in plants occurred was unknown until our dirigent protein (DP) discovery (from the Latin: *dirigere*, to guide or to align).

DP discovery began with the study of phenoxy-radical inter-molecular coupling control in lignan biosynthesis. DP physiological roles were later expanded to putative lignin-forming proteinaceous complexes containing DPs, and as gateway entry points to aromatic dimeric terpenoids. Some DP-like proteins are also gateway entry points to other plant phenol metabolic classes, such as pterocarpans, but do not involve inter-molecular radical coupling.

This chapter, while focusing on DPs involved in lignan biosynthesis, also addresses other known DP gateway entry points to distinct vascular plant phenol classes. Yet, it is important to stress that currently most members of the large DP multi-gene families in land plants have no known physiological/biochemical function. Such knowledge is essential, however, for gaining a full understanding of the biochemical (metabolic) diversity of DP gateway entry points in diverse plant phenol biosynthesis *in planta* (see Sections 4.3 and 4.4).

4.2 ROADMAP ENTRY POINTS TO LIGNANS AND OTHER PLANT PHENOLS

There has long been an ***enormous paucity in our biochemical knowledge of gateway entry points to the numerous and distinct plant phenol metabolic classes***. These include diverse and often higher molecular weight plant phenols, such as bioactive (oligomeric) lignans and neolignans, aromatic dimeric diterpenoids, pterocarpans, dimeric stilbenes, proanthocyanidins (condensed tannins), ellagitannins (hydrolysable tannins), as well as polymeric lignins and suberins (aromatic part), and many others. Together, they encompass not only tremendous structural diversity, but they can also be enormously abundant (i.e. together accounting for up to 40%–50% by weight of specific vascular plant tissues).

Structurally diverse lignans are typically phenylpropanoid dimers and higher oligomers, with physiological roles mainly in vascular plant defense (Vassão et al. 2010, Lewis and Davin 2022). Many lignans are optically active, with diverse coupling modes, although the 8–8′ inter-unit linkage, such as in (+)- and (−)-pinoresinols (**1a** and **1b**) (Figure 4.1), is perhaps most common.

The enormous Roadmap gap in our scientific knowledge needing filled, though, is in biochemically establishing the full nature of DP engendered biochemical gateway entry points, and subsequent downstream metabolic processes, i.e. affording various classes of (up to polymer size) plant-phenol-derived natural products *in planta*.

In this context, the *general biochemical mechanism(s)* of how distinct stereoselective or regiospecific phenoxy radical inter-molecular "coupling" in vascular plants stipulate(s) gateway entry points to a plethora of abundant vascular plant phenol natural products (such as lignans) is progressively yielding to enquiry. The resulting metabolic diversity so produced helps attain highly varied physiological functions *in planta*.

Such biochemical processes were also of critical evolutionary importance, as they ultimately led to emergence of terrestrial-based plant life (from aquatic forerunners), including initially the successful colonization of land which, in turn, led to the *circa* 391,000 vascular plant species in existence today.

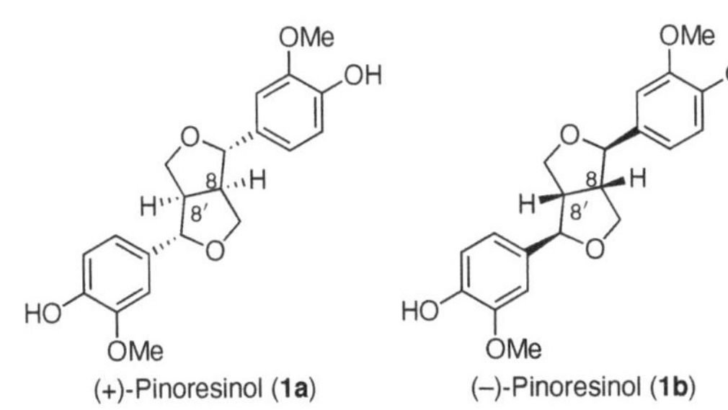

FIGURE 4.1 Examples of 8–8′-linked lignans, (+)- and (−)-pinoresinols (**1a** and **1b**).

4.2.1 Early Studies of Inter-molecular Phenoxy-Radical Coupling of Plant Phenols

Early studies of inter-molecular phenoxy-radical coupling *in vitro* initially focused solely on oxidative enzymes (e.g. peroxidases, laccases), their co-factors (e.g. H_2O_2 and O_2), and the coupling reactions and products obtained with various plant phenol substrates. Such coupling products formed *in vitro*, though frequently lacked either the stereoselectivity or regiospecificity observed *in planta*, as oxidases and their cofactors alone only gave products formed by so-called "random" (i.e. non-specific) radical coupling. Frequently, such assays gave mixtures of racemic products of differing regiochemistries that are not found in Nature *per se* including in those proportions (for a review, see Lewis and Davin 1998).

Nevertheless, these investigations helped provide foundational knowledge about those oxidative enzymes and their catalytic turnover mechanisms. On the other hand, they gave no useful insights into how differential control of the *outcomes* of phenoxy-radical coupling product formation occurred *in planta*. For the latter, there frequently appeared to be control of stereoselectivity to give one specific enantiomeric form and/or regiospecificity to afford specific regioisomers, whether dimeric or polymeric.

4.2.2 Dirigent Proteins

We thus hypothesized that there had to be a key component missing in control of stereoselective and regiospecific phenoxy-radical coupling *in vitro*, as clearly suggested, for example, from the optical activities of many lignans isolated from vascular plants. This key missing component was solved by our DP discovery (Davin et al. 1997) as the gateway entry point in 8–8′ lignan biosynthesis. Its discovery provided the first insights into how stereoselectivities and regiospecificities in phenoxy-radical coupling were stipulated *in vivo*. That is, coupling selectivity/specificity in 8–8′ lignan biosynthesis was only achieved *in vitro* when DPs were present in assays with oxidative enzymes, their co-factors, and substrates.

4.2.3 DP Multi-gene Families

There has been rapid progress in plant genome sequencing since 2000, when the first vascular plant genome sequence was reported (*Arabidopsis* Genome Initiative 2000). Genome sequencing, EST, and transcriptome databases/reports then led to recognition that DPs and dirigent-like proteins are in very large multi-gene sub-families *in planta* (Ralph et al. 2007, Corbin et al. 2018). In addition, Pfam (the protein families database) (Finn et al. 2016) is a widely used system of clustering protein sequences into families that are grouped by amino acid sequence similarity. All DPs are currently categorized into the single family, PFAM03018.

Our bioinformatics analysis of *circa* 800 DP sequences from the plant kingdom (i.e. including lycophytes, bryophytes, gymnosperms, and angiosperms) retrieved from the UniProt database led to construction of an unrooted phylogenetic tree using the Interactive Tree of Life (iTOL) (Letunic and Bork 2007) (Figure 4.2). This gave nine different families (DIR-a to DIR-i) and numerous sub-families (e.g. DIR-a_1 and DIR-a_2, not all shown in figure). A thorough BLAST (Basic Local Alignment Search Tool) search using *Arabidopsis* DP queries of unpublished plant genomic and transcriptomic data in the 1KP (1000 plants) project database (Matasci et al. 2014) also confirmed the presence of DP or dirigent-like proteins in bryophytes (mosses, hornworts, and liverworts), lycophytes, pteridophytes (ferns), and all gymnosperms and angiosperms.

Based on our analyses, however, DPs apparently evolved functionally during the early transition of plants to a dryland (desiccating) habitat. As examples, DP or DP-like genes are present in completed genome sequences of extant descendants of the earliest vascular plants, i.e. *Selaginella* (Banks et al. 2011), *Marchantia* (Bowman et al. 2017), and *Physcomitrella* (Rensing et al. 2008) (Figure 4.3). Of these, *Physcomitrella* seems to have only a few dirigent-like protein coding genes, *Marchantia* has about 40, and *Selaginella* even more, indicating early DP duplication and specialization. Higher plant genomes, though, typically have 20–50 dirigent-like protein coding genes.

Yet, it must be emphasized that currently only about 5%–10% of DP and dirigent-like protein gene functions are known at the physiological/biochemical level in all plant species harboring DPs. An example of an unknown DP biochemical function is a *Physcomitrella* dirigent-like protein that was reportedly expressed following fungal infection and may possibly be associated with cell wall modification (Reboledo et al. 2015). Another unknown DP biochemical function example is a dirigent associated with soybean pod opening (Funatsuki et al. 2014). Such observations, however, provide intriguing evidence in support of the growing importance of DPs in many distinct and diverse land plant processes involving plant phenol gateway entry points.

Evolutionarily, the transition of aquatic plants to a terrestrial habitat (Davin et al. 2008, Vassão et al. 2010) required significant biochemical pathway elaboration. This, in turn, was presumably facilitated via initial emergence of DP functions in controlling inter-molecular phenoxy radical-radical coupling processes. The latter were needed for vascular plant survival in a desiccating environment (suberins), for structural support to stand upright and as conduits for water and nutrient transport (lignins), and for plant defense (e.g. lignans). However, as angiosperms emerged, for example, other DP-engendered biochemical processes evolved as gateway entry points that were not inter-molecular radical coupling derived, i.e. such as to pterocarpans.

4.3 DP Roadmap to Diverse Plant Phenol Metabolic Classes

The Roadmap chronology of DP discovery, and additional DP characterization, as gateway entry points to distinct plant

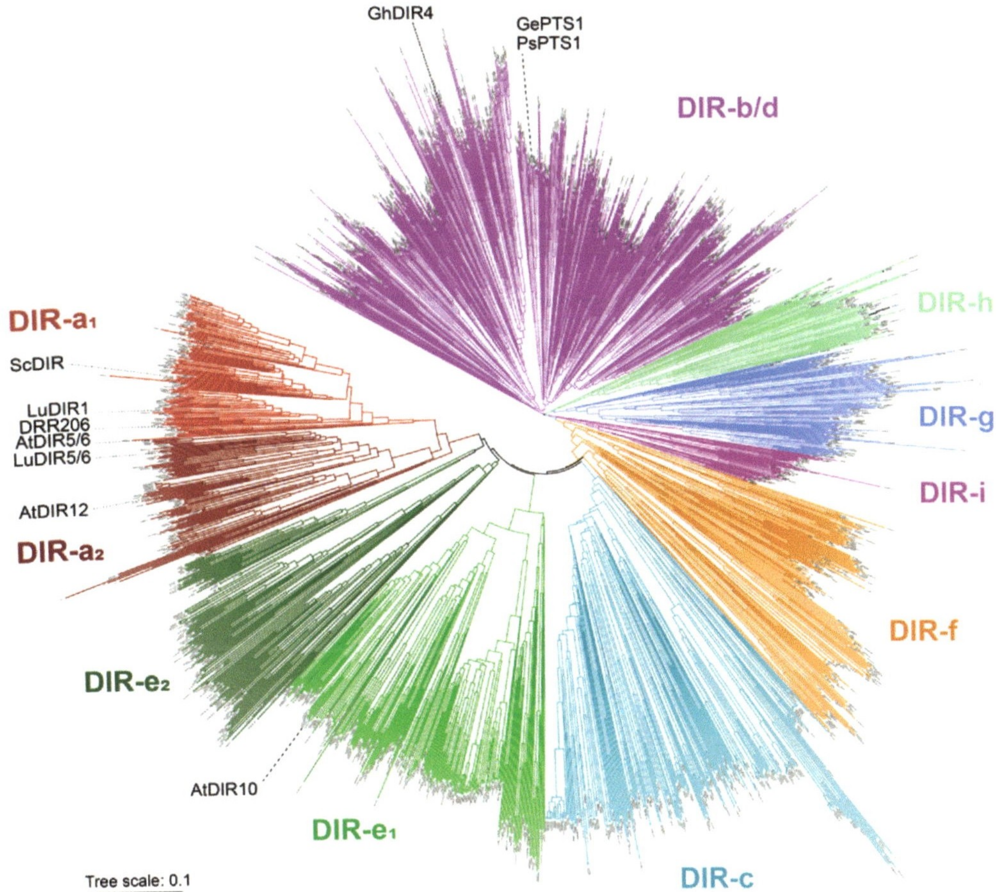

FIGURE 4.2 Unrooted phylogenetic tree of dirigent and dirigent-like protein family. The subfamily nomenclature of Ralph et al. (2007) is used. Dirigent proteins, whose functional characterization are known in the literature, are indicated (e.g. DRR206 in lignan formation).

phenol metabolic classes known to date is shown in Figure 4.4, i.e. to lignans, neolignans, aromatic diterpenoids, lignins, and pterocarpans, respectively (Umezawa et al. 1990, Davin et al. 1997, Gang et al. 1999, Xia et al. 2000, Halls and Lewis 2002, Halls et al. 2004, Davin and Lewis 2005b, Liu et al. 2008, Patten et al. 2008, Pickel et al. 2010, Vassão et al. 2010, Kim et al. 2012, Celoy 2013, Hosmani et al. 2013, Dalisay et al. 2015, Effenberger et al. 2015, Kim et al. 2015, Meng et al. 2020, Yonekura-Sakakibara et al. 2021). A brief summary of each of these distinct DP-engendered gateway entry points to different plant phenol metabolic classes is given below.

4.3.1 DP (Sub-family DIR-a) Roadmap to 8–8′ Lignans

Phenoxy radical-radical inter-molecular coupling control *in planta* was not considered by most researchers to involve anything other than free-radical intermediate formation (see Lewis and Davin 1998). Accordingly, an unproven yet popular paradigm emerged *circa* 60 years ago where only one electron (1e⁻) oxidation and subsequent non-enzymatic phenoxy radical-radical coupling was contemplated (see Lewis and Davin 1998); this remained so until our DP discovery (Davin et al. 1997).

Our discovery of DPs as a new protein class, and initially the gateway entry point in 8–8′ lignan biosynthesis, thus provided the first insight into how plant phenol-derived coupling/polymerization processes can be stereoselectively or regiospecifically controlled. For 8–8′ lignans (phenylpropanoid dimers), we demonstrated that either (+)- or (–)-pinoresinol-forming DPs, in the presence of a 1e⁻ oxidase (laccase or peroxidase), engendered stereoselective radical coupling of coniferyl alcohol (CA, **2**)-derived moieties to give (+)- or (–)-pinoresinols (**1a** or **1b**, Figure 4.5) depending upon DP type (Davin et al. 1997, Kim et al. 2002, Vassão et al. 2010, Kim et al. 2012, Dalisay et al. 2015, Seneviratne et al. 2015). The (+)- and (–)-pinoresinol-forming DPs (DRR206 and AtDIR6) are in the DIR-a_1 and DIR-a_2 sub-families (Figure 4.2).

These oxidative systems generated free radical CA˙, which binds to the DP(s), and whose differential orientation in the distinct DPs enabled 8–8′ coupling to occur either at the *si-si* face with concomitant intramolecular cyclization to afford (+)-pinoresinol (**1a**) in *Forsythia* (Davin et al. 1997), *Thuja plicata* (Kim et al. 2002), *Schizandra chinensis* (Kim et al. 2012), *Piper regnellii*, and pea (*Pisum sativum*) (Seneviratne et al. 2015), or *re-re* coupling to give the (–)-antipode (**1b**) in *Arabidopsis* (Vassão et al. 2010, Kim et al. 2012) and flax

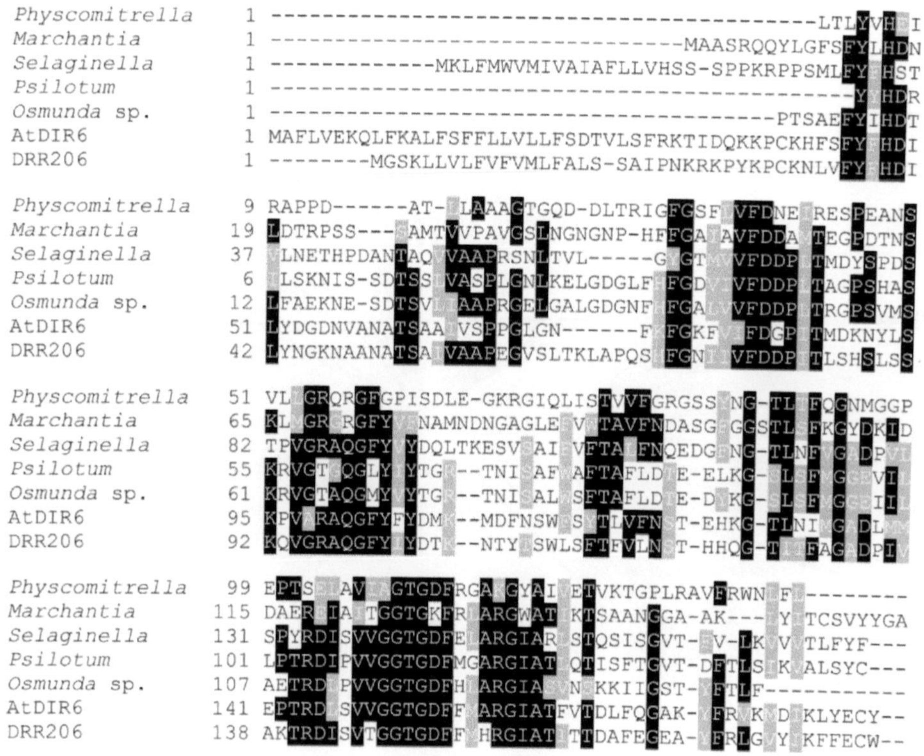

FIGURE 4.3 Comparison of DPs, from early land plants (so-called "primitive" plants) to that of angiosperm (+)- and (–)-pinoresinol forming DPs. The (–)-pinoresinol forming DP, AtDIR6 sequence was used as input for BLAST searches of genomes of early land plants, together with the (+)-pinoresinol forming DP, DRR206, from *Pisum sativum*. The published genomes of *Physcomitrella patens* (Rensing et al. 2008), *Marchantia polymorpha* (Bowman et al. 2017) and *Selaginella moellendorffii* (Banks et al. 2011) were searched in GenBank. Unpublished sequence data for the Eusporangiate and Leptosporangiate monilophytes *Psilotum nudum* and *Osmunda* sp. were retrieved by BLAST search of the 1KP initiative (Matasci et al. 2014). The best hit in each search was retrieved. Sequence IDs: XP_001752690.1 (*P. patens*), OAE21765.1 (*M. polymorpha*), XP_002981853.1 (*S. moellendorffii*), scaffold-QVMR-2004186 (*P. nudum*), and scaffold-UOMY-2005388 (*Osmunda* sp. gametophyte).

(*Linum usitatissimum*) (Dalisay et al. 2015). [AtDIR6, the *Arabidopsis* (–)-pinoresinol-forming DP (Vassão et al. 2010, Kim et al. 2012), was also reported independently by others (Pickel et al. 2010)].

The (+)-pinoresinol-forming pea DP (named here as DRR206 (Kim et al. 2015, Seneviratne et al. 2015)) is encoded by a gene of unknown function previously and described as *Disease Resistance Response-206, DRR206*). The latter's gene expression was earlier long known to be induced upon pea pod fungal exposure, but whose actual biochemical role was never established at that time (Fristensky et al. 1985, Riggleman et al. 1985, Fristensky et al. 1988). However, it was later predictably established to be yet another (+)-pinoresinol-forming DP (Seneviratne et al. 2015) based on our earlier work.

Their overall DP-mediated process here mechanistically involves three steps, namely radical formation (via laccase or peroxidase catalysis), DP-guided radical coupling to give a bis-quinone methide (bis-QM) via C–C bond formation, and intra-molecular cyclization (C–O bond formation) to give pinoresinol (**1**) (Figure 4.5).

(+)-Pinoresinol (**1a**) is also the biosynthetic pathway entry point to a plethora of medicinally important lignans, including (–)-podophyllotoxin (**3b**, Figure 4.6), a scaffold synthon to toposide, teniposide, and Etopophos, which are widely used in cancer treatment (Hartmann and Lipp 2006, Bailly 2012, Henkin et al. 2018).

Physiological roles of (–)-pinoresinol-forming DPs in *Arabidopsis* and flax: While AtDIR6 (and AtDIR5) from *Arabidopsis*-generated (–)-pinoresinol (**1b**) *in vitro*, we verified its role *in vivo* via AtDIR6 overexpression and RNAi strategies (Kim et al. 2012), i.e. proving AtDIR6 involvement in formation of (–)-pinoresinol (**1b**) and its downstream metabolites in *Arabidopsis*.

In flax (*Linum usitatissimum*) seed, the pinoresinol-derived secoisolariciresinol diglucoside [SDG (**5**), Figure 4.6)] released on mild alkali hydrolysis was composed of *circa* 99% (+)- and *circa* 1% (–)-secoisolariciresinols (**4a/b**) (Ford et al. 2001), suggesting presence of both (+)- and (–)-pinoresinol-forming DPs. We conducted a BLAST search of the flax genome (Wang et al. 2012), and identified 6 *DP* homologs with 45%–66% amino acid sequence identity to either (+)-pinoresinol-forming DP (ScDIR) or (–)-pinoresinol-forming DP (AtDIR6); LuDIR1 was found to be a (+)-pinoresinol-forming DP, whereas LuDIR6 and LuDIR7 were (–)-pinoresinol-forming DPs (Figure 4.5) (Dalisay et al. 2015).

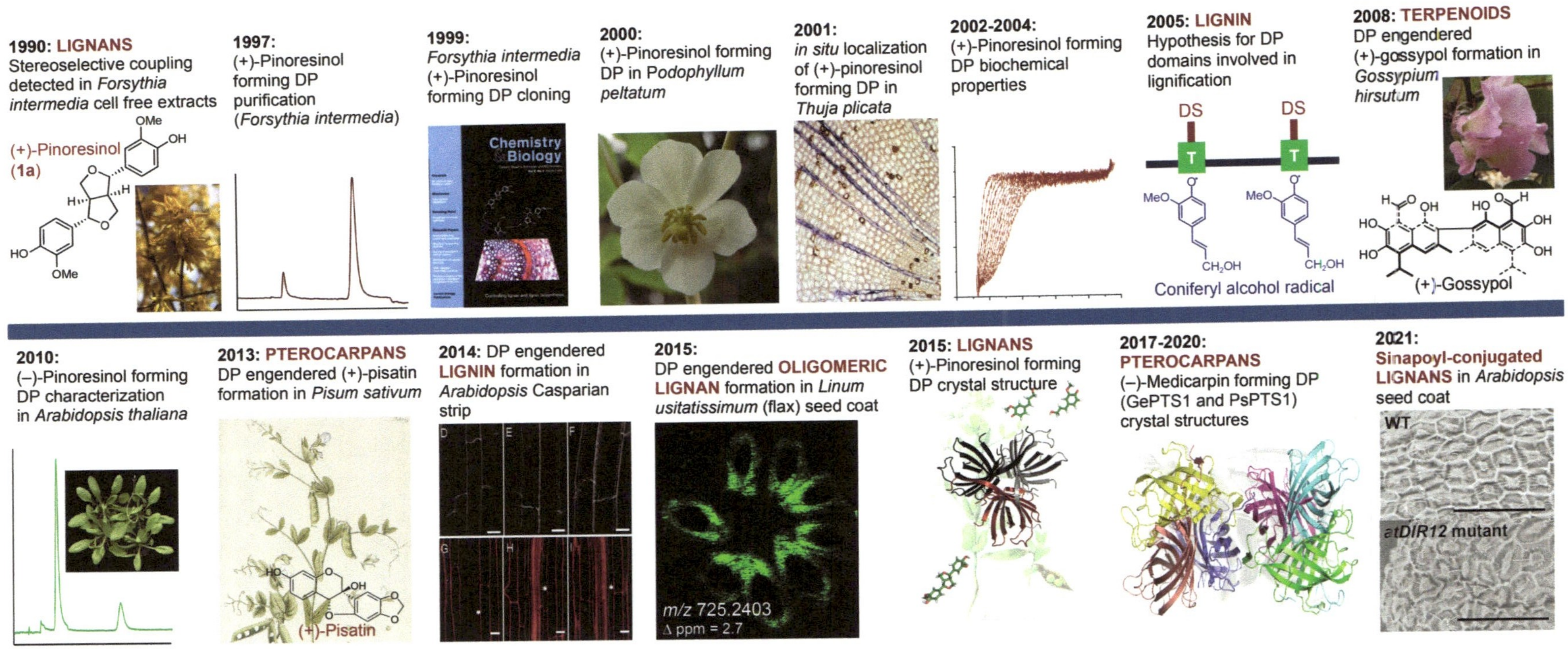

FIGURE 4.4 Current dirigent protein (DP) Roadmap to diverse plant phenols (Umezawa et al. 1990, Davin et al. 1997, Gang et al. 1999, Xia et al. 2000, Halls and Lewis 2002, Halls et al. 2004, Davin and Lewis 2005b, Liu et al. 2008, Patten et al. 2008, Pickel et al. 2010, Vassão et al. 2010, Kim et al. 2012, Celoy 2013, Hosmani et al. 2013, Dalisay et al. 2015, Effenberger et al. 2015, Kim et al. 2015, Meng et al. 2020, Yonekura-Sakakibara et al. 2021).

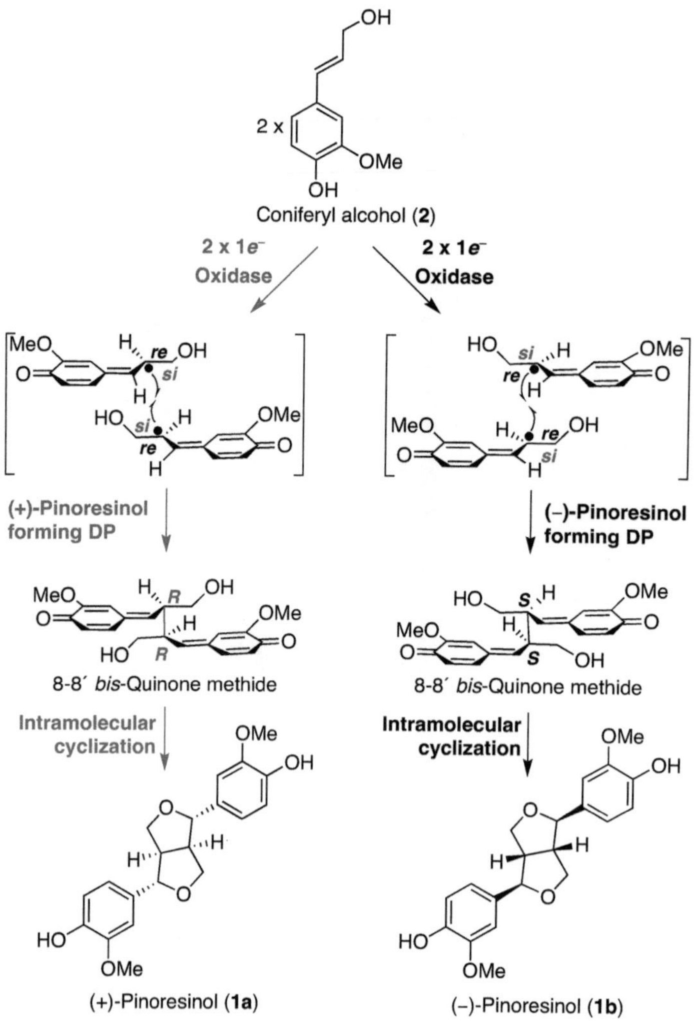

FIGURE 4.5 DP (sub-family DIR-a) Roadmap to 8–8′-lignans. Formation of (+)- or (–)-pinoresinol (**1a/1b**) from achiral coniferyl alcohol (**2**). 1e⁻ oxidation generates an intermediary free-radical mono-quinone methide (mono-QM) which either undergoes *si-si* or *re-re* coupling (C–C bond formation) to afford chiral 8–8′-bis-quinone methides (bis-QMs), depending on the DIR-a pinoresinol-forming DP to give, after intramolecular cyclization (C–O bond formation), (+)- or (–)-pinoresinol (**1a** or **1b**), respectively.

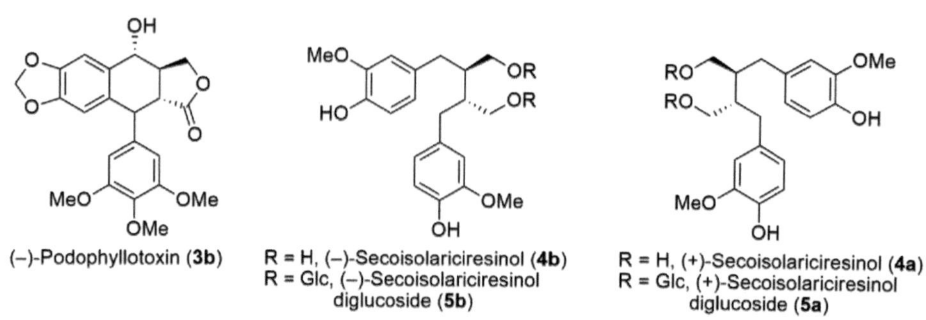

FIGURE 4.6 Medicinally important lignan, (–)-podophyllotoxin (**3b**), as well as (+)- and (–)-secoisolariciresinols (**4a** and **4b**) and their corresponding glucosides (**5a** and **5b**).

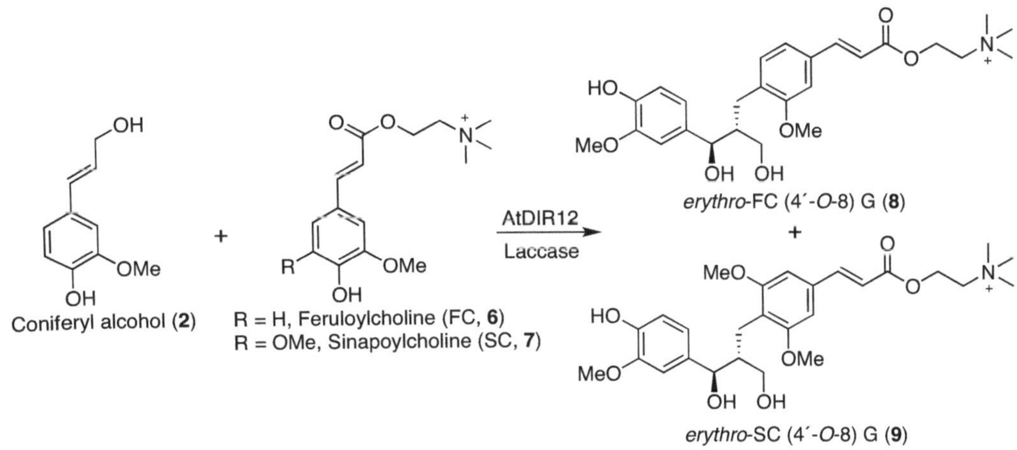

FIGURE 4.7 DP (sub-family DIR-a) Roadmap to 4′–*O*–8-lignans. Putative formation of feruloylcholine (FC)- and sinapoylcholine (SC)-conjugated lignans **8** and **9** by DIR-a AtDIR12 in presence of laccase.

4.3.2 DP (Sub-family DIR-a) Roadmap to 4′–*O*–8-Lignans

In a somewhat comparable manner, DIR-a sub-family member (AtDIR12) in seed coats of *Arabidopsis thaliana*, has been indirectly implicated in the biosynthesis of 4′–*O*–8-linked (neo)lignans, trivially named feruloylcholine (FC)- and sinapoylcholine (SC)-conjugated lignans **8** and **9**, respectively (Figure 4.7) (Yonekura-Sakakibara et al. 2021). It is proposed that these lignans may have a role in seed coat permeability.

This DP-engendered 4′–*O*–8 linkage observation is consistent with our earlier hypothesis that other DPs/dirigent-like proteins would permit entry into, for example, distinct lignan skeletal types (Lewis and Davin 1999). [Note that we use the nomenclature of 4′–*O*–8-linked lignans here, rather than 'neolignans'. The reasons for this nomenclature are summarized in Davin and Lewis (2022), i.e. where lignans (phenylpropanoid dimers to higher oligomeric lignans) are systematically differentiated on the basis of type of inter-unit linkages.]

Although not established *in vitro*, 4′–*O*–8 lignan formation is mechanistically contemplated to occur via initial coniferyl alcohol (**2**) 1e$^-$ (laccase) oxidation to afford the free radical mono-QM as above. This then undergoes subsequent DP-guided stereoselective 4′–*O*–8 coupling (C–O–C bond formation) with yet-to-be-determined feruloyl/sinapoyl (presumed radical) derivatives. Coupling of these radical species would give a mono-QM intermediate, which would then re-aromatize through hydroxylation at C-7 (not shown) to ultimately afford **8** and **9**. [It can also be speculated that the coniferyl alcohol radical binds first to this DP.]

4.3.3 DP (Sub-family DIR-b/d) Roadmap to Aromatic Dimeric Terpenoids

(+)-Gossypol (**11a**) is an aromatic diterpenoid imparting resistance against herbivorous insects and pathogens in cotton (*Gossypium hirsutum*) (Stipanovic et al. 2006). Its formation occurs via stereoselective coupling of its achiral precursor, hemigossypol (**10**) (Figure 4.8), through involvement of a (+)-gossypol-forming DP (Davin et al. 2008, Liu et al. 2008). The gene (GhDIR4) encoding its corresponding DP has been described (Effenberger et al. 2015, and our own unpublished work). Optical activity occurs via stereoselective coupling and restricted rotation of the C–C inter-unit biphenyl linkage in the (+)-gossypol product (**11a**) so formed.

The (+)-gossypol-forming DP (GhDIR4) is in the DIR-b/d sub-family. Its DP-mediated process is similar to that of DP-engendered 8–8′ lignan (pinoresinol) biosynthesis, i.e. it also involves three steps, these being oxidase-catalyzed radical and mono-QM formation, DP-guided radical coupling to give a bis-QM via C–C bond formation as before, with rearrangement and re-aromatization to give (+)-gossypol (**11a**).

4.3.4 DP (Sub-family DIR-b/d) Roadmap to Pterocarpans and Isoflavenes

Pterocarpans are important anti-fungal compounds (phytoalexins) in legumes (Ingham 1982, VanEtten et al. 1983). The first report of DP-like involvement in forming its pea (*Pisum sativum*) pterocarpans was from the laboratory of our late collaborator Hans von Etten in 2013 (Celoy 2013). Pterocarpan-forming dirigent-like proteins (pterocarpan synthases [PTSs]) were later reported in both licorice (*Glycyrrhiza echinata*) and pea. Both PTSs are in the very large DIR-b/d sub-family, together with a cluster of putative PTSs in other Leguminosae species.

The two PTS homologs from licorice and pea, named GePTS1 and PsPTS1, respectively, were individually expressed in *Escherichia coli*. Each purified recombinant protein enabled chiral pterocarpan formation *in vitro*, i.e. to (+)-medicarpin (**14a**) or (–)-medicarpin (**14b**). This occurs via intra-molecular cyclization (C–O bond formation) of chiral mono-QM intermediates generated from chiral isoflavonoid 7,2′-dihydroxy-4′-methoxy-isoflavanol (DMI) substrates (**12**) and (**13**) (Figure 4.9A) (Uchida et al. 2017, Meng et al. 2020).

For our pterocarpan-forming DP studies, the corresponding substrates (**12**) and (**13**) were synthesized, with the four

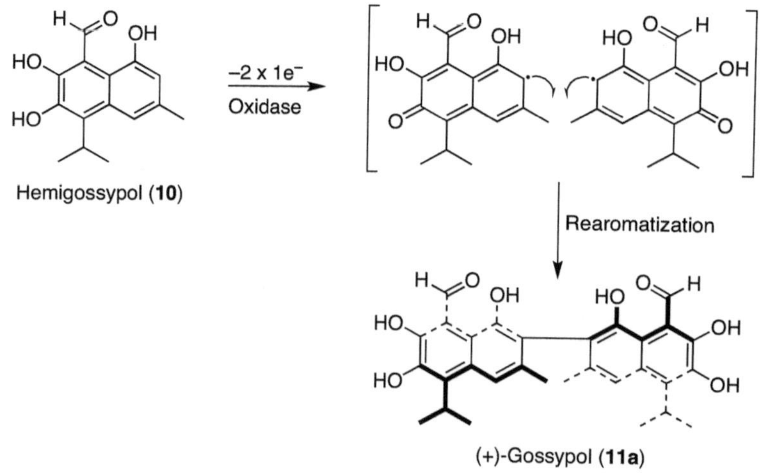

FIGURE 4.8 DP (sub-family DIR-b/d) Roadmap to aromatic dimeric terpenoids. (+)-Gossypol (**11a**) forming DIR-b/d DP (GhDIR4) whose action follows initial oxidase catalyzed 1e⁻ oxidation of achiral hemigossypol (**10**) to afford the proposed prochiral free-radical mono-quinone methide (mono-QM) intermediate; DP-guided stereoselective coupling then gives the chiral bis-quinone methide (bis-QM) derivative (C–C bond formation), re-aromatization of which generates (+)-gossypol (**11a**).

chiral forms of *cis*- and *trans*-DMI separated using chiral RU2 (Shiseido, Japan) column chromatography.

A second pea DIR-b/d dirigent-like protein homolog, PsPTS2, acts instead on 7,2-dihydroxy-4-methylenedioxyiso-flavanol (DMDI) substrates **15** and **16** *in vitro* (Celoy 2013, Uchida et al. 2020, Meng et al. manuscript in finalization), with the product being the achiral isoflavene DMDIF (**18**), which was reported earlier (Celoy and VanEtten 2014). Based on our comprehensive kinetic measurements, this occurs via conversion of **15** and **16** to first generate (–)- and (+)-maackiain (**17b**, **17a**), both of which are further processed to afford the isoflavene DMDI (**18**) (Figure 4.9B). Mechanistically, the latter conversion can be envisaged to occur via regeneration of the mono-QM intermediate as before via ring opening, but then with tautomerism at C3 and re-aromatization to give DMDIF (**18**) (Meng *et al.*, manuscript in finalization).

Interestingly, pea plants accumulate two phytoalexins (–)-maackiain (**17b**) and (+)-pisatin (**19a**) (Celoy and VanEtten 2014) that differ in their configurations at carbons 6a and 11a; this raises the possibility that achiral DMDIF (**18**) is further metabolized *in planta* to give (+)-pisatin (**19a**). If so, this presumably helps explain how phytoalexins **17** and **19** have different configurations at carbons 6a and 11a, respectively.

The DP-mediated transformations here thus represent another variation of the DP-mediated biochemical mechanisms (previously described in Figure 4.5). In Figure 4.9, an auxiliary oxidase is *not* needed to generate the mono-QM intermediate. Instead, these two DIR-b/d sub-family members generate mono-QMs, which then undergo either a C–O bond-forming process to give (**17**), or C3 hydride tautomerism to afford DMDIF (**18**).

4.3.5 DP (Sub-family DIR-e) Roadmap to Casparian Strip Lignins

DPs reportedly have a key role in ensuring that the pattern of lignin deposition, a plant cell wall polymerization process, in *Arabidopsis* root endodermis Casparian strip is uniform (Hosmani et al. 2013). DP involvement was proposed as it was observed that when expression of a DP gene (*AtDIR10*) was knocked out, orderly lignin deposition was reportedly interrupted. The DP controlling Casparian strip polymeric lignin deposition (AtDIR10) is in the DIR-e sub-family.

DP involvement in Casparian strip lignification is considered to be part of a hypothesized lignin-forming protein complex containing a Casparian strip domain protein (CASP) scaffold, DPs, various oxidative enzymes, and other enzymes affording co-factors, such as H_2O_2 (Hosmani et al. 2013). The proposed scaffold CASP membrane proteins are hypothesized to help organize bound DPs as part of a lignin-forming proteinaceous complex.

There are currently no *in vitro* studies identifying the precise biochemical step that the DIR-e DP enables in terms of the actual substrate(s) utilized and product(s) formed. Nevertheless, this report also apparently supports our earlier hypothesis of DPs and proteins harboring dirigent (substrate binding) sites/domains for control of lignan and lignin biosynthesis outcomes (Davin and Lewis 2005a, b).

4.3.6 DPs Are Absent in Algae/Cyanobacteria

Our bioinformatics analyses have indicated that DPs are absent in algae/cyanobacteria, as there were no comparable DP homologs in either their EST databases or available genome sequences (data not shown), i.e. DPs only evolved their functions during aquatic plant transition to land.

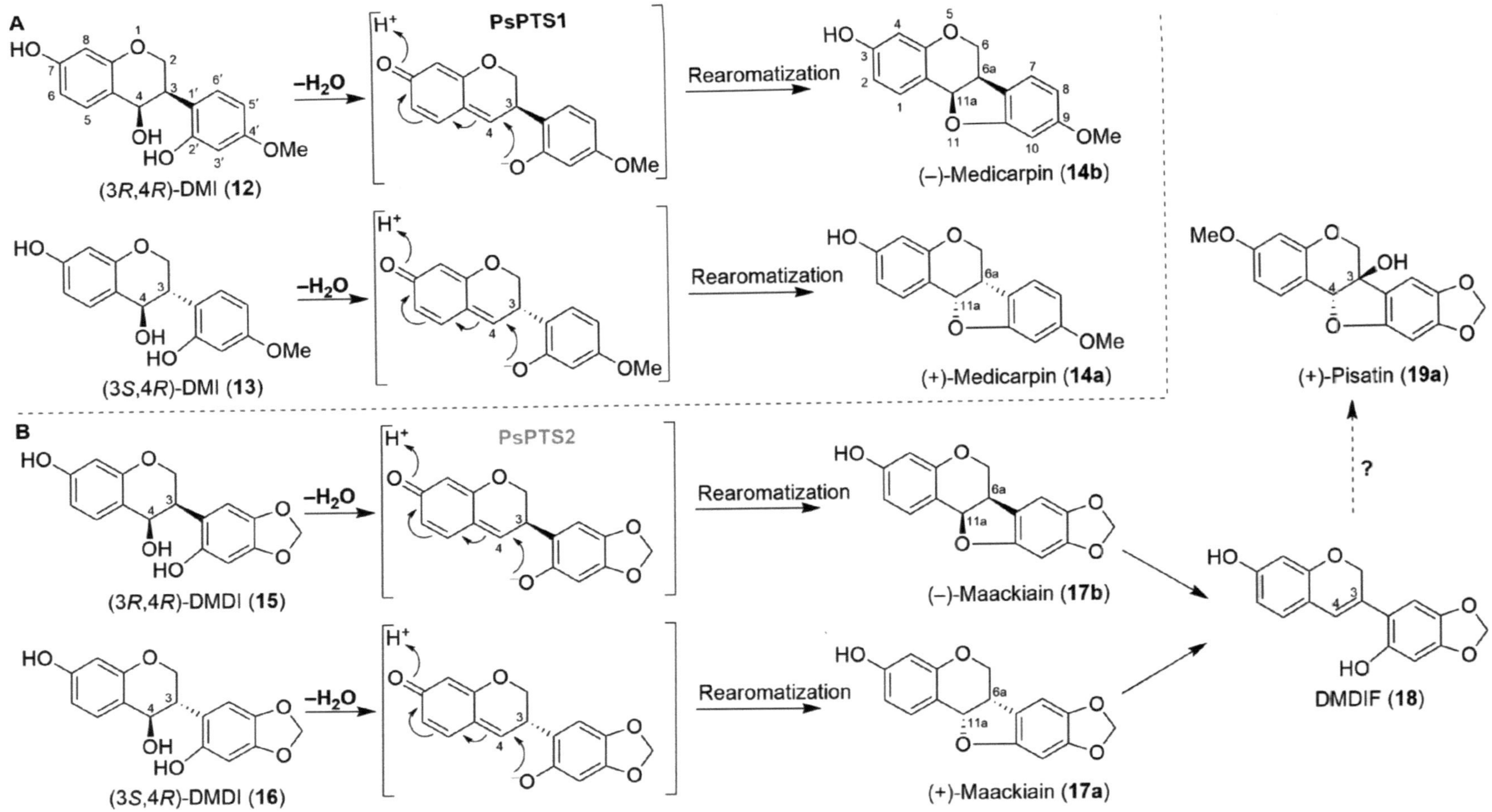

FIGURE 4.9 DP (sub-family DIR-b/d) Roadmap to pterocarpans and achiral flavonoids. **A.** Medicarpin-forming DIR-b/d subfamily member (PsPTS1) from chiral isoflavonoid substrates (3*R*,4*R*)-7,2′-dihydroxy-4′-methoxy-isoflavanol (DMI) (**12**) and (3*S*,4*R*)-DMI (**13**). Proposed biochemical mechanism involves mono-QM generation, intra-molecular cyclization and re-aromatization (C–O bond formation). **B.** 7,2′-dihydroxy-4′-methylenedioxyisoflav-3-ene (DMDIF, **18**) stepwise formation. Maackiain (**17**)-forming DIR-b/d (PsPTS2) from chiral isoflavonoids (3*R*,4*R*)-7,2′-dihydroxy-4′-methylenedioxyisoflavanol (DMDI) (**15**) and (3*S*,4*R*)-DMDI (**16**). Maackiain (**17**) is then converted into achiral 7,2′-dihydroxy-4′-methylenedioxyisoflav-3-ene (DMDIF, **18**). Note the change in numbering system from isoflavone to pterocarpan skeleta.

4.4 HYPOTHETICAL DP ROADMAP TO OTHER PLANT PHENOL METABOLIC CLASSES, SUCH AS PROANTHOCYANIDINS, STILBENOID AND ALLYLPHENOL/PROPENYLPHENOL DIMERS

There are numerous vascular plant phenol metabolic classes whose gateway entry points are unknown; this is perhaps better understood given that perhaps only 5%–10% of large multi-gene DP families are of known biochemical/physiological function. Three examples of unknown gateway entry points are to (oligomeric) proanthocyanidins, as well as to stilbenoid and allylphenol/propenylphenol dimers (and higher oligomers). These examples are discussed in this section to illustrate such (albeit hypothetical) DP Roadmap possibilities. There are many other such metabolic classes, where the gateway entry points to distinct plant phenols are unknown and where DPs can readily be hypothetically implicated.

4.4.1 Proposed DP Roadmap to Proanthocyanidins (PAs)

Proanthocyanidins, or PAs (from phenypropanoid-acetate dimers to higher oligomers/polymers), are widespread in vascular plants, and have roles in plant defense (de Colmenares et al. 1998, Debeaujon et al. 2000, Dixon et al. 2005) as well as in health-protecting properties (Sánchez-Moreno et al. 2003, Balalaie et al. 2018, Casanova-Martí et al. 2018). PA formation (oligomers/polymers) can hypothetically result via coupling of, for example, "catechin (**20**)/gallocatechin (**21**)" (Figure 4.10) chain extender (repeat) units.

In this context, we provisionally hypothesize that PA assembly is very similar to the DIR-b/d (dirigent-like protein) PTS biochemical mechanism, and thus may also involve a member of the DIR-b/d sub-family. Specifically, we hypothesize that biosynthetic chain extender (repeat) units for oligomer/polymer PA formation must be in an electrophilic/QM chemical form prior to coupling, and thus primed for nucleophilic attack by a PA terminal unit to produce PA dimers/trimers/polymers, etc. We envisage this occurs via action of a dirigent-like protein that we provisionally name here as a proanthocyanidin synthase/polymerase (PASP).

The precursors we hypothesize for PA formation are flavan-3,4-diols (e.g. **24–26**, Figure 4.10), which through H_2O loss could afford electrophilic (QM or carbocation) intermediates that can serve as chain extender (repeat) units (Figure 4.11). Such intermediates could react with chain terminal units, e.g. catechin (**20**), epicatechin (**22**), or flavan-3,4-diols (e.g. **24–26**) to afford dimers (e.g. **27**), higher oligomers, and polymers, as depicted in Figure 4.11, using leucocyanidin (**25**) and catechin (**20**) as examples. Analogously, epicatechin (**22**)/gallocatechin (**21**)/epigallocatechin (**23**) extender units could be formed from their corresponding flavan-3,4-diols (structures not shown). Yet, no proteins or enzymes are known to engender PA dimer/oligomer/polymer formation, although non-specific coupling reactions catalyzed by laccases/peroxidases, etc. *in vitro* have long been reported (López-Serrano and Ros Barceló 2001, 2002). The latter, however, do not result in PA formation.

If the terminal unit was a flavan-3,4-diol, as in **24–26**, then one question is whether it subsequently loses H_2O on further metabolism to generate corresponding mono-QM intermediates. If so, and depending upon the terminal flavan-3,4-diol potentially utilized (such as **25**), its resulting dehydration would give the corresponding QM intermediate, which could be reduced to afford PAs such as procyanidin B1 (**28**, Figure 4.12). This question, though, needs to be settled at the cell-free level, as does whether only PASPs are involved, or if additional proteins such as a reductase are required following putative H_2O loss in the terminal unit, i.e. if flavan-3,4-diols (e.g. **24–26**) are the initial terminal unit, rather than **20–23**.

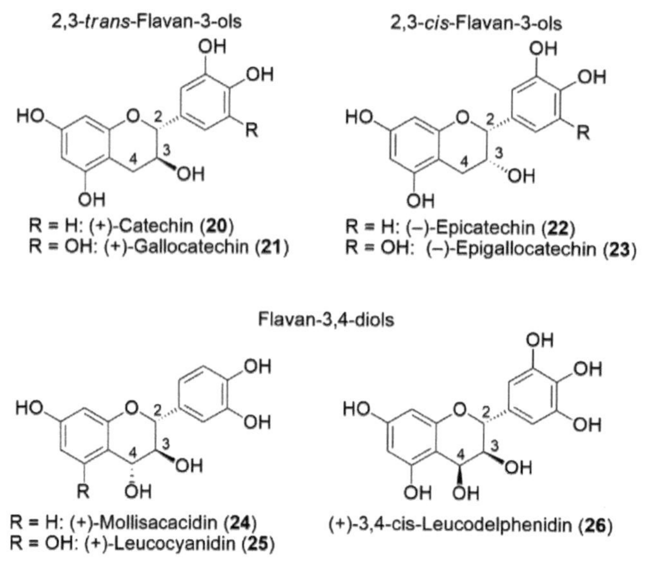

FIGURE 4.10 Flavonoid monomers **20 – 26**.

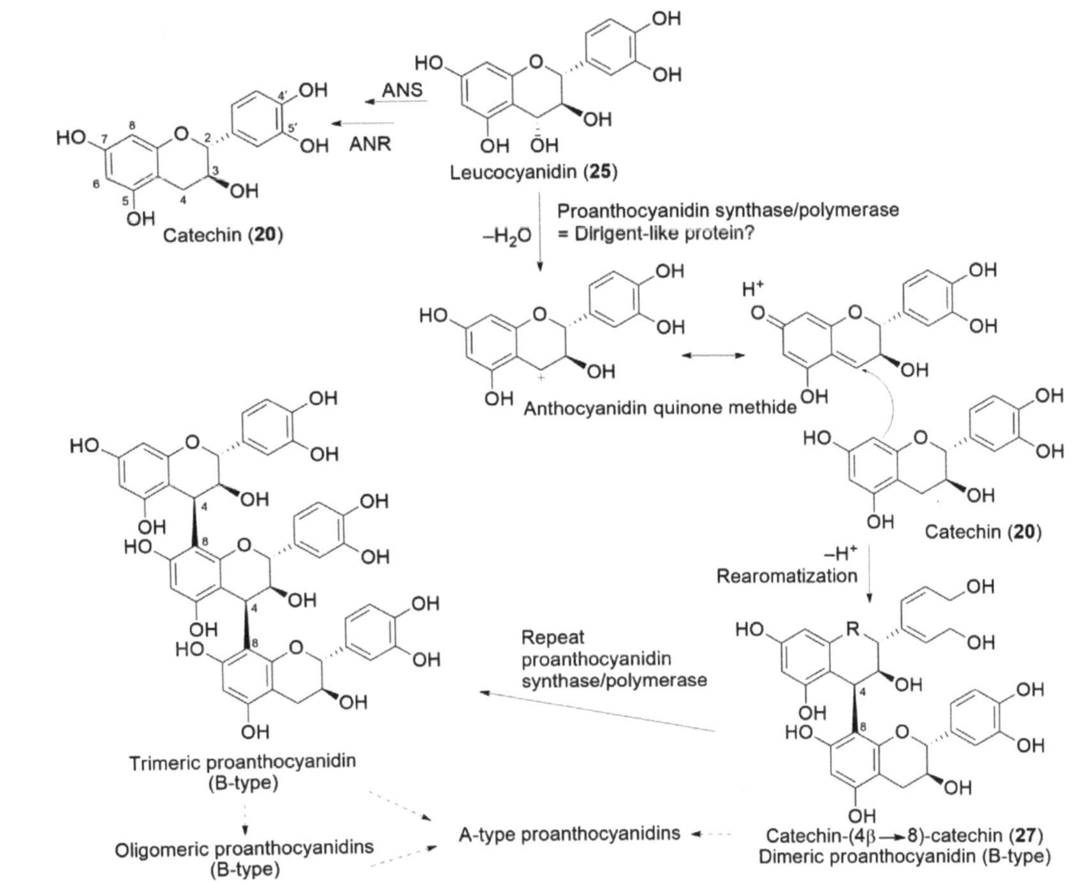

FIGURE 4.11 Hypothetical DP Roadmap to proanthocyanidins (PAs). Proposed mechanism of leucocyanidin (**25**) transformation by a putative proanthocyanidin synthase/polymerase (PASP) – hypothesized to be a paralog of a pterocarpan synthase (PTS). Coupling of terminal-unit catechin (**20**) to the anthocyanidin mono-QM or corresponding cation intermediate requires that a specific face of the quinone methide (either *re* or *si*) be attacked by catechin (**20**) in order to attain the regiospecificity and/or stereospecificity seen in proanthocyanidins. In other words, the substrates need to be precisely oriented for inter-molecular coupling (C–C bond formation), similar to how pinoresinol-forming DPs orient their bis-QM substrates. Note: Type A proanthocyanidins have 4→8 and 2→*O*→7 linkages, whereas Type B have 4→8 linkages. ANR: anthocyanidin reductase; ANS: anthocyanidin synthase.

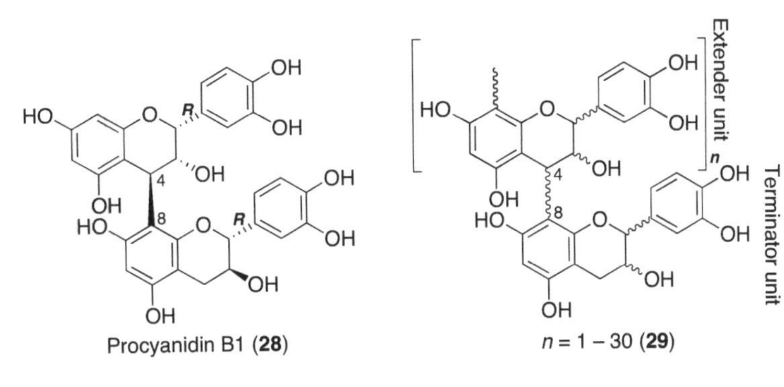

FIGURE 4.12 Procyanidin B1 (**28**), and generic oligomeric PA structures ($n = 1 - 30$, **29**, Ubillas et al. 1994).

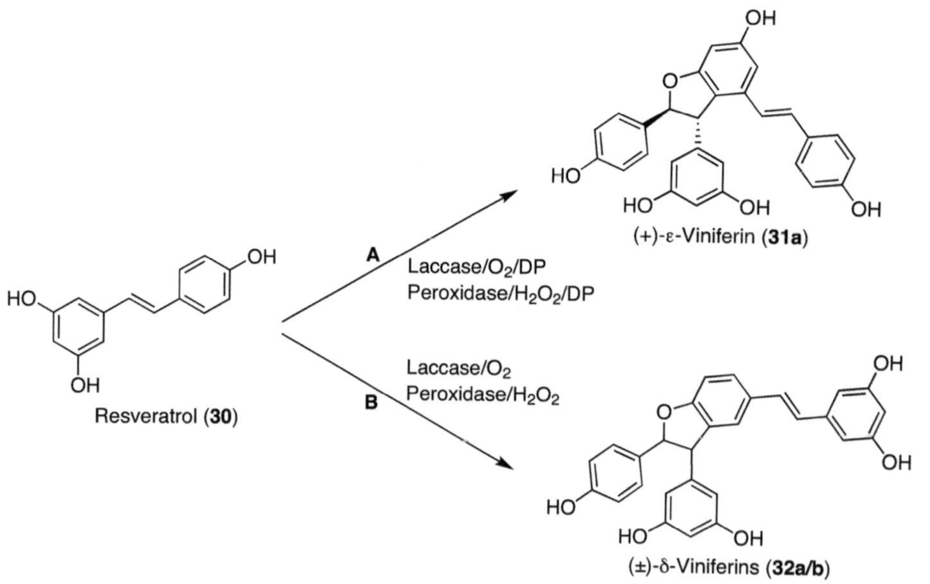

FIGURE 4.13 **A.** Hypothetical DP Roadmap to (+)-ε-viniferin (**31a**) from its achiral precursor, resveratrol (**30**), in *Vitis vinifera*. **B.** Formation of (±)-δ-viniferins (**32a/b**) from **30** by action of laccase/O_2 or peroxidase/H_2O_2.

How chain extension occurs with up to 30 extender (repeat) units (**29**, Ubillas et al. 1994, Figure 4.12) also needs resolution. Whether only PASPs are required for further chain-extending processes, or additional proteins/enzymes are needed must be determined. This oligomerization/polymerization (chain extension) can, however, be envisaged to continue via addition of ***quinone methide*** extender units as depicted in Figure 4.11. (However, we also cannot, at present, fully eliminate the possibility that an auxiliary oxidase or other enzymes/proteins may be *additionally* required for chain extension.)

4.4.2 Proposed DP Roadmap to Dimeric Stilbenoids

In addition to our previous predictions on the metabolic diversity of DPs as gateway entry points to distinct plant phenols (Lewis and Davin 1999, Vassão et al. 2010), there have been numerous deductions of DP involvement in dimeric stilbenoid formation (Pezet et al. 2004, Pickel and Schaller 2013, Borges 2014), e.g. to the resveratrol (**30**)-derived phytoalexin (+)-*trans*-ε-viniferin (**31a**, Figure 4.13A) in *Vitis vinifera* (Ito and Niwa 1996, Mattivi et al. 2011), or to (−)-*trans*-ε-viniferin (**31b**) in *Gnetum klossii* (Ali et al. 2003).

Conversely, both laccase/O_2 (Nicotra et al. 2004) and peroxidase/H_2O_2 (Langcake and Pryce 1977) catalysis of its presumed precursor resveratrol (**30**) *in vitro* form racemic *trans*-δ-viniferin (**32a/b**), rather than the naturally occurring (+)- or (−)-ε-viniferin (**31a** and **31b**) (Figure 4.13B). These observations have also led to consideration that DP sub-family members are instead involved *in planta*.

If correct, their formation can be envisaged to be analogous to that of pinoresinol formation, i.e. free radical formation catalyzed by an oxidase, DP-engendered stereoselective intermolecular radical-radical coupling (C–C bond formation), and intra-molecular cyclization (C–O–C bond formation).

4.4.3 Proposed DP Roadmap to Dimeric Allylphenols/Propenylphenols

There is also growing consideration for distinct DP families with different substrate recognition/binding properties affording other optically active (neo)lignan skeletal types (e.g. 8–2′, 8–5′, etc.), such as (−)-*cis*-blechnic acid (**33b**) in *Blechnum spicant* (Davin et al. 2003) and (−)-balanophonin (**34b**) in *Balanophora japonica* (Haruna et al. 1982), as well as the allylphenol-/propenylphenol-derived (+)-conocarpan (**35a**) and (−)-grandisin (**36b**) (Vassão et al. 2010) (Figure 4.14). Yet, while all of these gateway metabolic processes are envisaged

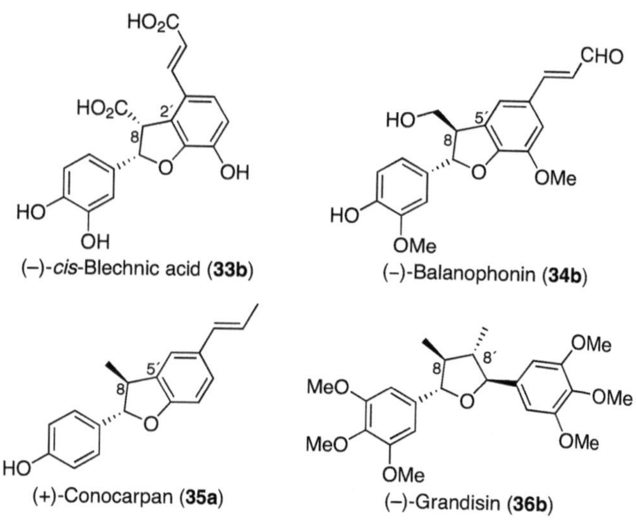

FIGURE 4.14 Optically active 8–2′-linked lignan, (−)-*cis*-blechnic acid (**33b**), from *Blechnum spicant* and 8–5′- and 8–8′-linked and allylphenol-/propenylphenol-derived lignans (**34b**, **35a**, and **36b**) from their corresponding monomers.

to utilize specific DPs, the putative DPs involved await discovery and characterization.

4.5 PROBING GENERAL CHARACTERISTICS OF DP BIOCHEMICAL MECHANISMS

Biochemical mechanistic investigations of DPs have led to much understanding of how different enantiomeric and regiospecific outcomes can be achieved at the molecular level. Furthermore, in 2015 we reported determination of the first (trimeric) DP lignan-forming crystal structure, and more recently in 2020, the first pterocarpan-forming DP-like protein structures. Progress in this area is described in the sections below.

4.5.1 Pinoresinol-Forming DPs, Heterologous Expression, and X-ray Crystallography

Using *Solanum peruvianum* (Kim et al. 2012) and *Pichia pastoris* expression systems and coniferyl alcohol (**2**) substrate, we characterized two distinct pinoresinol-forming DPs affording different 8–8′-linked lignan antipodes, (+)- or (–)-pinoresinols (**1a** or **1b**), *provided* 1e$^-$ oxidative capacity was present (e.g. from laccases, peroxidases, 1e$^-$ oxidants) (Figure 4.5). Both DPs need to be glycosylated post-translationally for their DP activities.

The (+)-pinoresinol-forming DP (DRR206) was the first 3D DP structure (apo-form) reported, and which was solved at 1.95 Å resolution (Kim et al. 2015) as a tightly-packed trimer. Its trimer axis was orientated along the 3-fold axis of a H32 space group (Figures 4.15A and 4.16A). In its trimer form, *circa* 2500 Å^2 of surface and over 30% total surface per monomer were buried.

By applying a Dali search (Holm and Rosenström 2010), several β-barrel proteins were distantly related to DRR206, the closest (~11% sequence similarity) being allene oxide cyclase (AOC) from *P. patens* (Schaller et al. 2005, Hofmann et al. 2006, Neumann et al. 2012). Both 8-stranded β-barrel proteins have a shear number 10, and can be superimposed with 2–3 Å *rmsd* over the entire barrel structure (Figure 4.15B).

Several highly conserved residues cluster around the putative substrate binding pocket including polar residues D40 and R141 (Figure 4.15C). Adjacent to D40, and on the opposite side of the β-sheet is H39, one of three highly conserved residues in an unusual cluster including T84 and S91 (not shown). These residues occur in or near an omega (Ω) loop between strands β2-β3 that fold back upon the barrel's exterior (Figures 4.15C and 4.16A). [An Ω loop is a non-regular secondary structure element characterized by a loop of varying structure with its N- and C-termini located on adjacent secondary structure elements, typically β-strands, close together in space (Fetrow 1995)]. Loops on the opposite end of the barrel (bottom) are more conserved, particularly the β2-β3 ("Ω") and β6-β7 loops. These may represent interaction sites for a DP-specific (per)oxidase, for example.

AtDIR6 (apo-form) was expressed in *P. pastoris,* with a preliminary diffraction data set collected at Stanford Synchrotron Radiation Lightsource (SSRL) beamline BL12-2. Its crystals diffracted to *circa* 3.6 Å resolution with unit cell dimensions of a = 105.46, b = 130.24, c = 83.21 in space group $P2_12_12$. This data gave a molecular replacement solution using search models derived from the DRR206 monomer and trimer, thereby indicating that AtDIR6 has two trimers in the asymmetric unit. Both trimers were arranged in the crystal structure such that their active site cavities faced towards each other. This may simply be an artifact of crystallization or it could suggest a mechanism by which free radical CA˙ in two juxtaposed AtDIR6 molecules could be brought together to initiate coupling.

The AtDIR6 structure was subsequently solved independently at 1.4 Å resolution (Gasper et al. 2016), using the DRR206 structure as the molecular replacement search model, and was similar to DRR206 as expected. That is, the reported

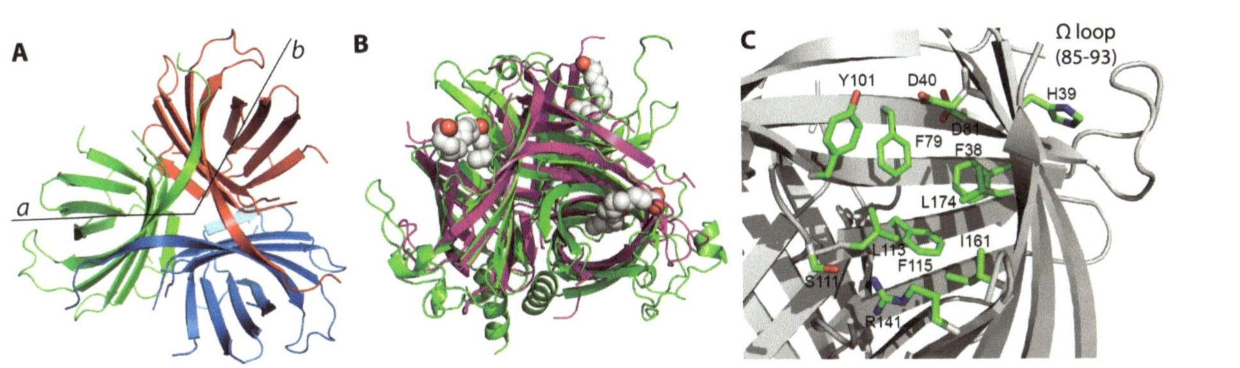

FIGURE 4.15 DIR-a DP in 8–8′-lignan biosynthesis **A.** DP DRR206 trimer viewed down 3-fold *c*-axis of H32 space group, with *a*- and *b*-axes indicated (Protein Data Bank [PDB] ID: 4REV). **B.** DRR206 (magenta) superposition with *Physcomitrella patens* allene oxide cyclase (AOC, PDB ID: 4H69, green, w/ligand, spheres), with *rmsd* ~ 2.1 Å over 102 residues. **C.** Several residues lining active site pocket are differentially conserved in (+)- or (–)-pinoresinol forming DPs, suggesting a role in determining substrate orientation. The Ω loop is labeled to indicate proximity to the backside of the active site. This research was originally published in the Journal of Biological Chemistry: Kim et al. (2015). Trimeric structure of (+)-pinoresinol forming dirigent protein at 1.95 Å resolution with three isolated active sites. *J. Biol. Chem.* 290:1308–1318. Copyright (2015) the American Society for Biochemistry and Molecular Biology.

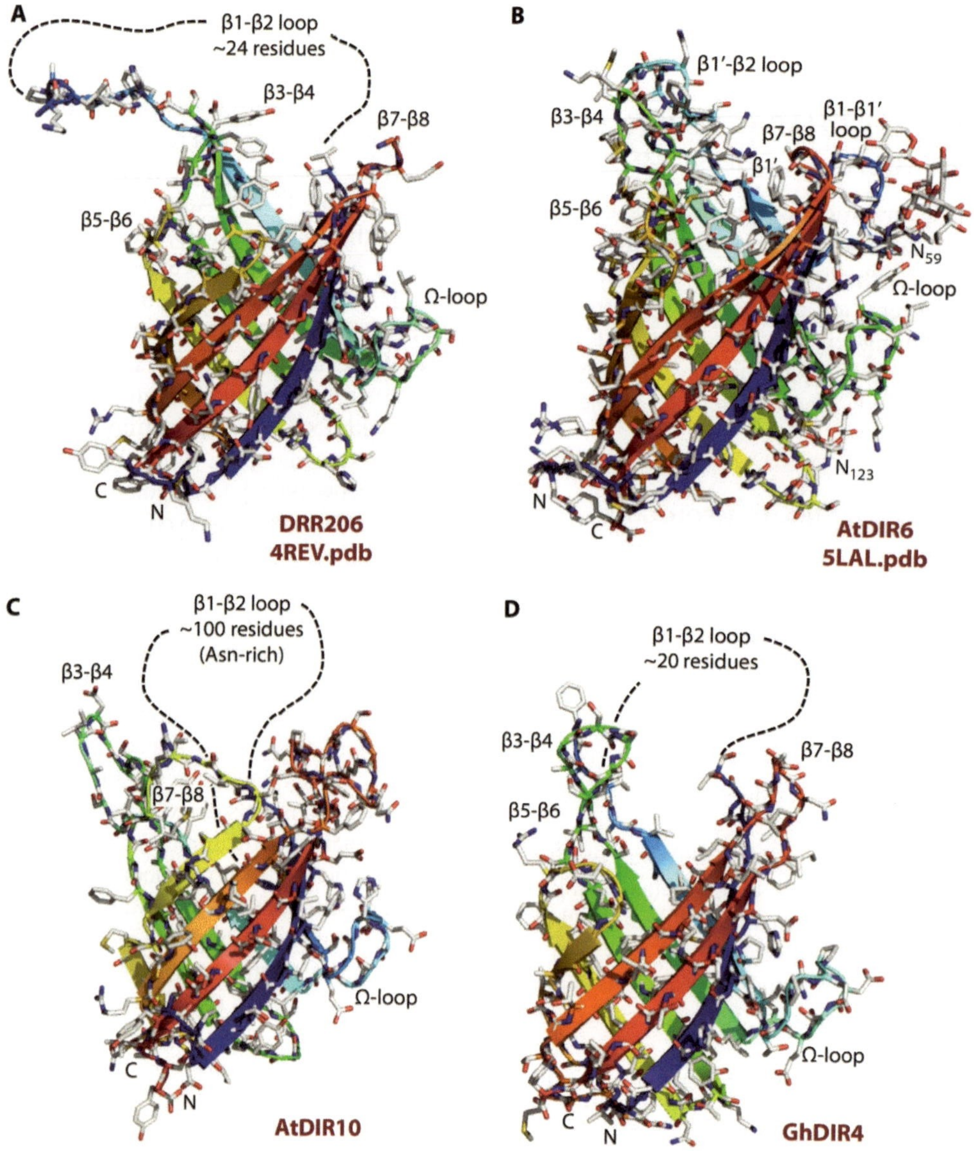

FIGURE 4.16 Sub-families DIR-a, DIR-b/d and DIR-e DPs in 8–8′-lignan, aromatic dimeric terpenoid and Casparian strip lignin biosynthesis. 3D structures of DRR206 (4REV, **A**) and AtDIR6 (5LAL, **B**). Homology models of AtDIR10 (**C**) and GhDIR4 (**D**). Adapted from Meng et al. 2020. This research (A, B and D) was originally published in the Journal of Biological Chemistry: Meng et al. (2020) Pterocarpan synthase (PTS) structures suggest a common quinone methide-stabilizing function in dirigent proteins and proteins with dirigent-like domains. *J. Biol. Chem.* 295:11584–11601. Copyright (2020) the American Society for Biochemistry and Molecular Biology.

trimer was structurally similar to the crystallographic trimer described for our reported DRR206, with backbone atom *rmsd* of 1.3 Å for the core β-barrel residues and having the same Ω loop bent upward and in contact with the exterior of the β-barrel. The register of the β-strands was also identical to that in our own homology model using the published DRR-206 structure (4REV, Figure 4.16A) as a template, together with the published sequence alignment; both our model and the AtDIR6 X-ray structure (5LAL, Figure 4.16B) had 1.7 Å *rmsd* for the core β-barrel residues (Gasper et al. 2016).

In that case, the protein crystallized in space group H3 with unit cell dimensions a = b = 101.09 Å, c = 90.24 Å. It contained two monomers in the asymmetric unit, and generated a trimer from each of the independent monomers. In this H3 crystal form, there did not appear to be the same alignment of putative active site cavities of one trimer with another crystallographically-generated trimer. The internal cavity in the AtDIR6 β-barrel can be separated into two putative binding pockets (A and B) and placement of free radical CA˙ in each pocket followed by simulated annealing energy minimization gave poses consistent with juxtaposition of the C8 and C8′ atoms, which could lead to coupling at the *re-re* face [to give ultimately (–)-pinoresinol (**1b**)] (Gasper et al. 2016).

This AtDIR6 structure is noteworthy in that it has a small additional β-strand that is formed by a segment of the large β1-β2 loop, forming two new smaller loops on either side of the additional strand (Figure 4.16B). This potentially indicates the existence of a conformational switch in this portion of the structure, modulated by association/dissociation of the small additional β-strand; for example, between an open or closed lid on the active site, as has been suggested, or some other function of the large loop related to surface binding, association with other proteins, or self-organization into separate soluble liquid phases or membrane-less organelles.

4.5.2 DP (Sub-families DIR-a, DIR-b/d and DIR-e) Homology Modelling Studies and Computational Docking

Figure 4.17 shows amino acid sequence alignment of the (+)-pinoresinol-forming DP DRR206, the (–)-pinoresinol-forming DP AtDIR6, the (+)-gossypol-forming DP GhDIR4 in *G. hirsutum*, and that reported to exercise control over lignification in *Arabidopsis* (AtDIR10). Alignment here showed that the β-strands forming the core β-barrel structure were largely conserved, suggesting homology modelling (see Figures 4.16C and D) was suitable to model and understand the putative active site and, in particular, the surrounding loops; those indicated as red bars (Figure 4.17) were noted to vary significantly in both length and sequence. We thus can hypothesize that the loops hold important roles, possibly helping confer substrate specificity. Interestingly, the large loop between β1-β2 (Figure 4.16) varied in length and sequence among different DP subfamilies and was provisionally likely to be conformationally flexible as it was missing in the DRR206 X-ray structure. Once again, this loop may possibly act as a lid on the active site that may control, for example, substrate binding or positioning for stereo- and regiocontrol of coupling, or it may mediate protein interactions in the proposed supramolecular complex.

However, short-lived intermediary radical substrate species, including their being bound and orientated on the DP in order to control the outcome of coupling, makes this somewhat mechanistically challenging to study biochemically. Indeed, in the current absence of a fully solved substrate/substrate analog bound 3D DP complex, and since the DRR206 and AtDIR6 putative substrate is reactive CA˙, the DP-substrate complex experimental investigation is not straightforward.

Accordingly, computational docking studies were employed using the DRR206 X-ray structure and models of both CA˙ and 8–8′ bis-QM, with the latter as a putative intermediate in pinoresinol (**1**) formation following (CA˙) radical-radical coupling (see Figure 4.5). Our initial docking studies with AUTODOCK (Morris et al. 2009) provisionally suggested that the binding site accommodated one CA˙ (Figure 4.18A), with dimensions appropriate for the long axis to span the distance from R141 to D40, with quinone oxygen and 9-OH moieties potentially interacting favorably with these charged residues. Preliminary docking studies using ICM-Pro, which allows for a fully flexible substrate and a partially flexible receptor, suggested that two CA˙ radicals can be accommodated in the DRR206 binding site, interacting with side chains Y101, D40, T110, and R141. Moreover, additional docking simulations with 8–8′ bis-QM and (–)-secoisolariciresinol (**4b**, Figure 4.6), which potentially mimics the 8–8′

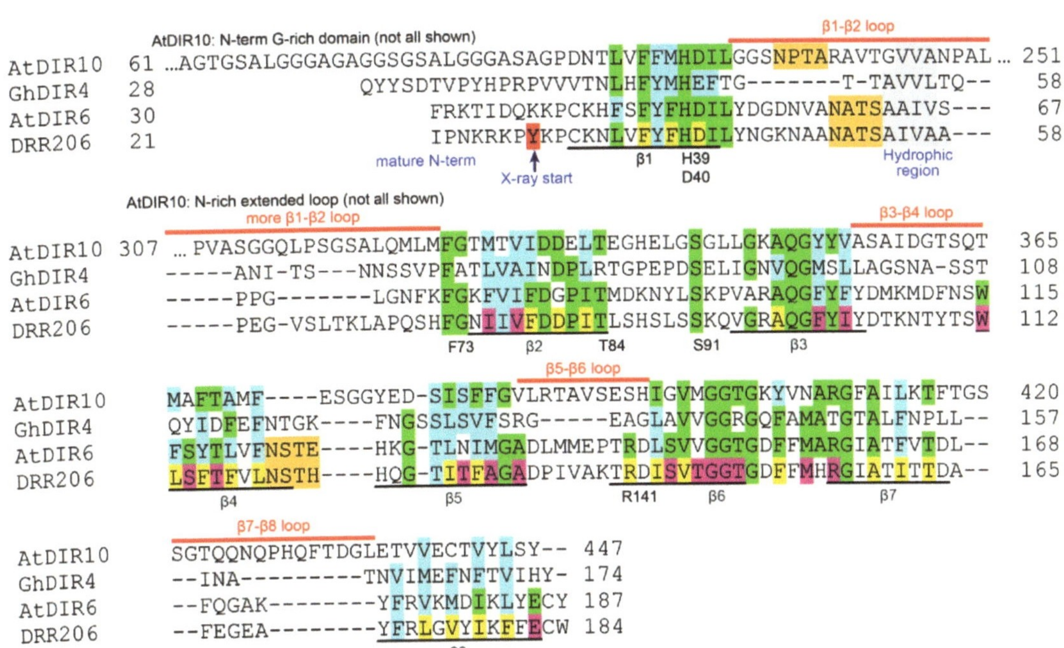

FIGURE 4.17 Sequence alignment of DIR-a, DIR-b/d and DIR-e sub-family members: AtDIR6 [(–)-pinoresinol (**1b**) forming DP], and DRR206 [(+)-pinoresinol (**1a**) forming DP], GhDIR4 [gossypol (**3**) forming DP] and AtDIR10 (Casparian strip lignin DP). For DRR206, yellow highlights core residues and pink indicates oligomer interface residues. Green and blue highlight residues identical or conserved (respectively) with respect to DRR206. Orange highlights known glycosylation sites, and grey indicate a small hydrophobic region present in the large β1-β2 loop from many sequences.

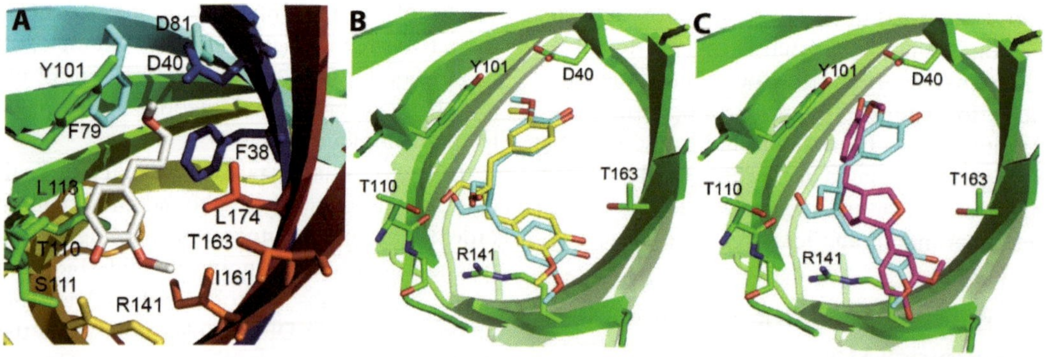

FIGURE 4.18 8–8′-lignan forming DIR-a DP [(+)-pinoresinol forming] substrate/substrate homolog docking. **A.** Docked substrate model *en route* to (+)-pinoresinol (**1a**) in putative DRR206 active site. CA˙ in *syn* configuration bound so that *re* face is exposed, this orientation couples with *re* face of another CA˙ to give 8–8′ bis-quinone methide (bis-QM) *R*,*R*-stereoisomer, the putative initial product from coupling of 2 CA˙. Preliminary molecular docking with ICM-Pro showing, in binding site of apo-DRR206, (–)-secoisolariciresinol (**4b**, yellow) and bis-QM (cyan) (**B**) and (+)-pinoresinol (**1a**, magenta) superimposed on bis-QM (cyan) (**C**). This research (A) was originally published in the Journal of Biological Chemistry: Kim et al. (2015). Trimeric structure of (+)-pinoresinol forming dirigent protein at 1.95 Å resolution with three isolated active sites. *J. Biol. Chem.* 290:1308–1318. Copyright (2015) the American Society for Biochemistry and Molecular Biology.

bis-QM, and (+)-pinoresinol (**1a**) may also be accommodated in the binding site. The most energetically favored docked structures for 8–8′ bis-QM and (–)-secoisolariciresinol (**4b**) were crescent-shaped conformations with the two quinone and phenolate rings, respectively, partially stacked against each other (Figure 4.18B). However, once the two fused tetra-hydrofuranyl rings formed to produce (+)-pinoresinol (**1a**), the molecule necessarily elongated (Figure 4.18C) and preliminary docking poses suggested a significant decrease in stabilization energy, which may be the trigger for product release.

The putative active site pocket in the published AtDIR6 crystal structure (Gasper et al. 2016) contained the same aforementioned conserved residues found also in DRR206, as well as substitutions that may direct the different stereochemical outcomes of the two DPs. The AtDIR6 structure was considered to have a binding pocket with a pronounced two-lobed character not recognized in DRR206. Modelling one CA˙ in each lobe of the pocket, and refinement and minimization by simulated annealing, gave results consistent with the hypothesis that two CA˙ bind and residues lining the pocket orient each substrate such that stereoselective 8–8' coupling favors production here of (–)-pinoresinol (**1b**). Presumably, an identical modelling exercise conducted with the DRR206 X-ray structure instead would indicate formation of the (+)-pinoresinol (**1a**) enantiomer.

4.5.3 Basis of Differing DP DIR-a Sub-family Stereoselectivities of (+)- and (–)-Pinoresinol-Forming DPs

Figure 4.19 summarizes results with differing recombinant DP amounts (from different plant species) in generating either **1a** or **1b** in enantiomeric excess (e. e.) in assays. We hypothesized that a bioinformatics analysis of our various (+)- and (–)-pinoresinol-forming DPs in different plant species would begin to shed light on how, for example, different stereoselectivities

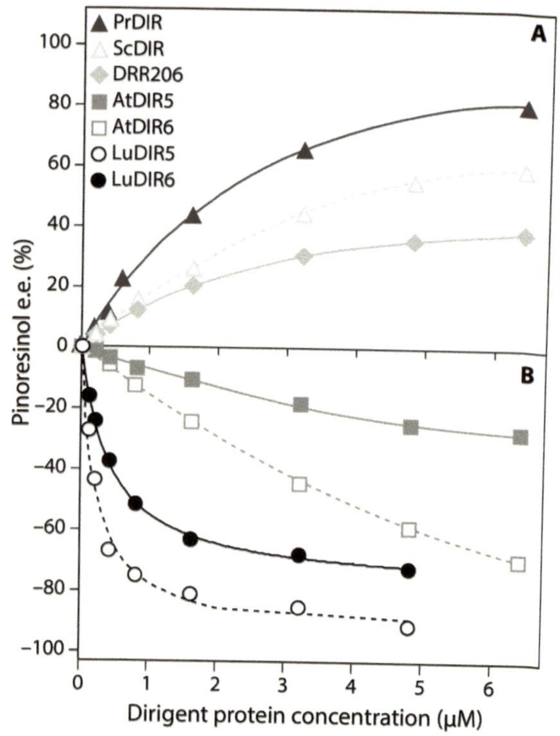

FIGURE 4.19 8–8′-lignan forming DIR-a DPs. Effect of varying (+)- and (–)-pinoresinol forming DP concentrations on formation of **A.** (+)-Pinoresinol (**1a**) and **B.** (–)-pinoresinol (**1b**) in enantiomeric excess (e.e.), respectively.

were achieved. Our analysis indicated 6 differentially conserved residues, 5 distributed in the "middle region" of the DPs, and 1 in the C-terminal region (Kim et al. 2012). Thus, it was instructive to ascertain whether domain swapping of an internal 49 amino acid "middle" region (Asn98 to Pro146 in ScDIR and Lys90 to Leu138 in AtDIR6, Figure 4.20A) affected stereoselectivity or substrate binding. Using a 3-step PCR approach (Grandori et al. 1997), a domain swapped ScDIR harboring the AtDIR6 49-amino-acid region gave a DP whose properties

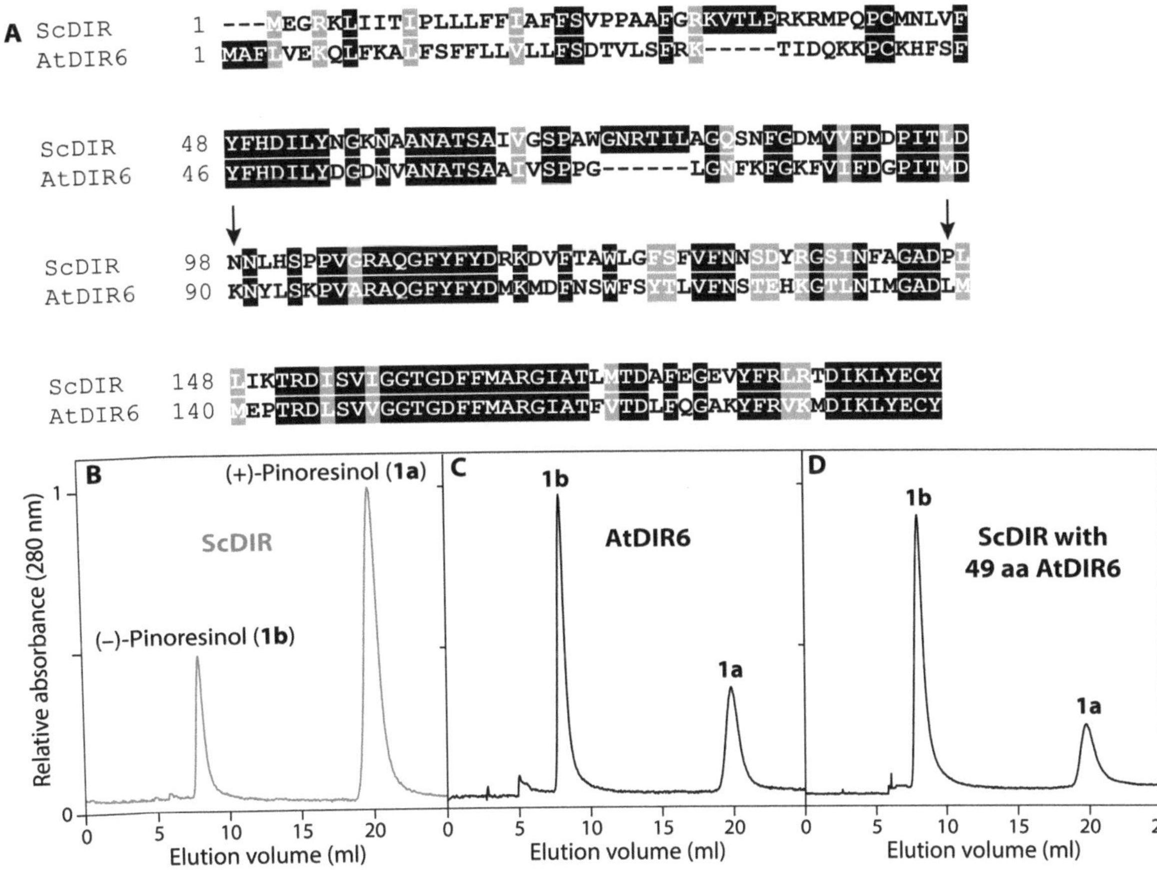

FIGURE 4.20 Effects of region swapping on 8–8′-lignan DIR-a (+)- and (–)-pinoresinol-forming DP stereoselectivity (Kim et al. 2012). **A.** Sequence alignment of ScDIR [(+)-pinoresinol forming DP] and AtDIR6 [(–)-pinoresinol forming DP] showing the 49 amino acid internal region used in domain swapping (Asn98 to Pro146 in ScDIR and Lys90 to Leu138 in AtDIR6 as indicated by arrows). **B–D.** Chiral HPLC analyses of pinoresinol (**1**) antipodes are shown under the assays conditions employed. Recombinant native *Schizandra chinensis* DP (ScDIR, **B**); recombinant native *Arabidopsis* DP (AtDIR6, **C**); and AtDIR6 49 amino acid internal region-B swapped into ScDIR (**D**). This research (B–D) was originally published in the Journal of Biological Chemistry: Kim et al. (2012). Opposite stereoselectivities of dirigent proteins in *Arabidopsis* and *Schizandra* species. *J. Biol. Chem.* 287:33957–33972. Copyright (2012) the American Society for Biochemistry and Molecular Biology.

completely reversed the original stereoselectivity, i.e. from a (+)-pinoresinol-forming DP to a (–)-pinoresinol-producing DP [assay conditions gave 50% enantiomeric excess (e. e.) of (–)-form (**1b**), Figure 4.20D] from 70% e.e. (+)-pinoresinol (**1a**) (Figure 4.20B) (Kim et al. 2012). That is, it was now functioning as recombinant AtDIR6 (Figure 4.20C). This demonstrated that key residues in the 49-amino-acid region differentially controlled stereoselectivity. Mutagenesis to place bulkier tryptophan residue side chains (V177W and F175W) in either of the lobes of the active site pocket in AtDIR6 reduced activity dramatically (Gasper et al. 2016).

4.5.4 Pterocarpan-Forming (and Related) DIR-b/d Sub-Family DPs, Heterologous Expression and X-Ray Crystallography

As indicated previously, pterocarpans are important plant phytoalexins (Ingham 1982, VanEtten et al. 1983). Their DP gateway entry points, pea (PsPTS1) and licorice (GePTS1) pterocarpan-forming dirigent-like proteins, were also crystallized with their X-ray structures determined at 1.5 and 2.6 Å, respectively. GePTS1 (Figure 4. 21A) and PsPTS1 (Figure 4.21B) exist as trimers (Meng et al. 2020), isostructural with the trimeric structures seen in DRR206 and AtDIR6. GePTS1 crystallizes as a dimer of trimers (Figure 4.21A).

The pterocarpan/isoflavene-forming dirigent-like protein PsPTS2 was crystallized in space group I222 with a hexamer in the asymmetric unit comprising two tightly packed trimers (Figure 4.21C, Meng et al., manuscript in finalization). Its trimeric structure is once again similar to the trimers in DRR206 and AtDIR6 and the two PTS1 enzymes, whereas the PsPTS2 hexamer is not isostructural with the GePTS1 hexamer. The N-termini of each PsPTS2 monomer crosses into the neighboring trimer and is involved in a domain-swapping event, whereby an additional β-strand from a monomer in one trimer is added to the β-barrel of a monomer in the second trimer (Figure 4.22A). Superposition of the monomers of PsPTS1 and GePTS1 onto one of the PsPTS2 monomers shows that

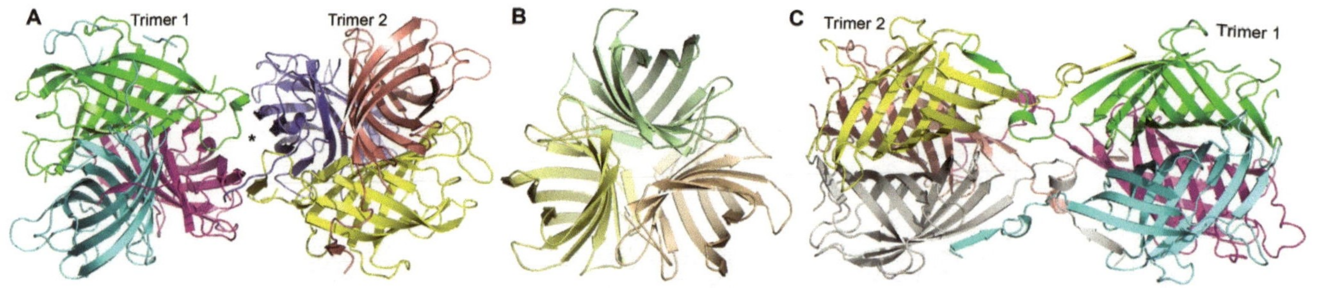

FIGURE 4.21 Ribbon representations of DIR-b/d sub-family DP (PST) structures. **A.** The biological unit of GePTS1 is a tightly packed trimer, and whose asymmetric unit contains a dimer of trimers related by a pseudo-two-fold axis perpendicular to the plane of the paper (Trimer 1: green, magenta, and cyan, and Trimer 2: yellow, pink and blue). **B.** PsPTS1 trimer. **C.** Ribbon representation of PsPTS2 hexamer showing two trimers (Trimer 1: yellow, pink and grey, and Trimer 2: green, magenta and cyan). This research (A) was originally published in the Journal of Biological Chemistry: Meng et al. (2020) Pterocarpan synthase (PTS) structures suggest a common quinone methide-stabilizing function in dirigent proteins and proteins with dirigent-like domains. *J. Biol. Chem.* 295:11584–11601. Copyright (2020) the American Society for Biochemistry and Molecular Biology.

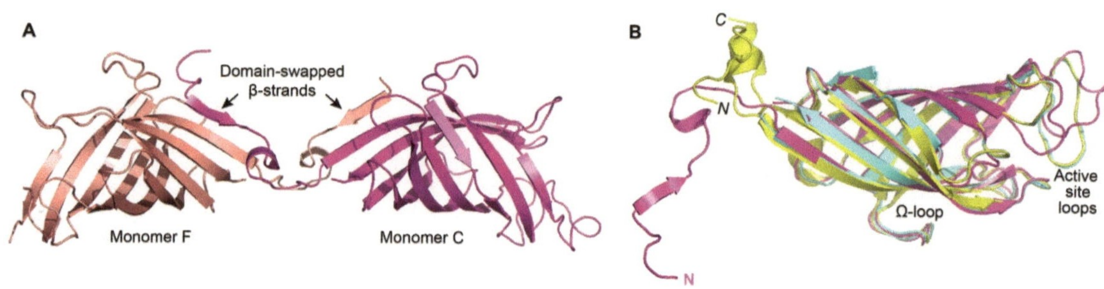

FIGURE 4.22 Dimer and monomer structure of DIR-b/d sub-family pterocarpan/flavene forming DP, PsPTS2. **A.** Domain-swapping between monomer C from Trimer 1 (magenta) and monomer F from Trimer 2 (pink). An extra β-strand from the N-terminus of each monomer is added to the β-barrel of the adjacent monomer. **B.** Superposition of PsPTS1 (cyan) and GePTS1 (yellow) onto monomer C of PsPTS2 (magenta). The N-terminus of PsPTS2 is labeled in magenta, and the N- and C-termini of GePTS1 are labeled in black.

the β-barrel, including the Ω-loop that projects from the barrel, is structurally conserved (Figure 4.22B).

A closer inspection of the putative active sites of the three enzymes shows a high level of conservation of the residues highlighted in DRR206 as playing a potential role in substrate orientation (Figure 4.15C), with the exception of the two juxtaposed phenylalanine residues in the central region of the pocket (F38 and F79 in DRR206, Figure 4.15C). At position 38 (DRR206 numbering), the two PTS1 enzymes have the canonical phenylalanine and PsPTS2 has a leucine, with position 79 (DRR206 numbering) being methionine in GePTS1 and PsPTS1 and once again leucine in PsPTS2. In the latter, these changes effectively make the central part of the active site pocket more open relative to the other enzymes. Structural changes are also evident in two loops at the opening of the PsPTS2 active site, relative to both PsPTS1 and GePTS1 (Meng et al., manuscript in finalization). The effect of the different sequence and loop conformation must await further structural studies.

4.6 FUTURE CONSIDERATIONS

Since the discovery of DPs in lignan biosynthesis (i.e. the (+)-pinoresinol-forming DP), DPs have been reported to be gateway entry points to diverse plant phenol metabolic classes as earlier predicted (Davin and Lewis 2005a, Davin et al. 2008). The DP Roadmap now includes gateway entry points to different lignan ('neolignan') types, polymeric lignins, aromatic diterpenoids, pterocarpans, and isoflavenes.

Clearly, from the large multigene DP and DP-like families in terrestrial plants, much more remains to be discovered as to their range of biochemical/physiological functions, i.e. affording other plant phenol metabolic classes and sub-classes. For example, DPs are hypothetically gateway entry points to (putative) proanthocyanidins, PAs, as well as dimeric/higher oligomeric stilbenoid and allylphenols/propenylphenols. Many more DP gateway entry points can be anticipated, e.g. to ellagitannins.

Also needed are additional comprehensive biochemical studies involved in fully establishing the factors in the DP active site(s) controlling DP stereoselectivity and/or regiospecificity. This includes both identifying the range of novel plant phenol DP Roadmap gateway entry points to not only distinct plant phenol metabolic classes, but also how higher oligomeric and polymeric entities are formed, e.g. such as hypothetically to oligo-lignans and proanthocyanidins.

A recent paper (Kim and Sattely 2021) characterizing formation of heterodimeric variants of pinoresinol (**1**) from coniferyl alcohol (**2**) homologs by a pinoresinol-forming DP from *Podophyllum hexandrum* reveals some additional features of the DP-guided enantioselective coupling reaction mechanism. In that report, endogenous substrate coniferyl alcohol (**2**) and a laccase were incubated *in vitro* with closely related non-canonical monolignols substituted at the 3- and/or 5-position of the aromatic ring in the presence or absence of the DP, i.e. to test substrate versatility. The product mixture in each case was analyzed for characteristics indicating specificity for one or both substrates: regiospecific 8–8′ coupling *versus* other coupling modes, enantiospecificity of the 8–8′ coupling, and homo- *versus* heterodimer formation in the presence of both substrates. Few of these non-canonical substrates formed 8–8′ homodimers, with the exception of 3-fluoroconiferyl alcohol, whereas some formed 8–8′ heterodimers with high enantioselectivity provided one monomer was coniferyl alcohol (**2**). For example, 8–8′ heterodimers were obtained between coniferyl alcohol (**2**) and variants where the methoxy group was replaced with a methyl or ethoxy group formed with high enantio- and regioselectivity. This suggests that the DP binds one monomer with high specificity for coniferyl alcohol (or 3-fluoroconiferyl alcohol), but can also accommodate hetero-coupling with high 8–8′ stereo- and regiospecificity, with some of the non-canonical monolignol substrates.

This observation is consistent with our proposed mechanism in which the pinoresinol-forming DP function is to bind and stabilize a coniferyl alcohol (**2**) QM radical with high specificity, positioning it for stereo- and regioselective 8–8′ coupling to a second propenylphenol radical within a single DP active site (Figure 4.5). The *P. hexandrum* DP apparently allows greater heterogeneity in the second radical substrate in hetero-dimer formation, at least in *in vitro* studies.

The range of heterogeneity that can be accommodated in the second radical substrate, including whether it needs to be a propenylphenol radical – or even a radical at all, which would allow for either initiation of coupling or polymerization by a radical chain mechanism – has probably not yet been fully explored for this and other pinoresinol-forming DPs. While these observations may not have physiological significance, they provide an opportunity to more deeply probe the biochemical mechanism in the future.

ACKNOWLEDGEMENTS

The authors thank the Chemical Sciences, Geosciences and Biosciences Division, US Department of Energy (DOE) Office of Basic Energy Sciences (DE-FG-0397ER20259) , as well as the Arthur and Katie Eisig-Tode Endowment. A portion of the research was performed using the Environmental Molecular Sciences Laboratory (EMSL), a DOE Office of Science User Facility sponsored by the Office of Biological and Environmental Research and located at Pacific Northwest National Laboratory (PNNL) in Richland, Washington. Use of the Stanford Synchrotron Radiation Lightsource (SSRL), SLAC National Accelerator Laboratory, is supported by the US DOE, Office of Science, Basic Energy Sciences (BES), under Contract No. DE-AC02–76SF00515. The SSRL Structural Molecular Biology Program is supported by the DOE Office of Biological and Environmental Research (BER), and by the National Institutes of Health, National Institute of General Medical Sciences (including P41GM103393). Mr. Kamal Seneviratne's assistance in obtaining AtDIR6 crystals for SLAC preliminary analysis (at 3.6 Å) is acknowledged.

REFERENCES

Ali, Z., T. Tanaka, I. Iliya, et al. 2003. Phenolic constituents of *Gnetum klossii*. *J. Nat. Prod.* 66:558–560.

Arabidopsis Genome Initiative, The. 2000. Analysis of the genome sequence of the flowering plant *Arabidopsis thaliana*. *Nature* 408:796–815.

Bailly, C. 2012. Contemporary challenges in the design of topoisomerase II inhibitors for cancer chemotherapy. *Chem. Rev.* 112:3611–3640.

Balalaie, A., M. B. Rezvani, and M. Mohammadi Basir. 2018. Dual function of proanthocyanidins as both MMP inhibitor and crosslinker in dentin biomodification: A literature review. *Dent. Mater. J.* 37:173–182.

Banks, J. A., T. Nishiyama, M. Hasebe, et al. 2011. The *Selaginella* genome identifies genetic changes associated with the evolution of vascular plants. *Science* 332:960–963.

Borges, A. F. G. 2014. The grapevine defences: Walking towards the "hidden" role of dirigent proteins in *Vitis vinifera* secondary metabolism. *PhD Thesis, Universidade Nova de Lisboa, Portugal. 192 pp.*

Bowman, J. L., T. Kohchi, K. T. Yamato, et al. 2017. Insights into land plant evolution garnered from the *Marchantia polymorpha* genome. *Cell* 171:287–304.

Casanova-Martí, À., J. Serrano, K. J. Portune, et al. 2018. Grape seed proanthocyanidins influence gut microbiota and enteroendocrine secretions in female rats. *Food Funct.* 9:1672–1682.

Celoy, R. M. 2013. (+)-Pisatin biosynthesis: From (−)-enantiomeric intermediates via an achiral isoflavene. *PhD Thesis, University of Arizona, 241 pp.*

Celoy, R. M., and H. D. VanEtten. 2014. (+)-Pisatin biosynthesis: From (−)-enantiomeric intermediates via an achiral 7,2′-dihydroxy-4′,5′-methylenedioxyisoflav-3-ene. *Phytochemistry* 98:120–127.

Corbin, C., S. Drouet, L. Markulin, et al. 2018. A genome-wide analysis of the flax (*Linum usitatissimum* L.) dirigent protein family: From gene identification and evolution to differential regulation. *Plant Mol. Biol.* 97:73–101.

Dalisay, D. S., K. W. Kim, C. Lee, et al. 2015. Dirigent protein-mediated lignan and cyanogenic glucoside formation in flax seed: Integrated omics and MALDI mass spectrometry imaging. *J. Nat. Prod.* 78:1231–1242.

Davin, L. B., H.-W. Wang, A. L. Crowell, et al. 1997. Stereoselective bimolecular phenoxy radical coupling by an auxiliary (dirigent) protein without an active center. *Science* 275:362–366.

Davin, L. B., C.-Z. Wang, G. L. Helms, and N. G. Lewis. 2003. [^{13}C]-Specific labeling of 8–2′ linked (−)-*cis*-blechnic, (−)-*trans*-blechnic and (−)-brainic acids in the fern *Blechnum spicant*. *Phytochemistry* 62:501–511.

Davin, L. B., and N. G. Lewis. 2005a. Dirigent phenoxy radical coupling: Advances and challenges. *Curr. Opin. Biotechnol.* 16:398–406.

Davin, L. B., and N. G. Lewis. 2005b. Lignin primary structures and dirigent sites. *Curr. Opin. Biotechnol.* 16:407–415.

Davin, L. B., M. Jourdes, A. M. Patten, et al. 2008. Dissection of lignin macromolecular configuration and assembly: Comparison to related biochemical processes in allyl/propenyl phenol and lignan biosynthesis *Nat. Prod. Rep.* 25:1015–1090.

Davin, L. B., and N. G. Lewis. 2022. Redefining lignan/neolignan nomenclature and diversity: A historical perspective. In *The Lignan Handbook* (N. G. Lewis and L. B. Davin eds.), Taylor & Francis, Boca Raton, London. pp 1–11

de Colmenares, N. G., J. R. Ramírez-Martínez, J. O. Aldana, et al. 1998. Isolation, characterisation and determination of biological activity of coffee proanthocyanidins. *J. Sci. Food Agric.* 77:368–372.

Debeaujon, I., K. M. Léon-Kloosterziel, and M. Koornneef. 2000. Influence of the testa on seed dormancy, germination, and longevity in *Arabidopsis. Plant Physiol.* 122:403–414.

Dixon, R. A., D.-Y. Xie, and S. B. Sharma. 2005. Proanthocyanidins – a final frontier in flavonoid research? *New Phytol.* 165:9–28.

Effenberger, I., B. Zhang, L. Li, et al. 2015. Dirigent proteins from cotton (*Gossypium* sp.) for the atropselective synthesis of gossypol. *Angew. Chem. Int. Ed.* 54:14660–14663.

Fetrow, J. S. 1995. Omega loops: Nonregular secondary structures significant in protein function and stability. *FASEB J.* 9:708–717.

Finn, R. D., P. Coggill, R. Y. Eberhardt, et al. 2016. The Pfam protein families database: Towards a more sustainable future. *Nucleic Acids Res.* 44:D279–D285.

Ford, J. D., K.-S. Huang, H.-B. Wang, L. B. Davin, and N. G. Lewis. 2001. Biosynthetic pathway to the cancer chemopreventive secoisolariciresinol diglucoside-hydroxymethyl glutaryl ester-linked lignan oligomers in flax (*Linum usitatissimum)* seed. *J. Nat. Prod.* 64:1388–1397.

Fristensky, B., R. C. Riggleman, W. Wagoner, and L. A. Hadwiger. 1985. Gene expression in susceptible and disease resistant interactions of peas induced with *Fusarium solani* pathogens and chitosan. *Physiol. Plant Pathol.* 27:15–28.

Fristensky, B., D. Horovitz, and L. A. Hadwiger. 1988. cDNA sequences for pea disease resistance response genes. *Plant Mol. Biol.* 11:713–715.

Funatsuki, H., M. Suzuki, A. Hirose, et al. 2014. Molecular basis of a shattering resistance boosting global dissemination of soybean. *Proc. Natl. Acad. Sci. USA* 111:17797–17802.

Gang, D. R., M. A. Costa, M. Fujita, et al. 1999. Regiochemical control of monolignol radical coupling: A new paradigm for lignin and lignan biosynthesis. *Chem. Biol.* 6:143–151.

Gasper, R., I. Effenberger, P. Kolesinski, et al. 2016. Dirigent protein mode of action revealed by the crystal structure of AtDIR6. *Plant Physiol.* 172:2165–2175.

Grandori, R., K. Struck, K. Giovanielli, and J. Carey. 1997. A three-step PCR protocol for construction of chimeric proteins. *Protein Eng.* 10:1099–1100.

Halls, S. C., and N. G. Lewis. 2002. Secondary and quaternary structures of the (+)-pinoresinol forming dirigent protein. *Biochemistry* 41:9455–9461.

Halls, S. C., L. B. Davin, D. M. Kramer, and N. G. Lewis. 2004. Kinetic study of coniferyl alcohol radical binding to the (+)-pinoresinol forming dirigent protein. *Biochemistry* 43:2587–2595.

Hartmann, J. T., and H.-P. Lipp. 2006. Camptothecin and podophyllotoxin derivatives: Inhibitors of topoisomerase I and II – Mechanisms of action, pharmacokinetics and toxicity profile. *Drug Saf.* 29:209–230.

Haruna, M., T. Koube, K. Ito, and H. Murata. 1982. Balanophonin, a new neo-lignan from *Balanophora japonica* Makino. *Chem. Pharm. Bull.* 30:1525–1527.

Henkin, J. M., Y. Ren, D. D. Soejarto, and A. D. Kinghorn. 2018. The search for anticancer agents from tropical plants. *Prog. Chem. Org. Nat. Prod.* 107:1–94.

Hofmann, E., P. Zerbe, and F. Schaller. 2006. The crystal structure of *Arabidopsis thaliana* allene oxide cyclase: Insights into the oxylipin cyclization reaction. *Plant Cell* 18:3201–3217.

Holm, L., and P. Rosenström. 2010. Dali server: Conservation mapping in 3D. *Nucleic Acids Res.* 38:W545–W549.

Hosmani, P. S., T. Kamiya, J. Danku, et al. 2013. Dirigent domain-containing protein is part of the machinery required for formation of the lignin-based Casparian strip in the root. *Proc. Natl. Acad. Sci. USA* 110:14498–14503.

Ingham, J. L. 1982. Phytoalexins. In *Phytoalexins from the Leguminosae* (J. A. Bailey and J. W. Mansfield eds.), Blackie and Son, Glasgow, UK. pp. 21–80.

Ito, J., and M. Niwa. 1996. Absolute structures of new hydroxystilbenoids, vitisin C and viniferal, from *Vitis vinifera* 'Kyohou'. *Tetrahedron* 52:9991–9998.

Kim, K.-W., S. G. A. Moinuddin, K. M. Atwell, et al. 2012. Opposite stereoselectivities of dirigent proteins in *Arabidopsis* and *Schizandra* species. *J. Biol. Chem.* 287:33957–33972.

Kim, K.-W., C. A. Smith, M. D. Daily, et al. 2015. Trimeric structure of (+)-pinoresinol-forming dirigent protein at 1.95 Å resolution with three isolated active sites. *J. Biol. Chem.* 290:1308–1318.

Kim, M. K., J.-H. Jeon, M. Fujita, L. B. Davin, and N. G. Lewis. 2002. The western red cedar (*Thuja plicata*) 8-8′ *DIRIGENT* family displays diverse expression patterns and conserved monolignol coupling specificity. *Plant Mol. Biol.* 49:199–214.

Kim, S. S., and E. S. Sattely. 2021. Dirigent proteins guide asymmetric heterocoupling for the synthesis of complex natural product analogues. *J. Am. Chem. Soc.* 143:5011–5021.

Langcake, P., and R. J. Pryce. 1977. Oxidative dimerisation of 4-hydroxystilbenes *in vitro*: Production of a grapevine phytoalexin mimic. *J. Chem. Soc., Chem. Commun.* 208–210.

Letunic, I., and P. Bork. 2007. Interactive Tree of Life (iTOL): An online tool for phylogenetic tree display and annotation. *Bioinformatics* 23:127–128.

Lewis, N. G., and L. B. Davin. 1998. The biochemical control of monolignol coupling and structure during lignan and lignin biosynthesis. In *Lignin and Lignan Biosynthesis* (N. G. Lewis and S. Sarkanen eds.), Vol. 697, ACS Symposium Series, Washington, DC. pp. 334–361.

Lewis, N. G., and L. B. Davin. 1999. Lignans: Biosynthesis and function. In *Comprehensive Natural Products Chemistry* (Sir D. H. R. Barton, K. Nakanishi and O. Meth-Cohn eds.), Vol. 1, Elsevier, Oxford. pp. 639–712.

Lewis, N. G., and L. B. Davin. 2022. *The Lignan Handbook.* Boca Raton, London: Taylor & Francis.

Liu, J., R. D. Stipanovic, A. A. Bell, L. S. Puckhaber, and C. W. Magill. 2008. Stereoselective coupling of hemigossypol to form (+)-gossypol in moco cotton is mediated by a dirigent protein. *Phytochemistry* 69:3038–3042.

López-Serrano, M., and A. Ros Barceló. 2001. Reversed-phase and size-exclusion chromatography as useful tools in the resolution of peroxidase-mediated (+)-catechin oxidation products. *J. Chromatogr. A* 919:267–273.

López-Serrano, M., and A. Ros Barceló. 2002. Comparative study of the products of the peroxidase-catalyzed and the polyphenoloxidase-catalyzed (+)-catechin oxidation. Their possible implications in strawberry (*Fragaria* × *ananassa*) browning reactions. *J. Agric. Food Chem.* 50:1218–1224.

Matasci, N., L.-H. Hung, Z. Yan, et al. 2014. Data access for the 1,000 plants (1KP) project. *GigaScience* 3:17.

Mattivi, F., U. Vrhovsek, G. Malacarne, et al. 2011. Profiling of resveratrol oligomers, important stress metabolites, accumulating in the leaves of hybrid *Vitis vinifera* (Merzling × Teroldego) genotypes infected with *Plasmopara viticola*. *J. Agric. Food Chem.* 59:5364–5375.

Meng, Q., S. G. A. Moinuddin, S.-J. Kim, et al. 2020. Pterocarpan synthase (PTS) structures suggest a common quinone methide-stabilizing function in dirigent proteins and proteins with dirigent-like domains. *J. Biol. Chem.* 295:11584–11601.

Morris, G. M., R. Huey, W. Lindstrom, et al. 2009. AutoDock4 and AutoDockTools4: Automated docking with selective receptor flexibility. *J. Comput. Chem.* 30:2785–2791.

Neumann, P., F. Brodhun, K. Sauer, et al. 2012. Crystal structures of *Physcomitrella patens* AOC1 and AOC2: Insights into the enzyme mechanism and differences in substrate specificity. *Plant Physiol.* 160:1251–1266.

Nicotra, S., M. R. Cramarossa, A. Mucci, et al. 2004. Biotransformation of resveratrol: Synthesis of *trans*-dehydrodimers catalyzed by laccases from *Myceliophthora thermophyla* and from *Trametes pubescens*. *Tetrahedron* 60:595–600.

Patten, A. M., L. B. Davin, and N. G. Lewis. 2008. Relationship of dirigent protein and 18s RNA transcript localization to heartwood formation in western red cedar. *Phytochemistry* 69:3032–3037.

Pezet, R., K. Gindro, O. Viret, and J. L. Spring. 2004. Glycosylation and oxidative dimerization of resveratrol are respectively associated to sensitivity and resistance of grapevine cultivars to downy mildew. *Physiol. Mol. Plant Pathol.* 65:297–303.

Pickel, B., M.-A. Constantin, J. Pfannstiel, et al. 2010. An enantiocomplementary dirigent protein for the enantioselective laccase-catalyzed oxidative coupling of phenols. *Angew. Chem. Int. Ed.* 49:202–204.

Pickel, B., J. Pfannstiel, A. Steudle, et al. 2012. A model of dirigent proteins derived from structural and functional similarities with allene oxide cyclase and lipocalins. *FEBS J.* 279:1980–1993.

Pickel, B., and A. Schaller. 2013. Dirigent proteins: molecular characteristics and potential biotechnological applications. *Appl. Microbiol. Biotechnol.* 97:8427–8438.

Ralph, S. G., S. Jancsik, and J. Bohlmann. 2007. Dirigent proteins in conifer defense II: Extended gene discovery, phylogeny, and constitutive and stress-induced gene expression in spruce (*Picea* spp.). *Phytochemistry* 68:1975–1991.

Reboledo, G., R. del Campo, A. Alvarez, et al. 2015. *Physcomitrella patens* activates defense responses against the pathogen *Colletotrichum gloeosporioides*. *Int. J. Mol. Sci.* 16:22280–22298.

Rensing, S. A., D. Lang, A. D. Zimmer, et al. 2008. The *Physcomitrella* genome reveals evolutionary insights into the conquest of land by plants. *Science* 319:64–69.

Riggleman, R. C., B. Fristensky, and L. A. Hadwiger. 1985. The disease resistance response in pea is associated with increased levels of specific mRNAs. *Plant Mol. Biol.* 4:81–86.

Sánchez-Moreno, C., G. Cao, B. Ou, and R. L. Prior. 2003. Anthocyanin and proanthocyanidin content in selected white and red wines. Oxygen radical absorbance capacity comparison with nontraditional wines obtained from highbush blueberry. *J. Agric. Food Chem.* 51:4889–4896.

Schaller, F., A. Schaller, and A. Stintzi. 2005. Biosynthesis and metabolism of jasmonates. *J. Plant Growth Regul.* 23:179–199.

Seneviratne, H. K., D. S. Dalisay, K.-W. Kim, et al. 2015. Non-host disease resistance response in pea (*Pisum sativum*) pods: Biochemical function of DRR206 and phytoalexin pathway localization. *Phytochemistry* 113:140–148.

Stipanovic, R. D., J. D. Lopez, M. K. Dowd, L. S. Puckhaber, and S. E. Duke. 2006. Effect of racemic and (+)- and (–)-gossypol on the survival and development of *Helicoverpa zea* larvae. *J. Chem. Ecol.* 32:959–968.

Ubillas, R., S. D. Jolad, R. C. Bruening, et al. 1994. SP-303, an antiviral oligomeric proanthocyanidin from the latex of *Croton lechleri* (Sangre de Drago). *Phytomedicine* 1:77–106.

Uchida, K., T. Akashi, and T. Aoki. 2017. The missing link in leguminous pterocarpan biosynthesis is a dirigent domain-containing protein with isoflavanol dehydratase activity. *Plant Cell Physiol.* 58:398–408.

Uchida, K., T. Aoki, H. Suzuki, and T. Akashi. 2020. Molecular cloning and biochemical characterization of isoflav-3-ene synthase, a key enzyme of the biosyntheses of (+)-pisatin and coumestrol. *Plant Biotechnol.* 37:301–310.

Umezawa, T., L. B. Davin, E. Yamamoto, D. G. I. Kingston, and N. G. Lewis. 1990. Lignan biosynthesis in *Forsythia* species. *J. Chem. Soc., Chem. Commun.* 1405–1408.

VanEtten, H. D., P. S. Matthews, and E. H. Mercer. 1983. (+)-Maackiain and (+)-medicarpin as phytoalexins in *Sophora japonica* and identification of the (–)-isomers by biotransformation. *Phytochemistry* 22:2291–2295.

Vassão, D. G., K.-W. Kim, L. B. Davin, and N. G. Lewis. 2010. Lignans (neolignans) and allyl/propenyl phenols: Biogenesis, structural biology, and biological/human health considerations. In *Comprehensive Natural Products II Chemistry and Biology* (L. Mander and H.-W. Lui eds.), Vol. 1, Elsevier, Oxford, UK. pp. 815–928.

Wang, Z., N. Hobson, L. Galindo, et al. 2012. The genome of flax (*Linum usitatissimum*) assembled *de novo* from short shotgun sequence reads. *Plant J.* 72:461–473.

Xia, Z.-Q., M. A. Costa, J. Proctor, L. B. Davin, and N. G. Lewis. 2000. Dirigent-mediated podophyllotoxin biosynthesis in *Linum flavum* and *Podophyllum peltatum*. *Phytochemistry* 55:537–549.

Yonekura-Sakakibara, K., M. Yamamura, F. Matsuda, et al. 2021. Seed-coat protective neolignans are produced by the dirigent protein AtDP1 and the laccase AtLAC5 in *Arabidopsis*. *Plant Cell* 33:129–152.

5 8–8′-Lignan Biosynthesis Roadmap

Laurence B. Davin and Norman G. Lewis

5.1 ROADMAP OVERVIEW

Much has been learned about 8–8′-lignan biochemical pathways, and of the enormous diversity of their chemical structures throughout the plant kingdom.

Progress made thus far in delineating the Roadmaps of their biochemical pathways in various plant species (many being economically important) is summarized in this chapter. This includes describing our current understanding of the various proteins, enzymes, and genes involved in either their general or species-specific plant lignan biosynthetic pathways.

First, we briefly address how the then unexplained mode of stereo-selective phenylpropanoid coupling was discovered *in planta* to afford optically active products and the entry point to the 8–8′-lignans. This includes what is known thus far concerning our discovery of dirigent proteins (DPs, see below) in stipulating the outcome of the 8–8′-linked lignan biosynthesis entry point. Evolution of DP physiological/biochemical functions seems to have occurred during the transition of aquatic plants to a terrestrial habitat.

DP discovery, however, ultimately led to recognition that the large DP multi-gene families *in planta* are involved in different branches of plant phenol metabolism, in addition to 8–8′-linked lignan biosynthesis. DPs are now known to stipulate the outcome of *both* inter-molecular coupling and intra-molecular cyclization in various plant phenol metabolic processes *in vivo* (Davin et al. 1997, Liu et al. 2008, Kim et al. 2012, Dalisay et al. 2015, Effenberger et al. 2015, Kim et al. 2015, Uchida et al. 2017, Davin et al. 2022, Meng et al. 2020, Yonekura-Sakakibara et al. 2021). Yet, in spite of these important advances, the bulk of the physiological functions of DPs remains to be urgently established.

5.2 8–8′-LIGNAN OPTICAL ACTIVITY ENIGMA: CONSIDERATIONS AND CAVEATS

At the beginning of our studies, how optically active lignans were formed was unknown, as all known monomeric (C_6C_3) coupling processes *in vitro* gave racemic products (Freudenberg 1959). It was uncertain whether the optical activity reported for many lignans from various plant species was introduced through stereoselective monomeric (C_6C_3) coupling. Alternatively, it could have resulted from non-specific coupling leading to racemic product formation, followed by enantiospecific post-coupling transformation of one or both of its two enantiomeric forms. If either occurred, this could result in detectable optical activity for the remaining products. Additionally, although many lignans are known to be optically active in various plant species, this does not by itself imply optical purity. Many lignan isolates from different plants and organs can contain mixtures of lignan enantiomers in differing amounts (Lewis and Davin 1999, Ford et al. 2001). Lignan enantiomeric compositions thus require chiral analysis to understand better the basis of their optical activity.

Our studies in identifying the biochemical entry point to the optically active 8–8′-linked lignans initially focused upon *Forsythia* shrubs (Oleaceae), a genus named after the Scottish botanist, William Forsyth (1737–1804). *Forsythia* species contain different 8–8′-linked lignans including (+)-pinoresinol (**1a**), (+)-*epi*-pinoresinol (**2a**) and its 4′-*O*-glucoside **3a**, (+)-phillygenin (**4a**), (+)-phillyrin (**5a**), (–)-matairesinol (**6b**) and its 4-*O*-glucoside **7b**, (–)-arctigenin (**8b**) and its 4-*O*-glucoside, (–)-arctiin (**9b**) (Chiba et al. 1978, Nishibe et al. 1988, Rahman et al. 1990) (Figure 5.1). We thus considered *Forsythia* sp. at the time to be an excellent choice to begin to investigate 8–8′-lignan biosynthetic pathways, and to explore and discover how lignan optical activity was established biochemically.

Our early studies resulted in the discovery that the (+)-pinoresinol (**1a**) enantiomer was the entry coupling product to various 8–8′-linked lignan skeleta, and whose downstream metabolism ultimately led to a variety of distinct structural types (Umezawa et al. 1990a, Umezawa et al. 1990b, Umezawa et al. 1991, Katayama et al. 1992, Katayama et al. 1993, Kato et al. 1997, Jiao et al. 1998, Xia et al. 2000). Distinct 8–8′-lignan structural types so produced included furofuran [e.g. (+)-sesamin (**10a**)], tetrahydrofuran [e.g. (+)-lariciresinol (**11a**)], dibenzylbutane [e.g. (–)-secoisolariciresinol (**12b**)], dibenzylbutyrolactone [e.g. (–)-matairesinol (**6b**)], aryltetrahydronaphthalene [e.g. (–)-6-methoxypodophyllotoxin (**13b**) and (–)-plicatic acid (**14b**)] lignans, as well as arylnaphthalene lignans, depending upon the species (Figure 5.2).

5.3 8–8′-FUROFURAN LIGNAN BIOSYNTHESIS ROADMAP IN PLANTA

This first section of the lignan Roadmap describes the biochemical steps involved in generating furofuran (8–8′-linked) lignans and their furofuran structural variants. In addition, Davin et al. (2022) summarize more fully the discoveries of DPs *in planta*, in terms of metabolic pathway diversity, structural biology, and evolutionary ramifications known thus far.

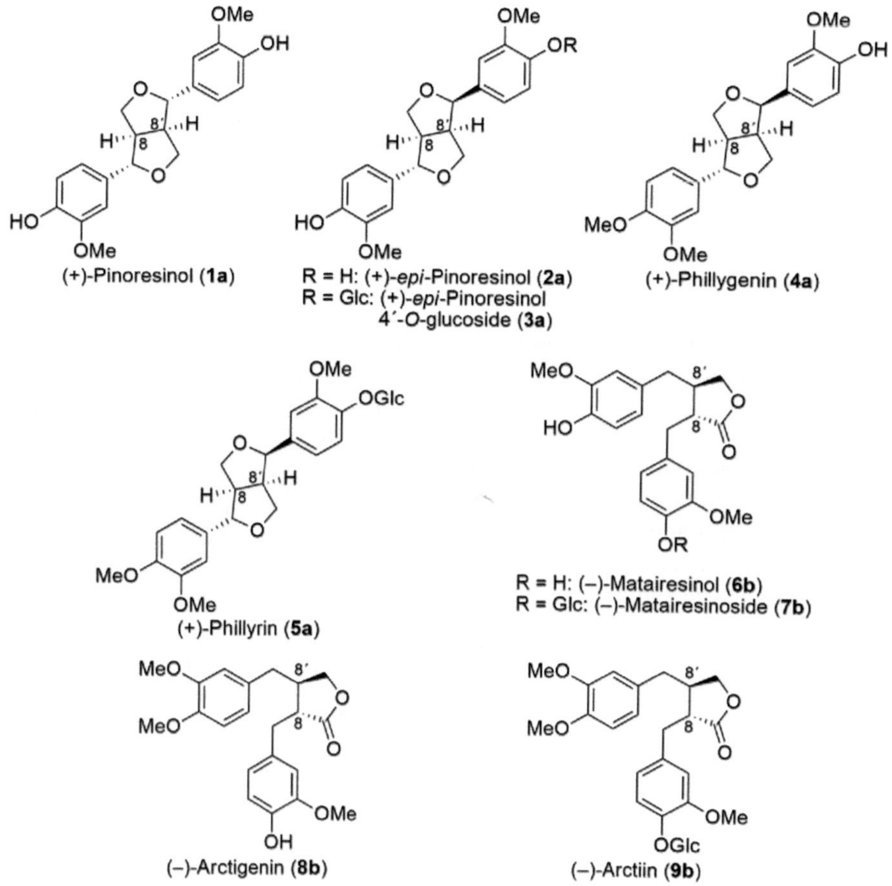

FIGURE 5.1 8–8′-Linked lignans from *Forsythia* species. "**a**" and "**b**" refer to the (+)- and (−)-enantiomeric forms, respectively.

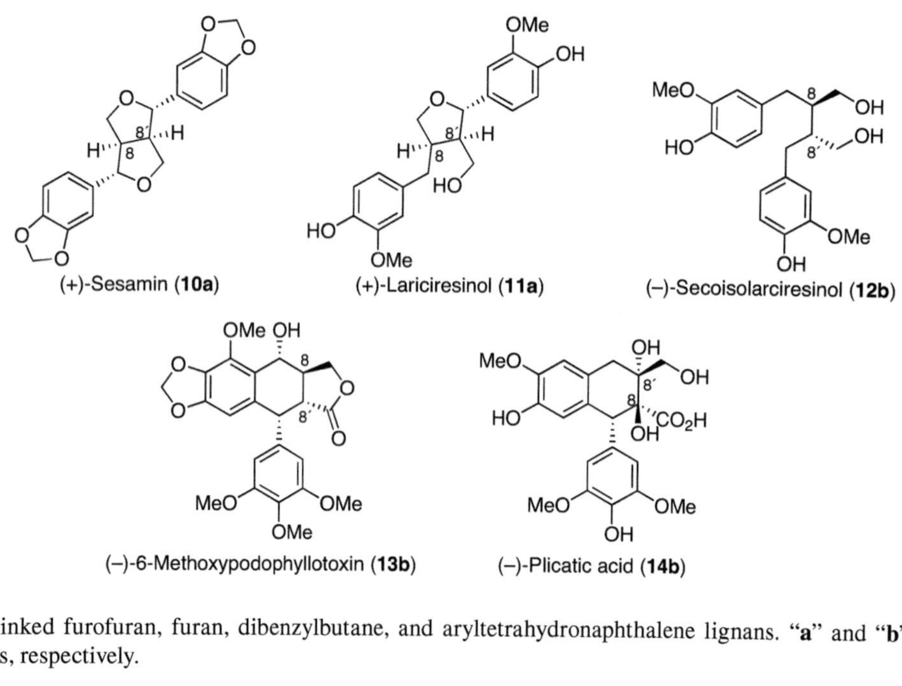

FIGURE 5.2 8–8′-Linked furofuran, furan, dibenzylbutane, and aryltetrahydronaphthalene lignans. "**a**" and "**b**" refer to the (+)- and (−)-enantiomeric forms, respectively.

5.3.1 (+)-Pinoresinol-Forming Dirigent Proteins and Encoding Genes

Initial detection of the first stereo-selective coupling of a monolignol, coniferyl alcohol (**15**, Figure 5.3), to afford (+)-pinoresinol (**1a**) in *Forsythia* was reported in 1992 (Davin et al. 1992). Additional studies using a putative cell-wall protein extract from its stems ultimately led to isolation of an ~78 kDa (+)-pinoresinol-forming trimeric protein, with subunit molecular weight of ~26 kDa. This protein was, however, inactive when incubated with coniferyl alcohol (**15**). However, it was able to convert coniferyl alcohol (**15**) into the 8–8′-linked furofuran (+)-pinoresinol (**1a**), provided that a one-electron oxidant or oxidase was also present in the assay mixture (Davin et al. 1997). Conversely, neither *p*-coumaryl (**16**) nor sinapyl (**17**) alcohols gave stereoselectively homo-coupled products in the *in vitro* assays. The protein was named as a dirigent protein, DP, from the Latin *dirigere*, to guide or align (Davin et al. 1997). The proposed biochemical mechanism for the DP-generated (+)-pinoresinol (**1a**) formation is shown in Figure 5.4A. This is envisaged to occur through *si-si* stereoselective coupling of the two free radical intermediates to afford the corresponding bis-quinone methide, which can then undergo intramolecular cyclization to give (+)-pinoresinol (**1a**).

On the other hand, if the DP was absent and a one-electron oxidant or an oxidase and its cofactor were incubated with coniferyl alcohol (**15**), only racemic dimeric product mixtures were obtained. Such well-known racemic products consisted of (±)-pinoresinols (**1a/b**), (±)-dehydrodiconiferyl alcohols (**18a/b**), and (±)-*erythro/threo* guaiacylglycerol 8–*O*–4′ coniferyl alcohol ethers (**19a/b**) due to non-regiospecific coupling (Figure 5.5).

Subsequent cloning and expression of the encoding gene for the (+)-pinoresinol-forming DP, Fi_DP1, showed that it encoded an ~18.3 kDa protein after removal of the secretory signal sequence, and not an ~ 26 kDa protein. Differences in estimated DP size were rationalized as due to the native DP being glycosylated by post-translational modification (Gang et al. 1999). Upon its heterologous expression using a eukaryotic system (baculovirus-infected *Drosophila* cells), an ~22 to 26 kDa glycosylated DP was obtained. This was also

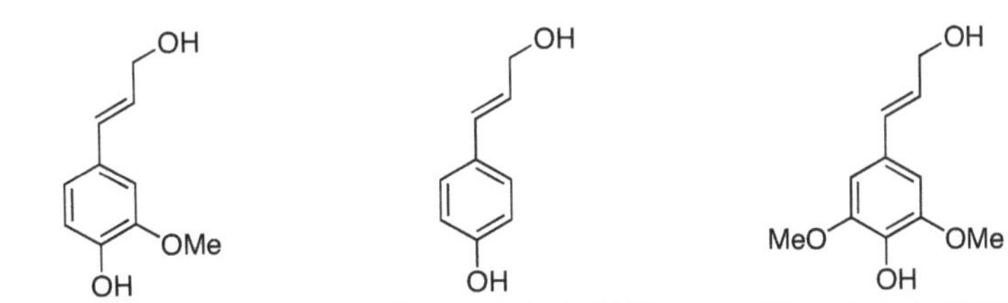

FIGURE 5.3 Monolignols **15–17**.

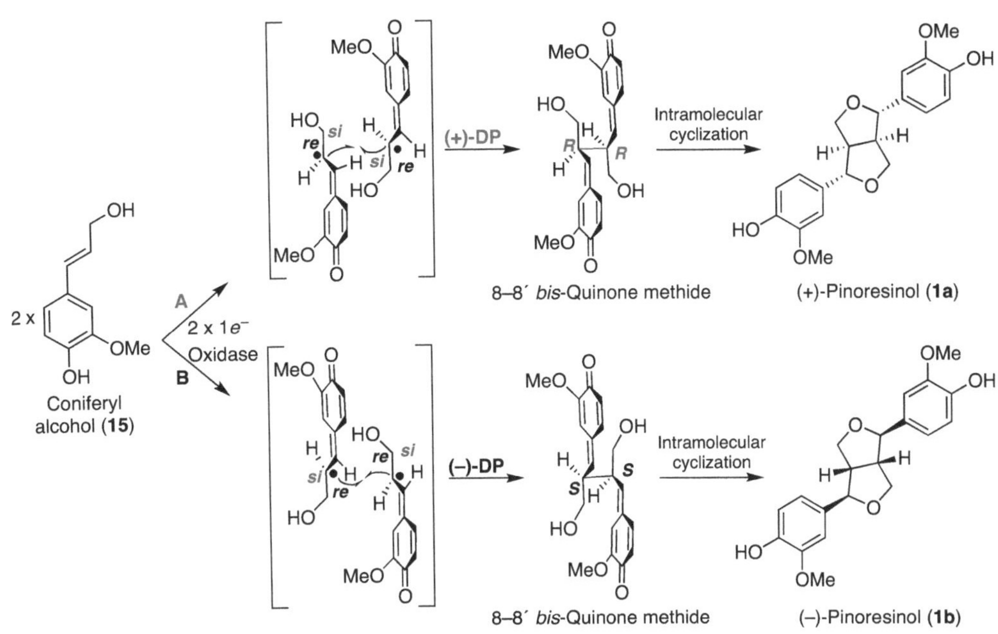

FIGURE 5.4 Proposed mechanism to afford 8–8′-linked furofurans, (+)- or (–)-pinoresinols (**1a** or **1b**), in DP-guided stereoselective coupling from coniferyl alcohol (**15**). **A.** (+)-Pinoresinol-forming DP. **B.** (–)-Pinoresinol-forming DP. "**a**" and "**b**" refer to the (+)- and (–)-enantiomeric forms, respectively.

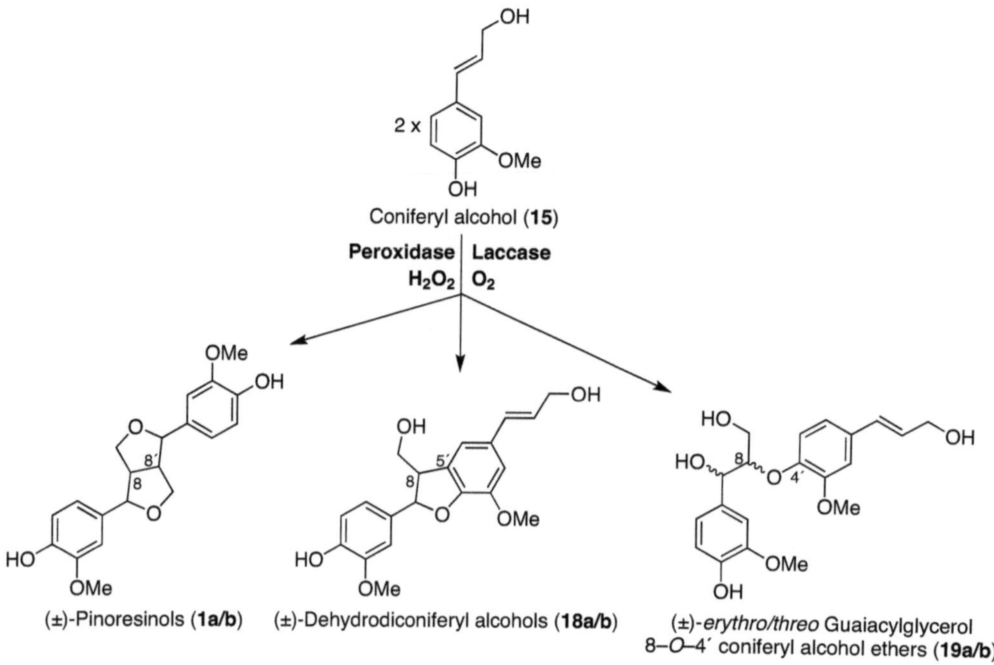

FIGURE 5.5 Initial dimeric (lignan) coupling products obtained *in vitro* from incubation of coniferyl alcohol (**15**) with an oxidase (peroxidase/laccase) in presence of H_2O_2 or O_2, respectively. "**a/b**" refers to racemates.

functionally capable of engendering stereoselective coupling of coniferyl alcohol (**15**) as previously stated, provided that a one-electron oxidase/oxidant was present, even though the glycosylation pattern was different (Gang et al. 1999).

Related DP studies were carried out with western red cedar (*Thuja plicata*). This species accumulates the 8–8′-linked aryltetrahydronaphthalene lignan (–)-plicatic acid (**14b**) and analogues thereof in its tissues, such as needles, stem, and bark (Gardner et al. 1966). We considered that these western red cedar lignans were also (+)-pinoresinol (**1a**) derived (Fujita et al. 1999). Nine DP genes (Tp_DIR1–Tp_DIR9) were obtained from this species, which have 72%–99.5% identity to each other (Kim et al. 2002), and encoded proteins of 180–183 amino acids with each having a predicted molecular mass of ~20 kDa including the signal peptide. Of these DPs, two (Tp_DIR5 and Tp_DIR8) engendered stereoselective coupling *in vitro* of coniferyl alcohol (**15**) to afford (+)-pinoresinol (**1a**) as previously discussed (Kim et al. 2002).

5.3.2 (–)-Pinoresinol-Forming Dirigent Proteins and Encoding Genes

Some plant species, such as *Daphne tangutica* (Zhuang et al. 1982), predominantly accumulate the opposite antipode (–)-pinoresinol (**1b**) in enantiomeric excess (e.e.). Additionally, flax (*Linum usitatissimum*) seed contains lignan-rich EtOH-H_2O solubles, which on mild alkali hydrolysis release secoisolariciresinol glucoside (SDG, **20**, Figure 5.6) (Ford et al. 2001). The latter, following incubation with a β-glucosidase, gave a secoisolariciresinol (**12**) isolate with a 99:1 enantiomeric composition of (+)- to (–)-antipodes, **12a** and **12b**. The preferential formation of (+)-secoisolariciresinol (**12a**), as described in Section 5.3.2.2, was considered to result from predominant formation of (–)-pinoresinol (**1b**) (see Figure 5.4B) and its subsequent downstream metabolism (Ford 2001, Teoh et al. 2003).

These findings allowed us to predict that (–)-pinoresinol-forming DPs were present in some species, and this was next established. DP assays with whole flaxseed crude enzyme preparations resulted in detection of (–)-pinoresinol (**1b**)-forming activity. Yet, all attempts to isolate a gene encoding a (–)-pinoresinol-forming DP were unsuccessful at that time; the primers employed only afforded those from (+)-pinoresinol-forming DP homologs (Ford 2001, Teoh et al. 2003).

5.3.2.1 *Arabidopsis thaliana* DPs

With *Arabidopsis*, however, there were two DP homologs, AtDIR5 and AtDIR6, which were closest in homology to the (+)-pinoresinol-forming DP from *Forsythia* (~61%–65% similarity and ~50%–52% identity, at the amino acid level) (Vassão et al. 2010, Kim et al. 2012). The corresponding recombinant proteins from these *Arabidopsis* homologs were next demonstrated to stereoselectively produce (–)-pinoresinol (**1b**) from coniferyl alcohol (**15**), in the presence of an oxidase/oxidant (see Figure 5.4B) (Vassão et al. 2010, Kim et al. 2012), an observation independently confirmed by others (Pickel et al. 2010). These studies established existence of distinct (+)- and (–)-pinoresinol (**1a** and **1b**)-forming DPs, as we had deduced (Lewis and Davin 1999) and reported (Vassão et al. 2010).

Additionally, overexpression of *AtDIR6* in *Arabidopsis* was carried out to further confirm its role *in vivo* (Kim et al. 2012). Roots of 4-week-old wild type and *AtDIR6* overexpressing

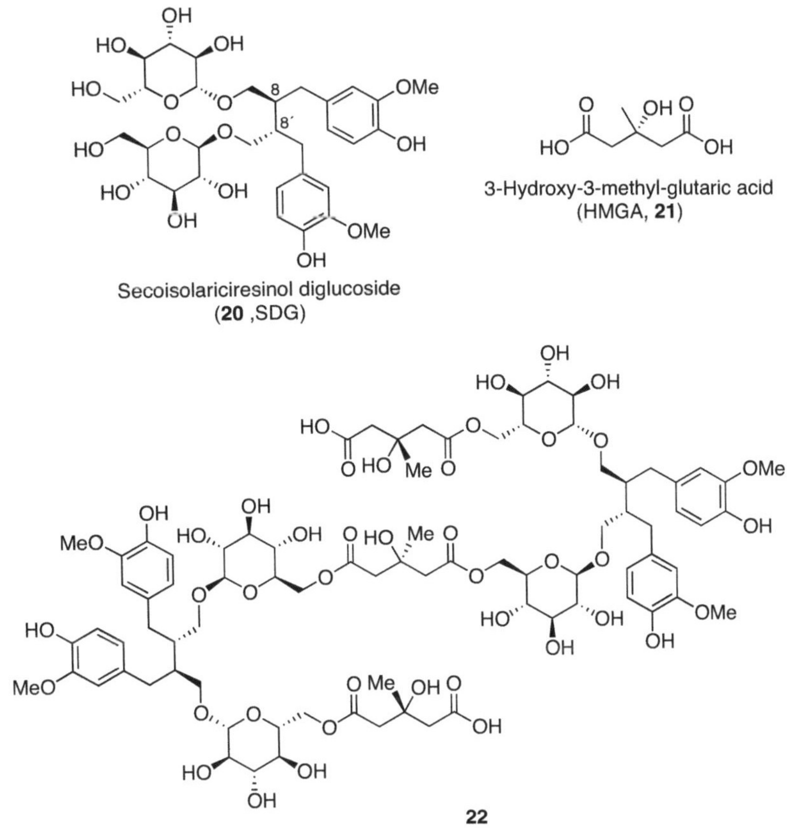

FIGURE 5.6 Flax seed 8–8′-linked secoisolariciresinol diglucoside (SDG, **20**), HMGA (**21**), and SDG "dimer" **22**.

lines were collected, with the pinoresinol (**1**) accumulating therein purified and submitted to chiral chromatography analysis. This established that the product was mainly (–)-pinoresinol (**1b**), and in an enantiomeric excess (~75% e.e., Figure 5.7B). For comparison, Figure 5.7A shows chiral separations of (+)- and (–)-pinoresinol enantiomers (**1a** and **1b**) in a racemic mixture.

Subsequent domain-swapping experiments using (+)- and (–)-pinoresinol-forming DPs from *Schizandra chinensis* and *Arabidopsis* were also carried out, with the corresponding recombinant DPs assayed *in vitro* to attempt to gain insight into how the different stereoselectivities were determined (Kim et al. 2012). These studies established that when the "center" region of the *Schizandra* (+)-pinoresinol-forming DP was swapped with the comparable region from the *Arabidopsis* (–)-pinoresinol-forming DP, the resulting chimeric DP was now (–)-pinoresinol forming. Thus, the region controlling the different stereoselectivities was identified.

5.3.2.2 Flax (*Linum usitatissimum*) DPs

In flaxseed, the dibenzylbutane lignan SDG (**20**) does not exist in free form, but is covalently linked and is found in various ester-linked oligomeric forms. Some are covalently attached to 3-hydroxy-3-methyl-glutaric acid (HMGA, **21**) (Ford et al. 2001, Kamal-Eldin et al. 2001), such as SDG-derived **22** (Figure 5.6). Interestingly, the SDG (**20**) released from such conjugates upon mild alkali treatment largely consists of the (+)-secoisolariciresinol (**12a**) enantiomer, in an ~99:1 ratio of (+) (**12a**) to (–) (**12b**) (Ford et al. 2001) as previously mentioned. The (+)-secoisolariciresinol (**12a**) was considered to be formed from (–)-pinoresinol (**1b**) via further downstream metabolism.

Conversely, flax leaf, stem, and floral tissues contain the dibenzylbutyrolactone lignan (–)-yatein (**23b**), (–)-hinokinin (**24b**), (–)-bursehernin (**25b**), (–)-*E*-anhydropodorhizol (**26b**), (–)-matairesinol dimethyl ether (**27b**), and (–)-thujaplicatin trimethyl ether (**28b**) (Figure 5.8) (von Heimendahl et al. 2005, Schmidt et al. 2006, Schmidt et al. 2008, Hemmati et al. 2010, Corbin et al. 2017). Other *Linum* species, such as *L. album*, also accumulate the aryltetrahydronaphthalene lignans (–)-podophyllotoxin (**29b**), (–)-6-methoxypodophyllotoxin (**13b**), and derivatives thereof (Federolf et al. 2007), whereas (–)-hinokinin (**24b**) is found in *L. corymbulosum* (Mohagheghzadeh et al. 2006) callus cultures. All of these lignans present in flax leaf, stem, and floral tissues and in other *Linum* species were considered by us to be (+)-pinoresinol (**1a**) derived, as described later.

With the flax genome sequence being reported (Wang et al. 2012), we next investigated the DP family in flax. We deduced that in the 44-member DP multi-gene family annotated (Corbin et al. 2018), six were likely to be pinoresinol-forming DPs (*LuDIR1* – *LuDIR6*). Of these, *LuDIR5* and *LuDIR6* were provisionally considered to be encoding (–)-pinoresinol-forming DPs having ~56%–57% identity at the amino acid

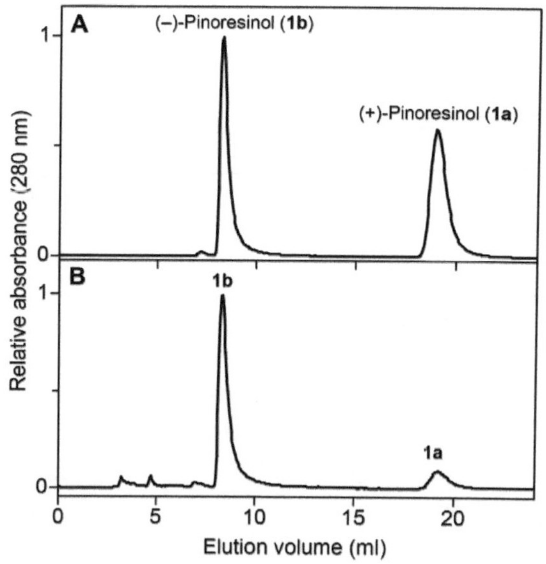

FIGURE 5.7 Chiral HPLC analyses of 8–8′-linked furofuran pinoresinol (**1**) enantiomeric compositions. **A.** Racemic pinoresinols (**1a** and **1b**). **B.** Pinoresinol (**1**) from root tissues of *A. thaliana* over-expressing *AtDIR6*, with the (–)-antipode (**1b**) being in 75% e.e. "**a**" and "**b**" refer to the (+)- and (–)-enantiomeric forms, respectively. This research was originally published in the Journal of Biological Chemistry: Kim et al. (2012). Opposite stereoselectivities of dirigent proteins in *Arabidopsis* and *Schizandra* species. *J. Biol. Chem.* 287:33957–33972. Copyright (2012) the American Society for Biochemistry and Molecular Biology.

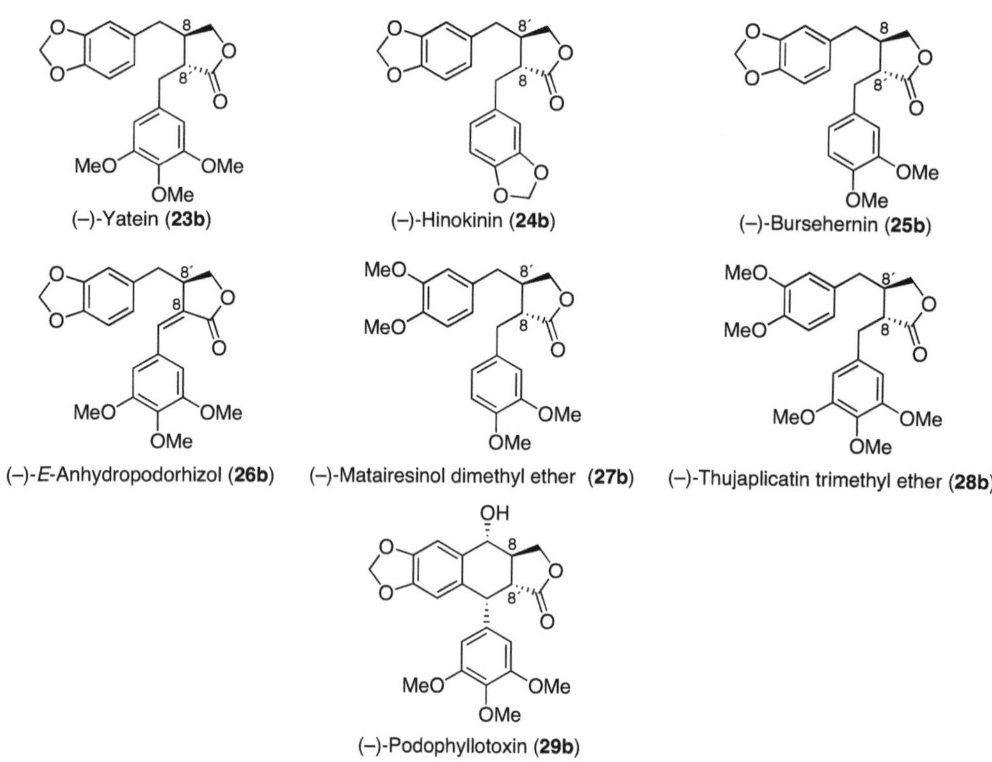

FIGURE 5.8 Various 8–8′-linked dibenzylbutyrolactone (**23b–28b**) and aryltetrahydronaphthalene (**29b**) lignans found in *Linum* species. "**b**" refers to the (–)-enantiomeric form.

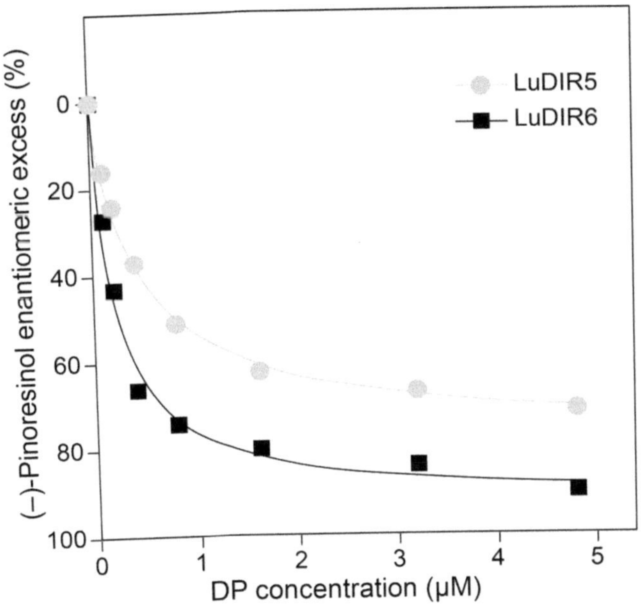

FIGURE 5.9 Preferential 8–8′-linked furofuran (–)-pinoresinol (**1b**) formation in enantiomeric excess (e.e.) using various concentrations of recombinant LuDIR5 and LuDIR6, respectively. This research was originally published in the Journal of Natural Products: Dalisay et al (2015). Dirigent protein-mediated lignan and cyanogenic glucoside formation in flax seed: Integrated omics and MALDI mass spectrometry imaging. *J. Nat. Prod.* 78:1231-1242. Copyright (2015) the American Chemical Society and the American Society of Pharmacognosy.

level to the (–)-pinoresinol-forming DP, AtDIR6. The others, *LuDIR1–LuDIR4*, with ~58%–66% identity at the amino acid level to the (+)-pinoresinol-forming DP, ScDIR, possibly encoded (+)-pinoresinol-forming DPs (Dalisay et al. 2015).

Our recombinant protein studies next established that *LuDIR5* and *LuDIR6* were (–)-pinoresinol-forming DPs (Figure 5.9) (Dalisay et al. 2015). Conversely, LuDIR1 had been previously shown to be a (+)-pinoresinol-forming DP (Ford 2001, Teoh et al. 2003). LuDIR2 to LuDIR4 were also hypothesized to be (+)-pinoresinol-forming DPs based on the high level of similarity between them and LuDIR1 (94%–96% amino acid identity) (Dalisay et al. 2015).

5.4 8–8′-ARYLDIHYDRONAPHTHALENE LIGNAN ROADMAP IN LIVERWORTS: ARE DPS INVOLVED?

Putative DPs engendering formation of other 8–8′-linked lignan types have been suggested (Lewis and Davin 1999, Vassão et al. 2010), such as to the aryldihydronaphthalene lignans, (–)-epiphyllic acid (**30b**, Figure 5.10) or its (+)-enantiomer (**30a**) in *Lophocolea heterophylla* and *Jamesoniella autumnalis* liverworts, respectively. While their intact cell cultures reportedly metabolized [8-^{2}H]-caffeic acid (**31**) into either lignan **30a** or **30b**, respectively, depending upon the species, their cell free extracts, when incubated with caffeic acid (**31**) and H_2O_2 *in vitro*, only gave racemic epiphyllic acid (**30a/30b**) under the conditions employed (Tazaki et al. 1999). To date, while it is hypothesized (Lewis and Davin 1999, Vassão et al. 2010) that a DP stipulates the outcome of stereoselective coupling of caffeic acid (**31**), or some other related precursor, to afford these liverwort lignans, this remains to be established.

5.5 8–8′-DIARYLTETRAHYDROFURAN, DIARYLBUTANE, AND ARYLTETRAHYDRONAPHTHALENE LIGNAN BIOSYNTHESIS ROADMAP IN THE CREOSOTE BUSH: POTENTIAL INVOLVEMENT OF DPS AND ENANTIOSPECIFIC POLYPHENOL OXIDASES

8–8′-Regiospecific coupling also affords various 9–9′-deoxygenated lignans in the creosote bush (*Larrea tridentata*), a desert shrub of 1–3 m height. Native Americans have long used this species in traditional medicine for treating more than 50 different ailments, including kidney and gallbladder stones. Its most abundant lignan, the diarylbutane lignan, nordihydroguairetic acid (NDGA, **32**, Figure 5.11A), and considered to be the main bioactive principle, can accumulate up to ~5%–10% of its leaves' dry weight (Arteaga et al. 2005). 9–9′-Deoxygenated lignans, such as the diaryltetrahydrofurans (–)-larreatricin (**33b**), (–)-8′-epi-larreatricin (**34b**), *meso*-3,3′-didemethoxynectandrin B (**35**), and (±)-3,3′-didemethoxyverrucosins (**36a/b**), are also present (Konno et al. 1990, Moinuddin et al. 2003), as are aryltetrahydronaphthalene

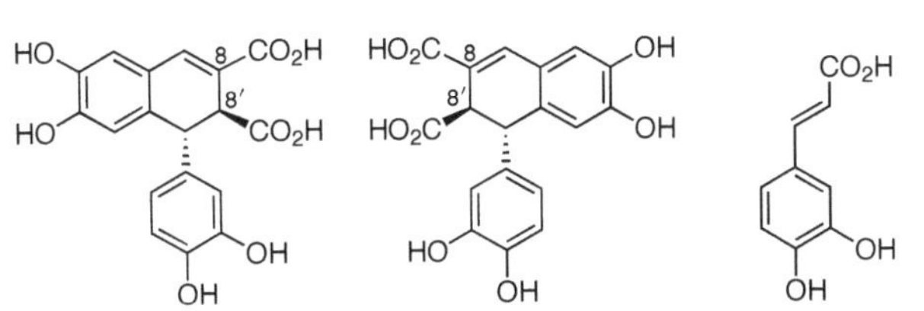

FIGURE 5.10 Aryldihydronaphthalene lignan biosynthesis: (–)- and (+)-Epiphyllic acid (**30b** and **30a**) from liverworts and their presumed precursor, caffeic acid (**31**). "**a**" and "**b**" refer to the (+)- and (–)-enantiomeric forms, respectively.

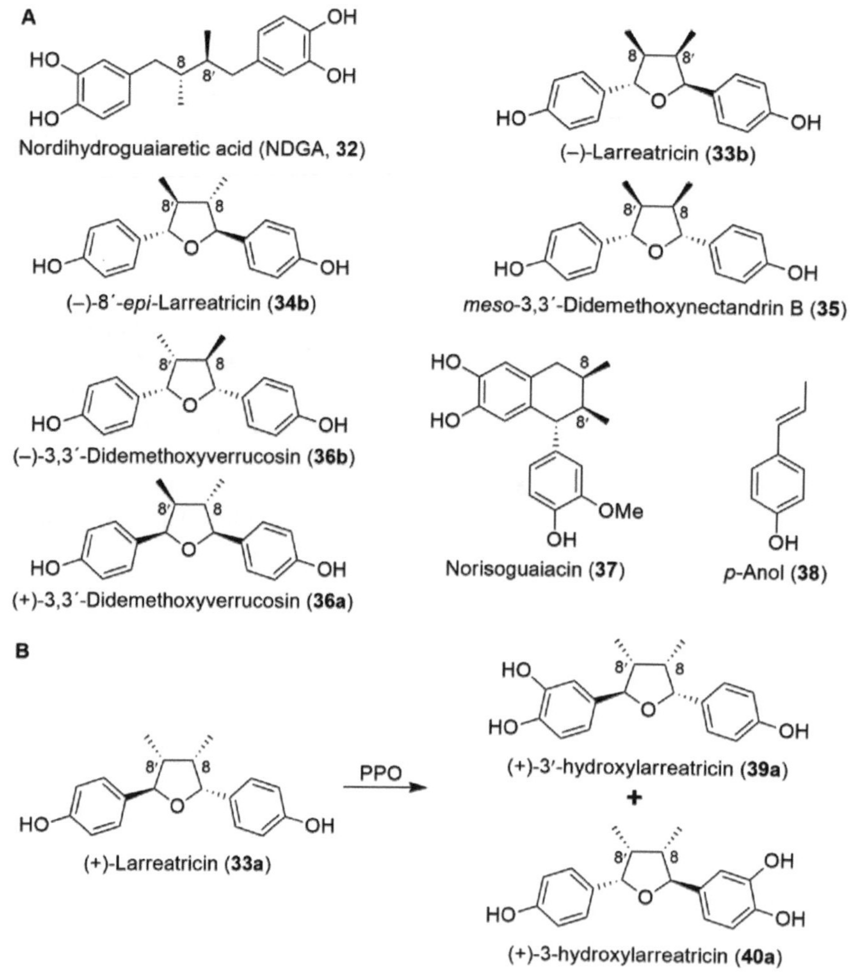

FIGURE 5.11 **A.** Creosote bush (*Larrea tridentata*) diarylbutane (**32**), diaryltetrahydrofuran (**33–36**), and aryltetrahydronaphthalene (**37**) lignans and putative allylphenol precursor, *p*-anol (**38**). **B.** Reaction catalyzed by *L. tridentata* polyphenol oxidase (PPO). "**a**" and "**b**" refer to the (+)- and (−)-enantiomeric forms, respectively.

lignans, e.g. the putative norlignan norisoguaiacin (**37**). The latter is considered as a "nor-lignan" as it lacks methylation of the hydroxyl group at C3.

Creosote bush 8–8′ linked lignans can be provisionally thought to be derived from allylphenol-/propenylphenol-coupling, e.g. from *p*-anol (**38**) to initially afford (+)- and (−)-larreatricins (**33a** and **33b**) and its possible tetrahydrofuran ring diastereoisomers (**34–36**). Yet, when *p*-anol (**38**) was incubated *in vitro* with oxidases/laccases, only non-regiospecific coupling products were obtained. Given the range of optical activities of lignans **33–36** in *Larrea*, their biosynthetic pathways are greatly in need of determination, and demonstration of the level of involvement of DP (or DP-like) proteinaceous control is occurring and with what substrates. The possibility also still exists that these natural products are derived from monolignol 8–8′ coupling, followed by subsequent metabolism to afford metabolites **32–36**.

In our previous studies, (+)-larreatricin (**33a**) [but not the (−)-antipode (**33b**)] was enantiospecifically hydroxylated *in vitro* by a *L. tridentata* polyphenol oxidase (PPO, Figure 5.11B) (Cho et al. 2003). This finding was in good agreement with the observed 92% e.e. of (−)-larreatricin (**33b**) *in planta* (Moinuddin et al. 2003).

The PPO effecting the C3′/C3 hydroxylation was purified ~1,700-fold to apparent homogeneity from *L. tridentata* twigs, to afford an ~43 kDa protein. This PPO preparation favored C3′ hydroxylation of (+)-larreatricin (**33a**) to give (+)-3′-hydroxylarreatricin (**39a**) in an ~7:1 ratio, relative to the C3-hydroxylated (+)-3-hydroxylarreatricin (**40a**) (Cho et al. 2003). This was the first report of an enantiospecific PPO. Its corresponding gene was obtained from a *L. tridentata* cDNA library. The predicted MW (66 kDa) of the corresponding protein was higher than that of the purified protein (~43 kDa), which is indicative of post-translational processing, a feature common to PPOs. Comparison of its deduced amino acid sequence to that of other PPOs showed that it had four histidine residues, His188, 197, 319, 323, and 353, presumed necessary for copper binding (Cho et al. 2003).

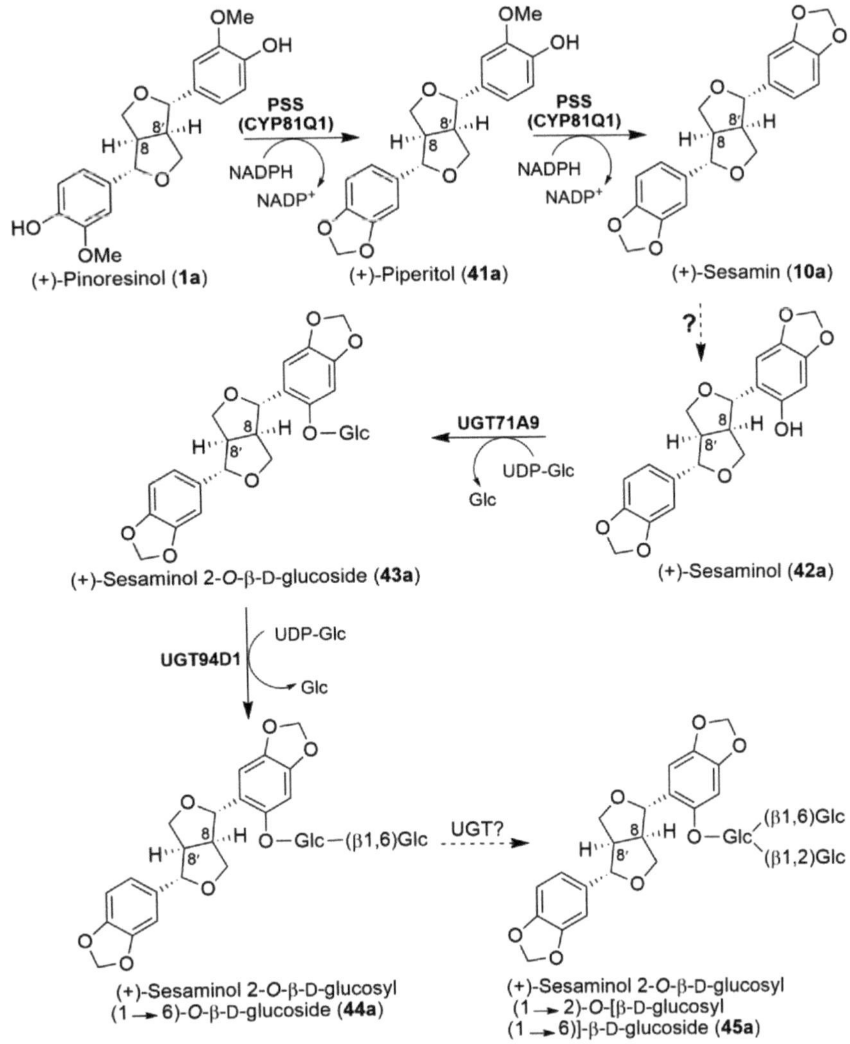

FIGURE 5.12 Proposed biosynthetic pathway to *Sesamum indicum* lignans. Legend: UGT?, indicates unknown glucosyltransferase at present. "**a**" refers to the (+)-antipode.

5.6 8–8′-FUROFURAN LIGNAN DERIVATIZATION ROADMAP IN *SESAMUM* SPECIES: METHYLENEDIOXY BRIDGE FORMATION, GLUCOSYLATION, AND OXYGEN INSERTIONS

8–8′-Furofuran lignan-derived transformations retaining the furofuran skeleton are largely encountered in sesame (*Sesamum indicum*) seeds, whose products are highly valued as both a source of antioxidant lignans and sesame oil (Fukuda et al. 1985, Osawa et al. 1985, Fukuda et al. 1986). The most abundant lignans reported from sesame seed are 8–8′-linked, with either one or two methylenedioxy bridges. They include (+)-sesamin (**10a**), (+)-piperitol (**41a**), (+)-sesaminol (**42a**) and the glucosides, (+)-sesaminol 2-*O*-mono- (**43a**), 2-*O*-di- (**44a**), and 2-*O*-tri- (**45a**) glucosides (Figure 5.12) (Fukuda et al. 1985, Katsuzaki et al. 1994). Two other lignans, (+)-sesamolin (**46a**) (Beroza 1955) and (+)-sesamolinol (**47a**) (Osawa et al. 1985) also have an unusual and rare acetal oxygen insertion between the furofuran moiety and a pendant aryl group (Figure 5.13).

5.6.1 Methylenedioxy Bridge Formation Roadmap

As described later, methylenedioxy bridge formation is quite common in 8–8′-lignan biosynthesis. Earlier studies in alkaloid biosynthesis showed enzymatic conversions leading to methylenedioxy bridge formation as being cytochrome P450 (CYP) catalyzed (Bauer and Zenk 1991). In a similar vein, much is now known about (+)-pinoresinol (**1a**) metabolism in sesame seed, following our initial discovery that the conversions into (+)-piperitol (**41a**) and (+)-sesamin (**10a**) were CYP catalyzed (Jiao et al. 1998). This was shown by administering (±)-[3,3′-$O^{14}CH_3$]-pinoresinols (**1a/1b**) to intact sesame seeds. Of these, only (+)-antipode (**1a**) was enantiospecifically metabolized into the sesame lignans **10a**, **41a**, and **46a**, and to different extents depending upon the seed maturation stage. The (–)-enantiomer **1b** was not incorporated (Jiao et al. 1998).

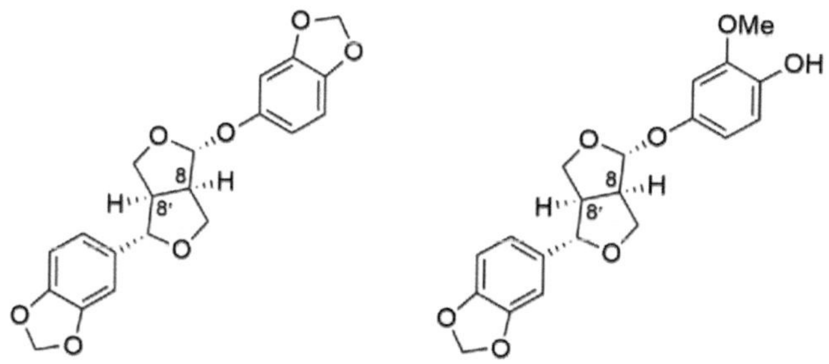

(+)-Sesamolin (**46a**) (+)-Sesamolinol (**47a**)

FIGURE 5.13 (+)-Sesamolin (**46a**) and (+)-sesamolinol (**47a**) from *Sesamum indicum*. "**a**" refers to the (+)-antipode.

Microsomal preparations were also obtained from the first and second stages of seed pod maturation, in order to probe whether these conversions were CYP catalyzed. Using (±)-[3,3′-$O^{14}CH_3$] pinoresinols (**1a/1b**) as potential substrates, only (+)-[3′-$O^{14}CH_2$, 3-$O^{14}CH_3$]-piperitol (**41a**) was formed when NADPH (1 mM) was present, but not the corresponding (–)-enantiomer (**41b**); the O_2-requiring, NADPH-dependent, CYP was named (+)-piperitol synthase (Jiao et al. 1998). Additionally, incubation of (+)-piperitol (**41a**) afforded (+)-sesamin (**10a**). Our studies thus demonstrated that both conversions were CYP catalyzed, as for alkaloid biosynthesis (Bauer and Zenk 1991).

Later work further confirmed these findings. A gene named *CYP81Q1* was obtained from a cDNA library of sesame seed actively synthesizing (+)-sesamin (**10a**) (Ono et al. 2006). When its corresponding recombinant protein CYP81Q1 was heterologously expressed in yeast, the resulting microsomal preparation was able to convert (+)-pinoresinol (**1a**) and (+)-piperitol (**41a**) into (+)-sesamin (**10a**), in the presence of NADPH (Figure 5.12). Apparent K_m values were determined of 10.2 and 11.7 μM for (+)-pinoresinol (**1a**) and (+)-piperitol (**41a**), respectively, these being in the expected range for CYPs. The enzyme was named (+)-piperitol/(+)-sesamin synthase, although apparently (–)-enantiomers **1b** and **41b** were not tested to establish true enantiospecificity. Nevertheless, these data confirmed our earlier findings (Jiao et al. 1998) that a CYP was involved. However, this microsomal preparation was unable to convert (+)-sesamolinol (**47a**) into (+)-sesamolin (**46a**), possibly suggesting another CYP as being responsible for formation of its additional methylenedioxy bridge. Nor was it able to use (+)-*epi*-pinoresinol (**2a**) as substrate (Ono et al. 2006).

Two related species, *S. radiatum* and *S. alatum*, were also found to harbor two other *CYP81Q1* homologous genes, *CYP81Q2* and *CYP81Q3*, respectively. In this regard, *S. radiatum* seeds accumulate (+)-sesamin (**10a**), whereas seeds of *S. alatum* accumulate (+)-*epi*-sesartemin A (**49a**), but not sesamin (**10a**). To our knowledge, (+)-*epi*-sesartemin A (**49a**) was first isolated and named in 1980 from *Artemisia absinthium* (Greger and Hofer 1980). Interestingly, Kamal-Eldin et al. (1992) sometime later described **49a** as being a new compound from *S. alatum*, naming it instead (+)-7′-*epi*-sesalatin.

Microsomal preparations harboring CYP81Q2, in the presence of NADPH, were able to convert (+)-pinoresinol (**1a**) into (+)-piperitol (**41a**) and (+)-sesamin (**10a**) (Ono et al. 2006), as for CYP81Q1. On the other hand, the preparation harboring CYP81Q3 formed (+)-pluviatilol (**48a**) from (+)-*epi*-pinoresinol (**2a**) (Figure 5.14) (Ono et al. 2018). However, it was inactive with either (+)- or (–)-pinoresinols (**1a** or **1b**). Interestingly, the name pluviatilol (**48**) came from its isolation from the Australian plant *Zanthoxylum pluviatile* (Corrie et al. 1970). Pluviatilol (**48**) can be provisionally envisaged to be a precursor to (+)-*epi*-sesartemin A (**49a**) via additional CYP catalyzed hydroxylations and *O*-methylations.

The reports of CYP81Q1 and CYP81Q2 apparently catalyzing dual methylenedioxy bridge formation differs from known CYPs in alkaloid metabolism. In the latter case, CYP719A2, obtained from *Eschscholtzia californica* cell cultures, converted (*S*)-cheilanthifoline (**50**) into (*S*)-stylopine (**51**); however, no analogous conversion of (*S*)-scoulerine

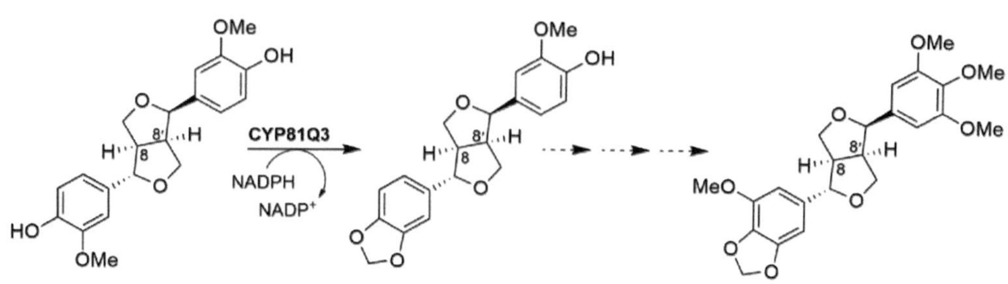

(+)-*epi*-Pinoresinol (**2a**) (+)-Pluviatilol (**48a**) (+)-*epi*-Sesartemin A (**49a**)

FIGURE 5.14 Proposed pathway to (+)-*epi*-sesartemin A (**49a**) in *Sesamum alatum*, where dotted line arrows indicate currently unknown genes encoding the CYPs C5 and C5′ hydroxylations, and *O*-methyltransferases potentially involved. "**a**" refers to the (+)-antipode.

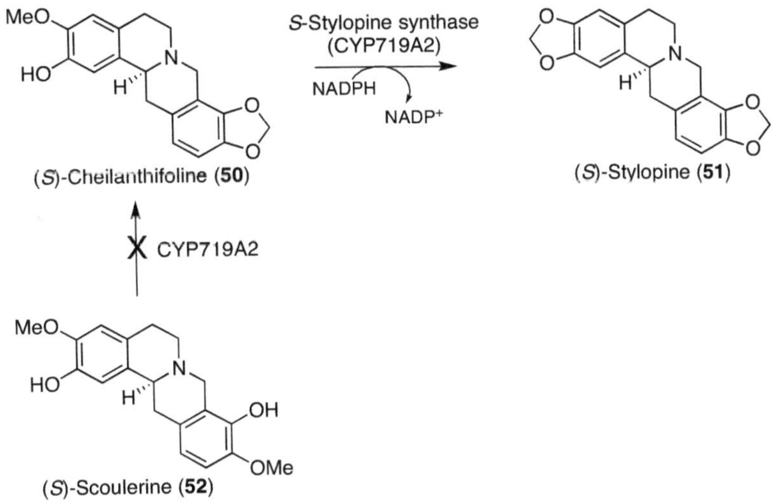

FIGURE 5.15 Isoquinoline alkaloid biosynthesis. (*S*)-cheilanthifoline (**50**) was converted into (*S*)-stylopine (**51**) by (*S*)-stylopine synthase (CYP719A2). (*S*)-Scoulerine (**52**) was not a CYP719A2 substrate for formation of (*S*)-cheilanthifoline (**50**).

(**52**) into (*S*)-cheilanthifoline (**50**) was observed (Figure 5.15) (Ikezawa et al. 2007).

In the future, resolution of the ternary structures of CYP81Q1/CYP81Q2, able to convert (+)-pinoresinol (**1a**) and (+)-piperitol (**41a**) into (+)-sesamin (**10a**), and of CYP81Q3, which converts (+)-*epi*-pinoresinol (**2a**) into (+)-pluviatilol (**48a**), would hopefully provide informative insight into the basis of their catalytic mechanisms, and how substrate specificity is effectuated. Of particular interest for CYP81Q1/CYP81Q2 would be determining whether the (+)-piperitol (**41a**) first synthesized is then released (or not) from their active site. If that occurred (i.e. with product released from the active site), this may explain why our previous study only detected (+)-piperitol (**41a**) formation when (+)-pinoresinol (**1a**) was incubated with a sesame seed microsomal preparation (Jiao et al. 1998). As one possibility, perhaps any released (+)-piperitol (**41a**) could have been in too small a concentration to compete with the much, much larger amounts of (+)-pinoresinol (**1a**) present in the assay mixture under saturating conditions.

Genome sequencing of an elite sesame cultivar named Zhongzhi No. 13 also confirmed the presence of the (+)-pinoresinol-forming DP and the downstream CYP (+)-piperitol/(+)-sesamin synthase (Wang et al. 2014). In addition, transgenic *Forsythia koreana* cell suspension cultures were obtained that had their pinoresinol/lariciresinol reductase (*PLR*) down-regulated using a RNAi strategy, as well as the sesame (+)-piperitol/(+)-sesamin synthase (*CYP81Q1*) overexpressed. The resulting transformed line had small amounts of sesamin (**10**) being produced (Kim et al. 2009).

5.6.2 Glucosylation Roadmap

Several sesame seed lignans are water soluble glucosylated compounds, the most abundant being (+)-sesaminol 2-*O*-triglucoside (**45a**) (Ryu et al. 1998, Moazzami et al. 2006). To investigate how their glucosides were possibly formed, a study directed towards identification of putative uridine diphosphate (UDP)-glucosyltransferases (UGTs) involved in formation of **45a** was carried out (Noguchi et al. 2008). Following two rounds of screening of the aforementioned sesame seed cDNA library using probes containing a well-conserved UGT sequence, 10 clones were obtained. Each of these were heterologously expressed in *E. coli* as His-tag fusion proteins, with two, UGT71A9 and UGT94D1, able to catalyze conversion of (+)-sesaminol (**42a**) into (+)-sesaminol 2-*O*-β-D-glucoside (**43a**), and the latter into (+)-sesaminol 2-*O*-β-D-glucosyl (1→6)-*O*-β-D-glucoside (**44a**), respectively, but not into the triglucoside **45a** (Figure 5.12). ^{1}H, ^{13}C, and 2D-NMR spectroscopic studies, and LC-MS analyses, confirmed the reaction product identities of the mono- and di-glucosides. The product formed by UGT71A9 catalysis had a *m/z* 555.1450 $[M + Na]^+$, indicating that one glucose had been added onto (+)-sesaminol (**42a**). Apparent K_m values obtained were 6.32 μM for (+)-sesaminol (**42a**) and 41.0 μM for UDP-glucose, with a k_{cat}/K_m value of 196,000 $M^{-1}\ s^{-1}$ for **42a**. UGT94D1 catalyzed formation of the diglucosylated product (**44a**) of *m/z* 717.2000 $[M + Na]^+$, with apparent K_m values of 77.0 μM for (+)-sesaminol 2-*O*-β-D-glucoside (**43a**) and 228.0 μM for UDP-glucose, but with a much lower k_{cat}/K_m of 11,700 $M^{-1}\ s^{-1}$ for **43a** as compared to UGT71A9. No UGT was identified that could catalyze the final β1→2 glucosylation to give (+)-sesaminol 2-*O*-triglucoside (**45a**) from **44a**.

To establish if the glucosyltransferase affording the monoglucoside sesaminol 2-*O*-β-D-glucoside (**43a**) is conserved in the sesame genome, the corresponding *S. alatum* and *S. radiatum* homologs of *UGT71A9*, namely *UGT71A8* and *UGT71A10* (with ~98% and 91% similarity to UGT71A9 at the amino acid level), were cloned. Both recombinant proteins, when heterologously expressed in *E. coli*, enzymatically converted (+)-sesaminol (**42a**) into (+)-sesaminol 2-*O*-β-D-glucoside (**43a**), providing good supporting evidence that the gene sequence is indeed conserved.

Sesamin (**10a**) levels in transformed *Forsythia koreana* cells were also modestly increased. To do this, a triple transformant was prepared that consisted of an RNA*i* construct against the pinoresinol (**1**) glucosylating enzyme, UGT71A18, (Ono et al. 2010), the gene encoding the CYP81Q1 that forms sesamin (**10a**), as well as a PLR-RNA*i* construct (Murata et al. 2015).

5.6.3 Oxygen Insertion and Rearrangement Roadmap

The insertion of oxygen to form the acetal bridge between the furofuran group and the aryl moiety during (+)-sesamolin (**46a**) formation was previously an unknown enzymatic conversion (Marchand et al. 1997), including the actual sequence of oxygenation and methylenedioxy bridge formation. Interestingly, (+)-sesamolin (**46a**) was known, though, to rearrange into (+)-sesaminol (**42a**), during acid-catalyzed processing of sesame oil, whereby the acetal bridge was considered to be cleaved to afford the presumed intermediate oxonium ion and sesamol (**53**), which can then undergo nucleophilic attack (Fukuda et al. 1986) (Figure 5.16A). Both samin (**54**) and sesamol (**53**) are also formed (Figure 5.16B), this being rationalized due to the trace amounts of water present (Fukuda et al. 1986).

It was previously considered that (+)-pinoresinol (**1a**), (+)-piperitol (**41a**), or (+)-sesamin (**10a**) could all potentially serve as substrates for the oxygen insertion step. Radiolabeled precursor administration experiments (Kato et al. 1997, Jiao et al. 1998) were carried out to investigate this question, but these studies were unable to distinguish between those possibilities at that time.

Nevertheless, while the enzymatic formation of (+)-sesamolin (**46a**) had not been characterized, a possible mechanism had been proposed earlier by us (Vassão et al. 2010) for the acetal bridge formation involving rearrangement of (+)-sesaminol (**42a**) to afford (+)-sesamolin (**46a**) (Figure 5.17). Recent studies (Murata et al. 2017) apparently confirmed our earlier hypothesis (Vassão et al. 2010), through the demonstration that both (+)-sesaminol (**42a**) and (+)-sesamolin (**46a**) could be catalytically produced from (+)-sesamin (**10a**) by a sesame CYP, named CYP92B14.

As before for the other enzymatic steps, structural biology and biochemical studies are needed to decipher how the substrates are actually bound, and how the various enzymatic

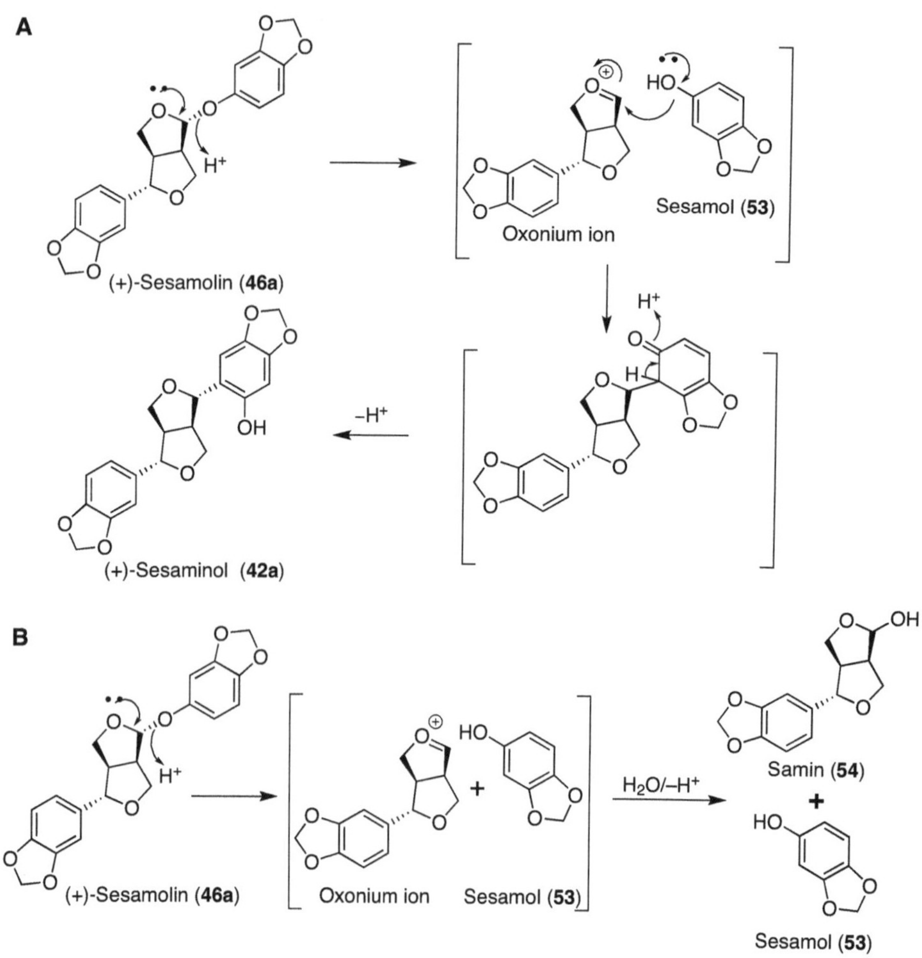

FIGURE 5.16 **A.** Proposed mechanism for rearrangement of (+)-sesamolin (**46a**) during acid catalyzed processing of sesame oil. **B.** Formation of sesamol (**53**) and samin (**54**) from (+)-sesamolin (**46a**) in presence of H_2O. "**a**" refers to the (+)-antipode.

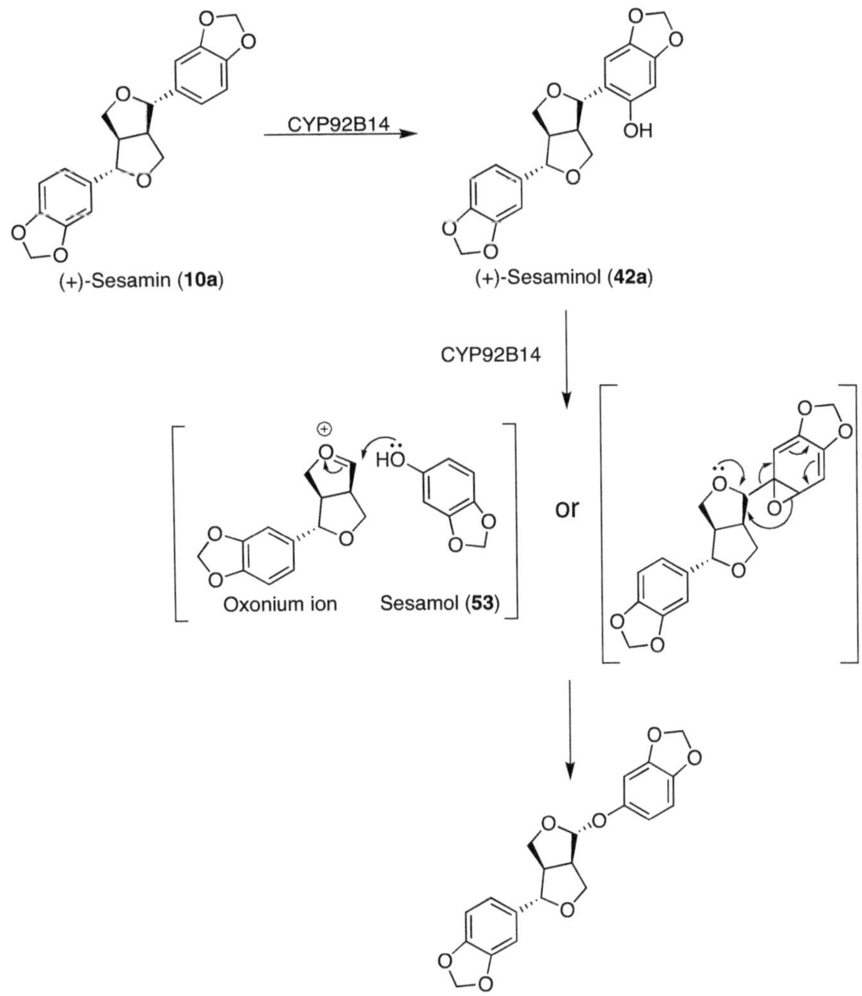

FIGURE 5.17 Sesamin (**10a**) cytochrome P450 (CYP92B14) hydroxylation to afford a mixture of (+)-sesaminol (**42a**) and its presumed rearrangement product (+)-sesamolin (**46a**), with possible intermediacy of oxonium ion and sesamol (**53**), and/or rearrangement as shown. "**a**" refers to the (+)-antipode.

transformations are effectuated mechanistically. It is also possible that the oxygen insertion occurs through an intermediate at carbons 5 and 6.

5.7 8–8′-FUROFURAN LIGNAN DOWNSTREAM METABOLISM ROADMAP: FURAN, DIBENZYLBUTANE AND DIBENZYLBUTYROLACTONE LIGNANS

Here, both the discoveries and subsequent studies of the enantiospecific transformations of pinoresinol (**1**) into three distinct 8–8′-linked metabolic classes are described, namely to the furanolignans [e.g. lariciresinol (**11**)], the dibenzylbutane lignans [e.g. secoisolariciresinol (**12**)], and the dibenzylbutyrolactone lignans [e.g. matairesinol (**6**)].

5.7.1 8–8′-Furofuran Reductase Roadmap: Pinoresinol/Lariciresinol Reductases (PLRs) and Pinoresinol Reductase (PLR homolog)

With different pinoresinol-forming DPs capable of stereoselectively affording either (+)- or (–)-pinoresinols (**1a** or **1b**), our attention had also been given to determine whether there were enzymes involved in subsequent downstream metabolism of the 8–8′-linked lignan entry point metabolite pinoresinol (**1**). This, in turn, led to the discovery of pinoresinol/lariciresinol reductases (PLRs) (Chu et al. 1993, Dinkova-Kostova et al. 1996). Indeed, PLRs as well as PLR homologs were found that catalyzed conversions of pinoresinol (**1**) into either lariciresinol (**11**), secoisolariciresinol (**12**), or both (Fujita et al. 1999, Ford 2001, Davin and Lewis 2003, Teoh et al. 2003, von Heimendahl et al. 2005, Hemmati et al. 2007, Bayindir et al. 2008, Nakatsubo et al. 2008, Corbin et al. 2017). Most of these studies were carried out with *Forsythia*, western red cedar (*T. plicata*), western hemlock (*Tsuga heterophylla*), *Linum*

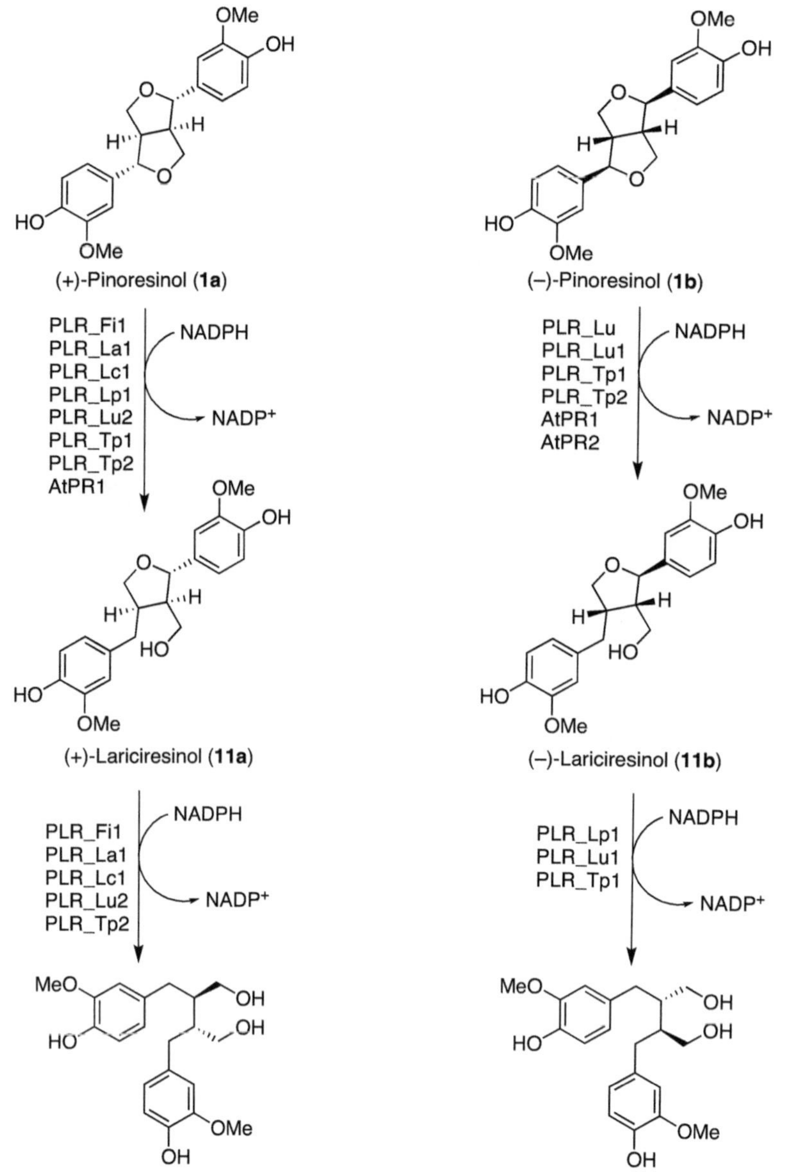

FIGURE 5.18 Reactions catalyzed *in vitro* by recombinant furofuran reductases, pinoresinol/ lariciresinol reductases (PLRs), pinoresinol reductases (PRs), and homologs thereof. At, *Arabidopsis thaliana*; Fi, *Forsythia intermedia*; La, *Linum album*; Lc, *L. corymbulosum*; Lp, *L. perenne*; Lu, *L. usitatissimum*; Tp, *Thuja plicata*. "**a**" and "**b**" refer to the (+)- and (–)-enantiomeric forms, respectively.

sp., and *A. thaliana* as summarized below. These advances, in turn, began to shed much needed additional light on the hitherto enigmatic differences in optical activities of various lignans from different plant species.

5.7.1.1 Discovery of *Forsythia* (+)-Pinoresinol/ (+)-Lariciresinol Reductase, PLR

PLR was discovered using *F. intermedia* (Chu et al. 1993, Dinkova-Kostova et al. 1996). The PLR so obtained catalyzed sequential NADPH-dependent enantioselective conversion of (+)-pinoresinol (**1a**) into (+)-lariciresinol (**11a**), and then the latter into (–)-secoisolariciresinol (**12b**), respectively (Figure 5.18) (Chu et al. 1993). Purified to apparent homogeneity, the *Forsythia* PLR gave apparent K_m, V_{max}, and k_{cat}/K_m values for **1a** and **11a** of 27/121 μM, 4.5/7.0 pkat μg^{-1} protein, and 5,800/2,000 M^{-1} s^{-1}, respectively (Dinkova-Kostova et al. 1996). Its encoding gene was subsequently cloned, and its recombinant protein PLR_Fi1 expressed in *E. coli*. The PLR_Fi1 monomer consisted of 312 amino acids with a molecular mass of 34.9 kDa (Dinkova-Kostova et al. 1996).

5.7.1.2 Gymnosperm PLR/PR Homologs of Differing Enantiospecificities

We also conducted investigations into how western red cedar (*Thuja plicata*) and western hemlock (*Tsuga heterophylla*) formed their lignan types, as exemplified by (–)-plicatic acid

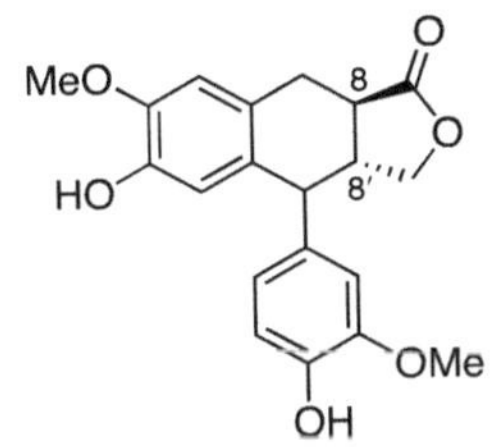

(–)-α-Conidendrin (**55b**)

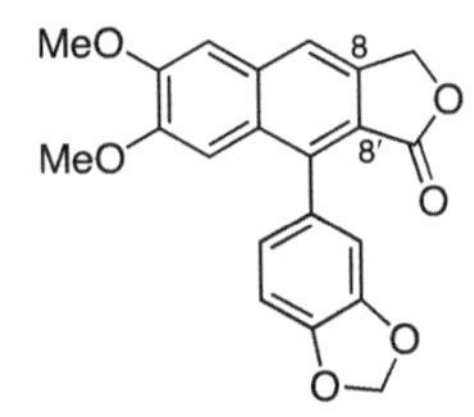

Justicidin B (**56**)

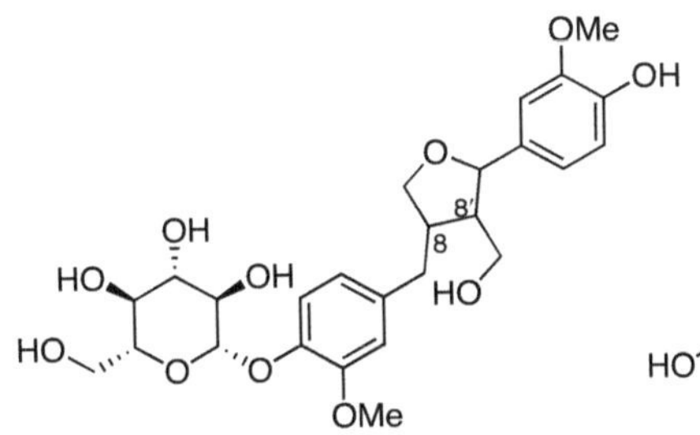

Lariciresinol-4-*O*-β-D-glucoside (**57**)

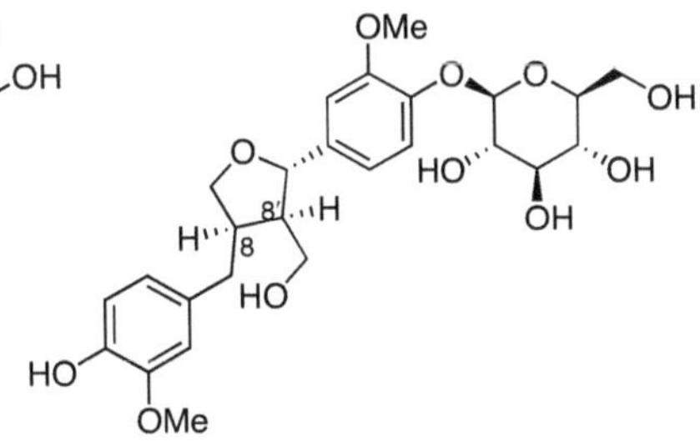

Lariciresinol-4′-*O*-β-D-glucoside (**58**)

FIGURE 5.19 (–)-α-Conidendrin (**55b**) from western hemlock (*Tsuga heterophylla*), justicidin B (**56**) from *Linum perenne*, and lariciresinol glucosides **57** and **58** from *Arabidopsis thaliana* roots. "**b**" refers to the (–)-antipode.

(**14b**) and (–)-α-conidendrin (**55b**, Figure 5.19), respectively. In this regard, PLR and PLR homolog cDNAs were isolated from these species, i.e. PLR_Tp1 to PLR_Tp4 and PLR_Th1/PLR_Th2, respectively, whose corresponding proteins had ~52%–61% identity and ~66%–79% similarity to the aforementioned PLR_Fi1 (Fujita et al. 1999).

Of these, the *E. coli* heterologously expressed recombinant western red cedar PLR_Tp2 and PLR_Tp1 (which shared >70% identity) had very different enantiospecificities (Hwang et al. 2020). Like the *Forsythia* PLR_Fi1, PLR_Tp2 efficiently converted (+)-pinoresinol (**1a**) into (–)-secoisolariciresinol (**12b**) via (+)-lariciresinol (**11a**), while also much more slowly reducing (–)-pinoresinol (**1b**) into (–)-lariciresinol (**11b**) (Figure 5.18). This was our first indication of distinct enantiospecific PLR and pinoresinol reductase (PR) activities, at least *in vitro*, in this one PLR/PR homolog isoform.

The PLR_Tp1 homolog, however, less efficiently catalyzed the NADPH-dependent conversion of (–)-pinoresinol (**1b**) into (–)-lariciresinol (**11b**) and **11b** into (+)-secoisolariciresinol (**12a**). It could also convert (+)-pinoresinol (**1a**) slowly into (+)-**11a**, i.e. also showing PR activity with this enantiomer (Figure 5.18). Kinetic analyses indicated that PLR_Tp2 was overall ~21- and 680-fold more catalytically active than PLR_Tp1 for **1a** and **1b**, respectively. The PLR_Tp2 enzymatic activity was consistent with its proposed role in (–)-plicatic acid (**14b**) biosynthesis.

5.7.1.3 *Linum* Species PLR and PR Homologs of Differing Enantiospecificities

PLRs and PLR homologs of differing enantiospecificities have been reported in various *Linum* species over the last 17 years or so. These include *L. usitatissimum* PLR_Lu (Ford 2001, Davin and Lewis 2003, Teoh et al. 2003), PLR_Lu1 and PLR_Lu2 (von Heimendahl et al. 2005, Hano et al. 2006, Hemmati et al. 2010, Corbin et al. 2017), *L. album* (PLR_La1) (von Heimendahl et al. 2005), the *L. corymbulosum* PLR_Lc1 (Bayindir et al. 2008), and the *L. perenne* PLR_Lp1 (Hemmati et al. 2007), respectively; their protein sequences ranged from ~60%–74% identity and ~78%–87% similarity to PLR_Fi1.

In our hands, the *E. coli* heterologously expressed flax PLR_Lu (Ford 2001, Davin and Lewis 2003, Teoh et al. 2003) was a (–)-pinoresinol reductase (PR homolog), whereas the PLR_Lu1 (von Heimendahl et al. 2005) catalyzed conversion of (–)-pinoresinol (**1b**) into (–)-lariciresinol (**11b**) and (+)-secoisolariciresinol (**12a**). Conversely, *PLR_Lu2* encoded a PLR that converted **1a** into **11a** and **12b** (Hemmati et al. 2010) (Figure 5.18).

To better understand the physiological roles of the distinct PLR/PR homologs, various tissues were subjected to reverse transcription polymerase chain reaction (RT-PCR) analysis. In this regard, transcripts of PLR_Lu1 were observed in developing seed coats at five different developmental stages (Hano et al. 2006), as well as in roots (Hemmati et al. 2010). This finding was consistent with (+)-secoisolariciresinol (**12a**) being produced from (–)-pinoresinol (**1b**) in seed coat tissues. By contrast, PLR_Lu2 transcripts were found in leaves, stems, and to a smaller extent in roots (Corbin et al. 2017), this being consistent with formation of the (–)-secoisolariciresinol (**12b**)-derived (–)-yatein (**23b**) in these tissues. PLR_Lu2 was also expressed in flax seed coats, but at a level five times lower compared to PLR_Lu1 (Corbin et al. 2017), i.e. perhaps partly explaining the 99:1 ratio of (+)- to (–)-secoisolariciresinol (**12a** and **12b**) in flaxseed (Ford et al. 2001). These data thus began to provide a biochemical explanation for the predominance of the (+)-secoisolariciresinol (**12a**) enantiomer in flaxseed, and of (–)-yatein (**23b**) in leaves and stems (Hano et al. 2006, Corbin et al. 2017).

Additional recombinant PLRs were obtained from *L. album* (PLR_La1) (von Heimendahl et al. 2005) and *L. corymbulosum* (PLR_Lc1) (Bayindir et al. 2008). These apparently also encoded PLRs able to convert (+)-pinoresinol (**1a**) into (–)-secoisolariciresinol (**12b**) (Figure 5.18). Their reported enantiospecific properties are consistent with the known optical activities of isolated *L. album* and *L. corymbulosum* lignans, respectively, as the former accumulates (–)-podophyllotoxin (**29b**), (–)-6-methoxypodophyllotoxin (**13b**), and derivatives thereof (Federolf et al. 2007), whereas the latter produces (–)-hinokinin (**24b**) (Mohagheghzadeh et al. 2006).

A PLR (PLR_Lp1) from *L. perenne*, which accumulates justicidin B (**56**, Figure 5.19) (Schmidt et al. 2007), gave unexpected findings (Hemmati et al. 2007), as shown from results of assays carried out with (±)-pinoresinols (**1a/1b**) in the presence of NADPH. Recombinant PLR_Lp1 reportedly utilized (+)-pinoresinol (**1a**) preferentially for the first reduction and (–)-lariciresinol (**11b**) for the second (Hemmati et al. 2007) (Figure 5.18). However, in order to perhaps better establish the actual substrate specificities and enantiospecificities in the future, it may require use of enantiomerically pure substrates, i.e. **1a**, **1b**, **11a**, and **11b**, rather than employing racemic mixtures, to obtain kinetic data.

5.7.1.4 *Isatis indigotica* PLRs

Chinese woad (*Isatis indigotica*) is very well known for its indigo dye and as a medicinal herb. It also reportedly accumulates lariciresinol (**11**) and SDG (**20**) (Xiao et al. 2015). A putative PLR gene (*IiPLR1*) was obtained from *I. indigotica* (Xiao et al. 2015). Its overexpression in hairy root cultures was carried out, resulting in higher amounts (~6.3-fold increase) of lariciresinol (**11**) relative to the non-transformed controls.

A limitation of the experimental design in that study was that no chiral analysis was carried out to establish the enantiomeric composition of the lariciresinol (**11**) formed. These researchers concluded though that IiPLR1 *in vitro* was able to convert racemic pinoresinols (**1a/1b**) into lariciresinol (**11**) and secoisolariciresinol (**12**), and racemic lariciresinols (**11a/11b**) into secoisolariciresinol (**12**), but with no experimental evidence provided to support these claims (Xiao et al. 2015). As a result, the lack of chiral analyses in this study gives findings of limited value until such analyses are carried out and reported.

5.7.1.5 *Arabidopsis* Pinoresinol Reductases (PRs)

Nakatsubo et al. (2008) later reported the existence of two additional PRs in *A. thaliana*, this being deduced following β-glucosidase treatment of wild-type *A. thaliana* roots whereby small amounts of (–)-lariciresinol (**11b**) in ~88% enantiomeric excess (e.e.) were released. cDNAs of the two *Arabidopsis* PLR homologs (*AtPR1* and *AtPR2*) were subsequently obtained, of ~59% identity and 75%–78% similarity to PLR_Fi1 at the amino acid level. Recombinant AtPR2 generated (–)-lariciresinol (**11b**) in 96% e.e., when incubated with (±)-pinoresinols (**1a/1b**) *in vitro*, whereas AtPrR1 gave the (+)-antipode **11a** in ~6% e.e. (Figure 5.18).

Of the two, AtPR1 was ~6 times more catalytically active *in vitro* with a k_{cat}/K_m ~11–16 times higher for (±)-pinoresinols (**1a/1b**) (12,800 and 19,000 M^{-1} s^{-1} for **1a** and **1b**, respectively), than AtPR2 for (–)-pinoresinol (**1b**) (1,170 M^{-1} s^{-1}). However, the (–)-lariciresinol (**11b**) predominance *in vivo* suggested that the catalytic properties of AtPrR2 played a major role.

T-DNA insertion mutants were obtained to investigate the question of which PR had the most dominant role *in vivo* as regards pinoresinol (**1**) metabolism. Homozygous lines were subsequently selected (i.e. *atprr1-1* and *atprr1-2* for AtPR1, and *atprr2* for AtPR2), as well as for the double mutant, *atprr1-1 atprr2*.

Small (and probably marginal) increases (~1.6- to 1.8-fold) in lariciresinol (**11**) contents in roots of single mutants, *atprr1-1, atprr1-2,* and *atprr2* were observed, as compared to the wild type line, with the (–)-lariciresinol (**11b**) being in e.e. of 94%, 96%, and 82%, respectively. In contrast, lariciresinol (**11**) was not detected in double mutant *atprr1-1 atprr2*. The latter instead accumulated (–)-pinoresinol (**1b**) in 74% e.e. [this being ~11 times higher than wild type levels of (±)-lariciresinols (**11a/11b**)]. Accumulation of (–)-pinoresinol (**1b**) in 74% e.e. was in agreement with the upstream DP-mediated (–)-pinoresinol (**1b**) formation as previously discussed (Section 5.3.2.1). These data indicated, when taken together, that the AtPR2 (*atprr2*) was most dominant as regards pinoresinol (**1**) metabolism.

5.7.1.6 *Arabidopsis* PRs Putatively Impacting Lignification: A Re-examination and Alternative Explanation

A fairly recent article made the unexpected claim that AtPR1 impacts lignin distribution in *Arabidopsis* (Zhao et al. 2015). That paper had as its focus, given the higher level of expression of AtPR1 in *Arabidopsis* stem rather than root tissues, an emphasis on lignin deposition.

The experimental design employed "mature stems from six-week-old plants" for analysis. Three lines were studied, namely *atprr1-1*, wild type, and an AtPR1 overexpression line. The surprising conclusion made, based on the evidence presented, was that AtPR1 impacts lignification. Other interpretations are considered more likely, as described in the following.

Lignan analyses: Nakatsubo et al. (2008) had earlier shown that recombinant AtPR1 can reduce both enantiomers of (±)-pinoresinols (**1a/1b**) *in vitro* to afford (±)-lariciresinols (**11a/11b**) in ~6% enantiomeric excess (e.e.) of (+)-antipode (**11a**). However, its corresponding *atprr1-1* mutant gave very similar amounts of lariciresinol (**11**) relative to the wild type, albeit with the (–)-antipode **11b** in slightly higher e.e. (94%) in the root tissue of the *atprr1-1* mutant versus ~88% in the wild type line. These data did not indicate that AtPR1 has any significant role in pinoresinol (**1**) metabolism, regardless of enantiomeric form.

The subsequent Zhao et al. (2015) study, with lignan levels expressed in μg g^{-1} dry weight sorbitol equivalents, gave expected results. A miniscule increase in pinoresinol (**1**) content in *atprrl-1* relative to the wild type (i.e. 0.19 *vs* 0.04 μg g^{-1}, respectively) was observed, whereas the corresponding overexpression AtPR1 line was essentially unchanged in pinoresinol (**1**) amount (0.03 μg g^{-1}). A shortcoming of the experimental design, though, was the absence of any chiral analysis to establish enantiomeric compositions.

Zhao et al. (2015) also indicated that no lariciresinol (**11**) could be detected in any of the stem samples. Yet, apparently, no analyses were carried out utilizing the root tissues as a control, which are known to accumulate lariciresinol (**11**) as its glucosides **57** and **58** (Figure 5.19) (Kim et al. 2012). That would have been a useful comparison to have made, i.e. in order to mitigate any possible shortcomings in the experimental protocol used by Zhao et al. (2015). Another shortcoming in their experimental design was that the extracts were not treated with β-glucosidase to establish if glucosylated lignans were present.

Lignin analysis: A comparative lignin analysis of "mature stems from six-week-old plants" was also carried out. Only two of the three lines were compared in the Zhao et al. (2015) study for this purpose, with two stems taken from *atprrl-1* and two from the wild type, respectively. The results obtained, however, are also open to alternative interpretations, given the experimental design.

Curiously, no actual lignin determinations in their study were done to establish relative lignin levels. Instead, thioacidolysis degradation was conducted, this being a chemical degradative process that cleaves 8–*O*–4′ inter-unit linkages in lignins releasing thioacidolysis derivatives that are derived from 8–*O*–4′ linked monolignols, *p*-coumaryl (**16**), coniferyl (**15**), and sinapyl (**17**) alcohols. The thioacidolysis products from each monolignol type are trivially designated H, G, and S, respectively (Corea et al. 2012).

At full maturation of wild type *Arabidopsis* stem growth and development, the amounts of thioacidolysis products are in the order of *circa* 300 μmol g^{-1} cell wall residue with a ~2:1 G:S ratio (Corea et al. 2012). This was established by carrying out stem analyses at weekly intervals of growth and development until senescence. In this way, Corea et al. (2012) established the changes in levels of thioacidolysis products and estimated lignin contents from the emergence of stem bolting to maturation. In that study, lignification began at about 3.5 weeks, and reached maturation of lignin deposition/growth from about 7–10 weeks growth.

The thioacidolysis product levels reported for the wild type line in Zhao et al. (2015) were ~170–200 μmol g^{-1} cell wall residue at six weeks, with an ~3.5:1 average G:S ratio. Such values are considerably lower than that for *Arabidopsis* stems at maturation (Corea et al. 2012). Clearly, the wild type stems in Zhao et al. (2015) study were still at a quite early growth/development stage, and quite far from maturity.

The *atprrl-1* plant line also had lower thioacidolysis product levels of circa 140 μmol g^{-1} cell wall residue with a G:S ratio on average of circa 4:1, but not far from the lower limits of 170 μmol g^{-1} in the wild type line above. In both the wild type and *atprrl-1* lines studied by Zhao et al. (2015), lignification in fibers (that increase the relative S levels during vascular apparatus maturation) were substantially behind those in wild type lines at maturity (as in Corea et al. 2012), i.e. reflecting the fact that the stem tissues in Zhao et al. (2015) were not at full maturation.

In sum, the single-point analyses, reported by Zhao et al. (2015), and their very limited experimental design with stem samples but not at maturation, do not allow for comparison of lignin deposition trends; i.e. it was not established whether there are differences in lignification at maturation in both lines, or whether there are other pleiotropic effects affecting growth, and so on.

Until time course studies are carried out as in Corea et al. (2012) to actual maturation of the wild type and *atprrl-1* mutant lines, the preliminary report of Zhao et al. (2015) must be viewed as simply a starting point for the needed experimental design to test their hypothesis, i.e. that AtPR1 can impact lignification. The data reported in the Zhao et al. (2015) study do not provide, however, proof for that assertion. In fact, the true physiological role as a so-called AtPR1 as such cannot even be considered established at this stage.

5.7.1.7 Structural Biology Roadmap of PLRs

PLR/PLR homolog structural biology studies provided needed insight into overall catalytic mechanisms, including determining hydride transfer stereospecificity, involvement of presumed enzyme-bound quinone methide intermediates, and biochemical basis for distinct enantiospecificities.

5.7.1.7.1 PLR Hydride Transfer Stereospecificity and Inversion of Product Configuration

Forsythia PLR is a type A reductase. With [4*R*-^{3}H]- and [4*S*-^{3}H]-NADPH as cofactors, the pro-*R* hydrogen on the NADPH nicotinamide ring was abstracted and transferred to form enzymatic products, (+)-lariciresinol (**11a**) and (–)-secoisolariciresinol (**12b**), respectively (Figure 5.20A), with the incoming hydride taking up the pro-*R* position (Chu et al. 1993). Additionally, the *Forsythia* PLR, when incubated with (±)-pinoresinols (**1a/1b**) and (±)-lariciresinols (**11a/11b**) and [4*R*-^{2}H]-NADPH, gave enzymatic products (+)-[7′*R*-^{2}H]-lariciresinol (**11a**) and (–)-[7*R*,7′*R*-^{2}H]-secoisolariciresinol (**12b**). Furthermore, with *Forsythia* PLR, (±)-[7,7′-$^{2}H_2$]-pinoresinols (**1a/1b**) and [7,7′-$^{2}H_3$]-lariciresinols (**11a/11b**) and unlabeled NADPH, NMR spectroscopic analyses established that the incoming hydride took up the pro-*R* position in the corresponding products (Figures 5.20D and 5.20E; for comparison, the corresponding ^{1}H NMR spectra of unlabeled **11** and **12** are shown in Figures 5.20B and 5.20C). That is, an "inversion" of configuration occurred at C7 for (+)-lariciresinol (**11a**) and at C7/C7′ for (–)-secoisolariciresinol (**12b**), with the *presumed* intermediates being enzyme-bound quinone methides (Figure 5.20A) (Chu et al. 1993, Dinkova-Kostova et al. 1996).

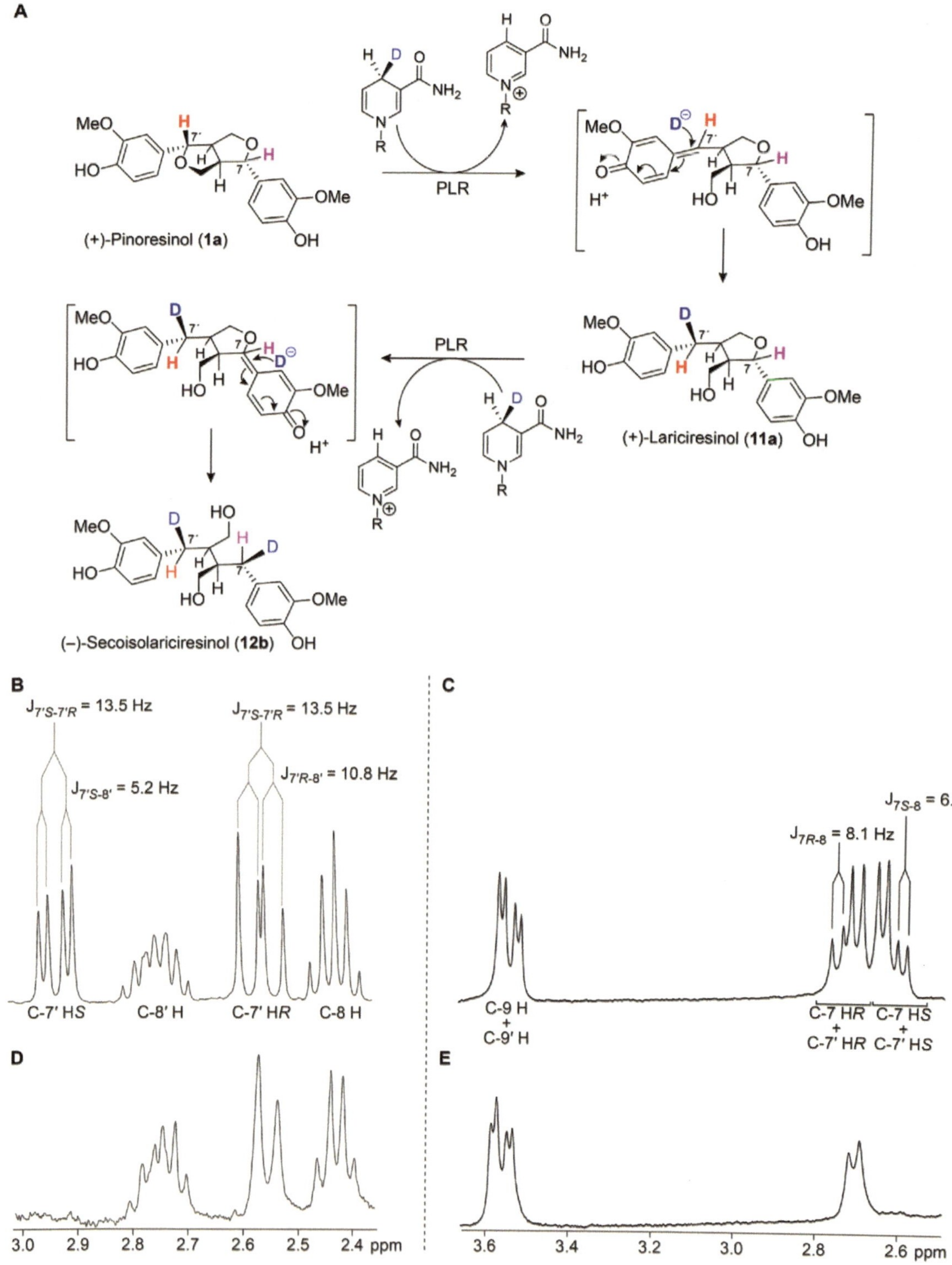

FIGURE 5.20 Pinoresinol lariciresinol reductase (PLR). **A.** Stereospecificity of hydride transfer during *F. intermedia* PLR catalysis resulting in proposed formation of quinone methide intermediate from (+)-pinoresinol (**1a**) and inversion of product configuration for (+)-lariciresinol (**11a**) and (−)-secoisolariciresinol (**12b**). **B–E.** Partial ^{1}H NMR spectra of lariciresinol (**11**) and secoisolariciresinol (**12**) showing regions for C-7′/C-8′/C-8 and C-7/C-7′/C-9 proton resonances, respectively. **B.** Synthetic (±)-lariciresinols (**11a/b**). **C.** Synthetic (±)-secoisolariciresinols (**12a/b**). **D.** Enzymatically synthesized (+)-[7,7′ *S*-$^2H_2$]lariciresinol (**11a**) obtained following incubation of PLR_Fi1 with (±)-[7,7′ $^2H_2$]pinoresinols (**1a/b**). **E.** Enzymatically synthesized (−)-[7,7′ *S*-$^2H_3$]secoisolariciresinol (**12a**) obtained following incubation of PLR_Fi1 with (±)-[7,7′ *S*-$^2H_3$]lariciresinols (**11a/b**). "**a**" and "**b**" refer to the (+)- and (−)-enantiomeric forms, respectively. This research was originally published in the Journal of Biological Chemistry: Chu et al. (1993). Stereospecificity of (+)-pinoresinol and (+)-lariciresinol reductases from *Forsythia intermedia*. *J. Biol. Chem.* 268:27026–27033. Copyright (1993) the American Society for Biochemistry and Molecular Biology.

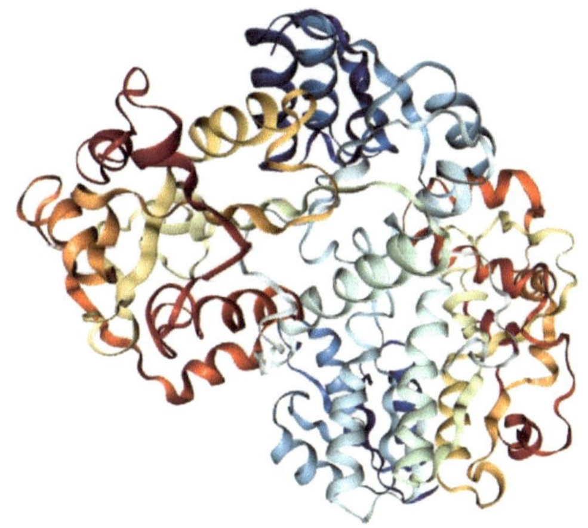

FIGURE 5.21 *Thuja plicata* PLR_Tp1 (PDB ID: 1QYD). Image created using NGL viewer (Rose et al. 2018).

5.7.1.7.2 *Structural Biology/Substrate Versatility Studies*

PLRs are members of the somewhat diverse short-chain dehydrogenase/reductase (SDR) superfamily (Min et al. 2003), with SDRs generally being about 250 amino acid residues in length and sharing many structural similarities. These include the conserved N-terminal nucleotide cofactor-binding sequence motif (GXXGXG, so-called Rossman fold) and the catalytic Lys residue.

The *T. plicata* PLR homolog (PLR_Tp1) had its X-ray crystal structure determined at 2.5 Å resolution (apo-form, 1QYD, Figure 5.21) and this was used for homology modelling of PLR_Tp2 (Min et al. 2003). PLR_Tp1 sedimentation equilibrium studies indicated it was a dimer with an apparent 69.9 ± 0.4 kDa molecular mass, whereas MALDI-TOF analysis gave a distinct monomer at 35,096 *m/z* and a broad less intense dimer at ~70,099 *m/z* (Min et al. 2003).

Both PLR_Tp2 and the PLR_Tp1 homolog have conserved nucleotide cofactor-binding N-terminal domains and presumed substrate binding C-terminal domains, with a deep cavity between the domains. The presumed highly conserved SDR catalytic lysine residue (Lys138) in the PLR_Tp1 active site was examined using site-directed mutagenesis, with the K138A mutant having its PLR activity abolished (Min et al. 2003).

In silico substrate/cofactor docking suggested that the general base Lys138 was in close proximity to the substrate phenolic group of (+)-pinoresinol (**1a**) to facilitate putative quinone methide intermediate formation (Chu et al. 1993) (Figure 5.22A). C7 hydride addition (from NADPH) then initially affords (+)-lariciresinol (**11a**), with the latter envisaged to leave the active site and then bind again via the furan group phenolic moiety adjacent to Lys138 as before. A second quinone methide can then be envisaged to be generated with hydride addition at C7′ affording (–)-secoisolariciresinol (**12b**). Additional indirect evidence for quinone methide intermediacy was also gained in the study of the western red cedar PLR homologs, PLR_Tp1 and PLR_Tp2, as ligballinol (**59**, Figure 5.22B), medioresinol (**60**) and syringaresinol (**61**) were processed as substrates, but sesamin (**10**) was not (Hwang et al. 2020). The latter, however, is unable to generate a quinone methide intermediate due to the presence of its two methylenedioxy bridges. These data were thus consistent with an apparent requirement for a free phenolic group in the substrate for quinone methide formation (Chu et al. 1993).

5.7.1.7.3 *PLR and PLR Homolog Enantiospecificity*

As PLR_Tp2 and PLR_Tp1 (see section 5.7.1.2) generally displayed opposite preferences as regards substrate enantiomer specificity, we carried out various site-directed mutagenesis studies (Hwang et al. 2020). Thus various "symmetrical" substitutions of amino acid residues (i.e. Leu128, Val268, Leu272, and Met305 in PLR_Tp1 and Met128, Gly267, Phe271, and Val304 in PLR_Tp2) were first selected. Leu128/Met305 (in PLR_Tp1) and Met128/Val304 (in PLR_Tp2) were not further considered as they were not associated with regions involved in substrate binding/catalysis (Min et al. 2003). Finally, Phe164 in PLR_Tp1 and Leu164 in PLR_Tp2 were also chosen as they could enhance substrate binding specificity (Min et al. 2003). Site-directed mutagenesis of Val268 to Gly268 in PLR_Tp1 enabled its enantiospecificity to go from (–)-pinoresinol (**1b**) as substrate to the (+)-antipode (**1a**), where the corresponding mutation (Gly267 to Val267 in PLR_Tp2) resulted only in attenuating significantly the ability to use (+)-pinoresinol (**1a**). Mutations of Phe164 to Leu164 or Leu272 to Phe271 in PLR_Tp1, resulted in proteins with no catalytic activity or with similar properties as the wild type PLR_Tp1 (Hwang et al. 2020).

5.7.1.8 Claimed Discovery of Pinoresinol Reductase Genes in Sphingomonads: A Re-assessment

Fukuhara et al. (2013) reported that resting cells of *Sphingobium* sp. strain SYK-6, when incubated with racemic pinoresinols (**1a/1b**) for 10 minutes, gave a single product believed to be racemic lariciresinols (**11a/11b**), based on retention time, MS fragmentation pattern, and comparison with an authentic standard. On the other hand, secoisolariciresinol (**12a/12b**) was not produced, even with a longer incubation period.

A gene named *pinZ*, was then obtained from the strain. This was considered to encode a PLR-like protein, albeit with very low homology (15-21% identity) to PLR_Fi1, PLR_La1, PLR_Lp1, PLR_Lu1, PLR_Tp1, and PLR_Tp2. Recombinant pinZ was able to completely metabolize racemic pinoresinols (**1a/1b**) to afford lariciresinol (**11**) and racemic lariciresinols (**11a/11b**) into secoisolariciresinol (**12**). Racemic syringaresinols (**61a/61b**) were also used as substrates but only underwent partial reduction. None of the products obtained were subjected to chiral analysis, a serious limitation of the study. Nor were kinetic parameters determined to establish efficacy relative to bona fide PLRs.

It is well known that many enzymes in nature can display substrate versatility, with different efficacies depending on the substrate tested. Given such versatility *in vitro*, reductions

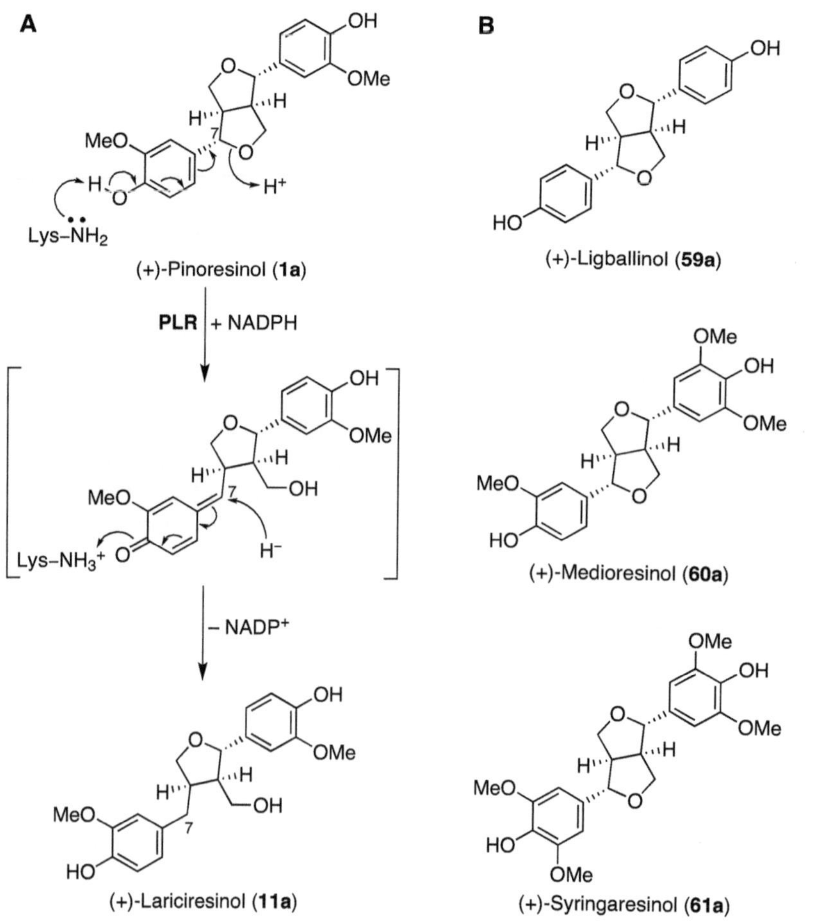

FIGURE 5.22 **A.** Proposed reaction mechanism for pinoresinol/lariciresinol reductases, PLR in active site. **B.** Structures of (+)-ligballinol (**59a**), (+)-medioresinol (**60a**) and (+)-syringaresinol (**61a**). "**a**" refers to the (+)-antipode.

such as these do not provide proof that the *in vivo* function of pinZ is that of a PLR proper. While interesting that a PLR-like activity was detected, no conclusions can be made, other than it is a PLR-like homolog of low percentage identity.

5.7.2 8–8′-Dibenzylbutane Lignan Derivatization Roadmap: Flaxseed Secoisolariciresinol Glucosylation and Hydroxymethylglutaric Acid Esterification

While the biochemical pathways to both (+)- and (–)-secoisolariciresinols (**12a** and **12b**) were determined, the product secoisolariciresinol (**12**) can also be further metabolized. In flaxseed, for example, the secoisolariciresinol (**12**) is essentially not found as such, but instead is covalently attached to other constituents.

Firstly, the secoisolariciresinol (**12**) can be glucosylated to afford the corresponding diglucoside **20**, with the aglycone being in a 99:1 ratio of (+)- to (–)-enantiomeric forms **12a** and **12b** upon deglycosylation with β-glucosidase (Ford et al. 2001).

SDG (**20**) is also essentially not detected as such, but it is further covalently modified *in planta*. In previous studies, the SDG (**20**) was found to be covalently attached to hydroxymethylglutaric acid (HMGA, **21**) moieties *in planta* to generate the esters such as 6a-HMG SDG (**62**) and 6a, 6a′-di-HMG SDG (**63**) (Figure 5.23) and **22** (Figure 5.6) (Ford et al. 2001). These components also do not exist as such *in planta* to any great extent, but undergo covalent modifications, although the precise chemical structures of the final products have not yet been fully determined (Moinuddin et al. 2019).

The following addresses though what is known thus far on secoisolariciresinol (**12**) covalent modifications in developing flax seed coats.

Ghose et al. (2014) reported that a flax glucosyltransferase (UGT74S1) was able to convert secoisolariciresinol (**12**) into SDG (**20**). Furthermore, the recombinant His-tag purified UGT74S1 (80 μg) gave a k_{cat} value of 0.89 sec^{-1}, with estimated apparent K_m values for secoisolariciresinol (**12**) and UDP-Glc of 79 and 1188 μM, respectively. *UGT74S1* mutants were also screened from an ethyl methane sulfonate (EMS) mutagenized flax population. From this, 13 homozygous missense mutations and two homozygous nonsense mutations were obtained (Fofana et al. 2017). These mutations reportedly altered SDG (**20**) profiles in flax seed, with SDG (**20**) being below the detection limit in seeds of homozygous

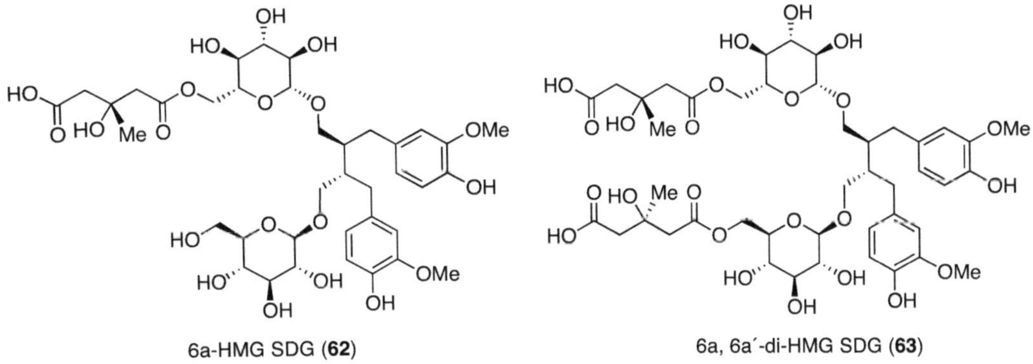

FIGURE 5.23 6a-HMG SDG (**62**) and 6a, 6a′-di-HMG SDG (**63**) from maturing whole flax seed tissue.

nonsense mutants from M3 and M4 generations, indicating UGT74S1 was required for SDG (**20**) biosynthesis.

While the HMGA derivatives **62** and **63** have been characterized (Ford et al. 2001), the specific enzymes responsible for their subsequent downstream conversions have not yet been described. On the other hand, HMGA (**21**) biosynthesis is very well-known, being a product derived from acetyl CoA (Rudney 1957).

5.7.3 8–8′-Dibenzylbutyrolactone Lignan Roadmap

5.7.3.1 Secoisolariciresinol Dehydrogenase (SDH) Discovery

SDH discovery began though studying lignan metabolism with *F. intermedia* cell-free extracts. Radiotracer/stable isotope studies with these showed that only (–)-secoisolariciresinol (**12b**) was enantiospecifically converted into (–)-matairesinol (**6b**) in the presence of NAD^+; the (+)-enantiomer (**12a**) was not converted into (+)-matairesinol (**6a**). Furthermore, when both (±)-[9,9′-^{3}H]- and (±)-[Ar-^{2}H]-secoisolariciresinols (**12a/12b**) were individually incubated in a similar manner, only (–)-[9-^{3}H]- and (–)-[Ar-^{2}H]-matairesinol (**6b**) were obtained. SDH was then purified (>6,000-fold) from *F. intermedia* stems, with the corresponding encoding gene (*SDH_Fi321*) cloned; its 831 base-pair cDNA encoded a 277-amino acid monomer. Functional recombinant SDH (SDH_Fi321) was expressed in *E. coli* (Xia et al. 2001) with catalysis occurring via intermediacy of (–)-lactol (**64b**) (Figure 5.24). On the other hand, (–)-lactol (**64b**) intermediacy was not detected when assaying native *F. intermedia* SDH.

A second SDH encoding gene (*SDH_Pp7*) was isolated from the (–)-podophyllotoxin (**29b**) producing species, *Podophyllum peltatum* (Xia et al. 2001). Following its heterologous recombinant protein expression as above in *E. coli*, purification to apparent homogeneity, and assays in presence of NAD^+ with (–)-lactol (**64b**) as substrate, apparent K_m, V_{max}

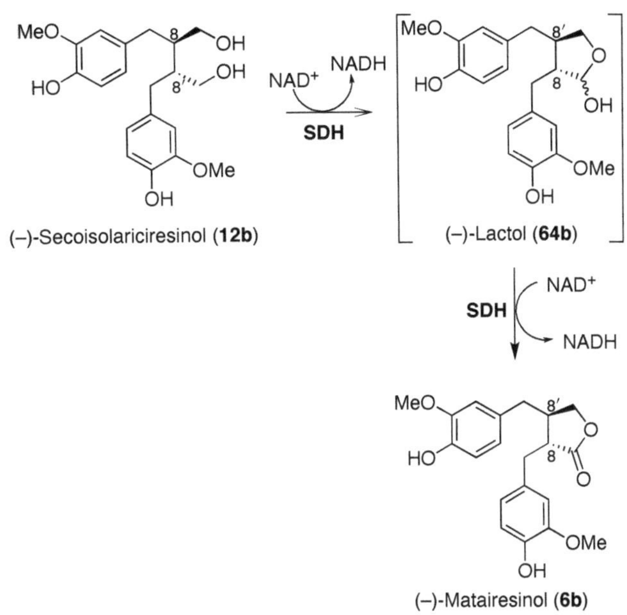

FIGURE 5.24 Reaction catalyzed by the dibenzylbutane lignan dehydrogenase, secoisolariciresinol dehydrogenase (SDH), through lactol (**64b**) intermediate. "**b**" refers to the (–)-antipode.

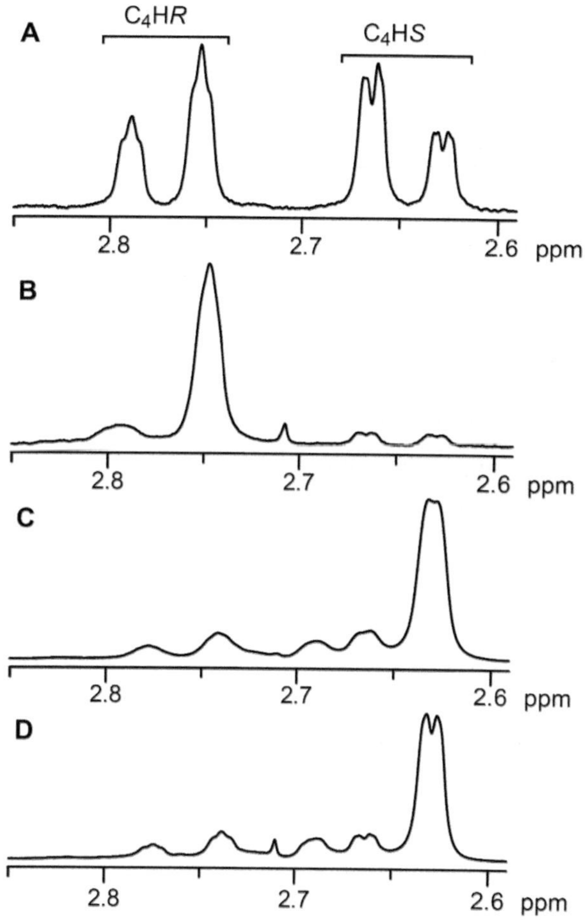

FIGURE 5.25 Partial ^{1}H NMR spectra of NADH showing spectroscopic regions for 4*S* and 4*R* protons at C-4 of the dihydropyridine ring. **A**. Natural abundance NADH. **B**. [4*S*-^{2}H]-NADH. **C**. [4*R*-^{2}H]-NADH. **D**. Enzymatically synthesized [4*R*-^{2}H]-NADH with SDH_Pp7 following incubation with (±)-secoisolariciresinols (**12a/b**) and [4-^{2}H]-NAD$^+$. This research was originally published in Organic and Biomolecular Chemistry: Moinuddin et al. (2006). Secoisolariciresinol dehydrogenase: Mode of catalysis and stereospecificity of hydride transfer in *Podophyllum peltatum*. *Org. Biomol. Chem.* 4:808–816. Copyright (2006) Royal Society of Chemistry.

and k_{cat}/K_m values of 160.2 μM, 118 pkat μg^{-1} protein, and 23,600 M^{-1} s^{-1} were obtained.

Hydride transfer stereospecificity from (–)-secoisolariciresinol (**12b**) to [4-^{2}H]-NAD$^+$ was also examined (Moinuddin et al. 2006). In order to do this, ^{1}H and heteronuclear multiple-quantum correlation (HMQC) nuclear magnetic resonance (NMR) spectroscopic analyses were carried out with natural-abundance (Figure 5.25A) and deuterated NADH (Figures 5.25B and 5.25C). These studies established that the 4*S*-proton was at ~δ2.65 ppm and the 4*R*-proton was at ~δ2.77 ppm in [4*R*-^{2}H]- and [4*S*-^{2}H]-NADH, respectively.

Subsequent incubation of (±)-secoisolariciresinols (**12a/12b**) with partially purified SDH and [4-^{2}H]-NAD$^+$ gave enzymatically generated (deuterated) NADH, which was purified and analyzed by ^{1}H (Figure 5.25D) and HMQC NMR spectroscopy. The broad doublet at δ 2.63 ppm ($J_{4S\text{-}5}$ 3.0 Hz), as well as lack of geminal coupling, revealed the deuterium to be in the 4*R* position. That is, the hydride abstracted from either (–)-secoisolariciresinol (**12b**) and/or (–)-lactol (**64b**) was added to the pro-*S* position of [4-^{2}H]-NAD$^+$, thereby generating [4*R*-^{2}H]-NADH, and showing SDH to be a type B dehydrogenase.

5.7.3.2 SDH Structural Biology Roadmap

P. peltatum SDH (SDH_Pp7) crystal structures were obtained in apo-form (2BGK), as well as with binary (2BGL) and ternary (2BGM, Figure 5.26A) complexes with NAD$^+$ and NAD$^+$/(–)-matairesinol (**6b**), at 1.6, 2.8 and 2.0 Å resolution, respectively (Youn et al. 2005). SDH, a member of the short-chain dehydrogenase (SDR) super-family, exists as a homo-tetramer in solution with each subunit having a single α,β domain structure. Its conserved Ser153–Tyr167–Lys171 (so-called catalytic triad) in the active site can apparently function (in combination with the cofactor ribose 2′-OH) as a putative proton-relay system (Figures 5.26B and 5.26C). The catalytic triad is considered responsible for cofactor binding, with Tyr167 also in close proximity to the alcohol functionality of (–)-secoisolariciresinol (**12b**) upon binding. Involvement of the catalytic triad was further supported by site-directed mutagenesis of residues Tyr167 and Lys171 to Ala, which abolished catalytic activity, and Ser153 to Ala, which displayed modest SDH activity (Moinuddin et al. 2006).

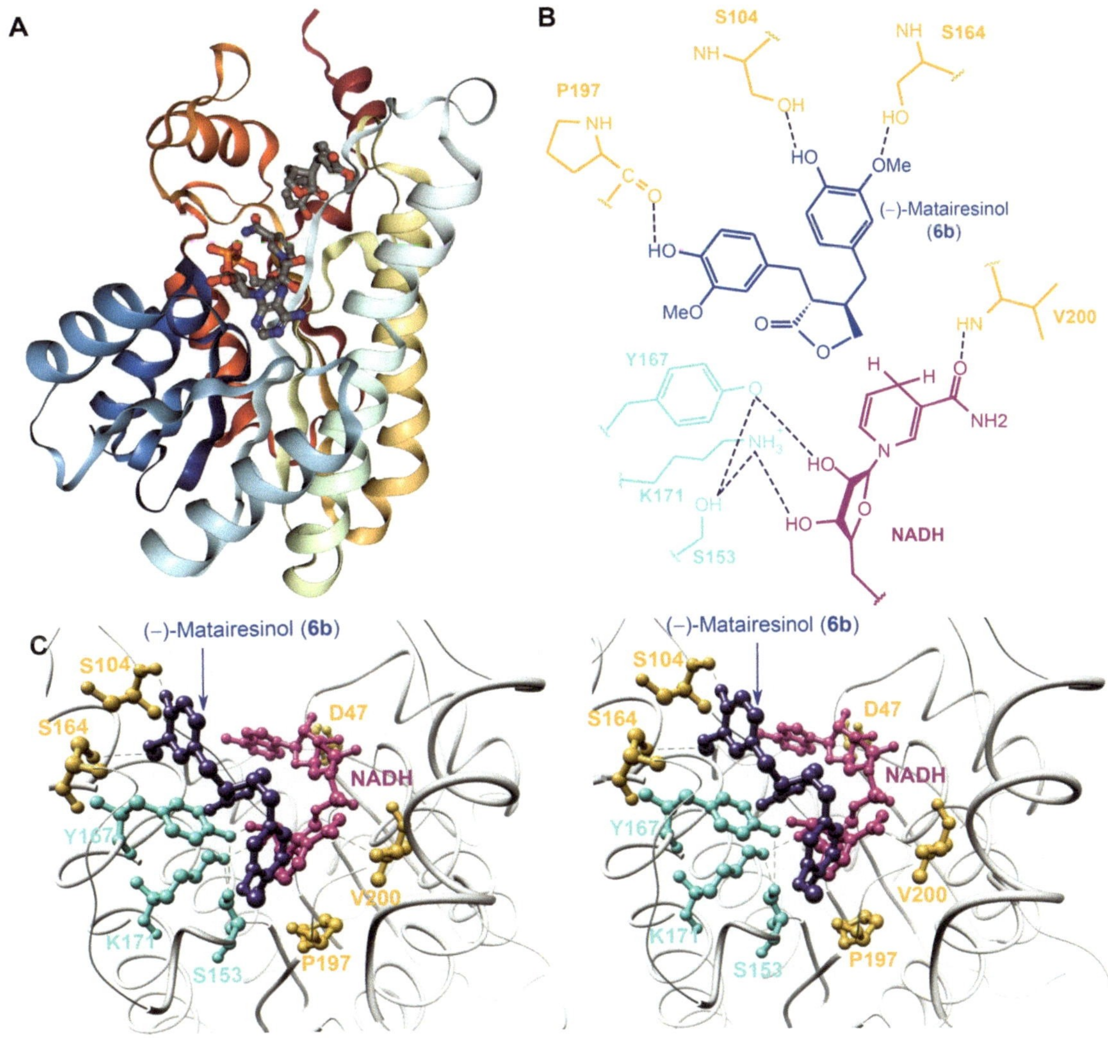

FIGURE 5.26 Secoisolariciresinol dehydrogenase structural biology and proposed catalytic mechanism. **A.** Ternary complex of SDH_Pp7 (2BGM) with NAD+ and (–)-matairesinol (**6b**). Image from the RCSB PDB (www.rcsb.org) of PDB ID 2BGM. **B.** The Tyr167 phenolic and Lys171-protonated amino groups are considered to be hydrogen-bonded to 2′- and 3′-OH of NADH, respectively. **C.** Structures of the substrate-binding pocket of SDH_Pp7 in ternary complex. Catalytic triad residues Ser153 (gold), Tyr167 (cyan), and Lys171 (cyan) are envisaged to form a hydrogen bond network with each other and water molecules. "**b**" refers to the (–)-antipode. This research was originally published in the Journal of Biological Chemistry: Youn et al. (2005). Crystal structures of apo-form and binary/ternary complexes of *Podophyllum* secoisolariciresinol dehydrogenase, an enzyme involved in formation of health-protecting and plant defense lignans. *J. Biol. Chem.* 280:12917–12926. Copyright (2005) the American Society for Biochemistry and Molecular Biology.

Entropically-favored NAD+ binding can be provisionally considered as stabilized by hydrogen bonding between cofactor ribose 2′- and 3′-OH groups and Tyr167 and Lys171, respectively (Figure 5.26B). This, in turn, would presumably favor Tyr167 deprotonation to afford the corresponding phenolate (which is further stabilized by the neighboring Ser153 hydroxyl group). Tyr167 was thus envisaged to function as a general catalytic base, abstracting (or perhaps more accurately influencing the local electronic environment of) the alcohol proton of (–)-secoisolariciresinol (**12b**), to facilitate intramolecular cyclization to form (–)-lactol (**64b**), i.e. with concerted C9-hydride transfer to the pro-*S* position of NAD+. The NADH is then released from the active site, with a second round of catalysis involving (–)-lactol (**64b**), i.e. formation of lactone (–)-matairesinol (**6b**), with concomitant hydride transfer to NAD+ as before.

5.7.3.3 SDH and Podophyllotoxin Biosynthesis Reports in *Podophyllum* Fungal Endophyte Isolates: Current Status?

More than a decade ago, Eyberger et al. (2006) reported that endophytic fungal isolates from *P. peltatum* produce podophyllotoxin (**29**) in amounts of *circa* 0.5 to 189 μg per liter during 4 weeks of culture growth. These isolates were reportedly *Phialocephala fortinii* Wang and Wilcos, and were coded as PPE5 and PPE7 strains, respectively.

A gene encoding an SDH (named *PPE7 SD*) was subsequently reported from PPE7, with the fungal isolate also renamed as *Phialocephala podophylli* (Arneaud and Porter 2015). These researchers also cloned the SDH encoding gene from *P. peltatum* that we had previously reported (Xia et al. 2001; GenBank Accession Number AF352734), and renamed it *PpSD*. The latter was identical to the *P. peltatum SDH*. On the other hand, the PPE7 SD only differed by four different amino acid residues in the entire amino acid sequence (at positions 49 55, 195, and 232). Nevertheless, the researchers asserted that *PPE7 SD* was a fungal gene, as attempts to obtain a *Rubisco* gene from the same fungal isolate were unsuccessful.

Somewhat bewilderingly, the researchers only heterologously expressed our previously described *P. peltatum* SDH in *Pichia* cell culture, and not the putative fungal SDH. Unsurprisingly, this recombinant *Podophyllum* species SDH was able to catalyze dehydrogenation of secoisolariciresinol (**12**) into matairesinol (**6**), in the presence of NAD^+, as had already been shown (Xia et al. 2001). The authors, however, did not carry out any chiral analyses of either substrates or products.

The researchers acknowledged correctly, however, that the genome sequencing of fungal isolate PPE7 is necessary to establish that its genome harbors not only SDH, but an entire podophyllotoxin (**29**) biosynthetic pathway. Until this is done, this report should be considered as in need of additional proof, such as genome sequencing, for an intact biochemical pathway to podophyllotoxin (**29**).

Somewhat comparable claims of fungal endophytes biosynthesizing structurally complex phytochemicals, such as Taxol™ (paclitaxel), have also been reported (Stierle et al. 1993, Zhang et al. 2009a, b), with evidence for same strongly disputed (Heinig et al. 2013). Yet, paclitaxel was also reported to occur in hazel (*Corylus avellana* L.) (Ottaggio et al. 2008), a species outside of the Taxaceae family. An endophytic fungus *Penicillium aurantiogriseum* NRRL 62431 from hazel also reportedly biosynthesizes Taxol (Yang et al. 2014).

The genome sequencing of the *P. aurantiogriseum* NRRL 62431 allowed these researchers to propose that Taxol biosynthetic pathway gene candidates were present in that organism, based on searches using 13 known Taxol biosynthetic gene sequences, which resulted in homologs with >30% identity (Yang et al. 2014). It was also concluded that this fungal pathway had evolved independently, and was not likely a consequence of horizontal gene transfer. Yet, none of the corresponding fungal proteins were apparently expressed to establish whether the genes identified truly have their putative functions, e.g. as a taxadiene synthase.

5.7.4 8–8′-Dibenzylbutyrolactone Lignan Derivatization Roadmap: Methylenedioxy Bridge Formation, Methylation, Glucosylation, and Hydroxylation

Based on known plant lignans retaining the 8–8′-dibenzylbutyrolactone skeleton, various subsequent conversions can occur throughout the plant kingdom. Such conversions frequently include, depending upon the plant species, regiospecific and enantiospecific derivatizations, such as methylenedioxy bridge formation, methylations, glucosylations, and hydroxylations. Currently, the methylenedioxy bridge-forming and methylation reactions are the best understood, in terms of the specific genes and enzymology involved. The advances known thus far are summarized below.

5.7.4.1 Methylenedioxy Bridge Formation

As for the furofuran 8–8′-lignans, various dibenzylbutyrolactone lignans with methylenedioxy bridges have been reported from different plant species. Examples include (–)-pluviatolide (**65b**, Figure 5.27) (Corrie et al. 1970), (–)-yatein (**23b**) (Erdtman and Harmatha 1979), and (–)-hinokinin (**24b**) (Yoshiki and Ishiguro 1933, Briggs 1935) (Figure 5.8). All three 8–8′-lignans are also present in *Anthriscus sylvestris* (Suzuki et al. 2002). As before, methylenedioxy bridge formation results via action of substrate-specific CYPs.

5.7.4.1.1 Pluviatolide

In our medicinal plant consortium sequencing project supported by the National Institutes of Health (NIH), transcriptomic and metabolomic analyses were carried out to potentially identify downstream (–)-matairesinol (**6b**) biosynthetic steps to the anticancer scaffold (–)-podophyllotoxin (**29b**) in *P. hexandrum* and *P. peltatum*. We had earlier deduced that the unknown enzymatic steps would include hydroxylations, regiospecific cyclization, methylations, and methylenedioxy bridge formation. However, the biochemical sequence of these putative enzymatic conversions was unknown. Nevertheless, as a first step, we put an emphasis on the CYP superfamily, as several of these putative reactions could potentially involve this enzyme class.

We first conducted transcriptome analysis and gene assemblies of *P. hexandrum* and *P. peltatum* at the whole tissue (rhizome, leaf and stem) level (Marques et al. 2013), together with targeted metabolite analyses (Cuthbertson et al. 2013). Mining of these data ultimately resulted in discovery of a previously unknown CYP, that we later trivially named pluviatolide synthase, and which was able to regiospecifically catalyze methylenedioxy bridge formation. Interestingly, of the two types of CYPs that we considered as possible methylenedioxy bridge-forming enzymes in each *Podophyllum* species, one type was of low similarity (32%) to (+)-piperitol/(+)-sesamin synthase in lignan biosynthesis, whereas the other had 63% similarity to (*S*)-canadine synthase in the alkaloids (Marques et al. 2013).

Specifically, the four potential CYPs (one homolog per family, from both *P. hexandrum* and *P. peltatum*) were heterologously expressed in yeast. Each of the resulting four microsomal preparations was assayed with potential biosynthetic intermediates to (–)-podophyllotoxin (**29b**). However, only two of the CYP719 homologs were catalytically active with the substrates employed. These were found able to

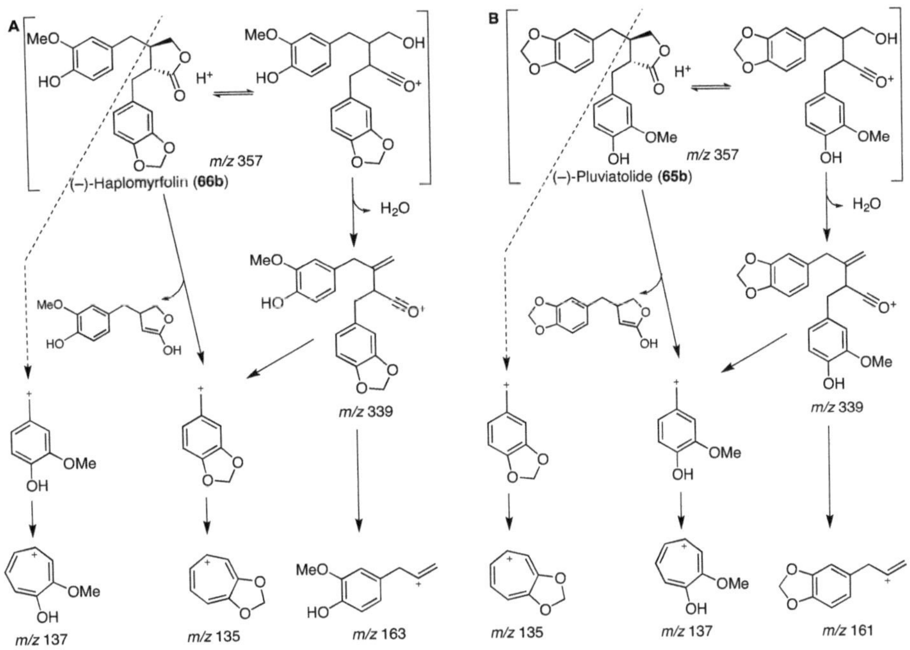

FIGURE 5.27 Fragmentation pattern of: **A.** (–)-haplomyrfolin (**66b**) and **B.** (–)-pluviatolide (**65b**). "**b**" refers to the (–)-antipode.

convert (–)-matairesinol (**6b**) into (–)-pluviatolide (**65b**), when NADPH was present in the assay mixture.

Initially, we had to distinguish whether the product formed was either (–)-pluviatolide (**65b**) or (–)-haplomyrfolin (**66b**). That it was the former was confirmed through analysis of the MS fragmentation pattern of the enzymatic product. The positive ion spectrum of the product had a base peak at *m/z* 339.1230 corresponding to $[M + H - H_2O]^+$ (calc. 339.1232), with much smaller molecular ion peaks in either protonated or Na^+ form, i.e. at *m/z* 357.1337 $[M + H]^+$ (calc. 357.1338) and *m/z* 379.1155 $[M + Na]^+$ (calc. 379.1157). The predicted fragmentation pattern unequivocally revealed on which of the two aromatic rings methylenedioxy bridge formation had occurred (Figure 5.27). Firstly, the fragment ion at *m/z* 135.0445 (calc. 135.0446) corresponded to the characteristic tropylium cation derivative of the aromatic ring housing the methylenedioxy group. Secondly, the fragment ion at *m/z* 137.0604 (calc. 137.0603) was due to the corresponding hydroxy-methoxy group substituted tropylium cation.

Thirdly, the most informative fragment ion was at *m/z* 161.0604. This enabled distinction between the two potential isobaric products, (–)-pluviatolide (**65b**) and (–)-haplomyrfolin (**66b**) (Figure 5.27). The allylbenzodioxole fragment (calc. 161.0603) detected showed that methylenedioxy bridge formation was at C3′–C4′ of (–)-matairesinol (**6b**), thereby forming (–)-pluviatolide (**65b**) (Figure 5.27B). Had the enzymatic product been (–)-haplomyrfolin (**66b**), this would have generated a 4-allyl-2-methoxyphenol fragment at *m/z* 163.0754 (Figure 5.27A). The latter was not observed.

The corresponding kinetic data for matairesinol (**6b**) was fairly similar for both CYPs, namely K_m 9.7 ± 2.2 μM and k_{cat} 14.9 ± 1.0 min^{-1} for the *P. hexandrum* CYP719A23, and K_m 5.8 ± 1.4 μM and k_{cat} 7.8 ± 0.4 min^{-1} for the *P. peltatum* CYP719A24. Interestingly, both CYPs catalyzing this conversion had a higher level of similarity (~60%) to the alkaloid methylenedioxy bridge-forming enzyme, than to the sesame lignan methylenedioxy bridge-modifying homolog.

5.7.4.1.2 Hinokinin

(–)-Hinokinin (**24b**) is perhaps best known as a trypanosomicidal phytochemical from *Piper cubeba* (Prabhu and Mulchandani 1985, de Souza et al. 2005). As indicated earlier, however, it is also found in both flax floral tissues and *L. corymbulosum* callus cultures. (–)-Hinokinin (**24b**) differs from (–)-pluviatolide (**65b**) though in having an additional methylenedioxy bridge at C3–C4, and thus could provisionally be considered derived from (–)-secoisolariciresinol (**12b**), (–)-matairesinol (**6b**) and (–)-pluviatolide (**65b**).

The only report as regards (–)-hinokinin (**24b**) biosynthesis focused on RNAi down-regulation of a putative PLR in *L. corymbulosum* hairy root cultures, which actively biosynthesize (–)-hinokinin (**24b**). Shoots of *L. corymbulosum* were transformed using *A. rhizogenes* harboring the construct *plr-Lc1-ihp*RNAi, for silencing its putative PLR_Lc1 (Bayindir et al. 2008). The resulting hairy root lines obtained could no longer biosynthesize (–)-hinokinin (**24b**). These data can be interpreted as secoisolariciresinol (**12**), matairesinol (**7**) and possibly (–)-pluviatolide (**65b**) biosynthesis being impaired.

While this interpretation is plausible, no metabolite analyses were reported to establish if any upstream metabolites accumulated, such as (+)-pinoresinol (**1a**). Nevertheless, these data were consistent with intermediacy of at least PLR and

SDH participation in the biosynthesis of (–)-hinokinin (**24b**) biosynthesis as previously proposed (Lewis and Davin 1999).

To our knowledge, there is currently no report as to how the methylenedioxy bridge at C3–C4 is formed at the gene/enzyme level, although it is certain to be a CYP, nor of the biochemical sequences of formation of its two methylenedioxy bridges.

5.7.4.1.3 Yatein

(–)-Yatein (**23b**) is found in flax leaf, stem and floral tissues (Schmidt et al. 2006, Schmidt et al. 2008), as well as in *Anthriscus sylvestris* (Suzuki 2002). It is also considered to be an intermediate en route to podophyllotoxin (**29b**) in *Podophyllum* species.

(–)-Yatein (**23b**) can be considered to be derived from (+)-pinoresinol (**1a**) via (–)-matairesinol (**6b**) and (–)-pluviatolide (**65b**). Possible biosynthetic pathways have been explored in three plant species, namely in flax, *A. sylvestris* and *Podophyllum* species.

Flax. These studies focused on the gene encoding the PLR (*LuPLR2*) needed to enantiospecifically reduce the (+)-pinoresinol antipode (**1a**) to afford (–)-secoisolariciresinol (**12b**) (Hemmati et al. 2010, Corbin et al. 2017). The studies focused on *PLR* gene expression, with *LuPLR2* expression being observed both in flax stems and leaves, this being consistent with the presence of the (–)-secoisolariciresinol (**12b**)-derived lignans **23–28** (Figure 5.8) in these tissues. However, neither the order of enzymatic reactions (C5′ hydroxylation, C4′ and C5′ *O*-methylation, and methylenedioxy bridge formation at C3–C4) leading to **23b** from **6b** has been reported in flax, nor have the specific genes or enzymes been identified other than the (+)-pinoresinol-forming DP (Dalisay et al. 2015) and LuPLR2 (Hemmati et al. 2010).

A. sylvestris. Biosynthesis of (–)-yatein (**23b**) in *A. sylvestris* was also preliminarily explored at the metabolite level using young shoots and ^{13}C- or ^{2}H-labeled precursors (Sakakibara et al. 2003). This plant species can form thujaplicatin (**67**), 5-methylthujaplicatin (**68**), 4,5-dimethylthujaplicatin (**69**), pluviatolide (**65**), 5′-hydroxypluviatolide (**70**), 4′-demethylyatein (**71**), bursehernin (**25**), 5′-demethylyatein (**72**), and yatein (**23**).

From the results obtained, two biosynthetic routes were considered possible *in vivo* (Figure 5.28). One set of biosynthetic conversions detected was the intact conversion of matairesinol (**6**) into bursehernin (**25**) via pluviatolide (**65**), whereas yatein (**23**) formation may be favored by a second route through hydroxylation of matairesinol (**6**) at C5 to initially afford thujaplicatin (**67**), which could then be regiospecifically methylated at the 5-OH functionality to generate 5-methylthujaplicatin (**68**). From the above feeding experiments, subsequent regiospecific methylation was then envisaged to occur at C-4 to afford 4,5-dimethylthujaplicatin (**69**), which may then be converted into yatein (**23**) through methylenedioxy bridge formation.

It will be informative to establish if these proposed transformations to yatein (**23**) can be demonstrated both *in vitro* using the corresponding recombinant proteins involved, and through genetic manipulations of the corresponding genes, and whether pluviatolide (**65**) is also involved in yatein (**23**) formation.

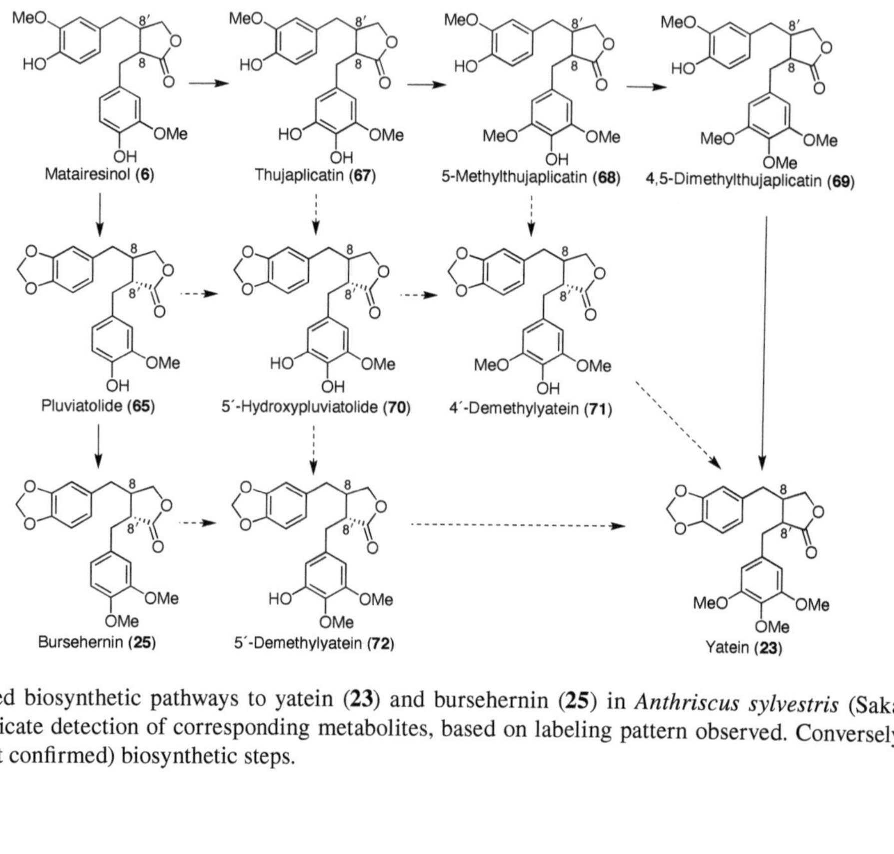

FIGURE 5.28 Proposed biosynthetic pathways to yatein (23) and bursehernin (25) in *Anthriscus sylvestris* (Sakakibara et al. 2003). Legend: Full arrows indicate detection of corresponding metabolites, based on labeling pattern observed. Conversely, dotted line arrows indicate possible (but not confirmed) biosynthetic steps.

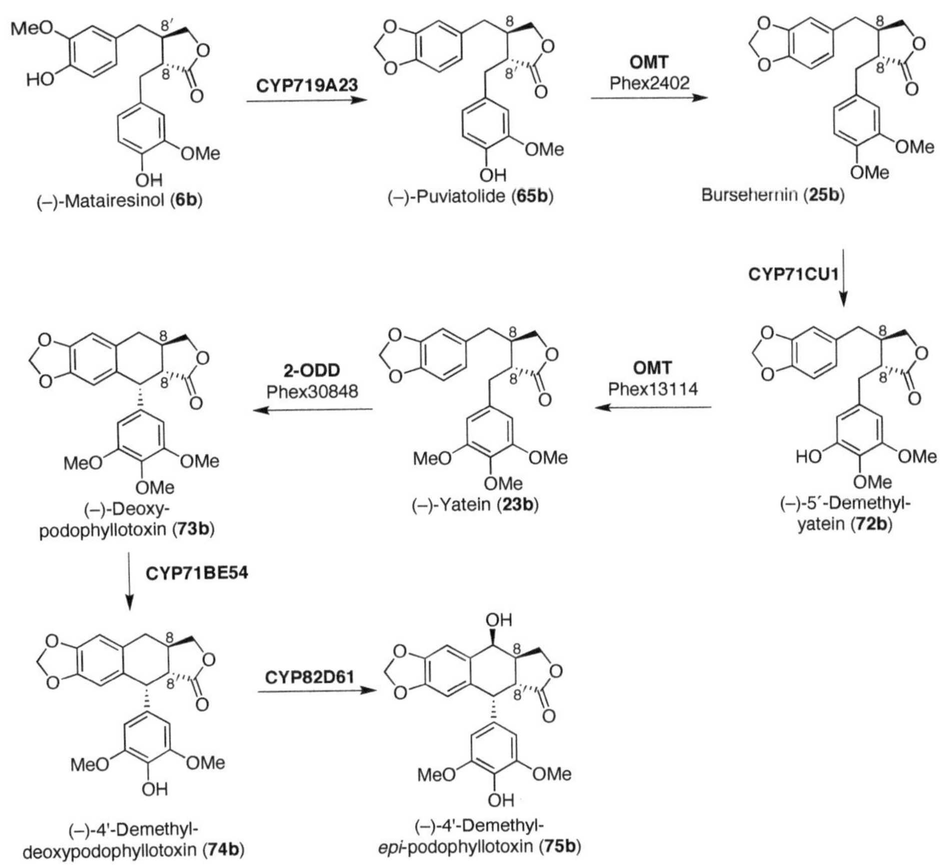

FIGURE 5.29 Proposed biosynthetic pathways to (–)-4′-demethyl-*epi*-podophyllotoxin (**75b**) as reported by Lau and Sattely (2015). "**b**" refers to the (–)-antipode.

***Podophyllum* sp.** Lau and Sattely (2015) reported a somewhat different metabolic pathway to (–)-yatein (**23b**) from (–)-pluviatolide (**65b**), using the aforementioned *P. hexandrum* transcriptomic data (Marques et al. 2013) and a combinatorial expression strategy in *Nicotiana benthamiana*. The biosynthetic steps described included a regiospecific methylation (Phex2402) to generate (–)-bursehernin (**25b**), which was then hydroxylated by CYP71CU1 to afford (–)-5′-demethylyatein (**72b**), the methylation of which by OMT (Phex13114) gives (–)-yatein (**23b**) (Figure 5.29).

5.7.4.2 Regiospecific *O*-methylations and Glucosylations

The enzymatic basis for observed regiospecific *O*-methylations and *O*-glucosylation of dibenzylbutyrolactone lignans has also been given attention in recent years.

While (–)-matairesinol (**6b**) is often found in different plant species, either regiospecific *O*-methylation, or *O*-glucosylation, or both, can occur *in planta*. For example, in safflower (*Carthamus tinctorius*) and *Forsythia* sp., *O*-methylated (–)-matairesinol (**6b**) derivatives include (–)-arctigenin (4′-*O*-methyl matairesinol, **8b**) (Chiba et al. 1978, Nishibe et al. 1988, Rahman et al. 1990, Sakakibara et al. 2007). Its 4-*O*-glucosylated derivative, (–)-matairesinoside (**7b**), has also been found in *Forsythia* sp. (Rahman et al. 1990). Additionally, (–)-arctiin (**9b**), which is derivatized via both regiospecific *O*-methylation and *O*-glucosylation, is found in both *C. tinctorius* (Umezawa et al. 2013) and *Forsythia* sp. (Nishibe et al. 1988). (–)-Isoarctigenin (4-*O*-methyl matairesinol), also known as (–)-buplerol (**76b**, Figure 5.30), has been isolated from *Bupleurum salicifolium* (Gonzalez et al. 1990).

In preliminary studies, when *F. intermedia* cell-free extracts were incubated with (±)-matairesinols (**6a/6b**), both (±)-arctigenins (**8a/8b**) and (±)-isoarctigenins (**76a/76b**) were obtained, with the naturally occurring substrate **6b** being slightly preferred (Ozawa et al. 1993). More recent studies, however, have focused on specific recombinant *O*-methyltransferases (OMTs) able to catalyze regiospecific conversions.

Using safflower, Umezawa et al. (2013) reported the presence of a regiospecific OMT, named CtMROMT. Its recombinant protein *in vitro* was able to *O*-methylate matairesinol (**6**) to afford arctigenin (**8**), but not isoarctigenin (**76**). It was later shown that only (–)-enantiomer **8b** was formed (Ragamustari et al. 2014) upon incubation with racemic **6a/6b**. Interestingly, CtMROMT could also efficiently convert the flavonoid apigenin (**77**) into acacetin (**78**) *in vitro* (Umezawa et al. 2013), although whether this has any physiological significance *in vivo* is perhaps improbable.

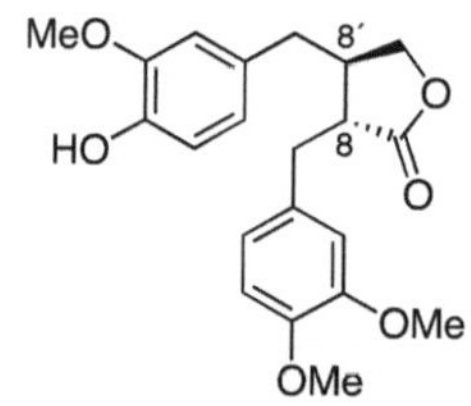

(−)-Isoarctigenin (**76b**)

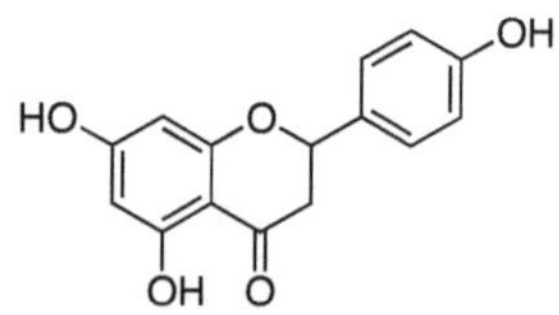

Apigenin (**77**)

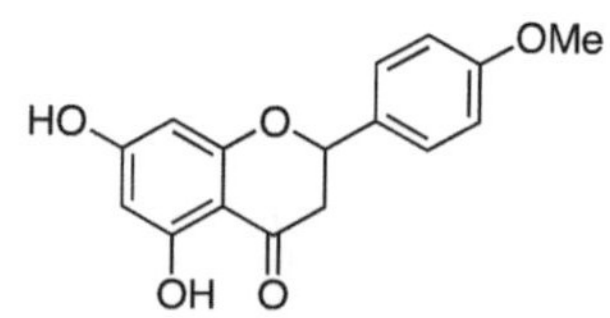

Acacetin (**78**)

FIGURE 5.30 (−)-Isoarctigenin (**76b**), and the flavonoids, apigenin (**77**) and acacetin (**78**). "**b**" refers to the (−)-antipode.

Two other regiospecific *MROMTs* were later isolated from *A. sylvestris* (*AsMROMT*) and *Forsythia koreana* (*FkMROMT*): Recombinant AsMROMT catalyzed formation of (−)-arctigenin (**8b**) from (−)-matairesinol (**6b**), whereas FkMROMT gave rise to (−)-isoarctigenin (**76b**) (Ragamustari et al. 2014).

In addition, matairesinol (**6**) in *A. sylvestris* is hydroxylated at C-5 to afford thujaplicatin (**67**), presumably via action of a CYP (Figure 5.28). The latter can then be sequentially regiospecifically methylated to form 5-methylthujaplicatin (**68**) and 4,5-dimethylthujaplicatin (**69**). The enzymology of this has been partly explained, as a cDNA was isolated from an *A. sylvestris* whose recombinant protein AsTJOMT was able to regiospecifically convert thujaplicatin (**67**) into 5-methylthujaplicatin (**68**), but not into dimethyl derivative (**69**). This OMT has a K_m of 3.8 μM for **67** and a k_{cat}/K_m of 1270 M^{-1} s^{-1} (Ragamustari et al. 2013).

There are no reports, however, to our knowledge, of the precise glucosyltransferase affording (−)-matairesinoside (**7b**).

5.7.4.3 Aryltetrahydronaphthalene and Arylnaphthalene Lignan Roadmap: Podophyllotoxin, 6-Methoxypodophyllotoxin and Justicidin B

Perhaps the best-known examples of these are (−)-podophyllotoxin (**29b**) and justicidin B (**56**). (−)-Podophyllotoxin (**29b**), an aryltetrahydronaphthalene lignan, commercially obtained from *Podophyllum* species, serves as one of the main plant natural product scaffolds in cancer treatment (Canel et al. 2000, Gordaliza et al. 2004). This scaffold is used to produce the semi-synthetic derivatives etoposide (**80**, Figure 5.31), teniposide (**81**), and Etopophos (**82**), which are widely and effectively utilized in chemotherapy (Loike 1982, Hande 1998). (−)-Podophyllotoxin (**29b**) also has a close homolog (−)-6-methoxypodophyllotoxin (**13b**) that is found in some *Linum* species (Xia et al. 2000, Federolf et al. 2007).

The arylnaphthalene lignans, justicidin B (**56**) and its methoxylated analog, justicidin A (**83**), from *Justicia hayatai* var. decumbens, also have potent bioactivity. As piscicidal agents, they have long been used by Taiwanese to catch fish over hundreds of years (Ohta et al. 1969). More recently,

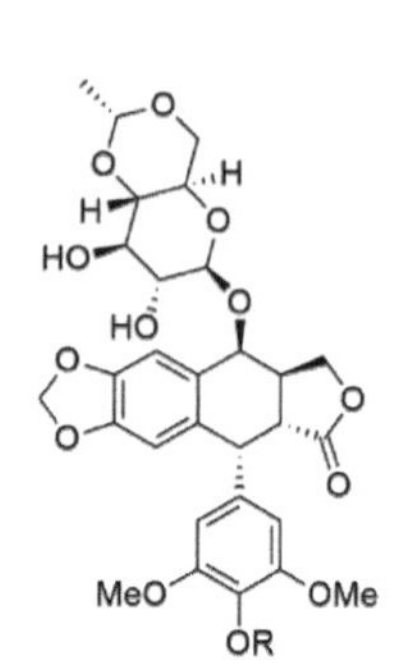

R = H, Etoposide (**80**)
R = PO(OH)$_2$, Etopophos (**82**)

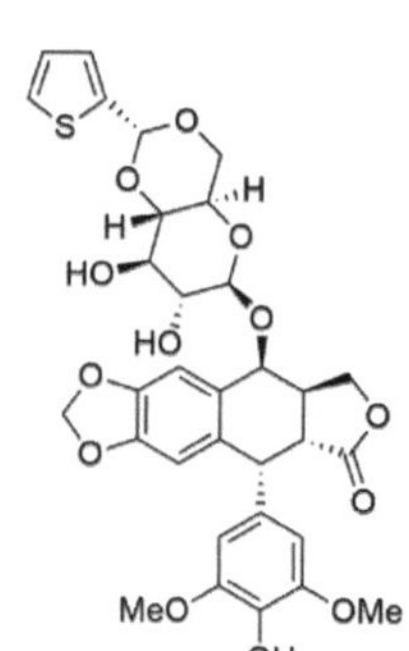

Teniposide (**81**)

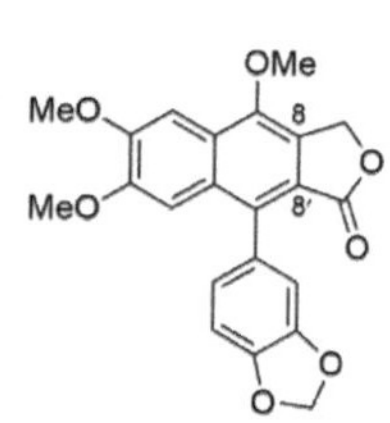

Justicidin A (**83**)

FIGURE 5.31 (−)-Podophyllotoxin (**29b**) semi-synthetic derivatives etoposide (**80**), teniposide (**81**), and Etopophos (**82**). Justicidin A (**83**) from *Justicia hayatai*. "**b**" refers to the (−)-antipode.

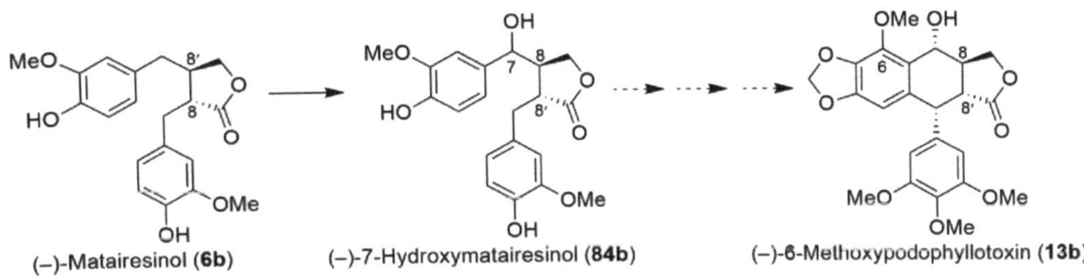

FIGURE 5.32 Proposed biosynthetic pathways to (–)-6-methoxypodophyllotoxin (**13b**) in *Linum flavum via* (–)-7-hydroxymatairesinol (**84b**) based on preliminary radiolabeling studies (Xia et al. 2000). "**b**" refers to the (–)-antipode.

justicidin B (**56**) has additionally shown promise as a potential trypanosomal agent (Hemmati and Seradj 2016).

Accordingly, the biosynthetic pathways to (–)-podophyllotoxin (**29b**) and homologs thereof, as well as to justicidin B (**56**), have attracted interest. The progress made to date is summarized in the following sections.

5.7.4.3.1 *Linum flavum* and *L. nodiflorum*

Using *L. flavum* root tissues, an active biosynthetic pathway to (–)-6-methoxypodophyllotoxin (**13b**) from (+)-pinoresinol (**1a**) to (–)-matairesinol (**6b**) was found to be operative (Xia et al. 2000) based on radiolabeled feeding studies. Uptake of matairesinol (**6**) by *L. flavum* root tissue, and its further metabolism for 6 hours, also resulted in its conversion to 7-hydroxymatairesinol (**84**) (Xia et al. 2000). Additionally, [7-^{3}H]-7-hydroxymatairesinol (**84**) was apparently intactly incorporated into (–)-6-methoxypodophyllotoxin (**13b**), suggesting **84** can also function as a pathway intermediate in *L. flavum* (Figure 5.32).

The biosynthetic pathways to both (–)-6-methoxypodophyllotoxin (**13b**) and β-peltatin A methyl ether (**86b**, Figure 5.33) have also been preliminarily explored using *L. flavum* cell suspension cultures. Based on a biochemical pathway to (–)-6-methoxypodophyllotoxin (**13b**) from (–)-deoxypodophyllotoxin (**73b**) proposed by Kranz and Petersen (2003), Molog et al. (2001) described detection of a putative deoxypodophyllotoxin 6-hydroxylase from crude microsomes of *L. flavum* cell suspension cultures. This preparation was able to catalyze formation of (–)-β-peltatin A (**85b**) from (–)-deoxypodophyllotoxin (**73b**) (Figure 5.33). Kinetic data gave apparent K_m's of 20 and 36 μM for **73b** and NADPH, respectively. Federolf et al. (2007) described similar findings using a crude *L. album* cell suspension culture microsomal preparation; the latter apparently also converted (–)-deoxypodophyllotoxin (**73b**) into (–)-β-peltatin A (**85b**) when NADPH was present (with apparent K_m's of 3 and 41 μM for **73b** and NADPH, respectively).

Additionally, a presumed regiospecific β-peltatin 6-*O*-methyltransferase was detected in cell suspension cultures of *Linum nodiflorum*. This preparation was able to catalyze conversion of β-peltatin A (**85b**) into β-peltatin A methylether (**86b**, Figure 5.33), with apparent K_m's estimated [using proteins recovered after $(NH_4)_2SO_4$ precipitation] of 40 and 15 μM for **85b** and S-adenosyl-methionine, respectively.

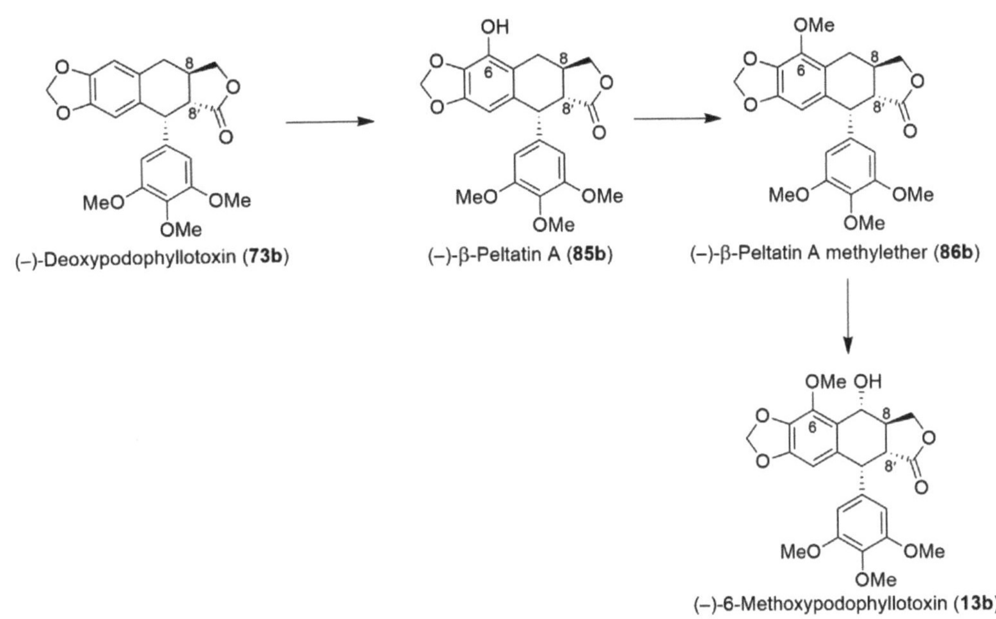

FIGURE 5.33 Proposed pathway to (–)-6-methoxypodophyllotoxin (**13b**) from (–)-deoxypodophyllotoxin (**73b**) (Molog et al. 2001, Federolf et al. 2007). "**b**" refers to the (–)-antipode.

Other potential substrates tested (i.e. matairesinol (**6**), pinoresinol (**1**), and podophyllotoxin (**29**)), were not methylated (Kranz and Petersen 2003).

While these preliminary studies give some possible insights into (–)-6-methoxypodophyllotoxin (**13b**) biosynthesis, additional studies are required to elucidate its complete pathway, i.e. in terms of the various hydroxylations, *O*-methylation, and methylenedioxy bridge-forming steps, as well as for aryltetrahydronaphthalene ring formation. In addition, the encoding genes and enzymes need to be identified in this species, and their biochemical roles proven both *in vitro* and *in planta*, as well as establishing whether there exists a single metabolic route or a metabolic grid.

5.7.4.3.2 *Podophyllum sp.*

By 2013, the biochemical pathway to (–)-podophyllotoxin (**29b**) had been elucidated to (–)-pluviatolide (**65b**) (Dinkova-Kostova et al. 1996, Davin et al. 1997, Fujita et al. 1999, Xia et al. 2001, Marques et al. 2013), with an 2-oxoglutarate/Fe(II)-dependent dioxygenase (2-ODD) generating the aryltetralin scaffold also having been obtained (Marques et al, unpublished results). Using the aforementioned *P. hexandrum* transcriptomic data (Marques et al. 2013) and combinatorial expression in *Nicotiana benthamiana*, Lau and Sattely (2015) further investigated some of the remaining biosynthetic steps to (–)-podophyllotoxin (**29b**). Their study identified six biosynthetic genes, not though to (–)-podophyllotoxin (**29b**) *per se*, but instead to (–)-4′-demethyl-*epi*-podophyllotoxin (**75b**, Figure 5.29). This epimer has the advantage of being more readily deployable for the semi-synthetic conversion to etoposide (**80**), teniposide (**81**), and Etopophos (**82**).

As described earlier for (–)-yatein (**23b**) biosynthesis in *Podophyllum* species (Lau and Sattely 2015), it was reportedly converted into (–)-deoxypodophyllotoxin (**73b**) via action of the above mentioned 2-oxoglutarate/Fe(II)-dependent dioxygenase (2-ODD), thereby generating the aryltetralin scaffold of **29b** between C7′ and C2. The subsequent conversions to (–)-4′-demethyl-*epi*-podophyllotoxin (**75b**) via (–)-4′-demethyldeoxypodophyllotoxin (**74b**) reportedly occurred via action of two CYPs (CYP71BE54 and CYP82D61), respectively.

While the overall pathway to (–)-podophyllotoxin (**29b**) is potentially close to completion (subject to verifying precise biochemical roles *in vitro* and *in planta*), one major obstacle nevertheless remained to be resolved. This was that of the production level of scaffold **75b**. When expressing DIR, PLR, SDH, CYP719A23, and the aforementioned OMTs and CYPs in tobacco leaves and infiltrating 100 μM (+)-pinoresinol (**1a**), only minuscule amounts (10.3 ng mg^{-1} of plant dry weight) of (–)-4′-demethyl-*epi*-podophyllotoxin (**75b**) were produced (Lau and Sattely 2015), when the last two CYPs were omitted, ~90 ng of deoxypodophyllotoxin (**73b**) per mg of plant dry weight were obtained. These levels have since been significantly increased to 0.71 mg g^{-1} of high purity (–)-deoxypodophyllotoxin (**73b**) (after column chromatography purification) (Schultz et al. 2019). Levels of (–)-4′-demethyl-*epi*-podophyllotoxin (**75b**) were, however, not reported.

5.7.4.3.3 *Justicia procumbens*

Justicidin B (**56**), isolated from *Justicia procumbens* (Ohta et al. 1969), is additionally found in various *Linum* species (Mohagheghzadeh et al. 2002, Vardapetyan et al. 2003, Koulman and Konuklugil 2004, Vasilev and Ionkova 2005a, Vasilev and Ionkova 2005b, Vasilev et al. 2006, Konuklugil et al. 2007, Schmidt et al. 2007). Justicidin B (**56**) has antiviral, antibacterial, piscicidal and antifungal activities (Ohta et al. 1969, Hemmati and Seradj 2016).

Unlike pluviatolide (**65b**), its methylenedioxy bridge is at C3′–C4′, i.e. on the pendant aryl ring. Little is yet understood as to how justicidin B (**56**) is biosynthesized, in terms of precise biochemical steps, and the genes and enzyme involved. Nevertheless, one preliminary study began examining how justicidin B (**56**) might be biosynthesized, but with the focus on putative PLRs. That study investigated if a putative *L. perenne* PLR (PLR_Lp1) was involved in its biosynthesis using an RNAi approach (Hemmati et al. 2007). To do this, *Agrobacterium rhizogenes* containing a vector with a RNAi construct for the PLR_Lp1 gene silencing was used for transformation of the *L. perenne* shoot culture. Various transformed hairy root cultures were obtained (six independent lines and control lines with and without empty constructs). These researchers reported that justicidin B (**56**) levels in the transformants were reduced to circa 24% of control levels, this being interpreted as due to partial suppression of PLR_Lp1 mRNA levels. This data thus provided preliminary experimental support to earlier predictions that arylnaphthalene lignans, such as justicidin B (**56**), were formed from pinoresinol (**1**) via action of PLR and other downstream enzymes (Lewis and Davin 1999).

As regards the remaining steps to form justicidin B (**56**), the biochemical sequence of methylenedioxy bridge formation, regiospecific methylation, ring closure at C7′–C2, and subsequent presumed dehydrogenation to afford the aryltetrahydronaphthalene skeleton needs to be established both *in vitro* and *in vivo*. Much more work is needed to establish the precise biochemical pathway at the precursor, gene, and enzyme levels.

ACKNOWLEDGEMENTS

The authors thank the Chemical Sciences, Geosciences and Biosciences Division, DOE Office of Basic Energy Sciences (DE-FG-0397ER20259), as well as the Arthur and Katie Eisig-Tode Endowment.

REFERENCES

Arneaud, S. L. B., and J. R. Porter. 2015. Investigation and expression of the secoisolariciresinol dehydrogenase gene involved in podophyllotoxin biosynthesis. *Mol. Biotechnol.* 57:961–973.

Arteaga, S., A. Andrade-Cetto, and R. Cárdenas. 2005. *Larrea tridentata* (creosote bush), an abundant plant of Mexican and US-American deserts and its metabolite nordihydroguaiaretic acid. *J. Ethnopharmacol.* 98:231–239.

Bauer, W., and M. H. Zenk. 1991. Two methylenedioxy bridge forming cytochrome P-450 dependent enzymes are involved in (*S*)-stylopine biosynthesis. *Phytochemistry* 30:2953–2961.

Bayindir, Ü., A. W. Alfermann, and E. Fuss. 2008. Hinokinin biosynthesis in *Linum corymbulosum* Reichenb. *Plant J.* 55:810–820.

Beroza, M. 1955. The structure of sesamolin and its stereochemical relation to sesamin, asarinin, and pinoresinol. *J. Am. Chem. Soc.* 77:3332–3334.

Briggs, L. H. 1935. Hinokinin as the enantiomorph of cubebinolide. *J. Am. Chem. Soc.* 57:1383–1384.

Canel, C., R. M. Moraes, F. E. Dayan, and D. Ferreira. 2000. Podophyllotoxin. *Phytochemistry* 54:115–120.

Chiba, M., S. Hisada, and S. Nishibe. 1978. Studies on the Chinese crude drug "Forsythiae Fructus." (III) On the constituents of fruits of *Forsythia viridissima* and *F. suspensa*. *Shoyakugaku Zasshi* 32:194–197.

Cho, M.-H., S. G. A. Moinuddin, G. L. Helms, et al. 2003. (+)-Larreatricin hydroxylase, an enantio-specific polyphenol oxidase from the creosote bush (*Larrea tridentata*). *Proc. Natl. Acad. Sci. USA* 100:10641–10646.

Chu, A., A. Dinkova, L. B. Davin, D. L. Bedgar, and N. G. Lewis. 1993. Stereospecificity of (+)-pinoresinol and (+)-lariciresinol reductases from *Forsythia intermedia*. *J. Biol. Chem.* 268:27026–27033.

Corbin, C., S. Drouet, I. Mateljak, et al. 2017. Functional characterization of the *pinoresinol-lariciresinol reductase-2* gene reveals its roles in yatein biosynthesis and flax defense response. *Planta* 246:405–420.

Corbin, C., S. Drouet, L. Markulin, et al. 2018. A genome-wide analysis of the flax (*Linum usitatissimum* L.) dirigent protein family: From gene identification and evolution to differential regulation. *Plant Mol. Biol.* 97:73–101.

Corea, O. R. A., C. Ki, C. L. Cardenas, et al. 2012. Arogenate dehydratase isoenzymes profoundly and differentially modulate carbon flux into lignins. *J. Biol. Chem.* 287:11446–11459.

Corrie, J. E. T., G. H. Green, E. Ritchie, and W. C. Taylor. 1970. Chemical constituents of Australian *Zanthoxylum* species. V. Constituents of *Z. pluviatile* Hartley; the structures of two new lignans. *Aust. J. Chem.* 23:133–145.

Cuthbertson, D. J., S. R. Johnson, J. Piljac-Žegarac, et al. 2013. Accurate mass–time tag library for LC/MS-based metabolite profiling of medicinal plants. *Phytochemistry* 91:187–197.

Dalisay, D. S., K.-W. Kim, C. Lee, et al. 2015. Dirigent protein-mediated lignan and cyanogenic glucoside formation in flax seed: Integrated omics and MALDI mass spectrometry imaging. *J. Nat. Prod.* 78:1231–1242.

Davin, L. B., D. L. Bedgar, T. Katayama, and N. G. Lewis. 1992. On the stereoselective synthesis of (+)-pinoresinol in *Forsythia suspensa* from its achiral precursor, coniferyl alcohol. *Phytochemistry* 31:3869–3874.

Davin, L. B., H.-B. Wang, A. L. Crowell, et al. 1997. Stereoselective bimolecular phenoxy radical coupling by an auxiliary (dirigent) protein without an active center. *Science* 275:362–367.

Davin, L. B., and N. G. Lewis. 2003. An historical perspective on lignan biosynthesis: Monolignol, allylphenol and hydroxycinnamic acid coupling and downstream metabolism. *Phytochem. Rev.* 2:257–288.

Davin, L. B., J. R. Cort, C. A. Smith, et al. 2022. Dirigent protein roadmap to lignans and other vascular plant phenol classes. In *The Lignan Handbook* (N. G. Lewis and L. B. Davin eds.), Taylor & Francis, Boca Raton, London. pp 57–77.

de Souza, V. A., R. da Silva, A. C. Pereira, et al. 2005. Trypanocidal activity of (–)-cubebin derivatives against free amastigote forms of *Trypanosoma cruzi*. *Bioorg. Med. Chem. Lett.* 15:303–307.

Dinkova-Kostova, A. T., D. R. Gang, L. B. Davin, et al. 1996. (+)-Pinoresinol/(+)-lariciresinol reductase from *Forsythia intermedia*. Protein purification, cDNA cloning, heterologous expression and comparison to isoflavone reductase. *J. Biol. Chem.* 271:29473–29482.

Effenberger, I., B. Zhang, L. Li, et al. 2015. Dirigent proteins from cotton (*Gossypium* sp.) for the atropselective synthesis of gossypol. *Angew. Chem. Int. Ed.* 54:14660–14663.

Erdtman, H., and J. Harmatha. 1979. The chemistry of the order Cupressales. Part 58. Phenolic and terpenoid heartwood constituents of *Libocedrus yateensis*. *Phytochemistry* 18:1495–1500.

Eyberger, A. L., R. Dondapati, and J. R. Porter. 2006. Endophyte fungal isolates from *Podophyllum peltatum* produce podophyllotoxin. *J. Nat. Prod.* 69:1121–1124.

Federolf, K., A. W. Alfermann, and E. Fuss. 2007. Aryltetralin-lignan formation in two different cell suspension cultures of *Linum album*: Deoxypodophyllotoxin 6-hydroxylase, a key enzyme for the formation of 6-methoxypodophyllotoxin. *Phytochemistry* 68:1397–1406.

Fofana, B., K. Ghose, A. Somalraju, et al. 2017. Induced mutagenesis in *UGT74S1* gene leads to stable new flax lines with altered secoisolariciresinol diglucoside (SDG) profiles. *Front. Plant Sci.* 8:1638.

Ford, J. D. 2001. Cancer chemopreventive flax seed lignans: Delineating the metabolic pathway(s) to the SDG-HMG ester-linked polymers. Ph.D. Thesis, Washington State University.

Ford, J. D., K.-S. Huang, H.-B. Wang, L. B. Davin, and N. G. Lewis. 2001. Biosynthetic pathway to the cancer chemopreventive secoisolariciresinol diglucoside-hydroxymethyl glutaryl ester-linked lignan oligomers in flax (*Linum usitatissimum*) seed. *J. Nat. Prod.* 64:1388–1397.

Freudenberg, K. 1959. Biosynthesis and constitution of lignin. *Nature* 183:1152–1155.

Fujita, M., D. R. Gang, L. B. Davin, and N. G. Lewis. 1999. Recombinant pinoresinol-lariciresinol reductases from western red cedar (*Thuja plicata*) catalyze opposite enantiospecific conversions. *J. Biol. Chem.* 274:618–627.

Fukuda, Y., T. Osawa, M. Namiki, and T. Ozaki. 1985. Studies on antioxidative substances in sesame seed. *Agric. Biol. Chem.* 49:301–306.

Fukuda, Y., M. Nagata, T. Osawa, and M. Namiki. 1986. A contribution of lignan analogues to antioxidative activity of refined unroasted sesame seed oil. *J. Am. Oil Chem. Soc.* 63:1027–1031.

Fukuhara, Y., N. Kamimura, M. Nakajima, et al. 2013. Discovery of pinoresinol reductase genes in sphingomonads. *Enzyme Microb. Technol.* 52:38–43.

Gang, D. R., M. A. Costa, M. Fujita, et al. 1999. Regiochemical control of monolignol radical coupling: A new paradigm for lignin and lignan biosynthesis. *Chem. Biol.* 6:143–151.

Gardner, J. A. F., E. P. Swan, S. A. Sutherland, and H. MacLean. 1966. The polyoxyphenols of western red cedar (*Thuja plicata* Donn) III. Structure of plicatic acid. *Can. J. Chem.* 44:52–58.

Ghose, K., K. Selvaraj, J. McCallum, et al. 2014. Identification and functional characterization of a flax UDP-glycosyltransferase glucosylating secoisolariciresinol (SECO) into secoisolariciresinol monoglucoside (SMG) and diglucoside (SDG). *BMC Plant Biol.* 14:82.

Gonzalez, A. G., R. Estevez-Reyes, and A. M. Estevez-Braun. 1990. Buplerol and guayarol, new lignans from the seeds of *Bupleurum salicifolium*. *J. Chem. Res., Synop.*:220–221.

Gordaliza, M., P. A. García, J. M. Miguel del Corral, M. A. Castro, and M. A. Gómez-Zurita. 2004. Podophyllotoxin: Distribution, sources, applications and new cytotoxic derivatives. *Toxicon* 44:441–459.

Greger, H., and O. Hofer. 1980. New unsymmetrically substituted tetrahydrofurofuran lignans from *Artemisia absinthium*. Assignment of the relative stereochemistry by lanthanide induced chemical shifts. *Tetrahedron* 36:3551–3558.

Hande, K. R. 1998. Etoposide: Four decades of development of a topoisomerase II inhibitor. *Eur. J. Cancer* 34:1514–1521.

Hano, C., I. Martin, O. Fliniaux, et al. 2006. *Pinoresinol-lariciresinol reductase* gene expression and secoisolariciresinol diglucoside accumulation in developing flax (*Linum usitatissimum*) seeds. *Planta* 224:1291–1301.

Heinig, U., S. Scholz, and S. Jennewein. 2013. Getting to the bottom of Taxol biosynthesis by fungi. *Fungal Diversity* 60:161–170.

Hemmati, S., T. J. Schmidt, and E. Fuss. 2007. (+)-Pinoresinol/(–)-lariciresinol reductase from *Linum perenne* Himmelszelt involved in the biosynthesis of justicidin B. *FEBS Lett.* 581:603–610.

Hemmati, S., C. B. I. von Heimendahl, M. Klaes, et al. 2010. Pinoresinol-lariciresinol reductases with opposite enantiospecificity determine the enantiomeric composition of lignans in the different organs of *Linum usitatissimum* L. *Planta Med.* 76:928–934.

Hemmati, S., and H. Seradj. 2016. Justicidin B: A promising bioactive lignan. *Molecules* 21:820.

Hwang, J. K., S. G. A. Moinuddin, L. B. Davin, and N. G. Lewis. 2020. Pinoresinol-lariciresinol reductase: Substrate versatility, enantiospecificity, and kinetic properties. *Chirality* 32:770–789.

Ikezawa, N., K. Iwasa, and F. Sato. 2007. Molecular cloning and characterization of methylenedioxy bridge-forming enzymes involved in stylopine biosynthesis in *Eschscholzia californica*. *FEBS J.* 274:1019–1035.

Jiao, Y., L. B. Davin, and N. G. Lewis. 1998. Furanofuran lignan metabolism as a function of seed maturation in *Sesamum indicum*: Methylenedioxy bridge formation. *Phytochemistry* 49:387–394.

Kamal-Eldin, A., and G. Yousif. 1992. A furofuran lignan from *Sesamum alatum*. *Phytochemistry* 31:2911–2912.

Kamal-Eldin, A., N. Peerlkamp, P. Johnsson, et al. 2001. An oligomer from flaxseed composed of secoisolariciresinoldiglucoside and 3-hydroxy-3-methyl glutaric acid residues. *Phytochemistry* 58:587–590.

Katayama, T., L. B. Davin, and N. G. Lewis. 1992. An extraordinary accumulation of (–)-pinoresinol in cell-free extracts of *Forsythia intermedia*: Evidence for enantiospecific reduction of (+)-pinoresinol. *Phytochemistry* 31:3875–3881.

Katayama, T., L. B. Davin, A. Chu, and N. G. Lewis. 1993. Novel benzylic ether reductions in lignan biogenesis in *Forsythia intermedia*. *Phytochemistry* 33:581–591.

Kato, M. J., A. Chu, L. B. Davin, and N. G. Lewis. 1997. Biosynthesis of antioxidant lignans in *Sesamum indicum* seeds. *Phytochemistry* 47:583–591.

Katsuzaki, H., S. Kawakishi, and T. Osawa. 1994. Sesaminol glucosides in sesame seeds. *Phytochemistry* 35:773–776.

Kim, H. J., E. Ono, K. Morimoto, et al. 2009. Metabolic engineering of lignan biosynthesis in *Forsythia* cell culture. *Plant Cell Physiol.* 50:2200–2209.

Kim, K.-W., S. G. A. Moinuddin, K. M. Atwell, et al. 2012. Opposite stereoselectivities of dirigent proteins in *Arabidopsis* and *Schizandra* species. *J. Biol. Chem.* 287:33957–33972.

Kim, K.-W., C. A. Smith, M. D. Daily, et al. 2015. Trimeric structure of (+)-pinoresinol forming dirigent protein at 1.95 Å resolution with three isolated active sites. *J. Biol. Chem.* 290:1308–1318.

Kim, M. K., J.-H. Jeon, M. Fujita, L. B. Davin, and N. G. Lewis. 2002. The western red cedar (*Thuja plicata*) 8–8′ *DIRIGENT* family displays diverse expression patterns and conserved monolignol coupling specificity. *Plant Mol. Biol.* 49:199–214.

Konno, C., Z.-Z. Lu, H.-Z. Xue, et al. 1990. Furanoid lignans from *Larrea tridentata*. *J. Nat. Prod.* 53:396–406.

Konuklugil, B., I. Ionkova, N. Vasilev, et al. 2007. Lignans from *Linum* species of sections *Syllinum* and *Linum*. *Nat. Prod. Res.* 21:1–6.

Koulman, A., and B. Konuklugil. 2004. Lignan profile of *Linum meletonis*. *Biochem. Syst. Ecol.* 32:91–93.

Kranz, K., and M. Petersen. 2003. β-peltatin 6-*O*-methyltransferase from suspension cultures of *Linum nodiflorum*. *Phytochemistry* 64:453–458.

Lau, W., and E. S. Sattely. 2015. Six enzymes from mayapple that complete the biosynthetic pathway to the etoposide aglycone. *Science* 349:1224–1228.

Lewis, N. G., and L. B. Davin. 1999. Lignans: Biosynthesis and function. In *Comprehensive Natural Products Chemistry. Vol. 1: Polyketides and Other Secondary Metabolites, Including Fatty Acids and Their Derivatives* (Sir D. H. R. Barton, K. Nakanishi and O. Meth-Cohn eds.). pp. 639–712.

Liu, J., R. D. Stipanovic, A. A. Bell, L. S. Puckhaber, and C. W. Magill. 2008. Stereoselective coupling of hemigossypol to form (+)-gossypol in moco cotton is mediated by a dirigent protein. *Phytochemistry* 69:3038–3042.

Loike, J. D. 1982. VP16-213 and podophyllotoxin. A study on the relationship between chemical structure and biological activity. *Cancer Chemother. Pharmacol.* 7:103–111.

Marchand, P. A., M. J. Kato, and N. G. Lewis. 1997. (+)-Episesaminone, a *Sesamum indicum* furofuran lignan. Isolation and hemisynthesis. *J. Nat. Prod.* 60:1189–1192.

Marques, J. V., K.-W. Kim, C. Lee, et al. 2013. Next generation sequencing in predicting gene-function in podophyllotoxin biosynthesis. *J. Biol. Chem.* 288:466–479.

Meng, Q., S. G. A. Moinuddin, S. J. Kim, et al. 2020. Pterocarpan synthase (PTS) structures suggest a common quinone methide-stabilizing function in dirigent proteins and proteins with dirigent-like domains. *J. Biol. Chem.* 295:11584–11601.

Min, T., H. Kasahara, D. L. Bedgar, et al. 2003. Crystal structures of pinoresinol-lariciresinol and phenylcoumaran benzylic ether reductases and their relationship to isoflavone reductases. *J. Biol. Chem.* 278:50714–50723.

Moazzami, A. A., R. E. Andersson, and A. Kamal-Eldin. 2006. HPLC analysis of sesaminol glucosides in sesame seeds. *J. Agric. Food Chem.* 54:633–638.

Mohagheghzadeh, A., T. J. Schmidt, and A. W. Alfermann. 2002. Arylnaphthalene lignans from *in vitro* cultures of *Linum austriacum*. *J. Nat. Prod.* 65:69–71.

Mohagheghzadeh, A., T. J. Schmidt, Ü. Bayindir, et al. 2006. Diarylbutyrolactone lignans from *Linum corymbulosum in vitro* cultures. *Planta Med.* 72:1165–1167.

Moinuddin, S. G. A., S. Hishiyama, M.-H. Cho, L. B. Davin, and N. G. Lewis. 2003. Synthesis and chiral HPLC analysis of the dibenzyltetrahydrofuran lignans, larreatricins, 8′-epi-larreatricins, 3,3′-didemethoxyverrucosins and

meso-3,3′-didemethoxynectandrin B in the creosote bush (*Larrea tridentata*): Evidence for regiospecific control of coupling. *Org. Biomol. Chem.* 1:2307–2313.
Moinuddin, S. G. A., B. Youn, D. L. Bedgar, et al. 2006. Secoisolariciresinol dehydrogenase: Mode of catalysis and stereospecificity of hydride transfer in *Podophyllum peltatum*. *Org. Biomol. Chem.* 4:808–816.
Moinuddin, S. G. A., J. R. Cort, C. A. Smith, et al. 2019. Linum lignan and associated biochemical pathways in human health and plant defense. In *Genetics and Genomics of Linum* (C. A. Cullis ed.), Springer, Cham, Switzerland. pp 167–193.
Molog, G. A., U. Empt, S. Kuhlmann, et al. 2001. Deoxypodophyllotoxin 6-hydroxylase, a cytochrome P450 monooxygenase from cell cultures of *Linum flavum* involved in the biosynthesis of cytotoxic lignans. *Planta* 214:288–294.
Murata, J., E. Matsumoto, K. Morimoto, T. Koyama, and H. Satake. 2015. Generation of triple-transgenic *Forsythia* cell cultures as a platform for the efficient, stable, and sustainable production of lignans. *PLoS One* 10:e0144519.
Murata, J., E. Ono, S. Yoroizuka, et al. 2017. Oxidative rearrangement of (+)-sesamin by CYP92B14 co-generates twin dietary lignans in sesame. *Nat. Commun.* 8:2155.
Nakatsubo, T., M. Mizutani, S. Suzuki, T. Hattori, and T. Umezawa. 2008. Characterization of *Arabidopsis thaliana* pinoresinol reductase, a new type of enzyme involved in lignan biosynthesis. *J. Biol. Chem.* 283:15550–15557.
Nishibe, S., A. Sakushima, S. Kitagawa, et al. 1988. Phenolic compounds from *Forsythia* leaves (III) On the comparison of constituents between hybrid and parents. *Shoyakugaku Zasshi* 42:324–328.
Noguchi, A., Y. Fukui, A. Iuchi-Okada, et al. 2008. Sequential glucosylation of a furofuran lignan, (+)-sesaminol, by *Sesamum indicum* UGT71A9 and UGT94D1 glucosyltransferases. *Plant J.* 54:415–427.
Ohta, K., Y.-L. Chen, S. Marumo, and K. Munakata. 1969. Studies on the piscicidal components of *Justicia hayatai* var. *decumbens*. *Agric. Biol. Chem.* 33:610–614.
Ono, E., M. Nakai, Y. Fukui, et al. 2006. Formation of two methylenedioxy bridges by a *Sesamum* CYP81Q protein yielding a furofuran lignan, (+)-sesamin. *Proc. Natl. Acad. Sci. USA* 103:10116–10121.
Ono, E., H. J. Kim, J. Murata, et al. 2010. Molecular and functional characterization of novel furofuran-class lignan glucosyltransferases from *Forsythia*. *Plant Biotechnol.* 27:317–324.
Ono, E., J. Murata, H. Toyonaga, et al. 2018. Formation of a methylenedioxy bridge in (+)-epipinoresinol by CYP81Q3 corroborates with diastereomeric specialization in sesame lignans. *Plant Cell Physiol.* 59:2278–2287.
Osawa, T., M. Nagata, M. Namiki, and Y. Fukuda. 1985. Sesamolinol, a novel antioxidant isolated from sesame seeds. *Agric. Biol. Chem.* 49:3351–3352.
Ottaggio, L., F. Bestoso, A. Armirotti, et al. 2008. Taxanes from shells and leaves of *Corylus avellana*. *J. Nat. Prod.* 71:58–60.
Ozawa, S., L. B. Davin, and N. G. Lewis. 1993. Formation of (−)-arctigenin in *Forsythia intermedia*. *Phytochemistry* 32:643–652.
Pickel, B., M.-A. Constantin, J. Pfannstiel, et al. 2010. An enantiocomplementary dirigent protein for the enantioselective laccase-catalyzed oxidative coupling of phenols. *Angew. Chem. Int. Ed.* 49:202–204.
Prabhu, B. R., and N. B. Mulchandani. 1985. Lignans from *Piper cubeba*. *Phytochemistry* 24:329–331.
Ragamustari, S. K., T. Nakatsubo, T. Hattori, et al. 2013. A novel *O*-methyltransferase involved in the first methylation step of yatein biosynthesis from matairesinol in *Anthriscus sylvestris*. *Plant Biotechnol.* 30:375–384.
Ragamustari, S. K., M. Yamamura, E. Ono, et al. 2014. Substrate-enantiomer selectivity of matairesinol *O*-methyltransferases. *Plant Biotechnol.* 31:257–267.
Rahman, M. M. A., P. M. Dewick, D. E. Jackson, and J. A. Lucas. 1990. Lignans of *Forsythia intermedia*. *Phytochemistry* 29:1971–1980.
Rose, A. S., A. R. Bradley, Y. Valasatava, et al. 2018. NGL viewer: Web-based molecular graphics for large complexes. *Bioinformatics* 34:3755–3758.
Rudney, H. 1957. Biosynthesis of β-hydroxy-β-methylglutaric acid. *J. Biol. Chem.* 227:363–377.
Ryu, S. N., C.-T. Ho, and T. Osawa. 1998. High performance liquid chromatographic determination of antioxidant lignan glycosides in some varieties of sesame. *J. Food Lipids* 5:17–28.
Sakakibara, N., S. Suzuki, T. Umezawa, and M. Shimada. 2003. Biosynthesis of yatein in *Anthriscus sylvestris*. *Org. Biomol. Chem.* 1:2474–2485.
Sakakibara, N., T. Nakatsubo, S. Suzuki, et al. 2007. Metabolic analysis of the cinnamate/monolignol pathway in *Carthamus tinctorius* seeds by a stable-isotope-dilution method. *Org. Biomol. Chem.* 5:802–815.
Schmidt, T. J., S. Hemmati, E. Fuss, and A. W. Alfermann. 2006. A combined HPLC-UV and HPLC-MS method for the identification of lignans and its application to the lignans of *Linum usitatissimum* L. and *L. bienne* Mill. *Phytochem. Anal.* 17:299–311.
Schmidt, T. J., S. Vößing, M. Klaes, and S. Grimme. 2007. An aryldihydronaphthalene lignan with a novel type of ring system and further new lignans from *Linum perenne* L. *Planta Med.* 73:1574–1580.
Schmidt, T. J., A. W. Alfermann, and E. Fuss. 2008. High-performance liquid chromatography/mass spectrometric identification of dibenzylbutyrolactone-type lignans: Insights into electrospray ionization tandem mass spectrometric fragmentation of lign-7-eno-9,9′-lactones and application to the lignans of *Linum usitatissimum* L. (common flax). *Rapid Commun. Mass Spectrom.* 22:3642–3650.
Schultz, B. J., S. Y. Kim, W. Lau, and E. S. Sattely. 2019. Total biosynthesis for milligram-scale production of etoposide intermediates in a plant chassis. *J. Am. Chem. Soc.* 141:19231–19235.
Stierle, A., G. Strobel, and D. Stierle. 1993. Taxol and taxane production by *Taxomyces andreanae*, an endophytic fungus of Pacific yew. *Science* 260:214–216.
Suzuki, S. 2002. Stereochemical diversity in lignan biosynthesis and establishment of norlignan biosynthetic pathway. *Wood Res.* 89:52–60.
Suzuki, S., N. Sakakibara, T. Umezawa, and M. Shimada. 2002. Survey and enzymatic formation of lignans of *Anthriscus sylvestris*. *J. Wood Sci.* 48:536–541.
Tazaki, H., T. Hayashida, F. Ishikawa, et al. 1999. Lignan biosynthesis in liverworts *Jamesoniella autumnalis* and *Lophocolea heterophylla*. *Tetrahedron Lett.* 40:101–104.
Teoh, K. H., J. D. Ford, M.-R. Kim, L. B. Davin, and N. G. Lewis. 2003. Delineating the metabolic pathway(s) to secoisolariciresinol diglucoside hydroxymethyl glutarate oligomers in flaxseed (*Linum usitatissimum*). In *Flaxseed in Human Nutrition* (L. U. Thompson and S. C. Cunnane eds.), 2nd Ed., AOCS Press, Champaign, IL. pp. 41–62.

Uchida, K., T. Akashi, and T. Aoki. 2017. The missing link in leguminous pterocarpan biosynthesis is a dirigent domain-containing protein with isoflavanol dehydratase activity. *Plant Cell Physiol.* 58:398–408.

Umezawa, T., L. B. Davin, and N. G. Lewis. 1990a. Formation of the lignan, (–)-secoisolariciresinol by cell-free extracts of *Forsythia intermedia. Biochem. Biophys. Res. Commun.* 171:1008–1014.

Umezawa, T., L. B. Davin, E. Yamamoto, D. G. I. Kingston, and N. G. Lewis. 1990b. Lignan biosynthesis in *Forsythia* species. *J. Chem. Soc., Chem. Commun.*:1405–1408.

Umezawa, T., L. B. Davin, and N. G. Lewis. 1991. Formation of lignans, (–)-secoisolariciresinol and (–)-matairesinol with *Forsythia intermedia* cell-free extracts. *J. Biol. Chem.* 266:10210–10217.

Umezawa, T., S. K. Ragamustari, T. Nakatsubo, et al. 2013. A lignan *O*-methyltransferase catalyzing the regioselective methylation of matairesinol in *Carthamus tinctorius. Plant Biotechnol.* 30:97–109.

Vardapetyan, H. R., A. B. Kirakosyan, A. A. Oganesyan, A. R. Penesyan, and W. A. Alfermann. 2003. Effect of various elicitors on lignan biosynthesis in callus cultures of *Linum austriacum. Russian J. Plant Physiol.* 50:297–300.

Vasilev, N., and I. Ionkova. 2005a. Lignan production by cell cultures of *Linum setaceum* and *Linum campanulatum. Pharm. Biol.* 43:509–511.

Vasilev, N., Elfahmi, R. Bos, et al. 2006. Production of justicidin B, a cytotoxic arylnaphthalene lignan from genetically transformed root cultures of *Linum leonii. J. Nat. Prod.* 69:1014–1017.

Vasilev, N. P., and I. Ionkova. 2005b. Cytotoxic activity of extracts from *Linum* cell cultures. *Fitoterapia* 76:50–53.

Vassão, D. G., K.-W. Kim, L. B. Davin, and N. G. Lewis. 2010. Lignans (neolignans) and allyl/propenyl phenols: Biogenesis, structural biology, and biological/human health considerations. In *Comprehensive Natural Products II Chemistry and Biology* (L. N. Mander and H.-W. Liu eds.), Vol. 1: *Structural Diversity I*, Elsevier, Oxford, UK. pp. 815–928.

von Heimendahl, C. B. I., K. M. Schäfer, P. Eklund, et al. 2005. Pinoresinol-lariciresinol reductases with different stereospecificity from *Linum album* and *Linum usitatissimum. Phytochemistry* 66:1254–1263.

Wang, L., S. Yu, C. Tong, et al. 2014. Genome sequencing of the high oil crop sesame provides insight into oil biosynthesis. *Genome Biol.* 15:R39.

Wang, Z., N. Hobson, L. Galindo, et al. 2012. The genome of flax (*Linum usitatissimum*) assembled *de novo* from short shotgun sequence reads. *Plant J.* 72:461–473.

Xia, Z.-Q., M. A. Costa, J. Proctor, L. B. Davin, and N. G. Lewis. 2000. Dirigent-mediated podophyllotoxin biosynthesis in *Linum flavum* and *Podophyllum peltatum. Phytochemistry* 55:537–549.

Xia, Z.-Q., M. A. Costa, H. C. Pélissier, L. B. Davin, and N. G. Lewis. 2001. Secoisolariciresinol dehydrogenase purification, cloning and functional expression: Implications for human health protection. *J. Biol. Chem.* 276:12614–12623.

Xiao, Y., Q. Ji, S. Gao, et al. 2015. Combined transcriptome and metabolite profiling reveals that IiPLR1 plays an important role in lariciresinol accumulation in *Isatis indigotica. J. Exp. Bot.* 66:6259–6271.

Yang, Y., H. Zhao, R. A. Barrero, et al. 2014. Genome sequencing and analysis of the paclitaxel-producing endophytic fungus *Penicillium aurantiogriseum* NRRL 62431. *BMC Genomics* 15:69.

Yonekura-Sakakibara, K., M. Yamamura, F. Matsuda, et al. 2021. Seed-coat protective neolignans are produced by the dirigent protein AtDP1 and the laccase AtLAC5 in Arabidopsis. *Plant Cell* 33:129–152.

Yoshiki, Y., and T. Ishiguro. 1933. Über die krystallisierten Bestandteile des Hinokiöls. *Yakugaku Zasshi* 53:73–151.

Youn, B., S. G. A. Moinuddin, L. B. Davin, N. G. Lewis, and C. Kang. 2005. Crystal structures of apo-form and binary/ternary complexes of *Podophyllum* secoisolariciresinol dehydrogenase, an enzyme involved in formation of health-protecting and plant defense lignans. *J. Biol. Chem.* 280:12917–12926.

Zhang, P., P.-P. Zhou, and L.-J. Yu. 2009a. An endophytic taxol-producing fungus from *Taxus* x *media, Aspergillus candidus* MD3. *FEMS Microbiol. Lett.* 293:155–159.

Zhang, P., P.-P. Zhou, and L.-J. Yu. 2009b. An endophytic taxol-producing fungus from *Taxus media, Cladosporium cladosporioides* MD2. *Curr. Microbiol.* 59:227–232.

Zhao, Q., Y. Zeng, Y. Yin, et al. 2015. Pinoresinol reductase 1 impacts lignin distribution during secondary cell wall biosynthesis in *Arabidopsis. Phytochemistry* 112:170–178.

Zhuang, L.-g., O. Seligmann, K. Jurcic, and H. Wagner. 1982. Constituents of *Daphne tangutica. Planta Med.* 45:172–176.

6 8–2′-, 8–3′-, 8–5′-, and 8–*O*–4′-Linked Lignan Biosynthesis Roadmap

Laurence B. Davin and Norman G. Lewis

In contrast to the 8–8′-linked lignan biosynthesis Roadmap, current understanding of the biosynthesis of 8–2′-, 8–3′-, 8–5′-, and 8–*O*–4′-linked lignans is not as well developed at present. Nevertheless, this contribution summarizes the current knowledge of what is known about their biosynthesis. Of these lignan sub-classes, the occurrence of 8–5′- and 8–*O*–4′-linked lignans in the plant kingdom appears to be near ubiquitous in vascular plants.

Each of these differentially linked lignan sub-classes are described separately in the following sections.

6.1 8–2′-Linked Lignan Biosynthesis Roadmap

The fern species, commonly known as deer fern or hard-fern (*Blechnum spicant* in the family Blechnaceae), accumulates various optically active 8–2′-linked lignans. Examples include (–)-(*E*/*Z*)-blechnic acids (**1b**/**2b**), and its derivative (–)-*E*-brainic acid (**3b**) (Figure 6.1A). These are structurally atypical of most other lignan sub-classes, in that the coupling mode (8–2′) is uncommon, and they also lack aromatic ring *O*-methylation functionalities. Based on their chemical structures, they can provisionally be considered as derived from stereoselective coupling of either *trans*- or *cis*- caffeic acids (**7**), although other possibilities cannot be excluded such as demethylation of OMe groups in the initial coupling product. The proposed mechanism of their formation, assuming a one-electron oxidation process is in effect and a hydroxycinnamic acid precursor, is shown in Figure 6.1B.

In work carried out to date, it was shown that (*E*/*Z*)-blechnic acid (**1b**/**2b**), and its derivative (–)-*E*-brainic acid (**3b**), were biosynthesized from [9-^{14}C]-cinnamic (**4**), [8-^{14}C]- and [8-^{13}C]-*p*-coumaric (**5**), and [8-^{13}C]-caffeic (**7**) acid precursors *in vivo*, as well as from [U-^{14}C]-, [1-^{13}C]-, [2-^{13}C]-, and [3-^{13}C]-Phe (**6**) (Davin et al. 2003). While provisionally these optically active lignans might be derived from dirigent protein (DP)-engendered coupling, with (–)-*Z*-blechnic acid (**2b**) perhaps being the initial coupling product (Figure 6.1B), this has not yet been established at either the gene or protein/enzyme level.

6.2 8–3′-Linked Lignan Biosynthesis Roadmap

The *Piper* genus, such as *Piper regnellii* in the Piperaceae, is well-known to contain various optically active lignans. Of these, *P. regnellii* leaf and root tissues accumulate optically active 8–3′-linked lignans, (+)-conocarpan (**8a**, Figure 6.2A), (–)-*epi*-conocarpan (**9b**), and fully conjugated derivatives thereof (e.g. **10**) (Sartorelli et al. 2001). These lignans, lacking any oxygenated functionality at C-9/C-9′, are provisionally considered derived from stereo-selective coupling of two *p*-anol (**11**) moieties (Figure 6.2B). Preliminary studies have been carried out to investigate this question, although again the nature of the proteins and enzymes involved in the coupling step has not yet been reported. Nevertheless, perhaps this is also a DP-mediated process, but whether the coupling *in vivo* employs an allyl/propenyl phenol or a monolignol has not been shown to date.

6.3 8–*O*–4′-Linked Lignan Biosynthesis Roadmap

While this type of inter-unit linkage is considered to be prevalent in lignin biopolymers, how the corresponding 8–*O*–4′-linked lignans are biosynthesized has also been studied. To date, such investigations have mainly focused on the small tree *Eucommia ulmoides,* a member of the family Eucommiaceae. Thus, using a presumed insoluble cell wall residue preparation from its stems, coniferyl alcohol (**12**) was converted into the optically active 8–*O*–4′-linked diastereomers (+)-*erythro*- and (–)-*threo*-guaiacylglycerol 8–*O*–4′ coniferyl alcohol ethers (**14a** and **14b**), respectively, in an ~3:2 ratio, provided H_2O_2 was present as a co-factor (Figure 6.3) (Katayama and Kado 1998).

In an analogous manner, when sinapyl alcohol (**13**) was employed as a substrate, both (+)-*erythro*- and (–)-*threo*-syringylglycerol 8–*O*–4′ sinapyl alcohol ethers (**15a** and **15b**, respectively), were produced in an ~10:1 ratio (Lourith et al. 2005a).

Furthermore, when young excised *E. ulmoides* shoots were provided either coniferyl (**12**) or sinapyl (**13**) alcohols, guaiacylglycerol 8–*O*–4′-sinapyl alcohol ether (**16**) was obtained in a *circa* 2.4:1 *erythro*/*threo* ratio (Lourith et al. 2005b).

As before, how this control over monolignol coupling occurs, including in also defining the stereo-chemistries at positions 7 and 8, is yet to be established. It is, however, tempting to speculate that DP protein(s) is (are) involved.

In various plant species, the dimeric 8–*O*–4′- linked lignans formed via initial coupling, can then undergo various modifications, such as side-chain reductions, glucosylations, etc. The most highly studied of these are the reductive side-chain modifications, such as that in (+)-ligraminol E (**17a,** Figure 6.4) (Kim et al. 2011). In this example, the C7′/ C8′ olefinic functionality is reduced, as well as there being no oxygenation at C7. This side-chain reductive step (such as at

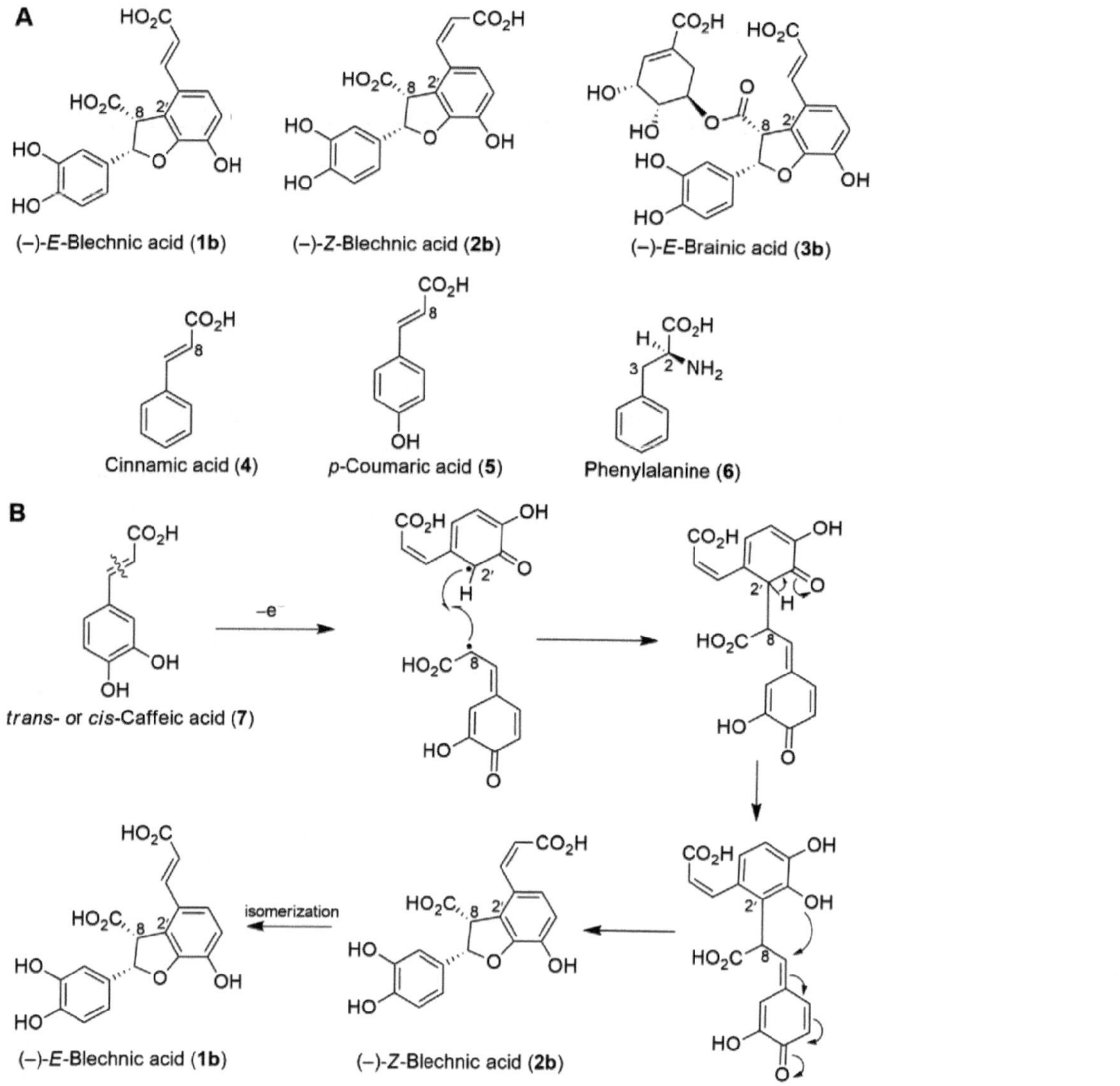

FIGURE 6.1 (–)-*E/Z*-Blechnic acid (**1b/2b**) putative biosynthesis. **A.** Structures of (–)-*E/Z*-blechnic acid (**1b/2b**), (–)-*E*-brainic acid (**3b**) and related compounds. **B.** Putative formation of (–)-*E/Z*-blechnic acids (**1b/2b**) from caffeic acid (**7**) precursor. "**b**" in compound numbers refers to the (–)-antipode.

C7′/ C8′) is quite commonly observed, not only in dimeric 8–*O*–4′-linked lignans, but also with various monomeric phenylpropanoid and dimeric 8–5′-linked lignans. What is known about this side-chain reductive process is summarized in the section below.

6.4 8–5′-Linked Lignan Biosynthesis Roadmap

8-5′-Linked lignans are apparently near ubiquitous throughout the vascular plant kingdom. To date, however, no DP has been reported in controlling 8–5′-coupling, such as of coniferyl alcohol (**12**) into dehydrodiconiferyl alcohol (**18a/b**).

Instead, most of the research work with this lignan subclass has focused upon various downstream modifications, which can include: (a) side-chain reductions such as to (+)-cedrusin (**19a**), and (+)-*trans*-dihydrodehydrodiconiferyl alcohol (**21a**) (Figure 6.5) (Agrawal et al. 1980, Kraus and Spiteller 1997), as well as analogous processes to various other monomeric and dimeric (side-chain-reduced) phenylpropanoids; (b) reduction of the dihydro-tetrafuran substructure in, for example, dehydrodiconiferyl alcohols (**18a/b**) to provide isodihydrodehydrodiconiferyl alcohols (**20a/b**); (c) glucosylation reactions, and (d) what appear to be de-methylation of methoxyl groups, e.g. (+)-cedrusin (**19a**).

Of these, the side-chain reductive modifications to various dihydrolignans (8–5′- and 8–*O*–4′- linked) and corresponding reduced monomers, are the most studied. Perhaps the earliest known example of this modification in terrestrial plants, from an evolutionary perspective, is from the fern (*Pteris vittata*), which accumulates the 8–5′-linked, side-chain-reduced, lignan **22b** (Satake et al. 1978).

Such side-chain modifications are also frequently found throughout the gymnosperms and the angiosperms. For example, in gymnosperm *Picea glauca* shoot and gall tissues, the

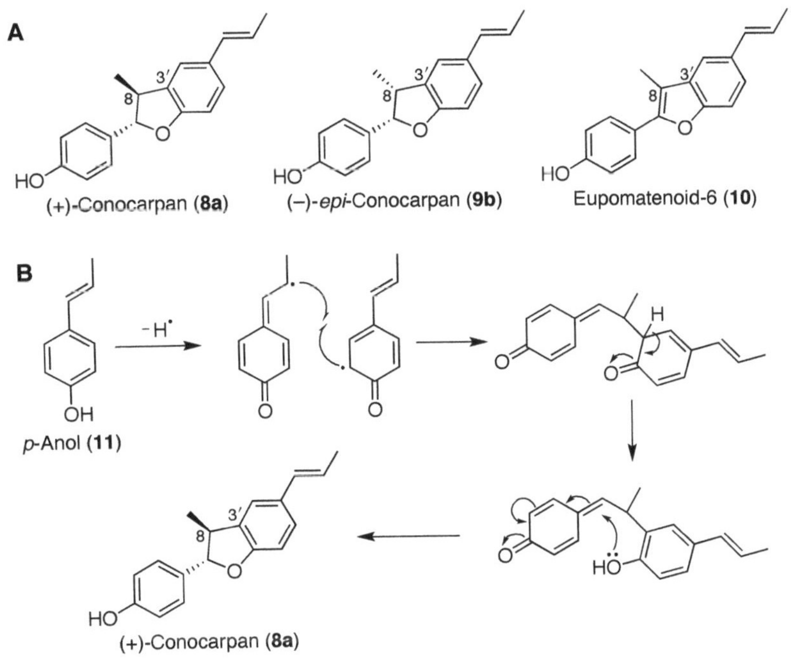

FIGURE 6.2 **A.** (+)-Conocarpan (**8a**), (–)-*epi*-conocarpan (**9b**) and eupomatenoid-6 (**10**). **B.** Putative biosynthesis of (+)-conocarpan (**8a**) from *p*-anol (**11**). "**a**" and "**b**" in compound numbers refer to the (+)- and (–)-antipodes, respectively.

presumed plant defense metabolites (+)-*trans*-dihydrodehydrodiconiferyl alcohol (**21a**), its demethylated homologue (+)-cedrusin (**19a**), dihydroconiferyl alcohol (**23**), dihydro-*p*-coumaric (**24**) acid, dihydroferulic acid (**26**), and their glucosides **25/27**, have been reported (Kraus and Spiteller 1997). Presence of (+)-cedrusin (**19a**) suggests the involvement of a demethylation step, perhaps being cytochrome P450 catalyzed.

6.4.1 Loblolly Pine and Putative *Arabidopsis* Phenylpropenal Double-Bond Reductases (PPDBRs)

Cell suspension cultures of loblolly pine (*Pinus taeda* in the Pinaceae) have long been known to produce "lignin-like" phenolic substances extracellularly when the cultures are elicited with 8% sucrose solution (Nose et al. 1995). During this process, accumulation of dihydrodehydrodiconiferyl alcohol (**21**) was observed to occur in the cell bathing medium together with a complex mixture of other plant phenolics.

This experimental system presented the opportunity of delineating how side-chain reduction occurred. This led to the finding that there was a nicotinamide adenine dinucleotide phosphate hydrogen (NADPH)-dependent conversion of (±)-dehydrodiconiferyl aldehydes (**28a/b**) and coniferyl aldehyde (**32**) to give the reduced products **29a/b** and **35** (Figure 6.6A), respectively. The corresponding protein was purified to apparent homogeneity, and its gene encoding an ~39 kDa protein was cloned, with the resulting functional recombinant protein, named PtPPDBR (*Pinus taeda* phenylpropenal double-bond reductase), expressed in *Escherichia coli* (Kasahara et al. 2004, Kasahara et al. 2006).

With the *Arabidopsis thaliana* genome having been sequenced in 2000, we deduced that there were also provisionally 11 PtPPDBR homologs in its genome, i.e. a putative 11-membered alkenal double-bond reductase (AtDBR) multigene family (Youn et al. 2006). This was deduced, based on ~63% amino acid homology and ~43% identity to PtPPDBR.

Each of the putative 11-membered *Arabidopsis* multi-gene DBR family was cloned, and their recombinant proteins individually expressed in *E. coli* as His-tag fusion proteins. Of these, only three had detectable catalytic activity with the substrates employed. Trivially named AtDBR1, 8 and 9, respectively, all three recombinant proteins were able to reduce the substrates, *p*-coumaryl (**30**) and coniferyl (**32**) aldehydes, in the presence of NADPH, to afford derivatives **33** and **35**. With both substrates tested *in vitro*, AtDBR1 had K_m, k_{cat}, and k_{cat}/K_m values of 0.53/0.41 mM, 2.82/0.43 s^{-1}, and 5360/1060 M^{-1} s^{-1} (Youn et al. 2006).

AtDBR1 less efficiently reduced caffeyl aldehyde (**31**) (Figure 6.6A), whereas other phenylpropanoid aldehydes, acids, and alcohols did not serve as potential substrates (Youn et al. 2006), i.e. 5-hydroxyconiferyl (**36**) and sinapyl (**37**) aldehydes, carboxylic acids (**5**, **7**, **38–40**), alcohols (**12**, **13**, **41–43**) (Figure 6.6B), and (±)-dehydrodiconiferyl aldehydes (**28a/b**) were not reduced under those conditions.

On the other hand, AtDBR1 could also reduce 4-hydroxy-(2*E*)-nonenal (4-HNE, **45**) to generate 4-hydroxynonanal (4-HNA, **46**, Figure 6.7) *in vitro* (Mano et al. 2002, Mano et al. 2005), but with a lower enzymatic efficacy (K_m of 0.28 mM, k_{cat} of 0.16 s^{-1} and k_{cat}/K_m of 600 M^{-1} s^{-1}) relative to *p*-coumaryl aldehyde (**30**) (Youn et al. 2006).

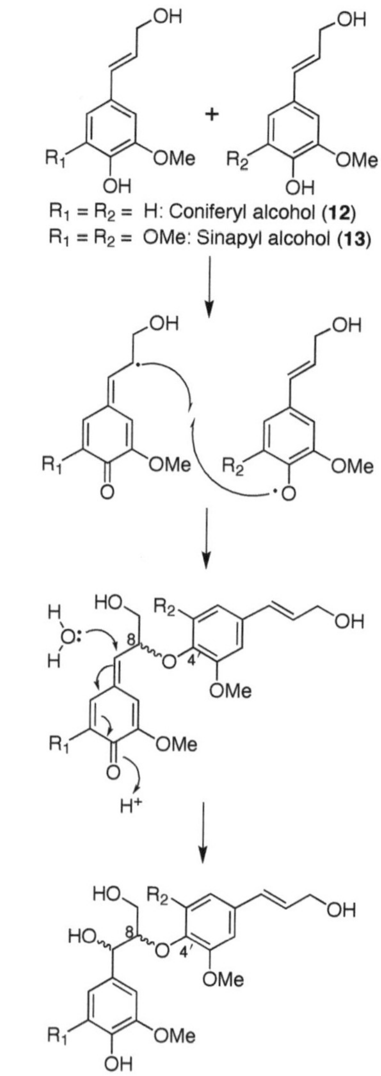

$R_1 = R_2 =$ H: Guaiacylglycerol 8-*O*-4′coniferyl alcohol ether (**14a/b**), *erythro:threo* ratio, 3:2
$R_1 = R_2 =$ OMe: Syringylglycerol 8-*O*-4′sinapyl alcohol ether (**15a/b**), *erythro:threo* ratio, 10:1
$R_1 =$ H, $R_2 =$ OMe: Guaiacylglycerol 8-*O*-4′sinapyl alcohol ether (**16a/b**), *erythro:threo* ratio, 2.4:1

FIGURE 6.3 Proposed biosynthesis of 8–*O*–4′ linked lignans in *Eucommia ulmoides.* "**a**" and "**b**" depict different enantiomeric and diastereomeric forms, with "**a**" for the (+)-forms and "**b**" for the (–)-forms.

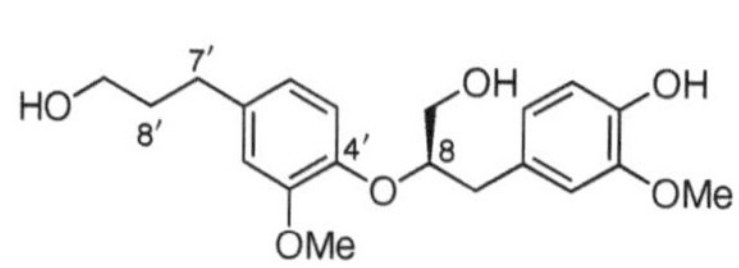

(+)-Ligraminol E (**17a**)

FIGURE 6.4 8–*O*–4′ Linked lignan, (+)-ligraminol E (**17a**) with C7′–C8′ reduced side-chain. "**a**" in compound number refers to the (+)-enantiomer.

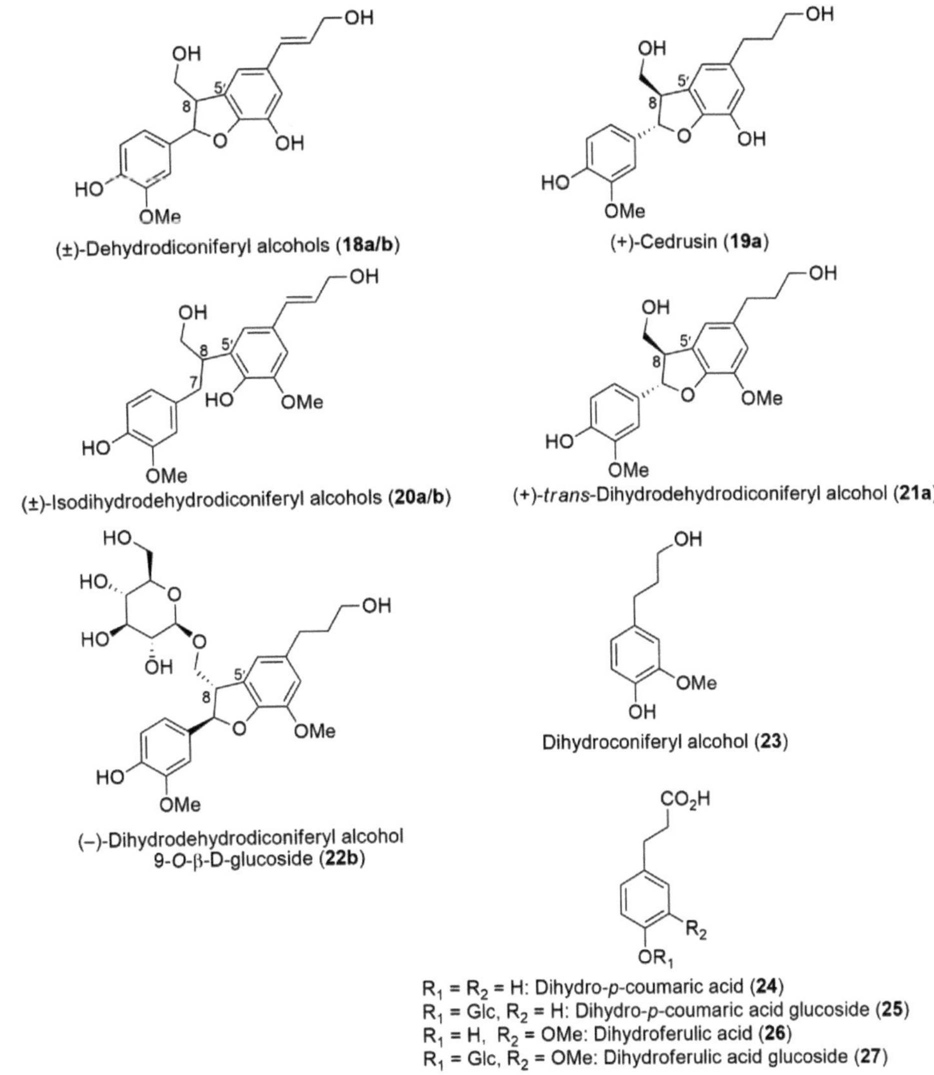

FIGURE 6.5 Representative 8–5′-linked lignans (**18** – **22**) and monomeric C_6C_3 (side-chain reduced) compounds **23** – **27**. "**a**" and "**b**" in compound numbers refer to the (+)- or (–)-enantiomers, respectively, whereas "**a/b**" together indicates racemate.

While these enzymatically catalyzed conversions were demonstrated *in vitro*, it is prudent to continue to caution that the true *Arabidopsis* AtDBR1, 8 and 9 physiological substrate(s) *in vivo* remained uncertain, as for all of the other members of this 11-membered multigene family.

These conversions, however, established the class of enzymes, and nature of substrates (α,β-unsaturated aldehydes), required for these reductive processes.

6.4.1.1 DBR Crystal Structure

Availability of the DBR recombinant proteins, nevertheless, allowed for analysis of their crystal structures and study of their catalytic mechanism. AtDBR1 was obtained in not only apo-form (2J3H) at 2.5 Å resolution, but also as binary (2J3I) and ternary (2J3J and 2J3K) complexes (2.8 Å resolution) with $NADP^+$ either in the absence or presence of putative physiological substrates, *p*-coumaryl aldehyde (**30**, Figure 6.8) and 4-HNE (**45**). A homodimer in open solution (Youn et al. 2006), and a member of the zinc-independent medium-chain dehydrogenase/reductase (MDR) family, it has the well-known nucleotide-binding Gly-rich motif. Its highly conserved MDR residue (Tyr260 in AtDBR1), presumed to function as a general acid during catalysis (see Figures 6.9A and 6.9C), was apparently H-bonded to the substrate aldehyde oxygen.

In mechanistic terms, double-bond migration could then be envisaged to afford an enolate/carbocation intermediate, with subsequent hydride addition (from NADPH) to the carbocation [at either C7 in *p*-coumaryl aldehyde (**30**) or C3 in 4-HNE (**45**)]. It can also be proposed that the carbon adjacent [C8 in *p*-coumaryl aldehyde (**30**) or C2 in 4-HNE (**45**)] is solvent protonated to regenerate the aldehyde functionality. On the other hand, hydride addition [at C7 of *p*-coumaryl aldehyde (**30**) or C3 of 4-HNE (**45**)] could allow for double bond migration and carbonyl oxygen protonation with H^+ from Tyr260 (Figures 6.9B and 6.9D), where the resulting enol product formed would undergo tautomerism to give aldehydes

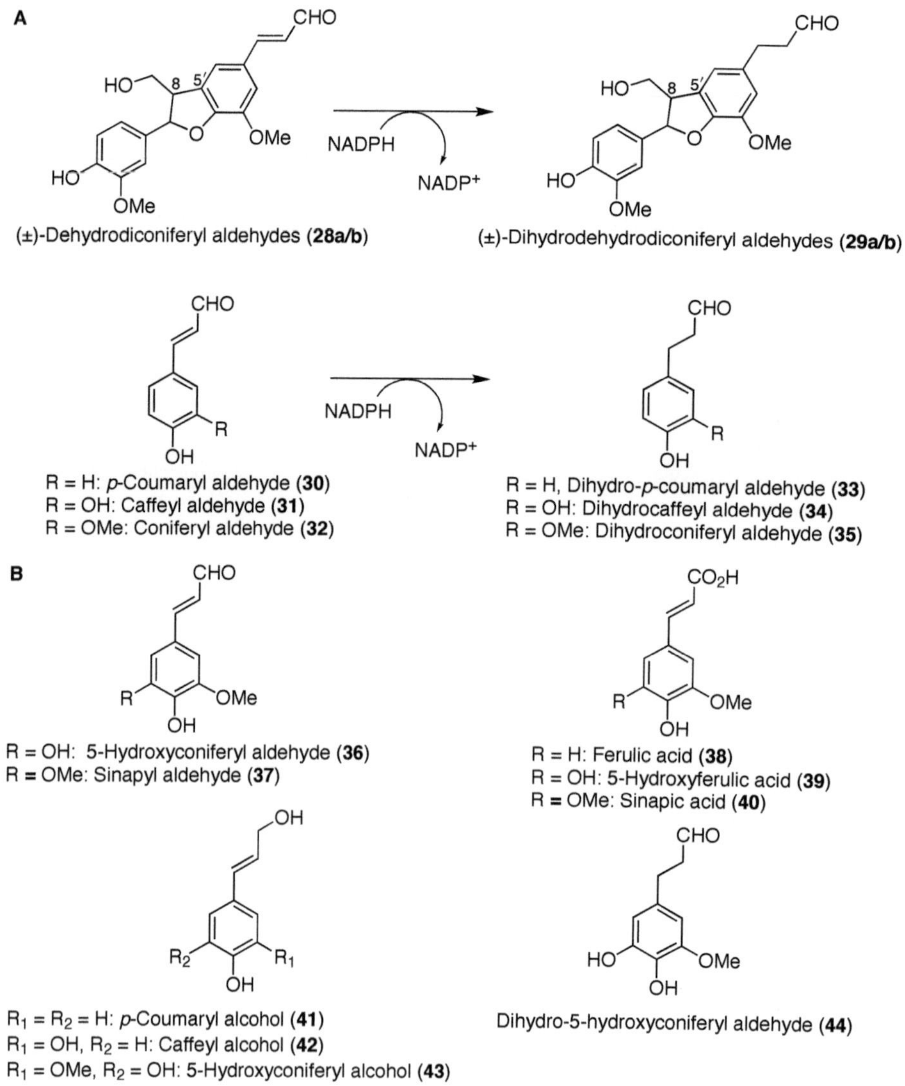

FIGURE 6.6 8–5′-Linked lignan and C_6C_3 aldehydes, acids and alcohols. **A.** Side-chain reduction of 8–5′-linked lignan and C_6C_3 aldehydes. **B.** Various monomeric C_6C_3 aldehydes, acids and alcohols. "**a/b**" depicts racemates.

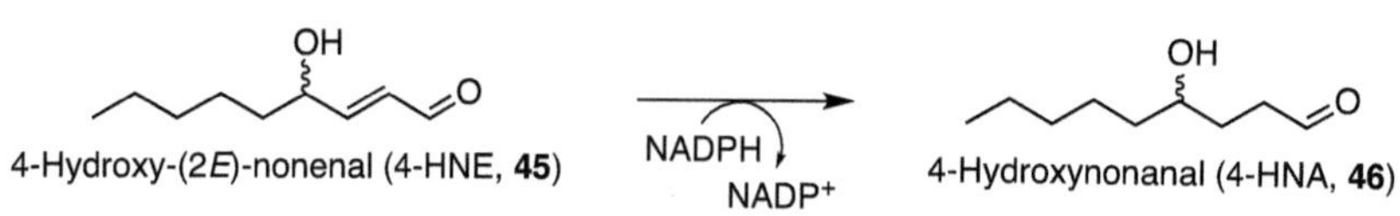

FIGURE 6.7 AtDBR1 catalyzed reduction of 4-hydroxy-(2*E*)-nonenal (4-HNE, **45**).

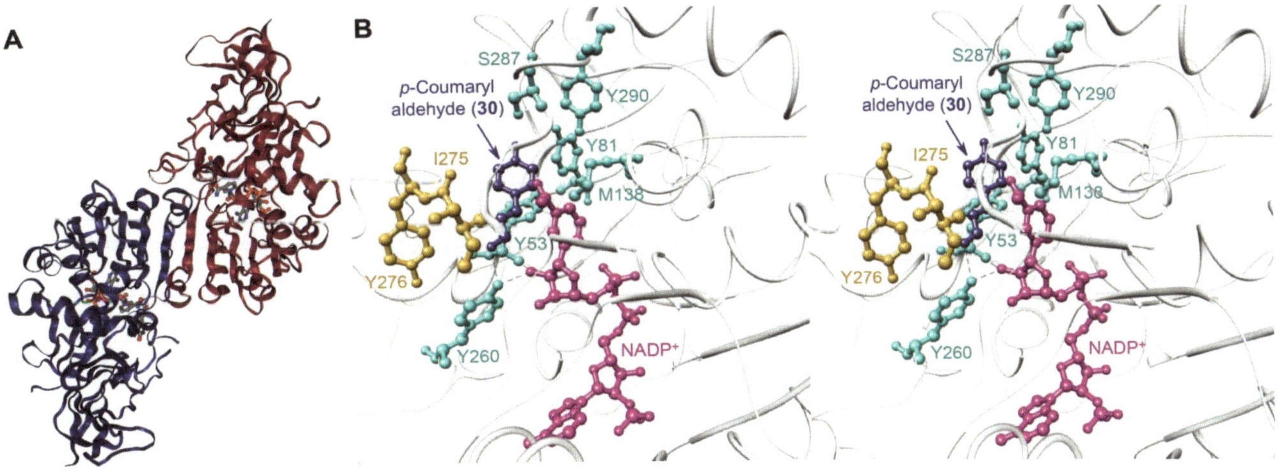

FIGURE 6.8 A. Crystal structure of AtDBR1 ternary complex with *p*-coumaryl aldehyde (**30**) (PDB: ID 2J3J). Image created using NGL viewer (Rose et al. 2018). **B**. Stereo-view of the substrate-binding pocket of AtDBR1 with NADP+ and *p*-coumaryl aldehyde (**30**). This research (B) was originally published in the Journal of Biological Chemistry: Youn et al. (2006). Mechanistic and structural studies of apoform, binary, and ternary complexes of the *Arabidopsis* alkenal double bond reductase At5g16970. *J. Biol. Chem.* 281:40076–40088. Copyright (2006) the American Society for Biochemistry and Molecular Biology.

33 or **46**. While the AtDBR1 crystal structure thus allows for homology modelling of the other DBRs, the overall mechanism is considered established from these studies.

6.4.1.2 DBRs in Liverworts

The liverworts are non-vascular land plants in the division Marchantiophyta, and at least two species, *Plagiochasma appendiculatum* (Wu et al. 2013) and *Marchantia paleacea* (Wu et al. 2018), have been reported to contain DBR homologs. In *P. appendiculatum,* these were trivially named *PaDBR1* and *PaDBR2,* whereas the two *M. paleacea* homologs obtained were called *MpDBR* and *MpMDBRL.*

The 2 *P. appendiculatum* DBR homologs, which differed by 2 amino acid residues, had relatively modest (~43% and ~57%) sequence identity to PtPPDBR and AtDBR1, respectively, as did the two *M. paleacea* DBR homologs with ~43/39% and ~59/43% identity to PtPPDBR and AtDBR1, respectively (Table 6.1). The PaDBR1 was able to catalyze the NADP(H)-dependent conversion of *p*-coumaryl (**30**) and caffeyl (**31**) aldehydes into propyl derivatives **33** and **34**, whereas PaDBR2 was able to use both these as well as aldehydes **32** and **36** to give the corresponding dihydroaldehydes **33–35** and **44** with higher catalytic efficiency as compared to PaDBR1.

Interestingly, site-directed mutation of one of the two different amino acid residues in PaDBR1, i.e. from Cys56 in PaDBR1 to Tyr (as in PaDBR2) gave a mutant, PaDBR1C56T, with the same substrate versatility as PaDBR2 (Wu et al. 2013). The two *M. paleacea* DBR homologues, MpDBR and MpMDBRL, also showed the same substrate specificity as PaDBR2. However, once again, whether these are the true physiological substrates for these liverwort DBRs *in vivo* has yet to be unambiguously established.

6.4.1.3 DBRs in Other Plant Species

Other DBRs have also been described in various metabolic pathways, i.e. (a) *Mentha piperita* (+)-pulegone reductase, which catalyzes reduction of the terpene pulegone into (–)-menthone and (+)-isomenthone (Ringer et al. 2003); (b) *Artemisia annua* artemisinic aldehyde reductase (AaDBR2), which converts the Δ11(13) double bond of the terpenoid artemisinic aldehyde into dihydroartemisinic aldehyde, but is inactive with artemisinic alcohol (Zhang et al. 2008); (c) a "generalized" DBR from *Nicotiana tabacum* (Mansell et al. 2013) able to act on various α,β-unsaturated aldehydes; and (d) raspberry (*Rubus idaeus*) ketone/zingerone synthase 1 (RiRZS1) with amino acid identities of ~48% to PtPPDBR and ~70% to AtDBR1 (Table 6.1). RiRZS1 reduces the C7–C8 double bond of the phenylpropanoid-polyketide pathway intermediates, 4-hydroxybenzalacetone (**47**) and 3-methoxy-4-hydroxybenzalacetone (**48**), to form raspberry ketone (**49**) and zingerone (**50**) (Figure 6.10A) (Koeduka et al. 2011), and d) an enoyl reductase (MdENRL3) from apple (*Malus* × *domestica*) able to convert *p*-coumaroyl CoA (**51**) into dihydro-*p*-coumaroyl CoA (**52**) (Figure 6.10B), an intermediate en route to phloridzin (Dare et al. 2013). MdENRL3 has, however, very low amino acid identities of ~15% to PtPPDBR and ~17% to AtDBR1, respectively.

All of these enzymes, though, have similar biochemical functions, in terms of substrates having a conjugated enone functionality (α,β-unsaturated aldehyde).

6.4.2 Phenylcoumaran Benzylic Ether Reductases (PCBERs) Roadmap

The Japanese gymnosperm *Cryptomeria japonica*, known as sugi in Japan, and as Japanese sugi pine or Japanese red-cedar in English, is a member of the Cupressaceae in the order Pinales. Its leaves have been reported to accumulate both the 8–5′-linked

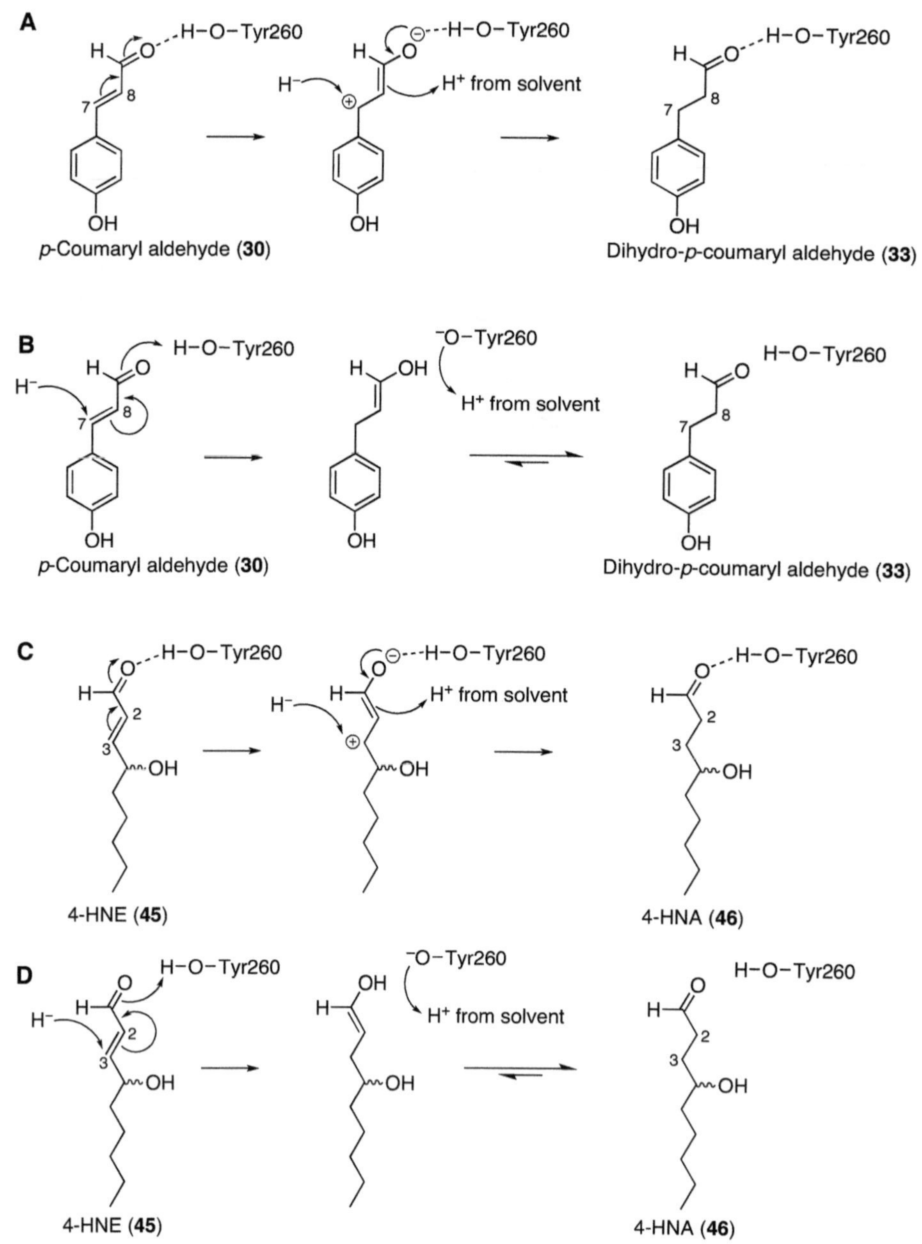

FIGURE 6.9 Proposed possible mechanisms for double bond reduction of *p*-coumaryl aldehyde (**30**) and HNE (**45**) by AtDBR1. **A** and **C.** Hydride addition to an enolate carbocation intermediate (Youn et al. 2006). **B** and **D.** Concerted hydride addition to give the enol form of the product, with subsequent tautomerization to afford the final aldehyde.

lignan (–)-*trans*-dihydrodehydrodiconiferyl alcohol triacetate (**53b**) and (–)-tetrahydrodehydrodiconiferyl alcohol tetraacetate (**54b**) (Figure 6.11), with the latter being a phenylcoumaran benzylic ether-reduced homolog (Su et al. 1995).

In addition to acylation, its formation required 7–*O*–4′ interunit linkage reduction, which we envisaged to possibly occur via a similar biochemical process to that of pinoresinol-lariciresinol reductase (PLR) through presumed quinone methide intermediacy with hydride attack at C7 (Davin and Lewis 2022).

It should be noted that compound **53b** was first reported as a *cis* isomer (Su et al. 1995), but this was later corrected by Wallis and co-workers (Li et al. 1997), who established that only the *trans* isomers were natural products, the earlier error being due to misinterpretation of spectroscopic data.

Accordingly, the *Forsythia* PLR cDNA was used to probe for PLR homologs in a *P. taeda* cell suspension culture cDNA library, this resulting in isolation of a PLR homolog (~45% amino acid identity to *Forsythia* PLR_Fi1). Its full-length cDNA encoded an 33.6 kDa protein, whose corresponding recombinant native protein was expressed in *E. coli* (Gang et al. 1999) and which was able to catalyze *in vitro* the NADPH-dependent reduction of (±)-dehydrodiconiferyl alcohols (**18a/b**), and (±)-dihydrodehydrodiconiferyl alcohols (**21a/b**), into the 7–*O*–4′-reduced products

TABLE 6.1
Identity (%) between double bond reductases (determined with Clustal Omega (Sievers et al. 2011))

	AtDBR1*	PtPPDBR	PaDBR1	PaDBR2	MpDBR	MpMDBRL	RiRZS1
PtPPDBR	44.7						
PaDBR1	57.1	43.1					
PaDBR2	57.1	43.4	99.4				
MpDBR	58.9	43.1	80.4	80.4			
MpMDBRL	42.7	39.0	41.4	42.0	43.8		
RiRZS1	69.9	48.3	56.0	56.0	58.9	45.0	
MdENRL3	16.9	14.6	19.1	19.1	16.7	16.2	17.9

* AtDBR1: *Arabidopsis thaliana* double-bond reductase 1; MdENRL3: *Malus* × *domestica* enoyl reductase-like 3; MpDBR: *Marchantia paleacea* double-bond reductase; MpMDBRL: *M. paleacea* microbial alkenal double-bond reductase-like; PaDBR1 and PaDBR2: *Plagiochasma appendiculatum* double-bond reductase 1 and 2; PtPPDBR: *Pinus taeda* phenylpropenal double-bond reductase; RiRZS1: *Rubus idaeus* raspberry ketone/zingerone synthase 1.

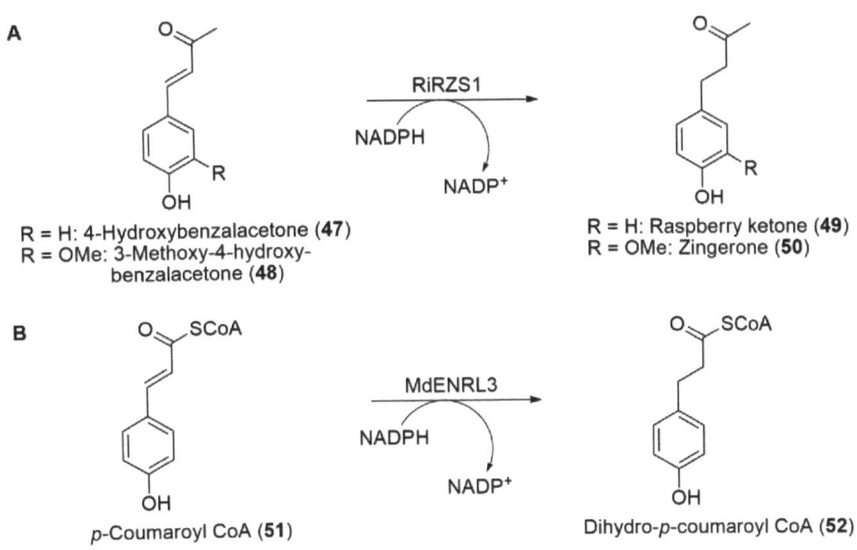

FIGURE 6.10 Miscellaneous reductive conversions catalyzed by PPDBR homologues.

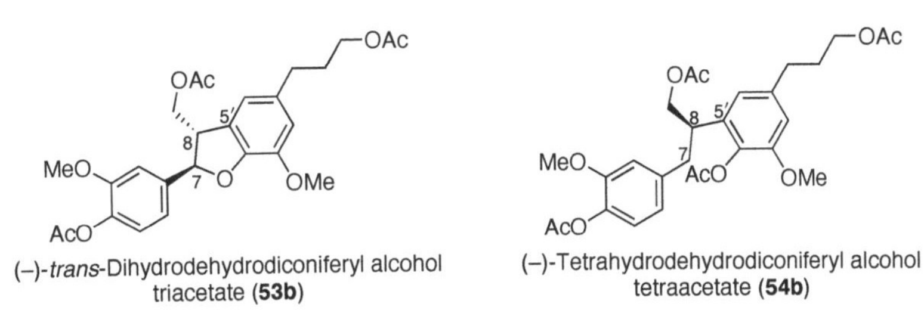

FIGURE 6.11 7–*O*–4′ reduction of 8–5′-linked lignan **53b**. "**b**" in compound numbers refers to the (–)-antipodes.

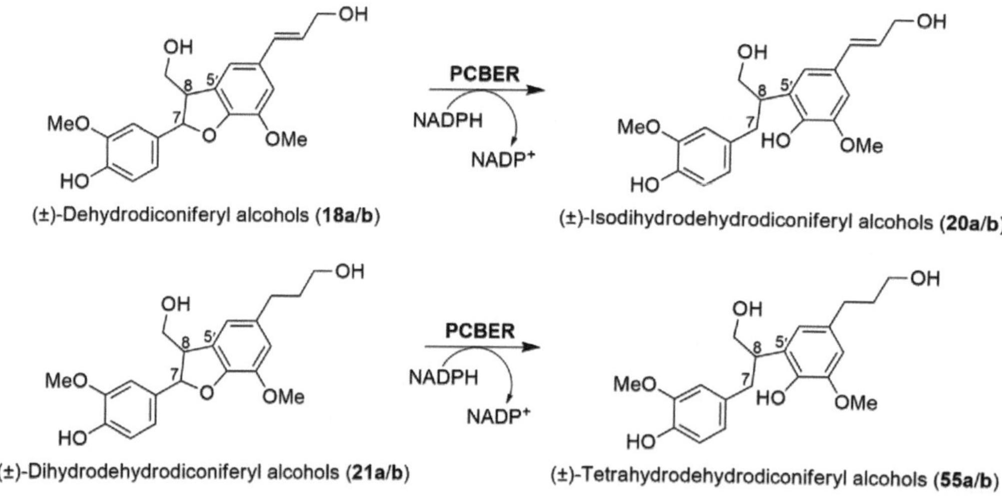

FIGURE 6.12 PCBER-catalyzed reduction of 8–5′-linked phenylcoumaran lignans. "**a/b**" in compound numbers depicts racemates.

(±)-isodihydrodehydrodiconiferyl alcohols (**20a/b**) and (±)-tetrahydrodehydrodiconiferyl alcohols (**55a/b**) (Figure 6.12), respectively.

The recombinant protein was *provisionally* designated as the phenylcoumaran benzylic ether reductase responsible, and was named PCBER_Pt1. This provisional naming was because kinetic parameters with (±)-dehydrodiconiferyl alcohols (**18a/b**) and (±)-dihydrodehydrodiconiferyl alcohols (**21a/b**) were very low for PCBER_Pt1 with K_m, V_{max}, and k_{cat}/K_m values of 0.61/1.95 mM, 0.029/0.016 pkat μg^{-1} protein, and 1.59/0.27 M^{-1} s^{-1}, respectively. Nevertheless, this enzymatic activity enabled demonstration of how phenylcoumarans could be converted into the corresponding 7–*O*–4′-reduced derivatives by enzymes of this class. PCBER_Pt1 was also established to be a type A reductase (as for PLR) (Gang et al. 1999).

At the same time, a PCBER from *Populus trichocarpa* (PCBER-Pop1) was also shown to catalyze the same conversions (Gang et al. 1999), as well as two homologues from *C. japonica* (PCBER_Cj79 and PCBER_Cj80) of ~83 and 87% similarity and ~72 and 80% identity to PCBER_Pt1, respectively (H. Kasahara, unpublished results).

Interestingly, recombinant PCBER_Pt1 differed from the enantiospecific PLRs, as it was able to regioselectively reduce both enantiomeric forms of the phenylcoumaran substrates tested (Gang et al. 1999).

Additional work with catalytically functional homologs of the above PCBERs (PCBER_Pt1 and PCBER_Pop1) from birch (*Betula*) and pear (*Pyrus communis*) established these enzymes to be cross-reactive allergens (eliciting Type 1 allergenic responses) in both birch and pear (Karamloo et al. 2001). Homologous allergens were also apparently present in apple, peach, orange, lychee fruit, strawberry, zucchini, and carrot.

In terms of catalytic properties, *we have repeatedly emphasized* that the kinetic parameters of these reductases are such as to question if they represent the true physiological function or functions or whether these are homologs of the true 7–*O*–4′ reductases that have been serendipitously obtained (Gang et al. 1999, Davin et al. 2008). However, they were not PLRs, as the PLR substrates proper, (±)-pinoresinols (**4a/b**), were not efficiently processed by either PCBER_Pt1 or PCBER_Pop1 (Gang et al. 1999).

This uncertainty as to which PCBER genes in poplar species are the true functional ones has continued, even though the reductive class of enzyme has been clearly established. Niculaes et al. (2014), in a metabolic profiling study, concluded that the

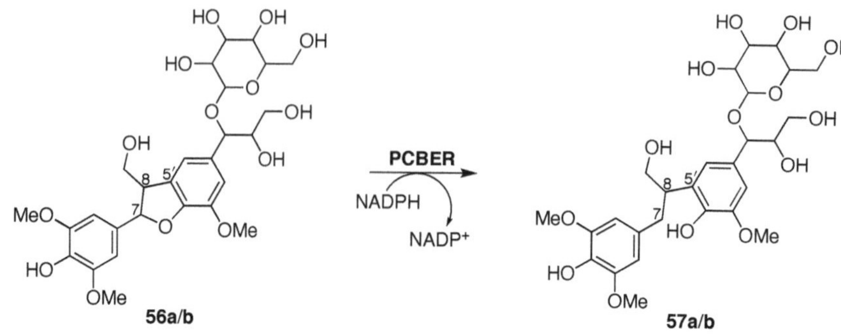

FIGURE 6.13 7–*O*–4′ reduction of 8–5′-linked lignans **56a/b**. (Nature of "hexose" not described (Niculaes et al. 2014).) "**a/b**" in compound numbers depicts racemates.

TABLE 6.2
Identity (%) between PCBERs (determined with Clustal Omega (Sievers et al. 2011))

	PCBER_Pt1*	AtPCBER1
AtPCBER1	61.9	
GbIRL1	80.0	63.6

* AtPCBER1: *Arabidopsis thaliana taeda* phenylcoumaran benzylic ether reductase 1; GbIRL1: *Ginkgo biloba* IFR-like protein 1; PCBER_Pt1: *Pinus taeda* phenylcoumaran benzylic ether reductase 1.

PCBER-Pop1 physiological substrate was an incompletely characterized hexosylated 8–5′-coupling product (**56a/b**) derived from sinapyl alcohol (**13**) and guaiacyl glycerol, to afford the corresponding 7–*O*–4′-reduced product **57a/b** (Figure 6.13) with this confirmed in an *in vitro* assay using recombinant protein. However, these researchers showed that other dimers with a phenylcoumaran moiety were also reduced. No study of enantiomeric composition of the products was carried out in these studies, however, of substrate or product, a serious flaw in experimental design. Moreover, as to be expected, down-regulation of this gene had no effect on lignin levels.

A PCBER homologue (~61.9% identity to PCBER_Pt1, Table 6.2) from *Arabidopsis thaliana* (the putative PCBER gene, At4g39230) was also cloned, and its recombinant protein *in vitro* was (as expected) able to reduce dehydrodiconiferyl alcohol **18** and (much more poorly) pinoresinol (Nuoendagula et al. 2016). This gene was both up- and down-regulated. None of the target lignans, e.g. dehydrodiconiferyl alcohol **18**, were detected, and lignin levels were unaltered. These authors concluded that "although AtPCBER1 can reduce dehydrodiconiferyl alcohol **18** as indicated, its natural substrate in *A. thaliana* has not been fully characterized."

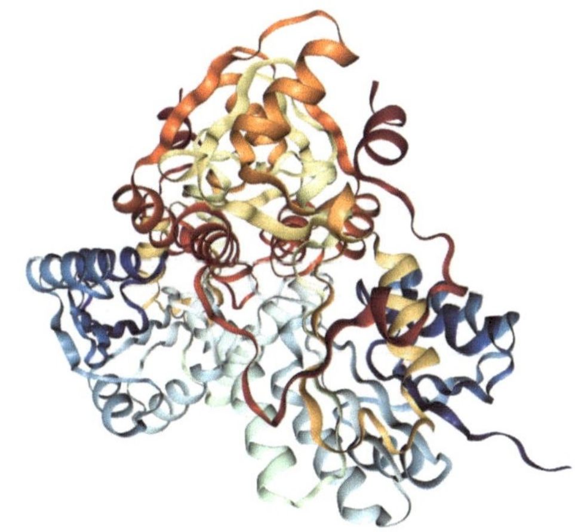

FIGURE 6.14 Schematic outlines of the dimeric forms of PCBER_Pt1 (PDB ID: 1QYC). Image created using NGL viewer (Rose et al. 2018).

Hua et al. (2013) also described a gene from *Ginkgo biloba* as being an isoflavone reductase (IFR)-like gene, with ~80.0% identity to PCBER_Pt1. We had described earlier that IFRs, PLRs, and PCBERs showed considerable homology to each other (Gang et al. 1999). As for PCBER_Pt1 and PCBER_Pop1, the *G. biloba* "IFR-like reductase" was able to (PDB ID: 1QYC). Image created using NGL viewer (Rose et al. 2018). educe dehydrodiconiferyl alcohol (**18**) and dihydrodehydrodiconiferyl alcohols (**21**) to afford the corresponding 7–*O*–4′-reduced products **20** and **55**. However, whether this is the true *G. biloba* physiological substrate also remains to be established.

PCBER Crystal Structure: PCBER_Pt1 crystals (1QYC) were obtained in apo-form (2.2 Å resolution) (Figure 6.14). With two domains separated by a cleft, a conserved nucleotide

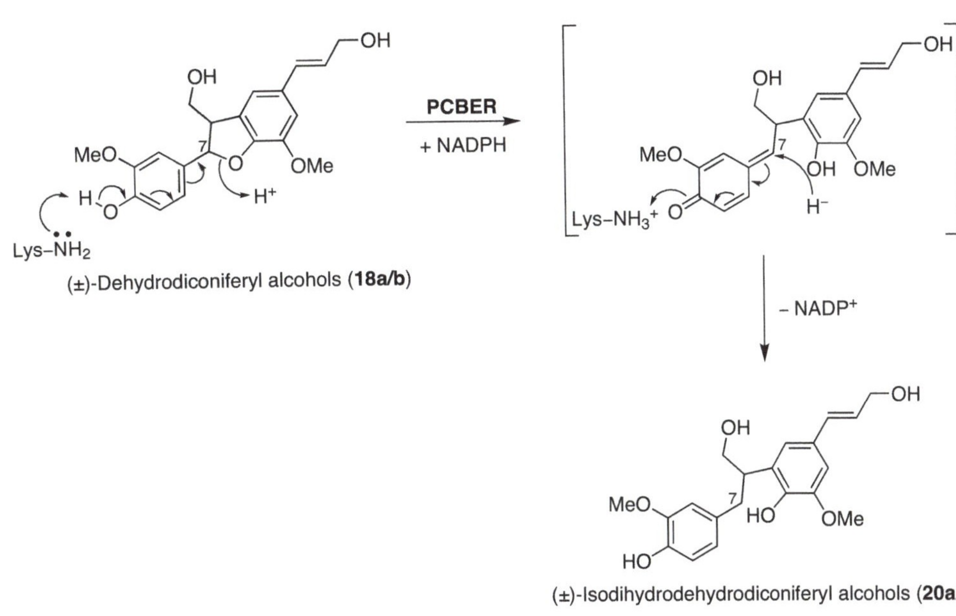

FIGURE 6.15 Proposed PCBER catalytical mechanism. "**a/b**" in compound numbers depicts racemates.

cofactor-binding Rossman fold motif, its highly conserved active site Lys134 catalytic residue generates a presumed quinone methide intermediacy, as for PLR, prior to reduction and product formation (Figure 6.15) (Min et al. 2003).

Acknowledgements

The authors thank the Chemical Sciences, Geosciences and Biosciences Division, DOE Office of Basic Energy Sciences (DE-FG-0397ER20259), as well as the Arthur and Katie Eisig-Tode Endowment.

References

Agrawal, P. K., S. K. Agarwal, and R. P. Rastogi. 1980. A new neolignan and other phenolic constituents from *Cedrus deodara*. *Phytochemistry* 19:1260–1261.

Dare, A. P., S. Tomes, J. M. Cooney, D. R. Greenwood, and R. P. Hellens. 2013. The role of enoyl reductase genes in phloridzin biosynthesis in apple. *Plant Physiol. Biochem.* 72:54–61.

Davin, L. B., C.-Z. Wang, G. L. Helms, and N. G. Lewis. 2003. [^{13}C]-Specific labeling of 8-2' linked (–)-*cis*-blechnic, (–)-*trans*-blechnic and (–)-brainic acids in the fern *Blechnum spicant*. *Phytochemistry* 62:501–511.

Davin, L. B., M. Jourdes, A. M. Patten, et al. 2008. Dissection of lignin macromolecular configuration and assembly: Comparison to related biochemical processes in allyl/propenyl phenol and lignan biosynthesis *Nat. Prod. Rep.* 25:1015–1090.

Davin, L. B., and N. G. Lewis. 2022. 8–8′-Lignan biosynthesis roadmap. In *The Lignan Handbook* (N. G. Lewis and L. B. Davin eds.), Taylor & Francis, Boca Raton, London. pp 79–112.

Gang, D. R., H. Kasahara, Z.-Q. Xia, et al. 1999. Evolution of plant defense mechanisms. Relationships of phenylcoumaran benzylic ether reductases to pinoresinol-lariciresinol and isoflavone reductases. *J. Biol. Chem.* 274:7516–7527.

Hua, C., L. Linling, X. Feng, et al. 2013. Expression patterns of an isoflavone reductase-like gene and its possible roles in secondary metabolism in *Ginkgo biloba*. *Plant Cell Rep.* 32:637–650.

Karamloo, F., A. Wangorsch, H. Kasahara, et al. 2001. Phenylcoumaran benzylic ether and isoflavonoid reductases are a new class of cross-reactive allergens in birch pollen, fruits and vegetables. *Eur. J. Biochem.* 268:5310–5320.

Kasahara, H., L. B. Davin, and N. G. Lewis. 2004. Patent. Aryl propenal double bond reductase. In United States Patent No. 6,703,229. Filed March 27, 2001. Issued March 9, 2004.

Kasahara, H., Y. Jiao, D. L. Bedgar, et al. 2006. *Pinus taeda* phenylpropenal double-bond reductase: Purification, cDNA cloning, heterologous expression in *Escherichia coli*, and subcellular localization in *P. taeda*. *Phytochemistry* 67:1765–1780.

Katayama, T., and Y. Kado. 1998. Formation of optically active neolignans from achiral coniferyl alcohol by cell-free extracts of *Eucommia ulmoides*. *J. Wood Sci.* 44:244–246.

Kim, K. H., H. K. Kim, S. U. Choi, et al. 2011. Bioactive lignans from the rhizomes of *Acorus gramineus*. *J. Nat. Prod.* 74:2187–2192.

Koeduka, T., B. Watanabe, S. Suzuki, et al. 2011. Characterization of raspberry ketone/zingerone synthase, catalyzing the alpha, beta-hydrogenation of phenylbutenones in raspberry fruits. *Biochem. Biophys. Res. Commun.* 412:104–108.

Kraus, C., and G. Spiteller. 1997. Comparison of phenolic compounds from galls and shoots of *Picea glauca*. *Phytochemistry* 44:59–67.

Li, S., T. Iliefski, K. Lundquist, and A. F. A. Wallis. 1997. Reassignment of relative stereochemistry at C-7 and C-8 in arylcoumaran neolignans. *Phytochemistry* 46:929–934.

Lourith, N., T. Katayama, K. Ishikawa, and T. Suzuki. 2005a. Biosynthesis of a syringyl 8-*O*-4′ neolignan in *Eucommia ulmoides*: Formation of syringylglycerol-8-*O*-4′-(sinapyl alcohol) ether from sinapyl alcohol. *J. Wood Sci.* 51:379–386.

Lourith, N., T. Katayama, and T. Suzuki. 2005b. Stereochemistry and biosynthesis of 8-*O*-4′ neolignans in *Eucommia ulmoides*: Diastereoselective formation of guaiacylglycerol-8-*O*-4′-(sinapyl alcohol) ether. *J. Wood Sci.* 51:370–378.

Mano, J., Y. Torii, S.-i. Hayashi, et al. 2002. The NADPH:quinone oxidoreductase P1-ζ-crystallin in *Arabidopsis* catalyzes the α,β-hydrogenation of 2-alkenals: Detoxication of the lipid peroxide-derived reactive aldehydes. *Plant Cell Physiol.* 43:1445–1455.

Mano, J., E. Belles-Boix, E. Babiychuk, et al. 2005. Protection against photooxidative injury of tobacco leaves by 2-alkenal reductase. Detoxication of lipid peroxide-derived reactive carbonyls. *Plant Physiol.* 139:1773–1783.

Mansell, D. J., H. S. Toogood, J. Waller, et al. 2013. Biocatalytic asymmetric alkene reduction: Crystal structure and characterization of a double bond reductase from *Nicotiana tabacum*. *ACS Catal.* 3:370–379.

Min, T., H. Kasahara, D. L. Bedgar, et al. 2003. Crystal structures of pinoresinol-lariciresinol and phenylcoumaran benzylic ether reductases and their relationship to isoflavone reductases. *J. Biol. Chem.* 278:50714–50723.

Niculaes, C., K. Morreel, H. Kim, et al. 2014. Phenylcoumaran benzylic ether reductase prevents accumulation of compounds formed under oxidative conditions in poplar xylem. *Plant Cell* 26:3775–3791.

Nose, M., M. A. Bernards, M. Furlan, et al. 1995. Towards the specification of consecutive steps in macromolecular lignin assembly. *Phytochemistry* 39:71–79.

Nuoendagula, N. Kamimura, T. Mori, et al. 2016. Expression and functional analyses of a putative phenylcoumaran benzylic ether reductase in *Arabidopsis thaliana*. *Plant Cell Rep.* 35:513–526.

Ringer, K. L., M. E. McConkey, E. M. Davis, G. W. Rushing, and R. Croteau. 2003. Monoterpene double-bond reductases of the (–)-menthol biosynthetic pathway: Isolation and characterization of cDNAs encoding (–)-isopiperitenone reductase and (+)-pulegone reductase of peppermint. *Arch. Biochem. Biophys.* 418:80–92.

Rose, A. S., A. R. Bradley, Y. Valasatava, et al. 2018. NGL viewer: Web-based molecular graphics for large complexes. *Bioinformatics* 34:3755–3758.

Sartorelli, P., P. J. C. Benevides, R. M. Ellensohn, et al. 2001. Enantioselective conversion of *p*-hydroxypropenylbenzene to (+)-conocarpan in *Piper regnellii*. *Plant Sci.* 161:1083–1088.

Satake, T., T. Murakami, Y. Saiki, and C.-M. Chen. 1978. Chemische Untersuchungen der Inhaltsstoffe von *Pteris vittata* L. *Chem. Pharm. Bull.* 26:1619–1622.

Sievers, F., A. Wilm, D. Dineen, et al. 2011. Fast, scalable generation of high-quality protein multiple sequence alignments using Clustal Omega. *Mol. Syst. Biol.* 7:539.

Su, W.-C., J.-M. Fang, and Y.-S. Cheng. 1995. Flavonoids and lignans from leaves of *Cryptomeria japonica*. *Phytochemistry* 40:563–566.

Wu, Y.-F., H.-B. Zheng, X.-Y. Liu, A.-X. Cheng, and H.-X. Lou. 2018. Molecular diversity of alkenal double bond reductases in the liverwort *Marchantia paleacea*. *Molecules* 23:e1630.

Wu, Y., Y. Cai, Y. Sun, et al. 2013. A single amino acid determines the catalytic efficiency of two alkenal double bond reductases produced by the liverwort *Plagiochasma appendiculatum*. *FEBS Lett.* 587:3122–3128.

Youn, B., S.-J. Kim, S. G. A. Moinuddin, et al. 2006. Mechanistic and structural studies of apoform, binary, and ternary complexes of the *Arabidopsis* alkenal double bond reductase At5g16970. *J. Biol. Chem.* 281:40076–40088.

Zhang, Y., K. H. Teoh, D. W. Reed, et al. 2008. The molecular cloning of artemisinic aldehyde Δ11(13) reductase and its role in glandular trichome-dependent biosynthesis of artemisinin in *Artemisia annua*. *J. Biol. Chem.* 283:21501–21508.

7 Allylphenol and Propenylphenol Biosynthesis and C9/C9′ Deoxygenation Roadmap

Laurence B. Davin, Sung-Jin Kim, Da Lu, and Norman G. Lewis

7.1 EARLY STUDIES

Many C_6C_3 (phenylpropanoid) monomers, lignans, norlignans, and others from "mixed" biochemical pathways lack an oxygenated functionality at C9, or at both C9/C9′ carbons (e.g. Figure 7.1 depicting various monomeric allylphenols and propenylphenols **1–10**). Indeed, numerous chemo-taxonomical investigations over several decades have described a myriad of such plant natural products, ranging from their presence in algae (at least putatively), hornworts, liverworts, ferns, gymnosperms, and angiosperms (Davin and Lewis 2022a). However, reports of C_6C_3 (phenylpropanoid) monomers lacking oxygenated functionalities can be frequently difficult to locate in the literature. This is because their presence is often recorded only in tabular form as minor constituents (frequently found in near trace amounts) of complex mixtures of various natural products, including terpenoids and the like.

7.1.1 EARLY CONSIDERATIONS OF ALLYLPHENOL AND PROPENYLPHENOL FORMATION *IN PLANTA*

During the 1950s, A. J. Birch's laboratory (of Birch reduction fame) at the Australian National University became intrigued as to how allylphenols and propenylphenols might be formed. They considered that a C9 oxygenated species was likely to be the precursor of C9 allylphenols and propenylphenols. To investigate this possibility, Birch and Slaytor (1956) discovered that 3,4-methylenedioxycinnamyl alcohol (**11**), when treated with $LiAlH_4$ in boiling diethyl ether in the presence of $AlCl_3$ (Lewis acid), gave the allylphenol and propenylphenol derivatives, safrole (**5**) and isosafrole (**10**) [~60% yield in an ~55:45 ratio] (Figure 7.2A). Conversely, no comparable conversion occurred with cinnamyl alcohol (**12**, Figure 7.2B). These findings suggested that a *para*-oxygenated aromatic ring moiety (not necessarily in free phenol form) was needed for C9 hydroxyl group displacement. From these studies, formation of safrole (**5**) and isosafrole (**10**) was mechanistically rationalized to occur via a resonance-stabilized allylic carbocation intermediate with subsequent terminal hydroxyl group loss and concomitant reduction via hydride addition.

Shortly thereafter, Birch (1963) proposed a modified means for displacement of the terminal C9 hydroxyl group (Figure 7.2C), whereby it was now activated for its more facile displacement. This was hypothesized as more likely to be the case, since in terpenoid biochemistry the formation and metabolism of isopentenyl pyrophosphates was known, with the pyrophosphate entity being displaced during further metabolism. This led Birch (1963) to propose that a monolignol-derived phosphate ester derivative (Figure 7.2C) was possibly formed, following which, an allylic hydride nucleophilic addition would result in displacement to generate the corresponding allylphenol.

Geissman and Crout (1969) further refined this contemplated biosynthetic pathway, where they envisaged that nucleophilic addition (of a hydride from NADH) to a (hydroxy)cinnamyl pyrophosphate could also explain how allylphenols and propenylphenols were formed. They considered two possibilities: Either a SN1 mechanism with formation and reduction of a delocalized allylic carbocation or an

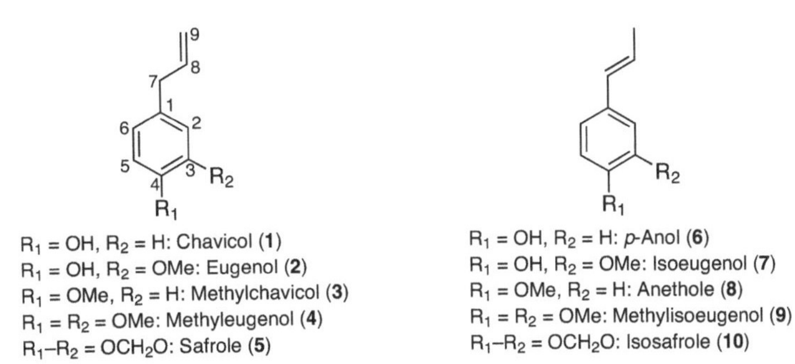

FIGURE 7.1 Allylphenols (**1–5**) and propenylphenols (**6–10**).

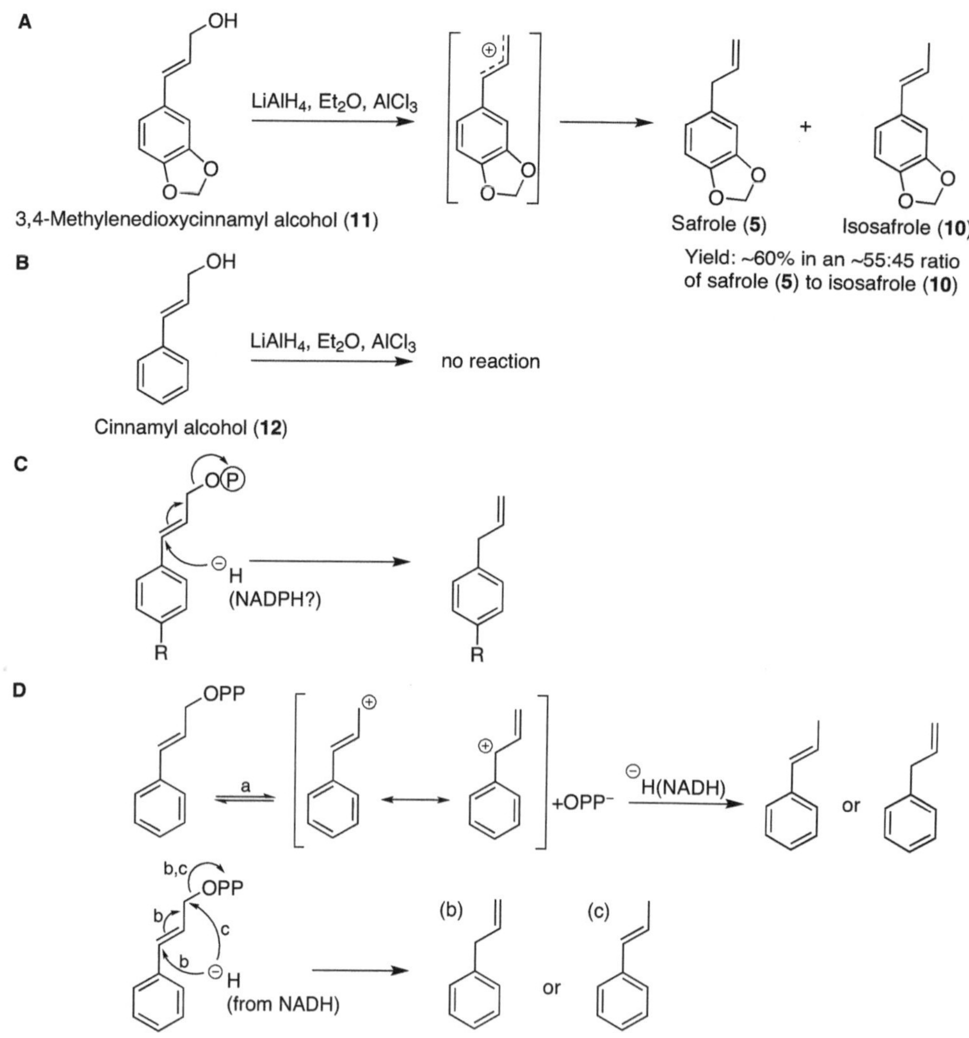

FIGURE 7.2 Proposed chemical and biochemical routes leading to allylphenols and propenylphenols prior to 1970. **A** and **B**. Birch and Slaytor (1956) chemical transformations. **C**. Birch's (1963) refinement to allylphenols via nucleophilic attack and phosphate displacement (R, not specified). **D**. Geissman and Crout (1969), **a**, S_N1 mechanism with hydride reduction, **b**, allylic and **c**, direct pyrophosphate nucleophilic displacement.

allylic or a direct pyrophosphate group displacement (Figure 7.2D).

Taken together, these studies then allowed for a hypothetical rationale as to how allylphenol and propenylphenol formation might occur via C9 hydroxyl group activation (e.g. with a good leaving group, such as an ester) and its subsequent displacement.

7.1.2 Early Radiolabel Tracer Studies: Phenylpropanoid Origins of Allylphenols and Propenylphenols

To begin to probe monomeric allylphenol and propenylphenol biosynthetic formation, Kaneko (1960a, b, 1961, 1962) carried out (preliminary) radiolabeling and chemical degradation experiments in fennel (*Foeniculum vulgare*) using potential precursors [2-^{14}C]-Phe (**13**, Figure 7.3), [Me-^{14}C]-Met (**15**), and [9-^{14}C]-*p*-coumaric acid (**17**), respectively. These were reportedly intactly incorporated into anethole (**8**), with the [2-^{14}C]-Phe (**13**) and [9-^{14}C]-*p*-coumaric acid (**17**) incorporations indicating phenylpropanoid origin and the [Me-^{14}C]-Met (**15**) accounting for the OMe in anethole (**8**). Later, Tressl and Drawert (1973) and Klischies et al. (1975) described allylphenol and propenylphenol formation in banana (*Musa* sp.) discs and basil (*Ocimum basilicum*). In banana, [1-^{14}C]-Phe (**13**) and [9-^{14}C]-caffeic acid (**18**) were individually incorporated into eugenol (**2**), methyleugenol (**4**), 5-methoxyeugenol (**23**), and elemicin (**24**), in agreement that they were phenylpropanoid derived. In the basil experiments, double-labeled [9-^{14}C, OC^{3}H$_3$]-ferulic acid glucoside (**19**), as well as [9-^{14}C, OC^{3}H$_3$]- and [9-^{14}C^{3}H$_2$OH]-coniferin (**22**) were also individually administered with the resulting [^{3}H/^{14}C]-eugenol (**2**) and methyleugenol (**4**) obtained having only minor changes in ^{3}H:^{14}C ratios from the administered

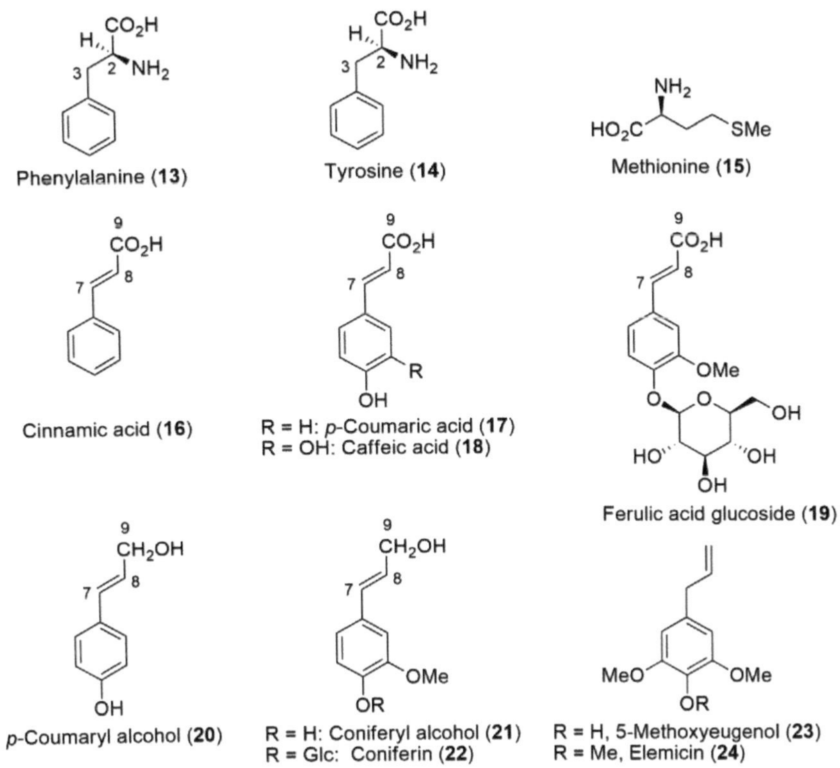

FIGURE 7.3 Probing monomeric allylphenol and propenylphenol formation with various potential biosynthetic precursors.

substrates. That the C9 in allylphenols **2** and **4** was retained from the precursor phenylpropanoid carbon skeleton was confirmed by chemical degradation studies. These studies were thus consistent with C9 retention during allylphenol and propenylphenol formation.

7.1.3 Another View: The Mistaken Belief that C9 Was Lost during Allylphenol and Propenylphenol Formation

During the 1970s, various reports appeared of putative C9 carbon loss during eugenol (**2**) biosynthesis in basil (Canonica et al. 1971, Manitto et al. 1974) and in cinnamon (*Cinnamomum zeylanicum*) (Senanayake et al. 1977). With basil, double-labeled Phe (**13**) (non-specifically tritiated, and also containing either 1-, 2- or 3-^{14}C radiolabels) precursor administrations were carried out, with the corresponding radiolabeled forms of eugenol (**2**) individually isolated. When the ^{14}C label was in the [1-^{14}C]-carboxylic acid group of Phe (**13**), these researchers reported that the eugenol (**2**) so obtained had a much higher ^{3}H:^{14}C radiolabel ratio (~20 times that of the ^{3}H:^{14}C precursor), suggesting that terminal C9 carbon loss had occurred during metabolism.

With cinnamon, the [1-^{14}C]-Phe (**13**) radiolabel was reportedly very poorly incorporated into eugenol (**2**, 0.003%), with marginally better incorporations using [2-^{14}C]- or [3-^{14}C]-Phe (**13**) (0.047 and 0.098%, respectively). It was also reported that [Me-^{14}C]-Met (**15**) was incorporated into eugenol (**2**, 0.1% incorporation), with the radiolabel believed to be at the C9 position (based on chemical degradation) (Senanayake et al. 1977), rather than in the OMe group. These data thus suggested to these researchers a methionine (**15**) origin for the C9. It is, however, well known that very low incorporation levels such as those reported (i.e. <0.1%) are best viewed with caution and should not be considered as definitive.

Nevertheless, a C9 origin (presumed to be derived from *S*-adenosyl methionine, SAM) in allylphenol and propenylphenol biosynthesis was still being considered in the early 2000s (Gang et al. 2001). Specifically, a *p*-coumaric acid (**17**) moiety was envisaged to be covalently attached to the methyl group from SAM to form cyclopropyl intermediate **25** (Figure 7.4), with the latter then thought to undergo C9 decarboxylation/ring opening to directly afford the allyl and propenyl side chains of chavicol (**1**) and *p*-anol (**6**). There is, however, no experimental evidence for this hypothetical transformation.

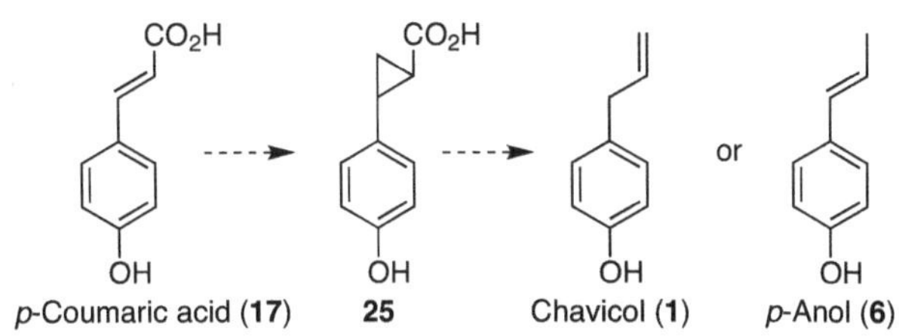

FIGURE 7.4 An earlier (incorrect) view of allylphenol and propenylphenol formation with C9 loss and SAM methylation.

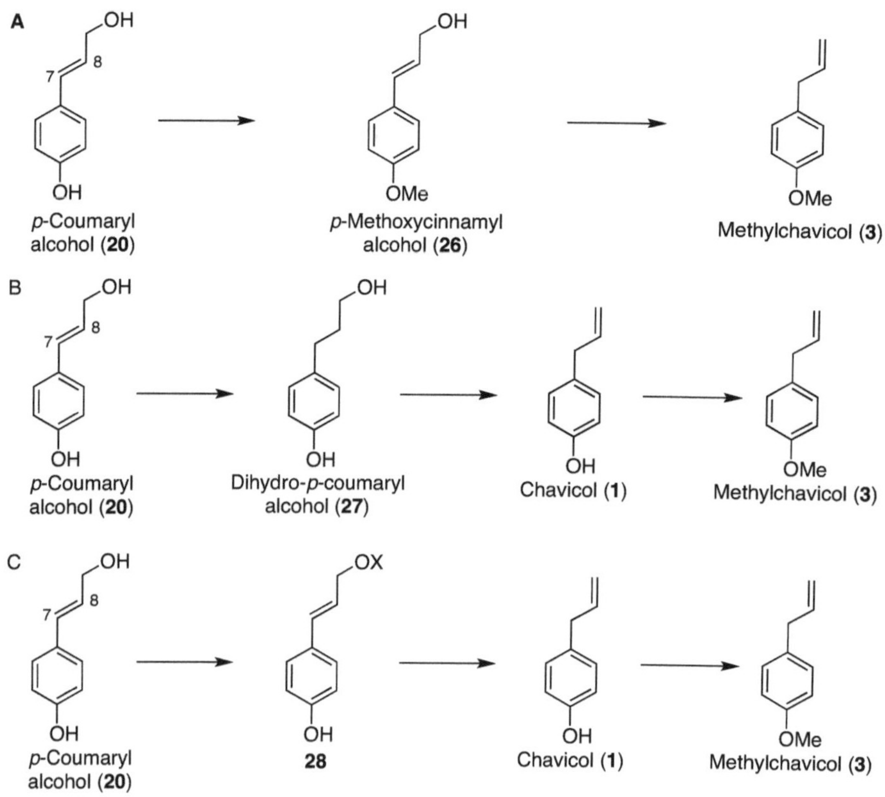

FIGURE 7.5 Potential biochemical pathways contemplated for methylchavicol (**3**) formation from *p*-coumaryl alcohol (**20**) (Vassão et al. 2006). X = Facile leaving group.

7.2 MONOLIGNOL ESTERS IN ALLYLPHENOL AND PROPENYLPHENOL BIOSYNTHESIS: CLUES FROM NORLIGNAN BIOSYNTHESIS

Our long-term interests in allylphenol and propenylphenol biosynthesis were particularly focused on basil, the creosote bush (*Larrea tridentata*), and various *Piper* sp., which can differentially harbor various allylphenols and propenylphenols in either monomeric, dimeric, or both forms, depending on the species. The Thai basil variety was of particular interest as it largely accumulates methylchavicol (**3**) in its glandular trichomes (Vassão et al. 2006), and was thus a good candidate in which to investigate monomeric allylphenol and propenylphenol formation. Our initial experiments demonstrated that young apical basil leaves (Vassão et al. 2006) were the most biosynthetically active for producing methylchavicol (**3**) *de novo*, with [U-^{14}C]-Phe (**13**), [8-^{14}C]-cinnamic acid (**16**) and [9-^{3}H]-*p*-coumaryl alcohol (**20**), but not [U-^{14}C]-Tyr (**14**), efficiently metabolized into methylchavicol (**3**) *in vivo* (~1%–4% radiolabel incorporation within 24 hours). On the other hand, with basil leaf cell-free extracts, neither unlabeled nor [9-^{3}H]-*p*-coumaryl alcohol (**20**) were metabolized into chavicol (**1**) *in vitro*, suggesting that additional enzymatic steps or then unknown co-factors were needed.

From these data, we hypothesized that there were three possible routes from *p*-coumaryl alcohol (**20**) to allylphenols and propenylphenols such as methylchavicol (**3**) *in planta*: (1) phenolic group *O*-methylation to generate *p*-methoxycinnamyl alcohol (**26**), prior to C9 deoxygenation (as proposed by Kaneko (1962); Figure 7.5A); (2) C7–C8 double bond reduction to initially afford dihydro-*p*-coumaryl alcohol (**27**) (i.e. analogous to the double bond reductase described in Chapter 6 (Davin and Lewis 2022b)) followed by loss of water to give chavicol (**1**) and then methylchavicol (**3**) after *O*-methylation (Figure 7.5B), and (3) C9 hydroxyl group activation via a *p*-coumaryl alcohol derivative (**28**) bearing a good leaving group, as proposed by Birch (1963) and Geissman and Crout (1969) (Figure 7.5C). Of these three distinct possibilities, routes 1 and 2 were readily eliminated following *in vivo* precursor administration and *in vitro* incubation experiments using cell-free enzyme preparations (Vassão et al. 2006).

With our focus next on potential C9 hydroxyl group activation, various monolignols were incubated with young basil leaf cell-free extracts with ATP, GTP, UTP, Glc-6-phosphate, and glutathione as possible activators at C9. However, in those experiments, no allylphenol and propenylphenol formation was detected. A potential clue to allylphenol and propenylphenol biosynthesis emerged at about the same time, however, in studies leading to clarification of how some types of norlignans were generated *in planta*; these were consistent with the need for activation of the C9 CH_2OH group to give a better leaving group. That is, the monolignol ester *p*-coumaryl coumarate (**29**) was discovered to be a norlignan biosynthesis

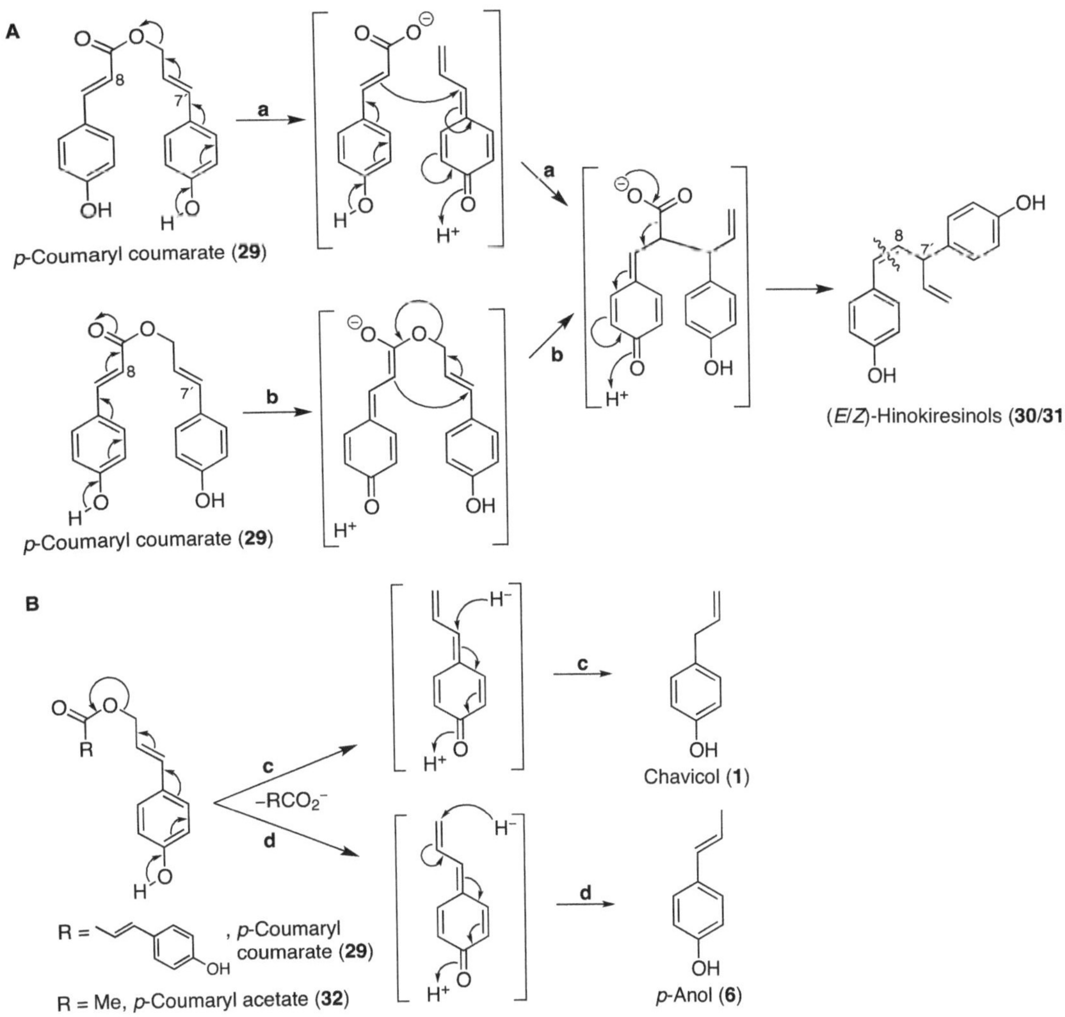

FIGURE 7.6 Proposed biosynthetic pathway mechanistic considerations leading to (**A**) norlignans, (*E*/*Z*)-hinokiresinol (**30**/**31**), and (**B**) chavicol (**1**)/*p*-anol (**6**).

substrate, affording *E*- and *Z*-hinokiresinols (**30** and **31**, Figure 7.6A) in *Asparagus officinalis* and *Cryptomeria japonica*, respectively (Suzuki et al. 2001, Suzuki et al. 2002, Suzuki et al. 2004, Davin and Lewis 2022d). Although this was an important finding, no enzymatic mechanism was proposed to account for these conversions.

We considered mechanistically how formation of **30** and **31** might occur with no additional co-factor (Figure 7.6A) (Vassão et al. 2006), and whether a somewhat comparable process could account for allylphenol and propenylphenol formation, i.e. through initial ester formation and subsequent displacement.

That is, based on the norlignan biosynthetic mechanism, we envisaged that formation of allylphenols and propenylphenols could potentially occur in an analogous manner, e.g. via an enzyme bound *p*-coumaryl alcohol ester (**29**), which could undergo C9 deoxygenation through ester displacement and reduction, such as to afford chavicol (**1**) and *p*-anol (**6**), respectively, and/or homologs thereof (Figure 7.6B). Such a mechanism, involving nucleophilic hydride addition, would be somewhat akin to that of the pinoresinol-lariciresinol reductase (PLR) we had discovered, and where formation of an intermediary quinone methide was invoked (Dinkova-Kostova et al. 1996, Vassão et al. 2010).

As for norlignan biosynthesis, a *p*-coumarate ester function was thus possibly a good leaving group, whereby nucleophilic hydride addition (from NADH/NADPH as cofactor) at either C7 or C9, or the generation of an intermediary quinone methide as for PLR, could provide the means to form isomeric allylphenols and propenylphenols.

To investigate this possibility, basil glandular trichome crude cell-free protein extracts were thus prepared, and these were found to be able to convert *p*-coumaryl coumarate (**29**) into chavicol (**1**) *in vitro*, in the presence of either NADPH or NADH (Vassão et al. 2006). Analogous experiments showed that other acyl esters, *p*-coumaryl acetate (**32**), and coniferyl acetate (**33**), afforded chavicol (**1**) and eugenol (**2**), respectively (Koeduka et al. 2006, Vassão et al. 2006). These substrate versatile conversions were also shown to occur with a creosote bush PLR homolog heterologously expressed and

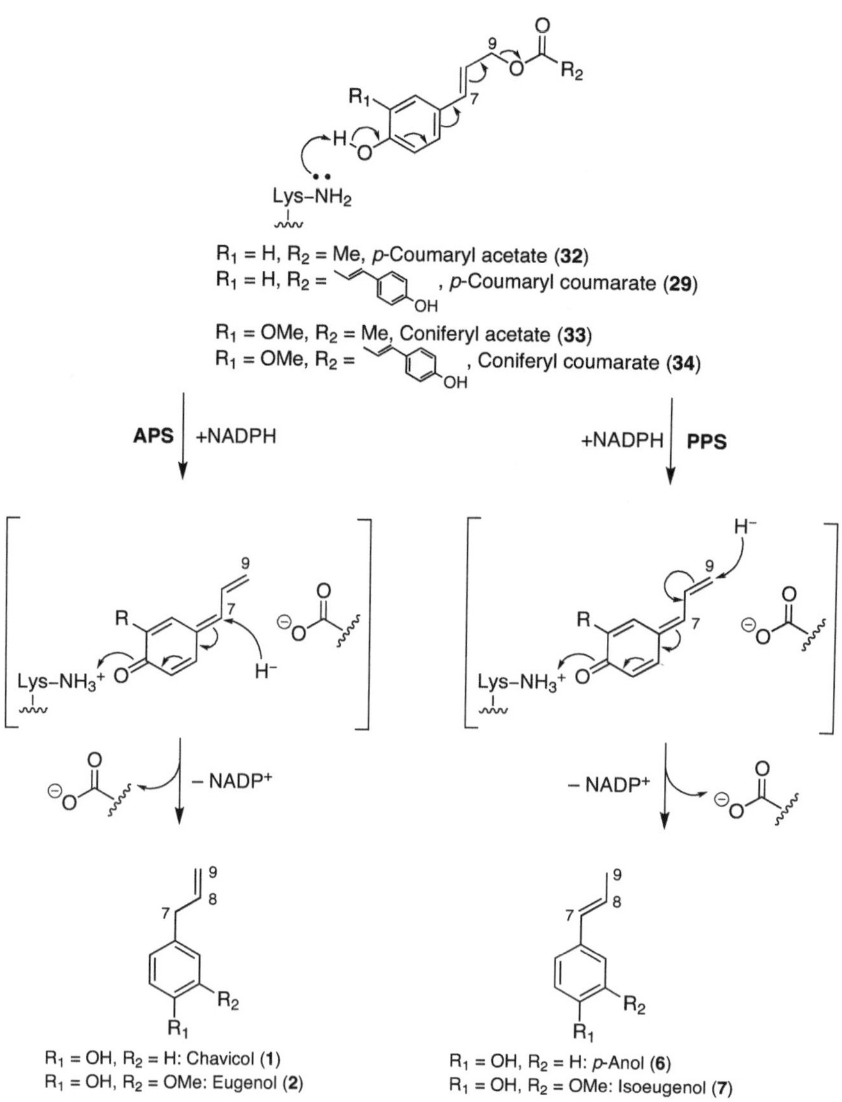

FIGURE 7.7 Allylphenol and propenylphenol (**1**, **2** and **6**, **7**) formation. Enzymatic mechanistic considerations from monolignol esters (**29**, **32–34**). APS: Allylphenol synthase, PPS: Propenylphenol synthase.

assayed with substrates (**29**, **32–34**) (Figure 7.7) (Vassão et al. 2007). Thus, allylphenol and propenylphenol formation was a two-step process of monolignol acylation and ester displacement in a NAD(P)H-dependent reduction conversion, presumably involving quinone methide intermediacy. Such reductive conversions were analogous to PLR, and the related homolog phenylcoumaran benzylic ether reductase (PCBER), etc. (Vassão et al. 2006, Vassão et al. 2007, Vassão et al. 2010).

7.3 GENES ENCODING ALLYLPHENOL AND PROPENYLPHENOL SYNTHASES

7.3.1 Identification of Genes Encoding Allylphenol and Propenylphenol Synthases

Isolation of the genes encoding allylphenol and propenylphenol synthases were obtained in an indirect way. Verdonk et al. (2005) had observed that suppression of a transcription factor (ODORANT1) in petunia (*Petunia hybrida*) resulted in upregulation of a PLR/PCBER/isoflavone reductase (IFR) homolog, but whose biochemical function was unknown. Various shikimate pathway-derived volatile benzenoid natural products decreased in levels, as did isoeugenol (**7**) amounts.

The full-length sequence of the PLR/PCBER/IFR homolog was obtained from a flower EST database, with the gene encoding a ~36 kDa peptide. Amino acid comparison gave sequence similarities/identities of ~59/40% to that of PLR_Tp2 from *Thuja plicata* (Fujita et al. 1999), ~62/42% for a PLR homolog from the creosote bush (Vassão et al. 2007), ~61/41% to the *Pinus taeda* PCBER (Gang et al. 1999), and ~59/37% to the IFR from *Medicago sativa* (Oommen et al. 1994). Scrutiny of the then basil trichome EST database also resulted in a basil homolog being obtained (based on sequence comparison with the petunia homolog above), with this gene encoding a 35.6 kDa protein (Koeduka et al. 2006).

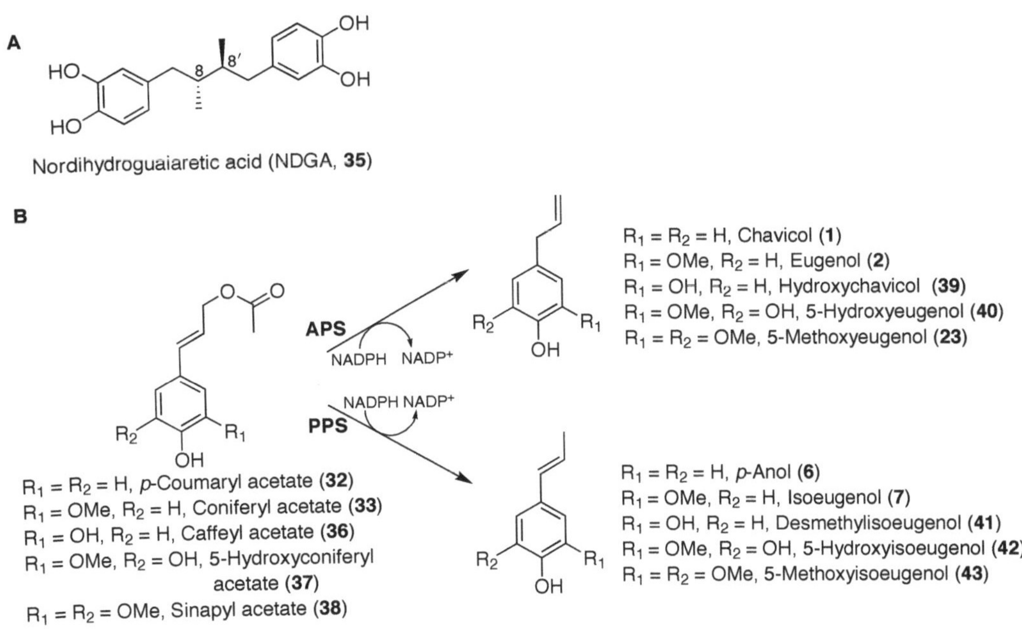

FIGURE 7.8 Allylphenol and propenylphenol formation in the creosote bush (*Larrea tridentata*). **A.** Nordihydroguaiaretic acid (NDGA, **35**). **B.** Substrate versatile APS and PPS transformations.

The creosote bush PLR homolog had been expressed in transformed *Escherichia coli* cells in 1999 (Lewis et al. 2015), this also being done with the corresponding petunia and basil homologs several years later. Each of their corresponding recombinant proteins were obtained and examined for their ability to form allylphenols or propenylphenols when incubated with either *p*-coumaryl or coniferyl acetate/*p*-coumarate esters (**29**, **32–34**) (Koeduka et al. 2006, Vassão et al. 2007).

All were found to harbor either chavicol (**1**)/eugenol (**2**) or *p*-anol (**6**)/isoeugenol (**7**) forming activities, thereby solving the long enigma as to how allylphenols and propenylphenols were actually biosynthetically produced. It must be emphasized though that, conceptually, Birch (1963) and Geissman and Crout (1969) had been prescient in their deductions that allylphenols and propenylphenols were derived from hydroxycinnamyl alcohols (monolignols) through C9 ester group activation. Importantly, they had hypothesized that an ester formation at C9 would afford a much better leaving group, with concomitant nucleophilic hydride addition at either C7 or C9 affording the desired products. While their original hypotheses contemplated phosphate or pyrophosphate as being the required moieties for ester linkages at C9, this turned out not to be the actual leaving groups needed. Nevertheless, their (bio)chemical mechanistic deductions were conceptually correct.

7.3.2 Creosote Bush Allylphenol Synthase (APS) and Propenylphenol Synthase (PPS) Kinetic Data

The creosote bush massively accumulates numerous 8–8′-linked lignans, e.g. nordihydroguaiaretic acid, NDGA (**35**, Figure 7.8A) (Konno et al. 1990, Moinuddin et al. 2003), lacking oxygenated functionalities at C9/C9′. However, it does not accumulate any significant levels of monomeric allylphenols and propenylphenols, such as chavicol (**1**)/eugenol (**2**) or *p*-anol (**6**)/isoeugenol (**7**), or C9 monomer derivatives thereof. The creosote bush PLR homolog mentioned above, though, was able to efficiently convert the five tested substrates, *p*-coumaryl coumarate (**29**), as well as *p*-coumaryl (**32**), coniferyl (**33**), caffeyl (**36**), 5-hydroxyconiferyl (**37**), and sinapyl (**38**) acetates, in an NADH/NADPH dependent manner, into the corresponding allylphenols (**1**, **2**, **39**, **40**, **23**; Figure 7.8B), respectively (Vassão et al. 2007, Kim et al. 2014). This was the first substrate versatile allylphenol synthase (APS) characterized, and was named LtAPS1 (*Larrea tridentata* allylphenol synthase). Subsequently, kinetic parameters were obtained using *p*-coumaryl (**32**), coniferyl (**33**), caffeyl (**36**), 5-hydroxyconiferyl (**37**), and sinapyl (**38**) acetates (Table 7.1) (Kim et al. 2014). While these indicated that *p*-coumaryl (**32**) and coniferyl (**33**) acetates were the most effective substrates with k_{cat}/K_m values of ~83,000 and 87,000 M^{-1} s^{-1}, respectively, the APS was clearly substrate versatile.

The creosote bush was also found to harbor a propenylphenol synthase, PPS (named LtPPS1) with ~44/57% identity/similarity to LtAPS1 (Tables 7.2 and 7.3). It too was substrate versatile and able to efficiently convert *p*-coumaryl (**32**), coniferyl (**33**), caffeyl (**36**), 5-hydroxyconiferyl (**37**), and sinapyl (**38**) acetates, in an NADH/NADPH dependent manner, into the corresponding propenylphenols (**6**, **7**, **41–43**), respectively, (Kim et al. 2014), i.e. the creosote bush harbors genes encoding proteins capable of forming either allylphenols or propenylphenols. Kinetic data established that coniferyl

TABLE 7.1
Kinetic parameters of LtAPS1 determined with *p*-coumaryl (32), coniferyl (33), caffeyl (36), 5-hydroxyconiferyl (37), and sinapyl (38) acetates (Kim et al. 2014).

Substrates	K_m (μM)	V_{max} (pkat/μg)	k_{cat} (s^{-1})	k_{cat}/K_m (M^{-1} s^{-1})
p-Coumaryl acetate (**32**)	212 ± 33	517.0 ± 24.3	17.6	83,000
Coniferyl acetate (**33**)	100 ± 19	253.9 ± 12.5	8.7	87,000
Caffeyl acetate (**36**)	107 ± 18	73.9 ± 3.5	2.5	23,400
5-Hydroxyconiferyl acetate (**37**)	236 ± 50	144.0 ± 8.4	4.9	20,800
Sinapyl acetate (**38**)	697 ± 37	557.7 ± 10.6	19.0	27,300

acetate (**33**) was the best substrate for LtPPS1 with a k_{cat}/K_m value of ~130,000 M^{-1} s^{-1} (Table 7.4)

It was instructive to ascertain whether the creosote bush LtAPS1 additionally displayed PLR, PCBER, or IFR activities. However, assays with their potential substrates, pinoresinol (**44**), (–) larreatricin (**45**), nectandrin B (**46**), licarin A (**47**), dehydrodiconiferyl alcohol (**48**), and the isoflavones genistein (**49**) and biochanin A (**50**) (Figure 7.9), gave at best marginal activity (Vassão et al. 2007). This indicated that LtAPS1 had no efficacious PLR, PCBER, or IFR activities, even though a later contribution incorrectly stated that LtAPS1 had not been assayed for PCBER activity (Koeduka et al. 2008).

7.3.3 APS and PPS in Piper regnellii

This *Piper* species accumulates monomeric allylphenols, such as myristicin (**51**), apiol (**52**), and dillapiol (**53**)], and the 8-5′ linked lignans (+)-conocarpan (**54a**), (–)-*epi*-conocarpan (**55b**), and eupomatenoid-6 (**56**) (Figure 7.10) (Benevides et al. 1999). In continuing our search for APS or PPS homologs in different species, three recombinant *P. regnellii* enzymes, trivially named PrAPS1, PrAPS2 (~80%/71% similarity/identity to LtAPS1), and PrPPS1 (~81%/69% similarity/identity to LtPPS1), were able to efficiently catalyze monomeric allylphenol and propenylphenol formation efficiently *in vitro* (S.-J. Kim, unpublished results).

7.3.4 Kinetic Parameters of APS and PPS in Other Species: Comparison to Creosote Bush APS and PPS

Koeduka et al. (2006) reported kinetic data for the basil and petunia APS and PPS homologs mentioned previously. Both were able to weakly convert coniferyl acetate (**33**) in the presence of NADPH or NADH, to give eugenol (**2**) or isoeugenol (**7**), respectively. These recombinant enzymes were named (a) ObEGS1 and PhEGS1 in basil, thereby depicting eugenol synthase activity and (b) PhIGS1 in petunia, for an isoeugenol synthase. Kinetic values for the basil (ObEGS1) and petunia (PhIGS1) enzymes, however, were much lower than those reported for LtAPS1, i.e. with apparent K_m's of 5.1 and 1.6 mM, V_{max} of ~20 and 7 pkat μg^{-1} protein, and k_{cat}/K_m of 136 and 160 M^{-1} s^{-1}, respectively. That is, their overall k_{cat}/K_m values for coniferyl acetate (**33**) were nearly two orders of magnitude lower than for the creosote bush LtAPS1, i.e. 87,000 M^{-1} s^{-1} for the latter, vs. 136 and 160 M^{-1} s^{-1} for the basil and petunia enzymes as we had noted (Vassão et al. 2006).

This very low activity for the basil/petunia kinetic data was recognized as problematic by Koeduka et al. (2006), and was rationalized as perhaps due to the true identity of the actual ester form employed *in vivo* being uncertain (i.e. *p*-coumarate (**17**), acetate, or some other ester).

The following year, the petunia PhIGS1 kinetic measurements were revised by the same researchers with no explanation/clarification (Koeduka et al. 2008), with a PhEGS1 homolog reported as well. The revised data were now much closer to our creosote bush LtAPS1 kinetic values (i.e. k_{cat}/K_m of 5,700 vs. 87,000 M^{-1} s^{-1} for coniferyl acetate (**33**) but still of lower efficacy by ~4 fold). The revised k_{cat}/K_m of 5,700 M^{-1} s^{-1} for PhIGS1 was also very different from the earlier petunia report of ~160 M^{-1} s^{-1}. The revised K_m value of 0.23 mM had decreased by a factor of 7, whereas the V_{max} increased by ~5 fold. Kinetic data for PhEGS1 were again lower than those of the creosote bush LtAPS1 and LtPPS1 when using coniferyl acetate (**33**) as substrate (i.e. k_{cat}/K_m of 2,400 M^{-1} s^{-1}).

No other kinetic data were reported with potential substrates, such as *p*-coumaryl acetate (**32**) and *p*-coumaryl coumarate (**29**). However, in our laboratory (D. G. Vassão, unpublished data) we found that the basil ObEGS1 could utilize these to give chavicol (**1**) with an efficacy similar to that obtained with coniferyl acetate (**33**). Yet, Louie et al. (2007) concluded that ObEGS1 was of "limited ability" as regards being able to use *p*-coumaryl acetate (**32**) as a potential substrate, in contrast to our unpublished experimental findings.

In terms of nomenclature, all of the APS and PPS enzymes investigated in our laboratory (from the creosote bush (Vassão et al. 2007), and basil (Vassão et al. 2006) were substrate versatile, rather than substrate specific. Yet, the basil enzyme was named as a eugenol synthase (ObEGS1), and the petunia enzymes as eugenol/isoeugenol synthase (PhEGS1/PhIGS1) (Koeduka et al. 2006). The basil and petunia are more accurately described as being substrate versatile, and thus APS and PPS nomenclatures are considered more appropriate.

Numerous other APS/PPS homologs were preliminarily characterized, and able to catalyze comparable monolignol ester reductions *in vitro*. These were reported in *Clarkia breweri* (CbIGS1, CbEGS1, CbEGS2) (Koeduka et al. 2008),

TABLE 7.2
Percent identity of characterized APS/PPS. Sequences were obtained from the NCBI protein sequence database. The BLOSUM62 matrix was applied for scoring.

	LtPPS1*	LtAPS1	LtAPS2	PrPPS1	PrAPS1	PrAPS2	CbIGS1	CbEGS1	CbEGS2	ObEGS1	PhIGS1	PhEGS1	PaAIS1	GoEGS1	GoEGS2	GcEGS1	GcEGS2	GdEGS1	GdEGS2	DcE(I)GS1	FaEGS1a	FaEGS1b
LtAPS1	44.2																					
LtAPS2	42.2	78.9																				
PrPPS1	69.2	44.8	42.9																			
PrAPS1	44.8	72.7	67.5	48.7																		
PrAPS2	47.1	70.1	66.9	46.1	72.4																	
CbIGS1	54.7	43.5	40.6	50.6	44.8	42.2																
CbEGS1	55.0	43.5	40.6	50.3	44.5	41.9	95.9															
CbEGS2	45.6	79.5	74.4	46.6	73.7	67.2	44.7	45.0														
ObEGS1	53.5	46.1	42.9	52.5	45.1	43.5	57.6	58.6	47.9													
PhIGS1	68.6	43.8	42.9	63.9	43.5	44.8	50.6	50.0	45.3	51.6												
PhEGS1	47.1	78.6	71.4	47.7	73.1	66.9	43.8	43.5	82.1	46.1	46.8											
PaAIS1	61.0	46.8	43.8	54.9	41.6	46.1	51.3	50.9	45.3	54.5	57.6	46.1										
GoEGS1	39.9	56.5	55.2	37.9	52.9	54.9	36.5	35.2	57.0	37.6	39.3	55.2	37.8									
GoEGS2	39.6	56.5	55.2	37.6	52.9	54.9	36.2	34.9	57.0	37.6	39.0	55.2	37.5	99.7								
GcEGS1	39.6	55.2	54.5	38.2	51.9	53.9	36.2	34.9	56.0	37.6	39.0	53.9	38.1	96.0	96.3							
GcEGS2	39.9	55.5	54.2	37.6	51.9	54.5	35.5	34.3	56.0	37.3	38.4	54.5	37.5	98.8	99.1	96.0						
GdEGS1	39.9	55.2	54.5	37.9	51.9	54.2	35.5	35.2	56.0	37.9	38.7	54.2	37.5	94.5	94.8	95.1	94.5					
GdEGS2	39.9	55.2	54.5	37.9	51.9	54.2	35.5	35.2	56.0	37.9	38.7	54.2	37.2	94.2	94.5	94.8	94.2	99.7				
DcE(I)GS1	55.3	44.8	43.5	53.0	39.3	42.5	50.9	50.6	44.7	56.1	54.7	44.2	70.6	36.6	36.3	35.9	36.3	35.3	35.0			
FaEGS1a	46.6	82.5	76.9	46.3	73.1	67.9	44.3	44.0	83.8	45.6	45.6	82.1	46.6	56.6	56.6	55.7	55.7	55.7	55.7	46.3		
FaEGS1b	46.3	82.8	76.9	45.6	73.7	68.5	44.0	43.7	83.2	45.3	45.6	82.8	46.6	56.3	56.3	55.3	55.3	55.3	55.3	46.0	98.1	
FaEGS2	44.3	71.4	67.5	43.0	65.9	64.9	40.5	40.1	71.8	41.1	41.1	69.8	43.0	51.8	51.8	50.5	50.8	51.1	50.8	39.8	71.5	71.5

* CbEGS1 and CbEGS2, *C. breweri* eugenol synthases 1 and 2; CbIGS1, *Clarkia breweri* isoeugenol synthase 1; DcE(I)GS1, *Daucus carota* eugenol/isoeugenol synthase 1; FaEGS1a, FaEGS1b, and FaEGS2, *Fragaria × ananassa* eugenol synthases 1a, 1b and 2. GcEGS1 and GcEGS2, *G. conopsea* eugenol synthases 1 and 2; GdEGS1 and GdEGS2, *G. densiflora* eugenol synthases 1 and 2; GoEGS1 and GoEGS2, *Gymnadenia odoratissima* eugenol synthases 1 and 2; LtAPS1 and LtAPS2, *L. tridentata* allylphenol synthases 1 and 2; LtPPS1, *Larrea tridentata* propenylphenol synthase 1; ObEGS1, *Ocimum basilicum* eugenol synthase 1; PaAIS1, *Pimpinella anisum* anol/isoeugenol synthase 1; PhEGS1, *Petunia hybrida* eugenol synthase 1; PhIGS1, *P. hybrida* isoeugenol synthase 1; PrAPS1 and PrAPS2, *Piper regnellii* allylphenol synthases 1 and 2; PrPPS1, *P. regnellii* propenylphenol synthase 1.

TABLE 7.3
Percent similarity of characterized APS/PPS. Sequences were obtained from the NCBI protein sequence database. The BLOSUM62 matrix was applied for scoring.

	LtPPS1*	LtAPS1	LtAPS2	PrPPS1	PrAPS1	PrAPS2	CbIGS1	CbEGS1	CbEGS2	ObEGS1	PhIGS1	PhEGS1	PaAIS1	GoEGS1	GoEGS2	GcEGS1	GcEGS2	GdEGS1	GdEGS2	DcE(I)GS1	FaEGS1a	FaEGS1b
LtAPS1	56.5																					
LtAPS2	57.1	86.4																				
PrPPS1	80.5	57.1	56.5																			
PrAPS1	59.7	79.2	78.9	61.7																		
PrAPS2	60.1	78.6	77.6	60.4	82.5																	
CbIGS1	68.9	56.8	55.8	64.8	57.8	57.5																
CbEGS1	68.9	57.5	56.5	64.8	58.1	57.8	97.2															
CbEGS2	59.5	85.1	82.5	57.9	82.1	78.2	58.3	59.2														
ObEGS1	67.2	58.1	55.8	67.2	58.8	57.8	70.1	70.7	60.5													
PhIGS1	78.0	57.8	57.1	74.6	56.5	57.8	65.4	64.8	57.9	65.9												
PhEGS1	59.4	83.8	79.9	58.4	81.2	78.6	56.5	57.1	88.3	58.4	57.1											
PaAIS1	71.4	59.4	58.4	68.3	56.8	60.1	67.3	67.0	59.9	68.8	68.4	59.4										
GoEGS1	51.3	67.2	67.5	50.8	64.6	64.3	50.9	50.9	68.3	52.2	52.0	67.2	51.7									
GoEGS2	50.9	67.2	67.5	50.5	64.6	64.3	50.6	50.6	68.3	52.2	51.7	67.2	51.4	99.7								
GcEGS1	50.9	66.2	67.2	50.8	64.0	63.0	50.3	50.3	67.6	51.9	52.0	66.2	52.0	96.6	96.9							
GcEGS2	50.9	66.9	67.2	50.5	64.3	64.0	50.3	50.3	68.0	52.2	51.7	66.9	51.4	99.4	99.7	96.6						
GdEGS1	51.3	66.6	67.5	50.8	64.3	63.6	50.3	50.6	68.0	52.5	52.0	66.9	52.3	95.7	96.0	96.0	95.7					
GdEGS2	51.3	66.6	67.5	50.8	64.3	63.6	50.3	50.6	68.0	52.5	52.0	66.9	52.0	95.4	95.7	95.7	95.4	99.7				
DcE(I)GS1	68.2	56.5	55.5	67.1	53.9	55.5	66.0	66.7	57.0	67.2	66.9	56.8	80.3	49.1	48.8	48.8	48.8	48.4	48.1			
FaEGS1a	60.2	86.7	83.1	57.9	81.8	78.6	57.9	58.6	90.0	59.5	58.9	87.7	60.5	68.9	68.9	68.3	68.6	68.6	68.6	58.6		
FaEGS1b	59.5	87.0	83.4	57.3	82.5	79.2	57.3	57.9	90.3	58.9	58.3	88.3	60.2	68.3	68.3	67.6	68.0	68.0	68.0	58.3	99.0	
FaEGS2	55.7	79.5	78.2	55.0	76.9	76.0	53.7	54.4	80.9	55.0	54.0	79.2	56.3	64.4	64.4	63.4	64.1	64.4	64.1	53.1	79.9	80.9

* For abbreviations, see Table 7.2.

TABLE 7.4
LtPPS1 kinetic parameters determined with *p*-coumaryl (32), coniferyl (33), caffeyl (36), 5-hydroxyconiferyl (37), and sinapyl (38) acetates (Kim et al. 2014).

Substrates	K_m (μM)	V_{max} (pkat/μg)	k_{cat} (s^{-1})	k_{cat}/K_m (M^{-1} s^{-1})
p-Coumaryl acetate (**32**)	99 ± 10	61.2 ± 2.3	2.2	22,200
Coniferyl acetate (**33**)	36 ± 4	103.0 ± 3.3	3.7	130,000
Caffeyl acetate (**36**)	33 ± 6	60.1 ± 3.3	2.2	66,700
5-Hydroxyconiferyl acetate (**37**)	76 ± 5	13.1 ± 0.3	0.5	6,600
Sinapyl acetate (**38**)	44 ± 4	34.7 ± 1.0	1.2	27,300

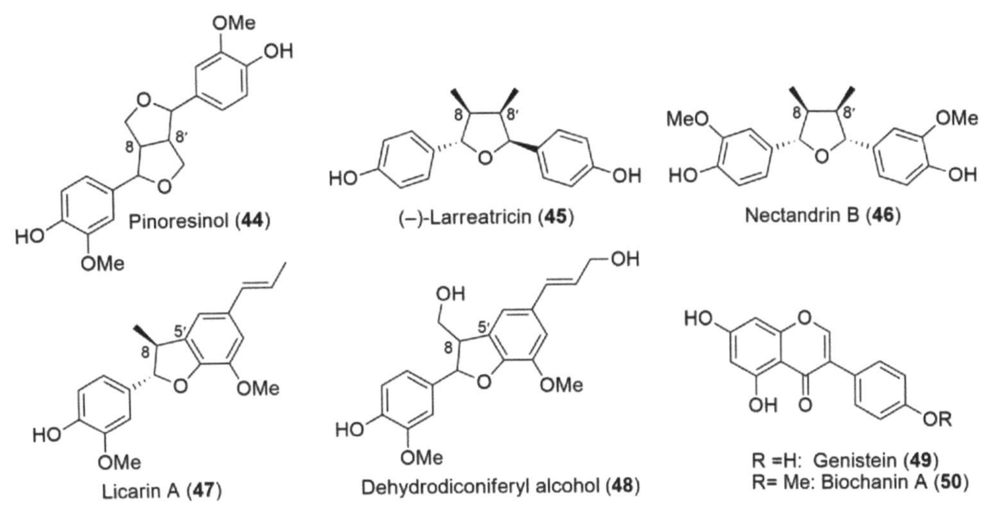

FIGURE 7.9 PLR (**44–46**), PCBER (**47, 48**) and IFR (**49, 50**) potential substrates tested with LtAPS1 (Vassão et al. 2006).

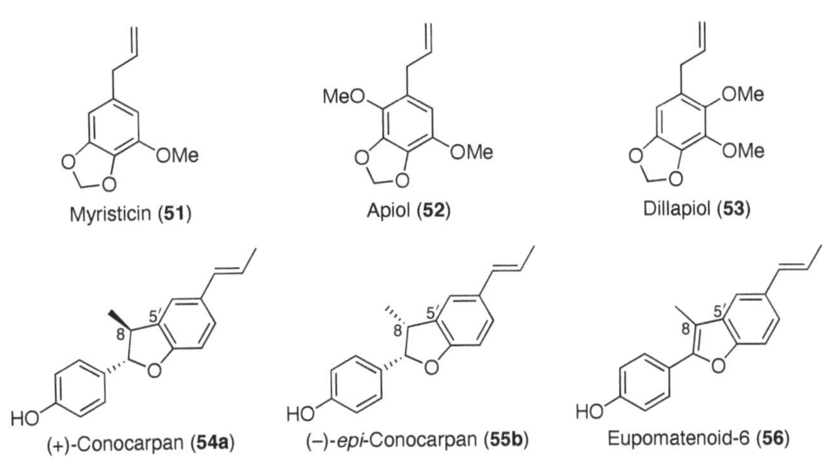

FIGURE 7.10 *Piper regnellii* metabolites (**51–56**). "**a**" and "**b**" in compounds refer to the (+)- and (–)-enantiomers, respectively.

as well as anise (*Pimpinella anisum*, PaAIS1) (Koeduka et al. 2009). Kinetic data for the *Clarkia* APS homologs (so-called CbEGS1 and CbEGS2) and PhEGS1 were again lower than those of the creosote bush LtAPS1 and LtPPS1 when using coniferyl acetate (**33**) as substrate, i.e. with k_{cat}/K_m values ranging from ~5-fold to more than an order of magnitude lower. The anise homolog, PaAIS1, also had a slow turnover compared to the creosote bush APS.

Achenes and receptacles of strawberry fruits produce eugenol (**2**). Interestingly, its amount is lower in cultivated strawberry (*Fragaria* × *ananassa*) as compared to the wild strawberry (*Fragaria vesca*) (Aragüez et al. 2013). Search of the strawberry EST database resulted in cloning of three CDNAs: *FaEGSla*, *FaEGSlb*, and *FaEGS2*. The corresponding recombinant proteins, with ~87%/82%, 87%/83%, and 80%/71% similarity/identity to LtAPS1, respectively, were heterologously expressed in *E. coli,* purified and assayed with coniferyl acetate (**33**) as substrate. FaEGSla and FaEGSlb catalyzed formation of eugenol (**2**), whereas FaEGS2 converted **33** into both eugenol (**2**) and isoeugenol (**7**). Kinetic parameters were also determined, with the data obtained for FaEGSla and FaEGSlb fitting two models depending on the coniferyl acetate (**33**) concentration, i.e. hyperbolic and sigmoidal. At lower coniferyl acetate (**33**) concentrations, K_m values were ~2040 and 770 μM, respectively, these being ~20 and 7 times higher as compared to LtAPS1. On the other hand, FaEGS2 displayed positive cooperativity for the production of both **2** and **7**, with similar affinity for coniferyl acetate (**33**), but at a five times higher velocity for eugenol (**2**) formation (Aragüez et al. 2013).

Additionally, cDNA encoding allylphenol synthases were isolated from three closely related orchid species, *Gymnadenia odoratissima* (*GoEGS1* and *GoEGS2*), *G. conopsea* (*GcEGS1* and *GcEGS2*), and *G. densiflora* (*GdEGS1* and *GdEGS2*) (Gupta et al. 2014), these having 67%/56% similarity to LtAPS1. The respective activities of the corresponding recombinant proteins were tested by expressing each individually in *E. coli* and growing the *E. coli* cells in the presence of coniferyl alcohol (**21**). Both *G. odoratissima* and *G. conopsea* EGS1 and EGS2 converted **21** into eugenol (**2**), whereas GdEGS1 and GdEGS2 produced trace amounts of isoeugenol (**7**) (Gupta et al. 2014). However, no kinetic data were obtained and only **21** was tested.

The aerial (leafy) part of carrot (*Daucus carota* subsp. *sativus*) accumulates terpenes, as well as allylphenols and propenylphenols, methyleugenol (**4**) and methylisoeugenol (**9**). Using genomic and transcriptomic data, a cDNA, *DcE(I) GS1* was isolated (Yahyaa et al. 2019). The corresponding recombinant protein, with ~68%/55% similarity/identity to LtPPS1, was heterologously expressed in *E. coli*, purified, and assayed. It catalyzed formation of both eugenol (**2**) and isoeugenol (**7**) in a 7:1 ratio. DcE(I)GS kinetic parameters determined with coniferyl acetate (**33**) as substrate gave an apparent K_m of ~250 μM, an apparent k_{cat} of 0.46 s^{-1}, and a k_{cat}/K_m of 1,840 M^{-1} s^{-1}. This k_{cat}/K_m value was ~70 and 47 times lower compared to that obtained for LtPPS1 and LtAPS1, respectively.

7.4 MONOLIGNOL ACYLTRANSFERASES

The initial studies (Koeduka et al. 2006, Vassão et al. 2006) in allylphenol and propenylphenol formation did not investigate into verifying the true chemical nature of the monolignol esters being formed *in vivo* (i.e. regarding their acyl moieties). Later though, Koeduka et al. (2008, 2009) essentially only tested coniferyl esters (e.g. **33**) for allylphenol and propenylphenol formation. This can be considered an oversimplification, given the substrate versatility observed in other studies.

Following identification that monolignol esters were allylphenol and propenylphenol synthase substrates, attention was directed towards identifying potential acyl transferases that would form the corresponding required esters. In this regard, the Clark laboratory (Dexter et al. 2007) reported identification of a cDNA that encoded a protein in the BAHD acyltransferase family, following an *in silico* analysis of a petunia petal EST database.

The BAHD superfamily name (St-Pierre and De Luca 2000) is an abbreviation for the first four enzymes characterized, these being: benzyl alcohol *O*-acetyltransferase (BEAT), anthocyanin *O*-hydroxycinnamoyl transferase (AHCT), anthranilate N-hydroxycinnamoyl/benzoyltransferase (HCBT), and deacetylvindoline 4-*O*-acetyltransferase (DAT), respectively. BAHD acyltransferases are implicated in biosynthetic pathways leading to several aroma and flavor compounds, even though amino acid identities among these enzyme homologues can vary enormously.

The corresponding putative acyltransferase gene in petunia, considered potentially involved in providing substrate(s) for the APS and PPS enzymes, was expressed primarily in petal limbs, and displayed a circadian rhythm profile that correlated well with floral volatile emission (i.e. the highest transcript levels were at night). Moreover, its RNAi-induced gene silencing led to a reduction in both level and emission of isoeugenol (**7**) [as well as reductions in other volatiles, such as 2-phenylethanol (**57**, Figure 7.11), phenylacetaldehyde (**58**), phenylethyl acetate (**59**), phenylethyl benzoate (**60**), and benzyl acetate (**61**)]. Conversely, an ~7-fold increase in coniferyl aldehyde (**62**) levels was also observed (Dexter et al. 2007).

The corresponding gene product was heterologously expressed in *E. coli*. Its recombinant protein efficiently acetylated coniferyl alcohol (**21**) at pH 7.5, with apparent K_m of 27.5 μM, k_{cat} of 0.81 s^{-1}, and k_{cat}/K_m of 29,450 M^{-1} s^{-1}). Other test substrates assayed, such as cinnamyl alcohol (**12**), sinapyl alcohol (**67**, Figure 7.12), octanol (**63**), and geraniol (**64**), were also processed albeit ~2–4 times slower. Enzymatic activities were 30- to 60-fold lower with *p*-coumaryl alcohol (**20**) and 2-phenylethanol (**57**) at comparable substrate concentrations. The acyltransferase was, however, very substrate versatile.

At pH 6.0, the enzyme with coniferyl alcohol (**21**) as substrate had a ~2-fold higher K_m (56.5 μM) and a ~2.5-fold higher k_{cat} (2.05 s^{-1}), which taken together was of similar overall

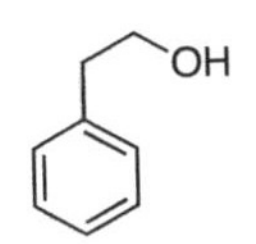
2-Phenylethanol (**57**)

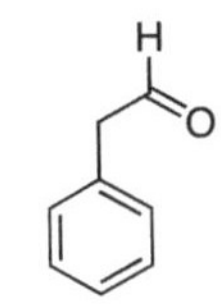
Phenylacetaldehyde (**58**)

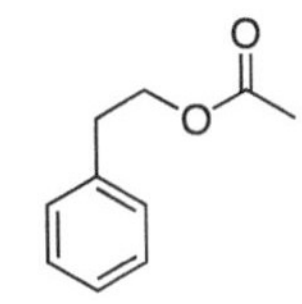
Phenylethyl acetate (**59**)

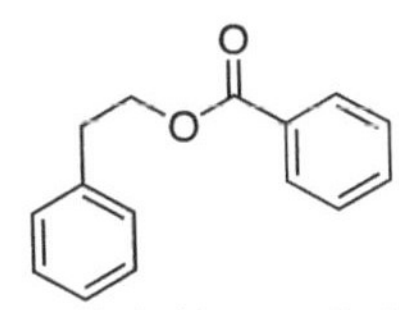
Phenylethyl benzoate (**60**)

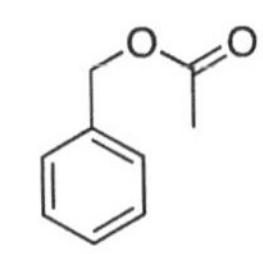
Benzyl acetate (**61**)

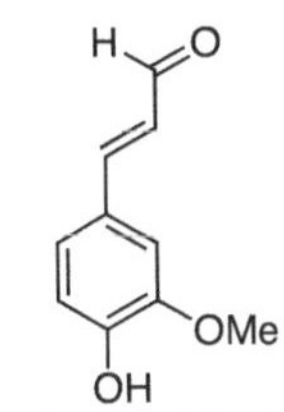
Coniferyl aldehyde (**62**)

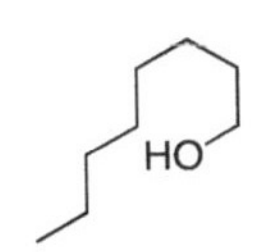
Octanol (**63**)

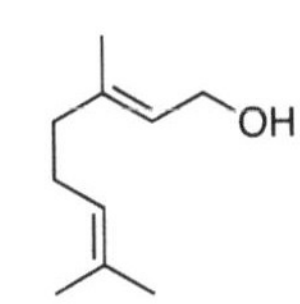
Geraniol (**64**)

FIGURE 7.11 Various volatiles formed in petunia (**57–61**), coniferyl aldehyde (**62**) and other potential acyl transferase substrates (**63** and **64**).

catalytic efficiency (k_{cat}/K_m of 36,280 M^{-1} s^{-1}). Albeit being very substrate versatile, it was named acetyl-CoA: coniferyl alcohol acetyltransferase (PhCAAT).

An analogous basil ObCAAT was also reported, this being able to acetylate *p*-coumaryl alcohol (**20**) *in vitro* as its preferred substrate, as well as coniferyl alcohol (**21**) and caffeyl alcohol (**65**) to give the corresponding acetates, *p*-coumaryl acetate (**32**), coniferyl acetate (**33**), and caffeyl acetate (**36**), respectively (Harrison and Gang 2006).

However, petunia also accumulates and emits benzoyl esters and phenylethanol derivatives. Using an RNAi strategy for the so-called PhCAAT, their levels were also adversely affected as noted above. Surprisingly, benzoyl-CoA and other potential substrates were apparently not tested with the petunia PhCAAT that is now implicated in allylphenol and propenylphenol biosynthesis. Nor were other putative substrates, such as *p*-coumaroyl- and feruloyl-CoAs, examined either. Alternate acyl donors were apparently also not tested during the preliminary characterization of the basil acyltransferase.

A gene encoding an acyltransferase, LtCAAT1, was isolated from *L. tridentata* (Kim et al. 2014). It encodes an ~51 kDa peptide with ~71%/57% amino acid similarity/identity to PhCAAT. Its pH optimum was determined to be ~4.8 with the five monolignols (**20, 21, 65–67**) tested (Figure 7.12). Kinetic data established that coniferyl alcohol (**21**) was the preferred substrate overall with k_{cat}/K_m of ~241,000 M^{-1} s^{-1}; *p*-coumaryl alcohol (**20**) was the next most effective (k_{cat}/K_m ~26,300 M^{-1} s^{-1}). The other three monolignols, **65–67**, were poorly utilized (Table 7.5). The k_{cat}/K_m for acetyl CoA, when measured with coniferyl alcohol (**21**), was determined to be ~203,000 M^{-1} s^{-1}. LtCAAT1 is thus fairly substrate versatile.

Given the observed substrate versatilities being noted, we thus consider that the depiction of a "true" *in vivo* substrate

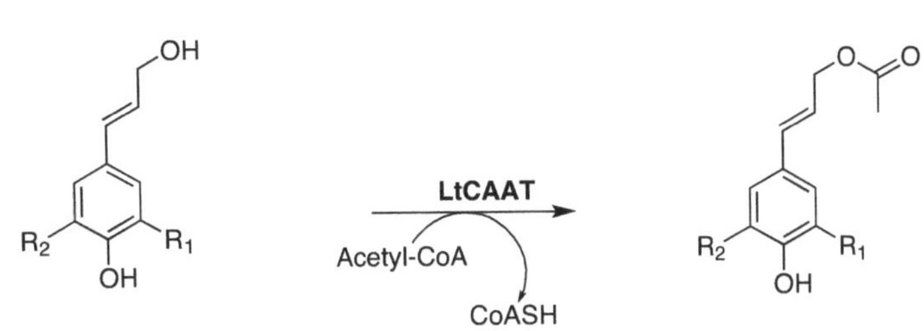

$R_1 = R_2 =$ H, *p*-Coumaryl alcohol (**20**)
$R_1 =$ OMe, $R_2 =$ H, Coniferyl alcohol (**21**)
$R_1 =$ OH, $R_2 =$ H, Caffeyl alcohol (**65**)
$R_1 =$ OMe, $R_2 =$ OH, 5-Hydroxyconiferyl alcohol (**66**)
$R_1 = R_2 =$ OMe, Sinapyl alcohol (**67**)

$R_1 = R_2 =$ H, *p*-Coumaryl acetate (**32**)
$R_1 =$ OMe, $R_2 =$ H, Coniferyl acetate (**33**)
$R_1 =$ OH, $R_2 =$ H, Caffeyl acetate (**36**)
$R_1 =$ OMe, $R_2 =$ OH, 5-Hydroxyconiferyl acetate (**37**)
$R_1 = R_2 =$ OMe, Sinapyl acetate (**38**)

FIGURE 7.12 Biosynthetic transformations catalyzed by the creosote bush (*Larrea tridentata*) acetyl-CoA: coniferyl alcohol acetyltransferase (LtCAAT).

TABLE 7.5
Kinetic parameters of LtCAAT1 determined *p*-coumaryl (20), coniferyl (21), caffeyl (65), 5-hydroxyconiferyl (66), and sinapyl (67) alcohols, as well as acetyl-CoA with coniferyl alcohol (21) (Kim et al. 2014).

Substrates	K_m (μM)	V_{max} (pkat/μg)		k_{cat} (s^{-1})	k_{cat}/K_m (M^{-1} s^{-1})
p-Coumaryl alcohol (**20**)	95 ± 6	48.5	± 1.2	2.5	26,300
Coniferyl alcohol (**21**)	174 ± 8.5	821.0	± 17.7	42.1	241,000
Caffeyl alcohol (**65**)	82 ± 8	10.3	± 0.4	0.5	6,100
5-Hydroxyconiferyl alcohol (**66**)	476 ± 27	81.7	± 2.8	4.2	8,800
Sinapyl alcohol (**67**)	302 ± 33	24.5	± 1.3	1.3	4,300
Acetyl-CoA [with coniferyl alcohol (**21**)]	133 ± 9	527.0	± 14.7	27	203,000

[e.g. coniferyl alcohol (**21**) and acetyl CoA] as being too restrictive at present, until further verification of the true substrate(s) is obtained.

7.5 APS/PPS, PCBER, PLR, AND IFR (-LIKE) MULTI-GENE FAMILY CONSIDERATIONS

Some of the greatest challenges in plant (and other) genomics are in identifying true biochemical/physiological functions and catalytic mechanisms, and in deducing/hypothesizing how such biochemical functions may have evolved. Plant genes belonging to APS/PPS, PCBER, PLR, and IFR-like multi-gene families are no different in that regard. Gene family members are often annotated function based on some approximate sequence similarity/identity level to *bona fide* members of a multi-gene family of truly known physiological/biochemical function.

In reality though, provisional gene annotations require validation in various mutually reinforcing ways to be truly rigorous. These include demonstrating: (a) comparable kinetic parameters *in vitro* of the candidate enzyme relative to the actual *bona fide* enzyme type; (b) that in addition to the annotated genes (and corresponding enzymes), the substrates/products are produced *in planta,* including whether they are co-localized (or not) to the same cell/tissue type; and (c) whether pathway enzyme-level manipulation (such as via overexpression or RNAi gene expression suppression) can alter the levels of pathway metabolites in the pathway of question. Yet, such rigor in experimental design is often lacking; see Koeduka et al. (2008) and Koeduka et al. (2009) as examples.

Addressed in the following, nevertheless, is what is currently known about the APS/PPS, PCBER, PLR, and IFR homologs.

7.6 APS/PPS, PLR, PCBER, AND IFR STRUCTURES AND MECHANISTIC STUDIES

APS/PPS (Vassão et al. 2007), PLR (Chu et al. 1993, Dinkova-Kostova et al. 1996, Fujita et al. 1999, Min et al. 2003), PCBER (Gang et al. 1999, Min et al. 2003), and IFR (Paiva et al. 1991, Paiva et al. 1994) are NADPH-dependent SDR (short-chain dehydrogenase/reductase) gene family members. Earlier PLR and PCBER X-ray crystal structure determinations, together with that of the modeled IFR (Min et al. 2003), provided important insights into presumed APS/PPS catalysis (Vassão et al. 2007). As for PLR/PCBER/IFR, the creosote bush LtAPS1 had a nucleotide-binding Rossman fold motif [residues 11–17 (GxxGxxG)] and a conserved catalytic Lys residue (Lys133 in LtAPS1) as a general base.

Its mutagenesis (i.e. the recombinant K133A mutant) resulted in no detectable APS catalytic activity (Vassão et al. 2007), with Lys133 viewed as abstracting the ester substrate's phenolic proton to afford the putative quinone methide intermediate via displacement of the corresponding ester functionality (Vassão et al. 2007) (Figure 7.7). Subsequent hydride addition at C7 from NAD(P)H with concomitant regeneration of the phenolic ring gave the corresponding allylphenol products, chavicol (**1**) and eugenol (**2**), depending on the substrates (**29**, **32**, **33**) tested (Vassão et al. 2007). Moreover, based on our previous studies, Tyr15 and Ile16 in LtAPS1 were hypothesized to interact with the NADPH pyrophosphate group, with Phe155 envisaged as stacked against the NADPH nicotinamide group. With LtPPS1, hydride addition to C9 of the putative quinone methide intermediate gave the conjugated propenylphenol products **6** and **7**. Thus, as previously proposed for PLR and PCBER (Davin and Lewis 2022c), these data allowed us to invoke involvement of quinone methides as enzyme-bound intermediates in APS and PPS catalysis.

X-ray crystal structures for the basil APS (ObEGS1) in its apo-form (Figure 7.13A) at ~1.8 Å, and in complexes with either $NADP^+$, NADPH, or $NADP^+$ and the inhibitor EMDF [(7*S*,8*S*)-ethyl-(7,8-methylene)-dihydroferulate, **68**, Figure 7.13B)] further confirmed our earlier study findings (Vassão et al. 2007). Interestingly though, the putative (but unproven) "true" coniferyl acetate (**33**) substrate was neither stably incorporated into the crystals nor readily modeled into the basil APS active site.

These X-ray structures (Louie et al. 2007), as had been anticipated in advance (Vassão et al. 2007), closely resembled those previously reported by ourselves for PLR and PCBER, IFR (Min et al. 2003, Davin and Lewis 2022b, c), and that modeled for LtAPS1. The basil APS also had a Rossman fold motif (residues 14–20 (GXXGXXG), *versus* residues 11–17

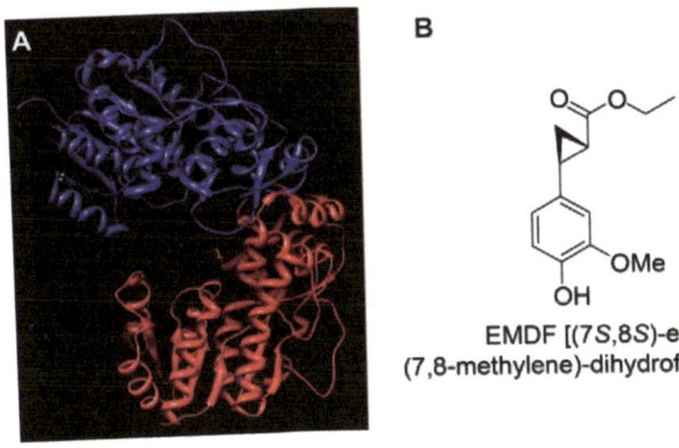

EMDF [(7*S*,8*S*)-ethyl-(7,8-methylene)-dihydroferulate (**68**)

FIGURE 7.13 A. Schematic outlines of the dimeric forms of *Ocimum basilicum* APS (EGS, PDB ID: 2QYS). Image created using NGL viewer (Rose et al. 2018). **B.** EMDF [(7*S*,8*S*)-ethyl-(7,8-methylene)-dihydroferulate (**68**).

in LtAPS1; Tyr and Ile at residues 18/19 interacting with the NADPH pyrophosphate group, *versus* at 15/16 for LtAPS1; Phe154 (Phe155 in LtAPS1), stacking against the NADPH nicotinamide group; and Lys132 (Lys133 in LtAPS1) for general base catalysis. Indeed, further confirmation of our earlier analysis (Vassão et al. 2007) was observed following mutation of the Lys132 residue to afford K132A (Louie et al. 2007), as this also abolished enzymatic activity. The X-ray data therefore fully confirmed our prior results from site-directed mutagenesis and homology modelling

PLRs, PCBERs, APSs, and PPSs from different plant species display considerable homology to each other, with similar catalytic mechanisms, i.e. involving transfer of a hydride onto a phenylpropanoid-derived side-chain substrate. It is not fully understood, though, to what extent these different enzyme types might display some degree of substrate versatility (degeneracy) as regards potential substrates or substrate homologs *in vitro,* even if there are massive differences in kinetic parameters. While monolignol esters **29**, **32**, and **33** are either smaller and/or of similar size than PLR substrates, they share some common structural features for potential binding and catalysis (e.g. a free phenolic group) – albeit even if less efficaciously processed by, for example, a PLR. A similar level of substrate versatility may be displayed by members of the PLR, PCBER, and IFR reductases. In contrast though, the APS/PPS enzymes might not necessarily be able to efficaciously utilize the larger (and less conformationally flexible) lignans as substrates, i.e. as productive binding and catalysis might be constrained as a result of greater steric hindrance with the larger substrates.

7.7 APS AND PPS TAXONOMICAL DIVERSITY

As indicated previously, APSs and PPSs are closely related to NADPH-dependent reductases in the phenylpropanoid/phenylpropanoid acetate/lignan pathways belonging to the "PIP" family. [The term "PIP" derives from PLR, isoflavone reductase (IFR) and PCBER (Vassão et al. 2008).] A phylogenetic analysis (Figure 7.14) indicates that these APSs/PPSs cluster together with PLRs, IFRs, and PCBERs as well as leucoanthocyanidin and pterocarpan reductases (LACRs and PTRs) (Letunic and Bork 2007). It must be cautioned, though, that while clustering occurs as depicted, true physiological/biochemical functions need to be demonstrated. Nevertheless, the following is known to date.

7.7.1 Creosote Bush

The creosote bush, as aforementioned, essentially exclusively and abundantly accumulates 8–8′-linked lignans, but apparently no other lignan types such as those with 8–5′ linkages, etc. Accordingly, the presence of both APS (LtAPS1 and LtAPS2) and PPS (LtPPS1) genes (of ~57%/43% similarity/identity to each other) in the creosote bush provides a biochemical explanation for both allylphenol and propenylphenol formation in this species. Yet interestingly, although APS genes cluster with various PCBER/PCBER homologs (Figure 7.14), only marginal (if any) PCBER activity was detected for LtAPS1 under the assay conditions deployed.

The creosote bush PPS (LtPPS1) has an amino acid sequence of ~78%/69% and ~71%/61% similarity/identity to petunia PPS (PhIGS1) and anise PPS (PaAIS1) homologs, respectively, with these grouping together as a clade (Figure 7.14). Such segregation may, though, not be indicative of independent evolution, as the creosote bush APS and PPS fall into different sub-clades. Many more *bona fide* plant PPS and APS enzymes than currently available thus need to be identified, and verified as such, to better consider evolutionary processes.

7.7.2 *Piper regnellii*

P. regnellii, as aforementioned, accumulates allylphenols myristicin (**51**), apiol (**52**), and dillapiol (**53**), as well as 8–8′-linked phenylpropene-type dimeric lignans, (+)-conocarpan (**54a**), (–)-*epi*-conocarpan (**55b**), and eupomatenoid-6 (**56**) (Figure 7.10) (Benevides et al. 1999). Their APS (PrAPS1 and PrAPS2) amino acid sequences (~80% similarity and ~71% identity to creosote bush LtAPS1), together with several putative PCBER and PCBER-like genes in other plant species, are within the same clade (Figure 7.14). Again, the presence of various putative PCBER gene family members in this clade further puts in question the nature of their true physiological role(s) in various organisms, as we have repeatedly cautioned (Gang et al. 1999). To our knowledge, *P. regnellii* lacks 8–3′/8–5′ linked lignans with 7–*O*–4′ inter-unit linkages reduced, so the presence of putative PCBER and PCBER-like homologs in the same clade may not be indicative of their actual biochemical functions *in planta*. Furthermore, PrAPS1 and PrAPS2 are unlikely to have PCBER functions *in vivo*, and presumably only serve in an APS function. This again reminds of potential limitations in only using phylogenetic analyses to attempt to predict potential biochemical function(s).

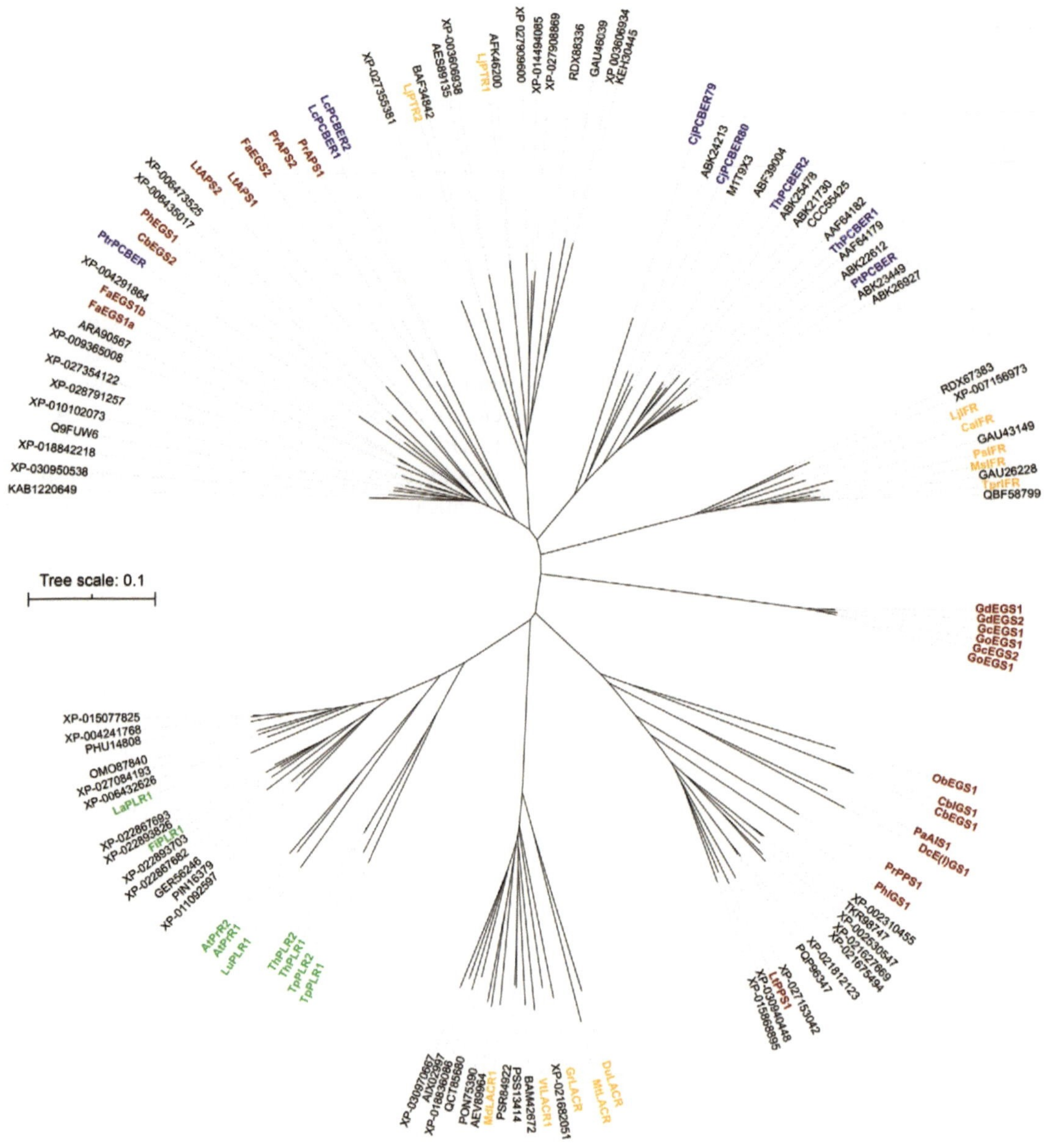

FIGURE 7.14 Unrooted phylogenetic tree of known and putative allylphenol and propenylphenol synthases (APS, PPS), pinoresinol-lariciresinol reductases (PLRs), phenylcoumaran benzylic ether reductases (PCBERs, isoflavone reductases (IFRs) and leucoanthocyanidin and pterocarpan reductases (LACRs and PTRs). Sequences were obtained from NCBI and a multiple sequence alignment was built with Clustal Omega (Madeira et al. 2019). The tree was rendered with iTOL (Letunic and Bork 2007). APSs and PPSs: *Clarkia breweri* [CbIGS1 (ABR24112), CbEGS1 (ABR24113) and CbEGS2 (ABR24114)], *Daucus carota* [*DcE(I)GS1* (XP_017241251), *Fragaria* x *ananassa* (FaEGS1a (AGV02006), FaEGS1b (AGV02007) and FaEGS2 (AGV02008), *Gymnadenia conopsea* [GcEGS1 (AKB11747) and GcEGS2 (AKB11748)]; *G. densiflora* [GdEGS1 (AKB11749) and GdEGS2 (AKB11750)], *G. odoratissima* [GoEGS1 (AKB11751) and GoEGS2 (AKB11752)], *Larrea tridentata* [LtAPS1 (AHA90804), LtAPS2 (AHA90805) and LtPPS1 (AHA90806)], *Ocimum basilicum* [ObEGS1 (ABD17321)], *Petunia hybrida* [PhIGS1 (ABD17322) and PhEGS1 (ABR24115)], *Pimpinella anisum* [PaAIS1 (ACL13526)] and *Piper regnellii* [PrAPS1 (AHA90807), PrAPS2 (AHA90808) and PrPPS1 (AHA90809)]. LACRs: *Desmodium uncinatum* [DuLACR (Q84V83)], *Gossypium raimondii* [GrLACR (CAI56324)], *Malus domestica* [MdLACR1 (AAZ79364)], *Medicago truncatula* [MtLACR (CAI56327)] and *Vitis vinifera* [VtLACR1 (CAI26309)]. IFRs: *Cicer arietinum* [CaIFR (Q00016)], *Lotus japonicus* [LjIFR (QBF58800)], *Medicago sativa* [MsIFR (AAC48976)], *Pisum sativum* [PsIFR (P52576)] and *Trifolium pratense* [TprIFR (QBF58799)]. PCBERs: *Cryptomeria japonica* [CjPCBER79 and CjPCBER80], *Linum corymbulosum* [LcPCBER1 (ACA60729) and LcPCBER2 (ACA60730)], *Pinus taeda* [PtPCBER (AAC32591)], *Populus trichocarpa* [PtrPCBER (CAA06706)] and *Tsuga heterophylla* (ThPCBER1 (AAF64176) and ThPCBER2 (AAF64177)]. PLRs: *Arabidopsis thaliana* [AtPrR1 (Q9FVQ6) and AtPrR2 (Q9SVP6)], *Forsythia intermedia* [FiPLR (AAC49608)], *Linum album* [LaPLR1 (Q4R0I0)], *Linum usitatissimum* [LuPLR1 (CAH60858)], *Thuja plicata* [TpPLR2 (Q9LD13) and TpPLR1 (Q9LD14)] and *Tsuga heterophylla* [ThPL1R (AAF64184) and ThPLR2 (AAF64185)]. PTRs: *Lotus japonicus* [LjPTR1 (BAF34841) and LjPTR2 (BAF34842)].

7.7.3 Basil

Basil glandular trichomes produce and accumulate chavicol (**1**) and eugenol (**2**). The gene encoding the basil APS (ObEGS1) currently falls into a sub-clade with the petunia PPS (PhIGS1), the creosote bush PPS (LtPPS1), and the *Clarkia* APS/PPS (CbEGS1 and CbIGS1) genes, but not the creosote bush APS (LtAPS1 and LtAPS2) (Figure 7.14). To date, more information is needed before any meaningful hypothesis can be made as regards biochemical pathway evolution, i.e. as to whether the evolutionary processes leading to APS activity in different plants were of either convergent or divergent nature.

7.7.4 *Pinus taeda*

Loblolly pine PtPCBER is in a separate clade from either APS or PPS gene families, according to Figure 7.14. While cognizant of its low kinetic parameters, PtPCBER (Gang et al. 1999) was isolated from a cell line harboring 8–5′-linked lignans, such as **48**, **69–70** (Figures 7.9 and 7.15) (Nose et al. 1995, Lewis and Davin 1999), and was able to catalyze 7–*O*–4′ reductions with 8–5′-linked lignans. PtPCBER is also expressed in vascular cambium/ray parenchyma cells, this being consistent with a lignan-related metabolic function. It is not known whether PtPCBER is bifunctional, i.e. also having APS or PPS properties *in vitro* and *in vivo*, in addition to the PCBER properties already documented. As Pinaceae members can contain very minor amounts of allylphenols and propenylphenols in their essential oils, this (dual) possibility cannot be excluded.

7.7.5 *Cryptomeria japonica*

C. japonica contains at least two PCBER genes (of ~85% and 72%–80% similarity and identity to that of the gene encoding PtPCBER) (Vassão et al. 2010) and biosynthesizes various 8–5′ linked lignans, including the 7–*O*–4′ reduced, tetra-acylated derivative **71** (Figure 7.15) (Su et al. 1995). *C. japonica* currently has no chemotaxonomic evidence for presence of allylphenols and propenylphenols. Formation of tetra-acetylated lignans, such as **71**, might involve intermediacy of monolignol esters (e.g. **33**), or alternatively acetylated post-coupling.

7.7.6 Petunia APS

The petunia APS (PhEGS1) was also examined for potential PCBER efficacy, but had marginal PCBER activity (Koeduka et al. 2008). This was not an unexpected finding given that there are currently no reports of 8–5′-linked lignans with 7–*O*–4′ inter-unit linkage-reduced metabolites from this species.

7.7.7 *Clarkia* APS

The *Clarkia* APS (CbEGS2) protein (Koeduka et al. 2008) falls into an APS and PCBER-like sub-clade (Figure 7.14). However, its APS activity was also an order of magnitude lower than the creosote bush APS (LtAPS1), i.e. indicative of fairly low catalytic activity overall. It was also investigated as to whether the *Clarkia* APS had PCBER activity, even though *Clarkia* flowers are not known to produce 7–*O*–4′-reduced 8–5′-linked lignan. It will be important to know if this is its biochemical function *in vivo*.

7.7.8 *Gymnadenia* APS

The six *Gymnadenia* APSs (Gupta et al. 2014) all fall in a separate clade between the PPS and IFR clades. Their biochemical/physiological roles await discovery.

7.7.9 *Fragaria* (Strawberry) APS

The three strawberry APSs all fall in the APS clade. That they are genuine eugenol synthases *in planta* was confirmed by the transient expression of the corresponding cDNAs in

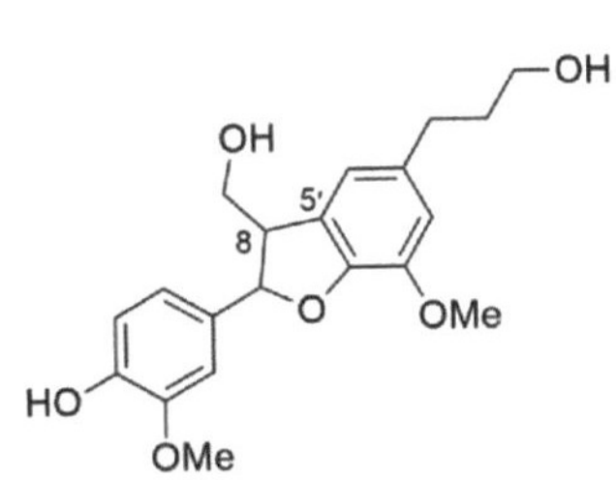

Dihydrodehydrodiconiferyl alcohol (**69**)

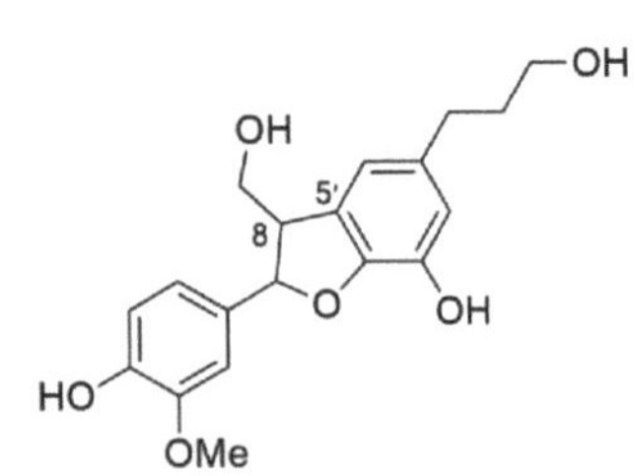

Cedrusin (**70**)

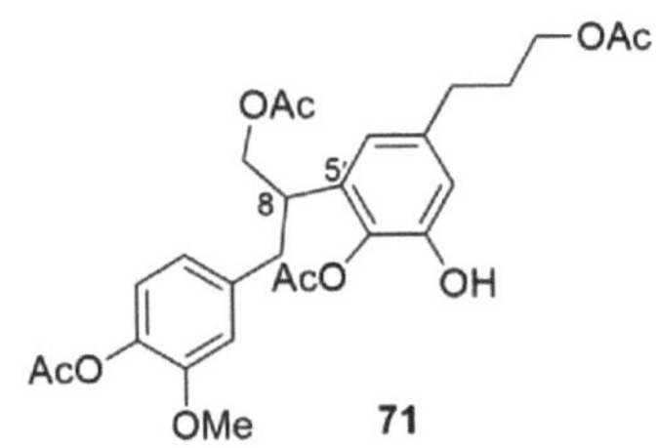

71

FIGURE 7.15 8–5′-Linked lignans **69–71**.

strawberry fruit and subsequent analysis of volatile analyses (Aragüez et al. 2013).

ACKNOWLEDGEMENTS

The authors thank the Chemical Sciences, Geosciences and Biosciences Division, DOE Office of Basic Energy Sciences (DE-FG-0397ER20259), as well as the Arthur and Katie Eisig-Tode Endowment.

REFERENCES

Aragüez, I., S. Osorio, T. Hoffmann, et al. 2013. Eugenol production in achenes and receptacles of strawberry fruits is catalyzed by synthases exhibiting distinct kinetics. *Plant Physiol.* 163:946–958.

Benevides, P. J. C., P. Sartorelli, and M. J. Kato. 1999. Phenylpropanoids and neolignans from *Piper regnellii*. *Phytochemistry* 52:339–343.

Birch, A. J., and M. Slaytor. 1956. Reduction of cinnamyl alcohols with aluminium chloride and lithium aluminium hydride. *Chem. Ind.*:1524.

Birch, A. J. 1963. Biosynthetic pathways. In *Chemical Plant Taxonomy* (T. Swain ed.), Academic Press, London, UK. pp. 141–166.

Canonica, L., P. Manitto, D. Monti, and M. Sanchez A. 1971. Biosynthesis of allylphenols in *Ocymum basilicum* L. *J. Chem. Soc., Chem. Commun.*:1108–1109.

Chu, A., A. Dinkova, L. B. Davin, D. L. Bedgar, and N. G. Lewis. 1993. Stereospecificity of (+)-pinoresinol and (+)-lariciresinol reductases from *Forsythia intermedia*. *J. Biol. Chem.* 268:27026–27033.

Davin, L. B., and N. G. Lewis. 2022a. Redefining lignan/neolignan nomenclature and diversity: A historical perspective. In *The Lignan Handbook* (N. G. Lewis and L. B. Davin eds.), Taylor & Francis, Boca Raton, London, pp 1–11.

Davin, L. B., and N. G. Lewis. 2022b. 8–2′-, 8–3′-, 8–5′-, and 8–*O*–4′-Linked lignan biosynthesis roadmap. In *The Lignan Handbook* (N. G. Lewis and L. B. Davin eds.), Taylor & Francis, Boca Raton, London, pp 113–125.

Davin, L. B., and N. G. Lewis. 2022c. 8–8′-Lignan biosynthesis roadmap. In *The Lignan Handbook M* (N. G. Lewis and L. B. Davin eds.), Taylor & Francis, Boca Raton, London, pp 79–112.

Davin, L. B., and N. G. Lewis. 2022d. 8–8′- and 8–7′-Norlignan biosynthesis roadmap. In *The Lignan Handbook* (N. G. Lewis and L. B. Davin eds.), Taylor & Francis, Boca Raton, London, pp 147–152.

Dexter, R., A. Qualley, C. M. Kish, et al. 2007. Characterization of a petunia acetyltransferase involved in the biosynthesis of the floral volatile isoeugenol. *Plant J.* 49:265–275.

Dinkova-Kostova, A. T., D. R. Gang, L. B. Davin, et al. 1996. (+)-Pinoresinol/(+)-lariciresinol reductase from *Forsythia intermedia*. Protein purification, cDNA cloning, heterologous expression and comparison to isoflavone reductase. *J. Biol. Chem.* 271:29473–29482.

Fujita, M., D. R. Gang, L. B. Davin, and N. G. Lewis. 1999. Recombinant pinoresinol-lariciresinol reductases from western red cedar (*Thuja plicata*) catalyze opposite enantiospecific conversions. *J. Biol. Chem.* 274:618–627.

Gang, D. R., H. Kasahara, Z.-Q. Xia, et al. 1999. Evolution of plant defense mechanisms. Relationships of phenylcoumaran benzylic ether reductases to pinoresinol-lariciresinol and isoflavone reductases. *J. Biol. Chem.* 274:7516–7527.

Gang, D. R., J. Wang, N. Dudareva, et al. 2001. An investigation of the storage and biosynthesis of phenylpropenes in sweet basil. *Plant Physiol.* 125:539–555.

Geissman, T. A., and D. H. G. Crout. 1969. Compounds derived from shikimic acid. In *Organic Chemistry of Secondary Metabolism* (T. A. Geissman and D. H. G. Crout eds.), Freeman, Cooper & Company, San Francisco, CA. pp. 137–166.

Gupta, A. K., I. Schauvinhold, E. Pichersky, and F. P. Schiestl. 2014. *Eugenol synthase* genes in floral scent variation in *Gymnadenia* species. *Funct. Integr. Genomics* 14:779–788.

Harrison, B., and D. R. Gang. 2006. Characterization of coniferyl acetate acetyl transferase from sweet basil (*Ocimum basilicum*). The 19th Rocky Mountain Regional Meeting, Tucson, AZ.

Kaneko, K. 1960a. Biogenetic studies of natural products. IV. Biosynthesis of anethole by *Foeniculum vulgare*. *Chem. Pharm. Bull.* 8:611–614.

Kaneko, K. 1960b. Biogenetic studies of natural products. V. Biosynthesis of anethole by *Foeniculum vulgare*. *Chem. Pharm. Bull.* 8:875–879.

Kaneko, K. 1961. Biogenetic studies of natural products. VI. Biosynthesis of anethole by *Foeniculum vulgare*. *Chem. Pharm. Bull.* 9:108–109.

Kaneko, K. 1962. Biogenetic studies of natural products. VIII. Biosynthesis of anethole by *Foeniculum vulgare*. *Chem. Pharm. Bull.* 10:1085–1087.

Kim, S.-J., D. G. Vassão, S. G. A. Moinuddin, et al. 2014. Allyl/propenyl phenol synthases from the creosote bush and engineering production of specialty/commodity chemicals, eugenol/isoeugenol, in *Escherichia coli*. *Arch. Biochem. Biophys.* 541:37–46.

Klischies, M., J. Stöckigt, and M. H. Zenk. 1975. Biosynthesis of the allylphenols eugenol and methyleugenol in *Ocymum basilicum* L. *J. Chem. Soc., Chem. Commun.*:879–880.

Koeduka, T., E. Fridman, D. R. Gang, et al. 2006. Eugenol and isoeugenol, characteristic aromatic constituents of spices, are biosynthesized via reduction of a coniferyl alcohol ester. *Proc. Natl. Acad. Sci. USA* 103:10128–10133.

Koeduka, T., G. V. Louie, I. Orlova, et al. 2008. The multiple phenylpropene synthases in both *Clarkia breweri* and *Petunia hybrida* represent two distinct protein lineages. *Plant J.* 54:362–374.

Koeduka, T., T. J. Baiga, J. P. Noel, and E. Pichersky. 2009. Biosynthesis of *t*-anethole in anise: Characterization of *t*-anol/isoeugenol synthase and an *O*-methyltransferase specific for a C7–C8 propenyl side chain. *Plant Physiol.* 149:384–394.

Konno, C., Z.-Z. Lu, H.-Z. Xue, et al. 1990. Furanoid lignans from *Larrea tridentata*. *J. Nat. Prod.* 53:396–406.

Letunic, I., and P. Bork. 2007. Interactive Tree of Life (iTOL): An online tool for phylogenetic tree display and annotation. *Bioinformatics* 23:127–128.

Lewis, N. G., and L. B. Davin. 1999. Lignans: Biosynthesis and function. In *Comprehensive Natural Products Chemistry* (Sir D. H. R. Barton, K. Nakanishi and O. Meth-Cohn eds.), Vol. 1, Elsevier, Oxford. pp. 639–712.

Lewis, N. G., L. B. Davin, S.-J. Kim, et al. 2015. Genes encoding chavicol/eugenol synthase from the creosote bush *Larrea tridentata*. No.: 9,131,648 B2, issued September 15, 2015. Washington State University, Pullman, WA (United States). http://www.osti.gov/scitech/servlets/purl/1215614.

Louie, G. V., T. J. Baiga, M. E. Bowman, et al. 2007. Structure and reaction mechanism of basil eugenol synthase. *PLoS One* 2:e993.

Madeira, F., Y. M. Park, J. Lee, et al. 2019. The EMBL-EBI search and sequence analysis tools APIs in 2019. *Nucleic Acids Res.* 47:W636-W641.

Manitto, P., D. Monti, and P. Gramatica. 1974. Biosynthesis of phenylpropanoid compounds. Part I. Biosynthesis of eugenol in *Ocimum basilicum* L. *J. Chem. Soc., Perkin Trans. 1*:1727–1731.

Min, T., H. Kasahara, D. L. Bedgar, et al. 2003. Crystal structures of pinoresinol-lariciresinol and phenylcoumaran benzylic ether reductases and their relationship to isoflavone reductases. *J. Biol. Chem.* 278:50714–50723.

Moinuddin, S. G. A., S. Hishiyama, M.-H. Cho, L. B. Davin, and N. G. Lewis. 2003. Synthesis and chiral HPLC analysis of the dibenzyltetrahydrofuran lignans, larreatricins, 8′-*epi*-larreatricins, 3,3′-didemethoxyverrucosins and *meso*-3,3′-didemethoxynectandrin B in the creosote bush (*Larrea tridentata*): Evidence for regiospecific control of coupling. *Org. Biomol. Chem.* 1:2307–2313.

Nose, M., M. A. Bernards, M. Furlan, et al. 1995. Towards the specification of consecutive steps in macrolecular lignin assembly. *Phytochemistry* 39:71–79.

Oommen, A., R. A. Dixon, and N. L. Paiva. 1994. The elicitor-inducible alfalfa isoflavone reductase promoter confers different patterns of developmental expression in homologous and heterologous transgenic plants. *Plant Cell* 6:1789–1803.

Paiva, N. L., R. Edwards, Y. Sun, G. Hrazdina, and R. A. Dixon. 1991. Stress responses in alfalfa (*Medicago sativa* L.) 11. Molecular cloning and expression of alfalfa isoflavone reductase, a key enzyme of isoflavonoid phytoalexin biosynthesis. *Plant Mol. Biol.* 17:653–667.

Paiva, N. L., Y. J. Sun, R. A. Dixon, H. D. VanEtten, and G. Hrazdina. 1994. Molecular cloning of isoflavone reductase from pea (*Pisum sativum* L.): Evidence for 3*R*-isoflavanone intermediate in (+)-pisatin biosynthesis. *Arch. Biochem. Biophys.* 312:501–510.

Rose, A. S., A. R. Bradley, Y. Valasatava, et al. 2018. NGL viewer: Web-based molecular graphics for large complexes. *Bioinformatics* 34:3755–3758.

Senanayake, U. M., R. B. H. Wills, and T. H. Lee. 1977. Biosynthesis of eugenol and cinnamic aldehyde in *Cinnamomum zeylanicum*. *Phytochemistry* 16:2032–2033.

St-Pierre, B., and V. De Luca. 2000. Evolution of acyltransferase genes: Origin and diversification of the BAHD superfamily of acyltransferases involved in secondary metabolism. In *Recent Advances in Phytochemistry* (J. T. Romeo, R. Ibrahim, L. Varin and V. De Luca eds.), Vol. 34, *Evolution of Metabolic Pathways*, Elsevier, Oxford, UK. pp. 285–315.

Su, W.-C., J.-M. Fang, and Y.-S. Cheng. 1995. Flavonoids and lignans from leaves of *Cryptomeria japonica*. *Phytochemistry* 40:563–566.

Suzuki, S., T. Umezawa, and M. Shimada. 2001. Norlignan biosynthesis in *Asparagus officinalis* L.: The norlignan originates from two non-identical phenylpropane units. *J. Chem. Soc., Perkin Trans. 1*:3252–3257.

Suzuki, S., T. Nakatsubo, T. Umezawa, and M. Shimada. 2002. First *in vitro* norlignan formation with *Asparagus officinalis* enzyme preparation. *J. Chem. Soc., Chem. Commun.*:1088–1089.

Suzuki, S., M. Yamamura, M. Shimada, and T. Umezawa. 2004. A heartwood norlignan, (*E*)-hinokiresinol, is formed from 4-coumaryl 4-coumarate by a *Cryptomeria japonica* enzyme preparation. *J. Chem. Soc., Chem. Commun.*:2838–2839.

Tressl, R., and F. Drawert. 1973. Biogenesis of banana volatiles. *J. Agric. Food Chem.* 21:560–565.

Vassão, D. G., D. R. Gang, T. Koeduka, et al. 2006. Chavicol formation in sweet basil (*Ocimum basilicum*): Cleavage of an esterified C9 hydroxyl group with NAD(P)H-dependent reduction. *Org. Biomol. Chem.* 4:2733–2744.

Vassão, D. G., S.-J. Kim, J. K. Milhollan, et al. 2007. A pinoresinol-lariciresinol reductase homologue from the creosote bush (*Larrea tridentata*) catalyzes the efficient *in vitro* conversion of *p*-coumaryl/coniferyl alcohol esters into the allylphenols chavicol/eugenol, but not the propenylphenols *p*-anol/isoeugenol. *Arch. Biochem. Biophys.* 465:209–218.

Vassão, D. G., L. B. Davin, and N. G. Lewis. 2008. Metabolic engineering of plant allyl/propenyl phenol and lignin pathways: Future potential for biofuels/bioenergy, polymer intermediates and specialty chemicals? In *Advances in Plant Biochemistry Molecular Biology* (H. J. Bohnert and H. T. Nguyen eds.), Vol. 1, Elsevier, Oxford. pp. 385–428.

Vassão, D. G., K.-W. Kim, L. B. Davin, and N. G. Lewis. 2010. Lignans (neolignans) and allyl/propenyl phenols: Biogenesis, structural biology, and biological/human health considerations. In *Comprehensive Natural Products II Chemistry and Biology* (L. Mander and H.-W. Liu eds.), Vol. 1: Structural Diversity I, Elsevier, Oxford, UK. pp. 815–928.

Verdonk, J. C., M. A. Haring, A. J. van Tunen, and R. C. Schuurink. 2005. *ODORANT1* regulates fragrance biosynthesis in petunia flowers. *Plant Cell* 17:1612–1624.

Yahyaa, M., A. Berim, B. Nawade, et al. 2019. Biosynthesis of methyleugenol and methylisoeugenol in *Daucus carota* leaves: Characterization of eugenol/isoeugenol synthase and *O*-methyltransferase. *Phytochemistry* 159:179–189.

8 8–8′- and 8–7′-Norlignan Biosynthesis Roadmap

Laurence B. Davin and Norman G. Lewis

As for other natural product classes (e.g. terpenoids, alkaloids) the prefix nor- that is used in lignan phytochemical descriptions depicts loss or removal of a single carbon functionality, potentially regardless of oxidation state, e.g. via loss of -CH_3, -CH_2, -CH_2X, or CH from the core natural product skeleton. There are, however, many structural variations of norlignan skeleta reported in the literature.

8.1 8–8′-NORLIGNAN ROADMAP

8–8′ Linked norlignans have been reported in the bryophytes (liverworts), as well as more commonly in gymnosperms and angiosperms. Liverworts, *Lepidozia incurvata*, *Chiloscyphus polyanthos*, *Jungermannia exsertifolia* spp. *cordifolia* (Cullmann et al. 1999), *Jamesoniella autumnalis* (Tazaki et al. 1995), and *Bazzania trilobata* (Martini et al. 1998), contain various 8–8′ linked aryl naphthalene and aryl-derived dihydronaphthalenes (e.g. compounds **1–6**, Figure 8.1). All of these lack a C9 functionality, perhaps via decarboxylation.

To our knowledge, the biosynthetic pathways to these metabolites have not been investigated to date. Interestingly, almost all of these bryophyte norlignans have catechol aromatic rings or rings apparently derived from same (i.e.

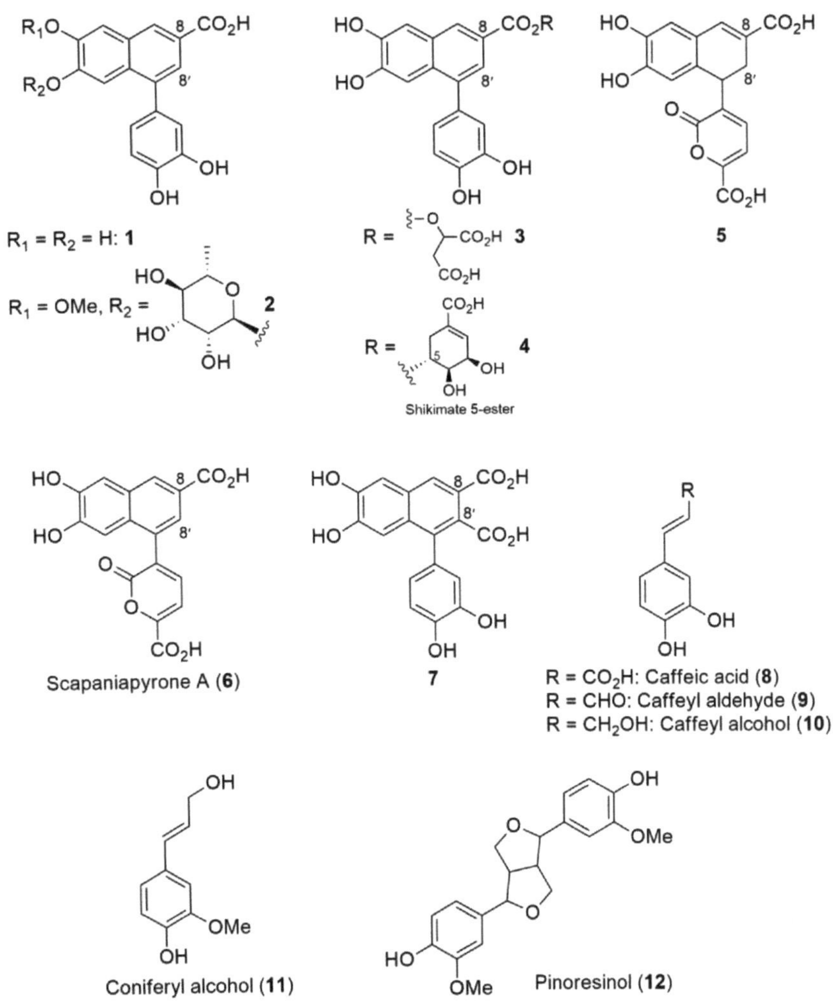

FIGURE 8.1 8–8′-Linked norlignans **1–6** and a lignan (**7**) found in bryophytes, and related metabolites **8–12**.

rather than having an *O*-methyl group functionality at the 3/3′-position, with one exception in compound **2**).

In those examples, however, the biosynthetic pathways to these 8–8′ norlignans may be somewhat similar to those of the monolignol-derived 8–8′ lignans (such as to arylnaphthalene lignans like **7** from *J. autumnalis* (Tazaki et al. 1995)), with the exception that a decarboxylation step is additionally involved. However, whether the initial coupling substrate to the bryophyte lignans is an hydroxycinnamic acid, aldehyde, or alcohol remains to be determined, i.e. caffeic acid (**8**), caffeyl aldehyde (**9**), or caffeyl alcohol (**10**) are all possibilities. Alternatively, these arylnaphthalenes (**1–4**, **6** ,**7**) and aryldihydronaphthalene (**5**) could be biosynthesized from coniferyl alcohol (**11**) through downstream pinoresinol (**12**) metabolism and subsequent demethylation.

8.2 8–7′-NORLIGNAN ROADMAP

Formation of 8–7′-linked norlignan skeleta cannot be explained, however, on the basis of the known 8–8′-lignan biosynthetic pathways (Davin and Lewis 2022). Examples of 8–7′-linked norlignans include (*E*)-hinokiresinol (**13**), agatharesinol (**15**), sequirin C (**16**), sequirin B (**18**), and cryptoresinol (**19**) (Figure 8.2) (Kai and Shimizu 1968a, b, Takahashi 1981, Takahashi et al. 1988) from *Cryptomeria japonica* heartwood, and (*Z*)-hinokiresinol (nyasol, **14**) from the angiosperm monocotyledon asparagus (*Asparagus officinalis*). Acetylenic alkylphenyl ethers, asparenydiol (**20**), asparenyol (**21**), and asparenyn (**22**), from asparagus have also been reported to be norlignans (see Section 8.2.3 for an alternative explanation) (Terada et al. 1995, Terada and Kamisako 1999).

8.2.1 Hinokiresinol Biosynthesis Roadmap

The 8–7′-linked norlignan biosynthetic pathway is now well understood using both *A. officinalis* and *C. japonica* as biological partners. (*E*/*Z*)-hinokiresinols (**13/14**) were initially considered to be formed from two non-identical phenylpropanoid moieties, *p*-coumaryl alcohol (**25**) and a *p*-coumaroyl derivative (i.e. CoA ester **24**, Figure 8.3) (Suzuki et al. 2001, Suzuki et al. 2002). Subsequently, a *p*-coumaryl coumarate (**26**) precursor (Suzuki et al. 2004) was implicated, based on various potential precursor labelling studies using *A. officinalis* crude cell-free extracts. From those initial deductions, the key discovery followed with isolation of hinokiresinol synthase (HRS) from *A. officinalis* elicitor-treated cells (Suzuki et al. 2007).

Interestingly, HRS consisted of two proteins (HRSα and HRSβ) of ~21 and 23 kDa, respectively, whose corresponding sequenced fragments allowed for cloning of their full-length cDNAs, with subsequent His-Tag recombinant protein (predicted masses of ~20.4 and 19.8 kDa, respectively) expression in *E. coli*. HRSα and HRSβ were ~49% identical to each other based on amino acid sequence. At that time, an *in silico* search gave several homologs of HRS in a clade (gene superfamily) annotated as either phloem protein 2 (PP2) or

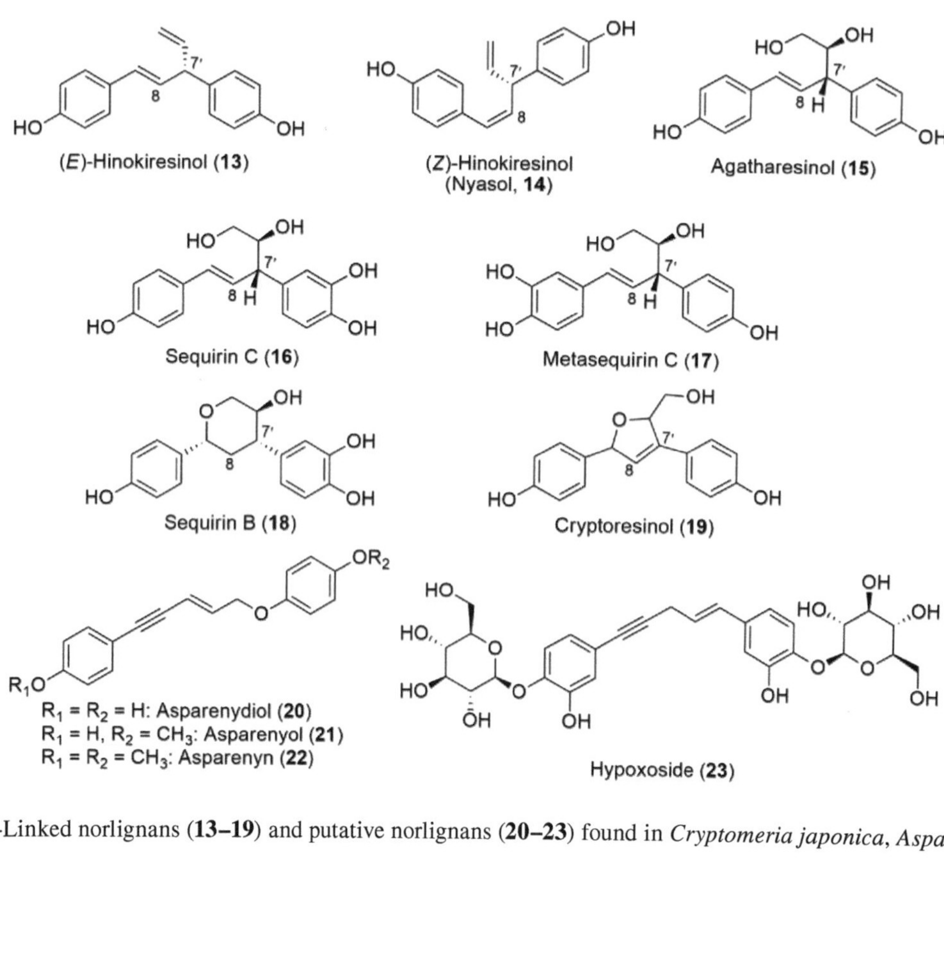

FIGURE 8.2 8–7′-Linked norlignans (**13–19**) and putative norlignans (**20–23**) found in *Cryptomeria japonica*, *Asparagus officinalis*, or *Hypoxis* spp.

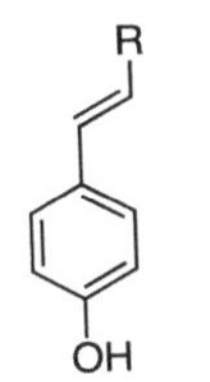

R= COSCoA: *p*-Coumaroyl CoA (**24**)
R= CH_2OH: *p*-Coumaryl alcohol (**25**)

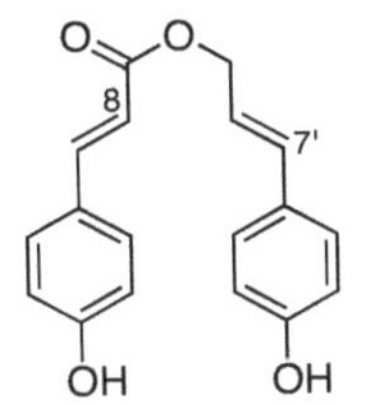

p-Coumaryl coumarate (**26**)

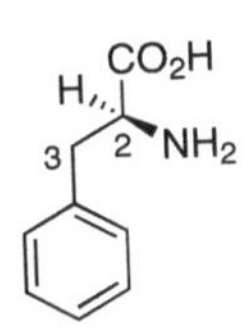

Phenylalanine (**27**)

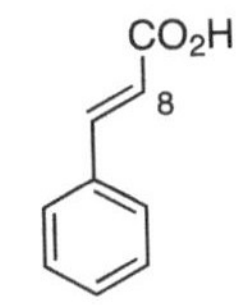

Cinnamic acid (**28**)

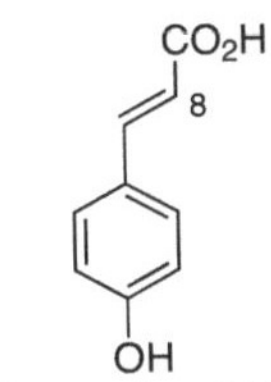

p-Coumaric acid (**29**)

FIGURE 8.3 Various monomeric C_6C_3 derivatives, *p*-coumaryl coumarate, and phenylalanine.

PP2-like, all of which were lacking precise physiological/biochemical function(s).

Individually, recombinant HRSα and HRSβ, with *p*-coumaryl coumarate (**26**) as a potential substrate, gave only (*E*)-hinokiresinol (**13**) as a product in both cases. On the other hand, asparagus plant tissues accumulated the *Z*-isomer **14**. This stereo-chemical incongruity was resolved when both HRSα and HRSβ (1:1 ratio) were assayed together. Under those conditions, the (*Z*)-isomer **14** was now produced.

Thus, *in vivo,* the asparagus HRS is apparently a heterodimer of HRSα and HRSβ, with kinetic parameters *in vitro* as follows: apparent K_m of 0.44 μM, V_{max} of ~0.075 pkat μg^{-1} protein, k_{cat} of ~1.5×10^{-3} s^{-1}, and k_{cat}/K_m of ~3,400 M^{-1} s^{-1} for *p*-coumaryl coumarate (**26**) (based on one active site per monomer).

From a mechanistic perspective, HRS catalysis involves both decarboxylation and 8–7′ bond-forming steps. Four possible mechanisms can be considered: an ester enolate Claisen rearrangement (Figure 8.4A); either a concerted (Figure 8.4D) or a two-step (Figure 8.4B) molecular rearrangement; or ester cleavage and bimolecular coupling (Figure 8.4C) (Vassão et al. 2006, Suzuki et al. 2007) Of these mechanisms, three implicated a putative quinone methide intermediate (Figure 8.4A–8.4C).

8.2.2 Agatharesinol and Sequirin C Biosynthesis Roadmap

Agatharesinol (**15**) and sequirin C (**16**) differ from (*E*)-hinokiresinol (**13**) by addition of two hydroxyl groups to the pendant olefinic group to afford agatharesinol (**15**), and an additional hydroxyl group on one of the phenyl rings for sequirin C (**16**). Both norlignans constitutively accumulate in *C. japonica* heartwood, with levels of agatharesinol (**15**) increasing from the sapwood side of the intermediate wood to the heartwood, whereas sequirin C (**16**) levels increased rapidly from intermediate wood tissue to heartwood tissue. Norlignan biosynthesis is also inducible: freshly felled tree sapwood slowly forms **15** when stored at room temperature (Ohashi et al. 1988).

Agatharesinol (**15**) biosynthesis was investigated using the *C. japonica* sapwood tissue, which could be induced to form norlignans (Imai et al. 2006a, b), although these investigations were not carried out at the gene or enzyme level. In contrast to the above-mentioned findings on (*E*)-hinokiresinol (**13**) biosynthesis, Imai et al. (2006a) reported that (*E*)-hinokiresinol (**13**) was not involved as an intermediate in agatharesinol (**15**) biosynthesis. In spite of the report by Imai et al. (2006a), however, this hypothesis was not unambiguously proven. This thus requires some further consideration of these studies.

These researchers, on administering various precursors to fresh sapwood sticks of *C. japonica*, had observed the following (Imai et al. 2006a, b): [ring-^{2}H]-Phe (**27**) and [ring-^{13}C]-(*E*)-cinnamic acid (**28**) (Figure 8.3) were both converted into agatharesinol (**15**) as to be expected. On the other hand, [9-$^2H_2$]-*p*-coumaryl alcohol (**25**) administration only afforded [9-$^2H_2$]-(*E*)-hinokiresinol (**13**), but not agatharesinol (**15**). This led these researchers to hypothesize that a different biosynthetic pathway from that giving (*E*)-hinokiresinol (**13**) was in place for agatharesinol (**15**) formation in the sapwood tissue.

However, other (simpler) explanations are possible. [Ring-^{2}H]-Phe (**27**) and [ring-^{13}C]-(*E*)-cinnamic acid (**28**) incorporations may have occurred as these were efficaciously transported to agatharesinol (**15**) biosynthesis site(s), whereas [9-$^2H_2$]-*p*-coumaryl alcohol (**25**) and [7-^{2}H]-(*E*)-hinokiresinol (**13**) may not have been.

In addition, how sequirin C (**16**) was biosynthesized was explored using microsomal fractions prepared from the heartwood side of *C. japonica* intermediate zone in the presence of NADPH/NADH (Imai et al. 2009). These microsomes, when incubated with [ring-^{13}C]-agatharesinol (**15**), afforded both sequirin C (**16**) and metasequirin C (**17**) under these conditions, the levels of which were enhanced upon FAD or FMN addition. Conversely, hydroxylation did not occur using microsomal preparations from sapwood or from the sapwood side of the intermediate zone.

The biosynthetic pathway to metasequirin C (**17**) would benefit in the future from determination of all of the genes and proteins involved in its biosynthetic pathway, i.e. to establish the actual conversions and sequence(s) of same.

8.2.3 Acetylenic Norlignans or Diarylpentanoids?

An additional variant reported as a member of the 8–7′ norlignan sub-group is that of the somewhat structurally unusual acetylenic norlignans present in *A. officinalis* and *Hypoxis* spp. Three of these contain an unusual oxygen insertion between the side chain and aromatic ring, such as with asparenydiol (**20**), asparenyol (**21**), and asparenyn (**22**), whereas hypoxoside (**23**) does not. Investigations directed towards

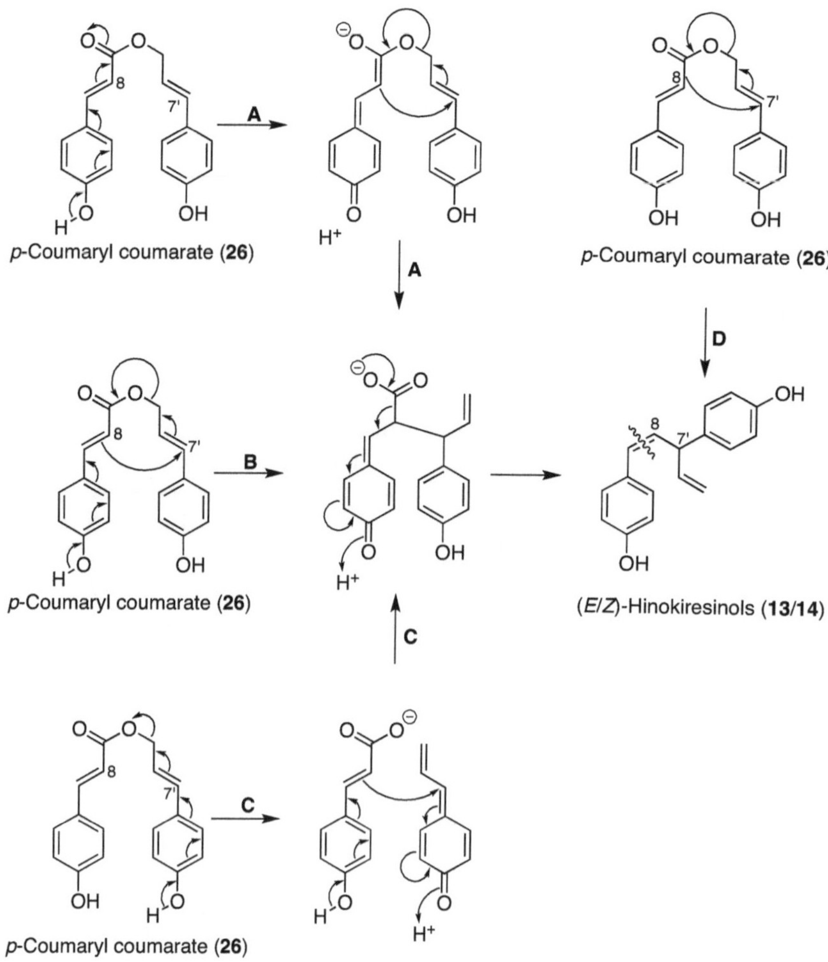

FIGURE 8.4 Proposed mechanisms for the formation of (*E/Z*)-hinokiresinols (**13/14**) from *p*-coumaryl coumarate (**26**). Ester enolate Claisen rearrangement (Route **A**), and two-step intramolecular rearrangement (Route **B**), involve formation of a quinone methide in the *p*-coumarate moiety of the substrate (Suzuki et al. 2007). Route **C** involves ester link cleavage with formation of a quinone methide in the *p*-coumaryl (alcohol)-derived moiety, followed by bimolecular coupling (Vassão et al. 2006) to give the same *p*-coumarate quinone methide intermediate as in Routes **A** and **B**. In all three proposed mechanisms (Routes **A**, **B**, and **C**), the *p*-coumarate quinone methide (center) undergoes decarboxylation with re-aromatization to give either (*E/Z*)-hinokiresinols (**13/14**). Route **D** involves direct intramolecular rearrangement, with concerted decarboxylation and C7′–C8 bond formation without intermediacy of a quinone methide (i.e. with no apparent participation of the phenolic rings) (Vassão et al. 2006).

understanding their biosynthetic pathways have thus far been quite preliminary.

More specifically, radiolabeling precursor studies using *Hypoxis* sp., with Phe (**27**), cinnamic acid (**28**), *p*-coumaric acid (**29**), caffeic acid (**8**) (Figures 8.1 and 8.3), and acetate, established these to be precursors. In addition, the corms were found to be the main sites of hypoxoside (**23**) biosynthesis, albeit with differing incorporation levels with the different precursors used (Bayley and van Staden 1990a, b, 1991).

Stable isotope-labeled precursors in *A. officinalis* also showed that asparenyn (**22**), asparenyol (**21**), and asparenydiol (**20**) skeleta were Phe (**27**) derived, based on results obtained using [ring-^{2}H]-, [3-^{2}H]-, [1-^{13}C]-, and [3-^{13}C]-Phe (**27**) (Terada et al. 1996, Terada and Kamisako 1999).

These researchers hypothesized a putative biosynthetic pathway (with no experimental proof) involving a hinokiresinol (**13/14**)-like skeleton and a *spiro* intermediate (Terada and Kamisako 1999). Moreover, neither enzymes nor genes have been reported as being involved in the formation of the unusual "oxygen-inserted" acetylenic norlignans **20–22**.

In the absence of experimental proof, a more plausible biochemical explanation is that the acetylenic products **20–23** are, in fact, diarylpentanoids with no biochemical relationship to norlignan biosynthesis. It can be speculated that, as for diarylheptanoid biosynthesis (Katsuyama et al. 2009a, b, Morita et al. 2010), entry into the diarylpentanoid metabolic class results instead from the action of specific polyketide synthases (PKSs).

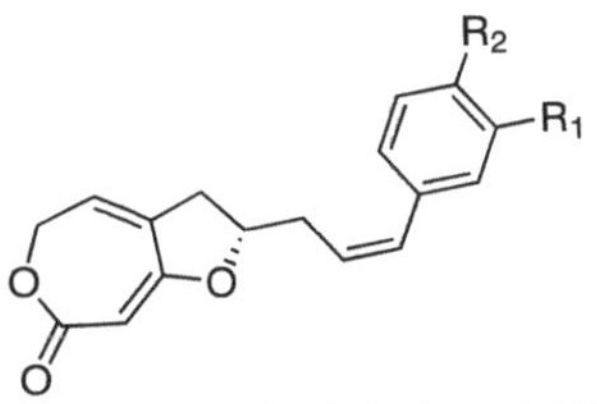

FIGURE 8.5 Capitulactones A–C (**30–32**) from *Curculigo capitulata.*

8.3 CAPITULACTONE NORLIGNAN ROADMAP

Li et al. (2019) reported the presence of three putative norlignans, capitulactones A–C (**30–32,** Figure 8.5) with an unusual 3,5-dihydro [2,3-*d*] oxepin-7(2*H*)-one scaffold from roots of *Curculigo capitulata,* a member of the Amaryllidaceae. Their biosynthetic pathway, however, is also currently unknown. Indeed, if these are true norlignans, then proof is needed at the (bio)chemical level to establish if they are norlignans (derived), with C9′ loss.

Alternatively, and presumably more likely, lactones (**30–32**) are of diarylpentanoid pathway origin, of which there are 20 reported thus far (Celebioglu et al. 2017). If this is the case, then entry to this metabolic class presumably also occurs via PKS enzymatic conversion with downstream metabolism, and they are not norlignans as reported. That is, their formation does not need to implicate involvement of the now established norlignan biosynthetic pathway (see Sections 8.2.1 and 8.2.2), as they may simply be formed via PKS action with subsequent downstream metabolism.

ACKNOWLEDGEMENTS

The authors thank the Chemical Sciences, Geosciences and Biosciences Division, DOE Office of Basic Energy Sciences (DE-FG-0397ER20259), as well as the Arthur and Katie Eisig-Tode Endowment.

REFERENCES

Bayley, A. D., and J. van Staden. 1990a. Is the corm the site of hypoxoside biosynthesis in *Hypoxis hemerocallidea*? *Plant Physiol. Biochem.* 28:691–695.

Bayley, A. D., and J. van Staden. 1990b. Precursors for hypoxoside in intact plants of *Hypoxis hemerocallidea. Plant Physiol. Biochem.* 28:697–702.

Bayley, A. D., and J. van Staden. 1991. Incorporation of various radiolabelled compounds into hypoxoside in intact plants of *Hypoxis hemerocallidea. Plant Physiol. Biochem.* 29:403–408.

Celebioglu, N., U. Ozgen, and H. Secen. 2017. Naturally occurring 1,5-diarylpentanoids: A review. *Org. Commun.* 10:250–258.

Cullmann, F., A. Schmidt, F. Schuld, M. L. Trennheuser, and H. Becker. 1999. Lignans from the liverworts *Lepidozia incurvata, Chiloscyphus polyanthos* and *Jungermannia exsertifolia* ssp. *cordifolia. Phytochemistry* 52:1647–1650.

Davin, L. B., and N. G. Lewis. 2022. 8–8′-Lignan biosynthesis roadmap. In *The Lignan Handbook* (N. G. Lewis and L. B. Davin eds.), Taylor & Francis, Boca Raton, London, pp 79-112.

Imai, T., M. Nomura, and K. Fukushima. 2006a. Evidence for involvement of the phenylpropanoid pathway in the biosynthesis of the norlignan agatharesinol. *J. Plant Physiol.* 163:483–487.

Imai, T., M. Nomura, Y. Matsushita, and K. Fukushima. 2006b. Hinokiresinol is not a precursor of agatharesinol in the norlignan biosynthetic pathway in Japanese cedar. *J. Plant Physiol.* 163:1221–1228.

Imai, T., K. Asai, M. Takino, and K. Fukushima. 2009. *In vitro* hydroxylation of a norlignan: From agatharesinol to sequirin C and metasequirin C with a microsomal preparation from *Cryptomeria japonica. Phytochem. Lett.* 2:196–200.

Kai, Y., and M. Shimizu. 1968a. On the phenolic constituents from *Cryptomeria japonica* D. Don. IV. The structure of sugiresinol. *Mokuzai Gakkaishi* 14:425–429.

Kai, Y., and M. Shimizu. 1968b. On the phenolic constituents from *Cryptomeria japonica* D. Don. V. The structure of hydroxysugiresinol. *Mokuzai Gakkaishi* 14:430–433.

Katsuyama, Y., T. Kita, N. Funa, and S. Horinouchi. 2009a. Curcuminoid biosynthesis by two type III polyketide synthases in the herb *Curcuma longa. J. Biol. Chem.* 284:11160–11170.

Katsuyama, Y., T. Kita, and S. Horinouchi. 2009b. Identification and characterization of multiple curcumin synthases from the herb *Curcuma longa. FEBS Lett.* 583:2799–2803.

Li, S., J.-H. Yu, Y.-Y. Fan, et al. 2019. Structural elucidation and total synthesis of three 9-norlignans from *Curculigo capitulata. J. Org. Chem.* 84:5195–5202.

Martini, U., J. Zapp, and H. Becker. 1998. Lignans from the liverwort *Bazzania trilobata. Phytochemistry* 49:1139–1146.

Morita, H., K. Wanibuchi, H. Nii, et al. 2010. Structural basis for the one-pot formation of the diarylheptanoid scaffold by curcuminoid synthase from *Oryza sativa. Proc. Natl. Acad. Sci. USA* 107:19778–19783.

Ohashi, H., T. Goto, T. Imai, and M. Yasue. 1988. Accumulation and distribution of extractives in the living stem of Sugi (*Cryptomeria japonica* D. Don.). *Res. Bull. Fac. Agr. Gifu Univ.* 53:301–314.

Suzuki, S., T. Umezawa, and M. Shimada. 2001. Norlignan biosynthesis in *Asparagus officinalis* L.: The norlignan originates from two non-identical phenylpropane units. *J. Chem. Soc., Perkin Trans. 1*:3252–3257.

Suzuki, S., T. Nakatsubo, T. Umezawa, and M. Shimada. 2002. First *in vitro* norlignan formation with *Asparagus officinalis* enzyme preparation. *J. Chem. Soc., Chem. Commun.*:1088–1089.

Suzuki, S., M. Yamamura, M. Shimada, and T. Umezawa. 2004. A heartwood norlignan, (*E*)-hinokiresinol, is formed from 4-coumaryl 4-coumarate by a *Cryptomeria japonica* enzyme preparation. *J. Chem. Soc., Chem. Commun.*:2838–2839.

Suzuki, S., M. Yamamura, T. Hattori, T. Nakatsubo, and T. Umezawa. 2007. The subunit composition of hinokiresinol synthase controls geometrical selectivity in norlignan formation. *Proc. Natl. Acad. Sci. USA* 104:21008–21013.

Takahashi, K. 1981. Heartwood phenols and their significance to color in *Cryptomeria japonica* D. Don. *Mokuzai Gakkaishi* 27:654–657.

Takahashi, K., M. Yasue, and K. Ogiyama. 1988. A norlignan, cryptoresinol, from the heartwood of *Cryptomeria japonica. Phytochemistry* 27:1550–1552.

Tazaki, H., K.-P. Adam, and H. Becker. 1995. Five lignan derivatives from *in vitro* cultures of the liverwort *Jamesoniella autumnalis. Phytochemistry* 40:1671–1675.

Terada, K., C. Honda, K. Suwa, et al. 1995. Acetylenic compounds isolated from cultured cells of *Asparagus officinalis. Chem. Pharm. Bull.* 43:564–566.

Terada, K., K. Suwa, S. Takeyama, C. Honda, and W. Kamisako. 1996. Biosynthesis of the acetylenic compounds in cultured cells of *Asparagus officinalis* from D- and ^{13}C-labelled phenylalanines. *Biol. Pharm. Bull.* 19:748–751.

Terada, K., and W. Kamisako. 1999. ^{1}H-NMR and mass spectral study of a D-enriched acetylenic norlignan, asparenyol, from cultured cells of *Asparagus officinalis. Biol. Pharm. Bull.* 22:561–566.

Vassão, D. G., D. R. Gang, T. Koeduka, et al. 2006. Chavicol formation in sweet basil (*Ocimum basilicum*): Cleavage of an esterified C9 hydroxyl group with NAD(P)H-dependent reduction. *Org. Biomol. Chem.* 4:2733–2744.

9 Roadmap to "Mixed Biochemical Pathway" Lignans

Laurence B. Davin and Norman G. Lewis

Little is currently known about how the various "mixed" biochemical pathway lignan products are actually formed, these being referred to, for example, as alkaloid lignans (e.g. Takemura et al. 1993), coumarinolignans and furanocoumarinolignans (e.g. Sajeli et al. 2006), flavonolignans (e.g. Sy-Cordero et al. 2010), stilbenolignans (e.g. Lee et al. 2001), terpenoid lignans (e.g. Konoshima et al. 1991), xantholignans (e.g. de M. Pinto et al. 1987), and diarylheptanoid- and diarylpentanoid-lignans (Huang et al. 2012, Liu et al. 2014), respectively. However, as discussed in greater detail in this chapter, whether some of these so-called "mixed" biochemical pathway metabolites should truly include the term "lignan" is worthy of scrutiny.

Nomenclature inconsistencies are illustrated by three examples of flavonolignans, namely (+)-pseudotsuganol (**1**) from Douglas-fir (*Pseudotsuga menziesii*) bark (Foo and Karchesy 1989), hydnowightin (**2**) from *Hydnocarpus wightiana* seed coats (Sharma et al. 1979), and (+)-silychristin (**3**) from *Silybum marianum* fruits (Wagner et al. 1978) (Figure 9.1).

(+)-Pseudotsuganol (**1**) and hydnowightin (**2**) can be considered true flavonolignans, as both contain *bona fide* lignan and flavonoid substructures. The former apparently results from coupling between lignan (+)-pinoresinol (**5**)-derived and flavonoid (+)-dihydroquercetin (**4**)-derived species. In a similar vein, hydnowightin (**2**) might be considered to result from coupling of luteolin (**6**) with dehydrodiconiferyl alcohol (**7**), with subsequent ring opening of the furano group. Yet, it could perhaps result from stepwise coupling of two coniferyl alcohol (**8**)-derived moieties onto luteolin (**6**), with this perhaps including dehydration (loss of H_2O at C7–C8) of an intermediate to afford **2**. However formed, there was no report of optical activity for hydnowightin (**2**).

By contrast, (+)-silychristin (**3**) from *S. marianum* fruits (Wagner et al. 1978) presumably results from coupling between (+)-dihydroquercetin (**4**) and coniferyl alcohol (**8**)-derived substrates. Compounds such as (+)-silychristin (**3**) lack a lignan skeleton proper, as only one C_6C_3 monolignol has been incorporated into the flavonoid structure. That is, there is no directly linked dimeric C_6C_3–C_6C_3 lignan product, but instead, there is a $C_6C_3C_6$–C_3C_6 coupled phenylpropanoid flavonoid product. The C_6C_3–C_6C_3 skeleton, though, is a lignan definition proper requirement (see Davin and Lewis 2022).

There are no reports of the actual entry points to these metabolically diverse "mixed" biochemical pathway plant products in the literature, i.e. in terms of whether there are either proteins and enzymes involved in coupling, or whether they are possibly artifacts from isolation and/or extract storage procedures used. Artifact formation is a possibility that cannot be excluded. This caution is given because many compounds, reported as plant natural products, are isolated from various plant tissues, sometimes after years of storage at ambient temperature, followed by lengthy extraction and isolation protocols, some of which can involve rather harsh chemical conditions. Artifact formation is a well-known phenomenon in plant phytochemical research.

Yet, regardless of how they are formed, the full natural product structural diversity of these "mixed biochemical pathway lignans" is likely very incomplete to this point, given that many plant species remain to be (exhaustively) studied for their phytochemical constituents. Nevertheless, even with this presumed paucity of knowledge currently, it will be important in the future to establish how these various natural product classes are precisely formed.

It needs to be emphasized, though, that it is not possible in this contribution to address all the examples of the various mixed biochemical pathway entries that can be found within this *Handbook*. Instead, just a few examples in various metabolic classes are selected for illustrative and discussion purposes in this chapter.

9.1 ALKALOID LIGNANS

There are various "mixed" biochemical pathway alkaloid lignans containing a single C_6C_3 unit, such as those with 8–*O*–4′ and benzodioxin-like substructures, e.g. *erythro*-canabisine H (**9**) from kenaf (*Hibiscus cannabinus*) bark (Seca et al. 2001), (–)-*erythro*-ipobscurine B (**11**) from *Ipomoea obscura* seeds (Jenett-Siems et al. 2003), and acrignine A (**13**) from *Citrus grandis* root tissue (Takemura et al. 1993), respectively (Figure 9.2).

These isolates were generally obtained in very small amounts. For example, *erythro*-canabisine H (**9**) was isolated in an amount of *circa* 53.8 mg (0.0036 % weight/weight, w/w) from *circa* 1.5 kg of air-dried kenaf bark. As no optical/specific rotation ($[\alpha]_D$) value was reported, it is unknown as to whether it was optically active or racemic. It is presumed to be possibly formed via coupling between feruloyl tyramine (**10**) and coniferyl alcohol (**8**)-derived radical species. Another report of *erythro*-canabisine H (**9**) came from a study of the constituents present in *Solanum nigrum* (Li et al. 2019), where 40 kg of its air-dried herb gave pure **9** (3.4 mg, 0.00001%

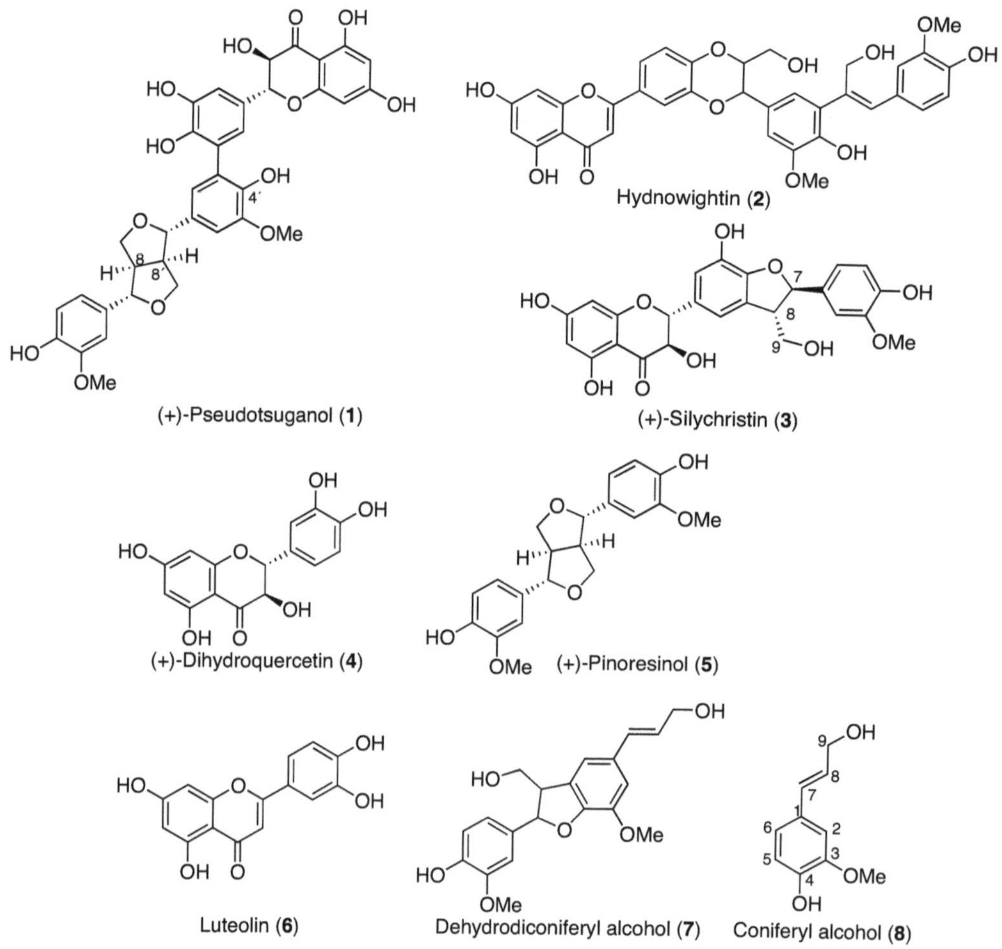

FIGURE 9.1 Some vascular plant "flavanolignans" (**1–3**) and their possible precursors (**4–8**).

w/w). This, however, was established to be a mixture of both enantiomeric forms, following chiral column separation, in roughly equimolar amounts.

(–)-*erythro*-Ipobscurine B (**11**) (*circa* 150 mg, 0.03% w/w) was obtained from 500 g seed of *I. obscura*. This alkaloid lignan can be provisionally envisaged to result from initial coupling of *p*-coumaroyl serotonin (**12**) and coniferyl alcohol (**8**) to ultimately afford the final product **11**. Given that the latter is optically active ($[\alpha]_D$ –35°, c 0.28, MeOH), its formation can provisionally be considered to be under proteinaceous control *in vivo*.

Racemic acrignine A (**13**) was isolated (in unknown amount) from *C. grandis* root tissue (Takemura et al. 1993). Its formation may possibly result from coupling of sinapyl alcohol (**14**) and citracridone III (**15**)-derived radical species. However, whether it is enzymatically generated *in vivo* (e.g. by action of either a peroxidase or laccase, both 1e⁻ oxidases) or is produced as an artifact is unknown.

Perhaps more importantly, neither *erythro*-canabisine H (**9**) (Seca et al. 2001), ipobscurine B (**11**) (Jenett-Siems et al. 2003), nor acrignine A (**13**) (Takemura et al. 1993) contain an actual lignan structure, i.e. none have a C_6C_3-C_6C_3 linkage, as per the definition requirement for lignans proper (see Davin and Lewis 2022).

Another alkaloid lignan is that of cannabisin O (**16**) from hemp (*Cannabis sativa*), this being obtained (1.72 mg, 0.00003% w/w) from 5.7 kg air-dried hemp seed (Yan et al. 2015). Cannabisin O (**16**) appears to be the result of two distinct forms of free radical coupling between feruloyl tyramine (**10**) moieties, and has an $[\alpha]_D$ +0.71° (c 0.07, MeOH). As previously stated, whether it is enzymatically generated or is an artifact cannot be gauged at this time. Although nitrogen containing, this metabolite, nevertheless, has two C_6C_3 lignan-like substructures, i.e. covalent linkages harboring an *O*-phenol linked feruloyltyramine (**10**)-derived moiety and an aryltetrahydrofuran-like substructure, respectively.

In addition, there are reports of cyclobutane alkaloid lignans, such as incarvillateine (**17**), obtained in an undisclosed amount from *Incarvillea sinensis* aerial tissue (18 kg) (Chi et al. 1990). Its formation can provisionally be considered to occur via a so-called head-to-tail 2 + 2 cycloaddition of two identical incarvine C (**18**) moieties (Chi et al. 1995). These are perhaps likely to be photochemically generated (Krauze-Baranowska 2002). Interestingly, in head-to-tail dimerization as here, there are various possible outcomes that can

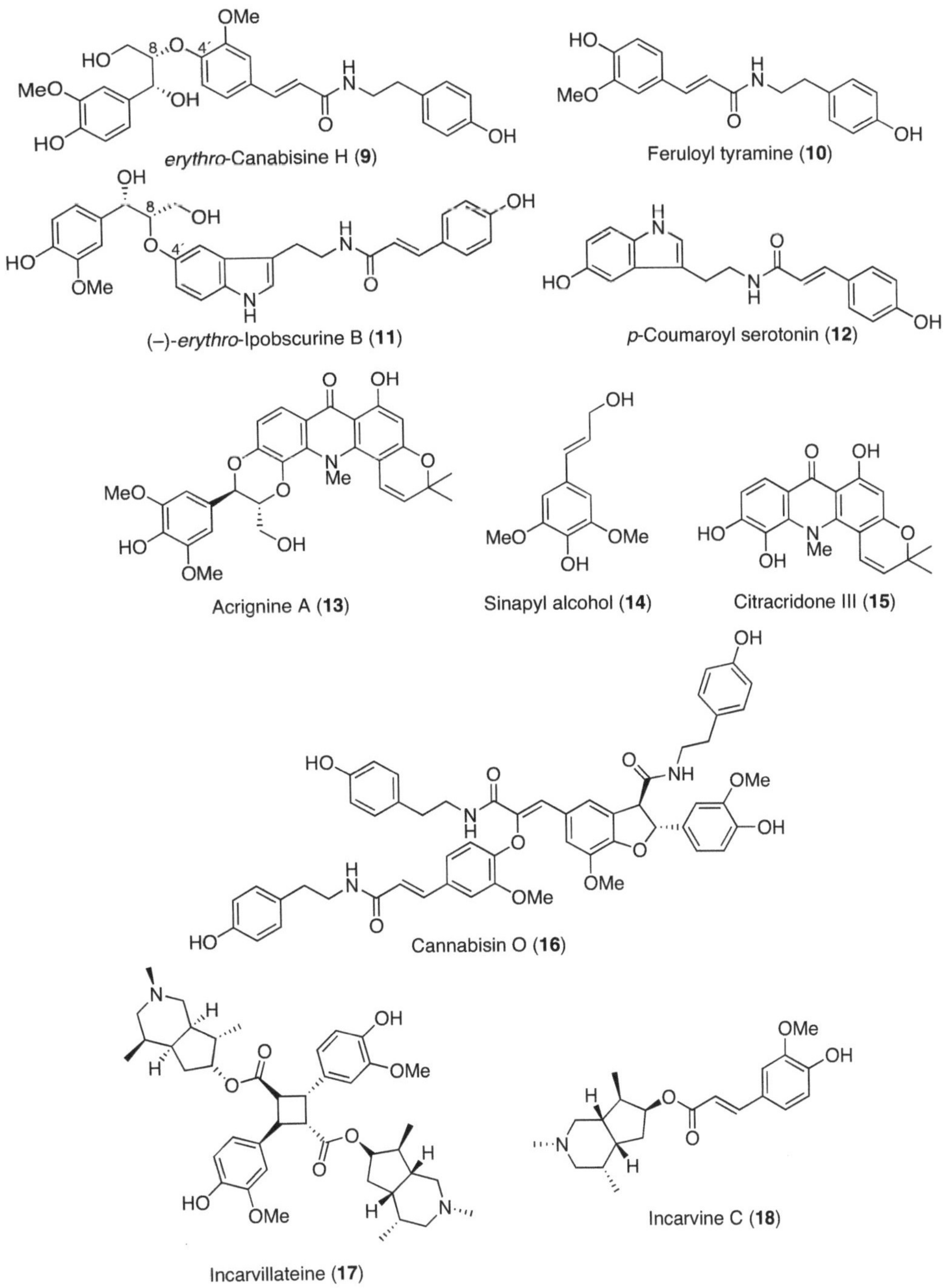

FIGURE 9.2 Vascular plant "alkaloid lignans" (**9**, **11**, **13**, **16**, and **17**) and some of their possible precursors (**10**, **12**, **14**, **15**, and **18**).

occur, namely *cis-cis*, *trans-cis*, *trans-trans*, and/or *cis-trans* coupling product formation. Such possibilities were not discussed. Nevertheless, its structure meets the criteria of being a lignan type proper, i.e. with direct C_6C_3–C_6C_3 linkages.

9.2 COUMARINO- AND FURANOCOUMARINO-LIGNANS

There are several reported examples of coumarinolignans and furanocoumarin lignans, such as the regiosiomers, cleomiscosin A (**19**) and cleomiscosin B (**21**) from *Cleome viscosa* seed (Ray et al. 1982), venkatasin (**20**) and hyosgerin (**22**) from *Hyoscyamus niger* seeds (Sajeli et al. 2006), as well as melicodin C (**24**) from aerial tissue of *Melicope denhamii* (Nakashima et al. 2012) (Figure 9.3).

Both coumarinolignans **19** and **21**, while differing in terms of regiospecificity of coupling, were racemic and both contain benzodioxin-like substructures. These can be considered to possibly result from coupling of coniferyl alcohol (**8**) with the catechol fraxetin (**23**) (Wessely and Demmer 1928, Tanaka

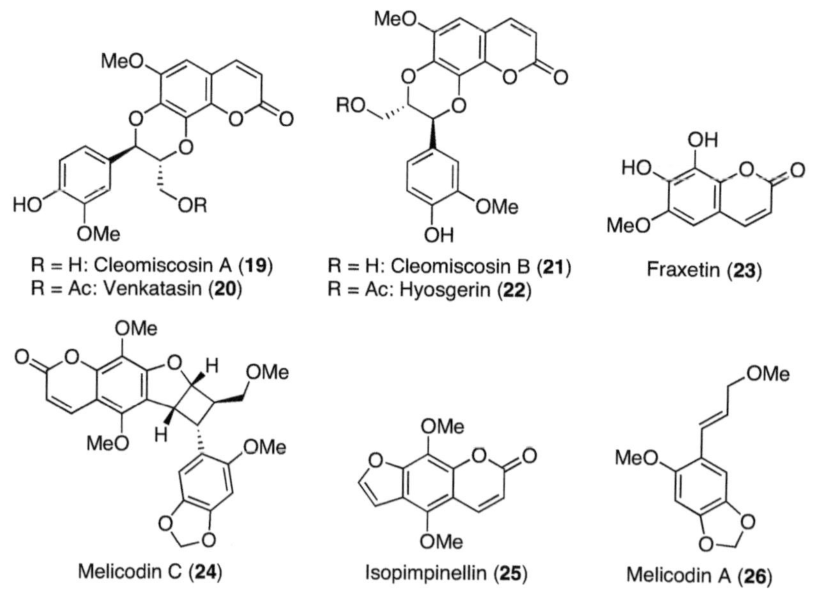

FIGURE 9.3 Vascular plant "coumarinolignans" (**19–22**), a "furano-coumarin lignan" (**24**), and some of the possible precursors (**23**, **25**, and **26**).

et al. 1986), in a manner similar to those described earlier. However, it is not possible at this time to distinguish whether they are true natural products or artifacts. Additionally, neither the amounts of plant seed tissues examined, nor the isolation protocols, nor the amounts isolated were reported.

Conversely, the closely related venkatasin (**20**) and hyosgerin (**22**) were both reported as optically active. (–)-Hyosgerin (**22**) (12.5 mg, 0.0003% w/w) and (–)-venkatasin (**20**) (9.1 mg, 0.0002% w/w) were obtained from 5 kg *H. niger* seeds. Both were optically active, ($[\alpha]_D$ –65.4°, c 0.38, $CHCl_3$) and ($[\alpha]_D$ +16.0°, c 0.09, $CHCl_3$), respectively, suggesting proteinaceous control over coupling (Sajeli et al. 2006).

Cyclobutane furanocoumarin lignan, melicodin C (**24**), was isolated (8 mg, 0.002% w/w) from aerial tissue of *M. denhamii* (490 g leaf tissue, air-dried) (Nakashima et al. 2012). It can be considered to possibly result from 2 + 2 cycloaddition of isopimpinellin (**25**) (Johns et al. 1968) and melicodin A (**26**) (Nakashima et al. 2012). It had no optical activity, indicating it was present in racemic form. Given that it was isolated from aerial tissue, it may have been formed photosynthetically.

All of these coumarinolignans lack a lignan C_6C_3–C_6C_3 structure proper.

9.3 FLAVANOLIGNANS

As known at this time, this is a much more diverse metabolic (mixed pathway) class than for the other "mixed metabolic pathway" lignans. Representatives known thus far can encompass dihydrochalcone, chalcone, flavonoid, and isoflavonoid skeleta linked to either monolignols, allyl/propenyl phenols, or lignans as shown, and in a variety of different inter-unit linkage types.

As for the previous types of "mixed biochemical pathway lignans", flavonolignans can be linked through, for example, 8–*O*–4′ and benzodioxin substructure-like inter-unit linkages, such as calquiquelignan E (**27,** Figure 9.4) (Chang et al. 2010), 5′-methoxyhydnocarpin D (**29**) (Stermitz et al. 2000), and alopecuroides A (**30**) (Zhang et al. 2007), as well as hydnowightin (**2**, Figure 9.1) (Parthasarathy et al. 1979, Sharma et al. 1979), respectively.

Calquiquelignan E (**27**) (4.8 mg, 0.00003% w/w) was obtained from air-dried stems (15.4 kg) of *Calamus quiquesetinervius* in optically active form ($[\alpha]_D$ +27°, c 0.48, MeOH) as the *threo* ester (Chang et al. 2010). [Its corresponding *erythro* ester (4.8 mg) was also isolated, with an $[\alpha]_D$ –58° (c 0.48, MeOH) (not shown)]. Their optical activities might provisionally suggest that the initial 8–*O*–4′-like linkage (of a *presumed p*-coumaryl alcohol (**28**)-derived species) was stereoselectively generated.

In contrast, racemic 5′-methoxyhydnocarpin D (**29**) (47 mg, 0.025% w/w) was isolated from dried, ground leaf tissue (3.32 kg, 0.0014% w/w) of *Berberis fremontii* (Stermitz et al. 2000), whereas alopecuroides A (**30**), $[\alpha]_D$ +4.44° (c 0.84, pyridine), was obtained from *Pueraria alopecuroides* plant tissue (isolated in unknown amount, and not specified as to tissues/organs analyzed) (Zhang et al. 2007). Nevertheless, both products appear to be possibly derived from initial coupling of coniferyl alcohol (**8**) radicals with either flavonoids, luteolin (**6**), selagin (**31**) or tricetin (**32**), or isoflavonoid, retusin (**33**), derived radical species. However, whether these were formed *in planta* as natural products, or were artifacts, cannot be gauged at this time.

If luteolin (**6**) was instead the substrate for coupling to ultimately afford 5′-methoxyhydnocarpin D (**29**), then a subsequent post-coupling methoxylation or hydroxylation/methylation step at C5′ to afford 5′-methoxyhydnocarpin D (**29**) would be required. Alternatively, selagin (5′-methoxyluteolin, **31**) [or tricetin (5′-hydroxyluteolin) (**32**) with subsequent

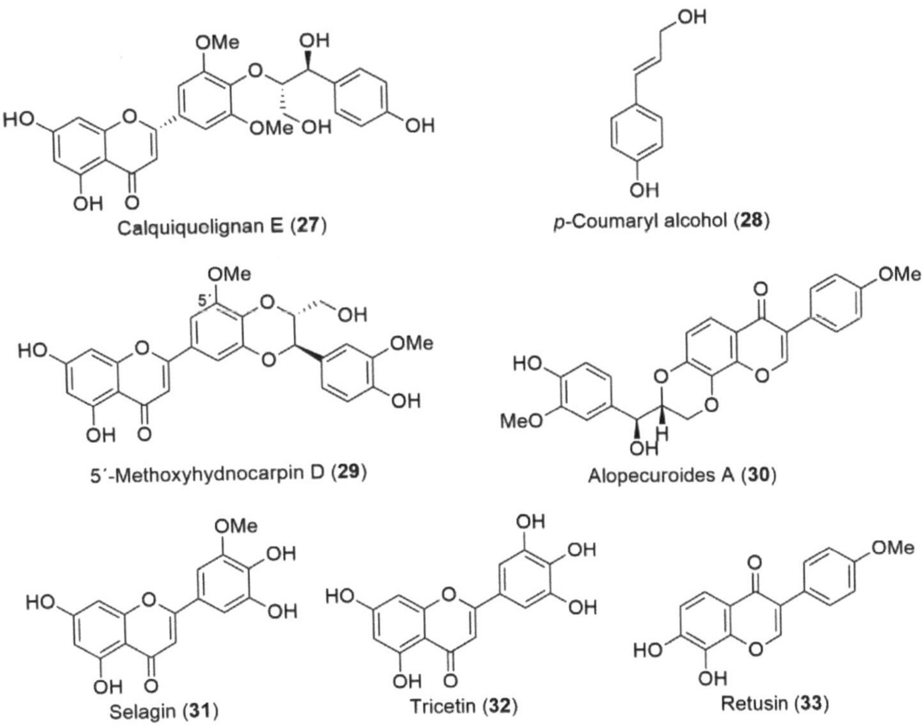

FIGURE 9.4 Some vascular plant "flavanolignans" (**27**, **29**, and **30**) and their possible precursors (**28**, **31–33**).

O-methylation at C5′] could serve as the substrate for coupling to afford 5′-methoxyhydnocarpin D (**29**).

Hydnowightin (**2**) differs from 5′-methoxyhydnocarpin D (**29**) in having an additional monolignol added, but no 5′-methoxy group. However, the report of its structure, following its isolation from *H. wightiana* seed coats (Parthasarathy et al. 1979), did not include either amount of plant material extracted, methods of isolation employed, amounts of hydnowightin (**2**) obtained, nor its $[\alpha]_D$ value. In terms of its formation as one possibility, hydnowightin (**2**) can be rationalized to possibly result from coupling between flavonoid luteolin (**6**) and two coniferyl alcohol (**8**)-derived moieties, respectively (Figure 9.5).

Other less likely possibilities are that either a dehydrodiconiferyl alcohol (**7**) lignan-derived intermediate would undergo coupling with luteolin (**6**) to afford the intermediate **34** (Figure 9.6), with subsequent opening of the dihydrofuran ring to afford the final product, or a corresponding ring-opened form (e.g. **35,** Figure 9.7) directly from same could also be used as the precursor, with subsequent loss of H_2O.

The available data, however, does not allow distinction as to whether either 5′-methoxyhydnocarpin D (**29**) or hydnowightin (**2**) formation occurs via radical coupling chemistry or not. Conversely, but perhaps also less likely, the corresponding *ortho*-di-quinones could be generated, with subsequent 4 + 2 Diels-Alder cycloadditions with the olefinic precursors to afford the corresponding products **29** or **2**.

Yet other examples of flavonolignans are (+)-silychristin (**3**, Figure 9.1) (Wagner et al. 1971, 1978), (+)-silydianin (**36**, Figure 9.8) (Abraham et al. 1970), and iryantherin A (**37**) (Garzon et al. 1987, Villamil et al. 1988). Of these, (+)-silychristin (**3**) ($[\alpha]_D$ +81.4°, pyridine) was isolated from *S. marianum* fruits (Wagner et al. 1971), although neither amounts of plant material, extraction or purification protocols utilized, nor amounts of **3** obtained, were reported. Its formation can be provisionally considered to result from coupling of a coniferyl alcohol (**8**)-derived species with (+)-dihydroquercetin (**4**), with subsequent intra-molecular cyclization. Given that there are three distinct chiral centers, it is not currently possible to ascertain if the coupling reaction is stereoselective.

For (+)-silydianin (**36**) ($[\alpha]_D$ +175°, acetone), isolated from fruits of *S. marianum* (Abraham et al. 1970), its formation also appears to involve a coniferyl alcohol (**8**)-derived species coupled to a luteolin (**6**) or a (+)-dihydroquercetin (**4**)-derived moiety. However, whether this occurs via a radical species or via a 4 + 2 hetero Diels–Alder cycloaddition remain to be established, although the former (based on current knowledge) is possibly more likely.

Racemic iryantherin A (**37**) was isolated from fruits (separated into aril, mesocarp, and kernel tissues) of *Iryanthera laevis* (Garzon et al. 1987, Villamil et al. 1988). The aril (495 g) and mesocarp (1,385 g) yielded 130 and 40 mg of iryantherin A (**37**), respectively, in 0.026 and 0.0029% w/w. Iryantherin A (**37**) has an unusual inter-unit-linkage type and can provisionally be considered to result from coupling of chavicol (**38**) with the 2′,4′-dihydroxy-4,6′-dimethoxy-dihydrochalcone (**39**) also present in *I. laevis* (Braz Filho et al. 1980). Its bonding pattern might suggest, if formed via radical coupling, that chavicol (**38**) is converted into a free radical species ("centered" at C7),

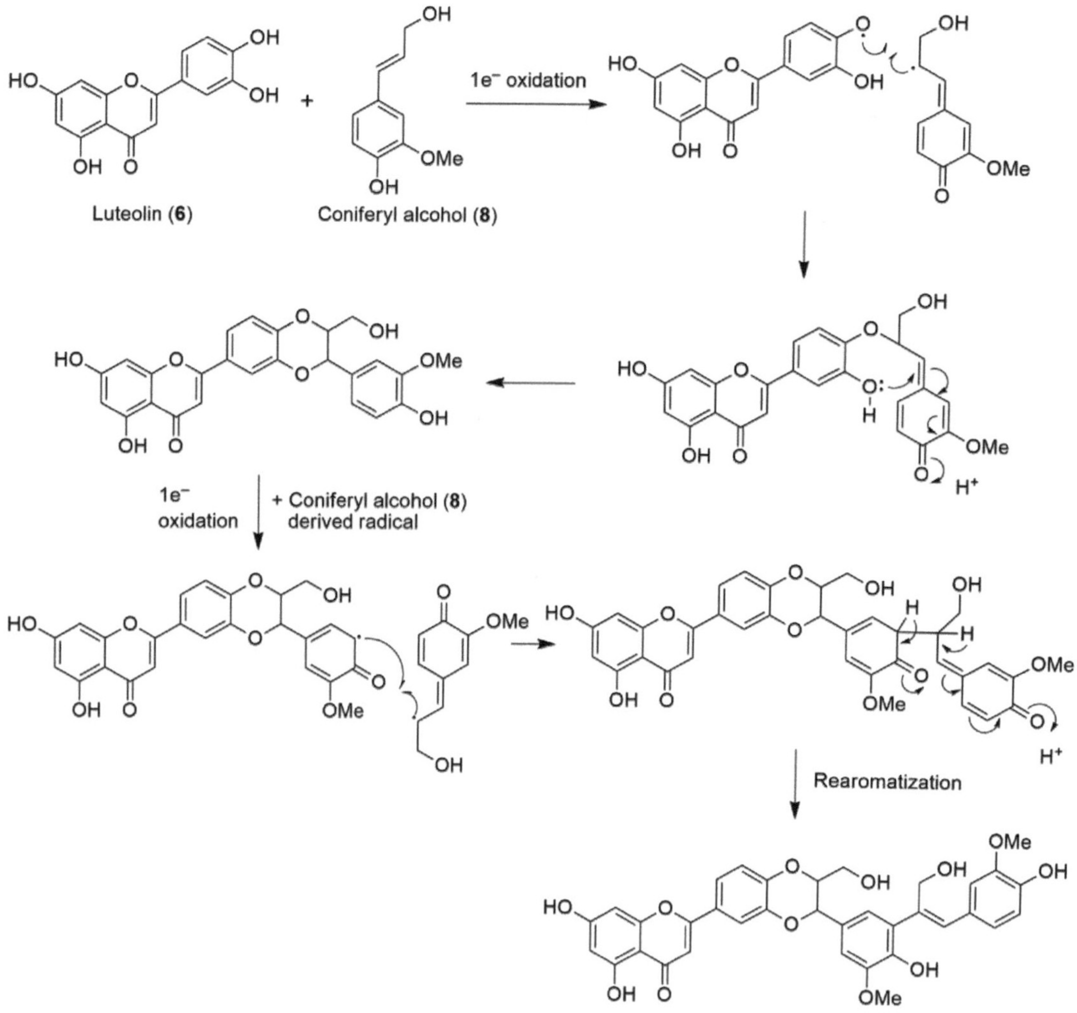

FIGURE 9.5 Possible route to hydnowightin (**2**) from luteolin (**6**) and coniferyl alcohol (**8**) via stepwise coupling.

which then couples with the radical species generated from **39**. However, this remains to be established.

9.4 STILBENOLIGNANS

The stilbenolignan, (–)-aiphanol (**40**, Figure 9.9), is apparently produced from coupling of a C_6C_2 stilbenoid [e.g. piceatannol (**41**)] with a C_6C_3 moiety, e.g. sinapyl alcohol (**14**).

(–)-Aiphanol (**40**, 6 mg, 0.0001% w/w) was obtained from 5.8 kg *Aiphanes aculeata* seeds, with an $[\alpha]_D$ –21.8° (c 0.13, MeOH) (Lee et al. 2001). This stilbenolignan and piceatannol (**41**) (250 mg, 0.0043% w/w) were isolated at the same time, with the latter having the requisite catechol functionality to allow for the benzodioxin-like substructure to be formed. Its structure was revised (to as shown) by Banwell et al. (2005).

Two other examples include the achiral salvianolic acid F (**42**, Figure 9.9) (Zhang et al. 2017) and tournefolic acid B (**43**) (Lin et al. 2002). The former (10 mg, 0.0002% w/w) was isolated from *Salvia miltiorrhiza* roots (5 kg). The latter was obtained from 20 kg of *Tournefortia sarmentosa* stems in an amount of 125 mg (0.0006% w/w), and could possibly be formed through intramolecular coupling of salvianolic acid F (**42**), the latter being a metabolite which co-occurs in its stems. These stilbenolignans all provisionally appear to be caffeic acid (**44**) derived.

Once again, whether **40**, **42**, and **43** are formed via radical coupling, or less likely that a Diels–Alder-like intermolecular cyclization reaction occurs to give **40** or **43**, is currently unknown.

None of these examples, however, contain a lignan (C_6C_3–C_6C_3) substructure proper, being instead C_6C_3–C_6C_2 coupled.

9.5 TERPENOID LIGNANS

Those reported thus far can contain either monomeric or dimeric (i.e. lignan) phenylpropanoid (C_6C_3) plant natural products. The monomeric phenylpropanoid (C_6C_3) terpenoid lignans can include those that are coniferyl (**8**) or sinapyl (**14**) alcohol derived, and linked via benzodioxin-like substructure linkages, e.g. brevitaxin (**45**, Figure 9.10) from the bark of

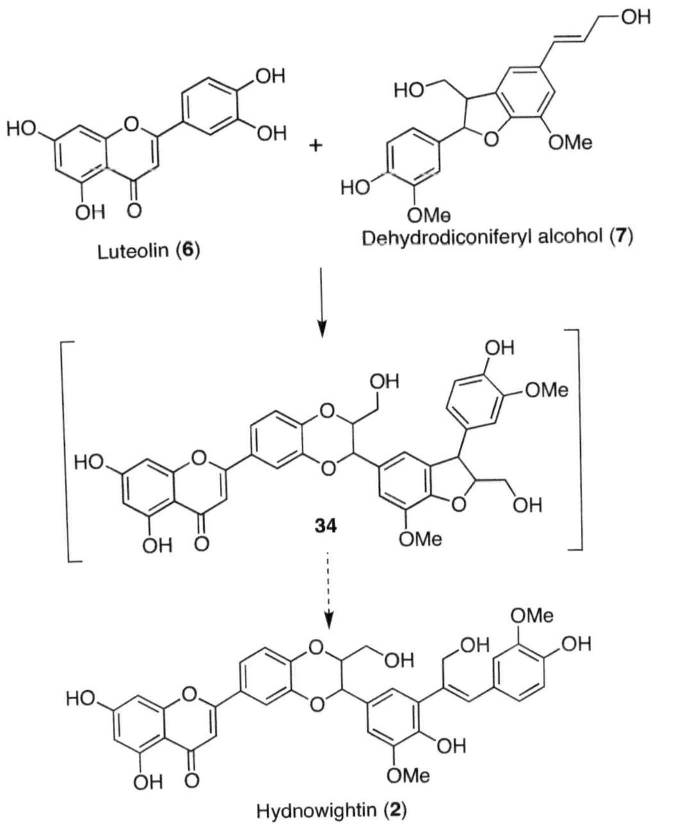

FIGURE 9.6 Possible route to hydnowightin (**2**) from luteolin (**6**) and dehydrodiconiferyl alcohol (**7**) directly, with subsequent ring opening of dihydrofurano ring.

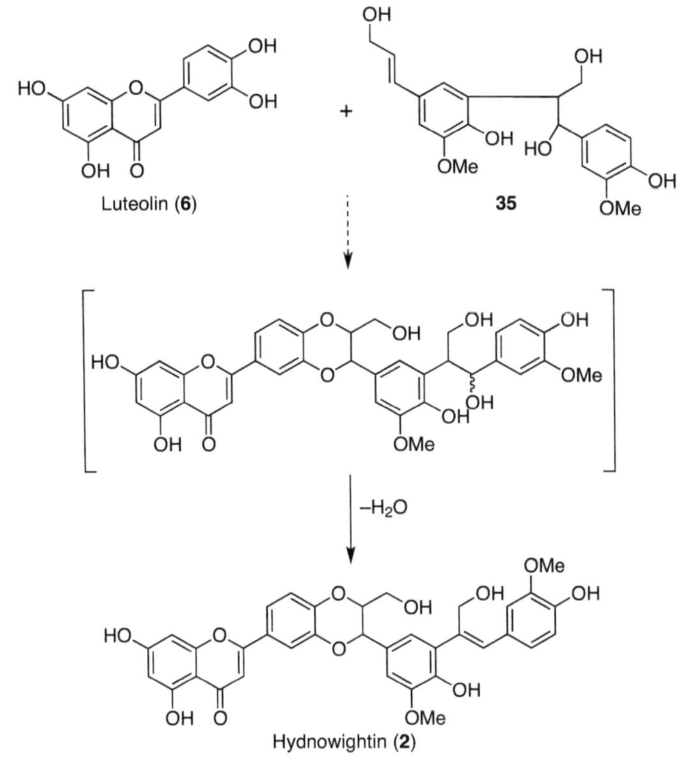

FIGURE 9.7 Possible route to hydnowightin (**2**) from luteolin (**6**) and coniferyl alcohol-derived dimer **35**.

Taxus brevifolia (Arslanian et al. 1995), which was isolated as a minor side-product during taxol (paclitaxel) isolation.

Brevitaxin (**45**), while containing the now rather common benzodioxin-like substructure (see earlier), displayed no optical activity. No details of the amount of bark extracted, or isolation conditions used to obtain it were provided. Another terpenoid lignan with a benzodioxin-like substructure is blepharodin (**46**) [9.3 mg, 0.0016% w/w, $[\alpha]_D$ −8.5° (c 0.82, $CHCl_3$)] isolated from the root bark (570 g) of *Maytenus magellanica* (Núñez et al. 2011). It may, however, only display optical activity because of its nor-triterpenoid component. Whether these are indeed true natural products produced enzymatically, or are isolation artifacts, is again an interesting question. While Núñez et al. (2011) described these as

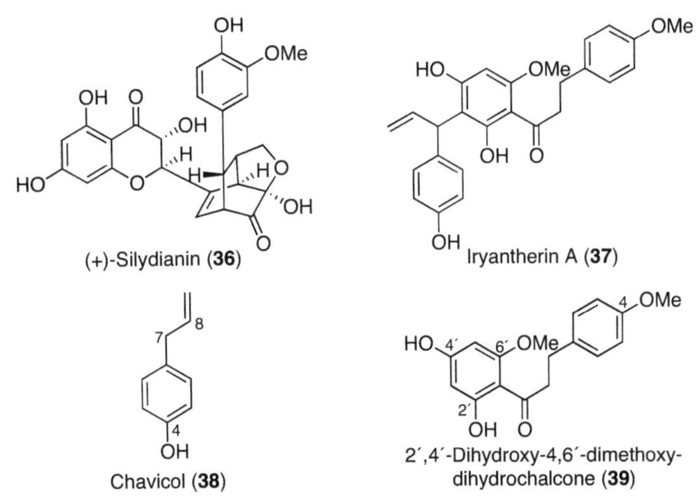

FIGURE 9.8 Vascular plant "flavanolignans", (+)-silydianin (**36**) and iryantherin A (**37**), and possible precursors (**38** and **39**) to iryantherin A (**37**).

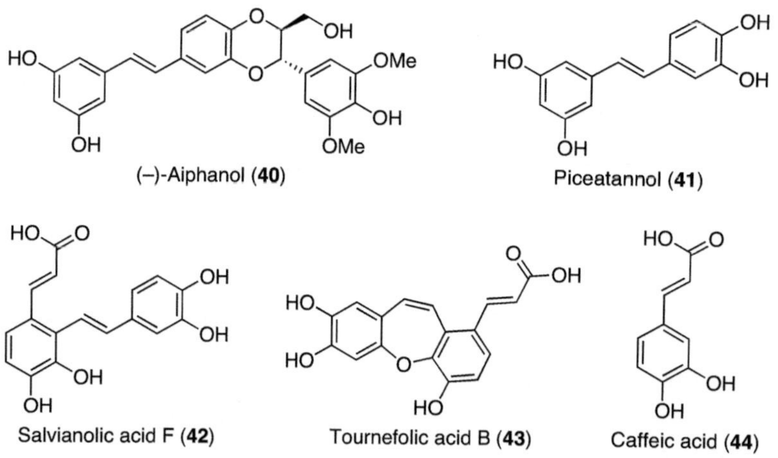

FIGURE 9.9 Stilbenolignans (–)-aiphanol (**40**), salvianolic acid F (**42**), and tournefolic acid B (**43**) and two possible precursors, piceatannol (**41**) and caffeic acid (**44**), respectively.

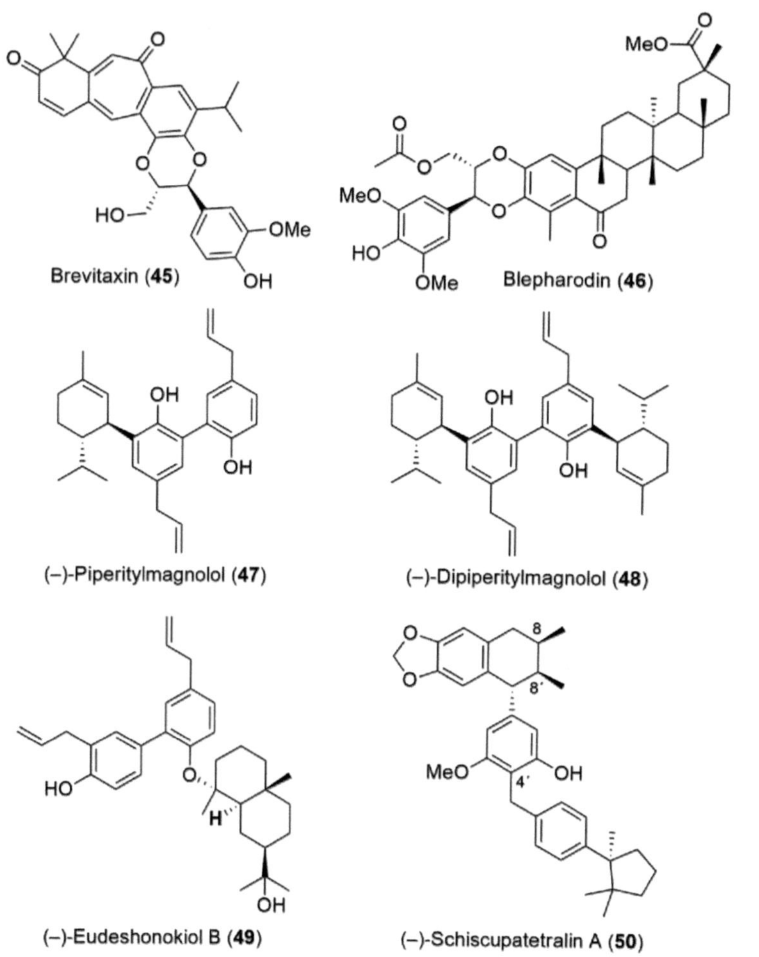

FIGURE 9.10 Various vascular plant terpenoid lignans (**45–50**).

hetero-Diels-Alder products, they could also be formed via radical coupling.

Other terpenoid lignan classes containing lignans proper that are covalently linked include those with either 1 or 2 monoterpenoid moieties or a diterpenoid C–C linked as shown, e.g. the biphenyl allylphenol lignans, (–)-piperitylmagnolol (**47**), also known as monoterpenylmagnolol, and (–)-dipiperitylmagnolol (**48**), as well as biphenyl ether lignans, e.g. (–)-eudeshonokiol B (**49**) (Figure 9.10). Other C–C linked terpenoid lignans include (–)-schiscupatetralin A (**50**), with this reportedly having an aryltetrahydronaphthalene lignan C–C bonded to an aromatic sesquiterpene-derived moiety as shown.

Of these, (–)-piperitylmagnolol (**47**) (Yahara et al. 1991) [or (–)-monoterpenylmagnolol (Konoshima et al. 1991)], isolates were obtained as colorless oils [890 mg, 0.018% w/w, $[\alpha]_D$ –146° (c 1.18, $CHCl_3$) and 780 mg, 0.036% w/w, $[\alpha]_D$ –138° (c 1.03, $CHCl_3$)] from *Magnolia officinalis* bark samples (5 and 2.2 kg), respectively, with (–)-dipiperitylmagnolol (**48**) [360 mg, 0.007% w/w, $[\alpha]_D$ –140° (c 0.31, $CHCl_3$)] also purified (Yahara et al. 1991). Another variant, (–)-eudeshonokiol B (**49**), was obtained [250 mg, 0.0025% w/w, $[\alpha]_D$ –71.7° (c 1.08, $CHCl_3$)] from purchased dried *Magnolia obovata* bark (10 kg) (Fukuyama et al. 1992). The aryltetrahydronaphthalene lignan, (–)-schiscupatetralin A (**50**) [12.2 mg, 0.00004% w/w, $[\alpha]_D$ –21° (c 1.22, MeOH)], was obtained from dried fruits of *Schizandra sphenanthera* (28 kg) (Tsai et al. 2014).

The biogenesis of these terpenoid lignans is unknown. However, both (–)-piperitylmagnolol (**47**) [(–)-monoterpenylmagnolol] and (–)-dipiperitylmagnolol (**48**) have very rare C–C linkage types between their terpenoid and lignan substructures. It will be of interest to identify both the terpenoid substrates and the enzymes/proteins responsible for these conversions, and establish how (–)-eudeshonokiol B (**49**) is biosynthesized.

The report of (–)-schiscupatetralin A (**50**) with four chiral centers is enigmatic. This is because it contains an aryltetrahydronaphthalene core that is reportedly C–C linked at C4′ to a sesquiterpene [presumed by these researchers to be (–)-γ-cuparenol (**51**, Figure 9.11) derived] (Tsai et al. 2014). Its proposed biosynthetic pathway also involves the putative lignan biosynthetic intermediate **52**, whose pendant aryl group has no phenolic OH at its C4′ position. Such a precursor, however, would be structurally highly unusual, as lignans are essentially always hydroxylated at that position. This needs to be established if correct or not as proposed.

In this context, recent literature contributions have revised many structures (Yang et al. 2015) from containing oxygenated functionalities at positions C3/C5 or C3′/C5′, but not C4/C4′, to have their functional groups revised to be at C3/C4 or C3′/C4′ instead as would be expected. As both the putative intermediate **52** reportedly has only oxygenated functionalities at C3/C5, it will be important to have it [and (–)-schiscupatetralin A (**50**)] re-examined to ensure the structures are correct.

9.6 XANTHOLIGNOIDS

Xanthones have the base skeleton (**53**, Figure 9.12). Xantholignans that have been reported include those linked to monolignols, such as sinapyl alcohol (**14**) in the xanthone (–)-gemixanthone A (**54**), harboring a benzodioxin-like substructure (Chung et al. 1999), and the xanthenone derivative, falloconvolin B (**55**) (Zhang et al. 2011), respectively.

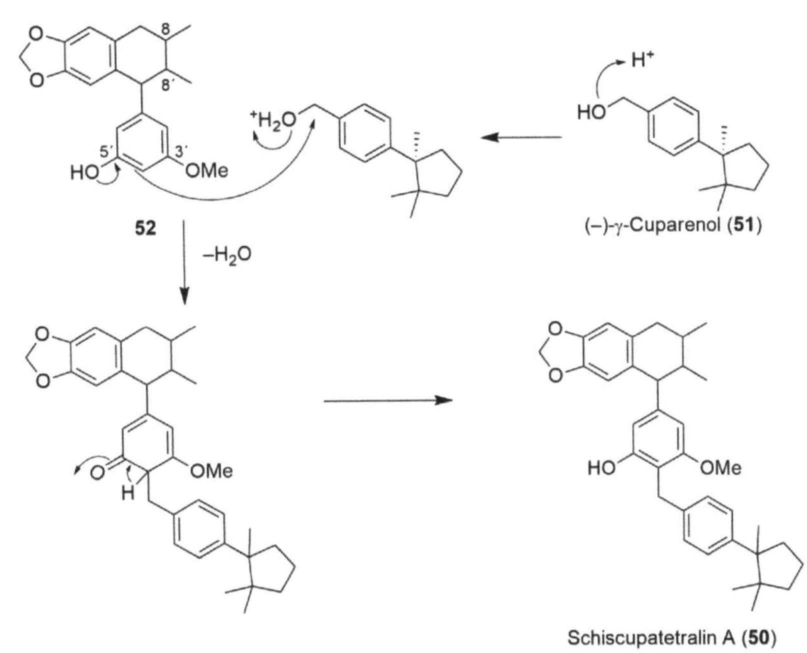

FIGURE 9.11 Proposed biosynthetic route to (–)-schiscupatetralin A (**50**) from putative aryltetrahydronaphthalene lignan **52** and (–)-γ-cuparenol (**51**) precursors as envisaged by Tsai et al. (2014).

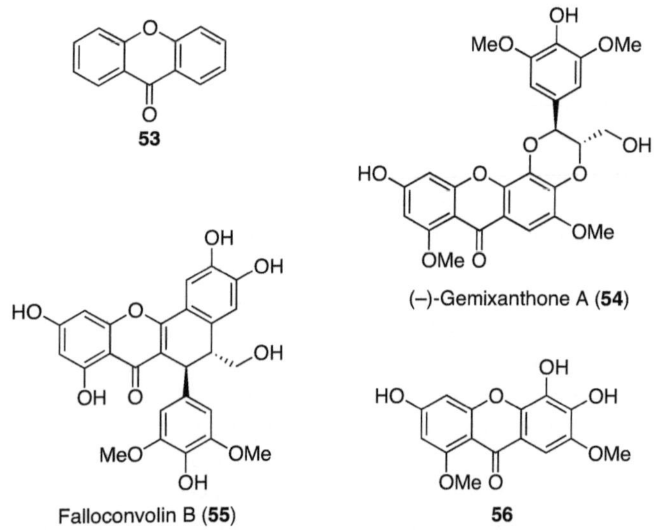

FIGURE 9.12 Xanthone base skeleton **53**, various xantholignoids (**54** and **55**) and possible biosynthetic precursor **56**.

(–)-Gemixanthone A (**54**) (10 mg, 0.0003% w/w) was isolated from *Hypericum geminiflorum* leaf tissue (3 kg) (Chung et al. 1999). It is, as for the mixed biochemical pathway lignans, most likely formed via radical coupling of the monolignol sinapyl alcohol (**14**) to a catechol xanthone, such as **56**, although a Diels-Alder adduct formation cannot be fully eliminated as another possibility.

Falloconvolin B (**55**) [25 mg, 0.00025% w/w, –20.5° (c 0.1, MeOH)] was obtained from aerial parts of *Fallopia convulvus* (10.1 kg) (Zhang et al. 2011), and can be provisionally considered to result via coupling of luteolin (**6**) with sinapyl alcohol (**14**).

None of these xantholignans contain lignan (C_6C_3–C_6C_3) substructures proper.

9.7 DIARYLHEPTANOID AND DIARYLPENTANOID LIGNANS

Two additional examples of mixed biochemical pathway products are those affording the rare diarylheptanoid and diarylpentanoid lignans (Figure 9.13). In this *Handbook* various entries for both diarylheptanoid and diarylpentanoid

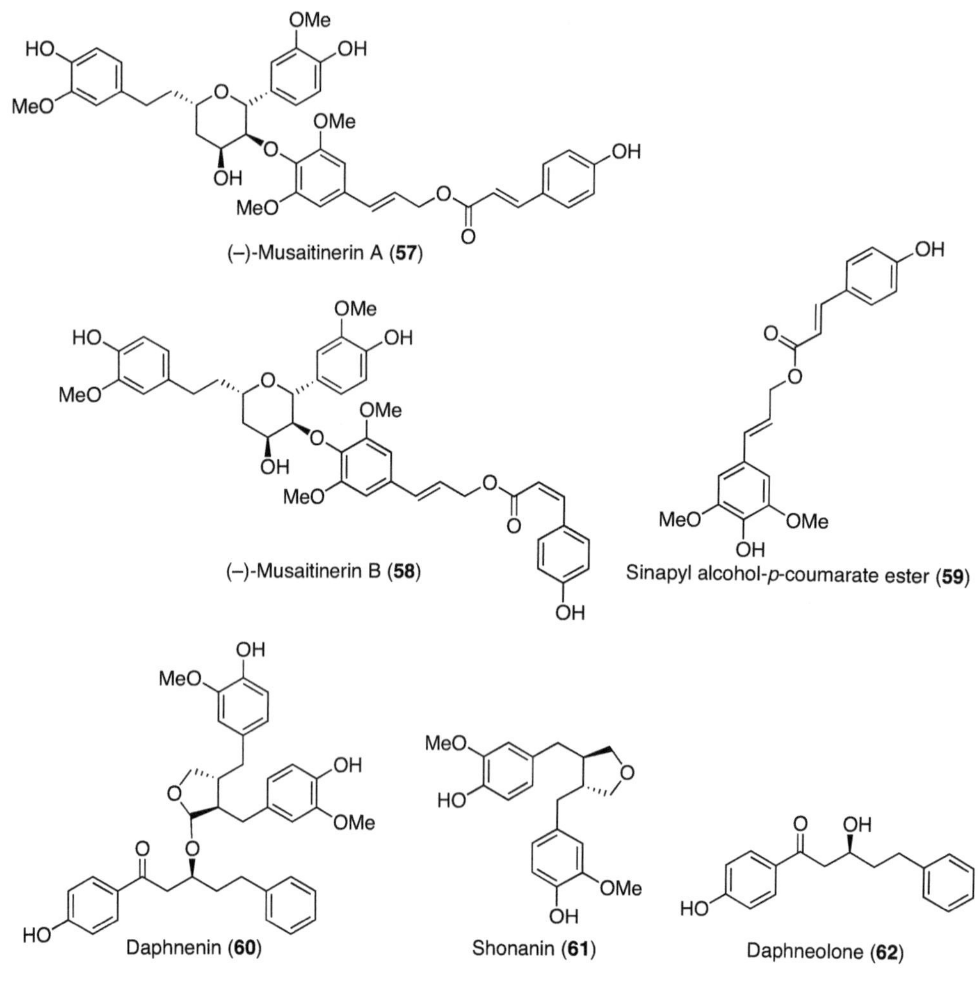

FIGURE 9.13 Rare diarylheptanoid- and diarylpentanoid-lignans (**57**, **58**, and **60**) and some possible biosynthetic precursors (**59**, **61**, and **62**).

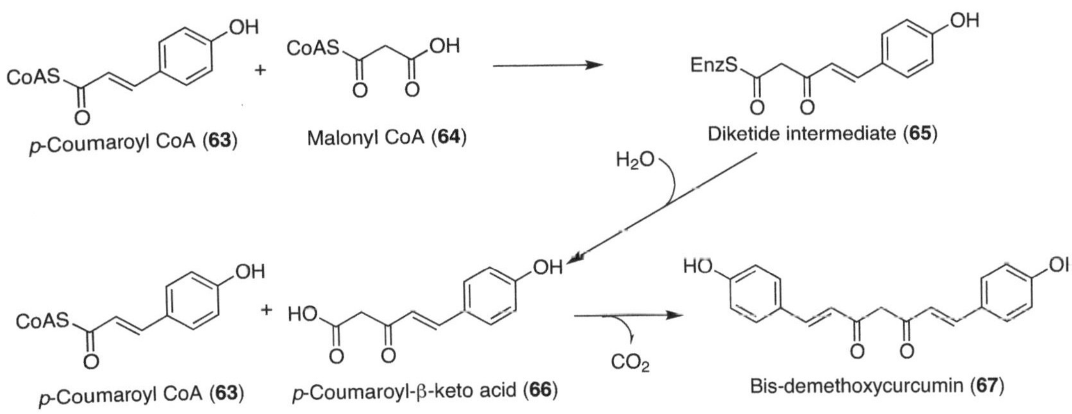

FIGURE 9.14 Curcuminoid synthase catalyzed conversion of *p*-coumaryl CoA (**63**) and malonyl CoA (**64**) to afford bis-demethoxycurcumin (**67**) (Morita et al. 2010).

metabolic classes are provided, but where only three plant products might be considered as either diarylheptanoid or diarylpentanoid lignans, i.e. in the same manner as for the other mixed biochemical pathways described earlier.

To our knowledge, the first diarylheptanoid lignans reported were (–)-musaitinerin A (**57**) $[\alpha]_D$ –36.3° (c 0.24, MeOH)) and (–)-musaitinerin B (**58**) $[\alpha]_D$ –35.7° (c 0.29, MeOH)), isolated from *Musa itinerans* fruits in 0.0009% and 0.0003% yield dry weight as both *E*- and *Z*-izomers (Liu et al. 2014). Their provisional inclusion as mixed biochemical pathway diarylheptanoid lignans results from the ether-linked sinapyl alcohol-*p*-coumarate ester (**59**) to a diarylheptanoid core.

In an analogous manner, there is only one diarylpentanoid lignan known thus far, namely daphnenin (**60**) $[\alpha]_D$ +71.6° (c 3.4, MeOH)) from *Daphne acutiloba* stems in 0.0001% yield dry weight (Huang et al. 2012). This metabolite is also ether linked, with the upper half of the molecule being anhydrose-coisolariciresinol (shonanin) (**61**) derived and the lower half being daphneolone (**62**).

Clearly, many of the reported diarylheptanoids and diarylpentanoids could potentially serve as substrates for mixed biochemical pathway diarylpentanoid- and diarylheptanoid-lignan formation *in planta*.

Diarylheptanoid biosynthesis has been studied, where it was established that entry into this metabolic class occurs via action of polyketide synthase-like enzymes (PKSs). For example, in rice (*Oryza sativa*), curcumin synthase, a type III PKS, catalyzes the one-pot formation of bis-demethoxycurcumin (**67**) from two 4-coumaroyl CoA (**63**) and one malonyl-CoA (**64**) molecules as shown in Figure 9.14 (Morita et al. 2010).

Similarly, entry into the diarylpentanoids can be envisaged to occur via action of somewhat comparable enzyme systems that afford entry into the diarylheptanoids.

On the other hand, nothing is currently known about the presumed biochemical pathways to these mixed biochemical pathway diarylheptanoid- and diarylpentanoid-lignans, including the substrates, proteins, and genes involved. This, of course, assumes that they are not artifacts of isolation.

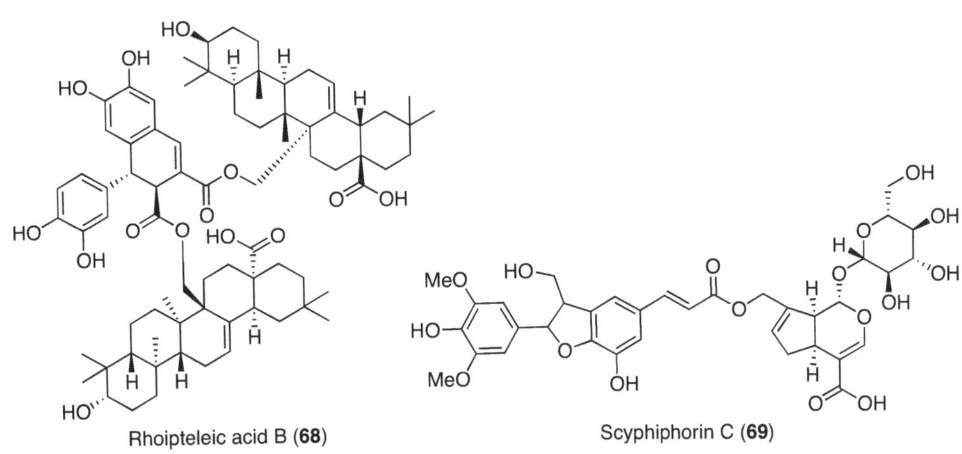

FIGURE 9.15 Miscellaneous lignan skeletal types ester linked to triterpene and iridoid moieties (**68** and **69**).

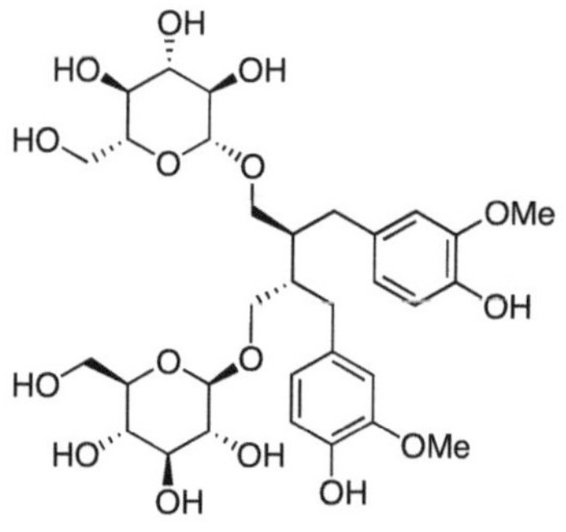

SDG (**70**)

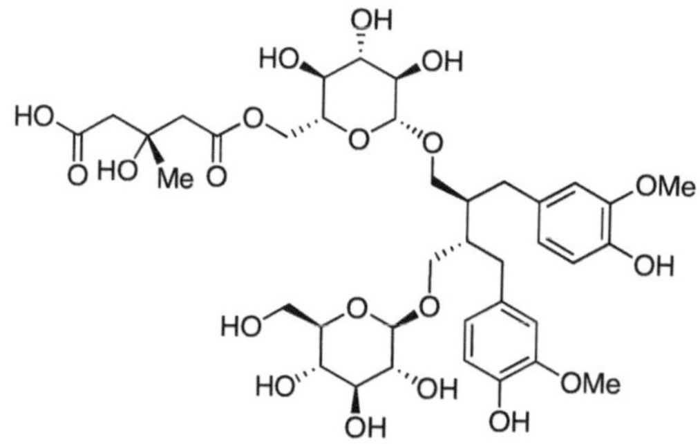

6a-HMG SDG (**71**)

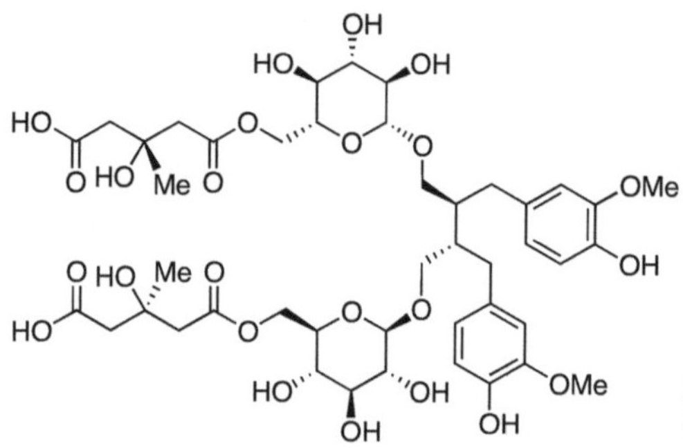

6a, 6a′-di-HMG SDG (**72**)

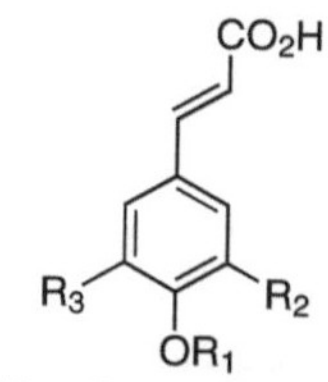

$R_1 = R_2 = R_3$ = H: *p*-Coumaric acid (**73**)
$R_1 = R_3$ = H, R_2 = OMe: Ferulic acid (**74**)
R_1 = H, $R_2 = R_3$ = OMe: Sinapic acid (**75**)
R_1 = Glc, $R_2 = R_3$ = H: *p*-Coumaric acid glucoside (**76**)
R_1 = Glc, R_2 = OH, R_3 = H: Caffeic acid glucoside (**77**)
R_1 = Glc, R_2 = OMe, R_3 = H: Ferulic acid glucoside (**78**)

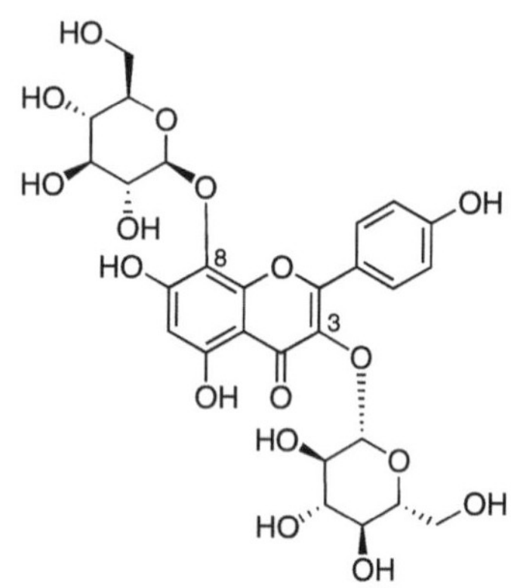

Herbacetin 3,8-*O*-diglucoside (**79**)

FIGURE 9.16 Lignan skeletal type ester linked hydroxymethylglutaric acid moieties (**71** and **72**), lignan secoisolariciresinol diglucoside (**70**), hydroxycinnamic acids (**73–75**), hydroxycinnamic acid glucosides (**76–78**), and herbacetin 3,8-*O*-diglucoside (**79**).

9.8 OTHER STRUCTURAL VARIANTS FOUND IN NATURE

In addition, there are many mixed biochemical pathway lignans (e.g. **68**, **69**, **71**, **72**) ester linked to various other natural product types. Examples here include lignans that are ester linked to triterpenes, an iridoid, and hydroxymethylglutaric acid (HMGA) moieties, respectively, such as rhoipteleic acid B (**68**) (Jiang et al. 1994), scyphiphorin C (**69**) (Tao et al. 2009), as well as secoisolariciresinol diglucoside (SDG, **70**), and its HMGA derivatives **71** and **72** (Ford et al. 2001), respectively (Figures 9.15 and 9.16).

Rhoipteleic acid B (**68**) [135 g, 0.03% w/w, $[\alpha]_D$ +49.4° (c 0.5, MeOH)] was obtained from 4.5 kg air-dried *Rhoiptelea chiliantha* bark (Jiang et al. 1994), whereas scyphiphorin C (**69**) [8 mg, 0.00008% w/w, $[\alpha]_D$ +28° (c 8.0, MeOH)] was purified from 10 kg stem bark of the Chinese mangrove (*Scyphiphora hydrophyllacea*) (Tao et al. 2009).

Rhoipteleic acid B (**68**) has an aryldihydronaphthalene lignan substructure that provisionally appears to be caffeic acid (**44**) derived, whereas the lignan substructure in scyphiphorin C (**69**) may be derived, for example, from dehydrodiconiferyl alcohol (**7**), or its methoxylated equivalent (not shown). In contrast, the flax (*Linum usitatissimum*) seed coat lignans containing SDG HMGA moieties covalently linked to other entities, such as hydroxycinnamic acids (**73–75**), hydroxycinnamic acid glucosides (**76–78**) and herbacetin 3,8-*O*-diglucoside (**79**) are present in much larger amounts (e.g. 3% w/w for just the SDG (**70**) component alone) (Ford et al. 2001). Not all of their biochemical pathways have been explored fully to date (Moinuddin et al. 2019).

9.9 CONCLUDING REMARKS

With a few exceptions [e.g. SDG (**70**) in flax seed coats], many if not most of the mixed biochemical pathway lignans are present in plant tissues in minuscule amounts, either in racemic or optically active forms. It should be noted that, for these very minor compounds, there is no definitive proof as to either the true biochemical or chemical mechanisms that are responsible for nearly all of their formation. Artifact formation in some cases is a possibility. Nor do we know which cell types these distinct metabolites accumulate *in planta*. Deciphering how these are indeed formed, and in which cell types (including subcellular locations), will be instructive in the future. Additionally, although a significant number of the "mixed-biochemical pathway lignans" are optically active, this does not establish optical purity, i.e. whether some isolates are enantiomerically pure or in enantiomeric excess. Future research needs to resolve this issue as well.

ACKNOWLEDGEMENTS

The authors thank the Chemical Sciences, Geosciences and Biosciences Division, DOE Office of Basic Energy Sciences (DE-FG-0397ER20259), as well as the Arthur and Katie Eisig-Tode Endowment.

REFERENCES

Abraham, D. J., S. Takagi, R. D. Rosenstein, et al. 1970. Structure of silydianin, an isomer of silymarin (silybin), by X-ray analysis. *Tetrahedron Lett.* 11:2675–2678.

Arslanian, R. L., D. T. Bailey, M. C. Kent, et al. 1995. Brevitaxin, a new diterpenolignan from the bark of *Taxus brevifolia*. *J. Nat. Prod.* 58:583–585.

Banwell, M. G., S. Chand, and G. P. Savage. 2005. An enantioselective total synthesis of the stilbenolignan (–)-aiphanol and the determination of its absolute stereochemistry. *Tetrahedron: Asymmetry* 16:1645–1654.

Braz Filho, R., M. S. Da Silva, and O. R. Gottlieb. 1980. The chemistry of Brazilian Myristicaceae. Part 14. Flavonoids from *Iryanthera laevis*. *Phytochemistry* 19:1195–1197.

Chang, C.-L., G.-J. Wang, L.-J. Zhang, et al. 2010. Cardiovascular protective flavonolignans and flavonoids from *Calamus quiquesetinervius*. *Phytochemistry* 71:271–279.

Chi, Y.-M., W.-M. Yan, and J.-S. Li. 1990. An alkaloid from *Incarvillea sinensis*. *Phytochemistry* 29:2376–2378.

Chi, Y.-M., F. Hashimoto, W.-M. Yan, and T. Nohara. 1995. Two alkaloids from *Incarvillea sinensis*. *Phytochemistry* 39:1485–1487.

Chung, M.-I., J.-R. Weng, M.-H. Lai, M.-H. Yen, and C.-N. Lin. 1999. A new chalcone, xanthones, and a xanthonolignoid from *Hypericum geminiflorum*. *J. Nat. Prod.* 62:1033–1035.

Davin, L. B., and N. G. Lewis. 2022. Redefining lignan/neolignan nomenclature and diversity: A historical perspective. In *The Lignan Handbook* (L. B. Davin and N. G. Lewis eds.), Taylor & Francis, Boca Raton, London, pp 1–11.

de M. Pinto, M. M., A. A. L. Mesquita, and O. R. Gottlieb. 1987. The chemistry of Brazilian Guttiferae. Part 40. Xanthonolignoids from *Kielmeyera coriacea*. *Phytochemistry* 26:2045–2048.

Foo, L. Y., and J. Karchesy. 1989. Pseudotsuganol, a biphenyl-linked pinoresinol-dihydroquercetin from Douglas-fir bark: Isolation of the first true flavonolignan. *J. Chem. Soc., Chem. Commun.*:217–219.

Ford, J. D., K.-S. Huang, H.-B. Wang, L. B. Davin, and N. G. Lewis. 2001. Biosynthetic pathway to the cancer chemopreventive secoisolariciresinol diglucoside-hydroxymethyl glutaryl ester-linked lignan oligomers in flax (*Linum usitatissimum*) seed. *J. Nat. Prod.* 64:1388–1397.

Fukuyama, Y., Y. Otoshi, K. Miyoshi, et al. 1992. Neurotrophic sesquiterpene-neolignans from *Magnolia obovata*: Structure and neurotrophic activity. *Tetrahedron* 48:377–392.

Garzon N, L., L. E. Cuca S, J. C. Martinez V, M. Yoshida, and O. R. Gottlieb. 1987. The chemistry of Colombian Myristicaceae. Part 5. Flavonolignoid from the fruit of *Iryanthera laevis*. *Phytochemistry* 26:2835–2837.

Huang, S. Z., X. J. Zhang, X. Y. Li, et al. 2012. Phenols with anti-HIV activity from *Daphne acutiloba*. *Planta Med.* 78:182–185.

Jenett-Siems, K., R. Weigl, M. Kaloga, J. Schulz, and E. Eich. 2003. Ipobscurines C and D: Macrolactam-type indole alkaloids from the seeds of *Ipomoea obscura*. *Phytochemistry* 62:1257–1263.

Jiang, Z.-H., T. Tanaka, and I. Kouno. 1994. Two diastereomeric triterpene-lignan esters having dimeric structure and their biosynthetically related triterpene caffeate from *Rhoiptelea chiliantha*. *Tetrahedron Lett.* 35:2031–2034.

Johns, S. R., J. A. Lamberton, J. R. Price, and A. A. Sioumis. 1968. Identification of coumarins isolated from *Lepiniopsis ternatensis* (Apocynaceae), *Pterocaulon sphacelatum* (Compositae), and *Melicope melanophloia* (Rutaceae). *Aust. J. Chem.* 21:3079–3080.

Konoshima, T., M. Kozuka, H. Tokuda, et al. 1991. Studies on inhibitors of skin tumor promotion. IX. Neolignans from *Magnolia officinalis*. *J. Nat. Prod.* 54:816–822.

Krauze-Baranowska, M. 2002. Truxillic and truxinic acids: Occurrence in plant kingdom. *Acta Pol. Pharm.* 59:403–410.

Lee, D., M. Cuendet, J. S. Vigo, et al. 2001. A novel cyclooxygenase-inhibitory stilbenolignan from the seeds of *Aiphanes aculeata*. *Org. Lett.* 3:2169–2171.

Li, C.-X., X.-Y. Song, W.-Y. Zhao, et al. 2019. Characterization of enantiomeric lignanamides from *Solanum nigrum* L. and their neuroprotective effects against MPP$^+$-induced SH-SY5Y cells injury. *Phytochemistry* 161:163–171.

Lin, Y.-L., Y.-Y. Chang, Y.-H. Kuo, and M.-S. Shiao. 2002. Anti-lipid-peroxidative principles from *Tournefortia sarmentosa*. *J. Nat. Prod.* 65:745–747.

Liu, F., Y. Zhang, Q.-Y. Sun, et al. 2014. Diarylheptanoids and phenylphenalenones from *Musa itinerans* fruits. *Phytochemistry* 103:171–177.

Moinuddin, S. G. A., J. R. Cort, C. A. Smith, et al. 2019. *Linum* lignan and associated biochemical pathways in human health and plant defense. In *Genetics and Genomics of Linum* (C. A. Cullis ed.), Springer, Cham. pp. 167–193.

Morita, H., K. Wanibuchi, H. Nii, et al. 2010. Structural basis for the one-pot formation of the diarylheptanoid scaffold by curcuminoid synthase from *Oryza sativa*. *Proc. Natl. Acad. Sci. USA* 107:19778–19783.

Nakashima, K.-i., M. Oyama, T. Ito, et al. 2012. Novel quinolinone alkaloids bearing a lignoid moiety and related constituents in the leaves of *Melicope denhamii*. *Tetrahedron* 68:2421–2428.

Núñez, M. J., M. L. Kennedy, I. A. Jimenez, and I. L. Bazzocchi. 2011. Uragogin and blepharodin, unprecedented hetero-Diels-Alder adducts from Celastraceae species. *Tetrahedron* 67:3030–3033.

Parthasarathy, M. R., K. R. Ranganathan, and D. K. Sharma. 1979. ^{13}C NMR of flavonolignans from *Hydnocarpus wightiana*. *Phytochemistry* 18:506–508.

Ray, A. B., S. K. Chattopadhyay, C. Konno, and H. Hikino. 1982. Structure of cleomiscosin B, a coumarino-lignoid of *Cleome viscosa* seeds. *Heterocycles* 19:19–22.

Sajeli, B., M. Sahai, R. Suessmuth, et al. 2006. Hyosgerin, a new optically active coumarinolignan, from the seeds of *Hyoscyamus niger*. *Chem. Pharm. Bull.* 54:538–541.

Seca, A. M. L., A. M. S. Silva, A. J. D. Silvestre, et al. 2001. Lignanamides and other phenolic constituents from the bark of kenaf (*Hibiscus cannabinus*). *Phytochemistry* 58:1219–1223.

Sharma, D. K., K. R. Ranganathan, M. R. Parthsarathy, B. Bhushan, and T. R. Seshadri. 1979. Flavonolignans from *Hydnocarpus wightiana*. *Planta Med.* 37:79–83.

Stermitz, F. R., J. Tawara-Matsuda, P. Lorenz, et al. 2000. 5′-Methoxyhydnocarpin-D and pheophorbide A: *Berberis* species components that potentiate berberine growth inhibition of resistant *Staphylococcus aureus*. *J. Nat. Prod.* 63:1146–1149.

Sy-Cordero, A., T. N. Graf, Y. Nakanishi, et al. 2010. Large-scale isolation of flavonolignans from *Silybum marianum* extract affords new minor constituents and preliminary structure-activity relationships. *Planta Med.* 76:644–647.

Takemura, Y., M. Abe, M. Ju-ichi, et al. 1993. Structure of acrignine-A, the first naturally occurring acridonolignoid from *Citrus* plants. *Chem. Pharm. Bull.* 41:406–407.

Tanaka, H., I. Kato, K. Ichino, and K. Ito. 1986. Coumarinolignoids, cleomiscosin A and cleomiscosin B, from *Aesculus turbinata*. *J. Nat. Prod.* 49:366–367.

Tao, S.-H., S.-H. Qi, S. Zhang, Z.-H. Xiao, and Q.-X. Li. 2009. Scyphiphorins C and D, two new iridoid glycosides from the Chinese mangrove *Scyphiphora hydrophyllacea*. *Heterocycles* 78:1557–1562.

Tsai, Y.-C., Y.-B. Cheng, I.-W. Lo, et al. 2014. Seven new sesquiterpenoids from the fruits of *Schisandra sphenanthera*. *Chem. Biodiversity* 11:1053–1068.

Villamil Z., E., L. E. Cuca S., and J. C. Martinez V. 1988. Partial study of the bark from *Iryanthera laevis*. *Spectroscopy* 6:157–166.

Wagner, H., O. Seligmann, L. Hoerhammer, M. Seitz, and J. Sonnenbichler. 1971. Structure of silychristin, a second silymarin isomer from *Silybum marianum*. *Tetrahedron Lett.* 12:1895–1899.

Wagner, H., V. M. Chari, M. Seitz, and I. Riess-Maurer. 1978. The structure of silychristin. A ^{13}C NMR study. *Tetrahedron Lett.* 19:381–384.

Wessely, F., and E. Demmer. 1928. Die Konstitution des Fraxetins. *Ber. Dtsch. Chem. Ges.* 61B:1279–1284.

Yahara, S., T. Nishiyori, A. Kohda, T. Nohara, and I. Nishioka. 1991. Isolation and characterization of phenolic compounds from Magnoliae Cortex produced in China. *Chem. Pharm. Bull.* 39:2024–2036.

Yan, X., J. Tang, C. dos Santos Passos, et al. 2015. Characterization of lignanamides from hemp (*Cannabis sativa* L.) seed and their antioxidant and acetylcholinesterase inhibitory activities. *J. Agric. Food Chem.* 63:10611–10619.

Yang, Y.-N., H. Zhu, Z. Chen, et al. 2015. NMR spectroscopic method for the assignment of 3,5-dioxygenated aromatic rings in natural products. *J. Nat. Prod.* 78:705–711.

Zhang, C.-F., J. Chen, L.-Q. Zhao, et al. 2011. Three new flavonoids from the active extract of *Fallopia convolvulus*. *J. Asian Nat. Prod. Res.* 13:136–142.

Zhang, J., Q. Jin, Y. Deng, et al. 2017. New depsides from the roots of *Salvia miltiorrhiza* and their radical-scavenging capacity and protective effects against H_2O_2-induced H9c2 cells. *Fitoterapia* 121:46–52.

Zhang, Z., G. M. Liu, Y. H. Wang, et al. 2007. A new isoflavonolignan from *Pueraria alopecuroides* Craib. *Chin. Chem. Lett.* 18:297–299.

10 Oligomeric Lignan Diversity *in Planta*
A Biochemical Roadmap for the Future

Syed G. A. Moinuddin, Laurence B. Davin, and Norman G. Lewis

10.1 INTRODUCTION

In terms of molecular size, the smallest lignans are C_6C_3 phenylpropanoid dimers. Of these, the biochemical pathways to 8–8′ linked lignans (C_6C_3 dimers), such as podophyllotoxin (**1b**, Figure 10.1), and its analogs are quite well understood (Vassão et al. 2010, Marques et al. 2013, Lau and Sattely 2015, Davin et al. 2022, Davin and Lewis 2022a, b, c).

Yet, there are many lignan skeletal types found in vascular plants whose formation – in terms of either coupling or downstream metabolism – remain poorly understood. Indeed, the enormous structural diversity found in the 'simplest' lignans (i.e. C_6C_3 phenylpropanoid dimers) points to the urgent need to explore how their different structural types (particularly those with different inter-unit linkages and additional structural rearrangements) are biochemically controlled.

An appreciation of this structural diversity of the 'simplest' lignans (i.e. C_6C_3 phenylpropanoid dimers) can readily be gleaned from the *Handbook* entries towards the end of these chapters. The fantastic level of structural diversity already known can be likened to that within the alkaloid families.

It is, however, not the intent of this contribution to focus upon the generation of these 'simplest' yet diverse lignan structures, even though this can be assumed to be a very productive line of scientific enquiry.

A purpose of this contribution is instead to attempt to potentially identify productive lines of lignan research for the future that address formation of higher oligomeric lignans. In so doing, the reader is reminded that oligomeric lignans of various skeletal types and molecular sizes can be found in different species, tissues, and organs. Some are, for example, found in heartwood of woody plant species. One example is that of the abundant 8–8′-linked lignan, plicatic acid (**2b**), and its higher oligomers (Gardner et al. 1966), which accumulate in western red cedar (*Thuja plicata*) heartwood (Johansson et al. 2000). In this species, lignan infusion from neighboring parenchyma cells into lignified cell walls represents an important part of its heartwood formation, and the durability that the heartwood has against opportunistic pathogens. On the other hand, what non-biochemical changes can occur to lignans in species such as western red cedar over timeframes spanning tens to hundreds to thousands of years during the plant's possible lifespan are not well understood.

There are numerous reports of oligomeric lignans throughout the plant kingdom. These, however, are generally present as mixtures of lignans (and other plant natural products) of varying molecular size, structural complexity, and amount, e.g. up to lignan pentamers (i.e. 10 × C_6C_3 monomers) and higher. Some are found in trace to near trace quantities, raising the additional uncertainty of whether they are true natural products or are artifacts of isolation or of tissue 'aging'. On the other hand, some others are more abundant. However, we emphasize that their relative amounts do not provide useful insight into whether they are either true natural products or are artifacts.

To classify lignans of varying size, we utilize the same nomenclature rationale as for terpene oligomers. Thus, one C_6C_3 unit is a phenylpropanoid monomer (but not a lignan), 2 C_6C_3 units linked together are lignans, 3 are sesqui-lignans, 4 are di-lignans (or bi-lignans), 5 are sester-lignans, 6 are tri-lignans, 7 are sesquar-lignans, 8 are tetra-lignans, 10 are penta-lignans, respectively, and so on.

The demarcation between oligomeric lignans and polymeric lignins with some researchers is currently blurred (Dima et al. 2015), both in chemical and physiological terms, but they can be considered separate natural product classes. Distinction can readily be made on the basis of whether they display polymeric properties (i.e. structural cell wall reinforcing lignins) *versus* oligomeric lignans. The latter are generally not considered as structural cell wall constituents, and are viewed as being largely involved in plant defense (Gang et al. 1998).

Oligomeric lignans in this contribution are described in order of decreasing size from the largest reported thus far, penta-lignans (10 C_6C_3 units linked together), to the sesquilignans (3 × C_6C_3 monomers covalently linked). It is important to note, however, that large oligomeric lignans are not

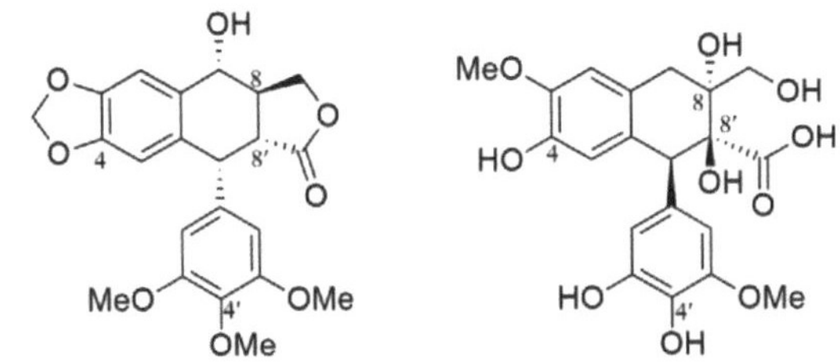

FIGURE 10.1 Aryltetrahydronaphthalene lignans, (–)-podophyllotoxin (**1b**) and (–)-plicatic acid (**2b**); "b" refers to the (–)-enantiomers.

to date very frequently reported. By far, the vast number of reports are the lignans themselves (2 × C_6C_3 monomers).

It needs to be emphasized that a critical literature review of oligomeric lignan occurrence in different plant species is frequently made difficult by the fragmented nature of many reports of a particular plant species' phytochemical constituents. Some literature contributors sprinkle a few new structures here and there into many papers on essentially the same topic. Such fragmentation of results has long been a practice discouraged by many scientific journals, i.e. that desire publishing a plant's phytochemical constituents as a comprehensive contribution.

10.2 PENTA- AND TETRA-LIGNANS

In contrast to the generalities above, a fairly comprehensive analysis of lignans was reported from *Clinopodium gracile* whole plant tissue (Aoshima et al. 2012). All of these are apparently (+)-rosmarinic acid (**3**), (–)-clinopodic acid J (**4**), and/or (+)-lithospermic acid (**5**) related, with the latter two containing 8′–2-inter-unit linkages (Figure 10.2). (This linkage type is also found in various lignans from other species; see the entries in this *Handbook*). The *C. gracile* lignans were present in an extract whose constituents ranged from penta-lignans to lignans (C_6C_3 dimers).

Specifically, the extract contained the 8′–2 linked penta-lignan (–)-clinopodic acid M (**6**, Figure 10.3), the corresponding tetra-lignans, (–)-clinopodic acid L (**7**), and (+)-clinopodic acid P (**8**), together with other lower molecular weight homologs/analogs, such as the tri-lignan (–)-clinopodic acid K (**10**, Figure 10.4). The "repeating units" in (–)-clinopodic acid M (**6**), (–)-clinopodic acid L (**7**), and (+)-clinopodic acid P (**8**) are somewhat structurally analogous to that of an aryl dihydrobenzofuran lignan (*cf.* dehydrodiconiferyl alcohol (**9**) as an example).

Interestingly, while lignans (**6–8**) and others were present in amounts ranging from 5 to 113 mg (isolated from 430 g powdered dried tissue, Table 10.1), both (–)-rosmarinic acid (**3**) and (–)-lithospermic acid (**5**) were much more abundant by comparison (*circa* 918 and 516 mg, respectively). That is, the lower molecular weight moieties, rosmarinic acid (**3**) and lithospermic acid (**5**), were by far the most dominant species, with the next two in abundance being the pentamer clinopodic acid M (**6**) and the trimer clinopodic acid K (**10**) (discussed below). The others were in much smaller amounts comparatively.

In terms of putative biosynthetic pathways, (–)-clinopodic acids L (**7**) and M (**6**) appear to be derived from coupling of either 4 or 5 rosmarinic acid (**3**) moieties, respectively. Coupling can be envisaged to occur sequentially by what may be regiospecific coupling of the caffeic acid (**11**) subunit of the rosmarinic acid (**3**) moiety with the 3-4-dihydroxyphenyllactic acid subunit of another rosmarinic acid (**3**) molecular species, i.e. to generate the corresponding quinone methide intermediates, which then re-aromatize sequentially via intramolecular 3–*O*–7′ bond formation as in (**6**), and so on. In contrast, formation of (+)-clinopodic acid P (**8**) seems to involve

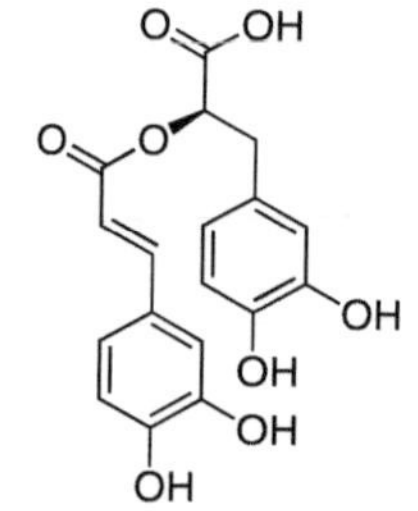

(+)-Rosmarinic acid (**3**)
(ester of caffeic acid and 3,4-dihydroxyphenyllactic acid)
$[\alpha]_D^{23}$ = +37.2° (c 2.92, MeOH)

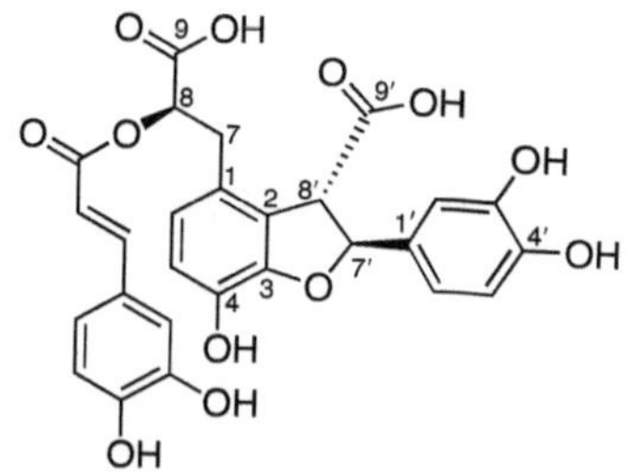

(–)-Clinopodic acid J (**4**)
$[\alpha]_D^{23}$ = –49.1° (c 1.08, MeOH)

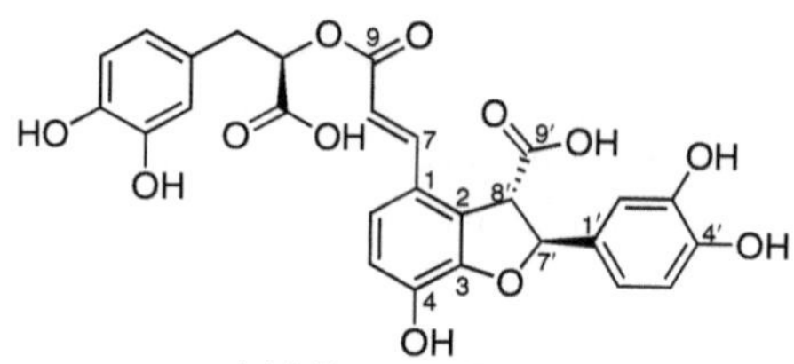

(+)-Lithospermic acid (**5**)
$[\alpha]_D^{22}$ = +56.5° (c 0.5, MeOH)

FIGURE 10.2 (+)-Rosmarinic acid (**3**), (–)-clinopodic acid J (**4**), and (+)-lithospermic acid (**5**) from *Clinopodium gracile* (Aoshima et al. 2012). (+)- and (–)- designations refer to the overall optical rotation, $[\alpha]_D$, reported in the literature.

stepwise coupling of three rosmarinic acid (**3**) moieties as discussed earlier, but with the fourth (terminal) unit being linked to the caffeic acid ester subunit of rosmarinic acid (**3**) instead (Figure 10.3).

Importantly, circular dichroism studies indicated that the dihydrobenzofuran rings in (–)-clinopodic acids L (**7**) and M (**6**) had the absolute stereo-chemistries shown. This, in turn, appears to provisionally suggest that the coupling reactions were stereoselective, but this needs to be experimentally established at the protein level. To date, only the genes and enzymes are known that enable formation of rosmarinic acid (**3**) (Petersen et al. 2009). Recent ^{13}C-labeling studies concluded, however, that rosmarinic acid (**3**) can undergo coupling, e.g. to form lithospermic acid (**5**) (Di et al. 2013). Nevertheless, neither genes nor proteins have yet been identified as involved in the formation of these higher molecular weight oligomers. This will be important to establish.

10.3 TRI-LIGNANS

Tri-lignans have been reported from various plant species; for example, the above-mentioned 8′-2–linked (–)-clinopodic acid K (**10**, Figure 10.4 and Table 10.1) from *C. gracile* (Aoshima et al. 2012), which is presumably derived from stereo-selective coupling of 3 rosmarinic acid (**3**) moieties as described previously. Structural variants have also been reported in other plant species, such as (+)-**12** from leaves of *Celastrus hindsii* (Ly et al. 2006), and (+)-yunnaneic acids A and B (**13** and **14**) from *Salvia yunnanensis* (Tanaka et al. 1996). Of these, tri-lignan (+)-**12** (167 mg, 0.24% dry weight) was isolated from

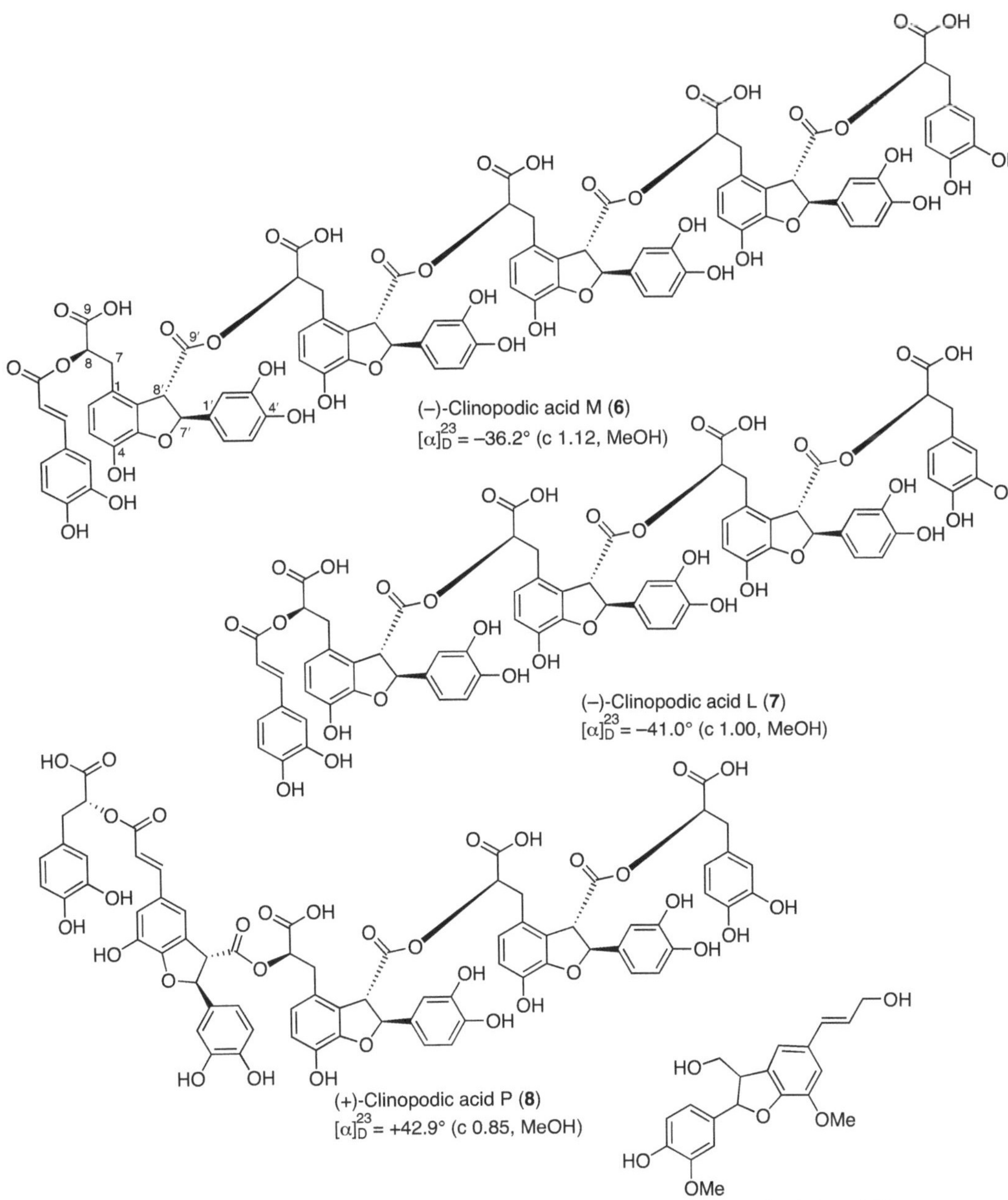

(−)-Clinopodic acid M (**6**)
$[\alpha]_D^{23} = -36.2°$ (c 1.12, MeOH)

(−)-Clinopodic acid L (**7**)
$[\alpha]_D^{23} = -41.0°$ (c 1.00, MeOH)

(+)-Clinopodic acid P (**8**)
$[\alpha]_D^{23} = +42.9°$ (c 0.85, MeOH)

Dehydrodiconiferyl alcohol (**9**)

FIGURE 10.3 Penta- and tetra-lignans, clinopodic acids M (**6**), L (**7**) and P (**8**) from *Clinopodium gracile* (Aoshima et al. 2012), and the dihydrobenzofuran lignan, dehydrodiconiferyl alcohol (**9**).

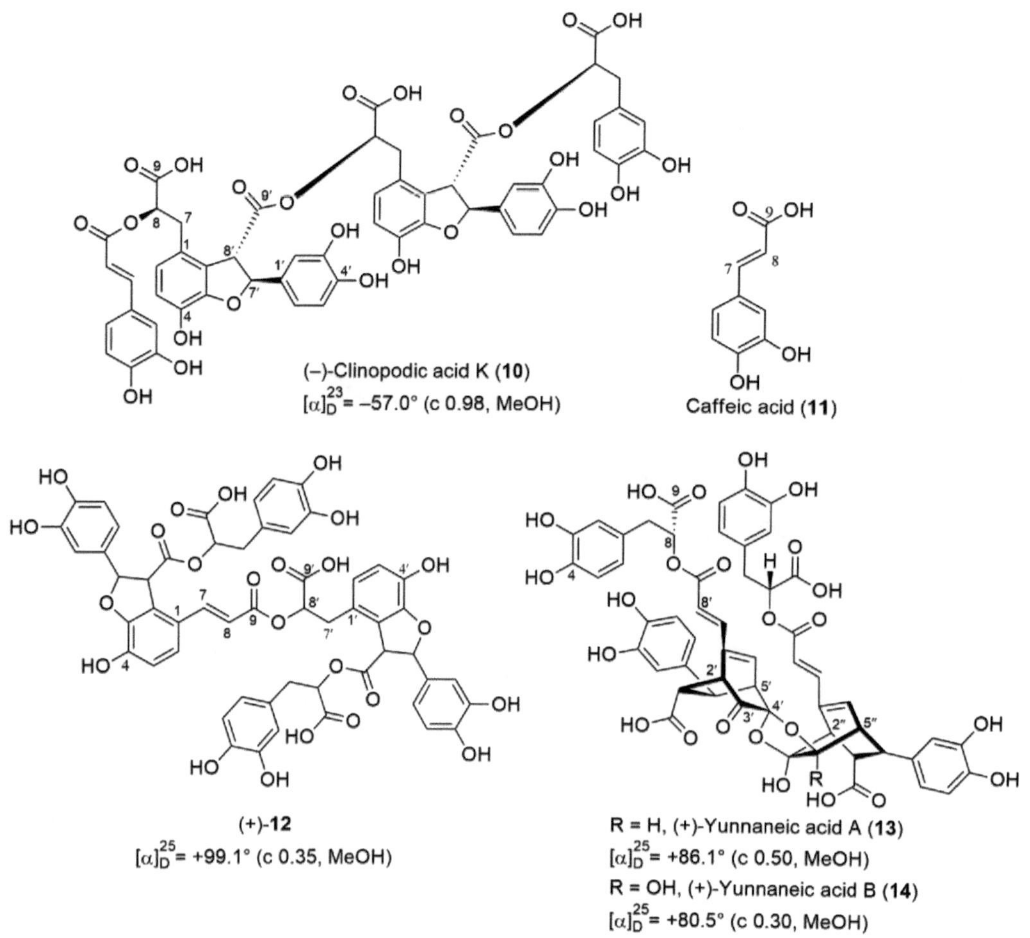

FIGURE 10.4 (−)-Clinopodic acid K (**10**) from *Clinopodium gracile* (Aoshima et al. 2012), (+)-tri-lignan **12** from *Celastrus hindsii* (Ly et al. 2006), and (+)-yunnaneic acids A and B (**13** and **14**) from *Salvia yunnanensis* (Tanaka et al. 1996), as well as caffeic acid (**11**).

70 g *C. hindsii* dried leaves (Ly et al. 2006), whereas (+)- yunnaneic acids A (**13**) (0.67 g, 0.07 % dry weight) and B (**14**) (3.8 g, 0.38 % dry weight) were obtained from 1 kg *S. yunnanensis* dried root tissue (Tanaka et al. 1996), respectively.

Their structures suggest they are also derived from coupling of rosmarinic acid (**3**). Notably, for (+)-yunnaneic acids A and B (**13** and **14**), two of their original aromatic rings have apparently undergone intramolecular (4 + 2) cycloaddition reactions with ester-linked caffeic acid (**11**)-derived C7–C8 side chains. Whether this occurs during chain extension or following tri-lignan formation is unknown. Furthermore, how their formation in each case occurs and differs biochemically, is also unknown presently. Assuming, however, this is not a photochemical reaction (e.g. during isolation), the enzymology here will be of considerable interest to resolve.

In *Cerbera* spp., a mixture of cerberalignans was reported from *Cerbera manghas* and *C. odollam* air-dried stems (Figures 10.5–10.7) (Abe et al. 1988a, b, 1989). This included

TABLE 10.1
Metabolites isolated from *Clinopodium gracile*[1] (Aoshima et al. 2012)

Compound isolated	Amount (mg)	(% dry weight)	Number of C_6C_3 units	Relative amounts
(+)-Rosmarinic acid (**3**)	918.0	0.2135	2	173
(+)-Lithospermic acid (**5**)	516.0	0.1200	3	97
(−)-Clinopodic acid J (**4**)	12.1	0.0028	3	2.3
(−)-Clinopodic acid K (**10**)	58.0	0.0135	6	10.9
(−)-Clinopodic acid L (**7**)	19.0	0.0044	8	3.6
(+)-Clinopodic acid P (**8**)	5.3	0.0012	8	1
(−)-Clinopodic acid M (**6**)	113.0	0.0263	10	21.3

1 Isolated from 430 g of dried aerial and subterranean parts.

a presumed coniferyl alcohol-derived tri-lignan, (–)-cerberalignan I (**30**, Figure 10.7) (Abe et al. 1989), as well as closely related sesqui-, bi-, sester-lignans, and lignans, albeit in vastly different amounts (Table 10.2). These included (–)-olivil (**15b**), (+)-cycloolivil (**16a**), and the cerberalignans **17–30** (Figures 10.5–10.7).

The tri-lignan (–)-cerberalignan I (**30**) was obtained in amounts of 24 and 4 mg from 6 kg *C. cerbera* and 1.1 kg *C. odollam* air-dried stems, these corresponding to *circa* 0.0004% dry weight. As one possibility, it may be derived directly from (–)-olivil (**15b**), which was also present in much higher amounts (3.21 g and 126 mg) from these two species (Abe et al. 1988a, b). If this assumption is correct, (–)-cerberalignan I (**30**) can be provisionally viewed as originating via coupling of (–)-olivil (**15b**) moieties linked both tail-to-tail (T–T) and head-to-tail (H–T) in this case. However, its very low amount (relative to (–)-olivil (**15b**)), and the different types of C5-C5′ coupling (T–T and H–T) observed raises the possibility (but not a proof) that these are artifacts of tissue 'aging' or isolation/storage protocols. This concern becomes more apparent (as discussed below), by (i) the diversity of (–)-olivil (**15b**)-derived sesqui-, bi-, and sester-lignans that can be linked head-to-head (H–H), tail-to-tail (T–T), or head-to-tail (H–T), and (ii) the presumed precursor (–)-olivil (**15b**) being present in large amount as a potential substrate for oxidative coupling during either "aging" of tissue or in isolation processes. Nevertheless, it cannot be ruled out that they are true natural products formed by protein/enzyme action. This, however, needs to be determined experimentally.

Other tri-lignans reported include those that are allylphenol-/propenylphenol-derived. An example is the racemic magnolianin (**31**, Figure 10.8) (2.2 g, 0.044%, dry weight) isolated from purchased dried *Magnolia obovata* bark tissue (5 kg) (Fukuyama et al. 1993, Matsuda et al. 2001). Magnolianin (**31**) is possibly formed via coupling of two obovatol (**32**)- and one magnolol (**33**)-derived moieties, respectively, these also being abundant lignans present in this species (**32**: 656 mg, 0.013%; **33**: 5.3 g, 0.106%, respectively). Again, while

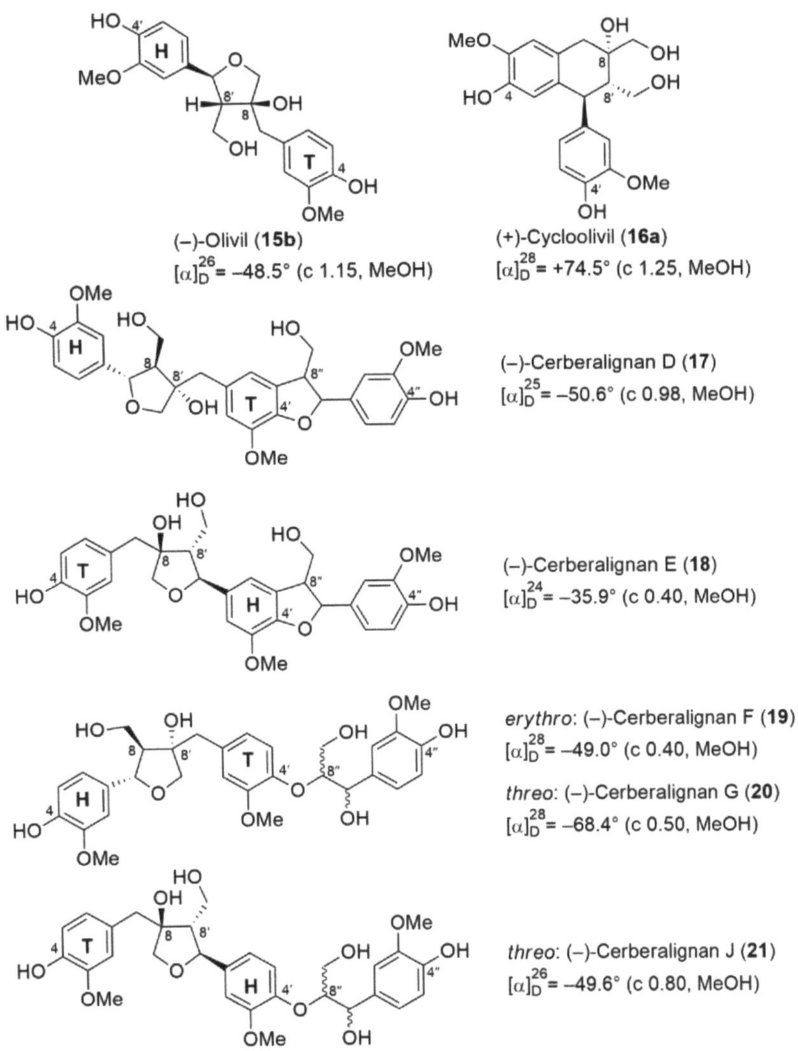

(–)-Olivil (**15b**) $[\alpha]_D^{26}$ = –48.5° (c 1.15, MeOH)

(+)-Cycloolivil (**16a**) $[\alpha]_D^{28}$ = +74.5° (c 1.25, MeOH)

(–)-Cerberalignan D (**17**) $[\alpha]_D^{25}$ = –50.6° (c 0.98, MeOH)

(–)-Cerberalignan E (**18**) $[\alpha]_D^{24}$ = –35.9° (c 0.40, MeOH)

erythro: (–)-Cerberalignan F (**19**) $[\alpha]_D^{28}$ = –49.0° (c 0.40, MeOH)

threo: (–)-Cerberalignan G (**20**) $[\alpha]_D^{28}$ = –68.4° (c 0.50, MeOH)

threo: (–)-Cerberalignan J (**21**) $[\alpha]_D^{26}$ = –49.6° (c 0.80, MeOH)

FIGURE 10.5 Lignans, (–)-olivil (**15b**) and (+)-cycloolivil (**16a**), and sesqui-lignans **17–21** from *Cerbera* spp. (Abe et al. 1988a, 1989). H = head, T = tail. "a" and "b" refer to the (+)- and (–)-enantiomeric forms, respectively.

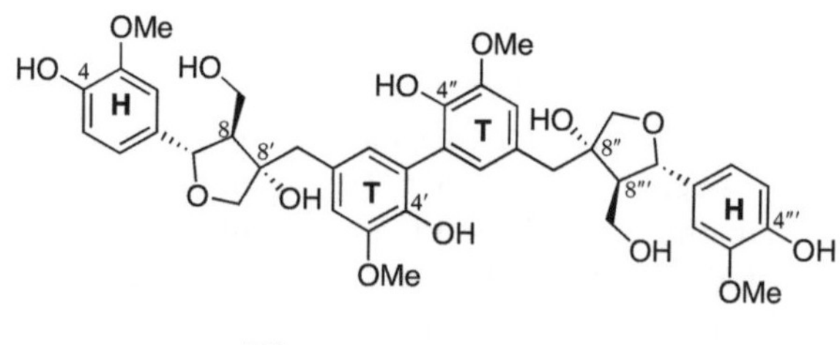

(−)-Cerberalignan A (**22**)
$[\alpha]_D^{29} = -76.0°$ (c 0.55, MeOH)

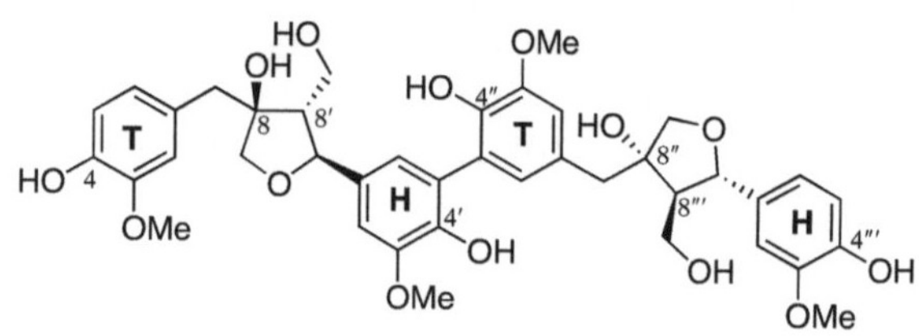

(−)-Cerberalignan B (**23**)
$[\alpha]_D^{28} = -75.4°$ (c 0.35, MeOH)

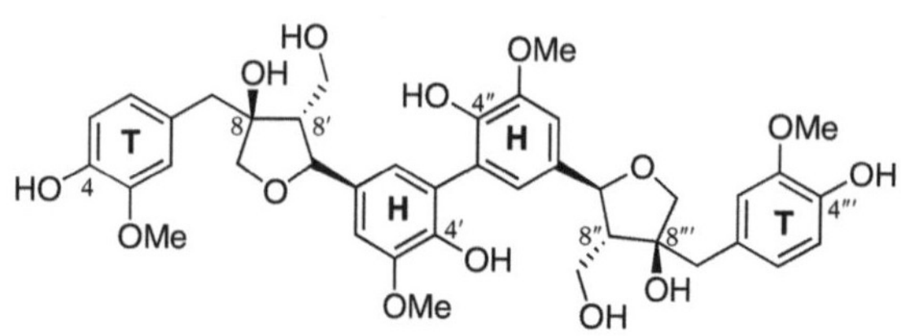

(−)-Cerberalignan C (**24**)
$[\alpha]_D^{27} = -57.3°$ (c 0.20, MeOH)

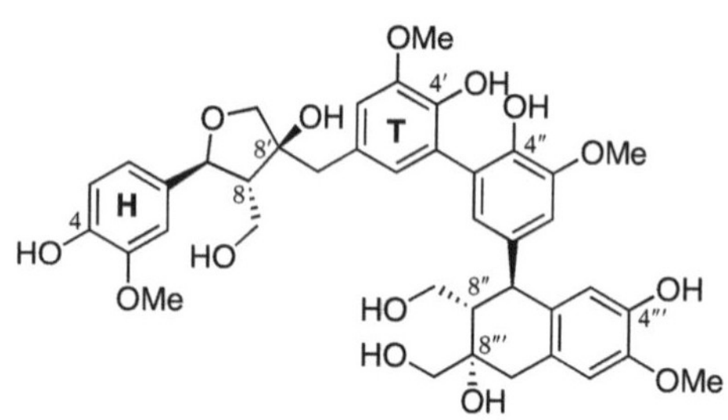

(+)-Cerberalignan K (**25**)
$[\alpha]_D^{32} = +19.1°$ (c 1.20, MeOH)

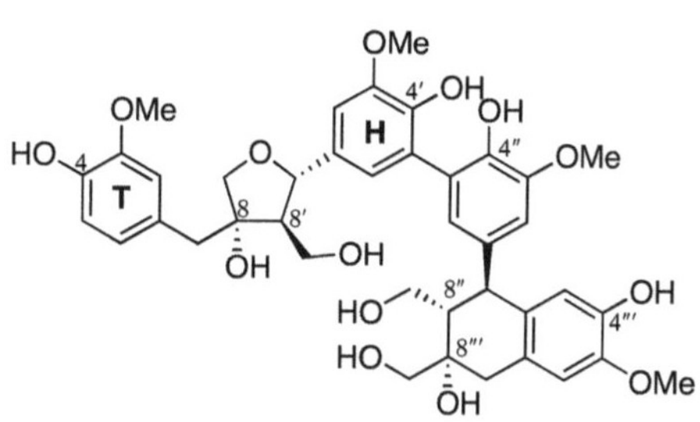

(+)-Cerberalignan L (**26**)
$[\alpha]_D^{32} = +36.1°$ (c 1.05, MeOH)

FIGURE 10.6 Di-lignans **22–26** from *Cerbera* spp. (Abe et al. 1988b, 1989). H = head, T = tail.

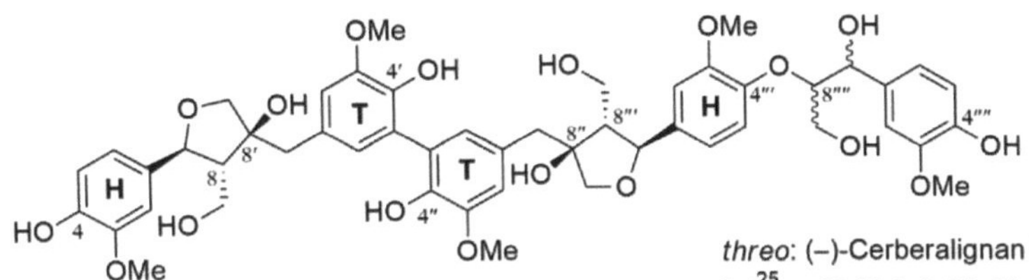

threo: (−)-Cerberalignan H (**27**)
$[\alpha]_D^{25} = -62.0°$ (c 0.39, MeOH)

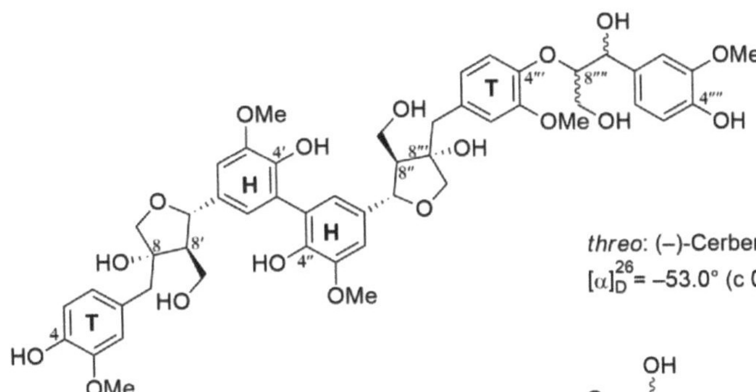

threo: (−)-Cerberalignan M (**28**)
$[\alpha]_D^{26} = -53.0°$ (c 0.5, MeOH)

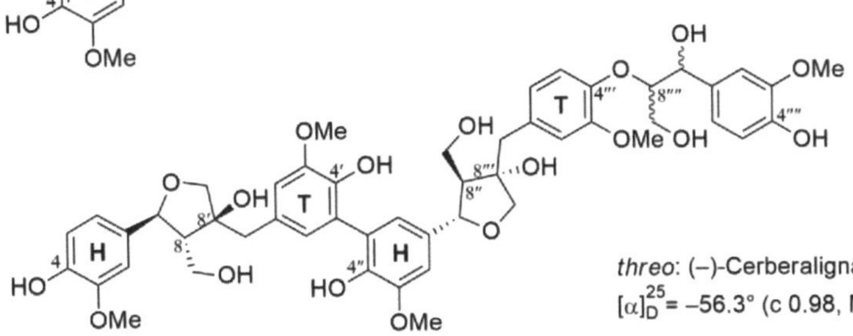

threo: (−)-Cerberalignan N (**29**)
$[\alpha]_D^{25} = -56.3°$ (c 0.98, MeOH)

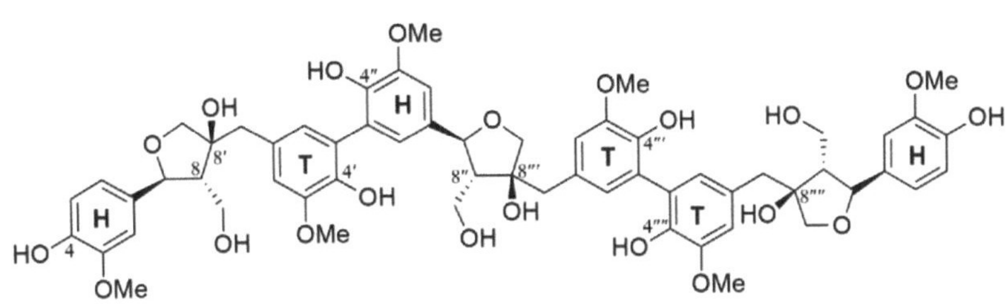

(−)-Cerberalignan I (**30**)
$[\alpha]_D^{25} = -76.7°$ (c 0.72, MeOH)

FIGURE 10.7 Sester- and tri-lignans **27–30** from *Cerbera* spp. (Abe et al. 1989). H = head, T = tail.

TABLE 10.2
Lignans and cerberalignans isolated from *Cerbera* species (Abe et al. 1988a, b, 1989)

Compound isolated	*Cerbera manghas*[1] Amount (mg)	(% dry weight)	*Cerbera odollam*[1] Amount (mg)	(% dry weight)	Number of C_6C_3 units
(–)-Olivil (**15b**)	3210	0.0535	126	0.0115	2
(+)-Cycloolivil (**16a**)	110	0.0018	4	0.0004	2
(–)-Cerberalignan D (**17**)	162	0.0027	7	0.0006	3
(–)-Cerberalignan E (**18**)	33	0.0006	3	0.0003	3
(–)-Cerberalignan F (**19**)	67	0.0011	4	0.0004	3
(–)-Cerberalignan G (**20**)	150	0.0025	9	0.0008	3
(–)-Cerberalignan J (**21**)	30	0.0005			3
(–)-Cerberalignan A (**22**)	70	0.0012	8	0.0007	4
(–)-Cerberalignan B (**23**)	230	0.0038	14	0.0013	4
(–)-Cerberalignan C (**24**)	90	0.0015	4	0.0004	4
(+)-Cerberalignan K (**25**)	26	0.0004			4
(+)-Cerberalignan L (**26**)	23	0.0004			4
(–)-Cerberalignan H (**27**)	30	0.0005	3	0.0003	5
(–)-Cerberalignan M (**28**)	15	0.0003			5
(–)-Cerberalignan N (**29**)	19	0.0003			5
(–)-Cerberalignan I (**30**)	24	0.0004	4	0.0004	6

[1] Isolated from dried stems: 6 kg (*C. manghas*) and 1.1 kg (*C. odollam*)

its formation presumably results from oxidative coupling, its low accumulation level, together with that of numerous other structurally related lignans, leaves open the possibility of their formation being artifacts of isolation, etc., rather than via biosynthesis proper. However, as before, they may indeed be *bona fide* natural products.

10.4 SESTER-LIGNANS

There are a few reports of sester-lignans. One is of herpepentol (**34**, Figure 10.9), isolated from seeds of *Herpetosporum caudigerum* (Kaouadji and Favre-Bonvin 1984), and present in unknown amount. It is one of the components in an extract from this species that also contains lower molecular weight lignans of related structures (see below). Herpepentol (**34**) has a structure with one 8–8′-like and three 8–5′-like inter-unit linkages, this perhaps suggesting that the 8–5′ inter-unit linkages were formed by sequential oxidative coupling of coniferyl alcohol (**35**)-derived species to an 8–8′-linked lignan (as one possibility only). Based on the structure of herpepentol (**34**), its 8–8′ lignan substructure could be lariciresinol (**36**) derived.

The "repeating unit" in herpepentol (**34**) is that of an aryl dihydrobenzofuran lignan (i.e. a dehydrodiconiferyl alcohol (**9**)-like derived) moiety. In this study, however, neither optical rotation nor circular dichroism data were reported, i.e. to provide any insight into whether the assembly processes were stereoselective.

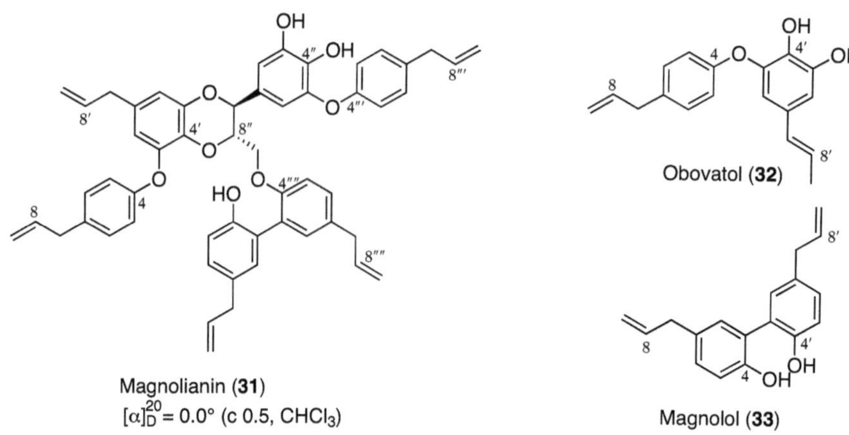

FIGURE 10.8 Tri-lignan, magnolianin (**31**), and possible lignan precursors (**32** and **33**) from *Magnolia obovata* (Fukuyama et al. 1993).

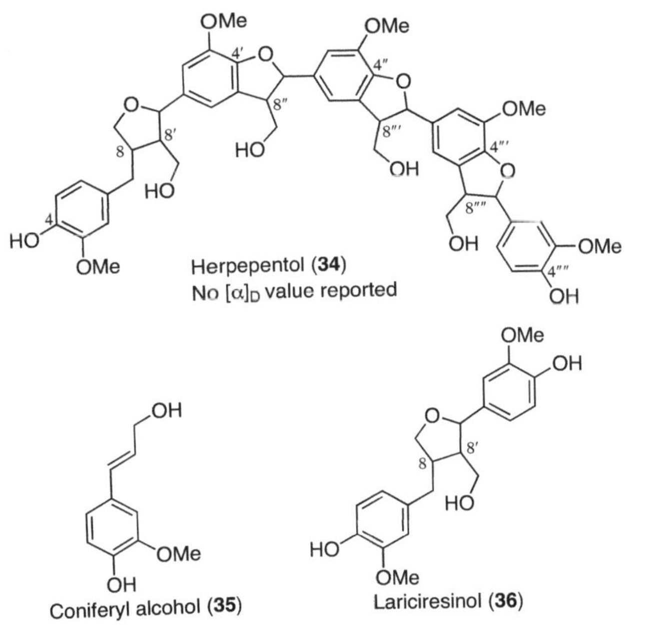

FIGURE 10.9 Sester-lignan, herpepentol (**34**), from *Herpetospermum caudigerum* (Kaouadji and Favre-Bonvin 1984), lignan lariciresinol (**36**), and coniferyl alcohol (**35**).

Sester-lignans, *threo*-(–)-cerberalignan H (**27**), *threo*-(–)-cerberalignan M (**28**), and *threo*-(–)-cerberalignan N (**29**) (Figure 10.7), were also reported from *C. manghas* stem tissue (Table 10.2) (Abe et al. 1989), with each having a coniferyl alcohol (**35**)-derived moiety aryl ether linked to two distinct di-lignan substructures; the latter were either linked tail-to-tail (T–T) as in (**27**), head-to-tail (H–H) as in **28**, or tail-to-head (T–H) as for **29**, via C5–C5′-like interunit linkages. They were isolated in amounts of 30 mg (0.0005% dry weight), 15 mg (0.0003%), and 19 mg (0.0003%), respectively, from air-dried stems. The "di-lignan cores" in **27–29** appear to be (–)-olivil (**15b**) derived, perhaps suggesting that coniferyl alcohol (**35**) is added last.

Two other sester-lignans, rhyncosides E (**37**) (5.1 mg, 0.0001% dry weight) and F (**38**) (4.7 mg, 0.0001% dry weight) (Figure 10.10), were isolated from Chinese mangrove tree (*Bruguiera sexangula* var *rhynchopetala*) air-dried stems (4.9 kg) (Bao et al. 2007); no optical activity data were reported for either compound. They are perhaps formed via coupling of coniferyl alcohol (**35**) with (–)-syringaresinol (**39b**)-derived precursors, with interunit linkages ultimately being 8–8′ [(–)-syringaresinol (**39b**) or (–)-pinoresinol (**40b**)] and 8–*O*–4′-like, respectively. Their biosynthetic pathway is currently unknown, but may possibly result from sequential addition of three coniferyl alcohol (**35**)-derived species to a syringaresinol (**39**)-derived species, but this needs to be established at the biochemical level. In that same study, three closely related di-lignans, hedyotisols A–C (**41–43**), were also obtained that differed by only having one coniferyl alcohol (**35**) moiety attached to each of the phenolic groups at C4′/C4″.

10.5 DI-LIGNANS

10.5.1 Di-lignans with "Core" Pinoresinol (40) or Syringaresinol (39) Substructures

There are various structurally diverse di-lignans that contain pinoresinol (**40**) or syringaresinol (**39**)-like subunits, such as herpetetradione (**44**) (Kaouadji and Pieraccini 1984); herpetetrone (**45**) (Kaouadji et al. 1987, Ma et al. 2019); 5‴-methoxyhedyotisol A (**46**) (Yang et al. 2007); and 4′,4″-di-*erythro*-*O*-(β-syringylglycerol)syringaresinol (**47**) (DellaGreca et al. 2007) (Figure 10.11).

Neither herpetetradione (**44**) nor herpetetrone (**45**), isolated from *H. caudigerum* seeds, in *circa* 0.005% and 0.0002% dry weight, respectively), had reports of any optical activity. Both have coniferyl alcohol (**35**)-derived moieties C-C linked to the C5′/C5″-like carbons of their pinoresinol (**40**) subunits. If pinoresinol (**40**) was the substrate for their formation, such coupling reactions might be anticipated to produce the corresponding aryl dihydrobenzofuran lignan substructures at either end of the pinoresinol-derived subunit, as one possibility for their formation.

Yet the aryl dihydrobenzofuran lignan substructure is only observed at the upper end of herpetetrone (**45**), whereas the lower end of the molecule has not undergone a similar ring closure reaction. This might suggest that upon coupling of the coniferyl alcohol (**35**)-derived moiety to the lower end at C5′, there is a re-aromatization step via reaction of an intermediate quinone methide with H_2O; the corresponding C7 alcohol so formed could then be oxidized to afford the ketone functionality as in **45**. In contrast, herpetetradione (**44**) formation has a keto-functionality at either end of the molecule.

Di-lignans, 5‴-methoxyhedyotisol A (**46**) and 4′,4″-di-*O*-(β-syringylglycerol)syringaresinol **47**, both have syringaresinol (**39**) subunits. 5‴-Methoxyhedyotisol A (**46**) was isolated from *Tarenna attenuata* (Yang et al. 2007), whereas compound **47** was a constituent of *Aptenia cordifolia* (DellaGreca et al. 2007). 5‴-Methoxyhedyotisol A (**46**) was obtained (3.7 mg, 0.00002% dry weight) from 17 kg dried whole plant material, with hedyotisol A (**41**, 32 mg), (+)-syringaresinol (**39a**, 20 mg), and (+)-pinoresinol (**40a**, 2 mg) also being present. In addition, compound **47** (4 mg, 0.00003% fresh weight) was recovered from 12 kg fresh leaf tissue, together with small amounts of syringaresinol (**39**, 3 mg) and pinoresinol (**40**, 2 mg). In a somewhat similar manner to that mentioned earlier, their formation might use syringaresinol (**39**) as the starting substrate for coupling at its C4′/C4″ OH groups with sinapyl (**48**) and coniferyl (**35**) alcohol moieties to give **46** or sinapyl alcohol (**48**) to afford **47**.

Another di-lignan is biplanispine A (**49**) (20 mg, 0.0001% dry weight), isolated from 17 kg *Zanthoxylum planispinum* air-dried bark; pinoresinol (**40**, 5 mg) was also isolated) (Yang et al. 2009). Assuming that the biplanispine A (**49**) structure is correct, its formation may occur through coupling of two lignan skeleta in a way whereby the C4′–*O*–C inter-unit linkage

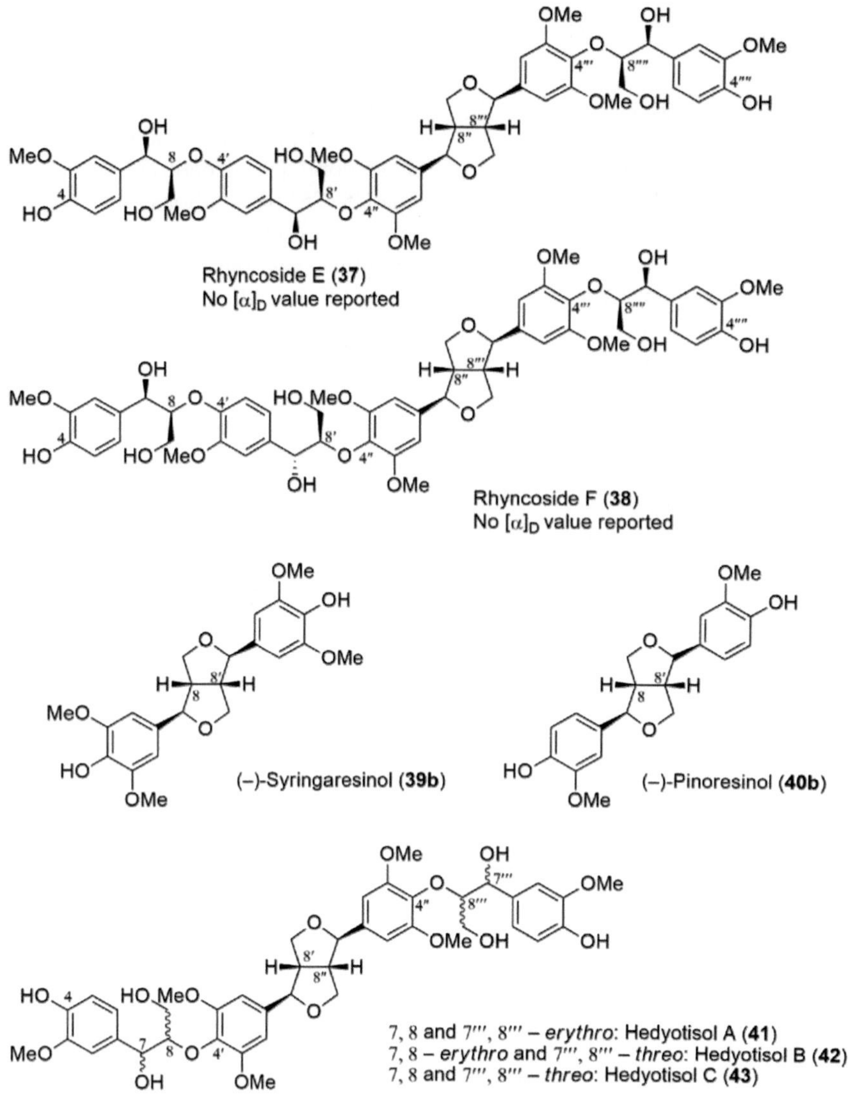

FIGURE 10.10 Rhyncosides E (**37**) and F (**38**) and hedyotisol A–C (**41–43**) from *Bruguiera sexangula* var. *rhynchopetala* (Bao et al. 2007) and lignans, (–)-syringaresinol (**39b**) and (–)-pinoresinol (**40b**). "b" refers to the (–)-enantiomers.

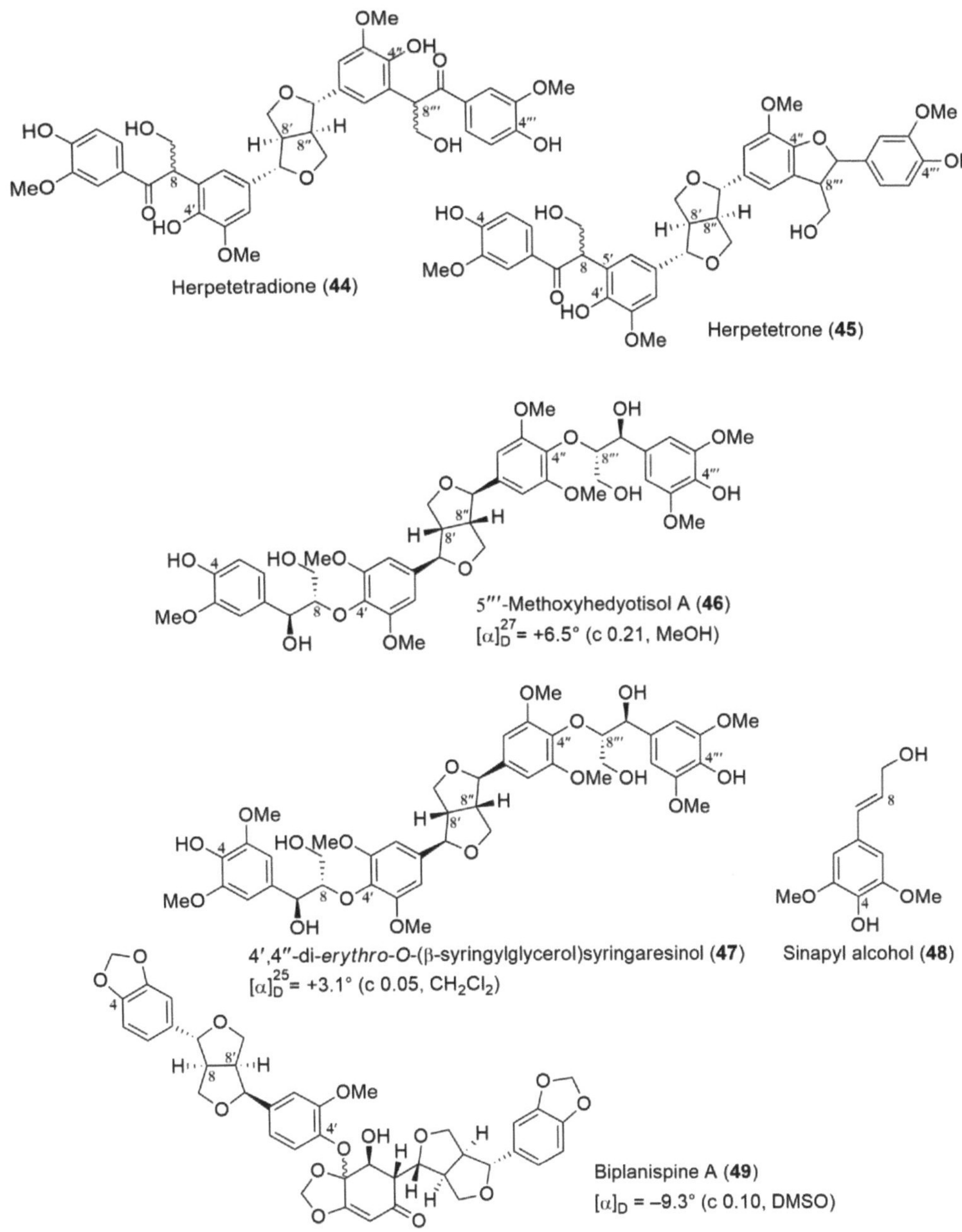

FIGURE 10.11 Di-lignans, herpetetradione (**44**) and herpetetrone (**45**) from *Herpetospermum caudigerum* (Kaouadji and Pieraccini 1984, Kaouadji et al. 1987); 5‴-methoxyhedyotisol A (**46**) from *Tarenna attenuata* (Yang et al. 2007), 4′,4″-di-*O*-(β-syringylglycerol)syringaresinol (**47**) from *Aptenia cordifolia* (DellaGreca et al. 2007) and biplanispine A (**49**) from *Zanthoxylum planispinum* (Yang et al. 2009), and monolignol, sinapyl alcohol (**48**).

is formed, with concomitant loss/modification of an aromatic ring to generate the unsaturated ketone derivative **49**.

It must be emphasized, though, that none of their biochemical pathways have been elucidated. These will be of interest to determine, particularly, how biplanispine A (**49**) is formed enzymatically.

10.5.2 Furano Di-lignans

In addition to the tri- and sester-cerberalignans described earlier, there are a number of di-lignans reported from *C. manghas* and *C. odollam* stem tissue (Abe et al. 1988a, b, 1989), i.e. (–)-cerberalignan A (**22**), (–)-cerberalignan B (**23**), (–)-cerberalignan C (**24**), (+)-cerberalignan K (**25**), and (+)-cerberalignan L (**26**) (Figure 10.6). These were isolated from 6 and 1.1 kg *C. manghas* and *C. odollam* stem tissue, respectively, as follows: (–)-cerberalignan A (**22**) (70 and 8 mg, 0.0012 and 0.0007% dry weight), (–)-cerberalignan B (**23**) (230 and 14 mg, 0.004 and 0.0013%), (–)-cerberalignan C (**24**) (90 and 4 mg, 0.002 and 0.0004%), whereas (+)-cerberalignan (K) (**25**) (26 mg, 0.0004% dry weight) and (+)-cerberalignan L (**26**) (23 mg, 0.0004%) were obtained from *C. manghas* stem tissue. Of these, cerberalignans A, B, and C (**22–24**) are linked tail-to-tail (T–T), head-to-tail (H–T), and head-to-head (H-H), respectively, and can provisionally be considered as formed via different regiocouplings of (–)-olivil (**15b**). Cerberalignans K and L (**25** and **26**), however, appear to be formed via coupling of (–)-olivil (**15b**) with (+)-cycloolivil (**16a**).

As before, their isolation procedures were extremely lengthy and the various lignans isolated were present together with much larger amounts of the presumed precursors, e.g., (–)-olivil (**15b**), raising the possibility that artifact formation cannot be discounted. Yet, as before, that they are indeed true natural products cannot be eliminated currently.

(–)-Larixsin (**50**, Figure 10.12) (9 mg, 0.00003% dry weight), obtained from 35 kg of *Larix olgensis* var. *koreana* air-dried bark (Yang et al. 2005), has a quite unusual structure. The upper half of the molecule appears to be derived from secoisolariciresinol (**51**), and the lower part perhaps from an analog/homolog thereof [e.g. (+)-neoolivil (**52a**)]. The nature of the inter-unit linkage, with a loss of an oxygen functionality from either the C9 or C9′ position suggests perhaps the involvement of a dehydration reaction to account for its formation. Interestingly, (–)-secoisolariciresinol (**51b**), (+)-lariciresinol (**36a**), and (+)-isolariciresinol (**54a**) were also isolated in amounts of 2, 6, and 4 grams, respectively, together with their 9′-*p*-coumarate derivatives, e.g. **53a** and **55a** (222 mg and 10.8 mg).

(+)-Syripinnalignin B (**56**, Figure 10.13) (23 mg, 0. 0008 % dry weight), isolated from 3 kg *Syringa pinnatifolia* var. *alashanensis* stem tissue (Wang et al. 2012), has yet another unusual furanolignan-like structure. The upper half of the molecule may also be derived from a secoisolariciresinol (**51**) homolog, which was also isolated from the same tissue (12 mg, 0.0004%) (Ao et al. 2012). The lower half of the molecule apparently stems from a 2 + 2 cycloaddition. Whether

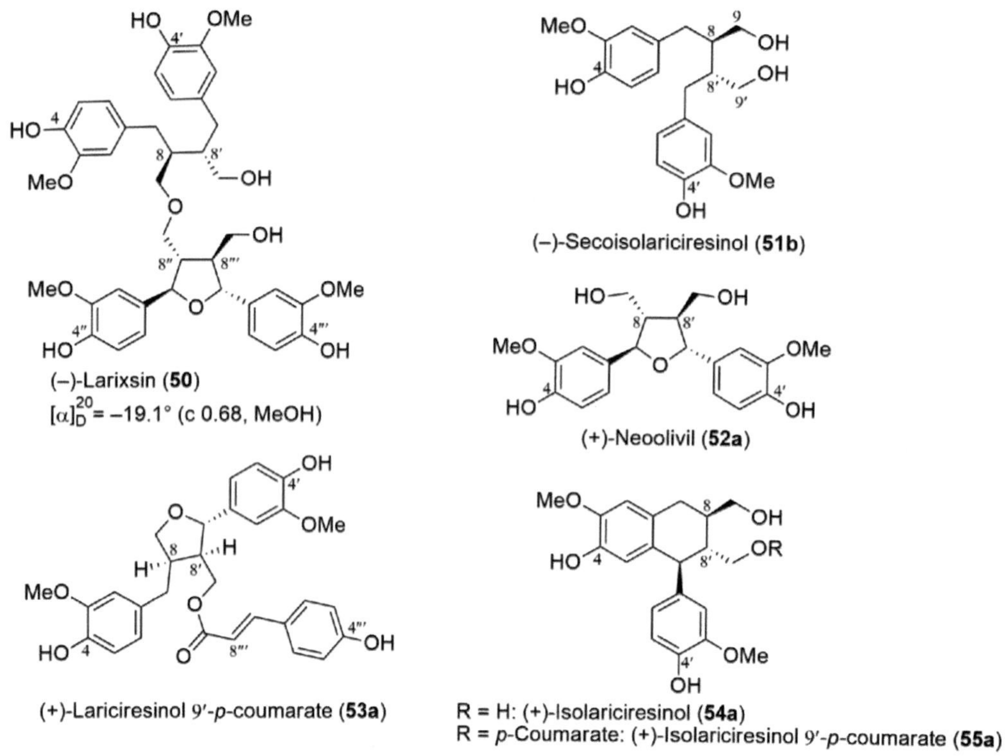

FIGURE 10.12 Di-lignan, (–)-larixsin (**50**) and lignans (–)-secoisolariciresinol (**51b**), (+)-neoolivil (**52a**), (+)-lariciresinol 9′-*p*-coumarate (**53a**), (+)-isolariciresinol (**54a**), and (+)-isolariciresinol 9′-*p*-coumarate (**55a**) isolated from *Larix olgensis* var. *koreana* (Yang et al. 2005). "a" and "b" refer to the (+)- and (–)-enantiomeric forms, respectively.

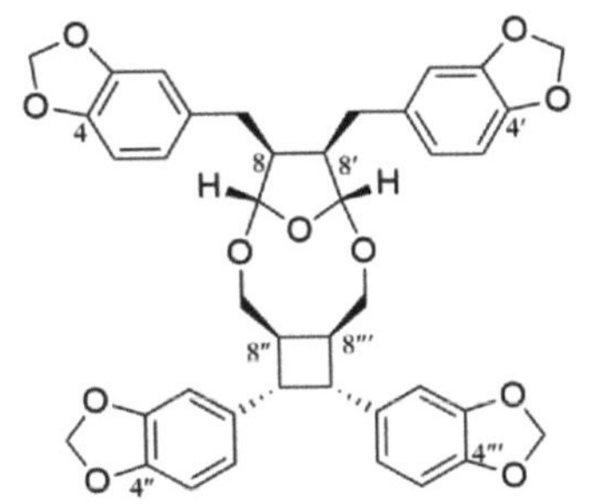

(+)-Syripinnalignin B (**56**)
$[\alpha]_D^{25}$ = +13.2° (c 0.10, MeOH)

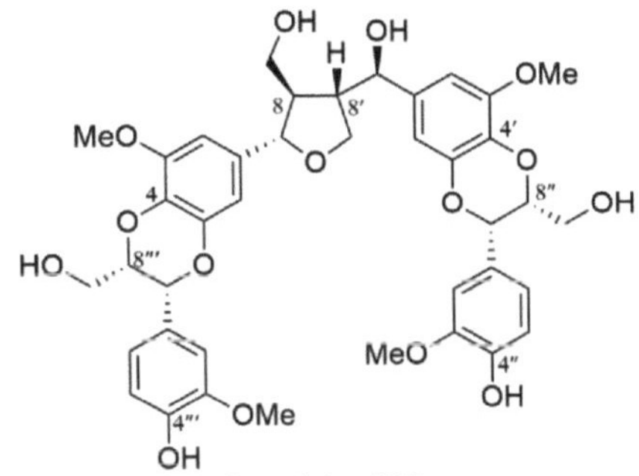

Cynarinine (**57**)
No $[\alpha]_D$ value reported

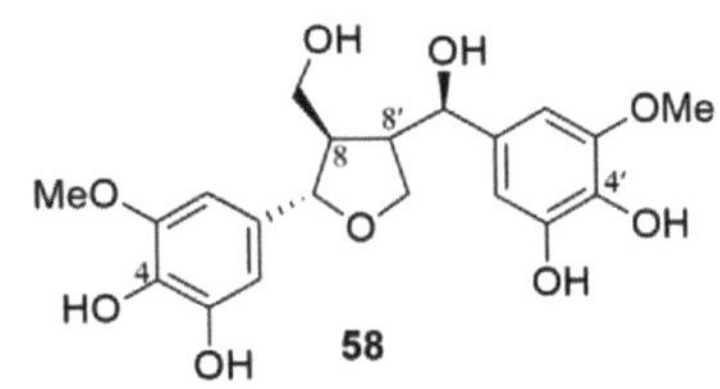

58

FIGURE 10.13 Di-lignans, (+)-syripinnalignin B (**56**) and cynarinine (**57**) isolated from *Syringa pinnatifolia* var. *alashanensis* (Wang et al. 2012) and *Cynara cardunculus* (Koubaa and Damak 2003), respectively, and possible precursor **58**.

this occurs via inter- or intra-molecular coupling cannot be gauged at this time. Given its presence in aerial tissue, the cyclobutane sub-structure could possibly result photochemically. However, this needs to be established as to whether this is either a biochemical or photochemical conversion.

Cynarinine (**57**) (36 mg, 0.013% dry weight), obtained from 270 g of dried, powdered *Cynara cardunculus* seeds (Koubaa and Damak 2003), reportedly has a 7′-hydroxyl group and two benzodioxin substructures at C4/C5 and C4′/C5′, respectively. Its formation could possibly be from a catechol lignan intermediate, such as **58**, that undergoes radical coupling with coniferyl alcohol (**35**)-derived species at either end of the molecule. No optical rotation data was described. There are, however, no reports of the putative catechol lignan **58** in the literature at present.

(–)-Bi-cubebin A (**59**, Figure 10.14) (3.9 and 7.7 mg, 0.0005 and 0.0031%) was isolated from 728.1 and 245.6 g underground *Aristolochia pubescens* and *A. lagesiana* tissues, respectively, together with (8*S*, 8′*R*, 9*S*)-cubebin (**60a**) (8.4 and 18.4 mg, 0.0012 and 0.0075%), (8*R*, 8′*R*, 9*R*)-cubebin (**60b**), and (8*R*, 8′*R*, 9*S*)-cubebin (**60c**), with **60b** and **60c** isolated as an inseparable mixture (2.8 and 4.8 mg, 0.0004 and 0.002%) (de Pascoli et al. 2006). It was also deduced that (–)-bi-cubebin A (**59**) was a dimer of **60b** and **60c**, perhaps

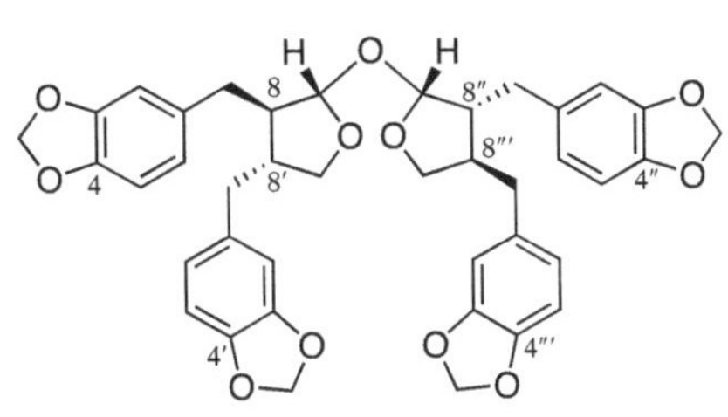

(–)-Bi-cubebin A (**59**)
$[\alpha]_D^{26}$ = –54.8° (c 0.36, $CHCl_3$)

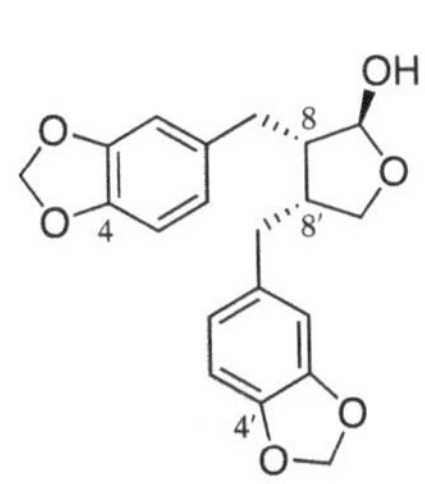

(8*S*,8′*R*,9*S*)-Cubebin (**60a**)

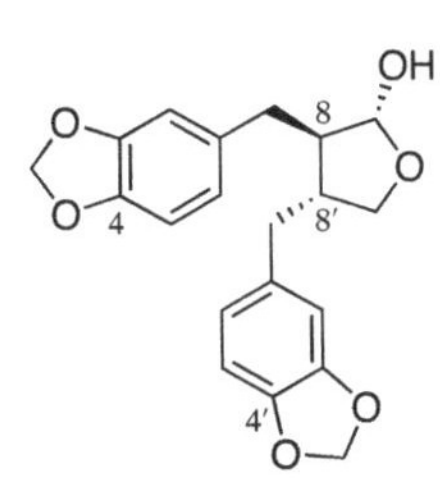

(8*R*,8′*R*,9*R*)-Cubebin (**60b**)

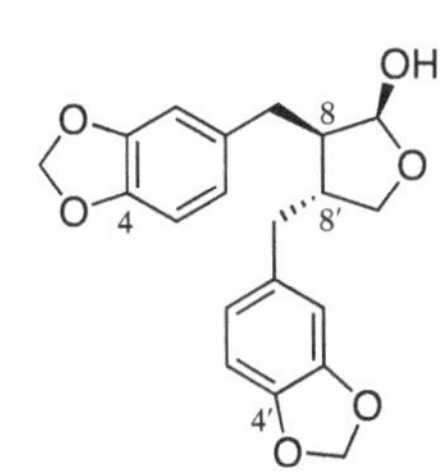

(8*R*,8′*R*,9*S*)-Cubebin (**60c**)

FIGURE 10.14 Di-lignan, (–)-bi-cubebin A (**59**), and cubebins **60a–60c** from *Aristolochia pubescens* and *A. lagesiana* (de Pascoli et al. 2006).

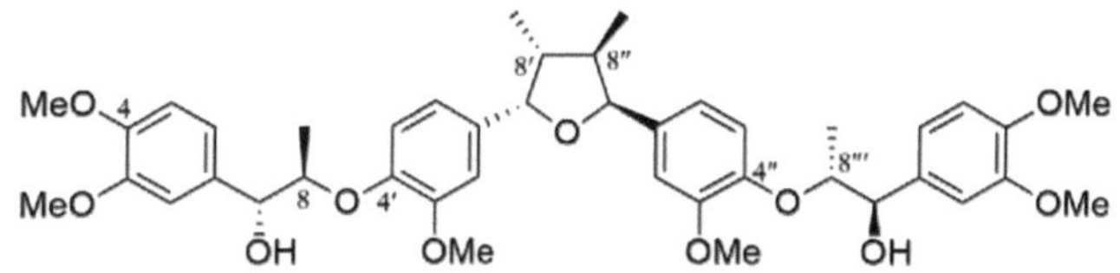

(–)-Manassantin A (**61**)
$[\alpha]_D^{25} = -102.1°$ (c 0.5, $CHCl_3$)

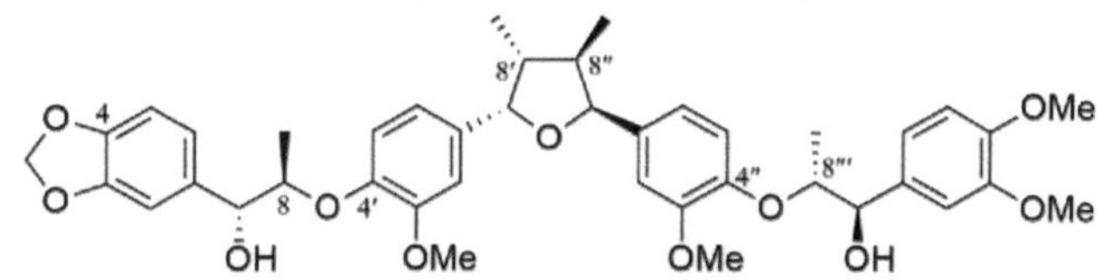

(–)-Manassantin B (**62**)
$[\alpha]_D^{25} = -99.8°$ (c 0.5, $CHCl_3$)

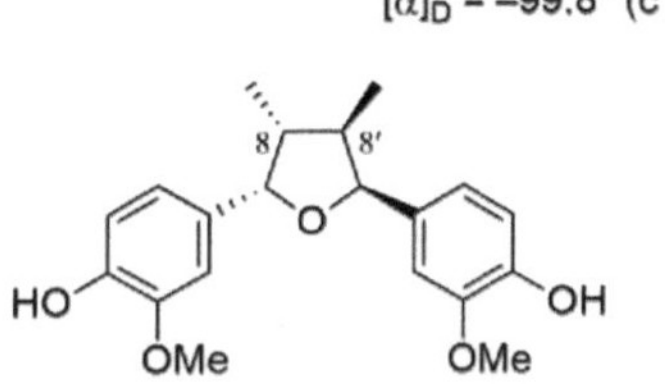

(–)-Saucernetin diol (**63b**)

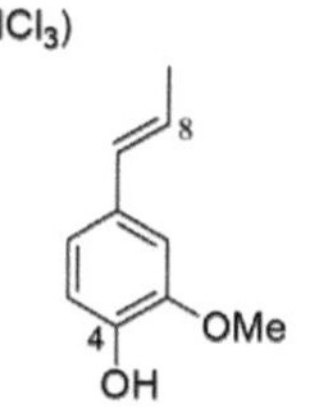

Isoeugenol (**64**)

FIGURE 10.15 Di-lignans, (–)-manassantin A (**61**) and B (**62**) from *Saururus chinensis* (Hwang et al. 2003, Cui et al. 2014) and possible precursor, (–)-saucernetin diol (**63b**), as well as isoeugenol (**64**). "b" refers to the (–)-enantiomers.

resulting from dehydration of **60b** and **60c** either chemically or biochemically.

Various furano di-lignans, such as (–)-manassantin A and B (**61** and **62**, Figure 10.15) (210 and 1090 mg, 0.0066 and 0.0341% dry weight), were obtained from 3.2 kg air-dried roots of *Saururus chinensis* (Hwang et al. 2003), together with numerous additional structurally related sesqui-lignans and lignans (Cui et al. 2014). (–)-Manassantin A (**61**) and B (**62**) may be derived from the furano-lignan (–)-saucernetin diol (**63b**), which is also present in *Saururus* sp. (Rao and Oruganty 1997, Qu et al. 2014), i.e. via radical coupling with isoeugenol (**64**), with re-aromatization through hydroxylation at C7/C7‴, and either *O*-methylation or methylenedioxy bridge formation.

None of the biosynthetic pathways to these final coupling products are known.

10.5.3 Dibenzylbutryolactone Di-lignans

There are various di-lignans reported that are apparently derived from (–)-matairesinol (**65b**)/(–)-arctigenin (**66b**, Figure 10.16) in *Arctium lappa* L, either from the seeds or from its fruit (the latter being described as Arctium fructus) (Ichihara et al. 1976, 1977, 1978, 1979, Umehara et al. 1993). The di-lignans reported include (–)-arctignan D (**75**), (–)-arctignan E (**76**), and (+)-lappaol F (**77**) (Figure 10.17), these being present together with a variety of sesqui-lignans (**67–74**, Figure 10.16).

Table 10.3 summarizes the reported levels of these metabolites from a 10-kg commercial purchase of Arctium Fructus. Based on the reported metabolites, it appears that (–)-matairesinol (**65b**) and (–)-arctigenin (**66b**) are the core structures from which the di-lignans (–)-arctignan D (**75**), (–)-arctignan E (**76**), and (+)-lappaol F (**77**) are formed, i.e. via addition of coniferyl alcohol (**35**)-derived moieties at the C5/C5′ positions as noted in other examples earlier.

Their quite high relative abundance (di-lignans, sesqui-lignans and lignans) is presumably such that they are almost certain to be natural products proper, and not artifacts. Formation of the di-lignans **75–77** can readily be explained on the basis of free-radical coupling between incoming coniferyl alcohol (**35**)- and (–)-matairesinol (**65b**)-derived species.

Di-lignans have also been reported from whole dried plant (1.5 kg) of *Saussurea conica* (Figure 10.18 and Table 10.4) (Fan et al. 2006), together with lignans **78–81**. The two di-lignans reported, (–)-conicaol A (**82**) and (–)-diarctigenin (**83**) (40 mg each, 0.003% dry weight), were apparently formed from inter-molecular coupling of either two (–)-matairesinol (**65b**) molecules to give (–)-conicaol A (**82**), or via intermolecular coupling of (–)-arctigenin (**66b**) and (–)-matairesinol (**65b**), respectively.

10.5.4 Diphenyl Di-lignans

The medicinal plant *Magnolia officinalis* has been extensively studied for its oligomeric lignans, apparently derived from chavicol (**84**, Figure 10.19) and its regio-isomer, 2-allylphenol (**85**). The plant tissues studied mainly focused upon the roots (Shih et al. 2013, Shih et al. 2016) and the bark (Magnoliae Cortex) (Yahara et al. 1991). The latter accumulate a broad range of di-lignans (**97–100**), and lignans (**86–96**). Table 10.5 summarizes the respective levels of the di-lignans (**97–100**) isolated, relative to the lignans (**86–96**) that were also present (Yahara et al. 1991).

In addition, study of its dried, powdered root tissue (5 kg) resulted in the isolation and structural elucidation of five di-lignans, (–)-houpulin A, B, D, K, and L (**101–105**, Figure 10.20 and Table 10.6) (Shih et al. 2013, 2016), these all being presumably derived from radical-radical coupling of magnolol (**33**)-derived species, respectively.

To our knowledge, there are no biochemical studies to date probing how these diphenyl di-lignans are formed.

Formation of (–)-houpulin A (**101**) and (–)-houpulin B (**102**) is quite interesting. The optical activity reported could only result from stereoselective coupling with restricted rotation around the di-phenyl-diphenyl linkages.

10.5.5 Aryl Dihydronaphthalene Di-lignans

Di-lignans, such as **106** and trilobatin E (**107**, Figure 10.21), from the liverwort *Bazzania trilobata* have been reported (Martini et al. 1998, Scher et al. 2003), as well as lignans (–)-epiphyllic acid (**108b**) and (–)-trilobatin D (**109b**). Di-lignan (**106**) is presumably formed via coupling of two (–)-epiphyllic acid (**108b**) moieties, whereas trilobatin E (**107**) may result from coupling of (–)-epiphyllic acid (**108b**) with

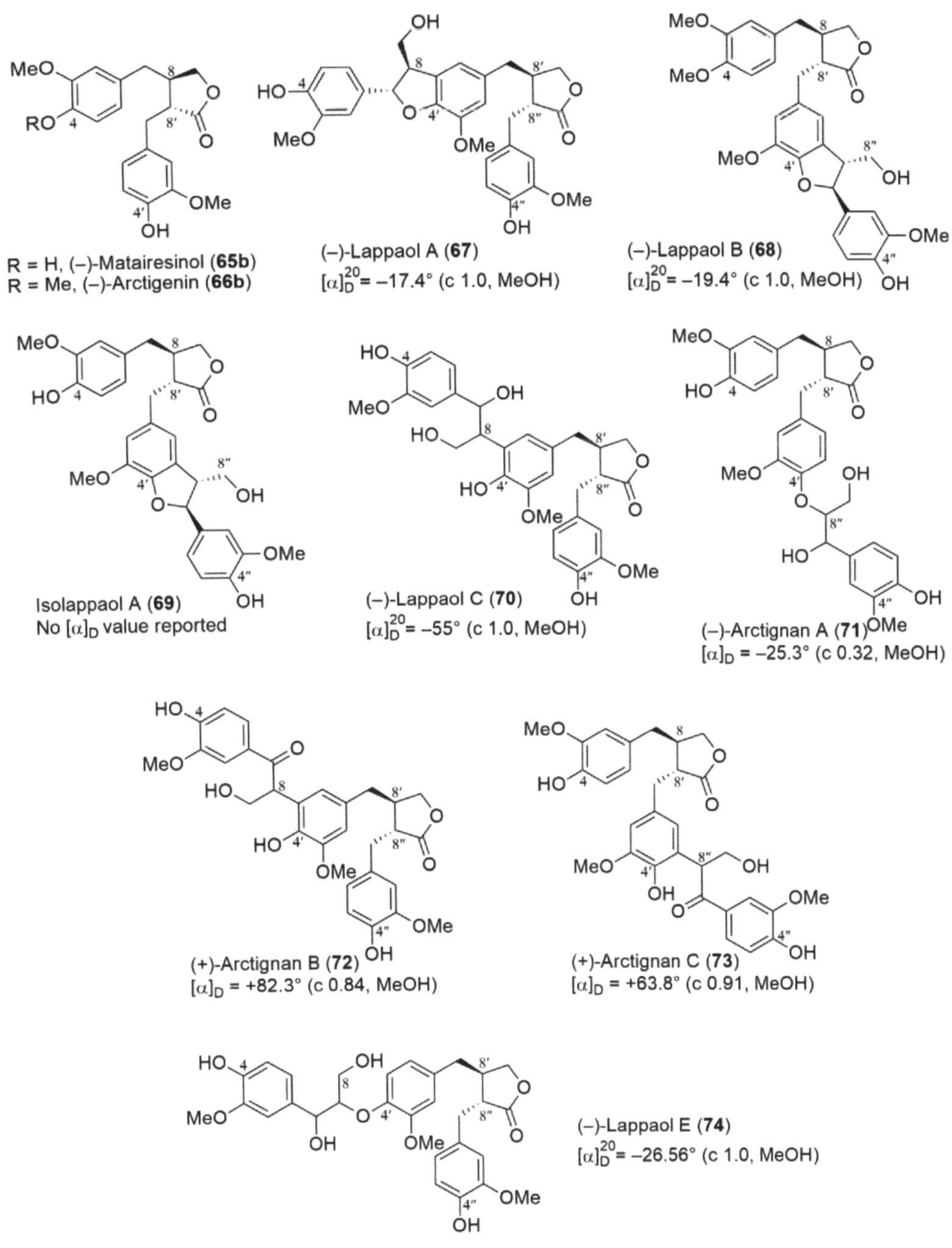

FIGURE 10.16 Dibenzylbutryolactone sesquilignans **67–74**, and lignans (–)-matairesinol (**65b**) and (–)-arctigenin (**66b**) from *Arctium lappa* L. (Ichihara et al. 1976, 1977, 1978, 1979, Umehara et al. 1993). "b" refers to the (–)-enantiomers.

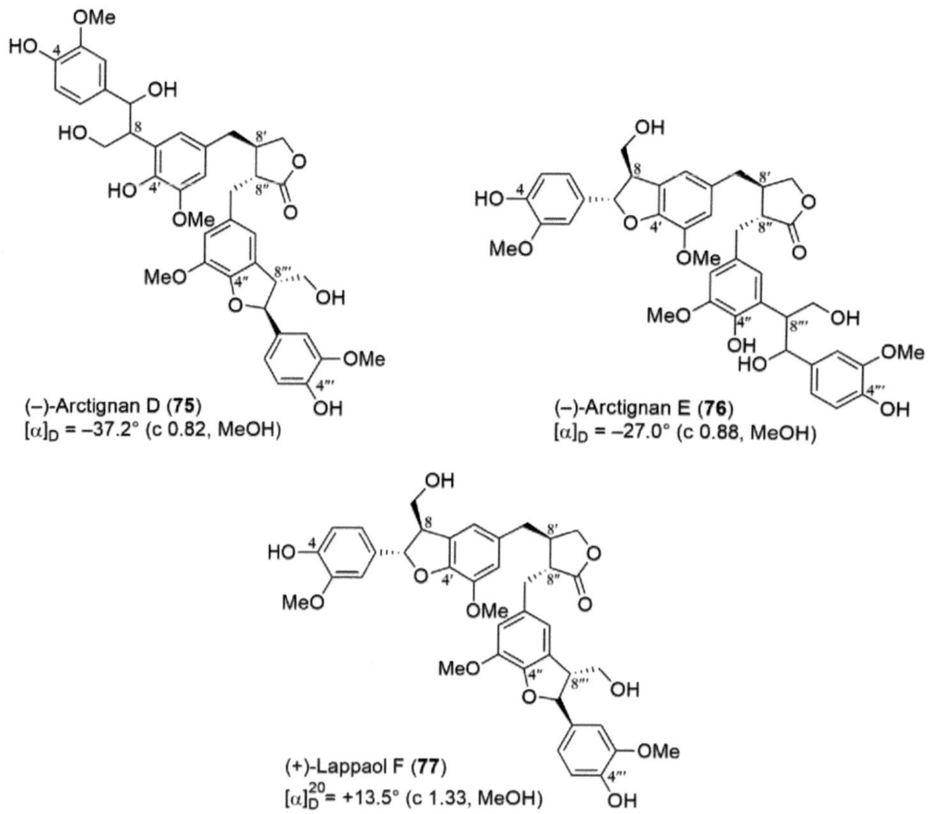

FIGURE 10.17 Dibenzylbutyrolactone di-lignans **75–77** from *Arctium lappa* L (Ichihara et al. 1976, 1977, 1978, 1979, Umehara et al. 1993).

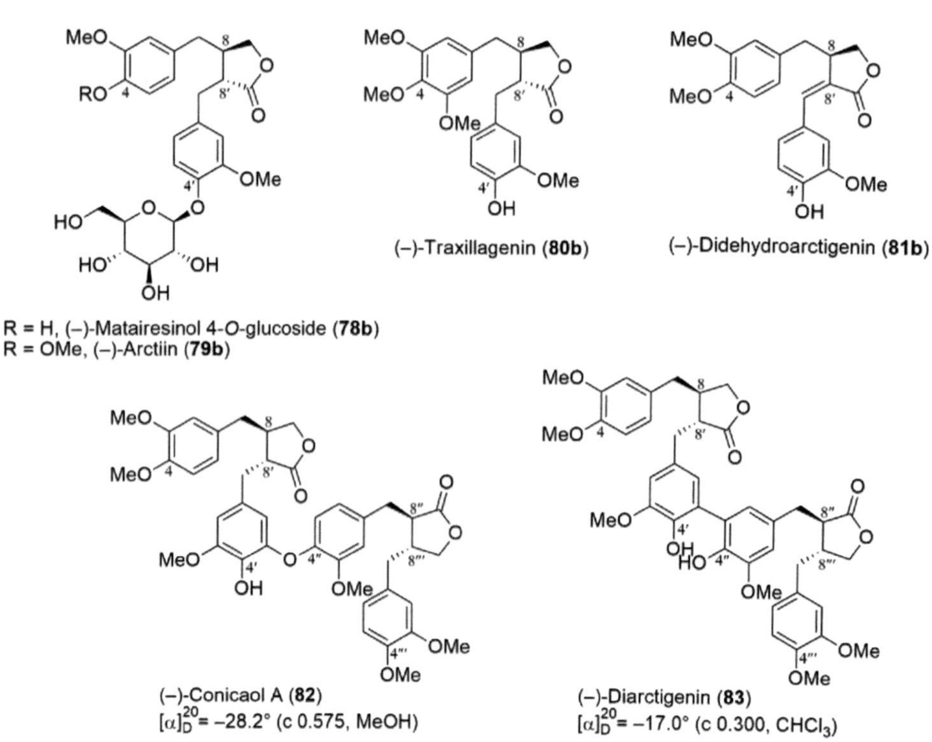

FIGURE 10.18 Dibenzylbutryolactone di-lignans (–)-conicaol A (**82**) and (–)-diarctigenin (**83**), as well as lignans **78b–81b** from *Saussurea conica* (Fan et al. 2006). "b" refers to the (–)-enantiomers.

TABLE 10.3
Lignans and di-lignans isolated from Arctium fructus[1] (Umehara et al. 1993)

Compound isolated	Amount (mg)	Amount (% dry weight)	Number of C_6C_3 units
(–)-Matairesinol (**65b**)	600	0.006	2
(–)-Arctigenin (**66b**)	3500	0.035	2
(–)-Lappaol A (**67**)	1000	0.010	3
(–)-Lappaol B (**68**)	700	0.007	3
Isolappaol A (**69**)	500	0.005	3
(–)-Lappaol C (**70**)	150	0.002	3
(–)-Arctignan A (**71**)	125	0.001	3
(+)-Arctignan B (**72**)	1500	0.015	3
(+)-Arctignan C (**73**)	1400	0.014	3
(–)-Lappaol E (**74**)	600	0.006	3
(–)-Arctignan D (**75**)	2500	0.025	4
(–)-Arctignan E (**76**)	2700	0.027	4
(+)-Lappaol F (**77**)	600	0.006	4

[1] Isolated from 10 kg of dried stems.

TABLE 10.4
Lignans and di-lignans isolated from *Saussurea conica*[1] **(Fan et al. 2006)**

Compound isolated	Amount (mg)	Amount (% dry weight)	Number of C_6C_3 units
(–)-Matairesinol (**65b**)	73	0.005	2
(–)-Arctigenin (**66b**)	3512	0.234	2
(–)-Matairesinol glucoside (**78b**)	260	0.017	2
(–)-Arctiin (**79b**)	480	0.032	2
(–)-Traxillagenin (**80b**)	15	0.001	2
(–)-Didehydroarctigenin (**81b**)	11	0.004	2
(–)-Conicaol A (**82**)	40	0.003	4
(–)-Diarctigenin (**83**)	40	0.003	4

[1] Isolated from 1.5 kg whole plant.

TABLE 10.5
Lignans and di-lignans isolated from Magnoliae Cortex[1] (Yahara et al. 1991)

Compound isolated	Amount (mg)	Amount (% dry weight)	Number of C_6C_3 units
Magnolol (**33**)	6000	0.1200	2
Honokiol (**86**)	1000	0.0200	2
Syringaresinol (**39**)	485	0.0097	2
Syringaresinol-glc (**87**)	98	0.0020	2
8-*O*-4 (**88**)	300	0.0060	2
Randainal (**89**)	130	0.0026	2
Magnaldehyde B (**90**)	120	0.0024	2
Magnaldehyde C (**91**)	15	0.0003	2
Magnolignan A (**92**)	15	0.0003	2
Magnolignan B (**93**)	200	0.0040	2
Magnolignan C (**94**)	35	0.0007	2
Magnolignan D (**95**)	38	0.0008	2
Magnolignan E (**96**)	29	0.0006	2
Magnolignan F (**97**)	96	0.0019	4
Magnolignan G (**98**)	153	0.0031	4
Magnolignan H (**99**)	115	0.0023	4
Magnolignan I (**100**)	120	0.0024	4

[1] Isolated from 5 kg bark.

TABLE 10.6
Lignans and di-lignans isolated from *Magnolia officinalis* **roots[1] (Shih et al. 2013, Shih et al. 2016)**

Compound isolated	Amount (mg)	Amount (% dry weight)	Number of C_6C_3 units
Honokiol (**86**)	19500	0.39000	2
(–)-Houpulin A (**101**)	17.4	0.00035	4
(–)-Houpulin B (**102**)	14.7	0.00029	4
(–)-Houpulin D (**103**)	5.8	0.00012	4
(–)-Houpulin K (**104**)	27.1	0.00054	4
(-)-Houpulin L (**105**)	1.6	0.00003	4

[1] Isolated from 5 kg bark.

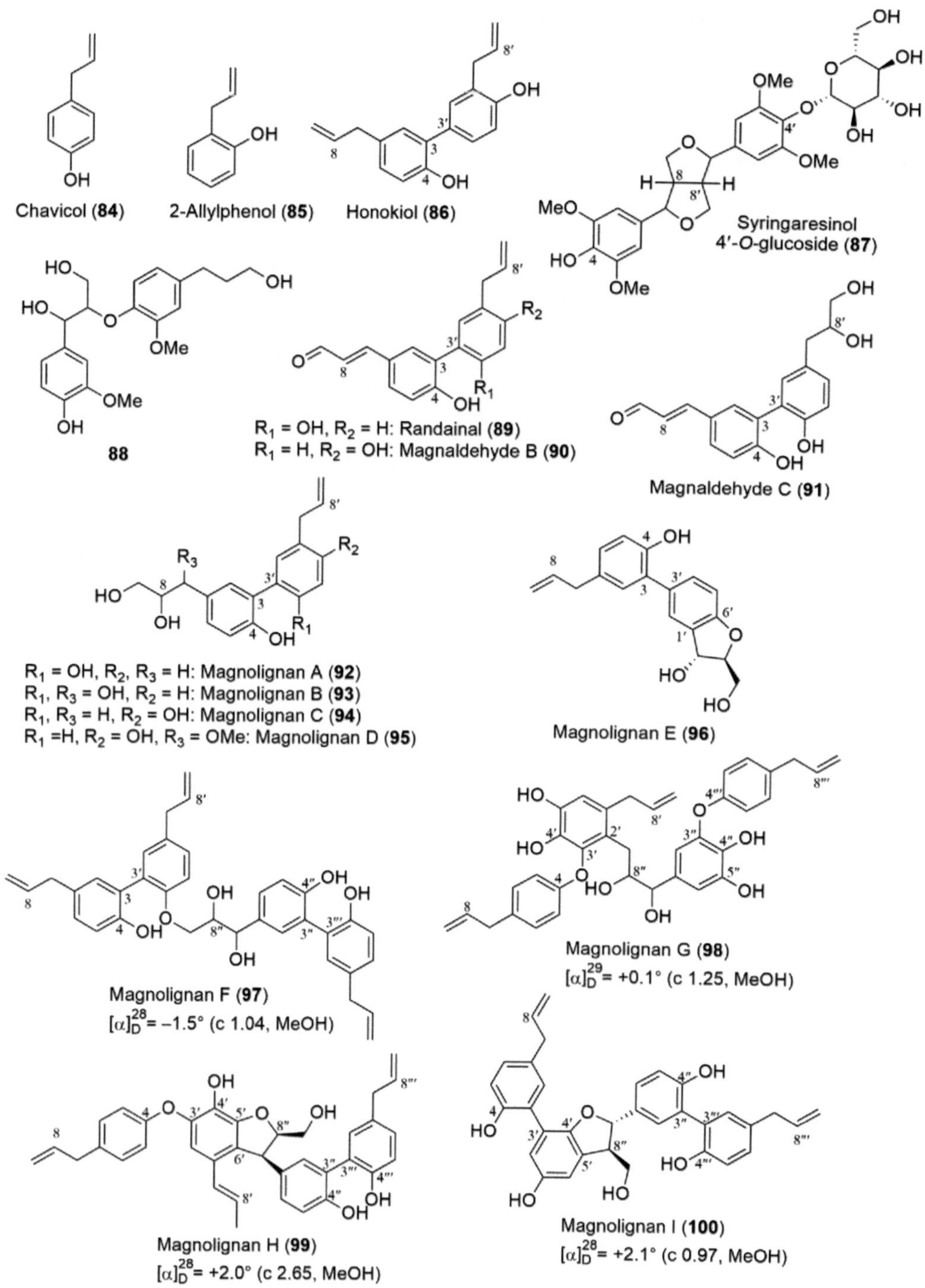

FIGURE 10.19 Allylphenols (**84** and **85**), lignans (**86–96**), and di-lignans (**97–100**) isolated from *Magnolia officinalis* (Yahara et al. 1991).

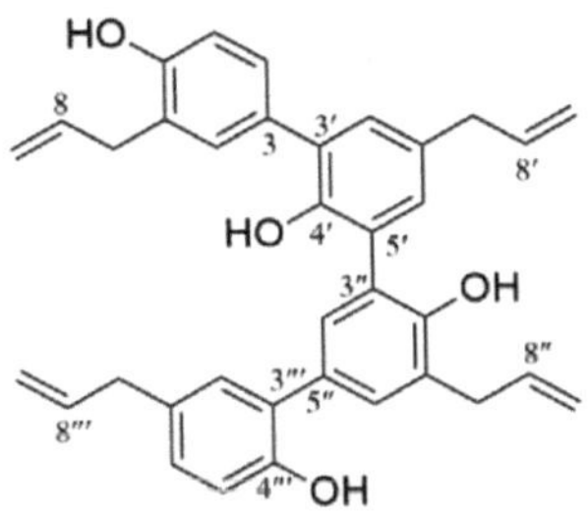

(–)-Houpulin A (**101**)
$[\alpha]_D^{25} = -7.88°$ (c 0.83, MeOH)

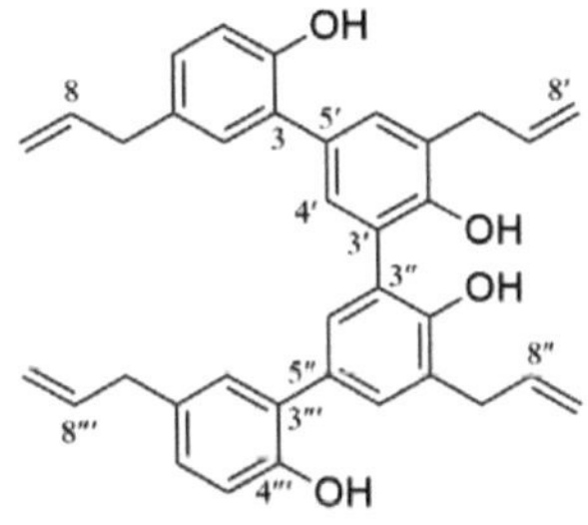

(–)-Houpulin B (**102**)
$[\alpha]_D^{25} = -7.98°$ (c 0.70, MeOH)

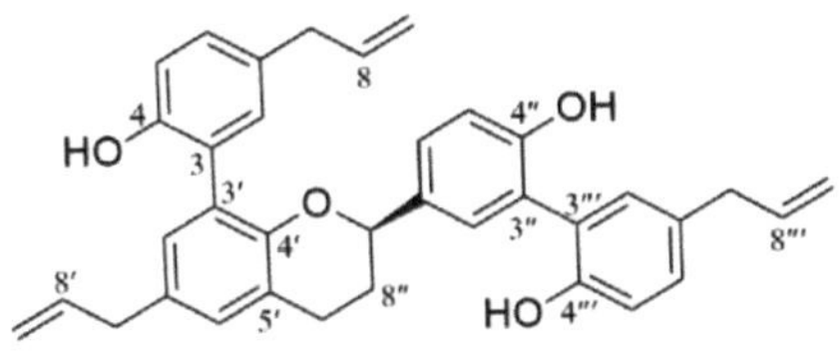

(–)-Houpulin D (**103**)
$[\alpha]_D^{25} = -57.64°$ (c 0.23, MeOH)

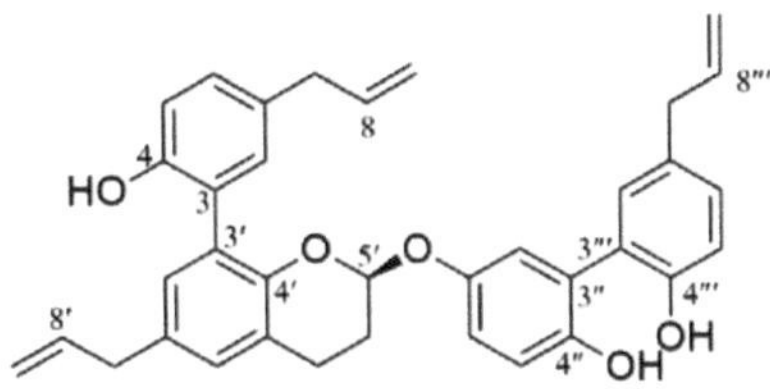

(–)-Houpulin K (**104**)
$[\alpha]_D^{25} = -8°$ (c 1.3, MeOH)

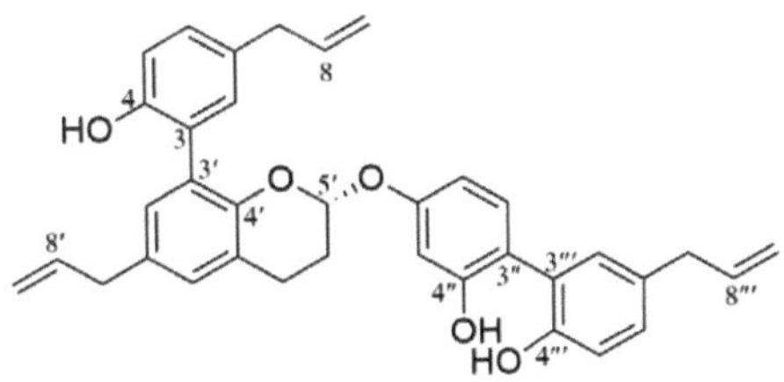

(–)-Houpulin L (**105**)
$[\alpha]_D^{25} = -72°$ (c 0.1, MeOH)

FIGURE 10.20 Di-lignans **101–105** from *Magnolia officinalis* (Shih et al. 2013, Shih et al. 2016).

(–)-trilobatin D (**109b**). These lignans are of particular interest given the absence of *O*-methylation on their aromatic rings.

There are no known biochemical studies on the formation of the aryl dihydronaphthalene lignans via lignan coupling processes.

10.5.6 Di-lignans with Multiple 8–*O*–4′-Like Linkages

Larch (*Larix leptolepis*) heartwood reportedly accumulates the di-lignan, leptolepisol E (**110**, Figure 10.22), and the sesqui-lignan, leptolepisol D (**111**) (Miki et al. 1979, 1980). In both cases, the lignans appear to be coniferyl alcohol (**35**) derived. While the different diastereomeric forms were not resolved in those studies, leptolepisol E (**110**) displayed a very small optical rotation.

Additionally, together with lignans syringaresinol (**39**), pinoresinol (**40**), and balanophonin (**112**), the sesquilignan buddlenol A (**113**), as well as the di-lignans, hedyotisol A (**41**), 5‴-methoxyhedyotisol (**46**), 4″-*O*-(8-guaiacylglycerol) buddlenol A (**114**), and 5″-methoxy-4″-*O*-(8-guaiacylglycerol) buddlenol A (**115**) from *T. attenuata* were reported (Yang et al. 2007).

However, as in all the previous examples, the actual biochemical or chemical processes involved in their coupling processes leading to these di-lignans needs to be determined.

10.6 SESQUI-LIGNANS

As indicated earlier, there were four sesqui-lignans **17–21** reported from *Cerbera* spp (Abe et al. 1988a, 1989), these being linked head-to-tail or tail-to-head.

There are also various reports of furanofuran sesqui-lignans, apparently derived from pinoresinol (**40**), or medioresinol (**117**, Figure 10.23), or syringaresinol (**39**), such as (+)-dracunculifoside R (**118**) from *Baccharis dracunculifolia* (Nagatani et al. 2002). Others include (+)-hedyotols C (**119**) and D (**120**) from *Hedyotis lawsoniae* leaves (Kikuchi et al. 1985); (–)-ficusesquilignan A (**121**) from *Ficus microcarpa* heartwood (Li and Kuo 2000); buddlenol C (**122**) from *Buddleja davidii* (Houghton 1985), as well as (+)-haedoxan A (**123**) and its congeners from *Phryma leptostachya* (Taniguchi et al. 1989). (+)-Haedoxan A (**123**) has the unusual

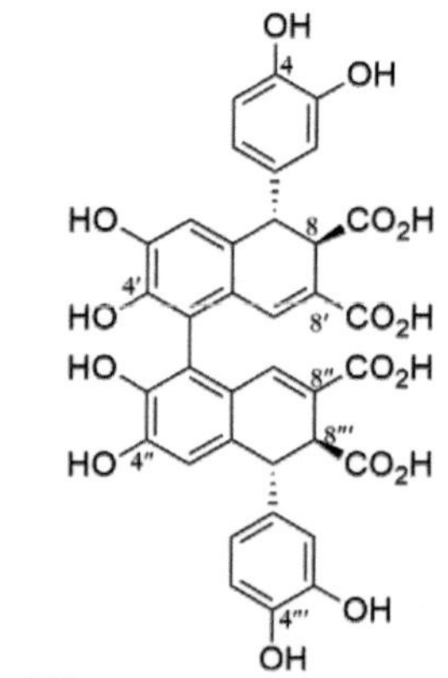

106
$[\alpha]_D^{20}$ = −106.1° (c 0.59, MeOH)

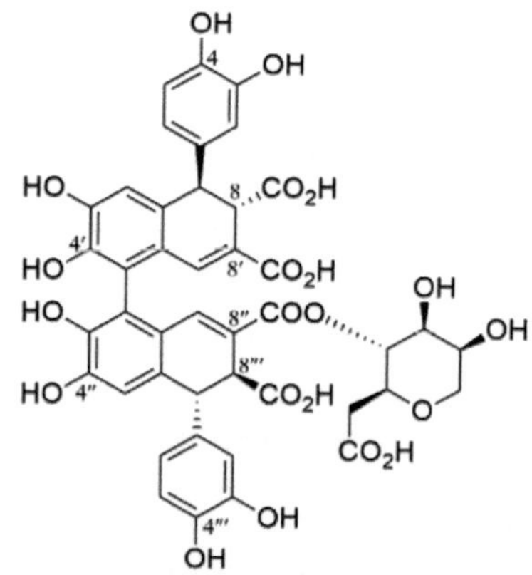

Trilobatin E (**107**)
No $[\alpha]_D$ value reported

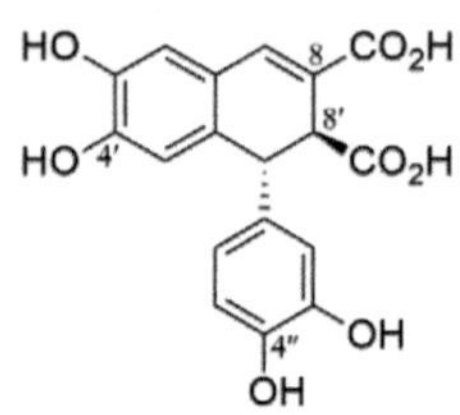

(−)-Epiphyllic acid (**108b**)

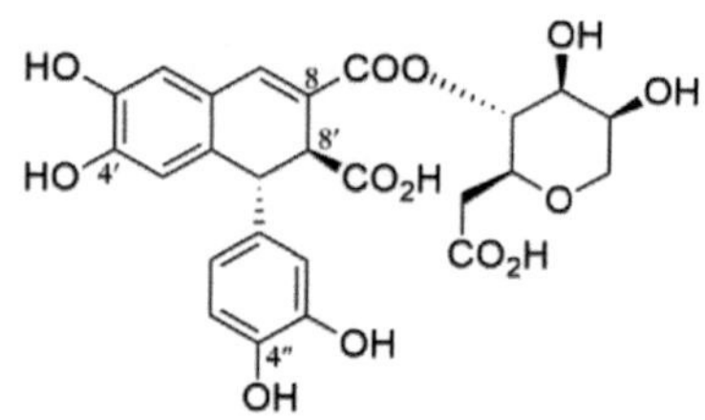

(−)-Trilobatin D (**109b**)
$[\alpha]_D^{20}$ = −146.7° (c 0.23, MeOH)

FIGURE 10.21 Di-lignans (**106** and **107**) and lignans (**108b** and **109b**) present in the liverwort *Bazzania trilobota* (Martini et al. 1998, Scher et al. 2003). "b" refers to the (–)-enantiomers.

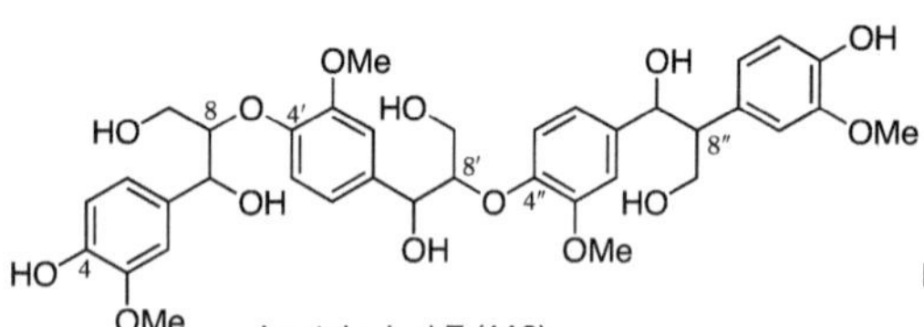

Leptolepisol E (**110**)
$[\alpha]_D^{28}$ = −3.2° (c 2.15, MeOH)

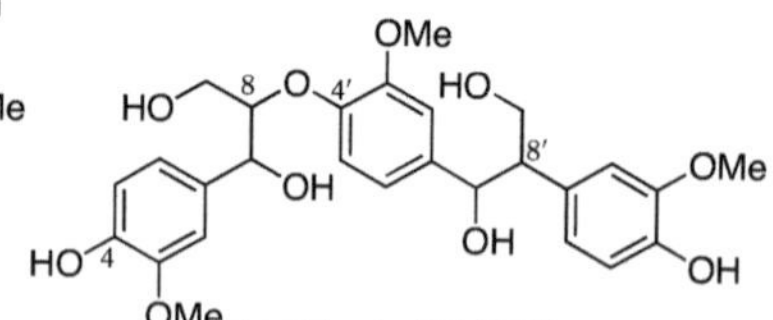

Leptolepisol D (**111**)
No $[\alpha]_D$ value reported

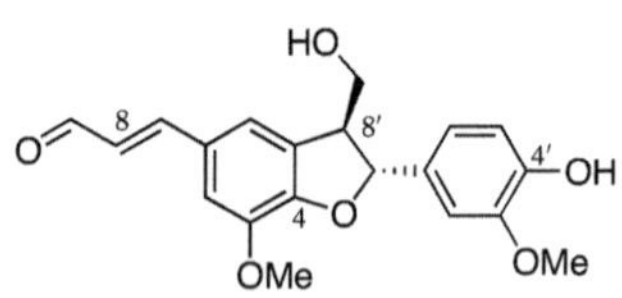

Balanophonin (**112**)

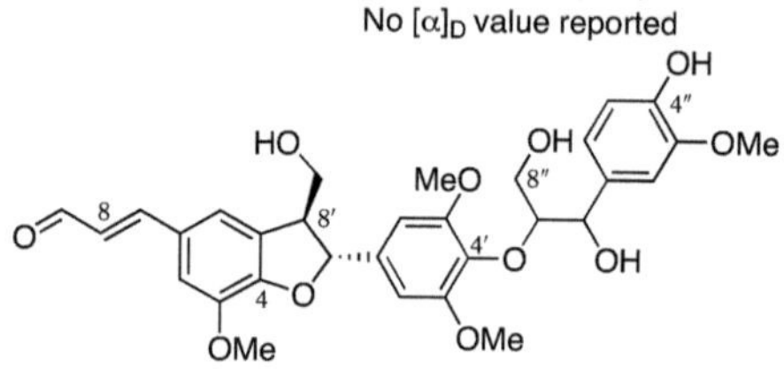

Buddlenol A (**113**)

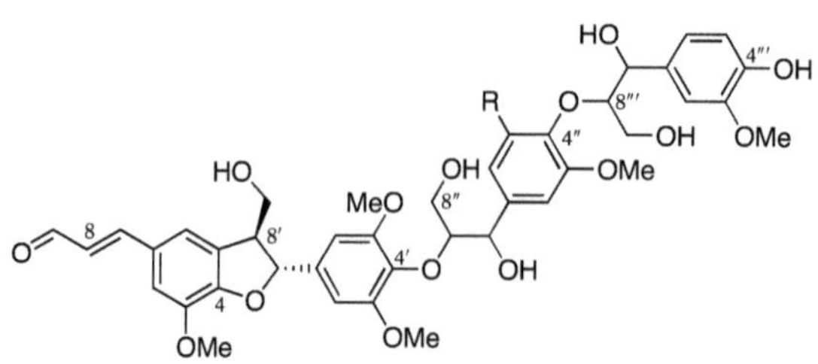

R = H: 4″-*O*-(8-guaiacylglycerol)buddlenol A (**114**)
R = OMe: 5″-Methoxy-4″-*O*-(8-guaiacylglycerol)buddlenol A (**115**)
114: $[\alpha]_D^{27}$ = +5.4° (c 0.34, MeOH)
115: $[\alpha]_D^{27}$ = +5.9° (c 0.38, MeOH)

FIGURE 10.22 Di-lignan **110** and sesqui-lignan **111** from larch (*Larix leptolepis*) (Miki et al. 1979, 1980) and lignan **112**, sesqui-lignan **113**, and di-lignans **114** and **115** from *Tarenna attenuata* (Yang et al. 2007).

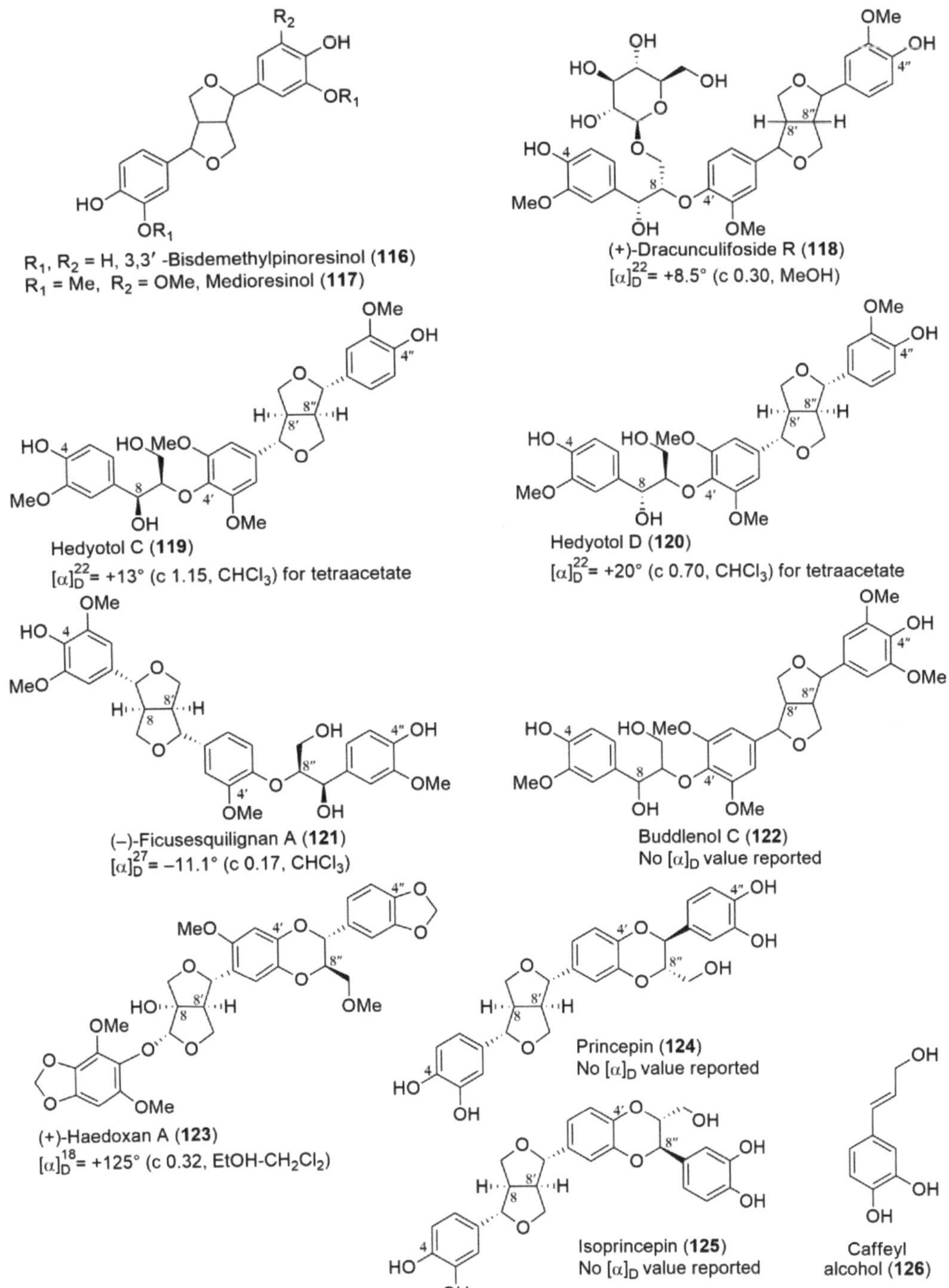

FIGURE 10.23 Monolignols, lignans and sesqui-lignans. Sesqui-lignans (**118–125**) from *Baccharis dracunculifolia* (Nagatani et al. 2002), *Hedyotis lawsoniae* (Kikuchi et al. 1985), *Ficus microcarpa* (Li and Kuo 2000), *Buddleja davidii* (Houghton 1985), *Phryma leptostachya* (Taniguchi et al. 1989) and *Joannesia princeps* (Waibel et al. 2003), lignans **116**, **117** and caffeyl alcohol (**126**).

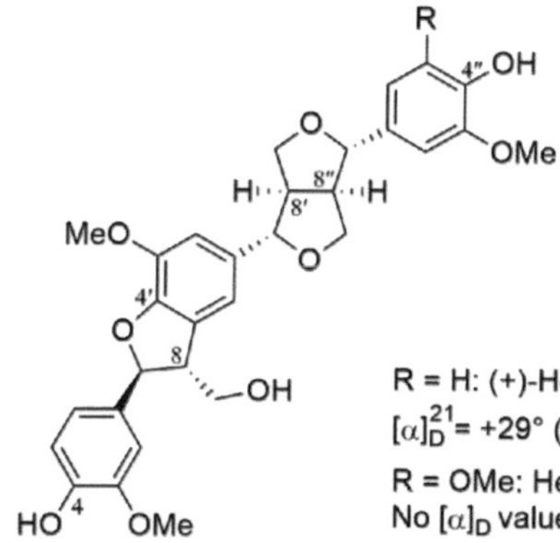

R = H: (+)-Hedyotol A (**127**)
$[\alpha]_D^{21}$ = +29° (c 1.0, $CHCl_3$) for triacetate
R = OMe: Hedyotol B (**128**)
No $[\alpha]_D$ value reported

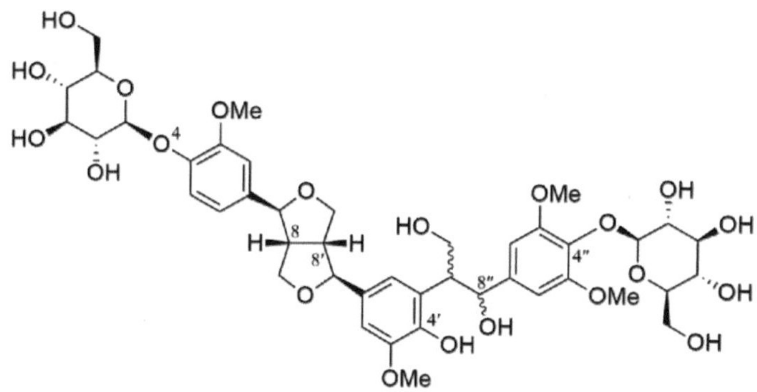

threo: Longifloroside C (**129**)
erythro: Longifloroside D (**130**)
No $[\alpha]_D$ value reported

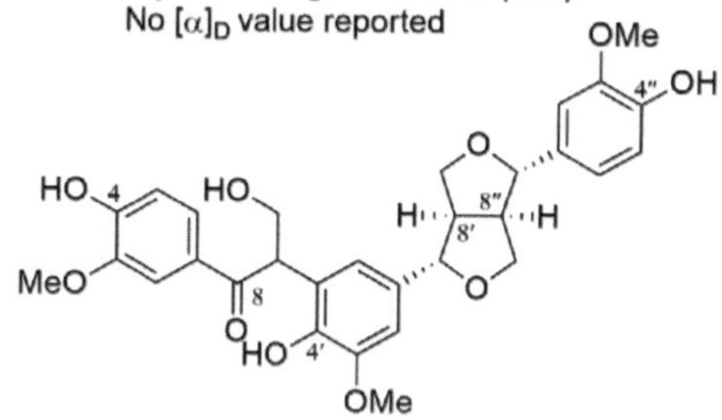

(+)-Herpetrione (**131**)
$[\alpha]_D$ = +140° (c 0.2, MeOH)

FIGURE 10.24 Sesqui-lignans (**127–131**) from *Hedyotis lawsoniae* (Kikuchi et al. 1985), *Pedicularis longiflora* (Wang and Jia 1997) and *Herpetospermum caudigerum* (Kaouadji and Favre-Bonvin 1983).

oxygen insertion between its furofuran rings and the aromatic ring moiety. Two other sesqui-lignans, both apparently derived from a 3,3′-bisdemethylpinoresinol (**116**) core were the regioisomeric benzodioxins, princepin (**124**) and isoprincepin (**125**), isolated from *Joannesia princeps* seeds (Waibel et al. 2003), these perhaps being caffeyl alcohol (**126**) derived.

Other variants have coniferyl (**35**) or sinapyl (**48**) alcohols linked to pinoresinol (**40**) or medioresinol (**117**) via 8-5′ linkages, such as in hedyotols A (**127**, Figure 10.24) and B (**128**) from *H. lawsoniae* (Kikuchi et al. 1985); longiflorosides C and D (**129** and **130**, mixture of *threo*- and *erythro*-isomers at C7″/C8″) from *Pedicularis longiflora* (Wang and Jia 1997); (+)-herpetrione (**131**) from *H. caudigerum* (Kaouadji and Favre-Bonvin 1983).

Future considerations: These examples represent a tiny sampling of the enormous structural diversity that lies within the oligomeric lignans, and of our current paucity in determining fully the biochemical implications thereof.

Yet, it is most likely that of the comprehensive lignan (including higher oligomer) entries in this *Handbook*, only those found to have important medicinal or plant defense properties will be first examined, i.e. as to how they are biochemically formed.

REFERENCES

Abe, F., T. Yamauchi, and A. S. C. Wan. 1988a. Sesqui- sester- and trilignans from stems of *Cerbera manghas* and *C. odollam*. *Phytochemistry* 27:3627–3631.

Abe, F., T. Yamauchi, and A. S. C. Wan. 1988b. Lignans related to olivil from genus *Cerbera*. *Chem. Pharm. Bull.* 36:795–799.

Abe, F., T. Yamauchi, and A. S. C. Wan. 1989. Cerberalignans J-N, oligolignans from *Cerbera manghas*. *Phytochemistry* 28:3473–3476.

Ao, W.-L.-J., X.-H. Bao, X.-L. Wu, and Q.-H. Wang. 2012. Lignans from *Syringa pinnatifolia* Hemsl. var. *alashanensis*. *J. Asian Nat. Prod. Res.* 14:396–400.

Aoshima, H., T. Miyase, and T. Warashina. 2012. Caffeic acid oligomers with hyaluronidase inhibitory activity from *Clinopodium gracile*. *Chem. Pharm. Bull.* 60:499–507.

Bao, S., Y. Ding, Z. Deng, P. Proksch, and W. Lin. 2007. Rhyncosides A-F, phenolic constituents from the Chinese mangrove plant *Bruguiera sexangula* var. *rhynchopetala*. *Chem. Pharm. Bull.* 55:1175–1180.

Cui, H., B. Xu, T. Wu, et al. 2014. Potential antiviral lignans from the roots of *Saururus chinensis* with activity against Epstein-Barr virus lytic replication. *J. Nat. Prod.* 77:100–110.

Davin, L. B., J. R. Cort, C. A. Smith, et al. 2022. Dirigent protein roadmap to lignans and other vascular plant phenol classes. In *The Lignan Handbook* (N. G. Lewis and L. B. Davin eds.), Taylor & Francis, Boca Raton, London, pp 57–77.

Davin, L. B., and N. G. Lewis. 2022a. 8–8′-Lignan biosynthesis roadmap. In *The Lignan Handbook* (N. G. Lewis and L. B. Davin eds.), Taylor & Francis, Boca Raton, London, pp 79–112.

Davin, L. B., and N. G. Lewis. 2022b. 8–2′-, 8–3′-, 8–5′- and 8–*O*–4′-Linked lignan biosynthesis roadmap. In *The Lignan Handbook* (N. G. Lewis and L. B. Davin eds.), Taylor & Francis, Boca Raton, London, pp 113–125.

Davin, L. B., and N. G. Lewis. 2022c. 8–8′- and 8–7′-Norlignan biosynthesis roadmap. In *The Lignan Handbook* (N. G. Lewis and L. B. Davin eds.), Taylor & Francis, Boca Raton, London, pp 147–152.

de Pascoli, I. C., I. R. Nascimento, and L. M. X. Lopes. 2006. Configurational analysis of cubebins and bicubebin from *Aristolochia lagesiana* and *Aristolochia pubescens*. *Phytochemistry* 67:735–742.

DellaGreca, M., A. Fiorentino, A. Izzo, et al. 2007. Phytotoxicity of secondary metabolites from *Aptenia cordifolia*. *Chem. Biodiversity* 4:118–128.

Di, P., L. Zhang, J. Chen, et al. 2013. ^{13}C Tracer reveals phenolic acids biosynthesis in hairy root cultures of *Salvia miltiorrhiza*. *ACS Chem. Biol.* 8:1537–1548.

Dima, O., K. Morreel, B. Vanholme, et al. 2015. Small glycosylated lignin oligomers are stored in *Arabidopsis* leaf vacuoles. *Plant Cell* 27:695–710.

Fan, C.-Q., X.-Z. Zhu, Z.-J. Zhan, et al. 2006. Lignans from *Saussurea conica* and their NO production suppressing activity. *Planta Med.* 72:590–595.

Fukuyama, Y., Y. Otoshi, K. Miyoshi, et al. 1993. Structure of magnolianin, a novel trilignan possessing potent 5-lipoxygenase inhibitory activity. *Tetrahedron Lett.* 34:1051–1054.

Gang, D. R., M. Fujita, L. B. Davin, and N. G. Lewis. 1998. The "abnormal lignins": Mapping heartwood formation through the lignan biosynthetic pathway. In *Lignin and Lignan Biosynthesis* (N. G. Lewis and S. Sarkanen eds.), Vol. 697, ACS Symposium Series, Washington, DC. pp. 389–421.

Gardner, J. A. F., E. P. Swan, S. A. Sutherland, and H. MacLean. 1966. The polyoxyphenols of western red cedar (*Thuja plicata* Donn) III. Structure of plicatic acid. *Can. J. Chem.* 44:52–58.

Houghton, P. J. 1985. Lignans and neolignans from *Buddleja davidii*. *Phytochemistry* 24:819–826.

Hwang, B. Y., J.-H. Lee, J. B. Nam, Y.-S. Hong, and J. J. Lee. 2003. Lignans from *Saururus chinensis* inhibiting the transcription factor NF-κB. *Phytochemistry* 64:765–771.

Ichihara, A., K. Oda, Y. Numata, and S. Sakamura. 1976. Lappaol A and B, novel lignans from *Arctium lappa* L. *Tetrahedron Lett.* 17:3961–3964.

Ichihara, A., Y. Numata, S. Kanai, and S. Sakamura. 1977. New sesquilignans from *Arctium lappa* L. The structure of lappaol C, D and E. *Agric. Biol. Chem.* 41:1813–1814.

Ichihara, A., S. Kanai, Y. Nakamura, and S. Sakamura. 1978. Structures of lappaol F and H, dilignans from *Arctium lappa* L. *Tetrahedron Lett.* 19:3035–3038.

Ichihara, A., Y. Nakamura, H. Kawagishi, and S. Sakamura. 1979. Stereochemistry of lappaol A and F. *Tetrahedron Lett.* 20:3735–3736.

Johansson, C. I., J. N. Saddler, and R. P. Beatson. 2000. Characterization of the polyphenolics related to the colour of western red cedar (*Thuja plicata* Donn) heartwood. *Holzforschung* 54:246–254.

Kaouadji, M., and J. Favre-Bonvin. 1983. Herpetrione, lignoide trimère isolé d' *Herpetospermum caudigerum*. *Tetrahedron Lett.* 24:5881–5884.

Kaouadji, M., and J. Favre-Bonvin. 1984. Herpepentol, nouveau lignoide pentamère isolé d' *Herpetospermum caudigerum* Wall. *Tetrahedron Lett.* 25:5137–5138.

Kaouadji, M., and E. Pieraccini. 1984. Herpetetradione, nouveau lignoide tétramère isolé d' *Herpetospermum caudigerum* Wall. *Tetrahedron Lett.* 25:5135–5136.

Kaouadji, M., J. Favre-Bonvin, F. Sarrazin, and D. Davoust. 1987. Herpetetrone, another tetrameric lignoid from *Herpetospermum caudigerum* seeds. *J. Nat. Prod.* 50:1089–1094.

Kikuchi, T., S. Matsuda, S. Kadota, and T. Tai. 1985. Studies on the constituents of medicinal and related plants in Sri Lanka. III. Novel sesquilignans from *Hedyotis lawsoniae*. *Chem. Pharm. Bull.* 33:1444–1451.

Koubaa, I., and M. Damak. 2003. A new dilignan from *Cynara cardunculus*. *Fitoterapia* 74:18–22.

Lau, W., and E. S. Sattely. 2015. Six enzymes from mayapple that complete the biosynthetic pathway to the etoposide aglycone. *Science* 349:1224–1228.

Li, Y.-C., and Y.-H. Kuo. 2000. Four new compounds, ficusal, ficusesquilignan A, B, and ficusolide diacetate from the heartwood of *Ficus microcarpa*. *Chem. Pharm. Bull.* 48:1862–1865.

Ly, T. N., M. Shimoyamada, and R. Yamauchi. 2006. Isolation and characterization of rosmarinic acid oligomers in *Celastrus hindsii* Benth leaves and their antioxidative activity. *J. Agric. Food Chem.* 54:3786–3793.

Ma, Y., H. Wang, R. Wang, et al. 2019. Cytotoxic lignans from the stems of *Herpetospermum pedunculosum*. *Phytochemistry* 164:102–110.

Marques, J. V., K.-W. Kim, C. Lee, et al. 2013. Next generation sequencing in predicting gene-function in podophyllotoxin biosynthesis. *J. Biol. Chem.* 288:466–479.

Martini, U., J. Zapp, and H. Becker. 1998. Lignans from the liverwort *Bazzania trilobata*. *Phytochemistry* 49:1139–1146.

Matsuda, H., T. Kageura, M. Oda, et al. 2001. Effects of constituents from the bark of *Magnolia obovata* on nitric oxide production in lipopolysaccharide-activated macrophages. *Chem. Pharm. Bull.* 49:716–720.

Miki, K., T. Sasaya, and A. Sakakibara. 1979. A new lignan consisting of four guaiacyl groups from *Larix leptolepis* Gord. *Mokuzai Gakkaishi* 25:678–679.

Miki, K., T. Sasaya, and A. Sakakibara. 1980. Lignans from heartwood of *Larix leptolepis* Gord. *Mokuzai Gakkaishi* 26:633–636.

Nagatani, Y., T. Warashina, and T. Noro. 2002. Studies on the constituents from the aerial part of *Baccharis dracunculifolia* DC. II. *Chem. Pharm. Bull.* 50:583–589.

Petersen, M., Y. Abdullah, J. Benner, et al. 2009. Evolution of rosmarinic acid biosynthesis. *Phytochemistry* 70:1663–1679.

Qu, W., J. Xue, F. H. Wu, and J. Y. Liang. 2014. Lignans from *Saururus chinensis* with antiplatelet aggregation and neuroprotective activities. *Chem. Nat. Compd.* 50:814–818.

Rao, K. V., and R. S. Oruganty. 1997. An improved method for the isolation of the lignan constituents of *Saururus cernuus* by reverse phase column chromatography. *J. Liq. Chromatogr. Relat. Technol.* 20:3121–3134.

Scher, J. M., J. Zapp, and H. Becker. 2003. Lignan derivatives from the liverwort *Bazzania trilobata*. *Phytochemistry* 62:769–777.

Shih, H.-C., T.-L. Hwang, H.-C. Chen, et al. 2013. Honokiol dimers and magnolol derivatives with new carbon skeletons from the roots of *Magnolia officinalis* and their inhibitory effects on superoxide anion generation and elastase release. *PLoS ONE* 8:e59502.

Shih, H.-C., P.-C. Kuo, S.-J. Wu, et al. 2016. Anti-inflammatory neolignans from the roots of *Magnolia officinalis*. *Bioorg. Med. Chem.* 24:1439–1445.

Tanaka, T., A. Nishimura, I. Kouno, G.-i. Nonaka, and T.-J. Young. 1996. Isolation and characterization of yunnaneic acids A-D, four novel caffeic acid metabolites from *Salvia yunnanensis*. *J. Nat. Prod.* 59:843–849.

Taniguchi, E., K. Imamura, F. Ishibashi, T. Matsui, and A. Nishio. 1989. Structure of the novel insecticidal sesquilignan, haedoxan A. *Agric. Biol. Chem.* 53:631–643.

Umehara, K., A. Sugawa, M. Kuroyanagi, A. Ueno, and T. Taki. 1993. Studies on differentiation-inducers from Arctium Fructus. *Chem. Pharm. Bull.* 41:1774–1779.

Vassão, D. G., K.-W. Kim, L. B. Davin, and N. G. Lewis. 2010. Lignans (neolignans) and allyl/propenyl phenols: Biogenesis, structural biology, and biological/human health considerations. In *Comprehensive Natural Products II Chemistry and Biology* (L. Manders and H.-W. Lui eds.), Vol. 1: Structural Diversity I, Elsevier, Oxford, UK. pp 815–928.

Waibel, R., G. Benirschke, M. Benirschke, and H. Achenbach. 2003. Sesquineolignans and other constituents from the seeds of *Joannesia princeps*. *Phytochemistry* 62:805–811.

Wang, C. Z., and Z. J. Jia. 1997. Neolignan glycosides from *Pedicularis longiflora*. *Planta Med.* 63:241–244.

Wang, Q.-H., W.-L.-J. Ao, Q. Wu, and X. Ying. 2012. Two new neolignans from the stems of *Syringa pinnatifolia* Hemsl. var. *alashanensis*. *Helv. Chim. Acta* 95:1680–1685.

Yahara, S., T. Nishiyori, A. Kohda, T. Nohara, and I. Nishioka. 1991. Isolation and characterization of phenolic compounds from Magnoliae Cortex produced in China. *Chem. Pharm. Bull.* 39:2024–2036.

Yang, B.-H., W.-D. Zhang, R.-H. Liu, et al. 2005. Lignans from bark of *Larix olgensis* var. *koreana*. *J. Nat. Prod.* 68:1175–1179.

Yang, G.-Z., Y. Hu, B. Yang, and Y. Chen. 2009. Lignans from the bark of *Zanthoxylum planispinum*. *Helv. Chim. Acta* 92:1657–1664.

Yang, X.-W., P.-J. Zhao, Y.-L. Ma, et al. 2007. Mixed lignan-neolignans from *Tarenna attenuata*. *J. Nat. Prod.* 70:521–525.

11 Lignan Chirality Roadmap

A Scientific Art in Danger of Being Lost?

Laurence B. Davin, Syed G. A. Moinuddin, Daneel Ferreira, and Norman G. Lewis

Plant natural products are frequently chiral, the word being derived from the Greek "*chir*" for hand. Many lignans are no exception. Yet, as for all chiral plant secondary metabolites, the unequivocal definition of the absolute configuration of lignans, especially those with multiple stereogenic centers, remains a challenging exercise.

Why is lignan chirality important to know? One reason is because, as for many natural product classes, differing lignan chiralities can massively affect biological, physiological, and medicinal/human health properties. The Roadmap herein thus addresses the importance of lignan chirality, and why gaining such knowledge can and does provide insights that could not be deduced otherwise. Representative examples of the needs for chiral lignan compound separations, and of the technologies available to determine chirality/absolute configuration, are also provided in this Roadmap As emphasized herein, this must not become a lost scientific art in either the lignan field or elsewhere, i.e. as regards full chemical compound identification.

11.1 ORIGINS OF CHIRALITY

Why and how many natural products (proteins, carbohydrates, RNA, lignans, etc.) emerged with specific chiral forms has no known full explanation that can be rationalized from our considerations of the beginnings of life to the present. This intriguing dilemma of how chirality emerged includes divergent (if not opposing) schools of thought from evolutionists, creationists, and agnostics alike (see Sarfati 1998, McCombs 2004, Singer 2014, Brazil 2015, for some popular, near layperson, articles).

Reports of chirality have not been restricted to Earth either, as meteorites have been reported to contain optically active amino acids, e.g. the Murchison meteorite (Engel and Nagy 1982). However, when these data were later re-examined/re-evaluated by others, the optical activity was considered to most probably result from terrestrial contamination (Bada et al. 1983). Yet later studies of "primitive" stony meteorites, known as carbonaceous chondrites, reportedly contained optically active α-methylamino acids (Pizzarello 2004).

In sum, how biological chirality might have come into effect has been the focus of numerous scholarly articles, including discussion of some of the factors that should perhaps be considered (e.g. Fuß 2009, Caglioti et al. 2011). The subject is that of an unfinished quest for an explanation, and likely to remain as such for the foreseeable future!

11.2 THE IMPORTANCE OF CHIRALITY IN LIVING SYSTEMS

Nevertheless, whatever the underlying cause (terrestrial or extraterrestrial) of homochirality in Nature, we live in a biological chiral world, with profound ramifications. For example, different forms of chiral natural products (e.g. distinct enantiomers) are well known to have different biological effects, such as in odor and in human health. A few examples here from the plant kingdom illustrate the importance of chirality in plant natural products as regards human health:

11.2.1 Chirality and Odor

The enantiomers of the monoterpenes, limonene (**1**) and carvone (**2**), have different odor properties. *R*-(+)-limonene (**1a**, Figure 11.1) has a citrus odor, whereas the *S*-(–)-form (**1b**) smells like turpentine or pine (Lis-Balchin et al. 1996). Carvone in its *S*-(+)-form (**2a**) has an odor characteristic of caraway seeds, with the *R*-(–)-enantiomer (**2b**) having a spearmint-like smell (Leitereg et al. 1971).

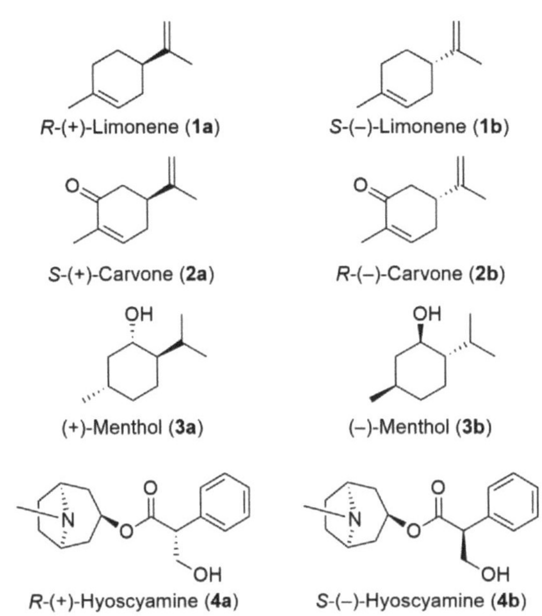

FIGURE 11.1 Limonene (**1**), carvone (**2**), and menthol (**3**) enantiomers and hyoscyamine (**4**) epimers. "**a**" and "**b**" refer to the (+)- and (–)-enantiomeric forms, respectively.

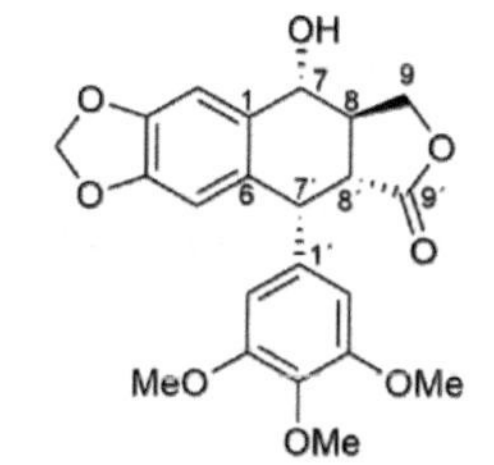

(–)-Podophyllotoxin (**5b**)

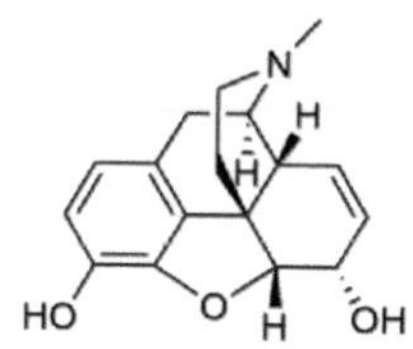

(–)-Morphine (**6b**)

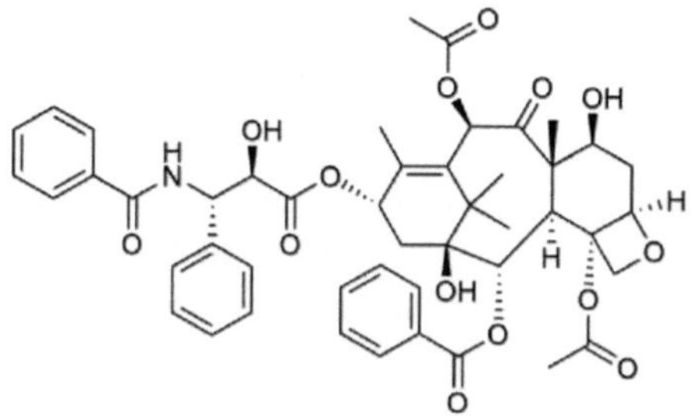

(–)-Paclitaxel (**7b**)

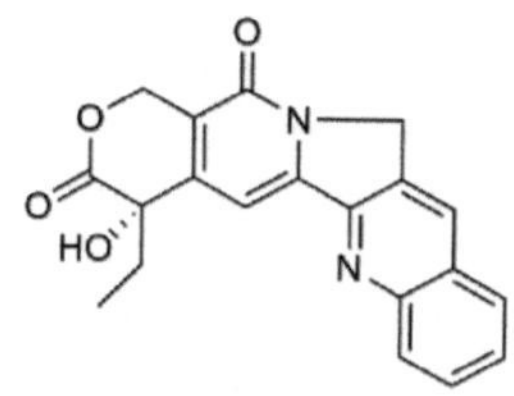

(+)-Camptothecin (**8a**)

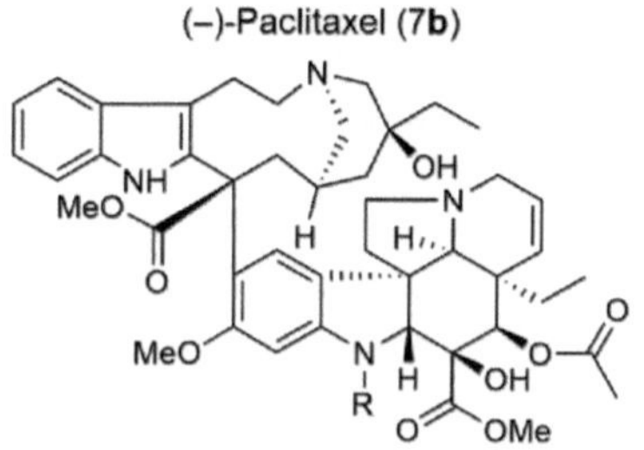

R = CH_3: (+)-Vinblastine (**9a**)
R = CHO: (+)-Vincristine (**10a**)

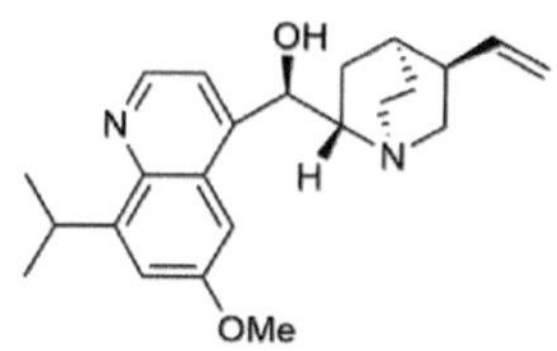

(–)-Quinine (**11b**)

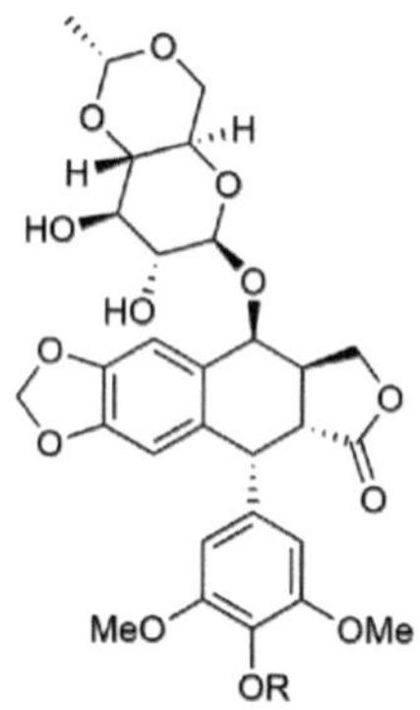

R = H, Etoposide (**12**)
R = $PO(OH)_2$, Etopophos (**13**)

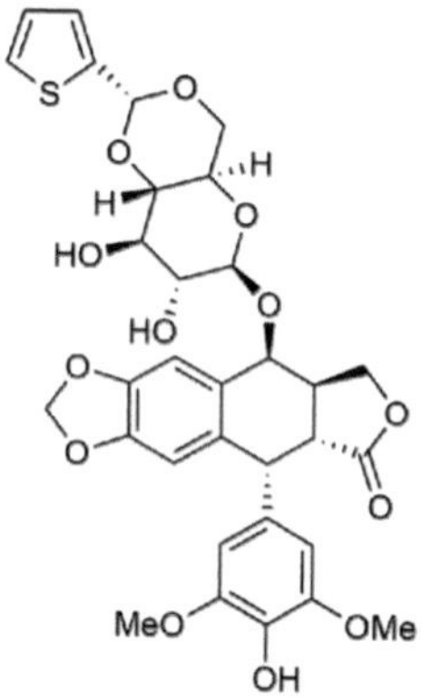

Teniposide (**14**)

FIGURE 11.2 Chiral natural products used in human health/disease treatment. "**a**" and "**b**" refer to the (+)- and (–)-enantiomeric forms, respectively.

11.2.2 Chirality and Human Health/Disease Treatment

Of the two enantiomers of limonene, *R*-(+)-limonene (**1a**) [*circa* 98.9% enantiomeric excess (e.e.)] was a more effective antibacterial against 25 *Listeria* strains, than its *S*-(–)-form (**1b**) (*circa* 73.5% e.e.) (Lis-Balchin et al. 1996). Additionally, menthol enantiomers (**3a** and **3b**) have different physiological cooling effects. The L-(–)-isomer (**3b**) elicits a much stronger physiological cooling sensation effect in humans than the corresponding D-(+)-form (**3a**) (Eccles et al. 1988, Jeffries and Waldron 2019). In addition, the tropane alkaloid hyoscyamine (also known as daturine) (**4**) has relatively potent analgesic properties in its *R*-(+)-form (**4a**), whereas its *S*-(–)-epimer form (**4b**) lacks comparable biological activity (Dei et al. 1997).

There are also many (generally structurally complex) chiral plant natural products that are massively used in human medicine. These include (–)-podophyllotoxin (**5b**, Figure 11.2), (–)-morphine (**6b**), (–)-paclitaxel (Taxol) (**7b**), (+)-camptothecin (**8a**) (marketed as its derivatives topotecan and ironotecan), vincristine/vinblastine (**9a/10a**), and (–)-quinine (**11b**). These examples alone further underscore the importance of stereochemistry/chirality in plant natural products, and our critical dependence on same.

In an example of lignans, (–)-podophyllotoxin (**5b**) exists in a specific chiral form with important medicinal benefits. That is, it is used in the treatment of venereal warts caused by *Papilloma* virus (Longstaff and von Krogh 2001), and also finds application as a scaffold in the semi-synthesis of optically active etoposide (**12**), etopophos (**13**), and teniposide (**14**) for cancer treatments (Stähelin 1970, 1973, Hande 1998).

11.2.3 Chirality of Lignans and Their Analyses: In Danger of Being a Lost Art in Science?

These selected examples of plant natural products (including lignans) begin to illustrate the significance that specific chiral plant natural product forms have, and why such knowledge on stereochemistry/chirality needs to be acquired. This recognition also underscores the necessity of not only fully determining plant natural product chemical structures (and biological activities/modes of action), but also their individual chiral properties (e.g. of distinct enantiomers and/or diastereomers) and their enantiomeric compositions.

In this regard, comprehensive phytochemical or natural product investigations of vascular plants leading to full structure determinations of isolated constituents requires the application of a number of interlocking physico-chemical techniques. These include ^{1}H and ^{13}C nuclear magnetic resonance (NMR) spectroscopic analyses (e.g. using the various pulse sequence programs, etc. now available), mass spectrometry (low and high resolution, including with ion mobility separations), chiral-phase column separations, UV, IR, X-ray crystallography, elemental analysis, melting point and/or boiling point determinations, as well as (most often) optical rotary dispersion (ORD) and electronic circular dichroism (ECD) measurements for chiral compounds. These comprehensive analyses, as well as others mentioned in this chapter, are required for complete and unambiguous determination of the chemical structures of isolated compound(s).

Yet, there is a disconcerting and growing trend in some scientific journals in current literature offerings to use only a small subset of said analytical techniques, and thus not attain full structural identification of isolated constituents. Of these, while mass spectrometry and gas chromatography (GC)/high-pressure liquid chromatography (HPLC) ("metabolomics") approaches have the advantage of being able to detect large numbers of constituents in complex plant mixtures/extracts, they also have most serious shortcomings.

Consequently, many such publications do not result in actual structure determinations, frequently reporting instead "tentative" structures where, for example, neither a sugar moiety attached to a putative aglycone is definitively identified, nor where bonding specifically occurs to the aglycone moiety either. Such analyses also frequently lack chiral-phase separations, and do not determine optical purities/specific rotations/ECD measurements of the metabolites being isolated, e.g. neither ORD/specific rotations nor ECD data are reported (see as an example, Dima et al. 2015). Indeed, the literature is being increasingly swamped with incomplete phytochemical compound studies and reporting of "tentative" structures. This is not to the benefit of the scientific enterprise, and experimental designs of this type can adversely affect the types of conclusions made from the work.

There are many methods to address chirality and absolute configurations, including X-ray diffraction, NMR spectroscopy, and stereoselective synthetic approaches. In addition to these, there are various *chiroptical methods* such as ECD (UV related), vibrational circular dichroism (VCD, IR related), Raman optical activity (ROA), and ORD available. Moreover, whereas ECD data interpretation was earlier stifled by reliance on empirical rules that were often misinterpreted, the advent of density functional chemistry (DFT) computational methods that permit the comparison of experimental and calculated ECD spectra has greatly decreased such reliance. It additionally led to a significant growth in VCD spectroscopy applications, with VCD offering the advantage over ECD in that it also applies to molecules devoid of a chromophore.

Additionally, various *chiral compound separation technologies* have emerged that deploy different solid packing material substrates used in chromatographic separations of enantiomers and diastereomers, and these have greatly impacted this branch of stereochemistry over the last several decades.

Yet, in the lignan classes of plant natural products (i.e. from C_6C_3 dimeric phenylpropanoids to higher oligomers), there is also now a growing absence of comprehensive chiral analyses, as evident in many of the physicochemical descriptions of the lignans reported to date in the dictionary contributions. Very often, there are neither enantiomeric compositions nor ECD data reported for isolated lignans, or the ORD data reported (including specific rotations) in general gives no definitive insight into the optical purity of a specific lignan.

Furthermore, ECD/specific rotation/ERD data of isolated lignans from a specific plant or a tissue do not necessarily inform as to whether the isolated lignan is optically pure or in enantiomeric excess (e.e.). As a further complication, different enantiomers of specific lignans can co-exist in the same tissue (including different enantiomers being in different compartments/cell types). Thus, what is obtained in such studies (of the overall tissue under investigation) is only the "averaged" ORD or ECD value. Such analyses therefore can come down to determining only if there is any optical rotation associated with the isolated compound(s) or not.

Accordingly, there is now a great need to determine the true enantiomeric compositions of isolated lignans, using chiral-phase column chromatography, these being done in conjunction with, for example, ORD or ECD analyses. There are many reasons for doing this, with perhaps the most important being that the biological activities of lignans (and many other natural product classes) can vary enormously depending upon the antipode in question.

A case in point of the importance of chiral analyses is that of how lignin *versus* lignan formation is envisaged to occur. The former, a vascular plant cell wall polymer, is viewed by some as being assembled in a "combinatorial" manner (i.e. essentially formed via random assembly and optically inactive) (Ralph et al. 2004, Ralph et al. 2008, Chen et al. 2012), whereas others consider lignin formation to be under proteinaceous control even if optically inactive (Davin and Lewis 2005, Davin et al. 2008, Hosmani et al. 2013). As described in Chapter 4, stereoselective lignan biosynthetic processes, on the other hand, are of documented proteinaceous control (Davin et al. 1997).

In this regard, analysis of vacuole metabolites in *Arabidopsis* leaves resulted in detection of putative oligomeric lignans that were viewed by these researchers as being somehow incorporated as "lignin oligomers" into the cell wall lignin (Dima et al. 2015), i.e. as part of the polymerization affording lignin in that cell (wall) type. The experimental design, however, in that study was not such as to either support or prove this claim.

This was because there was no compelling evidence that these small *lignan oligomers* were incorporated as such into cell wall lignin through polymerization. In fact, the body of scientific evidence in numerous other studies is exactly the opposite. A most serious limitation in the experimental design of the Dima et al. 2015 study was that no chiral analysis of these putative vacuole oligomeric lignans was carried out. This would have been very instructive to have done, because if they were in specific enantiomeric forms, these would presumably be oligomeric lignans used for plant defense (Corbin et al. 2018) rather than for lignification as claimed (Dima et al. 2015). Oligomeric lignans are well known to be deposited in various lignified cell types to help confer stability and durability against opportunistic pathogens (e.g. fungi, bacteria) (Hillis 1999). Until such analyses are done, and the putative vacuolar oligomeric lignans are unambiguously demonstrated to be involved in the polymerization process to give the lignin proper in the cell wall, the Dima et al. (2015) study must be viewed as deficient in its experimental design.

As an additional limitation to their experimental design, these "lignin oligomers" were neither purified nor their structures fully determined. (This included lack of chiral-phase separations and chiral analyses, including ORD/specific rotation and ECD measurements). Indeed, provisionally assuming these components were not *artifacts* of the isolation process, the lack of chiral analysis and the failure to determine the nature of sugar moieties and their positions of attachment cannot be viewed as full structural determination, excluding whether these were "lignin oligomers" or not. In contrast, these compounds are just as – if not more – likely to be lignan defense molecules until proven otherwise. In sum, their findings were only the beginning – not the completion – of the full characterization and determination of the metabolic fate of the putative oligolignans in *Arabidopsis* leaf vacuoles.

In contrast, the importance of the scientific advances made, and the technologies now available, in establishing and investigating lignan chirality are illustrated in the following examples.

11.3 CHIRAL-PHASE COLUMN CHROMATOGRAPHIC ANALYSES OF LIGNANS

Establishing the precise optical purity and enantiomeric compositions of various lignans has been an essential part of our studies including (a) determining the outcomes of both stereoselective and racemic monomeric C_6C_3 coupling leading to various dimeric lignans (Umezawa et al. 1990b, Davin et al. 1991, 1992, 1997, 2003, Cho et al. 2003, Moinuddin et al. 2003, Kim et al. 2012, Dalisay et al. 2015); (b) investigating the enantiospecificity of downstream lignan metabolism, such as with pinoresinol-lariciresinol reductase (PLR) (Katayama et al. 1992, Dinkova-Kostova et al. 1996, Fujita et al. 1999, Hwang et al. 2020), secoisolariciresinol dehydrogenase (Umezawa et al. 1991, Xia et al. 2001), phenylcoumaran benzylic ether reductase (PCBER) (Gang et al. 1999), and larreatricin 3′-hydroxylase (Cho et al. 2003); and (c) establishing the enantiomeric compositions of lignans in various plant species (Davin et al. 1991, Lewis and Davin 1999) or in genetically modified plants whose lignan contents and compositions have been altered (Kim et al. 2012). This can be best illustrated with brief examples.

11.3.1 (+)- AND (–)-PINORESINOL FORMING DIRIGENT PROTEINS

Our studies established that in the dirigent protein (DP) multigene families, two DP gene members in sub-family DIRa encoded proteins able to convert achiral coniferyl alcohol into either (+)- or (–)-pinoresinols (**15a** or **15b**), respectively, provided that a one-electron oxidase was also present (Davin et al. 1997, Kim et al. 2012, Dalisay et al. 2015).

Figure 11.3A shows the chiral-phase separation of a racemic mixture of the (+)- and (–)-antipodes of pinoresinol (**15a** and **15b**) using a chiral-phase column Chiralcel OD (Daicel, Chiral Technologies). The latter contains a polysaccharide-derived chiral stationary phase (CSP), this being sufficient to separate these enantiomers on the basis of their interaction with the derivatized polysaccharide matrix, *tris*-(3,5-dimethylphenylcarbamate).

This facile separation of the two enantiomers allowed for determination of pinoresinol enantiomeric composition in wild type (WT) *Arabidopsis thaliana* root tissue (Figure 11.3B), as well as from an *Arabidopsis* line overexpressing the (–)-pinoresinol-forming DP AtDIR6 (Figure 11.3C), and an *Arabidopsis* RNAi line where *AtDIR6* expression is greatly reduced (Figure 11.3D).

That is, to prove the physiological function of the AtDIR6 in *Arabidopsis*, genetically engineered *Arabidopsis* lines were

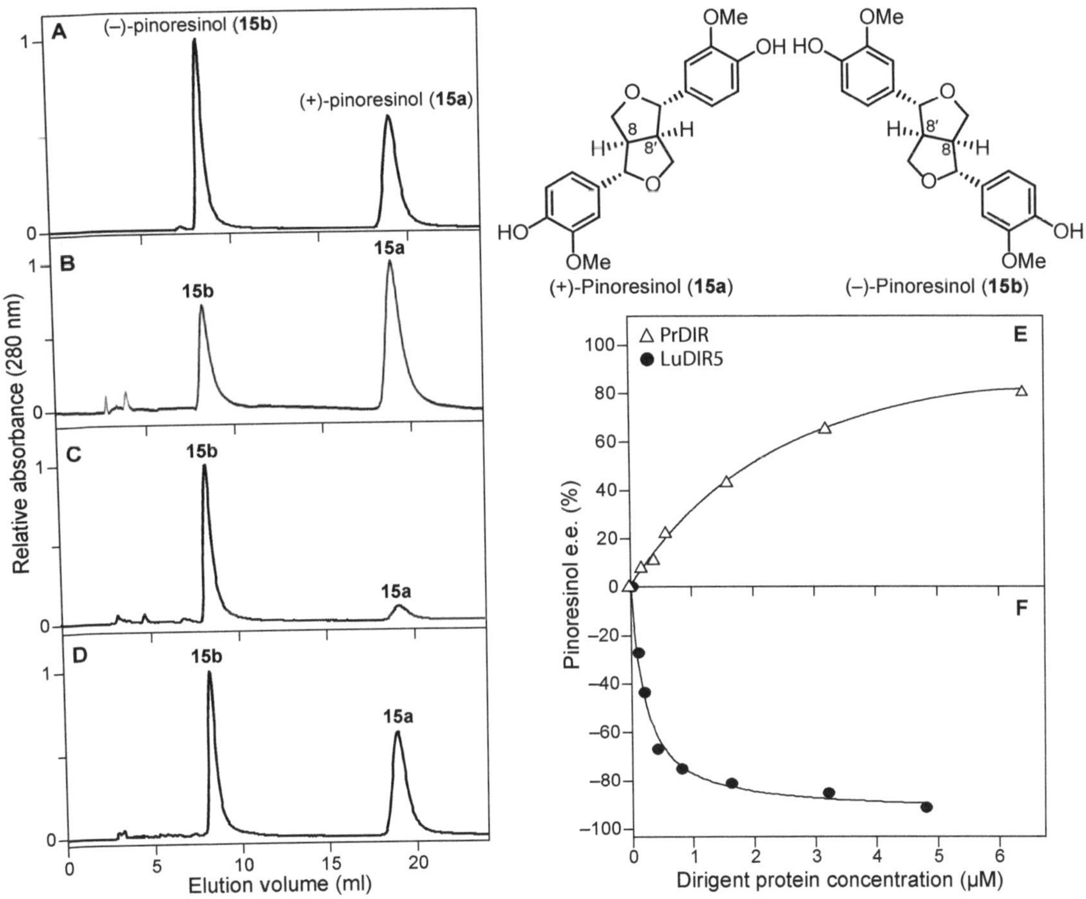

FIGURE 11.3 Chiral-phase HPLC analyses of pinoresinol (**15**) antipodes of racemic pinoresinols (**15a/b, A**); (**B–D**) pinoresinols (**15a/b**) isolated from root tissues of *Arabidopsis thaliana* WT and transgenic plant lines: WT (**B**), *Arabidopsis* overexpressing AtDIR6 line (**C**), and *Arabidopsis* RNAi transgenic line where *AtDIR6* mRNA expression RNAi was greatly reduced (**D**). The (+)- and (–)-enantiomers of pinoresinol (**15a/b**) were resolved on a Chiralcel OD (Daicel; 250 × 4.6 mm) eluted with hexanes–EtOH (1:1, flow rate: 0.8 ml min^{-1}). Effect of varying PrDIR (△, **E**, unpublished) and LuDIR5 (●, **F**) concentration on formation of (+)-pinoresinol (**15a**) and (–)-pinoresinol (**15b**) in enantiomeric excess (e.e.) in assays using coniferyl alcohol, in presence of either PrDIR or LuDIR5, provided a 1e oxidase or oxidant was present. Lu: *Linum usitatissimum*; Pr: *Piper regnellii*. "**a**" and "**b**" refer to the (+)- and (–)-enantiomeric forms, respectively.

This research (A–D) was originally published in the *Journal of Biological Chemistry*. Kim et al. (2012). Opposite stereoselectivities of dirigent proteins in *Arabidopsis* and *Schizandra* species. *J. Biol. Chem.* 287:33957–33972. Copyright (2012) the American Society for Biochemistry and Molecular Biology.

This research (F) was originally published in the *Journal of Natural Products*. Dalisay et al. (2015). Dirigent protein-mediated lignan and cyanogenic glucoside formation in flax seed: Integrated omics and MALDI mass spectrometry imaging. *J. Nat. Prod.* 78:1231–1242. Copyright (2015) the American Chemical Society and American Society of Pharmacognosy.

produced using overexpression and RNAi strategies (Kim et al. 2012). Chiral analysis of the pinoresinol (**15**) isolated from the WT *Arabidopsis* roots (72 ng/mg of dry weight) as discussed earlier showed that it was in 36% e.e. of the (+)-form (**15a**) (Figure 11.3B). However, overall levels of pinoresinol (**15**) substantially increased (585 ng/mg of dry weight root tissue) in roots of *Arabidopsis* plants overexpressing *AtDIR6*, this being in 75% e.e. of the (–)-form (**15b**) (Figure 11.3C) and further confirming the role of *AtDIR6* as a (–)-pinoresinol-forming DP. In the case of the RNAi transgenic lines, where *AtDIR6* expression was mostly repressed, levels of pinoresinol were slightly higher (89 ng/mg of dry weight) than in WT with an e.e of only 8% of the (+)- antipode (**15a**) (Figure 11.3D).

The same Chiralcel OD separations technology was used for rapid assays of DP stereoselectivity mentioned earlier, as illustrated in Figures 11.3E and 11.3F. As observed, increasing amounts of the (+)- and (–)-pinoresinol-forming DPs in the assays gave either (+)- or (–)-pinoresinols (**15a** or **15b**) in increasing enantiomeric purity levels (enantiomeric excess, e.e.).

In a related manner, this same chiral-phase separation technology was used to investigate the effect of domain swapping

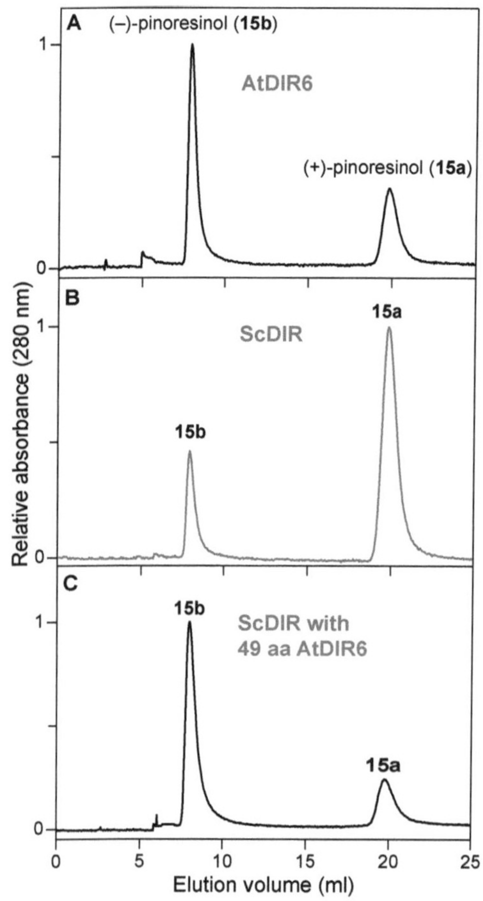

FIGURE 11.4 (+)- and (–)-Pinoresinol (**15a** and **15b**) chiral separations, and effects of region swapping on stereoselectivity of (+)-and (–)-pinoresinol-forming dirigent proteins. (A) Recombinant native ScDIR. (B) Recombinant native AtDIR6. (C) AtDIR6 49 internal amino acid region swapped into ScDIR. At: *Arabidopsis thaliana;* Sc: *Schizandra chinensis.*

This research was originally published in the *Journal of Biological Chemistry.* Kim et al. (2012). Opposite stereoselectivities of dirigent proteins in *Arabidopsis* and *Schizandra* species. *J. Biol. Chem.* 287:33957–33972. Copyright (2012) the American Society for Biochemistry and Molecular Biology.

in (+)- and (–)- pinoresinol-forming DPs. Domain swapping of a 49-amino-acid region of AtDIR6 (Figure 11.4A) into that of an ScDIR (Figure 11.4B) from *Schizandra chinensis* resulted in the (+)-pinoresinol-forming DP now displaying (–)-pinoresinol-forming capability (Figure 11.4C) (Kim et al. 2012) in assays using coniferyl alcohol as substrate, together with the DP and an 1e oxidase/oxidant. [Note: Assay conditions were not optimized.].

That is, the importance of conducting chiral-phase separations and their analyses was clearly demonstrated.

11.3.2 Pinoresinol-Lariciresinol Reductases

Pinoresinol-lariciresinol reductases (PLRs), or homologs thereof, are able to catalyze NADPH-dependent enantiospecific reductions of either (+)- or (–)-pinoresinols (**15a** or **15b**) to afford either the corresponding products, (+)- or (–)-lariciresinols (**16a** or **16b**, Figure 11.5), or (–)- or (+)-secoisolariciresinols (**21b** or **21a**, Figure 11.6) depending upon the PLR source (Davin and Lewis 2022). Chiral-phase separations of lariciresinol (**16a/b**) and secoisolariciresinol (**21a/b**) antipodes were achieved using Chiralcel OC and OD columns (Figures 11.5 and 11.7A), respectively (Umezawa et al. 1991, Dinkova-Kostova et al. 1996, Fujita et al. 1999). In the case of Chiralcel OC column chromatography, *tris* (phenylcarbamate) was coated to the polysaccharide matrix thus providing the basis for enantiomeric separation.

The substrate specificities of two PLRs (PLR_Tp1 and PLR_Tp2) from western red cedar (*Thuja plicata*) were also examined (Hwang et al. 2020). Different potential substrates were synthesized, namely (±)-ligballinols (**17a/b**), (±)-pinoresinols (**15a/b**), (±)-medioresinols (**18a/b**), and (±)-syringaresinols (**19a/b**) (Figure 11.6). Both of these distinct PLR homologs were able to readily catalyze different

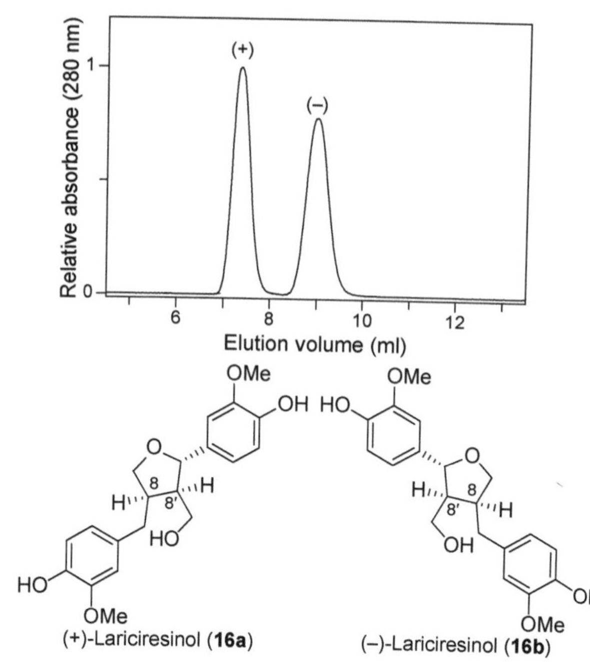

FIGURE 11.5 Chiral-phase HPLC analyses of racemic lariciresinols (**16a/b**). The (+)- and (–)-enantiomers of lariciresinol (**16a/b**) were resolved on a Chiralcel OC column (Daicel; 250 × 4.6 mm) eluted with hexanes–EtOH (1:4, flow rate: 0.3 ml min^{-1}).

This research was originally published in the *Journal of Biological Chemistry.* Chu et al. (1993). Stereospecificity of (+)-pinoresinol and (+)-lariciresinol reductases from *Forsythia intermedia. J. Biol. Chem.* 268:27026–27033. Copyright (1993) the American Society for Biochemistry and Molecular Biology.

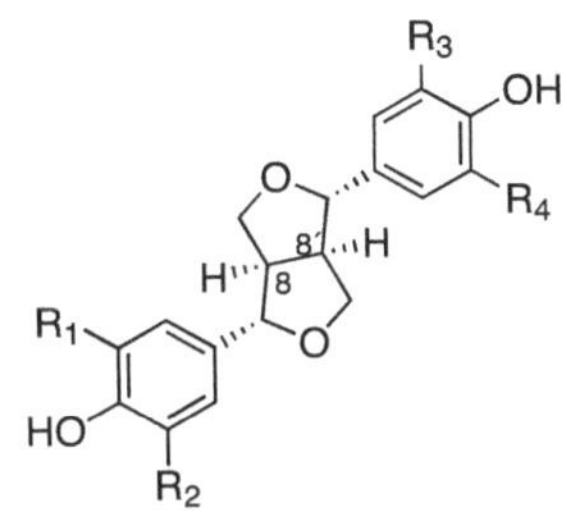

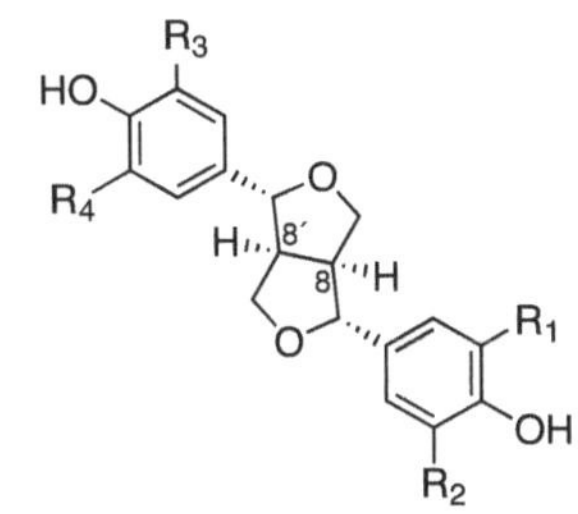

R_1, R_2, R_3, R_4 = H:	(+)-Ligballinol (**17a**)	(–)-Ligballinol (**17b**)
R_1 = H, R_2, R_3, R_4 = OMe:	(+)-Medioresinol (**18a**)	(–)-Medioresinol (**18b**)
R_1, R_2, R_3, R_4 = OMe:	(+)-Syringaresinol (**19a**)	(–)-Syringaresinol (**19b**)

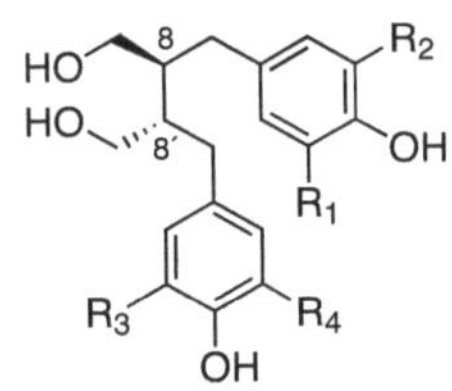

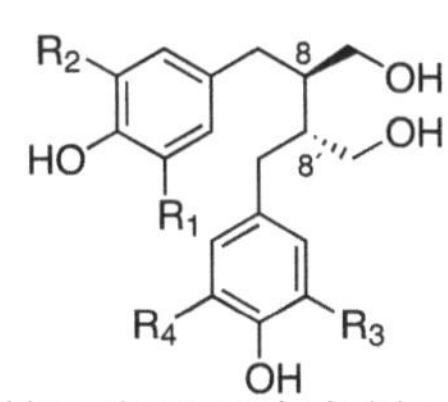

R_1, R_2, R_3, R_4 = H:	(+)-3,3′-Didemethoxysecoisolariciresinol (**20a**)	(–)-3,3′-Didemethoxysecoisolariciresinol (**20b**)
R_1, R_4 = H, R_2, R_3 = OMe:	(+)-Secoisolariciresinol (**21a**)	(–)-Secoisolariciresinol (**21b**)
R_1 = H, R_2, R_3, R_4 = OMe:	(+)-5′-Methoxysecoisolariciresinol (**22a**)	(–)-5′-Methoxysecoisolariciresinol (**22b**)
R_1, R_2, R_3, R_4 = OMe:	(+)-5,5′-Dimethoxysecoisolariciresinol (**23a**)	(–)-5,5′-Dimethoxysecoisolariciresinol (**23b**)

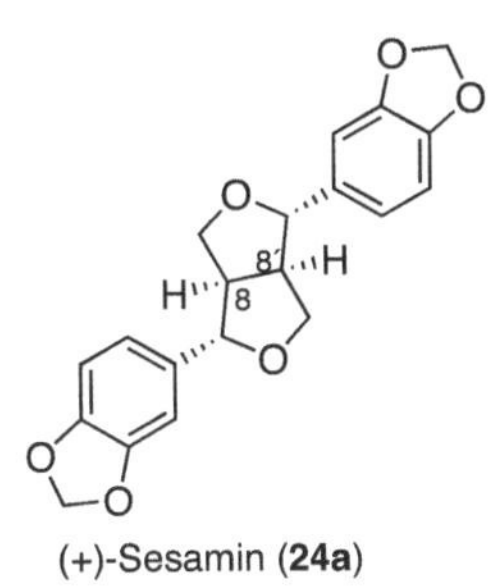

(+)-Sesamin (**24a**)

FIGURE 11.6 (+)- and (–)-Antipodes of lignans (**17a/b–23a/b**) and (+)-sesamin (**24a**). "**a**" and "**b**" refer to the (+)- and (–)-enantiomeric forms, respectively.

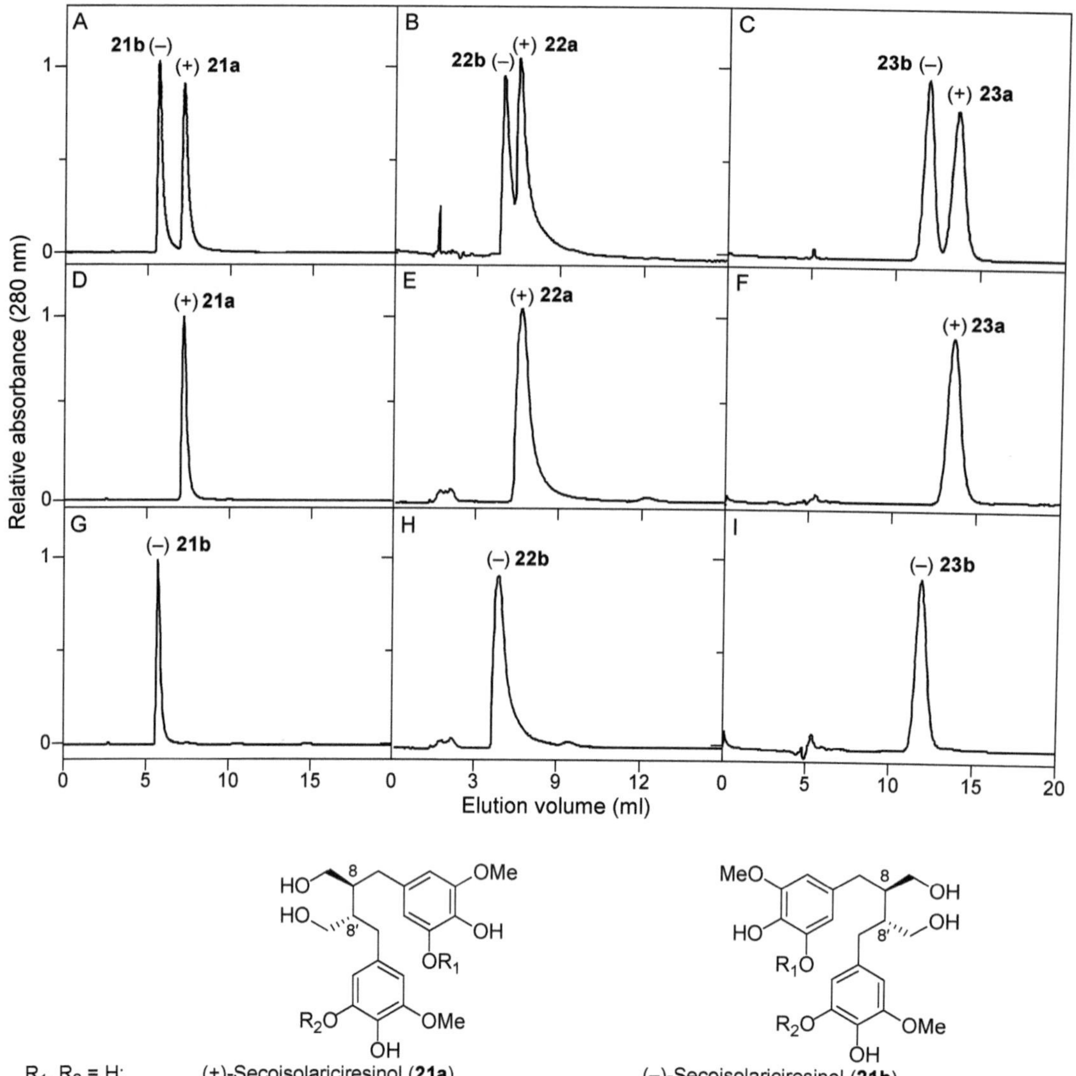

R_1, R_2 = H: (+)-Secoisolariciresinol (**21a**)
R_1 = H, R_2 = OMe: (+)-5′-Methoxysecoisolariciresinol (**22a**)
R_1, R_2 = OMe: (+)-5,5′-Dimethoxysecoisolariciresinol (**23a**)

(–)-Secoisolariciresinol (**21b**)
(–)-5′-Methoxysecoisolariciresinol (**22b**)
(–)-5,5′-Dimethoxysecoisolariciresinol (**23b**)

FIGURE 11.7 PLR enantiospecificity with different substrate analogues. Chiral-phase separations of synthetic racemic forms of (**A**) secoisolariciresinol (**21a/b**), (**B**) 5′-methoxysecoisolariciresinol (**22a/b**), and (**C**) 5,5′-dimethoxysecoisolariciresinol (**23a/b**). Upon incubation of PLR_Tp1 individually with (±)-pinoresinols (**15a/b**), (±)-medioresinols (**18a/b**), and (±)-syringaresinols (**19a/b**) in the presence of NADPH, (+)-secoisolariciresinol (**21a**, **D**), (+)-5′-methoxysecoisolariciresinol (**22a**, **E**), and (+)-5,5′-dimethoxysecoisolariciresinol (**23a**, **F**) were formed, respectively. The opposite antipodes, **21b** (**G**), **22b** (**H**), and **23b** (**I**) were generated when PLR_Tp2 was incubated with **7a/b**, **18a/b**, and **19a/b**. "**a**" and "**b**" refer to the (+)- and (–)-enantiomeric forms, respectively.

This research was originally published in Chirality: Hwang et al (2020). Pinoresinol-lariciresinol reductase: Substrate versatility, enantiospecificity, and kinetic properties. *Chirality* 32:770–789. Copyright (2020) with permission from John Wiley and Sons.

enantiospecific, NADP(H)-dependent reductions of three of these potential substrates (Figure 11.7).

That is, PLR_Tp1 catalyzed enantiospecific conversion of (–)-pinoresinol (**15b**) and homologs **18b** and **19b** into (+)-secoisolariciresinol (**21a**), (+)-5′-methoxysecoisolariciresinol (**22a**), and (+)-5,5′-dimethoxysecoisolariciresinol (**23a**), respectively (Figure 11.7 D–F).

PLR_Tp2, by contrast, catalyzed formation of (–)-secoisolariciresinol (**21b**), (–)-5′-methoxysecoisolariciresinol (**22b**) and (–)-5,5′-dimethoxysecoisolariciresinol (**23b**).

These studies were also extended to examining turnover efficacies of substrates. Both PLR homologs used pinoresinol (**15**) or medioresinol (**18**) as best substrates, with syringaresinol (**19**) being less effectively converted. In contrast, (±)-ligballinols (**17a/b**) served as a much poorer substrate and (+)-sesamin (**24a**) was not reduced. These data demonstrated that a free phenolic group was required for catalysis, with the presumed intermediates being ring-opened quinone methides as previously reported (Dinkova-Kostova et al. 1996).

These findings of different chiral substrate preferences with both PLR_Tp1 and PLR_Tp2 can also be envisaged as important for biotechnological manipulations in the future. That is, various lignan skeleta can be differentially "decorated" on their aromatic rings to afford the corresponding reduced products, e.g. different substrates (**15–19**) in optically active forms can be generated via substrate versatility of dirigent proteins.

11.3.3 Secoisolariciresinol Dehydrogenases and Chiral-Phase Separations

Secoisolariciresinol dehydrogenases (SDHs) from *Forsythia intermedia, Podophyllum hexandrum,* and *P. peltatum* catalyzed the enantiospecific NADP-dependent dehydrogenation of (–)-secoisolariciresinol (**21b**) into (–)-matairesinol (**25b**) (Umezawa et al. 1990a, 1991, Xia et al. 2001). Chiral-phase separation of (+)- and (–)-matairesinols (**25a** and **25b**) employed a Chiralcel OD column (Figure 11.8) (Umezawa et al. 1991, Chu et al. 1993).

11.3.4 Phenylcoumaran Benzylic Ether Reductase (PCBER), Phenylpropenal Double-Bond Reductase (PPDBR), and 8-5′ Linked Lignan Chiral-Phase Separations

Racemic forms of (±)-8-5′-linked phenylcoumaran lignans **26a/b–31a/b** were readily separated on either Chiralcel OF or Chirobiotic V (Astec) columns (Gang et al. 1999, Kasahara et al. 2006); in those cases, the Chiralcel OF coating was *tris* (4-chlorophenylcarbamate), whereas the Chirobiotic bonded phase was vancomycin.

These racemates were used to study PCBER and PPDBR, both isolated from loblolly pine (*Pinus taeda*) cell suspension cultures (Gang et al. 1999, Kasahara et al. 2006). PCBER regiospecifically converted **26a/b** and **28a/b** into the corresponding 7–*O*–4′ reduced lignans, **27a/b** and **29a/b** (Figure 11.9A), whereas PPDBR reduced the C7′–C8′ double bond of **30a/b** to form **31a/b** (Figure 11.9B). As an example, a chiral-phase separation of **28a** and **28b** is shown (Figure 11.10).

11.3.5 Larreatricin 3′-Hydroxylase

The creosote bush (*Larrea tridentata*) accumulates a number of 8–8′-linked lignans lacking an oxygen functionality at C9/C9′. In our studies of lignan biosynthesis in this species, a polyphenol oxidase (PPO) was isolated, with this being able to catalyze the enantiospecific oxidation of (+)-larreatricin (**32a**) to afford (+)-3′-hydroxylarreatricin (**33a**); the corresponding (–)-enantiomer (**32b**) did not function as a substrate (Cho et al. 2003, Moinuddin et al. 2003). This study was also greatly

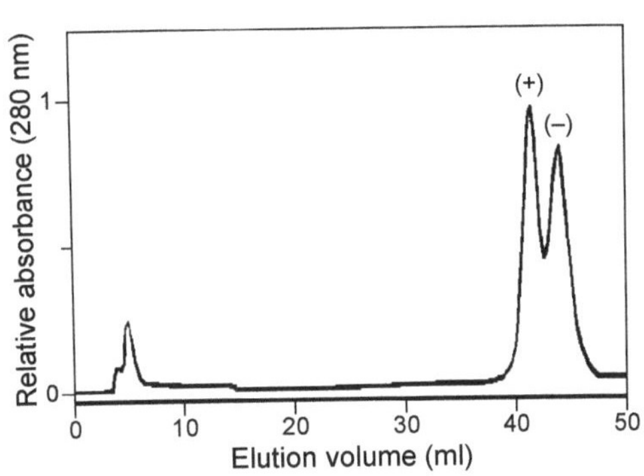

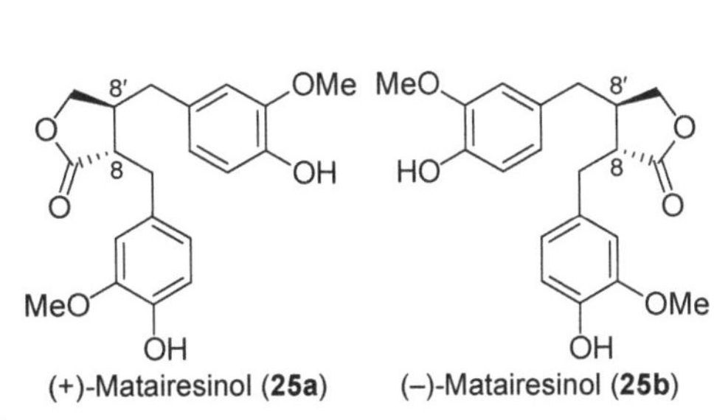

FIGURE 11.8 Chiral-phase HPLC analyses of racemic matairesinols (**25a/b**). The (+)- and (–)-enantiomers of matairesinol (**25a/b**) were resolved on a Chiralcel OD column (Daicel; 250 × 4.6 mm) eluted with 1% AcOH in hexanes–EtOH (85:15, flow rate: 1 ml min^{-1}), respectively. "**a**" and "**b**" refer to the (+)- and (–)-enantiomeric forms, respectively.

This research was originally published in the *Journal of Biological Chemistry*: Umezawa et al. (1991). Formation of lignans, (–)-secoisolariciresinol and (–)-matairesinol with *Forsythia intermedia* cell-free extracts. *J. Biol. Chem.* 266:10210–10217. Copyright (1991) the American Society for Biochemistry and Molecular Biology.

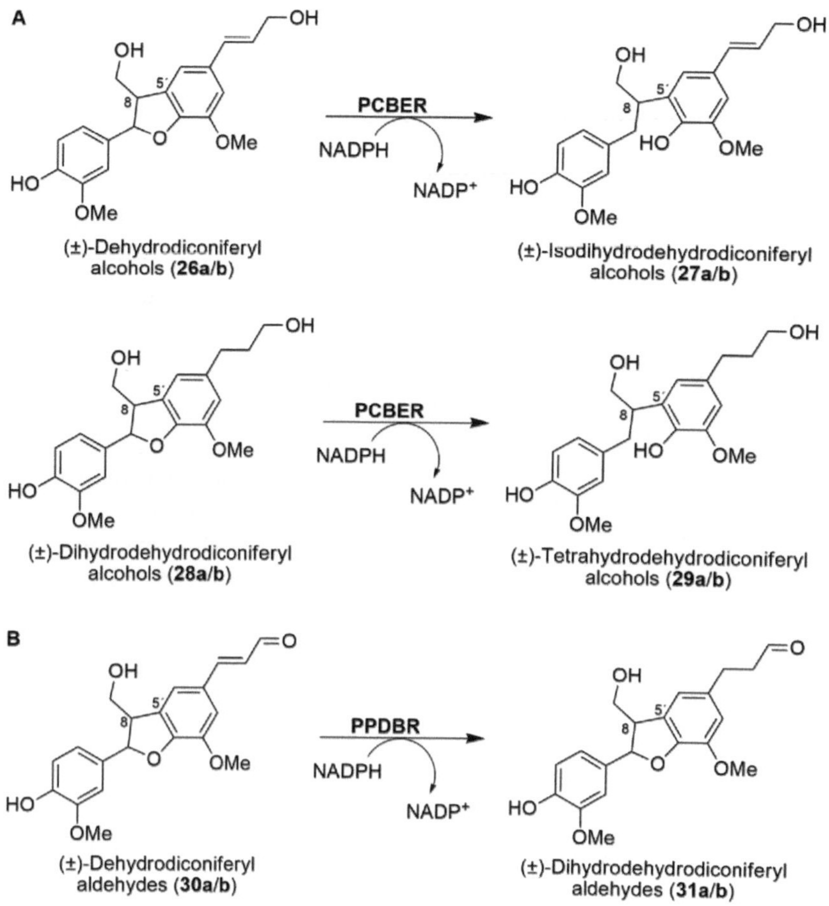

FIGURE 11.9 Reactions catalyzed by phenylcoumaran benzylic ether reductase (PCBER) and phenylpropenal double-bond reductase (PPDBR) (Gang et al. 1999, Kasahara et al. 2006). "**a/b**" refers to racemates.

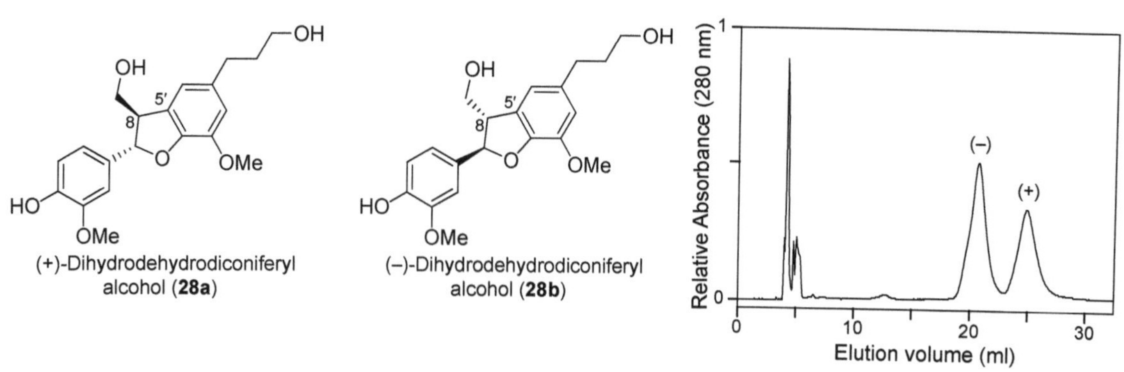

FIGURE 11.10 Chiral-phase HPLC analysis of (+)- and (–)-dihydrodehydrodiconiferyl alcohols (**28a** and **28b**). The (+)- and (–)-enantiomers **28a** and **28b** were resolved on a Chirobiotic V column eluted with an isocratic MeOH–NH_4NO_3 (1:9) solvent system at a flow rate of 1 ml min^{-1} with detection at 280 nm. "**a**" and "**b**" refer to the (+)- and (–)-enantiomeric forms, respectively.

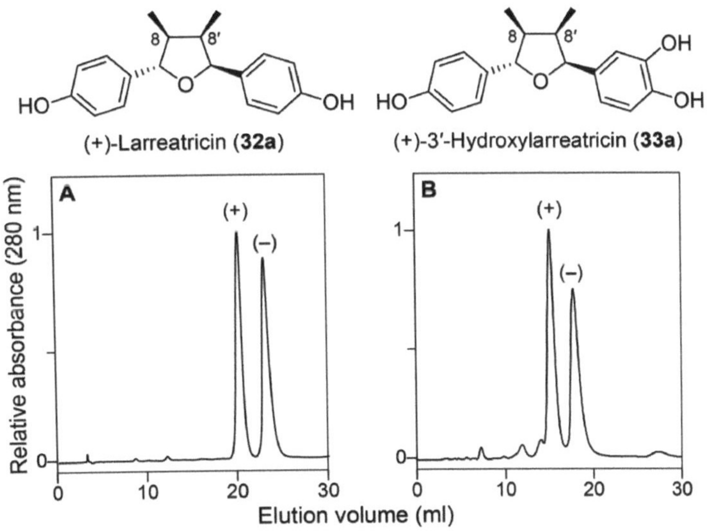

FIGURE 11.11 Chiral-phase HPLC analyses of: (A) (±)-larreatricins (**32a/b**) and (B) (±)-3′-hydroxylarreatricins (**33a/b**). Chiral separations used a Chirobiotic V column (Advanced Separation Technologies; 250 × 4.6 mm) eluted with hexanes–ethanol (88:12 for **32a/b** and 4:1 for **33a/b**; flow rate of 1 ml min^{-1}). "**a**" and "**b**" refer to the (+)- and (–)-enantiomeric forms, respectively.

This research was originally published in the *Proceedings of the National Academy of Sciences, USA*. Cho et al. (2003) (+)-Larreatricin hydroxylase, an enantio-specific polyphenol oxidase from the creosote bush (*Larrea tridentata*). *Proc. Natl. Acad. Sci. USA*. 2003; 100:10641–10646 Copyright (2003) National Academy of Sciences.

facilitated by the chiral-phase separation and analysis of the different chiral forms of substrate and product (Figure 11.11).

11.4 CIRCULAR DICHROISM ANALYSIS OF LIGNANS

The following examples serve to demonstrate the utility of both ECD (UV related), and VCD (IR related) spectroscopy to define the absolute configuration of different classes of naturally occurring and synthetically obtained lignans.

11.4.1 1,4-Benzodioxane Lignans

Enantioselective synthesis of a series of 1,4 benzodioxane lignans, (–)-eusiderin A (**34b**), (–)-eusiderin B (**35b**), (+)-eusiderin C (**36a**), and (–)-eusiderin M (**37b**) (Figure 11.12), and their subsequent ECD data analysis, indicated that analogues with an (8*R*) absolute configuration consistently exhibited a negative Cotton effect in the 230–260 nm region of the ECD spectrum (Pilkington and Barker 2012). The ECD spectra of (–)-eusiderin B (**35b**), its brominated analog (**38b**), and the brominated 7*R* homolog (**38a**), are shown in Figure 11.13. Combined with the ^{1}H NMR spectroscopic data, this also allowed for definition of the C-7 absolute configuration.

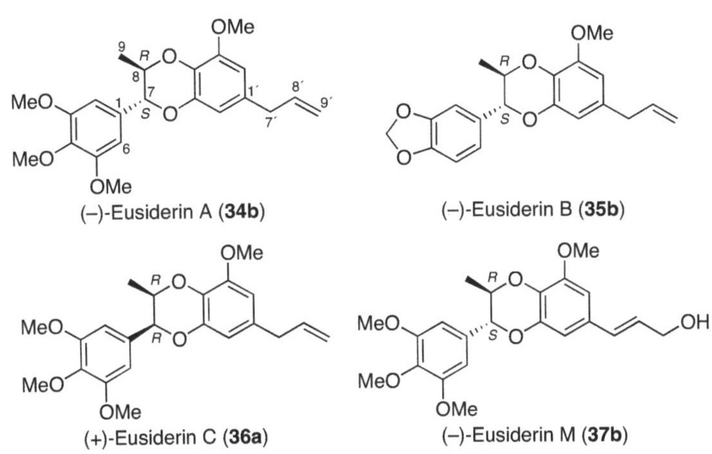

FIGURE 11.12 1,4-Benzodioxane lignans: (–)-eusiderins A (**34b**), B (**35b**), and M (**37b**), and (+)-eusiderin C (**36a**) (Pilkington and Barker 2012). "**a**" and "**b**" refer to the (+)- and (–)-enantiomeric forms, respectively.

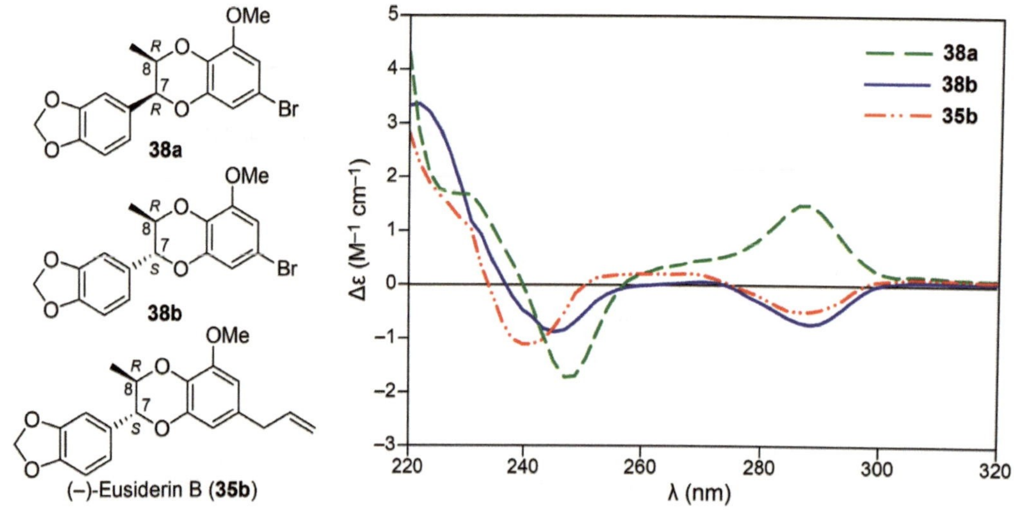

FIGURE 11.13 ECD spectra (MeOH) of **38a**, **38b** and (–)-eusiderin B (**35b**). "**a**" and "**b**" refer to the (+)- and (–)-enantiomeric forms, respectively.

Adapted with permission from Pilkington, L.I., and Barker, D. (2012). Asymmetric synthesis and CD investigation of the 1,4-benzodioxane lignans eusiderins A, B, C, G, L, and M. *J. Org. Chem.* 77:8156–8166. Copyright (2012) American Chemical Society.

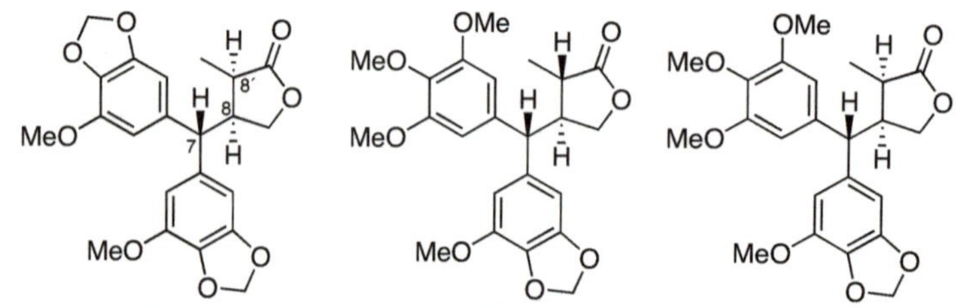

39 40 41

FIGURE 11.14 Seco-lignans from (**39–41**) *Peperomia blanda* (Felippe et al. 2012).

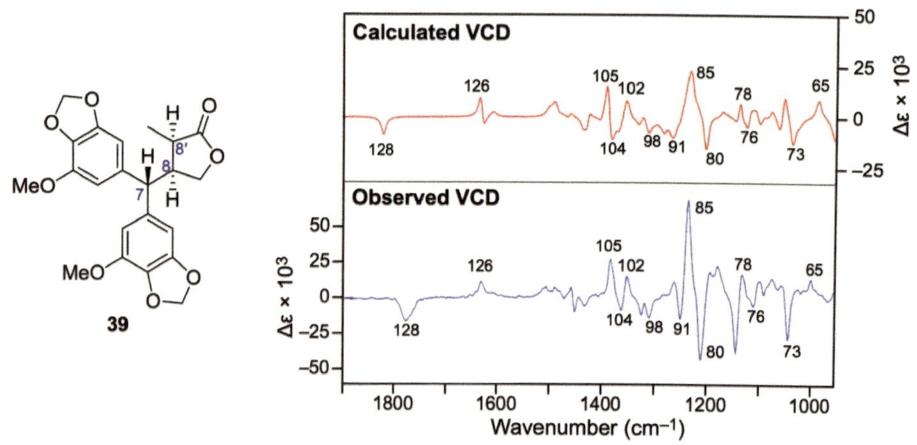

FIGURE 11.15 Comparison of the VCD spectra of the measured secolignan (–)-**39** from *Peperomia blanda* with the calculated [B3LYP/6-31G(d)] VCD of the Boltzmann average of the seven lowest-energy conformers of the corresponding (8*R*,8′*S*) **39**. The comparison establishes the absolute configuration of this molecule as (–)-(8*R*,8′*S*) **39**. Numbers represent fundamental vibrational modes.

Adapted with permission from Felippe et al. (2012). VCD to determine absolute configuration of natural product molecules: Secolignans from *Peperomia blanda*. *Org. Biomol. Chem.* 10:4208–4214. Copyright (2012) Royal Society of Chemistry.

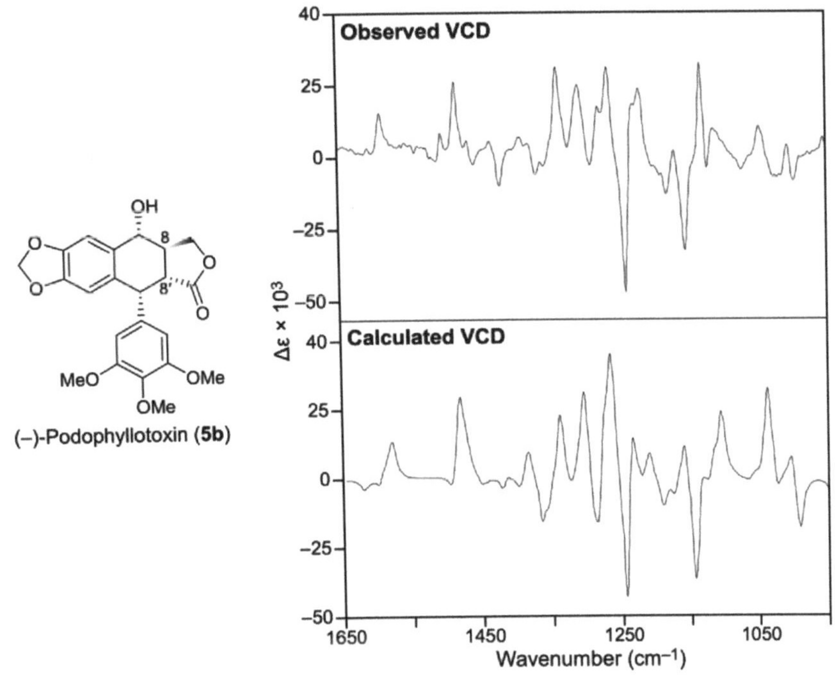

FIGURE 11.16 Comparison of the VCD spectra of the measured (–)-podophyllotoxin (**5b**) with the calculated B3LYP/DGDZVP DFT Boltzmann-weighted VCD. "**b**" refers to the (–)-enantiomeric form.

11.4.2 Seco-Lignans

Felippe et al. (2012) used both ECD and VCD data to assign the absolute configuration of three peperomin-type secolignans (**39–41**, Figure 11.14) isolated from *Peperomia blanda* (Felippe et al. 2012). Their results showed that ECD data alone were not sufficient to differentiate between C-7 diastereomers, a shortcoming that was resolved using VCD data (see Figure 11.15) as an example.

11.4.3 Aryltetralin Lactone-Type Lignans

(–)-Podophyllotoxin (**5b**), arguably currently the prime example of this class of lignans, and analogues thereof, have been subjected to both ECD (Sun et al. 2011) and VCD (Velázquez-Jiménez et al. 2011) spectroscopy for assignment of absolute configurations. The ECD data indicated that (–)-podophyllotoxin (**5b**) and its analogues with a (7′*R*) absolute configuration exhibited positive Cotton effects in the 280–290 nm region, and *vice versa* for those with a (7′*S*) absolute configuration.

Once such a conclusion as this is made, through-space NMR correlations may then be utilized to define the absolute configuration of the remaining stereocenters. VCD data were similarly applied to confirm the stereo-structures of podophyllotoxin-related lignans (see VCD data for (–)-podophyllotoxin (**5b**) in Figure 11.16).

11.4.4 Dibenzocyclooctadiene and Related Lignans

A series of dibenzocyclooctadiene lignans [e.g. (**42**)] and structural analogs [e.g. (**43**)] (Figure 11.17) have been identified in *Kadsura longipedunculata* (Wang et al. 2018, Liu et al. 2019). Compounds containing a biphenyl structural moiety possess ECD spectra reminiscent of this exciton-coupled chromophore (see Figure 11.18), while those of analogues related to (**43**) obviously lack this chromophore.

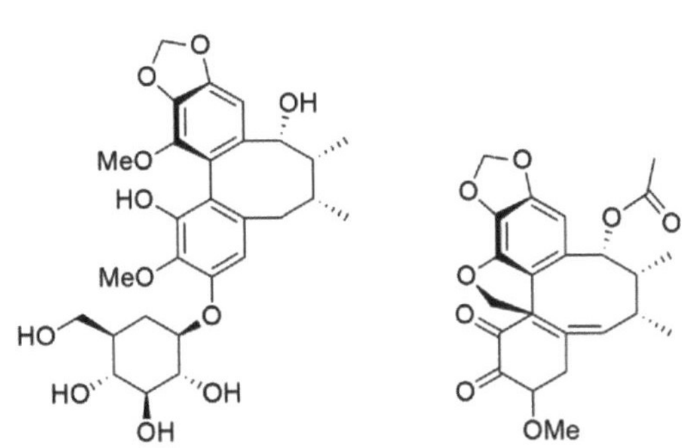

42 **43**

FIGURE 11.17 Dibenzocyclooctadienes **42** and **43** from *Kadsura longipedunculata* (Liu et al. 2019).

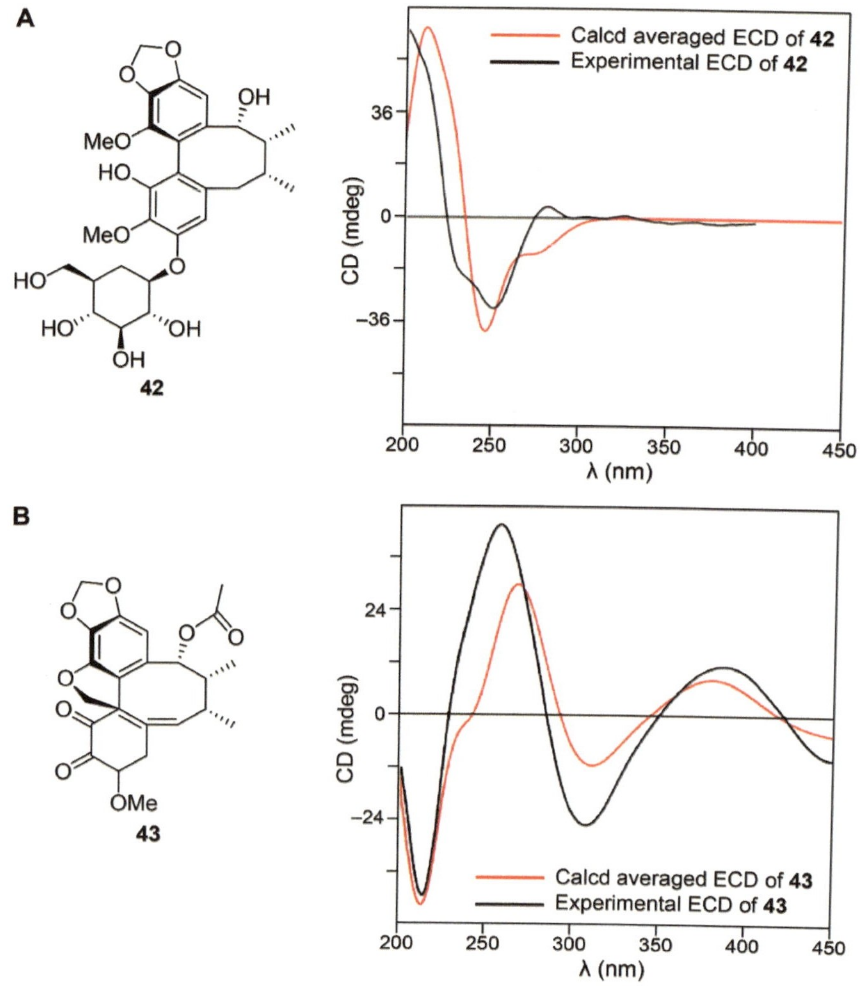

FIGURE 11.18 **A.** Experimental (black) and calculated (red) averaged Boltzmann-weighted ECD spectra of compounds **42** (A) and **43** (B) in MeOH. The σ-value (artificial line broadening) was set to 0.17 eV for **42** and 0.21 eV for **43**.

Adapted with permission from Liu et al. (2019) Hepatoprotective tetrahydrobenzocyclooctabenzofuranone lignans from *Kadsura longipedunculata*. *J. Nat. Prod.* 82:2842–2851. Copyright (2019) American Chemical Society and American Society of Pharmacognosy.

Their complex ECD data have recently been interpreted in terms of not only the absolute configurations of the biphenyl chromophore, but also of the remaining stereocenters.

11.5 CONCLUDING REMARKS

In sum, chiral analyses and chiral separations, including determinations of stereochemistry and absolute configurations, of lignans are arguably essential for all types of lignan research studies, e.g. such as for gaining a better understanding of the biological, physiological, and human health properties of members of the (oligo)lignan natural product class.

Indeed, our scientific capabilities are now such that the full power of EDC and VCD data can be fully utilized for unequivocal assignment of the absolute configurations of all classes of plant secondary metabolites, lignans included.

Moreover, the studies here can be viewed from a more positive perspective for the future. The purpose of this Lignan Chirality Roadmap is to encourage all researchers studying (oligo)lignans to benefit from such scientific advances and technologies that are available today to address lignan chirality questions. Such approaches can provide important information on not only chirality, but also as regards biological/physiological importance.

Routine application now and in the future of such technologies will thus help determine the important ramifications of specific stereogenic centers and chiral forms in lignans, as well as help allow for hypotheses and conclusions to be made from the perspective of the 3D world that we live in.

In this context, while focusing upon lignan chirality in this contribution, a thorough literature search will additionally reveal a wealth of information to assist the reader in solving such often difficult and complicated problems associated with stereogenic centers and chirality.

Yet, as noted herein, many research studies currently do not employ such approaches. This places the power of such analyses in danger to some extent of being a lost scientific art, if allowed to continue in this way in the literature.

ACKNOWLEDGEMENTS

The authors thank the Chemical Sciences, Geosciences and Biosciences Division, DOE Office of Basic Energy Sciences (DE-FG-0397ER20259), as well as the Arthur and Katie Eisig-Todc Endowment.

REFERENCES

Bada, J. L., J. R. Cronin, M.-S. Ho, et al. 1983. On the reported optical activity of amino acids in the Murchison meteorite. *Nature* 301:494–496.

Brazil, R. 2015. The origin of homochirality. Chemistry World. https://www.chemistryworld.com/features/the-origin-of-homochirality/9073.article#/ (accessed October 15, 2019).

Caglioti, L., K. Micskei, and G. Pályi. 2011. First molecules, biological chirality, origin(s) of life. *Chirality* 23:65–68.

Chen, F., Y. Tobimatsu, D. Havkin-Frenkel, R. A. Dixon, and J. Ralph. 2012. A polymer of caffeyl alcohol in plant seeds. *Proc. Natl. Acad. Sci. U. S. A.* 109:1772–1777.

Cho, M.-H., S. G. A. Moinuddin, G. L. Helms, et al. 2003. (+)-Larreatricin hydroxylase, an enantio-specific polyphenol oxidase from the creosote bush (*Larrea tridentata*). *Proc. Natl. Acad. Sci. USA* 100:10641–10646.

Chu, A., A. Dinkova, L. B. Davin, D. L. Bedgar, and N. G. Lewis. 1993. Stereospecificity of (+)-pinoresinol and (+)-lariciresinol reductases from *Forsythia intermedia*. *J. Biol. Chem.* 268:27026–27033.

Corbin, C., S. Drouet, L. Markulin, et al. 2018. A genome-wide analysis of the flax (*Linum usitatissimum* L.) dirigent protein family: From gene identification and evolution to differential regulation. *Plant Mol. Biol.* 97:73–101.

Dalisay, D. S., K.-W. Kim, C. Lee, et al. 2015. Dirigent protein-mediated lignan and cyanogenic glucoside formation in flax seed: Integrated omics and MALDI mass spectrometry imaging. *J. Nat. Prod.* 78:1231–1242.

Davin, L. B., T. Umezawa, and N. G. Lewis. 1991. Enantioselective separations in phytochemistry. In *Recent Advances in Phytochemistry* (N. K. Fischer, M. B. Isman and H. A. Stafford eds.), Vol. 25, Plenum Press, New York, NY. pp. 75–112.

Davin, L. B., D. L. Bedgar, T. Katayama, and N. G. Lewis. 1992. On the stereoselective synthesis of (+)-pinoresinol in *Forsythia suspensa* from its achiral precursor, coniferyl alcohol. *Phytochemistry* 31:3869–3874.

Davin, L. B., H.-W. Wang, A. L. Crowell, et al. 1997. Stereoselective bimolecular phenoxy radical coupling by an auxiliary (dirigent) protein without an active center. *Science* 275:362–366.

Davin, L. B., C.-Z. Wang, G. L. Helms, and N. G. Lewis. 2003. [^{13}C]-Specific labeling of 8–2′ linked (−)-*cis*-blechnic, (−)-*trans*-blechnic and (−)-brainic acids in the fern *Blechnum spicant*. *Phytochemistry* 62:501–511.

Davin, L. B., and N. G. Lewis. 2005. Lignin primary structures and dirigent sites. *Curr. Opin. Biotechnol.* 16:407–415.

Davin, L. B., M. Jourdes, A. M. Patten, et al. 2008. Dissection of lignin macromolecular configuration and assembly: Comparison to related biochemical processes in allyl/propenyl phenol and lignan biosynthesis *Nat. Prod. Rep.* 25:1015–1090.

Davin, L. B., and N. G. Lewis. 2022. 8–8′-Lignan biosynthesis roadmap. In *The Lignan Handbook* (N. G. Lewis and L. B. Davin eds.), Taylor & Francis, Boca Raton, London, pp 79–112.

Dei, S., A. Bartolini, C. Bellucci, et al. 1997. Differential analgesic activity of the enantiomers of atropine derivatives does not correlate with their muscarinic subtype selectivity. *Eur. J. Med. Chem.* 32:595–605.

Dima, O., K. Morreel, B. Vanholme, et al. 2015. Small glycosylated lignin oligomers are stored in *Arabidopsis* leaf vacuoles. *Plant Cell* 27:695–710.

Dinkova-Kostova, A. T., D. R. Gang, L. B. Davin, et al. 1996. (+)-Pinoresinol/(+)-lariciresinol reductase from *Forsythia intermedia*: Protein purification, cDNA cloning, heterologous expression and comparison to isoflavone reductase. *J. Biol. Chem.* 271:29473–29482.

Eccles, R., D. H. Griffiths, C. G. Newton, and N. S. Tolley. 1988. The effects of D and L isomers of menthol upon nasal sensation of airflow. *J. Laryngol. Otol.* 102:506–508.

Engel, M. H., and B. Nagy. 1982. Distribution and enantiomeric composition of amino acids in the Murchison meteorite. *Nature* 296:837–840.

Felippe, L. G., J. M. Batista, Jr., D. C. Baldoqui, et al. 2012. VCD to determine absolute configuration of natural product molecules: Secolignans from *Peperomia blanda*. *Org. Biomol. Chem.* 10:4208–4214.

Fujita, M., D. R. Gang, L. B. Davin, and N. G. Lewis. 1999. Recombinant pinoresinol-lariciresinol reductases from western red cedar (*Thuja plicata*) catalyze opposite enantiospecific conversions. *J. Biol. Chem.* 274:618–627.

Fuß, W. 2009. Does life originate from a single molecule? *Chirality* 21:299–304.

Gang, D. R., H. Kasahara, Z.-Q. Xia, et al. 1999. Evolution of plant defense mechanisms. Relationships of phenylcoumaran benzylic ether reductases to pinoresinol-lariciresinol and isoflavone reductases. *J. Biol. Chem.* 274:7516–7527.

Hande, K. R. 1998. Etoposide: Four decades of development of a topoisomerase II inhibitor. *Eur. J. Cancer* 34:1514–1521.

Hillis, W. E. 1999. The formation of heartwood and its extractives: An overview. In *Recent Advances in Phytochemistry* (J. T. Romeo ed.), Vol. 33, Kluwer Academic/Plenum Publishers, Boston, MA. pp. 215–253.

Hosmani, P. S., T. Kamiya, J. Danku, et al. 2013. Dirigent domain-containing protein is part of the machinery required for formation of the lignin-based Casparian strip in the root. *Proc. Natl. Acad. Sci. USA* 110:14498–14503.

Hwang, J. K., S. G. A. Moinuddin, L. B. Davin, and N. G. Lewis. 2020. Pinoresinol-lariciresinol reductase: Substrate versatility, enantiospecificity, and kinetic properties. *Chirality* 32:770–789.

Jeffries, O., and M. Waldron. 2019. The effects of menthol on exercise performance and thermal sensation: A meta-analysis. *J. Sci. Med. Sport* 22:707–715.

Kasahara, H., Y. Jiao, D. L. Bedgar, et al. 2006. *Pinus taeda* phenylpropenal double-bond reductase: Purification, cDNA cloning, heterologous expression in *Escherichia coli*, and subcellular localization in *P. taeda*. *Phytochemistry* 67:1765–1780.

Katayama, T., L. B. Davin, and N. G. Lewis. 1992. An extraordinary accumulation of (−)-pinoresinol in cell-free extracts of *Forsythia intermedia*: Evidence for enantiospecific reduction of (+)-pinoresinol. *Phytochemistry* 31:3875–3881.

Kim, K.-W., S. G. A. Moinuddin, K. M. Atwell, et al. 2012. Opposite stereoselectivities of dirigent proteins in *Arabidopsis* and *Schizandra* species. *J. Biol. Chem.* 287:33957–33972.

Leitereg, T. J., D. G. Guadagni, J. Harris, T. R. Mon, and R. Teranishi. 1971. Chemical and sensory data supporting the difference between the odors of the enantiomeric carvones. *J. Agric. Food. Chem.* 19:785–787.

Lewis, N. G., and L. B. Davin. 1999. Lignans: Biosynthesis and function. In *Comprehensive Natural Products Chemistry* (Sir D. H. R. Barton, K. Nakanishi and O. Meth-Cohn eds.), Vol. 1, Elsevier, Oxford. pp. 639–712.

Lis-Balchin, M., R. J. Ochocka, S. Deans, M. Asztemborska, and S. Hart. 1996. Bioactivity of the enantiomers of limonene. *Med. Sci. Res.* 24:309–310.

Liu, J., P. Pandey, X. Wang, et al. 2019. Hepatoprotective tetrahydrobenzocyclooctabenzofuranone lignans from *Kadsura longipedunculata. J. Nat. Prod.* 82:2842–2851.

Longstaff, E., and G. von Krogh. 2001. *Condyloma* eradication: Self-therapy with 0.15–0.5% podophyllotoxin versus 20–25% podophyllin preparations: An integrated safety assessment. *Regul. Toxicol. Pharmacol.* 33:117–137.

McCombs, C. 2004. Evolution hopes you don't know chemistry: The problem with chirality. Institute for Creation Research. https://www.icr.org/article/evolution-hopes-you-dont-know-chemistry-problem-wi/ (accessed October 10, 2019).

Moinuddin, S. G. A., S. Hishiyama, M.-H. Cho, L. B. Davin, and N. G. Lewis. 2003. Synthesis and chiral HPLC analysis of the dibenzyltetrahydrofuran lignans, larreatricins, 8′-*epi*-larreatricins, 3,3′-didemethoxyverrucosins and *meso*-3,3′-didemethoxynectandrin B in the creosote bush (*Larrea tridentata*): Evidence for regiospecific control of coupling. *Org. Biomol. Chem.* 1:2307–2313.

Pilkington, L. I., and D. Barker. 2012. Asymmetric synthesis and CD investigation of the 1,4-benzodioxane lignans eusiderins A, B, C, G, L, and M. *J. Org. Chem.* 77:8156–8166.

Pizzarello, S. 2004. Chemical evolution and meteorites: An update. *Origins Life Evol. Biosphere* 34:25–34.

Ralph, J., K. Lundquist, G. Brunow, et al. 2004. Lignins: Natural polymers from oxidative coupling of 4-hydroxyphenylpropanoids. *Phytochem. Rev.* 3:29–60.

Ralph, J., G. Brunow, P. J. Harris, et al. 2008. Lignification: Are lignins biosynthesized via simple combinatorial chemistry or via proteinaceous control and template replication? In *Recent Advances in Polyphenol Research* (F. Daayf and V. Lattanzio eds.), Vol. 1, Blackwell, Oxford, UK. pp. 36–66.

Sarfati, J. 1998. Origin of life: The chirality problem. *Journal of Creation* 12:263–266.

Singer, E. 2014. New twist found in the story of life's start. *Quanta Magazine.* https://www.quantamagazine.org/chiral-key-found-to-origin-of-life-20141126/ (accessed October 15, 2019).

Stähelin, H. 1970. 4′-Demethyl-epipodophyllotoxin thenylidene glucoside (VM 26), a *Podophyllum* compound with a new mechanism of action. *Eur. J. Cancer* 6:303–311.

Stähelin, H. 1973. Activity of a new glycosidic lignan derivative (VP 16-213) related to podophyllotoxin in experimental tumors. *Eur. J. Cancer* 9:215–221.

Sun, Y.-J., Z.-L. Li, H. Chen, et al. 2011. Three new cytotoxic aryltetralin lignans from *Sinopodophyllum emodi. Bioorg. Med. Chem. Lett.* 21:3794–3797.

Umezawa, T., L. B. Davin, and N. G. Lewis. 1990a. Formation of the lignan, (–)-secoisolariciresinol by cell-free extracts of *Forsythia intermedia. Biochem. Biophys. Res. Commun.* 171:1008–1014.

Umezawa, T., L. B. Davin, E. Yamamoto, D. G. I. Kingston, and N. G. Lewis. 1990b. Lignan biosynthesis in *Forsythia* species. *J. Chem. Soc., Chem. Commun.*:1405–1408.

Umezawa, T., L. B. Davin, and N. G. Lewis. 1991. Formation of lignans, (–)-secoisolariciresinol and (–)-matairesinol with *Forsythia intermedia* cell-free extracts. *J. Biol. Chem.* 266:10210–10217.

Velázquez-Jiménez, R., J. M. Torres-Valencia, C. M. Cerda-García-Rojas, et al. 2011. Absolute configuration of podophyllotoxin related lignans from *Bursera fagaroides* using vibrational circular dichroism. *Phytochemistry* 72:2237–2243.

Wang, X., J. Liu, P. Pandey, et al. 2018. Computationally assisted assignment of the kadsuraols, a class of chemopreventive agents for the control of liver cancer. *Org. Lett.* 20:5559–5563.

Xia, Z.-Q., M. A. Costa, H. C. Pélissier, L. B. Davin, and N. G. Lewis. 2001. Secoisolariciresinol dehydrogenase purification, cloning and functional expression: Implications for human health protection. *J. Biol. Chem.* 276:12614–12623.

Index

Introduction to Lignan Dictionary

COVERAGE

This dictionary section of the *Lignan Handbook* contains physical, chemical and structural data on 12,000 lignans. It is a subset of the *Dictionary of Natural Products* (DNP) database, an ongoing project that continuously reviews and abstracts the natural product literature. For this book project, the subset of DNP entries referring to lignans were carefully checked and reviewed and enhanced with a considerable amount of additional information relating to their natural occurrence. The intention is that the Handbook should contain data on every known lignan, and various checks carried out during the compilation process indicate that it comes very close to achieving this aim. In compiling this print edition the primary literature has been reviewed to mid-2021. The Editors would be grateful to learn of any accidental omissions.

DATA PRESENTATION AND ORGANISATION

The dictionary section of the Handbook is arranged alphabetically by entry name. Every entry is numbered to assist ready location.

Derivatives and Variants

Many compounds are included as derivatives of main entry compounds. A representative entry is shown in Figure 1.

Variants are used to describe the configuration of stereoisomers, e.g. (7*R*,7′*R*,8*S*)-form.

Chemical Names and Synonyms

The Handbook contains a wide range of synonyms which may be (a) those found in the primary literature, (b) Chemical Abstracts names, or (c) names added editorially to achieve as much consistency as possible with other closely related substances.

Synonym coverage is intended as comprehensive except in the case of drugs in clinical use, which may have a very large number of trade names for marketed formulations.

All important derivatives embedded within entries are named. If a compound cannot be located immediately in the main body of the entries, it is important to use the indexes.

Names which are known to be duplicated within the chemical literature (not necessarily within the *Lignan Handbook*), are marked with the sign ‡.

Nomenclature

It is expected that this handbook will be used by a wide variety of scientists, not all of whom will be specialist organic chemists. Therefore, compounds have been named so as to facilitate access to their factual data by keeping the nomenclature as simple as possible, whilst still adhering to good practice by IUPAC (the International Union of Pure and Applied Chemistry). A great deal of care has been taken to achieve this aim as near as possible.

1. There are many examples in the primary literature of compounds being named in ways which are violations of good IUPAC practice, e.g., where the substituents are ordered non-alphabetically. These are not reported in their incorrect form but have been corrected.
2. The number of trivial names used for acylating substituents has been kept to a minimum, but the following are used throughout the Handbook:
   Tigloyl = 2-methyl-2*E*-butenoyl
   Angeloyl = 2-methyl-2*Z*-butenoyl
   Cinnamoyl = 3-phenyl-2-propenoyl
3. Many other trivial appellations have from time to time appeared in the literature for other acyl groups. e.g.
   Senecioyl = 3-methyl-2-butenoyl,
   Feruloyl = 3-(4-hydroxy-3-methoxyphenyl)-2-propenoyl or 4-hydroxy-3-methoxycinnamoyl

The term **prenyl** for the common 3-methyl-2-butenyl substituent, $(H_3C)_2C{=}CHCH_2$-, is used throughout the Handbook. Several other names for this substituent have from time to time been used and appear in the primary literature including the following:

3-Methyl-2-butenyl (systematic)
γ, γ-Dimethylallyl
3,3-Dimethylallyl
3,3-Dimethylpropenyl
Dimethallyl
Isoprenyl
Isopentenyl
β,β-Dimethylacrylyl

CAS Registry Numbers

CAS Registry Numbers are identifying numbers allocated to each distinctly definable chemical substance indexed by the Chemical Abstracts Service since 1965 (plus retrospective allocation of numbers by CAS to earlier compounds). The numbers have no chemical significance, but they provide a label for each substance independent of any system of nomenclature.

Much effort has been expended to ensure that accurate CAS numbers are given in the Handbook for as many substances as possible.

Entry name → **2,7′-Cyclo-9,9′-epoxy-8,8′-lignan-3,3′,4,4′,5,5′-hexol** **C157** ← Entry number

Structural formula and stereochemical descriptor →

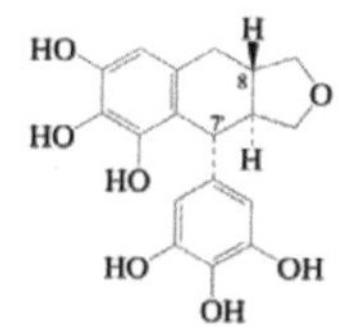

(7′R,8S,8′S)-form

9,9′-Epoxy-3,3′,4,4′,5,5′-hexahydroxy-2,7′-cyclolignan

Molecular formula → $C_{18}H_{18}O_7$ 346.335 ← Molecular weight

(7′R,8S,8′S)-form

Derivative heading → *Hexa-Me ether*

2,7′-Cyclo-9,9′-epoxy-3,3′,4,4′,5,5′-hexamethoxylignan. 9,9′-Epoxy-3,3′,4,4′,5,5′-hexamethoxy-2,7′-cyclolignan. ***Aglacin B*** ← Derivative synonyms

CAS registry number → [389628-93-7]

$C_{24}H_{30}O_7$ 430.497

Constit. of the stem bark of *Aglaia cordata*. Shows weak inhibition of human KB carcinoma cell line. ← Biological Source and other information

Needles (MeOH). Mp 138 - 139°. $[\alpha]_D^{20}$ +45 (c, 0.38 in $CHCl_3$). λ_{max} 212 (log ε 4.81) (MeOH) ← Physical data

Stereochemical descriptor → ***(7′S,8S,8′S)-form***

3,3′,5,5′-Tetra-Me ether

2,7′-Cyclo-9,9′-epoxy-3,3′,5,5′-tetramethoxy-8,8′-lignan-4,4′-diol. 9,9′-Epoxy-4,4′-dihydroxy-3,3′,5,5′-tetramethoxy-2,7′-cyclolignan. ***Gaultherin C***

[267665-74-7]

$C_{22}H_{26}O_7$ 402.443

Constit. of the roots of *Gaultheria yunnanensis*. Prisms (MeOH). Mp 185 - 186°. $[\alpha]_D^{21}$ +56 (c, 0.02 in MeOH). λ_{max} 219 (log ε 2.21); 276 (log ε 0.26) (MeOH).

Zhang, Z. *et al.*, *Heterocycles*, 2000, **53**, 675-679 (*Gaultherin C*).

Bibliographic references → Wang, B.-G. *et al.*, *J. Nat. Prod.*, 2001, **64**, 1521-1526 (*Aglacin B, cryst struct*). ← Reference tags

Xu, M. *et al.*, *Angew. Chem., Int. Ed.*, 2021, **60**, 16655-16660 (*Aglacin B, synth*) ← Reference tags

FIGURE 1

At the foot of the entry, immediately before the references, may be shown additional registry numbers. These are numbers which have been recognised by the editors as belonging to the entry concerned but which cannot be unequivocally assigned to any of the compounds covered by the entry. Their main use will be in helping those who need to carry out additional searches, especially online searches in the CAS or other databases, and who will be able to obtain additional hits using these numbers. Clearly, discretion is needed in their use for this purpose. Additional registry numbers may arise for a variety of reasons and many of them are labelled to indicate their attribution.

Further information on CAS number allocation policy can be obtained from CAS indexes or *The Organic Chemist's Desk Reference, Third Edition.*

Structural Formulae

Every attempt has been made to present the structures of chemical substances as accurately as possible according to best current practice and recommendations of IUPAC. As much consistency as possible has been aimed at between closely-related structures, for example, all sugars are shown as Haworth formulae, and whenever possible in complex structures the rings are oriented in the standard Haworth convention so that structural comparisons can be quickly made.

Stereochemical Conventions

Much attention has been paid to standardising the representations of chemical structures of related substances and to resolve uncertainties and irregularities in the published literature. Where the absolute configuration of a compound is known or can be inferred from this is indicated. Where only one stereoisomer is referred to in the text, the structural diagram indicates that stereoisomer. Wherever possible, stereostructures are described using the Cahn-Ingold-Prelog sequence-rule (*R,S*) and (*E,Z*) conventions.

Source

Taxonomic names and strain numbers for organisms given throughout are, in general, those given in the primary literature. Standardisation of minor orthographical variations has been carried out. The standard used is the *Catalogue of Life*, https://www.catalogueoflife.org/

Molecular Formula and Molecular Weight

The elements in the molecular formula are given according to the Hill convention (C, H, then other elements in alphabetical order). The molecular weights given are formula weights (or more strictly, molar masses in Daltons) and are rounded to three places in decimals. In the case of some high molecular mass substances, such as proteins, the value quoted may be that taken from an original literature source and may be an aggregate molar mass.

Physical Data

The Handbook gives the following physical characteristics of substances, when available; appearance, melting point, boiling point, optical rotation, density, refractive index, solubility, p*K*a.

Appearance

Organic compounds are considered to be colourless unless otherwise stated. Where the compound contains a chromophore which would be expected to lead to visible colour, but no colour is mentioned in the literature, the Handbook entry will mention this fact if it has been noticed by the contributor. An indication of crystal form and recrystallisation solvent is often given but these are imprecise items of data; most compounds can be crystallised from several solvent systems and the crystal form often varies. In the case of the small number of compounds where crystal behaviour has been intensively studied (e.g., pharmaceuticals), it is found that polymorphism is a very common phenomenon and there is no reason to believe that it is not widespread among organic compounds generally.

Melting Points and Boiling Points

The policy followed in the case of conflicting data is as follows:

(a) Where the literature melting points are closely similar, only one figure (the highest or most probable) is quoted
(b) Where two or more melting points are recorded and differ by several degrees (the most likely explanation being that one sample was impure) the lower figure is given in parentheses, thus Mp 139° (134–135°)
(c) Where quoted figures differ widely and some other explanation such as polymorphism or incorrect identity seems the most likely explanation, both figures are quoted without parentheses, thus Mp 142°, Mp 205–206°
(d) Known cases of polymorphism or double melting points are noted

Boiling point determination is less precise than that of melting points and conflicting boiling point data is not usually reported except when there appears to be a serious discrepancy between the different authors.

Optical Rotations

These are given wherever possible, and normally refer to what the contributor believes to be the best characterised sample of highest chemical and optical purity. Where available an indication of the optical purity (op) or enantiomeric excess (ee) of the sample measured follows the specific rotation value.

Spectroscopic Data

Many Handbook entries include ultraviolet spectra which are presented in the format:

[neutral] λ_{max} 198 (log ε 1.55); 224 (sh) (log ε 0.61); 241 (sh) (log ε 0.55) (H_2O) (Berdy)

where ε is the absorption coefficient for a given UV maxima value (λ_{max}). A description of the solvent conditions used, if reported in the literature, is listed at the beginning and end of the UV data in parentheses. All peak absorptions cited are maxima unless otherwise described, e.g., shoulder/inflection (sh) and end absorption (end). In addition, UV data may be followed by the term 'Berdy' or 'DEREP' indicating from which database the data originated. The absence of these terms implies that the data was abstracted from the primary literature.

Hazard and Toxicity Information

General

Toxicity and hazard information is highlighted by the symbol ► and has been selected to assist in risk assessments for experimental, manufacturing, and manipulative procedures with chemicals.

The Publishers cannot be held responsible for any inaccuracies in the reported information, neither does the omission of hazard data in the Handbook imply an absence of this data from the literature. Widely recognised hazards are included, however, and where possible key toxicity reviews are identified in the references. Further advice on the storage, handling and disposal of chemicals is given in *The Organic Chemist's Desk Reference, Third Edition.*

RTECS® Accession Numbers*

Many entries in this Handbook contain one or more RTECS® Accession Numbers. Possession of these numbers allows users to locate toxicity information on relevant substances from the NIOSH *Registry of Toxic Effects of Chemical Substances*, which is a compendium of toxicity data extracted from the scientific literature.

Bibliographic References

The selection of references is made with the aim of facilitating entry into the literature for the user who wishes to locate more detailed information about a particular compound. The contents of most references are indicated by reference tags (suffixes) indicating their content and in particular the stereoisomers and derivatives of the parent compound which they document. The number of references cited does not indicate the relative importance of a compound; one key recent citation may supersede several older ones.

Further References

Further useful information on a variety of topics concerned with the structure, description, stereochemistry, and nomenclature of organic compounds can be found in the *Organic Chemist's Desk Reference, Third Edition.*

INDEXES

There are two printed indexes

1. **Name Index** which lists every compound name and synonym in the dictionary section
2. **Type of Compound Index** listing all compounds given in the dictionary section, organised by structural type.

ABBREVIATIONS

The following is a selection of the most common Handbook abbreviations used:

Abbreviation	Name
[α]	specific rotation
abs config	absolute configuration
Ac	acetyl
acc	according
AcOH	acetic acid
Ac_2O	acetic anhydride
ACGIH	American Conference of Governmental Industrial Hygienists
ADI	Acceptable Daily Intake
alk	alkaline
amorph	amorphous
amt	amount
anal	analytical applications, analysis of detection
anhyd	anhydrous
aq	aqueous
BAN	British Approved Name
bibl	bibliography
biosynth	biosynthesis
Bp	boiling point
c	concentration
ca	(*circa*) about
cd	circular dichroism
CAS	Chemical Abstracts Service
chromatog	chromatography
clin	clinical
cmr	carbon (^{13}C) nuclear magnetic resonance
CNS	central nervous system
col	colour, coloration
coml	commercial(ly)
compd	compound

* RTECS® Accession Numbers are compiled and distributed by the National Institute for Occupational Safety and Health Service of the U.S. Department of Health and Human Services of The United States of America. All rights reserved (1996).

Abbreviation	Name
conc	concentrated
config	configuration
conformn	conformation
constit	constituent
cryst struct	X-ray crystal structure determination
d	density
dck	duck
dec	decomposes, decomposition
degradn	degradation
deriv(s)	derivative(s)
detn	detection, determination
dil	dilute, dilution
dimorph	dimorphic
diss	dissolves, dissolved
dist(n)	distil, distillation
DMF	dimethylformamide
DMSO	dimethyl sulfoxide
ee	enantiomeric excess
epr	electron paramagnetic (spin) resonance
equilib	equilibrium
esp	especially
Et	ethyl
Et_2O	diethyl ether
EtOAc	ethyl acetate
EtOH	ethanol
EtOH aq	aqueous ethanol
exp	experimental
FEMA	Flavor and Extract Manufacturers' Association
fl p	flash point
fluor	fluoresces, fluorescence
formn	formation
Fp	freezing point
g	gram
glc	gas liquid chromatography
Glc	β-D-glucopyranosyl
GRAS	Generally Recognised As Safe
ham	hamster
haz	hazard
hplc	high performance liquid chromatography
hydrol	hydrolyses, hydrolysed, hydrolysis
ihl	inhalation
ims	intramuscular
INN	International Nonproprietary Name
intermed	intermediate
ipr	intraperitoneal
ir	infra-red spectrum
isol(n)	isolation, isolated
isom	isomerism, isomers, isomerises
ivn	intravenous
JAN	Japanese Accepted Name
JMAF	Japanese Ministry for Agriculture, Forestry and Fisheries
LC	lethal concentration
LD	lethal dose: LD_{50}, a dose which is lethal to 50% of the animals tested
lit	literature
manuf	manufacturer, manufactured
max	maximum
Me	methyl
MeOH	methanol
Me_2CO	acetone
MEL	maximum exposure limit
metab	metabolite, metabolism
misc	miscible
mixt	mixture
mod	moderately
Mp	melting point
ms	mass spectrum
mus	mouse
n	index of refraction, e.g. n_D^{20} for 20° and sodium light
nmr	nuclear magnetic resonance spectrum (general)
obt	obtained
oc	open cup
occup	occupational
OES	Occupational Exposure Standard
op	optical purity
ord	optical rotatory dispersion
orl	oral
Ph	phenyl (C_6H_5)
pharmacol	pharmacology
pmr	proton (^1H) nuclear magnetic resonance
polarog	polarography
polym	polymerises, polymer
ppd	precipitated
ppm	parts per million
prob	probably
props	properties
purifn	purification
Py	pyridine
rbt	rabbit
redn	reduction
ref	reference
rel	relatively
resoln	resolution
rt	room temperature
scu	subcutaneous
sepd	separated
skn	skin
sl	slightly
sol	soluble
soln	solution
solv	solvent
sp	species (singular)
spar	sparingly
spp	species (plural)
ssp	subspecies
stereochem	stereochemistry, stereochemical(ly)
struct	structure
subl	sublimation, sublimes
substit	substituent
substn	substitution
synth	synthesis

Abbreviation	Name
tautom	tautomerism
THF	tetrahydrofuran
tlc	thin layer chromatography
TLV	Threshold Limit Value
tox	toxicity
unsatd	unsaturated
USAN	United States Adopted Name
uv	ultraviolet spectrum
v	very
var	variety
vis	visible
vol	volume

Reference Abbreviations

The following is a selection of the most common abbreviations used in reference tags, which appear in parentheses following the citation:

Abbreviation	Name
abs config	absolute configuration
activity	biological activity
anal	analysis
bibl	bibliography
biosynth	biosynthesis
cd	circular dichroism
chromatog	chromatography
cmr	^{13}C nuclear magnetic resonance spectrum
config	configuration
conformn	conformation
cryst struct	X-ray crystal structure determination
deriv(s)	derivative(s)
detn	determination, detection
dsc	differential scanning calorimetry
dta	differential thermal analysis
ed	electron diffraction
electrochem	electrochemistry, cyclic voltammetry
em	electron microscopy
epr	electron paramagnetic (spin) resonance spectrum
esca	electron spectroscopy for chemical analysis
exafs	extended X-ray diffraction fine structure
gc-ms	gas chromatography-mass spectrometry
glc	gas-liquid chromatography
haz	hazard
hplc	high performance liquid chromatography
ir	infrared spectrum
isol	isolation
isom	isomerism
manuf	manufacture
metab	metabolism
mineral	mineralogy
ms	mass spectrum
nmr	nuclear magnetic resonance spectrum
occur	occurrence
ord	optical rotatory dispersion
pe	photoelectron spectroscopy
pharmacol	pharmacology
photol	photolysis
pmr	proton (^{1}H) nuclear magnetic resonance spectrum
polarog	polarography
powder struct	X-ray powder structure determination
props	properties (chemical or physical)
Raman	Raman spectrum
resoln	resolution
rev	review
sepn	separation
soly	solubility
spectra	
struct	structure
synonyms	
synth	synthesis
tautom	tautomerism
tga	thermogravimetric analysis
theory	MO calculations etc.
tlc	thin layer chromatography
tox	toxicity
trans	transition(s)
use(s)	
uv	ultraviolet spectrum
uv-vis	ultraviolet visible spectrum

TYPE OF COMPOUND CLASSIFICATION

All lignans are classified under one or more codes according to structural type. Each structural type is assigned as a type of compound code, e.g., V.O.02500, V.O.08000.

The prefix V is specific to the *Dictionary of Natural Products*, and the median letter O refers to the lignan subset.

A list of the lignan type of compound codes used is as follows:

V.O.00200	Lignan monomers
V.O.00500	Simple dibenzylbutane 8,8′-lignans
V.O.01500	Saturated dibenzylbutyrolactone and -lactol 8,8′-lignans
V.O.02000	Unsaturated dibenzylbutyrolactone and -lactol 8,8′-lignans
V.O.02500	9,9′-Epoxyfuranoid 8,8′-lignans
V.O.03000	7,9′-Epoxyfuranoid 8,8′-lignans

V.O.03500	7,7′-Epoxyfuranoid 8,8′-lignans
V.O.04000	Simple furofuranoid 8,8′-lignans
V.O.04500	Side-chain oxygenated furofuranoid 8,8′-lignans
V.O.04700	Miscellaneous furofuranoid 8,8′-lignans
V.O.04800	1,2′-Cyclo-8,8′-lignans (Eupodienones)
V.O.05000	Simple 2,7′-cyclo-8,8′-lignans (arylnaphthalenes)
V.O.05500	2,7′-Cyclo-8,8′-lignans (side-chain oxygenated)
V.O.06000	2,7′-Cyclo-9,9′-epoxy-8,8′-lignans
V.O.06200	2,7′-Cyclo-8,8′-lignan-9,9′-olides
V.O.06300	2,7′-Cyclo-8,8′-lignan-9′,9-olides
V.O.06400	Miscellaneous naphthalenoid lignans
V.O.06700	7,7′/7,8′-Cyclolignans (cyclobutanes)
V.O.07500	2,2′-Cyclo-8,8′-lignans (dibenzocyclooctadienes)
V.O.08000	Norlignans
V.O.08500	Homolignans
V.O.09000	Secolignans
V.O.09500	Abeolignans
V.O.10000	8,8′-Sesquilignans
V.O.10250	Other sesquilignans
V.O.10750	Sesterlignans
V.O.12000	8,8′-Bilignans
V.O.12500	Other bilignans
V.O.16000	Flavonolignans
V.O.16500	Stilbenolignans
V.O.17000	Coumarinolignans
V.O.18000	Xanthonolignans
V.O.18500	Terpenoid lignans
V.O.19000	Benzodioxin-containing lignans
V.O.19500	Alkaloidal lignans
V.O.21800	1,8′-Lignans
V.O.23300	3,3′-Lignans
V.O.23800	3,8′-Lignans
V.O.27800	7,8′-Lignans
V.O.28900	8,9′-Lignans
V.O.29990	Other simple lignans
V.O.33700	3,7′-Cyclo-1,8′-lignans
V.O.35387	5,8′-Cyclo-3,7′-lignans
V.O.39990	Other cyclolignans
V.O.42700	2,7′-Epoxy-1,8′-lignans
V.O.42730	2,7′-Epoxy-3,8′-lignans
V.O.42800	2,8′-Epoxy-1,7′-lignans
V.O.43700	3,7′-Epoxy-2,8′-lignans
V.O.43740	3,7′-Epoxy-4,8′-lignans
V.O.44700	4,7′-Epoxy-3,8′-lignans
V.O.49990	Other simple epoxylignans
V.O.53700	3,7′-Epoxy-4,8′-oxylignans
V.O.54700	4,7′-Epoxy-3,8′-oxylignans
V.O.63300	3,3′-Oxylignans
V.O.63400	3,4′-Oxylignan
V.O.64400	4,4′-Oxylignans
V.O.64800	4,8′-Oxylignans
V.O.69900	Other simple oxylignans
V.O.70500	Rosmarinic/Salvianolic acid lignans
V.O.80000	Linear diarylheptanoids
V.O.80100	Diphenyl ether diarylheptanoids
V.O.80200	Biphenyl diarylheptanoids
V.O.80300	Tetrahydropyranoid diarylheptanoids
V.O.80500	Miscellaneous diarylheptanoids
V.O.99990	Lignans of unknown structure

FEEDBACK

Any comments and suggestions for inclusion may be sent to:

The Editors, The Handbook of Lignans
CRC Press/Taylor & Francis Group
18th Floor
240 Blackfriars Road
London SE1 8BU

Email: steve.walford@informa.com

THE IN HOUSE EDITORIAL TEAM

Dr I. Dager, CRC Press, London
Prof. R.A. Hill, University of Glasgow
Dr A.M. Hodgson, Consultant
Dr V.R.N. Munasinghe, Consultant
Mr A.D. Roberts, Consultant
Dr S.P. Walford, CRC Press, London
Dr J.G. Ward, Consultant

Additional data collected by contributors to the *CRC Chemical Database.*

Abiesol B A1

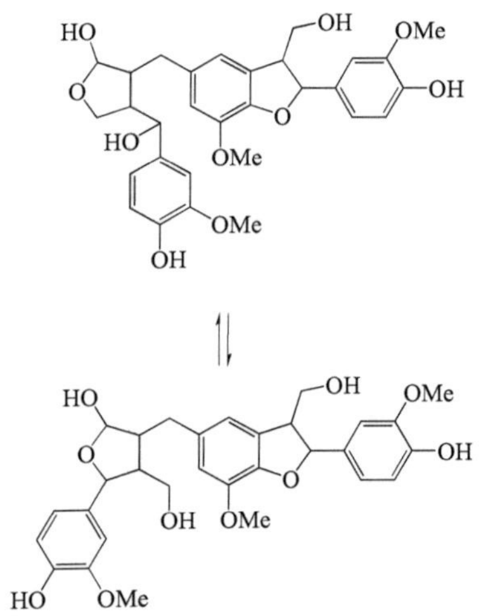

[136037-28-0]

$C_{30}H_{34}O_{10}$ 554.592

A hemiacetal in equilib. via the open-chain form. Isol. from wood of *Abies sachalinensis*. $[\alpha]^{23}_{D}$ -2.4 (c, 1.66 in MeOH).

Ozawa, S. *et al.*, *Mokuzai Gakkaishi*, 1991, **37**, 69-75 (*Abiesol B*).

Ozawa, S. *et al.*, *Stud. Nat. Prod. Chem.*, 1997, **20**, 613-657 (*rev*)

Acanthopanax A A2

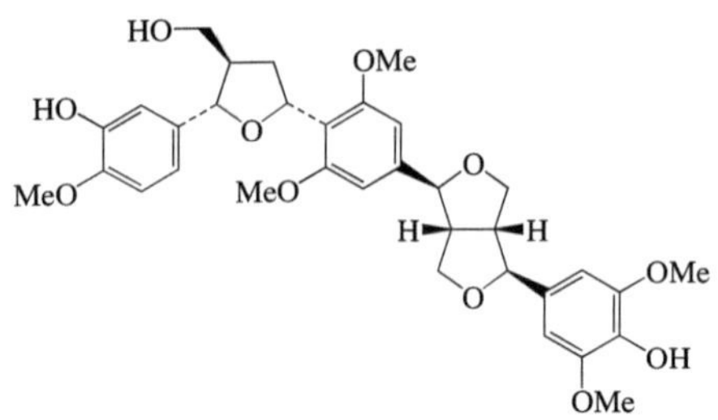

Absolute Configuration

$C_{34}H_{40}O_{11}$ 624.683

Constit. of the stems of *Acanthopanax senticosus*. Selective inhibitor of diacylglycerol acyltransferase 2. Powder. $[\alpha]^{25}_{D}$ -29.4 (c, 0.1 in MeOH). λ_{max} 214 (log ε 4.43); 247 (log ε 3.98); 289 (log ε 3.28) (MeOH).

Li, J.-L. *et al.*, *Fitoterapia*, 2016, **109**, 185-189 (*Acanthopanax A, cd, activity*)

Acanthotoxin A3

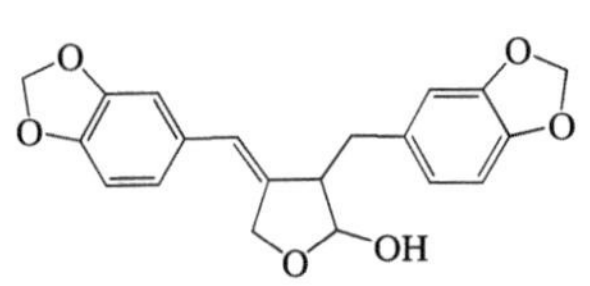

3-(1,3-Benzodioxol-5-ylmethyl)-4-(1,3-benzodioxol-5-ylmethylene) tetrahydro-2-furanol, CAS. *9,9′-Epoxy-3,4:3′,4′-bis(methylenedioxy)-8,8′-lign-7′-en-9-ol. 9-Hydroxy-3,4:3′,4′-bis(methylenedioxy)-9,9′-epoxylign-7′-ene*

[63831-09-4]

$C_{20}H_{18}O_6$ 354.358

Constit. of the stem bark of *Zanthoxylum acanthopodium*. Germination inhibitor. Mp 123°. $[\alpha]_{D}$ -133.7 ($CHCl_3$). λ_{max} 236 (ε 34600); 285 (ε 28200) (EtOH) (Berdy).

Roy, S. *et al.*, *Chem. Ind. (London)*, 1977, 231-232 (*Acanthotoxin, isol, reactivity*)

Acerogenin A A4

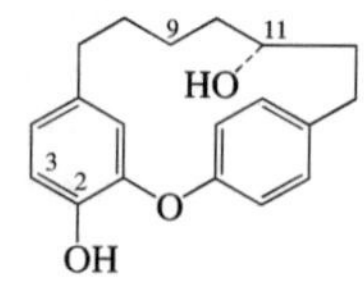

(*R*)-*form*

2-Oxatricyclo[13.2.2.1^{3,7}]eicosa-3,5,7(20),15,17,18-hexaene-4,12-diol, CAS

$C_{19}H_{22}O_3$ 298.382

Related to Acerogenin B, A5 and Galeon, G2

(R)-form [60503-28-8]

Isol. from stem bark of *Acer nikoense*. Antioxidant. Inhibitor of Na$^{\oplus}$-glucose co-transporter system. NO prodn. inhibitor. Mp 151 - 152°. $[\alpha]^{20}_{D}$ +57.3 (c, 0.8 in EtOH)

2-O-β-D-Glucopyranoside

Aceroside I

[65987-95-3]

$C_{25}H_{32}O_8$ 460.523

From *Acer nikoense*. Cryst. (Me_2CO). Mp 170 - 171°. $[\alpha]^{20}_{D}$ -7.7 (c, 1 in EtOH)

11-O-β-D-Glucopyranoside

Aceroside VI

[74390-28-6]

$C_{25}H_{32}O_8$ 460.523

Constit. of *Acer nikoense*. Prisms (EtOH). Mp 124 - 125°. $[\alpha]_{D}$ -69.3 (c, 1 in EtOH)

11-O-[β-D-Apiofuranosyl-(1→6)-β-D-glucopyranoside]

Aceroside III

[74390-29-7]

$C_{30}H_{40}O_{12}$ 592.638

Constit. of stem bark of *Acer nikoense*. Cryst. (MeOH aq.). Mp 138 - 141°. $[\alpha]_{D}$ -98.4 (c, 1.1 in EtOH)

11-O-[β-D-Apiofuranosyl-(1→6)-β-D-ribo-hexopyranos-3-ulosyl]

Aceroketoside

[497178-63-9]

$C_{30}H_{38}O_{12}$ 590.622

Constit. of the stem bark of *Acer nikoense*. Powder. $[\alpha]^{24}_{D}$ -82.2 (c, 0.2 in EtOH). λ_{max} 278 (log ε 3.41) (EtOH)

Di-Ac

[60432-35-1]

Needles (MeOH). Mp 116 - 118°. $[\alpha]^{20}_{D}$ +21.7 (c, 0.8 in EtOH)

2-Me ether

[60432-33-9]

Mp 124 - 125°. $[\alpha]^{20}_{D}$ +36.1

11-Ketone

Acerogenin C

[87425-32-9]

$C_{19}H_{20}O_3$ 296.366

Constit. of *Acer nikoense* and *Boswellia ovalifoliolata*. Exhibits antiproliferative activity. Needles (MeOH). Mp 116°. $[\alpha]^{20}_{D}$ 0 (c, 1.0 in EtOH)

11-Ketone, O-β-D-glucopyranoside

Aceroside IV

[74390-30-0]

$C_{25}H_{30}O_8$ 458.507

Constit. of stem bark of *Acer nikoense*. Needles (MeOH). Mp 153°. $[\alpha]^{16}_{D}$ -39.5 (c, 1.0 in EtOH)

7,8-Didehydro(E-), 11-ketone, 2-Me ether

Ovalifoliolatin B

[674292-89-8]

$C_{20}H_{20}O_3$ 308.377

Constit. of the stems of *Boswellia ovalifoliolata*. Semisolid. λ_{max} 208 (log ε 3.83); 260 (log ε 3.3) (MeOH)

9,10-Didehydro(E-), 11-ketone

Giffonin Q

[1966183-71-0]

$C_{19}H_{18}O_3$ 294.35

Constit. of the flowers of *Corylus avellana*. Amorph. solid

7,8,9,10-Tetradehydro(E,E-), 11-ketone

Giffonin R

[1966183-72-1]

$C_{19}H_{16}O_3$ 292.334

Constit. of the flowers of *Corylus avellana*. Amorph. solid

3-Hydroxy, 7,8,9,10,12,13-hexadehydro(7Z,9E,12Z-), 2-Me ether

Ostryopsitrienol

[1441932-11-1]
$C_{20}H_{18}O_4$ 322.36
Constit. of the stems of *Ostryopsis nobilis*. Cryst. (MeOH). $[\alpha]_D^{20}$ -595 (c, 0.02 in MeOH). Related to Galeon, G2.

12S-Hydroxy, 7,8-didehydro(E-), 11-ketone, 2-Me ether
Ovalifoliolatin A
[674292-88-7]
$C_{20}H_{20}O_4$ 324.376
Constit. of the stems of *Boswellia ovalifoliolata*. Semisolid. $[\alpha]_D^{25}$ -8.9 (c, 0.65 in MeOH). λ_{max} 207 (log ε 4.09); 261 (log ε 3.73) (MeOH)

13R-Hydroxy
2-Oxatricyclo[13.2.2.1^{3,7}]eicosa-3,5,7(20),15,17,18-hexaene-4,12,14-triol. ***Acerogenin F***
[150107-10-1]
$C_{19}H_{22}O_4$ 314.381
Constit. of *Acer nikoense*. Needles (MeOH). Mp 254 - 255°. $[\alpha]_D^{21}$ +127 (c, 0.5 in MeOH). Rel. stereochem.. λ_{max} 278 (log ε 3.40) (MeOH)

13S-Hydroxy
Acerogenin J
[150198-90-6]
$C_{19}H_{22}O_4$ 314.381
Constit. of *Acer nikoense*. Needles (MeOH). Mp 223 - 225°. $[\alpha]_D^{25}$ +65.2 (c, 0.5 in MeOH). Rel. stereochem.. λ_{max} 215 (log ε 4.14); 274 (log ε 3.43) (MeOH)

11-Deoxy, 8,9,10,11-tetradehydro(8E,10Z-)
2-Oxatricyclo[13.2.2.1^{3,7}]eicosa-3,5,7(20),9,11,15,17,18-octaen-4-ol. ***Tedarene A***
[1370415-61-4]
$C_{19}H_{18}O_2$ 278.351
Isol. from the sponge *Tedania ignis*. Amorph. solid. λ_{max} 257 (sh); 288 (sh) (MeOH)

3-Methoxy, 7,8,9,10-tetradehydro(E,E-), 11-ketone
Giffonin S
[1966183-73-2]
$C_{20}H_{18}O_4$ 322.36
Constit. of the flowers of *Corylus avellana*. Amorph. solid

(ξ)-form
7-Oxo
7-Oxoacerogenin A. ***Acerogenin H***
[150107-12-3]
$C_{19}H_{20}O_4$ 312.365
Constit. of *Acer nikoense*. Needles (EtOAc). Mp 245 - 246°. $[\alpha]_D^{20}$ -36.4 (c, 0.5 in MeOH). λ_{max} 232 (log ε 4.11); 279 (log ε 3.91); 307 (log ε 3.74); 354 (log ε 3.41) (EtOH). [acid] λ_{max} 250; 347 (EtOH/NaOAc aq.).

Nagai, M. *et al.*, *Chem. Pharm. Bull.*, 1978, **26**, 2805-2810 (*Aceroside I*).
Kubo, M. *et al.*, *Chem. Pharm. Bull.*, 1983, **31**, 1917-1922 (*Acerogenin A*).
Kubo, M. *et al.*, *Chem. Pharm. Bull.*, 1983, **31**, 1923-1928 (*Acerogenins A,C, Aceroside VI*).
Inoue, T. *et al.*, *Phytochemistry*, 1987, **26**, 1409-1411 (*biosynth*).
Nagumo, S. *et al.*, *Chem. Pharm. Bull.*, 1996, **44**, 1086-1089 (*Acerogenins F,H,J*).
Keserü, G.M. *et al.*, *Eur. J. Org. Chem.*, 1998, 521-524 (*synth*).
Gonzalez, G.I. *et al.*, *J. Org. Chem.*, 1999, **64**, 914-924 (*synth*).
Reddy, V.L.N. *et al.*, *Chem. Pharm. Bull.*, 2003, **51**, 1081-1084 (*Ovalifoliolatins A,B*).
Morikawa, T. *et al.*, *J. Nat. Prod.*, 2003, **66**, 86-91 (*Aceroketoside*).
Akihisa, T. *et al.*, *Chem. Pharm. Bull.*, 2006, **54**, 735-739 (*Aceroside I*).
Jeong, B.-S. *et al.*, *Eur. J. Org. Chem.*, 2007, 1338-1344 (*Acerogenin C, synth*).
Kumar, G.D.K. *et al.*, *Tetrahedron Lett.*, 2008, **49**, 2103-2105 (*Ovalifoliolatin B, synth*).
Morita, H. *et al.*, *Bioorg. Med. Chem. Lett.*, 2010, **20**, 1070-1074 (*Acerogenin A, conformn, activity*).
Akihisa, T. *et al.*, *Chem. Biodiversity*, 2012, **9**, 1475-1489 (*Acerogenin A, Acerosides I,III,IV*).
Costantino, V. *et al.*, *J. Org. Chem.*, 2012, **77**, 6377-6383 (*Tedarene A*).
Salih, M.Q. *et al.*, *Org. Lett.*, 2012, **14**, 4026-4029 (*synth*).
Zhang, Y.-X. *et al.*, *Chin. Chem. Lett.*, 2013, **24**, 512-514 (*Ostryopsitrienol*).
Pattawong, O. *Org. Biomol. Chem.*, 2014, **12**, 3303-3309 (*conformn, chirality*).
Lee, D.-S. *et al.*, *Molecules*, 2015, **20**, 12545-12557 (*Acerogenin A, activity*).
Masullo, M. *et al.*, *Phytochemistry*, 2016, **130**, 273-281 (*Giffonins Q,R,S*).
Maurent, K. *et al.*, *J. Nat. Prod.*, 2017, **80**, 1623-1630 (*Tedarene A, synth*).
Salih, M.Q. *et al.*, *Tetrahedron Lett.*, 2020, **61**, 151988:1-4 (*synth*)

Acerogenin B A5

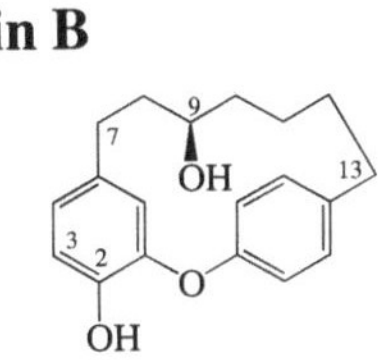
(R)-form

2-Oxatricyclo[13.2.2.1^{3,7}]eicosa-3,5,7(20),15,17,18-hexaene-4,10-diol,
CAS
$C_{19}H_{22}O_3$ 298.382
Closely related to Acerogenin A, A4 and Galeon, G2

(R)-form [497178-66-2]
Constit. of *Acer nikoense*. Inhibitor of Na⊕-glucose co-transporter system. NO prodn. inhibitor. Fine cryst. (MeOH). Mp 182 - 184.5°. $[\alpha]_D^{22}$ -95 (c, 0.1 in EtOH). λ_{max} 276 (log ε 3.32) (EtOH)

2-O-β-D-Glucopyranoside
Aceroside B₁
[497178-52-6]
[888460-23-9] (unspecified stereochem.)
$C_{25}H_{32}O_8$ 460.523
Constit. of the stem bark of *Acer nikoense*. Powder. $[\alpha]_D^{24}$ -94 (c, 0.4 in EtOH). λ_{max} 272 (log ε 3.24) (EtOH)

3-Hydroxy, 7,8-didehydro(Z-), 2-Me ether
7,8-Didehydro-3-hydroxy-2-O-methylacerogenin B
[177262-55-4]
$C_{20}H_{22}O_4$ 326.392
Constit. of *Betula platyphylla* var. *japonica*. Cryst. (EtOAc/hexane). Mp 186 - 187°. $[\alpha]_D$ +79 (c, 1 in $CHCl_3$). λ_{max} 223 (log ε 4.39); 258 (log ε 3.95) (MeOH)

(S)-form [497178-88-8]
Constit. of *Acer nikoense*. Fine cryst. (MeOH). Mp 181.3 - 183.4°. $[\alpha]_D^{22}$ +84.1 (c, 0.2 in EtOH). λ_{max} 276 (log ε 3.34) (EtOH)

2-O-β-D-Glucopyranoside
Aceroside B₂
[497178-62-8]
$C_{25}H_{32}O_8$ 460.523
Constit. of the stem bark of *Acer nikoense*. Powder. $[\alpha]_D^{24}$ +24.5 (c, 0.5 in EtOH). λ_{max} 273 (log ε 3.24) (EtOH)

(±)-form [74174-17-7]
Constit. of the stem bark of *Acer nikoense*. Mp 179°

Ketone
Acerogenin L
[177262-58-7]
$C_{19}H_{20}O_3$ 296.366
Constit. of *Acer nikoense*. Needles ($CHCl_3$/MeOH). Mp 188 - 190°. λ_{max} 206 (log ε 4.45); 276 (log ε 3.43) (EtOH)

(ξ)-form
13ξ-Hydroxy
2-Oxatricyclo[13.2.2.1^{3,7}]eicosa-3,5,7(20),15,17,18-hexaene-4,10,14-triol. ***Acerogenin I***
[150107-13-4]
$C_{19}H_{22}O_4$ 314.381

Constit. of *Acer nikoense*. Needles (MeOH aq.). Mp 214 - 215°. $[\alpha]_D^{20}$ -17.2 (c, 0.6 in MeOH). λ_{max} 213 (sh); 274 (log ε 3.40) (EtOH)

7-Oxo

7-Oxoacerogenin B. ***Acerogenin M*** [905705-54-6]

$C_{19}H_{20}O_4$ 312.365

Constit. of the stem bark of *Acer nikoense*. Amorph. solid. $[\alpha]_D^{25}$ +4 (c, 0.1 in MeOH). λ_{max} 213 (log ε 3.35); 231 (log ε 3.36); 278 (log ε 3.14); 303 (log ε 2.95) (MeOH).

Kubo, M. *Chem. Pharm. Bull.*, 1980, **28**, 1300-1303 (*Acerogenin B*).
Fuchino, H. *et al.*, *Chem. Pharm. Bull.*, 1996, **44**, 1033-1038 (*7,8-Didehydro-3-hydroxy-2-methylacerogenin B*).
Nagumo, S. *et al.*, *Chem. Pharm. Bull.*, 1996, **44**, 1086-1089 (*Acerogenins I,L*).
Gonzalez, G.I. *et al.*, *J. Org. Chem.*, 1999, **64**, 914-924 (*synth*).
Morikawa, T. *et al.*, *J. Nat. Prod.*, 2003, **66**, 86-91 (*Acerogenin B, Aceroside B_1, activity*).
Akihisa, T. *et al.*, *Chem. Pharm. Bull.*, 2006, **54**, 735-739 (*Acerogenin M*).
Morita, H. *et al.*, *Bioorg. Med. Chem. Lett.*, 2010, **20**, 1070-1094 (*Acerogenin B, activity*).
Pattawong, O. *et al.*, *Org. Biomol. Chem.*, 2014, **12**, 3303-3309 (*Acerogenin L, conformn, chirality*)

2-Acetyl-3-(3,4-dimethoxyphenyl)-3-hydroxy-2-methyl-5,6-(methylenedioxy)-1-indanone A6

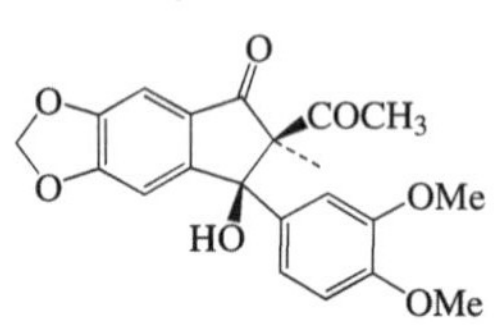

6-Acetyl-7-(3,4-dimethoxyphenyl)-6,7-dihydro-7-hydroxy-6-methyl-5H-indene[5,6-d]-1,3-dioxol-5-one, CAS [129684-16-8]

$C_{21}H_{20}O_7$ 384.384

Constit. of *Virola elongata* and *Virola sebifera*. Cryst. (Me_2CO). Mp 250 - 255°.

Lopes, L.M.X. *et al.*, *Phytochemistry*, 1984, **23**, 2021-2024 (*Virola sebifera constit*).
Kato, M.J. *et al.*, *Phytochemistry*, 1990, **29**, 1799-1810 (*Virola elongata constit*)

2-Acetyl-2′,4′,5-trihydroxy-3′-methoxy-4-(1-propenyl)benzophenone A7

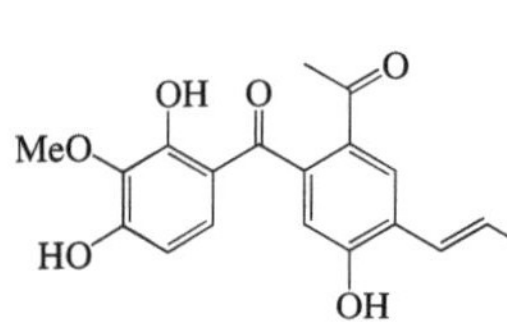

(*E*)-*form*

1-[2-(2,4-Dihydroxy-3-methoxybenzoyl)-4-hydroxy-5-(1-propenyl)phenyl] ethanone

$C_{19}H_{18}O_6$ 342.347

(E)-form

Penchinone A [2043024-05-9]

Constit. of *Penthorum chinense*. Exhibits protective activity against acetaminophen-induced damage to the HL7702 hepatocyte and selective cytotoxicity against human cancer ovarian cell line Hey *in vitro*. Pale yellow powder. λ_{max} 214 (log ε 4.05); 252 (log ε 4.23); 289 (log ε 4.11) (MeOH)

(Z)-form

Penchinone B [2043024-06-0]

Constit. of *Penthorum chinense*. Pale yellow cryst. (MeOH). λ_{max} 215 (log ε 4.14); 243 (log ε 4.21); 288 (log ε 4.17) (MeOH).

He, Y.-C. *et al.*, *RSC Adv.*, 2015, **5**, 76788-76794 (*Penchinones A,B, cryst struct, activity*).
Oh, Y. *et al.*, *J. Org. Chem.*, 2017, **82**, 11566-11572 (*synth*)

Acortatarinowin G A8

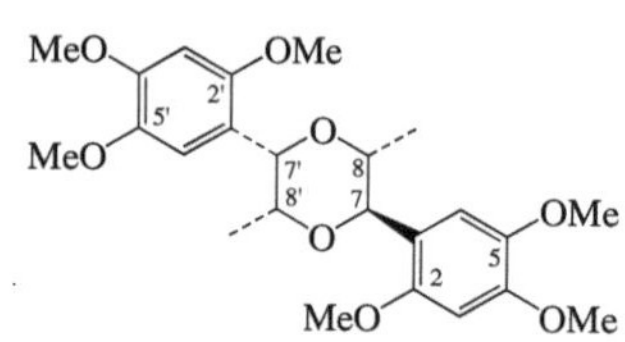

(*7R,7′S,8R,8′R*)-*form*

2,5-Dimethyl-3,6-bis(2,4,5-trimethoxyphenyl)-1,4-dioxane. 7′,8-Epoxy-2,2′,4,4′,5,5′-hexamethoxy-7,8′-oxylignan

$C_{24}H_{32}O_8$ 448.512

(7R,7′S,8R,8′R)-form

Constit. of the rhizomes of *Acorus tatarinowii*. Amorph. powder. $[\alpha]_D^{20}$ -28 (c, 0.1 in MeOH). Resolved by chiral hplc

(7S,7′R,8S,8′S)-form

Constit. of the rhizomes of *Acorus tatarinowii*. Amorph. powder. $[\alpha]_D^{20}$ +28 (c, 0.1 in MeOH). Resolved by chiral hplc

(7RS,7′SR,8RS,8′RS)-form

Constit. of the rhizomes of *Acorus tatarinowii*. Block cryst. (MeOH). Mp 186 - 187°. λ_{max} 205 (log ε 4.91); 230 (log ε 4.41); 291 (log ε 4.13) (MeOH).

Lu, Y. *et al.*, *Sci. Rep.*, 2016, **6**, 22909 (*Acortatarinowin G*)

Acortatarinowin H A9

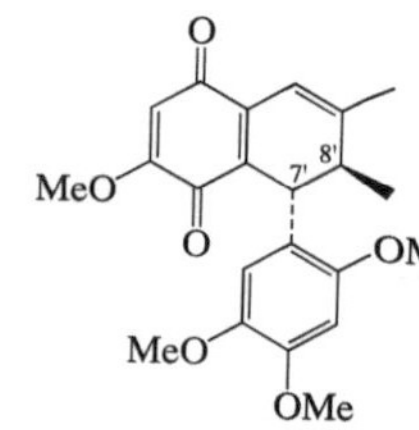

(*7′R,8′S*)-*form*

2,7′-Cyclo-2′,4,4′,5′-tetramethoxy-8,8′-lign-7-ene-3,6-dione

$C_{22}H_{24}O_6$ 384.428

(7′R,8′S)-form

Constit. of the rhizomes of *Acorus tatarinowii*. Amorph. brown-red powder. $[\alpha]_D^{20}$ +14 (c, 0.1 in CH_2Cl_2). Resolved by chiral hplc

(7′S,8′R)-form

Constit. of the rhizomes of *Acorus tatarinowii*. Amorph. brown-red powder. $[\alpha]_D^{20}$ -14 (c, 0.1 in CH_2Cl_2). Resolved by chiral hplc

(7′RS,8′SR)-form

Constit. of the rhizomes of *Acorus tatarinowii*. Amorph. brown-red powder. Resolved by chiral hplc. λ_{max} 208 (log ε 4.21); 243 (log ε 3.83); 290 (log ε 3.43); 315 (log ε 3.42) (MeOH).

Lu, Y. *et al.*, *Sci. Rep.*, 2016, **6**, 22909 (*Acortatarinowin H, abs config*)

Acorusin B A10

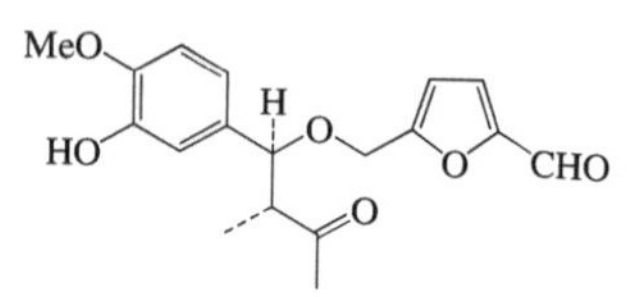

Relative Configuration

[2170741-43-0]

$C_{18}H_{20}O_6$ 332.352

Constit. of rhizomes of *Acorus tatarinowii*. Exhibits moderate cytotoxicity against the five cell lines NCI-H1650, HepG2, BGC-823, HCT-116 and MCF-7. Yellow solid. $[\alpha]_D^{20}$ +4.0 (c, 0.2 in MeOH). Hybrid-norlignan proposed by Ni *et al*, originating from a dibenzylbutane-type lignan which has undergone loss of benzyl via C7-C8 cleavage followed by coupling at C-7′ with a 5-hydroxy-2-furfuraldehyde moiety. λ_{max} 231 (log ε 3.47); 280 (log ε 3.58) (MeOH).

Ni, G. *et al.*, *Planta Med.*, 2016, **82**, 632-638 (*Acorusin B, activity*)

Acrignine A A11

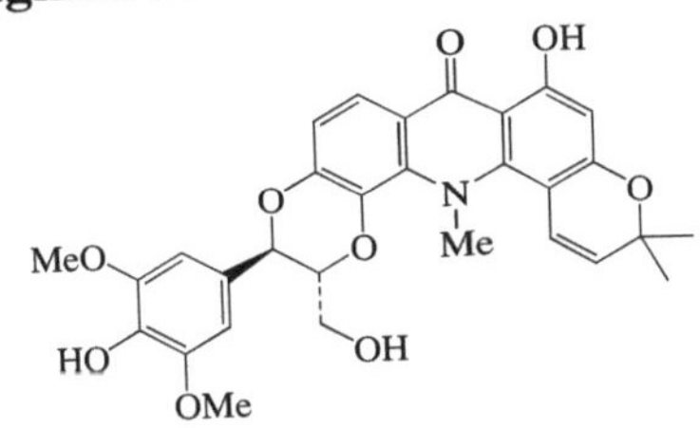

[149155-25-9]
$C_{30}H_{29}NO_9$ 547.56
Alkaloid from roots of *Citrus grandis* and *Citrus yuko*. Yellow prisms. Mp 177 - 179°. Racemic. λ_{max} 207; 236 (sh); 272; 316 (sh); 340; 400 (MeOH).

Takemura, Y. *et al.*, *Chem. Pharm. Bull.*, 1993, **41**, 406-407 (*Acrignine A, cryst struct*)

Actinidione A12

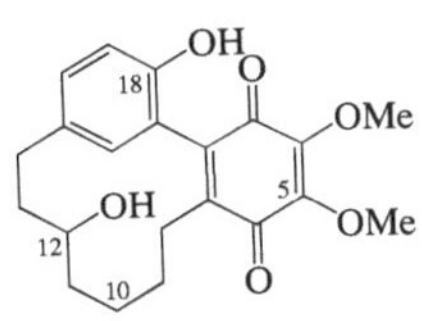

[902759-45-9]
$C_{21}H_{24}O_6$ 372.417
Constit. of *Clematoclethra actinidioides*, *Myrica nana* and *Myrica rubra*. Red gum. $[\alpha]^{20}_{D}$ -20.1 (c, 0.12 in MeOH). λ_{max} 229 (log ε 3.56); 278 (log ε 4.15); 382 (log ε 1.23) (MeOH)

O^5-De-Me, 18-Me ether
Myrubol
[1268690-01-2]
$C_{21}H_{24}O_6$ 372.417
Constit. of the roots of *Myrica rubra*. Red oil. $[\alpha]^{23}_{D}$ -18.5 (c, 0.35 in MeOH)

12-Deoxy, 10-oxo
Mangiferone
$C_{21}H_{22}O_6$ 370.401
Constit. of the bark of *Mangifera indica*. Red gum. $[\alpha]^{20}_{D}$ -22.5 (c, 0.35 in MeOH).

Guo, H. *et al.*, *Planta Med.*, 2006, **72**, 180-183 (*Actinidione*).
Li, G. *et al.*, *Nat. Prod. Res.*, 2011, **25**, 136-140 (*Myrubol*).
Yoshimura, M. *et al.*, *J. Nat. Prod.*, 2012, **75**, 1798-1802 (*Myrica rubra constit*).
Yu, X.-X. *et al.*, *Chin. Herb. Med.*, 2013, **5**, 320-322 (*Mangiferone*).
Massé, P. *et al.*, *J. Org. Chem.*, 2021, **86**, 3033-3040 (*synth*)

Aegicin A13

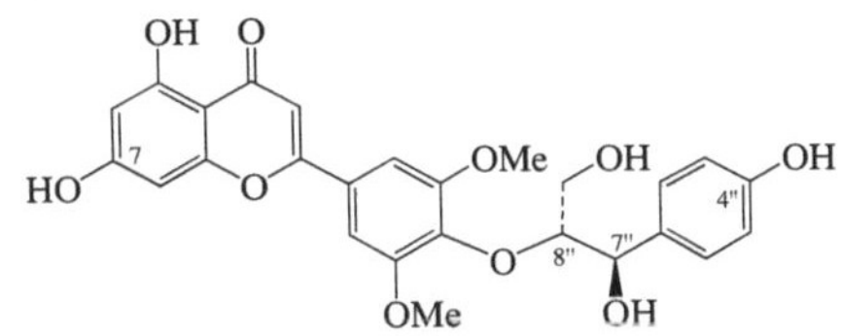

(7''*R*,8''*R*)-*form*

5,7-Dihydroxy-2-[4-[2-hydroxy-1-hydroxymethyl-2-(4-hydroxyphenyl) ethoxy]-3,5-dimethoxyphenyl]-4H-1-benzopyran-4-one, CAS. *Tricin 4'-O-(β-4-hydroxyphenylglyceryl) ether*
$C_{26}H_{24}O_{10}$ 496.468
Flavonolignan

(7''R,8''R)-form
7''-Me ether
[2025333-63-3]
$C_{27}H_{26}O_{10}$ 510.495
Constit. of the green grasses of *Elymus natans*. Amorph. yellow powder. $[\alpha]^{25}_{D}$ +26.4 (c, 0.1 in MeOH). λ_{max} 270; 340 (MeOH)

(7''R,8''S)-form
7''-Me ether
[2025333-65-5]
Constit. of the green grasses of *Elymus natans*. $[\alpha]^{25}_{D}$ -46.2 (c, 0.1 in MeOH). λ_{max} 270; 340 (MeOH)

(7''R*,8''R*)-form
Calquiquelignan E [1217897-50-1]
Constit. of the stems of *Calamus quiquesetinervius*. Exhibits moderate antiplatelet aggregation activity. Amorph. powder. Mp 247°. $[\alpha]^{25}_{D}$ +27 (c, 0.48 in MeOH). λ_{max} 201; 271; 329 (sh) (MeOH)

2S,3-Dihydro
Dihydrotricin 4'-O-(β-4-hydroxyphenylglyceryl) ether. ***Calquiquelignan C***
[1217897-49-8]
$C_{26}H_{26}O_{10}$ 498.484
Constit. of the stems of *Calamus quiquesetinervius*. Pale yellow powder. Mp 146°. $[\alpha]^{25}_{D}$ -16 (c, 0.06 in MeOH). λ_{max} 202; 230; 286; 330 (sh) (MeOH)

3''-Methoxy
Tricin 4'-O-(β-guaiacylglyceryl) ether. ***Salcolin A***
[369390-51-2]
$C_{27}H_{26}O_{11}$ 526.494
Constit. of *Avena sativa*, *Hordeum vulgare*, *Hyparrhenia hirta*, *Salsola collina*, *Sasa veitchii*, *Sorghum halepense*, *Zea mays* and *Zizania latifolia*. Potent platelet aggregation inhibitor and hydroxyl radical scavenger. Allelochemical. Active against cyanobacteria. Yellow solid. $[\alpha]^{24}_{D}$ -10 (c, 0.05 in MeOH) (Avena sativa constit). $[\alpha]^{24}_{D}$ +2.6 (c, 0.1 in MeOH) (Zea mays constit). λ_{max} 272; 287 (sh); 303 (sh); 334 (MeOH)

3''-Methoxy, 7-O-β-D-glucopyranoside
[462100-43-2]
$C_{33}H_{36}O_{16}$ 688.635
Constit. of the leaves of *Hyparrhenia hirta*. Yellow solid. λ_{max} 270; 287 (sh); 340 (MeOH)

3''-Methoxy, 7''-O-β-D-glucopyranoside
[1394179-23-7]
$C_{33}H_{36}O_{16}$ 688.635
Constit. of *Zizania latifolia*. Yellow powder (MeOH)

3''-Methoxy, 9''-Ac
[629646-28-2]
$C_{29}H_{28}O_{12}$ 568.531
Constit. of the leaves of *Sasa veitchii*. Amorph. yellow solid. $[\alpha]^{20}_{D}$ -48.5 (c, 0.11 in MeOH). λ_{max} 310 (log ε 4.06) (MeOH)

3''-Methoxy, 9''-O-(4-hydroxy-E-cinnamoyl)
[629646-41-9]
$C_{36}H_{32}O_{13}$ 672.639
Constit. of the leaves of *Sasa veitchii*. Amorph. yellow solid. $[\alpha]^{20}_{D}$ +52.1 (c, 0.48 in MeOH). λ_{max} 312 (log ε 4.58) (MeOH)

3''-Methoxy, 7''-Me ether
Salcolin C
[1394179-20-4]
$C_{28}H_{28}O_{11}$ 540.521
Constit. of *Oryza sativa*, *Zizania latifolia* and the roots of *Lepidium meyenii*. Inhibitor of NO prodn. in RAW 264.7 macrophage cells. Yellow powder (MeOH). $[\alpha]^{25}_{D}$ -0.9 (c, 1 in MeOH). λ_{max} 215; 272; 340 (MeOH)

3''-Methoxy, 7''-Me ether, 9''-Ac
[629646-35-1]
$C_{30}H_{30}O_{12}$ 582.558
Constit. of the leaves of *Sasa veitchii*. Amorph. yellow solid. $[\alpha]^{20}_{D}$ +5.8 (c, 0.41 in MeOH). λ_{max} 272 (log ε 4.39); 322 (log ε 4.4) (MeOH)

3''-Methoxy, 7''-Et ether
Quiquelignan C
[1215818-66-8]

$C_{29}H_{30}O_{11}$ 554.548
Constit. of the stems of *Calamus quiquesetinervius*. Antioxidant. Yellow powder. Mp 194°. $[\alpha]_D^{25}$ +7.1 (c, 0.28 in MeOH). λ_{max} 272; 335 (MeOH)

3″-Methoxy, 2S,3-dihydro Dihydrotricin 4′-O-(β-guaiacylglyceryl) ether. ***Calquiquelignan B***
[1217897-48-7]
$C_{27}H_{28}O_{11}$ 528.51
Constit. of the stems of *Calamus quiquesetinervius*. Exhibits moderate antiplatelet aggregation activity and hydroxy radical scavenger. Pale orange powder. Mp 118°. $[\alpha]_D^{25}$ +70 (c, 1.1 in MeOH). λ_{max} 203; 230; 288; 330 (sh) (MeOH)

3″-Methoxy, 2S,3-dihydro, 7″-Et ether
Quiquelignan A
[1215818-64-6]
$C_{29}H_{32}O_{11}$ 556.564
Constit. of the stems of *Calamus quiquesetinervius*. Antioxidant. Orange-yellow powder. Mp 145°. $[\alpha]_D$ -14.3 (c, 0.28 in MeOH). λ_{max} 212; 230; 286; 322 (MeOH)

(7″R*,8″S*)-form

Calquiquelignan D [65870-44-2]
Constit. of *Aegilops ovata*, *Calamus quiquesetinervius* and *Sinocalamus affinis* (preferred genus name *Dendrocalamus*). Exhibits moderate antiplatelet aggregation activity. Cryst. ($CHCl_3$/MeOH). Mp 235 - 236° (218°). $[\alpha]_D^{25}$ -58 (c, 0.48 in MeOH) (Calquiquelignan D). λ_{max} 201; 271; 333 (sh) (MeOH)

3″-Methoxy
Salcolin B
[369390-52-3]
Constit. of *Avena sativa*, *Hordeum vulgare*, *Hyparrhenia hirta*, *Lepidium meyenii*, *Oryza sativa*, *Salsola collina*, *Sasa veitchii*, *Sorghum halepense*, *Zea mays* and *Zizania latifolia*. Allelochemical. Active against cyanobacteria. Yellow solid. $[\alpha]_D^{24}$ -5 (c, 0.1 in MeOH) (Zea mays constit). $[\alpha]_D^{24}$ +15 (c, 0.05 in MeOH) (Avena sativa constit). λ_{max} 271; 288 (sh); 305 (sh); 335 (MeOH)

3″-Methoxy, 7-O-β-D-glucopyranoside
[462100-42-1]
Constit. of the leaves of *Hyparrhenia hirta*. Yellow solid. λ_{max} 270; 287 (sh); 340 (MeOH)

3″-Methoxy, 7″-O-β-D-glucopyranoside
[1394179-21-5]
$C_{33}H_{36}O_{16}$ 688.635
Constit. of *Zizania latifolia*. Yellow powder (MeOH)

3″-Methoxy, 9″-O-β-D-glucopyranoside
[1534387-44-4]
$C_{33}H_{36}O_{16}$ 688.635
Constit. of *Oryza sativa*. Yellow powder. $[\alpha]_D^{25}$ -19 (c, 0.24 in MeOH). λ_{max} 335 (MeOH)

3″-Methoxy, 9″-Ac
[629646-32-8]
Constit. of the leaves of *Sasa veitchii*. Amorph. yellow solid. $[\alpha]_D^{20}$ -70 (c, 0.15 in MeOH). λ_{max} 322 (log ε 3.94) (MeOH)

3″-Methoxy, 9″-O-(4-hydroxy-E-cinnamoyl)
[629646-44-2]
Constit. of the leaves of *Sasa veitchii*. Amorph. yellow solid. $[\alpha]_D^{20}$ -30.4 (c, 1.3 in dioxan). λ_{max} 310 (log ε 4.57) (MeOH)

3″-Methoxy, 7-Me ether, 5-O-β-D-glucopyranoside
$C_{34}H_{38}O_{16}$ 702.662
Constit. of the leaves of *Neosinocalamus affinis*. Yellow powder. $[\alpha]_D^{20}$ -35.5 (c, 0.11 in MeOH). λ_{max} 249 (log ε 3.9); 329 (log ε 4.1) (MeOH)

3″-Methoxy, 7″-Me ether
Salcolin D
[1979201-28-9]
$C_{28}H_{28}O_{11}$ 540.521
Constit. of *Zizania latifolia*. Inhibitor of LPS-induced NO prodn.. Yellow powder (MeOH)

3″-Methoxy, 7″-Me ether, 9″-Ac
[629646-37-3]
Constit. of the leaves of *Sasa veitchii*. Amorph. yellow solid. $[\alpha]_D^{20}$ -13.8 (c, 0.25 in MeOH). λ_{max} 271 (log ε 4.21); 335 (log ε 4.2) (MeOH)

3″-Methoxy, 2S,3-dihydro
Calquiquelignan A
[1217897-47-6]
$C_{27}H_{28}O_{11}$ 528.51
Constit. of the stems of *Calamus quiquesetinervius*. Exhibits moderate antiplatelet aggregation activity and hydroxy radical scavenger. Also shows vasodilatory props.. Pale yellow powder. Mp 204°. $[\alpha]_D^{25}$ -33 (c, 0.21 in MeOH). λ_{max} 203; 230; 288; 330 (sh) (MeOH)

3″-Methoxy, 2S,3-dihydro, 7″-Et ether, 9″-O-(4-hydroxybenzoyl)
Quiquelignan B
[1215818-65-7]
$C_{36}H_{36}O_{13}$ 676.671
Constit. of the stems of *Calamus quiquesetinervius*. Antioxidant. Pale yellow powder. Mp 116°. $[\alpha]_D^{25}$ -9.8 (c, 0.41 in MeOH). λ_{max} 216; 260; 275; 325 (MeOH).

Cooper, R. *et al.*, *Isr. J. Chem.*, 1977, **16**, 12-15 (*Aegicin*).
Syrchina, A.I. *et al.*, *Chem. Nat. Compd. (Engl. Transl.)*, 1992, **28**, 155-158 (*Salcolins A,B*).
Bouaziz, M. *et al.*, *Phytochemistry*, 2002, **60**, 515-520 (*Hyparrhenia hirta constits*).
Nakajima, Y. *et al.*, *Tetrahedron*, 2003, **59**, 8011-8015 (*Sasa veitchii constits*).
Wenzig, E. *et al.*, *J. Nat. Prod.*, 2005, **68**, 289-292 (*Salcolins A,B*).
Chang, C.-L. *et al.*, *Bioorg. Med. Chem.*, 2010, **18**, 518-525 (*Quiquelignans A,B,C, cd, activity*).
Chang, C.-L. *et al.*, *Phytochemistry*, 2010, **71**, 271-279 (*Calquiquelignans A-E, activity*).
Jeong, R.-H. *et al.*, *J. Korean Soc. Appl. Biol. Chem.*, 2011, **54**, 865-870 (*Salcolin C*).
Xiong, L. *et al.*, *J. Nat. Prod.*, 2011, **74**, 1188-1200 (*Sinocalamus affinis constits*).
Jeong, R.-H. *et al.*, *Chem. Nat. Compd. (Engl. Transl.)*, 2014, **49**, 1003-1005 (*Salcolin B 9″-glucoside*).
Xiao, X. *et al.*, *Environ. Microbiol.*, 2014, **16**, 1238-1251 (*Salcolins A,B, activity*).
Jung, Y.-J. *et al.*, *Arch. Pharmacal Res.*, 2015, **38**, 178-185 (*Salcolins A,B*).
Lee, S.-S. *et al.*, *Biosci., Biotechnol., Biochem.*, 2015, **79**, 700-706 (*Salcolins A-D, activity*).
Bai, N. *et al.*, *J. Agric. Food Chem.*, 2015, **63**, 2458-2463 (*Lepidium meyenii constits*).
Luo, G.-Y. *et al.*, *J. Asian Nat. Prod. Res.*, 2015, **17**, 248-255 (*Neosinocalamus affinis glucoside*).
Lee, S.-S. *et al.*, *Molecules*, 2015, **20**, 5616-5624 (*Salcolins A,B 7″-glucosides*).
Liu, Q. *et al.*, *J. Agric. Food Chem.*, 2017, **65**, 1320-1327 (*Elymus natans constits, cd*)

Aiphanol A14

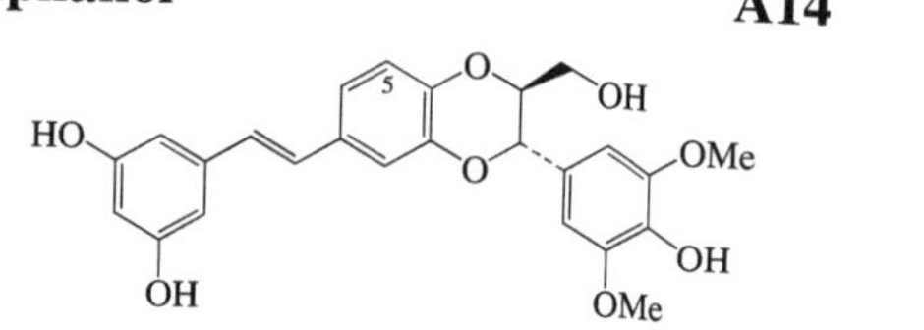

Absolute Configuration

[353482-13-0]
[578020-29-8] ((±)-form)
$C_{25}H_{24}O_8$ 452.459

Stilbenolignan. Related to Dendrocandin B, D16. Constit. of *Aiphanes aculeata* and *Smilax glabra*. COX-1 and COX-2 inhibitor. Brown powder. $[\alpha]_D^{20}$ -21.8 (c, 0.13 in MeOH). λ_{max} 233 (log ε 5.37); 322 (log ε 5.32) (MeOH)

5-Hydroxy

5-Hydroxyaiphanol

[1228276-51-4]

$C_{25}H_{24}O_9$ 468.458

Constit. of the seeds of *Syagrus romanzoffiana*. Exhibits moderate inhibitory activity against α-glucosidase. Amorph. yellow powder. Optically inactive. λ_{max} 289 (log ε 3.97); 321 (log ε 4.07) (MeOH).

Lee, D. *et al., Org. Lett.*, 2001, **3**, 2169-2171 (*Aiphanol, activity*).
Kuboki, A. *et al., Chem. Lett.*, 2003, **32**, 420-421 (*synth*).
Wang, X.L. *et al., Chin. Chem. Lett.*, 2004, **15**, 1036-1038 (*synth*).
Banwell, M.R. *et al., Tetrahedron: Asymmetry*, 2005, **16**, 1645-1654 (*synth, abs config*).
Lam, S.-H. *et al., Phytochemistry*, 2010, **71**, 792-797 (*5-Hydroxyaiphanol, activity*)

Alashinol H A15

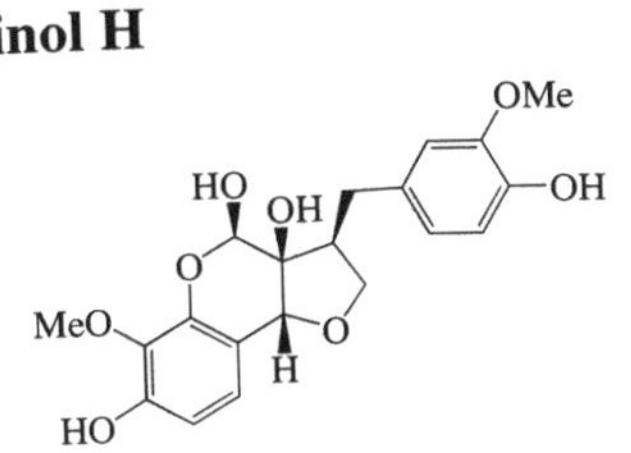

Absolute Configuration

2,9:7,9′-Diepoxy-3,3′-dimethoxy-8,8′-lignan-4,4′,8,9-tetrol.
4,4′,8,9-Tetrahydroxy-3,3′-dimethoxy-2,9:7,9′-diepoxylignan

[2247487-85-8]

$C_{20}H_{22}O_8$ 390.388

Constit. of the stem bark of *Syringa pinnatifolia*. Moderate NO prodn. inhibitor. Amorph. powder. $[\alpha]_D^{25}$ +14.8 (c, 0.1 in MeOH).

Su, G. *et al., Tetrahedron Lett.*, 2018, **59**, 1356-1359 (*Alashinol H, cd, activity*)

Aleuritin A16

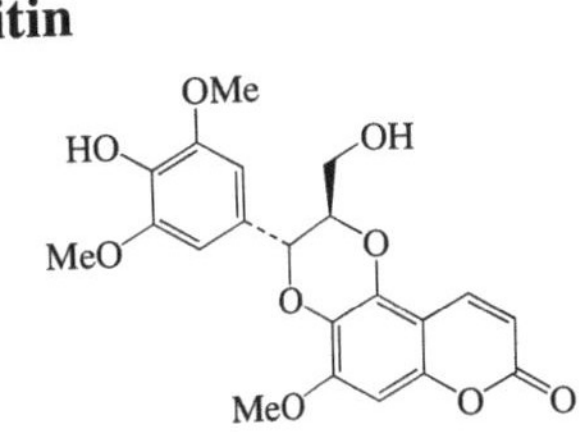

Relative Configuration

[124901-94-6]

$C_{21}H_{20}O_9$ 416.382

Constit. of the stems of *Aleurites fordii*. Cryst. (MeOH). Mp 238 - 239°. λ_{max} 240; 320 (MeOH).

Fozdar, B.I. *et al., Phytochemistry*, 1989, **28**, 2459-2461 (*Aleuritin*)

Alkaloid LC7 A17

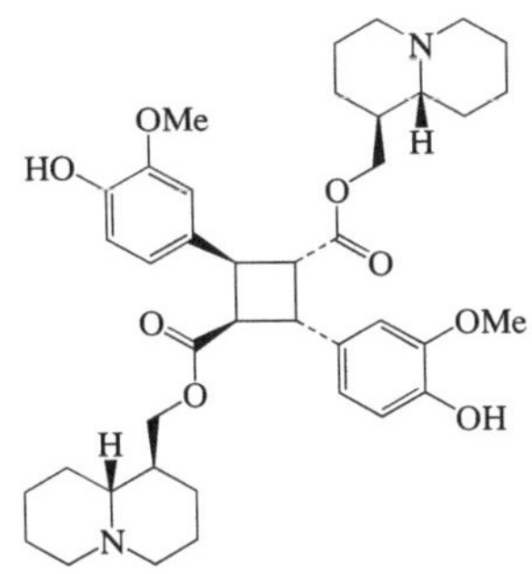

Probable structure

Alkaloid LV3‡

[71628-39-2]

$C_{40}H_{54}N_2O_8$ 690.878

Truxillate-type dimer of Epilupinyl *trans*-ferulate. Alkaloid from leaves of *Lupinus cosentinii*. Cryst. (MeOH/$CHCl_3$). Mp 228°. $[\alpha]_D^{25}$ +23 (c, 1 in $CHCl_3$/MeOH 1:1).

Beck, A.B. *et al., J. Nat. Prod.*, 1979, **42**, 385-398 (*Alkaloid LC 7*)

2-(4-Allyloxy-3-methoxyphenyl)-1-(3,4-methylenedioxyphenyl)-1-propanone A18

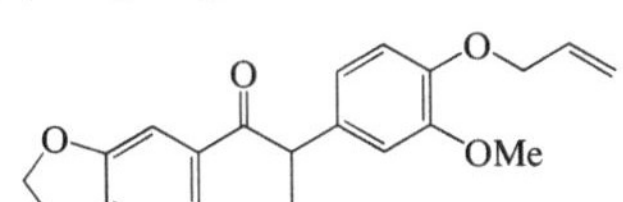

1-(1,3-Benzodioxol-5-yl)-2-[3-methoxy-4-(2-propenyloxy)phenyl]-1-propanone

[156665-58-6]

$C_{20}H_{20}O_5$ 340.375

Constit. of *Ocotea porosa*. Oil.

David, J.M. *et al., Phytochemistry*, 1994, **36**, 491-499 (*Ocotea porosa constit*)

Alnusdiol A19

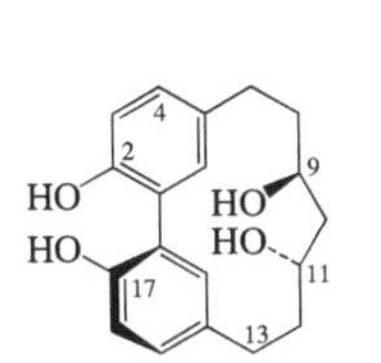

Absolute Configuration

Tricyclo[12.3.1.1^{2,6}]nonadeca-1(18),2,4,6(19),14,16-hexaene-3,9,11,17-tetrol, CAS. *3,5,11,15-Tetrahydroxy[7.0]metacyclophane*

[56973-51-4]

$C_{19}H_{22}O_4$ 314.381

Abs. config. not determined for derivs. Numbering systems vary. Related to Myricanol, M57. Constit. of *Alnus japonica*, *Betula maximowicziana* and *Casuarina junghuhniana*. Mp > 300° (293-298° dec.). $[\alpha]_D$ -46.7 (c, 0.52 in EtOH). λ_{max} 248 (sh); 302 (log ε 3.98) (EtOH)

Tetra-Ac

[56973-67-2]

Mp 159 - 162°

9-O-β-D-Glucopyranoside

Alnusdiol 9-glucoside

[181827-66-7]

$C_{25}H_{32}O_9$ 476.522

Constit. of *Betula maximowicziana*. Pale brown powder. $[\alpha]_D$ -26 (c, 0.8 in MeOH). λ_{max} 250 (log ε 4.05); 254 (log ε 4.03); 299 (log ε 3.77) (MeOH)

9-Ketone

Alnusonol

[52330-12-8]

$C_{19}H_{20}O_4$ 312.365

Constit. of *Alnus japonica*, *Alnus sieboldiana*, *Carpinus cordata* and *Myrica nana*. Mp 171 - 173°. $[\alpha]_D$ +41.7 (c, 0.2 in CHCl3)

9,11-Diketone

Alnusdione

[92533-50-1]

$C_{19}H_{18}O_4$ 310.349

Constit. of the wood of *Alnus sieboldiana*. α-Glucosidase inhibitor. Viscous liq.

10-Hydroxy, 9,11-diketone

10-Hydroxyalnusdione.

4-Hydroxy-3,5-alnusdione

[1622397-06-1]

$C_{19}H_{18}O_5$ 326.348

Constit. of the fruit of *Alnus japonica*. Exhibits anti-adipogenic activity mediated by down-regulation of PPARγ, C/EBPα and SREBP1c signaling. Brownish syrup. $[\alpha]_D^{25}$ -0.5 (c, 1 in MeOH)

11-Deoxy

Acerogenin K

[150107-14-5]

$C_{19}H_{22}O_3$ 298.382

Constit. of *Acer nikoense* and *Juglans regia*. Needles (MeOH). Mp 238 - 240°. Racemic

11-Deoxy, 9-O-β-D-glucopyranoside

Acerogenin K 9-glucoside

[1400570-81-1]

$C_{25}H_{32}O_8$ 460.523

Constit. of the bark of *Acer nikoense*. Needles (MeOH). Mp 166 - 169°. $[\alpha]_D^{25}$ -71.1 (c, 0.34 in EtOH). C-9 config. not determined. λ_{max} 217 (log ε 4.03); 256 (log ε 3.45); 297 (log ε 3.15) (MeOH)

11-Deoxy, 9-ketone

Acerogenin E. *Dihydroalnusone*

[52467-12-6]

$C_{19}H_{20}O_3$ 296.366

Constit. of *Acer nikoense*. Needles (MeOH); leaflets (EtOAc). Mp 240° (230-231°). Racemic

11-Deoxy, 9-ketone, 2-O-β-D-glucopyranoside

Aceroside XI

[150107-11-2]

$C_{25}H_{30}O_8$ 458.507

Constit. of *Acer nikoense*. Amorph. powder. $[\alpha]_D^{27}$ -17.7 (c, 1 in EtOH)

11-Deoxy, 3-hydroxy, 9-ketone, 2-Me ether

[181827-65-6]

$C_{20}H_{22}O_4$ 326.392

Constit. of *Betula maximowicziana*. Pale brown prisms (MeOH). Mp 167 - 170°. Racemic. λ_{max} 250 (log ε 4.02); 298 (log ε 3.73) (MeOH)

13-Oxo, 11-deoxy, 9-ketone, 2-Me ether

17-O-Methyl-7-oxoacerogenin E

[203642-80-2]

$C_{20}H_{20}O_4$ 324.376

Constit. of *Betula davurica*. Amorph. powder. Racemic. λ_{max} 250 (log ε 4.32); 283 (log ε 4.18) (MeOH)

13-Oxo, 4-methoxy, 11-deoxy, 9-ketone, 2-Me ether

15-Methoxy-17-O-methyl-7-oxo-acerogenin E

[203642-81-3]

$C_{21}H_{22}O_5$ 354.402

Constit. of *Betula davurica*. Needles (MeCN). Mp 223 - 224°. Racemic. λ_{max} 261 (log ε 4.42); 288 (log ε 4.15) (MeOH).

Nomura, M. *et al.*, *J. Chem. Soc., Chem. Commun.*, 1975, 316-317 (*Alnusdiol*).
Nomura, M. *et al.*, *Phytochemistry*, 1981, **20**, 1097-1104 (*Acerogenin E*).
Kaneda, N. *et al.*, *Phytochemistry*, 1990, **29**, 3366-3368 (*Alnusdiol, Casuarinondiol*).
Nagumo, S. *et al.*, *Chem. Pharm. Bull.*, 1993, **41**, 1255-1257 (*Aceroside XI*).
Inoue, T. *Yakugaku Zasshi*, 1993, **113**, 181-197 (*rev*).
Nagumo, S. *et al.*, *Chem. Pharm. Bull.*, 1996, **44**, 1086-1089 (*Acerogenin K*).
Fuchino, H. *et al.*, *Chem. Pharm. Bull.*, 1996, **44**, 1748-1758 (*Betula maximowicziana constits*).
Hanawa, F. *et al.*, *Phytochemistry*, 1997, **45**, 589-595 (*Betula maximowicziana constits, abs config, cryst struct*).
Fuchino, H. *et al.*, *Chem. Pharm. Bull.*, 1998, **46**, 166-168 (*Betula davurica constits*).
Akihisa, T. *et al.*, *Chem. Biodiversity*, 2012, **9**, 1475-1489 (*Acerogenin K 9-glucoside*).
Chiba, K. *et al.*, *J. Wood Chem. Technol.*, 2013, **33**, 44-51 (*Alnusdione*).
Ogura, T. *et al.*, *Tetrahedron*, 2013, **69**, 2807-2815 (*Acerogenins E,K, synth*).
Sung, S.H. *et al.*, *Bioorg. Med. Chem. Lett.*, 2015, **25**, 4648-4651 (*10-Hydroxyalnusdione*).
Jahng, Y. *et al.*, *Molecules*, 2018, **23**, 3107:1-42 (*rev, isol, synth, activity*)

Alnusone A20

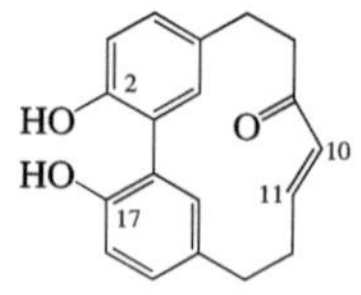

3,17-Dihydroxytricyclo[12.3.1.1^{2,6}]nonadeca-1(18),2,4,6(19),10,14,16-heptaen-9-one, CAS

$C_{19}H_{18}O_3$ 294.35

(E)-form [52330-11-7]

[52330-14-0] (di-Ac) [56438-63-2] (di-Me ether)

Constit. of *Alnus firma*, *Alnus japonica*, *Alnus sieboldiana* and *Corylus avellana*. Mp 253 - 255°. λ_{max} 217 (ε 22000); 239 (sh); 300 (ε 7200) (EtOH)

2-O-α-L-Arabinopyranoside, 17-O-β-D-glucopyranoside

Giffonin I

[1638542-54-7]

$C_{30}H_{36}O_{12}$ 588.606

Constit. of the leaves of *Corylus avellana*. Amorph. solid. $[\alpha]_D^{25}$ -4 (c, 0.1 in MeOH)

10,11-Epoxide

Alnusoxide

[52330-13-9]

$C_{19}H_{18}O_4$ 310.349

Constit. of the wood of *Alnus japonica*. Mp 266 - 269°.

Nomura, M. *et al.*, *J. Chem. Soc., Chem. Commun.*, 1974, 65-66 (*Alnusone, Alnusoxide*).
Semmelhack, M.F. *et al.*, *J. Am. Chem. Soc.*, 1981, **103**, 6460-6471 (*Alnusone, synth*).
Chiba, K. *et al.*, *J. Wood Chem. Technol.*, 2013, **33**, 44-51 (*Alnus sieboldiana constit*).
Masullo, M. *et al.*, *J. Nat. Prod.*, 2015, **78**, 17-25 (*Giffonin I*).
Masullo, M. *et al.*, *Phytochemistry*, 2016, **130**, 273-281 (*Corylus avellana constits*)

Alopecuroides A A21

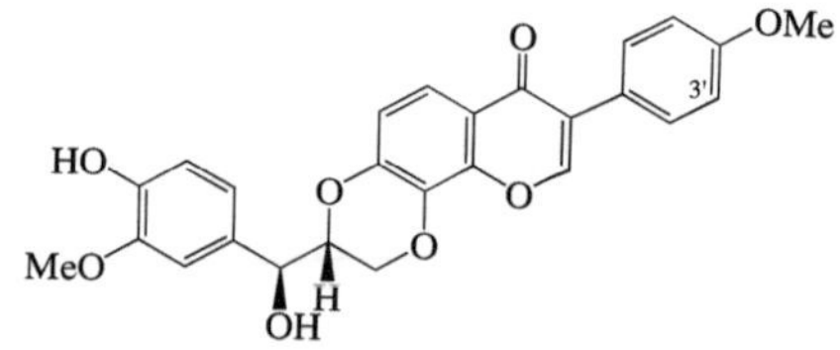

Relative Configuration

[949569-60-2]

$C_{26}H_{22}O_8$ 462.454

Flavonolignan. Constit. of *Pueraria alopecuroides*. Potential antifouling agent. Needles. Mp 225 - 227°. $[\alpha]_D^{27}$ +4.4 (c, 0.84 in Py). λ_{max} 207 (log ε 4.14); 217 (log ε 4.15); 260 (log ε 4.32); 379 (log ε 2.47) (MeOH/$CHCl_3$)

3′-Hydroxy

3′-Hydroxyalopecuroides A

[1238750-82-7]

$C_{26}H_{22}O_9$ 478.453

Constit. of the aerial parts of *Alhagi pseudalhagi*. Quinone reductase inducing agent. Powder. Racemic. λ_{max} 259 (MeOH).

Zhang, Z. *et al.*, *Chin. Chem. Lett.*, 2007, **18**, 297-299 (*Alopecuroides A*).
Pat. Coop. Treaty (WIPO), 2009, 115051 (*activity*).
Li, N. *et al.*, *Fitoterapia*, 2010, **81**, 1058-1061 (*3′-Hydroxyalopecuroides A*)

Alpinin A‡ A22

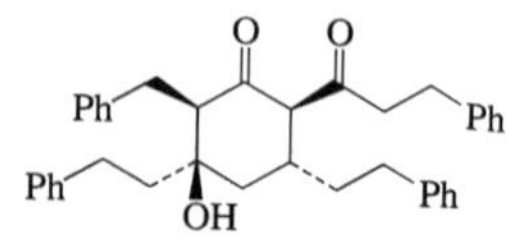

Relative Configuration

[1358917-48-2]

$C_{38}H_{40}O_3$ 544.735

Constit. of the rhizomes of *Alpinia officinarum*. Cryst.. $[\alpha]_D^{28}$ -66 (c, 0.01 in MeCN).

Liu, D. *et al.*, *Chin. Chem. Lett.*, 2012, **23**, 189-192 (*Alpinin A*)

Alpinin B‡ A23

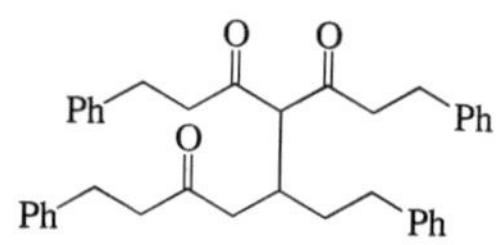

$C_{38}H_{40}O_3$ 544.735

Constit. of the rhizomes of *Alpinia officinarum*. Pale yellow oil.

Liu, D. *et al.*, *Chin. J. Nat. Med.*, 2014, **12**, 139-141 (*Alpinin B*)

Alpinin C A24

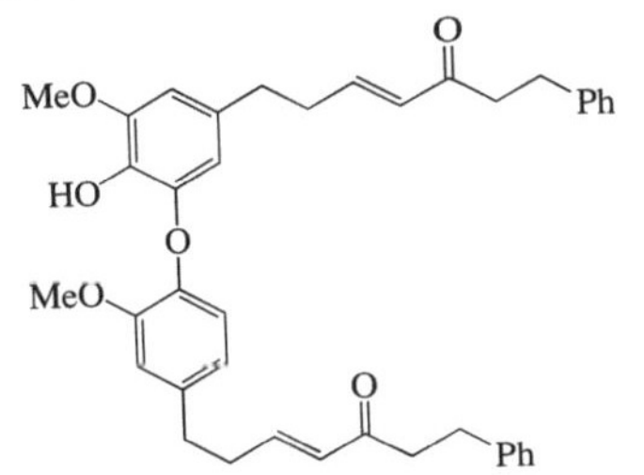

[1787232-96-5]
$C_{40}H_{42}O_6$ 618.77
Constit. of the rhizomes of *Alpinia officinarum*. Exhibits cytotoxic activity. Pale yellow oil. λ_{max} 208 (log ε 3.43) (MeOH).
Liu, D. *et al.*, *Fitoterapia*, 2014, **96**, 76-80 (*Alpinin C, activity*)

Alpinin E A25

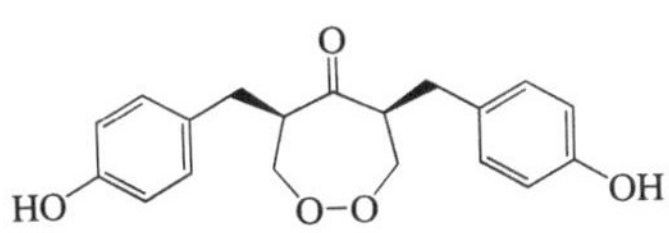

4,6-Bis(4-hydroxybenzyl)-1,2-dioxepan-5-one
[1997309-29-1]
$C_{19}H_{20}O_5$ 328.364
meso-Isomer. Constit. of the rhizomes of *Alpinia officinarum*. Cryst..
Liu, D. *et al.*, *Chem. Nat. Compd. (Engl. Transl.)*, 2016, **52**, 824-826 (*Alpinin E*)

Alpinisin A A26

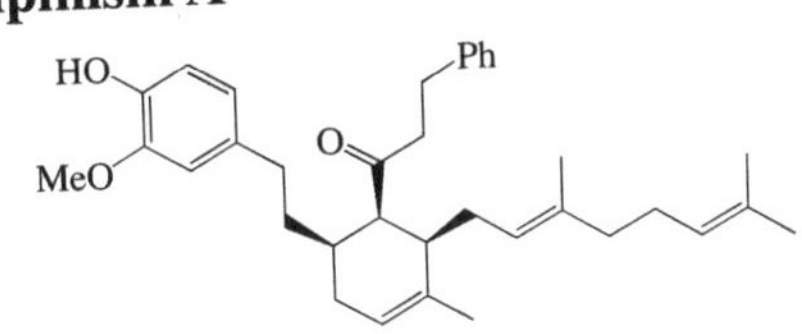

Relative Configuration

[2000239-46-1]
$C_{35}H_{46}O_3$ 514.750
Constit. of the rhizomes of *Alpinia officinarum*. Exhibits antitumour activity. Needles. $[\alpha]^{25}_{D}$ -19 (c, 0.01 in MeOH). λ_{max} 307 (log ε 4.08) (MeOH).
Wei, N. *et al.*, *Nat. Prod. Res.*, 2016, **30**, 2344-2349 (*Alpinisin A, activity*)

Alpinoid A A27

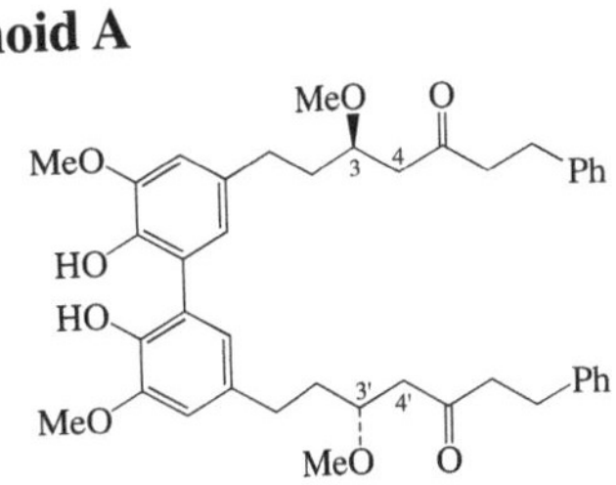

Absolute Configuration

[1005405-96-8]
$C_{42}H_{50}O_8$ 682.854
Constit. of the rhizomes of *Alpinia officinarum*. Oil. $[\alpha]^{25}_{D}$ -6.1 (c, 0.01 in $CHCl_3$). λ_{max} 211 (log ε 3.6); 220 (log ε 3.5); 283 (log ε 0.9) (MeOH)
3,3′-Didemethoxy, 3,3′,4,4′-tetradehydro(E,E-)
Alpinin D
[1787232-97-6]
$C_{40}H_{42}O_6$ 618.77
Constit. of the rhizomes of *Alpinia officinarum*. Pale yellow oil. λ_{max} 226 (log ε 3.23) (MeOH).
Sun, Y. *et al.*, *Helv. Chim. Acta*, 2008, **91**, 118-123 (*Alpinoid A*).
Liu, D. *et al.*, *Fitoterapia*, 2014, **96**, 76-80 (*Alpinin D*)

Amarusin A A28

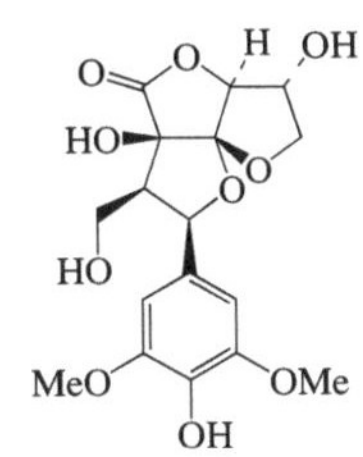

Absolute Configuration

[1620825-62-8]
$C_{17}H_{20}O_{10}$ 384.337
Constit. of the leaves of *Pleioblastus amarus*. Antioxidant. Amorph. powder. λ_{max} 210 (MeOH).
Sun, J. *et al.*, *Tetrahedron Lett.*, 2014, **55**, 4529-4531 (*Amarusin A, cd, activity*)

3-Amino-3-(3,4-dihydroxyphenyl)propanoic acid A29

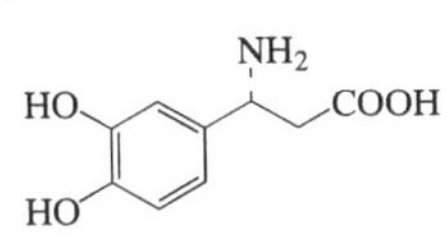

(*R*)-*form*

β-Amino-3,4-dihydroxybenzenepropanoic acid, CAS. *β-DOPA*
[174502-37-5]
$C_9H_{11}NO_4$ 197.19
(R)-form [220946-55-4]
[911224-87-8] (hydrochloride)
Present in the mushroom *Cortinarius violaceus* as the dark violet Fe^{III} complex responsible for the mushroom's colouring. Mp 194°. $[\alpha]^{23}_{D}$ -12.1 (c, 0.5 in 6*M* HCl)
(S)-form
Di-Me ether
β-Amino-3,4-dimethoxybenzenepropanoic acid. 3-Amino-3-(3,4-dimethoxyphenyl)propanoic acid
[696641-73-3]
$C_{11}H_{15}NO_4$ 225.244
Mp 224 - 229°. $[\alpha]^{25}_{D}$ +1.3 (c, 0.51 in H_2O)
Di-Me ether, hydrochloride
[491578-65-5]
Mp 177 - 181°. $[\alpha]^{25}_{D}$ +7.2 (c, 0.315 in H_2O) (>99% ee)
(±)-form
[26049-11-6] (hydrochloride)
Mp 186° (hydrochloride).
Ger. Pat., 1970, 1 931 240 (*synth*).
Von Nussbaum, F. *et al.*, *Angew. Chem., Int. Ed.*, 1998, **37**, 3292-3295 (*Cortinarius violaceus metab, synth*).
Spiteller, P. *et al.*, *Angew. Chem., Int. Ed.*, 2000, **39**, 2754-2756 (*biosynth*).
Davies, S.G. *et al.*, *Tetrahedron: Asymmetry*, 2006, **17**, 1793-1811 (*R-form, synth, pmr, cmr*).
Tassnádi, G. *et al.*, *Tetrahedron: Asymmetry*, 2008, **19**, 2072-2077 (*S-form, di-Me ether*)

2-Amino-3-(3,4-dihydroxyphenyl)-1-propanol A30

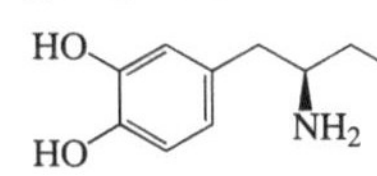

(*S*)-*form*

β-Amino-3,4-dihydroxybenzenepropanol. 4-(2-Amino-3-hydroxypropyl)-1,2-benzenediol. 3,4-Dihydroxyphenyalaninol. 3-Hydroxytyrosinol
$C_9H_{13}NO_3$ 183.207
(S)-form
N-(3,4-Dihydroxybenzoyl)
Nigerrimine B
[1607021-18-0]
$C_{16}H_{17}NO_6$ 319.313
Alkaloid from *Retiboletus nigerrimus*. Amorph. solid. Mp 239° dec.. $[\alpha]^{25}_{D}$ -52.2 (c, 10.6 in MeOH). λ_{max} 206 (sh); 256 (log ε 2.51); 285 (log ε 2.42) (MeOH)
(ξ)-form
3-Me ether, N-(4-hydroxy-3-methoxy-E-cinnamoyl)
Cinnaretamine
[1269005-81-3]
$C_{20}H_{23}NO_6$ 373.405
Alkaloid from *Cinnamomum reticulatum* and *Cinnamomum burmannii*. Amorph. powder. $[\alpha]^{25}_{D}$ +12.4 (c, 0.01 in MeOH). λ_{max} 220 (log ε 3.4); 290 (log ε 2.65); 316 (log ε 3.12) (MeCN)
3-Me ether, N-(4-hydroxy-3-methoxy-Z-cinnamoyl)
Cinnabutamine

[1360746-15-1]
$C_{20}H_{23}NO_6$ 373.405
Alkaloid from the stems of *Cinnamomum burmannii*. Amorph. powder. $[\alpha]_D^{25}$ +11.7 (c, 0.01 in MeOH). λ_{max} 220 (log ε 3.32); 290 (log ε 2.67); 320 (log ε 3.13) (MeCN).

Hong, Z.-L. *et al., Nat. Prod. Commun.*, 2011, **6**, 1297-1298 (*Cinnabutamine*).
Chen, C.-Y. *et al., Nat. Prod. Res.*, 2011, **25**, 26-30 (*Cinnaretamine*).
Gruber, G. *et al., Z. Naturforsch., B*, 2013, **68**, 675-682 (*Nigerrimine B*)

6′-Amino-3,3′-ligna-8,8′-dien-6-ol A31

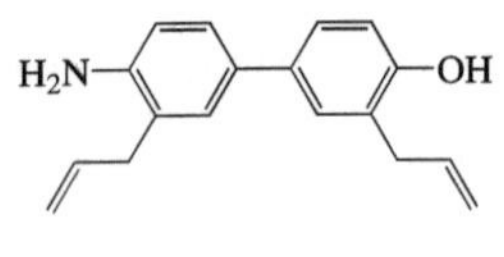

6-Amino-6′-hydroxy-3,3′-neoligna-8,8′-diene. 3,3′-Diallyl-4-amino-4′-hydroxybiphenyl
$C_{18}H_{19}NO$ 265.356
N-[2-(4-Hydroxyphenyl)ethyl]
[2137044-91-6]
$C_{26}H_{27}NO_2$ 385.507
Constit. of *Magnolia grandiflora*. Exhibits weak antiplasmodial activity. Oil. λ_{max} 219; 280 (MeCN aq.).

Latif, A. *et al., Chem. Biodiversity*, 2017, **14**, e1700209:1-7 (*Magnolia grandiflora constit, activity*)

7′-Amino-4,8′-oxylign-7-ene-3,3′,4′,5,5′,9,9′-heptol A32

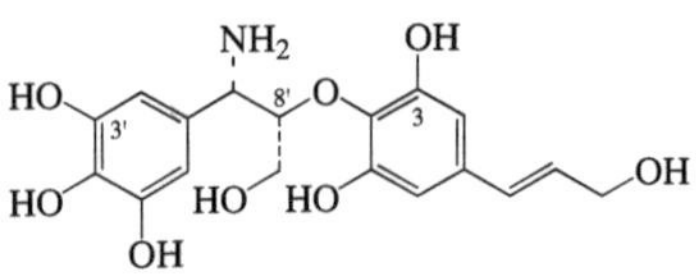

7′-Amino-3,3′,4′,5,5′,9,9′-heptahydroxy-4,8′-oxyneolign-7-ene
$C_{18}H_{21}NO_8$ 379.365
(7E,7′S,8′S)-form
N-(2-Hydroxyethyl), 3,3′,4′,5,5′-penta-Me ether
Hispidacine
[1640132-08-6]
$C_{25}H_{35}NO_9$ 493.553
Constit. of the stem bark of *Ficus hispida*. Moderate vasorelaxant. Light yellow oil. $[\alpha]_D^{25}$ +20 (c, 0.28 in $CHCl_3$). λ_{max} 228 (log ε 3.86); 271 (log ε 3.60) (EtOH).

Yap, V.A. *et al., Phytochemistry*, 2015, **109**, 96-102 (*Hispidacine, cd, activity*)

Anthelminthicol A A33

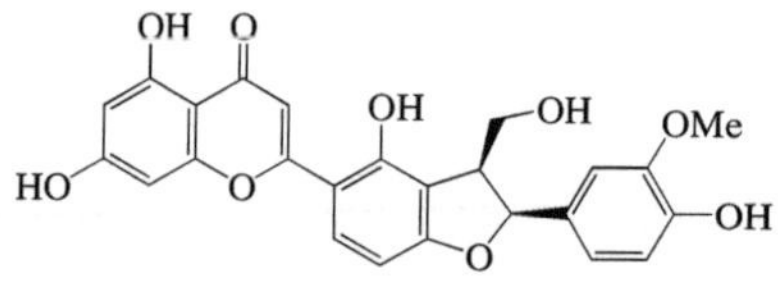

Relative Configuration

[1340493-70-0]
$C_{25}H_{20}O_9$ 464.426
Flavonolignan. Closely related to Isohydnocarpin, I38 and Silychristin, S61. Constit. of *Hydnocarpus anthelminthica* and *Inula falconeri*. Amorph. yellow powder. λ_{max} 281 (log ε 5.23); 339 (log ε 5.01); 385 (log ε 5.05) (MeOH).

Wang, J.-F. *et al., J. Asian Nat. Prod. Res.*, 2011, **13**, 80-83 (*Anthelminthicol A*)

Anthocerotonic acid A34

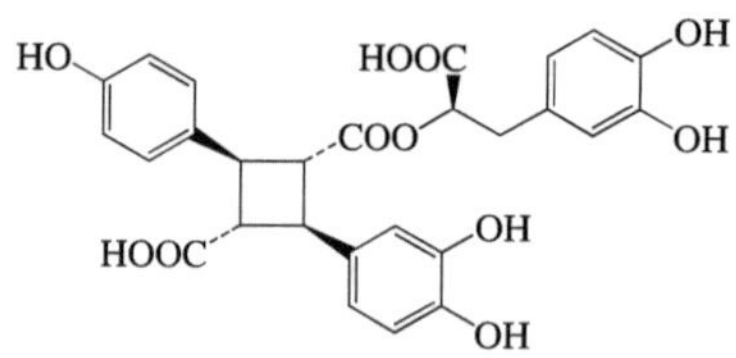

Absolute Configuration

Anthoceratonic acid
[130396-77-9]
$C_{27}H_{24}O_{11}$ 524.478
Isol. from the hornwort *Anthoceros punctatus*. $[\alpha]_D$ +4.4 (c, 1.1 in 5% AcOH aq.). First isol. of a lignan from a non-vascular plant. [acid] λ_{max} 226 (ε 18400); 282 (ε 6600) (5% AcOH) (Derep).

Takeda, R. *et al., Tetrahedron Lett.*, 1990, **31**, 4159-4162 (*Anthoceratonic acid, cd*)

Antiarisin A A35

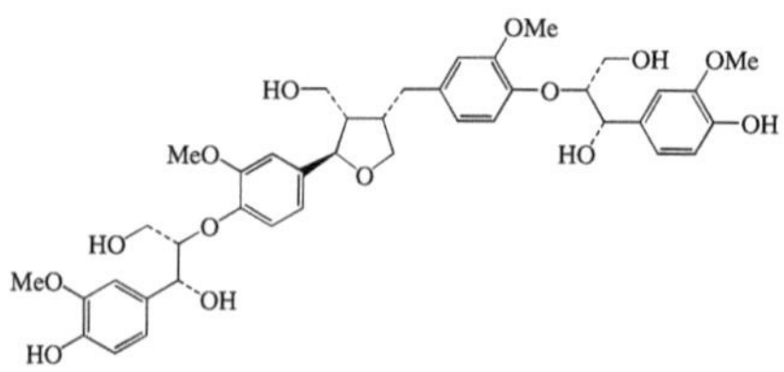

Relative Configuration

[1187831-93-1]
$C_{40}H_{48}O_{14}$ 752.81
Only the configs. of the chiral centres in each part of the molecule are known, not the relationship between them. Constit. of the stems of *Antiaris toxicaria*. Powder. $[\alpha]_D^{25}$ +54 (c, 0.1 in MeOH). λ_{max} 204 (log ε 3.95); 266 (log ε 3.21); 279 (log ε 2.13) (MeOH).

Jiang, M.-M. *et al., Planta Med.*, 2009, **75**, 340-345 (*Antiarisin A*)

Antidesmanin A A36

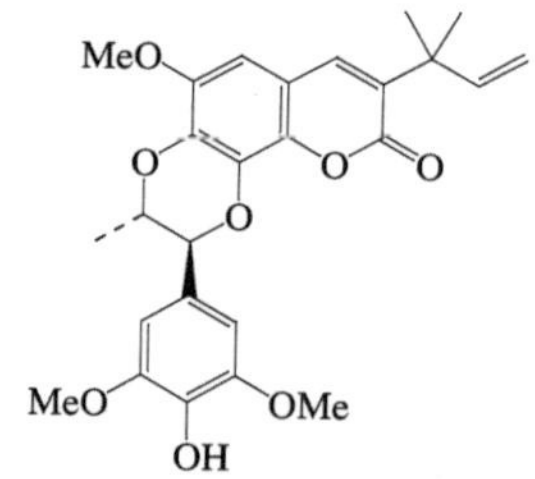

Relative Configuration

[842120-01-8]
$C_{26}H_{28}O_8$ 468.502
Constit. of the roots of *Antidesma pentandrum* var. *barbatum*. Shows weak cytotoxic activity. Oil. Racemic. λ_{max} 220 (log ε 5.01); 235 (sh); 328 (log ε 4.67) (MeOH). [base] λ_{max} 222 (log ε 5.02); 258 (log ε 4.83); 305 (sh) (MeOH/KOH).

Chen, Y.-C. *et al., Helv. Chim. Acta*, 2004, **87**, 2805-2811 (*Antidesmanin A, activity*)

Antidesmanin B A37

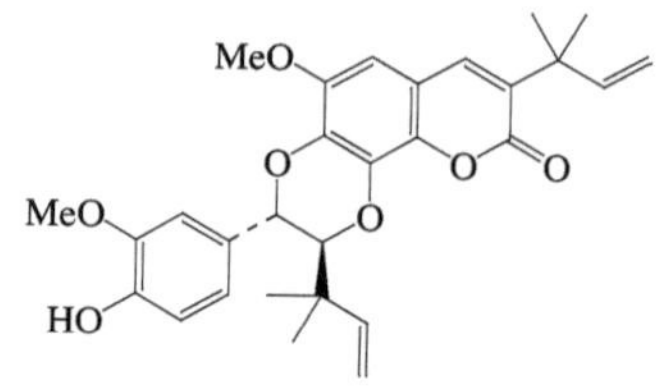

Relative Configuration

[842120-02-9]
$C_{29}H_{32}O_7$ 492.568
Constit. of the roots of *Antidesma pentandrum*. Shows weak cytotoxic activity. Yellowish oil. Racemic. λ_{max} 223 (log ε 5); 232 (log ε 5); 289 (sh); 331 (log ε 4.77) (MeOH). [base] λ_{max} 222 (log ε 5); 258 (log ε 4.83); 305 (sh); 340 (log ε 4.78) (MeOH/KOH).

Chen, Y.-C. *et al., Helv. Chim. Acta*, 2004, **87**, 2805-2811 (*Antidesmanin B, activity*)

Antidesmanin C A38

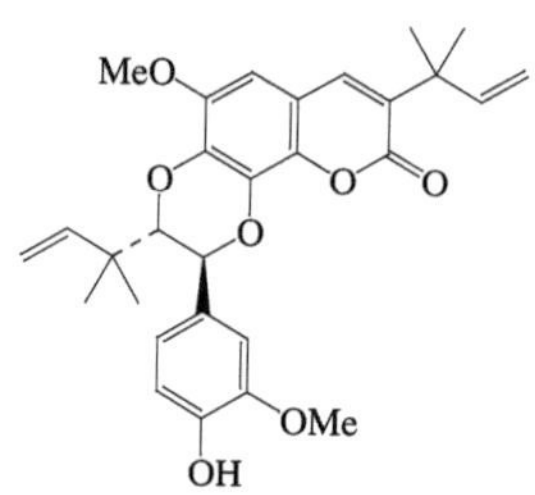

Relative Configuration

[842120-03-0]
$C_{29}H_{32}O_7$ 492.568
Constit. of the roots of *Antidesma pentandrum*. Shows weak cytotoxic activity. Oil. Racemic. Antidesmanin D

also isol. but struct. not fully elucidated. λ_{max} 219 (log ε 4.6); 235 (sh); 288 (log ε 3.98); 332 (log ε 4.21) (MeOH). [base] λ_{max} 219 (log ε 4.61); 258 (log ε 4.32); 302 (log ε 4.08); 348 (log ε 4.24) (MeOH/KOH).

Chen, Y.-C. *et al., Helv. Chim. Acta*, 2004, **87**, 2805-2811 (*Antidesmanin C, activity*)

Antidesmanin E A39

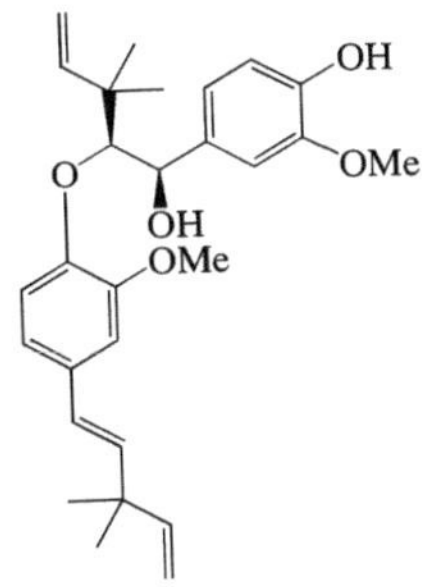

[1021300-68-4]
$C_{28}H_{36}O_5$ 452.591
Constit. of the roots of *Antidesma pentandrum* var. *barbatum*. Oil. Racemic. λ_{max} 210 (log ε 4.42); 266 (log ε 4.13); 306 (sh) (MeOH). [base] λ_{max} 210 (log ε 4.68); 258 (log ε 4.19); 298 (sh) (MeOH/KOH aq.)

threo-Diastereomer
Antidesmanin F
[1021300-69-5]
$C_{28}H_{36}O_5$ 452.591
Constit. of the roots of *Antidesma pentandrum* var. *barbatum*. Oil. Racemic. λ_{max} 210 (log ε 4.4); 268 (log ε 4.1); 306 (sh) (MeOH). [base] λ_{max} 210 (log ε 4.67); 262 (log ε 4.15); 294 (sh) (MeOH/KOH aq.).

Chen, Y.-C. *et al., J. Chin. Chem. Soc. (Taipei)*, 2007, **54**, 1325-1332 (*Antidesmanins E,F*)

Aphelandrine A40

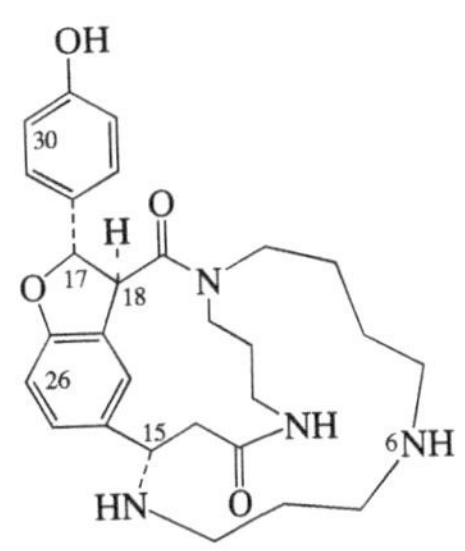

(+)-*form*

11-Epiephedradine A. 11-Epiorantine
[75363-58-5] (unspecified stereochem.)
$C_{28}H_{36}N_4O_4$ 492.62
(+)-form [69721-57-9]
Alkaloid from *Aphelandra squarrosa, Aphelandra tetragona, Encephalosphaera lasiandra* and *Premna integrifolia*. Plates (MeOH); needles (H_2O). Mp 263° dec.. [α] D +214 (c, 0.674 in $CHCl_3$/MeOH 9:1)

Perchlorate (1:2)
Cryst. (EtOH). Mp 252 - 253° dec.

N^6-Hydroxy
N^6-Hydroxyaphelandrine
[259813-07-5]
$C_{28}H_{36}N_4O_5$ 508.619
Alkaloid from *Aphelandra fuscopunctata*. Cryst.. Mp 186 - 190° dec.. $[\alpha]_D^{21}$ +125 (c, 0.18 in EtOH). λ_{max} 232; 280; 286 (EtOH)

N^6-Acetoxy
N^6-Acetoxyaphelandrine
[259813-08-6]
$C_{30}H_{38}N_4O_6$ 550.656
Alkaloid from *Aphelandra fuscopunctata*. Cryst.. Mp 176 - 184° dec.. $[\alpha]_D^{22}$ +192 (c, 0.2 in EtOH). λ_{max} 232; 280; 286 (EtOH)

17,18-Diepimer
Orantine. *Ephedradine A*
[71461-13-7]
$C_{28}H_{36}N_4O_4$ 492.62
Isol. from the crude drug ''Mao-kon'', made from the roots of *Ephedra* spp.. Hypotensive principle. Noncryst.; cryst. + $1H_2O$ (as dihydrochloride). Mp 222 - 225° (dihydrochloride). $[\alpha]_D^{25}$ -72 (c, 0.25 in $CHCl_3$/MeOH) (synthetic)

17,18-Diepimer, N,N,O-tri-Ac
[71657-56-2]
Mp 165 - 175°

17,18-Diepimer, Me ether
O-Methylorantine
[64366-67-2]
$C_{29}H_{38}N_4O_4$ 506.647
Alkaloid from *Aphelandra sinclairiana, Chaenorhinum minus* and *Chaenorhinum villosum*. Cryst. (C_6H_6). Mp 167° dec.. [α] D -84 (c, 0.88 in $CHCl_3$/MeOH)

17,18-Diepimer, 26-methoxy
Ephedradine D
[81349-26-0]
[81371-56-4] (dihydrobromide)
$C_{29}H_{38}N_4O_5$ 522.646
Isol. from the crude drug ''Mao-kon'', made from the roots of *Ephedra* spp.. Hypotensive principle. Noncryst.. Mp 219 - 221° (as dihydrobromide). [α] D -85.3 (H_2O) (dihydrobromide)

17,18-Diepimer, 30-methoxy
Ephedradine B
[71327-57-6]
[71295-26-6] (dihydrobromide)
$C_{29}H_{38}N_4O_5$ 522.646
Isol. from the crude drug ''Mao-kon'', made from the roots of *Ephedra* spp.. Hypotensive principle. Noncryst.; cryst. + $1H_2O$ (as dihydrobromide). Mp 219 - 221° (dihydrobromide). [α] D -101.5 (H_2O)

17,18-Diepimer, 30-methoxy, Me ether
Ephedradine C
[73276-37-6]
[73307-45-6] (dihydrobromide)
$C_{30}H_{40}N_4O_5$ 536.673
Isol. from the crude drug ''Mao-kon'', made from the roots of *Ephedra* spp.. Hypotensive principle. Noncryst.; cryst. + $1H_2O$ (as dihydrobromide). Mp 224 - 225° (dihydrobromide). [α] D -100.7 (H_2O)

(-)-form [133162-95-5]
Alkaloid from the whole plant of *Schweinfurthia papilionacea*. Mp 200° dec.. [α] D -13 (c, 0.3 in MeOH). This alkaloid, claimed as new and named as 11-Epiephedradine A or 11-Epiorantine, appears to be (-)-Aphelandrine of low opt. purity. However this requires confirmation.

Dätwyler, P. *et al., Helv. Chim. Acta*, 1978, **61**, 2646-2671 (*Aphelandrine*).
Dätwyler, P. *et al., Helv. Chim. Acta*, 1979, **62**, 2712-2723 (*Aphelandrine, Orantine, O-Methylorantine, abs config*).
Tamada, M. *et al., Heterocycles*, 1979, **12**, 783-786 (*Ephedradine B*).
Tamada, M. *et al., Tetrahedron Lett.*, 1979, 873-876 (*Ephedradine A, cryst struct, abs config*).
Konno, C. *et al., Heterocycles*, 1980, **14**, 295-298 (*Ephedradine C*).
Hikino, H. *et al., Heterocycles*, 1982, **17**, 155-158 (*Ephedradine D*).
Wasserman, H.M. *et al., J. Am. Chem. Soc.*, 1985, **107**, 519-521 (*O-Methylorantine, synth*).
Guggisberg, A. *et al., Helv. Chim. Acta*, 1986, **69**, 1012-1016 (*Aphelandrine, cryst struct, abs config*).
Ahmad, V.U. *et al., J. Nat. Prod.*, 1990, **53**, 1162-1167 (*11-Epiephedradine A*).
Papazoglou, G. *et al., Helv. Chim. Acta*, 1991, **74**, 565-571 (*Aphelandrine, biosynth*).
Youhnovski, N. *et al., Phytochemistry*, 1999, **52**, 1717-1723 (*N-Hydroxyaphelandrine, N-acetoxyaphelandrine*).
Nezbedova, L. *et al., Tetrahedron Lett.*, 2000, **41**, 7859-7862 (*Aphelandrine, Orantine, biosynth*).

Dimitrov, V. *et al.*, *Helv. Chim. Acta*, 2001, **84**, 2108-2118 (*Aphelandrine, Orantine, synth*).
Nezbedova, L. *et al.*, *Tetrahedron Lett.*, 2001, **42**, 4139-4141 (*Aphelandrine, synth*).
Kurosawa, W. *et al.*, *Tetrahedron*, 2004, **60**, 9615-9628 (*Ephedradine A, synth*)

Apteniol G A41

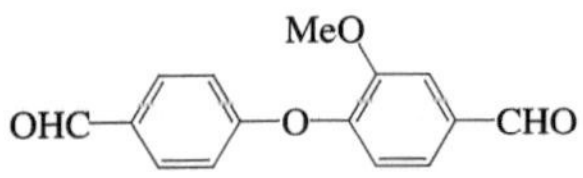

3-Methoxy-4,4′-oxydibenzaldehyde. 3-Methoxy-4,4′-oxybisbenzaldehyde. 4,4′-Diformyl-2-methoxydiphenyl ether. 3-Methoxy-8,8′,9,9′-tetranor-4,4′-oxylignan-7,7′-dial. 3-Methoxy-8,8′,9,9′-tetranor-7,7′-dioxo-4,4′-oxyneolignan
[408341-11-7]
$C_{15}H_{12}O_4$ 256.257
Struct. in doubt after synthetic studies in 2015. Constit. of the leaves of *Aptenia cordifolia*.

Greca, M.D. *et al.*, *Chem. Biodiversity*, 2007, **4**, 118-128 (*Apteniol G*).
Noshita, T. *et al.*, *Biosci., Biotechnol., Biochem.*, 2015, **79**, 1743-1749 (*synth*)

Araliangin A42

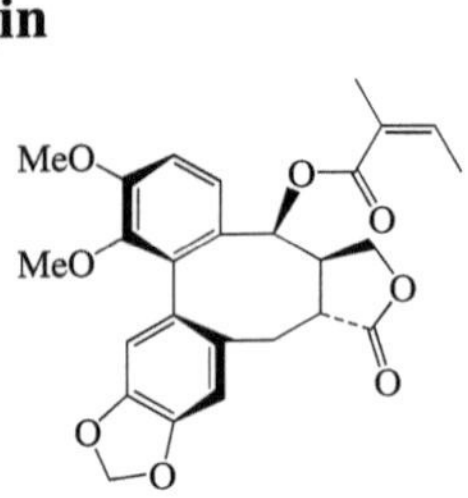

[85915-63-5]
$C_{26}H_{26}O_8$ 466.486
Constit. of *Steganotaenia araliacea*. Sol. MeOH, EtOH; poorly sol. H_2O, hexane. $[\alpha]^{22}_D$ -43 (c, 0.79 in $CHCl_3$). Similar struct. to Steganangin (see Steganol, S83). λ_{max} 255 (ε 10700); 287 (ε 5450) (EtOH) (Derep).

Taafrout, M. *et al.*, *Tetrahedron Lett.*, 1983, **24**, 197-200 (*isol, struct*).
Robin, J.P. *et al.*, *Tetrahedron Lett.*, 1986, **27**, 2871-2874 (*stereochem, anal*).
Wickramatne, D.B.M. *et al.*, *J. Nat. Prod.*, 1993, **56**, 2083-2090 (*isol*).
Baudoin, O. *et al.*, *Natural Products Chemistry*, 2003, **29**, 355-417 (*Araliangin, rev*)

Arctignan A A43

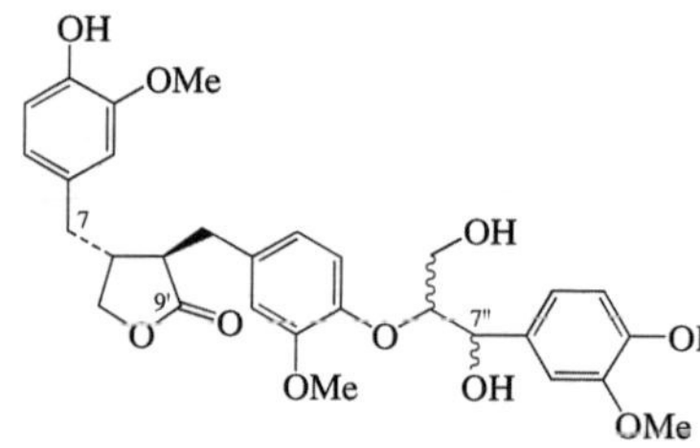

[155661-08-8]
$C_{30}H_{34}O_{10}$ 554.592
Not to be confused with Arctignan A, A44. Regioisomer of Lappaol E , L6. Constit. of the fruit of *Arctium lappa*. Amorph. powder. $[\alpha]_D$ -25.3 (c, 0.32 in MeOH)

7R-Hydroxy

7R-Hydroxyarctignan A
[845890-65-5]
$C_{30}H_{34}O_{11}$ 570.591
Constit. of *Picea abies*

7S-Hydroxy

7S-Hydroxyarctignan A
[845890-83-7]
$C_{30}H_{34}O_{11}$ 570.591
Constit. of *Picea abies*

7″-Deoxy

Sesquipinsapol A
[155969-62-3]
$C_{30}H_{34}O_9$ 538.593
Constit. of the wood of *Abies pinsapo* and *Pinus massoniana*. Syrup. $[\alpha]_D$ -8.3 (c, 1.3 in $CHCl_3$)

7″-Deoxy, 9′R-alcohol(lactol)

Sesquipinsapol C
[158204-47-8]
$C_{30}H_{36}O_9$ 540.609
Constit. of the wood of *Abies pinsapo*. Oil (as tetra-Ac). $[\alpha]^{25}_D$ -21.8 (c, 1.2 in $CHCl_3$) (tetra-Ac).

Umehara, K. *et al.*, *Chem. Pharm. Bull.*, 1993, **41**, 1774-1779 (*Arctignan A*).
Barrero, A.F. *et al.*, *Nat. Prod. Lett.*, 1993, **2**, 255-262 (*Sesquipinsapol A*).
Barrero, A.F. *et al.*, *J. Nat. Prod.*, 1994, **57**, 713-719 (*Sesquipinsapol C*).
Willfoer, S. *et al.*, *Holzforschung*, 2004, **58**, 345-354 (*7-Hydroxyarctignans*)

Arctignan A‡ A44

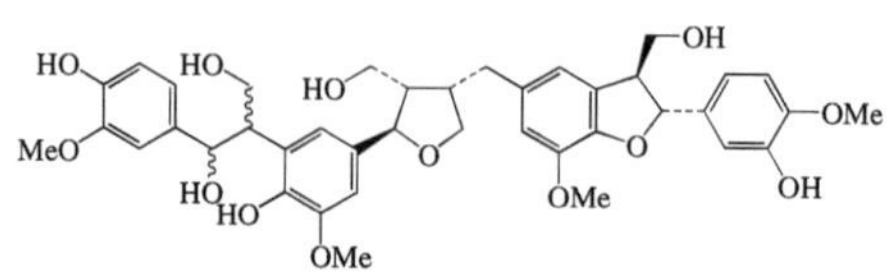

Relative Configuration

$C_{40}H_{46}O_{13}$ 734.795
Not to be confused with Arctignan A, A43. Constit. of the seeds of *Arctium lappa*. Oil. $[\alpha]^{26}_D$ -5.7 (c, 0.25 in MeOH). λ_{max} 282 (log ε 4.08) (MeOH).

Qin, Z.-B. *et al.*, *Phytochem. Lett.*, 2019, **34**, 43-49 (*Arctignan A*)

Arctignan B‡ A45

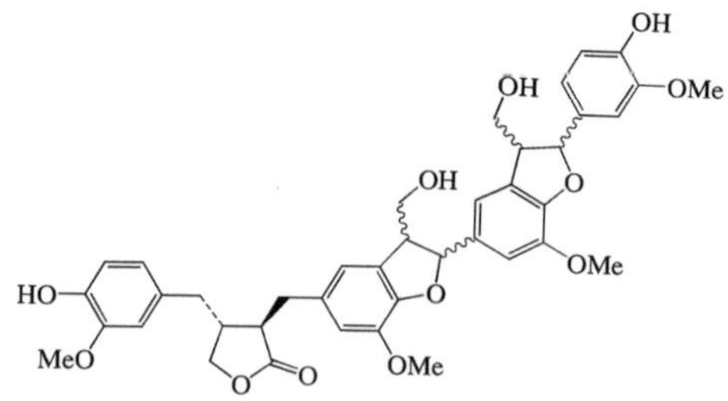

Absolute Configuration

$C_{40}H_{42}O_{12}$ 714.764
Not to be confused with Arctignan B, in L4. Constit. of the seeds of *Arctium lappa*. Oil. $[\alpha]^{26.4}_D$ +14.5 (c, 0.28 in MeOH). λ_{max} 205 (log ε 5.29); 283 (log ε 4.32) (MeOH).

Qin, Z.-B. *et al.*, *Phytochem. Lett.*, 2019, **34**, 43-49 (*Arctignan B, cd*)

Arctignan D A46

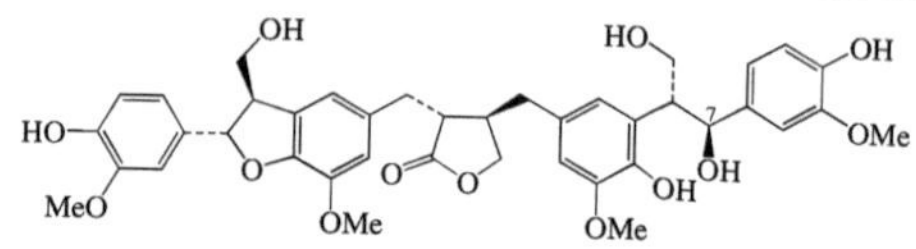

Relative Configuration

[155661-11-3]
$C_{40}H_{44}O_{13}$ 732.779
Related to Lappaol F, L7. Constit. of the fruit of *Arctium lappa*. Amorph. powder. $[\alpha]_D$ -37.2 (c, 0.82 in MeOH). λ_{max} 229 (sh); 282 (log ε 4.11) (MeOH)

7-Ketone

Arctignan F
[186541-65-1]
$C_{40}H_{42}O_{13}$ 730.763
Constit. of the fruit of *Arctium lappa*. Amorph. powder + 0.5 H_2O. $[\alpha]_D$ +82.9 (c, 0.6 in MeOH). λ_{max} 228 (log ε 3.63); 281 (log ε 4.26); 304 (log ε 3.99) (MeOH).

Umehara, K. *et al.*, *Chem. Pharm. Bull.*, 1993, **41**, 1774-1779 (*Arctignan D*).
Umehara, K. *et al.*, *Chem. Pharm. Bull.*, 1996, **44**, 2300-2304 (*Arctignan F*)

Arctignan E A47

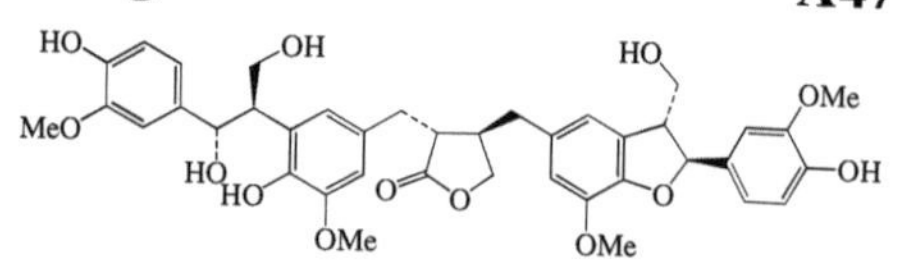

Relative Configuration

[155661-12-4]

$C_{40}H_{44}O_{13}$ 732.779
Related to Lappaol F, L7. Constit. of the fruit of *Arctium lappa*. Amorph. powder. [α] D -27 (c, 0.88 in MeOH).
Umehara, K. *et al.*, *Chem. Pharm. Bull.*, 1993, **41**, 1774-1779 (*Arctignan E*)

Arctignan G A48

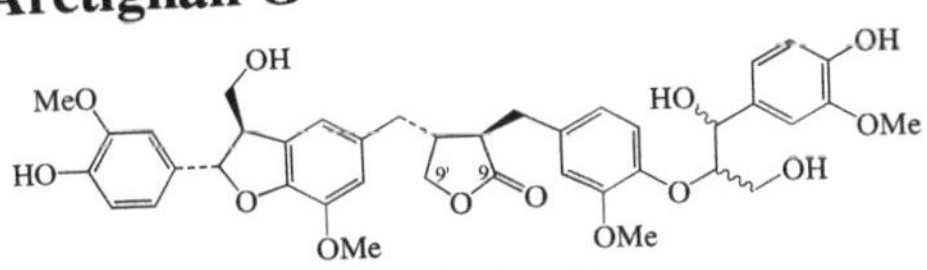

Absolute Configuration

[186541-89-9]
$C_{40}H_{44}O_{13}$ 732.779
Constit. of the fruit of *Arctium lappa*. Moderately potent differentiation inducer in mouse myeloid M1 cells. Yellow powder. [α] D +81.7 (c, 0.1 in MeOH). λ_{max} 231 (log ε 4.52); 281 (log ε 4.13) (MeOH)

9-Deoxo, 9′-oxo

Arctignan H
[186543-11-3]
$C_{40}H_{44}O_{13}$ 732.779
Constit. of the fruit of *Arctium lappa*. Moderately potent differentiation inducer in mouse myeloid M1 cells. Amorph. powder + ½H_2O. [α] D +145.2 (c, 0.06 in MeOH). λ_{max} 228 (log ε 4.65); 282 (log ε 4.3) (MeOH).

Umehara, K. *et al.*, *Chem. Pharm. Bull.*, 1996, **44**, 2300-2304 (*Arctignans G,H, cd, activity*)

Arctiidilactone A49

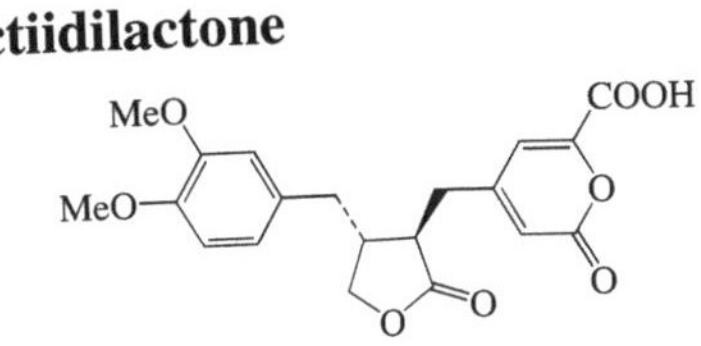

Absolute Configuration

4-[[4-(3,4-Dimethoxybenzyl)-2-oxo-tetrahydrofuran-3-yl]methyl]-2-oxo-2H-pyran-6-carboxylic acid
[1808085-00-8]
$C_{20}H_{20}O_8$ 388.372
Constit. of fruits of *Arctium lappa*. Exhibits strong antiinflammatory effect. Amorph. powder. $[\alpha]_D^{25}$ -28.9 (c, 0.26 in MeOH). λ_{max} 230 (log ε 3.51); 285 (log ε 3.29); 305 (log ε 3.22) (MeOH).
Yang, Y.-N. *et al.*, *J. Agric. Food Chem.*, 2015, **63**, 7958-7966 (*Arctiidilactone, activity, cd*)

Arctiisesquineolignan B A50

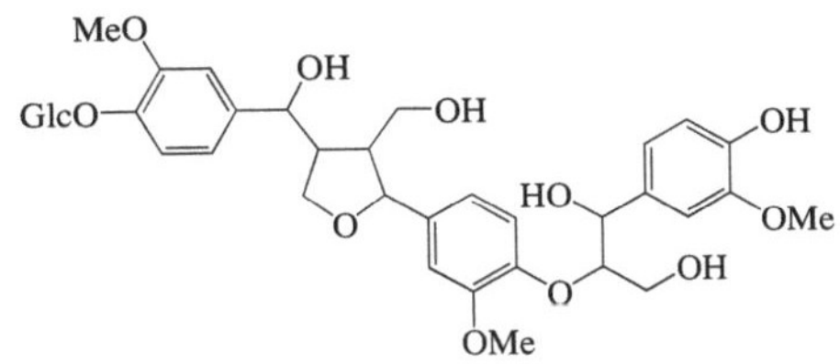

[2131752-70-8]
$C_{36}H_{46}O_{16}$ 734.748
Constit. of the fruit of *Arctium lappa*. Amorph. powder. $[\alpha]_D^{25}$ -10.9 (c, 0.16 in MeOH). λ_{max} 230 (log ε 3.12); 278 (log ε 2.69) (MeOH).
He, J. *et al.*, *J. Asian Nat. Prod. Res.*, 2016, **18**, 423-428 (*Arctiisesquineolignan B*)

Argenteane A51

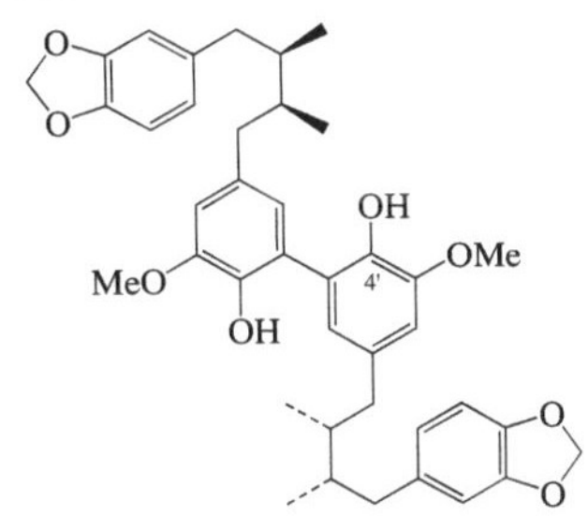

Relative Configuration

[474055-94-2]
$C_{40}H_{46}O_8$ 654.8
Constit. of *Myristica argentea*. Antioxidant. Oil. λ_{max} 220 (log ε 1.53); 250 (sh); 289 (log ε 1.04) (MeOH)

4′-Deoxy

Schibicolignan A
[1801516-65-3]
$C_{40}H_{46}O_7$ 638.801
Constit. of the stems of *Schisandra bicolor*. Yellow oil. $[\alpha]_D^{20}$ +4.28 (c, 0.35 in $CHCl_3$). λ_{max} 220 (log ε 3.91); 239 (log ε 3.85); 289 (log ε 3.73) (MeOH).

Filleur, F. *et al.*, *Nat. Prod. Lett.*, 2002, **16**, 1-7 (*Argenteane*).
Calliste, C.A. *et al.*, *Food Chem.*, 2010, **118**, 489-496 (*Argenteane, activity*).
Chen, Y. *et al.*, *Nat. Prod. Commun.*, 2013, **8**, 1121-1122 (*Schibicolignan A*)

Arillatose A A52

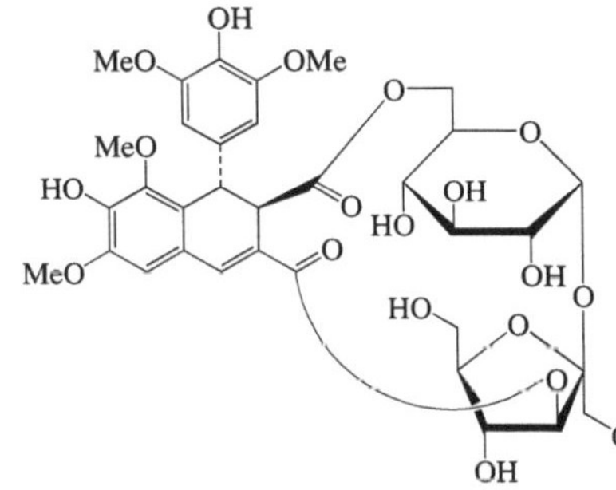

Absolute Configuration

[291746-00-4]
$C_{34}H_{40}O_{19}$ 752.675
Constit. of the roots of *Polygala arillata*. Amorph. powder. $[\alpha]_D^{27}$ +25.1 (c, 0.12 in MeOH). λ_{max} 205 (log ε 4.69); 250 (log ε 4.34); 340 (log ε 4.17) (MeOH).
Kobayashi, W. *et al.*, *J. Nat. Prod.*, 2000, **63**, 1066-1069 (*Arillatose A*)

Arisanschinin M A53

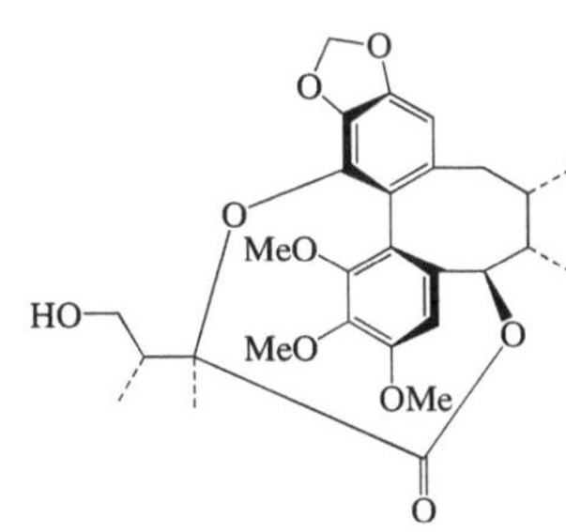

Absolute Configuration

[1420217-14-6]
$C_{28}H_{34}O_9$ 514.571
Constit. of the fruit of *Schisandra arisanensis*. Amorph. powder. $[\alpha]_D^{25}$ -13 (c, 0.1 in CH_2Cl_2). λ_{max} 230 (log ε 1.39); 256 (log ε 2.81); 285 (log ε 2.56) (MeOH).
Chen, Y.-C. *et al.*, *Bioorg. Med. Chem. Lett.*, 2013, **23**, 880-885 (*Arisanschinin M*)

Aristchamic C A54

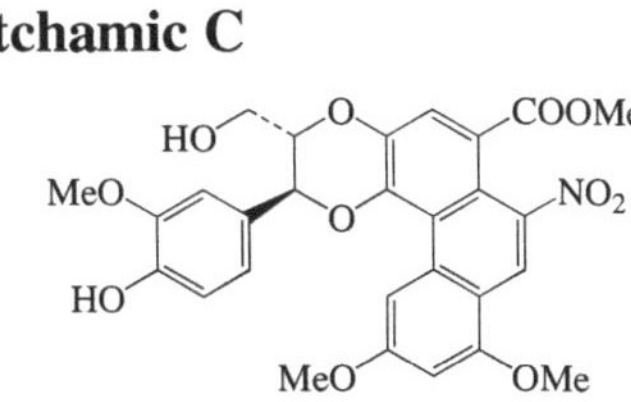

Absolute Configuration

[2247442-10-8]
$C_{28}H_{25}NO_{11}$ 551.504
Alkaloid from the rhizomes of *Aristolochia championii*. Yellow solid. $[\alpha]_D^{20}$ -21.6 (c, 0.04 in $CHCl_3$). Isol. as a mixt. with Aristchamic D and resolved by chiral hplc. λ_{max} 242 (log ε 4.82);

262 (log ε 4.78); 284 (log ε 4.55); 325 (log ε 4.28); 395 (log ε 4.16) (MeOH)

Enantiomer

Aristchamic D

[2247442-11-9]

$C_{28}H_{25}NO_{11}$ 551.504

Alkaloid from the rhizomes of *Aristolochia championii*. Yellow solid. $[\alpha]_D^{20}$ +19.7 (c, 0.04 in $CHCl_3$). Isol. as a mixt. with Aristchamic C and resolved by chiral hplc.

Wang, X. *et al.*, *Fitoterapia*, 2017, **118**, 63-68 (*Aristchamics C,D, cd*)

Aristolactam GI A55

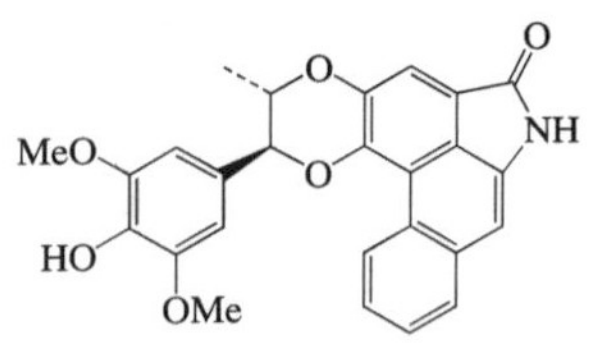

Absolute Configuration

Aristololactam G I

[1418191-51-1]

$C_{26}H_{21}NO_6$ 443.455

Alkaloid from *Fissistigma oldhamii*. Light yellow powder ($CHCl_3$/MeOH). $[\alpha]_D^{25}$ +6 (c, 0.2 in $CHCl_3$). λ_{max} 238 (log ε 4.45); 264 (log ε 4.45); 322 (log ε 4.09); 372 (log ε 3.87); 390 (log ε 3.87) (MeOH).

Ge, Y.-W. *et al.*, *Phytochemistry*, 2013, **86**, 201-207 (*Aristolactam GI*).

Luong, T.M. *et al.*, *J. Org. Chem.*, 2019, **84**, 5747-5756 (*synth*)

Artemicapin D A56

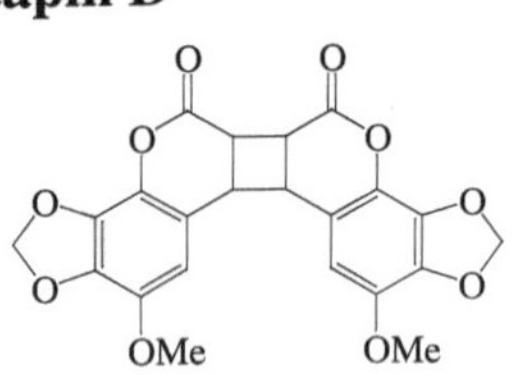

$C_{22}H_{16}O_{10}$ 440.36

Constit. of *Artemisia capillaris*. Needles ($CHCl_3$). Mp 138 - 139°. λ_{max} 222; 291 (sh) (MeOH).

Wu, T.-S. *et al.*, *Bioorg. Med. Chem.*, 2001, **9**, 77-83 (*Artemicapin D*)

Asatone A57

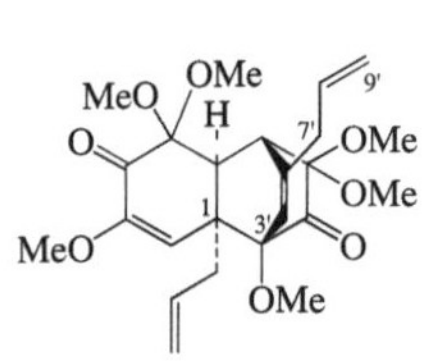

Relative Configuration

2,6'-Cyclo-1,2,5',6'-tetrahydro-3,3,3',5,5',5'-hexamethoxy-1,3'-ligna-8,8'-diene-4,4'(3H,3'H)-dione

[38451-63-7]

$C_{24}H_{32}O_8$ 448.512

Constit. of *Asarum taitonense*, *Heterotropa takaoi* and *Heterotropa aspera*. Exhibits antileukaemic activity in mice. Insect antifeedant. Cryst. (hexane). Mp 101 - 102°. Racemate. λ_{max} 229 (ε 5100); 278 (ε 5660) (MeOH) (Berdy)

$\Delta^{7'}$-Isomer

2,6'-Cyclo-1,2,5',6'-tetrahydro-3,3,3',5,5',5'-hexamethoxy-1,3'-ligna-7',8-diene-4,4'(3H,3'H)-dione. ***Isoasatone A***

[67451-73-4]

$C_{24}H_{32}O_8$ 448.512

Constit. of *Heterotropa takaoi*. Cryst. (Et_2O/hexane). Mp 122 - 124°. Racemate. Not to be confused with Isoasatone, I29.

$\Delta^{7'}$-Isomer, 9'-hydroxy

2,6'-Cyclo-1,2,5',6'-tetrahydro-9'-hydroxy-3,3,3',5,5',5'-hexamethoxy-1,3'-ligna-7',8-diene-4,4'(3H,3'H)-dione. ***Isoasatone B***

[67451-74-5]

$C_{24}H_{32}O_9$ 464.511

Constit. of *Heterotropa takaoi*. Cryst. (Et_2O/hexane). Mp 129 - 131°. Racemate.

Yamamura, S. *et al.*, *Bull. Chem. Soc. Jpn.*, 1976, **49**, 1940-1948 (*Asatone*).

Yamamura, S. *et al.*, *Bull. Chem. Soc. Jpn.*, 1982, **55**, 3573-3579 (*Isoasatones A,B*).

Honda, K. *et al.*, *J. Chem. Ecol.*, 1995, **21**, 1541-1548 (*Asatone, activity*).

Kürti, L. *et al.*, *Eur. J. Org. Chem.*, 1999, 2579-2581 (*synth*).

Chittimalla, S.K. *et al.*, *RSC Adv.*, 2015, **5**, 29391-29396 (*biosynth*)

Asparoffin B A58

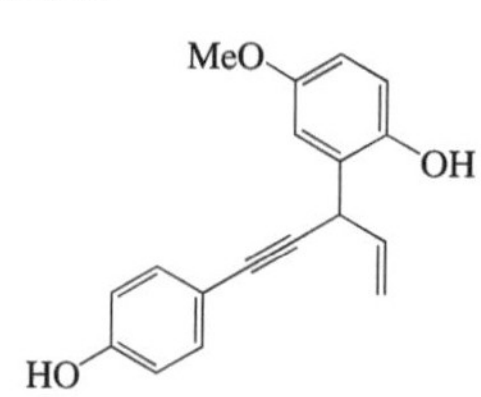

3-(2-Hydroxy-5-methoxyphenyl)-5-(4-hydroxyphenyl)-1-penten-4-yne. 5-Methoxy-9'-nor-7,8'-lign-8-en-7'-yne-2,4'-diol

[2065242-58-0]

$C_{18}H_{16}O_3$ 280.323

Similar to Hinokiresinol, H52. Constit. of the stems of *Asparagus officinalis*. Yellowish oil. Racemic. λ_{max} 278 (log ε 4.25); 293 (log ε 4.14) (MeOH).

Li, X.-M. *et al.*, *J. Asian Nat. Prod. Res.*, 2016, **18**, 344-348 (*Asparoffin B*)

Aspergillulactone A59

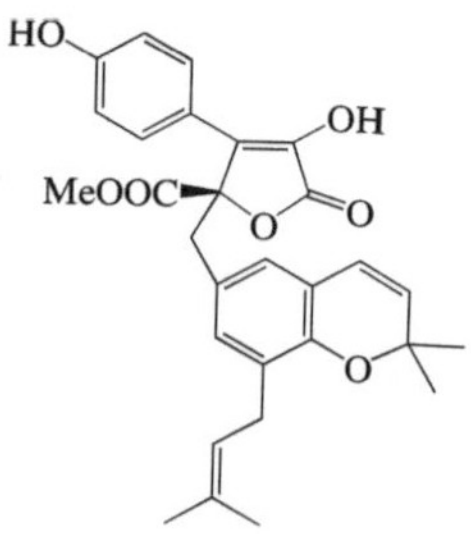

[1637408-24-2]

$C_{29}H_{30}O_7$ 490.552

Provisional prenylated lignan. Biosynth. pathway is unclear. Related to Butyrolactone I, B89 and Butyrolactone V, B92. Prod. by *Aspergillus* sp. PSU-RSPG185. Gum. $[\alpha]_D^{25}$ -12.8 (c, 1 in MeOH). λ_{max} 224 (log ε 3.23); 269 (log ε 2.68); 277 (log ε 2.73); 306 (log ε 2.92) (MeOH).

Rukachaisirikul, V. *et al.*, *J. Nat. Prod.*, 2014, **77**, 2375-2382 (*Aspergillulactone*)

Asperjinone A60

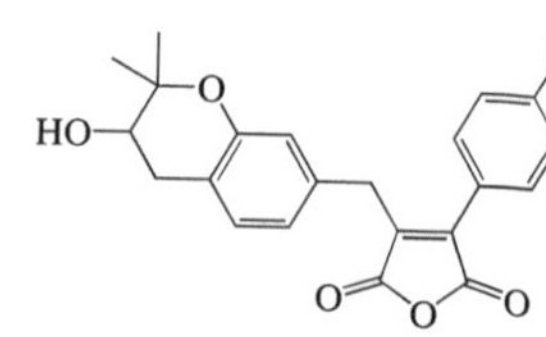

[1361320-14-0]

$C_{22}H_{20}O_6$ 380.396

Struct. revised in 2013. Prod. by *Aspergillus terreus* C9408-3. Light yellow oil.

Liao, W.-Y. *et al.*, *J. Nat. Prod.*, 2012, **75**, 630-635 (*Asperjinone*).

Elyashberg, M. *et al.*, *J. Nat. Prod.*, 2013, **76**, 113-116 (*struct*)

Aspernolide I‡ A61

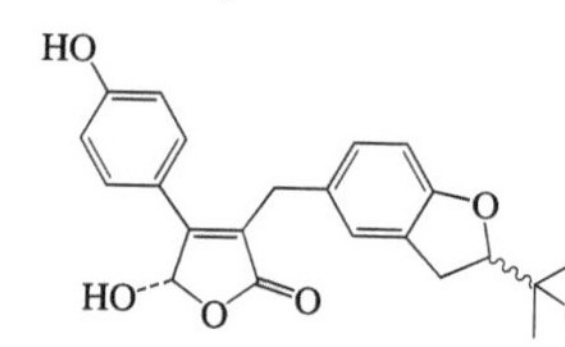

Absolute Configuration

$C_{22}H_{22}O_6$ 382.412

Prod. by *Aspergillus* sp. CBS-P-2. Antioxidant and cytotoxic agent. Light

brown gum. $[\alpha]_D^{20}$ -16.7 (c, 0.33 in MeOH). λ_{max} 300 (log ε 3.96) (MeOH).
An, X. *et al.*, *Molecules*, 2016, **21**, 1361:1-10 *(Aspernolide I, cd, activity)*

Aspernolide H‡ A62

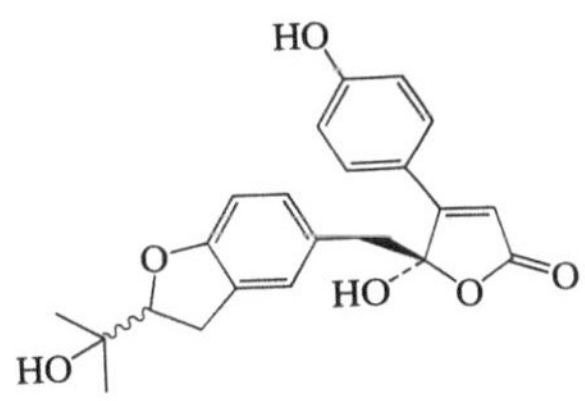

Absolute Configuration

$C_{22}H_{22}O_6$ 382.412
Related to Butyrolactone IV, B91. Prod. by *Aspergillus* sp. CBS-P-2. Antioxidant. Pale yellow gum. $[\alpha]_D^{20}$ -18.7 (c, 0.32 in MeOH). λ_{max} 202 (log ε 4.42) (MeOH).
An, X. *et al.*, *Molecules*, 2016, **21**, 1361:1-10 *(Aspernolide H, cd, activity)*

Asperphenol B A63

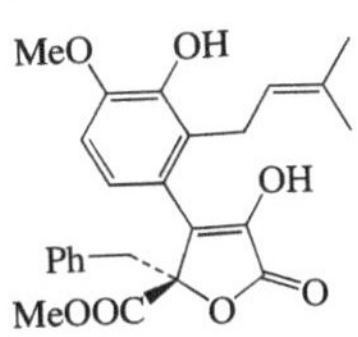

Absolute Configuration

[1675220-36-6]
$C_{25}H_{26}O_7$ 438.476
Provisional prenylated lignan. Biosynth. pathway is unclear. Related to Butyrolactone I, B89. Prod. by *Aspergillus versicolor*. Active against tobacco mosaic virus. Amorph. powder. $[\alpha]_D^{25}$ +79.8 (c, 0.2 in MeOH). λ_{max} 210 (log ε 4.32); 246 (log ε 3.51); 302 (log ε 3.74) (MeOH).
Ye, Y.-Q. *et al.*, *Bull. Korean Chem. Soc.*, 2014, **35**, 3059-3062 *(Asperphenol B, activity)*

Asperteretal A A64

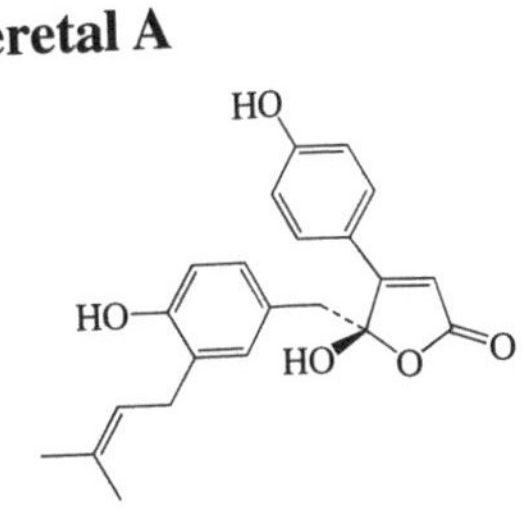

Absolute Configuration

5-Hydroxy-5-[[4-hydroxy-3-(3-methyl-2-butenyl)phenyl]methyl]-4-(4-hydroxyphenyl)-2(5H)-furanone, CAS. *4,4′,8′-Trihydroxy-9′-nor-3′-prenyl-7,8′-lign-7-en-9,8′-olide.* ***Butyrolactone IX***

[1853246-00-0]
$C_{22}H_{22}O_5$ 366.413
Prod. by *Aspergillus terreus* PR-P-2 and by the marine-derived *Paecilomyces variotii* EN-291. NO prodn. inhibitor. Oil. $[\alpha]_D^{25}$ +20 (c, 0.11 in MeOH). λ_{max} 203 (log ε 4.51); 302 (log ε 4.23) (MeOH).
Zhang, P. *et al.*, *Phytochem. Lett.*, 2015, **11**, 85-88 *(Butyrolactone IX)*.
Guo, F. *et al.*, *Fitoterapia*, 2016, **113**, 44-50 *(Asperteretal A, activity)*

Asperteretal B A65

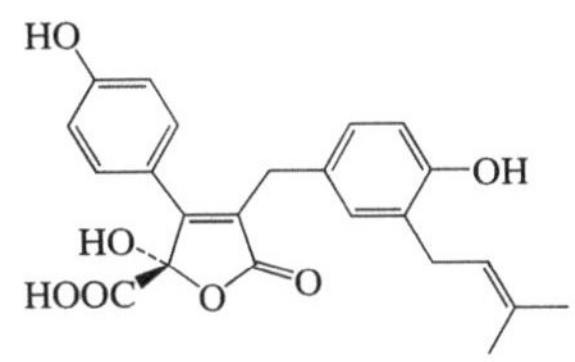

Absolute Configuration

4,4′,8-Trihydroxy-3′-prenyl-7,8′-lign-7(8′)-en-9′,8-olid-9-oic acid
[2131748-63-3]
$C_{23}H_{22}O_7$ 410.422
Related to Chromapedic acid, C70. Prod. by *Aspergillus terreus* PR-P-2. Oil. $[\alpha]_D^{20}$ -18.6 (c, 0.12 in MeOH). λ_{max} 224; 283 (MeOH).
Guo, F. *et al.*, *Fitoterapia*, 2016, **113**, 44-50 *(Asperteretal B)*

Asperteretal C A66

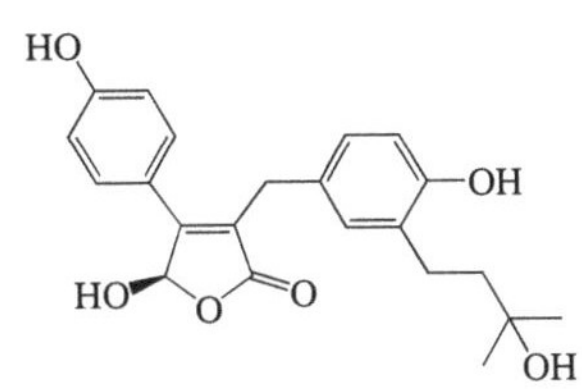

[2131748-65-5]
$C_{22}H_{24}O_6$ 384.428
Prod. by *Aspergillus terreus* PR-P-2. NO prodn. inhibitor. Oil. $[\alpha]_D^{20}$ -9 (c, 0.2 in MeOH). Described as the *R*-isomer but is probably a partial racemate. λ_{max} 225; 283; 305 (MeOH).
Guo, F. *et al.*, *Fitoterapia*, 2016, **113**, 44-50 *(Asperteretal C, activity)*

Asperterone A A67

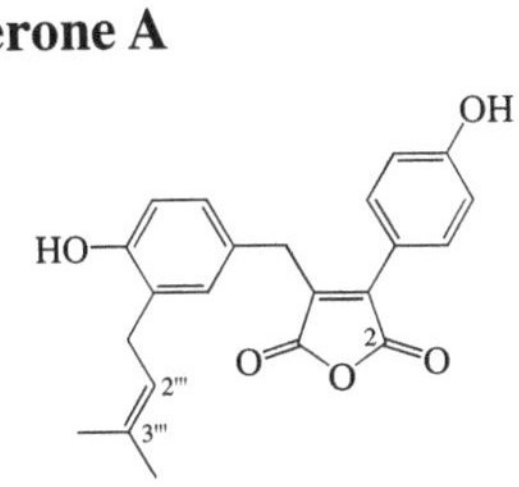

[1321528-92-0]
$C_{22}H_{20}O_5$ 364.397
Provisional prenylated norlignan. Related to 3-Benzyl-4-phenyl-2,5-furandione, B8. Prod. by *Aspergillus terreus* RCBC1002. Yellow gum. λ_{max} 205 (log ε 3.97) (MeOH)

2R-Alcohol
Aspernolide J‡
$C_{22}H_{22}O_5$ 366.413
Prod. by *Aspergillus* sp. CBS-P-2. Antioxidant. Light brown gum. $[\alpha]_D^{20}$ +2.6 (c, 1.4 in MeOH). λ_{max} 216 (log ε 4.08) (MeOH)

2‴,3‴-Dihydro, 3‴-hydroxy
Asperterone B
[1510806-57-1]
$C_{22}H_{22}O_6$ 382.412
Prod. by *Aspergillus terreus* MHL-P22. Exhibits moderate cytotoxic activity. Pale yellow gum. λ_{max} 217 (log ε 3.65); 282 (log ε 3.01) (MeOH).

Nuclear, P. *et al.*, *Chem. Pharm. Bull.*, 2010, **58**, 1221-1223 *(Asperterone A)*.
Gu, W. *et al.*, *Chem. Pharm. Bull.*, 2012, **60**, 1474-1477 *(Asperterone B)*.
An, X. *et al.*, *Molecules*, 2016, **21**, 1361:1-10 *(Aspernolide J, activity)*

Asperterone C A68

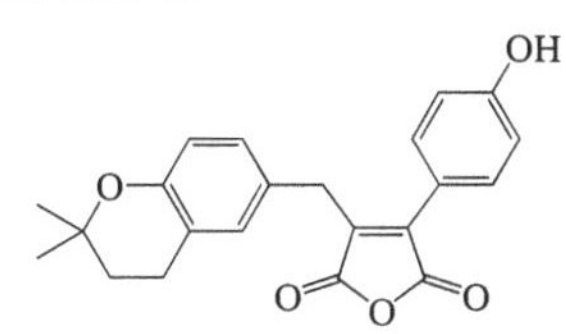

[1510806-58-2]
$C_{22}H_{20}O_5$ 364.397
Provisional prenylated norlignan. Related to 3-Benzyl-4-phenyl-2,5-furandione, B8. Prod. by *Aspergillus terreus* MHL-P22. Exhibits moderate cytotoxic activity. Pale yellow gum. λ_{max} 220 (log ε 4.08); 281 (log ε 3.12) (MeOH).
Gu, W. *et al.*, *Chem. Pharm. Bull.*, 2012, **60**, 1474-1477 *(Asperterone C, activity)*

Athrotaxin A69

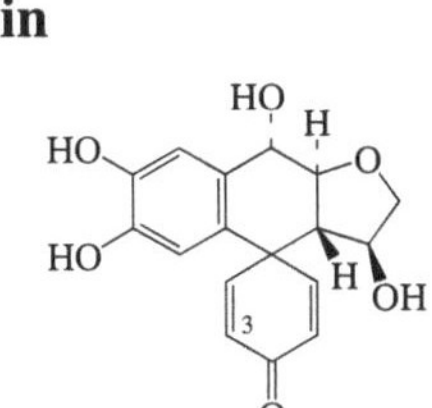

Absolute Configuration

7,8′-Cyclo-8′,9-epoxy-4′,5′,7′,8-tetrahydroxy-9′-nor-1,2′-lignan-4(1H)-one

[35124-85-7]
$C_{17}H_{16}O_6$ 316.309
Constit. of *Athrotaxis selaginoides* and *Metasequoia glyptostroboides*. Cryst. + H_2O (EtOH aq.). Mp 185° dec. (hydrate). Mp 285° dec. (anhyd.). $[\alpha]_D$ -225 (c, 0.23 in DMF aq.)

Di-Me ether

Mp 120 - 122°. $[\alpha]_D$ -257 (c, 0.12 in MeOH)

3-Hydroxy

7,8'-Cyclo-8',9-epoxy-3,4',5',7',8-pentahydroxy-9'-nor-1,2'-lignan-4(1H)-one. ***3'-Hydroxyathrotaxin***

[65560-08-9]
$C_{17}H_{16}O_7$ 332.308
Constit. of *Metasequoia glyptostroboides*.

Erdtman, H. *et al.*, *Acta Chem. Scand.*, 1960, **14**, 2161-2168 (*Arthrotaxin*).
Pilotti, A.-M. *Acta Crystallogr., Sect. B*, 1972, **28**, 2123-2128 (*cryst struct*).
Daniels, P. *et al.*, *J. Chem. Soc., Chem. Commun.*, 1972, 246-247 (*cryst struct*).
Enoki, A. *et al.*, *Mokuzai Gakkaishi*, 1978, **23**, 579-586 (*Athrotaxin*).
Enoki, A. *et al.*, *Mokuzai Gakkaishi*, 1978, **23**, 587-593 (*3'-Hydroxyathrotaxin*)

Austrafuran A A70

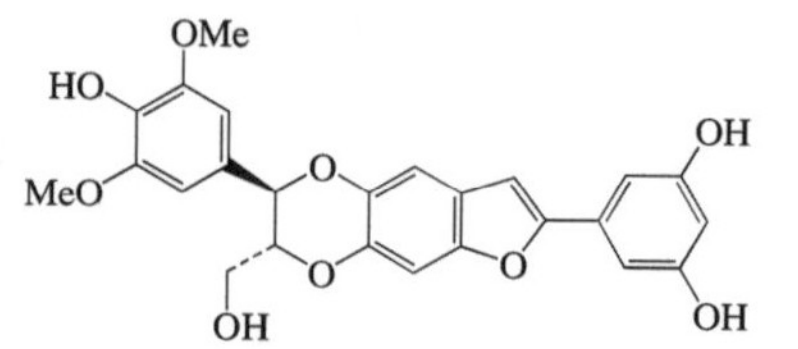

Relative Configuration

[1033529-12-2]
$C_{25}H_{22}O_9$ 466.442
Structure may need revision. Phenyl 3,5-dioxygenated substn. pattern in doubt as a result of work done by Yang *et al* in 2015. Constit. of the stems of *Morus australis*. Amorph. yellow powder. Mp 228 - 230°. λ_{max} 224 (log ε 3.74); 286 (log ε 3.36); 296 (log ε 3.38); 324 (log ε 3.69); 338 (log ε 3.66) (MeOH).

Zhang, Q.J. *et al.*, *Chin. Chem. Lett.*, 2008, **19**, 196-198 (*Austrafuran A*).
Yang, Y.-N. *et al.*, *J. Nat. Prod.*, 2015, **78**, 705-711 (*struct*)

Balanophonin A‡ B1

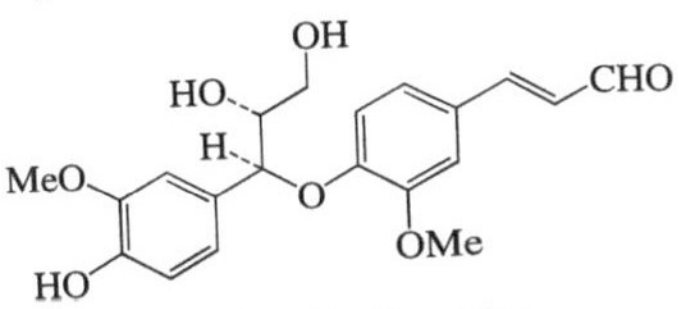

Absolute Configuration

4′,8′,9′-Trihydroxy-3,3′-dimethoxy-4,7′-oxylign-7-en-9-al. 4′,8′,9′-Trihydroxy-3,3′-dimethoxy-9-oxo-4,7′-oxyneolign-7-ene
[1482289-39-3]
$C_{20}H_{22}O_7$ 374.389
Not to be confused with Balanophonin, in E163. Constit. of the seeds of *Lithocarpus pachylepis*. Pale yellow oil. $[\alpha]_D^{20}$ -2.4 (c, 0.06 in MeOH). λ_{max} 212 (log ε 4.56); 256 (log ε 3.64); 290 (log ε 3.08) (MeOH).
Ma, G. *et al.*, *Phytochem. Lett.*, 2013, **6**, 152-155 (*Balanophonin A*)

Balanophonin C B2

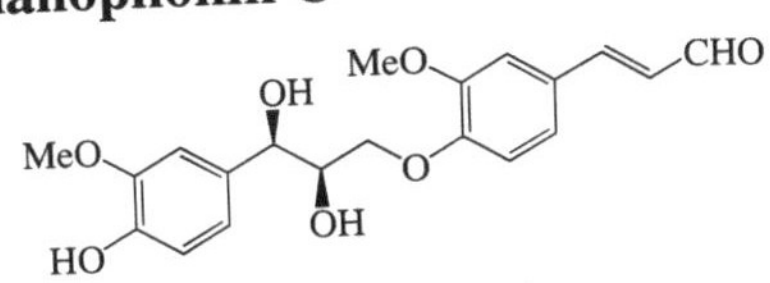

Relative Configuration

4′,7′,8′-Trihydroxy-3,3′-dimethoxy-4,9′-oxylign-7-en-9-al. 4′,7′,8′-Trihydroxy-3,3′-dimethoxy-9-oxo-4,9′-oxyneolign-7-ene
[1637672-74-2]
$C_{20}H_{22}O_7$ 374.389
Constit. of the seeds of *Lithocarpus pachylepis*. Amorph. yellow powder. $[\alpha]_D^{20}$ -0.2 (c, 0.1 in MeOH). λ_{max} 212 (log ε 4.16); 256 (log ε 3.78); 285 (log ε 3.15) (MeOH).
Xie, Y. *et al.*, *Molecules*, 2013, **18**, 10397-10403 (*Balanophonin C*)

Balanophonin E B3

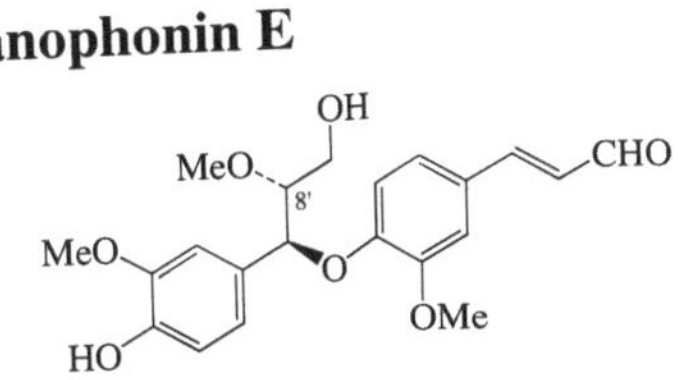

Relative Configuration

4′,9′-Dihydroxy-3,3′,8′-trimethoxy-4,7′-oxylign-7-en-9-al. 4′,9′-Dihydroxy-3,3′,8′-trimethoxy-9-oxo-4,7′-oxyneolign-7-ene
[1637672-76-4]
$C_{21}H_{24}O_7$ 388.416
Constit. of the seeds of *Lithocarpus pachylepis*. Amorph. yellow powder. $[\alpha]_D^{20}$ -0.36 (c, 0.08 in MeOH). λ_{max} 212 (log ε 4.04); 256 (log ε 3.61); 285 (log ε 3.23) (MeOH)

$O^{8'}$-De-Me, $O^{8'}$-Et

8′-Ethoxy-4′,9′-dihydroxy-3,3′-dimethoxy-4,7′-oxylign-7-en-9-al. 8′-Ethoxy-4′,9′-dihydroxy-3,3′-dimethoxy-9-oxo-4,7′-oxylign-7-ene.
Balanophonin D
[1637672-75-3]
$C_{22}H_{26}O_7$ 402.443
Constit. of the seeds of *Lithocarpus pachylepis*. Amorph. yellow powder. $[\alpha]_D^{20}$ -0.38 (c, 0.09 in MeOH). λ_{max} 212 (log ε 4.62); 256 (log ε 4.01); 285 (log ε 3.42) (MeOH).

Xie, Y. *et al.*, *Molecules*, 2013, **18**, 10397-10403 (*Balanophonins D,E*)

Bazzania acid B4

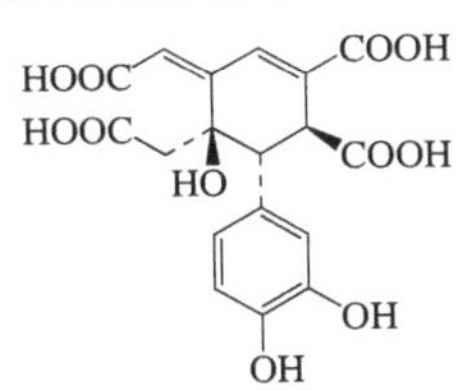

Relative Configuration

[217176-44-8]
$C_{18}H_{16}O_{11}$ 408.315
Seco-naphthalenoid lignan. Isol. from the liverwort *Bazzania trilobata*. Oil. $[\alpha]_D^{20}$ -16.5 (c, 0.2 in MeOH). λ_{max} 276 (MeOH).
Martini, U. *et al.*, *Phytochemistry*, 1998, **49**, 1139-1146 (*Bazzania acid*)

3-[2-(1,3-Benzodioxol-5-ylmethyl)-7-methoxy-5-benzofuranyl]-2-propenol B5

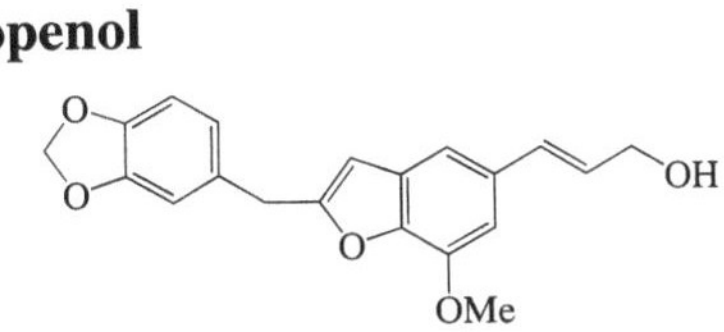

4,8′-Epoxy-5-methoxy-3′,4′-methylenedioxy-3,9′-lign-7-en-9-ol. 4,8′-Epoxy-9-hydroxy-5-methoxy-3′,4′-methylenedioxy-3,9′-neolign-7-ene
[86926-70-7]
$C_{20}H_{18}O_5$ 338.359
Constit. of the roots of *Heliopsis buphthalmoides*. Oil.
Bohlmann, F. *et al.*, *Liebigs Ann. Chem.*, 1983, 1202-1206 (*Heliopsis buphthalmoides constit*)

5-Benzyl-3-hydroxy-4-phenyl-2(5*H*)-furanone B6

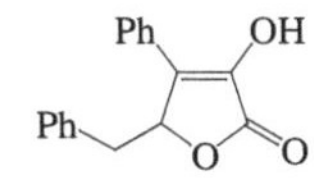

8-Hydroxy-9′-nor-7,8′-lign-7-en-9,8′-olide
$C_{17}H_{14}O_3$ 266.296
Provisional norlignan. Biosynth. pathway unclear. The nat. prods. were isol. as racemates or partial racemates. The isolate from *Phyllosticta* sp.was resolved by chiral hplc

(±)-form

Xenofuranone B [918631-18-2]
Prod. by *Aspergillus versicolor*, *Phyllosticta* sp. J13-2-12Y and *Xenorhabdus szentirmaii* DSM 16338. Solid. Mp 64°. $[\alpha]_D^{20}$ +19.5 (c, 0.002 in $CHCl_3$). Partial racemate. [acid] λ_{max} 207 (log ε 3.81); 286 (log ε 3.83) (MeOH/HCl). [neutral] λ_{max} 203 (log ε 3.92); 286 (log ε 3.83) (MeOH). [base] λ_{max} 208 (log ε 4.44); 321 (log ε 3.81) (MeOH/NaOH)

Me ether

5-Benzyl-3-methoxy-4-phenyl-2(5H)-furanone. 8-Methoxy-9′-nor-7,8′-lign-7-en-9,8′-olide.
Xenofuranone A
[918631-17-1]
$C_{18}H_{16}O_3$ 280.323
Prod. by *Xenorhabdus szentirmaii* DSM 16338. Antibacterial agent incl. *Clostridium difficile* NAP-1. Solid. Mp 77°. $[\alpha]_D^{20}$ +96 (c, 0.03 in $CHCl_3$). λ_{max} 204 (log ε 3.98); 282 (log ε 3.86) (MeOH). [base] λ_{max} 209 (log ε 4.48); 282 (log ε 3.84) (MeOH/NaOH). [acid] λ_{max} 207 (log ε 3.85); 282 (log ε 3.82) (MeOH/HCl).

Brachmann, A.O. *et al.*, *J. Nat. Prod.*, 2006, **69**, 1830-1832 (*Xenofuranones A,B, biosynth*).
Wang, H.-G. *et al.*, *Fitoterapia*, 2018, **124**, 86-91 (*Phyllosticta constits, cd, abs config*)

2-Benzyl-5-(2-phenylethyl)furan B7

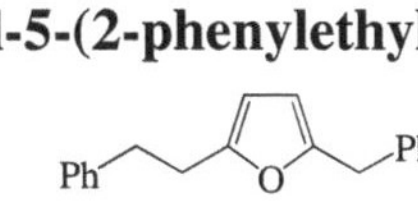

2-(2-Phenylethyl)-5-(phenylmethyl) furan
[26171-91-5]
$C_{19}H_{18}O$ 262.352
Constit. of the rhizomes of *Alpinia officinarum*

3″-Methoxy, 4″-hydroxy
2-(4-Hydroxy-3-methoxybenzyl)-5-(2-phenylethyl)furan. ***Alpinoid D***
[1041740-13-9]
$C_{20}H_{20}O_3$ 308.377
Constit. of the rhizomes of *Alpinia officinarum*. Yellow cryst.. Mp 70 - 72°. λ_{max} 202 (log ε 4.43); 227 (log ε 4.1); 279 (log ε 3.36) (MeOH).

Mavoungou-Gomes, L. *C. R. Seances Acad. Sci., Ser. C*, 1970, **270**, 750-753 (*synth*).
Sun, Y. *et al.*, *Planta Med.*, 2008, **74**, 427-431 (*Alpinoid D*).
An, N. *et al.*, *Food Chem.*, 2010, **119**, 513-517 (*Alpinoid D*).
Seçinti, H. *et al.*, *Helv. Chim. Acta*, 2015, **98**, 938-944 (*synth*)

3-Benzyl-4-phenyl-2,5-furandione B8

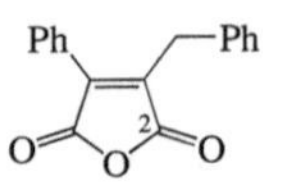

3-Phenyl-4-(phenylmethyl)-2,5-furandione, CAS. *Benzylphenylmaleic anhydride.* ***Dehydromicroperfuranone***
[65641-17-0]
$C_{17}H_{12}O_3$ 264.28
Provisional norlignan. Prod. by *Aspergillus nidulans* IFO 6398. Plant growth regulator. Needles (hexane). Mp 67 - 68°. λ_{max} 313 (ε 4800) (EtOH)

3′,3″,4′,4″-Tetrahydroxy
3-(3,4-Dihydroxybenzyl)-4-(3,4-dihydroxyphenyl)-2,5-furandione
[1708956-56-2]
$C_{17}H_{12}O_7$ 328.276
Constit. of *Leccinum chromapes*. Yellow solid. Oxidn. prod. of Chromapedic acid, C70. λ_{max} 207; 272; 384 (MeOH).

Hamasaki, T. *et al.*, *Agric. Biol. Chem.*, 1983, **47**, 891-892 (*Aspergillus nidulans metab*).
Gruber, G. *et al.*, *Z. Naturforsch., B*, 2014, **69**, 432-438 (*Leccinum chromapes constit*).
Roux, I. *et al.*, *ACS Synth. Biol.*, 2020, **9**, 1843-1854 (*Dehydromicroperfuranone, biosynth*)

Berchemin A B9

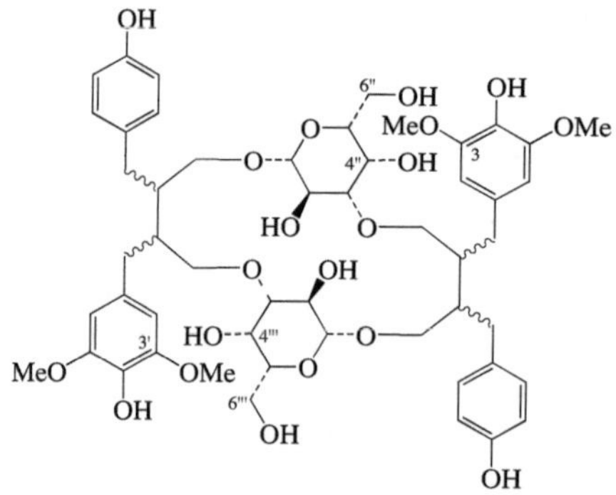

[791634-09-8]
$C_{52}H_{68}O_{20}$ 1013.096
Constit. of *Berchemia pakistanica*. Inhibitor of lipoxygenase and α-glucosidase. Yellow gummy solid. $[\alpha]^{27}_{D}$ +15.4 (c, 0.07 in MeOH). λ_{max} 228 (log ε 4.05); 279 (log ε 3.82) (MeOH)

4″,4‴-Diepimer, 3,3′-bis(demethoxy), 6″,6‴-dideoxy
Berchemin B
[791634-19-0]
$C_{50}H_{64}O_{16}$ 921.046
Constit. of *Berchemia pakistanica*. Moderate lipoxygenase inhibitor. Yellow gummy solid. $[\alpha]^{27}_{D}$ -9.3 (c, 0.15 in MeOH). λ_{max} 230 (log ε 3.91); 280 (log ε 3.79) (MeOH).

Mukhtar, N. *et al.*, *Helv. Chim. Acta*, 2004, **87**, 2050-2056 (*Berchemins A,B, activity*)

Bicubebin B10

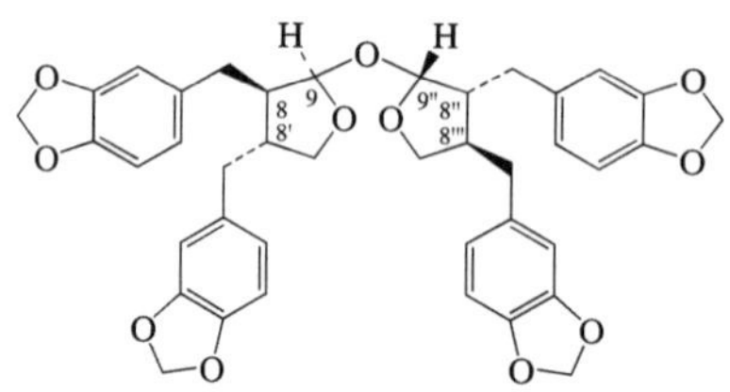

(8R,8′R,8″R,8‴R,9R,9″R)-form

$C_{40}H_{38}O_{11}$ 694.733
(8R,8′R,8″R,8‴R,9R,9″R)-form
Bicubebin C [2131170-31-3]
Mp 45 - 46°. $[\alpha]^{27}_{D}$ +76.9 (c, 0.16 in $CHCl_3$)

(8R,8′R,8″R,8‴R,9R,9″S)-form
Bicubebin A [885469-40-9]
Constit. of *Aristolochia lagesiana* and *Aristolochia pubescens*. Amorph. solid. $[\alpha]^{26}_{D}$ -54.8 (c, 0.36 in $CHCl_3$)

(8R,8′R,8″R,8‴R,9S,9″S)-form
Bicubebin B [898286-62-9]
Formerly assigned monomeric *cis*-Cubebin struct. in 2006. Struct. revised to a dimeric Cubebin in 2017. Constit. of *Aristolochia lagesiana* and *Aristolochia pubescens*. $[\alpha]^{26}_{D}$ -102.3 (c, 0.6 in $CHCl_3$).

De Pascoli, I.C. *et al.*, *Phytochemistry*, 2006, **67**, 735-742 (*Bicubebin A*).
Davidson, S.J. *et al.*, *Org. Lett.*, 2017, **19**, 5368-5371 (*Bicubebins A,B,C, synth*)

Bicycloillicinone asarone acetal B11

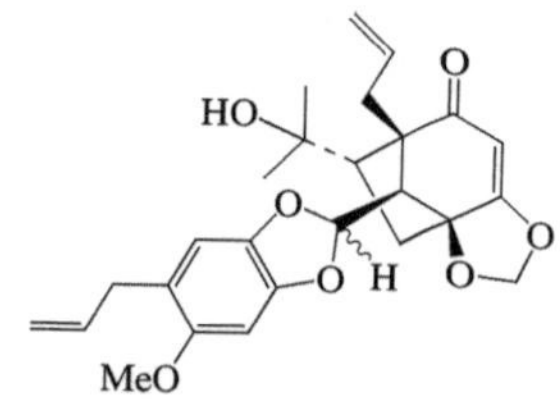

[193976-13-5]
$C_{26}H_{30}O_7$ 454.519
Constit. of *Illicium tashiroi*. Increaser of choline acetyltransferase (ChAT) activity. Amorph.. $[\alpha]_{D}$ -177.5 (c, 0.8 in $CHCl_3$).

Fukuyama, Y. *et al.*, *Planta Med.*, 1997, **63**, 275-277 (*Illicium tashiroi constit*)

Biginkgoside A B12

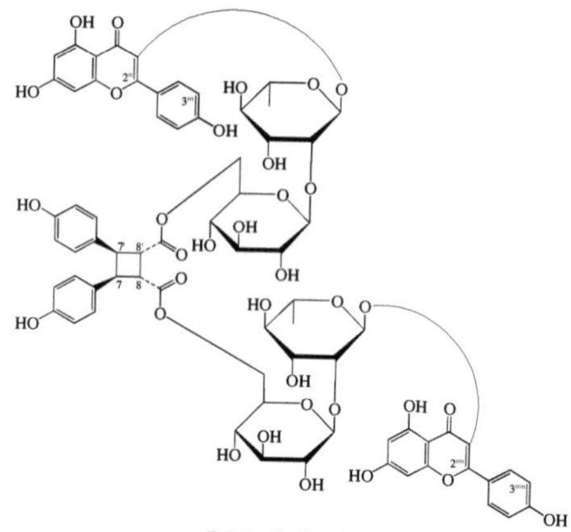

Relative Configuration

[1927027-84-6]
$C_{72}H_{72}O_{34}$ 1481.334
Constit. of the leaves of *Ginkgo biloba*. Pale yellow gum. $[\alpha]^{20}_{D}$ -71 (c, 0.2 in MeOH). λ_{max} 265 (log ε 4.47); 342 (log ε 4.29) (MeOH)

3‴,3⁗-Dihydroxy
Biginkgoside B
[1927027-85-7]
$C_{72}H_{72}O_{36}$ 1513.332
Constit. of the leaves of *Ginkgo biloba*. Pale yellow gum. $[\alpha]^{20}_{D}$ -67 (c, 0.2 in MeOH). λ_{max} 256 (log ε 4.41); 342 (log ε 4.26) (MeOH)

7′-Epimer
Biginkgoside G
[1927028-54-3]
$C_{72}H_{72}O_{34}$ 1481.334

Constit. of the leaves of *Ginkgo biloba*. Pale yellow gum. $[\alpha]_D^{20}$ -52 (c, 0.2 in MeOH). λ_{max} 265 (log ε 4.26); 342 (log ε 4.09) (MeOH)

7′,8-Diepimer

Biginkgoside C

[1927028-51-0]

$C_{72}H_{72}O_{34}$ 1481.334

Constit. of the leaves of *Ginkgo biloba*. Pale yellow gum. $[\alpha]_D^{20}$ -53 (c, 0.1 in MeOH). λ_{max} 265 (log ε 4.48); 342 (log ε 4.29) (MeOH)

7′,8′-Diepimer

Biginkgoside D

[1927028-52-1]

$C_{72}H_{72}O_{34}$ 1481.334

Constit. of the leaves of *Ginkgo biloba*. Pale yellow gum. $[\alpha]_D^{20}$ -71 (c, 0.1 in MeOH). λ_{max} 265 (log ε 4.59); 339 (log ε 4.40) (MeOH)

7′,8′-Diepimer, 3‴-hydroxy

Biginkgoside F

[1927027-86-8]

$C_{72}H_{72}O_{35}$ 1497.333

Constit. of the leaves of *Ginkgo biloba*. Pale yellow gum. $[\alpha]_D^{20}$ -65 (c, 0.1 in MeOH). λ_{max} 263 (log ε 4.56); 342 (log ε 4.37) (MeOH)

7′,8′-Diepimer, 3‴,3‴″-dihydroxy

Biginkgoside E

[1927028-53-2]

$C_{72}H_{72}O_{36}$ 1513.332

Constit. of the leaves of *Ginkgo biloba*. Exhibits inhibitory effect on NO prodn. in LPS-activated BV-2 cells. Pale yellow gum. $[\alpha]_D^{20}$ -58 (c, 0.1 in MeOH). λ_{max} 258 (log ε 4.40); 348 (log ε 4.21) (MeOH).

Ma, G.-L. *et al.*, *J. Nat. Prod.*, 2016, **79**, 1354-1364 (*Biginkgosides A-G, activity*)

Biginkgoside H B13

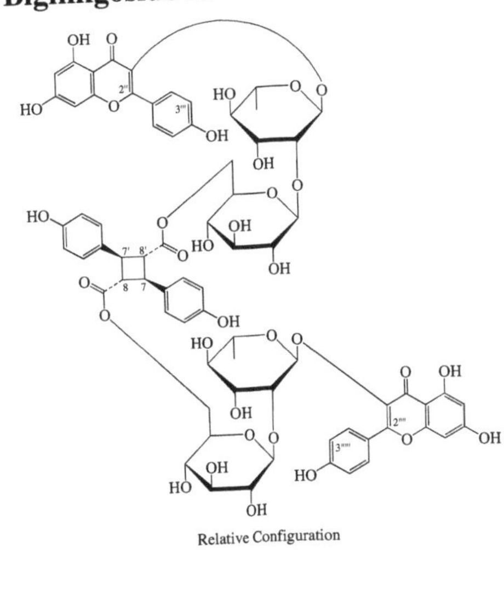

Relative Configuration

[1927027-87-9]

$C_{72}H_{72}O_{34}$ 1481.334

Constit. of the leaves of *Ginkgo biloba*. Exhibits inhibitory effect on NO prodn. in LPS-activated BV-2 cells. Pale yellow gum. $[\alpha]_D^{20}$ -75 (c, 0.2 in MeOH). λ_{max} 265 (log ε 4.33); 342 (log ε 4.14) (MeOH)

3‴,3‴″-Dihydroxy

Biginkgoside I

[1927027-88-0]

$C_{72}H_{72}O_{36}$ 1513.332

Constit. of the leaves of *Ginkgo biloba*. Pale yellow gum. λ_{max} 256 (log ε 4.38); 348 (log ε 4.23) (MeOH).

Ma, G.-L. *et al.*, *J. Nat. Prod.*, 2016, **79**, 1354-1364 (*Biginkgosides H,I, activity*)

5,5″-Bilariciresinol B14

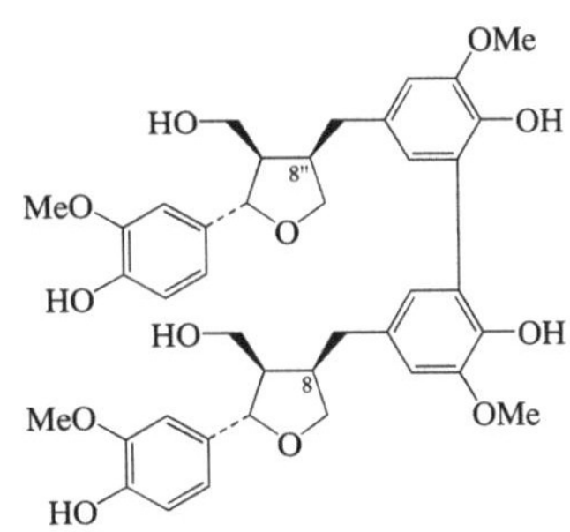

Relative Configuration

5,5″-Dilariciresinol

[1279013-31-8]

$C_{40}H_{46}O_{12}$ 718.796

Similar to Cerberalignan A, C44. Constit. of the leaves of *Mallotus philippensis*. Amorph. powder. $[\alpha]_D$ +26 (c, 0.5 in MeOH).

Mai, N.T. *et al.*, *Nat. Prod. Commun.*, 2010, **5**, 423-426 (*5,5″-Bilariciresinol*)

Biondinin A B15

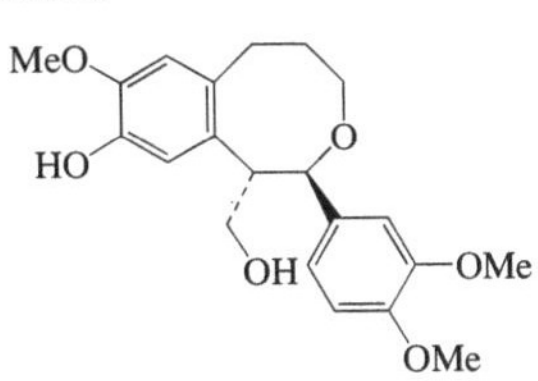

Relative Configuration

7′,9-Epoxy-3′,4′,5-trimethoxy-2,8′-lignan-4,9′-diol. 7′,9-Epoxy-4,9′-dihydroxy-3′,4′,5-trimethoxy-2,8′-neolignan

[142942-84-5]

$C_{21}H_{26}O_6$ 374.433

Constit. of *Magnolia biondii* and *Magnolia fargesii*. Antioxidant. Active against tobacco-mosaic virus. Oil.

Ma, Y.-L. *et al.*, *Phytochemistry*, 1996, **41**, 287-288 (*Magnolia biondii constit*).

Lee, J. *et al.*, *Chem. Pharm. Bull.*, 2009, **57**, 298-301 (*Magnolia fargesii constit, activity*).

Yang, Y.-K. *et al.*, *Asian J. Chem.*, 2013, **25**, 9610-9612 (*Magnolia fargesii constit, activity*)

Biphenylsalsonoid A B16

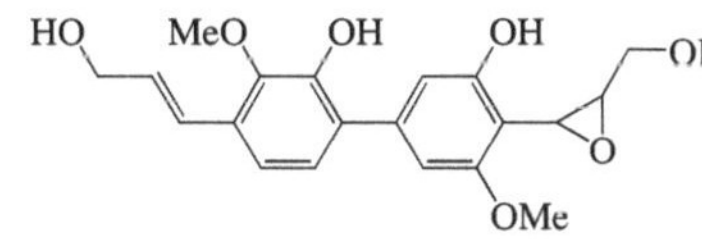

7,8-Epoxy-2′,6-dimethoxy-4,4′-lign-7′-ene-2,3′,9,9′-tetrol

[2412991-50-3]

$C_{20}H_{22}O_7$ 374.389

Constit. of the roots of *Salsola imbricata*. Exhibits antibacterial activity. Moderate antioxidant.

Oueslati, M.H. *et al.*, *Orient. J. Chem.*, 2017, **33**, 1871-1878 (*Biphenylsalsonoid A, activity*)

Biphenylsalsonoid B B17

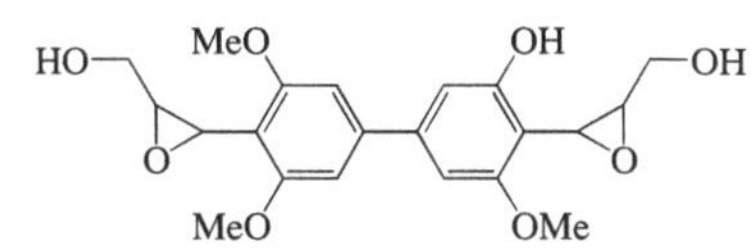

7,8:7′,8′-Diepoxy-2′,6,6′-trimethoxy-4,4′-lignan-2,9,9′-triol

[2412991-51-4]

$C_{21}H_{24}O_8$ 404.415

Constit. of the roots of *Salsola imbricata*. Exhibits antibacterial activity. Moderate antioxidant.

Oueslati, M.H. *et al.*, *Orient. J. Chem.*, 2017, **33**, 1871-1878 (*Biphenylsalsonoid B, activity*)

Bisabocurcumin B18

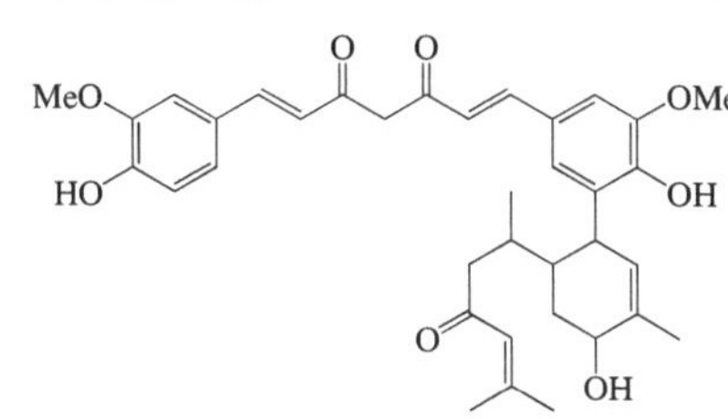

[1350290-10-6]

$C_{36}H_{42}O_8$ 602.724

Constit. of *Curcuma longa*. Yellow amorph. powder. $[\alpha]_D^{20}$ +0.8 (c, 0.45 in MeOH). λ_{max} 426 (log ε 4.36) (MeOH).

Xiao, Y.C. *et al.*, *Chin. Chem. Lett.*, 2011, **22**, 1457-1460 (*Bisabocurcumin*)

Bisabolocurcumin ether B19

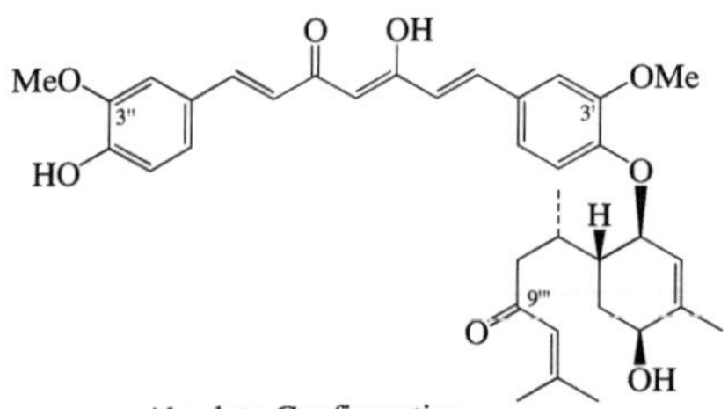

Absolute Configuration

[1363379-23-0]

$C_{36}H_{42}O_8$ 602.724

Hybrid of curcumins and bisabolanes. Enolised β-diketone. Constit. of the rhizomes of *Curcuma longa*. Displays cytotoxic activity against human cancer cell lines. Red cryst.. Mp 100 - 102°. $[\alpha]^{20}_{D}$ +0.6 (c, 0.6 in MeOH). $[\alpha]^{25}_{D}$ +3.5 (c, 0.01 in MeCN). λ_{max} 423 (log ε 4.72) (MeOH). [neutral] λ_{max} 419 (log ε 4.79) (MeCN)

3″-Demethoxy

Demethoxybisabolocurcumin ether

[1363379-22-9]

$C_{35}H_{40}O_7$ 572.698

Constit. of the rhizomes of *Curcuma longa*. Displays cytotoxic activity against human cancer cell lines. Amorph. yellow powder. $[\alpha]^{20}_{D}$ +0.5 (c, 0.45 in MeOH). $[\alpha]^{25}_{D}$ +3.1 (c, 0.01 in MeCN). λ_{max} 420 (log ε 4.78) (MeOH). [neutral] λ_{max} 417 (log ε 4.78) (MeCN)

Bis(demethoxy)

Didemethoxybisabolocurcumin ether

[1363379-25-2]

$C_{34}H_{38}O_6$ 542.672

Constit. of the rhizomes of *Curcuma longa*. Amorph. yellow powder. $[\alpha]^{20}_{D}$ +0.4 (c, 0.5 in MeOH). $[\alpha]^{25}_{D}$ +4.1 (c, 0.01 in MeCN). λ_{max} 241 (log ε 4.43); 416 (log ε 4.74) (MeOH). [neutral] λ_{max} 409 (log ε 4.35) (MeCN)

9‴-Deoxo

Terpecurcumin A

[1413592-69-4]

$C_{36}H_{44}O_7$ 588.741

Isol. from the rhizomes of *Curcuma longa*. Amorph. orange powder. $[\alpha]^{25}_{D}$ +2.9 (c, 0.01 in MeCN). λ_{max} 420 (log ε 4.74) (MeCN).

Xiao, Y.-C. *et al.*, *Helv. Chim. Acta*, 2012, **95**, 327-332 (*Bisabolocurcumin ethers*).
Lin, X. *et al.*, *J. Nat. Prod.*, 2012, **75**, 2121-2131 (*abs config, Terpecurcumin A, cryst struct*)

Bisavenanthramide B1 B20

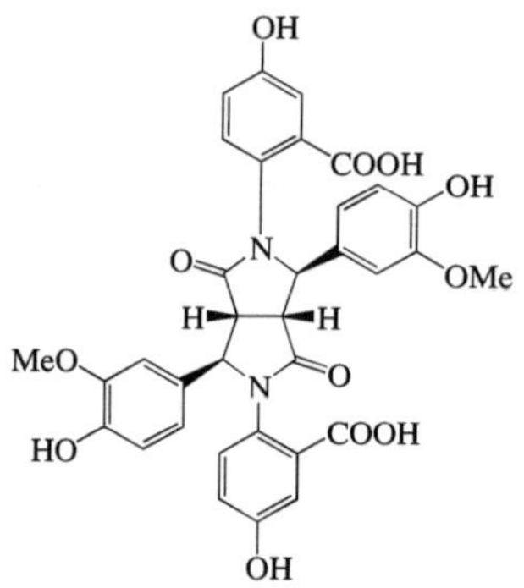

Relative Configuration

[939790-85-9]

$C_{34}H_{28}N_2O_{12}$ 656.600

Isol. from elicited oat leaves (*Avena sativa*). Amorph. solid. Racemic. λ_{max} 210 (log ε 4.83); 233 (sh) (log ε 4.44); 285 (log ε 4.04); 300 (sh) (log ε 3.84) (MeOH).

Okazaki, Y. *et al.*, *J. Org. Chem.*, 2007, **72**, 3830-3839 (*Bisavenanthramide B_1*)

Bisavenanthramide B2 B21

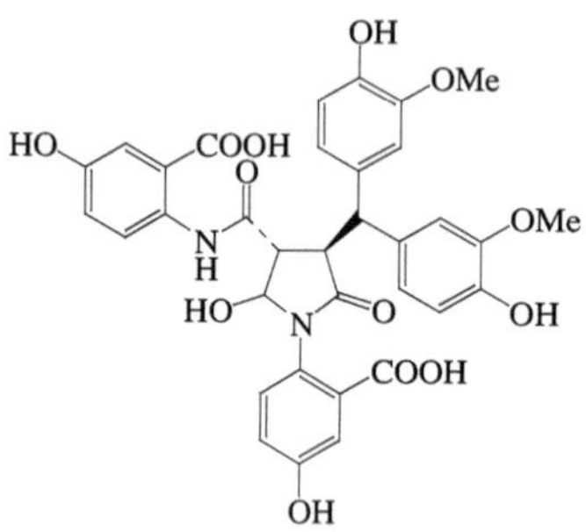

[939790-86-0]

$C_{34}H_{30}N_2O_{13}$ 674.615

Isol. from elicited oat leaves (*Avena sativa*). Yellowish solid. Racemic. λ_{max} 204 (log ε 5.04); 265 (log ε 4.23); 317 (log ε 3.07) (MeOH).

Okazaki, Y. *et al.*, *J. Org. Chem.*, 2007, **72**, 3830-3839 (*Bisavenanthramide B_2*)

Bisavenanthramide B5 B22

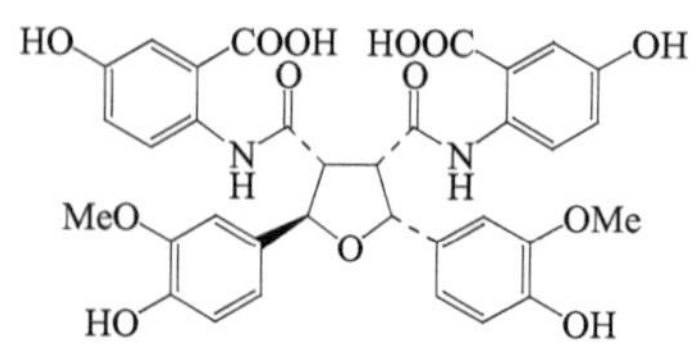

Relative Configuration

[939790-89-3]

$C_{34}H_{30}N_2O_{13}$ 674.615

Isol. from elicited oat leaves (*Avena sativa*). Yellow solid. λ_{max} 204 (log ε 4.79); 224 (log ε 4.51); 263 (log ε 4.19); 328 (log ε 3.75) (MeOH).

Okazaki, Y. *et al.*, *J. Org. Chem.*, 2007, **72**, 3830-3839 (*Bisavenanthramide B_5*)

Bisavenanthramide B6 B23

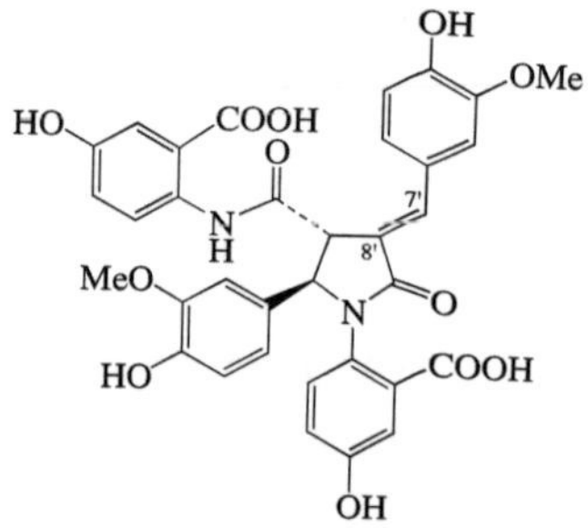

[719298-06-3]

$C_{34}H_{28}N_2O_{12}$ 656.600

Isol. from elicited oat leaves (*Avena sativa*). Phytoalexin. Racemic. λ_{max} 204 (log ε 4.61); 265 (log ε 4.31); 329 (log ε 4.45) (MeOH)

7′,8′α-Dihydro, 7′ξ-hydroxy (1)

Bisavenanthramide B_3

[939790-87-1]

$C_{34}H_{30}N_2O_{13}$ 674.615

Isol. from elicited oat leaves (*Avena sativa*). Yellowish solid. Racemic. C-7′ epimer of B_4. λ_{max} 205 (log ε 4.76); 225 (sh) (log ε 4.42); 267 (log ε 3.98); 319 (log ε 3.49) (MeOH)

7′,8′α-Dihydro, 7′ξ-hydroxy (2)

Bisavenanthramide B_4

[939790-88-2]

$C_{34}H_{30}N_2O_{13}$ 674.615

Isol. from elicited oat leaves (*Avena sativa*). Yellowish solid. Racemic. C-7′ epimer of B_3. λ_{max} 204 (log ε 4.87); 225 (sh) (log ε 4.59); 262 (log ε 4.19); 319 (log ε 3.67) (MeOH).

Okazaki, Y. *et al.*, *Tetrahedron*, 2004, **60**, 4765-4771 (*Bisavenanthramide B_6*).
Okazaki, Y. *et al.*, *J. Org. Chem.*, 2007, **72**, 3830-3839 (*Bisavenanthramides B_3,B_4*).
Di Maso, M.J. *et al.*, *Org. Lett.*, 2016, **18**, 1740-1743 (*synth*)

1,5-Bis(1,3-benzodioxol-5-yl)-tetrahydro-4*H*-furo[3,4-*d*][1,3]dioxane B24

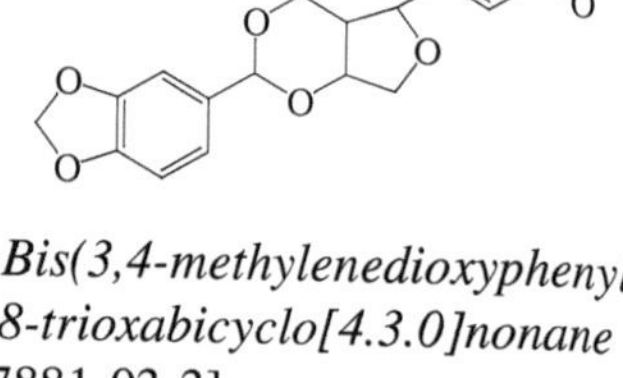

3,7-Bis(3,4-methylenedioxyphenyl)-2,4,8-trioxabicyclo[4.3.0]nonane

[877881-03-3]

$C_{20}H_{18}O_7$ 370.357
Isol. from the oil of *Camellia oleifera*.
Antioxidant. Cryst..

Lee, C.-P. *et al.*, *J. Agric. Food Chem.*, 2006, **54**, 779-784 (*Camellia oleifera constit, activity*)

2,3-Bis-O-[3-[1-carboxy-2-(3,4-dihydroxyphenyl)ethyl]-4,5-dihydroxycinnamoyl]tartaric acid B25

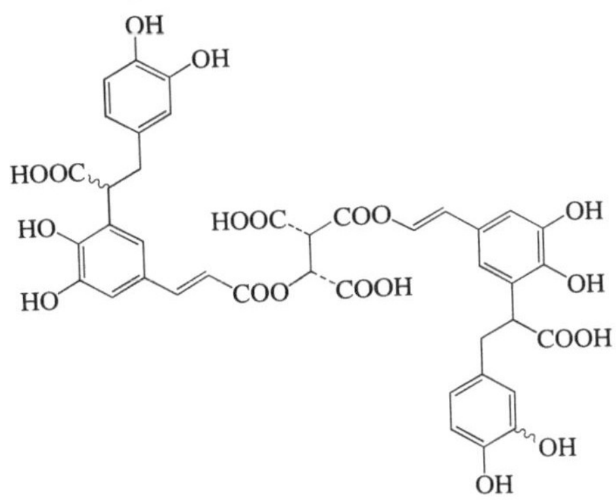

Absolute Configuration

[117869-66-6]
$C_{40}H_{34}O_{20}$ 834.692
Constit. of the leaves of *Echinacea pallida*. Amorph. solid

O-Deacyl, O-(3,4-dihydroxy-E-cinnamoyl)
[117841-80-2]
$C_{31}H_{26}O_{16}$ 654.533
Constit. of the leaves of *Echinacea pallida*. Amorph. solid.

Cheminat, A. *et al.*, *Phytochemistry*, 1988, **27**, 2787-2794 (*Echinacea pallida constits*)

Bis(deacyl)taiwankadsurin A B26

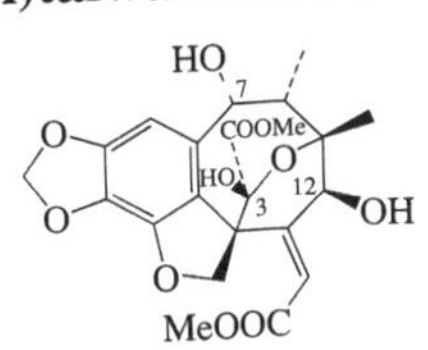

Relative Configuration

$C_{22}H_{24}O_{11}$ 464.423
Homolignan. CAS numbering shown

12-Angeloyl, 7-Ac
Taiwankadsurin D
[1594808-26-0]
$C_{29}H_{32}O_{13}$ 588.562
Constit. of *Kadsura philippinensis*. $[\alpha]^{26}_{D}$+57 (c, 0.5 in CH_2Cl_2). λ_{max} 225; 273 (MeOH)

7-Benzoyl, 12-Ac
Taiwankadsurin C
[870995-85-0]
$C_{31}H_{30}O_{13}$ 610.568
Constit. of *Kadsura philippinensis*. $[\alpha]^{26}_{D}$-30 (c, 0.7 in CH_2Cl_2). λ_{max} 246; 275; 282 (MeOH)

12-Benzoyl, 7-Ac
Taiwankadsurin A
[870995-83-8]
$C_{31}H_{30}O_{13}$ 610.568
Constit. of *Kadsura philippinensis*. Amorph. powder. $[\alpha]^{26}_{D}$+20 (c, 0.5 in CH_2Cl_2). λ_{max} 245; 276 (MeOH)

3-Epimer, 12-angeloyl, 7-Ac
Taiwankadsurin E
[1594808-27-1]
$C_{29}H_{32}O_{13}$ 588.562
Constit. of *Kadsura philippinensis*. $[\alpha]^{26}_{D}$-11 (c, 0.2 in CH_2Cl_2). λ_{max} 233; 276 (MeOH)

3-Epimer, 7,12-diangeloyl
Heteroclitin N
[1007566-04-2]
$C_{32}H_{36}O_{13}$ 628.627
Constit. of the stems of *Kadsura heteroclita*. Needles (Me_2CO/hexane). Mp 195 - 196°. $[\alpha]^{20}_{D}$+29 (c, 0.51 in MeOH). λ_{max} 220 (log ε 4.8); 275 (log ε 3.02) (MeOH)

3-Epimer, 12-benzoyl, 7-Ac
Taiwankadsurin B
[870995-84-9]
$C_{31}H_{30}O_{13}$ 610.568
Constit. of *Kadsura philippinensis*. $[\alpha]^{26}_{D}$+62 (c, 0.4 in CH_2Cl_2). λ_{max} 248; 274 (MeOH).

Shen, Y.-C. *et al.*, *Org. Lett.*, 2005, **7**, 5297-5300 (*Taiwankadsurins A,B,C*).
Xu, L.-J. *et al.*, *Helv. Chim. Acta*, 2008, **91**, 220-226 (*Heteroclitin N*).
Lin, Y.-C. *et al.*, *Molecules*, 2013, **18**, 6573-6583 (*Taiwankadsurins D,E*)

Bis(demethylene)sauchinone B27

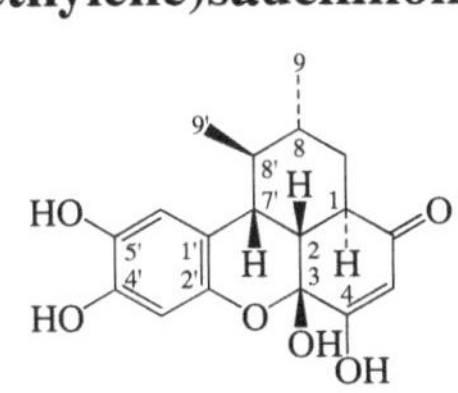

Relative Configuration

2,7′-Cyclo-2′,3-epoxy-2,3-dihydro-3,4,4′,5′-tetrahydroxy-8,8′-lignan-6(1H)-one
$C_{18}H_{20}O_6$ 332.352
Lignan numbering shown. Related to Hexakis(demethyl)polemannone C, H49. There are many errors in the lit. with respect to this series of compds.; care is therefore needed in their study

3,4-Methylene ether
Sauchinone C
[1686086-82-7]
$C_{19}H_{20}O_6$ 344.363
Constit. of *Saururus chinensis*. Cryst.. Mp 150 - 153°. $[\alpha]^{25}_{D}$-76 (c, 0.08 in MeOH). λ_{max} 241 (log ε 3.5); 297 (log ε 2.1) (MeOH)

4′,5′-Methylene, 4-Me ether
[1885114-26-0]
$C_{20}H_{22}O_6$ 358.39
Constit. of the roots of *Saururus chinensis*. Exhibits selective inhibitory activity against DGAT1. Amorph. brown powder. $[\alpha]^{25}_{D}$ -22.69 (c, 0.1 in $CHCl_3$). λ_{max} 253; 299 (MeOH)

Bis(methylene) ether
Sauchinone
[177931-17-8]
$C_{20}H_{20}O_6$ 356.374
Constit. of *Saururus chinensis*. Analgesic. Inhibits staurosporine-induced apoptosis in C6 rat glioma cells and NO prodn. in LPS-stimulated RAW264.7 cells. Attenutates CCl_4-induced toxicity in rat hepatocytes. Cryst. ($CHCl_3$/MeOH). Mp 224 - 226°. $[\alpha]_D$ +97.8 (c, 0.02 in $CHCl_3$) (prob. erroneous as there are 2 reported -ve values). $[\alpha]^{26}_{D}$-140 (c, 1 in $CHCl_3$). $[\alpha]^{25}_{D}$-96.2 (c, 1.7 in $CHCl_3$). λ_{max} 242 (log ε 4.25); 297 (log ε 3.72) (MeOH). [neutral] λ_{max} 253 (log ε 2.74); 300 (log ε 2.55) ($CHCl_3$)

1,7-Didehydro, bis(methylene) ether
Carpanone
[26430-30-8]
$C_{20}H_{18}O_6$ 354.358
Constit. of the bark of the carpano tree *Cinnamomum* sp.. Cryst.. Mp 210 - 212°. Racemate. λ_{max} 243 (log ε 3.93); 262 (log ε 4.04); 298 (log ε 3.71) (MeOH)

7′-Hydroxy, bis(methylene) ether
7′-Hydroxysauchinone. ***Sauchinone B***
[281655-25-2]
[1447990-21-7]
$C_{20}H_{20}O_7$ 372.373
Constit. of *Saururus chinensis*. Cryst.. Mp 155 - 158°. $[\alpha]^{25}_{D}$-45 (c, 0.08 in MeOH). λ_{max} 240 (log ε 3.47); 296 (log ε 2.98) (MeOH)

9′-Hydroxy, bis(methylene) ether
9′-Hydroxysauchinone. ***Sauchinone D***
[1186087-74-0]
$C_{20}H_{20}O_7$ 372.373
Constit. of *Saururus chinensis*. Inhibitor of NO prodn. in LPS-stimulated cells. Powder. $[\alpha]_D^{25}$ -33 (c, 0.01 in MeOH). λ_{max} 241 (log ε 2.96); 297 (log ε 2.52) (MeOH)

1-Epimer, bis(methylene) ether
1-Episauchinone.
Dihydrocarpanone
[26430-31-9]
$C_{20}H_{20}O_6$ 356.374
Constit. of *Saururus chinensis*. Mp 235 - 237°. $[\alpha]_D^{26}$ -100 (c, 0.1 in $CHCl_3$). The synonym 1′-Episauchinone has been used incorrectly elsewhere. λ_{max} 255 (log ε 2.51); 300 (log ε 2.32) ($CHCl_3$)

1-Epimer, bis(methylene) ether, configurational isomer
Carpananone
$C_{20}H_{20}O_6$ 356.374
Constit. of the bark of the carpano tree *Cinnamomum* sp.. Mp 227 - 229°. Thought to be a configurational isomer of dihydrocarpanone

2-Epimer, bis(methylene) ether
Sauchinone A
[26430-34-2]
$C_{20}H_{20}O_6$ 356.374
Constit. of *Saururus chinensis*. Amorph. powder. Mp 240 - 242°. $[\alpha]_D^{26}$ -80 (c, 0.1 in $CHCl_3$). λ_{max} 253 (log ε 2.54); 299 (log ε 2.12) ($CHCl_3$)

3-Epimer, 4′,5′-methylene, 4-Me ether
[1885114-24-8]
$C_{20}H_{22}O_6$ 358.39
Constit. of the roots of *Saururus chinensis*. Amorph. brown powder. $[\alpha]_D^{25}$ -16 (c, 0.1 in $CHCl_3$). λ_{max} 248; 299 (MeOH)

7′,8,8′-Triepimer, 1,7-didehydro, bis(methylene) ether (?)
Isocarpanone
[26430-33-1]
$C_{20}H_{18}O_6$ 354.358
Constit. of the bark of the carpano tree *Cinnamomum* sp.. Mp 216 - 217°.

Brophy, G.C. *et al.*, *Tetrahedron Lett.*, 1969, **10**, 5159-5162 (*Carpanone, Isocarpanone, Carpananone*).
Chapman, O.L. *et al.*, *J. Am. Chem. Soc.*, 1971, **93**, 6696-6698 (*Carpanone, synth*).
Matsumoto, M. *et al.*, *Tetrahedron Lett.*, 1981, **22**, 4437-4440 (*Carpanone, synth*).
Nicolaou, K.C. *et al.*, *Classics in Total Synthesis, Targets, Strategies, Methods*, VCH, 1996, 95-97 (*bibl, synth*).
Wang, E.-C. *et al.*, *Heterocycles*, 1996, **43**, 969-975 (*Sauchinone, cryst struct*).
Sung, S.H. *et al.*, *Biol. Pharm. Bull.*, 2000, **23**, 666-668 (*Sauchinone, hepatoprotective activity*).
Sung, S.H. *et al.*, *J. Nat. Prod.*, 2000, **63**, 1019-1021 (*1′-Episauchinone, Sauchinone A, activity*).
Baxendale, I.R. *et al.*, *J. Chem. Soc., Perkin Trans. 1*, 2002, 1850-1857 (*Carpanone, synth*).
Song, H. *et al.*, *Biol. Pharm. Bull.*, 2003, **26**, 1428-1430 (*Sauchinone, apoptosis inhibn*).
Hwang, B.Y. *et al.*, *Planta Med.*, 2003, **69**, 1096-1101 (*Sauchinone, activity*).
Seo, C-S. *et al.*, *Biol. Pharm. Bull.*, 2008, **31**, 523-526 (*Sauchinone*).
Daniels, R.N. *et al.*, *Org. Lett.*, 2008, **10**, 4097-4100 (*Carpanone, synth*).
Gao, X. *Planta Med.*, 2013, **79**, 1720-1723 (*Sauchinones B,C,D*).
Kim, Y.-J. *et al.*, *Biol. Pharm. Bull.*, 2015, **38**, 218-227 (*Sauchinone, pharmacokinet, metab*).
Li, N. *et al.*, *Fitoterapia*, 2015, **101**, 46-50 (*Saururus chinensis constits, activity*).
Cortés-Borda, D. *et al.*, *J. Org. Chem.*, 2018, **83**, 14286-14299 (*Carpanone, synth*).
Aka, E.C. *et al.*, *J. Org. Chem.*, 2019, **84**, 14101-14112 (*Carpanone, synth*)

5,5′-Bis[1-(3,4-dihydroxyphenyl)-1,2-dihydro-6,7-dihydroxy-2,3-naphthalenedicarboxylic acid] B28

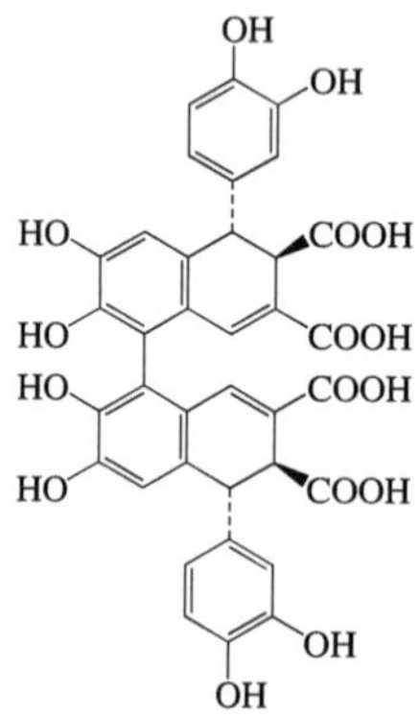

Relative Configuration

[217090-73-8]
$C_{36}H_{26}O_{16}$ 714.588
Dimer of Epiphyllic acid, in C339. Isol. from the liverwort *Bazzania trilobata*. $[\alpha]_D^{20}$ -106.1 (c, 0.6 in MeOH). λ_{max} 252; 329 (MeOH).

Martini, U. *et al.*, *Phytochemistry*, 1998, **49**, 1139-1146 (*Bazzania trilobata constit*)

1,7-Bis(3,4-dihydroxyphenyl)-2,5-heptanediol B29

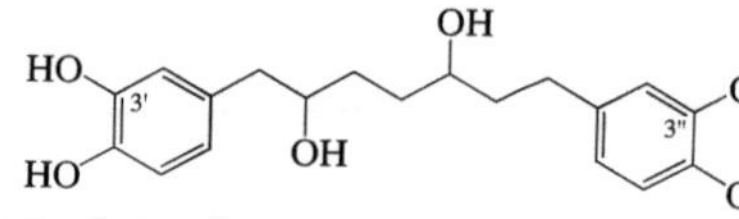

$C_{19}H_{24}O_6$ 348.395
(2ξ,5ξ)-form

3′,3″-Di-Me ether
1,7-Bis(4-hydroxy-3-methoxyphenyl)-2,5-heptanediol. Juglanol A

3′,3″-Di-Me ether, 5-O-β-D-xylopyranoside
Juglanol A 5-xyloside
[1198468-84-6]
$C_{26}H_{36}O_{10}$ 508.564
Constit. of the bark of *Juglans mandshurica* var. *sieboldiana*. $[\alpha]_D^{25}$ -25 (c, 0.1 in MeOH). λ_{max} 203 (log ε 4.74); 221 (log ε 4.23); 280 (log ε 3.88) (MeOH).

Machida, K. *et al.*, *J. Tohoku Pharm. Univ.*, 2008, **55**, 41-46 (*Juglanol A xyloside*)

1,7-Bis(3,4-dihydroxyphenyl)-3,5-heptanediol B30

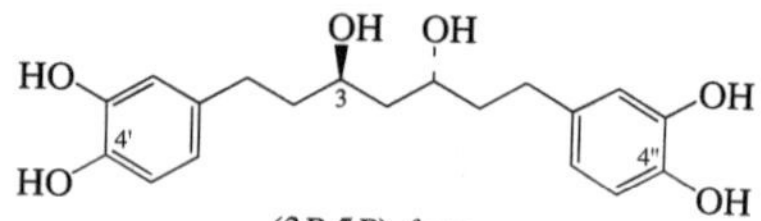

(3R,5R)-form

[93604-02-5]
$C_{19}H_{24}O_6$ 348.395
(3R,5R)-form [408324-01-6]
Constit. of the rhizomes of *Tacca chantrieri*. Viscous syrup. $[\alpha]_D^{23}$ +4 (c, 0.1 in MeOH). λ_{max} 283 (log ε 3.79) (MeOH)

3-O-β-D-Xylopyranoside
[883141-46-6]
$C_{24}H_{32}O_{10}$ 480.51
Constit. of the bark of *Alnus japonica*. Amorph. brown powder. $[\alpha]_D^{20}$ -6.3 (c, 1 in MeOH)

3-O-β-D-Glucopyranoside
[408324-05-0]
$C_{25}H_{34}O_{11}$ 510.536
Constit. of the rhizomes of *Tacca chantrieri*. Amorph. solid. $[\alpha]_D^{23}$ -8 (c, 0.1 in MeOH). λ_{max} 284 (MeOH)

3-Ac
[1269839-25-9]
$C_{21}H_{26}O_7$ 390.432
Constit. of the rhizomes of *Curcuma kwangsiensis*. Yellowish oil.

$[\alpha]^{25}_{D}$ +18.1 (c, 0.1 in MeOH). λ_{max} 283 (log ε 3.64) (MeOH)

3,5-Di-Ac

[1334172-42-7]

$C_{23}H_{28}O_8$ 432.469

Constit. of the rhizomes of *Curcuma kwangsiensis*

3′-Me ether

1-(3,4-Dihydroxyphenyl)-7-(4-hydroxy-3-methoxyphenyl)-3,5-heptanediol

[408324-13-0]

$C_{20}H_{26}O_6$ 362.422

Constit. of the rhizomes of *Curcuma kwangsiensis*

3′-Me ether, 3-O-β-D-glucopyranoside

[408324-06-1]

$C_{26}H_{36}O_{11}$ 524.563

Constit. of the rhizomes of *Tacca chantrieri*. Amorph. solid. $[\alpha]^{25}_{D}$ -16 (c, 0.1 in MeOH). λ_{max} 282 (log ε 3.45) (MeOH)

3′,3″-Di-Me ether, 3-O-β-D-glucopyranoside

[408324-08-3]

$C_{27}H_{38}O_{11}$ 538.59

Constit. of the rhizomes of *Tacca chantrieri*. Amorph. solid. $[\alpha]^{23}_{D}$ -18 (c, 0.1 in MeOH). λ_{max} 282 (log ε 3.7) (MeOH)

3′,3″-Di-Me ether, 4′-O-β-D-glucopyranoside

Dioscoroside E$_2$

[2250036-70-3]

$C_{27}H_{38}O_{11}$ 538.590

Constit. of the rhizomes of *Dioscorea septemloba*. Powder. $[\alpha]^{25}_{D}$ -27.7 (c, 0.87 in MeOH). λ_{max} 224 (log ε 4.43); 279 (log ε 4.05) (MeOH)

3-Ketone

1,7-Bis(3,4-dihydroxyphenyl)-5-hydroxy-3-heptanone.

Epihirsutanonol

[1198360-87-0]

$C_{19}H_{22}O_6$ 346.379

Constit. of the bark of *Alnus japonica*. Syrup. $[\alpha]^{20}_{D}$ +17 (c, 0.12 in MeOH). Enantiomer of Hirsutanonol below. λ_{max} 219 (log ε 3.49); 282 (log ε 3.04) (MeOH)

3-Ketone, 3″-Me ether

1-(3,4-Dihydroxyphenyl)-5-hydroxy-7-(4-hydroxy-3-methoxyphenyl)-3-heptanone

[1022082-07-0]

$C_{20}H_{24}O_6$ 360.406

Constit. of the rhizomes of *Alpinia officinarum*. Yellow oil. $[\alpha]^{20}_{D}$ -6.6 (c, 0.15 in $CHCl_3$)

3-Ketone, 5-Me ether

1,7-Bis(3,4-dihydroxyphenyl)-5-methoxy-3-heptanone.

Muricarpone A

[886226-14-8]

$C_{20}H_{24}O_6$ 360.406

Constit. of *Alnus japonica* and *Amomum muricarpum*. Syrup. $[\alpha]^{25}_{D}$ -7.1 (c, 0.88 in MeOH). λ_{max} 222 (log ε 4); 283 (log ε 3.73) (MeOH)

3-Ketone, 3′,3″-di-Me ether

5-Hydroxy-1,7-bis(4-hydroxy-3-methoxyphenyl)-3-heptanone.

5-Epihexahydrocurcumin

[100667-55-8]

[36062-05-2] (unspecified stereochem.) [93559-28-5] (deleted reg. number)

$C_{21}H_{26}O_6$ 374.433

Constit. of the rhizomes of *Alpinia officinarum*. Needles (C_6H_6). Mp 90 - 91°. $[\alpha]_{D}$ -10 (c, 0.28 in $CHCl_3$). λ_{max} 227 (log ε 4.36); 281 (log ε 4.05) (EtOH)

3,5-Diketone, 3′,3″-di-Me ether

1,7-Bis(4-hydroxy-3-methoxyphenyl)-3,5-heptanedione.

1,2,6,7-Tetrahydrocurcumin.

Tetrahydrocurcumin

[36062-04-1]

$C_{21}H_{24}O_6$ 372.417

Constit. of *Zingiber officinale*. Antioxidant for cosmetics. Cryst. (petrol or Et_2O/hexane). Mp 95 - 96°. Enolised β-diketone

(3R,5S)-form

Diarylcomosol IV [1801272-69-4]

Constit. of the rhizomes of *Curcuma comosa*. Pale yellow oil

3,5-Di-Ac

Diarylcomosol V

[1801272-70-7]

$C_{23}H_{28}O_8$ 432.469

Constit. of the rhizomes of *Curcuma comosa*. Pale yellow oil

3′-Me ether

[917078-13-8]

$C_{20}H_{26}O_6$ 362.422

Constit. of the rhizomes of *Zingiber ottensii*. Oil. $[\alpha]^{24}_{D}$ -1.3 (c, 0.86 in EtOH)

3′,3″-Di-Me ether

[135413-63-7]

Constit. of rhizomes of *Zingiber officinale*. Oil. λ_{max} 228 (log ε 3.81); 281 (log ε 3.48) (EtOH)

3′,3″-Di-Me ether, 3-Ac

[718638-70-1]

$C_{23}H_{30}O_7$ 418.486

Constit. of the rhizomes of *Zingiber officinale*. Oil. $[\alpha]^{24}_{D}$ +6 (c, 0.56 in $CHCl_3$)

3′,3″-Di-Me ether, 3,5-di-Ac

[135308-87-1]

Constit. of rhizomes of *Zingiber officinale*. Oil. λ_{max} 227 (sh); 281 (log ε 3.4) (EtOH)

(3S,5S)-form

3-O-β-D-Glucopyranoside

[1374348-29-4]

Constit. of the leaves of *Pyrostria major*. Amorph. solid. $[\alpha]^{25}_{D}$ -3.7 (c, 0.1 in MeOH). λ_{max} 227 (log ε 3.6); 281 (log ε 3.5) (MeOH)

3-O-Formyl, 5-O-β-D-glucopyranoside

[1374348-33-0]

$C_{26}H_{34}O_{12}$ 538.546

Constit. of the leaves of *Pyrostria major*. Amorph. solid. $[\alpha]^{25}_{D}$ -61 (c, 0.02 in MeOH). Stereochem. assigned tentatively. λ_{max} 222 (log ε 3.9); 281 (log ε 3.6) (MeOH)

3,5-Di-Ac

$C_{23}H_{28}O_8$ 432.469

Constit. of rhizomes of *Zingiber officinale*. Oil. $[\alpha]^{25}_{D}$ +1.5 (c, 1.3 in EtOH). λ_{max} 223 (log ε 4.07); 283 (log ε 3.7) (EtOH)

4′-Me ether, 5-O-β-D-glucopyranoside

[1374348-31-8]

$C_{26}H_{36}O_{11}$ 524.563

Constit. of the leaves of *Pyrostria major*. Amorph. solid. $[\alpha]^{25}_{D}$ -16 (c, 0.02 in MeOH). Stereochem. assigned tentatively. λ_{max} 221 (log ε 4.1); 282 (log ε 3.7) (MeOH)

3′,3″-Di-Me ether

1,7-Bis(4-hydroxy-3-methoxyphenyl)-3,5-heptanediol.

Octahydrocurcumin

[112494-41-4]

$C_{21}H_{28}O_6$ 376.449

Constit. of the rhizomes of *Zingiber officinale* and *Curcuma xanthorrhiza*. Needles. Mp 136°. $[\alpha]_{D}$ -18.5 (c, 0.26 in EtOH). λ_{max} 228 (log ε 3.81); 281 (log ε 3.48) (EtOH)

3′,3″-Di-Me ether, 3,5-di-Ac

[718638-64-3]

$C_{25}H_{32}O_8$ 460.523
Constit. of the rhizomes of *Zingiber officinale*. Oil. $[\alpha]_D^{26}$ +7 (c, 0.68 in $CHCl_3$)

3-Ketone
Hirsutanonol
[41137-86-4]
[91998-02-6]
$C_{19}H_{22}O_6$ 346.379
Constit. of *Alnus hirsuta*, *Alnus serrulatoides* and *Saxifraga tangutica*. Shows antibiotic activity. Inhibits nitric oxide synthase. Oil. $[\alpha]_D^{25}$ -1.7 (c, 0.78 in MeOH)

3-Ketone, 5-O-β-D-xylopyranoside
Oregonin
[55303-93-0]
$C_{24}H_{30}O_{10}$ 478.494
Constit. of *Alnus hirsuta*, *Alnus rubra*, other *Alnus* spp. and *Corylus maxima*. Nitric oxide synthase and cyclooxygenase-2 inhibitor. Antioxidant. Shows antibiotic and antimicrobial activities. Pale yellow oil. Mp 53 - 56° (as tetra-Me ether). $[\alpha]_D^{25}$ -21.5 (c, 0.3 in MeOH)

3-Ketone, 5-O-[2,3-bis-O-(3,4,5-trihydroxybenzoyl)-β-D-xylopyranoside]
2‴,3‴-Digalloyloregonin
[1781220-70-9]
$C_{38}H_{38}O_{18}$ 782.704
Constit. of the leaves of *Alnus sibirica*. Antioxidant. Dark yellow powder. Stereochem. not confirmed

3-Ketone, 5-O-[E-cinnamoyl-(→2)-β-D-xylopyranoside]
2‴-O-trans-Cinnamoyloregonin
[931117-60-1]
$C_{33}H_{36}O_{11}$ 608.64
Constit. of the leaves of *Alnus formosana*. $[\alpha]_D^{25}$ -9.1 (c, 1.1 in MeOH). λ_{max} 223 (log ε 4.13); 280 (log ε 2.64) (MeOH)

3-Ketone, 5-O-[4-hydroxy-E-cinnamoyl-(→2)-β-D-xylopyranoside]
2‴-O-trans-p-Coumaroyloregonin.
Oregonoyl A
[1198403-80-3]
$C_{33}H_{36}O_{12}$ 624.639
Constit. of the bark of *Alnus japonica*. Yellow syrup. $[\alpha]_D^{20}$ -9.6 (c, 0.32 in MeOH). λ_{max} 220 (log ε 3.22); 280 (log ε 2.94) (MeOH)

3-Ketone, 5-O-[4-hydroxy-3-methoxy-E-cinnamoyl-(→2)-β-D-xylopyranoside]
2‴-O-trans-Feruloyloregonin.
Oregonoyl B
[1198403-81-4]
$C_{34}H_{38}O_{13}$ 654.665
Constit. of the bark of *Alnus japonica*. Yellow syrup. $[\alpha]_D^{20}$ -12.4 (c, 0.34 in MeOH). λ_{max} 218 (log ε 3.4); 280 (log ε 3.14) (MeOH)

3-Ketone, 5-O-β-D-glucopyranoside
[93915-36-7]
$C_{25}H_{32}O_{11}$ 508.52
Constit. of *Alnus serrulatoides* and *Saxifraga tangutica*

3-Ketone, 5-O-(6-O-acetyl-β-D-glucopyranoside)
[1834610-14-8]
$C_{27}H_{34}O_{12}$ 550.557
Constit. of *Saxifraga tangutica*. Amorph. brown powder. $[\alpha]_D^{25}$ -6.2 (c, 0.22 in MeOH). λ_{max} 220 (log ε 4.07); 283 (log ε 3.72) (MeOH)

3-Ketone, 5-O-(6-O-benzoyl-β-D-glucopyranoside)
Oregonoside A
[213817-26-6]
$C_{32}H_{36}O_{12}$ 612.628
Constit. of the bark of *Alnus rubra*. Amorph. powder. $[\alpha]_D$ -7.5 (c, 0.07 in MeOH)

3-Ketone, 5-O-[4-hydroxy-3-methoxybenzoyl-(→6)-β-D-glucopyranoside]
Oregonoside B
[213817-27-7]
$C_{33}H_{38}O_{14}$ 658.653
Constit. of the bark of *Alnus rubra*. Amorph. powder. $[\alpha]_D$ -14.9 (c, 0.07 in MeOH)

3-Ketone, 5-O-[3,4,5-trihydroxybenzoyl-(→6)-β-D-glucopyranoside]
[211126-57-7]
$C_{32}H_{36}O_{15}$ 660.625
Constit. of *Alnus japonica*. Amorph. powder. $[\alpha]_D$ -18 (c, 1 in MeOH). λ_{max} 216 (log ε 4.95); 280 (log ε 4.52) (MeOH)

3-Ketone, 5-O-[4-hydroxy-E-cinnamoyl-(→6)-β-D-glucopyranoside]
[1527469-93-7]
$C_{34}H_{38}O_{13}$ 654.665
Constit. of bark of *Alnus glutinosa*. Exhibits considerable cytotoxicity and selectivity to NSCL carcinoma cells. $[\alpha]_D^{22}$ -7 (c, 1.0 in MeOH). λ_{max} 224 (sh) (log ε 4.51); 287 (log ε 4.37); 312 (log ε 4.31) (MeOH)

3-Ketone, 5-O-[3,4-dimethoxy-E-cinnamoyl-(→6)-β-D-glucopyranoside]
[1491159-91-1]
$C_{36}H_{42}O_{14}$ 698.718
Constit. of *Alnus glutinosa*. Exerts a chemoprotective effect on human lymphocytes DNA. Brown powder. $[\alpha]_D^{22}$ +4.0 (c, 1.0 in MeOH). λ_{max} 230 (log ε 4.15); 287 (log ε 4.02); 324 (log ε 4.02) (MeOH)

3-Ketone, 5-O-[3,4-dimethoxy-Z-cinnamoyl-(→6)-β-D-glucopyranoside]
[1527469-94-8]
$C_{36}H_{42}O_{14}$ 698.718
Constit. of bark of *Alnus glutinosa*. Brown gum. $[\alpha]_D^{22}$ -21 (c, 1.0 in MeOH). λ_{max} 222 (sh); 286 (log ε 4.11); 310 (log ε 4.04) (MeOH)

3-Ketone, 5-O-[3,4,5-trimethoxy-E-cinnamoyl-(→6)-β-D-glucopyranoside]
[1527469-95-9]
$C_{37}H_{44}O_{15}$ 728.744
Constit. of bark of *Alnus glutinosa*. Brown gum. $[\alpha]_D^{22}$ -21 (c, 1.0 in MeOH). λ_{max} 223 (log ε 4.44); 290 (log ε 4.19); 310 (log ε 4.14) (MeOH)

3-Ketone, 5-Me ether
5-O-Methylhirsutanonol
[876293-04-8]
$C_{20}H_{24}O_6$ 360.406
Constit. of the leaves of *Alnus japonica*. Viscous liq.. $[\alpha]_D$ +3.8 (c, 0.07 in MeOH). Enantiomer of Muricarpone A

3-Ketone, 3′,3″-di-Me ether
Hexahydrocurcumin
[39886-84-5]
$C_{21}H_{26}O_6$ 374.433
Constit. of the rhizomes of *Alpinia officinarum* and *Zingiber officinale*. Needles (C_6H_6). Mp 87°. $[\alpha]_D$ +9 ($CHCl_3$). λ_{max} 227 (log ε 4.36); 281 (log ε 4.05) (EtOH)

3-Ketone, 3′,3″-di-Me ether, 5-Ac
[718638-99-4]
$C_{23}H_{28}O_7$ 416.47
Constit. of the rhizomes of *Zingiber officinale*. Oil. $[\alpha]_D^{25}$ +3 (c, 0.6 in $CHCl_3$)

(3ξ,5ξ)-form
3,5-Di-Ac
[915039-08-6]

Constit. of *Zingiber officinale*. No phys. props. reported

3′-Me ether, 3,5-di-Ac

[138870-94-7]

$C_{24}H_{30}O_8$ 446.496

Constit. of rhizomes of *Zingiber officinale*. Oil. λ_{max} 225 (sh); 282 (log ε 3.6) (EtOH)

3-Ketone, 5-O-(1,5-anhydro-D-glucit-6-yl)

Hirsutoside

[91997-99-8]

$C_{25}H_{32}O_{10}$ 492.521

Isol. from green bark of *Alnus hirsuta* and *Alnus japonica*. Solid.

Roughley, P.J. *et al.*, *J. Chem. Soc., Perkin Trans. 1*, 1973, 2379-2388 (*Tetrahydrocurcumin, synth, pmr*).
Terazawa, M. *et al.*, *Mokuzai Gakkaishi*, 1973, **19**, 45-46 (*Hirsutanonol, isol, activity*).
Kato, N. *et al.*, *Chem. Pharm. Bull.*, 1984, **32**, 3323-3326 (*ketone, synth*).
Ohta, S. *et al.*, *J. Chem. Soc., Perkin Trans. 1*, 1984, 1635-1642 (*Alnus serrulatoid constits, isol, abs config, pmr, ms, struct*).
Terazawa, M. *et al.*, *Mokuzai Gakkaishi*, 1984, **30**, 587 , 601 (*Hirsutoside*).
Uehara, S. *et al.*, *Chem. Pharm. Bull.*, 1987, **35**, 3298-3304 (*Curcuma xanthorrhiza constits*).
Aoki, T. *et al.*, *Phytochemistry*, 1990, **29**, 3611-3614 (*Hirsutanonol, Oregonin, 3-ketone 5-glucoside 5S-form, occur, isol*).
Kikuzaki, H. *et al.*, *Chem. Pharm. Bull.*, 1991, **39**, 120-122 (*Hexahydrocurcumin, Zingiber officinale constits*).
Kikuzaki, H. *et al.*, *Phytochemistry*, 1991, **30**, 3647-3651 (*Zingiber officinale constits*).
Yamahara, J. *et al.*, *Yakugaku Zasshi*, 1992, **112**, 645-655 (*3′,3″-di-Me ethers*).
Saxena, G. *et al.*, *Int. J. Pharmacogn.*, 1995, **33**, 33-36 (*Oregonin, activity*).
Mazumder, A. *et al.*, *J. Med. Chem.*, 1997, **40**, 3057-3063 (*Tetrahydrocurcumin, synth*).
Wada, H. *et al.*, *Chem. Pharm. Bull.*, 1998, **46**, 1054-1055 (*ketone galloylglucoside*).
Gonzalez-Laredo, R.F. *et al.*, *J. Nat. Prod.*, 1998, **61**, 1292-1294 (*Oregonosides*).
Lee, M.-W. *et al.*, *Planta Med.*, 2000, **66**, 551-553 (*activity*).
Yokosuka, A. *et al.*, *J. Nat. Prod.*, 2002, **65**, 283-289 (*Tacca chantrieri constits*).
Girija, C.R. *et al.*, *Acta Crystallogr., Sect. C*, 2004, **60**, o611-o613 (*Tetrahydrocurcumin, cryst struct*).
Ma, J.P. *et al.*, *Chin. Chem. Lett.*, 2004, **15**, 1306-1308 (*3-ketone 3′,3″-di-Me Ac*).
Ma, J. *et al.*, *Phytochemistry*, 2004, **65**, 1137-1143 (*Zingiber officinale constits*).
Jolad, S.D. *et al.*, *Phytochemistry*, 2004, **65**, 1937-1954 (*Tetrahydrocurcumin, isol*).
Kuroyangi, M. *et al.*, *Chem. Pharm. Bull.*, 2005, **53**, 1519-1523 (*5-O-Methylhirsutanonol, Oregonin*).
Giang, P.M. *et al.*, *Chem. Pharm. Bull.*, 2005, **54**, 139-140 (*Muricarpone A*).
Kim, H.J. *et al.*, *Chin. Chem. Lett.*, 2005, **16**, 1337-1340 (*3-xyloside*).
Lee, W.S. *et al.*, *Planta Med.*, 2005, **71**, 295-299 (*Oregonin*).
Akiyama, K. *et al.*, *J. Nat. Prod.*, 2006, **69**, 1637-1640 (*3′-Me ether*).
Lee, S.-S. *et al.*, *Nat. Prod. Commun.*, 2006, **1**, 461-464 (*2‴-Cinnamoyloregonin*).
Ma, X. *et al.*, *Phytochemistry*, 2006, **67**, 2239-2255 (*3,5-di-Ac*).
An, N. *et al.*, *Fitoterapia*, 2008, **79**, 27-31 (*3-ketone 3″-Me ether*).
Tung, N.H. *et al.*, *Bioorg. Med. Chem. Lett.*, 2010, **20**, 1000-1003 (*Oregonoyls A,B*).
Tung, N.H. *et al.*, *J. Asian Nat. Prod. Res.*, 2010, **12**, 921-924 (*Epihirsutanonol*).
Li, J. *et al.*, *Bioorg. Med. Chem. Lett.*, 2011, **21**, 5363-5369 (*Curcuma kwangsiensis constits*).
Bhosale, S. *et al.*, *Tetrahedron Lett.*, 2011, **52**, 3397-3400 (*Octahydrocurcumin, synth*).
Beniddir, M.A. *et al.*, *Eur. J. Org. Chem.*, 2012, 1039-1046 (*Pyrostria major glucosides*).
Novakovic, M. *et al.*, *Planta Med.*, 2013, **79**, 499-505 (*Alnus glutinosa cinnamoylglucoside*).
Novakovic, M. *et al.*, *Phytochemistry*, 2014, **97**, 46-54 (*Alnus glutinosa constits, activity*).
Ponomarenko, J. *et al.*, *Phytochemistry*, 2014, **103**, 178-187 (*Oregonin, activity*).
Matsumoto, T. *et al.*, *J. Nat. Med.*, 2015, **69**, 142-147 (*Diarylcomosols IV,V*).
Aggarwal, B.B. *et al.*, *Molecules*, 2015, **20**, 185-205 (*Tetrahydrocurcumin, rev*).
Kim, M.H. *et al.*, *Nat. Prod. Res.*, 2016, **30**, 206-213 (*2‴,3‴-Digalloyloregonin*).
Dang, J. *et al.*, *Chem. Nat. Compd. (Engl. Transl.)*, 2017, **53**, 48-50 (*Saxifraga tangutica constits*).
Zhang, Y. *et al.*, *Fitoterapia*, 2017, **117**, 28-33 (*Dioscoroside E₂*)

1,7-Bis(3,4-dihydroxyphenyl)-1,3,5-heptanetriol B31

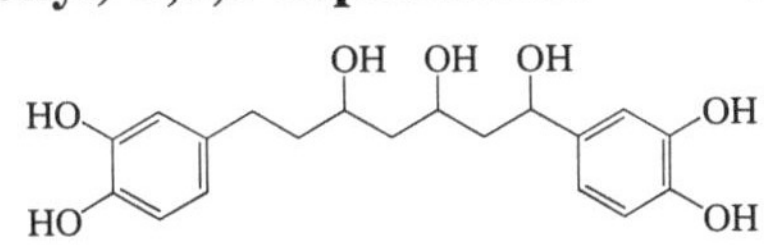

$C_{19}H_{24}O_7$ 364.394

(1ξ,3ξ,5ξ)-form

Curcusin A

Constit. of the rhizomes of *Curcuma singularis*. Soluble epoxide hydrolase inhibitor. Yellow oil. λ_{max} 281 (MeCN).

Cuong, N.M. *et al.*, *Bull. Korean Chem. Soc.*, 2017, **38**, 112-115 (*Curcusin A, activity*)

1,7-Bis(3,4-dihydroxyphenyl)-2,3,5-heptanetriol B32

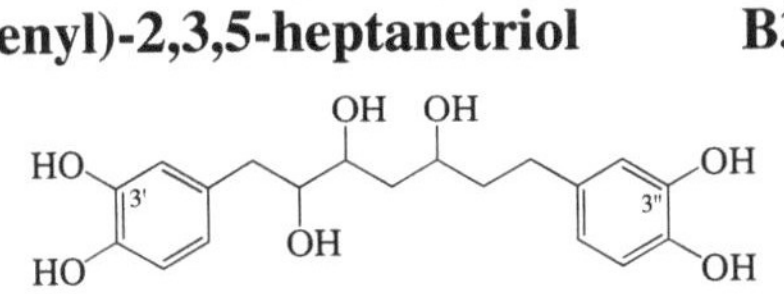

$C_{19}H_{24}O_7$ 364.394

3′,3″-Di-Me ether

1,7-Bis(4-hydroxy-3-methoxyphenyl)-2,3,5-heptanetriol

[882692-39-9]

$C_{21}H_{28}O_7$ 392.448

Constit. of *Juglans mandshurica*. Brown solid. $[\alpha]^{25}_{D}$ +15.9 (c, 0.19 in MeOH). λ_{max} 281 (log ε 3.76) (MeOH).

Li, G. *et al.*, *Bull. Korean Chem. Soc.*, 2005, **26**, 1878-1880 (*Juglans mandshurica constit*)

1,7-Bis(3,4-dihydroxyphenyl)-3-heptanol B33

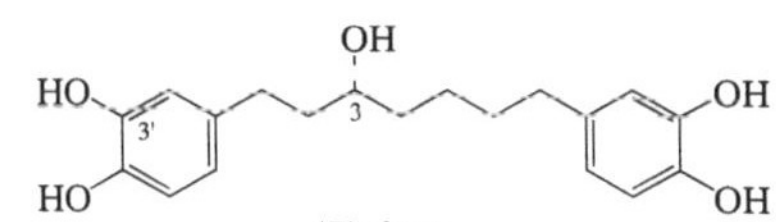

(R)-form

Rubranol. *3-Deoxohirsutanonol*

$C_{19}H_{24}O_5$ 332.396

(R)-form [211126-61-3]

Constit. of *Pinus flexilis*. Inhibitor of protein kinase C. Yellow gum. $[\alpha]_D$ -5 (c, 0.3 in MeOH)

3-O-β-D-Xylopyranoside

Rubranoside B

[229623-78-3]

$C_{24}H_{32}O_9$ 464.511

Constit. of *Alnus glutinosa*, *Alnus hirsuta* and *Alnus rubra*. Exhibits hepatoprotectant activity. $[\alpha]_D$ -28.3 (c, 0.3 in MeOH)

3-O-β-D-Glucopyranoside

Rubranoside A

[211126-58-8]

$C_{25}H_{34}O_{10}$ 494.537

Constit. of *Alnus glutinosa*, *Alnus japonica*, *Alnus rubra* and *Pinus flexilis*. Inhibitor of protein kinase C. Exhibits hepatoprotectant activity. Amorph. powder or yellow gum. $[\alpha]_D$ -25 (c, 1 in MeOH (-22)). λ_{max} 220 (log ε 4.19); 283 (log ε 3.86) (MeOH)

3-O-[3,4-Dimethoxy-E-cinnamoyl-(→6)-β-D-glucopyranoside]

[1527469-96-0]

$C_{36}H_{44}O_{13}$ 684.735

Constit. of bark of *Alnus glutinosa*. Brown gum. $[\alpha]^{22}_{D}$ -11 (c, 1.0 in MeOH). λ_{max} 220 (log ε 4.39); 232 (sh); 288 (log ε 4.14); 322 (log ε 4.11) (MeOH)

3-O-[3,4-Dimethoxy-Z-cinnamoyl-(→6)-β-D-glucopyranoside]

[1527469-97-1]

$C_{36}H_{44}O_{13}$ 684.735

Constit. of bark of *Alnus glutinosa*. Brown gum. $[\alpha]^{22}_{D}$ -20 (c, 1.0 in MeOH). λ_{max} 224 (sh); 288 (log ε 4.18); 310 (log ε 4.14) (MeOH)

3-O-[3,4,5-Trimethoxy-E-cinnamoyl-(→6)-β-D-glucopyranoside]
[1527469-98-2]
$C_{37}H_{46}O_{14}$ 714.761
Constit. of bark of *Alnus glutinosa*. Brown gum. $[\alpha]^{22}_{D}$ -40 (c, 1.0 in MeOH). λ_{max} 224 (sh); 289 (log ε 4.20); 310 (log ε 4.14) (MeOH)

3-O-[β-D-Apiofuranosyl-(1→6)-β-D-glucopyranoside]
Rubranoside D
[211126-60-2]
$C_{30}H_{42}O_{14}$ 626.652
Constit. of *Alnus japonica* and *Alnus rubra*. Amorph. powder. $[\alpha]_{D}$ -52 (c, 1 in MeOH). λ_{max} 220 (log ε 4.16); 283 (log ε 3.81) (MeOH)

3-O-[β-D-Glucopyranosyl-(1→3)-β-D-xylopyranoside]
Rubranoside C
[229623-79-4]
$C_{30}H_{42}O_{14}$ 626.652
Constit. of *Alnus hirsuta* and *Alnus rubra*. Exhibits hepatoprotectant activity. $[\alpha]_{D}$ -30.3 (c, 0.75 in MeOH)

3-Ketone
1,7-Bis(3,4-dihydroxyphenyl)-3-heptanone. ***Muricarpone B***
[886226-15-9]
$C_{19}H_{22}O_5$ 330.38
Constit. of *Alnus sibirica* and *Amomum muricarpum*. Also isol. from the nest of the wasp *Polistes* sp.. NO prodn. inhibitor in lipopolysaccharide-stimulated RAW 264.7 cells. Syrup

3-Ketone, 3′-Me ether
7-(3,4-Dihydroxyphenyl)-1-(4-hydroxy-3-methoxyphenyl)-3-heptanone. ***Dihydrogingerenone C***
[1376267-65-0]
$C_{20}H_{24}O_5$ 344.407
Constit. of the seeds of *Aframomum melegueta*. Yellow-brown oil. Misleading synonym. Not the dihydro deriv. of Gingerenone C. λ_{max} 215 (log ε 4.62); 280 (log ε 4.19) (MeOH)

(ξ)-form
3-Ac
[1083197-74-3]
$C_{21}H_{26}O_6$ 374.433
Constit. of the rhizomes of *Amomum muricarpum*. $[\alpha]_{D}$ -10.2 (c, 1.3 in MeOH).

Wada, H. *et al.*, *Chem. Pharm. Bull.*, 1998, **46**, 1054-1055 (*Alnus japonica constits*).
Lee, K.K. *et al.*, *J. Nat. Prod.*, 1998, **61**, 1407-1409 (*Pinus flexilis constit*).
Gonzalez-Laredo, R.F. *et al.*, *Nat. Prod. Lett.*, 1999, **13**, 75-80 (*Rubranosides A-D*).
Lee, M.W. *et al.*, *Arch. Pharmacal Res.*, 2000, **23**, 50-53 (*xyloside*).
Giang, P.M. *et al.*, *Chem. Pharm. Bull.*, 2006, **54**, 139-140 (*Muricarpone B*).
Park, D. *et al.*, *Chem. Pharm. Bull.*, 2010, **58**, 238-241 (*Rubranosides A,B,C, activity*).
He, J.-B. *et al.*, *Chem. Biodiversity*, 2011, **8**, 2270-2276 (*Muricarpone B, activity*).
El-Halawany, A.M. *et al.*, *Food Chem.*, 2012, **134**, 219-226 (*Dihydrogingerenone C*).
Giang, P.M. *et al.*, *Nat. Prod. Res.*, 2012, **26**, 1195-1200 (*Amomum muricarpum constit*).
Novakovic, M. *Planta Med.*, 2013, **79**, 499-505 (*Rubranosides A,B, activity*).
Novakovic, M. *et al.*, *Phytochemistry*, 2014, **97**, 46-54 (*Alnus glutinosa constits*).
Yin, J. *et al.*, *Nat. Prod. Sci.*, 2017, **23**, 146-150 (*Muricarpone B*)

1,7-Bis(3,4-dihydroxyphenyl)-1,4,6-heptatrien-3-one B34

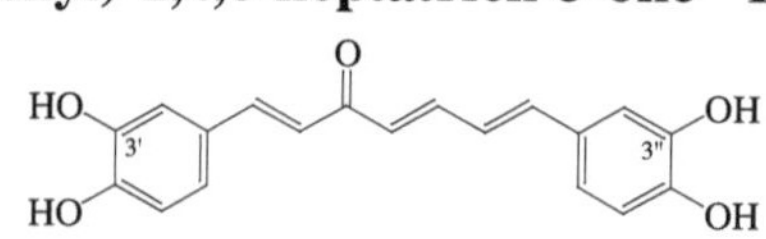

$C_{19}H_{16}O_5$ 324.332
3′,3″-Di-Me ether
1,7-Bis(4-hydroxy-3-methoxyphenyl)-1,4,6-heptatrien-3-one
[470466-02-5]
$C_{21}H_{20}O_5$ 352.386
Constit. of the rhizomes of *Curcuma longa*. Yellow powder. Mp 128 - 129°. λ_{max} 267 (log ε 3.54); 406 (log ε 3.84) (MeOH).

Park, S.-Y. *et al.*, *J. Nat. Prod.*, 2002, **65**, 1227-1231 (*di-Me ether*)

1,7-Bis(2,3-dihydroxyphenyl)-4-hepten-3-one B35

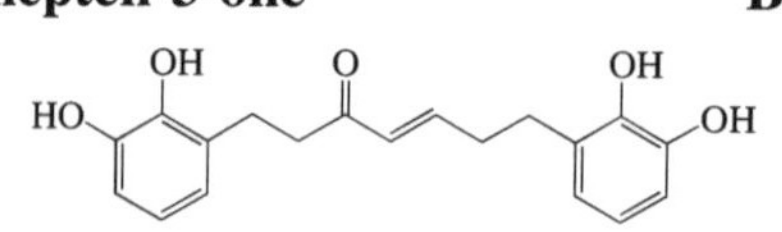

Nitidone B
$C_{19}H_{20}O_5$ 328.364
(E)-form
Constit. of *Alnus nitida*. λ_{max} 245 (log ε 3.22); 279 (log ε 3) (MeOH)

3′,3″-Dideoxy
1,7-Bis(2-hydroxyphenyl)-4-hepten-3-one. ***Nitidone A***
$C_{19}H_{20}O_3$ 296.366
Constit. of *Alnus nitida*. λ_{max} 245 (log ε 3.22); 279 (log ε 3) (MeOH).

Siddiqui, I.N. *et al.*, *Nat. Prod. Commun.*, 2010, **5**, 1787-1788 (*Nitidones A,B*)

1,7-Bis(3,4-dihydroxyphenyl)-4-hepten-3-one B36

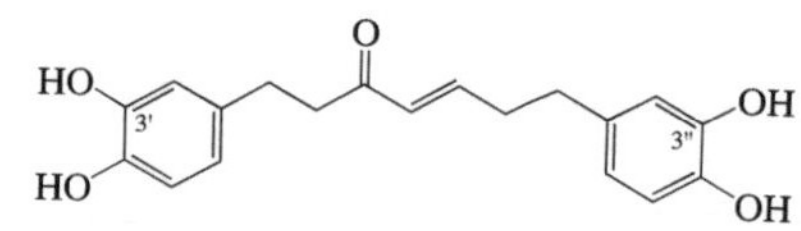

$C_{19}H_{20}O_5$ 328.364
(E)-form
Hirsutenone. Hirsutanone (incorr.) [41137-87-5]
Constit. of *Alnus hirsuta*, *Amomum muricarpum*, *Corylus maxima*, *Pinus flexilis*, *Renealmia alpinia* and *Viscum cruciatum*. Exhibits cytotoxic activity. Antioxidant. Antibacterial agent. Active against MRSA

3′-Me ether
7-(3,4-Dihydroxyphenyl)-1-(4-hydroxy-3-methoxyphenyl)-4-hepten-3-one
[138870-93-6]
$C_{20}H_{22}O_5$ 342.391
Constit. of the rhizomes of *Zingiber officinale*. Oil. λ_{max} 222 (log ε 4.2); 282 (log ε 3.7) (EtOH)

3′,3″-Di-Me ether
1,7-Bis(4-hydroxy-3-methoxyphenyl)-4-hepten-3-one. ***Gingerenone A***
[128700-97-0]
$C_{21}H_{24}O_5$ 356.418
Constit. of *Zingiber officinale*. Oil. λ_{max} 230 (log ε 4.27); 281 (log ε 3.85) (EtOH)

5′-Hydroxy, 3′-Me ether
1-(3,4-Dihydroxy-5-methoxyphenyl)-7-(3,4-dihydroxyphenyl)-4-hepten-3-one. ***Gingerenone D***
[1376699-70-5]
$C_{20}H_{22}O_6$ 358.390
Constit. of the seeds of *Aframomum melegueta*. Exhibits estrogen receptor binding activity. Yellow-brown oil. λ_{max} 215 (log ε 4.65); 280 (log ε 4.13) (MeOH)

5′-Methoxy, 3′,3″-di-Me ether
1-(4-Hydroxy-3,5-dimethoxyphenyl)-7-(4-hydroxy-3-methoxyphenyl)-4-hepten-3-one. ***Isogingerenone B***
[128700-99-2]
$C_{22}H_{26}O_6$ 386.444
Constit. of *Zingiber officinale*

5″-Methoxy, 3′,3″-di-Me ether
7-(4-Hydroxy-3,5-dimethoxyphenyl)-1-(4-hydroxy-3-methoxyphenyl)-4-hepten-3-one. ***Gingerenone B***

[128700-98-1]
$C_{22}H_{26}O_6$ 386.444
Constit. of *Zingiber officinale*.

Terazawa, M. *et al.*, *Mokuzai Gakkaishi*, 1973, **19**, 45-46 (*Hirsutenone*).
Endo, K. *et al.*, *Phytochemistry*, 1990, **29**, 797-799 (*Gingerenones A,B, Isogingerenone B*).
Kikuzaki, H. *et al.*, *Chem. Pharm. Bull.*, 1991, **39**, 120-122 (*Gingerenone A*).
Kikuzaki, H. *et al.*, *Phytochemistry*, 1991, **30**, 3647-3651 (*Zingiber officinale constit*).
Venkateswarlu, S. *et al.*, *Indian J. Chem., Sect. B*, 2001, **40**, 495-497 (*synth*).
Martin-Cordero, C. *et al.*, *Phytochemistry*, 2001, **58**, 567-569 (*Hirsutenone*).
Lee, W.S. *et al.*, *Planta Med.*, 2005, **71**, 295-299 (*Hirsutenone, activity*).
Jin, W.Y. *et al.*, *Arch. Pharmacal Res.*, 2007, **30**, 412-418 (*Hirsutenone*).
El-Halawany, A.M. *et al.*, *Food Chem.*, 2012, **134**, 219-226 (*Gingerenone D, activity*).
León-Gonzalez, A.J. *et al.*, *Phytomedicine*, 2014, **21**, 866-870 (*activity*).
Riethmüller, E. *et al.*, *J. Pharm. Biomed. Anal.*, 2015, **107**, 159-167 (*activity*)

1,7-Bis(3,4-dihydroxyphenyl)-5-oxo-3-heptanesulfonic acid B37

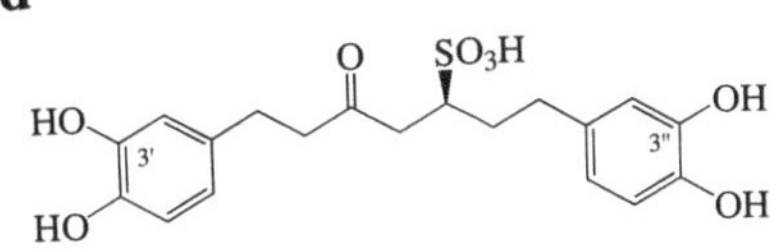

Shogasulfonic acid C
[532929-72-9]
$C_{19}H_{22}O_8S$ 410.437
Constit. of the rhizomes of *Zingiber officinale*. Pale yellow oil. $[\alpha]^{21}_{D}$ -5.6 (c, 0.25 in MeOH)

3'-Me ether
Shogasulfonic acid B
[532929-71-8]
$C_{20}H_{24}O_8S$ 424.464
Constit. of the rhizomes of *Zingiber officinale*. Pale green oil. $[\alpha]^{21}_{D}$ -1 (c, 1.6 in MeOH)

3',3"-Di-Me ether
1,7-Bis(4-hydroxy-3-methoxyphenyl)-5-oxo-3-heptanesulfonic acid. ***Shogasulfonic acid A***
[532929-70-7]
$C_{21}H_{26}O_8S$ 438.491
Constit. of the rhizomes of *Zingiber officinale*. Pale yellow powder. Mp 205° dec.. $[\alpha]^{21}_{D}$ -0.5 (c, 2 in MeOH)

5'-Hydroxy, 3',3"-di-Me ether
Shogasulfonic acid D
[532929-73-0]
$C_{21}H_{26}O_9S$ 454.49
Constit. of the rhizomes of *Zingiber officinale*. Pale yellow cryst.. Mp 154 - 158°. $[\alpha]^{21}_{D}$ -0.3 (c, 1 in MeOH).

Hori, Y. *et al.*, *Phytochemistry*, 2003, **62**, 613-617 (*Shogasulfonic acids A-D*).
Bencivenni, G. *et al.*, *Org. Biomol. Chem.*, 2020, **18**, 1091-1094 (*synth, abs config*)

2,3-Bis(3,4-dihydroxyphenyl)-1-propanol B38

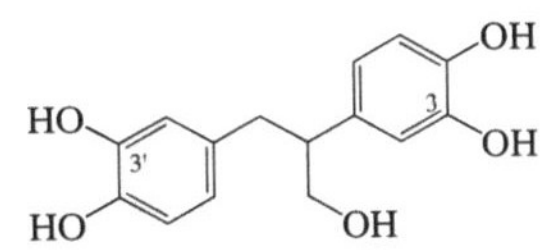

$C_{15}H_{16}O_5$ 276.288

(-)-form

3,3'-Di-Me ether
2,3-Bis(4-hydroxy-3-methoxyphenyl)-1-propanol
[908590-22-7]
$C_{17}H_{20}O_5$ 304.342
Constit. of the wood bark of *Durio zibethinus*. Amorph. solid. $[\alpha]^{24}_{D}$ -18.5 (c, 0.006 in MeOH).

Rudiyansyah, *et al.*, *J. Nat. Prod.*, 2006, **69**, 1218-1221 (*Durio zibethinus constit*)

2,3-Bis(3,4-dihydroxyphenyl)-2-propenal B39

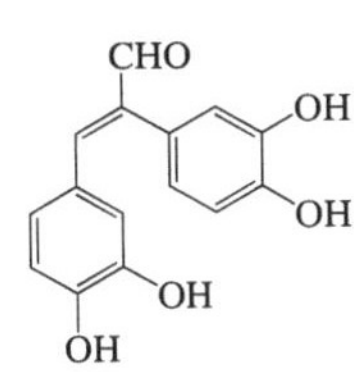

$C_{15}H_{12}O_5$ 272.256

(E)-form

3,3'-Di-Me ether
2,3-Bis(4-hydroxy-3-methoxyphenyl)-2-propenal
[1260667-95-5]
$C_{17}H_{16}O_5$ 300.31
Constit. of the roots of *Magnolia kachirachirai*. Mp 198 - 199°. λ_{max} 206 (log ε 4.42); 237 (sh) (log ε 4.16); 287 (sh) (log ε 3.93); 342 (log ε 4.25) (MeOH).

Chang, H.-S. *et al.*, *Chem. Biodiversity*, 2010, **7**, 2737-2747 (*3,3'-di-Me ether*)

1,2-Bis(3,4-dihydroxyphenyl)-2-propen-1-ol B40

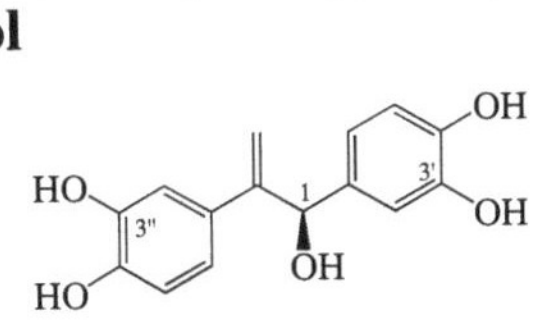

(*R*)-*form*

4,4'-(1-Hydroxy-2-propene-1,2-diyl)bis[1,2-benzenediol]
$C_{15}H_{14}O_5$ 274.272

(R)-form

1,3',3"-Tri-Me ether
[2452399-64-1]
$[\alpha]^{20}_{D}$ +65 (c, 0.1 in MeOH)

1-Et, 3',3"-di-Me ether
[2452399-62-9]
$[\alpha]^{20}_{D}$ +78 (c, 0.1 in MeOH)

(S)-form

1,3',3"-Tri-Me ether
[2452399-65-2]
$[\alpha]^{20}_{D}$ -70.5 (c, 0.1 in MeOH)

1-Et, 3',3"-di-Me ether
[2452399-63-0]
$[\alpha]^{20}_{D}$ -80.5 (c, 0.1 in MeOH)

(±)-form

1,3',3"-Tri-Me ether
1,2-Bis(4-hydroxy-3-methoxyphenyl)-1-methoxy-2-propene
[2452399-58-3]
$C_{18}H_{20}O_5$ 316.353
Constit. of the fruit of *Crataegus pinnatifida*. Oil. $[\alpha]^{20}_{D}$ +0.5 (c, 0.1 in MeOH). Partial racemate resolved by chiral hplc. λ_{max} 280 (log ε 0.15) (MeOH)

1-Et, 3',3"-di-Me ether
1-Ethoxy-1,2-bis(4-hydroxy-3-methoxyphenyl)-2-propene
[2452399-57-2]
$C_{19}H_{22}O_5$ 330.380
Constit. of the fruit of *Crataegus pinnatifida*. Oil. $[\alpha]^{20}_{D}$ +1.7 (c, 0.1 in MeOH). Partial racemate resolved by chiral hplc. λ_{max} 280 (log ε 0.49) (MeOH).

Zhao, P. *et al.*, *Bioorg. Chem.*, 2019, **93**, 103354:1-8 (*Crataegus pinnatifida constits*)

2,2-Bis(3,4-dihydroxyphenyl)-2H-pyran-3(6H)-one B41

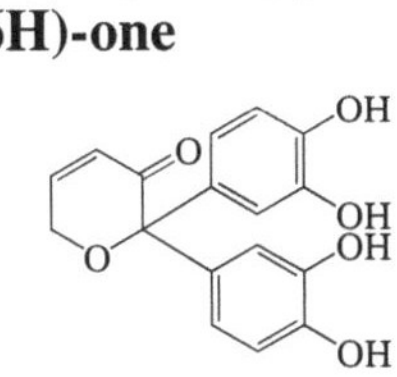

Breviscapin A‡
[1276541-08-2]
$C_{17}H_{14}O_6$ 314.293
Rearranged norlignan. Constit. of the rhizomes of *Curculigo breviscapa*. Black powder. λ_{max} 205 (log ε 4.6); 283 (log ε 3.87) (MeOH).

Zhu, C.-C. *et al.*, *Bull. Korean Chem. Soc.*, 2010, **31**, 224-226 (*Breviscapin A*)

4,5-Bis(3,4-dihydroxyphenyl)tetrahydro-3-furanol B42

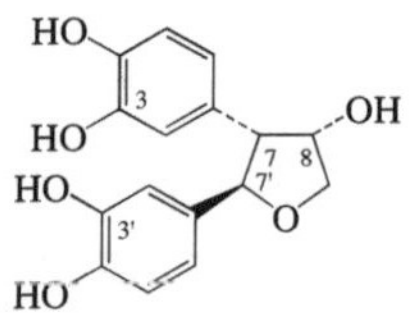

(7R,7'R,8S)-form

7',9-Epoxy-8',9'-dinor-7,7'-lignan-3,3',4,4',8-pentol. 7',9-Epoxy-3,3',4,4',8-pentahydroxy-8',9'-dinor-7,7'-neolignan

$C_{16}H_{16}O_6$ 304.298

Lignan numbering shown

(7R,7'R,8S)-form

3,3'-Di-Me ether

Tetrahydro-4,5-bis(4-hydroxy-3-methoxyphenyl)-3-furanol. 7',9-Epoxy-3,3'-dimethoxy-8',9'-dinor-7,7'-lignan-4,4',8-triol. 7',9-Epoxy-4,4',8-trihydroxy-3,3'-dimethoxy-8',9'-dinor-7,7'-lignan

[1954754-22-3]

$C_{18}H_{20}O_6$ 332.352

Constit. of fruit of *Melia toosendan*. Exhibits moderate agonistic activity on human melatonin receptor MT_1 on HEK293 cell line. Pale yellow oil. $[\alpha]_D^{20}$ -5.3 (c, 0.08 in MeOH). $[\alpha]_D^{20}$ -19 (c, 0.1 in MeOH). Mismatch in assignment of stereochem. between text and diag. (2019). λ_{max} 228 (log ε 4.26); 281 (log ε 3.88) (MeOH)

(7S,7'S,8R)-form

3,3'-Di-Me ether

[2452430-16-7]

$[\alpha]_D^{20}$ +17.2 (c, 0.1 in MeOH)

(7RS,7'RS,8SR)-form

3,3'-Di-Me ether

[2452430-15-6]

Constit. of the fruit of *Crataegus pinnatifida*. Oil. $[\alpha]_D^{20}$ -2.0 (c, 0.1 in MeOH). Partial racemate resolved by chiral hplc. λ_{max} 229 (log ε 0.21); 283 (log ε 0.1) (MeOH).

Wang, H. *et al.*, *Planta Med.*, 2015, **81**, 847-854 (*Melia toosendan constit*).
Zhao, P. *et al.*, *Bioorg. Chem.*, 2019, **93**, 103354:1-8 (*Crataegus pinnatifida constit, cd*)

6,6-Bis(3,4-dihydroxyphenyl)tetrahydro-2H-pyran-3-ol B43

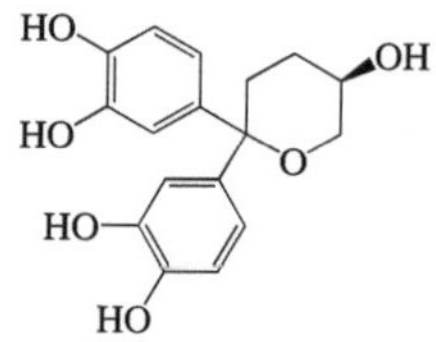

$C_{17}H_{18}O_6$ 318.325

Rearranged norlignan

(R)-form

3-O-β-D-Glucopyranoside

Breviscaside A

[1274654-79-3]

$C_{23}H_{28}O_{11}$ 480.466

Constit. of the rhizomes of *Curculigo breviscapa* and *Curculigo capitulata*. Powder. $[\alpha]_D^{25}$ -57 (c, 0.11 in MeOH). λ_{max} 190 (log ε 4.28); 206 (log ε 4.64); 289 (log ε 3.95) (MeOH).

Li, N. *et al.*, *Z. Naturforsch., B*, 2010, **65**, 79-82 (*Breviscaside A*)

Bishonokiol A B44

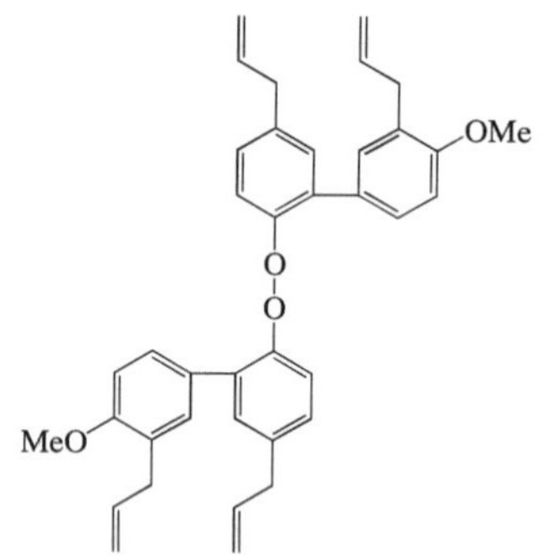

Bis(4'-methoxy-3',5-di-2-propenyl[1,1'-biphenyl]-2-yl)peroxide, CAS

[1629040-30-7]

$C_{38}H_{38}O_4$ 558.718

Constit. of seeds of *Magnolia grandiflora*. Exhibits potent antiproliferative activity against human cancer cell lines HepG2, MDA-MB-231, MDA-MB-435 and HNE1. Purple oil. $[\alpha]_D^{25}$ +1.7 (c, 0.1 in MeOH). λ_{max} 225 (log ε 2.76); 264 (log ε 2.77); 304 (log ε 2.95) (MeOH).

Li, H.-M. *et al.*, *Arch. Pharmacal Res.*, 2015, **38**, 1066-1071 (*Bishonokiol A, activity*)

Bis(4-hydroxycinnamyl) ether B45

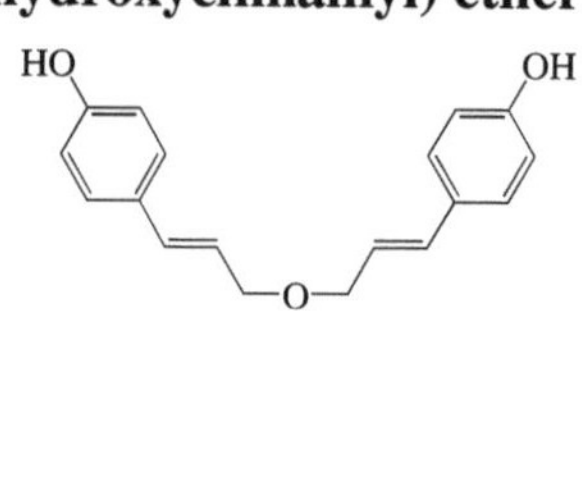

9,9'-Oxyligna-7,7'-diene-4,4'-diol. 4,4'-Dihydroxy-9,9'-oxyneoligna-7,7'-diene

$C_{18}H_{18}O_3$ 282.339

(E,E)-form

Di-Ac

Bis(4-acetoxycinnamyl) ether

[1179522-35-0]

$C_{22}H_{22}O_5$ 366.413

Constit. of the rhizomes of *Alpinia galanga*. Needles (petrol/EtOAc). Mp 108 - 109° (92.5°). λ_{max} 203 (log ε 4.2); 255 (log ε 3.9) (MeOH).

Zhu, X.-L. *et al.*, *Chin. J. Nat. Med.*, 2009, **7**, 19-20 (*Alpinia galanga constit*).
Zhao, L. *et al.*, *Chin. J. Nat. Med.*, 2012, **10**, 370-373 (*Alpinia galanga constit*)

1,2-Bis(4-hydroxy-3-methoxyphenyl)-1,3-propanediol, CAS B46

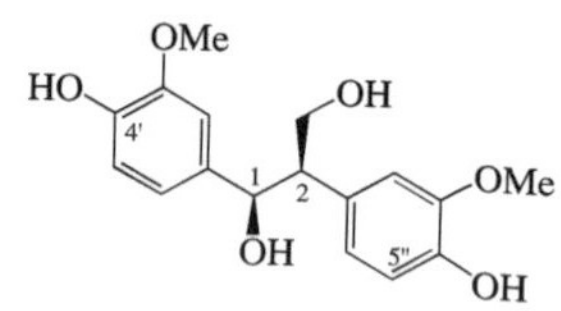

(1R,2R)-form

1,2-Diguaiacyl-1,3-propanediol

$C_{17}H_{20}O_6$ 320.341

(1R,2R)-form

1-Me ether

2,3-Bis(4-hydroxy-3-methoxyphenyl)-3-methoxy-1-propanol

[1414930-18-9]

$C_{18}H_{22}O_6$ 334.368

Constit. of *Phyllanthus glaucus*. Powder. $[\alpha]_D^{20}$ +30 (c, 0.18 in MeOH)

(1R,2S)-form [213612-02-3]

$[\alpha]_D^{25}$ -40.5 (c, 1.2 in MeOH)

1-O-β-D-Glucopyranoside

Hovetrichoside A

[213530-44-0]

$C_{23}H_{30}O_{11}$ 482.482

Constit. of the bark of *Hovenia trichocarpa*. Pale yellow powder. $[\alpha]_D^{25}$ -62.8 (c, 1.2 in MeOH). λ_{max} 210 (log ε 4.26); 228 (log ε 4.24); 280 (log ε 3.87) (MeOH)

3-Ac

[1620489-35-1]

$C_{19}H_{22}O_7$ 362.378

Constit. of the stems of *Rhododendron mariae*. Inhibitor of NO prodn. in LPS-activated C57BL6/J mouse macrophages. Amorph. powder. $[\alpha]_D^{20}$ +27.5 (c, 0.03 in MeOH). λ_{max} 230; 281 (MeOH)

1-Me ether
$C_{18}H_{22}O_6$ 334.368
Constit. of *Phyllanthus glaucus*. Powder. $[\alpha]^{20}_{D}$ -76.9 (c, 0.11 in MeOH)

(1S,2R)-form [213612-03-4]
$[\alpha]^{25}_{D}$ +41 (c, 0.7 in MeOH)

1-O-β-D-Glucopyranoside
Hovetrichoside B
[213530-45-1]
$C_{23}H_{30}O_{11}$ 482.482
Constit. of the bark of *Hovenia trichocarpa*. Pale yellow powder. $[\alpha]^{25}_{D}$ +24.1 (c, 0.7 in MeOH). λ_{max} 208 (log ε 4.01); 230 (log ε 3.86); 256 (log ε 3.45); 261 (log ε 3.39); 280 (log ε 3.46) (MeOH)

1-O-[α-L-Arabinofuranosyl-(1→6)-β-D-glucopyranoside]
Ichangoside
[1338457-41-2]
$C_{28}H_{38}O_{15}$ 614.597
Constit. of the leaves of *Viburnum ichangense*. Stress metabolite. Yellow gum. $[\alpha]^{20}_{D}$ -15 (c, 0.15 in MeOH). λ_{max} 210 (log ε 4.43); 260 (log ε 3.87); 280 (log ε 4.12) (MeOH)

4′-O-β-D-Glucopyranoside
[852866-15-0]
$C_{23}H_{30}O_{11}$ 482.482
Constit. of *Symplocos caudata* and *Tabebuia impetiginosa*. Powder. Mp 135 - 137°. $[\alpha]^{25}_{D}$ +24.6 (c, 0.11 in MeOH). λ_{max} 203 (log ε 4.66); 225 (sh); 278 (log ε 3.66) (MeOH)

1-Me ether
Constit. of *Phyllanthus glaucus*. $[\alpha]^{20}_{D}$ +79.5 (c, 0.13 in MeOH)

(1S,2S)-form [1643124-40-6]

1-Me ether
Constit. of *Phyllanthus glaucus*. Powder. $[\alpha]^{20}_{D}$ -30.8 (c, 0.22 in MeOH)

(1RS,2SR)-form

1-Me ether
[110995-87-4]
Constit. of the wood of *Zanthoxylum ailanthoides* and of *Irvingia malayana*. Two stereoisomers apparently isol., one dextrorotatory and one racemic, prob. *threo-*

(1R*,2R*)-form
[69887-40-7]

4′-O-β-D-Glucopyranoside
[886225-25-8]
$C_{23}H_{30}O_{11}$ 482.482
Constit. of the bark of *Tabebuia impetiginosa*. Amorph. powder. $[\alpha]^{22}_{D}$ -96 (c, 0.51 in MeOH). λ_{max} 202 (log ε 4.67); 223 (sh); 278 (log ε 3.66) (MeOH)

1-Et ether
3-Ethoxy-2,3-bis(4-hydroxy-3-methoxyphenyl)-1-propanol
[468094-64-6]
$C_{19}H_{24}O_6$ 348.395
Constit. of the roots of *Ficus beecheyana*. $[\alpha]^{25}_{D}$ +16 (c, 0.13 in $CHCl_3$). λ_{max} 280 (log ε 3.39) (MeOH)

1-Butyl ether
3-Butoxy-2,3-bis(4-hydroxy-3-methoxyphenyl)-1-propanol
[1416359-17-5]
$C_{21}H_{28}O_6$ 376.449
Constit. of the wood of *Casearia grewiifolia*. Amorph. solid. $[\alpha]^{25}_{D}$ +11.8 (c, 0.07 in $CHCl_3$). λ_{max} 280 (log ε 3.35) (MeOH)

(1R*,2S*)-form
[69887-41-8]

1-Et ether
[468094-65-7]
$C_{19}H_{24}O_6$ 348.395
Constit. of *Couepia ulei* and *Ficus beecheyana*. Amorph. powder or oil. λ_{max} 281 (log ε 3.42) (MeOH)

1-Butyl ether
[1416359-15-3]
$C_{21}H_{28}O_6$ 376.449
Constit. of the wood of *Casearia grewiifolia*. Amorph. solid. $[\alpha]^{25}_{D}$ -10.6 (c, 0.03 in $CHCl_3$). λ_{max} 279 (log ε 3.45) (MeOH)

(1ξ,2ξ)-form [4206-59-1]
Obt. by mild hydrol. of wood of *Fagus sylvatica*, *Larix leptolepis* and *Picea excelsa*. Prisms (EtOH aq.). Mp 143.5°. λ_{max} 282 (log ε 3.53) (EtOH)

5″-Methoxy
1-(4-Hydroxy-3-methoxyphenyl)-2-(4-hydroxy-3,5-dimethoxyphenyl)-1,3-propanediol
[4957-26-0]
$C_{18}H_{22}O_7$ 350.367
Obt. by mild hydrol. of lignin of *Fagus sylvatica*. Needles (MeOH aq.). Mp 192 - 193°.

Nimz, H. *Chem. Ber.*, 1966, **99**, 469-474 (*struct*).
Miki, K. *et al.*, *Phytochemistry*, 1980, **19**, 449 (*Larix leptolepis constit*).
Hsiao, J.-J. *et al.*, *Phytochemistry*, 1995, **39**, 899-902 (*1-Me ether*).
Mitsunaga, K. *et al.*, *Nat. Med. (Tokyo)*, 1996, **50**, 325-327 (*Irvingia malayana constit*).
Hsiao, J.-J. *et al.*, *Phytochemistry*, 1996, **41**, 1637 (*1-Me ether, erratum*).
Yoshikawa, K. *et al.*, *J. Nat. Prod.*, 1998, **61**, 1137-1139 (*Hovetrichosides A,B*).
Lee, T.-H. *et al.*, *J. Nat. Prod.*, 2002, **65**, 1497-1500 (*1-Et ethers*).
Jang, D.S. *et al.*, *Arch. Pharmacal Res.*, 2004, **27**, 169-172 (*Couepia ulei 1-Et ether*).
Jiang, J.S. *et al.*, *Chem. Pharm. Bull.*, 2005, **53**, 110-113 (*Symplocos 4′-glucoside*).
Warashina, T. *et al.*, *Chem. Pharm. Bull.*, 2006, **54**, 14-20 (*Tabebuia 4′-glucosides*).
Wu, B. *et al.*, *Helv. Chim. Acta*, 2011, **94**, 1677-1684 (*Ichangoside*).
Rayanil, K. *et al.*, *Phytochem. Lett.*, 2012, **5**, 59-62 (*Casearia 1-butyl ethers*).
Guo, Q. *et al.*, *J. Asian Nat. Prod. Res.*, 2014, **16**, 1-10 (*1R,2S-3-Ac, activity*).
Wu, Z. *et al.*, *Sci. Rep.*, 2016, **6**, 24809 (*Phyllanthus glaucus constits, abs config*)

3,3′-Bis-3-[(4-hydroxy-2-methoxyphenyl)-2-propenamide] B47

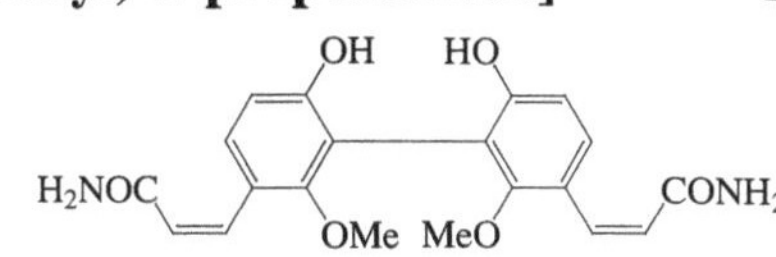

3,3′-(6,6′-Dihydroxy-2,2′-dimethoxy-[1,1′-biphenyl]-3,3′-diyl)bis[2-propenamide], CAS. *4,4′-Dihydroxy-2,2′-dimethoxy-3,3′-dicinnamide*
$C_{20}H_{20}N_2O_6$ 384.388

(Z,Z)-form [203524-81-6]
Constit. of *Sinularia flexibilis*.

Anjaneyulu, A.S.R. *et al.*, *Nat. Prod. Lett.*, 1997, **11**, 5-11 (*Sinularia flexibilis constit*)

2,4-Bis(4-hydroxyphenyl)-2-buten-1-ol B48

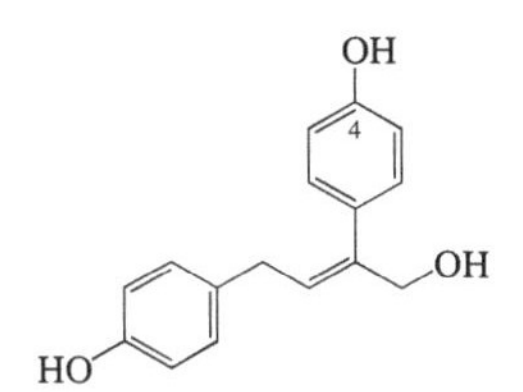

9,9′-Dinor-7,8′-lign-7(8′)-ene-4,4′,8-triol. 4,4′,8-Trihydroxy-9,9′-dinor-7,8′-neolign-7(8′)-ene
$C_{16}H_{16}O_3$ 256.301
Provisional norlignan. Biosynth. pathway unclear

(E)-form [2375168-59-3]
Constit. of *Caulophyllum robustum*. NO prodn. inhibitor. Amorph. powder

4-O-β-D-Glucopyranoside
Squadinorlignoside
[873077-46-4]
$C_{22}H_{26}O_8$ 418.442
Constit. of the stems of *Annona squamosa*. Syrup. λ_{max} 195; 225 (sh); 274 (MeCN).

Yang, Y.-L. *et al., Helv. Chim. Acta*, 2005, **88**, 2731-2737 (*Squadinorlignoside*).
Qin, B.-H. *et al., Chem. Nat. Compd. (Engl. Transl.)*, 2019, **55**, 465-468 (*Caulophyllum robustum constit, activity*)

4,5-Bis(4-hydroxyphenyl)-1,3-cyclopentanediol B49

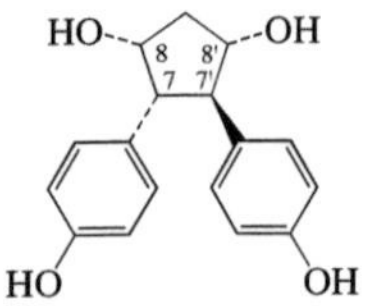

8,9'-Cyclo-9-nor-7,7'-lignan-4,4',8,8'-tetrol. 8,9'-Cyclo-4,4',8,8'-tetrahydroxy-9-nor-7,7'-neolignan

$C_{17}H_{18}O_4$ 286.327

Norlignan. Lignan numbering shown

(7R*,7'R*,8R*,8'S*)-form

Sonnerphenolic A [1771720-21-8]

Constit. of the mangrove plant *Sonneratia ovata*. Wax (EtOAc). $[\alpha]^{25}_{D}$ +3.8 (c, 0.37 in MeOH).

Nguyen, T.-H.-T. *et al., Bioorg. Med. Chem. Lett.*, 2015, **25**, 2366-2371 (*Sonnerphenolic A*)

1,7-Bis(4-hydroxyphenyl)-1,4-heptadiene B50

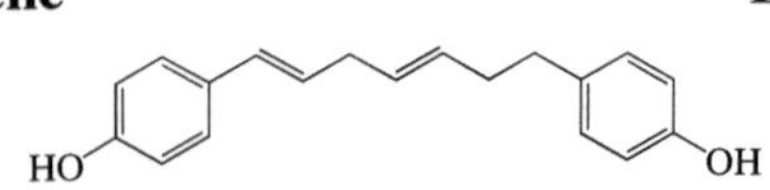

$C_{19}H_{20}O_2$ 280.367

(E,E)-form

Di-Me ether

1,7-Bis(4-methoxyphenyl)-1,4-heptadiene

[1563068-96-1]

$C_{21}H_{24}O_2$ 308.421

Constit.of *Ottelia alismoides*.

Hoye, T.R. *et al., Nat. Prod. Commun.*, 2013, **8**, 351-358 (*Ottelia alismoides constit*)

1,7-Bis(4-hydroxyphenyl)-3,5-heptanediol, CAS B51

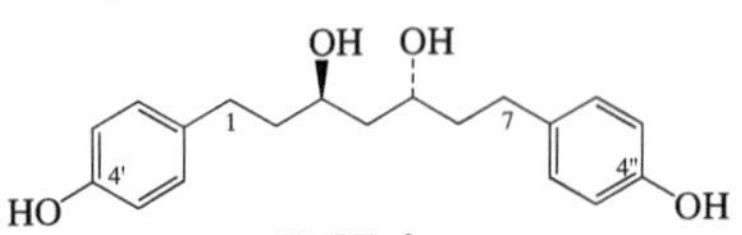

(3R,5R)-form

Hannokinol

[79120-40-4]

[56973-64-9] (unspecified stereochem.)

$C_{19}H_{24}O_4$ 316.397

Evidence for abs. config. of various isolates of Hannokinol and its derivs. is incomplete and sometimes contradictory

(3R,5R)-form [408324-76-5]

Constit. of *Amomum tsao-ko, Alnus hirsuta, Alpinia officinarum, Centrolobium robustum* and *Curcuma comosa*. $[\alpha]_D$ +11.4 (c, 0.38 in MeOH). λ_{max} 279 (log ε 3.5) (MeOH)

3-O-β-D-Glucopyranoside

[408324-09-4]

$C_{25}H_{34}O_9$ 478.538

Constit. of the rhizomes of *Tacca chantrieri*. Amorph. solid. $[\alpha]^{23}_{D}$ -8 (c, 0.1 in MeOH). λ_{max} 279 (log ε 3.44) (MeOH)

3-Ac

[1269839-24-8]

$C_{21}H_{26}O_5$ 358.434

Constit. of the rhizomes of *Curcuma kwangsiensis*. Yellowish oil. $[\alpha]^{25}_{D}$ +14.3 (c, 0.1 in MeOH). λ_{max} 279 (log ε 3.77) (MeOH)

3,5-Di-Ac

[1869012-35-0]

$C_{23}H_{28}O_6$ 400.471

Constit. of the rhizomes of *Curcuma kwangsiensis*. Yellow oil. $[\alpha]^{25}_{D}$ +11.8 (c, 0.2 in MeOH). λ_{max} 226; 272 (MeOH)

3-Ketone

5-Hydroxy-1,7-bis(4-hydroxyphenyl)-3-heptanone

[257301-36-3]

$C_{19}H_{22}O_4$ 314.381

Constit. of *Amomum muricarpum, Dioscorea villosa* and *Pellacalyx saccardianus*. Syrup. $[\alpha]^{25}_{D}$ +51.8 (c, 0.11 in MeOH)

Diketone

1,7-Bis(4-hydroxyphenyl)-3,5-heptanedione. Bisdemethoxytetrahydrocurcumin. ***Letestuianin C***

[113482-94-3]

$C_{19}H_{20}O_4$ 312.365

Constit. of the seeds of *Aframomum letestuianum*. Trypanocidal agent. Yellowish oil. λ_{max} 224 (log ε 2.4); 279 (log ε 3.4) (MeOH)

(3R,5S)-form

[79055-11-1]

Constit. of the fruit of *Amomum tsao-ko* and *Curcuma kwangsiensis*. Oil

5-O-β-D-Glucopyranoside

Betulaplatoside Ib

[213741-16-3]

$C_{25}H_{34}O_9$ 478.538

Constit. of the bark of *Betula platyphylla* var. *japonica*. Powder. $[\alpha]^{24}_{D}$ +4.4 (MeOH)

3-Ketone

Platyphyllonol. *Hannokinin. Platyphyllone*

[41137-85-3]

[79120-39-1]

$C_{19}H_{22}O_4$ 314.381

Constit. of *Alnus* spp., *Amomum tsao-ko, Betula platyphylla* and *Dioscorea villosa*. Active against KBNP-0028 (H9N2) influenza virus. Mp 127 - 129° (125-127°). $[\alpha]^{25}_{D}$ -1.8 (c, 1.02 in MeOH). Identity of Hannokinin and Platyphyllonol not stated but they have similar Mps. λ_{max} 225 (log ε 4.13) (EtOH)

3-Ketone, 5-O-β-D-xylopyranoside

[288141-04-8]

$C_{24}H_{30}O_8$ 446.496

Constit. of the bark of *Alnus rubra*. $[\alpha]^{28}_{D}$ -19.3 (c, 0.2 in MeOH)

3-Ketone, 5-O-β-D-glucopyranoside

Platyphylloside. *Platyphyllin*

[90803-80-8]

$C_{25}H_{32}O_9$ 476.522

Constit. of *Alnus glutinosa, Alnus viridis* and *Betula platyphylla*. Antioxidant. Chemo-protectant preventing damage by doxorubicin-induced cell death. Mp 62 - 64°. $[\alpha]^{25}_{D}$ -14.3 (c, 2.30 in MeOH). λ_{max} 224 (log ε 4.15); 279 (log ε 3.54) (EtOH)

3-Ketone, 5-O-[β-D-apiofuranosyl-(1→2)-β-D-glucopyranoside]

[146564-38-7]

$C_{30}H_{40}O_{13}$ 608.637

Constit. of *Betula pendula*. $[\alpha]_D$ -48 (c, 1 in EtOH)

3-Ketone, 5-O-[β-D-apiofuranosyl-(1→6)-β-D-glucopyranoside]

[146564-37-6]

$C_{30}H_{40}O_{13}$ 608.637

Constit. of *Alnus viridis* and *Betula pendula*. $[\alpha]_D$ -50 (c, 1 in EtOH)

3-Ketone, 5-O-[α-L-arabinofuranosyl-(1→6)-β-D-glucopyranoside]

Papyriferoside A

[956512-89-3]

$C_{30}H_{40}O_{13}$ 608.637

Constit. of the inner bark of *Betula papyrifera*. Amorph. powder. $[\alpha]^{25}_{D}$ -77.9 (c, 0.1 in MeOH). λ_{max} 224 (log ε 4.2); 279 (log ε 3.6) (MeOH)

3-Ketone, 5-O-[3,4,5-trihydroxybenzoyl-(→6)-β-D-glucopyranoside]

[1619924-73-0]

$C_{32}H_{36}O_{13}$ 628.627

Constit. of the bark of *Alnus viridis*. Amorph. reddish solid. $[\alpha]_D^{22}$ -38 (c, 0.1 in MeOH). λ_{max} 218 (log ε 4.01); 276 (log ε 3.78) (MeOH)

3-Ketone, 5-O-[4-hydroxy-E-cinnamoyl-(1→6)-β-D-glucopyranoside]

[1527469-89-1]

$C_{34}H_{38}O_{11}$ 622.667

Constit. of bark of *Alnus glutinosa*. Exhibits considerable cytotoxicity and selectivity to NSCL carcinoma cells. Red gum. $[\alpha]_D^{22}$ -32 (c, 1.0 in MeOH). λ_{max} 224 (log ε 4.58); 286 (log ε 4.38); 314 (log ε 4.41) (MeOH)

3-Ketone, 5-O-[3,4-dimethoxy-E-cinnamoyl-(1→6)-β-D-glucopyranoside]

[1527469-90-4]

$C_{36}H_{42}O_{12}$ 666.72

Constit. of bark of *Alnus glutinosa*. Yellow gum. $[\alpha]_D^{22}$ +12 (c, 2.0 in MeOH). λ_{max} 223 (sh); 246 (sh); 287 (log ε 4.10); 322 (log ε 4.13) (MeOH)

(3S,5S)-form [183184-32-9]

Constit. of seeds of *Alpinia blepharocalyx*. $[\alpha]_D$ -15 (c, 0.4 in MeOH)

5-O-β-D-Glucopyranoside

Betulaplatoside Ia

[213741-15-2]

$C_{25}H_{34}O_9$ 478.538

Constit. of the bark of *Betula platyphylla* var. *japonica*. Powder. $[\alpha]_D^{25}$ -11.5 (MeOH)

4′,4″-Di-Me ether, 5-O-β-D-glucopyranoside

$C_{27}H_{38}O_9$ 506.592

Constit. of the leaves of *Pyrostria major*. Amorph. solid. $[\alpha]_D^{25}$ -39 (c, 0.05 in MeOH). λ_{max} 224 (log ε 4.36); 278 (log ε 3.6) (MeOH)

3-Ketone, 5-Me ether

1,7-Bis(4-hydroxyphenyl)-5-methoxy-3-heptanone

[1622444-08-9]

$C_{20}H_{24}O_4$ 328.408

Constit. of the stem bark of *Alnus japonica*. Brown solid. $[\alpha]_D^{22}$ +21.6 (c, 1 in MeOH). λ_{max} 222 (log ε 4.35); 280 (log ε 3.96) (MeOH)

(3R*,5R*)-form

Isol. from woods of *Alnus hirsuta* and *Alnus japonica*. Plates. Mp 165 - 166° (154-156°). λ_{max} 203 (log ε 4.53); 226 (log ε 4.23); 281 (log ε 3.58) (EtOH). [base] λ_{max} 228 (log ε 4.19); 243 (log ε 4.22); 289 (log ε 3.64); 298 (log ε 3.64) (EtOH/NaOH)

(3ξ,5ξ)-form

3-O-[β-D-Apiofuranosyl-(1→6)-β-D-glucopyranoside]

[1619982-50-1]

$C_{30}H_{42}O_{13}$ 610.653

Constit. of the bark of *Alnus viridis*. Amorph. reddish solid. $[\alpha]_D^{22}$ -48 (c, 0.1 in MeOH). λ_{max} 226 (log ε 3.92); 278 (log ε 3.67) (MeOH).

Koga, T. *et al., Mokuzai Gakkaishi*, 1973, **19**, 47 ; 1984, **30**, 391-403 (*Platyphyllonol, Platyphyllin, Platyphylloside*).
Nomura, M. *et al., Phytochemistry*, 1981, **20**, 1097-1104 (*Hannokinol, Hannokinin*).
Ohta, S. *et al., Bull. Chem. Soc. Jpn.*, 1985, **58**, 2423-2424 (*Platyphylloside, abs config*).
Smite, E. *et al., Phytochemistry*, 1993, **32**, 365-369 (*glycosides*).
Matsuda, H. *et al., Bioorg. Med. Chem. Lett.*, 1998, **8**, 2939-2944 (*Betulaplatosides*).
Martin, T.S. *et al., J. Am. Oil Chem. Soc.*, 2000, **77**, 667-673 (*(+)-Hannokinol, meso-Hannokinol*).
Chen, J. *et al., Phytochemistry*, 2000, **53**, 971-973 (*3-ketone xyloside*).
Ali, M.S. *et al., J. Nat. Prod.*, 2001, **64**, 289-293 (*Alpinia blepharocalyx constits*).
Yokosuka, A. *et al., J. Nat. Prod.*, 2002, **65**, 283-289 (*3R,5R-3-glucoside*).
Kamnaing, P. *et al., J. Nat. Prod.*, 2003, **66**, 364-367 (*Letestuianin C*).
Mshvildadze, V. *et al., Phytochemistry*, 2007, **68**, 2531-2536 (*Papyriferoside A*).
Li, J. *et al., Bioorg. Med. Chem. Lett.*, 2011, **21**, 5363-5369 (*Curcuma kwangsiensis constits*).
Beniddir, M.A. *et al., Eur. J. Org. Chem.*, 2012, 1039-1046 (*Pyrostria major glucoside*).
Dong, S.-H. *et al., J. Nat. Prod.*, 2012, **75**, 2168-2177 (*Dioscorea villosa constits*).
Giang, P.M. *et al., Nat. Prod. Res.*, 2012, **26**, 1195-1200 (*Amomum muricarpum ketone*).
Reddy, P.R. *et al., Helv. Chim. Acta*, 2013, **96**, 289-295 (*synth*).
Dong, S.-H. *et al., J. Nat. Prod.*, 2013, **76**, 2005-2007 (*Dioscorea villosa constits, corrigendum*).
Novakovic, M. *et al., Chem. Biodiversity*, 2014, **11**, 872-885 (*Alnus viridus glycosides*).
Ibrahim, S.R.M. *et al., Nat. Prod. Res.*, 2014, **28**, 1765-1771 (*Alnus japonica ketone*).
Novakovic, M. *et al., Phytochemistry*, 2014, **97**, 46-54 (*Alnus glutinosa cinnamoylglucosides*).
Ponomarenko, J. *et al., Phytochemistry*, 2014, **103**, 178-187 (*Platyphylloside, activity*).
Chen, S.-D. *et al., Fitoterapia*, 2015, **102**, 67-73 (*Curcuma kwangsiensis di-Ac*).
Dinic, J. *et al., Fitoterapia*, 2015, **105**, 169-176 (*Alnus glutinosa constits, activity*).
Venkatesham, K. *et al., Helv. Chim. Acta*, 2015, **98**, 1307-1314 (*synth*).
Abed, S.A. *et al., Phytochem. Lett.*, 2015, **13**, 20-29 (*Pellacalyx saccardianus constit*).
Yadav, J.S. *et al., Tetrahedron Lett.*, 2015, **56**, 1360-1362 (*3R,5R-form, synth*)

1,7-Bis(4-hydroxyphenyl)-3-heptanol

B52

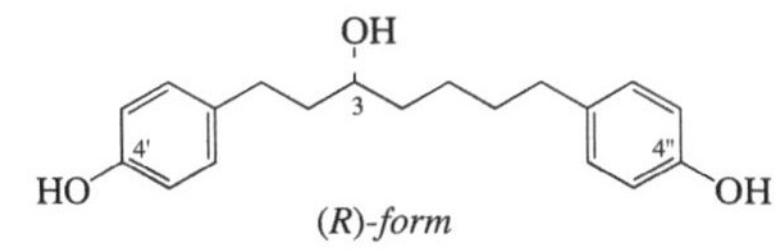

Centrolobol

[98119-94-9]

$C_{19}H_{24}O_3$ 300.398

(R)-form [30359-01-4]

Isol. from *Centrolobium paraense*, *Centrolobium sclerophyllum* and *Centrolobium tomentosum*. Needles. Mp 129 - 130°. $[\alpha]_D^{15}$ -8.3 (c, 0.34 in MeOH)

3-O-β-D-Glucopyranoside

Aceroside VII

[104109-45-7]

$C_{25}H_{34}O_8$ 462.539

Constit. of the bark of *Acer nikoense* and *Alnus viridis*. Cryst. powder. Mp 144 - 145°. $[\alpha]_D^{15}$ -28.4 (c, 0.8 in EtOH)

3-O-[β-D-Apiofuranosyl-(1→6)-β-D-glucopyranoside]

Aceroside VIII

[104109-46-8]

$C_{30}H_{42}O_{12}$ 594.654

Constit. of the bark of *Acer nikoense* and *Alnus viridis*. Film. $[\alpha]_D^{15}$ -64.3 (c, 0.6 in EtOH)

3-O-[α-L-Arabinofuranosyl-(1→6)-β-D-glucopyranoside]

[1619924-72-9]

$C_{30}H_{42}O_{12}$ 594.654

Constit. of the bark of *Alnus viridis*. Amorph. brown solid. $[\alpha]_D^{22}$ -91.1 (c, 0.09 in MeOH). λ_{max} 224 (log ε 3.88); 278 (log ε 3.57) (MeOH)

3-O-[β-D-Glucopyranosyl-(1→3)-β-D-xylopyranoside]

[1239976-63-6]

$C_{30}H_{42}O_{12}$ 594.654

Constit. of the stem bark of *Alnus hirsuta*. Amorph. brown powder. $[\alpha]_D^{26}$ -25.7 (c, 0.17 in MeOH)

3-O-[β-D-Apiofuranosyl-(1→2)-[β-D-apiofuranosyl-(1→6)]-β-D-glucopyranoside]

[146564-39-8]

$C_{35}H_{50}O_{16}$ 726.769

Constit. of the inner bark of *Betula pendula*. Amorph.. $[\alpha]_D$ -86.4 (c, 1.3 in EtOH)

4″-O-β-D-Glucopyranoside

Aceroside XIII

[156312-15-1]
$C_{25}H_{34}O_8$ 462.539
Constit. of *Acer triflorum*. Amorph. powder. [α] D -40.2 (c, 0.5 in MeOH)

4″-O-[β-D-Apiofuranosyl-(1→6)-β-D-glucopyranoside]
Aceroside XII
[156312-14-0]
$C_{30}H_{42}O_{12}$ 594.654
Constit. of *Acer triflorum*. Amorph. powder. [α] D -70.3 (c, 1 in MeOH)

6,7E-Didehydro
1,7-Bis(4-hydroxyphenyl)-6-hepten-3-ol
[1251830-59-7]
$C_{19}H_{22}O_3$ 298.382
Constit. of the rhizomes of *Curcuma kwangsiensis*. $[\alpha]_D^{25}$ +3.5 (c, 0.1 in MeOH). λ_{max} 226 (log ε 4.05); 280 (log ε 3.47) (MeOH)

3-Ketone
1,7-Bis(4-hydroxyphenyl)-3-heptanone. ***Acerogenin G***
[130233-83-9]
$C_{19}H_{22}O_3$ 298.382
Constit. of *Acer* spp. and *Curcuma kwangsiensis*. Amorph. powder

3-Ketone, 4″-O-β-D-glucopyranoside
Aceroside X
[130233-82-8]
$C_{25}H_{32}O_8$ 460.523
Constit. of *Acer griseum*, *Acer triflorum* and *Alnus hirsuta*. Shows cytotoxic activity. Needles. Mp 120 - 121°. $[\alpha]_D^{21}$ -29.7 (c, 1 in EtOH)

3-Ketone, 4″-O-[β-D-apiofuranosyl-(1→6)-β-D-glucopyranoside]
Aceroside IX
[130233-81-7]
$C_{30}H_{40}O_{12}$ 592.638
Constit. of *Acer griseum* and *Acer triflorum*. Amorph. powder. [α] D -63.5 (c, 1 in EtOH)

(S)-form [30359-04-7]
Isol. from *Centrolobium robustum* and an unidentified *Centrolobium* sp.. $[\alpha]_D^{25}$ +11. The occurrence of both enantiomers in different *C.* spp. is considered to be a genuine evolutionary trend

6,7E-Didehydro
[1251830-52-0]
Constit. of the rhizomes of *Curcuma kwangsiensis*. $[\alpha]_D^{25}$ -3.5 (c, 0.1 in MeOH).

Craveiro, A.A. *et al., Phytochemistry*, 1970, **9**, 1869-1875 (*Centrolobol*).
Jurd, L. *et al., Aust. J. Chem.*, 1984, **37**, 1127-1133 (*Centrolobol enantiomers, isol*).
Ohta, S. *et al., Bull. Chem. Soc. Jpn.*, 1985, **58**, 2423-2424 (*Centrolobol, abs config*).
Nagai, M. *et al., Chem. Pharm. Bull.*, 1986, **34**, 1056-1060 (*Centrolobol, abs config, Acerosides VII, VIII*).
Alegrio, L.V. *et al., Phytochemistry*, 1989, **28**, 2359-2362 (*Centrolobol, occur*).
Nagai, M. *et al., Chem. Pharm. Bull.*, 1990, **38**, 1506-1508 (*Acerogenin G, Acerosides IX,X*).
Smite, E. *et al., Phytochemistry*, 1993, **32**, 365-369 (*Betula pendula triglycoside*).
Shiratori, S. *et al., Chem. Pharm. Bull.*, 1994, **42**, 960-962 (*Acerosides XII,XIII*).
Park, D. *et al., Chem. Pharm. Bull.*, 2010, **58**, 238-241 (*3-glucosylxyloside*).
Li, J. *et al., J. Nat. Prod.*, 2010, **73**, 1667-1671 (*Curcuma kwangsiensis constits*).
Kashanna, J. *et al., Helv. Chim. Acta*, 2012, **95**, 1666-1671 (*synth*).
Ogura, T. *et al., Tetrahedron*, 2013, **69**, 2807-2815 (*Centrolobol, Acerogenin G, synth*).
Novaković, M. *et al., Chem. Biodiversity*, 2014, **11**, 872-885 (*Alnus viridis glycosides*)

1,7-Bis(4-hydroxyphenyl)-1,4,6-heptatrien-3-one B53

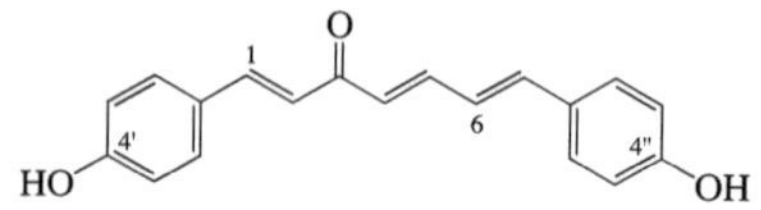

[470466-05-8]
$C_{19}H_{16}O_3$ 292.334

(all-E)-form [149732-52-5]
Constit. of the rhizomes of *Curcuma longa* and *Curcuma zedoaria*. Also from *Dioscorea opposita*, *Etlingera elatior* and *Smilax longifolia*. Antioxidant. Yellow powder. Mp 168 - 170°. λ_{max} 245 (log ε 4.06); 395 (log ε 4.58) (EtOH)

1,2-Dihydro
1,7-Bis(4-hydroxyphenyl)-4,6-heptadien-3-one
[332371-82-1]
$C_{19}H_{18}O_3$ 294.35
Constit. of *Alpinia blepharocalyx*, *Dioscorea opposita* and *Smilax longifolia*. COX-2 inhibitor. Amorph. yellow solid

6,7-Dihydro
1,7-Bis(4-hydroxyphenyl)-1,4-heptadien-3-one
[1412886-17-9]
$C_{19}H_{18}O_3$ 294.35
Constit. of *Viscum coloratum*. Exhibits antitumour activity

1,2,4,5-Tetrahydro
1,7-Bis(4-hydroxyphenyl)-6-hepten-3-one
[1251830-57-5]
$C_{19}H_{20}O_3$ 296.366
Constit. of the rhizomes of *Curcuma kwangsiensis*. Yellowish oil. λ_{max} 201 (log ε 4.86); 262 (log ε 3.05) (MeOH)

1,2,6,7-Tetrahydro
1,7-Bis(4-hydroxyphenyl)-4-hepten-3-one. ***Platyphyllenone***
[56973-65-0]
$C_{19}H_{20}O_3$ 296.366
Constit. of *Alnus japonica*, *Betula platyphylla* var. *japonica* and *Renealmia alpinia*. Oil

1,2,6,7-Tetrahydro, 6ξ-hydroxy
6-Hydroxy-1,7-bis(4-hydroxyphenyl)-4-hepten-3-one.
6-Hydroxyplatyphyllenone
[1772625-54-3]
$C_{19}H_{20}O_4$ 312.365
Constit. of *Betula platyphylla* var. *japonica*. Oil. λ_{max} 225 (log ε 4.2); 278 (log ε 3.67) (MeOH)

1,2,4,5,6,7-Hexahydro, 4″-Me ether
1-(4-Hydroxyphenyl)-7-(4-methoxyphenyl)-3-heptanone.
Renecernin
[1799553-42-6]
$C_{20}H_{24}O_3$ 312.409
Constit. of the leaves of *Renealmia cernua*. Amorph. powder. λ_{max} 224 (log ε 4.80); 278 (log ε 4.58) (MeOH)

1,2,4,5,6,7-Hexahydro, 4′-deoxy
7-(4-Hydroxyphenyl)-1-phenyl-3-heptanone
[100667-56-9]
$C_{19}H_{22}O_2$ 282.383
Constit. of the rhizomes of *Alpinia officinarum*. Oil.

Nomura, M. *et al., Phytochemistry*, 1981, **20**, 1097-1104 (*Platyphyllenone*).
Nakayama, R. *et al., Phytochemistry*, 1993, **33**, 501-502 (*Curcuma longa constit*).
Fuchino, H. *et al., Chem. Pharm. Bull.*, 1996, **44**, 1033-1038 (*Platyphyllenone, 6-Hydroxyplatyphyllenone*).
Ali, M.S. *et al., J. Nat. Prod.*, 2001, **64**, 289-293 (*Bis-4-hydroxyphenyl-4,6-heptadien-3-one*).
Jang, M.K. *et al., Arch. Pharmacal Res.*, 2004, **27**, 1220-1225 (*Curcuma zedoary constit*).
Mohamad, H. *et al., J. Nat. Prod.*, 2005, **68**, 285-288 (*Etlingera elatior constit*).
Li, W. *et al., Magn. Reson. Chem.*, 2009, **47**, 902-908 (*Curcuma longa constit*).
Li, J. *et al., J. Nat. Prod.*, 2010, **73**, 1667-1671 (*Bis-4-hydroxyphenyl-6-hepten-3-one*).
Zhao, Y.L. *et al., Z. Naturforsch., C*, 2012, **67**, 129-134 (*Bis-4-hydroxyphenyl-1,4-heptadien-3-one*).
Gilli, C. *et al., Biochem. Syst. Ecol.*, 2014, **56**, 178-184 (*Renecernin*).
Liu, D. *et al., Fitoterapia*, 2014, **96**, 76-80 (*7-(4-Hydroxyphenyl)-1-phenylheptanone*)

1,7-Bis(4-hydroxyphenyl)-2,4,6-heptatrien-1-one B54

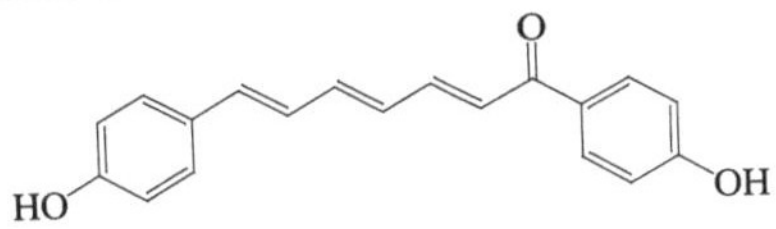

$C_{19}H_{16}O_3$ 292.334

(all-E)-form [849245-33-6]

Constit. of the rhizomes of *Etlingera elatior*. Antioxidant. Yellow powder. λ_{max} 395 (log ε 4.51) (MeOH).

Mohamad, H. *et al.*, *J. Nat. Prod.*, 2005, **68**, 285-288 (*Etlingera elatior constit, activity*)

1,7-Bis(4-hydroxyphenyl)-2-heptene B55

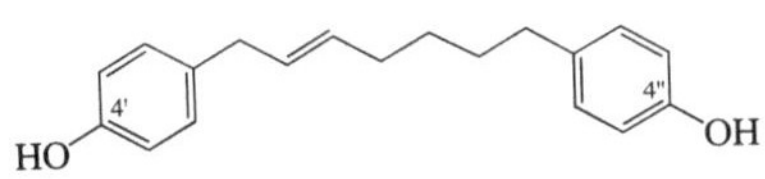

$C_{19}H_{22}O_2$ 282.383

(E)-form

4′-Me ether

7-(4-Hydroxyphenyl)-1-(4-methoxyphenyl)-2-heptene

[1220913-23-4]

$C_{20}H_{24}O_2$ 296.41

Constit. of the rhizomes of *Pleuranthodium racemigerum*. Cytotoxic. Oil. $[\alpha]^{25}_{D}$ -22.7 (c, 0.13 in MeOH). λ_{max} 203 (log ε 4.4); 225 (log ε 4.24); 278 (log ε 3.54) (MeOH).

Wohlmuth, H. *et al.*, *J. Nat. Prod.*, 2010, **73**, 743-746 (*Pleuranthodium racemigerum constit*).

Chang, C.-Y. *et al.*, *J. Chin. Chem. Soc. (Taipei)*, 2011, **58**, 286-289 (*4′-Me ether, synth*)

1,7-Bis(4-hydroxyphenyl)-1-heptene-3,5-diol B56

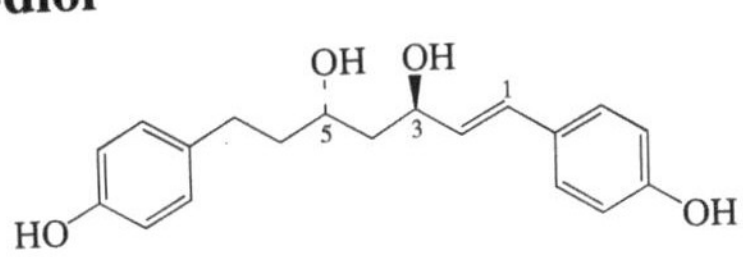

(1*E*,3*R*,5*S*)-*form*

$C_{19}H_{22}O_4$ 314.381

(1E,3R,5S)-form

3-Me ether

1,7-Bis(4-hydroxyphenyl)-5-methoxy-6-hepten-3-ol

[1426255-98-2]

$C_{20}H_{24}O_4$ 328.408

Constit. of twigs and leaves of *Viscum album*. Inhibits LPS-stimulated prodn. of TNF-α, IL-6 and IL-12p40 in bone marrow derived dendritic cells. Amorph. powder. $[\alpha]^{20}_{D}$ -10 (c, 0.1 in MeOH)

Diketone

1,7-Bis(4-hydroxyphenyl)-5-hydroxy-4,6-heptadien-3-one. 1,7-Bis(4-hydroxyphenyl)-1-heptene-3,5-dione. ***Bisdemethoxydihydrocurcumin***

[259795-67-0]

[470466-04-7] (unspecified stereochem.)

$C_{19}H_{18}O_4$ 310.349

Constit. of *Aframomum letestuianum, Alpinia blepharocalyx* and *Curcuma longa*. Platelet aggregation inhibitor and trypanocidal agent. Yellow needles (Et_2O/C_6H_6). Mp 148 - 149°. λ_{max} 215 (sh); 225 (sh); 362 (log ε 4.4) (MeOH)

(1E,3S,5S)-form

3-Me ether

[1426255-99-3]

$C_{20}H_{24}O_4$ 328.408

Constit. of the twigs and leaves of *Viscum album*. Exhibits moderate inhibitory activity against prodn. of TNF-α, IL-6 and IL-12p40. Amorph. powder. $[\alpha]^{20}_{D}$ -8 (c, 0.1 in MeOH)

(1E,5R)-form

3-Ketone

5-Hydroxy-1,7-bis(4-hydroxyphenyl)-1-hepten-3-one

[1415609-88-9]

$C_{19}H_{20}O_4$ 312.365

Constit. of *Dioscorea villosa*. Amorph. yellow powder. $[\alpha]^{22}_{D}$ +21.9 (c, 0.1 in MeOH). λ_{max} 213 (log ε 4.38); 327 (log ε 4.37) (MeOH)

(1E,5S)-form

3-Ketone

[1415609-89-0]

$C_{19}H_{20}O_4$ 312.365

Constit. of *Dioscorea villosa* and *Viscum album*. Amorph. yellow powder. $[\alpha]^{22}_{D}$ -18.1 (c, 0.1 in MeOH). λ_{max} 213 (log ε 4.38); 327 (log ε 4.37) (MeOH)

3-Ketone, 5-Me ether

1,7-Bis(4-hydroxyphenyl)-5-methoxy-1-hepten-3-one

[332371-81-0]

$C_{20}H_{22}O_4$ 326.392

Constit. of the seeds of *Alpinia blepharocalyx*. Amorph. yellow solid. $[\alpha]^{25}_{D}$ +17.5 (c, 0.13 in MeOH).

Dong, H. *et al.*, *J. Nat. Prod.*, 1998, **61**, 142-144 (*Bisdemethoxydihydrocurcumin*).

Ali, M.S. *et al.*, *J. Nat. Prod.*, 2001, **64**, 289-293 (*Alpinia blepharocalyx constit*).

Dong, S.-H. *et al.*, *J. Nat. Prod.*, 2012, **75**, 2168-2177 (*Dioscorea villosa constits*).

Nhiem, N.X. *et al.*, *J. Nat. Prod.*, 2013, **76**, 495-502 (*Viscum album constits, activity*)

2,3-Bis(3-hydroxyphenyl)-1-propanol B57

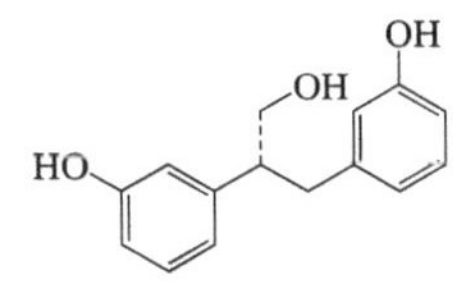

3-Hydroxy-β-(3-hydroxyphenyl) benzenepropanol, CAS

$C_{15}H_{16}O_3$ 244.29

(R)-form [1658414-06-2]

Obt. from faeces of *Trogopterus xanthipes*. Light brown oil. $[\alpha]^{25}_{D}$ -15.4 (c, 1.0 in EtOH).

Baek, S. *et al.*, *Beilstein J. Org. Chem.*, 2014, **10**, 2955-2962 (*Trogopterus xanthipes constit, abs config*)

2,3-Bis(4-hydroxyphenyl)-2-propenal B58

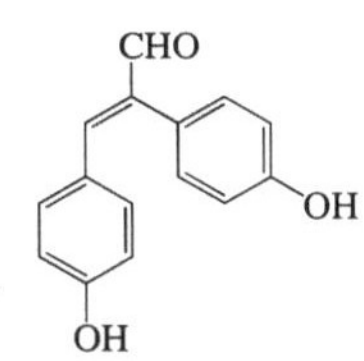

$C_{15}H_{12}O_3$ 240.258

(E)-form [2183518-19-4]

Constit. of the seeds of *Machilus yunnanensis*. Amorph. yellow powder. $[\alpha]^{22}_{D}$ -3.2 (c, 0.52 in MeOH). Opt. rotn. unaccounted for. λ_{max} 231 (log ε 3.96); 329 (log ε 4.05) (MeOH).

Long, H. *et al.*, *J. Asian Nat. Prod. Res.*, 2016, **18**, 952-958 (*Machilus yunnanensis constit*)

1,1′-[[2,4-Bis(6-methoxy-1,3-benzodioxol-5-yl)-1,3-cyclobutanediyl]dicarbonyl]bispyrrolidine, CAS B59

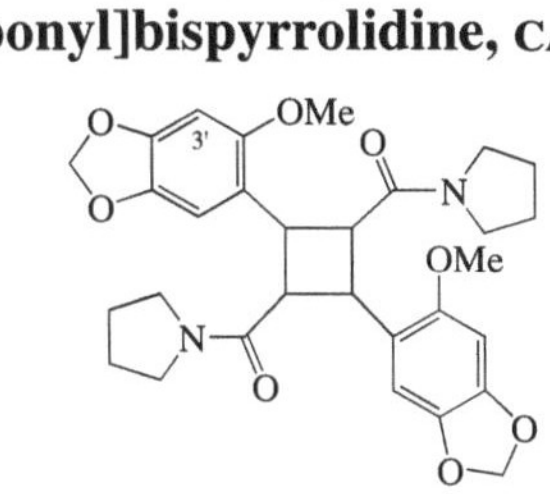

2,4-Bis(5-methoxy-1,3-benzodioxol-6-yl)-1,3-cyclobutanedicarboxylic acid bispyrrolidide

[173866-77-8]

$C_{30}H_{34}N_2O_8$ 550.608

Alkaloid from leaves and stems of *Piper peepuloides*. Also obt. by irradiation (sunlight) of Peepuloidine. Mp 258°. λ_{max} 240; 304; 330 (MeOH)

3′-Methoxy

1,1′-[[2-(6,7-Dimethoxy-1,3-benzodioxol-5-yl)-4-(6-methoxy-1,3-benzodioxol-5-yl)-1,3-cyclobutanediyl] dicarbonyl]bispyrrolidine, CAS. *4-(4,5-Dimethoxy-1,3-benzodioxol-6-yl)-2-(5-methoxy-1,3-benzodioxol-6-yl)-1,3-cyclobutanedicarboxylic acid bispyrrolidide*

[173866-76-7]

$C_{31}H_{36}N_2O_9$ 580.634

Alkaloid from leaves and stems of *Piper peepuloides*. Mp 265°. λ_{max} 240; 305; 333 (MeOH).

Dhar, K.L. *et al.*, *Fitoterapia*, 1995, **66**, 390-392 (*Piper peepuloides constits*)

1,6-Bis(4-methoxyphenyl)-1,5-hexadiene B60

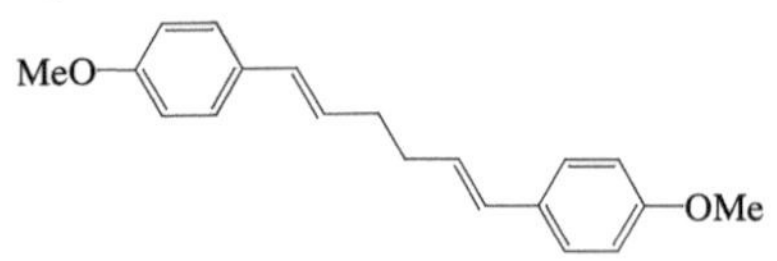

1,1′-(1,5-Hexadiene-1,6-diyl)bis[4-methoxybenzene], CAS. *4,4′-Dimethoxy-9,9′-ligna-7,7′-diene*

[106175-42-2]

$C_{20}H_{22}O_2$ 294.394

(E,E)-form

Ocimin [72448-90-9]

Isol. from the essential oil of *Ocimum americanum*. Fluorescent flakes (EtOAc). Mp 170 - 171°.

Thappa, R.K. *et al.*, *Phytochemistry*, 1979, **18**, 1242 (*Ocimin*).

Desai, D.G. *et al.*, *Indian J. Chem., Sect. B*, 1982, **21**, 491-492 (*synth*).

Khasnis, D.S. *et al.*, *Org. Prep. Proced. Int.*, 1991, **23**, 126-127 (*synth*).

Ornelas, M.A. *et al.*, *Synth. Commun.*, 1999, **29**, 1401-1403 (*synth*).

Kadam, A.J. *et al.*, *Indian J. Chem., Sect. B*, 2000, **39**, 628 (*synth*).

Flores, M. *et al.*, *Rev. Iberoam. Cienc.*, 2015, **2**, 31-37 (*synth*)

3,4:3′,4′-Bis(methylenedioxy)lignan B61

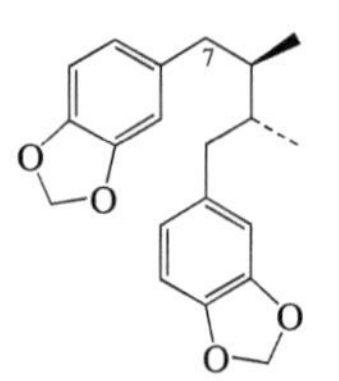

(8*R*,8′*R*)-*form*

5,5′-(2,3-Dimethyl-1,4-butanediyl)bis-1,3-benzodioxole, CAS. *2,3-Bis(3,4-methylenedioxybenzyl)butane*. *2,3-Dimethyl-1,4-dipiperonylbutane*

$C_{20}H_{22}O_4$ 326.392

Lignan numbering shown

(8R,8′R)-form

Austrobailignan 5 [55890-23-8] [188436-66-0]

Constit. of *Austrobaileya scandens*, *Myristica argentea*, *Saururus cernuus*, *Talauma ovata* and *Myristica fragrans*. Exhibits moderate antitumour activity against human HL-60, stomach cancer BGC-823, breast cancer MDA-MB-435 cells. Prisms (hexane). Mp 43 - 44°. Bp 100° at 0.02 mm. $[\alpha]^{23}_{D}$ -40 (c, 0.01 in MeOH)

(8R,8′S)-form

Machilin A. 8′-Epiaustrobailignan 5 [110269-50-6]

Constit. of *Machilus thunbergii*. Needles. Mp 48 - 50°

(8RS,8′RS)-form [68964-81-8]

Constit. of *Myristica dactyloides*. Oil.

Murphy, S.T. *et al.*, *Aust. J. Chem.*, 1975, **28**, 81-90 (*Austrobaileya scandens constit, struct*).

Biftu, T. *et al.*, *J. Chem. Soc., Perkin Trans. 1*, 1978, 1147-1150 (*8RS,8′RS-form, synth*).

Mahalanabis, K.K. *et al.*, *Tetrahedron Lett.*, 1982, **23**, 3975-3978 (*8RS,8′RS-form, synth*).

Shimomura, H. *et al.*, *Phytochemistry*, 1987, **26**, 1513-1515 (*Machilin A*).

Chattopadhyay, S.K. *et al.*, *Tetrahedron*, 1987, **43**, 669-678 (*Saururus cernuus constit*).

Rao, K.V. *et al.*, *J. Nat. Prod.*, 1990, **53**, 212-215 (*Saururus cernuus constit*).

Herath, H.M.T.B. *et al.*, *Phytochemistry*, 1997, **44**, 699-703 (*Myristica dactyloides constit*).

Filleur, F. *et al.*, *Planta Med.*, 2001, **67**, 700-704 (*Myristica argentea constit, struct*).

Stefanello, M.E.A. *et al.*, *Fitoterapia*, 2002, **73**, 135-139 (*Talauma ovata constit*).

Kwon, H.S. *et al.*, *Bioorg. Med. Chem. Lett.*, 2008, **18**, 194-198 (*Myristica fragrans constit*).

Xia, Y. *et al.*, *J. Serb. Chem. Soc.*, 2010, **75**, 1325-1335 (*Austrobailignan 5, activity*)

3,4:3′,4′-Bis(methylenedioxy)-8,8′-lignan-9,9′-olide B62

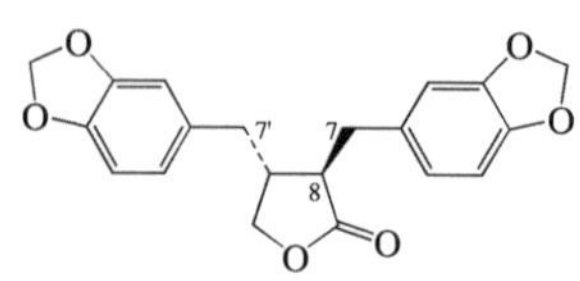

(8*R*,8′*R*)-*form*

3,4-Bis(1,3-benzodioxol-5-ylmethyl)dihydro-2(3H)-furanone, CAS

$C_{20}H_{18}O_6$ 354.358

(8R,8′R)-form

Hinokinin. Cubebinolide [26543-89-5]

Constit. of *Aristolochia indica*, *Chamaecyparis obtusa*, *Linum corymbulosum*, *Virola sebifera*, *Juniperus chinensis*, *Anthriscus sylvestris*, *Piper cubeba* and many other plants. Exhibits cytotoxic activity and antitrypanosomal props.. Plates (EtOH or C_6H_6). Mp 64 - 65°. $[\alpha]^{16}_{D}$-33.7 ($CHCl_3$) (-25). Cubebinolide was erroneously descr. as the (+)-enantiomer of Hinokinin (1933). They were the same enantiomer. However the (+)-enantiomer has since been isol.. λ_{max} 235 (ε 6760); 287 (ε 6760) (MeOH) (Berdy)

(8S,8′R)-form

Isohinokinin [26543-90-8]

Alkaline epimerisation prod. of Hinokinin. Leaflets (MeOH) or needles. Mp 116 - 117°. $[\alpha]^{21}_{D}$+107 (c, 1.03 in $CHCl_3$) (38)

(8S,8′S)-form

Heliobuphthalmin lactone.

(+)-Hinokinin [580-73-4]

Constit. of *Phyllanthus urinaria* and *Heliopsis buphthalmoides*. $[\alpha]^{25}_{D}$+38 (c, 1 in $CHCl_3$). λ_{max} 226 (sh) (log ε 4.25); 285 (log ε 3.95) (MeCN)

(8RS,8′RS)-form [24563-02-8]

Mp 108°.

Haworth, R.D. *et al.*, *J. Chem. Soc.*, 1938, 1985-1989 (*synth*).

Haworth, R.D. *et al.*, *J. Chem. Soc.*, 1950, 71-72 (*synth, struct*).

Burden, R.S. *et al.*, *J. Chem. Soc. C*, 1969, 693-701 (*ord*).

Karrer, W. *Konstitution und Vorkommen der Organischen Pflanzenstoffe, 2nd edn., Birkhäuser Verlag*, 1976-1985, no.1166 (*bibl*).

Asano, Y. *et al.*, *Bull. Chem. Soc. Jpn.*, 1976, **49**, 3232-3235 (*synth*).

Wenkert, E. *et al.*, *Phytochemistry*, 1976, **15**, 1547-1551 (*cmr*).

Mahalanabis, K.K. *et al.*, *Tetrahedron Lett.*, 1982, **23**, 3975-3978 (*synth*).

Lopes, L.M.X. *et al.*, *Phytochemistry*, 1983, **22**, 1516-1518 (*Virola sebifera constit*).

Jakupovic, J. *et al.*, *Planta Med.*, 1986, **52**, 18-20 (*Heliopsis buphthalmoides constit*).

Belletire, J.L. *et al.*, *J. Org. Chem.*, 1987, **52**, 2549-2555 (*synth*).

Rehnberg, N. *et al.*, *J. Org. Chem.*, 1990, **55**, 4340-4349 (*synth*).

Yoda, H. *et al.*, *Tetrahedron Lett.*, 1990, **31**, 7623-7626 (*synth*).

Lu, X. *et al.*, *Synlett*, 1993, 68-70 (*Isohinokinin, synth*).

Suarez, A. *et al.*, *Synth. Commun.*, 1993, **23**, 1991-2001 (*synth*).

Honda, T. *et al.*, *J. Chem. Soc., Perkin Trans. 1*, 1994, 1043-1046 (*synth*).

van Oeveren, A. *et al.*, *J. Org. Chem.*, 1994, **59**, 5999-6007 (*synth*).

Ikeda, R. *et al.*, *Chem. Pharm. Bull.*, 1998, **46**, 875-878 (*Anthriscus sylvestris constit*).

Xia, Y.-M. *et al., Chin. J. Chem.*, 2003, **21**, 1540-1542 (*synth*).
Chang, C.-C. *et al., Phytochemistry*, 2003, **63**, 825-833 (*Heliobuphthalmin lactone*).
Enders, D. *et al., Z. Naturforsch., B*, 2007, **62**, 117-120 (*synth, pmr, cmr*).
Bayindir, Ü. *et al., Plant J.*, 2008, **55**, 810-820 (*biosynth*).
Marcotullio, M.C. *et al., Molecules*, 2014, **19**, 14862-14878 (*Hinokinin, rev, biosynth, activity*).
Nguyen, V.T. *et al., Phytochem. Lett.*, 2014, **7**, 182-185 (*Heliobuphthalmin lactone, activity*).
Zhou, Q.-L. *Nat. Prod. Bioprospect.*, 2015, **5**, 255-261 (*synth*)

3,4:3′,4′-Bis(methylenedioxy)-8,8′-lign-7-en-9,9′-olide B63

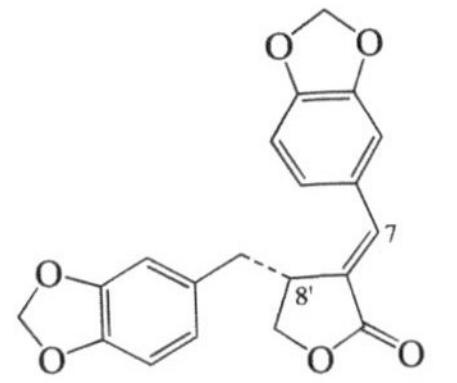

(7E,8′R)-form

4-(1,3-Benzodioxol-5-ylmethyl)-3-(1,3-benzodioxol-5-ylmethylene) dihydro-2(3H)-furanone, CAS. *Dihydro-4-(3,4-methylenedioxybenzyl)-3-(3,4-methylenedioxybenzylidene)-2(3H)-furanone*

$C_{20}H_{16}O_6$ 352.342

(7E,8′R)-form

Savinin. *Hibalactone*. *Taiwanin B*
[493-95-8]

Constit. of *Juniperus sabina* and *Juniperus conferta* needles and the bark of *Chloroxylon swietenia* and *Zanthoxylum nitidum*. Also present in the heartwood of *Taiwania cryptomerioides* and in *Chamaecyparis obtusa*. Shows antifungal activity and moderate antiparasitic activity against *Trypanosoma cruzi*. Inhibitor of SARS-CoV 3CL protease. Cryst. (C_6H_6 or EtOH or $CHCl_3$/EtOH). Mp 146 - 148° (127°, 139-142°). Mp 139 - 142°. Mp 127°. $[\alpha]^{21}_{D}$-87 (c, 0.95 in $CHCl_3$)

(7E,8′S)-form

Isogadain [26607-96-5]

Constit. of *Jatropha gossypifolia*. Cryst. (C_6H_6)

(7Z,8′R)-form

Isohibalactone [22554-63-8]

Constit. of *Juniperus chinensis* and *Hypoestes purpurea*. Cryst. (C_6H_6). Mp 147 - 149° (145-146°). $[\alpha]^{25}_{D}$-155 (c, 0.85 in $CHCl_3$)

(7Z,8′S)-form

Gadain [15914-41-7]

Isol. from seeds, roots and stems of *Jatropha gossypifolia*. Cryst. (C_6H_6). Mp 145°. $[\alpha]^{25}_{D}$+86 ($CHCl_3$). λ_{max} 235 (ε 13183); 293 (ε 11228); 337 (ε 15849) (EtOH)

(7E,8′RS)-form [24562-97-8]

Cryst. (C_6H_6 or EtOH). Mp 136 - 138°

(7Z,8′RS)-form [97233-05-1]

Cryst. (MeOH). Mp 154 - 155°.

Hartwell, J.L. *et al., J. Am. Chem. Soc.*, 1953, **75**, 235-236 (*Savinin*).
Schrecker, A.W. *et al., J. Am. Chem. Soc.*, 1954, **76**, 4896-4899 (*Savinin, uv, ir*).
Lin, Y.-T. *et al., J. Chin. Chem. Soc. (Taipei)*, 1963, **10**, 163-165 (*Savinin*).
Burden, R.S. *et al., J. Chem. Soc. C*, 1969, 693-701 (*cd*).
Doi, K. *et al., Phytochemistry*, 1972, **11**, 1175 (*Savinin*).
Bhide, K.S. *et al., Indian J. Chem., Sect. B*, 1977, **15**, 440-444 (*Savinin*).
Banerji, J. *et al., Phytochemistry*, 1984, **23**, 2323-2327 (*Gadain*).
Banerji, J. *et al., Heterocycles*, 1985, **23**, 661-665 (*Gadain, synth*).
Banerji, J. *Indian J. Chem., Sect. B*, 1987, 972-973 (*Gadain, Isogadain, pmr*).
Prashad, M. *et al., Tetrahedron Lett.*, 1989, **30**, 2877-2880 (*synth*).
Shieh, H.-L. *et al., J. Org. Chem.*, 1990, **55**, 5139-5145 (*Savinin, cryst struct, conformn*).
Lee, E. *et al., J. Chem. Soc., Chem. Commun.*, 1991, 1314-1315 (*synth*).
Fang, J.-M. *et al., Phytochemistry*, 1992, **31**, 3659-3661 (*Isohibalactone*).
Honda, T. *et al., J. Chem. Soc., Perkin Trans. 1*, 1994, 1043-1046 (*synth*).
Das, B. *et al., Planta Med.*, 1996, **62**, 90 (*Isogadain*).
Charlton, J.L. *et al., Can. J. Chem.*, 1997, **75**, 1076-1083 (*Savinin, synth*).
Rossi, R. *et al., Synthesis*, 1997, 1061-1065 (*synth, ms, ir*).
Chang, S.-T. *et al., Holzforschung*, 1999, **53**, 487-490 (*Savinin, activity*).
Bastos, J.K. *et al., Planta Med.*, 1999, **65**, 541-544 (*Savinin, activity*).
Shen, C-C. *et al., J. Nat. Prod.*, 2004, **67**, 1947-1949 (*Hypoestes pururea constit*).
Da Silva, R. *et al., Magn. Reson. Chem.*, 2005, **43**, 966-969 (*Savinin, pmr, cmr*).
Chyu, C-F. *et al., Helv. Chim. Acta*, 2007, **90**, 738-747 (*Taiwania cryptomerioides constit*).
Wen, C-C. *et al., J. Med. Chem.*, 2007, **50**, 4087-4095 (*Savinin, activity*).
Kuroyanagi, M. *et al., Chem. Pharm. Bull.*, 2008, **56**, 60-63 (*Chamaecyparis obtusa constit*).
Yang, C-H. *et al., Chem. Biodiversity*, 2009, **6**, 846-857 (*Zanthoxylum nitidum constit*).
Lloyd, M.G. *et al., Org. Biomol. Chem.*, 2016, **14**, 8971-8988 (*Gadain, Savinin, synth*)

4′-O-5″-Bisnortrachelogenin B64

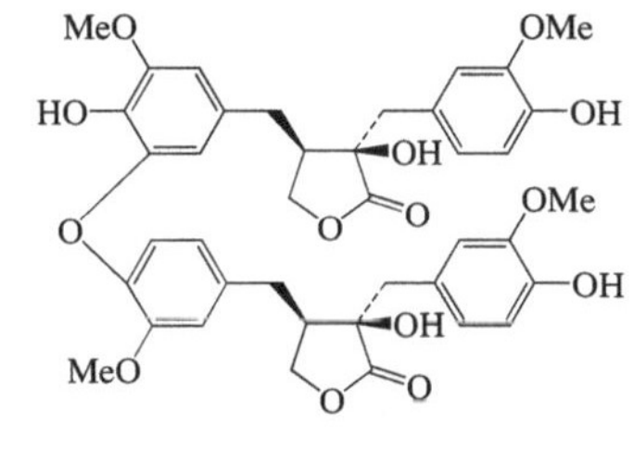

Absolute Configuration

4′-O-5″-Dinortrachelogenin
[1437802-62-4]

$C_{40}H_{42}O_{14}$ 746.762

Constit. of the roots of *Wikstroemia indica*. Powder (MeOH). Mp 188 - 190°. $[\alpha]^{20}_{D}$+29.5 (c, 0.9 in MeOH).

Wang, G.-C. *et al., Chem. Pharm. Bull.*, 2012, **60**, 920-923 (*Bisnortrachelogenin*)

5,5′-Bisnortrachelogenin B65

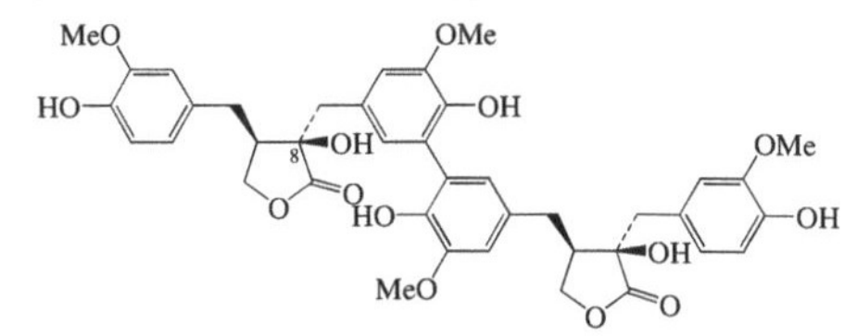

Absolute Configuration

5,5′-Biswikstromol. *5,5′-Dinortrachelogenin*
[870480-57-2]

$C_{40}H_{42}O_{14}$ 746.762

Not to be confused with 5,5-Bisnortrachelogenin, B67 and 5′,5′-Bisnortrachelogenin, B66. Constit. of the roots of *Wikstroemia indica*. Light yellow oil. $[\alpha]^{23}_{D}$+68.1 (c, 0.42 in MeOH). λ_{max} 280 (log ε 4.22) (MeOH)

8-Deoxy

8-Deoxy-5,5′-bisnortrachelogenin. *5-(5-Matairesinol)nortrachelogenin*
[1437802-60-2]

$C_{40}H_{42}O_{13}$ 730.763

Constit. of the roots of *Wikstroemia indica*. Powder (MeOH). Mp 179 - 181°. $[\alpha]^{20}_{D}$+41.2 (c, 1.1 in MeOH)

8,8-Dideoxy

5,5′-Bismatairesinol

$C_{40}H_{42}O_{12}$ 714.764

Constit. of the roots of *Wikstroemia indica*. Gum. $[\alpha]^{20}_{D}$+29.8 (c, 0.09 in MeOH). λ_{max} 230 (log ε 3.04); 282 (log ε 2.68) (MeOH).

Wang, L.-Y. *et al., Chem. Pharm. Bull.*, 2005, **53**, 1348-1351 (*5,5′-Bisnortrachelogenin*).
Wang, G.-C. *et al., Chem. Pharm. Bull.*, 2012, **60**, 920-923 (*8-Deoxy-5,5′-bisnortrachelogenin*).

Chang, H. *et al.*, *Fitoterapia*, 2017, **121**, 31-37 (*5,5′-Bismatairesinol*)

5′,5′-Bisnortrachelogenin B66

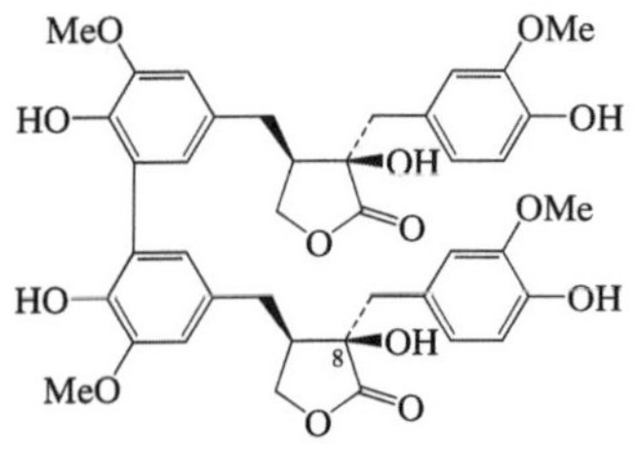

Absolute Configuration

5′,5′-Biswikstromol. 5′,5′-Dinortrachelogenin
[870480-56-1]
$C_{40}H_{42}O_{14}$ 746.762
Not to be confused with 5,5-Bisnortrachelogenin, B67 and 5,5′-Bisnortrachelogenin, B66. Constit. of the roots of *Wikstroemia indica.* Inhibits NO prodn.. Light yellow oil. $[\alpha]_D^{23}$ +55 (c, 0.1 in MeOH). λ_{max} 205 (log ε 4.98); 284 (log ε 4.07) (MeOH)

8-Deoxy
8-Deoxy-5′,5′-bisnortrachelogenin. *5′-(5′-Matairesinol) nortrachelogenin*
[1437802-61-3]
$C_{40}H_{42}O_{13}$ 730.763
Constit. of the roots of *Wikstroemia indica.* Powder (MeOH). Mp 204 - 206°. $[\alpha]_D^{20}$ +29.5 (c, 1.1 in MeOH)

8,8′-Dideoxy
5′,5′-Bismatairesinol
$C_{40}H_{42}O_{12}$ 714.764
Constit. of the roots of *Wikstroemia indica.* Gum. $[\alpha]_D^{20}$ +32.6 (c, 0.1 in MeOH). λ_{max} 230 (log ε 3.41); 282 (log ε 3.08) (MeOH).

Wang, L.-Y. *et al.*, *Chem. Pharm. Bull.*, 2005, **53**, 1348-1351 (*5′,5′-Bisnortrachelogenin*).
Wang, G.-C. *et al.*, *Chem. Pharm. Bull.*, 2012, **60**, 920-923 (*8-Deoxy-5′,5′-bisnortrachelogenin*).
Chang, H. *et al.*, *Fitoterapia*, 2017, **121**, 31-37 (*5′,5′-Bismatairesinol*)

5,5-Bisnortrachelogenin B67

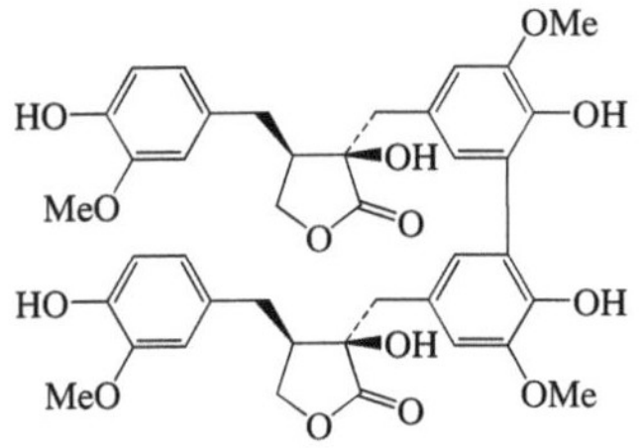

Absolute Configuration

5,5-Biwikstromol. 5,5-Biswikstromol. 5,5-Dinortrachelogenin
[1437802-59-9]
$C_{40}H_{42}O_{14}$ 746.762
Not to be confused with 5,5′-Bisnortrachelogenin, B67 and 5′,5′-Bisnortrachelogenin, B65. Related to Diarctigenin, D23. Constit. of the roots of *Wikstroemia indica.* Powder (MeOH). Mp 200 - 202°. $[\alpha]_D^{20}$ +13.9 (c, 1 in MeOH).

Wang, G.-C. *et al.*, *Chem. Pharm. Bull.*, 2012, **60**, 920-923 (*5,5-Bisnortrachelogenin*)

Bispicropodophyllin B68

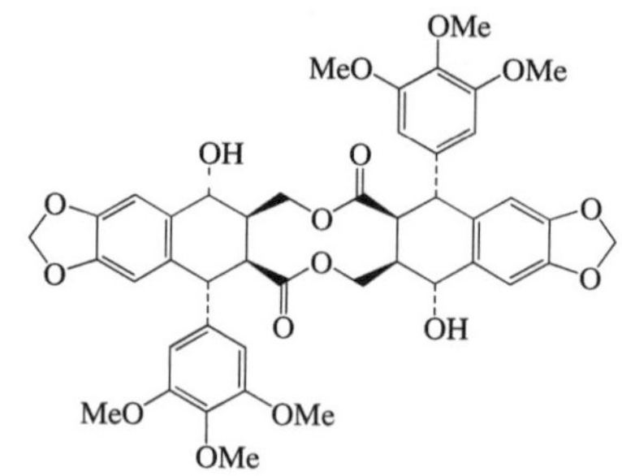

Relative Configuration

$C_{44}H_{44}O_{16}$ 828.82

Di-O-β-D-glucopyranoside
Bis(picropodophyllin glucoside)
[619267-10-6]
$C_{56}H_{64}O_{26}$ 1153.102
Constit. of *Withania coagulans.* Amorph. powder. $[\alpha]_D$ -20.5 (c, 0.1 in Py). λ_{max} 206 (log ε 4.13); 240 (log ε 3.91); 280 (log ε 3.76); 324 (log ε 3.34) (MeOH).

Nur-e-Alam, M. *et al.*, *Helv. Chim. Acta*, 2003, **86**, 607-614 (*Withania coagulans constit*)

1,7-Bis(3,4,5-trihydroxyphenyl)-3,5-heptanediol B69

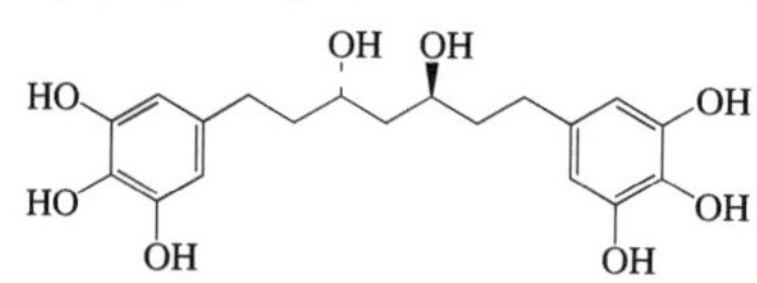

$C_{19}H_{24}O_8$ 380.393
(3S,5S)-form

3′,3″,4′,4″,5′,5″-Hexa-Me ether
1,7-Bis(3,4,5-trimethoxyphenyl)-3,5-heptanediol
[1346752-72-4]
Pale yellow oil. $[\alpha]_D^{23}$ +2.8 (c, 0.5 in $CHCl_3$). λ_{max} 230 (log ε 5); 270 (log ε 4.03) (EtOH)

3′,3″,4′,4″,5′,5″-Hexa-Me ether, 3,5-di-Ac
[1347742-64-6]
$C_{29}H_{40}O_{10}$ 548.629
Constit. of the rhizomes of *Zingiber mekongense.* Pale yellow oil. $[\alpha]_D^{26}$ +4.9 (c, 0.5 in $CHCl_3$). λ_{max} 230 (sh) (log ε 4.93); 269 (log ε 4.06) (EtOH).

Chareonkla, A. *et al.*, *Fitoterapia*, 2011, **82**, 534-538 (*Zingiber mekongense constit*)

1,7-Bis(3,4,5-trihydroxyphenyl)-3-heptanol B70

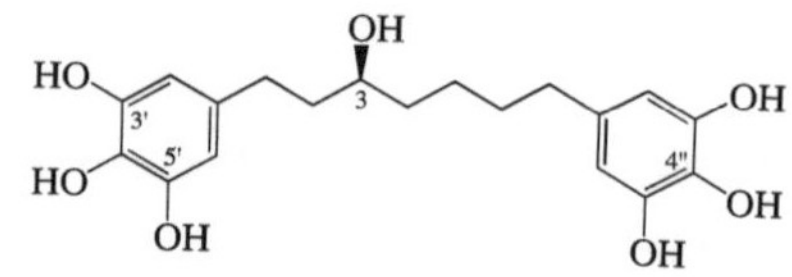

$C_{19}H_{24}O_7$ 364.394
(S)-form

3′,5′-Di-Me ether, 3-Ac
[1870838-65-5]
$C_{23}H_{30}O_8$ 434.485
Constit. of *Aframomum melegueta.* Yellow-brown oil. $[\alpha]_D^{25}$ +36.5 (c, 0.1 in MeOH).

El-Halawany, A.M. *et al.*, *Sci. Rep.*, 2014, **4**, 5880:1-9 (*Aframomum melegueta constit*)

1,6-Bis(3,4,5-trihydroxyphenyl)-1,6-hexanedione B71

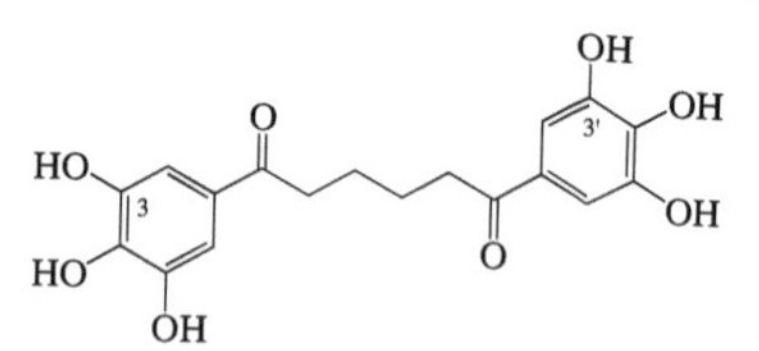

3,3′,4,4′,5,5′-Hexahydroxy-9,9′-lignan-7,7′-dione. 3,3′,4,4′,5,5′-Hexahydroxy-7,7′-dioxo-9,9′-neolignan
$C_{18}H_{18}O_8$ 362.334

3,3′,5,5′-Tetra-Me ether
1,6-Bis(4-hydroxy-3,5-dimethoxyphenyl)-1,6-hexanedione. 4,4′-Dihydroxy-3,3′,5,5′-tetramethoxy-9,9′-lignan-7,7′-dione
[2260774-74-9]
$C_{22}H_{26}O_8$ 418.442
Constit. of the stems of *Eleutherococcus senticosus.* Diacylglycerol acyltransferase 1 (DGAT1) inhibitor. Pale yellow powder. $[\alpha]_D^{25}$ -3.2 (c, 0.1 in MeOH). λ_{max} 233 (log ε 3.45); 276 (log ε 3.04) (MeOH)

3′-Deoxy, 3,5,5′-tri-Me ether
1-(4-Hydroxy-3,5-dimethoxyphenyl)-6-(4-hydroxy-3-methoxyphenyl)-1,6-hexanedione. 4,4′-Dihydroxy-3,3′,5-trimethoxy-9,9′-lignan-7,7′-dione

[2260774-75-0]
$C_{21}H_{24}O_7$ 388.416
Constit. of the stems of *Eleutherococcus senticosus*. Diacylglycerol acyltransferase 1 (DGAT1) inhibitor. Powder. $[\alpha]_D^{25}$ -4.4 (c, 0.1 in MeOH). λ_{max} 233 (log ε 3.61); 273 (log ε 3.08) (MeOH).

Li, B.-B. *et al.*, *Phytochem. Lett.*, 2017, **20**, 123-127 (*Eleutherococcus senticosus constits, activity*)

1,2-Bis(3,4,5-trihydroxyphenyl)propane B72

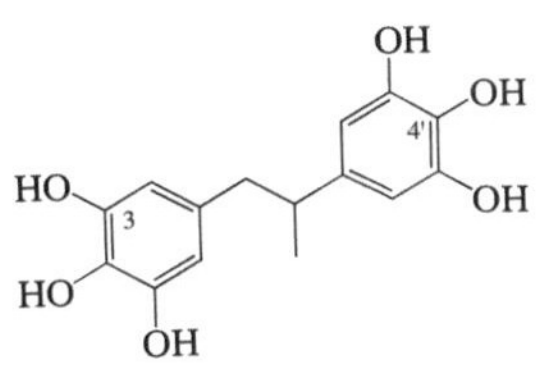

8,9,9′-Trinor-7,7′-lignan-3,3′,4,4′,5,5′-hexol. 3,3′,4,4′,5,5′-Hexahydroxy-8,9,9′-trinor-7,7′-neolignan
$C_{15}H_{16}O_6$ 292.287
Lignan numbering shown
(ξ)-form
4′-O-(2-Propenyl), penta-Me ether
1,3-Dimethoxy-5-[1-methyl-2-(3,4,5-trimethoxyphenyl)ethyl]-2-(2-propenyloxy)benzene.
Aurein A
[61243-90-1]
$C_{23}H_{30}O_6$ 402.487
Constit. of *Licaria aurea* and *Magnolia saulangiana*. Cryst. (C_6H_6/hexane). Mp 80 - 81°. λ_{max} 235 (sh); 270 (ε 1200) (EtOH)
4′-O-(2-Propenyl), 3,4-methylene, 3′,5,5′-tri-Me ether
Aurein B
[139906-82-4]
$C_{22}H_{26}O_6$ 386.444
Constit. of *Licaria aurea*. Oil.

Gottlieb, O.R. *et al.*, *Phytochemistry*, 1976, **15**, 1289-1291 (*Aurein A*).
Wenkert, E. *et al.*, *Phytochemistry*, 1976, **15**, 1547-1551 (*Aurein A, cmr*).
Marques, M.O.M. *et al.*, *Phytochemistry*, 1992, **31**, 360-361 (*Aurein B*)

Blepharocalyxin A B73

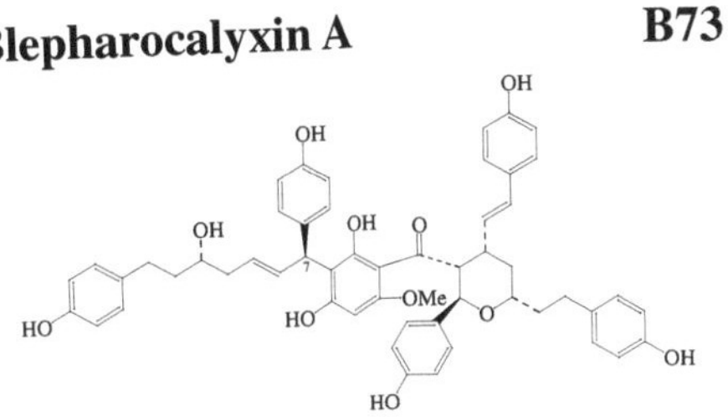

Absolute configuration

[183137-69-1]
$C_{54}H_{54}O_{11}$ 879.015
Constit. of the seeds of *Alpinia blepharocalyx*. Immunoregulator. Inhibits nitric oxide production in murine macrophages. Pale yellow amorph. solid. $[\alpha]_D$ -56.4 (c, 0.2 in MeOH)
7-Epimer
Blepharocalyxin B
[183254-62-8]
$C_{54}H_{54}O_{11}$ 879.015
Constit. of the seeds of *Alpinia blepharocalyx*. Immunoregulator. Inhibits nitric oxide production in murine macrophages. Pale yellow amorph. solid. $[\alpha]_D$ -97.7 (c, 0.2 in MeOH).

Kadota, S. *et al.*, *Tetrahedron Lett.*, 1996, **37**, 7283-7286 (*Blepharocalyxins A,B, activity*).
Prasain, J.K. *et al.*, *J. Nat. Prod.*, 1998, **61**, 212-216 (*Blepharocalyxins A,B, struct*).
Ali, M.S. *et al.*, *Biol. Pharm. Bull.*, 2001, **24**, 525-528 (*Blepharocalyxins A,B, activity*)

Blepharocalyxin C B74

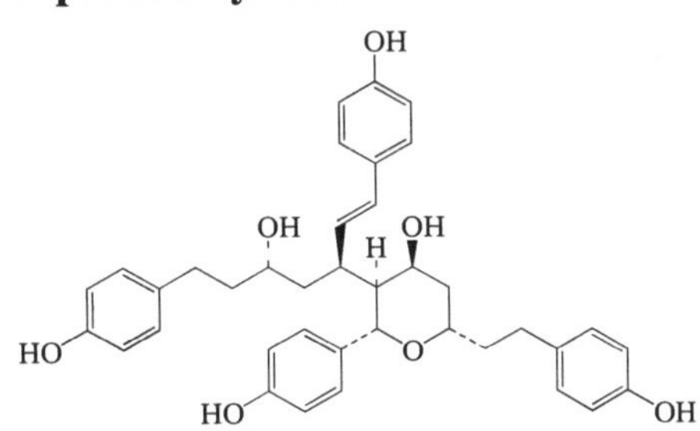

Absolute Configuration

[294855-51-9]
$C_{38}H_{42}O_7$ 610.747
Constit. of the seeds of *Alpinia blepharocalyx*. Amorph. light yellow solid. $[\alpha]_D^{25}$ +63.5 (c, 0.03 in MeOH).

Tezuka, Y. *et al.*, *Tetrahedron Lett.*, 2000, **41**, 5903-5907 (*Blepharocalyxin C, activity*).
Ali, M.S. *et al.*, *J. Nat. Prod.*, 2001, **64**, 491-496 (*Blepharocalyxin C, activity*)

Blepharocalyxin D B75

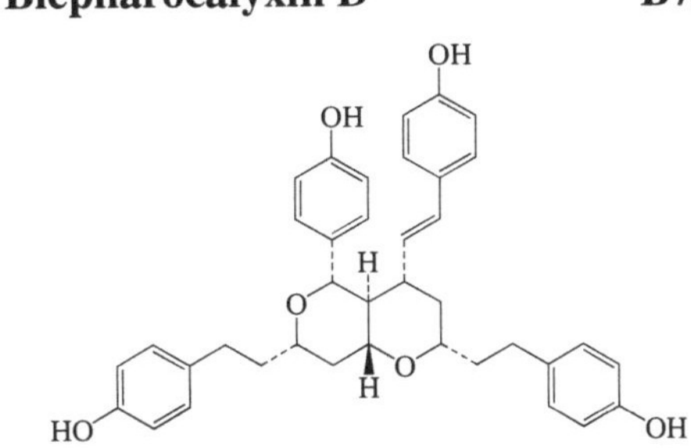

Absolute Configuration

[294855-52-0]
$C_{38}H_{40}O_6$ 592.732
Constit. of the seeds of *Alpinia blepharocalyx*. Antiproliferative agent. Light yellow amorph. powder. $[\alpha]_D^{25}$ +18.5 (c, 0.02 in MeOH).

Tezuka, Y. *et al.*, *Tetrahedron Lett.*, 2000, **41**, 5903-5907 (*Blepharocalyxin D, activity*).
Ali, M.S. *et al.*, *J. Nat. Prod.*, 2001, **64**, 491-496 (*Blepharocalyxin D, activity*).
Ko, H.M. *et al.*, *Tetrahedron*, 2007, **63**, 5797-5805 (*synth*).
Cons, B.D. *et al.*, *Org. Lett.*, 2013, **15**, 2046-2049 (*synth*)

Blepharocalyxin E B76

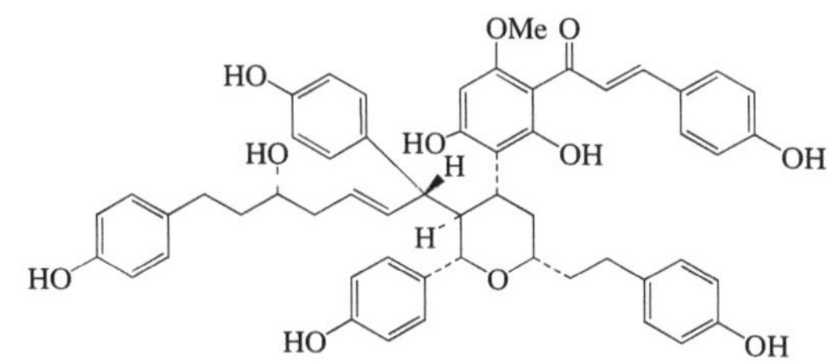

Absolute Configuration

[294855-53-1]
$C_{54}H_{54}O_{11}$ 879.015
Constit. of the seeds of *Alpinia blepharocalyx*. Antiproliferative agent. Amorph. light yellow powder. $[\alpha]_D^{25}$ +145.5 (c, 0.02 in MeOH).

Tezuka, Y. *et al.*, *Tetrahedron Lett.*, 2000, **41**, 5903-5907 (*Blepharocalyxin E, activity*).
Ali, M.S. *et al.*, *J. Nat. Prod.*, 2001, **64**, 491-496 (*Blepharocalyxin E, activity*)

Blepharodin B77

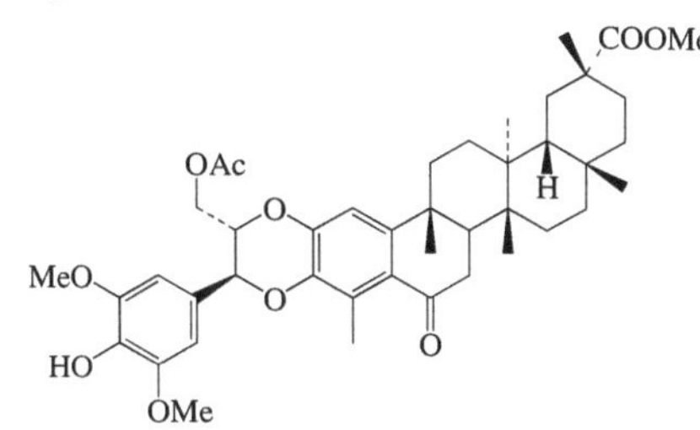

[1308252-39-2]
$C_{43}H_{56}O_{10}$ 732.911
Constit. of *Maytenus magellanica*. Glass. $[\alpha]_D^{20}$ -8.5 (c, 0.82 in $CHCl_3$). λ_{max} 236 (log ε 4.02); 279 (log ε 3.77); 311 (log ε 3.34) (EtOH).

Núñez, M.J. *et al.*, *Tetrahedron*, 2011, **67**, 3030-3033 (*Blepharodin*)

Bonaspectin D B78

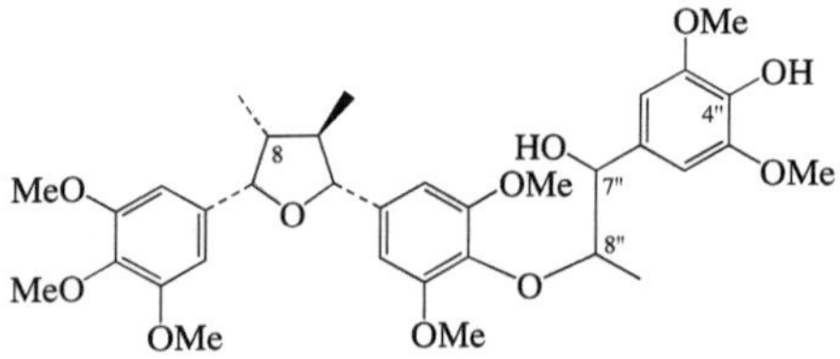

Relative Configuration

$C_{34}H_{44}O_{11}$ 628.715
Sesquilignan. Related to Saucerneol, in T17. Parent not isol.

4″-O-β-D-Glucopyranoside
[261949-32-0]
$C_{40}H_{54}O_{16}$ 790.856
Constit. of *Bonamia spectabilis*. Antiplasmodial agent. Oil. $[\alpha]_D^{20}$ +15.3 (c, 0.3 in $CHCl_3$). C-7″, C-8″ has *erythro*-config.. λ_{max} 217; 270 (MeOH)

4″-Me ether
Bonaspectin E
[479199-13-8]
$C_{35}H_{46}O_{11}$ 642.742
Constit. of *Bonamia spectabilis*. Oil. C-7″, C-8″ has *erythro*-config.

4″-Me ether, 7″-Ac
Bonaspectin G
[479199-15-0]
$C_{37}H_{48}O_{12}$ 684.779
Constit. of *Bonamia spectabilis*. Oil. C-7″, C-8″ has *erythro*-config.

7″-Deoxy, 4″-O-β-D-glucopyranoside
Bonaspectin C 4″-β-D-glucoside
[261949-31-9]
$C_{40}H_{54}O_{15}$ 774.857
Constit. of *Bonamia spectabilis*. Antiplasmodial agent. Oil. $[\alpha]_D^{20}$ +12.4 (c, 0.23 in $CHCl_3$). C-8″ config. not determined. λ_{max} 217; 239 (sh); 271 (MeOH)

7″-Deoxy, 4″-Me ether
Bonaspectin B
[261949-30-8]
$C_{35}H_{46}O_{10}$ 626.743
Constit. of *Bonamia spectabilis*. Oil. $[\alpha]_D^{20}$+11.5 (c, 0.2 in $CHCl_3$). C-8″ config. not determined. λ_{max} 210; 236 (sh); 270 (MeOH)

8-Epimer, 4″-Me ether
Bonaspectin F
[479199-14-9]
$C_{35}H_{46}O_{11}$ 642.742
Constit. of *Bonamia spectabilis*. Oil. C-7″, C-8″ has *erythro*-config.

8-Epimer, 4″-Me ether, 7″ -Ac
Bonaspectin H
[479199-16-1]
$C_{37}H_{48}O_{12}$ 684.779
Constit. of *Bonamia spectabilis*. Oil. C-7″, C-8″ has *erythro*-config.

8-Epimer, 7″-deoxy, 4″-Me ether
Bonaspectin A
[261949-29-5]
$C_{35}H_{46}O_{10}$ 626.743
Constit. of *Bonamia spectabilis*. Oil. C-8″ config. not determined. λ_{max} 210; 236 (sh); 270 (MeOH).

Tofern, B. *et al.*, *Phytochemistry*, 2000, **53**, 119-128 (*Bonaspectins A-D*).
Kraft, C. *et al.*, *Phytochemistry*, 2002, **60**, 167-173 (*Bonaspectins A,E-H, activity*)

Brainin A B79

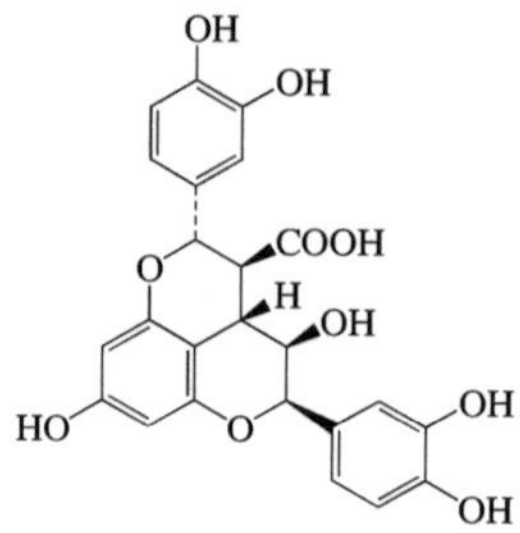

Relative Configuration

[1289526-20-0]
$C_{24}H_{20}O_{10}$ 468.414
Constit. of the rhizomes of *Brainea insignis*. Amorph. yellow powder. Mp 185 - 187°. $[\alpha]_D^{20}$+45.6 (c, 0.2 in MeOH).

Yang, M.H. *et al.*, *Chin. Chem. Lett.*, 2011, **22**, 455-457 (*Brainin A*)

Breviscapin B B80

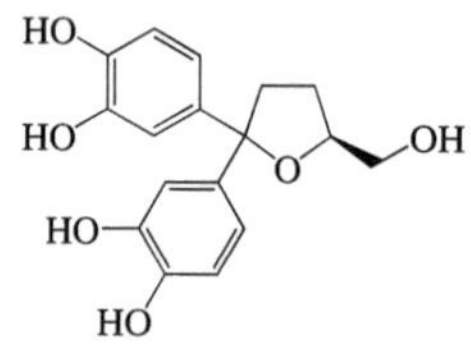

4,4′-[Dihydro-5-(hydroxymethyl)-2(3H)-furanylidene]bis[1,2-benzenediol], CAS. *5,5-Bis(3,4-dihydroxyphenyl)dihydro-2-furanmethanol*
$C_{17}H_{18}O_6$ 318.325
Rearranged norlignan
(S)-form [1274654-80-6]
Constit. of the rhizomes of *Curculigo breviscapa*. Powder. $[\alpha]_D^{27}$-7.1 (c, 0.22 in MeOH). λ_{max} 195 (log ε 4.19); 206 (log ε 4.59); 288 (log ε 3.85) (MeOH).

Li, N. *et al.*, *Z. Naturforsch., B*, 2010, **65**, 79-82 (*Breviscapin B*)

Brevitaxin B81

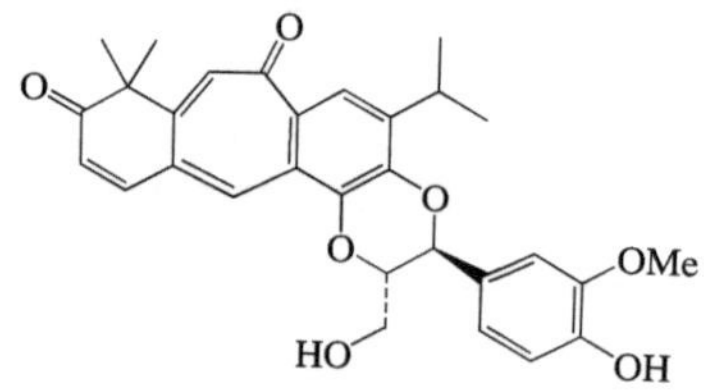

Relative Configuration

[164416-32-4]
$C_{30}H_{30}O_7$ 502.563
Constit. of *Amentotaxus formosana*, *Taxus brevifolia*, *Taxus chinensis* and *Taxus yunnanensis*. Exhibits cytotoxic activity. Yellow powder. Mp 252 - 254°. Mp 280°. $[\alpha]_D^{20}$0 (c, 1 in CH_2Cl_2). λ_{max} 317 (ε 19400) (MeOH).

Arslanian, R.L. *et al.*, *J. Nat. Prod.*, 1995, **58**, 583-585 (*Brevitaxin*).
Liang, J.-Y. *et al.*, *Phytochemistry*, 1998, **47**, 69-72 (*Taxus chinensis constit*).
Chen, H.-L. *et al.*, *Fitoterapia*, 2011, **82**, 219-224 (*Amentotaxus formosana constit, activity*)

Buddlenol B B82

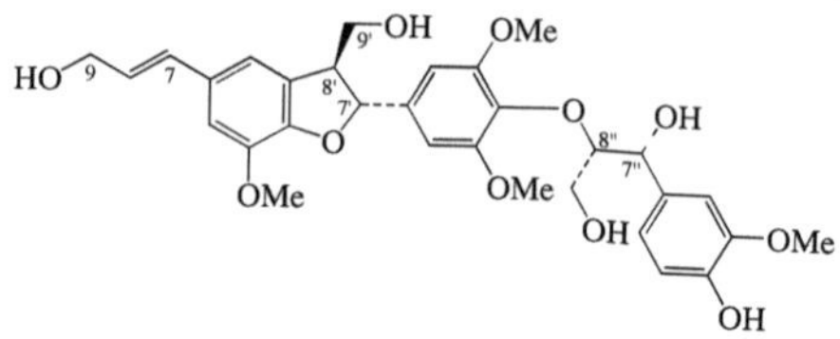

(7E,7′R,7″R,8′S,8″S)-form

[97399-80-9]
[172823-69-7] (unspecified stereochem.) [828916-86-5] (*threo*-Buddlenol B) [828916-87-6] (*erythro*-Buddlenol B) [844637-85-0] (7*E*,7′ξ,7″*R**,8′ξ,8″*R**-form) [844637-86-1] (7*E*,7′ξ,7″*R**,8′ξ,8″*S**-form) [1857283-12-5] (unspecified stereochem.)
$C_{31}H_{36}O_{11}$ 584.618
Buddlenol B may be used as a general synonym for all compds. with this planar struct. since many isolations are not accompanied by stereochem. data. Some compds., isol. more recently, have had absolute and/or relative configs. assigned and these are listed separately below. Buddlenol B usually assigned (7′*R**,8′*S**)-stereochem. with either *threo*-(7″*R**,8″*R**)- or *erythro*-(7″*R**,8″*S**)-stereochem.. Constit. of *Buddleja davidii*, *Campylotropis hirtella*, *Cassia occidentalis*, *Echinacea purpurea* and *Rosa laevigata*

9-Aldehyde
Buddlenol A
[97399-78-5]

[263383-54-6] (*erythro*-Buddlenol A) [263383-55-7] (*threo*-Buddlenol A) [844637-87-2] (7*E*,7′ξ,7″*R**,8′ξ,8″*R**-form) [844637-88-3] (7*E*,7′ξ,7″*R**,8′ξ,8″*S**-form) [1857283-13-6] (7*E*,7′*R*,7″ξ, 8′*S*,8″ξ-form)
$C_{31}H_{34}O_{11}$ 582.602
Constit. of *Buddleja davidii, Crataegus pinnatifida, Firmiana simplex, Hevea brasiliensis, Picrasma quassioides* and *Tarenna attenuata*. NO prodn. inhibitor. Antioxidant. Mp 102 - 104°. Similarly for Buddlenol B, this also has many isol. without stereochem. data. λ_{max} 225 (ε 81280); 276 (ε 6310); 334 (ε 16600) (MeOH)

7,8-Dihydro
see Dihydrobuddlenol B, D112

(7E,7′R,7″R,8′S,8″S)-form
(-)-Jatrointelignan B [1842427-56-8]
Constit. of *Jatropha integerrima*. Oil. $[\alpha]^{20}_{D}$ -42 (c, 0.12 in $CHCl_3$). λ_{max} 225 (log ε 4.61); 276 (log ε 4.14) (MeOH)

9-Me ether
Pinnatifidanin A
[1571064-75-9]
$C_{32}H_{38}O_{11}$ 598.645
Constit. of *Crataegus pinnatifida*. Yellow oil

(7E,7′R,7″S,8′S,8″S)-form
(-)-Jatrointelignan A [1842427-55-7]
Constit. of *Jatropha integerrima*. Oil. $[\alpha]^{20}_{D}$ -13 (c, 0.19 in $CHCl_3$). λ_{max} 225 (log ε 4.46); 276 (log ε 4.0) (MeOH)

9-Me ether
Pinnatifidanin AI
[1715014-20-2]
$C_{32}H_{38}O_{11}$ 598.645
Constit. of the seeds of *Crataegus pinnatifida*. Yellow oil

(7E,7′S,7″R,8′R,8″R)-form
(+)-Jatrointelignan A [1696388-90-5]
Constit. of *Coix lachryma-jobi* and *Jatropha integerrima*. Amorph. powder. $[\alpha]^{25}_{D}$ +27 (c, 0.1 in $CHCl_3$). λ_{max} 225 (log ε 4.46); 276 (log ε 4.0) (MeOH)

(7E,7′S,7″S,8′R,8″R)-form
(+)-Jatrointelignan B [1696388-89-2]
Constit. of *Coix lachryma-jobi* and *Jatropha integerrima*. Amorph. powder. $[\alpha]^{20}_{D}$ +38 (c, 0.15 in $CHCl_3$). λ_{max} 225 (log ε 4.61); 276 (log ε 4.14) (MeOH)

(7E,7′R*,7″R*,8′R*,8″S*)-form
9-Me ether
9-O-Methyl-cis-cis-buddlenol B. 9-Methoxy-cis-cis-buddlenol B
$C_{32}H_{38}O_{11}$ 598.645
Constit. of *Aglaia odorata* var. *microphyllina*. Pale yellow oil. $[\alpha]^{25}_{D}$ +0.5 (c, 0.1 in MeOH)

(7E,7′R*,7″R*,8′S*,8″R*)-form
9′-O-β-D-Glucopyranoside
Alangisesquin B
[212007-43-7]
$C_{37}H_{46}O_{16}$ 746.759
Constit. of *Alangium premnifolium*. Powder. $[\alpha]^{24}_{D}$ -14.3 (c, 0.2 in MeOH). λ_{max} 225 (sh); 275 (log ε 4.09); 284 (sh) (MeOH)

(7E,7′R*,7″R*,8′S*,8″S*)-form
[613684-55-2]
Constit. of *Brassica fruticulosa*

9′-O-β-D-Glucopyranoside
Alangisesquin A
[212007-42-6]
$C_{37}H_{46}O_{16}$ 746.759
Constit. of the stems of *Alangium premnifolium*. Pale yellow powder. $[\alpha]_D$ -29 (c, 0.9 in MeOH). λ_{max} 224 (sh); 276 (log ε 4.26); 284 (sh) (MeOH).

Houghton, P.J. *et al.*, *Phytochemistry*, 1985, **24**, 819-826 (*Buddlenols A,B*).
Kijima, K. *et al.*, *Phytochemistry*, 1998, **48**, 669-676 (*Alangisesquins A,B*).
Cutillo, F. *et al.*, *J. Agric. Food Chem.*, 2003, **51**, 6165-6172 (*Brassica fruticulosa lignan*).
Yang, X.-W. *et al.*, *J. Nat. Prod.*, 2007, **70**, 521-525 (*Buddlenol A, activity*).
Huang, X.-X. *et al.*, *Biochem. Syst. Ecol.*, 2013, **48**, 1-5 (*Pinnatifidanin AI*).
Li, L.-Z. *et al.*, *Chin. J. Nat. Med.*, 2013, **11**, 411-414 (*Buddlenol A*).
Huang, X.-X. *et al.*, *Biochem. Syst. Ecol.*, 2014, **54**, 208-212 (*Pinnatifidanin A*).
Choi, G. *et al.*, *Chem. Biodiversity*, 2014, **12**, 380-387 (*Coix lachryma-jobi constits*).
Liu, S. *et al.*, *Nat. Prod. Commun.*, 2014, **9**, 7-8 (*9-O-Methyl-cis-cis-buddlenol B*).
Zhu, J.-Y. *et al.*, *RSC Adv.*, 2015, **5**, 12202-12208 (*Jatrointelignans A,B*).
Woo, K.W. *et al.*, *Bioorg. Med. Chem. Lett.*, 2016, **26**, 730-733 (*Buddlenol A*)

Buddlenol E B83

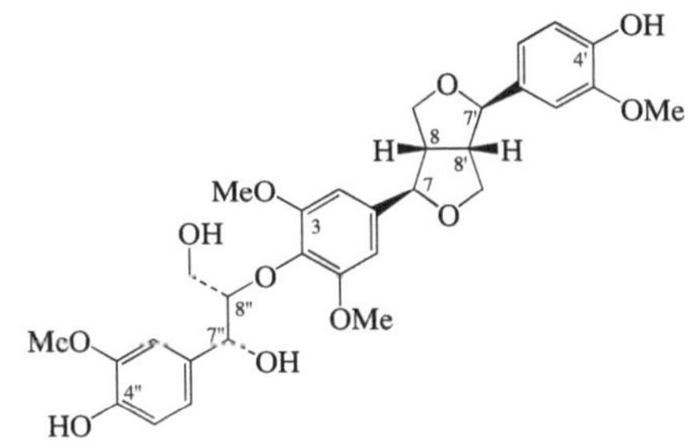

(7*R*,7′*R*,7″*R*,8*S*,8′*S*,8″*S*)-form

3,3′,3″,5-Tetramethoxy-4,8″-oxy-8,8′-sesquilignan-4′,4″,7″,9″-tetrol. 4′,4″,7″,9″-Tetrahydroxy-3,3′,3″,5-tetramethoxy-4,8″-oxy-8,8′-sesquineolignan
[97399-82-1]
[828916-88-7] (stereoisomer)
[1001329-65-2] (stereoisomer)
[1001329-67-4] (stereoisomer)
$C_{31}H_{36}O_{11}$ 584.618
Buddlenol E may be used as a general synonym for all compds. with this planar struct. since many older isolations are not accompanied by stereochem. data. Many compds. isol. more recently have had abs. and/or rel. configs. assigned and these are listed separately below. Prob. exists as various mixts. of (7*R**,7′*R**,7″*R**,8*S**,8′*S**,8″*R**)- and (7*R**,7′*R**,7″*R**,8*S**,8′*S**,8″*S**)-diastereoisomers. Constit. of *Acer saccharum, Buddleja davidii, Euryale ferox, Firmiana simplex, Helianthus annuus* and *Tarenna attenuata*

5′-Methoxy
3,3′,3″,5,5′-Pentamethoxy-4,8″-oxy-8,8′-sesquilignan-4′,4″,7″,9″-tetrol. 4′,4″,7″,9″-Tetrahydroxy-3,3′,3″,5,5′-pentamethoxy-4,8″-oxy-8,8′-sesquilignan.
Buddlenol C
[97465-73-1]
$C_{32}H_{38}O_{12}$ 614.644
Constit. of *Aquilaria sinensis, Buddleja davidii, Cassia occidentalis, Chenopodium album, Picrasma quassioides* and *Tarenna attenuata*. Exhibits antiinflammatory activity

5′-Methoxy, 4″-O-β-D-glucopyranoside
Buddlenol C 4″-glucoside
[209864-41-5]
$C_{38}H_{48}O_{17}$ 776.785
Constit. of *Scutellaria baicalensis*

5″-Methoxy
3,3′,3″,5,5″-Pentamethoxy-4,8″-oxy-8,8′-sesquilignan-4′,4″,7″,9″-tetrol. 4′,4″,7″,9″-Tetrahydroxy-3,3′,3″,5,5″-pentamethoxy-4,8″-oxy-8,8′-sesquilignan. **Buddlenol F**
[97399-83-2]
$C_{32}H_{38}O_{12}$ 614.644
Constit. of *Buddleja davidii*

5′,5″-Dimethoxy
3,3′,3″,5,5′,5″-Hexamethoxy-4,8″-oxy-8,8′-sesquilignan-4′,4″,7″,9″-tetrol. 4′,4″,7″,9″-Tetrahydroxy-3,3′,3″,5,5′,5″-hexamethoxy-4,8″-oxy-8,8′-sesquilignan. **Buddlenol D**
[97465-75-3]
$C_{33}H_{40}O_{13}$ 644.67
Constit. of *Buddleja davidii* and *Tarenna attenuatum*

5′,5″-Dimethoxy, 4′-O-β-D-glucopyranoside
Albizzioside C
[160896-41-3]
$C_{39}H_{50}O_{18}$ 806.811
Constit. of *Albizzia myriophylla*. Oil. $[\alpha]_D^{18}$ -0.9 (c, 0.4 in MeOH)

5′,5″-Dimethoxy, 4″-O-β-D-glucopyranoside
[212010-66-7]
$C_{39}H_{50}O_{18}$ 806.811
Constit. of *Aster auriculatus*. Powder + $3H_2O$. Mp 121 - 123°. $[\alpha]_D^{31}$ +15.8 (c, 0.065 in MeOH). Rel. config. determined

5′,5″-Dimethoxy, 4′,4″-di-O-β-D-glucopyranoside
[109825-41-4]
[115588-15-3]
$C_{45}H_{60}O_{23}$ 968.952
Constit. of the bark of *Eucommia ulmoides*. Amorph. powder. $[\alpha]_D^{26}$ -11.1 (c, 0.14 in MeOH). λ_{max} 228; 270; 279 (MeOH)

5′,5″-Dimethoxy, 4′-O-[β-D-apiofuranosyl-(1→2)-β-D-glucopyranoside]
Albizzioside B
[160896-40-2]
$C_{44}H_{58}O_{22}$ 938.926
Constit. of *Albizzia myriophylla*. Powder. $[\alpha]_D^{18}$ -2.8 (c, 1.2 in MeOH)

(7R,7′R,7″R,8S,8′S,8″S)-form
[1292294-23-5]
Constit. of *Acanthopanax senticosus*, *Ilex pubescens* and *Sinocalamus affinis*. Amorph. powder. $[\alpha]_D^{20}$ -4 (c, 0.1 in MeCN). λ_{max} 233 (log ε 3.71); 280 (log ε 3.12) (MeOH)

8R-Hydroxy
3,3′,3″,5-Tetramethoxy-4,8″-oxy-8,8′-sesquilignan-4′,4″,7″,8,9″-pentol. 4′,4″,7″,8,9″-Pentahydroxy-3,3′,3″,5-tetramethoxy-4,8″-oxy-8,8′-sesquilignan. **Firmianol B**
[1864954-49-3]
$C_{31}H_{36}O_{12}$ 600.617
Constit. of the stems of *Firmiana simplex*. Gum. $[\alpha]_D^{25}$ +8.8 (MeOH). C-8 has *R*-config. due to change in Cahn-Ingold-Prelog priorities

5′-Methoxy
[1292294-25-7]
$C_{32}H_{38}O_{12}$ 614.644
Constit. of *Sinocalamus affinis*. Amorph. powder. $[\alpha]_D^{20}$ -3 (c, 0.05 in MeCN). λ_{max} 205 (log ε 4.38); 235 (log ε 3.71); 277 (log ε 3.09) (MeOH)

5′,5″-Dimethoxy
[1292294-26-8]
$C_{33}H_{40}O_{13}$ 644.67
Constit. of *Sinocalamus affinis*. Amorph. powder. $[\alpha]_D^{20}$ -3 (c, 0.05 in MeCN). λ_{max} 206 (log ε 4.29); 238 (log ε 3.59); 277 (log ε 2.93) (MeOH)

(7R,7′R,7″S,8S,8′S,8″R)-form
Constit. of *Ilex pubescens*. Yellow oil. λ_{max} 230 (log ε 2.66); 257 (log ε 2.74) (MeOH)

4′-O-β-D-Glucopyranoside
Samsesquinoside
[1621072-59-0]
$C_{37}H_{46}O_{16}$ 746.759
Constit. of *Sambucus williamsii*. Pale yellow powder. $[\alpha]_D^{26}$ -36.7 (c, 0.6 in MeOH). λ_{max} 205 (log ε 5.0); 228 (log ε 4.41); 280 (log ε 3.84) (MeOH)

(7R,7′R,7″S,8S,8′S,8″S)-form
[1292294-24-6]
Constit. of *Acanthopanax senticosus* and *Sinocalamus affinis*. Amorph. powder. $[\alpha]_D$ -4.8 (c, 0.1 in MeCN). λ_{max} 232 (log ε 3.73); 280 (log ε 3.20) (MeOH)

3′-O-De-Me, 4′-Me ether
3,3″,4′,5-Tetramethoxy-4,8″-oxy-8,8′-sesquilignan-3′,4″,7″,9″-tetrol. 3′,4″,7″,9″-Tetrahydroxy-3,3″,4′,5-tetramethoxy-4,8″-oxy-8,8′-sesquilignan
$C_{31}H_{36}O_{11}$ 584.618
Constit. of *Acanthopanax senticosus*. Powder. $[\alpha]_D^{25}$ -4.4 (c, 0.1 in MeOH). λ_{max} 233 (log ε 3.61); 273 (log ε 3.08) (MeOH)

3″-O-De-Me, 4″-Me ether
3,3′,4″,5-Tetramethoxy-4,8″-oxy-8,8′-sesquilignan-3″,4′,7″,9″-tetrol. 3″,4′,7″,9″-Tetrahydroxy-3,3′,4″,5-tetramethoxy-4,8″-oxy-8,8′-sesquilignan
$C_{31}H_{36}O_{11}$ 584.618
Constit. of *Acanthopanax senticosus*. Pale yellow powder. $[\alpha]_D^{25}$ -4.6 (c, 0.1 in MeOH). λ_{max} 236 (log ε 3.65); 278 (log ε 3.10) (MeOH)

5′-Methoxy
[931114-00-0]
$C_{32}H_{38}O_{12}$ 614.644
Constit. of *Acanthopanax senticosus* and *Sinocalamus affinis*. Amorph. powder. $[\alpha]_D^{20}$ -6 (c, 0.1 in MeOH). λ_{max} 206 (log ε 4.21); 238 (log ε 3.50); 277 (log ε 2.81) (MeOH)

5′-Methoxy, 3″-O-de-Me, 4″-Me ether
3,3′,4″,5,5′-Pentamethoxy-4,8″-oxy-8,8′-sesquilignan-3″,4′,7″,9″-tetrol. 3″,4′,7″,9″-Tetrahydroxy-3,3′,4″,5,5′-pentamethoxy-4,8″-oxy-8,8′-sesquilignan
$C_{32}H_{38}O_{12}$ 614.644
Constit. of *Acanthopanax senticosus*. Pale yellow powder. $[\alpha]_D^{25}$ -3.2 (c, 0.1 in MeOH). λ_{max} 233 (log ε 3.45); 276 (log ε 3.04) (MeOH)

5′,5″-Dimethoxy
[1292294-27-9]
$C_{33}H_{40}O_{13}$ 644.67
Constit. of *Sinocalamus affinis*. Amorph. powder. $[\alpha]_D^{20}$ -5 (c, 0.1 in MeCN). λ_{max} 207 (log ε 4.22); 240 (log ε 3.47); 274 (log ε 2.65) (MeOH)

(7S,7′S,7″R,8R,8′R,8″R)-form
Ecdysanol D. Hedyotol D [97465-80-0]
Constit. of *Hedyotis lawsoniae* and *Urceola rosea*. Exhibits antiinflammatory activity. Amorph. powder. $[\alpha]_D^{25}$ +13 (c, 0.1 in MeOH). Ecdysanol D abs. config. assigned; however that of Hedyotol D is only probable and may be a mixt. of (7″*R*,8″*R*)- and (7″*R*,8″*S*)-stereoisomers.

4″-O-β-D-Glucopyranoside
Hedyotol D 4″-glucoside
[209864-40-4]
$C_{37}H_{46}O_{16}$ 746.759

Constit. of *Scutellaria baicalensis*
7″-O-β-D-Glucopyranoside
$C_{37}H_{46}O_{16}$ 746.759
Constit. of *Helicteres angustifolia*. Amorph. powder. $[\alpha]_D^{25}$ -13.5 (c, 0.3 in MeOH). Abs. config. assigned. λ_{max} 214 (log ε 3.93); 280 (log ε 3.22) (MeOH)
(7S,7′S,7″S,8R,8′R,8″R)-form
Ecdysanol C. Hedyotol C [97465-79-7]
Constit. of *Euonymus alatus*, *Hedyotis lawsoniae*, *Sinocalamus affinis* and *Urceola rosea*. Exhibits antiinflammatory activity. Amorph. powder. $[\alpha]_D^{25}$ +13.9 (c, 0.1 in MeOH). Ecdysanol C abs. config. assigned; however that of Hedyotol C is only probable and may be a mixt. of (7″*R*,8″*R*)- and (7″*R*,8″*S*)-stereoisomers.
4″-O-β-D-Glucopyranoside
Hedyotol C 4″-glucoside
[209864-39-1]
$C_{37}H_{46}O_{16}$ 746.759
Constit. of *Scutellaria baicalensis*
7″-O-β-D-Glucopyranoside
$C_{37}H_{46}O_{16}$ 746.759
Constit. of *Helicteres angustifolia*. Amorph. powder. $[\alpha]_D^{25}$ +14.2 (c, 0.4 in MeOH). λ_{max} 218 (log ε 3.83); 280 (log ε 3.15) (MeOH)
4′,4″-Di-O-β-D-glucopyranoside
Hedyotol C 4′,4″-diglucoside
[107668-75-7]
[114884-47-8]
$C_{43}H_{56}O_{21}$ 908.9
Constit. of the bark of *Eucommia ulmoides*. Component of Du Zhong. Amorph. powder. $[\alpha]_D^{28}$ -11.5 (c, 0.35 in MeOH). λ_{max} 232; 271 (MeOH)
3-Demethoxy, 5′-methoxy, 4′-Me ether
3′,3″,4′,5,5′-Pentamethoxy-4,8″-oxy-8,8′-sesquilignan-4″,7″,9″-triol. 4″,7″,9″-Trihydroxy-3′,3″,4′,5,5′-pentamethoxy-4,8″-oxy-8,8′-sesquilignan.
Colocasinol A
[1224161-89-0]
$C_{32}H_{38}O_{11}$ 598.645
Constit. of the tuber bark of *Colocasia antiquorum* var. *esculenta*. Yellowish gum. $[\alpha]_D^{25}$ -12.4 (c, 0.1 in MeOH). λ_{max} 230 (log ε 3.7); 281 (log ε 3.3) (MeOH)
3-Demethoxy, 5′,5″-dimethoxy, 4′,4″-di-Me ether
3′,3″,4′,4″,5,5′,5″-Heptamethoxy-4,8″-oxy-8,8′-sesquilignan-7″,9″-diol. 7″,9″-Dihydroxy-3′,3″,4′,4″,5,5′,5″-heptamethoxy-4,8″-oxy-8,8′-sesquilignan.
Euonymolin A
[2030376-02-2]
$C_{34}H_{42}O_{12}$ 642.698
Constit. of the twigs of *Euonymus alatus*. Exhibits cytotoxicity against SK-MEL-2 cell line and inhibits NO prodn. in LPS-activated BV-2 cells. Gum. $[\alpha]_D^{25}$ -43.4 (c, 0.3 in MeOH)
(7R*,7′R*,7″R*,8S*,8′S*,8″R*)-form
5′-Methoxy, 9″-Me ether
3,3′,3″,5,5′,9″-Hexamethoxy-4,8″-oxy-8,8′-sesquilignan-4′,4″,7″-triol. 4′,4″,7″-Trihydroxy-3,3′,3″,5,5′,9″-hexamethoxy-4,8″-oxy-8,8′-sesquilignan
[931109-33-0]
$C_{33}H_{40}O_{12}$ 628.671
Constit. of *Chenopodium album*. $[\alpha]_D^{25}$ +13 (c, 0.07 in $CHCl_3$). λ_{max} 220 (log ε 4.52); 257 (log ε 3.90) ($CHCl_3$)
5′,5″-Dimethoxy, 9″-Me ether
3,3′,3″,5,5′,5″,9″-Heptamethoxy-4,8″-oxy-8,8′-sesquilignan-4′,4″,7″-triol. 4′,4″,7″-Trihydroxy-3,3′,3″,5,5′,5″,9″-heptamethoxy-4,8″-oxy-8,8′-sesquilignan
[931109-34-1]
$C_{34}H_{42}O_{13}$ 658.697
Constit. of *Chenopodium album*. $[\alpha]_D^{25}$ +18 (c, 0.04 in $CHCl_3$). λ_{max} 226 (log ε 4.60); 265 (log ε 3.00) ($CHCl_3$)
3-Demethoxy, 5″-methoxy
3′,3″,5,5″-Tetramethoxy-4,8″-oxy-8,8′-sesquilignan-4′,4″,7″,9″-tetrol. 4′,4″,7″,9″-Tetrahydroxy-3′,3″,5,5″-tetramethoxy-4,8″-oxy-8,8′-sesquilignan.
Kandelisesquilignan A
[1572636-35-1]
$C_{31}H_{36}O_{11}$ 584.618
Constit. of *Kandelia obovata*. Antioxidant. Oil. $[\alpha]_D^{20}$ -6.8 (c, 1.0 in MeOH). λ_{max} 233; 276 (MeOH)
(7R*,7′R*,7″R*,8S*,8′S*,8″S*)-form
7″-O-β-D-Glucopyranoside
[2193053-00-6]
$C_{37}H_{46}O_{16}$ 746.759
Constit. of the stems of *Erycibe obtusifolia*. Powder. $[\alpha]_D^{20}$ -72.8 (c, 0.05 in MeOH). λ_{max} 278 (no solvent reported)
5″-Methoxy, 4′,4″-di-O-β-D-glucopyranoside
[1370732-74-3]
$C_{44}H_{58}O_{22}$ 938.926
Constit. of the leaves of *Viburnum plicatum* var. *plicatum*. Amorph. powder. $[\alpha]_D^{25}$ -14 (c, 0.2 in MeOH). λ_{max} 211 (log ε 4.6); 230 (sh); 274 (log ε 3.5) (MeOH)
5′,5″-Dimethoxy, 7″-O-β-D-glucopyranoside
[1646332-24-2]
$C_{39}H_{50}O_{18}$ 806.811
Constit. of *Erycibe hainanensis*. Shows moderate hepatoprotective activity. Powder. $[\alpha]_D^{20}$ -25.4 (c, 0.19 in MeOH). λ_{max} 215 (log ε 4.81); 238 (sh); 271 (log ε 3.70) (MeOH)
3-Demethoxy, 7″-O-β-D-glucopyranoside
Mananthoside K
[1069137-65-0]
$C_{36}H_{44}O_{15}$ 716.733
Constit. of *Mananthes patentiflora*. Amorph. solid (MeOH). $[\alpha]_D^{25}$ +2 (c, 0.34 in MeOH). λ_{max} 205 (log ε 4.7); 230 (log ε 4.3); 280 (log ε 3.89); 333 (log ε 2.81); 352 (log ε 2.66); 360 (log ε 2.77) (MeOH)
3-Demethoxy, 9″-O-β-D-glucopyranoside
Dracunculifoside R
[454204-34-3]
$C_{36}H_{44}O_{15}$ 716.733
Constit. of the aerial parts of *Baccharis dracunculifolia*. Amorph. powder. $[\alpha]_D^{22}$ +8.5 (c, 0.3 in MeOH). λ_{max} 204 (log ε 4.88); 231 (log ε 4.31); 279 (log ε 3.88) (MeOH)
3-Demethoxy, 5′-methoxy
***Ficusesquilignan A**. Ficusequilignan A*
[321991-56-4]
$C_{31}H_{36}O_{11}$ 584.618
Constit. of *Alyxia schlechteri*, *Berberis koreana*, *Dysoxylum densiflorum*, *Ficus microcarpa*, *Rhododendron alutaceum*, *Tarenna attenuata* and *Toona sinensis*. Exhibits antifungal and cytotoxic activities. NO prodn. inhibitor. Pale yellow oil. $[\alpha]_D^{27}$ -11.1 (c, 0.17 in

$CHCl_3$). λ_{max} 235 (log ε 3.87); 280 (log ε 3.5) (MeOH)

3-Demethoxy, 5′-methoxy, 5-O-de-Me

Xanthiumnolic B

[2226004-45-9]

$C_{30}H_{34}O_{11}$ 570.591

Constit. of the fruit of *Xanthium sibiricum*. Moderate NO prodn. inhibitor. Amorph. powder. $[\alpha]_D^{20}$ -16.5 (c, 0.1 in MeOH). Assignment of stereochem. unclear. λ_{max} 235 (log ε 3.79) (MeOH)

3-Demethoxy, 5′,5″-dimethoxy

3′,3″,5,5′,5″-Pentamethoxy-4,8″-oxy-8,8′-sesquilignan-4′,4″,7″,9″-tetrol. 4′,4″,7″,9″-Tetrahydroxy-3′,3″,5,5′,5″-pentamethoxy-4,8″-oxy-8,8′-sesquilignan.

Kandelisesquilignan B

[1573023-79-6]

$C_{32}H_{38}O_{12}$ 614.644

Constit. of *Kandelia obovata*. Antioxidant. Oil. $[\alpha]_D^{20}$-5.3 (c, 1.0 in MeOH). λ_{max} 235; 276 (MeOH)

(all-ξ)-form

5″-Methoxy, 7″-deoxy, 4′,4″-di-O-β-D-glucopyranoside

Hedysalignan A

[847392-05-6]

$C_{44}H_{58}O_{21}$ 922.927

Constit. of *Hedysarum polybotrys*. Mp 160 - 164°. $[\alpha]_D^{25}$+4 (c, 0.22 in MeOH).

Kikuchi, T. *et al.*, *Chem. Pharm. Bull.*, 1985, **33**, 1444-1451 (*Hedyotols C,D*).
Houghton, P.J. *et al.*, *Phytochemistry*, 1985, **24**, 819-826 (*Buddlenols C-F*).
Deyama, T. *et al.*, *Chem. Pharm. Bull.*, 1986, **34**, 4933-4938 (*Hedyotol C 4,4″-diglucoside*).
Deyama, T. *et al.*, *Chem. Pharm. Bull.*, 1987, **35**, 1803-1807 (*Eucommia ulmoides diglucoside*).
Ito, A. *et al.*, *Phytochemistry*, 1994, **37**, 1455-1458 (*Albizziosides B,C*).
Miyaichi, Y. *et al.*, *Nat. Med. (Tokyo)*, 1998, **52**, 82-86 (*Hedyotols C,D 4″-glucosides*).
Wang, C.Z. *et al.*, *Phytochemistry*, 1998, **48**, 711-717 (*Aster auriculatus lignan*).
Li, Y.-C. *et al.*, *Chem. Pharm. Bull.*, 2000, **48**, 1862-1865 (*Ficusesquilignan A*).
Nagatani, Y. *et al.*, *Chem. Pharm. Bull.*, 2002, **50**, 583-589 (*Dracunculifoside R*).
Hai, L.-Q. *et al.*, *Yaoxue Xuebao*, 2003, **38**, 592-595 (*Hedysalignan A*).
Cutillo, F. *et al.*, *Phytochem. Anal.*, 2006, **17**, 344-349 (*Chenopodium album constits*).
Tian, J.-M. *et al.*, *J. Asian Nat. Prod. Res.*, 2008, **10**, 228-232 (*Mananthoside K*).
Kim, K.H. *et al.*, *J. Agric. Food Chem.*, 2010, **58**, 4779-4785 (*Colocasinol A, cd*).
Li, L. *et al.*, *J. Agric. Food Chem.*, 2011, **59**, 7708-7716 (*maple syrup constit*).
Xiong, L. *et al.*, *J. Nat. Prod.*, 2011, **74**, 1188-1200 (*Sinocalamus affinis constits, abs config*).
Katagiri, S. *et al.*, *Nat. Prod. Commun.*, 2011, **6**, 1901-1904 (*Viburnum plicatum diglucoside*).
Li, H.-Z. *et al.*, *Arch. Pharmacal Res.*, 2012, **35**, 1887-1893 (*Ficusesquilignan A*).
Nan, H. *et al.*, *Heterocycles*, 2013, **87**, 1093-1098 (*Kandelisesquilignans A,B, activity*).
Xiao, H.-H. *et al.*, *Fitoterapia*, 2014, **94**, 29-35 (*Samsesquinoside*).
Feng, Z.-M. *et al.*, *Phytochem. Lett.*, 2014, **9**, 163-167 (*Erycibe hainanensis constit, activity*).
Sribuhom, T. *et al.*, *Phytochem. Lett.*, 2015, **11**, 80-84 (*Ficusesquilignan A, activity*).
Woo, K.W. *et al.*, *Bioorg. Med. Chem. Lett.*, 2016, **26**, 730-733 (*Firmianol B, Buddlenol E*).
Lee, S. *et al.*, *Chem. Biodiversity*, 2016, **13**, 1391-1396 (*Euonymolin A, activity*).
Li, J.-L. *et al.*, *Fitoterapia*, 2016, **109**, 185-189 (*Acanthopanax senticosus constits, activity*).
Xiong, H. *et al.*, *J. Nat. Med.*, 2016, **70**, 673-678 (*Ilex pubescens constits*).
Yin, X. *et al.*, *Molecules*, 2016, **21**, 1506:1-9 (*Helicteres angustifolia 7″-glucosides*).
Feng, Z.-M. *et al.*, *Phytochem. Lett.*, 2016, **17**, 45-49 (*Erycibe obtusifolia constit*).
Jiang, H. *et al.*, *Fitoterapia*, 2017, **117**, 11-15 (*Xanthiumnolic B, activity*).
Dong, D.-D. *et al.*, *Fitoterapia*, 2019, **134**, 96-100 (*Ecdysanols C,D, activity*)

Bulbovaginatin B84

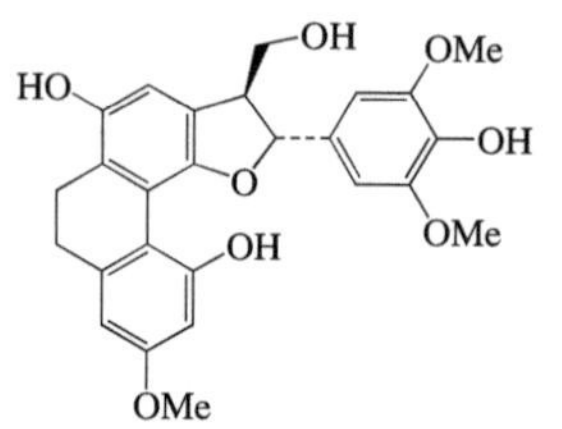

Relative Configuration

$C_{26}H_{26}O_8$ 466.486

Constit. of *Bulbophyllum vaginatum* and *Dendrobium moniliforme*. Gum. $[\alpha]_D^{27}$ -3.5 (c, 0.85 in $CHCl_3$). λ_{max} 230 (sh) (log ε 3.12); 278 (log ε 4.85); 300 (log ε 4.01); 318 (log ε 3.77) (EtOH).

Leong, Y.-W. *et al.*, *J. Nat. Prod.*, 2004, **67**, 1601-1603 (*Bulbovaginatin*)

Burmanicumol A B85

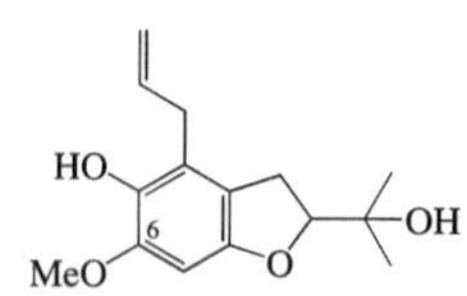

2,3-Dihydro-5-hydroxy-6-methoxy-α,α-dimethyl-4-(2-propenyl)-2-benzofuranmethanol

[2052959-91-6]

$C_{15}H_{20}O_4$ 264.321

Prenylated lignan monomer. Constit. of the stem bark of *Illicium burmanicum*. Oil. $[\alpha]_D^{24}$-9.5 (c, 0.72 in $CHCl_3$)

6-O-De-Me, 6-Et ether

Burmanicumol B

[2052959-92-7]

$C_{16}H_{22}O_4$ 278.348

Constit. of the stem bark of *Illicium burmanicum*. Oil. $[\alpha]_D^{24}$-8.7 (c,0.15 in $CHCl_3$).

Huang, D. *et al.*, *Fitoterapia*, 2015, **107**, 22-28 (*Burmanicumols A,B*)

Burmanicumol D B86

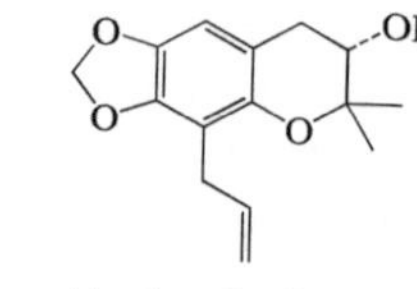

Absolute Configuration

7,8-Dihydro-6,6-dimethyl-4-(2-propenyl)-6H-1,3-dioxolo[4,5-g][1]benzopyran-7-ol. 8-Allyl-3,4-dihydro-3-hydroxy-6,7-(methylenedioxy)-2,2-dimethyl-2H-1-benzopyran

[2052959-94-9]

$C_{15}H_{18}O_4$ 262.305

Prenylated lignan monomer. Constit. of the stem bark of *Illicium burmanicum*. Oil. $[\alpha]_D^{24}$-8.1 (c, 0.29 in $CHCl_3$).

Huang, D. *et al.*, *Fitoterapia*, 2015, **107**, 22-28 (*Burmanicumol D*)

Burmaniol A B87

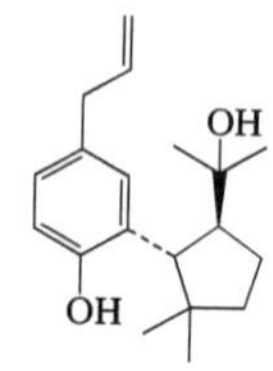

Absolute Configuration

[2250441-60-0]

$C_{19}H_{28}O_2$ 288.431

Constit. of the branches and leaves of *Illicium burmanicum*. Exhibits cytotoxicity against A549 and HCT116 cells. Wax. $[\alpha]_D^{20}$+114 (c, 0.1 in MeOH).

Tian, X. *et al.*, *RSC Adv.*, 2017, **7**, 10849-10854 (*Burmaniol A, activity*)

Burmaniol B B88

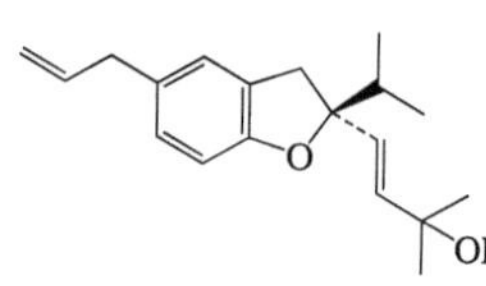

Absolute Configuration

[2250441-61-1]

$C_{19}H_{26}O_2$ 286.415

Prenylated lignan monomer. Constit. of the branches and leaves of *Illicium*

burmanicum. Exhibits cytotoxicity against A549 and HCT116 cells. Oil. $[\alpha]_D^{20}$ -11.7 (c, 0.3 in MeOH).
Tian, X. *et al.*, *RSC Adv.*, 2017, **7**, 10849-10854 (*Burmaniol B, activity*)

Butyrolactone I B89

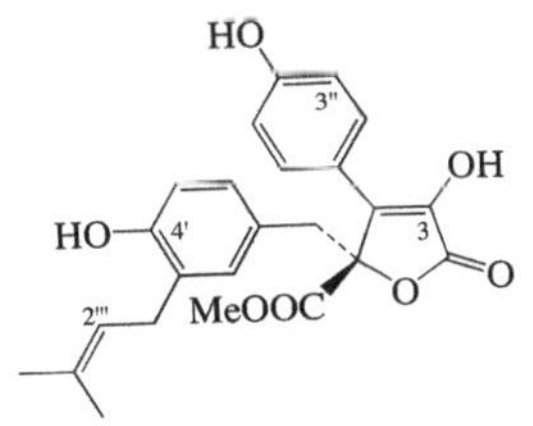

Absolute Configuration

F06Z-509A
[87414-49-1]
$C_{24}H_{24}O_7$ 424.449
Provisional prenylated lignan. Biosynthetic pathway unclear. Prod. by terrestrial and marine-derived *Aspergillus terreus* (strains IFO 8835, DRCC 152, HKI0499 and XWC21-10). Also from a marine-derived *Paecilomyces variotii* EN-291. Cyclin dependent kinase inhibitor. Exhibits cytotoxic activity. α-Glucosidase inhibitor. Quorum sensing molecule for *A. terreus*. Shows anti-neuroinflammatory props. Cryst. (CH_2Cl_2). Mp 95° dec.. $[\alpha]_D^{17}$ +86 (c, 0.5 in EtOH). λ_{max} 317 (log ε 4.4) (MeOH)

Et ester analogue
Butyrolactone VII
[1338563-92-0]
$C_{25}H_{26}O_7$ 438.476
Prod. by *Aspergillus terreus* BCC 4651. Amorph. solid. Mp 124 - 125°. $[\alpha]_D^{27}$ +54 (c, 0.08 in MeOH). λ_{max} 227 (log ε 4.08); 303 (log ε 4.17) (MeOH)

3-Sulfate
[1012860-64-8]
$C_{24}H_{24}O_{10}S$ 504.506
Prod. by *Aspergillus terreus* HKI0499 and a marine-derived *Pseudoalteromonas* sp. strain 06204B. Exhibits cytotoxic activity. Oil. $[\alpha]_D^{23}$ +68.3 (c, 0.1 in MeOH). λ_{max} 311 (log ε 4.4) (MeOH)

4'-Sulfate
[1012860-65-9]
$C_{24}H_{24}O_{10}S$ 504.506
Prod. by *Aspergillus terreus* (HKI0499). Oil. $[\alpha]_D^{23}$ +28.2 (c, 0.1 in MeOH). λ_{max} 312 (log ε 4.4) (MeOH)

3-Me ether
3-O-Methylbutyrolactone I. ***Aspernolide C‡***
[1815632-13-3]
$C_{25}H_{26}O_7$ 438.476
Prod. by the marine-derived *Aspergillus* sp. XS-20090B15, *Aspergillus iizukae* and *Aspergillus versicolor*. Exhibits antibacterial activity against *Staphylococcus aureus*. Oil. $[\alpha]_D^{25}$ +75.2 (c, 0.2 in MeOH). λ_{max} 318 (log ε 4.2) (MeOH)

4''-Me ether
Versicolactone D‡
[1803200-18-1]
$C_{25}H_{26}O_7$ 438.476
Prod. by *Aspergillus versicolor* KJ801852. Amorph. powder. $[\alpha]_D^{24}$ +82.6 (c, 0.2 in MeOH). λ_{max} 210 (log ε 4.38); 244 (log ε 3.56); 305 (log ε 3.72) (MeOH)

3-Et ether
3-O-Ethylbutyrolactone I. ***Aspernolide F***
[2042321-44-6]
$C_{26}H_{28}O_7$ 452.503
Prod. by *Aspergillus terreus*. Exhibits activity against *Cryptococcus neoformans* and mild activity against MRSA. Yellow gum. $[\alpha]_D$ +39 (c, 0.25 in MeOH). λ_{max} 216 (log ε 4.01); 254 (log ε 3.68); 299 (log ε 0.52) (MeOH)

2''',3'''-Epoxide
Butyrolactone III. *F06Z-509C*
[87414-45-7]
$C_{24}H_{24}O_8$ 440.448
Prod. by *Aspergillus terreus* IFO 4100 and a marine-derived *Aspergillus terreus* Y10. Cytotoxic. Yellowish mass (CH_2Cl_2). Mp 70°. $[\alpha]_D^{25}$ +80 (c, 0.7 in EtOH)

1'''-Hydroxy, 2''',3'''-epoxide
7''-Hydroxybutyrolactone III
[1550903-37-1]
$C_{24}H_{24}O_9$ 456.447
Prod. by a mangrove-derived *Aspergillus terreus* A8-4. Yellow powder. Mp 127 - 128°. λ_{max} 308 (MeOH)

2'-Hydroxy, 4'-Me ether
Terrephenol B
[1643389-21-2]
$C_{25}H_{26}O_8$ 454.475
Prod. by *Aspergillus terreus* and *Penicillium oxalicum* 0403. Yellow gum. $[\alpha]_D^{25}$ +74.3 (c, 0.2 in MeOH). λ_{max} 220 (log ε 4.38); 248 (log ε 3.42); 305 (log ε 3.72) (MeOH)

3''-Hydroxy, 3-Me ether
Methylbutyrolactone III
[1863110-12-6]
$C_{25}H_{26}O_8$ 454.475
Prod. of *Aspergillus flavipes* PJ03-11. α-Glucosidase inhibitor. Brown-yellow oil. $[\alpha]_D$ +55 (c, 0.12 in MeOH). Misleading synonym

4'''-Hydroxy(E-)
Aspernolide D‡
[1940169-16-3]
$C_{24}H_{24}O_8$ 440.448
Prod. by *Aspergillus versicolor*. Amorph. powder. $[\alpha]_D^{25}$ +78.6 (c, 0.2 in MeOH). λ_{max} 218 (log ε 4.10); 250 (log ε 3.58); 295 (log ε 3.38) (MeOH)

4''-Deoxy
Versicolactone B‡
[1803200-16-9]
$C_{24}H_{24}O_6$ 408.45
Prod. by *Aspergillus versicolor* KJ801852 and the marine-derived *Aspergillus terreus* strains MF972904 and MF993118. Exhibits strong cytotoxic activity. Amorph. powder. $[\alpha]_D^{24}$ +70.2 (c, 0.2 in MeOH). λ_{max} 220 (log ε 4.18); 255 (log ε 3.52); 300 (log ε 2.26) (MeOH)

4''-Deoxy, 4'-Me ether
Versicolactone C‡
[1803200-17-0]
$C_{25}H_{26}O_6$ 422.477
Prod. by *Aspergillus versicolor* KJ801852. Amorph. powder. $[\alpha]_D^{25}$ +75.4 (c, 0.2 in MeOH). λ_{max} 210 (log ε 4.29); 246 (log ε 3.52); 300 (log ε 3.72) (MeOH)

5'-Methoxy
Terrephenol A
[1643389-20-1]
$C_{25}H_{26}O_8$ 454.475
Prod. by *Aspergillus terreus*. Yellow gum. $[\alpha]_D^{25}$ +68.4 (c, 0.2 in MeOH). λ_{max} 220 (log ε 4.27); 250 (log ε 3.48); 306 (log ε 3.78) (MeOH)

5'-Methoxy, 4''-Me ether
Terrelactone A‡. *Terrephenol C*
[1675220-34-4]
$C_{26}H_{28}O_8$ 468.502
Prod. by *Aspergillus terreus* and *Penicillium oxalicum* 0403. Shows high anti-tobacco mosaic virus activity. Yellow gum. $[\alpha]_D^{25}$ +65.8

(c, 0.2 in MeOH). λ_{max} 210 (log ε 4.36); 252 (log ε 3.53); 308 (log ε 3.73) (MeOH)

2‴,3‴-Dihydro, 3‴-hydroxy

Aspernolide B. Terrelactone‡

[1141433-18-2]

$C_{24}H_{26}O_8$ 442.464

Prod. by marine-derived *Aspergillus terreus* and *Penicillium oxalicum* 0403. Light brown or yellow oil. $[\alpha]_D^{28}$ +48.3 (c, 0.29 in MeOH) (Aspernolide B). $[\alpha]_D^{25}$ +71 (c, 0.5 in $CHCl_3$) (Terrelactone A). λ_{max} 240; 303 (MeOH) (Aspernolide B). [neutral] λ_{max} 231 (log ε 4); 310 (log ε 4.4) (MeOH) (Terrelactone A)

2‴,3‴-Dihydro, 2‴S,3‴-dihydroxy

Aspernolide D‡

[1321528-89-5]

$C_{24}H_{26}O_9$ 458.463

Prod. by *Aspergillus terreus* RCBC1002 and the marine-derived *Eurotium rubrum* SH-823. Light yellow gum. $[\alpha]_D^{20}$ +40 (c, 0.1 in MeOH). Structure revised in 2018. λ_{max} 290 (log ε 4.14) (MeOH)

2‴,3‴-Dihydro, 2‴R,3‴-dihydroxy, 3-Me ether

Aspernolide H‡

[2170730-37-5]

$C_{25}H_{28}O_9$ 472.490

Prod. by *Aspergillus iizukae* IFB-XS5-1 and *Aspergillus* sp. CPCC 400735. Yellow gum. $[\alpha]_D^{25}$ +48.8 (c, 0.11 in MeOH). λ_{max} 222 (log ε 4.23); 309 (log ε 4.31) (MeOH)

2‴,3‴-Dihydro, 2‴S,3‴-dihydroxy, 3-Me ether

Aspernolide I‡

[2170730-38-6]

$C_{25}H_{28}O_9$ 472.490

Prod. by *Aspergillus iizukae* IFB-XS5-1 and *Aspergillus* sp. CPCC 400735. Yellow gum. $[\alpha]_D^{25}$ +46.8 (c, 0.1 in MeOH). λ_{max} 222 (log ε 4.11); 309 (log ε 4.18) (MeOH)

2‴,3‴-Dihydro, 2‴S,3‴-dihydroxy, 3-Et ether

Aspernolide G

[2042321-50-4]

$C_{26}H_{30}O_9$ 486.517

Prod. by *Aspergillus terreus*. Yellow gum. $[\alpha]_D$ +68 (c, 0.01 in MeOH). Struct. revised in 2018. λ_{max} 235 (log ε 3.85); 307 (log ε 3.24) (MeOH)

2‴,3‴-Dihydro, 3″,3‴-dihydroxy, 3-Me ether

Aspernolide J‡

[2171509-99-0]

$C_{25}H_{28}O_9$ 472.490

Prod. by *Aspergillus iizukae* IFB-XS5-1. Yellow gum. $[\alpha]_D^{25}$ +39.5 (c, 0.11 in MeOH). λ_{max} 205 (log ε 4.15); 327 (log ε 3.80) (MeOH)

2‴,3‴-Dihydro, 3‴-methoxy

Aspernolide C‡

[1141433-19-3]

$C_{25}H_{28}O_8$ 456.491

Isol. from extracts of a marine-derived *Aspergillus terreus*. Light brown syrup. Artifact.

Nitta, K. *et al.*, *Chem. Pharm. Bull.*, 1983, **31**, 1528-1533 (*Butyrolactone III, synth, biosynth*).
Kitagawa, M. *et al.*, *Oncogene*, 1994, **9**, 2549-2557 (*Butyrolactone I, activity*).
Rao, K.V. *et al.*, *Chem. Pharm. Bull.*, 2000, **48**, 559-562 (*Butyrolactone I, isol, cmr, activity*).
Cazar, M.E. *et al.*, *World J. Microbiol. Biotechnol.*, 2005, **21**, 1067-1075 (*Butyrolactone I, activity*).
Niu, X. *et al.*, *J. Nat. Prod.*, 2008, **71**, 689-692 (*sulfates*).
Parvatkar, R.R. *Phytochemistry*, 2009, **70**, 128-132 (*Aspernolides B,C*).
Nuclear, P. *et al.*, *Chem. Pharm. Bull.*, 2010, **58**, 1221-1223 (*Aspernolide D*).
Haritakun, R. *et al.*, *Chem. Pharm. Bull.*, 2010, **58**, 1545-1548 (*Butyrolactone VII*).
Wang, F.-S. *et al.*, *J. Chin. Pharm. Sci.*, 2010, **19**, 251-255 (*F06Z-509A, F06Z-509C*).
Wang, Y. *et al.*, *Mar. Drugs*, 2011, **9**, 1368-1378 (*Terrelactone*).
Shen, Y. *et al.*, *Chem. Pharm. Bull.*, 2012, **60**, 1437-1441 (*7″-Hydroxybutyrolactone III*).
Nagia, M.M.S. *et al.*, *Org. Med. Chem. Lett.*, 2012, **2**, No.9 (*Butyrolactone I*).
Raina, S. *et al.*, *Process Biochem.*, 2012, **47**, 843-852 (*activity*).
Ye, Y.-Q. *et al.*, *Heterocycles*, 2014, **89**, 2177-2182 (*Terrephenols A,B, activity*).
Zhou, K. *et al.*, *Asian J. Chem.*, 2015, **27**, 3532-3534 (*Terrephenol C, activity*).
Chen, M. *et al.*, *Chem. Biodiversity*, 2015, **12**, 1398-1406 (*3-O-Methylbutyrolactone I*).
Zhou, M. *et al.*, *J. Braz. Chem. Soc.*, 2015, **26**, 545-549 (*Aspernolides C,D*).
Yuan, L. *et al.*, *Nat. Prod. Res.*, 2015, **29**, 1914-1919 (*Terrelactone A, activity*).
Ibrahim, S.R.M. *et al.*, *Phytochem. Lett.*, 2015, **14**, 84-90 (*Aspernolide F, activity*).
Zhou, M. *et al.*, *Planta Med.*, 2015, **81**, 235-240 (*Versicolactones B,C,D, cryst struct*).
Zhang, L.-H. *et al.*, *Bioorg. Med. Chem. Lett.*, 2016, **26**, 346-350 (*Methylbutyrolactone III, activity*).
Li, L.-J. *et al.*, *Phytochem. Lett.*, 2016, **16**, 134-140 (*Aspernolides H,I,J*).
Ibrahim, S.R.M. *et al.*, *Nat. Prod. Commun.*, 2017, **12**, 791-800 (*rev, occur, biosynth, activity*).
Qi, C. *et al.*, *Bioorg. Med. Chem.*, 2018, **26**, 5903-5910 (*Versicolactone B, cryst struct, activity*).
Zhang, Y.Y. *et al.*, *Mar. Drugs*, 2018, **16**, 202:1-13 (*isol, activity*).
Wu, W. *et al.*, *FASEB J.*, 2019, **33**, 12616-12629 (*activity*).
Lin, X. *et al.*, *Nat. Prod. Res.*, 2019, **33**, 494-499 (*3-sulfate, activity*)

Butyrolactone II B90

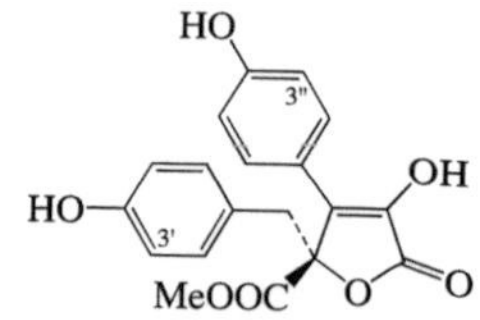

Absolute Configuration

F06Z-509B

[87414-44-6]

$C_{19}H_{16}O_7$ 356.33

Provisional lignan. Biosynth. pathway unclear. Prod. by *Aspergillus terreus* (strains DRCC 152 and IFO 4100) and *Retiboletus retipes*. Cytotoxic. Plates (CH_2Cl_2). Mp 94 - 96°. $[\alpha]_D^{25}$ +85 (c, 1.0 in EtOH)

3-Me ether

3-O-Methylbutyrolactone II

[1471279-33-0]

$C_{20}H_{18}O_7$ 370.357

Prod. by a marine-derived *Aspergillus* sp.. Oil. $[\alpha]_D^{25}$ +38 (c, 0.2 in MeOH). λ_{max} 316 (log ε 4.2) (MeOH)

3-Et ether

Terrenolide S

[1813541-93-3]

$C_{21}H_{20}O_7$ 384.384

Prod. by *Aspergillus terreus*. Yellow gum. $[\alpha]_D$ +63.6 (c, 0.05 in MeOH). λ_{max} 213 (log ε 3.98); 249 (log ε 3.62); 296 (log ε 0.55) (MeOH)

3′-Formyl

3′-Formylbutyrolactone II

[1397705-03-1]

$C_{20}H_{16}O_8$ 384.34

Prod. by a marine-derived *Aspergillus terreus*. β-Glucuronidase inhibitor. Pale yellow powder. Mp 86 - 88°. $[\alpha]_D^{25}$ +128 (c, 0.05 in MeOH). λ_{max} 203; 224; 258; 306 (MeOH)

4′-Ac

Versicolactone F

[1964496-87-4]

$C_{21}H_{18}O_8$ 398.367

Prod. by *Aspergillus versicolor*. Amorph. powder. $[\alpha]_D^{25}$ +65.8 (c, 0.2 in MeOH). λ_{max} 210 (log ε 4.11); 239 (log ε 3.50); 285 (log ε 3.72) (MeOH)

4'-Deoxy, 3-Me ether

Versicolactone E

[1964496-86-3]

$C_{20}H_{18}O_6$ 354.358

Prod. by *Aspergillus versicolor*. Amorph. powder. $[\alpha]_D^{25}$ +68.4 (c, 0.2 in MeOH). λ_{max} 210 (log ε 4.05); 235 (log ε 3.41); 286 (log ε 3.68) (MeOH)

4'-Deoxy, 4"-Me ether

Asperphenol A

$C_{20}H_{18}O_6$ 354.358

Prod. by *Aspergillus versicolor*. Amorph. powder. $[\alpha]_D^{25}$ +63.2 (c, 0.2 in MeOH). λ_{max} 210 (log ε 4.18); 236 (log ε 3.38); 287 (log ε 3.62) (MeOH)

4'-Deoxy, 3"-hydroxy, 4"-Me ether

Asperphenol C

$C_{20}H_{18}O_7$ 370.357

Prod. by *Aspergillus versicolor*. Amorph. powder. $[\alpha]_D^{25}$ +62.3 (c, 0.2 in MeOH). λ_{max} 215 (log ε 4.06); 242 (log ε 3.46); 285 (log ε 3.68) (MeOH).

Nitta, K. *et al.*, *Chem. Pharm. Bull.*, 1983, **31**, 1528-1533 (*Butyrolactone II, synth, biosynth*).
Rao, K.V. *et al.*, *Chem. Pharm. Bull.*, 2000, **48**, 559-562 (*Butyrolactone II, activity*).
Justus, K. *et al.*, *Eur. J. Org. Chem.*, 2007, 5560-5572 (*synth, biosynth*).
Wang, F.-S. *et al.*, *J. Chin. Pharm. Sci.*, 2010, **19**, 251-255 (*F06Z-509B*).
Haroon, M.H. *et al.*, *Nat. Prod. Res.*, 2013, **27**, 1060-1066 (*3-Formylbutyrolactone II*).
Guo, C.-J. *et al.*, *Org. Lett.*, 2013, **15**, 3562-3565 (*biosynth*).
Ye, Y.-Q. *et al.*, *Bull. Korean Chem. Soc.*, 2014, **35**, 3059-3062 (*Asperphenol A*).
Li, Y. *et al.*, *Asian J. Chem.*, 2015, **27**, 3529-3531 (*Asperphenol C, activity*).
Chen, M. *et al.*, *Chem. Biodiversity*, 2015, **12**, 1398-1406 (*3-O-Methylbutyrolactone II*).
Zhou, K. *et al.*, *Chem. Nat. Compd. (Engl. Transl.)*, 2016, **52**, 591-594 (*Versicolactones E,F*).
Elkhayat, E.S. *et al.*, *Nat. Prod. Res.*, 2016, **30**, 814-820 (*Terrenolide S*)

Butyrolactone IV B91

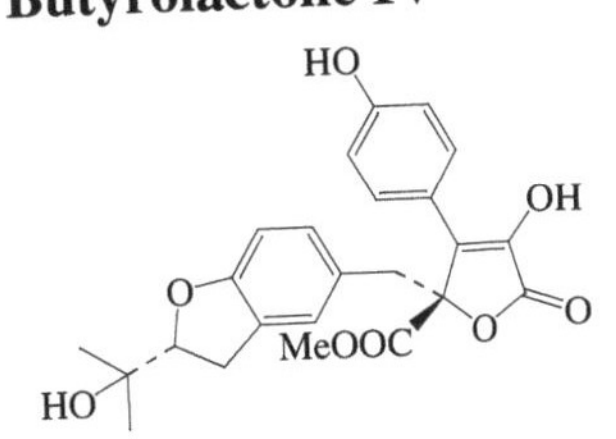

Absolute Configuration

[271587-58-7]

$C_{24}H_{24}O_8$ 440.448

Provisional prenylated lignan. Biosynth. pathway unclear. Prod. by *Aspergillus terreus* strains BCC 4651, DRCC 152, MF993118, RCBC1002, SCSGAF0162 and a marine-derived *Paecilomyces variotii* EN-291. Cryst. ($CHCl_3$/EtOAc). Mp 110 - 112°. $[\alpha]_D^{27}$ +92 (c, 0.7 in EtOH). λ_{max} 204; 227; 309 (MeOH).

Rao, K.V. *et al.*, *Chem. Pharm. Bull.*, 2000, **48**, 559-562 (*Butyrolactone IV, cryst struct*)

Butyrolactone V B92

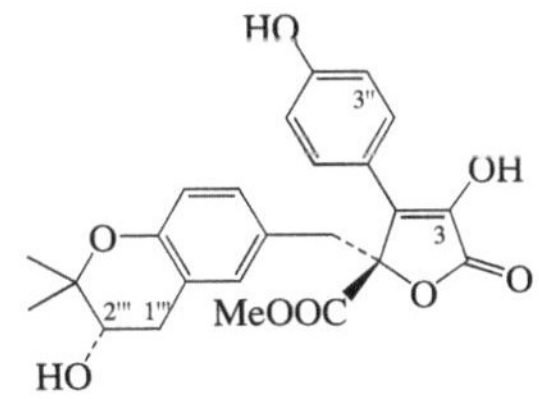

Absolute Configuration

Butyrolactone VI

[1151509-01-1]

[1338563-91-9] (Butyrolactone VI)

$C_{24}H_{24}O_8$ 440.448

Provisional prenylated lignan. Biosynth. pathway unclear. Structure revision of Butyrolactone VI in 2018 suggested identity with Butyrolactone V, however the stereochemistry at C-2‴ was not discussed. Prod. by *Aspergillus* sp. F1, *Aspergillus terreus* BCC 4651, *Penicillium expansum* 091006 and the marine-derived *Aspergillus terreus* 2P-22 and *Paecilomyces variotii* EN-291. Antiplasmodial agent. Cryst.. Mp 136 - 137° (Butyrolactone VI). Mp 139 - 140° (Butyrolactone V). $[\alpha]_D^{27}$ +71 (c, 0.1 in MeOH) (Butyrolactone V). $[\alpha]_D^{26}$ +79 (c, 0.09 in MeOH) (Butyrolactone VI). λ_{max} 229 (log ε 3.99); 306 (log ε 4.15) (MeOH)

3-Me ether

Aspernolide K

[2170730-39-7]

$C_{25}H_{26}O_8$ 454.475

Prod. by *Aspergillus iizukae* IFB-XS5-1 and *Aspergillus* sp. CPCC 400735

3"-Hydroxy, 3-Me ether

[913690-68-3]

$C_{25}H_{26}O_9$ 470.474

Prod. by *Aspergillus iizukae* IFB-XS5-1

2‴-Deoxy

Aspernolide A

[65372-43-2]

$C_{24}H_{24}O_7$ 424.449

Prod. by various terrestrial and marine-derived *Aspergillus terreus*. Also from *Paecilomyces variotii* EN-291 and *Penicillium oxalicum* 0403. Soybean lipoxygenase inhibitor. DPPH radical-scavenging agent. Gummy solid. $[\alpha]_D^{28}$ +88.7 (c, 0.58 in $CHCl_3$). λ_{max} 240; 303 (MeOH)

2‴-Deoxy, 1′",2′"-didehydro

Aspernolide E

[1443332-65-7]

$C_{24}H_{22}O_7$ 422.433

Prod. by a marine-derived *Aspergillus terreus* SCSGAF0162. Amorph. yellow powder. $[\alpha]_D^{30}$ +100 (c, 0.002 in MeOH). λ_{max} 228 (log ε 3.47); 325 (log ε 4.13) (MeOH).

Lin, T. *et al.*, *Nat. Prod. Res.*, 2009, **23**, 77-85 (*Butyrolactone V*).
Parvatkar, R.R. *et al.*, *Phytochemistry*, 2009, **70**, 128-132 (*Aspernolide A*).
Sugiyama, Y. *et al.*, *Biosci., Biotechnol., Biochem.*, 2010, **74**, 881-883 (*Aspernolide A, activity*).
Haritakun, R. *et al.*, *Chem. Pharm. Bull.*, 2010, **58**, 1545-1548 (*Butyrolactones VI,V, abs config, activity*).
San-Martin, A. *et al.*, *J. Chil. Chem. Soc.*, 2011, **56**, 625-627 (*Butyrolactone VI*).
Nagia, M.M.S. *et al.*, *Org. Med. Chem. Lett.*, 2012, **2**, No.9 (*Butyrolactone V*).
He, F. *et al.*, *J. Nat. Prod.*, 2013, **76**, 1182-1186 (*Aspernolide E*).
Ibrahim, S.R.M. *et al.*, *Phytochem. Lett.*, 2015, **14**, 84-90 (*Butyrolactone VI*).
Li, L.-J. *et al.*, *Phytochem. Lett.*, 2016, **16**, 134-140 (*Aspergillus iizukae constits*).
Ibrahim, S.R.M. *et al.*, *Nat. Prod. Commun.*, 2017, **12**, 791-800 (*rev, occur, biosynth, activity*).
Liu, M. *et al.*, *RSC Adv.*, 2018, **8**, 13040-13047 (*Butyrolactone VI, struct revision*)

Butyrolactone VIII B93

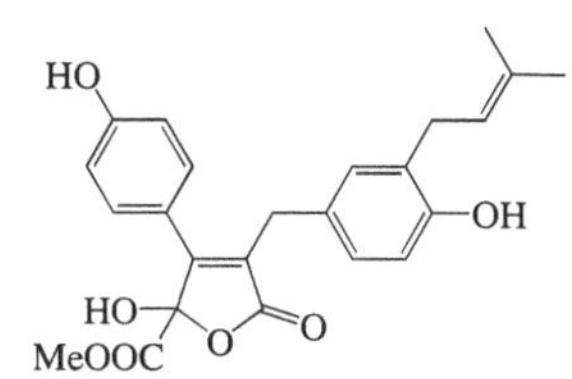

[1818288-01-5]

$C_{24}H_{24}O_7$ 424.449

Provisional lignan. Biosynth. pathway unclear. Prod. by a marine-derived *Aspergillus terreus* MXH-23. Pale yellow oil. $[\alpha]_D^{25}$ -11.2 (c, 0.1 in MeOH). λ_{max} 210 (log ε 4.45); 307 (log ε 4.27) (MeOH).

Ma, X. *et al.*, *J. Ocean Univ. China*, 2014, **13**, 1067-1070 (*Butyrolactone VIII*)

4-*O*-Cadinylangolensin C1

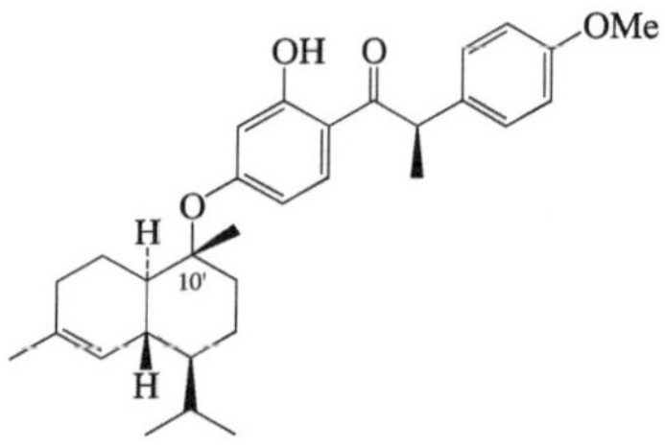

(10′*R*)-*form*

$C_{31}H_{40}O_4$ 476.657

(10′R)-form

4-*O*-α-Cadinylangolensin [75917-91-8]

Constit. of *Pterocarpus angolensis* heartwood. Needles (EtOH). Mp 136°

(10′S)-form

4-*O*-*T*-Cadinylangolensin [75872-84-3]

Constit. of *Pterocarpus angolensis* heartwood. Noncryst.. Mp 44 - 46°.

Bezuidenhoudt, B.C.B. *et al.*, *J. Chem. Soc., Perkin Trans. 1*, 1980, 2179-2183 (*Pterocarpus angolensis constits, cryst struct, cd*)

6-*O*-Caffeoylglucose C2

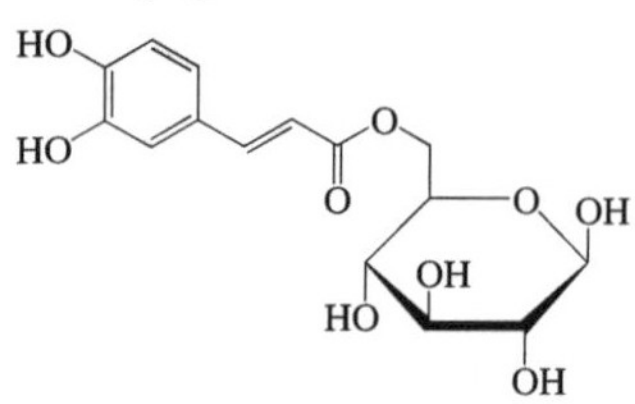

β-D-Pyranose-*form*

6-O-(3,4-Dihydroxycinnamoyl)glucose

[10066-92-9]

$C_{15}H_{18}O_9$ 342.3

D-*form* [114297-66-4]

Isol. from bark of *Prunus grayana* and *Prunus buergeriana* and from *Lavandula stoechas*. Amorph. pale orange powder. $[\alpha]_D^{25}$ +26.8 (c, 0.74 in MeOH)

D-Pyranose-form

3-O-β-D-Glucopyranosyl, 4′-O-[E-cinnamoyl-(→6)-β-D-glucopyranoside]

Susaroyside B

[2414074-29-4]

$C_{36}H_{44}O_{20}$ 796.728

Constit. of the fruit of *Forsythia suspensa*. Amorph. yellow powder. $[\alpha]_D^{20}$ -55.4 (c, 0.1 in MeOH). λ_{max} 205 (log ε 4.24); 282 (log ε 3.99) (MeOH)

3-O-β-D-Glucopyranosyl, 4′-O-[4-hydroxy-E-cinnamoyl-(→6)-β-D-glucopyranoside]

Susaroyside C

[2414074-30-7]

$C_{36}H_{44}O_{21}$ 812.727

Constit. of the fruit of *Forsythia suspensa*. Amorph. yellow powder. $[\alpha]_D^{20}$ -79.8 (c, 0.1 in MeOH). λ_{max} 206 (log ε 4.30); 312 (log ε 4.21) (MeOH)

4′-O-[4-Hydroxy-E-cinnamoyl-(→6)-β-D-glucopyranoside]

Susaroyside A

[2414074-28-3]

$C_{30}H_{34}O_{16}$ 650.586

Constit. of the fruit of *Forsythia suspensa*. Exhibits significant antiinflammatory activity. Amorph. yellow powder. $[\alpha]_D^{20}$ -48 (c, 0.1 in MeOH). λ_{max} 312 (log ε 4.38) (MeOH)

α-D-Pyranose-form

1-O-β-D-Fructofuranosyl, 4′-O-[4-hydroxy-E-cinnamoyl-(→6)-β-D-glucopyranoside]

Susaroyside E

[2414074-32-9]

$C_{36}H_{44}O_{21}$ 812.727

Constit. of the fruit of *Forsythia suspensa*. Amorph. yellow powder. $[\alpha]_D^{20}$ -171.5 (c, 0.1 in MeOH). λ_{max} 228 (log ε 4.6); 313 (log ε 4.63) (MeOH)

β-D-Pyranose-form

Et glycoside

[1610369-07-7]

$C_{17}H_{22}O_9$ 370.354

Constit. of *Artemisia capillaris*. Brown gum. $[\alpha]_D^{27}$ -25.8 (c, 0.12 in MeOH). λ_{max} 216 (log ε 4.27); 329 (log ε 4.26) (MeOH)

4-Formyl-2-methoxyphenyl glycoside, 3′-Me ether

6-O-Feruloyl-1-O-vanilloyl-β-D-glucopyranoside

[1141892-36-5]

$C_{24}H_{26}O_{11}$ 490.461

Flavour constit. of cured vanilla beans (*Vanilla plantifolia*). No phys. props. reported. λ_{max} 247; 275; 311 (MeCN)

1-O-(4-Hydroxycinnamoyl)

6-Caffeoyl-1-coumaroylglucose

[114217-03-7]

$C_{24}H_{24}O_{11}$ 488.445

Isol. from bark of *Prunus buergeriana*. Pale yellow powder. $[\alpha]_D^{25}$ +12.1 (c, 0.61 in MeOH)

1,3-Bis-O-(3,4,5-trihydroxybenzoyl)

6-O-trans-Caffeoyl-1,3-di-O-galloyl-β-D-glucoside

$C_{29}H_{26}O_{17}$ 646.51

Constit. of the leaves of *Juglans sigillata*. Amorph. yellowish powder. Mp 193 - 195°. $[\alpha]_D^{20}$ +10.5 (c, 0.5 in MeOH). λ_{max} 245; 300; 332 (no solvent reported)

3′-Me ether

6-O-Feruloyl-β-D-glucopyranoside

[10066-93-0]

$C_{16}H_{20}O_9$ 356.327

Constit. of the leaves of *Eucalyptus maideni*. Mp 218° (198-201° dec.). $[\alpha]_D^{18}$ +30 (c, 0.3 in MeOH). λ_{max} 238 (log ε 4.07); 327 (log ε 4.2) (MeOH)

2,3,4-Trihydroxy-3-methylbutyl glycoside, 3′-Me ether

[257939-69-8]

$C_{21}H_{30}O_{12}$ 474.459

Constit. of *Salvia officinalis* leaf. Powder

2,3,4,5-Tetrahydroxyhexyl glycoside

[151778-89-1]

$C_{21}H_{30}O_{13}$ 490.458

Constit. of the leaves of *Viburnum dilatatum*.

Watanabe, S. *et al.*, *Agric. Biol. Chem.*, 1966, **30**, 420-422 (*synth*).
Birkofer, L. *et al.*, *Justus Liebigs Ann. Chem.*, 1966, **699**, 223-231 (*synth*).
Shimomura, H. *et al.*, *Phytochemistry*, 1987, **26**, 249-251 ; 1988, **27**, 641-644 (*isol, pmr*).
Machida, K. *et al.*, *CA*, 1994, **120**, 4693x (*Viburnum dilatatum constits*).
Wang, M. *et al.*, *J. Agric. Food Chem.*, 2000, **48**, 235-238 (*Me ether trihydroxymethylbutyl glycoside*).
Schwarz, B. *et al.*, *J. Agric. Food Chem.*, 2009, **57**, 3729-3737 (*vanilloyl glycoside 3′-Me ether*).
Tian, L.-W. *et al.*, *Helv. Chim. Acta*, 2010, **93**, 2194-2202 (*6-feruloylglucose*).
Zhao, Y. *et al.*, *Fitoterapia*, 2014, **95**, 187-193 (*Et glycoside*).
Si, C.-L. *et al.*, *Chem. Nat. Compd. (Engl. Transl.)*, 2016, **52**, 1008-1010 (*1,3-digalloyl*).
Shao, S.-Y. *et al.*, *Bioorg. Med. Chem. Lett.*, 2019, **29**, 126635:1-5 (*Susaroysides A,B,C,E, activity*)

1-*O*-Caffeoylglycerol C3

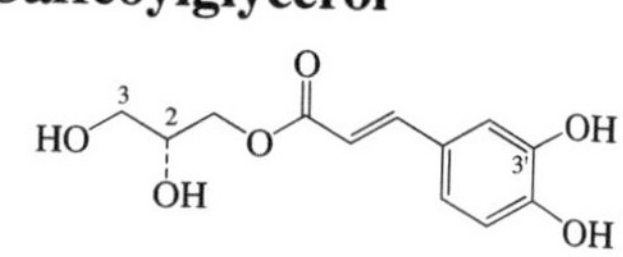

(*R,E*)-*form*

1-O-(3,4-Dihydroxycinnamoyl)glycerol

$C_{12}H_{14}O_6$ 254.238

(R,E)-form

3-O-β-D-Glucopyranoside

3-O-(3,4-Dihydroxycinnamoyl)-1-O-glucosyl-sn-glycerol.

Epiregaloside C [120665-04-5]

$C_{18}H_{24}O_{11}$ 416.379

Constit. of *Lilium pardarinum*. Obt. as a mixture with Regaloside C

3'-Me ether, 3-O-β-D-glucopyranoside

3-O-Feruloyl-1-O-glucosyl-sn-glycerol. ***Epiregaloside F***

[120665-05-6]

$C_{19}H_{26}O_{11}$ 430.406

Constit. of *Lilium auratum*. Obt. as a mixture with Regaloside F

(S,E)-form [138812-55-2]

Constit. of *Juncus effusus*, *Lilium lancifolium* and a Spanish propolis. $[\alpha]_D$ +10.5 (MeOH)

2-O-β-D-Glucopyranoside

1-O-(3,4-Dihydroxycinnamoyl)-2-O-glucosyl-sn-glycerol. ***Regaloside K***

[138772-00-6]

$C_{18}H_{24}O_{11}$ 416.379

Constit. of *Lilium mackliniae*. Pale yellow powder. $[\alpha]_D^{28}$ -27.4 (c, 0.2 in MeOH). λ_{max} 240 (sh); 300 (sh); 330 (log ε 4.33) (MeOH)

3-O-β-D-Glucopyranoside

1-O-(3,4-Dihydroxycinnamoyl)-3-O-glucosyl-sn-glycerol. ***Regaloside C***

[117591-85-2]

$C_{18}H_{24}O_{11}$ 416.379

Isol. from bulbs of *Lilium henryi* and *Lilium pardarinum*. Pale yellow amorph. powder. $[\alpha]_D^{27}$ -11.3 (c, 0.62 in MeOH)

3-Ac, 2-O-β-D-glucopyranoside

Regaloside E

[123134-21-4]

$C_{20}H_{26}O_{12}$ 458.416

Constit. of *Lilium longiflorum*

3'-Me ether

1-O-(4-Hydroxy-3-methoxycinnamoyl)glycerol. ***1-O-Feruloylglycerol***

[120601-69-6]

$C_{13}H_{16}O_6$ 268.265

Constit. of *Juncus effusus* and *Lilium brownii*

3'-Me ether, 2-O-β-D-glucopyranoside

1-O-Feruloyl-2-O-glucosyl-sn-glycerol. ***Regaloside G***

[120601-64-1]

$C_{19}H_{26}O_{11}$ 430.406

Isol. from *Lilium auratum*. Pale yellow amorph. powder. $[\alpha]_D^{23}$ -19.3 (c, 0.14 in MeOH)

3'-Me ether, 3-O-β-D-glucopyranoside

1-O-Feruloyl-3-O-glucosyl-sn-glycerol. ***Regaloside F***

[120601-65-2]

$C_{19}H_{26}O_{11}$ 430.406

Constit. of *Lilium lancifolium*

(±)-(E)-form

3'-Me ether

[108026-20-6]

Constit. of *Lilium brownii*, *Lilium auratum* and *Lilium mackliniae*

(ξ)-(E)-form

3'-Me ether, 3-pentacosanoyl

1-O-Feruloyl-3-O-pentacosanoylglycerol. ***Niloticol***

[110037-32-6]

$C_{38}H_{64}O_7$ 632.923

Constit. of the roots of *Tamarix nilotica*. Cryst. (Me_2CO) (as di-Ac). Mp 68° (di-Ac)

3'-Me ether, 3-O-(26-hydroxyhexacosanoyl)

1-O-Feruloyl-3-O-(26-hydroxyhexacosanoyl)glycerol

[1251855-60-3]

$C_{39}H_{66}O_8$ 662.949

Constit. of the roots of *Incarvillea mairei*.

Shimomura, H. *et al.*, *Chem. Pharm. Bull.*, 1987, **36**, 2430-2446 (*Regaloside C*).

Shimomura, H. *et al.*, *Chem. Pharm. Bull.*, 1987, **36**, 4841-4848 (*Lilium pardarinum constits*).

Shimomura, H. *et al.*, *Phytochemistry*, 1987, **26**, 844-845 (*Lilium auratum constits*).

Barakat, H.H. *et al.*, *Phytochemistry*, 1987, **26**, 1837-1838 (*Niloticol*).

Mimaki, Y. *et al.*, *Chem. Pharm. Bull.*, 1990, **38**, 3055-3059 (*Lilium brownii constits*).

Sashida, Y. *et al.*, *Chem. Pharm. Bull.*, 1991, **39**, 2362-2368 (*Regaloside K*).

Garcia-Viguera, C. *et al.*, *Z. Naturforsch., C*, 1992, **47**, 634-637 (*propolis constit*).

Greca, M.D. *et al.*, *Nat. Prod. Lett.*, 1998, **12**, 263-270 (*Juncus effusus constits*).

Huang, Z.S. *et al.*, *Zhongguo Zhongyao Zazhi*, 2009, **34**, 1672-1675 (*Incarvillea mairei constit*)

1-*O*-Caffeoylquinic acid C4

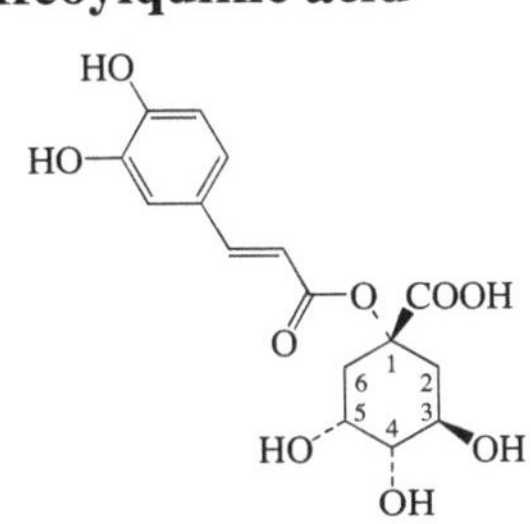

1-(3,4-Dihydroxycinnamoyl)quinic acid

[1241-87-8]

$C_{16}H_{18}O_9$ 354.311

Constit. of coffee and artichoke leaves (*Cynara scolymus*). Noncryst.. $[\alpha]_D^{15}$ -8.3 (c, 3 in H_2O)

(1'→3)-Lactone

Cryst.. Mp 205 - 208°. $[\alpha]_D^{25}$ -17 (c, 2 in EtOH)

4-Deoxy

1-O-Caffeoyl-4-deoxyquinic acid

[153444-59-8]

$C_{16}H_{18}O_8$ 338.312

Isol. from foliage of *Arachis paraguariensis*. Inhibits development of *Spodoptera litura* larvae. No phys. props. reported.

Scarpati, M.L. *et al.*, *Ann. Chim. (Rome)*, 1958, **48**, 997-1006 (*lactone, isol*).

Waiss, A.C. *et al.*, *Chem. Ind. (London)*, 1964, 1984-1985 (*pmr*).

Litinenko, V.I. *et al.*, *Planta Med.*, 1975, **27**, 372 (*rev*).

Stevenson, P.C. *et al.*, *J. Chem. Ecol.*, 1993, **19**, 2917-2933 (*4-deoxy, isol, activity*).

Sefkow, M. *et al.*, *Eur. J. Org. Chem.*, 2001, 2735-2742 (*synth*).

Schütz, K. *et al.*, *J. Agric. Food Chem.*, 2004, **52**, 4090-4096 (*artichoke constits*)

3-*O*-Caffeoylquinic acid C5

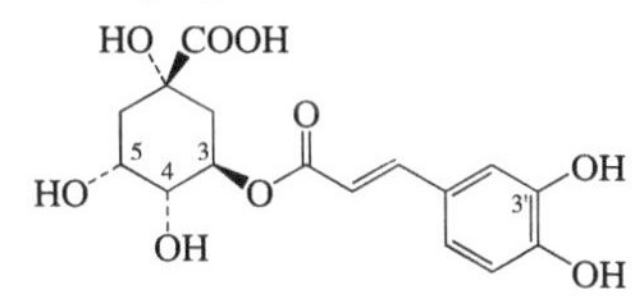

(*E*)-*form*

3-(3,4-Dihydroxycinnamoyl)quinic acid. ***Chlorogenic acid.*** *Caffeylquinic acid. Caffetannic acid. Helianthic acid. Igasuric acid*

[327-97-9]

$C_{16}H_{18}O_9$ 354.311

GU8480000

(E)-form [202650-88-2]

Constit. of many plants including *Chrozophora*, *Cinchona*, *Scabiosa*, *Valeriana*, *Senecio*, *Baccharis* and *Hypericum* spp., isol. originally from Liberian coffee. Also from *Arachis paraguaiensis*, *Baccharis* spp., *Eriobotrya japonica*, *Persea americana*, *Cynara scolymus*, *Cydonia oblonga* and many other plants. Insecticide. Inhibits development of *Spodoptera litura* larvae. Inhibitor of glucose 6-phosphate translocase. Ion chelator. Shows antiviral activity (HSV-1). Shows antioxidant activity and scavenging effect on DPPH radical. Needles + ½ H_2O. Mp 208°. $[\alpha]_D$ -35.2 (H_2O)

4'-O-β-D-Glucopyranoside

$C_{22}H_{28}O_{14}$ 516.452

Constit. of the flower buds of *Lonicera japonica*. Amorph. powder. $[\alpha]_D^{20}$ -74.2 (c, 0.1 in H_2O). λ_{max} 212 (sh); 228 (sh); 289 (log ε 4.12); 319 (log ε 4.06) (MeOH)

Me ester
Methyl chlorogenate
[29708-87-0]
$C_{17}H_{20}O_9$ 368.338
Constit. of *Eremochloa ophiuroides, Eriobotrya japonica, Manilkara zapota, Lonicera japonica, Phyllostachys edulis* and *Viburnum cylindricum*. Amorph. powder. $[\alpha]_D^{26}$ -31.7 (c, 0.12 in MeOH). Was earlier misassigned as the 1-Me ether. λ_{max} 218 (sh); 244 (log ε 4.0); 297 (sh); 330 (log ε 4.1) (MeOH)

Butyl ester
[132741-56-1]
$C_{20}H_{26}O_9$ 410.419
Constit. of *Spondias mombin*. Antiviral agent

Pentyl ester
[1384192-74-8]
$C_{21}H_{28}O_9$ 424.446
Constit. of *Anthocephalus chinensis*

5-Ac
5-O-Acetyl-3-O-caffeoylquinic acid
[548474-69-7]
$C_{18}H_{20}O_{10}$ 396.348
Constit. of the fruit of *Solanum melongena*

Penta-Ac
Needles (EtOH aq.). Mp 181°

4-Benzoyl
Mumeic acid A
[1446205-98-6]
$C_{23}H_{22}O_{10}$ 458.419
Constit. of the flower buds of *Prunus mume*. Inhibits melanogenesis. Amorph. powder. $[\alpha]_D^{15}$ -82.9 (c, 0.22 in MeOH). λ_{max} 232 (log ε 4.28); 329 (log ε 4.41) (MeOH)

4-Benzoyl, Me ester
[1443678-34-9]
$C_{24}H_{24}O_{10}$ 472.446
Constit. of the flower buds of *Prunus mume*. Inhibits melanogenesis. Amorph. powder. $[\alpha]_D^{15}$ -11.9 (c, 0.81 in MeOH). λ_{max} 232 (log ε 4.20); 328 (log ε 4.48) (MeOH)

4-O-(4-Hydroxybenzoyl)
Castusic acid
[1136303-47-3]
$C_{23}H_{22}O_{11}$ 474.418
Constit. of flowers of *Vitex agnus-castus*. Amorph. powder. λ_{max} 214 (sh); 253 (log ε 3.81); 330 (log ε 3.76) (MeOH)

4-O-(3,4,5-Trihydroxybenzoyl)
4-O-Galloylchlorogenic acid
[574747-59-4]
$C_{23}H_{22}O_{13}$ 506.416
Constit. of the fruit of *Manilkara zapota*. Antioxidant. Yellow powder. Mp > 300°. $[\alpha]_D^{25}$ -65.3 (c, 0.0007 in MeOH). Dec. starts at 180°

4-O-(3,4,5-Trihydroxybenzoyl), Me ester
[574747-58-3]
$C_{24}H_{24}O_{13}$ 520.443
Constit. of the fruit of *Manilkara zapota*. Antioxidant. Yellow powder. Mp 183 - 185°. $[\alpha]_D^{25}$ -137.4 (c, 0.005 in MeOH)

4-O-(4-Hydroxy-3-methoxybenzoyl), Me ester
[1207645-18-8]
$C_{25}H_{26}O_{12}$ 518.471
Constit. of *Erycibe hainanensis*. Powder. $[\alpha]_D^{20}$ -108.2 (c, 0.04 in MeOH). λ_{max} 254 (log ε 4.16); 298 (log ε 4.28); 330 (log ε 4.26) (MeOH)

4-O-(4-Hydroxy-3,5-dimethoxybenzoyl)
3-O-Caffeoyl-4-O-syringoylquinic acid. ***4-O-Syringoylchlorogenic acid***
[1207645-15-5]
$C_{25}H_{26}O_{13}$ 534.47
Constit. of *Erycibe hainanensis*. Powder. $[\alpha]_D^{20}$ -106.7 (c, 0.05 in MeOH). λ_{max} 249 (log ε 4.1); 289 (log ε 4.32); 331 (log ε 4.25) (MeOH)

4-O-(4-Hydroxy-3,5-dimethoxybenzoyl), Me ester
[1820924-09-1]
$C_{26}H_{28}O_{13}$ 548.497
Constit. of *Erycibe obtusifolia*. Exhibits weak *in vitro* inhibition of respiratory syncytial virus (RSV). Yellow oil. $[\alpha]_D^{17}$ -89.4 (c, 0.29 in MeOH). λ_{max} 218 (log ε 4.34); 293 (log ε 4.11); 332 (log ε 4.06) (MeOH)

5-O-(4-Hydroxy-3,5-dimethoxybenzoyl)
3-O-Caffeoyl-5-O-syringoylquinic acid. ***5-O-Syringoylchlorogenic acid***
[1207645-16-6]
$C_{25}H_{26}O_{13}$ 534.47
Constit. of *Erycibe hainanensis*. Powder. $[\alpha]_D^{20}$ -96.4 (c, 0.05 in MeOH). λ_{max} 249 (log ε 4.11); 290 (log ε 4.34); 330 (log ε 4.26) (MeOH)

5-O-(4-Hydroxy-3,5-dimethoxybenzoyl), Me ester
[946599-45-7]
$C_{26}H_{28}O_{13}$ 548.497
Constit. of the roots of *Erycibe obtusifolia*. Amorph. powder. Mp 139 - 141°. $[\alpha]_D^{25}$ -78 (c, 0.05 in MeOH). λ_{max} 287 (log ε 4.34); 330 (log ε 4.26) (MeOH)

5-Malonyl
3-O-Caffeoyl-5-O-malonylquinic acid
[1325713-70-9]
$C_{19}H_{20}O_{12}$ 440.357
Constit. of the fruit of *Solanum melongena*. Amorph. powder. $[\alpha]_D^{20}$ -48.3 (c, 0.0003 in MeOH). λ_{max} 218 (log ε 3.69); 300 (sh); 328 (log ε 3.72) (MeOH)

1-Me ether
3-O-Caffeoyl-1-O-methylquinic acid
[374728-77-5]
$C_{17}H_{20}O_9$ 368.338
Constit. of *Phyllostachys edulis*. Powder. Mp 202 - 203°. $[\alpha]_D^{25}$ -16.9 (c, 0.35 in H_2O). λ_{max} 245 (sh); 300; 329 (MeOH)

3'-Me ether
3-O-Feruloylquinic acid
$C_{17}H_{20}O_9$ 368.338
Constit. of coffee beans, *Lycopersicon esculentum, Helianthus annuus, Swietenia mahagoni, Nicotiana tabacum* and *Phellodendron amurense* bark. Protease inhibitor. Mp 196 - 197°. $[\alpha]_D^{28}$ -42.8 (EtOH). λ_{max} 325 (ε 19200) (EtOH) (Berdy)

3'-Me ether, 4'-O-β-D-glucopyranoside
3-O-(4-O-Glucosylferuloyl)quinic acid. *GPFQ*
[438544-16-2]
$C_{23}H_{30}O_{14}$ 530.479
Constit. of the roots and rhizomes of *Hydrastis canadensis*. Off-white powder. $[\alpha]_D^{19}$ -53 (c, 0.02 in MeOH). $[\alpha]_D^{16}$ -58 (c, 0.02 in H_2O). λ_{max} 217 (log ε 4.32); 232 (log ε 4.31); 292 (log ε 4.21); 320 (log ε 4.28) (MeOH)

3'-Me ether, Me ester
[154418-15-2]
$C_{18}H_{22}O_9$ 382.365

Constit. of *Phellodendron amurense* bark. Powder. Mp 78 - 81°. [α] D -29.6 (c, 1 in MeOH)

3′-Me ether, butyl ester

Hycandinic acid ester 1

$C_{21}H_{28}O_9$ 424.446

Constit. of a commercial root sample of *Hydrastis canadensis*. Oil. $[\alpha]^{24}_{D}$ -38.6 (c, 0.14 in $CHCl_3$). λ_{max} 223 (log ε 4.96); 298 (log ε 5); 340 (log ε 5.02) (EtOH)

3′-Me ether, eicosyl ester

[1458723-51-7]

$C_{37}H_{60}O_9$ 648.878

Constit. of the roots of *Acacia senegal*

4-Me ether

3-O-Caffeoyl-4-O-methylquinic acid

[374728-76-4]

$C_{17}H_{20}O_9$ 368.338

Constit. of *Phyllostachys edulis*. Powder. Mp 200 - 201°. $[\alpha]^{25}_{D}$ -47.2 (c, 0.2 in EtOH). Misnamed in ref.. λ_{max} 244 (sh); 296; 328 (MeOH)

5′-Methoxy, 3′-Me ether

3-O-Sinapoylquinic acid

[53905-81-0]

$C_{18}H_{22}O_{10}$ 398.364

Isol. from *Antitoxicum sibiricum* (preferred genus name *Vincetoxicum*). [α] D -35 (MeOH)

3′,4′-Quinone

Chlorogenoquinone

[24321-18-4]

$C_{16}H_{16}O_9$ 352.295

Unstable prod. from the oxidn. of 3-*O*-Caffeoylquinic acid by *o*-diphenyl oxidase. Forms coloured adducts with amino acids and peptides, implicated in food discolouration. No phys. props. reported

4-Epimer

4-Epichlorogenic acid. *1-Epineochlorogenic acid*

$C_{16}H_{18}O_9$ 354.311

Constit. of the fruit of *Sorbus domestica*. Amorph. powder. $[\alpha]^{25}_{D}$ -4.4 (c, 0.2 in MeOH)

4-Epimer, (1″→4)-lactone

3-O-Caffeoyl-epi-δ-quinide

[1036742-36-5]

$C_{16}H_{16}O_8$ 336.296

5-Epimer

3-O-Caffeoyl-muco-quinic acid

[214680-17-8]

$C_{16}H_{18}O_9$ 354.311

Constit. of *Asimina triloba*. Oviposition stimulant for the zebra swallowtail butterfly *Eurytides marcellus*. Powder. λ_{max} 225; 325 (MeOH)

5-Epimer, (1″→5)-lactone

3-O-Caffeoyl-muco-γ-quinide

[1036742-39-8]

$C_{16}H_{16}O_8$ 336.296

(Z)-form [15016-60-1]

Constit. of the fruit of *Solanum melongena*.

Aldrich Library of FT-IR Spectra, 1st edn., 1985, **2**, 288C (*ir*).
Aldrich Library of ¹³C and ¹H FT-NMR Spectra, 1992, **2**, 1235C (*nmr*).
Fischer, H.O.L. *et al.*, *Ber. Dtsch. Chem. Ges.*, 1932, **65**, 1037-1040 (*3-Caffeoylquinic acid*).
Corse, J. *et al.*, *Tetrahedron*, 1962, **18**, 1207-1210 (*3-Feruloylquinic acid*).
Waiss, A.C. *et al.*, *Chem. Ind. (London)*, 1964, 1984-1985 (*pmr*).
Pierpoint, W.S. *et al.*, *Biochem. J.*, 1966, **98**, 567-580 (*Chlorogenoquinone*).
Pierpoint, W.S. *et al.*, *Biochem. J.*, 1969, **112**, 609-616 (*Chlorogenoquinone*).
Shatokhina, R.K. *et al.*, *Chem. Nat. Compd. (Engl. Transl.)*, 1974, **10**, 525 (*3-O-Sinapoylquinic acid*).
Kelley, C.J. *et al.*, *J. Org. Chem.*, 1976, **41**, 449-455 (*cmr*).
Karrer, W. *Konstitution und Vorkommen der Organischen Pflanzenstoffe, 2nd edn., Birkhäuser Verlag*, 1976-1985, no. 990 (*occur*).
Nagels, L. *et al.*, *Phytochemistry*, 1976, **15**, 703-706 (*biosynth*).
Herrmann, K. *Fortschr. Chem. Org. Naturst.*, 1978, **35**, 73-132 (*rev*).
Nishizawa, M. *et al.*, *Chem. Pharm. Bull.*, 1987, **35**, 2133-2135 (*3-Sinapoylquinic acid*).
Corthout, J. *et al.*, *Phytochemistry*, 1992, **31**, 1979-1981 (*butyl ester*).
Stevenson, P.C. *et al.*, *J. Chem. Ecol.*, 1993, **19**, 2917-2933 (*Arachis paraguaiensis constit, activity*).
Ida, Y. *et al.*, *Phytochemistry*, 1994, **35**, 209-215 (*3-Feruloylquinic acid*).
Fuchs, C. *et al.*, *J. Mass Spectrom.*, 1996, **31**, 602-608 (*hplc, ms*).
Gentry, E.J. *et al.*, *J. Nat. Prod.*, 1998, **61**, 1187-1193 (*Hycandinic acid ester 1*).
Haribal, M. *et al.*, *Phytochemistry*, 1998, **49**, 103-108 (*muco isomer*).
De Almeida, A.P. *et al.*, *Phytother. Res.*, 1998, **12**, 562-567 (*activity*).
Jung, H.A. *et al.*, *Arch. Pharmacal Res.*, 1999, **22**, 213-218 (*Eriobotrya japonica constit, activity*).
Lin, L.-C. *et al.*, *J. Nat. Prod.*, 1999, **62**, 405-408 (*cmr*).
Pauli, G.F. *et al.*, *Magn. Reson. Chem.*, 1999, **37**, 827-836 (*pmr, cmr*).
Sefkow, M. *Eur. J. Org. Chem.*, 2001, 1137-1141 (*synth*).
Kweon, M.-H. *et al.*, *J. Agric. Food Chem.*, 2001, **49**, 4646-4655 (*1-Me ether, 4-Me ether*).
Whitaker, B.D. *et al.*, *J. Agric. Food Chem.*, 2003, **51**, 3448-3454 (*5-Acetyl-3-caffeoylquinic acid*).
Ma, J. *et al.*, *J. Nat. Prod.*, 2003, **66**, 983-986 (*4-Galloylchlorogenic acid*).
McNamara, C.E. *et al.*, *J. Nat. Prod.*, 2004, **67**, 1818-1822 (*GPFQ*).
Zhu, X.-D. *et al.*, *Helv. Chim. Acta*, 2005, **88**, 339-342 (*Me ester*).
Liu, J. *et al.*, *Phytochemistry*, 2007, **68**, 1775-1780 (*Erycibe obtusifolia ester*).
Frank, O. *et al.*, *J. Agric. Food Chem.*, 2008, **56**, 9581-9586 (*roasted coffee constits*).
Lee, E.J. *et al.*, *Food Chem.*, 2010, **120**, 134-139 (*Me ester*).
Song, S. *et al.*, *J. Nat. Prod.*, 2010, **73**, 177-184 (*Erycibe hainanensis constits*).
Ma, C. *et al.*, *J. Agric. Food Chem.*, 2011, **59**, 9645-9651 (*3-Caffeoyl-5-malonylquinic acid*).
Dahmani-Hamzaoui, N. *et al.*, *Helv. Chim. Acta*, 2012, **95**, 564-576 (*Atrtemisia herba-alba constit, pmr, cmr*).
Jain, R. *et al.*, *J. Pharm. Res.*, 2012, **5**, 4934-4935 (*Acacia senegal eicosyl ester*).
Mishra, D.P. *et al.*, *Trends Carbohydr. Res.*, 2012, **4**, 42-46 (*pentyl ester*).
Nakamura, S. *et al.*, *Phytochemistry*, 2013, **92**, 128-136 (*Prunus mume constits*).
Fan, L. *et al.*, *Planta Med.*, 2013, **79**, 1558-1564 (*4-syringoyl Me ester, activity*).
Yu, Y. *et al.*, *Acta Pharm. Sin. B*, 2015, **5**, 210-214 (*Lonicera 4′-glucoside*).
Forino, M. *et al.*, *Food Chem.*, 2015, **178**, 306-310 (*4-Epichlorogenic acid*).
Kirmizibekmez, H. *et al.*, *Helv. Chim. Acta*, 2016, **99**, 518-522 (*Castusic acid*).
Santana-Gálvez, J. *et al.*, *Molecules*, 2017, **22**, 358:1-21 (*rev, activity, props*).
Lewis, R.J. *Sax's Dangerous Properties of Industrial Materials, 8th edn., Van Nostrand Reinhold*, 1992, CHK175

4-*O*-Caffeoylquinic acid C6

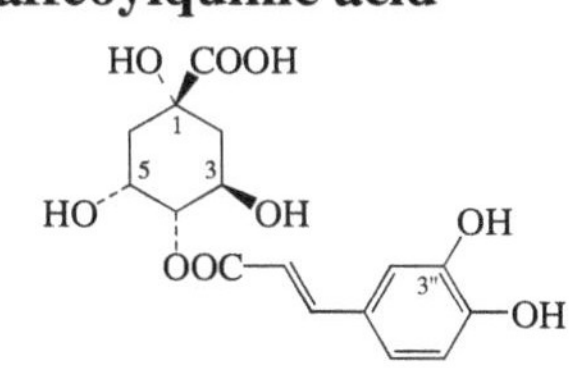

4-(3,4-Dihydroxycinnamoyl)quinic acid.

Cryptochlorogenic acid

[905-99-7]

$C_{16}H_{18}O_9$ 354.311

Constit. of *Coffea* sp., *Cynara scolymus* heads, *Cydonia oblonga*, *Helianthus* sp., *Ipomoea batatas* tubers and *Prunus domestica*. Mp 182 - 184°. [α] D -75 (-54) (H_2O). λ_{max} 225 (log ε 2.73); 247 (log ε 2.65); 280 (log ε 2.57); 328 (log ε 2.79) (MeOH)

3″-O-α-D-Glucopyranoside

[1401352-63-3]

$C_{22}H_{28}O_{14}$ 516.452

Isol. from the leaves of *Moringa oleifera*. Amorph. powder. [α] D +73.5 (c, 0.85 in MeOH)

4″-O-α-D-Glucopyranoside

[1401352-62-2]

$C_{22}H_{28}O_{14}$ 516.452

Isol. from the leaves of *Moringa oleifera*. Amorph. powder. [α]D +82.5 (c, 0.89 in MeOH)

4″-O-β-D-Glucopyranoside

$C_{22}H_{28}O_{14}$ 516.452

Constit. of the flower buds of *Lonicera japonica*. Prod. by a mangrove-derived *Penicillium citrinum*. Exhibits cytotoxic activity. Amorph. powder. $[\alpha]_D^{20}$ -78.6 (c, 0.1 in H_2O). λ_{max} 212 (sh); 228 (sh); 236 (sh); 291 (log ε 4.28); 317 (sh) (MeOH)

Me ester

[123372-74-7]

$C_{17}H_{20}O_9$ 368.338

Constit. of *Viburnum cylindricum* and *Viburnum dilatatum*. Amorph. powder. $[\alpha]_D^{25}$ -33.6 (c, 0.004 in MeOH). λ_{max} 216; 329 (MeOH)

Butyl ester

[1315322-56-5]

$C_{20}H_{26}O_9$ 410.419

Constit. of the flowers of *Hemerocallis fulva*. Amorph. powder. $[\alpha]_D^{25}$ -8.1 (c, 0.15 in MeOH). λ_{max} 220 (log ε 3.82); 233 (log ε 3.78); 243 (log ε 3.78); 295 (log ε 3.89); 329 (log ε 3.98) (MeOH)

5-Ac

5-O-Acetyl-4-O-caffeoylquinic acid

[548474-70-0]

$C_{18}H_{20}O_{10}$ 396.348

Constit. of the fruit of *Solanum melongena*

5-O-(4-Hydroxy-3,5-dimethoxybenzoyl)

4-O-Caffeoyl-5-O-syringoylquinic acid

[1207645-17-7]

$C_{25}H_{26}O_{13}$ 534.47

Constit. of *Erycibe hainanensis*. Powder. $[\alpha]_D^{20}$ -83.5 (c, 0.05 in MeOH). λ_{max} 248 (log ε 4.1); 290 (log ε 4.32); 330 (log ε 4.25) (MeOH)

5-O-(4-Hydroxy-3,5-dimethoxybenzoyl), Me ester

[1819972-59-2]

$C_{26}H_{28}O_{13}$ 548.497

Constit. of *Erycibe obtusifolia*. Exhibits weak *in vitro* inhibition of respiratory syncytial virus (RSV). Yellow oil. $[\alpha]_D^{17}$ -32.8 (c, 0.46 in MeOH). λ_{max} 218 (log ε 3.95); 290 (log ε 3.76); 333 (log ε 3.61) (MeOH)

5-O-(4-Hydroxy-3,5-dimethoxy-E-cinnamoyl), Me ester

[1820924-08-0]

$C_{28}H_{30}O_{13}$ 574.535

Constit. of *Erycibe obtusifolia*. Exhibits weak *in vitro* inhibition of respiratory syncytial virus (RSV). Yellow oil. $[\alpha]_D^{17}$ -44.5 (c, 0.2 in MeOH). λ_{max} 204 (log ε 4.01); 221 (log ε 4.07); 242 (log ε 4.03); 328 (log ε 4.26) (MeOH)

3-Malonyl

4-O-Caffeoyl-3-O-malonylquinic acid

[1325713-71-0]

$C_{19}H_{20}O_{12}$ 440.357

Constit. of the fruit of *Solanum melongena*. Amorph. powder. $[\alpha]_D^{20}$ -42.9 (c, 0.0001 in MeOH). λ_{max} 217 (log ε 3.59); 300 (sh); 329 (log ε 3.6) (MeOH)

3-Me ether

4-O-Caffeoyl-3-O-methylquinic acid

[1160614-90-3]

$C_{17}H_{20}O_9$ 368.338

Constit. of *Pteris multifida*. Amorph. powder. [α]D -43.1 (c, 0.1 in MeOH)

3″-Me ether, 5-O-(4-hydroxy-3,5-dimethoxybenzoyl), Me ester

[2068019-22-5]

$C_{27}H_{30}O_{13}$ 562.524

Constit. of the roots and stems of *Erycibe obtusfolia*. Yellow oil. $[\alpha]_D^{25}$ -68.3 (c, 0.13 in MeOH). λ_{max} 216 (log ε 4.45); 288 (log ε 4.15); 327 (log ε 4.10) (MeOH)

(1′→3)-Lactone

4-O-Caffeoyl-γ-quinide

[157048-84-5]

$C_{16}H_{16}O_8$ 336.296

Constit. of roasted coffee

(1′→3)-Lactone, 3″-Me ether

4-O-Feruloyl-γ-quinide

[157048-83-4]

$C_{17}H_{18}O_8$ 350.323

Constit. of roasted coffee.

Scarpati, M.L. *et al.*, *Tetrahedron Lett.*, 1963, **4**, 1147-1150 (*Cynara scolymus constit, Coffea constit*).
Machida, K. *et al.*, *Phytochemistry*, 1992, **31**, 3654-3656 (*Me ester*).
Fuchs, C. *et al.*, *J. Mass Spectrom.*, 1996, **31**, 602-698 (*hplc, ms*).
Nakatani, N. *et al.*, *J. Agric. Food Chem.*, 2000, **48**, 5512-5516 (*Prunus domestica constit*).
Sefkow, M. *et al.*, *Eur. J. Org. Chem.*, 2001, 2735-2742 (*synth*).
Whitaker, B.D. *et al.*, *J. Agric. Food Chem.*, 2003, **51**, 3448-3454 (*5-Acetyl-4-caffeoylquinic acid*).
Silva, B.M. *et al.*, *J. Agric. Food Chem.*, 2004, **52**, 4705-4712 (*Cydonia oblonga constit*).
Zhu, X.-D. *et al.*, *Helv. Chim. Acta*, 2005, **88**, 339-342 (*Me ester*).
Farah, A. *et al.*, *J. Agric. Food Chem.*, 2005, **53**, 1505-1513 (*coffee γ-quinides*).
Dini, I. *et al.*, *J. Agric. Food Chem.*, 2006, **54**, 8733-8737 (*Ipomoea batatas constit*).
Harinantenaina, L. *et al.*, *J. Nat. Med.*, 2008, **62**, 452-455 (*3-Me ether*).
Song, S. *et al.*, *J. Nat. Prod.*, 2010, **73**, 177-184 (*4-Caffeoyl-5-syringoylquinic acid*).
Lin, Y.-L. *et al.*, *J. Agric. Food Chem.*, 2011, **59**, 8789-8795 (*butyl ester*).
Ma, C. *et al.*, *J. Agric. Food Chem.*, 2011, **59**, 9645-9651 (*4-Caffeoyl-3-malonylquinic acid*).
Kashiwada, Y. *et al.*, *J. Nat. Med.*, 2012, **66**, 217-221 (*α-glucosides*).
Fan, L. *et al.*, *Planta Med.*, 2013, **79**, 1558-1564 (*Erycibe obtusifolia constits, activity*).
Yu, Y. *et al.*, *Acta Pharm. Sin. B*, 2015, **5**, 210-214 (*Lonicera 4″-glucoside*).
Liu, J. *et al.*, *Nat. Prod. Commun.*, 2015, **10**, 1203-1205 (*Penicillium 4″-glucoside, activity*).
Fan, L. *et al.*, *Phytochem. Lett.*, 2015, **14**, 185-189 (*Erycibe obtusifolia constit*)

5-*O*-Caffeoylquinic acid C7

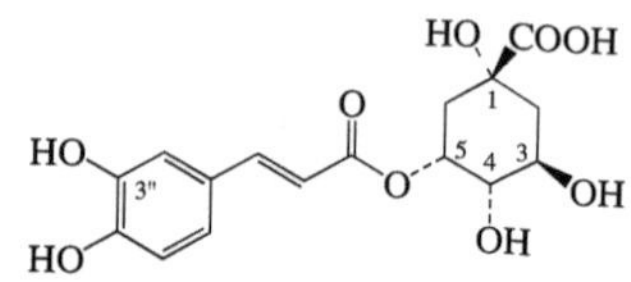

5-(3,4-Dihydroxycinnamoyl)quinic acid.
Neochlorogenic acid

[906-33-2]

[534-61-2]

$C_{16}H_{18}O_9$ 354.311

(E)-form [342811-68-1]

Constit. *Cynara scolymus*, *Cydonia oblonga*, *Aster scaber*, *Arachis paraguariensis*, *Prunus domestica*, *Hydrangea macrophylla* and many other plants. First isol. from peaches (*Prunus persica*).. Shows HIV-1 inhibitory activity; also antiviral and antihepatotoxic activities.Aldose reductase inhibitor. Inhibits development of *Spodoptera litura* larvae. Mp 177 - 179° dec.. [α]D -5.4 (EtOH aq.). [α]D -24.8 (c, 0.49 in MeOH). Forms a co-pigment with Myrtillin and Al^{3+} to give a supramolecular metal complex found in *Hydrangea macrophylla*

NH_4 salt

[1536184-42-5]

$C_{16}H_{21}NO_9$ 371.342

Constit. of *Hemerocallis fulva* var. *sempervirens*. Powder. λ_{max} 197 (log ε 4.59); 242 (log ε 4.18);

302 (log ε 4.19); 324 (log ε 4.22) (MeOH)

4-Hydroxyphenylethylamine salt (1:1)
[1536184-39-0]
$C_{24}H_{29}NO_{10}$ 491.493
Constit. of *Hemerocallis fulva* var. *sempervirens*. Powder. λ_{max} 202 (log ε 4.26); 221 (log ε 4.31); 244 (log ε 3.99); 300 (log ε 4.11) (no solvent reported)

4″-O-β-D-Glucopyranoside
$C_{22}H_{28}O_{14}$ 516.452
Constit. of the flower buds of *Lonicera japonica*. Amorph. powder. $[\alpha]^{20}_{D}$ -71.7 (c, 0.15 in H_2O). λ_{max} 214 (log ε 4.54); 237 (sh); 291 (log ε 4.54); 315 (log ε 4.48) (MeOH)

Me ester
[123410-65-1]
$C_{17}H_{20}O_9$ 368.338
Constit. of *Viburnum cylindricum* and *Viburnum dilatatum*. Amorph. powder. $[\alpha]^{25}_{D}$ +7.3 (c, 0.003 in MeOH). λ_{max} 213; 328 (MeOH)

Butyl ester
[409361-64-4]
$C_{20}H_{26}O_9$ 410.419
Constit. of *Erigeron breviscapus*. Mp 121 - 122°. $[\alpha]^{20}_{D}$ -27.8 (MeOH)

(1′→3)-Lactone
Neochlorogenic γ-lactone. 5-O-Caffeoyl-γ-quinide
[157089-01-5]
$C_{16}H_{16}O_8$ 336.296
Constit. of *Betula platyphylla*. Powder (EtOAc). Mp 236 - 238°. $[\alpha]^{21}_{D}$ +48.5 (10 min) → +11.4 (40 min) (c, 1.7 in EtOH). Data corresponds to synthetic material

3″-Me ether
5-O-Feruloylquinic acid
No phys. props. reported

3″-Me ether, Me ester
[154461-64-0]
$C_{18}H_{22}O_9$ 382.365
Constit. of *Phellodendron amurense*. Cryst.. Mp 173 - 175°. $[\alpha]_D$ +17.7 (c, 1 in MeOH)

3″-Me ether, butyl ester
Hycandinic acid ester 2
$C_{21}H_{28}O_9$ 424.446
Constit. of the commercial root sample of *Hydrastis canadensis*. Oil. $[\alpha]^{24}_{D}$ -29.6 (c, 0.25 in $CHCl_3$). λ_{max} 239 (log ε 4.82); 296 (log ε 4.8) (EtOH)

3″-Me ether, (1′→3)-lactone
5-O-Feruloyl-1,3-quinide
[157048-81-2]
$C_{17}H_{18}O_8$ 350.323

Tri-Me ether
Cryst.. Mp 83 - 85°

(Z)-form [32719-11-2]
Isol. from *Gleditsia triacanthes* and *Lonicera henryi*. Mp 217 - 218°. $[\alpha]_D$ +6.6 (c, 2 in MeOH). [acid] λ_{max} 316 (HCOOH/MeOH aq.).

Corse, J. *et al.*, *Nature (London)*, 1953, **172**, 771-772 (*Prunus persica constit*).
Scarpati, M.L. *et al.*, *Tetrahedron Lett.*, 1963, **4**, 1147-1150 (*struct*).
Ruveda, E.A. *et al.*, *Chem. Ind. (London)*, 1964, 239-240 (*lactone, synth*).
Waiss, A.C. *et al.*, *Chem. Ind. (London)*, 1964, 1984-1985 (*pmr*).
Haslam, E. *et al.*, *J. Chem. Soc.*, 1964, 2137-2146 (*synth*).
Weinges, K. *et al.*, *Justus Liebigs Ann. Chem.*, 1966, **691**, 181-185 (*Z-form*).
Machida, K. *et al.*, *Phytochemistry*, 1992, **31**, 3654-3656 (*Me ester*).
Stevenson, P.C. *et al.*, *J. Chem. Ecol.*, 1993, **19**, 2917-2933 (*activity*).
Ida, Y. *et al.*, *Phytochemistry*, 1994, **35**, 209-215 (*5-Feruloylquinic acid*).
Fuchs, C. *et al.*, *J. Mass Spectrom.*, 1996, **31**, 602-608 (*hplc, ms*).
Gentry, E.J. *et al.*, *J. Nat. Prod.*, 1998, **61**, 1187-1193 (*Hycandinic acid ester 2*).
Pauli, G.F. *et al.*, *Magn. Reson. Chem.*, 1999, **37**, 827-836 (*pmr, cmr*).
Kwon, H.C. *et al.*, *Chem. Pharm. Bull.*, 2000, **48**, 1796-1798 (*Aster scaber constit, activity*).
Nakatani, N. *et al.*, *J. Agric. Food Chem.*, 2000, **48**, 5512-5516 (*Prunus domestica constit*).
Wang, S.J. *et al.*, *Chin. Chem. Lett.*, 2001, **12**, 509-510 (*Neochlorogenic γ-lactone*).
Sefkow, M. *et al.*, *Eur. J. Org. Chem.*, 2001, 2735-2742 (*synth*).
Zhang, W.-D. *et al.*, *Yaoxue Xuebao*, 2001, **36**, 360-363 (*butyl ester*).
Silva, B.M. *et al.*, *J. Agric. Food Chem.*, 2004, **52**, 4705-4712 (*coffee constit*).
Zhu, X.-D. *et al.*, *Helv. Chim. Acta*, 2005, **88**, 339-342 (*Me ester*).
Farah, A. *et al.*, *J. Agric. Food Chem.*, 2005, **53**, 1505-1513 (*lactone, 3′-Me ether lactone, isol*).
Kondo, T. *et al.*, *Tetrahedron Lett.*, 2005, **46**, 6645-6649 (*Hydrangea supramolecular pigment*).
Toyama-Kato, Y. *et al.*, *Heterocycles*, 2007, **72**, 239-254 (*Hydrangea supramolecular pigment*).
Ogawa, Y. *et al.*, *Chem. Nat. Compd. (Engl. Transl.)*, 2013, **49**, 991-995 (*Hemerocallis fulva salts*).
Liu, J. *et al.*, *Chem. Pharm. Bull.*, 2013, **61**, 655-661 (*Hydrangea macrophylla constit, activity*).
Jaiswal, R. *et al.*, *Phytochemistry*, 2014, **108**, 252-263 (*Z-form, Lonicera henryi constit, activity*).
Yu, Y. *et al.*, *Acta Pharm. Sin. B*, 2015, **5**, 210-214 (*Lonicera 4″-glucoside*)

3-*O*-Caffeoylshikimic acid C8

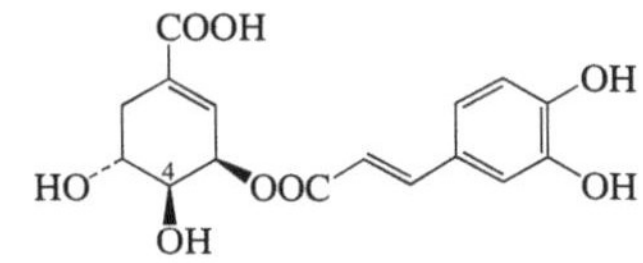

Neodattelic acid. ***Neodactylifric acid***
[6082-44-6]
$C_{16}H_{16}O_8$ 336.296
Constit. of *Phoenix dactylifera*. Cryst. (H_2O). Mp 224 - 225° dec.. $[\alpha]^{25}_{D}$ -124 (c, 1.16 in EtOH)

4-O-β-D-Glucopyranoside
[1781220-71-0]
$C_{22}H_{26}O_{13}$ 498.437
Constit. of the leaves of *Alnus sibirica*. Pale yellow powder. $[\alpha]_D$ -149.7 (c, 0.01 in MeOH).

Maier, V.P. *et al.*, *Biochem. Biophys. Res. Commun.*, 1964, **14**, 124-128 (*Neodactylifric acid*).
Harborne, J.B. *et al.*, *Phytochemistry*, 1974, **13**, 1557-1559 (*occur*).
Herrmann, K. *Fortschr. Chem. Org. Naturst.*, 1978, **35**, 73-132 (*rev*).
Kim, M.H. *et al.*, *Nat. Prod. Res.*, 2016, **30**, 206-213 (*4-glucoside*)

4-*O*-Caffeoylshikimic acid C9

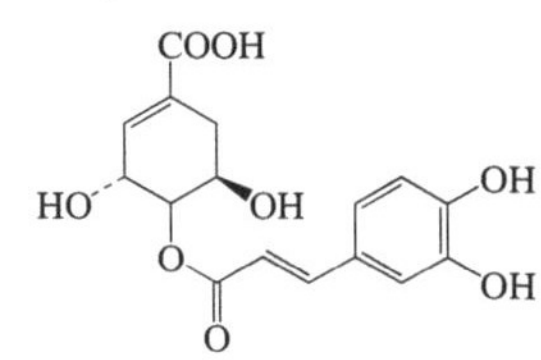

Isodactylifric acid. *Isodattelic acid*
[180842-65-3]
$C_{16}H_{16}O_8$ 336.296
Isol. from *Phoenix dactylifera*

3″-Deoxy
4-O-p-Coumaroylshikimic acid
[196496-50-1]
$C_{16}H_{16}O_7$ 320.297
Constit. of *Dicksonia antarctica*. Solid.

Maier, V.P. *et al.*, *Biochem. Biophys. Res. Commun.*, 1964, **14**, 124-128 (*Isodactylifric acid*).
Saito, T. *et al.*, *Biosci., Biotechnol., Biochem.*, 1997, **61**, 1397-1398 (*p-Coumaroylshikimic acid*).
Habib, H.M. *et al.*, *J. Sci. Food Agric.*, 2014, **94**, 1084-1089 (*Phoenix dactylifera constit, ms*)

5-*O*-Caffeoylshikimic acid C10

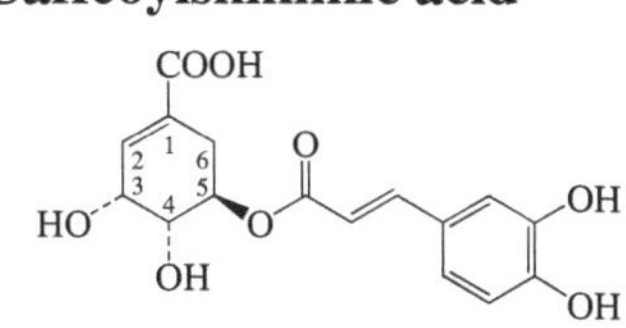

Dattelic acid. *Dactylifric acid*
[73263-62-4]

$C_{16}H_{16}O_8$ 336.296
Isol. from several ferns incl. *Blechnum orentale*, *Equisetum arvense*, *Phymatodes scolopendria* (preferred genus name *Dipteris*), *Pteridium aquilinum* and *Woodwardia* spp. Also from unripe dates *Phoenix dactylifera*. Mp 300° (dec. at 200°) (224-225°). [α] D -135 (c, 0.1 in MeOH).
Maier, V.P. *et al., Biochem. Biophys. Res. Commun.*, 1964, **14**, 124-128 (*Dactylifric acid*).
Herrmann, K. *Fortschr. Chem. Org. Naturst.*, 1978, **35**, 73-132 (*rev*).
Veit, M. *et al., Phytochemistry*, 1992, **31**, 3483-3485 (*Equisetum arvense constit*).
Mansouri, A. *et al., Food Chem.*, 2005, **89**, 411-420 (*Phoenix dactylifera constit*).
Habib, H.M. *et al., J. Sci. Food Agric.*, 2014, **94**, 1084-1089 (*Phoenix dactylifera constit, ms*)

Calebin A C11

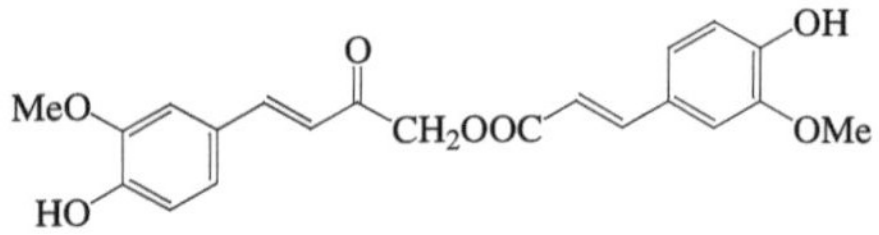

[336784-82-8]
$C_{21}H_{20}O_7$ 384.384
Constit. of the rhizomes of *Curcuma longa*. Protects PC12 cells from β-Amyloid insult. Pale yellow powder. Mp 138 - 139°. λ_{max} 219 (log ε 3.42); 240 (log ε 3.44); 339 (log ε 3.73) (MeOH).
Kim, D.S.H.L. *et al., Bioorg. Med. Chem. Lett.*, 2001, **11**, 2541-2543 (*synth*).
Park, S.-Y. *et al., J. Nat. Prod.*, 2002, **65**, 1227-1231 (*Calebin A*)

Calyxin I C12

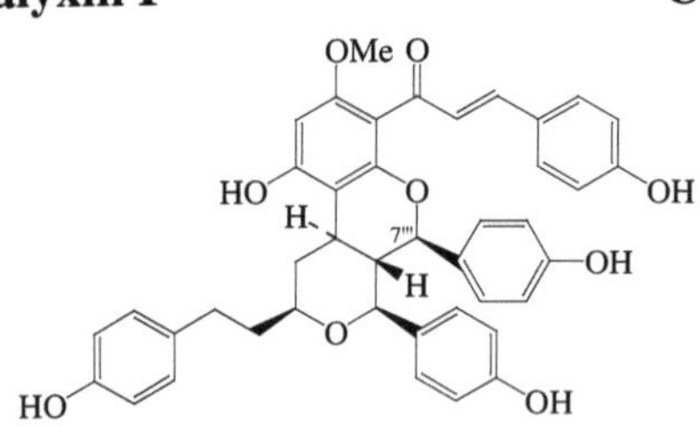

[252060-65-4]
$C_{42}H_{38}O_9$ 686.757
Constit. of the seeds of *Alpinia blepharocalyx*. Antiproliferative agent. Active against human HT-1080 fibrosarcoma and murine colon 26-L5 carcinoma. Amorph. light yellow solid. $[\alpha]^{25}_{D}$ -16.4 (c, 0.05 in MeOH)

7'''-Epimer

Epicalyxin I
[332877-80-2]
$C_{42}H_{38}O_9$ 686.757
Constit. of the seeds of *Alpinia blepharocalyx*. Antiproliferative agent. Active against human HT-1080 fibrosarcoma cells. Amorph. pale yellow solid. $[\alpha]^{25}_{D}$ +28.3 (c, 0.02 in MeOH).
Gewali, M.B. *et al., Org. Lett.*, 1999, **1**, 1733-1736 (*Calyxin I*).
Tezuka, Y. *et al., J. Nat. Prod.*, 2001, **64**, 208-213 (*Calyxin I, Epicalyxin I*)

Calyxin B C13

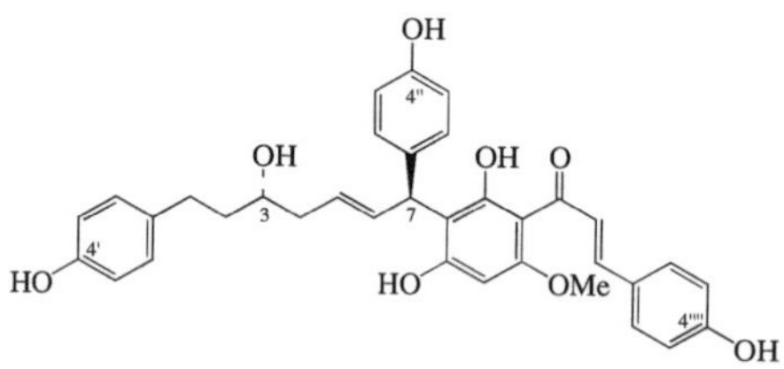

Absolute Configuration

[164991-53-1]
$C_{35}H_{34}O_8$ 582.649
Constit. of the seeds of *Alpinia blepharocalyx*. Amorph. light yellow solid. [α] D -24.7 (c, 0.4 in MeOH)

4'-Deoxy

Calyxin H
[202596-22-3]
$C_{35}H_{34}O_7$ 566.65
Constit. of the seeds of *Alpinia blepharocalyx*. Amorph. pale yellow solid. [α] D -4.7 (c, 0.2 in MeOH)

4',4'''-Dideoxy

Alpinnanin B
[872422-19-0]
$C_{35}H_{34}O_6$ 550.651
Constit. of the rhizomes of *Alpinia pinnanensis*. Pale yellow powder. $[\alpha]^{25}_{D}$ -39.3 (c, 0.28 in MeOH)

4',4'',4'''-Trideoxy

Katsumain A
[1221251-94-0]
$C_{35}H_{34}O_5$ 534.652
Constit. of the seeds of *Alpinia katsumadai*. Pale yellow solid. $[\alpha]^{22}_{D}$ -68.2 (c, 0.45 in MeOH)

3-Epimer, 4'''-deoxy

7-Epikatsumain C
[1337550-60-3]
$C_{35}H_{34}O_7$ 566.65
Constit. of the seeds of *Alpinia katsumadai*. Amorph. yellow solid. $[\alpha]^{25}_{D}$ -10.1 (c, 0.1 in MeOH). λ_{max} 216 (log ε 4.8); 288 (log ε 4.5); 347 (log ε 4.7) (MeOH)

3-Epimer, 4',4'''-dideoxy

Alpinnanin A
[872422-18-9]
$C_{35}H_{34}O_6$ 550.651
Constit. of the rhizomes of *Alpinia pinnanensis*. Pale yellow powder. $[\alpha]^{25}_{D}$ -33.3 (c, 0.21 in MeOH)

7-Epimer

Epicalyxin B
[164991-54-2]
$C_{35}H_{34}O_8$ 582.649
Constit. of the seeds of *Alpinia blepharocalyx*. Exhibits antiproliferative props. against human fibrosarcoma and murine colon cancer cells. Amorph. light yellow solid. [α] D +11.5 (c, 0.5 in MeOH)

7-Epimer, 4'-deoxy

Epicalyxin H
[202596-23-4]
$C_{35}H_{34}O_7$ 566.65
Constit. of the seeds of *Alpinia blepharocalyx* and *Alpinia katsumadai*. Exhibits moderate antiproliferative activity against human fibrosarcoma and murine colon carcinoma cells. Amorph. pale yellow solid. $[\alpha]^{25}_{D}$ +11.6 (c, 0.2 in MeOH)

7-Epimer, 4',4'''-dideoxy

ent-Alpinnanin A
[1337550-62-5]
$C_{35}H_{34}O_6$ 550.651
Constit. of the seeds of *Alpinia katsumadai*. Amorph. yellow solid. $[\alpha]^{25}_{D}$ +5.8 (c, 0.07 in MeOH). λ_{max} 217 (log ε 4.9); 289 (log ε 4.5); 351 (log ε 4.8) (MeOH)

7-Epimer, 4',4'',4'''-trideoxy

Katsumain B
[1221251-95-1]
$C_{35}H_{34}O_5$ 534.652
Constit. of the seeds of *Alpinia katsumadai*. Pale yellow solid. $[\alpha]^{22}_{D}$ +102 (c, 0.5 in MeOH)

3,7-Diepimer, 4'-deoxy

ent-Calyxin H
[1337550-63-6]
$C_{35}H_{34}O_7$ 566.65
Constit. of the seeds of *Alpinia katsumadai*. Amorph. yellow solid. $[\alpha]^{25}_{D}$ +22.5 (c, 0.1 in MeOH). λ_{max} 228 (log ε 4.7); 371 (log ε 4.8) (MeOH)

3,7-Diepimer, 4'''-deoxy

Katsumain C
[1337550-59-0]
$C_{35}H_{34}O_7$ 566.65
Constit. of the seeds of *Alpinia katsumadai*. Amorph. yellow solid. $[\alpha]^{25}_{D}$ +9.4 (c, 0.2 in MeOH). λ_{max}

217 (log ε 4.9); 288 (log ε 4.5); 348 (log ε 4.7) (MeOH)

3,7-Diepimer, 4′,4‴′-dideoxy

***ent*-Alpinnanin B**

[1337550-61-4]

$C_{35}H_{34}O_6$ 550.651

Constit. of the seeds of *Alpinia katsumadai*. Amorph. yellow solid. $[\alpha]^{25}_{D}$ +7 (c, 0.05 in MeOH). λ_{max} 217 (log ε 4.9); 289 (log ε 4.5); 350 (log ε 4.8) (MeOH).

Kadota, S. *et al.*, *Chem. Pharm. Bull.*, 1994, **42**, 2647-2649 (*Calyxin B, Epicalyxin B*).
Prasain, J.K. *et al.*, *Tetrahedron*, 1997, **53**, 7833-7842 (*Alpinia blepharocalyx constits, abs config*).
Prasain, J.K. *et al.*, *J. Nat. Prod.*, 1998, **61**, 212-216 (*Calyxin H, Epicalyxin H*).
Ali, M.S. *et al.*, *Biol. Pharm. Bull.*, 2001, **24**, 525-528 (*Epicalyxins B,H, activity*).
Giang, P.M. *et al.*, *Chem. Pharm. Bull.*, 2005, **53**, 1335-1337 (*Alpinnanins A,B*).
Li, Y.-Y. *et al.*, *Helv. Chim. Acta*, 2010, **93**, 382-388 (*Katsumains A,B*).
Nam, J.-W. *et al.*, *J. Nat. Prod.*, 2011, **74**, 2109-2115 (*Alpinia katsumadai constits*)

Calyxin C C14

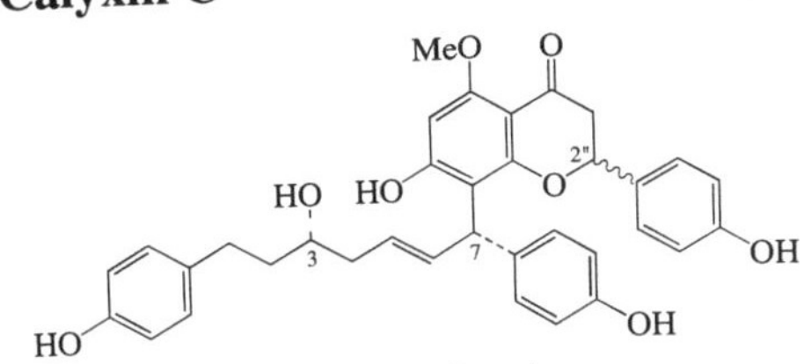

Absolute Configuration

[193816-82-9]

$C_{35}H_{34}O_8$ 582.649

Constit. of *Alpinia blepharocalyx*. Shows mod. antiproliferative activity against human and murine tumour cell lines. Inhibitor of NO prodn. in activated murine macrophages. Pale yellow amorph. solid. $[\alpha]_D$ -55.1 (c, 0.24 in MeOH)

2″-Epimer

Epicalyxin C

[193816-85-2]

$C_{35}H_{34}O_8$ 582.649

Constit. of *Alpinia blepharocalyx*. Shows mod. antiproliferative activity against human and murine tumour cell lines. Inhibitor of NO prodn. in activated murine macrophages. Pale yellow amorph. solid. $[\alpha]_D$ -38.9 (c, 0.8 in MeOH)

7-Epimer

Calyxin D

[193816-86-3]

$C_{35}H_{34}O_8$ 582.649

Constit. of *Alpinia blepharocalyx*. Shows mod. antiproliferative activity against human and murine tumour cell lines. Inhibitor of NO prodn. in activated murine macrophages. Pale yellow amorph. solid. $[\alpha]_D$ +43 (c, 0.4 in MeOH)

2″,7-Diepimer

Epicalyxin D

[193816-93-2]

$C_{35}H_{34}O_8$ 582.649

Constit. of *Alpinia blepharocalyx*. Shows mod. antiproliferative activity against human and murine tumour cell lines. Inhibitor of NO prodn. in activated murine macrophages. Pale yellow amorph. solid. $[\alpha]_D$ +26.6 (c, 0.45 in MeOH).

Prasain, J.K. *et al.*, *Tetrahedron*, 1997, **53**, 7833-7842 (*Calyxins C,D, Epicalyxins C,D, activity*).
Prasain, J.K. *et al.*, *Biol. Pharm. Bull.*, 1998, **21**, 371-374 (*Calyxins C,D, Epicalyxins C,D: activity*).
Ali, M.A. *et al.*, *Biol. Pharm. Bull.*, 2001, **24**, 525-528 (*antiproliferative activity*)

Calyxin E C15

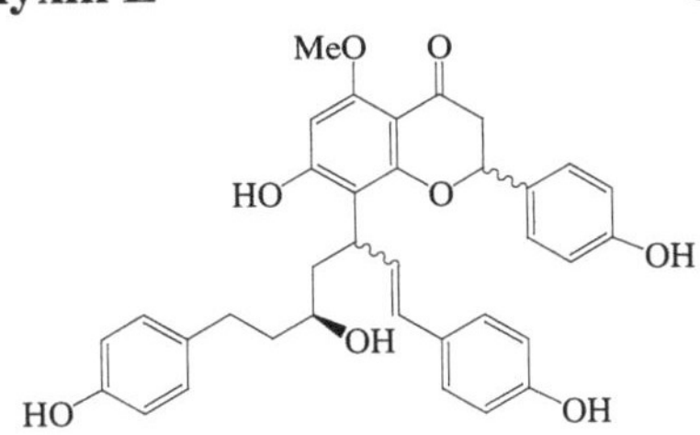

[205313-08-2]

$C_{35}H_{34}O_8$ 582.649

Constit. of *Alpinia blepharocalyx*. Inhibits NO prodn. in activated murine macrophages. Pale yellow solid. $[\alpha]^{25}_{D}$ +10.3 (c, 0.3 in MeOH).

Prasain, J.K. *et al.*, *Biol. Pharm. Bull.*, 1998, **21**, 371-374 (*NO prodn inhibitor*).
Prasain, J.K. *et al.*, *J. Chem. Res., Synop.*, 1998, 22-23 (*Calyxin E*)

Calyxin F C16

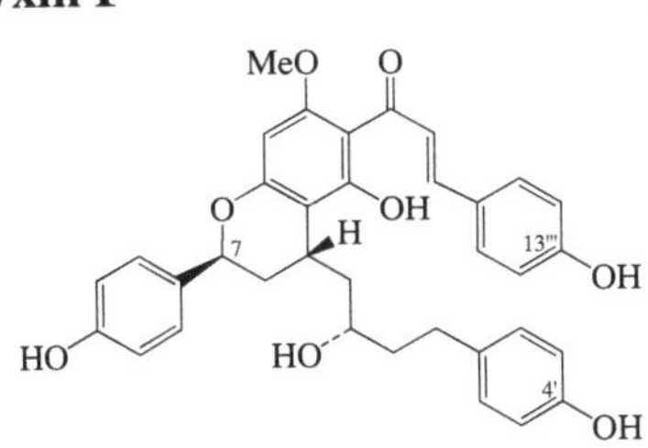

[205313-09-3]

$C_{35}H_{34}O_8$ 582.649

Struct. revised in 2006. Constit. of the seeds of *Alpinia blepharocalyx*. Pale yellow solid. $[\alpha]^{25}_{D}$ +5.7 (c, 0.3 in MeOH). $[\alpha]^{25}_{D}$ +16.3 (c, 0.18 in MeOH)

6S-Hydroxy

6-Hydroxycalyxin F

[205234-22-6]

$C_{35}H_{34}O_9$ 598.648

Constit. of *Alpinia blepharocalyx*. Pale yellow solid. $[\alpha]^{25}_{D}$ -9 (c, 0.4 in MeOH). Struct. requires revision

13‴-Deoxy

Katsumain F

[1486465-13-7]

$C_{35}H_{34}O_7$ 566.65

Constit. of *Alpinia katsumadai*. Amorph. yellow powder. $[\alpha]^{25}_{D}$ +15.91 (c, 0.2 in MeOH). λ_{max} 229 (log ε 4.7); 371 (log ε 4.8) (MeOH)

5,7-Diepimer

Epicalyxin F

[960618-53-5]

$C_{35}H_{34}O_8$ 582.649

Constit. of the seeds of *Alpinia blepharocalyx*. Cytotoxic. Active against human HT1080 sarcoma and murine 26-L5 carcinoma. The most active member of the group. Amorph. light yellow solid. $[\alpha]^{25}_{D}$ +103.1 (c, 0.05 in MeOH). $[\alpha]^{25}_{D}$ +13.2 (c, 0.2 in MeOH). Struct. finally confirmed in 2007

5,7-Diepimer, 4′-deoxy

Katsumain G

[1486465-14-8]

$C_{35}H_{34}O_7$ 566.65

Constit. of *Alpinia katsumadai*. Amorph. yellow powder. $[\alpha]^{25}_{D}$ +24.56 (c, 0.06 in MeOH). λ_{max} 229 (log ε 4.7); 372 (log ε 4.7) (MeOH)

5,7-Diepimer, 13‴-deoxy

Katsumain E

[1486465-12-6]

$C_{35}H_{34}O_7$ 566.65

Constit. of *Alpinia katsumadai*. Amorph. yellow powder. $[\alpha]^{25}_{D}$ +21.43 (c, 0.2 in MeOH). λ_{max} 228 (log ε 4.6); 371 (log ε 4.7) (MeOH).

Prasain, J.K. *et al.*, *J. Chem. Res., Synop.*, 1998, 22-23 (*Calyxin F, 6-hydroxycalyxin F, struct*).
Gewali, M.B. *et al.*, *Org. Lett.*, 1999, **1**, 1733-1736 (*Epicalyxin F, Calyxin F*).
Ali, M.S. *et al.*, *Biol. Pharm. Bull.*, 2001, **24**, 525-528 (*activity*).
Tezuka, Y. *et al.*, *J. Nat. Prod.*, 2001, **64**, 208-213 (*Epicalyxin F*).
Tian, X. *et al.*, *J. Org. Chem.*, 2006, **71**, 3176-3183 (*synth, struct*).

Tian, X. *et al.*, *Org. Lett.*, 2007, **9**, 4955-4958 (*Epicalyxin F, Calyxin F, revised struct, synth, abs config*).
Nam, J.-W. *et al.*, *Helv. Chim. Acta*, 2013, **96**, 1670-1680 (*Katsumains E,F,G*)

Calyxin G C17

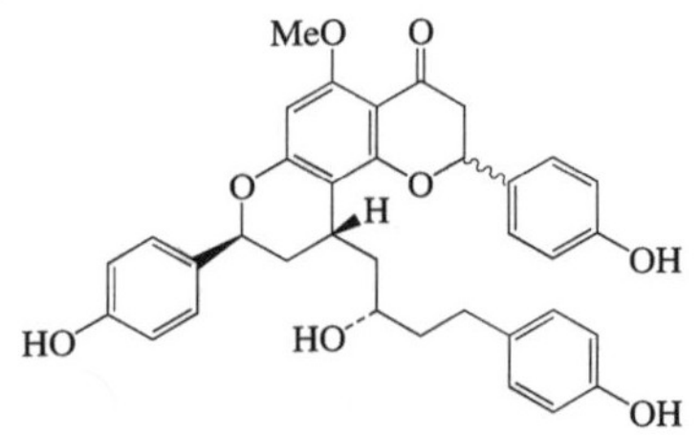

[205313-10-6]
$C_{35}H_{34}O_8$ 582.649
Struct. revised in 2006. Constit. of *Alpinia blepharocalyx*. Isol. as a mixt. with its epimer

Epimer

Epicalyxin G
[205313-11-7]
$C_{35}H_{34}O_8$ 582.649
Constit. of *Alpinia blepharocalyx*

Stereoisomer(?)

Calyxin K‡
$C_{35}H_{34}O_8$ 582.649
Constit. of the seeds of *Alpinia blepharocalyx*. Amorph. pale yellow solid. $[\alpha]_D^{25}$ +35.5 (c, 0.06 in MeOH). Struct. requires revision

Stereoisomer(?)

Epicalyxin K‡
$C_{35}H_{34}O_8$ 582.649
Constit. of the seeds of *Alpinia blepharocalyx*. Amorph. pale yellow solid. $[\alpha]_D^{25}$ -17 (c, 0.08 in MeOH). Struct. requires revision.

Prasain, J.K. *et al.*, *J. Chem. Res., Synop.*, 1998, 22-23 (*Calyxin G, Epicalyxin G, struct*).
Tezuka, Y. *et al.*, *J. Nat. Prod.*, 2001, **64**, 208-213 (*Alpinia blepharocalyx constits, struct*).
Tian, X. *et al.*, *J. Org. Chem.*, 2006, **71**, 3176-3183 (*Calyxin G, Epicalyxin G, struct, synth*)

Calyxin J C18

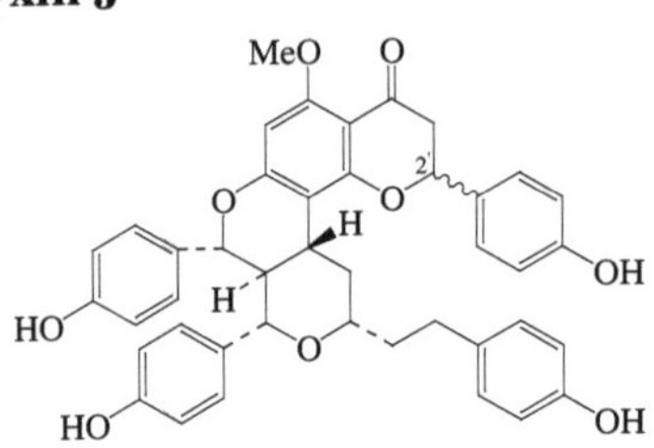

[332877-81-3]
$C_{42}H_{38}O_9$ 686.757
Constit. of the seeds of *Alpinia blepharocalyx*. Amorph. pale yellow solid. $[\alpha]_D^{25}$ +99.2 (c, 0.18 in MeOH)

2′-Epimer

Epicalyxin J
[332877-82-4]
$C_{42}H_{38}O_9$ 686.757
Constit. of the seeds of *Alpinia blepharocalyx*. Shows potent antiproliferative activity against H-1080 fibrosarcoma cells. Amorph. pale yellow solid.

Ali, M.S. *et al.*, *Biol. Pharm. Bull.*, 2001, **24**, 525-528 (*Epicalyxin J: antiproliferative activity*).
Tezuka, Y. *et al.*, *J. Nat. Prod.*, 2001, **64**, 208-213 (*Calyxin J, Epicalyxin J, struct, abs config, cytotox*)

Calyxin L C19

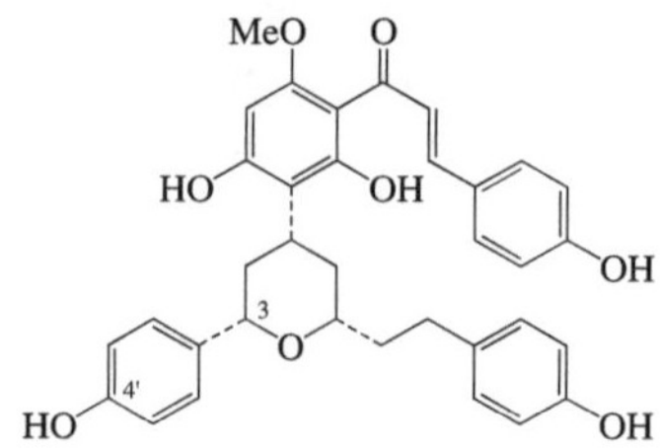

Absolute Configuration

[252060-62-1]
$C_{35}H_{34}O_8$ 582.649
Struct. reported in 2001 as Epicalyxin F. Struct. revised in 2006. Constit. of the seeds of *Alpinia blepharocalyx*. Amorph. pale yellow solid. $[\alpha]_D^{25}$ +77.1 (c, 0.05 in MeOH). $[\alpha]_D^{25}$ +103.1 (c, 0.05 in MeOH)

3-Epimer, 4′-deoxy

Katsumain D
[1486465-11-5]
$C_{35}H_{34}O_7$ 566.65
Constit. of *Alpinia katsumadai*. Yellow powder. $[\alpha]_D^{25}$ +12.33 (c, 1.0 in MeOH). λ_{max} 228 (log ε 4.6); 371 (log ε 4.7) (MeOH).

Tezuka, Y. *et al.*, *J. Nat. Prod.*, 2001, **64**, 208-213 (*Calyxin L*).
Tian, X. *et al.*, *J. Org. Chem.*, 2006, **71**, 3176-3183 (*synth, revised struct*).
Nam, J.-W. *et al.*, *Helv. Chim. Acta*, 2013, **96**, 1670-1680 (*Katsumain D*)

Calyxin M C20

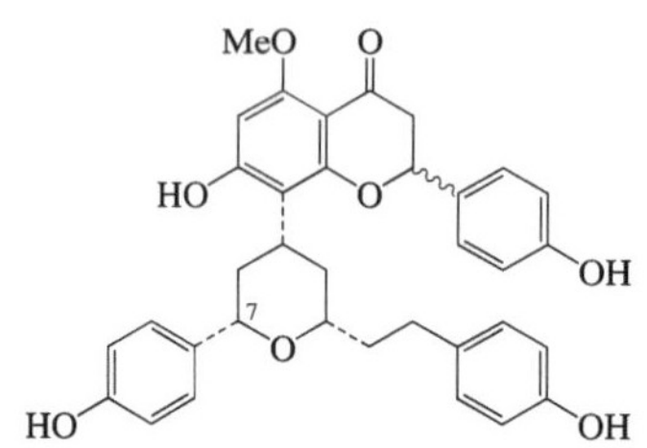

Absolute Configuration

Calyxin K‡
$C_{35}H_{34}O_8$ 582.649
Struct. revised in 2006. Constit. of the seeds of *Alpinia blepharocalyx*. Amorph. yellow solid. $[\alpha]_D^{25}$ +35.5 (c, 0.06 in MeOH). Obt. as a mixt. with its epimer. Confusing nomenclature in earlier lit.; this pair of epimeric structs. reported in 2001 as Calyxin K and Epicalyxin K

7-Epimer

Epicalyxin M. Epicalyxin K‡
$C_{35}H_{34}O_8$ 582.649
Constit. of the seeds of *Alpinia blepharocalyx*. $[\alpha]_D$ -17 (c, 0.085 in MeOH). Obt. as a mixt. with its epimer.

Tezuka, Y. *et al.*, *J. Nat. Prod.*, 2001, **64**, 208-213 (*Calyxin K, Epicalyxin K*).
Tian, X. *et al.*, *J. Org. Chem.*, 2006, **71**, 3176-3183 (*Calyxin M, Epicalyxin M, struct, synth*)

Calyxin W C21

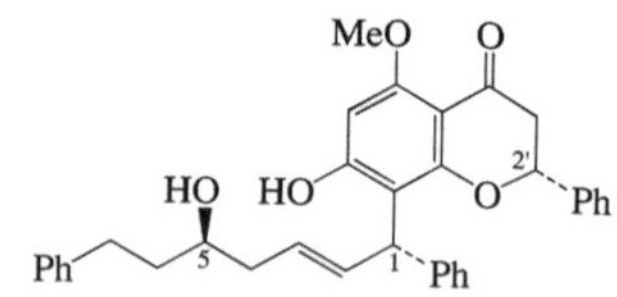

Absolute Configuration

[2267995-21-9]
$C_{35}H_{34}O_5$ 534.652
Constit. of the seeds of *Alpinia katsumadai*. Exhibits moderate antiproliferative activity. Amorph. pale yellow powder. $[\alpha]_D^{25}$ -41.4 (c, 0.1 in MeOH). λ_{max} 289 (log ε 4.1) (MeOH)

2′-Epimer

ent-Calyxin T
[2267995-19-5]
$C_{35}H_{34}O_5$ 534.652
Constit. of the seeds of *Alpinia katsumadai*. Exhibits moderate antiproliferative activity. Amorph. pale yellow powder. $[\alpha]_D^{25}$ -91.1 (c, 0.11 in MeOH). λ_{max} 206 (log ε 4.62); 289 (log ε 4.16) (MeOH)

5-Epimer

ent-Calyxin U
$C_{35}H_{34}O_5$ 534.652
Constit. of the seeds of *Alpinia katsumadai*. Exhibits moderate antiproliferative activity. Amorph. pale yellow powder. $[\alpha]_D^{25}$ -26.9 (c, 0.11 in MeOH). λ_{max} 207 (log ε 4.60); 289 (log ε 4.18) (MeOH)

1,2′-Diepimer

Calyxin U
[2267995-20-8]
$C_{35}H_{34}O_5$ 534.652

Constit. of the seeds of *Alpinia katsumadai*. Exhibits moderate antiproliferative activity. Amorph. pale yellow powder. $[\alpha]_D^{25}$ +42.3 (c, 0.11 in MeOH). λ_{max} 207 (log ε 4.57); 289 (log ε 4.17) (MeOH)

1,5-Diepimer

Calyxin T

[2267995-18-4]

$C_{35}H_{34}O_5$ 534.652

Constit. of the seeds of *Alpinia katsumadai*. Exhibits strong antiproliferative activity. Needles (MeOH). Mp 262 - 264°. $[\alpha]_D^{25}$ +165.5 (c, 0.11 in MeOH). λ_{max} 207 (log ε 4.43); 289 (log ε 3.99) (MeOH)

2′,5-Diepimer

Calyxin V

[2267995-22-0]

$C_{35}H_{34}O_5$ 534.652

Constit. of the seeds of *Alpinia katsumadai*. Exhibits moderate antiproliferative activity. Amorph. pale yellow powder. $[\alpha]_D^{25}$ -116.2 (c, 0.11 in MeOH). λ_{max} 287 (log ε 4.18) (MeOH).

Wang, X.-B. *et al.*, *Fitoterapia*, 2017, **119**, 121-129 (*Calyxins T-W, ent-Calyxins T,U, cryst struct*)

Canangafruticoside A cyclolignan diester C22

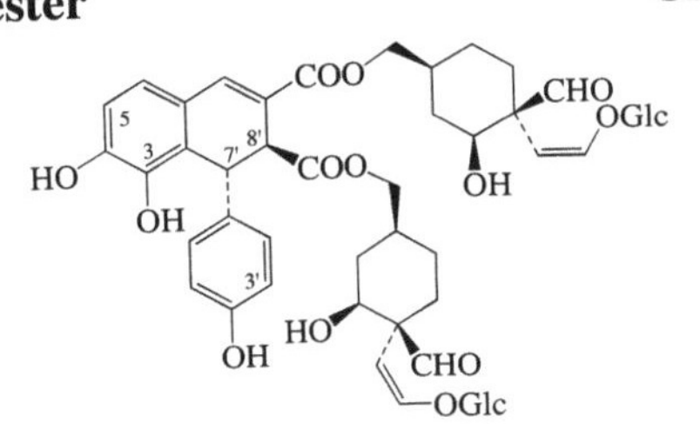

Relative Configuration

[1245700-99-5]

$C_{50}H_{62}O_{23}$ 1031.023

Lignan numbering shown. Constit. of *Cananga odorata* var. *fruticosa*. Amorph. powder. $[\alpha]_D^{25}$ +78.3 (c, 1.13 in MeOH)

3-Deoxy, 3′-hydroxy, 7′,8′-didehydro

[2252179-83-0]

$C_{50}H_{60}O_{23}$ 1029.007

Constit. of the leaves of *Cananga odorata* var. *odorata*. Amorph. powder. $[\alpha]_D^{25}$ +12.8 (c, 1 in MeOH). λ_{max} 251 (log ε 4.16); 286 (log ε 3.95); 309 (log ε 3.90); 327 (log ε 3.83); 350 (log ε 3.41) (MeOH)

3-Deoxy, 5-hydroxy, 7′,8′-didehydro

[2252179-82-9]

$C_{50}H_{60}O_{23}$ 1029.007

Constit. of the leaves of *Cananga odorata* var. *odorata*. Amorph. powder. $[\alpha]_D^{25}$ +17.5 (c, 0.7 in MeOH). λ_{max} 227 (log ε 4.15); 255 (log ε 4.26); 286 (log ε 3.93); 311 (log ε 3.84); 330 (log ε 3.79) (MeOH).

Nagashima, J. *et al.*, *Phytochemistry*, 2010, **71**, 1564-1572 (*Cananga odorata fruticosa constit*).
Katsui, H. *et al.*, *Chem. Pharm. Bull.*, 2017, **65**, 97-101 (*Cananga odorata odorata constits*)

Canangafruticoside E dimer C23

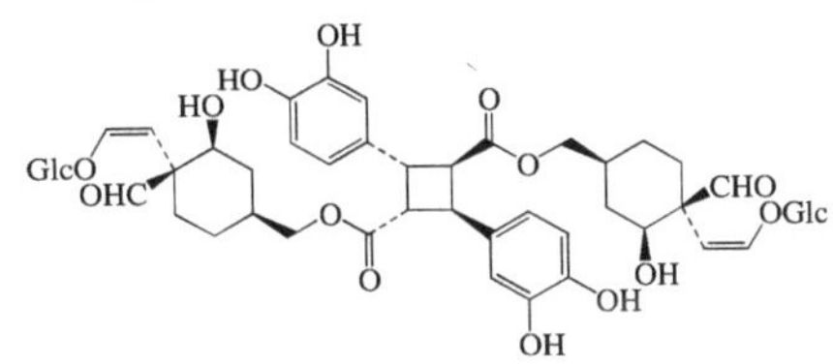

Relative Configuration

[2252179-84-1]

$C_{50}H_{64}O_{24}$ 1049.038

Constit. of the leaves of *Cananga odorata* var. *odorata*. Amorph. powder. $[\alpha]_D^{25}$ +13.5 (c, 0.8 in MeOH). λ_{max} 227 (log ε 4.02); 253 (log ε 3.78); 286 (log ε 3.76) (MeOH).

Katsui, H. *et al.*, *Chem. Pharm. Bull.*, 2017, **65**, 97-101 (*Cananga odorata constit*)

Canangalignan I C24

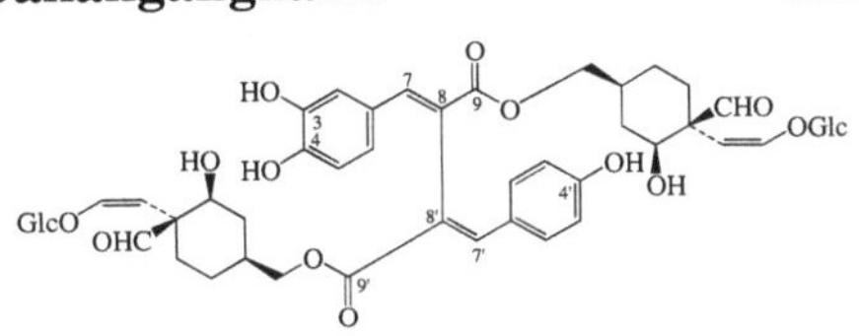

Absolute Configuration

[1579211-70-3]

$C_{50}H_{62}O_{23}$ 1031.023

Constit. of flower buds of *Cananga odorata*. Amorph. powder. $[\alpha]_D^{25}$ +25 (c, 0.6 in MeOH). λ_{max} 200 (log ε 4.20); 314 (log ε 4.22) (MeOH).

Matsumoto, T. *et al.*, *J. Nat. Prod.*, 2014, **77**, 990-999 (*Canangalignan I*)

Canangalignan II C25

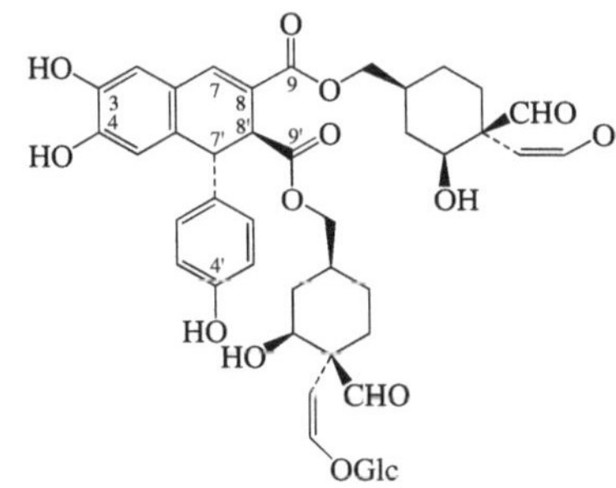

Absolute Configuration

[1579211-72-5]

$C_{50}H_{62}O_{23}$ 1031.023

Constit. of flower buds of *Cananga odorata*. Amorph. powder. $[\alpha]_D^{25}$ +33 (c, 0.5 in MeOH). λ_{max} 226 (log ε 4.20); 331 (log ε 4.33) (MeOH).

Matsumoto, T. *et al.*, *J. Nat. Prod.*, 2014, **77**, 990-999 (*Canangalignan II, abs config*)

Candinol B C26

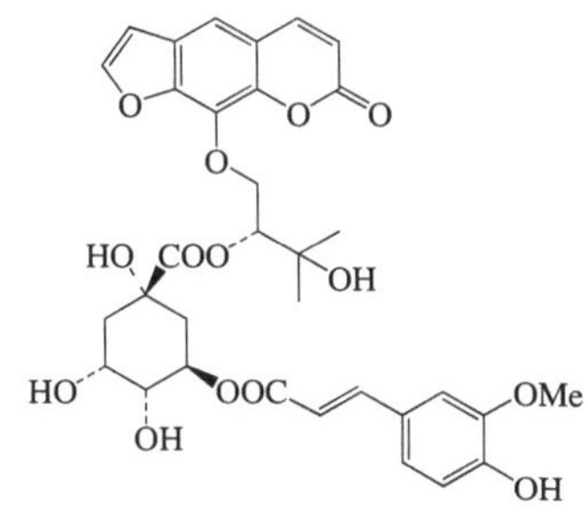

[1301167-51-0]

$C_{33}H_{34}O_{14}$ 654.621

Ester of Heraclenol with Me ether of 3-*O*-Caffeoylquinic acid, C5. Constit. of the roots of *Heracleum candicans*. Viscous yellowish oil. $[\alpha]_D$ +18.9 (c, 0.44 in MeOH). λ_{max} 200 (log ε 4.41); 215 (log ε 4.39); 244 (log ε 4.3); 300 (log ε 4.18); 323 (sh) (log ε 4.16) (MeOH).

Inoue, A. *et al.*, *J. Nat. Med.*, 2010, **64**, 175-181 (*Candinol B*)

Canellin C C27

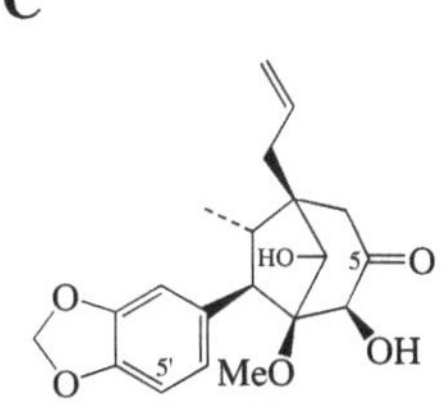

Relative Configuration

3,7′-Cyclo-1,2,3,6-tetrahydro-2,4-dihydroxy-3-methoxy-3′,4′-methylenedioxy-1,8′-lign-8-en-5(4H)-one

[54835-72-2]

$C_{20}H_{24}O_6$ 360.406
Lignan numbering shown. Constit. of *Licaria canella*, *Ocotea catharinensis* and *Endlicheria* spp.. Inhibitor of DNA polymerase β-lyase. Oil

5′-Methoxy

3,7′-Cyclo-1,2,3,6-tetrahydro-2,4-dihydroxy-3,5′-dimethoxy-3′,4′-methylenedioxy-1,8′-lign-8-en-5(4H)-one. ***5′-Methoxycanellin C***
[65703-35-7]
$C_{21}H_{26}O_7$ 390.432
Constit. of *Aniba williamsii* and *Ocotea catharinensis*. Oil. λ_{max} 237 (sh); 274 (ε 2050) (MeOH)

5-Deoxo, 5α-methoxy

3,7′-Cyclo-1,2,3,4,5,6-hexahydro-3,5-dimethoxy-3′,4′-methylene-dioxy-1,8′-lign-8-ene-2,4-diol.3,7′-Cy-clo-1,2,3,4,5,6-hexahydro-2,4-dihy-droxy-3,5-dimethoxy-3′,4′-methy-lenedioxy-1,8′-lign-8-ene. ***Canellin A***
[54835-74-4]
$C_{21}H_{28}O_6$ 376.449
Constit. of *Licaria canella*, *Ocotea aciphylla* and *Endlicheria* spp.. Inhibitor of DNA polymerase β-lyase. Cryst. (C_6H_6). Mp 144.5 - 145°.

Giesbrecht, A.M. *et al.*, *Phytochemistry*, 1974, **13**, 2285-2293 (*Canellins A,C*).
De Alvarenga, M.A. *et al.*, *Phytochemistry*, 1977, **16**, 1801-1804 (*5′-Methoxycanellin C*).
Haraguchi, M. *et al.*, *Phytochemistry*, 1983, **22**, 561-564 (*Ocotea catharinensis constits*).
Rodrigues, M. de M. *et al.*, *Phytochemistry*, 1984, **23**, 667-669 (*Canellin A, cryst struct*).
Chaturvedula, V.S.P. *et al.*, *J. Nat. Prod.*, 2004, **67**, 964-967 (*Endlicheria constits, activity*)

Cannabisin E C28

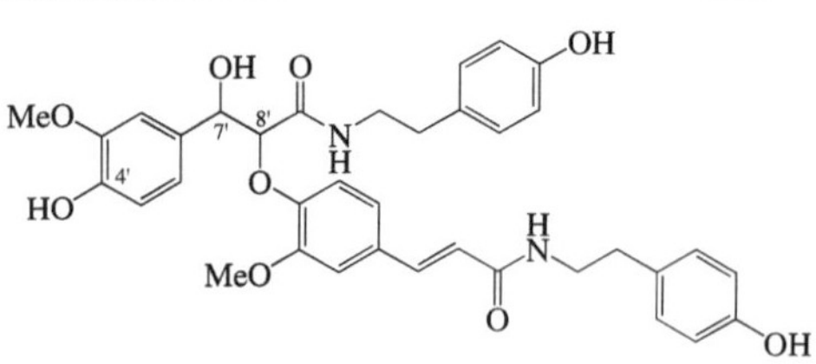

[163136-16-1]
$C_{36}H_{38}N_2O_9$ 642.705
Constit. of *Alocasia cucullata*, *Cannabis sativa*, *Datura metel*, *Hyoscyamus niger*, and *Nicandra physaloides*. Germination inhibitor of *Allium fistulosum*. Amorph. powder. λ_{max} 226 (log ε 4.16); 283 (log ε 4.43); 313 (log ε 4.31) (MeOH)

7′-Deoxy, 7′,8′-didehydro(Z-)

Cannabisin F
[163136-19-4]
$C_{36}H_{36}N_2O_8$ 624.69
Constit of *Alocasia cucullata*, *Cannabis sativa* and *Mitrephora thorelli*. Amorph. powder. λ_{max} 287 (log ε 4.55); 320 (log ε 4.57) (MeOH)

7′-Deoxy, 7′,8′-didehydro(Z-), 3′-O-de-Me, 4′-Me ether
[1374216-67-7]
$C_{36}H_{36}N_2O_8$ 624.69
Constit. of *Peperomia tetraphylla*. Amorph. powder. λ_{max} 280 (log ε 4.50); 309 (log ε 4.55) (MeOH)

Didemethoxy, 7′-deoxy, 7′,8′-didehy-dro(Z-)

Melongenamide C
[1676050-26-2]
$C_{34}H_{32}N_2O_6$ 564.638
Constit. of the roots of *Solanum melongena*. NO prodn. inhibitor. Amorph. powder. λ_{max} 218 (sh); 287; 310 (sh) (MeOH).

Sakakibara, I. *et al.*, *Phytochemistry*, 1995, **38**, 1003-1007 (*Cannabisins E,F*).
Li, Y.-Z. *et al.*, *Chem. Biodiversity*, 2012, **9**, 769-776 (*Peperomia tetraphylla constit*).
Zhang, W.-N. *et al.*, *J. Agric. Food Chem.*, 2012, **60**, 1682-1687 (*Hyoscyamus niger constit, activity*).
Xia, Y.-M. *et al.*, *Chem. Pap.*, 2014, **68**, 384-391 (*Cannabisin F, synth*).
Sun, J. *et al.*, *Fitoterapia*, 2014, **98**, 110-116 (*Melongenamide C, activity*)

Cannabisin K C29

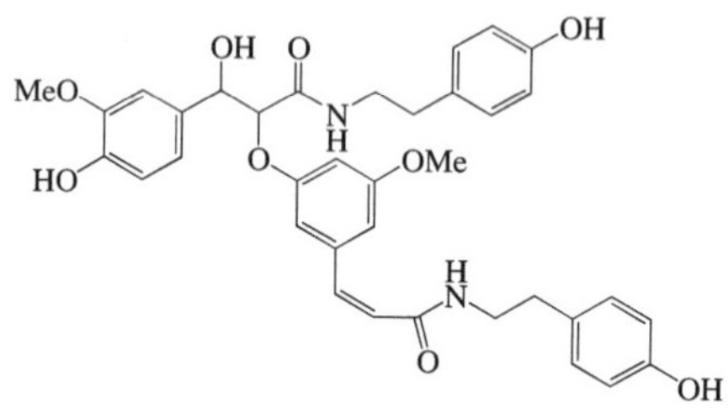

[1360434-15-6]
$C_{36}H_{38}N_2O_9$ 642.705
Constit. of the seeds of *Hyoscyamus niger*. Amorph. powder. $[\alpha]^{25}_{D}$ +7.3 (c, 0.15 in MeOH). λ_{max} 253 (log ε 4.4); 278 (log ε 4.0); 310 (log ε 3.8) (MeOH).

Zhang, W.-N. *et al.*, *J. Agric. Food Chem.*, 2012, **60**, 1682-1687 (*Cannabisin K*)

Cannabisin L C30

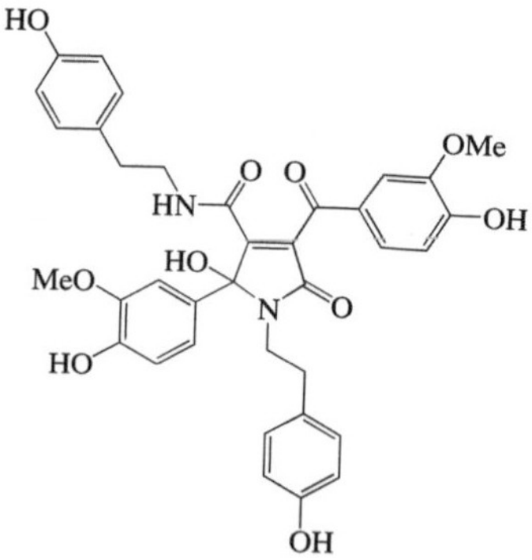

[1360434-16-7]
$C_{36}H_{34}N_2O_{10}$ 654.672
Constit. of *Alocasia cucullata* and *Hyoscyamus niger*. Amorph. yellow powder. $[\alpha]^{25}_{D}$ +13.5 (c, 0.08 in MeOH). λ_{max} 224 (log ε 4.0); 280 (log ε 3.9); 311 (log ε 3.8) (MeOH).

Zhang, W.-N. *et al.*, *J. Agric. Food Chem.*, 2012, **60**, 1682-1687 (*Cannabisin L*).
Zhang, N. *et al.*, *Chem. Nat. Compd. (Engl. Transl.)*, 2016, **52**, 963-965 (*Alocasia cucullata constit*)

Capitulactone A C31

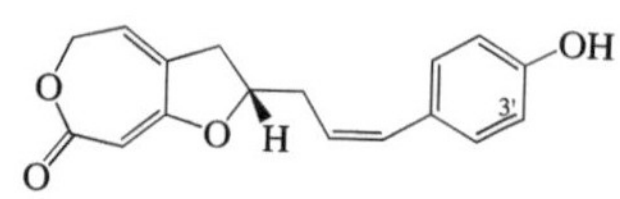

Absolute Configuration

3,5-Dihydro-2-[3-(4-hydroxyphenyl)-2-propenyl]furo[2,3-d]oxepin-7(2H)-one, CAS
[2345640-84-6]
$C_{17}H_{16}O_4$ 284.311
Constit. of the roots of *Curculigo capitulata*. Amorph. powder. $[\alpha]^{20}_{D}$ -10.7 (c, 0.14 in MeOH)

4′-Me ether

Capitulactone B
[2345639-85-0]
$C_{18}H_{18}O_4$ 298.338
Constit. of the roots of *Curculigo capitulata*. Amorph. powder. $[\alpha]^{20}_{D}$ -13.6 (c, 0.14 in MeOH)

3′-Hydroxy, 4′-Me ether

Capitulactone C
[2345668-51-9]
$C_{18}H_{18}O_5$ 314.337
Constit. of the roots of *Curculigo capitulata*. Amorph. powder. $[\alpha]^{20}_{D}$ -8.5 (c, 0.15 in MeOH).

Li, S. *et al.*, *J. Org. Chem.*, 2019, **84**, 5195-5202 (*Capitulactones A,B,C, synth, cd*)

Capituloside B C32

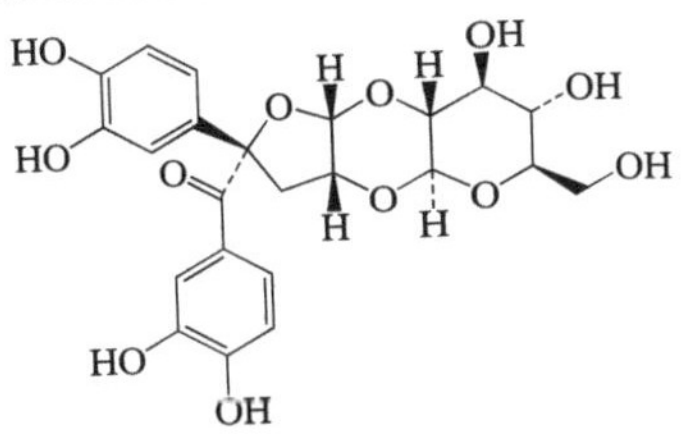

Absolute Configuration

[1282498-98-9]
$C_{23}H_{24}O_{12}$ 492.433
Norlignan glycoside related to Curcapicycloside. Constit. of the rhizomes of *Curculigo capitulata*. Powder. $[\alpha]_D^{21}$ -82.1 (c, 0.12 in MeOH). λ_{max} 204 (log ε 4.35); 232 (log ε 3.97); 284 (log ε 3.79); 314 (log ε 3.69) (MeOH).

Di, L. *et al.*, *Bull. Korean Chem. Soc.*, 2010, **31**, 2999-3002 (*Capituloside B*)

6-Carboxy-2,2′-cyclo-8,8′-lignan-3,3′,4,4′,5,5′,7′-heptol C33

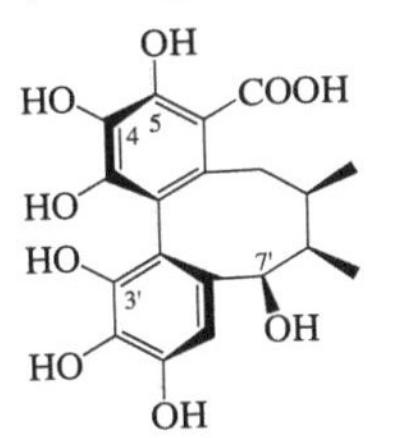

6-Carboxy-3,3′,4,4′,5,5′,7′-heptahydroxy-2,2′-cyclolignan
$C_{19}H_{20}O_9$ 392.36
(7′R,8R,8′R,S_{axial})-form
4′,5′-Methylene, 3,3′,4-tri-Me ether
6-Carboxy-2,2′-cyclo-3,3′,4-trimethoxy-4′,5′-methylenedioxy-8,8′-lignan-5,7′-diol. 6-Carboxy-5,7′-dihydroxy-3,3′,4-trimethoxy-4′,5′-methylenedioxy-2,2′-cyclolignan. ***Neglignan B***
[1441710-07-1]
$C_{23}H_{26}O_9$ 446.452
Isol. from the stems of *Schisandra neglecta*. Displays anti-HIV-1 activity. Yellow gum. $[\alpha]_D^{24}$ +23.8 (c, 0.2 in MeOH). λ_{max} 210 (log ε 4.58); 250 (log ε 3.64); 320 (log ε 0.68) (MeOH)
4′,5′-Methylene, 3,3′,4,5-tetra-Me ether
6-Carboxy-2,2′-cyclo-3,3′,4,5-tetramethoxy-4′,5′-methylenedioxy-8,8′-lignan-7′-ol. 6-Carboxy-7′-hydroxy-3,3′,4,5-tetraethoxy-4′,5′-methylenedioxy-2,2′-cyclolignan. ***Neglignan A***
[1441710-00-4]
$C_{24}H_{28}O_9$ 460.479
Isol. from the stems of *Schisandra neglecta*. Yellow gum. $[\alpha]_D^{25}$ +15.6 (c, 0.2 in MeOH). λ_{max} 210 (log ε 4.62); 248 (log ε 3.72); 320 (log ε 0.64) (MeOH).

Gao, X.-M. *et al.*, *J. Nat. Prod.*, 2013, **76**, 1052-1057 (*Neglignans A,B*)

3-(2-Carboxy-3-hydroxyphenyl)-2-propenoic acid C34

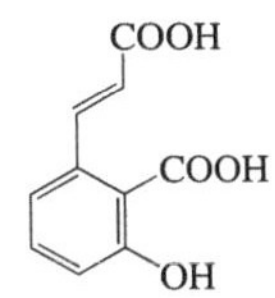

2-(2-Carboxyethenyl)-6-hydroxybenzoic acid, CAS. *2-Carboxy-3-hydroxycinnamic acid*
$C_{10}H_8O_5$ 208.169
(E)-form [73274-92-7]
Prod. by *Mycosphaerella fijiensis* IMI 105378.

Stierle, A.A. *et al.*, *Experientia*, 1991, **47**, 853-859 (*Mycosphaerella fijiensis constit*)

Carpinontriol B C35

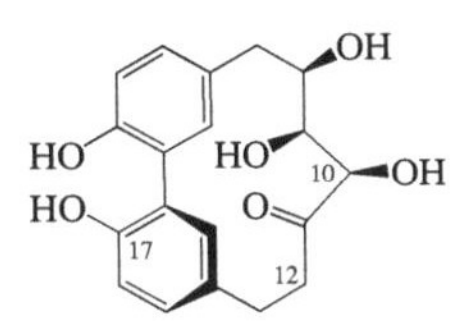

Absolute Configuration

[473451-73-9]
$C_{19}H_{20}O_6$ 344.363
Constit. of *Carpinus cordata*, *Carpinus turczaninowii* and *Corylus avellana*. Antioxidant. Needles. Mp 232 - 235°. $[\alpha]_D^{28}$ +65.9 (c, 0.73 in MeOH)
17-O-β-D-Glucopyranoside
Giffonin T
[2098225-04-6]
$C_{25}H_{30}O_{11}$ 506.504
Constit. of the green leafy covers of *Corylus avellana*. Amorph. solid. $[\alpha]_D^{25}$ -6 (c, 0.1 in MeOH)
10-Deoxy, 11-alcohol
Ostryopsitriol
[1441932-18-8]
$C_{19}H_{22}O_5$ 330.380
Constit. of *Ostryopsis nobilis*. Amorph. powder. $[\alpha]_D^{20}$ -34 (c, 0.18 in MeOH). Stereochem. unknown
10-Deoxy, 12ξ-hydroxy
Carpinontriol A*. 9-Dehydrobetulatetraol A*
[473451-72-8]
$C_{19}H_{20}O_6$ 344.363
Constit. of *Carpinus cordata*, *Carpinus turczaninowii* and *Betula pendula*. Antioxidant. Needles or amorph. powder. Mp 157 - 159°. $[\alpha]_D^{28}$ +12.7 (c, 0.6 in MeOH). Abs. stereochem. not studied
10-Deoxy, 12ξ-hydroxy, 11-alcohol
Tricyclo[12.3.1.1^{2,6}]nonadeca-1(18),2,4,6(19),14,16-hexaene-3,8,9,11,12,17-hexol.
Betulatetraol
[148335-60-8]
$C_{19}H_{22}O_6$ 346.379
Constit. of *Alnus japonica*, *Betula platyphylla* var. *latifolia* and *Ostryopsis nobilis*. Amorph. powder. $[\alpha]_D^{19}$ -8.6 (c, 1.1 in MeOH). Stereochem. unknown.

Lee, M.-W. *et al.*, *Arch. Pharmacal Res.*, 1992, **15**, 211-214 (*Betulatetraol*).
Lee, J.S. *et al.*, *J. Nat. Prod.*, 2002, **65**, 1367-1370 (*Carpinontriols A,B*).
Zhang, Y.-X. *et al.*, *Chin. Chem. Lett.*, 2013, **24**, 512-514 (*Ostryopsitriol*).
Ko, H.N. *et al.*, *J. Korean Chem. Soc.*, 2013, **57**, 520-523 (*Carpinontriols A,B, activity*).
Feng, M.-M. *et al.*, *Zhongcaoyao*, 2013, **44**, 2650-2656 (*Betulatetraol*).
Tsvetkov, D.E. *et al.*, *Russ. Chem. Bull. (Engl. Transl.)*, 2015, **64**, 1413-1418 (*9-Dehydrobetulatetraol A*).
Cerulli, A. *et al.*, *J. Nat. Prod.*, 2017, **80**, 1703-1713 (*Carpinontriol B, Giffonin T, cd, abs config*)

Cartorimin C36

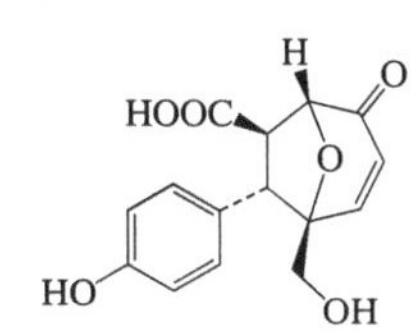

Relative Configuration

[289720-93-0]
$C_{15}H_{14}O_6$ 290.271
Related to Acorusin C in H104. Constit. of *Carthamus tinctorius*. Prisms (MeOH). Mp 206 - 207°. $[\alpha]_D^{25}$ -2.6 (c, 0.005 in MeOH). λ_{max} 225 (log ε 4.04) (MeOH).

Yin, H.-B. *et al.*, *J. Nat. Prod.*, 2000, **63**, 1164-1165 (*Cartorimin*).
Snider, B.B. *et al.*, *Tetrahedron*, 2006, **62**, 5171-5177 (*synth*)

Casearialignan E C37

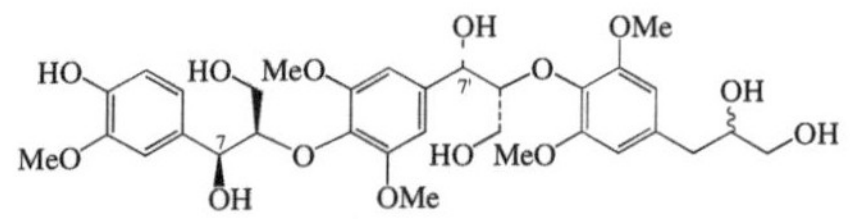

Absolute Configuration

[1217193-10-6]

$C_{32}H_{42}O_{14}$ 650.674

Similar to Leptolepisol B, L18. Constit. of the leaves of *Casearia sylvestris*. Gum. $[\alpha]_D^{20}$ +10.8 (c, 0.09 in MeOH). λ_{max} 230 (log ε 3.54); 278 (log ε 3.08) (MeOH)

7,7′-Diepimer

Casearialignan F

[1217193-11-7]

$C_{32}H_{42}O_{14}$ 650.674

Constit. of the leaves of *Casearia sylvestris*. Gum. $[\alpha]_D^{20}$ -2.3 (c, 0.08 in MeOH). λ_{max} 231 (log ε 3.5); 279 (log ε 3.12) (MeOH).

Wang, W. *et al.*, *Helv. Chim. Acta*, 2010, **93**, 139-146 (*Casearialignans E,F*)

Cassumunarin A C38

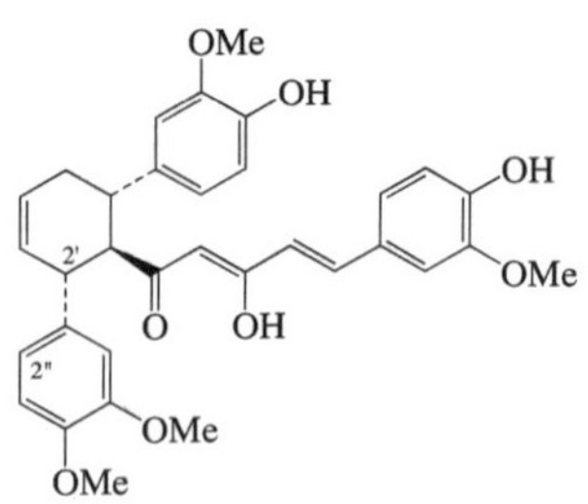

Relative Configuration

[155518-35-7]

$C_{33}H_{34}O_8$ 558.627

Constit. of the rhizomes of *Zingiber cassumunar*. Antioxidant. Yellow powder. $[\alpha]_D^{24}$ +8.9 (c, 1 in $CHCl_3$). λ_{max} 223; 262; 371 (MeOH) (Derep)

2′-Epimer

Cassumunarin B

[155551-80-7]

$C_{33}H_{34}O_8$ 558.627

Constit. of the rhizomes of *Zingiber cassumunar*. Antioxidant. $[\alpha]_D^{24}$ -2.5 (c, 1 in $CHCl_3$). λ_{max} 223; 262; 371 (MeOH) (Derep)

2′-Epimer, 2″-methoxy

Cassumunarin C

[155518-36-8]

$C_{34}H_{36}O_9$ 588.653

Constit. of the rhizomes of *Zingiber cassumunar*. Antioxidant. λ_{max} 227; 283; 368 (MeOH) (Derep).

Jitoe, A. *et al.*, *Tetrahedron Lett.*, 1994, **35**, 981-984 (*Cassumunarins A,B,C, activity*).

Masuda, T. *et al.*, *J. Am. Oil Chem. Soc.*, 1995, **72**, 1053-1057 (*Cassumunarins A,B,C, activity*)

Cassumunin A C39

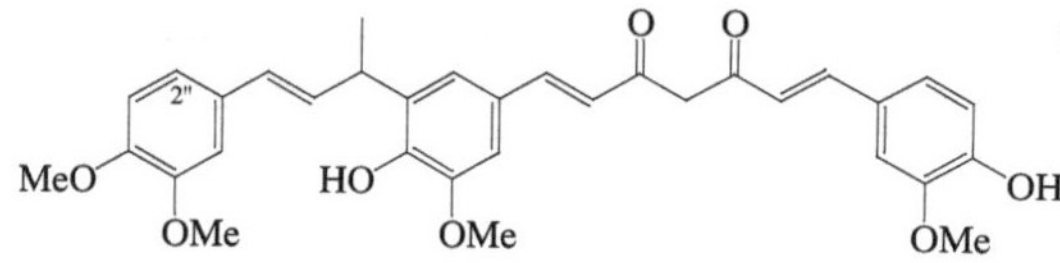

[146763-90-8]

$C_{33}H_{34}O_8$ 558.627

Related to Curcumin, C130. Constit. of the rhizomes of *Zingiber cassumunar*. Antioxidant and antiinflammatory agent. Amorph. yellow powder. $[\alpha]_D^{24}$ -11 (c, 1 in $CHCl_3$). λ_{max} 260; 297 (sh); 426 (MeOH)

2″-Methoxy

Cassumunin B

[146763-91-9]

$C_{34}H_{36}O_9$ 588.653

Constit. of the rhizomes of *Zingiber cassumunar*. Antioxidant and antiinflammatory agent. Yellow powder. $[\alpha]_D$ -4 (c, 1 in $CHCl_3$). λ_{max} 260; 314; 326 (MeOH).

Masuda, T. *et al.*, *J. Agric. Food Chem.*, 1994, **42**, 1850-1856 (*Cassumunins A,B, activity*).

Masuda, T. *et al.*, *J. Nat. Prod.*, 1998, **61**, 609-613 (*synth, pmr*)

Cassumunin C C40

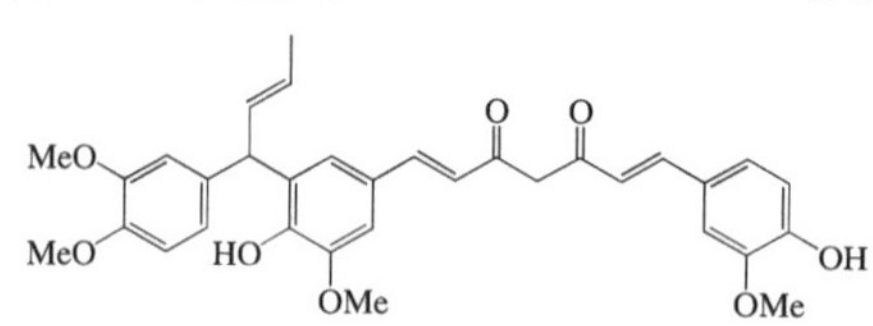

[146763-92-0]

$C_{33}H_{34}O_8$ 558.627

Related to Curcumin, C130. Constit. of the rhizomes of *Zingiber cassumunar*. Antioxidant and antiinflammatory agent. Yellow powder. $[\alpha]_D$ -13 (c, 0.4 in $CHCl_3$).

Masuda, T. *et al.*, *Chem. Lett.*, 1993, 189-192 (*Cassumunin C, activity*).

Masuda, T. *et al.*, *J. Agric. Food Chem.*, 1994, **42**, 1850-1856 (*Cassumunin C, activity, cd*).

Dai, Y. *et al.*, *Synthesis*, 2014, **46**, 3041-3046 (*synth*)

Casuarinondiol C41

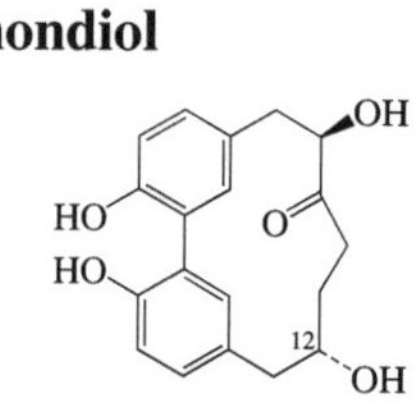

[132923-25-2]

$C_{19}H_{20}O_5$ 328.364

Constit. of *Carpinus cordata*, *Casuarina junghuhniana*, *Ostryopsis nobilis* and *Scyphiphora hydrophyllacea*. Cryst. (MeOH). Mp 243 - 244°. $[\alpha]_D^{20}$ +10 (c, 0.2 in EtOH)

Stereoisomer

[909205-37-4]

$C_{19}H_{20}O_5$ 328.364

Constit. of the wood of *Betula pendula*. Amorph. powder

Stereoisomer, 12-ketone

[1883800-71-2]

$C_{19}H_{18}O_5$ 326.348

Constit. of the wood of *Betula pendula*. Amorph. powder.

Kaneda, N. *et al.*, *Phytochemistry*, 1990, **29**, 3366-3368 (*Casuarinondiol*).

Feng, M.-M. *et al.*, *Zhongcaoyao*, 2013, **44**, 2650-2656 (*Casuarinondiol*).

Tsvetkov, D.E. *et al.*, *Russ. Chem. Bull. (Engl. Transl.)*, 2015, **64**, 1413-1418 (*Betula pendula constits*)

Catunaregin C42

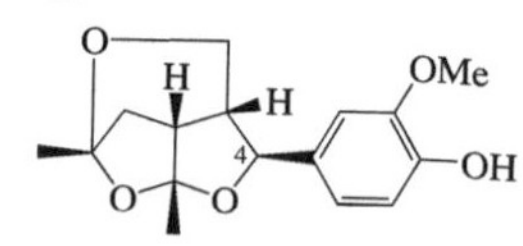

Relative Configuration

[1190713-93-9]

$C_{16}H_{20}O_5$ 292.331

Constit. of the stem bark of the mangrove *Catunaregam spinosa*. Angiogenesis inhibitor. Amorph. solid. $[\alpha]_D^{20}$ +1.4 (c, 0.8 in MeOH). λ_{max} 198 (log ε 3.16); 218 (log ε 3.54); 227 (log ε 3.72); 279 (log ε 3.19) (MeOH)

4-Epimer

Epicatunaregin

[1190713-95-1]

$C_{16}H_{20}O_5$ 292.331

Constit. of the stem bark of *Catunaregam spinosa*. Amorph. solid. Optically inactive. λ_{max} 220 (log ε 3.9); 223 (log ε 3.97); 278 (log ε 3.39) (MeOH).

Gao, G.-C. *et al.*, *Helv. Chim. Acta*, 2010, **93**, 339-344 (*Catunaregin, Epicatunaregin*).

Liu, J.-X. *et al.*, *Mar. Drugs*, 2014, **12**, 2790-2801 (*isol, activity*).

Abe, H. *et al.*, *Org. Chem. Front.*, 2016, **3**, 1084-1086 (*synth*).

Abe, H. *et al.*, *Eur. J. Org. Chem.*, 2018, 1655-1664 (*synth*)

Cerberalignan I C43

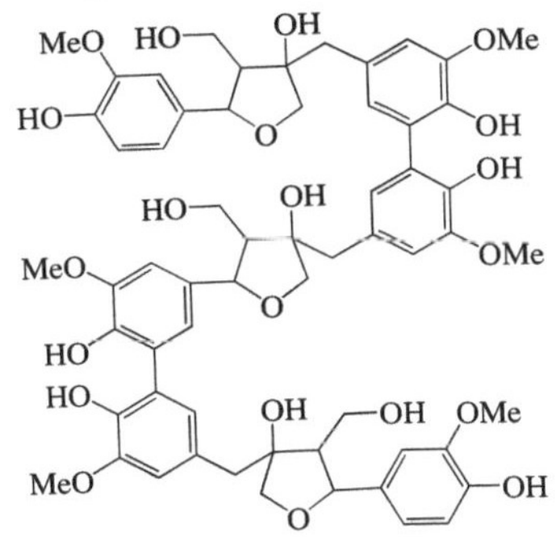

[119420-39-2]

$C_{60}H_{68}O_{21}$ 1125.183

Constit. of *Cerbera manghas* and *Cerbera odollam*. Solid. $[\alpha]_D^{25}$ -76.7 (c, 0.72 in MeOH).

Abe, F. *et al.*, *Phytochemistry*, 1988, **27**, 3627-3631 (*Cerberalignan I*)

Cerberalignan A C44

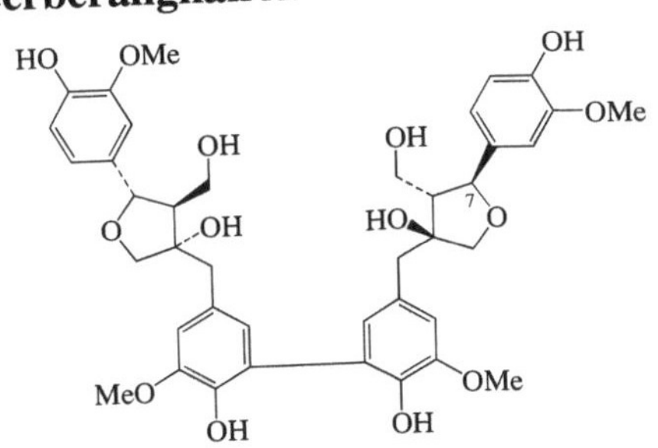

Relative Configuration

5,5″-Biolivil. 5,5″-Diolivil

[119478-09-0]

[114449-09-1] (5,5″-Biolivil)

$C_{40}H_{46}O_{14}$ 750.794

Similar to 5,5″-Bilariciresinol, B14. Isol. from *Cerbera manghas* and *Cerbera odollam*. Solid. $[\alpha]_D^{29}$ -76 (c, 0.55 in MeOH). λ_{max} 222 (ε 39800); 282 (ε 11500) (MeOH).

Abe, F. *et al.*, *Chem. Pharm. Bull.*, 1988, **36**, 795-799 (*Cerberalignan A*)

Cerberalignan B C45

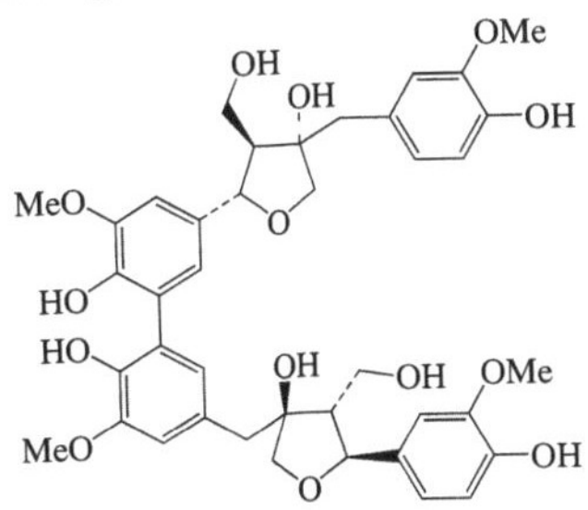

Relative Configuration

5,5‴-Biolivil. 5,5‴-Diolivil

[114435-23-3]

$C_{40}H_{46}O_{14}$ 750.794

Isol. from *Cerbera manghas* and *Cerbera odollam*. Solid. $[\alpha]_D^{20}$ -75.4 (c, 0.35 in MeOH). λ_{max} 223 (ε 46000); 282 (ε 12000) (MeOH).

Abe, F. *et al.*, *Chem. Pharm. Bull.*, 1988, **36**, 795-799 (*Cerberalignan B*)

Cerberalignan C C46

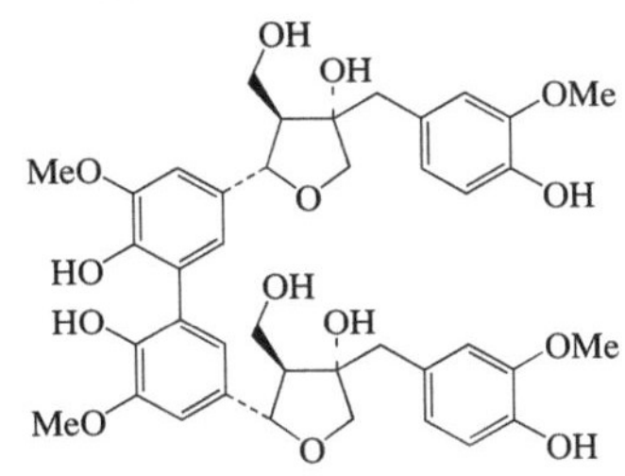

Relative Configuration

5′,5‴-Biolivil. 5′,5‴-Diolivil

[114435-24-4]

$C_{40}H_{46}O_{14}$ 750.794

Isol. from *Cerbera manghas* and *Cerbera odollam*. Solid. $[\alpha]_D^{27}$ -57.3 (c, 0.20 in MeOH). λ_{max} 223 (ε 49000); 282 (ε 12000) (MeOH).

Abe, F. *et al.*, *Chem. Pharm. Bull.*, 1988, **36**, 795-799 (*Cerberalignan C*)

Cerberalignan D C47

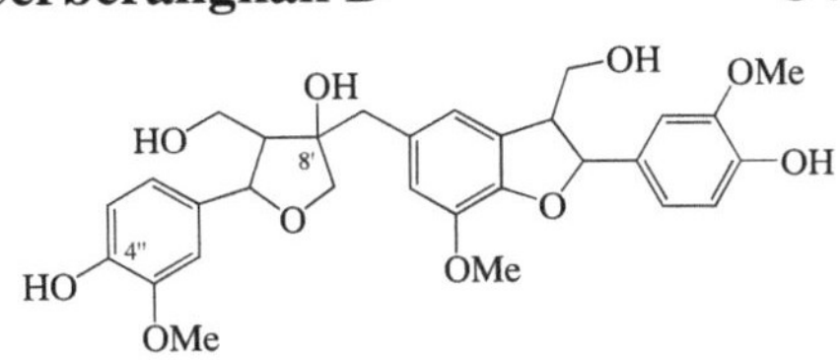

[119403-22-4]

$C_{30}H_{34}O_{10}$ 554.592

Constit. of *Cerbera manghas* and *Cerbera odollam*. Solid. $[\alpha]_D^{25}$ -50.6 (c, 0.98 in MeOH). λ_{max} 228 (log ε 4.37); 280 (log ε 3.99); 286 (log ε 3.96) (MeOH)

8′-Deoxy

Abiesol A

[136037-58-6]

$C_{30}H_{34}O_9$ 538.593

Isol. from the wood of *Abies sachalinensis*. Amorph.. $[\alpha]_D^{22}$ +15 (c, 1.04 in MeOH)

8′-Deoxy, 4″-O-β-D-glucopyranoside

Capselloside

[2118981-52-3]

$C_{36}H_{44}O_{14}$ 700.734

Constit. of *Capsella bursa-pastoris*. Brownish gum. $[\alpha]_D^{25}$ -34 (c, 0.05 in MeOH). Possesses (7*R*,7″*S*,8*S*,8′*S*,8″*R*)-config.. λ_{max} 214 (log ε 3.1); 228 (log ε 3.3); 280 (log ε 1.4) (MeOH).

Abe, F. *et al.*, *Phytochemistry*, 1988, **27**, 3627-3631 (*Cerberalignan D*).

Ozawa, S. *et al.*, *Mokuzai Gakkaishi*, 1991, **37**, 69-75 (*Abiesol A*).

Cha, J.M. *et al.*, *Molecules*, 2017, **22**, 1023:1-8 (*Capselloside, cd*)

Cerberalignan E C48

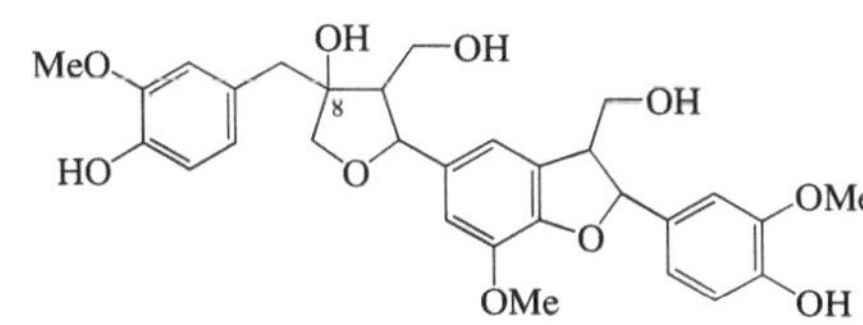

[119403-23-5]

$C_{30}H_{34}O_{10}$ 554.592

Constit. of *Cerbera manghas* and *Cerbera odollam*. Solid. $[\alpha]_D^{24}$ -35.9 (c, 0.4 in MeOH). λ_{max} 228 (log ε 4.35); 280 (log ε 3.96); 286 (log ε 3.92) (MeOH)

8-Deoxy

Herpetriol

[73140-41-7]

$C_{30}H_{34}O_9$ 538.593

Constit. of *Herpetospermum caudigerum*. Mp 110°. $[\alpha]_D^{20}$ +72 (c, 0.38 in MeOH). λ_{max} 214; 230; 278 (ε 11000); 285 (sh) (MeOH).

Kaouadji, M. *et al.*, *Z. Naturforsch., C*, 1979, **34**, 1129-1132 (*Herpetriol*).

Abe, F. *et al.*, *Phytochemistry*, 1988, **27**, 3627-3631 (*Cerberalignan E*)

Cerberalignan F C49

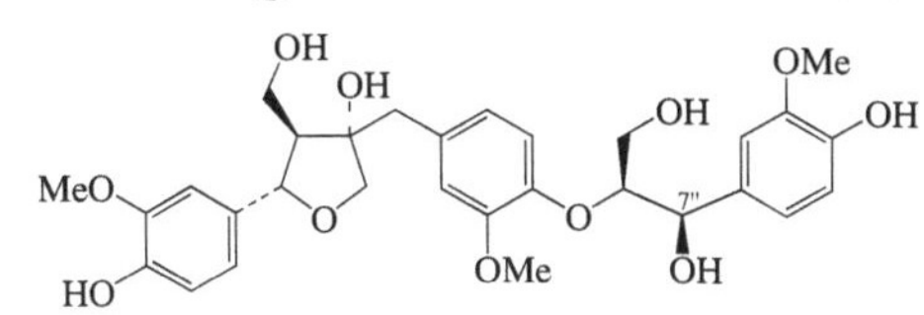

Relative Configuration

Guaiacylglycerylolivil

[119420-37-0]

$C_{30}H_{36}O_{11}$ 572.607

Similar to Ehletianol C, E3. Constit. of *Cerbera manghas* and *Cerbera odollam*. Solid. $[\alpha]_D^{28}$ -49 (c, 0.4 in MeOH)

7″- or 8″-Epimer

Cerberalignan G

[119478-88-5]

$C_{30}H_{36}O_{11}$ 572.607

Constit. of *Cerbera manghas* and *Cerbera odollam*. Solid. $[\alpha]_D^{28}$ -68.4 (c, 0.5 in MeOH).

Abe, F. *et al.*, *Phytochemistry*, 1988, **27**, 3627-3631 (*Cerberalignans F,G*)

Cerberalignan H C50

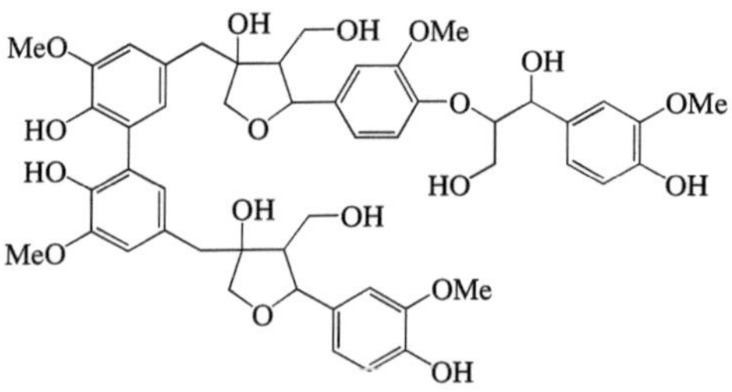

[119420-38-1]

$C_{50}H_{58}O_{18}$ 946.996

Constit. of *Cerbera manghas* and *Cerbera odollam*. Solid. $[\alpha]^{25}_{D}$ -62 (c, 0.39 in MeOH).

Abe, F. *et al.*, *Phytochemistry*, 1988, **27**, 3627-3631 (*Cerberalignan H*)

Cerberalignan J C51

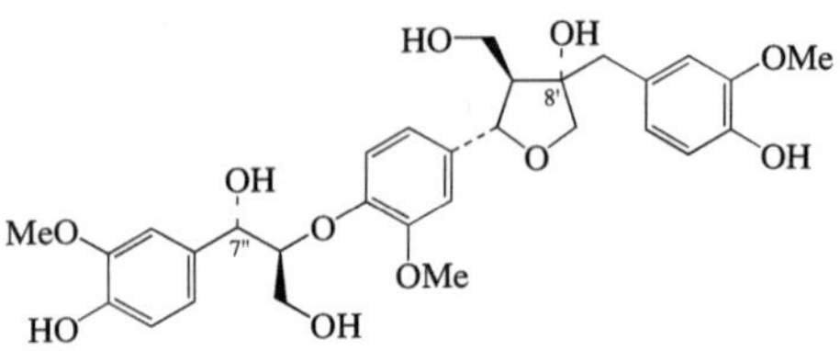

Relative Configuration

Guaiacylglycosylolivil

[126199-50-6]

$C_{30}H_{36}O_{11}$ 572.607

The rel. configs. of C-7″/C-8″ were not determined and are based on those of the deoxy deriv. Similar to Seslignanoccidentaliol A, S52. Constit. of *Cerbera manghas*. Solid. $[\alpha]^{26}_{D}$ -49.6 (c, 0.8 in MeOH)

8′-Deoxy

Guaiacylglycerol β-O-4-lariciresinol ether. ***8-Deoxycerberalignan J***

[613683-99-1]

$C_{30}H_{36}O_{10}$ 556.608

Constit. of *Brassica fruticulosa*. Plant germination inhibitor. $[\alpha]^{25}_{D}$ -3.4 (c, 2 in MeOH).

Abe, F. *et al.*, *Phytochemistry*, 1989, **28**, 3473-3476 (*Cerberalignan J*).

Cutillo, F. *et al.*, *J. Agric. Food Chem.*, 2003, **51**, 6165-6172 (*8-Deoxycerberalignan J, activity*)

Cerberalignan K C52

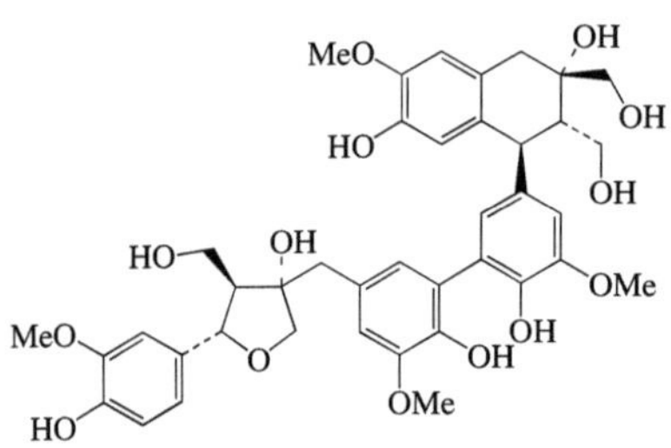

Relative Configuration

(5‴-Olivyl)-5′-cycloolivil

[126176-87-2]

$C_{40}H_{46}O_{14}$ 750.794

Constit. of *Cerbera manghas*. Solid. $[\alpha]^{32}_{D}$ +19.1 (c, 1.2 in MeOH).

Abe, F. *et al.*, *Phytochemistry*, 1989, **28**, 3473-3476 (*Cerberalignan K*)

Cerberalignan L C53

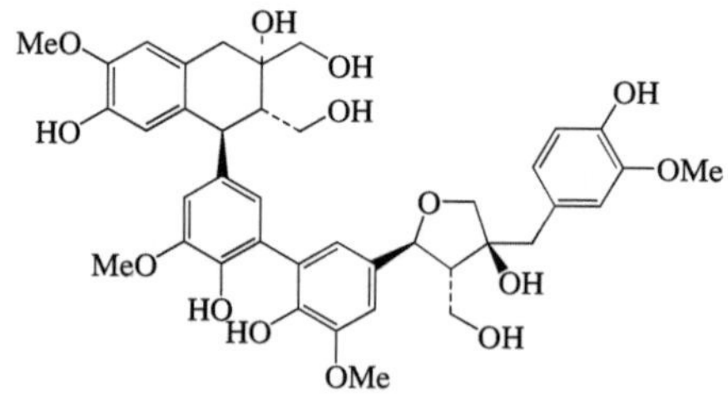

Relative Configuration

(5″-Olivyl)-5′-cycloolivil

[126176-88-3]

$C_{40}H_{46}O_{14}$ 750.794

Isol. from *Cerbera manghas*. Solid. $[\alpha]^{32}_{D}$ +36.1 (c, 1.05 in MeOH). λ_{max} 220 (log ε 4.80); 283 (log ε 4.20) (MeOH).

Abe, F. *et al.*, *Phytochemistry*, 1989, **28**, 3473-3476 (*Cerberalignan L*)

Cerberalignan M C54

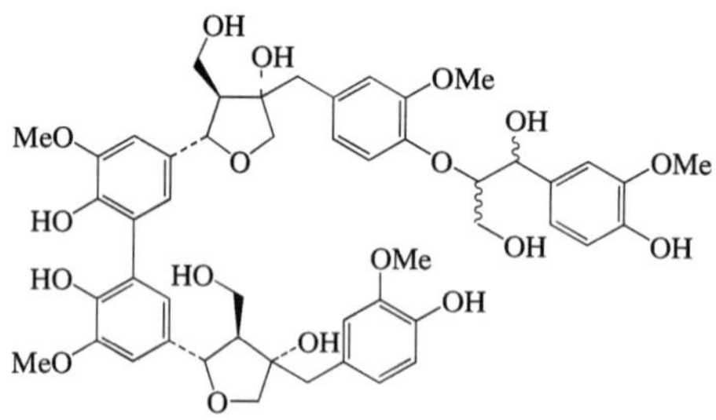

Relative Configuration

[126176-89-4]

$C_{50}H_{58}O_{18}$ 946.996

Constit. of *Cerbera manghas*. Solid. $[\alpha]^{26}_{D}$ -53 (c, 0.5 in MeOH).

Abe, F. *et al.*, *Phytochemistry*, 1989, **28**, 3473-3476 (*Cerberalignan M*)

Cerberalignan N C55

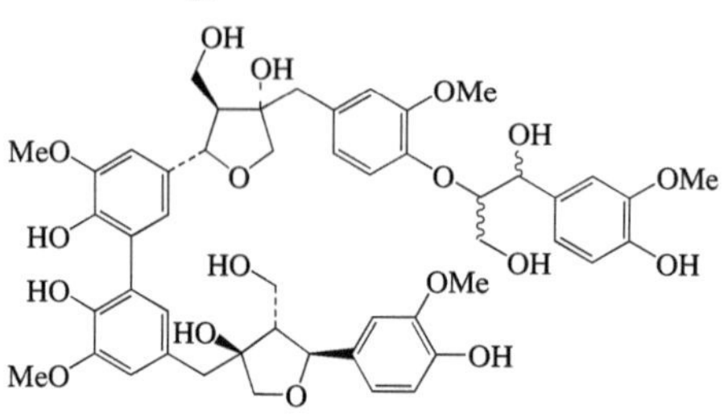

Relative Configuration

[126176-90-7]

$C_{50}H_{58}O_{18}$ 946.996

Constit. of *Cerbera manghas*. Solid. $[\alpha]^{25}_{D}$ -56.3 (c, 0.98 in MeOH).

Abe, F. *et al.*, *Phytochemistry*, 1989, **28**, 3473-3476 (*Cerberalignan N*)

Chamaecypanone C C56

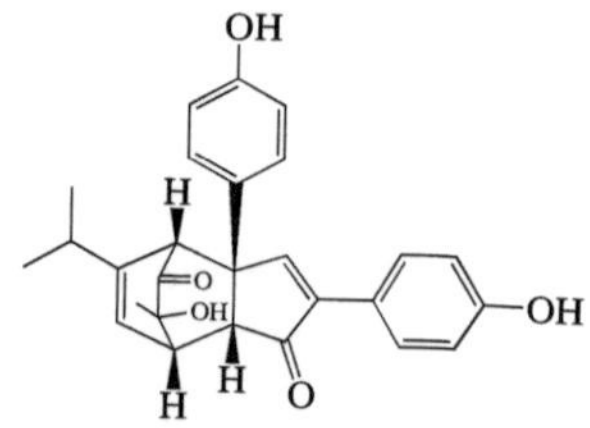

Absolute Configuration

[929195-26-6]

$C_{27}H_{26}O_5$ 430.5

Constit. of the heartwood of *Chamaecyparis obtusa* var. *formosana*. Microtubule inhibitor. Cytotoxic to human nasopharyngeal (HONE-1), oral epidermoid (KB) and gastric (TSGH) carcinoma cells. Amorph. solid. $[\alpha]^{23}_{D}$ +175.7 (c, 0.85 in MeOH). λ_{max} 227 (log ε 4.32); 282 (log ε 3.93); 302 (sh) (MeOH).

Chien, S.-C. *et al.*, *Tetrahedron Lett.*, 2007, **48**, 1567-1569 (*Chamaecypanone C*).

Dong, S. *et al.*, *Angew. Chem., Int. Ed.*, 2009, **48**, 1494-1497 (*synth*).

Hsieh, C.-C. *et al.*, *Biochem. Pharmacol.*, 2010, **79**, 1261-1271 (*activity*)

Cherinonaine C57

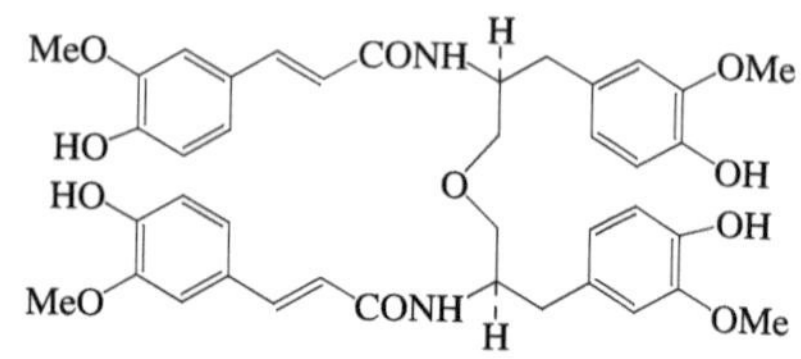

[202589-82-0]

$C_{40}H_{44}N_2O_{11}$ 728.795

Alkaloid from the stems of *Annona cherimola*. Amorph. powder. Opt. inactive (*meso*-compd.). λ_{max} 220; 290; 317 (no solvent reported).

Chen, C.-Y. *et al.*, *Tetrahedron Lett.*, 1998, **39**, 407-410 (*Cherinonaine*)

Chilianthin A C58

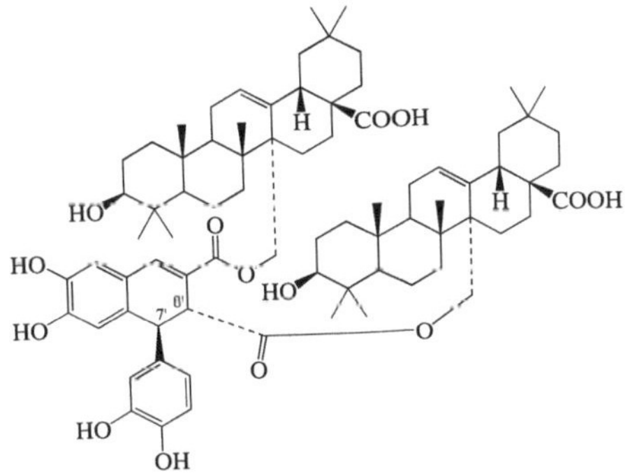

Rhoipteleic acid A

[155748-94-0]

$C_{78}H_{106}O_{14}$ 1267.692

Constit. of *Rhoiptelea chiliantha*. Amorph. powder. $[\alpha]_D$ +86.9 (c, 0.5 in MeOH). Diester of 3,27-Dihydroxy-12-oleanen-28-oic acid and 2,7′-Cyclo-3′,4,4′,5-tetrahydroxy-8,8′-lign-7-ene-9,9′-dioic acid, C339

7′,8′-Diepimer

Chilianthin B. Rhoipteleic acid B [155661-13-5]

$C_{78}H_{106}O_{14}$ 1267.692

Constit. of *Rhoiptelea chiliantha*. Amorph. powder. $[\alpha]_D$ +49.4 (c, 0.5 in MeOH).

Jiang, Z.-H. *et al.*, *Chem. Pharm. Bull.*, 1996, **44**, 1669-1675 (*Chilianthins A,B*)

Chilianthin C C59

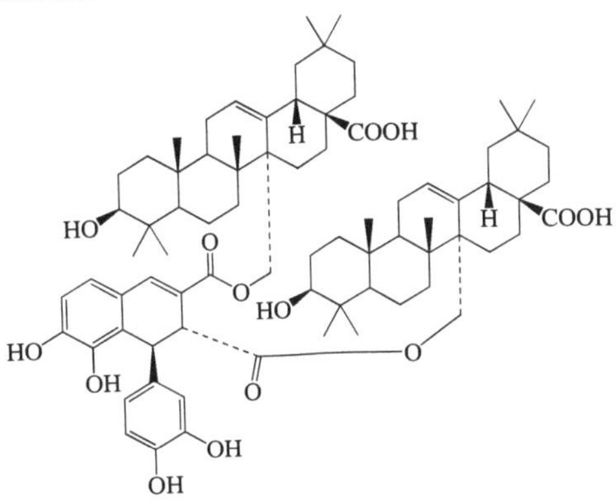

[182161-86-0]

$C_{78}H_{106}O_{14}$ 1267.692

Constit. of *Rhoiptelea chiliantha*. Amorph. powder. $[\alpha]^{20}_{D}$ +60.8 (c, 0.3 in MeOH).

Jiang, Z.-H. *et al.*, *Chem. Pharm. Bull.*, 1996, **44**, 1669-1675 (*Chilianthin C*)

Chilianthin D C60

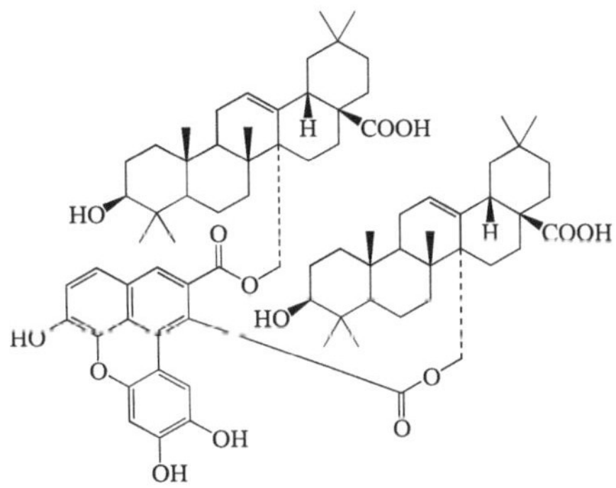

[182161-88-2]

$C_{78}H_{102}O_{14}$ 1263.66

Constit. of *Rhoiptelea chiliantha*. Yellow powder. $[\alpha]^{24}_{D}$ +14.3 (c, 0.4 in MeOH).

Jiang, Z.-H. *et al.*, *Chem. Pharm. Bull.*, 1996, **44**, 1669-1675 (*Chilianthin D*)

Chilianthin E C61

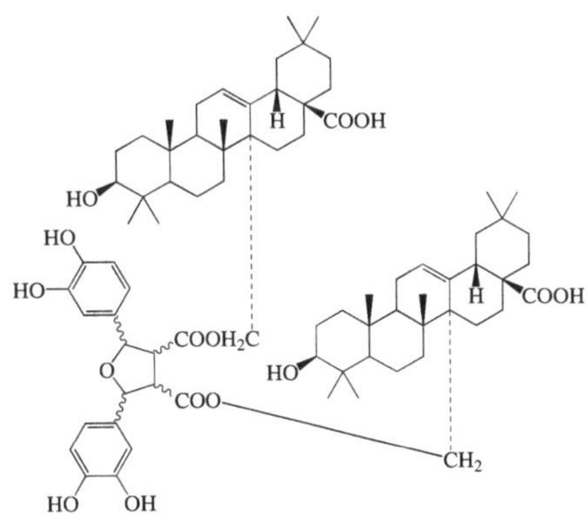

[182161-89-3]

$C_{78}H_{108}O_{15}$ 1285.707

Constit. of *Rhoiptelea chiliantha*. Amorph. powder. $[\alpha]^{15}_{D}$ +97.1 (c, 1 in MeOH). Mixt. of diastereoisomers.

Jiang, Z.-H. *et al.*, *Chem. Pharm. Bull.*, 1996, **44**, 1669-1675 (*Chilanthin E*)

Chimarrhinin C62

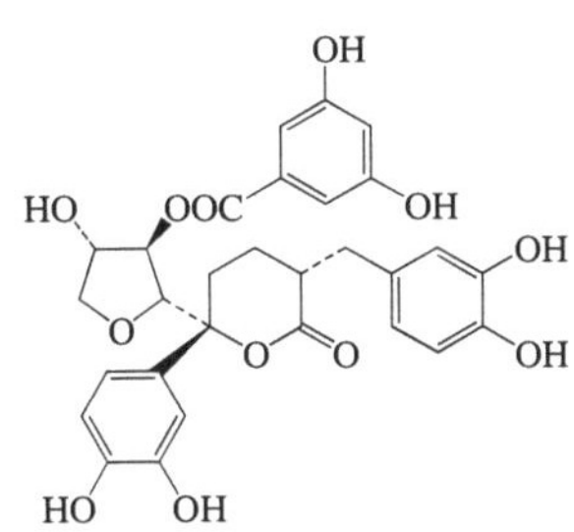

Relative Configuration

[1276543-21-5]

$C_{29}H_{28}O_{12}$ 568.531

Structure may need revision. Phenyl 3,5-dioxygenated substn. pattern in doubt as a result of work done by Yang *et al* in 2015. Constit. of the leaves of *Chimarrhis turbinata*. Antioxidant. Pale brown powder. $[\alpha]^{25}_{D}$ -204 (c, 0.01 in MeOH).

Cardoso, C.L. *et al.*, *J. Nat. Prod.*, 2011, **74**, 487-491 (*Chimarrhinin, activity*).

Yang, Y.-N. *et al.*, *J. Nat. Prod.*, 2015, **78**, 705-711 (*pmr, struct*)

9-Chloro-7,9′-epoxy-4′,5-dimethoxy-8,8′-lignan-3,3′,8-triol C63

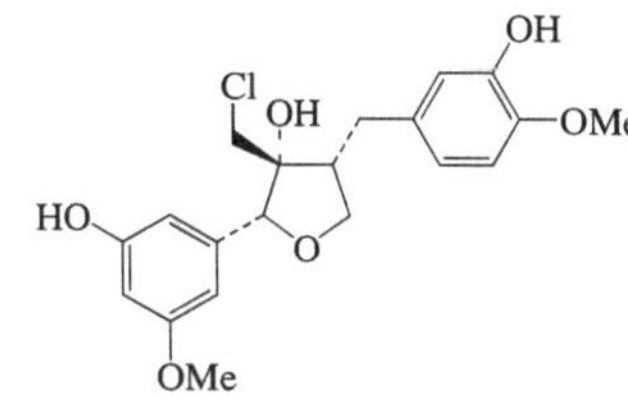

9-Chloro-3,3′,8-trihydroxy-4′,5-dimethoxy-7,9′-epoxylignan

$C_{20}H_{23}ClO_6$ 394.848

Structures in this series may need revision. Phenyl 3,5-dioxygenated substn. pattern in doubt as a result of work done by Yang *et al* in 2015

(7R*,8R*,8′R*)-form [1206464-63-2]

Constit. of *Viburnum foetidum* var. *foetidum*. Amorph. powder. $[\alpha]^{23}_{D}$ -116.7 (c, 0.15 in MeOH). λ_{max} 204 (log ε 3.2); 231 (log ε 2.8); 282 (log ε 3.7) (MeOH).

Chen, X.-Q. *et al.*, *Chem. Pharm. Bull.*, 2009, **57**, 1129-1131 (*Viburnum foetidum constit*).

Yang, Y.-N. *et al.*, *J. Nat. Prod.*, 2015, **78**, 705-711 (*pmr, struct*)

1-*O*-Chlorogenoylchlorogenic acid C64

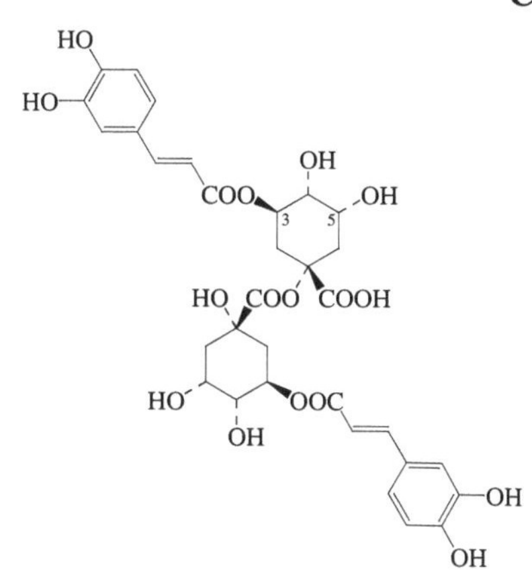

[393782-12-2]

$C_{32}H_{34}O_{17}$ 690.607

Constit. of *Cussonia barteri*. $[\alpha]^{25}_{D}$ -2 (c, 0.5 in MeOH). λ_{max} 206 (ε 27600); 218 (ε 23000); 236 (ε 15100); 244 (ε 14200); 302 (ε 15000); 330 (ε 18700) (MeOH)

3-Decaffeoyl, 5-O-caffeoyl

1-O-Chlorogenoylneochlorogenic acid

[393782-13-3]

$C_{32}H_{34}O_{17}$ 690.607

Constit. of *Cussonia barteri*. $[\alpha]^{25}_{D}$ -3 (c, 0.3 in MeOH). λ_{max} 206 (ε 32200); 218 (ε 27500); 234 (ε 17800); 244 (ε 16600); 302 (ε 18100); 330 (ε 22400) (MeOH).

Papajewski, S. *et al.*, *Planta Med.*, 2001, **67**, 732-736 (*Cussonia barteri constits*)

3-Chloro-3-phenyl-1,2-propanediol C65

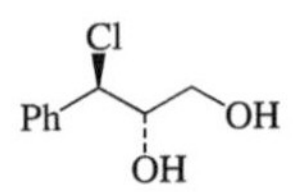

$C_9H_{11}ClO_2$ 186.635

(2S,3R)-form

1-Benzoyl

3-Chloro-2-hydroxy-3-phenylpropyl benzoate

$C_{16}H_{15}ClO_3$ 290.743

Constit. of *Desmos pedunculosus*. Oil. $[\alpha]^{21}_{D}$ -15.4 (c, 0.1 in $CHCl_3$).

Clement, J.A. *et al.*, *Phytochem. Lett.*, 2017, **22**, 117-121 (*Desmos pedunculosus constit*)

3-Chloro-6-(1-propenyl)-1,2,4-benzenetriol C66

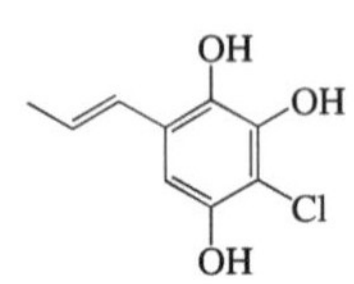

$C_9H_9ClO_3$ 200.618

(E)-form

2-Me ether

3-Chloro-2-methoxy-6-(1-propenyl)-1,4-benzenediol

$C_{10}H_{11}ClO_3$ 214.645

2-Me ether, 1-O-β-D-glucopyranoside

Przewatangoside A

[2260925-94-6]

$C_{16}H_{21}ClO_8$ 376.786

Constit. of *Przewalskia tangutica*. Amorph. powder. $[\alpha]^{20}_{D}$ +21 (c, 0.5 in MeOH).

Zhao, M. *et al.*, *Phytochem. Lett.*, 2017, **20**, 168-171 (*Przewatangoside A*)

5-Chloro-6-propyl-1,2,4-benzenetriol C67

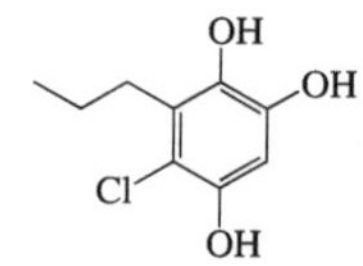

$C_9H_{11}ClO_3$ 202.634

4-Me ether, 1-O-β-D-glucopyranoside

Globosumoside A

[1469997-27-0]

$C_{16}H_{23}ClO_8$ 378.802

Prod. by *Chaetomium elatum* No. 89-1-3-1. Also isol. from *Przewalskia tangutica*. Needles (MeOH). λ_{max} 204 (log ε 4.32); 286 (log ε 3.39) (MeOH)

4-Me ether, 1-O-(6-O-acetyl-β-D-glucopyranoside)

Globosumoside B

[1469997-28-1]

$C_{18}H_{25}ClO_9$ 420.839

Prod. by *Chaetomium elatum* No. 89-1-3-1. Amorph. powder. λ_{max} 206 (log ε 4.7); 284 (log ε 3.88) (MeOH).

Chen, G.-D. *et al.*, *Planta Med.*, 2012, **78**, 1683-1689 (*Globosumosides A,B*)

Ocotea catharinensis $C_{42}H_{44}O_{12}$ Cyclolignan dimer C68

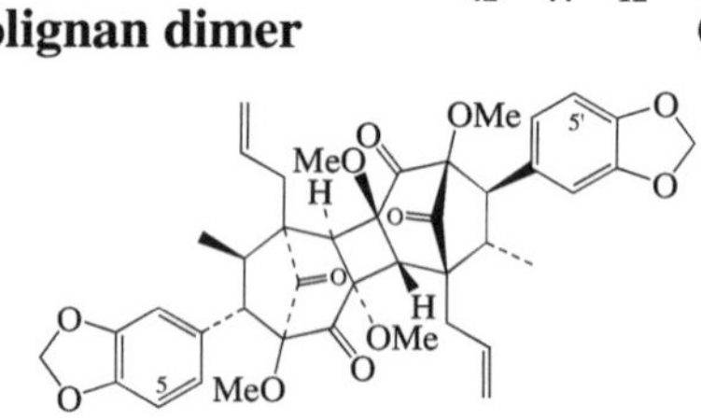

Relative Configuration

[1178974-90-7]

$C_{42}H_{44}O_{12}$ 740.802

Constit. of the leaves of *Ocotea catharinensis*. Viscous oil. $[\alpha]^{21}_{D}$ +220 (c, 0.07 in MeOH)

5,5′-Dimethoxy

Ocotea catharinensis $C_{44}H_{48}O_{14}$ ***Cyclolignan dimer***

[1178974-89-4]

$C_{44}H_{48}O_{14}$ 800.854

Constit. of the leaves of *Ocotea catharinensis*. Viscous oil. $[\alpha]^{21}_{D}$ +57 (c, 0.62 in MeOH).

Funasaki, M. *et al.*, *J. Braz. Chem. Soc.*, 2009, **20**, 853-859 (*Ocotea catharinensis dimers*)

Chokorin C69

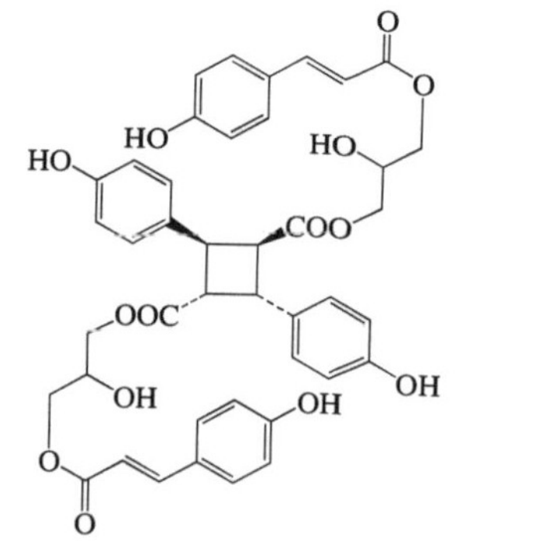

[115812-55-0]

$C_{42}H_{40}O_{14}$ 768.768

Constit. of stromata of *Epichloe typhina* on *Phleum pratense*. Needles ($CHCl_3$/MeOH). Mp 226 - 228°.

Koshino, H. *et al.*, *Phytochemistry*, 1988, **27**, 1333-1338 (*Chokorin, struct*)

Chromapedic acid C70

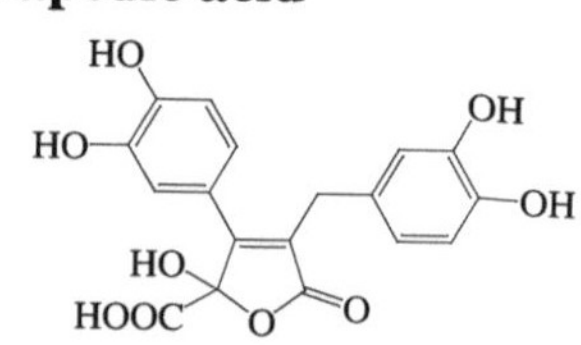

3-(3,4-Dihydroxyphenyl)-4-[(3,4-dihydroxyphenyl)methy]-2,5-dihydro-2-hydroxy-5-oxo-2-furancarboxylic acid, CAS. *3,3′,4,4′,8-Pentahydroxy-7,8′-lign-7(8′)-en-9′,8-olid-9-oic acid*

[1704729-38-3]

$C_{18}H_{14}O_9$ 374.301

Provisional lignan. Isol. from *Leccinum chromapes*. Yellowish oil. $[\alpha]^{20}_{D}$ +4 (c, 1.6 in MeCN). λ_{max} 211 (log ε 3.49); 249 (log ε 3.22); 296 (log ε 3.18); 304 (log ε 3.19); 323 (log ε 3.19) (MeOH).

Gruber, G. *et al.*, *Z. Naturforsch., B*, 2014, **69**, 432-438 (*Chromapedic acid*)

Chrysophyllin A C71

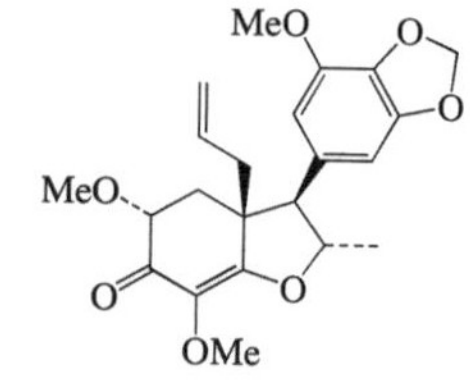

Relative Configuration

2,8′-Epoxy-5,6-dihydro-3,3′,5-trimethoxy-4′,5′-methylenedioxy-1,7′-lign-8-en-4(1H)-one

[85538-70-1]

$C_{22}H_{26}O_7$ 402.443

Constit. of the trunk wood of *Licaria chrysophylla*. Cryst. (C_6H_6). Mp 183

- 185°. λ_{max} 258 (sh); 273 (ε 20100) (MeOH).

Ferreira, Z.S. *et al.*, *Phytochemistry*, 1982, **21**, 2756-2758 (*Chrysophyllin A*)

Chrysophyllin B C72

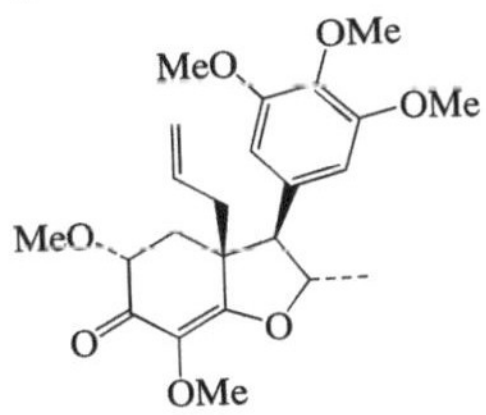

Relative Configuration

3,3a,4,5-Tetrahydro-5,7-dimethoxy-2-methyl-3a-(2-propenyl)-3-(3,4,5-trimethoxyphenyl)-6(2H)-benzofuranone, CAS. *2,8′-Epoxy-5,6-dihydro-3,3′,4′,5,5′-pentamethoxy-1,7′-lign-8-en-4(1H)-one*

[85541-04-4]

$C_{23}H_{30}O_7$ 418.486

Constit. of the trunk wood of *Licaria chrysophylla.*

Ferreira, Z.S. *et al.*, *Phytochemistry*, 1982, **21**, 2756-2758 (*Chrysophyllin B*)

Chrysophyllon IIIA C73

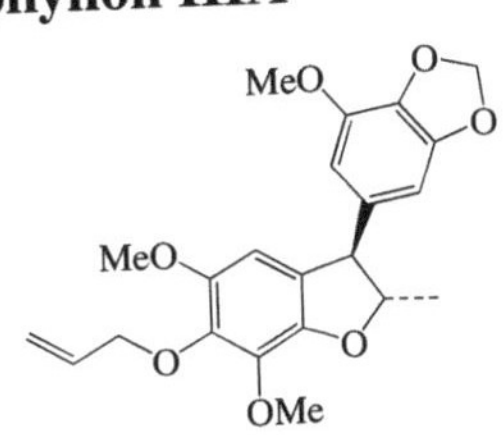

[106897-13-6]

$C_{22}H_{24}O_7$ 400.427

Constit. of *Licaria chrysophylla*. Oil.

Lopes, M.N. *et al.*, *Phytochemistry*, 1986, **25**, 2609-2612 (*Chrysophyllon IIIA*)

Chrysophyllon IIIB C74

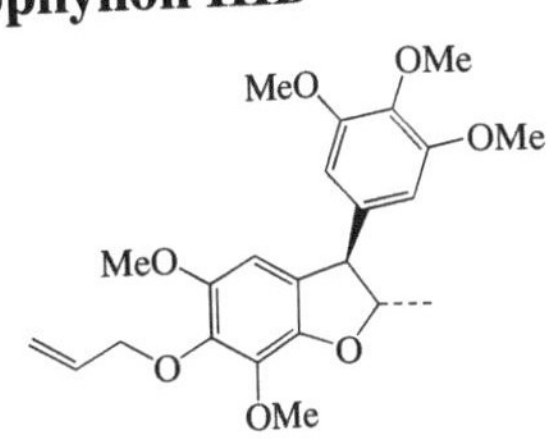

[106897-14-7]

$C_{23}H_{28}O_7$ 416.47

Constit. of *Licaria chrysophylla*. Oil. λ_{max} 294 (ε 1000) (MeOH)

O-Depropenyl

Chrysophyllon VIB

[1454880-25-1]

$C_{20}H_{24}O_7$ 376.405

Constit. of *Licaria chrysophylla*.

Lopes, M.N. *et al.*, *Phytochemistry*, 1986, **25**, 2609-2612 (*Chrysophyllon IIIB*).

Bezerra, A. de M. *et al.*, *Quim. Nova*, 2012, **35**, 2226-2228 (*Chrysophyllon VIB*)

Chushizisin H C75

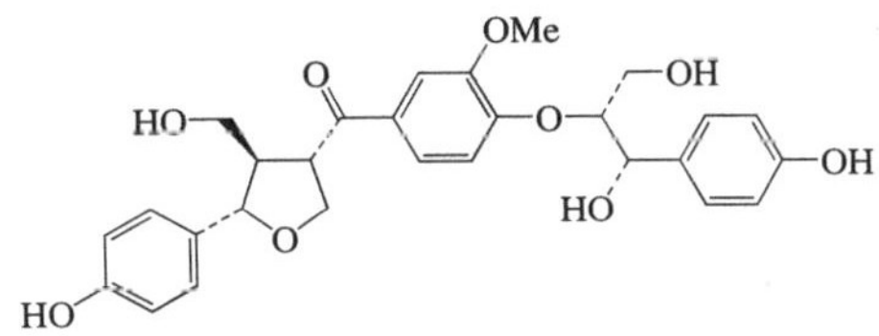

Relative Configuration

[1142925-37-8]

$C_{28}H_{30}O_9$ 510.539

Only the configs. of the chiral centres in each half of the molecule are known, not the relationship between the halves. Related to Ehletianol C, E3. Constit. of the fruit of *Broussonetia papyrifera*. Oil. λ_{max} 202 (log ε 4.49); 226 (log ε 4.44); 277 (log ε 4.09); 306 (log ε 3.89) (MeOH)

Stereoisomer

[1248570-22-0]

$C_{28}H_{30}O_9$ 510.539

Constit. of the fruit of *Broussonetia papyrifera*. Amorph. solid. $[\alpha]^{22}_{D}$ -2 (c, 0.1 in MeOH). λ_{max} 201 (log ε 4.42); 226 (log ε 4.25); 277 (log ε 3.92); 306 (log ε 3.74) (MeOH).

Mei, R.-Q. *et al.*, *J. Nat. Prod.*, 2009, **72**, 621-625 (*Chushizisin H*).

Zhou, X.-J. *et al.*, *J. Asian Nat. Prod. Res.*, 2010, **12**, 399-406 (*stereoisomer*)

Cinerin C C76

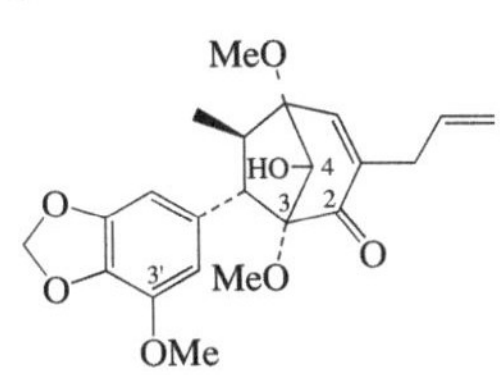

Probable Absolute Configuration

5,8′-Cyclo-4,5-dihydro-4-hydroxy-3,3′,5-trimethoxy-4′,5′-methylenedioxy-3,7′-lign-8-en-2(3H)-one

[1166328-49-9]

$C_{22}H_{26}O_7$ 402.443

Lignan numbering shown. Abs. configs. in this series are uncertain. Related to Nectamazin C, N1. Constit. of the leaves of *Pleurothyrium cinereum*. PAF antagonist. Potent COX-2 inhibitor. Cryst. (Me_2CO). Mp 176 - 177°. $[\alpha]^{25}_{D}$ -13.2 (c, 0.28 in $CHCl_3$). λ_{max} 215 (log ε 4.55); 245 (log ε 4.12); 285 (log ε 3.4) (MeOH)

4-Ketone

5,8′-Cyclo-3,3′,5-trimethoxy-4′,5′-methylenedioxy-3,7′-lign-8-ene-2,4(3H,5H)-dione. ***Cinerin A***

[1166328-44-4]

$C_{22}H_{24}O_7$ 400.427

Constit. of the leaves of *Pleurothyrium cinereum*. Yellowish oil. $[\alpha]^{25}_{D}$ -18.8 (c, 0.23 in $CHCl_3$). λ_{max} 209 (log ε 4.22); 223 (log ε 4.07); 275 (log ε 3.65) (MeOH)

4-Ketone, 2α-alcohol

5,8′-Cyclo-2,3-dihydro-2-hydroxy-3,3′,5-trimethoxy-4′,5′-methylenedioxy-3,7′-lign-8-en-4(3H)-one. ***Cinerin D***

[1166328-51-3]

$C_{22}H_{26}O_7$ 402.443

Constit. of the leaves of *Pleurothyrium cinereum*. Needles (hexane). Mp 104 - 105°. $[\alpha]^{25}_{D}$ -10.3 (c, 0.2 in $CHCl_3$). λ_{max} 210 (log ε 4.72); 236 (log ε 3.88); 289 (log ε 3.52) (MeOH)

Stereoisomer (1) (?)

Macrophyllin‡

[55127-81-6]

$C_{22}H_{26}O_7$ 402.443

Constit. of the trunkwood of *Licaria macrophylla*. Cryst. ($CHCl_3$/MeOH). Mp 175 - 178°. Struct. not certain. May be identical with Cinerin C, but no opt. rotn. was reported and the alternative 6′-oxo struct. could not be ruled out. λ_{max} 215 (ε 53100); 235 (sh); 270 (sh) (EtOH)

Stereoisomer (2) (?)

[1186228-71-6]

$C_{22}H_{26}O_7$ 402.443

Constit. of the fruit of *Ocotea heterochroma*. Needles (Me_2CO). Mp 184 - 185°. λ_{max} 213 (log ε 4.72); 243 (log ε 4.27); 283 (log ε 3.51) (MeOH).

Franca, N.C. *et al.*, *Phytochemistry*, 1974, **13**, 2839-2842 (*Macrophyllin*).

Agrawal, P.K. *et al.*, *Magn. Reson. Chem.*, 1985, **23**, 389-418 (*cmr*).

Coy, E.D. *et al.*, *Bioorg. Med. Chem. Lett.*, 2009, **19**, 6922-6925 (*Cinerin C, activity*).

Cuca, L.E. *et al.*, *Chem. Nat. Compd. (Engl. Transl.)*, 2009, **45**, 179-181 (*Ocotea heterochroma constit*).

Coy, E.D. *et al.*, *J. Nat. Prod.*, 2009, **72**, 1245-1248 (*Cinerins A,C,D*).

Coy, E.D. *et al.*, *Org. Biomol. Chem.*, 2010, **8**, 2003-2005 (*Cinerin C, synth*).

Coy-Barrera, E.D. *et al.*, *Acta Crystallogr., Sect. C*, 2012, **68**, o320-o322 (*Cinerin C, cryst struct*)

Cinncassin C C77

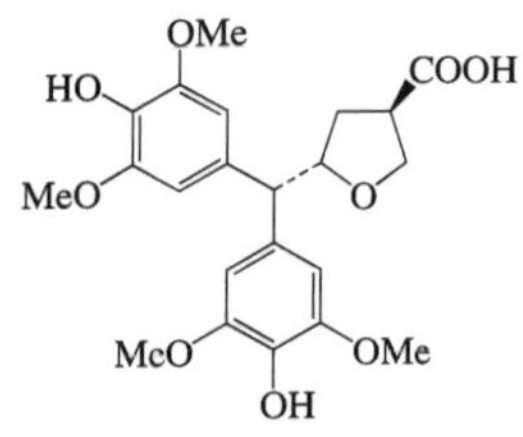
Absolute Configuration

Tetrahydro-5-[bis(4-hydroxy-3,5-dimethoxyphenyl)methyl]-3-furancarboxylic acid
$C_{22}H_{26}O_9$ 434.441
Constit. of the bark of *Cinnamomum cassia*. Amorph. powder. $[\alpha]_D^{20}$ -27 (c, 0.1 in MeOH). λ_{max} 204 (log ε 2.63); 225 (log ε 2.26); 280 (log ε 1.75) (MeOH).
He, S. *et al.*, *Fitoterapia*, 2016, **112**, 153-160 (*Cinncassin C, cd*)

Cinncassin D C78

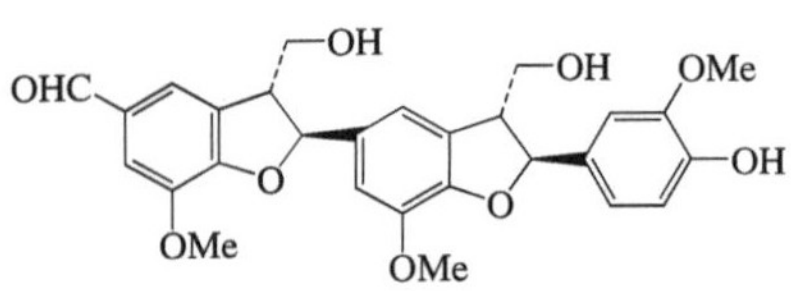
Absolute Configuration

$C_{28}H_{28}O_9$ 508.523
Constit. of the bark of *Cinnamomum cassia*. Amorph. powder. $[\alpha]_D^{20}$ -36 (c, 0.1 in MeOH). λ_{max} 207 (log ε 2.65); 235 (log ε 2.23); 289 (log ε 1.98); 307 (log ε 1.88) (MeOH).
He, S. *et al.*, *Fitoterapia*, 2016, **112**, 153-160 (*Cinncassin D, cd*)

Cinncassin L C79

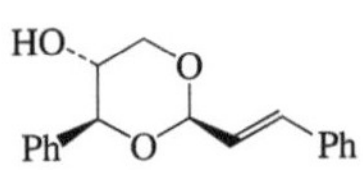
Absolute Configuration

4-Phenyl-2-(2-phenylethenyl)-1,3-dioxan-5-ol. 9,9′-Epoxy-7,9′-oxylign-7′-en-8-ol. 9,9′-Epoxy-8-hydroxy-7,9′-oxyneolign-7′-ene. Cinnacassin L
[2232873-51-5]
$C_{18}H_{18}O_3$ 282.339
Constit. of the twigs of *Cinnamomum cassia*. Amorph. solid. $[\alpha]_D^{25}$ -7 (c, 0.6 in MeOH). λ_{max} 205 (log ε 4.33); 215 (log ε 4.25); 245 (log ε 4.05) (MeOH).
Liu, X. *et al.*, *J. Nat. Prod.*, 2018, **81**, 1333-1342 (*Cinncassin L, cd*)

Claulansine K C80

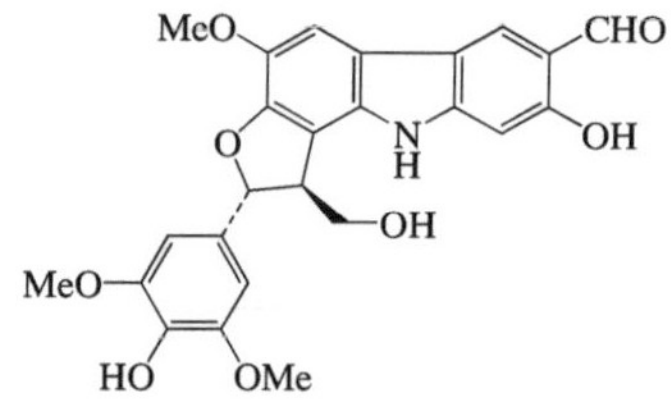
Relative Configuration

Claulansine L
[1808205-50-6]
[1894227-44-1] (Claulansine L)
$C_{25}H_{23}NO_8$ 465.458
Alkaloid from the stems and peel of *Clausena lansium*. α-Glucosidase inhibitor. Yellow powder. $[\alpha]_D^{22}$ -3.8 (c, 0.5 in MeOH). Claulansine K is probably a partial racemate whereas Claulansine L was isol. as a racemic mixt. of (7′*R*,8′*S*)- and (7′*S*,8′*R*)-forms. λ_{max} 205 (log ε 0.72); 309 (log ε 0.31) (MeOH).
Deng, H.-D. *et al.*, *J. Asian Nat. Prod. Res.*, 2014, **16**, 1024-1028 (*Claulansine K*).
Du, Y.-Q. *et al.*, *Fitoterapia*, 2015, **103**, 122-128 (*Claulansine L*)

Clausenaglycoside B C81

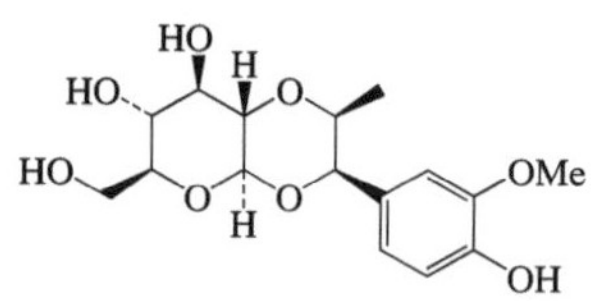
Absolute Configuration

$C_{16}H_{22}O_8$ 342.344
Regioisomer of Hemidesmoside A, in J5. Constit. of *Clausena dunniana* var. *robusta*. Yellow oil. $[\alpha]_D^{20}$ +18.9 (c, 0.1 in MeOH).
Guo, Y. *et al.*, *Phytochem. Lett.*, 2017, **22**, 138-141 (*Clausenaglycoside B*)

Cleomiscosin A C82

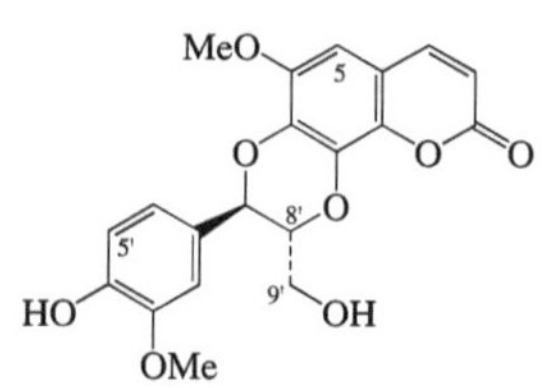

[76948-72-6]
[906624-70-2] (unspecified stereochem) [1482512-83-3]
(7′R,8′R)-form
$C_{20}H_{18}O_8$ 386.356
Coumarinolignan and regioisomer of Cleomiscosin B, C83. Variable enantiochem. shown by Cleomiscosin A and its relatives. It has been isol. as the predominant (*R*,*R*)-diastereoisomer shown here, and also as the (*S*,*S*)-form and the racemate. Other nat. prods. covered by this entry have not been fully investigated. Constit. of *Acer nikoense*, *Aesculus turbinata*, *Ailanthus altissima*, *Annona squamosa*, *Brucea javanica*, *Cleome viscosa*, *Hyoscyamus niger*, *Soulamea soulameoides*, *Matayba arborescens*, *Xanthium sibiricum* and many other plant spp.. Tyrosinase and phosphodiesterase inhibitor. Antioxidant. Shows antiinflammatory, hepatoprotectant and vasorelaxant props.. Cryst.. Mp 261° (245-247°). $[\alpha]_D^{20}$ +22.5 (c, 0.1 in MeOH) (7′*R*,8′*R*-form). $[\alpha]_D^{20}$ -24.7 (c, 0.1 in MeOH) (7′*S*,8′*S*-form). λ_{max} 211 (log ε 3.38); 231 (log ε 1.30); 287 (log ε 0.34); 321 (log ε 0.53) (MeOH)

9′-Ac

Durantin A. Venkatasin
[102070-29-1]
$C_{22}H_{20}O_9$ 428.393
Constit. of *Jatropha gossypifolia*, *Hyoscyamus niger* and *Duranta repens*. Phosphodiesterase inhibitor. Mp 113 - 115°. Mp 120 - 121° (racemate). $[\alpha]_D^{25}$ +16 (c, 0.09 in $CHCl_3$). λ_{max} 322 (log ε 3.63) (EtOH)

4′-O-E-Cinnamoyl

4′-O-trans-Cinnamoylcleomiscosin A
[1207663-67-9]
$C_{29}H_{24}O_9$ 516.502
Constit. of the roots of *Terminalia tropophylla*. Amorph. solid. $[\alpha]_D^{25}$ -77 (c, 0.1 in MeOH). λ_{max} 211 (log ε 3.7); 284 (log ε 3.2); 352 (sh) (MeOH)

9′-O-(4-Hydroxy-E-cinnamoyl)

Repenin A
[1174169-13-1]
$C_{29}H_{24}O_{10}$ 532.501
Constit. of *Duranta repens*. Antioxidant. Light yellow powder. Mp 191 - 194° dec.. Racemate. λ_{max} 286; 328 (no solvent reported)

9′-O-(3,4-Dihydroxy-E-cinnamoyl)

Repenin C
[1174169-15-3]
$C_{29}H_{24}O_{11}$ 548.5
Constit. of *Duranta repens*. Antioxidant. Light yellow powder. Mp 206 - 209° dec.. Racemate

9′-O-(4-Hydroxy-3-methoxy-E-cinnamoyl)
Repenin B
[1174169-14-2]
$C_{30}H_{26}O_{11}$ 562.527
Constit. of *Duranta repens*. Antioxidant. Light yellow solid. Mp 199 - 201° dec.

4′-Me ether
4′-O-Methylcleomiscosin A
[90706-60-8]
$C_{21}H_{20}O_8$ 400.383
Constit. of the seeds of *Hyoscyamus niger*. Cryst.. Mp 205 - 210°. Racemate. λ_{max} 288 (sh); 327 (MeOH)

4′-Me ether, 9′-Ac
Durantin B
[90706-61-9]
$C_{23}H_{22}O_9$ 442.42
Constit. of *Duranta repens*. Creamy yellow amorph. powder. λ_{max} 289 (log ε 3.78); 329 (log ε 4.03) (MeOH)

3′-O-De-Me, 4′-Me ether
Cleomiscosin E
[1581730-22-4]
$C_{20}H_{18}O_8$ 386.356
Constit. of the seeds of *Brucea javanica*. Exhibits potent antiinflammatory activity by inhibiting NO prodn.. Pale yellow powder. $[\alpha]_D^{21}$ -2.3 (c, 0.08 in MeOH). Assigned (7′*S*,8′*S*)-stereochem.. λ_{max} 204 (log ε 4.63); 233 (log ε 4.24); 324 (log ε 3.90) (MeOH)

5′-Hydroxy
5′-Hydroxycleomiscosin A. 5″-Demethylaquillochin
[305364-91-4]
$C_{20}H_{18}O_9$ 402.355
Constit. of *Acer saccharum* and *Mallotus apelta*. Needles. Mp 245 - 246°

5′-Hydroxy, 6-O-de-Me
Malloapelin A
[1109272-09-4]
$C_{19}H_{16}O_9$ 388.328
Constit. of the roots of *Mallotus apelta*. Amorph. powder. Mp 258 - 259°. Racemate. λ_{max} 237; 252; 322 (MeOH)

9′-Deoxy
Propacin
[90475-47-1]
[94062-48-3] (*trans*-(±)-form)
[121700-28-5] (*trans*-form)
$C_{20}H_{18}O_7$ 370.357
Constit. of *Jatropha gossypifolia*, *Melochia chamaedrys*, *Mondia whitei* and *Protium opacum*. Shows moderate reversal of multi-drug resistance in *S. aureus*. Cryst. ($CHCl_3$/MeOH). Mp 238 - 241° (226-228°). λ_{max} 231 (ε 33000); 257 (sh); 286 (ε 7800); 323 (ε 13950) (MeOH)

9′-Deoxy, 6-O-de-Me
Jatrocin A
[664305-65-1]
$C_{19}H_{16}O_7$ 356.33
Constit. of *Jatropha gossypifolia*. Viscous gum. Racemate

9′-Deoxy, 5-methoxy
5-Methoxypropacin
[883133-64-0]
$C_{21}H_{20}O_8$ 400.383
Constit. of the wood of *Protium unifoliolatum*. Amorph. solid

5′-Methoxy
Aquillochin. Cleomiscosin C
[84575-10-0]
[121587-17-5] *trans*-form
$C_{21}H_{20}O_9$ 416.382
Constit. of *Acer saccharum*, *Annona squamosa*, *Aquilaria agallocha*, *Brucea javanica*, *Cleome viscosa*, *Cleome icosandra*, *Rhododendron colletianum* and *Xanthium sibiricum*. Exhibits antiinflammatory activity. Potent tyrosinase inhibitor and also inhibits lipid peroxidation. Needles (MeOH/EtOAc). Sol. MeOH, $CHCl_3$; poorly sol. H_2O. Mp 255°. λ_{max} 232 (ε 25600); 280 (ε 4160); 322 (ε 12160) (MeOH) (Berdy). [base] λ_{max} 258; 328 (MeOH/NaOH) (Berdy)

5′-Methoxy, 4′-O-β-D-glucopyranoside
Yinxiancaoside C
[1159908-74-3]
$C_{27}H_{30}O_{14}$ 578.523
Constit. of *Chloranthus japonicus*. Amorph. powder. Mp 230 - 233°. $[\alpha]_D^{20}$ -19.1 (c, 0.1 in MeOH). λ_{max} 210 (log ε 2); 326 (log ε 0.8) (MeOH)

5′-Methoxy, 9′-Ac
Durantin C
[586968-92-5]
$C_{23}H_{22}O_{10}$ 458.419
Constit. of *Duranta repens*. Phosphodiesterase inhibitor. Creamy yellow amorph. solid. Racemate. λ_{max} 233 (log ε 3.28); 323 (log ε 4.1) (MeOH)

5′-Methoxy, 9′-O-(3,4-dihydroxybenzoyl)
Repenin D
[1174169-16-4]
$C_{28}H_{24}O_{12}$ 552.488
Constit. of *Duranta repens*. Shows antioxidant activity. Light yellow solid. Mp 225 - 227° dec.. Racemate

5′-Methoxy, 9′-deoxy
9′-Deoxycleomiscosin C. *5′-Methoxypropacin. Jatrocin B*
[851724-27-1]
[664305-66-2] (Jatrocin B)
$C_{21}H_{20}O_8$ 400.383
Constit. of *Hibiscus syriacus*, *Jatropha gossypifolia*, *Mondia whitei* and *Xanthium sibiricum*. Inhibits lipid peroxidation. Pale yellow solid. Mp 240 - 244°. $[\alpha]_D$ -6.4 (c, 0.7 in $CHCl_3$/MeOH). λ_{max} 212 (log ε 4.67); 233 (sh); 322 (log ε 3.91) (MeOH)

5′-Methoxy, 9′-deoxy, 4′-Me ether
4′-O-Methyljatrocin B
[1169705-33-2]
$C_{22}H_{22}O_8$ 414.41
Constit. of *Eurycorymbus cavaleriei*

6-Demethoxy
Demethoxyisodaphneticin
[1174499-52-5]
$C_{19}H_{16}O_7$ 356.33
Constit. of *Daphne giraldii*

6-Demethoxy, 5′-methoxy
Isodaphneticin
[477770-78-8]
$C_{20}H_{18}O_8$ 386.356
Constit. of the leaves of *Daphne gnidium* and *Daphne pedunculata*. Powder. Mp 248 - 249°. $[\alpha]_D$ +11 (c, 1.4 in $CHCl_3$) (synthetic). λ_{max} 260 (log ε 0.8); 320 (log ε 1.1) (MeOH)

5-Chloro, 9′-deoxy
5-Chloropropacin
[851724-28-2]
$C_{20}H_{17}ClO_7$ 404.799
Constit. of the roots of *Mondia whitei*. Amorph. solid

8′-Epimer
8′-Epicleomiscosin A
[848366-79-0]
$C_{20}H_{18}O_8$ 386.356
Constit. of *Rhododendron collettianum*. Tyrosinase inhibitor. Amorph.

powder. $[\alpha]_D^{25}$ +15.5 (c, 0.18 in Py). Possesses 7′R,8′S-config.. λ_{max} 214 (log ε 2.04); 219 (log ε 2.53); 329 (log ε 2.09) (MeOH).

Zoghbi, M.D.G.B. *et al., Phytochemistry*, 1981, **20**, 180 (*Propacin*).
Bhandari, P. *et al., Phytochemistry*, 1982, **21**, 2147-2149 (*Cleomiscosin C*).
Arnoldi, A. *et al., Heterocycles*, 1984, **22**, 1537-1544 (*Cleomiscosins A,C, synth*).
Arisawa, M. *et al., J. Nat. Prod.*, 1984, **47**, 300-307 (*Cleomiscosin A, isol*).
Tanaka, H. *et al., Chem. Pharm. Bull.*, 1985, **33**, 3218-3223 (*Cleomiscosin A, synth*).
Ray, A.B. *et al., Tetrahedron*, 1985, **41**, 209-214 (*synth*).
Tanaka, H. *et al., Chem. Pharm. Bull.*, 1988, **36**, 1738-1743 (*Propacin, synth*).
Tanaka, H. *et al., Chem. Pharm. Bull.*, 1988, **36**, 3833-3837 (*Cleomiscosin C, synth*).
Tanaka, H. *et al., Heterocycles*, 1988, **27**, 2651-2658 (*Cleomiscosins A,C, synth*).
Lin, L.-J. *et al., J. Chem. Res., Miniprint*, 1988, 3052-3080 (*synth*).
Luyengi, L. *et al., Phytochemistry*, 1996, **43**, 409-412 (*isol*).
Das, B. *et al., Nat. Prod. Lett.*, 1999, **13**, 293-297 (*Venkatasin*).
Cheng, X.F. *et al., Fitoterapia*, 2000, **71**, 341-342 (*5″-Demethylaquillochin*).
Yun, B.S. *et al., J. Nat. Prod.*, 2001, **64**, 1238-1240 (*9′-Deoxycleomiscosin C, activity*).
Cottiglia, F. *et al., Magn. Reson. Chem.*, 2002, **40**, 551-553 (*Isodaphneticin*).
Ren, X. *et al., Tetrahedron: Asymmetry*, 2002, **13**, 1799-1804 (*Isodaphneticin, synth*).
Iqbal, K. *et al., Heterocycles*, 2003, **60**, 151-157 (*Durantins A,B,C, Cleomiscosin A, activity*).
Das, R. *et al., Int. J. Chem. Sci.*, 2003, **1**, 159-164 (*Jatrocins A,B*).
Ahmad, V.I. *et al., Chem. Pharm. Bull.*, 2004, **52**, 1458-1461 (*8′-Epicleomiscosin A*).
Liao, S.-G. *et al., Helv. Chim. Acta*, 2005, **88**, 2873-2878 (*Demethoxyisodaphneticin*).
Patnam, R. *et al., Phytochemistry*, 2005, **66**, 683-686 (*Mondia whitei constits*).
Marquez, B. *et al., Phytochemistry*, 2005, **66**, 1804-1811 (*Propacin, activity*).
Sajeli, B. *et al., Chem. Pharm. Bull.*, 2006, **54**, 538-541 (*Cleomiscosin A, Venkatasin, isol, cryst struct*).
Magalhaes, A. *et al., Nat. Prod. Res.*, 2006, **20**, 43-46 (*5-Methoxypropacin*).
Iizuka, T. *et al., Biol. Pharm. Bull.*, 2007, **30**, 1164-1166 (*Acer nikoense constits, activity*).
Xu, J.-F. *et al., Chem. Biodiversity*, 2008, **5**, 591-597 (*Malloapelin A*).
Kuboki, A. *et al., Tetrahedron Lett.*, 2008, **49**, 4516-4518 (*Cleomiscosin C, synth*).
Kuang, H.-X. *et al., Arch. Pharmacal Res.*, 2009, **32**, 329-334 (*Yinxiancaoside C*).
Ahmad, N. *et al., Bioorg. Med. Chem. Lett.*, 2009, **19**, 3521-3524 (*Repenins A-D*).
Cheng, L. *et al., J. Chromatogr., A*, 2009, **1216**, 4859-4867 (*Cleomiscosins A,C, Jatrocin B, ms*).
Begum, S. *et al., Fitoterapia*, 2010, **81**, 178-184 (*4′-O-Methylcleomiscosin A, activity*).
Cao, S. *et al., Phytochemistry*, 2010, **71**, 95-99 (*4′-Cinnamoylcleomiscosin A*).
Begum, S.A. *et al., Prog. Chem. Org. Nat. Prod.*, 2010, **93**, 1-28 (*rev, struct, activity*).
Meena, A. *et al., Chem. Biol. Drug Des.*, 2011, **78**, 567-579 (*antiinflammatory activity*).
Yoshikawa, K. *et al., J. Nat. Med.*, 2011, **65**, 191-193 (*Acer saccharum constits*).
Kan, S. *et al., Nat. Prod. Res.*, 2011, **25**, 1243-1249 (*Cleomiscosins A,C, Xanthium constits*).
Akbar, E. *et al., Asian J. Chem.*, 2012, **24**, 4831-4842 (*rev*).
Yang, J. *et al., Helv. Chim. Acta*, 2014, **97**, 278-282 (*Brucea javanica constits*).
Yan, Z.-Y. *et al., Fitoterapia*, 2018, **130**, 234-240 (*Cleomiscosin A, resoln, cd*).
Begum, A.S. *et al., Top. Curr. Chem.*, 2018, **376**, 34 (*rev, chem, pharmacol*)

Cleomiscosin B C83

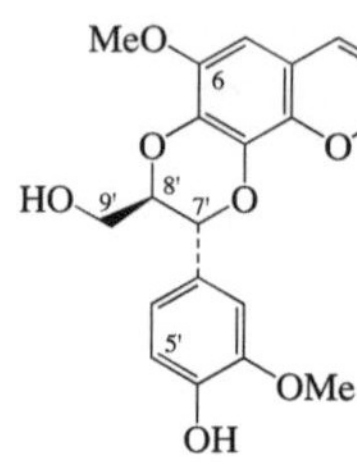

[76985-93-8]
[121587-24-4] (*trans*-form)
[1137653-03-2] (*S,S*-form)
$C_{20}H_{18}O_8$ 386.356
Coumarinolignan and regioisomer of Cleomiscosin A, C82. Variable enantiochem. shown by Cleomiscosin B and its relatives. Cleomiscosin B predominantly isol. as a racemate, but Hyosgerin had the (*R,R*)- abs. config. illus. The other nat. prods. incl. in this entry have not been fully investigated. Constit. of *Annona squamosa, Cleome viscosa, Erycibe obtusifolia, Hyoscyamus niger, Mallotus apelta, Salsola laricifolia* and other plant spp.. Shows hepatoprotectant activity. Cryst.. Mp 274°. $[\alpha]_D^{25}$ -50 (c, 0.3 in MeOH) (synthetic (*R,R*)-form). λ_{max} 285; 328 (EtOH) (Berdy)

9′-Ac
Hyosgerin
[897929-32-7]
$C_{22}H_{20}O_9$ 428.393
Constit. of the seeds of *Hyoscyamus niger*. Yellow cryst. (MeOH). Mp 198 - 199°. $[\alpha]_D^{25}$ -65.4 (c, 0.38 in $CHCl_3$) (100% o.p.). λ_{max} 232 (log ε 4.29); 287 (log ε 3.71); 322 (log ε 3.87) (MeOH)

6-O-De-Me
6-O-Demethylcleomiscosin B
[1864843-80-0]
$C_{19}H_{16}O_8$ 372.329
Constit. of the aerial parts of *Euphorbia antiquorum*. Amorph. solid. $[\alpha]_D^{23.3}$ -24 (c, 0.15 in MeOH). Possesses (7′S,8′S)- abs. config. Lit. contains an error in the diag.. λ_{max} 274 (log ε 4.17); 334 (log ε 4.10) (no solvent reported)

5′-Hydroxy
5′-Hydroxycleomiscosin B. ***Malloapelin C***
[1109272-11-8]
[1245817-31-5] (5′-Hydroxycleomiscosin B)
$C_{20}H_{18}O_9$ 402.355
Constit. *Eurycorymbus cavaleriei, Excoecaria acerifolia* and *Mallotus apelta*. Shows quinone reductase induction and moderate cytotoxic activity. Amorph. powder. Mp 275 - 276° (Malloapelin C). $[\alpha]_D^{25}$ -0.7 (c, 0.1 in MeOH) (5′-Hydroxycleomiscosin B). Malloapelin C was racemic. λ_{max} 228; 254; 327 (MeOH)

5′-Hydroxy, O⁶-de-Me
Malloapelin B
[1109272-10-7]
$C_{19}H_{16}O_9$ 388.328
Constit. of the roots of *Mallotus apelta*. Amorph. powder. Mp 285 - 286°. Racemate. λ_{max} 232; 254; 323 (MeOH)

4′-Deoxy, 5′-hydroxy
Hemidesmin 2
[143050-12-8]
$C_{20}H_{18}O_8$ 386.356
Constit. of *Hemidesmus indicus* and *Jatropha curcas*. Amorph.. Mp 245 - 250°. Stereochem. not determined

9′-Deoxy
Isopropacin. ***Jatrophin‡***
[118141-23-4]
[90523-83-4] (*S,S*-form)
[92077-88-8] (*trans*-(±)-form)
[113085-00-0] (unspecified stereochem.) [121700-72-9] (*trans*-form)
$C_{20}H_{18}O_7$ 370.357
Isol. from *Jatropha glandulifera* and *Jatropha curcas*. Needles (EtOAc/petrol). Mp 245 - 248°. $[\alpha]_D$ -56

5′-Methoxy
Cleomiscosin D
[94062-49-4]
$C_{21}H_{20}O_9$ 416.382

Constit. of *Cleome viscosa*, *Hibiscus syriacus*, *Paederia scandens*, *Salsola laricifolia* and *Zanthoxylum avicennae*. Shows antiinflammatory activity. Needles (MeOH/EtOAc). Mp 243 - 244°

5′-Methoxy, 4′-Me ether

4′-O-Methylcleomiscosin D

[1002753-44-7]

$C_{22}H_{22}O_9$ 430.409

Constit. of the stems of *Zanthoxylum avicennae*. Shows antiinflammatory activity. Needles ($CHCl_3$/MeOH). Mp 252 - 254°. $[\alpha]^{25}_{D}$ -15.2 (c, 0.11 in $CHCl_3$). λ_{max} 325 (log ε 4.26) (MeOH)

5′-Methoxy, O⁶-de-Me

Daphnecin

[1266611-47-5]

$C_{20}H_{18}O_9$ 402.355

Constit. of *Daphne mucronata*. Pale yellow solid. $[\alpha]^{25}_{D}$ +2 (c, 0.3 in MeOH). λ_{max} 324 (MeOH)

5′-Methoxy, O^6-de-Me, 4′-Me ether

Hemidesmin 1

[143050-11-7]

$C_{21}H_{20}O_9$ 416.382

Constit. of the roots of *Hemidesmus indicus*. Cryst. (EtOAc/MeOH). Mp 232 - 234°. Stereochem. not determined

6-Demethoxy, 5′-methoxy

Daphneticin

[83327-22-4]

$C_{20}H_{18}O_8$ 386.356

Constit. of *Daphne oleoides*, *Daphne tangutica* and *Excoecaria acerifolia*. Cryst. (MeOH/Me_2CO). Mp 235 - 238°. $[\alpha]^{25}_{D}$ -8 (c, 1.1 in $CHCl_3$) (synthetic). λ_{max} 242 (ε 9170); 260 (ε 8892); 317 (ε 11230) (MeOH) (Berdy)

6-Demethoxy, 5′-methoxy, 4′-O-β-D-glucopyranoside

Daphneticin glucoside

[231287-98-2]

[1073173-27-9] (*S,S*-form)

$C_{26}H_{28}O_{13}$ 548.497

Constit. of *Daphne oleoides*. Needles (MeOH). Mp 254 - 255°. $[\alpha]_D$ +23.5 (c, 0.1 in DMSO).

Zhuang, L.-G. *et al.*, *Phytochemistry*, 1983, **22**, 617-619 (*Daphneticin*).
Arnoldi, A. *et al.*, *Heterocycles*, 1984, **22**, 1537-1544 (*Cleomiscosin B, synth*).
Parthasarathy, M.R. *et al.*, *Indian J. Chem., Sect. B*, 1984, **23**, 1105 (*synth*).
Parthasarathy, M.R. *et al.*, *Phytochemistry*, 1984, **23**, 867-869 (*Jatrophin*).
Tanaka, H. *et al.*, *Heterocycles*, 1985, **23**, 1991-197 (*Cleomiscosin B, synth*).
Ray, A.B. *et al.*, *Tetrahedron*, 1985, **41**, 209-214 (*Cleomiscosin B, activity*).
Tanaka, H. *et al.*, *Chem. Pharm. Bull.*, 1988, **36**, 3833-3837 (*Cleomiscosin D, synth*).
Lin, L.-J. *et al.*, *J. Chem. Res., Miniprint*, 1988, 3052-3080 (*synth*).
Kumar, S. *et al.*, *Phytochemistry*, 1988, **27**, 636-638 (*Cleomiscosin D*).
Chen, M. *et al.*, *Zhiwu Xuebao*, 1988, **30**, 308 (*Jatrophin*).
Das, P.C. *et al.*, *Indian J. Chem., Sect. B*, 1992, **31**, 342-345 (*Hemidesmins 1,2*).
Ullah, N. *et al.*, *Phytochemistry*, 1999, **51**, 103-105 (*Daphneticin glucoside*).
Ren, X. *et al.*, *Tetrahedron: Asymmetry*, 2002, **13**, 1799-1804 (*Daphneticin, synth*).
Sajeli, B. *et al.*, *Chem. Pharm. Bull.*, 2006, **54**, 538-541 (*Hyosgerin, abs config*).
Xu, J.-F. *et al.*, *Chem. Biodiversity*, 2008, **5**, 591-597 (*Malloapelins B,C*).
Chen, J.-J. *et al.*, *J. Nat. Prod.*, 2008, **71**, 212-217 (*4′-O-Methylcleomiscosin D, activity*).
Ma, Z. *et al.*, *Fitoterapia*, 2009, **80**, 320-326 (*5′-Hydroxycleomiscosin B*).
Rasool, M.A. *et al.*, *J. Asian Nat. Prod. Res.*, 2010, **12**, 324-327 (*Daphnecin*).
Meena, A. *et al.*, *Chem. Biol. Drug Des.*, 2011, **78**, 567-579 (*activity*).
Akbar, E. *et al.*, *Asian J. Chem.*, 2012, **24**, 4831-4842 (*rev*).
Wu, X.-D. *et al.*, *J. Asian Nat. Prod. Res.*, 2013, **15**, 151-157 (*Excoecaria acerifolia constits, activity*).
Hu, J. *et al.*, *Chem. Nat. Compd. (Engl. Transl.)*, 2016, **52**, 48-51 (*6-O-Demethylcleomiscosin B*)

Cleosandrin C84

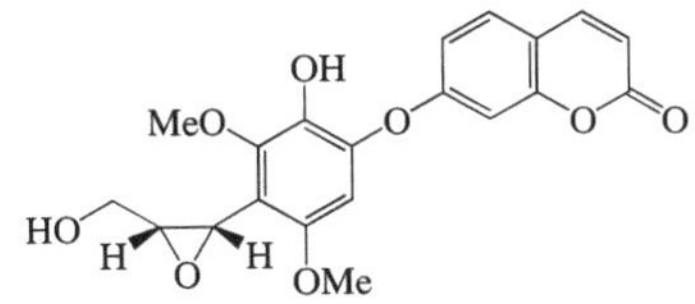

Relative Configuration

[74342-09-9]

$C_{20}H_{18}O_8$ 386.356

Struct. illus. in doubt. Cleosandrin may be the same or similar to Cleomiscosin A, C82. Constit. of the seeds of *Cleome icosandra* and *Cleome viscosa*.

Ramchandran, A.G. *Indian J. Chem., Sect. B*, 1979, **17**, 438-440 (*Cleosandrin*).
Ray, A.B. *et al.*, *Tetrahedron Lett.*, 1980, **21**, 4477-4480 (*struct*)

Cleroden A C85

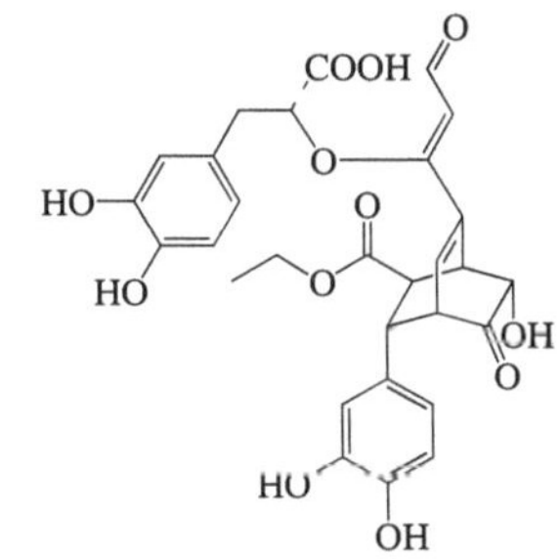

Absolute Configuration

$C_{29}H_{28}O_{12}$ 568.531

Related to Yunnaneic acid D, Y2. Constit. of *Clerodendranthus spicatus*. Exhibits moderate inhibitory activity on LPS-induced NO prodn.. Amorph. powder. $[\alpha]^{20}_{D}$ +237.6 (c, 0.17 in MeOH). λ_{max} 275 (log ε 3.28) (MeOH).

Ma, G.-X. *et al.*, *Fitoterapia*, 2015, **105**, 61-65 (*Cleroden A, activity, cd*)

Cleroden D C86

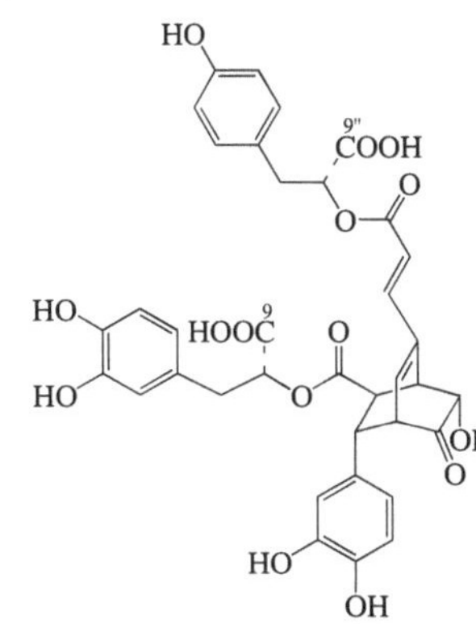

Absolute Configuration

$C_{36}H_{32}O_{15}$ 704.637

Closely related to Helisterculin B in C87 and Yunnaneic acid D, Y2. Constit. of *Clerodendranthus spicatus*. Exhibits significant inhibitory activity on LPS-induced NO prodn.. Amorph. powder. $[\alpha]^{20}_{D}$ +145.6 (c, 0.09 in MeOH). λ_{max} 275 (log ε 3.02) (MeOH)

9-Me ester

Cleroden C

$C_{37}H_{34}O_{15}$ 718.664

Constit. of *Clerodendranthus spicatus*. Exhibits moderate inhibitory activity on LPS-induced NO prodn.. Amorph. powder. $[\alpha]^{20}_{D}$ +174.1 (c, 0.14 in MeOH). λ_{max} 275 (log ε 3.25) (MeOH)

9″-Me ester

Cleroden B

$C_{37}H_{34}O_{15}$ 718.664

Constit. of *Clerodendranthus spicatus*. Exhibits moderate inhibitory activity on LPS-induced NO prodn.. Amorph. powder. $[\alpha]^{20}_{D}$ +145.7 (c, 0.12 in MeOH). λ_{max} 277 (log ε 3.45) (MeOH).

Ma, G.-X. *et al.*, *Fitoterapia*, 2015, **105**, 61-65 (*Clerodens B,C,D, activity, cd*)

Cleroden E C87

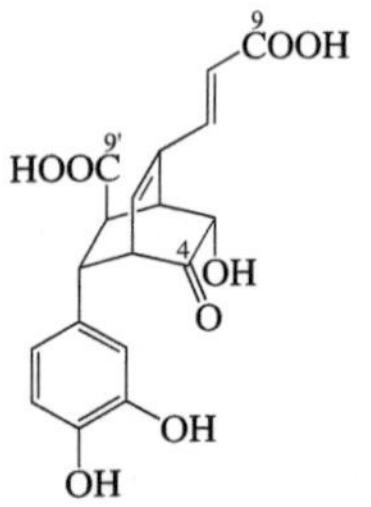

Absolute Configuration

5,7'-Cyclo-3,3',4'-trihydroxy-4-oxo-2,8'-ligna-1(6),7-dien-9,9'-dioic acid
$C_{18}H_{16}O_8$ 360.318
Caffeic acid dimer. Related to Yunnaneic acid D, Y2. Constit. of the aerial parts of *Clerodendranthus spicatus*. Exhibits moderate antibacterial activity against drug resistant bacteria *in vitro*. Amorph. brown powder. $[\alpha]^{25}_{D}$+378 (c, 0.5 in MeOH). λ_{max} 269 (log ε 4.22) (MeOH)

9-[1R-Carboxy-2-(3,4-dihydroxyphenyl)ethyl] ester
Cleroden F
$C_{27}H_{24}O_{12}$ 540.477
Constit. of the aerial parts of *Clerodendranthus spicatus*. Amorph. brown powder. $[\alpha]^{25}_{D}$+93 (c, 0.5 in MeOH). λ_{max} 272 (log ε 4.19) (MeOH)

9-[1R-Carboxy-2-(3,4-dihydroxyphenyl)ethyl] ester, 9'-Me ester
Cleroden G
$C_{28}H_{26}O_{12}$ 554.504
Constit. of the aerial parts of *Clerodendranthus spicatus*. Amorph. brown powder. $[\alpha]^{25}_{D}$+182 (c, 0.5 in MeOH). λ_{max} 271 (log ε 4.17) (MeOH)

9-[2-(3,4-Dihydroxyphenyl)-1R-(methoxycarbonyl)ethyl] ester
Cleroden H
$C_{28}H_{26}O_{12}$ 554.504
Constit. of the aerial parts of *Clerodendranthus spicatus*. Amorph. brown powder. $[\alpha]^{25}_{D}$+378 (c, 0.5 in MeOH). λ_{max} 274 (log ε 4.22) (MeOH)

9-[2-(3,4-Dihydroxyphenyl)-1R-(methoxycarbonyl)ethyl] ester, 9'-Me ester
Helisterculin A
[1115210-87-1]
[223436-38-2]
$C_{29}H_{28}O_{12}$ 568.531
Constit. of the fruit of *Helicteres isora*. Amorph. yellow solid. $[\alpha]^{26}_{D}$ +239.2 (c, 0.75 in MeOH). λ_{max} 276 (log ε 4.28) (MeOH)

9'-[1R-Carboxy-2-(3,4-dihydroxyphenyl)ethyl] ester
Cleroden I
$C_{27}H_{24}O_{12}$ 540.477
Constit. of the aerial parts of *Clerodendranthus spicatus*. Amorph. brown powder. $[\alpha]^{25}_{D}$+15 (c, 0.5 in MeOH). λ_{max} 269 (log ε 4.09) (MeOH)

9'-[2-(3,4-Dihydroxyphenyl)-1R-(methoxycarbonyl)ethyl] ester
Cleroden J
$C_{28}H_{26}O_{12}$ 554.504
Constit. of the aerial parts of *Clerodendranthus spicatus*. Amorph. brown powder. $[\alpha]^{25}_{D}$+139 (c, 0.5 in MeOH). λ_{max} 274 (log ε 4.07) (MeOH)

9,9'-Bis[1R-carboxy-2-(3,4-dihydroxyphenyl)ethyl] ester
Helisterculin C
$C_{36}H_{32}O_{16}$ 720.636
Constit. of *Clerodendranthus spicatus*. Amorph. brown powder. λ_{max} 272 (log ε 4.50) (MeOH)

9'-[2-(3,4-Dihydroxyphenyl)-1R-(methoxycarbonyl)ethyl], 9-[1R-carboxy-2-(3,4-dihydroxyphenyl)ethyl] ester
Helisterculin D
$C_{37}H_{34}O_{16}$ 734.663
Constit. of *Clerodendranthus spicatus*. Amorph. yellowish powder. $[\alpha]^{25}_{D}$+92 (c, 0.5 in MeOH). λ_{max} 271 (log ε 4.53) (MeOH)

4S-Alcohol, 9,9'-bis[2-(3,4-dihydroxyphenyl)-1R-(methoxycarbonyl)ethyl] ester
Helisterculin B
[223436-39-3]
$C_{38}H_{38}O_{16}$ 750.706
Constit. of the fruit of *Helicteres isora*. Amorph. yellow solid. $[\alpha]^{26}_{D}$ +51.9 (c, 1.5 in MeOH). λ_{max} 279 (log ε 4.36) (MeOH).

Tezuka, Y. *et al.*, *Helv. Chim. Acta*, 1999, **82**, 408-417 (*Helisterculins A,B*).
Snyder, S.A. *et al.*, *J. Am. Chem. Soc.*, 2009, **131**, 1745-1752 (*Helisterculin A, synth*).
Li, Q. *et al.*, *Fitoterapia*, 2016, **114**, 110-114 (*Clerodens E-J, cd, activity*).
Zhou, H.-C. *et al.*, *J. Asian Nat. Prod. Res.*, 2017, **19**, 974-980 (*Helisterculins C,D*)

Clinopodic acid I C88

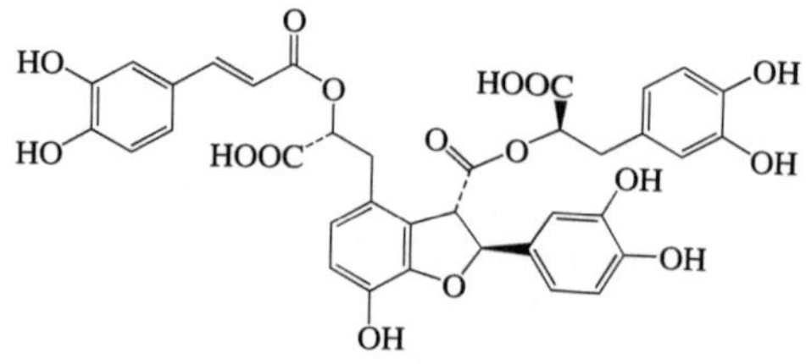

Absolute Configuration

[1174909-60-4]
$C_{36}H_{30}O_{16}$ 718.62
Constit. of *Clinopodium chinense* var. *parviflorum* and *Satureja biflora*. Amorph. solid. $[\alpha]^{23}_{D}$-76.6 (c, 0.55 in MeOH). λ_{max} 231 (sh); 289 (log ε 4.21); 331 (log ε 4.17) (MeOH)

Stereoisomer
[894784-31-7]
$C_{36}H_{30}O_{16}$ 718.62
Constit. of the leaves of *Celastrus hindsii*. Pale yellow-green wax + $4H_2O$. $[\alpha]^{25}_{D}$+78.3 (c, 0.3 in MeOH). Stereochem. undetermined. λ_{max} 287 (log ε 4.39); 331 (log ε 4.33) (MeOH).

Ly, T.N. *et al.*, *J. Agric. Food Chem.*, 2006, **54**, 3786-3793 (*Celastrus hindsii constit*).
Murata, T. *et al.*, *J. Nat. Prod.*, 2009, **72**, 1379-1384 (*Clinopodic acid I*)

Clinopodic acid J C89

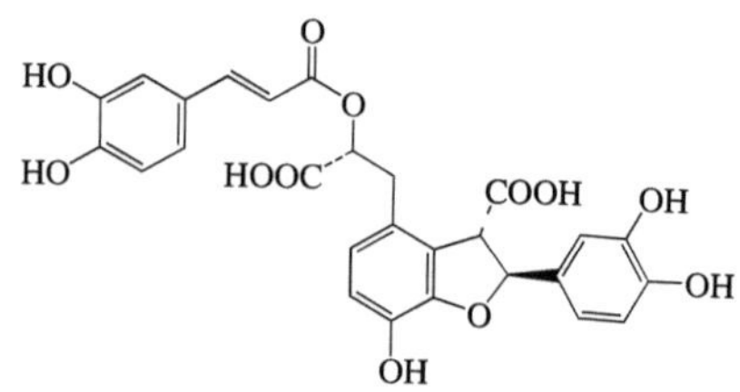

Absolute Configuration

[1427521-24-1]
$C_{27}H_{22}O_{12}$ 538.461
Isol. from *Clinopodium gracile*. Amorph. solid. $[\alpha]^{23}_{D}$-49.1 (c, 1.08 in MeOH). λ_{max} 290 (log ε 4.12); 329 (log ε 4.12) (MeOH).

Aoshima, H. *et al.*, *Chem. Pharm. Bull.*, 2012, **60**, 499-507 (*Clinopodic acid J*)

Clinopodic acid K C90

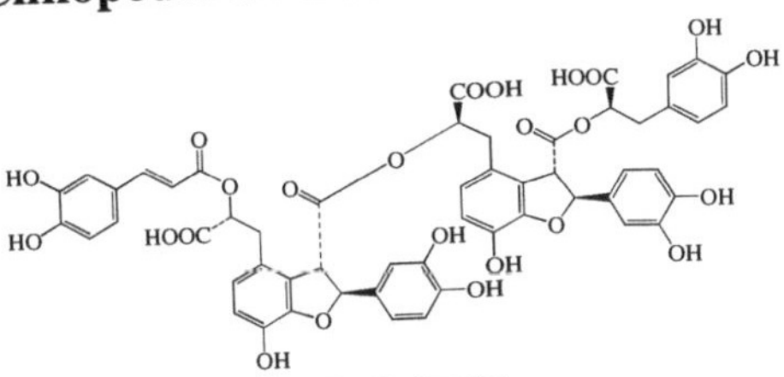

Absolute Configuration

[1427521-25-2]
$C_{54}H_{44}O_{24}$ 1076.922
Constit. of *Clinopodium gracile* and *Satureja biflora*. Amorph. solid. $[\alpha]^{23}_{D}$ -57.0 (c, 0.98 in MeOH). λ_{max} 287 (log ε 4.34); 329 (log ε 4.19) (MeOH).
Aoshima, H. *et al.*, *Chem. Pharm. Bull.*, 2012, **60**, 499-507 (*Clinopodic acid K*)

Clinopodic acid L C91

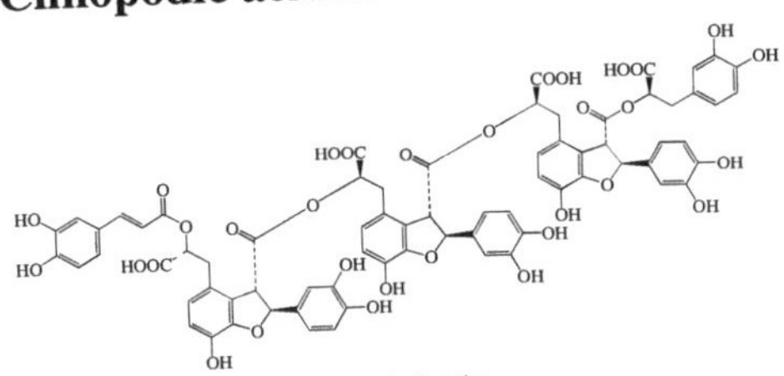

Absolute Configuration

[1427521-26-3]
$C_{72}H_{58}O_{32}$ 1435.224
Constit. of *Clinopodium gracile*. Shows hyaluronidase inhibitory activity. Amorph. solid. $[\alpha]^{23}_{D}$ -41.0 (c, 1.0 in MeOH). λ_{max} 287 (log ε 4.36); 329 (log ε 4.15) (MeOH).
Aoshima, H. *et al.*, *Chem. Pharm. Bull.*, 2012, **60**, 499-507 (*Clinopodic acid L, activity*)

Clinopodic acid M C92

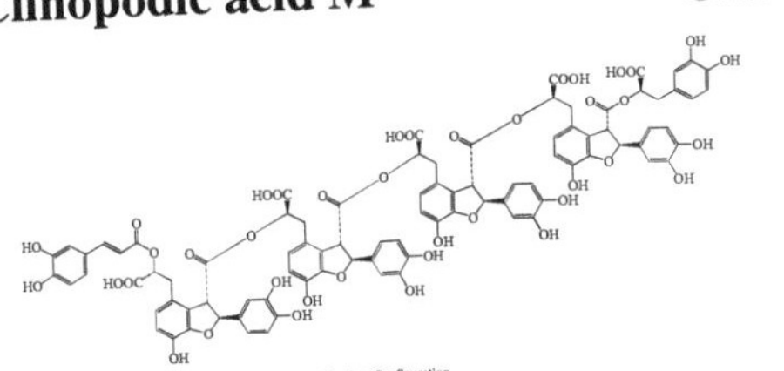

Absolute Configuration

[1427521-27-4]
$C_{90}H_{72}O_{40}$ 1793.526
Isol. from *Clinopodium gracile*. Displays significant hyaluronidase inhibitory activity. Amorph. solid. $[\alpha]^{23}_{D}$ -36.2 (c, 1.12 in MeOH). λ_{max} 286 (log ε 4.44); 330 (log ε 4.15) (MeOH).
Aoshima, H. *et al.*, *Chem. Pharm. Bull.*, 2012, **60**, 499-507 (*Clinopodic acid M*)

Clinopodic acid O C93

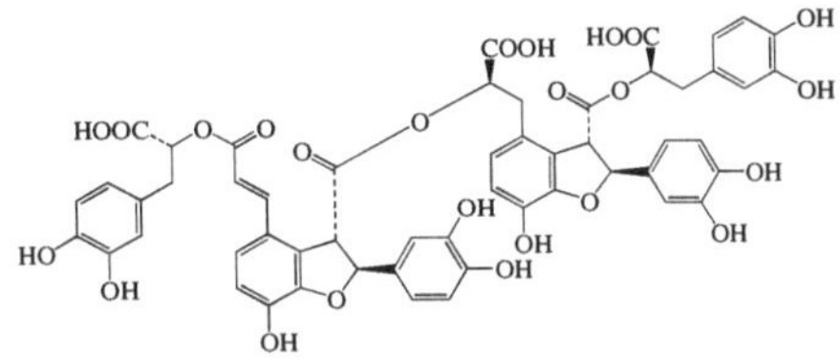

Absolute Configuration

[1427521-29-6]
$C_{54}H_{44}O_{24}$ 1076.922
Isol. from *Clinopodium gracile* and *Satureja biflora*. Amorph. solid. $[\alpha]^{23}_{D}$ +33.5 (c, 1.0 in MeOH). λ_{max} 287 (log ε 4.22); 331 (log ε 4.13) (MeOH).
Aoshima, H. *et al.*, *Chem. Pharm. Bull.*, 2012, **60**, 499-507 (*Clinopodic acid O*)

Clinopodic acid P C94

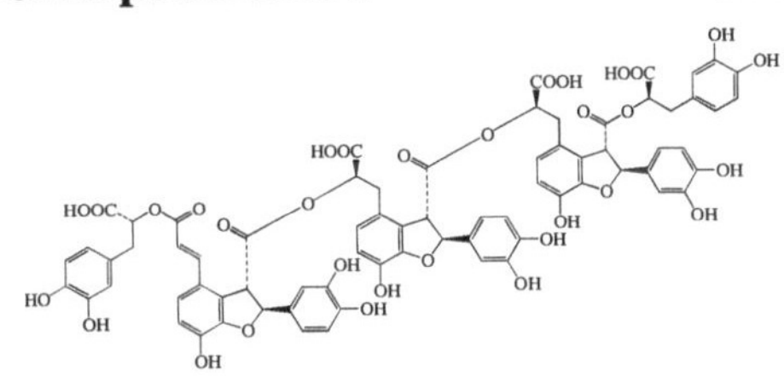

Absolute Configuration

[1427521-30-9]
$C_{72}H_{58}O_{32}$ 1435.224
Isol. from *Clinopodium gracile* and *Satureja biflora*. Displays moderate hyaluronidase inhibitory activity. Amorph. solid. $[\alpha]^{23}_{D}$ +42.9 (c, 0.85 in MeOH). λ_{max} 286 (log ε 4.35); 330 (log ε 4.17) (MeOH).
Aoshima, H. *et al.*, *Chem. Pharm. Bull.*, 2012, **60**, 499-507 (*Clinopodic acid P, activity*)

Clovanedunnianol C95

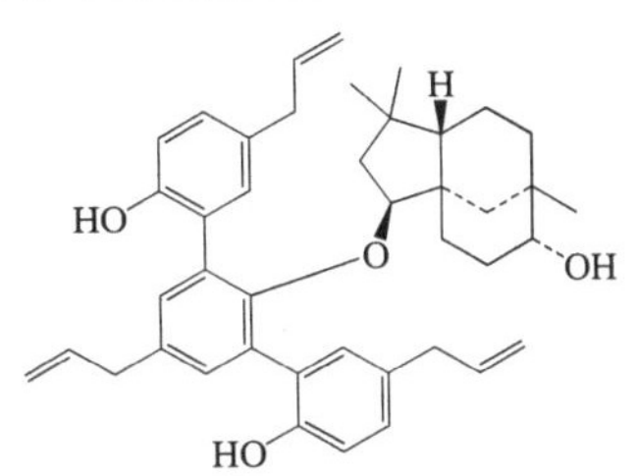

Relative Configuration

[1582261-58-2]
$C_{42}H_{50}O_{4}$ 618.858
Constit. of *Illicium simonsii*. Moderate ACE inhibitor. Gum. $[\alpha]^{22}_{D}$ -8.2 (c, 0.2 in MeOH). λ_{max} 217 (log ε 4.04); 290 (log ε 3.3) (MeOH).
Dong, C. *et al.*, *Planta Med.*, 2013, **79**, 338-347 (*Clovanedunnianol*)

Clovanemagnolol C96

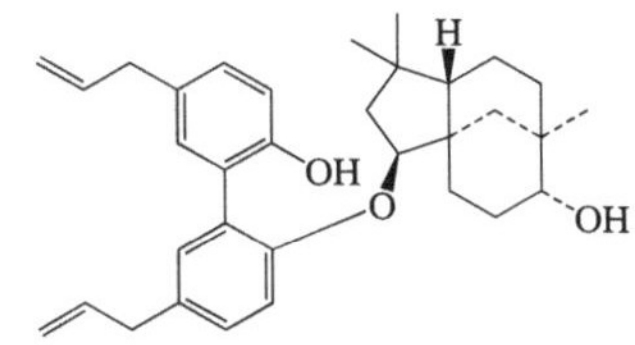

Relative Configuration

[130756-35-3]
$C_{33}H_{42}O_{3}$ 486.696
Constit. of bark of *Magnolia obovata*. Shows neurotrophic activity. Oil. $[\alpha]_{D}$ +21 (c, 1.5 in $CHCl_3$). λ_{max} 204 (ε 4600); 208 (ε 41000); 286 (ε 5800) (EtOH) (Derep).
Fukuyama, Y. *et al.*, *Tetrahedron*, 1992, **48**, 377-392 (*Clovanemagnolol*).
Khaing, Z. *et al.*, *Bioorg. Med. Chem. Lett.*, 2011, **21**, 4808-4812 (*activity*).
Cheng, X. *et al.*, *Org. Biomol. Chem.*, 2012, **10**, 383-393 (*synth, activity*)

Cochinchin C97

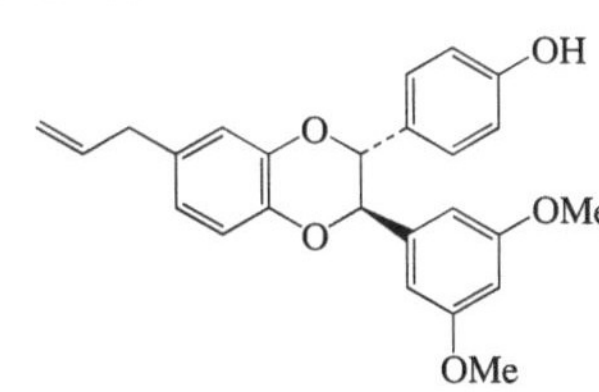

Relative Configuration

[796039-33-3]
$C_{25}H_{24}O_{5}$ 404.462
Constit. of the resin of *Dracaena cochinchinensis*. $[\alpha]_{D}$ -30.2 (c, 0.005 in Me_2CO).
He, L. *et al.*, *Chin. J. Chem.*, 2004, **22**, 867-869 (*Cochinchin*)

Codonopiloneolignanin A C98

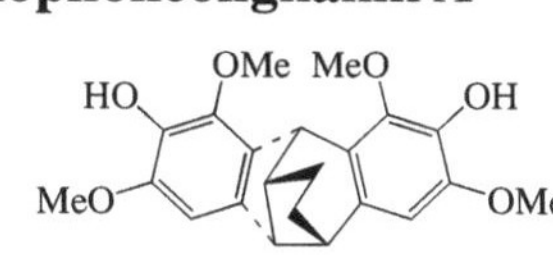

Absolute Configuration

[1865780-13-7]
$C_{22}H_{24}O_{6}$ 384.428
Constit. of the roots of *Codonopsis pilosula*. Amorph. yellow powder. $[\alpha]^{20}_{D}$ +19.3 (c, 0.06 in MeOH). λ_{max} 203 (log ε 4.72); 288 (log ε 3.57) (MeOH).
Jiang, Y.-P. *et al.*, *Chin. Chem. Lett.*, 2016, **27**, 55-58 (*Codonopiloneolignanin A*).
Gao, Z. *et al.*, *Org. Lett.*, 2021, **23**, 5684-5688 (*synth*)

Coelobillardin C99

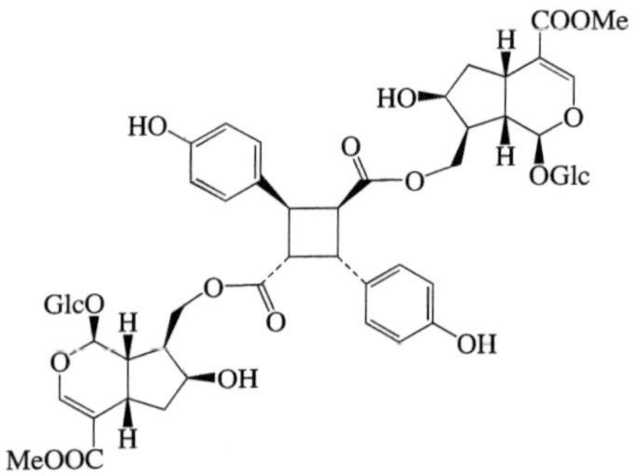

[452962-76-4]

$C_{52}H_{64}O_{26}$ 1105.058

Constit. of *Coelospermum billardieri*. [α] D -55.7 (c, 1 in MeOH). λ_{max} 235 (log ε 4.46); 280 (log ε 3.27) (MeOH).

Magiatis, P. *et al.*, *Phytochemistry*, 2002, **60**, 415-418 (*Coelobillardin*)

Coenzyme A, CAS C100

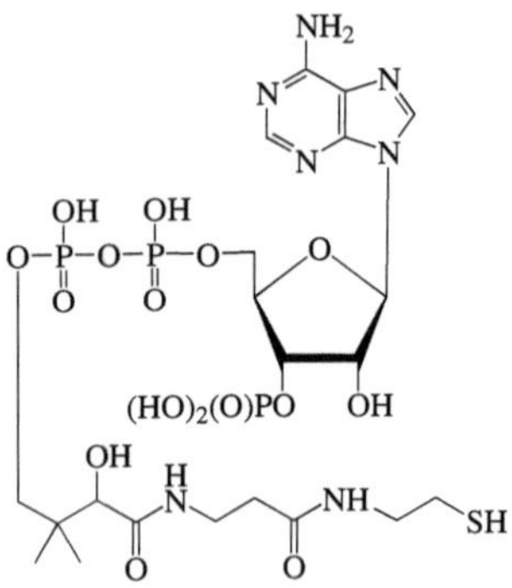

Adenosine 5′-(trihydrogen diphosphate) 3′-(dihydrogen phosphate) 5′-[3-hydroxy-4-[[3-[(2-mercaptoethyl) amino]-3-oxopropyl]amino]-2,2-dimethyl-4-oxobutyl] ester. Aluzime. CoA. Coalip. Lucina

[85-61-0]

[31416-98-5] [18439-24-2] (tri-Li salt)

[55672-92-9] (Na salt)

$C_{21}H_{36}N_7O_{16}P_3S$ 767.533

Constit. of many microorganisms. A cofactor in enzymatic acetyl transfer reactions. *S*-Acyl derivs (see below) are biochemically important in acyl transfer reactions. They are powerful inhibitors of *de novo* fatty acid biosynthesis. Powder. Sol. H_2O. Inactivated by phosphatase. Oxid. by air to inactive disulfide. λ_{max} 260 (ε 16800) (H_2O)

S-Ac

Acetylcoenzyme A. Acetyl-CoA

[72-89-9]

[75520-41-1]

$C_{23}H_{38}N_7O_{17}P_3S$ 809.57

The most important metabolic acyl transfer agent. In aerobic organisms the Ac group is formed by degradn. of carbohydrates, aminoacids and fats; in anaerobic organisms by carbonylation of Me groups attached to tetrahydrofolic acid

S-Propanoyl

Propionylcoenzyme A. Propionyl-CoA

[317-66-8]

$C_{24}H_{40}N_7O_{17}P_3S$ 823.597

Formed by catabolism of some amino acids and by β-oxidation of odd-chain fatty acids. Enters the tricarboxylic acid cycle in humans via Methylmalonylcoenzyme A and Succinylcoenzyme A.

S-Butanoyl

[2140-48-9]

$C_{25}H_{42}N_7O_{17}P_3S$ 837.624

S-(2E-Butenoyl)

Crotonylcoenzyme A

[992-67-6]

$C_{25}H_{40}N_7O_{17}P_3S$ 835.608

S-Hexanoyl

[5060-32-2]

$C_{27}H_{46}N_7O_{17}P_3S$ 865.678

S-Octanoyl

[1264-52-4]

$C_{29}H_{50}N_7O_{17}P_3S$ 893.732

S-Decanoyl

[1264-57-9]

$C_{31}H_{54}N_7O_{17}P_3S$ 921.786

S-Dodecanoyl

[6244-92-4]

$C_{33}H_{58}N_7O_{17}P_3S$ 949.84

S-Tetradecanoyl

[3130-72-1]

$C_{35}H_{62}N_7O_{17}P_3S$ 977.894

S-Hexadecanoyl

Palmitoylcoenzyme A

[1763-10-6]

$C_{37}H_{66}N_7O_{17}P_3S$ 1005.948

S-Octadecanoyl

Stearoylcoenzyme A

[362-66-3]

$C_{39}H_{70}N_7O_{17}P_3S$ 1034.002

S-(9Z-Octadecenoyl)

Oleylcoenzyme A

[1716-06-9]

$C_{39}H_{68}N_7O_{17}P_3S$ 1031.986

S-(9Z,12Z-Octadecadienoyl)

[6709-57-5]

[59042-85-2] (unspecified stereochem.)

$C_{39}H_{66}N_7O_{17}P_3S$ 1029.97

S-(5Z,8Z,11Z,14Z-Eicosatetraenoyl)

[17046-56-9]

$C_{41}H_{66}N_7O_{17}P_3S$ 1053.992

S-(3-Methyl-2-butenoyl)

β-Methylcrotonylcoenzyme A

[6712-03-4]

$C_{26}H_{42}N_7O_{17}P_3S$ 849.635

S-(3-Oxobutanoyl)

Acetoacetylcoenzyme A

[1420-36-6]

$C_{25}H_{40}N_7O_{18}P_3S$ 851.607

S-(Carboxyacetyl)

Malonylcoenzyme A. 5-(Hydrogen propanedioate)coenzyme A, CAS

[524-14-1]

$C_{24}H_{38}N_7O_{19}P_3S$ 853.579

Involved in control of hepatic fatty acid synthesis and oxidation. Acts as a metabolic coupling factor in pancreatic β-cell signal transduction mechanisms linking insulin secretion and fatty acid metabolism. λ_{max} 236 (H_2O)

S-(Methoxycarbonylacetyl)

Methylmalonylcoenzyme A. Methylmalonyl-CoA

[1264-45-5]

$C_{25}H_{40}N_7O_{19}P_3S$ 867.606

Intermed. in the pathway from Propionylcoenzyme A to Succinylcoenzyme A.

S-(3-Carboxypropanoyl)

Succinylcoenzyme A. Succinyl-CoA

[604-98-8]

$C_{25}H_{40}N_7O_{19}P_3S$ 867.606

Participates in the tricarboxylic acid cycle in humans, as an intermed. between α-ketoglutarate and succinate. Also formed in the pathway from Propionylcoenzyme A.

S-(4-Carboxy-3-hydroxy-3-methylbutanoyl)

β-Hydroxy-β-methylglutarylcoenzyme A. HMG-CoA

[1553-55-5]

$C_{27}H_{44}N_7O_{20}P_3S$ 911.659

Biochemical precursor of mevalonic acid

S-(4-Hydroxycinnamoyl)

p-Coumaroylcoenzyme A

[30802-00-7]

$C_{30}H_{42}N_7O_{18}P_3S$ 913.678

Intermed. in biosynth. of lignans and C_9 lignan precursors

S-(3,4-Dihydroxycinnamoyl)

Caffeoylcoenzyme A

[53034-79-0]

[1021913-49-4] (*E*-form)

$C_{30}H_{42}N_7O_{19}P_3S$ 929.677

Intermed. in biosynth. of lignans and C_9 lignan precursors

S-(4-Hydroxy-3-methoxycinnamoyl)

Feruloylcoenzyme A

[30802-02-9]

[142185-30-6] (*E*-form)

$C_{31}H_{44}N_7O_{19}P_3S$ 943.704

Intermed. in biosynth. of lignans and C_9 lignan precursors

S-(3,4-Dihydroxy-5-methoxycinnamoyl)

5-Hydroxyferuloylcoenzyme A

[59862-14-5]

$C_{31}H_{44}N_7O_{20}P_3S$ 959.703

Intermed. in biosynth. of lignans and C_9 lignan precursors

S-(4-Hydroxy-3,5-dimethoxycinnamoyl)

Sinapoylcoenzyme A

[54429-80-0]

$C_{32}H_{46}N_7O_{20}P_3S$ 973.73

Intermed. in biosynth. of lignans and C_9 lignan precursors.

Lipmann, F. *et al., J. Biol. Chem.*, 1950, **186**, 235-243 (*purifn*).
Wilson, I.B. *J. Am. Chem. Soc.*, 1952, **74**, 3205-3206 (*synth*).
Lipmann, F. *Bacteriol. Rev.*, 1953, **17**, 1-16 (*rev*).
Kaufman, S. *et al., J. Biol. Chem.*, 1953, **203**, 869-888 (*synth*).
Simon, E.J. *et al., J. Am. Chem. Soc.*, 1953, **75**, 2520-2520 (*Succinylcoenzyme A*).
Baddiley, J. *Adv. Enzymol. Relat. Subj. Biochem.*, 1955, **16**, 1-21 (*rev*).
Hilz, H. *et al., Biochem. Z.*, 1958, **329**, 476-489 (*HMG-CoA, CrotonylCoA, β-Methylcrotonyl-CoA*).
Lynen, F. *et al., Biochem. Z.*, 1958, **330**, 269-295 (*HMG-CoA, AcetoacetylCoA*).
Goldman, P. *et al., J. Biol. Chem.*, 1961, **236**, 2620-2623 (*PalmitoylCoA, OleylCoA*).
Moffatt, J.G. *et al., J. Am. Chem. Soc.*, 1961, **83**, 663-675 (*synth*).
Eggerer, H. *et al., Biochem. Z.*, 1962, **335**, 540-547 (*MaleoylCoA*).
Kloss, R.A. *et al., Biochim. Biophys. Acta*, 1963, **70**, 90-91 (*MaleoylCoA*).
Bonner, W.M. *et al., J. Biol. Chem.*, 1972, **247**, 3123-3133 (*StearoylCoA*).
Pullman, M.E. *Anal. Biochem.*, 1973, **54**, 188-198 (*chromatog*).
Lee, C-H. *et al., J. Am. Chem. Soc.*, 1975, **97**, 1225-1236 (*conformn*).
Wilson, G.E. *et al., J. Am. Chem. Soc.*, 1975, **97**, 2907-2908 (*pmr*).
Wenganmeyer, H. *et al., Eur. J. Biochem.*, 1976, **65**, 529-536 (*CoumaroylCoA, CaffeoylCoA, FeruloylCoA, 5-HydroxyferuloylCoA, SinapoylCoA*).
Pucci, P. *et al., Clin. Ter. (Rome)*, 1977, **82**, 161-188 (*rev*).
Baker, F.C. *et al., Anal. Biochem.*, 1979, **94**, 417-424 (*anal, purifn, esters*).
Kawaguchi, A. *et al., J. Biochem. (Tokyo)*, 1981, **89**, 337-339 (*acyl derivs, synth, hplc, rev*).
Shimizu, S. *et al., FEBS Lett.*, 1983, **151**, 303-306 (*synth*).
Robishaw, J.D. *et al., Am. J. Physiol.*, 1985, **248**, E1-E9 (*metab, rev*).
Prentki, M. *et al., Diabetes*, 1996, **45**, 273-283 (*MaleoylCoA, rev*).
Ducasse-Cabanot, S. *et al., Antimicrob. Agents Chemother.*, 2004, **48**, 242-249 (*Acetoacetyl-CoA*).
Encyclopedia of Biological Chemistry, (eds. Lennarz, W.J. *et al*), Elsevier, 2004, **3**, 447-451 (*PropionylCoA, rev*).
Martindale, The Complete Drug Reference, 36th edn., Pharmaceutical Press, 2009, 2285.
Arnedo, M. *et al., J. Lipid Res.*, 2012, **53**, 2046-2056 (*HMG-CoA*)

Combretastatin D_2 C101

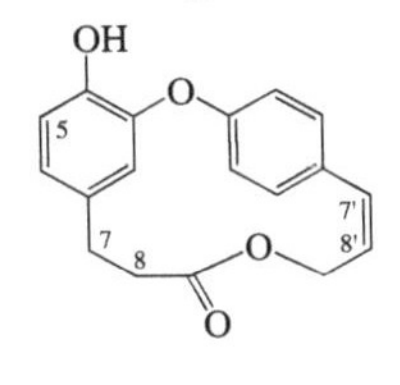

4-Hydroxy-3,4′-oxylign-7′-en-9,9′-olide

[126191-23-9]

$C_{18}H_{16}O_4$ 296.322

Constit. of *Combretum caffrum* and *Celastrus hindsii*. Cytostatic agent. Needles (Me_2CO/hexane). Mp 148 - 151°. λ_{max} 235 (ε 7300); 274 (ε 2260); 339 (ε 1050) (MeOH)

7′R,8′S-Epoxide

7′,8′-Epoxy-4-hydroxy-3,4′-oxylignan-9,9′-olide. ***Combretastatin D$_1$***

[117709-78-1]

$C_{18}H_{16}O_5$ 312.321

Constit. of *Combretum caffrum*. Needles (Me_2CO/hexane). Sol. MeOH, $CHCl_3$; poorly sol. H_2O, hexane. Mp 180 - 181°. $[\alpha]_D$ -100 (c, 0.015 in $CHCl_3$). λ_{max} 224 (ε 15200); 278 (ε 3070) ($CHCl_3$)

7′,8′-Dihydro

4-Hydroxy-3,4′-oxylignan-9,9′-olide. Dihydrocombretastatin D$_2$. ***Combretastatin D$_4$. Corniculatolide A***

[126191-24-0]

$C_{18}H_{18}O_4$ 298.338

Constit. of *Aegiceras corniculatum*, *Getonia floribunda* and *Lumnitzera littorea*. Amorph. solid. Mp 155.4 - 156.3° (Combretastatin D_4). Mp 175 - 177° (Corniculatolide A). λ_{max} 239 (log ε 3.81); 276 (log ε 3.51) ($CHCl_3$)

7′,8′-Dihydro, 4-Me ether

4-Methoxy-3,4′-oxylignan-9,9′-olide. ***11-O-Methylcorniculatolide A.*** *11-O-Methylcombretastatin D$_4$*

[1358550-64-7]

$C_{19}H_{20}O_4$ 312.365

Constit. of the mangrove plant *Aegiceras corniculatum*. Needles (Me_2CO). Mp 142 - 145°. λ_{max} 278 (ε 2480) (MeOH)

5-Hydroxy, 7′,8′-dihydro, 4-Me ether

5-Hydroxy-4-methoxy-3,4′-oxylignan-9,9′-olide. ***12-Hydroxy-11-O-methylcorniculatolide A***

[1358550-67-0]

$C_{19}H_{20}O_5$ 328.364

Constit. of *Aegiceras corniculatum* and *Lumnitzera littorea*. Needles (hexane/Me_2CO). Mp 244 - 245°. λ_{max} 275 (ε 1795) (MeOH)

5-Hydroxy, 7,8-didehydro, 4-Me ether

5-Hydroxy-4-methoxy-3,4′-oxyligna-7,7′-dien-9,9′-olide. ***Combretastatin D$_3$***

[862695-36-1]

$C_{19}H_{16}O_5$ 324.332

Constit. of the stems of *Getonia floribunda* and *Celastrus hindsii*. Cytotoxic. Shows weak antitubercular props.. Needles. Mp 212.6 - 213.1°. λ_{max} 239 (log ε 4.39); 288 (log ε 4.17) ($CHCl_3$)

5-Hydroxy, 7,8-didehydro, 7′,8′-dihydro, 4-Me ether

5-Hydroxy-4-methoxy-3,4′-oxylign-7-en-9,9′-olide. ***Hindsiilactone A***

[1624588-15-3]

$C_{19}H_{18}O_5$ 326.348

Constit. of the stems of *Celastrus hindsii*. Exhibits moderate cytotoxic activity. Cryst.. λ_{max} 208 (log ε 3.33); 278 (log ε 2.83) (MeOH).

Singh, S.B. *et al., J. Org. Chem.*, 1990, **55**, 2797-2800 (*Combretastatin D$_2$*).
Deshpande, V.H. *et al., Tetrahedron Lett.*, 1992, **33**, 4213-4216 (*Combretastatin D$_2$, synth*).
Rychnovsky, S.D. *et al., J. Org. Chem.*, 1994, **59**, 5414-5418 (*Combretastatins D$_1$,D$_2$, synth*).
Couladouros, E.A. *et al., Tetrahedron Lett.*, 1994, **35**, 4409-4412 (*Combretastatin D$_2$, synth*).
Rychnovsky, S.D. *et al., Tetrahedron Lett.*, 1994, **35**, 8927-8930 (*Combretastatin D$_1$, synth, abs config*).
Couladouros, E.A. *et al., Tetrahedron Lett.*, 1995, **36**, 9369-9372 (*Combretastatin D$_1$, synth*).
Gangakhedkar, K.K. *Synth. Commun.*, 1996, **26**, 1887-1896 (*Combretastatin D$_2$, synth*).
Vongvanich, N. *et al., Planta Med.*, 2005, **71**, 191-193 (*Combretastatins D$_3$,D$_4$*).
Cousin, D. *et al., Org. Biomol. Chem.*, 2006, **4**, 54-62 (*synth*).
Uno, K. *et al., Heterocycles*, 2008, **75**, 291-296 (*Combretastatin D$_4$, synth*).
Pettit, G.R. *et al., J. Nat. Prod.*, 2009, **72**, 876-883 (*Combretastatins D$_2$,D$_4$, synth*).
Ponnapalli, M.G. *et al., J. Nat. Prod.*, 2012, **75**, 275-279 (*Corniculatolides A*).

Hu, X.-Q. *et al., Phytochem. Lett.*, 2014, **7**, 169-172 (*Celastrus hindsii constits, Hindsiilactone A, activity, cryst struct*).
Harras, M. *et al., J. Org. Chem.*, 2017, **82**, 579-587 (*Combretastatins D_1,D_2,D_4, synth*)

Commiphorin C102

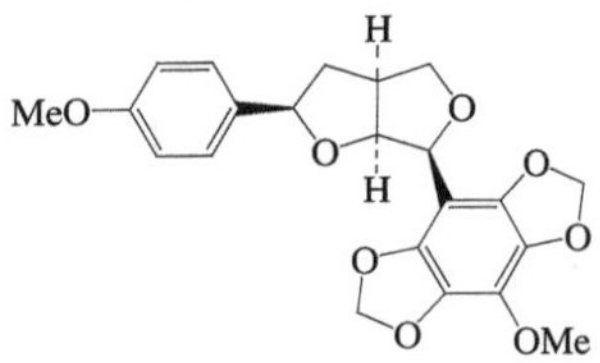

Relative Configuration

[895518-74-8]
$C_{22}H_{22}O_8$ 414.41
Constit. of the resin of *Commiphora mukul*. Exhibits antibacterial activity. Amorph. solid. $[\alpha]^{25}_D$ +97.3 (c, 0.1 in $CHCl_3$). λ_{max} 245 (ε 5370); 276 (ε 5370) ($CHCl_3$).
Sultana, N. *et al., Z. Naturforsch., B*, 2005, **60**, 1202-1206 (*Commiphorin, activity*)

Conicaol A C103

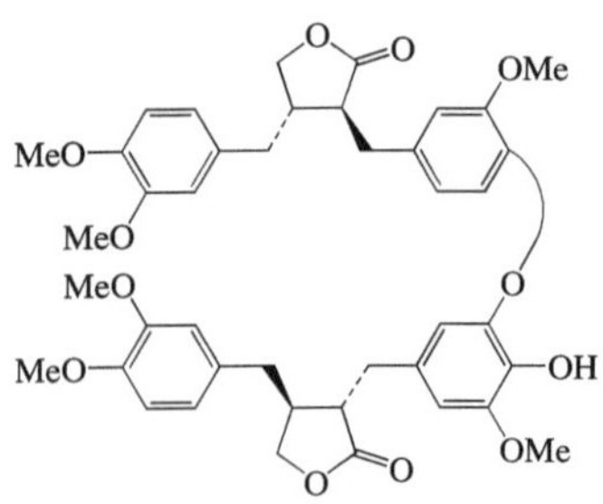

Relative Configuration

[916888-03-4]
$C_{42}H_{46}O_{12}$ 742.818
Constit. of *Saussurea conica*. Amorph. powder. $[\alpha]^{20}_D$ -28.2 (c, 0.57 in MeOH). λ_{max} 280 (log ε 4.08); 331 (log ε 3.3) (MeOH).
Fan, C.-Q. *et al., Planta Med.*, 2006, **72**, 590-595 (*Conicaol A*)

α-Conidendrin guaiacylglyceryl ether C104

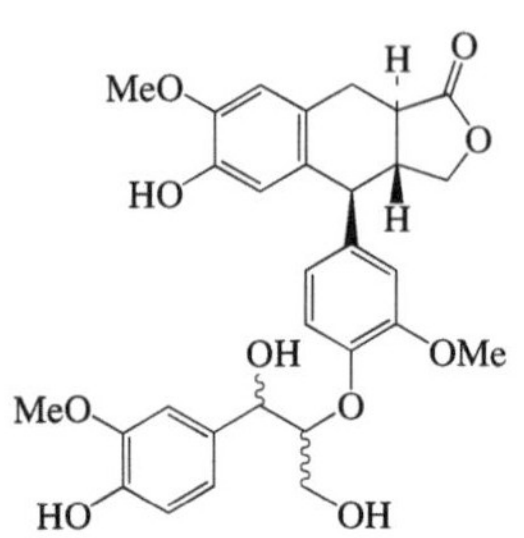

2,7′-Cyclo-4,4″,7″,9″-tetrahydroxy-3′,3″,5-trimethoxy-4′,8″-oxy-8,8′-sesquilignan-9,9′-olide
$C_{30}H_{32}O_{10}$ 552.576

(7′S,7″ξ,8R,8′R,8″ξ)-form
[845890-69-9]
Constit. of *Picea abies*.
Willfoer, S. *et al., Holzforschung*, 2004, **58**, 345-354 (*Picea abies constit*)

Conoliferine C105

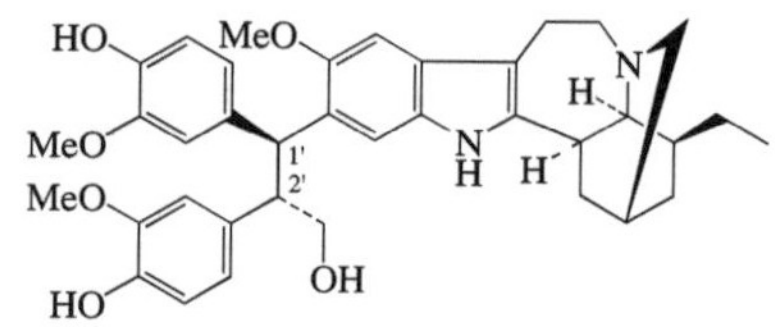

Absolute Configuration

[1169792-92-0]
$C_{37}H_{44}N_2O_6$ 612.767
Alkaloid from the stem bark of *Tabernaemontana corymbosa*. Isol. as inseparable mixt. with Isoconoliferine.
λ_{max} 229; 287; 301 (MeOH)
1′,2′-Diepimer
Isoconoliferine
[1169792-94-2]
$C_{37}H_{44}N_2O_6$ 612.767
Alkaloid from the stem bark of *Tabernaemontana corymbosa*.
Lim, K.-H. *et al., Tetrahedron Lett.*, 2009, **50**, 3756-3759 (*Conoliferine, Isoconoliferine*)

Cornigerin C106

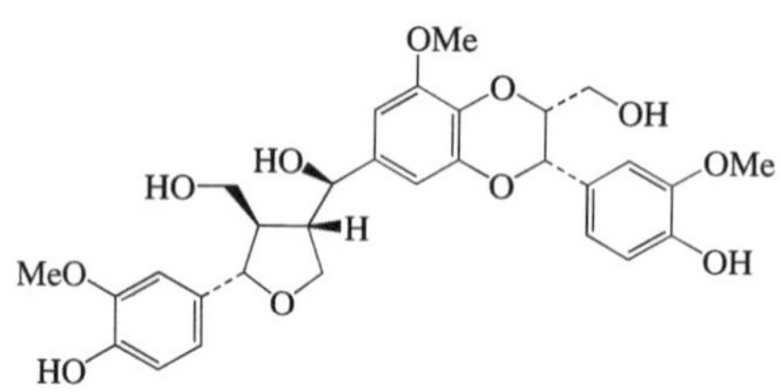

Relative Configuration

$C_{30}H_{34}O_{11}$ 570.591
Related to Cynarinin, C356. Constit. of the aerial parts of *Cynara cornigera*. λ_{max} 280 (MeOH).
El Sohafy, S.M. *et al., Fitoterapia*, 2016, **115**, 101-105 (*Cornigerin*)

Coumanolignan C107

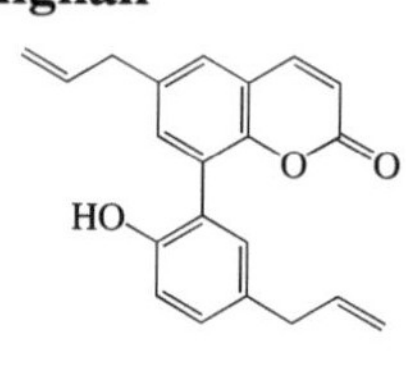

8-[2-Hydroxy-5-(2-propenyl)phenyl]-6-(2-propenyl)-2H-1-benzopyran-2-one
[1033749-97-1]
$C_{21}H_{18}O_3$ 318.372
Constit. of the stem bark of *Magnolia obovata*. Amorph. powder. Mp 156 - 158°. λ_{max} 240 (log ε 4.5); 290 (log ε 3.9) (no solvent reported).
Youn, U.J. *et al., Chem. Pharm. Bull.*, 2008, **56**, 115-117 (*Coumanolignan*)

Kadsura heteroclita Coumarinolignan C108

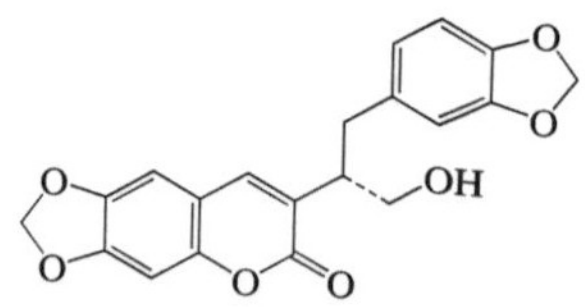

Absolute Configuration

7-[2-(1,3-Benzodioxol-5-yl)-1-(hydroxymethyl)ethyl]-6H-1,3-dioxolo[4,5-g][1]benzopyran-6-one, CAS
[1649427-44-0]
$C_{20}H_{16}O_7$ 368.341
Constit. of the stems of *Kadsura heteroclita*. Exhibits anti-HBV activity against HBeAg and HBsAg, and moderate antifibrotic and neuroprotective activity. Flaky cryst. ($CHCl_3$/MeOH). $[\alpha]^{20}_D$ +135.6 ($CHCl_3$/MeOH).
Su, W. *et al., Bioorg. Med. Chem. Lett.*, 2015, **25**, 1506-1508 (*Coumarinolignan, cryst struct, activity*)

1-*O*-*p*-Coumaroylglucose C109

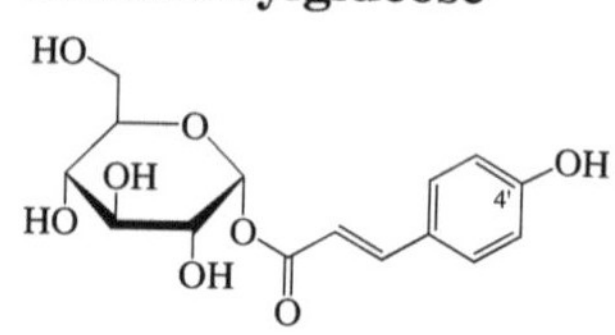

α-D-*form*

Glucose 1-[3-(4-hydroxyphenyl)-2-propenoate], CAS. *1-O-p-Hydroxycinnamoylglucose*
$C_{15}H_{18}O_8$ 326.301
α-D-*Pyranose-form* [1964461-48-0]
Constit. of the needles of *Cedrus deodara*. Needles
β-D-*Pyranose-form* [7139-64-2]
Isol. from many plants, e.g. *Solanum, Primula, Antirrhinum, Fragaria* spp.. Platelets (H_2O). Mp 212°. $[\alpha]^{20}_D$ -7.95 (50% MeOH aq.)
2-O-Sulfate
[85784-00-5]
$C_{15}H_{18}O_{11}S$ 406.358
Isol. from *Adiantum capillus-veneris*. Isol. as mixt. with the 6-sulfate. CAS no. refers to open chain compd.
3-O-Sulfate
$C_{15}H_{18}O_{11}S$ 406.358

Isol. from *Asplenium septentrionale*

6-O-Sulfate

[58115-23-4]

$C_{15}H_{18}O_{11}S$ 406.358

Isol. from *Adiantum capillus-veneris*, as mixt. with the 2-sulfate, and from *Pteridium aquilinum*. CAS no. refers to open chain compd.

3-Ac

[1583276-04-3]

$C_{17}H_{20}O_9$ 368.338

Isol. from *Solidago canadensis*. Displays moderate hypolipidaemic effects in HFD-fed hamsters. Amorph. powder. $[\alpha]_D^{23}$ +60.11 (c, 0.02 in MeOH). λ_{max} 228 (log ε 3.75); 315 (log ε 4.05) (MeOH)

3-O-(3,4,5-Trihydroxybenzoyl)

1-O-p-Coumaroyl-3-O-galloyl-β-D-glucopyranose

[1195785-60-4]

$C_{22}H_{22}O_{12}$ 478.406

Constit. of the rhizomes of *Balanophora harlandii*. Amorph. yellow powder. $[\alpha]_D^{25}$ -22.5 (c, 3.8 in MeOH). λ_{max} 193 (ε 17780); 201 (ε 23410); 316 (ε 17100) (MeOH)

3,4,6-Tris-O-(3,4,5-trihydroxybenzoyl)

1-O-p-Coumaroyl-3,4,6-trigalloyl-β-D-glucopyranose

[1219501-92-4]

$C_{36}H_{30}O_{20}$ 782.616

Constit. of the rhizomes of *Balanophora fungosa*. Yellow powder. $[\alpha]_D^{24}$ -9.2 (c, 0.3 in MeOH). λ_{max} 218 (log ε 4.15) (no solvent reported)

2′-Hydroxy, 4′-Me ether

[145227-80-1]

$C_{16}H_{20}O_9$ 356.327

Constit. of *Matricaria chamomilla*

3′-Hydroxy

1-O-Caffeoylglucose. Phoeniceoside

[13080-40-5]

[14364-08-0]

$C_{15}H_{18}O_9$ 342.300

Present in many plants, e.g. *Solanum, Primula, Cestrum, Raphanus, Begonia* etc. spp. Found in 31% of 204 plant spp. investigated. Amorph.. Mp 96 - 98°. $[\alpha]_D$ -47 (EtOH)

3′-Hydroxy, 2-O-sulfate

[81148-98-3]

$C_{15}H_{18}O_{12}S$ 422.357

Isol. from fresh fronds of *Ceterach officinarum*. Isol. as mixt. with other isomeric sulfates

3′-Hydroxy, 3-O-sulfate

[81148-97-2]

$C_{15}H_{18}O_{12}S$ 422.357

Isol. from fresh fronds of *Ceterach officinarum*. Isol. as mixt. with other isomeric sulfates

3′-Hydroxy, 4-O-sulfate

[81148-99-4]

$C_{15}H_{18}O_{12}S$ 422.357

Isol. from fresh fronds of *Ceterach officinarum*. Isol. as mixt. with other isomeric sulfates

3′-Hydroxy, 6-O-sulfate

[58115-22-3]

$C_{15}H_{18}O_{12}S$ 422.357

Isol. from *Adiantum* spp., *Ceterach officinarum* and *Pteridium aquilinum*

3′-Hydroxy, 3-O-α-L-rhamnopyranosyl

1-O-Caffeoylrungiose

[845736-11-0]

$C_{21}H_{28}O_{13}$ 488.442

Constit. of *Aeginetia indica*. Amorph. powder. $[\alpha]_D^{22}$ -54.5 (c, 0.25 in MeOH)

3′-Hydroxy, 3-O-(3,4,5-trihydroxybenzoyl)

1-O-Caffeoyl-3-O-galloyl-β-D-glucopyranose

[359819-42-4]

$C_{22}H_{22}O_{13}$ 494.405

Constit. of *Balanophora japonica*. Amorph. yellow powder. $[\alpha]_D^{15}$ -48.9 (c, 0.8 in MeOH)

3′-Hydroxy, 3,4-bis(3,4,5-trihydroxybenzoyl)

1-O-Caffeoyl-3,4-di-O-galloyl-β-D-glucopyranose

[359819-46-8]

$C_{29}H_{26}O_{17}$ 646.51

Constit. of *Balanophora japonica*. Amorph. yellow powder + 2½ H_2O. $[\alpha]_D^{15}$ -74.9 (c, 0.4 in MeOH)

3′-Hydroxy, 4,6-bis(3,4,5-trihydroxybenzoyl)

1-O-Caffeoyl-4,6-di-O-galloyl-β-D-glucopyranose

[359819-47-9]

$C_{29}H_{26}O_{17}$ 646.51

Constit. of *Balanophora japonica*. Amorph. yellow powder + $3H_2O$. $[\alpha]_D^{15}$ -177.8 (c, 0.2 in MeOH)

3′-Hydroxy, 3,4,6-tris-O-(3,4,5-trihydroxybenzoyl)

1-O-Caffeoyl-3,4,6-trigalloyl-β-D-glucopyranose

[1219501-93-5]

$C_{36}H_{30}O_{21}$ 798.615

Constit. of the rhizomes of *Balanophora fungosa*. Yellow powder. $[\alpha]_D^{24}$ -23.2 (c, 0.3 in MeOH). λ_{max} 219 (log ε 6.01); 293 (log ε 5.67); 330 (log ε 5.43) (EtOH)

3′-Hydroxy, 6-O-(3,4-dihydroxyphenylethyl) ether, 3-O-α-L-rhamnopyranosyl

Barleriside B

[1206462-83-0]

$C_{29}H_{36}O_{15}$ 624.592

Constit. of *Barleria acanthoides*. Gummy solid. $[\alpha]_D^{25}$ -45.7 (c, 0.001 in MeOH). λ_{max} 223 (log ε 4.27); 245 (log ε 4.1); 290 (log ε 4.2); 332 (log ε 4.25) (no solvent reported)

3′-(β-D-Glucopyranosyloxy)

[62018-87-5]

[133318-51-1] (*E*-form)

$C_{21}H_{28}O_{14}$ 504.441

Constit. of *Balanophora japonica* and *Lantana hybrida*. Amorph. yellow powder. $[\alpha]_D^{15}$ -438.4 (c, 0.5 in MeOH)

3′-Methoxy

1-O-Feruloylglucose

[7196-71-6]

$C_{16}H_{20}O_9$ 356.327

Present in *Solanum, Antirrhinum, Raphanus, Begonia* etc. spp.. Short prisms (EtOH). Mp 123 - 126°. $[\alpha]_D$ -42.1 (c, 0.8 in MeOH). λ_{max} 204 (log ε 3.7); 239 (log ε 3.4); 287 (log ε 3.7) (MeOH)

3′,5′-Dimethoxy

1-O-Sinapoylglucose*. Raphanusol B*

[78185-48-5]

$C_{17}H_{22}O_{10}$ 386.353

Present in *Rheum rhaponticum, Brassica oleracea* and *Raphanus sativus*. Also from *Lilium speciosum* and *Pelargonium zonale*. Antioxidant. Pale yellow needles. Mp 123 - 124°. $[\alpha]_D$ -61 (c, 1.8 in MeOH). λ_{max} 227 (log ε 3.9); 239 (log ε 3.9); 331 (log ε 3) (MeOH).

Birkofer, L. *et al., Naturwissenschaften*, 1960, **47**, 469-470 (*Feruloylglucose, isol, synth*).

Harborne, J.B. *et al., Biochem. J.*, 1961, **81**, 242-250 (*occur*).

Birkofer, L. *et al., Z. Naturforsch., B*, 1961, **16**, 249-251 (*occur, synth*).

Asen, S. *et al.*, *Phytochemistry*, 1962, **1**, 169-174 (*occur*).
Harborne, J.B. *et al.*, *Phytochemistry*, 1964, **3**, 421-428 (*occur*).
Cooper-Driver, G. *et al.*, *Phytochemistry*, 1975, **14**, 2506-2507 (*6-sulfate*).
Litvinenko, V.I. *et al.*, *Planta Med.*, 1975, **27**, 372-380 (*Feruloylglucose, rev*).
Imperato, F. *et al.*, *Phytochemistry*, 1976, **15**, 1786 (*3′-glucosyloxy*).
Herrmann, K. *Fortschr. Chem. Org. Naturst.*, 1978, **35**, 73-132 (*rev*).
Imperato, F. *et al.*, *Chem. Ind. (London)*, 1981, 691-692 (*Ceterach officinarum sulfates*).
Pauli, G.F. *et al.*, *Phytochemistry*, 1995, **38**, 1245-1250 (*Sinapoylglucose*).
Jiang, Z.-H. *et al.*, *Chem. Pharm. Bull.*, 2001, **49**, 887-892 (*3′-glucosyloxy, caffeoylgalloylglucoses*).
Takaya, Y. *et al.*, *J. Agric. Food Chem.*, 2003, **51**, 8061-8066 (*Feruloylglucose, Sinapoylglucose*).
Ho, J.-C. *et al.*, *J. Chin. Chem. Soc. (Taipei)*, 2004, **51**, 1073-1076 (*1-Caffeoylrungiose*).
Galland, S. *et al.*, *J. Agric. Food Chem.*, 2007, **55**, 7573-7579 (*1-Caffeoylglucose, synth, pmr, cmr*).
Panthama, N. *et al.*, *Chem. Pharm. Bull.*, 2009, **57**, 1352-1355 (*Balanophora trigalloylglucoses*).
Wang, W. *et al.*, *Helv. Chim. Acta*, 2009, **92**, 1817-1822 (*1-Coumaroyl-3-galloylglucose*).
Karim, A. *et al.*, *J. Enzyme Inhib. Med. Chem.*, 2009, **24**, 1332-1335 (*Barleriside B*).
Zhang, J.M. *et al.*, *Chem. Nat. Compd. (Engl. Transl.)*, 2011, **47**, 272-274 (*Cedrus deodara constit*).
Huang, Y. *et al.*, *J. Asian Nat. Prod. Res.*, 2013, **15**, 319-324 (*Solidago canadensis constit*)

2-*O*-*p*-Coumaroylglucose C110

2-O-(4-Hydroxycinnamoyl)glucose
$C_{15}H_{18}O_8$ 326.301
D-***form*** [110351-44-5]
Isol. from several Japanese sp. of *Ajuga* and from various commercial rhubarbs (*Rheum* sp.). Needles (H_2O). Mp 204 - 206°. $[\alpha]^{27}_{D}$ +32.8 (c, 0.3 in MeOH). $[\alpha]^{20}_{D}$ +27.3 → +51.8 (c, 1 in MeOH)
6-O-Sulfate
$C_{15}H_{18}O_{11}S$ 406.358
Isol. from fern *Asplenium fontanum* var. *obovatum*
β-D-***Pyranose-form*** [1382338-59-1]
1,4-Bis(3,4,5-trihydroxybenzoyl)
2-O-p-Coumaroyl-1,4-digalloyl-β-D-glucopyranose
$C_{29}H_{26}O_{16}$ 630.511
Constit. of *Balanophora polyandra*. Light yellow powder. $[\alpha]^{20}_{D}$ -0.7 (c, 0.1 in Me_2CO).

Imperato, F. *Chem. Ind. (London)*, 1981, 691-692 (*sulfate*).
Shimomura, H. *et al.*, *Phytochemistry*, 1987, **26**, 1981-1983 (*Ajuga constit*).
Kashiwada, Y. *et al.*, *Phytochemistry*, 1988, **27**, 1473-1477 (*Rheum constit*).
Wang, Y. *et al.*, *Acta Pharm. Sin. B*, 2013, **3**, 46-50 (*Balanophora polyandra constit*)

3-*O*-*p*-Coumaroylglucose C111

$C_{15}H_{18}O_8$ 326.301
β-D-***Pyranose-form***
3′-Hydroxy
3-O-Caffeoyl-β-D-glucopyranose
[359819-50-4]
$C_{15}H_{18}O_9$ 342.3
Constit. of the rhizomes of *Balanophora harlandii*. Amorph. yellow powder
3′-Hydroxy, 1,4-bis(3,4,5-trihydroxybenzoyl)
3-O-Caffeoyl-1,4-di-O-galloyl-β-D-glucopyranose
[359819-45-7]
$C_{29}H_{26}O_{17}$ 646.51
Constit. of *Balanophora japonica*. Amorph. yellow powder + 2½ H_2O. $[\alpha]^{15}_{D}$ -123.7 (c, 0.7 in MeOH).

Jiang, Z.-H. *et al.*, *Chem. Pharm. Bull.*, 2001, **49**, 887-892 (*3-Caffeoyl-1,4-digalloylglucose*).
Wang, W. *et al.*, *Helv. Chim. Acta*, 2009, **92**, 1817-1822 (*3-Caffeoylglucose*)

4-*O*-*p*-Coumaroylglucose C112

Glucose 4-[3-(4-hydroxyphenyl)-2-propenoate], CAS. *4-O-(4-Hydroxycinnamoyl)glucose*
[91465-76-8]
$C_{15}H_{18}O_8$ 326.301
Isol. from ferns *Plagiogyria euphlebia* and *Microlepia speluncae*, radish seedlings and from the petals of *Antirrhinum majus*. Needles (MeOH). Mp 228 - 230°. $[\alpha]_D$ +21 (c, 0.8 in Py)
2-Ac
[91432-05-2]
$C_{17}H_{20}O_9$ 368.338
Isol. from *Microlepia speluncae*. Prisms (MeOH). Mp 196 - 198°. $[\alpha]^{22}_{D}$ -34.2 (c, 0.47 in MeOH)
Penta-Ac
Oil. $[\alpha]^{23}_{D}$ +16 (c, 0.5 in $CHCl_3$)
3′-Hydroxy
4-O-trans-Caffeoylglucose
[1419169-01-9]
$C_{15}H_{18}O_9$ 342.3
Constit. of the aerial parts of *Lygodium japonicum*. Amorph. powder. $[\alpha]^{25}_{D}$ +14.9 (c, 0.6 in MeOH). λ_{max} 217 (log ε 3.97); 243 (sh) (log ε 3.81); 304 (log ε 3.8); 330 (log ε 3.99) (MeOH)
3′-Hydroxy, 1,6-bis(3,4,5-trihydroxybenzoyl)
4-O-Caffeoyl-1,6-digalloyl-β-D-glucopyranose. ***Balapolyphorin B***
[931426-08-3]
$C_{29}H_{26}O_{17}$ 646.51
Constit. of *Balanophora polyandra*. Pale brown powder. $[\alpha]^{21}_{D}$ +55 (c, 0.5 in MeOH). λ_{max} 204 (log ε 3.94); 250 (log ε 3.2); 255 (log ε 3.2); 261 (log ε 3.05) (MeOH).

Kuraishi, T. *et al.*, *Chem. Pharm. Bull.*, 1984, **32**, 1998-2000 (*Plagiogyria euphlebia constit*).
Wang, K.-J. *et al.*, *Chem. Biodiversity*, 2006, **3**, 1317-1324 (*Balapolyphorin B*).
Duan, Y.-H. *et al.*, *J. Asian Nat. Prod. Res.*, 2012, **14**, 286-292 (*4-Caffeoylglucose*)

6-*O*-*p*-Coumaroylglucose C113

6-O-(4-Hydroxycinnamoyl)glucose
$C_{15}H_{18}O_8$ 326.301
D-***form*** [114297-65-3]
Isol. from various commercial rhubarbs (*Rheum* sp.), from bark of *Prunus buergeriana* and grape pomace, *Vitis vinifera*. α-Glucosidase inhibitor. Needles (H_2O). Mp 184 - 186°. $[\alpha]^{24}_{D}$ +30.1 (c, 0.72 in MeOH). $[\alpha]_D$ +32 → +39 (80% dioxan aq.). λ_{max} 313 (EtOH)
3′,5′-Dimethoxy
6-O-Sinapoylglucose
[149998-27-6] (anomer)
[149998-28-7] (anomer)
$C_{17}H_{22}O_{10}$ 386.353
Constit. of the roots of *Cynanchum hancockianum*. Powder (MeOH). Mp 129 - 131°. Obt. as an anomeric mixt.
D-***Pyranose-form***
3″-O-β-D-Glucopyranosyl, 4′-O-[4-hydroxy-E-cinnamoyl-(→6)-β-D-glucopyranoside]
Susaroyside D
[2414074-31-8]
$C_{36}H_{44}O_{20}$ 796.728
Constit. of the fruit of *Forsythia suspensa*. Amorph. yellow powder. $[\alpha]^{20}_{D}$ -61.5 (c, 0.1 in MeOH). λ_{max} 205 (log ε 4.46); 223 (log ε 4.33); 299 (log ε 4.48) (MeOH)
α-D-***Pyranose-form*** [912279-95-9] [25545-14-6] [698393-98-5]

Constit. of *Eucalyptus citriodora* and *Picrorhiza scrophulariifolia*

β-D-Pyranose-form

Constit. of the stems of *Picrorhiza scrophulariifolia*

1,2-Bis-O-(3,4,5-trihydroxybenzoyl)

6-O-p-Coumaroyl-1,2-digalloyl-glucose

$C_{29}H_{26}O_{16}$ 630.511

Isol. from various commercial rhubarbs (*Rheum* spp.). Also from *Eucalyptus citriodora*. Amorph. powder + $3H_2O$. $[\alpha]^{21}_{D}$ -68.7 (c, 0.95 in Me_2CO)

3′-Hydroxy, 1,3,4-tris-O-(3,4,5-trihydroxybenzoyl)

6-O-Caffeoyl-1,3,4-tri-O-galloyl-β-D-glucopyranose

[359819-48-0]

$C_{36}H_{30}O_{21}$ 798.615

Constit. of *Balanophora japonica*. Amorph. yellow powder + 2½H_2O. $[\alpha]^{15}_{D}$ +14.9 (c, 0.6 in MeOH).

Watanabe, S. *et al.*, *Agric. Biol. Chem.*, 1966, **30**, 420-422 (*synth*).
Shimomura, H. *et al.*, *Phytochemistry*, 1988, **27**, 641-644 (*Prunus constit*).
Kashiwada, Y. *et al.*, *Phytochemistry*, 1988, **27**, 1473-1477 (*rhubarb constits, cmr, pmr*).
Lou, H. *et al.*, *Phytochemistry*, 1993, **32**, 1283-1286 (*6-Sinapoylglucose*).
Jiang, Z.-H. *et al.*, *Chem. Pharm. Bull.*, 2001, **49**, 887-892 (*6-Caffeoyl-1,3,4-trigalloylglucose*).
Huang, S.-X. *et al.*, *Helv. Chim. Acta*, 2004, **87**, 598-604 (*isol, pmr, cmr*).
Freitas, M.O. *et al.*, *Quim. Nova*, 2007, **30**, 1926-1929 (*6-Coumaroyl-1,2-digalloylglucose*).
Shao, S.-Y. *et al.*, *Bioorg. Med. Chem. Lett.*, 2019, **29**, 126635:1-5 (*Susaroyside D*)

1-*O*-Coumaroylglycerol C114

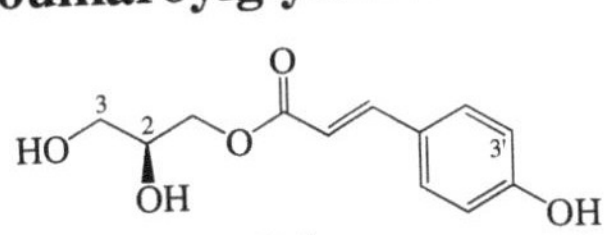

(S,E)-form

1-O-(4-Hydroxycinnamoyl) glycerol. 2,3-Dihydroxyprop-1-yl 3-(4-hydroxyphenyl)-2-propenoate

[63529-09-9]

[106055-11-2] [122045-55-0]

$C_{12}H_{14}O_5$ 238.239

(S,E)-form [114420-68-7]

Constit. of *Juncus effusus* and *Lilium* spp.. Cryst. ($CHCl_3$/MeOH). Mp 118 - 120°. $[\alpha]^{20}_{D}$ +12 (c, 0.1 in MeOH). λ_{max} 208 (ε 11000); 225 (ε 12000); 300 (ε 14800) (EtOH)

2-O-β-D-Glucopyranoside

2-O-Glucosyl-1-O-p-hydroxycinnamoyl-sn-glycerol. ***Regaloside D***

[120601-66-3]

$C_{18}H_{24}O_{10}$ 400.38

Isol. from various lily spp., *Lilium longiflorum*, *Lilium pardarinum*, *Lilium auratum* and *Lilium brownii*. Pale yellow amorph. powder. $[\alpha]^{27}_{D}$ -24.6 (c, 0.39 in MeOH)

3-O-β-D-Glucopyranoside

3-O-Glucosyl-1-O-p-hydroxycinnamoyl-sn-glycerol. ***Regaloside A***

[114420-66-5]

$C_{18}H_{24}O_{10}$ 400.38

Acylated glycerol glycoside. Isol. from the bulbs of *Lilium henryi*, *Lilium lancifolium*, *Lilium regale*, *Lilium speciosum*, *Lilium pardarinum*, *Lilium auratum* and *Fritillaria camtschatcensis*. Pale yellow amorph. powder. $[\alpha]^{25}_{D}$ -17 (c, 1.0 in MeOH) (-14.1). Bitter taste

2,3-O-Isopropylidene

1-O-p-Coumaroyl-2,3-isopropylideneglycerol. ***Juncusyl ester A***

[173357-20-5]

$C_{15}H_{18}O_5$ 278.304

Constit. of *Juncus effusus*. Cryst. (petrol/EtOAc). Mp 121 - 123°. $[\alpha]^{22}_{D}$ +6.6 (c, 0.6 in MeOH). λ_{max} 211 (ε 1300); 228 (ε 1740); 313 (ε 3310) (EtOH)

3-Ac, 2-O-β-D-glucopyranoside

Regaloside B

[114420-67-6]

$C_{20}H_{26}O_{11}$ 442.417

Isol. from bulbs of *Lilium longiflorum*, *Lilium regale* and *Lilium brownii*. Pale yellow amorph. powder. $[\alpha]^{15}_{D}$ -21.2 (c, 1.0 in MeOH). Bitter taste

(R,E)-form

3-O-β-D-Glucopyranoside

1-O-Glucosyl-3-O-p-hydroxycinnamoyl-sn-glycerol. ***Epiregaloside A***

[120665-03-4]

$C_{18}H_{24}O_{10}$ 400.38

Constit. of *Lilium auratum* and *Lilium pardarinum*. Pale yellow powder. $[\alpha]^{26}_{D}$ -12.1 (c, 0.6 in MeOH)

(±)-(E)-form [108026-22-8]

Constit. of *Lilium auratum*, *Lilium henryi* and *Lilium brownii*. Powder. λ_{max} 228 (log ε 4.14); 302 (sh); 313 (log ε 4.36) (MeOH)

(ξ)-(E)-form

3′,5′-Dimethoxy

1-Sinapoylglycerol. Glycerol sinapate. ***Yimunol A***

[858251-57-7]

$C_{14}H_{18}O_7$ 298.291

Constit. of *Leonurus sibiricus* var. *albiflora*. Yellow oil.

Shimomura, H. *et al.*, *Phytochemistry*, 1987, **26**, 844-845 (*Lilium auratum constits*).
Shimomura, H. *et al.*, *Chem. Pharm. Bull.*, 1988, **36**, 4841-4848 (*Epiregaloside A*).
Shimomura, H. *et al.*, *Phytochemistry*, 1988, **27**, 451-454 (*Lilium regale constits*).
Mimaki, Y. *et al.*, *Phytochemistry*, 1989, **28**, 3453-3458 (*Lilium auratum constits*).
Mimaki, Y. *et al.*, *Chem. Pharm. Bull.*, 1990, **38**, 3055-3059 (*Lilium brownii constits*).
Mimaki, Y. *et al.*, *Phytochemistry*, 1990, **29**, 2267-2271 (*Lilium brownii constits*).
Mimaki, Y. *et al.*, *Phytochemistry*, 1991, **30**, 937-940 (*Lilium speciosum constits*).
Jin Dong-Zhe, *et al.*, *Phytochemistry*, 1996, **41**, 545-547 (*Juncusyl ester A*).
Greca, M.D. *et al.*, *Nat. Prod. Lett.*, 1998, **12**, 263-270 (*Juncus effusus constits*).
Pan, S.-M. *et al.*, *Chin. Pharm. J. (Taipei)*, 2006, **58**, 35-40 (*Yimunol A*)

2-*O*-Coumaroylglycerol C115

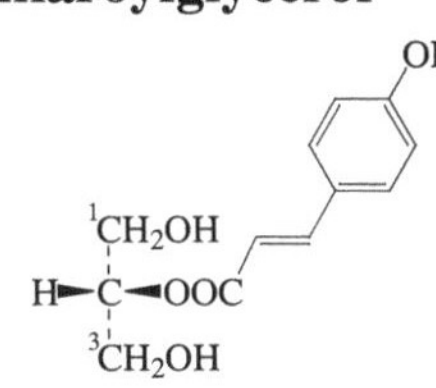

2-O-(4-Hydroxycinnamoyl)glycerol. ***Juncusyl ester B****. 1,3-Dihydroxyprop-2-yl 3-(4-hydroxyphenyl)-2-propenoate*

$C_{12}H_{14}O_5$ 238.239

The parent compd. is achiral

(E)-form [142942-87-8]

Constit. of *Juncus effusus*. Cryst. ($CHCl_3$/MeOH). Mp 110 - 114°. λ_{max} 211 (ε 6460); 312 (ε 5250) (EtOH)

1-O-β-D-Glucopyranoside

Regaloside H

[126239-77-8]

$C_{18}H_{24}O_{10}$ 400.38

Constit. of *Lilium auratum* var. *platyphyllum*. Amorph. powder. $[\alpha]^{28}_{D}$ -32 (c, 0.58 in MeOH)

3-Ac, 1-O-β-D-glucopyranoside

Regaloside I

[126239-78-9]

$C_{20}H_{26}O_{11}$ 442.417

Constit. of *Lilium auratum* var. *platyphyllum*. Amorph. powder. $[\alpha]_{D}$ -11.1 (c, 0.51 in MeOH)

4′-Me ether

2-O-(4-Methoxycinnamoyl) glycerol

[220852-58-4]

$C_{13}H_{16}O_5$ 252.266

Constit. of *Juncus effusus*

3″-Hydroxy, 1-O-β-D-glucopyranoside

Regaloside L

[138797-43-0]

$C_{18}H_{24}O_{11}$ 416.379

Constit. of *Lilium mackliniae*. Pale yellow powder. $[\alpha]_D^{28}$ -29.1 (c, 0.5 in MeOH). λ_{max} 243 (sh) (log ε 3.74); 302 (sh); 325 (log ε 4.02) (MeOH).

Mimaki, Y. *et al.*, *Phytochemistry*, 1989, **28**, 3453-3458 (*Regalosides H,I*).
Sashida, Y. *et al.*, *Chem. Pharm. Bull.*, 1991, **39**, 2362-2368 (*Regaloside L*).
Jin, D.-Z. *et al.*, *Phytochemistry*, 1996, **41**, 545-547 (*Juncusyl ester B*).
Greca, M.D. *et al.*, *Nat. Prod. Lett.*, 1998, **12**, 263-270 (*4′-Me ether*)

3-*O*-*p*-Coumaroylquinic acid C116

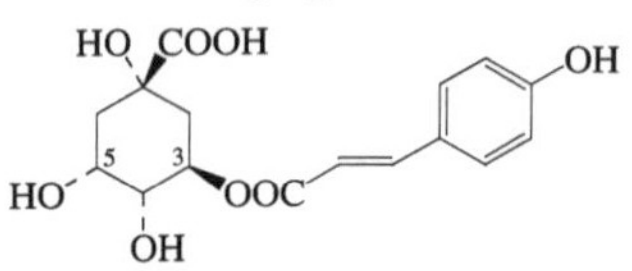

3-(4-Hydroxycinnamoyl)quinic acid, CAS. *3-pCoQA*

[1899-30-5]

$C_{16}H_{18}O_8$ 338.312

Found in *Humulus lupulus*, *Ilex glabra*, *Gynura divaricata*, *Malus* spp. and other plant sources. Needles (H_2O). Mp 252 - 255° dec. (247-248°). $[\alpha]_D^{20}$ -53.6 (c, 1.04 in MeOH)

Et ester

[1443678-33-8]

$C_{18}H_{22}O_8$ 366.366

Constit. of the flower buds of *Prunus mume*. Inhibits melanogenesis. Amorph. powder. $[\alpha]_D^{15}$ -8.9 (c, 0.29 in MeOH). λ_{max} 228 (log ε 4.34); 314 (log ε 4.50) (MeOH)

Butyl ester

[420795-38-6]

$C_{20}H_{26}O_8$ 394.42

Constit. of *Araucaria cunninghamii* and *Chaenomeles sinensis*

1-Me ether

[160400-61-3]

Needles (H_2O). Mp 248°. $[\alpha]_D^{22}$ -47.5 (c, 0.5 in EtOH)

Z-Isomer, butyl ester

[1361544-15-1]

$C_{20}H_{26}O_8$ 394.42

Constit. of *Araucaria cunninghamii*. Powder. $[\alpha]_D^{25}$ -40 (c, 0.1 in MeOH). λ_{max} 228 (log ε 3.66); 313 (log ε 3.93) (MeOH).

Williams, A.H. *Chem. Ind. (London)*, 1958, 1200 (*apple constit*).
Haslam, E. *et al.*, *J. Chem. Soc.*, 1961, 5153-5156 (*synth*).
Hemmerle, H. *et al.*, *J. Med. Chem.*, 1997, **40**, 137-145 (*synth*).
Kenji, A. *et al.*, *Nat. Med. (Tokyo)*, 2001, **55**, 255-257 (*butyl ester*).
Chen, J. *et al.*, *Planta Med.*, 2011, **77**, 485-488 (*butyl esters*).
Nakamura, S. *et al.*, *Phytochemistry*, 2013, **92**, 128-136 (*Et ester*).
Chen, J. *et al.*, *Fitoterapia*, 2014, **99**, 1-6 (*Gynura divaricata constit*).
Tanaka, Y. *et al.*, *J. Agric. Food Chem.*, 2014, **62**, 2198-2206 (*Humulus lupulus constit*).
Jaiswal, R. *et al.*, *Phytochemistry*, 2014, **106**, 141-155 (*Ilex glabra constit*).
Ortiz, A.L.G. *et al.*, *Tetrahedron: Asymmetry*, 2017, **28**, 419-427 (*synth*)

4-*O*-*p*-Coumaroylquinic acid C117

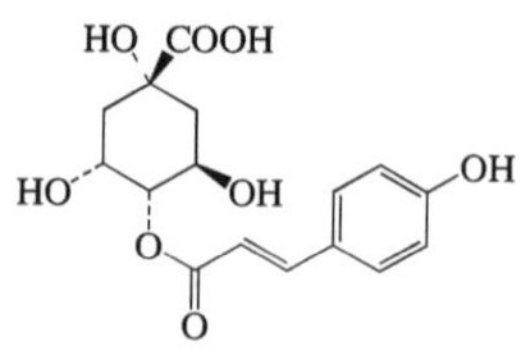

4-(4-Hydroxycinnamoyl)quinic acid

[53539-37-0]

$C_{16}H_{18}O_8$ 338.312

Constit. of the leaves of *Juglans regia* and *Malus domestica*. Plates (H_2O). Mp 192 - 193° dec.. $[\alpha]_D^{20}$ -47.3 (c, 1.4 in MeOH)

Me ester

[864467-51-6]

$C_{17}H_{20}O_8$ 352.339

Constit. of *Vittaria anguste-elongata*. Syrup. $[\alpha]_D$ -107.4 (c, 0.03 in MeOH). λ_{max} 210 (log ε 3.96); 314 (log ε 4.12) (MeOH)

1-Me ether

Plates. Mp 193 - 194°. $[\alpha]_D^{21}$ -44.1 (c, 1.0 in EtOH)

(1′→3)-Lactone

4-O-p-Coumaroyl-1,3-quinide

[157048-82-3]

$C_{16}H_{16}O_7$ 320.297

Constit. of roasted coffee beans. Also numbered 1,5-quinide.

Haslam, E. *et al.*, *J. Chem. Soc.*, 1964, 2137-2146 (*synth*).
Stopar, M. *et al.*, *J. Agric. Food Chem.*, 2002, **50**, 1643-1646 (*Malus domestica constit*).
Farah, A. *et al.*, *J. Agric. Food Chem.*, 2005, **53**, 1505-1513 (*lactone*).
Wu, P.-L. *et al.*, *J. Nat. Prod.*, 2005, **68**, 1180-1184 (*Me ester*).
Pereira, J.A. *et al.*, *Food Chem. Toxicol.*, 2007, **45**, 2287-2295 (*Juglans regia constit*).
Pereira, J.A. *et al.*, *Food Chem. Toxicol.*, 2008, **46**, 2326-2331 (*Juglans regia constit*).
Ortiz, A.L.G. *et al.*, *Tetrahedron: Asymmetry*, 2017, **28**, 419-427 (*synth*)

5-*O*-*p*-Coumaroylquinic acid C118

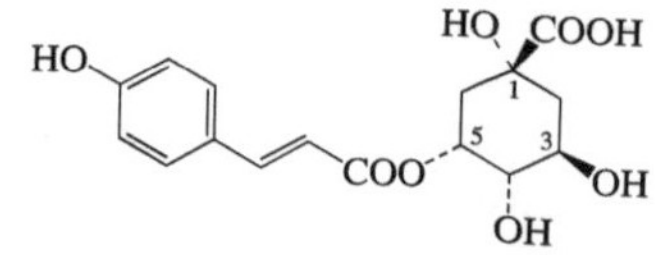

5-(4-Hydroxycinnamoyl)quinic acid

[87099-71-6]

$C_{16}H_{18}O_8$ 338.312

Constit. of *Galium fagetorum*. Found in *Hydrangea macrophylla* as a supramolecular copigment with Myrtillin and Al^{3+}. Prisms (H_2O). Mp 194°. $[\alpha]_D^{19}$ -5.6 (c, 0.6 in MeOH)

1′→3-Lactone

5-O-p-Coumaroyl-1,3-quinide

[157048-80-1]

$C_{16}H_{16}O_7$ 320.297

Constit. of roasted coffee beans.

Haslam, E. *et al.*, *J. Chem. Soc.*, 1964, 2137-2146 (*synth, struct*).
Bogaevskii, A.K. *et al.*, *Chem. Nat. Compd. (Engl. Transl.)*, 1970, **6**, 761-762 (*occur*).
Farah, A. *et al.*, *J. Agric. Food Chem.*, 2005, **53**, 1505-1513 (*lactone, isol*).
Kondo, T. *et al.*, *Tetrahedron Lett.*, 2005, **46**, 6645-6649 (*Hydrangea supramolecular pigment*).
Toyama-Kato, Y. *et al.*, *Heterocycles*, 2007, **72**, 239-254 (*Hydrangea supramolecular pigment*).
Ortiz, A.L.G. *et al.*, *Tetrahedron: Asymmetry*, 2017, **28**, 419-427 (*synth*)

Crassifogenin C C119

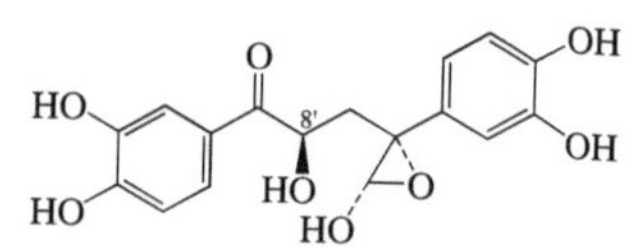

Relative Configuration

7,8-Epoxy-3,3′,4,4′,8,8′-hexahydroxy-9-nor-7,9′-lignan-7′-one

[1093958-73-6]

$C_{17}H_{16}O_8$ 348.307

Constit. of *Curculigo capitulata* and *Curculigo crassifolia*. Antioxidant. Powder. λ_{max} 206 (log ε 4.46); 230 (log ε 4.14); 281 (log ε 3.94); 310 (log ε 3.79) (MeOH)

8′-O-β-D-Glucopyranoside

Breviscaside B

[1257411-16-7]

$C_{23}H_{26}O_{13}$ 510.448

Constit. of *Curculigo breviscapa*, *Curculigo capitulata* and *Curculigo glabrescens*. Powder. $[\alpha]_D^{25}$ -15.3 (c, 0.3 in MeOH). Stereochem. unconfirmed. λ_{max} 204 (log ε 4.57); 290 (log ε 3.97) (MeOH)

8′-Epimer

Isocrassifogenin C

$C_{17}H_{16}O_8$ 348.307

Constit. of the rhizomes of *Curculigo crassifolia*. Isol. as a mixt. with Crassifogenin C.

Wang, K.J. *et al., Arch. Pharmacal Res.*, 2008, **31**, 1313-1316 (*Crassifogenin C, activity*).
Di, L. *et al., Bull. Korean Chem. Soc.*, 2010, **31**, 2999-3002 (*Breviscaside B, Crassifogenin C*).
Li, N. *et al., Fitoterapia*, 2010, **81**, 528-531 (*Breviscaside B*)

Crassifoside E C120

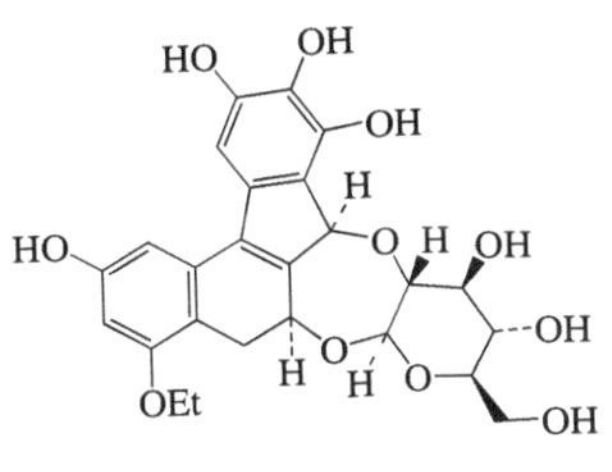

Absolute Configuration

[865758-27-6]

$C_{25}H_{26}O_{11}$ 502.472

Constit. of the rhizomes of *Curculigo crassifolia*. ACE inhibitor. Brown powder. $[\alpha]_D^{24}$ +23.1 (c, 0.2 in MeOH). λ_{max} 224 (log ε 4.42); 258 (log ε 4.75); 294 (log ε 4.46); 347 (log ε 3.31) (MeOH)

4-Deethoxy, 4-chloro

Crassifoside H

[1232140-08-7]

$C_{23}H_{21}ClO_{10}$ 492.861

Constit. of the rhizomes of *Curculigo glabrescens* and *Curculigo sinensis*. Antioxidant. Black powder. $[\alpha]_D^{28}$ +20.9 (c, 0.17 in MeOH). Stereochem. not necessarily the same as Crassifoside E. λ_{max} 203 (log ε 4.22); 224 (log ε 4.11); 261 (log ε 4.33); 296 (log ε 4.11) (MeOH).

Li, N. *et al., Tetrahedron Lett.*, 2005, **46**, 6445-6447 (*Crassifoside E*).
Zhu, C.-C. *et al., Z. Naturforsch., B*, 2009, **64**, 1077-1080 (*Crassifoside H*)

Crassifoside F C121

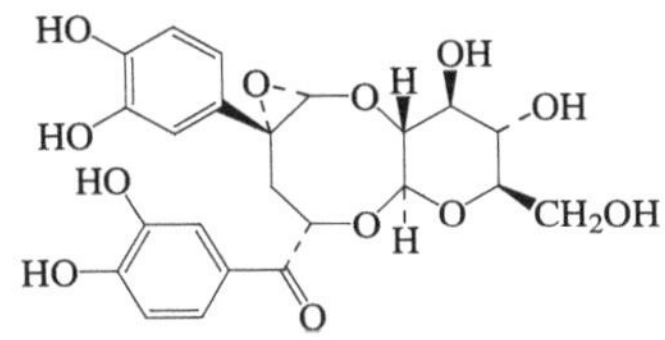

Absolute Configuration

[865758-28-7]

$C_{23}H_{24}O_{12}$ 492.433

Constit. of the rhizomes of *Curculigo crassifolia*. ACE inhibitor. Powder. $[\alpha]_D^{28}$ +112.9 (c, 0.2 in MeOH). λ_{max} 196 (log ε 4.09); 206 (log ε 4.5); 232 (log ε 4.18); 283 (log ε 3.99); 315 (log ε 3.86) (MeOH).

Li, N. *et al., Tetrahedron Lett.*, 2005, **46**, 6445-6447 (*Crassifoside F, activity*).
Wang, K.-J. *et al., Arch. Pharmacal Res.*, 2008, **31**, 1313-1316 (*activity*)

Cryptoresinol C122

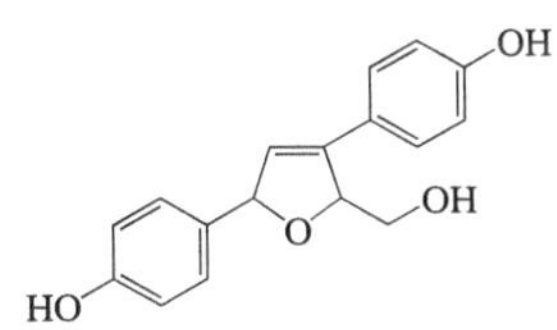

2,5-Dihydro-3,5-bis(4-hydroxyphenyl)-2-furanmethanol, CAS. *7′,8-Epoxy-9′-nor-7,8′-lign-7(8′)-ene-4,4′,9-triol. 7′,8-Epoxy-4,4′,9-trihydroxy-9′-nor-7,8′-neolign-7(8′)-ene*

[115713-11-6]

$C_{17}H_{16}O_4$ 284.311

Constit. of *Araucaria angustifolia, Cryptomeria japonica* and *Taxodium ascendens*. Pink needles (Et_2O/MeOH or MeOH aq.). Mp 253 - 255° (230-232°). $[\alpha]_D^{25}$ -170.4 (c, 1.25 in MeOH). λ_{max} 265 (log ε 4.26) (MeOH).

Takahashi, K. *et al., Phytochemistry*, 1988, **27**, 1550-1552 (*Cryptoresinol*).
Ohashi, H. *et al., Phytochemistry*, 1992, **31**, 1371-1374 (*Aurucaria angustifolia constit*).
Zhang, Y.M. *et al., Fitoterapia*, 2009, **80**, 361-363 (*Taxodium ascendens constit*)

Cuminiresinol C123

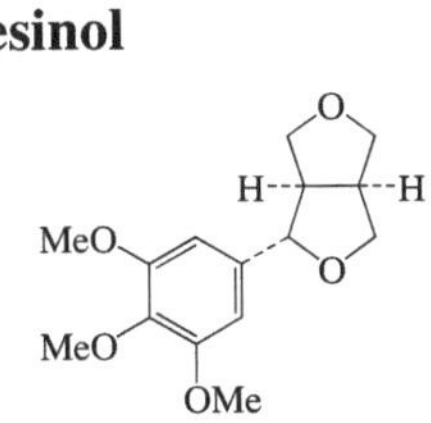

Relative Configuration

2-(3,4,5-Trimethoxyphenyl)-3,7-dioxabicyclo[3.3.0]octane

[1151670-45-9]

$C_{15}H_{20}O_5$ 280.32

Norlignan. Constit. of the stem bark of *Syzygium cumini*. Pale yellow cryst. (Me_2CO). Mp 260 - 261°.

Mir, Q.Y. *et al., Nat. Prod. Res.*, 2009, **23**, 422-430 (*Cuminiresinol*)

Cuneataside F‡ C124

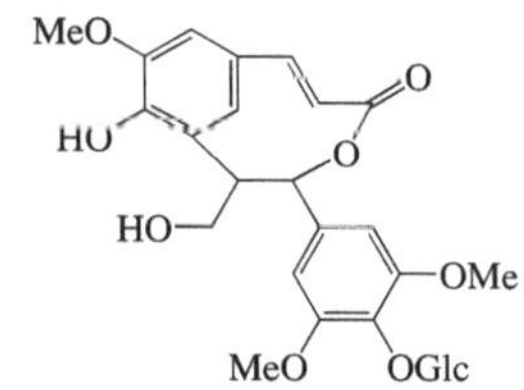

[1217890-08-8]

$C_{27}H_{32}O_{13}$ 564.540

Constit. of the stems of *Sargentodoxa cuneata*. Amorph. powder. λ_{max} 222; 313 (no solvent reported)

Aglycone

4,4′,9′-Trihydroxy-3′,5,5′-trimethoxy-3,8′-lign-7-en-9,7′-olide.

Chen, Z.X. *et al., Chin. Chem. Lett.*, 2009, **20**, 1339-1341 (*Cuneataside F*)

Cuneifolin‡ C125

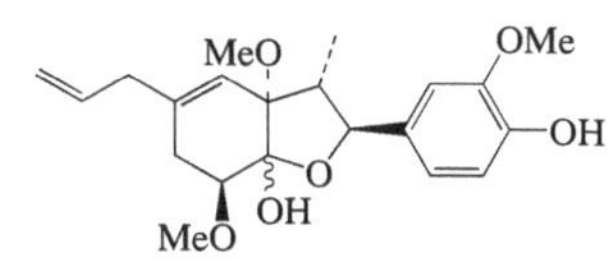

Relative Configuration

4,7′-Epoxy-3,4,5,6-tetrahydro-3,3′,5-trimethoxy-3,8′-lign-8-ene-4,4′-diol. 4,7′-Epoxy-3,4,5,6-tetrahydro-4,4′-dihydroxy-3,3′,5-trimethoxy-3,8′-neolign-8-ene

[176686-23-0]

$C_{21}H_{28}O_6$ 376.449

Constit. of *Piper cuneifolium*. Needles. Mp 120 - 121°. $[\alpha]_D^{20}$ -14 (c, 0.1 in $CHCl_3$). λ_{max} 239; 291 (MeOH).

Abdallah, O.M. *et al., Pharmazie*, 1996, **51**, 187-188 (*Cuneifolin*)

Curcapital C126

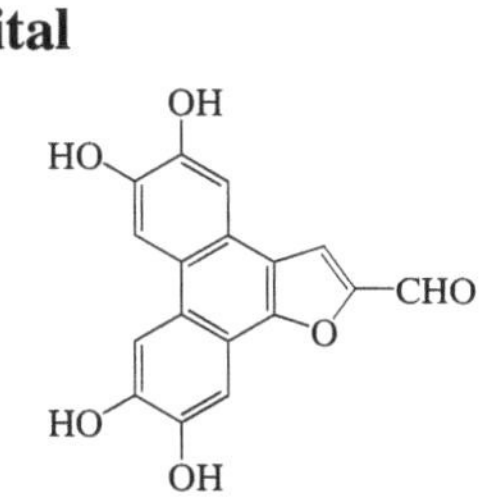

$C_{17}H_{10}O_6$ 310.261

Constit. of *Arundina gramnifolia, Curculigo capitulata* and *Curculigo crassifolia*. Amorph. orange solid. λ_{max} 221 (log ε 4.38); 250 (log ε 4.63); 276

(log ε 4.55); 309 (sh) (log ε 4.15); 376 (log ε 4.25) (MeOH).
Chang, W.-L. *et al.*, *J. Nat. Prod.*, 1999, **62**, 734-739 (*Curcapital*)

Curcumaromin A C127

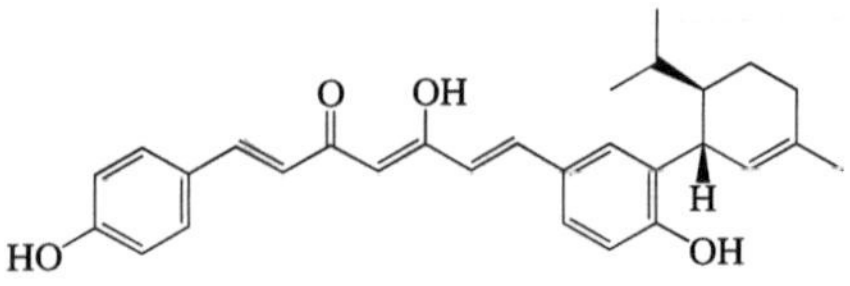

Relative Configuration

[1810034-38-8]
$C_{29}H_{32}O_4$ 444.571
Constit. of *Curcuma aromatica.* Orange powder. $[\alpha]^{17}_{D}$ -237 (c, 0.1 in MeOH). λ_{max} 423 (MeOH).
Qin, X.-D. *et al.*, *Helv. Chim. Acta*, 2015, **98**, 1325-1331 (*Curcumaromin A*)

Curcumaromin B C128

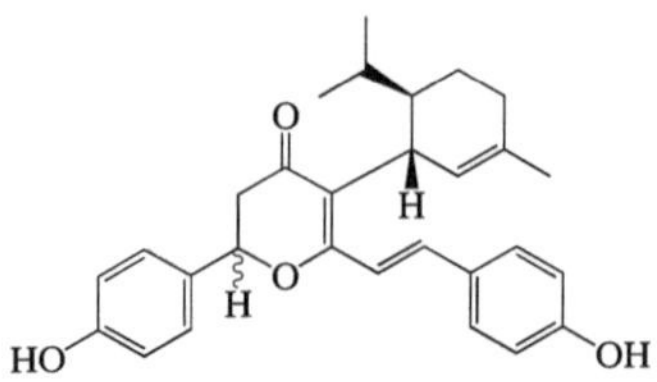

Relative Configuration

[1810034-39-9]
$C_{29}H_{32}O_4$ 444.571
Constit. of *Curcuma aromatica.* Yellow powder. $[\alpha]_D$ +17 (c, 0.1 in MeOH). λ_{max} 370 (MeOH).
Qin, X.-D. *et al.*, *Helv. Chim. Acta*, 2015, **98**, 1325-1331 (*Curcumaromin B*)

Curcumaromin C C129

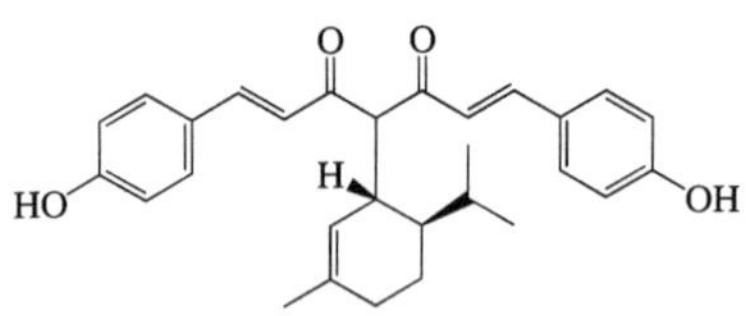

Relative Configuration

[1810034-40-2]
$C_{29}H_{32}O_4$ 444.571
Constit. of *Curcuma aromatica.* Yellow powder. $[\alpha]^{18}_{D}$ -149.7 (c, 0.1 in MeOH). λ_{max} 346 (MeOH).
Qin, X.-D. *et al.*, *Helv. Chim. Acta*, 2015, **98**, 1325-1331 (*Curcumaromin C*)

Curcumin C130

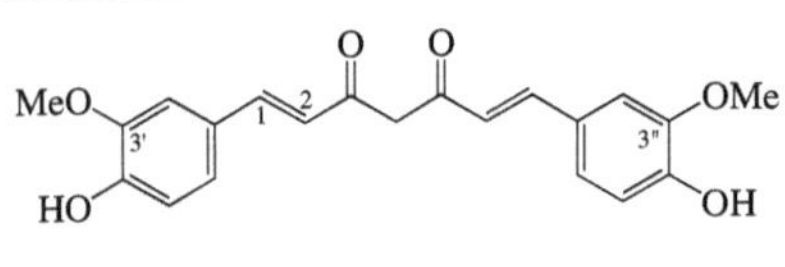

1,7-Bis(4-hydroxy-3-methoxyphenyl)-1,6-heptadiene-3,5-dione, CAS. *Diferuloylmethane. Curcumin I. Curcuma yellow. E100. C.I. Natural Yellow 3*
[458-37-7]
$C_{21}H_{20}O_6$ 368.385
Exists in enolised β-diketone form. Constit. of *Curcuma longa, Curcuma xanthorrhiza, Curcuma zedoaria* and other *Curcuma* spp. Also from *Clutia lanceolata.* Natural colouring matter used extensively in Indian curries etc. Used as 0.1% soln. in AcOH for photometric detn. of B (λ_{max} 550 nm, ε 180000) and of Ni (indirectly). Antiedemic, antiinflammatory agent. Bile stimulant showing hepatoprotective props. Possesses bactericidal and fungicidal props. Antitumour agent functioning through apoptosis; induces global DNA hypomethylation in leukaemia cells. Nutraceutical. Shows antioxidant activity. Acetylcholinesterase inhibitor. Orange prisms. Mp 183°. Log P 2.05 (calc). λ_{max} 420 (MeOH)
MI5230000

Di-Ac
Cryst.. Mp 170 - 171°

Dibenzoyl
Cryst.. Mp 210°

O-De-Me
1-(3,4-Dihydroxyphenyl)-7-(4-hydroxy-3-methoxyphenyl)-1,6-heptadiene-3,5-dione. ***De-O-methylcurcumin***
[149732-51-4]
$C_{20}H_{18}O_6$ 354.358
Constit. of the rhizomes of *Curcuma domestica.* Orange powder

1,2-Dihydro
5-Hydroxy-1,7-bis(4-hydroxy-3-methoxyphenyl)-4,6-heptadien-3-one, CAS. ***Dihydrocurcumin****. Letestuianin B*
[76474-56-1]
$C_{21}H_{22}O_6$ 370.401
Constit. of *Curcuma longa* and the seeds of *Aframomum letestuianum.* Yellow cryst. + ½ H_2O. Mp 179 - 180°. Enolised β-diketone. λ_{max} 250 (log ε 4.1); 260 (sh); 280 (log ε 3.84); 305 (log ε 3.73); 375 (log ε 4.39) (EtOH)

5′-Hydroxy
1-(3,4-Dihydroxy-5-methoxyphenyl)-7-(4-hydroxy-3-methoxyphenyl)-1,6-heptadien-3,5-dione. 5′-Hydroxycurcumin. ***Curcumalongin C***
[1375177-99-3]
$C_{21}H_{20}O_7$ 384.384
Constit. of the rhizomes of *Curcuma longa.* Amorph. yellow powder. λ_{max} 264 (log ε 3.74); 422 (log ε 4.27) (MeOH)

5′-Methoxy
1-(4-Hydroxy-3,5-dimethoxyphenyl)-7-(4-hydroxy-3-methoxypheny)-1,6-heptadiene-3,5-dione, CAS. ***5′-Methoxycurcumin***
$C_{22}H_{22}O_7$ 398.411
Constit. of *Curcuma xanthorrhiza.* Yellow powder. Mp 145 - 146°

Demethoxy
1-(4-Hydroxy-3-methoxyphenyl)-7-(4-hydroxyphenyl)-1,6-heptadiene-3,5-dione, CAS. *p-Hydroxycinnamoylferuloylmethane.* ***Demethoxycurcumin. Curcumin II***
[22608-11-3]
$C_{20}H_{18}O_5$ 338.359
Isol. from *Curcuma zedoaria, Curcuma longa* and *Curcuma xanthorrhiza.* Shows antioxidant activity. Orange-yellow powder. Mp 168°

3′-Demethoxy, 3″-O-de-Me
1-(3,4-Dihydroxyphenyl)-7-(4-hydroxyphenyl)-1,6-heptadiene-3,5-dione
[869789-09-3]
$C_{19}H_{16}O_5$ 324.332
Constit. of the rhizomes of *Curcuma longa.* Yellow powder

3′-Demethoxy, 1,2-dihydro
5-Hydroxy-7-(4-hydroxy-3-methoxyphenyl)-1-(4-hydroxyphenyl)-4,6-heptadien-3-one. 1-(4-Hydroxy-3-methoxyphenyl)-7-(4-hydroxyphenyl)-1-heptene-3,5-dione
[1198806-65-3]
$C_{20}H_{20}O_5$ 340.375
Constit. of *Curcuma longa.* Yellow powder. Mp 145.5°. λ_{max} 368 (log ε 4.08) (EtOH)

3″-Demethoxy, 1,2-dihydro
5-Hydroxy-1-(4-hydroxy-3-methoxyphenyl)-7-(4-hydroxyphenyl)-4,6-heptadien-3-one. 7-(4-Hydroxy-3-methoxyphenyl)-1-(4-hydroxyphenyl)-1-heptene-3,5-dione. ***Letestuianin A***
[502623-93-0]

$C_{20}H_{20}O_5$ 340.375
Constit. of *Curcuma longa* and the seeds of *Aframomum letestuianum*. Yellowish oil. Enolised β-diketone. λ_{max} 283 (log ε 3.9); 380 (log ε 3.2) (MeOH)

Bis(demethoxy)
1,7-Bis(4-hydroxyphenyl)-1,6-heptadiene-3,5-dione, CAS. *p,p'-Dihydroxydicinnamoylmethane*. ***Bisdemethoxycurcumin. Curcumin III***
[33171-05-0]
[22608-12-4] [24939-16-0] (unspecified stereo)
$C_{19}H_{16}O_4$ 308.333
Isol. from *Curcuma zedoaria*, *Curcuma longa* and *Curcuma aromatica*. Absent in some *Curcuma* spp.. Antitumour agent. Shows antioxidant activity. Yellow plates + H_2O (EtOH). Mp 224°

Bis(demethoxy), 1,2-dihydro
5-Hydroxy-1,7-bis(4-hydroxyphenyl)-4,6-heptadien-3-one, CAS
[207792-17-4]
$C_{19}H_{18}O_4$ 310.349
Constit. of the rhizomes of *Curcuma longa*. Yellow powder.

Aldrich Library of FT-IR Spectra, 1st edn., 1985, **2**, 52C (*Curcumin, ir*).
Aldrich Library of ^{13}C *and* ^{1}H *FT-NMR Spectra*, 1992, **2**, 800B (*Curcumin, nmr*).
Lampe, V. *Ber. Dtsch. Chem. Ges.*, 1918, **51**, 1347-1355 (*Curcumin, synth*).
Srinivasan, K.R. *J. Pharm. Pharmacol.*, 1953, **5**, 448-457 (*Curcuma longa constits, chromat.*).
Spicer, G.S. *et al.*, *Anal. Chim. Acta*, 1958, **18**, 231-239 (*Curcumin, use, detn, B*).
Umland, F. *et al.*, *Fresenius' Z. Anal. Chem.*, 1963, **197**, 151-160 (*Curcumin, use, detn, Ni*).
Jentzsch, K. *et al.*, *Sci. Pharm.*, 1968, **36**, 251-260 (*Curcumin*).
Kuroyagi, M. *et al.*, *Yakugaku Zasshi*, 1970, **90**, 1467-1470 (*Curcumin, isol, pmr*).
Dyrssen, D.W. *et al.*, *Anal. Chim. Acta*, 1972, **60**, 139-151 (*Curcumin, use, detn, B*).
Roughley, P.J. *et al.*, *J. Chem. Soc., Perkin Trans. 1*, 1973, 2379-2388 (*biosynth*).
Karig, F. *Dtsch. Apoth. -Ztg.*, 1975, **115**, 325-328 (*Cinnamon bark constits, chromatog*).
Quint, P. *et al.*, *Fresenius' Z. Anal. Chem.*, 1977, **285**, 356-358 (*Curcumin, use, detn, B*).
Kashima, C. *et al.*, *Heterocycles*, 1977, **7**, 241-246 (*Curcumin, synth*).
Wahlstrom, B. *et al.*, *Acta Pharmacol. Toxicol.*, 1978, **43**, 86-92 (*Curcumin, metab*).
Holder, G.M. *et al.*, *Xenobiotica*, 1978, **8**, 761-768 (*metab*).
Quint, P. *et al.*, *Fresenius' Z. Anal. Chem.*, 1979, **295**, 269-270 (*use*).
Ravindranath, V. *et al.*, *Phytochemistry*, 1980, **19**, 2031-2032 (*Dihydrocurcumin, struct*).
Tønnesen, H.H. *et al.*, *Acta Chem. Scand., Ser. B*, 1982, **36**, 475-479 (*Curcumin, cryst struct, bibl*).
Cheng, K.L. *et al.*, *Handbook of Organic Analytical Reagents*, CRC Press, Boca Raton, 1982, 511 (*Curcumin, use*).
Jitoe, A. *et al.*, *J. Agric. Food Chem.*, 1992, **40**, 1337-1340 (*Curcuminoids, occur*).
Masuda, T. *et al.*, *Phytochemistry*, 1992, **31**, 3645-3647 (*5'-Methoxycurcumin, struct*).
Nakayama, R. *et al.*, *Phytochemistry*, 1993, **33**, 501-502 (*O-de-Me, struct*).
Encyclopedia of Food and Color Additives, (ed. Burdock, G.A.), CRC Press , 1997, 2856-2858 (*Curcumin*).
Dong, H. *et al.*, *J. Nat. Prod.*, 1998, **61**, 142-144 (*Curcumin, struct*).
Syu, W.-J. *et al.*, *J. Nat. Prod.*, 1998, **61**, 1531-1534 (*Curcumin, isol, synth*).
Khar, A. *et al.*, *FEBS Lett.*, 1999, **445**, 165-168 (*Curcumin, antitumour props*).
Huang, J. *et al.*, *Chem. Pharm. Bull.*, 2000, **48**, 1228-1229 (*Curcumin, activity*).
Kamnaing, P. *et al.*, *J. Nat. Prod.*, 2003, **66**, 364-367 (*Letestuianins, struct*).
Mague, J.T. *et al.*, *Acta Crystallogr., Sect. C*, 2004, **60**, o608-o610 (*di-Ac, cryst struct*).
Gafner, S. *et al.*, *Phytochemistry*, 2004, **65**, 2849-2859 (*Curcumin, activity*).
Weber, W.M. *et al.*, *Bioorg. Med. Chem.*, 2005, **13**, 3811-3820 (*sar*).
Parimita, S.P. *et al.*, *Acta Crystallogr., Sect. E*, 2007, **63**, o860-o862 (*Curcumin, cryst struct*).
Payton, F. *et al.*, *J. Nat. Prod.*, 2007, **70**, 143-146 (*Curcumin, pmr, cmr*).
Goel, A. *et al.*, *Biochem. Pharmacol.*, 2008, **75**, 787-809 (*rev*).
Simoni, D. *et al.*, *Bioorg. Med. Chem. Lett.*, 2008, **18**, 845-849 (*antitumour activity*).
Kita, T. *et al.*, *Biosci., Biotechnol., Biochem.*, 2008, **72**, 1789-1798 (*Curcuminoides, biosynth, struct, cmr*).
Benassi, R. *et al.*, *J. Mol. Struct.*, 2008, **892**, 168-176 (*pmr, cmr, uv, ft-ir*).
Liu, Z. *et al.*, *Bioorg. Med. Chem. Lett.*, 2009, **19**, 706-709 (*activity*).
Kita, T. *et al.*, *Biosci., Biotechnol., Biochem.*, 2009, **73**, 1113-1117 (*Curcuminoides, desmethoxycurcumin, Dihydrocurcumins, isol*).
Uddin, S. *et al.*, *Expert Opin. Invest. Drugs*, 2009, **18**, 57-67 (*Curcumin, pharmacol*).
Li, W. *et al.*, *Magn. Reson. Chem.*, 2009, **47**, 902-908 (*O-de-Me, demethoxy O-de-Me, struct*).
Sanphui, P. *et al.*, *Chem. Commun.*, 2011, **47**, 5013-5015 (*Curcumin, polymorphism*).
Esatbeyoglu, T. *et al.*, *Angew. Chem., Int. Ed.*, 2012, **51**, 5308-5332 (*rev*).
Gupta, S.C. *et al.*, *Clin. Exp. Pharmacol. Physiol.*, 2012, **39**, 283-299 (*rev*).
Dao, T.T. *et al.*, *Food Chem.*, 2012, **134**, 21-28 (*Curcumalongin C*).
Zingg, J.-M. *et al.*, *BioFactors*, 2013, **39 (Special Issue)**, 1-149 (*revs*).
Prasad, S. *et al.*, *Biotechnol. Adv.*, 2014, **32**, 1053-1064 (*rev*).
Marchiani, A. *et al.*, *Curr. Med. Chem.*, 2014, **21**, 204-222 (*rev*).
Priyadarsini, K.I. *Molecules*, 2014, **19**, 20091-20112 (*rev, chem, props*).
Lestari, M.L.A.D. *et al.*, *Profiles Drug Subst., Excipients, Relat. Methodol.*, 2014, **39**, 113-204 (*rev*).
Salem, M. *et al.*, *RSC Adv.*, 2014, **4**, 10815-10829 (*rev, chem, activity*).
Perrone, D. *et al.*, *Exp. Ther. Med.*, 2015, **10**, 1615-1623 (*rev, biol, activity*).
Aggarwal, B.B. *et al.*, *Molecules*, 2015, **20**, 185-205 (*rev*).
Pulido-Moran, M. *et al.*, *Molecules*, 2016, **21**, 264:1-22 (*rev, activity, props*).
Parveen, M. *et al.*, *Nat. Prod. Bioprospect.*, 2016, **6**, 267-278 (*Clutia lanceolata constit, cryst struct, activity*).
Nelson, K.M. *et al.*, *ACS Med. Chem. Lett.*, 2017, **8**, 467-470 (*rev*).
Teymouri, M. *et al.*, *BioFactors*, 2017, **43**, 331-346 (*rev, activity, props*).
Mirzaei, H. *et al.*, *Biomed. Pharmacother.*, 2017, **85**, 102-112 (*rev, pharmacol, pharmacokinet*).
Kunnumakkara, A.B. *et al.*, *Br. J. Pharmacol.*, 2017, **174**, 1325-1348 (*rev, activity, props*).
Kocaadam, B. *et al.*, *Crit. Rev. Food Sci. Nutr.*, 2017, **57**, 2889-2895 (*rev, chem, activity*).
Di Martino, R.M. *et al.*, *Expert Opin. Ther. Pat.*, 2017, **27**, 579-590 (*rev, patents*).
Nelson, K.M. *et al.*, *J. Med. Chem.*, 2017, **60**, 1620-1637 (*rev, biol, activity*).
Stanic, Z *Plant Foods Hum. Nutr.*, 2017, **72**, 1-12 (*rev, activity*).
Imran, M. *et al.*, *Crit. Rev. Food Sci. Nutr.*, 2018, **58**, 1271-1293 (*rev, activity, props*).
Tsuda, T. *Food Funct.*, 2018, **9**, 705-714 (*rev, activity, props*).
Ramezani, M. *et al.*, *J. Cell Physiol.*, 2018, **233**, 880-887 (*Bisdemethoxycurcumin, rev, activity*).
Rauf, A. *et al.*, *Trends Food Sci. Technol.*, 2018, **74**, 33-45 (*rev, activity, props*).
Salehi, B. *et al.*, *Eur. J. Med. Chem.*, 2019, **163**, 527-545 (*rev, activity, use*).
Zheng, D, *et al.*, *Chem. Biodiversity*, 2020, **17**, e2000171:1-14 (*rev, activity*)

Curcumin dimer 1 C131

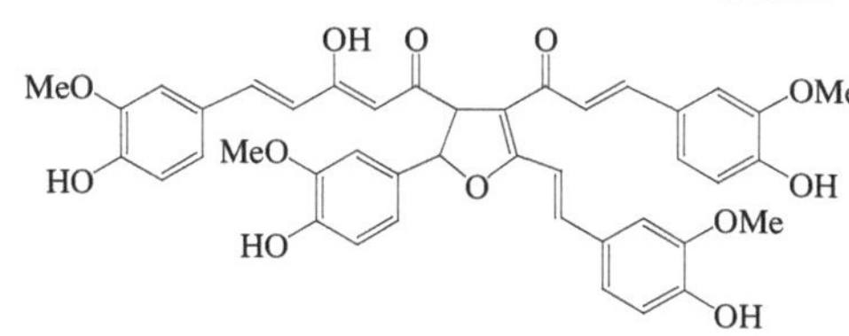

[220155-32-8]
$C_{42}H_{38}O_{12}$ 734.754
Dimer of Curcumin, C130. Constit. of the rhizomes of *Curcuma longa*. Yellow powder.

Masuda, T. *et al.*, *J. Agric. Food Chem.*, 2002, **50**, 2524-2530 (*isol, pmr*)

Curcumin dimer 2 C132

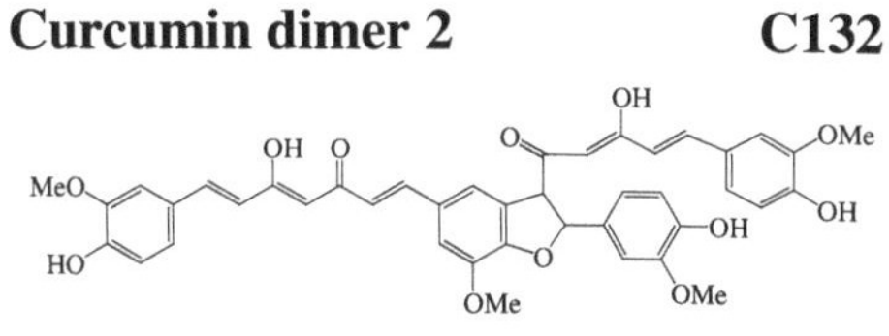

[431888-25-4]
$C_{42}H_{38}O_{12}$ 734.754
Dimer of Curcumin, C130. Constit. of the rhizomes of *Curcuma longa*. Yellow solid.

Masuda, T. *et al.*, *J. Agric. Food Chem.*, 2002, **50**, 2524-2530 (*isol, pmr, cmr*)

Curcumin dimer 3 C133

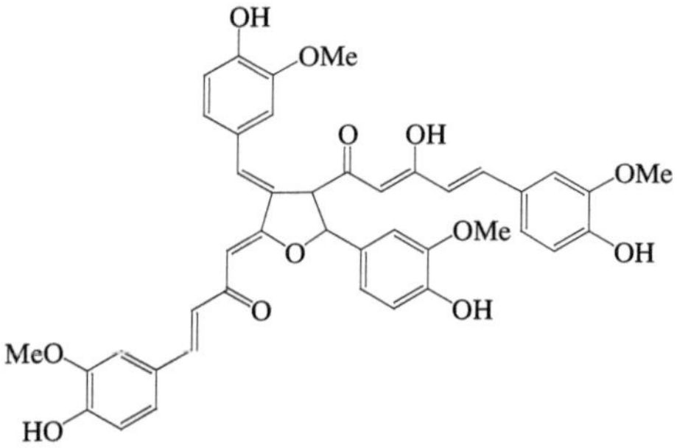

[431888-26-5]

$C_{42}H_{38}O_{12}$ 734.754

Dimer of Curcumin, C130. Constit. of the rhizomes of *Curcuma longa*. Yellow powder.

Masuda, T. *et al.*, *J. Agric. Food Chem.*, 2002, **50**, 2524-2530 (*isol, pmr, cmr*)

Curcusin B C134

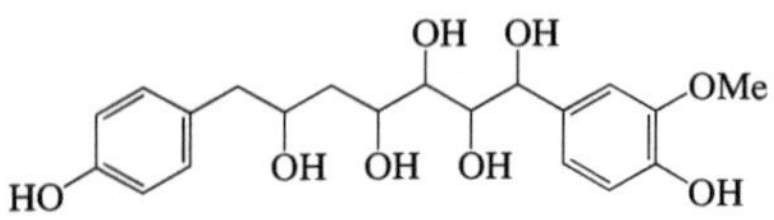

1-(4-Hydroxy-3-methoxyphenyl)-7-(4-hydroxyphenyl)-1,2,3,4,6-heptanepentol

$C_{20}H_{26}O_8$ 394.420

Constit. of the rhizomes of *Curcuma singularis*. Soluble epoxide hydrolase inhibitor. Amorph. powder. λ_{max} 225; 280 (MeCN).

Cuong, N.M. *et al.*, *Bull. Korean Chem. Soc.*, 2017, **38**, 112-115 (*Curcusin B, activity*)

Curcuterpene G C135

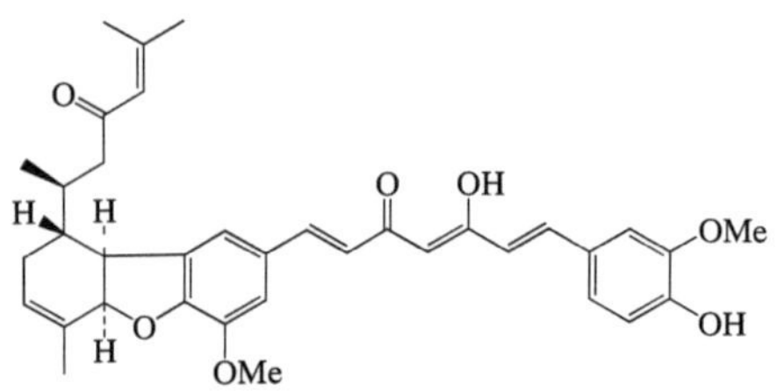

Relative Configuration

$C_{36}H_{40}O_7$ 584.709

Constit. of the rhizomes of *Curcuma longa*. Exhibits cytotoxic activity. Mp 93 - 98°. $[\alpha]^{25}_{D}$ -219 (c, 0.15 in MeOH). λ_{max} 426 (MeOH).

Li, R. *et al.*, *Zhongcaoyao*, 2016, **47**, 2418-2423 (*Curcuterpene G, activity*)

8-*O*-4/8-5-Cyclic dehydrotriferuloyltyramine C136

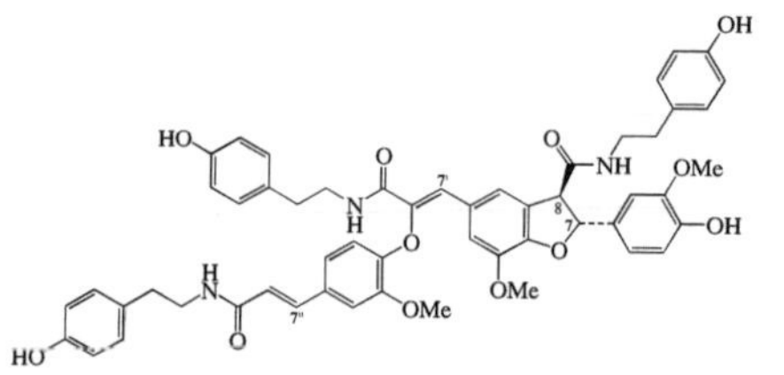

(7R*,7'Z,7''E,8R*)-form

$C_{54}H_{53}N_3O_{12}$ 936.027

Related to Grossamide, G29

(7R*,7'Z,7''E,8R*)-form

Cannabisin O [1831134-15-6]

Constit. of the seeds of *Cannabis sativa*. Amorph. powder. $[\alpha]^{20}_{D}$ +0.71 (c, 0.07 in MeOH). λ_{max} 225 (log ε 4.78); 287 (log ε 4.48); 320 (log ε 4.46) (MeOH)

(7ξ,7'Z,7''E,8ξ)-form [1260102-08-6]

Isol. from common scab lesions of *Solanum tuberosum*. Oil.

King, R.R. *et al.*, *Phytochemistry*, 2010, **71**, 2187-2189 (*Solanum tuberosum constit*).

Yan, X. *et al.*, *J. Agric. Food Chem.*, 2015, **63**, 10611-10619 (*Cannabisin O*)

2,7'-Cyclo-3',4':4,5-bis(methylenedioxy)-1,7-seco-8,8'-lignan-9,9'-olide C137

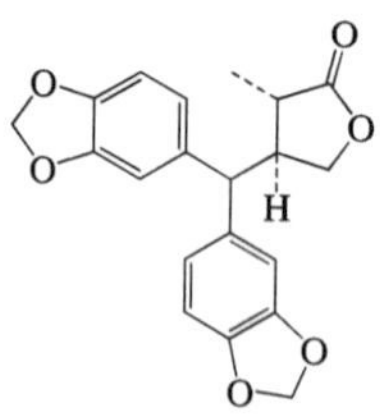

Absolute Configuration

4-[Bis(1,3-benzodioxol-5-yl)methyl] dihydro-3-methyl-2(3H)-furanone, CAS. *3',4':4,5-Bis(methylenedioxy)-1,7-seco-2,7'-cyclolignan-9,9'-olide*. ***Peperomin D***

[184587-37-9]

$C_{20}H_{18}O_6$ 354.358

Secolignan. Constit. of *Peperomia glabella*. Cryst. (heptane). Mp 122 - 124°. $[\alpha]_D$ +23 (c, 0.1 in $CHCl_3$) (+18.4).

Delle Monache, F. *et al.*, *Phytochemistry*, 1996, **43**, 1097-1098 (*Peperomin D*).

Sibi, M.P. *et al.*, *Can. J. Chem.*, 2001, **79**, 1546-1555 (*synth*)

Cyclocurcumin C138

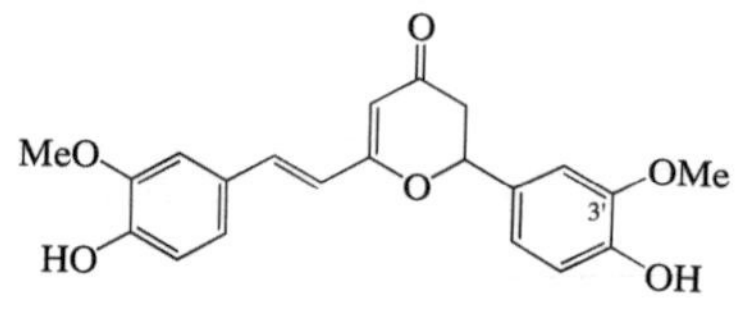

5,6-Dihydro-2-[2-(4-hydroxy-3-methylphenyl)ethenyl]-6-(4-hydroxy-3-methoxyphenyl)-4H-pyran-4-one

[153127-42-5]

$C_{21}H_{20}O_6$ 368.385

Constit. of the rhizome of *Curcuma longa*. Inhibitor of cell proliferation. Exhibits anti-vasoconstrictive activity. Antioxidant. Yellow gum. λ_{max} 233 (log ε 4.4); 262 (log ε 4.28); 373 (log ε 4.89) (EtOH)

3'-Demethoxy

3'-Demethoxycyclocurcumin

[1188320-05-9]

$C_{20}H_{18}O_5$ 338.359

Constit. of the tubers of *Curcuma xanthorrhiza*. Antibacterial agent. Amber oil. Racemic

Bis(demethoxy)

Didemethoxycyclocurcumin

[1042441-12-2]

$C_{19}H_{16}O_4$ 308.333

Constit. of the roots of *Curcuma longa*. Brownish-yellow needles (MeOH). Mp 138 - 139°. λ_{max} 277; 284; 360 (EtOH).

Kiuchi, F. *et al.*, *Chem. Pharm. Bull.*, 1993, **41**, 1640-1643 (*Cyclocurcumin*).

Wang, L.-Y. *et al.*, *Biochem. Syst. Ecol.*, 2008, **36**, 476-480 (*Didemethoxycyclocurcumin*).

Yamada, K. *et al.*, *Biosci., Biotechnol., Biochem.*, 2009, **73**, 776-780 (*3'-Demethoxycyclocurcumin*).

Adhikary, R. *et al.*, *J. Phys. Chem. B*, 2011, **115**, 10707-10714 (*photochem, fluorescence*).

Kim, K. *et al.*, *J. Nat. Prod.*, 2017, **80**, 196-200 (*activity*).

Li, Y. *et al.*, *New J. Chem.*, 2018, **42**, 12698-12705 (*activity*).

Marazzi, M. *et al.*, *Phys. Chem. Chem. Phys.*, 2020, **22**, 4749-4757 (*photochem*)

3,8'-Cyclo-2,3-dihydro-3,3',4,4',5'-pentahydroxy-1,7'-lign-8-en-6(1*H*)-one C139

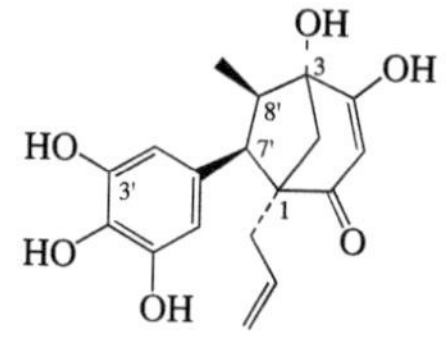

(1*R*,3*R*,7'*R*,8'*R*)-form

4,5-Dihydroxy-6-methyl-1-(2-propenyl)-7-(3,4,5-trihydroxyphenyl)bicyclo[3.2.1]oct-3-en-2-one
$C_{18}H_{20}O_6$ 332.352
(1R,3R,7′R,8′R)-form
3,4-Methylene, 3′,4′,5′-tri-Me ether
3,8′-Cyclo-2,3-dihydro-3′,4′,5′-trimethoxy-3,4-methylenedioxy-1,7′-lign-8-en-6(1H)-one
[1651214-82-2]
$C_{22}H_{26}O_6$ 386.444
Constit. of *Ocotea cymosa*. Exhibits antiplasmodial activity. Oil. $[\alpha]_D^{25}$+101 (c, 0.02 in MeOH). λ_{max} 217; 250 (MeOH aq.)
3,4:3′,4′-Bis(methylene), 5′-Me ether
3,8′-Cyclo-2,3-dihydro-3′-methoxy-3,4:4′,5′-bis(methylenedioxy)-1,7′-lign-8-en-6(1H)-one. ***Ocobullenone***
[149990-50-1]
$C_{21}H_{22}O_6$ 370.401
Constit. of *Ocotea bullata* and *Ocotea cymosa*. 5-Lipoxygenase inhibitor. Needles. Mp 151°. $[\alpha]_D^{25}$ +204 (c, 0.16 in $CHCl_3$)
(1R,3R,7′R,8′S)-form
3,4:3′,4′-Bis(methylene), 5′-Me ether
Isoocobullenone
[165306-72-9]
$C_{21}H_{22}O_6$ 370.401
Constit. of the bark of *Ocotea bullata*. Needles (CH_2Cl_2/hexane). Mp 142 - 144°. $[\alpha]_D^{23}$+122.8 (c, 0.002 in $CHCl_3$)
(1S,3S,7′R,8′S)-form
3,4-Methylene, 3′,4′,5′-tri-Me ether
[1651214-81-1]
$C_{22}H_{26}O_6$ 386.444
Constit. of *Ocotea cymosa*. Oil. $[\alpha]_D^{25}$-51 (c, 0.02 in MeOH). λ_{max} 217; 250 (MeOH aq.)
3,4:3′,4′-Bis(methylene) ether
3,8′-Cyclo-2,3-dihydro-3′-hydroxy-3,4:4′,5′-bis(methylenedioxy)-1,7′-lign-8-en-6(1H)-one.
Demethylsibyllenone
[1651214-80-0]
$C_{20}H_{20}O_6$ 356.374
Constit. of *Ocotea cymosa*. Exhibits antiplasmodial activity. Amorph. powder. $[\alpha]_D^{25}$+6 (c, 0.01 in MeOH). λ_{max} 205 (log ε 3.31); 240 (log ε 2.7); 300 (log ε 1.9) (MeOH)
3′-Deoxy, 3,4:4′,5′-bis(methylene) ether
3,8′-Cyclo-2,3-dihydro-3,4:3′,4′-bis(methylenedioxy)-1,7′-lign-8-en-6(1H)-one. ***Demethoxysibyllenone***
[1651214-79-7]
$C_{20}H_{20}O_5$ 340.375
Constit. of *Ocotea cymosa*. Exhibits antiplasmodial activity. Oil. $[\alpha]_D^{25}$-144 (c, 0.1 in MeOH). λ_{max} 240; 290 (MeOH aq.)
(1RS,3RS,7′SR,8′RS)-form
3,4:3′,4′-Bis(methylene), 5′-Me ether
Sibyllenone
[299175-09-0]
$C_{21}H_{22}O_6$ 370.401
Constit. of *Ocotea bullata* and *Ocotea cymosa*. 5-Lipoxygenase inhibitor. Cryst.. Mp 160°.

Sehlapelo, B.M. *et al.*, *Phytochemistry*, 1993, **32**, 1352-1353 (*Ocobullenone*).
Drewes, S.E. *et al.*, *Phytochemistry*, 1995, **38**, 1505-1508 (*Isoocobullenone*).
Zshocke, S. *et al.*, *Phytochemistry*, 2000, **54**, 591-595 (*Ocobullenone, Isoocobullenone, Sibyllenone, cryst struct*).
Rakotondraibe, L.H. *et al.*, *J. Nat. Prod.*, 2015, **78**, 431-440 (*Ocotea cymosa constits, activity*)

5,8′-Cyclo-4,5-dihydro-3′,4,4′,5,5′-pentahydroxy-3,7′-lign-8-en-2(3*H*)-one C140

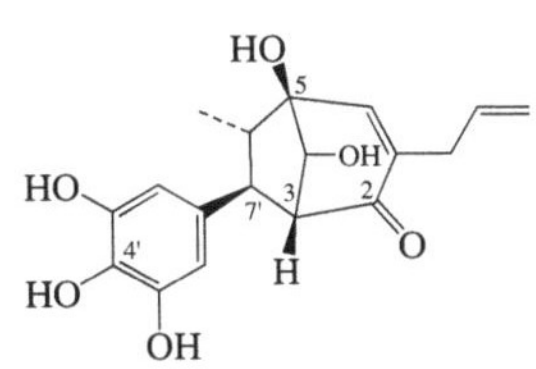

(3R,4S,5S,7′S,8′S)-form

$C_{18}H_{20}O_6$ 332.352
(3R,4S,5S,7′S,8′S)-form
3′-Deoxy, 4′,5,5′-tri-Me ether, 4-Ac
Piperulin A*. Puberulin A‡*
[164268-04-6]
$C_{23}H_{28}O_6$ 400.471
Constit. of *Piper puberulum*. PAF receptor antagonist. Cubes (Et_2O). Mp 130 - 131°. $[\alpha]_D$ -28.1 (c, 5.1 in $CHCl_3$). λ_{max} 231 (log ε 4.05); 280 (log ε 3.39) (MeOH)
3′-Deoxy, 4′,5′-methylene, 5-Me ether, 4-Ac
Piperulin B*. Puberulin B*
[164454-37-9]
$C_{22}H_{24}O_6$ 384.428
Constit. of *Piper puberulum*. Amorph. solid. $[\alpha]_D$ -33.9 (c, 3.2 in $CHCl_3$). λ_{max} 233 (log ε 3.92); 288 (log ε 3.52) (MeOH)
(3S,4R,5R,7′R,8′R)-form
3′,4′-Methylene, 5,5′-di-Me ether
5,8′-Cyclo-4,5-dihydro-4-hydroxy-3′,5-dimethoxy-4′,5′-methylenedioxy-3,7′-lign-8-en-2(3H)-one.
Cinerin B
[1166328-46-6]
$C_{21}H_{24}O_6$ 372.417
Constit. of the leaves of *Pleurothyrium cinereum*. Oil. $[\alpha]_D^{25}$ -13.2 (c, 0.3 in $CHCl_3$). λ_{max} 214 (log ε 4.48); 248 (log ε 4.17); 288 (log ε 3.4) (MeOH)
3′-Deoxy, 4′,5,5′-tri-Me ether
5,8′-Cyclo-4,5-dihydro-4-hydroxy-3′,4′,5-trimethoxy-3,7′-lign-8-en-2(3H)-one. ***Kadsurenin C***
[145722-88-9]
$C_{21}H_{26}O_5$ 358.434
Constit. of *Piper kadsura*. PAF antagonist. Oil. $[\alpha]_D^{18}$-24 (c, 0.07 in $CHCl_3$)
3′-Deoxy, 4′,5,5′-tri-Me ether, 4-Ac
Kadsurenin L
[149438-61-9]
$C_{23}H_{28}O_6$ 400.471
Constit. of *Piper kadsura*. PAF antagonist. $[\alpha]_D^{18}$-36.9 ($CHCl_3$)
3′-Deoxy, 4′,5′-methylene, 5-Me ether
5,8′-Cyclo-4,5-dihydro-4-hydroxy-5-methoxy-3′,4′-methylenedioxy-3,7′-lign-8-en-2(3H)-one.
Kadsurenin B
[145701-13-9]
$C_{20}H_{22}O_5$ 342.391
Constit. of *Piper kadsura*. PAF receptor antagonist. Oil. $[\alpha]_D^{18}$-29.5 (c, 0.11 in $CHCl_3$)
3′-Deoxy, 4′,5′-methylene, 5-Me ether, 4-Ac
Kadsurenin A
[145701-12-8]
$C_{22}H_{24}O_6$ 384.428
Constit. of *Piper kadsura*. Cryst.. Mp 82 - 84°
(3S,4S,5R,7′R,8′R)-form
3′,4′,5,5′-Tetra-Me ether
5,8′-Cyclo-4,5-dihydro-4-hydroxy-3′,4′,5,5′-tetramethoxy-3,7′-lign-8-en-2(3H)-one. ***Ocophyllol C***
[1189119-92-3]
$C_{22}H_{28}O_6$ 388.46
Constit. of the leaves of *Ocotea macrophylla*. Moderate PAF antagonist. Needles. Mp 188 - 190°. $[\alpha]_D^{25}$-29.8 (c, 0.63 in $CHCl_3$)

3′-Deoxy, 4′,5,5′-tri-Me ether
Ocophyllol B
[1189119-91-2]
$C_{21}H_{26}O_5$ 358.434
Constit. of the leaves of *Ocotea macrophylla*. Moderate PAF antagonist. Oil. $[\alpha]^{25}_{D}$ -25.3 (c, 0.37 in $CHCl_3$)

3′-Deoxy, 4′,5′-methylene, 5-Me ether
Ocophyllol A
[1189119-90-1]
$C_{20}H_{22}O_5$ 342.391
Constit. of *Ocotea macrophylla*. Moderate PAF antagonist. Oil. $[\alpha]^{25}_{D}$ -27.4 (c, 0.41 in $CHCl_3$)

3′-Deoxy, 4-ketone, 5,5′-di-Me ether
5,8′-Cyclo-4′-hydroxy-3′,5-dimethoxy-3,7′-lign-8-ene-2,4(3H,5H)-dione. ***Kadsurenin K***
[149560-83-8]
$C_{20}H_{22}O_5$ 342.391
Constit. of *Piper kadsura*. PAF receptor antagonist. Oil. $[\alpha]^{18}_{D}$ -54.6 (c, 0.05 in $CHCl_3$). Possesses (3*R*,5*R*,7′*R*,8′*R*)-config. due to change in Cahn-Ingold-Prelog priorities

3′-Deoxy, 4-ketone, 4′,5,5′-tri-Me ether
5,8′-Cyclo-3′,4′,5-trimethoxy-3,7′-lign-8-ene-2,4(3H,5H)-dione. ***Denudadione A***
[149560-81-6]
[245648-17-3]
$C_{21}H_{24}O_5$ 356.418
Constit. of *Magnolia biondii* and *Magnolia denudata*. Possesses (3*R*,5*R*,7′*R*,8′*R*)-config. due to change in Cahn-Ingold-Prelog priorities

3′-Deoxy, 4-ketone, 4′,5′-methylene, 5-Me ether
5,8′-Cyclo-5-methoxy-3′,4′-methylenedioxy-3,7′-lign-8-ene-2,4(3H,5H)-dione. ***Denudadione B***
[288260-35-5]
$C_{20}H_{20}O_5$ 340.375
Constit. of *Magnolia denudata*. Moderate NO prodn. inhibitor. Possesses (3*R*,5*R*,7′*R*,8′*R*)-config. due to change in Cahn-Ingold-Prelog priorities

(3R,5S,7′R,8′R)-form

4-Ketone, 3′,4′,5,5′-tetra-Me ether
5,8′-Cyclo-3′,4′,5,5′-tetramethoxy-3,7′-lign-8-ene-2,4(3H,5H)-dione. 6-Ketomacrophyllin B. ***2-Oxomacrophyllin B***
[74984-56-8]
$C_{22}H_{26}O_6$ 386.444
Constit. of *Pleurothyrium cinereum*. Oil. Possesses (3*S*,5*S*,7′*R*,8′*R*)-config. due to change in Cahn-Ingold-Prelog priorities

4-Ketone, 2R-alcohol, 3′,4′,5,5′-tetra-Me ether
5,8′-Cyclo-2,3-dihydro-2-hydroxy-3′,4′,5,5′-tetramethoxy-3,7′-lign-8-en-4(5H)-one. ***Macrophyllin B‡***
[74944-98-2]
$C_{22}H_{28}O_6$ 388.46
Constit. of *Nectandra amazonicum* and *Pleurothyrium cinereum*. Moderate, selective inhibitor of COX-2 and PAF. Oil. Possesses (2*R*,3*S*,5*S*,7′*R*,8′*R*)-config. due to change in Cahn-Ingold-Prelog priorities

3′-Deoxy, 4-ketone, 5,5′-di-Me ether
Liliflodione
[87402-81-1]
$C_{20}H_{22}O_5$ 342.391
Constit. of *Magnolia liliflora*. Oil. $[\alpha]_{D}$ +254.4 (c, 1.07 in MeOH). Possesses (3*S*,5*S*,7′*R*,8′*R*)-config. due to change in Cahn-Ingold-Prelog priorities.

Braz Filho, R. *et al.*, *Phytochemistry*, 1980, **19**, 659-662 (*Macrophyllin B, 2-Oxomacrophyllin B*).
Iida, T. *et al.*, *Phytochemistry*, 1983, **22**, 763-766 (*Liliflodione*).
Rodrigues, M. de M. *et al.*, *Phytochemistry*, 1984, **23**, 667-669 (*2-Oxomacrophyllin B, cryst struct*).
Han, G. *et al.*, *Chin. Chem. Lett.*, 1992, **3**, 521-524 (*Kadsurenins A,B,C, activity*).
Ma, Y. *et al.*, *Yaoxue Xuebao*, 1993, **28**, 207-211 (*Piper kadsura constits, activity*).
Zhang, S.-X. *et al.*, *J. Nat. Prod.*, 1995, **58**, 540-547 (*Piperulins A,B, cryst struct*).
Zhang, S.-X. *et al.*, *J. Nat. Prod.*, 1996, **59**, 1002 (*nomenclature*).
Horne, D.A. *et al.*, *Tetrahedron Lett.*, 1999, **40**, 5443-5447 (*Liliflodione, synth*).
Kuroyanagi, M. *et al.*, *Chem. Pharm. Bull.*, 2000, **48**, 832-837 (*Denudadiones A,B, cd, activity*).
Wang, M. *et al.*, *J. Org. Chem.*, 2002, **67**, 5405-5407 (*Kadsurenins C,L, synth*).
Jiang, R.W. *et al.*, *J. Mol. Struct.*, 2003, **654**, 177-182 (*Kadsurenin C, cryst struct*).
Coy-Barrera, E.D. *et al.*, *Biochem. Syst. Ecol.*, 2008, **36**, 674-677 (*Pleurothyrium cinereum constit*).
Noshita, T. *et al.*, *Biosci., Biotechnol., Biochem.*, 2008, **72**, 2775-2778 (*Denudadione B, activity*).
Coy, E.D. *et al.*, *Bioorg. Med. Chem. Lett.*, 2009, **19**, 6922-6925 (*Macrophyllin B, 2-Oxomacrophyllin B, activity*).
Coy, E.D. *et al.*, *J. Nat. Prod.*, 2009, **72**, 1245-1248 (*Cinerin B, cd, activity*).
Coy-Barrera, E.D. *et al.*, *Phytochemistry*, 2009, **70**, 1309-1314 (*Ocophyllols A,B,C, cd, activity*).
Nguyen, T.T.M. *et al.*, *Nat. Prod. Sci.*, 2017, **23**, 119-124 (*Denudadione A*)

2,7′-Cyclo-4,4′-dihydroxy-3′,5-dimethoxy-8,9-dinor-7,8′-lign-7′-en-9′-al C141

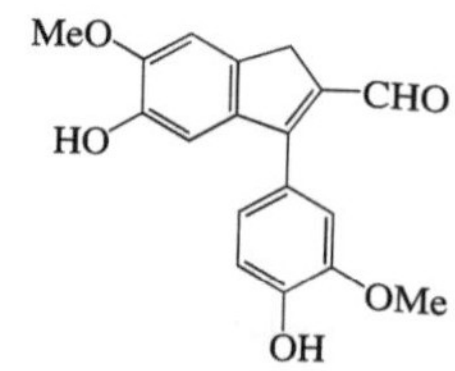

5-Hydroxy-3-(4-hydroxy-3-methoxyphenyl)-6-methoxy-1H-indene-2-carboxaldehyde. ***Vitexdoin F‡***
$C_{18}H_{16}O_5$ 312.321
Isol. from the seeds of *Vitex negundo*. Displays antiinflammatory and antiosteoporotic activity. Amorph. yellow powder. Not to be confused with Vitexdoin F in C181. λ_{max} 223; 264; 364 (MeOH).

Zheng, C.-J. *et al.*, *Fitoterapia*, 2014, **93**, 31-38 (*Vitexdoin F*)

2,7′-Cyclo-4,4′-dihydroxy-3′,5-dimethoxy-9-nor-8,8′-ligna-7,7′-dien-9′-al C142

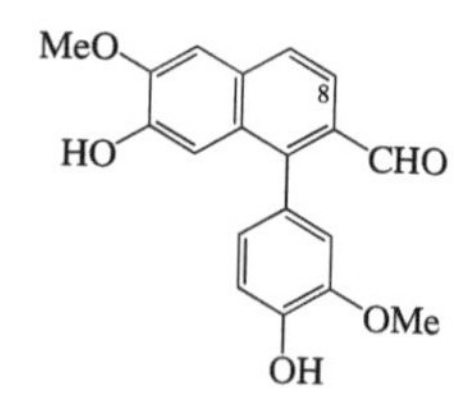

4,4′-Dihydroxy-3′,5-dimethoxy-9-nor-2,7′-cycloligna-7,7′-dien-9′-al. 7-Hydroxy-1-(4-hydroxy-3-methoxyphenyl)-6-methoxy-2-naphthalenecarboxaldehyde. ***Vitexdoin C***
[1186021-80-6]
$C_{19}H_{16}O_5$ 324.332
Norlignan. Lignan numbering shown. Constit. of the seeds of *Vitex negundo*. Amorph. yellow powder. λ_{max} 206; 266; 316 (MeOH)

8-Hydroxy
2,7′-Cyclo-4,4′,8-trihydroxy-3′,5-dimethoxy-9-nor-8,8′-ligna-7,7′-dien-9′-al. 4,4′,8-Trihydroxy-3′,5-dimethoxy-9-nor-2,7′-cycloligna-7,7′-dien-9′-al. 3,7-Dihydroxy-1-(4-hydroxy-3-methoxyphenyl)-6-methoxy-2-naph-

thalenecarboxaldehyde. ***Vitexdoin D***
[1186021-81-7]
$C_{19}H_{16}O_6$ 340.331
Constit. of the seeds of *Vitex negundo*. Potent LPS-induced NO prodn. inhibitor. Amorph. yellow powder. λ_{max} 203; 265; 333 (MeOH).

Zheng, C.-J. *et al.*, *J. Nat. Prod.*, 2009, **72**, 1627-1630 (*Vitexdoins C,D*)

7,7′-Cyclo-4,4′-dihydroxy-8,8′-lignan-9,9′-dioic acid C143

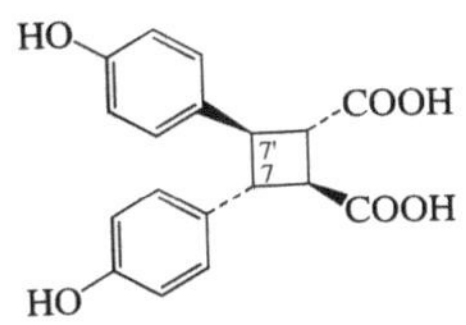

(7α,7′β,8β,8′α)-*form*

3,4-Bis(4-hydroxyphenyl)-1,2-cyclobutanedicarboxylic acid. 4′,4″-Dihydroxytruxinic acid
$C_{18}H_{16}O_6$ 328.32
Photodimer of 3-(4-Hydroxyphenyl)-2-propenoic acid, H126

(7α,7′β,8β,8′α)-form
Mono-Me ester
Anisumic acid
[2095621-37-5]
$C_{19}H_{18}O_6$ 342.347
Constit. of *Clausena anisum-olens*. Powder. $[\alpha]^{21}_{D}$ +13 (c, 3 in MeOH). λ_{max} 205; 230; 279; 360 (MeOH)

(all-ξ)-form [133097-99-1]
Constit. of plant cell walls. Prod. by *Streptomyces* sp. YIM 67086. Exhibits antifungal activity.

Ford, C.W. *et al.*, *J. Sci. Food Agric.*, 1989, **46**, 301-310 (*occur*).
Yang, X. *et al.*, *Nat. Prod. Res.*, 2015, **29**, 331-335 (*Streptomyces metab*).
Wang, Y.-S. *et al.*, *J. Nat. Prod.*, 2017, **80**, 798-804 (*Anisumic acid*)

7′,8-Cyclo-4,4′-dihydroxy-7,8′-lignan-9,9′-dioic acid C144

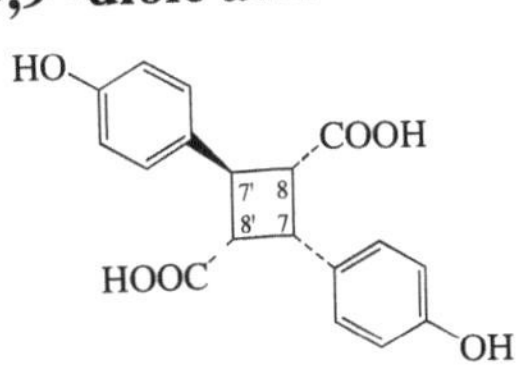

(7α,7′β,8α,8′α)-*form*

2,4-Bis(4-hydroxyphenyl)-1,3-cyclobutanedicarboxylic acid, CAS
[133098-00-7]
$C_{18}H_{16}O_6$ 328.32

(7α,7′β,8α,8′α)-form
Bis-L-tyrosine amide
Abrusamide C
$C_{36}H_{34}N_2O_{10}$ 654.672
Isol. from *Abrus mollis*. λ_{max} 226; 275 (MeOH)

(7α,7′β,8α,8′β)-form
4,4-Dihydroxy-α-truxillic acid. ***Thesinic acid*** [490-18-6]
Component of plant cell walls. Mp > 300°. Artifact, formed *in situ* by the action of light on 3-(4-Hydroxyphenyl)-2-propenoic acid, H126

Bis-L-tyrosine amide
Abrusamide A
[1422043-31-9]
$C_{36}H_{34}N_2O_{10}$ 654.672
Isol. from *Abrus mollis*. Shows significant hepatoprotective activity. Amorph. solid. $[\alpha]^{20}_{D}$ +44.6 (c, 0.086 in MeOH). λ_{max} 226; 278 (MeOH)

(7β,7′β,8α,8′α)-form
Bis-L-tyrosine amide
Abrusamide B
[1422043-32-0]
$C_{36}H_{34}N_2O_{10}$ 654.672
Isol. from *Abrus mollis*. Hepatoprotectant. Amorph. light brown solid. $[\alpha]^{20}_{D}$ -32.5 (c, 0.081 in MeOH). λ_{max} 226; 278 (MeOH).

Arendaruk, A.P. *J. Gen. Chem. USSR (Engl. Transl.)*, 1960, **30**, 507-510 (*Thesinic acid, partial struct*).
Hartley, R.D. *et al.*, *Phytochemistry*, 1988, **27**, 349-351 (*isol*).
Koshino, H. *et al.*, *Phytochemistry*, 1988, **27**, 1333-1338 (*synth, ms, pmr*).
Turner, L.B. *Phytochemistry*, 1993, **33**, 791-796 (*occur*).
Yuan, X. *et al.*, *Phytochem. Lett.*, 2014, **7**, 137-142 (*Abrusamides A,B*).
Yuan, X.-J. *et al.*, *Yaoxue Xuebao*, 2016, **51**, 1595-1599 (*Abrusamide C*)

2,7′-Cyclo-4′,4″-dihydroxy-3′,3″,5-trimethoxy-4,8″-oxy-8,8′-sesquiligna-7,7″-diene-9,9′,9″-trioic acid C145

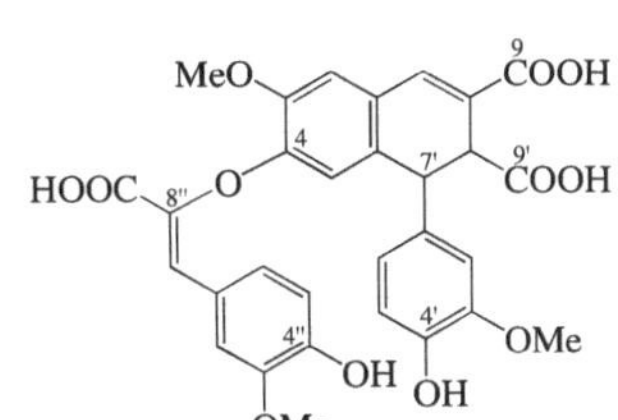

Ferulic acid-(8-O-4)-4,4′-dihydroxy-3′,5-dimethoxy-2,7′-cyclolign-7-ene-9,9′-dioic acid
[848649-29-6]
$C_{30}H_{26}O_{12}$ 578.526
Lignan numbering shown. Constit. of maize bran (*Zea mays*).

Funk, C. *et al.*, *Phytochemistry*, 2005, **66**, 363-371 (*Zea mays constit*)

2,7′-Cyclo-4,5-dimethoxy-3′,4′-methylenedioxy-8,9-dinor-7,8′-lign-7′-en-7-one C146

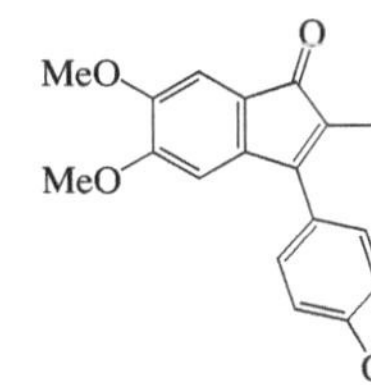

5,6-Dimethoxy-2-methyl-3-(3,4-methylenedioxyphenyl)indenone. 2,7′-Cyclo-4,5-dimethoxy-3′,4′-methylenedioxy-8,9-dinor-7-oxo-7,8′-neolign-7′-ene
[213842-67-2]
[93446-23-2]
$C_{19}H_{16}O_5$ 324.332
Struct. revised in 1998. Cmr still shows inconsistencies. Constit. of *Virola elongata* and *Virola sebifera*. Red solid. Mp 214 - 216°. λ_{max} 265 (ε 32400); 340 (ε 7600) (MeOH)

7′,8′-Dihydro(cis-)
2,7′-Cyclo-4,5-dimethoxy-3′,4′-methylenedioxy-8,9-dinor-7,8′-lignan-7-one. 2,7′-Cyclo-4,5-dimethoxy-3′,4′-methylenedioxy-8,9-dinor-7-oxo-7,8′-lignan
[93446-24-3]
$C_{19}H_{18}O_5$ 326.348
Constit. of *Virola sebifera*. Struct. uncertain in view of the revision of the parent indenone.

Lopes, L.M.X. *et al.*, *Phytochemistry*, 1984, **23**, 2021-2024 (*Virola sebifera constits*).
Kato, M.J. *et al.*, *Phytochemistry*, 1990, **29**, 1799-1810 (*Virola elongata constit*).
Harrowven, D.C. *et al.*, *Tetrahedron Lett.*, 1998, **39**, 6757-6760 (*struct, synth*)

2,7′-Cyclo-6,9′-epoxy-1,6-dihydro-3′,4,4′,5-tetrahydroxy-1,8′-lign-8-en-3(2*H*)-one C147

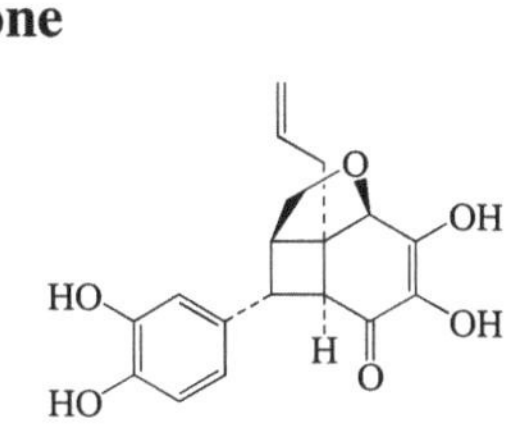

Relative Configuration

$C_{18}H_{18}O_6$ 330.336

Tetra-Me ether

2,7′-Cyclo-6,9′-epoxy-1,6-dihydro-3′,4,4′,5-tetramethoxy-1,8′-lign-8-en-3(2H)-one. ***Madangone A***

$C_{22}H_{26}O_6$ 386.444

Constit. of the stem bark of *Beilschmiedia madang*. Exhibits antioxidant activity and moderate inhibitory activity against COX-2. Amorph. solid. $[\alpha]_D^{20}$ +0.29 (c, 1.2 in $CHCl_3$). λ_{max} 231 (log ε 1.48); 285 (log ε 1.24) (MeOH)

3′,4′-Methylene, 4,5-di-Me ether

2,7′-Cyclo-6,9′-epoxy-1,6-dihydro-4,5-dimethoxy-3′,4′-methylenedioxy-1,8′-lign-8-en-3(2H)-one.

Kunstlerone

[1340473-43-9]

$C_{21}H_{22}O_6$ 370.401

Constit. of *Beilschmiedia kunstleri* and *Beilschmiedia madang*. Antioxidant. Amorph. solid. $[\alpha]_D^{25}$ +163.6 (c, 0.002 in MeOH). λ_{max} 332 (log ε 4); 404 (log ε 1.36) (MeOH).

Mollataghi, A. *et al., Molecules*, 2011, **16**, 6582-6590 (*Kunstlerone*).

Salleh, W.M.N.H.W. *et al., Phytochem. Lett.*, 2016, **15**, 168-173 (*Madangone A, activity*)

2,7′-Cyclo-7,9′-epoxy-3′,5-dimethoxy-8,8′-lignan-4,4′,9-triol C148

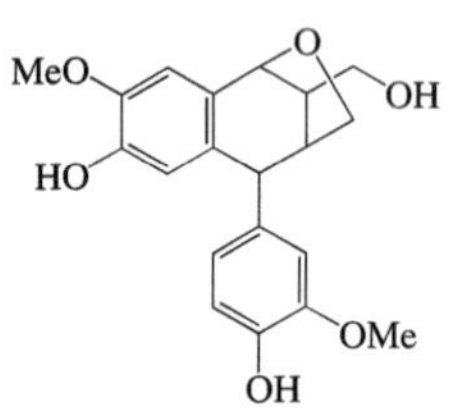

7,9′-Epoxy-4,4′,9-trihydroxy-3′,5-dimethoxy-2,7′-cyclolignan

[131653-18-4]

$C_{20}H_{22}O_6$ 358.390

Constit. of the roots of *Taxus wallichiana*. Solid.

Abe, F. *et al., Chem. Pharm. Bull.*, 1990, **38**, 2143-2145 (*synth*).

Dang, P.H. *et al., J. Nat. Prod.*, 2017, **80**, 1876-1882 (*Taxus wallichiana constit*)

2,7′-Cyclo-2′,9-epoxy-8,8′-lignan-3,3′,4,4′,5,5′,9′-heptol C149

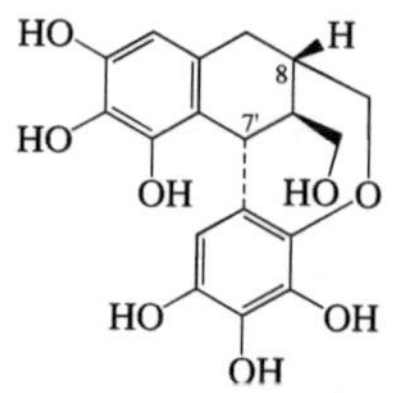

$C_{18}H_{18}O_8$ 362.334

(7′R,8S,8′S)-form

3,3′,5,5′-Tetra-Me ether

2,7′-Cyclo-2′,9-epoxy-3,3′,5,5′-tetramethoxy-8,8′-lignan-4,4′,9′-triol. 2′,9-Epoxy-4,4′,9′-trihydroxy-3,3′,5,5′-tetramethoxy-2,7′-cyclolignan

3,3′,5,5′-Tetra-Me ether, 9′-O-β-D-xylopyranoside

Gaultheroside G

$C_{27}H_{34}O_{12}$ 550.557

Constit. of *Gaultheria yunnanensis*. Solid. $[\alpha]_D^{17}$ +36 (c, 1.5 in MeOH). 2′,9-Epoxynaphthalenoid lignan related to Ovafolinin B, O16. λ_{max} 211 (log ε 1.68); 268 (log ε 0.27) ($CHCl_3$).

Gao, Y.-P. *et al., J. Asian Nat. Prod. Res.*, 2014, **16**, 724-729 (*Gaultheroside G*)

2,7′-Cyclo-2′,9′-epoxy-8,8′-lignan-3,3′,4,4′,5,5′,9-heptol C150

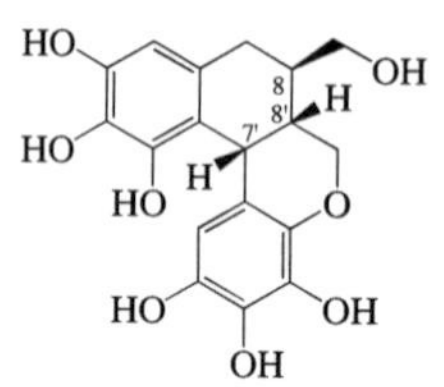

(7′R,8R,8′R)-form

$C_{18}H_{18}O_8$ 362.334

The nat. prod. Pithecellobiumin B was isol. as a racemate and resolved by chiral hplc

(7′R,8R,8′R)-form

3,3′,5,5′-Tetra-Me ether

(−)-Pithecellobiumin B

[2225884-70-6]

$C_{22}H_{26}O_8$ 418.442

Inhibits the formn. of $A\beta_{42}$ aggregation. Yellow oil. $[\alpha]_D^{20}$ -7 (c, 0.1 in MeOH)

(7′S,8S,8′S)-form

3,3′,5,5′-Tetra-Me ether

(+)-Pithecellobiumin B

[2225891-16-5]

Inhibits the formn. of $A\beta_{42}$ aggregation. Yellow oil. $[\alpha]_D^{20}$ +5 (c, 0.1 in MeOH)

(7′RS,8SR,8′SR)-form

3,3′,5,5′-Tetra-Me ether

2,7′-Cyclo-2′,9′-epoxy-3,3′,5,5′-tetramethoxy-8,8′-lignan-4,4′,9-triol.

Pithecellobiumin B

$C_{22}H_{26}O_8$ 418.442

Constit. of the twigs and leaves of *Pithecellobium clyperia*. Yellow oil. $[\alpha]_D^{20}$ -2 (c, 0.1 in MeOH). λ_{max} 225 (log ε 4.73); 282 (log ε 3.11) (MeOH).

Wang, Y.-X. *et al., Bioorg. Chem.*, 2018, **77**, 579-585 (*Pithecellobiumin B, activity*)

2,7′-Cyclo-8,8′-epoxy-8,8′-lignan-3′,4,4′,5,5′,6,7-heptol C151

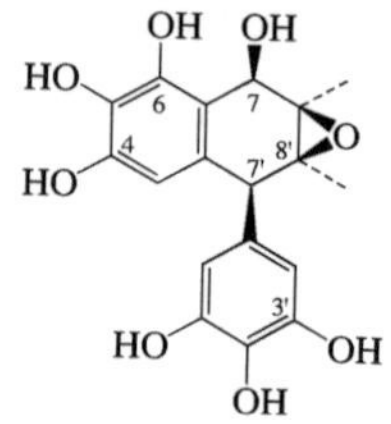

8,8′-Epoxy-3′,4,4′,5,5′,6,7-heptahydroxy-2,7′-cyclolignan

$C_{18}H_{18}O_8$ 362.334

(7R*,7′S*,8R*,8′S*)-form

3′,4,5′,6-Tetra-Me ether

2,7′-Cyclo-8,8′-epoxy-3′,4,5′,6-tetramethoxy-8,8′-lignan-4′,5,7-triol. 8,8′-Epoxy-4′,5,7-trihydroxy-3′,4,5′,6-tetramethoxy-2,7′-cyclolignan.

Sacidumlignan C

[848986-17-4]

$C_{22}H_{26}O_8$ 418.442

Constit. of the whole plant of *Sarcostemma acidum*. Amorph. powder. $[\alpha]_D^{20}$ -72 (c, 0.5 in Me_2CO). λ_{max} 204 (log ε 4.97); 274 (log ε 3.53) (MeOH).

Gan, L.-S. *et al., J. Nat. Prod.*, 2005, **68**, 221-225 (*Sacidumlignan C*)

2,7′-Cyclo-9,9′-epoxy-8,8′-lignan-3,3′,4,4′,5,5′,7-heptol C152

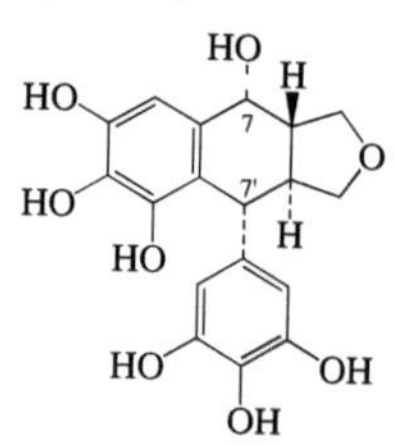

(7R,7′R,8S,8′S)-form

9,9′-Epoxy-3,3′,4,4′,5,5′,7-Heptahydroxy-2,7′-cyclolignan
$C_{18}H_{18}O_8$ 362.334
(7R,7′R,8S,8′S)-form
3,3′,4,4′,5,5′-Hexa-Me ether, 7-Ac
Aglacin A
[389628-92-6]
$C_{26}H_{32}O_9$ 488.533
Constit. of the stem bark of *Aglaia cordata*. Weak inhibitor of human KB carcinoma cell line. Needles (MeOH). Mp 155 - 156°. $[\alpha]_D^{20}$ +55.6 (c, 0.5 in $CHCl_3$). λ_{max} 212 (log ε 4.92) (MeOH)
(7R,7′S,8R,8′R)-form
3,3′,4,4′,5,5′-Hexa-Me ether
2,7′-Cyclo-9,9′-epoxy-3,3′,4,4′,5,5′-hexamethoxy-8,8′-lignan-7-ol. 9,9′-Epoxy-7-hydroxy-3,3′,4,4′,5,5′-hexamethoxy-2,7′-cyclolignan. ***Aglacin E***
[477529-36-5]
$C_{24}H_{30}O_8$ 446.496
Constit. of the stem bark of *Aglaia cordata*. Waxy solid. $[\alpha]_D^{20}$+17 (c, 0.69 in $CHCl_3$)
(7S,7′S,8R,8′R)-form
3,3′,4,4′,5,5′-Hexa-Me ether
Aglacin F
[477529-37-6]
$C_{24}H_{30}O_8$ 446.496
Constit. of the stem bark of *Aglaia cordata*. Waxy solid. $[\alpha]_D^{20}$-1 (c, 0.41 in $CHCl_3$)
(7R*,7′R*,8R*,8′R*)-form
7-Ketone, 3,3′,4,4′,5,5′-hexa-Me ether
2,7′-Cyclo-9,9′-epoxy-3,3′,4,4′,5,5′-hexamethoxy-8,8′-lignan-7-one. 9,9′-Epoxy-3,3′,4,4′,5,5′-hexamethoxy-7-oxo-2,7′-cyclolignan.
Aglacin D
[389628-95-9]
$C_{24}H_{28}O_8$ 444.48
Constit. of the stem bark of *Aglaia cordata*. Waxy solid (MeOH). $[\alpha]_D^{20}$-87.2 (c, 0.58 in $CHCl_3$). λ_{max} 212 (log ε 4.69); 229 (sh); 282 (log ε 4.05) (MeOH).

Wang, B.-G. *et al.*, *J. Nat. Prod.*, 2001, **64**, 1521-1526 (*Aglacins A,D,E,F, cryst struct*).
Wang, B.-G. *et al.*, *Tetrahedron Lett.*, 2002, **43**, 5783-5787 (*Aglacins E,F, abs config*).
Xu, M. *et al.*, *Angew. Chem., Int. Ed.*, 2021, **60**, 16655-16660 (*Aglacins A,E, synth*)

2,2′-Cyclo-7,7′-epoxy-8,8′-lignan-3,3′,4,4′,5,5′-hexol C153

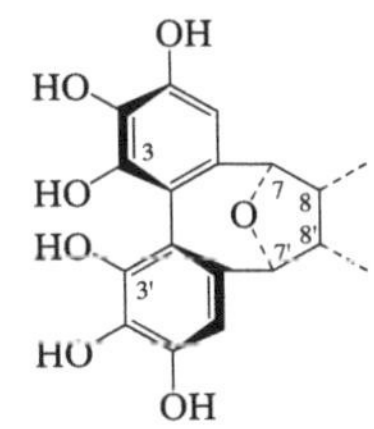

7,7′-Epoxy-3,3′,4,4′,5,5′-hexahydroxy-2,2′-cyclolignan
$C_{18}H_{18}O_7$ 346.335
Derivs. have a potential axis of symmetry. Stereochem. of several derivs. not certain. Compds. readily atropisomerise
(7R,7′S,8R,8′S,S$_{axial}$)-form
3,3′,4,4′,5-Penta-Me ether
2,2′-Cyclo-7,7′-epoxy-3,3′,4,4′,5-pentamethoxy-8,8′-lignan-5′-ol. 7,7′-Epoxy-5′-hydroxy-3,3′,4,4′,5-pentamethoxy-2,2′-cyclolignan.
Propinquain I
[1426059-95-1]
$C_{23}H_{28}O_7$ 416.47
Isol. from the roots of *Schisandra propinqua* ssp. *sinensis*. Amorph. yellow powder ($CHCl_3$). $[\alpha]_D^{20}$-1.2 (c, 0.61 in CH_2Cl_2). Appears to be a mixt. of atropisomers. λ_{max} 220 (log ε 4.58); 243 (log ε 4.34); 284 (log ε 4.17) (MeOH)
Hexa-Me ether
2,2′-Cyclo-7,7′-epoxy-3,3′,4,4′,5,5′-hexamethoxylignan. 7,7′-Epoxy-3,3′,4,4′,5,5′-hexamethoxy-2,2′-cyclolignan.
Kadsulignan N
[163564-58-7]
$C_{24}H_{30}O_7$ 430.497
Constit. of the seeds of *Kadsura coccinea*. Cryst. (MeOH). Mp 113 - 115°. Racemic. λ_{max} 225 (log ε 4.99); 258 (sh); 283 (log ε 4.73) (MeOH)
4,5-Methylene, 3,3′,4′-tri-Me ether
2,2′-Cyclo-7,7′-epoxy-3,3′,4′-trimethoxy-4,5-methylenedioxy-8,8′-lignan-5′-ol. 7,7′-Epoxy-5′-hydroxy-3,3′,4′-trimethoxy-4,5-methylenedioxy-2,2′-cyclolignan. ***Kadsulignan M***
[163564-59-8]
$C_{22}H_{24}O_7$ 400.427
Constit. of *Kadsura coccinea*. Shows anti-HIV activity. Cryst. (MeOH). Mp 199 - 201°. $[\alpha]_D$ +17.5 (c, 0.6 in MeOH). Atropisomerises, the major stereoisomer appears to have the S$_{axial}$-config.. λ_{max} 220 (ε 89125); 286 (ε 19054) (MeOH) (Berdy)
4,5-Methylene, 3,3′,4′,5′-tetra-Me ether
2,2′-Cyclo-7,7′-epoxy-3,3′,4′,5′-tetramethoxy-4,5-methylenedioxylignan. 7,7′-Epoxy-3,3′,4′,5′-tetramethoxy-4,5-methylenedioxy-2,2′-cyclolignan. ***Kadsulignan L***. *Neokadsuranin*
[163660-06-8]
[115181-68-5]
$C_{23}H_{26}O_7$ 414.454
Constit. of *Kadsura coccinea* and *Clerodendron inerme*. Needles (MeOH); cryst. (Et_2O). Mp 149 - 150°. $[\alpha]_D$+12.2 (c, 3.3 in MeOH) (Kadsulignan L, initial). Atropisomerises. Kadsulignan L appears to be the S_{axial} isomer and Neokadsuranin the R_{axial} isomer (revised from the initial assignment) however they equilibrate in soln.. λ_{max} 228 (log ε 4.91); 262 (sh); 288 (log ε 4.58) (MeOH)
4,5:4′,5′-Bis(methylene), 3,3′-di-Me ether
2,2′-Cyclo-7,7′-epoxy-3,3′-dimethoxy-4,5:4′,5′-bis(methylenedioxy)lignan. 7,7′-Epoxy-3,3′-dimethoxy-4,5:4′,5′-bis(methylenedioxy)-2,2′-cyclolignan
[81345-36-0]
$C_{22}H_{22}O_7$ 398.411
Constit. of *Clerodendron inerme*. Cryst. (EtOH). Mp 175 - 177°. Racemic. Incorrectly stated in the lit. to be a *meso*-compd..

Spencer, G.F. *et al.*, *Phytochemistry*, 1981, **20**, 2757-2759 (*Clerodendron inerme constits*).
Lian-niang, L. *et al.*, *Planta Med.*, 1988, **54**, 45-46 (*Neokadsuranin*).
Liu, J.S. *et al.*, *Phytochemistry*, 1995, **38**, 241-245 (*Kadsulignans L,M,N, abs config, activity*).
Jin, M.N. *et al.*, *Planta Med.*, 2012, **78**, 807-813 (*Propinquain I*)

2,2′-Cyclo-7,7′-epoxy-8,8′-lignan-3,4,4′,5,5′,6′-hexol C154

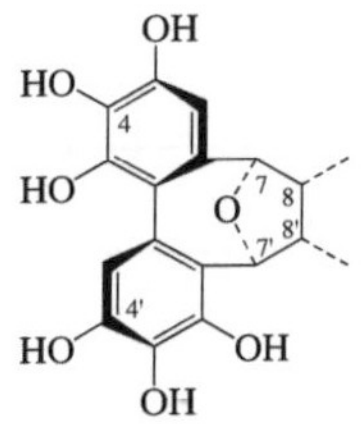

(7*R*,7′*S*,8*R*,8′*S*,S_{axial})-*form*

7,7′-Epoxy-3,4,4′,5,5′,6′-hexahydroxy-2,2′-cyclolignan

$C_{18}H_{18}O_7$ 346.335

Gymnothelignans L and M reported to have R_{axial}-config. but struct. drawn with S_{axial}-config.

(7R,7′S,8R,8′S,S$_{axial}$)-form

4,5-Methylene, 3,4′,6′-tri-Me ether

Gymnothelignan L

[1401068-88-9]

$C_{22}H_{24}O_7$ 400.427

Constit. of *Gymnotheca chinensis*. Yellow powder. $[\alpha]_D^{20}$ -2 (c, 0.2 in Me_2CO). λ_{max} 307 (MeOH)

(7R,7′S,8S,8′S,S$_{axial}$)-form

4,5-Methylene, 3,4′,6′-tri-Me ether

Gymnothelignan M

[1401068-89-0]

$C_{22}H_{24}O_7$ 400.427

Constit. of *Gymnotheca chinensis*. Yellow powder. $[\alpha]_D^{20}$ +8 (c, 0.03 in MeOH). λ_{max} 300 (MeOH)

(7R*,7′S*,8R*,8′S*,S*$_{axial}$)-form

3,4,4′,5,6′-Penta-Me ether

Gymnothelignan V

[1629581-98-1]

$C_{23}H_{28}O_7$ 416.47

Constit. of *Gymnotheca chinensis*. Exhibits moderate cytotoxicity against HCT116, HCT15, A549, MCF-7 and HepG2 cancer cell lines. Yellow gum. $[\alpha]_D^{20}$ -19 (c, 0.31 in MeOH). λ_{max} 294 (log ε 3.86) (MeOH).

He, D. *et al.*, *J. Org. Chem.*, 2012, **77**, 8435-8443 (*Gymnothelignans L,M*).
Xiao, S.-J. *et al.*, *Phytochem. Lett.*, 2014, **8**, 38-40 (*Gymnothelignan V*).
Soorukram, D. *et al.*, *J. Org. Chem.*, 2018, **83**, 4173-4179 (*synth*).
Chen, P. *et al.*, *Org. Chem. Front.*, 2018, **5**, 1124-1128 (*synth*)

2,2′-Cyclo-7,9′-epoxy-8,8′-lignan-3,3′,4,4′,5,5′-hexol C155

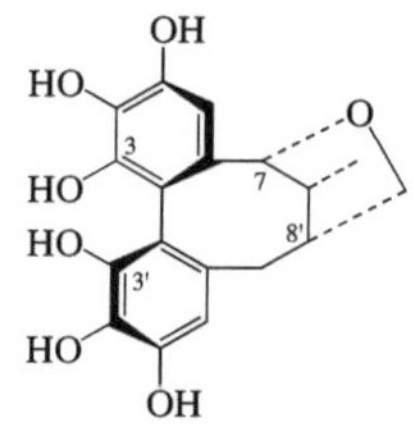

(7*R*,8*R*,8′*R*,R_{axial})-*form*

7,9′-Epoxy-3,3′,4,4′,5,5′-hexahydroxy-2,2′-cyclolignan

$C_{18}H_{18}O_7$ 346.335

Stereochem. of the Pyramidatins A-H (isol. in 2000) was reviewed and revised in 2010. Prev. publ. data on the Pyramidatins contained errors in stereochem. However the 2010 paper also contains errors in assignment of Pyramidatin names

(7R,8R,8′R,R$_{axial}$)-form

3,4,4′,5,5′-Penta-Me ether

2,2′-Cyclo-7,9′-epoxy-3,4,4′,5,5′-pentamethoxy-8,8′-lignan-3′-ol. 7,9′-Epoxy-3′-hydroxy-3,4,4′,5,5′-pentamethoxy-2,2′-cyclolignan.

3′-Hydroxy-3,4,4′,5,5′-pentamethoxypyramidatin. *Pyramidatin B*

[325827-84-7]

$C_{23}H_{28}O_7$ 416.47

Constit. of *Magnolia fraseri* and *Magnolia pyramidata*. Cryst.. Referred to both as Pyramidatin B and Pyramidatin C in the Schühly paper. λ_{max} 217; 224; 258; 292 (sh) (MeCN aq.)

Hexa-Me ether

2,2′-Cyclo-7,9′-epoxy-3,3′,4,4′,5,5′-hexamethoxylignan. 7,9′-Epoxy-3,3′,4,4′,5,5′-hexamethoxy-2,2′-cyclolignan.

3,3′,4,4′,5,5′-Hexamethoxypyramidatin

[325827-82-5]

$C_{24}H_{30}O_7$ 430.497

Constit. of *Magnolia fraseri* and *Magnolia pyramidata*. Yellowish oil. $[\alpha]_D^{21}$ -11 (c, 1.5 in MeOH). λ_{max} 219 (log ε 2.97); 255 (log ε 2.55); 272 (sh) (MeOH)

4,5-Methylene, 3,3′,4′,5′-tetra-Me ether

2,2′-Cyclo-7,9′-epoxy-3,3′,4′,5′-tetramethoxy-4,5-methylenedioxylignan. 7,9′-Epoxy-3,3′,4′,5′-tetramethoxy-4,5-methylenedioxy-2,2′-cyclolignan.

3,3′,4′,5′-Tetramethoxy-4,5-methylenedioxypyramidatin

[1254174-82-7]

$C_{23}H_{26}O_7$ 414.454

Constit. of *Magnolia fraseri*, *Magnolia pyramidata* and *Talauma gloriensis*. Cryst.. Mp 95 - 97°. $[\alpha]_D^{21}$ -41 (c, 2.3 in MeOH). λ_{max} 213; 217; 226; 260 (sh); 288 (sh) (MeCN aq.)

4′,5′-Methylene, 3,3′,4,5-tetra-Me ether

2,2′-Cyclo-7,9′-epoxy-3,3′,4,5-tetramethoxy-4′,5′-methylenedioxylignan. 7,9′-Epoxy-3,3′,4,5-tetramethoxy-4′,5′-methylenedioxy-2,2′-cyclolignan.

3,3′,4,5-Tetramethoxy-4′,5′-methylenedioxypyramidatin

[1254174-81-6]

$C_{23}H_{26}O_7$ 414.454

Constit. of *Talauma gloriensis*. Cryst.. λ_{max} 218; 226; 257; 292 (sh) (MeCN aq.)

4,5:4′,5′-Bis(methylene), 3,3′-di-Me ether

2,2′-Cyclo-7,9′-epoxy-3,3′-dimethoxy-4,5:4′,5′-bis(methylenedioxy)lignan. 7,9′-Epoxy-3,3′-dimethoxy-4,5:4′,5′-bis(methylenedioxy)-2,2′-cyclolignan.

3,3′-Dimethoxy-4,5:4′,5′-bis(methylenedioxy)pyramidatin

[325827-86-9]

$C_{22}H_{22}O_7$ 398.411

Constit. of *Magnolia fraseri* and *Talauma gloriensis*. Amorph. solid. Mp 168 - 169°. $[\alpha]_D^{21}$ -73 (c, 0.68 in CH_2Cl_2). λ_{max} 220 (log ε 2.98); 254 (log ε 2.42); 264 (sh) (MeOH)

(7R,8R,8′R,S$_{axial}$)-form

3,3′,4,4′,5-Penta-Me ether

2,2′-Cyclo-7,9′-epoxy-3,3′,4,4′,5-pentamethoxy-8,8′-lignan-5′-ol. 7,9′-Epoxy-5′-hydroxy-3,3′,4,4′,5-pentamethoxy-2,2′-cyclolignan.

5′-Hydroxy-3,3′,4,4′,5-pentamethoxypyramidatin. *Pyramidatin E*

[325827-27-8]

$C_{23}H_{28}O_7$ 416.47

Constit. of *Magnolia pyramidata*. Amorph. solid. λ_{max} 214 (log ε 4.68); 258 (log ε 3.84); 291 (log ε 3.28) (MeOH)

3,4,4′,5,5′-Penta-Me ether
Pyramidatin C. *Pyramidatin B (incorr.)*
[325827-87-0]
$C_{23}H_{28}O_7$ 416.47
Constit. of *Magnolia pyramidata*. Yellow oil. Referred to as both Pyramidatin B and Pyramidatin C in the 2010 paper. λ_{max} 216 (log ε 2.85); 254 (log ε 2.37); 295 (sh) (MeOH)
3′,4,4′,5,5′-Penta-Me ether
2,2′-Cyclo-7,9′-epoxy-3′,4,4′,5,5′-pentamethoxy-8,8′-lignan-3-ol. 7,9′-Epoxy-3-hydroxy-3′,4,4′,5,5′-pentamethoxy-2,2′-cyclolignan. ***3-Hydroxy-3′,4,4′,5,5′-pentamethoxypyramidatin***. *Pyramidatin D*
[1254174-79-2]
$C_{23}H_{28}O_7$ 416.47
Constit. of *Magnolia pyramidata*. Yellow solid. λ_{max} 217; 224; 256; 294 (sh) (MeCN aq.)
Hexa-Me ether
Pyramidatin A
[1254174-78-1]
$C_{24}H_{30}O_7$ 430.497
Constit. of *Magnolia fraseri* and *Magnolia pyramidata*. Needles. Mp 148 - 149°. $[\alpha]^{21}_{D}$ -20 (c, 3.4 in MeOH). λ_{max} 218 (log ε 3.02); 253 (log ε 2.49); 293 (sh) (MeOH)
4,5-Methylene, 3,3′,4′,5′-tetra-Me ether
[1254064-61-3]
$C_{23}H_{26}O_7$ 414.454
Constit. of *Magnolia fraseri* and *Talauma gloriensis*. Cryst.. Mp 126°. $[\alpha]^{21}_{D}$ -18 (c, 0.61 in CH_2Cl_2). λ_{max} 217 (log ε 3.02); 255 (log ε 2.48); 293 (sh) (MeOH)
4′,5′-Methylene, 3′,4,5-tri-Me ether
2,2′-Cyclo-7,9′-epoxy-3′,4,5-trimethoxy-4′,5′-methylenedioxy-8,8′-lignan-3-ol. 7,9′-Epoxy-3-hydroxy-3′,4,5-trimethoxy-4′,5′-methylenedioxy-2,2′-cyclolignan. ***3-Hydroxy-3′,4,5-trimethoxy-4′,5′-methylenedioxypyramidatin***. *Pyramidatin F*
[325827-88-1]
$C_{22}H_{24}O_7$ 400.427
Constit. of *Magnolia pyramidata*. Yellowish solid. λ_{max} 218; 226; 258; 292 (MeCN aq.)
4′,5′-Methylene, 3,3′,4,5-tetra-Me ether
Pyramidatin G
[325827-89-2]
$C_{23}H_{26}O_7$ 414.454
Constit. of *Magnolia fraseri*. Amorph. solid. Mp 150 - 152°. λ_{max} 218; 225; 255; 294 (sh) (MeCN aq.)
4,5:4′,5′-Bis(methylene), 3,3′-di-Me ether
Pyramidatin H
[1254174-80-5]
$C_{22}H_{22}O_7$ 398.411
Constit. of *Magnolia fraseri*, *Magnolia pyramidata* and *Talauma gloriensis*. Amorph. powder. Mp 206°. $[\alpha]^{21}_{D}$ -50 (c, 0.6 in MeOH). λ_{max} 224 (log ε 3.02); 257 (log ε 2.47); 296 (sh) (MeOH).

Song, Q. *et al.*, *Phytochemistry*, 2000, **55**, 653-661 (*Pyramidatins A-H*).
Schühly, W. *et al.*, *Phytochemistry*, 2010, **71**, 1787-1795 (*Pyramidatins, cd, cryst struct, abs config*)

2,2′-Cyclo-8,8′-epoxy-8,8′-lignan-3,3′,4,4′,5,5′-hexol C156

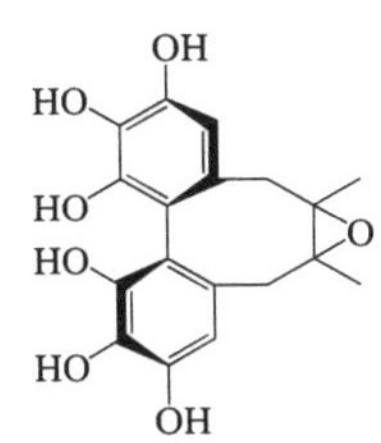

8,8′-Epoxy-3,3′,4,4′,5,5′-hexahydroxy-2,2′-cyclolignan
$C_{18}H_{18}O_7$ 346.335
C-8 and C-8′ are not chiral centers
(R_{axial})-form
Hexa-Me ether
Schisanchinin C
[1613761-70-8]
$C_{24}H_{30}O_7$ 430.497
Isol. from *Schisandra chinensis*. Yellowish oil. $[\alpha]^{23}_{D}$ +8.7 (c, 0.01 in MeOH). λ_{max} 218 (log ε 3.79) (MeOH).

Hu, D. *et al.*, *Phytochemistry*, 2014, **104**, 72-78 (*Schisanchinin C*)

2,7′-Cyclo-9,9′-epoxy-8,8′-lignan-3,3′,4,4′,5,5′-hexol C157

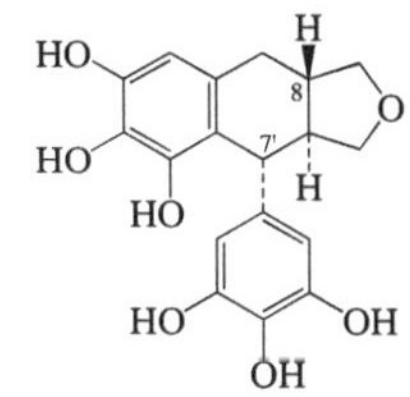

(7′R,8S,8′S)-form

9,9′-Epoxy-3,3′,4,4′,5,5′-hexahydroxy-2,7′-cyclolignan
$C_{18}H_{18}O_7$ 346.335
(7′R,8S,8′S)-form
Hexa-Me ether
2,7′-Cyclo-9,9′-epoxy-3,3′,4,4′,5,5′-hexamethoxylignan. 9,9′-Epoxy-3,3′,4,4′,5,5′-hexamethoxy-2,7′-cyclolignan. ***Aglacin B***
[389628-93-7]
$C_{24}H_{30}O_7$ 430.497
Constit. of the stem bark of *Aglaia cordata*. Shows weak inhibition of human KB carcinoma cell line. Needles (MeOH). Mp 138 - 139°. $[\alpha]^{20}_{D}$ +45 (c, 0.38 in $CHCl_3$). λ_{max} 212 (log ε 4.81) (MeOH)
(7′S,8S,8′S)-form
3,3′,5,5′-Tetra-Me ether
2,7′-Cyclo-9,9′-epoxy-3,3′,5,5′-tetramethoxy-8,8′-lignan-4,4′-diol. 9,9′-Epoxy-4,4′-dihydroxy-3,3′,5,5′-tetramethoxy-2,7′-cyclolignan. ***Gaultherin C***
[267665-74-7]
$C_{22}H_{26}O_7$ 402.443
Constit. of the roots of *Gaultheria yunnanensis*. Prisms (MeOH). Mp 185 - 186°. $[\alpha]^{21}_{D}$ +56 (c, 0.02 in MeOH). λ_{max} 219 (log ε 2.21); 276 (log ε 0.26) (MeOH).

Zhang, Z. *et al.*, *Heterocycles*, 2000, **53**, 675-679 (*Gaultherin C*).
Wang, B.-G. *et al.*, *J. Nat. Prod.*, 2001, **64**, 1521-1526 (*Aglacin B, cryst struct*).
Xu, M. *et al.*, *Angew. Chem., Int. Ed.*, 2021, **60**, 16655-16660 (*Aglacin B, synth*)

2,7′-Cyclo-9,9′-epoxy-8,8′-lignan-3′,4,4′,5,8,9-hexol C158

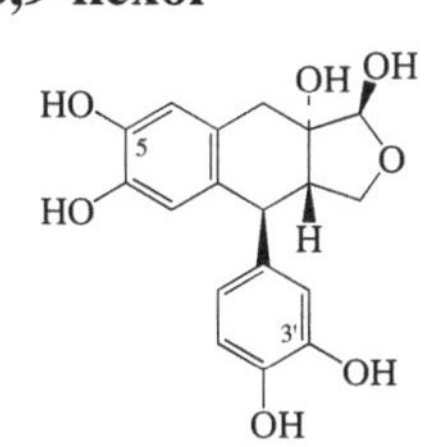

9,9′-Epoxy-3′,4,4′,5,8,9-hexahydroxy-2,7′-cyclolignan

$C_{18}H_{18}O_7$ 346.335

3′,5-Di-Me ether

2,7′-Cyclo-9,9′-epoxy-3′,5-dimethoxy-8,8′-lignan-4,4′,8,9-tetrol. 9,9′-Epoxy-4,4′,8,9-tetrahydroxy-3′,5-dimethoxy-2,7′-cyclolignan. ***Africanal‡***

[73340-43-9]

$C_{20}H_{22}O_7$ 374.389

Constit. of *Olea europaea* ssp. *africana*. $[\alpha]^{25}_{D}$ +221 (c, 0.24 in MeOH).

Viviers, P.M. *et al., Tetrahedron Lett.*, 1979, **20**, 3773-3776 (*Olea africana constit, abs config*).

Tsakamoto, H. *et al., Phytochemistry*, 1984, **23**, 2839-2841 (*Olea europaea africana constit*)

7′,9-Cyclo-9,9′-epoxy-2,8′-lignan-3′,4,4′,5,7,9-hexol C159

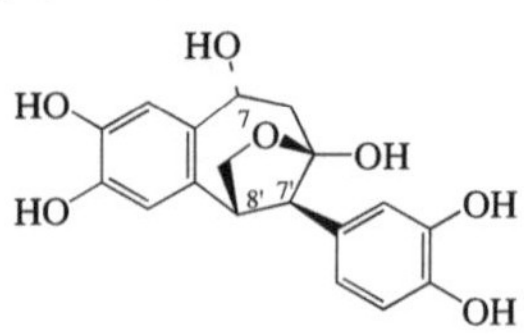

(*7R,7′R,8′R,9R*)-*form*

11-(3,4-Dihydroxyphenyl)-2,4,5,6-tetrahydro-1H-1,4-methanobenzo[d]oxocine-4,6,8,9-tetrol. 7′,9-Cyclo-9,9′-epoxy-3′,4,4′,5,7,9-hexahydroxy-2,8′-neolignan

$C_{18}H_{18}O_7$ 346.335

The nat. prods. Verniciasin A and 7′-*O*-Methylverniciasin A were isol. as racemates and resolved by chiral hplc

(7R,7′R,8′R,9R)-form

(*7R,7′R,8′R,9R*)-Verniciasin A

[2158209-50-6]

NO prodn. inhibitor

7-Me ether

(7R,7′R,8′R,9R)-7-O-Methylverniciasin A. (7R,7′R,8′R,9R)-7-Methoxyverniciasin A (incorr.)

[2158209-52-8]

NO prodn. inhibitor

(7S,7′S,8′S,9S)-form

(*7S,7′S,8′S,9S*)-Verniciasin A

[2158209-51-7]

NO prodn. inhibitor

7-Me ether

(7S,7′S,8′S,9S)-7-O-Methylverniciasin A

[2158209-53-9]

NO prodn. inhibitor

(7RS,7′RS,8′RS,9RS)-form

Verniciasin A [2159057-29-9]

Constit. of the seeds of *Vernicia fordii*. Yellow oil

7-Me ether

7-O-Methylverniciasin A. 7-Methoxyverniciasin A

[2159057-30-2]

$C_{19}H_{20}O_7$ 360.362

Constit. of the seeds of *Vernicia fordii*. NO prodn. inhibitor. Yellow oil.

Zhao, W.-H. *et al., Sci. Rep.*, 2017, **7**, 13608:1-10 (*Verniciasin A, 7′-O-Methylverniciasin A, cd, activity*)

2,2′-Cyclo-8,9-epoxy-8,8′-lignan-3,3′,4,4′,5,5′,7,7′-octol C160

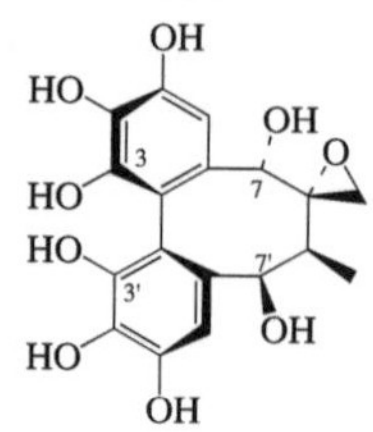

8,9-Epoxy-3,3′,4,4′,5,5′,7,7′-octahydroxy-2,2′-cyclolignan

$C_{18}H_{18}O_9$ 378.333

(7S,7′R,8S,8′S,S_axial)-form

3′,4,4′,5-Tetra-Me ether, 7-benzoyl, 7′-angeloyl

Renchangianin D

[770733-58-9]

$C_{34}H_{36}O_{11}$ 620.651

Constit. of the stems of *Kadsura renchangiana*. Brown powder. $[\alpha]^{28}_{D}$ -14.7 (c, 0.65 in MeOH). λ_{max} 221 (log ε 3.99) (MeOH)

4′,5′-Methylene, 3′,4,5-tri-Me ether, 7,7′-diangeloyl

Kadsuphilol C

[947256-78-2]

$C_{32}H_{36}O_{11}$ 596.629

Constit. of *Kadsura heteroclita* and *Kadsura philippinensis*. Amorph. powder. $[\alpha]^{25}_{D}$ +52 (c, 0.01 in CH_2Cl_2). λ_{max} 219 (log ε 3.68) (MeOH)

4′,5′-Methylene, 3′,4,5-tri-Me ether, 7-benzoyl, 7′-angeloyl

Kadheterin B

$C_{34}H_{34}O_{11}$ 618.635

Constit. of *Kadsura heteroclita*. Amorph. powder. $[\alpha]^{21}_{D}$ -34.3 (c, 0.11 in MeOH). λ_{max} 222 (log ε 4.74); 280 (log ε 3.63) (MeOH).

Chen, M. *et al., Helv. Chim. Acta*, 2004, **87**, 1368-1376 (*Renchangianin D*).

Shen, Y.-C. *et al., J. Nat. Prod.*, 2007, **70**, 1139-1145 (*Kadsuphilol C*).

Luo, Y.-Q. *et al., Fitoterapia*, 2017, **119**, 150-157 (*Kadheterin B*)

2,7′-Cyclo-9,9′-epoxy-8,8′-lignan-3,3′,4,4′,5,5′,7,9′-octol C161

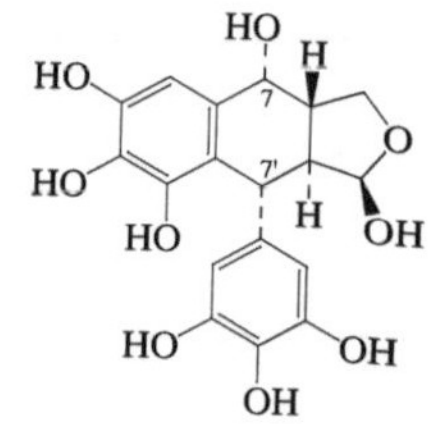

9,9′-Epoxy-3,3′,4,4′,5,5′,7,9′-octahydroxy-2,7′-cyclolignan

$C_{18}H_{18}O_9$ 378.333

(7R*,7′R*,8S*,8′S*,9′R*)-form

3,3′,4,4′,5,5′-Hexa-Me ether

2,7′-Cyclo-9,9′-epoxy-3,3′,4,4′,5,5′-hexamethoxy-8,8′-lignan-7,9′-diol. 9,9′-Epoxy-7,9′-dihydroxy-3,3′,4,4′,5,5′-hexamethoxy-2,7′-cyclolignan

3,3′,4,4′,5,5′-Hexa-Me ether, 7-Ac

Aglacin I

[686744-61-6]

$C_{26}H_{32}O_{10}$ 504.532

Constit. of the stem bark of *Aglaia cordata*. Obt. as a mixt. with Aglacin J, in C196

Wang, B.-G. *et al., J. Nat. Prod.*, 2004, **67**, 682-684 (*Aglacin I*)

2,2′-Cyclo-7,7′-epoxy-8,8′-lignan-3,4,4′,5,5′-pentol C162

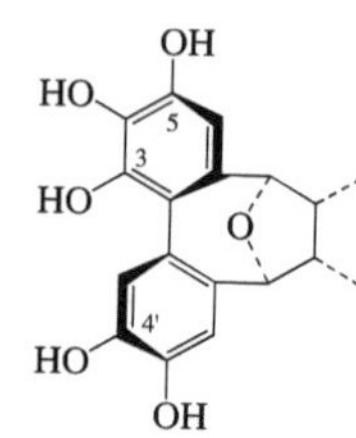

(*7R*,7′S*,8R*,8′S*,S*axial*)-*form*

7,7′-Epoxy-3,4,4′,5,5′-pentahydroxy-2,2′-cyclolignan

$C_{18}H_{18}O_6$ 330.336

(7R*,7′S*,8R*,8′S*,S*_axial)-form

4,5-Methylene, 3,4′-di-Me ether

2,2′-Cyclo-7,7′-epoxy-3,4′-dimethoxy-4,5-methylenedioxy-8,8′-lignan-5′-ol. 7,7′-Epoxy-5′-hydroxy-3,4′-dimethoxy-4,5-methylenedioxy-2,2′-cyclolignan. ***Gymnothelignan W***

[1621958-62-0]

$C_{21}H_{22}O_6$ 370.401

Constit. of *Gymnotheca chinensis*. Exhibits weak cytotoxicity against

A549 cancer cell line. Powder. $[\alpha]^{20}_{D}$ -8 (c, 0.20 in MeOH). λ_{max} 295 (log ε 3.76) (MeOH)

Stereoisomer 4,5-methylene, 3,4′-di-Me ether

Manglisin B

$C_{21}H_{22}O_6$ 370.401

Constit. of carpels of *Manglietiastrum sinicum*. Exhibits moderate antimicrobial activity against *Staphylococcus* aureus and MRSA types 82, 92, 98 and 331. Oil. $[\alpha]^{16}_{D}$ +4 (c, 0.2 in MeOH). λ_{max} 195 (log ε 4.05); 219 (log ε 4.29); 291 (log ε 3.92) (MeOH).

Ding, J.-Y. *et al.*, *J. Nat. Prod.*, 2014, **77**, 1800-1805 (*Manglisin B*).
Xiao, S.-J. *et al.*, *Phytochem. Lett.*, 2014, **8**, 38-40 (*Gymnothelignan W*)

2,7′-Cyclo-9,9′-epoxy-8,8′-lignan-3,3′,4,4′,5-pentol C163

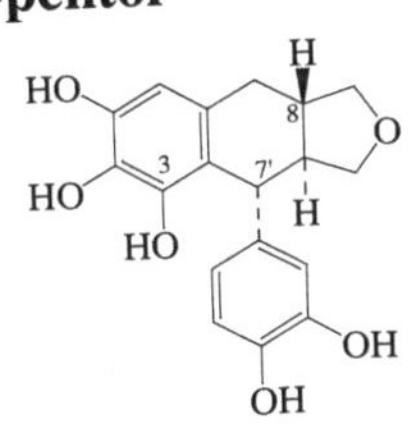

9,9′-Epoxy-3,3′,4,4′,5-pentahydroxy-2,7′-cyclolignan

$C_{18}H_{18}O_6$ 330.336

(7′R*,8S*,8′S*)-form

Penta-Me ether

2,7′-Cyclo-9,9′-epoxy-3,3′,4,4′,5-pentamethoxylignan. 9,9′-Epoxy-3,3′,4,4′,5-pentamethoxy-2,7′-cyclolignan. ***Aglacin C***

[389628-94-8]

$C_{23}H_{28}O_6$ 400.471

Constit. of the stem bark of *Aglaia cordata*. Waxy solid. $[\alpha]^{20}_{D}$ +25.6 (c, 0.6 in $CHCl_3$). λ_{max} 211 (log ε 4.67) (MeOH).

Wang, B.-G. *et al.*, *J. Nat. Prod.*, 2001, **64**, 1521-1526 (*Aglacin C*)

2,7′-Cyclo-9,9′-epoxy-8,8′-lignan-3′,4,4′,5,5′-pentol C164

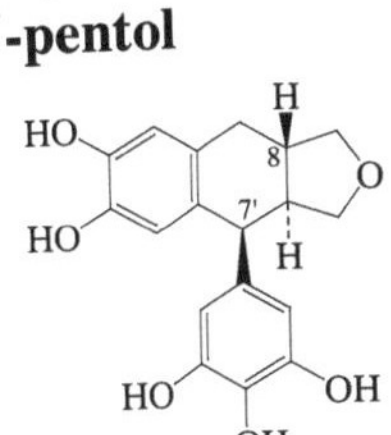

9,9′-Epoxy-3′,4,4′,5,5′-pentahydroxy-2,7′-cyclolignan

$C_{18}H_{18}O_6$ 330.336

(7′S,8S,8′S)-form

3′,5,5′-Tri-Me ether

2,7′-Cyclo-9,9′-epoxy-3′,5,5′-trimethoxy-8,8′-lignan-4,4′-diol. 9,9′-Epoxy-4,4′-dihydroxy-3′,5,5′-trimethoxy-2,7′-cyclolignan. ***Gaultherin D***

[267665-75-8]

$C_{21}H_{24}O_6$ 372.417

Constit. of the roots of *Gaultheria yunnanensis*. Prisms (MeOH). Mp 196 - 197°. $[\alpha]^{21}_{D}$ +55 (c, 0.04 in MeOH). λ_{max} 218 (log ε 2.18); 278 (log ε 0.3) (MeOH).

Zhang, Z. *et al.*, *Heterocycles*, 2000, **53**, 675-679 (*Gaultherin D*)

2,7′-Cyclo-9,9′-epoxy-8,8′-lignan-3′,4,4′,5,7-pentol C165

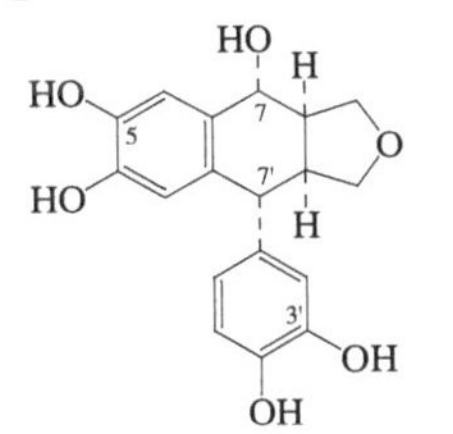

9,9′-Epoxy-3′,4,4′,5,7-pentahydroxy-2,7′-cyclolignan

$C_{18}H_{18}O_6$ 330.336

(7R*,7′R*,8R*,8′S*)-form

3′,5-Di-Me ether

2,7′-Cyclo-9,9′-epoxy-3′,5-dimethoxy-8,8′-lignan-4,4′,7-triol. 9,9′-Epoxy-4,4′,7-trihydroxy-3′,5-dimethoxy-2,7′-cyclolignan. ***Lanceolin‡***

[957468-36-9]

$C_{20}H_{22}O_6$ 358.39

Constit. of the heartwood of *Cunninghamia lanceolata*. Oil (as tri-Ac). $[\alpha]^{26}_{D}$ +17.2 (c, 0.2 in $CHCl_3$) (tri-Ac).

Lee, T.-H. *et al.*, *Biosci., Biotechnol., Biochem.*, 2007, **71**, 2075-2078 (*Lanceolin*)

2,7′-Cyclo-9,9′-epoxy-8,8′-lignan-3′,4,4′,5,9-pentol C166

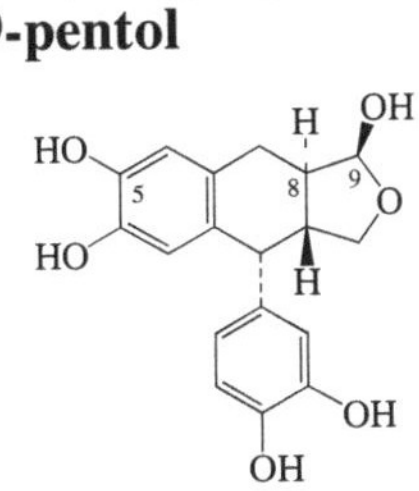

(7′R,8R,8′R,9S)-form

9,9′-Epoxy-3′,4,4′,5,9-pentahydroxy-2,7′-cyclolignan

$C_{18}H_{18}O_6$ 330.336

Derivs. readily epimerise at C-9. 9-Configs. given are tentative

(7′R,8R,8′R,9S)-form

3′,5,9-Tri-Me ether

7′-Epitsugacetal

[2105814-21-7]

$C_{21}H_{24}O_6$ 372.417

Constit. of the roots of *Taxus wallichiana*. Amorph. solid. $[\alpha]^{25}_{D}$ +48.5 (c, 0.01 in MeOH)

(7′S,8R,8′R,9R)-form

3′,5-Di-Me ether

2,7′-Cyclo-9,9′-epoxy-3′,5-dimethoxy-8,8′-lignan-4,4′,9-triol. 9,9′-Epoxy-4,4′,9-trihydroxy-3′,5-dimethoxy-2,7′-cyclolignan. ***Todolactol B***

[111956-89-9]

$C_{20}H_{22}O_6$ 358.39

Constit. of *Abies sachalinensis* and *Pinus armandii*. Mp 181 - 182°. $[\alpha]^{23}_{D}$ +38.8 (c, 0.33 in Me_2CO). 9-Config. appears uncertain

3′,5,9-Tri-Me ether

2,7′-Cyclo-9,9′-epoxy-3′,5,9-trimethoxy-8,8′-lignan-4,4′-diol. 9,9′-Epoxy-4,4′-dihydroxy-3′,5,9-trimethoxy-2,7′-cyclolignan. α-Conidendral cyclic methyl acetal. ***Formosanol. Todolactol C***

[71724-99-7]

$C_{21}H_{24}O_6$ 372.417

Constit. of *Abies sachalinensis*, *Dacrydium intermedium* and *Juniperus formosana*. Plates (MeOH aq. or CH_2Cl_2/hexane). Mp 229° (225-227°). $[\alpha]^{20}_{D}$ -71 (c, 1.0 in $CHCl_3$). Readily isomerises at C-9. The 9-epimer has been synthesised

9-Et, 3′,5-di-Me ether

2,7′-Cyclo-9,9′-epoxy-9-ethoxy-3′,5-dimethoxy-8,8′-lignan-4,4′-diol. 9,9′-Epoxy-9-ethoxy-4,4′-dihydroxy-3′,5-dimethoxy-2,7′-cyclolignan. ***Todolactol D***

[114924-88-8]

$C_{22}H_{26}O_6$ 386.444

Isol. from *Dacrydium intermedium* and *Abies sachalinensis*. Prisms. Mp 187 - 191°. $[\alpha]^{20}_{D}$ +30.1. Poss. artifact

(7′S,8R,8′R,9S)-form

3′,5,9-Tri-Me ether

Tsugacetal

[101312-79-2]

$C_{21}H_{24}O_6$ 372.417
Constit. of *Dacrydium intermedium*, *Cunninghamia konishii*, *Taxus wallichiana* and *Tsuga chinensis* var. *formosana*,. Larviicide. Cryst. (EtOH or EtOAc/hexane). Mp 188 - 190°. $[\alpha]_D$ +68.3 (c, 0.9 in Me_2CO). λ_{max} 212 (ε 15000); 283 (ε 6900) (EtOH)

3′,5,9-Tri-Me ether, di-Ac
Mp 201°. $[\alpha]_D^{20}$ -76.3 (c, 0.5 in $CHCl_3$)

(7′R*,8S*,8′S*,9R*)-form

9-Butyl, 3′,5-di-Me ether
9-Butoxy-2,7′-cyclo-9,9′-epoxy-3′,5-dimethoxy-8,8′-lignan-4,4′-diol. 9-Butoxy-9,9′-epoxy-4,4′-dihydroxy-3′,5-dimethoxy-2,7′-cyclolignan. ***9′-Butoxytsugacetal***
[1391860-63-1]
$C_{24}H_{30}O_6$ 414.498
Isol. from *Pseudolarix kaempferi*. Shows antimicrobial and cytotoxic activities. Amorph. powder. $[\alpha]_D^{20}$ +70 (c, 0.15 in $CHCl_3$). λ_{max} 204 (log ε 4.68); 283 (log ε 3.78) (MeOH).

Cambie, R.C. *et al.*, *Aust. J. Chem.*, 1979, **32**, 2741-2751 (*Dacrydium intermedium constits*).
Cambie, R.C. *et al.*, *Aust. J. Chem.*, 1985, **38**, 1631-1645 (*Dacrydium intermedium constits, struct*).
Fang, J.-M. *et al.*, *Phytochemistry*, 1985, **24**, 1363-1365 (*Tsugacetal, cryst struct*).
Dhal, R. *et al.*, *Tetrahedron*, 1986, **42**, 2005-2016 (*α-Conidendral, synth*).
Ozawa, S. *et al.*, *Mokuzai Gakkaishi*, 1987, **33**, 747-747 (*Abies sachalinensis constits*).
Ogawa, S. *et al.*, *Mokuzai Gakkaishi*, 1988, **34**, 169-175 (*Abies sachalinensis constits*).
Ozawa, S. *et al.*, *Mokuzai Gakkaishi*, 1988, **34**, 851-857 (*Abies sachalinensis constits*).
Fang, J.-M. *et al.*, *J. Chin. Chem. Soc. (Taipei)*, 1989, **36**, 483-485 (*Pinus armandii constit*).
Kuo, Y.-H. *et al.*, *Chem. Pharm. Bull.*, 1990, **38**, 3195-3201 (*Juniperus formosana constits*).
Castro, M.A. *et al.*, *Phytochemistry*, 1996, **41**, 995-1011 (*Todolactols, occur, rev*).
Ozawa, S. *et al.*, *Stud. Nat. Prod. Chem.*, 1997, **20**, 613-657 (*Todolactol B, struct, rev*).
He, W.-J. *et al.*, *Planta Med.*, 2011, **77**, 1924-1931 (*9′-Butoxytsugacetal*).
Dang, P.H. *et al.*, *J. Nat. Prod.*, 2017, **80**, 1876-1882 (*Taxus wallichiana constits*)

2,9-Cyclo-9,9′-epoxy-8,8′-lignan-3′,4,4′,5-tetrol C167

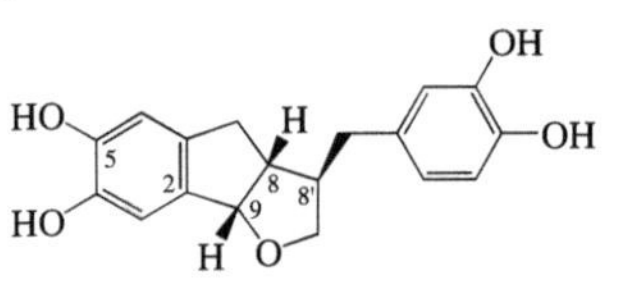

9,9′-Epoxy-3′,4,4′,5-tetrahydroxy-2,9-cyclolignan
$C_{18}H_{18}O_5$ 314.337

(8R*,8′R*,9S*)-form

3′,5-Di-Me ether
2,9-Cyclo-9,9′-epoxy-3′,5-dimethoxy-8,8′-lignan-4,4′-diol
[1803041-26-0]
$C_{20}H_{22}O_5$ 342.391
Constit. of the stems of *Syringa pinnatifolia*. Needles (MeOH aq.). $[\alpha]_D^{25}$ +12.8 (c, 0.1 in MeOH). λ_{max} 263 (log ε 3.86) (MeOH).

Wang, Q. *et al.*, *Nat. Prod. Res.*, 2016, **30**, 796-799 (*Syringa pinnatifolia constit*)

2,2′-Cyclo-7,7′-epoxy-1,7-seco-8,8′-lignan-3,3′,4,4′,5,5′,7-heptol C168

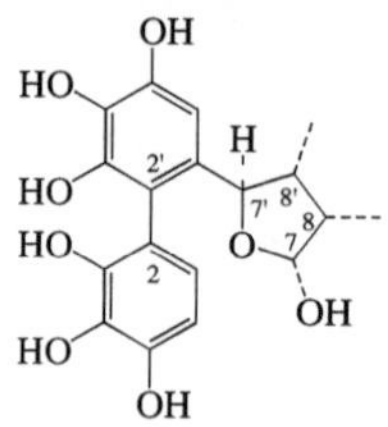

7,7′-Epoxy-3,3′,4,4′,5,5′,7-heptahydroxy-1,7-seco-2,2′-cyclolignan
$C_{18}H_{20}O_8$ 364.350

(7R*,7′R*,8S*,8′R*)-form

3,3′,4′,5,5′-Penta-Me ether
2,2′-Cyclo-7,7′-epoxy-3,3′,4′,5,5′-pentamethoxy-1,7-seco-8,8′-lignan-4,7-diol. 7,7′-Epoxy-4,7-dihydroxy-3,3′,4′,5,5′-pentamethoxy-1,7-seco-2,2′-cyclolignan.
Gymnothelignan Y_3
$C_{23}H_{30}O_8$ 434.485
Constit. of *Gymnotheca involucrata*.

Zhang, P.-Z. *et al.*, *Zhongcaoyao*, 2017, **48**, 1509-1512 (*Gymnothelignan Y_3*)

2,2′-Cyclo-7,7′-epoxy-1,7-seco-8,8′-lignan-3′,4,4′,5,5′,6,7-heptol C169

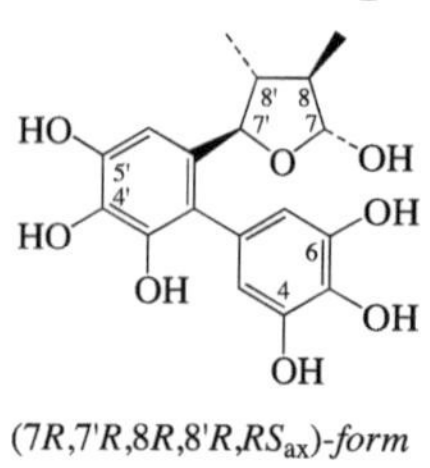

$(7R,7'R,8R,8'R,RS_{ax})$-*form*

7,7′-Epoxy-3′,4,4′,5,5′,6,7-heptahydroxy-1,7-seco-2,2′-cyclolignan
$C_{18}H_{20}O_8$ 364.350

$(7R,7'R,8R,8'R,RS_{axial})$-form

3′,4,4′,5′,6,7-Hexa-Me ether
2,2′-Cyclo-7,7′-epoxy-3′,4,4′,5′,6,7-hexamethoxy-1,7-seco-8,8′-lignan-5-ol. 7,7′-Epoxy-5-hydroxy-3′,4,4′,5′,6,7-hexamethoxy-1,7-seco-2,2′-cyclolignan.
Gymnothelignan E
[1401068-82-3]
$C_{24}H_{32}O_8$ 448.512
Constit. of *Gymnotheca chinensis*. Powder. $[\alpha]_D^{20}$ -7 (c, 0.35 in Me_2CO). λ_{max} 277 (MeOH)

4′,5′-Methylene, 3′,4,6,7-tetra-Me ether
2,2′-Cyclo-7,7′-epoxy-3′,4,6,7-tetramethoxy-4′,5′-methylenedioxy-1,7-seco-8,8′-lignan-5-ol. 7,7′-Epoxy-5-hydroxy-3′,4,6,7-tetramethoxy-4′,5′-methylenedioxy-1,7-seco-2,2′-cyclolignan.
Gymnothelignan G
[1401068-84-5]
$C_{23}H_{28}O_8$ 432.469
Constit. of *Gymnotheca chinensis*. Powder. $[\alpha]_D^{20}$ -10 (c, 0.25 in Me_2CO). λ_{max} 296 (MeOH)

$(7R,7'R,8S,8'S,RS_{axial})$-form

4′,5′-Methylene, 3′,4,6,7-tetra-Me ether
Gymnothelignan H
[1401068-85-6]
$C_{23}H_{28}O_8$ 432.469
Constit. of *Gymnotheca chinensis*. Cryst.. $[\alpha]_D^{20}$ -80 (c, 0.03 in Me_2CO). λ_{max} 296 (MeOH)

$(7S,7'R,8S,8'R,RS_{axial})$-form

3′,4,4′,5′,6-Penta-Me ether
2,2′-Cyclo-7,7′-epoxy-3′,4,4′,5′,6-pentamethoxy-1,7-seco-8,8′-lignan-5,7-diol. 7,7′-Epoxy-5,7-dihydroxy-3′,4,4′,5′,6-pentamethoxy-1,7-seco-2,2′-cyclolignan.
Gymnothelignan J
[1401068-86-7]
$C_{23}H_{30}O_8$ 434.485
Constit. of *Gymnotheca chinensis*. Yellowish powder. $[\alpha]_D^{20}$ -4 (c, 0.1 in Me_2CO). λ_{max} 298 (MeOH)

3′,4,4′,5′,6,7-Hexa-Me ether
Gymnothelignan D
[1401068-81-2]
$C_{24}H_{32}O_8$ 448.512
Constit. of *Gymnotheca chinensis*. Cryst.. Mp 133 - 134°. $[\alpha]_D^{20}$ -12 (c, 0.03 in MeOH). λ_{max} 276 (MeOH)

4′,5′-Methylene, 3′,4,6-tri-Me ether
2,2′-Cyclo-7,7′-epoxy-3′,4,6-trimethoxy-4′,5′-methylenedioxy-1,7-seco-8,8′-lignan-5,7-diol. 7,7′-Epoxy-5,7-dihydroxy-3′,4,6-trimethoxy-4′,5′-methylenedioxy-1,7-seco-2,2′-cyclolignan.
Gymnothelignan I
[1401068-96-9]
$C_{22}H_{26}O_8$ 418.442
Constit. of *Gymnotheca chinensis*. Yellowish powder. $[\alpha]^{20}_{D}$ -8 (c, 0.1 in Me_2CO). λ_{max} 302 (MeOH)

4′,5′-Methylene, 3′,4,6,7-tetra-Me ether
Gymnothelignan F
[1401068-83-4]
$C_{23}H_{28}O_8$ 432.469
Constit. of *Gymnotheca chinensis*. Yellowish powder. $[\alpha]^{20}_{D}$ -2 (c, 0.2 in Me_2CO). λ_{max} 287 (MeOH)

7-Ketone(lactone), hexa-Me ether
Gymnothelignan X$_5$
[2172644-48-1]
$C_{24}H_{30}O_8$ 446.496
Constit. of *Gymnotheca involucrata*. Powder. $[\alpha]^{20}_{D}$ -58 (c, 0.24 in MeOH). λ_{max} 292 (log ε 3.47) (MeOH)

7-Ketone(lactone), 4′,5′-methylene, tetra-Me ether
2,2′-Cyclo-7,7′-epoxy-3′,4,5,6-tetramethoxy-4′,5′-methylenedioxy-1,7-seco-8,8′-lignan-7-one. 7,7′-Epoxy-3′,4,5,6-tetramethoxy-4′,5′-methylenedioxy-7-oxo-1,7-seco-2,2′-cyclolignan. **Gymnothelignan X$_6$**
[2172644-49-2]
$C_{23}H_{26}O_8$ 430.453
Constit. of *Gymnotheca involucrata*. Powder. $[\alpha]^{20}_{D}$ -86 (c, 0.11 in MeOH). λ_{max} 292 (log ε 3.47) (MeOH)

(7′R,8R,8′S,RS$_{axial}$)-form
7-Ketone(lactone)
2,2′-Cyclo-7,7′-epoxy-3′,4,4′,5,5′,6-hexahydroxy-1,7-seco-8,8′-lignan-7-one. 7,7′-Epoxy-3′,4,4′,5,5′,6-hexahydroxy-7-oxo-1,7-seco-2,2′-cyclolignan

7-Ketone(lactone), 4′,5′-methylene, tetra-Me ether
Eupomatilone 4
[139068-10-3]
$C_{23}H_{26}O_8$ 430.453
Constit. of *Eupomatia bennettii*. Gum. $[\alpha]_D$ +36.4 (c, 0.5 in $CHCl_3$). Mixt. of atropisomers. λ_{max} 214 (ε 43100); 288 (ε 2850) (MeOH)

8,9-Didehydro, 7-ketone(lactone), hexa-Me ether
2,2′-Cyclo-7,7′-epoxy-3′,4,4′,5,5′,6-hexamethoxy-1,7-seco-8,8′-lign-8-en-7-one. 7,7′-Epoxy-3′,4,4′,5,5′,6-hexamethoxy-7-oxo-1,7-seco-2,2′-cyclolign-8-ene. **Eupomatilone 2**
[139006-76-1]
$C_{24}H_{28}O_8$ 444.48
Constit. of *Eupomatia bennettii*. Exhibits activity against human breast MDA-MB-435 cells. Gum. $[\alpha]_D$ +3.3 (c, 0.5 in $CHCl_3$). Mixt. of atropisomers. λ_{max} 212 (ε 50110); 273 (ε 6710) (MeOH)

8,9-Didehydro, 7-ketone(lactone), 4′,5′-methylene, tetra-Me ether
2,2′-Cyclo-7,7′-epoxy-3′,4,5,6-tetramethoxy-4′,5′-methylenedioxy-1,7-seco-8,8′-lign-8-en-7-one.7,7′-Epoxy-3′,4,5,6-tetramethoxy-4′,5′-methylenedioxy-7-oxo-1,7-seco-2,2′-cyclolign-8-ene. **Eupomatilone 1**
[139006-75-0]
$C_{23}H_{24}O_8$ 428.437
Constit. of *Eupomatia bennettii*. Gum. $[\alpha]_D$ -10 (c, 0.5 in $CHCl_3$). Mixt. of atropisomers. λ_{max} 212 (ε 50120); 271 (ε 6610); 341 (ε 790) (MeOH)

(7′S,8R,8′R,RS$_{axial}$)-form
7-Ketone(lactone), 4′,5′-methylene, tetra-Me ether
Eupomatilone 3
[139006-77-2]
$C_{23}H_{26}O_8$ 430.453
Constit. of *Eupomatia bennettii*. Gum. $[\alpha]_D$ -40 (c, 0.5 in $CHCl_3$). Mixt. of atropisomers. λ_{max} 214 (ε 45480); 280 (ε 3320) (MeOH)

(7R*,7′R*,8R*,8′R*,RS$_{axial}$)-form
3′,4,4′,5′,7-Penta-Me ether
Gymnothelignan Y$_4$
$C_{23}H_{30}O_8$ 434.485
Constit. of *Gymnotheca involucrata*.

Carroll, A.R. *et al., Aust. J. Chem.*, 1991, **44**, 1705-1714 (*Eupomatilones 1-6*).
Coleman, R.S. *et al., Org. Lett.*, 2004, **6**, 4025-4028 (*Eupomatilones 4,6, synth*).
Rainka, M.P. *et al., Angew. Chem., Int. Ed.*, 2005, **44**, 6177-6180 (*Eupomatilone 3, synth*).
Kabalka, G.W. *et al., Tetrahedron Lett.*, 2005, **46**, 7325-7328 (*Eupomatilone 2, synth*).
Johnson, J.B. *et al., Angew. Chem., Int. Ed.*, 2007, **46**, 4514-4518 (*Eupomatilone 4, synth*).
Mitra, S. *et al., J. Org. Chem.*, 2007, **72**, 8724-8736 (*Eupomatilones 1,2,4,5, synth*).
He, D. *et al., J. Org. Chem.*, 2012, **77**, 8435-8443 (*Gymnothelignans D-J*).
Hirokawa, Y. *et al., Eur. J. Org. Chem.*, 2013, 721-727 (*Eupomatilones 1,2, synth*).
Xiao, S.-J. *et al., Planta Med.*, 2016, **82**, 723-728 (*Gymnothelignans X$_5$,X$_6$*).
Zhang, P.-Z. *et al., Zhongcaoyao*, 2017, **48**, 1509-1512 (*Gymnothelignan Y$_4$*).
Choi, H. *et al., Org. Lett.*, 2019, **21**, 7857-7862 (*Eupomatilones 2,5,6, synth*)

2,7′-Cyclo-9,9′-epoxy-1,7-seco-8,8′-lignan-3′,4,4′,5,5′,6,9-heptol C170

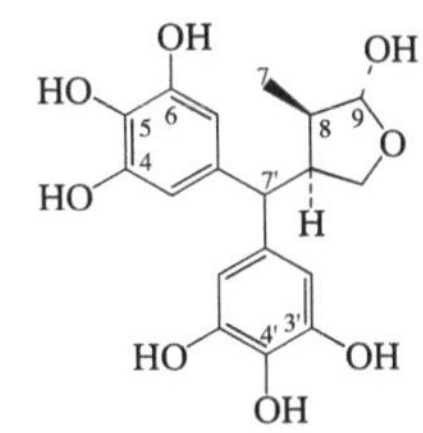

(8R*,8′S*,9R*)-form

9,9′-Epoxy-3′,4,4′,5,5′,6,9-heptahydroxy-1,7-seco-2,7′-cyclolignan
$C_{18}H_{20}O_8$ 364.35
Lignan numbering shown. Care needed. Free rotation of the aryl rings in the secolignan struct. means that numbering round the rings is ambiguous, and the two aryl rings also become equivalent. Numbering of substits. given here is to give lowest locant numbers according to standard nomenclature rules. Note that in lignan numbering, given here, the numbering of the two aryl rings is different although they are constitutionally identical. In unsymmetrically substd. compds. C-7′ is an additional chiral centre

(8R*,8′S*,9R*)-form
3′,4′,4,5-Bis(methylene), 5′,6-di-Me ether
2,7′-Cyclo-9,9′-epoxy-3′,4-dimethoxy-4′,5′:5,6-bis(methylenedioxy)-1,7-seco-8,8′-lignan-9-ol. 9,9′-Epoxy-9-hydroxy-3′,4-dimethoxy-4′,5′:5,6-bis(methylenedioxy)-1,7-seco-2,7′-cyclolignan
[885109-65-9]
$C_{22}H_{24}O_8$ 416.426
Constit. of *Peperomia dindygulensis*. Pale yellow gum. $[\alpha]^{20}_{D}$ -110.7 (c, 0.03 in $CHCl_3$). λ_{max} 216; 249; 280 (MeOH)

(8R,8′S)-form

9-Ketone(lactone)

Dihydro-3-methyl-4-[bis(3,4,5-trihydroxyphenyl)methyl]-2(3H)-furanone. 2,7′-Cyclo-3′,4,4′,5,5′,6-hexahydroxy-1,7-seco-8,8′-lignan-9,9′-olide. 3′,4,4′,5,5′,6-Hexahydroxy-1,7-seco-2,7′-cyclolignan-9,9′-olide

9-Ketone(lactone), 3′,4′-methylene, 4,5,5′,6-tetra-Me ether(7′S-)

*2,7′-Cyclo-3′,4,5,6-tetramethoxy-4′,5′-methylenedioxy-1,7-seco-8,8′-lignan-9,9′-olide. 3′,4,5,6-Tetramethoxy-4′,5′-methylenedioxy-1,7-seco-2,7′-cyclolignan-9,9′-olide. cis-***Peperomin B**

[1379746-43-6]

$C_{23}H_{26}O_8$ 430.453

Constit. of the aerial parts of *Peperomia blanda*. Pale yellow oil. $[\alpha]_D^{21}$ -55.3 (c, 0.49 in $CHCl_3$). λ_{max} 224; 247; 275 (MeOH)

9-Ketone(lactone), 3′,4′;4,5-bis(methylene), 5′,6-di-Me ether

*2,7′-Cyclo-3′,4-dimethoxy-4′,5′:5,6-bis(methylenedioxy)-1,7-seco-8,8′-lignan-9,9′-olide. 3′,4-Dimethoxy-4′,5′:5,6-bis(methylenedioxy)-1,7-seco-2,7′-cyclolignan-9,9′-olide. cis-***Peperomin A**

[1383481-91-1]

$C_{22}H_{22}O_8$ 414.41

Constit. of the aerial parts of *Peperomia blanda*. Pale yellow oil. $[\alpha]_D^{21}$ -54.7 (c, 0.14 in $CHCl_3$). λ_{max} 220; 248; 276 (MeOH)

9-Ketone(lactone), 7,8-didehydro, 3′,4,4′,5,5′-penta-Me ether

$C_{23}H_{26}O_8$ 430.453

Constit. of *Peperomia dindygulensis*. Exhibits significant inhibitory activity against JAK-STAT signalling pathways. Amorph. powder. $[\alpha]_D^{25}$ +15.7 (c, 0.6 in MeOH)

9-Ketone(lactone), 7,8-didehydro, 3′,4,4′,5′,6-penta-Me ether

$C_{23}H_{26}O_8$ 430.453

Constit. of *Peperomia dindygulensis*. Exhibits significant inhibitory activity against JAK-STAT signalling pathways. Amorph. powder. $[\alpha]_D^{25}$ -34 (c, 0.1 in MeOH)

(8S,8′S)-form

9-Ketone(lactone), 3′,4,4′,5′-tetra-Me ether

2,7′-Cyclo-4,5-dihydroxy-3′,4′,5′,6-tetramethoxy-1,7-seco-8,8′-lignan-9,9′-olide. 4,5-Dihydroxy-3′,4′,5′,6-tetramethoxy-1,7-seco-2,7′-cyclolignan-9,9′-olide

[933773-69-4]

$C_{22}H_{26}O_8$ 418.442

Constit. of *Peperomia heyneana*. Gum. $[\alpha]_D^{20}$ +40 (c, 0.27 in MeOH). λ_{max} 247; 276 (MeOH aq.)

9-Ketone(lactone), 3′,4,4′,5,5′-penta-Me ether

2,7′-Cyclo-4-hydroxy-3′,4′,5,5′,6-pentamethoxy-1,7-seco-8,8′-lignan-9,9′-olide. 4-Hydroxy-3′,4′,5,5′,6-pentamethoxy-1,7-seco-2,7′-cyclolignan-9,9′-olide

[885109-60-4]

$C_{23}H_{28}O_8$ 432.469

Constit. of *Peperomia dindygulensis*. Pale yellow gum. $[\alpha]_D^{20}$ +14.5 (c, 0.27 in $CHCl_3$). λ_{max} 212; 246; 279 (MeOH)

9-Ketone(lactone), 3′,4,4′,5′,6-penta-Me ether

2,7′-Cyclo-5-hydroxy-3′,4,4′,5′,6-pentamethoxy-1,7-seco-8,8′-lignan-9,9′-olide. 5-Hydroxy-3′,4,4′,5′,6-pentamethoxy-1,7-seco-2,7′-cyclolignan-9,9′-olide

[885109-58-0]

$C_{23}H_{28}O_8$ 432.469

Constit. of *Peperomia dindygulensis*. Pale yellow gum. $[\alpha]_D^{20}$ +41.5 (c, 0.44 in $CHCl_3$). λ_{max} 212; 245; 274 (MeOH)

9-Ketone(lactone), hexa-Me ether

2,7′-Cyclo-3′,4,4′,5,5′,6-hexamethoxy-1,7-seco-8,8′-lignan-9,9′-olide. 3′,4,4′,5,5′,6-Hexamethoxy-1,7-seco-2,7′-cyclolignan-9,9′-olide. **Peperomin C**

[129341-10-2]

$C_{24}H_{30}O_8$ 446.496

Constit. of *Peperomia japonica* and *Peperomia dindygulensis*. Enhances Ca accumulation in human MDR 2780 ovarian cancer cells. Cryst. (MeOH or EtOAc/hexane). Mp 158 - 160° (149-150°). $[\alpha]_D^{27}$ +42.7 (c, 0.059 in $CHCl_3$) (+27.3)

9-Ketone(lactone), 3′,4′-methylene, 4,5-di-Me ether

$C_{21}H_{22}O_8$ 402.399

Constit. of *Peperomia dindygulensis*. Amorph. powder. $[\alpha]_D^{25}$ +28.6 (c, 0.5 in MeOH)

9-Ketone(lactone), 3′,4′-methylene, 5,5′-di-Me ether

$C_{21}H_{22}O_8$ 402.399

Constit. of *Peperomia dindygulensis*. Amorph. powder. $[\alpha]_D^{25}$ +27.3 (c, 0.44 in MeOH)

9-Ketone(lactone), 3′,4′-methylene, 4,5′,6-tri-Me ether

2,7′-Cyclo-5-hydroxy-3′,4,6-trimethoxy-4′,5′-methylenedioxy-1,7-seco-8,8′-lignan-9,9′-olide. 5-Hydroxy-3′,4,6-trimethoxy-4′,5′-methylenedioxy-1,7-seco-2,7′-cyclolignan-9,9′-olide

$C_{22}H_{24}O_8$ 416.426

Constit. of *Peperomia dindygulensis*. Pale yellow gum. $[\alpha]_D^{20}$ +62 (c, 0.13 in $CHCl_3$). λ_{max} 212; 245; 283 (MeOH)

9-Ketone(lactone), 3′,4′-methylene, 4,5,6-tri-Me ether

2,7′-Cyclo-3′-hydroxy-4,5,6-trimethoxy-4′,5′-methylenedioxy-1,7-seco-8,8′-lignan-9,9′-olide. 3′-Hydroxy-4,5,6-trimethoxy-4′,5′-methylenedioxy-1,7-seco-2,7′-cyclolignan-9,9′-olide

[933773-70-7]

$C_{22}H_{24}O_8$ 416.426

Constit. of *Peperomia heyneana*. Gum. $[\alpha]_D^{20}$ +35.4 (c, 0.13 in MeOH). λ_{max} 250; 278 (MeOH aq.)

9-Ketone(lactone), 3′,4′-methylene, 4,5,5′,6-tetra-Me ether (7′S-)

Peperomin B

[129011-96-7]

$C_{23}H_{26}O_8$ 430.453

Constit. of *Peperomia blanda*, *Peperomia japonica* and *Peperomia dindygulensis*. Inhibitor of tube formn. in human umbilical vein endothelial cells. Cryst. (MeOH aq.). Mp 143 - 145°. $[\alpha]_D^{27}$ +28.9 (c, 0.444 in $CHCl_3$)

9-Ketone(lactone), 3′,4′:4,5-bis(methylene), 6-Me ether

2,7′-Cyclo-3′-hydroxy-4-methoxy-4′,5′:5,6-bis(methylenedioxy)-1,7-seco-8,8′-lignan-9,9′-olide. 3′-Hydroxy-4-methoxy-4′,5′:5,6-bis(methylenedioxy)-1,7-seco-2,7′-cyclolignan-9,9′-olide. **Peperomin H**

[1311291-66-3]
$C_{21}H_{20}O_8$ 400.383
Constit. of *Peperomia dindygulensis*. Powder. $[\alpha]^{23}_{D}$ +23.8 (c, 0.05 in $CHCl_3$). C-7′ config. not determined. λ_{max} 250 (log ε 3.74); 276 (log ε 3.23) (MeOH)

9-Ketone(lactone), 3′,4′:4,5-bis(methylene), 5′,6-di-Me ether
Peperomin A
[129341-09-9]
$C_{22}H_{22}O_8$ 414.41
Constit. of *Peperomia japonica* and *Peperomia dindygulensis*. Cryst. (MeOH aq.). Mp 143 - 145°. $[\alpha]^{27}_{D}$ +20.6 (c, 0.136 in $CHCl_3$)

7,8-Didehydro, 9-ketone(lactone), 3′,4′-methylene, 4,5′,6-tri-Me ether
2,7′-Cyclo-5-hydroxy-3′,4,6-trimethoxy-4′,5′-methylenedioxy-1,7-seco-8,8′-lign-7-en-9,9′-olide. 5-Hydroxy-3′,4,6-trimethoxy-4′,5′-methylenedioxy-1,7-seco-2,7′-cyclolign-7-en-9,9′-olide. ***7,8-Didehydropeperomin B***
[885109-64-8]
$C_{22}H_{22}O_8$ 414.41
Constit. of *Peperomia dindygulensis*. Antiinflammatory agent. Pale yellow gum. $[\alpha]^{20}_{D}$ +20.8 (c, 0.17 in $CHCl_3$). Called 2,6-didehydro- in the lit.. λ_{max} 213; 250; 280 (MeOH)

7,8-Didehydro, 9-ketone(lactone), 3′,4′:4,5-bis(methylene), 5′,6-di-Me ether
2,7′-Cyclo-3′,4-dimethoxy-4′,5′:5,6-bis(methylenedioxy)-1,7-seco-8,8′-lign-7-en-9,9′-olide. 3′,4-Dimethoxy-4′,5′:5,6-bis(methylenedioxy)-1,7-seco-2,7′-cyclolign-7-en-9,9′-olide. ***Peperomin E***
[220736-94-7]
$C_{22}H_{20}O_8$ 412.394
Constit. of *Peperomia dindygulensis*. Antiinflammatory agent. Inhibitor of tube formn. in human umbilical vein endothelial cells. Mp 140°. $[\alpha]_D$ +2.7 (c, 0.7 in $CHCl_3$). λ_{max} 221 (log ε 4.32); 254 (log ε 4.04); 280 (log ε 3.65) (no solvent reported)

(8R*,8′R*)-form
9-Ketone(lactone), 3′,4′-methylene, 4,5,5′-tri-Me ether
2,7′-Cyclo-4-hydroxy-3′,5,6-trimethoxy-4′,5′-methylenedioxy-1,7-seco-8,8′-lignan-9,9′-olide. 4-Hydroxy-3′,5,6-trimethoxy-4′,5′-methylenedioxy-1,7-seco-2,7′-cyclolignan-9,9′-olide
$C_{22}H_{24}O_8$ 416.426
Constit. of *Peperomia pellucida*. Shows weak suppressant activity against human HL-60 cancer cells. Gum. $[\alpha]^{20}_{D}$ +19.5 (c, 0.61 in Me_2CO). 7′-Config. not determined. λ_{max} 247; 277 (MeOH aq.)

7,8-Didehydro, 9-ketone(lactone), 3′,4′-methylene, 4,5,5′,6-tetra-Me ether
2,7′-Cyclo-3′,4,5,6-tetramethoxy-4′,5′-methylenedioxy-1,7-seco-8,8′-lign-7-en-9,9′-olide. 3′,4,5,6-Tetramethoxy-4′,5′-methylenedioxy-1,7-seco-2,7′-cyclolign-7-en-9,9′-olide
[249633-03-2]
$C_{23}H_{24}O_8$ 428.437
Constit. of *Peperomia pellucida*. Shows antiproliferative activity against human breast cancer MFC-7, cervical cancer HeLa and leukaemia HL-60 cells. Gum. $[\alpha]^{20}_{D}$ -5 (c, 0.18 in Me_2CO). λ_{max} 248; 280 (MeOH aq.).

Chen, C.-M. *et al.*, *Heterocycles*, 1989, **29**, 411-414 (*Peperomins A,B,C*).
Lee, T.-J. *et al.*, *Acta Crystallogr., Sect. C*, 1990, **46**, 338-340 (*Peperomin B, cryst struct*).
Cruz-Almanza, R. *et al.*, *Heterocycles*, 1992, **34**, 2323-2330 (*Peperomin C, synth*).
Govindachari, T.R. *et al.*, *Phytochemistry*, 1998, **49**, 2129-2131 (*Peperomin E*).
Sibi, M.P. *et al.*, *Can. J. Chem.*, 2001, **79**, 1546-1555 (*Peperomins A,C, synth*).
Xu, S. *et al.*, *J. Nat. Prod.*, 2006, **69**, 247-250 (*Peperomia pellucida constits*).
Wu, J.L. *et al.*, *J. Nat. Prod.*, 2006, **69**, 790-794 (*Peperomia dindygulensis constits, activity*).
Zhang, G.-L. *et al.*, *J. Nat. Prod.*, 2007, **70**, 662-664 (*Peperomia heyneana lignans*).
Tsutsui, C. *et al.*, *Bioorg. Med. Chem. Lett.*, 2009, **19**, 4084-4087 (*Peperomin E, Didehydropeperomin B, activity*).
Lin, M.-G. *et al.*, *Chem. Biodiversity*, 2011, **8**, 862-871 (*Peperomin H, activity*).
Felippe, L.G. *et al.*, *Phytochem. Lett.*, 2011, **4**, 245-249 (*cis-Peperomin A*).
Felippe, L.G. *et al.*, *Org. Biomol. Chem.*, 2012, **10**, 4208-4214 (*vcd, abs config, cis-Peperomin B*).
Soorukram, D. *et al.*, *Tetrahedron*, 2014, **70**, 7577-7583 (*Peperomin C, synth*).
Lloyd, M.G. *et al.*, *Org. Biomol. Chem.*, 2016, **14**, 8971-8988 (*Peperomin E, synth*).
Sun, R.-F. *et al.*, *Fitoterapia*, 2017, **122**, 80-84 (*Peperomia dindygulensis constits, activity*)

2,2′-Cyclo-7,7′-epoxy-1,7-seco-8,8′-lignan-3′,4,4′,5,5′,7-hexol C171

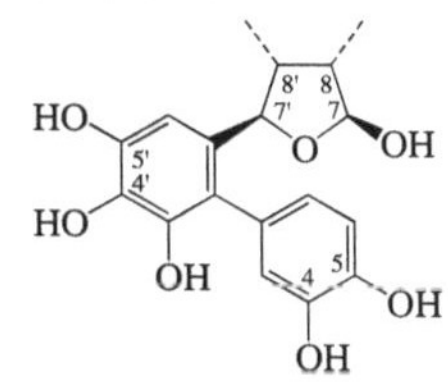

(7S,7′R,8S,8′R,RS_{ax})-form

7,7′-Epoxy-3′,4,4′,5,5′,7-hexahydroxy-1,7-seco-2,2′-cyclolignan
$C_{18}H_{20}O_7$ 348.351
(7S,7′R,8S,8′R,RS$_{axial}$)-form
3′,4,4′,5′-Tetra-Me ether
2,2′-Cyclo-7,7′-epoxy-3′,4,4′,5′-tetramethoxy-1,7-seco-8,8′-lignan-5,7-diol. 7,7′-Epoxy-5,7-dihydroxy-3′,4,4′,5′-tetramethoxy-1,7-seco-2,2′-cyclolignan. ***Gymnothelignan Y_1***
[2095034-77-6]
$C_{22}H_{28}O_7$ 404.459
Constit. of *Gymnotheca chinensis*. Pale yellow gum. $[\alpha]^{20}_{D}$ -45 (c, 0.19 in MeOH). λ_{max} 289 (MeOH)

4′,5′-Methylene, 3′,4-di-Me ether
2,2′-Cyclo-7,7′-epoxy-3′,4-dimethoxy-4′,5′-methylenedioxy-1,7-seco-8,8′-lignan-5,7-diol. 7,7′-Epoxy-5,7-dihydroxy-3′,4-dimethoxy-4′,5′-methylenedioxy-1,7-seco-2,2′-cyclolignan. ***Gymnothelignan Y_2***
[2095034-78-7]
$C_{21}H_{24}O_7$ 388.416
Constit. of *Gymnotheca chinensis*. Light yellow gum. $[\alpha]^{20}_{D}$ -49 (c, 0.46 in MeOH). λ_{max} 288 (MeOH)

7-Ketone(lactone), 3′,4,4′,5′-tetra-Me ether
2,2′-Cyclo-7,7′-epoxy-5-hydroxy-3′,4,4′,5′-tetramethoxy-1,7-seco-8,8′-lignan-7-one. 7,7′-Epoxy-5-hydroxy-3′,4,4′,5′-tetramethoxy-7-oxo-1,7-seco-2,2′-cyclolignan. ***Gymnothelignan X_1***
[2095034-75-4]
$C_{22}H_{26}O_7$ 402.443
Constit. of *Gymnotheca chinensis*. Exhibits moderate cytotoxic activity. Light yellow gum. $[\alpha]^{20}_{D}$ -39 (c, 0.44 in MeOH). λ_{max} 287 (MeOH)

7-Ketone(lactone), 4′,5′-methylene, 3′,4-di-Me ether
2,2′-Cyclo-7,7′-epoxy-5-hydroxy-3′,4-dimethoxy-4′,5′-methylene-

dioxy-1,7-seco-8,8′-lignan-7-one. 7,7′-Epoxy-5-hydroxy-3′,4-dimethoxy-4′,5′-methylenedioxy-7-oxo-1,7-seco-2,2′-cyclolignan. **Gymnothelignan X_2**
[2095034-76-5]
$C_{21}H_{22}O_7$ 386.400
Constit. of *Gymnotheca chinensis*. Light yellow gum. $[\alpha]_D^{20}$ -56 (c, 0.34 in MeOH). λ_{max} 288 (MeOH)

7-Ketone(lactone), 4′,5′-methylene, tri-Me ether
2,2′-Cyclo-7,7′-epoxy-3′,4,5-trimethoxy-4′,5′-methylenedioxy-1,7-seco-8,8′-lignan-7-one. 7,7′-Epoxy-3′,4,5-trimethoxy-4′,5′-methylenedioxy-7-oxo-1,7-seco-2,2′-cyclolignan. **Gymnothelignan X_3**
[2172780-49-1]
$C_{22}H_{24}O_7$ 400.427
Constit. of *Gymnotheca involucrata*. Powder. $[\alpha]_D^{20}$ -38 (c, 0.1 in MeOH). λ_{max} 289 (log ε 3.57) (MeOH)

(7′R,8R,8′S,RS$_{axial}$)-form
7-Ketone(lactone)
2,2′-Cyclo-7,7′-epoxy-3′,4,4′,5,5′-pentahydroxy-1,7-seco-8,8′-lignan-7-one. 7,7′-Epoxy-3′,4,4′,5,5′-pentahydroxy-7-oxo-1,7-seco-2,2′-cyclolignan

7-Ketone(lactone), 4′,5′-methylene, tri-Me ether
Eupomatilone 7
[139006-80-7]
$C_{22}H_{24}O_7$ 400.427
Constit. of *Eupomatia bennettii*. Gum. $[\alpha]_D$ +35.8 (c, 0.1 in $CHCl_3$). Mixt. of atropisomers

(7′R,8S,8′S,RS$_{axial}$)-form
7-Ketone(lactone), 4,5-methylene, tri-Me ether
2,2′-Cyclo-7,7′-epoxy-3′,4′,5′-trimethoxy-4,5-methylenedioxy-1,7-seco-8,8′-lignan-7-one. 7,7′-Epoxy-3′,4′,5′-trimethoxy-4,5-methylenedioxy-7-oxo-1,7-seco-2,2′-cyclolignan. **Eupomatilone 6**
[139006-78-3]
$C_{22}H_{24}O_7$ 400.427
Constit. of *Eupomatia bennettii*. Gum. $[\alpha]_D$ -25.6 (c, 0.5 in $CHCl_3$). Mixt. of atropisomers. Stereochem. revised in 2006

7-Ketone(lactone), 4′,5′-methylene, tri-Me ether
Gymnothelignan X_4
[2172780-50-4]
$C_{22}H_{24}O_7$ 400.427
Constit. of *Gymnotheca involucrata*. Exhibits weak antifungal activity. Powder. $[\alpha]_D^{20}$ -86 (c, 0.11 in MeOH). λ_{max} 292 (log ε 3.47) (MeOH)

8,9-Didehydro, 7-ketone(lactone), 4,5-methylene, tri-Me ether
2,2′-Cyclo-7,7′-epoxy-3′,4′,5′-trimethoxy-4,5-methylenedioxy-1,7-seco-8,8′-lign-8-en-7-one. 7,7′-Epoxy-3′,4′,5′-trimethoxy-4,5-methylenedioxy-7-oxo-1,7-seco-2,2′-cyclolign-8-ene. **Eupomatilone 5**
[139006-79-4]
$C_{22}H_{22}O_7$ 398.411
Constit. of *Eupomatia bennettii*. Gum. $[\alpha]_D$ +6.5 (c, 0.5 in $CHCl_3$). $[\alpha]_D^{24}$ +26.4 (c, 1.04 in $CHCl_3$). Mixt. of atropisomers.

Carroll, A.R. *et al.*, *Aust. J. Chem.*, 1991, **44**, 1705-1714 (*Eupomatilones 5,6,7*).
Gurjar, M.K. *et al.*, *J. Org. Chem.*, 2005, **70**, 9658-9661 (*Eupomatilone 6, synth, abs config*).
Kabalka, G.W. *et al.*, *Tetrahedron Lett.*, 2005, **46**, 7325-7328 (*Eupomatilone 5, synth*).
Johnson, J.B. *et al.*, *Angew. Chem., Int. Ed.*, 2007, **46**, 4514-4518 (*Eupomatilones 6,7, synth*).
Mitra, S. *et al.*, *J. Org. Chem.*, 2007, **72**, 8724-8736 (*Eupomatilone 5, synth*).
He, D. *et al.*, *J. Org. Chem.*, 2012, **77**, 8435-8443 (*Eupomatilone 6, config*).
Hirokawa, Y. *et al.*, *Eur. J. Org. Chem.*, 2013, 721-727 (*Eupomatilone 5, synth*).
Yodwaree, S. *et al.*, *Org. Biomol. Chem.*, 2014, **12**, 6885-6894 (*Eupomatilone 6, epimers, synth*).
Shiji, X. *et al.*, *Youji Huaxue*, 2014, **34**, 1677-1681 (*Gymnothelignans X_1,X_2,Y_1,Y_2*).
Xiao, S.-J. *et al.*, *Planta Med.*, 2016, **82**, 723-728 (*Gymnothelignans X_3,X_4*)

2,2′-Cyclo-7,7′-epoxy-1,7-seco-8,8′-lignan-3′,4,4′,5,5′,7-hexol C172

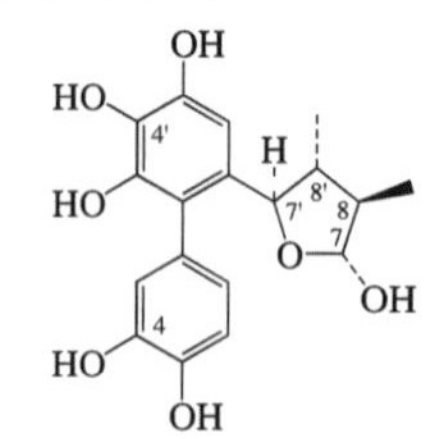

(7R*,7′R*,8R*,8′R*)-form

4-(3,4-Dihydroxyphenyl)-5-(tetrahydro-5-hydroxy-3,4-dimethyl-2-furanyl)-1,2,3-benzenetriol. 7,7′-Epoxy-3′,4,4′,5,5′,7-hexahydroxy-1,7-seco-2,2′-cyclolignan
$C_{18}H_{20}O_7$ 348.351
Secolignan

(7R*,7′R*,8R*,8′R*)-form
3′,4,4′,5′,7-Penta-Me ether
2,2′-Cyclo-7,7′-epoxy-3′,4,4′,5′,7-pentamethoxy-1,7-seco-8,8′-lignan-5-ol. 7,7′-Epoxy-5-hydroxy-3′,4,4′,5′,7-pentamethoxy-1,7-seco-2,2′-cyclolignan.
Gymnothelignan S
[1608094-33-2]
$C_{23}H_{30}O_7$ 418.486
Constit. of *Gymnotheca chinensis*. Powder. $[\alpha]_D^{20}$ -17 (c, 0.036 in MeOH). Mixt. of atropisomers. λ_{max} 266 (log ε 3.43) (MeOH)

4′,5′-Methylene, 3′,4,7-tri-Me ether
2,2′-Cyclo-7,7′-epoxy-3′,4,7-trimethoxy-4′,5′-methylenedioxy-1,7-seco-8,8′-lignan-5-ol. 7,7′-Epoxy-5-hydroxy-3′,4,7-trimethoxy-4′,5′-methylenedioxy-1,7-seco-2,2′-cyclolignan. **Gymnothelignan P**
[1608094-30-9]
$C_{22}H_{26}O_7$ 402.443
Constit. of *Gymnotheca chinensis*. Powder. $[\alpha]_D^{20}$ -90 (c, 0.04 in MeOH). Mixt. of atropisomers. λ_{max} 267 (log ε 3.69) (MeOH)

(7R*,7′R*,8S*,8′S*)-form
4′,5′-Methylene, 3′,4,7-tri-Me ether
Gymnothelignan R
[1608094-32-1]
$C_{22}H_{26}O_7$ 402.443
Constit. of *Gymnotheca chinensis*. Powder. $[\alpha]_D^{20}$ -97 (c, 0.041 in MeOH). Mixt. of atropisomers. λ_{max} 267 (log ε 3.60) (MeOH)

(7R*,7′S*,8R*,8′S*)-form
4′,5′-Methylene, 3′,4,7-tri-Me ether
Gymnothelignan Q
[1608094-31-0]
$C_{22}H_{26}O_7$ 402.443
Constit. of *Gymnotheca chinensis*. Powder. $[\alpha]_D^{20}$ -64 (c, 0.048 in MeOH). Mixt. of atropisomers. λ_{max} 267 (log ε 3.60) (MeOH).

Xiao, S. *et al.*, *Helv. Chim. Acta*, 2014, **97**, 499-506 (*Gymnothelignans P-S*)

2,7′-Cyclo-9,9′-epoxy-1,7-seco-8,8′-lignan-3′,4,4′,5,5′,6,7,9-octol C173

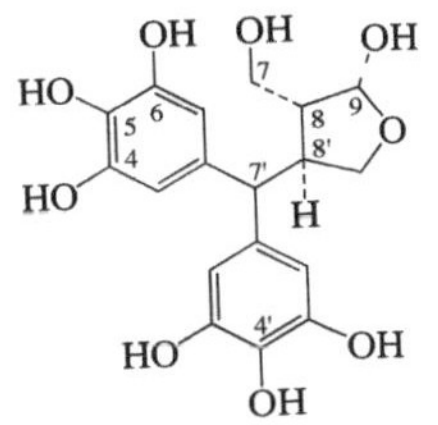

(8R,8'S,9R)-form

Tetrahydro-2-hydroxy-4-[bis(3,4,5-trihydroxyphenyl) methyl]-3-furanmethanol. 3′,4,4′,5,5′,6,7,9-Octahydroxy-9,9′-epoxy-1,7-seco-2,7′-cyclolignan

$C_{18}H_{20}O_9$ 380.349

Secolignan. Care needed with numbering; the two aromatic rings in the parent compd. are equivalent

(8R,8′S,9R)-form

3′,4′:4,5-Bis(methylene), 5′,6-di-Me ether, 7-Ac

Peperomin F

[220736-95-8]

$C_{24}H_{26}O_{10}$ 474.462

Constit. of *Peperomia dindygulensis*. Mp 170°. [α] D -41.2 (c, 0.17 in MeOH). Species given as Peperomia dindigulensis in ref.. λ_{max} 253 (log ε 2.91); 278 (log ε 2.63) (no solvent reported)

9-Ketone(lactone), 3′,4′-methylene, 4,5′,6-tri-Me ether

$C_{22}H_{24}O_9$ 432.425

Constit. of *Peperomia dindygulensis*. Amorph. powder. $[\alpha]^{25}_{D}$ -93.2 (c, 0.14 in MeOH)

(8R*,8′R*,9ξ)-form

3′,4′-Methylene, 4,5′,6-tri-Me ether, 7-Ac

$C_{24}H_{28}O_{10}$ 476.478

Constit. of *Peperomia dindygulensis*. Possesses negative opt. rotn.

9-Ketone(lactone)

2,7′-Cyclo-3′,4,4′,5,5′,6,7-heptahydroxy-1,7-seco-8,8′-lignan-9,9′-olide

9-Ketone(lactone), 3′,4,4′,5′-tetra-Me ether, 7-Ac

[1588563-99-8]

$C_{24}H_{28}O_{10}$ 476.478

Constit. of *Peperomia dindygulensis*. Possesses negative opt. rotn.

9-Ketone(lactone), 3′,4′-methylene, 4,5′,6-tri-Me ether, 7-Ac

[885109-63-7]

$C_{24}H_{26}O_{10}$ 474.462

Constit. of *Peperomia dindygulensis*. Moderate inhibitor of lung tumour VA-13 and lung fibroblast WI-38 cells. Pale yellow gum. $[\alpha]^{20}_{D}$ -44 (c, 0.14 in $CHCl_3$). λ_{max} 214; 250; 280 (MeOH)

9-Ketone(lactone), 3′,4′-methylene, 4,5,5′,6-tetra-Me ether

2,7′-Cyclo-7-hydroxy-3′,4,5,6-tetramethoxy-4′,5′-methylenedioxy-1,7-seco-8,8′-lignan-9,9′-olide. 7-Hydroxy-3′,4,5,6-tetramethoxy-4′,5′-methylenedioxy-1,7-seco-2,7′-cyclolignan-9,9′-olide

[885109-62-6]

$C_{23}H_{26}O_9$ 446.452

Constit. of *Peperomia dindygulensis*. Pale yellow gum. $[\alpha]^{20}_{D}$ -75.4 (c, 0.18 in $CHCl_3$). λ_{max} 212; 249; 281 (MeOH)

9-Ketone(lactone), 3′,4′:4,5-bis(methylene), 5′,6-di-Me ether

2,7′-Cyclo-7-hydroxy-3,3′-dimethoxy-4,5:4′,5′-bis(methylenedioxy)-1,7-seco-8,8′-lignan-9,9′-olide. 7-Hydroxy-3,3′-dimethoxy-4,5:4′,5′-bis(methylenedioxy)-1,7-seco-2,7′-cyclolignan-9,9′-olide

[885109-61-5]

$C_{22}H_{22}O_9$ 430.409

Constit. of *Peperomia dindygulensis*. Pale yellow gum. $[\alpha]^{20}_{D}$ -42.2 (c, 1.12 in MeOH). λ_{max} 212; 250; 283 (MeOH)

(8R*,8′S*,9R*)-form

3′,4′-Methylene, 4,5,5′,6-tetra-Me ether, 7-Ac

[933773-71-8]

$C_{25}H_{30}O_{10}$ 490.505

Constit. of *Peperomia heyneana*. Gum. [α] D -68.6 (c, 0.18 in MeOH). C-7′ is a chiral centre. λ_{max} 250; 280 (MeOH).

Govindachari, T.R. *et al.*, *Phytochemistry*, 1998, **49**, 2129-2131 (*Peperomin F*).

Wu, J.-L. *et al.*, *J. Nat. Prod.*, 2006, **69**, 790-794 (*Peperomia dindygulensis constits*).

Zhang, G.L. *et al.*, *J. Nat. Prod.*, 2007, **70**, 662-664 (*Peperomia heyneana lignan*).

Wang, X.-Z. *et al.*, *J. Pharm. Biomed. Anal.*, 2014, **94**, 1-11 (*Peperomia dindygulensis constits*).

Sun, R.-F. *et al.*, *Fitoterapia*, 2017, **122**, 80-84 (*Peperomia dindygulensis constit*)

2,7′-Cyclo-9,9′-epoxy-1,7-seco-8,8′-lignan-4′,5,7,9-tetrol C174

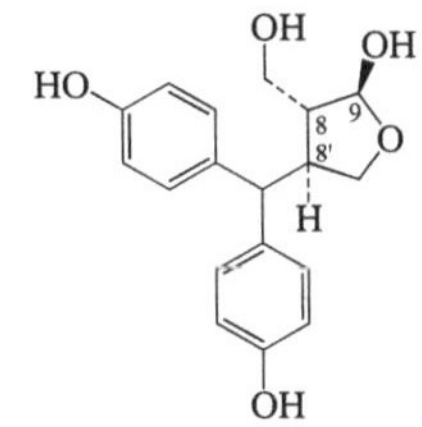

Tetrahydro-4-[bis(4-hydroxyphenyl) methyl]-3-hydroxymethyl-2-furanol. 4′,5,7,9-Tetrahydroxy-9,9′-epoxy-1,7-seco-2,7′-cyclolignan

$C_{18}H_{20}O_5$ 316.353

(8R*,8′S*,9S*)-form

9-Me ether

2,7′-Cyclo-9,9′-epoxy-9-methoxy-1,7-seco-8,8′-lignan-4′,5,7-triol. 9,9′-Epoxy-4′,5,7-trihydroxy-9-methoxy-1,7-seco-2,7′-cyclolignan

[1228379-88-1]

$C_{19}H_{22}O_5$ 330.38

Secolignan. Constit. of *Selaginella sinensis*. Amorph. solid. $[\alpha]^{20}_{D}$ -91 (c, 0.12 in MeOH). λ_{max} 204; 237; 277 (MeOH)

9-Ketone(lactone)

2,7′-Cyclo-4′,5,7-trihydroxy-1,7-seco-8,8′-lignan-9,9′-olide. Dihydro-3-(hydroxymethyl)-4-[bis(4-hydroxyphenyl) methyl]-2(3H)-furanone

[1228379-87-0]

$C_{18}H_{18}O_5$ 314.337

Constit. of *Selaginella sinensis*. Amorph. solid. $[\alpha]^{20}_{D}$ +49 (c, 0.23 in MeOH). λ_{max} 205; 233; 281 (MeOH).

Feng, W.-S. *et al.*, *J. Asian Nat. Prod. Res.*, 2009, **11**, 658-662 (*Selaginella sinensis constits*)

2,7′-Cyclo-2′,3-epoxy-4,4′,5′-trihydroxy-9′-nor-8,8′-ligna-7,7′-dien-9-oic acid C175

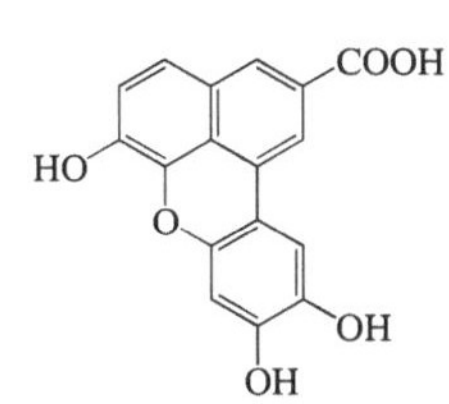

6,9,10-Trihydroxybenzo[k,l]xanthene-2-carboxylic acid. Menthalignan

[1622067-73-5]

$C_{17}H_{10}O_6$ 310.261

Constit. of *Mentha haplocalyx*. Mp > 300°.

Xu, L.-Y. *et al.*, *Zhongcaoyao*, 2013, **44**, 2798-2802 (*Menthalignan*)

2,7′-Cyclo-3′,4,4′,5,5′,6,7-heptahydroxy-8,8′-ligna-7,7′-dien-9′,9-olide C176

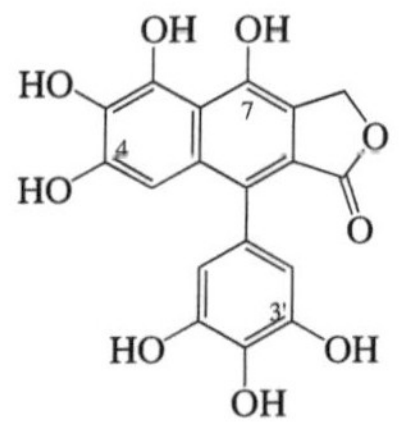

3′,4,4′,5,5′,6,7-Heptahydroxy-2,7′-cycloligna-8,8′-ligna-7,7′-dien-9′,9-olide

$C_{18}H_{12}O_9$ 372.285

4,5-Methylene, 3′,4′,5′,6-tetra-Me ether

2,7′-Cyclo-7-hydroxy-3′,4′,5′,6-tetramethoxy-4,5-methylenedioxy-8,8′-ligna-7,7′-dien-9′,9-olide. 7-Hydroxy-3′,4′,5′,6-tetramethoxy-4,5-methylenedioxy-2,7′-cycloligna-7,7′-dien-9′,9-olide. ***6-Methoxydehydropodophyllotoxin****. 6-Methoxytetradehydropodophyllotoxin*

[151703-06-9]

$C_{23}H_{20}O_9$ 440.404

Constit. of *Hyptis verticillata*. Cytotoxic. Needles. Mp > 225° dec.. λ_{max} 223 (log ε 4.52); 268 (log ε 4.36); 325 (log ε 3.68); 357 (log ε 3.38) (MeOH).

Novelo, M. *et al.*, *J. Nat. Prod.*, 1993, **56**, 1728-1736 (*6-Methoxydehydropodophyllotoxin*)

2,7′-Cyclo-3′,4,4′,5,5′,6,9-heptahydroxy-8,8′-lignan-9′-oic acid C177

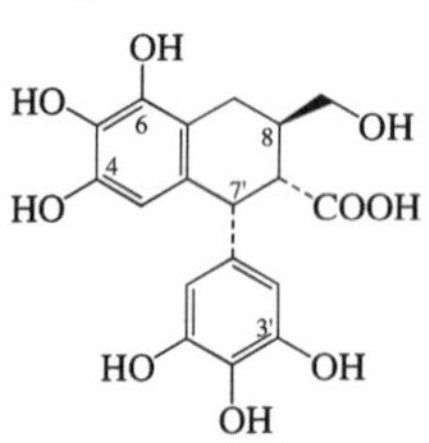

(7′R,8R,8′R)-form

3′,4,4′,5,5′,6,9-Heptahydroxy-2,7′-cyclolignan-9′-oic acid

$C_{18}H_{18}O_9$ 378.333

(7′R,8R,8′R)-form

4,5-Methylene, 3′,4′,5′,6-tetra-Me ether

2,7′-Cyclo-9-hydroxy-3′,4′,5′,6-tetramethoxy-4,5-methylenedioxy-8,8′-lignan-9′-oic acid. 9-Hydroxy-3′,4′,5′,6-tetramethoxy-4,5-methylenedioxy-2,7′-cyclolignan-9′-oic acid

$C_{23}H_{26}O_9$ 446.452

4,5-Methylene, 3′,4′,5′,6-tetra-Me ether, Me ester

Sabinaperin A

[1391976-65-0]

$C_{24}H_{28}O_9$ 460.479

Isol. from epigeal part of *Juniperus sabina*. Solid. λ_{max} 204 (ε 16000); 283 (ε 1000) (MeOH)

(7′R,8R,8′S)-form

4,5-Methylene, 3′,4′,5′,6-tetra-Me ether, Me ester

Sabinaperin B

[1391976-66-1]

$C_{24}H_{28}O_9$ 460.479

Isol. from epigeal part of *Juniperus sabina*. Solid. λ_{max} 203 (ε 29700); 280 (ε 1200) (MeOH).

Janar, J. *et al.*, *Heterocycles*, 2012, **84**, 1259-1263 (*Sabinaperins A,B*)

2,7′-Cyclo-3′,4,4′,5,5′,7,9-heptahydroxy-8,8′-lignan-9′-oic acid C178

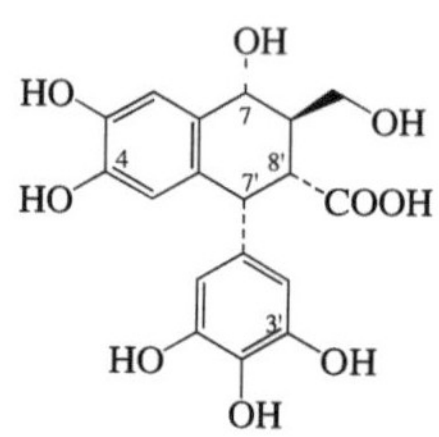

(7R,7′R,8R,8′R)-form

3′,4,4′,5,5′,7,9-Heptahydroxy-2,7′-cyclolignan-9′-oic acid

[75171-58-3] (7*R**,7′*S**,8*S**,8′*S**-Me ester) [122871-76-5] (7*R**,7′*S**,8*R**,8′*S**-Me ester)

$C_{18}H_{18}O_9$ 378.333

(7R,7′R,8R,8′R)-form

(9′→9)-Lactone

2,7′-Cyclo-3′,4,4′,5,5′,7-hexahydroxy-8,8′-lignan-9′,9-olide, C199

4,5-Methylene, 3′,4′,5′-tri-Me ether

2,7′-Cyclo-7,9-dihydroxy-3′,4′,5′-trimethoxy-4,5-methylenedioxy-8,8′-lignan-9′-oic acid. 7,9-Dihydroxy-3′,4′,5′-trimethoxy-4,5-methylenedioxy-2,7′-cyclolignan-9′-oic acid. ***Podophyllic acid****. Toxinic acid. Podophyllinic acid. Podophyllotoxinic acid*

[1853-37-8]

[518-48-9] (7*R*,7′*R*,8*S*,8′*S*-form) [79355-65-0] (7*R**,7′*S**,8*S**,8′*S**-form) [122871-78-7] (7*R**,7′*S**,8*R**,8′*R**-form) [122871-80-1] (7*R**,7′*S**,8*S**,8′*R**-form)

$C_{22}H_{24}O_9$ 432.425

Constit. of *Juniperus thurifera*. Antimitotic agent. Derivs. show cytotoxic activity. Cryst. + $3H_2O$ (EtOH aq.). Mp 163 - 165° dec. (synthetic). $[\alpha]^{25}_{D}$ -96 (synthetic). Nat. prod. isol. as di-Ac, Me ester. Numerous synthetic stereoisomers and their derivs. studied

TP8750000

4,5-Methylene, 3′,4′,5′-tri-Me ether, N-ethylhydrazide

Mitopodozide*. NSC 72274*

[1508-45-8]

$C_{24}H_{30}N_2O_8$ 474.51

Antineoplastic agent. Mp 100°. $[\alpha]_{D}$ -149 (c, 0.5 in $CHCl_3$). Log P -2.13 (calc)

TP8800000. LD_{50} (rat, ivn) 140 mg/kg

4,5-Methylene, 3′,4′,5′-tri-Me ether, di-Ac, Me ester

[139955-57-0]

$C_{27}H_{30}O_{11}$ 530.526

Oil. $[\alpha]^{23}_{D}$ -164 (c, 0.8 in $CHCl_3$). λ_{max} 207 (ε 34600); 288 (ε 3000) (MeOH) (Berdy)

(7S,7′R,8R,8′R)-form

4,5-Methylene, 3′,4′,5′-tri-Me ether

Epipodophyllic acid

4,5-Methylene, 3′,4′,5′-tri-Me ether, Me ester, 7-O-[β-D-glucopyranosyl-(1→6)-β-D-glucopyranoside]

[386267-12-5]

$C_{35}H_{46}O_{19}$ 770.734

Constit. of *Sinopodophyllum emodi*. Powder. $[\alpha]^{15}_{D}$ -120 (c, 0.34 in MeOH). λ_{max} 210; 280 (MeOH).

Gensler, W.J. *et al.*, *J. Am. Chem. Soc.*, 1954, **76**, 315-316 (*(±)-Podophyllic acid*).

Gensler, W.J. *et al.*, *Chem. Pharm. Bull.*, 1956, **21**, 261 (*Podophyllic acid, cryst struct*).

Fr. Pat., 1962, 1 312 (*Mitopodozide*).

Stamm, O. *et al.*, *Cancer (Hoboken, NJ, U.S.)*, 1965, **18**, 1096 (*Mitopodozide, activity*).

Hulbert, P.B. *et al.*, *J. Chem. Res., Synop.*, 1981, 27 (*cd, uv*).

San Feliciano, A. *et al.*, *Phytochemistry*, 1992, **31**, 267-270 (*Juniperus thurifera constit*).

Zhao, C. *et al.*, *Chem. Pharm. Bull.*, 2001, **49**, 1474-1476 (*Sinopodophyllum emodi constit*)

2,7′-Cyclo-3,3′,4,4′,5,5′,7-heptahydroxy-8,8′-lignan-9′,9-olide C179

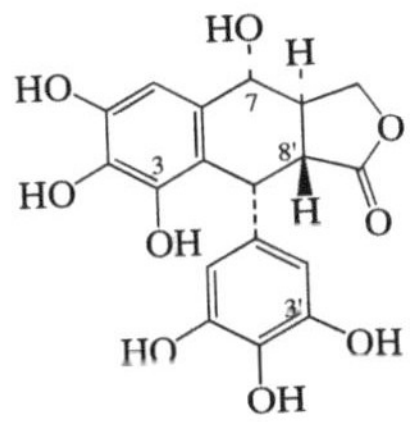

(7R,7′R,8R,8′R)-form

3,3′,4,4′,5,5′,7-Heptahydroxy-2,7′-cyclolignan-9′,9-olide
$C_{18}H_{16}O_9$ 376.317
(7R,7′R,8R,8′R)-form
4,5-Methylene, 3,3′,4′,5′-tetra-Me ether
2,7′-Cyclo-7-hydroxy-3,3′,4′,5′-tetramethoxy-4,5-methylenedioxy-8,8′-lignan-9′,9-olide. 7-Hydroxy-3,3′,4′,5′-tetramethoxy-4,5-methylenedioxy-2,7′-cyclolignan-9′,9-olide.
3-Methoxypodophyllotoxin
$C_{23}H_{24}O_9$ 444.436
Constit. of the seeds of *Hernandia sonora*. Amorph. solid. $[\alpha]_D$ -41.8 (c, 1.4 in $CHCl_3$). Called 5-methoxy- in the lit.
4,5-Methylene, 3,3′,4′,5′-tetra-Me ether, 7-Ac
3-Methoxypodophyllotoxin acetate
[225669-50-1]
$C_{25}H_{26}O_{10}$ 486.473
Constit. of the seeds of *Hernandia sonora*. Amorph. solid. $[\alpha]_D$ -63.5 (c, 2.1 in $CHCl_3$)
(7R,7′R,8R,8′S)-form
4,5-Methylene, 3,3′,4′,5′-tetra-Me ether, 7-Ac
7-O-Acetyl-3-methoxypicropodophyllin
[445042-93-3]
$C_{25}H_{26}O_{10}$ 486.473
Constit. of the twigs of *Hernandia ovigera*. Amorph.. $[\alpha]^{20}_D$+51 (c, 0.07 in MeCN). λ_{max} 213 (log ε 4) (MeCN).

Udino, L. *et al.*, *Planta Med.*, 1999, **65**, 279-281 (*3-Methoxypodophyllotoxin, 3-Methoxypodophyllotoxin acetate*).
Gu, J.-Q. *et al.*, *J. Nat. Prod.*, 2002, **65**, 1065-1068 (*7-Acetyl-3-methoxypicropodophyllin*)

2,7′-Cyclo-3′,4,4′,5,5′,6,7-heptahydroxy-8,8′-lignan-9′,9-olide C180

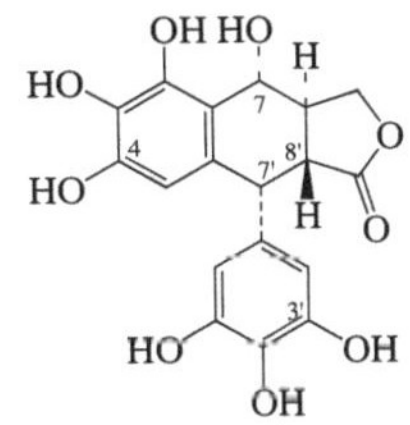

(7R,7′R,8R,8′R)-form

3′,4,4′,5,5′,6,7-Heptahydroxy-2,7′-cyclolignan-9′,9-olide
$C_{18}H_{16}O_9$ 376.317
(7R,7′R,8R,8′R)-form
4,5-Methylene, 3′,5′,6-tri-Me ether
2,7′-Cyclo-4′,7-dihydroxy-3′,5′,6-trimethoxy-4,5-methylenedioxy-8,8′-lignan-9′,9-olide. 4′,7-Dihydroxy-3′,5′,6-trimethoxy-4,5-methylenedioxy-2,7′-cyclolignan-9′,9-olide.
4′-De-O-methyl-6-methoxypodophyllotoxin
[139047-84-0]
$C_{22}H_{22}O_9$ 430.409
Constit. of *Linum cariense* and *Linum flavum*
4,5-Methylene, 3′,4′,5′,6-tetra-Me ether
2,7′-Cyclo-7-hydroxy-3′,4′,5′,6-tetramethoxy-4,5-methylenedioxy-8,8′-lignan-9′,9-olide. 7-Hydroxy-3′,4′,5′,6-tetramethoxy-4,5-methylenedioxy-2,7′-cyclolignan-9′,9-olide. ***6-Methoxypodophyllotoxin***. *6-MPTOX*
[128443-52-7]
$C_{23}H_{24}O_9$ 444.436
Isol. from *Juniperus sabina*, *Linum flavum*, *Linum cariense*, *Linum perenne*, *Hernandia sonora* and *Libocedrus chevalieri*. Cytotoxic against human KB carcinoma cells. Cryst. (hexane/CH_2Cl_2). Mp 203°. $[\alpha]_D$ -129.5 (c, 1 in $CHCl_3$). Wrongly descr. as β-Peltatin A methyl ether in one ref.. λ_{max} 219 (ε 35000); 279 (ε 2400) (MeOH) (Berdy)
4,5-Methylene, 3′,4′,5′,6-tetra-Me ether, 7-O-β-D-glucopyranoside
[119188-52-2]
$C_{29}H_{34}O_{14}$ 606.577
Isol. from *Linum capitatum* and *Linum flavum*. Cytotoxic
4,5-Methylene, 3′,4′,5′,6-tetra-Me ether, 7-Ac
7-O-Acetyl-6-methoxypodophyllotoxin
[128443-54-9]
$C_{25}H_{26}O_{10}$ 486.473
Constit. of *Linum flavum*, *Linum capitatum* and *Hernandia sonora*. Antitumour agent
(7R,7′R,8R,8′S)-form
4,5-Methylene, 3′,4′,5′,6-tetra-Me ether
6-Methoxy-8′-epipodophyllotoxin.
6-Methoxypicropodophyllin
[128443-51-6]
Constit. of *Juniperus sabina*, *Linum cariense* and *Linum perenne*. $[\alpha]^{24}_D$ -73.2 (c, 0.36 in $CHCl_3$). λ_{max} 221 (ε 24800); 279 (ε 2100); 291 (ε 1700) (EtOH)
(7S,7′R,8R,8′R)-form
4,5-Methylene, 3′,4′,5′,6-tetra-Me ether
6-Methoxy-7-epipodophyllotoxin
[951208-76-7]
Constit. of the bark of *Libocedrus chevalieri*. Cytotoxic against human epidermoid KB cells. Tubulin polymerisation inhibitor. Amorph. yellowish powder. $[\alpha]^{25}_D$ -29 (c, 0.75 in $CHCl_3$)
(7S,7′R,8R,8′S)-form
4,5-Methylene, 3′,4′,5′,6-tetra-Me ether
6-Methoxyepipicropodophyllin
[128397-02-4]
Constit. of *Juniperus sabina*. Moderate antineoplastic and antiviral agent. Cryst. (hexane/CH_2Cl_2). Mp 228°. $[\alpha]^{24}_D$-2.8 (c, 0.5 in $CHCl_3$). λ_{max} 213 (ε 58000); 279 (ε 2900); 290 (ε 2100) (EtOH).

Berlin, J. *et al.*, *J. Nat. Prod.*, 1986, **49**, 435-439 (*6-Methoxypodophyllotoxin*).
Berlin, J. *et al.*, *Planta Med.*, 1988, **54**, 204-206 (*6-Methoxypodophyllotoxin, activity*).
San Feliciano, A. *et al.*, *Phytochemistry*, 1990, **29**, 1335-1338 (*Juniperus sabina constits*).
Broomhead, A.J. *et al.*, *Phytochemistry*, 1990, **29**, 3839-3844 (*Linum constits*).
Wichers, H.J. *et al.*, *Phytochemistry*, 1991, **30**, 3601-3604 (*biosynth, pmr, hplc*).
Van Uden, W. *et al.*, *J. Nat. Prod.*, 1992, **55**, 102-110 (*Linum flavum constit, pmr, cmr*).
Udino, L. *et al.*, *Planta Med.*, 1999, **65**, 279-281 (*Hernandia sonora constits*).
Konuklugil, B. *Chem. Nat. Compd. (Engl. Transl.)*, 2005, **41**, 306-307 (*Linum cariense constit*).
Zhang, Y.-J. *et al.*, *J. Nat. Prod.*, 2007, **70**, 1368-1370 (*6-Methoxyepipodophyllotoxin*).

Mohagheghzadeh, A. *et al.*, *Z. Naturforsch., C*, 2007, **62**, 43-49 (*6-Methoxypodophyllotoxin, isol*).
Jullian-Pawlicki, N. *et al.*, *Phytochem. Anal.*, 2015, **26**, 310-319 (*Linum perenne constits, abs config*)

2,7′-Cyclo-3,3′,4,4′,5,5′,9′-heptahydroxy-8,8′-lign-7-en-9-al C181

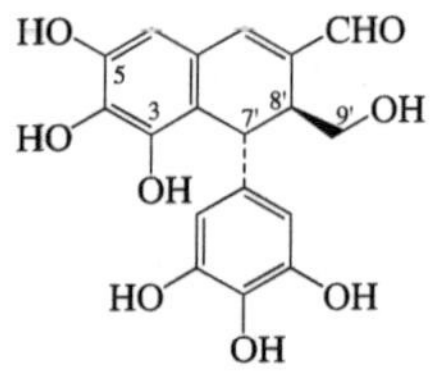

(7′R,8′S)-form

3,3′,4,4′,5,5′,9′-Heptahydroxy-9-oxo-2,7′-cyclolign-7-ene
$C_{18}H_{16}O_8$ 360.318
(7′R,8′S)-form
3,3′,5,5′-Tetra-Me ether, 9′-O-β-D-glucopyranoside
[1428904-01-1]
$C_{28}H_{34}O_{13}$ 578.567
Constit. of *Eurya japonica*. Amorph. yellow solid. $[\alpha]^{25}_{D}$ -5.9 (c, 0.2 in MeOH). λ_{max} 253 (log ε 3.84); 270 (sh); 320 (sh); 354 (log ε 3.61) (MeOH)
(7′S,8′R)-form
3,3′,5-Tri-Me ether
2,7′-Cyclo-3′,4,4′,9′-tetrahydroxy-3,5,5′-trimethoxy-8,8′-lign-7-en-9-al.
3′,4,4′,9′-Tetrahydroxy-3,5,5′-trimethoxy-9-oxo-2,7′-cyclolign-7-ene. ***Vitexdoin F‡***
[1693721-46-8]
$C_{21}H_{22}O_8$ 402.399
Constit. of the seeds of *Vitex negundo* var. *cannabifolia*. Amorph. yellow powder. $[\alpha]^{16}_{D}$ -73.9 (c, 0.6 in MeOH). Not to be confused with 2,7′-Cyclo-4,4′-dihydroxy-3′,5-dimethoxy-8,9-dinor-7,8′-lign-7′-en-9′-al, C141. λ_{max} 203; 254; 359 (MeOH)
3,3′,5,5′-Tetra-Me ether, 9′-O-β-D-glucopyranoside
[1428904-00-0]
Constit. of *Eurya japonica*. Amorph. yellow solid. $[\alpha]_{D}$ +64.9 (c, 0.2 in MeOH). λ_{max} 254 (log ε 3.87); 270 (sh); 354 (log ε 3.66); 356 (sh) (MeOH)
(7′R*,8′R*)-form
3,3′,4,4′,5,5′-Hexa-Me ether
2,7′-Cyclo-9′-hydroxy-3,3′,4,4′,5,5′-hexamethoxy-8,8′-lign-7-en-9-al.
9′-Hydroxy-3,3′,4,4′,5,5′-hexamethoxy-2,7′-cyclolign-7-en-9-al.
Aglacin G
[477529-38-7]
$C_{24}H_{28}O_8$ 444.48
Constit. of the stem bark of *Aglaia cordata*. Waxy solid. $[\alpha]^{20}_{D}$ -42 (c, 0.78 in $CHCl_3$)
(7′R*,8′S*)-form
3,3′,5,5′-Tetra-Me ether
2,7′-Cyclo-4,4′,9′-trihydroxy-3,3′,5,5′-tetramethoxy-8,8′-lign-7-en-9-al.
4,4′,9′-Trihydroxy-3,3′,5,5′-tetramethoxy-2,7′-cyclolign-7-en-9-al.
Ovafolinin E
[1252008-45-9]
$C_{22}H_{24}O_8$ 416.426
Constit. of the wood of *Lyonia ovalifolia* and the rhizomes of *Sinomenium acutum*. Significantly decreases prodn. of NO in LPS-induced BV-2 cells. Amorph. solid or yellowish gum. $[\alpha]^{25}_{D}$ +105.7 (c, 0.16 in MeOH). $[\alpha]_{D}$ +28 (c, 0.1 in MeOH). λ_{max} 290 (log ε 3.85); 354 (log ε 4.01) (MeOH). [neutral] λ_{max} 260 (log ε 3.6); 350 (log ε 3.3) (MeOH).

Wang, B.-G. *et al.*, *Tetrahedron Lett.*, 2002, **43**, 5783-5787 (*Aglacin G*).
Kashima, K. *et al.*, *Chem. Pharm. Bull.*, 2010, **58**, 191-194 (*Ovafolinin E*).
Kuo, L.-M.Y. *et al.*, *J. Nat. Prod.*, 2013, **76**, 580-587 (*Eurya japonica constits, cd*).
Lou, Z.-H. *et al.*, *J. Asian Nat. Prod. Res.*, 2014, **16**, 963-969 (*Vitexdoin F*).
Lee, S.Y. *et al.*, *Phytochem. Lett.*, 2016, **17**, 79-84 (*Sinomenium acutum constit, activity*)

2,7′-Cyclo-3,3′,4,4′,5,5′,9′-heptahydroxy-8,8′-lign-7-en-9-oic acid C182

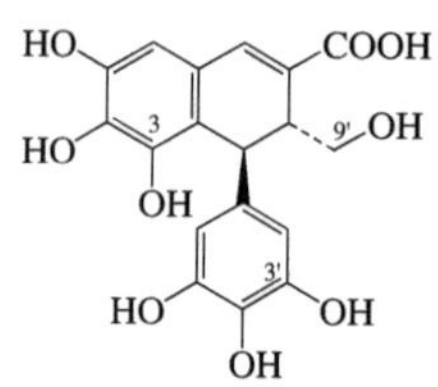

(7′S,8′R)-form

3,3′,4,4′,5,5′,9′-Heptahydroxy-2,7′-cyclolign-7-en-9-oic acid
$C_{18}H_{16}O_9$ 376.317
(7′S,8′R)-form
3,3′,5,5′-Tetra-Me ether, [(4-hydroxyphenyl)ethyl]amide
Majusamide B
$C_{30}H_{33}NO_9$ 551.592
Constit. of *Chelidonium majus*. Amorph. powder. $[\alpha]^{20}_{D}$ -34 (c, 0.1 in MeOH). λ_{max} 205 (log ε 4.6); 225 (log ε 4.6); 266 (log ε 4.6) (MeOH)
(7′R*,8′R*)-form
3,3′,5,5′-Tetra-Me ether, 9N-(4-acetamidobutyl)amide
Lycichinenamide
$C_{28}H_{36}N_2O_9$ 544.601
Constit. of the root bark of *Lycium chinense*. NO prodn. inhibitor. Amorph. powder. $[\alpha]^{25}_{D}$ +0.2 (c, 0.16 in MeOH). λ_{max} 206; 326 (MeOH)
(7′R*,8′S*)-form
3,3′,5,5′-Tetra-Me ether, 9-[(4-hydroxyphenyl)ethyl]amide
Lyciumamide H
$C_{30}H_{33}NO_9$ 551.592
Constit. of root bark of *Lycium chinense*. Amorph. powder. λ_{max} 223 (log ε 3.93); 308 (log ε 3.42) (MeOH)
9′-Carboxylic acid, 3,3′,5-tri-Me ether
2,7′-Cyclo-3′,4,4′-trihydroxy-3,5,5′-trimethoxy-8,8′-lign-7-en-9,9′-dioic acid.
3′,4,4′-Trihydroxy-3,5,5′-trimethoxy-2,7′-cyclolign-7-en-9,9′-dioic acid
9′-Carboxylic acid, 3,3′,5-tri-Me ether, bis[(4-hydroxyphenyl)ethyl] amide
Flavifloramide A
[1440789-26-3]
$C_{37}H_{38}N_2O_{10}$ 670.715
Alkaloid from *Piper flaviflorum*. Powder. $[\alpha]^{25}_{D}$ -13 (c, 0.1 in MeOH). λ_{max} 213 (log ε 4.82); 245 (log ε 4.44); 293 (log ε 4.03); 330 (log ε 4.16) (MeOH)
9′-Carboxylic acid, 3,3′,5,5′-tetra-Me ether, bis[(4-hydroxyphenyl)ethyl] amide (1)
[200135-05-3]
$C_{38}H_{40}N_2O_{10}$ 684.742
Alkaloid from *Commelina communis* and *Porcelia macrocarpa*. Amorph. powder. $[\alpha]^{25}_{D}$ -20 (c, 0.06 in MeOH). λ_{max} 210 (log ε 4.87);

240 (log ε 4.5); 282 (log ε 4.11); 325 (log ε 4.23) (MeOH)

9′-Carboxylic acid, 3,3′,5,5′-tetra-Me ether, bis[(4-hydroxyphenyl)ethyl] amide (2)

Flavifloramide B

[1440789-29-6]

$C_{38}H_{40}N_2O_{10}$ 684.742

Alkaloid from *Piper flaviflorum*. Powder. $[\alpha]_D^{25}$ +19.0 (c, 0.1 in MeOH). Has the enantiomeric config. to Flavifloramide A and appears to be the enantiomer of the *C. communis/ P. macrocarpa* alkaloid. λ_{max} 215 (log ε 4.80); 246 (log ε 4.41); 294 (log ε 4.01); 324 (log ε 4.15) (MeOH)

9′-Carboxylic acid, 3,3′,5,5′-tetra-Me ether, 9-[(4-hydroxyphenyl)ethyl] amide, 9′-[(4-acetylamino)butyl] amide

Lyciumamide I

$C_{36}H_{43}N_3O_{10}$ 677.751

Constit. of root bark of *Lycium chinense*. Amorph. powder. λ_{max} 206 (log ε 4.46); 244 (log ε 4.07); 328 (log ε 3.73) (MeOH)

9′-Carboxylic acid, 3,3′,5,5′-tetra-Me ether, 9′-[(4-hydroxyphenyl)ethyl] amide, 9-[(4-acetylamino)butyl] amide

Lyciumamide J

Constit. of root bark of *Lycium chinense*. Exhibits potent antihyperlipidaemic activity. Amorph. powder. λ_{max} 211 (log ε 4.49); 325 (log ε 3.46) (MeOH)

***(7′RS,8′RS)*-form**

3,3′,5,5′-Tetra-Me ether

2,7′-Cyclo-4,4′,9′-trihydroxy-3,3′,5,5′-tetramethoxy-8,8′-lign-7-en-9-oic acid. 4,4′,9′-Trihydroxy-3,3′,5,5′-tetramethoxy-2,7′-cyclolign-7-en-9-oic acid. ***Thomasic acid***

[21880-64-8]

$C_{22}H_{24}O_9$ 432.425

Constit. of *Ulmus* spp., esp. *Ulmus thomasii*. Flat needles (H_2O). Mp 232 - 234°. Racemic. Stereochem. revised to *trans-* soon after first isolation. λ_{max} 246 (ε 22600); 331 (ε 13500 Very pH dependent) (EtOH)

3,3′,5,5′-Tetra-Me ether, Me ester

[50870-02-5]

Needles (MeOH aq.). Mp 229°

3,3′,5,5′-Tetra-Me ether, tri-Ac

[18014-14-7]

Needles (C_6H_6/petrol). Mp 197.5 - 198°

9′-Carboxylic acid, 3,3′,5,5′-tetra-Me ether

2,7′-Cyclo-4,4′-dihydroxy-3,3′,5,5′-tetramethoxy-8,8′-lign-7-en-9,9′-dioic acid. 4,4′-Dihydroxy-3,3′,5,5′-tetramethoxy-2,7′-cyclolign-7-en-9,9′-dioic acid. ***Thomasidioic acid***

[26350-60-7]

$C_{22}H_{22}O_{10}$ 446.408

Occurs in *Ulmus* spp. Found in cereal grains. Amorph.. Mp 210 - 250°

9′-Carboxylic acid, hexa-Me ether, di-Me ester

[51019-84-2]

Cryst. (MeOH). Mp 147 - 148°.

Seikel, M.K. *et al.*, *Tetrahedron*, 1968, **24**, 1475-1488 (*Thomasic acid*).
Wallis, A.F.A. *Tetrahedron Lett.*, 1968, **9**, 5287-5288 (*stereochem*).
Hostettler, F.D. *et al.*, *Tetrahedron*, 1969, **25**, 2325-2327 (*Ulmus thomasii constits*).
Rowe, J.W. *et al.*, *Phytochemistry*, 1972, **11**, 2513-2517 (*Thomasic acid, Thomasidioic acid, occur*).
Wallis, A.F.A. *Aust. J. Chem.*, 1973, **26**, 1571-1576 (*Thomasidioic acid, synth*).
Ahmed, R. *et al.*, *Tetrahedron*, 1973, **29**, 3753-3759 (*Thomasic acid, synth*).
Chaves, M.H. *et al.*, *Phytochemistry*, 1997, **46**, 879-881 (*bisphenylethylamide*).
Charlton, J.L. *et al.*, *Tetrahedron Lett.*, 1997, **38**, 7311-7312 (*Thomasidioic acid, synth*).
Niwa, T. *et al.*, *Bioorg. Med. Chem. Lett.*, 2002, **12**, 963-965 (*Thomasidioic acid, synth*).
Bunzel, M. *et al.*, *J. Agric. Food Chem.*, 2003, **51**, 1427-1434 (*Thomasidioic acid, biosynth*).
Yang, Q. *et al.*, *Biochem. Syst. Ecol.*, 2007, **35**, 621-623 (*bisphenylethylamide*).
Zoia, L. *et al.*, *Molecules*, 2008, **13**, 129-148 (*Thomasidioic acid, synth*).
Wu, Y. *et al.*, *Helv. Chim. Acta*, 2013, **96**, 951-955 (*Flavifloramides A,B*).
Chen, H. *et al.*, *Fitoterapia*, 2017, **122**, 119-125 (*Lyciumamides H,I,J, activity*).
Huang, H.-C. *et al.*, *Chem. Nat. Compd. (Engl. Transl.)*, 2019, **55**, 1002-1006 (*Lycichinenamide, Lyciumamide H, activity*).
Huang, X.-Y. *et al.*, *Fitoterapia*, 2019, **139**, 104359:1-6 (*Majusamide B, cd*)

2,7′-Cyclo-3′,4,4′,5,5′,7-hexahydroxy-8,8′-ligna-7,7′-dien-9′-oic acid

C183

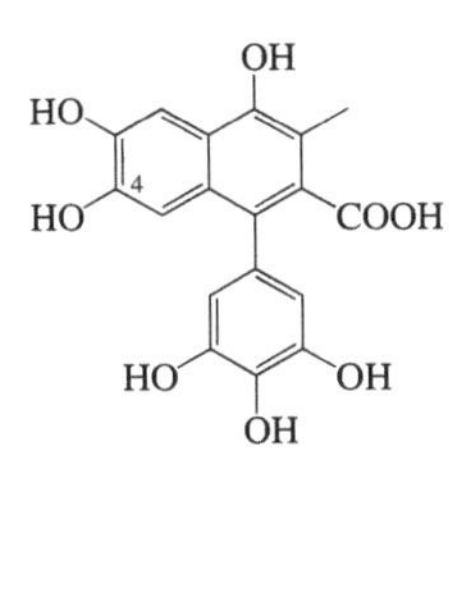

4,6,7-Trihydroxy-3-methyl-1-(3,4,5-trihydroxyphenyl)-2-naphthalenecarboxylic acid. 3′,4,4′,5,5′,7-Hexahydroxy-2,7′-cycloligna-7,7′-dien-9′-oic acid

$C_{18}H_{14}O_8$ 358.302

4,5-Methylene, 3′,4′,5′-tri-Me ether

2,7′-Cyclo-7-hydroxy-3′,4′,5′-trimethoxy-4,5-methylenedioxy-8,8′-ligna-7,7′-dien-9′-oic acid. 7-Hydroxy-3′,4′,5′-trimethoxy-4,5-methylenedioxy-2,7′-cycloligna-7,7′-dien-9′-oic acid.

Junaphtoic acid

[138935-77-0]

$C_{22}H_{20}O_8$ 412.394

Isol. from *Juniperus sabina*. Oil. λ_{max} 212 (ε 24000); 255 (ε 21200); 295 (ε 4500) (EtOH).

San Feliciano, A. *et al.*, *Phytochemistry*, 1991, **30**, 3483-3485 (*Junaphtoic acid*).
San Feliciano, A. *et al.*, *An. Quim.*, 1992, **88**, 256-259 (*Junaphtoic acid*)

2,7′-Cyclo-2′,3,4,4′,5′,7-hexahydroxy-8,8′-ligna-7,7′-dien-9′,9-olide

C184

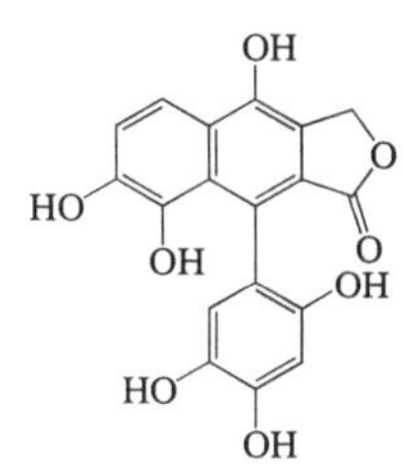

2′,3,4,4′,5′,7-Hexahydroxy-2,7′-cycloligna-7,7′-dien-9′,9-olide

$C_{18}H_{12}O_8$ 356.286

3,4:4′,5′-Bis(methylene) ether

2,7′-Cyclo-2′,7-dihydroxy-3,4:4′,5′-bis(methylenedioxy)-8,8′-ligna-7,7′-dien-9′,9-olide. 2′,7-Dihydroxy-3,4:4′,5′-bis(methylenedioxy)-2,7′-cycloligna-7,7′-dien-9′,9-olide. ***2′-Hydroxyjustirumalin.*** *2′,7-Dihydroxyretrohelioxanthin*

[1176949-73-7]

$C_{20}H_{12}O_8$ 380.308

Constit. of the leaves of *Acanthus mollis*. Pale yellow cryst.. Mp 251 - 253°. $[\alpha]_D^{23}$ +156 (c, 0.01 in MeOH). Exhibits opt. rotation due to hindered rotation. Has R_{axial}-config.. λ_{max} 252 (log ε 4.67); 313 (log ε 3.91) (MeOH)

3,4:4′,5′-Bis(methylene) ether, 2′-O-(6-deoxy-β-D-glucopyranoside)
[1176949-76-0]
$C_{26}H_{22}O_{12}$ 526.45
Constit. of the leaves of *Acanthus mollis*. Pale yellow powder. $[\alpha]_D^{20}$ +164 (c, 0.02 in MeOH). *R*-config.

3,4:4′,5′-Bis(methylene) ether, 7-O-α-L-rhamnopyranoside
[1176949-75-9]
$C_{26}H_{22}O_{12}$ 526.45
Constit. of the leaves of *Acanthus mollis*. Pale yellow powder. $[\alpha]_D^{20}$ +165 (c, 0.02 in MeOH). *R*-config.

3,4:4′,5′-Bis(methylene) ether, 2′-O-(6-deoxy-β-D-glucopyranoside), 7-O-[β-D-apiofuranosyl-(1→2)-[α-L-rhamnopyranosyl-(1→3)] [β-D-xylopyranosyl-(1→6)]-β-D-glucopyranoside]
[1176949-77-1]
$C_{48}H_{58}O_{29}$ 1098.963
Constit. of the leaves of *Acanthus mollis*. Pale yellow powder. $[\alpha]_D^{20}$ +160 (c, 0.02 in MeOH). *R*-config..

Rezanka, T. *et al.*, *Phytochemistry*, 2009, **70**, 1049-1054 (*Acanthus mollis constits*)

2,7′-Cyclo-2′,4,4′,5,5′,6-hexahydroxy-8,8′-ligna-7,7′-dien-9,9′-olide C185

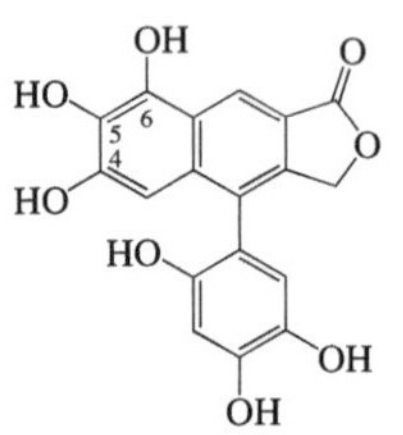

2′,4,4′,5,5′,6-Hexahydroxy-2,7′-cycloligna-7,7′-dien-9,9′-olide
$C_{18}H_{12}O_8$ 356.286

4,5:4′,5′-Bis(methylene), 6-Me ether
2,7′-Cyclo-2′-hydroxy-6-methoxy-4,5:4′,5′-bis(methylenedioxy)-8,8′-ligna-7,7′-dien-9,9′-olide. 2′-Hydroxy-6-methoxy-4,5:4′,5′-bis(methylenedioxy)-2,7′-cycloligna-7,7′-dien-9,9′-olide. ***Prostalidin A***
[73461-17-3]
$C_{21}H_{14}O_8$ 394.335
Isol. from whole plant of *Justicia prostrata*. Shows antidepressant activity. Mp 303 - 305°. λ_{max} 230 (log ε 4.93); 255 (log ε 4.86); 303 (sh); 310 (log ε 3.28); 315 (log ε 3.25); 345 (sh) (MeOH)

4,5:4′,5′-Bis(methylene), 2′,6-di-Me ether
2,7′-Cyclo-2′,6-dimethoxy-4,5:4′,5′-bis(methylenedioxy)-8,8′-ligna-7,7′-dien-9,9′-olide. 2′,6-Dimethoxy-4,5:4′,5′-bis(methylenedioxy)-2,7′-cycloligna-7,7′-dien-9,9′-olide. ***Prostalidin B***
[73428-15-6]
$C_{22}H_{16}O_8$ 408.362
Isol. from *Justicia prostrata*, also by methylation of Prostratin A. Mp 245 - 247°. λ_{max} 235 (log ε 4.86); 258 (log ε 4.59); 293 (sh); 310 (log ε 3.82); 316 (log ε 3.84); 350 (sh) (no solvent reported).

Ghosal, S. *et al.*, *Chem. Ind. (London)*, 1979, 854-855 (*Prostalidins A,B*)

2,7′-Cyclo-2′,4,4′,5,5′,7-hexahydroxy-8,8′-ligna-7,7′-dien-9,9′-olide C186

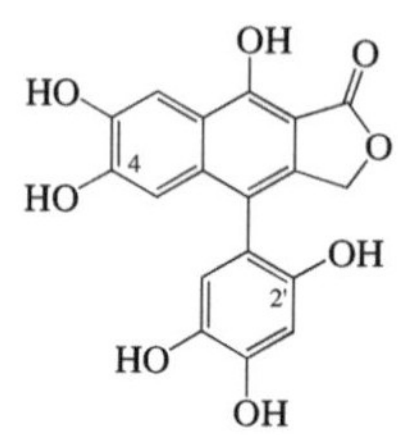

2′,4,4′,5,5′,7-Hexahydroxy-2,7′-cycloligna-7,7′-dien-9,9′-olide
$C_{18}H_{12}O_8$ 356.286

4′,5′-Methylene, 4,5,7-tri-Me ether
2,7′-Cyclo-2′-hydroxy-4,5,7-trimethoxy-4′,5′-methylenedioxy-8,8′-ligna-7,7′-dien-9,9′-olide. 2′-Hydroxy-4,5,7-trimethoxy-4′,5′-methylenedioxy-2,7′-cycloligna-7,7′-dien-9,9′-olide. ***6′-Hydroxyjusticidin C***
[904328-26-3]
$C_{22}H_{18}O_8$ 410.378
Constit. of aerial parts of *Justicia procumbens*. Pale yellow powder

4′,5′-Methylene, 4,5,7-tri-Me ether, 2′-O-β-D-glucopyranoside
Justicidinoside A
[177912-21-9]
$C_{28}H_{28}O_{13}$ 572.519
Constit. of *Justicia procumbens*. Powder. Mp 160° dec.. $[\alpha]_D^{25}$ +2.4 (c, 0.5 in MeOH). λ_{max} 257 (log ε 4.66); 304 (log ε 4.12); 356 (log ε 3.7) (MeOH).

Asano, J. *et al.*, *Phytochemistry*, 1996, **42**, 713-717 (*Justicidinoside A*).
Yang, M. *et al.*, *Magn. Reson. Chem.*, 2006, **44**, 727-730 (*6′-Hydroxyjusticidin C*).
Zhou, P. *et al.*, *Molecules*, 2015, **20**, 7048-7058 (*6′-Hydroxyjusticidin C, isol, purifn*)

2,7′-Cyclo-2′,4,4′,5,5′,7-hexahydroxy-8,8′-ligna-7,7′-dien-9′,9-olide C187

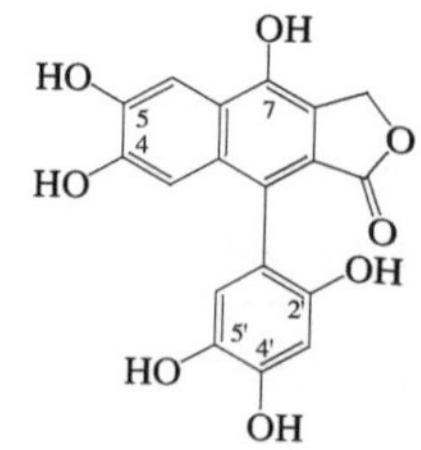

2′,4,4′,5,5′,7-Hexahydroxy-2,7′-cycloligna-7,7′-dien-9′,9-olide
$C_{18}H_{12}O_8$ 356.286

4′,5′-Methylene, 4,5,7-tri-Me ether
2,7′-Cyclo-2′-hydroxy-4,5,7-trimethoxy-4′,5′-methylenedioxy-8,8′-ligna-7,7′-dien-9′,9-olide. 2′-Hydroxy-4,5,7-trimethoxy-4′,5′-methylenedioxy-2,7′-cyclolignan-9′,9-olide. ***6′-Hydroxyjusticidin A****. Justicidin G‡. JR6*
[145971-08-0]
$C_{22}H_{18}O_8$ 410.378
Constit. of *Justicia procumbens*. Cytotoxic. Induces apoptosis in human bladder cancer EJ cells. Pale yellow powder or solid (DMSO aq.). Mp 260°. λ_{max} 203; 256; 307 (solvent not reported)

4′,5′-Methylene, 4,5,7-tri-Me ether, 2′-O-β-D-glucopyranoside
Justicidinoside B
[177912-22-0]
$C_{28}H_{28}O_{13}$ 572.519
Constit. of *Justicia procumbens* var. *leucantha* and *Justicia prostrata*. Powder. Mp 164 - 165° dec.. $[\alpha]_D^{25}$ -8.3 (c, 0.05 in MeOH). λ_{max} 234 (log ε 4.39); 262 (log ε 4.71); 304 (log ε 4.14); 352 (log ε 3.73) (MeOH).

Japan. Pat., 1992, 211 609 (*Justicidin G*).
Asano, J. *et al.*, *Phytochemistry*, 1996, **42**, 713-717 (*Justicidinoside B*).
Yang, M. *et al.*, *Magn. Reson. Chem.*, 2006, **44**, 727-730 (*6′-Hydroxyjusticidin A*).
Pandey, D.P. *et al.*, *Asian J. Chem.*, 2011, **23**, 2125-2127 (*Justicia prostrata constit*).
Xiong, L. *et al.*, *J. Asian Nat. Prod. Res.*, 2012, **14**, 322-326 (*6′-Hydroxyjusticidin A, synth*).
Wang, S.-P. *et al.*, *J. Asian Nat. Prod. Res.*, 2013, **15**, 644-649 (*Justicidinoside B, synth*).
Jin, H. *et al.*, *Fitoterapia*, 2014, **94**, 70-76 (*6′-Hydroxyjusticidin A*)

2,7′-Cyclo-3,3′,4′,5,6,7-hexahydroxy-8,8′-ligna-7,7′-dien-9′,9-olide C188

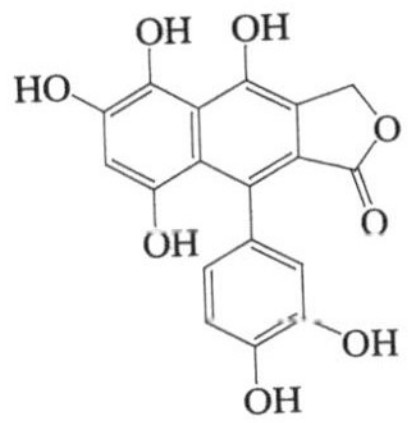

3,3′,4′,5,6,7-Hexahydroxy-2,7′-cyclolignа-7,7′-dien-9′,9-olide
$C_{18}H_{12}O_8$ 356.286
3′,4′-Methylene, 3,5,7-tri-Me ether
2,7′-Cyclo-6-hydroxy-3,5,7-trimethoxy-3′,4′-methylenedioxy-8,8′-ligna-7,7′-dien-9′,9-olide. 6-Hydroxy-3,5,7-trimethoxy-3′,4′-methylenedioxy-2,7′-cyclolignа-7,7′-dien-9′,9-olide. ***Justicinol‡***
[857049-63-9]
$C_{22}H_{18}O_8$ 410.378
Constit. of the leaves of *Justicia patentiflora*. Cytotoxic against human KB and MCF-7 breast carcinoma cell lines. Powder. Not to be confused with Justicinol, in C283. λ_{max} 202 (log ε 4.46); 217 (log ε 4.23); 231 (log ε 4.25); 268 (log ε 4.48); 310 (log ε 3.69); 362 (log ε 3.49) (MeOH).

Susplugas, S. *et al.*, *J. Nat. Prod.*, 2005, **68**, 734-738 (*Justicinol*)

2,7′-Cyclo-3′,4,4′,5,5′,6-hexahydroxy-8,8′-ligna-7,7′-dien-9′,9-olide C189

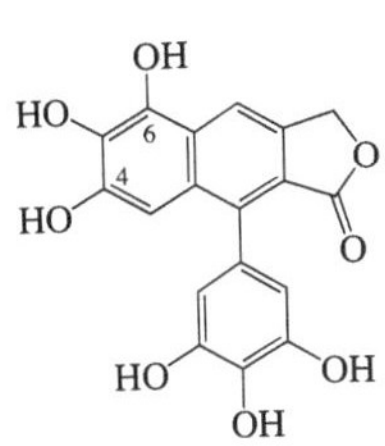

3′,4,4′,5,5′,6-Hexahydroxy-2,7′-cyclolignа-7,7′-dien-9′,9-olide
$C_{18}H_{12}O_8$ 356.286
4,5-Methylene, 3′,4′,5′,6-tetra-Me ether
2,7′-Cyclo-3′,4′,5′,6-tetramethoxy-4,5-methylenedioxy-8,8′-ligna-7,7′-dien-9′,9-olide. 3′,4′,5′,6-Tetramethoxy-4,5-methylenedioxy-2,7′-cyclolignа-7,7′-dien-9′,9-olide. ***Dehydro-β-peltatin methyl ether***
[15656-72-1]
$C_{23}H_{20}O_8$ 424.405
Constit. of *Hyptis verticillata*. Mp 266 - 268°

Novelo, M. *et al.*, *J. Nat. Prod.*, 1993, **56**, 1728-1736 (*Dehydro-β-peltatin methyl ether*)

2,7′-Cyclo-3′,4,4′,5,5′,7-hexahydroxy-8,8′-ligna-7,7′-dien-9′,9-olide C190

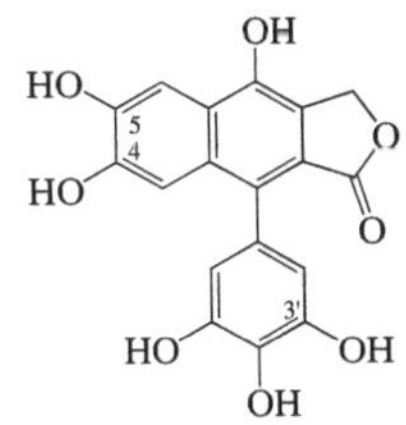

3′,4,4′,5,5′,7-Hexahydroxy-2,7′-cyclolignа-7,7′-dien-9′,9-olide
$C_{18}H_{12}O_8$ 356.286
3′,5-Di-Me ether
2,7′-Cyclo-3′,4,4′,7-tetrahydroxy-5,5′-dimethoxy-8,8′-ligna-7,7′-dien-9′,9-olide. 3′,4,4′,7-Tetrahydroxy-5,5′-dimethoxy-2,7′-cyclolignа-7,7′-dien-9′,9-olide. ***Plicatinaphthol***
[22127-07-7]
$C_{20}H_{16}O_8$ 384.34
Isol. from *Thuja plicata*. Cryst. (EtOH). Mp 260 - 270° dec.. Blue fluor. in soln.. λ_{max} 263 (log ε 4.59); 312 (log ε 3.89); 323 (log ε 3.88); 365 (log ε 3.68) (EtOH)
4,5-Methylene, 3′,5′-di-Me ether
2,7′-Cyclo-4′,7-dihydroxy-3′,5′-dimethoxy-4,5-methylenedioxy-8,8′-ligna-7,7′-dien-9′,9-olide. 4′,7-Dihydroxy-3′,5′-dimethoxy-4,5-methylenedioxy-2,7′-cyclolignа-7,7′-dien-9′,9-olide. ***4′-Demethyldehydropodophyllotoxin***
[117669-31-5]
$C_{21}H_{16}O_8$ 396.351
Constit. of the leaves of *Podophyllum hexandrum*. Amorph. powder. λ_{max} 205 (ε 34800); 229 (ε 22400); 264 (ε 29000); 314 (ε 6780); 357 (ε 400) (EtOH) (Berdy)
4,5-Methylene, 3′,4′,5′-tri-Me ether
2,7′-Cyclo-7-hydroxy-3′,4′,5′-trimethoxy-4,5-methylenedioxy-8,8′-ligna-7,7′-dien-9′,9-olide. 7-Hydroxy-3′,4′,5′-trimethoxy-4,5-methylenedioxy-2,7′-cyclolignа-7,7′-dien-9′,9-olide. ***Dehydropodophyllotoxin***. *7,7′,8,8′-Tetradehydropodophyllotoxin*
[42123-27-3]
$C_{22}H_{18}O_8$ 410.378
Constit. of *Podophyllum peltatum*, *Hernandia ovigera* and *Hyptis verticillata*. Anitiinflammatory, toxic to brine shrimp. Needles (EtOH). Mp 272 - 274°. λ_{max} 213 (ε 49000); 224 (ε 32360); 262 (ε 44670); 322 (ε 10470); 358 (ε 5130) (MeOH) (Berdy). [neutral] λ_{max} 206 (ε 26964); 225 (ε 15850); 263 (ε 25200); 322 (ε 6300); 356 (ε 2500) (EtOH) (Berdy)
4,5-Methylene, 3′,4′,5′-tri-Me ether, 7-O-β-D-glucopyranoside
Dehydropodophyllotoxin 7-glucoside
[1798290-31-9]
$C_{28}H_{28}O_{13}$ 572.519
Constit. of the leaves of *Cleistanthus boivinianus*. Exhibits modest antiproliferative activity against A2780 ovarian cancer cell line. Amorph. powder. $[\alpha]^{25}_{D}$ -14 (c, 0.22 in MeOH). λ_{max} 260 (log ε 4.7); 315 (log ε 4.0); 351 (log ε 3.69) (MeOH).

Kofod, H. *et al.*, *Acta Chem. Scand.*, 1954, **8**, 1296-1297 (*Dehydropodophyllotoxin*).
MacLean, H. *et al.*, *Can. J. Chem.*, 1969, **47**, 457-460 (*Plicatinaphthol*).
Tanoguchi, M. *et al.*, *Chem. Pharm. Bull.*, 1987, **35**, 4162-4168 (*Tetradehydropodophyllotoxin*).
Kuhnt, M. *et al.*, *Phytochemistry*, 1994, **36**, 485-489 (*Dehydropodophyllotoxin*).
Atta-ur-Rahman, *et al.*, *Phytochemistry*, 1995, **40**, 427-431 (*4′-Demethyldehydropodophyllotoxin*).
Liu, Y. *et al.*, *J. Nat. Prod.*, 2015, **78**, 1543-1547 (*Dehydropodophyllotoxin 7-glucoside, activity*)

2,7′-Cyclo-3′,4,4′,5,6,7-hexahydroxy-8,8′-ligna-7,7′-dien-9′,9-olide C191

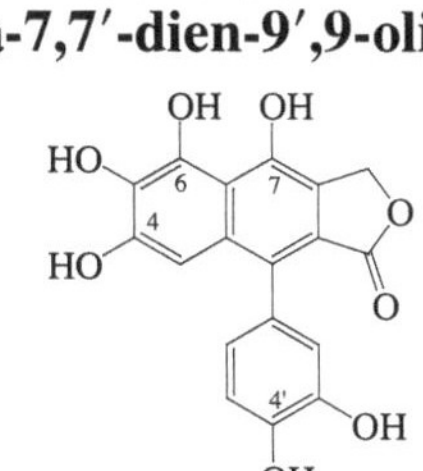

3′,4,4′,5,6,7-Hexahydroxy-2,7′-cyclolignа-7,7′-dien-9′,9-olide
$C_{18}H_{12}O_8$ 356.286
3′,4′-Methylene, 4,5,6-tri-Me ether
2,7′-Cyclo-7-hydroxy-4,5,6-trimethoxy-3′,4′-methylenedioxy-8,8′-ligna-7,7′-dien-9′,9-olide. 7-Hydroxy-4,5,6-trimethoxy-3′,4′-meth-

ylenedioxy-2,7′-cycloligna-7,7′-dien-9′,9-olide. ***6-Methoxydiphyllin***
$C_{22}H_{18}O_8$ 410.378
Constit. of Kenyan propolis. Amorph. yellow powder

3′,4′-Methylene, 4,5,7-tri-Me ether
2,7′-Cyclo-6-hydroxy-4,5,7-trimethoxy-3′,4′-methylenedioxy-8,8′-ligna-7,7′-dien-9′,9-olide. 6-Hydroxy-4,5,7-trimethoxy-3′,4′-methylenedioxy-2,7′-cycloligna-7,7′-dien-9′,9-olide. ***5-Hydroxyjusticidin A***
[880884-23-1]
$C_{22}H_{18}O_8$ 410.378
Constit. of *Mananthes patentiflora*. Yellow powder. Mp 211 - 213°. λ_{max} 203 (log ε 4.66); 268 (log ε 4.53); 310 (log ε 3.69); 370 (log ε 3.58) (MeOH)

3′,4′-Methylene, 4,5,6,7-tetra-Me ether
2,7′-Cyclo-4,5,6,7-tetramethoxy-3′,4′-methylenedioxy-8,8′-ligna-7,7′-dien-9′,9-olide. 4,5,6,7-Tetramethoxy-3′,4′-methylenedioxy-2,7′-cycloligna-7,7′-dien-9′,9-olide. ***5-Methoxyjusticidin A***
[205505-62-0]
$C_{23}H_{20}O_8$ 424.405
Constit. of *Protium unifoliolatum* and *Monsonia angustifolia*. Potential antidementia agent. Flakes (MeOH). Mp 193°. λ_{max} 236 (log ε 5.02); 264 (log ε 5.28); 368 (log ε 4.31) (EtOH).

Siani, A.C. *et al.*, *J. Nat. Prod.*, 1998, **61**, 796-797 (*5-Methoxyjusticidin A*).
Tian, J. *et al.*, *Helv. Chim. Acta*, 2006, **89**, 291-298 (*5-Hydroxyjusticidin A*).
Petrova, A. *et al.*, *Fitoterapia*, 2010, **81**, 509-514 (*6-Methoxydiphyllin*).
Eur. Pat., 2014, 2 712 615 (*Monsonia angustifolia constit, activity*)

2,7′-Cyclo-3′,4,4′,5,6,9-hexahydroxy-8,8′-ligna-7,7′-dien-9′,9-olide C192

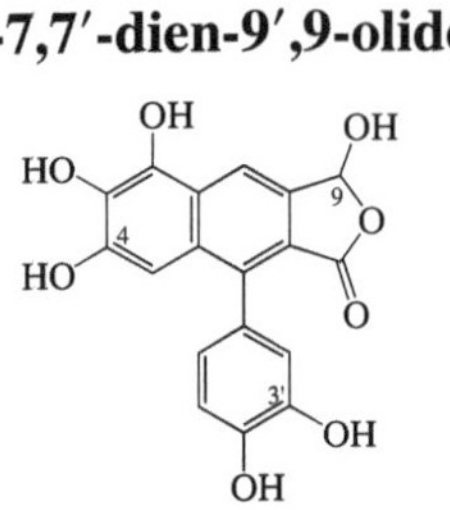

3′,4,4′,5,6,9-Hexahydroxy-2,7′-cycloligna-7,7′-dien-9′,9-olide
$C_{18}H_{12}O_8$ 356.286

(ξ)-form
3′,4′-Methylene, 4,5,6,9-tetra-Me ether
2,7′-Cyclo-4,5,6,9-tetramethoxy-3′,4′-methylenedioxy-8,8′-ligna-7,7′-dien-9′,9-olide. 4,5,6,9-Tetramethoxy-3′,4′-methylenedioxy-2,7′-cycloligna-7,7′-dien-9′,9-olide. ***Phyllamyricin D***
[184094-77-7]
$C_{23}H_{20}O_8$ 424.405
Constit. of *Phyllanthus myrtifolius*. λ_{max} 231 (log ε 4.6); 262 (log ε 4.93); 290 (log ε 4.11); 360 (log ε 3.64) (MeOH).

Lee, S.-S. *et al.*, *J. Nat. Prod.*, 1996, **59**, 1061-1065 (*Phyllamyricin D*).
Wang, C-Y. *et al.*, *Phytochem. Anal.*, 2011, **22**, 352-360 (*Phyllamyricin D, pmr, ms*)

2,7′-Cyclo-3′,4,4′,5,7,9-hexahydroxy-8,8′-ligna-7,7′-dien-9′,9-olide C193

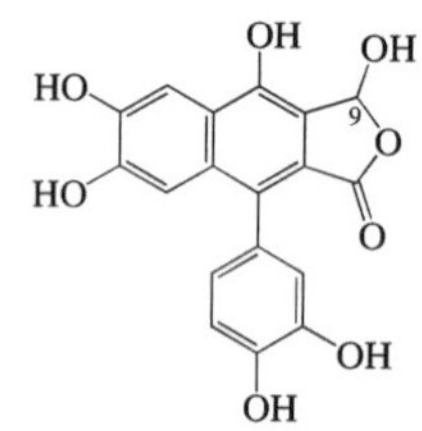

3′,4,4′,5,7,9-Hexahydroxy-2,7′-cycloligna-7,7′-dien-9′,9-olide
$C_{18}H_{12}O_8$ 356.286

(ξ)-form
3′,4′-Methylene, 4,5,9-tri-Me ether
2,7′-Cyclo-7-hydroxy-4,5,9-trimethoxy-3′,4′-methylenedioxy-8,8′-ligna-7,7′-dien-9′,9-olide. 7-Hydroxy-4,5,9-trimethoxy-3′,4′-methylenedioxy-2,7′-cycloligna-7,7′-dien-9′,9-olide. ***Cleistanone‡****. 9-Methoxydiphyllin*
[660394-57-0]
$C_{22}H_{18}O_8$ 410.378
Constit. of the aerial parts of *Cleistanthus collinus*. Cytotoxic against MT_2 cell line. Pale yellow cryst. (MeOH). Mp 217 - 218°. $[\alpha]^{25}_{D}$ +4 (c, 1.5 in MeOH)

3′,4′-Methylene, 4,5,7,9-tetra-Me ether
2,7′-Cyclo-4,5,7,9-tetramethoxy-3′,4′-methylenedioxy-8,8′-ligna-7,7′-dien-9′,9-olide. 4,5,7,9-Tetramethoxy-3′,4′-methylenedioxy-2,7′-cycloligna-7,7′-dien-9′,9-olide. ***Justicidin P***
[95585-91-4]
[86012-93-3]
$C_{23}H_{20}O_8$ 424.405
Constit. of *Justicia extensa*. Shows insecticidal props.. Cryst.. Mp 208 - 210°. λ_{max} 264; 295; 309; 350 (EtOH) (Derep).

Wang, C.-L.J. *et al.*, *J. Org. Chem.*, 1983, **48**, 2555-2557 (*Justicidin P*).
Ramesh, C. *et al.*, *Chem. Pharm. Bull.*, 2003, **51**, 1299-1300 (*Cleistanone*)

7,7′-Cyclo-3,3′,4,4′,5,5′-hexahydroxy-8,8′-lignan-9,9′-dioic acid C194

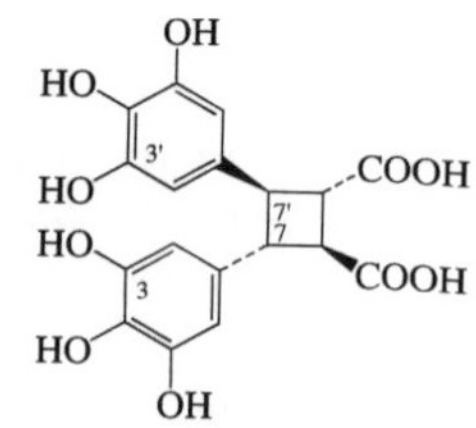

(7*R*,7′*R*,8*S*,8′*S*)-*form*

3,4-Bis(3,4,5-trihydroxyphenyl)-1,2-cyclobutanedicarboxylic acid. 3′,3″,4′,4″,5′,5″-Hexahydroxytruxinic acid
$C_{18}H_{16}O_{10}$ 392.316

(7R,7′R,8S,8′S)-form
3,3′,5,5′-Tetra-Me ether, di-Me ester
(+)-Isatiscycloneolignan A
[2412569-89-0]
Exhibits moderate neuroprotective activity. $[\alpha]^{20}_{D}$ +30.3 (c, 0.1 in MeOH)

(7S,7′S,8R,8′R)-form
3,3′,5,5′-Tetra-Me ether, di-Me ester
(-)-Isatiscycloneolignan A
[2412569-90-3]
Exhibits moderate neuroprotective activity. $[\alpha]^{20}_{D}$ -29 (c, 0.1 in MeOH)

(7RS,7′RS,8SR,8′SR)-form
3,3′,5,5′-Tetra-Me ether
3,4-Bis(4-hydroxy-3,5-dimethoxyphenyl)-1,2-cyclobutanedicarboxylic acid. 7,7′-Cyclo-4,4′-dihydroxy-3,3′,5,5′-tetramethoxy-8,8′-lignan-9,9′-dioic acid

3,3′,5,5′-Tetra-Me ether, di-Me ester
Isatiscycloneolignan A
[2412569-91-4]
$C_{24}H_{28}O_{10}$ 476.478
Constit. of *Isatis indigotica*. Yellow solid. $[\alpha]^{20}_{D}$ -3.6 (c, 0.1 in MeOH). Partial racemate resolved by chiral hplc. λ_{max} 229 (log ε 4.67); 255 (log ε 4.19) (MeOH).

Xi, Y.-F. *et al.*, *Bioorg. Chem.*, 2019, **88**, 102926:1-6 (*Isatiscycloneolignans A, activity*)

2,7′-Cyclo-2′,4,4′,5,5′,8′-hexahydroxy-8,8′-lignan-9′,9-olide C195

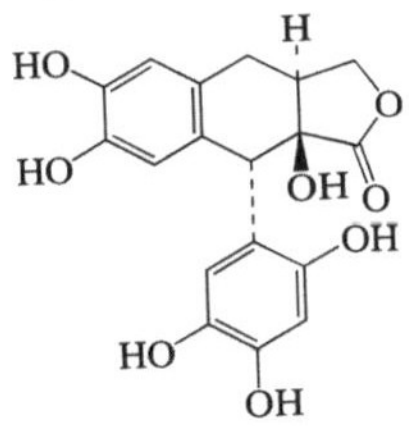

2′,4,4′,5,5′,8′-Hexahydroxy-2,7′-cyclolignan-9′,9-olide
$C_{18}H_{16}O_8$ 360.318
(7′S,8S,8′S)-form
4,5:4′,5′-Bis(methylene), 2′-Me ether
2,7′-Cyclo-8′-hydroxy-2′-methoxy-4,5:4′,5′-bis(methylenedioxy)-8,8′-lignan-9′,9-olide. 8′-Hydroxy-2′-methoxy-4,5:4′,5′-bis(methylenedioxy)-2,7′-cyclolignan-9′,9-olide
[768400-88-0]
$C_{21}H_{18}O_8$ 398.367
Constit. of the aerial parts of *Justicia heterocarpa*. Cryst.. Mp 215°.

Al-Juaid, S.S. *et al.*, *Chem. Pharm. Bull.*, 2004, **52**, 507-509 (*Justicia heterocarpa constit, cryst struct*)

2,7′-Cyclo-3,3′,4,4′,5,5′-hexahydroxy-8,8′-lignan-9′,9-olide C196

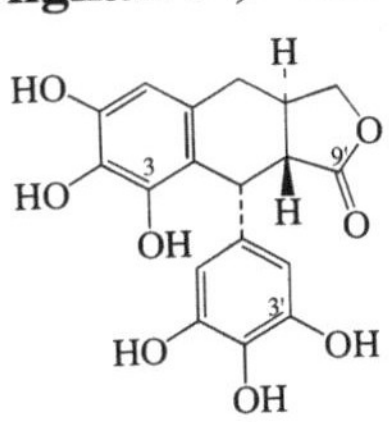

(7′R,8R,8′R)-form

3,3′,4,4′,5,5′-Hexahydroxy-2,7′-cyclolignan-9′,9-olide
$C_{18}H_{16}O_8$ 360.318
(7′R,8R,8′R)-form
4,5-Methylene, 3,3′,4′,5′-tetra-Me ether
2,7′-Cyclo-3,3′,4′,5′-tetramethoxy-4,5-methylenedioxy-8,8′-lignan-9′,9-olide. 3,3′,4′,5′-Tetramethoxy-4,5-methylenedioxy-2,7′-cyclolignan-9′,9-olide. ***Hernandin‡****. 3-Methoxydeoxypodophyllotoxin*
[94425-97-5]
$C_{23}H_{24}O_8$ 428.437
Constit. of seeds of *Hernandia ovigera* and *Hernandia sonora*. Needles (EtOH). Mp 210 - 213°. $[\alpha]_D^{24}$ -70 (c, 0.7 in $CHCl_3$). λ_{max} 204 (ε 58344); 262 (ε 58921); 318 (ε 9743) (EtOH)
(7′R*,8S*,8′S*)-form
9′R-Lactol, 3,3′,4,4′,5,5′-hexa-Me ether*
2,7′-Cyclo-9,9′-epoxy-3,3′,4,4′,5,5′-hexamethoxy-8,8′-lignan-9′-ol. 9,9′-Epoxy-9′-hydroxy-3,3′,4,4′,5,5′-hexamethoxy-2,7′-cyclolignan. ***Aglacin J***
[686744-62-7]
$C_{24}H_{30}O_8$ 446.496
Constit. of the stem bark of *Aglaia cordata*. Obt. as a mixt. with Aglacin I, in C161

Yamaguchi, H. *et al.*, *Chem. Pharm. Bull.*, 1982, **30**, 3212-3218 (*Hernandin, cryst struct*).
Tanoguchi, M. *et al.*, *Chem. Pharm. Bull.*, 1989, **37**, 68-72 (*Hernandin, synth*).
Tanoguchi, M. *et al.*, *Chem. Pharm. Bull.*, 1991, **39**, 1873-1876 (*Hernandin, cd*).
Udino, L. *et al.*, *Planta Med.*, 1999, **65**, 279-281 (*Hernandia sonora constit*).
Wang, B.-G. *et al.*, *J. Nat. Prod.*, 2004, **67**, 682-684 (*Aglacin J*)

2,7′-Cyclo-3,3′,4,4′,5,8′-hexahydroxy-8,8′-lignan-9′,9-olide C197

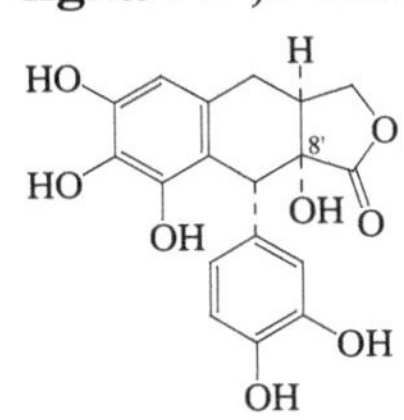

3,3′,4,4′,5,8′-Hexahydroxy-2,7′-cyclolignan-9′,9-olide
$C_{18}H_{16}O_8$ 360.318
(7′R*,8S*,8′R*)-form
3′,4′-Methylene, 3,4,5-tri-Me ether
2,7′-Cyclo-8′-hydroxy-3,4,5-trimethoxy-3′,4′-methylenedioxy-8,8′-lignan-9′,9-olide. 8′-Hydroxy-3,4,5-trimethoxy-3′,4′-methylenedioxy-2,7′-cyclolignan-9′,9-olide
3′,4′-Methylene, 3,4,5-tri-Me ether, 8′-Ac
Erlangerin B
[465535-62-0]
$C_{24}H_{24}O_9$ 456.447
Constit. of the resin of *Commiphora erlangeriana*. Cryst. (MeOH/CH_2Cl_2). Mp 205 - 207°. $[\alpha]_D^{22}$ +131 (c, 0.15 in $CHCl_3$). λ_{max} 205 (log ε 2.67); 283 (log ε 1.62) (EtOH).

Dekebo, A. *et al.*, *J. Nat. Prod.*, 2002, **65**, 1252-1257 (*Erlangerin B, cryst struct*)

2,7′-Cyclo-3′,4,4′,5,5′,6-hexahydroxy-8,8′-lignan-9′,9-olide C198

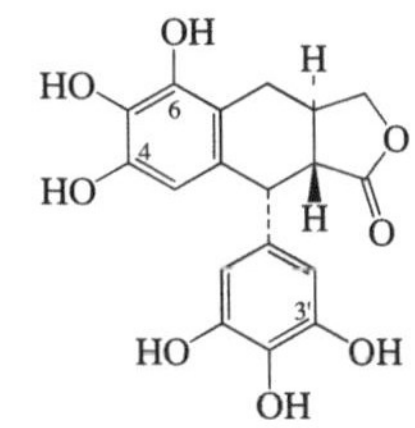

(7′R,8R,8′R)-form

3′,4,4′,5,5′,6-Hexahydroxy-2,7′-cyclolignan-9′,9-olide
$C_{18}H_{16}O_8$ 360.318
(7′R,8R,8′R)-form
4,5-Methylene, 3′,5′-di-Me ether
2,7′-Cyclo-4′,6-dihydroxy-3′,5′-dimethoxy-4,5-methylenedioxy-8,8′-lignan-9′,9-olide. 4′,6-Dihydroxy-3′,5′-dimethoxy-4,5-methylenedioxy-2,7′-cyclolignan-9′,9-olide. ***α-Peltatin****. α-Peltatin A*
[568-53-6]
$C_{21}H_{20}O_8$ 400.383
Isol. from *Podophyllum peltatum*, *Podophyllum versipelle*, *Asarum canadense*, *Diphylleia cymosa*, *Amanoa oblongifolia*, *Linum capitatum*, *Anthriscus sylvestris*, *Eriope latifolia* and others. Antimitotic, antineoplastic agent. Shows antiviral activity. Flat prisms (EtOH). Mp 242 - 246°. $[\alpha]_D^{20}$ -124.5 (c, 0.5 in $CHCl_3$). Log P 0.31 (calc). λ_{max} 272 (EtOH) (Berdy)
LV2476800
4,5-Methylene, 3′,5′-di-Me ether, 6-O-β-D-glucopyranoside
[11024-58-1]
$C_{27}H_{30}O_{13}$ 562.524
From *Podophyllum peltatum*, *Podophyllum versipelle* and *Linum capitatum*. Shows antimitotic activity. Prisms (Me_2CO). Poorly sol. hexane. Mp 168 - 171° dec.. $[\alpha]_D^{20}$-128.9 (c, 0.5 in MeOH). λ_{max} 288 (EtOH) (Berdy)
4,5-Methylene, 3′,5′-di-Me ether, di-Ac
Prisms (EtOH). Mp 233 - 234°. $[\alpha]_D^{20}$-113.2 (c, 0.5 in $CHCl_3$)
4,5-Methylene, 3′,4′,5′-tri-Me ether
2,7′-Cyclo-6-hydroxy-3′,4′,5′-trimethoxy-4,5-methylenedioxy-8,8′-lignan-9′,9-olide. 6-Hydroxy-3′,4′,5′-trime-

thoxy-4,5-methylenedioxy-2,7′-cyclolignan-9′,9-olide. ***β-Peltatin.***
β-Peltatin A. NSC 24819
[518-29-6]
$C_{22}H_{22}O_8$ 414.41
From *Podophyllum peltatum.*
Antimitotic, antineoplastic agent.
Prisms (EtOH). Mp 231 - 238°.
$[\alpha]_D^{20}$-115 (c, 1 in EtOH). Log P 0.24 (calc)
RY0542000

4,5-Methylene, 3′,4′,5′-tri-Me ether, 6-O-β-D-glucopyranoside
[11024-59-2]
$C_{28}H_{32}O_{13}$ 576.551
From *Podophyllum peltatum, Podophyllum versipelle* and *Linum capitatum*. Shows antimitotic activity. Amorph.. Sol. MeOH, $CHCl_3$; poorly sol. H_2O, hexane. Mp 156 - 159°. $[\alpha]_D^{20}$-123 (c, 0.5 in MeOH). λ_{max} 289 (EtOH) (Berdy)

4,5-Methylene, 3′,4′,5′,6-tetra-Me ether
2,7′-Cyclo-3′,4′,5′,6-tetramethoxy-4,5-methylenedioxy-8,8′-lignan-9′,9-olide.
3′,4′,5′,6-Tetramethoxy-4,5-methylenedioxy-2,7′-cyclolignan-9′,9-olide. ***β-Peltatin methyl ether.*** *β-Peltatin A methyl ether*
[23978-65-6]
$C_{23}H_{24}O_8$ 428.437
From *Bursera fagaroides*. Antineoplastic and cytotoxic agent. Prisms (EtOH). Mp 124 - 126°. $[\alpha]_D^{20}$-116 (c, 1.006 in $CHCl_3$). Log P 0.78 (calc)

3′,4′:4,5-Bis(methylene), 5′,6-di-Me ether
2,7′-Cyclo-3′,6-dimethoxy-4,5:4′,5′-bis(methylenedioxy)-8,8′-lignan-9′,9-olide.
Austrobailignan 9
[1941213-45-1]
$C_{22}H_{20}O_8$ 412.394
Constit. of *Austrobaileya scandens*. Exhibits potent antiproliferative activity against human prostate cancer PC3 cells. Amorph. powder. $[\alpha]_D^{25}$-62 (c, 0.05 in MeOH). λ_{max} 208 (log ε 4.2); 285 (log ε 3.1) (MeOH)

(7′R,8R,8′S)-form
4,5-Methylene, 3′,4′,5′,6-tetra-Me ether
β-Peltatin B methyl ether
[38943-35-0]
$C_{23}H_{24}O_8$ 428.437
Constit. of *Juniperus sabina*. Cryst. (MeOH). Mp 184°. $[\alpha]_D^{24}$+3.3 (c, 0.45 in $CHCl_3$).

Hartwell, J.L. *et al., J. Am. Chem. Soc.*, 1950, **72**, 246-253 (*β-Peltatin*).
Stoll, A. *et al., J. Am. Chem. Soc.*, 1955, **77**, 1710-1711 (*glucosides*).
Von Wartburg, A. *et al., Helv. Chim. Acta*, 1957, **40**, 1331-1357 (*Peltatin glucosides*).
Pelter, A. *J. Chem. Soc. C*, 1968, 74-79 (*ms*).
Bianchi, E. *et al., Tetrahedron Lett.*, 1969, 2759-2762 (*β-Peltatin methyl ether*).
Markkanen, T. *et al., Drugs Exp. Clin. Res.*, 1981, **7**, 711-718 (*activity*).
Kamil, W.M. *et al., Phytochemistry*, 1986, **25**, 2089-2092 (*biosynth*).
Ayres, D.C. *et al., Lignans, Chem. Biol. and Clinical Props.*, Cambridge University Press, 1990, (*rev*).
San Feliciano, A. *et al., Phytochemistry*, 1990, **29**, 1335-1338 (*β-Peltatin B methyl ether*).
Broomhead, A.J. *et al., Phytochemistry*, 1990, **29**, 3839-3844 (*Linum capitatum constits*).
Kashima, T. *et al., Chem. Pharm. Bull.*, 1991, **39**, 192-194 (*β-Peltatin methyl ether, synth*).
Eich, E. *et al., J. Med. Chem.*, 1996, **39**, 86-95 (*activity*).
Charlton, J.L. *J. Nat. Prod.*, 1998, **61**, 1447-1451 (*activity, pmr*).
Santos, E.O. *et al., Nat. Prod. Res.*, 2011, **25**, 1450-1453 (*Eriope latifolia constit*).
Hendrawati, O. *et al., Phytochemistry*, 2011, **72**, 2172-2179 (*Anthriscus sylvestris constit*).
Rojas-Sepulveda, A.M. *et al., Molecules*, 2012, **17**, 9506-9519 (*β-Peltatin methyl ether*).
Tran, T.D. *et al., J. Nat. Prod.*, 2016, **79**, 1514-1523 (*Austrobailignan 9, activity*)

2,7′-Cyclo-3′,4,4′,5,5′,7-hexahydroxy-8,8′-lignan-9′,9-olide C199

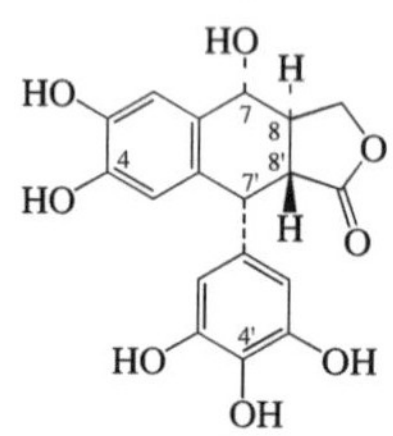

(7R,7′R,8R,8′R)-form

3′,4,4′,5,5′,7-Hexahydroxy-2,7′-cyclolignan-9′,9-olide
$C_{18}H_{16}O_8$ 360.318
Many syntheses of Podophyllotoxin and its analogues reported. Only those since 2000 given here

(7R,7′R,8R,8′R)-form
3′,5′-Di-Me ether
2,7′-Cyclo-4,4′,5,7-tetrahydroxy-3′,5′-dimethoxy-8,8′-lignan-9′,9-olide.
4,4′,5,7-Tetrahydroxy-3′,5′-dimethoxy-2,7′-cyclolignan-9′,9-olide.
4′-Demethyl-4,5-demethylenepodophyllotoxin
[138456-90-3]
$C_{20}H_{20}O_8$ 388.372
Constit. of the roots and rhizomes of *Sinopodophyllum emodi*. Inhibitor of HeLa cells. Cryst. (EtOAc). Mp 208 - 211° dec.. $[\alpha]_D^{25}$-101 (c, 0.25 in Me_2CO)

3′,4′,5′-Tri-Me ether
2,7′-Cyclo-4,5,7-trihydroxy-3′,4′,5′-trimethoxy-8,8′-lignan-9′,9-olide.
4,5,7-Trihydroxy-3′,4′,5′-trimethoxy-2,7′-cyclolignan-9′,9-olide.
4,5-Demethylenepodophyllotoxin
[1174-97-6]
$C_{21}H_{22}O_8$ 402.399
Constit. of the roots and rhizomes of *Sinopodophyllum emodi*. Cytotoxic against human MEL-28, HT-29 and A-549 and mouse P-388 cancer cells. Weak insecticide. Cryst. (EtOAc or EtOH aq.). Mp 226 - 228°. $[\alpha]_D^{25}$-120 (c, 0.5 in EtOH). λ_{max} 212 (ε 29100); 285 (ε 2900) (EtOH)

3′,4,4′,5,5′-Penta-Me ether
2,7′-Cyclo-7-hydroxy-3′,4,4′,5,5′-pentamethoxy-8,8′-lignan-9′,9-olide.
7-Hydroxy-3′,4,4′,5,5′-pentamethoxy-2,7′-cyclolignan-9′,9-olide.
Sikkimotoxin
[18651-67-7]
$C_{23}H_{26}O_8$ 430.453
Constit. of *Podophyllum sikkimensis*. Shows anti-HIV activity and moderate insecticidal activity. Cryst.. Mp 125 - 130°. λ_{max} 229 (ε 18800); 281 (ε 3810) (MeOH)

4,5-Methylene, 3′-Me ether
2,7′-Cyclo-3′,4′,7-trihydroxy-5′-methoxy-4,5-methylenedioxy-8,8′-lignan-9′,9-olide.
3′,4′,7-Trihydroxy-5′-methoxy-4,5-methylenedioxy-2,7′-cyclolignan-9′,9-olide. ***3′,4′-Di-O-demethylpodophyllotoxin***
[105662-24-6]
$C_{20}H_{18}O_8$ 386.356
Constit. of *Dysosma versipellis*. Mp 219 - 221°

4,5-Methylene, 3′-Me ether, 7-O-β-D-glucopyranoside
3′,4′-Di-O-demethylpodophyllotoxin 7-glucoside
[1269169-18-7]
$C_{26}H_{28}O_{13}$ 548.497

Constit. of the roots and rhizomes of *Sinopodophyllum emodi*. Cytotoxic. Powder. λ_{max} 207; 282 (no solvent reported)

4,5-Methylene, 3′,4′-di-Me ether
2,7′-Cyclo-3′,7-dihydroxy-4′,5′-dimethoxy-4,5-methylenedioxy-8,8′-lignan-9′,9-olide. 3′,7-Dihydroxy-4′,5′-dimethoxy-4,5-methylenedioxy-2,7′-cyclolignan-9′,9-olide. ***3′-Demethylpodophyllotoxin***
[54747-66-9]
$C_{21}H_{20}O_8$ 400.383
Isol. from whole plant of *Linum album*. Amorph. pale yellow solid. Mp 85 - 105°

4,5-Methylene, 3′,5′-di-Me ether
2,7′-Cyclo-4′,7-dihydroxy-3′,5′-dimethoxy-4,5-methylenedioxy-8,8′-lignan-9′,9-olide. 4′,7-Dihydroxy-3′,5′-dimethoxy-4,5-methylenedioxy-2,7′-cyclolignan-9′,9-olide. ***4′-Demethylpodophyllotoxin***
[40505-27-9]
$C_{21}H_{20}O_8$ 400.383
Isol. from *Podophyllum emodi*, *Podophyllum polygama* and other *Podophyllum* spp., also root of *Dysosma majorense*. Cryst. (EtOH). Mp 250 - 251.6°. $[\alpha]_D^{20}$ -130 ($CHCl_3$). λ_{max} 206 (ε 29500); 240 (ε 6310); 289 (ε 2000); 290 (MeOH) (Berdy)
LV2476100

4,5-Methylene, 3′,5′-di-Me ether, 7-O-β-D-glucopyranoside
[40505-30-4]
$C_{27}H_{30}O_{13}$ 562.524
Isol. from *Podophyllum emodi*. Powder. Mp 165 - 170°. $[\alpha]_D^{20}$-75 (H_2O)

4,5-Methylene, 3′,5′-di-Me ether, 4′-Ac
4′-Acetyl-4′-demethylpodophyllotoxin
[1314092-86-8]
$C_{23}H_{22}O_9$ 442.42
Constit. of the roots and rhizomes of *Sinopodophyllum emodi*. Cytotoxic against KB and HeLa cells. Needles ($CHCl_3$). Mp 231 - 233°. $[\alpha]_D$ -98.1 (c, 0.8 in $CHCl_3$). λ_{max} 208 (log ε 1.8); 292 (log ε 0.1) (MeOH)

4,5-Methylene, 3′,4′,5′-tri-Me ether
2,7′-Cyclo-7-hydroxy-3′,4′,5′-trimethoxy-4,5-methylenedioxy-8,8′-lignan-9′,9-olide. 7-Hydroxy-3′,4′,5′-trimethoxy-4,5-methylenedioxy-2,7′-cyclolignan-9′,9-olide. ***Podophyllotoxin***. *Podofilox. Condyline. Condylox. Warticon. NSC 24818*
[518-28-5]
[77519-37-0]
(7*R**,7′*R**,8*R**,8′*R**-form)
$C_{22}H_{22}O_8$ 414.41
Constit. of *Podophyllum emodi*, *Podophyllum hexandrum*, *Podophyllum peltatum* and other *Podophyllum* spp. Also from *Diphylleia grayi*, *Callitris drummondii*, *Juniperus* spp., *Linum flavum* and others. Anticancer agent. Cell division inhibitor. Inhibitor of human DNA topoisomerase II. Antimitotic agent. Used topically for treatment of external anogenital warts and plantar warts. Also shows ichthyotoxic, insecticidal and phytotoxic props.. Cryst. (Me_2CO aq. or MeOH). Sol. H_2O. Mp 117 - 118°. $[\alpha]_D^{14}$-101.3 (EtOH). Log P -0.84 (calc). 4 Crystal modifications reported. Mps up to 188-189° have been reported for carefully dried samples. λ_{max} 292 (MeOH) (Berdy). [neutral] λ_{max} 287 (ε 4300); 291 (ε 4350) (EtOH) (Berdy)
LV2500000. Severe eye, skin and respiratory tract irritant. Severe adverse effects reported if swallowed incl. gastrointestinal, CNS, hepatotoxic, nephrotoxic, haemopoietic effects and development of peripheral and autonomic neuropathies. LD_{50} (rat, skn) 500 mg/kg. LD_{50} (mus, orl) 100 mg/kg

4,5-Methylene, 3′,4′,5′-tri-Me ether, 7-O-(6-O-acetyl-α-D-glucopyranoside)
[917085-35-9]
$C_{30}H_{34}O_{14}$ 618.588
Constit. of *Podophyllum hexandrum*

4,5-Methylene, 3′,4′,5′-tri-Me ether, 7-O-β-D-glucopyranoside
[16481-54-2]
$C_{28}H_{32}O_{13}$ 576.551
Isol. from *Podophyllum emodi*, *Libocedrus chevalieri*, *Sinopodophyllum emodi* and from plants and cell cultures of *Linum flavum*. Antineoplastic agent. Amorph. solid. Mp 152 - 153°. $[\alpha]_D^{20}$-65 (c, 0.5 in H_2O). $[\alpha]_D$ -75 (c, 0.6 in MeOH). $[\alpha]_D^{20}$-117 (c, 0.67 in Py). Log P -3.13 (uncertain value) (calc). λ_{max} 287; 291 (EtOH)

4,5-Methylene, 3′,4′,5′-tri-Me ether, 7-O-(6-O-acetyl-β-D-glucopyranoside)
Sinolignan A
[917085-34-8]
$C_{30}H_{34}O_{14}$ 618.588
Constit. of *Podophyllum hexandrum* and *Sinopodophyllum emodi*. Amorph. powder (MeOH). $[\alpha]_D^{25}$ -79.1 (c, 0.5 in MeOH). λ_{max} 208 (log ε 1.7); 288 (log ε 0.2) (MeOH)

4,5-Methylene, 3′,4′,5′-tri-Me ether, 7-Ac
[1180-34-3]
Cryst. (MeOH). Mp 209.5 - 210.5°. $[\alpha]_D^{20}$-143 (c, 1 in $CHCl_3$)

4,5-Methylene, 3′,4′,5′-tri-Me ether, 7-angeloyl
Angeloylpodophyllotoxin
[229309-98-2]
$C_{27}H_{28}O_9$ 496.512
Constit. of the roots of *Anthriscus sylvestris*

7-Ketone, 4,5-methylene, 3′,5′-di-Me ether
2,7′-Cyclo-4′-hydroxy-3′,5′-dimethoxy-4,5-methylenedioxy-7-oxo-8,8′-lignan-9′,9-olide. 4′-Hydroxy-3′,5′-dimethoxy-4,5-methylenedioxy-7-oxo-2,7′-cyclolignan-9′,9-olide. ***4′-O-Demethylpodophyllotoxone***
[93780-84-8]
$C_{21}H_{18}O_8$ 398.367
Constit. of *Podophyllum hexandrum*, *Podophyllum emodi*, *Podophyllum peltatum*, *Podophyllum pleianthum* and *Dysosma majorense*. Antitumour agent. Cryst. (EtOH). Sol. MeOH, EtOAc; poorly sol. H_2O. Mp 173 - 185°. λ_{max} 273 (EtOH)

7-Ketone, 4,5-methylene, 3′,4′,5′-tri-Me ether
2,7′-Cyclo-3′,4′,5′-trimethoxy-4,5-methylenedioxy-7-oxo-8,8′-lignan-9′,9-olide. 3′,4′,5′-Trimethoxy-4,5-methylenedioxy-7-oxo-2,7′-cyclolignan-9′,9-olide. ***Podophyllotoxone***

[477-49-6]
[86594-36-7]
Constit. of *Diphylleia* spp., *Dysosma* spp., *Anthriscus sylvestris* and *Podophyllum* spp. such as *Podophyllum hexandrum* and *Podophyllum peltatum*. Antineoplastic agent. Needles (EtOH). Mp 191 - 192°. $[\alpha]^{25}_{D}$-128 (c, 1.4 in $CHCl_3$). λ_{max} 235 (ε 15488); 276 (ε 7413); 317 (ε 6310) (EtOH)

(7R,7′R,8R,8′S)-form

4,5-Methylene, 3′,5′-di-Me ether,
4′-Demethylpicropodophyllotoxin.
4′-Demethylpicropodophyllin
[112066-16-7]
$C_{21}H_{20}O_8$ 400.383
Constit. of *Sinopodophyllum emodi*

4,5-Methylene, 3′,5′-di-Me ether, 7-O-β-D-glucopyranoside
4′-Demethylpicropodophyllotoxin 7-glucoside
[7598-48-3]
$C_{27}H_{30}O_{13}$ 562.524
Constit. of *Sinopodophyllum emodi*. Powder. $[\alpha]^{29}_{D}$-5.2 (c, 0.25 in MeOH). λ_{max} 208; 287 (MeOH)

4,5-Methylene, 3′,5′-di-Me ether, 7-O-[β-D-glucopyranosyl-(1→6)-β-D-glucopyranoside]
Sinolignan B
[1314092-87-9]
$C_{33}H_{40}O_{18}$ 724.665
Constit. of the roots and rhizomes of *Sinopodophyllum emodi*. Amorph. powder (MeOH). $[\alpha]^{25}_{D}$ -40 (c, 0.2 in MeOH). λ_{max} 208 (log ε 2.1); 284 (log ε 0.3) (MeOH)

4,5-Methylene, 3′,4′,5′-tri-Me ether
Picropodophyllin.
Picropodophyllotoxin
[477-47-4]
$C_{22}H_{22}O_8$ 414.41
Isol. from *Podophyllum emodi*, *Diphylleia grayi*, *Dysosma majorense* and others. ERK activator, inducer of ERK phosphorylation. Needles (MeOH or C_6H_6). Mp 226 - 227°. $[\alpha]_D$ +5.3 ($CHCl_3$). λ_{max} 205 (ε 60200); 240 (ε 9590); 290 (ε 5130) (MeOH) (Berdy)
TJ8750000

4,5-Methylene, 3′,4′,5′-tri-Me ether, 7-O-β-D-glucopyranoside
[1400-92-6]
$C_{28}H_{32}O_{13}$ 576.551
Isol. from *Podophyllum emodi* and *Sinopodophyllum emodi*. Needles (H_2O). Mp 237 - 238°. $[\alpha]^{26}_{D}$-91 (c, 0.08 in EtOH)

4,5-Methylene, 3′,4′,5′-tri-Me ether, 7-O-[β-D-glucopyranosyl-(1→6)-β-D-glucopyranoside]
Picropodophyllin gentiobioside
[619296-83-2]
$C_{34}H_{42}O_{18}$ 738.692
Constit. of the roots and rhizomes of *Sinopodophyllum emodi*. Needles. $[\alpha]^{29}_{D}$-46 (c, 0.6 in MeOH). λ_{max} 208; 287 (MeOH)

4,5-Methylene, 3′,4′,5′-tri-Me ether, 7-Ac
Acetylpicropodophyllin
[38491-90-6]
$C_{24}H_{24}O_9$ 456.447
Constit. of the twigs of *Hernandia ovigera*. Amorph. powder. Mp 212 - 214°. $[\alpha]_D$ +44.9 (c, 0.1 in MeCN). λ_{max} 213 (log ε 4.1) (MeCN)

4,5-Methylene, 7-Et, 3′,4′,5′-tri-Me ether
2,7′-Cyclo-7-ethoxy-3′,4′,5′-trimethoxy-4,5-methylenedioxy-8,8′-lignan-9′,9-olide. 7-Ethoxy-3′,4′,5′-trimethoxy-4,5-methylenedioxy-2,7′-cyclolignan-9′,9-olide.
O-Ethylpicropodophyllin
[106709-55-1]
$C_{24}H_{26}O_8$ 442.464
Constit. of *Diphylleia sinensis*

7-Ketone, 4,5-methylene, 3′,4′,5′-tri-Me ether
Picropodophyllone.
Picropodophyllotoxone
[477-48-5]
[64937-82-2] ((±)-form)
Constit. of *Diphylleia* spp., *Dysosma* spp., *Juniperus sabina* and *Podophyllum* spp. such as *Podophyllum hexandrum* and *Podophyllum versipelle*. Antifungal and insecticidal agent. Cryst.. Mp 153 - 154°. $[\alpha]^{25}_{D}$-164 (c, 0.34 in $CHCl_3$) (-142). λ_{max} 206 (ε 57500); 240 (ε 29450); 280 (ε 9400); 324 (ε 8320) (EtOH) (Berdy)

(7R,7′R,8S,8′R)-form

4,5-Methylene, 3′,4′,5′-tri-Me ether
Epiisopicropodophyllin
[55568-80-4]
Semisynthetic. The name Epiisopicropodophyllin is inconsistent owing to misassignment of configs. in 1963

7-Ketone, 4,5-methylene, 3′,5′-di-Me ether
4′-O-Demethylisopicropodophyllone
[93780-85-9]
$C_{21}H_{18}O_8$ 398.367
Constit. of *Haplophyllum hexandrum*, *Haplophyllum peltatum*, *Haplophyllum pleianthum*, *Podophyllum pleianthum*, *Podophyllum hexandrum* and *Podophyllum peltatum*. Antitumour agent

7-Ketone, 4,5-methylene, 3′,4′,5′-tri-Me ether
Isopicropodophyllone
[55515-07-6]
Constit. of *Diphylleia* spp., *Dysosma* and *Podophyllum* spp.. Cryst. (EtOH). Mp 171 - 174°. $[\alpha]_D$ -288 (c, 1.1 in $CHCl_3$). λ_{max} 237 (ε 25000); 275 (ε 10000); 323 (ε 7900) (MeOH) (Berdy). [neutral] λ_{max} 235 (ε 23100); 259 (ε 10600); 318 (ε 6500) (EtOH) (Berdy)

(7R,7′R,8S,8′S)-form

4,5-Methylene, 3′,4′,5′-tri-Me ether
Epiisopodophyllotoxin
[55568-79-1]
$C_{22}H_{22}O_8$ 414.41
Semisynthetic. The name Epiisopodophyllotoxin is inconsistent owing to misassignment of configs. in 1963

(7S,7′R,8R,8′R)-form

4,5-Methylene, 3′,5′-di-Me ether
4′-Demethylepipodophyllotoxin
[6559-91-7]
$C_{21}H_{20}O_8$ 400.383
Antimitotic agent. Cryst. (MeOH). Mp 228 - 230°. $[\alpha]^{20}_{D}$-69.8 ($CHCl_3$)
LV2476200

4,5-Methylene, 3′,5′-di-Me ether, 7-O-β-D-glucopyranoside
4′-Demethylepipodophyllotoxin 7-glucoside
[23363-35-1]
$C_{27}H_{30}O_{13}$ 562.524
Constit. of *Sinopodophyllum emodi*. Needles. $[\alpha]^{15}_{D}$-29.3 (c, 0.63 in MeOH)

4,5-Methylene, 3′,4′,5′-tri-Me ether
Epipodophyllotoxin
[4375-07-9]
Semisynthetic. Topoisomerase II inhibitor. Needles (EtOH aq.). Mp

159.4 - 161.2°. $[\alpha]^{20}_{D}$ -75 (c, 1 in $CHCl_3$)
LV2520000

(7S,7′R,8R,8′S)-form

4,5-Methylene, 3′,4′,5′-tri-Me ether
Epipicropodophyllin
[4375-06-8]
$C_{22}H_{22}O_8$ 414.41
Semisynthetic

(7S,7′R,8S,8′R)-form

4,5-Methylene, 3′,4′,5′-tri-Me ether
Isopicropodophyllin
[60660-50-6]
Synthetic, obt. for the first time in 1989 as racemate

(7S,7′R,8S,8′S)-form

4,5-Methylene, 3′,5′-di-Me ether
4′-Demethylisopodophyllotoxin
[96554-39-1]
$C_{21}H_{20}O_8$ 400.383
Isol. from roots of *Dysosma versipellis* var. *tomentosa*. Needles. Mp 256 - 258°. $[\alpha]^{22}_{D}$ -6.6 (c, 0.0073 in Py)

4,5-Methylene, 3′,4′,5′-tri-Me ether
Isopodophyllotoxin
[26540-82-9]
Semisynthetic

4,5-Methylene, 3′,4′,5′-tri-Me ether, 7-O-[β-D-glucopyranosyl-(1→6)-β-D-glucopyranoside]
Isopodophyllotoxin 7-gentiobioside
[354527-09-6]
$C_{34}H_{42}O_{18}$ 738.692
Constit. of *Sinopodophyllum emodi*. Needles. $[\alpha]^{29}_{D}$ -45.7 (c, 0.6 in MeOH). λ_{max} 208; 287 (MeOH)

7-Ketone, 4,5-methylene, 3′,4′,5′-tri-Me ether
Isopodophyllotoxone
[60660-51-7]
[64550-42-1]
Mp 206 - 208°. $[\alpha]_{D}$ -63.8 (c, 0.6 in $CHCl_3$).

Aldrich Library of FT-IR Spectra, 1st edn., 1985, **2**, 320A (*ir*).
Aldrich Library of ^{13}C and ^{1}H FT-NMR Spectra, 1992, **2**, 1304A (*nmr*).
Hartwell, J.L. *et al.*, *J. Am. Chem. Soc.*, 1951, **73**, 2909-2916 (*Podophyllotoxin, Picropodophyllin, struct*).
Chatterjee, R. *et al.*, *J. Am. Pharm. Assoc.*, 1952, **41**, 415-419 (*Podophyllum sikkimensis constit*).
Nadkarni, M.V. *et al.*, *J. Am. Chem. Soc.*, 1953, **75**, 1308-1312 (*4′-Demethylpodophyllotoxin, Picropodophyllin glucoside*).
Stoll, A. *et al.*, *J. Am. Chem. Soc.*, 1954, **76**, 3103-3104 (*Podophyllotoxin, Podophyllotoxin glucoside*).
Schrecker, A.W. *et al.*, *J. Org. Chem.*, 1956, **21**, 288-291 (*Podophyllotoxin, polymorphism*).
Schrecker, A.W. *et al.*, *J. Org. Chem.*, 1956, **21**, 381-382 (*abs config*).
Gensler, W.J. *et al.*, *J. Am. Chem. Soc.*, 1960, **82**, 6074-6081 (*Podophyllotoxone, Picropodophyllone*).
Schreier, E. *et al.*, *Helv. Chim. Acta*, 1964, **47**, 1529-1555 (*Sikkimotoxin*).
Duffield, A.M. *J. Heterocycl. Chem.*, 1967, **4**, 16-22 (*ms*).
Ayres, D.C. *et al.*, *J. Chem. Soc., Perkin Trans. 1*, 1972, 1343-1350 (*pmr*).
Petcher, T.J. *et al.*, *J. Chem. Soc., Perkin Trans. 2*, 1973, 288-292 (*cryst struct*).
Weiss, S.G. *et al.*, *J. Pharm. Sci.*, 1975, **64**, 95-98 (*3′-Demethylpodophyllotoxin*).
Chang, F.C. *et al.*, *Phytochemistry*, 1975, **14**, 1440 (*Podophyllum pleianthum constits, uv, pmr*).
Karrer, W. *Konstitution und Vorkommen der Organischen Pflanzenstoffe, 2nd edn., Birkhäuser Verlag*, 1976-1985, no.1182 (*Podophyllotoxin, bibl*).
Nambi, A.V. *et al.*, *J. Org. Chem.*, 1977, **42**, 246-249 (*Isopicropodophyllin, Isopodophyllotoxin, Isopodophyllotoxone, synth*).
Hulbert, P.B. *et al.*, *J. Chem. Res., Miniprint*, 1981, 401-449 (*Sikkimotoxin, uv*).
Ayres, D.C. *et al.*, *J. Chem. Soc., Perkin Trans. 1*, 1981, 2134-2136 (*Podophyllotoxin, biosynth*).
Dewick, P.M. *et al.*, *Phytochemistry*, 1981, **20**, 2277-2280 (*Podophyllotoxone*).
Jackson, D.E. *et al.*, *Phytochemistry*, 1984, **23**, 1029-1036 (*4,5-Demethylenepodophyllotoxin, struct*).
Jackson, D.E. *et al.*, *Phytochemistry*, 1984, **23**, 1037-1042 (*Podophyllotoxin, biosynth*).
Jackson, D.E. *et al.*, *Phytochemistry*, 1984, **23**, 1147-1152 (*Podophyllotoxone, Isopicropodophyllone, 4′-O-Demethylpodophyllotoxone, 4′-O-Demethylisopicropodophyllone*).
Inamori, Y. *et al.*, *Chem. Pharm. Bull.*, 1986, **34**, 3928-3932 (*activity, rev*).
Kamil, W.M. *et al.*, *Phytochemistry*, 1986, **25**, 2093-2102 (*Podophyllotoxin, biosynth*).
Takano, S. *et al.*, *Heterocycles*, 1987, **25**, 69-73 (*Sikkimotoxin, isol, synth*).
Thurston, L.S. *et al.*, *J. Med. Chem.*, 1989, **32**, 604-608 (*Podophyllotoxin, pharmacol*).
Andersen, K.V. *et al.*, *J. Chem. Soc., Perkin Trans. 2*, 1990, 1871-1879 (*Podophyllotoxin, cryst struct*).
Ayres, D.C. *et al.*, *Lignans, Chem. Biol. and Clinical Props.*, Cambridge, 1990, (*rev*).
Broomhead, A.J. *et al.*, *Phytochemistry*, 1990, **29**, 3839-3844 (*Podophyllotoxin 7-glucoside*).
Yu, P.Z. *et al.*, *J. Nat. Prod.*, 1991, **54**, 1422-1424 (*4′-Demethylisopodophyllotoxin*).
Eich, E. *et al.*, *Planta Med.*, suppl., 1991, **57**, A7 (*Epipodophyllotoxin, activity*).
Sicheri, F.V. *et al.*, *Acta Crystallogr., Sect. C*, 1992, **48**, 747-749 (*4′-Demethylepipodophyllotoxin, cryst struct*).
Wang, Z.-Q. *et al.*, *J. Med. Chem.*, 1992, **35**, 871-877 (*4′-Demethyl-4,5-demethylenepodophyllotoxin, synth*).
Ward, R.S. *Synthesis*, 1992, 719-730 (*rev, synth, bibl*).
Martindale, The Extra Pharmacopoeia, 30th edn., Pharmaceutical Press, 1993, 765.
Sackett, D.L. *Pharmacol. Ther.*, 1993, **59**, 163-228 (*rev, activity*).
Ma, C. *et al.*, *Yaoxue Xuebao*, 1993, **28**, 690-694 (*O-Ethylpicropodophyllin*).
Lim, Y.-H. *et al.*, *Arch. Pharmacal Res.*, 1999, **22**, 208-212 (*Angeloylpodophyllotoxin*).
Berkowitz, D.B. *et al.*, *J. Org. Chem.*, 2000, **65**, 847-860 (*Podophyllotoxin, synth, bibl*).
Canel, C. *et al.*, *Phytochemistry*, 2000, **54**, 115-120 (*rev*).
Zhao, C. *et al.*, *Chem. Pharm. Bull.*, 2001, **49**, 773-775 (*Isopodophyllotoxin 7-gentiobioside, 4′-Demethylpicropodophyllotoxin 7-glucoside*).
Zhao, C. *et al.*, *Chem. Pharm. Bull.*, 2001, **49**, 1474-1476 (*Sinopodophyllum emodi glycosides*).
Gordaliza, M. *et al.*, *Farmaco*, 2001, **56**, 297-304 (*Podophyllotoxone, activity*).
Rajnikant, G. *et al.*, *Indian J. Phys., A*, 2002, **76**, 171-173 (*Podophyllotoxone, cryst struct*).
Gu, J.-Q. *et al.*, *J. Nat. Prod.*, 2002, **65**, 1065-1068 (*Acetylpicropodophyllin*).
Engelhardt, U. *et al.*, *Angew. Chem., Int. Ed.*, 2003, **42**, 2487-2489 (*Epipodophyllotoxin, synth*).
Zhao, C. *et al.*, *Chem. Pharm. Bull.*, 2003, **51**, 255-261 (*Sinopodophyllum emodi glycosides, cryst struct*).
Castro, A. *et al.*, *Eur. J. Med. Chem.*, 2003, **38**, 65-74 (*synth, activity*).
Zhu, X-K. *et al.*, *Bioorg. Med. Chem.*, 2004, **12**, 4267-4273 (*Sikkimotoxin, activity*).
Puri, S.C. *et al.*, *J. Chromatogr. Sci.*, 2006, **44**, 239-243 (*7-(6-acetylglucopyranosides)*).
López-Pèrez, J.L. *et al.*, *Tetrahedron*, 2006, **62**, 2370-2379 (*Podophyllotoxone, Isopicropodophyllone, struct*).
Lei, M. *et al.*, *Tetrahedron*, 2006, **62**, 8928-8932 (*Isopodophyllotoxone, props*).
Lin, Y.-Q. *et al.*, *Curr. Bioact. Compd.*, 2007, **3**, 37-66 (*rev*).
Stadler, D. *et al.*, *Angew. Chem., Int. Ed.*, 2008, **47**, 7557-7559 (*Podophyllotoxin, synth*).
Saitoh, T. *et al.*, *Bioorg. Med. Chem.*, 2008, **16**, 5815-5825 (*activity*).
Vasilcanu, R. *et al.*, *Mol. Pharmacol.*, 2008, **73**, 930-939 (*Picropodophyllin, activity*).
Wu, Y. *et al.*, *Org. Lett.*, 2009, **11**, 597-600 (*(±)-Podophyllotoxin, synth*).
Sun, Y.-J. *et al.*, *Bioorg. Med. Chem. Lett.*, 2011, **21**, 3794-3797 (*Sinopodophyllum emodi constits*).
Zhao, C.Q. *et al.*, *Chin. Chem. Lett.*, 2011, **25**, 181-184 (*3′,4′-Di-O-demethylpodophyllotoxin 7-glucoside*).
Lv, M. *et al.*, *Mini-Rev. Med. Chem.*, 2011, **11**, 901-909 (*Podophyllotoxin, rev*).
Liu, Y-Q. *et al.*, *Nat. Prod. Res.*, 2011, **25**, 1570-1576 (*Sikkimotoxin, activity*).
Velázquez-Jimènez, R. *et al.*, *Phytochemistry*, 2011, **72**, 2237-2243 (*vcd, abs config*).
Jiang, F. *et al.*, *Zhongcaoyao*, 2011, **42**, 634-639 (*3′,4′-Di-O-demethylpodophyllotoxin*).
Guerram, M. *et al.*, *Chin. J. Nat. Med.*, 2012, **10**, 161-169 (*Podophyllotoxin, rev*).
Wang, Y. *et al.*, *J. Agric. Food Chem.*, 2012, **60**, 8435-8443 (*Picropodophyllone, activity*).
He, S. *et al.*, *J. Agric. Food Chem.*, 2013, **61**, 618-625 (*4,5-Demethylenepodophyllotoxin, activity*).
Marques, J.V. *et al.*, *J. Biol. Chem.*, 2013, **288**, 466-479 (*Podophyllotoxin, biosynth*).
Takahashi, M. *et al.*, *J. Org. Chem.*, 2013, **78**, 3250-3261 (*Podophyllotoxin, synth*).

Ting, C.P. *et al.*, *Angew. Chem., Int. Ed.*, 2014, **53**, 3115-3119 (*Podophyllotoxin, synth*).
Medrado, H.H.S. *et al.*, *Quim. Nova*, 2015, **38**, 243-258 (*Podophyllotoxin, rev*).
Lisiecki, K. *et al.*, *Org. Biomol. Chem.*, 2016, **14**, 460-469 (*Podophyllotoxin, synth*).
Saumen, S. *et al.*, *Org. Lett.*, 2017, **19**, 6530-6533 (*Podophyllotoxin, synth*).
Xiao, J. *et al.*, *Org. Lett.*, 2018, **20**, 1651-1654 (*synth*).
Li, J. *et al.*, *Angew. Chem., Int. Ed.*, 2019, **58**, 11657-11660 (*synth*).
Chang, W.-c. *et al.*, *Org. Lett.*, 2019, **21**, 228-232 (*biosynth*).
Ting, C.P. *et al.*, *Tetrahedron*, 2019, **75**, 3299-3308 (*Podophyllotoxin, synth*).
Lewis, R.J. *Sax's Dangerous Properties of Industrial Materials, 11th edn., J. Wiley*, 2004, EBB600

2,7′-Cyclo-3′,4,4′,5,5′,8′-hexahydroxy-8,8′-lignan-9′,9-olide C200

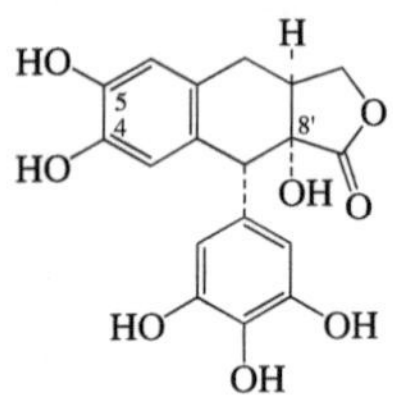

(7′R,8S,8′R)-form

3′,4,4′,5,5′,8′-Hexahydroxy-2,7′-cyclolignan-9′,9-olide
$C_{18}H_{16}O_8$ 360.318

(7′R,8S,8′R)-form

4,5-Methylene, 3′,5′-di-Me ether
2,7′-Cyclo-4′,8′-dihydroxy-3′,5′-dimethoxy-4,5-methylenedioxy-8,8′-lignan-9′,9-olide. 4′,8′-Dihydroxy-3′,5′-dimethoxy-4,5-methylenedioxy-2,7′-cyclolignan-9′,9-olide. ***8′-Hydroxy-4′-demethyldeoxypodophyllotoxin***
[157201-75-7]
$C_{21}H_{20}O_8$ 400.383
Constit. of *Casearia clarkei*. Needles ($CHCl_3$). Mp 255 - 256°. $[\alpha]_D$ -239 (c, 0.05 in $CHCl_3$). Called 2-Hydroxy-4′-demethyldesoxypodophyllotoxin in the lit.. λ_{max} 207 (log ε 4.22); 245 (sh); 294 (log ε 3.22) (EtOH)

(7′R*,8S*,8′R*)-form

4,5-Methylene, 3′,4′,5′-tri-Me ether
2,7′-Cyclo-8′-hydroxy-3′,4′,5′-trimethoxy-4,5-methylenedioxy-8,8′-lignan-9′,9-olide. 8′-Hydroxy-3′,4′,5′-trimethoxy-4,5-methylenedioxy-2,7′-cyclolignan-9′,9-olide

4,5-Methylene, 3′,4′,5′-tri-Me ether, 8′-Ac
Erlangerin C
[465535-63-1]
$C_{24}H_{24}O_9$ 456.447
Constit. of the resin of *Commiphora erlangeriana*. Cryst. ($CHCl_3$/MeOH). Mp 224 - 226°. $[\alpha]_D^{22}$+35 (c, 0.34 in $CHCl_3$). λ_{max} 205 (log ε 2.71); 292 (log ε 1.68) (EtOH).

Shaari, K. *et al.*, *J. Nat. Prod.*, 1994, **57**, 720-724 (*8′-Hydroxy-4′-demethyldeoxypodophyllotoxin*).
Dekebo, A. *et al.*, *J. Nat. Prod.*, 2002, **65**, 1252-1257 (*Erlangerin C*)

2,7′-Cyclo-3′,4,4′,5,5′,8-hexahydroxy-8,8′-lignan-9′,9-olide C201

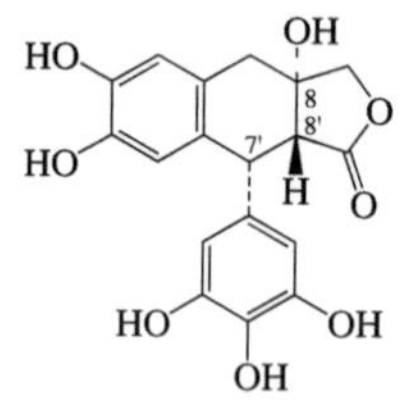

3′,4,4′,5,5′,8-Hexahydroxy-2,7′-cyclolignan-9′,9-olide
$C_{18}H_{16}O_8$ 360.318

(7′R,8S,8′R)-form

4,5-Methylene, 3′,4′,5′-tri-Me ether, 8-O-β-D-glucopyranoside
[1798290-30-8]
$C_{28}H_{32}O_{13}$ 576.551
Constit. of the leaves of *Cleistanthus boivinianus*. Exhibits potent antiproliferative activity against A2780 ovarian cancer cell line. Amorph. powder. $[\alpha]_D^{25}$+120 (c, 0.1 in MeOH). λ_{max} 208 (log ε 4.4); 245 (log ε 3.7); 291 (log ε 3.19) (MeOH).

Liu, Y. *et al.*, *J. Nat. Prod.*, 2015, **78**, 1543-1547 (*Cleistanthus boivinianus constit, activity*)

2,7′-Cyclo-3′,4,4′,5,6,7-hexahydroxy-8,8′-lignan-9′,9-olide C202

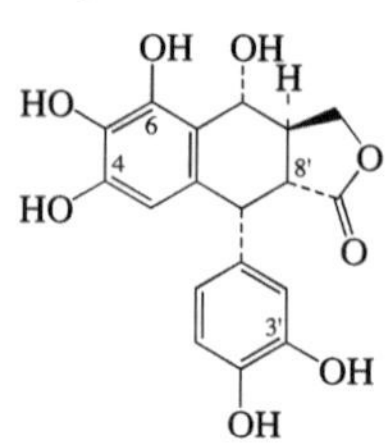

(7R,7′R,8R,8′R)-form

3′,4,4′,5,6,7-Hexahydroxy-2,7′-cyclolignan-9′,9-olide
$C_{18}H_{16}O_8$ 360.318

(7R,7′R,8R,8′R)-form

4,5-Methylene, 3′,4′,6-tri-Me ether
2,7′-Cyclo-7-hydroxy-3′,4′,6-trimethoxy-4,5-methylenedioxy-8,8′-lignan-9′,9-olide. 7-Hydroxy-3′,4′,6-trimethoxy-4,5-methylenedioxy-2,7′-cyclolignan-9′,9-olide. ***5′-Demethoxy-6-methoxypodophyllotoxin***
[138773-22-5]
$C_{22}H_{22}O_8$ 414.41
Constit. of *Linum cariense*, *Linum album* and *Linum flavum*

3′,4′:4,5-Bis(methylene), 6-Me ether
2,7′-Cyclo-7-hydroxy-6-methoxy-3′,4′:4,5-bis(methylenedioxy)-8,8′-lignan-9′,9-olide. 7-Hydroxy-6-methoxy-3′,4′:4,5-bis(methylenedioxy)-2,7′-cyclolignan-9′,9-olide. ***Cleistantoxin***
[1397199-07-3]
$C_{21}H_{18}O_8$ 398.367
Isol. from fruits of *Cleistanthus indochinensis*. Cytotoxic against human KB and MCF-7 cancer cells. Microcryst. solid (Me_2CO). Mp 195 - 196°. $[\alpha]_D^{30}$-148 (c, 0.5 in $CHCl_3$). λ_{max} 207 (log ε 3.39); 240 (log ε 4.18); 287 (log ε 3.83) ($CHCl_3$)

3′,4′:4,5-Bis(methylene), 6-Me ether, 7-O-β-D-glucopyranoside
Cleisindoside D
[1821136-72-4]
$C_{27}H_{28}O_{13}$ 560.508
Constit. of the fruits of *Cleistanthus indochinensis*. Cryst. (Me_2CO/MeOH). Mp 261 - 262°. $[\alpha]_D^{30}$-188 (c, 0.5 in MeOH)

3′,4′:4,5-Bis(methylene), 6-Me ether, 7-O-(6-O-acetyl-β-D-glucopyranoside)
Cleisindoside E
[1821136-73-5]
$C_{29}H_{30}O_{14}$ 602.545
Constit. of the fruit of *Cleistanthus indochinensis*. Cryst. (Me_2CO/MeOH). Mp 265 - 266°. $[\alpha]_D^{30}$-196 (c, 0.5 in MeOH). λ_{max} 212 (log ε 3.59); 240 (log ε 4.19); 288 (log ε 3.82) (MeOH)

(7R,7′R,8R,8′S)-form

4,5-Methylene, 3′,4′,6-tri-Me ether
5′-Demethoxy-6-methoxypicropodophyllin
[944382-47-2]
$C_{22}H_{22}O_8$ 414.41
Constit. of *Linum cariense*.

Wichers, H.J. *et al.*, *Phytochemistry*, 1991, **30**, 3601-3604 (*Linum flavum constit*).
Smollny, T. *et al.*, *Phytochemistry*, 1998, **48**, 975-979 (*Linum album constit*).
Konuklugil, B. *et al.*, *Nat. Prod. Res.*, 2007, **21**, 1-6 (*Linum cariense constits*).
Thanh, V.T.T. *J. Nat. Prod.*, 2012, **75**, 1578-1583 (*Cleistantoxin*).
Thanh, V.T.T. *et al.*, *Planta Med.*, 2014, **80**, 695-702 (*Cleisindosides D,E*)

2,2′-Cyclo-4,4′,5,5′,6′,7′-hexahydroxy-8,8′-lignan-7-one C203

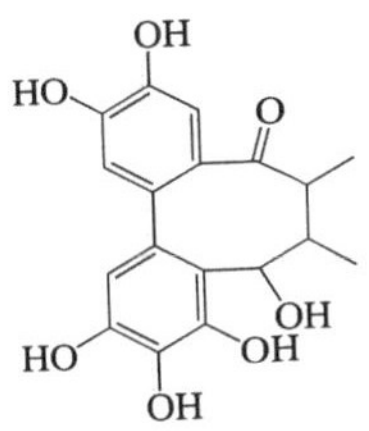

4,4′,5,5′,6,7-Hexahydroxy-7′-oxo-2,2′-cyclolignan
$C_{18}H_{18}O_7$ 346.335
4,4′,5,5′,6′-Penta-Me ether
2,2′-Cyclo-7′-hydroxy-4,4′,5,5′,6′-pentamethoxy-8,8′-lignan-7-one. 7-Hydroxy-4,4′,5,5′,6-pentamethoxy-7′-oxo-2,2′-cyclolignan. ***Saururine A***
$C_{23}H_{28}O_7$ 416.47
Constit. of *Saururus chinensis*. Powder. Stereochemistry not determined.

Liu, J. *et al.*, *Chem. Nat. Compd. (Engl. Transl.)*, 2010, **46**, 631-633 (*Saururine A*)

2,7′-Cyclo-3′,4,4′,5,7′,8-hexahydroxy-8,8′-lignan-7-one C204

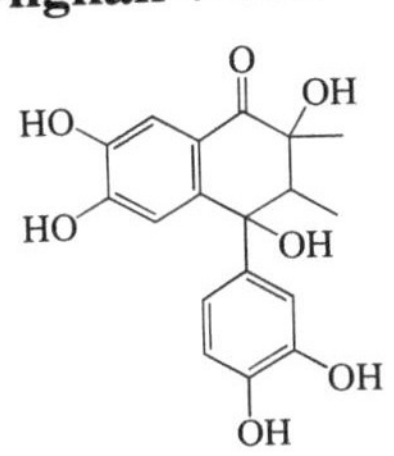

3′,4,4′,5,7′,8-Hexahydroxy-7-oxo-2,7′-cyclolignan
$C_{18}H_{18}O_7$ 346.335
(7′ξ,8ξ,8′ξ)-form
3′,4′-Methylene, 4,5-di-Me ether
2,7′-Cyclo-7′,8-dihydroxy-4,5-dimethoxy-3′,4′-methylenedioxy-8,8′-lignan-7-one. 7′,8-Dihydroxy-4,5-dimethoxy-3′,4′-methylenedioxy-7-oxo-2,7′-cyclolignan
[93446-19-6]
$C_{21}H_{22}O_7$ 386.4
Constit. of the fruit of *Virola sebifera*. Cryst. (MeOH). Mp 81 - 85°. λ_{max} 235 (ε 24200); 275 (ε 11550); 315 (ε 7200) (MeOH)
4,5-Methylene, 3′,4′-di-Me ether
2,7′-Cyclo-7′,8-dihydroxy-3′,4′-dimethoxy-4,5-methylenedioxy-8,8′-lignan-7-one. 7′,8-Dihydroxy-3′,4′-dimethoxy-4,5-methylenedioxy-7-oxo-2,7′-cyclolignan
[93471-89-7]
$C_{21}H_{22}O_7$ 386.4
Constit. of the fruit of *Virola sebifera*. Amorph. solid. λ_{max} 235 (ε 24150); 248 (ε 11600); 254 (ε 11000); 260 (ε 11400); 277 (ε 12750); 310 (ε 5950) (MeOH).

Lopes, L.M.X. *et al.*, *Phytochemistry*, 1984, **23**, 2021-2024 (*Virola sebifera constits*)

2,7′-Cyclo-3,3′,4,4′,5,9′-hexahydroxy-8,8′-lign-7-en-9-oic acid C205

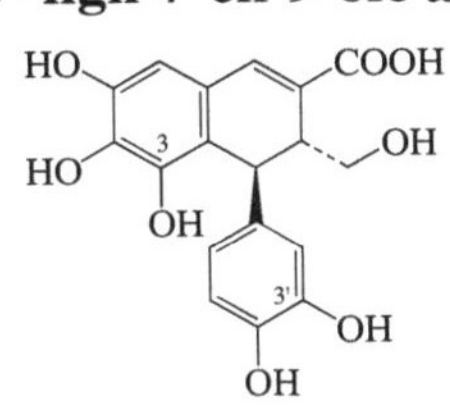

3,3′,4,4′,5,9′-Hexahydroxy-2,7′-cyclolign-7-en-9-oic acid
$C_{18}H_{16}O_8$ 360.318
(7′S,8′R)-form
3,3′,5-Tri-Me ether, [(4-hydroxyphenyl)ethyl]amide
Majusamide A
$C_{29}H_{31}NO_8$ 521.566
Constit. of *Chelidonium majus*. Amorph. powder. $[\alpha]^{20}_{D}$ -40.2 (c, 0.12 in MeOH). λ_{max} 218 (log ε 4.6); 265 (log ε 4.6) (MeOH).

Huang, X.-Y. *et al.*, *Fitoterapia*, 2019, **139**, 104359:1-6 (*Majusamide A*)

2,7′-Cyclo-3′,4,4′,5,5′,6-hexahydroxy-8,8′-lign-8(8′)-en-9′,9-olide C206

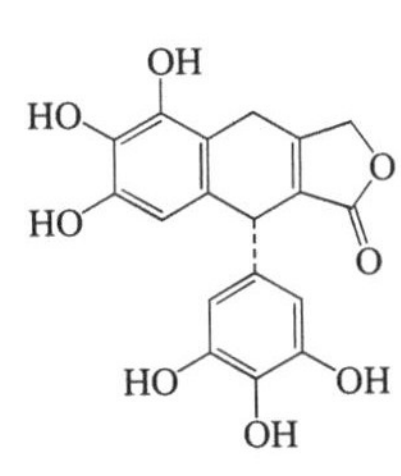

3′,4,4′,5,5′,6-Hexahydroxy-2,7′-cyclolign-8(8′)-en-9′,9-olide
$C_{18}H_{14}O_8$ 358.302
(R)-form
4,5-Methylene, 3′,4′,5′,6-tetra-Me ether
2,7′-Cyclo-3′,4′,5′,6-tetramethoxy-4,5-methylenedioxy-8,8′-lign-8(8′)-en-9′,9-olide. 3′,4′,5′,6-Tetramethoxy-4,5-methylenedioxy-2,7′-cyclolign-8(8′)-en-9′,9-olide. ***Hyptoside***
[1443626-75-2]
$C_{23}H_{22}O_8$ 426.421
Constit. of the aerial parts of *Hyptis verticillata*. Inhibitor of an adult T-cell leukaemia cell, S1T. Amorph. solid. $[\alpha]^{25}_{D}$ -63 (c, 0.05 in CH_2Cl_2). Assigned unconfirmed struct. revision; analogous with Hyptinin, C312 in 2017. Sign of opt. rotn. may be incorrect. λ_{max} 275 (log ε 3.53); 331 (log ε 3.21) (CH_2Cl_2).

Hamada, T. *et al.*, *Molecules*, 2012, **17**, 9931-9938 (*Hyptoside, activity*).
Maeda, K. *et al.*, *J. Nat. Prod.*, 2017, **80**, 1446-1449 (*struct*)

2,7′-Cyclo-3,3′,4,4′,5,5′-hexahydroxy-9′-nor-8,8′-ligna-7,7′-dien-9-oic acid C207

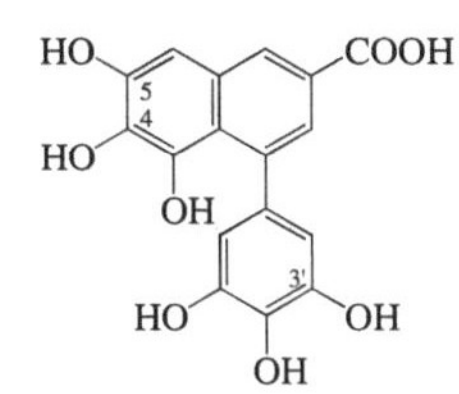

3,3′,4,4′,5,5′-Hexahydroxy-9′-nor-2,7′-cycloligna-7,7′-dien-9-oic acid. 5,6,7-Trihydroxy-4-(3,4,5-trihydroxyphenyl)-2-naphthalenecarboxylic acid
$C_{17}H_{12}O_8$ 344.275
Lignan numbering shown
3,3′,5,5′-Tetra-Me ether
2,7′-Cyclo-4,4′-dihydroxy-3,3′,5,5′-tetramethoxy-9′-nor-8,8′-ligna-7,7′-dien-9-oic acid. ***Roseochelin A***
[2170691-53-7]
$C_{21}H_{20}O_8$ 400.383
Prod. by *Phaeobacter inhibens* DSM17395. Cryst. (MeOH)
3′,4,5,5′-Tetra-Me ether
2,7′-Cyclo-3,4′-dihydroxy-3′,4,5,5′-tetramethoxy-9′-nor-8,8′-ligna-7,7′-dien-9-oic acid. 3,4′-Dihydroxy-3′,4,5,5′-tetramethoxy-9′-nor-2,7′-cycloligna-7,7′-dien-9-oic acid. ***Gardenianan A***

[1318790-63-4]
$C_{21}H_{20}O_8$ 400.383
Constit. of the fruit of *Gardenia jasminoides*
6-(Methylthio), 3,3′,5′-tri-Me ether
2,7′-Cyclo-4,4′,5-trihydroxy-3,3′,5′-trimethoxy-6-(methylthio)-9′-nor-8,8′-ligna-7,7′-dien-9-oic acid. ***Roseochelin B***
[2170691-54-8]
$C_{21}H_{20}O_8S$ 432.443
Prod. by *Phaeobacter inhibens* DSM17395. Exhibits antialgal activity. Cryst. (MeOH).
Yu, Y. *et al.*, *Zhongcaoyao*, 2010, **41**, 509-514 (*Gardenianan A*).
Wang, R. *et al.*, *Org. Lett.*, 2017, **19**, 5138-5141 (*Roseochelins A,B, biosynth, cryst struct, activity*)

2,7′-Cyclo-3′,4,4′,5,7,9-hexahydroxy-1,7-seco-8,8′-lignan-9′-oic acid C208

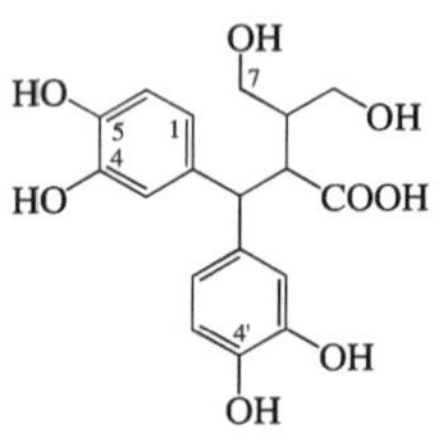

2-[Bis(3,4-dihydroxyphenyl)methyl]-4-hydroxy-3-(hydroxymethyl)butanoic acid. 3′,4,4′,5,7,9-Hexahydroxy-2,7′-cyclo-1,7-secolignan-9′-oic acid
$C_{18}H_{20}O_8$ 364.35
Secolignans
4,4′-Di-Me ether
2,7′-Cyclo-3′,5,7,9-tetrahydroxy-4,4′-dimethoxy-1,7-seco-8,8′-lignan-9′-oic acid. ***Marphenol E***
[1420791-99-6]
$C_{20}H_{24}O_8$ 392.404
Constit. of the fruits of *Schisandra wilsoniana*. Yellow gum. $[\alpha]^{25}_{D}$ +19.6 (c, 0.18 in MeOH). Reference is made to determination of the relative config., but the molecule has only one chiral centre. The abs. config. was not determined. λ_{max} 205 (log ε 4.92); 282 (log ε 4.18); 326 (log ε 2.25) (MeOH)
3′,4′-Methylene, 4-Me ether
2,7′-Cyclo-5,7,9-trihydroxy-4-methoxy-3′,4′-methylenedioxy-1,7-seco-8,8′-lignan-9′-oic acid. ***Marphenol F***
[1420792-00-2]
$C_{20}H_{22}O_8$ 390.388
Constit. of the fruits of *Schisandra wilsoniana*. Yellow gum. $[\alpha]^{25}_{D}$ +26.8 (c, 0.20 in MeOH). λ_{max} 208 (log ε 4.82); 282 (log ε 4.22); 325 (log ε 2.18) (MeOH).
Yang, G.-Y. *et al.*, *J. Nat. Prod.*, 2013, **76**, 250-255 (*Marphenols E,F*)

2,7′-Cyclo-2′,4,4′,5′,7,8-hexamethoxy-8,8′-lignan-3,6-dione C209

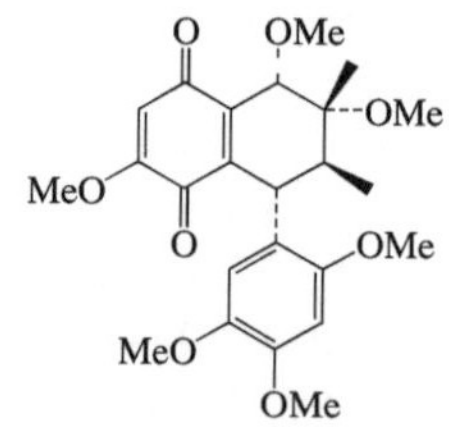

Absolute Configuration

2′,4,4′,5′,7,8-Hexamethoxy-2,7′-cyclolignan-3,6-dione
[1458672-48-4]
$C_{24}H_{30}O_8$ 446.496
Constit. of the rhizomes of *Acorus gramineus*. Brown-red gum. $[\alpha]^{25}_{D}$+9 (c, 0.25 in $CHCl_3$). λ_{max} 280 (log ε 3.4) (MeOH).
Lee, S.Y. *et al.*, *Biosci., Biotechnol., Biochem.*, 2013, **77**, 276-280 (*Acorus gramineus constit*)

2,7′-Cyclo-9′-hydroxy-3,3′,4,4′,5,5′-hexamethoxy-9-nor-8,8′-lignan-7-one C210

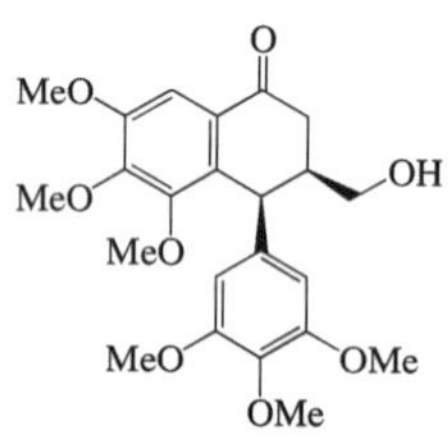

Relative Configuration

9′-Hydroxy-3,3′,4,4′,5,5′-hexamethoxy-9-nor-7-oxo-2,7′-cyclolignan. ***Aglacin H***
[477529-39-8]
$C_{23}H_{28}O_8$ 432.469
Constit. of the stem bark of *Aglaia cordata*. Waxy solid. $[\alpha]^{20}_{D}$ -40 (c, 0.57 in $CHCl_3$).
Wang, B.-G. *et al.*, *Tetrahedron Lett.*, 2002, **43**, 5783-5787 (*Aglacin H*)

2,7′-Cyclo-4-hydroxy-5-methoxy-3′,4′-methylenedioxy-7′,8′-seco-8,8′-lignan-7′,8′-dione C211

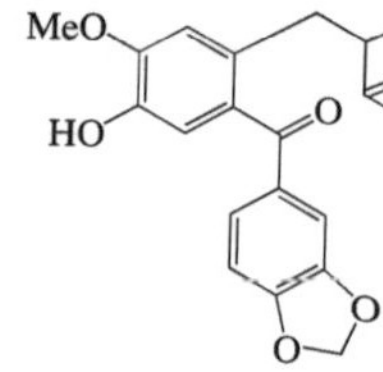

4-[[2-(1,3-Benzodioxol-5-yl)carbonyl]-4-hydroxy-5-methoxyphenyl]-3-methyl-2-butanone, CAS. *4-Hydroxy-5-methoxy-3′,4′-methylenedioxy-7′,8′-dioxo-7′,8′-seco-2,7′-cyclolignan*
$C_{20}H_{20}O_6$ 356.374
(ξ)-form [188894-38-4]
Constit. of *Virola* aff. *pavonis*. Viscous oil. $[\alpha]_D$ +25 (c, 0.1 in $CHCl_3$). λ_{max} 235 (log ε 4.44); 280 (log ε 4.01); 320 (log ε 4.27) (MeOH)
3′,4′-Demethylene, 3′-Me ether ***Machilupin A***
$C_{20}H_{22}O_6$ 358.390
Constit. of *Machilus philippinensis*. Syrup. $[\alpha]^{25}_{D}$ +28 (c, 0.03 in MeOH). λ_{max} 235 (log ε 3.5); 285 (log ε 3.2); 316 (log ε 3.2) (MeOH).
Martinez, J.C. *et al.*, *Phytochemistry*, 1997, **44**, 1179-1182 (*Virola pavonis constit*).
Tai, S.-H. *et al.*, *RSC Adv.*, 2019, **9**, 21616-21625 (*Machilupin A*)

2,7′-Cyclo-8,8′-ligna-7,7′-diene-4,4′-diol C212

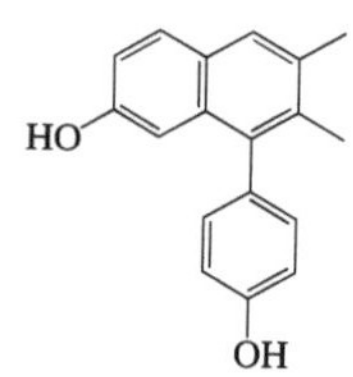

8-(4-Hydroxyphenyl)-6,7-dimethyl-2-naphthalenol. 4,4′-Dihydroxy-2,7′-cycloligna-7,7′-diene
$C_{18}H_{16}O_2$ 264.324
4′-Me ether
2,7′-Cyclo-4′-methoxy-8,8′-ligna-7,7′-dien-4-ol. 8-(4-Methoxyphenyl)-6,7-dimethyl-2-naphthalenol, CAS. *4-Hydroxy-4′-methoxy-2,7′-cycloligna-7,7′-diene.* ***Megislignan***
[473699-86-4]
$C_{19}H_{18}O_2$ 278.351
Constit. of the stem bark of *Iryanthera megistophylla*. Acetyl

cholinesterase inhibitor. Amorph. powder. Mp 189 - 190°. $[\alpha]_D^{20}$ -18 (c, 1 in $CHCl_3$). Opt. rotn. unaccounted for. λ_{max} 247 (log ε 4.42); 284 (log ε 4.05); 336 (log ε 3.62) ($CHCl_3$).

Ming, D.S. *et al.*, *J. Nat. Prod.*, 2002, **65**, 1412-1416 (*Meglslignan*)

2,7′-Cyclo-8,8′-ligna-7,7′-diene-3,4,4′,5,7,9,9′-heptol C213

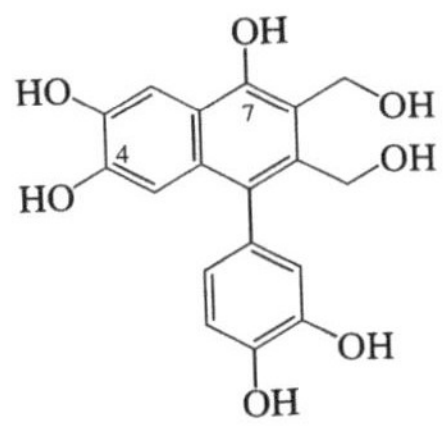

3,4,4′,5,7,9,9′-Heptahydroxy-2,7′-cycloligna-7,7′-diene
$C_{18}H_{16}O_7$ 344.319

3′,4′-Methylene, 4,5-di-Me ether
2,7′-Cyclo-4,5-dimethoxy-3′,4′-methylenedioxy-8,8′-ligna-7,7′-diene-7,9,9′-triol. 7,9,9′-Trihydroxy-4,5-dimethoxy-3′,4′-methylenedioxy-2,7′-cycloligna-7,7′-diene

3′,4′-Methylene, 4,5-di-Me ether, 7-O-β-D-glucopyranoside
Phyllamyricoside B
[184094-81-3]
$C_{27}H_{30}O_{12}$ 546.525
Constit. of *Phyllanthus myrtifolius*. Amorph. solid. Mp 170 - 175°. λ_{max} 243 (log ε 4.63); 249 (log ε 4.65); 287 (log ε 3.86); 331 (log ε 3.26) (MeOH).

Lee, S.-S. *et al.*, *J. Nat. Prod.*, 1996, **59**, 1061-1065 (*Phyllamyricoside B*)

2,7′-Cyclo-8,8′-ligna-7,7′-diene-3′,4,4′,5,5′,6-hexol C214

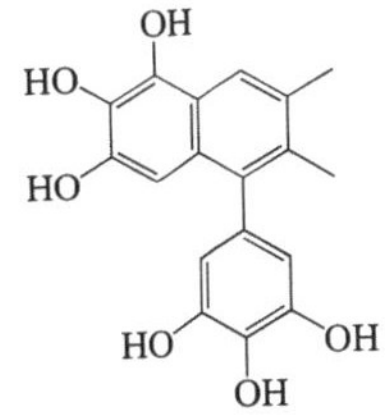

6,7-Dimethyl-5-(3,4,5-trihydroxyphenyl)-1,2,3-naphthalenetriol, CAS. *3′,4,4′,5,5′,6-Hexahydroxy-2,7′-cycloligna-7,7′-diene*
$C_{18}H_{16}O_6$ 328.32

3′,4,5′,6-Tetra-Me ether
2,7′-Cyclo-3′,4,5′,6-tetramethoxy-8,8′-ligna-7,7′-diene-4′,5-diol. 4′,5-Dihydroxy-3′,4,5′,6-tetramethoxy-2,7′-cycloligna-7,7′-diene.
Sacidumlignan A
[848986-16-3]
$C_{22}H_{24}O_6$ 384.428
Constit. of *Sarcostemma acidum*. Shows moderate antibacterial activity against Staphylococci. Needles (MeOH); cryst. (Me_2CO). λ_{max} 205 (log ε 4.81); 239 (log ε 4.94); 290 (log ε 3.97); 331 (log ε 3.38) (MeOH).

Gan, L.-S. *et al.*, *J. Nat. Prod.*, 2005, **68**, 221-225 (*Sacidumlignan A*).
Zhang, J.-J. *et al.*, *Org. Biomol. Chem.*, 2013, **11**, 2498-2513 (*Sacidumlignan A*).
Reddel, J.C.T. *et al.*, *Chem. Sci.*, 2017, **8**, 2156-2160 (*synth*)

2,7′-Cyclo-8,8′-ligna-7,7′-diene-3′,4,4′,5,7,9-hexol C215

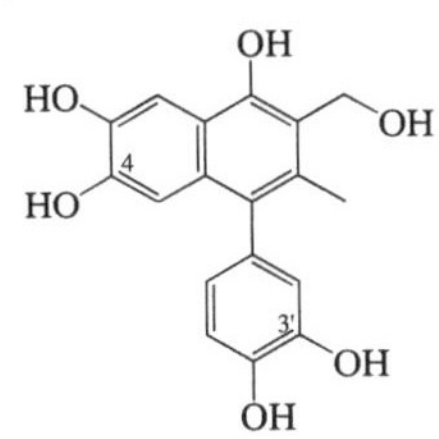

3′,4,4′,5,7,9-Hexahydroxy-2,7′-cycloligna-7,7′-diene
$C_{18}H_{16}O_6$ 328.32

3′,4,5-Tri-Me ether
2,7′-Cyclo-3′,4,5-trimethoxy-8,8′-ligna-7,7′-diene-4′,7,9-triol. 4′,7,9-Trihydroxy-3′,4,5-trimethoxy-2,7′-cycloligna-7,7′-diene

3′,4,5-Tri-Me ether, 7-O-β-D-glucopyranoside
Phyllamyricoside C
[184094-82-4]
$C_{27}H_{32}O_{11}$ 532.542
Constit. of *Phyllanthus myrtifolius*. Amorph. solid. Mp 159 - 163°. λ_{max} 243 (log ε 4.75); 247 (log ε 4.74); 290 (log ε 3.98); 331 (log ε 3.32) (MeOH)

3′,4′-Methylene, 4,5-di-Me ether
2,7′-Cyclo-4,5-dimethoxy-3′,4′-methylenedioxy-8,8′-ligna-7,7′-diene-7,9-diol. 7,9-Dihydroxy-4,5-dimethoxy-3′,4′-methylenedioxy-2,7′-cycloligna-7,7′-diene

3′,4′-Methylene, 4,5-di-Me ether, 7-O-β-D-glucopyranoside
Phyllamyricoside A
[184094-80-2]
$C_{27}H_{30}O_{11}$ 530.526
Constit. of *Phyllanthus myrtifolius*. Increases the activity of HIV-1 reverse transcriptase. λ_{max} 245 (ε 63100); 250 (ε 63100); 289 (ε 10960); 330 (ε 1995) (MeOH) (Berdy).

Lee, S-S. *et al.*, *J. Nat. Prod.*, 1998, **59**, 1061-1065 (*Phyllamyricosides A,C*)

2,7′-Cyclo-8,8′-ligna-7,7′-diene-3′,4,4′,5,9,9′-hexol C216

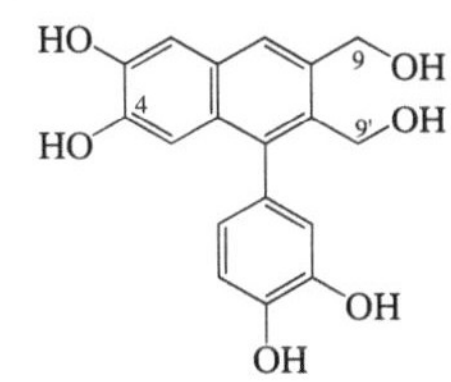

3′,4,4′,5,9,9′-Hexahydroxy-2,7′-cycloligna-7,7′-diene
$C_{18}H_{16}O_6$ 328.32

3′,5-Di-Me ether
2,7′-Cyclo-3′,5-dimethoxy-8,8′-ligna-7,7′-diene-4,4′,9,9′-tetrol. 4,4′,9,9′-Tetrahydroxy-3′,5-dimethoxy-2,7′-cycloligna-7,7′-diene.
Alashinol D
[2226514-60-7]
$C_{20}H_{20}O_6$ 356.374
Constit. of the stem bark of *Syringa pinnatifolia*. Inhibits LPS-induced NO prodn.. Powder. λ_{max} 242 (log ε 4.82); 283 (log ε 4.04); 316 (log ε 3.56); 331 (log ε 3.59) (MeOH)

4,5-Methylene, 3′,4′-di-Me ether
1-(3,4-Dimethoxyphenyl)-2,3-bis(hydroxymethyl)-6,7-methylenedioxynaphthalene. 2,7′-Cyclo-3′,4′-dimethoxy-4,5-methylenedioxy-8,8′-ligna-7,7′-diene-9,9′-diol. 9,9′-Dihydroxy-3′,4′-dimethoxy-4,5-methylenedioxy-2,7′-cycloligna-7,7′-diene
[42923-60-4]
$C_{21}H_{20}O_6$ 368.385
Constit. of *Jatropha gossypifolia*. Cryst. (C_6H_6/petrol). Mp 184°. Mp 210 - 211° (synthetic). λ_{max} 247 (ε 54954); 290 (ε 9818); 332 (ε 2692) (EtOH)

9′-Aldehyde, 3′,4′-methylene, 4,5,9-tri-Me ether
6,7-Dimethoxy-3-(methoxymethyl)-1-(3,4-methylenedioxyphenyl)-2-naphthalenecarboxaldehyde. 2,7′-Cyclo-4,5,9-trimethoxy-3′,4′-methylenedioxy-8,8′-ligna-7,7′-dien-9′-al. 4,5,9-Trimethoxy-3′,4′-methylenedioxy-2,7′-cycloligna-7,7′-dien-9′-al. ***Phyllamyricin B***
[164124-33-8]
$C_{22}H_{20}O_6$ 380.396
Constit. of *Phyllanthus myrtifolius*. Strongly inhibits HIV reverse transcriptase. Cryst. (MeOH/$CHCl_3$). Mp 140 - 141°

9′-Carboxylic acid, 3′,4′-methylene, 4,5-di-Me ether
2,7′-Cyclo-9-hydroxy-4,5-dimethoxy-3′,4′-methylenedioxy-8,8′-ligna-7,7′-dien-9′-oic acid. 9-Hydroxy-4,5-dimethoxy-3′,4′-methylenedioxy-2,7′-cycloligna-7,7′-dien-9′-oic acid

9′-Carboxylic acid, 3′,4′-methylene, 4,5-di-Me ether, 9-Ac, β-D-glucopyranosyl ester
Phyllanthostatin A
[119767-19-0]
$C_{29}H_{30}O_{13}$ 586.546
Constit. of *Phyllanthus acuminatus* and *Phyllanthus anisolobus*. Cytostatic agent. Amorph. solid. Sol. MeOH. Mp 127 - 130°. $[\alpha]_D^{26}$ +24.5 (c, 1.5 in CH_2Cl_2). Slowly forms Justicidin B, in C332 in soln.. λ_{max} 248; 288 (MeOH) (Berdy)

9,9′-Dicarboxylic acid
1-(3,4-Dihydroxyphenyl)-6,7-dihydroxy-2,3-naphthalenedicarboxylic acid, CAS. *2,7′-Cyclo-3′,4,4′,5-tetrahydroxy-8,8′-ligna-7,7′-diene-9,9′-dioic acid. 3′,4,4′,5-Tetrahydroxy-2,7′-cycloligna-7,7′-diene-9,9′-dioic acid*
[171901-70-5]
$C_{18}H_{12}O_8$ 356.286
Isol. from the liverwort *Jamesoniella autumnalis*. Light yellow powder. λ_{max} 224 (ε 19953); 252 (log ε 41687); 307 (ε 14791) (MeOH)

9,9′-Dicarboxylic acid, bis[(4-hydroxyphenyl)ethyl]amide
Cannabisin A
[130508-46-2]
[130508-47-3] (hexa-Ac)
$C_{34}H_{30}N_2O_8$ 594.62
Constit. of the fruit of *Cannabis sativa*. Amorph. powder; needles (as hexa-Ac). Mp 204 - 207° (hexa-Ac). λ_{max} 257 (log ε 4.79) (EtOH)

9,9′-Dicarboxylic acid, 4,5-methylene, 3′,4′-di-Me ether, anhydride
1-(3,4-Dimethoxyphenyl)-6,7-methylenedioxy-2,3-naphthalenic anhydride
[42923-53-5]
$C_{21}H_{14}O_7$ 378.336
Constit. of *Cleistanthus collinus* and *Jatropha gossypifolia*. Deep yellow cryst. ($CHCl_3$/MeOH). Mp 272 - 274°.

Anjaneyulu, A. *et al.*, *Indian J. Chem., Sect. B*, 1986, **25**, 589-595 (*Jatropha gossypifolia constits, synth*).
Pettit, G.R. *et al.*, *J. Nat. Prod.*, 1988, **51**, 1104-1112 (*Phyllanthostatin A*).
Das, B. *et al.*, *Phytochemistry*, 1988, **27**, 3684-3686 (*Jatropha gossypifolia constit*).
Sakakibara, I. *et al.*, *Phytochemistry*, 1991, **30**, 3013-3016 (*Cannabisin A*).
Bachmann, T.L. *et al.*, *Phytochemistry*, 1993, **33**, 189-191 (*Phyllanthus anisolobus constit*).
Lin, M.-T. *et al.*, *J. Nat. Prod.*, 1995, **58**, 244-249 (*Phyllamyricin B*).
Tazaki, H. *et al.*, *Phytochemistry*, 1995, **40**, 1671-1675 (*Jamesoniella autumnalis constit*).
Satyanarayana, P. *et al.*, *Indian J. Chem., Sect. B*, 1998, **37**, 717-719 (*Cleistanthus collinus naphthalenic anhydride*).
Kim, H.J. *et al.*, *J. Nat. Prod.*, 1998, **61**, 145-148 (*Phyllamyricin B, activity*).
Wang, C.-Y. *et al.*, *Phytochem. Anal.*, 2011, **22**, 352-360 (*Phyllamyricin B, pmr, hplc*).
Su, G. *et al.*, *Fitoterapia*, 2016, **114**, 63-68 (*Alashinol D, activity*)

2,7′-Cyclo-8,8′-ligna-7,7′-diene-3′,4,4′,5,5′-pentol C217

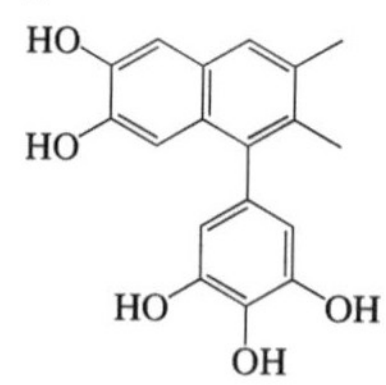

6,7-Dimethyl-5-(3,4,5-trihydroxyphenyl)-2,3-naphthalenediol. 3′,4,4′,5,5′-Pentahydroxy-2,7′-cycloligna-7,7′-diene
$C_{18}H_{16}O_5$ 312.321

3′,4,4′,5′-Tetra-Me ether
2,7′-Cyclo-3′,4,4′,5′-tetramethoxy-8,8′-lignan-5-ol. 5-Hydroxy-3,4,4′,5′-tetramethoxy-2,7′-cyclolignan. ***Daphnelignan B***
[1306747-70-5]
$C_{22}H_{24}O_5$ 368.429
Constit. of *Daphne acutiloba* leaves and stems. Shows anti-HIV props.. Cryst. (MeOH aq.). λ_{max} 205 (log ε 6.12); 238 (log ε 4.87); 287 (log ε 3.16) (MeOH).

Cao, J.-L. *et al.*, *Asian J. Chem.*, 2010, **22**, 6509-6512 (*Daphnelignan B*)

2,7′-Cyclo-8,8′-ligna-7,7′-diene-3,4,4′,5,6-pentol C218

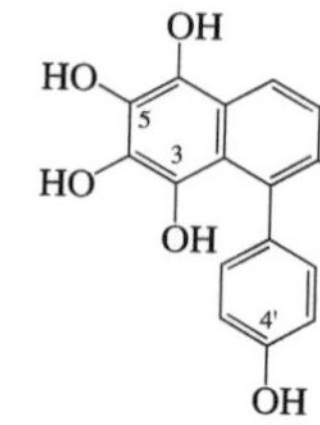

5-(4-Hydroxyphenyl)-6,7-dimethyl-1,2,3,4-naphthalenetetrol. 3,4,4′,5,6-Pentahydroxy-2,7′-cycloligna-7,7′-diene
$C_{18}H_{16}O_5$ 312.321
Lignan numbering shown

4,5-Methylene, 3,6-di-Me ether
2,7′-Cyclo-3,6-dimethoxy-4,5-methylenedioxy-8,8′-ligna-7,7′-dien-4′-ol. 4′-Hydroxy-3,6-dimethoxy-4,5-methylenedioxy-2,7′-cycloligna-7,7′-diene. ***Pycnanthulignene D***
[1207532-19-1]
$C_{21}H_{20}O_5$ 352.386
Constit. of the roots of *Pycnanthus angolensis*. Brown cryst. (C_6H_6). Mp 164.5 - 165°.

Nono, E.C.N. *et al.*, *J. Nat. Prod.*, 2010, **73**, 213-216 (*Pycnanthulignene D*)

2,7′-Cyclo-8,8′-ligna-7,7′-diene-3′,4,4′,5,9-pentol C219

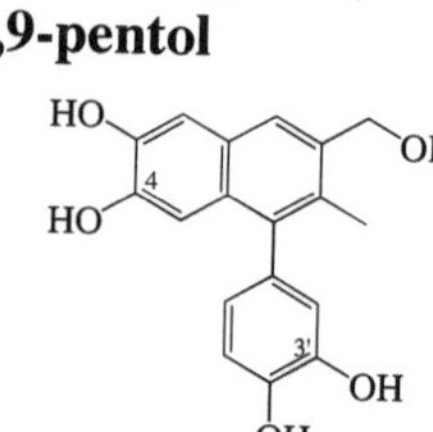

3′,4,4′,5,9-Pentahydroxy-2,7′-cycloligna-7,7′-diene
$C_{18}H_{16}O_5$ 312.321

3′,4′-Methylene, 4,5-di-Me ether
2,7′-Cyclo-4,5-dimethoxy-3′,4′-methylenedioxy-8,8′-ligna-7,7′-dien-9-ol. 9-Hydroxy-4,5-dimethoxy-3′,4′-methylenedioxy-2,7′-cycloligna-7,7′-diene. ***Phyllamyricin F***

[184094-79-9]
$C_{21}H_{20}O_5$ 352.386
Isol. from *Phyllanthus myrtifolius*. λ_{max} 244 (log ε 5); 290 (log ε 4.54); 313 (log ε 3.87); 328 (log ε 3.96) (MeOH).

Lee, S.-S. *et al.*, *J. Nat. Prod.*, 1996, **59**, 1061-1065 (*Phyllamyricin F*)

2,9′-Cyclo-8,8′-ligna-7,8′-diene-3′,4,4′,5,9-pentol C220

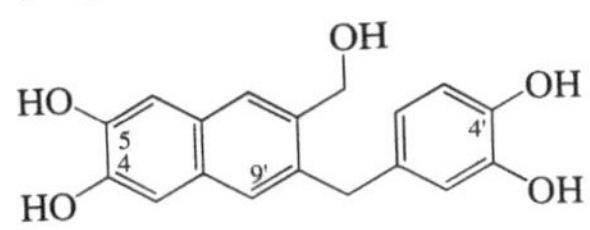

3′,4,4′,5,9-Pentahydroxy-2,9′-cycloligna-7,8′-diene
$C_{18}H_{16}O_5$ 312.321

9-Aldehyde, 3′,5-di-Me ether
2,9′-Cyclo-4,4′-dihydroxy-3′,5-dimethoxy-8,8′-ligna-7,8′-dien-9-al. 4,4′-Dihydroxy-3′,5-dimethoxy-2,9′-cycloligna-7,8′-dien-9-al
[1801457-37-3]
$C_{20}H_{18}O_5$ 338.359
Constit. of the stems of *Syringa pinnatifolia*. Needles (MeOH aq.). λ_{max} 267 (log ε 4.25) (MeOH).

Wang, Q. *et al.*, *Nat. Prod. Res.*, 2016, **30**, 85-88 (*Syringa pinnatifolia constit*)

2,7′-Cyclo-8,8′-ligna-7,7′-diene-3′,4,4′,5-tetrol C221

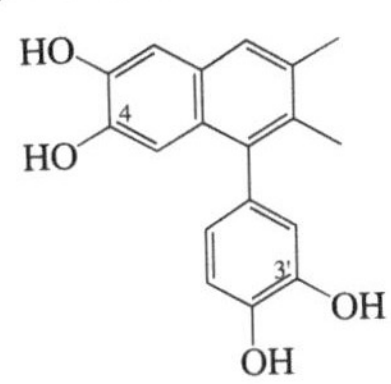

5-(3,4-Dihydroxyphenyl)-6,7-dimethyl-2,3-naphthalenediol. 3′,4,4′,5-Tetrahydroxy-2,7′-cycloligna-7,7′-diene
$C_{18}H_{16}O_4$ 296.322

3′,4-Di-Me ether
2,7′-Cyclo-3′,4-dimethoxy-8,8′-ligna-7,7′-diene-4′,5-diol. 4′,5-Dihydroxy-3′,4-dimethoxy-2,7′-cycloligna-7,7′-diene. ***Cinnamophilin A‡. Furfuracin***
[1217184-44-5]
$C_{20}H_{20}O_4$ 324.376
Constit. of *Cinnamomum philippinense* and *Knema furfuracea*. Cryst.. Mp 205 - 207°. Not to be confused with Cinnamophilin, in L68. λ_{max} 236 (log ε 4.97); 287 (log ε 4.18); 324 (log ε 3.65) (MeOH). [neutral] λ_{max} 235 (log ε 3.02); 280 (log ε 1.90) (MeCN)

3′,5-Di-Me ether
2,7′-Cyclo-3′,5-dimethoxy-8,8′-ligna-7,7′-diene-4,4′-diol. 4,4′-Dihydroxy-3′,5-dimethoxy-2,7′-cycloligna-7,7′-diene. ***Dehydroguaiaretic acid***
[20601-86-9]
$C_{20}H_{20}O_4$ 324.376
Constit. of *Combretum alfredii, Guaiacum officinale* and *Knema furfuracea*. Gum. Mp 209 - 215°

3′,4′-Methylene, 5-Me ether
2,7′-Cyclo-5-methoxy-3′,4′-methylenedioxy-8,8′-ligna-7,7′-dien-4-ol. 4-Hydroxy-5-methoxy-3′,4′-methylenedioxy-2,7′-cycloligna-7,7′-diene. ***4-O-Demethyldehydroisogalcatin***
[129931-98-2]
$C_{20}H_{18}O_4$ 322.36
Constit. of the bark of *Virola calophylla*. Cryst. (EtOAc/hexane). Mp 168 - 169°. λ_{max} 239 (ε 44850); 290 (ε 8400); 318 (ε 2950); 333 (ε 3950) (MeOH)

3′,4′-Methylene, 4,5-di-Me ether
2,7′-Cyclo-4,5-dimethoxy-3′,4′-methylenedioxy-7,7′-lignadiene. 4,5-Dimethoxy-3′,4′-methylenedioxy-2,7′-cycloligna-7,7′-diene. ***Tetradehydroisogalcatin****. Tetradehydroisootobain (incorr.)*
$C_{21}H_{20}O_4$ 336.387
Constit. of *Virola calophylla* and *Virola sebifera*. Yellow cryst. (MeOH). Mp 188 - 190°.

Wallace, R. *et al.*, *J. Chem. Soc.*, 1963, 1445-1449 (*synth*).
King, F.E. *et al.*, *J. Chem. Soc.*, 1964, 4011-4024 (*Guaiacum officinale constit, synth*).
Aulin-Erdtman, G. *et al.*, *Acta Chem. Scand.*, 1968, **22**, 1187-1209 (*Dehydroguaiaretic acid*).
Lopes, L.M.X. *et al.*, *Phytochemistry*, 1984, **23**, 2647-2652 (*Virola sebifera constit*).
Konno, C. *et al.*, *J. Nat. Prod.*, 1989, **52**, 1113-1117 (*Tetradehydroisogalcatin*).
Pinto, M.M.M. *et al.*, *Phytochemistry*, 1990, **29**, 1985-1988 (*Knema furfuracea constits*).
Martinez V., J.C. *et al.*, *Phytochemistry*, 1990, **29**, 2655-2657 (*Virola calophylla constits*).
Rangkaew, N. *et al.*, *Fitoterapia*, 2009, **80**, 377-379 (*Furfuracin*).
Chen, C.-Y. *et al.*, *Chem. Nat. Compd. (Engl. Transl.)*, 2011, **47**, 519-520 (*Cinnamophilin A*)

2,7′-Cyclo-8,8′-ligna-7,7′-diene-4,4′,5-triol C222

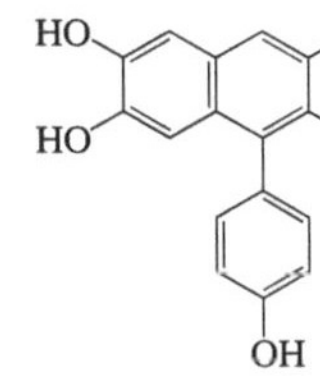

5-(4-Hydroxyphenyl)-6,7-dimethyl-2,3-naphthalenediol. 4,4′,5-Trihydroxy-2,7′-cycloligna-7,7′-diene
[124962-10-3]
$C_{18}H_{16}O_3$ 280.323
Isol. from *Larrea tridentata*. Mp 127 - 128°

4′-Me ether
5-(4-Methoxyphenyl)-6,7-dimethyl-2,3-naphthalenediol. 2,7′-Cyclo-4′-methoxy-8,8′-ligna-7,7′-diene-4,5-diol. 4,5-Dihydroxy-4′-methoxy-2,7′-cycloligna-7,7′-diene
[2085301-54-6]
$C_{19}H_{18}O_3$ 294.350
Constit. of *Combretum alfredii*. Exhibits strong antibacterial activity. Amorph. powder

4,5-Methylene, 4′-Me ether
2,7′-Cyclo-4′-methoxy-4,5-methylenedioxy-7,7′-lignadiene. 4′-Methoxy-4,5-methylenedioxy-2,7′-cycloligna-7,7′-diene. ***Pycnanthulignene C***
[1207532-18-0]
$C_{20}H_{18}O_3$ 306.361
Constit. of the roots of *Pycnanthus angolensis*. Shows significant antimicrobial activity against various drug-resistant pathogens. Powder ($CHCl_3$). Mp 158 - 160°.

Konno, C. *et al.*, *J. Nat. Prod.*, 1989, **52**, 1113-1117 (*Larrea tridentata constit*).
San Feliciano, A. *et al.*, *Magn. Reson. Chem.*, 1993, **31**, 868-875 (*cmr*).
Nono, E.C.N. *et al.*, *J. Nat. Prod.*, 2010, **73**, 213-216 (*Pycnanthulignene C*).
Reddel, J.C.T. *et al.*, *Chem. Sci.*, 2017, **8**, 2156-2160 (*synth*).
Bai, M. *et al.*, *Nat. Prod. Res.*, 2017, **31**, 1022-1027 (*Combretum alfredii constit, activity*)

7′,8-Cyclo-7,8′-lignan-9,9′-dioic acid C223

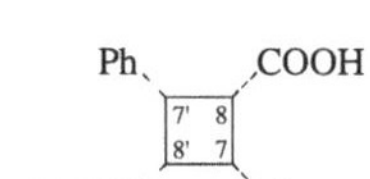

(7α,7′α,8α,8′α)-*form*

2,4-Diphenyl-1,3-cyclobutanedicarboxylic acid, CAS.
Truxillic acid
[4462-95-7]
$C_{18}H_{16}O_4$ 296.322
All five possible stereoisomeric forms known. All are *meso-*. Solid
(7α,7′α,8α,8′α)-form
η-Truxillic acid. *peri*-Truxillic acid
[528-39-2]
Cryst. (AcOH/C_6H_6). Sol. EtOH; insol. Et_2O, C_6H_6. Mp 266° dec.. KOH fusion → ε-Truxillic acid
Mono-Me ester
$C_{19}H_{18}O_4$ 310.349
Cryst. (MeOH). Mp 192°
Di-Me ester
Cryst. (MeOH). Mp 104 - 105°
Monoamide
η-Truxillamic acid
$C_{18}H_{17}NO_3$ 295.338
Prisms (AcOH/C_6H_6). Spar. sol. AcOH; insol. EtOH. Mp 256.5°
Anhydride
$C_{18}H_{14}O_3$ 278.307
Prisms (AcOH). Mp 285°
(7α,7′α,8α,8′β)-form
epi-Truxillic acid [552-20-5]
Cryst. (EtOH aq. or C_6H_6/AcOH). V. spar. sol. Et_2O, C_6H_6. Mp 285 - 287°. Forms ε-Truxillic acid at Mp. Forms two monomethyl esters
Di-Me ester
Mp 111 - 112°
Diamide
$C_{18}H_{18}N_2O_2$ 294.354
Needles (EtOH aq.). Mp 252 - 253°
(7α,7′α,8β,8′β)-form
ε-Truxillic acid. β-Cocaic acid
[528-38-1]
Needles (EtOH aq. or Et_2O). Mp 192°
Mono-Me ester
Prisms (MeOH). Mp 131°
Di-Me ester
Prisms (EtOH aq.). Mp 64°
Dichloride
Prisms (petrol). Mp 106 - 107°
Monoamide
ε-Truxillamic acid
Needles (EtOH). Sol. EtOH, hot H_2O; insol. C_6H_6. Mp 213°
(7α,7′β,8α,8′α)-form
γ-Truxillic acid [490-19-7]
Needles (EtOH aq.). Spar. sol. hot H_2O. Mp 228°. KOH fusion → α- and ε-Truxillic acids
Mono-Me ester
Needles (EtOH aq.). Insol. H_2O, petrol. Mp 183.5 - 184°. Resolvable
Di-Me ester
Needles (MeOH aq.). Sol. MeOH, AcOH. Mp 126°
Monoamide
γ-Truxillamic acid
Prisms (AcOH aq.). Mp 240°. Resolvable
Anhydride
Needles ($CHCl_3$/EtOH). Mp 191°
(7α,7′β,8α,8′β)-form
α-*Truxillic acid*. γ-Isatropic acid. Cocaic acid. Gratissimic acid [490-20-0]
Constit. of *Halimium verticillatum*. Component of α-Truxilline, T124. Isol from illicit cocaine. Needles (EtOH aq.); cryst. + 2MeOH (MeOH). Mp 285° (274°). Does not form anhydride. Sublimes unchanged *in vacuo*. KOH fusion → ε-Truxillic acid
Mono-Me ester
Needles. Mp 195°
Di-Me ester
Gratissimin
[56586-34-6]
$C_{20}H_{20}O_4$ 324.376
Obt. from *Ocimum gratissimum*. Plates or needles (MeOH). Spar. sol. MeOH, AcOH. Mp 174°. Bp *ca.* 330°
Dichloride
[37597-30-1]
$C_{18}H_{14}Cl_2O_2$ 333.208
Prisms + ½C_6H_6 (C_6H_6/petrol). Mp 125°
Monoamide
α-Truxillamic acid
Needles (AcOH). Sol. EtOH, AcOH; insol. Et_2O, petrol. Mp 261°.

Criegee, R. *et al.*, *Chem. Ber.*, 1960, **93**, 2521-2524 (*synth*).
Cohen, M.D. *et al.*, *J. Chem. Soc.*, 1964, 2000-2013 (*α-trixillic acid*).
Qudrat-i-Khuda, M. *et al.*, *Sci. Res. (Dacca)*, 1964, **1**, 217-222 (*Gratissimic acid, Gratissimin*).
Qudrat-i-Khuda, M. *et al.*, *Sci. Res. (Dacca)*, 1965, **2**, 8-10 (*Gratissimic acid, Gratissimin*).
Rodd's Chem. Carbon Compd., 2nd edn., 1967, **2A**, 72-103 (*rev*).
Carter, R.O. *et al.*, *Appl. Spectrosc.*, 1972, **26**, 378-381 (*ir*).
Montaudo, G. *et al.*, *Org. Magn. Reson.*, 1974, **6**, 534-536 (*pmr*).
Caccamese, S. *et al.*, *Org. Mass Spectrom.*, 1974, **9**, 1114-1123 (*ms*).
Lurie, I.S. *et al.*, *J. Chromatogr.*, 1987, **405**, 273-281 (*cocaine constits*).
Urones, J.G. *et al.*, *Phytochemistry*, 1994, **36**, 529-530 (*Ocimum gratissimum constit, cryst struct*).
Khan, M. *et al.*, *J. Am. Chem. Soc.*, 2008, **130**, 1741-1748 (*synth*).
Cañellas, S. *et al.*, *J. Am. Chem. Soc.*, 2018, **140**, 17349-17355 (*7α,7′α,8α,8′β-form, synth, ir, pmr, cmr*)

2,7′-Cyclo-8,8′-lignan-4,4′-diol C224

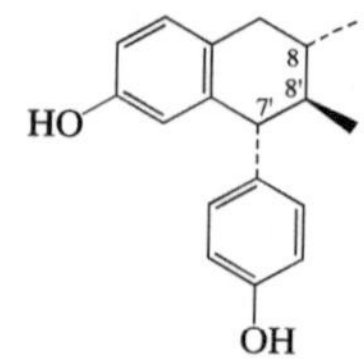

5,6,7,8-Tetrahydro-8-(4-hydroxyphenyl)-6,7-dimethyl-2-naphthalenol, CAS.
4,4′-Dihydroxy-2,7′-cyclolignan
$C_{18}H_{20}O_2$ 268.356
Lignan numbering shown
(7′R*,8S*,8′R*)-form [88662-13-9]
Constit. of the fruit of *Iryanthera grandis* and of *Iryanthera lancifolia*. Shows estrogenic activity. Cryst. (hexane/MeOH). Mp 138 - 140°. λ_{max} 225 (ε 18300); 290 (ε 4250) (MeOH).

Vieira, P.C. *et al.*, *Phytochemistry*, 1983, **22**, 2281-2286 (*Iryanthera grandis constit*).
Lopes, N.P. *et al.*, *Phytochemistry*, 1998, **49**, 1405-1410 (*Iryanthera lancifolia constit*).
Mesa-Siverio, D. *et al.*, *Bioorg. Med. Chem.*, 2008, **16**, 3387-3394 (*activity*)

2,2′-Cyclo-8,8′-lignan-3,3′,4,4′,5,5′,7-heptol C225

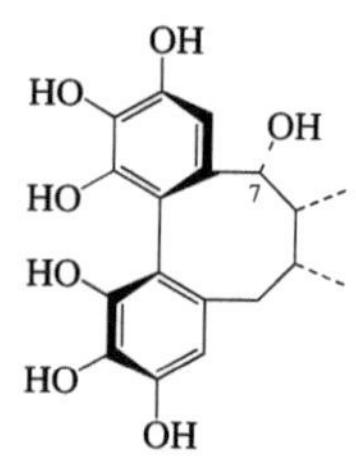

(*7R,8R,8′R,R*$_{axial}$)*-form*

3,3′,4,4′,5,5′,7-Heptahydroxy-2,2′-cyclolignan
$C_{18}H_{20}O_7$ 348.351
(7R,8R,8′R,R$_{axial}$)-form
3,3′,4,4′-Tetra-Me ether, 7-benzoyl
Longipedlignan K
[2376337-93-6]
$C_{29}H_{32}O_8$ 508.567
Constit. of the roots of *Kadsura longipedunculata*. Amorph. yellow

powder. $[\alpha]_D^{20}$ +8 (c, 0.5 in MeOH). λ_{max} 220 (log ε 1.64) (MeOH)

4,5-Methylene, 3,4′,5′-tri-Me ether, 7-Ac

Schisantherin O. *Schizantherin O*
[150132-87-9]
$C_{24}H_{28}O_8$ 444.48
Constit. of the seeds of *Kadsura coccinea*. Cryst. (MeOH). Mp 194 - 195°. $[\alpha]_D$ +13.4 (c, 0.72 in MeOH)

4,5-Methylene, 3,4′,5′-tri-Me ether, 7-benzoyl

Longipedunin C
[886441-74-3]
$C_{29}H_{30}O_8$ 506.551
Constit. of *Kadsura longipedunculata*. Amorph. powder. $[\alpha]_D$ +25.4 (c, 0.2 in MeOH). λ_{max} 225 (log ε 4.15); 254 (log ε 3.48); 280 (log ε 3) (MeOH)

4,5-Methylene, 3,3′,4′,5′-tetra-Me ether, 7-Ac

Ananolignan A
[1280213-59-3]
$C_{25}H_{30}O_8$ 458.507
Constit. of the seeds of *Kadsura ananosma*. Solid. $[\alpha]_D^{26}$ +68.1 (c, 0.17 in $CHCl_3$). λ_{max} 199 (log ε 3.64); 208 (log ε 3.64); 213 (log ε 3.64); 219 (log ε 3.66); 226 (log ε 3.68); 230 (log ε 3.67); 241 (log ε 3.99) ($CHCl_3$)

4′,5′-Methylene, 3,3′,4-tri-Me ether, 7-E-cinnamoyl

Tiegusanin L
[1156503-63-7]
$C_{31}H_{32}O_8$ 532.589
Constit. of *Schisandra propinqua* var. *sinensis*. Amorph. yellow solid. $[\alpha]_D^{27}$ -68.2 (c, 0.4 in $CHCl_3$). λ_{max} 202 (log ε 3.84); 206 (log ε 3.84); 221 (log ε 3.85); 229 (log ε 3.86); 236 (log ε 3.88); 280 (log ε 4.02) ($CHCl_3$)

4,5:4′,5′-Bis(methylene), 3,3′-di-Me ether

2,2′-Cyclo-3,3′-dimethoxy-4,5:4′,5′-bis(methylenedioxy)-8,8′-lignan-7-ol. 7-Hydroxy-3,3′-dimethoxy-4,5:4′,5′-bis(methylenedioxy)-2,2′-cyclolignan.

Schisanlignaol D
[144606-85-9]

[143729-19-5] [143729-20-8] [143729-21-9]

$C_{22}H_{24}O_7$ 400.427
Constit. of *Schisandra viridis*. Mp 120 - 122°. $[\alpha]_D^{17}$ -34.2 (c, 0.36 in $CHCl_3$). λ_{max} 228 (log ε 4.48); 260 (log ε 3.96); 284 (log ε 3.64) (MeOH)

7-Ketone, 3,3′,4,4′,5′-penta-Me ether

2,2′-Cyclo-5-hydroxy-3,3′,4,4′,5′-pentamethoxy-8,8′-lignan-7-one. 5-Hydroxy-3,3′,4,4′,5′-pentamethoxy-7-oxo-2,2′-cyclolignan.

Schisanlignone B
[135459-86-8]
$C_{23}H_{28}O_7$ 416.47
Constit. of the seeds of a Chinese *Kadsura* sp.. Shows antileukaemic activity agains P-388 cells. Mp 151 - 152°. $[\alpha]_D^{19.5}$ -36.9 (c, 0.2355 in $CHCl_3$). λ_{max} 208 (log ε 4.65); 224 (log ε 4.56); 245 (log ε 4.43); 270 (log ε 4.15); 323 (log ε 3.75) (EtOH)

7-Ketone, 3,3′,4,4′,5,5′-hexa-Me ether

2,2′-Cyclo-3,3′,4,4′,5,5′-hexamethoxy-8,8′-lignan-7-one. 3,3′,4,4′,5,5′-Hexamethoxy-7-oxo-2,2′-cyclolignan.

Schisanlignone A
[135557-67-4]
$C_{24}H_{30}O_7$ 430.497
Constit. of the seeds of a Chinese *Kadsura* sp.. Shows antileukaemic activity against P-388 cells. Mp 104 - 105°. $[\alpha]_D^{24}$ -74.8 (c, 0.210 in $CHCl_3$). λ_{max} 207 (log ε 4.64); 225 (log ε 4.56); 243 (log ε 4.38); 270 (log ε 4.11); 325 (log ε 3.68) (EtOH)

7-Ketone, 4′,5′-methylene, 3,3′,4,5-tetra-Me ether

2,2′-Cyclo-3,3′,4,5-tetramethoxy-4′,5′-methylenedioxy-8,8′-lignan-7-one. 3,3′,4,5-Tetramethoxy-4′,5′-methylenedioxy-7-oxo-2,2′-cyclolignan.

Schisanlignone C
[144606-83-7]
$C_{23}H_{26}O_7$ 414.454
Constit. of *Schisandra viridis*. Mp 139 - 140°. $[\alpha]_D^{17}$ -38.2 (c, 0.39 in $CHCl_3$). Abs. config. of biaryl system uncertain. Drawn as R_{axial} in the lit, but appears to be stated to have opposite config. to Gomisin O, which is *R*-. The opt. rotns. are certainly opposite. λ_{max} 216 (log ε 4.39); 228 (log ε 4.45); 250 (log ε 4.16); 272 (log ε 3.59); 320 (log ε 3.12) (MeOH)

7-Ketone, 4,5:4′,5′-bis(methylene), 3,3′-di-Me ether

2,2′-Cyclo-3,3′-dimethoxy-4,5:4′,5′-bis(methylenedioxy)-8,8′-lignan-7-one. 3,3′-Dimethoxy-4,5:4′,5′-bis(methylenedioxy)-7-oxo-2,2′-cyclolignan.

Schisanlignone D
[144606-84-8]
$C_{22}H_{22}O_7$ 398.411
Constit. of *Schisandra viridis*. Mp 175 - 176°. $[\alpha]_D^{17}$ -45.6 (c, 0.44 in $CHCl_3$). λ_{max} 218 (log ε 4.39); 236 (log ε 4.44); 294 (log ε 3.97) (MeOH)

(7R,8R,8′R,S_{axial})-form

3,4,4′,5′-Tetra-Me ether

2,2′-Cyclo-3,4,4′,5′-tetramethoxy-8,8′-lignan-3′,5,7-triol. 3′,5,7-Trihydroxy-3,4,4′,5′-tetramethoxy-2,2′-cyclolignan. **Deangeloylschizantherin F**
[117047-91-3]
$C_{22}H_{28}O_7$ 404.459
Constit. of the fruit of *Kadsura japonica*. Prisms (Me_2CO). Mp 195 - 196°. $[\alpha]_D^{24}$ -12.5 (c, 0.9 in dioxan)

3,4,4′,5′-Tetra-Me ether, 3′-O-(2ξ-methylbutanoyl)

Kadsuralignan L
[1351336-81-6]
$C_{27}H_{36}O_8$ 488.577
Constit. of the rhizomes of *Kadsura coccinea*. Needles. $[\alpha]_D^{20}$ -61 (c, 0.25 in MeOH). λ_{max} 220 (log ε 4.11); 259 (log ε 3.0); 291 (log ε 3.05) (MeOH)

3,4,4′,5′-Tetra-Me ether, 3′-angeloyl

Schizantherin F. *Schisantherin F*
[117047-88-8]
$C_{27}H_{34}O_8$ 486.561
Isol. from roots of *Kadsura* sp.. Mp 169 - 171°. $[\alpha]_D$ +23.3 ($CHCl_3$)

3,3′,4,4′,5′-Penta-Me ether

2,2′-Cyclo-3,3′,4,4′,5′-pentamethoxy-8,8′-lignan-5,7-diol. 5,7-Dihydroxy-3,3′,4,4′,5′-pentamethoxy-2,2′-cyclolignan. **Propinquain H**
[1426059-47-3]
$C_{23}H_{30}O_7$ 418.486
Isol. from the roots of *Schisandra propinqua* ssp. *sinensis*. Amorph. yellow powder ($CHCl_3$). $[\alpha]_D^{20}$ -36.5 (c, 0.32 in $CHCl_3$). λ_{max} 214 (log ε 4.70); 246 (log ε 4.08) (MeOH)

3,4,4′,5,5′-Penta-Me ether
2,2′-Cyclo-3,4,4′,5,5′-pentamethoxy-8,8′-lignan-3′,7-diol. 3′,7-Dihydroxy-3,4,4′,5,5′-pentamethoxy-2,2′-cyclolignan. Binankadsurin B
[117047-93-5]
$C_{23}H_{30}O_7$ 418.486
Powder. $[\alpha]_D^{27}$ -32.8 (c, 2 in $CHCl_3$). Not yet found as a nat. prod.

3,4,4′,5,5′-Penta-Me ether, 7-Ac
7-Acetylbinankadsurin B
[167114-93-4]
$C_{25}H_{32}O_8$ 460.523
Constit. of the fruit of *Kadsura japonica*. Powder. $[\alpha]_D^{28}$ +7.7 (c, 1.3 in $CHCl_3$)

3,4,4′,5,5′-Penta-Me ether, 3′-O-(2-methylpropanoyl)
Tiegusanin B
[1156503-53-5]
$C_{27}H_{36}O_8$ 488.577
Constit. of *Schisandra propinqua* var. *sinensis*. Powder. $[\alpha]_D^{26}$ +24 (c, 0.12 in $CHCl_3$). λ_{max} 198 (log ε 3.48); 201 (log ε 3.47); 210 (log ε 3.46); 223 (log ε 3.53); 227 (log ε 3.54); 233 (log ε 3.53); 241 (log ε 3.87) ($CHCl_3$)

3,4,4′,5,5′-Penta-Me ether, 3′-O-(2-methylbutanoyl)
Tiegusanin A
[1156503-52-4]
$C_{28}H_{38}O_8$ 502.604
Constit. of *Schisandra propinqua* var. *sinensis*. Powder. λ_{max} 200 (log ε 3.49); 208 (log ε 3.48); 227 (log ε 3.58); 242 (log ε 3.9) ($CHCl_3$)

3,4,4′,5,5′-Penta-Me ether, 3′-angeloyl
3′-Angeloylbinankadsurin B. *Methylschisantherin F*
[117047-89-9]
$C_{28}H_{36}O_8$ 500.588
Constit. of *Schisandra propinqua* var. *sinensis*

3,4,4′,5,5′-Penta-Me ether, 7-angeloyl
7-Angeloylbinankadsurin B
[167114-92-3]
$C_{28}H_{36}O_8$ 500.588
Constit. of the fruit of *Kadsura japonica*. Plates (hexane/Et_2O). Mp 146.5 - 148°. $[\alpha]_D^{25}$ +33.6 (c, 1 in $CHCl_3$)

3′,4,4′,5,5′-Penta-Me ether, 3-benzoyl
Tiegusanin C
[1156503-54-6]
$C_{30}H_{34}O_8$ 522.594
Constit. of *Schisandra propinqua* var. *sinensis*. Powder. $[\alpha]_D^{28}$ -57.9 (c, 0.22 in $CHCl_3$). λ_{max} 195 (log ε 3.58); 199 (log ε 3.57); 201 (log ε 3.57); 205 (log ε 3.56); 208 (log ε 3.56); 224 (log ε 3.64); 229 (log ε 3.65); 241 (log ε 4.1) ($CHCl_3$)

Hexa-Me ether, 7-Ac
Kadsufolin B
[1300063-41-5]
$C_{26}H_{34}O_8$ 474.55
Constit. of *Kadsura oblongifolia*. Needles (petrol/Me_2CO). Mp 104 - 105°. $[\alpha]_D^{25}$ -89 (c, 1 in MeOH). λ_{max} 215 (log ε 4.73); 250 (sh); 286 (sh) (MeOH)

Hexa-Me ether, 7-angeloyl
Kadsufolin A
[1300063-38-0]
$C_{29}H_{38}O_8$ 514.615
Constit. of *Kadsura oblongifolia*. Cytotoxic. Needles (petrol/Me_2CO). Mp 107 - 108°. $[\alpha]_D^{25}$ -10 (c, 1.4 in MeOH). λ_{max} 216 (log ε 4.64); 252 (sh); 287 (sh) (MeOH)

Hexa-Me ether, 7-benzoyl
Schizanrin L
[874472-16-9]
$C_{31}H_{36}O_8$ 536.621
Constit. of *Kadsura japonica*. Mp 89 - 90°. $[\alpha]_D^{28}$ -100 (c, 0.1 in MeOH)

4,5-Methylene, 3,4′-di-Me ether, 5′-O-β-D-glucopyranoside
Longipedunculatin D
[2408237-12-5]
$C_{27}H_{34}O_{12}$ 550.557
Constit. of the roots of *Kadsura longipedunculata*. Amorph. yellow powder. $[\alpha]_D^{20}$ -28 (c, 0.2 in MeOH). λ_{max} 218 (log ε 2.57); 255 (log ε 1.92); 286 (log ε 1.54) (MeOH)

4,5-Methylene, 4′,5′-di-Me ether, 7-angeloyl
Kadsutherin A
[903573-40-0]
$C_{26}H_{30}O_8$ 470.518
Constit. of *Kadsura* spp.. Yellow powder. $[\alpha]_D^{25}$ +236.7 (c, 0.02 in MeOH). λ_{max} 220 (log ε 4.67); 280 (sh) (MeOH)

4,5-Methylene, 3,3′,4′-tri-Me ether
2,2′-Cyclo-3,3′,4′-trimethoxy-4,5-methylenedioxy-8,8′-lignan-5′,7-diol. 5′,7-Dihydroxy-3,3′,4′-trimethoxy-4,5-methylenedioxy-2,2′-cyclolignan. **Kadsuralignan A**
[913237-64-6]
$C_{22}H_{26}O_7$ 402.443
Constit. of the rhizomes of *Kadsura coccinea*. Needles (hexane). Mp 157 - 158°. λ_{max} 220 (log ε 3.9); 255 (sh); 289 (sh) (MeOH)

4,5-Methylene, 3,3′,4′-tri-Me ether, 7-Ac
Schilancifolignan C
[1310536-21-0]
$C_{24}H_{28}O_8$ 444.48
Constit. of *Schisandra lancifolia*. Yellow gum. $[\alpha]_D^{24}$ +20.2 (c, 0.2 in MeOH). λ_{max} 210 (log ε 5.32); 248 (log ε 3.35); 320 (log ε 1.26) (MeOH)

4,5-Methylene, 3,3′,4′-tri-Me ether, 7-O-(2-methylpropanoyl)
Neglignan G
[1441710-42-4]
$C_{26}H_{32}O_8$ 472.534
Constit. of *Schisandra neglecta*. Yellow gum. $[\alpha]_D^{24}$ +11.8 (c, 0.2 in MeOH). λ_{max} 210 (log ε 4.76); 246 (log ε 3.85); 320 (log ε 1.22) (MeOH)

4,5-Methylene, 3,4′,5′-tri-Me ether
2,2′-Cyclo-3,4′,5′-trimethoxy-4,5-methylenedioxy-8,8′-lignan-3′,7-diol. 3′,7-Dihydroxy-3,4′,5′-trimethoxy-4,5-methylenedioxy-2,2′-cyclolignan. **Binankadsurin A**
[77165-79-8]
$C_{22}H_{26}O_7$ 402.443
Isol. from stems of *Kadsura longipedunculata*. Plant extracts are active against arthritis and intestinal ulcers. Prisms (Et_2O/hexane). Mp 153 - 155° (145-146°). $[\alpha]_D^{14}$ -27.4 (c, 0.09 in $CHCl_3$) (-20.7)

4,5-Methylene, 3,4′,5′-tri-Me ether, 7-Ac
Acetylbinankadsurin A. *Norkadsurin*
[77174-33-5]
$C_{24}H_{28}O_8$ 444.48
Occurs in *Kadsura japonica*. Needles (Me_2CO/EtOH). Mp 188 - 192°. $[\alpha]_D^{24}$ 0 ($CHCl_3$)

4,5-Methylene, 3,4′,5′-tri-Me ether, 7-propanoyl
7-Propanoylbinankadsurin A. **Longipedunin B**
[886441-73-2]
$C_{25}H_{30}O_8$ 458.507

Constit. of *Kadsura longipedunculata*. Amorph. powder. [α] D +8.3 (c, 0.1 in MeOH). λ_{max} 218 (log ε 3.8); 254 (log ε 3.18); 280 (log ε 2.81) (MeOH)

4,5-Methylene, 3,4′,5′-tri-Me ether, 7-O-(2-methylpropanoyl)
7-Isobutyrylbinankadsurin A. ***Lanciphenol B***
[135541-45-6]
$C_{26}H_{32}O_8$ 472.534
Constit. of *Kadsura longipedunculata* and *Schisandra lancifolia*. Cryst. (Et_2O). Mp 152 - 153°. $[\alpha]_D^{16}$ +21 (c, 0.07 in $CHCl_3$)

4,5-Methylene, 3,4′,5′-tri-Me ether, 7-butanoyl
7-Butyrylbinankadsurin A
[135459-84-6]
$C_{26}H_{32}O_8$ 472.534
Constit. of *Kadsura* sp.

4,5-Methylene, 3,4′,5′-tri-Me ether, 3′-O-(2-methylbutanoyl)
Kadsuralignan J
[960511-18-6]
$C_{27}H_{34}O_8$ 486.561
Constit. of the rhizomes of *Kadsura coccinea*. Amorph. powder. $[\alpha]_D^{25}$ -59 (c, 0.25 in MeOH). λ_{max} 220 (log ε 4.7) (MeOH)

4,5-Methylene, 3,4′,5′-tri-Me ether, 7-O-(3-methylbutanoyl)
7-Isovaleroylbinankadsurin A
[135541-44-5]
$C_{27}H_{34}O_8$ 486.561
Constit. of *Kadsura longipedunculata*. Cryst. (Et_2O). Mp 126 - 128°. $[\alpha]_D^{15}$ +33.6 (c, 0.06 in $CHCl_3$)

4,5-Methylene, 3,4′,5′-tri-Me ether, 7-hexanoyl
7-Capreoylbinankadsurin A
[77165-81-2]
$C_{28}H_{36}O_8$ 500.588
Constit. of *Kadsura japonica*. Prisms. Mp 83 - 87°. $[\alpha]_D^{25}$ +14.7 (c, 2.73 in $CHCl_3$)

4,5-Methylene, 3,4′,5′-tri-Me ether, 3′-angeloyl
Kadsuralignan I
[960511-17-5]
$C_{27}H_{32}O_8$ 484.545
Constit. of the rhizomes of *Kadsura coccinea*. Amorph. powder. $[\alpha]_D^{25}$ -23 (c, 0.25 in MeOH). λ_{max} 220 (log ε 4.77) (MeOH)

4,5-Methylene, 3,4′,5′-tri-Me ether, 7-angeloyl
7-Angeloylbinankadsurin A
[77165-80-1]
$C_{27}H_{32}O_8$ 484.545
Constit. of *Kadsura japonica*. Prisms (Et_2O/hexane). Mp 121 - 122°. $[\alpha]_D^{25}$ +61.5 (c, 2.26 in $CHCl_3$)

4,5-Methylene, 3,4′,5′-tri-Me ether, 7-tigloyl
Propinquain G
$C_{27}H_{32}O_8$ 484.545
Isol. from the roots of *Schisandra propinqua* ssp. *sinensis*. Amorph. yellow powder ($CHCl_3$). $[\alpha]_D^{20}$ -38.8 (c, 0.39 in CH_2Cl_2). λ_{max} 251 (log ε 4.05); 281 (log ε 3.69) (MeOH)

4,5-Methylene, 3,4′,5′-tri-Me ether, 7-O-(4,4-dimethoxy-2-methyl-2Z-butenoyl)
Kadsuralignan G
[1433964-42-1]
$C_{29}H_{36}O_{10}$ 544.597
Constit. of the rhizomes of *Kadsura coccinea*. Needles. $[\alpha]_D^{20}$ -43 (c, 0.25 in MeOH). λ_{max} 220 (log ε 4.21); 242 (log ε 3.07); 290 (log ε 3.15) (MeOH)

4,5-Methylene, 3,4′,5′-tri-Me ether, 3′-benzoyl
Kadsuralignan K
[960511-19-7]
$C_{29}H_{30}O_8$ 506.551
Constit. of the rhizomes of *Kadsura coccinea*. Amorph. powder. $[\alpha]_D^{25}$ -37 (c, 0.25 in MeOH). λ_{max} 219 (log ε 4.63) (MeOH)

4,5-Methylene, 3,4′,5′-tri-Me ether, 7-benzoyl
7-Benzoylbinankadsurin A
[135541-43-4]
$C_{29}H_{30}O_8$ 506.551
Constit. of *Kadsura longipedunculata*. Amorph.. $[\alpha]_D^{22}$ -71.3 (c, 0.13 in $CHCl_3$)

4,5-Methylene, 3,4′,5′-tri-Me ether, 7-O-E-cinnamoyl
Longipedunin A. *7-Cinnamoylbinankadsurin A. 3′-O-Demethylkadsuphilin A*
[886441-72-1]
$C_{31}H_{32}O_8$ 532.589
Constit. of *Kadsura longipedunculata* and *Kadsura philippinensis*. Prisms (MeOH) or pale yellow powder. Mp 176 - 178°. [α] D +52.2 (c, 0.2 in MeOH). $[\alpha]_D^{28}$ -49 (c, 1 in CH_2Cl_2). The opposite reported opt. rotns. for the two isolates are presumably a solvent effect. λ_{max} 222 (log ε 4.19); 258 (log ε 3.91); 278 (log ε 3.91) (MeOH). [neutral] λ_{max} 225 (log ε 4.93); 260 (log ε 4.46); 280 (log ε 3.75) (MeOH)

4,5-Methylene, 3,3′,4′,5′-tetra-Me ether
2,2′-Cyclo-3,3′,4′,5′-tetramethoxy-4,5-methylenedioxy-8,8′-lignan-7-ol. 7-Hydroxy-3,3′,4′,5′-tetramethoxy-4,5-methylenedioxy-2,2′-cyclolignan. ***Yunnankadsurin B***
[51670-42-9]
$C_{23}H_{28}O_7$ 416.47
Constit of *Kadsura* spp.. Yellow powder. $[\alpha]_D^{28}$ -41.8 (c, 0.02 in MeOH). λ_{max} 215 (log ε 4.2); 250 (sh); 280 (sh) (MeOH)

4,5-Methylene, 3,3′,4′,5′-tetra-Me ether, 7-Ac
Kadsurin‡
[51670-40-7]
$C_{25}H_{30}O_8$ 458.507
Constit. of stems of *Kadsura japonica*. Cryst.. Mp 157 - 158°. $[\alpha]_D^{25}$ -39 (c, 0.13 in $CHCl_3$)

4,5-Methylene, 3,3′,4′,5′-tetra-Me ether, 7-propanoyl
Kadsutherin B
[903573-41-1]
$C_{26}H_{32}O_8$ 472.534
Constit. of *Kadsura* spp.. Powder. $[\alpha]_D^{25}$ +239 (c, 0.01 in MeOH). λ_{max} 218 (log ε 4.46); 253 (sh); 281 (sh) (MeOH)

4,5-Methylene, 3,3′,4′,5′-tetra-Me ether, 7-O-(2-methylbutanoyl)
Heteroclitin A
[140369-75-1]
$C_{28}H_{36}O_8$ 500.588
Constit. of the stems of *Kadsura heteroclita*. Amorph. powder. $[\alpha]_D^{25}$ +29.2 (c, 0.66 in MeOH)

4,5-Methylene, 3,3′,4′,5′-tetra-Me ether, 7-angeloyl
Heteroclitin B
[140461-47-8]
$C_{28}H_{34}O_8$ 498.572
Constit. of the stems of *Kadsura heteroclita*. Amorph. powder. $[\alpha]_D^{25}$ +58.3 (c, 3.92 in MeOH)

4,5-Methylene, 3,3′,4′,5′-tetra-Me ether, 7-tigloyl
Heteroclitin C
[140460-42-0]
$C_{28}H_{34}O_8$ 498.572

Constit. of the stems of *Kadsura heteroclita*. Amorph. powder. $[\alpha]_D^{25}$ -12 (c, 3.61 in MeOH)

4,5-Methylene, 3,3′,4′,5′-tetra-Me ether, 7-O-E-cinnamoyl

Kadsuphilin A

[887770-02-7]

$C_{32}H_{34}O_8$ 546.616

Constit. of *Kadsura philippinensis* and *Kadsura coccinea*. Pale yellow powder. $[\alpha]_D^{28}$ -12 (c, 1 in CH_2Cl_2). λ_{max} 236 (log ε 4.92); 260 (log ε 4.25); 283 (log ε 3.64) (MeOH)

4,5-Methylene, 3,3′,4′,5′-tetra-Me ether, 7-Z-cinnamoyl

Polysperlignan F

[1360440-61-4]

$C_{32}H_{34}O_8$ 546.616

Isol. from the stems of *Kadsura polysperma*. Yellow oil. $[\alpha]_D^{17}$ +380 (c, 0.10 in MeOH). λ_{max} 218 (log ε 4.26); 256 (log ε 3.81) (MeOH)

4,5-Methylene, 3,3′,4′,7-tetra-Me ether

2,2′-Cyclo-3,3′,4′,7-tetramethoxy-4,5-methylenedioxy-8,8′-lignan-5′-ol. 5′-Hydroxy-3,3′,4′,7-tetramethoxy-4,5-methylenedioxy-2,2′-cyclolignan. **Schilancifolignan B**

[1310536-20-9]

$C_{23}H_{28}O_7$ 416.47

Constit. of *Schisandra lancifolia*. Yellow gum. $[\alpha]_D^{23}$ +18.5 (c, 0.22 in MeOH). λ_{max} 210 (log ε 5.02); 245 (log ε 3.42); 320 (log ε 0.61) (MeOH)

4,5-Methylene, penta-Me ether

2,2′-Cyclo-3,3′,4′,5′,7-pentamethoxy-4,5-methylenedioxylignan. 3,3′,4′,5′,7-Pentamethoxy-4,5-methylenedioxy-2,2′-cyclolignan. **Schilancifolignan A**

[1310536-19-6]

$C_{24}H_{30}O_7$ 430.497

Constit. of *Schisandra lancifolia*. Yellow gum. $[\alpha]_D^{22}$ +22.5 (c, 0.24 in MeOH). λ_{max} 210 (log ε 5.06); 243 (log ε 3.51); 320 (log ε 0.58) (MeOH)

4,5:4′,5′-Bis(methylene), 3-Me ether, 3′-tigloyl

Kadsurain A

[1445972-66-6]

$C_{26}H_{28}O_8$ 468.502

Constit. of roots of *Kadsura coccinea*. Exhibits significant antiproliferative effect on HCT116 and A549 tumour cell lines. Amorph. powder. Mp 108 - 110°. $[\alpha]_D^{20}$ -36.8 (c, 0.1 in $CHCl_3$). λ_{max} 220 (log ε 4.77) (MeOH)

4,5:4′,5′-Bis(methylene), 3-Me ether, 3′-tigloyl, 7-Ac

Kadsurain B

[1445972-67-7]

$C_{28}H_{30}O_9$ 510.539

Constit. of roots of *Kadsura coccinea*. Exhibits significant antiproliferative effect on HCT116 and A549 tumour cell lines. Amorph. powder. Mp 132 - 134°. $[\alpha]_D^{20}$ -36 (c, 0.1 in $CHCl_3$). λ_{max} 220 (log ε 4.77) (MeOH)

4,5:4′,5′-Bis(methylene), 3,7-di-Me ether, 3′-tigloyl

Kadsurain C

[1445972-68-8]

$C_{27}H_{30}O_8$ 482.529

Constit. of roots of *Kadsura coccinea*. Exhibits significant antiproliferative effect on HCT116 and A549 tumour cell lines. Amorph. powder. Mp 134 - 136°. $[\alpha]_D^{20}$ -20 (c, 0.1 in $CHCl_3$). λ_{max} 237 (log ε 4.73) (MeOH)

7-Ketone, 3,4,4′,5,5′-penta-Me ether

2,2′-Cyclo-3′-hydroxy-3,4,4′,5,5′-pentamethoxy-8,8′-lignan-7-one. 3′-Hydroxy-3,4,4′,5,5′-pentamethoxy-7-oxo-2,2′-cyclolignan. ***Rubriflorin B‡***

[835875-96-2]

$C_{23}H_{28}O_7$ 416.47

Constit. of the stems of *Schisandra rubriflora*. Yellowish powder ($CHCl_3$). Mp 175 - 176°. $[\alpha]_D^{19}$ -32.7 (c, 0.15 in $CHCl_3$). λ_{max} 222 (log ε 4.8); 254 (log ε 4.25); 285 (log ε 3.98) ($CHCl_3$)

7-Ketone, 4,5-methylene, 3,4′,5′-tri-Me ether, 3′-benzoyl

Schisanlignone E. *Schizanlignone E*

[143625-36-9]

$C_{29}H_{28}O_8$ 504.535

Constit. of *Schisandra viridis*

(7R,8R,8′S,R$_{axial}$)-form

4′,5′-Methylene, 3,3′,4-tri-Me ether, 7-E-cinnamoyl

Kadsufolin C

[1300063-42-6]

$C_{31}H_{32}O_8$ 532.589

Constit. of *Kadsura oblongifolia*. Mp 90 - 92°. $[\alpha]_D^{25}$ +33.5 (c, 0.3 in MeOH). λ_{max} 216 (log ε 4.69); 250 (sh); 277 (sh) (MeOH)

(7R,8R,8′S,S$_{axial}$)-form

3,4,4′,5,5′-Penta-Me ether, 3′-angeloyl

Propinquanin F

[897929-00-9]

$C_{28}H_{36}O_8$ 500.588

Constit. of the stems of *Schisandra propinqua*. Needles. Mp 109°. $[\alpha]_D^{20}$ +8.5 (c, 1.48 in $CHCl_3$). λ_{max} 214 (log ε 3.97); 257 (sh); 282 (sh) (EtOH)

4′,5′-Methylene, 3,3′,4-tri-Me ether, 7-Ac

Arisanschinin B

[1333378-31-6]

$C_{24}H_{28}O_8$ 444.48

Constit. of *Schisandra arisanensis*. Amorph. solid. $[\alpha]_D^{25}$ +17.5 (c, 0.2 in MeOH). λ_{max} 232 (log ε 4.13); 256 (log ε 3.82); 280 (log ε 3.2) (MeOH)

4′,5′-Methylene, 3,3′,4,5-tetra-Me ether

2,2′-Cyclo-3,3′,4,5-tetramethoxy-4′,5′-methylenedioxy-8,8′-lignan-7-ol. 7-Hydroxy-3,3′,4,5-tetramethoxy-4′,5′-methylenedioxy-2,2′-cyclolignan. **Gomisin O**

[72960-22-6]

$C_{23}H_{28}O_7$ 416.47

Constit. of fruits of *Schisandra chinensis* and of *Schisandra neglecta* and *Schisandra sphenanthera*. Cryst. (Et_2O/hexane). Mp 146.5 - 147.5°. $[\alpha]_D^{24}$ -30.1 (c, 1.06 in $CHCl_3$)

4′,5′-Methylene, 3,3′,4,5-tetra-Me ether, 7-Ac

Rubschisantherin. *Acetylgomisin O*

[102637-03-6]

$C_{25}H_{30}O_8$ 458.507

Isol. from *Schizandra rubrifolia*. Amorph. powder. $[\alpha]_D^{18}$ -69.1 (EtOH)

4′,5′-Methylene, 3,3′,4,5-tetra-Me ether, 7-angeloyl

Angeloylgomisin O

[83864-69-1]

$C_{28}H_{34}O_8$ 498.572

Isol. from *Schizandra chinensis*. Amorph. powder. $[\alpha]_D^{24}$ +47.1 (c, 1.36 in $CHCl_3$)

4′,5′-Methylene, 3,3′,4,5-tetra-Me ether, 7-tigloyl
Tigloylgomisin O
[130855-74-2]
$C_{28}H_{34}O_8$ 498.572
Isol. from *Schizandra sphenanthera*

4′,5′-Methylene, 3,3′,4,5-tetra-Me ether, 7-benzoyl
Benzoylgomisin O
[130783-32-3]
$C_{30}H_{32}O_8$ 520.578
Isol. from *Schizandra chinensis* and *Schizandra sphenanthera*. Amorph. powder. $[\alpha]_D^{25}$ -41.4 (c, 1 in $CHCl_3$)

4′,5′-Methylene, 3,4,5,7-tetra-Me ether
2,2′-Cyclo-3,4,5,7-tetramethoxy-4′,5′-methylenedioxy-8,8′-lignan-3′-ol. 3′-Hydroxy-3,4,5,7-tetramethoxy-4′,5′-methylenedioxy-2,2′-cyclolignan. **Wilsonilignan C**
[1313513-96-0]
$C_{23}H_{28}O_7$ 416.47
Constit. of the fruit of *Schisandra wilsoniana*. Yellow gum. $[\alpha]_D^{24}$ +20.2 (c, 0.2 in MeOH). λ_{max} 210 (log ε 5.32); 248 (log ε 3.35); 320 (log ε 1.26) (MeOH)

4′,5′-Methylene, 3,3′,4,5,7-penta-Me ether
2,2′-Cyclo-3,3′,4,5,7-pentamethoxy-4′,5′-methylenedioxylignan. 3,3′,4,5,7-Pentamethoxy-4′,5′-methylenedioxy-2,2′-cyclolignan. **O-Methylgomisin O**
[1276654-07-9]
$C_{24}H_{30}O_7$ 430.497
Constit. of the fruit of *Schisandra sphenanthera*. Amorph. powder. $[\alpha]_D^{27}$ +9.6 (c, 0.16 in $CHCl_3$)

7-Ketone, 4′,5′-methylene, 3,3′,4-tri-Me ether
2,2′-Cyclo-5-hydroxy-3,3′,4-trimethoxy-4′,5′-methylenedioxy-8,8′-lignan-7-one. 5-Hydroxy-3,3′,4-trimethoxy-4′,5′-methylenedioxy-7-oxo-2,2′-cyclolignan. **Shizanrin M**. *Yunnankadsurin A*
[868227-32-1]
$C_{22}H_{24}O_7$ 400.427
Constit. of *Kadsura japonica* and other *Kadsura* spp.. Yellow solid. Mp 230 - 231°. $[\alpha]_D^{28}$ -190 (c, 0.1 in MeOH). λ_{max} 208 (log ε 4.25); 250 (sh); 278 (sh); 320 (log ε 3.19) (MeOH)

(7R,8S,8′S,S$_{axial}$)-form
3,3′,4,4′-Tetra-Me ether
2,2′-Cyclo-3,3′,4,4′-tetramethoxy-8,8′-lignan-5,5′,7-triol. 5,5′,7-Trihydroxy-3,3′,4,4′-tetramethoxy-2,2′-cyclolignan. **Marlignan D**
[1228647-25-3]
$C_{22}H_{28}O_7$ 404.459
Constit. of *Schisandra wilsoniana*. Yellow gum. $[\alpha]_D^{25}$ +42.8 (c, 0.28 in MeOH). λ_{max} 210 (log ε 5.84); 241 (log ε 3.52); 326 (log ε 0.52); 345 (log ε 0.32) (MeOH)

3,3′,4,4′-Tetra-Me ether, 7-Ac
Marlignan E
[1228647-26-4]
$C_{24}H_{30}O_8$ 446.496
Constit. of *Schisandra wilsoniana*. Yellow gum. $[\alpha]_D^{23}$ +52 (c, 0.25 in MeOH). λ_{max} 205 (log ε 5.22); 242 (log ε 4.28); 285 (log ε 2.9); 309 (log ε 1.18); 360 (log ε 1.85) (MeOH)

3,3′,4,4′-Tetra-Me ether, 7-benzoyl
Marlignan G
[1228647-28-6]
$C_{29}H_{32}O_8$ 508.567
Constit. of *Schisandra wilsoniana*. Yellow gum. $[\alpha]_D^{24}$ +35.2 (c, 0.21 in MeOH). λ_{max} 203 (log ε 5.22); 243 (log ε 4.22); 281 (log ε 2.18); 374 (log ε 2.42); 395 (log ε 1.87) (MeOH)

4,4′,5,5′-Tetra-Me ether
2,2′-Cyclo-4,4′,5,5′-tetramethoxy-8,8′-lignan-3,3′,7-triol. 3,3′,7-Trihydroxy-4,4′,5,5′-tetramethoxy-2,2′-cyclolignan. **Marlignan N‡**
[1310041-39-4]
$C_{22}H_{28}O_7$ 404.459
Constit. of the fruits of *Schisandra wilsoniana*. Shows anti-HIV-1 activity. Yellow gum. $[\alpha]_D^{25}$ -18.6 (c, 0.22 in MeOH). λ_{max} 212 (log ε 4.86); 240 (log ε 3.72); 324 (log ε 0.78) (MeOH)

3,3′,4,4′,5-Penta-Me ether
2,2′-Cyclo-3,3′,4,4′,5-pentamethoxy-8,8′-lignan-5′,7-diol. 5′,7-Dihydroxy-3,3′,4,4′,5-pentamethoxy-2,2′-cyclolignan. **Gomisin U**
[135095-46-4]
$C_{23}H_{30}O_7$ 418.486
Isol. from *Schizandra sphenanthera*

3,3′,4,4′,5-Penta-Me ether, 7-benzoyl
7-Benzoylgomisin U
[130783-33-4]
$C_{30}H_{34}O_8$ 522.594
Isol. from *Schizandra sphenanthera*. Needles. $[\alpha]_D$ -61.6 ($CHCl_3$)

3,3′,4,4′,7-Penta-Me ether
2,2′-Cyclo-3,3′,4,4′,7-pentamethoxy-8,8′-lignan-5,5′-diol. 5,5′-Dihydroxy-3,3′,4,4′,7-pentamethoxy-2,2′-cyclolignan. **Marlignan F**
[1228647-27-5]
$C_{23}H_{30}O_7$ 418.486
Constit. of *Schisandra wilsoniana*. Yellow gum. $[\alpha]_D^{22}$ +33.2 (MeOH). λ_{max} 205 (log ε 5.64); 245 (log ε 3.28); 329 (log ε 0.64); 348 (log ε 0.48) (MeOH)

3,3′,4′,5,5′-Penta-Me ether
2,2′-Cyclo-3,3′,4′,5,5′-pentamethoxy-8,8′-lignan-4,7-diol. 4,7-Dihydroxy-3,3′,4′,5,5′-pentamethoxy-2,2′-cyclolignan. **Marlignan M‡**
[1420791-92-9]
$C_{23}H_{30}O_7$ 418.486
Constit. of the fruits of *Schisandra wilsoniana*. Shows anti-HIV-1 activity. Yellow gum. $[\alpha]_D^{24}$ +22.8 (c, 0.20 in MeOH). λ_{max} 211 (log ε 4.76); 242 (log ε 3.68); 322 (log ε 1.82) (MeOH)

3,3′,4,4′,5,5′-Hexa-Me ether, 7-Ac
Marlignan I
[1228647-30-0]
$C_{26}H_{34}O_8$ 474.55
Constit. of *Schisandra wilsoniana*. Yellow gum. $[\alpha]_D^{23}$ +38.4 (c, 0.32 in MeOH). λ_{max} 206 (log ε 5.88); 242 (log ε 3.62); 328 (log ε 0.61); 346 (log ε 0.47) (MeOH)

3,3′,4,4′,5,7-Hexa-Me ether, 5′-Ac
Rubrilignan B
[1354630-94-6]
$C_{26}H_{34}O_8$ 474.55
Constit. of the fruit of *Schisandra rubriflora*. Yellow gum. $[\alpha]_D^{25}$ -27.6 (c, 0.25 in MeOH). λ_{max} 210 (log ε 4.26); 252 (log ε 3.38); 316 (log ε 0.88) (MeOH)

3,3′,4,4′,5′,7-Hexa-Me ether, 5-Ac
Rubrilignan A
[1354757-49-5]
$C_{26}H_{34}O_8$ 474.55
Constit. of the fruit of *Schisandra rubriflora*. Yellow gum. $[\alpha]_D^{25}$ -36.5 (c, 0.25 in MeOH). λ_{max} 212 (log ε

4.16); 252 (log ε 3.36); 318 (log ε 0.94) (MeOH)

7-Et, 3,3′,4,4′-tetra-Me ether
Neglectalignan C. **Neglectaphenol C**
[1686087-24-0]
$C_{24}H_{32}O_7$ 432.513
Constit. of the stems of *Schisandra neglecta*. Exhibits moderate anti-HIV activity. Yellow gum. $[\alpha]_D^{24.9}$+28.6 (c, 0.2 in MeOH). λ_{max} 205 (log ε 4.24); 240 (log ε 3.81); 329 (log ε 0.86) (MeOH)

7-Et, 3,3′,4,4′,5,5′-hexa-Me ether
Marlignan J
[1228647-31-1]
$C_{26}H_{36}O_7$ 460.567
Constit. of *Schisandra wilsoniana*. Yellow gum. $[\alpha]_D^{24}$+35.5 (c, 0.16 in MeOH). λ_{max} 205 (log ε 5.64); 240 (log ε 3.6); 329 (log ε 0.64); 348 (log ε 0.48) (MeOH)

7-O-(2-Methylpropyl), 3,3′,4,4′,5′-penta-Me ether
Lanciphenol A. *Neglignan F*
[1441710-35-5]
[1702301-10-7] (Lanciphenol A)
$C_{27}H_{38}O_7$ 474.594
Constit. of *Schisandra lancifolia* and *Schisandra neglecta*. Exhibits anti-HIV activity. Yellow gum. $[\alpha]_D^{25}$+33.6 (c, 0.2 in MeOH). λ_{max} 210 (log ε 4.80); 240 (log ε 3.62); 320 (log ε 0.89) (MeOH)

7-O-(2-Methylpropyl), 3,3′,4,4′,5,5′-hexa-Me ether
Neglignan E
[1441710-28-6]
$C_{28}H_{40}O_7$ 488.621
Constit. of *Schisandra neglecta*. Yellow gum. $[\alpha]_D^{25}$+38.2 (c, 0.2 in MeOH). λ_{max} 210 (log ε 4.87); 241 (log ε 3.65); 320 (log ε 0.68) (MeOH)

4,5-Methylene, 3,3′,4′-tri-Me ether
Schisanwilsonin O
[1610942-19-2]
$C_{22}H_{26}O_7$ 402.443
Constit. of *Schisandra wilsoniana*. Amorph. powder. $[\alpha]_D^{25}$+42 (c, 0.02 in MeOH). λ_{max} 213 (log ε 4.23); 257 (sh); 288 (sh) (MeOH)

4,5-Methylene, 3′,4′,5′-tri-Me ether, 7-O-(2-methylbutanoyl)
Schizanrin B
[343326-63-6]
$C_{27}H_{34}O_8$ 486.561
Constit. of *Kadsura matsudai*. Amorph. light yellow powder. $[\alpha]_D$ -37.5 (c, 0.1 in $CHCl_3$)

4,5-Methylene, 3′,4′,5′-tri-Me ether, 7-hexanoyl
Schizanrin C
[343326-64-7]
$C_{28}H_{36}O_8$ 500.588
Constit. of *Kadsura matsudai*. Amorph. light yellow powder. $[\alpha]_D$ -13.5 (c, 0.14 in $CHCl_3$)

4,5-Methylene, 3′,4′,5′-tri-Me ether, 7-tigloyl
Schizanrin D
[343326-65-8]
$C_{27}H_{32}O_8$ 484.545
Constit. of *Kadsura matsudai*. Amorph. light yellow powder

4,5-Methylene, 3′,4′,5′-tri-Me ether, 7-benzoyl
Schizanrin E
[343326-66-9]
$C_{29}H_{30}O_8$ 506.551
Constit. of *Kadsura matsudai*. Amorph. light yellow powder. $[\alpha]_D$ -325 (c, 0.1 in $CHCl_3$). Doubt concerning magnitude of opt. rotn.

4,5-Methylene, 3′,4′,5′-tri-Me ether, 7-O-E-cinnamoyl
Schizanrin A
[248257-23-0]
$C_{31}H_{32}O_8$ 532.589
Constit. of *Schizandra arisanensis*. Prisms. Mp 173 - 175°. $[\alpha]_D$ +48 (c, 0.1 in $CHCl_3$)

4,5-Methylene, 3,3′,4′,5′-tetra-Me ether
Isogomisin O
[83916-76-1]
Amorph. powder. $[\alpha]_D^{24}$-27.3 (c, 1.10 in $CHCl_3$)

4,5-Methylene, 3,3′,4′,5′-tetra-Me ether, 7-angeloyl
7-Angeloylisogomisin O. *Schisanwilsonin N*
[83864-70-4]
$C_{28}H_{34}O_8$ 498.572
Constit. of fruits of *Schizandra chinensis* and of *Schizandra wilsoniana*. Prisms (Et_2O/hexane) or amorph. powder. Mp 122 - 123°. $[\alpha]_D^{24}$+52.2 (c, 1.80 in $CHCl_3$) (7-Angeloylgomisin O). $[\alpha]_D^{25}$+168 (c, 0.01 in MeOH) (Schisanwilsonin N)

4,5-Methylene, 3,3′,4′,5′-tetra-Me ether, 7-tigloyl
Schisanwilsonin M
[1610942-17-0]
$C_{28}H_{34}O_8$ 498.572
Constit. of *Schisandra wilsoniana*. Amorph. powder. $[\alpha]_D^{25}$+91 (c, 0.01 in MeOH). λ_{max} 216 (log ε 4.65); 255 (sh); 286 (sh) (MeOH)

4,5-Methylene, 3,3′,4′,5′-tetra-Me ether, 7-benzoyl
7-Benzoylisogomisin O
[83864-71-5]
$C_{30}H_{32}O_8$ 520.578
Isol. from *Schizandra chinensis*. Inhibits ACAT. Amorph. powder. $[\alpha]_D^{24}$-14.7 (c, 1.70 in $CHCl_3$)

4,5-Methylene, 3,3′,4′,7-tetra-Me ether
5′-O-Demethylwuweilignan I
[1040392-75-3]
$C_{23}H_{28}O_7$ 416.47
Constit. of *Schisandra rubriflora*. Oil. $[\alpha]_D^{26}$+8.3 (c, 0.1 in $CHCl_3$)

4,5-Methylene, 3,4′,5′,7-tetra-Me ether
2,2′-Cyclo-3,4′,5′,7-tetramethoxy-4,5-methylenedioxy-8,8′-lignan-3′-ol. 3′-Hydroxy-3,4′,5′,7-tetramethoxy-4,5-methylenedioxy-2,2′-cyclolignan. ***Marlignan R***
[1420791-95-2]
$C_{23}H_{28}O_7$ 416.47
Constit. of the fruits of *Schisandra wilsoniana*. Shows anti-HIV-1 activity. Yellow gum. $[\alpha]_D^{24}$+43.8. λ_{max} 210 (log ε 4.82); 246 (log ε 3.79); 286 (log ε 2.64); 329 (log ε 1.68) (MeOH)

4,5-Methylene, 3′,4′,5′,7-tetra-Me ether
2,2′-Cyclo-3′,4′,5′,7-tetramethoxy-4,5-methylenedioxy-8,8′-lignan-3-ol. 3-Hydroxy-3′,4′,5′,7-tetramethoxy-4,5-methylenedioxy-2,2′-cyclolignan. ***Marlignan S***
[1420791-96-3]
$C_{23}H_{28}O_7$ 416.47
Constit. of the fruits of *Schisandra wilsoniana*. Shows anti-HIV-1 activity. Yellow gum. $[\alpha]_D^{24}$+45.8. λ_{max} 210 (log ε 4.74); 245 (log ε 3.83); 280 (log ε 2.54); 325 (log ε 1.74) (MeOH)

4,5-Methylene, penta-Me ether
Methylisogomisin O. *Wuweilignan I*
[1023744-70-8]
$C_{24}H_{30}O_7$ 430.497
Constit. of *Schizandra propinqua* var. *propinqua*. Prisms. Mp 148 - 150°. $[\alpha]_D^{28}$ +46.7 (c, 0.4 in MeOH). λ_{max} 217 (log ε 4.51) (MeOH)

4′,5′-Methylene, 3,3′,4-tri-Me ether
2,2′-Cyclo-3,3′,4-trimethoxy-4′,5′-methylenedioxy-8,8′-lignan-5,7-diol. 5,7-Dihydroxy-3,3′,4-trimethoxy-4′,5′-methylenedioxy-2,2′-cyclolignan.
Marlignan O‡
$C_{22}H_{26}O_7$ 402.443
Constit. of the fruits of *Schisandra wilsoniana*. Shows anti-HIV-1 activity. Yellow gum. $[\alpha]_D^{24}$ +38.4 (c, 0.22 in MeOH). λ_{max} 212 (log ε 4.68); 240 (log ε 3.75); 330 (log ε 1.74) (MeOH)

4,5:4′,5′-Bis(methylene), 3,3′-di-Me ether
Gomisin R
[83864-72-6]
$C_{22}H_{24}O_7$ 400.427
Constit. of fruits of *Schizandra chinensis*. Amorph. powder. $[\alpha]_D^{24}$ -21.1 (c, 1.94 in $CHCl_3$)

4,5:4′,5′-Bis(methylene), 3,3′-di-Me ether, 7-angeloyl
Angeloylgomisin R
[128324-09-4]
$C_{27}H_{30}O_8$ 482.529
Constit. of *Kadsura interior* and *Kadsura longipedunculata*. Needles (EtOH). Mp 160 - 162°. $[\alpha]_D^{29}$ +66 (c, 0.06 in $CHCl_3$). λ_{max} 218 (log ε 4.72); 255 (log ε 4.32); 280 (log ε 3.72) (EtOH)

4,5:4′,5′-Bis(methylene), 3,3′-di-Me ether, 7-benzoyl
Interiotherin A. *Benzoylgomisin R*
[181701-06-4]
$C_{29}H_{28}O_8$ 504.535
Constit. of *Kadsura interior*. Shows anti-HIV activity. Prisms (MeOH). Mp 153 - 154°. $[\alpha]_D$ -53.4 (c, 2.2 in $CHCl_3$). λ_{max} 222 (log ε 4.84); 255 (log ε 4.1); 278 (log ε 3.79) (MeOH)

4,5:4′,5′-Bis(methylene), 3,7-di-Me ether
2,2′-Cyclo-3,7-dimethoxy-4,5:4′,5′-bis(methylenedioxy)-8,8′-lignan-3′-ol. 3′-Hydroxy-3,7-dimethoxy-4,5:4′,5′-bis(methylenedioxy)-2,2′-cyclolignan. ***Marlignan P***
[1420791-93-0]
$C_{22}H_{24}O_7$ 400.427
Constit. of the fruits of *Schisandra wilsoniana*. Shows anti-HIV-1 activity. Yellow gum. $[\alpha]_D^{23}$ -28.5 (c, 0.22 in MeOH). λ_{max} 210 (log ε 4.86); 252 (log ε 3.81); 325 (log ε 2.44); 340 (log ε 1.86) (MeOH)

4,5:4′,5′-Bis(methylene), 3′,7-di-Me ether
2,2′-Cyclo-3′,7-dimethoxy-4,5:4′,5′-bis(methylenedioxy)-8,8′-lignan-3-ol. 3-Hydroxy-3′,7-dimethoxy-4,5:4′,5′-bis(methylenedioxy)-2,2′-cyclolignan.
Marlignan Q
[1420791-94-1]
$C_{22}H_{24}O_7$ 400.427
Constit. of the fruits of *Schisandra wilsoniana*. Shows anti-HIV-1 activity. Yellow gum. $[\alpha]_D^{24}$ +38.6 (c, 0.25 in MeOH). λ_{max} 210 (log ε 4.73); 250 (log ε 3.66); 285 (log ε 2.54); 325 (log ε 2.08); 340 (log ε 1.55) (MeOH)

4,5:4′,5′-Bis(methylene), 3,3′,7-tri-Me ether
2,2′-Cyclo-3,3′,7-trimethoxy-4,5:4′,5′-bis(methylenedioxy)lignan. 3,3′,7-Trimethoxy-4,5:4′,5′-bis(methylenedioxy)-2,2′-cyclolignan.
Methylgomisin R
[1039730-63-6]
$C_{23}H_{26}O_7$ 414.454
Constit. of *Schisandra rubriflora*. Amorph. yellow powder. $[\alpha]_D^{27}$ +36.4 (c, 0.82 in $CHCl_3$). λ_{max} 215 (log ε 4.65); 257 (sh); 282 (log ε 3.58) (MeOH)

4,5:4′,5′-Bis(methylene), 7-Et, 3,3′-di-Me ether
2,2′-Cyclo-7-ethoxy-3,3′-dimethoxy-4,5:4′,5′-bis(methylenedioxy)lignan. 7-Ethoxy-3,3′-dimethoxy-4,5:4′,5′-bis(methylenedioxy)-2,2′-cyclolignan.
Schisanwilsonin L
[1610531-89-9]
$C_{24}H_{28}O_7$ 428.481
Constit. of *Schisandra wilsoniana*. Amorph. powder. $[\alpha]_D^{25}$ +38 (c, 0.02 in MeOH). λ_{max} 214 (log ε 4.51); 253 (sh); 282 (sh) (MeOH)

7-Ketone, 3,3′,4,4′-tetra-Me ether
2,2′-Cyclo-5,5′-dihydroxy-3,3′,4,4′-tetramethoxy-8,8′-lignan-7-one. 5,5′-Dihydroxy-3,3′,4,4′-tetramethoxy-7-oxo-2,2′-cyclolignan.
Marlignan H
[1228647-29-7]
$C_{22}H_{26}O_7$ 402.443
Constit. of *Schisandra wilsoniana*. Yellow gum. $[\alpha]_D^{25}$ +26.5 (c, 0.28 in MeOH). λ_{max} 203 (log ε 5.48); 240 (log ε 4.15); 280 (log ε 2.25); 370 (log ε 2.26) (MeOH)

7-Ketone, 4,5-methylene, 3′,4′,5′-tri-Me ether
2,2′-Cyclo-3-hydroxy-3′,4′,5-trimethoxy-4,5-methylenedioxy-8,8′-lignan-7-one. 3-Hydroxy-3′,4′,5-trimethoxy-4,5-methylenedioxy-7-oxo-2,2′-cyclolignan.
Schizanrin N
[874472-17-0]
$C_{22}H_{24}O_7$ 400.427
Constit. of *Kadsura japonica*. Mp 65 - 67°. $[\alpha]_D^{28}$ -180 (c, 0.1 in MeOH)

(7S,8R,8′R,R_{axial})-form

3,4′,5,5′-Tetra-Me ether, 7-E-cinnamoyl
Kadsuphilol D
[947256-79-3]
$C_{31}H_{34}O_8$ 534.605
Constit. of *Kadsura philippinensis*. Amorph. powder. $[\alpha]_D^{25}$ +89 (c, 0.01 in CH_2Cl_2). λ_{max} 213 (log ε 3.35); 280 (log ε 3.01) (MeOH)

4,5-Methylene, 3,4′,5′-tri-Me ether
Kadsuphilol A
[947256-76-0]
$C_{22}H_{26}O_7$ 402.443
Constit. of *Kadsura philippinensis*. Amorph. powder. $[\alpha]_D^{25}$ +15 (c, 0.01 in CH_2Cl_2). λ_{max} 216 (log ε 3.86); 254 (log ε 3.25) (MeOH)

4,5-Methylene, 3,4′,5′-tri-Me ether, 3′-Ac
Polysperlignan K
[1360440-86-3]
$C_{24}H_{28}O_8$ 444.48
Isol. from the stems of *Kadsura polysperma*. Yellow oil. $[\alpha]_D^{17}$ -2.2 (c, 0.10 in $CHCl_3$). λ_{max} 193 (log ε 2.89); 241 (log ε 3.47) ($CHCl_3$)

(7S,8R,8′R,S_{axial})-form

4,5-Methylene, 3,3′,4′-tri-Me ether

Kadsuralignan E

[906081-86-5]

$C_{22}H_{26}O_7$ 402.443

Constit. of the rhizomes of *Kadsura coccinea*. Needles (hexane). Mp 157 - 158°. λ_{max} 220 (log ε 3.9); 255 (sh); 289 (sh) (MeOH)

(7S,8R,8′S,S_{axial})-form

4,5-Methylene, 3′,4′,5′-tri-Me ether, 7-Ac

Kadsumarin A

[246236-13-5]

$C_{24}H_{28}O_8$ 444.48

Constit. of *Kadsura matsudai*, *Kadsura phillipinensis*, *Schizandra arisanensis* and *Schizandra rubriflora*. Pale pink powder. Mp 175 - 178°. [α] D +13 (c, 0.1 in $CHCl_3$)

4′,5′-Methylene, 3,3′,4-tri-Me ether

Kadsuphilin B

[887770-03-8]

$C_{22}H_{26}O_7$ 402.443

Constit. of *Kadsura philippinensis* and *Kadsura coccinea*. Pale yellow powder. Mp 144 - 147°. $[\alpha]^{28}_{D}$ +51 (c, 1 in CH_2Cl_2) (+20). λ_{max} 244 (log ε 4.46); 251 (log ε 4.12); 284 (log ε 3.54) (MeOH)

4′,5′-Methylene, 3,3′,4-tri-Me ether, 7-benzoyl

Arisanschinin A. *Kadsufolin D*

[1300063-43-7]

$C_{29}H_{30}O_8$ 506.551

Constit. of *Schisandra arisanensis* and *Kadsura oblongifolia*. Cytotoxic against A549, DU145, KB and HCT-8 tumour cell lines. Amorph. solid or powder. Mp 91 - 93° (Kadsufolin D). $[\alpha]^{25}_{D}$ -19.5 (c, 0.2 in MeOH) (Kadsufolin D). $[\alpha]^{25}_{D}$ +42.7 (c, 0.3 in MeOH). λ_{max} 230 (log ε 4.46); 256 (log ε 3.92); 280 (log ε 3.69) (MeOH) (Kadsufolin D). [neutral] λ_{max} 221 (log ε 4.77); 255 (sh); 280 (sh) (MeOH)

4′,5′-Methylene, 3,4,5-tri-Me ether

2,2′-Cyclo-3,4,5-trimethoxy-4′,5′-methylenedioxy-8,8′-lignan-3′,7-diol. 3′,7-Dihydroxy-3,4,5-trimethoxy-4′,5′-methylenedioxy-2,2′-cyclolignan.

Wilsonilignan A

[1313513-94-8]

$C_{22}H_{26}O_7$ 402.443

Constit. of the fruit of *Schisandra wilsoniana*. Yellow gum. $[\alpha]^{23}_{D}$ -58.2 (c, 0.28 in MeOH). λ_{max} 215 (log ε 4.64); 253 (log ε 3.84); 314 (log ε 0.62) (MeOH)

4′,5′-Methylene, 3′,4,5-tri-Me ether

2,2′-Cyclo-3′,4,5-trimethoxy-4′,5′-methylenedioxy-8,8′-lignan-3,7-diol. 3,7-Dihydroxy-3′,4,5-trimethoxy-4′,5′-methylenedioxy-2,2′-cyclolignan.

Wilsonilignan B

[1313513-95-9]

$C_{22}H_{26}O_7$ 402.443

Constit. of the fruit of *Schisandra wilsoniana*. Yellow gum. $[\alpha]^{23}_{D}$ -59.3 (c, 0.25 in MeOH). λ_{max} 210 (log ε 4.72); 252 (log ε 3.63); 314 (log ε 0.53) (MeOH)

4′,5′-Methylene, 3,3′,4,5-tetra-Me ether

Epigomisin O. *Gomisin E‡*

[73036-31-4]

$C_{23}H_{28}O_7$ 416.47

Isol. from *Schizandra chinensis*. Amorph. powder. $[\alpha]^{23}_{D}$ -66.7 (c, 1.50 in $CHCl_3$)

(7S,8S,8′S,S_{axial})-form

3,3′,4,4′-Tetra-Me ether

Schisandlignan D

[1610338-28-7]

$C_{22}H_{28}O_7$ 404.459

Constit. of rattans of *Schisandra sphenanthera*. Oil. λ_{max} 210 (log ε 3.55); 250 (log ε 3.00); 290 (log ε 2.70) (MeOH)

3,3′,4,4′,5-Penta-Me ether

Gomisin S

[119239-49-5]

$C_{23}H_{30}O_7$ 418.486

Isol. from *Schizandra chinensis*. Prisms. Mp 172 - 176°. $[\alpha]^{23}_{D}$ -63 (c, 0.49 in $CHCl_3$)

4′,5′-Methylene, 3,3′,5-tri-Me ether

Neglectalignan D. ***Neglectaphenol D***

[1686087-88-6]

$C_{22}H_{26}O_7$ 402.443

Constit. of the stems of *Schisandra neglecta*. Exhibits moderate anti-HIV activity. Yellow gum. $[\alpha]^{25}_{D}$ -45.2 (c, 0.25 in MeOH)

4,5:4′,5′-Bis(methylene), 3,3′-di-Me ether

7-Epigomisin R

[852211-14-4]

$C_{22}H_{24}O_7$ 400.427

Constit. of *Kadsura philippinensis*. Pale yellow powder. $[\alpha]^{28}_{D}$ +22 (c, 1 in CH_2Cl_2). λ_{max} 237 (log ε 4.73); 261 (log ε 4.36); 281 (log ε 3.78) (MeOH)

4,5:4′,5′-Bis(methylene), 3,3′-di-Me ether, 7-Ac

Acetyl-7-epigomisin R

[1257391-37-9]

$C_{24}H_{26}O_8$ 442.464

Constit. of the roots of *Kadsura coccinea*. Cryst. (MeOH). Mp 146 - 147°. $[\alpha]^{25}_{D}$ +85 (c, 0.25 in MeOH)

(8S,8′R,S_{ax})-form

7-Ketone, hexa-Me ether

Schisantherin S

[2230512-49-7]

$C_{24}H_{30}O_7$ 430.497

Constit. of the roots of *Kadsura coccinea*. Powder

7-Ketone, 4,5:4′,5′-bis(methylene), 3,3′-di-Me ether

Schisantherin R

[2230512-48-6]

$C_{22}H_{22}O_7$ 398.411

Constit. of the roots of *Kadsura coccinea*. Powder.

Chen, Y.-P. *et al.*, *Bull. Chem. Soc. Jpn.*, 1977, **50**, 1824-1826 (*Kadsurin*).

Merric, M. *et al.*, *J. Am. Chem. Soc.*, 1977, **99**, 7673-7678 (*Kadsurin, synth*).

Ikeya, Y. *et al.*, *Chem. Pharm. Bull.*, 1979, **27**, 2695-2709 (*Gomisin O, Epigomisin O*).

Ookawa, N. *et al.*, *Chem. Pharm. Bull.*, 1981, **29**, 123-127 (*Acetylbinankadsurin A, Angeloylbinankadsurin A, Capreoylbinankadsurin A*).

Ikeya, Y. *et al.*, *Chem. Pharm. Bull.*, 1982, **30**, 3202-3206 (*Angeloylgomisin O, Angeloylisogomisin O, Benzoylisogomisin O, Gomisin R*).

Tan, R, *et al.*, *Planta Med.*, 1984, **50**, 414-417 (*Angeloylgomisin R*).

Wang, H.-J. *et al.*, *Yaoxue Xuebao*, 1985, **20**, 832-841 (*Rubschisantherin*).

Ikeya, Y. *et al.*, *Chem. Pharm. Bull.*, 1988, **36**, 3974-3979 (*Gomisin S*).

Liu, J.-S. *et al.*, *Huaxue Xuebao*, 1988, **46**, 460-464 (*Schizantherin F*).

Liu, J-S. *et al.*, *Huaxue Xuebao*, 1991, **49**, 412-416 (*Schisanlactones A,B, Butyrylbinankadsurin A*).

Ikeya, Y. *et al.*, *Phytochemistry*, 1991, **30**, 975-980 (*Gomisin U, Benzoylgomisin U, Tigloylgomisin O*).

Li, L.-N. *et al.*, *Planta Med.*, 1991, **57**, 169-171 (*Kadsura longipedunculata lignans*).

Luo, G. *et al.*, *Huaxue Xuebao*, 1992, **50**, 515-520 (*Schisanlignone E*).

Luo, G. *et al.*, *Huaxue Xuebao*, 1992, **50**, 620-624 (*Schisandra viridis constits*).

Chen, D.-F. *et al.*, *Phytochemistry*, 1992, **31**, 629-632 (*Heteroclitins*).

Liu, J.-S. *et al.*, *Phytochemistry*, 1993, **32**, 1293-1296 (*Schisantherin O*).

Chen, C.-C. *et al.*, *J. Nat. Prod.*, 1994, **57**, 1164-1165 (*Benzoylgomisin O*).
Ookawa, N. *et al.*, *Phytochemistry*, 1995, **39**, 1187-1191 (*Deangeloylschizantherin F, Binankadsurin B derivs*).
Chen, D.-F. *et al.*, *J. Nat. Prod.*, 1996, **59**, 1066-1068 (*Interiotherin A*).
Kuo, Y.-H. *et al.*, *Chem. Pharm. Bull.*, 1999, **47**, 1047-1048 (*Kadsumarin A*).
Kuo, Y.-H. *et al.*, *J. Org. Chem.*, 1999, **64**, 7023-7027 (*Schizanrin A*).
Kwon, B.M. *et al.*, *Planta Med.*, 1999, **65**, 74-76 (*activity*).
Kuo, Y.-H. *et al.*, *J. Nat. Prod.*, 2001, **64**, 487-490, 1608 (*Schizanrins B-E*).
Li, L. *et al.*, *Helv. Chim. Acta*, 2004, **87**, 2943-2947 (*Schisandra rubriflora constits*).
Jia, Z. *et al.*, *Helv. Chim. Acta*, 2005, **88**, 2288-2293 (*Yunnankadsurins A,B*).
Coleman, R.S. *et al.*, *J. Org. Chem.*, 2005, **70**, 8932-8941 (*Gomisin O, Interiotherin A, synth*).
Kuo, Y.-H. *et al.*, *Planta Med.*, 2005, **71**, 646-653 (*Schizanrins L,N*).
Kuo, Y.-H. *et al.*, *Planta Med.*, 2005, **71**, 646-653 (*Shizanrin M*).
Sun, Q.-Z. *et al.*, *Chem. Pharm. Bull.*, 2006, **54**, 129-132 (*Longipedunins A-C*).
Xu, L.J. *et al.*, *Chem. Pharm. Bull.*, 2006, **54**, 542-545 (*Propinquanin F*).
Li, H.-R. *et al.*, *Chem. Pharm. Bull.*, 2006, **54**, 1022-1025 (*Kadsuralignan A*).
Lu, Y. *et al.*, *Helv. Chim. Acta*, 2006, **89**, 895-901 (*Kadsutherins*).
Li, H. *et al.*, *Heterocycles*, 2006, **68**, 1259-1265 (*Kadsuralignan E*).
Shen, Y.-C. *et al.*, *J. Nat. Prod.*, 2006, **69**, 963-966 (*Kadsura phillipinensis constits*).
Lei, C. *et al.*, *Chem. Pharm. Bull.*, 2007, **55**, 1281-1283 (*Methylisogomisin O*).
Shen, Y.-C. *et al.*, *J. Nat. Prod.*, 2007, **70**, 1139-1145 (*Kadsuphilols A,D*).
Li, H. *et al.*, *J. Nat. Prod.*, 2007, **70**, 1999-2002 (*Kadsuralignans I,J,K*).
Li, H.-M. *et al.*, *Helv. Chim. Acta*, 2008, **91**, 1053-1062 (*5′-O-Demethylwuweilignan I, Methylgomisin R*).
Li, X.-N. *et al.*, *J. Nat. Prod.*, 2009, **72**, 1133-1141 (*Tiegusanins A,B,C,L, 3′-Angeloylbinankadsurin B*).
Ban, N.K. *et al.*, *Planta Med.*, 2009, **75**, 1253-1257 (*Acetyl-7-epigomisin R*).
Yang, G.-Y. *et al.*, *Chem. Pharm. Bull.*, 2010, **58**, 734-737 (*Schilancifolignans A-C*).
Yang, G.-Y. *et al.*, *J. Asian Nat. Prod. Res.*, 2010, **12**, 470-476 (*Wilsonilignans A,B,C*).
Yang, G.-Y. *et al.*, *J. Nat. Prod.*, 2010, **73**, 915-919 (*Marlignans D-J*).
Liu, C.-C. *et al.*, *Planta Med.*, 2010, **76**, 1605-1610 (*Arisanschinins A,B*).
Ren, R. *et al.*, *Z. Naturforsch., B*, 2010, **65**, 211-218 (*O-Methylgomisin O*).
Huang, Z. *et al.*, *Helv. Chim. Acta*, 2011, **94**, 519-527 (*Kadsufolins A-D*).
Mu, H.-X. *et al.*, *J. Asian Nat. Prod. Res.*, 2011, **13**, 393-399 (*Rubrilignans A,B*).
Duan, Y-X. *et al.*, *J. Asian Nat. Prod. Res.*, 2011, **13**, 592-598 (*Schisandra neglecta constits*).
Hu, W. *et al.*, *J. Asian Nat. Prod. Res.*, 2012, **14**, 364-369 (*Kadsuralignans G,L*).
Dong, K. *et al.*, *J. Nat. Prod.*, 2012, **75**, 249-256 (*Polysperlignans F,K*).
Jin, M.N. *et al.*, *Planta Med.*, 2012, **78**, 807-813 (*Propinquains G,H*).
Liang, C-Q. *et al.*, *Chem. Pharm. Bull.*, 2013, **61**, 96-100 (*Schisandra sphenenthera constits*).
Li, Y.-F. *et al.*, *J. Asian Nat. Prod. Res.*, 2013, **15**, 934-940 (*Schisandlignan D*).
Gao, X.-M. *et al.*, *J. Braz. Chem. Soc.*, 2013, **24**, 2021-2027 (*Neglectaphenols C,D, activity*).
Yang, G.-Y. *et al.*, *J. Nat. Prod.*, 2013, **76**, 250-255 (*Marlignans M-S*).
Gao, X.-M. *et al.*, *J. Nat. Prod.*, 2013, **76**, 1052-1057 (*Neglignans E,F,G*).
Ma, W.-H. *et al.*, *Planta Med.*, 2013, **79**, 1051-1055 (*Schisanwilsonins K-O*).
Zhao, Q.-J. *et al.*, *Arch. Pharmacal Res.*, 2014, **37**, 1375-1379 (*Kadsurains A,B,C*).
Li, Y.-K. *et al.*, *Asian J. Chem.*, 2014, **26**, 224-226 (*Lanciphenol A, activity*).
Yang, H.-Y. *et al.*, *Asian J. Chem.*, 2014, **26**, 3152-3154 (*Lanciphenol B*).
Xu, L. *et al.*, *Chem. Nat. Compd. (Engl. Transl.)*, 2018, **54**, 242-244 (*Schisantherins R,S*).
Liu, J. *et al.*, *J. Nat. Prod.*, 2019, **82**, 2842-2851 (*Longipedunculatin D, Longipedlignan K, cd*)

2,2′-Cyclo-8,8′-lignan-3,3′,4,4′,5,5′,8-heptol C226

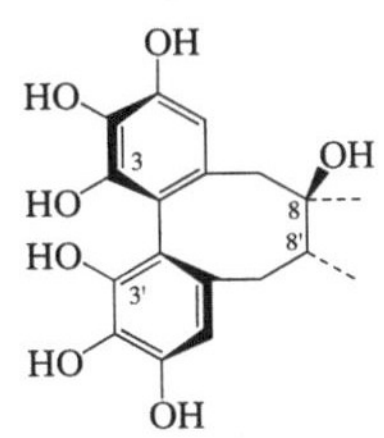

($8R,8'R,S_{axial}$)-*form*

3,3′,4,4′,5,5′,8-Heptahydroxy-2,2′-cyclolignan

$C_{18}H_{20}O_7$ 348.351

Axial abs. config. of Schisandrin revised in 2014 on spectroscopic (VCD) evidence. If the reassignment is correct it is not clear if this might also apply to other related lignans. Abs. configs. in this entry are given according to the most recent lit. assignment for each compd.

($8R,8'R,S_{axial}$)-form

3,3′,4,4′-Tetra-Me ether, 5-O-β-D-glucopyranoside

[2014382-50-2]

$C_{28}H_{38}O_{12}$ 566.6

Constit. of the stems of *Schisandra chinensis*. Amorph. powder. $[\alpha]_D^{20}$ +6 (c, 0.4 in MeOH). λ_{max} 218 (log ε 3.65); 258 (sh) (MeOH)

($8R,8'S,R_{axial}$)-form

3,3′,4,4′,5,5′-Hexa-Me ether

2,2′-Cyclo-3,3′,4,4′,5,5′-hexamethoxy-8,8′-lignan-8-ol. 8-Hydroxy-3,3′,4,4′,5,5′-hexamethoxy-2,2′-cyclolignan. ***Isoschisandrin. Isoschizandrin***

[114422-18-3]

$C_{24}H_{32}O_7$ 432.513

Constit. of the fruit of *Schisandra chinensis*. Antiulcer agent. Amorph. powder. $[\alpha]_D^{25}$+92 (c, 1.22 in $CHCl_3$). $[\alpha]_D^{24}$+110 (c, 0.4 in $CHCl_3$) (synthetic). Genus given as Schizandra in ref.

4′,5′-Methylene, 3,4,5-tri-Me ether, 3′-tigloyl

3′-Tigloylschinlignan D

[2089242-48-6]

$C_{27}H_{32}O_8$ 484.545

Constit. of the fruit of *Schisandra chinensis*. Amorph. solid. $[\alpha]_D^{20}$ +9.7 (c, 0.2 in $CHCl_3$). λ_{max} 258 (log ε 3.76); 283 (log ε 3.58) (MeOH)

($8S,8'R,R_{axial}$)-form

3,3′,4,4′,5,5′-Hexa-Me ether

Schisandrin*. Schisandrol A.* ***Schizandrin****. Schizandrol A. Wuweizi alcohol A. Wuweizichun A*

[7432-28-2]

$C_{24}H_{32}O_7$ 432.513

Constit. of *Schisandra chinensis*. Microsomal cytochrome P 450 inducer. Antihepatotoxic agent. Free radical scavenger. Mp 128 - 129°. $[\alpha]_D$+91.5 (c, 0.1 in EtOH). $[\alpha]_D^{22}$+88.7 ($CHCl_3$). See note above on abs. configs. Racemate reported from *Schisandra chinensis*. λ_{max} 217 (ε 79400); 246 (ε 19500); 251 (ε 20000); 256 (sh); 288 (sh) (EtOH) (Derep) HP1626000

($8S,8'S,R_{axial}$)-form

3,3′,4,4′-Tetra-Me ether

2,2′-Cyclo-3,3′,4,4′-tetramethoxy-8,8′-lignan-5,5′,8-triol. 5,5′,8-Trihydroxy-3,3′,4,4′-tetramethoxy-2,2′-cyclolignan. ***Marlignan L***

[1228647-33-3]

$C_{22}H_{28}O_7$ 404.459

Constit. of *Schisandra wilsoniana*. Yellow gum. $[\alpha]_D^{25}$+35.2 (c, 0.24 in MeOH). λ_{max} 210 (log ε 5.06); 239 (log ε 3.48); 330 (log ε 0.58); 348 (log ε 0.27) (MeOH)

3,4,4′,5-Tetra-Me ether

2,2′-Cyclo-3,4,4′,5-tetramethoxy-8,8′-lignan-3′,5′,8-triol. 3′,5′,8-Trihydroxy-3,4,4′,5-tetramethoxy-2,2′-cyclolignan. ***Marlignan K***

[1228647-32-2]

$C_{22}H_{28}O_7$ 404.459
Constit. of *Schisandra wilsoniana*. Yellow gum. $[\alpha]_D^{25}$ +42.2 (c, 0.18 in MeOH). λ_{max} 208 (log ε 5.48); 240 (log ε 3.22); 325 (log ε 0.52); 345 (log ε 0.31) (MeOH)

3,3′,4,4′,5-Penta-Me ether
2,2′-Cyclo-3,3′,4,4′,5-pentamethoxy-8,8′-lignan-5′,8-diol. 5′,8-Dihydroxy-3,3′,4,4′,5-pentamethoxy-2,2′-cyclolignan. ***Gomisin T***
[119139-66-1]
$C_{23}H_{30}O_7$ 418.486
Isol. from *Schisandra chinensis*. Powder. $[\alpha]_D^{23}$ +60 (c, 0.50 in $CHCl_3$)

3,3′,4,4′,5′-Penta-Me ether, 5-O-β-D-glucopyranoside
Schisandroside B
[2059120-55-5]
$C_{29}H_{40}O_{12}$ 580.627
Constit. of the roots of *Schisandra chinensis*. Amorph. powder. $[\alpha]_D^{25}$ +6.17 (c, 0.3 in MeOH)

3,4,4′,5,5′-Penta-Me ether
2,2′-Cyclo-3,4,4′,5,5′-pentamethoxy-8,8′-lignan-3′,8-diol. 3′,8-Dihydroxy-3,4,4′,5,5′-pentamethoxy-2,2′-cyclolignan. ***Gomisin H****. Norschizandrin*
[66056-20-0]
$C_{23}H_{30}O_7$ 418.486
Isol. from fruit of *Schisandra chinensis*. Plates (hexane/Et_2O). Mp 144 - 145°. $[\alpha]_D^{26}$ +41.2 (c, 1.0 in $CHCl_3$)

3,4,4′,5,5′-Penta-Me ether, 3′-angeloyl
Angeloylgomisin H*. Micrantherin A*
[66056-22-2]
[877994-30-4] (Micrantherin A)
$C_{28}H_{36}O_8$ 500.588
Constit. of *Kadsura longipedunculata, Schisandra chinensis* and *Schisandra micrantha*. Component of Wu Wei Zi. Amorph. powder. $[\alpha]_D^{22}$ +19.4 (c, 1.6 in $CHCl_3$) (Angeloylgomisin H). $[\alpha]_D^{18}$ +16.2 (c, 0.45 in $CHCl_3$) (Micrantherin A). Identity of Angeloylgomisin H and Micrantherin A not confirmed, but the reported opt. rotns. are similar. λ_{max} 215 (log ε 4.72); 248 (sh); 286 (sh) (MeOH)

3,4,4′,5,5′-Penta-Me ether, 3′-tigloyl
Tigloylgomisin H
[66069-55-4]
$C_{28}H_{36}O_8$ 500.588
Constit. of the fruit of *Schisandra chinensis*. Component of Wu Wei Zi. Amorph. powder. $[\alpha]_D^{23}$ +67.7 (c, 1.3 in $CHCl_3$). λ_{max} 215 (log ε 4.75); 248 (sh); 285 (sh) (MeOH)

3,4,4′,5,5′-Penta-Me ether, 3′-benzoyl
Benzoylgomisin H
[66056-23-3]
$C_{30}H_{34}O_8$ 522.594
Constit. of the fruit of *Schisandra chinensis*. Component of Wu Wei Zi. Amorph. powder. $[\alpha]_D^{24}$ +96.8 (c, 1.2 in $CHCl_3$). λ_{max} 217 (log ε 4.66); 250 (sh); 282 (sh) (MeOH)

3,4,4′,5,5′-Penta-Me ether, 3′-E-cinnamoyl
Neglschisandrin C
[1042663-75-1]
$C_{32}H_{36}O_8$ 548.632
Constit. of the stems of *Schisandra neglecta*. Powder. $[\alpha]_D^{22}$ +72.1 (c, 0.48 in MeOH). λ_{max} 217 (log ε 4.44); 252 (log ε 4.05); 280 (log ε 3.23) (MeOH)

3′,4,4′,5,5′-Penta-Me ether
Schinlignan F
[1714123-65-5]
$C_{23}H_{30}O_7$ 418.486
Constit. of the fruit of *Schisandra chinensis*. Amorph. powder. $[\alpha]_D^{22.8}$ +22.3 (c, 0.17 in MeOH). λ_{max} 212 (log ε 4.44); 250 (log ε 3.86); 282 (log ε 3.64) (MeOH)

4′,5′-Methylene, 3,3′,4-tri-Me ether, 5-O-β-D-glucopyranoside
Schisandroside A
[2059120-54-4]
$C_{28}H_{36}O_{12}$ 564.584
Constit. of the roots of *Schisandra chinensis*. Amorph. powder. $[\alpha]_D^{25}$ +12.25 (c, 0.2 in MeOH). λ_{max} 218 (log ε 4.69); 257 (log ε 4.20); 292 (log ε 3.52) (MeOH)

4′,5′-Methylene, 3,4,5-tri-Me ether, 3′-angeloyl
Schinlignan D
[1714123-63-3]
$C_{27}H_{32}O_8$ 484.545
Constit. of the fruit of *Schisandra chinensis*. Amorph. powder. $[\alpha]_D^{23.2}$ +13.4 (c, 0.26 in MeOH). λ_{max} 212 (log ε 4.4); 251 (log ε 3.84); 286 (log ε 3.21) (MeOH)

4′,5′-Methylene ether, 3′,4,5-tri-Me, 3-angeloyl
Schinlignan E
[1714123-64-4]
$C_{27}H_{32}O_8$ 484.545
Constit. of the fruit of *Schisandra chinensis*. Amorph. powder. $[\alpha]_D^{23}$ -31.8 (c, 0.14 in MeOH). λ_{max} 212 (log ε 4.42); 250 (log ε 3.80); 282 (log ε 3.33) (MeOH)

4′,5′-Methylene, 3,3′,4,5-tetra-Me ether
2,2′-Cyclo-3,3′,4,5-tetramethoxy-4′,5′-methylenedioxy-8,8′-lignan-8-ol. 8-Hydroxy-3,3′,4,5-tetramethoxy-4′,5′-methylenedioxy-2,2′-cyclolignan. ***Gomisin A****. Wuweizisu B. Wuweizichun B. Schisandrol B. Schizandrol B. Wuweizi alcohol B. Besigomisin. TJN 101*
[58546-54-6]
[61281-39-8]
$C_{23}H_{28}O_7$ 416.47
Constit. of fruit of *Schisandra chinensis*. Antineoplastic agent showing apoptotic activity through caspase-7 cleavage in colon carcinoma cells. Hepatoprotective agent. Vasodilator. Component of chinese traditional medicines showing hepatoprotective activity. Mp 88 - 89° (54-56°). $[\alpha]_D$ +67.9. $[\alpha]_D^{22}$ +60.8 (c, 0.58 in $CHCl_3$). λ_{max} 217 (ε 79400); 246 (ε 19500); 251 (ε 20000); 256 (sh); 288 (sh) (EtOH) (Derep)
DE8396900

(8S,8′R,S_{axial})-form

3′,4,4′,5,5′-Penta-Me ether, 3-angeloyl
Schinlignan C
[1714123-62-2]
$C_{28}H_{36}O_8$ 500.588
Constit. of the fruit of *Schisandra chinensis*. Amorph. powder. $[\alpha]_D^{23.2}$ +13.4 (c, 0.26 in MeOH). λ_{max} 212 (log ε 4.4); 251 (log ε 3.84); 286 (log ε 3.21) (MeOH).

Kochetkov, N.K. *et al., Tetrahedron Lett.*, 1961, **2**, 730-734 (*Schizandrin*).
Ikeya, Y. *et al., Chem. Pharm. Bull.*, 1978, **26**, 328-331 (*Angeloylgomisin H, Tigloylgomisin H, Benzoylgomisin H*).
Ghera, E. *et al., J. Chem. Soc., Chem. Commun.*, 1978, 480-481 (*Schizandrin, synth*).
Ikeya, Y. *et al., Chem. Pharm. Bull.*, 1979, **27**, 1383-1394 (*Gomisin A, Schizandrin, abs config*).
Ikeya, Y. *et al., Chem. Pharm. Bull.*, 1979, **27**, 1576-1582 (*Gomisin H, Angeloylgomisin H, Tigloylgomisin H, Benzoylgomisin H*).
Tan , R. *et al., Planta Med.*, 1986, **52**, 49-51 (*Wuweizisu B*).

Ikeya, Y. *et al.*, *Chem. Pharm. Bull.*, 1988, **36**, 3974-3979 (*Gomisin T*).
Ikeya, Y. *et al.*, *Phytochemistry*, 1988, **27**, 569-573 (*Isoschizandrin*).
Warshawsky, A.M. *et al.*, *J. Am. Chem. Soc.*, 1990, **112**, 8090-8099 (*Schizandrin, Isoschizandrin, synth, abs config*).
Mizoguchi, Y. *et al.*, *Planta Med.*, 1991, **57**, 320-324 (*Gomisin A, activity*).
Takeya, T. *et al.*, *Chem. Pharm. Bull.*, 1994, **42**, 430-437 (*Gomisin A, Schizandrin, synth*).
Takeya, T. *et al.*, *Chem. Pharm. Bull.*, 1994, **42**, 438-442 (*Isoschizandrin, synth*).
Tanaka, M. *et al.*, *J. Org. Chem.*, 1995, **60**, 4339-4352 (*Gomisin A, Schizandrin, Isoschizandrin, synth*).
Molander, G.A. *et al.*, *J. Org. Chem.*, 2003, **68**, 9533-9540 (*Isoschizandrin, synth*).
Li, R.-T. *et al.*, *Zhongguo Tianran Yaowu*, 2005, **3**, 208-212 (*Micrantherin A*).
Jang, J.-C. *et al.*, *Synth. Commun.*, 2006, **36**, 2259-2268 (*Schizandrin, isol, synth*).
Chen, M. *et al.*, *Molecules*, 2008, **13**, 1148-1155 (*Neglschisandrin C*).
Yang, G.Y. *et al.*, *J. Nat. Prod.*, 2010, **73**, 915-919 (*Marlignans K,L*).
Hwang, D. *et al.*, *Bioorg. Med. Chem. Lett.*, 2011, **21**, 6054-6057 (*Gomisin A, activity*).
He, P. *et al.*, *Tetrahedron Lett.*, 2014, **55**, 2965-2968 (*Schizandrin, vcd, abs config*).
Kim, H.M. *et al.*, *Chem. Pharm. Bull.*, 2015, **63**, 746-751 (*Schisandrosides A,B*).
Xue, Y. *et al.*, *Phytochemistry*, 2015, **116**, 253-261 (*Schinlignans C-F*).
Yang, B.-Y. *et al.*, *Chem. Biodiversity*, 2016, **13**, 1118-1125 (*Schisandra chinensis 5-glucoside*).
Szopa, A. *et al.*, *Phytochem. Rev.*, 2017, **16**, 195-218 (*Schisandra chinensis constits, rev*).
Pel, P. *et al.*, *Phytochemistry*, 2017, **136**, 119-124 (*3′-Tigloylschinlignan D*)

2,7′-Cyclo-8,8′-lignan-2′,3,4,4′,5′,6,7-heptol C227

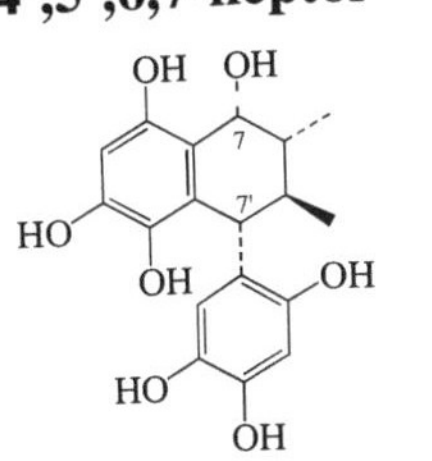

(7R,7′R,8R,8′S)-form

1,2,3,4-Tetrahydro-2,3-dimethyl-4-(2,4,5-trihydroxyphenyl)-1,5,6,8-naphthalenetetrol. 2′,3,4,4′,5′,6,7-Heptahydroxy-2,7′-cyclolignan

$C_{18}H_{20}O_7$ 348.351

(7R,7′R,8R,8′S)-form

Hepta-Me ether

1,2,3,4-Tetrahydro-1,5,6,8-tetramethoxy-2,3-dimethyl-4-(2,4,5-trimethoxyphenyl) naphthalene. 2′,3,4,4′,5′,6,7-Heptamethoxy-2,7′-cyclolignan. ***(-)-Asarolignan G***

$C_{25}H_{34}O_7$ 446.54

Constit. of the rhizomes of *Acorus tatarinowii*. Yellowish oil. $[\alpha]_D^{23}$ -106 (c, 0.5 in MeOH). Obt. by chiral hplc. λ_{max} 207 (log ε 5.44); 235 (log ε 4.90); 289 (log ε 4.52) (MeOH)

(7S,7′S,8S,8′R)-form

Hepta-Me ether

(+)-Asarolignan G

$C_{25}H_{34}O_7$ 446.54

Constit. of the rhizomes of *Acorus tatarinowii*. Yellowish oil. $[\alpha]_D^{23}$ +102 (c, 0.5 in MeOH). Obt. by chiral hplc.

Qin, D.-P. *et al.*, *RSC Adv.*, 2017, **7**, 8512-8520 (*Asarolignans G, cd*)

2,7′-Cyclo-8,8′-lignan-3,3′,4,4′,5,5′,7-heptol C228

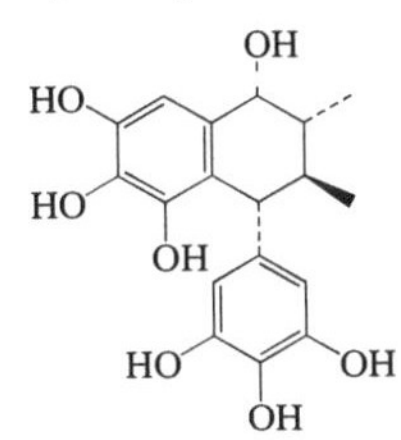

3,3′,4,4′,5,5′,7-Heptahydroxy-2,7′-cyclolignan

$C_{18}H_{20}O_7$ 348.351

(7R,7′R,8R,8′S)-form

3,3′,5,5′,7-Penta-Me ether

2,7′-Cyclo-3,3′,5,5′,7-pentamethoxy-8,8′-lignan-4,4′-diol. 4,4′-Dihydroxy-3,3′,5,5′,7-pentamethoxy-2,7′-cyclolignan

$C_{23}H_{30}O_7$ 418.486

Constit. of the roots of *Codonopsis pilosula*. $[\alpha]_D^{20}$ -13.5 (c, 0.06 in MeOH). λ_{max} 279 (log ε 3.37) (MeOH).

Jiang, Y.-P. *et al.*, *Yaoxue Xuebao*, 2016, **51**, 616-625 (*Codonopsis pilosula constit, cd*)

2,7′-Cyclo-8,8′-lignan-3,3′,4,4′,5,9,9′-heptol C229

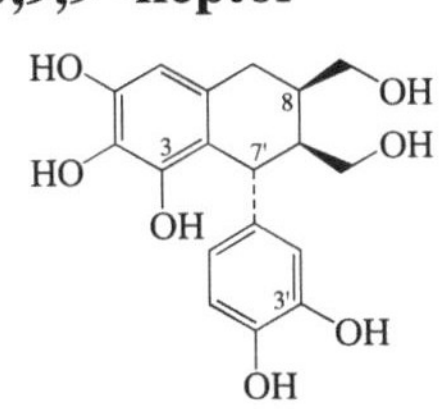

(7′R,8R,8′S)-form

3,3′,4,4′,5,9,9′-Heptahydroxy-2,7′-cyclolignan

$C_{18}H_{20}O_7$ 348.351

(7′R,8R,8′S)-form

4,5-Methylene, 3,3′,4′,9,9′-penta-Me ether

Nirtetralin B

[1312543-36-4]

$C_{24}H_{30}O_7$ 430.497

Constit. of *Phyllanthus niruri*. Cryst. (MeOH). $[\alpha]_D^{25}$+7.5 (c, 8.8 in Me_2CO). λ_{max} 241 (log ε 4.0); 281 (log ε 3.74) (MeOH)

(7′R,8S,8′S)-form

3,3′,5-Tri-Me ether, 9′-O-β-D-glucopyranoside

$C_{27}H_{36}O_{12}$ 552.573

Isol. from the stem bark of *Stemmadenia minima*. $[\alpha]_D$ -4

3,4-Methylene, 3′,4′,5,9′-tetra-Me ether

2,7′-Cyclo-3′,4′,5,9′-tetramethoxy-3,4-methylenedioxy-8,8′-lignan-9-ol. 9-Hydroxy-3′,4′,5,9′-tetramethoxy-3,4-methylenedioxy-2,7′-cyclolignan

3,4-Methylene, 3′,4′,5,9′-tetra-Me ether, 9-O-[β-D-apiofuranosyl-(1→6)-β-D-glucopyranoside]

[884856-00-2]

$C_{34}H_{46}O_{16}$ 710.726

Constit. of the roots of *Rodgersia sambucifolia*. Amorph. powder. Mp 228 - 230°. $[\alpha]_D^{20}$-35.2 (c, 0.43 in MeOH). Named as a glycoside of Hypophyllanthin rather than 9-O-demethylhypophyllanthin. λ_{max} 210; 228; 272 (no solvent reported)

4,5-Methylene, 3,3′,4′,9,9′-penta-Me ether

2,7′-Cyclo-3,3′,4′,9,9′-pentamethoxy-4,5-methylenedioxylignan. 3,3′,4′,9,9′-Pentamethoxy-4,5-methylenedioxy-2,7′-cyclolignan. ***Nirtetralin***

[50656-78-5]

$C_{24}H_{30}O_7$ 430.497

Constit. of *Phyllanthus niruri*, *Phyllanthus urinaria* and *Rodgersia sambucifolia*. Endothelin receptor antagonist. Cryst. (hexane). Mp 55°. $[\alpha]_D$+14.39 (c, 1.39 in $CHCl_3$). λ_{max} 201 (ε 8183); 229 (ε 21877); 281 (ε 7079) (MeCN)

(7′S,8R,8′R)-form

3,3′-Di-Me ether

2,7′-Cyclo-3,3′-dimethoxy-8,8′-lignan-4,4′,5,9,9′-pentol. 4,4′,5,9,9′-Pentahydroxy-3,3′-di-

methoxy-2,7′-cyclolignan.
Lonicerinol
[1277182-37-2]
$C_{20}H_{24}O_7$ 376.405
Constit. of the caulis of *Lonicera japonica*

3,3′,5-Tri-Me ether
2,7′-Cyclo-3,3′,5-trimethoxy-8,8′-lignan-4,4′,9,9′-tetrol. 4,4′,9,9′-Tetrahydroxy-3,3′,5-trimethoxy-2,7′-cyclolignan.
3-Methoxyisolariciresinol
[124681-08-9]
$C_{21}H_{26}O_7$ 390.432
Isol. from *Justicia tranquebariensis*. Cryst. (C_6H_6/MeOH). Mp 163 - 164°. $[\alpha]_D$ +61.7 (c, 0.66 in Me_2CO)

3,3′,5-Tri-Me ether, 4-O-β-D-glucopyranoside
[1355468-08-4]
$C_{27}H_{36}O_{12}$ 552.573
Constit. of the roots of *Rhodiola crenulata*. Amorph. powder. $[\alpha]_D^{20}$ -21 (c, 0.02 in MeOH). λ_{max} 238; 282 (MeOH)

3,3′,5-Tri-Me ether, 9′-O-β-D-xylopyranoside
Tiliamuroside A
[1446929-06-1]
$C_{26}H_{34}O_{11}$ 522.547
Constit. of *Tilia amurensis*. Gum. $[\alpha]_D^{25}$ +24.8 (c, 0.18 in MeOH). λ_{max} 228 (log ε 4.2); 284 (log ε 3.3) (MeOH)

3,3′,5-Tri-Me ether, 9′-O-β-D-glucopyranoside
Isol. from stem bark of *Stemmadenia minima*. $[\alpha]_D$ +15

3,3′,5-Tri-Me ether, 9′-O-α-L-rhamnopyranoside
[756835-77-5]
$C_{27}H_{36}O_{11}$ 536.574
Constit. of the aerial parts of *Vitis thunbergii*. Amorph. solid. $[\alpha]_D^{25}$ +13.9 (c, 0.1 in MeOH). λ_{max} 238 (log ε 3.72); 283 (log ε 3.43) (MeOH)

(7′S,8R,8′S)-form

3,3′,5-Tri-Me ether, 9-O-β-D-xylopyranoside
[1235577-49-7]
$C_{26}H_{34}O_{11}$ 522.547
Constit. of the stems of *Litsea glutinosa*. Light yellow powder. $[\alpha]_D^{22}$ +13.9 (c, 0.1 in MeOH). λ_{max} 238; 284 (MeOH)

3,3′,5-Tri-Me ether, 9′-O-β-D-glucopyranoside
[872993-35-6]
Constit. of the stems of *Annona squamosa* and rhizomes of *Clematis manshurica*. Weak inhibitor of cyclooxygenase 2. Syrup. $[\alpha]_D^{22}$ +130.8 (c, 0.03 in MeCN). λ_{max} 201; 232 (sh); 282 (MeCN)

4,5-Methylene, 3,3′,4′,9,9′-penta-Me ether
Neonirtetralin. ***Nirtetralin A***
[471913-06-1]
$C_{24}H_{30}O_7$ 430.497
Constit. of *Phyllanthus niruri* and *Phyllanthus urinaria*. Shows significant cytotoxicity against CHO and J774 cell lines. Oil. Renamed as Nirtetralin A

(7′R*,8S*,8′S*)-form

3,4-Methylene, 3′,4′,5,9,9′-penta-Me ether
2,7′-Cyclo-3′,4′,5,9,9′-pentamethoxy-3,4-methylenedioxylignan. 3′,4′,5,9,9′-Pentamethoxy-3,4-methylenedioxy-2,7′-cyclolignan. ***Hypophyllanthin***
[33676-00-5]
$C_{24}H_{30}O_7$ 430.497
Constit. of *Phyllanthus niruri*, *Phyllanthus amarus* and *Phyllanthus urinaria*. Exhibits moderate cytotoxicity against CHO and J774 cancer cell lines. Antihyperuricaemic agent. Long needles (MeOH or petrol). Mp 128°. $[\alpha]_D^{25}$ +6 (c, 1 in $CHCl_3$). Several incorrect structs. originally proposed. The current struct. dates from 1980. λ_{max} 213 (ε 44670); 231 (ε 20400); 280 (ε 4460) (EtOH) (Berdy).

Subba Rao, G. *et al.*, *Tetrahedron Lett.*, 1971, **12**, 3175-3178 (*Hypophllanthin, bibl*).
Anjaneyulu, A.S.R. *et al.*, *Tetrahedron*, 1973, **29**, 1291-1298 (*Nirtetralin*).
Bhadbade, M.M. *et al.*, *Tetrahedron Lett.*, 1980, **21**, 3097-3098 (*Hypophyllanthin, cryst struct*).
Schneiders, G.E. *et al.*, *J. Chem. Soc., Perkin Trans. 1*, 1982, 999-1003 (*Hypophyllanthin, struct*).
Raju, G.V.S. *et al.*, *Indian J. Chem., Sect. B*, 1989, **28**, 558-561 (*3-Methoxyisolariciresinol*).
Achenbach, H. *et al.*, *Planta Med.*, 1992, **58**, 270-272 (*Stemmadenia glucosides*).
Somanabandhu, A. *et al.*, *J. Nat. Prod.*, 1993, **56**, 233-239 (*Hypophyllanthin, pmr, cmr*).
Hussain, R.A. *et al.*, *J. Nat. Prod.*, 1995, **58**, 1515-1520 (*Nirtetralin, props*).
Wei, W-X. *et al.*, *Bull. Korean Chem. Soc.*, 2002, **23**, 896-898 (*Nirtetralin A*).
Chang, C.-C. *et al.*, *Phytochemistry*, 2003, **63**, 825-833 (*Nirtetralin, Hypophyllanthin, cd, abs config*).
He, Y.-H. *et al.*, *Heterocycles*, 2004, **63**, 871-877 (*Vitis rhamnoside*).
Yang, Y.-L. *et al.*, *Helv. Chim. Acta*, 2005, **88**, 2731-2737 (*Annona squamosa glucoside*).
Hu, H-B. *et al.*, *Indian J. Chem., Sect. B*, 2005, **44**, 2399-2403 (*Rodgersia sambucifolia constits*).
Murugaiyah, V. *et al.*, *Planta Med.*, 2006, **72**, 1262-1267 (*Hypophyllanthin, activity*).
Shi, S-P. *et al.*, *Z. Naturforsch., B*, 2006, **61**, 1299-1303 (*Clematis manshurica constit, activity*).
Pan, J.-Y. *et al.*, *Helv. Chim. Acta*, 2010, **93**, 951-957 (*Litsea glutinosa xyloside*).
Yean, M.H. *et al.*, *Nat. Prod. Sci.*, 2010, **16**, 15-19 (*Lonicerinol*).
Kim, K.H. *et al.*, *Food Chem. Toxicol.*, 2012, **50**, 3680-3686 (*Tiliamuroside A*).
Yang, Y.-N. *et al.*, *J. Agric. Food Chem.*, 2012, **60**, 964-972 (*Rhodiola crenulata glucoside*).
Wei, W. *et al.*, *Phytother. Res.*, 2012, **26**, 964-968 (*Nirtetralin B, cryst struct*).
Van Thanh, N. *et al.*, *Phytochem. Lett.*, 2014, **7**, 182-185 (*Nirtetralin A, activity*).
Nguyen, V.T. *et al.*, *Phytochem. Lett.*, 2014, **7**, 182-185 (*activity*)

2,7′-Cyclo-8,8′-lignan-3′,4,4′,5,5′,9,9′-heptol C230

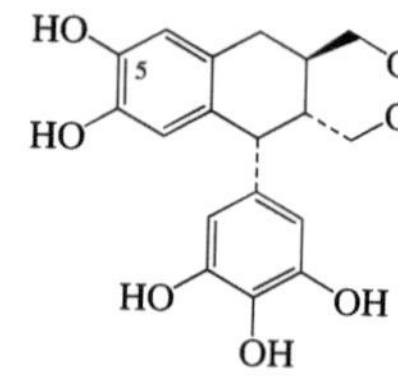

(7′R,8R,8′R)-form

3′,4,4′,5,5′,9,9′-Heptahydroxy-2,7′-cyclolignan
$C_{18}H_{20}O_7$ 348.351

(7′R,8R,8′R)-form

3′,4,5,5′-Tetra-Me ether, 9-O-(3,4-di-O-methyl-β-D-glucopyranoside)
[1894199-71-3]
$C_{30}H_{42}O_{12}$ 594.654
Constit. of the bark of *Aglaia exima*. Waxy solid. $[\alpha]_D^{20}$ +31.7 (c, 0.02 in MeOH). Stereochem. is uncertain. λ_{max} 284 (log ε 4.39); 327 (log ε 4.93) (MeOH)

(7′R,8S,8′R)-form

3′,5,5′-Tri-Me ether, 9′-O-α-L-rhamnopyranoside
[2185026-84-8]
$C_{27}H_{36}O_{11}$ 536.574
Constit. of aerial parts of *Lespedeza cuneata*. Amorph. powder. $[\alpha]_D^{20}$ -72.8 (c, 0.17 in MeOH). λ_{max} 284 (log ε 3.12) (MeOH)

9′-Carboxylic acid, 4,5-methylene, 3′,4′,5′-tri-Me ether
2,7′-Cyclo-9-hydroxy-3′,4′,5′-trimethoxy-4,5-methylenedioxy-8,8′-lignan-9′-oic acid. 9-Hydroxy-3′,4′,5′-trimethoxy-4,5-methylenedioxy-2,7′-cyclolignan-9′-oic acid. ***Deoxypodophyllic acid***
[38943-33-8]
$C_{22}H_{24}O_8$ 416.426
Isol. from *Thujopsis dolabrata*. Antineoplastic agent. Cryst. (MeOH). Mp 171 - 172°. $[\alpha]^{24}_{D}$ -158.5 (Py). Log P 0.5 (calc)

9′-Carboxylic acid, 4,5-methylene, 3′,4′,5′-tri-Me ether, Me ester
Methyl deoxypodophyllotoxinate
[1643-12-5]
$C_{23}H_{26}O_8$ 430.453
Co-occurs with the parent acid in *Juniperus thurifera*. Oil. $[\alpha]^{23}_{D}$ -72.5 (c, 0.1 in EtOH). λ_{max} 215 (ε 20215); 289 (ε 3750) (MeOH) (Berdy)

9′-Carboxylic acid, 4,5-methylene, 3′,4′,5′-tri-Me ether, β-D-glucopyranosyl ester
[59092-95-4]
$C_{28}H_{34}O_{13}$ 578.567
Isol. from roots of *Podophyllum peltatum* and *Podophyllum emodi*. Powder (Me_2CO/Et_2O). Mp 123 - 128°. $[\alpha]^{22}_{D}$ -151.5 (c, 0.575 in MeOH)

(7′R,8S,8′S)-form

3′,5,5′-Tri-Me ether
2,7′-Cyclo-3′,5,5′-trimethoxy-8,8′-lignan-4,4′,9,9′-tetrol. 4,4′,9,9′-Tetrahydroxy-3′,5,5′-trimethoxy-2,7′-cyclolignan. ***5′-Methoxyisolariciresinol***
[143236-07-1]
$C_{21}H_{26}O_7$ 390.432
Constit. of *Gaultheria yunnanensis* and *Sinocalamus affinis*. Cryst. (MeOH). Mp 175 - 178°. $[\alpha]^{25}_{D}$ -46.4 (MeOH)

3′,5,5′-Tri-Me ether, 9′-O-α-L-rhamnopyranoside
[1819968-34-7]
$C_{27}H_{36}O_{11}$ 536.574
Constit. of the aerial parts of *Lespedeza cuneata*. Amorph. solid. $[\alpha]^{25}_{D}$ -160.9 (c, 0.08 in MeOH). λ_{max} 206 (log ε 4.74); 284 (log ε 3.56) (MeOH)

(7′S,8R,8′R)-form

3′,5,5′-Tri-Me ether, 4-O-β-D-glucopyranoside
[1355468-11-9]
$C_{27}H_{36}O_{12}$ 552.573
Constit. of the roots of *Rhodiola crenulata*. Amorph. powder. $[\alpha]^{20}_{D}$ -49.5 (c, 0.01 in MeOH). λ_{max} 234; 282 (MeOH)

3′,5,5′-Tri-Me ether, 9′-O-α-L-arabinopyranoside
[903559-38-6]
$C_{26}H_{34}O_{11}$ 522.547
Constit. of the bark of *Walsura yunnanensis*. Antioxidant. Needles (MeOH). Mp 206 - 208°. $[\alpha]^{28}_{D}$ +24.7 (c, 0.44 in MeOH). λ_{max} 207 (log ε 4.12); 283 (log ε 3.84) (MeOH)

3′,5,5′-Tri-Me ether, 9′-O-α-L-rhamnopyranoside
[132923-04-7]
Constit. of *Quercus glauca*. Yellow solid. Mp 146°. $[\alpha]^{25}_{D}$ +10.0 (c, 1.0 in MeOH)

3′,5,5′-Tri-Me ether, 9′-O-[β-D-apiofuranosyl-(1→2)-β-D-glucopyranoside]
[1913248-36-8]
$C_{32}H_{44}O_{16}$ 684.688
Constit. of the branches of *Tabebuia chrysotricha*. Exhibits moderate DPPH radical scavenging activity. Amorph. powder. $[\alpha]^{24}_{D}$ -26.7 (c, 0.13 in MeOH). λ_{max} 213 (log ε 4.23); 280 (log ε 3.86); 332 (log ε 3.63) (MeOH)

3′,5,5′-Tri-Me ether, 9,9′-di-Ac
Gaultherin A
[243468-37-3]
$C_{25}H_{30}O_9$ 474.506
Isol. from roots of *Gaultheria yunnanensis*. Powder (Me_2CO aq.). Mp 159 - 160°. $[\alpha]^{23}_{D}$ +30 (c, 0.08 in MeOH). λ_{max} 225 (log ε 2.68); 283 (log ε 0.91) (MeOH)

4,5-Methylene, 3′,5′-di-Me ether
2,7′-Cyclo-3′,5′-dimethoxy-4,5-methylenedioxy-8,8′-lignan-4′,9,9′-triol. 4′,9,9′-Trihydroxy-3′,5′-dimethoxy-4,5-methylenedioxy-2,7′-cyclolignan

4,5-Methylene, 3′,5′-di-Me ether, 9,9′-di-Ac
Rostellulin A
[1019848-65-7]
$C_{25}H_{28}O_9$ 472.49
Constit. of *Rostellularia procumbens*. Weak antibacterial agent. Amorph. powder. Mp 175 - 177°. $[\alpha]^{19}_{D}$ +43.9 (c, 0.11 in $CHCl_3$). Abs. config. not confirmed. λ_{max} 241 (log ε 3.85); 293 (log ε 3.45) ($CHCl_3$).

Kuhn, M. *et al.*, *Helv. Chim. Acta*, 1963, **46**, 2127-2139 (*Deoxypodophyllic acid glucosyl ester*).
Pelter, A. *J. Chem. Soc. C*, 1968, 74-79 (*Deoxypodophyllic acid, ms*).
Akahori, A. *et al.*, *Chem. Pharm. Bull.*, 1972, **20**, 1150-1155 (*Thujopsis dolabrata constit*).
Vecchietti, V. *et al.*, *Phytochemistry*, 1979, **18**, 1847-1849 (*5′-Methoxyisolariciresinol*).
Loriot, M. *et al.*, *Bull. Soc. Chim. Fr.*, 1985, 871-875 (*5′-Methoxyisolariciresinol, synth*).
Feliciano, A.S. *et al.*, *Phytochemistry*, 1992, **31**, 267-270 (*Methyl deoxypodophyllotoxinate*).
Zhang, Z. *et al.*, *Zhongcaoyao*, 1998, **29**, 508-511 (*5′-Methoxyisolariciresinol*).
Zhang, Z. *et al.*, *Phytochemistry*, 1999, **51**, 469-472 (*Gaultherin A*).
Luo, X.-D. *et al.*, *Chem. Biodiversity*, 2006, **3**, 224-230 (*Walsura arabinoside*).
Zhang, Y. *et al.*, *Planta Med.*, 2007, **73**, 1596-1599 (*Rostellulin A*).
Xiong, L. *et al.*, *J. Nat. Prod.*, 2011, **74**, 1188-1200 (*Sinocalamus affinis constit*).
Shen, C.-C. *et al.*, *Chem. Pharm. Bull.*, 2012, **60**, 924-929 (*Quercus glauca constit*).
Yang, Y.-N. *et al.*, *J. Agric. Food Chem.*, 2012, **60**, 964-972 (*Rhodiola crenulata glucoside*).
Takahashi, S. *et al.*, *Am. J. Plant Sci.*, 2015, **6**, 676-684 (*Tabebuia chrysotricha constit, cd, activity*).
Zhang, C.-F. *et al.*, *J. Asian Nat. Prod. Res.*, 2016, **18**, 913-920 (*Lespedeza cuneata constit*).
Sianturi, J. *et al.*, *Nat. Prod. Res.*, 2016, **30**, 2204-2208 (*Aglaia exima constit*).
Zhou, J. *et al.*, *Phytochemistry*, 2016, **121**, 58-64 (*Lespedeza cuneata constit*)

2,7′-Cyclo-8,8′-lignan-3′,4,4′,5,8,9,9′-heptol C231

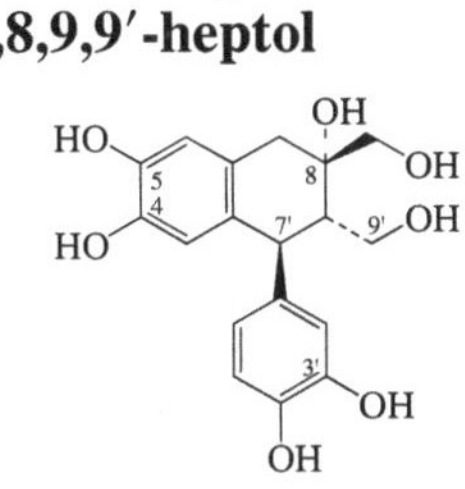

3′,4,4′,5,8,9,9′-Heptahydroxy-2,7′-cyclolignan
$C_{18}H_{20}O_7$ 348.351

(7′S,8S,8′S)-form

3′,5-Di-Me ether
2,7′-Cyclo-3′,5-dimethoxy-8,8′-lignan-4,4′,8,9,9′-pentol. 4,4′,8,9,9′-Pentahydroxy-3′,5-dimethoxy-2,7′-cyclolignan. ***Cycloolivil.*** *Isoolivil*
[3064-05-9]
$C_{20}H_{24}O_7$ 376.405

Isol. from *Olea cunninghamii*, *Gmelina asiatica*, *Carissa spinarum*, *Jasminum giraldii* and *Gymnelaea cunninghamii*. Shows moderate anti-DPPH free radical activity. Mp 171 - 173°. $[\alpha]_D^{28}$ +74.5 (c, 1.2 in MeOH). λ_{max} 231 (ε 4898) (EtOH)

3′,5-Di-Me ether, 4-O-β-D-glucopyranoside

[151228-11-4]

$C_{26}H_{34}O_{12}$ 538.546

Constit. of *Osmanthus asiaticus*. Amorph. powder. $[\alpha]_D^{21}$ -19.6 (c, 0.2 in MeOH). λ_{max} 230 (sh); 280 (log ε 3.8) (MeOH)

3′,5-Di-Me ether, 4′-O-β-D-glucopyranoside

[878807-15-9]

$C_{26}H_{34}O_{12}$ 538.546

Constit. of *Stereospermum cylindricum*. Amorph. powder. $[\alpha]_D^{24}$ +17.6 (c, 1.6 in MeOH)

3′,5-Di-Me ether, 9-O-β-D-xylopyranoside

[1403967-11-2]

$C_{25}H_{32}O_{11}$ 508.52

Constit. of *Cathaya argyrophylla*. Amorph. solid. $[\alpha]_D^{20}$ +26.7 (c, 0.1 in MeOH). λ_{max} 203 (log ε 4.8); 284 (log ε 3.85) (MeOH)

3′,5-Di-Me ether, 9-O-[β-D-apiofuranosyl-(1→6)-β-D-glucopyranoside]

Yemuoside YM$_6$

[122127-69-9]

$C_{31}H_{42}O_{16}$ 670.661

Isol. from *Stauntonia chinensis*. Powder. Mp 143 - 145°. $[\alpha]_D^{23}$ -21.2 (c, 0.10 in MeOH)

3′,5-Di-Me ether, 9′-O-β-D-xylopyranoside

[1355468-13-1]

$C_{25}H_{32}O_{11}$ 508.52

Constit. of the roots of *Rhodiola crenulata*. Amorph. powder. $[\alpha]_D^{20}$ +43.5 (c, 0.02 in MeOH). λ_{max} 233; 285 (MeOH)

3′,5-Di-Me ether, 9′-O-β-D-glucopyranoside

[1105067-02-4]

$C_{26}H_{34}O_{12}$ 538.546

Constit. of the roots of *Urtica triangularis*. Amorph. powder. $[\alpha]_D^{20}$ +26.1 (c, 0.2 in MeOH)

3′,5-Di-Me ether, tetra-Ac

[56193-50-1]

Cryst.. Mp 150°. $[\alpha]_D$ +3.8 (c, 1.0 in $CHCl_3$)

3′,5-Di-Me ether, 8,9-cyclic carbonate

Cycloolivil carbonate

[705259-92-3]

$C_{21}H_{22}O_8$ 402.399

Constit. of the roots of *Strychnos guianensis*. Cryst.. Mp 118 - 120°. $[\alpha]_D^{25}$ +68 (c, 0.05 in MeOH)

9,9′-Methylene, 3′,5-di-Me ether

2,7′-Cyclo-3′,5-dimethoxy-9,9′-methylenedioxy-8,8′-lignan-4,4′,8-triol. 4,4′,8-Trihydroxy-3′,5-dimethoxy-9,9′-methylenedioxy-2,7′-cyclolignan.

9,9′-Methylenecycloolivil. *Cycloolivil formaldehyde condensate*

[1264084-92-5]

$C_{21}H_{24}O_7$ 388.416

Constit. of *Epimedium brevicornum*. Amorph. powder. $[\alpha]_D^{25}$ +128.4 (c, 0.1 in $CHCl_3$)

8,9-Isopropylidene, 3′,5-di-Me ether

Cycloolivil 8,9-acetonide

[152242-97-2]

$C_{23}H_{28}O_7$ 416.47

Constit. of the trunkwood of *Tabebuia incana*. Cryst. (CH_2Cl_2/hexane). Mp 194 - 195°. Probable artifact. λ_{max} 288 (log ε 4.37) (EtOH aq.)

9-Carboxylic acid, 3′,5-di-Me ether

2,7′-Cyclo-4,4′,8,9′-tetrahydroxy-3′,5-dimethoxy-8,8′-lignan-9-oic acid. 4,4′,8,9′-Tetrahydroxy-3′,5-dimethoxy-2,7′-cyclolignan-9-oic acid. 8-Hydroxy-α-conidendric acid

9-Carboxylic acid, 3′,5-di-Me ether, Me ester

[189204-40-8]

$C_{21}H_{24}O_8$ 404.415

Constit. of *Tsuga heterophylla*. Amorph. powder. $[\alpha]_D^{25}$ +65.4 (c, 2 in MeOH). λ_{max} 209 (log ε 4.67); 231 (sh) (log ε 4.24); 285 (log ε 3.89) (MeOH)

9-Carboxylic acid, 3′,5-di-Me ether, 9,9′-lactone

2,7′-Cyclo-4,4′,8-trihydroxy-3′,5-dimethoxy-8,8′-lignan-9,9′-olide.

8-Hydroxy-α-conidendrin

[189204-39-5]

$C_{20}H_{20}O_7$ 372.373

Constit. of *Tsuga heterophylla*. Pale yellow powder. $[\alpha]_D^{25}$ -55.4 (c, 0.6 in MeOH). λ_{max} 210 (log ε 4.71); 229 (sh) (log ε 4.25); 284 (log ε 3.89) (MeOH)

9,9′-Dicarboxylic acid

2,7′-Cyclo-3′,4,4′,5,8-pentahydroxy-8,8′-lignan-9,9′-dioic acid

9,9′-Dicarboxylic acid, di-Me ester

[1357560-24-7]

$C_{20}H_{20}O_9$ 404.371

Constit. of the roots of *Pulsatilla koreana*. Amorph. brown powder. $[\alpha]_D^{25}$ +28 (c, 0.06 in MeOH). λ_{max} 248 (log ε 2.04); 267 (log ε 0.82) (EtOH).

Ayres, D.C. *et al.*, *J. Chem. Soc.*, 1965, 3586-3589 (*Cycloolivil*).
Manerikar, S.V. *et al.*, *Indian J. Chem., Sect. B*, 1979, **18**, 182-183 (*Cycloolivil, synth*).
Abe, F. *et al.*, *Chem. Pharm. Bull.*, 1988, **36**, 795-799 (*Cycloolivil*).
Wang, H.B. *et al.*, *J. Nat. Prod.*, 1989, **52**, 342-345 (*Yemuoside YM$_6$*).
Sugiyama, M. *et al.*, *Phytochemistry*, 1993, **33**, 1215-1219 (*Cycloolivil 4-glucoside*).
De Oliveira, A.B. *et al.*, *Phytochemistry*, 1993, **34**, 1409-1412 (*Cycloolivil acetonide*).
Moritani, Y. *et al.*, *J. Chem. Soc., Perkin Trans. 1*, 1996, 2747-2753 (*Cycloolivil, synth*).
Kawamura, F. *et al.*, *Phytochemistry*, 1997, **44**, 1351-1357 (*Tsuga heterophylla constits*).
Kawamura, F. *et al.*, *Phytochemistry*, 2000, **54**, 439-444 (*Tsuga heterophylla constits, pmr*).
Pinheiro, M.L.B. *et al.*, *Quim. Nova*, 2004, **27**, 188-192 (*Cycloolivil carbonate*).
Kanchanapoom, T. *et al.*, *Phytochemistry*, 2006, **67**, 516-520 (*Cycloolivil 4′-glucoside*).
Yan, X.-G. *et al.*, *Chem. Pharm. Bull.*, 2008, **56**, 1463-1465 (*Cycloolivil 9′-glucoside*).
Li, H.Z. *et al.*, *Chin. Chem. Lett.*, 2011, **22**, 85-87 (*9,9′-Methylenecycloolivil*).
Cuong, T.D. *et al.*, *Planta Med.*, 2011, **77**, 66-69 (*Pulsatilla koreana constit*).
Yang, Y.-N. *et al.*, *J. Agric. Food Chem.*, 2012, **60**, 964-972 (*Cycloolivil 9′-xyloside*).
He, W.J. *et al.*, *Phytochemistry*, 2012, **83**, 63-69 (*Cycloolivil 9-xyloside*).
Wangteeraprasert, R. *et al.*, *Phytother. Res.*, 2012, **26**, 1496-1499 (*Cycloolivil, activity*).
Yue, Z. *et al.*, *Molecules*, 2013, **18**, 4766-4775 (*Jasminum giraldii constit*).
Vakiti, J.R. *et al.*, *Org. Lett.*, 2020, **22**, 3345-3350 (*Cycloolivil, synth, abs config*)

2,7′-Cyclo-8,8′-lignan-3′,4,4′,5,8′,9,9′-heptol C232

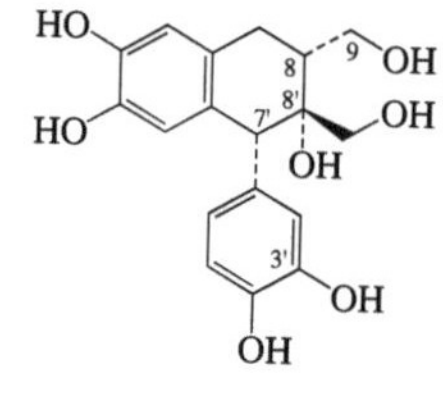

3′,4,4′,5,8′,9,9′-Heptahydroxy-2,7′-cyclolignan

$C_{18}H_{20}O_7$ 348.351

(7′R,8R*,8′R*)-form*
9-Carboxylic acid, 3′-Me ether
2,7′-Cyclo-4,4′,5,8′,9′-pentahydroxy-3′-methoxy-8,8′-lignan-9-oic acid. ***Daphnretusic acid***
[1617497-00-3]
$C_{19}H_{20}O_8$ 376.361
Constit. of *Daphne retusa*. Amorph. powder. Mp 164 - 166°. $[\alpha]_D$ -78 (c, 1.02 in MeOH). λ_{max} 206 (log ε 4.16); 226 (log ε 4.56); 278 (log ε 3.92) (MeOH).
Mansoor, F. *et al.*, *J. Asian Nat. Prod. Res.*, 2014, **16**, 210-215 (*Daphnretusic acid*)

2,2′-Cyclo-8,8′-lignan-3,3′,4,4′,5,5′-hexol C233

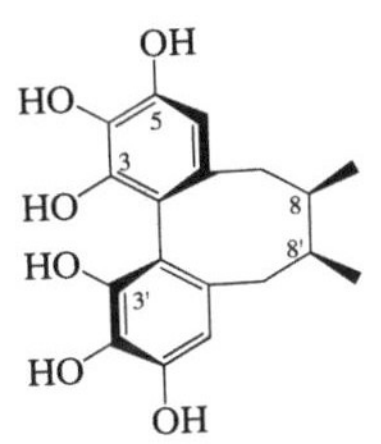

(8*R*,8′*S*,*R*axial)-*form*

5,6,7,8-Tetrahydro-6,7-dimethyldibenzo[a,c]cyclooctene-1,2,3,10,11,12-hexol, CAS. *3,3′,4,4′,5,5′-Hexahydroxy-2,2′-cyclolignan*
$C_{18}H_{20}O_6$ 332.352
Parent compd. shows symmetry such that for example (8*R*,8′*S*,R_{axial}) ≡ (8*S*,8′*R*,R_{axial}) by rotn. about the x-axis. Compds. here are referred to the (8*R*)-label, with stereoisomers of derivs. distinguished by different numbering around the primed and unprimed rings. Different numbering systems have been used

(8R,8′S,R$_{axial}$)-form
3,4,4′,5′-Tetra-Me ether
2,2′-Cyclo-3,4,4′,5′-tetramethoxy-8,8′-lignan-3′,5-diol. 3′,5-Dihydroxy-3,4,4′,5′-tetramethoxy-2,2′-cyclolignan. ***Arisanschinin F***
[1181216-78-3]
$C_{22}H_{28}O_6$ 388.46
Constit. of the fruit of *Schisandra arisanensis*. Amorph. powder. $[\alpha]^{25}_D$+13.3 (c, 0.1 in CH_2Cl_2). Stereochem. at C-8 and C-8′ stated to be relative, i.e. could be the 3′,4,4′,5-tetra-Me ether. λ_{max} 242 (log ε 3.85); 278 (log ε 3.65) (MeOH)

3,4,4′,5′-Tetra-Me ether, 3′-angeloyl
Schisanchinin B
[1613761-69-5]
$C_{27}H_{34}O_7$ 470.562
Constit. of fruit of *Schisandra chinensis*. Potent inhibitor of NO release by LPS-activated microgelia cells. Yellowish oil. $[\alpha]^{23}_D$+1.3 (c, 0.05 in MeOH). λ_{max} 213 (log ε 3.01) (MeOH)

3,3′,4,4′,5-Penta-Me ether
2,2′-Cyclo-3,3′,4,4′,5-pentamethoxy-8,8′-lignan-5′-ol. 5′-Hydroxy-3,3′,4,4′,5-pentamethoxy-2,2′-cyclolignan. ***Gomisin K$_2$***
[75684-44-5]
$C_{23}H_{30}O_6$ 402.487
Isol. from fruit of *Schisandra chinensis*. Amorph. powder. $[\alpha]^{24}_D$ +81.7 ($CHCl_3$). Genus given as Schizandra in ref.. λ_{max} 216 (log ε 4.74); 250 (log ε 4.21); 276 (sh); 285 (log ε 3.46) (EtOH)

3,4,4′,5,5′-Penta-Me ether
2,2′-Cyclo-3,4,4′,5,5′-pentamethoxy-8,8′-lignan-3′-ol. 3′-Hydroxy-3,4,4′,5,5′-pentamethoxy-2,2′-cyclolignan. ***Gomisin K$_3$****. Schizanhenol. Schisanhenol*
[69363-14-0]
$C_{23}H_{30}O_6$ 402.487
Isol. from *Schisandra henryi*, *Schisandra chinensis* and *Schisandra rubriflora*. Needles (Et_2O/hexane). Mp 100 - 101°. $[\alpha]^{23}_D$ +60.8 ($CHCl_3$). Abs. config. correlation of Gomisin K_3 and Schizanhenol not formally secured. λ_{max} 219 (log ε 4.71); 251 (sh); 277 (log ε 3.58); 283 (sh) (EtOH)

3,4,4′,5,5′-Penta-Me ether, 3′-angeloyl
Angeloylgomisin K$_3$
[1023744-69-5]
$C_{28}H_{36}O_7$ 484.589
Constit. of *Schisandra propinqua* var. *propinqua* and *Schisandra rubriflora*. Yellowish powder. $[\alpha]^{19}_D$ +16.5 (c, 0.12 in MeOH). λ_{max} 193 (log ε 4.39); 215 (log ε 4.71) (MeOH)

3,4,4′,5,5′-Penta-Me ether, 3′-tigloyl
Tigloylgomisin K$_3$
[1040392-74-2]
$C_{28}H_{36}O_7$ 484.589
Constit. of *Schisandra rubriflora*. Yellowish oil. $[\alpha]^{27}_D$+71.7 (c, 0.33 in $CHCl_3$). Misleading synonym

3,4,4′,5,5′-Penta-Me ether, 3′-benzoyl
Neglschisandrin D
[849411-43-4]
$C_{30}H_{34}O_7$ 506.595
Constit. of the stems of *Schisandra neglecta*. Powder. $[\alpha]^{22}_D$+79.6 (c, 1.3 in MeOH). λ_{max} 226 (log ε 4.42); 250 (log ε 4.12); 279 (log ε 3.71) (MeOH)

3,4,4′,5,5′-Penta-Me ether, 3′-E-cinnamoyl
Neglschisandrin A
[1015009-45-6]
$C_{32}H_{36}O_7$ 532.633
Constit. of *Schisandra neglecta*. Powder. $[\alpha]_D$ +40.2 (c, 0.53 in MeOH). λ_{max} 220 (log ε 4.8); 251 (log ε 4.48); 280 (log ε 4.47) (MeOH)

3′,4,4′,5,5′-Penta-Me ether
2,2′-Cyclo-3′,4,4′,5,5′-pentamethoxy-8,8′-lignan-3-ol. 3-Hydroxy-3′,4,4′,5,5′-pentamethoxy-2,2′-cyclolignan. ***Schisanchinin D***
[1614245-13-4]
$C_{23}H_{30}O_6$ 402.487
Constit. of fruits of *Schisandra chinensis*. Yellowish oil. $[\alpha]^{23}_D$+5.9 (c, 0.04 in MeOH). MF incorrectly given as $C_{27}H_{34}O_7$ in the experimental section. λ_{max} 214 (log ε 2.12) (MeOH)

Hexa-Me ether
2,2′-Cyclo-3,3′,4,4′,5,5′-hexamethoxylignan. 3,3′,4,4′,5,5′-Hexamethoxy-2,2′-cyclolignan. ***Deoxyschizandrin****. Schizandrin A. Schisandrin A. Wuweizisu A. Dimethylgomisin J. Desoxyschizandrin*
[61281-38-7]
$C_{24}H_{32}O_6$ 416.514
Isol. from kernels of *Schisandra chinensis* and *Schisandra rubriflora*. Cryst. (MeOH). Mp 114 - 116°. $[\alpha]^{20}_D$+107 (c, 5.7 in $CHCl_3$)
HP1595000

4,5-Methylene, 3′,4′,5′-tri-Me ether
2,2′-Cyclo-3′,4′,5′-trimethoxy-4,5-methylenedioxy-8,8′-lignan-3-ol. 3-Hydroxy-3′,4′,5′-trimethoxy-4,5-methylenedioxy-2,2′-cyclolignan. ***Gomisin M$_2$***
[82425-45-4]
$C_{22}H_{26}O_6$ 386.444
Isol. from *Kadsura coccinea* and *Kadsura chinensis*. Prisms (Et_2O/

hexane). Mp 159 - 160°. $[\alpha]_D^{24}$ +54.1

4,5-Methylene, 3,4′,5′-tri-Me ether, 3′-angeloyl

Angeloylgomisin M₁
[95152-96-8]
$C_{27}H_{32}O_7$ 468.546
Constit. of *Kadsura longipedunculata*. Amorph. powder. λ_{max} 217 (log ε 4.63); 250 (sh); 278 (log ε 3.39) (EtOH)

4,5-Methylene, 3,3′,4′,5′-tetra-Me ether

2,2′-Cyclo-3,3′,4′,5′-tetramethoxy-4,5-methylenedioxylignan. 3,3′,4′,5′-Tetramethoxy-4,5-methylenedioxy-2,2′-cyclolignan. ***Kadsuranin****. (+)-Gomisin N*
[713102-92-2]
$C_{23}H_{28}O_6$ 400.471
Isol. from roots and stems of *Kadsura longipedunculata*. Prisms (EtOH). Mp 104 - 106°. $[\alpha]_D^{19}$ +98.9 (c, 0.216 in $CHCl_3$)

4′,5′-Methylene, 3′,4,5-tri-Me ether

2,2′-Cyclo-3′,4,5-trimethoxy-4′,5′-methylenedioxy-8,8′-lignan-3-ol. 3-Hydroxy-3′,4,5-trimethoxy-4′,5′-methylenedioxy-2,2′-cyclolignan. ***Schizanhenol B****. Schisanhenol B*
[102681-52-7]
$C_{22}H_{26}O_6$ 386.444
Isol. from *Schisandra henryi* and *Schisandra rubriflora*. Cryst. (petrol/Et_2O). Mp 144 - 146°

4′,5′-Methylene, 3,4,5-tri-Me ether, 3′-angeloyl

Kadsutherin
[99481-39-7]
$C_{27}H_{32}O_7$ 468.546
Isol. from rootstem of *Kadsura coccinea*. Amorph. powder. $[\alpha]_D^{25}$ +40 (MeOH)

4′,5′-Methylene, 3,3′,4,5-tetra-Me ether

2,2′-Cyclo-3,3′,4,5-tetramethoxy-4′,5′-methylenedioxylignan. 3,3′,4,5-Tetramethoxy-4′,5′-methylenedioxy-2,2′-cyclolignan. Isokadsuranin. ***Schisandrin B****. Schizandrin B. (+)-γ-Schisandrin. Wuweizisu B*
[82467-52-5]
[61281-37-6]
$C_{23}H_{28}O_6$ 400.471
Constit. of *Kadsura coccinea* and *Schisandra chinensis*. Exhibits anticancer, antiinflammatory and antiviral activities. Antioxidant. Cryst. (Et_2O/hexane). Mp 105 - 107°. $[\alpha]_D^{21}$ +81.9 ($CHCl_3$)

3′,4,4′,5,5′-Penta-Me ether, 3-angeloyl

Schinlignan G*. 3-O-Angeloylschisanchinin D*
[1714123-66-6]
$C_{28}H_{36}O_7$ 484.589
Constit. of the fruit of *Schisandra chinensis*. Exhibits potent antihepatitis B virus activity. Amorph. powder. $[\alpha]_D^{23.5}$ +9.4 (c, 0.38 in MeOH). λ_{max} 212 (log ε 4.43); 252 (log ε 3.78); 284 (log ε 3.19) (MeOH)

(8R,8′S,S$_{axial}$)-form

3,4,4′-Tri-Me ether

2,2′-Cyclo-3,4,4′-trimethoxy-8,8′-lignan-3′,5,5′-triol. 3′,5,5′-Trihydroxy-3,4,4′-trimethoxy-2,2′-cyclolignan. ***Marlignan A***
[1228647-22-0]
$C_{21}H_{26}O_6$ 374.433
Constit. of *Schisandra wilsoniana*. Yellow gum. $[\alpha]_D^{24}$ +42.3 (c, 0.21 in MeOH). λ_{max} 210 (log ε 5.22); 241 (log ε 3.28); 326 (log ε 0.45) (MeOH)

3,4,4′-Tri-Me ether, 5′-Ac

Neglectalignan A*. Neglectaphenol A*
[1686087-56-8]
$C_{23}H_{28}O_7$ 416.47
Constit. of the stems of *Schisandra neglecta*. Exhibits moderate anti-HIV activity. Yellow gum. $[\alpha]_D^{24.5}$ +38.2 (c, 0.25 in MeOH). λ_{max} 210 (log ε 4.18); 245 (log ε 3.26); 326 (log ε 1.22) (MeOH)

4,4′,5′-Tri-Me ether

2,2′-Cyclo-4,4′,5′-trimethoxy-8,8′-lignan-3,3′,5-triol. 3,3′,5-Trihydroxy-4,4′,5′-trimethoxy-2,2′-cyclolignan. ***Marlignan B***
[1228647-23-1]
$C_{21}H_{26}O_6$ 374.433
Constit. of *Schisandra wilsoniana*. Yellow gum. $[\alpha]_D^{24}$ +38.5

4,4′,5′-Tri-Me ether, 5-Ac

Neglectalignan B*. Neglectaphenol B*
[1686086-85-0]
$C_{23}H_{28}O_7$ 416.47
Constit. of the stems of *Schisandra neglecta*. Exhibits moderate anti-HIV activity. Yellow gum. $[\alpha]_D^{24.8}$ +32.1 (c, 0.25 in MeOH). λ_{max} 210 (log ε 4.22); 245 (log ε 3.18); 326 (log ε 0.97) (MeOH)

4,5,5′-Tri-Me ether

2,2′-Cyclo-4,5,5′-trimethoxy-8,8′-lignan-3,3′,4′-triol. 3,3′,4′-Trihydroxy-4,5,5′-trimethoxy-2,2′-cyclolignan. ***Marlignan C***
[1228647-24-2]
$C_{21}H_{26}O_6$ 374.433
Constit. of *Schisandra wilsoniana*. Yellow gum. $[\alpha]_D^{25}$ +35.8 (c, 0.18 in MeOH). λ_{max} 210 (log ε 5.33); 240 (log ε 3.3); 328 (log ε 0.43) (MeOH)

3,3′,4,4′-Tetra-Me ether

2,2′-Cyclo-3,3′,4,4′-tetramethoxy-8,8′-lignan-5,5′-diol. 5,5′-Dihydroxy-3,3′,4,4′-tetramethoxy-2,2′-cyclolignan. ***Gomisin J***
[66280-25-9]
$C_{22}H_{28}O_6$ 388.46
Isol. from fruits of *Schisandra chinensis* and from *Kadsura coccinea*. $Ca^{2\oplus}$ antagonist enhancing arterial blood flow. Inhibits ACAT. Mp 149 - 150°. $[\alpha]_D^{24}$ -43.9 (c, 1.0 in $CHCl_3$). λ_{max} 214 (log ε 4.7); 248 (log ε 4.15); 276 (log ε 3.53) (EtOH)

3,3′,4,4′-Tetra-Me ether, 5,5′-di-O-β-D-glucopyranoside
[2014382-49-9]
$C_{34}H_{48}O_{16}$ 712.742
Constit. of the stems of *Schisandra chinensis*. Amorph. powder. $[\alpha]_D^{20}$ +9.6 (c, 0.5 in MeOH). λ_{max} 222 (log ε 3.65); 258 (sh) (MeOH)

3,3′,4,5′-Tetra-Me ether

2,2′-Cyclo-3,3′,4,5′-tetramethoxy-8,8′-lignan-4′,5-diol. 4′,5-Dihydroxy-3,3′,4,5′-tetramethoxy-2,2′-cyclolignan. ***Schineglignan A***
[1366275-19-5]
$C_{22}H_{28}O_6$ 388.46
Constit. of *Schisandra neglecta*. Shows moderate anti-HIV-1 activity. Yellow gum. $[\alpha]_D^{23}$ +17.3 (c, 0.22 in MeOH). λ_{max} 216 (log ε 4.97); 249 (log ε 3.47); 322 (log ε 1.22) (MeOH)

3,4,4′,5′-Tetra-Me ether
Rubrisandrin A
[919289-30-8]
$C_{22}H_{28}O_6$ 388.46
Constit. of *Schisandra rubriflora*. Rubrisandrin A was a 2:1 mixt. with the 3′,4,4′,5-tetra-Me positional isomer

3′,4,4′,5′-Tetra-Me ether
2,2′-Cyclo-3′,4,4′,5′-tetramethoxy-8,8′-lignan-3,5-diol. 3,5-Dihydroxy-3′,4,4′,5′-tetramethoxy-2,2′-cyclolignan. ***Schineglignan B***
[1367085-49-1]
$C_{22}H_{28}O_6$ 388.46
Constit. of *Schisandra neglecta*. Shows moderate anti-HIV-1 activity. Yellow gum. $[\alpha]^{23}_{D}$ -26.7 (c, 0.26 in MeOH). λ_{max} 215 (log ε 5.08); 248 (log ε 3.68); 329 (log ε 0.97); 348 (log ε 0.69) (MeOH)

4,4′,5,5′-Tetra-Me ether
2,2′-Cyclo-4,4′,5,5′-tetramethoxy-8,8′-lignan-3,3′-diol. 3,3′-Dihydroxy-4,4′,5,5′-tetramethoxy-2,2′-cyclolignan. ***Rubrisandrin B***. *Rubrifloralignan A*
[887767-48-8]
$C_{22}H_{28}O_6$ 388.46
Constit. of the fruit of *Schisandra rubriflora*. Amorph. solid. $[\alpha]^{22}_{D}$ +33.7 (c, 0.1 in MeOH). λ_{max} 217 (log ε 4.93); 268 (log ε 4.06) (MeOH)

3,3′,4,4′,5-Penta-Me ether
Gomisin K₁
[75629-20-8]
$C_{23}H_{30}O_6$ 402.487
Isol. from fruit of *Schisandra chinensis*. Prisms (Et_2O/hexane). Mp 99 - 101°. $[\alpha]^{23}_{D}$ -96.7 ($CHCl_3$)

4,5-Methylene, 3,3′,4′-tri-Me ether
2,2′-Cyclo-3,3′,4′-trimethoxy-4,5-methylenedioxy-8,8′-lignan-5′-ol. 5′-Hydroxy-3,3′,4′-trimethoxy-4,5-methylenedioxy-2,2′-cyclolignan. ***Gomisin L₂***
[82425-44-3]
$C_{22}H_{26}O_6$ 386.444
Isol. from *Schisandra chinensis*. Inhibits ACAT. Amorph. powder. $[\alpha]^{24}_{D}$ -98.1 (c, 0.52 in $CHCl_3$)

4,5-Methylene, 3,4′,5′-tri-Me ether
2,2′-Cyclo-3,4′,5′-trimethoxy-4,5-methylenedioxy-8,8′-lignan-3′-ol. 3′-Hydroxy-3,4′,5′-trimethoxy-4,5-methylenedioxy-2,2′-cyclolignan. ***Gomisin L₁***
[82425-43-2]
$C_{22}H_{26}O_6$ 386.444
Isol. from *Schisandra chinensis*. Prisms (Et_2O/hexane). Mp 194 - 196°. $[\alpha]^{24}_{D}$ -53.5 (c, 1.70 in $CHCl_3$)

4,5-Methylene, 3,3′,4′,5′-tetra-Me ether
Rubschisandrin. *(-)-γ-Schisandrin*
[82467-51-4]
$C_{23}H_{28}O_6$ 400.471
Constit. of the kernels of *Schisandra rubiflora* and seed oil of *Schisandra chinensis*. Prisms (MeOH). Mp 126.5 - 128° (118-119°). $[\alpha]^{17}_{D}$ -32 (EtOH). λ_{max} 220 (log ε 4.81); 245 (log ε 4.25); 285 (log ε 3.98) (EtOH)

4′,5′-Methylene, 3,3′,4-tri-Me ether
2,2′-Cyclo-3,3′,4-trimethoxy-4′,5′-methylenedioxy-8,8′-lignan-5-ol. 5-Hydroxy-3,3′,4-trimethoxy-4′,5′-methylenedioxy-2,2′-cyclolignan. ***Schisanbicolorin A***
[1350802-13-9]
$C_{22}H_{26}O_6$ 386.444
Constit. of the stems of *Schisandra bicolor*. Powder. $[\alpha]^{25}_{D}$ -23.6 (c, 1.3 in MeOH). λ_{max} 241 (log ε 4.28) (MeOH)

4′,5′-Methylene, 3,3′,4,5-tetra-Me ether
Gomisin N
[69176-52-9]
$C_{23}H_{28}O_6$ 400.471
Isol. from *Schisandra chinensis*. Inhibitor of ACAT and chitin synthase II. Shows insecticidal props.. Prisms (Et_2O/hexane). Mp 105 - 107°. $[\alpha]^{23}_{D}$ -84.7 (c, 2.17 in $CHCl_3$)

4,5:4′,5′-Bis(methylene), 3,3′-di-Me ether
2,2′-Cyclo-3,3′-dimethoxy-4,5:4′,5′-bis(methylenedioxy)lignan. 3,3′-Dimethoxy-4,5:4′,5′-bis(methylenedioxy)-2,2′-cyclolignan. ***Wuweizisu C***. *Schizandrin C. Schisandrin C*
[61301-33-5]
$C_{22}H_{24}O_6$ 384.428
Isol. from kernels of *Schisandra chinensis*. Inhibits ACAT. Chitin synthase II inhibitor. Cryst. (Me_2CO/MeOH). Mp 122 - 123°. $[\alpha]^{22}_{D}$ -54.9 (c, 1 in $CHCl_3$). $[\alpha]_D$ -68 ($CHCl_3$)
GX9500000

(8S,8′R,R$_{axial}$)-form
4,5-Methylene, 3,3′,4′-tri-Me ether
Neglschisandrin E
[1464719-44-5]
$C_{22}H_{26}O_6$ 386.444
Constit. of the stems of *Schisandra neglecta*. Amorph. powder. $[\alpha]^{22}_{D}$ +26.9 (c, 0.26 in MeOH). λ_{max} 216 (log ε 4.67); 250 (log ε 4.22); 280 (log ε 3.97) (MeOH)

(8S,8′S,S$_{axial}$)-form
4,5-Methylene, 3,3′-di-Me ether, 5′-O-β-D-glucopyranoside
Schisandroside C
[2059120-56-6]
$C_{27}H_{34}O_{11}$ 534.558
Constit. of the roots of *Schisandra chinensis*. Amorph. powder. $[\alpha]^{25}_{D}$ -25.68 (c, 0.25 in MeOH). λ_{max} 219 (log ε 4.91); 254 (log ε 4.28); 281 (log ε 4.01) (MeOH)

(8R*,8′S*,RS$_{axial}$)-form
3,4,4′,5′-Tetra-Me ether
Arisanschinin G
[1181216-79-4]
$C_{22}H_{28}O_6$ 388.46
Constit. of *Schisandra arisanensis*. Pale yellow oil. $[\alpha]^{25}_{D}$ -86.7 (c, 0.1 in CH_2Cl_2). Has S_{axial}-config.. λ_{max} 221 (log ε 2.75); 250 (log ε 4.03) (MeOH)

4,5-Methylene, 3,4′,5′-tri-Me ether
Gomisin M₁
[82467-50-3]
$C_{22}H_{26}O_6$ 386.444
Isol. from *Kadsura coccinea* and *Schisandra chinensis*. Needles (Et_2O/hexane). Mp 116 - 119°. No CD absorption between 200 and 400 nm.

Kochetkov, N.K. *et al., J. Gen. Chem. USSR (Engl. Transl.)*, 1961, **31**, 3218-3224 (*Deoxyschisandrin*).
Kochetkov, N.K. *et al., Tetrahedron Lett.*, 1962, **3**, 361-363 (*Deoxyschizandrin, struct, synth*).
Chen, Y.-Y. *et al., Sci. Sin. (Engl. Edn.)*, 1976, **19**, 276-290 (*Deoxyschizandrin*).
Liu, C.S. *Sci. Sin. (Engl. Edn.)*, 1976, **21**, 483-502 (*γ-Schizandrin*).
Ghera, E. *et al., Tetrahedron Lett.*, 1977, **18**, 463-466 (*Deoxyschizandrin, synth*).
Ikeya, Y. *et al., Chem. Pharm. Bull.*, 1978, **27**, 1583-1588 (*Gomisin J, Deoxyschizandrin*).
Liu, C.S. *et al., CA*, 1979, **90**, 103711 (*Schizanhenol*).

Ikeya, Y. *et al.*, *Chem. Pharm. Bull.*, 1979, **27**, 2695-2709 (*Gomisin N*).
Biftu, J. *et al.*, *J. Chem. Soc., Perkin Trans. 1*, 1979, 2276-2281 (*Deoxyschizandrin, synth*).
Ikeya, Y. *Chem. Pharm. Bull.*, 1980, **28**, 2422-2427 (*Gomisins K_1,K_2,K_3*).
Schneiders, G.E. *et al.*, *J. Org. Chem.*, 1981, **46**, 2969-2971 (*Wuweizisu C, synth, struct*).
Ikeya, Y. *et al.*, *Chem. Pharm. Bull.*, 1982, **30**, 132-139 (*Gomisins L_1,L_2,M_1,M_2, isol, γ-Schizandrin, synth, config*).
Ikeya, Y. *et al.*, *Chem. Pharm. Bull.*, 1982, **30**, 3207-3211 (*Wuweizisu C, abs config*).
Hikino, H. *et al.*, *Planta Med.*, 1984, **50**, 213-218 (*activity*).
Tan, R. *et al.*, *Planta Med.*, 1984, **50**, 414-417 (*Kadsuranin, Angeloylgomisin M_1*).
Takeya, T. *et al.*, *Chem. Pharm. Bull.*, 1985, **33**, 3599-3607 (*Wuweizisu C, Deoxyschizandrin, synth*).
Li, L.-N. *et al.*, *Planta Med.*, 1985, **51**, 297-300 (*Kadsura coccinea constits*).
Kiso, Y. *et al.*, *Planta Med.*, 1985, **53**, 331-334 (*Wuweizisu C, activity*).
Wang, H.-J. *et al.*, *Yaoxue Xuebao*, 1985, **20**, 832-841 (*Schisandra rubriflora constits*).
Suekawa, M. *et al.*, *Yakugaku Zasshi*, 1987, **107**, 720-726 (*Gomisin J, activity*).
Carroll, A.R. *et al.*, *Aust. J. Chem.*, 1994, **47**, 1579-1589 (*Deoxyschizandrin, synth*).
Tanaka, M. *et al.*, *Tetrahedron*, 1995, **51**, 11693-11702 (*Wuweizisu C, (-)-Gomisin J, (-)-Gomisin N, (±)-γ-Schizandrin, synth*).
Wu, A. *et al.*, *Chin. Sci. Bull.*, 1997, **42**, 995-998 (*synth*).
Marek, J. *et al.*, *Acta Crystallogr., Sect. C*, 1998, **54**, 1548-1550 (*Gomisin N, cryst struct*).
Miyazawa, M. *et al.*, *Nat. Prod. Lett.*, 1998, **12**, 175-180 (*Gomisins B,N*).
Kwon, B.M. *et al.*, *Planta Med.*, 1999, **65**, 74-76 (*activity*).
Hwang, E.I. *et al.*, *Yakhak Hoechi*, 1999, **43**, 509-515 (*Wuweizisu C, activity*).
Chen, M. *et al.*, *J. Nat. Prod.*, 2006, **69**, 1697-1701 (*Rubrisandrins A,B*).
Tian, R.R. *et al.*, *Zhongguo Tianran Yaowu*, 2006, **4**, 40-44 (*Rubrifloralignan A*).
Lei, C. *et al.*, *Chem. Pharm. Bull.*, 2007, **55**, 1281-1283 (*Angeloylgomisin K_3*).
Li, H.-M. *et al.*, *Helv. Chim. Acta*, 2008, **91**, 1053-1062 (*Tigloylgomisin K_3*).
Chen, M. *et al.*, *Molecules*, 2008, **13**, 548-555 (*Neglschisandrin A*).
Chen, M. *et al.*, *Molecules*, 2008, **13**, 1148-1155 (*Neglschisandrin D*).
Cheng, Y.-B. *et al.*, *J. Nat. Prod.*, 2009, **72**, 1663-1668 (*Arisanschinins F,G*).
Li, Y. *et al.*, *Synthesis*, 2009, 3383-3390 (*synth*).
Ma, W.-H. *et al.*, *Biochem. Syst. Ecol.*, 2010, **38**, 1156-1159 (*Schisanbicolorin A*).
Yang, G.-Y. *et al.*, *J. Nat. Prod.*, 2010, **73**, 915-919 (*Marlignans A,B,C*).
Duan, Y.-X. *et al.*, *J. Asian Nat. Prod. Res.*, 2011, **13**, 592-598 (*Schineglignans A,B*).
Zheng, S. *et al.*, *Chem. - Eur. J.*, 2012, **18**, 3193-3198 (*Deoxyschizandrin, synth*).
Gao, X.-M. *et al.*, *J. Braz. Chem. Soc.*, 2013, **24**, 2021-2027 (*Neglectaphenols A,B, activity*).
Chen, M. *et al.*, *Molecules*, 2013, **18**, 2297-2306 (*Neglschisandrin E, cd*).
Hu, D. *et al.*, *Phytochemistry*, 2014, **104**, 72-78 (*Schisanchinins B,D*).
Kim, H.M. *et al.*, *Chem. Pharm. Bull.*, 2015, **63**, 746-751 (*Schisandroside C*).
Xue, Y. *et al.*, *Phytochemistry*, 2015, **116**, 253-261 (*Schinlignan G, activity*).
Yang, B.-Y. *et al.*, *Chem. Biodiversity*, 2016, **13**, 1118-1125 (*Schisandra chinensis 5,5′-diglucoside*).
Szopa, A. *et al.*, *Phytochem. Rev.*, 2017, **16**, 195-218 (*Schisandra chinensis constits, rev*)

2,7′-Cyclo-8,8′-lignan-3,3′,4,4′,5,5′-hexol C234

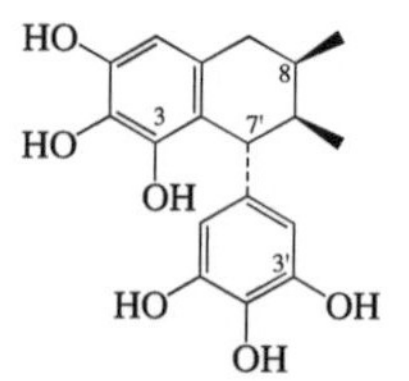

(7′R,8R,8′R)-form

5,6,7,8-Tetrahydro-6,7-dimethyl-8-(3,4,5-trihydroxyphenyl)-1,2,3-naphthalenetriol.
3,3′,4,4′,5,5′-Hexahydroxy-2,7′-cyclolignan
$C_{18}H_{20}O_6$ 332.352

(7′R,8R,8′R)-form

3′,4,4′,5-Tetra-Me ether
2,7′-Cyclo-4,4′,5,5′-tetramethoxy-8,8′-lignan-3,3′-diol. ***Heilaohusu E***
$C_{22}H_{28}O_6$ 388.460
Constit. of the roots of *Kadsura coccinea*. Yellow gel. $[\alpha]_D^{25}$ +48 (c, 0.09 in MeOH). λ_{max} 241 (log ε 4.59) (MeOH)

3′,4′-Methylene, 4,5,5′-tri-Me ether
2,7′-Cyclo-3′,4,5-trimethoxy-4′,5′-methylenedioxy-8,8′-lignan-3-ol.
3-Hydroxy-3′,4,5-trimethoxy-4′,5′-methylenedioxy-2,7′-cyclolignan. ***Kadsuralignan H***
[960511-16-4]
$C_{22}H_{26}O_6$ 386.444
Constit. of the rhizomes of *Kadsura coccinea*. Powder. $[\alpha]_D^{25}$ +49.1 (c, 0.25 in MeOH). λ_{max} 219 (log ε 4.29); 270 (log ε 3.28) (MeOH)

(7′S,8R,8′R)-form

4,5-Methylene, 3,3′,4′-tri-Me ether
2,7′-Cyclo-3,4′,5′-trimethoxy-4,5-methylenedioxy-8,8′-lignan-3′-ol.
3′-Hydroxy-3,4′,5′-trimethoxy-4,5-methylenedioxy-2,7′-cyclolignan. ***Sinensisin C***
[1401673-26-4]
$C_{22}H_{26}O_6$ 386.444
Constit. of *Schisandra propinqua* var. *sinensis*. Amorph. powder. $[\alpha]_D^{22}$ -11 (c, 0.2 in $CHCl_3$). λ_{max} 241 (log ε 3.76); 280 (log ε 3.15) ($CHCl_3$)

(7′S,8R,8′S)-form

3′,4′:4,5-Bis(methylene), 3,5′-di-Me ether
2,7′-Cyclo-3,3′-dimethoxy-4,5:4′,5′-bis(methylenedioxy)-8,8′-lignan.
3,3′-Dimethoxy-4,5:4′,5′-bis(methylenedioxy)-2,7′-cyclolignan. ***Kadsurindutin C***
[1149351-53-0]
$C_{22}H_{24}O_6$ 384.428
Constit. of the stems of *Kadsura induta*. Yellow powder. $[\alpha]_D^{25}$ -36 (c, 0.01 in MeOH). λ_{max} 211 (log ε 4.66); 282 (sh) (MeOH)

(7′R*,8R*,8′R*)-form

3,3′,5-Tri-Me ether
2,7′-Cyclo-3,5,5′-trimethoxy-8,8′-lignan-3′,4,4′-triol.
3′,4,4′-Trihydroxy-3,5,5′-trimethoxy-2,7′-cyclolignan. ***Ocholignan A***
[858518-79-3]
$C_{21}H_{26}O_6$ 374.433
Constit. of the leaves of *Scyphocephalium ochocoa*. Shows antibacterial activity against MRSA

(7′R*,8S*,8′R*)-form

3,3′,5-Tri-Me ether
Ocholignan B
[858518-80-6]
$C_{21}H_{26}O_6$ 374.433
Constit. of the leaves of *Scyphocephalium ochocoa*.

Hu, J.-F. *et al.*, *Phytochemistry*, 2005, **66**, 1077-1082 (*Ocholignans A,B*).
Li, H. *et al.*, *J. Nat. Prod.*, 2007, **70**, 1999-2002 (*Kadsuralignan H*).
Ma, W. *et al.*, *Helv. Chim. Acta*, 2009, **92**, 709-715 (*Kadsurindutin C*).
Li, X.-N. *et al.*, *Fitoterapia*, 2012, **83**, 249-252 (*Sinensisin C*).
Yang, Y. *et al.*, *Fitoterapia*, 2019, **139**, 104368:1-6 (*Heilaohusu E, cd*)

2,7′-Cyclo-8,8′-lignan-3,3′,4,4′,5,7′-hexol C235

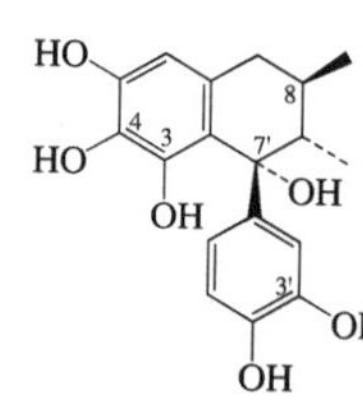

(7′R,8R,8′S)-form

3,3′,4,4′,5,7′-Hexahydroxy-2,7′-cyclolignan

$C_{18}H_{20}O_6$ 332.352
(7′R,8R,8′S)-form
3,4:3′,4′-Bis(methylene), 5-Me ether
2,7′-Cyclo-5-methoxy-3,4:3′,4′-bis(methylenedioxy)-8,8′-lignan-7′-ol. 7′-Hydroxy-5-methoxy-3,4:3′,4′-bis(methylenedioxy)-2,7′-cyclolignan. ***Hydroxymethoxyotobain***
[98819-57-9]
$C_{21}H_{22}O_6$ 370.401
Isol. from leaves and bark of *Virola calophylloidea*. Oil. λ_{max} 239 (ε 7300); 292 (ε 4500); 308 (ε 3450) (MeOH)
(7′S,8R,8′R)-form
3,4:3′,4′-Bis(methylene), 5-Me ether
[98767-34-1]
Isol. from leaves and bark of *Virola calophylloidea*. Cryst. (MeOH). Mp 66 - 67°. λ_{max} 240 (ε 7150); 290 (ε 5500); 308 (ε 3500) (MeOH).

Martinez, J.C. *et al.*, *Phytochemistry*, 1985, **24**, 1867-1868 (*Virola calophylloidea constits*)

2,7′-Cyclo-8,8′-lignan-3′,4,4′,5,7′,8′-hexol C236

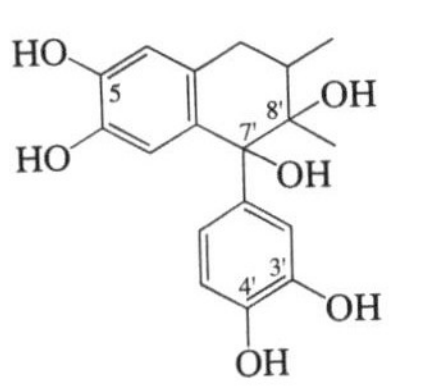

3′,4,4′,5,7′,8′-Hexahydroxy-2,7′-cyclolignan
$C_{18}H_{20}O_6$ 332.352
3′,4′-Methylene, 5-Me ether
2,7′-Cyclo-5-methoxy-3′,4′-methylenedioxy-8,8′-lignan-4,7′,8′-triol. 4,7′,8′-Trihydroxy-5-methoxy-3′,4′-methylenedioxy-2,7′-cyclolignan
[188894-40-8]
$C_{20}H_{22}O_6$ 358.39
Constit. of *Virola pavonis*. Viscous oil. $[\alpha]_D$ -28.5 (c, 0.2 in $CHCl_3$). λ_{max} 230 (log ε 3.93); 282 (log ε 3.8) (MeOH).

Martinez, V., J.C. *et al.*, *Phytochemistry*, 1997, **44**, 1179-1182 (*Virola pavonis constit*)

2,7′-Cyclo-8,8′-lignan-3′,4,4′,5,9,9′-hexol C237

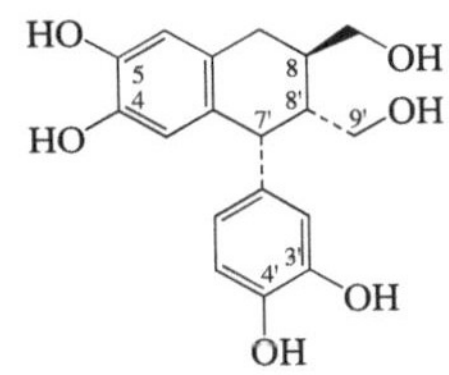

(7′R,8R,8′R)-form

3′,4,4′,5,9,9′-Hexahydroxy-2,7′-cyclolignan
$C_{18}H_{20}O_6$ 332.352
(7′R,8R,8′R)-form
3′,5-Di-Me ether
2,7′-Cyclo-3′,5-dimethoxy-8,8′-lignan-4,4′,9,9′-tetrol. 4,4′,9,9′-Tetrahydroxy-3′,5-dimethoxy-2,7′-cyclolignan. ***(-)-Burselignan***
[871711-74-9]
$C_{20}H_{24}O_6$ 360.406
Constit. of *Bursera tonkinensis, Ecdysanthera rosea, Isatis indigotica, Kadsura heteroclita, Styrax perkinsae* and *Tripterygium wilfordii*. Powder. $[\alpha]^{20}_D$ -64.2 (c, 0.01 in MeOH). λ_{max} 222 (log ε 4); 284 (log ε 3.5) (MeOH)
3′,5-Di-Me ether, 9′-O-α-L-arabinopyranoside
Burselignan 9′-α-L-arabinopyranoside
$C_{25}H_{32}O_{10}$ 492.521
Constit. of the pericarp of *Litchi chinensis*. Amorph. powder. $[\alpha]^{20}_D$ -99.8 (c, 0.43 in MeOH). λ_{max} 284 (log ε 3.55) (MeOH)
3′,5-Di-Me ether, 9′-O-β-D-xylopyranoside
Tiliamuroside B
[1446929-57-2]
$C_{25}H_{32}O_{10}$ 492.521
Constit. of *Tilia amurensis*. Exhibits significant cytotoxic activity. Gum. $[\alpha]^{25}_D$ -75.3 (c, 0.35 in MeOH). λ_{max} 223 (log ε 4.0); 283 (log ε 3.5) (MeOH)
3′,5-Di-Me ether, 9′-O-β-D-glucopyranoside
Densispicoside A
[1028335-67-2]
$C_{26}H_{34}O_{11}$ 522.547
Constit. of the aerial parts of *Pedicularis densispica*. Amorph. powder. $[\alpha]^{20}_D$ -155.7 (c, 0.15 in MeOH). λ_{max} 206 (log ε 4.74) (MeOH)
3′,5-Di-Me ether, 9′-O-[β-D-apiofuranosyl-(1→6)-β-D-glucopyranoside]
$C_{31}H_{42}O_{15}$ 654.662
Constit. of the branches of *Alangium kurzii*. Pale brown oil. $[\alpha]^{20}_D$ -2.35 (c, 0.25 in MeOH)
3′,5-Di-Me ether, 9′-O-[β-D-xylopyranosyl-(1→6)-β-D-glucopyranoside]
$C_{31}H_{42}O_{15}$ 654.662
Constit. of the branches of *Alangium kurzii*. Pale yellow oil. $[\alpha]^{20}_D$ -2.3 (c, 0.23 in MeOH)
(7′R,8S,8′R)-form
3′,5-Di-Me ether, 4-O-β-D-glucopyranoside
Sargentodoside A
[1868129-12-7]
$C_{26}H_{34}O_{11}$ 522.547
Constit. of *Sargentodoxa cuneata*. Amorph. yellow powder. $[\alpha]^{25}_D$ -31.4 (c, 0.17 in MeOH). λ_{max} 282 (log ε 3.86) (MeOH)
(7′R,8S,8′S)-form
3′,4′-Di-Me ether, 4-O-β-D-glucopyranoside
Sargentodoside C
[1868129-14-9]
$C_{26}H_{34}O_{11}$ 522.547
Constit. of *Sargentodoxa cuneata*. Amorph. yellow powder. $[\alpha]^{25}_D$ -6.9 (c, 0.17 in MeOH)
3′,5-Di-Me ether
ent-Isolariciresinol
[110268-37-6]
$C_{20}H_{24}O_6$ 360.406
Constit. of *Reseda suffruticosa, Berberis koreana* and *Wikstroemia glabra*. Moderately cytotoxic against various human cancer cell lines. Cryst. ($CHCl_3$/MeOH). Mp 152 - 154°. $[\alpha]^{25}_D$ -53.8 (c, 1.47 in MeOH). $[\alpha]^{22}_D$ -48.6 (c, 0.4 in Me_2CO)
3′,5-Di-Me ether, 4-O-β-D-glucopyranoside
[365493-99-8]
$C_{26}H_{34}O_{11}$ 522.547
Constit. of *Balanophora japonica* and *Cynomorium songaricum*. Pale yellow amorph. powder. $[\alpha]^{15}_D$ -33 (c, 0.2 in MeOH)
3′,5-Di-Me ether, 9-O-β-D-glucopyranoside
[1206882-32-7]
$C_{26}H_{34}O_{11}$ 522.547
Constit. of the leaves of *Glochidion rubrum*. Amorph. powder. $[\alpha]^{24}_D$

-48.6 (c, 0.29 in MeOH). λ_{max} 209 (log ε 3.82); 280 (log ε 3.23) (MeOH)

3′,5-Di-Me ether, 9′-O-α-L-arabinopyranoside

[903559-37-5]

$C_{25}H_{32}O_{10}$ 492.521

Constit. of the bark of *Walsura yunnanensis*. Antioxidant. Cryst. (MeOH). Mp 142 - 144°. $[\alpha]^{18}_{D}$ -50.4 (c, 0.3 in MeOH). λ_{max} 206 (log ε 4.18); 284 (log ε 3.89) (MeOH)

3′,5-Di-Me ether, 9′-O-β-D-xylopyranoside

Saracoside

[80734-72-1]

$C_{25}H_{32}O_{10}$ 492.521

Constit. of *Picea abies* and *Saraca indica*. Inhibitor of DNA topoisomerase IB. Amorph.. $[\alpha]^{22}_{D}$ -55.1 (c, 0.9 in MeOH)

3′,5-Di-Me ether, 9′-O-α-L-rhamnopyranoside

[1819968-33-6]

$C_{26}H_{34}O_{10}$ 506.548

Constit. of the aerial parts of *Lespedeza cuneata*. Exhibits ability to activate the transcription of xbp1 promoter. Amorph. solid. $[\alpha]^{25}_{D}$ -156.8 (c, 0.28 in MeOH). λ_{max} 206 (log ε 4.78); 284 (log ε 3.88) (MeOH)

3′,5-Di-Me ether, 9′-O-β-D-glucopyranoside

[143236-04-8]

$C_{26}H_{34}O_{11}$ 522.547

Constit. of *Stemmadenia minima*. $[\alpha]_{D}$ -32 (c, 0.11 in MeOH). λ_{max} 233 (sh); 283 (log ε 3.55) (MeOH)

3′,5-Di-Me ether, 9′-O-[3-hydroxy-4-methoxybenzoyl-(→6)-β-D-glucopyranoside]

Tarennanoside F

[1201192-84-8]

$C_{34}H_{40}O_{14}$ 672.68

Constit. of *Tarenna attenuata*. Amorph. powder. $[\alpha]^{19}_{D}$ -44.2 (c, 0.92 in MeOH). λ_{max} 265 (log ε 4.04); 287 (log ε 4.03) (MeOH)

3′,5-Di-Me ether, 9′-O-[4-hydroxy-3-methoxybenzoyl-(→6)-β-D-glucopyranoside]

Tarennanoside D

[1201192-82-6]

$C_{34}H_{40}O_{14}$ 672.68

Constit. of *Tarenna attenuata*. Amorph. powder. $[\alpha]^{19}_{D}$ -37.3 (c, 0.25 in MeOH). λ_{max} 264 (log ε 3.7); 288 (log ε 3.69) (MeOH)

3′,5-Di-Me ether, 9′-O-[β-D-apiofuranosyl-(1→2)-β-D-glucopyranoside]

[610754-91-1]

$C_{31}H_{42}O_{15}$ 654.662

Constit. of the aerial parts of *Sauropus androgynus*. Amorph. powder. $[\alpha]^{28}_{D}$ -48 (c, 1.7 in MeOH). λ_{max} 215 (log ε 4.15); 282 (log ε 3.74) (MeOH)

3′,5-Di-Me ether, 9-O-[α-L-rhamnopyranosyl-(1→6)-β-D-glucopyranoside]

Flacoside B

[1204224-35-0]

$C_{32}H_{44}O_{15}$ 668.689

Constit. of *Flacourtia ramontchi*. Syrup. $[\alpha]^{20}_{D}$ -2.5 (c, 0.08 in DMSO). λ_{max} 219 (log ε 4.62); 284 (log ε 3.95) (MeOH)

Hexa-Me ether

2,7′-Cyclo-3′,4,4′,5,9,9′-hexamethoxylignan. 3′,4,4′,5,9,9′-Hexamethoxy-2,7′-cyclolignan.

Phyltetralin

[123048-17-9]

[50656-79-6] (unspecified stereochem.) [78215-55-1] ((±)-form)

Constit. of *Phyllanthus niruri* and *Phyllanthus urinaria*. Mp 110 - 111°. $[\alpha]^{25}_{D}$ +40 (c, 0.5 in $CHCl_3$). λ_{max} 232 (sh); 282 (log ε 3.85) (MeCN)

3′,4′-Methylene, 4,5,9,9′-tetra-Me ether

2,7′-Cyclo-4,5,9,9′-tetramethoxy-3′,4′-methylenedioxylignan. 4,5,9,9′-Tetramethoxy-3′,4′-methylenedioxy-2,7′-cyclolignan.

Lintetralin

[73231-44-4]

$C_{23}H_{28}O_6$ 400.471

Isol. from *Phyllanthus* spp.. Needles (hexane). Mp 87 - 88°. $[\alpha]^{25}_{D}$ -7 (c, 1 in $CHCl_3$). Revised abs. config.. λ_{max} 225 (sh); 275 (log ε 3.92) (MeCN)

4,5-Methylene, 3′,4′,9,9′-tetra-Me ether

2,7′-Cyclo-3′,4′,9,9′-tetramethoxy-4,5-methylenedioxylignan. 3′,4′,9,9′-Tetramethoxy-4,5-methylenedioxy-2,7′-cyclolignan.

Isolintetralin

[145459-30-9]

$C_{23}H_{28}O_6$ 400.471

Isol. from *Phyllanthus niruri* and *Phyllanthus urinaria*. Powder. $[\alpha]^{25}_{D}$ +46.2 (c, 0.2 in $CHCl_3$). Revised abs. config.. λ_{max} 226 (sh); 283 (log ε 3.85) (MeCN)

3′,4′:4,5-Bis(methylene), 9,9′-di-Me ether

2,7′-Cyclo-9,9′-dimethoxy-3′,4′:4,5-bis(methylenedioxy)lignan. 9,9′-Dimethoxy-3′,4′:4,5-bis(methylenedioxy)-2,7′-cyclolignan.

Urinatetralin

$C_{22}H_{24}O_6$ 384.428

Constit. of *Phyllanthus niruri* and *Phyllanthus urinaria*. Amorph. solid. $[\alpha]^{25}_{D}$ +7 (c, 1 in $CHCl_3$). λ_{max} 232 (sh); 287 (log ε 3.91) (MeCN)

(7′S,8R,8′R)-form

Shows potent cytotoxic activity

3′-Me ether

2,7′-Cyclo-3′-methoxy-8,8′-lignan-4,4′,5,9,9′-pentol. 4,4′,5,9,9′-Pentahydroxy-3′-methoxy-2,7′-cyclolignan

[349150-66-9]

$C_{19}H_{22}O_6$ 346.379

Constit. of the heartwood of *Taxus baccata*. Pinkish powder. $[\alpha]^{22}_{D}$ -77 (c, 0.41 in MeOH)

5-Me ether

2,7′-Cyclo-5-methoxy-8,8′-lignan-3′,4,4′,9,9′-pentol. 3′,4,4′,9,9′-Pentahydroxy-5-methoxy-2,7′-cyclolignan. ***Isotaxiresinol***

[26194-57-0]

$C_{19}H_{22}O_6$ 346.379

Constit. of *Taxus baccata*, *Taxus cuspidata*, *Taxus meieri*, *Podocarpus spicata*, *Fitzroya cupressoides* and *Justicia* sp.. Cryst. (AcOH aq.). Mp 171°. Opt. activity not recorded. λ_{max} 227; 285 (MeOH) (Berdy)

3′,4′-Di-Me ether, 4-O-β-D-glucopyranoside

Sargentodoside B

[1868129-13-8]

$C_{26}H_{34}O_{11}$ 522.547

Constit. of *Sargentodoxa cuneata*. Amorph. yellow powder. $[\alpha]^{25}_{D}$ +72.5 (c, 0.17 in MeOH). λ_{max} 282 (log ε 3.86) (MeOH)

3′,5-Di-Me ether

(+)-Isolariciresinol. *α-Conidendryl alcohol*

[548-29-8]

$C_{20}H_{24}O_6$ 360.406

Constit. of *Acer saccharum*, *Fitzroya cupressoides*, *Justicia tranquebariensis*, *Picea excelsa*, *Pinus massoniana* and *Taxus baccata*. Needles ($CHCl_3$/MeOH). Mp 157 - 158°. $[\alpha]^{22}_{D}$ +69.5 (c, 1.19 in Me_2CO)

3′,5-Di-Me ether, 4-O-β-D-glucopyranoside
Isolariciresinol 4-glucoside. ***Woorenoside XI***
[195873-45-1]
$C_{26}H_{34}O_{11}$ 522.547
Constit. of *Coptis japonica* var. *dissecta* and *Vitis thunbergii*. Amorph. solid. $[\alpha]_{D}$ -29.3 (c, 2.7 in MeOH). λ_{max} 240 (log ε 3.93); 283 (log ε 3.59) (EtOH)

3′,5-Di-Me ether, 4-O-β-D-glucopyranoside, 9′-O-β-D-xylopyranoside
Isolariciresinol 4-glucoside 9′-xyloside
[756835-81-1]
$C_{31}H_{42}O_{15}$ 654.662
Constit. of the aerial parts of *Vitis thunbergii*. Amorph. solid. $[\alpha]^{25}_{D}$ -19.1 (c, 0.2 in MeOH). λ_{max} 235 (log ε 4.05); 285 (log ε 3.68) (MeOH)

3′,5-Di-Me ether, 4′-O-β-D-glucopyranoside
Isolariciresinol 4′-glucoside
[145748-07-8]
$C_{26}H_{34}O_{11}$ 522.547
Constit. of Riesling wine. Amorph. solid

3′,5-Di-Me ether, 9-O-α-L-arabinopyranoside
Isolariciresinol 9-O-α-L-arabinopyranoside. **Bridelioside**
[1013938-23-2]
$C_{25}H_{32}O_{10}$ 492.521
Constit. of the leaves of *Bridelia glauca* f. *balansae*. Amorph. powder. $[\alpha]^{20}_{D}$ +23 (c, 0.61 in MeOH). λ_{max} 216 (log ε 4.25); 284 (log ε 3.77) (MeOH)

3′,5-Di-Me ether, 9-O-β-D-glucopyranoside
Isolariciresinol 9-glucoside
[63358-11-2]
[62080-76-6]
$C_{26}H_{34}O_{11}$ 522.547
Constit. of *Pinus massoniana* and *Pinus sylvestris*. Amorph.. $[\alpha]^{20}_{D}$ +8 (c, 0.9 in EtOH)

3′,5-Di-Me ether, 9′-O-α-L-arabinofuranoside
Isolariciresinol 9′-O-α-L-arabinofuranoside
[63358-10-1]
[62080-75-5]
$C_{25}H_{32}O_{10}$ 492.521
Constit. of *Pinus massoniana* and *Pinus sylvestris*. Amorph.. $[\alpha]^{20}_{D}$ -36.8 (c, 1 in EtOH)

3′,5-Di-Me ether, 9′-O-α-L-arabinopyranoside
(+)-Isolariciresinol 9′-α-L-arabinopyranoside
$C_{25}H_{32}O_{10}$ 492.521
Constit. of the pericarp of *Litchi chinensis*. Amorph. powder. $[\alpha]^{20}_{D}$ +40 (c, 0.45 in MeOH). λ_{max} 284 (log ε 4.0) (MeOH)

3′,5-Di-Me ether, 9-O-β-D-xylopyranoside
Difengpioside C
[2041284-99-3]
$C_{25}H_{32}O_{10}$ 492.521
Constit. of the stem bark of *Illicium difengpi*. Amorph. powder. $[\alpha]^{20}_{D}$ -32 (c, 0.2 in MeOH). λ_{max} 281 (log ε 3.64) (MeOH)

3′,5-Di-Me ether, 9′-O-β-D-xylopyranoside
Schizandriside
[71222-06-5]
$C_{25}H_{32}O_{10}$ 492.521
Constit. of *Illicium difengpi*, *Litchi chinensis*, *Schizandra nigra*, *Pinus massoniana*, *Picea abies* and *Tilia amurensis*. Exhibits significant cytotoxic activity. Cryst.. Mp 225 - 226°. $[\alpha]^{27}_{D}$ +28.3 (c, 0.52 in EtOH)

3′,5-Di-Me ether, 9′-O-β-D-fucopyranoside
[2185026-83-7]
$C_{26}H_{34}O_{10}$ 506.548
Constit. of aerial parts of *Lespedeza cuneata*. Amorph. powder. $[\alpha]^{20}_{D}$ +18.9 (c, 0.1 in MeOH). λ_{max} 284 (log ε 2.22) (MeOH)

3′,5-Di-Me ether, 9′-O-α-L-rhamnopyranoside
Isolariciresinol 9′-rhamnoside. ***Aviculin***. *Chestnutlignanoside*
[156765-33-2]
$C_{26}H_{34}O_{10}$ 506.548
Constit. of *Castanea mollissima*, *Trigonostemon heterophyllus* and *Polygonum aviculare*. Amorph. powder. Mp 155 - 157°. Mp 211 - 214°. $[\alpha]^{25}_{D}$ +38.6 (c, 0.32 in MeOH)

3′,5-Di-Me ether, 9′-O-β-D-glucopyranoside
Isolariciresinol 9′-glucoside
[63358-12-3]
$C_{26}H_{34}O_{11}$ 522.547
Constit. of *Myrsine seguinii*, *Pinellia ternata* and *Pinus sylvestris*. Needles (MeOH). Mp 135 - 138°. $[\alpha]^{23}_{D}$ +45.4 (c, 0.8 in MeOH)

3′,5-Di-Me ether, 9′-O-[3-hydroxy-4-methoxybenzoyl-(→6)-β-D-glucopyranoside]
Tarennanoside E
[1201192-83-7]
$C_{34}H_{40}O_{14}$ 672.68
Constit. of *Tarenna attenuata*. Amorph. powder. $[\alpha]^{19}_{D}$ +9.5 (c, 0.35 in MeOH). λ_{max} 264 (log ε 4.02); 286 (log ε 4) (MeOH)

3′,5-Di-Me ether, 9′-O-[4-hydroxy-3-methoxybenzoyl-(→6)-β-D-glucopyranoside]
Tarennanoside C
[1201192-80-4]
$C_{34}H_{40}O_{14}$ 672.68
Constit. of *Tarenna attenuata*. Amorph. powder. $[\alpha]^{20}_{D}$ +15.7 (c, 0.89 in MeOH). λ_{max} 266 (log ε 3.94); 287 (log ε 3.93) (MeOH)

3′,5-Di-Me ether, 9′-O-[4-hydroxy-3,5-dimethoxybenzoyl-(→6)-β-D-glucopyranoside]
Tarennanoside G
[1201192-85-9]
$C_{35}H_{42}O_{15}$ 702.706
Constit. of *Tarenna attenuata*. Amorph. powder. $[\alpha]^{24}_{D}$ +22.4 (c, 0.62 in MeOH). λ_{max} 282 (log ε 4.16) (MeOH)

3′,5-Di-Me ether, 9′-O-[4-hydroxy-E-cinnamoyl-(→6)-β-D-glucopyranoside]
[1819968-31-4]
$C_{35}H_{40}O_{13}$ 668.692
Constit. of the aerial parts of *Lespedeza cuneata*. Exhibits ability to activate the transcription of xbp1 promoter. Amorph. solid. $[\alpha]^{25}_{D}$ +24.8 (c, 0.25 in MeOH). λ_{max} 206 (log ε 4.17); 289 (log ε 3.69); 315 (log ε 4.22) (MeOH)

3′,5-Di-Me ether, 9′-O-[4-hydroxy-Z-cinnamoyl-(→6)-β-D-glucopyranoside]
[1819968-32-5]
$C_{35}H_{40}O_{13}$ 668.692
Constit. of the aerial parts of *Lespedeza cuneata*. Amorph. solid. $[\alpha]_D^{25}$ +53.9 (c, 0.07 in MeOH). λ_{max} 206 (log ε 4.89); 290 (log ε 4.37); 314 (log ε 4.34) (MeOH)

3′,5-Di-Me ether, 9′-O-[β-D-apiofuranosyl-(1→2)-β-D-glucopyranoside]
[1913248-35-7]
$C_{31}H_{42}O_{15}$ 654.662
Constit. of branches of *Tabebuia chrysotricha*. Exhibits moderate DPPH radical scavenging activity. Amorph. powder. $[\alpha]_D^{24}$ -27.8 (c, 0.43 in MeOH). λ_{max} 277 (log ε 3.88); 331 (log ε 3.74) (no solvent reported)

3′,5-Di-Me ether, 9′-O-[α-L-rhamnopyranosyl-(1→6)-β-D-glucopyranoside]
Isolariciresinol 9′-rutinoside
[1161572-52-6]
$C_{32}H_{44}O_{15}$ 668.689
Constit. of the branches of *Dipteronia dyeriana*. Powder. $[\alpha]_D^{25}$ +3.9 (c, 0.43 in MeOH). λ_{max} 283 (log ε 3.77) (MeOH)

3′,5-Di-Me ether, 9′-O-[β-D-glucopyranosyl-(1→2)-β-D-glucopyranoside]
Saposide A
[1686144-39-7]
$C_{32}H_{44}O_{16}$ 684.688
Constit. of the sap of *Acer saccharum*. Amorph. powder. $[\alpha]_D^{22}$ +9.2 (c, 0.6 in MeOH). λ_{max} 230 (log ε 4.06); 283 (log ε 3.76) (MeOH)

3′,5-Di-Me ether, 9′-O-β-D-glucopyranoside, 9-O-β-D-apiofuranoside
Cyclolignanyingoside
$C_{31}H_{42}O_{15}$ 654.662
Constit. of the tubers of *Pinellia ternata*. Antioxidant. Yellow oil. $[\alpha]_D^{25}$ -43.1 (c, 0.1 in MeOH). λ_{max} 210 (log ε 0.94); 283 (log ε 0.17) (MeOH)

3′,5-Di-Me ether, 9′-O-sulfate
Isolariciresinol 9′-O-sulfate
[448949-57-3]
$C_{20}H_{24}O_9S$ 440.463
Constit. of *Polygonum cuspidatum* and *Myrsine seguinii*. Cryst. (MeOH). Mp 185 - 195° dec.. $[\alpha]_D$ +34 (c, 0.11 in MeOH) (as Na salt). λ_{max} 209 (log ε 4.5); 229 (sh); 284 (log ε 3.75) (MeOH)

3′,5-Di-Me ether, 9′-Ac
9′-O-Acetylisolariciresinol
[877997-55-2]
$C_{22}H_{26}O_7$ 402.443
Constit. of *Ephedra viridis* and *Leea aequata*. Yellow syrup. $[\alpha]_D^{28}$ +5.2 (c, 0.5 in $CHCl_3$)

3′,5-Di-Me ether, 9,9′-di-Ac
Yunnanensin A
[1426545-05-2]
$C_{24}H_{28}O_8$ 444.48
Constit. of *Lagerstroemia indica* and *Parakmeria yunnanensis*. Yellow oil. $[\alpha]_D^{23}$ -6.8 (c, 0.75 in MeOH). λ_{max} 203 (log ε 3.31); 284 (log ε 2.4) (MeOH)

3′,5-Di-Me ether, 9-O-(5-hydroxy-4-oxopentanoyl)
Clemomanshurinane B
[943441-38-1]
$C_{25}H_{30}O_9$ 474.506
Constit. of the roots and rhizomes of *Clematis manshurica*. Cyclooxygenase 2 inhibitor. Mp 158°. $[\alpha]_D^{20}$ +36.7 (c, 0.35 in MeOH). λ_{max} 240 (log ε 4.21) (MeOH)

3′,5-Di-Me ether, 9′-O-(5-hydroxy-4-oxopentanoyl)
Clemomanshurinane A
[943441-37-0]
$C_{25}H_{30}O_9$ 474.506
Constit. of the roots and rhizomes of *Clematis manshurica*. Cyclooxygenase 2 inhibitor. Mp 153°. $[\alpha]_D^{20}$ +39.5 (c, 0.38 in MeOH). λ_{max} 241 (log ε 4.18) (MeOH)

3′,5-Di-Me ether, 9′-benzoyl, 9-Ac
$C_{29}H_{30}O_8$ 506.551
Constit. of *Tripterygium wilfordii*. NO prodn. inhibitor. Yellow oil. $[\alpha]_D^{25}$ +45.3 (c, 0.1 in MeOH). λ_{max} 228 (log ε 2.89); 283 (log ε 2.33) (MeOH)

3′,5-Di-Me ether, 9′-O-(2-hydroxybenzoyl)
(+)-Isolariciresinol 9′-salicylate
[1624270-03-6]
$C_{27}H_{28}O_8$ 480.513
Constit. of *Gaultheria yunnanensis*. Brown oil. $[\alpha]_D^{17}$ +10 (c, 0.75 in MeOH). λ_{max} 259 (log ε 0.98); 285 (log ε 2.12) (MeOH)

3′,5-Di-Me ether, 9′-O-(4-hydroxy-E-cinnamoyl)
9′-O-p-Coumaroyl-(+)-isolariciresinol
[864378-12-1]
$C_{29}H_{30}O_8$ 506.551
Constit. of the bark of *Larix olgensis* var. *koreana*. Powder (MeOH). Mp 125 - 127°. $[\alpha]_D^{20}$ +86.5 (c, 0.48 in MeOH)

3′,5-Di-Me ether, 9′-O-(4-hydroxy-3-methoxy-E-cinnamoyl)
9′-O-trans-Feruloyl-(+)-isolariciresinol
[1353749-58-2]
$C_{30}H_{32}O_9$ 536.577
Constit. of *Abies nephrolepis*. NO prodn. inhibitor. Yellow powder. $[\alpha]_D^{24}$ +46 (c, 0.2 in MeOH). λ_{max} 288 (log ε 5.03); 326 (log ε 5.02) (MeOH)

4,5-Di-Me ether
2,7′-Cyclo-4,5-dimethoxy-8,8′-lignan-3′,4′,9,9′-tetrol. 3′,4′,9,9′-Tetrahydroxy-4,5-dimethoxy-2,7′-cyclolignan
[23141-17-5]
$C_{20}H_{24}O_6$ 360.406
Constit. of *Chirita longgangensis* var. *hongyao* and *Fitzroya cupressoides*. Needles ($CHCl_3$/MeOH). Mp 181 - 182° (172-173°). $[\alpha]_D^{24}$ +34.9 (c, 1.49 in Me_2CO)

3′,4′,5-Tri-Me ether
2,7′-Cyclo-3′,4′,5-trimethoxy-8,8′-lignan-4,9,9′-triol. 4,9,9′-Trihydroxy-3′,4′,5-trimethoxy-2,7′-cyclolignan. ***Isolariciresinol 4′-methyl ether***
[67560-72-9]
$C_{21}H_{26}O_6$ 374.433
Constit. of *Araucaria angustifolia*. Cryst. (MeOH/Me_2CO). Mp 188 - 190°. $[\alpha]_D^{25}$ +16 (c, 1 in MeOH)

3′,4,5-Tri-Me ether, 9′-O-β-D-xylopyranoside
Bryophylluside.
4-Methoxyschizandriside
[1627340-71-9]
$C_{26}H_{34}O_{10}$ 506.548
Constit. of *Bryophyllum pinnatum*. Mp 210 - 212°. $[\alpha]_D^{25}$ +13.2 (c, 3.15 in MeOH). λ_{max} 201 (log ε 4.29); 220 (log ε 4.53); 237 (log ε 4.58); 256 (log ε 4.70) (MeCN)

3′,4,4′,5-Tetra-Me ether
2,7′-Cyclo-3′,4,4′,5-tetramethoxy-8,8′-lignan-9,9′-diol. 9,9′-Dihydroxy-3′,4,4′,5-tetramethoxy-2,7′-cyclolignan
Mp 161 - 164°. $[\alpha]_D$ +15.8 (c, 1.79 in $CHCl_3$)

9′-O-(1-Hydroxy-1-methylethyl), 5-Me ether

2,7′-Cyclo-9′-(1-hydroxyisopropoxy)-5-methoxy-8,8′-lignan-3′,4,4′,9-tetrol. 3′,4,4′,9-Tetrahydroxy-9′-(1-hydroxyisopropoxy)-5-methoxy-2,7′-cyclolignan

[349150-65-8]

$C_{22}H_{28}O_7$ 404.459

Constit. of the heartwood of *Taxus baccata*. Yellowish cryst.. Mp 138°. $[\alpha]_D^{22}$ -110 (c, 1.1 in MeOH)

(7′S,8S,8′S)-form

3′,5-Di-Me ether, 9′-O-α-L-rhamnopyranoside

[1819968-35-8]

$C_{26}H_{34}O_{10}$ 506.548

Constit. of the aerial parts of *Lespedeza cuneata*. Exhibits ability to activate the transcription of xbp1 promoter. Amorph. solid. $[\alpha]_D^{25}$ +81.5 (c, 0.1 in MeOH). λ_{max} 206 (log ε 4.67); 284 (log ε 3.72) (MeOH)

3′,5-Di-Me ether, 9′-O-β-D-glucopyranoside

[2185026-82-6]

$C_{26}H_{34}O_{11}$ 522.547

Constit. of aerial parts of *Lespedeza cuneata*. Amorph. powder. $[\alpha]_D^{20}$ +85.5 (c, 0.12 in MeOH). λ_{max} 284 (log ε 2.19) (MeOH)

(7′R*,8R*,8′S*)-form

3′,4′-Di-Me ether

2,7′-Cyclo-3′,4′-dimethoxy-8,8′-lignan-4,5,9,9′-tetrol. 4,5,9,9′-Tetrahydroxy-3′,4′-dimethoxy-2,7′-cyclolignan. ***Torreyinol***

[109237-35-6]

$C_{20}H_{24}O_6$ 360.406

Constit. of *Torreya jackii*. Identified by ms only

(7′R*,8S*,8′S*)-form

3′,4,5-Tri-Me ether

Scaphopetalone

[533925-59-6]

$C_{21}H_{26}O_6$ 374.433

Constit. of the stem bark of *Scaphopetalum thonneri*. Brown viscous oil. λ_{max} 219 (log ε 2.32); 274 (log ε 0.45) (MeOH).

King, F.E. *et al.*, *J. Chem. Soc.*, 1952, 17-24 (*Isotaxiresinol*).

Weinges, K. *Chem. Ber.*, 1961, **94**, 2522-2533 (*Isolariciresinol*).

Erdtman, H. *et al.*, *Acta Chem. Scand.*, 1969, **23**, 2021-2024 (*Fitzroya cupressoides constits*).

Anjaneyulu, A.S.R. *et al.*, *Tetrahedron*, 1973, **29**, 1291-1298 (*Phyltetralin*).

Popoff, T. *et al.*, *Acta Chem. Scand., Ser. B*, 1977, **31**, 329-337 (*Isolariciresinol glycosides*).

Stevenson, R. *Tetrahedron*, 1977, **33**, 2913-2917 (*Phyltetralin, struct*).

Fonseca, S.F. *et al.*, *Phytochemistry*, 1978, **17**, 499-502 (*Isolariciresinol, Isolariciresinol 4′-Me ether, cmr*).

Takani, M. *et al.*, *Chem. Pharm. Bull.*, 1979, **27**, 1422-1425 (*Schizandriside*).

Ganeshepure, P.A. *et al.*, *J. Chem. Soc., Perkin Trans. 1*, 1981, 1681-1684 (*Lintetralin, Phyltetralin, synth*).

Lundgren, L.N. *et al.*, *Phytochemistry*, 1981, **20**, 1967-1969 (*Picea abies constit*).

Lundgren, L.N. *et al.*, *Acta Chem. Scand., Ser. B*, 1985, **39**, 241-248 (*Pinus massoniana constits*).

Charlton, J.L. *et al.*, *J. Org. Chem.*, 1986, **51**, 3490-3493 (*Isolariciresinol dimethyl ether, synth*).

Chen, R. *et al.*, *CA*, 1987, **107**, 55719z (*Torreyinol*).

Urones, J.G. *et al.*, *Phytochemistry*, 1987, **26**, 1540-1541 (*ent-Isolariciresinol*).

Meyers, A.I. *J. Am. Chem. Soc.*, 1988, **110**, 4611-4624 (*Phyltetralin, synth, abs config*).

Sashida, Y. *et al.*, *Chem. Pharm. Bull.*, 1989, **37**, 3301-3303 (*Schizandriside*).

Raju, G.V.S. *et al.*, *Indian J. Chem., Sect. B*, 1989, **28**, 558-561 (*Justicia tranquebariensis constit*).

Marinos, V.A. *et al.*, *Phytochemistry*, 1992, **31**, 4307-4312 (*Isolariciresinol 4′-glucoside*).

Achenbach, H. *et al.*, *Planta Med.*, 1992, **58**, 270-272 (*Stemmadenia minima constit*).

Huang, Y.L. *et al.*, *Planta Med.*, 1992, **58**, 473-474 (*Isolintetralin*).

Kim, H.J. *et al.*, *J. Nat. Prod.*, 1994, **57**, 581-586 (*Aviculin*).

Coltart, D.M. *et al.*, *Can. J. Chem.*, 1996, **74**, 88-94 (*Isolariciresinol dimethyl ether, synth*).

Shen, Y.-C. *et al.*, *Chin. Pharm. J. (Taipei)*, 1997, **49**, 285-296 (*isol, activity*).

Yoshikawa, K. *et al.*, *Nat. Med. (Tokyo)*, 1997, **51**, 244-248 (*Woorenoside XI*).

Zhong, X.-N. *et al.*, *Phytochemistry*, 1998, **49**, 1777-1778 (*Isolariciresinol 9′-sulfate*).

Jiang, Z.-H. *et al.*, *Chem. Pharm. Bull.*, 2001, **49**, 1036-1038 (*3′,5-di-Me 4-glucoside*).

Xiao, K. *et al.*, *Chem. Pharm. Bull.*, 2002, **50**, 605-608 (*Isolariciresinol 9′-sulfate*).

Enders, D. *et al.*, *Chirality*, 2003, **15**, 510-513 (*Lintetralin, struct, synth*).

Garzino, F. *et al.*, *Eur. J. Org. Chem.*, 2003, 1410-1414 (*Phyltetralin, synth*).

Erdemoglu, N. *et al.*, *J. Mol. Struct.*, 2003, **655**, 459-466 (*Taxus baccata constits*).

Vardamides, J.C. *et al.*, *Phytochemistry*, 2003, **62**, 647-650 (*Scaphopetalone*).

Chang, C.-C. *et al.*, *Phytochemistry*, 2003, **63**, 825-833 (*Urinatetralin, Lintetralin, Isolintetralin*).

Kanchanapoom, T. *et al.*, *Phytochemistry*, 2003, **63**, 985-988 (*Sauropus androgynus glycoside*).

He, Y.-H. *et al.*, *Heterocycles*, 2004, **63**, 871-877 (*Vitis thunbergii diglycoside*).

Tang, W.-Z. *et al.*, *Yaoxue Xuebao*, 2004, **39**, 531-533 (*Chestnutlignanoside*).

Yang, B. *et al.*, *J. Nat. Prod.*, 2005, **68**, 1175-1179 (*9′-Coumaroylisolariciresinol*).

Jutiviboonsuk, A. *et al.*, *Phytochemistry*, 2005, **66**, 2745-2751 (*Burselignan*).

Pullela, S.V. *et al.*, *Planta Med.*, 2005, **71**, 787-789 (*9′-O-Acetylisolariciresinol*).

Luo, X.-D. *et al.*, *Chem. Biodiversity*, 2006, **3**, 224-230 (*Walsura yunnanensis lignan*).

Elfahmi, *et al.*, *J. Nat. Prod.*, 2006, **69**, 55-58 (*Urinatetralin*).

Shi, S.-P. *et al.*, *Z. Naturforsch., B*, 2006, **61**, 1299-1303 (*Clemomanshurinanes A,B*).

Sueyoshi, E. *et al.*, *J. Nat. Med.*, 2007, **61**, 468-471 (*Bridelioside*).

Guo, R. *et al.*, *Chin. Chem. Lett.*, 2008, **19**, 1215-1217 (*Isolariciresinol 9′-rutinoside*).

Latté, K.P. *et al.*, *Phytochemistry*, 2008, **69**, 820-826 (*Isolariciresinol 9′-glucoside*).

Yang, X.-W. *et al.*, *Chem. Biodiversity*, 2009, **6**, 540-550 (*Tarennanosides C-G*).

Cai, W.-H. *et al.*, *J. Nat. Med.*, 2009, **63**, 408-411 (*Glochidion rubrum constit*).

Chai, X.-Y. *et al.*, *Planta Med.*, 2009, **75**, 1246-1252 (*Flacoside B*).

Zhu, X. *et al.*, *Fitoterapia*, 2010, **81**, 906-909 (*Burselignan, Ecdysanthera rosea constit*).

Kim, K.H. *et al.*, *Bioorg. Med. Chem. Lett.*, 2011, **21**, 2270-2273 (*ent-Isolariciresinol, Berberis koreana constit, activity*).

Li, Y.-L. *et al.*, *Chem. Biodiversity*, 2011, **8**, 2299-2309 (*9′-Feruloylisolariciresinol*).

Wang, M.-Y. *et al.*, *Yaoxue Xuebao*, 2011, **46**, 179-182 (*Chirita longgangensis constit*).

Chu, H.B. *et al.*, *Z. Naturforsch., B*, 2011, **66**, 641-646 (*Densispicoside A*).

Kim, K.H. *et al.*, *Food Chem. Toxicol.*, 2012, **50**, 3680-3686 (*Tiliamuroside B, activity*).

Mukherjee, T. *et al.*, *Nat. Prod. Commun.*, 2012, **7**, 767-769 (*Saracoside, activity*).

Pat. Coop. Treaty (WIPO), 2012, 2012 021 981 (*Acer saccharum constit*).

Shang, S.-Z. *et al.*, *Arch. Pharmacal Res.*, 2013, **36**, 1223-1230 (*Yunnanensin A*).

Yoshikawa, K. *et al.*, *Molecules*, 2013, **18**, 9641-9649 (*Saposide A*).

Ye, Y.-Y. *et al.*, *Zhongcaoyao*, 2013, **44**, 2642-2646 (*Bryophylluside*).

Li, Y.-X. *et al.*, *J. Asian Nat. Prod. Res.*, 2014, **16**, 549-553 (*Trigonostemon heterophyllus constit, Aviculin*).

Gao, Y.-P. *et al.*, *J. Asian Nat. Prod. Res.*, 2014, **16**, 724-729 (*(+)-Isolariciresinol 9′-salicylate*).

Wen, L. *et al.*, *J. Funct. Foods*, 2014, **8**, 26-34 (*Schizandriside, Litchi constit*).

Ma, Q. *et al.*, *J. Funct. Foods*, 2014, **9**, 156-161 (*Litchi 9′-arabinosides*).

Takahashi, S. *et al.*, *Am. J. Plant Sci.*, 2015, **6**, 676-684 (*Tabebuia chrysotricha constit, cd, activity*).

Zeng, X. *et al.*, *Fitoterapia*, 2015, **101**, 153-161 (*Sargentodosides A,B,C, cd*).

Wu, Y.-Y. *et al.*, *J. Asian Nat. Prod. Res.*, 2015, **17**, 1097-1103 (*Cyclolignanyingoside, activity*).

Zhang, D. *et al.*, *Zhongcaoyao*, 2015, **46**, 2209-2211 (*Yunnanensin A*).

Zhang, C.-F. *et al.*, *J. Asian Nat. Prod. Res.*, 2016, **18**, 913-920 (*Lespedeza cuneata constits*).

Pan, Z.-H. *et al.*, *Molecules*, 2016, **21**, 607:1-8 (*Difengpioside C*).

Zhou, J. *et al.*, *Phytochemistry*, 2016, **121**, 58-64 (*Lespedeza cuneata constits, cd, activity*).

Ren, J. *et al.*, *Fitoterapia*, 2017, **121**, 152-158 (*Alangium kurzii constits*).

Chen, F. *et al.*, *Fitoterapia*, 2018, **129**, 133-137 (*Tripterygium wilfordii constit, activity*).

Sampei, M. *et al.*, *J. Nat. Med.*, 2018, **72**, 651-654 (*Isolariciresinol, Saracoside, Schizandriside, synth*)

2,7′-Cyclo-8,8′-lignan-3′,4,4′,5′,9,9′-hexol C238

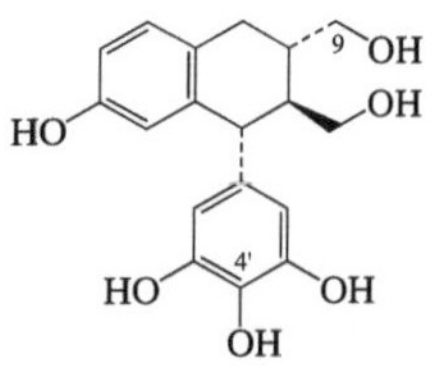

3′,4,4′,5′,9,9′-Hexahydroxy-2,7′-cyclolignan
$C_{18}H_{20}O_6$ 332.352
(7′R,8S,8′S)-form
3′,5′-Di-Me ether
2,7′-Cyclo-3′,5′-dimethoxy-8,8′-lignan-4,4′,9,9′-tetrol. 4,4′,9,9′-Tetrahydroxy-3′,5′-dimethoxy-2,7′-cyclolignan
3′,5′-Di-Me ether, 9-O-β-D-glucopyranoside
[1010453-76-5]
$C_{26}H_{34}O_{11}$ 522.547
Constit. of aerial parts of *Pelargonium reniforme*. Amorph. powder. $[\alpha]^{20}_{D}$ -56 (c, 0.06 in MeOH).

Latté, K.P. *et al.*, *Phytochemistry*, 2008, **69**, 820-826 (*Pelargonium reniforme constit*)

2,2′-Cyclo-8,8′-lignan-3,3′,4,4′,5,5′,7,7′,8-nonol C239

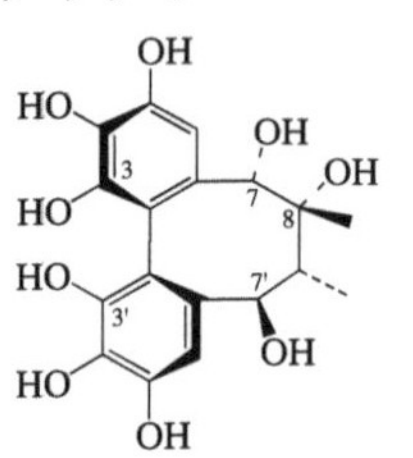

(7S,7′R,8R,8′R,S_{axial})-*form*

3,3′,4,4′,5,5′,7,7′,8-Nonahydroxy-2,2′-cyclolignan
$C_{18}H_{20}O_9$ 380.349
(7S,7′R,8R,8′R,S$_{axial}$)-form
4′,5′-Methylene, 3,3′,4,5-tetra-Me ether, 7′-E-cinnamoyl, 7-Ac
Arisanschinin C
[1333378-32-7]
$C_{34}H_{36}O_{11}$ 620.651
Constit. of *Schisandra arisanensis*. Amorph. solid. $[\alpha]^{25}_{D}$ -53 (c, 0.2 in MeOH). λ_{max} 223 (log ε 4.16); 260 (log ε 3.83); 279 (log ε 3.22) (MeOH)

(7S,7′R,8R,8′S,S$_{axial}$)-form
4′,5′-Methylene, 3′,4,5-tri-Me ether, 7,7′-diangeloyl
Kadheterin C
$C_{32}H_{38}O_{11}$ 598.645
Constit. of *Kadsura heteroclita*. Amorph. powder. $[\alpha]^{26}_{D}$ +125.4 (c, 0.13 in MeOH). λ_{max} 220 (log ε 4.69); 288 (log ε 3.57); 365 (log ε 2.07) (MeOH)
4′,5′-Methylene, 3′,4,5-tri-Me ether, 7-benzoyl, 7′-angeloyl
Kadheterin D
$C_{34}H_{36}O_{11}$ 620.651
Constit. of *Kadsura heteroclita*. Amorph. powder. $[\alpha]^{22}_{D}$ -31.3 (c, 0.1 in MeOH). λ_{max} 206 (log ε 4.76); 222 (log ε 4.89); 283 (log ε 3.79) (MeOH)
4′,5′-Methylene, 3′,4,5-tri-Me ether, 7′-benzoyl, 7-angeloyl
Kadheterin E
$C_{34}H_{36}O_{11}$ 620.651
Constit. of *Kadsura heteroclita*. Amorph. powder. $[\alpha]^{22}_{D}$ -28.7 (c, 0.1 in MeOH). λ_{max} 206 (log ε 4.79); 222 (log ε 4.93); 283 (log ε 3.84) (MeOH)

(7S,7′R,8S,8′R,S$_{axial}$)-form
4,5-Methylene, 3,4′,5′-tri-Me ether, 7-Ac
Kadsuphilin D
[942925-09-9]
$C_{24}H_{28}O_{10}$ 476.478
Constit. of *Kadsura philippinensis*. Yellowish powder. $[\alpha]^{30}_{D}$ +9.2 (c, 0.2 in CH_2Cl_2). λ_{max} 215 (log ε 4.4); 258 (log ε 3.73); 285 (log ε 3.49) (MeOH)
4,5-Methylene, 3,4′,5′-tri-Me ether, 7-angeloyl
Kadsuphilin H
[1017242-92-0]
$C_{27}H_{32}O_{10}$ 516.543
Constit. of *Kadsura philippinensis*. Amorph. powder. $[\alpha]^{25}_{D}$ -12 (c, 0.3 in CH_2Cl_2). λ_{max} 215 (log ε 4.31) (MeOH)
4,5-Methylene, 3,4′,5′-tri-Me ether, 7-benzoyl
Kadsuphilin F
[942925-11-3]
$C_{29}H_{30}O_{10}$ 538.549
Constit. of *Kadsura philippinensis*. Yellowish powder. $[\alpha]^{30}_{D}$ -21.6 (c, 0.2 in CH_2Cl_2). λ_{max} 221 (log ε 3.97); 257 (log ε 3.04); 286 (log ε 3) (MeOH)
4′,5′-Methylene, 3′,4,5-tri-Me ether, 7-angeloyl, 7′-propanoyl
Kadsuphilol S
[1275533-57-7]
$C_{30}H_{36}O_{11}$ 572.607
Isol. from *Kadsura philippinensis*. Amorph. yellow powder. $[\alpha]^{25}_{D}$ -61 (c, 0.9 in MeOH). λ_{max} 252 (log ε 3.79); 291 (log ε 3.48) (MeOH)
4′,5′-Methylene, 3′,4,5-tri-Me ether, 7,7′-diangeloyl
Kadsuphilol R
$C_{32}H_{38}O_{11}$ 598.645
Isol. from *Kadsura philippinensis*. Amorph. yellow powder. $[\alpha]^{25}_{D}$ -75 (c, 0.6 in MeOH). λ_{max} 227 (log ε 3.98); 249 (log ε 3.77); 293 (log ε 3.68) (MeOH)
4′,5′-Methylene, 3′,4,5-tri-Me ether, 7-benzoyl, 7′-angeloyl
Kadsuphilol T
[1275533-59-9]
$C_{34}H_{36}O_{11}$ 620.651
Isol. from *Kadsura philippinensis*. Amorph. yellow powder. $[\alpha]^{25}_{D}$ -26 (c, 0.3 in MeOH). λ_{max} 238 (log ε 3.83); 252 (log ε 3.71) (MeOH)

(7S,7′R,8S,8′S,S$_{axial}$)-form
3′,4,4′,5-Tetra-Me ether, 7-benzoyl, 7′-hexanoyl
Propinquanin B
[908011-54-1]
$C_{35}H_{42}O_{11}$ 638.71
Constit. of the stems of *Schisandra propinqua*. Cytotoxic against human HL-60 and Hep-G2 tumour cell lines. Needles. Mp 90 - 91°. $[\alpha]_{D}$ -129.7 (c, 0.94 in $CHCl_3$). λ_{max} 220 (log ε 4.07); 254 (sh); 284 (sh) (EtOH)
4,4′,5,5′-Tetra-Me-ether, 7-benzoyl, 7′-Ac
Renchangianin A
[770733-53-4]
$C_{31}H_{34}O_{11}$ 582.602
Constit. of the stems of *Kadsura renchangiana*. Needles (MeOH). Mp 231 - 232°. $[\alpha]^{28}_{D}$ -177.9 (c, 0.89 in MeOH). λ_{max} 222 (log ε 4.11); 274 (sh); 284 (sh) (MeOH)
4,4′,5,5′-Tetra-Me ether, 7-benzoyl, 7′-angeloyl
Renchangianin B
[770733-54-5]
$C_{34}H_{38}O_{11}$ 622.667
Constit. of the stems of *Kadsura renchangiana*. Powder. $[\alpha]^{28}_{D}$ -136.1

(c, 0.31 in MeOH). λ_{max} 226 (log ε 4.24); 273 (sh); 283 (sh) (MeOH)

3,3′,4,4′,5′-Penta-Me ether, 7-angeloyl, 7′-propanoyl

Propinquain F

[1426058-82-3]

$C_{31}H_{40}O_{11}$ 588.65

Isol. from the roots of *Schisandra propinqua* ssp. *sinensis*. Amorph. yellow powder. $[\alpha]^{20}_{D}$ -31.2 (c, 0.26 in CH_2Cl_2). λ_{max} 217 (log ε 4.66); 258 (log ε 4.13) (MeOH)

3,3′,4,4′,5-Penta-Me ether, 7-tigloyl, 7′-hexanoyl

Propinquanin A

[908011-53-0]

$C_{34}H_{46}O_{11}$ 630.731

Constit. of the stems of *Schisandra propinqua*. Yellowish powder. Mp 70 - 72°. $[\alpha]_{D}$ -65 (c, 0.2 in $CHCl_3$). λ_{max} 221 (log ε 4.22); 258 (sh); 284 (sh) (EtOH)

3,3′,4,4′,5′-Penta-Me ether, 7′-tigloyl, 7-angeloyl

Propinquain E

[1426059-09-7]

$C_{33}H_{42}O_{11}$ 614.688

Isol. from the roots of *Schisandra propinqua* ssp. *sinensis*. Amorph. yellow powder. $[\alpha]^{20}_{D}$-22.5 (c, 0.89 in $CHCl_3$). λ_{max} 218 (log ε 4.73); 246 (log ε 4.23) (MeOH)

3,3′,4,4′,5-Penta-Me ether, 7-benzoyl, 7′-angeloyl

Intermedin A‡

[1046340-30-0]

$C_{35}H_{40}O_{11}$ 636.694

Constit. of *Schisandra propinqua* var. *intermedia*

3,3′,4,4′,5-Penta-Me ether, 7-benzoyl, 7′-hexanoyl

Schizantherin I. *Schisantherin I*

[117073-94-6]

$C_{36}H_{44}O_{11}$ 652.737

Isol. from roots of *Kadsura* sp.. Mp 164 - 166°. $[\alpha]_{D}$ -139 ($CHCl_3$)

3′,4,4′,5,5′-Penta-Me ether, 7-angeloyl, 7′-Ac

Kadsuphilol J

[1141889-63-5]

$C_{30}H_{38}O_{11}$ 574.623

Constit. of *Kadsura philippinensis*. Amorph. yellow powder. $[\alpha]^{25}_{D}$ -10.2 (c, 2 in CH_2Cl_2). λ_{max} 224 (log ε 4.65); 255 (log ε 4.45); 286 (log ε 4.16) (MeOH)

3′,4,4′,5,5′-Penta-Me ether, 7-benzoyl, 7′-Ac

Kadsuphilin E

[942925-10-2]

$C_{32}H_{36}O_{11}$ 596.629

Constit. of *Kadsura philippinensis*. Yellowish powder. $[\alpha]^{30}_{D}$-59.7 (c, 0.6 in CH_2Cl_2). λ_{max} 220 (log ε 5); 255 (log ε 4.81); 282 (log ε 3.93) (MeOH)

3′,4,4′,5,5′-Penta-Me ether, 7-E-cinnamoyl, 7′-Ac

Kadangustin C

[1013633-21-0]

$C_{34}H_{38}O_{11}$ 622.667

Constit. of the stems of *Kadsura angustifolia*. Amorph. powder. $[\alpha]^{17}_{D}$-53.7 (c, 0.09 in MeOH). λ_{max} 218 (log ε 5.01); 254 (log ε 4.57); 280 (log ε 4.62); 361 (log ε 3.13) (MeOH)

3,3′,4,4′,5,5′-Hexa-Me ether, 7-benzoyl, 7′-Ac

Schizanrin H

[697228-89-0]

$C_{33}H_{38}O_{11}$ 610.656

Constit. of the dried stems of *Kadsura matsudai*. Needles. Mp 193 - 195°

4,5-Methylene, 4′,5′-di-Me ether, 7,7′-diangeloyl

Propinquanin E

[137656-52-1]

$C_{31}H_{36}O_{11}$ 584.618

Constit. of the stems of *Schisandra propinqua*. Needles. Mp 110 - 112°. $[\alpha]^{20}_{D}$+34 (c, 0.22 in $CHCl_3$). λ_{max} 219 (log ε 4.05); 255 (sh); 283 (sh) (EtOH)

4′,5′-Methylene, 3,4,5-tri-Me ether, 7-angeloyl, 7′-Ac

Schizantherin G. *Schisantherin G*

[117047-86-6]

$C_{29}H_{34}O_{11}$ 558.58

Isol. from roots of *Kadsura* sp.. Mp 229 - 231°. $[\alpha]_{D}$ +12.6 ($CHCl_3$)

4′,5′-Methylene, 3,4,5-tri-Me ether, 7-angeloyl, 7′-propanoyl

Propinquanin C

[908011-55-2]

$C_{30}H_{36}O_{11}$ 572.607

Constit. of the stems of *Schisandra propinqua*. Needles. Mp 87 - 90°. $[\alpha]_{D}$ +24 (c, 0.19 in $CHCl_3$). λ_{max} 221 (log ε 3.97); 257 (sh); 284 (sh) (EtOH)

4′,5′-Methylene, 3,4,5-tri-Me ether, 7,7′-diangeloyl

Schizanrin J

[874472-14-7]

$C_{32}H_{38}O_{11}$ 598.645

Constit. of the stems of *Kadsura japonica*. Light yellow solid. Mp 102 - 103°. $[\alpha]^{28}_{D}$-344.8 (c, 0.1 in MeOH)

4′,5′-Methylene, 3,4,5-tri-Me ether, 7′-benzoyl, 7-Ac

Heteroclitalignan A

[911835-14-8]

$C_{31}H_{32}O_{11}$ 580.586

Constit. of the stems of *Kadsura heteroclita*. Yellow cryst. (MeOH). Mp 103 - 104°. $[\alpha]^{20}_{D}$-58 (c, 0.2 in MeOH). λ_{max} 223 (log ε 4.63) (MeOH)

4′,5′-Methylene, 3,4,5-tri-Me ether, 7,7′-dibenzoyl

Schizanrin I

[874472-13-6]

$C_{36}H_{34}O_{11}$ 642.657

Constit. of the stems of *Kadsura japonica*. Yellow solid. Mp 109 - 110°. $[\alpha]^{28}_{D}$-344.8 (c, 0.1 in MeOH)

4′,5′-Methylene, 3,4,5-tri-Me ether, 7′-E-cinnamoyl, 7-Ac

Kadsuphilol L

[1141889-65-7]

$C_{33}H_{34}O_{11}$ 606.624

Constit. of *Kadsura philippinensis*. Amorph. yellow powder. $[\alpha]^{25}_{D}$-3.2 (c, 0.25 in CH_2Cl_2). λ_{max} 219 (log ε 4.67); 281 (log ε 4.16) (MeOH)

4′,5′-Methylene, 3′,4,5-tri-Me ether, 7-angeloyl, 7′-Ac

Schizanrin G

[180864-19-1]

$C_{29}H_{34}O_{11}$ 558.58

Constit. of the dried stems of *Kadsura matsudai* and of *Kadsura phillipinensis*. Needles. Mp 179 - 182°

4′,5′-Methylene, 3′,4,5-tri-Me ether, 7-angeloyl, 7′-O-(2ξ-methylbutanoyl)

Kadheterin H

$C_{32}H_{40}O_{11}$ 600.661

Constit. of *Kadsura heteroclita*. Amorph. powder. $[\alpha]^{26}_{D}$+36.5 (c, 0.14 in MeOH). λ_{max} 222 (log ε 4.76); 289 (log ε 3.72); 416 (log ε 2.32) (MeOH)

4′,5′-Methylene, 3′,4,5-tri-Me ether, 7′-angeloyl, 7-O-(2-methylpropanoyl)
Kadheterin G
$C_{31}H_{38}O_{11}$ 586.634
Constit. of *Kadsura heteroclita*. Amorph. powder. $[\alpha]_D^{22}$ +47.1 (c, 0.12 in MeOH). λ_{max} 222 (log ε 4.68); 289 (log ε 3.63); 429 (log ε 2.20) (MeOH)

4′,5′-Methylene, 3′,4,5-tri-Me ether, 7,7′-diangeloyl
Schizantherin H. *Schisantherin H*
[117047-87-7]
$C_{32}H_{38}O_{11}$ 598.645
Isol. from roots of *Kadsura* sp.. Mp 188 - 190°. $[\alpha]_D$ +90.8 ($CHCl_3$)

4′,5′-Methylene, 3′,4,5-tri-Me ether, 7-benzoyl
Kadsuphilol K
[1141889-64-6]
$C_{29}H_{30}O_{10}$ 538.549
Constit. of *Kadsura philippinensis*. Amorph. yellow powder. $[\alpha]_D^{25}$ -14 (c, 0.2 in CH_2Cl_2). λ_{max} 223 (log ε 4.37); 257 (log ε 3.3); 288 (log ε 2.54) (MeOH)

4′,5′-Methylene, 3′,4,5-tri-Me ether, 7-benzoyl, 7′-Ac
Kadsuphilin C
[180864-20-4]
$C_{31}H_{32}O_{11}$ 580.586
Constit. of *Kadsura philippinensis*. Yellowish powder. $[\alpha]_D^{30}$ -5.6 (c, 1 in CH_2Cl_2). λ_{max} 219 (log ε 4.76); 257 (log ε 3.96); 280 (log ε 3.78) (MeOH)

4′,5′-Methylene, 3′,4,5-tri-Me ether, 7,7′-dibenzoyl
Kadheterin F
$C_{36}H_{34}O_{11}$ 642.657
Constit. of *Kadsura heteroclita*. Amorph. powder. $[\alpha]_D^{22}$ -155.2 (c, 0.1 in MeOH). λ_{max} 225 (log ε 4.76); 280 (log ε 3.72) (MeOH)

4′,5′-Methylene, 3,3′,4,5-tetra-Me ether, 7,7′-di-Ac
Kadsuralignan B
[913183-88-7]
$C_{27}H_{32}O_{11}$ 532.542
Constit. of the rhizomes of *Kadsura coccinea*. Needles (hexane). Mp 140 - 142°. λ_{max} 230 (log ε 3.88); 255 (sh); 280 (sh) (MeOH)

4′,5′-Methylene, 3,3′,4,5-tetra-Me ether, 7-angeloyl, 7′-Ac
Kadsurarin
[51670-41-8]
$C_{30}H_{36}O_{11}$ 572.607
Constit. of *Kadsura japonica*, *Kadsura matsudai* and *Kadsura heteroclita*. Cryst. (EtOAc). Mp 255 - 256° (sealed tube). $[\alpha]_D^{25}$ -65 (c, 0.1 in $CHCl_3$)

4′,5′-Methylene, 3,3′,4,5-tetra-Me ether, 7-angeloyl, 7′-propanoyl
Heteroclitalignan B
[911835-15-9]
$C_{31}H_{38}O_{11}$ 586.634
Constit. of the stems of *Kadsura heteroclita*. Powder (MeOH). Mp 134 - 136°. $[\alpha]_D^{20}$ +42.5 (c, 0.08 in MeOH). λ_{max} 220 (log ε 4.82) (MeOH)

4′,5′-Methylene, 3,3′,4,5-tetra-Me ether, 7-benzoyl, 7′-Ac
Schizanrin F
[697228-88-9]
$C_{32}H_{34}O_{11}$ 594.613
Constit. of the dried stems of *Kadsura matsudai*. Amorph. solid. Mp 141 - 143°

4′,5′-Methylene, 3,3′,4,5-tetra-Me ether, 7-benzoyl, 7′-propanoyl
Schizanrin O
$C_{33}H_{36}O_{11}$ 608.64
Constit. of leaves of *Kadsura induta*. Amorph. powder. $[\alpha]_D^{25}$ -56 (c, 0.1 in $CHCl_3$)

4′,5′-Methylene, 3,3′,4,5-tetra-Me ether, 7-benzoyl, 7′-O-(2-methylbutanoyl)
Schizantherin J. *Schisantherin J*
[135432-28-9]
$C_{35}H_{40}O_{11}$ 636.694
Constit. of the seeds of *Kadsura longipedunculata*

4′,5′-Methylene, 3,3′,4,5-tetra-Me ether, 7-benzoyl, 7′-pentanoyl
Kadsulignan O‡
[1430817-27-8]
$C_{35}H_{40}O_{11}$ 636.694
Constit. of *Kadsura coccinea*

4′,5′-Methylene, 3,3′,4,5-tetra-Me ether, 7-benzoyl, 7′-hexanoyl
Tiegusanin E
[1156503-56-8]
$C_{36}H_{42}O_{11}$ 650.721
Constit. of *Schisandra propinqua* var. *sinensis*. Pale yellow powder. $[\alpha]_D^{26}$ -108.4 (c, 0.63 in $CHCl_3$). λ_{max} 212 (log ε 3.53); 219 (log ε 3.56); 226 (log ε 3.57); 231 (log ε 3.57); 241 (log ε 4.05) ($CHCl_3$)

4′,5′-Methylene, 3,3′,4,5-tetra-Me ether, 7′-benzoyl, 7-Ac
Heteroclitalignan D
[137637-52-6]
$C_{32}H_{34}O_{11}$ 594.613
Constit. of the stems of *Kadsura heteroclita*. Powder (MeOH). Mp 79 - 80°. $[\alpha]_D^{20}$ -55 (c, 0.13 in MeOH). λ_{max} 223 (log ε 4.73) (MeOH)

4′,5′-Methylene, 3,3′,4,5-tetra-Me ether, 7′-benzoyl, 7-angeloyl
$C_{35}H_{38}O_{11}$ 634.678
Constit. of *Kadsura heteroclita*. Amorph. powder. $[\alpha]_D^{21}$ -63.1 (c, 0.11 in MeOH). λ_{max} 222 (log ε 4.74); 255 (log ε 4.02); 280 (log ε 3.57) (MeOH)

4′,5′-Methylene, 3,3′,4,5-tetra-Me ether, 7,7′-dibenzoyl
Tiegusanin D
[1156503-55-7]
$C_{37}H_{36}O_{11}$ 656.684
Constit. of *Schisandra propinqua* var. *sinensis*. Amorph. powder. $[\alpha]_D^{21}$ -200 (c, 0.22 in $CHCl_3$). λ_{max} 200 (log ε 3.72); 207 (log ε 3.7); 223 (log ε 3.73); 226 (log ε 3.74); 241 (log ε 4.24) ($CHCl_3$)

4′,5′-Methylene, 3,3′,4,5-tetra-Me ether, 7-E-cinnamoyl, 7′-benzoyl
Tiegusanin F
[1156503-57-9]
$C_{39}H_{38}O_{11}$ 682.722
Constit. of *Schisandra propinqua* var. *sinensis*. Powder. $[\alpha]_D^{21}$ -180.6 (c, 0.25 in $CHCl_3$). λ_{max} 195 (log ε 3.96); 198 (log ε 3.96); 201 (log ε 3.95); 221 (log ε 3.95); 227 (log ε 3.96); 232 (log ε 3.96); 240 (log ε 4.15); 282 (log ε 4.04); 355 (log ε 1.92); 399 (log ε 1.85) ($CHCl_3$)

4′,5′-Methylene, 3,3′,4,5-tetra-Me ether, 7′-E-cinnamoyl, 7-angeloyl
Tiegusanin G
[1156503-58-0]
$C_{37}H_{40}O_{11}$ 660.716
Constit. of *Schisandra propinqua* var. *sinensis*. Shows anti-HIV-1 activity. Powder. $[\alpha]_D^{22}$ -47.5 (c, 0.2 in $CHCl_3$). λ_{max} 192 (log ε 3.9); 194 (log ε 3.9); 200 (log ε 3.88); 205 (log ε 3.88); 210 (log ε 3.88); 240 (log ε 4.13); 262 (log ε 3.97); 283 (log ε 3.97) ($CHCl_3$)

4,5:4′,5′-Bis(methylene), 3,3′-di-Me ether, 7-angeloyl
Kadsurindutin B

[1001644-52-5]
$C_{27}H_{30}O_{10}$ 514.527
Constit. of the stems of *Kadsura induta*. Powder. $[\alpha]^{25}_{D}$ +38.6 (c, 0.007 in MeOH). λ_{max} 222 (log ε 4.11); 274 (sh); 284 (sh) (MeOH)

4,5:4′,5′-Bis(methylene), 3,3′-di-Me ether, 7-angeloyl, 7′-Ac
Kadsurindutin A
[1001644-51-4]
$C_{29}H_{32}O_{11}$ 556.564
Constit. of the stems of *Kadsura induta* and of *Kadsura coccinea*. Powder. $[\alpha]^{25}_{D}$ +86 (c, 0.005 in MeOH). λ_{max} 222 (log ε 4.24); 257 (sh); 284 (sh) (MeOH)

7′-Ketone, 4′,5′-methylene, 3,3′,4,5-tetra-Me ether, 7-angeloyl
Schizanrin K
[874472-15-8]
$C_{28}H_{32}O_{10}$ 528.554
Constit. of the stems of *Kadsura japonica*. Yellow solid. Mp 63 - 64°. $[\alpha]^{28}_{D}$ -30 (c, 0.1 in MeOH)

(7ξ,7′ξ,8ξ,8′ξ,ξ$_{axial}$)-form

4′,5′-Methylene, 3′,4,5-tri-Me ether, 7-benzoyl, 7′-Ac
Heteroclitin Q
$C_{31}H_{32}O_{11}$ 580.586
Constit. of the stems of *Kadsura heteroclita*. Yellow gum. $[\alpha]_{D}$ +80.4 (c, 1.5 in $CHCl_3$). λ_{max} 221 (log ε 4.45); 257 (log ε 3.57); 280 (log ε 3.8) (MeOH)

4′,5′-Methylene, 3′,4,5-tri-Me ether, 7′-benzoyl, 7-angeloyl
Heteroclitin P
[1007566-06-4]
$C_{34}H_{36}O_{11}$ 620.651
Constit. of the stems of *Kadsura heteroclita*. Needles. Mp 195 - 197°. $[\alpha]^{20}_{D}$ -110 (c, 0.09 in $CHCl_3$).

Chen, Y.-P. *et al.*, *Bull. Chem. Soc. Jpn.*, 1977, **50**, 1824-1826 (*Kadsurarin*).
Liu, J. *et al.*, *Huaxue Xuebao*, 1988, **46**, 465-471 (*Schizantherins G,H,I*).
Liu, J. *et al.*, *Huaxue Xuebao*, 1991, **49**, 308-312 (*Schizantherin J*).
Wu, M.D. *et al.*, *Chem. Pharm. Bull.*, 2003, **51**, 1233-1236 (*Schizanrins F,G,H*).
Chen, M. *et al.*, *Helv. Chim. Acta*, 2004, **87**, 1368-1376 (*Renchangianins A,B*).
Kuo, Y-H. *et al.*, *Bioorg. Med. Chem.*, 2005, **13**, 1555-1561 (*Kadsurarin*).
Kuo, Y.-H. *et al.*, *Planta Med.*, 2005, **71**, 646-653 (*Schizanrins I,J,K*).
Xu, L.-J. *et al.*, *Chem. Pharm. Bull.*, 2006, **54**, 542-545 (*Propinquanin E*).
Li, H.-R. *et al.*, *Chem. Pharm. Bull.*, 2006, **54**, 1022-1025 (*Kadsuralignan B*).
Xu, L.-J. *et al.*, *Planta Med.*, 2006, **72**, 169-174 (*Propinquanins A,B,C*).
Wang, W. *et al.*, *Planta Med.*, 2006, **72**, 284-288 (*Heteroclitalignans A,B,D, Kadsurarin*).
Ma, W. *et al.*, *Chem. Biodiversity*, 2007, **4**, 966-972 (*Kadsurindutins A,B*).
Shen, Y.-C. *et al.*, *Chem. Pharm. Bull.*, 2007, **55**, 280-283 (*Kadsuphilins C-F, Schisanrin G*).
Li, H.-M. *et al.*, *Arch. Pharmacal Res.*, 2008, **31**, 684-687 (*Intermedin A*).
Xu, L.-J. *et al.*, *Helv. Chim. Acta*, 2008, **91**, 220-226 (*Heteroclitins P,Q*).
Shen, Y.C. *et al.*, *Helv. Chim. Acta*, 2008, **91**, 483-494 (*Kadsuphilin H*).
Gao, X.-M. *et al.*, *J. Nat. Prod.*, 2008, **71**, 558-563 (*Kadangustin C*).
Li, X.-N. *et al.*, *J. Nat. Prod.*, 2009, **72**, 1133-1141 (*Tiegusanins D-G*).
Liu, H.-T. *et al.*, *Magn. Reson. Chem.*, 2009, **47**, 609-612 (*Kadsurarin*).
Shen, Y.-C. *et al.*, *Phytochemistry*, 2009, **70**, 114-120 (*Kadsuphilols J,K,L*).
Liu, C.-C. *et al.*, *Planta Med.*, 2010, **76**, 1605-1610 (*Arisanschinin C*).
Cheng, Y.-B. *et al.*, *Helv. Chim. Acta*, 2011, **94**, 148-158 (*Kadsuphilols R,S,T*).
Jin, M.N. *et al.*, *Planta Med.*, 2012, **78**, 807-813 (*Propinquains E,F*).
Shu, Y.-Z. *et al.*, *Zhongcaoyao*, 2012, **43**, 428-431 (*Kadsulignan O*).
Minh, P.T.H. *et al.*, *Bull. Korean Chem. Soc.*, 2014, **35**, 1859-1862 (*Schizanrin O*).
Fang, L. *et al.*, *Phytochem. Lett.*, 2014, **9**, 158-162 (*Kadsurindutin A*).
Luo, Y.-G. *et al.*, *Fitoterapia*, 2017, **119**, 150-157 (*Kadheterins C-H*)

2,2′-Cyclo-8,8′-lignan-3,3′,4,4′,5,5′,7,8,9-nonol C240

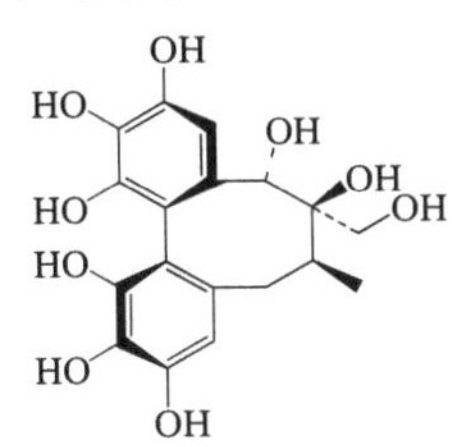

3,3′,4,4′,5,5′,7,8,9-Nonahydroxy-2,2′-cyclolignan
$C_{18}H_{20}O_9$ 380.349

(7S,8R,8′S,R$_{axial}$)-form

3,3′,4,4′,5,5′-Hexa-Me ether, 7-angeloyl
9-Hydroxyangeloylgomisin Q
[1791438-94-2]
$C_{29}H_{38}O_{10}$ 546.613
Constit. of leaves and stems of *Schisandra chinensis*. Yellow gum. $[\alpha]^{26}_{D}$ -14 (c, 0.1 in MeOH). λ_{max} 217 (log ε 4.0) (MeOH).

Shi, Y. *et al.*, *Chin. J. Chem.*, 2014, **32**, 734-740 (*9-Hydroxyangeloylgomisin Q*)

2,7′-Cyclo-8,8′-lignan-3′,4,4′,5,5′,8,8′,9,9′-nonol C241

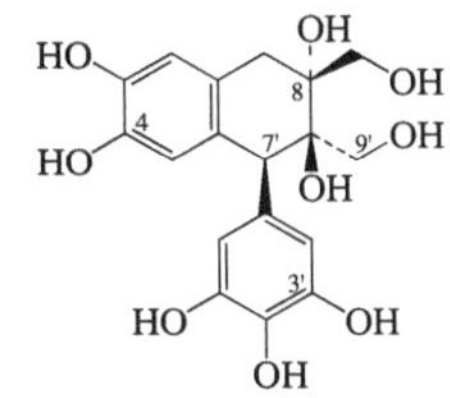

(7′S,8R,8′R)-form

3′,4,4′,5,5′,8,8′,9,9′-Nonahydroxy-2,7′-cyclolignan
$C_{18}H_{20}O_9$ 380.349

(7′S,8R,8′R)-form

9′-Carboxylic acid, 3′,5-di-Me ether
2,7′-Cyclo-3′,4,4′,8,8′,9-hexahydroxy-5,5′-dimethoxy-8,8′-lignan-9′-oic acid. ***Plicatic acid***
[16462-65-0]
[7531-92-2] (unspecified stereochem.)
$C_{20}H_{22}O_{10}$ 422.386
Isol. from *Thuja plicata*. Amorph. tan powder. $[\alpha]^{20}_{D}$ -18 (c, 0.04 in H_2O) (-9.99). Has 8′*S*-config. (change of Cahn-Ingold-Prelog priorities). λ_{max} 283 (ε 4210) (MeOH)
TP3400000

9′-Carboxylic acid, 9′,9-lactone, 3′,5-di-Me ether
2,7′-Cyclo-3′,4,4′,8,8′-pentahydroxy-5,5′-dimethoxy-8,8′-lignan-9′,9-olide. Plicatic acid lactone. ***Plicatin‡***
[16462-67-2]
$C_{20}H_{20}O_9$ 404.371
Isol. from heartwood of *Thuja plicata*. λ_{max} 283 (ε 4630) (MeOH)

(7′R*,8S*,8′S*)-form

9′-Carboxylic acid, 9′,9-lactone, 4,5-methylene, 3′,4′,5′-tri-Me ether
2,7′-Cyclo-8,8′-dihydroxy-3′,4′,5′-trimethoxy-4,5-methylenedioxy-8,8′-lignan-9′,9-olide

9′-Carboxylic acid, 9′,9-lactone, 4,5-methylene, 3′,4′,5′-tri-Me ether, 8′-Ac
Erlangerin D
[465535-64-2]
$C_{24}H_{24}O_{10}$ 472.446
Constit. of the resin of *Commiphora erlangeriana*. Cytotoxic. Cryst. (MeOH/$CHCl_3$). Mp 244 - 246°. Has 8′*R**-config. (change of Cahn-Ingold-Prelog priorities). λ_{max} 211 (log ε 2.82); 290 (log ε 2.03) (EtOH).

Gardner, J.A.F. *et al.*, *Can. J. Chem.*, 1959, **37**, 1703-1709 (*Plicatic acid, Plicatic acid lactone*).
Gardner, J.A.F. *et al.*, *Can. J. Chem.*, 1960, **38**, 2387-2394 (*Plicatic acid*).
Gardner, J.A.F. *et al.*, *Can. J. Chem.*, 1966, **44**, 52-58 (*Plicatic acid, cryst struct*).
Swan, R.J. *et al.*, *Can. J. Chem.*, 1967, **45**, 319-324 (*Plicatic acid, ord, abs config*).
Hulbert, P.B. *et al.*, *J. Chem. Res., Miniprint*, 1981, 401-449 (*Plicatic acid, Plicatin, uv*).
Dekebo, A. *et al.*, *J. Nat. Prod.*, 2002, **65**, 1252-1257 (*Erlangerin D, cryst struct*).
Habtemariam, S. *Toxicon*, 2003, **41**, 723-727 (*Erlangerin D, activity*).
Sun, B.-F. *et al.*, *J. Am. Chem. Soc.*, 2009, **131**, 10384-10385 (*(-)-Plicatic acid, synth*)

2,2′-Cyclo-8,8′-lignan-3,3′,4,4′,5,5′,7,7′-octol C242

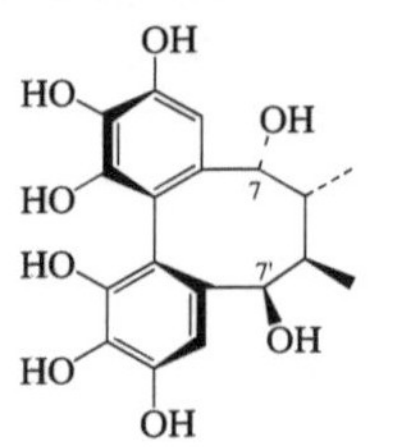

(7*R*,7′*R*,8*R*,8′*R*,*S*axial)-*form*

3,3′,4,4′,5,5′,7,7′-Octahydroxy-2,2′-cyclolignan

$C_{18}H_{20}O_8$ 364.35

Care needed with numbering and stereodescriptors owing to the unusual symmetry props. of the parent. All derivs. are numbered as shown in the diag. Dibenzocyclooctadiene lignan. Isol. from the stems of *Kadsura polysperma*

(7R,7′R,8R,8′R,S_{axial})-form

3,4,4′,5′-Tetra-Me ether, 3′-angeloyl, 7′-Ac

Propinquanin D

[908011-56-3]

$C_{29}H_{36}O_{10}$ 544.597

Constit. of the stems of *Schisandra propinqua*. Needles. Mp 123 - 125°. $[\alpha]_D$ +161 (c, 0.2 in $CHCl_3$). λ_{max} 222 (log ε 3.66); 255 (sh); 284 (sh) (EtOH)

4,5-Methylene, 3,3′,4′,5′-tetra-Me ether, 7′-angeloyl, 7-propanoyl

Tiegusanin I

[1156503-60-4]

$C_{31}H_{38}O_{10}$ 570.635

Constit. of *Schisandra propinqua* var. *sinensis*. Powder. $[\alpha]^{22}_D$ +55.9 (c, 0.44 in $CHCl_3$). λ_{max} 197 (log ε 3.56); 227 (log ε 3.6); 241 (log ε 4.04) ($CHCl_3$)

4,5-Methylene, 3,3′,4′,5′-tetra-Me ether, 7′-angeloyl, 7-benzoyl

Tiegusanin H

[1156503-59-1]

$C_{35}H_{38}O_{10}$ 618.679

Constit. of *Schisandra propinqua* var. *sinensis*. Amorph. powder. $[\alpha]^{27}_D$ -2.7 (c, 0.18 in $CHCl_3$). λ_{max} 197 (log ε 3.62); 214 (log ε 3.62); 222 (log ε 3.66); 225 (log ε 3.66); 230 (log ε 3.67); 241 (log ε 4.12) ($CHCl_3$)

4,5-Methylene, 3,3′,4′,5′-tetra-Me ether, 7,7′-ditigloyl

Tiegusanin J

[1156503-61-5]

$C_{33}H_{40}O_{10}$ 596.673

Constit. of *Schisandra propinqua* var. *sinensis*. Powder. $[\alpha]^{27}_D$ -3.9 (c, 0.26 in $CHCl_3$). λ_{max} 192 (log ε 3.61); 195 (log ε 3.6); 199 (log ε 3.59); 214 (log ε 3.58); 226 (log ε 3.62); 241 (log ε 4.03) ($CHCl_3$)

(7R,7′R,8R,8′S,S_{axial})-form

3,4,4′,5′-Tetra-Me ether, 7′-benzoyl, 7-Ac

Kadsuphilin I

[1017242-93-1]

$C_{31}H_{34}O_{10}$ 566.603

Constit. of *Kadsura philippinensis*. Amorph. powder. $[\alpha]^{25}_D$ -1 (c, 5 in CH_2Cl_2). λ_{max} 231 (log ε 4.63); 283 (log ε 3.91) (MeOH)

4,4′,5,5′-Tetra-Me ether, 7′-E-cinnamoyl, 7-angeloyl

Renchangianin C

[770733-56-7]

$C_{36}H_{40}O_{10}$ 632.706

Constit. of the stems of *Kadsura renchangiana*. Yellow powder. $[\alpha]^{28}_D$ +50.4 (c, 1.1 in MeOH). λ_{max} 222 (log ε 3.9); 254 (log ε 3.48); 275 (log ε 3.52) (MeOH)

3,3′,4,4′,5-Penta-Me ether, 7,7′-di-Ac

Ananonin K

[1314021-86-7]

$C_{27}H_{34}O_{10}$ 518.559

Isol. from the seeds of *Kadsura ananosma*. Solid. $[\alpha]^{27}_D$ +60.6 (c, 0.19 in $CHCl_3$). λ_{max} 215 (log ε 3.6); 221 (log ε 3.64); 241 (log ε 3.93) ($CHCl_3$)

3,3′,4,4′,5-Penta-Me ether, 7′-angeloyl, 7-Ac

Ananonin L

[1314021-88-9]

$C_{30}H_{38}O_{10}$ 558.624

Isol. from the seeds of *Kadsura ananosma*. Solid. $[\alpha]^{27}_D$ +58.3 (c, 0.18 in $CHCl_3$). λ_{max} 224 (log ε 3.62); 228 (log ε 3.62); 241 (log ε 3.92) ($CHCl_3$)

3,3′,4,4′,5-Penta-Me ether, 7′-benzoyl

Ananonin M

[1314021-89-0]

$C_{30}H_{34}O_9$ 538.593

Isol. from the seeds of *Kadsura ananosma*. Moderately active neuroprotective agent. Solid. $[\alpha]^{26}_D$ -50.3 (c, 0.22 in $CHCl_3$). λ_{max} 204 (log ε 3.74); 224 (log ε 3.69); 230 (log ε 3.7); 241 (log ε 4.13) ($CHCl_3$)

3,3′,4,4′,5-Penta-Me ether, 7′-benzoyl, 7-Ac

Ananonin N

[1314021-91-4]

$C_{32}H_{36}O_{10}$ 580.63

Isol. from the seeds of *Kadsura ananosma*. Solid. $[\alpha]^{28}_D$ -52.1 (c, 0.18 in $CHCl_3$). λ_{max} 198 (log ε 3.72); 219 (log ε 3.62); 241 (log ε 4.08) ($CHCl_3$)

3,4,4′,5,5′-Penta-Me ether, 7-Ac

Kadsuphilin K

[1013633-19-6]

$C_{25}H_{32}O_9$ 476.522

Constit. of *Kadsura philippinensis*. Amorph. powder. $[\alpha]^{25}_D$ -19 (c, 0.1 in CH_2Cl_2). λ_{max} 228 (log ε 4.52); 256 (log ε 3.85); 287 (log ε 3.42) (MeOH)

3,4,4′,5,5′-Penta-Me ether, 7′-Ac

Kadsuphilol I

[1141889-62-4]

$C_{25}H_{32}O_9$ 476.522

Constit. of *Kadsura philippinensis*. Amorph. yellow powder. $[\alpha]^{25}_D$ -22.6 (c, 0.6 in CH_2Cl_2). λ_{max} 222 (log ε 4.62); 254 (log ε 4.21); 287 (log ε 3.92) (MeOH)

3,4,4′,5,5′-Penta-Me ether, 3′-angeloyl, 7′-Ac

Schizantherin K. *Schisantherin K*

[143625-35-8]

$C_{30}H_{38}O_{10}$ 558.624

Constit. of *Schisandra viridis*

3,4,4′,5,5′-Penta-Me ether, 7′-angeloyl, 7-Ac

Kadangustin F

[1013633-24-3]

$C_{30}H_{38}O_{10}$ 558.624

Constit. of the stems of *Kadsura angustifolia*. Pale yellow powder. $[\alpha]^{16}_D$ +92 (c, 0.09 in MeOH). λ_{max}

214 (log ε 4.99); 344 (log ε 2.92); 380 (log ε 2.9) (MeOH)

3,4,4′,5,5′-Penta-Me ether, 3′,7′-diangeloyl

Polysperlignan I

[1360440-76-1]

$C_{33}H_{42}O_{10}$ 598.689

Isol. from the stems of *Kadsura polysperma*. Solid. Mp 65 - 66°. $[\alpha]_D^{17}$ +137.9 (c, 0.29 in MeOH). λ_{max} 193 (log ε 4.05); 216 (log ε 4.4) (MeOH)

3,4,4′,5,5′-Penta-Me ether, 7′-tigloyl, 7-Ac

Kadangustin G

[1013633-25-4]

$C_{30}H_{38}O_{10}$ 558.624

Constit. of the stems of *Kadsura angustifolia*. Pale yellow powder. $[\alpha]_D^{25}$ +16 (c, 0.21 in MeOH). λ_{max} 214 (log ε 4.96); 344 (log ε 2.93); 380 (log ε 2.89) (MeOH)

3,4,4′,5,5′-Penta-Me ether, 7′-tigloyl, 3′-angeloyl

Polysperlignan J

[1360440-81-8]

$C_{33}H_{42}O_{10}$ 598.689

Isol. from the stems of *Kadsura polysperma*. Yellow gum. $[\alpha]_D^{23}$ +39.6 (c, 0.10 in $CHCl_3$). λ_{max} 241 (log ε 3.59) ($CHCl_3$)

3,4,4′,5,5′-Penta-Me ether, 7′-benzoyl, 7-Ac

Kadangustin E. Kadsuphilol Q

[1013633-23-2]

$C_{32}H_{36}O_{10}$ 580.63

Constit. of the stems of *Kadsura angustifolia*; also from *Kadsura philippinensis*. Pale yellow powder. $[\alpha]_D^{25}$ -31 (c, 0.4 in MeOH). λ_{max} 217 (log ε 4.96); 281 (log ε 3.98); 376 (log ε 2.76); 397 (log ε 2.6) (MeOH). [neutral] λ_{max} 229 (log ε 3.88); 258 (log ε 3.51); 282 (log ε 3.27) (MeOH)

3,4,4′,5,5′-Penta-Me ether, 7′-E-cinnamoyl, 7-Ac

Kadangustin D

[1013633-22-1]

$C_{34}H_{38}O_{10}$ 606.668

Constit. of the stems of *Kadsura angustifolia*. Amorph. powder. $[\alpha]_D^{18}$ -110.1 (c, 0.05 in MeOH). λ_{max} 217 (log ε 5.04); 254 (log ε 4.57); 280 (log ε 4.66); 347 (log ε 3.12) (MeOH)

3,3′,4,4′,5,5′-Hexa-Me ether, 7,7′-di-Ac

Angustifolin D

[225109-09-1]

$C_{28}H_{36}O_{10}$ 532.586

Constit. of *Kadsura angustifolia*. Amorph. powder. $[\alpha]_D^{19}$ 0 (c, 0.018 in MeOH). Only the rel. config. appears to have been determined but it is probably S_{axial} as determined for the other natural lignans. λ_{max} 219 (log ε 4.64); 255 (sh); 284 (sh) (MeOH)

4,5-Methylene, 3,4′,5′-tri-Me ether, 7-Ac

Kadsuphilin G

[1017242-91-9]

$C_{24}H_{28}O_9$ 460.479

Constit. of *Kadsura philippinensis*. Amorph. yellow powder. $[\alpha]_D^{25}$ -10 (c, 0.8 in CH_2Cl_2). λ_{max} 222 (log ε 4.61); 257 (log ε 3.96); 288 (log ε 3.63) (MeOH)

4,5-Methylene, 3,4′,5′-tri-Me ether, 3′,7′-diangeloyl

Polysperlignan G

[1360440-66-9]

$C_{32}H_{38}O_{10}$ 582.646

Isol. from the stems of *Kadsura polysperma*. Cryst.. Mp 129 - 130°. $[\alpha]_D^{23}$ +132.6 (c, 0.10 in $CHCl_3$). λ_{max} 241 (log ε 3.69) ($CHCl_3$)

4,5-Methylene, 3,4′,5′-tri-Me ether, 7′-tigloyl, 3′-angeloyl

Polysperlignan H

[1360440-71-6]

$C_{32}H_{38}O_{10}$ 582.646

Isol. from the stems of *Kadsura polysperma*. Solid. Mp 129 - 130°. $[\alpha]_D^{24}$ +49.0 (c, 0.09 in MeOH). λ_{max} 198 (log ε 3.41); 212 (log ε 3.97); 290 (log ε 2.62) (MeOH)

4,5-Methylene, 3′,4′,5′-tri-Me ether, 7′-angeloyl, 7-Ac

Ananonin H

[1314021-81-2]

$C_{29}H_{34}O_{10}$ 542.581

Isol. from the seeds of *Kadsura ananosma*. Solid. $[\alpha]_D^{27}$ +49.3 (c, 0.24 in $CHCl_3$). λ_{max} 194 (log ε 3.78); 220 (log ε 3.78); 241 (log ε 4.1) ($CHCl_3$)

4,5-Methylene, 3′,4′,5′-tri-Me ether, 7′-angeloyl, 7-propanoyl

Ananonin I

[1314021-82-3]

$C_{30}H_{36}O_{10}$ 556.608

Isol. from the seeds of *Kadsura ananosma*. Solid. $[\alpha]_D^{28}$ +35.7 (c, 0.18 in $CHCl_3$). λ_{max} 198 (log ε 3.7); 200 (log ε 3.69); 241 (log ε 4.05) ($CHCl_3$)

4,5-Methylene, 3′,4′,5′-tri-Me ether, 7′-angeloyl, 7-O-(2-methylbutanoyl)

Ananonin J

[1314021-84-5]

$C_{32}H_{40}O_{10}$ 584.662

Isol. from the seeds of *Kadsura ananosma*. Solid. $[\alpha]_D^{27}$ +69.9 (c, 0.19 in $CHCl_3$). λ_{max} 198 (log ε 3.7); 223 (log ε 3.71); 241 (log ε 4.06) ($CHCl_3$)

4,5-Methylene, 3,4′,5′-tri-Me ether, 7-benzoyl, 7′-Ac

Kadsuphilol B

[947256-77-1]

$C_{31}H_{32}O_{10}$ 564.587

Constit. of *Kadsura philippinensis*. Amorph. powder. $[\alpha]_D^{25}$ +11 (c, 0.01 in CH_2Cl_2). λ_{max} 220 (log ε 3.6) (MeOH)

4,5-Methylene, 3,4′,5′-tri-Me ether, 7′-benzoyl, 7-Ac

Kadsuphilol P

[1275533-55-5]

$C_{31}H_{32}O_{10}$ 564.587

Isol. from *Kadsura philippinensis*. Amorph. yellowish powder. $[\alpha]_D^{25}$ -21 (c, 0.1 in MeOH). λ_{max} 225 (log ε 3.64); 259 (log ε 3.4); 284 (log ε 3.18) (MeOH)

4,5-Methylene, 3,4′,5′-tri-Me ether, 7-E-cinnamoyl, 7′-angeloyl

Tiegusanin K

[1156503-62-6]

$C_{36}H_{38}O_{10}$ 630.69

Constit. of *Schisandra propinqua* var. *sinensis*. Powder. $[\alpha]_D^{22}$ +14.6 (c, 0.1 in $CHCl_3$). λ_{max} 197 (log ε 3.92); 219 (log ε 3.92); 230 (log ε 3.94); 240 (log ε 4.08); 261 (log ε 4); 281 (log ε 4.05) ($CHCl_3$)

4,5-Methylene, 3,3′,4′,5′-tetra-Me ether

Kadangustin L

[1415922-02-9]

$C_{23}H_{28}O_8$ 432.469

Constit. of the stems of *Kadsura angustifolia*. Pale yellow powder. $[\alpha]_D^{17}$ -15.6 (c, 0.55 in MeOH). λ_{max} 217 (log ε 4.32) (MeOH)

4,5-Methylene, 3,3′,4′,5′-tetra-Me ether, 7′-Ac

Ananolignan E

[1280213-63-9]
$C_{25}H_{30}O_9$ 474.506
Constit. of the seeds of *Kadsura ananosma*. Solid. $[\alpha]^{26}_{D}$ +58.5 (c, 0.22 in $CHCl_3$). λ_{max} 192 (log ε 3.58); 204 (log ε 3.58); 226 (log ε 3.51); 242 (log ε 3.95) ($CHCl_3$)

4,5-Methylene, 3,3′,4′,5′-tetra-Me ether, 7,7′-di-Ac
Ananolignan F
[1280213-64-0]
$C_{27}H_{32}O_{10}$ 516.543
Constit. of the seeds of *Kadsura ananosma*. Shows significant neuroprotective effects in a SH-SY5Y neuroblastoma cell line *in vitro* assay. Solid. $[\alpha]^{29}_{D}$ +74.3 (c, 0.21 in $CHCl_3$). λ_{max} 199 (log ε 3.66); 226 (log ε 3.55); 241 (log ε 4.03) ($CHCl_3$)

4,5-Methylene, 3,3′,4′,5′-tetra-Me ether, 7-propanoyl, 7′-Ac
Ananolignan G
[1280213-65-1]
$C_{28}H_{34}O_{10}$ 530.57
Constit. of the seeds of *Kadsura ananosma*. Solid. $[\alpha]^{27}_{D}$ +76.1 (c, 0.17 in $CHCl_3$). λ_{max} 199 (log ε 3.65); 204 (log ε 3.65); 231 (log ε 3.68); 241 (log ε 4) ($CHCl_3$)

4,5-Methylene, 3,3′,4′,5′-tetra-Me ether, 7-O-(2-methylpropanoyl), 7′-Ac
Ananolignan H
[1280213-66-2]
$C_{29}H_{36}O_{10}$ 544.597
Constit. of the seeds of *Kadsura ananosma*. Solid. $[\alpha]^{27}_{D}$ +90.5 (c, 0.18 in $CHCl_3$). λ_{max} 193 (log ε 3.68); 199 (log ε 3.67); 241 (log ε 4.05) ($CHCl_3$)

4,5-Methylene, 3,3′,4′,5′-tetra-Me ether, 7-butanoyl, 7′-Ac
Ananolignan I
[1280213-67-3]
$C_{29}H_{36}O_{10}$ 544.597
Constit. of the seeds of *Kadsura ananosma*. Solid. $[\alpha]^{28}_{D}$ +57.3 (c, 0.16 in $CHCl_3$). λ_{max} 192 (log ε 3.68); 194 (log ε 3.67); 241 (log ε 4.01) ($CHCl_3$)

4,5-Methylene, 3,3′,4′,5′-tetra-Me ether, 7-O-(3-methylbutanoyl), 7′-Ac
Ananolignan J
[1280213-68-4]
$C_{30}H_{38}O_{10}$ 558.624
Constit. of the seeds of *Kadsura ananosma*. Solid. $[\alpha]^{27}_{D}$ +103.3 (c, 0.19 in $CHCl_3$). λ_{max} 196 (log ε 3.61); 209 (log ε 3.6); 232 (log ε 3.64); 241 (log ε 3.97) ($CHCl_3$)

4,5-Methylene, 3,3′,4′,5′-tetra-Me ether, 7′-angeloyl, 7-Ac
***Interiotherin C**. Rubriflorin A‡*
[460090-65-7]
$C_{30}H_{36}O_{10}$ 556.608
Constit. of the stems of *Kadsura interior*, *Schisandra rubriflora* and *Schisandra neglecta*. Needles (MeOH or $CHCl_3$). Mp 179 - 181° (176-178°). $[\alpha]_{D}$ +127.7 (c, 1.17 in $CHCl_3$). $[\alpha]^{26}_{D}$ +63 (c, 0.25 in $CHCl_3$). λ_{max} 217 (log ε 5.9); 260 (sh); 285 (sh) (MeOH). [neutral] λ_{max} 218 (log ε 4.6); 253 (log ε 4.15); 284 (log ε 3.78) ($CHCl_3$)

4,5-Methylene, 3,3′,4′,5′-tetra-Me ether, 7′-angeloyl, 7-O-(2-methylpropanoyl)
Ananolignan M
[1280213-71-9]
$C_{32}H_{40}O_{10}$ 584.662
Constit. of the seeds of *Kadsura ananosma*. Solid. $[\alpha]^{27}_{D}$ +77.8 (c, 0.16 in $CHCl_3$). λ_{max} 194 (log ε 3.7); 202 (log ε 3.68); 222 (log ε 3.59); 241 (log ε 4.1) ($CHCl_3$)

4,5-Methylene, 3,3′,4′,5′-tetra-Me ether, 7′-angeloyl, 7-butanoyl
Ananolignan N
[1280213-72-0]
$C_{32}H_{40}O_{10}$ 584.662
Constit. of the seeds of *Kadsura ananosma*. Solid. $[\alpha]^{27}_{D}$ +64.8 (c, 0.21 in $CHCl_3$). λ_{max} 198 (log ε 3.71); 210 (log ε 3.71); 241 (log ε 4.09) ($CHCl_3$)

4,5-Methylene, 3,3′,4′,5′-tetra-Me ether, 7′-angeloyl, 7-O-(2-methylbutanoyl)
Ananonin G
[1314021-79-8]
$C_{33}H_{42}O_{10}$ 598.689
Isol. from the seeds of *Kadsura ananosma*. Solid. $[\alpha]^{29}_{D}$ +75.1 (c, 0.19 in $CHCl_3$). λ_{max} 200 (log ε 3.62); 223 (log ε 3.46); 231 (log ε 3.49); 241 (log ε 4) ($CHCl_3$)

4,5-Methylene, 3,3′,4′,5′-tetra-Me ether, 7,7′-diangeloyl
Polysperlignan A
[1360440-45-4]
$C_{33}H_{40}O_{10}$ 596.673
Isol. from the stems of *Kadsura polysperma*. Moderate neuroprotective agent. Cryst.. Mp 156 - 157°. $[\alpha]^{17}_{D}$ +270.1 (c, 0.07 in MeOH). λ_{max} 191 (log ε 4.08); 219 (log ε 4.44) (MeOH)

4,5-Methylene, 3,3′,4′,5′-tetra-Me ether, 7-tigloyl, 7′-Ac
Polysperlignan B
[1360440-48-7]
$C_{30}H_{36}O_{10}$ 556.608
Isol. from the stems of *Kadsura polysperma*. Solid. Mp 86 - 88°. $[\alpha]^{27}_{D}$ +9.4 (c, 0.17 in $CHCl_3$). λ_{max} 212 (log ε 3.71); 215 (log ε 3.73); 223 (log ε 3.74); 241 (log ε 4.07) ($CHCl_3$)

4,5-Methylene, 3,3′,4′,5′-tetra-Me ether, 7-tigloyl, 7′-angeloyl
Polysperlignan D
[1360440-54-5]
$C_{33}H_{40}O_{10}$ 596.673
Isol. from the stems of *Kadsura polysperma*. Moderate neuroprotectant. Yellow oil. $[\alpha]^{27}_{D}$ +1.5 (c, 0.16 in $CHCl_3$). λ_{max} 194 (log ε 3.63); 201 (log ε 3.61); 203 (log ε 3.60); 218 (log ε 3.45); 221 (log ε 3.45); 227 (log ε 3.47); 231 (log ε 3.48); 241 (log ε 4.00) ($CHCl_3$)

4,5-Methylene, 3,3′,4′,5′-tetra-Me ether, 7′-tigloyl
Ananosin A‡
[402942-26-1]
$C_{28}H_{34}O_9$ 514.571
Constit. of the stem bark of *Kadsura ananosma*. Cryst.. Mp 132 - 134°. $[\alpha]_{D}$ -16.5 (c, 0.6 in $CHCl_3$). λ_{max} 225 (log ε 4.78); 288 (log ε 3.43) (MeOH)

4,5-Methylene, 3,3′,4′,5′-tetra-Me ether, 7′-tigloyl, 7-Ac
Ananolignan L
[1280213-70-8]
$C_{30}H_{36}O_{10}$ 556.608
Constit. of the seeds of *Kadsura ananosma*. Shows significant neuroprotective effects in a SH-SY5Y neuroblastoma cell line *in vitro* assay. Solid. $[\alpha]^{29}_{D}$ -22.6 (c, 0.19 in $CHCl_3$). λ_{max} 195 (log ε 3.63); 202 (log ε 3.61); 241 (log ε 4.01) ($CHCl_3$)

4,5-Methylene, 3,3′,4′,5′-tetra-Me ether, 7-benzoyl, 7′-Ac
Ananolignan K
[1280213-69-5]
$C_{32}H_{34}O_{10}$ 578.614

Constit. of the seeds of *Kadsura ananosma*. Solid. $[\alpha]^{27}_{D}$ +1.6 (c, 0.40 in $CHCl_3$). λ_{max} 196 (log ε 3.59); 218 (log ε 3.58); 224 (log ε 3.59); 241 (log ε 3.95) ($CHCl_3$)

4,5-Methylene, 3,3′,4′,5′-tetra-Me ether, 7′-benzoyl

Ananonin A

[1314021-67-4]

$C_{30}H_{32}O_9$ 536.577

Isol. from the seeds of *Kadsura ananosma*. Solid. $[\alpha]^{27}_{D}$ -40.3 (c, 0.24 in $CHCl_3$). λ_{max} 214 (log ε 3.75); 221 (log ε 3.77); 241 (log ε 4.14) ($CHCl_3$)

4,5-Methylene, 3,3′,4′,5′-tetra-Me ether, 7′-benzoyl, 7-Ac

Arisanschinin D. *Ananonin B*

[1314021-69-6]

$C_{32}H_{34}O_{10}$ 578.614

Constit. of *Schisandra arisanensis* and the seeds of *Kadsura ananosma*. Amorph. solid. $[\alpha]^{25}_{D}$ -22.5 (c, 0.2 in MeOH). $[\alpha]^{29}_{D}$ -48.8 (c, 0.18 in $CHCl_3$). λ_{max} 229 (log ε 4.3); 255 (log ε 3.65); 279 (log ε 3.43) (MeOH). [neutral] λ_{max} 199 (log ε 3.71); 203 (log ε 3.7); 225 (log ε 3.58); 241 (log ε 4.15) ($CHCl_3$)

4,5-Methylene, 3,3′,4′,5′-tetra-Me-ether, 7′-benzoyl, 7-propanoyl

Ananonin C

[1314021-70-9]

$C_{33}H_{36}O_{10}$ 592.641

Isol. from the seeds of *Kadsura ananosma*. Solid. $[\alpha]^{28}_{D}$ -41 (c, 0.16 in $CHCl_3$). λ_{max} 199 (log ε 3.72); 219 (log ε 3.58); 241 (log ε 4.1) ($CHCl_3$)

4,5-Methylene, 3,3′,4′,5′-tetra-Me ether, 7′-benzoyl, 7-O-(2-methylpropanoyl)

Ananonin E

[1314021-74-3]

$C_{34}H_{38}O_{10}$ 606.668

Isol. from the seeds of *Kadsura ananosma*. Solid. $[\alpha]^{27}_{D}$ -21.7 (c, 0.36 in $CHCl_3$). λ_{max} 202 (log ε 3.55); 215 (log ε 3.55); 226 (log ε 3.58); 241 (log ε 3.95) ($CHCl_3$)

4,5-Methylene, 3,3′,4′,5′-tetra-Me ether, 7′-benzoyl, 7-butanoyl

Ananonin D

[1314021-72-1]

$C_{34}H_{38}O_{10}$ 606.668

Isol. from the seeds of *Kadsura ananosma*. Solid. $[\alpha]^{26}_{D}$ -18.1 (c, 0.26 in $CHCl_3$). λ_{max} 195 (log ε 3.56); 222 (log ε 3.57); 241 (log ε 3.95) ($CHCl_3$)

4,5-Methylene, 3,3′,4′,5′-tetra-Me ether, 7′-benzoyl, 7-O-(2-methylbutanoyl)

Ananonin F

[1314021-76-5]

$C_{35}H_{40}O_{10}$ 620.695

Isol. from the seeds of *Kadsura ananosma*. Solid. $[\alpha]^{27}_{D}$ -5.5 (c, 0.18 in $CHCl_3$). λ_{max} 197 (log ε 3.72); 220 (log ε 3.75); 241 (log ε 4.1) ($CHCl_3$)

4,5-Methylene, 3,3′,4′,5′-tetra-Me ether, 7′-benzoyl, 7-tigloyl

Polysperlignan C

[1360440-50-1]

$C_{35}H_{38}O_{10}$ 618.679

Isol. from the stems of *Kadsura polysperma*. Solid. Mp 69 - 70°. $[\alpha]^{17}_{D}$ -133.3 (c, 0.07 in MeOH). λ_{max} 218 (log ε 4.27) (MeOH)

4,5-Methylene, 3,3′,4′,5′-tetra-Me ether, 7′-E-cinnamoyl, 7-angeloyl

Polysperlignan E

[1360440-57-8]

$C_{37}H_{40}O_{10}$ 644.717

Isol. from the stems of *Kadsura polysperma*. Moderate neuroprotectant. Solid. Mp 132 - 134°. $[\alpha]^{17}_{D}$ -1.7 (c, 0.08 in MeOH). λ_{max} 218 (log ε 4.47); 277 (log ε 4.01) (MeOH)

4′,5′-Methylene, 3,4,5-tri-Me ether, 7-angeloyl, 7′-tigloyl

3-O-Demethylpolysperlignan D

[1667747-83-2]

$C_{32}H_{38}O_{10}$ 582.646

Constit. of the roots of *Kadsura coccinea*. Yellow oil. $[\alpha]^{20}_{D}$ +40.5 (c, 0.37 in MeOH)

4,5:4′,5′-Bis(methylene), 3,3′-di-Me ether, 7-Ac

Kadangustin B

[1013633-20-9]

$C_{24}H_{26}O_9$ 458.463

Constit. of the stems of *Kadsura angustifolia*. Pale yellow solid. $[\alpha]^{23}_{D}$ +56.1 (c, 0.27 in MeOH). λ_{max} 217 (log ε 5); 369 (log ε 3.3) (MeOH)

4,5:4′,5′-Bis(methylene), 3,3′-di-Me ether, 7′-benzoyl

Angustifolin C

[211632-08-5]

$C_{29}H_{28}O_9$ 520.534

Constit. of *Kadsura angustifolia*. Gum. $[\alpha]^{25}_{D}$ -14.5 (c, 0.28 in MeOH). λ_{max} 223 (log ε 4.78); 256 (sh); 283 (sh) (MeOH)

4,5:4′,5′-Bis(methylene), 3,3′-di-Me ether, 7′-benzoyl, 7-Ac

***Angustifolin B*‡**

[211632-06-3]

$C_{31}H_{30}O_{10}$ 562.571

Constit. of *Kadsura angustifolia*. Gum. $[\alpha]^{25}_{D}$ -3 (c, 0.36 in MeOH). λ_{max} 224 (log ε 4.76); 257 (sh); 282 (sh) (MeOH)

4,5:4′,5′-Bis(methylene), 3,3′-di-Me ether, 7,7′-dibenzoyl

Angustifolin A

[211632-04-1]

$C_{36}H_{32}O_{10}$ 624.642

Constit. of *Kadsura angustifolia*. Gum. $[\alpha]^{25}_{D}$ -43.9 (c, 0.37 in MeOH). λ_{max} 228 (log ε 4.59); 256 (sh); 284 (sh) (MeOH)

7-Ketone

2,2′-Cyclo-3,3′,4,4′,5,5′,7′-heptahydroxy-8,8′-lignan-7-one. 3,3′,4,4′,5,5′,7′-Heptahydroxy-7-oxo-2,2′-cyclolignan

7-Ketone, 4,5:4′,5′-bis(methylene), 3,3′-di-Me ether

Schizantherin Q. *Schisantherin Q. Kadsurindutin H*

[163565-72-8]

$C_{22}H_{22}O_8$ 414.41

Constit. of *Kadsura coccinea*, *Kadsura induta* and *Kadsura angustifolia*. Cryst. (MeOH). Mp 173 - 175°. $[\alpha]^{25}_{D}$ +12.1 (c, 0.1 in MeOH). λ_{max} 215 (log ε 4.52); 240 (log ε 4.44); 292 (log ε 3.98); 330 (sh) (MeOH)

(7R,7′S,8R,8′S,S$_{axial}$)-form

4,5-Methylene, 3,3′,4′,5′-tetra-Me ether

Ananolignan C

[1280213-61-7]

$C_{23}H_{28}O_8$ 432.469

Constit. of the seeds of *Kadsura ananosma*. Solid. $[\alpha]^{27}_{D}$ -35.5 (c, 0.17 in $CHCl_3$). λ_{max} 198 (log ε 3.66); 205 (log ε 3.66); 210 (log ε 3.67); 222 (log ε 3.68); 241 (log ε 3.99) ($CHCl_3$)

4,5-Methylene, 3,3′,4′,5′-tetra-Me ether, 7-Ac

Ananolignan D

[1280213-62-8]

$C_{25}H_{30}O_9$ 474.506

Constit. of the seeds of *Kadsura ananosma*. Solid. $[\alpha]^{27}_{D}$ -26.6 (c, 0.2 in $CHCl_3$). λ_{max} 214 (log ε 3.71); 224 (log ε 3.73); 241 (log ε 4.08) ($CHCl_3$)

4,5:4′,5′-Bis(methylene), 3,3′-di-Me ether

Schizantherin P. Schisantherin P

[164203-85-4]

$C_{22}H_{24}O_8$ 416.426

Constit. of the seeds of *Kadsura coccinea* and stems of *Kadsura angustifolia*. Cryst. (MeOH). Mp 107 - 108°. $[\alpha]_D$ +8 (c, 1.3 in MeOH). λ_{max} 214 (log ε 4.56); 252 (sh); 282 (log ε 3.53) (MeOH)

4,5:4′,5′-Bis(methylene), 3,3′-di-Me ether, 7′-angeloyl

Schizantherin L. Schisantherin L

[149990-51-2]

[150133-29-2]

$C_{27}H_{30}O_9$ 498.528

Constit. of the seeds of *Kadsura coccinea* and stems of *Kadsura angustifolia*. Needles (Et_2O). Mp 147 - 148°. $[\alpha]_D$ +97.1 (c, 0.5 in MeOH)

4,5:4′,5′-Bis(methylene), 3,3′-di-Me ether, 7′-angeloyl, 7-Ac

Acetylschizantherin L. Acetylschisantherin L

[149998-51-6]

$C_{29}H_{32}O_{10}$ 540.565

Constit. of the seeds of *Kadsura coccinea*. Pale yellow gum. $[\alpha]_D$ +79.9 (c, 0.55 in MeOH)

4,5:4′,5′-Bis(methylene), 3,3′-di-Me ether, 7,7′-diangeloyl

Schizantherin N. Schisantherin N

[150132-86-8]

$C_{32}H_{36}O_{10}$ 580.63

Constit. of the seeds of *Kadsura coccinea*

4,5:4′.5′-Bis(methylene), 3,3′-di-Me ether, 7′-angeloyl, 7-tigloyl

Schizantherin M. Schisantherin M

[149990-52-3]

$C_{32}H_{36}O_{10}$ 580.63

Constit. of the seeds of *Kadsura coccinea*. Cryst. (MeOH). Mp 154 - 155°. $[\alpha]_D$ +51.2 (c, 1.6 in MeOH)

(7R*,7′R*,8R*,8′S*,S*$_{axial}$)-form

3,4,4′,5,5′-Penta-Me ether, 7-Ac

Kadangustin A

$C_{25}H_{32}O_9$ 476.522

Constit. of the stems of *Kadsura angustifolia*. Amorph. powder. $[\alpha]^{23}_{D}$ +56 (c, 0.95 in MeOH). λ_{max} 210 (log ε 5.01); 284 (log ε 3.9); 309 (log ε 2.62); 360 (log ε 2.75) (MeOH)

(7′S,8S,8′S,R$_{axial}$)-form

7-Ketone, 4′,5′-methylene, 3,3′,4,5-tetra-Me ether, 7′-Ac

Ananolignan B

[1280213-60-6]

$C_{25}H_{28}O_9$ 472.49

Constit. of the seeds of *Kadsura ananosma*. Solid. $[\alpha]^{27}_{D}$ +47.8 (c, 0.19 in $CHCl_3$). λ_{max} 205 (log ε 3.8); 220 (log ε 3.82); 227 (log ε 3.83); 241 (log ε 4.06) ($CHCl_3$).

Luo, G. *et al.*, *Huaxue Xuebao*, 1992, **50**, 515-520 (*Schisantherin K*).

Liu, J.-S. *et al.*, *Phytochemistry*, 1993, **32**, 1293-1296 (*Schisantherins L,M,N, Acetylschizantherin L*).

Liu, J-S. *et al.*, *Phytochemistry*, 1995, **38**, 1009-1011 (*Schisantherins P,Q, abs config*).

Chen, Y.G. *et al.*, *J. Asian Nat. Prod. Res.*, 1998, **1**, 125-131 (*Angustifolin D*).

Chen, Y.G. *et al.*, *Phytochemistry*, 1998, **48**, 1059-1062 (*Angustifolins A,B,C*).

Chen, Y-G. *et al.*, *Phytochemistry*, 2001, **58**, 1277-1280 (*Ananosin A*).

Chen, D.-F. *et al.*, *J. Nat. Prod.*, 2002, **65**, 1242-1245 (*Interiotherin C*).

Chen, M. *et al.*, *Helv. Chim. Acta*, 2004, **87**, 1368-1376 (*Renchangianin C*).

Li, L. *et al.*, *Helv. Chim. Acta*, 2004, **87**, 2943-2947 (*Rubriflorin A*).

Xu, L.-J. *et al.*, *Planta Med.*, 2006, **72**, 169-174 (*Propinquanin D*).

Shen, Y.-C. *et al.*, *J. Nat. Prod.*, 2007, **70**, 1139-1145 (*Kadsuphilol B*).

Shen, Y.-C. *et al.*, *Helv. Chim. Acta*, 2008, **91**, 483-494 (*Kadsuphilins G,I,K*).

Gao, X.-M. *et al.*, *J. Nat. Prod.*, 2008, **71**, 558-563 (*Kadangustins A,B,D,E,F,G, Schisantherins L,P,Q*).

Ma, W. *et al.*, *Helv. Chim. Acta*, 2009, **92**, 709-715 (*Kadsurindutin H*).

Li, X.-N. *et al.*, *J. Nat. Prod.*, 2009, **72**, 1133-1141 (*Tiegusanins H-K*).

Shen, Y.-C. *et al.*, *Phytochemistry*, 2009, **70**, 114-120 (*Kadsuphilol I*).

Liu, C.-C. *et al.*, *Planta Med.*, 2010, **76**, 1605-1610 (*Arisanschinin D*).

Cheng, Y.-B. *et al.*, *Helv. Chim. Acta*, 2011, **94**, 148-158 (*Kadsuphilols P,Q*).

Yang, J.-H. *et al.*, *Tetrahedron*, 2011, **67**, 4498-4504 (*Ananonins A-N*).

Gao, X.-M. *et al.*, *J. Asian Nat. Prod. Res.*, 2012, **14**, 129-134 (*Kadangustin L*).

Dong, K. *et al.*, *J. Nat. Prod.*, 2012, **75**, 249-256 (*Polysperlignans A-E,G-J*).

Gao, X-M. *et al.*, *J. Nat. Prod.*, 2013, **76**, 1052-1057 (*Interiotherin C*).

Fang, L. *et al.*, *Phytochem. Lett.*, 2014, **9**, 158-162 (*3-O-Demethylpolysperlignan D*)

2,2′-Cyclo-8,8′-lignan-3,3′,4,4′,5,5′,7,8-octol C243

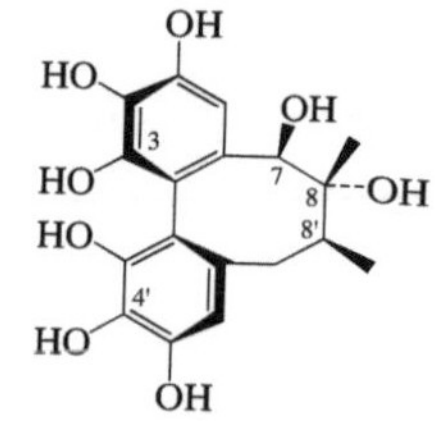

(*7R,8R,8′S,S*$_{axial}$)-*form*

3,3′,4,4′,5,5′,7,8-Octahydroxy-2,2′-cyclolignan

$C_{18}H_{20}O_8$ 364.35

(7R,8R,8′S,S$_{axial}$)-form

4′,5′-Methylene, 3,3′,4,5-tetra-Me ether

2,2′-Cyclo-3,3′,4,5-tetramethoxy-4′,5′-methylenedioxy-8,8′-lignan-7,8-diol. 7,8-Dihydroxy-3,3′,4,5-tetramethoxy-4′,5′-methylenedioxy-2,2′-cyclolignan. ***Gomisin P****. Schizandiol. Schisandiol*

[69980-01-4]

[58546-59-1] [128440-05-1]

$C_{23}H_{28}O_8$ 432.469

Needles. Mp 114 - 116°. $[\alpha]_D$ -94.3 (c, 0.53 in $CHCl_3$). No stereochem. given for Schisandiol but it appears to be identical to Gomisin P

4′,5′-Methylene, 3,3′,4,5-tetra-Me ether, 7-angeloyl

Angeloylgomisin P. *Schizantherin C. Schisantherin C*

[64938-51-8]

$C_{28}H_{34}O_9$ 514.571

Isol. from fruits of *Schizandra chinensis* and *Schizandra phenanthera*. Antiproliferative agent; cell cycle progression inhibitor in cancer cells. $[\alpha]_D$ -98.5

4′,5′-Methylene, 3,3′,4,5-tetra-Me ether, 7-tigloyl

Tigloylgomisin P

[69176-51-8]

$C_{28}H_{34}O_9$ 514.571

Isol. from *Schizandra chinensis*. Amorph. powder. $[\alpha]_D$ -64.2. $[\alpha]^{22}_{D}$ -64.2 (c, 0.463 in $CHCl_3$)

4′,5′-Methylene, 3,3′,4,5-tetra-Me ether, 7-benzoyl

Benzoylgomisin P

[129445-43-8]

$C_{30}H_{32}O_9$ 536.577

Isol. from fruit of *Schizandra sphenanthera*. Shows antitussive, tonic and sedative props.. Amorph. powder. $[\alpha]_D^{24}$ -22.1 (c, 0.4 in $CHCl_3$)

(7S,8R,8′R,S_{axial})-form

3,3′,4,4′,5-Penta-Me ether, 7-angeloyl

Arisanschinin J

[1181216-82-9]

$C_{28}H_{36}O_9$ 516.587

Constit. of the fruit of *Schisandra arisanensis*. Pale yellow oil. $[\alpha]_D^{23}$ -30 (c, 0.1 in CH_2Cl_2). λ_{max} 226 (log ε 3.95) (MeOH)

4,5-Methylene, 3,3′-di-Me ether, 7-benzoyl

Schisphenlignan C

[1477666-07-1]

$C_{28}H_{28}O_9$ 508.523

Isol. from *Schisandra sphenanthera*. Amorph. powder. $[\alpha]_D^{26}$ -120.5 (c, 0.26 in MeOH). λ_{max} 222 (log ε 4.13) (MeOH)

4,5-Methylene, 3,3′,4′-tri-Me ether, 7-benzoyl

Schisphenlignan B

[1477666-05-9]

$C_{29}H_{30}O_9$ 522.55

Isol. from *Schisandra sphenanthera*. Amorph. powder. $[\alpha]_D^{23}$ -245.7 (c, 0.16 in MeOH). λ_{max} 221 (log ε 4.31) (MeOH)

4,5-Methylene, 3,4′,5′-tri-Me ether, 7-benzoyl

Schisphenlignan A

[1477666-04-8]

$C_{29}H_{30}O_9$ 522.55

Isol. from *Schisandra sphenanthera*. Amorph. powder. $[\alpha]_D^{27}$ -113.5 (c, 0.17 in $CHCl_3$). λ_{max} 240 (log ε 3.60) ($CHCl_3$)

4′,5′-Methylene, 3,3′,4-tri-Me ether, 7-benzoyl

Schisphenlignan L

$C_{29}H_{30}O_9$ 522.55

Constit. of the fruit of *Schisandra sphenanthera*. Exhibits neuroprotective effect. Powder. $[\alpha]_D^{25}$ -38.8 (c, 0.1 in MeOH). λ_{max} 222 (log ε 4.32) (MeOH)

4′,5′-Methylene, 3′,4,5-tri-Me ether, 7-angeloyl

Arisanschinin I

[1181216-81-8]

$C_{27}H_{32}O_9$ 500.544

Constit. of the fruit of *Schisandra arisanensis*. Pale yellow oil. $[\alpha]_D^{23}$ -25 (c, 0.1 in CH_2Cl_2). λ_{max} 220 (log ε 3.96) (MeOH)

(7S,8R,8′S,S_{axial})-form

3,3′,4,4′,5-Penta-Me ether, 7-tigloyl

Schisanwilsonin D

[1133723-85-9]

$C_{28}H_{36}O_9$ 516.587

Constit. of the fruit of *Schisandra wilsoniana*. Powder. $[\alpha]_D^{22}$ -53.5 (c, 0.02 in MeOH). λ_{max} 214 (log ε 4.78) (MeOH)

3,3′,4,4′,5-Penta-Me ether, 7-benzoyl

Schisanwilsonin G

[1189056-63-0]

$C_{30}H_{34}O_9$ 538.593

Constit. of the fruit of *Schisandra wilsoniana*. Powder. $[\alpha]_D^{22}$ -58.5 (c, 0.02 in MeOH). λ_{max} 217 (log ε 4.76) (MeOH)

4′,5′-Methylene, 3,3′,4,5-tetra-Me ether, 7-Ac

Schisanwilsonin F

[1189056-62-9]

$C_{25}H_{30}O_9$ 474.506

Constit. of the fruit of *Schisandra wilsoniana*. Powder. $[\alpha]_D^{22}$ -32 (c, 0.04 in MeOH). λ_{max} 219 (log ε 4.34); 290 (sh) (MeOH)

4′,5′-Methylene, 3,3′,4,5-tetra-Me ether, 7-O-(2-methylpropanoyl)

Schisanwilsonin E

[1189056-61-8]

$C_{27}H_{34}O_9$ 502.56

Constit. of the fruit of *Schisandra wilsoniana*. Powder. $[\alpha]_D^{22}$ +20 (c, 0.02 in MeOH). λ_{max} 211 (log ε 4.65); 290 (sh) (MeOH)

4′,5′-Methylene, 3,3′,4,5-tetra-Me ether, 7-angeloyl

Schisanwilsonin A

[181228-78-4]

$C_{28}H_{34}O_9$ 514.571

Constit. of the fruit of *Schisandra wilsoniana*. Powder. $[\alpha]_D^{22}$ -17.5 (c, 0.04 in MeOH). λ_{max} 221 (log ε 4.18); 290 (log ε 3.27) (MeOH)

4′,5′-Methylene, 3,3′,4,5-tetra-Me ether, 7-tigloyl

Arisanschinin L. Schisanwilsonin B

[1181216-84-1]

$C_{28}H_{34}O_9$ 514.571

Constit. of the fruits of *Schisandra arisanensis* and *Schisandra wilsoniana*. Powder. $[\alpha]_D^{23}$ -73 (c, 0.1 in CH_2Cl_2). λ_{max} 219 (log ε 3.98) (MeOH)

4′,5′-Methylene, 3,3′,4,5-tetra-Me ether, 7-benzoyl

Arisanschinin K. Schisanwilsonin C

[1181216-83-0]

$C_{30}H_{32}O_9$ 536.577

Constit. of the fruits of *Schisandra arisanensis* and *Schisandra wilsoniana*. Powder. $[\alpha]_D^{22}$ -70.5 (c, 0.04 in CH_2Cl_2). λ_{max} 217 (log ε 4.78) (MeOH)

(7S,8S,8′S,S_{axial})-form

3,3′,4,4′,5-Penta-Me ether, 7-tigloyl

Schisphenin A

[1420217-16-8]

$C_{28}H_{36}O_9$ 516.587

Constit. of the fruit of *Schisandra sphenanthera*. Amorph. powder. $[\alpha]_D^{25}$ +125 (c, 0.1 in CH_2Cl_2). λ_{max} 232 (log ε 2.77); 254 (log ε 3.17); 285 (log ε 3.05) (MeOH)

3,3′,4,4′,5-Penta-Me ether, 7-benzoyl

Schizantherin E. Schisantherin E

[64917-83-5]

$C_{30}H_{34}O_9$ 538.593

Isol. from *Schizandra sphenanthera*

3,3′,4,4′,5,5′-Hexa-Me ether

2,2′-Cyclo-3,3′,4,4′,5,5′-hexamethoxy-8,8′-lignan-7,8-diol. 7,8-Dihydroxy-3,3′,4,4′,5,5′-hexamethoxy-2,2′-cyclolignan. ***Gomisin Q***

[66096-74-0]

$C_{24}H_{32}O_8$ 448.512

Mp 191 - 193°. $[\alpha]_D^{22}$ -106 (c, 1.24 in $CHCl_3$)

3,3′,4,4′,5,5′-Hexa-Me ether, 7-angeloyl

Angeloylgomisin Q

[72561-28-5]

$C_{29}H_{38}O_9$ 530.614

Isol. from fruit of *Schizandra chinensis*. Mp 82.5 - 83.5°. $[\alpha]_D^{24}$ -26.4 (c, 1.1 in $CHCl_3$)

3,3′,4,4′,5,5′-Hexa-Me ether, 7-benzoyl

Benzoylgomisin Q

[129385-73-5]

$C_{31}H_{36}O_9$ 552.62

Isol. from fruit of *Schizandra sphenanthera*. $[\alpha]_D^{23}$ -118 (c, 0.9 in $CHCl_3$)

3,3′,4,4′,5,5′-Hexa-Me ether, 7-O-(3-pyridinecarbonyl)

Nicotinoylgomisin Q

$C_{30}H_{35}NO_9$ 553.608

Constit. of leaves and stems of *Schisandra chinensis*. Exhibits anti HIV-1 activity. Yellow gum. $[\alpha]_D^{26}$

-78 (c, 0.09 in MeOH). λ_{max} 217 (log ε 3.78); 250 (log ε 3.23) (no solvent reported)

4,5-Methylene, 3,4′,5′-tri-Me ether, 7-angeloyl
Arisanschinin H
[1181216-80-7]
$C_{27}H_{32}O_9$ 500.544
Constit. of the fruit of *Schisandra arisanensis*. Amorph. powder. $[\alpha]_D^{25}$ +16.7 (c, 0.1 in CH_2Cl_2). λ_{max} 238 (log ε 3.96) (MeOH)

4,5-Methylene, 3′,4′,5′-tri-Me ether, 7-angeloyl
Schirubrisin D
[1366232-11-2]
$C_{27}H_{32}O_9$ 500.544
Constit. of the stems of *Schisandra rubriflora*. Pale yellow oil. $[\alpha]_D^{19}$ +6.4 (c, 0.15 in MeOH). λ_{max} 220 (log ε 4.28) (MeOH)

4,5-Methylene, 3,3′,4′,5′-tetra-Me ether, 7-angeloyl
Gomisin F
[62956-47-2]
$C_{28}H_{34}O_9$ 514.571
Isol. from fruits of *Schizandra chinensis*. Needles. Mp 119.5 - 120.5°. $[\alpha]_D^{23}$ 0 (c, 0.40 in $CHCl_3$). λ_{max} 218 (ε 94500); 257 (sh); 292 (sh) (EtOH) (Derep)

4,5-Methylene, 3,3′,4′,5′-tetra-Me ether, 7-benzoyl
Gomisin G
[62956-48-3]
$C_{30}H_{32}O_9$ 536.577
Isol. from *Schizandra chinensis*. Prisms (Me_2CO/Et_2O). Mp 97 - 98°. $[\alpha]_D^{25}$ -126 (c, 0.427 in $CHCl_3$). λ_{max} 220 (ε 57500); 253 (sh); 284 (sh) (EtOH) (Derep)

4′,5′-Methylene, 3,3′,4-tri-Me ether, 7-angeloyl
Schirubrisin C
[1366232-10-1]
$C_{27}H_{32}O_9$ 500.544
Constit. of the stems of *Schisandra rubriflora*. Pale yellow powder. $[\alpha]_D^{18}$ -29.1 (c, 0.07 in MeOH). λ_{max} 218 (log ε 4.28) (MeOH)

4′,5′-Methylene, 3,3′,4-tri-Me ether, 7-angeloyl, 5-O-β-D-glucopyranoside
Schisandroside D
[2059120-57-7]
$C_{33}H_{42}O_{14}$ 662.685
Constit. of the roots of *Schisandra chinensis*. Amorph. powder. $[\alpha]_D^{25}$ -36.4 (c, 0.2 in MeOH). λ_{max} 220 (log ε 4.86); 257 (log ε 4.23); 292 (log ε 3.72) (MeOH)

4′,5′-Methylene, 3′,4,5-tri-Me ether, 7-angeloyl
Schirubrisin A
$C_{27}H_{32}O_9$ 500.544
Constit. of the stems of *Schisandra rubriflora*. Pale yellow oil. $[\alpha]_D^{19}$ -25.8 (c, 0.12 in MeOH). λ_{max} 220 (log ε 4.37) (MeOH)

4′,5′-Methylene, 3′,4,5-tri-Me ether, 7-benzoyl
Schirubrisin B
[1366232-09-8]
$C_{29}H_{30}O_9$ 522.55
Constit. of the stems of *Schisandra rubriflora*. Pale yellow powder. $[\alpha]_D^{19}$ -106.5 (c, 0.21 in MeOH). λ_{max} 223 (log ε 4.26) (MeOH)

4′,5′-Methylene, 3,3′,4,5-tetra-Me ether, 7-angeloyl
Gomisin B. *Schizantherin B. Schisantherin B. Wuweizi ester B*
[58546-55-7]
$C_{28}H_{34}O_9$ 514.571
Isol. from *Schizandra chinensis* and *Schizandra sphenanthera*. Antihepatitic agent. Shows insecticidal props.. Cryst.. Mp 95 - 97°. $[\alpha]_D^{23}$ -27.6 (c, 0.5 in CH_2Cl_2). λ_{max} 218 (ε 94500); 257 (sh); 292 (sh) (EtOH) (Derep)

4′,5′-Methylene, 3,3′,4,5-tetra-Me ether, 7-benzoyl
Gomisin C. *Schisantherin A*. ***Schizantherin A***. *Wuweizi ester A*
[58546-56-8]
$C_{30}H_{32}O_9$ 536.577
Isol. from *Schizandra chinensis* and *Schizandra sphenanthera*. Cryst.. λ_{max} 220 (ε 57500); 253 (sh); 284 (sh) (EtOH) (Derep)

4,5:4′,5′-Bis(methylene), 3,3′-di-Me ether, 7-benzoyl
Schizantherin D. *Schisantherin D*
[64917-82-4]
$C_{29}H_{28}O_9$ 520.534
Isol. from *Schizandra sphenanthera, Schizandra chinensis* and *Kadsura interior*. Exhibits potent anti-HIV activity. Inhibits ACAT. Mp 108 - 110°. $[\alpha]_D$ -180. $[\alpha]_D^{25}$ -98 ($CHCl_3$)

4,5:4′,5′-Bis(methylene), 3,3′-di-Me ether, 7-angeloyl
Interiotherin B
[181701-07-5]
$C_{27}H_{30}O_9$ 498.528
Constit. of *Kadsura interior*. HIV reverse transcriptase inhibitor. Prisms (MeOH). Mp 96 - 98.5°. $[\alpha]_D$ +50.8 (c, 4.2 in $CHCl_3$). λ_{max} 221 (log ε 4.82); 256 (log ε 4.07); 278 (log ε 3.63) (MeOH)

7-Ketone, 4,5-methylene, 3,4′-di-Me ether
2,2′-Cyclo-3′,5′,8-trihydroxy-3,4′-dimethoxy-4,5-methylenedioxy-8,8′-lignan-7-one. 3′,5′,8-Trihydroxy-3,4′-dimethoxy-4,5-methylenedioxy-7-oxo-2,2′-cyclolignan.
Kadoblongifolin B
[1174173-74-0]
$C_{21}H_{22}O_8$ 402.399
Constit. of the stems of *Kadsura oblongifolia*. Powder. $[\alpha]_D^{20}$ -59.4 (c, 0.27 in MeOH). λ_{max} 220 (log ε 4.48); 251 (log ε 4.09); 274 (log ε 3.66) (MeOH)

7-Ketone, 4,5-methylene, 3,4′,5′-tri-Me ether
2,2′-Cyclo-3′,8-dihydroxy-3,4′,5′-trimethoxy-4,5-methylenedioxy-8,8′-lignan-7-one. 3′,8-Dihydroxy-3,4′,5′-trimethoxy-4,5-methylenedioxy-7-oxo-2,2′-cyclolignan.
Kadoblongifolin A
[1174173-73-9]
$C_{22}H_{24}O_8$ 416.426
Constit. of the stems of *Kadsura oblongifolia*. Powder. $[\alpha]_D^{20}$ -47.3 (c, 0.05 in MeOH). λ_{max} 221 (log ε 4.44); 251 (log ε 4.08); 274 (log ε 3.6) (MeOH)

7-Ketone, 4,5-methylene, 3,3′,4′,5′-tetra-Me ether
2,2′-Cyclo-8-hydroxy-3,3′,4′,5′-tetramethoxy-4,5-methylenedioxy-8,8′-lignan-7-one. ***Kadoblongifolin C***
[62956-33-6]
$C_{23}H_{26}O_8$ 430.453
Constit. of the stems of *Kadsura oblongifolia*. Powder

9-Chloro, 4′,5′-methylene, 3,3′,4,5-tetra-Me ether, 7-angeloyl
9-Chlorogomisin B. *9-Chloroschizantherin B*
[1186227-88-2]
$C_{28}H_{33}ClO_9$ 549.013
Constit. of the fruit of *Schisandra sphenanthera*. $[\alpha]_D^{27}$ -29.7 (c, 0.12 in $CHCl_3$)

(8R,8′R,S$_{axial}$)-form

8-Hydroperoxide, 7-ketone, 4,5-methylene, 3,3′,4′,5′-tetra-Me ether

Tiegusanin M

[1156503-64-8]

$C_{23}H_{26}O_9$ 446.452

Constit. of *Schisandra propinqua* var. *sinensis*. Brown gum. $[\alpha]^{21}_{D}$ -83.3 (c, 0.07 in $CHCl_3$). λ_{max} 197 (log ε 3.49); 207 (log ε 3.48); 222 (log ε 3.52); 241 (log ε 3.82) ($CHCl_3$).

Liu, C.S. *et al., Sci. Sin. (Engl. Edn.)*, 1978, **21**, 483-502 (*Schizantherin D*).
Ikeya, Y. *Chem. Pharm. Bull.*, 1979, **27**, 1383-1394 (*Gomisins B,C,F,G*).
Ikeya, Y. *et al., Chem. Pharm. Bull.*, 1979, **27**, 2536-2538 (*Angeloylgomisin Q*).
Ikeya, Y. *et al., Chem. Pharm. Bull.*, 1980, **28**, 3357-3361 (*Angeloylgomisin P, Tigloylgomisin P*).
Ikeya, Y. *et al., Chem. Pharm. Bull.*, 1982, **30**, 3207-3211 (*Schisantherin D*).
Hikino, H. *et al., Planta Med.*, 1984, **50**, 213-218 (*pharmacol*).
Ikeya, Y. *et al., Chem. Pharm. Bull.*, 1990, **38**, 1408-1411 (*Schisantherins A-E*).
Chen, D.-F. *et al., J. Nat. Prod.*, 1996, **59**, 1066-1068 (*Interiotherin B*).
Miyazawa, M. *et al., Nat. Prod. Lett.*, 1998, **12**, 175-180 (*Gomisin B, pmr, cmr*).
Kwon, B.M. *et al., Planta Med.*, 1999, **65**, 74-76 (*activity*).
Min, H.Y. *et al., Bioorg. Med. Chem. Lett.*, 2008, **18**, 523-526 (*Schisantherin C, activity*).
Ma, W.-H. *et al., Bioorg. Med. Chem. Lett.*, 2009, **19**, 4958-4962 (*Schisanwilsonins A-G*).
Li, X.-N. *et al., J. Nat. Prod.*, 2009, **72**, 1133-1141 (*Tiegusanin M*).
Cheng, Y.-B. *et al., J. Nat. Prod.*, 2009, **72**, 1663-1668 (*Arisanschinins H-L*).
Liu, H.-T. *et al., Magn. Reson. Chem.*, 2009, **47**, 609-612 (*Kadoblongifolins A,B,C*).
Ren, R. *et al., Z. Naturforsch., B*, 2010, **65**, 211-218 (*9-Chlorogomisin B*).
Luo, X. *et al., Chin. J. Nat. Med.*, 2011, **9**, 167-172 (*Schirubrisins A-D*).
Chen, Y.-C. *et al., Bioorg. Med. Chem. Lett.*, 2013, **23**, 880-885 (*Arisanschinin N, Schisphenin A*).
Liang, C.-Q. *et al., Chem. Pharm. Bull.*, 2013, **61**, 96-100 (*Schisphenlignans A,B,C*).
Liu, H.-W. *et al., Eur. J. Med. Chem.*, 2013, **59**, 265-273 (*Schizantherin A, cd, cryst struct*).
Shi, Y. *et al., Chin. J. Chem.*, 2014, **32**, 734-740 (*Nicotinoylgomisin Q*).
Kim, H.M. *et al., Chem. Pharm. Bull.*, 2015, **63**, 746-751 (*Schisandroside D*).
Liu, Y. *et al., Fitoterapia*, 2017, **116**, 10-16 (*Schisphenlignan L, activity*).
Szopa, A. *et al., Phytochem. Rev.*, 2017, **16**, 195-218 (*Schisandra chinensis constits, rev*)

2,2′-Cyclo-8,8′-lignan-3,3′,4,4′,5,5′,7,8′-octol C244

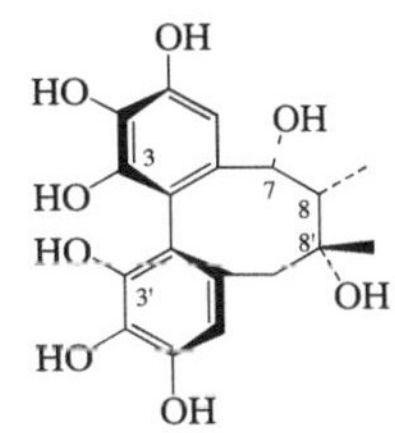

(7R,8S,8′R,S$_{axial}$)-form

3,3′,4,4′,5,5′,7,8′-Octahydroxy-2,2′-cyclolignan

$C_{18}H_{20}O_8$ 364.350

(7R,8S,8′R,S$_{ax}$)-form

4,5-Methylene, 3,4′,5′-tri-Me ether

3,4′,5′-Trimethoxy-4,5-methylenedioxy-2,2′-cyclo-8,8′-lignan-3′,7,8′-triol. 3′,7,8′-Trihydroxy-3,4′,5′-trimethoxy-4,5-methylenedioxy-2,2′-cyclolignan

4,5-Methylene, 3,4′,5′-tri-Me ether, 7-angeloyl

Longipedlignan E

[2229859-81-6]

$C_{27}H_{32}O_9$ 500.544

Constit. of the roots of *Kadsura longipedunculata*. Amorph. yellow powder. $[\alpha]^{20}_{D}$+16 (c, 0.1 in MeOH). λ_{max} 220 (log ε 1.05) (MeOH)

4,5-Methylene, 3,4′,5′-tri-Me ether, 7-benzoyl

Longipedlignan A

[2229859-79-2]

$C_{29}H_{30}O_9$ 522.550

Constit. of the roots of *Kadsura longipedunculata*. Amorph. yellow powder. $[\alpha]^{20}_{D}$-11 (c, 0.2 in MeOH). λ_{max} 233 (log ε 1.94) (MeOH)

4,5-Methylene, 3,4′,5′-tri-Me ether, 7-E-cinnamoyl

Longipedlignan C

[2229859-80-5]

$C_{31}H_{32}O_9$ 548.588

Constit. of the roots of *Kadsura longipedunculata*. Amorph. yellow powder. $[\alpha]^{20}_{D}$-40 (c, 0.5 in MeOH). λ_{max} 217 (log ε 2.23); 259 (log ε 2.33); 279 (log ε 1.86) (MeOH)

(7R,8S,8′S,S$_{ax}$)-form

4,5-Methylene, 3,4′,5′-tri-Me ether, 7-benzoyl

Longipedlignan B

[2229860-15-3]

$C_{29}H_{30}O_9$ 522.550

Constit. of the roots of *Kadsura longipedunculata*. Amorph. yellow powder. $[\alpha]^{20}_{D}$-15 (c, 0.1 in MeOH). λ_{max} 210 (log ε 2.22); 222 (log ε 2.23); 259 (log ε 1.54); 283 (log ε 1.23) (MeOH)

4,5-Methylene, 3,4′,5′-tri-Me ether, 7-E-cinnamoyl

Longipedlignan D

[2229860-16-4]

$C_{31}H_{32}O_9$ 548.588

Constit. of the roots of *Kadsura longipedunculata*. Amorph. yellow powder. $[\alpha]^{20}_{D}$-67 (c, 0.5 in MeOH). λ_{max} 217 (log ε 4.75); 259 (log ε 4.36); 279 (log ε 4.37) (MeOH).

Liu, J. *et al., J. Nat. Prod.*, 2018, **81**, 846-857 (*Longipedlignans A-E*)

2,7′-Cyclo-8,8′-lignan-3,3′,4,4′,5,5′,7,7′-octol C245

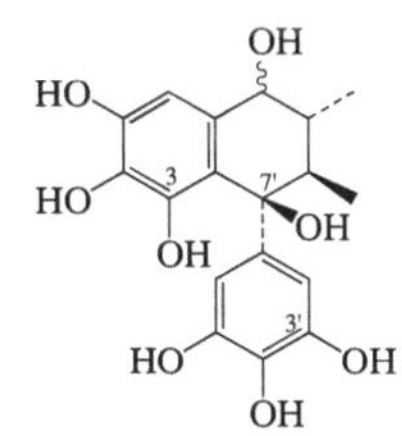

(7ξ,7′S,8R,8′R)-form

3,3′,4,4′,5,5′,7,7′-Octahydroxy-2,7′-cyclolignan

$C_{18}H_{20}O_8$ 364.35

(7ξ,7′S,8R,8′R)-form

7′-Et, 3′,4,4′,5,5′-penta-Me ether

2,7′-Cyclo-7′-ethoxy-3′,4,4′,5,5′-pentamethoxy-8,8′-lignan-3,7-diol. 7′-Ethoxy-3,7-dihydroxy-3′,4,4′,5,5′-pentamethoxy-2,7′-cyclolignan

[1187888-72-7]

$C_{25}H_{34}O_8$ 462.539

Constit. of the wood of *Eusideroxylon zwageri*

(7ξ,7′S,8S,8′R)-form

7′-Et, 3′,4,4′,5,5′-penta-Me ether

[1187888-74-9]

Constit. of the wood of *Eusideroxylon zwageri*.

Yoosu, S. *et al., Chem. Nat. Compd. (Engl. Transl.)*, 2009, **45**, 356-359 (*Eusideroxylon zwageri constits*)

2,7′-Cyclo-8,8′-lignan-3,3′,4,4′,5,5′,9,9′-octol **C246**

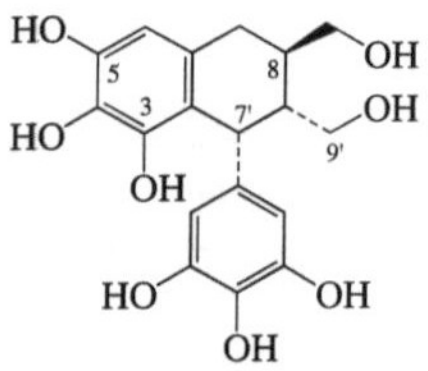

(7′R,8R,8′R)-form

3,3′,4,4′,5,5′,9,9′-Octahydroxy-2,7′-cyclolignan
[3737-93-7]
$C_{18}H_{20}O_8$ 364.35
Lignan numbering shown

(7′R,8R,8′R)-form

3,3′,5,5′-Tetra-Me ether
2,7′-Cyclo-3,3′,5,5′-tetramethoxy-8,8′-lignan-4,4′,9,9′-tetrol. 4,4′,9,9′-Tetrahydroxy-3,3′,5,5′-tetramethoxy-2,7′-cyclolignan.
7′-Epilyoniresinol
[894079-07-3]
$C_{22}H_{28}O_8$ 420.458
Constit. of *Euonymus alatus* and *Toona sinensis*. Moderate inhibitor of NO prodn. in LPS-stimulated RAW264.7 cells. Amorph. solid. λ_{max} 232 (log ε 4); 274 (log ε 3.2) (MeOH)

3,3′,5,5′-Tetra-Me ether, 9′-O-α-L-arabinopyranoside
Gaultheroside A
[233686-33-4]
$C_{27}H_{36}O_{12}$ 552.573
Constit. of *Gaultheria yunnanensis*. Mp 154 - 155°. λ_{max} 209; 278 (MeOH)

3,3′,5,5′-Tetra-Me ether, 9′-O-β-D-glucopyranoside
[162681-51-8]
$C_{28}H_{38}O_{13}$ 582.599
Constit. of the bark of *Aegle marmelos* and silver birch and leaves of *Laurus nobilis*. Amorph. solid. $[\alpha]^{26}_{D}$ -144.8 (c, 0.42 in MeOH)

3,3′,5,5′-Tetra-Me ether, 4,9′-di-O-β-D-glucopyranoside
[1432625-93-8]
$C_{34}H_{48}O_{18}$ 744.74
Constit. of *Bambusa tuldoides*. Amorph. powder. $[\alpha]_D$ -73.1 (c, 0.46 in MeOH). λ_{max} 275 (MeOH)

(7′R,8R,8′S)-form

3,3′,5,5′-Tetra-Me ether, 9′-O-[β-D-xylopyranosyl-(1→6)-β-D-glucopyranoside]
$C_{33}H_{46}O_{17}$ 714.714
Constit. of the branches of *Alangium kurzii*. Yellow oil. $[\alpha]^{20}_{D}$ -2.65 (c, 0.38 in MeOH)

(7′R,8S,8′R)-form

3,3′,5,5′-Tetra-Me ether, 4-O-β-D-glucopyranoside
[1014977-24-2]
$C_{28}H_{38}O_{13}$ 582.599
Constit. of the roots of *Capparis tenera*. Pale yellow powder. $[\alpha]^{25}_{D}$ -1.2 (c, 0.08 in MeOH). λ_{max} 215 (log ε 1.96); 274 (log ε 0.21) (MeOH)

3,3′,5,5′-Tetra-Me ether, 9′-O-α-L-rhamnopyranoside
Racemiside
[1181372-30-4]
$C_{28}H_{38}O_{12}$ 566.6
Constit. of *Bruguiera gymnorrhiza* and *Cotoneaster racemiflora*. Amorph. solid. $[\alpha]^{27}_{D}$ -33.7 (c, 0.7 in MeOH) (*B. gymnorrhiza*). $[\alpha]^{20}_{D}$ +4.8 (c, 0.0006 in MeOH) (*C. racemiflora*). Stereochem. of the isolates has not been shown to be identical. Only *B. gymnorrhiza* isolate has abs. config. assigned. λ_{max} 206 (log ε 3.8); 240 (log ε 4.02); 276 (log ε 3.9) (MeOH)

3,3′,5,5′-Tetra-Me ether, 9′-O-β-D-glucopyranoside
[1175149-18-4]
$C_{28}H_{38}O_{13}$ 582.599
Constit. of the stems of *Carissa carandas*. Amorph. yellow solid. $[\alpha]^{20}_{D}$ -46.9 (c, 0.04 in MeOH). λ_{max} 224 (log ε 4.32); 278 (log ε 3.64) (MeOH)

4,5-Methylene, 3,3′,5′-tri-Me ether
2,7′-Cyclo-3,3′,5′-trimethoxy-4,5-methylenedioxy-8,8′-lignan-4′,9,9′-triol. 4′,9,9′-Trihydroxy-3,3′,5′-trimethoxy-4,5-methylenedioxy-2,7′-cyclolignan

4,5-Methylene, 3,3′,5′-tri-Me ether, 9,9′-di-Ac
[929692-68-2]
$C_{26}H_{30}O_{10}$ 502.516
Constit. of *Peperomia duclouxii*. Gum. $[\alpha]^{25}_{D}$ +28.7 (c, 0.03 in $CHCl_3$). λ_{max} 242 (log ε 3.46); 281 (log ε 2.98) ($CHCl_3$)

(7′R,8S,8′S)-form

3,3′,5,5′-Tetra-Me ether
(-)-Lyoniresinol. *Pygeoresinol*
[31768-94-2]
$C_{22}H_{28}O_8$ 420.458
Aglycone from *Enkianthus nudipes*. Cryst. (H_2O or MeOH). Mp 171 - 172° (202-204°). $[\alpha]^{31}_{D}$ -53.2 (c, 2.01 in Me_2CO). $[\alpha]_D$ -5.24 (MeOH)

3,3′,5,5′-Tetra-Me ether, 9-O-β-D-glucopyranoside
[162613-63-0]
$C_{28}H_{38}O_{13}$ 582.599
Constit. of *Tabernaemontana cymosa*. Oil. $[\alpha]_D$ -23 (c, 0.3 in MeOH)

3,3′,5,5′-Tetra-Me ether, 9′-O-β-D-xylopyranoside
Nudiposide
[62058-46-2]
$C_{27}H_{36}O_{12}$ 552.573
Isol. from *Quercus petraea*, *Betula pendula* and *Enkianthus nudipes*. Mp 175 - 178° (157-160°). $[\alpha]^{29}_{D}$ -67 (c, 1.4 in EtOH)

3,3′,5,5′-Tetra-Me ether, 9′-O-α-L-rhamnopyranoside
(-)-Lyoniresinol 9′-rhamnoside
[27193-56-2]
$C_{28}H_{38}O_{12}$ 566.6
Constit. of the bark of *Trichilia estipulata*. Viscous oil (as hexa-Ac). $[\alpha]^{20}_{D}$ -11.7 (c, 0.03 in $CHCl_3$) (hexa-Ac). λ_{max} 203 (ε 52990); 275 (ε 1040) (MeOH) (hexa-Ac)

3,3′,5,5′-Tetra-Me ether, 9′-O-β-D-glucopyranoside
[143236-02-6]
$C_{28}H_{38}O_{13}$ 582.599
Isol. from bark of *Stemmadenia minima* and from *Tabernaemontana cymosa*. $[\alpha]_D$ -110

3,3′,5,5′-Tetra-Me ether, 9′-O-[3,4,5-trihydroxybenzoyl-(→6)-β-D-glucopyranoside]
(-)-Quercoresinoside B
[2021245-86-1]
$C_{35}H_{42}O_{17}$ 734.704
Obt. from the extract of toasted *Quercus petraea* heartwood. Useful as a bitter standard in wines. Amorph. powder. $[\alpha]^{20}_{D}$ -26 (c, 0.05 in MeOH)

3,3′,5,5′-Tetra-Me ether, 9′-O-[4-hydroxy-3,5-dimethoxybenzoyl-(→6)-β-D-glucopyranoside]
Tarennanoside B
[1201192-78-0]
$C_{37}H_{46}O_{17}$ 762.758
Constit. of *Tarenna attenuata*. Antioxidant. Amorph. powder. $[\alpha]_D^{20}$ -40.9 (c, 0.75 in MeOH). λ_{max} 277 (log ε 4.05) (MeOH)

3,3′,5,5′-Tetra-Me ether, 4,9′-di-O-β-D-glucopyranoside
[1432625-94-9]
Constit. of *Bambusa tuldoides*. Amorph. powder. $[\alpha]_D$ -15.5 (c, 2.0 in MeOH). λ_{max} 275 (MeOH)

3,3′,5,5′-Tetra-Me ether, 9′-sulfate
(-)-Lyoniresinol 9′-sulfate
[721882-34-4]
[448949-56-2]
$C_{22}H_{28}O_{11}S$ 500.515
Constit. of the roots of *Polygonum cuspidatum*. Amorph. powder (as Na salt). $[\alpha]_D^{25}$ -17 (c, 0.13 in MeOH). λ_{max} 205; 238 (sh); 282 (MeOH)

3,3′,5,5′-Tetra-Me ether, 9-O-(3,4-dihydroxybenzoyl), 9′-O-β-D-xylopyranoside
9-O-(3,4-Dihydroxybenzoyl) nudiposide
[1559068-07-3]
$C_{34}H_{40}O_{15}$ 688.679
Constit. of the bark of *Tetrapterys mucronata*. Amorph. powder. $[\alpha]_D^{25}$ -4.1 (c, 0.1 in MeOH). λ_{max} 206 (log ε 3.43); 264 (log ε 2.57); 300 (sh) (MeOH)

3,3′,5,5′-Tetra-Me ether, 9-O-(3,4,5-trihydroxybenzoyl), 9′-O-β-D-xylopyranoside
9-O-(3,4,5-Trihydroxybenzoyl) nudiposide. **9-Galloylnudiposide**
[1622396-84-2]
$C_{34}H_{40}O_{16}$ 704.678
Constit. of the wood of *Quercus petraea*. Amorph. powder. $[\alpha]_D^{20}$ -47.3 (c, 0.35 in MeOH)

(7′S,8R,8′R)-form
3,3′,5,5′-Tetra-Me ether
(+)-Lyoniresinol*. Dimethoxyisolariciresinol*
[14464-90-5]
$C_{22}H_{28}O_8$ 420.458
Isol. from *Tripetaleia paniculata*, *Cotoneaster adpressus*, *Quercus petraea* and *Justicia tranquebariensis*. Mp 115 - 116°. Mp 165 - 167° (dimorph.). $[\alpha]_D$ +52 (EtOH). $[\alpha]_D$ +13.3 (c, 0.32 in MeOH)

3,3′,5,5′-Tetra-Me ether, 4-O-α-L-arabinofuranoside
(+)-Lyoniresinol 4-O-α-L-arabinofuranoside
$C_{27}H_{36}O_{12}$ 552.573
Constit. of *Achillea millefolium*. Brown powder. $[\alpha]_D^{24}$ +18.1 (c, 0.32 in MeOH)

3,3′,5,5′-Tetra-Me ether, 4-O-β-D-glucopyranoside
[374928-35-5]
$C_{28}H_{38}O_{13}$ 582.599
Constit. of *Antidesma membranaceum*. λ_{max} 205 (log ε 4.84); 275 (log ε 3.5) (MeOH)

3,3′,5,5′-Tetra-Me ether, 4,4′-di-O-β-D-glucopyranoside
(+)-Lyoniresinol 4,4′-diglucoside
[853913-74-3]
$C_{34}H_{48}O_{18}$ 744.74
Constit. of *Indigofera hetrantha*. Lipoxygenase inhibitor. Gummy solid. $[\alpha]_D^{25}$ +18.2 (c, 0.21 in MeOH). λ_{max} 206 (log ε 4.15); 242 (log ε 2.41); 278 (log ε 2.61) (MeOH)

3,3′,5,5′-Tetra-Me ether, 4-O-β-D-glucopyranoside, 9′-O-β-D-xylopyranoside
[2407648-37-5]
$C_{33}H_{46}O_{17}$ 714.714
Constit. of *Betula schmidtii*. Gum. $[\alpha]_D^{25}$ +13.2 (c, 0.1 in MeOH). λ_{max} 208 (log ε 4.13); 230 (sh); 276 (log ε 3.82) (MeOH)

3,3′,5,5′-Tetra-Me ether, 4,9′-di-O-β-D-glucopyranoside
[142542-81-2]
$C_{34}H_{48}O_{18}$ 744.74
Isol. from the stem bark of *Albizzia julibrissin*. Powder. $[\alpha]_D^{25}$ -21.2 (c, 0.12 in MeOH)

3,3′,5,5′-Tetra-Me ether, 4-O-[4-hydroxy-3-methoxy-E-cinnamoyl-(→6)-β-D-glucopyranoside]
Rhusemialin B
[946852-23-9]
$C_{38}H_{46}O_{16}$ 758.77
Constit. of the roots of *Rhus semialata*. Powder. $[\alpha]_D^{27}$ -9.4 (c, 1.1 in Me_2CO). λ_{max} 219 (log ε 4.33); 233 (sh); 282 (log ε 4.03); 301 (log ε 4); 331 (log ε 4.13) (MeOH)

3,3′,5,5′-Tetra-Me ether, 4-O-[4-hydroxy-3-methoxy-E-cinnamoyl-(→4)-[4-hydroxy-3-methoxy-E-cinnamoyl-(→6)]-β-D-glucopyranoside]
Rhusemialin C
[946852-24-0]
$C_{48}H_{54}O_{19}$ 934.941
Constit. of the roots of *Rhus semialata*. Powder. $[\alpha]_D^{27}$ -3.1 (c, 1.9 in Me_2CO)

3,3′,5,5′-Tetra-Me ether, 4′-O-β-D-glucopyranoside
[1014977-25-3]
$C_{28}H_{38}O_{13}$ 582.599
Constit. of the roots of *Capparis tenera*. Pale yellow powder. $[\alpha]_D^{25}$ -6.7 (c, 0.12 in MeOH). λ_{max} 213 (log ε 1.96); 278 (log ε 0.15) (MeOH)

3,3′,5,5′-Tetra-Me ether, 9-O-α-D-galactopyranoside, 9′-O-β-D-glucopyranoside
[786712-60-5]
$C_{34}H_{48}O_{18}$ 744.74
Constit. of the aerial parts of *Acanthus ilicifolius*. Amorph. powder. $[\alpha]_D^{25}$ +40 (c, 0.6 in MeOH)

3,3′,5,5′-Tetra-Me ether, 9-O-β-D-glucopyranoside
[213667-46-0]
$C_{28}H_{38}O_{13}$ 582.599
Constit. of the bark of *Aegle marmelos*. Amorph. solid. $[\alpha]_D^{27}$ -51.7 (c, 0.41 in MeOH)

3,3′,5,5′-Tetra-Me ether, 9-O-[β-D-glucopyranosyl-(1→4)-β-D-glucopyranoside]
[142542-80-1]
$C_{34}H_{48}O_{18}$ 744.74
From *Albizzia julibrissin*. Powder. $[\alpha]_D^{25}$ +0.4 (c, 0.47 in MeOH)

3,3′,5,5′-Tetra-Me ether, 9′-O-α-L-arabinofuranoside
Linderanoside A
[1809032-77-6]
$C_{27}H_{36}O_{12}$ 552.573
Constit. of trunk of *Lindera glauca*. Exhibits moderate to weak cytotoxicity against A498 cells. Amorph. gum. $[\alpha]_D^{25}$ +20 (c, 0.3 in MeOH). λ_{max} 228 (log ε 4.1); 284 (log ε 3.2) (MeOH)

3,3′,5,5′-Tetra-Me ether, 9′-O-β-D-xylopyranoside
Lyoniside*. Lyonoside. Pygeoside*
[34425-25-7]
[75587-29-0] (Pygeoside)

$C_{27}H_{36}O_{12}$ 552.573
Isol. from *Lyonia ovalifolia*, *Alnus glutinosa*, *Cotoneaster adpressus*, *Quercus petraea*, *Sorbus* spp., *Tetrapterys mucronata*, *Ficus ischnopodia* and *Vaccinium myrtillus*. Shows antifungal, antioxidant and antifungal activities. Mp 124 - 125°. Mp 165 - 165° (dimorph.). $[\alpha]_D$ +37.5 (c, 1.1 in MeOH). Stereochem. of Pygeoside not confirmed

3,3′,5,5′-Tetra-Me ether, 9′-O-α-L-rhamnopyranoside
[73745-47-8]
$C_{28}H_{38}O_{12}$ 566.6
Obt. from heartwood of *Ulmus thomasii* and from *Cotoneaster adpressus*. Amorph.

3,3′,5,5′-Tetra-Me ether, 9′-O-β-D-galactopyranoside
Amaranthoside
[903574-30-1]
$C_{28}H_{38}O_{13}$ 582.599
Constit. of *Amaranthus spinosus*

3,3′,5,5′-Tetra-Me ether, 9′-O-β-D-glucopyranoside
[87585-32-8]
$C_{28}H_{38}O_{13}$ 582.599
Isol. from bark of *Cinnamomum cassia* and from *Quercus petraea*. Amorph. powder. Mp 178 - 180°. $[\alpha]_D$ +22.4 (c, 1.01 in MeOH)

3,3′,5,5′-Tetra-Me ether, 9′-O-[3,4,5-trihydroxybenzoyl-(→6)-β-D-glucopyranoside]
(+)-Quercoresinoside A
[2021245-85-0]
$C_{35}H_{42}O_{17}$ 734.704
Obt. from the extract of toasted *Quercus petraea* heartwood. Useful as a bitter standard in wines. Amorph. powder. $[\alpha]_D^{20}$ +10 (c, 0.05 in MeOH)

3,3′,5,5′-Tetra-Me ether, 9′-O-[2-hydroxy-5-methoxybenzoyl-(→6)-β-D-glucopyranoside]
Tarennanoside H
[1201192-86-0]
$C_{36}H_{44}O_{16}$ 732.732
Constit. of *Tarenna attenuata*. Amorph. powder. $[\alpha]_D^{25}$ +22 (c, 0.43 in MeOH). λ_{max} 263 (log ε 3.12); 328 (log ε 3.95) (MeOH)

3,3′,5,5′-Tetra-Me ether, 9′-O-[4-hydroxy-3-methoxybenzoyl-(→6)-β-D-glucopyranoside]
Fernandoside
[366492-00-4]
$C_{36}H_{44}O_{16}$ 732.732
Constit. of *Fernandoa adenophylla*. Amorph. powder. $[\alpha]_D^{26}$ +25 (c, 1.24 in MeOH). λ_{max} 250; 362 (MeOH)

3,3′,5,5′-Tetra-Me ether, 9′-O-[4-hydroxy-3,5-dimethoxybenzoyl-(→2)-β-D-glucopyranoside]
Acanfolioside
[339295-49-7]
$C_{37}H_{46}O_{17}$ 762.758
Constit. of *Acanthus ilicifolius*. Amorph. powder. $[\alpha]_D^{22}$ +28.3 (c, 2.7 in MeOH)

3,3′,5,5′-Tetra-Me ether, 9′-O-[4-hydroxy-3,5-dimethoxybenzoyl-(→6)-β-D-glucopyranoside]
Tarennanoside A
[1174139-79-7]
$C_{37}H_{46}O_{17}$ 762.758
Constit. of *Tarenna attenuata* and the stems of *Callicarpa japonica*. Antioxidant. Amorph. powder. $[\alpha]_D^{31}$ +14.8 (c, 0.7 in MeOH). $[\alpha]_D^{25}$ +29.1 (c, 0.52 in MeOH). λ_{max} 277 (log ε 4.03) (MeOH)

3,3′,5,5′-Tetra-Me ether, 9′-O-[4-hydroxy-3,5-dimethoxy-E-cinnamoyl-(→6)-β-D-glucopyranoside]
Hazaleanin B
[147742-23-2]
$C_{39}H_{48}O_{17}$ 788.796
Constit. of the bark of *Fagara rhetza*, fruit of *Gardenia jasminoides* and Zhuyeqing tea, *Camellia sinensis*. Amorph. solid. $[\alpha]_D^{26}$ +7.5 (c, 1.1 in MeOH). $[\alpha]_D^{25}$ +32 (c, 0.5 in MeOH). λ_{max} 272 (ε 6300); 320 (ε 11500) (MeOH)

3,3′,5,5′-Tetra-Me ether, 9′-O-[β-D-apiofuranosyl-(1→2)-β-D-glucopyranoside]
[833486-03-6]
$C_{33}H_{46}O_{17}$ 714.714
Constit. of the roots of *Strobilanthes cusia*. Amorph. powder. $[\alpha]_D^{25}$ -2.5 (c, 0.12 in MeOH)

3,3′,5,5′-Tetra-Me ether, 9′-O-[β-D-xylopyranosyl-(1→6)-β-D-glucopyranoside]
Alangilignoside A
[193153-51-4]
$C_{33}H_{46}O_{17}$ 714.714
Constit. of *Alangium premnifolium*. Amorph. powder. $[\alpha]_D$ +23.4 (c, 0.9 in MeOH). λ_{max} 209 (log ε 4.75); 238 (log ε 4.08); 276 (log ε 3.55) (MeOH)

3,3′,5,5′-Tetra-Me ether, 9′-O-[α-D-galactopyranosyl-(1→6)-β-D-glucopyranoside]
[786712-59-2]
$C_{34}H_{48}O_{18}$ 744.74
Constit. of the aerial parts of *Acanthus ilicifolius*. Amorph. powder. $[\alpha]_D^{25}$ +36.5 (c, 0.8 in MeOH)

3,3′,5,5′-Tetra-Me ether, 9′-O-[β-D-glucopyranosyl-(1→3)-β-D-glucopyranoside]
Pteleifoside G
[1415247-15-2]
$C_{34}H_{48}O_{18}$ 744.74
Constit. of the roots and rhizomes of *Melicope pteleifolia*. Amorph. powder. $[\alpha]_D^{25}$ +50.5 (c, 0.05 in MeOH)

3,3′,5,5′-Tetra-Me ether, di-Ac
Gaultherin B
[111515-96-9]
$C_{26}H_{32}O_{10}$ 504.532
Constit. of *Gaultheria yunnanensis*. Powder (Me_2CO aq.). Mp 120 - 121° (99-100°,88-89°). $[\alpha]_D^{23}$ +40 (c, 0.05 in MeOH). λ_{max} 218 (log ε 2.34); 274 (log ε 0.45) (MeOH)

3,3′,5,5′-Tetra-Me ether, 9-O-(4-hydroxy-3,5-dimethoxybenzoyl), 9′-O-[4-hydroxy-3,5-dimethoxybenzoyl-(→2)-β-D-glucopyranoside]
Strychnoside
[845738-88-7]
$C_{46}H_{54}O_{21}$ 942.917
Constit. of the stems of *Strychnos vanprukii*. Amorph. powder. $[\alpha]_D^{25}$ -8.4 (c, 0.1 in MeOH). λ_{max} 206; 277 (MeOH)

3,3′,5,5′-Tetra-Me ether, 9′-O-(4-hydroxy-3,5-dimethoxybenzoyl), 9-O-β-D-glucopyranoside
Vanprukoside
[845738-87-6]
$C_{37}H_{46}O_{17}$ 762.758
Constit. of the stems of *Strychnos vanprukii*. Amorph. powder. $[\alpha]_D^{25}$ -3.2 (c, 0.1 in MeOH). λ_{max} 208; 277 (MeOH)

3,3′,5,5′-Tetra-Me ether, 9-O-(3,4,5-trihydroxybenzoyl), 9′-O-β-D-xylopyranoside
9-O-(3,4,5-Trihydroxybenzoyl) lyoniside. ***9-Galloyllyoniside***
[1567408-69-8]

$C_{34}H_{40}O_{16}$ 704.678
Constit. of *Quercus mongolica* var. *crispula*. Amorph. yellow powder. $[\alpha]^{27}_{D}$ +15.3 (c, 0.12 in MeOH). λ_{max} 277 (ε 14500) (MeOH)

3,3′,5,5′-Tetra-Me ether, 9′-O-(3,4-dihydroxy-E-cinnamoyl)
9′-O-trans-Caffeoyllyoniresinol.
Rhusemialin A
[946852-22-8]
$C_{31}H_{34}O_{11}$ 582.602
Constit. of the roots of *Rhus semialata*. Powder. $[\alpha]^{27}_{D}$ -1.6 (c, 3.05 in Me_2CO). λ_{max} 216 (log ε 4.3); 243 (sh); 287 (sh); 302 (sh); 330 (log ε 4.19) (MeOH)

3,3′,5,5′-Tetra-Me ether, 9-O-(4-hydroxy-3-methoxy-E-cinnamoyl)
9-O-Feruloyl-(+)-lyoniresinol
[1290544-76-1]
$C_{32}H_{36}O_{11}$ 596.629
Constit. of the heartwood of *Cinnamomum osmophloeum*. Yellow solid. $[\alpha]^{22}_{D}$ +10 (c, 0.3 in MeOH). λ_{max} 196 (log ε 4.8); 206 (log ε 4.03); 235 (log ε 4.55); 285 (log ε 4.28); 325 (log ε 4.36) (MeOH)

3,3′,5,5′-Tetra-Me ether, 9,9′-bis-O-(4-hydroxy-3-methoxy-E-cinnamoyl)
9,9′-Di-O-feruloyl-(+)-lyoniresinol
[1290544-77-2]
$C_{42}H_{44}O_{14}$ 772.8
Constit. of the heartwood of *Cinnamomum osmophloeum*. Cytotoxic. Yellow solid. $[\alpha]^{22}_{D}$ +63.3 (c, 0.15 in MeOH). λ_{max} 194 (log ε 4.98); 206 (log ε 5.23); 235 (log ε 4.81); 285 (log ε 4.58); 325 (log ε 3.75) (MeOH)

3,3′,4′,5,5′-Penta-Me ether
2,7′-Cyclo-3,3′,4′,5,5′-pentamethoxy-8,8′-lignan-4,9,9′-triol.
4,9,9′-Trihydroxy-3,3′,4′,9,9′-pentamethoxy-2,7′-cyclolignan

3,3′,4′,5,5′-Penta-Me ether, 4-O-β-D-glucopyranoside
[374928-36-6]
$C_{29}H_{40}O_{13}$ 596.626
Constit. of *Antidesma membranaceum*. λ_{max} 205 (log ε 4.63); 273 (log ε 3.18) (MeOH)

3,3′,4,4′,5,5′-Hexa-Me ether
[7333-11-1]
Cryst.. Mp 163 - 164°. [α] D +45 ($CHCl_3$)

9,9′-Isopropylidene, 3,3′,5,5′-tetra-Me ether
(+)-Lyoniresinol 9,9′-acetonide.
Brugunin A
[1000779-06-5]
$C_{25}H_{32}O_8$ 460.523
Constit. of the mangrove *Bruguiera gymnorrhiza*. Amorph. solid. $[\alpha]^{20}_{D}$ +7.6 (c, 0.19 in $CHCl_3$). λ_{max} 279 (log ε 3.52) (MeOH). [acid] λ_{max} 276 (log ε 3.55) (MeOH/HCl). [base] λ_{max} 251 (log ε 3.74) (MeOH/NaOH)

(7′S,8R,8′S)-form

3,3′,5,5′-Tetra-Me ether, 9′-O-β-D-glucopyranoside
[873077-48-6]
$C_{28}H_{38}O_{13}$ 582.599
Constit. of the stems of *Annona squamosa* and *Carissa carandas*

3,3′,5,5′-Tetra-Me ether, 9′-O-[β-D-xylopyranosyl-(1→6)-β-D-glucopyranoside]
$C_{33}H_{46}O_{17}$ 714.714
Constit. of the branches of *Alangium kurzii*. Pale yellow oil. $[\alpha]^{20}_{D}$ -6.35 (c, 0.25 in MeOH)

(7′S,8S,8′R)-form

3,3′,5,5′-Tetra-Me ether
Epilyoniresinol
[73745-49-0]
Isol. from *Grevillea robusta*. Pinkish prisms (EtOAc/C_6H_6). Mp 198 - 200°. [α] D -28.9 (MeOH)

3,3′,5,5′-Tetra-Me ether, 9′-O-β-D-glucopyranoside
[73706-67-9]
$C_{28}H_{38}O_{13}$ 582.599
Isol. from *Grevillea robusta*

3′,4′,5,5′-Tetra-Me ether
2,7′-Cyclo-3′,4′,5,5′-tetramethoxy-8,8′-lignan-3,4,9,9′-tetrol.
3,4,9,9′-Tetrahydroxy-3′,4′,5,5′-tetramethoxy-2,7′-cyclolignan.
Lingueresinol
[129502-58-5]
$C_{22}H_{28}O_8$ 420.458
Isol. from stem bark of *Persea lingue*. Cryst. (MeOH). Mp 179 - 183°. [α] D +35.9 (c, 1.255 in MeOH). Tentative struct.

(7′S,8S,8′S)-form

3,3′,5,5′-Tetra-Me ether, 9′-O-β-D-xylopyranoside
Linderanoside B
[1809032-79-8]
$C_{27}H_{36}O_{12}$ 552.573
Constit. of trunk of *Lindera glauca*. Exhibits moderate to weak cytotoxicity against A498 cells. Amorph. gum. $[\alpha]^{25}_{D}$ +108 (c, 0.35 in MeOH). λ_{max} 225 (log ε 4.0); 284 (log ε 3.1) (MeOH)

(7′RS,8RS,8′RS)-form

3,3′,5,5′-Tetra-Me ether
Polystachyol
[936020-33-6]
$C_{22}H_{28}O_8$ 420.458
Constit. of *Aphanamixis polystachya*, *Euonymus acanthocarpus* and *Sorbus lanata*. Amorph. solid

(7′RS,8SR,8′SR)-form

3,3′,5,5′-Tetra-Me ether
(±)-Lyoniresinol
[23212-64-8]
$C_{22}H_{28}O_8$ 420.458
Isol. from *Prunus jamasakura*, *Quercus petraea* and *Ulmus* spp.. Cryst. ($CHCl_3$). Mp 193 - 194°

3,3′,5,5′-Tetra-Me ether, di-Ac
Cryst.. Mp 103°

3,3′,5,5′-Tetra-Me ether, 9-O-(3,4,5-trihydroxybenzoyl)
9-Galloyllyoniresinol.
Quercoresinol
[1622862-36-5]
$C_{29}H_{32}O_{12}$ 572.563
Constit. of the wood of *Quercus petraea*. $[\alpha]^{20}_{D}$ +5 (c, 1.2 in MeOH). Partial racemate

3,3′,4,4′,5,5′-Hexa-Me ether
[23212-66-0]
Cryst. (Et_2O). Mp 162 - 164°.

Freudenberg, K. *et al.*, *Tetrahedron Lett.*, no.17, 1959, **1**, 19-22 (*Dimethoxyisolariciresinol, 3,3′,5,5′-tetra-Me ether 9-xyloside*).
Weinges, K. *Chem. Ber.*, 1961, **94**, 2522-2533 (*Dimethoxyisolariciresinol xyloside*).
Yasue, M. *et al.*, *Yakugaku Zasshi*, 1961, **81**, 529-532 (*Lyoniside*).
Arya, V.P. *et al.*, *Acta Chem. Scand.*, 1962, **16**, 518-520 (*Lyoniside, struct, props*).
Narnsimbachari, N. *et al.*, *Can. J. Chem.*, 1962, **40**, 1118-1123 (*Lyoniside, polymorphs*).
Klyne, W. *et al.*, *J. Chem. Soc. C*, 1966, 893-896 (*ord*).
Hostettler, F.D. *et al.*, *Tetrahedron*, 1969, **25**, 2325-2337 (*Ulmus thomasii constit*).
Rowe, J.W. *et al.*, *Phytochemistry*, 1972, **11**, 2513-2517 (*Ulmus constits*).
Sakakibura, J. *et al.*, *Yakugaku Zasshi*, 1974, **94**, 1377-1388 (*Lyoniside, abs config*).
Ogawa, M. *et al.*, *Chem. Pharm. Bull.*, 1976, **24**, 2102-2105 (*Nudiposide*).
Mehta, M. *et al.*, *Indian J. Chem., Sect. B*, 1976, **14**, 127-128 (*Lyoniresinol dimethyl ether, synth*).
Vecchietti, V. *et al.*, *Phytochemistry*, 1979, **18**, 1847-1849 (*Lyoniresinol, Lyoniside, cmr*).

Chandel, R.S. *et al.*, *Indian J. Chem., Sect. B*, 1980, **19**, 279-282 (*Pygeoside*).
Varma, M. *et al.*, *Z. Naturforsch., C*, 1980, **35**, 344-345 (*Epilyoniresinol*).
Hulbert, P.B. *et al.*, *J. Chem. Res., Synop.*, 1981, 27 (*cd, uv*).
Miyamura, M. *et al.*, *Phytochemistry*, 1983, **22**, 215-218 (*Cinnamomum cassia constit*).
Dada, G. *et al.*, *J. Nat. Prod.*, 1989, **52**, 1327-1330 (*Quercus petraea glycosides*).
Yoshinari, K. *et al.*, *Phytochemistry*, 1990, **29**, 1675-1678 ((±)-*Lyoniresinol*).
Sepulveda-Boza, S. *et al.*, *Phytochemistry*, 1990, **29**, 2357-2358 (*Lingueresinol*).
Higuchi, H. *et al.*, *Chem. Pharm. Bull.*, 1992, **40**, 534-535 (*Albizzia julibrissin glucosides*).
Shibuya, H. *et al.*, *Chem. Pharm. Bull.*, 1992, **40**, 2639-2646 (*Hazaleanin B*).
Achenbach, R.S. *et al.*, *Planta Med.*, 1992, **58**, 270-272 (*Stemmadenia minima glucoside*).
Ohashi, K. *et al.*, *Chem. Pharm. Bull.*, 1994, **42**, 1924-1926 (*Aegle marmelos constits*).
Achenbach, H. *et al.*, *Phytochemistry*, 1997, **45**, 325-335 (*Tabernaemontana cymosa constits*).
Yuasa, K. *et al.*, *Phytochemistry*, 1997, **45**, 611-615 (*Alangilignoside A*).
Cortez, D.A.G. *et al.*, *Nat. Prod. Lett.*, 1998, **11**, 255-262 ((-)-*Lyoniresinol rhamnoside*).
Zhang, Z. *et al.*, *Phytochemistry*, 1999, **51**, 469-472 (*Gaultherin B*).
Zhang, Z. *et al.*, *Yaoxue Xuebao*, 1999, **34**, 128-131 (*Gaultheroside A*).
Buske, A. *et al.*, *Eur. J. Org. Chem.*, 2001, 3537-3543 (*Antidesma 4-glucosides*).
Kanchanapoom, T. *et al.*, *Phytochemistry*, 2001, **56**, 369-372 ; 57, 1245 - 1248 (*Acanfolioside, Fernandoside*).
Xiao, K. *et al.*, *Chem. Pharm. Bull.*, 2002, **50**, 605-608 (*Lyoniresinol 9′-sulfate*).
Tanaka, T. *et al.*, *Chem. Pharm. Bull.*, 2004, **52**, 1242-1245 ((+)-*Lyoniresinol 3-apiosylglucoside*).
Thongphasuk, P. *et al.*, *Fitoterapia*, 2004, **75**, 623-628 (*Strychnoside, Vanprukoside*).
De Marino, S. *et al.*, *J. Agric. Food Chem.*, 2004, **52**, 7525-7531 (*Lyoniside*).
Wu, J. *et al.*, *Z. Naturforsch., B*, 2004, **59**, 341-344 (*Acanthus ilicifolius glycosides*).
Aziz-ur-Rehman, *et al.*, *Chem. Pharm. Bull.*, 2005, **53**, 263-266 (*Lyoniresinol 4,4′-diglucoside*).
Takemoto, M. *et al.*, *Chem. Pharm. Bull.*, 2006, **54**, 226-229 (*Lyoniresinol, synth*).
Sadhu, S.K. *et al.*, *J. Nat. Med.*, 2006, **60**, 258-260 (*Polystachyol*).
Azhar-ul-Haq, *et al.*, *Pol. J. Chem.*, 2006, **80**, 259-263 (*Amaranthoside*).
Su, D.-M. *et al.*, *Chem. Biodiversity*, 2007, **4**, 2852-2862 (*Capparis tenera glucosides*).
Ouyang, M.-A. *et al.*, *Chem. Pharm. Bull.*, 2007, **55**, 804-807 (*Rhusemialins A,B,C*).
Han, L. *et al.*, *J. Asian Nat. Prod. Res.*, 2007, **9**, 327-331 (*Brugunin A*).
Li, N. *et al.*, *J. Nat. Prod.*, 2007, **70**, 544-548 (*Peperomia duclouxii constit*).
Rahman, M.D.A. *et al.*, *J. Wood Sci.*, 2007, **53**, 161-167 (*Lyoniresinol, Lyoniside*).
Yang, X.-W. *et al.*, *Chem. Biodiversity*, 2009, **6**, 540-550 (*Tarennanosides A,B,H*).
Wangteeraprasert, R. *et al.*, *Helv. Chim. Acta*, 2009, **92**, 1217-1223 (*Carissa carandas glucosides*).
Khan, S. *et al.*, *J. Asian Nat. Prod. Res.*, 2009, **11**, 44-48 (*Racemiside*).
Ono, M. *et al.*, *J. Nat. Med.*, 2009, **63**, 86-90 (*Callicarpa japonica constit*).
Chen, T.-H. *et al.*, *Planta Med.*, 2010, **76**, 613-619 (*9-Feruloyllyoniresinol, 9,9′-Diferuloyllyoniresinol*).
Jeong, E.J. *et al.*, *Bioorg. Med. Chem. Lett.*, 2011, **21**, 2283-2286 (*7′-Epilyoniresinol, activity*).
Szakiel, A. *Phytochem. Lett.*, 2011, **4**, 138-143 (*Lyoniside, bilberry, activity*).
Zhu, J.X. *et al.*, *Arch. Pharmacal Res.*, 2012, **35**, 1739-1747 (*Polystachyol*).
Zhang, Y. *et al.*, *Carbohydr. Res.*, 2012, **361**, 114-119 (*Pteleifoside G*).
Yu, Y. *et al.*, *Fitoterapia*, 2012, **83**, 563-567 (*Hazaleanin B*).
Sun, J. *et al.*, *J. Agric. Food Chem.*, 2013, **61**, 4556-4562 (*Bambusa tuldoides constits*).
Omar, M. *et al.*, *Phytochem. Lett.*, 2013, **6**, 486-490 (*Quercus mongolica constit*).
Gao, Y.-P. *et al.*, *J. Asian Nat. Prod. Res.*, 2014, **16**, 724-729 (*Gaultheria yunnanensis constits*).
Queiroz, M.M.F. *et al.*, *J. Nat. Prod.*, 2014, **28**, 1052-1057 (*9-Dihydroxybenzoylnudiposide*).
Xu, Z.-S. *et al.*, *Nat. Prod. Res.*, 2014, **77**, 650-656 (*Ficus ischnopoda constit*).
Gao, H.-Y. *et al.*, *Phytochem. Lett.*, 2014, **7**, 150-155 (*Zhuyeqing liquor constit*).
Suh, W.S. *et al.*, *Helv. Chim. Acta*, 2015, **98**, 1087-1094 (*Linderanosides A,B, activity, cd*).
Panyadee, A. *et al.*, *Phytochem. Lett.*, 2015, **11**, 215-219 (*Racemiside, abs config*).
Marchal, A. *et al.*, *Tetrahedron*, 2015, **71**, 3148-3156 (*Quercoresinol, Lyoniside, Nudiposide, 9-Galloylnudiposide*).
Sindt, L. *et al.*, *J. Nat. Prod.*, 2016, **79**, 2432-2438 (*Quercoresinosides A,B, use*).
Ren, J. *et al.*, *Fitoterapia*, 2017, **121**, 152-158 (*Alangium kurzii constits*).
Liu, J. *et al.*, *Phytochem. Lett.*, 2017, **22**, 247-250 ((+)-*Lyoniresinol 4-arabinoside*).
Park, K.J. *et al.*, *Phytochemistry*, 2019, **167**, 112085:1-8 (*Betula schmidtii constit*)

2,7′-Cyclo-8,8′-lignan-3,3′,4,4′,5,6,9,9′-octol C247

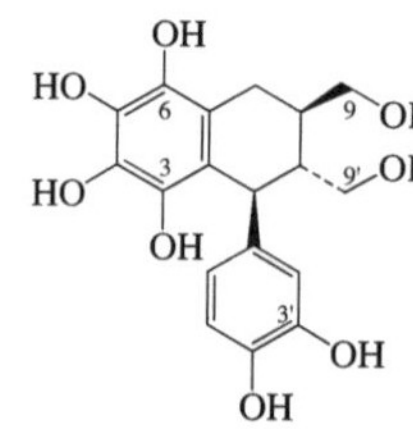

3,3′,4,4′,5,6,9,9′-Octahydroxy-2,7′-cyclolignan

$C_{18}H_{20}O_8$ 364.35

(7′S,8R,8′R)-form

3,4-Methylene, 3′,4′,5,6,9′-penta-Me ether

2,7′-Cyclo-3′,4′,5,6,9′-pentamethoxy-3,4-methylenedioxy-8,8′-lignan-9-ol.
9-Hydroxy-3′,4′,5,6,9′-pentamethoxy-3,4-methylenedioxy-2,7′-cyclolignan

$C_{24}H_{30}O_8$ 446.496

3,4-Methylene, 3′,4′,5,6,9′-penta-Me ether, 9-O-[β-D-apiofuranosyl-(1→6)-β-D-glucopyranoside]

[884856-01-3]

$C_{35}H_{48}O_{17}$ 740.752

Constit. of the roots of *Rodgersia sambucifolia*. Amorph. powder. Mp 234 - 236°. $[\alpha]_D^{20}$ -37.6 (c, 0.41 in MeOH). Shown with the opposite abs. config. to Hypophyllanthin, in C229. This is prob. erroneous. λ_{max} 223; 267; 282 (no solvent reported).

Hu, H.-B. *et al.*, *Indian J. Chem., Sect. B*, 2005, **44**, 2399-2403 (*Rodgersia glycoside*)

2,7′-Cyclo-8,8′-lignan-3,3′,4,4′,5,8,9,9′-octol C248

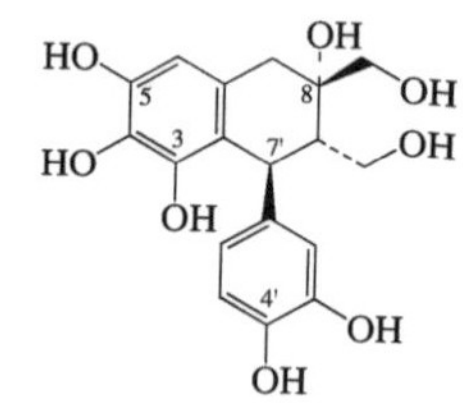

3,3′,4,4′,5,8,9,9′-Octahydroxy-2,7′-cyclolignan

$C_{18}H_{20}O_8$ 364.35

(7′S,8S,8′S)-form

3,3′,5-Tri-Me ether

2,7′-Cyclo-3,3′,5-trimethoxy-8,8′-lignan-4,4′,8,9,9′-pentol.
4,4′,8,9,9′-Pentahydroxy-3,3′,5-trimethoxy-2,7′-cyclolignan

3,3′,5-Tri-Me ether, 4-O-β-D-glucopyranoside

[1355468-10-8]

$C_{27}H_{36}O_{13}$ 568.572

Constit. of the roots of *Rhodiola crenulata*. Amorph. powder. $[\alpha]_D^{20}$ -5.4 (c, 0.03 in MeOH). λ_{max} 237; 282 (MeOH).

Yang, Y.-N. *et al.*, *J. Agric. Food Chem.*, 2012, **60**, 964-972 (*Rhodiola crenulata glucoside*)

2,7′-Cyclo-8,8′-lignan-3′,4,4′,5,5′,7,9,9′-octol C249

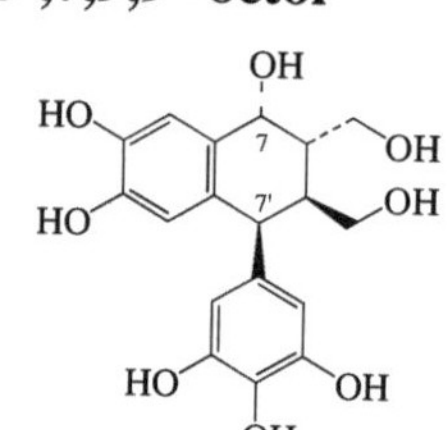

3′,4,4′,5,5′,7,9,9′-Octahydroxy-2,7′-cyclolignan

$C_{18}H_{20}O_8$ 364.350

(7R,7′S*,8S*,8′S*)-form*

9,9′-Dicarboxylic acid, 3′,4,5,5′-tetra-Me ether

2,7′-Cyclo-4′,7-dihydroxy-3′,4,5,5′-tetramethoxy-8,8′-lignan-9,9′-dioic acid. 4′,7-Dihydroxy-3′,4,5,5′-tetramethoxy-2,7′-cyclolignan-9,9′-dioic acid

9,9′-Dicarboxylic acid, 3′,4,5,5′-tetra-Me ether, bis[(4-hydroxyphenyl) ethyl]amide

Foveolatamide

$C_{38}H_{42}N_2O_{10}$ 686.758

Alkaloid from the stems of *Ficus faveolata*. Exhibits antibacterial activity. Yellow powder. $[\alpha]_D^{20}$ +2.8 (c, 0.06 in MeOH). Possesses (7*R**,7′*S**,8*R**,8′*S**)-config. due to change in Cahn-Ingold-Prelog priorities. λ_{max} 277 (log ε 7.0); 325 (log ε 1.2) (MeOH).

Meerungrueang, W. *et al., Nat. Prod. Commun.*, 2016, **11**, 91-94 (*Foveolatamide, activity*)

2,7′-Cyclo-8,8′-lignan-3′,4,4′,5,8,8′,9,9′-octol C250

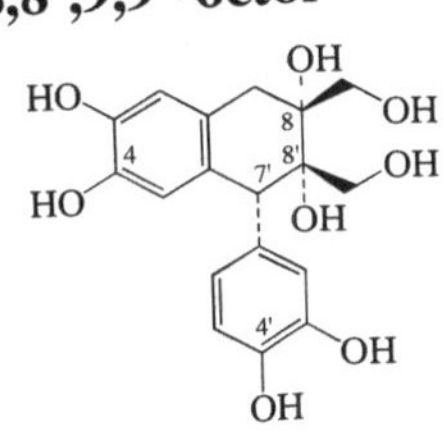

(7′R,8R,8′S)-form

3′,4,4′,5,8,8′,9,9′-Octahydroxy-2,7′-cyclolignan

$C_{18}H_{20}O_8$ 364.35

(7′R,8R,8′S)-form

3′,5-Di-Me ether

2,7′-Cyclo-3′,5-dimethoxy-8,8′-lignan-4,4′,8,8′,9,9′-hexol. 4,4′,8,8′,9,9′-Hexahydroxy-3′,5-dimethoxy-2,7′-cyclolignan. ***Dolomiaeasin B***

[1421227-94-2]

$C_{20}H_{24}O_8$ 392.404

Constit. of the roots of *Dolomiaea souliei*. $[\alpha]_D^{20}$ -16.3 (c, 0.24 in MeOH). λ_{max} 210 (log ε 4.8); 284 (log ε 3.99) (MeOH)

(7′S,8R,8′R)-form

3′,5-Di-Me ether

Dolomiaeasin A

[1421227-93-1]

$C_{20}H_{24}O_8$ 392.404

Constit. of the roots of *Dolomiaea souliei*. $[\alpha]_D^{20}$ -4 (c, 0.22 in MeOH). λ_{max} 207 (log ε 4.46); 283 (log ε 3.6) (MeOH).

Wei, H. *et al., Molecules*, 2012, **17**, 5544-5549 (*Dolomiaeasins A,B*)

2,7′-Cyclo-8,8′-lignan-3,3′,4,4′,5-pentol C251

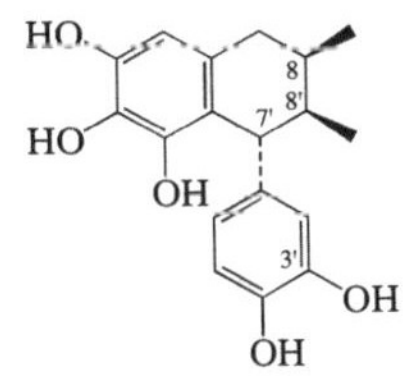

8-(3,4-Dihydroxyphenyl)-5,6,7,8-tetrahydro-6,7-dimethyl-1,2,3-naphthalenetriol. 3,3′,4,4′,5-Pentahydroxy-2,7′-cyclolignan

$C_{18}H_{20}O_5$ 316.353

(7′R,8R,8′R)-form

3,3′,5-Tri-Me ether

2,7′-Cyclo-3,3′,5-trimethoxy-8,8′-lignan-4,4′-diol. 4,4′-Dihydroxy-3,3′,5-trimethoxy-2,7′-cyclolignan

[1196997-41-7]

$C_{21}H_{26}O_5$ 358.434

Constit. of the bark of *Machilus wangchiana*. Moderate inhibitor of PAF-induced β-glucuronidase release. Amorph. powder. $[\alpha]_D^{20}$ -13 (c, 0.03 in $CHCl_3$). λ_{max} 203 (log ε 4.72); 231 (log ε 4.15); 281 (log ε 3.71) (MeOH).

Cheng, W. *et al., J. Nat. Prod.*, 2009, **72**, 2145-2152 (*Machilus wangchiana constit*).

Chumsri, N. *et al., Tetrahedron Lett.*, 2021, **66**, 152827:1-5 (*synth, abs config*)

2,7′-Cyclo-8,8′-lignan-3,3′,4,4′,6-pentol C252

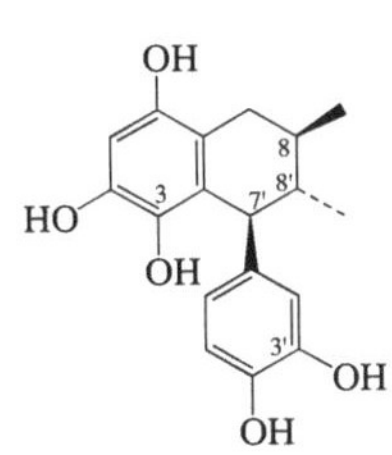

Absolute Configuration

8-(3,4-Dihydroxyphenyl)-5,6,7,8-tetrahydro-6,7-dimethyl-1,2,4-naphthalenetriol. 3,3′,4,4′,6-Pentahydroxy-2,7′-cyclolignan

$C_{18}H_{20}O_5$ 316.353

(7′S,8R,8′S)-form

3,4:3′,4′-Bis(methylene) ether

2,7′-Cyclo-3,4:3′,4′-bis(methylenedioxy)-8,8′-lignan-6-ol. 6-Hydroxy-3,4:3′,4′-bis(methylenedioxy)-2,7′-cyclolignan. ***Saurunarin***

[1532522-59-0]

$C_{20}H_{20}O_5$ 340.375

Constit. of *Saururus chinensis*. Amorph. brown powder. $[\alpha]_D^{62}$ -9.0 (c, 0.5 in MeOH). λ_{max} 241 (log ε 3.0); 289 (log ε 2.9) (MeOH).

Tsai, W.-J. *et al., J. Nat. Prod.*, 2014, **77**, 125-131 (*Saurunarin*)

2,7′-Cyclo-8,8′-lignan-3,3′,4,4′,7-pentol C253

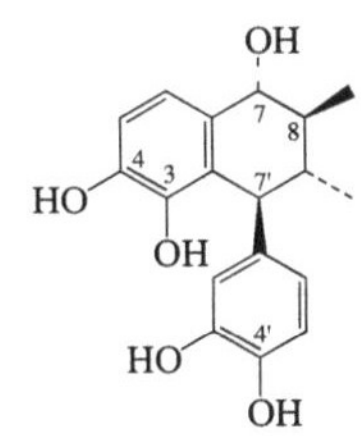

(7R,7′S,8S,8′R)-form

3,3′,4,4′,7-Pentahydroxy-2,7′-cyclolignan

$C_{18}H_{20}O_5$ 316.353

(7R,7′S,8S,8′R)-form

3,4:3′,4′-Bis(methylene) ether

2,7′-Cyclo-3,4:3′,4′-bis(methylenedioxy)-8,8′-lignan-7-ol. 7-Hydroxy-3,4:3′,4′-bis(methylenedioxy)-2,7′-cyclolignan. ***Oleiferin E****. Dihydrootobanone*

[125074-10-4]

$C_{20}H_{20}O_5$ 340.375

Constit. of the leaves of *Virola oleifera*. Needles (EtOAc). Mp 217 - 219°. $[\alpha]_D^{24}$ -6.4 (c, 0.23 in $CHCl_3$). Probable abs. config. based on the abs. config. of Otobanone. Shown in the ref. with the opposite abs. config.. λ_{max} 232 (log ε 3.9); 285 (log ε 3.7) (MeOH)

(7′R,8R,8′S)-form

7-Ketone, 3,4:3′,4′-bis(methylene) ether

2,7′-Cyclo-3,4:3′,4′-bis(methylenedioxy)-8,8′-lignan-7-one. 3,4:3′,4′-Bis(methylenedioxy)-7-oxo-2,7′-cyclolignan. ***Otobanone****. 7-Oxootobain*

[34426-79-4]

$C_{20}H_{18}O_5$ 338.359

Constit. of *Chamaecyparis obtusa* var. *formosana*, *Cordia exaltata*,

Myristica cagayanesis, *Myristica simarum* and *Myristica fragrans*. Cryst.. Mp 176°. [α] D -27.1 (c, 0.7 in $CHCl_3$)

(7′R,8S,8′S)-form

7-Ketone, 3,4:3′,4′-bis(methylene) ether

Isootobanone

[875392-13-5]

$C_{20}H_{18}O_5$ 338.359

Constit. of the heartwood of *Chamaecyparis obtusa* var. *formosana*. Amorph.. $[\alpha]_D^{26}$ -55.9 (c, 0.72 in $CHCl_3$). λ_{max} 235 (log ε 4.19); 285 (log ε 3.85); 312 (log ε 3.67) (MeOH).

Kuo, Y.-H. *et al.*, *Chem. Pharm. Bull.*, 1989, **37**, 2310-2312 (*Myristica cagayanesis constit*).
Fernandes, A.M.A. *et al.*, *Phytochemistry*, 1993, **32**, 1567-1572 (*Oleiferin E*).
Kuo, Y.-H. *et al.*, *Tetrahedron Lett.*, 2001, **42**, 6731-6735 (*Isootobanone, Otobanone, abs config*).
Nogueira, T.B. *et al.*, *Molecules*, 2013, **18**, 11086-11099 (*Cordia exaltata constit*)

2,7′-Cyclo-8,8′-lignan-3,3′,4,4′,7′-pentol C254

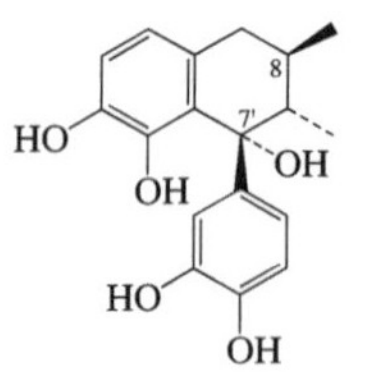

3,3′,4,4′,7′-Pentahydroxy-2,7′-cyclolignan

$C_{18}H_{20}O_5$ 316.353

(7′R,8R,8′S)-form

3,4:3′,4′-Bis(methylene) ether

2,7′-Cyclo-3,4:3′,4′-bis(methylenedioxy)-8,8′-lignan-7′-ol. 7′-Hydroxy-3,4:3′,4′-bis(methylenedioxy)-2,7′-cyclolignan.

Hydroxyotobain

[10240-17-2]

$C_{20}H_{20}O_5$ 340.375

Constit. of *Myristica otoba*, *Osteophloeum platyspermum* and *Virola cuspidata*. Prisms (CH_2Cl_2/petrol). Mp 116 - 117° (113-114°). [α] D -28 (c, 0.8 in $CHCl_3$). Abs. config. revised in 1982. λ_{max} 235 (ε 10500); 288 (ε 7980) (MeOH).

Gilchrist, T. *et al.*, *J. Chem. Soc.*, 1962, 1780-1786 (*struct*).
Wallace, R. *et al.*, *J. Chem. Soc.*, 1963, 1445-1449 (*Myristica otoba constit*).
Blair, G.E. *et al.*, *Phytochemistry*, 1969, **8**, 497-500 (*Virola cuspidata constit*).
Hulbert, P.M. *et al.*, *J. Chem. Res., Miniprint*, 1981, 401-449 (*uv*).
Lopes, L.M.X. *et al.*, *Phytochemistry*, 1982, **21**, 751-755 (*cd, abs config*)

2,7′-Cyclo-8,8′-lignan-3′,4,4′,5,5′-pentol C255

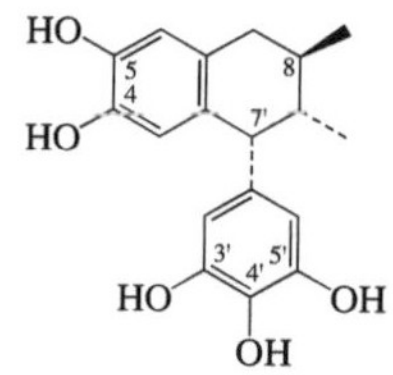

(7′R,8R,8′S)-form

5,6,7,8-Tetrahydro-5-(3,4,5-trihydroxyphenyl)-6,7-dimethyl-2,3-naphthalenediol. 3′,4,4′,5,5′-Pentahydroxy-2,7′-cyclolignan

$C_{18}H_{20}O_5$ 316.353

Lignan numbering shown

(7′R,8R,8′S)-form

4,5-Methylene, 3′,4′,5′-tri-Me ether

2,7′-Cyclo-3′,4′,5′-trimethoxy-4,5-methylenedioxy-8,8′-lignan. 3′,4′,5′-Trimethoxy-4,5-methylenedioxy-2,7′-cyclolignan.

Austrobailignan 4

[55924-17-9]

$C_{22}H_{26}O_5$ 370.445

Constit. of *Austrobaileya scandens*. Needles (MeOH). Mp 118 - 120°. $[\alpha]_D^{20}$ -124 (c, 2.4 in $CHCl_3$)

3′,4′:4,5-Bis(methylene), 5′-Me ether

2,7′-Cyclo-3′-methoxy-4,5:4′,5′-bis(methylenedioxy)-8,8′-lignan. 3′-Methoxy-4,5:4′,5′-bis(methylenedioxy)-2,7′-cyclolignan.

Austrobailignan 3

[55890-22-7]

$C_{21}H_{22}O_5$ 354.402

Constit. of *Austrobaileya scandens*. Rods (MeOH or petrol). Mp 87 - 90°. $[\alpha]_D^{20}$ -120 (c, 2.5 in $CHCl_3$)

(7′S,8R,8′S)-form

3′,5-Di-Me ether, 5′-O-β-D-glucopyranoside

$C_{26}H_{34}O_{10}$ 506.548

Constit. of roots and stems of *Kadsura coccinea*. Powder

3′,5,5′-Tri-Me ether

2,7′-Cyclo-3′,5,5′-trimethoxy-8,8′-lignan-4,4′-diol. 4,4′-Dihydroxy-3′,5,5′-trimethoxy-2,7′-cyclolignan. ***Kadsuralignan C***

[913183-89-8]

$C_{21}H_{26}O_5$ 358.434

Constit. of the rhizomes of *Kadsura coccinea* and leaves and stems of *Kadsura longipedunculata*. Moderate inhibitor of NO bioprodn.. Needles (hexane). Mp 209 - 210°. λ_{max} 215 (log ε 4); 240 (sh); 288 (log ε 3.5) (MeOH)

(7′S,8S,8′S)-form

3′,5-Di-Me ether

2,7′-Cyclo-5,5′-dimethoxy-8,8′-lignan-3′,4,4′-triol. 3′,4,4′-Trihydroxy-5,5′-dimethoxy-2,7′-cyclolignan

[1310453-60-1]

$C_{20}H_{24}O_5$ 344.407

Constit. of the bark of *Machilus robusta*. Amorph. powder. $[\alpha]_D^{20}$ +66.6 (c, 0.12 in $CHCl_3$). λ_{max} 204 (log ε 4.86); 237 (sh); 284 (log ε 3.63) (MeOH)

3′,5,5′-Tri-Me ether

[1310453-62-3]

$C_{21}H_{26}O_5$ 358.434

Constit. of the bark of *Machilus robusta*. Inhibitor of galactosamine-induced hepatocyte damage. Amorph. powder. $[\alpha]_D^{20}$ +97 (c, 0.1 in $CHCl_3$). λ_{max} 205 (log ε 4.35); 283 (log ε 3.28) (MeOH).

Murphy, S.T. *et al.*, *Aust. J. Chem.*, 1975, **28**, 81-90 (*Austrobailignans 3,4*).
Li, H.-R. *et al.*, *Chem. Pharm. Bull.*, 2006, **54**, 1022-1025 (*Kadsuralignan C*).
Pu, J-X. *et al.*, *Chem. Pharm. Bull.*, 2008, **56**, 1143-1146 (*Kadsura longipedunculata constit*).
Li, Y. *et al.*, *J. Nat. Prod.*, 2011, **74**, 1444-1452 (*Machilus robusta constits*).
Yeon, J.-H. *et al.*, *Chin. J. Nat. Med.*, 2014, **12**, 782-285 (*Kadsura coccinea constit*).
Chumsri, N. *et al.*, *Tetrahedron Lett.*, 2021, **66**, 152827:1-5 (*synth, abs config*)

2,7′-Cyclo-8,8′-lignan-3′,4,4′,5,7-pentol C256

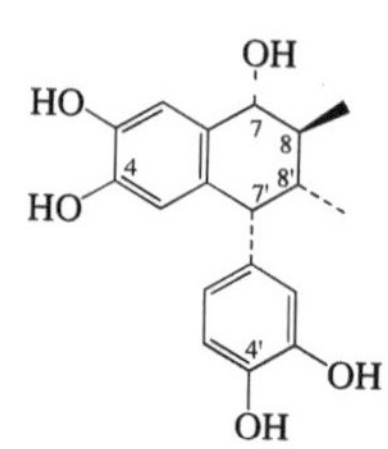

(7R,7′R,8S,8′R)-form

3′,4,4′,5,7-Pentahydroxy-2,7′-cyclolignan

$C_{18}H_{20}O_5$ 316.353

(7R,7′R,8S,8′R)-form

3′,4′-Methylene, 4,5-di-Me ether

2,7′-Cyclo-4,5-dimethoxy-3′,4′-methylenedioxy-8,8′-lignan-7-ol. 7-Hydroxy-4,5-dimethoxy-3′,4′-methylenedioxy-2,7′-cyclolignan. ***Aristotetraol***

[113952-99-1]

$C_{21}H_{24}O_5$ 356.418

Constit. of the roots of *Aristolochia chilensis*. Amorph.. [α] D -149 (c, 1.5 in $CHCl_3$). λ_{max} 233 (log ε 4.05); 285 (log ε 3.80) (MeOH)

7-Ketone

2,7′-Cyclo-3′,4,4′,5-tetrahydroxy-8,8′-lignan-7-one. 3′,4,4′,5-Tetrahydroxy-7-oxo-2,7′-cyclolignan

7-Ketone, 3′,4′,5-tri-Me ether

2,7′-Cyclo-4-hydroxy-3′,4′,5-trimethoxy-8,8′-lignan-7-one. 4-Hydroxy-3′,4′,5-trimethoxy-7-oxo-2,7′-cyclolignan. ***8,8′-Epiholostylone****. 8,8′-Diepiholostylone*

[694489-73-1]

$C_{21}H_{24}O_5$ 356.418

Constit. of the roots of *Holostylis reniformis*. Amorph. yellow solid. $[\alpha]^{25}_{D}$-171.6 (c, 0.32 in $CHCl_3$). λ_{max} 213 (log ε 4.3); 234 (log ε 4.3); 279 (log ε 4.1); 314 (log ε 3.8) (MeOH)

7-Ketone, 4,4′,5-tri-Me ether

2,7′-Cyclo-3′-hydroxy-4,4′,5-trimethoxy-8,8′-lignan-7-one. 3′-Hydroxy-4,4′,5-trimethoxy-7-oxo-2,7′-cyclolignan. ***Aristochilone***

[113952-98-0]

$C_{21}H_{24}O_5$ 356.418

Constit. of *Aristolochia chilensis*. Amorph.. [α] D -134 (c, 1.62 in $CHCl_3$). λ_{max} 233 (log ε 4.32); 278 (log ε 4.08); 314 (log ε 3.77) (MeOH)

7-Ketone, tetra-Me ether

2,7′-Cyclo-3′,4,4′,5-tetramethoxy-8,8′-lignan-7-one. 3′,4,4′,5-Tetramethoxy-7-oxo-2,7′-cyclolignan. ***Aristoligone***

[114029-72-0]

$C_{22}H_{26}O_5$ 370.445

Constit. of *Aristolochia chilensis* and *Holostylis reniformis*. Amorph. yellow solid. $[\alpha]^{25}_{D}$-206.4 (c, 0.72 in $CHCl_3$). λ_{max} 232 (log ε 4.35); 276 (log ε 4.06); 313 (log ε 3.74) (MeOH)

7-Ketone, 3′,4′-methylene, 4,5-di-Me ether

2,7′-Cyclo-4,5-dimethoxy-3′,4′-methylenedioxy-8,8′-lignan-7-one. 4,5-Dimethoxy-3′,4′-methylenedioxy-7-oxo-2,7′-cyclolignan. ***Aristotetralone***

[111188-75-1]

$C_{21}H_{22}O_5$ 354.402

Constit. of *Aristolochia chilensis* and *Holostylis reniformis*. Amorph. yellow solid. [α] D -164.3 (c, 1.54 in $CHCl_3$). λ_{max} 233 (log ε 4.33); 280 (log ε 4) (MeOH)

7-Ketone, 3′,4′:4,5-bis(methylene) ether

2,7′-Cyclo-3′,4′:4,5-bis(methylenedioxy)-8,8′-lignan-7-one. 3′,4′:4,5-Bis(methylenedioxy)-7-oxo-2,7′-cyclolignan. ***Cagayanone A***

[887501-29-3]

$C_{20}H_{18}O_5$ 338.359

Constit. of *Holostylis reniformis* and *Litsea pedunculata*. Amorph. yellow solid. $[\alpha]^{25}_{D}$-27.4 (c, 0.78 in $CHCl_3$). λ_{max} 235 (log ε 3.9); 278 (log ε 3.5); 320 (log ε 3.4) (MeOH)

(7R,7′R,8S,8′S)-form

3′,4′,5-Tri-Me ether

2,7′-Cyclo-3′,4′,5-trimethoxy-8,8′-lignan-4,7-diol. 4,7-Dihydroxy-3′,4′,5-trimethoxy-2,7′-cyclolignan

$C_{21}H_{26}O_5$ 358.434

Constit. of roots of *Holostylis reniformis*. Exhibits moderate antiplasmodial activity against *Plasmodium falciparum*. Yellow oil. $[\alpha]^{25}_{D}$-38 (c, 0.7 in $CHCl_3$)

3′,4,4′,5-Tetra-Me ether

2,7′-Cyclo-3′,4,4′,5-tetramethoxy-8,8′-lignan-7-ol. 7-Hydroxy-3′,4,4′,5-tetramethoxy-2,7′-cyclolignan. ***Holostylol A***

[887501-31-7]

$C_{22}H_{28}O_5$ 372.461

Constit. of the roots of *Holostylis reniformis*. Amorph. yellow solid. $[\alpha]^{25}_{D}$-36 (c, 0.8 in $CHCl_3$). λ_{max} 213 (log ε 3.9); 231 (log ε 3.7); 280 (log ε 3.3) (MeOH)

3′,4′,5,7-Tetra-Me ether

2,7′-Cyclo-3′,4′,5,7-tetramethoxy-8,8′-lignan-4-ol. 4-Hydroxy-3′,4′,5,7-tetramethoxy-2,7′-cyclolignan

$C_{22}H_{28}O_5$ 372.461

Constit. of roots of *Holostylis reniformis*. Yellow oil. $[\alpha]^{25}_{D}$-26 (c, 0.6 in $CHCl_3$)

7-Ketone, 3′,4-di-Me ether

2,7′-Cyclo-4′,5-dihydroxy-3′,4-dimethoxy-8,8′-lignan-7-one. 4′,5-Dihydroxy-3′,4-dimethoxy-7-oxo-2,7′-cyclolignan. Arisantetralone A. ***Wulignan A$_1$***

[117047-76-4]

$C_{20}H_{22}O_5$ 342.391

Isol. from the fruits of *Schisandra arisanensis*, *Schisandra henryi* and *Schisandra sphenenthera*. Pale yellow prisms. Mp 186 - 187°. $[\alpha]^{25}_{D}$-35 (c, 0.1 in CH_2Cl_2). λ_{max} 209 (log ε 3.71); 233 (log ε 3.65); 277 (log ε 3.36); 317 (log ε 3.02) (MeOH)

7-Ketone, 3′,5-di-Me ether

2,7′-Cyclo-4,4′-dihydroxy-3′,5-dimethoxy-8,8′-lignan-7-one. 4,4′-Dihydroxy-3′,5-dimethoxy-7-oxo-2,7′-cyclolignan. ***Arisantetralone B***

[1181216-77-2]

$C_{20}H_{22}O_5$ 342.391

Constit. of the fruit of *Schisandra arisanensis*. Amorph. powder. $[\alpha]^{25}_{D}$-35 (c, 0.1 in CH_2Cl_2). λ_{max} 204 (log ε 3.69); 234 (log ε 3.32); 281 (log ε 3.08) (MeOH)

7-Ketone, 3′,4,4′-tri-Me ether

2,7′-Cyclo-5-hydroxy-3′,4,4′-trimethoxy-8,8′-lignan-7-one. 5-Hydroxy-3′,4,4′-trimethoxy-7-oxo-2,7′-cyclolignan. ***Schisandrone****. Arisantetralone C*

[98619-25-1]

$C_{21}H_{24}O_5$ 356.418

Constit. of *Schisandra arisanensis* and *Schisandra sphenanthera*. Cryst. (Et_2O/hexane). Mp 180 - 181° (176-177°). $[\alpha]^{17}_{D}$-32.5 (c, 0.12 in $CHCl_3$). λ_{max} 205 (log ε 4.3); 232 (log ε 4.31); 274 (log ε 4.06); 315 (log ε 3.72) (EtOH)

7-Ketone, 3′,4,5-tri-Me ether

2,7′-Cyclo-4′-hydroxy-3′,4,5-trimethoxy-8,8′-lignan-7-one. 4′-Hydroxy-3′,4,5-trime-

thoxy-7-oxo-2,7′-cyclolignan.
Holostyligone
[887501-28-2]
$C_{21}H_{24}O_5$ 356.418
Constit. of the roots of *Holostylis reniformis*. Amorph. yellow solid. $[\alpha]_D^{25}$ -27 (c, 1.2 in $CHCl_3$)

7-Ketone, 3′,4′,5-tri-Me ether
8-Epiholostylone
[694489-75-3]
$C_{21}H_{24}O_5$ 356.418
Constit. of the roots of *Holostylis reniformis*. Yellow cryst.. Mp 169 - 172°. $[\alpha]_D^{25}$ -40.6 (c, 1.2 in $CHCl_3$). λ_{max} 210 (log ε 4.9); 235 (log ε 4.9); 278 (log ε 4.6); 314 (log ε 4.4) (MeOH)

7-Ketone, tetra-Me ether
8′-Epiaristoligone
[117404-43-0]
$C_{22}H_{26}O_5$ 370.445
Constit. of *Holostylis reniformis* and *Phoebe minutiflora*. Yellow cryst.. Mp 130 - 132°. $[\alpha]_D^{25}$ -64.3 (c, 1 in $CHCl_3$). λ_{max} 214 (log ε 3.9); 234 (log ε 4); 275 (log ε 3.8); 313 (log ε 3.6) (MeOH)

7-Ketone, 3′,4′-methylene, 4-Me ether
Schisphentetralone A
[1311164-16-5]
$C_{20}H_{20}O_5$ 340.375
Constit. of the fruit of *Schisandra sphenanthera*. Exhibits antiinflammatory activity. $[\alpha]_D^{25}$ -16.5 (c, 0.1 in CH_2Cl_2). λ_{max} 233 (log ε 2.95); 278 (log ε 2.68) (MeOH)

7-Ketone, 3′,4′-methylene, 4,5-di-Me ether
8′-Epiaristotetralone
$C_{21}H_{22}O_5$ 354.402
Constit. of *Phoebe minutiflora*. Amorph. yellow solid. $[\alpha]_D^{24}$ -42 (c, 0.1 in MeOH). λ_{max} 230 (log ε 4.5); 277 (log ε 4.2); 314 (log ε 3.85) (MeOH)

7-Ketone, 4,5-methylene, 3′-Me ether
2,7′-Cyclo-4′-hydroxy-3′-methoxy-4,5-methylenedioxy-8,8′-lignan-7-one.
4′-Hydroxy-3′-methoxy-4,5-methylenedioxy-7-oxo-2,7′-cyclolignan.
Enshicine
[93710-74-8]
$C_{20}H_{20}O_5$ 340.375
Constit. of *Schisandra henryi*. Cryst. (MeOH aq.). Mp 146 - 147°. $[\alpha]_D^{23}$ -52.3 (c, 0.13 in $CHCl_3$). λ_{max} 210 (log ε 4.08); 236 (log ε 4.12); 279 (log ε 3.73); 320 (log ε 3.55) (MeOH)

7-Ketone, 4,5-methylene, 3′,4′-di-Me ether
2,7′-Cyclo-3′,4′-dimethoxy-4,5-methylenedioxy-8,8′-lignan-7-one.
3′,4′-Dimethoxy-4,5-methylenedioxy-7-oxo-2,7′-cyclolignan.
4′-O-Methylenshicine
[694489-76-4]
$C_{21}H_{22}O_5$ 354.402
Constit. of the roots of *Holostylis reniformis*. Amorph. yellow solid. $[\alpha]_D^{25}$ -47.1 (c, 1 in $CHCl_3$). λ_{max} 213 (log ε 4.7); 233 (log ε 4.9); 275 (log ε 4.7); 311 (log ε 4.4) (MeOH)

7-Ketone, 3′,4′:4,5-bis(methylene) ether
Cagayanone B
[887501-30-6]
$C_{20}H_{18}O_5$ 338.359
Constit. of the roots of *Holostylis reniformis*. Amorph. yellow solid. $[\alpha]_D^{25}$ -33 (c, 0.45 in $CHCl_3$). λ_{max} 240 (log ε 3.2); 277 (log ε 3.2); 324 (log ε 3.2) (MeOH)

(7′R,8R,8′R)-form
7-Ketone, tetra-Me ether
Aristosynone
[114029-73-1]
$C_{22}H_{26}O_5$ 370.445
Constit. of the roots of *Aristolochia chilensis*

(7′R,8R,8′S)-form
7-Ketone, 3′,4-di-Me ether
8-Epiwulignan A₁
[117047-78-6]
$C_{20}H_{22}O_5$ 342.391
Isol. from the fruit of *Schisandra henryi*

7-Ketone, 3′,5-di-Me ether
Wulignan A₂
[117047-77-5]
$C_{20}H_{22}O_5$ 342.391
Constit. of *Holostylis reniformis* and *Schisandra henryi*. Amorph. yellow solid. $[\alpha]_D^{14}$ -61.2 (c, 0.17 in $CHCl_3$) (-55)

7-Ketone, 3′,4,4′-tri-Me ether
8-Epischizandrone. *8-Epischisandrone. Arisantetralone D*
[98619-26-2]
$C_{21}H_{24}O_5$ 356.418
Constit. of *Holostylis reniformis*, *Schisandra henryi* and fruit of *Kadsura longipedunculata*. Amorph. yellow solid. $[\alpha]_D^{25}$ -24 (c, 1.2 in $CHCl_3$). $[\alpha]_D^{14}$ +5.5 (c, 1.05 in $CHCl_3$)

7-Ketone, 3′,4′,5-tri-Me ether
Holostylone
[694489-74-2]
$C_{21}H_{24}O_5$ 356.418
Constit. of the roots of *Holostylis reniformis*. Amorph. yellow solid. $[\alpha]_D^{25}$ -27.4 (c, 0.3 in $CHCl_3$). λ_{max} 240 (log ε 4); 278 (log ε 4); 312 (log ε 3.9) (MeOH)

7-Ketone, tetra-Me ether
8,8′-Epiaristoligone.
8,8′-Diepiaristoligone
[117404-44-1]
$C_{22}H_{26}O_5$ 370.445
Constit. of the roots of *Holostylis reniformis*. Yellow cryst.. Mp 146.5 - 149.3°. $[\alpha]_D^{25}$ -30 (c, 1.3 in $CHCl_3$). λ_{max} 212 (log ε 4.4); 232 (log ε 4.5); 274 (log ε 4.2); 308 (log ε 4) (MeOH)

7-Ketone, 3′,4′-methylene, 4,5-di-Me ether
[82427-60-9]
Constit. of *Cordia exaltata*, *Virola elongata* and *Virola sebifera*. Needles. Mp 131° (128.2-130.4°). $[\alpha]_D^{25}$ -44.7 (c, 0.25 in $CHCl_3$). Abs. config. revised in 1990

7-Ketone, 4,5-methylene, 3′-Me ether
8-Epienshicine
[115939-80-5]
$C_{20}H_{20}O_5$ 340.375
Constit. of *Schisandra propinqua* var. *sinensis*. Active against leukaemia P-388 virus *in vitro*. Mp 163 - 164°. $[\alpha]_D^{11}$ -18.2 ($CHCl_3$). λ_{max} 234 (ε 25120); 276 (ε 9660); 318 (ε 7250) (EtOH) (Berdy)

7-Ketone, 3′,4′:4,5-bis(methylene) ether
Cagayanone. *7-Oxocagayanin*
[124961-79-1]
$C_{20}H_{18}O_5$ 338.359
Isol. from *Myristica cagayanensis*. Needles. Mp 236 - 238°. $[\alpha]_D^{23}$ -27.6 (c, 1.0 in $CHCl_3$). λ_{max} 235 (log ε 4.4); 275 (log ε 4.12) (MeOH)

(7′ξ,8ξ,8′ξ)-form
7-Ketone, 3′,4′-methylene, 5-Me ether
2,7′-Cyclo-4-hydroxy-5-methoxy-3′,4′-methylenedioxy-8,8′-lignan-7-one.
4-Hydroxy-5-methoxy-3′,4′-methylenedioxy-7-oxo-2,7′-cyclolignan

[188894-36-2]

$C_{20}H_{20}O_5$ 340.375

Constit. of *Virola* aff. *pavonis*. Viscous oil. λ_{max} 228 (log ε 4.34); 275 (log ε 4.02); 308 (log ε 3.85) (MeOH).

Liu, J.-S. *et al., Phytochemistry*, 1984, **23**, 1143-1145 (*Enshicine*).
Lian-niang, L. *et al., Planta Med.*, 1985, 217-219 (*Schisandrone*).
Urzúa, A. *et al., Phytochemistry*, 1987, **26**, 2414-2415 (*Aristotetralone*).
Liu, J.-S. *Huaxue Xuebao*, 1988, **46**, 345-348 (*Epienshicine*).
Liu, J.-S. *et al., Huaxue Xuebao*, 1988, **46**, 483-488 (*Wulignans, Epischisandrone*).
Urzúa, A. *et al., J. Nat. Prod.*, 1988, **51**, 117-121 (*Aristochilone, Aristoligone, Aristosynone, Aristotetraol*).
Kuo, Y.-H. *et al., Chem. Pharm. Bull.*, 1989, **37**, 2310-2312 (*Cagayanone*).
Kato, M.J. *et al., Phytochemistry*, 1990, **29**, 1799-1810 (*Virola elongata constit*).
Martinez V., J.C. *et al., Phytochemistry*, 1997, **44**, 1179-1182 (*Virola aff. pavonis constit*).
Da Silva, T. *et al., Phytochemistry*, 2004, **65**, 751-759 (*Holostylis reniformis constits*).
Ku, Y.-L. *et al., Nat. Prod. Res.*, 2006, **20**, 1199-1206 (*8′-Epiaristotetralone, 8′-Epiaristoligone*).
Da Silva, T. *et al., Phytochemistry*, 2006, **67**, 929-937 (*Holostylis reniformis constits*).
Liu, H-T. *et al., Chem. Pharm. Bull.*, 2009, **57**, 405-407 (*Schisandra henryi constits*).
Cheng, Y.-B. *et al., J. Nat. Prod.*, 2009, **72**, 1663-1668 (*Arisantetralones A-D*).
Messiano, G.B. *et al., Phytochemistry*, 2009, **70**, 590-596 (*biosynth*).
Huang, H.-C. *et al., Food Chem.*, 2011, **128**, 348-357 (*Schisphentetralone A, activity*).
Zaugg, J. *et al., Phytochemistry*, 2011, **72**, 2385-2395 (*Kadsura lonipedunculata constit*).
Nogueira, T.B. de S. de S. *et al., Molecules*, 2013, **18**, 11086-11099 (*Cordia exaltata constit*).
Reddel, J.C.T. *et al., Angew. Chem., Int. Ed.*, 2014, **53**, 1395-1398 (*synth*).
Tsai, Y-C. *et al., Chem. Biodiversity*, 2014, **11**, 1053-1068 (*Schisandra sphenanthera constits*).
Pereira, M.D.P. *et al., Phytochem. Lett.*, 2015, **13**, 200-205 (*Holostylis reniformis constits, activity*)

2,7′-Cyclo-8,8′-lignan-3′,4,4′,5,8-pentol C257

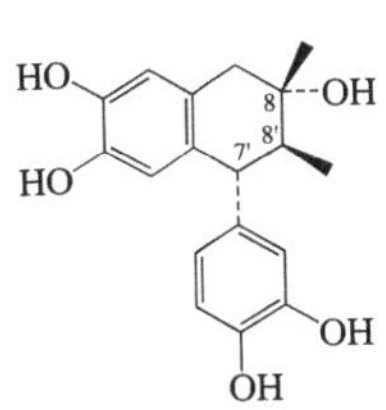

3′,4,4′,5,8-Pentahydroxy-2,7′-cyclolignan

$C_{18}H_{20}O_5$ 316.353

(7′R,8S,8′S)-form

3′,5-Di-Me ether

2,7′-Cyclo-3′,5-dimethoxy-8,8′-lignan-4,4′,8-triol. ***8-Hydroxyisoguaiacin***

[1220531-14-5]

$C_{20}H_{24}O_5$ 344.407

Constit. of *Laguncularia racemosa*. Oil. $[\alpha]^{20}_{D}$ -92 (c, 0.5 in MeOH). λ_{max} 282 (MeOH).

Shi, C. *et al., Phytochemistry*, 2010, **71**, 435-442, 1025 (*8-Hydroxyisoguaiacin*)

2,7′-Cyclo-8,8′-lignan-3′,4,4′,5,9′-pentol C258

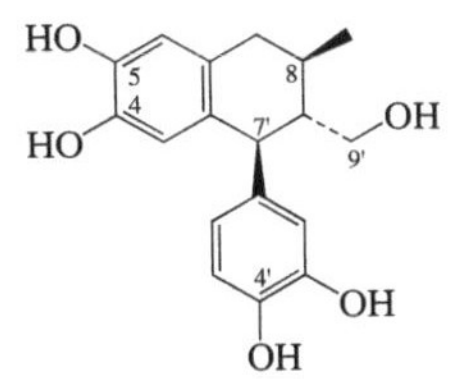

3′,4,4′,5,9′-Pentahydroxy-2,7′-cyclolignan

$C_{18}H_{20}O_5$ 316.353

Lignan numbering shown

(7′S,8R,8′S)-form

3′,4,5-Tri-Me ether

2,7′-Cyclo-3′,4,5-trimethoxy-8,8′-lignan-4′,9′-diol. 4′,9′-Dihydroxy-3′,4,5-trimethoxy-2,7′-cyclolignan

3′,4,5-Tri-Me ether, 9′-O-β-D-glucopyranoside

[208840-79-3]

$C_{27}H_{36}O_{10}$ 520.575

Constit. of the bark of *Trichilia estipulata*. Oil (as penta-Ac). $[\alpha]^{20}_{D}$ +8.5 (c, 0.02 in $CHCl_3$) (penta-Ac). λ_{max} 201 (ε 41380); 277 (ε 2480) (MeOH) (penta-Ac).

Cortez, D.A.G. *et al., Nat. Prod. Lett.*, 1998, **11**, 255-262 (*Trichilia estipulata constit*)

2,2′-Cyclo-8,8′-lignan-4,5,9,9′-tetrol C259

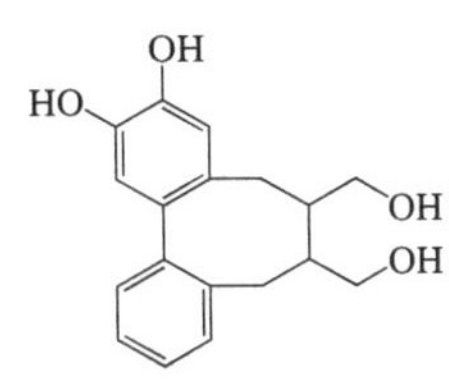

4,5,9,9′-Tetrahydroxy-2,2′-cyclolignan

$C_{18}H_{20}O_4$ 300.354

$(8\xi,8'\xi,\xi_{axial})$-form

4,5-Di-Me ether, 9′-O-β-D-glucopyranoside

[1699543-64-0]

$C_{26}H_{34}O_9$ 490.549

Constit. of *Oplopanax horridus*. Amorph. powder. $[\alpha]^{20}_{D}$ +18.1 (c, 0.5 in MeOH). λ_{max} 208 (log ε 0.78); 216 (log ε 1.43); 228 (log ε 1.88) (MeOH).

Huang, W.-H. *et al., Helv. Chim. Acta*, 2015, **98**, 201-209 (*Oplopanax glucoside*)

2,7′-Cyclo-8,8′-lignan-3,3′,4,4′-tetrol C260

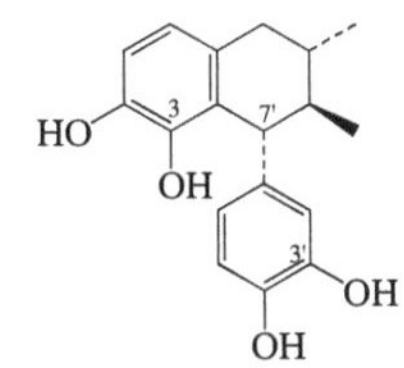

8-(3,4-Dihydroxyphenyl)-5,6,7,8-tetrahydro-6,7-dimethyl-1,2-naphthalenediol. 3,3′,4,4′-Tetrahydroxy-2,7′-cyclolignan

$C_{18}H_{20}O_4$ 300.354

(7′R,8S,8′R)-form

3,4:3′,4′-Bis(methylene) ether

2,7′-Cyclo-3,4:3′,4′-bis(methylenedioxy)lignan. 3,4:3′,4′-Bis(methylenedioxy)-2,7′-cyclolignan. ***Otobain***. *Otobite*

[3738-01-0]

$C_{20}H_{20}O_4$ 324.376

Isol. from fruits of *Myristica otoba* and *Myristica simarum*. Cryst. (petrol). Mp 136 - 137°. $[\alpha]_D$ -43 (c, 0.8 in $CHCl_3$). λ_{max} 235 (ε 9850); 288 (ε 7020) (MeOH).

Gilchrist, T. *et al., J. Chem. Soc.*, 1962, 1780-1786 (*Otobain*).
Wallis, R. *et al., J. Chem. Soc.*, 1963, 1445-1449 (*pmr*).
Klyne, W. *et al., J. Chem. Soc. C*, 1966, 893-896 (*abs config*).
Maclean, I. *et al., J. Chem. Soc. C*, 1966, 1717-1719 (*Otobain, synth*).
McMurry, T.B.H. *et al., Tetrahedron Lett.*, 1966, 975-977 (*Otobain, synth*).
Kuo, Y.H. *et al., Experientia*, 1976, **32**, 828-829 (*Myristica simarum constit*)

2,7′-Cyclo-8,8′-lignan-3′,4,4′,5-tetrol C261

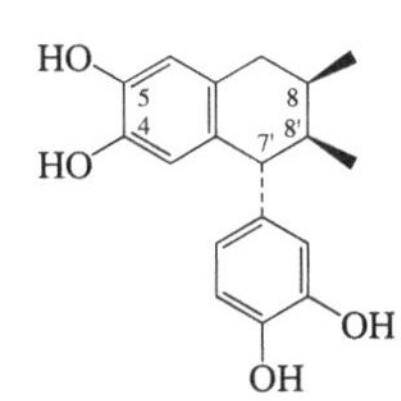

(7′R,8R,8′R)-form

5-(3,4-Dihydroxyphenyl)-5,6,7,8-tetrahydro-6,7-dimethyl-2,3-

naphthalenediol. 3′,4,4′,5-Tetrahydroxy-2,7′-cyclolignan
$C_{18}H_{20}O_4$ 300.354
Lignan numbering shown. There are various errors and uncertainties concerning abs. configs. and notes are given throughout the entry
(7′R,8R,8′R)-form
Di-O-demethylisoguaiacin. *3′-O-*Demethylnorisoguaiacin [125074-14-8]
Constit. of *Larrea nitida* and *Larrea tridentata*. Potent estrogenic signalling molecule. Mp 85 - 87°. CD $[\theta]_{207}$ +2700 (c, 0.017 in MeOH). λ_{max} 200 (log ε 4.79); 228 (sh); 285 (log ε 3.85) (MeOH)
3′-Me ether
2,7′-Cyclo-3′-methoxy-8,8′-lignan-4,4′,5-triol. 4,4′,5-Trihydroxy-3′-methoxy-2,7′-cyclolignan. ***Norisoguaiacin***. *Desoxydemethylenechicanine*
[50376-42-6]
$C_{19}H_{22}O_4$ 314.381
Constit. of *Larrea divaricata* and *Larrea tridentata*. Inhibitor of *Streptococcus* sp, *Bacillus subtilis* and *Pseudomonas aeruginosa*. Potent estrogenic signalling molecule. Mp 153 - 155° (148-149°). Desoxydemethylenechicanine was a semisynthetic prod. obt. by transformation of Chicanine (Liu *et al*, 1984). However (1) the abs. config. of the Chicanine enantiomers was revised in 2011 and (2) the transformation involves a ring opening rearrangement. Therefore opt. rotn. has not been assigned here. λ_{max} 200 (log ε 4.5); 224 (sh) (log ε 4.16); 282 (log ε 3.47) (MeOH)
3′,4-Di-Me ether
2,7′-Cyclo-3′,4-dimethoxy-8,8′-lignan-4′,5-diol. 4′,5-Dihydroxy-3′,4-dimethoxy-2,7′-cyclolignan
[852360-84-0]
$C_{20}H_{24}O_4$ 328.408
Constit. of the flowering tops of *Larrea tridentata*. Brown gum. $[\alpha]_D^{25}$ -19.2 (c, 0.5 in $CHCl_3$). λ_{max} 213; 280 (MeCN)
3′,5-Di-Me ether
2,7′-Cyclo-3′,5-dimethoxy-8,8′-lignan-4,4′-diol. 3′,5-Dimethoxy-4,4′-dihydroxy-2,7′-cyclolignan. ***(-)-Isoguaiacin***
[78341-26-1]
$C_{20}H_{24}O_4$ 328.408
Isol. from *Guaiacum officinale*, *Machilus thunbergii* and *Calyptranthes pallens*. Inhibits platelet aggregation. Shows neuroprotective and antiosteoporotic activity. Hepatoprotectant. Cryst.. Mp 195 - 197°. $[\alpha]_D^{25}$ -32 (c, 0.5 in $CHCl_3$). $[\alpha]_D^{20}$ -14 (c, 0.1 in MeOH). The (+)-enantiomer also occurs with a much greater opt. rotn. reported and these may be may be partial racemates
3′,4,5-Tri-Me ether
2,7′-Cyclo-3′,4,5-trimethoxy-8,8′-lignan-4′-ol. 4′-Hydroxy-3′,4,5-trimethoxy-2,7′-cyclolignan. ***Sinensisin A***
[1401673-22-0]
$C_{21}H_{26}O_4$ 342.435
Constit. of *Schisandra propinqua* var. *sinensis*. Amorph. solid. $[\alpha]_D^{22}$ -5.9 (c, 0.12 in $CHCl_3$). λ_{max} 240 (log ε 3.22); 283 (log ε 2.99) ($CHCl_3$)
Tetra-Me ether
2,7′-Cyclo-3′,4,4′,5-tetramethoxylignan. 3′,4,4′,5-Tetramethoxy-2,7′-cyclolignan. (-)-Isogalbulin. Isoguaiacin dimethyl ether. Dimethylisoguaiacin
[54548-00-4]
Semisynthetic. Shows antiosteoporotic activity. Needles (MeOH). Mp 100 - 101°. $[\alpha]_D^{26}$ -46 (c, 0.105 in $CHCl_3$) (-35.8)
3′,4′-Methylene, 4,5-di-Me ether
2,7′-Cyclo-4,5-dimethoxy-3′,4′-methylenedioxylignan. 4,5-Dimethoxy-3′,4′-methylenedioxy-2,7′-cyclolignan. ***Isogalcatin B***
[1174682-81-5]
$C_{21}H_{24}O_4$ 340.419
Semisynthetic, by acid catalysis of Magnovatin B in L68. Constit. of the bark of *Machilus thunbergii*. Shows moderate neuroprotectant activity. Amorph. powder. $[\alpha]_D^{25}$ -21.1 (c, 1.2 in $CHCl_3$) (natural). $[\alpha]_D^{25}$ -11.7 (c, 0.1 in $CHCl_3$) (semisynthetic). λ_{max} 208 (log ε 4.6); 235 (log ε 3.9); 289 (log ε 3.7) (MeOH) (semisynthetic)
4,5-Methylene, 3′-Me ether
2,7′-Cyclo-3′-methoxy-4,5-methylenedioxy-8,8′-lignan-4′-ol. 4′-Hydroxy-3′-methoxy-4,5-methylenedioxy-2,7′-cyclolignan. Desoxyenshicine
[93780-83-7]
Semisynthetic, by reductive deoxygenation of Enshicine in C256. Needles (MeOH). Mp 121 - 122°. $[\alpha]_D^{23}$ -33.3 (c, 0.09 in MeOH)
Bis(methylene) ether
2,7′-Cyclo-3′,4′:4,5-bis(methylenedioxy)lignan. 3′,4′:4,5-Bis(methylenedioxy)-2,7′-cyclolignan
[1174682-79-1]
Semisynthetic, by acid-catalysed cyclisation of Magnovatin A in L68. Oil. $[\alpha]_D^{25}$ -48 (c, 0.1 in $CHCl_3$). λ_{max} 210 (log ε 4.5); 244 (log ε 3.8); 294 (log ε 3.7) (MeOH)
(7′S,8R,8′R)-form
3′,4,4′-Tri-Me ether
2,7′-Cyclo-3′,4,4′-trimethoxy-8,8′-lignan-5-ol. 5-Hydroxy-3′,4,4′-trimethoxy-2,7′-cyclolignan. ***Sinensisin B***
[1401673-24-2]
$C_{21}H_{26}O_4$ 342.435
Constit. of *Schisandra propinqua* var. *sinensis*. Gum. $[\alpha]_D^{22}$ +4.7 (c, 0.18 in $CHCl_3$). λ_{max} 240 (log ε 3.76); 283 (log ε 3.57) ($CHCl_3$)
(7′S,8R,8′S)-form
3′,5-Di-Me ether
Guaiacin
[36531-08-5]
$C_{20}H_{24}O_4$ 328.408
Constit. of *Machilus edulis*, *Guaiacum officinale*, *Osteophloeum platyspermum*, *Saururus cernuus*, *Virola carinata*, *Virola surinamensis*, *Magnolia kachirachirai* and *Myristica fragrans*. Shows neuroprotectant and antiosteoporotic activity. Cryst. (C_6H_6). Mp 204 - 206° (198-200°). $[\alpha]_D^{25}$ +46 ($CHCl_3$) (+33.5)
3′,4′,5-Tri-Me ether
2,7′-Cyclo-3′,4′,5-trimethoxy-8,8′-lignan-4-ol. 4-Hydroxy-3′,4′,5-trimethoxy-2,7′-cyclolignan
[207384-95-0]
$C_{21}H_{26}O_4$ 342.435
Constit. of *Virola oleifera*. Amorph. solid. $[\alpha]_D^{21}$ -14.2 (c, 0.4 in $CHCl_3$)
Tetra-Me ether
Galbulin
[521-54-0]
[4915-80-4] ((±)-Galbulin)
[59573-54-5] ((+)-Galbulin)

$C_{22}H_{28}O_4$ 356.462
Occurs in *Himantandra baccata* (preferred genus name *Galbulimima*), *Virola surinamensis* and other *Virola* spp.. Needles (EtOH). Mp 135° (130-132°). $[\alpha]^{25}_D$ -43.5 (c, 0.38 in $CHCl_3$) (Kato *et al*). $[\alpha]^{25}_D$ +8 ($CHCl_3$) (Majumder *et al*). The (*S,R,S*)-abs config. derives from Majumder *et al* (1972). Hong *et al* (2012) report (*R,S,R*)- with the opposite abs. config. but it is not clear whether two enantiomers were isol.

3′,4′-Methylene, 5-Me ether
2,7′-Cyclo-5-methoxy-3′,4′-methylenedioxy-8,8′-lignan-4-ol. 4-Hydroxy-5-methoxy-3′,4′-methylenedioxy-2,7′-cyclolignan.
Otobaphenol
[10240-16-1]
$C_{20}H_{22}O_4$ 326.392
Isol. from fruit of *Myristica otoba* and from *Osteophloeum platyspermum* and *Kadsura longipedunculata*. Cryst. (MeOH aq. or CH_2Cl_2/petrol). Mp 134 - 136°. $[\alpha]_D$ +42 (c, 1.5 in $CHCl_3$). λ_{max} 233 (ε 11200); 288 (ε 7750) (MeOH)

3′,4′-Methylene, 4,5-di-Me ether
Isogalcatin. *Isootobain‡*
[10240-19-4]
[24150-38-7] ((+)-Isogalcatin)
$C_{21}H_{24}O_4$ 340.419
Constit. of the fruit oil of *Myristica otoba* and bark of *Virola calophylla* (no stereochem. assigned). Also by methylation of Otobaphenol. Needles or prisms (EtOH). Mp 106 - 108°. $[\alpha]_D$ +5.2 (c, 1.424 in $CHCl_3$) (+2). The name Isootobain is misleading. Kato (1990) reports (-)-Isogalcatin ($[\alpha]_D$ -10.9) drawn with the (*S,R,S*)-config. but stated in the experimental section to be *R,S,R*-

4,5-Methylene, 3′-Me ether
Isootobaphenol
[59985-35-2]
$C_{20}H_{22}O_4$ 326.392
Constit. of *Virola carinata*, *Schisandra glaucescens* and *Myristica fragrans*. Shows moderate LDL antioxidant activity. Cryst. (MeOH/hexane). Mp 130 - 132°. $[\alpha]^{20}_D$ -4 (MeOH)

4,5-Methylene, 3′-Me ether, Ac
[59985-36-3]
Mp 125 - 128°

4,5-Methylene, 3′,4′-di-Me ether
2,7′-Cyclo-3′,4′-dimethoxy-4,5-methylenedioxy-8,8′-lignan. 3′,4′-Dimethoxy-4,5-methylenedioxy-2,7′-cyclolignan.
Galcatin
[4892-34-6]
$C_{21}H_{24}O_4$ 340.419
Isol. from *Himantandra baccata*, *Virola carinata* and *Schisandra glaucescens*. Cryst. (MeOH). Mp 117 - 118°. $[\alpha]^{20}_D$ -8.8 (c, 2 in $CHCl_3$)

Bis(methylene) ether
Cagayanin. *Isootobain‡*
[99096-51-2]
$C_{20}H_{20}O_4$ 324.376
Prod. by acid-catalysed cyclisation of 3,4:3′,4′-Bis(methylenedioxy) lignan, B61. Isol. from *Myristica cagayanensis*, *Virola surinamensis* and *Iryanthera lancifolia*. Needles. Mp 122 - 123°. $[\alpha]^{20}_D$ -33.5 (c, 1.0 in $CHCl_3$). The name Isootobain was also incorr. applied to Otobaphenol in 1963

(7′S,8S,8′S)-form

5-Me ether
2,7′-Cyclo-5-methoxy-8,8′-lignan-3′,4,4′-triol. 3′,4,4′-Trihydroxy-5-methoxy-2,7′-cyclolignan
[1310453-61-2]
$C_{19}H_{22}O_4$ 314.381
Constit. of the bark of *Machilus robusta*. Amorph. powder. $[\alpha]^{20}_D$ +11.8 (c, 0.07 in $CHCl_3$). λ_{max} 204 (log ε 4.32); 232 (log ε 3.8); 286 (log ε 3.46) (MeOH)

3′,5-Di-Me ether
(+)-Isoguaiacin
$C_{20}H_{24}O_4$ 328.408
Iso. from *Manglietiastrum sinicum*. $[\alpha]^{26}_D$ +89.7 (c, 0.39 in $CHCl_3$)

Tetra-Me ether
(+)-Isogalbulin
[94667-55-7]
$C_{22}H_{28}O_4$ 356.462
Constit. of the fruit of *Virola sebifera*. Cryst. (MeOH). Mp 87.8 - 89.6°. Mp 120 - 123°. λ_{max} 220 (ε 3500); 285 (ε 2300) (MeOH).

Hughes, G.K. *et al.*, *Aust. J. Chem.*, 1957, **7**, 104-112 (*Galcatin, Galbulin*).
Birch, A.J. *J. Chem. Soc.*, 1958, 4471-4476 (*(-)-Galbulin, (-)-Galcatin, abs config*).
Traverso, G. *et al.*, *Farmaco, Ed. Sci.*, 1961, **16**, 457-471 (*Galbulin, synth*).
Adjangba, M.S. *Bull. Soc. Chim. Fr.*, 1962, 1970-1977 (*Galcatin, synth*).
Adjangba, M.S. *Bull. Soc. Chim. Fr.*, 1963, 1942-1950 (*Galbulin, synth*).
Wallace, R. *et al.*, *J. Chem. Soc.*, 1963, 1445-1449 (*Isootobain*).
King, F.E. *et al.*, *J. Chem. Soc.*, 1964, 4011-4024 (*Guaiacin, Isoguaiacin*).
Klyne, W. *et al.*, *J. Chem. Soc. C*, 1966, 893-896 (*(+)-Otobaphenol, Isogalcatin, Galbulin, abs config*).
Kohen, F. *et al.*, *J. Chem. Soc. C*, 1966, 1775-1780 (*Otobaphenol*).
Pelter, A. *J. Chem. Soc. C*, 1968, 74-79 (*ms*).
Majumder, P.L. *et al.*, *Phytochemistry*, 1972, **11**, 811-814 (*(+)-Guaiacin, (+)-Galbulin*).
Gisvold, O. *et al.*, *J. Pharm. Sci.*, 1974, **63**, 1905-1907 (*Norisoguaiacin*).
Gottlieb, O.R. *et al.*, *Phytochemistry*, 1976, **15**, 773-774 (*Isootobaphenol*).
Fonseca, S.F. *et al.*, *Phytochemistry*, 1979, **18**, 1703-1708 (*cmr*).
Mpango, G.P. *et al.*, *Tetrahedron Lett.*, 1980, **21**, 4827-4830 (*Galcatin, Isogalcatin, synth*).
Liu, J.-S. *et al.*, *Can. J. Chem.*, 1981, **59**, 1680-1684 (*(+)-Galcatin, Desoxydemethylenechicanine, synth, ord, abs config*).
Liu, J-S. *et al.*, *Phytochemistry*, 1984, **23**, 1143-1146 (*Desoxyenshicine, Desoxydemethylenechicanine*).
Ito, K. *et al.*, *Phytochemistry*, 1984, **23**, 2643-2646 (*Magnolia kachirachirai constit*).
Lopes, L.M.X. *et al.*, *Phytochemistry*, 1984, **23**, 2647-2652 (*Isogalbulin, ord, cd, abs config*).
Kuo, Y.-H. *et al.*, *J. Chin. Chem. Soc. (Taipei)*, 1985, **32**, 177-178 (*Cagayanin*).
Hattori, M. *et al.*, *Chem. Pharm. Bull.*, 1988, **36**, 648-653 (*Cagayanin, synth*).
Konno, C. *et al.*, *J. Nat. Prod.*, 1989, **52**, 1113-1117 (*Di-O-demethylisoguaiacin*).
Kato, M. *et al.*, *Phytochemistry*, 1990, **29**, 1799-1810 (*Virola elongata constits*).
Fernandes, A.M.A.P. *et al.*, *Phytochemistry*, 1994, **36**, 533-534 (*(-)-Galcatin, (-)-Cagayanin, abs config, bibl*).
Sartorelli, P. *et al.*, *Phytochemistry*, 1998, **47**, 1003-1006 (*Virola oleifera constit*).
Lopes, N. *et al.*, *Phytochemistry*, 1999, **51**, 29-33 (*Virola surinamensis constits*).
Kasatkin, A.N. *et al.*, *J. Org. Chem.*, 2000, **65**, 3236-3238 (*Galbulin, Isogalbulin, synth, cmr*).
Uck, Y. *et al.*, *J. Pharm. Pharmacol.*, 2000, **52**, 1163-1169 (*Isoguaiacin, hepatoprotective activity*).
Wang, B.-G. *et al.*, *Planta Med.*, 2000, **66**, 511-515 (*Isoguaiacin, activity*).
Datta, P.K. *et al.*, *J. Org. Chem.*, 2001, **66**, 8606-8611 (*Cagayanin, Galbulin, synth, bibl*).
Ma, C.J. *et al.*, *Planta Med.*, 2004, **70**, 79-80 (*Guaiacin, Isoguaiacin, neuroprotective activity*).
Lambert, J.D. *et al.*, *Phytochemistry*, 2005, **66**, 811-815 (*Larrea tridentata 3′,4-di-Me ether*).
Lee, M.K. *et al.*, *Biol. Pharm. Bull.*, 2007, **30**, 814-817 (*Guaiacin, Isoguaiacin, antiosteoporetic activity*).
Kwon, H.S. *et al.*, *Bioorg. Med. Chem. Lett.*, 2008, **18**, 194-198 (*Myristica fragrans constits, antioxidant activity*).

Ma, C.J. *et al.*, *J. Enzyme Inhib. Med. Chem.*, 2009, **24**, 1117-1121 (*Isogalcatin B, activity*).
Barros, L.F.L. *et al.*, *J. Nat. Prod.*, 2009, **72**, 1529-1532 (*3′,4′-methylene 4,5-di-Me ether, bismethylene ether, synth, cd, abs config*).
Messiano, G.B. *et al.*, *J. Nat. Prod.*, 2010, **73**, 1933-1937 (*(-)-Isogalbulin, synth*).
Li, Y. *et al.*, *J. Nat. Prod.*, 2011, **74**, 1444-1452 (*Machilus robusta constit*).
Hong, B.-C. *et al.*, *Chem. Commun.*, 2012, **48**, 2385-2387 (*(+)-Galbulin, synth, abs config*).
Li, X.-N. *et al.*, *Fitoterapia*, 2012, **83**, 249-252 (*Sinensisins A,B*).
Schmidt, T.J. *et al.*, *Phytochem. Lett.*, 2012, **5**, 632-638 (*Larrea tridentata constits, abs config*).
Peng, Y. *et al.*, *Org. Biomol. Chem.*, 2013, **11**, 7574-7586 (*Galbulin, Isogalbulin, Cagayanin, synth*).
Reddel, J.C.T. *et al.*, *Angew. Chem., Int. Ed.*, 2014, **53**, 1395-1398 (*Galcatin, synth*).
Yu, H-Y. *et al.*, *J. Nat. Prod.*, 2014, **77**, 1311-1320 (*Schisandra glaucescens constits*).
Li, X. *et al.*, *Tetrahedron Lett.*, 2014, **55**, 6324-6327 (*Galbulin, Isogalbulin, synth*).
Jeong, S.-Y. *et al.*, *Bull. Korean Chem. Soc.*, 2015, **36**, 2254-2259 (*Larrea nitida constits, activity*).
Clausen, F. *et al.*, *Org. Lett.*, 2020, **22**, 6780-6783 (*Galbulin, synth*)

7,7′-Cyclo-8,8′-lignan-3,3′,4,4′-tetrol
C262

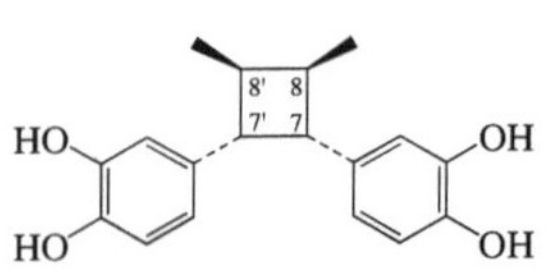

(7α,7′α,8β,8′β)-*form*

1,2-Bis(3,4-dihydroxyphenyl)-3,4-dimethylcyclobutane.
3,3′,4,4′-Tetrahydroxy-7,7′-cyclolignan
$C_{18}H_{20}O_4$ 300.354
(7α,7′α,8β,8′β)-form
3′,3′-Di-Me ether
7,7′-Cyclo-3,3′-dimethoxy-8,8′-lignan-4,4′-diol. 4,4′-Dihydroxy-3,3′-dimethoxy-7,7′-cyclolignan. ***Endiandrin B***
[1140478-56-3]
$C_{20}H_{24}O_4$ 328.408
Constit. of roots of *Endiandra anthropophagorum*. Exhibits moderate cytotoxicity against lung carcinoma. Needles. Mp 145 - 147°. Opt. inactive (*meso-*). λ_{max} 233 (log ε 3.78); 284 (log ε 3.44) (MeOH)
Tetra-Me ether
3,3′,4,4′-Tetramethoxy-7,7′-cyclolignan. ***Cinbalansan***
[58045-93-5]
$C_{22}H_{28}O_4$ 356.462
Constit. of the leaves of *Cinnamomum balansae*. Shows moderate cytotoxicity against lung carcinoma A 549 cells. Cryst. (pentane). Mp 64 - 64.2°. Opt. inactive (*meso-*)
(7β,7′α,8α,8′β)-form
3,3′-Di-Me ether
Endiandrin A
[946065-33-4]
$C_{20}H_{24}O_4$ 328.408
Constit. of the roots of *Endiandra anthropophagorum*. Glucocorticoid receptor binder. Shows moderate cytotoxicity against lung carcinoma. Oil. $[\alpha]_D^{22}$ -51 (c, 0.19 in $CHCl_3$). λ_{max} 231 (log ε 3.99); 283 (log ε 3.63) (MeOH).

Marshall, W.R. *et al.*, *J. Am. Chem. Soc.*, 1989, **111**, 1749-1754 (*Cinbalansan, synth*).
Cuong, N.M. *et al.*, *Nat. Prod. Lett.*, 2001, **15**, 331-338 (*Cinbalansan*).
Davis, R.A. *et al.*, *J. Nat. Prod.*, 2007, **70**, 1118-1121 (*Endiandrin A*).
Davis, R.A. *et al.*, *Bioorg. Med. Chem.*, 2009, **17**, 1387-1392 (*Endiandrins A,B, cryst struct, activity*).
Davis, R.A. *et al.*, *J. Med. Chem.*, 2011, **54**, 1682-1692 (*Endiandrin A, activity*).
Barnes, T.H. *et al.*, *Chem. Commun.*, 2020, **56**, 8747-8749 (*synth*)

2,7′-Cyclo-8,8′-lignan-3,4,4′-triol
C263

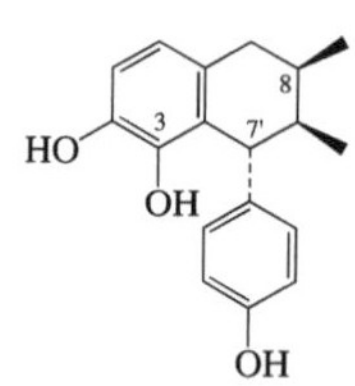

5,6,7,8-Tetrahydro-8-(4-hydroxyphenyl)-6,7-dimethyl-1,2-naphthalenediol, CAS. *1,2,3,4-Tetrahydro-7,8-dihydroxy-1-(4-hydroxyphenyl)-2,3-dimethylnaphthalene. 3,4,4′-Trihydroxy-2,7′-cyclolignan*
$C_{18}H_{20}O_3$ 284.355
Lignan numbering shown
(7′R,8R,8′R)-form [171204-45-8]
Constit. of *Larrea tridentata*. $[\alpha]_D^{25}$ +118 (c, 0.1 in MeOH).

Gnabre, J. *et al.*, *Tetrahedron*, 1995, **51**, 12203-12210 (*Larrea tridentata constit*)

2,7′-Cyclo-8,8′-lignan-4,4′,5-triol
C264

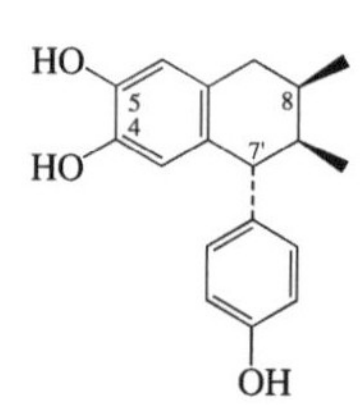

(7′*R**,8*R**,8′*R**)-*form*

5,6,7,8-Tetrahydro-5-(4-hydroxyphenyl)-6,7-dimethyl-2,3-naphthalenediol, CAS. *4,4′,5-Trihydroxy-2,7′-cyclolignan*
$C_{18}H_{20}O_3$ 284.355
(7′R*,8R*,8′R*)-form
3′-Demethoxy-5-O-demethylisoguaiacin. Nor-3′-demethoxyisoguaiacin [71113-16-1]
Isol. from leaves and twigs of *Larrea tridentata*. Shows antibacterial activity against *Staphylococcus aureus, Enterococcus faecalis, Escherichia coli, Enterobacter cloacae* and *Mycobacterium tuberculosis*. Fertility regulator. Cryst. (CH_2Cl_2/hexane). Mp 165 - 166°. Called 3′-Demethoxy-6-*O*-demethylisoguaiacin in the lit. (different numbering system)
4-Me ether
2,7′-Cyclo-4-methoxy-8,8′-lignan-4′,5-diol. 4′,5-Dihydroxy-4-methoxy-2,7′-cyclolignan. ***3′-Demethoxyisoguaiacin***
[54473-22-2]
$C_{19}H_{22}O_3$ 298.382
Isol. from *Larrea divaricata*. Needles (AcOH aq.). Mp 96 - 98°
(7′R*,8S*,8′R*)-form
4,5-Methylene ether
2,7′-Cyclo-4,5-methylenedioxy-8,8′-lignan-4′-ol. 4′-Hydroxy-4,5-methylenedioxy-2,7′-cyclolignan. ***Attenuol***
[66761-07-7]
$C_{19}H_{20}O_3$ 296.366
Constit. of *Knema attenuata*. Needles ($CHCl_3$/hexane). Mp 160 - 161°. $[\alpha]_D^{25}$ -20.52 (c, 1.16 in $CHCl_3$). CD spectrum reported but abs. config. not assigned. λ_{max} 227 (ε 11220); 285 (ε 6310); 294 (ε 5370) (CH_2Cl_2)
(7RS,8SR,8′RS)-form [72029-72-2]
Cryst. (CH_2Cl_2/hexane). Mp 165 - 166°.

Gisvold, O. *J. Pharm. Sci.*, 1974, **63**, 1905-1907 (*3′-Demethoxyisoguaiacin*).
Brown, E. *et al.*, *Tetrahedron Lett.*, 1979, **20**, 1389-1392 (*Attenuol, synth*).
Joshi, B.S. *et al.*, *Tetrahedron*, 1979, **35**, 1665-1671 (*Attenuol*).
Loriot, M. *et al.*, *Tetrahedron*, 1983, **39**, 2795-2798 (*Attenuol, synth*).
Konno, C. *et al.*, *J. Nat. Prod.*, 1989, **52**, 1113-1117 (*3′-Demethoxy-6-O-dimethylisoguaiacin*).
Favela-Hernández, J.M.J. *et al.*, *Molecules*, 2015, **20**, 12450-12458 (*3′-Demethoxy-6-O-demethylisoguaicin, activity*)

2,2′-Cyclo-8,8′-lign-8-ene-3,3′,4,4′,5,5′,7-heptol C265

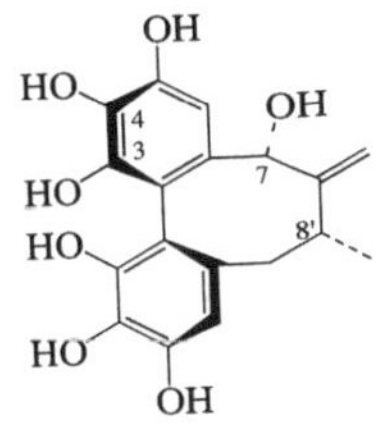

(7R,8′R,S_{axial})-form

3,3′,4,4′,5,5′,7-Heptahydroxy-2,2′-cyclolign-8-ene

$C_{18}H_{18}O_7$ 346.335

(7R,8′R,S_{axial})-form

4′,5′-Methylene, 3,3′,4,5-tetra-Me ether, 7-angeloyl

Schiviridin A

[1826796-41-1]

$C_{28}H_{32}O_8$ 496.556

Constit. of the fruit of *Schisandra viridis*. Amorph. powder. $[\alpha]_D^{20}$ +5.82 (c, 0.22 in MeOH). λ_{max} 220 (log ε 3.96) (MeOH)

4,5:4′,5′-Bis(methylene), 3,3′-di-Me ether, 7-angeloyl

Schisandrene A

[1040392-76-4]

$C_{27}H_{28}O_8$ 480.513

Constit. of *Schisandra rubriflora*. Amorph. powder. $[\alpha]_D^{27}$ +67.8 (c, 0.12 in Me_2CO). λ_{max} 219 (log ε 4.69); 267 (sh); 282 (log ε 3.53) (MeOH)

4,5:4′,5′-Bis(methylene), 3,3′-di-Me ether, 7-benzoyl

Schisandrene. Schizandrene

[874364-30-4]

$C_{29}H_{26}O_8$ 502.519

Constit. of the fruit of *Schisandra chinensis*. Yellow oil. $[\alpha]_D^{25}$ -13 (c, 0.51 in $CHCl_3$). λ_{max} 244 (log ε 1.89) (MeOH)

(7R,8′S,S_{axial})-form

4,5-Methylene, 3,3′,4′,5′-tetra-Me ether, 7-benzoyl

Schisanchinin A

[1613761-68-4]

$C_{30}H_{30}O_8$ 518.562

Constit. of fruit of *Schisandra chinensis*. Moderate inhibitor of NO release in LPS-activated microglial cells. Yellowish oil. $[\alpha]_D^{23}$ -3.6 (c, 0.03 in MeOH). λ_{max} 213 (log ε 3.42) (MeOH)

4′,5′-Methylene, 3,3′,4,5-tetra-Me ether, 8α,9-epoxide, 7-angeloyl

Schisanwilsonin K

[1610530-95-4]

$C_{28}H_{32}O_9$ 512.555

Constit. of *Schisandra wilsoniana*. Amorph. powder. $[\alpha]_D^{25}$ +22.5 (c, 0.02 in MeOH). λ_{max} 219 (log ε 4.38); 257 (sh); 290 (sh) (MeOH)

4′,5′-Methylene, 3,3′,4,5-tetra-Me ether, 8α,9-epoxide, 7-benzoyl

Schisandrin A_1. ***Schisphenlignan D***

[1422349-64-1]

$C_{30}H_{30}O_9$ 534.561

Isol. from *Schisandra sphenanthera*. Amorph. powder or needles (MeOH). Mp 118 - 120°. $[\alpha]_D^{14}$ -188.9 (c, 0.15 in $CHCl_3$). $[\alpha]_D^{20}$ -38.6 (c, 0.01 in MeOH) (Schisandrin A_1). λ_{max} 240 (log ε 3.68) ($CHCl_3$). [neutral] λ_{max} 223 (log ε 0.63) (MeOH) (Schisandrin A_1).

Choi, Y.-W. *et al., J. Nat. Prod.*, 2006, **69**, 356-359 (*Schisandrene*).

Li, H.-M. *et al., Helv. Chim. Acta*, 2008, **91**, 1053-1062 (*Schisandrene A*).

Liang, C.Q. *et al., Chem. Pharm. Bull.*, 2013, **61**, 96-100 (*Schisphenlignan D*).

Liu, H.-W. *et al., Eur. J. Med. Chem.*, 2013, **59**, 265-273 (*Schisandrin A_1, cd, cryst struct*).

Ma, W-H. *et al., Planta Med.*, 2013, **79**, 1051-1055 (*Schisanwilsonin K*).

Hu, D. *et al., Phytochemistry*, 2014, **104**, 72-78 (*Schisanchinin A*).

Tian, T. *et al., Chem. Nat. Compd. (Engl. Transl.)*, 2015, **51**, 1046-1048 (*Schiviridin A*)

2,7′-Cyclo-8,8′-lign-7-ene-3,3′,4,4′,5,9,9′-heptol C266

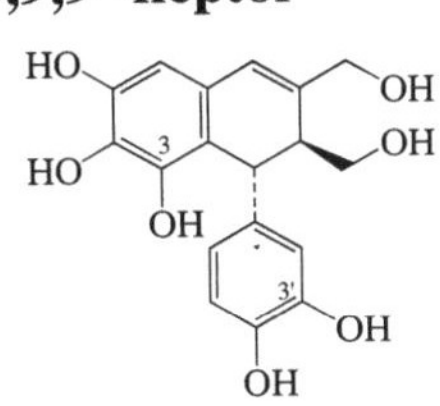

(7′R,8′S)-form

3,3′,4,4′,5,9,9′-Heptahydroxy-2,7′-cyclolign-7-ene

$C_{18}H_{18}O_7$ 346.335

(7′R,8′S)-form

3,3′,4,4′,5-Penta-Me ether

2,7′-Cyclo-3,3′,4,4′,5-pentamethoxy-8,8′-lign-7-ene-9,9′-diol. 9,9′-Dihydroxy-3,3′,4,4′,5-pentamethoxy-2,7′-cyclolign-7-ene.

Magnoliadiol

[177602-19-6]

$C_{23}H_{28}O_7$ 416.47

Constit. of *Magnolia fargesii*. Pale yellow oil. $[\alpha]_D$ +62.9 (c, 2 in $CHCl_3$)

(7′S,8′R)-form

9-Aldehyde, 3,3′,5-tri-Me ether

2,7′-Cyclo-4,4′,9′-trihydroxy-3,3′,5-trimethoxy-8,8′-lign-7-en-9-al

9-Aldehyde, 3,3′,5-tri-Me ether, 9′-O-β-D-glucopyranoside

Vitexnegheteroin F. Vitecannaside C

[2135569-93-4]

$C_{27}H_{32}O_{12}$ 548.541

Constit. of the seeds and the aerial parts of *Vitex negundo* var. *heterophylla*. Exhibits ABTS radical scavenging activity. Yellowish oil. $[\alpha]_D^{20}$ -126.3 (c, 0.1 in MeOH) (Vitexnegheteroin F). $[\alpha]_D^{20}$ -91.6 (c, 0.19 in MeOH) (Vitecannaside C). λ_{max} 255 (MeOH) (Vitexnegheteroin F). [neutral] λ_{max} 254 (log ε 2.74); 358 (log ε 2.66) (MeOH) (Vitecannaside C).

Miyazawa, M. *et al., Phytochemistry*, 1996, **42**, 531-533 (*Magnoliadiol*).

Hu, P. *et al., Fitoterapia*, 2016, **111**, 147-153 (*Vitexnegheteroin F, activity*).

Nie, X.-F. *et al., J. Asian Nat. Prod. Res.*, 2016, **18**, 656-661 (*Vitecannaside C*)

2,2′-Cyclo-8,8′-lign-7-ene-3,3′,4,4′,5,5′-hexol C267

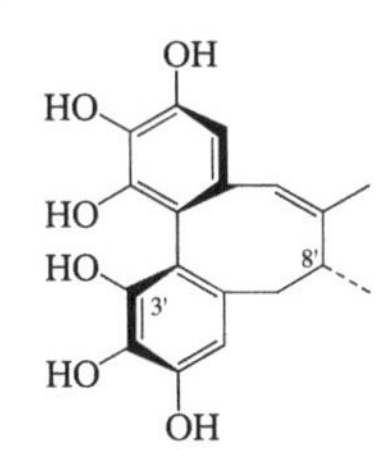

3,3′,4,4′,5,5′-Hexahydroxy-2,2′-cyclolign-7-ene

$C_{18}H_{18}O_6$ 330.336

(8′R,R_{axial})-form

3,4,4′,5,5′-Penta-Me ether, 3′-angeloyl

Neglschisandrin F

[1464719-45-6]

$C_{28}H_{34}O_7$ 482.573

Constit. of the stems of *Schisandra neglecta*. Weakly cytotoxic against human colorectal (HCT-8) and human lung (A549) carcinoma cell lines. Amorph. powder. $[\alpha]_D^{22}$ +16.9 (c, 0.31 in MeOH). λ_{max} 216 (log ε 4.78); 250 (log ε 4.37); 279 (log ε 3.97) (MeOH).

Chen, M. *et al., Molecules*, 2013, **18**, 2297-2306 (*Neglschisandrin F, cd, activity*)

2,2′-Cyclo-8,8′-lign-8-ene-3,3′,4,4′,5,5′-hexol C268

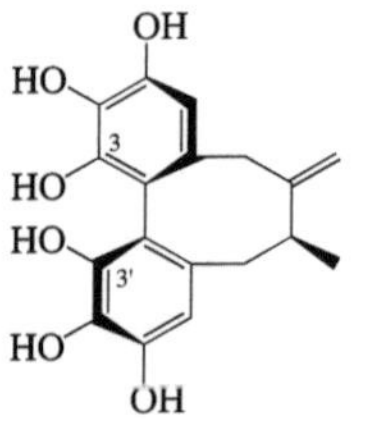

3,3′,4,4′,5,5′-Hexahydroxy-2,2′-cyclolign-8-ene

$C_{18}H_{18}O_6$ 330.336

(8′S,R$_{axial}$)-form

3,4,4′,5,5′-Penta-Me ether, 3′-angeloyl

Neglschisandrin B

[1015009-48-9]

$C_{28}H_{34}O_7$ 482.573

Constit. of the stems of *Schisandra neglecta*. Powder. $[\alpha]_D^{22}$ +17.9 (c, 0.11 in MeOH). λ_{max} 216 (log ε 4.21); 248 (log ε 3.77); 280 (log ε 3.26) (MeOH).

Chen, M. *et al., Molecules*, 2008, **13**, 548-555 (*Neglschisandrin B*)

2,7′-Cyclo-8,8′-lign-7-ene-2′,3,4,4′,5′,6-hexol C269

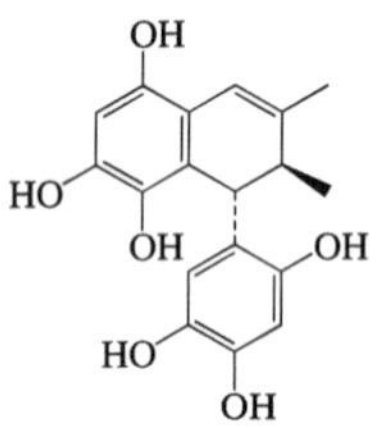

5,6-Dihydro-6,7-dimethyl-5-(2,4,5-trihydroxyphenyl)-1,3,4-naphthalenetriol.

2′,3,4,4′,5′,6-Hexahydroxy-2,7′-cyclolign-7-ene

$C_{18}H_{18}O_6$ 330.336

(7′RS,8′SR)-form

Hexa-Me ether

2,7′-Cyclo-2′,3,4,4′,5′,6-hexamethoxy-8,8′-lign-7-ene. 2′,3,4,4′,5′,6-Hexamethoxy-2,7′-cyclolign-7-ene. ***Magnoshinin****. MSI*

[86702-02-5]

$C_{24}H_{30}O_6$ 414.498

Constit. of *Magnolia salicifolia* and rhizomes of *Acorus gramineus*. Shows antiinflammatory activity. Weak inhibitor of various cancer cell lines. Moderate inhibitor of interleukin-stimulated tube formation. Cryst. (Et_2O or diisopropyl ether). Mp 113.5 - 115° (107-109°). λ_{max} 232 (ε 20893); 269 (ε 16218); 278 (ε 16596); 310 (ε 8912); 320 (ε 8318) (MeOH) (Berdy).

Kikuchi, T. *et al., Chem. Pharm. Bull.*, 1983, **31**, 1112-1114 (*Magnolia salicifolia constit*).
Kadota, S. *et al., Tetrahedron Lett.*, 1987, **28**, 2857-2860 (*Magnoshinin, synth, activity*).
Yoshida, S. *et al., Synlett*, 1994, 895-898 (*Magnoshinin, synth*).
Kobayashi, S. *et al., Biol. Pharm. Bull.*, 1996, **19**, 1304-1306 (*activity*).
Yvon, B.L. *et al., Synthesis*, 2001, 1556-1560 (*Magnoshinin, synth, cmr, ms*).
Lee, S.Y. *et al., Biosci., Biotechnol., Biochem.*, 2013, **77**, 276-280 (*Acorus gramineus constit, activity*)

2,7′-Cyclo-8,8′-lign-7-ene-3′,4,4′,5,5′,6-hexol C270

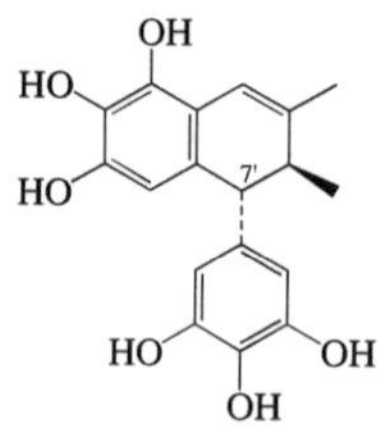

5,6-Dihydro-6,7-dimethyl-5-(3,4,5-trihydroxyphenyl)-1,2,3-naphthalenetriol.

3′,4,4′,5,5′,6-Hexahydroxy-2,7′-cyclolign-7-ene

$C_{18}H_{18}O_6$ 330.336

(7′R*,8′S*)-form

3′,4,5′,6-Tetra-Me ether

2,7′-Cyclo-3′,4,5′,6-tetramethoxy-8,8′-lign-7-ene-4′,5-diol. 4′,5-Dihydroxy-3′,4,5′,6-tetramethoxy-2,7′-cyclolign-7-ene. ***Sacidumlignan B****. Orthosilignin*

[781658-08-0]

$C_{22}H_{26}O_6$ 386.444

Constit. of *Orthosiphon wulfenioides* and *Sarcostemma acidum*. Amorph. powder. $[\alpha]_D^{20}$ -116 (c, 1.44 in Me_2CO). λ_{max} 204 (log ε 4.69); 279 (log ε 4.06) (MeOH).

Xiang, W. *et al., Yunnan Zhiwu Yanjiu*, 2002, **24**, 535-538 (*Orthosilignin*).
Gan, L.-S. *et al., J. Nat. Prod.*, 2005, **68**, 221-225 (*Sacidumlignan B*).
Rout, J.K. *et al., J. Org. Chem.*, 2012, **77**, 1566-1571 (*synth*).
Reddel, J.C.T. *et al., Chem. Sci.*, 2017, **8**, 2156-2160 (*synth*).
Luo, Z.-B. *et al., Org. Biomol. Chem.*, 2020, **18**, 2054-2057 (*synth*)

2,7′-Cyclo-8,8′-lign-7-ene-3′,4,4′,5,9,9′-hexol C271

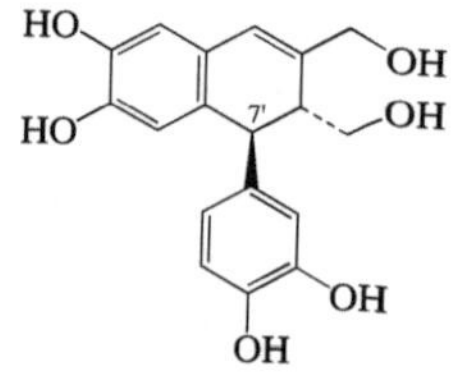

(7′S,8′R)-form

3′,4,4′,5,9,9′-Hexahydroxy-2,7′-cyclolign-7-ene

$C_{18}H_{18}O_6$ 330.336

(7′S,8′R)-form

9-Aldehyde, 3′,5-di-Me ether

2,7′-Cyclo-4,4′,9′-trihydroxy-3′,5-dimethoxy-8,8′-lign-7-en-9-al. 4,4′,9′-Trihydroxy-3′,5-dimethoxy-2,7′-cyclolign-7-en-9-al. 7,8,9,9-Tetradehydroisolariciresinol. ***Vitexin 1****. VB1*

[357645-16-0]

$C_{20}H_{20}O_6$ 356.374

Constit. of the seeds of *Vitex negundo*. Antitumour agent. Cryst. (MeOH). Mp 126 - 127°. $[\alpha]_D^{26}$ -176 (c, 0.005 in MeOH). λ_{max} 255 (log ε 4.31); 359 (log ε 4.24) (MeOH). [base] λ_{max} 279 (log ε 4.2); 427 (log ε 4.43) (MeOH/NaOH)

9-Aldehyde, 3′,4,4′,5-tetra-Me ether

2,7′-Cyclo-9′-hydroxy-3′,4,4′,5-tetramethoxy-8,8′-lign-7-en-9-al. ***Vitexin 6****. VB6*

[357418-62-3]

$C_{22}H_{24}O_6$ 384.428

Constit. of the seeds of *Vitex negundo*. Antitumour agent

(7′R*,8′S*)-form

3′,5-Di-Me ether

2,7′-Cyclo-3′,5-dimethoxy-8,8′-lign-7-ene-4,4′,9,9′-tetrol. 4,4′,9,9′-Tetrahydroxy-3′,5-dimethoxy-2,7′-cyclolign-7-ene.

Negundin B

[844692-51-9]

$C_{20}H_{22}O_6$ 358.39

Constit. of the roots of *Vitex negundo*. Potent lipoxygenase inhibitor. Amorph. solid. $[\alpha]_D^{26}$ -56 (c, 0.11 in MeOH). λ_{max} 222 (log ε 4.7); 283 (log ε 4.34) (MeOH)

9-Aldehyde, 3′-Me ether

2,7′-Cyclo-4,4′,5,9′-tetrahydroxy-3′-methoxy-8,8′-lign-7-en-9-al. 4,4′,5,9′-Tetrahydroxy-3′-methoxy-2,7′-cyclolign-7-en-9-al. ***Vitexdoin A***

[1186021-77-1]
$C_{19}H_{18}O_6$ 342.347
Constit. of the seeds of *Vitex negundo*. Amorph. yellow powder. $[\alpha]_D^{15}$ -156 (c, 0.3 in MeOH). λ_{max} 203; 258; 369 (MeOH)

9-Aldehyde, 3′-Me ether, 9′-O-β-D-glucopyranoside
Vitexnegheteroin E
$C_{25}H_{28}O_{11}$ 504.488
Constit. of the seeds of *Vitex negundo* var. *heterophylla*. Exhibits ABTS radical scavenging activity and inhibitory activity on LPS-induced NO prodn.. Yellowish oil. $[\alpha]_D^{20}$ -99.8 (c, 1.10 in MeOH). λ_{max} 259 (MeOH)

9-Aldehyde, 3′,5-di-Me ether, 9′-O-β-D-glucopyranoside
Vitecannaside B
[944711-14-2]
$C_{26}H_{30}O_{11}$ 518.515
Constit. of the fruit of *Vitex cannabifolia*. Amorph. powder. $[\alpha]_D^{13}$ -139.1 (c, 0.5 in MeOH)

9-Aldehyde, 3′,5-di-Me ether, 9′-hexadecanoyl
Vitexdoin G
$C_{36}H_{50}O_7$ 594.789
Isol. from seeds of *Vitex negundo*. Antiinflammatory and antiosteoporotic agent. Syrup. $[\alpha]_D^{20}$ -140 (c, 0.2 in MeOH). λ_{max} 223; 256; 358 (MeOH)

9-Carboxylic acid, 3′,5-di-Me ether, 9′-benzoyl, [(4-hydroxyphenyl) ethyl]amide
Acorusin A
$C_{35}H_{33}NO_8$ 595.648
Constit. of the rhizomes of *Acorus tatarinowii*. Yellow solid. Racemate. λ_{max} 227 (log ε 3.95); 283 (log ε 3.46); 328 (log ε 3.53) (MeOH).

Chawla, A.S. *et al.*, *Phytochemistry*, 1992, **31**, 4378-4379 (*Vitex negundo constit*).
Azhar-ul-Haq, *et al.*, *Chem. Pharm. Bull.*, 2004, **52**, 1269-1272 (*Negundin B*).
Ono, M. *et al.*, *J. Nat. Prod.*, 2004, **67**, 2073-2075 (*Vitex negundo constit*).
Yamasaki, T. *et al.*, *J. Nat. Med.*, 2008, **62**, 47-51 (*Vitecannaside B*).
Zhou, Y. *et al.*, *Clin. Cancer Res.*, 2009, **15**, 5161-5169 (*VB1, activity*).
Zheng, C.-J. *et al.*, *J. Nat. Prod.*, 2009, **72**, 1627-1630 (*Vitexdoin A, activity*).
Zhou, J. *et al.*, *Anti-Cancer Drugs*, 2013, **24**, 928-936 (*Vitexin 6, activity*).
Zheng, C.-J. *et al.*, *Fitoterapia*, 2014, **93**, 31-38 (*Vitexdoin G*).
Hu, P. *et al.*, *Fitoterapia*, 2016, **111**, 147-153 (*Vitexnegheteroin E, activity*).
Ni, G. *et al.*, *Planta Med.*, 2016, **82**, 632-638 (*Acorusin A*)

2,2′-Cyclo-8,8′-lign-8-ene-3,3′,4,4′,5,5′,7,7′-octol C272

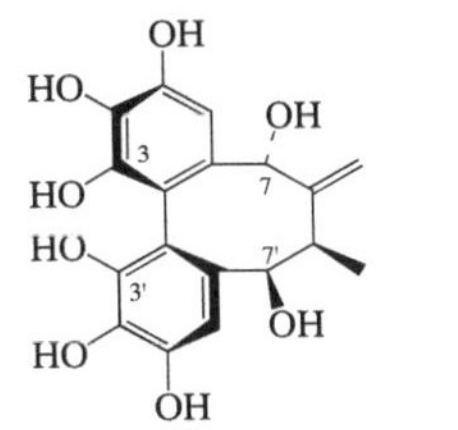

3,3′,4,4′,5,5′,7,7′-Octahydroxy-2,2′-cyclolign-8-ene
$C_{18}H_{18}O_8$ 362.334
(7R,7′R,8′R,S_{axial})-form

4′,5′-Methylene, 3′,4,5-tri-Me ether, 7,7′-diangeloyl
Kadheterin A
$C_{32}H_{36}O_{10}$ 580.630
Constit. of the stems of *Kadsura heteroclita*. Exhibits moderate cytotoxic activity. Needles (MeOH/$CHCl_3$). Mp 186 - 187°. $[\alpha]_D^{26}$ +109.7 (c, 0.12 in MeOH). λ_{max} 220 (log ε 4.74); 288 (log ε 3.72); 499 (log ε 2.51) (MeOH).

Luo, Y.-Q. *et al.*, *Fitoterapia*, 2017, **119**, 150-157 (*Kadheterin A, cryst struct, activity*)

2,7′-Cyclo-8,8′-lign-7-ene-3,3′,4,4′,5,5′,9,9′-octol C273

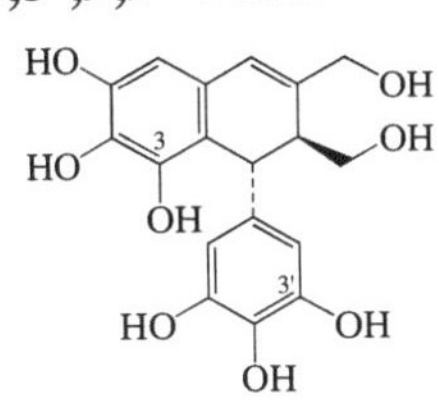

3,3′,4,4′,5,5′,9,9′-Octahydroxy-2,7′-cyclolign-7-ene
$C_{18}H_{18}O_8$ 362.334
(7′R,8′S)-form

3,3′,5,5′,9-Penta-Me ether
2,7′-Cyclo-3,3′,5,5′,9-pentamethoxy-8,8′-lign-7-ene-4,4′,9′-triol. 4,4′,9′-Trihydroxy-3,3′,5,5′,9-pentamethoxy-2,7′-cyclolign-7-ene
$C_{23}H_{28}O_8$ 432.469
Constit. of the rhizomes of *Sinomenium acutum*. Significantly decreases the prodn. of NO in LPS-induced BV-2 cells. Yellowish gum. $[\alpha]_D$ +15.8 (c, 0.06 in MeOH). λ_{max} 220 (log ε 4.14); 255 (log ε 3.88); 356 (log ε 3.56) (MeOH).

Lee, S.Y. *et al.*, *Phytochem. Lett.*, 2016, **17**, 79-84 (*Sinomenium acutum constit, activity*)

2,7′-Cyclo-8,8′-lign-7′-ene-3,3′,4,4′,5-pentol C274

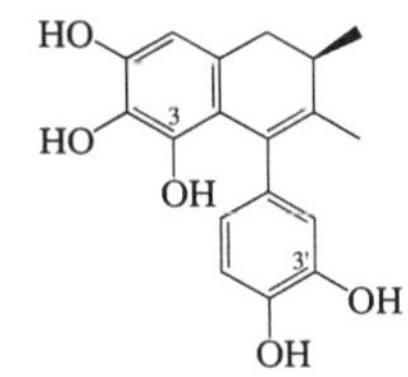

8-(3,4-Dihydroxyphenyl)-5,6-dihydro-6,7-dimethyl-1,2,3-naphthalenetriol. 3,3′,4,4′,5-Pentahydroxy-2,7′-cyclolign-7′-ene
$C_{18}H_{18}O_5$ 314.337
(R)-form

3,4:3′,4′-Bis(methylene), 5-Me ether
2,7′-Cyclo-5-methoxy-3,4:3′,4′-bis(methylenedioxy)-7′-lignene. 5-Methoxy-3,4:3′,4′-bis(methylenedioxy)-2,7′-cyclolign-7′-ene
[98767-35-2]
$C_{21}H_{20}O_5$ 352.386
Constit. of *Virola calophylloidea*. Cryst. (MeOH). Mp 152°. $[\alpha]_D^{20}$ +51.7 (c, 0.79 in $CHCl_3$). Probable abs. config. Poss. artifact. λ_{max} 279 (ε 1200); 292 (ε 1050); 332 (ε 300) (MeOH).

Martinez, V.J.C. *et al.*, *Phytochemistry*, 1985, **24**, 1867-1868 (*Virola calophylloidea constit*)

2,7′-Cyclo-8,8′-lign-8′-ene-3′,4,4′,5,9-pentol C275

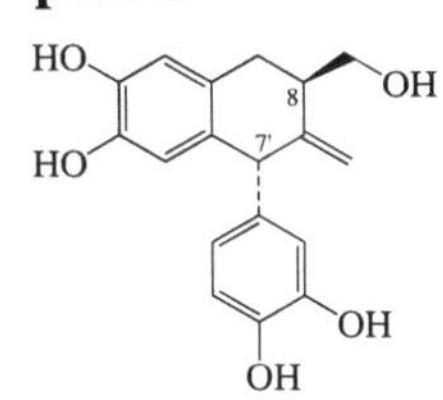

3′,4,4′,5,9-Pentahydroxy-2,7′-cyclolign-8′-ene
$C_{18}H_{18}O_5$ 314.337
(7′R,8R)-form

3′,5-Di-Me ether
2,7′-Cyclo-3′,5-dimethoxy-8,8′-lign-8′-ene-4,4′,9-triol. 4,4′,9-Trihydroxy-3′,5-dimethoxy-2,7′-cyclolign-8′-ene. ***Alashinol A***
[2226511-92-6]
$C_{20}H_{22}O_5$ 342.391
Constit. of the stem bark of *Syringa pinnatifolia*. Inhibits LPS-induced

NO prodn. and prodn. of inflammatory mediators TNF-α and IL-6. Exhibits neuroprotective effect against glutamate-induced injury in PC12 line. Powder. $[\alpha]_D^{25}$ -54.4 (c, 0.1 in MeOH). λ_{max} 206 (log ε 4.41); 230 (log ε 1.90); 283 (log ε 0.75) (MeOH).

Su, G. *et al.*, *Fitoterapia*, 2016, **114**, 63-68 (*Alashinol A, activity*)

2,7′-Cyclo-8,8′-lign-7-ene-3′,4,4′,5-tetrol C276

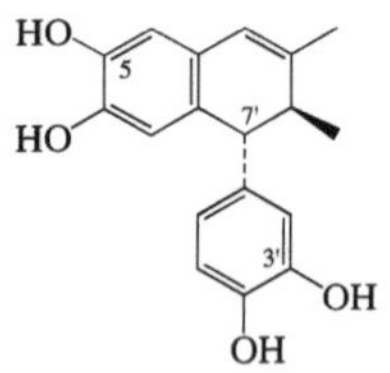

(7′R,8′S)-form

5-(3,4-Dihydroxyphenyl)-5,6-dihydro-6,7-dimethyl-2,3-naphthalenediol. 3′,4,4′,5-Tetrahydroxy-2,7′-cyclolign-7-ene

$C_{18}H_{18}O_4$ 298.338

(7′R,8′S)-form

Tetra-Me ether

2,7′-Cyclo-3,4,4′,5-tetramethoxy-7-lignene. 3,4,4′,5-Tetramethoxy-2,7′-cyclolign-7-ene. ***(-)-Cyclogalgravin***

[887501-32-8]

$C_{22}H_{26}O_4$ 354.446

Constit. of the roots of *Holostylis reniformis*. Amorph. yellow solid (MeOH or Et_2O). Mp 89 - 91°. $[\alpha]_D^{20}$ -128 (c, 0.2 in $CHCl_3$) (-105.9). λ_{max} 241 (log ε 3.7); 280 (log ε 3.4) (MeOH)

(7′S,8′R)-form

3′,5-Di-Me ether

2,7′-Cyclo-3′,5-dimethoxy-8,8′-lign-7-ene-4,4′-diol. 4,4′-Dihydroxy-3′,5-dimethoxy-2,7′-cyclolign-7-ene. ***Myrisfragransin***

[135962-21-9]

[180470-31-9]

$C_{20}H_{22}O_4$ 326.392

Isol. from *Myristica fragrans*. Powder. $[\alpha]_D$ +44.8 (c, 1.0 in $CHCl_3$). Stereochem. ambiguously shown in the ref. and stated to be (*S,S*-). Methylation gave (+)-Cyclogalgravin but the opt. rotn. did not correspond

Tetra-Me ether

Cyclogalgravin

[73366-01-5]

$C_{22}H_{26}O_4$ 354.446

Isol. from heartwood knots of *Araucaria angustifolia*. Mp 89 - 90°. $[\alpha]_D$ +135.5 (c, 0.28 in $CHCl_3$) (+23.7)

(7′R*,8′S*)-form

3′,4′-Methylene, 4,5-di-Me ether

2,7′-Cyclo-4,5-dimethoxy-3′,4′-methylenedioxy-7-lignene. 4,5-Dimethoxy-3′,4′-methylenedioxy-2,7′-cyclolign-7-ene. ***Pycnanthulignene B***

[1207532-17-9]

$C_{21}H_{22}O_4$ 338.403

Constit. of the roots of *Pycnanthus angolensis*. Yellow viscous oil. $[\alpha]_D^{20}$ +118.5 (c, 0.34 in EtOH).

Traverso, G. *et al.*, *Farmaco, Ed. Sci.*, 1961, **16**, 445-452 (*tetra-Me ether, synth*).
Fonseca, S. *et al.*, *Phytochemistry*, 1979, **18**, 1703-1708 (*Cyclogalgravin*).
Miyazawa, M. *et al.*, *Nat. Prod. Lett.*, 1996, **8**, 25-26 (*Myrisfragransin*).
Yvon, B.L. *et al.*, *Synthesis*, 2001, 1556-1560 (*Cyclogalgravin, synth, pmr, ms*).
Da Silva, T. *et al.*, *Phytochemistry*, 2006, **67**, 929-937 (*(-)-Cyclogalgravin*).
Nono, E.C.N. *et al.*, *J. Nat. Prod.*, 2010, **73**, 213-216 (*Pycnanthulignene B*).
Rye, C.E. *et al.*, *J. Org. Chem.*, 2011, **76**, 6636-6648 (*synth*).
Reddel, J.C.T. *et al.*, *Angew. Chem., Int. Ed.*, 2014, **53**, 1395-1398 (*synth*).
Sakuma, D. *et al.*, *Chem. Lett.*, 2014, **43**, 610-611 (*synth*).
Reddel, J.C.T. *et al.*, *Chem. Sci.*, 2017, **8**, 2156-2160 (*synth*)

2,7′-Cyclo-8,8′-lign-7′-ene-3′,4,4′,5-tetrol C277

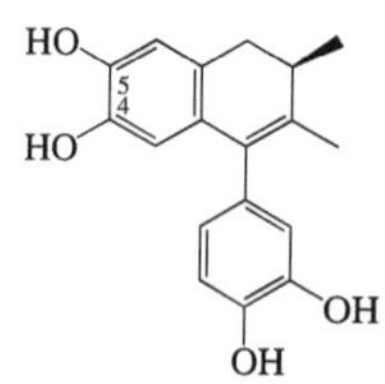

8-(3,4-Dihydroxyphenyl)-5,6-dihydro-6,7-dimethyl-2,3-naphthalenediol. 3′,4,4′,5-Tetrahydroxy-2,7′-cyclolign-7′-ene

$C_{18}H_{18}O_4$ 298.338

(R)-form

3′,4′-Methylene, 4,5-di-Me ether

2,7′-Cyclo-4,5-dimethoxy-3′,4′-methylenedioxy-7′-lignene. 4,5-Dimethoxy-3′,4′-methylenedioxy-2,7′-cyclolign-7′-ene. ***Dehydroisogalcatin***

[129931-97-1]

$C_{21}H_{22}O_4$ 338.403

Isol. from bark of *Virola calophylla*. Cryst. (EtOAc/hexane). Mp 62 - 63°. $[\alpha]_D$ +53.2 (c, 0.7 in $CHCl_3$). λ_{max} 238 (ε 11250); 287 (ε 3750); 330 (ε 1150) (MeOH).

Martinez, J.C. *et al.*, *Phytochemistry*, 1990, **29**, 2655-2657 (*Virola calophylla constit*)

2,7′-Cyclo-8,8′-lign-7-ene-4,4′,5-triol C278

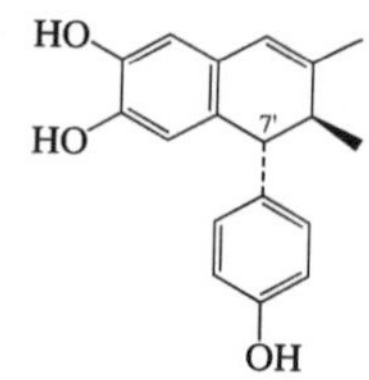

5,6-Dihydro-5-(4-hydroxyphenyl)-6,7-dimethyl-2,3-naphthalenediol. 4,4′,5-Trihydroxy-2,7′-cyclolign-7-ene

$C_{18}H_{18}O_3$ 282.339

(7′R*,8′S*)-form

4,5-Di-Me ether

2,7′-Cyclo-4,5-dimethoxy-8,8′-lign-7-en-4′-ol. 4′-Hydroxy-4,5-dimethoxy-2,7′-cyclolign-7-ene.

Pycnanthulignene A

[1207532-16-8]

$C_{20}H_{22}O_3$ 310.393

Constit. of the roots of *Pycnanthus angolensis*. Shows antimicrobial activity against a range of drug-resistant pathogens. Powder ($CHCl_3$). Mp 141 - 143°. $[\alpha]_D^{20}$ +33.8 (c, 0.31 in EtOH).

Nono, E.C.N. *et al.*, *J. Nat. Prod.*, 2010, **73**, 213-216 (*Pycnanthulignene A*).
Rye, C.E. *et al.*, *J. Org. Chem.*, 2011, **76**, 6636-6648 (*synth*).
Kuete, V. *et al.*, *Nat. Prod. Res.*, 2011, **25**, 432-443 (*Pycnanthulignene A, activity*)

2,7′-Cyclo-9′-nor-8,8′-ligna-7,7′-diene-3′,4,4′,5,9-pentol C279

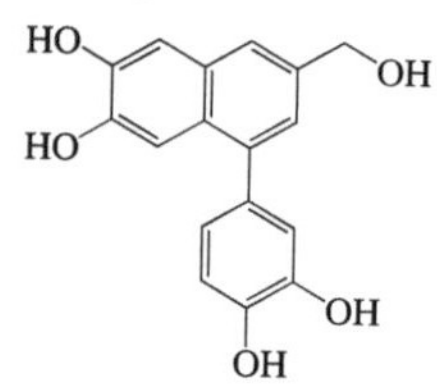

4-(3,4-Dihydroxyphenyl)-6,7-dihydroxy-2-naphthalenemethanol. 3′,4,4′,5,9-Pentahydroxy-9′-nor-2,7′-cycloligna-7,7′-diene

$C_{17}H_{14}O_5$ 298.294

3′,5-Di-Me ether
3′,5-Dimethoxy-9′-nor-2,7′-cyclo-ligna-7,7′-diene-4,4′,9-triol. ***Noralashinol C***
[1472708-67-0]
$C_{19}H_{18}O_5$ 326.348
Constit. of the stem bark of *Syringa pinnatifolia*. Powder. λ_{max} 240 (log ε 5.49); 288 (log ε 1.17) (MeOH).

Zhang, R.-F. *et al.*, *J. Asian Nat. Prod. Res.*, 2017, **19**, 416-422 (*Noralashinol C*)

2,7′-Cyclo-3,3′,4,4′,5,7,8,8′-octahydroxy-8,8′-lignan-9′,9-olide C280

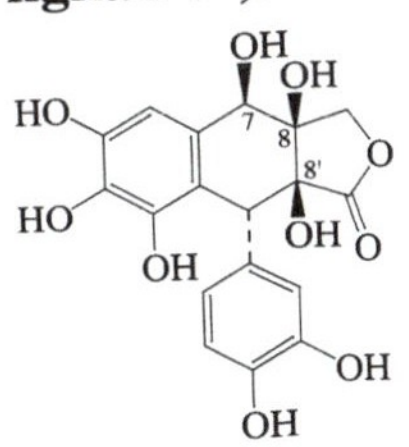

3,3′,4,4′,5,7,8,8′-Octahydroxy-2,7′-cyclolignan-9′,9-olide
$C_{18}H_{16}O_{10}$ 392.316
(7R*,7′R*,8R*,8′S*)-form
3′,4′-Methylene, 3,4,5-tri-Me ether
2,7′-Cyclo-7,8,8′-trihydroxy-3,4,5-trimethoxy-3′,4′-methylenedioxy-8,8′-lignan-9′,9-olide. 7,8,8′-Trihydroxy-3,4,5-trimethoxy-3′,4′-methylenedioxy-2,7′-cyclolignan-9′,9-olide
$C_{22}H_{22}O_{10}$ 446.408
3′,4′-Methylene, 3,4,5-tri-Me ether, 7-angeloyl
Erlangerin A
[465535-61-9]
$C_{27}H_{28}O_{11}$ 528.51
Constit. of the resin of *Commiphora erlangeriana*. Cryst. ($CHCl_3$/hexane). Mp 76 - 78°. $[\alpha]_D^{22}$ +55 (c, 0.2 in $CHCl_3$). λ_{max} 208 (log ε 2.86); 223 (log ε 2.88); 284 (log ε 2.32) (EtOH).

Dekebo, A. *et al.*, *J. Nat. Prod.*, 2002, **65**, 1252-1257 (*Erlangerin A, cryst struct*)

2,7′-Cyclo-2′,4,4′,5,5′-pentahydroxy-8,8′-ligna-7,7′-dien-9,9′-olide C281

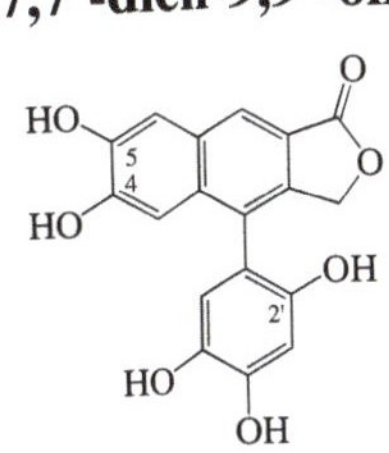

2′,4,4′,5,5′-Pentahydroxy-2,7′-cycloligna-7,7′-dien-9,9′-olide
$C_{18}H_{12}O_7$ 340.287
4′,5′-Methylene, 2′,4,5-tri-Me ether
2,7′-Cyclo-2′,4,5-trimethoxy-4′,5′-methylenedioxy-8,8′-ligna-7,7′-dien-9,9′-olide. 2′,4,5-Trimethoxy-4′,5′-methylenedioxy-2,7′-cycloligna-7,7′-dien-9,9′-olide. ***Procumphthalide A***
[775342-84-2]
$C_{22}H_{18}O_7$ 394.379
Constit. of *Justicia procumbens* and *Justicia prostrata*. Amorph. powder. λ_{max} 213 (log ε 3.35); 250 (log ε 3.23); 310 (sh) (log ε 2.69) (MeOH)
4,5:4′,5′-Bis(methylene) ether
2,7′-Cyclo-2′-hydroxy-4,5:4′,5′-bis(methylenedioxy)-8,8′-ligna-7,7′-dien-9,9′-olide. 2′-Hydroxy-4,5:4′,5′-bis(methylenedioxy)-2,7′-cycloligna-7,7′-dien-9,9′-olide. ***Prostalidin C***
[73428-14-5]
$C_{20}H_{12}O_7$ 364.309
Isol. from *Justicia prostrata*. Mp 263 - 265°.

Ghosal, S. *et al.*, *Chem. Ind. (London)*, 1979, 854-855 (*Prostalidin C*).
Weng, J.-R. *et al.*, *Arch. Pharm. (Weinheim, Ger.)*, 2004, **337**, 207-212 (*Procumphthalide A*).
Pandey, D.P. *et al.*, *Asian J. Chem.*, 2011, **23**, 2125-2127 (*Justicia procumbens constit*)

2,7′-Cyclo-2′,4,4′,5,5′-pentahydroxy-8,8′-ligna-7,7′-dien-9′,9-olide C282

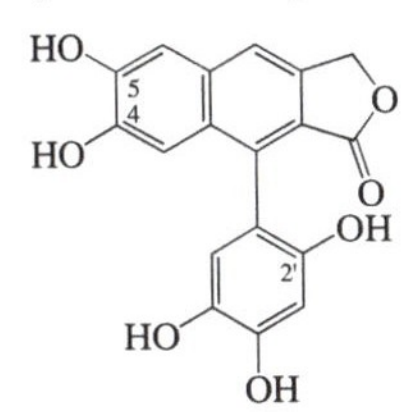

2′,4,4′,5,5′-Pentahydroxy-2,7′-cycloligna-7,7′-dien-9′,9-olide
$C_{18}H_{12}O_7$ 340.287
4′,5′-Methylene, 4,5-di-Me ether
2,7′-Cyclo-2′-hydroxy-4,5-dimethoxy-4′,5′-methylenedioxy-8,8′-ligna-7,7′-dien-9′,9-olide. 2′-Hydroxy-4,5-dimethoxy-4′,5′-methylenedioxy-2,7′-cycloligna-7,7′-dien-9′,9-olide. ***Justicidin H.*** *6′-Hydroxyjusticidin B*
[145726-39-2]
$C_{21}H_{16}O_7$ 380.352
Constit. of *Justicia procumbens*. Pale yellow powder
4′,5′-Methylene, 4,5-di-Me ether, 2′-O-β-D-glucopyranoside
Justicidinoside C
[177912-23-1]
$C_{27}H_{26}O_{12}$ 542.493
Constit. of aerial parts of *Justicia procumbens*. Powder. Mp 166° dec. $[\alpha]_D^{25}$ -15.2 (c,0.5 in MeOH). λ_{max} 258 (log ε 4.68); 300 (log ε 4.04); 347 (log ε 3.64) (MeOH).

Japan. Pat., 1992, 211 609 (*Justicidin H*).
Asano, J. *et al.*, *Phytochemistry*, 1996, **42**, 713-717 (*Justicidinoside C*).
Yang, M. *et al.*, *Magn. Reson. Chem.*, 2006, **44**, 727-730 (*Justicidin H*)

2,7′-Cyclo-3,3′,4,4′,5′-pentahydroxy-8,8′-ligna-7,7′-dien-9,9′-olide C283

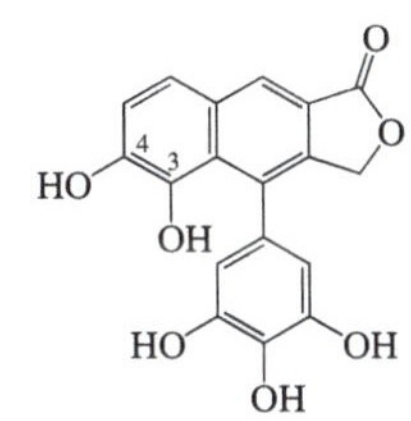

3,3′,4,4′,5′-Pentahydroxy-2,7′-cycloligna-7,7′-dien-9,9′-olide
$C_{18}H_{12}O_7$ 340.287
3,4:3′,4′-Bis(methylene) ether
2,7′-Cyclo-3′-hydroxy-3,4:4′,5′-bis(methylenedioxy)-8,8′-ligna-7,7′-dien-9,9′-olide. 3′-Hydroxy-3,4:4′,5′-bis(methylenedioxy)-2,7′-cycloligna-7,7′-dien-9,9′-olide. ***Justicinol‡***
[75340-41-9]
$C_{20}H_{12}O_7$ 364.309
Constit. of the leaves of *Justicia flava*. Shows mild CNS motor effects. Yellow prisms. Mp 250 - 252°. Not to be confused with Justicinol, C188

Olaniyi, A.A. *et al.*, *J. Nat. Prod.*, 1980, **43**, 482-486 (*Justicinol*)

2,7′-Cyclo-3,3′,4,4′,5-pentahydroxy-8,8′-ligna-7,7′-dien-9,9′-olide C284

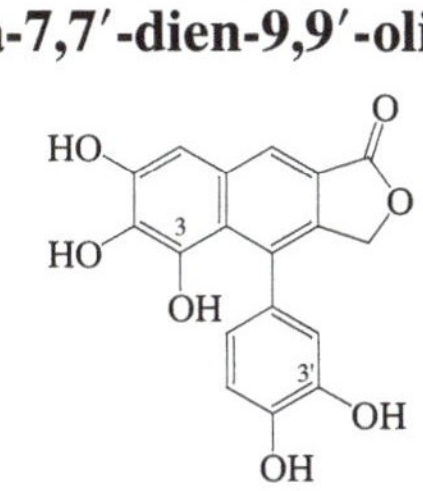

3,3′,4,4′,5-Pentahydroxy-2,7′-cycloligna-7,7′-dien-9,9′-olide
$C_{18}H_{12}O_7$ 340.287

3′,4′-Methylene, 3,4,5-tri-Me ether
2,7′-Cyclo-3,4,5-trimethoxy-3′,4′-methylenedioxy-8,8′-ligna-7,7′-dien-9,9′-olide. 3,4,5-Trimethoxy-3′,4′-methylenedioxy-2,7′-cyclolignа-7,7′-dien-9,9′-olide. ***Chaihunaphthone****. NSC 283809*
[6258-43-1]
$C_{22}H_{18}O_7$ 394.379
Constit. of the roots of *Bupleurum scorzonerifolium*. Needles. Mp 164 - 166°. λ_{max} 228 (log ε 4.4); 258 (log ε 4.48); 318 (log ε 3.84) ($CHCl_3$).

San Feliciano, A. *et al.*, *Arch. Pharm. (Weinheim, Ger.)*, 1993, **326**, 421-426 (*synth*).
Chang, W.-L. *et al.*, *Phytochemistry*, 2003, **64**, 1375-1379 (*Chaihunaphthone*).
Eur. Pat., 2005, 1 527 784 (Buddhist Tzu Chi General Hospital)(*Bupleurum scorzonerifolium constit*)

2,7′-Cyclo-3,3′,4,4′,7-pentahydroxy-8,8′-ligna-7,7′-dien-9,9′-olide C285

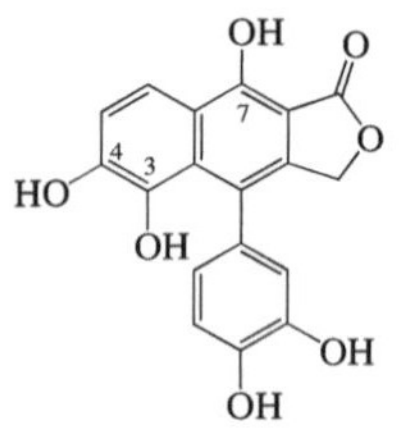

3,3′,4,4′,7-Pentahydroxy-2,7′-cyclolignа-7,7′-dien-9,9′-olide
$C_{18}H_{12}O_7$ 340.287
3,4:3′,4′-Bis(methylene), 7-Me ether
2,7′-Cyclo-7-methoxy-3,4:3′,4′-bis(methylenedioxy)-8,8′-ligna-7,7′-dien-9,9′-olide. 7-Methoxy-3,4:3′,4′-bis(methylenedioxy)-2,7′-cyclolignа-7,7′-dien-9,9′-olide. ***Lignan J1****. 7-Methoxyhelioxanthin*
[131189-73-6]
$C_{21}H_{14}O_7$ 378.336
Constit. of *Justicia hyssopifolia* and *Justicia procumbens*. Cryst. (EtOAc/hexane). Mp 208 - 209°. λ_{max} 248; 264; 358 (EtOH).

Trujillo, J.M. *et al.*, *Phytochemistry*, 1990, **29**, 2991-2993 (*Lignan J1*).
Zhou, P. *et al.*, *Molecules*, 2015, **20**, 7048-7058 (*Lignan J1, isol, purifn*)

2,7′-Cyclo-3,3′,4,4′,7-pentahydroxy-8,8′-ligna-7,7′-dien-9′,9-olide C286

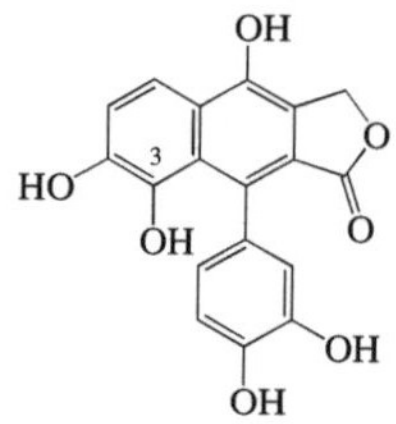

3,3′,4,4′,7-Pentahydroxy-2,7′-cyclolignа-7,7′-dien-9′,9-olide
$C_{18}H_{12}O_7$ 340.287
3′,4′-Methylene, 7-Me ether
2,7′-Cyclo-3,4-dihydroxy-7-methoxy-3′,4′-methylenedioxy-8,8′-ligna-7,7′-dien-9′,9-olide. 3,4-Dihydroxy-7-methoxy-3′,4′-methylenedioxy-2,7′-cyclolignа-7,7′-dien-9′,9-olide. ***3-De-O-methylorosunol***
[81992-86-1]
$C_{20}H_{14}O_7$ 366.325
Constit. of the roots of *Justicia flava*. Yellow needles (MeOH/ $CHCl_3$). Mp 158 - 160°
3,4-Methylene, 3′,4′-di-Me ether
2,7′-Cyclo-7-hydroxy-3′,4′-dimethoxy-3,4-methylenedioxy-8,8′-ligna-7,7′-dien-9′,9-olide. 7-Hydroxy-3′,4′-dimethoxy-3,4-methylenedioxy-2,7′-cyclolignа-7,7′-dien-9′,9-olide
[104756-96-9]
$C_{21}H_{16}O_7$ 380.352
Constit. of *Bupleurum marginatum*. Yellow cryst.. Mp 259 - 260°. λ_{max} 204 (log ε 4.82); 254 (log ε 4.85); 307 (log ε 2.97) (MeOH)
3′,4′-Methylene, 3,7-di-Me ether
2,7′-Cyclo-4-hydroxy-3,7-dimethoxy-3′,4′-methylenedioxy-8,8′-ligna-7,7′-dien-9′,9-olide. 4-Hydroxy-3,7-dimethoxy-3′,4′-methylenedioxy-2,7′-cyclolignа-7,7′-dien-9′,9-olide. ***Orosunol***
[82012-44-0]
$C_{21}H_{16}O_7$ 380.352
Constit. of the roots of *Justicia flava*. Yellow prisms (MeOH/ $CHCl_3$). Mp 172 - 173°
3,4:3′,4′-Bis(methylene) ether
2,7′-Cyclo-7-hydroxy-3,4:3′,4′-bis(methylenedioxy)-8,8′-ligna-7,7′-dien-9′,9-olide. 7-Hydroxy-3,4:3′,4′-bis(methylenedioxy)-2,7′-cyclolignа-7,7′-dien-9′,9-olide. ***Justirumalin***
[247591-64-6]
$C_{20}H_{12}O_7$ 364.309
Constit. of *Justicia neesii*. Shows weak to moderate antiviral activity. Yellow solid. Mp 258 - 260°. λ_{max} 254; 318 (MeOH).

Olaniyi, A.A. *Planta Med.*, 1982, **44**, 154-156 (*Orosunol, 3-Demethylorosunol*).
Rajasekhar, D. *et al.*, *Indian J. Chem., Sect. B*, 1999, **38**, 713-717 (*Justirumalin*).
Yeo, H. *et al.*, *J. Med. Chem.*, 2005, **48**, 534-546 (*Justirumalin, activity*).
Liu, Y. *et al.*, *Helv. Chim. Acta*, 2008, **91**, 2316-2320 (*Bupleurum marginatum constit*)

2,7′-Cyclo-3,3′,4,4′,9′-pentahydroxy-8,8′-ligna-7,7′-dien-9,9′-olide C287

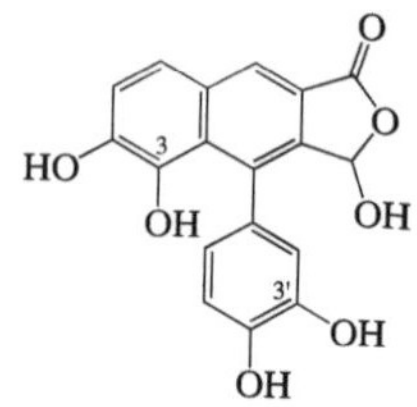

3,3′,4,4′,9′-Pentahydroxy-2,7′-cyclolignа-7,7′-dien-9,9′-olide
$C_{18}H_{12}O_7$ 340.287
(9′ξ)-form
3,4:3′,4′-Bis(methylene ether)
2,7′-Cyclo-9′-hydroxy-3,4:3′,4′-bis(methylenedioxy)-8,8′-ligna-7,7′-dien-9,9′-olide. 9′-Hydroxy-3,4:3′,4′-bis(methylenedioxy)-2,7′-cyclolignа-7,7′-dien-9,9′-olide. ***Jusmicranthin***
[247591-61-3]
$C_{20}H_{12}O_7$ 364.309
Constit. of *Justicia neesii* and *Justicia purpurea*. Yellow cryst. (EtOAc/petrol) (as Me ether). Mp 196 - 198°. λ_{max} 265; 357 (MeOH) (Me ether).

Rajasekhar, D. *et al.*, *Indian J. Chem., Sect. B*, 1999, **38**, 713-717 (*Jusmicranthin*).
Rajasekhar, D. *Fitoterapia*, 2000, **71**, 598-599 (*Justicia neesii constit*).
Kavitha, J. *et al.*, *J. Nat. Prod.*, 2003, **66**, 1113-1115 (*Justicia purpurea constit*)

2,7′-Cyclo-3′,4,4′,5,5′-pentahydroxy-8,8′-ligna-7,7′-dien-9′,9-olide C288

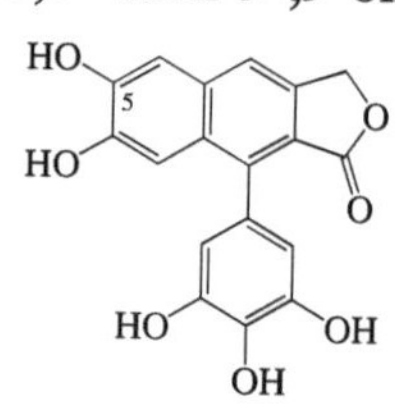

3′,4,4′,5,5′-Pentahydroxy-2,7′-cyclolignа-7,7′-dien-9′,9-olide

$C_{18}H_{12}O_7$ 340.287

3′,5-Di-Me ether

2,7′-Cyclo-3′,4,4′-trihydroxy-5,5′-dimethoxy-8,8′-ligna-7,7′-dien-9′,9-olide. 3′,4,4′-Trihydroxy-5,5′-dimethoxy-2,7′-cycloligna-7,7′-dien-9′,9-olide. ***Plicatinaphthalene***

[26560-82-7]

$C_{20}H_{16}O_7$ 368.341

Constit. of *Thuja plicata*. Cryst. (EtOH or CH_2Cl_2). Mp 157.4° (from CH_2Cl_2). Mp 248 - 254° dec. (from EtOH)

3′,5-Di-Me ether, tri-Ac

[26560-84-9]

Mp 221 - 222°

4,5-Methylene, 3′,4′,5′-tri-Me ether

2,7′-Cyclo-3′,4′,5′-trimethoxy-4,5-methylenedioxy-8,8′-ligna-7,7′-dien-9′,9-olide. 3′,4′,5′-Trimethoxy-4,5-methylenedioxy-2,7′-cycloligna-7,7′-dien-9′,9-olide. ***Dehydrodesoxypodophyllotoxin***

[6258-41-9]

$C_{22}H_{18}O_7$ 394.379

Constit. of seeds of *Hernandia ovigera*. Cryst. (EtOH). Mp 270 - 275°

3′,4′:4,5-Bismethylene, 5′-Me ether

2,7′-Cyclo-3′-methoxy-4,5:4′,5′-bis(methylenedioxy)-8,8′-ligna-7,7′-dien-9′,9-olide. 3′-Methoxy-4,5:4′,5′-bis(methylenedioxy)-2,7′-cycloligna-7,7′-dien-9′,9-olide. ***Koelreuterin 1***

[162616-74-2]

$C_{21}H_{14}O_7$ 378.336

Constit. of *Koelreuteria henryi*. Cytotoxic agent. Needles (hexane/EtOAc). Mp 270 - 272°. λ_{max} 261 (ε 23850); 309 (ε 19219); 350 (ε 7375) ($CHCl_3$) (Berdy).

MacLean, H. *et al.*, *Can. J. Chem.*, 1969, **47**, 4495-4498 (*Plicatinaphthalene*).

Yamaguchi, H. *et al.*, *Chem. Pharm. Bull.*, 1982, **30**, 3212-3218 (*Dehydrodesoxypodophyllotoxin*).

Song, Y.-N. *et al.*, *J. Nat. Prod.*, 1994, **57**, 1670-1674 (*Koelreuterin 1*)

2,7′-Cyclo-3′,4,4′,5,6-pentahydroxy-8,8′-ligna-7,7′-dien-9,9′-olide C289

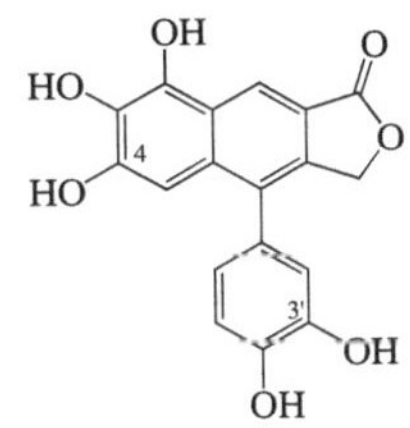

3′,4,4′,5,6-Pentahydroxy-2,7′-cycloligna-7,7′-dien-9,9′-olide

$C_{18}H_{12}O_7$ 340.287

3′,4′-Methylene, 4,5,6-tri-Me ether

2,7′-Cyclo-4,5,6-trimethoxy-3′,4′-methylenedioxy-8,8′-ligna-7,7′-dien-9,9′-olide. 4,5,6-Trimethoxy-3′,4′-methylenedioxy-2,7′-cycloligna-7,7′-dien-9,9′-olide. ***Phyllamyricin A***

[164124-32-7]

$C_{22}H_{18}O_7$ 394.379

Constit. of *Phyllanthus myrtifolius*. Cryst. (MeOH/$CHCl_3$). Mp 231 - 232°. λ_{max} 228 (log ε 4.36); 259 (log ε 4.69); 295 (log ε 3.79); 326 (log ε 3.94); 354 (sh) (MeOH)

3′,4′:4,5-Bis(methylene), 6-Me ether

2,7′-Cyclo-6-methoxy-3′,4′:4,5-bis(methylenedioxy)-8,8′-ligna-7,7′-dien-9,9′-olide. 6-Methoxy-3′,4′:4,5-bis(methylenedioxy)-2,7′-cycloligna-7,7′-dien-9,9′-olide. ***Justicidin G‡***

[364779-86-2]

$C_{21}H_{14}O_7$ 378.336

Constit. of *Justicia neesii*. Cryst. (EtOAc/petrol). Mp 218 - 220°. λ_{max} 226; 260; 321 (MeOH).

Lin, M.-T. *et al.*, *J. Nat. Prod.*, 1995, **58**, 244-249 (*Phyllamyricin A*).

Gopalaiah, K. *et al.*, *Indian J. Chem., Sect. B*, 2001, **40**, 596-600 (*Justicidin G*).

Kudoh, T. *et al.*, *Synlett*, 2013, **24**, 1509-1512 (*Phyllamyricin A, synth*)

2,7′-Cyclo-3′,4,4′,5,6-pentahydroxy-8,8′-ligna-7,7′-dien-9′,9-olide C290

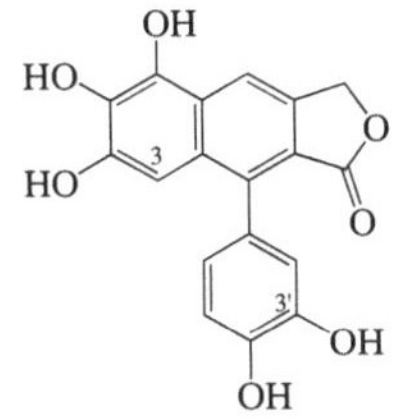

3′,4,4′,5,6-Pentahydroxy-2,7′-cycloligna-7,7′-dien-9′,9-olide

$C_{18}H_{12}O_7$ 340.287

Lignan numbering shown

3′,4′-Methylene, 4,5,6-tri-Me ether

2,7′-Cyclo-4,5,6-trimethoxy-3′,4′-methylenedioxy-8,8′-ligna-7,7′-dien-9′,9-olide. 4,5,6-Trimethoxy-3′,4′-methylenedioxy-2,7′-cycloligna-7,7′-dien-9′,9-olide. ***Phyllamyricin C***

[164124-34-9]

$C_{22}H_{18}O_7$ 394.379

Constit. of *Phyllanthus myrtifolius* and Kenyan propolis. Mp 190 - 191°. Incorrectly named as Phyllamycin C by Kudoh *et al* (2013). λ_{max} 230 (log ε 4.37); 259 (log ε 4.64); 290 (log ε 3.93); 316 (sh); 361 (log ε 3.62) (MeOH).

Lin, M.-T. *et al.*, *J. Nat. Prod.*, 1995, **58**, 244-249 (*Phyllamyricin C*).

Petrova, A. *et al.*, *Fitoterapia*, 2010, **81**, 509-514 (*propolis constit*).

Kudoh, T. *et al.*, *Synlett*, 2013, **24**, 1509-1512 (*Phyllamyricin C, synth*)

2,7′-Cyclo-3′,4,4′,5,7-pentahydroxy-8,8′-ligna-7,7′-dien-9,9′-olide C291

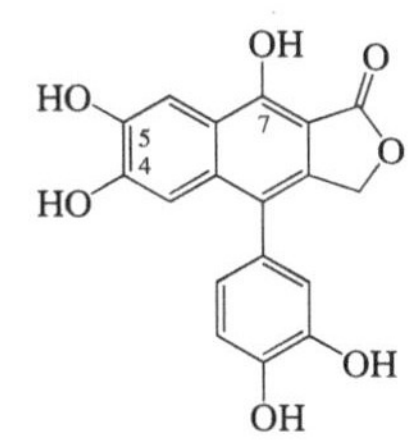

3′,4,4′,5,7-Pentahydroxy-2,7′-cycloligna-7,7′-dien-9,9′-olide

$C_{18}H_{12}O_7$ 340.287

3′,4′-Methylene, 4,7-di-Me ether

2,7′-Cyclo-5-hydroxy-4,7-dimethoxy-3′,4′-methylenedioxy-8,8′-ligna-7,7′-dien-9,9′-olide. 5-Hydroxy-4,7-dimethoxy-3′,4′-methylenedioxy-2,7′-cycloligna-7,7′-dien-9,9′-olide. ***Neojusticin C***

[1020151-87-4]

$C_{21}H_{16}O_7$ 380.352

Constit. of *Justicia procumbens*. Amorph. powder

3′,4′-Methylene, 4,5,7-tri-Me ether

2,7′-Cyclo-4,5,7-trimethoxy-3′,4′-methylenedioxy-8,8′-ligna-7,7′-dien-9,9′-olide. 4,5,7-Trimethoxy-3′,4′-methylenedioxy-2,7′-cycloligna-7,7′-dien-9,9′-olide. ***Justicidin C****. Neojusticin B*

[17803-12-2]

$C_{22}H_{18}O_7$ 394.379

Isol. from *Justicia procumbens*, *Justicia simplex* and *Cneoridium dumosum*. Piscicide. Shows antibacterial activity. Cryst. (EtOAc). Mp 266°

3′,4′:4,5-Bis(methylene), 7-Me ether

2,7′-Cyclo-7-methoxy-3′,4′:4,5-bis(methylenedioxy)-8,8′-ligna-7,7′-dien-9,9′-olide. 7-Methoxy-3′,4′:4,5-bis(methylenedioxy)-2,7′-cycloligna-7,7′-dien-9,9′-olide. ***Justicidin D****. Neojusticin A*

[27041-98-1]

$C_{21}H_{14}O_7$ 378.336

Isol. from *Justicia procumbens* and *Rostellularia procumbens*. Piscicide. Cryst. (EtOAc). Mp 273 - 275°.

Dreyer, D.L. *et al., Phytochemistry*, 1969, **8**, 1499-1501 (*Cneoridium dumosum constit*).
Ohta, K. *et al., Tetrahedron Lett.*, 1970, **11**, 923-925 (*Justicidins C,D*).
Okigawa, M. *et al., Tetrahedron*, 1970, **26**, 4301-4305 (*Neojusticins A,B*).
Horii, Z. *et al., Chem. Pharm. Bull.*, 1977, **25**, 1803-1808 (*Justicidin D, synth*).
Sastry, K.V. *et al., Indian J. Chem., Sect. B*, 1979, **7**, 415-416 (*Justicia simplex constit*).
Kobayashi, K. *et al., J. Chem. Soc., Perkin Trans. 1*, 1997, 443-446 (*Neojusticin A,B*).
Yang, M. *et al., Magn. Reson. Chem.*, 2006, **44**, 727-730 (*Justicidin D, pmr, cmr*).
Zhang, Y. *et al., Planta Med.*, 2007, **73**, 1596-1599 (*Justicidin C, activity*).
Liu, G. *et al., Magn. Reson. Chem.*, 2008, **46**, 283-286 (*Neojusticin C*).
Jin, H. *et al., Fitoterapia*, 2014, **94**, 70-76 (*Justicia procumbens constits*).
Kim, T. *et al., Eur. J. Org. Chem.*, 2017, 1704-1712 (*synth*)

2,7′-Cyclo-3′,4,4′,5,7-pentahydroxy-8,8′-ligna-7,7′-dien-9′,9-olide C292

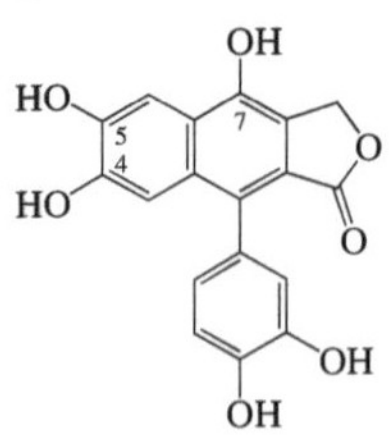

3′,4,4′,5,7-Pentahydroxy-2,7′-cycloligna-7,7′-dien-9′,9-olide

$C_{18}H_{12}O_7$ 340.287

5-Me ether

2,7′-Cyclo-3′,4,4′,7-tetrahydroxy-5-methoxy-8,8′-ligna-7,7′-dien-9′,9-olide. 3′,4,4′,7-Tetrahydroxy-5-methoxy-2,7′-cycloligna-7,7′-dien-9′,9-olide

5-Me ether, 7-O-β-D-apiofuranoside

Haploborin

[355118-52-4]

$C_{24}H_{22}O_{11}$ 486.429

Constit. of *Haplophyllum suaveolens*. Amorph. powder. λ_{max} 225; 262; 292; 315; 360 (MeOH)

4,5-Di-Me ether

2,7′-Cyclo-3′,4′,7-trihydroxy-4,5-dimethoxy-8,8′-ligna-7,7′-dien-9′,9-olide. 3′,4′,7-Trihydroxy-4,5-dimethoxy-2,7′-cycloligna-7,7′-dien-9′,9-olide. ***Taiwanin H***

[102040-04-0]

$C_{20}H_{16}O_7$ 368.341

Constit. of *Taiwania cryptomerioides*

4,5,7-Tri-Me ether, 4′-O-β-D-glucopyranoside

Procumbenoside K

[2243471-11-4]

$C_{27}H_{28}O_{12}$ 544.509

Constit. of *Justicia procumbens*. Amorph. powder (MeOH). $[\alpha]_D^{20}$ -10 (c, 0.05 in DMSO)

3′,4,5,7-Tetra-Me ether

2,7′-Cyclo-4′-hydroxy-3′,4,5,7-tetramethoxy-8,8′-ligna-7,7′-dien-9′,9-olide. 4′-Hydroxy-3′,4,5,7-tetramethoxy-2,7′-cycloligna-7,7′-dien-9′,9-olide. ***Cilinaphthalide A***

[238423-03-5]

$C_{22}H_{20}O_7$ 396.395

Constit. of *Justicia ciliata* and *Rostellularia procumbens*. Shows moderate antibacterial activity against *E.coli*. Cryst.. Mp 221 - 222°. λ_{max} 250 (log ε 4.05); 260 (log ε 4.06); 265 (log ε 4.05); 275 (log ε 4.04); 315 (log ε 3.98); 350 (log ε 3.75) ($CHCl_3$)

4,4′,5,7-Tetra-Me ether

2,7′-Cyclo-3′-hydroxy-4,4′,5,7-tetramethoxy-8,8′-ligna-7,7′-dien-9′,9-olide. 3′-Hydroxy-4,4′,5,7-tetramethoxy-2,7′-cycloligna-7,7′-dien-9′,9-olide. ***Pronaphthalide A***

[1624617-93-1]

$C_{22}H_{20}O_7$ 396.395

Constit. of *Justicia procumbens*. Displays cytotoxic activity. Amorph. powder. $[\alpha]_D^{20}$ +27.4 (c, 0.05 in MeOH). λ_{max} 322 (log ε 3.65); 364 (log ε 3.65) (MeOH)

Penta-Me ether

2,7′-Cyclo-3′,4,4′,5,7-pentamethoxy-8,8′-ligna-7,7′-dien-9′,9-olide. 3′,4,4′,5,7-Pentamethoxy-2,7′-cycloligna-7,7′-dien-9′,9-olide. ***Cilinaphthalide B***

[102040-06-2]

$C_{23}H_{22}O_7$ 410.422

Constit. of *Justicia ciliata* and *Justicia procumbens*. Shows antiplatelet activity. Cryst.. Mp 202.5 - 203.5°. λ_{max} 260 (log ε 3.58); 290 (log ε 3.57); 305 (log ε 3.57); 325 (log ε 3.56); 350 (log ε 3.56) ($CHCl_3$)

3′,4′-Methylene, 4-Me ether

2,7′-Cyclo-5,7-dihydroxy-4-methoxy-3′,4′-methylenedioxy-8,8′-ligna-7,7′-dien-9′,9-olide. 5,7-Dihydroxy-4-methoxy-3′,4′-methylenedioxy-2,7′-cycloligna-7,7′-dien-9′,9-olide

$C_{20}H_{14}O_7$ 366.325

3′,4′-Methylene, 4-Me ether, 5-O-α-L-arabinopyranoside, 7-O-β-D-glucopyranoside

Reticulatuside B

[1479020-43-3]

$C_{31}H_{32}O_{16}$ 660.581

Constit. of *Phyllanthus reticulatus*. Amorph. yellow powder. $[\alpha]_D^{25}$ -22 (c, 0.04 in MeOH). λ_{max} 259 (log ε 4.69) (MeOH)

3′,4′-Methylene, 4-Me ether, 7-O-[β-D-apiofuranosyl-(1→2)-β-D-apiofuranoside]

Mananthoside F

[881422-02-2]

$C_{30}H_{30}O_{15}$ 630.555

Constit. of *Mananthes patentiflora*. Amorph. powder. Mp 158 - 160°. $[\alpha]_D^{24}$ -111.1 (c, 0.74 in MeOH). λ_{max} 203 (log ε 4.75); 224 (log ε 4.49); 263 (log ε 4.67); 290 (log ε 3.96); 323 (log ε 4.03) (MeOH)

3′,4′-Methylene, 5-Me ether

2,7′-Cyclo-4,7-dihydroxy-5-methoxy-3′,4′-methylenedioxy-8,8′-ligna-7,7′-dien-9′,9-olide. 4,7-Dihydroxy-5-methoxy-3′,4′-methylenedioxy-2,7′-cycloligna-7,7′-dien-9′,9-olide. ***Haplomyrtin***

[105037-92-1]

$C_{20}H_{14}O_7$ 366.325

Isol. from *Haplophyllum myrtifolium*, *Haplophyllum telephioides* and *Phyllanthus oligospermus*. Amorph. powder

3′,4′-Methylene, 5-Me ether, 7-O-β-D-apiofuranoside

Haplomyrtoside

[177912-20-8]
$C_{25}H_{22}O_{11}$ 498.44
Constit. of *Haplophyllum cappadocicum*. Amorph. solid. [α] D -52.8 (c, 0.1 in MeOH)

3′,4′-Methylene, 5-Me ether, 7-O-β-D-glucopyranoside
Procumbenoside I
[2243471-10-3]
$C_{26}H_{24}O_{12}$ 528.466
Constit. of *Justicia procumbens*. Amorph. powder (MeOH). $[\alpha]_D^{20}$ -38 (c, 0.05 in MeOH)

3′,4′-Methylene, 4,5-di-Me ether
see Diphyllin, D296

3′,4′-Methylene, 4,5-di-Me ether, 7-O-[β-D-glucopyranosyl-(1→4)-2-O-acetyl-6-deoxy-β-D-glucopyranoside]
Justiprocumin B
[1422256-20-9]
$C_{35}H_{38}O_{17}$ 730.672
Constit. of *Justicia gendarussa*. Potent inhibitor against drug resistant HIV isolates NRTI (nucleoside reverse transcriptase inhibitor) and NNRTI (non-nucleoside reverse transcriptase inhibitor). Powder. $[\alpha]_D^{20}$ -32 (c, 0.74 in MeOH). λ_{max} 227 (log ε 0.63); 262 (log ε 1.41); 294 (log ε 0.29); 315 (log ε 0.31); 356 (log ε 0.14) (MeOH)

3′,4′-Methylene, 4,5-di-Me ether, 7-O-[β-D-glucopyranosyl-(1→4)-3-O-acetyl-6-deoxy-β-D-glucopyranoside]
Justiprocumin A
[1422256-19-6]
$C_{35}H_{38}O_{17}$ 730.672
Constit. of *Justicia gendarussa*. Powder. $[\alpha]_D^{20}$ -26.2 (c, 0.55 in MeOH). λ_{max} 225 (log ε 0.41); 262 (log ε 1.01); 294 (log ε 0.19); 315 (log ε 0.21); 356 (log ε 0.07) (MeOH)

3′,4′-Methylene, 5,7-di-Me ether
2,7′-Cyclo-4-hydroxy-5,7-dimethoxy-3′,4′-methylenedioxy-8,8′-ligna-7,7′-dien-9′,9-olide. 4-Hydroxy-5,7-dimethoxy-3′,4′-methylenedioxy-2,7′-cycloligna-7,7′-dien-9′,9-olide. **Phyllanthuspermin A**. *Phyllanthusmin A*
[914639-24-0]
$C_{21}H_{16}O_7$ 380.352
Constit. of *Phyllanthus oligospermus* and *Phyllanthus poilanei*. Shows significant cytotoxicity against KB and P-388 cancer cell lines. Needles (MeOH). Mp 283 - 284°. λ_{max} 230 (log ε 4.29); 263 (log ε 4.61); 292 (log ε 3.93); 310 (sh); 356 (log ε 3.64) (MeOH)

3′,4′-Methylene, 4,5,7-tri-Me ether
2,7′-Cyclo-4,5,7-trimethoxy-3′,4′-methylenedioxy-8,8′-ligna-7,7′-dien-9′,9-olide. 4,5,7-Trimethoxy-3′,4′-methylenedioxy-2,7′-cycloligna-7,7′-dien-9′,9-olide. **Justicidin A**
[25001-57-4]
$C_{22}H_{18}O_7$ 394.379
Isol. from *Justicia hayati*, *Justicia procumbens*, *Haplophyllum tuberculatum*, *Phyllanthus myrtifolius* and *Justicia ciliata*. Cytotoxic, piscicide. Mp 291°. λ_{max} 264; 295; 309; 350 (EtOH) (Derep). [neutral] λ_{max} 265 (ε 54280); 295 (ε 11520); 315 (ε 11520); 355 (ε 5280) (EtOH) (Berdy)

4,5-Methylene, 7-Me ether
2,7′-Cyclo-3′,4′-dihydroxy-7-methoxy-4,5-methylenedioxy-8,8′-ligna-7,7′-dien-9′,9-olide. 3′,4′-Dihydroxy-7-methoxy-4,5-methylenedioxy-2,7′-cycloligna-7,7′-dien-9′,9-olide

4,5-Methylene, 7-Me ether, 3′-O-β-D-glucopyranoside
Justalakonin
[602331-39-5]
$C_{26}H_{24}O_{12}$ 528.466
Constit. of *Justicia purpurea*. Amorph. solid. Mp 205 - 207°. $[\alpha]_D^{25}$ -31 (c, 0.25 in MeOH)

4,5-Methylene, 3′,4′-di-Me ether
2,7′-Cyclo-7-hydroxy-3′,4′-dimethoxy-4,5-methylenedioxy-8,8′-ligna-7,7′-dien-9′,9-olide. 7-Hydroxy-3′,4′-dimethoxy-4,5-methylenedioxy-2,7′-cycloligna-7,7′-dien-9′,9-olide. **Chinensinaphthol**. *Isodiphyllin*
[53965-06-3]
$C_{21}H_{16}O_7$ 380.352
Constit. of *Polygala chinensis*. Solid (EtOAc). Mp 285 - 286° (256°, 265-270°)

4,5-Methylene, 3′,4′-di-Me ether, 7-O-β-D-glucopyranoside
Procumbenoside D
[1020151-89-6]
$C_{27}H_{26}O_{12}$ 542.493
Constit. of *Justicia procumbens*. Amorph. powder

4,5-Methylene, 3′,7-di-Me ether
2,7′-Cyclo-4′-hydroxy-3′,7-dimethoxy-4,5-methylenedioxy-8,8′-ligna-7,7′-dien-9′,9-olide. 4′-Hydroxy-3′,7-dimethoxy-4,5-methylenedioxy-2,7′-cycloligna-7,7′-dien-9′,9-olide.
4′-De-O-methyl-O-methylchinensinaphthol
[184579-23-5]
$C_{21}H_{16}O_7$ 380.352
Constit. of *Justicia procumbens*. Prisms (MeOH). Mp 240 - 242°

4,5-Methylene, 3′,4′,7-tri-Me ether
2,7′-Cyclo-3′,4′,7-trimethoxy-4,5-methylenedioxy-8,8′-ligna-7,7′-dien-9′,9-olide. 3′,4′,7-Trimethoxy-4,5-methylenedioxy-2,7′-cycloligna-7,7′-dien-9′,9-olide.
O-Methylchinensinaphthol
[53965-11-0]
$C_{22}H_{18}O_7$ 394.379
Constit. of *Polygala chinensis*. Cryst. ($CHCl_3$/MeOH). Mp 257 - 258°

3′,4′:4,5-Bis(methylene) ether
2,7′-Cyclo-7-hydroxy-3′,4′:4,5-bis(methylenedioxy)-8,8′-ligna-7,7′-dien-9′,9-olide. 7-Hydroxy-3′,4′:4,5-bis(methylenedioxy)-2,7′-cycloligna-7,7′-dien-9′,9-olide. **Taiwanin E**
[22743-05-1]
$C_{20}H_{12}O_7$ 364.309
Constit. of the heartwood of *Taiwania cryptomeroides* and of *Cleistanthus collinus* and *Justicia patentiflora*. Shows cytotoxic activity. Pale yellow needles (MeOH). Mp 292°. Mp 304 - 306°. Struct. revised in 1969. λ_{max} 250 (ε 19953); 257 (ε 12589); 303 (ε 39811); 357 (ε 7943) (EtOH)

3′,4′:4,5-Bis(methylene) ether, 7-O-β-D-glucopyranoside
Procumbenoside C
[1020151-88-5]
$C_{26}H_{22}O_{12}$ 526.45
Constit. of *Justicia procumbens*. Amorph. powder

3′,4′:4,5-Bis(methylene) ether, 7-O-(3,4-di-O-methyl-β-D-xylopyranoside)
O-(3,4-Di-O-methyl-β-D-xylopyranosyl)taiwanin E

[81410-46-0]
$C_{27}H_{24}O_{11}$ 524.478
Constit. of the heartwood of *Cleistanthus collinus*. Needles (MeOH). Mp 210 - 212°. $[\alpha]_D$ +11. β-Anomer not confirmed

3′,4′:4,5-Bis(methylene), 7-Me ether
2,7′-Cyclo-7-methoxy-3′,4′:4,5-bis(methylenedioxy)-8,8′-ligna-7,7′-dien-9′,9-olide. 7-Methoxy-3′,4′:4,5-bis(methylenedioxy)-2,7′cycloligna-7,7′-dien-9′,9-olide. ***Justicidin F****. Taiwanin E methyl ether*
[30403-00-0]
$C_{21}H_{14}O_7$ 378.336
Constit. of *Justicia procumbens* and *Justicia neesii*. Mp 227 - 230°. λ_{max} 223; 267 (MeOH).

Horii, Z. *et al.*, *Tetrahedron Lett.*, 1969, **10**, 1079-1081 (*Taiwanin E*).
Okigawa, M. *et al.*, *Tetrahedron*, 1970, **26**, 4301-4305 (*Taiwanin E methyl ether, Justicidin A*).
Arnold, B.J. *et al.*, *J. Chem. Soc., Perkin Trans. 1*, 1973, 1266-1270 (*Taiwanin E, synth*).
Momose, T. *et al.*, *Chem. Pharm. Bull.*, 1978, **26**, 3195-3198 (*Taiwanin E, synth*).
González, A.G. *et al.*, *Tetrahedron*, 1978, **34**, 1011-1013 (*Isodiphyllin, synth*).
Anjaneyulu, A.S.R. *et al.*, *Tetrahedron*, 1981, **37**, 3641-3652 (*Cleistanthus collinus constits*).
Kuo, Y.H. *et al.*, *J. Chin. Chem. Soc. (Taipei)*, 1985, **32**, 381-383 (*Taiwanin H*).
Fukamiya, N. *et al.*, *J. Nat. Prod.*, 1986, **49**, 348-350 (*Justicidin A, activity*).
Evcim, U. *et al.*, *Phytochemistry*, 1986, **25**, 1949-1951 (*Haplomyrtin*).
Harrowven, D.C. *Tetrahedron*, 1993, **49**, 9039-9048 (*Chinensinaphthol, synth*).
Ulubelen, A. *et al.*, *Phytochemistry*, 1994, **35**, 1600-1601 (*Haplophyllum telephioides constit*).
Chen, C.C. *et al.*, *J. Nat. Prod.*, 1996, **59**, 1149-1150 (*4′-De-O-methyl-O-methylchinensinaphthol*).
Gözler, B. *et al.*, *Phytochemistry*, 1996, **42**, 689-693 (*Haplomyrtoside*).
Al-Abed, Y. *et al.*, *Phytochemistry*, 1998, **49**, 1779-1781 (*Taiwanin H*).
Day, S.-H. *et al.*, *J. Nat. Prod.*, 1999, **62**, 1056-1058 (*Cilinaphthalides*).
Chang, S.-T. *et al.*, *Phytochemistry*, 2000, **55**, 227-232 (*Taiwanin E, activity*).
Gopalaiah, K. *et al.*, *Indian J. Chem., Sect. B*, 2001, **40**, 596-600 (*Justicidin F, uv*).
Ivanova, A. *et al.*, *Z. Naturforsch., C*, 2001, **56**, 329-333 (*Haploborin*).
Kavitha, J. *et al.*, *J. Nat. Prod.*, 2003, **66**, 1113-1115 (*Justalakonin*).
Sato, Y. *et al.*, *Angew. Chem., Int. Ed.*, 2004, **43**, 2436-2440 (*Taiwanin E, synth*).
Weng, J-R. *et al.*, *Arch. Pharm. (Weinheim, Ger.)*, 2004, **337**, 207-212 (*Cilinaphthalide B, activity*).
Susplugas, S. *et al.*, *J. Nat. Prod.*, 2005, **68**, 734-738 (*Justicia patentiflora constit*).
Wu, S.-J. *et al.*, *Chem. Pharm. Bull.*, 2006, **54**, 1223-1225 (*Phyllanthuspermin A, Haplomyrtin, activity*).
Tian, J. *et al.*, *Helv. Chim. Acta*, 2006, **89**, 291-298 (*Mananthoside F*).
Yang, M. *et al.*, *Magn. Reson. Chem.*, 2006, **44**, 727-730 (*Chinensinaphthol, pmr, cmr*).
Zhang, Y. *et al.*, *Planta Med.*, 2007, **73**, 1596-1599 (*Cilinaphthalide A, activity*).
Liu, G. *et al.*, *Magn. Reson. Chem.*, 2008, **46**, 283-286 (*Procumbenosides C,D*).
Ma, J.-X. *et al.*, *J. Asian Nat. Prod. Res.*, 2012, **14**, 1073-1077 (*Reticulatuside B*).
Jin, H. *et al.*, *Fitoterapia*, 2014, **94**, 70-76 (*Pronaphthalide A*).
Naresh, G. *et al.*, *Org. Lett.*, 2015, **17**, 3446-3449 (*Justicidin A, Taiwanin E, synth*).
Kim, T. *et al.*, *Eur. J. Org. Chem.*, 2017, 1704-1712 (*synth*).
Jin, H. *et al.*, *J. Asian Nat. Prod. Res.*, 2017, **19**, 1-8 (*Procumbenosides I,K*).
Zhang, H.-J. *et al.*, *Phytochemistry*, 2017, **136**, 94-100 (*Justiprocumins A,B, activity*)

2,7′-Cyclo-3,4,4′,5,9-pentahydroxy-8,8′-ligna-7,7′-dien-9′,9-olide C293

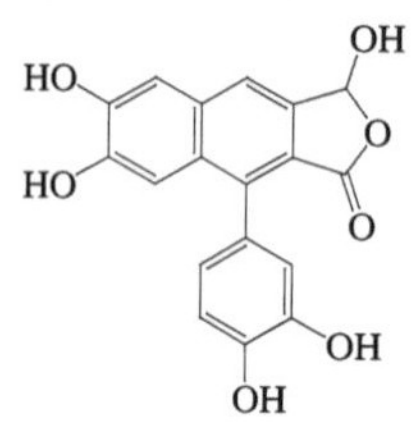

3,4,4′,5,9-Pentahydroxy-2,7′-cycloligna-7,7′-dien-9′,9-olide
$C_{18}H_{12}O_7$ 340.287
(9ξ)-form

3′,4′-Methylene, 4,5-di-Me ether
2,7′-Cyclo-9-hydroxy-4,5-dimethoxy-3′,4′-methylenedioxy-8,8′-ligna-7,7′-dien-9′,9-olide. 9-Hydroxy-4,5-dimethoxy-3′,4′-methylenedioxy-2,7′-cycloligna-7,7′-dien-9′,9-olide. ***Piscatorin***
[623943-73-7]
$C_{21}H_{16}O_7$ 380.352
Constit. of the aerial parts of *Phyllanthus piscatorum*. Piscicidal agent. Amorph. powder. Mp 247°. $[\alpha]_D^{20}$ -3 (c, 0.1 in $CHCl_3$). λ_{max} 263 ($CHCl_3$)

3′,4′-Methylene, 4,5,9-tri-Me ether
2,7′-Cyclo-4,5,9-trimethoxy-3′,4′-methylenedioxy-8,8′-ligna-7,7′-dien-9′,9-olide. 4,5,9-Trimethoxy-3′,4′-methylenedioxy-2,7′-cycloligna-7,7′-dien-9′,9-olide. ***Phyllamyricin E***
[184094-78-8]
$C_{22}H_{18}O_7$ 394.379
Constit. of *Phyllanthus myrtifolius*. Amorph. solid. Mp 95 - 99°. $[\alpha]_D^{20}$ -1.3 (c, 0.7 in MeOH). λ_{max} 260 (log ε 4.62); 291 (log ε 3.74) (MeOH).

Lee, S.-S. *et al.*, *J. Nat. Prod.*, 1996, **59**, 1061-1065 (*Phyllamyricin E*).
Gertsch, J. *et al.*, *Planta Med.*, 2003, **69**, 420-424 (*Piscatorin*).
Wang, C-Y. *et al.*, *Phytochem. Anal.*, 2011, **22**, 352-360 (*Phyllamyricin E, pmr, ms*)

2,7′-Cyclo-3′,4,4′,5,9′-pentahydroxy-8,8′-ligna-7,7′-dien-9,9′-olide C294

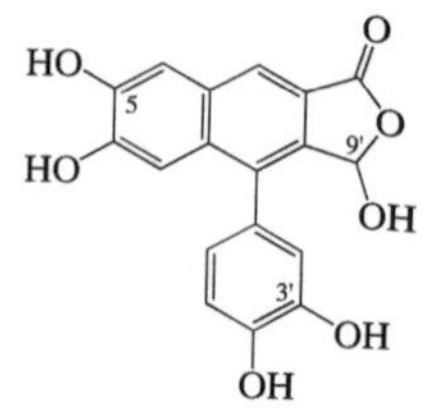

3′,4,4′,5,9′-Pentahydroxy-2,7′-cycloligna-7,7′-dien-9,9′-olide
(9′ξ)-form

9′-Et, 3′,5-di-Me ether
2,7′-Cyclo-9′-ethoxy-4,4′-dihydroxy-3′,5-dimethoxy-8,8′-ligna-7,7′-dien-9,9′-olide. 9′-Ethoxy-4,4′-dihydroxy-3′,5-dimethoxy-2,7′-cycloligna-7,7′-dien-9,9′-olide. ***Vitexdoin I***
$C_{22}H_{20}O_7$ 396.395
Isol. from seeds of *Vitex negundo*. Amorph. powder. $[\alpha]_D^{20}$ -14 (c, 0.2 in MeOH). Partial racemate. λ_{max} 262 (MeOH).

Zheng, C.-J. *et al.*, *Fitoterapia*, 2014, **93**, 31-38 (*Vitexdoin I*)

2,7′-Cyclo-3′,4,4′,5,5′-pentahydroxy-8,8′-lignan-9′-oic acid C295

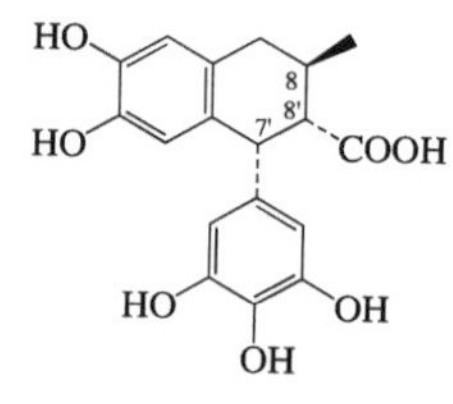

3′,4,4′,5,5′-Pentahydroxy-2,7′-cyclolignan-9′-oic acid
$C_{18}H_{18}O_7$ 346.335
(7′R*,8R*,8′S*)-form

3′,4′:4,5-Bis(methylene), 5′-Me ether
2,7′-Cyclo-3′-methoxy-4,5:4′,5′-bis(methylenedioxy)-8,8′-lignan-9′-oic acid. ***Austrobailignan 8***
[1941213-44-0]
$C_{21}H_{20}O_7$ 384.384
Constit. of *Austrobaileya scandens*. Exhibits moderate antiproliferative

activity against human prostate cancer PC3 cells. Amorph. powder. $[\alpha]^{25}_{D}$ +33 (c, 0.02 in MeOH). λ_{max} 214 (log ε 4.3); 245 (log ε 3.2); 282 (log ε 3.0) (MeOH).

Tran, T.D. *et al.*, *J. Nat. Prod.*, 2016, **79**, 1514-1523 (*Austrobailignan 8, activity*)

2,7′-Cyclo-3′,4,4′,5,9-pentahydroxy-8,8′-lignan-9′-oic acid C296

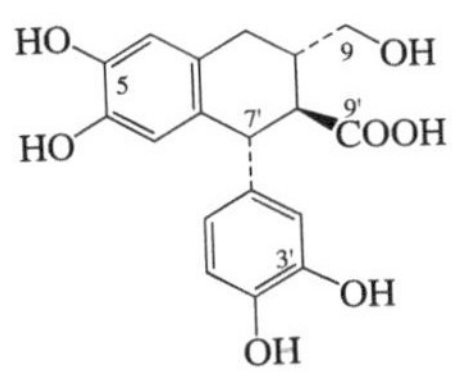

(7′R,8S,8′S)-form

1-(3,4-Dihydroxyphenyl)-1,2,3,4-tetrahydro-6,7-dihydroxy-3-(hydroxymethyl)-2-naphthalenecarboxylic acid.
3′,4,4′,5,9-Pentahydroxy-2,7′-cyclolignan-9′-oic acid
$C_{18}H_{18}O_7$ 346.335

(7′R,8S,8′S)-form

3′,5-Di-Me ether
2,7′-Cyclo-4,4′,9-trihydroxy-3′,5-dimethoxy-8,8′-lignan-9′-oic acid. ***(-)-Isolariciresinol-9′-carboxylic acid***

3′,5-Di-Me ether, Me ester
[915096-21-8]
$C_{21}H_{24}O_7$ 388.416
Constit. of the roots of *Smilax bockii*. Cryst. (MeOH). Mp 109 - 111°. $[\alpha]^{21}_{D}$ -46.9 (c, 0.1 in MeOH)

(7′S,8R,8′R)-form [2408952-93-0]
Constit. of the roots of *Hypochaeris radicata*. Yellow gum. $[\alpha]^{20}_{D}$ +68 (c, 0.9 in MeOH). Possible artifact.

Xu, J. *et al.*, *Pharmazie*, 2006, **61**, 812 (*Smilax bockii constit*).
Shulha, O. *et al.*, *Phytochemistry*, 2019, **165**, 112047:1-12 (*Hypochaeris radicata constit, cd*)

2,7′-Cyclo-3,3′,4,4′,5-pentahydroxy-8,8′-lignan-9′,9-olide C297

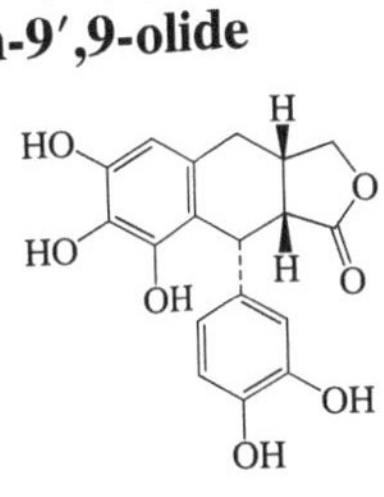

3,3′,4,4′,5-Pentahydroxy-2,7′-cyclolignan-9′,9-olide
$C_{18}H_{16}O_7$ 344.319

(7′R*,8S*,8′R*)-form

3′,4′-Methylene, 3,4,5-tri-Me ether
2,7′-Cyclo-3,4,5-trimethoxy-3′,4′-methylenedioxy-8,8′-lignan-9′,9-olide.
3,4,5-Trimethoxy-3′,4′-methylenedioxy-2,7′-cyclolignan-9′,9-olide. ***Marginatoxin***
[1422536-56-8]
$C_{22}H_{22}O_7$ 398.411
Constit. of the aerial parts of *Bupleurum marginatum*. Cytotoxic against human HeLa and HepG2 cancer cells. Cryst. (hexane).

Ashour, M.L. *et al.*, *Phytochem. Lett.*, 2012, **5**, 387-392 (*Marginatoxin*)

2,7′-Cyclo-3′,4,4′,5,5′-pentahydroxy-8,8′-lignan-9′,9-olide C298

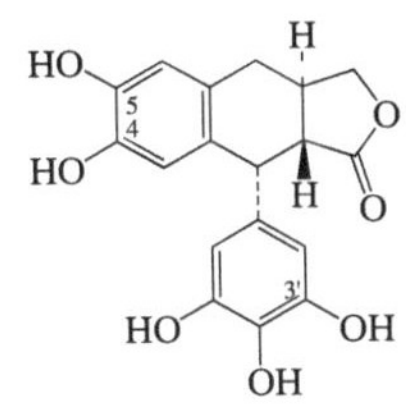

(7′R,8R,8′R)-form

3′,4,4′,5,5′-Pentahydroxy-2,7′-cyclolignan-9′,9-olide
$C_{18}H_{16}O_7$ 344.319

(7′R,8R,8′R)-form

3′,4′,5′-Tri-Me ether
2,7′-Cyclo-4,5-dihydroxy-3′,4′,5′-trimethoxy-8,8′-lignan-9′,9-olide.
4,5-Dihydroxy-3′,4′,5′-trimethoxy-2,7′-cyclolignan-9′,9-olide.
4,5-Demethylene-7-desoxypodophyllotoxin
[1173-42-8]
$C_{21}H_{22}O_7$ 386.4
Isol. from seeds of *Hernandia ovigera*. Needles (EtOH). Mp 238 - 241°. $[\alpha]^{20}_{D}$ -113 (c, 0.086 in $CHCl_3$)

4,5-Methylene, 3′,5′-di-Me ether
2,7′-Cyclo-4′-hydroxy-3′,5′-dimethoxy-4,5-methylenedioxy-8,8′-lignan-9′,9-olide.
4′-Hydroxy-3′,5′-dimethoxy-4,5-methylenedioxy-2,7′-cyclolignan-9′,9-olide.
4′-Demethyldeoxypodophyllotoxin
[3590-93-0]
$C_{21}H_{20}O_7$ 384.384
Isol. from *Amanoa oblongifolia*, *Hyptis verticillata*, *Polygala macradenia* and *Polygala paenea*. Exhibits cytotoxic and tumour inhibitory activity. Cryst.. Mp 244 - 249° (subl. at 215°). $[\alpha]^{20}_{D}$ -122.6 (c, 0.555 in $CHCl_3$). λ_{max} 282 (ε 4340); 291 (ε 4070) (EtOH) (Berdy)
LV2476000

4,5-Methylene, 3′,5′-di-Me ether, 4′-O-β-glucopyranoside
4′-Demethyldeoxypodophyllotoxin 4′-glucoside
[78178-42-4]
$C_{27}H_{30}O_{12}$ 546.525
Isol. from *Diphylleia cymosa*, *Juniperus communis*, *Podophyllum emodi* and *Podophyllum peltatum*. Amorph. solid + ½H_2O. Mp 146 - 158°. $[\alpha]_D$ -77 (MeOH). λ_{max} 287; 292 (sh) (EtOH)

4,5-Methylene, 3′,4′,5′-tri-Me ether
2,7′-Cyclo-3′,4′,5′-trimethoxy-4,5-methylenedioxy-8,8′-lignan-9′,9-olide.
3′,4′,5′-Trimethoxy-4,5-methylenedioxy-2,7′-cyclolignan-9′,9-olide.
Deoxypodophyllotoxin. *Desoxypodophyllotoxin. Hernandion. Anthricin. Silicicolin*
[19186-35-7]
$C_{22}H_{22}O_7$ 398.411
Isol. from *Bursera microphylla*, *Libocedrus bidwillii*, *Anthriscus sylvestris*, *Hernandia ovigera*, *Juniperus silicicola*, *Pulsatilla koreana*, *Linum album*, *Podophyllum* spp. and others. Cytotoxic, antineoplastic agent having an inhibitory effect on the tube-like formation of human umbilical vein on endothelial cells. Also causes G_2/M cell arrest and apoptosis in HeLa cells. Cryst. (EtOH). Sol. MeOH, Et_2O. Mp 170 - 171°. $[\alpha]^{19}_{D}$ -119 ($CHCl_3$). Log P 0.8 (calc). Readily epimerised by base to Deoxypicropodophyllin. λ_{max} 291 (ε 6200); 294 (ε 4776) (EtOH) (Berdy)
TP9002800

3′,4′:4,5-Bis(methylene), 5′-Me ether
2,7′-Cyclo-3′-methoxy-4,5:4′,5′-bis(methylenedioxy)-8,8′-lignan-9′,9-olide.
3′-Methoxy-4,5:4′,5′-bis(methylenedioxy)-2,7′-cyclolignan-9′,9-olide. ***Austrobailignan 1***
[55955-07-2]
$C_{21}H_{18}O_7$ 382.368

Constit. of *Austrobaileya scandens* and *Koelreuteria henryi*. Needles (MeOH). Mp 183 - 185°. $[\alpha]_D^{24}$ -110 (c, 1.2 in $CHCl_3$)

(7′R,8R,8′S)-form

4,5-Methylene, 3′,5′-di-Me ether

4′-Demethyldeoxypicropodophyllotoxin

[157242-05-2]

$C_{21}H_{20}O_7$ 384.384

Constit. of *Casearia clarkei*. Amorph. solid. Mp 174 - 175°. $[\alpha]_D$ +37 (c, 0.05 in $CHCl_3$)

4,5-Methylene, 3′,4′,5′-tri-Me ether

Deoxypicropodophyllin. *Cicutin. Silicicolin B*

[24150-39-8]

$C_{22}H_{22}O_7$ 398.411

Isol. from *Thuja plicata, Hyptis verticillata, Austrobaileya scandeus, Juniperus sabina, Bursera microphylla, Thujopsis dolabrata, Austrobaileya scandeus, Cicuta maculata* and others. Toxic to brine shrimps; cytotoxic agent. Cryst. (EtOH aq.). Sol. MeOH, $CHCl_3$; poorly sol. H_2O, hexane. Mp 171 - 172°. $[\alpha]_D^{21}$ +32 ($CHCl_3$). Poss. artifact of epimerisation. λ_{max} 204 (ε 63000); 293 (ε 7760) (EtOH) (Berdy)

3′,4′:4,5-Bis(methylene), 5′-Me ether

Austrobailignan 2

[55890-20-5]

$C_{21}H_{18}O_7$ 382.368

Isol. from *Austrobaileya scandens* and *Koelreuteria henryi*. Needles (MeOH). Mp 217 - 219°. $[\alpha]_D^{20}$ +88 (c, 2.5 in $CHCl_3$). Prob. artifact. λ_{max} 244 (ε 7387); 289 (ε 5096) ($CHCl_3$) (Berdy)

(7′R,8S,8′S)-form

4,5-Methylene, 3′,4′,5′-tri-Me ether

Isodeoxypodophyllotoxin

[17187-81-4]

$C_{22}H_{22}O_7$ 398.411

Constit. of *Hyptis verticillata*. Needles. Mp 256 - 258° (246-248°). $[\alpha]_D$ -38.5 (c, 0.4 in $CHCl_3$)

(7′S,8R,8′S)-form

4,5-Methylene, 3′,4′,5′-tri-Me ether

Isopicrodeoxypodophyllotoxin

[83023-41-0]

$C_{22}H_{22}O_7$ 398.411

Constit. of *Thuja occidentalis*. Oil. $[\alpha]_D^{20}$ +12.5 (c, 0.06 in $CHCl_3$). λ_{max} 217 (log ε 4); 241 (log ε 3.5); 291 (log ε 3.2) (MeOH).

Hartwell, J.L. *et al., J. Am. Chem. Soc.*, 1953, **75**, 2138-2140 (*Deoxypodophyllotoxin, synth*).
Hartwell, J.L. *et al., J. Am. Chem. Soc.*, 1954, **76**, 4034-4035 (*Hernandion, Cicutin*).
Polonsky, J. *et al., Bull. Soc. Chim. Fr.*, 1962, 1722-1726 (*Polygala paenea constit*).
Von Wartburg, A. *et al., Helv. Chim. Acta*, 1964, **47**, 1203-1210 (*4′-Demethyldeoxypodophyllotoxin 4′-glucoside*).
Bianchi, E. *et al., J. Pharm. Sci.*, 1968, **57**, 696-697 (*Bursea microphylla constits, activity*).
German, V.F. *J. Pharm. Sci.*, 1971, **60**, 649-650 (*Hyptis verticillata constit*).
Klemm, L.H. *et al., J. Org. Chem.*, 1971, **36**, 3740-3743 (*Deoxypicropodophyllin, synth*).
Quon, H.H. *et al., Bi-Mon. Res. Notes - Can., For. Serv.*, 1972, **28**, 23-24 (*Deoxypicropodophyllin*).
Akahori, A. *et al., Chem. Pharm. Bull.*, 1972, **20**, 1150-1155 (*Thujopsis dolabrata constit*).
Murphy, S.T. *et al., Aust. J. Chem.*, 1975, **28**, 81-90 (*Austrobaileya scandens constits*).
Russell, G.B. *Phytochemistry*, 1975, **14**, 2708 (*Libocedrus bidwillii constit*).
Hoffmann, J.J. *et al., J. Pharm. Sci.*, 1977, **66**, 586-587 (*4′-Demethyldeoxypodophyllotoxin, isol, activity*).
Yamaguchi, H. *et al., Chem. Pharm. Bull.*, 1982, **30**, 3212-3218 (*Hernandia ovigera constits*).
Robin, J.-P. *et al., Tetrahedron*, 1982, **38**, 3667-3671 (*Isodeoxypodophyllotoxin, synth*).
Tanoguchi, M. *et al., Chem. Pharm. Bull.*, 1987, **35**, 4162-4168 (*Hernandia ovigera constits*).
Pullockaran, A.J. *et al., J. Nat. Prod.*, 1989, **52**, 1290-1295 (*Deoxypodophyllotoxin, pmr*).
Brown, E. *et al., Tetrahedron*, 1989, **45**, 141-154 (*Deoxypicropodophyllin, synth*).
Feliciano, A.S. *et al., Phytochemistry*, 1990, **29**, 1335-1338 (*Juniperus sabina constit*).
Morimoto, T. *et al., Tetrahedron Lett.*, 1990, **31**, 261-264 (*Deoxypodophyllotoxin, synth*).
Kashima, T. *et al., Chem. Pharm. Bull.*, 1991, **39**, 192-194 (*Deoxypodophyllotoxin, synth*).
Ito, C. *et al., Chem. Pharm. Bull.*, 1992, **40**, 1318-1321 (*4,5-Demethylene-7-desoxypodophyllotoxin*).
Novelo, M. *et al., J. Nat. Prod.*, 1993, **56**, 1728-1736 (*Isodeoxypodophyllotoxin*).
Shaari, K. *et al., J. Nat. Prod.*, 1994, **57**, 720-724 (*Demethyldeoxypicropodophyllotoxin*).
Song, Y.N. *et al., J. Nat. Prod.*, 1994, **57**, 1670-1674 (*Koelreuteria henryi constits*).
Middel, O. *J. Med. Chem.*, 1995, **38**, 2112-2118 (*Deoxypicropodophyllin, synth*).
Bogucki, D.E. *et al., J. Org. Chem.*, 1995, **60**, 588-593 (*Deoxypodophyllotoxin, synth*).
Van Uden, W. *et al., J. Nat. Prod.*, 1997, **60**, 401-403 (*Deoxypodophyllotoxin, isol*).
Chang, L.C. *et al., J. Nat. Prod.*, 2000, **63**, 1235-1238 (*Isopicrodeoxypodophyllotoxin, struct*).
Kim, Y. *et al., Planta Med.*, 2002, **68**, 271-274 (*Pulsatilla koreana constit, activity*).
Kolly-Kovač, T. *et al., Synthesis*, 2005, 1459-1466 (*Deoxypodophyllotoxin, synth*).
Federolf, K. *et al., Phytochemistry*, 2007, **68**, 1397-1406 (*Linum album constit*).
Julsing, M.K. *et al., Eur. J. Med. Chem.*, 2008, **43**, 1171-1179 (*Deoxypodophyllotoxin, metab*).
Yong, Y. *et al., Bioorg. Med. Chem. Lett.*, 2009, **19**, 4367-4371 (*Deoxypodophyllotoxin, activity*).
Lee, T.H. *et al., J. Nat. Med.*, 2009, **63**, 209-214 (*Koelreuteria henryi constits*).
Xu, H. *et al., Bioorg. Med. Chem. Lett.*, 2010, **20**, 2500-2502 (*Deoxypodophyllotoxin, activity, sar*).
Soon, Y.S. *Cancer Lett.*, 2010, **287**, 231-239 (*Deoxypodophyllotoxin, activity*).
Iida, N. *et al., Chem. Pharm. Bull.*, 2010, **58**, 742-746 (*Juniperus communis constit*).
Velázquez-Jimènez, R. *et al., Phytochemistry*, 2011, **72**, 2237-2243 (*Deoxypodophyllotoxin, abs config*).
Khaled, M. *et al., J. Ethnopharmacol.*, 2013, **149**, 24-34 (*Deoxypodophyllotoxin, rev*).
Saavedra, D.I. *et al., J. Org. Chem.*, 2018, **83**, 2018-2026 (*Deoxypodophyllotoxin, synth*)

2,7′-Cyclo-3′,4,4′,5,5′-pentahydroxy-8,8′-lignan-9,9′-olide C299

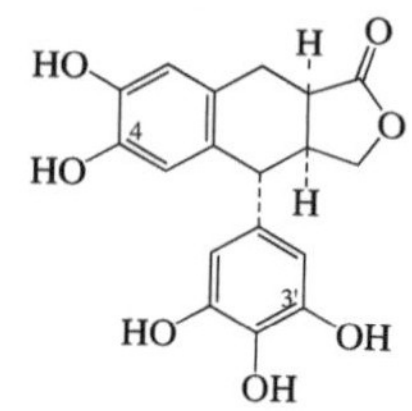

3′,4,4′,5,5′-Pentahydroxy-2,7′-cyclolignan-9,9′-olide

$C_{18}H_{16}O_7$ 344.319

(7′R,8R,8′S)-form

4,5-Methylene, 3′,4′,5′-tri-Me ether

2,7′-Cyclo-3′,4′,5′-trimethoxy-4,5-methylenedioxy-8,8′-lignan-9,9′-olide. 3′,4′,5′-Trimethoxy-4,5-methylenedioxy-2,7′-cyclolignan-9,9′-olide.

Formosalactone

[148906-40-5]

$C_{22}H_{22}O_7$ 398.411

Constit. of the bark of *Juniperus formosana* var. *concolor* and leaves, bark and wood of *Juniperus brevifolia*. Light yellow liq.. $[\alpha]_D^{20}$ -35 (c, 0.3 in $CHCl_3$).

Kuo, Y.-H. *et al., Heterocycles*, 1993, **36**, 529-535 (*Formosalactone*).
Moujir, L. *et al., Planta Med.*, 2008, **74**, 751-753 (*Juniperus brevifolia constit*)

2,7′-Cyclo-3′,4,4′,5,6-pentahydroxy-8,8′-lignan-9′,9-olide C300

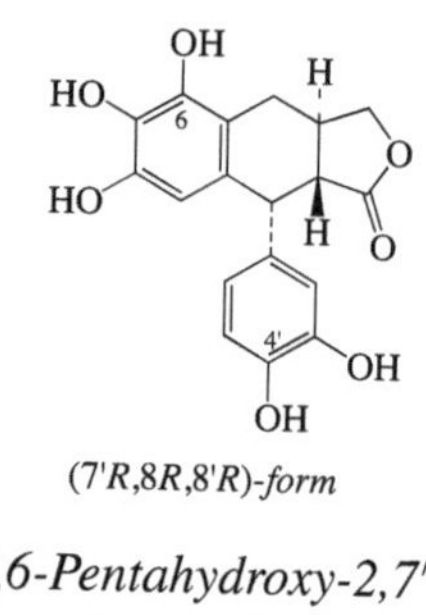

(7′R,8R,8′R)-form

3′,4,4′,5,6-Pentahydroxy-2,7′-cyclolignan-9′,9-olide

$C_{18}H_{16}O_7$ 344.319

(7′R,8R,8′R)-form

4,5-Methylene, 3′,4′-di-Me ether

2,7′-Cyclo-6-hydroxy-3′,4′-dimethoxy-4,5-methylenedioxy-8,8′-lignan-9′,9-olide. 6-Hydroxy-3′,4′-dimethoxy-4,5-methylenedioxy-2,7′-cyclolignan-9′,9-olide. ***5′-Demethoxy-β-peltatin A***

4,5-Methylene, 3′,4′-di-Me ether, 6-O-β-D-glucopyranoside

[312610-05-2]

$C_{27}H_{30}O_{12}$ 546.525

Constit. of *Bridelia ferruginea.* Gum. $[\alpha]_D$ -101 (c, 0.1 in MeOH). λ_{max} 210 (log ε 4.76); 230 (sh); 287 (log ε 3.88) (MeOH)

4,5-Methylene, 3′,4′,6-tri-Me ether

2,7′-Cyclo-3′,4′,6-trimethoxy-4,5-methylenedioxy-8,8′-lignan-9′,9-olide. 3′,4′,6-Trimethoxy-4,5-methylenedioxy-2,7′-cyclolignan-9′,9-olide

[32970-80-2]

[23978-66-7]

$C_{22}H_{22}O_7$ 398.411

Constit. of *Bursera fagaroides.* Antitumour agent. Cryst. (MeOH/Me_2CO). Mp 142 - 143°. Mp 167 - 167.5°. Mp 182 - 185° (trimorph.). $[\alpha]^{24}_{D}$ -146. Log P 1.35 (calc)

3′,4′:4,5-Bis(methylene), 6-Me ether

2,7′-Cyclo-6-methoxy-3′,4′:4,5-bis(methylenedioxy)-8,8′-lignan-9′,9-olide. 6-Methoxy-3′,4′:4,5-bis(methylenedioxy)-2,7′-cyclolignan-9′,9-olide. **Picroburseranin**. Picrobursenin

[1810049-49-0]

$C_{21}H_{18}O_7$ 382.368

Constit. of the fruits of *Cleistanthus indochinensis.* Potent cytotoxic agent against oral epidermoid carcinoma cells. Cryst. (Me_2CO). Mp 186 - 188°. $[\alpha]^{30}_{D}$ -100.5 (c, 0.55 in $CHCl_3$). λ_{max} 228 (log ε 2.19); 286 (log ε 1.77) (MeOH)

(7′R,8R,8′S)-form

3′,4′:4,5-Bis(methylene), 6-Me ether

Burseranin

[852920-06-0]

$C_{21}H_{18}O_7$ 382.368

Constit. of the stems of *Bursera graveolens.* Inhibitor of human HT 1080 fibrosarcoma cells. Waxy solid. $[\alpha]_D$ +32 (c, 0.27 in $CHCl_3$). λ_{max} 206 (ε 25200); 287 (ε 5100) (MeOH).

Bianchi, E. *et al., Tetrahedron Lett.*, 1969, 2759-2762 (*Bursera fagaroides ether*).

Bates, R.B. *et al., J. Org. Chem.*, 1972, **37**, 562-565 (*cryst struct*).

Rashid, M.A. *et al., Nat. Prod. Lett.*, 2000, **14**, 285-292 (*Bridelia ferruginea constit*).

Nakanishi, T. *et al., Chem. Pharm. Bull.*, 2005, **53**, 229-231 (*Burseranin*).

Thanh, V.T.T. *et al., Planta Med.*, 2014, **80**, 695-702 (*Picroburseranin, cd, activity*)

2,7′-Cyclo-3′,4,4′,5,7-pentahydroxy-8,8′-lignan-9′,9-olide C301

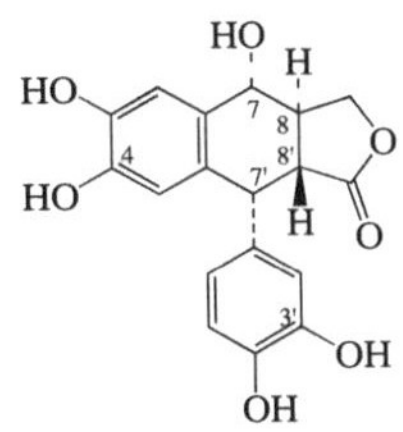

(7R,7′R,8R,8′R)-form

3′,4,4′,5,7-Pentahydroxy-2,7′-cyclolignan-9′,9-olide

$C_{18}H_{16}O_7$ 344.319

(7R,7′R,8R,8′R)-form

4,5-Methylene, 3′-Me ether

2,7′-Cyclo-4′,7-dihydroxy-3′-methoxy-4,5-methylenedioxy-8,8′-lignan-9′,9-olide. 4′,7-Dihydroxy-3′-methoxy-4,5-methylenedioxy-2,7′-cyclolignan-9′,9-olide. NC 370

[139022-95-0]

$C_{20}H_{18}O_7$ 370.357

Constit. of *Linum flavum*

(7R,7′R,8R,8′S)-form

3′,4′:4,5-Bis(methylene) ether

2,7′-Cyclo-7-hydroxy-3′,4′:4,5-bis(methylenedioxy)-8,8′-lignan-9′,9-olide. 7-Hydroxy-3′,4′:4,5-bis(methylenedioxy)-2,7′-cyclolignan-9′,9-olide.

7-Hydroxypicropolygamin

[1810049-73-0]

$C_{20}H_{16}O_7$ 368.341

Constit. of the fruit of *Cleistanthus indochinensis.* Cryst. (Me_2CO). Mp 259 - 260°. $[\alpha]^{30}_{D}$ +14 (c, 0.5 in $CHCl_3$). λ_{max} 215 (log ε 3.38); 240 (log ε 4.11); 291 (log ε 4.05) ($CHCl_3$)

(7S,7′R,8R,8′R)-form

3′,4′:4,5-Bis(methylene) ether, 7-O-β-D-glucopyranoside

Cleisindoside A

[1821136-69-9]

$C_{26}H_{26}O_{12}$ 530.482

Constit. of the fruit of *Cleistanthus indochinensis.* Cryst. (Me_2CO/MeOH). Mp 261 - 262°. $[\alpha]^{30}_{D}$ -50 (c, 0.3 in MeOH). λ_{max} 215 (log ε 4.32); 288 (log ε 3.98) (MeOH)

3′,4′:4,5-Bis(methylene) ether, 7-O-[β-D-apiofuranosyl-(1→6)-β-D-glucopyranoside]

Cleisindoside B

[1821136-70-2]

$C_{31}H_{34}O_{16}$ 662.597

Constit. of the fruit of *Cleistanthus indochinensis.* Cryst. (Me_2CO/MeOH). Mp 248 - 249°. $[\alpha]^{30}_{D}$ -52 (c, 1.0 in MeOH). λ_{max} 215 (log ε 4.36); 238 (log ε 3.96); 289 (log ε 3.96) (MeOH)

3′,4′:4,5-Bis(methylene) ether, 7-O-[β-D-xylopyranosyl-(1→6)-β-D-glucopyranoside]

Cleisindoside C

[1821136-71-3]

$C_{31}H_{34}O_{16}$ 662.597

Constit. of the fruit of *Cleistanthus indochinensis.* Cryst. (Me_2CO/MeOH). Mp 278 - 279°. $[\alpha]^{30}_{D}$ -65 (c, 0.9 in MeOH). λ_{max} 216 (log ε 3.86); 235 (log ε 3.15); 287 (log ε 3.12) (MeOH).

Wichers, H.J. *et al., Phytochemistry*, 1991, **30**, 3601-3604 (*Linum flavum constit*).

Thanh, V.T.T. *et al., Planta Med.*, 2014, **80**, 695-702 (*Cleisindosides A,B,C, 7-Hydroxypicropolygamin*)

1,2′-Cyclo-3,3′,4′,5′,7′-pentahydroxy-8,8′-lignan-4(1*H*)-one C302

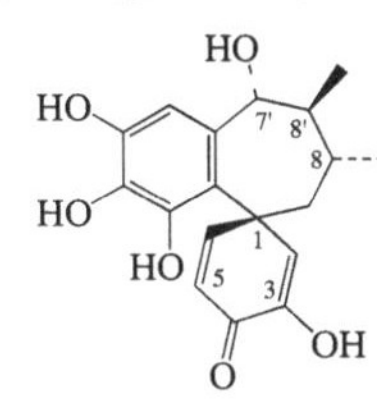

Absolute Configuration

$C_{18}H_{20}O_6$ 332.352

(1S,7′R,8R,8′S)-form

4′,5′-Methylene, 3,3′-di-Me ether

1,2′-Cyclo-7′-hydroxy-3,3′-dimethoxy-4′,5′-methylenedioxy-8,8′-lignan-4(1H)-one

4′,5′-Methylene, 3,3′-di-Me ether, 7′-Ac

Eupodienone 7

[73020-06-1]

$C_{23}H_{26}O_7$ 414.454

Constit. of *Eupomatia laurina.* Gum. $[\alpha]_D$ +58 (c, 0.5 in $CHCl_3$).

λ_{max} 217 (sh); 248 (ε 12600); 278 (sh) ($CHCl_3$)

4',5'-Methylene, 3,3'-di-Me ether, 7'-benzoyl

Eupodienone 6

[73020-05-0]

$C_{28}H_{28}O_7$ 476.525

Constit. of *Eupomatia laurina*. Gum. [α] D -25 (c, 0.5 in $CHCl_3$). λ_{max} 221 (ε 46200); 250 (sh); 280 (sh) ($CHCl_3$)

5-Methoxy, 3,3',4',5'-tetra-Me ether

1,2'-Cyclo-7'-hydroxy-3,3',4',5,5'-pentamethoxy-8,8'-lignan-4(1H)-one. ***Eupodienone 2***

[39350-63-5]

$C_{23}H_{30}O_7$ 418.486

Constit. of the bark of *Eupomatia laurina*. Needles (C_6H_6/petrol). Mp 207 - 209°. [α] D -34 (c, 1.5 in $CHCl_3$)

5-Methoxy, 3,3',4',5'-tetra-Me ether, 7'-Ac

Eupodienone 1

[75539-66-1]

$C_{25}H_{32}O_8$ 460.523

Constit. of the bark of *Eupomatia laurina*. Needles (Et_2O). Mp 172 - 173°. [α] D +26 (c, 0.5 in $CHCl_3$)

5-Methoxy, 3,3',4',5'-tetra-Me ether, 7'-benzoyl

Eupodienone 3

[39350-66-8]

$C_{30}H_{34}O_8$ 522.594

Constit. of the bark of *Eupomatia laurina*. Needles (MeOAc/hexane). Mp 170 - 172°

5-Methoxy, 7'-Et, 3,3',4',5'-tetra-Me ether

7'-O-Ethyleupodienone 2

[75539-71-8]

$C_{25}H_{34}O_7$ 446.54

Constit. of *Eupomatia laurina*. Needles. Mp 215 - 218°. [α] D -34 ($CHCl_3$). Probable artifact

5-Methoxy, 4',5'-methylene, 3,3'-di-Me ether, 7'-Ac

Eupodienone 5

[73020-04-9]

$C_{24}H_{28}O_8$ 444.48

Constit. of *Eupomatia laurina*. Gum. [α] D +29 (c, 0.5 in $CHCl_3$). λ_{max} 221 (ε 41000); 251 (ε 8500); 282 (ε 10400) ($CHCl_3$)

5-Methoxy, 4',5'-methylene, 3,3'-di-Me ether, 7'-benzoyl

Eupodienone 4

[73020-03-8]

$C_{29}H_{30}O_8$ 506.551

Constit. of *Eupomatia laurina*. Gum. [α] D -65 (c, 0.5 in $CHCl_3$). λ_{max} 222 (ε 38400); 236 (sh); 279 (ε 9200) ($CHCl_3$).

Bowden, B.F. *et al.*, *Aust. J. Chem.*, 1980, **33**, 1823-1831 (*Eupodienones 1,2,3*).

Read, R.W. *et al.*, *Aust. J. Chem.*, 1981, **34**, 1125-1134 (*Eupodienones 4-7*)

2,7'-Cyclo-2',3,3',4,4'-pentahydroxy-8,8'-lignan-7-one C303

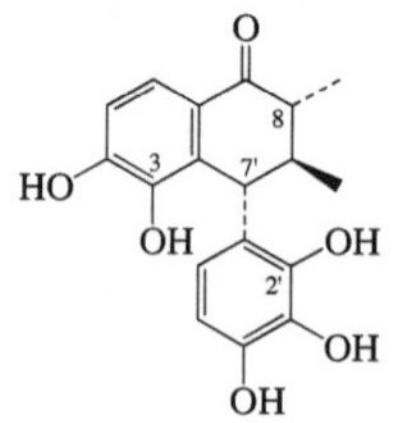

2',3,3',4,4'-Pentahydroxy-7-oxo-2,7'-cyclolignan

$C_{18}H_{18}O_6$ 330.336

(7'R*,8R*,8'S*)-form

3,4:3',4'-Bis(methylene) ether

2,7'-Cyclo-2'-hydroxy-3,4:3',4'-bis(methylenedioxy)-8,8'-lignan-7-one. 2'-Hydroxy-3,4:3',4'-bis(methylenedioxy)-7-oxo-2,7'-cyclolignan

[488735-14-4]

$C_{20}H_{18}O_6$ 354.358

Constit. of the fruit of *Virola sebifera*. Solid. Mp 167 - 168°. [α] D -29.5 (c, 0.1 in MeOH). Exhibits atropisomerism.

Rezende, K.R. *et al.*, *Phytochemistry*, 2002, **61**, 427-432 (*Virola sebifera constit*)

2,7'-Cyclo-3,3',4,4',7'-pentahydroxy-8,8'-lignan-7-one C304

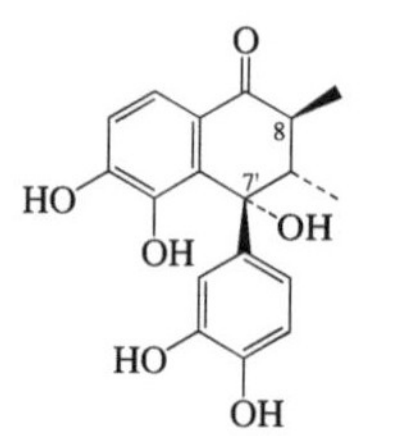

3,3',4,4',7'-Pentahydroxy-7-oxo-2,7'-cyclolignan

$C_{18}H_{18}O_6$ 330.336

(7'R,8S,8'S)-form

3,4:3',4'-Bis(methylene) ether

2,7'-Cyclo-7'-hydroxy-3,4:3',4'-bis(methylenedioxy)-8,8'-lignan-7-one. 7'-Hydroxy-3,4:3',4'-bis(methylenedioxy)-7-oxo-2,7'-cyclolignan. ***Hydroxyoxootobain***

[82467-97-8]

$C_{20}H_{18}O_6$ 354.358

Isol. from seeds of *Virola sebifera* and *Virola elongata*. Mp 86 - 87°. Mp 115 - 117°. Abs. config. revised in 1982. λ_{max} 234 (ε 21150); 282 (ε 9900); 300 (sh); 315 (ε 6300) (MeOH).

Lopes, L.M.X. *et al.*, *Phytochemistry*, 1982, **21**, 751-755 (*Viola sebifera constit, cd, abs config*).

Martinez, J.C. *et al.*, *Phytochemistry*, 1985, **24**, 1612-1614 (*Virola elongata constit*)

2,7'-Cyclo-3',4,4',5,7'-pentahydroxy-8,8'-lignan-7-one C305

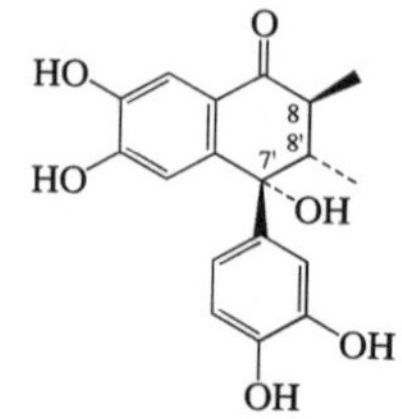

(7'R,8S,8'S)-form

3',4,4',5,7'-Pentahydroxy-2,7'-cyclolignan-7-one

$C_{18}H_{18}O_6$ 330.336

(7'R,8S,8'S)-form

3',4'-Methylene, 4,5-di-Me ether

2,7'-Cyclo-7'-hydroxy-4,5-dimethoxy-3',4'-methylenedioxy-8,8'-lignan-7-one. 7'-Hydroxy-4,5-dimethoxy-3',4'-methylenedioxy-2,7'-cyclolignan-7-one

[82427-59-6]

$C_{21}H_{22}O_6$ 370.401

Constit. of *Virola elongata* and *Virola sebifera*. Cryst. (MeOH). Mp 167 - 171° (*V. elongata*). Mp 177 - 180° (*V. sebifera*). $[\alpha]^{25}_{D}$ -80 (c, 0.01 in MeOH). Abs. config. revised in 1990

8,8'-Didehydro

2,7'-Cyclo-3',4,4',5,7'-pentahydroxy-8,8'-lign-8(8')-en-7-one. 3',4,4',5,7'-Pentahydroxy-2,7'-cyclolign-8(8')-en-7-one

8,8'-Didehydro, 3',4'-methylene, 4,5-di-Me ether

2,7'-Cyclo-7'-hydroxy-4,5-dimethoxy-3',4'-methylenedioxy-8,8'-lign-8(8')-en-7-one. 7'-Hydroxy-4,5-dimethoxy-3',4'-methylenedioxy-2,7'-cyclolign-8(8')-en-7-one

[93446-21-0]

$C_{21}H_{20}O_6$ 368.385

Constit. of the fruit of *Virola sebifera*. Yellow cryst. (MeOH). Mp 215 - 217°. Config. not determined. λ_{max} 245 (ε 33500); 289 (ε 11600) (MeOH).

Lopes, L.M.X. *et al.*, *Phytochemistry*, 1982, **21**, 751-755 (*Virola sebifera constit*).
Lopes, L.M.X. *et al.*, *Phytochemistry*, 1984, **23**, 2021-2024 (*Virola sebifera constit*).
Kato, M.J. *et al.*, *Phytochemistry*, 1990, **29**, 1799-1810 (*Virola elongata constit, abs config*)

2,7′-Cyclo-3′,4,4′,5,8-pentahydroxy-8,8′-lignan-7-one C306

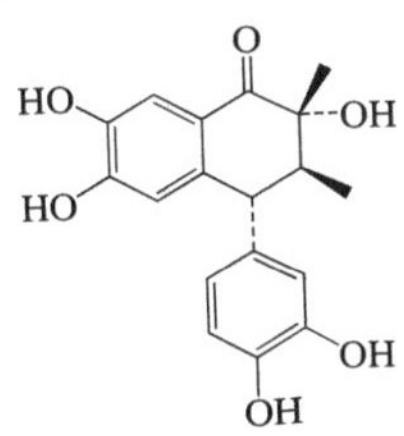

(7′R,8R,8′S)-form

3′,4,4′,5,8-Pentahydroxy-7-oxo-2,7′-cyclolignan
$C_{18}H_{18}O_6$ 330.336
(7′R,8R,8′S)-form
3′,4,4′,5-Tetra-Me ether
2,7′-Cyclo-8-hydroxy-3′,4,4′,5-tetramethoxy-8,8′-lignan-7-one. 8-Hydroxy-3′,4,4′,5-tetramethoxy-7-oxo-2,7′-cyclolignan.
8-Hydroxy-8′-epiaristoligone
[887304-88-3]
$C_{22}H_{26}O_6$ 386.444
Constit. of the roots of *Holostylis reniformis*. Amorph. yellow solid. $[\alpha]^{25}_{D}$ -14 (c, 1.3 in $CHCl_3$). λ_{max} 212 (ε 7943); 235 (ε 10000); 277 (ε 5012); 310 (ε 2512) (MeOH)
(7′R*,8R*,8′R*)-form
3′,4′-Methylene, 4,5-di-Me ether
2,7′-Cyclo-8-hydroxy-4,5-dimethoxy-3′,4′-methylenedioxy-8,8′-lignan-7-one. 8-Hydroxy-4,5-dimethoxy-3′,4′-methylenedioxy-7-oxo-2,7′-cyclolignan.
Hydroxyaristotetralone
[113953-01-8]
$C_{21}H_{22}O_6$ 370.401
Constit. of *Aristolochia chilensis*. Amorph.. [α] D -62 (c, 0.23 in $CHCl_3$)
3′,4′-Methylene, 4,5-di-Me ether, 8-Ac
Acetoxyaristotetralone
[113953-00-7]
$C_{23}H_{24}O_7$ 412.438
Constit. of *Aristolochia chilensis*. Amorph.. [α] D -62 (c, 0.58 in $CHCl_3$)
3′,4′:4,5-Bis(methylene) ether
2,7′-Cyclo-8-hydroxy-3′,4′:4,5-bis(methylenedioxy)-8,8′-lignan-7-one. 8-Hydroxy-3′,4′:4,5-bis(methylenedioxy)-7-oxo-2,7′-cyclolignan.
Pedunculin A‡
[1229002-94-1]
$C_{20}H_{18}O_6$ 354.358
Constit. of *Litsea pedunculata*. Cryst.. Mp 168 - 170°. $[\alpha]^{25}_{D}$ -111.2 (c, 0.22 in $CHCl_3$). λ_{max} 295 (log ε 3.5); 317 (log ε 3.8) (MeOH).

Urzúa, A. *et al.*, *J. Nat. Prod.*, 1988, **51**, 117-121 (*Hydroxyaristotetralone, Acetoxyaristotetralone*).
Da Silva, T. *et al.*, *Phytochemistry*, 2006, **67**, 929-937 (*8-Hydroxy-8′-epiaristoligone*).
Wang, L. *et al.*, *J. Asian Nat. Prod. Res.*, 2009, **11**, 1028-1031 (*Pedunculin A*)

2,7′-Cyclo-3,3′,4,4′,9′-pentahydroxy-8,8′-lign-7-en-9-al C307

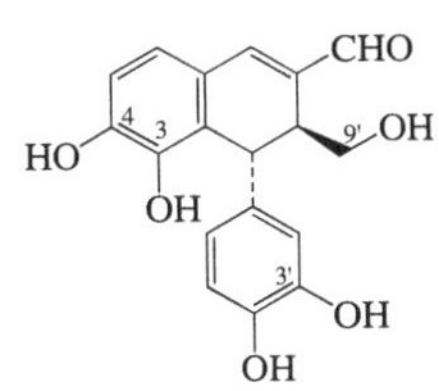

3,3′,4,4′,9′-Pentahydroxy-2,7′-cyclolign-7-en-9-al. 3,3′,4,4′,9′-Pentahydroxy-9-oxo-2,7′-cyclolign-7-ene
$C_{18}H_{16}O_6$ 328.32
(7′R*,8′S*)-form
3,3′-Di-Me ether
2,7′-Cyclo-4,4′,9′-trihydroxy-3,3′-dimethoxy-8,8′-lign-7-en-9-al. 4,4′,9′-Trihydroxy-3,3′-dimethoxy-2,7′-cyclolign-7-en-9-al. 4,4′,9′-Trihydroxy-3,3′-dimethoxy-9-oxo-2,7′-cyclolign-7-ene. ***Vitedoin A***. *Vitexin 2. VB2*
[819861-40-0]
$C_{20}H_{20}O_6$ 356.374
Constit. of the seeds of *Vitex negundo*. Antioxidant. Moderate inhibitor of human COC1 ovarian cancer cell line. Amorph. powder. $[\alpha]^{13}_{D}$ -79.4 (c, 3.2 in MeOH). λ_{max} 221; 288; 341 (MeOH)
3,3′-Di-Me ether, 9′-O-β-D-glucopyranoside
Vitecannaside A
[1018447-94-3]
$C_{26}H_{30}O_{11}$ 518.515
Constit. of the fruit of *Vitex cannabifolia*. Amorph. powder. $[\alpha]^{13}_{D}$ -90.6 (c, 0.8 in MeOH)
3,3′-Di-Me ether, 9′-hexadecanoyl
Vitexdoin H
$C_{36}H_{50}O_7$ 594.789
Isol. from seeds of *Vitex negundo*. Shows antiinflammatory and antiosteoporetic activity. Syrup. $[\alpha]^{20}_{D}$ -75 (c, 0.2 in MeOH). λ_{max} 223; 283; 338 (MeOH).

Ono, M. *et al.*, *J. Nat. Prod.*, 2004, **67**, 2073-2075 (*Vitedoin A*).
Eur. Pat., 2008, 1 972 611 (Dietetech Investments Ltd.)(*Vitedoin A, activity*).
Yamasaki, T. *et al.*, *J. Nat. Med.*, 2008, **62**, 47-51 (*Vitecannaside A*).
Zheng, C.-J. *et al.*, *Fitoterapia*, 2014, **93**, 31-38 (*Vitexdoin H*)

3′,8-Cyclo-4,5,5′,7,9-pentahydroxy-2,4′-lign-7′-en-9′-al C308

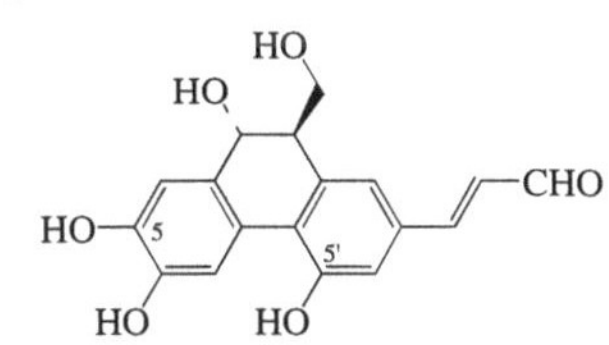

3-[9,10-Dihydro-4,6,7,9-tetrahydroxy-10-(hydroxymethyl)-2-phenanthrenyl]-2-propenal
$C_{18}H_{16}O_6$ 328.32
(7R*,7′E,8R*)-form
5,5′-Di-Me ether
3′,8-Cyclo-4,7,9-trihydroxy-5,5′-dimethoxy-2,4′-lign-7′-en-9′-al
[352281-90-4]
$C_{20}H_{20}O_6$ 356.374
Constit. of *Trema orientalis*. Light yellow solid. $[\alpha]^{20}_{D}$ +0.2 (c, 0.56 in $CHCl_3$). λ_{max} 219 (log ε 4.5); 237 (sh); 258 (sh); 278 (sh); 311 (sh); 343 (log ε 4.3) (MeOH).

Dijoux-Franca, M.-G. *et al.*, *J. Nat. Prod.*, 2001, **64**, 832-835 (*Trema orientalis constit*)

3′,8-Cyclo-5,5′,6′,7,9-pentahydroxy-2,4′-lign-7′-en-9′-al C309

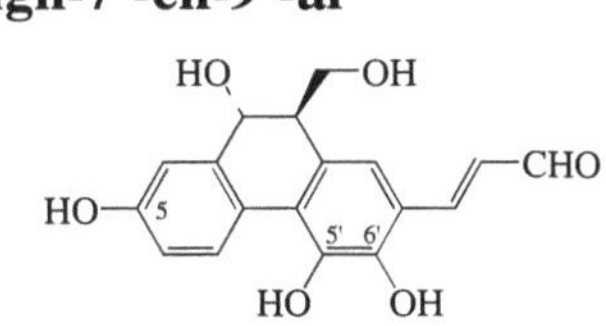

3-[9,10-Dihydro-3,4,7,9-tetrahydroxy-10-(hydroxymethyl)-2-phenanthrenyl]-2-propenal. 3′,8-Cyclo-5,5′,6′,7,9-pentahydroxy-9′-oxo-2,4′-neolign-7′-ene
$C_{18}H_{16}O_6$ 328.32

(7R*,7′E,8R*)-form

5,6′-Di-Me ether

3′,8-Cyclo-5′,7,9-trihydroxy-5,6′-dimethoxy-2,4′-lign-7′-en-9′-al. 3′,8-Cyclo-5′,7,9-trihydroxy-5,6′-dimethoxy-9′-oxo-2,4′-lign-7′-ene

5,6′-Di-Me ether, 5′-O-β-D-glucopyranoside

[352281-91-5]

$C_{26}H_{30}O_{11}$ 518.515

Constit. of *Trema orientalis*. Resin. $[\alpha]_D^{20}$ +0.8 (c, 0.25 in MeOH). λ_{max} 219 (log ε 5.5); 237 (sh); 258 (sh); 278 (sh); 311 (sh); 343 (log ε 4.7) (MeOH).

Dijoux-Franca, M.G. *et al., J. Nat. Prod.*, 2001, **64**, 832-835 (*Trema orientalis constit*)

2,7′-Cyclo-3,3′,4,4′,5-pentahydroxy-8,8′-lign-7-ene-9,9′-dioic acid C310

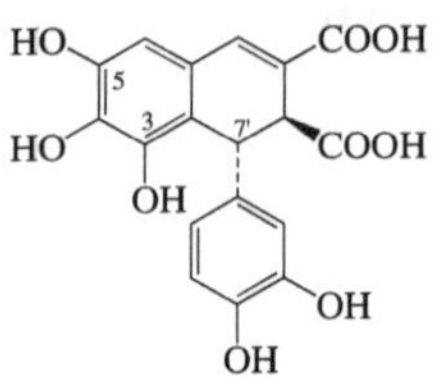

(7′R,8′S)-form

3,3′,4,4′,5-Pentahydroxy-2,7′-cyclolign-7-ene-9,9′-dioic acid

$C_{18}H_{14}O_9$ 374.301

(7′R,8′S)-form

3,5-Di-Me ether

2,7′-Cyclo-3′,4,4′-trihydroxy-3,5-dimethoxy-8,8′-lign-7-ene-9,9′-dioic acid. 3′,4,4′-Trihydroxy-3,5-dimethoxy-2,7′-cyclolign-7-ene-9,9′-dioic acid

3,5-Di-Me ether, bis[2-(4-hydroxyphenyl)ethylamide]

[200135-06-4]

$C_{36}H_{36}N_2O_9$ 640.689

Alkaloid from *Porcelia macrocarpa*. Amorph. powder. $[\alpha]_D^{25}$ -12 (c, 0.08 in MeOH). λ_{max} 249 (log ε 4.49); 283 (log ε 4.22); 332 (log ε 4.24) (EtOH)

(7′R*,8′R*)-form

3,3′,5-Tri-Me ether, 9′-Me 9-(α-D-glucopyranosyl-(1→2)-β-D-fructofuranosid-6-yl) diester

Echiumin D

[2227328-50-7]

$C_{34}H_{42}O_{19}$ 754.691

Constit. of *Echium angustifolium*. Exhibits cytotoxic activity. Amorph. yellow powder. $[\alpha]_D^{20}$ +71 (c, 0.05 in MeOH). λ_{max} 282 (ε 152); 316 (ε 173); 332 (ε 158) (MeOH)

3′-Dexoy, 3,5-di-Me ether, 9′-Me 9-(α-D-glucopyranosyl-(1→2)-β-D-fructofuranosid-6-yl) diester

Echiumin C

[2227328-46-1]

$C_{33}H_{40}O_{18}$ 724.665

Constit. of *Echium angustifolium*. Exhibits weak cytotoxic activity. Amorph. yellow powder. $[\alpha]_D^{20}$ +67 (c, 0.07 in MeOH). λ_{max} 282 (ε 143); 326 (ε 216); 346 (ε 183) (MeOH)

(7′R*,8′S*)-form

3,3′,5-Tri-Me ether

2,7′-Cyclo-4,4′-dihydroxy-3,3′,5-trimethoxy-8,8′-lign-7-ene-9,9′-dioic acid. 4,4′-Dihydroxy-3,3′,5-trimethoxy-2,7′-cyclolign-7-ene-9,9′-dioic acid

3,3′,5-Tri-Me ether, bis[2-(4-hydroxyphenyl)ethylamide]

Melongenamide A

[1676050-20-6]

$C_{37}H_{38}N_2O_9$ 654.716

Alkaloid from roots of *Solanum melongena*. Green amorph. powder. $[\alpha]_D^{21}$ -22 (c, 0.1 in MeOH). λ_{max} 230 (log ε 4.26); 285 (log ε 4.15); 326 (log ε 4.51) (MeOH)

(7′RS,8′SR)-form

3,5-Di-Me ether, 9′-[2-(4-hydroxyphenyl)ethyl]amide, 9-[4-(acetylamino)butyl]amide

[1416474-65-1]

$C_{34}H_{39}N_3O_9$ 633.698

Isol. from *Lycium chinense*. Yellow gum. λ_{max} 250 (log ε 3.48); 303 (log ε 4.42) (MeOH).

Chaves, M.H. *et al., Phytochemistry*, 1997, **46**, 879-881 (*Porcelia macrocarpa constit*).
Zhang, J.-X. *et al., J. Nat. Prod.*, 2013, **76**, 51-58 (*Lycium chinense constit*).
Sun, J. *et al., Fitoterapia*, 2014, **98**, 110-116 (*Melongenamide A*).
El-Rokh, A.R. *et al., Phytochemistry*, 2018, **149**, 155-160 (*Echiumins C,D, activity*)

2,7′-Cyclo-3,3′,4,4′,7-pentahydroxy-8,8′-lign-8(8′)-en-9′,9-olide C311

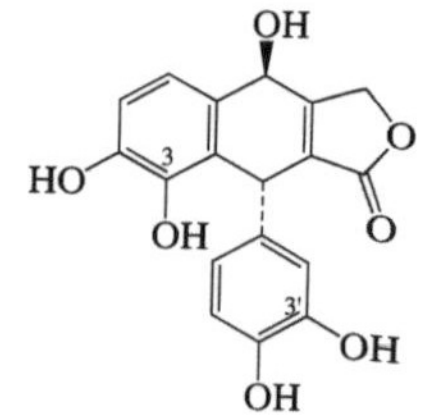

3,3′,4,4′,7-Pentahydroxy-2,7′-cyclolign-8(8′)-en-9′,9-olide

$C_{18}H_{14}O_7$ 342.303

(7R,7′R)-form

3,4:3′,4′-Bis(methylene) ether

2,7′-Cyclo-7-hydroxy-3,4:3′,4′-bis(methylenedioxy)-8,8′-lign-8(8′)-en-9′,9-olide. 7-Hydroxy-3,4:3′,4′-bis(methylenedioxy)-2,7′-cyclolign-8(8′)-en-9′,9-olide. ***Jusneesiinol***

[216955-78-1]

$C_{20}H_{14}O_7$ 366.325

Constit. of *Justicia neesii*. Cryst. (petrol/EtOAc). Mp 220 - 222°. $[\alpha]_D^{25}$ +218.4 (c, 0.5 in Me_2CO). λ_{max} 286 (log ε 3.95) (MeOH).

Rajasekhar, D. *et al., Tetrahedron*, 1998, **54**, 13227-13236 (*Jusneesiinol*)

2,7′-Cyclo-3′,4,4′,5,5′-pentahydroxy-8,8′-lign-8(8′)-en-9′,9-olide C312

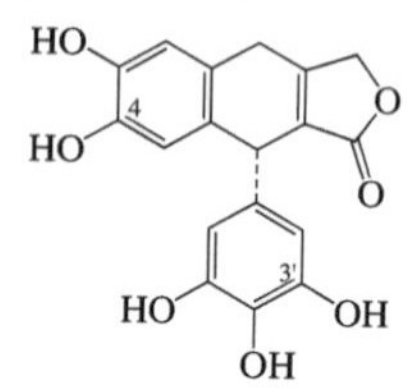

(R)-form

3′,4,4′,5,5′-Pentahydroxy-2,7′-cyclolign-8(8′)-en-9′,9-olide

$C_{18}H_{14}O_7$ 342.303

(R)-form

4,5-Methylene, 3′,4′,5′-tri-Me ether

2,7′-Cyclo-3′,4′,5′-trimethoxy-4,5-methylenedioxy-8,8′-lign-8(8′)-en-9′,9-olide. 3′,4′,5′-Trimethoxy-4,5-methylenedioxy-2,7′-cyclolign-8(8′)-en-9′,9-olide. ***β-Apopicropodophyllin. Hyptinin***

[477-52-1]

[518-32-1] (α-Apopicropodophyllin) [668-01-9] (γ-Apopicropodopyllin) [156764-88-4] (Hyptinin)

$C_{22}H_{20}O_7$ 396.395

Obt. by action of base on the α-isomer. Constit. of *Goldfussia*

psilostachys, *Hyptis verticillata* and *Micranthemum umbrosum*.
Cytotoxic agent. Inhibitor of an adult T-cell leukaemia cell line, S1T. Crayfish feeding deterrent.
Cryst. (petrol). Mp 215 - 217° (212-214°). $[\alpha]_D^{24}$ +63 (c, 1 in $CHCl_3$). $[\alpha]_D^{25}$ +77 ($CHCl_3$) (Hyptinin). α-Apopicropodophyllin, obt. by acid-cat. dehydration of Picropodophyllin, and γ-Apopicropodophyllin, obt. by base-catalysed isom. of the β-isomer, are semisynthetic isomers (Δ^7- and $\Delta^{7'}$- respectively). Structure of Hyptinin revised in 2017. λ_{max} 250 (ε 6300); 301 (ε 4000); 317 (ε 500) (MeOH) (Berdy)

(ξ)-form

3′,5-Di-Me ether

2,7′-Cyclo-3′,4,4′-trihydroxy-5,5′-dimethoxy-8,8′-lign-8(8′)-en-9′,9-olide. 3′,4,4′-Trihydroxy-5,5′-dimethoxy-2,7′-cyclolign-8(8′)-en-9′,9-olide. ***β-Apoplicatitoxin***

[39993-02-7]

$C_{20}H_{18}O_7$ 370.357

Constit. of *Thuja plicata*. No opt. rotn. reported. λ_{max} 284 (log ε 3.75) (EtOH)

Penta-Me ether

[41565-17-7]

Cryst. (MeOH). Mp 220 - 222°. λ_{max} 282 (log ε 3.61) (EtOH).

Schrecker, A.W. *et al.*, *J. Am. Chem. Soc.*, 1952, **74**, 5676-5683 (*Apopicropodophyllins*).
Schrecker, A.W. *et al.*, *J. Am. Chem. Soc.*, 1953, **75**, 5916-5924 (*β-Apopicropodophyllin, abs config*).
Ayres, D.C. *et al.*, *J. Chem. Soc. C*, 1971, 1111-1114 (*cd*).
MacDonald, B.F. *et al.*, *Can. J. Chem.*, 1973, **51**, 482-485 (*β-Apoplicatitoxin*).
Novelo, M. *et al.*, *J. Nat. Prod.*, 1993, **56**, 1728-1736 (*Hyptis vericillata constit*).
Kuhnt, M. *et al.*, *Phytochemistry*, 1994, **36**, 485-489 (*Hyptinin*).
Kennedy-Smith, J.J. *et al.*, *Org. Lett.*, 2004, **6**, 1325-1327 (*β-Apopicropodophyllin, synth*).
Parker, J.D. *et al.*, *J. Chem. Ecol.*, 2006, **32**, 815-833 (*Micranthemum umbrosum constit*).
Lane, A.L. *et al.*, *Phytochemistry*, 2006, **67**, 1224-1231 (*activity*).
Da Silva, R. *et al.*, *Magn. Reson. Chem.*, 2009, **47**, 523-526 (*β-Apopicropodophyllin, pmr, cmr*).
Hamada, T. *et al.*, *Molecules*, 2012, **17**, 9931-9938 (*Hyptinin, activity*).
Maeda, K. *et al.*, *J. Nat. Prod.*, 2017, **80**, 1446-1449 (*Hyptinin, synth, struct*)

2,7′-Cyclo-3′,4,4′,5,8-pentahydroxy-8,8′-lign-7′-en-7-one C313

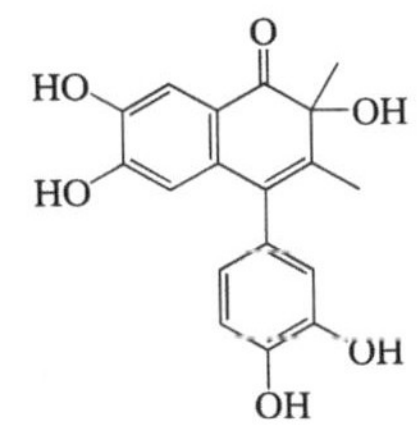

3′,4,4′,5,8-Pentahydroxy-7-oxo-2,7′-cyclolign-7′-ene

$C_{18}H_{16}O_6$ 328.32

(8ξ)-form

3′,4′-Methylene, 4,5-di-Me ether

2,7′-Cyclo-8-hydroxy-4,5-dimethoxy-3′,4′-methylenedioxy-8,8′-lign-7′-en-7-one. 8-Hydroxy-4,5-dimethoxy-3′,4′-methylenedioxy-7-oxo-2,7′-cyclolign-7′-ene

[93446-20-9]

$C_{21}H_{20}O_6$ 368.385

Constit. of the fruit of *Virola sebifera*. Amorph. solid. λ_{max} 237 (ε 15700); 265 (ε 6950); 284 (ε 7650); 315 (ε 4700) (MeOH).

Lopes, L.M.X. *et al.*, *Phytochemistry*, 1984, **23**, 2021-2024 (*Virola sebifera constit, cd*)

2,7′-Cyclo-3,3′,4,4′,8′-pentahydroxy-9′-nor-8,8′-ligna-7,7′-dien-9-al C314

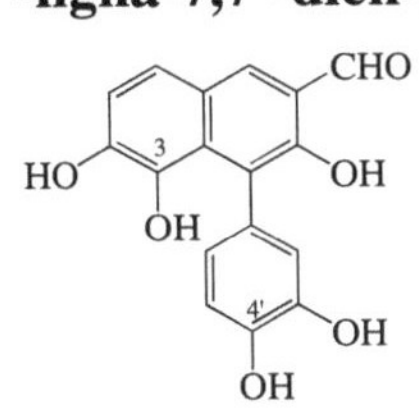

$C_{17}H_{12}O_6$ 312.277

Norlignan. Lignan numbering shown

3,3′,4′-Tri-Me ether

4-(3,4-Dimethoxyphenyl)-3,6-dihydroxy-5-methoxy-2-naphthalenecarboxaldehyde, CAS. *2,7′-Cyclo-4,8′-dihydroxy-3,3′,4′-trimethoxy-9′-nor-8,8′-ligna-7,7′-dien-9-al. 4,8′-Dihydroxy-3,3′,4′-trimethoxy-9′-nor-2,7′-cycloligna-7,7′-dien-9-al.* ***Vitrofolal B***

[261903-76-8]

$C_{20}H_{18}O_6$ 354.358

Constit. of the roots of *Vitex rotundifolia*. Amorph. yellow powder. λ_{max} 206 (log ε 4.3); 283 (log ε 4.2); 338 (log ε 3.8) (MeOH).

Kawazoe, K. *et al.*, *Phytochemistry*, 1999, **52**, 1657-1659 (*Vitrofolal B*)

2,7′-Cyclo-3′,4,4′,5,7-pentahydroxy-9′-nor-8,8′-ligna-7,7′-dien-9-al C315

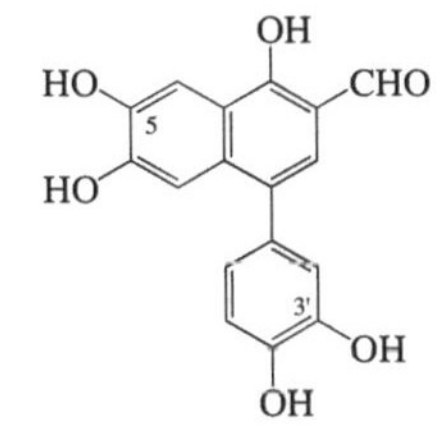

$C_{17}H_{12}O_6$ 312.277

3′,5-Di-Me ether

1,6-Dihydroxy-4-(4-hydroxy-3-methoxyphenyl)-7-methoxy-2-naphthalenecarboxaldehyde. 2,7′-Cyclo-4,4′,7-trihydroxy-3′,5-dimethoxy-9′-nor-8,8′-ligna-7,7′-dien-9-al. 4,4′,7-Trihydroxy-3′,5-dimethoxy-9′-nor-2,7′-cycloligna-7,7′-dien-9-al. ***Vitexdoin E***

[1186021-82-8]

$C_{19}H_{16}O_6$ 340.331

Constit. of the seeds of *Vitex negundo*. NO prodn. inhibitor. Amorph. yellow powder. λ_{max} 233; 278; 333 (MeOH).

Zheng, C.-J. *et al.*, *J. Nat. Prod.*, 2009, **72**, 1627-1630 (*Vitexdoin E*)

2,7′-Cyclo-3′,4,4′,5,8′-pentahydroxy-9′-nor-8,8′-ligna-7,7′-dien-9-oic acid C316

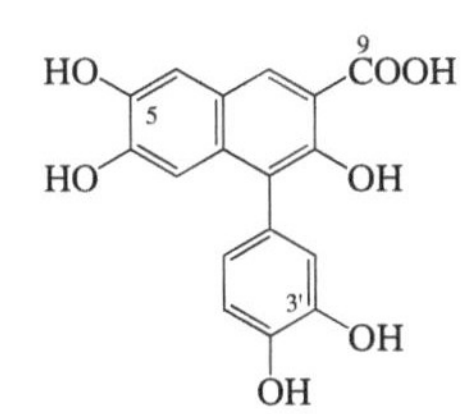

$C_{17}H_{12}O_7$ 328.276

3′,5-Di-Me ether

2,7′-Cyclo-4,4′,8′-trihydroxy-3′,5-dimethoxy-9′-nor-8,8′-ligna-7,7′-dien-9-oic acid. ***Vitexnegheteroin G***

$C_{19}H_{16}O_7$ 356.330

Constit. of the seeds of *Vitex negundo* var. *heterophylla*. Amorph. yellow powder. λ_{max} 255 (MeOH)

9-Aldehyde, 3′,5-di-Me ether

3,6-Dihydroxy-4-(4-hydroxy-3-methoxyphenyl)-7-methoxy-2-naphthalenecarboxaldehyde. 2,7′-Cyclo-4,4′,8′-trihydroxy-3′,5-dimethoxy-9′-nor-8,8′-ligna-7,7′-dien-9-al. 4,4′,8′-Trihydroxy-3′,5-dimethoxy-9′-nor-2,7′-cycloligna-7,7′-dien-9-al. ***Vitrofolal F***

[345902-01-4]
$C_{19}H_{16}O_6$ 340.331
Constit. of the roots of *Vitex rotundifolia* and the seeds of *Vitex negundo*. Amorph. yellow solid. λ_{max} 239 (log ε 4.3); 269 (log ε 4.5) (dioxan).

Kawazoe, K. *et al.*, *J. Nat. Prod.*, 2001, **64**, 588-591 (*Vitrofolal F*).
Zheng, C.-J. *et al.*, *J. Nat. Prod.*, 2009, **72**, 1627-1630 (*Vitrofolal F*).
Hu, P. *et al.*, *Fitoterapia*, 2016, **111**, 147-153 (*Vitexnegheteroin G*)

2,7′-Cyclo-3′,4,4′,5,5′-pentahydroxy-7-oxo-8,8′-lign-8-en-9′-oic acid C317

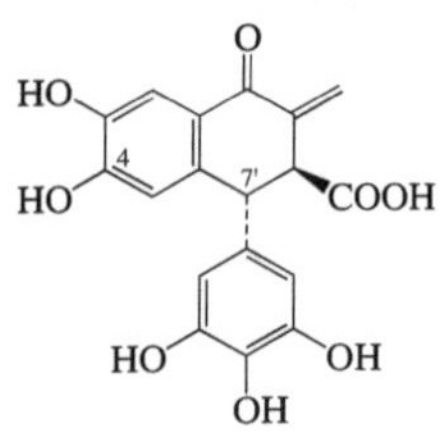

3′,4,4′,5,5′-Pentahydroxy-7-oxo-2,7′-cyclolign-8-en-9′-oic acid
$C_{18}H_{14}O_8$ 358.302
(7′R,8′S)-form
4,5-Methylene, 3′,4′,5′-tri-Me ether
2,7′-Cyclo-3′,4′,5′-trimethoxy-4,5-methylenedioxy-7-oxo-8,8′-lign-8-en-9′-oic acid. 3′,4′,5′-Trimethoxy-4,5-methylenedioxy-7-oxo-2,7′-cyclolign-8-en-9′-oic acid. ***Thuriferic acid***
[120090-82-6]
$C_{22}H_{20}O_8$ 412.394
Isol. from leaves of *Juniperus thurifera*. Solid (Et_2O) or oil. Mp 92 - 96°. $[\alpha]_D^{23}$ -179.6 ($CHCl_3$). λ_{max} 205 (ε 35200); 235 (ε 16000); 273 (ε 4900); 326 (ε 4600) (EtOH).

San Feliciano, A. *et al.*, *Tetrahedron*, 1988, **44**, 7255-7260 (*Thuriferic acid*).
Lopez-Perez, J. *et al.*, *Tetrahedron*, 1995, **51**, 6343-6348 (*abs config, conformn*).
Lopez-Perez, J. *et al.*, *Tetrahedron*, 1996, **52**, 4903-4910 (*stereochem*).
Pohmakotr, M. *et al.*, *Tetrahedron*, 2005, **61**, 5311-5321 (*synth*).
Koprowski, M. *et al.*, *Org. Biomol. Chem.*, 2016, **14**, 1822-1830 (*synth*)

2,7′-Cyclo-3′,4,4′,5,7-pentahydroxy-1,7-seco-8,8′-lignan-9,9′-olide C318

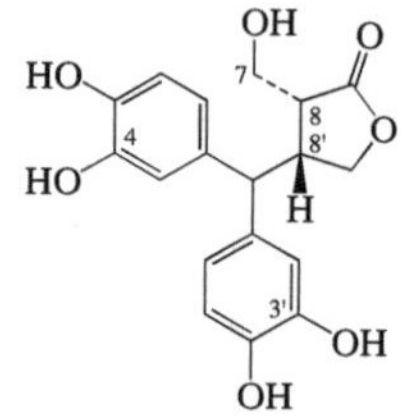

(8*R*,8′*R*)-*form*

4-[Bis(3,4-dihydroxyphenyl)methyl] dihydro-3-(hydroxymethyl)-2(3H)-furanone. 3′,4,4′,5,7-Pentahydroxy-1,7-seco-2,7′-cyclolignan-9,9′-olide
$C_{18}H_{18}O_7$ 346.335
Secolignan. Lignan numbering shown. Care needed however; free rotation around the (2→7′) bond in the secolignan struct. means the numbering round the ring is ambiguous and the two aryl rings also become equivalent
(8R,8′R)-form
Urticin B [2075707-02-5]
Constit. of the rhizomes of *Urtica fissa*. Exhibits antiinflammatory activity. Needles. $[\alpha]_D^{20}$ -2.8 (c, 0.1 in MeOH). λ_{max} 240; 282 (MeOH)
3,3′,4,4′-Tetra-Me ether, 7-O-β-D-glucopyranoside
Urticin A
[2075707-03-6]
$C_{28}H_{36}O_{12}$ 564.584
Constit. of the rhizomes of *Urtica fissa*. Exhibits antiinflammatory activity. Amorph. powder. $[\alpha]_D^{20}$ -9.2 (c, 0.4 in MeOH). λ_{max} 240; 282 (MeOH)
(8R*,8′R*)-form
3′,4-Di-Me ether
Urticol
[1160701-42-7]
$C_{20}H_{22}O_7$ 374.389
Constit. of the roots of *Urtica mairei*. Amorph. powder (Me_2CO). Mp 113 - 114°. $[\alpha]_D^{20}$ +18.5 (c, 1.35 in MeOH). Symmetrically substituted, two chiral centres
3′,4-Di-Me ether, 4′-O-β-D-glucopyranoside
Urticaside A
[1259568-20-1]
$C_{26}H_{32}O_{12}$ 536.53
Constit. of the roots of *Urtica triangularis*. Pale yellow powder
3′,4-Di-Me ether, 7-O-β-D-glucopyranoside
[1162717-95-4]
$C_{26}H_{32}O_{12}$ 536.53
Constit. of the roots of *Urtica fissa*. Amorph. powder. $[\alpha]_D^{20}$ -18.8 (c, 0.48 in MeOH)
3′,4,4′-Tri-Me ether, 7-O-β-D-glucopyranoside
[1427290-50-3]
$C_{27}H_{34}O_{12}$ 550.557
Constit. of the roots of *Urtica fissa*. Pale yellow powder (MeOH). Mp 131 - 133°. $[\alpha]_D^{20}$ -11.7 (c, 0.9 in MeOH). Unsymmetrically substituted, 7′ is a centre of chirality. Config. at this centre not determined
(8R*,8′S*)-form
3′,4-Di-Me ether, 4′-O-β-D-glucopyranoside
Urticaside B
[1259568-21-2]
$C_{26}H_{32}O_{12}$ 536.53
Constit. of the roots of *Urtica triangularis*. Pale yellow powder
3′,4-Di-Me ether, 7-O-β-D-glucopyranoside
[1162717-97-6]
Constit. of the roots of *Urtica fissa*. Amorph. powder. $[\alpha]_D^{20}$ -39.1 (c, 1.15 in MeOH)
3′,4,4′-Tri-Me ether, 7-O-β-D-glucopyranoside
[1231709-00-4]
Constit. of the roots of *Urtica fissa*. Pale yellow powder (MeOH). Mp 128 - 130°. $[\alpha]_D^{20}$ -32.4 (c, 0.7 in MeOH)
3,3′,4,4′-Tetra-Me ether, 7-O-β-D-glucopyranoside
[1228313-00-5]
$C_{28}H_{36}O_{12}$ 564.584
Constit. of the roots of *Urtica fissa*. Pale yellow powder (MeOH). Mp 120 - 122°. $[\alpha]_D^{20}$ -34.6 (c, 2.4 in MeOH).

Wang, W. *et al.*, *Chin. Chem. Lett.*, 2008, **19**, 1212-1214 (*Urticol*).
Ji, B.-Q. *et al.*, *Helv. Chim. Acta*, 2009, **92**, 997-1001 (*3′,4-di-Me ether 7-glucosides*).
Feng, B. *et al.*, *Fitoterapia*, 2010, **81**, 607-609 (*Urticasides A,B*).
Feng, B. *et al.*, *J. Nat. Med.*, 2012, **66**, 562-565 (*tri-Me ether 7-glucosides, tetra-Me ether 7-glucoside*).
Feng, X. *et al.*, *J. Nat. Med.*, 2017, **71**, 553-557 (*Urticins A,B, activity*)

2,7′-Cyclo-3′,4,4′,5,7-pentahydroxy-1,7-seco-8,8′-lign-8(8′)-en-9′,9-olide C319

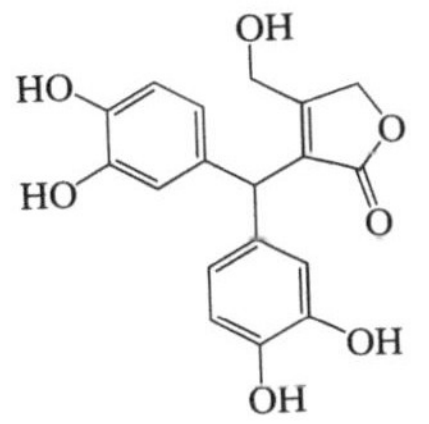

3′,4,4′,5,7-Pentahydroxy-1,7-seco-2,7′-cyclolign-8(8′)-en-9′,9-olide
$C_{18}H_{16}O_7$ 344.319
Secolignan. Achiral compd.

3′,4′:4,5-Bis(methylene) ether
2,7′-Cyclo-7-hydroxy-3′,4′:4,5-bis(methylenedioxy)-1,7-seco-8,8′-lign-8(8′)-en-9′,9-olide. 7-Hydroxy-3′,4′:4,5-bis(methylenedioxy)-1,7-seco-2,7′-cyclolign-8(8′)-en-9′,9-olide. ***Juspurpudin***
[600728-48-1]
$C_{20}H_{16}O_7$ 368.341
Constit. of *Justicia procumbens*. Cytotoxic agent. Amorph. yellow powder. $[\alpha]^{20}_{D}$ -1.5 (c, 0.065 in MeOH). λ_{max} 289 (log ε 3.95) (MeOH)

3′,4′:4,5-Bis(methylene) ether, O-β-D-glucopyranoside
Juspurpurin
[602331-38-4]
$C_{26}H_{26}O_{12}$ 530.482
Constit. of *Justicia purpurea*. Oil. $[\alpha]^{25}_{D}$ -17.5 (c, 0.75 in Me_2CO).

Kavitha, J. *et al.*, *J. Nat. Prod.*, 2003, **66**, 1113-1115 (*Juspurpurin*).
Jin, H. *et al.*, *Fitoterapia*, 2014, **94**, 70-76 (*Juspurpudin*)

2,7′-Cyclo-1,7-seco-8,8′-lignan-3′,4,4′,5,5′,6,9-heptol C320

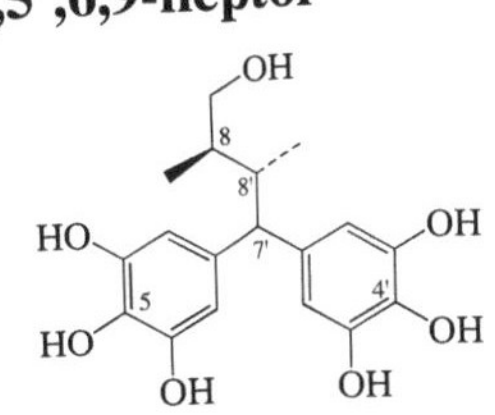

2,3-Dimethyl-4,4-bis(3,4,5-trihydroxyphenyl)-1-butanol.
3′,4,4′,5,5′,6,9-Heptahydroxy-1,7-seco-2,7′-cyclolignan
$C_{18}H_{22}O_7$ 350.367
Secolignan. Lignan numbering shown. Care needed however; free rotation of the phenyl groups means the numbering around the rings is ambiguous and the two aryl rings also become equivalent

(8R*,8′R*)-form

3′,4′-Methylene, 4,5,5′,6-tetra-Me ether
2,7′-Cyclo-3′,4,5,6-tetramethoxy-4′,5′-methylenedioxy-1,7-seco-8,8′-lignan-9-ol. 9-Hydroxy-3′,4,5,6-tetramethoxy-4′,5′-methylenedioxy-1,7-seco-2,7′-cyclolignan. ***Schilancifolignan D****. Schineglphenol A*
[1276015-36-1]
$C_{23}H_{30}O_7$ 418.486
Constit. of *Schisandra lancifolia* and *Schisandra chinensis*. Antioxidant. Exhibits strong cytotoxic and weak anti-HIV activities. Pale yellow solid. $[\alpha]^{16}_{D}$ +8.4 (c, 0.18 in EtOH). Stereochem. identity of Schilancifolignan D and Schineglphenol A not confirmed. λ_{max} 205 (log ε 4.95); 282 (log ε 4.15); 326 (log ε 2.58) (MeOH)

9-Deoxy, 3′,4′-methylene, 4,5,5′,6-tetra-Me ether
2,7′-Cyclo-3′,4,5,6-tetramethoxy-4′,5′-methylenedioxy-1,7-secolignan. 3′,4,5,6-Tetramethoxy-4′,5′-methylenedioxy-1,7-seco-2,7′-cyclolignan. ***Schilancifolignan E***
[1346122-62-0]
$C_{23}H_{30}O_6$ 402.487
Constit. of *Schisandra lancifolia*. Exhibits strong cytotoxic and weak anti-HIV activities. Yellow gum. $[\alpha]^{22}_{D}$ +9.7 (c, 0.15 in MeOH). λ_{max} 205 (log ε 5.01); 282 (log ε 4.11); 326 (log ε 2.65) (MeOH).

Xue, J.-J. *et al.*, *J. Asian Nat. Prod. Res.*, 2011, **13**, 492-497 (*Schilancifolignans D,E, activity*).
Chinese Pat., 2013, 101 982 467 (*Schineglphenol A, activity*).
Davidson, S.J. *et al.*, *Phytochem. Lett.*, 2015, **14**, 138-142 (*synth, pmr, cmr, config*)

2,7′-Cyclo-1,7-seco-8,8′-lignan-3′,4,4′,5,7,9,9′-heptol C321

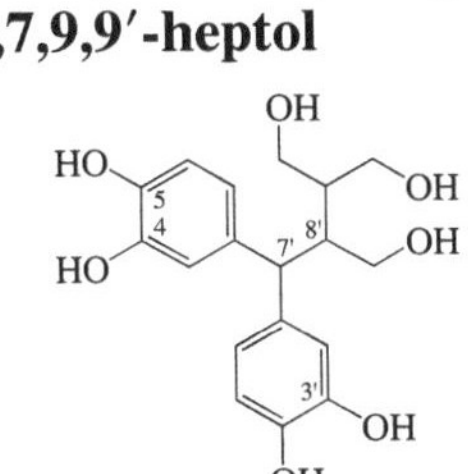

2-[Bis(3,4-dihydroxyphenyl)methyl]-3-(hydroxymethyl)-1,4-butanediol.
3′,4,4′,5,7,9,9′-Heptahydroxy-1,7-seco-2,7′-cyclolignan
$C_{18}H_{22}O_7$ 350.367
Secolignan. Lignan numbering shown. The parent compd. and symmetrically substituted derivs. have only one chiral centre

3′,4-Di-Me ether
2,7′-Cyclo-3′,4-dimethoxy-1,7-seco-8,8′-lignan-4′,5,7,9,9′-pentol. 4′,5,7,9,9′-Pentahydroxy-3′,4-dimethoxy-1,7-seco-2,7′-cyclolignan

3′,4-Di-Me ether, 4′-O-β-D-glucopyranoside
Rehmaneolignan C
[2001061-02-3]
$C_{26}H_{36}O_{12}$ 540.562
Constit. of the roots of *Rehmannia glutinosa*. Pale yellow powder. $[\alpha]^{20}_{D}$ -35.6 (c, 0.14 in MeOH). 2 Chiral centres. Configs. not determined. λ_{max} 278 (MeOH)

3′,5-Di-Me ether
2,7′-Cyclo-3′,5-dimethoxy-1,7-seco-8,8′-lignan-4,4′,7,9,9′-pentol. 4,4′,7,9,9′-Pentahydroxy-3′,5-dimethoxy-1,7-seco-2,7′-cyclolignan. ***Daphneresinol***
[1073342-08-1]
[1149745-72-1] (*Quercus glauca* constit.)
$C_{20}H_{26}O_7$ 378.421
Constit. of the stem bark of *Daphne feddei* and leaves of *Quercus glauca*. Shows antioxidant props.. Pale yellow oil or amorph. powder. Mp 122°. $[\alpha]^{20}_{D}$ -26 (c, 0.11 in MeOH). $[\alpha]^{20}_{D}$ -11 (c, 0.001 in MeOH). One chiral centre. Abs. configs. not determined. Identity of samples not established. λ_{max} 281 (MeOH)

3′,4,4′-Tri-Me ether
2,7′-Cyclo-3′,4,4′-trimethoxy-1,7-seco-8,8′-lignan-5,7,9,9′-tetrol. 5,7,9,9′-Tetrahydroxy-3′,4,4′-trimethoxy-1,7-seco-2,7′-cyclolignan. ***Feddeiphenol B***
[1383847-71-9]
$C_{21}H_{28}O_7$ 392.448
Constit. of the leaves and stems of *Daphne feddei*. Exhibits strong cytotoxic and weak anti-HIV activities. Pale yellow gum. $[\alpha]^{24}_{D}$ -25.4 (c, 0.22 in MeOH). Two chiral centres. Has (7′*R**,8′*R**)-config.. λ_{max} 210 (log

ε 4.86); 280 (log ε 4.21); 324 (log ε 2.52) (MeOH)

3′,4,4′,5-Tetra-Me ether

2,7′-Cyclo-3′,4,4′,5-tetramethoxy-1,7-seco-8,8′-lignan-7,9,9′-triol. 7,9,9′-Trihydroxy-3′,4,4′,5-tetramethoxy-1,7-seco-2,7′-cyclolignan.

Feddeiphenol A

[1383847-70-8]

$C_{22}H_{30}O_7$ 406.475

Constit. of *Daphne feddei* and *Magnolia fargesii*. Exhibits strong cytotoxic and weak anti-HIV activities. Pale yellow gum. [α]$_D^{25}$ -28.5 (c, 0.21 in MeOH). One chiral centre. Abs. config. not determined. λ_{max} 210 (log ε 4.82); 278 (log ε 4.15); 324 (log ε 2.47) (MeOH)

3′,4′-Methylene, 4,5,9,9′-tetra-Me ether

2,7′-Cyclo-4,5,9,9′-tetramethoxy-3′,4′-methylenedioxy-1,7-seco-8,8′-lignan-7-ol. 7-Hydroxy-4,5,9,9′-tetramethoxy-3′,4′-methylenedioxy-1,7-seco-2,7′-cyclolignan.

Seco-7-hydroxylintetralin

[113851-64-2]

$C_{23}H_{30}O_7$ 418.486

Isol. from leaves of *Phyllanthus niruri*. Oil. [α]$_D^{28}$ +11.5 ($CHCl_3$). 3 Chiral centres, has (7′*R**,8*S**,8′*R**)-config. Called seco-4-hydroxy- in the lit..

Satyanarayana, P. *et al.*, *J. Nat. Prod.*, 1988, **51**, 44-49 (*Seco-7-hydroxylintetralin*).

Ko, R.K. *et al.*, *Bull. Korean Chem. Soc.*, 2008, **29**, 2531-2532 (*Quercus glauca constit*).

Liang, S. *et al.*, *J. Nat. Prod.*, 2008, **71**, 1902-1905 (*Daphneresinol*).

Hu, Q.-F. *et al.*, *Chem. Pharm. Bull.*, 2011, **59**, 1421-1424 (*Feddeiphenols A,B, activity*).

Li, M. *et al.*, *Molecules*, 2015, **20**, 15192-15201 (*Rehmaneolignan C*)

2,7′-Cyclo-1,7-seco-8,8′-lignan-3′,4,4′,5,5′,6,7,9,9′-nonanol C322

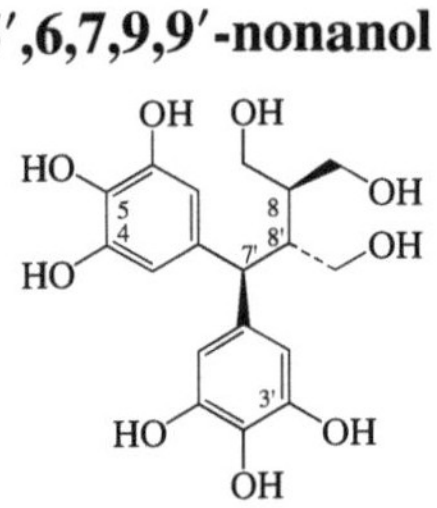

Relative Configuration

3′,4,4′,5,5′,6,7,9,9′-Nonahydroxy-1,7-seco-2,7′-cyclolignan

$C_{18}H_{22}O_9$ 382.365

3′,4′-Methylene, 4,5,5′,6-tetra-Me ether

Peperomialin

$C_{23}H_{30}O_9$ 450.484

Constit. of *Peperomia japonica*. [α]$_D^{27}$ -26.7 (c, 0.15 in MeOH). λ_{max} 208 (log ε 4.70); 254 (log ε 4.34); 275 (log ε 4.14) (MeOH)

3′,4′-Methylene, 4,5,5′,6-tetra-Me ether, 7-Ac

7-O-Acetylpeperomialin

$C_{25}H_{32}O_{10}$ 492.521

Constit. of *Peperomia japonica*. Amorph. yellow solid. [α]$_D^{26}$ -33.3 (c, 0.35 in MeOH). λ_{max} 225 (log ε 4.01); 250 (log ε 3.86); 269 (log ε 3.53) (MeOH).

Nagashima, K. *et al.*, *Phytochem. Lett.*, 2018, **27**, 30-35 (*Peperomialin, 7-Acetylpeperomialin*)

2,7′-Cyclo-1,7-seco-8,8′-lignan-3′,4,4′,5,7-pentol C323

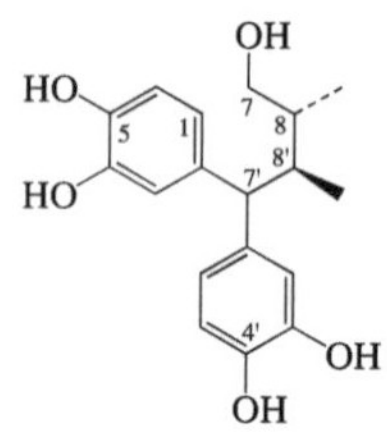

(8*R*,8′*S*)-*form*

4,4-Bis(3,4-dihydroxyphenyl)-2,3-dimethyl-1-butanol. 3′,4,4′,5,7-Pentahydroxy-1,7-seco-2,7′-cyclolignan

$C_{18}H_{22}O_5$ 318.369

Secolignan. Lignan numbering shown. Care needed however; free rotation of the phenyl groups means the numbering around the rings is ambiguous and the two aryl rings also become equivalent

(8R,8′S)-form

3′,4-Di-Me ether, 7-Ac

8-Epihenricine B

[1801690-32-3]

$C_{22}H_{28}O_6$ 388.46

Constit. of the roots of *Machilus obovatifolia*. Exhibits ABTS cation radical scavenging activity. Yellowish oil. [α]$_D^{25}$ -68.6 (c, 0.1 in $CHCl_3$). λ_{max} 208 (log ε 4.52); 238 (log ε 4.01); 282 (log ε 3.76) (MeOH)

3′,5-Di-Me ether

2,7′-Cyclo-3′,5-dimethoxy-1,7-seco-8,8′-lignan-4,4′,7-triol. ***Machilupin C***

$C_{20}H_{26}O_5$ 346.423

Constit. of *Machilus philippinensis*. Powder. Mp 132 - 134°. [α]$_D^{25}$ +26 (c, 0.04 in MeOH). λ_{max} 281 (log ε 2.6) (MeOH)

3′,4,4′,5-Tetra-Me ether

(-)-Kadangustin J

[1316841-28-7]

Oil which solidifies on standing. [α]$_D$ -20.7 (c, 1.19 in MeOH)

(8S,8′R)-form

3′,4,4′,5-Tetra-Me ether

4,4-Bis(3,4-dimethoxyphenyl)-2,3-dimethyl-1-butanol. 2,7′-Cyclo-3′,4,4′,5-tetramethoxy-1,7-seco-8,8′-lignan-7-ol. 7-Hydroxy-3′,4,4′,5-tetramethoxy-1,7-seco-2,7′-cyclolignan.

Kadangustin J

[1013633-28-7]

$C_{22}H_{30}O_5$ 374.477

Constit. of the stems of *Kadsura angustifolia*. Pale yellow solid. [α]$_D^{16}$ +4.9 (c, 0.17 in MeOH). λ_{max} 205 (log ε 5.02); 281 (log ε 4.08) (MeOH)

(8S,8′S)-form

3′,4-Di-Me ether, 7-Ac

Henricine B

[1158301-83-7]

$C_{22}H_{28}O_6$ 388.46

Constit. of the stems of *Schisandra henryi*. Yellow gum. [α]$_D^{20}$ -4 (c, 0.1 in $CHCl_3$). λ_{max} 207 (log ε 4.49); 239 (log ε 3.94); 282 (log ε 3.68) (MeOH)

(8R*,8′R*)-form

3′,5-Di-Me ether

Constit. of roots of *Kadsura longipedunculata*. [α]$_D^{20}$ -30 (c, 0.1 in MeOH)

3′,4,4′,5-Tetra-Me ether

Constit. of roots of *Kadsura longipedunculata*. [α]$_D^{20}$ -10 (c, 0.05 in MeOH)

3′,4′-Methylene, 4,5-di-Me ether

4-(3,4-Dimethoxyphenyl)-4-(3,4-methylenedioxyphenyl)-2,3-dimethyl-1-butanol. 2,7′-Cyclo-3′,4′-dimethoxy-4,5-methylenedioxy-1,7-seco-8,8′-lignan-7-ol. 2,7′-Cyclo-7-hydroxy-3′,4′-dimethoxy-4,5-methylenedioxy-1,7-secolignan

[94513-80-1]

$C_{21}H_{26}O_5$ 358.434

Constit. of *Virola elongata* and *Virola sebifera*. Oil. $[\alpha]_D^{25}$ -5 (c, 0.01 in $CHCl_3$)

3′,4′-Methylene, 4,5-di-Me ether, 7-Ac

[129684-17-9]

$C_{23}H_{28}O_6$ 400.471

Constit. of *Virola elongata*. Oil. $[\alpha]_D^{25}$ -6 (c, 0.08 in $CHCl_3$)

(8ξ,8′ξ)-form

3′,4-Di-Me ether

4,4-Bis(4-hydroxy-3-methoxyphenyl)-2,3-dimethyl-1-butanol. 2,7′-Cyclo-3′,4-dimethoxy-1,7-seco-8,8′-lignan-4′,5,7-triol. 4′,5,7-Trihydroxy-3′,4-dimethoxy-1,7-seco-2,7′-cyclolignan. ***Cinnaburmanin A***

[913643-31-9]

$C_{20}H_{26}O_5$ 346.423

Constit. of *Cinnamomum burmanii* and *Schisandra propinqua*. Cytotoxic agent. Powder. $[\alpha]_D^{20}$ -13 (c, 0.162 in $CHCl_3$). λ_{max} 206 (log ε 4.27); 245 (sh); 282 (sh) (EtOH). [neutral] λ_{max} 235 (log ε 3.10); 280 (log ε 1.90) (MeCN)

3′,4,5-Tri-Me ether

4-(3,4-Dimethoxy-phenyl)-4-(4-hydroxy-3-methoxyphenyl)-2,3-dimethyl-1-butanol. 2,7′-Cyclo-3′,4,5-trimethoxy-1,7-seco-8,8′-lignan-4′,7-diol. 4′,7-Dihydroxy-3′,4,5-trimethoxy-1,7-seco-2,7′-cyclolignan. ***Kadangustin K***

[1013633-29-8]

$C_{21}H_{28}O_5$ 360.45

Constit. of the stems of *Kadsura angustifolia*. Amorph. solid. Error in CA struct.. λ_{max} 205 (log ε 4.92); 281 (log ε 4.01); 321 (log ε 2.85); 362 (log ε 2.4) (MeOH)

3′,4,5-Tri-Me ether, 7-Ac

Propinquain K

[1427163-89-0]

$C_{23}H_{30}O_6$ 402.487

Isol. from the roots of *Schisandra propinqua* ssp. *sinensis* and the fruit of *Schisandra sphenanthera*. Yellow oil. $[\alpha]_D^{20}$ -4.5 (c, 0.67 in CH_2Cl_2). $[\alpha]_D^{20}$ -20 (c, 1.1 in Me_2CO). λ_{max} 210 (log ε 4.52); 225 (log ε 4.30); 245 (log ε 4.21); 278 (log ε 3.79) (MeOH)

3′,4′-Methylene, 4-Me ether

Marphenol G

[1389397-49-2]

$C_{20}H_{24}O_5$ 344.407

Constit. of the leaves of *Schisandra wilsoniana*. Pale yellow gum. $[\alpha]_D^{25}$ +18.9 (c, 0.2 in MeOH). λ_{max} 280 (log ε 4.18); 325 (log ε 2.28) (MeOH)

7-Deoxy, 3′,4-di-Me ether

2,7′-Cyclo-3′,4-dimethoxy-1,7-seco-8,8′-lignan-4′,5-diol. 4′,5-Dihydroxy-3′,4-dimethoxy-1,7-seco-2,7′-cyclolignan. ***Daphnelignan C***

[1318774-06-9]

$C_{20}H_{26}O_4$ 330.424

Constit. of the leaves of *Daphne feddei*. Shows potential cytotoxic and weak anti-HIV activities. Pale yellow solid. $[\alpha]_D^{23}$ +11.5 (c, 0.18 in MeOH). λ_{max} 205 (log ε 4.92); 278 (log ε 4.12); 322 (log ε 2.47) (MeOH).

Lopes, L.M.X. *et al.*, *Phytochemistry*, 1984, **23**, 2647-2652 (*Virola sebifera constit*).

Kato, M.J. *et al.*, *Phytochemistry*, 1990, **29**, 1799-1810 (*Virola elongata constits*).

Xu, L.-J. *et al.*, *J. Integ. Plant Biol.*, 2006, **48**, 1493-1497 (*3′,4-di-Me ether*).

Gao, X.-M. *et al.*, *J. Nat. Prod.*, 2008, **71**, 558-563 (*Kadangustins J,K*).

Liu, H.-T. *et al.*, *Chem. Pharm. Bull.*, 2009, **57**, 405-407 (*Henricine B*).

Wan, L. *et al.*, *Asian J. Chem.*, 2010, **22**, 8162-8166 (*Daphnelignan C, activity*).

Yang, G.-Y. *et al.*, *Chem. Pharm. Bull.*, 2011, **59**, 1344-1347 (*Marphenol G*).

Rye, C.E. *et al.*, *J. Org. Chem.*, 2011, **76**, 6636-6648 (*(-)-Kadangustin J, synth*).

Jin, M.N. *et al.*, *Planta Med.*, 2012, **78**, 807-813 (*Propinquain K*).

Peng, Y. *et al.*, *Org. Biomol. Chem.*, 2013, **11**, 7574-7586 (*(±)-Kadangustin J, synth*).

Song, Q.-Y. *et al.*, *Phytochem. Lett.*, 2013, **6**, 174-178 (*Schisandra sphenanthera constit*).

Davidson, S.J. *et al.*, *Phytochem. Lett.*, 2015, **14**, 138-142 (*synth, pmr, cmr, config*).

Yuan, L.T. *et al.*, *Chem. Nat. Compd. (Engl. Transl.)*, 2017, **53**, 623-625 (*Cinnaburmanin A*).

Qi, X.-Z. *et al.*, *Zhongcaoyao*, 2017, **48**, 2164-2171 (*Kadsura longipedunculata constits*).

Tai, S.-H. *et al.*, *RSC Adv.*, 2019, **9**, 21616-21625 (*Machilupin C, cd*)

2,7′-Cyclo-1,7-seco-8,8′-lign-7-ene-3′,4,4′,5,9,9′-hexol C324

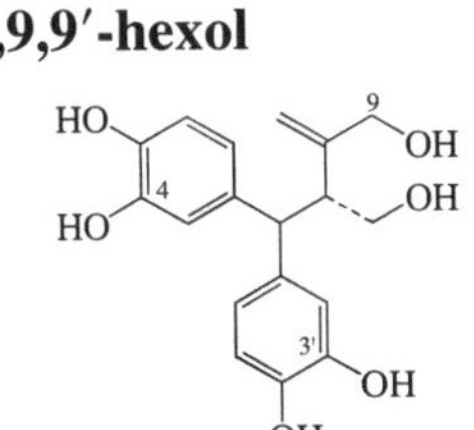

3′,4,4′,5,9,9′-Hexahydroxy-1,7-seco-2,7′-cyclolign-7-ene

$C_{18}H_{20}O_6$ 332.352

(R)-form

9-Aldehyde, 3′,4-di-Me ether, 9′-Ac

9′-Acetoxy-2,7′-cyclo-4′,5-dihydroxy-3′,4-dimethoxy-1,7-seco-8,8′-lign-7-en-9-al. ***Lucidenal***

[1564313-95-6]

$C_{22}H_{24}O_7$ 400.427

Constit. of *Echinochloa utilis* and *Feroniella lucida*. Shows cytotoxicity against various cancer cell lines. Viscous brown liq.. $[\alpha]_D^{24}$ -26.7 (c, 0.33 in MeOH). λ_{max} 244 (log ε 2.75); 281 (log ε 2.74) ($CHCl_3$).

Sriyatep, T. *et al.*, *Tetrahedron*, 2014, **70**, 1773-1779 (*Lucidenal*)

3,7′-Cyclo-1,2,3,4-tetrahydro-1,8′-lign-8-ene-2,3,3′,4,4′,5-hexol C325

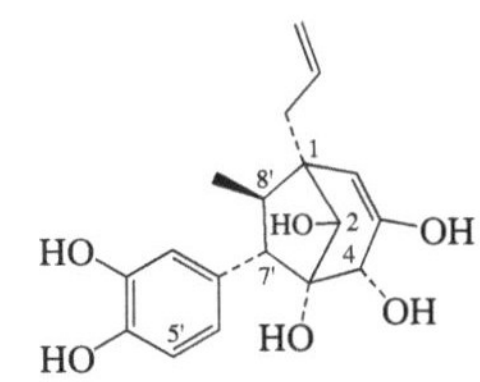

(1*S*,2*S*,3*R*,4*S*,7′*S*,8′*R*)-*form*

3,7′-Cyclo-1,2,3,4-tetrahydro-2,3,3′,4,4′,5-hexahydroxy-1,8′-neolign-8-ene

$C_{18}H_{22}O_6$ 334.368

(1S,2S,3R,4S,7′S,8′R)-form

4-Ketone, 3′,4′-methylene ether

3,7′-Cyclo-1,2-dihydro-2,3,5-trihydroxy-3′,4′-methylenedioxy-1,8′-lign-8-en-4(3H)-one

[716343-75-8]

$C_{19}H_{20}O_6$ 344.363

Constit. of the roots of *Endlicheria* aff.. Viscous oil. $[\alpha]_D^{25}$ -18.8 (c, 0.46 in $CHCl_3$). λ_{max} 232 (log ε 3.24); 273 (log ε 3.45) (MeOH)

4-Ketone, 3′,4′-methylene, 3-Me ether

3,7′-Cyclo-1,2-dihydro-2,5-dihydroxy-3-methoxy-3′,4′-methylenedioxy-1,8′-lign-8-en-4(3H)-one

[716343-78-1]

$C_{20}H_{22}O_6$ 358.39

Constit. of *Ocotea catharinensis* and *Endlicheria* aff.. Prisms. Mp 142 - 144°. λ_{max} 233 (ε 4750); 272 (ε 4450) (MeOH)

4-Ketone, 3′,4′-methylene, 5-Me ether

3,7′-Cyclo-1,2-dihydro-2,3-dihydroxy-5-methoxy-3′,4′-methylenedioxy-1,8′-lign-8-en-4(3H)-one

[716343-73-6]

$C_{20}H_{22}O_6$ 358.39

Constit. of the roots of *Endlicheria* aff.. Viscous oil. $[\alpha]_D^{25}$ -42.2 (c, 0.62 in $CHCl_3$). λ_{max} 235 (log ε 3.24); 267 (log ε 4.28) (MeOH)

4-Ketone, 3′,4′-methylene, 3,5-di-Me ether
3,7′-Cyclo-1,2-dihydro-2-hydroxy-3,5-dimethoxy-3′,4′-methylenedioxy-1,8′-lign-8-en-4(3H)-one. ***3′-Methoxyguianin***
[61240-31-1]
$C_{21}H_{24}O_6$ 372.417
Constit. of an *Aniba* sp. and *Ocotea aciphylla*. Viscous oil. λ_{max} 234 (log ε 3.96); 263 (log ε 3.82) (MeOH)

(1R*,3S*,4R*,7′R*,8′S*)-form

2-Ketone, 3′,4′-methylene, 3,5-di-Me ether
3,7′-Cyclo-3,4-dihydro-4-hydroxy-3,5-dimethoxy-3′,4′-methylenedioxy-1,8′-lign-8-en-2(1H)-one
[199916-62-6]
$C_{21}H_{24}O_6$ 372.417
Constit. of *Ocotea catharinensis*. Viscous oil. $[\alpha]_D^{21}$ +6 (c, 1 in MeOH). Possesses (1*R**,3*R**,4*R**,7′*R**,8′*S**)-config. due to change in Cahn-Ingold-Prelog priorities

2,4-Diketone, 3′,4′-methylene, 3,5-di-Me ether
3,7′-Cyclo-3,5-dimethoxy-3′,4′-methylenedioxy-1,8′-lign-8-ene-2,4(1H,3H)-dione
[1178974-87-2]
$C_{21}H_{22}O_6$ 370.401
Constit. of the leaves of *Ocotea catharinensis*. Viscous oil. $[\alpha]_D^{21}$ -18 (c, 3.75 in MeOH)

5′-Methoxy, 2-ketone, 3′,4′-methylene, 3,5-di-Me ether
3,7′-Cyclo-3,4-dihydro-4-hydroxy-3,3′,5-trimethoxy-4′,5′-methylenedioxy-1,8′-lign-8-en-2(1H)-one
[199916-63-7]
$C_{22}H_{26}O_7$ 402.443
Constit. of *Licaria brasiliensis*, *Ocotea costulatum* and *Virola surinamensis*. Cryst. (MeOH). Mp 170 - 172°. Possesses (1*R**,3*R**,4*R**,7′*R**,8′*S**)-config. due to change in Cahn-Ingold-Prelog priorities

5′-Methoxy, 2,4-diketone, 3′,4′-methylene, 3,5-di-Me ether
3,7′-Cyclo-3,3′,5-trimethoxy-4′,5′-methylenedioxy-1,8′-lign-8-ene-2,4(1H,3H)-dione
[65527-41-5]
$C_{22}H_{24}O_7$ 400.427
Constit. of *Aniba simulans* and *Ocotea catharinensis*. Viscous oil. $[\alpha]_D^{21}$ -29 (c, 4.5 in MeOH)

(1R*,3S*,4S*,7′R*,8′S*)-form

2-Ketone, 3′,4′-methylene, 3,5-di-Me ether
[199916-64-8]
$C_{21}H_{24}O_6$ 372.417
Constit. of *Ocotea catharinensis*. Viscous oil. $[\alpha]_D^{21}$ -10 (c, 1.91 in MeOH). Possesses (1*R**,3*R**,4*S**,7′*R**,8′*S**)-config. due to change in Cahn-Ingold-Prelog priorities.

Fernandes, J.B. *et al.*, *Phytochemistry*, 1976, **15**, 1033-1036 (*3′-Methoxyguianin*).
Ishige, M. *et al.*, *Phytochemistry*, 1991, **30**, 4121-4128 (*Ocotea catharinensis constit*).
Lordello, A.L. *et al.*, *Phytochemistry*, 1997, **46**, 741-744 (*Ocotea catharinensis constits*).
Chaturvedula, V.S.P. *et al.*, *J. Nat. Prod.*, 2004, **67**, 964-967 (*Endlicheria aff. constits*).
Funasaki, M. *et al.*, *J. Braz. Chem. Soc.*, 2009, **20**, 853-859 (*Ocotea catharinensis constits*)

3,7′-Cyclo-1,2,3,4-tetrahydro-1,8′-lign-8-ene-2,3′,4,4′,5-pentol C326

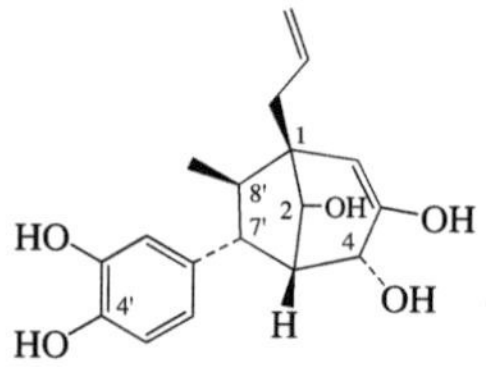

(1*R*,2*R*,3*R*,4*R*,7′*R*,8′*R*)-*form*

7-(3,4-Dihydroxyphenyl)-6-methyl-5-(2-propenyl)bicyclo[3.2.1]oct-3-ene-2,3,8-triol. 3,7′-Cyclo-1,2,3,4-tetrahydro-2,3′,4,4′,5-pentahydroxy-1,8′-neolign-8-ene
$C_{18}H_{22}O_5$ 318.369
Parent not known

(1R,2R,3R,4R,7′R,8′R)-form

4-Ketone, 3′,4′-methylene, 5-Me ether
3,7′-Cyclo-1,2-dihydro-2-hydroxy-5-methoxy-3′,4′-methylenedioxy-1,8′-lign-8-en-4(3H)-one. ***2-Epiguianin***
[1188415-72-6]
[156767-72-5] (relative config.)
$C_{20}H_{22}O_5$ 342.391
Constit. of *Licaria puchury-major*, *Ocotea macrophylla* and *Ocotea porosa*. COX-1 inhibitor. Viscous oil. λ_{max} 234 (ε 8900); 274 (ε 8600) (MeOH)

(1R,2S,3R,7′R,8′R)-form

4-Ketone, 3′,4′,5-tri-Me ether
3,7′-Cyclo-1,2-dihydro-2-hydroxy-3′,4′,5-trimethoxy-1,8′-lign-8-en-4(3H)-one. ***Kadsurenin I***
[145553-01-1]
$C_{21}H_{26}O_5$ 358.434
Constit. of *Piper kadsura*. Oil. $[\alpha]_D^{18}$ +31.5 ($CHCl_3$)

4-Ketone, 3′,4′,5-tri-Me ether, 2-Ac
[145553-02-2]
$C_{23}H_{28}O_6$ 400.471
Constit. of *Piper kadsura*. Cryst.. $[\alpha]_D^{18}$ +54.5 ($CHCl_3$)

4-Ketone, 3′,4′-methylene, 5-Me ether
Guianin
[50939-40-7]
$C_{20}H_{22}O_5$ 342.391
Constit. of *Aniba guianensis* and *Ocotea porosa*. Cryst. (C_6H_6/$CHCl_3$). Mp 189 - 190°. λ_{max} 235 (ε 8600); 265 (ε 10300) (EtOH)

(1S,2R,3S,7′R,8′R)-form

4-Ketone, 3′,5-di-Me ether
3,7′-Cyclo-1,2-dihydro-2,4′-dihydroxy-3′,5-dimethoxy-1,8′-lign-8-en-4(3H)-one. ***Kadsurenin E***
[149596-53-2]
$C_{20}H_{24}O_5$ 344.407
Constit. of *Piper kadsura*

4-Ketone, 3′,4′,5-tri-Me ether
Kadsurenin G
[149572-30-5]
[140669-87-0]
$C_{21}H_{26}O_5$ 358.434
Constit. of *Piper kadsura*. Oil. $[\alpha]_D^{18}$ -166.7 (c, 0.1 in $CHCl_3$)

4-Ketone, 3′,4′,5-tri-Me ether, 2-Ac
Kadsurenin H
[140669-88-1]
$C_{23}H_{28}O_6$ 400.471
Constit. of *Piper kadsura*. Cryst.. $[\alpha]_D^{18}$ -111 (c, 0.01 in $CHCl_3$)

(1R*,2R*,3S*,4R*,7′R*,8′R*)-form

5S,6-Dihydro, 3′,4′-methylene, 5-Me ether
3,7′-Cyclo-1,2,3,4,5,6-hexahydro-5-methoxy-3′,4′-methylenedioxy-1,8′-lign-8-ene-2,4-diol. 3,7′-Cyclo-1,2,3,4,5,6-hexahydro-2,4-dihydroxy-5-methoxy-3′,4′-methylenedioxy-1,8′-lign-8-ene
[156665-55-3]
$C_{20}H_{26}O_5$ 346.423

Constit. of the bark of *Ocotea porosa*. Oil

(1R*,2S*,3R*,4R*,7′S*,8′S*)-form

3′,4′-Methylene, 5-Me ether

3,7′-Cyclo-1,2,3,4-tetrahydro-5-methoxy-3′,4′-methylenedioxy-1,8′-lign-8-ene-2,4-diol. 3,7′-Cyclo-1,2,3,4-tetrahydro-2,4-dihydroxy-5-methoxy-3′,4′-methylenedioxy-1,8′-lign-8-ene

[156767-79-2]

$C_{20}H_{24}O_5$ 344.407

Constit. of the bark of *Ocotea porosa*. Oil

(1R*,2S*,3R*,4S*,7′S*,8′S*)-form

3′,4′-Methylene, 5-Me ether

[156665-54-2]

Constit. of the bark of *Ocotea porosa*. Oil

(1R*,2R*,3R*,7′R*,8′R*)-form

4-Ketone, 3′,4′-methylene, 5-Me ether, 2-Ac

[156767-78-1]

$C_{22}H_{24}O_6$ 384.428

Constit. of the bark of *Ocotea porosa*. Oil

(1R*,2S*,3R*,7′R*,8′R*)-form

4-Ketone, 3′,4′,5-tri-Me ether

Piperol A

[195259-21-3]

$C_{21}H_{26}O_5$ 358.434

Constit. of *Piper betle*. PAF receptor antagonist

(1R*,2S*,3R*,7′S*,8′S*)-form

4-Ketone, 3′,5-di-Me ether, 2-Ac

Piperbetol. *Piperulin C. Puberulin C*

[195259-20-2]

[164268-05-7] (Puberulin C)

$C_{22}H_{26}O_6$ 386.444

Constit. of *Piper betle* and *Piper puberulum*. PAF receptor antagonist. Needles (Et_2O). Mp 205 - 205.5°. $[\alpha]_D$ +5.8 (c, 1.2 in $CHCl_3$). λ_{max} 231 (log ε 3.84); 259 (log ε 3.70) (MeOH)

4-Ketone, 3′,4′,5-tri-Me ether

Piperol B

[184973-38-4]

$C_{21}H_{26}O_5$ 358.434

Constit. of *Nectandra grandiflora* and *Piper betle*. PAF receptor antagonist

4-Ketone, 3′,4′,5-tri-Me ether, 2-Ac

Acetylpiperol B. ***Methylpiperbetol***

[195202-51-8]

$C_{23}H_{28}O_6$ 400.471

Constit. of *Piper betle*. PAF receptor antagonist

4-Ketone, 3′,4′-methylene, 5-Me ether

Puberulin D

[156767-71-4]

$C_{20}H_{22}O_5$ 342.391

Constit. of *Ocotea porosa* and *Piper betle*. Exhibits synergistic antibacterial activity combined with Norfloxacin. Abs. config. has been determined

4-Ketone, 3′,4′-methylene, 5-Me ether, 2-Ac

2-O-Acetylpuberulin D

[156767-77-0]

$C_{22}H_{24}O_6$ 384.428

Constit. of the leaves of *Ocotea porosa*. Oil

(1R*,3R*,4R*,7′R*,8′S*)-form

2-Ketone, 3′,4′,5-tri-Me ether

3,7′-Cyclo-3,4-dihydro-4-hydroxy-3′,4′,5-trimethoxy-1,8′-lign-8-en-2(1H)-one. ***Hancinol***

[108864-50-2]

$C_{21}H_{26}O_5$ 358.434

Constit. of *Piper hancei*. Incorrect struct. has been widely reported. Possesses (1*R**,3*S**,4*R**,7′*R**,8′*S**)-config. due to change in Cahn-Ingold-Prelog priorities

(1R*,3R*,4R*,7′S*,8′S*)-form

2-Ketone, 3′,4′-methylene, 5-Me ether

3,7′-Cyclo-3,4-dihydro-4-hydroxy-5-methoxy-3′,4′-methylenedioxy-1,8′-lign-8-en-2(1H)-one. ***Clarkinol***

[156768-14-8]

$C_{20}H_{22}O_5$ 342.391

Constit. of *Piper clarkii*. Brown oil. $[\alpha]^{22}_D$ +52.8 (c, 0.5 in MeOH). Possesses (1*R**,3*S**,4*R**,7′*S**,8′*S**)-config. due to change in Cahn-Ingold-Prelog priorities. λ_{max} 215; 230 (sh); 282 (MeOH)

(1R*,3R*,4S*,7′S*,8′S*)-form

2-Ketone, 3′,4′-methylene, 5-Me ether

4-Epiclarkinol

[156665-56-4]

$C_{20}H_{22}O_5$ 342.391

Constit. of *Ocotea porosa*. Oil. Possesses (1*R**,3*S**,4*S**,7′*S**,8′*S**)-config. due to change in Cahn-Ingold-Prelog priorities

(1R*,3R*,7′R*,8′S*)-form

2,4-Diketone, 3′,4′-methylene, 5-Me ether

3,7′-Cyclo-5-methoxy-3′,4′-methylenedioxy-1,8′-lign-8-ene-2,4(1H,3H)-dione. ***Denudadione C***

[288260-44-6]

$C_{20}H_{20}O_5$ 340.375

Constit. of *Magnolia denudata*

(1R*,3R*,7′S*,8′S*)-form

2,4-Diketone, 3′,4′,5-tri-Me ether

3,7′-Cyclo-3′,4′,5-trimethoxy-1,8′-lign-8-ene-2,4(1H,3H)-dione. ***2-Oxopiperol B***. *2-Ketopiperol B. Kadsurenin D*

[873435-38-2]

[140669-89-2] (Kadsurenin D)

$C_{21}H_{24}O_5$ 356.418

Constit. of *Nectandra grandiflora* and *Piper kadsura*. Light brown oil. Identity of 2-Oxopiperol B and Kadsurenin D is uncertain and they may be diastereoisomers

2,4-Diketone, 3′,4′-methylene, 5-Me ether

4-Ketoclarkinol

[167613-00-5]

$C_{20}H_{20}O_5$ 340.375

Constit. of *Piper clarkii*. Oil. $[\alpha]^{22}_D$ +97.4 (c, 0.6 in $CHCl_3$)

5′-Methoxy, 2,4-diketone, 3′,4′-methylene, 5-Me ether

3,7′-Cyclo-3′,5-dimethoxy-4′,5′-methylenedioxy-1,8′-lign-8-ene-2,4(1H,3H)-dione

[65527-40-4]

$C_{21}H_{22}O_6$ 370.401

Constit. of *Aniba simulans*. Cryst. (Et_2O). Mp 147 - 148°.

Von Bülow, M.V. *et al.*, *Phytochemistry*, 1973, **12**, 1805-1808 (*Guianin*).

Büchi, G. *et al.*, *J. Am. Chem. Soc.*, 1977, **99**, 8073-8075 (*Guianin, synth*).

De Alvarenga, M.A. *et al.*, *Phytochemistry*, 1977, **16**, 1801-1804 (*Aniba simulans constits*).

Li, S. *et al.*, *Yaoxue Xuebao*, 1987, **22**, 196-202 (*Hancinol*).

De Carvalho, M.G. *et al.*, *Phytochemistry*, 1988, **27**, 2319-2323 (*Guianin, 2-Epiguianin*).

Da Silva, W.D. *et al.*, *Phytochemistry*, 1989, **28**, 661-662 (*Ocotea costulatum constit*).

Ma, Y. *et al.*, *Chin. Chem. Lett.*, 1991, **2**, 623-626 (*Kadsurenins G,H, cryst struct*).

Han, G. *et al.*, *Beijing Yike Daxue Xuebao*, 1992, **24**, 347-350 (*Kadsurenin E*).

Ma, Y. *et al.*, *Chin. Chem. Lett.*, 1992, **3**, 635-636 (*Kadsurenins I,J*).

Engler, T.A. *et al.*, *J. Org. Chem.*, 1994, **59**, 6588-6599 (*Guianin, synth*).

David, J.M. *et al.*, *Phytochemistry*, 1994, **36**, 491-499 (*Ocotea porosa constits*).

Jensen, S. *et al.*, *Phytochemistry*, 1994, **36**, 789-792 (*Clarkinol*).

Zhang, S.-X. *et al.*, *J. Nat. Prod.*, 1995, **58**, 540-547 (*Puberulin C*).
Prasad, A.K. *et al.*, *Phytochemistry*, 1995, **39**, 655-658 (*Piper clarkii 2,4-diketone*).
Zhang, S.-X. *et al.*, *J. Nat. Prod.*, 1996, **59**, 1002 (*Puberulin C, erratum*).
Lordello, A.L. *et al.*, *Phytochemistry*, 1997, **46**, 741-744 (*Ocotea catharinensis constits*).
Zeng, H.-W. *et al.*, *Planta Med.*, 1997, **63**, 296-298 (*Piperols A,B, Piperbetol, Methylpiperbetol, activity*).
Kuroyanagi, M. *et al.*, *Chem. Pharm. Bull.*, 2000, **48**, 832-837 (*Denudadione C*).
Jiang, R.W. *et al.*, *J. Mol. Struct.*, 2003, **654**, 177-182 (*Kadsurenin H, cryst struct*).
Ribeiro, A.B. *et al.*, *J. Braz. Chem. Soc.*, 2005, **16**, 526-530 (*2-Oxopiperol B*).
Coy, E.D. *et al.*, *Bioorg. Med. Chem. Lett.*, 2009, **19**, 6922-6925 (*2-Epiguianin, activity*).
Funasaki, M. *et al.*, *J. Braz. Chem. Soc.*, 2009, **20**, 853-859 (*Ocotea catharinensis constits*).
Sun, Z.-L. *et al.*, *RSC Adv.*, 2016, **6**, 43518-43525 (*Puberulin D, cd, activity*)

5,7′-Cyclo-2,3,4,5-tetrahydro-3′,4,4′-trihydroxy-2,8′-lign-7-en-9′,3-olid-9-oic acid C327

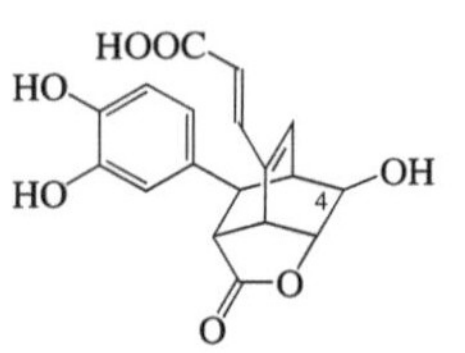

Relative Configuration

Rufescenolide C‡
$C_{18}H_{16}O_7$ 344.319
Constit. of *Cordia americana*. Antioxidant. Yellow oil. $[\alpha]_D^{25}$ +141.4 (c, 0.145 in MeOH). λ_{max} 218 (ε 3855); 259 (ε 5659) (MeOH)

Me ester
Rufescenolide
[1367349-25-4]
$C_{19}H_{18}O_7$ 358.346
Constit. of the stems of *Cordia rufescens*. Amorph. solid. Mp 176° dec.. $[\alpha]_D^{25}$ +15 (c, 0.5 in MeOH)

[1R-Carboxy-2-(3,4-dihydroxyphenyl) ethyl] ester
Rufescenolide D
[1609652-16-5]
$C_{27}H_{24}O_{11}$ 524.478
Constit. of *Cordia americana*. Antioxidant. Brown oil. $[\alpha]_D^{25}$ +5.2 (c, 0.13 in MeOH). λ_{max} 215 (ε 14430); 270 (ε 17954) (MeOH)

[2-(3,4-Dihydroxyphenyl)-1R-(methoxycarbonyl)ethyl] ester
Rufescenolide B‡
$C_{28}H_{26}O_{11}$ 538.505
Constit. of *Cordia americana*. Antioxidant. Brown oil. $[\alpha]_D^{25}$ +71.2 (c, 0.015 in MeOH). λ_{max} 213 (ε 17366); 275 (ε 24028) (MeOH)

4-O-[3-(3,4-Dihydroxyphenyl)-2ξ-hydroxypropanoyl]
[1927045-90-6]
$C_{27}H_{24}O_{11}$ 524.478
Constit. of *Cordia leucocephala*. Brown resin. $[\alpha]_D^{25}$ +310.4 (c, 0.45 in MeOH).

Do Vale, A.E. *et al.*, *Phytochemistry*, 2012, **76**, 158-161 (*Rufescenolide*).
Fernandez, L.R. *et al.*, *Planta Med.*, 2013, **79**, 1724-1729 (*Rufescenolides B,C,D*).
Griffith, D.R. *et al.*, *J. Org. Chem.*, 2014, **49**, 88-105 (*Rufescenolide, synth*).
Diniz, J.C. *et al.*, *J. Braz. Chem. Soc.*, 2015, **26**, 325-330 (*Cordia leucocephala constit*)

2,7′-Cyclo-2′,4,4′,5′-tetrahydroxy-8,8′-ligna-7,7′-diene-3,6-dione C328

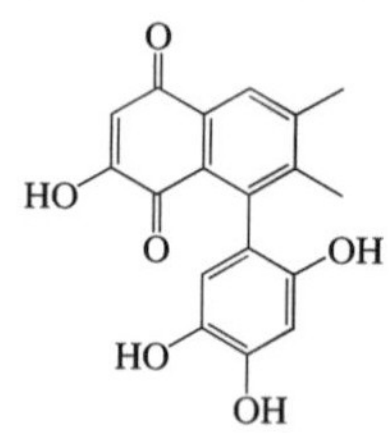

2′,4,4′,5′-Tetrahydroxy-2,7′-cycloligna-7,7′-diene-3,6-dione
$C_{18}H_{14}O_6$ 326.304

Tetra-Me ether
2,7′-Cyclo-2′,4,4′,5′-tetramethoxy-8,8′-ligna-7,7′-diene-3,6-dione. 2′,4,4′,5′-Tetramethoxy-2,7′-cycloligna-7,7′-diene-3,6-dione
[1458672-49-5]
$C_{22}H_{22}O_6$ 382.412
Constit. of the rhizomes of *Acorus gramineus*. Brown-red gum. $[\alpha]_D^{25}$ +30.5 (c, 0.15 in $CHCl_3$). Opt. rotn. presumably due to atropisomerism, but no comment made. λ_{max} 253 (log ε 3.4); 293 (log ε 3.5) (MeOH).

Lee, S.Y. *et al.*, *Biosci., Biotechnol., Biochem.*, 2013, **77**, 276-280 (*Acorus gramineus constit*)

2,7′-Cyclo-3,3′,4,4′-tetrahydroxy-8,8′-ligna-7,7′-dien-9,9′-olide C329

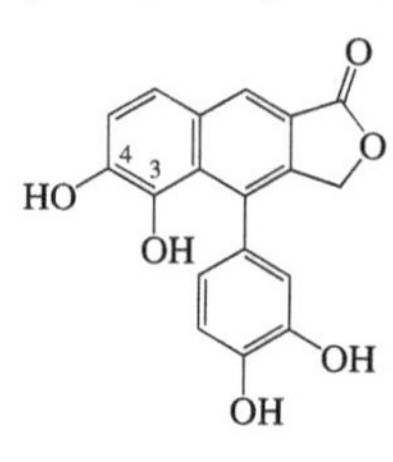

3,3′,4,4′-Tetrahydroxy-2,7′-cycloligna-7,7′-dien-9,9′-olide
$C_{18}H_{12}O_6$ 324.288

3,3′,4′-Tri-Me ether
2,7′-Cyclo-4-hydroxy-3,3′,4′-trimethoxy-8,8′-ligna-7,7′-dien-9,9′-olide. 4-Hydroxy-3,3′,4′-trimethoxy-2,7′-cycloligna-7,7′-dien-9,9′-olide
[345298-50-2]
$C_{21}H_{18}O_6$ 366.369
Constit. of the roots of *Vitex rotundifolia*. Amorph. brown solid. λ_{max} 260 (log ε 4.4); 329 (log ε 3.6) (dioxan)

3′,4′-Methylene ether
2,7′-Cyclo-3,4-dihydroxy-3′,4′-methylenedioxy-8,8′-ligna-7,7′-dien-9,9′-olide. 3,4-Dihydroxy-3′,4′-methylenedioxy-2,7′-cycloligna-7,7′-dien-9,9′-olide
Mp 123 - 125°

3′,4′-Methylene ether, 3-O-β-D-glucopyranoside
Elenoside. *Lignan J2*
[131189-74-7]
$C_{25}H_{22}O_{11}$ 498.44
Isol. from leaves of *Justicia hyssopifolia*. Exhibits sedative effects. Shows strong cytotoxic activity to human cancer cell lines. Mp 232 - 236°. $[\alpha]_D^{25}$ +91.5 (MeOH). LD_{50} (mus, ipr) 305 mg/kg

3,4:3′,4′-Bis(methylene) ether
2,7′-Cyclo-3,4:3′,4′-bis(methylenedioxy)-8,8′-ligna-7,7′-dien-9,9′-olide. 3,4:3′,4′-Bis(methylenedioxy)-2,7′-cycloligna-7,7′-dien-9,9′-olide. ***Helioxanthin***
[18920-47-3]
$C_{20}H_{12}O_6$ 348.31
Lignan from the roots of *Heliopsis scabra*, heartwood of *Taiwania cryptomeroides* and *Justicia neesii*. Inhibitor of interleukin-induced MIP-1β production. Suppresses gene expression of human hepatitis B virus. Yellow cryst. with strong blue fluor.. Mp 240 - 241°.

Burden, R.S. *et al.*, *J. Chem. Soc. C*, 1969, 693-701 (*Helioxanthin*).
Holmes, T.L. *et al.*, *J. Chem. Soc. C*, 1971, 2091-2094 (*Helioxanthin, synth*).
Stevenson, R. *et al.*, *J. Nat. Prod.*, 1989, **52**, 367-375 (*Helioxanthin, synth*).

Trujillo, J.M. *et al.*, *Phytochemistry*, 1990, **29**, 2991-2993 (*Elenoside*).
Rajasekhar, D. *et al.*, *Indian J. Chem., Sect. B*, 1999, **38**, 713-717 (*Justicia neesii constit*).
Navarro, E. *J. Nat. Prod.*, 2001, **64**, 134-135 (*Elenoside, activity*).
Kawazoe, K. *et al.*, *J. Nat. Prod.*, 2001, **64**, 588-591 (*3,3′,4′-tri-Me ether*).
Tseng, P-C. *et al.*, *Biochem. Pharmacol.*, 2008, **76**, 1121-1133 (*Taiwania cryptomeroides constit, activity*).
Janmanchi, D. *et al.*, *Bioorg. Med. Chem.*, 2013, **21**, 2163-2176 (*Helioxanthin, activity*).
Kao, T.-T. *et al.*, *J. Org. Chem.*, 2015, **80**, 6708-6714 (*Helioxanthin, synth*)

2,7′-Cyclo-3,3′,4,4′-tetrahydroxy-8,8′-ligna-7,7′-dien-9′,9-olide C330

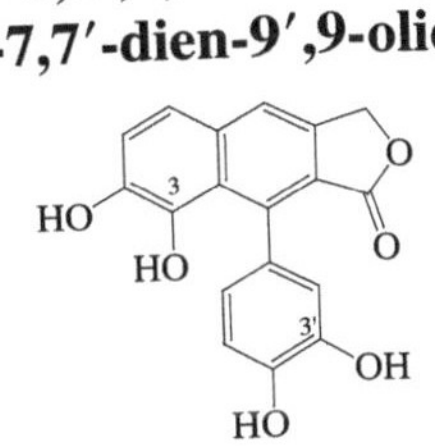

3,3′,4,4′-Tetrahydroxy-2,7′-cycloligna-7,7′-dien-9′,9-olide
$C_{18}H_{12}O_6$ 324.288

3,4-Methylene, 3′-Me ether
2,7′-Cyclo-4′-hydroxy-3′-methoxy-3,4-methylenedioxy-8,8′-ligna-7,7′-dien-9′,9-olide. 4′-Hydroxy-3′-methoxy-3,4-methylenedioxy-2,7′-cycloligna-7,7′-dien-9′,9-olide
[1126088-86-5]
$C_{20}H_{14}O_6$ 350.326
Constit. of *Bupleurum marginatum* aerial parts. Shows weak activity against hepatitis B. Green cryst.. Mp 244 - 246°. λ_{max} 204 (log ε 4.54); 259 (log ε 4.52) (MeOH)

3′,4′-Methylene, 3,4-di-Me ether
2,7′-Cyclo-3,4-dimethoxy-3′,4′-methylenedioxy-8,8′-ligna-7,7′-dien-9′,9-olide. 3,4-Dimethoxy-3′,4′-methylenedioxy-2,7′-cycloligna-7,7′-dien-9′,9-olide. ***Isojusticidin B***
[400887-35-6]
$C_{21}H_{16}O_6$ 364.353
Constit. of root cultures of *Linum austriacum* and aerial parts of *Linum perenne*. Mp 190 - 192° (synthetic). λ_{max} 252; 290 (MeOH)

3,4:3′,4′-Bis(methylene) ether
2,7′-Cyclo-3,4:3′,4′-bis(methylenedioxy)-8,8′-ligna-7,7′-dien-9′,9-olide. 3,4:3′,4′-Bis(methylenedioxy)-2,7′-cycloligna-7,7′-dien-9′,9-olide. ***Retrohelioxanthin***
[32502-02-6]
$C_{20}H_{12}O_6$ 348.31
Constit. of *Justicia neesii* and aerial parts of *Linum perenne*. Antihepatitis B agent. Cryst. (MeOH). Mp 264 - 268° (263-270°).

Holmes, T.L. *et al.*, *J. Chem. Soc. C*, 1971, 2091-2094 (*Retrohelioxanthin, synth*).
Stevenson, R. *et al.*, *J. Nat. Prod.*, 1989, **52**, 367-375 (*Retrohelioxanthin, synth*).
Rajasekhar, D. *et al.*, *Indian J. Chem., Sect. B*, 1999, **38**, 713-717 (*Justicia neesii constit*).
Mohagheghzadeh, A. *et al.*, *J. Nat. Prod.*, 2002, **65**, 69-71 (*Isojusticidin B*).
Schmidt, T.J. *et al.*, *Planta Med.*, 2007, **73**, 1574-1580 (*Linum perenne constits, activity*).
Liu, Y. *et al.*, *Helv. Chim. Acta*, 2008, **91**, 2316-2320 (*Bupleurum marginatum constit*).
Janmanchi, D. *et al.*, *Bioorg. Med. Chem.*, 2013, **21**, 2163-2176 (*3,4-methylene-3′-Me ether, synth, activity*).
Kocsis, L.S. *et al.*, *Org. Lett.*, 2014, **16**, 4158-4161 (*Isojusticidin B, synth*)

2,7′-Cyclo-3′,4,4′,5-tetrahydroxy-8,8′-ligna-7,7′-dien-9,9′-olide C331

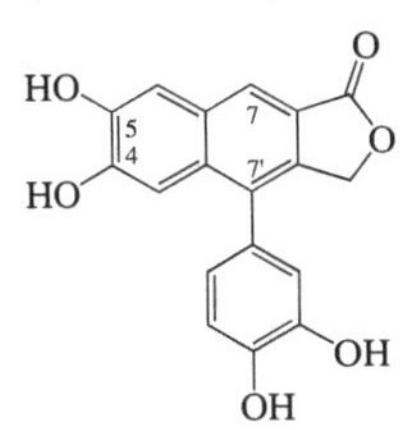

3′,4,4′,5-Tetrahydroxy-2,7′-cycloligna-7,7′-dien-9,9′-olide
$C_{18}H_{12}O_6$ 324.288
Lignan numbering shown

3′-Me ether
2,7′-Cyclo-4,4′,5-trihydroxy-3′-methoxy-8,8′-ligna-7,7′-dien-9,9′-olide. 4,4′,5-Trihydroxy-3′-methoxy-2,7′-cycloligna-7,7′-dien-9,9′-olide. ***Vitexdoin G***
[2135569-94-5]
$C_{19}H_{14}O_6$ 338.315
Constit. of *Vitex negundo*. Amorph. pale yellow powder. λ_{max} 223 (log ε 2.95); 260 (log ε 3.09); 329 (log ε 2.49); 386 (sh) (MeOH)

3′,5-Di-Me ether
2,7′-Cyclo-4,4′-dihydroxy-3′,5-dimethoxy-8,8′-ligna-7,7′-dien-9,9′-olide. 4,4′-Dihydroxy-3′,5-dimethoxy-2,7′-cycloligna-7,7′-dien-9,9′-olide. ***7,7′,8,8′-Tetradehydroconidendrin***. *Detetrahydroconidendrin*
[19463-47-9]
$C_{20}H_{16}O_6$ 352.342
Isol. from the heartwood of *Juniperus formosana* and seeds of *Vitex negundo*. Antibacterial agent active against MRSA. Antioxidant. Mp 254°. λ_{max} 258 (ε 56234); 322 (ε 15488) (MeOH)

4,4′-Di-Me ether
2,7′-Cyclo-3′,5-dihydroxy-4,4′-dimethoxy-8,8′-ligna-7,7′-dien-9,9′-olide. 3′,5-Dihydroxy-4,4′-dimethoxy-2,7′-cycloligna-7,7′-dien-9,9′-olide
[1243200-53-4]
$C_{20}H_{16}O_6$ 352.342
Constit. of *Liatris elegans*. Oil. λ_{max} 220 (log ε 3.56); 259 (log ε 3.86); 321 (log ε 4.2) (MeOH)

3′,4′,5-Tri-Me ether
2,7′-Cyclo-4-hydroxy-3′,4′,5-trimethoxy-8,8′-ligna-7,7′-dien-9,9′-olide. 4-Hydroxy-3′,4′,5-trimethoxy-2,7′-cycloligna-7,7′-dien-9,9′-olide
[345298-51-3]
$C_{21}H_{18}O_6$ 366.369
Constit. of the roots of *Vitex rotundifolia*. Yellowish oil. λ_{max} 257 (log ε 4.6); 319 (log ε 4) (dioxan)

3′,4′-Methylene, 4,5-di-Me ether
2,7′-Cyclo-4,5-dimethoxy-3′,4′-methylenedioxy-8,8′-ligna-7,7′-dien-9,9′-olide. 4,5-Dimethoxy-3′,4′-methylenedioxy-2,7′-cycloligna-7,7′-dien-9,9′-olide. ***Retrojusticidin B***
[82001-16-9]
$C_{21}H_{16}O_6$ 364.353
Isol. from *Phyllanthus myrtifolius*. Strongly inhibits HIV reverse transcriptase. Cryst. (MeOH/$CHCl_3$). Mp 223 - 224° (218°)

4,5-Methylene ether
2,7′-Cyclo-3′,4′-dihydroxy-4,5-methylenedioxy-8,8′-ligna-7,7′-dien-9,9′-olide. 3′,4′-Dihydroxy-4,5-methylenedioxy-2,7′-cycloligna-7,7′-dien-9,9′-olide. ***Prostalidin D***
[327022-81-1]
$C_{19}H_{12}O_6$ 336.299
Constit. of *Justicia diffusa* var. *prostrata*. Cryst.. Mp 230 - 232°. λ_{max} 265; 280; 300; 328; 340 (MeOH)

4,5-Methylene, 3′-Me ether
2,7′-Cyclo-4′-hydroxy-3′-methoxy-4,5-methylenedioxy-

8,8′-ligna-7,7′-dien-9,9′-olide.
4′-Hydroxy-3′-methoxy-4,5-methylenedioxy-2,7′-cyclolignа-7,7′-dien-9,9′-olide.
4′-O-Demethylretrochinensin
[849832-84-4]
$C_{20}H_{14}O_6$ 350.326
Constit. of the stems of *Jatropha gossypifolia*

4,5-Methylene, 3′,4′-di-Me ether
2,7′-Cyclo-3′,4′-dimethoxy-4,5-methylenedioxy-8,8′-ligna-7,7′-dien-9,9′-olide.
3′,4′-Dimethoxy-4,5-methylenedioxy-2,7′-cyclolignа-7,7′-dien-9,9′-olide. ***Retrochinensin***
[5707-96-0]
$C_{21}H_{16}O_6$ 364.353
Isol. from whole plant of *Justicia prostrata*. Needles. Mp 234°

3′,4′:4,5-Bis(methylene) ether
2,7′-Cyclo-3′,4′:4,5-bis(methylenedioxy)-8,8′-ligna-7,7′-dien-9,9′-olide.
3′,4′:4,5-Bis(methylenedioxy)-2,7′-cyclolignа-7,7′-dien-9,9′-olide. ***Justicidin E***
[27792-97-8]
$C_{20}H_{12}O_6$ 348.31
Isol. from *Justicia procumbens*, *Hypoestes purpurea* and *Justicia obiculata*. Piscicide. Pale yellow cryst. (MeOH). Mp 270 - 271°. λ_{max} 244 (ε 33800); 300 (ε 9500) (CH_2Cl_2)

4′-O-[3-(4-Hydroxy-3-methoxyphenyl)propyl], 4,5-methylene, 3′-Me ether
Taxutone
[1252004-92-4]
$C_{30}H_{26}O_8$ 514.53
Constit. of the needles of *Taxus baccata*.

Horii, Z. *et al.*, *Chem. Pharm. Bull.*, 1977, **25**, 1803-1808 (*Justicidin E, synth*).
Ghosal, S. *et al.*, *Chem. Ind. (London)*, 1979, 854-855 (*Retrochinensin*).
Ghosal, S. *et al.*, *J. Chem. Soc., Chem. Commun.*, 1979, 165-166 (*Retrochinensin, synth*).
Pelter, A. *et al.*, *Tetrahedron Lett.*, 1982, **23**, 571-572 (*Retrochinensin, Tetradehydroconidendrin, synth*).
Takano, S. *et al.*, *Tetrahedron Lett.*, 1985, **26**, 1659-1660 (*Justicidin E, Retrochinensin, synth*).
Anjaneyulu, A.S.R. *et al.*, *Indian J. Chem., Sect. B*, 1986, **25**, 955-956 (*Retrochinensin, Justicidin B, synth*).
Anjaneyulu, A.S.R. *et al.*, *Indian J. Chem., Sect. B*, 1987, **26**, 305-307 (*Retrochinensin, synth*).
Stevenson, R. *et al.*, *J. Nat. Prod.*, 1989, **52**, 367-375 (*Justicidin E, synth*).
Kuo, Y.-H. *et al.*, *Chem. Pharm. Bull.*, 1990, **38**, 3195-3201 (*Detetrahydroconidendrin*).
Seko, S. *et al.*, *Tetrahedron Lett.*, 1990, **31**, 6883-6886 (*Justicidin E, synth*).
Anastas, P.T. *et al.*, *J. Nat. Prod.*, 1991, **54**, 1687-1691 (*Retrochinensin, synth*).
Lin, M.-T. *et al.*, *J. Nat. Prod.*, 1995, **58**, 244-249 (*Retrojusticidin B*).
Tanabe, Y. *et al.*, *J. Chem. Soc., Perkin Trans. 1*, 1996, 2157-2165 (*Justicidin E, synth*).
Kim, H.J. *et al.*, *J. Nat. Prod.*, 1998, **61**, 145-148 (*Retrojusticidin B, activity*).
Rajasekhar, D. *et al.*, *J. Asian Nat. Prod. Res.*, 2000, **2**, 289-300 (*Prostalidin D, synth*).
Ramesh, B. *et al.*, *Indian J. Chem., Sect. B*, 2001, **40**, 864-866 (*Justicia obiculata constit*).
Kawazoe, K. *et al.*, *J. Nat. Prod.*, 2001, **64**, 588-591 (*3′,5-di-Me ether, 3′,4′,5-tri-Me ether, activity*).
Harrowven, D.C. *et al.*, *Tetrahedron Lett.*, 2001, **42**, 6973-6975 (*Retrojusticidin B, synth*).
Flanagan, S.R. *et al.*, *Tetrahedron*, 2002, **58**, 5989-6001 (*Justicidin E, Retrojusticidin B, Retrochinensin, synth*).
Das, R. *et al.*, *Indian J. Heterocycl. Chem.*, 2004, **14**, 169-170 (*4′-O-Demethylretrochinensin*).
Shen, C-C. *et al.*, *J. Nat. Prod.*, 2004, **67**, 1947-1949 (*Hypoestes purpurea constit*).
Ono, M. *et al.*, *J. Nat. Prod.*, 2004, **67**, 2073-2075 (*Vitex negundo constit, activity*).
Das, B. *et al.*, *Indian J. Heterocycl. Chem.*, 2009, **18**, 407-408 (*Taxutone*).
Zhang, Z. *et al.*, *Nat. Prod. Res.*, 2010, **24**, 1079-1085 (*Liatris elegans constit*).
Patel, R.M. *et al.*, *Org. Lett.*, 2013, **15**, 14-17 (*Retrojusticidin B, synth*).
Kao, T.-T. *et al.*, *J. Org. Chem.*, 2015, **80**, 6708-6714 (*Justicidin E, synth*).
Nie, X.-F. *et al.*, *J. Asian Nat. Prod. Res.*, 2016, **18**, 656-661 (*Vitexdoin G*)

2,7′-Cyclo-3′,4,4′,5-tetrahydroxy-8,8′-ligna-7,7′-dien-9′,9-olide C332

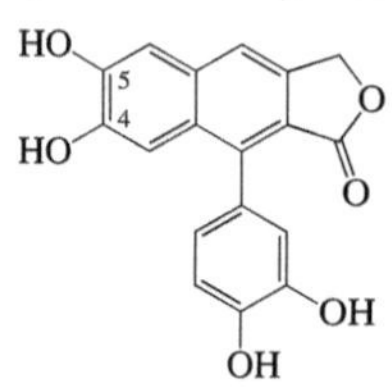

3′,4,4′,5-Tetrahydroxy-2,7′-cyclolignа-7,7′-dien-9′,9-olide
$C_{18}H_{12}O_6$ 324.288

4,5-Di-Me ether, 4′-O-β-D-glucopyranoside
Procumbenoside L
[2243471-12-5]
$C_{26}H_{26}O_{11}$ 514.483
Constit. of *Justicia procumbens*. Amorph. powder (MeOH). $[\alpha]_D^{20}$ -44.6 (c, 0.065 in DMSO)

3′,4′-Methylene, 4-Me ether
2,7′-Cyclo-5-hydroxy-4-methoxy-3′,4′-methylenedioxy-8,8′-ligna-7,7′-dien-9′,9-olide.
5-Hydroxy-4-methoxy-3′,4′-methylenedioxy-2,7′-cyclolignа-7,7′-dien-9′,9-olide. ***Daurinol***
[79862-78-5]
$C_{20}H_{14}O_6$ 350.326
Isol. from *Haplophyllum dauricum*, *Haplophyllum cappodocicum* and *Daphne acutiloba*. Mp 256 - 257°. λ_{max} 224 (ε 23988); 259 (ε 50119); 322 (ε 10233) (MeOH)

3′,4′-Methylene, 4-Me ether, 5-O-β-D-glucopyranoside
Daurinol glucoside
[129950-40-9]
$C_{26}H_{24}O_{11}$ 512.467
Constit. of *Haplophyllum buxbaumii*

3′,4′-Methylene, 5-Me ether
2,7′-Cyclo-4-hydroxy-5-methoxy-3′,4′-methylenedioxy-8,8′-ligna-7,7′-dien-9′,9-olide.
4-Hydroxy-5-methoxy-3′,4′-methylenedioxy-2,7′-cyclolignа-7,7′-dien-9′,9-olide. ***Isodaurinol***
[79862-77-4]
$C_{20}H_{14}O_6$ 350.326
Constit. of *Haplophyllum cappadocicum*

3′,4′-Methylene, 4,5-di-Me ether
2,7′-Cyclo-4,5-dimethoxy-3′,4′-methylenedioxy-8,8′-ligna-7,7′-dien-9′,9-olide.
4,5-Dimethoxy-3′,4′-methylenedioxy-2,7′-cyclolignа-7,7′-dien-9′,9-olide. ***Justicidin B.***
Dehydrocollinusin
[17951-19-8]
[96400-37-2] (Justicin B incorr.)
$C_{21}H_{16}O_6$ 364.353
Constit. of *Justicia hayati*, *Justicia procumbens*, *Haplophyllum dauricum*, *Phyllanthus piscatorum* and many other plants. Also prod. by a marine-derived *Nocardia* sp. ALAA 2000. Exhibits antineoplastic, antifungal and piscicidal activities. Mp 240°. λ_{max} 260 (ε 71290); 295 (ε 12150); 310 (ε 12150); 350 (ε 6125) (EtOH) (Berdy)

4,5-Methylene, 3′-Me ether
2,7′-Cyclo-4′-hydroxy-3′-methoxy-4,5-methylenedioxy-8,8′-ligna-7,7′-dien-9′,9-olide.
4′-Hydroxy-3′-methoxy-4,5-methylenedioxy-2,7′-cyclolignа-7,7′-dien-9′,9-olide.
4′-De-O-methylchinensin
[1126088-85-4]
$C_{20}H_{14}O_6$ 350.326

Constit. of *Bupleurum marginatum*. Needles. Mp 244 - 246°. λ_{max} 258 (log ε 4.42); 288 (log ε 3.88) (MeOH/$CHCl_3$)

4,5-Methylene, 3′,4′-di-Me ether

2,7′-Cyclo-3′,4′-dimethoxy-4,5-methylenedioxy-8,8′-ligna-7,7′-dien-9′,9-olide. 3′,4′-Dimethoxy-4,5-methylenedioxy-2,7′-cycloligna-7,7′-dien-9′,9-olide. ***Chinensin‡****. 4-Deoxyisodiphyllin*

[31888-76-3]

$C_{21}H_{16}O_6$ 364.353

Isol. from *Polygala chinensis* and *Bupleurum* spp.. Cryst. (Me_2CO/petrol or MeOH/$CHCl_3$). Mp 227°. λ_{max} 250 (ε 46600); 300 (ε 10600); 340 (ε 5500) (CH_2Cl_2)

3′,4′:4,5-Bis(methylene) ether

2,7′-Cyclo-3′,4′:4,5-bis(methylenedioxy)-8,8′-ligna-7,7′-dien-9′,9-olide. 3′,4′:4,5-Bis(methylenedioxy)-2,7′-cycloligna-7,7′-dien-9′,9-olide. ***Taiwanin C***

[14944-34-4]

$C_{20}H_{12}O_6$ 348.31

Isol. from heartwood of *Taiwania cryptomeroides* and from *Justicia procumbens*. Shows antifungal activity. Inhibitor of cyclooxygenases 1 and 2. Mp 276°

4-O-(3-Methyl-2-butenyl), 3′,4′-methylene, 5-Me ether

2,7′-Cyclo-5-methoxy-3′,4′-methylenedioxy-4-prenyloxy-8,8′-ligna-7,7′-dien-9′,9-olide. 5-Methoxy-3′,4′-methylenedioxy-4-prenyloxy-2,7′-cycloligna-7,7′-dien-9′,9-olide. ***4-O-Prenylisodaurinol****. 4-O-(3-Methyl-2-butenyl) isodaurinol*

[664344-47-2]

$C_{25}H_{22}O_6$ 418.445

Constit. of *Haplophyllum myrtifolium*. Amorph. solid. λ_{max} 209 (log ε 4.27); 223 (log ε 4.21); 257 (log ε 4.32); 293 (log ε 3.82); 311 (log ε 3.79); 348 (sh) (MeOH).

Ghosal, S. *et al.*, *Phytochemistry*, 1974, **13**, 2281-2284 (*Chinensin*).
González, A.G. *et al.*, *Tetrahedron*, 1978, **34**, 1011-1013 (*4-Deoxyisodiphyllin, synth*).
Batsuren, D. *et al.*, *Chem. Nat. Compd. (Engl. Transl.)*, 1981, **17**, 223-225 (*Daurinol, Justicidin B*).
Abdullaev, N.D. *Chem. Nat. Compd. (Engl. Transl.)*, 1987, **23**, 63-74 (*cmr*).
Stevenson, R. *et al.*, *J. Nat. Prod.*, 1989, **52**, 367-375 (*Taiwanin C, synth*).
Al-Abed, Y. *et al.*, *Phytochemistry*, 1990, **29**, 2659-2661 (*Daurinol glucoside*).
Seko, S. *et al.*, *Tetrahedron Lett.*, 1990, **31**, 6883-6886 (*Taiwanin C, synth*).
Anastas, P.T. *et al.*, *J. Nat. Prod.*, 1991, **54**, 1687-1691 (*Daurinol, synth*).
Goezler, B. *et al.*, *Phytochemistry*, 1992, **31**, 2473-2475 (*Daurinol, Isodaurinol*).
Kamal, A. *et al.*, *Tetrahedron Lett.*, 1994, **35**, 3879-3882 (*Justicidin B, synth*).
Ogiku, T. *et al.*, *J. Org. Chem.*, 1995, **60**, 4585-4590 (*Taiwanin C, synth*).
Tanabe, Y. *et al.*, *J. Chem. Soc., Perkin Trans. 1*, 1996, 2157 (*Taiwanin C, synth*).
Chang, S.-T. *et al.*, *Holzforschung*, 1999, **53**, 487-490 (*Taiwanin C, isol, activity*).
Ban, H.S. *et al.*, *Biochem. Pharmacol.*, 2002, **64**, 1345-1354 (*Taiwanin C, activity*).
Flanagan, S.R. *et al.*, *Tetrahedron*, 2002, **58**, 5989-6001 (*Justicidin B, Taiwanin C, Chinensin, synth*).
Sağlam, H. *et al.*, *Fitoterapia*, 2003, **74**, 564-569 (*4-Prenylisodaurinol*).
Gertsch, J. *et al.*, *Planta Med.*, 2003, **69**, 420-424 (*Justicidin B, activity*).
Sato, Y. *et al.*, *Angew. Chem., Int. Ed.*, 2004, **43**, 2436-2440 (*Taiwanin C, synth*).
Vasilev, N. *et al.*, *J. Nat. Prod.*, 2006, **69**, 1014-1017 (*Justicidin B, bibl*).
Yang, M. *et al.*, *Magn. Reson. Chem.*, 2006, **44**, 727-730 (*Taiwanin C, pmr, cmr*).
Da Silva, R. *et al.*, *Magn. Reson. Chem.*, 2007, **45**, 902-904 (*Justicidin B, Taiwanin C, pmr, cmr*).
Liu, Y. *et al.*, *Helv. Chim. Acta*, 2008, **91**, 2316-2320 (*4′-De-O-methylchinensin*).
Cao, J-L. *et al.*, *Asian J. Chem.*, 2010, **22**, 6509-6512 (*Daphne acutiloba constit*).
Patel, R.M. *et al.*, *Org. Lett.*, 2013, **15**, 14-17 (*synth*).
Jin, H. *et al.*, *Fitoterapia*, 2014, **94**, 70-76 (*Justicidin B, Taiwanin C*).
Zhou, P. *et al.*, *Molecules*, 2015, **20**, 7048-7058 (*Justicidin B, isol, purifn*).
Hayat, F. *et al.*, *Tetrahedron*, 2015, **71**, 2945-2950 (*Daurinol, Justicidin B, synth*).
Hemmati, S. *et al.*, *Molecules*, 2016, **21**, 820:1-20 (*Justicidin B, rev, synth, occur, activity*).
Jin, H. *et al.*, *J. Asian Nat. Prod. Res.*, 2017, **19**, 1-8 (*Procumbenoside L*)

7,7′-Cyclo-3,3′,4,4′-tetrahydroxy-8,8′-lignan-9,9′-dioic acid C333

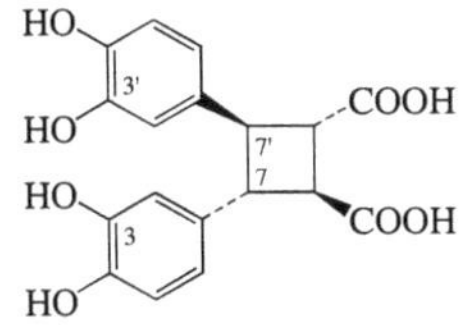

(7α,7′β,8β,8′α)-*form*

3,4-Bis(3,4-dihydroxyphenyl)-1,2-cyclobutanedicarboxylic acid. 3′,3″,4′,4″-Tetrahydroxytruxinic acid

$C_{18}H_{16}O_8$ 360.318

(7α,7′β,8β,8′α)-form

Mono-Me ester

[1917362-03-8]

$C_{19}H_{18}O_8$ 374.345

Constit. of the aerial parts of *Lysimachia clethroides*. Exhibits moderate activity against aldose reductase. Solid. $[\alpha]_D$ -82.8 (c, 0.11 in MeOH). λ_{max} 285 (log ε 3.72) (MeOH)

Di-Me ester

[1383572-08-4]

$C_{20}H_{20}O_8$ 388.372

Constit. of the leaves of *Vitex quinata*. Light yellow gum. $[\alpha]_D^{20}$ +5.4 (c, 0.1 in MeOH). λ_{max} 215 (log ε 4.07); 345 (log ε 3.88) (MeOH)

3,3′-Di-Me ether, 9,9′-O-[α-D-glucopyranosyl-(1→2)-β-D-fructofuranose-3,6-diyl] ester

Coixlachryside A

$C_{32}H_{38}O_{17}$ 694.639

Constit. of the roots of *Coix lachryma-jobi*. Pale yellow syrup. $[\alpha]_D^{25}$+9.9 (c, 0.6 in MeOH). λ_{max} 235 (log ε 4.08); 285 (log ε 3.96) (MeOH)

(7β,7′β,8α,8′α)-form

Diester with 3-OH of (-)-Quinic acid

Eucommicin A

[1845774-14-2]

$C_{32}H_{36}O_{18}$ 708.622

Constit. of the leaves of *Eucommia ulmoides*. Exhibits selective anticancer stem cell activity and inhibits tumour sphere formn. by iCSCL cells. Amorph. solid. $[\alpha]_D^{20}$ -30.2 (c, 0.136 in MeOH).

Deng, Y. *et al.*, *Phytochem. Lett.*, 2011, **4**, 213-217 (*di-Me ester*).
Liang, D. *et al.*, *Phytochem. Lett.*, 2015, **11**, 116-119 (*Lysimachia clethroides constit*).
Hong, S.S. *et al.*, *Phytochem. Lett.*, 2016, **17**, 152-157 (*Coixlachryside A*).
Fujiwara, A. *et al.*, *Phytochemistry*, 2016, **122**, 139-145 (*Eucommicin A, activity*)

7′,8-Cyclo-3,3′,4,4′-tetrahydroxy-7,8′-lignan-9,9′-dioic acid C334

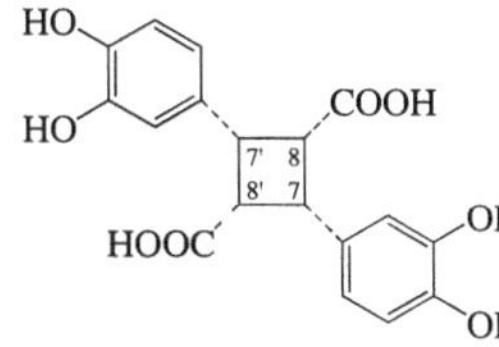

2,4-Bis(3,4-dihydroxyphenyl)-1,3-cyclobutanedicarboxylic acid. 3,3′,4,4′-Tetrahydroxytruxillic acid

$C_{18}H_{16}O_8$ 360.318
(7α,7′α,8α,8′α)-form
A *meso*-stereoisomer
Di-Me ester
[501012-85-7]
$C_{20}H_{20}O_8$ 388.372
Constit. of the whole plant of *Bidens parviflora*. PGE_2 prodn. inhibitor. Needles. Mp 220°. λ_{max} 236 (log ε 4.5); 284 (log ε 4.36) (MeOH)
Bis(2-hydroxyphenyl-β-D-glucopyranosid-6-yl) ester
Dodegranoside B
[1198215-16-5]
$C_{42}H_{44}O_{20}$ 868.794
Constit. of the leaves of *Dodecadenia grandiflora*. Amorph. brown powder. $[\alpha]_D^{26}$ -35.3 (c, 0.03 in MeOH). λ_{max} 214; 278 (MeOH).

Wang, N. *et al.*, *Phytochemistry*, 2003, **62**, 741-746 (*Bidens parviflora constit*).
Kumar, M. *et al.*, *Phytochemistry*, 2009, **70**, 1448-1455 (*Dodegranoside B*)

7,8′-Cyclo-3′,4,4′,5-tetrahydroxy-2,7′-lignan-9,9′-dioic acid C335

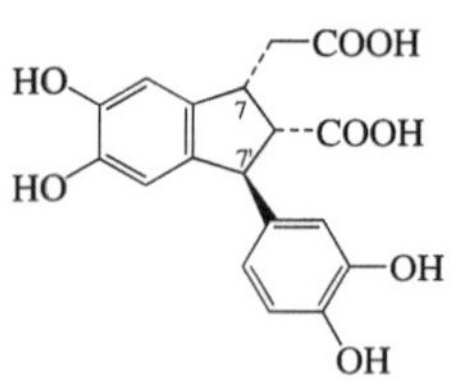

(7*R**,7′*R**,8′*R**)-form

2-Carboxy-3-(3,4-dihydroxyphenyl)-2,3-dihydro-5,6-dihydroxy-1H-indene-1-acetic acid
$C_{18}H_{16}O_8$ 360.318
(7R*,7′R*,8′R*)-form
Di-Me ester
[191280-20-3]
$C_{20}H_{20}O_8$ 388.372
Constit. of the seeds of *Helianthus annuus*. Gum. $[\alpha]_D^{24}$ +93.6 (c, 0.25 in MeOH)
(7R*,7′S*,8′S*)-form
Di-Me ester
[191280-19-0]
Constit. of the seeds of *Helianthus annuus*. Gum. $[\alpha]_D^{24}$ +58 (c, 0.25 in MeOH).

Kato, T. *et al.*, *Nat. Prod. Lett.*, 1997, **9**, 161-165 (*Helianthus annuus constits*)

2,7′-Cyclo-3,3′,4,4′-tetrahydroxy-8,8′-lignan-9′,9-olide C336

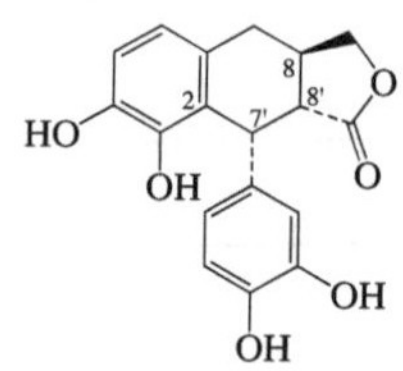

3,3′,4,4′-Tetrahydroxy-2,7′-cyclolignan-9′,9-olide
$C_{18}H_{16}O_6$ 328.32
Lignan numbering shown
(7′R,8R,8′R)-form
3′,4′-Methylene, 3,4-di-Me ether
2,7′-Cyclo-3,4-dimethoxy-3′,4′-methylenedioxy-8,8′-lignan-9′,9-olide.
3,4-Dimethoxy-3′,4′-methylenedioxy-2,7′-cyclolignan-9′,9-olide.
Polygamatin
[71640-48-7]
$C_{21}H_{20}O_6$ 368.385
Isol. from whole plant of *Polygala polygama*. Prisms. Mp 184 - 186°. $[\alpha]_D^{24}$ -115 (c, 0.1 in $CHCl_3$). λ_{max} 286 (ε 7000) (EtOH) (Berdy).

Hokanson, G.C. *J. Nat. Prod.*, 1979, **42**, 378-384 (*Polygamatin*)

2,7′-Cyclo-3′,4,4′,5-tetrahydroxy-8,8′-lignan-9′,9-olide C337

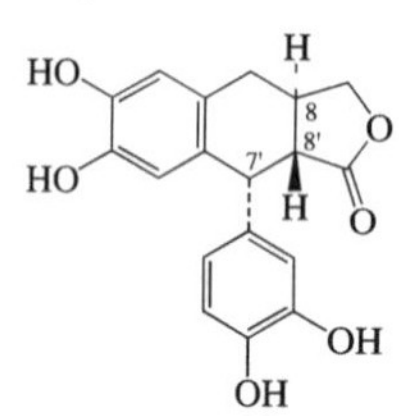

(7*R*,8*R*,8′*R*)-form

3′,4,4′,5-Tetrahydroxy-2,7′-cyclolignan-9′,9-olide
$C_{18}H_{16}O_6$ 328.32
(7′R,8R,8′R)-form
3′,4′-Methylene, 4,5-di-Me ether
2,7′-Cyclo-4,5-dimethoxy-3′,4′-methylenedioxy-8,8′-lignan-9′,9-olide.
4,5-Dimethoxy-3′,4′-methylenedioxy-2,7′-cyclolignan-9′,9-olide.
Tetrahydrojusticidin B
$C_{21}H_{20}O_6$ 368.385
Constit. of Kenyan propolis. Amorph. yellow powder. $[\alpha]_D^{21}$ -46 (c, 1.1 in $CHCl_3$)
3′,4′:4,5-Bis(methylene) ether
2,7′-Cyclo-3′,4′:4,5-bis(methylenedioxy)-8,8′-lignan-9′,9-olide.
3′,4′;4,5-Bis(methylenedioxy)-2,7′-cyclolignan-9′,9-olide.
Polygamain
[71640-49-8]
$C_{20}H_{16}O_6$ 352.342
Isol. from *Polygala polygama*, *Commiphora incisa*, tubers of *Alisma orientalis* and aerial parts of *Haplophyllum tuberculatum*. Shows antibacterial activity. Potent antiproliferative and microtubule depolymerising agent. Needles (MeOH). Mp 187 - 188°. $[\alpha]_D^{22}$ -140 (c, 0.18 in $CHCl_3$). λ_{max} 237 (log ε 3.95); 287 (log ε 3.98) (MeOH). [neutral] λ_{max} 243 (ε 8150); 288 (ε 9300) ($CHCl_3$) (Berdy)
(7′R,8R,8′S)-form
3′,4′:4,5-Bis(methylene) ether
Picropolygamain
[98819-09-1]
$C_{20}H_{16}O_6$ 352.342
Constit. of *Bursera graveolens*, *Bursera simaruba* and *Commiphora incisa*. Shows cytotoxic props.. Needles (Et_2O). Mp 171 - 173°. $[\alpha]_D$ +65 (c, 0.95 in $CHCl_3$). Incorrectly called Burseranin in one 2012 ref.. λ_{max} 206 (ε 21500); 287 (ε 11600) (MeOH)
(7′R,8S,8′S)-form
3′,4′:4,5-Bis(methylene) ether
Dehydroheliobuphthalmin lactone
[103001-07-6]
$C_{20}H_{16}O_6$ 352.342
Constit. of *Heliopsis buphthalmoides*. Mp 262°. Stereochem. given is tentative
(7′S,8R,8′R)-form
Constit. of the roots of *Hypochaeris radicata*. Yellow gum. $[\alpha]_D^{20}$ +65.1 (c, 0.35 in MeOH)
3′,4′:4,5-Bis(methylene) ether
7β-Polygamain
[85761-47-3]
$C_{20}H_{16}O_6$ 352.342
Constit. of *Haplophyllum ptilostylum*. HIV reverse transcriptase inhibitor. $[\alpha]_D$ -94 (c, 1.2 in $CHCl_3$). Called 1β-Polygamain in the lit., different numbering system. λ_{max} 235 (log ε 4.2); 288 (log ε 3.8) (MeOH).

Brown, E. *et al.*, *C. R. Seances Acad. Sci., Ser. 2*, 1982, **295**, 563-566 (*synth*).
Provan, G.J. *et al.*, *Planta Med.*, 1985, **51**, 271-272 (*Picropolygamain*).
Jakupovic, J. *et al.*, *Planta Med.*, 1986, **52**, 18-20 (*Dehydroheliobuphthalmin lactone*).

Sheriha, G.M. *et al.*, *Phytochemistry*, 1987, **26**, 3339-3341 (*Polygamain*).
Peraza-Sanchez, S.R. *et al.*, *J. Nat. Prod.*, 1992, **55**, 1768-1771 (*Picropolygamain, activity*).
Ulubelen, A. *et al.*, *Phytochemistry*, 1995, **39**, 417-422 (7β-*Polygamain*).
Da Silva, R. *et al.*, *Magn. Reson. Chem.*, 2004, **42**, 985-989 (*Polygamain, pmr, cmr*).
Nakanishi, T. *et al.*, *Chem. Pharm. Bull.*, 2005, **53**, 229-231 (*Picropolygamain, cd, activity*).
Kolly-Kovač, T. *et al.*, *Synthesis*, 2005, 1459-1466 (*synth, 7β-Polygamain, pmr, cmr, ms*).
Petrova, A. *et al.*, *Fitoterapia*, 2010, **81**, 509-514 (*Tetrahydrojusticidin B*).
Peng, J. *et al.*, *J. Nat. Prod.*, 2012, **75**, 494-496 (*Polygamain, activity*).
Qiang, Y. *et al.*, *Chem. Nat. Compd. (Engl. Transl.)*, 2014, **49**, 1143-1145 (*Alisma orientalis constit*).
Shulha, O. *et al.*, *Phytochemistry*, 2019, **165**, 112047:1-12 (*Hypochaeris radicata constit, cd*)

2,7′-Cyclo-3′,4,4′,5-tetrahydroxy-8,8′-lignan-9,9′-olide C338

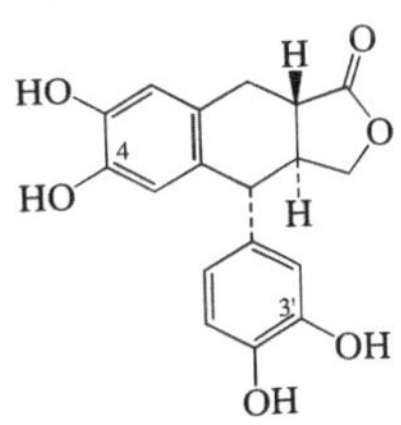

(7′R,8S,8′S)-form

3′,4,4′,5-Tetrahydroxy-2,7′-cyclolignan-9,9′-olide
$C_{18}H_{16}O_6$ 328.32
(7′R,8S,8′S)-form
α-Norconidendrin
3′,5-Di-Me ether
2,7′-Cyclo-4,4′-dihydroxy-3′,5-dimethoxy-8,8′-lignan-9,9′-olide. 4,4′-Dihydroxy-3′,5-dimethoxy-2,7′-cyclolignan-9,9′-olide. ***α-Conidendrin****. Conidendrin. Sulfite liquors lactone. Tsugalactone. Tsugaresinol*
[85699-62-3]
$C_{20}H_{20}O_6$ 356.374
Constit. of numerous conifers incl. *Tsuga sieboldii*, other *Tsuga* spp., *Trachelospermum asiaticum*, *Picea excelsa*, *Podocarpus spicata*, *Larix lyallii*, *Taxus mairei*, *Taxus wallichiana* and *Abies amabilis*. Isol. from spent sulfite liquor, e.g. from aspen wood paper manuf. (*Populus* spp.).. Shows potent cytotoxic activity. Cryst. (EtOH). Mp 254 - 255°. $[\alpha]_D^{20}$ -54.5 (Me_2CO). β-Conidendrin is a semisynthetic epimerisation product. λ_{max} 216; 284 (MeOH) (Berdy)
Tetra-Me ether
α-Dimethylconidendrin
[25925-39-7]
Mp 179 - 180°. $[\alpha]_D^{18}$ -99.4 (Me_2CO)
(7′S,8R,8′R)-form
3′,4,5-Tri-Me ether
2,7′-Cyclo-4′-hydroxy-3′,4,5-trimethoxy-8,8′-lignan-9,9′-olide. 4′-Hydroxy-3′,4,5-trimethoxy-2,7′-cyclolignan-9,9′-olide
[2055706-32-4]
$C_{21}H_{22}O_6$ 370.401
Constit. of the bark of *Abies beshanzuensis*. Amorph. yellow powder. $[\alpha]_D^{20}$ -20 (c, 0.05 in MeOH). λ_{max} 236 (log ε 4.69); 282 (log ε 4.39) (MeOH).

Haworth, R.D. *et al.*, *J. Chem. Soc.*, 1935, 636-644 (*α-Conidendrin, struct, bibl*).
Hearon, W.M. *et al.*, *J. Am. Chem. Soc.*, 1951, **73**, 4005-4007 (*β-Conidendrin*).
Schrecker, A.W. *et al.*, *J. Am. Chem. Soc.*, 1955, **77**, 432-437 (*abs config*).
Pearl, I.A. *et al.*, *J. Org. Chem.*, 1961, **26**, 546-550 (*Populus constit*).
Freudenberg, K. *et al.*, *Tetrahedron*, 1961, **15**, 115-128 (*nomencl*).
Barton, G.M. *et al.*, *J. Org. Chem.*, 1962, **27**, 322-323 (*Abies constit, Tsuga constit*).
Karrer, W. *Konstitution und Vorkommen der Organischen Pflanzenstoffe, 2nd edn., Birkhäuser Verlag*, 1976-1985, no. 1175 (*α-Conidendrin, bibl, occur*).
Miller, R.W. *et al.*, *J. Nat. Prod.*, 1982, **45**, 78-82 (*Taxus wallichiana constit*).
Nabi, Y. *et al.*, *Chem. Lett.*, 1983, 1543-1546 (*α-Conidendrin, synth*).
Dhal, R. *et al.*, *Tetrahedron*, 1986, **42**, 2005-2016 (*(±)-β-Conidendrin, synth*).
Boissin, P. *et al.*, *Tetrahedron*, 1992, **48**, 687-694 (*(-)-α-Conidendrin, synth*).
Shen, Y.-C. *et al.*, *Chin. Pharm. J. (Taipei)*, 1997, **49**, 285-296 (*Taxus mairei constit, activity*).
Stomberg, R. *et al.*, *Z. Kristallogr. - New Cryst. Struct.*, 1999, **214**, 403-404 (*α-Conidendrin, cryst struct*).
Fischer, J. *et al.*, *Org. Lett.*, 2004, **6**, 1345-1348 (*(-)-α-Conidendrin, synth*).
Hu, C.-L. *et al.*, *Nat. Prod. Res.*, 2017, **31**, 1358-1364 (*Abies beshanzuensis constit*).
Hartmann, P. *et al.*, *J. Org. Chem.*, 2019, **84**, 5831-5837 (*(-)-α-Conidendrin, synth*)

2,7′-Cyclo-3′,4,4′,5-tetrahydroxy-8,8′-lign-7-ene-9,9′-dioic acid C339

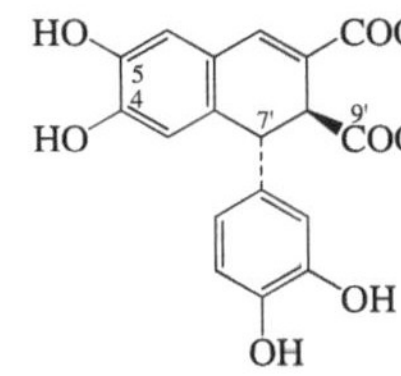

(7′R,8′S)-form

1-(3,4-Dihydroxyphenyl)-1,2-dihydro-6,7-dihydroxy-2,3-naphthalenedicarboxylic acid, CAS. *3′,4,4′,5-Tetrahydroxy-2,7′-cyclolign-7-ene-9,9′-dioic acid*
$C_{18}H_{14}O_8$ 358.302
Lignan numbering shown
(7′R,8′S)-form
Epiphyllic acid [152367-34-5]
Constit. of *Pellia epiphylla*, *Lepicolea ochroleuca*, *Lophocolea heterophylla*, *Lepidozia incurvata* and *Bazzania trilobata*. Antiinflammatory agent. Inhibitor of human leukocyte elastase. $[\alpha]_D^{20}$ -131 (MeOH)
9′-Me ester
[152367-35-6]
$C_{19}H_{16}O_8$ 372.329
Isol. from *Pellia epiphylla*. $[\alpha]_D^{20}$ -112.1 (MeOH)
9′-Me ester, 4-O-β-D-glucopyranoside
$C_{25}H_{26}O_{13}$ 534.47
Constit. of *Lepicolea ochroleuca*. $[\alpha]_D^{20}$ -124.1 (c, 0.66 in MeOH). Poss. artifact
9-O-(1S,2-Dicarboxyethyl) ester
2-O-Epiphylloylmalic acid
[423170-80-3]
$C_{22}H_{18}O_{12}$ 474.374
Constit. of the liverwort *Lophocolea heterophylla*. Syrup. $[\alpha]_D$ -61 (c, 1.4 in MeOH). λ_{max} 205 (log ε 4.14); 222 (log ε 3.95); 253 (log ε 3.9); 310 (log ε 3.54); 337 (log ε 3.57) (MeOH)
Bis(1R,2S,3-tricarboxypropyl) ester
Diisocitryl epiphyllate. ***Sevanol***
[1401564-66-6]
$C_{30}H_{26}O_{20}$ 706.518
Constit. of *Thymus armeniacus*. Inhibitor of ASIC3 channel current. Exhibits analgesic and antiinflammatory activities. [acid] λ_{max} 254 (ε 4860) (0.1% TFA)
9′-[1R-Carboxy-2-(3,4-dihydroxyphenyl)ethyl] ester
Salvianolic acid R
[1497207-97-2]

$C_{27}H_{22}O_{12}$ 538.461
Constit. of *Origanum dictamnus*. Abs. config. drawn incorrectly in paper and is therefore in doubt

9,9′-Bis[1R-carboxy-2-(3,4-dihydroxyphenyl)ethyl] ester
see Rabdosiin, R1

9-(Shikim-5-yl) ester
[152367-36-7]
$C_{25}H_{22}O_{12}$ 514.439
Isol. from *Pellia epiphylla*. $[\alpha]^{20}_{D}$ -122.2 (MeOH).

9′-(Shikim-5-yl) ester, 4-O-β-D-glucopyranoside
[261366-29-4]
$C_{31}H_{32}O_{17}$ 676.58
Constit. of *Lepicolea ochroleuca*. $[\alpha]^{20}_{D}$-137.4 (c, 0.82 in MeOH)

9,9′-Bis(shikim-5-yl) ester
[261366-31-8]
$C_{32}H_{30}O_{16}$ 670.576
Constit. of *Lepicolea ochroleuca*. $[\alpha]^{20}_{D}$-243.6 (c, 0.7 in MeOH)

9-(2,3,4,5,6,7-Hexahydroxyheptyl) ester, 9′-(shikim-5-yl) ester, 4-O-β-D-glucopyranoside
[261377-87-1]
$C_{38}H_{46}O_{23}$ 870.763
Constit. of *Lepicolea ochroleuca*. $[\alpha]_{D}$ -119.1 (c, 0.38 in MeOH)

3′,5-Di-Me ether
2,7′-Cyclo-4,4′-dihydroxy-3′,5-dimethoxy-8,8′-lign-7-ene-9,9′-dioic acid. *4,4′-Dihydroxy-3′,5-dimethoxy-2,7′-cyclolign-7-ene-9,9′-dioic acid. 6,7′-Cyclo-8,8′-diferulic acid*
[160097-31-4]
$C_{20}H_{18}O_{8}$ 386.356
Constit. of various grass cell walls. Off-white solid. Mp 175 - 180°. *trans*-Form. Abs. config. not detd.

3′,4,4′,5-Tetra-Me ether
2,7′-Cyclo-3′,4,4′,5-tetramethoxy-8,8′-lign-7-ene-9,9′-dioic acid. 3′,4,4′,5-Tetramethoxy-2,7′-cyclolign-7-ene-9,9′-dioic acid
[119152-56-6]
$C_{22}H_{22}O_{8}$ 414.41
Obt. from Rabdosiin, R1 by methylation and hydrol.. Isol. from *Pellia epiphylla*. $[\alpha]^{25}_{D}$-202 (c, 1.1 in $CHCl_3$)

(7′RS,8′RS)-form

3′,4-Di-Me ether, bis[2-(4-hydroxyphenyl)ethylamide]
[879223-64-0]
$C_{36}H_{36}N_{2}O_{8}$ 624.69
Alkaloid from the leaves of *Aptenia cordifolia*. Amorph. powder. λ_{max} 223 (log ε 3.9); 284 (log ε 1.2); 333 (log ε 1.1) (MeOH)

3′,4′-Di-Me ether, bis[2-(4-hydroxyphenyl)ethylamide]
Sargentodognan F
[1859917-98-8]
$C_{36}H_{36}N_{2}O_{8}$ 624.69
Constit. of the stems of *Sargentodoxa cuneata*. Yellow needles. λ_{max} 224 (log ε 4.49); 283 (log ε 4.05); 324 (log ε 3.99) (MeOH)

(7′RS,8′SR)-form

Bis[2-(4-hydroxyphenyl)ethyl]amide
Cannabisin B
[144506-17-2]
$C_{34}H_{32}N_{2}O_{8}$ 596.636
Isol. from fruits of *Cannabis sativa* and aerial parts of *Tetragonia tetragonoides*. Amorph. powder. λ_{max} 250 (log ε 4.32); 277 (log ε 4.00); 337 (log ε 4.02) (MeOH)

5-Me ether, bis[2-(4-hydroxyphenyl) ethyl]amide
Cannabisin C
[144506-18-3]
$C_{35}H_{34}N_{2}O_{8}$ 610.663
Isol. from fruits of *Cannabis sativa*. Amorph. powder. λ_{max} 248 (log ε 4.38); 286 (log ε 4.13); 328 (log ε 4.17) (MeOH)

3′,5-Di-Me ether, bis[2-(4-hydroxyphenyl)ethyl]amide
Cannabisin D
[144506-19-4]
$C_{36}H_{36}N_{2}O_{8}$ 624.69
Isol. from fruits of *Cannabis sativa* and seeds of *Hyoscyamus niger*. Needles (MeOH). Mp 165 - 168°. λ_{max} 246 (log ε 4.36); 285 (log ε 4.04); 336 (log ε 4.09) (MeOH)

(7′SR,8′SR)-form

3′,4′-Di-Me ether, bis[2-(4-hydroxyphenyl)ethylamide]
Sargentodognan G
$C_{36}H_{36}N_{2}O_{8}$ 624.69
Constit. of the stems of *Sargentodoxa cuneata*. Exhibits antibacterial activities against *Staphylococcus aureus* ATCC 25923. Yellow needles. λ_{max} 224 (log ε 4.49); 283 (log ε 4.05); 324 (log ε 3.99) (MeOH)

(7′ξ,8′ξ)-form [130752-21-5]
Obt. by oxidn. of caffeic acid by plant enzymes or $KMnO_4$, present in various plant materials. Exhibits antigonadotrophic activity. Further characterisation desirable

9-[1R-Carboxy-2-(3,4-dihydroxyphenyl)ethyl] ester
Globoidnan B
[147058-15-9]
$C_{27}H_{22}O_{12}$ 538.461
Constit. of *Lobostemon fruticosus, Origanum glandulosum, Pulmonaria officinalis* and *Symphytum officinale*. Exhibits moderate activity against β-amyloid aggregation. Brown powder. $[\alpha]^{21}_{D}$+114.9 (c, 0.2 in MeOH). λ_{max} 236 (log ε 3.23); 370 (log ε 4.26) (MeOH).

Agata, I. *et al.*, *Chem. Pharm. Bull.*, 1988, **36**, 3223-3225 (*tetra-Me ether*).
Nahrstedt, A. *et al.*, *Planta Med.*, 1990, **56**, 395-398 (*synth*).
Sakakibara, I. *et al.*, *Phytochemistry*, 1992, **31**, 3219-3223 (*Cannabisins B,C,D*).
Cullmann, F. *et al.*, *Phytochemistry*, 1993, **34**, 831-834 (*Pellia epiphylla constits*).
Ralph, J. *et al.*, *J. Chem. Soc., Perkin Trans. 1*, 1994, 3485-3498 (*occur*).
Martin, U. *et al.*, *Phytochemistry*, 1998, **49**, 1139-1146 (*Bazzania trilobata constit*).
Cullmann, F. *et al.*, *Phytochemistry*, 1999, **52**, 1647-1650 (*Lepidozia incurvata constit*).
Cullmann, F. *et al.*, *Phytochemistry*, 1999, **52**, 1651-1656 (*Lepicolea ochroleuca constits*).
Tazaki, H. *et al.*, *Tetrahedron Lett.*, 1999, **40**, 101-104 (*biosynth*).
Melzig, M.F. *et al.*, *Pharmazie*, 2001, **56**, 967-970 (*activity*).
Tazaki, H. *et al.*, *Biosci., Biotechnol., Biochem.*, 2002, **66**, 255-261 (*Lophocolea heterophylla constits*).
Ma, C-Y. *et al.*, *J. Nat. Prod.*, 2002, **65**, 206-209 (*Hyoscyamus niger constit*).
Della Greca, M. *et al.*, *Tetrahedron*, 2006, **62**, 2877-2882 (*Aptenia cordifolia constit*).
Lee, K.H. *et al.*, *Heterocycles*, 2008, **75**, 419-426 (*Tetragonia tetragonoides constit*).
Zhang, W-N. *et al.*, *J. Agric. Food Chem.*, 2012, **60**, 1682-1687 (*Hyoscyamus niger constit*).
Dubinnyl, M.A. *et al.*, *J. Biol. Chem.*, 2012, **287**, 32993-33000 (*Sevanol, activity*).
Exarchou, V. *et al.*, *Phytochem. Lett.*, 2013, **6**, 46-52 (*Salvianolic acid R*).
Basli, A. *et al.*, *Rec. Nat. Prod.*, 2014, **8**, 208-216 (*Globoidnan B, activity*).
Zeng, X. *et al.*, *Fitoterapia*, 2015, **101**, 153-161 (*Sargentodognans F,G*).
Xia, Y. *et al.*, *J. Chem. Res.*, 2015, **39**, 606-608 (*Cannabisin B, synth*).
Osmakov, D.I. *et al.*, *Russ. J. Bioorg. Chem. (Engl. Transl.)*, 2015, **41**, 543-547 (*Sevanol, activity*).
Krzyzanowska-Kowalczyk, J. *et al.*, *Molecules*, 2018, **23**, 2277:1-32 (*Globoidnan B, cd*).
Belozerova, O.A. *et al.*, *Tetrahedron*, 2018, **74**, 1449-1453 (*Sevanol, synth*).
Trifan, A. *et al.*, *J. Ethnopharmacol.*, 2020, **262**, 113169 (*Globoidnan B*)

7,8′-Cyclo-3′,4,4′,5-tetrahydroxy-2,7′-lign-7-ene-9,9′-dioic acid C340

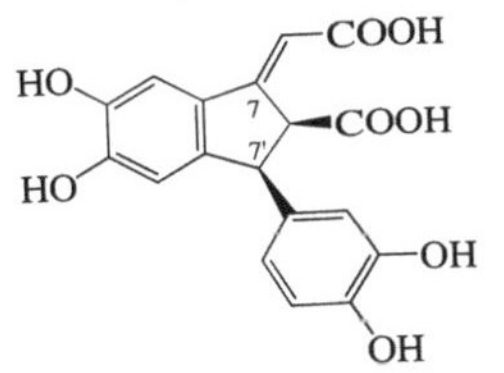

$C_{18}H_{14}O_8$ 358.302
(7E,7′S,8′S)-form
Bis[2-(4-hydroxyphenyl)ethylamide]
Limoniastramide
[1482416-61-4]
$C_{34}H_{32}N_2O_8$ 596.636
Constit. of *Limoniastrum guyonianum*. Antioxidant. Powder. $[\alpha]_D^{20}$ -7.5 (c, 0.07 in MeOH). Abs. config. not certain
3′,5-Di-Me ether
7,8′-Cyclo-4,4′-dihydroxy-3′,5-dimethoxy-2,7′-lign-7-ene-9,9′-dioic acid
3′,5-Di-Me ether, bis[2-(4-hydroxyphenyl)ethylamide]
Lyciumamide D
[2230089-59-3]
$C_{36}H_{36}N_2O_8$ 624.690
Alkaloid from *Polygonum chinense*. Yellow powder.

Trabelsi, N. *et al.*, *Food Chem.*, 2014, **146**, 466-471 (*Limoniastramide, activity*).
Zheng, H.-C. *et al.*, *Bioorg. Med. Chem. Lett.*, 2018, **28**, 1495-1500 (*Lyciumamide D*)

2,7′-Cyclo-3,3′,4,4′-tetrahydroxy-8,8′-lign-7′-en-9′,9-olide C341

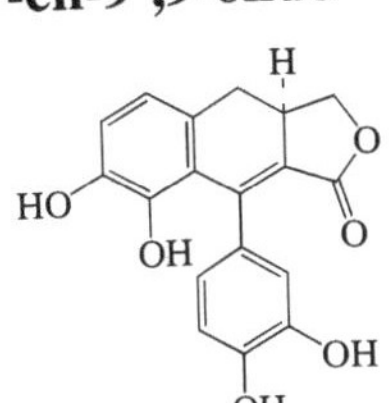

3,3′,4,4′-Tetrahydroxy-2,7′-cyclolign-7′-en-9′,9-olide
$C_{18}H_{14}O_6$ 326.304
(R)-form
3′,4′-Methylene, 3,4-di-Me ether
2,7′-Cyclo-3,4-dimethoxy-3′,4′-methylenedioxy-8,8′-lign-7′-en-9′,9-olide. 3,4-Dimethoxy-3′,4′-methylenedioxy-2,7′-cyclolign-7′-en-9′,9-olide.
7,8-Dihydroisojusticidin B
[1008094-31-2]
$C_{21}H_{18}O_6$ 366.369
Constit. of aerial parts of *Linum perenne*. Amorph. solid. $[\alpha]_D^{20}$ -17 (c, 0.06 in $CHCl_3$). λ_{max} 225 (sh); 288 (log ε 3.73); 350 (sh) (MeOH)
3,4:3′,4′-Bis(methylene) ether
2,7′-Cyclo-3,4:3′,4′-bis(methylenedioxy)-8,8′-lign-7′-en-9′,9-olide. 3,4:3′,4′-Bis(methylenedioxy)-2,7′-cyclolign-7′-en-9′,9-olide.
7,8-Dihydroretrohelioxanthin
[1008094-32-3]
$C_{20}H_{14}O_6$ 350.326
Constit. of *Linum perenne*. Amorph. solid. $[\alpha]_D^{20}$ -17 (c, 0.09 in $CHCl_3$). λ_{max} 237 (sh); 253 (sh); 296 (log ε 3.74); 350 (sh) (MeOH).

Schmidt, T.J. *et al.*, *Planta Med.*, 2007, **73**, 1574-1580 (*Linum perenne constits*)

2,7′-Cyclo-3,3′,4,4′-tetrahydroxy-8,8′-lign-8(8′)-en-9′,9-olide C342

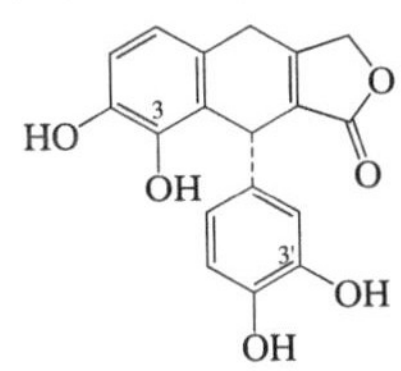

3,3′,4,4′-Tetrahydroxy-2,7′-cyclolign-8(8′)-en-9′,9-olide
$C_{18}H_{14}O_6$ 326.304
(R)-form
3′,4′-Methylene, 3,4-di-Me ether
2,7′-Cyclo-3,4-dimethoxy-3′,4′-methylenedioxy-8,8′-lign-8(8′)-en-9′,9-olide. 3,4-Dimethoxy-3′,4′-methylenedioxy-2,7′-cyclolign-8(8′)-en-9′,9-olide.
β-Apopolygamatin
[71640-47-6]
[68970-48-9] (unspecified stereochem.)
$C_{21}H_{18}O_6$ 366.369
Constit. of *Polygala polygama*. Needles (MeOH). Mp 170 - 172°. $[\alpha]_D^{23}$ -75 (c, 0.2 in $CHCl_3$). Struct. revised in 1991. λ_{max} 293 (ε 9600) (EtOH) (Berdy)
3,4:3′,4′-Bis(methylene) ether
2,7′-Cyclo-3,4:3′,4′-bis(methylenedioxy)-8,8′-lign-8(8′)-en-9′,9-olide. 3,4:3′,4′-Bis(methylenedioxy)-2,7′-cyclolign-8(8′)-en-9′,9-olide. ***Jusneesiin.*** *7,7′-Dihydroretrohelioxanthin*
[216955-77-0]
$C_{20}H_{14}O_6$ 350.326
Constit. of *Justicia neesii*. Amorph. solid (petrol/EtOAc). Mp 207 - 209°. $[\alpha]_D^{25}$+8 (c, 0.1 in Me_2CO). λ_{max} 288 (log ε 3.88) (MeOH).

Momose, T. *et al.*, *Chem. Pharm. Bull.*, 1978, **26**, 3195-3198 (*β-Apopolygamatin, synth*).
Hokanson, G.C. *J. Nat. Prod.*, 1979, **42**, 378-384 (*β-Apopolygamatin*).
Stevenson, R. *et al.*, *J. Nat. Prod.*, 1991, **54**, 310-314 (*β-Apopolygamatin, struct*).
Rajasekhar, D. *et al.*, *Tetrahedron*, 1998, **54**, 13227-13236 (*Jusneesiin*)

2,7′-Cyclo-3′,4,4′,5-tetrahydroxy-8,8′-lign-7′-en-9′,9-olide C343

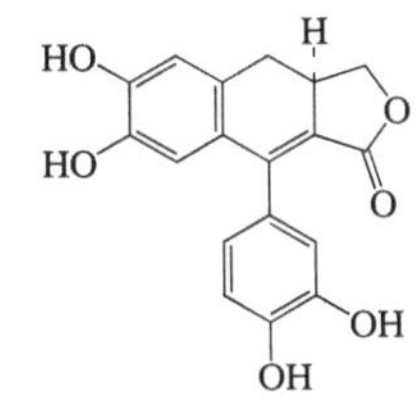

(R)-form

3′,4,4′,5-Tetrahydroxy-2,7′-cyclolign-7′-en-9′,9-olide
$C_{18}H_{14}O_6$ 326.304
(R)-form
3′,4′-Methylene, 4,5-di-Me ether
2,7′-Cyclo-4,5-dimethoxy-3′,4′-methylenedioxy-8,8′-lign-7′-en-9′,9-olide. 4,5-Dimethoxy-3′,4′-methylenedioxy-2,7′-cyclolign-7′-en-9′,9-olide.
Collinusin
[17990-72-6]
$C_{21}H_{18}O_6$ 366.369
Constit. of the leaves of *Cleistanthus collinus*. Mp 196° (192.5-193.5°). $[\alpha]_D$ +132.48 (c, 2.04 in $CHCl_3$)
4,5-Methylene ether, 3′-O-β-D-glucopyranoside
Cleisindoside F
[1821136-74-6]
$C_{25}H_{24}O_{11}$ 500.456
Constit. of the fruit of *Cleistanthus indochinensis*. Cryst. (Me_2CO/MeOH). Mp 261 - 262°. λ_{max} 250 (log ε 3.71); 348 (log ε 3.33) (MeOH)
3′,4′:4,5-Bis(methylene) ether
2,7′-Cyclo-3′,4′:4,5-bis(methylenedioxy)-8,8′-lign-7′-en-9′,9-olide. 3′,4′:4,5-Bis(methylenedioxy)-2,7′-cyclolign-7′-en-9′,9-olide
[1008094-33-4]
$C_{20}H_{14}O_6$ 350.326
Constit. of aerial parts of *Linum perenne*. Amorph. solid. $[\alpha]_D^{20}$ +25 (c, 0.04 in $CHCl_3$). λ_{max} 227 (sh);

252 (sh); 283 (log ε 3.38); 350 (sh) (MeOH)

(ξ)-form

3′,4′:4,5-Bis(methylene) ether

7,8-Dihydrotaiwanin C

[81410-42-6]

$C_{20}H_{14}O_6$ 350.326

Constit. of the heartwood of *Cleistanthus collinus*. Cryst. (C_6H_6). Mp 199 - 200°.

Govindachari, T.R. *et al.*, *Tetrahedron*, 1969, **25**, 2815-2821 (*Collinusin*).
Block, E. *J. Org. Chem.*, 1971, **36**, 3453-3455 (*Collinusin, synth*).
Anjaneyulu, A.S.R. *et al.*, *Tetrahedron*, 1981, **37**, 3641-3652 (*Cleistanthus collinus constit*).
Morimoto, T. *et al.*, *Tetrahedron*, 1993, **49**, 1793-1806 (*Collinusin, synth*).
Schmidt, T.J. *et al.*, *Planta Med.*, 2007, **73**, 1574-1580 (*Linum perenne constit*).
Kocsis, L.S. *et al.*, *Org. Lett.*, 2014, **16**, 4158-4161 (*Dihydrotaiwanin C, synth*).
Thanh, V.T.T. *et al.*, *Planta Med.*, 2014, **80**, 695-702 (*Cleisindoside F*)

2,7′-Cyclo-3′,4,4′,5-tetrahydroxy-8,8′-lign-8(8′)-en-9′,9-olide C344

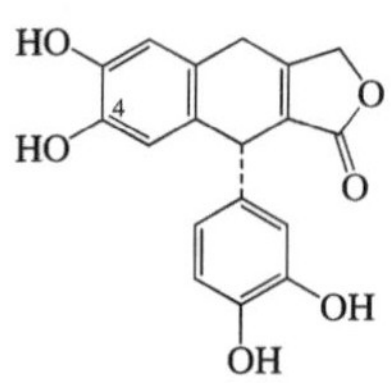

(*R*)-*form*

3′,4,4′,5-Tetrahydroxy-2,7′-cyclolign-8(8′)-en-9′,9-olide

(R)-form

3′,4′:4,5-Bis(methylene) ether

2,7′-Cyclo-3′,4′:4,5-bis(methylenedioxy)-8,8′-lign-8(8′)-en-9′,9-olide. 3′,4′:4,5-Bis(methylenedioxy)-2,7′-cyclolign-8(8′)-en-9′,9-olide.

7,7′R-Dihydrotaiwanin C

[216955-79-2]

$C_{20}H_{14}O_6$ 350.326

Constit. of *Justicia neesii*. Cryst. ($CHCl_3$). Mp 248 - 250°. $[\alpha]^{25}_{D}$ +107.4 (c, 0.23 in $CHCl_3$)

(S)-form

3′,4′:4,5-Bis(methylene) ether

7,7′S-Dihydrotaiwanin C

[1148131-30-9]

Constit. of the heartwood of *Chamaecyparis formosensis*. Cytotoxic. Powder. Mp 245 - 246°. $[\alpha]^{25}_{D}$ -100 (c, 0.4 in $CHCl_3$). λ_{max} 292 (log ε 3.47) (MeOH).

Rajasekhar, D. *et al.*, *Tetrahedron*, 1998, **54**, 13227-13236 (*7,7′R-Dihydrotaiwanin C*).
Chen, T.-H. *et al.*, *Planta Med.*, 2008, **74**, 1806-1811 (*7,7′S-Dihydrotaiwanin C*)

2,7′-Cyclo-3′,4,4′,5-tetrahydroxy-8,8′-lign-8(8′)-en-9,9′-olide C345

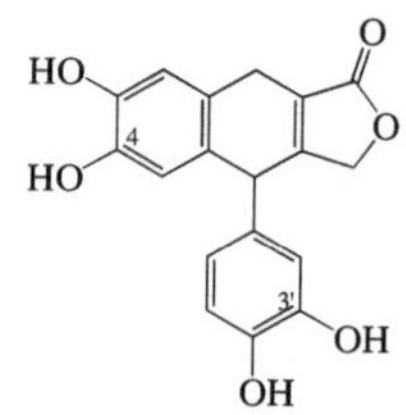

3′,4,4′,5-Tetrahydroxy-2,7′-cyclolign-8(8′)-en-9,9′-olide

$C_{18}H_{14}O_6$ 326.304

(±)-form

3′,4′-Methylene, 5-Me ether

2,7′-Cyclo-4-hydroxy-5-methoxy-3′,4′-methylenedioxy-8,8′-lign-8(8′)-en-9,9′-olide. 4-Hydroxy-5-methoxy-3′,4′-methylenedioxy-2,7′-cyclolign-8(8′)-en-9,9′-olide. ***Konyanin***

[90902-18-4]

$C_{20}H_{16}O_6$ 352.342

Constit. of *Haplophyllum vulcanicum*. Needles (MeOH). Mp 263 - 265°

(ξ)-form

4,5-Methylene, 3′,4′-di-Me ether

2,7′-Cyclo-3′,4′-dimethoxy-4,5-methylenedioxy-8,8′-lign-8(8′)-en-9,9′-olide. 3′,4′-Dimethoxy-4,5-methylenedioxy-2,7′-cyclolign-8(8′)-en-9,9′-olide.

7,7′-Dihydroretrochinensin

[174793-29-4]

$C_{21}H_{18}O_6$ 366.369

Constit. of *Bupleurum handiense*. Viscous oil. $[\alpha]^{25}_{D}$ +4 (c, 0.3 in $CHCl_3$).

Gözler, T. *et al.*, *Tetrahedron*, 1984, **40**, 1145-1150 (*Konyanin*).
Lopez, H. *et al.*, *J. Nat. Prod.*, 1996, **59**, 493-494 (*7,7′-Dihydroretrochinensin*)

2,7′-Cyclo-3,3′,4,4′-tetrahydroxy-9′-nor-8,8′-ligna-7,7′-dien-9-al C346

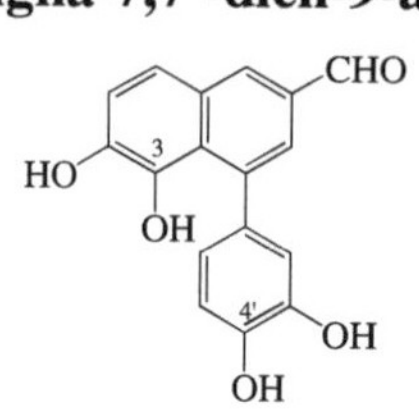

$C_{17}H_{12}O_5$ 296.278

3,3′-Di-Me ether

6-Hydroxy-4-(4-hydroxy-3-methoxyphenyl)-5-methoxy-2-naphthalenecarboxaldehyde. 2,7′-Cyclo-4,4′-dihydroxy-3,3′-dimethoxy-9′-nor-8,8′-ligna-7,7′-dien-9-al. 4,4′-Dihydroxy-3,3′-dimethoxy-9′-nor-2,7′-cycloligna-7,7′-dien-9-al.

Vitexdoin B

[1186021-79-3]

$C_{19}H_{16}O_5$ 324.332

Constit. of the seeds of *Vitex negundo*. Amorph. yellow powder. λ_{max} 268; 330 (MeOH)

3,3′,4′-Tri-Me ether

4-(3,4-Dimethoxyphenyl)-6-hydroxy-5-methoxy-2-naphthalenecarboxaldehyde. 2,7′-Cyclo-4-hydroxy-3,3′,4′-trimethoxy-9′-nor-8,8′-ligna-7,7′-dien-9-al. 4-Hydroxy-3,3′,4′-trimethoxy-9′-nor-2,7′-cycloligna-7,7′-dien-9-al.

Vitrofolal A

[261903-75-7]

$C_{20}H_{18}O_5$ 338.359

Constit. of the roots of *Vitex rotundifolia*. Amorph. yellow powder. λ_{max} 208 (log ε 4.6); 224 (log ε 4.5); 269 (log ε 4.5); 329 (log ε 4) (MeOH).

Kawazoe, K. *et al.*, *Phytochemistry*, 1999, **52**, 1657-1659 (*Vitrofolal A*).
Zheng, C.-J. *et al.*, *J. Nat. Prod.*, 2009, **72**, 1627-1630 (*Vitexdoin B*)

2,7′-Cyclo-3′,4,4′,5-tetrahydroxy-9′-nor-8,8′-ligna-7,7′-dien-9-oic acid C347

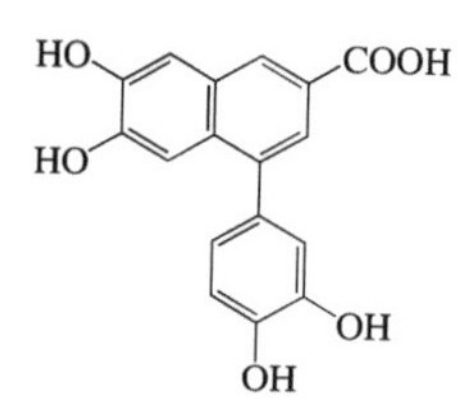

4-(3,4-Dihydroxyphenyl)-6,7-dihydroxy-2-naphthalenecarboxylic acid, CAS. *3′,4,4′,5-Tetrahydroxy-9′-nor-2,7′-cycloligna-7,7′-dien-9-oic acid*

[121242-02-2]

$C_{17}H_{12}O_6$ 312.277

Constit. of *Pellia epiphylla*, *Bazzania trilobata* and other liverworts. Cryst. (MeOH). Mp 325 - 327° dec.. λ_{max} 243 (log ε 4.25); 293 (sh); 312 (log ε 4) (MeOH)

(1,2-Dicarboxyethyl) ester

[261361-61-9]

$C_{21}H_{16}O_{10}$ 428.349

Constit. of the liverwort *Chiloscyphus polyanthos*. $[\alpha]^{20}_{D}$ -7.5 (c, 0.28 in MeOH)

Shikim-5-yl ester

[261361-62-0]

$C_{24}H_{20}O_{10}$ 468.414

Constit. of the liverwort *Jungermannia exsertifolia* ssp. *cordifolia*. $[\alpha]^{20}_{D}$ -14 (c, 0.88 in MeOH)

4-Me ether, 5-O-α-L-rhamnopyranoside

[261361-63-1]

$C_{24}H_{24}O_{10}$ 472.446

Constit. of the liverwort *Lepidozia incurvata*. $[\alpha]^{20}_{D}$ -72 (c, 0.18 in MeOH)

9-Aldehyde, 3′,5-di-Me ether

6-Hydroxy-4-(4-hydroxy-3-methoxyphenyl)-7-methoxy-2-naphthalenecarboxaldehyde. 2,7′-Cyclo-4,4′-dihydroxy-3′,5-dimethoxy-9′-nor-8,8′-ligna-7,7′-dien-9-al. 4,4′-Dihydroxy-3′,5-dimethoxy-9′-nor-2,7′-cycloligna-7,7′-dien-9-al. ***Vitrofolal E***

[345902-02-5]

$C_{19}H_{16}O_5$ 324.332

Constit. of *Syringa pinnatifolia*, *Vitex rotundifolia* and *Vitex negundo*. Amorph. yellow solid. λ_{max} 267 (log ε 4.5); 320 (log ε 4.1) (dioxan)

9-Aldehyde, di-Me acetal, 3′,5-di-Me ether

2,7′-Cyclo-3′,5,9,9-tetramethoxy-9′-nor-8,8′-ligna-7,7′-diene-4,4′-diol. ***Noralashinol A***

[1884709-66-3]

$C_{21}H_{22}O_6$ 370.401

Constit. of *Syringa pinnatifolia* var. *alashanensis*. Powder. Misleading synonym. λ_{max} 246 (log ε 0.83); 320 (log ε 0.32) (MeOH).

Rischmann, M. *et al.*, *Phytochemistry*, 1989, **28**, 867-869 (*Pellia epiphylla constit, synth*).
Martini, U. *et al.*, *Phytochemistry*, 1998, **49**, 1139-1146 (*Bazzania trilobata constit*).
Cullmann, F. *et al.*, *Phytochemistry*, 1999, **52**, 1647-1650 (*esters, Me ether rhamnoside*).
Zheng, C.-J. *et al.*, *J. Nat. Prod.*, 2009, **72**, 1627-1630 (*Vitrofolal E*).
Su, G. *et al.*, *Nat. Prod. Res.*, 2016, **30**, 2149-2153 (*Noralashinol A*)

2,7′-Cyclo-3′,4,4′,5-tetrahydroxy-1,7-seco-8,8′-lign-7-en-9,9′-olide C348

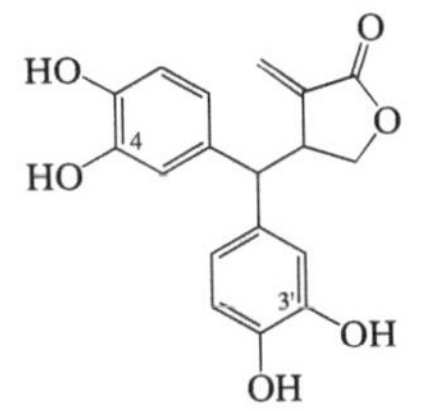

4-[Bis(3,4-dihydroxyphenyl)methyl]dihydro-3-methylene-2(3H)-furanone. 3′,4,4′,5-Tetrahydroxy-1,7-seco-2,7′-cyclolign-7-en-9,9′-olide

$C_{18}H_{16}O_6$ 328.32

(ξ)-form

3′,4-Di-Me ether

2,7′-Cyclo-4′,5-dihydroxy-3′,4-dimethoxy-1,7-seco-8,8′-lign-7-en-9,9′-olide. 4′,5-Dihydroxy-3′,4-dimethoxy-1,7-seco-2,7′-cyclolign-7-en-9,9′-olide.

Urticene

[1160701-43-8]

$C_{20}H_{20}O_6$ 356.374

Constit. of the roots of *Urtica mairei*. Amorph. powder (MeOH). Mp 108 - 110°. $[\alpha]^{20}_{D}$ -27.3 (c, 0.33 in MeOH).

Wang, W. *et al.*, *Chin. Chem. Lett.*, 2008, **19**, 1212-1214 (*Urticene*)

2,7′-Cyclo-3′,4′,5′-trihydroxy-8,8′-lignan-9′,9-olide C349

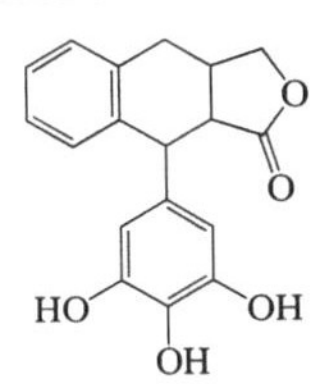

3′,4′,5′-Trihydroxy-2,7′-cyclolignan-9′,9-olide

$C_{18}H_{16}O_5$ 312.321

Tri-Me ether

3a,4,9,9a-Tetrahydro-9-(3,4,5-trimethoxyphenyl)naphtho[2,3-c]furan-1(3H)-one. 2,7′-Cyclo-3′,4′,5′-trimethoxy-8,8′-lignan-9′,9-olide. 3′,4′,5′-Trimethoxy-2,7′-cyclolignan-9′,9-olide

$C_{21}H_{22}O_5$ 354.402

Isol. from seed oil of *Hernandia ovigera*. Unusual substitution pattern lacking A-ring oxygen substituents. May deserve reinvestigating.

Hata, C. *Nippon Kagaku Kaishi*, 1942, **63**, 1540-1544 (*Hernandia ovigera constit*)

2′,8-Cyclo-3,4,4′-trihydroxy-2,7′-lign-7-en-9,3′-olid-9′-oic acid C350

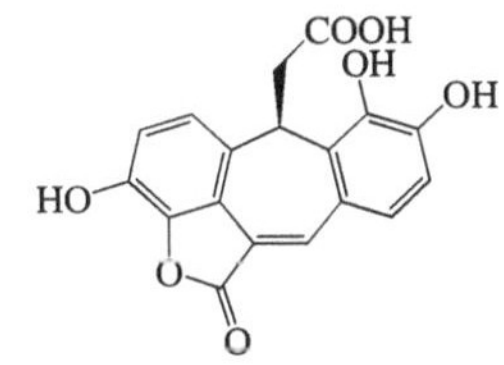

(R)-form

2-(3,7,8-Trihydroxy-1-oxo-1,6-dihydrobenzo[5,6]cyclohepta[1,2,3-cd]benzofuran-6-yl)acetic acid

$C_{18}H_{12}O_7$ 340.287

(R)-form

Miltiolignanolide C [2043953-85-9]

Constit. of the roots of *Salvia miltiorrhiza*. Exhibits antioxidant activity with respect to ABTS, and is a moderate inhibitor of NO prodn. in LPS-induced microglial cells. Moderately increases the viability of H9C2 cells. Amorph. yellow powder

Me ester

Miltiolignanolide E

[2043953-87-1]

$C_{19}H_{14}O_7$ 354.314

Constit. of the roots of *Salvia miltiorrhiza*. Inhibitor of NO prodn. in LPS-induced microglial cells. Amorph. yellow powder

(S)-form

Miltiolignanolide D [2043953-86-0]

Constit. of the roots of *Salvia miltiorrhiza*. Moderate inhibitor of NO prodn. in LPS-induced microglial cells. Moderately increases the viability of H9C2 cells. Amorph. yellow powder

Me ester

Miltiolignanolide F

[2043953-88-2]

$C_{19}H_{14}O_7$ 354.314

Constit. of the roots of *Salvia miltiorrhiza*. Inhibitor of NO prodn. in LPS-induced microglial cells. Amorph. yellow powder.

Li, L.-Z. *et al.*, *Org. Biomol. Chem.*, 2016, **14**, 10050-10057 (*Miltiolignanolides C-F, abs config, activity*)

2′,8-Cyclo-3,4,4′-trihydroxy-9′-nor-2,7′-lign-7-en-9,3′-olide C351

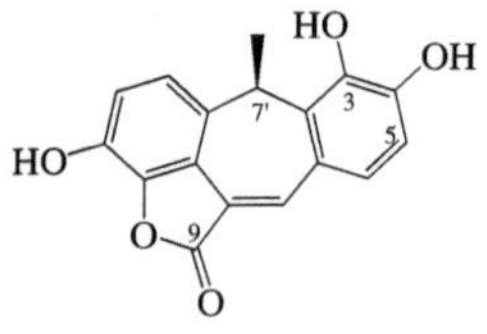

(R)-form

3,7,8-Trihydroxy-6-methylbenzo[5,6]cyclohepta[1,2,3-cd]benzofuran-1(6H)-one
$C_{17}H_{12}O_5$ 296.278
(R)-form
Miltiolignanolide A [2043953-83-7]
Constit. of the roots of *Salvia miltiorrhiza*. Significantly increases the viability of H9C2 cells from H_2O_2-induced cell death. Moderate inhibitor of NO prodn. in LPS-induced microglial cells. Amorph. yellow powder
(S)-form
Miltiolignanolide B [2043953-84-8]
Constit. of the roots of *Salvia miltiorrhiza*. Significantly increases the viability of H9C2 cells from H_2O_2-induced cell death. Moderate inhibitor of NO prodn. in LPS-induced microglial cells. Amorph. yellow powder
(ξ)-form
3-Deoxy, 5-hydroxy
2′,8-Cyclo-4,4′,5-trihydroxy-9′-nor-2,7′-lign-7-en-9,3′-olide.
Miltiolignanolide G
[2043953-89-3]
$C_{17}H_{12}O_5$ 296.278
Constit. of the roots of *Salvia miltiorrhiza*. Amorph. yellow powder.

Li, L.-Z. *et al.*, *Org. Biomol. Chem.*, 2016, **14**, 10050-10057 (*Miltiolignanolides A,B,G, abs config, activity*)

2,7′-Cyclo-4,4′,4″-trihydroxy-5,5′,5″-trimethoxy-3′,3″:8,8′-sesquiligna-7,7″-diene-9,9′,9″-trioic acid C352

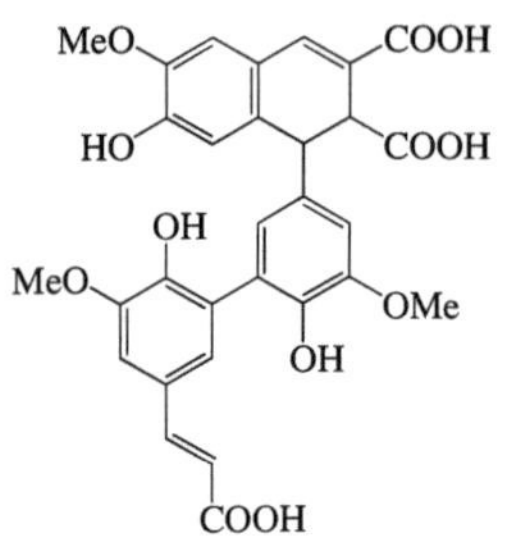

Ferulic acid-(5′-5″)-4,4′-dihydroxy-3′,5-dimethoxy-2,7′-cyclolign-7-ene-9,9′-dioic acid. 8-8(Cyclic)/5-5-TriFA
[1995908-21-8]
$C_{30}H_{26}O_{12}$ 578.526
Identified in alkaline hydrolysates of insol. fibres from maize, wheat and sugar beet.

Waterstraat, M. *et al.*, *J. Agric. Food Chem.*, 2016, **64**, 7244-7250 (*8-8(Cyclic)/5-5-Tri-FA*)

2,7′-Cyclo-3′,3″,5-trimethoxy-4′,8″-oxy-8,8′-sesquiligpnan-4,4″,7″-triol C353

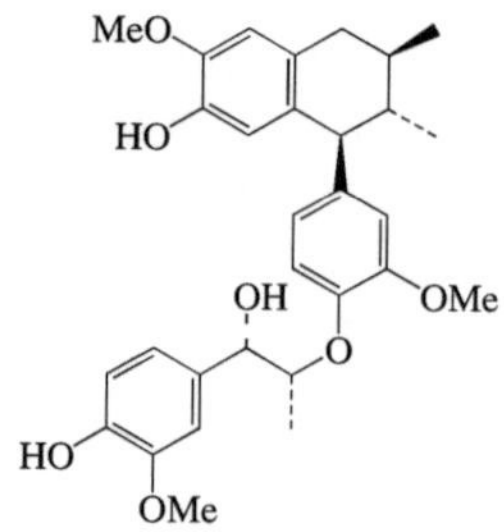

2,7′-Cyclo-4,4″,7″-trihydroxy-3′,3″,5-trimethoxy-4′,8″-oxysesquineolignan
$C_{30}H_{36}O_7$ 508.611
(7′S,7″S,8R,8′S,8″R)-form [1310453-57-6]
Isol. from the bark of *Machilus robusta*. Amorph. solid. $[\alpha]^{20}_{D}$ -40.5 (c, 0.01 in MeOH). λ_{max} 204 (log ε 4.43); 232 (log ε 3.52); 280 (log ε 3.13) (MeOH).

Li, Y. *et al.*, *J. Nat. Prod.*, 2011, **74**, 1444-1452 (*Machilus robusta constit*)

Cymodiene C354

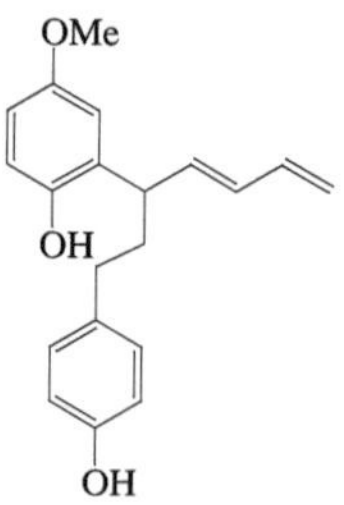

7-(4-Hydroxyphenyl)-5-(2-hydroxy-5-methoxyphenyl)-1,3-heptadiene
[852381-02-3]
$C_{20}H_{22}O_3$ 310.393
Constit. of the sea grass *Cymodocea nodosa*. Oil. $[\alpha]_D$ -17 (c, 0.1 in CH_2Cl_2). λ_{max} 245 (log ε 2.92); 260 (log ε 2.16); 286 (log ε 2.73) (CH_2Cl_2).

Kontiza, I. *et al.*, *Tetrahedron Lett.*, 2005, **46**, 2845-2847 (*Cymodiene*)

Cymodienol C355

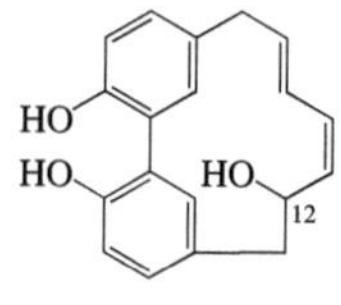

[852381-01-2]
$C_{19}H_{18}O_3$ 294.35
Constit. of the sea grass *Cymodocea nodosa*. Cytotoxic. Greenish oil. $[\alpha]_D$ -181 (c, 0.5 in CH_2Cl_2). λ_{max} 279 (log ε 1.75); 300 (log ε 1.84); 386 (log ε 1.26) (CH_2Cl_2)
12-Deoxy
12-Deoxycymodienol
[1012852-14-0]
$C_{19}H_{18}O_2$ 278.351
Constit. of *Cymodocea nodosa* and *Zostera marina*. Oil. λ_{max} 254 (log ε 0.39); 295 (log ε 0.1) (CH_2Cl_2).

Kontiza, I. *et al.*, *Tetrahedron Lett.*, 2005, **46**, 2845-2847 (*Cymodienol*).
Kontiza, I. *et al.*, *Tetrahedron*, 2008, **64**, 1696-1702 (*12-Deoxycymodienol*).
Li, Y. *et al.*, *Tetrahedron Lett.*, 2019, **60**, 150930:1-3 (*Zostera marina constit*)

Cynarinin C356

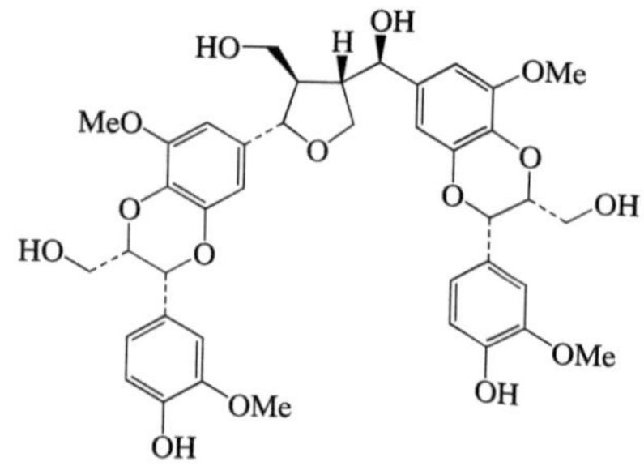

Relative Configuration

[618907-44-1]
$C_{40}H_{44}O_{15}$ 764.777
Bilignan. Constit. of the seeds of *Cynara cardunculus*. Amorph. solid. λ_{max} 242; 280 (EtOH).

Koubaa, I. *et al.*, *Fitoterapia*, 2003, **74**, 18-22 (*Cynarinin*)
$[\alpha]^{23.3}_{D}$

Daglesioside I D1

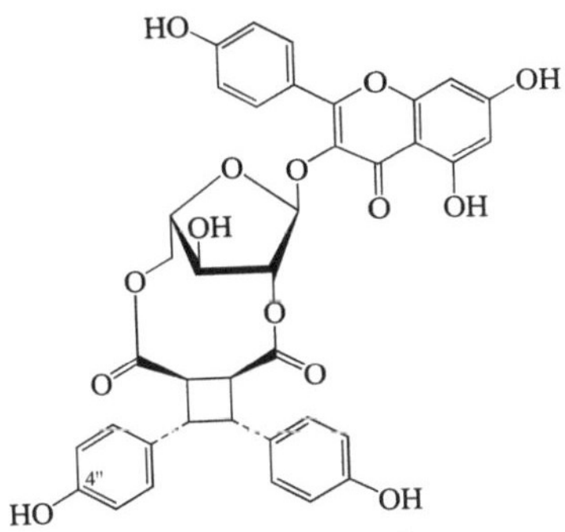

Absolute Configuration

3-[[2,5-O-[[3,4-Bis(4-hydroxyphenyl)-1,2-cyclobutanediyl] dicarbonyl]-α-L-arabinofuranosyl] oxy]-5,7-dihydroxy-2-(4-hydroxyphenyl)-4H-1-benzopyran-4-one, CAS

[606137-06-8]

$C_{38}H_{30}O_{14}$ 710.644

Constit. of the needles of *Pseudotsuga menziesii*. Powder. λ_{max} 267; 295 (sh); 348 (MeOH). [acid] λ_{max} 276; 316 (sh); 349; 399 (MeOH/$AlCl_3$). [base] λ_{max} 274; 335 (sh); 403 (MeOH/NaOH)

4''-Deoxy

Daglesioside II

[606137-67-1]

$C_{38}H_{30}O_{13}$ 694.645

Constit. of the needles of *Pseudotsuga menziesii*. Powder. λ_{max} 267; 293 (sh); 354 (MeOH). [acid] λ_{max} 276; 316 (sh); 347; 403 (MeOH/$AlCl_3$). [base] λ_{max} 274; 336 (sh); 400 (MeOH/NaOH).

Krauze-Baranowska, M. *et al.*, *Z. Naturforsch., C*, 2013, **68**, 87-96 (*Daglesiosides I,II*)

Danshensuan C D2

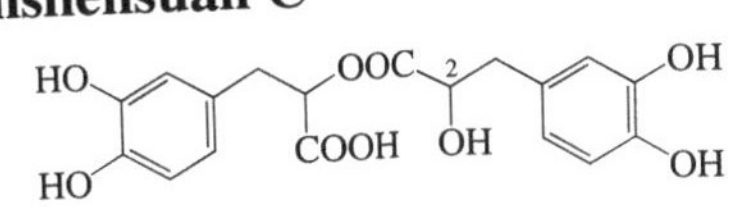

[80944-40-7]

$C_{18}H_{18}O_9$ 378.333

Constit. of *Salvia miltiorrhiza*

2-O-(3,4-Dihydroxy-E-cinnamoyl), Me ester

Salpalaestinin

[1562218-05-6]

$C_{28}H_{26}O_{12}$ 554.504

Constit. of *Salvia palaestina*. Potent antioxidant. Amorph. brown powder. $[\alpha]_D^{25}$ +1.9 (c, 0.0004 in MeOH). λ_{max} 290; 350 (MeOH).

Chen, Z. *et al.*, *Yaoxue Tongbao*, 1981, **16**, 24-25 (*Danshensuan C*).
Al-Qudah, M.A. *et al.*, *Phytochemistry*, 2014, **99**, 115-120 (*Salpalaestinin, activity*)

Daphnauranin B D3

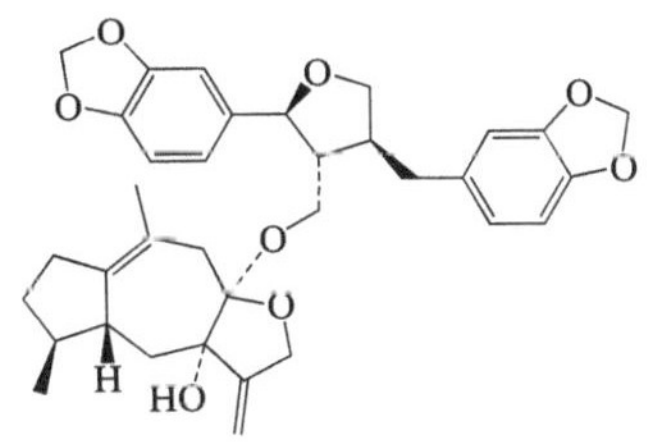

[2425647-44-3]

$C_{35}H_{40}O_8$ 588.697

Constit. of *Daphne aurantiaca*. Oil. $[\alpha]_D^{18}$ -43.3 (c, 0.091 in MeOH). λ_{max} 236 (log ε 3.98); 287 (log ε 3.95) (MeOH).

Huang, S.-Z. *et al.*, *Fitoterapia*, 2017, **122**, 11-15 (*Daphnauranin B*)

Daphnenin D4

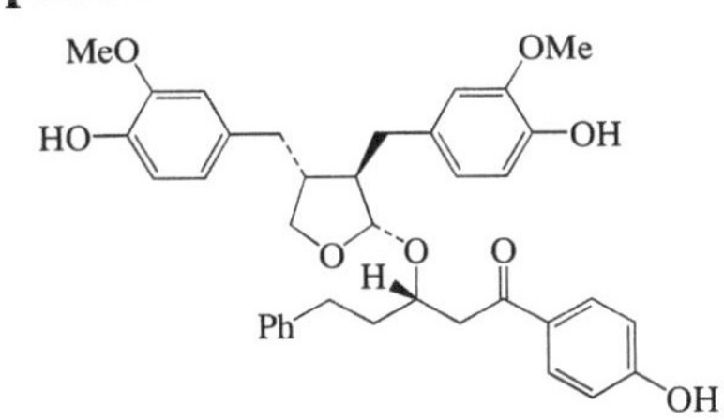

[1395922-31-2]

$C_{37}H_{40}O_8$ 612.719

Abs. config. is known only of the acyclic chiral centre. The other 3 centres are relative only. Constit. of the stems of *Daphne acutiloba*. Shows anti-HIV activity. Amorph. powder. $[\alpha]_D^{24}$ +71.6 (c, 3.4 in MeOH). λ_{max} 281 (log ε 4.2) (MeOH).

Huang, S.Z. *et al.*, *Planta Med.*, 2012, **78**, 182-185 (*Daphnenin, activity*)

Deacylinteriorin D5

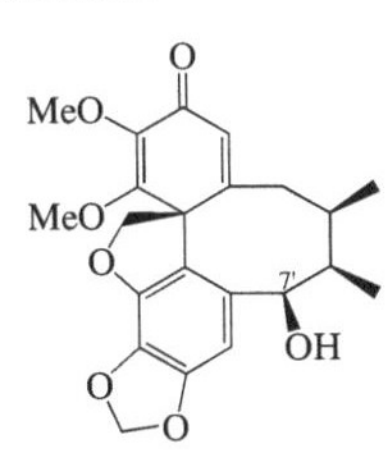

Absolute Configuration

7'-Hydroxy-3,4-dimethoxy-4',5'-methylenedioxy-2,2'-cyclo-2a-homo,3'-epoxy-8,8'-lignan-5(2H)-one

$C_{22}H_{24}O_7$ 400.427

Stereochem. of derivs. not fully assigned except for Kadsutherin D

Ac

Interiorin C

[133360-39-1]

$C_{24}H_{26}O_8$ 442.464

Constit. of *Kadsura interior*. Stereochem. not determined

2-Methylpropenoyl

Kadsutherin D

[1173933-01-1]

$C_{26}H_{28}O_8$ 468.502

Constit. of the stems of *Kadsura* spp.. Powder (EtOAc/petrol). $[\alpha]_D^{22}$ -89.3 (c, 0.03 in MeOH)

Angeloyl

Interiorin

[119139-55-8]

$C_{27}H_{30}O_8$ 482.529

Isol. from aerial parts of *Kadsura heteroclita* and *Kadsura interior*. Cryst. (Et_2O). Mp 157 - 159°. $[\alpha]_D^{20}$ -50 (c, 1.0 in MeOH). Rel. stereochem.

Benzoyl

Interiorin D

[133360-40-4]

$C_{29}H_{28}O_8$ 504.535

Constit. of *Kadsura interior*. Stereochem. not determined

Stereoisomer, angeloyl

Interiorin B. *Isointeriorin*

[133442-66-7]

$C_{27}H_{30}O_8$ 482.529

Constit. of *Kadsura interior*. Stereochem. not determined.

Shide, L. *et al.*, *Planta Med.*, 1988, **54**, 440-443 (*Interiorin, cryst struct*).
Ding, Z. *et al.*, *Huaxue Xuebao*, 1990, **48**, 1075-1079 (*Isointeriorin, Interiorin, Interiorins C,D*).
Chen, D.-F. *et al.*, *Phytochemistry*, 1992, **31**, 629-632 (*Kadsura heteroclita constit*).
Lu, Y. *et al.*, *Nat. Prod. Res.*, 2008, **22**, 1344-1349 (*Kadsutherin D*)

Deacylkadsulignan C D6

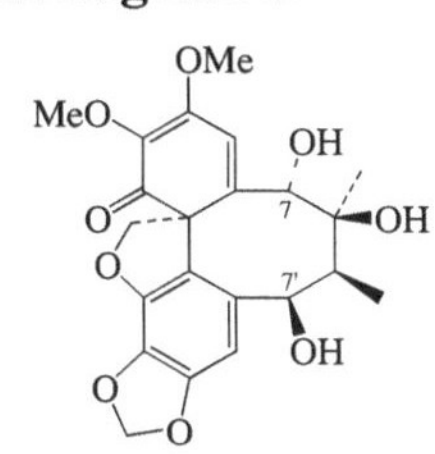

Absolute Configuration

7,7',8-Trihydroxy-4,5-dimethoxy-4',5'-methylenedioxy-2,2'-cyclo-2a-homo-3'-epoxy-8,8'-lignan-3(2H)-one

$C_{22}H_{24}O_9$ 432.425

7-Angeloyl, 7'-Ac

Kadsuphilol M

[1141889-66-8]

$C_{29}H_{32}O_{11}$ 556.564

Constit. of *Kadsura philippinensis*. Amorph. yellow powder. $[\alpha]_D^{25}$ -4.6 (c, 0.6 in CH_2Cl_2). λ_{max} 222 (log ε 4.68); 253 (log ε 4.1); 283 (log ε 3.53) (MeOH)

7,7′-Diangeloyl

Kadsulignan D

[137637-50-4]

$C_{32}H_{36}O_{11}$ 596.629

Isol. from *Kadsura longipedunculata*. Active against leukaemia P388. Mp 160 - 161°. $[\alpha]_D^{10}$ -1.2 (c, 2.14 in $CHCl_3$). Rel. stereochem.

7-Benzoyl, 7′-Ac

Heteroclitin J‡

[936620-28-9]

$C_{31}H_{30}O_{11}$ 578.57

Constit. of the stems of *Kadsura heteroclita*. Yellow gum. $[\alpha]_D^{20}$ -60.5 (c, 0.16 in MeOH). λ_{max} 220 (log ε 4.21); 281 (log ε 3.12); 326 (log ε 3.15) (MeOH)

7-Benzoyl, 7′-angeloyl

***Kadsuphilol E**. Heteroclitin I‡*

[936620-27-8]

$C_{34}H_{34}O_{11}$ 618.635

Constit. of *Kadsura philippinensis* and the stems of *Kadsura heteroclita*. Yellow needles. Mp 192 - 194° (Heteroclitin I). $[\alpha]_D^{20}$ -122.3 (c, 1.1 in MeOH) (Heteroclitin I). $[\alpha]_D^{25}$ -120 (c, 0.01 in CH_2Cl_2) (Kadsuphilol E). λ_{max} 219 (log ε 4.75); 275 (sh); 325 (log ε 3.41) (MeOH) (Heteroclitin I)

7′-Benzoyl, 7-Ac

Kadsulignan C

[137637-49-1]

$C_{31}H_{30}O_{11}$ 578.57

Isol. from the root bark of *Kadsura longipedunculata*. Active against leukaemia P388. Mp 244 - 246°. $[\alpha]_D^{17}$ -86.3 (c, 0.45 in $CHCl_3$). Rel. stereochem.

7′-Benzoyl, 7-angeloyl

***Kadsuphilol F**. Heteroclitin L*

[936620-30-3]

$C_{34}H_{34}O_{11}$ 618.635

Constit. of *Kadsura philippinensis* and the stems of *Kadsura heteroclita*. Yellow needles. Mp 128 - 130° (Heteroclitin L). $[\alpha]_D^{25}$ -29 (c, 0.01 in CH_2Cl_2) (Kadsuphilol F). $[\alpha]_D^{20}$ -28 (c, 1 in MeOH) (Heteroclitin L). λ_{max} 219 (log ε 3.47) (MeOH) (Kadsuphilol F). [neutral] λ_{max} 221 (log ε 4.65); 278 (log ε 3.11); 327 (log ε 3.24) (MeOH) (Heteroclitin L)

7,7′-Dibenzoyl

Heteroclitin K

[936620-29-0]

$C_{36}H_{32}O_{11}$ 640.641

Constit. of the stems of *Kadsura heteroclita*. Yellow needles. Mp 195 - 198°. $[\alpha]_D^{20}$ -161.7 (c, 1 in MeOH). λ_{max} 220 (log ε 4.56); 280 (log ε 3.21); 325 (log ε 3.29) (MeOH)

7′-E-Cinnamoyl, 7-Ac

***Kadsulignan P**. Kadsuphilol U*

[1275533-61-3]

$C_{33}H_{32}O_{11}$ 604.608

Constit. of the stems of *Kadsura oblongifolia* and *Kadsura philippinensis*. Yellowish needles. $[\alpha]_D^{20}$ -30 (c, 0.2 in MeOH). $[\alpha]_D^{25}$ -48 (c, 0.2 in MeOH). λ_{max} 221 (log ε 3.35); 271 (log ε 3.06) (MeOH). [neutral] λ_{max} 223 (log ε 4.02); 285 (log ε 3.85); 335 (log ε 3.69) (MeOH)

7-Deoxy, 7′-benzoyl

Longipedlignan G

[2229860-17-5]

$C_{29}H_{28}O_9$ 520.534

Constit. of the roots of *Kadsura longipedunculata*. Amorph. yellow powder. $[\alpha]_D^{20}$ -16 (c, 0.1 in MeOH). λ_{max} 220 (log ε 1.69); 320 (log ε 0.95) (MeOH)

7-Deoxy, 7′-E-cinnamoyl

Longipedlignan N

[2408242-34-0]

$C_{31}H_{30}O_9$ 546.572

Constit. of the roots of *Kadsura longipedunculata*. Amorph. yellow powder. $[\alpha]_D^{20}$ -19 (c, 0.1 in MeOH). λ_{max} 220 (log ε 1.73); 281 (log ε 1.22) (MeOH)

8-Epimer, 7-deoxy, 7′-benzoyl

Longipedlignan F

[2229859-82-7]

$C_{29}H_{28}O_9$ 520.534

Constit. of the roots of *Kadsura longipedunculata*. Exhibits moderate hepatoprotective activity. Amorph. yellow powder. $[\alpha]_D^{20}$ +46 (c, 0.1 in MeOH). λ_{max} 221 (log ε 4.64); 325 (log ε 3.43) (MeOH)

8-Epimer, 7-deoxy, 7′-E-cinnamoyl

Longipedlignan M

[2377873-31-7]

$C_{31}H_{30}O_9$ 546.572

Constit. of the roots of *Kadsura longipedunculata*. NO prodn. inhibitor. Amorph. yellow powder. $[\alpha]_D^{20}$ -5.6 (c, 0.4 in MeOH). λ_{max} 219 (log ε 1.75); 280 (log ε 1.23) (MeOH)

8′-Epimer, 8-deoxy, 7′-Ac

Longipedlignan H

[2229859-83-8]

$C_{24}H_{26}O_9$ 458.463

Constit. of the roots of *Kadsura longipedunculata*. Amorph. yellow powder. $[\alpha]_D^{20}$ +35 (c, 0.1 in MeOH). λ_{max} 220 (log ε 1.95); 275 (log ε 0.56); 320 (log ε 0.58) (MeOH)

8′-Epimer, 8-deoxy, 7′-O-(2-methylpropanoyl)

Longipedlignan I

[2229859-84-9]

$C_{26}H_{30}O_9$ 486.517

Constit. of the roots of *Kadsura longipedunculata*. Amorph. yellow powder. $[\alpha]_D^{20}$ +23 (c, 0.1 in MeOH). λ_{max} 219 (log ε 1.37); 327 (log ε 0.14) (MeOH)

8′-Epimer, 8-deoxy, 7′-E-cinnamoyl

Longipedlignan J

[2229859-85-0]

$C_{31}H_{30}O_9$ 546.572

Constit. of the roots of *Kadsura longipedunculata*. Amorph. yellow powder. $[\alpha]_D^{20}$ +6.2 (c, 0.5 in MeOH). λ_{max} 218 (log ε 1.55); 271 (log ε 1.12) (MeOH).

Liu, J.-S. *et al.*, *Can. J. Chem.*, 1991, **69**, 1403-1407 (*Kadsulignans C,D*).

Xu, L.-J. *et al.*, *Heterocycles*, 2007, **71**, 941-947 (*Heteroclitins I-L*).

Shen, Y.-C. *et al.*, *J. Nat. Prod.*, 2007, **70**, 1139-1145 (*Kadsuphilols E,F*).

Shen, Y.-C. *et al.*, *Phytochemistry*, 2009, **70**, 114-120 (*Kadsuphilol M, Heteroclitin J*).

Liu, H. *et al.*, *Fitoterapia*, 2011, **82**, 731-734 (*Kadsulignan P*).

Cheng, Y.-B. *et al.*, *Helv. Chim. Acta*, 2011, **94**, 148-158 (*Kadsuphilol U*).

Liu, J. *et al.*, *J. Nat. Prod.*, 2018, **81**, 846-857 (*Longipedlignans F-J, activity*).

Liu, J. *et al.*, *J. Nat. Prod.*, 2019, **82**, 2842-2851 (*Longipedlignans M,N, cd, activity*)

Deacylkadsulignan E D7

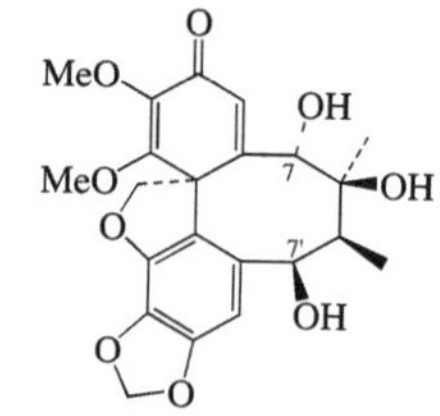

Absolute Configuration

7,7′,8-Trihydroxy-3,4-dimethoxy-4′,5′-methylenedioxy-2,2′-cyclo-2a-homo,3′-epoxy-8,8′-lignan-5(2H)-one
$C_{22}H_{24}O_9$ 432.425
7-Angeloyl, 7′-propanoyl
Heteroclitalignan C
[911052-16-9]
$C_{30}H_{34}O_{11}$ 570.591
Constit. of the stems of *Kadsura heteroclita*. Powder (MeOH). Mp 138 - 140°. $[\alpha]_D^{20}$ -10 (c, 0.11 in MeOH). λ_{max} 220 (log ε 4.75); 250 (sh) (MeOH)
7-Angeloyl, 7′-Ac
Kadsulignan F
[142674-81-5]
$C_{29}H_{32}O_{11}$ 556.564
Constit. of the root bark of *Kadsura longipedunculata*. Cryst. ($CHCl_3$/petrol). Mp 223 - 225°. $[\alpha]_D^{20}$ -8.1 (c, 1.52 in $CHCl_3$). λ_{max} 226 (log ε 4.5); 248 (sh); 306 (log ε 3.66) (MeOH)
7,7′-Diangeloyl
Kadsulignan G
[142674-82-6]
$C_{32}H_{36}O_{11}$ 596.629
Constit. of the root bark of *Kadsura longipedunculata*. Cryst. ($CHCl_3$/petrol). Mp 269 - 271°. $[\alpha]_D^{16}$ -36.1 (c, 1.02 in $CHCl_3$). λ_{max} 228 (log ε 4.52); 309 (log ε 3.53) (MeOH)
7′-Benzoyl, 7-Ac
Kadsulignan E
[142674-80-4]
$C_{31}H_{30}O_{11}$ 578.57
Constit. of the root bark of *Kadsura longipedunculata*. Cryst. ($CHCl_3$/petrol). Mp 281 - 283°. $[\alpha]_D^{15}$ -199.7 (c, 0.32 in $CHCl_3$). λ_{max} 221 (log ε 4.73); 280 (log ε 3.65); 310 (log ε 3.65) (MeOH)
7′-E-Cinnamoyl, 7-Ac
Kadsulignan O‡. *Kadsuphilol V*
[1275533-63-5]
$C_{33}H_{32}O_{11}$ 604.608
Constit. of the stems of *Kadsura oblongifolia* and *Kadsura philippinensis*. Yellowish needles. $[\alpha]_D^{20}$ -91.7 (c, 0.63 in MeOH). $[\alpha]_D^{25}$ -31 (c, 0.1 in MeOH). λ_{max} 221 (log ε 3.61); 283 (log ε 3.19) (MeOH). [neutral] λ_{max} 225 (log ε 3.94); 288 (log ε 3.63); 297 (log ε 3.28) (MeOH).

Liu, J.-S. *et al.*, *Phytochemistry*, 1992, **31**, 957-960 (*Kadsulignans E,F,G*).
Wang, W. *et al.*, *Planta Med.*, 2006, **72**, 284-288 (*Heteroclitalignan C*).
Liu, H. *et al.*, *Fitoterapia*, 2011, **82**, 731-734 (*Kadsulignan O*).
Cheng, Y.-B. *et al.*, *Helv. Chim. Acta*, 2011, **94**, 148-158 (*Kadsuphilol V*)

Deacyltaiwanschirin D8

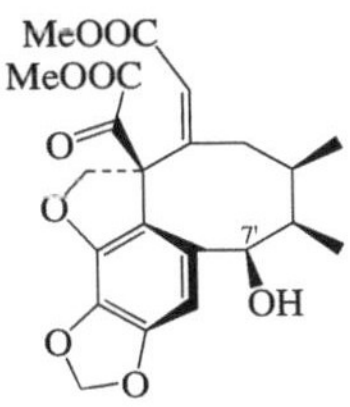

Deangeloylheteroclitin F
$C_{22}H_{24}O_9$ 432.425
4,5-Seco-homo-2,2′-cyclolignans. Abs. configs. not fully determined but almost certainly as shown
7′-Ac
Taiwanschirin B
[248257-21-8]
$C_{24}H_{26}O_{10}$ 474.462
Constit. of the stems of *Schizandra arisanensis* and *Kadsura japonica*. Prisms. Mp 190 - 193°. $[\alpha]_D$ +344 (c, 0.1 in $CHCl_3$)
7′-O-(2-Methylbutanoyl)
Taiwanschirin A
[248257-20-7]
$C_{27}H_{32}O_{10}$ 516.543
Constit. of the stems of *Schizandra arisanensis* and *Kadsura japonica*. Prisms. Mp 147 - 150°. $[\alpha]_D$ +415 (c, 0.1 in $CHCl_3$)
7′-Angeloyl
Heteroclitin F
[144049-67-2]
$C_{27}H_{30}O_{10}$ 514.527
Constit. of the stems of *Kadsura heteroclita* and *Kadsura japonica*. Powder. Mp 87 - 89°. $[\alpha]_D^{25}$ +261.2 (c, 0.33 in MeOH). λ_{max} 216 (ε 35800) (MeOH)
7′-Hexanoyl
Taiwanschirin D
[321992-41-0]
$C_{28}H_{34}O_{10}$ 530.57
Constit. of the stems of *Kadsura matsudai*. Cytotoxic agent against human hepatoma hepa-3B cells. Light yellow powder. Mp 47 - 50°. $[\alpha]_D$ +315 (c, 0.3 in CH_2Cl_2)
7′-Benzoyl
Taiwanschirin C
[248257-22-9]
$C_{29}H_{28}O_{10}$ 536.533
Constit. of the stems of *Schizandra arisanensis* and *Kadsura japonica*. Cytotoxic agent against hepatoma hepa-3B cells. Prisms. Mp 86 - 89°. $[\alpha]_D$ +128 (c, 0.1 in $CHCl_3$).

Yang, X.-W. *et al.*, *Chem. Pharm. Bull.*, 1992, **40**, 1510-1516 (*Heteroclitin F*).
Kuo, Y.-H. *et al.*, *J. Org. Chem.*, 1999, **64**, 7023-7027 (*Taiwanschirins A,B,C*).
Li, S.-Y. *et al.*, *Chem. Pharm. Bull.*, 2000, **48**, 1992-1993 (*Taiwanschirin D*).
Kuo, Y.-H. *et al.*, *Planta Med.*, 2005, **71**, 646-653 (*Kadsura japonica constits*)

Deangeloylinteriotherin D D9

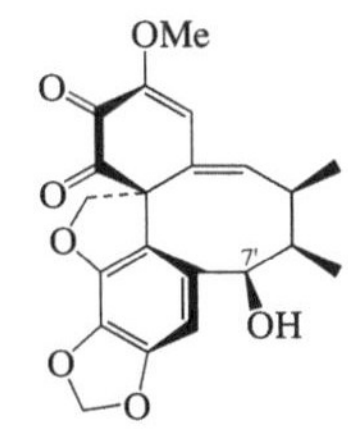

Absolute Configuration

$C_{21}H_{20}O_7$ 384.384
7′-Angeloyl
Interiotherin D
[460090-66-8]
$C_{26}H_{26}O_8$ 466.486
Constit. of the stems of *Kadsura interior*. Yellow prisms (MeOH). Mp 148 - 151°. $[\alpha]_D$ -271.2 (c, 1.18 in $CHCl_3$). λ_{max} 214 (log ε 5.73); 240 (sh); 379 (sh) (MeOH).

Chen, D.-F. *et al.*, *J. Nat. Prod.*, 2002, **65**, 1242-1245 (*Interiotherin D*)

Debenzoylkadsutherin C D10

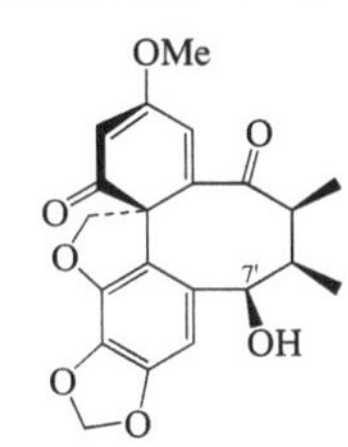

Absolute Configuration

$C_{21}H_{20}O_7$ 384.384
7′-Benzoyl
Kadsutherin C
[903573-42-2]
$C_{28}H_{24}O_8$ 488.492
Constit. of *Kadsura* spp.. Yellow powder. $[\alpha]_D^{25}$ -50.3 (c, 0.04 in MeOH). λ_{max} 216 (log ε 4.28); 374 (log ε 2.88) (MeOH).

Bai, H. *et al.*, *Helv. Chim. Acta*, 2006, **89**, 884-894 (*Kadsutherin C*)

Decarboxyflavipesolide C D11

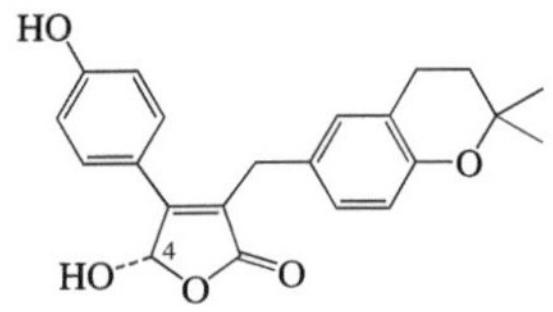

$C_{22}H_{22}O_5$ 366.413

(S)-form

4-Me ether

Decarboxy-4-O-methylflavipesolide C

[2190517-51-0]

$C_{23}H_{24}O_5$ 380.440

Prod. by a marine-derived *Aspergillus terreus* OUCMDZ-2739. Brown-green oil. $[\alpha]^{25}_{D}$ -18 (c, 0.1 in CH_2Cl_2). λ_{max} 221 (log ε 4.05); 247 (log ε 3.33); 285 (log ε 3.76) (MeOH).

Sun, K. *et al.*, *Tetrahedron*, 2018, **74**, 83-87 (*Decarboxy-4-O-methylflavipesolide C*)

De(methoxycarbonyl)aspernolide A D12

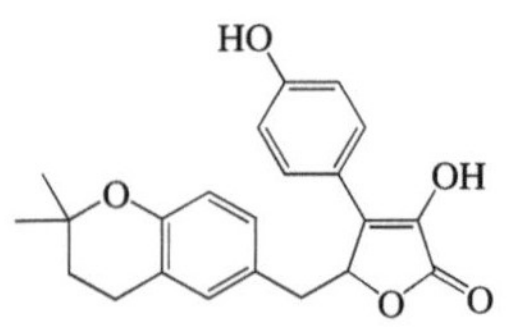

5-[(3,4-Dihydro-2,2-dimethyl-2H-1-benzopyran-6-yl)methyl]-3-hydroxy-4-(4-hydroxyphenyl)-2(5H)-furanone, CAS [65372-45-4]

$C_{22}H_{22}O_5$ 366.413

Related to Butyrolactone V, B92. Prod. by a marine-derived *Aspergillus terreus* PT06-2. α-Glucosidase inhibitor. Mp 229 - 231°. λ_{max} 223 (log ε 4.22); 291 (log ε 4.27); 305 (log ε 4.31) (EtOH).

Kiriyama, N. *et al.*, *Chem. Pharm. Bull.*, 1977, **25**, 2593-2601 (*synth*).
Wang, Y. *et al.*, *Mar. Drugs*, 2011, **9**, 1368-1378 (*Aspergillus terreus metab*).
Sun, Y. *et al.*, *Bioorg. Med. Chem. Lett.*, 2018, **28**, 315-318 (*isol, activity*)

O-Demethylcentrolobin D13

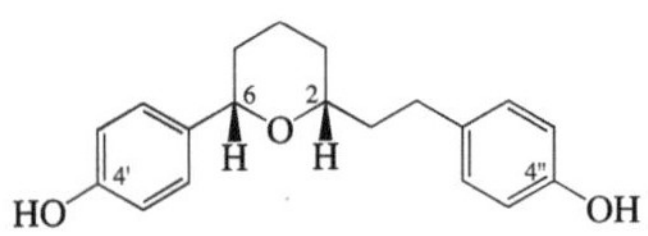

(2*R*,6*S*)-*form*

4-[Tetrahydro-6-[2-(4-hydroxyphenyl)ethyl]-2H-pyran-2-yl]phenol

$C_{19}H_{22}O_3$ 298.382

(2R,6S)-form

(-)-De-O-methylcentrolobin

[30359-00-3]

Isol. from heartwood of *Centrolobium paraense*, *Centrolobium robustum* and *Centrolobium tomentosum*. Needles (Me_2CO/C_6H_6). Mp 246 - 248° (183°). $[\alpha]^{25}_{D}$ -95.1 (c, 0.9 in MeOH)

4'-Me ether

4-[2-[Tetrahydro-6-(4-methoxyphenyl)-2H-pyran-2-yl]ethyl]phenol. (-)-Centrolobin. (-)-Centrolobine

[30358-99-7]

$C_{20}H_{24}O_3$ 312.409

Constit. of *Brosimum potabile*, *Centrolobium paraense*, *Centrolobium robustum* and *Centrolobium tomentosum*. $[\alpha]^{25}_{D}$ -92.2 (c, 1 in $CHCl_3$)

4,5-Didehydro

[227289-51-2]

$C_{19}H_{20}O_3$ 296.366

Constit. of *Alpinia blepharocalyx*. Needles ($CHCl_3$/MeOH). Mp 150 - 152°. $[\alpha]^{25}_{D}$ -135.5 (c, 0.2 in MeOH). Possesses (2*S*,6*S*)-config. owing to change in *R*,*S*-priorities. Incorrect stereochem. shown in 1999 ref.

(2S,6R)-form

(+)-De-O-methylcentrolobin

[30359-03-6]

Constit. of the wood of *Centrolobium tomentosum*. Needles (MeOH/C_6H_6). Mp 246 - 248° (183°). $[\alpha]^{25}_{D}$ +99.2 (c, 1 in MeOH)

4'-Me ether

(+)-Centrolobin. (+)-Centrolobine

[30359-02-5]

$C_{20}H_{24}O_3$ 312.409

Constit. of the wood of *Centrolobium robustum* and *Centrolobium tomentosum*. Antibacterial agent. Cryst.. Mp 87°. $[\alpha]^{20}_{D}$ +97 (c, 0.4 in $CHCl_3$)

4R,5S-Dihydroxy

4R,5S-Dihydroxy-O-demethylcentrolobin

[344362-68-1]

$C_{19}H_{22}O_5$ 330.380

Constit. of the seeds of *Alpinia blepharocalyx*. Amorph. yellow solid

4S,5R-Dihydroxy

4S,5R-Dihydroxy-O-demethylcentrolobin

[344362-67-0]

$C_{19}H_{22}O_5$ 330.380

Constit. of the seeds of *Alpinia blepharocalyx*. Amorph. yellow solid

4S,5S-Dihydroxy

4S,5S-Dihydroxy-O-demethylcentrolobin

[344362-66-9]

$C_{19}H_{22}O_5$ 330.380

Constit. of the seeds of *Alpinia blepharocalyx*. Amorph. yellow solid. $[\alpha]^{25}_{D}$ +28.5 (c, 0.04 in MeOH)

(2S,6S)-form

4,5-Didehydro

5,6-Didehydrode-O-methylcentrolobin

[344362-65-8]

$C_{19}H_{20}O_3$ 296.366

Constit. of the seeds of *Alpinia blepharocalyx*. Amorph. powder. $[\alpha]^{25}_{D}$ -12.3 (c, 0.33 in MeOH). Most prob. config. is (2*R*,6*S*) consequent of a change in *R*,*S*-priorities (see Rogano et al 2010 paper)

4',4''-Dideoxy, 4S,5R-dihydroxy

Tetrahydro-2-phenyl-6-(2-phenylethyl)-2H-pyran-3,4-diol. ***4,5-Dihydroxy-4'-demethoxy-4''-deoxycentrolobin***

$C_{19}H_{22}O_3$ 298.382

Constit. of the seeds of *Alpinia katsumadai*. Amorph. light yellow solid. $[\alpha]^{25}_{D}$ -30.2 (c, 0.1 in MeOH). λ_{max} 207 (log ε 3.63); 258 (log ε 0.28) (MeOH)

(2RS,6SR)-form

4'-Me ether

(±)-Centrolobin. (±)-Centrolobine

[1484-91-9]

Oil. Bp 190° at 0.03 mm.

De Albuquerque, I.L. *et al.*, *Gazz. Chim. Ital.*, 1964, **94**, 287-295 (*(+)-Centrolobin*).
Galeffi, C. *et al.*, *Gazz. Chim. Ital.*, 1965, **95**, 95-100 (*(±)-Centrolobin, synth*).
Craveiro, A.A. *et al.*, *Phytochemistry*, 1970, **9**, 1869-1875 (*Centrolobium robustum constits, Centrolobium tomentosum constits*).
Jurd, L. *et al.*, *Aust. J. Chem.*, 1984, **37**, 1127-1133 (*(-)-De-O-methylcentrolobin*).
Nagal, M. *et al.*, *Chem. Pharm. Bull.*, 1986, **34**, 1056-1060 (*(-)-Centrolobin, abs config*).
Araujo, C.A.C. *et al.*, *Phytochemistry*, 1998, **49**, 751-754 (*Demethylcentrolobin*).
Prasain, J.K. *et al.*, *Planta Med.*, 1999, **65**, 196 (*Didehydro-O-demethylcentrolobin*).
Alcântara, A.F. de C. *et al.*, *Fitoterapia*, 2000, **71**, 613-615 (*Brosimum potabile constit*).
Prasain, J.K. *et al.*, *Planta Med.*, 2000, **66**, 590 (*Didehydro-O-demethylcentrolobin, erratum*).
Ali, M.S. *et al.*, *J. Nat. Prod.*, 2001, **64**, 491-496 (*4,5-Dihydroxy-O-demethylcentrolobins*).

Colobert, F. *et al.*, *Org. Lett.*, 2002, **4**, 1723-1725 (*(-)-Centrolobin, synth*).
Marumoto, S. *et al.*, *Org. Lett.*, 2002, **4**, 3919-3922 (*(-)-Centrolobin, synth*).
Evans, P.A. *et al.*, *Org. Lett.*, 2003, **5**, 3883-3885 (*(-)-Centrolobin, synth*).
Lee, E. *et al.*, *Bull. Korean Chem. Soc.*, 2004, **25**, 1609-1610 (*(-)-Centrolobin, synth*).
Boulard, L. *et al.*, *Tetrahedron Lett.*, 2004, **45**, 6603-6605 (*(-)-Centrolobin, synth*).
Clarke, P.A. *et al.*, *Tetrahedron*, 2005, **61**, 5433-5438 (*(±)-Centrolobin, synth*).
Jennings, M.P. *et al.*, *Tetrahedron Lett.*, 2005, **46**, 2021-2024 (*(-)-Centrolobin, synth*).
Chandrasekhar, S. *et al.*, *Tetrahedron Lett.*, 2005, **46**, 6651-6653 (*(-)-Centrolobin, synth*).
Prasad, K.R. *et al.*, *Tetrahedron*, 2007, **63**, 1089-1092 (*(-)-Centrolobin, synth*).
Takeuchi, T. *et al.*, *Tetrahedron Lett.*, 2008, **49**, 6462-6465 (*(-)-Centrolobin, synth*).
Zhou, H. *et al.*, *Tetrahedron Lett.*, 2009, **50**, 4368-4371 (*(±)-Centrolobin, synth*).
Rogano, F. *et al.*, *Helv. Chim. Acta*, 2010, **93**, 1281-1298 (*(+)-Centrolobin, (-)-Centrolobin, synth, abs config*).
Rogano, F. *et al.*, *Helv. Chim. Acta*, 2010, **93**, 1299-1312 (*Centrolobin regioisomers, synth, abs config*).
Mohapatra, D.K. *et al.*, *Heterocycles*, 2010, **80**, 219-227 (*(-)-Centrolobin, synth*).
Fuwa, H. *et al.*, *Heterocycles*, 2010, **82**, 641-647 (*(±)-Centrolobin, synth*).
Chaladaj, W. *et al.*, *J. Org. Chem.*, 2010, **75**, 1740-1743 (*(-)-Centrolobin, synth*).
Reddy, C.R. *et al.*, *Tetrahedron: Asymmetry*, 2010, **21**, 103-105 (*(+)-Centrolobin, synth*).
Fujioka, H. *et al.*, *Angew. Chem., Int. Ed.*, 2011, **50**, 12232-12235 (*(+)-Centrolobin, synth*).
Jeong, Y. *et al.*, *Org. Biomol. Chem.*, 2011, **9**, 374-378 (*(±)-Centrolobin, synth*).
Xie, J.-H. *et al.*, *Org. Lett.*, 2012, **14**, 4758-4761 (*(-)-Centrolobin, synth*).
Sudarshan, K. *et al.*, *Eur. J. Org. Chem.*, 2013, 2298-2302 (*(+)-Centrolobin, synth*).
Zeng, J. *et al.*, *Chem. - Eur. J.*, 2014, **20**, 405-409 (*(±)-Centrolobin, synth*).
Latif, M. *et al.*, *J. Org. Chem.*, 2015, **80**, 3315-3320 (*(-)-Centrolobin, synth*).
Nagarjuna, B. *et al.*, *Tetrahedron Lett.*, 2015, **56**, 4916-4918 (*Centrolobin diastereoisomers, synth*).
Gharpure, S.J. *et al.*, *Org. Lett.*, 2017, **19**, 6534-6537 (*(±)-Centrolobin, synth*).
Yang, C.-S. *et al.*, *Phytochem. Lett.*, 2019, **30**, 190-193 (*Alpinia katsumadai constit*).
Schmidt, J.P. *et al.*, *Angew. Chem., Int. Ed.*, 2020, **59**, 23485-23490 (*(-)-Centrolobin, synth*)

3′-*O*-Demethyldenudanolide B D14

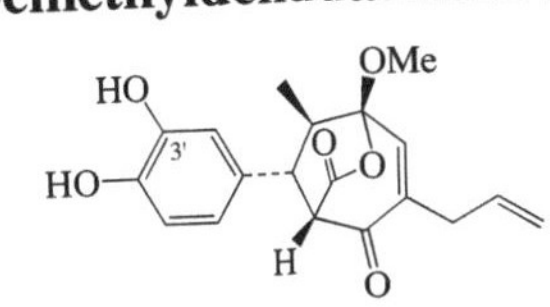

Absolute Configuration

$C_{19}H_{20}O_6$ 344.363
Parent not known
3′-Me ether
Denudanolide B
[288259-82-5]
$C_{20}H_{22}O_6$ 358.39
Constit. of *Magnolia denudata*
3′,4′-Di-Me ether
Denudanolide C
[288259-86-9]
$C_{21}H_{24}O_6$ 372.417
Constit. of *Magnolia denudata*
3′,4′-Methylene ether
Denudanolide A
[288259-72-3]
$C_{20}H_{20}O_6$ 356.374
Constit. of *Magnolia denudata.* Exhibits moderate antiplatelet-activating props.
5′-Methoxy, 3′,4′-di-Me ether
Denudanolide D
[288259-90-5]
$C_{22}H_{26}O_7$ 402.443
Constit. of *Magnolia denudata.*

Kuroyanagi, M. *et al.*, *Chem. Pharm. Bull.*, 2000, **48**, 832-837 (*Denudanolides A-D, cd, activity*)

De-*O*-methyliryantherin D D15

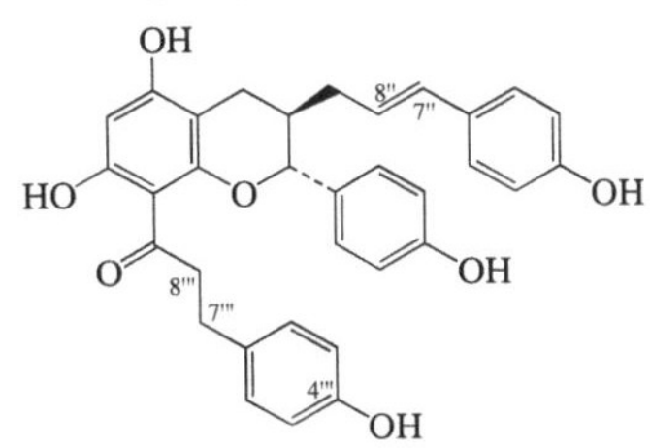

$C_{33}H_{30}O_7$ 538.596
Parent not isol.. All derivs., as racemates, show antibacterial activity against *Staphylococcus aureus*. However, their (2*S*,3*S*)-(-)-enantiomers are significantly more active. This group of compds. have been isol. as racemates, separated by chiral hplc, and had their abs. configs. determined
(2RS,3RS,7″E)-form
4‴-Me ether
Iryantherin D
[132396-74-8]
[1774358-35-8] ((+)-form)
[1774358-39-2] ((-)-form)
[1774358-68-7] ((±)-form)
$C_{34}H_{32}O_7$ 552.623
Constit. of *Iryanthera laevis*, *Iryanthera ulei* and *Populus balsamifera.* Yellow powder. $[\alpha]_D$ -2.4 (c, 0.3 in Me_2CO) (*S,S*-form). λ_{max} 202 (log ε 4.52); 225 (log ε 4.31); 264 (log ε 4.19); 285 (log ε 4.09); 292 (log ε 4.09); 330 (log ε 3.53) (MeOH)
5-Me ether
Balsacone K
[1774358-43-8]
[1774358-36-9] ((+)-form)
[1774358-40-5] ((-)-form)
$C_{34}H_{32}O_7$ 552.623
Constit. of *Populus balsamifera.* Yellow powder. $[\alpha]^{25}_D$-1 (c, 2.5 in Me_2CO) (*S,S*-form). λ_{max} 205 (log ε 4.77); 225 (log ε 4.65); 271 (log ε 4.51); 285 (log ε 4.46); 330 (log ε 3.79) (MeOH)
4‴,5-Di-Me ether
Iryantherin E
[132396-75-9]
$C_{35}H_{34}O_7$ 566.65
Constit. of *Iryanthera laevis* and *Iryanthera ulei.* λ_{max} 269 (ε 4150); 294 (ε 4100) (MeOH)
7‴,8‴-Didehydro(E-), 5-Me ether
Balsacone M
[1774358-38-1]
$C_{34}H_{30}O_7$ 550.607
Constit. of *Populus balsamifera.* Yellow powder. λ_{max} 204 (log ε 4.78); 259 (log ε 4.32); 371 (log ε 4.47) (MeOH)
4‴-Deoxy
Balsacone J
[1774358-42-7]
[1774341-32-0] ((-)-form)
[1774358-34-7] ((+)-form)
$C_{33}H_{30}O_6$ 522.597
Constit. of *Populus balsamifera.* Yellow powder. $[\alpha]^{22}_D$+8.6 (c, 2.5 in Me_2CO) (*R,R*-form). λ_{max} 205 (log ε 4.60); 226 (log ε 4.43); 264 (log ε 4.22); 293 (log ε 4.15); 330 (log ε 3.67) (MeOH)
4‴-Deoxy, 7‴,8‴-didehydro(E-)
Balsacone L
[1774358-44-9]
[1774358-37-0] ((+)-form)
[1774358-41-6] ((-)-form)
$C_{33}H_{28}O_6$ 520.581
Constit. of *Populus balsamifera.* Yellow powder. $[\alpha]^{22}_D$+50.4 (c, 1 in Me_2CO) (*R,R*-form). λ_{max} 204 (log ε 4.65); 262 (log ε 4.32); 310 (log ε 4.24); 349 (log ε 4.11) (MeOH).

Conserva, L.M. *et al.*, *Phytochemistry*, 1990, **29**, 3911-3918 (*Iryantherins D,E*).
Simard, F. *et al.*, *J. Nat. Prod.*, 2015, **78**, 1147-1153 (*Balsacones J-M, Iryantherin D, cd, activity*)

Dendrocandin B D16

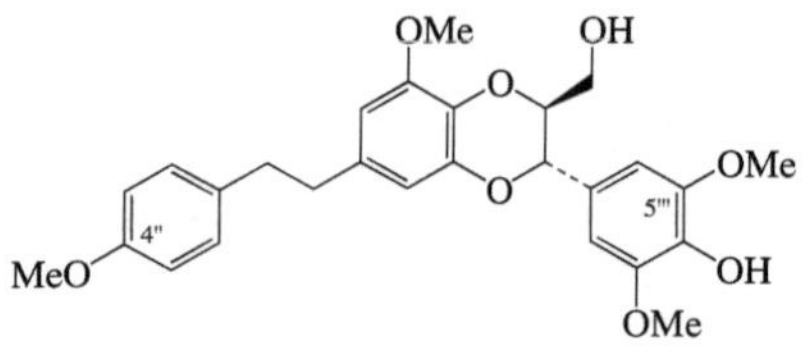

Absolute Configuration

[1104820-02-1]

$C_{27}H_{30}O_8$ 482.529

Stilbenolignan. Related to Aiphanol, A14. Constit. of *Dendrobium candidum* and *Dendrobium officinale*. Moderate neurite outgrowth promotor. Powder. $[\alpha]^{20}_D$ -4.6 (c, 0.11 in $CHCl_3$). λ_{max} 238 (log ε 4.3); 273 (log ε 3.8) ($CHCl_3$)

4″-O-De-Me

Dendrocandin U

[1922084-93-2]

$C_{26}H_{28}O_8$ 468.502

Constit. of the stems of *Dendrobium officinale*. Weak neurite outgrowth inhibitor. Powder. $[\alpha]^{20}_D$ -4.7 (c, 0.22 in MeOH). λ_{max} 209; 275 (MeOH)

3′,4″-Di-O-de-Me

Dendrocandin O

[1820612-31-4]

$C_{25}H_{26}O_8$ 454.475

Constit. of the stems of *Dendrobium candidum*. Yellow syrup. Racemate. λ_{max} 279 (log ε 3.6) (MeOH)

3″-Methoxy, 4″-O-de-Me

Dendrocandin T

[1922084-92-1]

$C_{27}H_{30}O_9$ 498.528

Constit. of the stems of *Dendrobium officinale*. Weak neurite outgrowth inhibitor. Oil. $[\alpha]_D$ -4.4 (c, 0.26 in MeOH). λ_{max} 208; 277 (MeOH)

5‴-Demethoxy

Dendrocandin V

$C_{26}H_{28}O_7$ 452.503

Constit. of *Dendrobium wardianum*. Exhibits cytotoxic activity. Light yellow oil. $[\alpha]^{20}_D$ -14.7 (c, 0.12 in MeOH). λ_{max} 223 (log ε 4.3); 278 (log ε 3.5) (MeOH).

Li, Y. *et al.*, *Chem. Pharm. Bull.*, 2008, **56**, 1477-1479 (*Dendrocandin B*).
Li, Y. *et al.*, *J. Asian Nat. Prod. Res.*, 2014, **16**, 1035-1043 (*Dendrocandin O*).
Yang, L. *et al.*, *J. Asian Nat. Prod. Res.*, 2015, **17**, 125-131 (*Dendrocandins B,T,U, activity*).
Zhang, C. *et al.*, *Fitoterapia*, 2017, **122**, 76-79 (*Dendrocandin V, activity*)

Dendrocandin M D17

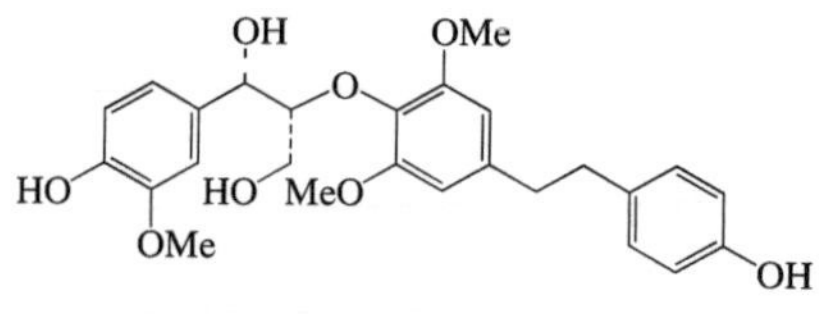

Absolute Configuration

[1820612-29-0]

$C_{26}H_{30}O_8$ 470.518

Stilbenolignan. Constit. of the stems of *Dendrobium candidum*. Yellow syrup. $[\alpha]^{20}_D$ -18.9 (c, 0.1 in MeOH). λ_{max} 279 (log ε 3.6) (MeOH).

Li, Y. *et al.*, *J. Asian Nat. Prod. Res.*, 2014, **16**, 1035-1043 (*Dendrocandin M*)

Dendrocandin N D18

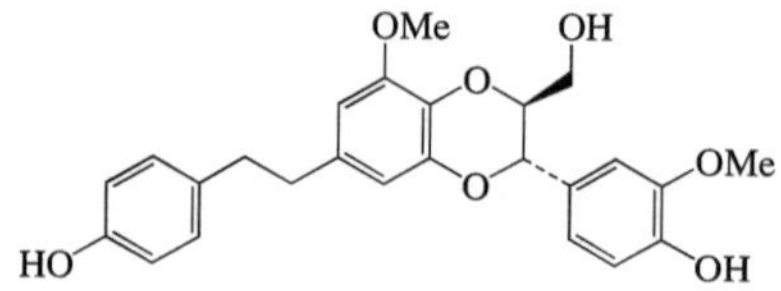

[1820612-30-3]

$C_{25}H_{26}O_7$ 438.476

Constit. of the stems of *Dendrobium candidum*. Yellow syrup. Racemate. λ_{max} 279 (log ε 3.7) (MeOH).

Li, Y. *et al.*, *J. Asian Nat. Prod. Res.*, 2014, **16**, 1035-1043 (*Dendrocandin N*)

Dendrocandin P_1 D19

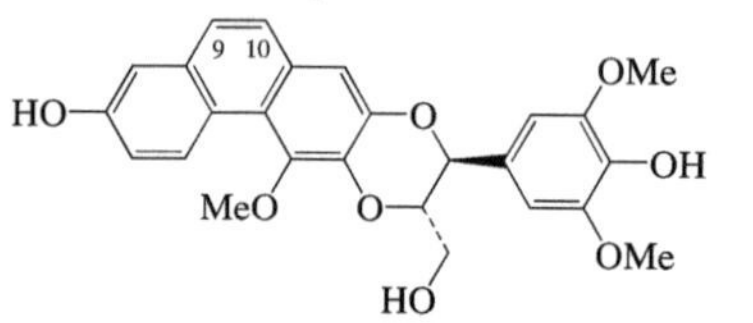

Absolute Configuration

$C_{26}H_{24}O_8$ 464.470

Constit. of the stems of *Dendrobium officinale*. Exhibits cytotoxic activity. Amorph. yellow solid. $[\alpha]^{20}_D$ +58 (c, 0.2 in MeOH). λ_{max} 225 (log ε 0.55); 264 (log ε 1.11) (MeOH)

9,10-Dihydro

Dendrocandin P_2

$C_{26}H_{26}O_8$ 466.486

Constit. of the stems of *Dendrobium officinale*. Exhibits cytotoxic activity. Amorph. solid. $[\alpha]^{20}_D$ +62 (c, 0.45 in MeOH).

Zhao, G.-Y. *et al.*, *J. Nat. Med.*, 2018, **72**, 246-251 (*Dendrocandins P_1,P_2, cd, activity*)

Denudatone D20

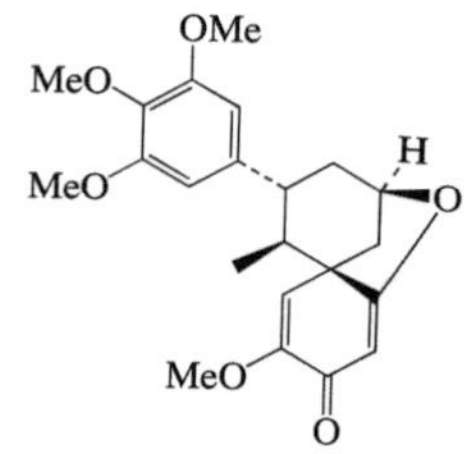

Relative Configuration

7′,9-Cyclo-2,8-epoxy-3′,4′,5,5′-tetramethoxy-1,8′-lignan-4(1H)-one.

Magliflloenone. *Magliflonenone*

[82427-77-8]

$C_{22}H_{26}O_6$ 386.444

Similar to Futoenone, F17. Constit. of *Magnolia biondii*, *Magnolia denudata* and *Magnolia liliflora*. Cryst. (C_6H_6 or $CHCl_3$/petrol). Mp 232° (Magliflloenone). Mp 238 - 240° (Denudatone). $[\alpha]_D$ -90.1 (c, 0.08 in $CHCl_3$) (Magliflloenone). $[\alpha]_D$ -78.3 (c, 0.4 in $CHCl_3$) (Denudatone). λ_{max} 258 (log ε 4.32) (EtOH).

Talapatra, B. *et al.*, *Phytochemistry*, 1982, **21**, 747-750 (*Magliflloenone*).
Iida, T. *et al.*, *Phytochemistry*, 1982, **21**, 2939-2941 (*Denudatone*).
Iida, T. *et al.*, *Phytochemistry*, 1983, **22**, 763-766 (*Denudatone*).
Nguyen, T.T.M. *et al.*, *Nat. Prod. Sci.*, 2017, **23**, 119-124 (*Magliflloenone*)

Deoxycalyxin A D21

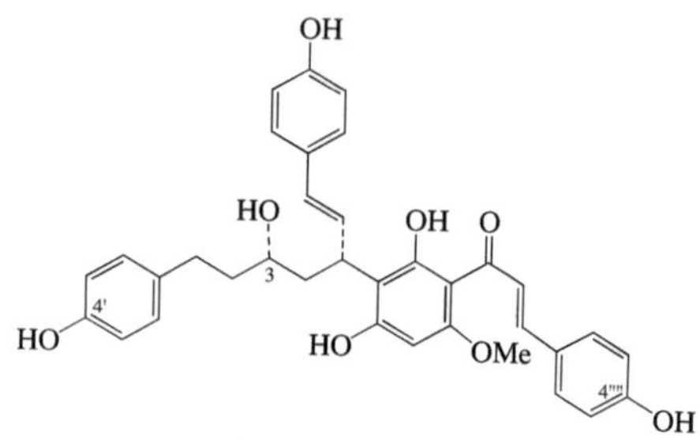

Absolute Configuration

[332396-25-5]

$C_{35}H_{34}O_8$ 582.649

Constit. of the seeds of *Alpinia blepharocalyx*. Exhibits antiproliferative activity. Amorph. pale yellow solid. $[\alpha]^{25}_D$ +147.9 (c, 0.03 in MeOH)

3-Hydroperoxide

Calyxin A

[164991-52-0]

$C_{35}H_{34}O_9$ 598.648

Constit. of the seeds of *Alpinia blepharocalyx*. Amorph. light yellow solid. $[\alpha]_D$ -58.9 (c, 0.1 in MeOH)

4′,4‴′-Dideoxy
Alpinnanin C
[872422-20-3]
$C_{35}H_{34}O_6$ 550.651
Constit. of the rhizomes of *Alpinia pinnanensis* and seeds of *Alpinia blepharocalyx*. Exhibits antiproliferative effects. Pale yellow powder. $[\alpha]^{25}_{D}$ +42.9 (c, 0.14 in MeOH).
Kadota, S. *et al., Chem. Pharm. Bull.*, 1994, **42**, 2647-2649 (*Calyxin A, pmr, cmr*).
Prasain, J.K. *et al., J. Chem. Res., Synop.*, 1998, 22-23 (*Calyxin A, struct*).
Gewali, M.B. *et al., Org. Lett.*, 1999, **1**, 1733-1736 (*Calyxin A, activity*).
Ali, M.S. *et al., Biol. Pharm. Bull.*, 2001, **24**, 525-528 (*activity*).
Tezuka, Y. *et al., J. Nat. Prod.*, 2001, **64**, 208-213 (*Deoxycalyxin A, Calyxin A, activity*).
Giang, P.M. *et al., Chem. Pharm. Bull.*, 2005, **53**, 1335-1337 (*Alpinnanin C*)

Descuraic acid D22

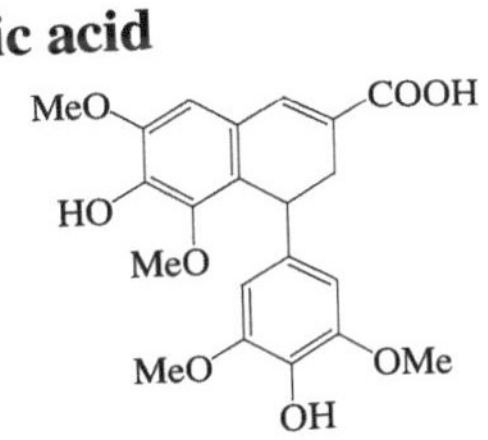

3,4-Dihydro-6-hydroxy-4-(4-hydroxy-3,5-dimethoxyphenyl)-5,7-dimethoxy-2-naphthalenecarboxylic acid. 2,7′-Cyclo-4,4′-dihydroxy-3,3′,5,5′-tetramethoxy-9′-nor-8,8′-lign-7-en-9-oic acid
$C_{21}H_{22}O_8$ 402.399
Norlignan
(-)-form [921914-53-6]
Constit. of the seeds of *Descurainia sophia*. Needles (MeOH). Mp 211 - 213°. $[\alpha]^{20}_{D}$ -3.6 (c, 0.01 in MeOH).
Sun, K. *et al., Nat. Prod. Res.*, 2006, **20**, 519-522 (*Descuraic acid*)

Diarctigenin D23

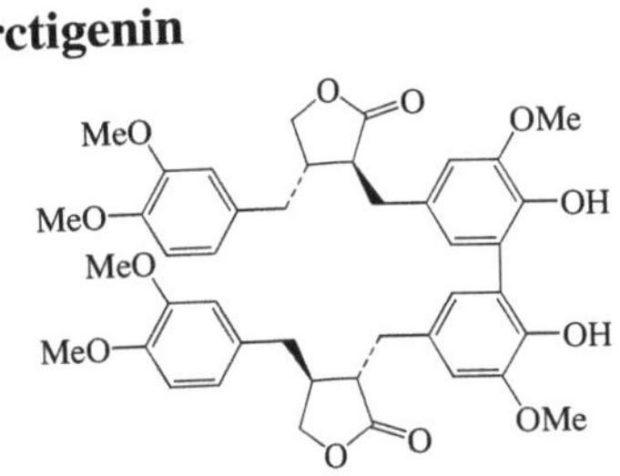

Relative Configuration

5,5-Biarctigenin
[160433-41-0]
$C_{42}H_{46}O_{12}$ 742.818
Similar to 5,5-Bisnortrachelogenin, B67. Constit. of *Arctium lappa* and *Saussurea conica*. NO production inhibitor. Antiinflammatory agent. Amorph. solid. Mp 88 - 92°. $[\alpha]^{24}_{D}$ -17 (c, 0.3 in $CHCl_3$). λ_{max} 240 (ε 39800); 285 (ε 15850) (EtOH).
Han, B.H. *et al., Phytochemistry*, 1994, **37**, 1161-1163 (*Diarctigenin*).
Fan, C.-Q. *et al., Planta Med.*, 2006, **72**, 590-595 (*Saussurea conica constit, activity*).
Park, S.Y. *et al., Chem. Pharm. Bull.*, 2007, **55**, 150-152 (*Arctium lappa constit, activity*).
Kuo, P.-C. *et al., Tetrahedron Lett.*, 2013, **54**, 6955-6958 (*synth*)

Diasarone 1 D24

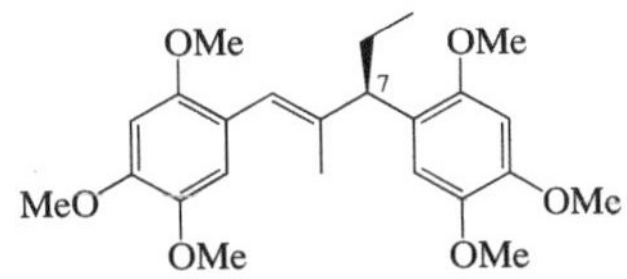

(R,E)-form

1,1′-(3-Ethyl-2-methyl-1-propene-1,3-diyl)bis[2,4,5-trimethoxybenzene]. 2-Methyl-1,3-bis(2,4,5-trimethoxyphenyl)-1-pentene. 2,2′,4,4′,5,5′-Hexamethoxy-7,8′-lign-7′-ene
[130766-51-7]
$C_{24}H_{32}O_6$ 416.514
Originally formulated as a cyclobutane
(R,E)-form
Tatarinan S [1776969-79-9]
Constit. of the roots of *Acorus tatarinowii*. Exhibits protective effect against β-amyloid toxicity. Cryst. (Me_2CO/petrol)
(ξ,E)-form [80423-94-5]
Isol. from *Asarum europaeum* and *Acorus tatarinowii*. Exhibits protective effect against β-amyloid toxicity. Cryst. (MeOH). Mp 96 - 97°. Isomerises to γ-Diasarone, D25. Also obt. by irradiation of Asarone. λ_{max} 295 (ε 11230) (MeOH).
Lemini, C. *et al., Org. Prep. Proced. Int.*, 1981, **13**, 374-378 (*synth*).
Bohlmann, F. *et al., Arch. Pharm. (Weinheim, Ger.)*, 1982, **315**, 474-476 (*bibl*).
Lemini, C. *et al., Acta Crystallogr., Sect. C*, 1990, **46**, 1542-1545 (*cryst struct*).
Luo, X.-H. *et al., Fitoterapia*, 2016, **108**, 5-8 (*Tatarinan S, cryst struct, activity*)

γ-Diasarone D25

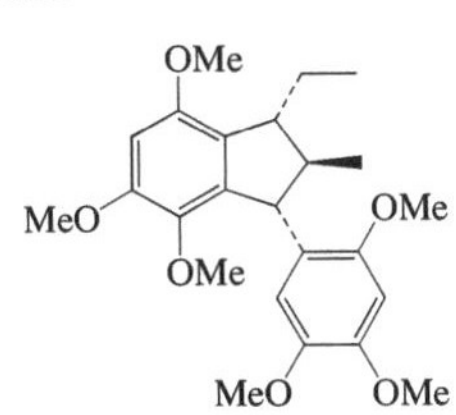

Relative Configuration

1-Ethyl-2,3-dihydro-4,5,7-trimethoxy-2-methyl-3-(2,4,5-trimethoxyphenyl)-1H-indene, CAS. *7,8′-Cyclo-2′,3,4,4′,5′,6-hexamethoxy-2,7′-lignan.* ***Diasarone 2***
[80434-33-9]
[101910-59-2] (unspecified stereochem.) [130746-43-9] (deleted)
[137490-15-4] (*all-S*-form)
$C_{24}H_{32}O_6$ 416.514
Constit. of *Acorus calamus* and *Asarum europaeum*. Prisms. Mp 99 - 100° (84°).
Lemini, C. *et al., Org. Prep. Proced. Int.*, 1981, **13**, 374-378 (*synth*).
Bohlmann, F. *et al., Arch. Pharm. (Weinheim, Ger.)*, 1982, **315**, 474-476 (*Asarum europaeum constit*).
Saxena, D.B. *Phytochemistry*, 1986, **25**, 553-555 (*Acorus calamus constit*).
Lemini, C. *et al., Acta Crystallogr., Sect. C*, 1990, **46**, 1542-1545 (*cryst struct*).
Al-Farhan, E. *et al., J. Chem. Res., Synop.*, 1992, 36-37 (*synth, struct*).
Alesso, E. *et al., ARKIVOC*, 2003, **x**, 283-297 (*synth*).
Lantano, B. *et al., Tetrahedron*, 2008, **64**, 4090-4102 (*synth*)

1,5-Di-*O*-[3,4-bis(3,4-dihydroxyphenyl)-1,2-cyclobutanedicarbonyl]-3,4-di-*O*-caffeoylquinic acid D26

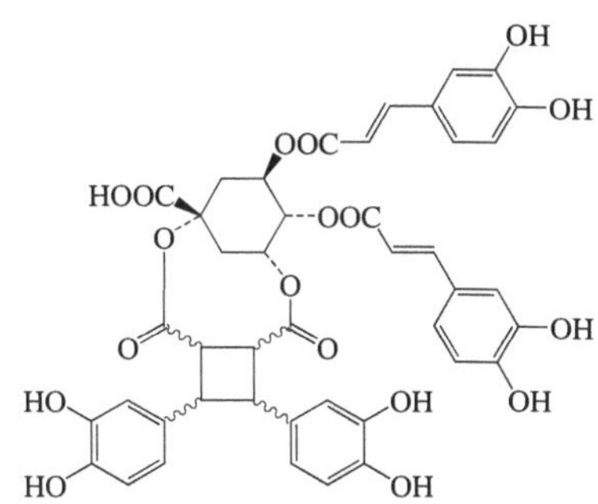

[158364-85-3] (stereoisomer)
[158414-91-6] (stereoisomer)
$C_{43}H_{36}O_{18}$ 840.743
Isol. as mixture of stereoisomers. Constit. of *Pluchea symphytifolia*. Shows moderate anthelmintic activity. Amorph. yellowish-white powder. $[\alpha]^{20}_{D}$ -175 (c, 2 in MeOH). λ_{max} 233 (ε 17783); 292 (ε 19498); 329 (ε 23442) (MeOH).
Scholz, E. *et al., Planta Med.*, 1994, **60**, 360-364 (*Pluchea symphytifolia constit*)

1,5-*O*-Dicaffeoyl-3-(2-*O*-caffeoyl-1-*O*-methyl-4-maloyl)quinic acid D27

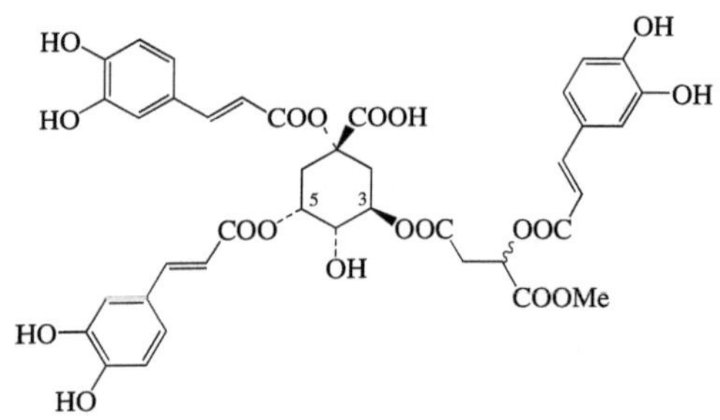

Absolute Configuration

$C_{39}H_{36}O_{19}$ 808.698
Numbering systems vary. Positions 3 and 5 frequently interchanged. Isol. from cultured cells of *Saussurea involucrata*. Amorph. powder. Mp > 250°. $[\alpha]^{20}_{D}$ +26.6 (c, 0.3 in MeOH).

Zou, X.-W. *et al., Chin. J. Nat. Med.*, 2015, **13**, 295-298 (*Saussurea involucrata constit*)

1,2-Dicaffeoylglucose D28

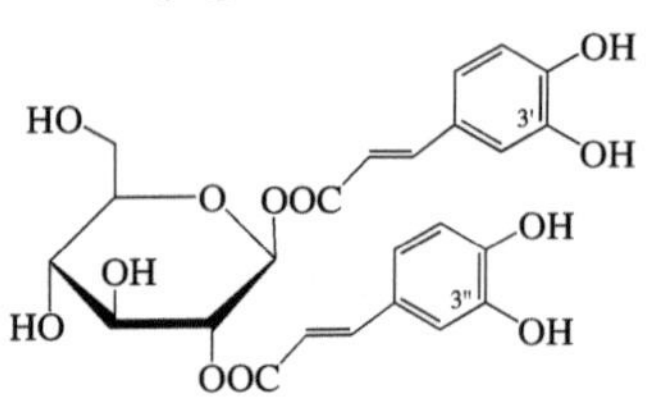

$C_{24}H_{24}O_{12}$ 504.444
β-D-***Pyranose-form*** [359819-37-7]
Constit. of *Balanophora japonica* and *Spiraea prunifolia* var. *simpliciflora*. Amorph. yellow powder +1½ H_2O. $[\alpha]^{25}_{D}$ -15.3 (c, 0.7 in MeOH)

3′-Deoxy
2-O-Caffeoyl-1-O-p-coumaroyl-β-D-glucopyranose
[359819-39-9]
$C_{24}H_{24}O_{11}$ 488.445
Constit. of *Balanophora japonica*. Amorph. yellow powder + ½H_2O. $[\alpha]^{15}_{D}$ -268.6 (c, 0.8 in MeOH)

3″-Deoxy
1-O-Caffeoyl-2-O-p-coumaroyl-β-D-glucopyranose
[1771793-33-9]
$C_{24}H_{24}O_{11}$ 488.445
Constit. of the twigs of *Spiraea prunifolia* var. *simpliciflora*. Yellow gum. $[\alpha]^{25}_{D}$ -14 (c, 0.1 in MeOH). λ_{max} 227; 314 (MeOH).

Jiang, Z.-H *et al., Chem. Pharm. Bull.*, 2001, **49**, 887-892 (*Balanophora japonica constits*).
Jang, S.W. *et al., Arch. Pharmacal Res.*, 2015, **38**, 1943-1951 (*Spiraea prunifolia constits*)

1,4-Dicaffeoylglucose D29

$C_{24}H_{24}O_{12}$ 504.444
β-D-***Pyranose-(E,E)-form***
[1615694-34-2]
Constit. of the roots of *Aruncus sylvester*. Amorph. solid (MeOH). $[\alpha]^{20}_{D}$ +29.4 (c, 0.51 in DMSO). λ_{max} 247 (log ε 4.35); 302 (sh); 334 (log ε 4.51) (MeOH).

Que, M. *et al., J. Asian Nat. Prod. Res.*, 2014, **16**, 158-162 (*Aruncus sylvester constit*)

3,6-Dicaffeoylglucose D30

[610787-37-6]
$C_{24}H_{24}O_{12}$ 504.444
Constit. of the aerial parts of *Rubus sanctus*. Amorph. powder. Obt. as anomeric mixt.. λ_{max} 250 (sh); 295; 334 (MeOH).

Hussein, S.A.M. *et al., Phytochemistry*, 2003, **63**, 905-911 (*Rubus sanctus constit*)

1,3-Di-*O*-caffeoylquinic acid D31

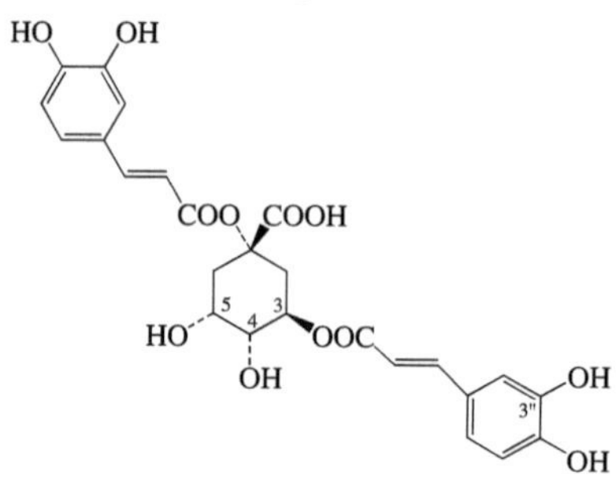

1,3-Bis(3,4-dihydroxycinnamoyl)quinic acid
[879293-21-7]
$C_{25}H_{24}O_{12}$ 516.455
Care needed with numbering. Positions 3 and 5 frequently interchanged. The numbering shown is the usual scheme. Isol. from *Arnica* spp., *Arctium lappa* and *Ipomoea batatas* tubers. Yellow powder. λ_{max} 225 (log ε 2.95); 247 (log ε 2.81); 280 (log ε 2.75); 343 (log ε 2.76) (MeOH)

5-O-[(Methoxycarbonyl)carbonyl]
1,3-Di-O-trans-caffeoyl-5-methoxalylquinic acid
[1290614-27-5]
$C_{28}H_{26}O_{15}$ 602.501
Constit. of the flowers of *Arnica montana*

4-Succinoyl
1,3-Di-O-caffeoyl-4-O-succinoylquinic acid
[168002-97-9]
$C_{29}H_{28}O_{15}$ 616.528
Constit. of *Arctium lappa*. Oil. $[\alpha]_D$ +41 (c, 0.3 in MeOH). Referred to as 1,5-dicaffeoyl in the ref.

5-Succinoyl
1,3-Di-O-caffeoyl-5-O-succinoylquinic acid
[168002-96-8]
$C_{29}H_{28}O_{15}$ 616.528
Constit. of *Arctium lappa*. Oil. $[\alpha]_D$ +117 (c, 0.3 in MeOH). Descr. as 1,5-dicaffeoyl-3-succinoyl in the ref.

4,5-Disuccinoyl
1,3-Di-O-caffeoyl-4,5-di-O-succinoylquinic acid
[168002-98-0]
$C_{33}H_{32}O_{18}$ 716.601
Constit. of *Arctium lappa*. Oil. $[\alpha]_D$ +45 (MeOH). Called 1,5-dicaffeoyl-3,4-disuccinoyl in the ref.

4-O-(3-Carboxy-2E-propenoyl)
1,3-Di-O-trans-caffeoyl-4-fumaroylquinic acid
[1290614-33-3]
$C_{29}H_{26}O_{15}$ 614.512
Constit. of the flowers of *Arnica montana*

5-O-(3-Carboxy-2E-propenoyl)
1,3-Di-O-trans-caffeoyl-5-fumaroylquinic acid
[1290614-34-4]
$C_{29}H_{26}O_{15}$ 614.512
Constit. of the flowers of *Arnica montana*

5-O-(3-Carboxy-3-hydroxypropanoyl)
MDCQA
[1373335-93-3]
$C_{29}H_{28}O_{16}$ 632.527
Obt. from cultured cells of *Saussurea involucrata*. Exhibits DPPH and ABTS radical scavenging activity. Amorph. powder. Mp 205 - 206°. $[\alpha]^{20}_{D}$ +28 (c, 0.2 in MeOH). λ_{max} 245 (log ε 4.02); 300 (sh); 330 (log ε 4.31) (MeOH)

3″-Me ether
1-trans-Caffeoyl-3-trans-feruloylquinic acid
[1290614-26-4]
$C_{26}H_{26}O_{12}$ 530.482
Constit. of the flowers of *Arnica montana*

1-Epimer
1,3-Di-O-caffeoylepiquinic acid
[879293-19-3]
$C_{25}H_{24}O_{12}$ 516.455

Constit. of the flowers of *Chrysanthemum morifolium*. Mp 178 - 180°. $[\alpha]_D^{21}$ +183.7 (c, 1.94 in MeOH). λ_{max} 219 (log ε 4.08); 245 (log ε 3.63); 269 (sh); 330 (log ε 4.57) (MeOH)

3,4-Diepimer, Me ester

Evolvoid B

[947701-08-8]

$C_{26}H_{26}O_{12}$ 530.482

Constit. of *Evolvulus alsinoides*. Amorph. powder. $[\alpha]_D^{29}$ -50.8 (c, 0.04 in $CHCl_3$). λ_{max} 211; 273; 327 (MeOH).

Merfort, I. *Phytochemistry*, 1992, **31**, 2111-2113 (*Arnica constits*).
Maruta, Y. *et al.*, *J. Agric. Food Chem.*, 1995, **43**, 2592-2595 (*Arctium lappa constits*).
Kim, H.J. *et al.*, *Planta Med.*, 2005, **71**, 867-870 (*1,3-Dicaffeoylepiquinic acid*).
Dini, I. *et al.*, *J. Agric. Food Chem.*, 2006, **54**, 8733-8737 (*Ipomoea batatas constit*).
Gupta, P. *et al.*, *Chem. Pharm. Bull.*, 2007, **55**, 771-775 (*Evolvoid B*).
Jaiswal, R. *et al.*, *J. Agric. Food Chem.*, 2011, **59**, 4033-4039 (*Arnica montana constits*).
Zou, X. *et al.*, *Phytochem. Lett.*, 2014, **7**, 133-136 (*MDCQA, activity*)

1,4-Di-*O*-caffeoylquinic acid D32

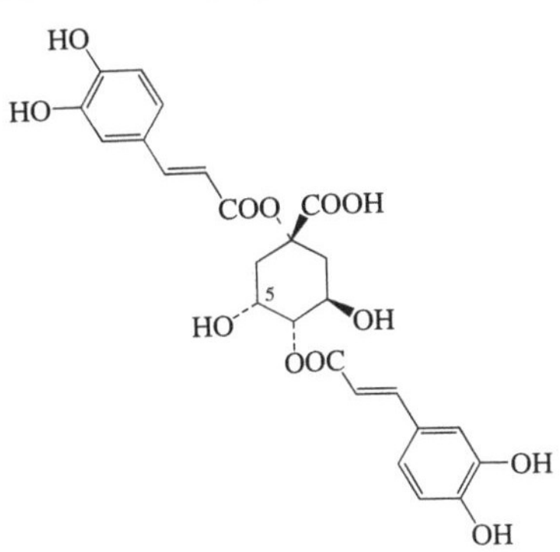

NSC 91529. 1,4-Bis(3,4-dihydroxycinnamoyl)quinic acid

[1182-34-9]

$C_{25}H_{24}O_{12}$ 516.455

Constit. of *Cynara cardunculus* ssp. *scolymus*, *Sempervivum ruthenicum*, *Helichrysum italicum* and *Achillea cartilaginea*. Mp 228 - 231° (218-220°). $[\alpha]_D^{18}$ -30 (c, 0.5 in MeOH)

7′,7″,8′,8″-Tetrahydro, 5-O-(4-hydroxy-3-methoxy-E-cinnamoyl)

1,4-Bis-O-(dihydrocaffeoyl)-5-O-feruloylquinic acid.

Feruloylpodospermic acid B

[944747-08-4]

$C_{35}H_{36}O_{15}$ 696.658

Constit. of *Scorzonera divaricata*. Amorph. solid. $[\alpha]_D^{20}$ -89.6 (c, 0.05 in MeOH). Tentative struct. assigned. λ_{max} 220; 330 (MeOH aq.)

4,5-Diepimer

[1401523-58-7]

$C_{25}H_{24}O_{12}$ 516.455

Constit. of *Erigeron bonariensis*. Pale yellow gummy solid. $[\alpha]_D^{29}$ -22 (c, 0.3 in MeOH). λ_{max} 220; 330 (MeOH).

Panizzi, L. *et al.*, *Gazz. Chim. Ital.*, 1965, **95**, 71-82 (*synth*).
Gumenyuk, L.A. *Chem. Nat. Compd. (Engl. Transl.)*, 1971, **7**, 503 (*Sempervivum ruthenicum constit*).
Zapesochnaya, G.G. *et al.*, *Chem. Nat. Compd. (Engl. Transl.)*, 1992, **28**, 40-44 (*Helichrysum italicum constit, Achillea cartilaginea constit*).
Tsevegsuren, N. *et al.*, *J. Nat. Prod.*, 2007, **70**, 962-967 (*Feruloylpodospermic acid B*).
Lattanzio, V. *et al.*, *J. Funct. Foods*, 2009, **1**, 131-144 (*artichoke constit*).
Li, F. *et al.*, *Free Radical Res.*, 2012, **46**, 286-294 (*ms*).
Zahoor, A. *et al.*, *Helv. Chim. Acta*, 2012, **95**, 1613-1622 (*Erigeron bonariensis diepimer*)

1,5-Di-*O*-caffeoylquinic acid D33

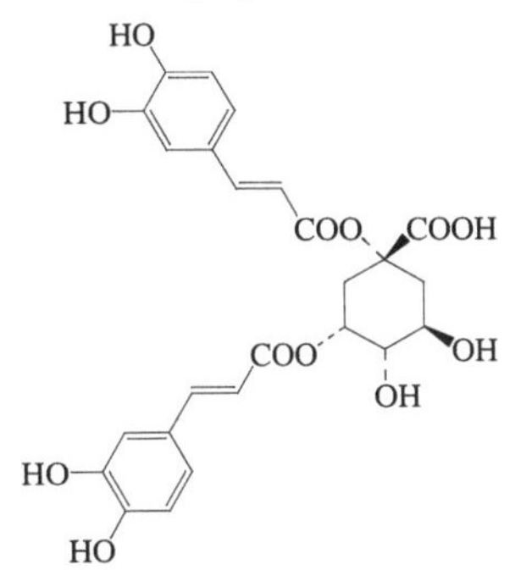

1,5-Bis(3,4-dihydroxycinnamoyl)quinic acid. ***Cynarine,*** *INN. Plemocil. Cinarcaf. Cynarex. Cinarina. Cynarin. Listrocol. Nivellipid. CYN*

[1884-24-8] (unspecified stereochem.)

$C_{25}H_{24}O_{12}$ 516.455

Can also be numbered 1,3-dicaffeoyl. 1,5- is more usual. Has been confused in CAS with the 1,4-isomer

(E,E)-form [30964-13-7]

Isol. from *Cynara scolymus*, *Cynara cardunculus*, *Ferula elaeochytris*, *Artemisia vulgaris*, *Rhus typhina*, *Senecio nemorensis*, *Cirsium arvense* and other plants. Specific inhibitor of human HIV-1 integrase. Choleretic agent. Shows hepatoprotective activity *in vitro*. Moderate protectant against jacaraca (snake) venom. Cryst.. Mp 227 - 228°. $[\alpha]_D^{25}$ -59 (c, 2 in MeOH). λ_{max} 326 (E1%/1cm 616) (MeOH) (Berdy)

▶ GD8970000. LD_{50} (mus, ipr) 1900 mg/kg

Hexa-Ac

Cryst.. Mp 168 - 172°

4-O-[(Methoxycarbonyl)carbonyl]

1,5-Di-O-trans-caffeoyl-4-methoxalylquinic acid

[1290614-28-6]

$C_{28}H_{26}O_{15}$ 602.501

Constit. of the flowers of *Arnica montana*

3′,3″-Di-Me ether

1,5-Di-O-feruloylquinic acid

[325172-03-0]

$C_{27}H_{28}O_{12}$ 544.509

Constit. of the roots of nutrient-deprived *Brachiaria decumbens* and *Brachiaria ruziziensis* and the pollen of *Calendula officinalis*. Mushroom tyrosinase inhibitor. Pale yellow powder. Called 1,3-diferuloyl in the ref.. λ_{max} 295 (sh); 330 (4.26); 388 (sh) (H_2O)

4′,4″-Di-Me ether

1,5-Diisoferuloylquinic acid

[1629858-73-6]

$C_{27}H_{28}O_{12}$ 544.509

Constit. of pollen of *Calendula officinalis*. Exhibits inhibitory activity against mushroom tyrosinase. λ_{max} 212 (sh); 225; 297; 310 (MeOH).

Panizzi, L. *et al.*, *Nature (London)*, 1954, **174**, 1062 (*Cynarine*).
Preziosi, P. *et al.*, *Arch. Int. Pharmacodyn. Ther.*, 1958, **117**, 63-80 (*pharmacol*).
Antona, A. *et al.*, *Clin. Ter. (Rome)*, 1971, **58**, 109-116 (*metab*).
Horman, I. *et al.*, *J. Agric. Food Chem.*, 1984, **32**, 538-540 (*struct, pmr, conformn*).
Sakushima, A. *et al.*, *Phytochemistry*, 1985, **24**, 325-328 (*ms*).
McDougall, B. *Antimicrob. Agents Chemother.*, 1998, **42**, 140-146 (*activity*).
Wenzl, P. *et al.*, *Phytochemistry*, 2000, **55**, 389-395 (*Diferuloylquinic acid*).
Mors, W.B. *et al.*, *Phytochemistry*, 2000, **55**, 627-642 (*activity*).
Slanina, J. *et al.*, *Tetrahedron Lett.*, 2001, **42**, 3383-3385 (*synth, cmr*).
Alkhatib, R. *et al.*, *Phytochemistry*, 2008, **69**, 2979-2983 (*Ferula elaeochytris constit*).
Lattanzio, V. *et al.*, *J. Funct. Foods*, 2009, **1**, 131-144 (*Cynara scolymus constit*).
Jaiswal, R. *et al.*, *J. Agric. Food Chem.*, 2011, **59**, 4033-4039 (*4-methoxalyl*).
Olennikov, D.N. *et al.*, *Chem. Nat. Compd. (Engl. Transl.)*, 2014, **50**, 589-593 (*1,5-Diferuloylquinic acid, 1,5-Diisoferuloylquinic acid, activity*)

3,4-Di-*O*-caffeoylquinic acid D34

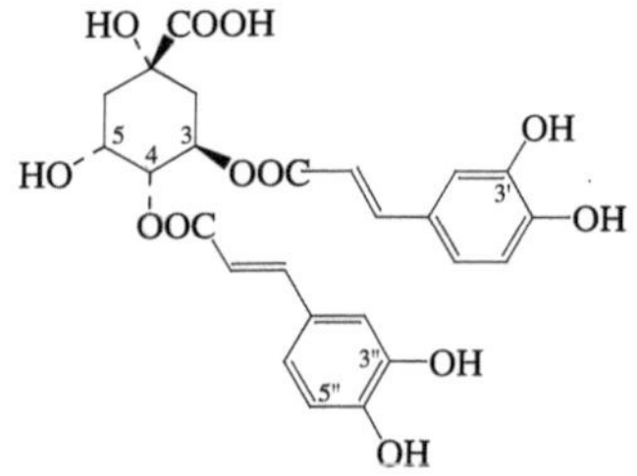

Isochlorogenic acid B
[14534-61-3]
[2271-12-7]
$C_{25}H_{24}O_{12}$ 516.455
Numbering systems vary. Isol. from coffee and maté and *Cynara cardunculus* ssp. *scolymus*, *Gynura divaricata*, *Artemisia herba-alba*, *Chrysothamnus paniculatus*, *Dichrocephala bicolor*, *Lonicera japonica*, *Solidago canadensis* and others. Neuroprotectant. Moderate antioxidant. Powder. Mp 234 - 238° dec.. $[\alpha]^{28}_{D}$ -225 (MeOH)

Me ester
Macroantoin F. *Macranthoin F*
[114637-83-1]
$C_{26}H_{26}O_{12}$ 530.482
Constit. of *Gynura divaricata*, *Lonicera japonica* and *Lonicera macranthoides*. Light yellow powder. Mp 115 - 116°

Et ester
[1802332-38-2]
$C_{27}H_{28}O_{12}$ 544.509
Constit. of *Gynura divaricata*

Butyl ester
[488128-33-2]
$C_{29}H_{32}O_{12}$ 572.563
Constit. of *Isertia pittieri*. λ_{max} 242 (log ε 4.29); 300 (sh); 331 (log ε 4.52) (MeOH)

1,5-Bis-O-(4-hydroxy-E-cinnamoyl)
3,4-Dicaffeoyl-1,5-di-p-coumaroylquinic acid
[854049-13-1]
$C_{43}H_{36}O_{16}$ 808.745
Constit. of the leaves of *Ipomoea pes-caprae*. Amorph. solid. $[\alpha]^{22}_{D}$ -329 (c, 0.9 in MeOH). λ_{max} 221 (log ε 4.7); 234 (log ε 4.6); 299 (sh); 316 (log ε 4.9) (MeOH)

1-O-[(Methoxycarbonyl)carbonyl]
3,4-Di-O-trans-caffeoyl-1-methoxalylquinic acid
[1290614-30-0]
$C_{28}H_{26}O_{15}$ 602.501
Constit. of the flowers of *Arnica montana*

5-O-(3-Hydroxy-3-methylglutaroyl)
[103744-76-9]
$C_{31}H_{32}O_{16}$ 660.581
Isol. from fruit of *Gardenia jasminoides*. Lipoxygenase inhibitor. $[\alpha]^{18}_{D}$ -169.8 (c, 0.95 in MeOH)

3′-Me ether
4-O-Caffeoyl-3-O-feruloylquinic acid
[96990-65-7]
[102851-10-5]
$C_{26}H_{26}O_{12}$ 530.482
Constit. of coffee beans

3″-Me ether
3-Caffeoyl-4-feruloylquinic acid
[125132-81-2]
$C_{26}H_{26}O_{12}$ 530.482
Constit. of the peel *Citrus reticulata*. Amorph. yellow solid. $[\alpha]^{29}_{D}$ -24.7 (c, 0.1 in MeOH). λ_{max} 206; 216; 233; 328 (EtOH)

3″-Me ether, Me ester
[345642-66-2]
$C_{27}H_{28}O_{12}$ 544.509
Constit. of the peel of *Citrus reticulata*. Amorph. solid. $[\alpha]^{29}_{D}$ -171 (c, 0.1 in MeOH). λ_{max} 219; 236; 328 (EtOH)

7″,8″-Dihydro
3-O-Caffeoyl-4-O-(dihydrocaffeoyl) quinic acid. ***Tungtungmadic acid***
[883238-44-6]
$C_{25}H_{26}O_{12}$ 518.471
Constit. of *Salicornia herbacea*. Antioxidant

5″-Methoxy, 3″-Me ether
3-O-Caffeoyl-4-O-sinapoylquinic acid
[110241-35-5]
$C_{27}H_{28}O_{13}$ 560.508
Constit. of Gardenia fructus (*Gardenia jasminoides*). Lipoxygenase inhibitor. Amorph. powder. $[\alpha]^{20}_{D}$ -252 (c, 0.54 in MeOH). Log P -0.19 (calc). λ_{max} 220 (ε 31800); 239 (ε 28800); 328 (ε 40300) (MeOH)

5″-Methoxy, 3″-Me ether, Me ester
[879627-81-3]
$C_{28}H_{30}O_{13}$ 574.535
Constit. of the fruit of *Gardenia jasminoides*. Mp 130 - 132°. $[\alpha]^{19}_{D}$ -238.7 (c, 1.45 in MeOH). λ_{max} 239 (log ε 1.74); 330 (log ε 4.32) (MeOH)

5″-Methoxy, 3″-Me ether, Et ester
$C_{29}H_{32}O_{13}$ 588.562
Constit. of the fruit of *Gardenia jasminoides*. Mp 129 - 130°. $[\alpha]^{20}_{D}$ -225.9 (c, 0.29 in MeOH). λ_{max} 242 (log ε 2.36); 331 (log ε 4.34) (MeOH).

Scarpati, M.L. *Tetrahedron Lett.*, 1964, **5**, 2851-2853 (*Isochlorogenic acid B*).
Timmermann, B.N. *et al.*, *J. Nat. Prod.*, 1983, **46**, 365-368 (*Chrysothamnus paniculatus constit, pmr, cmr, ms*).
Iwahashi, H. *et al.*, *Phytochemistry*, 1985, **24**, 630-632 (*Caffeoylferuloylquinic acid*).
Nishizawa, M. *et al.*, *Chem. Pharm. Bull.*, 1986, **34**, 1419-1421 (*3-hydroxy-3-methylglutaroyl*).
Nishizawa, M. *et al.*, *Chem. Pharm. Bull.*, 1987, **35**, 2133-2135 (*3-Caffeoyl-4-sinapoylquinic acid*).
Chen, M. *et al.*, *Yaoxue Xuebao*, 1994, **29**, 617-620 (*Macroantoin F*).
Lin, L.-C. *et al.*, *J. Nat. Prod.*, 1999, **62**, 405-408 (*Dichrocephala bicolor constit, pmr, cmr*).
Choi, E.M. *et al.*, *Phytochemistry*, 2001, **56**, 733-739 (*3-Caffeoyl-4-feruloylquinic acid*).
Um, B.H. *et al.*, *Fitoterapia*, 2002, **73**, 550-552 (*butyl ester*).
Chung, Y.C. *et al.*, *Arch. Pharmacal Res.*, 2005, **28**, 1122-1126 (*Tungtungmadic acid*).
Teramachi, F. *et al.*, *J. Nat. Prod.*, 2005, **68**, 794-796 (*Dicaffeoyl-1,6-dicoumaroylquinic acid*).
Kim, H.J. *et al.*, *J. Nat. Prod.*, 2006, **69**, 600-603 (*3-Caffeoyl-4-sinapoylquinic esters*).
Nakajima, Y. *et al.*, *Life Sci.*, 2007, **80**, 370-377 (*activity*).
Bradette-Herbert, M-E. *et al.*, *Chem. Pharm. Bull.*, 2008, **56**, 82-84 (*Solidago canadensis constit*).
Lattanzio, V. *et al.*, *J. Funct. Foods*, 2009, **1**, 131-144 (*Cynara cardunculus constit*).
Jaiswal, R. *et al.*, *J. Agric. Food Chem.*, 2011, **59**, 4033-4039 (*1-methoxalyl*).
Dahmani-Hamzaoui, N. *et al.*, *Helv. Chim. Acta*, 2012, **95**, 564-576 (*Artemisia herba-alba constit, pmr, cmr, activity*).
Chen, J. *et al.*, *Fitoterapia*, 2014, **99**, 1-6 (*Gynura divaricata constits*).
Giang, P.M. *et al.*, *Chem. Nat. Compd. (Engl. Transl.)*, 2017, **53**, 674-677 (*cd, abs config*)

3,5-Di-*O*-caffeoylquinic acid D35

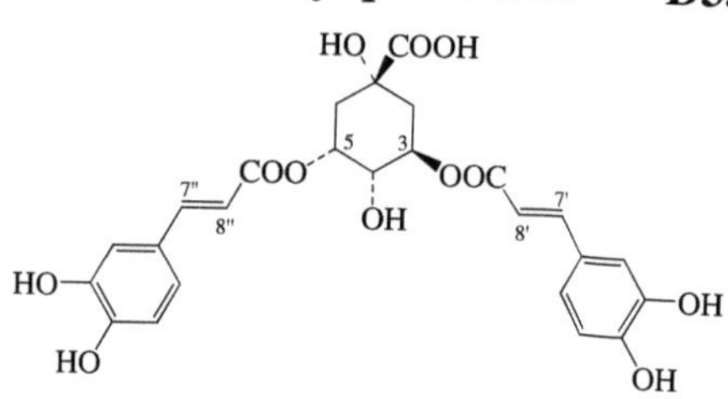

3,5-Bis(3,4-dihydroxycinnamoyl)quinic acid. ***Isochlorogenic acid A***
[2450-53-5]
$C_{25}H_{24}O_{12}$ 516.455
Care needed with numbering. Positions 3 and 5 frequently interchanged. Isol. from plant sources incl. *Cynara*

scolymus, Cydonia oblonga, Vaccinium arctostaphylos, Aster scaber, Ipomoea batatas, Gynura divaricata, Artemisia herba-alba, Lonicera japonica, Pyrus pyrifolia, Chrysothamnus paniculatus, Ferula lutea, Lactuca tuberosa and Brazilain propolis. Active against HIV-1 integrase. Shows antibacterial, antiviral and antihepatotoxic activities. Moderate antioxidant. Prevents lesion formation by *Alternaria alternata* on pear leaves. Cryst.. Mp 170 - 172° dec.. $[\alpha]_D$ -198 (MeOH). λ_{max} 225 (log ε 3.08); 247 (log ε 3.02); 280 (log ε 2.98); 343 (log ε 3.03) (MeOH)

Me ester

Macroantoin G. *Macranthoin G*
[159934-13-1]
$C_{26}H_{26}O_{12}$ 530.482
Constit. of *Erigeron breviscapus, Gynura divaricata* and *Lonicera macranthoides*. Light yellow powder. Mp 123 - 124° (118-119°). $[\alpha]^{25}_D$ -81 (c, 1 in MeOH)

Butyl ester
[329281-01-8]
$C_{29}H_{32}O_{12}$ 572.563
Constit. of *Lonicera japonica*

1-O-(2-Methylpropanoyl)
3,5-Dicaffeoyl-1-isobutyrylquinic acid
[239133-36-9]
$C_{29}H_{30}O_{13}$ 586.546
Constit. of the flowerheads of *Buphthalmum salicifolia*. Amorph. pale yellow powder. λ_{max} 242; 300 (sh); 328 (MeOH)

1-O-[3-Carboxy-3-(3,4-dihydroxy-E-cinnamoyloxy)propanoyl]
MTCQA
[1623074-63-4]
$C_{38}H_{34}O_{19}$ 794.671
Obt. from cultured cells of *Saussurea involucrata*. Exhibits DPPH and ABTS radical scavenging activity. Amorph. powder. Mp 233 - 234°. $[\alpha]^{20}_D$ +81.76 (c, 0.17 in MeOH)

4-O-(4-Carboxy-3-hydroxy-3-methylbutanoyl)
3,5-Dicaffeoyl-4-O-(3-hydroxy-3-methylglutaroyl)quinic acid
[114637-82-0]
$C_{31}H_{32}O_{16}$ 660.581
Constit. of the fruit of *Gardenia jasminoides*. Component of Zhi Zi. Amorph. powder + ½H_2O. $[\alpha]^{20}_D$ -170 (c, 0.95 in MeOH). λ_{max} 216 (ε 26600); 242 (ε 17600); 304 (sh); 329 (ε 30800) (MeOH)

4-O-(4-Carboxy-3-hydroxy-3-methylbutanoyl), Me ester
[879627-86-8]
$C_{32}H_{34}O_{16}$ 674.608
Constit. of the fruit of *Gardenia jasminoides*. Mp 165 - 168° dec.. $[\alpha]^{20}_D$ -129.6 (c, 0.3 in MeOH). λ_{max} 243 (log ε 1.95); 269 (sh); 332 (log ε 4.42) (MeOH)

4-O-(4-Hydroxy-E-cinnamoyl)
3,5-Dicaffeoyl-4-p-coumaroylquinic acid
[854049-02-8]
$C_{34}H_{30}O_{14}$ 662.6
Constit. of the leaves of *Ipomoea pes-caprae*. Amorph. solid. $[\alpha]^{22}_D$ -378 (c, 1.2 in MeOH). λ_{max} 219 (log ε 4.8); 232 (log ε 4.7); 300 (sh); 323 (log ε 4.9) (MeOH)

4-O-[(Methoxycarbonyl)carbonyl]
3,5-Di-O-trans-caffeoyl-4-methoxalylquinic acid
[1290614-29-7]
$C_{28}H_{26}O_{15}$ 602.501
Constit. of the flowers of *Arnica montana*

1,4-Bis-O-[(methoxycarbonyl)carbonyl]
3,5-Di-O-trans-caffeoyl-1,4-bis(-methoxalyl)quinic acid
[1290614-35-5]
$C_{31}H_{28}O_{18}$ 688.547
Constit. of the flowers of *Arnica montana*

4-Malonoyl
3,5-Dicaffeoyl-4-malonoylquinic acid. ***Irbic acid***
[956712-44-0]
$C_{28}H_{26}O_{15}$ 602.501
Isol. from cell cultures of *Centella asiatica*. Antioxidant. Amorph. powder. $[\alpha]^{28}_D$ -220 (c, 0.18 in MeOH). λ_{max} 245 (sh); 299; 327 (MeOH). [base] λ_{max} 265 (sh); 309; 373 (MeOH/KOH)

4-O-Succinoyl
3,5-Dicaffeoyl-4-succinoylquinic acid
[179761-31-0]
$C_{29}H_{28}O_{15}$ 616.528
Constit. of *Chrysanthemum coronarium*. Antioxidant

4-O-(4-Hydroxyphenylacetyl)
[953422-42-9]
$C_{33}H_{30}O_{14}$ 650.589
Constit. of *Hubertia tomentosa*

4-O-(1-Hydroxy-4-oxo-2,5-cyclohexadien-1-acetyl)
[953422-40-7]
$C_{33}H_{30}O_{15}$ 666.588
Constit. of *Hubertia ambavilla* and *Hubertia tomentosa*. Amorph. powder. $[\alpha]^{25}_D$ -151 (c, 0.2 in MeOH)

1-Me ether
[1360540-39-1]
$C_{26}H_{26}O_{12}$ 530.482
Isol. from *Inula crithmoides*. Displays hepatoprotective activity. Amorph. yellow powder. λ_{max} 208; 330 (MeOH)

1-Me ether, Me ester
[409361-62-2]
$C_{27}H_{28}O_{12}$ 544.509
Constit. of *Erigeron breviscapus*. Mp 132 - 134°. $[\alpha]^{20}_D$ -34.7 (MeOH)

3″-Me ether, 4-O-[(methoxycarbonyl)carbonyl]
3-O-trans-Caffeoyl-4-methoxalyl-5-O-trans-feruloylquinic acid
[1290614-32-2]
$C_{29}H_{28}O_{15}$ 616.528
Constit. of the flowers of *Arnica montana*

5′-Methoxy, 3′-Me ether, Me ester
3-O-Caffeoyl-5-O-sinapoylquinic acid methyl ester
[879627-83-5]
$C_{28}H_{30}O_{13}$ 574.535
Constit. of the fruit of *Gardenia jasminoides*. Mp 140 - 144° dec.. $[\alpha]^{21}_D$ -148 (c, 0.3 in MeOH). λ_{max} 242 (log ε 2.68); 330 (log ε 4.78) (MeOH)

5′-Methoxy, 3′-Me ether, Et ester
[879627-84-6]
$C_{29}H_{32}O_{13}$ 588.562
Constit. of the fruit of *Gardenia jasminoides*. Mp 140 - 142° dec.. $[\alpha]^{21}_D$ -133.7 (c, 0.32 in MeOH). λ_{max} 238 (log ε 2.06); 331 (log ε 4.7) (MeOH)

7′,8′-Dihydro
[1919029-54-1]
$C_{25}H_{26}O_{12}$ 518.471
Constit. of *Salicornia herbacea*. Antioxidant. Amorph. powder. $[\alpha]^{25}_D$ -2.3 (c, 0.1 in MeOH)

7′,8′-Dihydro, Me ester
[2232186-44-4]
$C_{26}H_{28}O_{12}$ 532.498
Constit. of *Salicornia herbacea*. Antioxidant. Amorph. powder. $[\alpha]^{25}_D$ -12.8 (c, 0.1 in MeOH)

7′,7″,8′,8″-Tetrahydro, Me ester
[2232186-46-6]
$C_{26}H_{30}O_{12}$ 534.514
Constit. of *Salicornia herbacea*. Antioxidant. Amorph. powder. $[\alpha]^{25}_{D}$ -12.4 (c, 0.1 in MeOH)

7′Z-Isomer
3-O-cis-Caffeoyl-5-O-trans-caffeoylquinic acid
[1291058-39-3]
$C_{25}H_{24}O_{12}$ 516.455
Constit. of the flowers of *Arnica montana*

7″Z-Isomer
5-O-cis-Caffeoyl-3-O-trans-caffeoylquinic acid
[1290614-25-3]
$C_{25}H_{24}O_{12}$ 516.455
Constit. of the flowers of *Arnica montana*

1-Epimer
3,5-Di-O-caffeoylepiquinic acid
[879305-14-3]
$C_{25}H_{24}O_{12}$ 516.455
Constit. of *Ainsliaea acerifolia* and *Chrysanthemum morifolium*. Antioxidant. Mp 184 - 187° dec.. $[\alpha]^{21}_{D}$ -166 (c, 0.27 in MeOH). λ_{max} 218 (log ε 3.81); 243 (log ε 2.91); 297 (sh); 328 (log ε 4.59) (MeOH)

1-Epimer, Me ester
[2173001-24-4]
$C_{26}H_{26}O_{12}$ 530.482
Constit. of *Ainsliaea acerifolia*. Amorph. yellow powder. $[\alpha]^{20}_{D}$ -120 (c, 0.2 in MeOH). λ_{max} 296 (log ε 4.07); 330 (log ε 4.27) (MeOH)

3-Epimer
3,5-Di-O-caffeoyl-muco-quinic acid
[314295-64-2]
$C_{25}H_{24}O_{12}$ 516.455
Constit. of *Aster scaber*. Active against HIV-1 integrase. Yellowish gum. $[\alpha]_{D}$ -153.8 (c, 0.78 in MeOH). λ_{max} 219 (log ε 4.43); 244 (log ε 4.28); 300 (sh); 328 (log ε 4.52) (EtOH).

Haslam, E. *et al.*, *J. Chem. Soc.*, 1964, 2137-2146 (*coffee constit*).
Scarpati, M.L. *et al.*, *Tetrahedron Lett.*, 1964, **5**, 2851-2853 (*Isochlorogenic acid A*).
Corse, J. *et al.*, *Phytochemistry*, 1965, **4**, 527-529 (*uv, pmr*).
Nichiforescu, E.A. *et al.*, *Khim. Prir. Soedin.*, 1970, **7**, 546 (*Vaccinium arctostaphylos constit*).
Nichiforescu, E.A. *et al.*, *Plant. Med. Phytother.*, 1970, **4**, 56-62 (*Cynara scolymus constit*).
Timmermann, B.N. *et al.*, *J. Nat. Prod.*, 1983, **46**, 365-368 (*Chrysothamnus paniculatus constit, pmr, cmr, ms*).
Nishizawa, M. *et al.*, *Chem. Pharm. Bull.*, 1988, **36**, 87-95 (*4-3-hydroxy-3-methylglutarate*).
Chen, M. *et al.*, *Yaoxue Xuebao*, 1994, **29**, 617-620 (*Macroantoin G*).
Tatefuji, T. *et al.*, *Biol. Pharm. Bull.*, 1996, **19**, 966-970 (*propolis constit*).
Chudo, Y. *et al.*, *J. Agric. Food Chem.*, 1996, **44**, 2037-2039 (*4-succinate*).
Kodoma, M. *et al.*, *Phytochemistry*, 1998, **47**, 371-373 (*Pyrus pyrifolia constit, activity*).
Miketova, P. *et al.*, *J. Mass Spectrom.*, 1999, **34**, 1240-1252 (*ms*).
Heilmann, J. *et al.*, *Phytochemistry*, 1999, **51**, 713-718 (*Dicaffeoyl-1-isobutyrylquinic acid*).
Kwon, H.C. *et al.*, *Chem. Pharm. Bull.*, 2000, **48**, 1796-1798 (*3,5-Dicaffeoylmucoquinic acid*).
Peng, L.Y. *et al.*, *Fitoterapia*, 2000, **71**, 713-715 (*butyl ester*).
Zhang, W.-D. *et al.*, *J. Asian Nat. Prod. Res.*, 2000, **2**, 283-288 (*Macroantoin G*).
Zhang, W.-D. *et al.*, *Yaoxue Xuebao*, 2001, **36**, 360-363 (*1-Me ether Me ester*).
Silva, B.M. *et al.*, *J. Agric. Food Chem.*, 2004, **52**, 4705-4712 (*Cydonia oblonga constit*).
Teramachi, F. *et al.*, *J. Nat. Prod.*, 2005, **68**, 794-796 (*Dicaffeoyl-4-coumaroylquinic acid*).
Kim, H.J. *et al.*, *Planta Med.*, 2005, **71**, 871-876 (*3,5-Dicaffeoylepiquinic acid*).
Dini, I. *et al.*, *J. Agric. Food Chem.*, 2006, **54**, 8733-8737 (*Ipomoea batatas constit*).
Kim, H.J. *et al.*, *J. Nat. Prod.*, 2006, **69**, 600-603 (*4-O-(4-carboxy-3-hydroxy-3-methylbutanoyl) Me ester, 5′-methoxy 3′-Me ether Et ester*).
Sprogøe, K. *et al.*, *J. Nat. Prod.*, 2007, **70**, 1472-1477 (*Hubertia constits*).
Antognoni, F. *et al.*, *Fitoterapia*, 2011, **82**, 950-954 (*Irbic acid*).
Jaiswal, R. *et al.*, *J. Agric. Food Chem.*, 2011, **59**, 4033-4039 (*Arnica montana constits*).
Raheem, K.S. *et al.*, *Tetrahedron Lett.*, 2011, **52**, 7175-7177 (*synth*).
Ela, M.A.A. *et al.*, *Helv. Chim. Acta*, 2012, **95**, 61-66 (*1-Me ether*).
Dahmani-Hamzaoui, N. *et al.*, *Helv. Chim. Acta*, 2012, **95**, 564-576 (*Artemisia herba-alba constit*).
Xiong, J. *et al.*, *Food Chem.*, 2013, **138**, 327-333 (*activity*).
Stojakowska, A. *et al.*, *Food Chem.*, 2013, **138**, 1250-1255 (*Lactuca tuberosa constit*).
Chen, J. *et al.*, *Fitoterapia*, 2014, **99**, 1-6 (*Gynura divaricata constits*).
Znati, M. *et al.*, *Molecules*, 2014, **19**, 16959-16975 (*Ferula lutea constit*).
Zou, X. *et al.*, *Phytochem. Lett.*, 2014, **7**, 133-136 (*MTCQA, activity*).
Cho, J.-Y. *et al.*, *Molecules*, 2016, **21**, 1097:1-13 (*Salicornia herbacea constits, activity*).
Kim, T. *et al.*, *Phytochem. Lett.*, 2016, **16**, 146-151 (*Ainsliaea acerifolia ester*).
Ge, L. *et al.*, *RSC Adv.*, 2018, **8**, 35374-35385 (*Lonicera japonica constit*)

4,5-Di-*O*-caffeoylquinic acid D36

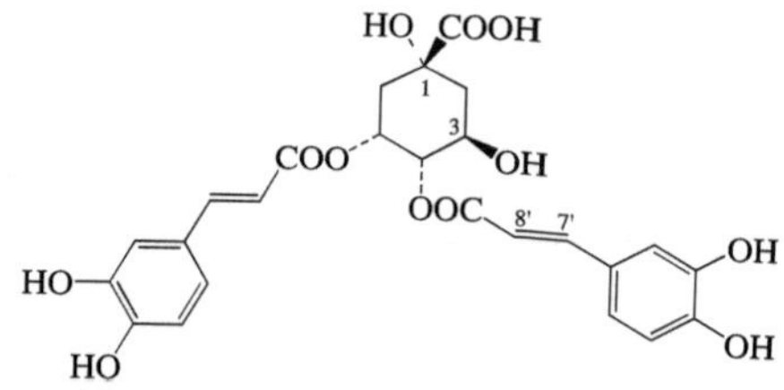

4,5-Bis(3,4-dihydroxycinnamoyl)quinic acid. ***Isochlorogenic acid C***
[57378-72-0]
$C_{25}H_{24}O_{12}$ 516.455
Numbering systems vary. Isol. from *Aster scaber*, *Gynura divaricata*, *Artemisia princeps*, *Artemisia herba-alba*, *Chrysothamnus paniculatus*, *Dichrocephala bicolor*, *Lonicera japonica* and *Pluchea symphytifolia*, coffee, Brazilian propolis and maté. Active against HIV-1 integrase. Shows antiviral and antihepatotoxic activities. Immunomodulator. Moderate antioxidant. Cryst.. Mp 140° dec.. $[\alpha]^{28}_{D}$ -257.8 (c, 0.46 in MeOH) (-199.6)

Me ester
[188742-80-5]
[170717-20-1]
$C_{26}H_{26}O_{12}$ 530.482
Constit. of *Erigeron breviscapus*, *Gynura divaricata*, *Lonicera japonica* and *Tessaria integrifolia*. Antioxidant

1-Me ether
[1360540-40-4]
$C_{26}H_{26}O_{12}$ 530.482
Isol. from *Inula crithmoides*. Displays hepatoprotective activity. Amorph. yellow powder. λ_{max} 208; 332 (MeOH)

3′-Me ether, 3-O-[(methoxycarbonyl) carbonyl]
5-O-trans-Caffeoyl-3-methoxalyl-4-O-trans-feruloylquinic acid
[1290614-31-1]
$C_{29}H_{28}O_{15}$ 616.528
Constit. of the flowers of *Arnica montana*

(1′→3)-Lactone
4,5-Di-O-caffeoyl-1,3-quinide
[140895-65-4]
$C_{25}H_{22}O_{11}$ 498.440
Constit. of roasted coffee beans

7′,8′-Dihydro, Me ester
[2232186-45-5]
$C_{26}H_{28}O_{12}$ 532.498

Constit. of *Salicornia herbacea*. Antioxidant. Amorph. powder. $[\alpha]_D^{25}$ -13.6 (c, 0.1 in MeOH).

Scarpati, M.L. *et al.*, *Tetrahedron Lett.*, 1964, **5**, 2851-2853 (*Isochlorogenic acid C*).
Timmermann, B.N. *et al.*, *J. Nat. Prod.*, 1983, **46**, 365-368 (*Chrysothamnus paniculatus constit*).
Tatefuji, T. *et al.*, *Biol. Pharm. Bull.*, 1996, **19**, 966-970 (*propolis constit*).
Miketova, P. *et al.*, *J. Mass Spectrom.*, 1999, **34**, 1240-1252 (*ms*).
Lin, L.C. *et al.*, *J. Nat. Prod.*, 1999, **62**, 405-408 (*Dichrocephala bicoilor constit*).
Kwon, H.C. *et al.*, *Chem. Pharm. Bull.*, 2000, **48**, 1796-1798 (*Aster scaber constit, activity*).
Zhang, W.-D. *et al.*, *J. Asian Nat. Prod. Res.*, 2000, **2**, 283-288 (*Me ester*).
Farah, A. *et al.*, *J. Agric. Food Chem.*, 2005, **53**, 1505-1513 (*(1'→3)-lactone*).
Jaiswal, R. *et al.*, *J. Agric. Food Chem.*, 2011, **59**, 4033-4039 (*Arnica montana constits*).
Ela, M.A.A. *et al.*, *Helv. Chim. Acta*, 2012, **95**, 61-66 (*1-Me ether*).
Dahmani-Hamzaoui, N. *et al.*, *Helv. Chim. Acta*, 2012, **95**, 564-576 (*Atremisia herba-alba constit, activity*).
Chen, J. *et al.*, *Fitoterapia*, 2014, **99**, 1-6 (*Gynura divaricata constits*).
Cho, J.-Y. *et al.*, *Molecules*, 2016, **21**, 1097:1-13 (*Salicornia herbacea constit, activity*)

3,5-Dichloro-6-(1-propenyl)-1,2,4-benzenetriol D37

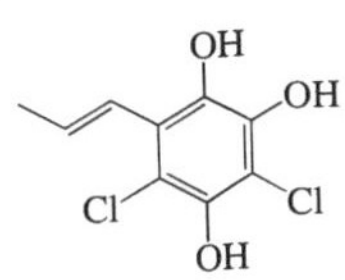

$C_9H_8Cl_2O_3$ 235.060

(E)-form

2-Me ether

3,5-Dichloro-2-methoxy-6-(1-propenyl)-1,4-benzenediol

2-Me ether, 1-O-β-D-glucopyranoside

Przewatangoside B

[2260925-95-7]

$C_{16}H_{20}Cl_2O_8$ 411.228

Constit. of *Przewalskia tangutica*. Amorph. powder. $[\alpha]_D^{20}$ +13 (c, 0.1 in MeOH)

2,4-Di-Me ether

3,5-Dichloro-2,4-dimethoxy-6-(1-propenyl)phenol

2,4-Di-Me ether, 1-O-β-D-glucopyranoside

Przewatangoside C

[2260925-96-8]

$C_{17}H_{22}Cl_2O_8$ 425.255

Constit. of *Przewalskia tangutica*. Exhibits weak cytotoxic activity. Amorph. powder. $[\alpha]_D^{20}$ +17 (c, 0.1 in MeOH).

Zhao, M. *et al.*, *Phytochem. Lett.*, 2017, **20**, 168-171 (*Przewatangosides B,C, activity*)

5′,5′-Dicinnamophilin D38

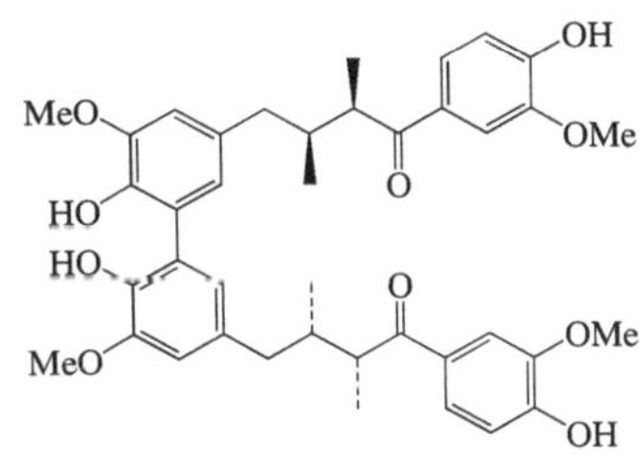

Absolute Configuration

5′,5′-Bis(4,4′-dihydroxy-3,3′-dimethoxy-8,8′-lignan-7-one).

5′,5′-Biscinnamophilin

$C_{40}H_{46}O_{10}$ 686.798

Constit. of the stems of *Machilus philippinensis*. Syrup. $[\alpha]_D^{25}$ +32 (c, 0.2 in $CHCl_3$). λ_{max} 227 (log ε 3.5); 280 (log ε 2.9) (MeOH).

Tai, S.-H. *et al.*, *RSC Adv.*, 2019, **9**, 21616-21625 (*5′,5′-Dicinnamophilin*)

1,4-Di-*p*-coumaroylglucose D39

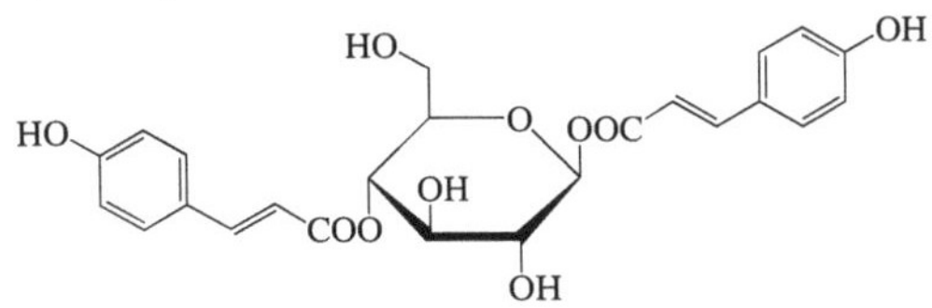

1,4-Bis-O-(4-hydroxycinnamoyl)glucose

$C_{24}H_{24}O_{10}$ 472.446

β-D-Pyranose (E,E-)-form

[75799-16-5]

Isol. from fern *Dennstaedtia scandens*. Needles (MeOH). Mp 235 - 236°. $[\alpha]_D^{25}$ -7.1 (c, 0.4 in MeOH).

Tanaka, N. *et al.*, *Chem. Pharm. Bull.*, 1980, **28**, 2843-2845 (*Denstaedtia scandens constit*)

1,6-Di-*p*-coumaroylglucose D40

1,6-Bis-O-(4-hydroxycinnamoyl)glucose

$C_{24}H_{24}O_{10}$ 472.446

β-D-Pyranose-(E,E)-form

[365460-31-7] (penta-Ac)

Constit. of the leaves of *Catalpa ovata*.

Machida, K. *et al.*, *Nat. Med. (Tokyo)*, 2001, **55**, 64-67 (*Catalpa ovata constit*)

1,2-Di-*p*-coumaroylglycerol D41

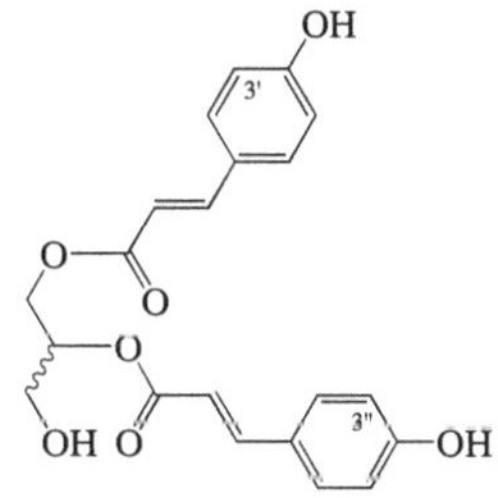

1,2-Bis-O-(4-hydroxycinnamoyl)glycerol

$C_{21}H_{20}O_7$ 384.384

(E,E)-form [115788-10-8]

Isol. from *Epichloe typhina*

3′-Methoxy

2-O-p-Coumaroyl-1-O-feruloylglycerol

[126240-21-9]

$C_{22}H_{22}O_8$ 414.41

Constit. of *Lilium auratum*.

Racemic

3″-Methoxy

1-O-p-Coumaroyl-2-O-feruloylglycerol

[126240-26-4]

$C_{22}H_{22}O_8$ 414.41

Constit. of *Lilium auratum*.

Racemic

3′,3″-Dimethoxy

1,2-Di-O-feruloylglycerol

[126240-20-8]

$C_{23}H_{24}O_9$ 444.436

Constit. of *Lilium auratum*, *Juncus effusus* and *Smilax excelsa*. Powder. Racemic.

Shimomura, H. *et al.*, *Phytochemistry*, 1987, **26**, 844-845 (*Lilium auratum constits*).
Koshino, H. *et al.*, *Phytochemistry*, 1988, **27**, 1333-1338 (*Epichloe typhina constit*).
Dellagreca, M. *et al.*, *Nat. Prod. Lett.*, 1998, **12**, 263-270 (*Juncus effusus constit*).
Ivanova, A. *et al.*, *Chem. Nat. Compd. (Engl. Transl.)*, 2010, **46**, 295-297 (*Smilax excelsa constit*)

1,3-Di-*O*-*p*-coumaroylglycerol D42

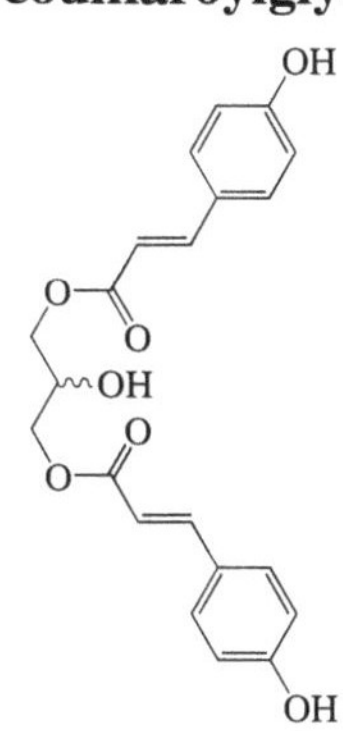

1,3-Bis-O-(4-hydroxycinnamoyl)glycerol

$C_{21}H_{20}O_7$ 384.384

***(E,E)*-form** [115788-09-5]
Constit. of *Aegilops ovata*, *Sparganium stoloniferum* and *Juncus effusus*. Isol. from *Epichloe typhina*. Amorph. solid. Mp 194 - 196° (178-184°)

2-Ac
2-O-Acetyl-1,3-di-O-p-coumaroylglycerol. ***Lasiocarpin A***
[65556-57-2]
$C_{23}H_{22}O_8$ 426.421
Constit. of *Populus lasiocarpa*. Cryst. (AcOH aq.). Mp 121 - 122°. The *E,E*-config. was not specifically proven for the Lasiocarpins

3′-Hydroxy
1-O-Caffeoyl-3-O-p-coumaroylglycerol
[125564-68-3]
$C_{21}H_{20}O_8$ 400.383
Constit. of the fruit of *Ananas comosus*

3′-Hydroxy, 2-Ac
2-O-Acetyl-1-O-trans-caffeoyl-3-O-p-coumaroylglycerol. ***Lasiocarpin B***
[65526-89-8]
$C_{23}H_{22}O_9$ 442.42
Constit. of *Populus lasiocarpa*. Yellow oil. *E,E*-Config. not explicitly proven

3′,3″-Dihydroxy
1,3-Di-O-trans-caffeoylglycerol. ***Ananasate***
[123134-22-5]
$C_{21}H_{20}O_9$ 416.382
Constit. of *Ananas comosus*, *Paepalanthus microphyllus* and *Tillandsia streptocarpa*. Powder. Mp 157°

3′,3″-Dihydroxy, 2-Ac
2-O-Acetyl-1,3-di-O-trans-caffeoylglycerol. ***Lasiocarpin C***
[65526-90-1]
$C_{23}H_{22}O_{10}$ 458.419
Constit. of *Populus lasiocarpa*. Cryst. (Me_2CO). Mp 160 - 162°. *E,E*-Config. not explicitly proven

3′-Methoxy
1-O-p-Coumaroyl-3-O-feruloylglycerol
[108026-21-7]
$C_{22}H_{22}O_8$ 414.41
Constit. of *Aegilops ovata*, *Lilium* spp. and the fruit of *Ananas comosus* (pineapple). Pale yellow powder. $[\alpha]^{25}_{D}$ -0.9 (c, 1.3 in MeOH)

3′-Methoxy, 3″-hydroxy
1-O-Caffeoyl-3-O-feruloylglycerol
[910101-38-1]
$C_{22}H_{22}O_9$ 430.409
Constit. of the fruit of *Ananas comosus*

3′-Methoxy, 3″,5″-dihydroxy
1-O-Feruloyl-3-O-(5-hydroxycaffeoyl)glycerol. *1-O-Feruloyl-3-O-(3,4,5-trihydroxycinnamoyl)glycerol*
[953420-02-5]
$C_{22}H_{22}O_{10}$ 446.408
Constit. of the fruit of *Ananas comosus*

3′,3″-Dimethoxy
1,3-Di-O-feruloylglycerol
[83008-46-2]
$C_{23}H_{24}O_9$ 444.436
Constit. of *Aegilops ovata*, *Lilium* spp. and *Sparganium stoloniferum*. Amorph. solid. Mp 47 - 56°

3′,3″-Dimethoxy, 5′-hydroxy
1-O-Feruloyl-3-O-(5-hydroxyferuloyl)glycerol. *1-O-(3,4-Dihydroxy-5-methoxycinnamoyl)-3-O-feruloylglycerol*
[953420-01-4]
$C_{23}H_{24}O_{10}$ 460.435
Constit. of the fruit of *Ananas comosus*

***(E,Z)*-form**
3′-Methoxy
[1166972-55-9]
$C_{22}H_{22}O_8$ 414.41
Constit. of the rhizomes of *Sparganium stoloniferum*. Yellowish solid. $[\alpha]^{25}_{D}$ -1.2 (c, 0.5 in MeOH). λ_{max} 230; 291; 312 (MeOH).

Asakawa, Y. *et al.*, *Phytochemistry*, 1977, **16**, 1791-1795 (*Lasiocarpins A-C*).
Cooper, R. *et al.*, *Phytochemistry*, 1978, **17**, 1673-1675 (*Aegilops ovata constit*).
Ori, K. *et al.*, *Phytochemistry*, 1992, **31**, 2767-2775 (*3′-methoxy, bibl*).
Delaporte, R.H. *et al.*, *Biochem. Syst. Ecol.*, 2006, **34**, 599-602 (*Ananasate*).
Ma, C. *et al.*, *J. Chromatogr., A*, 2007, **1165**, 39-44 (*pineapple cinnamates*).
Xiong, Y. *et al.*, *Arch. Pharmacal Res.*, 2009, **32**, 717-720 (*1-trans-Coumaroyl-3-cis-feruloylglycerol*).
Sun, S. *et al.*, *J. Am. Oil Chem. Soc.*, 2014, **91**, 1339-1345 (*Ananasate, synth*).
Sun, S. *et al.*, *J. Agric. Food Chem.*, 2015, **63**, 7261-7269 (*3′,3″-dimethoxy, synth*).
Ma, T. *et al.*, *Molecules*, 2017, **22**, 506:1-17 (*isol, activity*)

Di-*O*-demethylwikstrocoumarin A D43

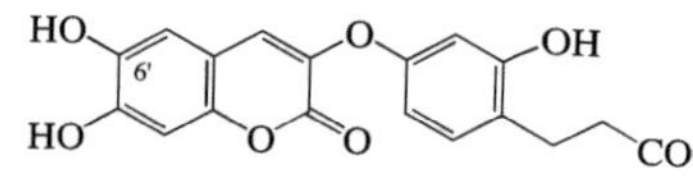

4-[(6,7-Dihydroxy-2-oxo-2H-1-benzopyran-3-yl)oxy]-2-hydroxybenzenepropanoic acid
$C_{18}H_{14}O_8$ 358.302

6′-Me ether
[2102024-17-7]
$C_{19}H_{16}O_8$ 372.329
Constit. of *Daphne oleoides*. No phys. props. reported

6′-Me ether, 2-O-β-D-glucopyranoside
[2102024-16-6]
$C_{25}H_{26}O_{13}$ 534.470
Constit. of *Daphne oleoides*. No phys. props. reported

6′-Me ether, Me ester
Wikstrocoumarin A
[1169690-86-1]
$C_{20}H_{18}O_8$ 386.356
Constit. of *Wikstroemia indica*. Needles ($CHCl_3$). λ_{max} 263; 346 (MeOH)

6′-Me ether, Et ester
[1448666-09-8]
$C_{21}H_{20}O_8$ 400.383
Constit. of *Daphne pedunculata*. Amorph. powder. Mp 193 - 194°.

Chen, Y. *et al.*, *Chin. Chem. Lett.*, 2009, **20**, 592-594 (*Wikstrocoumarin A*).
Du, J.-L. *et al.*, *Chem. Nat. Compd. (Engl. Transl.)*, 2013, **49**, 426-427 (*Daphne pedunculata Et ester*).
Venditti, A. *et al.*, *Chem. Biodiversity*, 2017, **14**, e1700072:1-8 (*Daphne oleoides constits*)

Didymochlaenone A D44

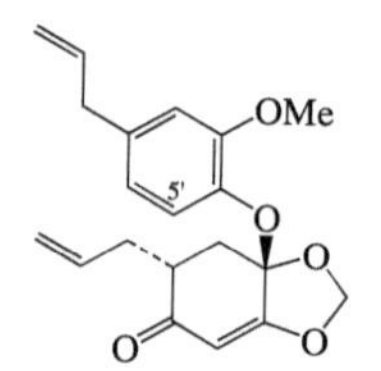

Absolute Configuration

1,2-Dihydro-3′-methoxy-3,4-methylenedioxy-3,4′-oxyligna-8,8′-dien-6(3H)-one
[877659-97-7]
$C_{20}H_{22}O_5$ 342.391
Constit. of the roots of *Didymochlaena truncatula*. Oil. $[\alpha]^{22}_{D}$ +112.5 (c, 0.08 in EtOH). λ_{max} 246 (log ε 3.97); 283 (log ε 3.46) (EtOH)

5′-Methoxy

1,2-Dihydro-3′,5′-dimethoxy-3,4-methylenedioxy-3,4′-oxyligna-8,8′-dien-6(3H)-one. ***Didymochlaenone C***

[1651214-78-6]

$C_{21}H_{24}O_6$ 372.417

Constit. of the stems of *Ocotea cymosa*. Oil. $[\alpha]^{25}_{D}$ -119 (c, 0.2 in MeOH). λ_{max} 246; 280 (MeOH aq.).

Cao, S. *et al., J. Nat. Prod.*, 2006, **69**, 284-286 (*Didymochlaenone A*).
Rakotondraibe, L.H. *et al., J. Nat. Prod.*, 2015, **78**, 431-440 (*Didymochlaenone C*)

7,9′:7′,9-Diepoxy-3,4:3′,4′-bis(methylenedioxy)-8,8′-lignan D45

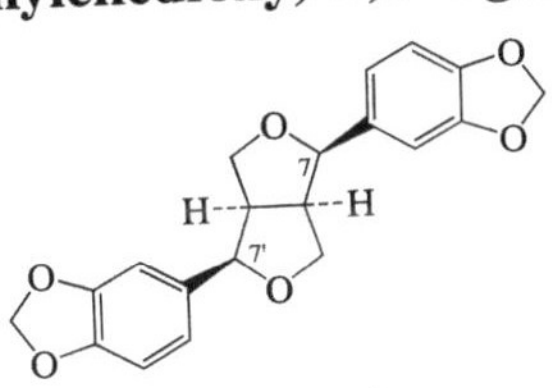

(7R,7′R,8R,8′R)-form

5,5′-(Tetrahydro-1H,3H-furo[3,4-c]furan-1,4-diyl)bis-1,3-benzodioxole, CAS. *2,6-Bis(3,4-methylenedioxyphenyl)-3,7-dioxabicyclo[3.3.0]octane. 3,4:3′,4′-Bis(methylenedioxy)-7,9′:7′,9-diepoxylignan*

$C_{20}H_{18}O_6$ 354.358

Lignan numbering shown. Log P 1.9 (calc)

(7R,7′R,8R,8′R)-form

Diasesamin. Epiasarinin [551-30-4]

Obt. by isom. of Asarinin. Constit. of the stems of *Cinnamomum camphora*. Powder. Mp 168 - 171°. $[\alpha]^{21}_{D}$+214 (c, 0.3 $CHCl_3$). λ_{max} 217 (log ε 4.10); 240 (log ε 4.14); 282 (log ε 4.06) (MeCN)

(7R,7′R,8S,8′S)-form

(-)-Sesamin [13079-95-3]

Formed by heating (-)-Asarinin with EtOH/HCl. Isol. from *Magnolia campellii, Magnolia mutabilis, Zanthoxylum tingoassuiba, Zanthoxylum senegalensis, Fagara viridis, Piper* sp., *Amanoa oblongifolia, Paraaristolochia flos-avis, Sideritis canariensis, Acanthopanax sessiliflorus* and others. Antisickling agent. Moderate inhibitor of LPS-induced NO prodn.. Sol. MeOH, Et_2O; poorly sol. H_2O. Mp 122 - 124°. $[\alpha]^{17}_{D}$-68 (c, 0.5 in $CHCl_3$). λ_{max} 236 (ε 9330); 286 (ε 8130) (MeOH) (Berdy). [neutral] λ_{max} 205; 237; 288 (EtOH) (Berdy)

(7R,7′S,8R,8′R)-form

(+)-Asarinin. (+)-Episesamin. Isosesamin. Xanthoxylin S [133-03-9]

Constit. of *Acronychia muelleri, Lindera obtusiloba, Asarum sieboldii, Zanthoxylum carolianicum, Asarum* sp., *Piper* spp.. Antitubercular and antiinflammatory agent. Shows antitumour props. Inhibitor of superoxide generation in neutrophils, and of fMLP-induced elastase release. Cryst. (EtOH or Me_2CO). Sol. MeOH, Et_2O; poorly sol. H_2O. Mp 121 - 122°. $[\alpha]^{18}_{D}$+121.7 ($CHCl_3$). Log P 1.9 (calc). Also formed by isomerisation of (+)-Sesamin. λ_{max} 230 (ε 8900); 287 (ε 8500) (MeOH) (Berdy)

(7S,7′S,8R,8′R)-form

(+)-Sesamin. Pseudocubebin. Fagarol [607-80-7]

Widespread. Isol. from *Sideritis canariensis, Commiphora mukul, Zanthoxylum* spp., *Artemisia gorgonum, Chrysanthemum cinerariaefolium, Asarum sieboldii, Piper longum* and *Talauma hodgsoni*. Constit. of sesame oil. Present in the Rutaceae, Piperaceae, Scrophulariaceae, Compositae and Ginkgoaceae. Synergist for pyrethrum insecticides. Antiinflammatory and antiplasmodial agent. Mitigator of oxidative stress in endothelial cells. Inhibitor of P-glycoprotein mediated efflux from LS-180V cells, and of superoxide generation in neutrophils. Inhibits fMLP-induced elastase release, and phagocycte oxidative burst. Metalloproteinase inhibitor. Needles (EtOH). Mp 123 - 124°. $[\alpha]^{20}_{D}$+78.4 ($CHCl_3$)

(7S,7′R,8S,8′S)-form

(-)-Asarinin. (-)-Episesamin [133-04-0]

Isol. from *Asarum, Boronia* and *Zanthoxylum* spp.. Shows antibacterial activity. Cryst. (EtOH). Mp 121°. $[\alpha]^{18}_{D}$-122 (c, 3.54 in $CHCl_3$). Acid isom. → (-)-Sesamin

(7RS,7′RS,8SR,8′SR)-form

(±)-Sesamin [81602-22-4]

Constit. of *Zanthoxylum senegalense, Zanthoxylum alatum, Coleonema pulchellum* and *Fagara xanthoxyloides*. Cryst. (EtOH). Mp 129 - 130°

(7RS,7′SR,8RS,8′RS)-form

[105661-97-0]

Constit. of the leaves of *Zanthoxylum alatum*. Mp 134 - 135°.

Davenport, J.B. *et al., Aust. J. Chem.*, 1954, **7**, 384-386 (*Asarinin*).
Carnmalm, B. *et al., Acta Chem. Scand.*, 1955, **9**, 1111-1118 (*(±)-form, isol*).
Beroza, M. *et al., J. Am. Chem. Soc.*, 1956, **78**, 1242-1247 (*(±)-Sesamin, (±)-Asarinin, synth*).
Beroza, M. *J. Am. Chem. Soc.*, 1956, **78**, 5082-5084 (*Epiasarinin, synth*).
Ramaswamy, A.S. *Naturwissenschaften*, 1957, **44**, 380 (*Sesamin, antitubercular activity*).
Freudenberg, K. *et al., Chem. Ber.*, 1961, **94**, 851-862 (*config*).
Takahashi, K. *et al., Chem. Pharm. Bull.*, 1966, **14**, 641-647 (*pmr, config*).
Pelter, A. *J. Chem. Soc. C*, 1967, 1376-1380 (*ms*).
Krajniak, R.E. *et al., Aust. J. Chem.*, 1973, **26**, 687-689 (*Sesamin, isol*).
Abe, F. *et al., Chem. Pharm. Bull.*, 1974, **22**, 2650-2655 (*(-)-Asarinin, (-)-Sesamin*).
Talapatra, B. *et al., Phytochemistry*, 1975, **14**, 589-590 (*(-)-form, isol*).
Dutta, C.P. *et al., Phytochemistry*, 1975, **14**, 2090-2091 (*(+)-Sesamin, Piper longum constit*).
Karrer, W. *Konstitution und Vorkommen der Organischen Pflanzenstoffe, 2nd edn., Birkhäuser Verlag*, 1976-1985, no. 1154 (*occur*).
Pelter, A. *et al., Tetrahedron*, 1976, **32**, 2783-2788 (*struct, pmr, cmr*).
Anjaneyulu, A.S.R. *et al., Tetrahedron*, 1977, **33**, 133-143 (*pmr*).
Pelter, A. *et al., Tetrahedron Lett.*, 1977, **18**, 4137-4140 (*Sesamin, Episesamin, cmr*).
Pelter, A. *et al., Tetrahedron Lett.*, 1978, **19**, 1509-1512 (*Sesamin, pmr, cmr*).
Orito, K. *et al., Tetrahedron Lett.*, 1991, **32**, 5999-6002 (*(±)-Sesamin, synth*).
Stevens, D.R. *et al., J. Chem. Soc., Perkin Trans. 1*, 1992, 185-190 (*synth*).
Takano, S. *et al., Synlett*, 1993, 785-787 (*Asarinin, synth*).
Il'in, S.G. *et al., Chem. Nat. Compd. (Engl. Transl.)*, 1994, **30**, 525-526 (*Asarinin, cryst struct*).
Samizu, K. *et al., Chem. Lett.*, 1995, 543-544 (*(+)-Sesamin, (-)-Asarinin, synth*).
Suginone, H. *et al., J. Org. Chem.*, 1995, **60**, 3052-3064 (*synth, pmr*).
Takasaki, M. *et al., Biol. Pharm. Bull.*, 1997, **20**, 776-780 (*Asarinin, antitumour activity*).
Ahsan, M. *et al., Fitoterapia*, 1997, **68**, 464-465 (*(-)-Episesamin, antibacterial activity*).
Kato, M.J. *et al., Phytochemistry*, 1998, **47**, 583-591 (*biosynth*).
Christov, R. *et al., Fitoterapia*, 1999, **70**, 89-92 (*cmr*).
Rana, K.K. *et al., Tetrahedron Lett.*, 2000, **41**, 9337-9338 (*(±)-Sesamin, synth*).
Brown, R.C.D. *et al., J. Org. Chem.*, 2001, **66**, 6719-6728 (*Asarinin, synth*).
Roy, S.C. *et al., J. Org. Chem.*, 2002, **67**, 3242-3248 (*(±)-Sesamin, synth*).
Aldous, D.J. *et al., Org. Lett.*, 2002, **4**, 1159-1162 (*(±)-Epiasarinin, synth*).
Enders, D. *et al., Synthesis*, 2002, 512-522 (*(-)-form, synth, ir, pmr, cmr*).
Li, C.-Y. *et al., J. Nat. Prod.*, 2005, **68**, 1622-1624 (*Sesamin, Asarinin, cryst struct*).
Banerjee, B. *et al., Synthesis*, 2005, 2913-2919 (*(-)-form, synth, ir, pmr, cmr*).

Hsieh, T.J. *et al.*, *Nat. Prod. Commun.*, 2006, **1**, 21-25 (*Diasesamin, Sesamin, Episesamin*).
Kim, J.-C. *et al.*, *Tetrahedron: Asymmetry*, 2006, **17**, 3-6 (*(+)-form, (-)-form, synth*).
Seo, S.-M. *et al.*, *Chem. Nat. Compd. (Engl. Transl.)*, 2008, **44**, 419-423 (*(+)-Sesamin*).
Lee, W.-J. *et al.*, *J. Agric. Food Chem.*, 2009, **57**, 11406-11417 (*Sesamin, endothelial cell protectant*).
Mbaze, L.M. *et al.*, *Phytochemistry*, 2009, **70**, 1442-1447 (*(+)-Sesamin, oxidative burst suppression*).
Okura, T. *et al.*, *Biol. Pharm. Bull.*, 2010, **33**, 255-259 (*(+)-Sesamin, glycoprotein efflux inhibitor*).
Chen, J.-J. *et al.*, *Food Chem.*, 2011, **125**, 282-287 (*(+)-Sesamin, (+)-Episesamin, superoxide elastase inhibitor*).
Ortet, R. *et al.*, *J. Ethnopharmacol.*, 2011, **138**, 637-640 (*(+)-Sesamin, antiplasmodial activity*).
Trowitzsch-Kienast, W. *et al.*, *Z. Naturforsch., C*, 2011, **66**, 460-466 (*Episesamin, abs config*).
Quang, T.H. *et al.*, *Bioorg. Med. Chem. Lett.*, 2012, **22**, 2527-2533 (*Asarum sieboldii constits, antiinflammatory activity*).
Negi, D.S. *et al.*, *Acta Crystallogr., Sect. C*, 2013, **69**, 87-89 (*Asarinin, Sesamin, cryst struct, bibl*).
Dar, A.A. *et al.*, *Bioorg. Chem.*, 2013, **50**, 1-10 (*Sesamin, rev, activity, biosynth, occur*).
Lee, D.-Y. *et al.*, *Molecules*, 2013, **18**, 41-49 (*(-)-Sesamin, NO prodn inhibitor*).
Ishikawa, R. *et al.*, *Chem. Lett.*, 2014, **43**, 1572-1574 (*(+)-Sesamin, synth*).
Inai, M. *et al.*, *Synthesis*, 2015, **47**, 3518-3521 (*(+)-Sesamin, synth*).
Li, K. *et al.*, *Fitoterapia*, 2019, **134**, 135-140 (*Asarinin, cryst struct*).
Hajra, S. *et al.*, *Tetrahedron*, 2020, **76**, 131483:1-6 (*(+)-Sesamin, synth*)

7,9′:7′,9-Diepoxy-5,5′-diethyl-8,8′-lignan-3,3′,4,4′-tetrol D46

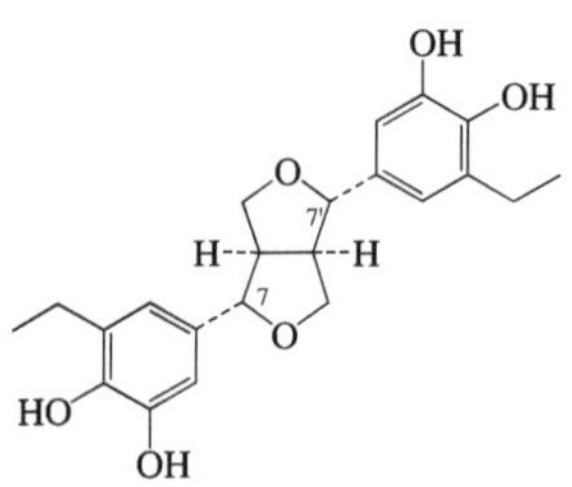

2,6-Bis(3-ethyl-4,5-dihydroxyphenyl)-3,7-dioxabicyclo[3.3.0]octane. 3,3′-Diethyl-4,4′,5,5′-tetrahydroxy-7,9′:7′,9-diepoxylignan
$C_{22}H_{26}O_6$ 386.444
(7S,7′S,8R,8′R)-form
3,3′-Di-Me ether
7,9′:7′,9-Diepoxy-5,5′-diethyl-3,3′-dimethoxy-8,8′-lignan-4,4′-diol. 2,6-Bis(3-ethyl-4-hydroxy-5-methoxyphenyl)-3,7-dioxabicyclo[3.3.0]octane. 3,3′-Diethyl-4,4′-dihydroxy-5,5′-dimethoxy-7,9′:7′,9-diepoxylignan
[334699-29-5]
$C_{24}H_{30}O_6$ 414.498
Constit. of *Daphne oleoides*. Amorph. powder. Mp 151 - 153°. $[\alpha]_D^{20}$ +30.5 (c, 0.1 in $CHCl_3$). λ_{max} 225; 272 (CH_2Cl_2).

Ullah, N. *et al.*, *Helv. Chim. Acta*, 2001, **84**, 157-162 (*Daphne oleoides constit*)

7,9′:7′,9-Diepoxy-3,3′-dimethoxy-8,8′-lignan-4,4′-diol D47

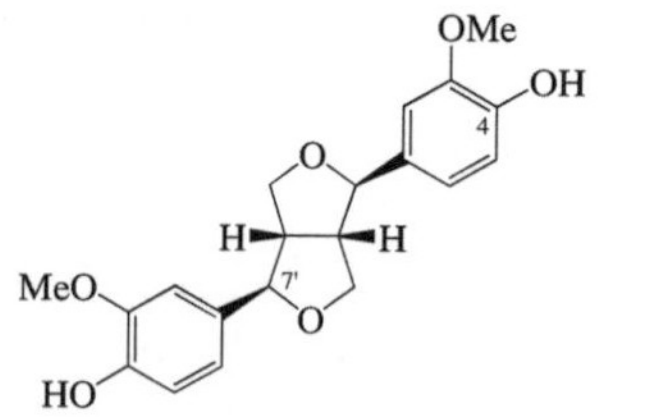

(7R,7′R,8S,8′S)-form

4,4′-(Tetrahydro-1H,3H-furo[3,4-c]furan-1,4-diyl)bis[2-methoxyphenol], CAS. 2,6-Bis-(4-hydroxy-3-methoxyphenyl)-3,7-dioxabicyclo[3.3.0]octane, CAS. 4,4′-Dihydroxy-3,3′-dimethoxy-7,9′:7′,9-diepoxylignan
[23171-10-0] (D-Pinoresinol)
$C_{20}H_{22}O_6$ 358.390
(7R,7′R,8S,8′S)-form
(-)-Pinoresinol [81446-29-9]
Isol. from *Balanophora japonica*, *Lonicera insulata*, *Berberis koreana* and *Festuca argentina*. Shows moderate cytotoxicity against XF-498 cells. Inhibitor of NO prodn. in LPS-activated BV-Z cells. $[\alpha]_D^{23}$ -63.5 (c, 1.97 in MeOH)
4-O-β-D-Glucopyranoside
[41607-20-9]
$C_{26}H_{32}O_{11}$ 520.531
Constit. of *Festuca argentina*, *Berberis koreana* and *Symplocos lucida*. $[\alpha]_D$ -82.3 (MeOH). Genus name given as Simplocos
Di-O-β-D-glucopyranoside
[153061-63-3]
$C_{32}H_{42}O_{16}$ 682.672
Constit. of *Daphne pseudo-mezereum*. Mp 240 - 248°. $[\alpha]_D^{28}$ -57.3 (c, 0.89 in H_2O)
4-O-[4-Hydroxy-3-methoxy-E-cinnamoyl-(→6)-β-D-glucopyranoside]
Pinoresinol 4-(6-trans-feruloylglucoside)
[947172-56-7]
$C_{36}H_{40}O_{14}$ 696.702
Constit. of the roots of *Rhus javanica* var. *roxburghiana*. Powder. $[\alpha]_D^{27}$ -14 (c, 0.43 in Me_2CO). λ_{max} 220 (log ε 4.38); 231 (sh); 283 (log ε 4.08); 301 (log ε 4); 329 (log ε 4.12) (MeOH)
4-O-[3,6-Bis(4-hydroxy-3-methoxy-E-cinnamoyl)-β-D-glucopyranoside]
Pinoresinol 4-(3,6-di-trans-feruloylglucoside)
[947172-57-8]
$C_{46}H_{48}O_{17}$ 872.873
Constit. of the roots of *Rhus javanica* var. *roxburghiana*. Powder. $[\alpha]_D^{27}$ -15.5 (c, 1.1 in Me_2CO). λ_{max} 222 (log ε 4.35); 284 (log ε 4.11); 300 (log ε 4.02); 331 (log ε 4.14) (MeOH)
4-O-[4,6-Bis(4-hydroxy-3-methoxy-E-cinnamoyl)-β-D-glucopyranoside]
Pinoresinol 4-(4,6-di-trans-feruloylglucoside)
[947172-58-9]
$C_{46}H_{48}O_{17}$ 872.873
Constit. of the roots of *Rhus javanica* var. *roxburghiana*. Powder. $[\alpha]_D^{27}$ -6.9 (c, 0.45 in Me_2CO). λ_{max} 219 (log ε 4.28); 234 (log ε 4.02); 284 (log ε 4.06); 302 (log ε 4.04); 331 (log ε 4.16) (MeOH)
Di-Me ether
see 7,9′:7′,9-Diepoxy-3,3′,4,4′-tetramethoxy-8,8′-lignan, D104
4-O-(3-Methyl-2-butenyl)
4-O-Prenyl-(-)-pinoresinol. ***Boropinan***
[299422-22-3]
Constit. of the roots of *Boronia pinnata*. Oil. $[\alpha]_D$ -54 (c, 0.12 in $CHCl_3$). λ_{max} 206; 231; 280 (MeOH)
Bis-O-(3-methyl-2-butenyl)
Di-O-(3-methyl-2-butenyl)-(-)-pinoresinol. *Di-O-prenyl-(-)-pinoresinol*
$C_{30}H_{38}O_6$ 494.628
Constit. of the bark of *Zanthoxylum simulans*. Solid. $[\alpha]_D^{26.8}$ -32.1 (c, 0.3 in $CHCl_3$). λ_{max} 241 (log ε 3.37); 282 (log ε 3.13) ($CHCl_3$)
4-O-(3,7-Dimethyl-2E,6-octadienyl)
4-O-Geranyl-(-)-pinoresinol. ***Planispin B***
[1202693-36-4]
$C_{30}H_{38}O_6$ 494.628
Constit. of the bark of *Zanthoxylum armatum* and *Zanthoxylum*

planispinum. Oil. [α] D -5.4 (c, 0.95 in Me_2CO)
(7R,7′S,8R,8′R)-form
(+)-Epipinoresinol [24404-50-0]
Isol. from *Picea excelsa*, *Asarum sieboldii*, *Forsythia intermedia* and *Streptomyces* sp.. Cryst. (EtOH). Sol. MeOH, $CHCl_3$, C_6H_6, EtOAc; poorly sol. H_2O. Mp 137 - 138°. $[\alpha]^{25}_{D}$+130 (Me_2CO). [α] D +110 (c, 5.0 in MeOH). λ_{max} 232; 282 (MeOH) (Berdy)
4-O-β-D-Glucopyranoside
(+)-Epipinoresinol glucoside
[74983-66-7]
$C_{26}H_{32}O_{11}$ 520.531
Constit. of *Anisacanthus virgularis*, *Forsythia intermedia* and *Forsythia suspensa*. Cryst. (Me_2CO). Mp 192 - 194°. $[\alpha]^{22}_{D}$+242.4 (c, 0.33 in MeOH). λ_{max} 228; 279; 310 (MeOH)
4,4′-Di-O-β-D-glucopyranoside
(+)-Epipinoresinol diglucoside
[1426550-88-0]
$C_{32}H_{42}O_{16}$ 682.672
Constit. of *Sinopodophyllum emodi*. Amorph. powder. $[\alpha]^{25}_{D}$-25 (c, 0.3 in MeOH). λ_{max} 208 (log ε 2.46); 228 (log ε 1.1); 278 (log ε 0.36) (MeOH)
4-O-[α-L-Rhamnopyranosyl-(1→6)-β-D-glucopyranoside]
Versicoside
[101391-02-0]
$C_{32}H_{42}O_{15}$ 666.673
Constit. of *Haplophyllum versicolor*. Cryst. (Me_2CO). Mp 223 - 224°. $[\alpha]^{20}_{D}$-8.8 (c, 0.61 in EtOH)
(7R,7′S,8S,8′S)-form
(-)-Epipinoresinol. Symplocosigenol
[10061-38-8]
Constit. of *Asiasarum heterotropoides*, *Balanophora abbreviata* and *Zanthoxylum planispinum*. Moderately potent inhibitor of LPS-induced iNOS expression in RAW 264.7 cells. Mp 141 - 141.5°. $[\alpha]^{17}_{D}$-118.9 (MeOH)
4-O-β-D-Glucopyranoside
Symplocosin. *(-)-Epipinoresinol glucoside*
[11042-30-1]
$C_{26}H_{32}O_{11}$ 520.531
Isol. from *Symplocos lucida*. Mp 171 - 171.5°. $[\alpha]^{11}_{D}$-44.9

4-O-(3-Methyl-2-butenyl)
Planispin A
[1202761-42-9]
$C_{25}H_{30}O_6$ 426.509
Constit. of the bark of *Zanthoxylum armatum* and *Zanthoxylum planispinum*. Oil. [α] D -22.3 (c, 0.35 in Me_2CO)
(7S,7′S,8R,8′R)-form
(+)-Pinoresinol [487-36-5]
Constit. of various *Picea*, *Pinus* and *Abies* spp. Also from *Daphne tangutica*, *Eucommia ulmoides*, *Forsythia suspensa*, *Forsythia koreana*, *Streptomyces* sp., *Himatanthus sucuuba*, *Acanthopanax sessiliflorus* and *Nardostachys jatamansi*. Antinociceptive agent. Moderately active inhibitor of oestradiol response. Inhibitor of NO prodn. RAW 264.7 cells. Potent antifungal agent against *Candida albicans*. Cryst. (EtOH aq.). Sol. MeOH, Et_2O; poorly sol. H_2O. Mp 122°. $[\alpha]^{21}_{D}$+84.4 (c, 5.0 in Me_2CO). λ_{max} 232 (ε 15500); 282 (ε 6300) (MeOH) (Berdy). [neutral] λ_{max} 230 (ε 14800); 280 (ε 6620) (EtOH) (Berdy)
4-O-β-D-Apiofuranoside, 4′-O-β-D-glucopyranoside
[1374324-58-9]
$C_{31}H_{40}O_{15}$ 652.646
Constit. of *Broussonetia papyrifera*. Shows inhibitory activity against HepG-2 cells. Amorph. yellow powder. $[\alpha]^{25}_{D}$-53.6 (c, 1.11 in MeOH)
4-O-α-D-Galactopyranoside
[332901-08-3]
$C_{26}H_{32}O_{11}$ 520.531
Constit. of *Daphne oleoides*. Amorph. powder. Abs. config. not certain
4-O-β-D-Glucopyranoside
[69251-96-3]
$C_{26}H_{32}O_{11}$ 520.531
Occurs in fruits of *Forsythia suspensa* and *Forsythia koreana*
Di-O-β-D-glucopyranoside
[63902-38-5]
$C_{32}H_{42}O_{16}$ 682.672
Constit. of the bark of *Eucommia ulmoides*. Mp 221 - 230°. $[\alpha]^{25}_{D}$ -27.3 (c, 0.54 in H_2O)
4-O-[3,4,5-Trihydroxybenzoyl-(→6)-β-D-glucopyranoside]
Pinoresinol 4-(6-galloylglucoside)
[1177291-33-6]

$C_{33}H_{36}O_{15}$ 672.636
Constit. of the leaves of *Macaranga tanarius*. Amorph. powder. $[\alpha]^{25}_{D}$+18.1 (c, 0.44 in MeOH). λ_{max} 218 (log ε 4.42); 276 (log ε 4.05) (MeOH)
4-O-[3,4-Dihydroxy-E-cinnamoyl-(→6)-β-D-glucopyranoside]
Pinoresinol 4-O-(6-trans-caffeoylglucoside)
[449142-26-1]
$C_{35}H_{38}O_{14}$ 682.675
Constit. of the aerial parts of *Baccharis dracunculifolia*. Amorph. powder. $[\alpha]^{22}_{D}$+39 (c, 0.69 in MeOH). λ_{max} 203 (log ε 4.95); 204 (log ε 4.91); 221 (log ε 4.38); 230 (sh); 284 (log ε 4.07); 300 (log ε 4.03); 330 (log ε 4.16) (MeOH)
4-O-[β-D-Apiofuranosyl-(1→2)-β-D-glucopyranoside]
[123564-49-8]
$C_{31}H_{40}O_{15}$ 652.646
Constit. of the leaves of *Parsonsia laevigata*. Solid. $[\alpha]^{25}_{D}$-47.5 (c, 0.7 in MeOH)
4-O-[α-L-Rhamnopyranosyl-(1→2)-β-D-glucopyranoside]
Hibiscuside
[250724-23-3]
$C_{32}H_{42}O_{15}$ 666.673
Constit. of *Urtica laetevirens* and the root bark of *Hibiscus syriacus*. Antioxidant. Amorph. powder (MeOH). Mp 143 - 145°. $[\alpha]^{20}_{D}$+8.5 (c, 0.01 in MeOH aq.). Confusion in the lit. over whether this is a rhamnosylglucoside or a glucosylrhamnoside. The former is correct. λ_{max} 215 (sh); 229 (log ε 1.79); 279 (log ε 0.78) (MeOH aq.)
4-O-[β-D-Glucopyranosyl-(1→2)-[β-D-glucopyranosyl-(1→6)]-β-D-glucopyranoside]
KP3
[150035-90-8]
$C_{38}H_{52}O_{21}$ 844.813
Constit. of sesame seeds. Antioxidant. $[\alpha]^{25}_{D}$-12 (c, 0.14 in H_2O)
4-O-Sulfate
Pinoresinol 4-sulfate
[928045-60-7]
$C_{20}H_{22}O_9S$ 438.447
Constit. of the roots of *Frankenia thymifolia*. Pale yellow cryst.. Mp 147°. $[\alpha]^{22}_{D}$+95 (c, 0.4 in MeOH). λ_{max} 230; 275 (MeOH)

Di-Ac
[32971-25-8]
$C_{24}H_{26}O_8$ 442.464
Prod. by *Pinus* sp.. Mp 166 - 167.5°. $[\alpha]_D^{20}$ +50.9 (c, 2.5 in Me_2CO)

Bis-O-(3-methylbutanoyl)
Pinoresinol diisovalerate
[81915-92-6]
$C_{30}H_{38}O_8$ 526.626
Constit. of *Callilepis salicifolia*. Gum

4-O-(3-Methyl-2-butenyl)
O-(3-Methyl-2-butenyl)pinoresinol. *4-O-Prenyl-(+)-pinoresinol*
[230307-52-5]
$C_{25}H_{30}O_6$ 426.509
Constit. of the fruit of *Zanthoxylum integrifoliolum*. Oil. $[\alpha]_D^{25}$ +35.8 (c, 0.065 in $CHCl_3$)

Bis-O-(3-methyl-2-butenyl)
Di-O-(3-methyl-2-butenyl)-(+)-pinoresinol.
Di-O-prenyl-(+)-pinoresinol
[230307-51-4]
$C_{30}H_{38}O_6$ 494.628
Constit. of the fruit of *Zanthoxylum integrifoliolum*. Oil. $[\alpha]_D^{25}$ +41.6 (c, 0.13 in $CHCl_3$). λ_{max} 234 (log ε 5.4); 281 (log ε 5.93) (EtOH)

4,4′-Diangeloyl
4,4′-Diangeloyl-(+)-pinoresinol
[1428768-21-1]
$C_{30}H_{34}O_8$ 522.594
Lignan isol. from the aerial parts of *Rudbeckia laciniata*. Gum. $[\alpha]_D^{25}$ +14.0 (c, 0.17 in $CHCl_3$). λ_{max} 216 (log ε 4.0); 276 (log ε 3.9) (MeOH).

Erdtman, H. *Justus Liebigs Ann. Chem.*, 1935, **516**, 162-176 (*(+)-Pinoresinol, di-Ac*).
Haworth, R.D. *J. Chem. Soc.*, 1942, 448-456 (*struct*).
Nishida, K. *et al.*, *Nippon Rin Gakkaishi*, 1951, **33**, 235-239 (*Symplocosigenol, struct*).
Nishida, K. *et al.*, *Nippon Rin Gakkaishi*, 1951, **33**, 269-272 (*Symplocosin, Symplocosigenol*).
Freudenberg, K. *Chem. Ber.*, 1953, **86**, 1157-1166 (*synth*).
Gurevich, A.I. *et al.*, *Dokl. Akad. Nauk SSSR*, 1960, **131**, 833-836 (*Symplocosin, abs config*).
Weinges, K. *Tetrahedron Lett.*, 1960, **No. 20**, 1-2 (*Pinoresinol, Epipinoresinol*).
Freudenberg, K. *et al.*, *Tetrahedron Lett.*, 1960, **No. 20**, 3-6 (*Pinoresinol, abs config*).
Freudenberg, K. *et al.*, *Chem. Ber.*, 1961, **94**, 851-862 (*Pinoresinol, abs config*).
Weinges, K. *et al.*, *Chem. Ber.*, 1961, **94**, 2522-2533 (*(+)-Pinoresinol, (+)-Epipinoresinol, stereochem*).
Kratzl, K. *et al.*, *Monatsh. Chem.*, 1963, **94**, 434-439 (*(±)-Pinoresinol, synth*).
Ludwig, C.H. *et al.*, *J. Am. Chem. Soc.*, 1964, **86**, 1186-1196 (*pmr*).
Pelter, A. *J. Chem. Soc. C*, 1967, 1376-1380 (*ms*).
Duffield, A.M. *J. Heterocycl. Chem.*, 1967, **4**, 16-22 (*Pinoresinol, ms*).
Sih, C.J. *et al.*, *J. Am. Chem. Soc.*, 1976, **98**, 5412-5413 (*Pinoresinol diglucoside*).
Chiba, M. *et al.*, *Phytochemistry*, 1980, **19**, 335-336 (*Symplocosin*).
Bohlmann, F. *et al.*, *Phytochemistry*, 1982, **21**, 139-142 (*diisovalerate*).
Tsukamoto, H. *et al.*, *Tennen Yuki Kagobutsu Toronkai Koen Yoshishu*, 1983, **26**, 181-188 (*glucosides*).
Batirov, E.Kh. *et al.*, *Chem. Nat. Compd. (Engl. Transl.)*, 1985, **21**, 584-587 (*Versicoside*).
Abe, F. *et al.*, *Phytochemistry*, 1989, **28**, 1737-1741 (*4-apiosylglucoside*).
Rahman, M.M.A. *et al.*, *Phytochemistry*, 1990, **29**, 1971-1980 (*(+)-Epipinoresinol, glucoside, occur*).
Nabeta, K. *et al.*, *Phytochemistry*, 1991, **30**, 3591-3593 (*biosynth*).
Davin, L.B. *et al.*, *Phytochemistry*, 1992, **31**, 3869-3874 (*(+)-Pinoresinol, biosynth*).
Katayama, T. *et al.*, *Phytochemistry*, 1992, **31**, 3875-3881 (*(-)-Pinoresinol, biosynth*).
Katsuzaki, H. *et al.*, *Heterocycles*, 1993, **36**, 933-936 (*KP3*).
Konishi, T. *et al.*, *Yakugaku Zasshi*, 1993, **113**, 670-675 (*diglucoside*).
Maruyama, J. *et al.*, *Heterocycles*, 1994, **37**, 839-845 (*(±)-Pinoresinol, synth*).
Casabuono, A.C. *et al.*, *Phytochemistry*, 1994, **35**, 479-483 (*(-)-Pinoresinol*).
Chen, I.S. *et al.*, *J. Nat. Prod.*, 1999, **62**, 833-837 (*O-Prenylpinoresinols*).
Lee, S.-J. *et al.*, *Planta Med.*, 1999, **65**, 658-660 (*Hibiscuside*).
Ito, C. *J. Nat. Prod.*, 2000, **63**, 1344-1348 (*Boropinan*).
Riaz, M. *et al.*, *Z. Naturforsch., B*, 2000, **55**, 1216-1222 (*4-galactoside*).
Lee, K. *et al.*, *Planta Med.*, 2001, **67**, 99-102 (*Lonicera insulata constit*).
Nagatani, Y. *et al.*, *Chem. Pharm. Bull.*, 2002, **50**, 583-589 (*6-caffeoylglucoside*).
Roy, S.C. *et al.*, *J. Org. Chem.*, 2002, **67**, 3242-3248 (*(±)-Pinoresinol, synth*).
El-Domiaty, M.M. *et al.*, *Pharm. Biol.*, 2002, **40**, 96-102 (*Epipinoresinol 4-glucoside*).
Li, N. *et al.*, *Yunnan Zhiwu Yanjiu*, 2003, **25**, 711-715 (*Curculigo capitulata glucoside*).
Hosokawa, A. *et al.*, *Chem. Pharm. Bull.*, 2004, **52**, 1265-1267 (*(-)-Epipinoresinol, iNOS expression inhibitor*).
Harkat, H. *et al.*, *Biochem. Syst. Ecol.*, 2007, **35**, 176-179 (*Pinoresinol 4-sulfate*).
Ouyang, M.-A. *et al.*, *J. Agric. Food Chem.*, 2007, **55**, 6460-6465 (*4-feruloylglucosides*).
Hu, Y. *et al.*, *Bull. Korean Chem. Soc.*, 2009, **30**, 1884-1886 (*Planispins A,B*).
Zhou, Y. *et al.*, *J. Nat. Med.*, 2009, **63**, 100-101 (*2-rhamnosylglucoside*).
Matsunami, K. *et al.*, *Phytochemistry*, 2009, **70**, 1277-1285 (*4-6-galloylglucoside*).
Hwang, B. *et al.*, *Molecules*, 2010, **15**, 3507-3516 (*(+)-Pinoresinol, antifungal activity*).
Kim, K.-H. *et al.*, *Bioorg. Med. Chem. Lett.*, 2011, **21**, 2270-2273 (*Berberis koreana constits, cytotoxicity, NO prodn inhibitor*).
Aehle, E. *et al.*, *Phytochemistry*, 2011, **72**, 2396-2405 (*Pinoresinol, estradiol response inhibitor*).
Sun, Y.-J. *et al.*, *Planta Med.*, 2012, **78**, 480-484 (*Epipinoresinol diglucoside*).
Ran, X-K. *et al.*, *Chin. J. Nat. Med.*, 2013, **11**, 269-273 (*Broussonetia papyrifera glycoside*).
Lee, S.Y. *et al.*, *Helv. Chim. Acta*, 2013, **96**, 320-325 (*4,4′-Diangeloyl-(+)-pinoresinol*).
Lee, D.-Y. *et al.*, *Molecules*, 2013, **18**, 41-49 (*Acanthopanax sessiliflorus constit, NO prodn inhibitor*).
Albertson, A.K.F. *et al.*, *Angew. Chem., Int. Ed.*, 2015, **54**, 2204-2208 (*Pinoresinol, synth*).
Peng, C.-Y. *et al.*, *J. Asian Nat. Prod. Res.*, 2015, **17**, 232-238 (*Di-O-prenyl-(-)-pinoresinol*).
Zhang, W. *et al.*, *Molecules*, 2018, **23**, 617:1-10 (*Planispins A,B*)

7,7′:9,9′-Diepoxy-3,3′-dimethoxy-8,8′-lignan-4,4′,8-triol D48

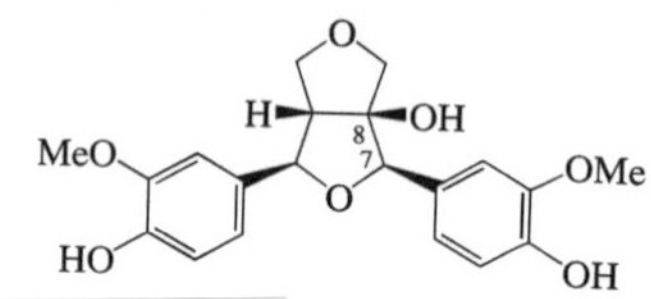

Relative Configuration

1-Hydroxy-2,4-bis(4-hydroxy-3-methoxyphenyl)-3,7-dioxabicyclo[3.3.0]octane. 4,4′,8-Trihydroxy-3,3′-dimethoxy-7,7′:9,9′-diepoxylignan
$C_{20}H_{22}O_7$ 374.389

(7R*,7′R*,8S*,8′S*)-form
[408513-62-2]
Constit. of the aerial parts of *Saussurea medusa*. Weak immunosuppressant. Amorph. powder. $[\alpha]_D^{25}$ +20.7 (c, 1.4 in MeOH). λ_{max} 231 (log ε 4.14); 280 (log ε 3.71) (MeOH).

Duan, H. *et al.*, *Phytochemistry*, 2002, **59**, 85-90 (*Saussurea medusa constit, activity*)

7,9′:7′,9-Diepoxy-5′-ethyl-8,8′-lignan-3,3′,4,4′,5-pentol D49

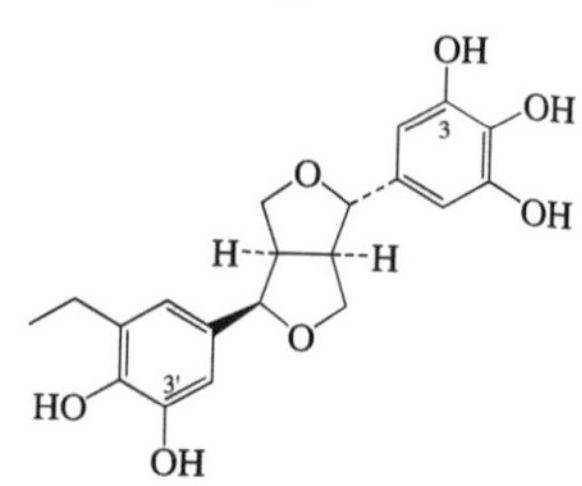

2-(3-Ethyl-4,5-dihydroxyphenyl)-6-(3,4,5-trihydroxyphenyl)-3,7-dioxabicyclo[3.3.0]octane. 3-Ethyl-3′,4,4′,5,5′-pentahydroxy-7,9′:7′,9-diepoxylignan
$C_{20}H_{22}O_7$ 374.389

(7S,7′R,8R,8′R)-form
3,3′,5-Tri-Me ether
7,9′:7′,9-Diepoxy-5′-ethyl-8,8′-lignan-4,4′-diol. 3-Ethyl-4,4′-dihydroxy-3′,5,5′-trimethoxy-7,9′:7′,9-diepoxylignan
[334699-28-4]
$C_{23}H_{28}O_7$ 416.47
Constit. of *Daphne oleoides*. Amorph. powder. Mp 162 - 163°. $[\alpha]^{20}_{D}$ +36.5 (c, 0.1 in $CHCl_3$). λ_{max} 220; 275 (CH_2Cl_2).

Ullah, N. *et al.*, *Helv. Chim. Acta*, 2001, **84**, 157-162 (*Daphne oleoides constit*)

7,9′:9,9*a*-Diepoxy-9*a*-homo-8,8′-lignan-3,3′,4,4′,8,8′-hexol D50

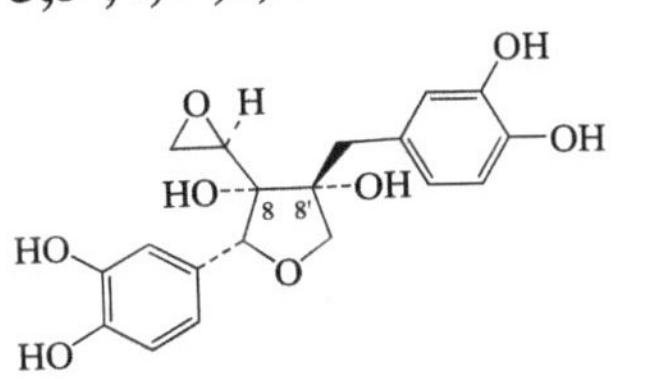

$C_{19}H_{20}O_8$ 376.361
(7R,8R,8′R,9R)-form
3,3′-Di-Me ether
7,9′:9,9a-Diepoxy-9a-homo-3,3′-dimethoxy-8,8′-lignan-4,4′,8,8′-tetrol
$C_{21}H_{24}O_8$ 404.415
Constit. of *Dolomiaea berardioidea*. Yellow needles. $[\alpha]^{20}_{D}$ -25.2 (c, 0.1 in MeOH).

Huang, Y.-Y. *et al.*, *Phytochem. Lett.*, 2019, **31**, 125-130 (*Dolomiaea berardioidea constit, cryst struct*)

7,9′:7′,9-Diepoxy-8,8′-lignan-4,4′-diol D51

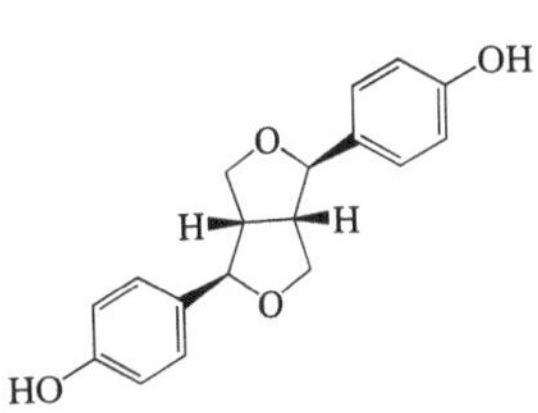

(7R,7′R,8S,8′S)-form

4,4′-(Tetrahydro-1H,3H-furo[3,4-c]furan-1,4-diyl)bisphenol, CAS. *2,6-Bis-(4-hydroxyphenyl)-3,7-dioxabicyclo[3.3.0]octane. 4,4′-Dihydroxy-7,9′:7′,9-diepoxylignan*
$C_{18}H_{18}O_4$ 298.338
(7R,7′R,8S,8′S)-form
(-)-Ligballinol [55332-75-7]
Constit. of *Ecballium elaterium*, *Gynostemma pentaphyllum*, *Carya cathayensis*, *Zanthoxylum wutaiense* and *Podocarpus fleuryi*. Needles ($CHCl_3$). Mp 264 - 266°. $[\alpha]_D$ -7.1 (c, 1.61 in MeOH)
(7R,7′S,8R,8′R)-form
(-)-Idaeusinol B [2376889-70-0]
Exhibits weak neuroprotective activity. Oil. $[\alpha]^{20}_{D}$ -36.3 (c, 0.14 in MeOH)
(7S,7′R,8S,8′S)-form
(+)-Idaeusinol B [2376889-69-7]
Oil. $[\alpha]^{20}_{D}$ +38.1 (c, 0.14 in MeOH)
(7S,7′S,8R,8′R)-form
(+)-Ligballinol [88082-92-2]
Stress metab. of cell cultures of *Vigna angularis*, also from *Mappianthus iodoides*. Needles. $[\alpha]^{28}_{D}$ +25.3 (c, 0.06 in MeOH). λ_{max} 227; 276; 283 (sh) (MeOH)
(7RS,7′SR,8RS,8′RS)-form
Idaeusinol B
Constit. of the fruit of *Rubus ideaus*. Light oil. $[\alpha]^{20}_{D}$ -0.3 (c, 0.1 in MeOH). Partial racemate resolved by chiral hplc. λ_{max} 226 (log ε 4.16); 277 (log ε 3.45) (MeOH).

Rao, M.M. *et al.*, *Tetrahedron*, 1974, **30**, 3309-3313 (*Ecballium elaterium constit*).
Kobayashi, M. *et al.*, *Phytochemistry*, 1983, **22**, 1257-1261 (*Vigna angularis constit*).
Wang, X.W. *et al.*, *Chin. Chem. Lett.*, 2009, **20**, 589-591 (*Gynostemma pentaphyllum constit, cryst struct*).
Zhou, L. *et al.*, *Phytochemistry*, 2019, **164**, 122-129 (*Idaeusinol B, cd, activity*)

7,9′:7′,9-Diepoxy-8,8′-lignan-2,2′,3,3′,4,4′,9-heptol D52

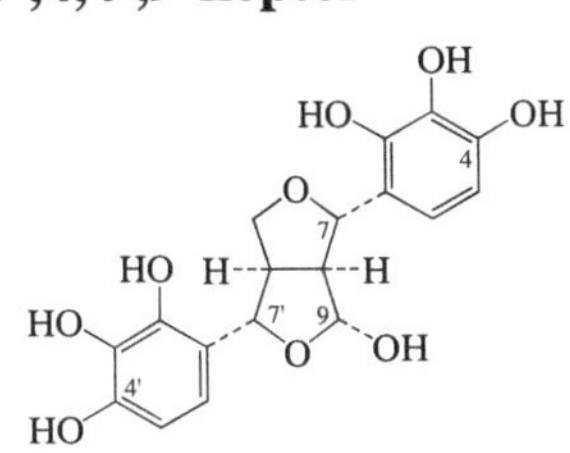

2,2′,3,3′,4,4′,9-Heptahydroxy-7,9′:7′,9-diepoxylignan
$C_{18}H_{18}O_9$ 378.333
(7S,7′S,8S,8′R,9R)-form
3,4:3′,4′-Bis(methylene), 9-Me ether
7,9′:7′,9-Diepoxy-9-methoxy-3,4:3′,4′-bis(methylenedioxy)-8,8′-lignan-2,2′-diol. 2,2′-Dihydroxy-9-methoxy-3,4:3′,4′-bis(methylenedioxy)-7,9′:7′,9-diepoxylignan. ***2,2′-Dihydroxy-9-methoxysesamin****. 9-Methoxysesamin-2,2′-diol*
[878000-25-0]
$C_{21}H_{20}O_9$ 416.382
Constit. of the leaves of *Callicarpa furfuraceae*. Oil. $[\alpha]^{27}_{D}$ -39 (c, 0.08 in MeOH).

Shao, Y. *et al.*, *Helv. Chim. Acta*, 2006, **89**, 64-72 (*2,2′-Dihydroxy-9-methoxysesamin*)

7,9′:7′,9-Diepoxy-8,8′-lignan-2,2′,4,4′,5,5′,8-heptol D53

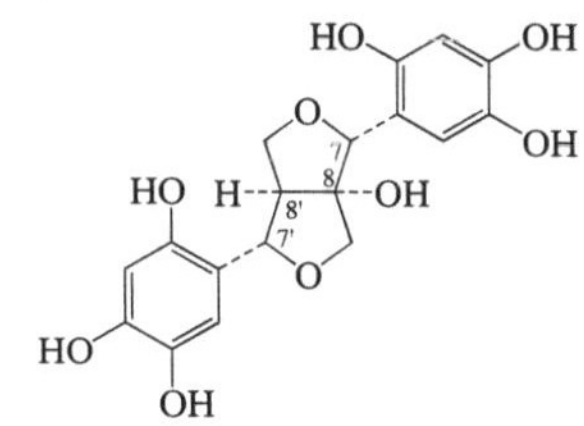

2,2′,4,4′,5,5′,8-Heptahydroxy-7,9′:7′,9-diepoxylignan
$C_{18}H_{18}O_9$ 378.333
(7R,7′S,8S,8′R)-form
4,5:4′,5′-Bis(methylene), 2,2′-di-Me ether
7,9′:7′,9-Diepoxy-2,2′-dimethoxy-4,5:4′,5′-bis(methylenedioxy)-8,8′-lignan-8-ol. 8-Hydroxy-2,2′-dimethoxy-4,5:4′,5′-bis(methylenedioxy)-7,9′:7′,9-diepoxylignan
4,5:4′,5′-Bis(methylene), 2,2′-di-Me ether, 8-Ac
Phrymarin I
[303155-52-4]
$C_{24}H_{24}O_{10}$ 472.446
Constit. of *Phryma leptostachya*. $[\alpha]^{27}_{D}$ +28.2 (c, 1 in $CHCl_3$) (synthetic).

Ishibashi, F. *et al.*, *Biosci., Biotechnol., Biochem.*, 2001, **65**, 29-34 (*Phrymarin I, synth, abs config*)

7,9′:7′,9-Diepoxy-8,8′-lignan-2,3,3′,4,4′,5,5′-heptol D54

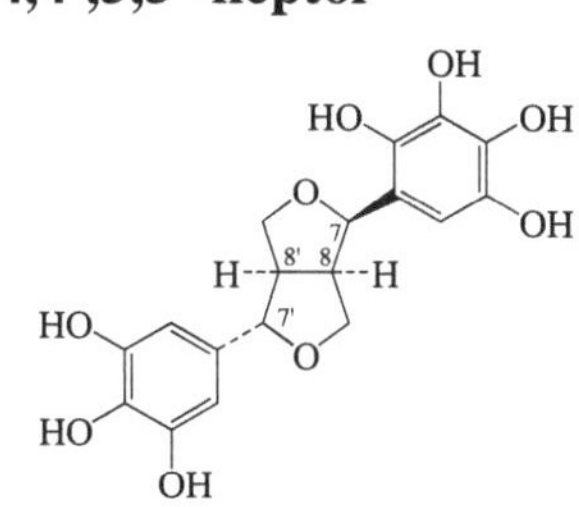

(7R,7′S,8R,8′R)-form

2,3,3′,4,4′,5,5′-Heptahydroxy-7,9′:7′,9-diepoxylignan
$C_{18}H_{18}O_9$ 378.333
(7R,7′S,8R,8′R)-form
3′,4′-Methylene, 3,4,5,5′-tetra-Me ether
7,9′:7′,9-Diepoxy-3,3′,4,5-tetramethoxy-4′,5′-

methylenedioxy-8,8′-lignan-2-ol. 2-Hydroxy-3,3′,4,5-tetramethoxy-4′,5′-methylenedioxy-7,9′:7′,9-diepoxylignan

3′,4′-Methylene, 3,4,5,5′-tetra-Me ether, 2-O-β-D-glucopyranoside

7-Epiterminaloside D

[1906937-18-5]

$C_{29}H_{36}O_{14}$ 608.593

Constit. of the leaves of *Terminalia citrina*. Amorph. powder. $[\alpha]_D^{25}$ +119.7 (c, 0.2 in MeOH). λ_{max} 216 (log ε 4.41); 277 (log ε 3.66) (MeOH)

(7S,7′S,8R,8′R)-form

3,3′,4,4′,5,5′-Hexa-Me ether, 2-O-β-D-glucopyranoside

Terminaloside G

[1906937-21-0]

$C_{30}H_{40}O_{14}$ 624.636

Constit. of the leaves of *Terminalia citrina*. Exhibits inhibitory effects on MCF-7 and T47D estrogen-responsive human breast cancer cell lines. Amorph. pale yellow powder. $[\alpha]_D^{25}$ +42.3 (c, 0.2 in MeOH). λ_{max} 214 (log ε 4.40); 277 (log ε 3.52) (MeOH)

3′,4′-Methylene, 3,4,5′-tri-Me ether, 2-O-β-D-glucopyranoside

Terminaloside C

[1906937-16-3]

$C_{28}H_{34}O_{14}$ 594.566

Constit. of the leaves of *Terminalia citrina*. Exhibits antiestrogenic activity against MCF-7 cells. Amorph. yellow powder. $[\alpha]_D^{25}$ +28.1 (c, 0.2 in MeOH). λ_{max} 219 (log ε 4.21); 281 (log ε 3.67) (MeOH)

3′,4′-Methylene, 3,4,5,5′-tetra-Me ether, 2-O-β-D-glucopyranoside

Terminaloside D

[1906937-17-4]

$C_{29}H_{36}O_{14}$ 608.593

Constit. of the leaves of *Terminalia citrina*. Amorph. pale yellow powder. $[\alpha]_D^{25}$ +45.4 (c, 0.2 in MeOH). λ_{max} 219 (log ε 4.25); 279 (log ε 3.61) (MeOH).

Muhit, M.A. *et al., J. Nat. Prod.*, 2016, **79**, 1298-1307 (*Terminalosides C,D,G, 2-Epiterminaloside D, activity*)

7,9′:7′,9-Diepoxy-8,8′-lignan-2,3,3′,4,4′,5,6-heptol D55

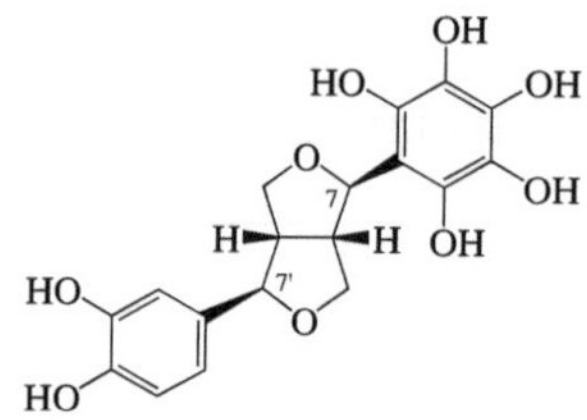

2,3,3′,4,4′,5,6-Heptahydroxy-7,9′:7′,9-diepoxylignan

$C_{18}H_{18}O_9$ 378.333

(7R*,7′R*,8S*,8′S*)-form

3,4:3′,4′-Bis(methylene), 2,5,6-tri-Me ether

7,9′:7′,9-Diepoxy-2,3,6-trimethoxy-3′,4′:4,5-bis(methylenedioxy)lignan. 2,3,6-Trimethoxy-3′,4′:4,5-bis(methylenedioxy)-7,9′:7′,9-diepoxylignan. ***Ecbolin A***

[509076-79-3]

$C_{23}H_{24}O_9$ 444.436

Constit. of the roots of *Ecbolium linneanum*. Needles. Mp 118 - 120°. λ_{max} 232; 280 (EtOH).

Venkataraman, R. *et al., Phytochemistry*, 2002, **61**, 963-966 (*Ecbolin A, cryst struct*)

7,9′:7′,9-Diepoxy-8,8′-lignan-2,3,3′,4,4′,8,8′-heptol D56

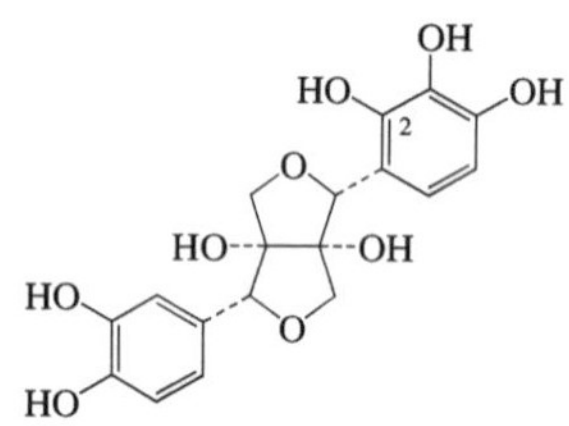

2,3,3′,4,4′,8,8′-Heptahydroxy-7,9′:7′,9-diepoxylignan

$C_{18}H_{18}O_9$ 378.333

(7R,7′R,8S,8′S)-form

3,4:3′,4′-Bis(methylene), 2-Me ether

7,9′:7′,9-Diepoxy-2-methoxy-3,4:3′,4′-bis(methylenedioxy)-8,8′-lignan-8,8′-diol. 8,8′-Dihydroxy-2-methoxy-3,4:3′,4′-bis(methylenedioxy)-7,9′:7′,9-diepoxylignan. ***Palatiferin B***

[1361308-21-5]

$C_{21}H_{20}O_9$ 416.382

Constit. of the roots of *Pseuderanthemum palatiferum*. Moderate cytotoxic agent against KB and HepG2 cells. Light yellow cryst.. Mp 167 - 169°. $[\alpha]_D^{30}$ -53 (c, 0.28 in $CHCl_3$). λ_{max} 213 (log ε 4.3); 284 (log ε 3.6) (MeOH).

Mai, H.D.T. *et al., Planta Med.*, 2011, **77**, 951-954 (*Palatiferin B*)

7,9′:7′,9-Diepoxy-8,8′-lignan-3,3′,4,4′,5,5′,7-heptol D57

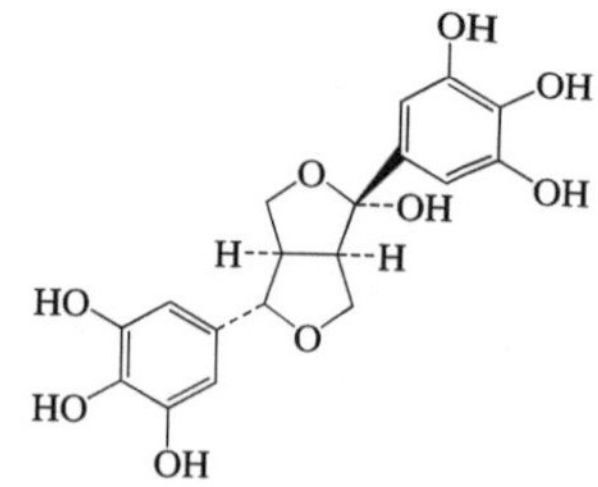

3,3′,4,4′,5,5′,7-Heptahydroxy-7,9′:7′,9-diepoxylignan

$C_{18}H_{18}O_9$ 378.333

(7R*,7′S*,8R*,8′R*)-form

Hepta-Me ether

7,9′:7′,9-Diepoxy-3,3′,4,4′,5,5′,7-heptamethoxylignan. 3,3′,4,4′,5,5′,7-Heptamethoxy-7,9′:7′,9-diepoxylignan. ***Caruilignan A***

[331960-70-4]

Constit. of aerial parts of *Artemisia caruifolia* and *Artemisia sieversiana*. Moderately cytotoxic to Meth-A cells. Amorph. powder. $[\alpha]_D^{27}$ +86 (c, 0.93 in $CHCl_3$) (+61.6). λ_{max} 243 (ε 7300); 270 (ε 1600) ($CHCl_3$).

Ma, C.-M. *et al., Chem. Pharm. Bull.*, 2001, **49**, 183-187 (*Caruilignan A*).
Mori, N. *et al., Synthesis*, 2006, 400-404 (*synth*).
Cui, J. *et al., Chem. Nat. Compd. (Engl. Transl.)*, 2020, **56**, 1120-1121 (*Caruilignan A*)

7,9′:7′,9-Diepoxy-8,8′-lignan-3,3′,4,4′,5,5′,8-heptol D58

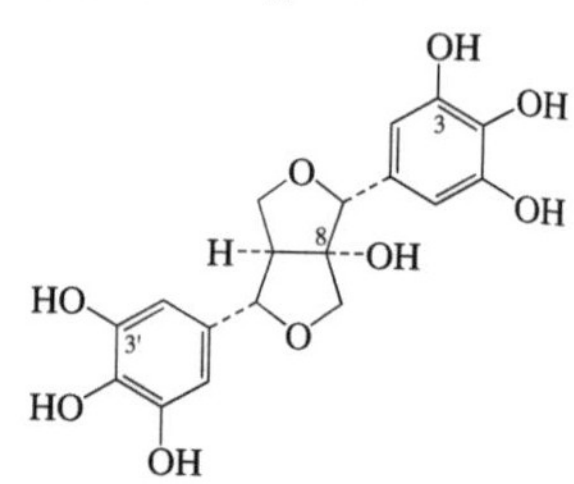

3,3′,4,4′,5,5′,8-Heptahydroxy-7,9′:7′,9-diepoxylignan

$C_{18}H_{18}O_9$ 378.333

(7R,7′S,8S,8′R)-form

3,3′,5,5′-Tetra-Me ether

7,9′:7′,9-Diepoxy-3,3′,5,5′-tetramethoxy-8,8′-lignan-4,4′,8-triol. 4,4′,8-Trihydroxy-3,3′,5,5′-tetramethoxy-7,9′:7′,9-diepoxy-

lignan. ***1-Hydroxysyringaresinol***. *8-Hydroxysyringaresinol*
[89199-95-1]
$C_{22}H_{26}O_9$ 434.441
Isol. from *Fraxinus mandshurica* var. *japonica*, *Fraxinus japonica*, *Mappianthus iodoides* and *Pseuderanthemum carruthersii* var. *atropurpureum*. Cryst. powder. Mp 95 - 97°. $[\alpha]_D^{26}$ +24.6 (c, 0.12 in $CHCl_3$)
3,3′,5,5′-Tetra-Me ether, 4-O-β-D-glucopyranoside
Ussuriensislignan B
[2231301-59-8]
$C_{28}H_{36}O_{14}$ 596.582
Constit. of fruit of *Pyrus ussuriensis*. Amorph. powder. $[\alpha]_D^{20}$ -3.7 (c, 0.27 in MeOH). λ_{max} 210 (log ε 5.74); 274 (log ε 4.46) (MeOH)
3,3′,5,5′-Tetra-Me ether, 4′-O-β-D-glucopyranoside
Ussuriensislignan A
[2231301-58-7]
$C_{28}H_{36}O_{14}$ 596.582
Constit. of fruit of *Pyrus ussuriensis*. Amorph. powder. $[\alpha]_D^{20}$ -5.26 (c, 0.57 in MeOH). λ_{max} 210 (log ε 5.42); 272 (log ε 3.12) (MeOH)
3,3′,5,5′-Tetra-Me ether, 8-O-β-D-glucopyranoside
Staunoside C
[153980-86-0]
$C_{28}H_{36}O_{14}$ 596.582
Constit. of *Stauntonia hexaphylla* and roots of *Sideritis pullulans*. Amorph. powder. Mp 101 - 104°. $[\alpha]_D^{20}$ -21.7 (c, 0.66 in MeOH). λ_{max} 240 (ε 16982); 262 (ε 5129) (EtOH).

Tsukamoto, H. *et al.*, *Chem. Pharm. Bull.*, 1984, **32**, 4482-4489 (*1-Hydroxysyringaresinol*).
Wang, H.-B. *et al.*, *Phytochemistry*, 1993, **34**, 1621-1624 (*Staunoside C*).
Faiella, L. *et al.*, *Phytochemistry*, 2014, **106**, 164-170 (*Sideritis pullulans constit*).
Zhao, Z.-Q. *et al.*, *J. Asian Nat. Prod. Res.*, 2016, **18**, 1151-1157 (*Ussuriensislignans A,B*)

7,9′:7′,9-Diepoxy-8,8′-lignan-3,3′,4,4′,5,5′,9-heptol D59

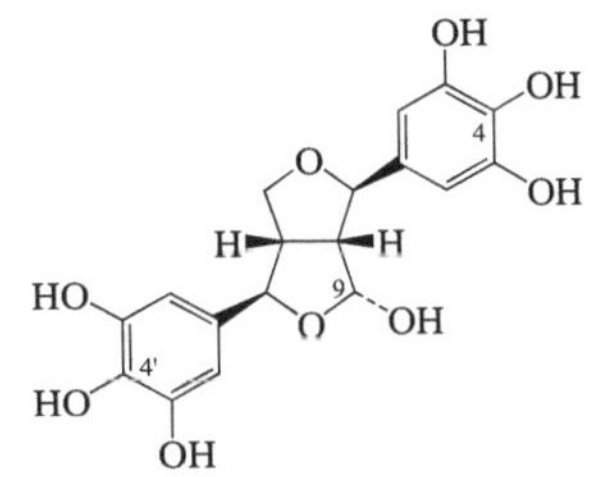

(7R,7′R,8R,8′S,9R)-form

3,3′,4,4′,5,5′,9-Heptahydroxy-7,9′:7′,9-diepoxylignan
$C_{18}H_{18}O_9$ 378.333
(7R,7′R,8R,8′S,9R)-form
3,3′,5,5′-Tetra-Me ether, 9-O-β-D-glucopyranoside
[2042682-90-4]
$C_{28}H_{36}O_{14}$ 596.582
Constit. of rhizomes of *Atractylodes lancea*. Amorph. powder. $[\alpha]_D^{20}$ -88.2 (c, 0.07 in MeOH). λ_{max} 208 (log ε 4.51); 240 (log ε 3.87); 279 (log ε 3.48) (MeOH)
(7S,7′S,8R,8′S,9R)-form
3,3′,5,5′-Tetra-Me ether, 9-O-β-D-glucopyranoside
[2042682-89-1]
Constit. of rhizomes of *Atractylodes lancea*. Amorph. powder. $[\alpha]_D^{20}$ -11.3 (c, 0.11 in MeOH). λ_{max} 209 (log ε 4.81); 242 (log ε 4.03); 272 (log ε 3.17) (MeOH)
(7S,7′S,8S,8′R,9R)-form
3,3′,5,5′-Tetra-Me ether
7,9′:7′,9-Diepoxy-3,3′,5,5′-tetramethoxy-8,8′-lignan-4,4′,9-triol. 4,4′,9-Trihydroxy-3,3′,5,5′-tetramethoxy-7,9′:7′,9-diepoxylignan
[869884-50-4]
$C_{22}H_{26}O_9$ 434.441
Constit. of callus cultures of *Cichorium intybus*
3,3′,5,5′-Tetra-Me ether, 4′-O-β-D-glucopyranoside
[869884-52-6]
$C_{28}H_{36}O_{14}$ 596.582
Constit. of callus cultures of *Cichorium intybus*
(7S,7′S,8S,8′R,9S)-form
3,3′,5,5′-Tetra-Me ether
[869884-51-5]
Constit. of callus cultures of *Cichorium intybus*
3,3′,5,5′-Tetra-Me ether, 4′-O-β-D-glucopyranoside
[869884-53-7]
Constit. of callus cultures of *Cichorium intybus*.

Malarz, J. *et al.*, *Plant Cell Rep.*, 2005, **24**, 246-249 (*Cichorium intybus constits*).
Xu, K. *et al.*, *Bioorg. Med. Chem. Lett.*, 2016, **26**, 5187-5192 (*Atractylodes lancea constits*)

7,9′:7′,9-Diepoxy-8,8′-lignan-3,3′,4,4′,5,8,8′-heptol D60

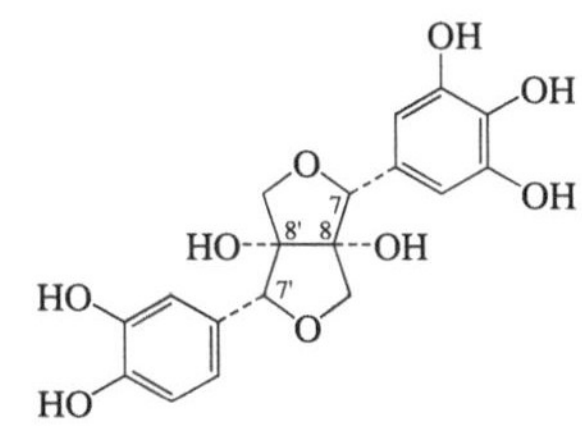

(7R,7′R,8S,8′S)-form

3,3′,4,4′,5,8,8′-Heptahydroxy-7,9′:7′,9-diepoxylignan
$C_{18}H_{18}O_9$ 378.333
(7R,7′R,8S,8′S)-form
3,3′,5-Tri-Me ether
7,9′:7′,9-Diepoxy-3,3′,5-trimethoxy-8,8′-lignan-4,4′,8,8′-tetrol. 4,4′,8,8′-Tetrahydroxy-3,3′,5-trimethoxy-7,9′:7′,9-diepoxylignan.
5-Methoxyprinsepiol
[1593092-57-9]
$C_{21}H_{24}O_9$ 420.414
Isol. from the roots of *Vladimiria muliensis*. Amorph. powder. $[\alpha]_D^{20}$ -36 (c, 1.1 in MeOH). λ_{max} 205 (log ε 3.82); 228 (log ε 3.48); 273 (log ε 3.22) (MeOH)
3,3′,5-Tri-Me ether, 4-O-β-D-glucopyranoside
[1593092-59-1]
$C_{27}H_{34}O_{14}$ 582.555
Isol. from the roots of *Vladimiria muliensis*. Oil. $[\alpha]_D^{20}$ -50 (c, 1.0 in MeOH). λ_{max} 209 (log ε 3.76); 225 (log ε 3.48); 273 (log ε 3.18) (MeOH)
(7R,7′S,8S,8′S)-form
3,3′,5-Tri-Me ether
Dipsalignan C
[1841146-34-6]
$C_{21}H_{24}O_9$ 420.414
Constit. of the roots of *Dipsacus asper*. Amorph. powder. $[\alpha]_D^{20}$ +35.8 (c, 0.06 in MeOH). λ_{max} 204; 231; 280 (MeOH)

(7S,7′S,8R,8′R)-form

3,3′,5-Tri-Me ether

Dipsalignan D

[1841146-35-7]

$C_{21}H_{24}O_9$ 420.414

Constit. of the roots of *Dipsacus asper*. Amorph. powder. $[\alpha]_D^{20}$ +71.2 (c, 0.05 in MeOH). λ_{max} 208; 230; 279 (MeOH).

Chen, J.-J. *et al.*, *Planta Med.*, 2013, **79**, 1470-1473 (*Vladimiria muliensis constits*).

Sun, X. *et al.*, *Molecules*, 2015, **20**, 2165-2175 (*Dipsalignans C,D*)

7,7′:7,8-Diepoxy-8,8′-lignan-3,3′,4,4′,5,5′-hexol D61

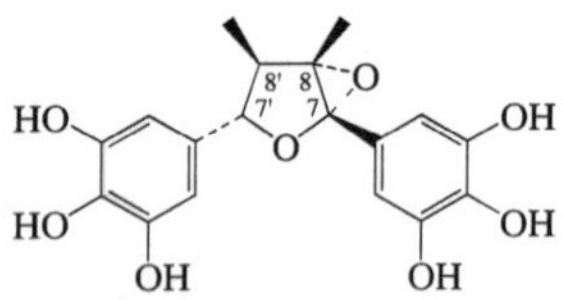

3,3′,4,4′,5,5′-Hexahydroxy-7,7′:7,8-diepoxylignan

$C_{18}H_{18}O_8$ 362.334

(7S,7′S,8S,8′R)-form

Hexa-Me ether

7,7′:7,8-Diepoxy-3,3′,4,4′,5,5′-hexamethoxylignan. 3,3′,4,4′,5,5′-Hexamethoxy-7,7′:7,8-diepoxylignan. ***4,5-Epoxybeilschmin B***

[1000269-48-6]

$C_{24}H_{30}O_8$ 446.496

Constit. of the leaves of *Beilschmiedia tsangii*. Oil. $[\alpha]_D^{24}$ +24.6 (c, 0.12 in $CHCl_3$). λ_{max} 219 (log ε 4.48); 270 (log ε 3.61) (MeOH)

3,4-Methylene, 3′,4′,5,5′-tetra-Me ether

7,7′:7,8-Diepoxy-3′,4′,5,5′-tetramethoxy-3,4-methylenedioxylignan. 3′,4′,5,5′-Tetramethoxy-3,4-methylenedioxy-7,7′:7,8-diepoxylignan. ***4,5-Epoxybeilschmin A***

[1000269-47-5]

$C_{23}H_{26}O_8$ 430.453

Constit. of the leaves of *Beilschmiedia tsangii*. Oil. $[\alpha]_D^{24}$ +22.6 (c, 0.12 in $CHCl_3$). λ_{max} 216 (log ε 4.42); 275 (log ε 4.16) (MeOH)

3,4:3′,4′-Bis(methylene), 5,5′-di-Me ether

7,7′:7,8-Diepoxy-3,3′-dimethoxy-4,5:4′,5′-bis(methylenedioxy) lignan. 3,3′-dimethoxy-4,5:4′,5′-bis(methylenedioxy)-7,7′:7,8-diepoxylignan. ***Beilschmin D***

[1000272-34-3]

$C_{22}H_{22}O_8$ 414.41

Constit. of the leaves of *Beilschmiedia tsangii*. Oil. $[\alpha]_D^{25}$ +23.8 (c, 0.1 in $CHCl_3$). λ_{max} 217 (log ε 4.41); 278 (log ε 3.53) (MeOH).

Chen, J.-J. *et al.*, *Planta Med.*, 2007, **73**, 567-571 (*Epoxybeilschmins A,B, Beilschmin D*)

7,7′:8,8′-Diepoxy-8,8′-lignan-3,3′,4,4′,9,9′-hexol D62

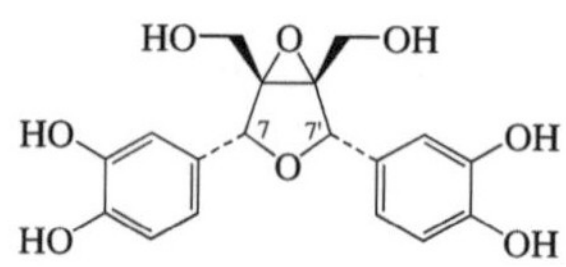

3,3′,4,4′,9,9′-Hexahydroxy-7,7′:8,8′-diepoxylignan

$C_{18}H_{18}O_8$ 362.334

(7R*,7′S*,8S*,8′R*)-form

3,3′-Di-Me ether

7,7′:8,8′-Diepoxy-3,3′-dimethoxy-8,8′-lignan-4,4′,9,9′-tetrol. ***Schisanpropinin***

$C_{20}H_{22}O_8$ 390.388

Constit. of *Schisandra propinqua* var. *propinqua*. Yellowish solid (MeOH). $[\alpha]_D^{20}$ -31.2 (c, 0.1 in MeOH). λ_{max} 205 (log ε 4.96); 286 (log ε 4.06) (MeOH).

Liu, M. *et al.*, *Nat. Prod. Bioprospect.*, 2017, **7**, 257-262 (*Schisanpropinin*)

7,9′:7′,9-Diepoxy-8,8′-lignan-2,2′,3,3′,4,4′-hexol D63

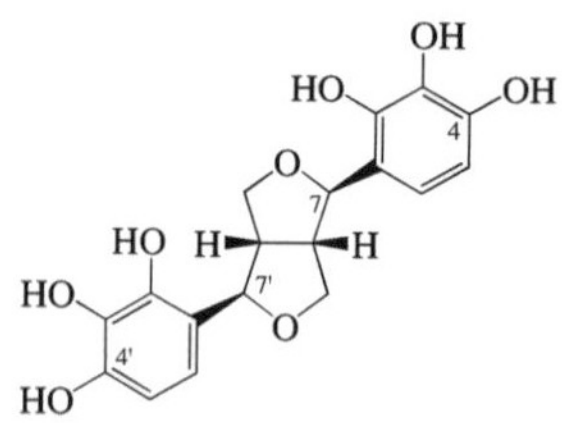

(7R,7′R,8S,8′S)-form

2,2′,3,3′,4,4′-Hexahydroxy-7,9′:7′,9-diepoxylignan

$C_{18}H_{18}O_8$ 362.334

(7R,7′R,8S,8′S)-form

3,4:3′,4′-Bis(methylene) ether

7,9′:7′,9-Diepoxy-3,4:3′,4′-bis(methylenedioxy)-8,8′-lignan-2,2′-diol. 2,2′-Dihydroxy-3,4:3′,4′-bis(methylenedioxy)-7,9′:7′,9-diepoxylignan. (-)-Sesamin-2,2′-diol

[1152441-87-6]

$C_{20}H_{18}O_8$ 386.356

Constit. of the aerial parts of *Isodon japonicus*. Amorph. powder. $[\alpha]_D^{25}$ -30 (c, 0.03 in MeOH). λ_{max} 202; 237; 274 (MeOH)

(7S,7′S,8R,8′R)-form

3,4:3′,4′-Bis(methylene) ether

2,2′-Dihydroxysesamin. *Sesamin-2,2′-diol*

[878000-24-9]

$C_{20}H_{18}O_8$ 386.356

Constit. of the leaves of *Callicarpa furfuraceae*. Amorph. powder. $[\alpha]_D^{27}$ +39.1 (c, 0.06 in $CHCl_3$).

Shao, Y. *et al.*, *Helv. Chim. Acta*, 2006, **89**, 64-72 (*2,2′-Dihydroxysesamin*).

Hong, S.S. *et al.*, *Arch. Pharmacal Res.*, 2009, **32**, 501-504 (*Isodon japonicus constit*)

7,9′:7′,9-Diepoxy-8,8′-lignan-2,3,3′,4,4′,5′-hexol D64

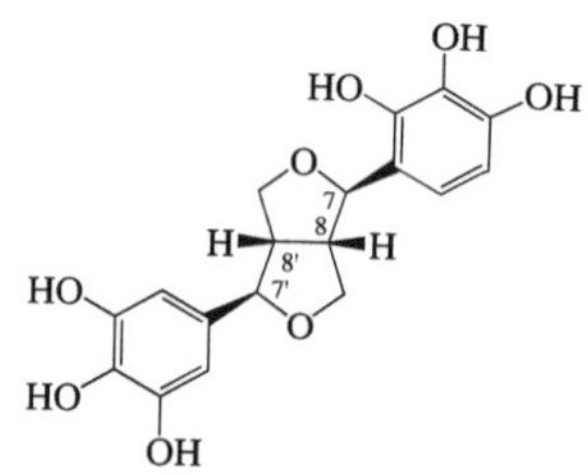

2,3,3′,4,4′,5′-Hexahydroxy-7,9′:7′,9-diepoxylignan

$C_{18}H_{18}O_8$ 362.334

(7R*,7′R*,8S*,8′S*)-form

3,4:3′,4′-Bis(methylene), 2,5′-di-Me ether

7,9′:7′,9-Diepoxy-2,3′-dimethoxy-3,4:4′,5′-bis(methylenedioxy)-8,8′-lignan. 2,3′-dimethoxy-3,4:4′,5′-bis(methylenedioxy)-7,9′:7′,9-diepoxylignan.

2,5′-Dimethoxysesamin

[1416429-63-4]

$C_{22}H_{22}O_8$ 414.41

Constit. of the root bark of *Leucophyllum frutescens*. Exhibits moderate antimycobacterial activity. Brown oil.

Alanis-Garza, B. *et al.*, *Nat. Prod. Commun.*, 2012, **7**, 597-598 (*2,5′-Dimethoxysesamin*)

7,9′:7′,9-Diepoxy-8,8′-lignan-2,3,3′,4,4′,5-hexol D65

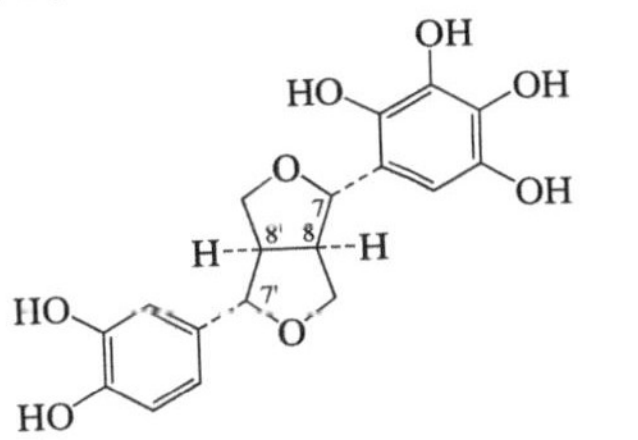

2,3,3′,4,4′,5-Hexahydroxy-7,9′:7′,9-diepoxylignan
$C_{18}H_{18}O_8$ 362.334
(7S,7′S,8R,8′R)-form
3′,4′-Methylene, 3,4-di-Me ether
7,9′:7′,9-Diepoxy-3,4-dimethoxy-3′,4′-methylenedioxy-8,8′-lignan-2,5-diol. 2,5-Dihydroxy-3,4-dimethoxy-3′,4′-methylenedioxy-7,9′:7′,9-diepoxylignan
3′,4′-Methylene, 3,4-di-Me ether, 2-O-β-D-glucopyranoside
Terminaloside A
[1906937-14-1]
$C_{27}H_{32}O_{13}$ 564.54
Constit. of the leaves of *Terminalia citrina*. Amorph. pale yellow powder. $[\alpha]^{25}_{D}$ +3.7 (c, 0.2 in MeOH). λ_{max} 212 (log ε 4.17); 283 (log ε 3.82) (MeOH)
3′,4′-Methylene, 3,4,5-tri-Me ether, 2-O-β-D-glucopyranoside
Terminaloside B
[1906937-15-2]
$C_{28}H_{34}O_{13}$ 578.567
Constit. of the leaves of *Terminalia citrina*. Exhibits inhibitory effects on MCF-7 and T47D estrogen-responsive human breast cancer cell lines. Amorph. pale yellow powder. $[\alpha]^{25}_{D}$ +38.3 (c, 0.2 in MeOH). λ_{max} 226 (log ε 4.08); 282 (log ε 3.84) (MeOH).

Muhit, M.A. *et al., J. Nat. Prod.*, 2016, **79**, 1298-1307 (*Terminalosides A,B, activity*)

7,9′:7′,9-Diepoxy-8,8′-lignan-2,3,3′,4,4′,8′-hexol D66

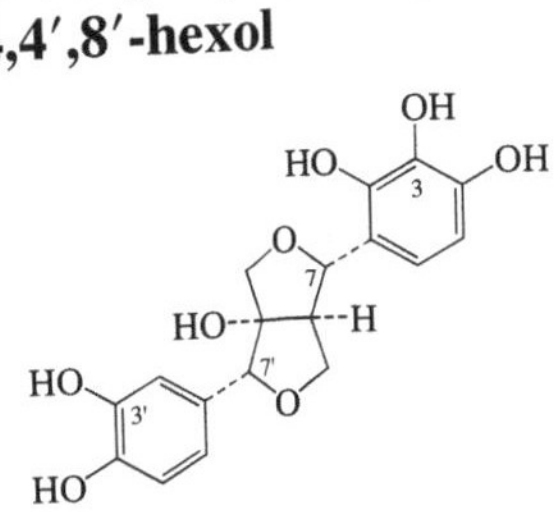

2,3,3′,4,4′,8′-Hexahydroxy-7,9′:7′,9-diepoxylignan
$C_{18}H_{18}O_8$ 362.334
(7S,7′R,8R,8′S)-form
3,4:3′,4′-Bis(methylene), 2-Me ether
7,9′:7′,9-Diepoxy-2′-methoxy-3,4:3′,4′-bis(methylenedioxy)-8,8′-lignan-8-ol. 8′-Hydroxy-2-methoxy-3,4:3′,4′-bis(methylenedioxy)-7,9′:7′,9-diepoxylignan. ***Palatiferin A***
[1361308-20-4]
$C_{21}H_{20}O_8$ 400.383
Constit. of the roots of *Pseuderanthemum palatiferum*. Shows moderate cytotoxicity against HepG2 cells. Light yellow cryst.. Mp 103 - 104°. $[\alpha]^{30}_{D}$ +28.2 (c, 0.71 in $CHCl_3$). Probable abs. config. based on opt. rotn. comparisons. λ_{max} 215 (log ε 3.8); 284 (log ε 3.2) (MeOH).

Mai, H.D.T. *et al., Planta Med.*, 2011, **77**, 951-954 (*Palatiferin A*)

7,9′:7′,9-Diepoxy-8,8′-lignan-2,3′,4,4′,5,8′-hexol D67

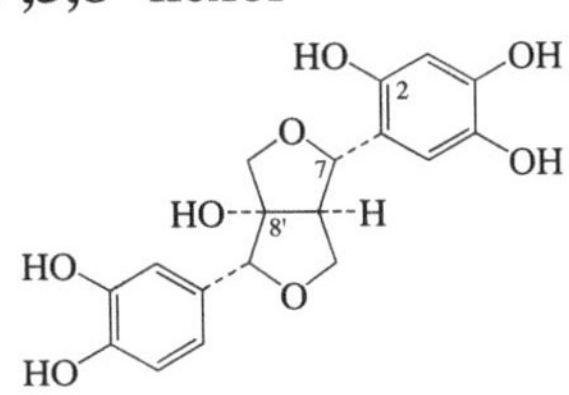

2,3′,4,4′,5,8′-Hexahydroxy-7,9′:7′,9-diepoxylignan
$C_{18}H_{18}O_8$ 362.334
(7S,7′R,8R,8′S)-form
3′,4′:4,5-Bis(methylene), 2-Me ether
7,9′:7′,9-Diepoxy-2′-methoxy-3,4:4′,5′-bis(methylenedioxy)-8,8′-lignan-8-ol. 8′-Hydroxy-2-methoxy-3′,4′:4,5-bis(methylenedioxy)-7,9′:7′,9-diepoxylignan
3′,4′:4,5-Bis(methylene), 2-Me ether, 8′-Ac
Phrymarin II
[331947-56-9]
$C_{23}H_{22}O_9$ 442.42
Constit. of *Phryma leptostachya* and *Isodon lophanthoides* var. *gerardianus*. $[\alpha]^{27}_{D}$ +8.9 (c, 0.38 in $CHCl_3$). $[\alpha]^{22}_{D}$ +17.68 (c, 0.7 in $CHCl_3$). λ_{max} 205 (log ε 4.18); 240 (log ε 3.36); 288 (log ε 3.32) ($CHCl_3$).

Jiang, B. *et al., Zhiwu Xuebao*, 2000, **42**, 95-97 (*Isodon lophanthoides var gerardianus constit*).
Ishibashi, F. *et al., Biosci., Biotechnol., Biochem.*, 2001, **65**, 29-34 (*Phrymarin II, synth, abs config*)

7,9′:7′,9-Diepoxy-8,8′-lignan-3,3′,4,4′,5,5′-hexol D68

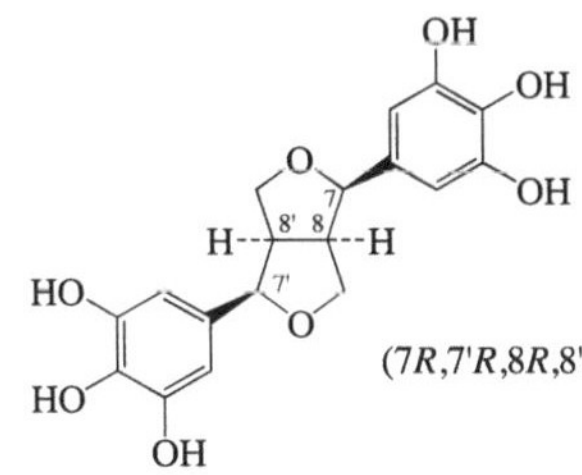

(7R,7′R,8R,8′R)-form

Tetrahydro-1,4-bis(3,4,5-trihydroxyphenyl)-1H,3H-furo[3,4-c]furan. 2,6-Bis(3,4,5-trihydroxyphenyl)-3,7-dioxabicyclo[3.3.0]octane. 3,3′,4,4′,5,5′-Hexahydroxy-7,9′:7′,9-diepoxylignan
$C_{18}H_{18}O_8$ 362.334
(7R,7′R,8R,8′R)-form
3,3′,5,5′-Tetra-Me ether
7,9′:7′,9-Diepoxy-3,3′,5,5′-tetramethoxy-8,8′-lignan-4,4′-diol. 4,4′-Dihydroxy-3,3′,5,5′-tetramethoxy-7,9′:7′,9-diepoxylignan. ***Diasyringaresinol***
[551-29-1]
$C_{22}H_{26}O_8$ 418.442
Isol. from *Macropiper excelsum* and other woods. Also from *Annona cherimola* and *Bupleurum salicifolium*. Mp 185 - 186° (174-176°). $[\alpha]_D$ +48.9 ($CHCl_3$). λ_{max} 217 (ε 16500); 237 (sh); 273 (ε 2090) (MeOH) (Derep)
Hexa-Me ether
7,9′:7′,9-Diepoxy-3,3′,4,4′,5,5′-hexamethoxy-8,8′-lignan. 3,3′,4,4′,5,5′-Hexamethoxy-7,9′:7′,9-diepoxylignan. ***Diayangambin***. *Lirioresinol C dimethyl ether*
[21453-68-9]
$C_{24}H_{30}O_8$ 446.496
Isol. from *Macropiper excelsum, Liriodendron tulipifera, Artemisia absinthinum, Piper arborescens* and resin of *Commiphora wightii*. Cryst. ($CHCl_3$). Mp 151 - 152° (145-147.5°). $[\alpha]_D$ +284 (c, 0.1 in $CHCl_3$)
3,4-Methylene, 3′,4′,5,5′-tetra-Me ether
7,9′:7′,9-Diepoxy-3,3′,4,5-tetramethoxy-4′,5′-methylenedioxy-8,8′-lignan.

3,3′,4,5-Tetramethoxy-4′,5′-methylenedioxy-7,9′:7′,9-diepoxylignan. ***Diasesartemin***
[77449-33-3]
$C_{23}H_{26}O_8$ 430.453
Isol. from roots of *Artemisia absinthium* and from *Hernandia ovigera* and *Commiphora wightii* resin. α-Glucosidase inhibitor. Mp 102 - 104°. $[\alpha]_D$ +315

(7R,7′R,8S,8′S)-form

3,3′,4,4′-Tetra-Me ether, 5,5′-di-Ac
Schproplignan C
[1242247-94-4]
$C_{26}H_{30}O_{10}$ 502.516
Constit. of the fruit of *Schisandra propinqua* var. *propinqua*. Exhibits anti-HIV activity. Amorph. solid. $[\alpha]_D^{24}$ +6 (c, 0.12 in MeOH). Relative config. only. λ_{max} 208 (log ε 5.96); 282 (log ε 4.88) (MeOH)

3,3′,5,5′-Tetra-Me ether
(-)-Syringaresinol
[6216-81-5]
$C_{22}H_{26}O_8$ 418.442
Isol. from *Swertia chirata*, *Annona montana*, *Liriodendron tulipifera*, *Berberis koreana*, *Pharbitis nil* and *Gymnosporia trigyna*. Hepatoprotective agent, DNA strand scissioning agent, platelet aggregation inhibitor. Promotes neurite outgrowth in PC12h cells. Shows activity against SK-MEL-2 and XF 498 cancer cell lines. Inhibitor of NO prodn. in LPS-stimulated BV-2 cells. Prisms. Mp 185 - 186° (175-176°). $[\alpha]_D$ -5.2 ($CHCl_3$). $[\alpha]_D^{20}$ -50.6 (c, 0.32 in $CHCl_3$) (resolved material). λ_{max} 206; 220; 271 (MeOH) (Berdy)

3,3′,5,5′-Tetra-Me ether, 4-O-β-D-glucopyranoside
[137038-13-2]
$C_{28}H_{36}O_{13}$ 580.583
Constit. of *Albizzia julibrissin*, *Pharbitis nil*, *Aster auriculatus* and *Magnolia coco*. $[\alpha]_D$ -15 (c, 0.2 in MeOH)

3,3′,5,5′-Tetra-Me ether, 4-O-(6-O-acetyl-β-D-glucopyranoside)
Pharsyringaresinol
[1393346-97-8]
$C_{30}H_{38}O_{14}$ 622.62
Constit. of the seeds of *Pharbitis nil*. Gum. $[\alpha]_D^{25}$ -18 (c, 0.1 in MeOH). λ_{max} 230 (log ε 3.82); 280 (log ε 3.4) (MeOH)

3,3′,5,5′-Tetra-Me ether, 4-O-[4,6-bis(4-hydroxy-3-methoxy-E-cinnamoyl)-β-D-glucopyranoside]
[947172-59-0]
$C_{48}H_{52}O_{19}$ 932.925
Constit. of the roots of *Rhus javanica* var. *roxburghiana*. Inhibitor of tobacco mosaic virus replication. Powder. $[\alpha]_D^{27}$ -6.5 (c, 0.51 in Me_2CO). λ_{max} 224 (log ε 4.4); 232 (log ε 4.01); 288 (log ε 4.1); 301 (log ε 4.03); 330 (log ε 4.1) (MeOH)

3,3′,5,5′-Tetra-Me ether, di-O-β-D-glucopyranoside
Acanthoside D. *Eleutheroside E. (-)-Syringaresinol β-D-diglucoside*
[96038-87-8]
[66791-77-3] ((-)-Syringaresinol β-D-diglucoside)
$C_{34}H_{46}O_{18}$ 742.724
Isol. from *Eleutherococcus senticosus*, *Eucommia ulmoides*, *Ligustrum japonicum*, *Albizzia julibrissin* and *Viscum album*. Needles. Mp 269 - 270°. $[\alpha]_D^{22}$ -18.5 (c, 0.54 in 50% EtOH aq.)

3,3′,5,5′-Tetra-Me ether, 4-O-[β-D-apiofuranosyl-(1→2)-β-D-glucopyranoside]
[136997-64-3]
$C_{33}H_{44}O_{17}$ 712.698
Isol. from stem bark of *Albizzia julibrissin*. Shows sedative props.. Pale yellow powder. $[\alpha]_D$ -2.6 (MeOH)

3,3′,5,5′-Tetra-Me ether, 4-O-[β-D-apiofuranosyl-(1→2)-β-D-glucopyranoside], 4′-O-β-D-glucopyranoside
[136997-65-4]
$C_{39}H_{54}O_{22}$ 874.839
Isol. from stem bark of *Albizzia julibrissin*. Shows sedative props.. Pale yellow powder. $[\alpha]_D$ -23.6 (MeOH)

3,3′,5,5′-Tetra-Me ether, di-O-[β-D-apiofuranosyl-(1→2)-β-D-glucopyranoside]
[136997-66-5]
$C_{44}H_{62}O_{26}$ 1006.954
Isol. from stem bark of *Albizzia julibrissin*. Shows sedative props.. Pale yellow powder. $[\alpha]_D$ -30.7 (MeOH)

3,3′,4,5,5′-Penta-Me ether
7,9′:7′,9-Diepoxy-3,3′,4′,5,5′-pentamethoxy-8,8′-lignan-4-ol. 4-Hydroxy-3,3′,4′,5,5′-pentamethoxy-7,9′,7′,9-diepoxylignan
[1290633-29-2]
$C_{23}H_{28}O_8$ 432.469
Constit. of the stems of *Sinocalamus affinis*. Amorph. powder. $[\alpha]_D^{20}$ -45.8 (c, 0.03 in MeOH). λ_{max} 207 (log ε 4.27); 236 (log ε 3.55); 272 (log ε 2.86) (MeOH)

3,4-Methylene, 3′,5-di-Me ether
7,9′:7′,9-Diepoxy-3′,5-dimethoxy-4′,5′-methylenedioxy-8,8′-lignan-3,4-diol.
3,4-Dihydroxy-3′,5-dimethoxy-4′,5′-methylenedioxy-7,9′:7′,9-diepoxylignan
[929692-65-9]
$C_{21}H_{22}O_8$ 402.399
Constit. of *Peperomia duclouxii*. Inhibitor of VA-13 and HepG2 cancer cell lines; inhibitor of Ca accumulation in multidrug resistant human ovarian cancer. Gum. $[\alpha]_D^{25}$ -32.8 (c, 0.5 in $CHCl_3$). λ_{max} 246 (log ε 4.2); 275 (log ε 3.93) ($CHCl_3$)

3,4-Methylene, 3′,5,5′-tri-Me ether
7,9′:7′,9-Diepoxy-3,3′,5-trimethoxy-4′,5′-methylenedioxy-8,8′-lignan-4-ol. 4-Hydroxy-3,3′,5-trimethoxy-4′,5′-methylenedioxy-7,9′:7′,9-diepoxylignan
[929692-64-8]
$C_{22}H_{24}O_8$ 416.426
Constit. of *Peperomia duclouxii*. Gum. $[\alpha]_D^{25}$ -32.8 (c, 0.09 in $CHCl_3$). λ_{max} 243 (log ε 3.4); 272 (log ε 3.9) ($CHCl_3$)

3,4:3′,4′-Bis(methylene), 5,5′-di-Me ether
7,9′:7′,9-Diepoxy-3,3′-dimethoxy-4,5:4′,5′-bis(methylenedioxy)-8,8′-lignan. 3,3′-dimethoxy-4,5:4′,5′-bis(methylenedioxy)-7,9′:7′,9-diepoxylignan. 5,5′-Dimethoxysesamin
[929692-63-7]
$C_{22}H_{22}O_8$ 414.41
Constit. of *Podolepis rugata* and *Peperomia duclouxii*. Gum. $[\alpha]_D^{25}$ -28.5 (c, 0.3 in $CHCl_3$). λ_{max} 244 (log ε 2.9); 276 (log ε 3.91) ($CHCl_3$)

(7R,7′S,8R,8′R)-form

3,3′-Di-Me ether, 4-O-β-D-glucopyranoside

Brachyside A

$C_{26}H_{32}O_{13}$ 552.529

Constit. of aerial parts of *Caragana brachyantha*. Moderate antioxidant. Yellow gummy solid. $[\alpha]_D^{25}$ +23 (c, 0.3 in MeOH)

3,3′-Di-Me ether, 4-O-[β-D-glucopyranosyl-(1→6)-β-D-glucopyranoside]

Brachyside B

$C_{32}H_{42}O_{18}$ 714.67

Constit. of aerial parts of *Caragana brachyantha*. Weak antioxidant. Yellow gummy solid. $[\alpha]_D^{25}$+35 (c, 0.2 in MeOH)

3,3′,5,5′-Tetra-Me ether

(+)-Episyringaresinol. *Lirioresinol A. Symplicosigenol*

[21453-71-4]

$C_{22}H_{26}O_8$ 418.442

Isol. from *Liriodendron tulipifera, Artemisia absinthium, Persea japonica, Aralia bipinnata, Annona cherimola, Pseuderanthemum carruthersii* and *Magnolia grandiflora*. Prod. of degradation of birch lignin. Cryst. ($CHCl_3$/MeOH). Mp 210 - 211°. $[\alpha]_D$ +127 ($CHCl_3$). λ_{max} 217 (ε 16500); 237 (sh); 273 (ε 2090) (MeOH) (Derep)

3,3′,5,5′-Tetra-Me ether, 4-O-β-D-glucopyranoside

Tortoside A

[190655-16-4]

$C_{28}H_{36}O_{13}$ 580.583

Constit. of Acanthopanax koreanum, *Ancistrocladus cochinchinensis, Ilex pubescens* and *Pedicularis torta*. Amorph. powder. $[\alpha]_D^{15}$+28.3 (c, 0.3 in MeOH)

3,3′,5,5′-Tetra-Me ether, 4′-O-β-D-glucopyranoside

Eleutheroside E$_2$

[390407-21-3]

$C_{28}H_{36}O_{13}$ 580.583

Constit. of the roots of *Eleutherococcus senticosus*. $[\alpha]_D^{25}$+22 (c, 0.73 in Py). λ_{max} 207 (log ε 4); 238 (log ε 3.32) (MeOH)

3,3′,5,5′-Tetra-Me ether, di-O-β-D-glucopyranoside

Eleutheroside D

[79484-75-6]

$C_{34}H_{46}O_{18}$ 742.724

Constit. of *Eleutherococcus senticosus* and *Sargentodoxa cuneata*

3,3′,4′,5,5′-Penta-Me ether

4′-O-Demethylepiyangambin

[188564-12-7]

$C_{23}H_{28}O_8$ 432.469

Constit. of *Ocotea duckei*

Hexa-Me ether

7-Epiyangambin

[24192-64-1]

$C_{24}H_{30}O_8$ 446.496

Constit. of *Virola elongata, Achillea holosericea, Hernandia nymphaefolia* and *Artemisia* spp.. Modulator of intracellular calcium. Cryst. (Et_2O). Mp 119 - 120°. $[\alpha]_D^{25}$ +122 ($CHCl_3$). λ_{max} 210 (log ε 4.8); 238 (sh); 272 (log ε 3.51) (MeOH)

3,4-Methylene, 3′,4′,5,5′-tetra-Me ether

Episesartemin A

[77449-31-1]

$C_{23}H_{26}O_8$ 430.453

Isol. from roots of *Artemisia absinthium*. Mp 112 - 114°. $[\alpha]_D$ +115

3,4:3′,4′-Bis(methylene), 5,5′-di-Me ether

Epiexcelsin

[51020-09-8]

$C_{22}H_{22}O_8$ 414.41

Constit. of *Macropiper excelsum* and *Litsea verticillata*. Mp 122 - 123°. $[\alpha]_D^{18}$+131 ($CHCl_3$). λ_{max} 242 (ε 11000); 275 (ε 2800) (EtOH)

(7R,7′S,8S,8′S)-form

3,3′,5,5′-Tetra-Me ether

(-)-Episyringaresinol

[6216-82-6]

$C_{22}H_{26}O_8$ 418.442

Isol. from *Nandina domestica*. Mp 208 - 211°

3,3′,5,5′-Tetra-Me ether, 4-O-[β-D-apiofuranosyl-(1→2)-β-D-glucopyranoside]

Albizzioside A

[161104-16-1]

$C_{33}H_{44}O_{17}$ 712.698

Constit. of *Albizzia myriophylla*. Oil. $[\alpha]_D^{18}$-3.7 (c, 1.3 in MeOH)

(7S,7′R,8R,8′R)-form

3,3′,5,5′-Tetra-Me ether, 4-O-[4-hydroxy-3,5-dimethoxybenzoyl-(→5)-β-D-apiofuranosyl-(1→6)-β-D-glucopyranoside]

Obtusifoside D

[1431790-53-2]

$C_{42}H_{52}O_{21}$ 892.857

Constit. of stems of *Erycibe obtusifolia*. Powder. Mp 161 - 163°. $[\alpha]_D^{20}$-45.5 (c, 0.05 in MeOH)

3,3′,5,5′-Tetra-Me ether, 4-O-[4-hydroxy-3,5-dimethoxy-E-cinnamoyl-(→5)-β-D-apiofuranosyl-(1→6)-β-D-glucopyranoside]

Obtusifoside E

[1431790-54-3]

$C_{44}H_{54}O_{21}$ 918.895

Constit. of stems of *Erycibe obtusifolia*. Yellow powder. Mp 156 - 158°. $[\alpha]_D^{20}$-59.3 (c, 0.05 in MeOH)

3,4-Methylene, 3′,4′,5,5′-tetra-Me ether

Episesartemin B. ***2-Episesalatin***. *Episesartemin*

[77449-32-2]

$C_{23}H_{26}O_8$ 430.453

Constit. of *Achillea holosericea, Artemisia absinthium, Sesamum alatum* and *Virola elongata*. Cryst. (Et_2O). Mp 114 - 115°. $[\alpha]_D$ +127. Greger (1980) calls this Episesartemin B. MacRae (1985) calls it Episesartemin

(7S,7′S,8R,8′R)-form

3,3′-Di-Me ether

7,9′:7′,9-Diepoxy-5,5′-dimethoxy-8,8′-lignan-3,3′,4,4′-tetrol. 3,3′,4,4′-Tetrahydroxy-5,5′-dimethoxy-7,9′:7′,9-diepoxylignan.

5,5′-Dihydroxy-(+)-pinoresinol

$C_{20}H_{22}O_8$ 390.388

Prod. from Pu-erh ripe leaves of *Camellia sinensis*. Amorph. powder. $[\alpha]_D^{20}$+12.6 (c, 0.1 in MeOH). λ_{max} 208 (log ε 4.90) (MeOH)

3,3′-Di-Me ether, 4-O-β-D-glucopyranoside

Tortoside C

[190655-18-6]

$C_{26}H_{32}O_{13}$ 552.529

Constit. of *Pedicularis torta*. Amorph. powder. $[\alpha]_D^{15}$-8.9 (c, 0.4 in MeOH)

3,3′,5′-Tri-Me ether

7,9′:7′,9-Diepoxy-3′,5,5′-trimethoxy-8,8′-lignan-3,4,4′-triol.

Cinncassin F. *Cinnacassin F*

[2232873-45-7]

$C_{21}H_{24}O_8$ 404.415

Constit. of *Cinnamomum cassia*. Yellow oil. $[\alpha]_D^{25}$ +6 (c, 1.2 in MeOH). λ_{max} 205 (log ε 4.31); 235 (log ε 4.32); 275 (log ε 3.47) (MeOH)

3,3',5-Tri-Me ether, 4'-O-β-D-glucopyranoside

Viscoloratin

[1018989-85-9]

$C_{27}H_{34}O_{13}$ 566.556

Constit. of the stems of *Viscum coloratum*

3,3',5,5'-Tetra-Me ether

(+)-Syringaresinol. *Lirioresinol B*

[21453-69-0]

[19046-36-7] (all-*S*-form)

$C_{22}H_{26}O_8$ 418.442

Constit. of *Annona cherimola, Allamanda neriifolia, Pseuderanthemum carruthersii, Euonymus alatus, Liriodendron tulipifera* and *Tectona grandis*. Also prod. by *Actinomadura miaoliensis* BCRC 16873. Promotes neurite outgrowth in PC12h cells. Inhibitor of LPS induced NO prodn. Shows antiinflammatory and antinociceptive effects. Plant growth inhibitor. Cryst.. Mp 187.5 - 189.5° (175°). $[\alpha]_D$ +62.2 ($CHCl_3$) (+53.3). Syringaresinol of unreported abs. config. also isol. from a number of other spp. Lirioresinol C isol. by Dickey (1958) shown to be impure Lirioresinol B by Briggs *et al* (1968). λ_{max} 217 (ε 16500); 237 (sh); 273 (ε 2090) (MeOH) (Derep). [neutral] λ_{max} 239 (ε 10900); 273 (ε 2800) (MeOH) (Berdy)

3,3',5,5'-Tetra-Me ether, 4-O-β-D-glucopyranoside

Acanthoside B. *Eleutheroside E₁. (+)-Syringaresinol β-D-glucoside*

[7374-79-0]

$C_{28}H_{36}O_{13}$ 580.583

Constit. of *Acanthopanax senticosus, Epimedium grandiflorum, Magnolia grandiflora, Scrophularia scopolii, Scrophularia albida-colchica, Sargentodoxa cuneata, Syringa velutina* and other plants. Needles (MeOH). Mp 187 - 191° (174-177°). $[\alpha]_D^{25}$ -24 (c, 1.98 in MeOH). Some cmr values revised in 2004

3,3',5,5'-Tetra-Me ether, 4-O-β-D-glucopyranoside, 4'-O-α-L-rhamnopyranoside

Dioscoroside H₂

[2291325-07-8]

$C_{34}H_{46}O_{17}$ 726.725

Constit. of the rhizomes of *Dioscorea septemloba*. Powder. $[\alpha]_D^{25}$ -54.9 (c, 0.85 in MeOH). λ_{max} 235 (log ε 4.17); 255 (log ε 3.94); 280 (sh); 317 (sh) (MeOH)

3,3',5,5'-Tetra-Me ether, di-O-β-D-glucopyranoside

Liriodendrin

[573-44-4]

$C_{34}H_{46}O_{18}$ 742.724

Isol. from bark of yellow poplar *Liriodendron tulipifera*, from *Salvadora persica, Eleutherococcus senticosus, Allamanda neriifolia, Epimedium grandiflorum, Kalopanax pictus, Ducrosia ismaelis, Sargentodoxa cuneata* and *Acanthopanax senticosus*. Enhances β-endorphin level. Immunostimulatory, antiosteoporotic agent, shows antiinflammatory and antinociceptive effects. Cryst. (EtOH). Mp 269 - 270° (235-240°). $[\alpha]_D^{18}$ -18.5 (c, 0.2 in Py). $[\alpha]_D^{28}$ -15.9 (c, 0.32 in MeOH). $[\alpha]_D^{23}$ -48.5 (c, 1.02 in Py). λ_{max} 268 (ε 1970) (H_2O) (Derep)

3,3',5,5'-Tetra-Me ether, 4-O-[β-D-apiofuranosyl-(1→6)-β-D-glucopyranoside]

Obtusifoside A

[1429784-29-1]

$C_{33}H_{44}O_{17}$ 712.698

Constit. of the stems of *Erycibe obtusifolia*. Exhibits hepatoprotective activity. Powder. Mp 150 - 152°. $[\alpha]_D^{20}$ -62.9 (c, 0.05 in MeOH)

3,3',5,5'-Tetra-Me ether, 4-O-[4-hydroxy-3,5-dimethoxybenzoyl-(→5)-β-D-apiofuranosyl-(1→6)-β-D-glucopyranoside]

Obtusifoside B

[1431790-51-0]

$C_{42}H_{52}O_{21}$ 892.857

Constit. of stems of *Erycibe obtusifolia*. Powder. Mp 130 - 132°. $[\alpha]_D^{20}$ -38.2 (c, 0.05 in MeOH)

3,3',5,5'-Tetra-Me ether, 4-O-[4-hydroxy-3,5-dimethoxy-E-cinnamoyl-(→5)-β-D-apiofuranosyl-(1→6)-β-D-glucopyranoside]

Obtusifoside C

[1431790-52-1]

$C_{44}H_{54}O_{21}$ 918.895

Constit. of stems of *Erycibe obtusifolia*. Yellow powder. Mp 133 - 135°. $[\alpha]_D^{20}$ -57.9 (c, 0.05 in MeOH)

3,3',5,5'-Tetra-Me ether, 4-O-[β-D-glucopyranosyl-(1→6)-β-D-glucopyranoside]

Syringaresinol 4-gentiobioside

[639857-59-3]

$C_{34}H_{46}O_{18}$ 742.724

Constit. of the dried rhizomes of *Smilax glabra*. Amorph. powder. $[\alpha]_D^{24}$ -10.7 (c, 0.55 in MeOH). λ_{max} 210 (log ε 4.67); 271 (log ε 3.44) (MeOH)

3,3',5,5'-Tetra-Me ether, 4,4'-diangeloyl

4,4'-Diangeloyl-(+)-syringaresinol

[1428768-23-3]

$C_{32}H_{38}O_{10}$ 582.646

Isol. from the aerial parts of *Rudbeckia laciniata*. Gum. $[\alpha]_D^{25}$ +71.0 (c, 0.15 in MeOH). λ_{max} 216 (log ε 4.0); 272 (log ε 3.9) (MeOH)

3,3',4,4',5-Penta-Me ether

7,9':7',9-Diepoxy-3',4,4',5,5'-pentamethoxy-8,8'-lignan-3-ol. 3-Hydroxy-3',4,4',5,5'-pentamethoxy-7,9':7',9-diepoxylignan

[1126087-60-2]

$C_{23}H_{28}O_8$ 432.469

Constit. of the bark of *Magnolia kobus*. Cryst.. Mp 116 - 118°. $[\alpha]_D^{20}$ +42.6 (c, 0.1 in $CHCl_3$). λ_{max} 235 (log ε 3.9); 273 (log ε 3.6) ($CHCl_3$)

3,3',4,5,5'-Penta-Me ether, 4'-O-β-D-glucopyranoside

Icariside E₇

[135743-07-6]

$C_{29}H_{38}O_{13}$ 594.61

Constit. of *Epimedium sagittatum*. Needles (MeOH). Mp 88 - 89.5°. $[\alpha]_D^{23}$ -25 (c, 0.14 in MeOH)

Hexa-Me ether

Yangambin. *Di-O-methyllirioresinol B. Di-O-methylsyringaresinol*

[13060-14-5]

$C_{24}H_{30}O_8$ 446.496

Constit. of *Piper guineense, Eremophila glabra, Liriodendron tulipfera, Micropiper excelsum* (preferred genus name *Peperomia*), *Ocotea duckei, Virola elongata, Achillea holosericea, Rudbeckia maxima, Rudbeckia nitida* and *Rudbeckia scabrifolia*. Also prod. by *Actinomadura miaoliensis* BCRC 16873. Selective platelet aggregation inhibitor. PAF receptor antagonist. Needles (EtOH); prisms (MeOH). Mp 125.5 - 126.5° (122-123°). $[\alpha]_D$ +46.6 (c, 6.18 in $CHCl_3$). λ_{max} 207 (ε 61700); 230 (sh); 268 (ε 14500) (MeOH) (Derep)

3,4-Methylene, 3′,4′,5-tri-Me ether, 5′-O-β-D-glucopyranoside

Terminaloside E

[1906939-25-0]

$C_{28}H_{34}O_{13}$ 578.567

Constit. of the leaves of *Terminalia citrina*. Exhibits inhibitory activity against T47D cell line. Amorph. pale yellow powder. $[\alpha]_D^{25}$ +38.1 (c, 0.2 in MeOH). λ_{max} 217 (log ε 4.21); 273 (log ε 3.45) (MeOH)

3,4-Methylene, 3′,4′,5,5′-tetra-Me ether

(+)-Sesartemin. *Sesalatin*

[77394-27-5]

$C_{23}H_{26}O_8$ 430.453

Constit. of *Artemisia absinthium, Artemisia arborescens, Achillea cretica, Achillea holosericea, Beilschmiedia pulverulenta* and *Virola elongata*. Mp 115 - 116°. $[\alpha]_D^{25}$ +50 ($CHCl_3$)

3,4:3′,4′-Bis(methylene), 5,5′-di-Me ether

(+)-Excelsin‡

[50886-57-2]

$C_{22}H_{22}O_8$ 414.41

Constit. of *Beilschmiedia pulverulenta* and *Macropiper excelsum*. Solid. Mp 122 - 123°. $[\alpha]_D^{25}$ +124.9 (c, 0.02 in $CHCl_3$). λ_{max} 242 (ε 10500); 275 (ε 2900) (EtOH)

(7R*,7′R*,8S*,8′S*)-form

3,3′,5,5′-Tetra-Me ether, 4′-O-β-D-allopyranoside

[2368241-83-0]

$C_{28}H_{36}O_{13}$ 580.583

Constit. of the rhizomes of *Cimicifuga dahurica*. Brownish-yellow gum. $[\alpha]_D^{25}$ -34.6 (no solvent reported). λ_{max} 210 (log ε 4.93) (MeOH)

(7RS,7′RS,8SR,8′SR)-form

3,3′,5,5′-Tetra-Me ether

(±)-Syringaresinol

[1177-14-6]

$C_{22}H_{26}O_8$ 418.442

Isol. from *Fagus sylvatica, Quercus rubra, Fraxinus mandshurica, Epimedium koreanum, Magnolia officinalis, Capparis flavicans* and *Machilus thunbergii*. Inhibitor of estrogen-enhanced cell proliferation *in vitro*. Cryst. (EtOH aq.). Mp 174 - 175°

3,4:3′,4′-Bis(methylene), 5,5′-di-Me ether

(±)-Excelsin

[82373-97-5]

$C_{22}H_{22}O_8$ 414.41

Isol. from *Vismia guaramirangae* and *Amorphophallus konjac*. Amorph.. The isoln. of the (±)-form as a nat. prod. is doubtful.

Dickey, E.E. *J. Org. Chem.*, 1958, **23**, 179-184 (*Lirioresinols*).
Jefferies, P.R. *et al.*, *Aust. J. Chem.*, 1961, **14**, 175-176 (*Lirioresinol B di-Me ether*).
Nimz, H. *et al.*, *Chem. Ber.*, 1965, **98**, 538-539 (*(±)-Syringaresinol*).
Hänsel, R. *et al.*, *Z. Naturforsch., B*, 1966, **21**, 530-535 (*Yangambin*).
Ovodov, Y.S. *et al.*, *Chem. Nat. Compd. (Engl. Transl.)*, 1967, **3**, 53-54 (*Eleutherosides D,E*).
Briggs, L.H. *et al.*, *J. Chem. Soc. C*, 1968, 3042-3045 (*Lirioresinol C di-Me ether, Lirioresinols A,B*).
Seikel, M.F. *et al.*, *Phytochemistry*, 1971, **10**, 2249-2251 (*Syringaresinol, isol*).
Erickson, M. *et al.*, *Acta Chem. Scand.*, 1972, **26**, 3085-3096 (*(±)-Syringaresinol, synth*).
Sudo, K. *et al.*, *Mokuzai Gakkaishi*, 1973, **19**, 165-169 (*(±)-Syringaresinol*).
Russell, G.B. *et al.*, *Phytochemistry*, 1973, **12**, 1799-1803 (*Excelsin, Epiexcelsin*).
Kunimoto, J.-I. *et al.*, *Yakugaku Zasshi*, 1975, **95**, 445-447 (*(-)-Episyringaresinol, struct*).
Bryan, R.F. *et al.*, *J. Chem. Soc., Perkin Trans. 2*, 1976, 341-345 (*Syringaresinol, cryst struct, bibl*).
Chen, C.-L. *et al.*, *Phytochemistry*, 1976, **15**, 547-550 (*Liriodendron tulipfera constits*).
Greger, H. *et al.*, *Tetrahedron*, 1980, **36**, 3551-3558 (*Artemisia absinthium constits*).
Camele, G. *et al.*, *Phytochemistry*, 1982, **21**, 417-419 (*5,5′-Dimethoxysesamin*).
MacRae, W.D. *et al.*, *Phytochemistry*, 1985, **24**, 561-566 (*Virola elongata constits*).
Bytheway, I.R. *et al.*, *Aust. J. Chem.*, 1987, **40**, 1913-1917 (*Di-O-methylsyringaresinol, abs config, cryst struct*).
Miyase, T. *et al.*, *Chem. Pharm. Bull.*, 1987, **35**, 3713-3719 (*Epimedium grandiflorum constits*).
Hofer, O. *et al.*, *Monatsh. Chem.*, 1988, **119**, 1143-1153 (*Diasesartemin, Episesartemin B, cryst struct*).
Abe, F. *et al.*, *Phytochemistry*, 1988, **27**, 575-577 (*Allamanda neriifolia constits*).
Jaensch, M. *et al.*, *Phytochemistry*, 1989, **28**, 3497-3501 (*Podolepis rugata constit*).
Nishibe, S. *et al.*, *Chem. Pharm. Bull.*, 1990, **38**, 1763-1765 (*(+)-Syringaresinol glucoside, diglucoside, activity*).
Kinjo, J. *et al.*, *Chem. Pharm. Bull.*, 1991, **39**, 1623-1625 (*(-)-Syringaresinol glycosides*).
Matsushita, H. *et al.*, *Phytochemistry*, 1991, **30**, 2025-2027 (*Icariside E_7*).
Vermes, B. *et al.*, *Phytochemistry*, 1991, **30**, 3087-3089 (*Syringaresinol glycosides, synth, struct*).
Macias, F.A. *et al.*, *Acta Crystallogr., Sect. C*, 1992, **48**, 2240-2244 (*Yangambin, cryst struct*).
Hofer, O. *et al.*, *Monatsh. Chem.*, 1992, **123**, 105-120 (*Sesartemin, Diasesartemin, Episesartemin B, conformn*).
Kamal-Eldin, A. *et al.*, *Phytochemistry*, 1992, **31**, 2911-2912 (*2-Episesalitin*).
Chakravarty, A.K. *et al.*, *Indian J. Chem., Sect. B*, 1994, **33**, 405-408 (*(-)-Syringaresinol*).
Ito, A. *et al.*, *Phytochemistry*, 1994, **37**, 1455-1458 (*Albizzioside A*).
Suginome, H. *et al.*, *J. Org. Chem.*, 1995, **60**, 3052-3064 (*Yangambin, synth*).
Castro-Faria-Neto, H.C. *et al.*, *Planta Med.*, 1995, **61**, 101-105 (*Yangambin, PAF receptor antagonist*).
Castro-Faria-Neto, H.C. *et al.*, *Planta Med.*, 1995, **61**, 106-112 (*Yangambin, activity*).
Morais, L.C.S.L. *et al.*, *Fitoterapia*, 1996, **67**, 557 (*4′-O-Demethylepiyangambin*).
Wang, C.Z. *et al.*, *Phytochemistry*, 1997, **45**, 159-166 (*Tortosides*).
Chen, C.-Y. *et al.*, *J. Chin. Chem. Soc. (Taipei)*, 1998, **45**, 629-634 (*Annona cherimola constits*).
Yu, H.J. *et al.*, *J. Chin. Chem. Soc. (Taipei)*, 1998, **45**, 773-778 (*(-)-Syringaresinol glucoside, spectra*).
Chang, F.-R. *et al.*, *J. Nat. Prod.*, 1998, **61**, 863-866 (*Diasyringaresinol*).
Wang, C.Z. *et al.*, *Phytochemistry*, 1998, **48**, 711-717 (*Aster auriculatus constit*).
Christov, R. *et al.*, *Fitoterapia*, 1999, **70**, 89-92 (*cmr*).
Das, B. *et al.*, *Fitoterapia*, 1999, **70**, 101-102 (*Syringaresinol, pmr*).
Jesus-Morais, C.M.Q. *et al.*, *Planta Med.*, 2000, **66**, 211-216 (*Yangambin, selective PAF antagonist*).
Li, X.-C. *et al.*, *Planta Med.*, 2001, **67**, 776-778 (*Eleutheroside E_2*).
Chiba, K. *et al.*, *Biol. Pharm. Bull.*, 2002, **25**, 791-793 (*(±)-Syringaresinol, resoln, neurite outgrowth promotion*).
Chao, Y.Y. *et al.*, *Eur. J. Pharmacol.*, 2002, **443**, 31-38 (*7-Epiyangambin, activity*).
Ahmed, A.A. *et al.*, *Phytochemistry*, 2002, **59**, 851-856 (*Achillea holosericea constits*).
Yuan, J. *et al.*, *Heterocycles*, 2003, **60**, 1633-1637 (*4-gentiobioside*).
Jung, H.-J. *et al.*, *Planta Med.*, 2003, **69**, 610-616 (*Syringaresinol, Liriodendrin, activity*).
Shahat, A.A. *et al.*, *Fitoterapia*, 2004, **75**, 771-773 (*Syringaresinol glucoside*).
Mori, N. *et al.*, *Synthesis*, 2006, 400-404 (*Yangambin, synth, ir, pmr, cmr*).
Yin, J. *et al.*, *Asian J. Tradit. Med.*, 2007, **2**, 58-60 (*Viscoloratin*).

Ouyang, M.-A. *et al.*, *J. Agric. Food Chem.*, 2007, **55**, 6460-6465 (*4-4,6-diferuloylglucoside, TMV replication inhibitor*).
Li, N. *et al.*, *J. Nat. Prod.*, 2007, **70**, 544-548 (*Peperomia duclouxii constits, activity*).
Lee, H.-J. *et al.*, *Helv. Chim. Acta*, 2008, **91**, 2361-2366 (*Magnolia kobus penta-Me ether*).
Luecha, P. *et al.*, *J. Nat. Prod.*, 2009, **72**, 1954-1959 (*(±)-Syringaresinol, cell proliferation inhibitor*).
Fan, P. *et al.*, *Asian J. Chem.*, 2010, **22**, 1705-1708 (*Schproplignan C, activity*).
Kim, K.H. *et al.*, *Bioorg. Med. Chem. Lett.*, 2011, **21**, 2270-2273 (*(-)-Syringaresinol, activity*).
Jeong, E.J. *et al.*, *Bioorg. Med. Chem. Lett.*, 2011, **21**, 2283-2286 (*(+)-Syringaresinol, activity*).
Quang, T.H. *et al.*, *Bull. Korean Chem. Soc.*, 2011, **32**, 4049-4054 (*Liriodendrin, activity*).
Kim, K.H. *et al.*, *Chem. Pharm. Bull.*, 2011, **59**, 1425-1429 (*Pharbitis nil constits*).
Xiong, L. *et al.*, *J. Nat. Prod.*, 2011, **74**, 1188-1200 (*Sinocalamus affinis constit*).
Lacret, R. *et al.*, *Phytochem. Lett.*, 2012, **5**, 382-386 (*(+)-Syringaresinol, activity*).
Liu, Z.-Z. *et al.*, *Carbohydr. Res.*, 2013, **372**, 47-54 (*Obtusifosides A-E, activity*).
Cheng, M.-J. *et al.*, *Chem. Nat. Compd. (Engl. Transl.)*, 2013, **49**, 226-228 (*Actinomadura miaoliensis constits*).
Lee, S.Y. *et al.*, *Helv. Chim. Acta*, 2013, **96**, 320-325 (*4,4′-Diangeloyl-(+)-syringaresinol*).
El-Mekkawy, S. *et al.*, *Nat. Prod. Res.*, 2013, **27**, 146-154 (*Commiphora wightii constits, activity*).
Morgan, A.M.A. *et al.*, *Bioorg. Med. Chem. Lett.*, 2014, **24**, 3434-3439 (*Liriodendrin, activity*).
Syed, M.K. *et al.*, *Eur. J. Org. Chem.*, 2014, 5549-5556 (*synth*).
Parveen, S. *et al.*, *J. Chem. Res.*, 2014, **38**, 46-49 (*Brachysides A,B, activity*).
Matsuo, Y. *et al.*, *Chem. Pharm. Bull.*, 2016, **64**, 918-923 (*5,5′-Dihydroxy-(+)-pinoresinol*).
Zhang, Y. *et al.*, *Fitoterapia*, 2016, **115**, 165-172 (*Dioscoroside H_2*).
Muhit, M.A. *et al.*, *J. Nat. Prod.*, 2016, **79**, 1298-1307 (*Terminaloside E, activity*).
Salleh, W.M.N.H.W. *et al.*, *Nat. Prod. Sci.*, 2016, **22**, 225-230 (*(+)-Excelsin, (+)-Sesartemin, activity*).
Liu, X. *et al.*, *J. Nat. Prod.*, 2018, **81**, 1333-1342 (*Cinncassin F*).
Lu, Q. *et al.*, *Bioorg. Med. Chem. Lett.*, 2019, **29**, 1774-1778 (*tetra-Me ether 4′-alloside*)

7,9′:7′,9-Diepoxy-8,8′-lignan-3,3′,4,4′,5,7-hexol D69

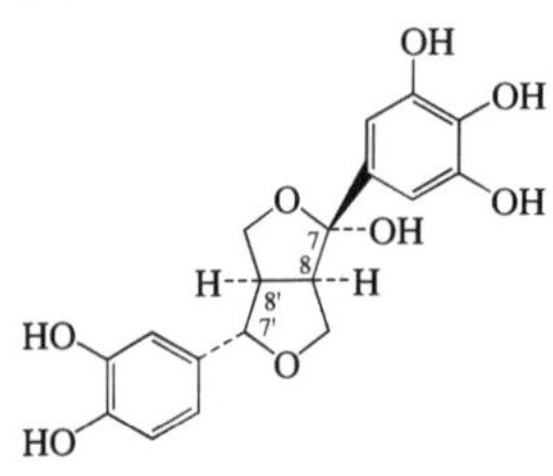

3,3′,4,4′,5,7-Hexahydroxy-7,9′:7′,9-diepoxylignan
$C_{18}H_{18}O_8$ 362.334

(7R*,7′S*,8R*,8′R*)-form
3,3′,5,7-Tetra-Me ether
7,9′:7′,9-Diepoxy-3,3′,5,7-tetramethoxy-8,8′-lignan-4,4′-diol. 4,4′-Dihydroxy-3,3′,5,7-tetramethoxy-7,9′:7′,9-diepoxylignan. ***7-Methoxy-7-epimedioresinol***
[1412398-54-9]
$C_{22}H_{26}O_8$ 418.442
Constit. of *Chromolaena odorata*. Cryst. (Me_2CO). Mp 205 - 206°. $[\alpha]^{20}_{D}$ +90 (c, 0.01 in MeOH). λ_{max} 233; 280 (MeOH)
3′,4′-Methylene, 3,4,5,7-tetra-Me ether
7,9′:7′,9-Diepoxy-3,4,5,7-tetramethoxy-3′,4′-methylenedioxylignan. 3,4,5,7-Tetramethoxy-3′,4′-methylenedioxy-7,9′:7′,9-diepoxylignan. ***Caruilignan B***
[331960-71-5]
$C_{23}H_{26}O_8$ 430.453
Constit. of aerial parts of *Artemisia caruifolia*. Shows moderate cytotoxicity against Meth-A cells. Amorph. powder. $[\alpha]^{24}_{D}$ +78.8 (c, 0.51 in $CHCl_3$). λ_{max} 242 (ε 9900); 283 (ε 5300) ($CHCl_3$).

Ma, C.-M. *et al.*, *Chem. Pharm. Bull.*, 2001, **49**, 183-187 (*Caruilignan B*).
Zhang, M.-L. *et al.*, *J. Nat. Prod.*, 2012, **75**, 2076-2081 (*7-Methoxy-7-epimedioresinol*)

7,9′:7′,9-Diepoxy-8,8′-lignan-3,3′,4,4′,5,8-hexol D70

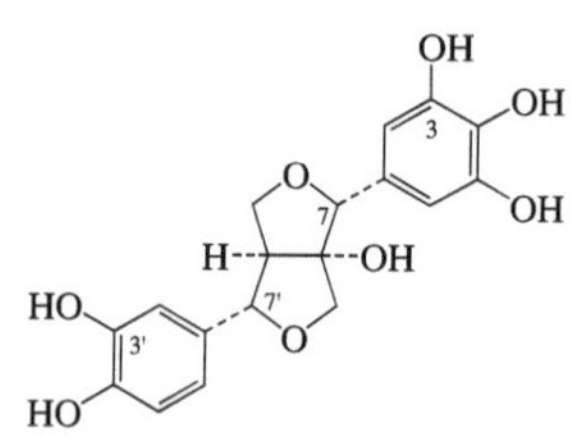

(7R,7′S,8S,8′R)-form

3,3′,4,4′,5,8-Hexahydroxy-7,9′:7′,9-diepoxylignan
$C_{18}H_{18}O_8$ 362.334

(7R,7′S,8S,8′R)-form
3,3′,5-Tri-Me ether
7,9′:7′,9-Diepoxy-3,3′,5-trimethoxy-8,8′-lignan-4,4′,8-triol. 4,4′,8-Trihydroxy-3,3′,5-trimethoxy-7,9′:7′,9-diepoxylignan. ***Fraxiresinol. Pseuderesinol***
[89199-97-3]
[1451255-62-1] (Pseuderesinol)
$C_{21}H_{24}O_8$ 404.415
Constit. of *Fraxinus mandshurica* var. *japonica*, *Fraxinus japonica* and *Salvia santolinifolia*. Also from *Pseuderanthemum carruthersii* var. *atropurpureum*. Lipoxygenase inhibitor. Exhibits moderate cytotoxic activity. Amorph. powder. $[\alpha]^{25}_{D}$ +29.3 (c, 0.61 in $CHCl_3$) (Fraxiresinol). $[\alpha]^{20}_{D}$ -488 (c, 0.4 in $CHCl_3$) (Pseuderesinol). Structure of Pseuderesinol revised by research publ. in 2015. λ_{max} 232 (ε 12590); 280 (ε 3470) (EtOH)
3,3′,5-Tri-Me ether, 4′-O-β-D-glucopyranoside
Fraxiresinol 4′-β-D-glucoside
[745835-63-6]
$C_{27}H_{34}O_{13}$ 566.556
Constit. of the roots of *Valeriana prionophylla*. Amorph. yellow solid. $[\alpha]^{25}_{D}$ -15.3 (c, 1 in MeOH). λ_{max} 232; 280 (MeOH)
3,3′,5-Tri-Me ether, 8-O-β-D-glucopyranoside
Fraxiresinol 8-β-D-glucoside
[89199-94-0]
$C_{27}H_{34}O_{13}$ 566.556
Constit. of *Olea europeae* and *Stauntonia hexaphylla*. Amorph. powder. Mp 125 - 126° (119-122°). $[\alpha]^{20}_{D}$ -19.4 (c, 0.53 in MeOH). λ_{max} 232 (log ε 4.2); 279 (log ε 3.89) (EtOH)

(7S,7′S,8S,8′R)-form
3,3′,5-Tri-Me ether
Dipsalignan A
[1841146-32-4]
$C_{21}H_{24}O_8$ 404.415
Constit. of the roots of *Dipsacus asper*. Amorph. powder. $[\alpha]^{20}_{D}$ +28.2 (c, 0.06 in MeOH). Stereochem. revised in 2018. λ_{max} 211; 236; 280 (MeOH).

Tsukamoto, H. *et al.*, *Chem. Pharm. Bull.*, 1984, **32**, 4482-4489 (*Fraxiresinol*).
Tsukamoto, H. *et al.*, *Chem. Pharm. Bull.*, 1985, **33**, 1232-1241 (*Fraxiresinol 8-glucoside*).
Wang, H.B. *et al.*, *Phytochemistry*, 1993, **34**, 1621-1624 (*Fraxiresinol 8-glucoside*).
Piccinelli, A.L. *et al.*, *J. Nat. Prod.*, 2004, **67**, 1135-1140 (*Fraxiresinol 4′-glucoside*).
Mehmood, S. *et al.*, *Pol. J. Chem.*, 2008, **82**, 571-575 (*Salvia santolinifolia constit, activity*).
Vo, T.N. *et al.*, *Chem. Pharm. Bull.*, 2012, **60**, 1125-1133 (*Pseuderesinol, activity*).
Sun, X. *et al.*, *Molecules*, 2015, **20**, 2165-2175 (*Dipsalignan A*).
Shao, S.-Y. *et al.*, *J. Nat. Prod.*, 2018, **81**, 1023-1028 (*Pseuderesinol, Dipsalignan A, struct*)

7,9′:7′,9-Diepoxy-8,8′-lignan-3,3′,4,4′,5,9-hexol D71

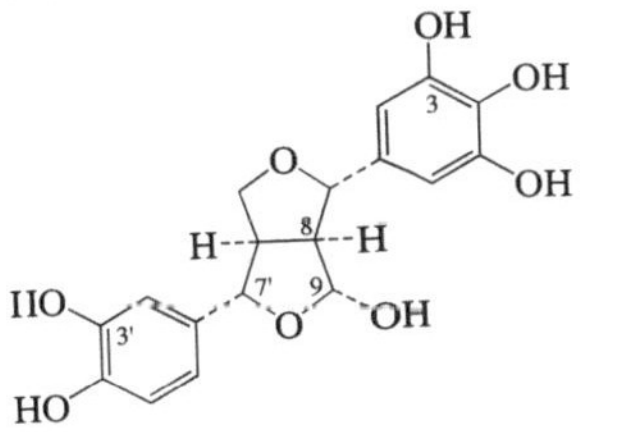

3,3′,4,4′,5,9-Hexahydroxy-7,9′:7′,9-diepoxylignan
$C_{18}H_{18}O_8$ 362.334
(7S,7′S,8S,8′R,9R)-form
3,3′,5-Tri-Me ether
7,9′:7′,9-Diepoxy-3,3′,5-trimethoxy-8,8′-lignan-4,4′,9-triol. 4,4′,9-Trihydroxy-3,3′,5-trimethoxy-7,9′:7′,9-diepoxylignan. ***9α-Hydroxymedioresinol***
[114558-78-0]
$C_{21}H_{24}O_8$ 404.415
Constit. of *Allamanda neriifolia*. Prisms (MeOH). Mp 210 - 213°. $[\alpha]_D^{18}$ +44 (c, 0.49 in MeOH).

Abe, F. *et al.*, *Phytochemistry*, 1988, **27**, 575-577 (*9α-Hydroxymedioresinol*)

7,9′:7′,9-Diepoxy-8,8′-lignan-3,3′,4,4′,5,9′-hexol D72

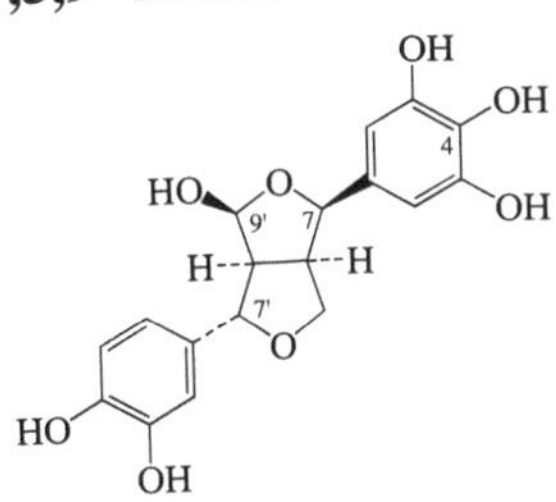

3,3′,4,4′,5,9′-Hexahydroxy-7,9′:7′,9-diepoxylignan
$C_{18}H_{18}O_8$ 362.334
(7R,7′S,8R,8′S,9′S)-form
3,3′,5-Tri-Me ether
7,9′:7′,9-Diepoxy-3,3′,5-trimethoxy-8,8′-lignan-4,4′,9′-triol. 4,4′,9′-Trihydroxy-3,3′,5-trimethoxy-7,9′:7′,9-diepoxylignan
3,3′,5-Tri-Me ether, 9′-O-β-D-glucopyranoside
Lactuberin B
[1423043-26-8]
$C_{27}H_{34}O_{13}$ 566.556
Constit. of the tubers of *Lactuca tuberosa*. Amorph. solid.

Stojakowska, A. *et al.*, *Food Chem.*, 2013, **138**, 1250-1255 (*Lactuberin B*)

7,9′:7′,9-Diepoxy-8,8′-lignan-3,3′,4,4′,7,8-hexol D73

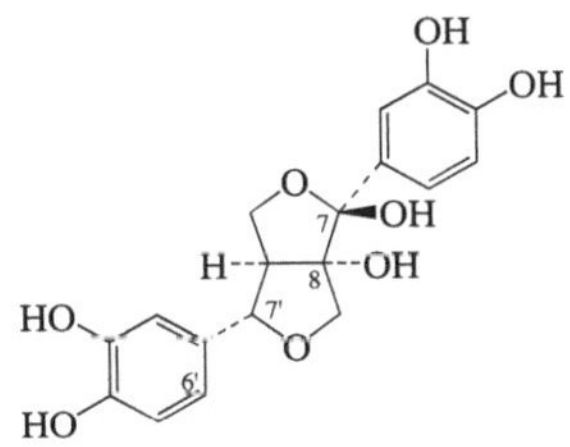

(7*R*,7′*S*,8*S*,8′*R*)-*form*

3,3′,4,4′,7,8-Hexahydroxy-7,9′:7′,9-diepoxylignan
$C_{18}H_{18}O_8$ 362.334
(7R,7′S,8S,8′R)-form
3,4:3′,4′-Bis(methylene) ether
7,9′:7′,9-Diepoxy-3,4:3′,4′-bis(methylenedioxy)-8,8′-lignan-7,8-diol. 7,8-Dihydroxy-3,4:3′,4′-bis(methylenedioxy)-7,9′:7′,9-diepoxylignan. ***Isoarboreol***
[54826-94-7]
$C_{20}H_{18}O_8$ 386.356
Constit. of the heartwood of *Gmelina arborea*. Mp 162°. $[\alpha]_D$ +39 (c, 1 in $CHCl_3$)
6′-Bromo, 3,4:3′,4′-bis(methylene) ether
6′-Bromoisoarboreol
[58976-43-5]
$C_{20}H_{17}BrO_8$ 465.252
Constit. of *Gmelina arborea*. Mp 190°. $[\alpha]_D$ -22
(7S,7′S,8S,8′R)-form
3,4:3′,4′-Bis(methylene) ether
Arboreol
[54868-71-2]
$C_{20}H_{18}O_8$ 386.356
Constit. of the heartwood of *Gmelina arborea* and aerial parts of *Rehmannia chingii*. Needles (C_6H_6). Mp 134 - 136°. Mp 162°. $[\alpha]_D^{30}$ +70 ($CHCl_3$)
3,4:3′,4′-Bis(methylene), 7-Me ether
7,9′:7′,9-Diepoxy-7-methoxy-3,4:3′,4′-bis(methylenedioxy)-8,8′-lignan-8-ol. 8-Hydroxy-7-methoxy-3,4:3′,4′-bis(methylenedioxy)-7,9′:7′,9-diepoxylignan. ***Methylarboreol***
[37589-50-7]
$C_{21}H_{20}O_8$ 400.383
Constit. of *Gmelina arborea*. Needles (MeOH). Mp 134°. $[\alpha]_D^{30}$ +84.5 ($CHCl_3$)
3,4:3′,4′-Bis(methylene), 7-Et ether
7,9′:7′,9-Diepoxy-7-ethoxy-3,4:3′,4′-bis(methylenedioxy)-8,8′-lignan-8-ol. 7-Ethoxy-8-hydroxy-3,4:3′,4′-bis(methylenedioxy)-7,9′:7′,9-diepoxylignan
$C_{22}H_{22}O_8$ 414.41
Constit. of *Gmelina arborea*. Needles (EtOAc). Mp 164°. $[\alpha]_D^{30}$ +136 ($CHCl_3$).

Anjaneyulu, A.S.R. *et al.*, *Tetrahedron*, 1975, **31**, 1277-1285 (*Arboreol, Isoarboreol*).
Anjaneyulu, A.S.R. *et al.*, *Tetrahedron*, 1977, **33**, 133-143 (*6′-Bromoisoarboreol*).
Zheng, C-J. *et al.*, *Chem. Nat. Compd. (Engl. Transl.)*, 2014, **50**, 560-561 (*Rehmannia chingii constit*)

7,9′:7′,9-Diepoxy-8,8′-lignan-3,3′,4,4′,8,8′-hexol D74

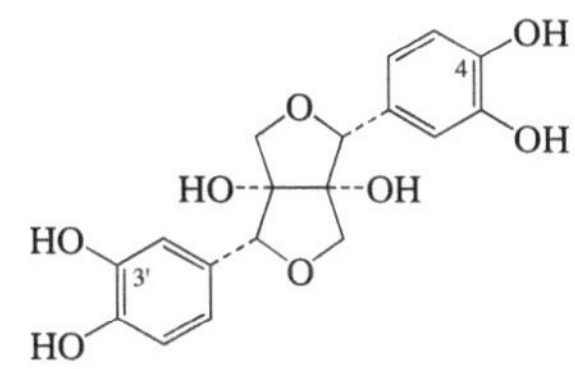

(7*R*,7′*R*,8*S*,8′*S*)-*form*

2,6-Bis(3,4-dihydroxyphenyl)-3,7-dioxabicyclo[3.3.0]octane-1,5-diol. 3,3′,4,4′,8,8′-Hexahydroxy-7,9′:7′,9-diepoxylignan
$C_{18}H_{18}O_8$ 362.334
Lignan numbering shown
(7R,7′R,8S,8′S)-form
8-O-α-L-Rhamnopyranoside
[1321997-66-3]
$C_{24}H_{28}O_{12}$ 508.476
Constit. of the seed pods of *Bauhinia retusa*. Yellow cryst.. Mp 195°. $[\alpha]_D$ -26.4 (c, 1 in MeOH). λ_{max} 233; 278 (no solvent reported)
3,3′-Di-Me ether
7,9′:7′,9-Diepoxy-3,3′-dimethoxy-8,8′-lignan-4,4′,8,8′-tetrol. 4,4′,8,8′-Tetrahydroxy-3,3′-dimethoxy-7,9′:7′,9-diepoxylignan.
Prinsepiol
[82667-99-0]
$C_{20}H_{22}O_8$ 390.388
Constit. of *Prinsepia utilis*, *Schisandra chinensis*, *Vladimiria muliensis*, *Styrax japonica* and *Valeriana prionophylla*. Weak vasodilator. Cryst. (EtOAc/petrol). Mp 191 - 192°. $[\alpha]_D$ -18.41 (EtOAc). λ_{max} 234; 280 (MeOH)
3,3′-Di-Me ether, 4-O-β-D-glucopyranoside
Ambrosidin
[536745-95-6]

$C_{26}H_{32}O_{13}$ 552.529
Constit. of the roots of *Cephalaria ambrosioides*, *Saussurea lappa* and *Valeriana prionophylla*. Amorph. powder. $[\alpha]_D^{26}$ -41.2 (c, 1 in MeOH). λ_{max} 204 (log ε 6.87); 225 (log ε 7.5) (MeOH). [neutral] λ_{max} 233 (log ε 3.86); 278 (log ε 3.72) (MeOH)

3,3′-Di-Me ether, 8-Ac
8-O-Acetylprinsepiol
[1924602-42-5]
$C_{22}H_{24}O_9$ 432.425
Constit. of *Helianthemum sessiliflorum*. Oil. $[\alpha]_D^{22}$ -22.5 (c, 0.27 in EtOAc)

3,4-Methylene ether
7,9′:7′,9-Diepoxy-3′,4′-methylenedioxy-8,8′-lignan-3,4,8,8′-tetrol. 3,4,8,8′-Tetrahydroxy-3′,4′-methylenedioxy-7,9′:7′,9-diepoxylignan

3,4-Methylene ether, 3′-O-β-D-glucopyranoside
Khainaoside A
[1193330-72-1]
$C_{25}H_{28}O_{13}$ 536.486
Constit. of the leaves of *Vitex glabrata*. Amorph. brown powder. $[\alpha]_D^{25}$ -40 (c, 0.2 in MeOH). λ_{max} 230 (log ε 4.26); 282 (log ε 3.96); 317 (sh) (MeOH)

3,4:3′,4′-Bis(methylene) ether
7,9′:7′,9-Diepoxy-3,4:3′,4′-bis(methylenedioxy)-8,8′-lignan-8,8′-diol. 8,8′-Dihydroxy-3,4:3′,4′-bis(methylenedioxy)-7,9′:7′,9-diepoxylignan. ***Wodeshiol.*** *8,8′-Dihydroxysesamin. Kigeliol*
[58007-97-9]
$C_{20}H_{18}O_8$ 386.356
Lignan from *Cleistanthus collinus* and wood of *Kigelia pinnata*. Needles. Mp 153 - 154° (150-151°). $[\alpha]_D^{24}$ -34.6 ($CHCl_3$)

(7R,7′S,8S,8′S)-form

3,3′-Di-Me ether
Dipsalignan B
[1841146-33-5]
$C_{20}H_{22}O_8$ 390.388
Constit. of the roots of *Dipsacus asper*. Amorph. powder. $[\alpha]_D^{20}$ +21.6 (c, 0.07 in $CHCl_3$). λ_{max} 204; 231; 280 (MeOH)

3,4:3′,4′-Bis(methylene) ether
Vitexkarinol
[1533418-56-2]
$C_{20}H_{18}O_8$ 386.356
Constit. of *Vitex leptobotrys*. Gum. $[\alpha]_D^{25}$ +32.8 (c, 4.8 in $CHCl_3$). λ_{max} 236 (log ε 3.85); 286 (log ε 3.77) (MeOH).

Inoue, K. *et al.*, *Phytochemistry*, 1981, **20**, 2271-2276 (*Kigelia pinnata constit*).
Anjaneyulu, A.S.R. *et al.*, *Tetrahedron*, 1981, **37**, 3641-3652 (*Wodeshiol, struct*).
Kilidhar, S.B. *et al.*, *Phytochemistry*, 1982, **21**, 796-797 (*Prinsepiol*).
Han, X. *et al.*, *Org. Lett.*, 1999, **1**, 1871-1872 (*Wodeshiol, synth*).
Pasi, S. *et al.*, *Nat. Prod. Lett.*, 2002, **16**, 365-370 (*Ambrosidin*).
Piccinelli, A.L. *et al.*, *J. Nat. Prod.*, 2004, **67**, 1135-1140 (*Prinsepiol, activity*).
Xu, Y.-Q. *Acta Crystallogr., Sect. E*, 2005, **61**, o2559-o2560 (*Prinsepiol, cryst struct*).
Li, S. *et al.*, *J. Asian Nat. Prod. Res.*, 2006, **8**, 281-286 (*Ambrosidin*).
Luecha, P. *J. Nat. Prod.*, 2009, **72**, 1954-1959 (*Khainaoside A*).
Wu, N. *et al.*, *Asian J. Chem.*, 2011, **23**, 931-932 (*Styrax japonica constit*).
Semwal, S. *et al.*, *Chin. Chem. Lett.*, 2011, **22**, 1081-1083 (*8-rhamnoside*).
Chen, J-J. *et al.*, *Planta Med.*, 2013, **79**, 1470-1473 (*Vladimiria muliensis constit*).
Shi, Y. *et al.*, *Chin. J. Chem.*, 2014, **32**, 734-740 (*Schisandra chinensis constit*).
Pan, W. *et al.*, *J. Nat. Prod.*, 2014, **77**, 663-667 (*Vitexkarinol*).
Sun, X. *et al.*, *Molecules*, 2015, **20**, 2165-2175 (*Dipsalignan B*).
Benabdelaziz, I. *et al.*, *Rec. Nat. Prod.*, 2015, **9**, 342-348 (*8-Acetylprinsepiol*)

7,9′:7′,9-Diepoxy-8,8′-lignan-3,3′,4,4′,8,9′-hexol D75

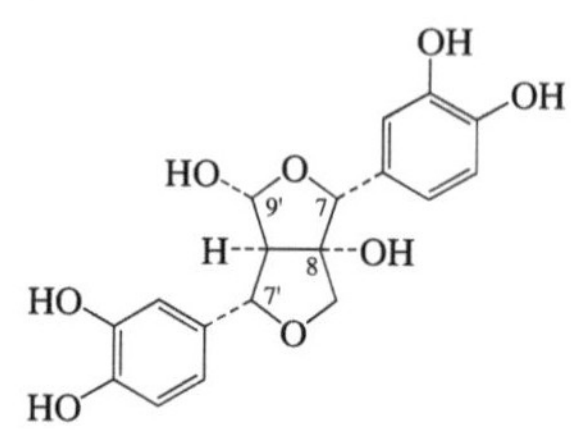

(7R,7′S,8S,8′R,9′R)-form

3,3′,4,4′,8,9′-Hexahydroxy-7,9′:7′,9-diepoxylignan
$C_{18}H_{18}O_8$ 362.334

(7R,7′S,8S,8′R,9′R)-form

3,4:3′,4′-Bis(methylene) ether
7,9′:7′,9-Diepoxy-3,4:3′,4′-bis(methylenedioxy)-8,8′-lignan-8,9′-diol. 8,9′-Dihydroxy-3,4:3′,4′-bis(methylenedioxy)-7,9′:7′,9-diepoxylignan. ***Gummadiol***
[57684-87-4]
$C_{20}H_{18}O_8$ 386.356
Constit. of *Gmelina arborea*. Mp 130°. $[\alpha]_D$ +32

3,4:3′,4′-Bis(methylene) ether, di-Ac
Mp 140°. $[\alpha]_D$ +31

(7R,7′S,8S,8′R,9′S)-form

3,4:3′,4′-Bis(methylene) ether, 9′-O-β-D-glucopyranoside
[63398-40-3]
$C_{26}H_{28}O_{13}$ 548.497
Constit. of *Gmelina arborea* and *Premna corymbosa*. Amorph. powder. Mp 95 - 97°. $[\alpha]_D$ +44.8 (c, 0.79 in MeOH). Config. at C-9′ is tentative.

Anjaneyulu, A.S.R. *et al.*, *Tetrahedron*, 1977, **33**, 133-143 (*Gmelina arborea constits*).
Pelter, A. *et al.*, *Tetrahedron Lett.*, 1979, **20**, 2275-2278 (*synth*).
Yuasa, K. *et al.*, *J. Nat. Prod.*, 1993, **56**, 1695-1699 (*pmr, cmr, glucoside*)

7,9′:7′,9-Diepoxy-8,8′-lignan-3,3′,4,4′,8,9-hexol D76

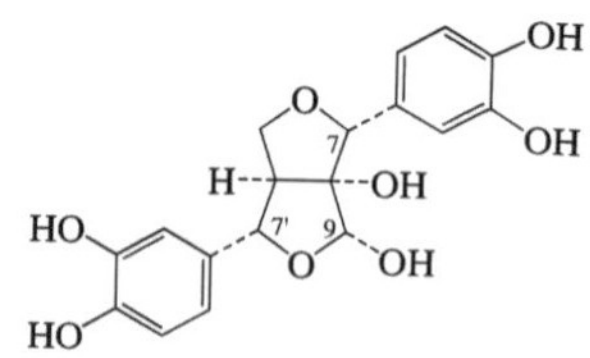

3,3′,4,4′,8,9-Hexahydroxy-7,9′:7′,9-diepoxylignan
$C_{18}H_{18}O_8$ 362.334

(7R,7′S,8R,8′R,9R)-form

3,4:3′,4′-Bis(methylene) ether
7,9′:7′,9-Diepoxy-3,4:3′,4′-bis(methylenedioxy)-8,8′-lignan-8,9-diol. 8,9-Dihydroxy-3,4:3′,4′-bis(methylenedioxy)-7,9′:7′,9-diepoxylignan. ***Premnan B***
$C_{20}H_{18}O_8$ 386.356
Constit. of the wood of *Premna serratifolia*. Increases melanin prodn.. Amorph. solid. $[\alpha]_D^{25}$ +1.6 (c, 0.1 in MeOH). λ_{max} 245 (log ε 3.70); 288 (log ε 3.78) (MeOH).

Woo, S.-Y. *et al.*, *Fitoterapia*, 2019, **133**, 35-42 (*Premnan B, activity*)

7,9′:7′,9-Diepoxy-8,8′-lignan-3,3′,4,4′,9,9′-hexol D77

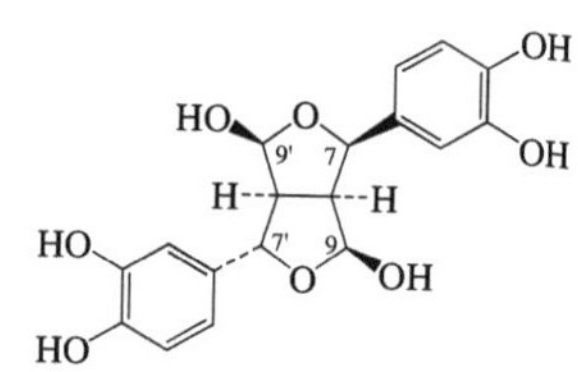

(7R,7′S,8S,8′S,9S,9′S)-form

4,6-Dihydroxy-2,6-bis(3,4-dihydroxyphenyl)-3,7-dioxabicyclo[3.3.0]octane.

3,3′,4,4′,9,9′-Hexahydroxy-7,9′:7′,9-diepoxylignan
$C_{18}H_{18}O_8$ 362.334
(7R,7′S,8S,8′S,9S,9′S)-form
3,3′-Di-Me ether
7,9′:7′,9-Diepoxy-3,3′-dimethoxy-8,8′-lignan-4,4′,9,9′-tetrol. 4,4′,9,9′-Tetrahydroxy-3,3′-dimethoxy-7,9′:7′,9-diepoxylignan. ***9,9′-Dihydroxyepipinoresinol***
[869799-75-7]
$C_{20}H_{22}O_8$ 390.388
Constit. of the heartwood of *Tsuga dumosa*. Mp 128 - 130°. $[\alpha]^{27}_{D}$ +36.8 (c, 0.29 in MeOH)
(7S,7′S,8S,8′S,9R,9′R)-form
3,3′-Di-Me ether, 9-O-β-D-glucopyranoside
Lactucaside
[541547-02-8]
$C_{26}H_{32}O_{13}$ 552.529
Constit. of *Lactuca indica* and *Lactuca sibirica*. Shows antidiabetic activity. Off-white powder. $[\alpha]^{25}_{D}$ -44.3 (c, 0.4 in MeOH)
3,3′-Di-Me ether, 9-O-[4-hydroxyphenylacetyl-(→6)-β-D-glucopyranoside]
6″-O-(4-Hydroxyphenylacetyl) lactucaside
[1101831-89-3]
$C_{34}H_{38}O_{15}$ 686.663
Constit. of *Lactuca sibirica*. Gum
3,3′-Di-Me ether, 9-O-[4-methoxyphenylacetyl-(→6)-β-D-glucopyranoside]
6″-O-(4-Methoxyphenylacetyl) lactucaside
[1101831-88-2]
$C_{35}H_{40}O_{15}$ 700.69
Constit. of *Lactuca sibirica*. Gum. $[\alpha]^{20}_{D}$-21.8 (c, 0.7 in MeOH)
3,3′:4,4′-Tetra-Me ether
7,9′:7′,9-Diepoxy-3,3′,4,4′-tetramethoxy-8,8′-lignan-9,9′-diol. ***9,9′-Dihydroxy-3,3′,4,4′-tetramethoxy-7,9′:7′,9-diepoxylignan***
[111614-78-9]
$C_{22}H_{26}O_8$ 418.442
Constit. of *Eremophila dalyana*. Cryst. (EtOH/Me_2CO). Mp 207 - 209°. $[\alpha]_D$ -16.9 (c, 0.2 in MeOH)
(7R*,7′R*,8R*,8′R*,9ξ,9′ξ)-form
3,4:3′,4′-Bis(methylene) ether
7,9′:7′,9-Diepoxy-3,4:3′,4′-bis(methylenedioxy)-8,8′-lignan-9,9′-diol. 9,9′-Dihydroxy-3,4:3′,4′-bis(methylenedioxy)-7,9′:7′,9-diepoxylignan.
4,8-Dihydroxysesamin
[63398-39-0]
$C_{20}H_{18}O_8$ 386.356
Constit. of *Gmelina arborea*. Needles (C_6H_6). Mp 150°. $[\alpha]_D$ -14.4
3,4:3′,4′-Bis(methylene) ether, di-Ac
[63398-44-7]
$C_{24}H_{22}O_{10}$ 470.43
Needles (MeOH). Mp 158°
9-Ketone(lactone), 3,4:3′,4′-bis(methylene) ether
7,9′-Epoxy-3,4:3′,4′-bis(methylenedioxy)-8,8′-lignan-9,7′-olide. ***Vitelignin A***
[1383440-30-9]
$C_{20}H_{16}O_8$ 384.34
Constit. of the seeds of *Vitex negundo*. Moderate antifungal agent. Syrup. $[\alpha]^{25}_{D}$+18.2 (c, 0.26 in $CHCl_3$). Probable abs. config.. λ_{max} 236 (log ε 4.12); 282 (log ε 3.69) (MeOH).

Anjaneyulu, A.S.R. *et al.*, *Tetrahedron*, 1977, **33**, 133-143 (*4,8-Dihydroxysesamin*).
Pelter, A. *et al.*, *J. Chem. Soc., Perkin Trans. 1*, 1982, 175-181 (*synth*).
Ghisalberti, E.L. *et al.*, *Aust. J. Chem.*, 1987, **40**, 405-411 (*tetra-Me ether*).
Hou, C.C. *et al.*, *J. Nat. Prod.*, 2003, **66**, 625-629 (*Lactucaside*).
Zhao, Y.-X. *et al.*, *Yunnan Zhiwu Yanjiu*, 2004, **26**, 229-233 (*9,9′-Dihydroxyepipinoresinol*).
Kisiel, W. *et al.*, *Fitoterapia*, 2008, **79**, 241-244 (*phenylacetyllactucasides*).
Zheng, C.-J. *et al.*, *Phytochem. Lett.*, 2011, **4**, 298-300 (*Vitelignin A*)

7,9′:7′,9-Diepoxy-8,8′-lignan-3,3′,4,5,5′,8-hexol D78

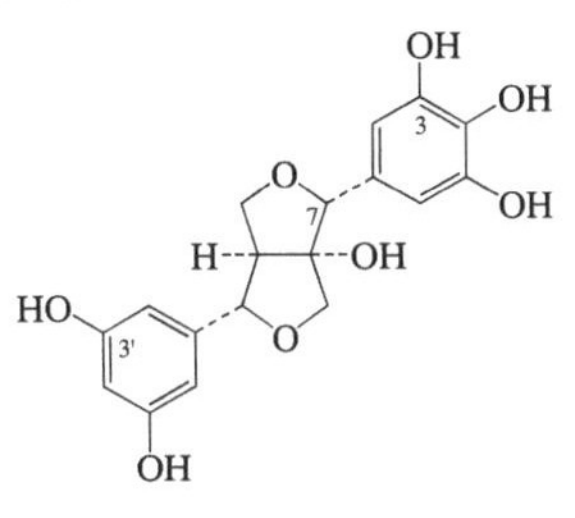

3,3′,4,5,5′,8-Hexahydroxy-7,9′:7′,9-diepoxylignan
$C_{18}H_{18}O_8$ 362.334
Structures in this series may need revision. Phenyl 3,5-dioxygenated substn. pattern in doubt as a result of work done by Yang *et al* in 2015
(7R,7′S,8S,8′R)-form
3,3′,5-Tri-Me ether
7,9′:7′,9-Diepoxy-3′,5,5′-trimethoxy-8,8′-lignan-3,4′,8′-triol. 3′,4,8-Trihydroxy-3,5,5′-trimethoxy-7,9′:7′,9-diepoxylignan.
Melodinin B
$C_{21}H_{24}O_8$ 404.415
Constit. of the stems of *Melodinus suaveolens*. Exhibits antiinflammatory activity. Amorph. powder. $[\alpha]^{25}_{D}$+8 (c, 0.05 in MeOH). Structure formerly assigned to Pseuderesinol but revised by research in 2015.

Yang, Y.-N. *et al.*, *J. Nat. Prod.*, 2015, **78**, 705-711 (*pmr, struct*).
Shao, S.-Y. *et al.*, *J. Nat. Prod.*, 2018, **81**, 1023-1028 (*config*).
Fang, L. *et al.*, *Phytochem. Lett.*, 2018, **26**, 134-137 (*Melodinin B, activity*)

7,9′:7′,9-Diepoxy-8,8′-lignan-2,2′,3,3′,4,4′,6,6′,8-nonol D79

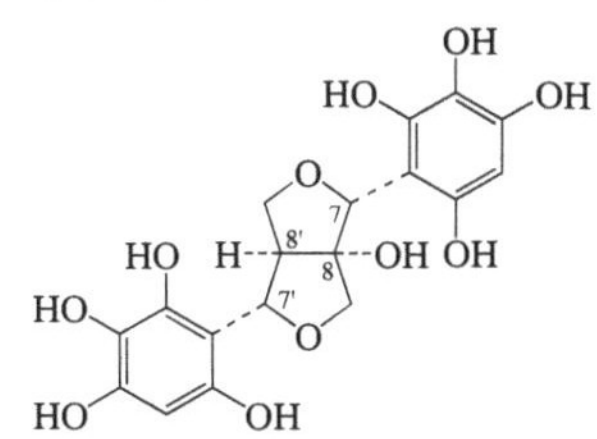

2,2′,3,3′,4,4′,6,6′,8-Nonahydroxy-7,9′:7′,9-diepoxylignan
$C_{18}H_{18}O_{11}$ 410.331
(7R*,7′S*,8S*,8′R*)-form
3,4:3′,4′-Bis(methylene), 2,2′,6,6′-tetra-Me ether
7,9′:7′,9-Diepoxy-2,2′,6,6′-tetramethoxy-3,4:3′,4′-bis(methylenedioxy)-8,8′-lignan-8-ol. 8-Hydroxy-2,2′,6,6′-tetramethoxy-3,4:3′,4′-bis(methylenedioxy)-7,9′:7′,9-diepoxylignan. Leptostachyol
[35942-44-0]
$C_{24}H_{26}O_{11}$ 490.461
Cryst.. Mp 134 - 136°
3,4:3′,4′-Bis(methylene)-2,2′,6,6′-tetra-Me ether, 8-Ac
Leptostachyol acetate
[35770-58-2]
$C_{26}H_{28}O_{12}$ 532.498
Constit. the roots of *Phryma leptostachya* var. *asiatica*. Larvicidal agent. Mp 178 - 179°. λ_{max} 236 (log ε 4.2); 295 (log ε 4.0) (EtOH).

Taniguchi, E. *et al.*, *Tetrahedron Lett.*, 1972, **13**, 653-656 (*Leptostachyol acetate*).

Park, I.-K. *et al.*, *J. Agric. Food Chem.*, 2005, **53**, 969-972 (*activity*)

7,9′:7′,9-Diepoxy-8,8′-lignan-2,2′,3,3′,4,4′,5,5′-octol **D80**

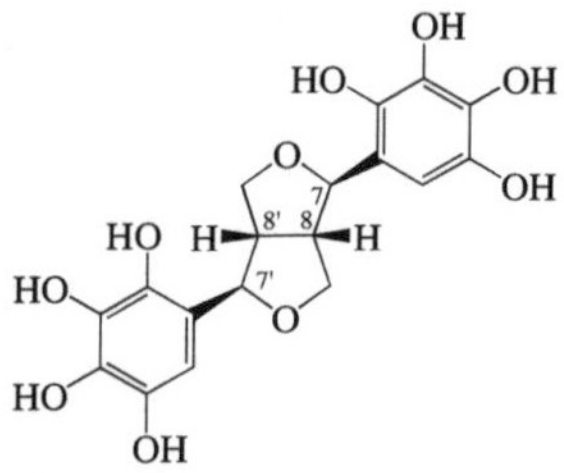

(7R,7′R,8S,8′S)-form

2,2′,3,3′,4,4′,5,5′-Octahydroxy-7,9′:7′,9-diepoxylignan

$C_{18}H_{18}O_{10}$ 394.332

(7R,7′R,8S,8′S)-form

2,2′,3,3′,4,4′-Hexa-Me ether, 5-O-β-glucopyranoside

Terminaloside I

[1906937-23-2]

$C_{30}H_{40}O_{15}$ 640.635

Constit. of the leaves of *Terminalia citrina*. Exhibits antiestrogenic activity against MCF-7 cells. Amorph. pale yellow powder. $[\alpha]^{25}_{D}$ +8 (c, 0.2 in MeOH). λ_{max} 219 (log ε 4.2); 280 (log ε 3.63) (MeOH)

2′,3,3′,4,4′,5′-Hexa-Me ether, 2-O-β-D-glucopyranoside

Terminaloside H

[1906937-22-1]

$C_{30}H_{40}O_{15}$ 640.635

Constit. of the leaves of *Terminalia citrina*. Amorph. pale yellow powder. $[\alpha]^{25}_{D}$ +7.6 (c, 0.2 in MeOH). λ_{max} 215 (log ε 4.23); 281 (log ε 3.63) (MeOH)

2,2′,3,3′,4,4′,5′-Hepta-Me ether, 5-O-β-D-glucopyranoside

Terminaloside J

[1906937-24-3]

$C_{31}H_{42}O_{15}$ 654.662

Constit. of the leaves of *Terminalia citrina*. Amorph. yellow powder. $[\alpha]^{25}_{D}$ +51.2 (c, 0.2 in MeOH). λ_{max} 212 (log ε 4.6); 279 (log ε 3.77) (MeOH)

(7S,7′R,8R,8′R)-form

3,3′,4,4′,5,5′-Hexa-Me ether, 2-O-β-D-glucopyranoside

6-Epiterminaloside K

[1906937-26-5]

$C_{30}H_{40}O_{15}$ 640.635

Constit. of the leaves of *Terminalia citrina*. Exhibits antiestrogenic activity against MCF-7 cells. Amorph. pale yellow powder. $[\alpha]^{25}_{D}$ +52.3 (c, 0.2 in MeOH). λ_{max} 214 (log ε 4.24); 283 (log ε 3.76) (MeOH)

(7S,7′S,8R,8′R)-form

3,3′,4,4′,5,5′-Hexa-Me ether, 2-O-β-D-glucopyranoside

Terminaloside K

[1906937-25-4]

$C_{30}H_{40}O_{15}$ 640.635

Constit. of the leaves of *Terminalia citrina*. Amorph. yellow powder. $[\alpha]^{25}_{D}$ +32.3 (c, 0.2 in MeOH). λ_{max} 214 (log ε 4.29); 283 (log ε 3.77) (MeOH)

4′,5′-Methylene, 2′,3,3′,4-tetra-Me ether, 2-O-β-D-glucopyranoside

Terminaloside F

[1906937-20-9]

$C_{29}H_{36}O_{15}$ 624.592

Constit. of the leaves of *Terminalia citrina*. Exhibits antiestrogenic activity against MCF-7 cells. Amorph. pale yellow powder. $[\alpha]^{25}_{D}$ +40.6 (c, 0.2 in MeOH). λ_{max} 218 (log ε 4.26); 283 (log ε 3.69) (MeOH).

Muhit, M.A. *et al.*, *J. Nat. Prod.*, 2016, **79**, 1298-1307 (*Terminalosides F,H-K, 6-Epiterminaloside K, activity*)

7,9′:7′,9-Diepoxy-8,8′-lignan-2,2′,3,4,4′,5′,6,8′-octol **D81**

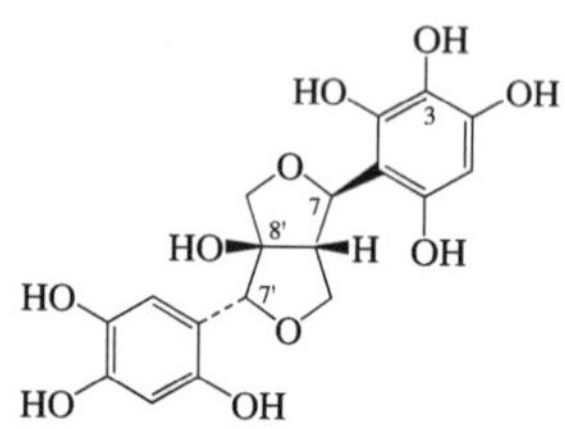

2,2′,3,4,4′,5′,6,8′-Octahydroxy-7,9′:7′,9-diepoxylignan

$C_{18}H_{18}O_{10}$ 394.332

(7R*,7′R*,8S*,8′R*)-form

3,4:4′,5′-Bis(methylene), 2,2′,6-tri-Me ether

7,9′:7′,9-Diepoxy-2,2′,6′-trimethoxy-3′,4′:4,5-bis(methylenedioxy)-8,8′-lignan-8-ol. 8′-Hydroxy-2,2′,6-trimethoxy-3,4:4′,5′-bis(methylenedioxy)-7,9′:7′,9-diepoxylignan. Epiphrymarol

[123513-88-2]

$C_{23}H_{24}O_{10}$ 460.435

3,4:4′,5′-Bis(methylene), 2,2′,6-tri-Me ether, 8′-Ac

[139405-55-3]

$C_{25}H_{26}O_{11}$ 502.472

Constit. of the roots of *Phryma leptostachya* var. *asiatica*.

Taniguchi, E. *et al.*, *Tennen Yuki Kagobutsu Toronkai Koen Yoshishu*, 1988, **30**, 1-8 (*Epiphrymarol*).
Park, I.-K. *et al.*, *J. Agric. Food Chem.*, 2005, **53**, 969-972 (*Phryma leptostachya constit*)

7,9′:7′,9-Diepoxy-8,8′-lignan-3,3′,4,4′,5,5′,8,8′-octol **D82**

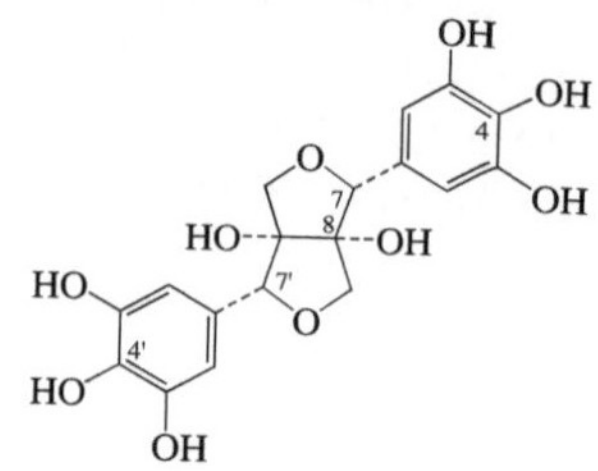

3,3′,4,4′,5,5′,8,8′-Octahydroxy-7,9′:7′,9-diepoxylignan

$C_{18}H_{18}O_{10}$ 394.332

(7R,7′R,8S,8′S)-form

3,3′,5,5′-Tetra-Me ether

7,9′:7′,9-Diepoxy-3,3′,5,5′-tetramethoxy-8,8′-lignan-4,4′,8,8′-tetrol. 4,4′,8,8′-Tetrahydroxy-3,3′,5,5′-tetramethoxy-7,9′:7′,9-diepoxylignan

3,3′,5,5′-Tetra-Me ether, 4-O-β-D-glucopyranoside

Aketrilignoside A

[1020108-68-2]

$C_{28}H_{36}O_{15}$ 612.581

Constit. of the stems of *Akebia trifoliata*. Powder. $[\alpha]^{20}_{D}$ -34 (c, 0.12 in MeOH). λ_{max} 214 (log ε 4.1); 270 (log ε 3.2) (MeOH).

Guan, S.-H. *et al.*, *Magn. Reson. Chem.*, 2008, **46**, 186-190 (*Aketrilignoside A*)

7,9′:7′,9-Diepoxy-8,8′-lignan-2,3,3′,4,4′-pentol **D83**

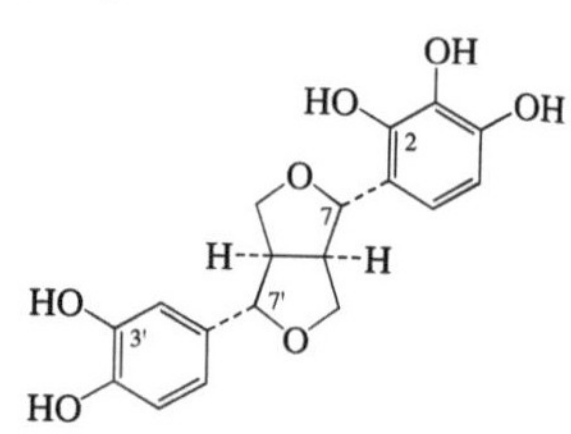

2,3,3′,4,4′-Pentahydroxy-7,9′:7′,9-diepoxylignan

$C_{18}H_{18}O_7$ 346.335

Lignan numbering shown

(7S,7′S,8R,8′R)-form

3,4-Methylene, 2,3′-di-Me ether

7,9′:7′,9-Diepoxy-2′,3-dimethoxy-3′,4′-methylenedioxy-8,8′-lignan-4-ol. 4′-Hydroxy-2,3′-dimethoxy-3,4-methylenedioxy-7,9′:7′,9-diepoxylignan

[501412-94-8]

$C_{21}H_{22}O_7$ 386.4

Constit. of *Leucophyllum ambiguum*. Calmodulin inhibitor. Oil. $[\alpha]_D$ +74 (c, 1 in $CHCl_3$). λ_{max} 240 (log ε 4.2); 279 (log ε 4) (EtOH)

3,4-Methylene, 2,3′,4′-tri-Me ether

7,9′:7′,9-Diepoxy-2,3′,4′-trimethoxy-3,4-methylenedioxylignan. 2,3′,4′-Trimethoxy-3,4-methylenedioxy-7,9′:7′,9-diepoxylignan. ***2′-Methoxykobusin****. 2′-Methoxyspinescin*

[501412-93-7]

$C_{22}H_{24}O_7$ 400.427

Constit. of *Leucophyllum ambiguum*. Calmodulin inhibitor. Needles. Mp 90 - 92°. $[\alpha]_D$ +104 (c, 1 in $CHCl_3$). λ_{max} 229 (log ε 4.1); 278 (log ε 4) (EtOH)

3,4:3′,4′-Bis(methylene) ether

7,9′:7′,9-Diepoxy-3,4:3′,4′-bis(methylenedioxy)-8,8′-lignan-2-ol. 2-Hydroxy-3,4:3′,4′-bis(methylenedioxy)-7,9′:7′,9-diepoxylignan. ***2-Hydroxysesamin****. Sesamin-2-ol*

[878000-23-8]

$C_{20}H_{18}O_7$ 370.357

Constit. of the leaves of *Callicarpa furfuraceae*. Amorph. powder. $[\alpha]^{27}_D$ -4 (c, 0.09 in $CHCl_3$).

Rojas, S. *et al.*, *J. Nat. Prod.*, 2003, **66**, 221-224 (*Leucophyllum ambiguum constits*).
Shao, Y. *et al.*, *Helv. Chim. Acta*, 2006, **89**, 64-72 (*2-Hydroxysesamin*)

7,9′:7′,9-Diepoxy-8,8′-lignan-2,3′,4,4′,5-pentol D84

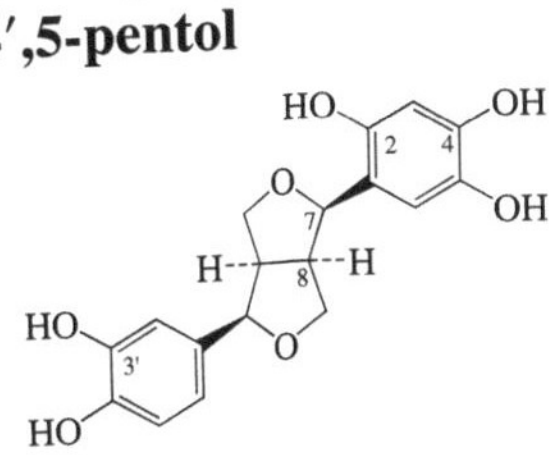

(7R,7′R,8R,8′R)-form

2,3′,4,4′,5-Pentahydroxy-7,9′:7′,9-diepoxylignan

$C_{18}H_{18}O_7$ 346.335

Lignan numbering shown

(7R,7′R,8R,8′R)-form

3′,4′:4,5-Bis(methylene) ether

7,9′:7′,9-Diepoxy-3′,4′:4,5-bis(methylenedioxy)-8,8′-lignan-2-ol. 2-Hydroxy-3′,4′:4,5-bis(methylenedioxy)-7,9′:7′,9-diepoxylignan. 7,7′-Diepisesaminol. Diasesaminol

Mp 194 - 196°. $[\alpha]_D$ +325.9 (c, 0.1 in $CHCl_3$)

(7R,7′S,8R,8′R)-form

4,5-Methylene, 3′-Me ether

7,9′:7′,9-Diepoxy-3′-methoxy-4,5-methylenedioxy-8,8′-lignan-2,4′-diol. 2,4′-Dihydroxy-3′-methoxy-4,5-methylenedioxy-7,9′:7′,9-diepoxylignan. ***6-Hydroxy-7-epipiperitol***

$C_{20}H_{20}O_7$ 372.373

Constit. of the perisperm of *Sesamum indicum*. Amorph. yellow solid. $[\alpha]_D$ +53 (c, 0.2 in MeOH)

3′,4′:4,5-Bis(methylene) ether

7-Episesaminol

[104319-96-2]

Prod. during industrial bleaching of unroasted sesame oil. Yellow prisms ($CHCl_3$/MeOH) (as 3,5-dinitrobenzoyl). Mp 169 - 170° (3,5-dinitrobenzoyl). $[\alpha]_D$ +101.5 (c, 1 in $CHCl_3$) (dinitrobenzoyl)

3′,4′:4,5-Bis(methylene) ether, 2-O-β-D-glucopyranoside

Cuscutoside C

[1260542-48-0]

$C_{26}H_{28}O_{12}$ 532.498

Constit. of the leaves of *Eucalyptus maidenii* and the seeds of *Cuscuta chinensis*. Amorph. powder. Mp 112 - 113°. $[\alpha]^{27}_D$ -37.3 (c, 0.3 in MeOH). $[\alpha]^{20}_D$ -73.5 (c, 0.07 in MeOH). λ_{max} 203 (log ε 4.66); 236 (log ε 3.98); 290 (log ε 3.91) (MeOH). [neutral] λ_{max} 237; 291 (MeOH)

3′,4′:4,5-Bis(methylene) ether, 2-O-[β-D-xylopyranosyl-(1→6)-β-D-glucopyranoside]

Cuscutoside B

[164176-39-0]

$C_{31}H_{36}O_{16}$ 664.613

Constit. of the fruit of *Cuscuta chinensis*. Pale yellow powder. $[\alpha]^{28}_D$ -7.2 (c, 0.36 in MeOH)

3′,4′:4,5-Bis(methylene) ether, 2-O-[β-D-apiofuranosyl-(1→2)-β-D-glucopyranoside]

Cuscutoside A

[164176-40-3]

$C_{31}H_{36}O_{16}$ 664.613

Constit. of the fruit of Chinese dodder *Cuscuta chinensis*. Pale yellow powder. $[\alpha]^{28}_D$ -25.1 (c, 0.48 in MeOH aq.)

3′,4′:4,5-Bis(methylene) ether, 2-O-[β-D-glucopyranosyl-(1→6)-β-D-glucopyranoside]

Neocuscutoside C

[339164-15-7]

$C_{32}H_{38}O_{17}$ 694.639

Constit. of the seeds of *Cuscuta chinensis*. Powder. $[\alpha]^{20}_D$ -12.5 (c, 0.4 in Py)

3′,4′:4,5-Bis(methylene) ether, 2-O-[β-D-apiofuranosyl-(1→2)-[β-D-glucopyranosyl-(1→6)]-β-D-glucopyranoside]

Cuscutoside D

[1311963-71-9]

$C_{37}H_{46}O_{21}$ 826.754

Constit. of the seeds of *Cuscuta chinensis*. Amorph. powder. Mp 153 - 154°. $[\alpha]^{20}_D$ -99 (c, 0.07 in MeOH). λ_{max} 237; 291 (MeOH)

3′,4′:4,5-Bis(methylene) ether, 2-O-[α-L-arabinopyranosyl-(1→6)-[β-D-apiofuranosyl-(1→2)]-β-D-glucopyranoside]

Neocuscutoside B

[339193-40-7]

$C_{36}H_{44}O_{20}$ 796.728

Constit. of the seeds of *Cuscuta chinensis*. Powder. $[\alpha]^{20}_D$ -20 (c, 0.5 in Py)

3′,4′:4,5-Bis(methylene) ether, 2-O-[β-D-glucopyranosyl-(1→4)-β-D-apiofuranosyl-(1→2)-β-D-glucopyranoside]

Neocuscutoside A

[339192-58-4]

$C_{37}H_{46}O_{21}$ 826.754

Constit. of the seeds of *Cuscuta chinensis*. Powder. $[\alpha]^{20}_D$ -19.8 (c, 0.1 in Py)

(7S,7′R,8R,8′R)-form

3′,4′:4,5-Bis(methylene) ether

7′-Episesaminol

Prod. during industrial bleaching of unroasted sesame oil. Mp 160°. $[\alpha]_D$ +138.2 (c, 1 in $CHCl_3$)

(7S,7′S,8R,8′R)-form

4,5-Methylene, 3′-Me ether

6-Hydroxypiperitol

[156941-78-5]

$C_{20}H_{20}O_7$ 372.373

Prod. by *Streptomyces* sp. IT-44. Syrup

3',4':4,5-Bis(methylene) ether

Sesaminol. *Justisolin*

[74061-79-3]

[105616-55-5] [112531-05-2]

$C_{20}H_{18}O_7$ 370.357

Prod. during industrial bleaching of unroasted sesame oil. Isol. from the whole plant of *Justicia simplex* and *Sesamum indicum*. CNS stimulant, antioxidant. Protectant of low-density lipoprotein from oxidative damage. Mp 146 - 148°. $[\alpha]_D^{28}$ +42.7 (c, 0.21 in $CHCl_3$)

3',4':4,5-Bis(methylene) ether, 2-O-β-D-glucopyranoside

[153512-13-1]

$C_{26}H_{28}O_{12}$ 532.498

Constit. of *Sesamum indicum*. $[\alpha]_D^{25}$ -6.24 (c, 1.3 in MeOH)

3',4':4,5-Bis(methylene) ether, 2-O-[β-D-glucopyranosyl-(1→2)-β-D-glucopyranoside]

[157469-82-4]

$C_{32}H_{38}O_{17}$ 694.639

Constit. of *Sesamum indicum*. $[\alpha]_D^{25}$ -25.3 (c, 0.2 in MeOH)

3',4':4,5-Bis(methylene) ether, 2-O-[β-D-glucopyranosyl-(1→2)-[β-D-glucopyranosyl-(1→6)]-β-D-glucopyranoside]

Sesaminol triglucoside

[157469-83-5]

$C_{38}H_{48}O_{22}$ 856.78

Constit. of *Sesamum indicum*. $[\alpha]_D^{25}$ -9.75 (c, 0.3 in H_2O)

3',4':4,5-Bis(methylene), 2-Me ether

7,9':7',9-Diepoxy-2-methoxy-3',4':4,5-bis(methylenedioxy) lignan. 2-Methoxy-3',4':4,5-bis(methylenedioxy)-7,9':7',9-diepoxylignan. ***Sesangolin***

[100268-57-3]

$C_{21}H_{20}O_7$ 384.384

Constit. of *Sesamum angolense*. Shows high synergistic activity for pyrethrum. Cryst. (MeOH). Mp 101° (87-88°). $[\alpha]_D^{23}$ +48.5 (c, 0.54 in $CHCl_3$).

Jones, W.A. *et al., J. Org. Chem.*, 1962, **27**, 3232-3235 (*Sesangolin*).

Birch, A.J. *et al., J. Chem. Soc. C*, 1967, 1968-1972 (*Sesangolin, config*).

Ghosal, S. *et al., Phytochemistry*, 1980, **19**, 332-334 (*Sesaminol*).

Fukuda, Y. *et al., Heterocycles*, 1986, **24**, 923-926 (*sesame oil constits*).

Nagata, M. *et al., Agric. Biol. Chem.*, 1987, **51**, 1285-1289 (*Sesaminol, cryst struct, diastereomers*).

Chiung, Y.-M. *et al., J. Antibiot.*, 1994, **47**, 487-491 (*6-Hydroxypiperitol*).

Katsuzaki, H. *et al., Phytochemistry*, 1994, **35**, 773-776 (*Sesaminol glycosides*).

Yahara, S. *et al., Phytochemistry*, 1994, **37**, 1755-1757 (*Cuscutosides A,B*).

Kang, M-H. *et al., Life Sci.*, 2000, **66**, 161-172 (*Sesaminol, activity*).

Xiang, S.-X. *et al., Chin. J. Chem.*, 2001, **19**, 282-285 (*Neocuscutosides A,B,C*).

Ohtsuki, T. *et al., Biosci., Biotechnol., Biochem.*, 2003, **67**, 2304-2306 (*Sesaminol glucosides, synth*).

Moazzami, A.A. *et al., J. Agric. Food Chem.*, 2006, **54**, 633-638 (*Sesaminol glycosides*).

Grougnet, R. *et al., J. Agric. Food Chem.*, 2006, **54**, 7570-7574 (*6-Hydroxy-7-epipiperitol*).

Jan, K.C. *et al., J. Agric. Food Chem.*, 2009, **57**, 6101-6106 (*Sesaminol triglucoside, metab*).

Mochizuki, M. *et al., J. Agric. Food Chem.*, 2009, **57**, 10429-10434 (*Sesaminol, metab, 6-Hydroxypiperitol*).

Tian, L.-W. *et al., Helv. Chim. Acta*, 2010, **93**, 2194-2202 (*7-Episesaminol 2-glucoside*).

He, X.-H. *et al., J. Asian Nat. Prod. Res.*, 2010, **12**, 934-939 (*Cuscutosides C,D*).

Dar, A.A. *et al., Bioorg. Chem.*, 2013, **50**, 1-10 (*Sesaminol, rev, activity, biosynth*).

Ishikawa, R. *et al., Chem. Lett.*, 2014, **43**, 1572-1574 (*Sesaminol*)

7,9':7',9-Diepoxy-8,8'-lignan-3,3',4,4',5-pentol D85

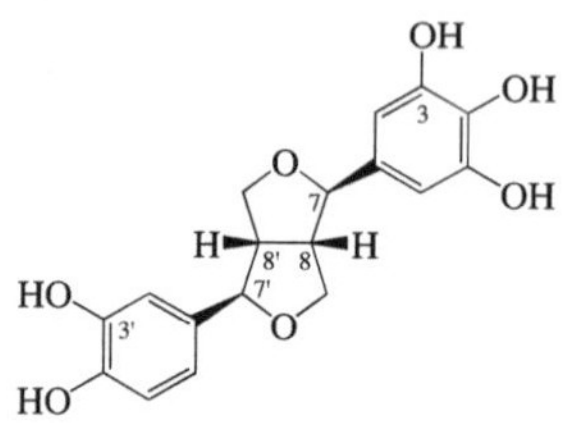

(7R,7'R,8S,8'S)-form

3,3',4,4',5-Pentahydroxy-7,9':7',9-diepoxylignan

$C_{18}H_{18}O_7$ 346.335

Numbering can differ confusingly between the primed and unprimed lignan residues

(7R,7'R,8S,8'S)-form

3,3',5-Tri-Me ether

7,9':7',9-Diepoxy-3,3',5-trimethoxy-8,8'-lignan-4,4'-diol. 4,4'-Dihydroxy-3,3',5-trimethoxy-7,9':7',9-diepoxylignan.

(-)-Medioresinol

[125638-34-8]

$C_{21}H_{24}O_7$ 388.416

Isol. from *Dirca occidentalis*, *Goniothalamus dumontetii* and *Dendrobium loddigesii*. Potent inhibitor of protein kinases, DYRK1A and CDK1/Cyclin B, implicated in some cancers and neurodegenerative disorders. Inhibitor of NO prodn. in RAW 264.7 macrophage-like cells. Sol. MeOH, $CHCl_3$. Mp 155 - 160°. $[\alpha]_D^{25}$ -17 (c. 0.14 in $CHCl_3$). λ_{max} 232 (ε 38000); 280 (ε 13500) (EtOH) (Berdy)

Penta-Me ether

7,9':7',9-Diepoxy-3,3',4,4',5-pentamethoxylignan. 3,3',4,4',5-Pentamethoxy-7,9':7',9-diepoxylignan. ***(-)-Magnolin***

$C_{23}H_{28}O_7$ 416.47

Constit. of the bark of *Zanthoxylum simulans*. Oil. $[\alpha]_D^{23.5}$ -15.3 (c, 0.1 in $CHCl_3$). λ_{max} 204 (log ε 3.71); 276 (log ε 2.53) (hexane)

(7R,7'S,8R,8'R)-form

3,3',4,5-Tetra-Me ether

7,9':7',9-Diepoxy-3,3',4',5'-tetramethoxy-8,8'-lignan-4-ol. 4'-Hydroxy-3,3',4,5-tetramethoxy-7,9':7',9-diepoxylignan. ***Praderin***

[116481-69-7]

$C_{22}H_{26}O_7$ 402.443

Constit. of *Lindera praecox* and *Hernandia peltata*. Prisms (MeOH). Mp 155 - 156°. $[\alpha]_D$ +113.8 (c, 0.05 in $CHCl_3$)

Penta-Me ether

Epimagnolin

[41689-51-4]

$C_{23}H_{28}O_7$ 416.47

Constit. of *Hernandia ovigera*, *Hernandia nymphaeifolia* and *Commiphora wightii*. Shows anti-platelet aggregation activity. Mp 84°. $[\alpha]_D^{25}$ +112 (c, 0.6 in $CHCl_3$). Config. corrected in 1977

3,4-Methylene, 3',4',5-tri-Me ether

7,9':7',9-Diepoxy-3,3',4'-trimethoxy-4,5-methylenedioxylignan. 3,3',4'-Trimethoxy-4,5-methylenedioxy-7,9':7',9-diepoxylignan. ***5'-Demethoxyepisesartemin A***

[873867-94-8]

$C_{22}H_{24}O_7$ 400.427

Constit. of the leaves of *Piper fimbriulatum*. Needles (MeOH). Mp 134 - 136°. $[\alpha]_D$ +115.2 (c, 0.24 in $CHCl_3$). λ_{max} 204 (log ε 5.39); 233 (log ε 4.54); 278 (log ε 4.16) (EtOH)

4,5-Methylene, 3,3'-di-Me ether

Epimethoxypiperitol

$C_{21}H_{22}O_7$ 386.400
Constit. of *Eremophila macdonnellii*. Exhibits moderate cytotoxicity. Powder. $[\alpha]_D^{23}$ +97.1 (c, 0.51 in $CHCl_3$). λ_{max} 205 (log ε 4.90); 231 (log ε 4.29); 280 (log ε 3.81) (MeOH)

(7R,7′S,8S,8′S)-form

3,3′,5-Tri-Me ether

Asarinin B
[2211144-29-3]
$C_{21}H_{24}O_7$ 388.416
Constit. of the roots and rhizomes of *Asarum heterotropoides*. Exhibits antiinflammatory activity. Amorph. powder. $[\alpha]_D^{20}$ -126.7 (c, 0.6 in $CHCl_3$). λ_{max} 241 (log ε 3.86); 280 (log ε 3.58) ($CHCl_3$)

3-O-(3,7-Dimethyl-2E,6-octadienyl), 3′,4′:4,5-bis(methylene) ether
7,9′:7′,9-Diepoxy-3-geranyl-3′,4′:4,5-bis(methylenedioxy) lignan. **Armatumin**
[1376098-91-7]
$C_{30}H_{34}O_7$ 506.595
Constit. of *Zanthoxylum armatum*. Gum. $[\alpha]_D^{24}$ -6.5 (c, 0.58 in $CHCl_3$). λ_{max} 210; 233; 286 (MeOH)

(7S,7′R,8R,8′R)-form

3,3′,4,5-Tetra-Me ether

4′-Demethylepimagnolin A
[155419-01-5]
$C_{22}H_{26}O_7$ 402.443
Constit. of *Ocotea duckei*. Oil. $[\alpha]_D^{23}$ +70.1 (c, 1 in $CHCl_3$)

Penta-Me ether

Epimagnolin A
[143235-93-2]
$C_{23}H_{28}O_7$ 416.47
Constit. of *Magnolia fargesii*, *Artemisia gorgonum*, *Achillea holosericea* and *Rollinia mucosa*. Insect growth inhibitor. iNOS and COX-2 expression inhibitor; shows significant antiplasmodial activity. Yellowish oil. $[\alpha]_D$ +105.2 (c, 3.5 in $CHCl_3$)

3′,4′-Methylene, 3,5-di-Me ether
7,9′:7′,9-Diepoxy-3,5-dimethoxy-3′,4′-methylenedioxy-8,8′-lignan-4-ol. 4-Hydroxy-3,5-dimethoxy-3′,4′-methylenedioxy-7,9′;7′,9-diepoxylignan. **5-Methoxypluviatilol**
[1365563-04-7]
$C_{21}H_{22}O_7$ 386.4
Constit. of *Asarum sieboldii* and *Asarum heteropoides*. PPARs transcription activator. Amorph. powder. $[\alpha]_D^{25}$ -115.3 (c, 0.1 in CH_2Cl_2). λ_{max} 258 (log ε 3.08) (MeOH)

3′,4′-Methylene, 3,4,5-tri-Me ether
7,9′:7′,9-Diepoxy-3,4,5-trimethoxy-3′,4′-methylenedioxylignan. 3,4,5-Trimethoxy-3′,4′-methylenedioxy-7,9′:7′,9-diepoxylignan. **Epiaschantin**. *Epiashantin*
[41689-50-3]
$C_{22}H_{24}O_7$ 400.427
Constit. of *Hernandia ovigera*, *Hernandia nymphaeifolia*, *Artemisia absinthinum* and *Artemisia sieversiana*. Shows anti-platelet aggregation activity. Moderate antimicrobial agent. Mp 123°. $[\alpha]_D^{19}$ +114 (c, 0.5 in $CHCl_3$). λ_{max} 229 (ε 13500); 285 (ε 4560) (EtOH)

3,4:3′,4′-Bis(methylene), 5-Me ether
7,9′:7′,9-Diepoxy-3-methoxy-3′,4′:4,5-bis(methylenedioxy) lignan. 3-Methoxy-3′,4′:4,5-bis(methylenedioxy)-7,9′:7′,9-diepoxylignan. **5′-Demethoxyepiexcelsin**. *7′-Epi-3-methoxysesamin*
[496790-13-7]
$C_{21}H_{20}O_7$ 384.384
Constit of *Litsea verticillata*. Shows anti-HIV activity. Cryst. (Me_2CO). $[\alpha]_D^{20}$ +116.3 (c, 1.35 in $CHCl_3$)

(7S,7′S,8R,8′R)-form [1151670-47-1]
Constit. of the stem bark of *Syzygium cumini*. Yellowish cryst. ($CHCl_3$). Mp 245 - 246°. λ_{max} 240 (log ε 6.1); 360 (log ε 5.9) (MeOH)

3-Me ether
7,9′:7′,9-Diepoxy-5-methoxy-8,8′-lignan-3,3′,4,4′-tetrol. 3,3′,4,4′-Tetrahydroxy-5-methoxy-7,9′:7′,9-diepoxylignan. **Jatrophasin A**
[1342302-29-7]
$C_{19}H_{20}O_7$ 360.362
Constit. of the seeds of *Jatropha curcas*. Antioxidant. Brown powder. $[\alpha]_D^{25}$ +6.1 (c, 0.1 in MeOH). λ_{max} 220 (log ε 4.09); 232 (log ε 3.96); 254 (log ε 3.4); 279 (log ε 3.44) (MeOH)

3,3′-Di-Me ether
7,9′:7′,9-Diepoxy-3′,5-dimethoxy-8,8′-lignan-3,4,4′-triol. 3,4,4′-Trihydroxy-3′,5-dimethoxy-7,9′:7′,9-diepoxylignan. 5-Hydroxy-(+)-pinoresinol. ***(+)-Neomedioresinol***
Constit. of *Balanophora involucrate* and *Callitris preissii*. Moderate α-glucosidase inhibitor. $[\alpha]_D$ +39.6 (c, 0.05 in $CHCl_3$)

3,3′-Di-Me ether, 4,4′-di-O-β-D-glucopyranoside
Neomedioresinol 4,4′-diglucoside
[1334387-43-7]
$C_{32}H_{42}O_{17}$ 698.671
Constit. of the leaves of *Viburnum plicatum*. Needles. Mp 175 - 176°. $[\alpha]_D^{25}$ +82 (c, 0.12 in MeOH). λ_{max} 221 (log ε 4.5); 273 (log ε 3.7) (MeOH)

3,3′,5-Tri-Me ether

(+)-Medioresinol
[40957-99-1]
$C_{21}H_{24}O_7$ 388.416
Constit. of *Allamanda neriifolia*, *Viscum album*, *Tectona grandis* and *Justicia tranquebariensis*. Plant growth inhibitor. Inhibitor of LPS-stimulated prodn. of IL-12p40. Prisms (MeOH). Mp 170 - 172°. $[\alpha]_D^{28}$ +77.7 (c, 0.69 in MeOH)

3,3′,5-Tri-Me ether, 4-O-β-D-glucopyranoside

Eucommin A
[99633-12-2]
$C_{27}H_{34}O_{12}$ 550.557
Isol. from *Eucommia ulmoides* and *Cistanche tuberosa*. Mp 163°. $[\alpha]_D^{18}$ +20.4 (c, 2.4 in MeOH)

3,3′,5-Tri-Me ether, 4′-O-β-D-glucopyranoside

Isoeucommin A
[134859-90-8]
$C_{27}H_{34}O_{12}$ 550.557
Isol. from *Osmanthus asiaticus*, *Cistanche tubulosa*, *Vitex glabrata* and *Festuca argentina*. Mp 110 - 113°. $[\alpha]_D^{23}$ 0 (c, 0.6 in MeOH). $[\alpha]_D$ +3.1 (c, 0.1 in Me_2CO)

3,3′,5-Tri-Me ether, 4,4′-di-O-β-D-glucopyranoside
[88142-63-6]
$C_{33}H_{44}O_{17}$ 712.698
Isol. from *Eucommia ulmoides*, *Acanthopanax senticosus* and *Dendropanax dentiger*. Needles (EtOH). Mp 222°. $[\alpha]_D^{22}$ -9.1 (c, 0.1 in Py)

3,3′,5-Tri-Me ether, 4,4′-diangeloyl

4,4′-Diangeloyl-(+)-medioresinol

[1428768-22-2]
$C_{31}H_{36}O_9$ 552.62
Isol. from the aerial parts of *Rudbeckia laciniata*. Gum. $[\alpha]^{25}_{D}$ +52 (c, 0.15 in $CHCl_3$). λ_{max} 223 (log ε 4.0); 275 (log ε 4.1) (MeOH)

3,3′,4,4′-Tetra-Me ether
7,9′:7′,9-Diepoxy-3′,4,4′,5-tetramethoxy-8,8′-lignan-3-ol. 3-Hydroxy-3′,4,4′,5-tetramethoxy-7,9′,7′,9-diepoxylignan. ***4′-De-O-methylmagnolin***
[145213-47-4]
$C_{22}H_{26}O_7$ 402.443
Constit. of *Magnolia fargesii*. Oil. $[\alpha]^{23}_{D}$ +58.1 (c, 2.75 in $CHCl_3$)

3,3′,4,5-Tetra-Me ether
7-Epipraderin
$C_{22}H_{26}O_7$ 402.443
Constit. of *Artemisia arborescens*. Oil. $[\alpha]_D$ +18 (c, 0.9 in $CHCl_3$)

Penta-Me ether
(+)-Magnolin
[31008-18-1]
$C_{23}H_{28}O_7$ 416.47
Constit. of *Achillea holosericea*, *Artemisia gorgonum*, *Magnolia fargesii* and *Pseuderanthemum carruthersii*. Moderate cytotoxic agent against MCF-7 cancer cell line. Mp 96 - 97°. $[\alpha]_D$ +55.7 ($CHCl_3$)

3,4-Methylene, 3′,5-di-Me ether
7,9′:7′,9-Diepoxy-3,3′-dimethoxy-4′,5′-methylenedioxy-8,8′-lignan-4-ol. 4′-Hydroxy-3,3′-dimethoxy-4,5-methylenedioxy-7,9′:7′,9-diepoxylignan. ***5-Methoxypiperitol***
[106874-16-2]
$C_{21}H_{22}O_7$ 386.4
Constit. of *Nectandra turbacensis*. Viscous oil. $[\alpha]_D$ +59 (c, 1 in $CHCl_3$). λ_{max} 230 (ε 11900); 280 (ε 4400) (MeOH)

3′,4′-Methylene, 3,4-di-Me ether
7,9′:7′,9-Diepoxy-4,5-dimethoxy-3′,4′-methylenedioxy-8,8′-lignan-3-ol. 3-Hydroxy-4,5-dimethoxy-3′,4′-methylenedioxy-7,9′:7′,9-diepoxylignan. ***3-O-Demethylaschantin***. *5′-Hydroxy-4′-O-methylpiperitol. 3′-O-Demethylaschantin*
[1126087-62-4]
[1151670-49-3] (relative stereochem.)
$C_{21}H_{22}O_7$ 386.4
Constit. of the bark of *Magnolia kobus* and stem bark of *Syzygium cumini*. Cryst. or pale yellow cryst. ($CHCl_3$/petrol). Mp 210° (129-131°). $[\alpha]^{20}_{D}$ +56.2 (c, 0.14 in $CHCl_3$). λ_{max} 248 (log ε 4.1); 276 (log ε 3.7) ($CHCl_3$)

3′,4′-Methylene, 3,5-di-Me ether
4-O-Demethylaschantin
[929692-66-0]
$C_{21}H_{22}O_7$ 386.4
Constit. of *Peperomia duclouxii*. Gum. $[\alpha]^{25}_{D}$ +62.6 (c, 0.02 in $CHCl_3$). λ_{max} 244 (log ε 3.46); 284 (log ε 3.42) ($CHCl_3$)

3′,4′-Methylene, 3,4,5-tri-Me ether
Aschantin. *Ashantin*
[13060-15-6]
$C_{22}H_{24}O_7$ 400.427
Constit. of the fruits of Congo cubebs (*Piper* spp.). Also from *Magnolia fargesii*, *Artemisia gorgonum* and *Achillea holosericea*. Potent superoxide scavenger. Shows antiplasmodial activity. Mp 123°. $[\alpha]^{18}_{D}$ +104 (c, 0.40 in $CHCl_3$) (+65)

3,4:3′,4′-Bis(methylene), 5-Me ether
5-Methoxysesamin. *Demethoxyexcelsin*
[50696-38-3]
$C_{21}H_{20}O_7$ 384.384
Isol. from *Macropiper excelsum*, *Artemisia arborescens*, *Artemisia sieversiana* and *Podolepis rugata*. Oil or cryst.. Mp 101 - 102°. $[\alpha]^{18}_{D}$ +54.2 ($CHCl_3$)

3,4:3′,4′-Bis(methylene) ether, 5-O-[β-D-glucopyranosyl-(1→2)-[β-D-glucopyranosyl-(1→6)]-β-D-glucopyranoside]
[1627497-44-2]
$C_{38}H_{48}O_{22}$ 856.78
Isol. from *Tabebuia argentea* and *Catalpa bignonioides*. Amorph. yellow powder. $[\alpha]_D$ -14 (c, 0.1 in MeOH).

Hänsel, R. *et al.*, *Arch. Pharm. Ber. Dtsch. Pharm. Ges.*, 1969, **302**, 940-942 (*Aschantin*).
Kakisawa, H. *et al.*, *Phytochemistry*, 1972, **11**, 2289-2293 (*Magnolin*).
Russell, G.B. *et al.*, *Phytochemistry*, 1973, **12**, 1799-1803 (*5-Methoxysesamin*).
Nishino, C. *et al.*, *Tetrahedron Lett.*, 1973, **14**, 335-338 (*Epiaschantin, Epimagnolin*).
Pelter, A. *et al.*, *Tetrahedron Lett.*, 1977, **47**, 4137-4140 (*Epiaschantin, Epimagnolin, config*).
Greger, H. *et al.*, *Tetrahedron*, 1980, **36**, 3551-3558 (*Aschantin, Epiaschantin, config*).
Deyama, T. *Chem. Pharm. Bull.*, 1983, **31**, 2993-2997 (*(+)-Medioresinol diglucoside*).
Badawi, M.M. *et al.*, *J. Pharm. Sci.*, 1983, **72**, 1285-1287 (*(-)-Medioresinol, isol*).
Deyama, T. *et al.*, *Chem. Pharm. Bull.*, 1985, **33**, 3651-3657 (*Eucommin A, struct*).
De Carvalho, M.G. *et al.*, *Phytochemistry*, 1987, **26**, 265-267 (*5-Methoxypiperitol*).
Ichino, K. *et al.*, *Phytochemistry*, 1988, **27**, 1906-1907 (*Praderin*).
Subba Raju, G.V. *et al.*, *Indian J. Chem., Sect. B*, 1989, **28**, 558-561 (*(+)-Medioresinol*).
Sugiyama, M. *et al.*, *Chem. Pharm. Bull.*, 1991, **39**, 483-485 (*Isoeucommin A*).
Miyazawa, M. *et al.*, *Phytochemistry*, 1992, **31**, 3666-3668 (*4′-De-O-methylmagnolin*).
Casabuono, A.C. *et al.*, *Phytochemistry*, 1994, **35**, 479-483 (*Isoeucommin A*).
Miyazawa, M. *et al.*, *Phytochemistry*, 1994, **35**, 611-613 (*Epimagnolin A, insect growth inhibitor*).
Marco, J.A. *et al.*, *Phytochemistry*, 1997, **44**, 1133-1137 (*7-Epipraderin*).
Morais, L.C.S.L. *et al.*, *Fitoterapia*, 1998, **69**, 91-92 (*4′-Demethylepimagnolin A*).
Tan, R.X. *et al.*, *Phytochemistry*, 1998, **49**, 157-161 (*5-Methoxysesamin*).
Christov, R. *et al.*, *Fitoterapia*, 1999, **70**, 89-92 (*Aschantin, cmr*).
Chen, J.-J. *et al.*, *Planta Med.*, 2000, **66**, 251-256 (*Epiaschantin, Epimagnolin, anti-platelet aggregation activity*).
Brown, R.C.D. *et al.*, *J. Org. Chem.*, 2001, **66**, 6719-6728 (*(±)-Epimagnolin A, synth*).
Hoang, V.D. *et al.*, *Phytochemistry*, 2002, **59**, 325-329 (*5′-Demethoxyepiexcelsin*).
Ahmed, A.A. *et al.*, *Phytochemistry*, 2002, **59**, 851-856 (*Achillea holosericea constits*).
Enders, D. *et al.*, *Synthesis*, 2002, 515-522 (*Aschantin, synth, ir, pmr, cmr*).
Estrada-Reyes, R. *et al.*, *Z. Naturforsch., C*, 2002, **57**, 29-31 (*Epimagnolin A, pmr, cmr*).
Pettit, G.R. *et al.*, *J. Nat. Prod.*, 2004, **67**, 214-220 (*Praderin, cryst struct, abs config*).
Solis, P.N. *et al.*, *Pharm. Biol.*, 2005, **43**, 378-381 (*5′-Demethoxyepisesartemin A*).
Suleimenov, F.M. *et al.*, *Chem. Nat. Compd. (Engl. Transl.)*, 2007, **43**, 232-233 (*Epiaschantin, antimicrobial activity*).
Lee, J. *et al.*, *Chem. Pharm. Bull.*, 2007, **55**, 137-139 (*Aschantin, superoxide radical scavenger*).
Li, N. *et al.*, *J. Nat. Prod.*, 2007, **70**, 544-548 (*4-De-O-methylaschantin*).
Lee, H.J. *et al.*, *Helv. Chim. Acta*, 2008, **91**, 2361-2366 (*3-O-Demethylaschantin*).
Mir, Q.Y. *et al.*, *Nat. Prod. Res.*, 2009, **23**, 422-430 (*Syzygium cumini constits*).
Ito, M. *et al.*, *Chem. Pharm. Bull.*, 2010, **58**, 628-633 (*(-)-Medioresinol, NO prodn inhibitor*).
Li, L. *et al.*, *Chin. Herb. Med.*, 2010, **2**, 245-247 (*Jatrophasin A*).
Ortet, R. *et al.*, *J. Ethnopharmacol.*, 2011, **138**, 637-640 (*Artemisia gorgonum constits, antiplasmodial activity*).
Kikuchi, M. *et al.*, *J. Nat. Med.*, 2011, **65**, 202-205 (*Neomedioresinol diglucoside*).
Quang, T.H. *et al.*, *Bioorg. Med. Chem. Lett.*, 2012, **22**, 2527-2533 (*5-Methoxypluviatilol, PPARs transcription activator*).
Zheng, L.-P. *et al.*, *Chem. Nat. Compd. (Engl. Transl.)*, 2012, **48**, 883-885 (*(+)-Medioresinol diglucoside, spectra*).
Vo, T.N. *et al.*, *Chem. Pharm. Bull.*, 2012, **60**, 1125-1133 (*(+)-Magnolin, cytotoxicity*).
Guo, T. *et al.*, *Nat. Prod. Res.*, 2012, **26**, 859-864 (*Armatumin*).

Lacret, R. *et al.*, *Phytochem. Lett.*, 2012, **5**, 382-386 (*(+)-Medioresinol, plant growth inhibitor*).
Lee, S.Y. *et al.*, *Helv. Chim. Acta*, 2013, **96**, 320-325 (*4,4′-Diangeloyl-(+)-medioresinol*).
Nhiem, N.X. *et al.*, *J. Nat. Prod.*, 2013, **76**, 495-502 (*(+)-Medioresinol, IL-12p40 prodn inhibitor*).
Marti, G. *et al.*, *Molecules*, 2013, **18**, 3018-3027 (*(-)-Medioresinol, DYRK1A inhibitor, CDK1/ Cyclin B inhibitor*).
De Abreu, M.B. *et al.*, *Phytochem. Lett.*, 2014, **7**, 85-88 (*bismethylene ether triglucoside*).
Peng, C.-Y. *et al.*, *J. Asian Nat. Prod. Res.*, 2015, **17**, 232-238 (*(-)-Magnolin*).
Wei, J. *et al.*, *Fitoterapia*, 2017, **121**, 129-135 (*(+)-Neomedioresinol, activity*).
Jing, Y. *et al.*, *Molecules*, 2017, **22**, 125:1-20 (*Asarinin B, activity*).
Tan, Y.P. *et al.*, *Fitoterapia*, 2018, **126**, 93-97 (*Epimethoxypiperitol, activity*).
Hajra, S. *et al.*, *Tetrahedron*, 2020, **76**, 131483:1-6 (*(+)-Aschantin, synth*)

7,9′:7′,9-Diepoxy-8,8′-lignan-3,3′,4,4′,7-pentol D86

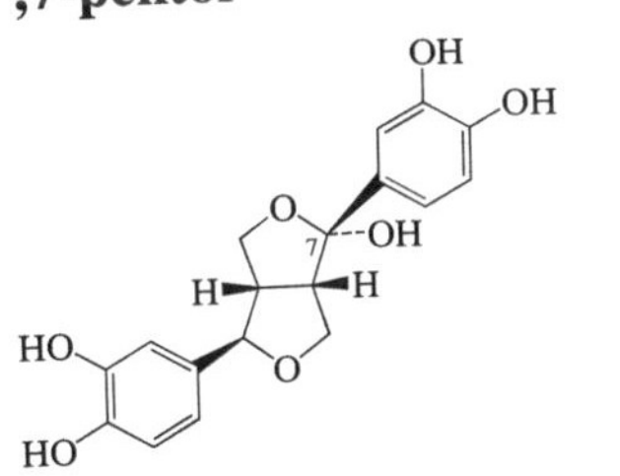

3,3′,4,4′,7-Pentahydroxy-7,9′:7′,9-diepoxylignan
$C_{18}H_{18}O_7$ 346.335
(7R*,7′R*,8S*,8′S*)-form
3,3′,7-Tri-Me ether
7,9′:7′,9-Diepoxy-3,3′,7-trimethoxy-8,8′-lignan-4,4′-diol. 4,4′-Dihydroxy-3,3′,7-trimethoxy-7,9′:7′,9-diepoxylignan. ***7-Methoxypinoresinol***
[805230-73-3]
$C_{21}H_{24}O_7$ 388.416
Constit. of *Boschniakia himalaica*. Powder. Mp 174 - 175°. $[\alpha]^{24}_{D}$ +153.8 (c, 0.01 in Py).

Liu, L.H. *et al.*, *Chin. Chem. Lett.*, 2004, **15**, 43-45 (*7-Methoxypinoresinol*)

7,9′:7′,9-Diepoxy-8,8′-lignan-3,3′,4,4′,8-pentol D87

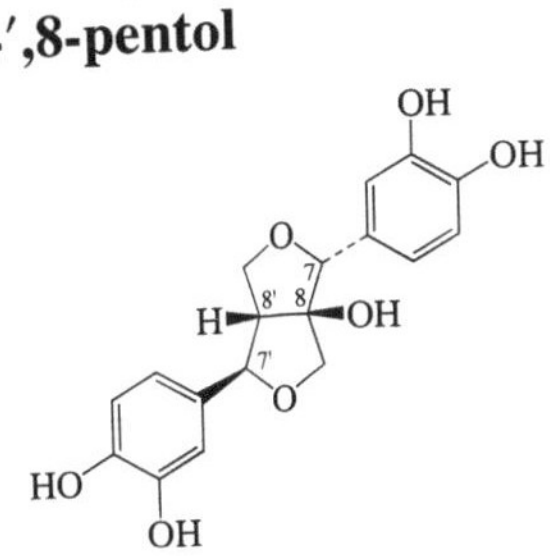

(7R,7′R,8R,8′S)-form

3,3′,4,4′,8-Pentahydroxy-7,9′:7′,9-diepoxylignan
$C_{18}H_{18}O_7$ 346.335
Numbering systems vary
(7R,7′R,8R,8′S)-form
3,4:3′,4′-Bis(methylene) ether
7,9′:7′,9-Diepoxy-3,4:3′,4′-bis(methylenedioxy)-8,8′-lignan-8-ol. 8-Hydroxy-3,4:3′,4′-bis(methylenedioxy)-7,9′:7′,9-diepoxylignan. **Neopaulownin**
[121123-80-6]
$C_{20}H_{18}O_7$ 370.357
Constit. of *Vitex leptobotrys*
(7R,7′R,8S,8′R)-form
3,3′-Di-Me ether, 4-O-β-D-glucopyranoside
8-Hydroxyepipinoresinol 4-glucoside
[1637239-82-7]
$C_{26}H_{32}O_{12}$ 536.53
Constit. of fruit of *Forsythia suspensa*. Exhibits cytotoxicity against human cancer cell lines (A549, Colo-205, Hep-3B, HL-60 and KB cells). Amorph. yellow powder (MeOH). $[\alpha]^{25}_{D}$-8.1 (c, 0.05 in MeOH). λ_{max} 207; 227; 278 (MeOH)
3,3′-Di-Me ether, 4′-O-β-D-glucopyranoside
Forsythialanside E
$C_{26}H_{32}O_{12}$ 536.53
Constit. of fruit of *Forsythia suspensa*. Yellow gum. $[\alpha]^{26}_{D}$+24.9 (c, 0.3 in MeOH). λ_{max} 226 (log ε 4.21); 277 (log ε 3.79) (MeOH)
3,3′,4′-Tri-Me ether, 8-O-β-D-glucopyranoside
[1355451-19-2]
$C_{27}H_{34}O_{12}$ 550.557
Constit. of the flowers of *Osmanthus fragrans* var. *aurantiacus*. Amorph. powder (MeOH). $[\alpha]_D$ +45 (c, 0.5 in MeOH)
3,3′,4,4′-Tetra-Me ether
7,9′:7′,9-Diepoxy-3,3′,4,4′-tetramethoxy-8,8′-lignan-8-ol. 8-Hydroxy-3,3′,4,4′-tetramethoxy-7,9′:7′,9-diepoxylignan. **Gmelinol**
[469-28-3]
$C_{22}H_{26}O_7$ 402.443
Constit. of the wood of *Gmelina leichhardtii*. Plates or prisms (H_2O or EtOH). Mp 124° +117.1. $[\alpha]_D$ +123.3 ($CHCl_3$). λ_{max} 231 (ε 9120); 279 (log ε 3467) (EtOH)
3,3′,4,4′-Tetra-Me ether, 8-Ac
Prisms (EtOH). Mp 118°
3,4-Methylene ether
7,9′:7′,9-Diepoxy-3′,4′-methylenedioxy-8,8′-lignan-3,4,8′-triol. 3,4,8′-Trihydroxy-3′,4′-methylenedioxy-7,9′:7′,9-diepoxylignan
[1627500-20-2]
$C_{19}H_{18}O_7$ 358.346
Constit. of *Callicarpa nudiflora*. Antioxidant. Amorph. yellow powder. $[\alpha]^{22}_{D}$+72 (c, 0.10 in $CHCl_3$). λ_{max} 230 (log ε 4.11); 283 (log ε 3.96) (MeOH)
(7R,7′S,8S,8′R)-form
3,3′-Di-Me ether
7,9′:7′,9-Diepoxy-3,3′-dimethoxy-8,8′-lignan-4,4′,8-triol. 4,4′,8-Trihydroxy-3,3′-dimethoxy-7,9′:7′,9-diepoxylignan. ***8-Hydroxypinoresinol***. *1-Hydroxypinoresinol*
[81426-17-7]
$C_{20}H_{22}O_7$ 374.389
Constit. of bark of *Olea europaea* and roots of *Nardostachys jatamansi*. Also from *Valeriana officinalis*. Cryst. powder. Mp 183 - 185°. $[\alpha]^{15}_{D}$+39 (c, 0.65 in EtOH)
3,3′-Di-Me ether, 4-O-β-D-glucopyranoside
[72896-51-6]
$C_{26}H_{32}O_{12}$ 536.53
Constit. of *Olea europaea*, *Olea africana*, *Valeriana officinalis* and *Fraxinus mandshurica*. Mp 127 - 129°. $[\alpha]^{23}_{D}$-9.3 (c, 0.42 in MeOH)
3,3′-Di-Me ether, 4′-O-β-D-glucopyranoside
[102582-69-4]
$C_{26}H_{32}O_{12}$ 536.53
Constit. of *Eucommia ulmoides* and *Valeriana officinalis*. Amorph.. $[\alpha]^{18}_{D}$-28.8 (c, 1 in MeOH). λ_{max} 228; 278 (MeOH)
3,3′-Di-Me ether, 8-O-β-D-glucopyranoside
[81495-71-8]
$C_{26}H_{32}O_{12}$ 536.53
Constit. of *Olea europaea*, *Olea africana* and *Salvia scapiformis*. Amorph. powder. Mp 179 - 183°. $[\alpha]^{23}_{D}$-17.5 (c, 0.38 in EtOH)
3,3′-Di-Me ether, 8-O-[4-hydroxybenzoyl-(→6)-β-D-glucopyranoside]
[1700655-32-8]

$C_{33}H_{36}O_{14}$ 656.637
Constit. of *Salvia scapiformis*. Gum. $[\alpha]_D^{22}$ +4.3 (c, 0.16 in MeOH). λ_{max} 207 (log ε 4.82); 234 (log ε 4.26); 259 (log ε 4.21) (MeOH)

3,3′-Di-Me ether, 4,4′-di-O-β-D-glucopyranoside
[112747-99-6]
$C_{32}H_{42}O_{17}$ 698.671
Constit. of the bark of *Eucommia ulmoides*

3,3′-Di-Me ether, 8-Ac
8-Acetoxypinoresinol
[81426-14-4]
$C_{22}H_{24}O_8$ 416.426
Constit. of *Olea europaea*. Amorph. powder. $[\alpha]_D^{21}$ +31.4 (c, 0.99 in EtOH). Called 1-acetoxy in the lit.

3,3′-Di-Me ether, 8-Ac, 4-O-β-D-glucopyranoside
[72896-49-2]
$C_{28}H_{34}O_{13}$ 578.567
Constit. of bark of *Olea europaea* and *Olea africana*. Needles. Mp 183.5 - 185°. $[\alpha]_D^{22}$ +7.9 (c, 1.0 in EtOH)

3,3′,4′-Tri-Me ether
7,9′:7′,9-Diepoxy-3,3′,4′-trimethoxy-8,8′-lignan-4,8-diol. 4,8-Dihydroxy-3,3′,4′-trimethoxy-7,9′:7′,9-diepoxylignan
[81426-18-8]
$C_{21}H_{24}O_7$ 388.416
Constit. of *Olea europaea*. Amorph. powder. $[\alpha]_D^{23}$ +37.9 (c, 0.81 in $CHCl_3$)

3,3′,4′-Tri-Me ether, 8-Ac
[81426-15-5]
$C_{23}H_{26}O_8$ 430.453
Constit. of *Olea europaea*. Amorph. powder. $[\alpha]_D^{23}$ +29.9 (c, 1.14 in $CHCl_3$)

3,3′,4′-Tri-Me ether, 8-Ac, 4-O-β-D-glucopyranoside
[72896-50-5]
$C_{29}H_{36}O_{13}$ 592.594
Constit. of *Olea europaea* and *Olea africana*. Amorph. powder. $[\alpha]_D^{20}$ +9.1 (c, 1.6 in EtOH)

3,3′,4,4′-Tetra-Me ether
Isogmelinol
[597-01-3]
Prisms (xylene). Mp 149°. $[\alpha]_D^{15}$ +30 ($CHCl_3$)

3,4-Methylene ether, 3′-O-[4-hydroxybenzoyl-(→6)-β-D-glucopyranoside]
Isocannabilignin
[2082741-80-6]
$C_{32}H_{32}O_{14}$ 640.594
Constit. of the leaves of *Vitex negundo* var. *cannabifolia*. Gum. $[\alpha]_D^{21}$ -18 (c, 0.1 in MeOH). λ_{max} 206 (log ε 2.50); 232 (log ε 2.00); 259 (log ε 2.03) (MeOH)

3′,4′-Methylene ether, 3-O-[4-hydroxybenzoyl-(→6)-β-D-glucopyranoside]
Cannabilignin
[2082741-79-3]
$C_{32}H_{32}O_{14}$ 640.594
Constit. of the leaves of *Vitex negundo* var. *cannabifolia*. Gum. $[\alpha]_D^{21}$ -34 (c, 0.1 in MeOH). λ_{max} 207 (log ε 2.50); 230 (log ε 2.09); 259 (log ε 2.10) (MeOH)

3,4:3′,4′-Bis(methylene) ether
Paulownin
[13040-46-5]
$C_{20}H_{18}O_7$ 370.357
Constit. of *Paulownia tomentosa*, *Tabebuia rosea*, *Stereospermum acuminatissimum*, *Kigelia pinnata* and *Gmelina arborea*. Cryst. (MeOH or EtOAc/hexane). Mp 106 - 107°. $[\alpha]_D^{30}$ +29 (c, 1 in $CHCl_3$)

3,4:3′,4′-Bis(methylene) ether, 8-Ac
8-Acetylpaulownin
[28071-36-5]
$C_{22}H_{20}O_8$ 412.394
Constit. of *Isodon lophanthoides* var. *gerardianus*. Cubic cryst.. Mp 168 - 169.3° (144-145°). $[\alpha]_D^{22}$ +34.4 (c, 0.47 in $CHCl_3$)

(7S,7′R,8S,8′R)-form

3,3′-Di-Me ether
8-Hydroxy-7,7′-diepipinoresinol. *1-Hydroxy-2,6-bisepinoresinol*
[767340-96-5]
$C_{20}H_{22}O_7$ 374.389
Constit. of the roots of *Valeriana laxiflora*. Amorph. solid. $[\alpha]_D^{22}$ +21.8 (c, 0.1 in MeOH). Stereochem. revised in 2018. λ_{max} 210 (log ε 4.62); 235 (log ε 4.33); 280 (log ε 4.11) (MeOH)

3,4-Methylene ether
[1627500-19-9]
Constit. of *Callicarpa nudiflora*. Amorph. yellow powder. $[\alpha]_D^{22}$ +19.6 (c, 0.10 in MeOH). λ_{max} 232 (log ε 3.36); 284 (log ε 3.19) (MeOH)

3′,4′-Methylene ether
7,9′:7′,9-Diepoxy-3′,4′-methylenedioxy-8,8′-lignan-3,4,8-triol. 3,4,8-Trihydroxy-3′,4′-methylenedioxy-7,9′:7′,9-diepoxylignan
[1627500-21-3]
$C_{19}H_{18}O_7$ 358.346
Constit. of *Callicarpa nudiflora*. Antioxidant. Amorph. yellow powder. $[\alpha]_D^{22}$ +20.4 (c, 0.10 in MeOH). λ_{max} 230 (log ε 3.63); 285 (log ε 3.45) (MeOH)

(7S,7′S,8S,8′R)-form

3,3′,4,4′-Tetra-Me ether
Neogmelinol
[17669-45-3]
Acid isom. prod. of Gmelinol. Plates (MeOH). Mp 163 - 164°. $[\alpha]_D^{25}$ +60. Originally assigned an isomeric struct.

3,4:3′,4′-Bis(methylene) ether
Isopaulownin
[10590-41-7]
Acid-cat. epimerisation prod. of Paulownin. Mp 132°. $[\alpha]_D$ +127 (c, 0.80 in $CHCl_3$)

(7RS,7′SR,8SR,8′RS)-form

3′,4′-Methylene, 3-Me ether
7,9′:7′,9-Diepoxy-3-methoxy-3′,4′-methylenedioxy-8,8′-lignan-4,8-diol. 4,8-Dihydroxy-3-methoxy-3′,4′-methylenedioxy-7,9′:7′,9-diepoxylignan
[1660050-48-5]
$C_{20}H_{20}O_7$ 372.373
Constit. of the inner bark of *Tabebuia avellanedae*. Oil. $[\alpha]_D$ +0.71 (c, 1.9 in MeOH). λ_{max} 232 (log ε 1.39); 284 (log ε 1.10) (MeOH).

Birch, A.J. *et al.*, *J. Chem. Soc.*, 1964, 2705-2708 (*Gmelinol*).
Takahashi, K. *et al.*, *Chem. Pharm. Bull.*, 1966, **14**, 641-647 (*Paulownin*).
Pelter, A. *J. Chem. Soc. C*, 1967, 1376-1380 (*ms*).
Birch, A.J. *et al.*, *J. Chem. Soc. C*, 1967, 1968-1972 (*Neogmelinol*).
Takahashi, K. *et al.*, *Chem. Pharm. Bull.*, 1970, **18**, 421-428 (*Isopaulownin*).
Chiba, M. *et al.*, *Chem. Pharm. Bull.*, 1979, **27**, 2868-2873 (*8-Acetoxypinoresinol*).
Tsukamoto, H. *et al.*, *Chem. Pharm. Bull.*, 1984, **32**, 2730-2735 (*Hydroxypinoresinol*).
Tsukamoto, H. *et al.*, *Chem. Pharm. Bull.*, 1985, **33**, 1232-1241 (*Olea europaea glucosides, Gmelinol*).

Deyama, T. *et al.*, *Chem. Pharm. Bull.*, 1986, **34**, 523-527 (*Eucommia ulmoides 4-glucoside*).
Deyama, T. *et al.*, *CA*, 1988, **108**, 91694s (*Eucommia ulmoides diglucoside*).
Kraus, G.A. *et al.*, *J. Am. Chem. Soc.*, 1990, **112**, 3464-3466 (*Paulownin, synth*).
Macias, F.A. *et al.*, *Acta Crystallogr., Sect. C*, 1992, **48**, 2240-2244 (*Hydroxypinoresinol, cryst struct*).
Adhikara, S. *et al.*, *Tetrahedron Lett.*, 1992, **33**, 6025-6026 (*Paulownin, Isogmelinol, synth*).
Roy, S.C. *et al.*, *Tetrahedron*, 1993, **49**, 8415-8422 (*Isogmelinol, synth*).
Okazaki, M. *et al.*, *Biosci., Biotechnol., Biochem.*, 1997, **61**, 743-745 (*Paulownin, synth*).
Joshi, R.K. *et al.*, *Fitoterapia*, 1998, **69**, 275-276 (*Hydroxypinoresinol, pmr, cmr*).
Jiang, B. *et al.*, *Zhiwu Xuebao*, 2000, **42**, 95-97 (*8-Acetylpaulownin*).
Ishibashi, F. *et al.*, *Biosci., Biotechnol., Biochem.*, 2001, **65**, 29-34 (*Paulownin, synth*).
Schumacher, B. *et al.*, *J. Nat. Prod.*, 2002, **65**, 1479-1485 (*Valeriana officinalis constits*).
Gu, J.-Q. *et al.*, *Planta Med.*, 2004, **70**, 509-514 (*8-Hydroxy-7,7′-diepipinoresinol*).
Pohmakotr, M. *et al.*, *J. Org. Chem.*, 2006, **71**, 386-389 (*Gmelinol, synth*).
Angle, S.R. *et al.*, *J. Org. Chem.*, 2008, **73**, 6268-6278 (*Paulownin, synth*).
Singh, P. *et al.*, *J. Indian Chem. Soc.*, 2010, **87**, 1403-1407 (*Kigelia pinnata constit*).
Lee, D.-G. *et al.*, *Arch. Pharmacal Res.*, 2011, **34**, 2029-2035 (*Osmanthus fragrans 8-glucoside*).
Sob, S.V.T. *et al.*, *J. Asian Nat. Prod. Res.*, 2011, **13**, 1128-1134 (*Stereospermum acuminatissimum constit*).
Sichaem, J. *et al.*, *Fitoterapia*, 2012, **83**, 1456-1459 (*Tabebuia rosea constit*).
Lai, Y. *et al.*, *Molecules*, 2013, **18**, 11377-11383 (*Salvia scapiformis constits*).
Chang, L. *et al.*, *Chin. J. Nat. Med.*, 2014, **12**, 697-699 (*Forsythialanside E*).
Zhang, L. *et al.*, *Heterocycles*, 2014, **89**, 731-738 (*Tabebuia avellanedae constit*).
Yan, X.-J. *et al.*, *J. Asian Nat. Prod. Res.*, 2014, **16**, 602-610 (*8-Hydroxyepipinoresinol, isol, activity*).
Pan, W. *et al.*, *J. Nat. Prod.*, 2014, **77**, 663-667 (*Neopaulownin*).
Luo, Y-H. *et al.*, *Phytochem. Lett.*, 2014, **7**, 194-197 (*Callicarpa nudiflorum constits*).
Li, Y.-T. *et al.*, *Nat. Prod. Res.*, 2017, **31**, 918-924 (*Cannabilignin, Isocannabilignin, cd*).
Shao, S.-Y. *et al.*, *J. Nat. Prod.*, 2018, **81**, 1023-1028 (*Gmelinol, Hydroxypinoresinol, 8-Hydroxy-7,7′-diepipinoresinol, pmr*)

7,9′:7′,9-Diepoxy-8,8′-lignan-3,3′,4,4′,9-pentol D88

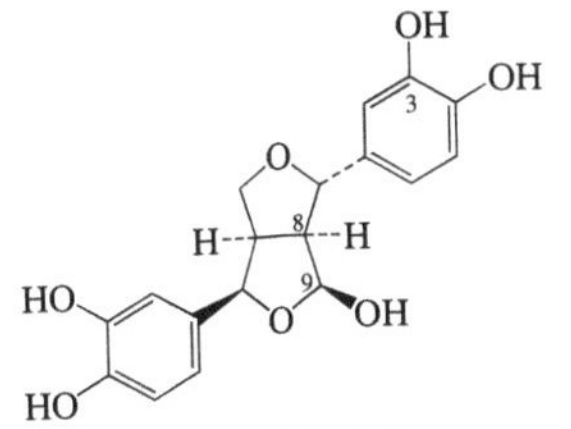

(7S,7′R,8S,8′R,9S)-form

3,3′,4,4′,9-Pentahydroxy-7,9′:7′,9-diepoxylignan
$C_{18}H_{18}O_7$ 346.335
Lignan numbering shown
(7S,7′R,8S,8′R,9S)-form
3,3′-Di-Me ether, 9-O-β-D-glucopyranoside
Lactuberin A
[1423043-69-9]
$C_{26}H_{32}O_{12}$ 536.53
Constit. of the tubers of *Lactuca tuberosa*. Amorph. solid. $[\alpha]_D^{23}$ +84.9 (c, 1.06 in MeOH)
(7S,7′S,8S,8′R,9R)-form
3,3′-Di-Me ether
7,9′:7′,9-Diepoxy-3,3′-dimethoxy-8,8′-lignan-4,4′,9-triol. 4,4′,9-Trihydroxy-3,3′-dimethoxy-7,9′:7′,9-diepoxylignan.
9α-Hydroxypinoresinol
[114580-49-3]
$C_{20}H_{22}O_7$ 374.389
Constit. of *Allamanda neriifolia* and *Ligularia kanaitzensis*. Amorph.. $[\alpha]_D^{26}$+55 (c, 0.95 in MeOH)
3,3′-Di-Me ether, 4-O-β-D-glucopyranoside
Petaslignolide A. *9-Hydroxypinoresinol 4-glucoside*
[870700-90-6]
$C_{26}H_{32}O_{12}$ 536.53
Constit. of the leaves of *Petasites japonicus*. Antioxidant. Neuroprotective agent. Amorph. powder (MeOH). $[\alpha]_D$ -10 (c, 0.23 in MeOH). λ_{max} 204 (log ε 4.73); 229 (log ε 4.19); 279 (log ε 3.75) (MeOH)
3,3′-Di-Me ether, 9-O-β-D-glucopyranoside
[903520-83-2]
$C_{26}H_{32}O_{12}$ 536.53
Constit. of *Ligularia virgaurea* ssp. *oligocephala*. Amorph. powder. $[\alpha]_D^{29}$-14 (c, 0.7 in MeOH)
3,3′-Di-Me ether, 9-angeloyl
9-Angeloyloxypinoresinol
[851309-69-8]
$C_{25}H_{28}O_8$ 456.491
Constit. of *Ligularia kanaitzensis*. Inhibitor of HIV-1 RT. Gum. $[\alpha]_D^{25}$ +26.3 (c, 0.42 in $CHCl_3$)
3,4′-Di-Me ether, 3′,4-diangeloyl
Crellisin A
[170663-63-5]
$C_{30}H_{34}O_9$ 538.593
Constit. of *Cremanthodium ellisii*. Gum. $[\alpha]_D^{20}$+3.7 (c, 0.4 in $CHCl_3$). λ_{max} 242 (ε 16982); 280 (ε 13183) ($CHCl_3$)
3,3′,9-Tri-Me ether
9-Methoxypinoresinol
[2102394-06-7]
$C_{21}H_{24}O_7$ 388.416
Constit. of *Calotropis gigantea*. Exhibits potent cytotoxic activity. Amorph. solid. Abs. config. assumed
3,3′,4,4′-Tetra-Me ether
7,9′:7′,9-Diepoxy-3,3′,4,4′-tetramethoxy-8,8′-lignan-9-ol. 9-Hydroxy-3,3′,4,4′-tetramethoxy-7,9′:7′,9-diepoxylignan
[767348-26-5]
$C_{22}H_{26}O_7$ 402.443
Constit. of the roots of *Echinops giganteus* var. *lelyi*. Powder (EtOAc). Mp 171 - 173°. $[\alpha]_D^{22}$ +29.5 (c, 0.45 in $CHCl_3$). λ_{max} 203 (log ε 4.96); 205 (log ε 4.86); 232 (log ε 4.18); 279 (log ε 3.65) (EtOH)
3,3′,4,4′-Tetra-Me ether, 9-O-(3-methylbutanoyl)
[767348-24-3]
$C_{27}H_{34}O_8$ 486.561
Constit. of the roots of *Echinops giganteus* var. *lelyi*. Powder (MeOH). Mp 94 - 95°. $[\alpha]_D^{22}$+9.7 (c, 0.35 in $CHCl_3$). λ_{max} 201 (log ε 4.82); 232 (log ε 4.14); 279 (log ε 3.73) (EtOH)
3,4-Methylene, 3′,4′-di-Me ether
7,9′:7′,9-Diepoxy-3′,4′-dimethoxy-3,4-methylenedioxy-8,8′-lignan-9-ol. 9-Hydroxy-3′,4′-dimethoxy-3,4-methylenedioxy-7,9′:7′,9-diepoxylignan. ***9-Hydroxykobusin***
[929601-14-9]
$C_{21}H_{22}O_7$ 386.4
Constit. of *Geranium thunbergii*. Oil. $[\alpha]_D^{24}$-42 (c, 0.63 in MeOH). λ_{max} 232; 284 (MeOH)
(7S,7′S,8S,8′R,9S)-form
3,3′-Di-Me ether
9β-Hydroxypinoresinol
[96738-89-5]
$C_{20}H_{22}O_7$ 374.389
Constit. of *Lonicera hypoleuca*. λ_{max} 237 (log ε 4.58); 281 (log ε 4.19) (MeOH)
3,3′-Di-Me ether, 9-O-β-D-glucopyranoside
[1311450-09-5]

Constit. of *Lactuca tuberosa* and *Scorzonera judaica*. Amorph. powder. $[\alpha]_D^{25}$ +26 (c, 0.1 in MeOH). Uncertain stereochem. in ref.. λ_{max} 231 (log ε 3.8); 281 (log ε 4.05) (MeOH)

(7S,7′S,8S,8′R,9ξ)-form

3,4:3′,4′-Bis(methylene) ether

7,9′:7′,9-Diepoxy-3,4:3′,4′-bis(methylenedioxy)-8,8′-lignan-9-ol. 9-Hydroxy-3,4:3′,4′-bis(methylenedioxy)-7,9′:7′,9-diepoxylignan. ***Aptosimol. 9-Hydroxysesamin***

[61254-18-0]

[2081945-54-0]

((7*S*,7′*S*,8*S*,8′*R*,9*R*)-form)

$C_{20}H_{18}O_7$ 370.357

Constit. of *Aptosimum spinescens* and *Cuscuta auralis*. Needles. Mp 165° (158°). $[\alpha]_D^{20}$ +34 (c, 2 in Me_2CO). $[\alpha]_D$ +58.4. Incorrect struct. originally assigned. Identity of Aptosimol and 9-Hydroxysesamin not establ. 9-Configs. variable (acetals)

(7R,7′S,8S,8′R,9ξ)(?)-form

3,4:3′,4′-Bis(methylene) ether

Cinnamonol

[58262-61-6]

$C_{20}H_{18}O_7$ 370.357

Isol. from leaves of *Cinnamomum camphora*.

Takaoka, D. *et al., Nippon Kagaku Kaishi*, 1975, 2192-2196 (*Cinnamonol*).

Brieskorn, C.H. *et al., Tetrahedron Lett.*, 1976, **17**, 2221-2114 (*Aptosimol*).

Anjaneyulu, A.S.R. *et al., Tetrahedron*, 1977, **33**, 133-143 (*9-Hydroxysesamin*).

Cooper, R. *et al., Tetrahedron*, 1979, **35**, 861-868 (*struct*).

Pelter, A. *et al., J. Chem. Soc., Perkin Trans. 1*, 1982, 183-190 (*Aptosimol, pmr*).

Khan, K.A. *et al., Phytochemistry*, 1985, **24**, 628-630 (*9β-Hydroxypinoresinol*).

Abe, F. *et al., Phytochemistry*, 1988, **27**, 575-577 (*9α-Hydroxypinoresinol*).

Stevens, D.R. *et al., J. Chem. Soc., Perkin Trans. 1*, 1992, 185-190 (*synth*).

Yang, L. *et al., Indian J. Chem., Sect. B*, 1995, **34**, 975-977 (*Crellisin A*).

Tene, M. *et al., Phytochemistry*, 2004, **65**, 2101-2105 (*Echinops giganteus constits*).

Min, B.-S. *et al., Arch. Pharmacal Res.*, 2005, **28**, 1023-1026 (*Petaslignolide A*).

Cui, H.S. *et al., J. Agric. Food Chem.*, 2005, **53**, 8526-8532 (*Petaslignolide A, activity*).

Li, Y.-S. *et al., Nat. Prod. Res.*, 2005, **19**, 125-129 (*9-Angeloyloxypinoresinol*).

Liu, Q.-H. *et al., Arch. Pharmacal Res.*, 2006, **29**, 1109-1113 (*9-Hydroxykobusin*).

Li, Y. *et al., Helv. Chim. Acta*, 2006, **89**, 870-875 (*3,3′-di-Me 9-glucoside*).

Bader, A. *et al., J. Nat. Prod.*, 2011, **74**, 1421-1426 (*Scorzonera judaica glucoside*).

Stojakowska, A. *et al., Food Chem.*, 2013, **138**, 1250-1255 (*Lactuberin A*).

Nguyen, K.D.H. *et al., Bioorg. Med. Chem. Lett.*, 2017, **27**, 2902-2906 (*9-Methoxypinoresinol, activity*).

Rho, T. *et al., J. Sep. Sci.*, 2018, **41**, 2169-2177 (*9-Hydroxysesamin*)

7,9′:7′,9-Diepoxy-8,8′-lignan-3,3′,4,5,5′-pentol D89

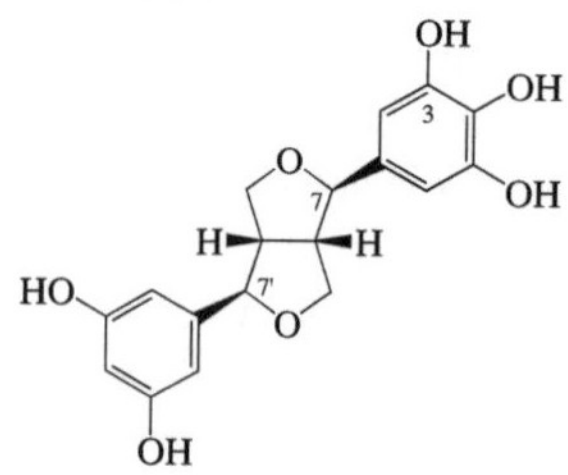

(7R,7′R,8S,8′S)-form

3,3′,4,5,5′-Pentahydroxy-7,9′:7′,9-diepoxylignan

$C_{18}H_{18}O_7$ 346.335

Structures in this series may need revision. Phenyl 3,5-dioxygenated substn. pattern in doubt as a result of work done by Yang *et al* in 2015

(7R,7′R,8S,8′S)-form

Penta-Me ether

(-)-Acortatarinowin F

[1807543-37-8]

$C_{23}H_{28}O_7$ 416.47

Constit. of rhizomes of *Acorus tatarinowii*. Exhibits weak inhibitory activity against NO prodn. in activated macrophages. Amorph. powder. $[\alpha]_D^{20}$ -44 (c, 0.1 in MeOH). Resolved by chiral hplc

(7S,7′R,8R,8′R)-form

Penta-Me ether

3,3′,4,5,5′-Pentamethoxy-7,9′:7′,9-diepoxylignan. ***Epimagnolin B***

[1134188-26-3]

$C_{23}H_{28}O_7$ 416.47

Constit. of the flower buds of *Magnolia fargesii*. Yellow oil. $[\alpha]_D$ +92.9 (c, 0.9 in $CHCl_3$). λ_{max} 207 (log ε 4.84); 274 (log ε 3.85) ($CHCl_3$)

(7S,7′S,8R,8′R)-form

Penta-Me ether

(+)-Acortatarinowin F

[1807543-36-7]

$C_{23}H_{28}O_7$ 416.47

Constit. of rhizomes of *Acorus tatarinowii*. Amorph. powder. $[\alpha]_D^{20}$ +37 (c, 0.1 in MeOH). Resolved by chiral hplc

(7RS,7′RS,8SR,8′SR)-form

Penta-Me ether

Acortatarinowin F

[1807543-42-5]

$C_{23}H_{28}O_7$ 416.47

Constit. of rhizomes of *Acorus tatarinowii*. Amorph. powder. $[\alpha]_D^{20}$ -20 (c, 0.11 in MeOH). Partial racemate. Resolved by chiral hplc. λ_{max} 207 (log ε 4.76); 232 (log ε 4.10); 276 (log ε 3.36) (MeOH).

Kim, J.Y. *et al., Bioorg. Med. Chem. Lett.*, 2009, **19**, 937-940 (*Epimagnolin B*).

Yang, Y.-N. *et al., J. Nat. Prod.*, 2015, **78**, 705-711 (*pmr, struct*).

Lu, Y. *et al., J. Nat. Prod.*, 2015, **78**, 2205-2214 (*Acortatarinowin F, activity*)

7,9′:7′,9-Diepoxy-8,8′-lignan-3,3′,4,5′,8-pentol D90

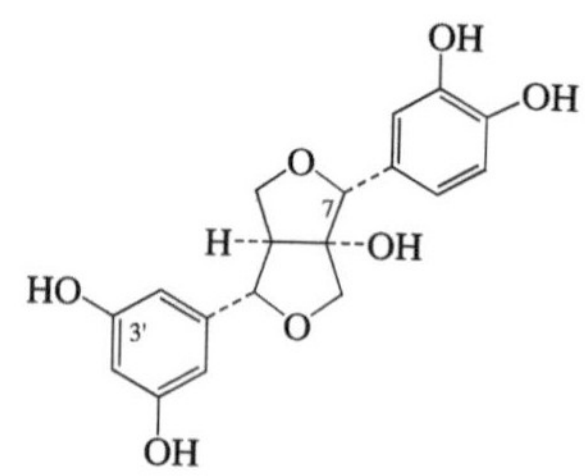

Absolute Configuration

3,3′,4,5′,8-Pentahydroxy-7,9′:7′,9-diepoxylignan

$C_{18}H_{18}O_7$ 346.335

(7R,7′S,8S,8′R)-form

3,3′-Di-Me ether

7,9′:7′,9-Diepoxy-3′,5-dimethoxy-8,8′-lignan-3,4′,8′-triol. 3′,4,8-Trihydroxy-3,5′-dimethoxy-7,9′:7′,9-diepoxylignan. ***Melodinin A***

$C_{20}H_{22}O_7$ 374.389

Constit. of the stems of *Melodinus suaveolens*. Exhibits moderate antiinflammatory activity. Amorph. powder. $[\alpha]_D^{25}$ +16 (c, 0.05 in MeOH).

Fang, L. *et al., Phytochem. Lett.*, 2018, **26**, 134-137 (*Melodinin A, activity*)

7,7′:7,8-Diepoxy-8,8′-lignan-3,3′,4,4′-tetrol D91

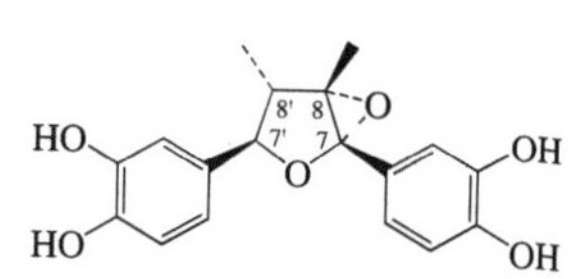

3,3′,4,4′-Tetrahydroxy-7,7′:7,8-diepoxylignan

$C_{18}H_{18}O_6$ 330.336

(7S,7′R,8S,8′S)-form

3,4-Methylene, 3′-Me ether

7,7′:7′,8′-Diepoxy-3-methoxy-3′,4′-methylenedioxy-8,8′-lignan-4-ol. 4-Hydroxy-3-methoxy-3′,4′-methylenedioxy-7,7′:7′,8′-diepoxylignan

[1886973-83-6]

$C_{20}H_{20}O_6$ 356.374

Constit. of the roots of *Schisandra sphenanthera*. Antioxidant. Yellow oil. $[\alpha]_D^{25}$ +40 (c, 0.1 in MeOH). λ_{max} 229 (log ε 0.18) (EtOH)

3,4-Methylene, 3′,4′-di-Me ether

7,7′:7,8-Diepoxy-3′,4′-dimethoxy-3,4-methylenedioxy-8,8′-lignan. 3′,4′-dimethoxy-3,4-methylenedioxy-7,7′:7,8-diepoxylignan

[1886973-82-5]

$C_{21}H_{22}O_6$ 370.401

Constit. of the roots of *Schisandra sphenanthera*. Yellow oil. $[\alpha]_D^{25}$ +20 (c, 0.1 in MeOH). λ_{max} 227 (log ε 0.32) (EtOH)

3,4:3′,4′-Bis(methylene) ether

7,7′:7,8-Diepoxy-3,4:3′,4′-bis(methylenedioxy)-8,8′-lignan. 3,4:3′,4′-Bis(methylenedioxy)-7,7′:7,8-diepoxylignan. ***7,8-Epoxyzuonin A***

[475598-86-8]

$C_{20}H_{18}O_6$ 354.358

Constit. of *Chamaecyparis obtusa* var. *formosana* and *Schisandra sphenanthera*. Liq.. $[\alpha]_D^{25}$ -14.4 (c, 0.31 in $CHCl_3$). Sl. misleading synonym; config. is not same as Zuonin A. λ_{max} 237 (log ε 3.52); 287 (log ε 3.43) (MeOH).

Kuo, Y.-H. *et al.*, *Chem. Pharm. Bull.*, 2002, **50**, 978-980 (*7,8-Epoxyzuonin A*).

Jiang, K. *et al.*, *Fitoterapia*, 2015, **103**, 63-70 (*Schisandra sphenanthera constits*)

7,9′:7′,9-Diepoxy-8,8′-lignan-2,2′,4,4′-tetrol D92

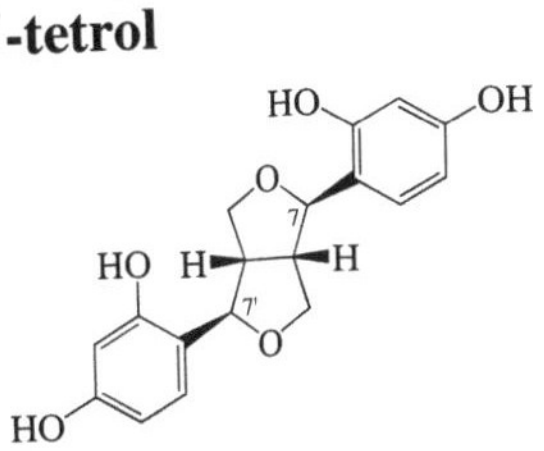

2,2′,4,4′-Tetrahydroxy-7,9′:7′,9-diepoxylignan

$C_{18}H_{18}O_6$ 330.336

(7R,7′R*,8S*,8′S*)-form*

2,2′-Di-Me ether

7,9′:7′,9-Diepoxy-2,2′-dimethoxy-8,8′-lignan-4,4′-diol. 4,4′-Dihydroxy-2,2′-dimethoxy-7,9′:7′,9-diepoxylignan

2,2′-Di-Me ether, 4-O-β-D-glucopyranoside

Strangusin A

[432551-47-8]

$C_{26}H_{32}O_{11}$ 520.531

Constit. of *Serratula strangulata*. Gum. λ_{max} 242 (log ε 4.23); 280 (log ε 4.12) (no solvent reported).

Dai, J.Q. *et al.*, *Chin. Chem. Lett.*, 2002, **13**, 143-146 (*Strangusin A*)

7,9′:7′,9-Diepoxy-8,8′-lignan-3,3′,4,4′-tetrol D93

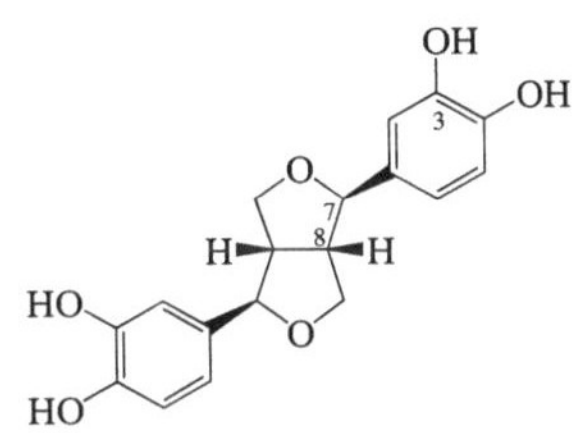

(7R,7′R,8S,8′S)-form

3,3′,4,4′-Tetrahydroxy-7,9′:7′,9-diepoxylignan

$C_{18}H_{18}O_6$ 330.336

(7R,7′R,8S,8′S)-form

3,3′,4-Tri-Me ether

7,9′:7′,9-Diepoxy-3,3′,4′-trimethoxy-8,8′-lignan-4-ol. 4-Hydroxy-3,3′,4′-trimethoxy-7,9′:7′,9-diepoxylignan.

O-Methylpinoresinol

[95404-51-6]

$C_{21}H_{24}O_6$ 372.417

Constit. of the roots of *Stellaria chamaejasme*. Yellow powder. $[\alpha]_D^{30}$ -58 (c, 0.05 in EtOH). λ_{max} 223 (log ε 3.75); 279 (log ε 4.15) (MeOH)

(7R,7′S,8R,8′R)-form

3-Me ether

7,9′:7′,9-Diepoxy-3′-methoxy-8,8′-lignan-3,4,4′-triol. 3,4,4′-Trihydroxy-3′-methoxy-7,9′:7′,9-diepoxylignan.

3′-O-Demethylepipinoresinol

[252333-73-6]

$C_{19}H_{20}O_6$ 344.363

Constit. of *Taxus mairei*. Gum. $[\alpha]_D^{24}$ +95.3 (c, 1.9 in $CHCl_3$)

3,4-Di-Me ether

7,9′:7′,9-Diepoxy-3′,4′-dimethoxy-8,8′-lignan-3,4-diol. 3,4-Dihydroxy-3′,4′-dimethoxy-7,9′:7′,9-diepoxylignan. ***Lantibetin***

[1096706-66-9]

$C_{20}H_{22}O_6$ 358.39

Constit. of *Lancea tibetica*. Amorph. powder (MeOH). $[\alpha]_D^{24}$ +92.5 (c, 0.19 in MeOH). λ_{max} 206 (log ε 4.38); 231 (log ε 3.96); 281 (log ε 4.08) (MeOH)

3,3′,4-Tri-Me ether

(+)-Phillygenin*. Forsythigenol. Phyllygenin. Sylvatesmin. Forsythigenin*

[487-39-8]

$C_{21}H_{24}O_6$ 372.417

Isol. from fruit of *Forsythia suspensa*, *Forsythia intermedia* and other *Forsythia* spp.. Also from *Phillyrea* spp., *Lancea tibetica*, *Osmanthus fragrans* and *Piper sylvaticum*. Antitumour agent. Antioxidant. Mp 133 - 134°. $[\alpha]_D^{22}$ +120 (c, 0.04 in MeOH). $[\alpha]_D^{20}$ +158 (c, 0.5 in MeOH)

3,3′,4-Tri-Me ether, 4′-O-β-D-glucopyranoside

Phillyrin*. Forsythin. Phillyroside. Chionanthin*

[487-41-2]

$C_{27}H_{34}O_{11}$ 534.558

Isol. from *Forsythia suspensa*, *Forsythia intermedia*, *Brassica rapa*, *Chionanthus virginica*, *Phillyrea latifolia* and *Phillyrea angustifolia*. Prod. by *Colletotrichum gloeosporioides* on *Forsythia* spp.. Mp 154 - 155° (146-148°). Mp 184 - 185° (dimorph.). $[\alpha]_D^{21}$ +46.9 (c, 0.25 in MeOH)

3,3′,4-Tri-Me ether, 4′-O-[β-D-xylopyranosyl-(1→6)-β-D-glucopyranoside]

Lantibeside

[98891-87-3]

$C_{32}H_{42}O_{15}$ 666.673

Isol. from *Lancea tibetica*. Mp 214°

3,3′,4-Tri-Me ether, 4′-O-[α-L-rhamnopyranosyl-(1→6)-β-D-glucopyranoside]

6″-O-α-L-Rhamnopyranosylphillyrin

[1660155-66-7]

$C_{33}H_{44}O_{15}$ 680.7

Constit. of *Pinus banksiana*. Amorph. yellow powder. Mp 169

- 171°. $[\alpha]_D^{20}$ +47.3 (c, 2 in MeOH). λ_{max} 235; 280 (MeOH)

3,3′,4-Tri-Me ether, Ac

Mp 122.5 - 123°. $[\alpha]_D^{25}$ +98.3 (c, 0.29 in MeOH)

(7S,7′R,8S,8′S)-form

3,3′,4-Tri-Me ether

(-)-Phillygenin

[23397-47-9]

$C_{21}H_{24}O_6$ 372.417

Constit. of *Pararistolochia flos-avis*. Cryst. (EtOAc/hexane). Mp 128 - 130°. $[\alpha]_D^{20}$ -114 (c, 0.5 in $CHCl_3$)

(7S,7′S,8R,8′R)-form

Bisdemethyl-(+)-pinoresinol

[340167-81-9]

Constit. of *Morinda citrifolia* and *Syzygium cumini*. LDL oxidn. inhibitor. Cryst. ($CHCl_3$/petrol). Mp 260 - 261°. $[\alpha]_D^{24}$ -3.4 (c, 0.3 in MeOH). λ_{max} 210 (log ε 4.4); 223 (log ε 4.12); 283 (log ε 3.86) (MeOH)

3,3′-Di-Me ether

see 7,9′:7′,9-Diepoxy-3,3′-dimethoxy-8,8′-lignan-4,4′-diol, D47

3,4-Di-Me ether

[237066-55-6]

$C_{20}H_{22}O_6$ 358.39

Constit. of *Melicope hayesii*. Amorph. solid. $[\alpha]_D$ +29 (c, 0.05 in MeOH). λ_{max} 206; 217; 279 (MeOH)

4,4′-Di-Me ether

7,9′:7′,9-Diepoxy-4,4′-dimethoxy-8,8′-lignan-3,3′-diol. 3,3′-Dihydroxy-4,4′-dimethoxy-7,9′:7′,9-diepoxylignan. ***Clemaphenol A***

[362606-60-8]

$C_{20}H_{22}O_6$ 358.390

Constit. of *Clematis chinensis*, *Clausena lansium*, *Cinnamomum kotoense*, *Daphne genkwa* and *Melia toosendan*. Oil. $[\alpha]_D^{25}$ +72 (c, 0.03 in $CHCl_3$). Probably isol. as a mixt. of diastereoisomers in most cases

4,4′-Di-Me ether, 3-O-β-D-glucopyranoside

[869013-35-4]

$C_{26}H_{32}O_{11}$ 520.531

Constit. of the rhizomes of *Curculigo capitulata*

3,3′,4-Tri-Me ether

Pinoresinol monomethyl ether

[29395-87-7]

$C_{21}H_{24}O_6$ 372.417

Constit. of *Forsythia japonica*, *Forsythia giraldiana*, *Torreya yunnanensis*, *Inula hookeri* and *Zanthoxylum planispinum*. Amorph. powder. $[\alpha]_D^{23}$ +65.6 (c, 1.9 in $CHCl_3$)

3,3′,4-Tri-Me ether, 4′-O-β-D-glucopyranoside

[74957-57-6]

$C_{27}H_{34}O_{11}$ 534.558

Constit. of *Forsythia suspensa*, *Forsythia japonica*, *Forsythia giraldiana* and *Torreya yunnanensis*. Needles (EtOH). Mp 148 - 150°. $[\alpha]_D^{23}$ +8.7 (c, 1.1 in MeOH)

Tetra-Me ether

7,9′:7′,9-Diepoxy-3,3′,4,4′-tetramethoxy-8,8′-lignan, D104

Methylene ether

7,9′:7′,9-Diepoxy-3′,4′-methylenedioxy-8,8′-lignan-3,4-diol, D102

Bis(methylene) ether

7,9′:7′,9-Diepoxy-3,4:3′,4′-bis(methylenedioxy)-8,8′-lignan, D45

(7RS,7′RS,8SR,8′SR)-form

Didemethyl-(±)-pinoresinol

Constit. of *Capparis flavicans* and *Joannesia princeps*. Brown cryst. (MeOH/$CHCl_3$). Mp 106 - 109°. λ_{max} 208 (log ε 4.88); 224 (log ε 4.57); 282 (log ε 4.29) (MeOH)

(7R*,7′R*,8R*,8′S*)-form

3,3′,4′-Tri-Me ether, 4-O-[E-cinnamoyl-(→2)-α-L-rhamnopyranoside]

[2170257-34-6]

$C_{36}H_{40}O_{11}$ 648.705

Constit. of the needles of *Pinus banksiana*. Amorph. powder. Mp 199 - 201°. $[\alpha]_D^{20}$ -47.9 (c, 0.5 in MeOH). λ_{max} 235; 285 (MeOH)

4-O-(3-Methyl-2-butenyl), 3-Me ether

Zanthonin

$C_{24}H_{28}O_6$ 412.482

Constit. of the leaves of *Zanthoxylum armatum*. Oil. $[\alpha]_D^{25}$ -116 (c, 0.12 in MeOH). λ_{max} 211 (log ε 3.34); 231 (log ε 2.82); 281 (log ε 1.14) (MeOH)

4-O-(3,7-Dimethyl-2E,6-octadienyl), 3-Me ether

Zanthoxylumin B

[1822319-65-2]

$C_{29}H_{36}O_6$ 480.601

Constit. of the bark of *Zanthoxylum simulans*. Solid. $[\alpha]_D^{18.6}$ -43.7 (c, 0.11 in MeOH). λ_{max} 228 (log ε 3.43); 281 (log ε 2.95) (MeOH).

Karrer, W. *Konstitution und Vorkommen der Organischen Pflanzenstoffe, 2nd edn., Birkhäuser Verlag*, 1976-1985, no. 1151 , no. 1152 (*Phillygenin, occur*).

Chiba, M. *et al., Phytochemistry*, 1980, **19**, 335-336 (*Phillygenin, Phillyrin, cmr*).

Banerji, A. *et al., J. Nat. Prod.*, 1982, **45**, 672-675 (*Sylvatesmin*).

Zheng, R. *et al., Zhiwu Xuebao*, 1985, **27**, 402-406 (*Lantibeside*).

Sun, N.-J. *et al., Phytochemistry*, 1987, **26**, 3051-3053 (*(-)-Phillygenin*).

Kitagawa, S. *et al., Chem. Pharm. Bull.*, 1988, **36**, 3667-3670 (*Phillygenin, Phillyrin, Pinoresinol monomethyl ether, Pinoresinol monomethyl ether 4′-glucoside, occur*).

Rahman, M.M.A. *et al., Phytochemistry*, 1990, **29**, 1841-1846 (*Phillygenin, Phillyrin, biosynth*).

Rahman, M.M.A. *et al., Phytochemistry*, 1990, **29**, 1971-1980 (*Phillygenin, Phillyrin*).

Luo, S.Q. *et al., Phytochemistry*, 1993, **33**, 193-196 (*Phillyrin*).

Yang, S.-J. *et al., J. Chin. Chem. Soc. (Taipei)*, 1999, **46**, 811-820 (*3′-O-Demethylepipinoresinol*).

Latip, J. *et al., Phytochemistry*, 1999, **51**, 107-110 (*3,4-di-Me ether*).

Su, B. *et al., Planta Med.*, 1999, **65**, 558-561 (*Sylvatesmin, activity*).

Chen, C.-C. *et al., Planta Med.*, 1999, **65**, 709-711 (*Phillygenin, activity*).

He, M. *et al., J. Chin. Pharm. Sci.*, 2001, **10**, 180-182 (*Clemaphenol A*).

Xu, Z.-H. *et al., Yaoxue Xuebao*, 2001, **36**, 668-671 (*O-Methylpinoresinol*).

Waibel, R. *et al., Phytochemistry*, 2003, **62**, 805-811 (*(±)-Didemethylpinoresinol*).

Li, N. *et al., Yunnan Zhiwu Yanjiu*, 2003, **25**, 711-715 (*Curculigo capitulata constit*).

Kamiya, K. *et al., J. Agric. Food Chem.*, 2004, **52**, 5843-5848 (*Didemethylpinoresinol, activity*).

Yin, R.-J. *et al., Acta Crystallogr., Sect. E*, 2006, **62**, 4606-4608 (*Phillyrin, cryst struct*).

Chang, M.-J. *et al., Biosci., Biotechnol., Biochem.*, 2008, **72**, 2750-2755 (*Phillygenin, activity*).

Li, T. *et al., Planta Med.*, 2008, **74**, 1391-1396 (*Lantibetin*).

Mir, Q.Y. *et al., Nat. Prod. Res.*, 2009, **23**, 422-430 (*Bisdemethylpinoresinol*).

Zhang, Q. *et al., Fitoterapia*, 2012, **83**, 1500-1505 (*Colletotrichum gloeosporioides constit*).

Hung, C.-Y. *et al., Molecules*, 2012, **17**, 10724-10737 (*(+)-Phillygenin*).

Wu, Q. *et al., Chem. Nat. Compd. (Engl. Transl.)*, 2013, **49**, 852-856 (*Phillyrin, spectra*).

Ye, L.-H. *et al., Eur. J. Drug Metab. Pharmacokinet.*, 2013, **38**, 201-207 (*Phillygenin, hplc, detn, pharmacokinetics*).

Si, C.-L. *et al., Holzforschung*, 2013, **67**, 357-363 (*6″-Rhamnopyranosylphillyrin*).

Yang, M. *et al., Yingyong Yu Huanjing Shengwu Xuebao*, 2014, **20**, 245-248 (*Clemaphenol A*).

Peng, C.-Y. *et al., J. Asian Nat. Prod. Res.*, 2015, **17**, 232-238 (*Zanthoxylumin B*).

Liu, C.-Y. *et al.*, *Chem. Nat. Compd. (Engl. Transl.)*, 2017, **53**, 1020-1024 (*Pinus banksiana constit*).
Bhatt, V. *et al.*, *Nat. Prod. Commun.*, 2017, **12**, 99-100 (*Zanthonin*)

7,9′:7′,9-Diepoxy-8,8′-lignan-3,3′,4,5′-tetrol D94

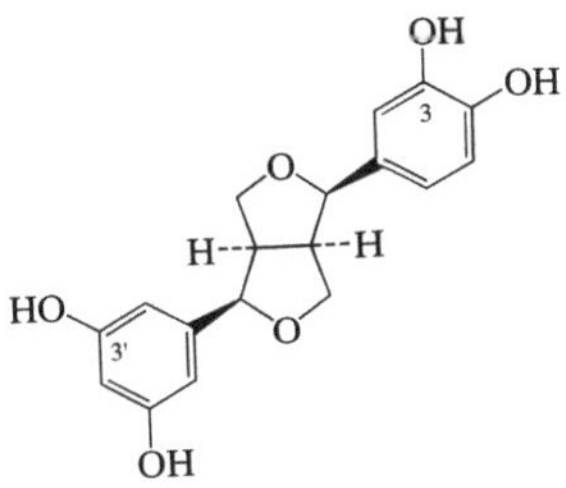

(7R,7′R,8R,8′R)-form

3,3′,4,5′-Tetrahydroxy-7,9′:7′,9-diepoxylignan
$C_{18}H_{18}O_6$ 330.336
Structs. in this series may need revision. Phenyl 3,5-dioxygenated substn. pattern in doubt as a result of work done by Yang *et al* in 2015

(7R,7′R,8R,8′R)-form
3,3′-Di-Me ether
7,9′:7′,9-Diepoxy-3,5′-dimethoxy-8,8′-lignan-3′,4-diol. 3′,4-Dihydroxy-3,5′-dimethoxy-7,9′:7′,9-diepoxylignan. **Inovonoid C**
$C_{20}H_{22}O_6$ 358.390
Constit. of *Balanophora involucrata*. Moderate α-glucosidase inhibitor. Amorph. powder. $[\alpha]_D^{25}$ -10.3 (c, 0.1 in MeOH). λ_{max} 229 (log ε 3.02); 281 (log ε 2.28) (MeOH)

(7R,7′R,8S,8′S)-form
3,3′-Di-Me ether, 4-O-β-D-glucopyranoside
Koreanaside A
[2231630-68-3]
$C_{26}H_{32}O_{11}$ 520.531
Constit. of the flowers of *Forsythia koreana*. Antioxidant. Pale yellow needles (MeOH). Mp 78 - 79°. $[\alpha]_D$ -9.2 (c, 0.2 in MeOH)

(7S,7′S,8R,8′R)-form
3,3′,5′-Tri-Me ether
7,9′:7′,9-Diepoxy-3,3′,5′-trimethoxy-8,8′-lignan-4-ol. 4-Hydroxy-3,3′,5′-trimethoxy-7,9′:7′,9-diepoxylignan. **Pandanusphenol A**
[1610891-88-7]
$C_{21}H_{24}O_6$ 372.417
Constit. of the fruit of *Pandanus tectorius*. Amorph. solid. $[\alpha]_D$ +12.5 (c, 0.1 in MeOH). λ_{max} 242 (log ε 1.96); 282 (log ε 1.52) (MeOH)

(7S,7′S,8S,8′S)-form
3,3′,5′-Tri-Me ether
Bombasinol A
[1413874-28-8]
$C_{21}H_{24}O_6$ 372.417
Constit. of the roots of *Bombax ceiba*. Shows activity against hepatitis B. Yellow oil. $[\alpha]_D^{30}$ +1.5 (c, 0.08 in MeOH). λ_{max} 209 (log ε 2.91); 258 (log ε 0.39); 278 (log ε 0.51) (MeOH)

(7R*,7′S*,8R*,8′R*)-form
3,3′,4,5′-Tetra-Me ether
7,9′:7′,9-Diepoxy-3,3′,4,5′-tetramethoxylignan. 3,3′,4,5′-Tetramethoxy-7,9′:7′,9-diepoxylignan
[1394179-95-3]
$C_{22}H_{26}O_6$ 386.444
Isol. from the flower buds of *Magnolia liliflora*. Amorph. solid. $[\alpha]_D^{20}$ -16.4 (c, 0.06 in MeOH). λ_{max} 231 (log ε 3.41); 284 (log ε 3.93) (MeOH)

3,4-Methylene, 3′,5′-di-Me ether
7,9′:7′,9-Diepoxy-3′,5′-dimethoxy-3,4-methylenedioxylignan. 3′,5′-dimethoxy-3,4-methylenedioxy-7,9′:7′,9-diepoxylignan
[1394179-90-8]
$C_{21}H_{22}O_6$ 370.401
Isol. from the flower buds of *Magnolia liliflora*. Oil. $[\alpha]_D^{20}$ -30 (c, 0.15 in $CHCl_3$). λ_{max} 239 (log ε 2.89); 297 (log ε 3.92) (MeOH)

(7R*,7′S*,8S*,8′S*)-form
3,4-Methylene, 3′,5′-di-Me ether
[1394179-91-9]
Isol. from the flower buds of *Magnolia liliflora*. Shows strong antigenotoxic activity. Cryst.. Mp 133 - 134°. $[\alpha]_D^{20}$ +11.8 (c, 0.12 in $CHCl_3$). λ_{max} 257 (log ε 2.91); 296 (log ε 3.86) ($CHCl_3$).

Wang, W.-S. *et al.*, *Planta Med.*, 2012, **78**, 141-147 (*Magnolia liliflora constits*).
Wang, G.K. *et al.*, *Nat. Prod. Res.*, 2013, **27**, 1348-1352 (*Bombasinol A*).
Zhang, X. *et al.*, *Rec. Nat. Prod.*, 2013, **7**, 359-362 (*Pandanusphenol A*).
Yang, Y.-N. *et al.*, *J. Nat. Prod.*, 2015, **78**, 705-711 (*pmr, struct*).
Wei, J. *et al.*, *Fitoterapia*, 2017, **121**, 129-135 (*Inovonoid C, activity*).
Lee, Y.-G. *et al.*, *Chem. Biodiversity*, 2018, **15**, e1800026:1-8 (*Koreanaside A, activity*)

7,9′:7′,9-Diepoxy-8,8′-lignan-3,3′,5,5′-tetrol D95

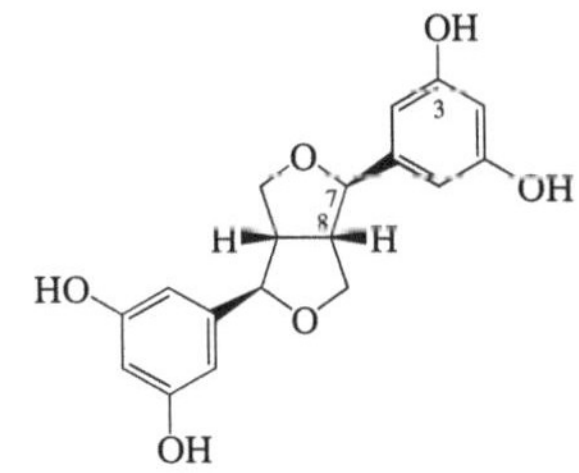

(7R*,7′R*,8S*,8′S*)-form

3,3′,5,5′-Tetrahydroxy-7,9′:7′,9-diepoxylignan
$C_{18}H_{18}O_6$ 330.336
Structures in this series may need revision. Phenyl 3,5-dioxygenated substn. pattern in doubt as a result of work done by Yang *et al* in 2015

(7R*,7′R*,8S*,8′S*)-form
3,3′,5,5′-Tetra-Me ether
7,9′:7′,9-Diepoxy-3,3′,5,5′-tetramethoxylignan. 3,3′,5,5′-Tetramethoxy-7,9′:7′,9-diepoxylignan. **Eudesmin A**
[1355016-87-3]
$C_{22}H_{26}O_6$ 386.444
Constit. of *Haplophyllum villosum*. Needles (MeOH). Mp 92 - 94°. λ_{max} 242 (log ε 1.88); 281 (log ε 1.36) ($CHCl_3$)

(7R*,7′S*,8S*,8′S*)-form
3,3′,5,5′-Tetra-Me ether
[1394179-92-0]
$C_{22}H_{26}O_6$ 386.444
Isol. from the flower buds of *Magnolia liliflora*. Amorph. solid. $[\alpha]_D^{20}$ -43.3 (c, 0.03 in MeOH). λ_{max} 239 (log ε 3.40); 280 (log ε 3.97) (MeOH).

Parhoodeh, P. *et al.*, *Molecules*, 2011, **16**, 2268-2273 (*Eudesmin A*).
Wang, W.-S. *et al.*, *Planta Med.*, 2012, **78**, 141-147 (*Magnolia liliflora constit*).
Yang, Y.-N. *et al.*, *J. Nat. Prod.*, 2015, **78**, 705-711 (*pmr, struct*)

7,9′:7′,9-Diepoxy-8,8′-lignan-3,4,4′,5-tetrol D96

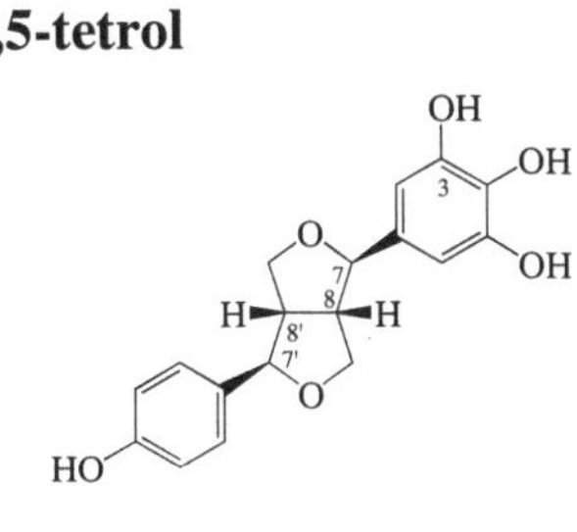

3,4,4′,5-Tetrahydroxy-7,9′:7′,9-diepoxylignan
$C_{18}H_{18}O_6$ 330.336
(7R*,7′R*,8S*,8′S*)-form
3,5-Di-Me ether
7,9′:7′,9-Diepoxy-3,5-dimethoxy-8,8′-lignan-4,4′-diol. ***Diphyllin A***
$C_{20}H_{22}O_6$ 358.390
Constit. of the roots of *Diphylleia sinensis*. Amorph. powder. $[\alpha]_D^{20}$ +17.3 (c, 0.17 in MeOH). Not to be confused with Diphyllin, D296
Tetra-Me ether
7,9′:7′,9-Diepoxy-3,4,4′,5-tetramethoxy-8,8′-lignan. ***Isoeudesmin***
$C_{22}H_{26}O_6$ 386.444
Constit. of the flower buds of *Magnolia biondii*. Exhibits antiallergic activities. Yellowish wax. $[\alpha]_D^{25}$ +70 (c, 0.1 in MeOH). λ_{max} 207; 232; 273 (MeOH).

Zhang, X. *et al., RSC Adv.*, 2017, **7**, 34236-34243 (*Isoeudesmin, activity*).
Sun, C. *et al., Zhongcaoyao*, 2019, **50**, 5193-5197 (*Diphyllin A*)

7,9′:7′,9-Diepoxy-8,8′-lignan-3,4,4′,8′-tetrol D97

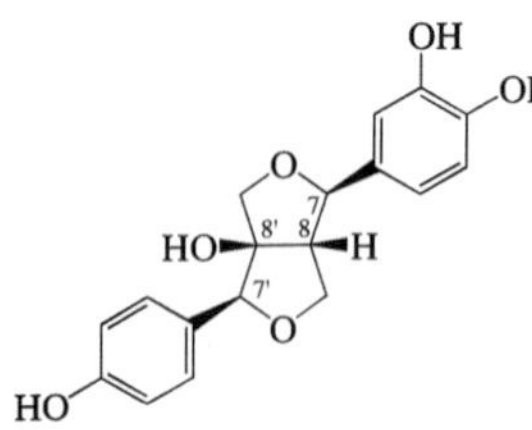

3,4,4′,8′-Tetrahydroxy-7,9′:7′,9-diepoxylignan
$C_{18}H_{18}O_6$ 330.336
(7R,7′S,8S,8′R)-form
3,4-Methylene ether
7,9′:7′,9-Diepoxy-3,4-methylenedioxy-8,8′-lignan-4′,8-diol. 4′,8-Dihydroxy-3,4-methylenedioxy-7,9′:7′,9-diepoxylignan. ***Firmianol A***
[1864954-48-2]
$C_{19}H_{18}O_6$ 342.347
Constit. of the stems of *Firmiana simplex*. Gum. $[\alpha]_D^{25}$ +18 (MeOH).

Woo, K.Y. *et al., Bioorg. Med. Chem. Lett.*, 2016, **26**, 730-733 (*Firmianol A*)

7,9′:7′,9-Diepoxy-8,8′-lignan-3,4,4′,8-tetrol D98

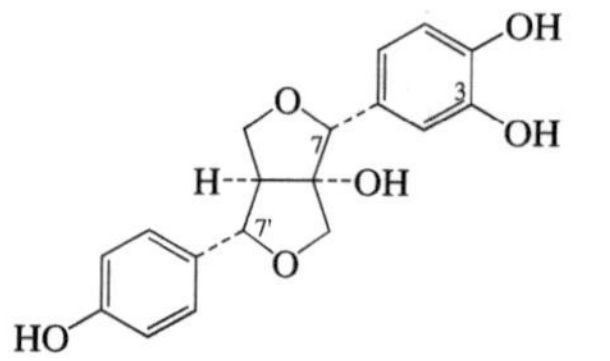

3,4,4′,8-Tetrahydroxy-7,9′:7′,9-diepoxylignan
$C_{18}H_{18}O_6$ 330.336
(7R,7′S,8S,8′R)-form
3-Me ether
7,9′:7′,9-Diepoxy-3-methoxy-8,8′-lignan-4,4′,8-triol. ***Mappianthin B***
$C_{19}H_{20}O_6$ 344.363
Constit. of *Mappianthus iodoides*. NO prodn. inhibitor. Amorph. powder. $[\alpha]_D^{20}$ +35.8 (c, 0.13 in MeOH). λ_{max} 231 (log ε 4.28); 280 (log ε 3.69) (MeOH).

Jiang, Z.-H. *et al., Phytochem. Lett.*, 2017, **22**, 194-196 (*Mappianthin B, activity*)

7,9′:8,9-Diepoxy-8,8′-lignan-3,3′,4,4′-tetrol D99

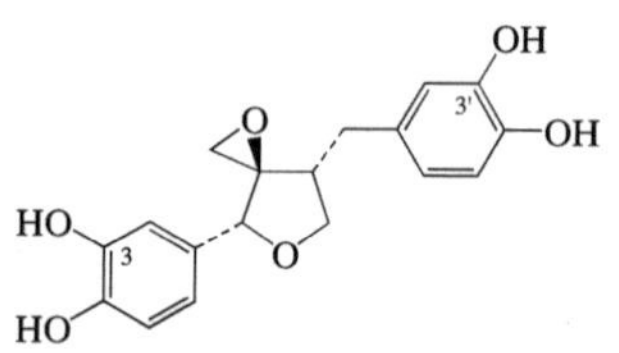

(7R*,8R*,8′R*)-form

3,3′,4,4′-Tetrahydroxy-7,9′:8,9-diepoxylignan. 8,9-Epoxy-3,3′,4,4′-tetrahydroxy-7,9′-epoxylignan
$C_{18}H_{18}O_6$ 330.336
(7R*,8R*,8′R*)-form
3,3′-Di-Me ether
7,9′:8,9-Diepoxy-3,3′-dimethoxy-8,8′-lignan-4,4′-diol. 4,4′-Dihydroxy-3,3′-dimethoxy-7,9′:8,9-diepoxylignan
[1206464-62-1]
$C_{20}H_{22}O_6$ 358.39
Constit. of *Viburnum foetidum* var. *foetidum*. Amorph. powder. $[\alpha]_D^{23}$ -64.7 (c, 0.1 in MeOH). λ_{max} 204 (log ε 3.3); 231 (log ε 2.9); 282 (log ε 4) (MeOH)
(7R*,8S*,8′S*)-form
3′,4′-Methylene, 3-Me ether
7,9′:8,9-Diepoxy-3-methoxy-3′,4′-methylenedioxy-8,8′-lignan-4-ol. 4-Hydroxy-3-methoxy-3′,4′-methylenedioxy-7,9′:8,9-diepoxylignan
[1345841-55-5]
$C_{20}H_{20}O_6$ 356.374
Constit. of *Phillyrea angustifolia*. Plant growth inhibitor. $[\alpha]_D^{25}$ -22 (c, 0.12 in MeOH). Descr. as 8*R*-config. in the text of the paper, but the diag. shows 8*S*-.

Chen, X.-Q. *et al., Chem. Pharm. Bull.*, 2009, **57**, 1129-1131 (*Viburnum foetidum 3,3′-di-Me ether*).
DellaGreca, M. *et al., Phytochem. Lett.*, 2011, **4**, 118-121 (*Phillyrea angustifolia constit*)

7,9′:7′,9-Diepoxy-8,8′-lignan-3,4,4′-triol D100

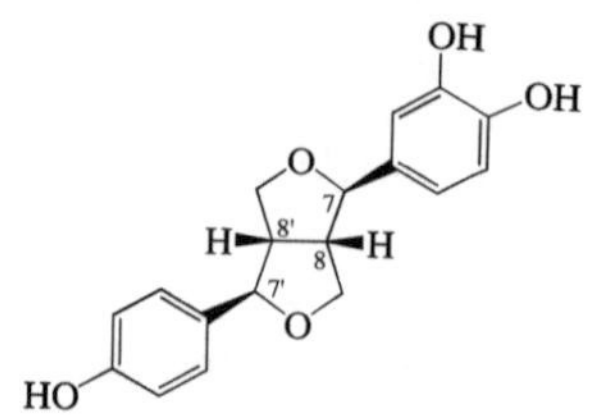

(7R,7′R,8S,8′S)-form

2-(3,4-Dihydroxyphenyl)-6-(4-hydroxyphenyl)-3,7-dioxabicyclo[3.3.0]octane. 3,4,4′-Trihydroxy-7,9′:7′,9-diepoxylignan
$C_{18}H_{18}O_5$ 314.337
(7R,7′R,8S,8′S)-form
3-Me ether
7,9′:7′,9-Diepoxy-3-methoxy-8,8′-lignan-4,4′-diol. 4,4′-Dihydroxy-3-methoxy-7,9′:7′,9-diepoxylignan. ***Demethoxy-(-)-pinoresinol***
[832735-37-2]
$C_{19}H_{20}O_5$ 328.364
Constit. of *Balanophora abbreviata*. Inhibitor of LPS-induced iNOS expression *in vitro*. Amorph. solid. $[\alpha]_D^{23}$ -66.9 (c, 0.28 in $CHCl_3$). λ_{max} 228 (log ε 5.55); 279 (log ε 1.58) (EtOH)
3-Me ether, 4-Ac
Smallanlignan A
[1345430-89-8]
$C_{21}H_{22}O_6$ 370.401
Constit. of *Smallanthus sonchifolius*. Exhibits antioxidant activity. Amorph. solid. $[\alpha]_D^{23}$ -55 (c, 0.28 in $CHCl_3$). λ_{max} 226 (log ε 5.18); 282 (log ε 1.61) ($CHCl_3$)
(7R,7′S,8S,8′S)-form
3-Me ether
[154966-95-7]
Constit. of *Festuca argentina*. $[\alpha]_D$ -110 (c, 0.03 in $CHCl_3$)

(7S,7′S,8R,8′R)-form
3-Me ether
Demethoxy-(+)-pinoresinol
[137172-13-5]
$C_{19}H_{20}O_5$ 328.364
Isol. from *Mikania saltensis*. Gum
Tri-Me ether
7,9′:7′,9-Diepoxy-3,4,4′-trimethoxylignan. 3,4,4′-Trimethoxy-7,9′:7′,9-diepoxylignan. **Membrin**
[148057-46-9]
$C_{21}H_{24}O_5$ 356.418
Constit. of the seeds of *Rollinia membranacea*. [α] D +52.3 (c, 0.08 in MeOH)
3,4-Methylene ether
7,9′:7′,9-Diepoxy-3′,4′-methylenedioxy-8,8-lignan-4-ol. 4-Hydroxy-3′,4′-methylenedioxy-7,9′:7′,9-diepoxylignan. **Beechenol**
[845739-57-3]
$C_{19}H_{18}O_5$ 326.348
Constit. of the stems of *Zanthoxylum beecheyanum*. Oil. $[\alpha]_D^{25}$ +35.9 (c, 0.05 in $CHCl_3$). λ_{max} 229 (log ε 3.91); 284 (log ε 3.58) (MeOH)
(7R,7′S*,8R*,8′R*)-form*
Tri-Me ether
7-Epimembrin
[425407-76-7]
$C_{21}H_{24}O_5$ 356.418
Constit. of the leaves of *Rollinia mucosa*. Cryst.. Mp 128 - 130°. $[\alpha]_D^{25}$ +134.6 (c, 1.2 in $CHCl_3$).

Cuenca, M. del R., *et al.*, *J. Nat. Prod.*, 1991, **54**, 1162-1164 (*Demethoxy-(+)-pinoresinol*).
Saez, J. *et al.*, *J. Nat. Prod.*, 1993, **56**, 351-356 (*Membrin*).
Casabuono, A.C. *et al.*, *Phytochemistry*, 1994, **35**, 479-483 (*Festuca argentina constit*).
Wirth, T. *et al.*, *Liebigs Ann./Recl.*, 1997, 1155-1158 (*(+)-Membrin, synth*).
Estrada-Reyes, R. *et al.*, *Z. Naturforsch., C*, 2002, **57**, 29-31 (*7-Epimembrin, Membrin*).
Hosokawa, A. *et al.*, *Chem. Pharm. Bull.*, 2004, **52**, 1265-1267 (*Demethoxy-(-)-pinoresinol*).
Cheng, M.-J. *et al.*, *J. Chin. Chem. Soc. (Taipei)*, 2004, **51**, 1065-1072 (*Beechenol*).
Xue, J.-J. *et al.*, *Asian J. Chem.*, 2011, **23**, 933-934 (*Smallanlignan A, cd, activity*)

7,9′:7′,9-Diepoxy-8,8′-lignan-4,4′,8-triol D101

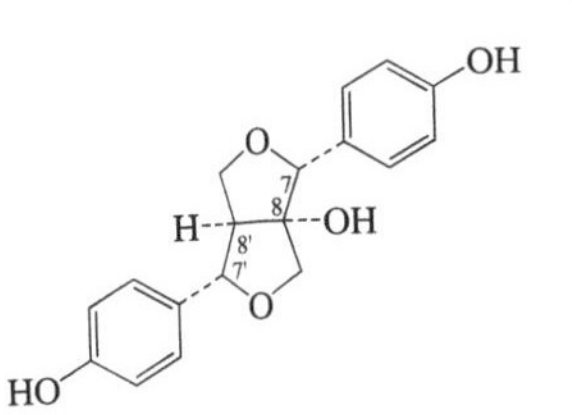

4,4′,8-Trihydroxy-7,9′:7′,9-diepoxylignan
$C_{18}H_{18}O_5$ 314.337
(7R,7′S,8S,8′R)-form
Mappianthin A
Constit. of *Mappianthus iodoides*. NO prodn. inhibitor. Amorph. powder. $[\alpha]_D^{20}$ +32.6 (c, 0.12 in MeOH). λ_{max} 232 (log ε 4.31); 283 (log ε 3.72) (MeOH).

Jiang, Z.-H. *et al.*, *Phytochem. Lett.*, 2017, **22**, 194-196 (*Mappianthin A, activity*)

7,9′:7′,9-Diepoxy-3′,4′-methylenedioxy-8,8′-lignan-3,4-diol D102

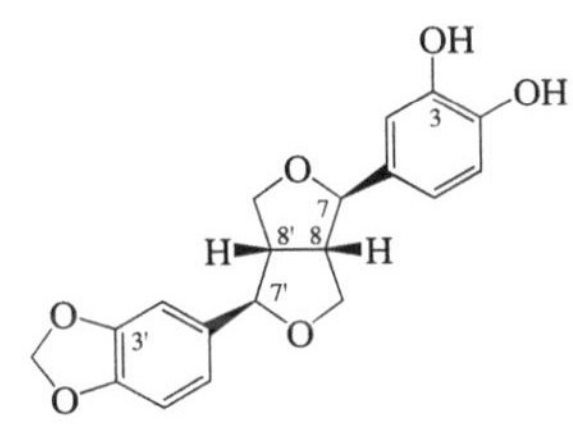

(7R,7′R,8S,8′S)-form

4-[4-(1,3-Benzodioxol-5-yl)tetrahydro-1H,3H-furo[3,4-c]furan-1-yl]-1,2-benzenediol. 2-(3,4-Dihydroxyphenyl)-6-(3,4-methylenedioxyphenyl)-3,7-dioxabicyclo[3.3.0]octane. 3,4-Dihydroxy-3′,4′-methylenedioxy-7,9′:7′,9-diepoxylignan
$C_{19}H_{18}O_6$ 342.347
Lignan numbering shown. Not all abs. configs. are firmly established and the allocation of trivial names to various stereoisomers varies considerably
(7R,7′R,8S,8′S)-form
4-O-β-D-Glucopyranoside
Demethyl-(-)-piperitol 4-glucoside
[1260542-47-9]
$C_{25}H_{28}O_{11}$ 504.488
Constit. of the leaves of *Eucalyptus maidenii*. Amorph. powder. $[\alpha]_D^{28}$ -72.9 (c, 0.5 in MeOH). λ_{max} 204 (log ε 4.75); 282 (log ε 3.89); 376 (log ε 3.34) (MeOH)
3-Me ether
7,9′:7′,9-Diepoxy-3-methoxy-3′,4′-methylenedioxy-8,8′-lignan-4-ol. 4-Hydroxy-3-methoxy-3′,4′-methylenedioxy-7,9′:7′,9-diepoxylignan. ***(-)-Piperitol***
[54983-96-9]
$C_{20}H_{20}O_6$ 356.374
Obt. from *Zanthoxylum piperitum*. Syrup. $[\alpha]_D^{21}$ -66.3 (c, 2.5 in $CHCl_3$). Genus name given as Xanthoxylum in ref.. λ_{max} 233 (log ε 4.11); 285 (log ε 3.87) (MeOH)
Di-Me ether
7,9′:7′,9-Diepoxy-3,4-dimethoxy-3′,4′-methylenedioxy-8,8′-lignan. 3,4-dimethoxy-3′,4′-methylenedioxy-7,9′:7′,9-diepoxylignan. (-)-Kobusin.
(-)-Methylpiperitol
[189066-84-0]
$C_{21}H_{22}O_6$ 370.401
Isol. from *Samadera* aff. *bidwillii* and *Geranium eriostemon*. Shows moderate cytotoxicity *in vitro*. Oil or cryst.. Mp 74°. [α] D -21 (c, 0.38 in $CHCl_3$). $[\alpha]_D^{25}$ -71.5 (c, 0.4 in $CHCl_3$) (synthetic)
(7R,7′S,8R,8′R)-form
Di-Me ether
(+)-Fargesin*. Planinin. Methylpluviatilol*
[68296-27-5]
[78737-48-1] (deleted)
$C_{21}H_{22}O_6$ 370.401
Constit. of *Artemisia absinthium, Magnolia fargesii, Magnolia pterocarpa, Magnolia coco, Virola elongata* and *Zanthoxylum armatum*. Platelet-activating factor inhibitor. Cryst.. Mp 135 - 136°. $[\alpha]_D^{20}$ +122 (c, 1 in $CHCl_3$) (+78). One ref. shows with the opposite abs. config.
(7R,7′S,8S,8′S)-form
3-Me ether
(-)-Xanthoxylol
[54983-95-8]
$C_{20}H_{20}O_6$ 356.374
Isol. from *Zanthoxylum piperitum* and *Zanthoxylum aeanthapodium*. Pillars (C_6H_6/hexane). Mp 140 - 142°. $[\alpha]_D^{16}$ -117 (c, 0.5 in $CHCl_3$). λ_{max} 233 (log ε 4.06); 286 (log ε 3.82) (MeOH)
4-O-(3-Methyl-2-butenyl), 3-Me ether
7,9′:7′,9-Diepoxy-3-methoxy-3′,4′-methylenedioxy-4-prenyloxy-8,8′-lignan. ***Xanthoxylol γ,γ-dimethylallyl ether***
[54615-53-1]
$C_{25}H_{28}O_6$ 424.493
Isol. from *Zanthoxylum piperitum*. Oil. $[\alpha]_D^{19}$ -86.5 (c, 4.08 in $CHCl_3$). λ_{max} 234 (log ε 4.28); 284 (log ε 3.97) (MeOH)
(7S,7′R,8R,8′R)-form
3-Me ether
(+)-Xanthoxylol

[92620-81-0]
$C_{20}H_{20}O_6$ 356.374
Isol. from leaves of *Juniperus chinensis*. Mp 140 - 142°. $[\alpha]_D^{25}$ +122 (c, 5.8 in $CHCl_3$). The abs. config. of the Xanthoxylol enantiomers appears to be uncertain

3-Me ether, 4-O-β-D-glucopyranoside
Lantibeside D
[1096706-63-6]
$C_{26}H_{30}O_{11}$ 518.515
Constit. of *Androya decaryi* and *Lancea tibetica*. Amorph. powder (MeOH). $[\alpha]_D^{24}$ +33.5 (c, 0.1 in MeOH). λ_{max} 204 (log ε 4.46); 230 (log ε 3.86); 282 (log ε 3.59) (MeOH)

3-Me ether, 4-O-[β-D-xylopyranosyl-(1→6)-β-D-glucopyranoside]
Lantibeside C
[1096706-35-2]
$C_{31}H_{38}O_{15}$ 650.63
Constit. of *Lancea tibetica*. Amorph. powder (MeOH). $[\alpha]_D^{24}$ +13.1 (c, 0.07 in MeOH). λ_{max} 204 (log ε 4.5); 230 (log ε 3.86); 282 (log ε 3.58) (MeOH)

4-Me ether
7,9′:7′,9-Diepoxy-4-methoxy-3′,4′-methylenedioxy-8,8′-lignan-3-ol. 3-Hydroxy-4-methoxy-3′,4′-methylenedioxy-7,9′:7′,9-diepoxylignan. ***Horsfieldin***
[85922-40-3]
$C_{20}H_{20}O_6$ 356.374
Constit. of *Horsfieldia iryaghedhi*, *Zanthoxylum planispinum* and *Daphne pedunculata*. Cryst.. Mp 164 - 165°. $[\alpha]_D$ +38 ($CHCl_3$)

Di-Me ether
Neofargesin.
(+)-Methylxanthoxylol
[112572-28-8]
[189066-86-2] (deleted)
$C_{21}H_{22}O_6$ 370.401
Constit. of *Samadera* sp. nov. aff. *biwillii* and *Eremophila calorhabdos*. Prisms (petrol/Me_2CO). Mp 124 - 125°. $[\alpha]_D$ +112 (c, 0.4 in $CHCl_3$)

(7S,7′R,8S,8′S)-form

3-Me ether
Pluviatilol
[28115-67-5]
$C_{20}H_{20}O_6$ 356.374
Isol. from *Zanthoxylum pluviatile*, *Asarum heterotropoides* and *Lindera obtusiloba*. Cytotoxic agent. Cryst. (C_6H_6/petrol). Mp 161.5 - 162.5°. $[\alpha]_D$ -136 ($CHCl_3$). Stereochem. not certain, some confusion in the lit.

3-Me ether, 4-O-β-D-glucopyranoside
Styraxjaponoside C
[944147-14-2]
$C_{26}H_{30}O_{11}$ 518.515
Constit. of the stem bark of *Styrax japonica*. Yellowish gum. $[\alpha]_D^{23}$ -30.9 (c, 0.06 in MeOH). λ_{max} 233 (log ε 4.12); 284 (log ε 3.85) (MeOH)

4-O-(3-Methyl-2-butenyl), 3-Me ether
O-Prenylpluviatilol. *Pluviatilol 3,3-dimethylallyl ether*
[85994-79-2]
Isol. from *Zanthoxylum ailanthoides* and *Zanthoxylum podocarpum*. Cryst.. Mp 105 - 107°

(7S,7′S,8R,8′R)-form
Demethylpiperitol [170713-44-7]
Constit. of the leaves of *Apollonias barbujana*. Metab. of Sesamin in rat liver. Shows antioxidant props.. Amorph. powder. $[\alpha]_D^{25}$ +46.3 (c, 0.3 in $CHCl_3$)

3-Me ether
(+)-Piperitol‡. *Regiomontanin*
[52151-92-5]
$C_{20}H_{20}O_6$ 356.374
Constit. of *Aptosimum spinescens*, *Asarum sieboldii*, *Magnolia stellata*, *Nectandra turbacensis* and *Zanthoxylum petiolare*. Metab. of Sesamin in rat liver. Potent vasodilator. Activator of PPARs transcriptionn. Antioxidant. Mp 79°. $[\alpha]_D^{20}$ +76.2 (c, 2 in $CHCl_3$)

3-Me ether, 4-O-β-D-glucopyranoside
Simplexoside
[74061-78-2]
$C_{26}H_{30}O_{11}$ 518.515
Isol. from whole plant of *Justicia simplex*. CNS depressant. Straw-coloured cryst. +1H_2O. Mp 205 - 210°. $[\alpha]_D$ -12.5 (c, 0.34 in MeOH)

3-Me ether, 4-O-[β-D-xylopyranosyl-(1→6)-β-D-glucopyranoside]
Lantibeside B. *Tenuifloroside*
[1096706-64-7]
$C_{31}H_{38}O_{15}$ 650.63
Constit. of *Lancea tibetica* and *Castilleja tenuiflora*. Amorph. powder (MeOH). $[\alpha]_D^{24}$ -5.1 (c, 0.13 in MeOH). $[\alpha]_D^{28}$ +10 (c, 1.0 in H_2O) (Tenuifloroside). λ_{max} 204 (log ε 4.48); 231 (log ε 3.81); 282 (log ε 3.56) (MeOH). [neutral] λ_{max} 201; 228; 282 (MeCN aq.) (Tenuifloroside)

Di-Me ether
Spinescin. *Kobusin‡. Desmethoxyaschantin. O-Methylpiperitol*
[36150-23-9]
[55781-78-7] (deleted CAS no.)
$C_{21}H_{22}O_6$ 370.401
Constit. of *Aptosimum spinescens*, Geranium thunbergii, *Magnolia kobus*, *Magnolia stellata*, sassafras root and other sources. Inhibits lipopolysaccharide stimulated NO prodn. and expression of NO synthase. Cryst. or oil. Sol. MeOH, $CHCl_3$. Mp 55°. Mp 75°. $[\alpha]_D$ +58.6 (c, 3.33 in $CHCl_3$). $[\alpha]_D^{20}$ +73.6 ($CHCl_3$). $[\alpha]_D$ +272. λ_{max} 232; 284 (MeOH) (Berdy). [neutral] λ_{max} 233 (ε 12950); 284 (ε 7480) (EtOH) (Berdy)

4-O-(3-Methyl-2-butenyl), 3-Me ether
O-Prenyl-(+)-piperitol. *(+)-Piperitol γ,γ-dimethylallyl ether*
[157659-20-6]
$C_{25}H_{28}O_6$ 424.493
Constit. of *Zanthoxylum tetraspermum*, *Zanthoxylum petiolare* and *Zanthoxylum nitidum*. Needles ($CHCl_3$). Mp 60°. $[\alpha]_D^{25}$ +25 ($CHCl_3$)

(7S,7′S,8S,8′S)-form

Di-Me ether
[128820-38-2]
$C_{21}H_{22}O_6$ 370.401
Constit. of *Zanthoxylum* sp.

(7R*,7′R*,8S*,8′S*)-form

4-O-(3-Methyl-2-butenyl), 3-Me ether
O-Prenyl-(-)-piperitol. *(-)-Piperitol γ,γ-dimethylallyl ether*
[54631-93-5]
Constit. of *Zanthoxylum piperitum*. Needles. Mp 58 - 59°. $[\alpha]_D^{20}$ -45.7 (c, 5.4 in $CHCl_3$). λ_{max} 234 (log ε 4.26); 284 (log ε 3.95) (MeOH)

(7RS,7′RS,8SR,8′SR)-form

Di-Me ether
Epifargesin
[63087-38-7]
[62839-87-6] (deleted)
Constit. of *Fleischmannia pycnocephala*. Oil. One isolate was definitely racemic, the earlier isolate had no reported opt. rotn.

(7RS,7′SR,8RS,8′RS)-form

Di-Me ether

***Fargesin**. Methyl pluviatilol*

[31008-19-2]

[55821-75-5] (deleted)

[187604-37-1] (deleted)

$C_{21}H_{22}O_6$ 370.401

Isol. from flower buds of *Magnolia fargesii* and stem bark of *Zanthoxylum acanthopodium*. Plates (C_6H_6/petrol). One isolate was definitely racemic, the earlier isolate had no reported opt. rotn.

(7ξ,7′ξ,8ξ,8′ξ)-form

4-O-(2-Hydroxy-3-methyl-3-butenyl), 3-Me ether

O-(2-Hydroxy-3-methyl-3-butenyl) piperitol

[157563-38-7]

$C_{25}H_{28}O_7$ 440.492

Constit. of *Zanthoxylum petiolare*. Viscous oil. $[\alpha]_D$ +80.9 (c, 0.0004 in $CHCl_3$).

Corrie, J.E.T. *et al., Aust. J. Chem.*, 1970, **23**, 133-145 (*Pluviatilol*).

Höke, M. *et al., Arch. Pharm. (Weinheim, Ger.)*, 1972, **305**, 33-39 (*Desmethoxyaschantin*).

Kakisawa, H. *et al., Phytochemistry*, 1972, **11**, 2289-2293 (*Fargesin*).

Abe, F. *et al., Chem. Pharm. Bull.*, 1974, **22**, 2650-2655 (*Zanthoxylum piperitum constits*).

Kamaido, T. *Agric. Biol. Chem.*, 1975, **39**, 833-836 (*Kobusin*).

Rao, E.V. *et al., Curr. Sci.*, 1975, **44**, 228-230 (*(±)-Fargesin*).

Brieskorn, C.H. *et al., Tetrahedron Lett.*, 1976, **17**, 2221-2224 (*Piperitol, Spinescin*).

Pelter, A. *et al., Tetrahedron*, 1976, **32**, 2783-2788 (*Pluviatilol, Methyl pluviatilol, Xanthoxylol, struct, stereochem*).

Bohlmann, F. *et al., Chem. Ber.*, 1977, **110**, 301-314 (*(±)-Epifargesin*).

Pelter, A. *et al., Tetrahedron Lett.*, 1977, **18**, 4137-4140 (*cmr, stereochem*).

Vaquette, J. *et al., Planta Med.*, 1979, **35**, 42-47 (*Piperitol γ,γ-dimethylallyl ether*).

Ghosal, S. *et al., Phytochemistry*, 1980, **19**, 332-334 (*Simplexoside*).

De Carvalho, M.G. *et al., Phytochemistry*, 1980, **26**, 265-267 (*Piperitol, isol, activity*).

Hofer, O. *et al., Tetrahedron*, 1981, **37**, 1181-1186 (*cd, abs config*).

Iida, T. *et al., Phytochemistry*, 1982, **21**, 673-675 (*Kobusin*).

Gunatilaka, A.L.L. *et al., Phytochemistry*, 1982, **21**, 2719-2723 (*Horsfieldin*).

Iida, T. *et al., Phytochemistry*, 1983, **22**, 211-214 (*Magnolia stellata constit*).

Ishii, H. *et al., Yakugaku Zasshi*, 1983, **103**, 279-292 (*O-Prenylpluviatilol*).

Pelter, A. *et al., J. Chem. Soc., Perkin Trans. 1*, 1985, 587-594 (*synth*).

Bytheway, I.R. *et al., Aust. J. Chem.*, 1987, **40**, 1913-1917 (*(+)-Methylxanthoxylol*).

Chen, J. *et al., Yaoxue Xuebao*, 1988, **23**, 422-425 (*Planinin*).

Marcos, M. *et al., Planta Med.*, 1990, **56**, 89-91 (*Zanthoxylum constit*).

Stevens, D.R. *et al., J. Chem. Soc., Perkin Trans. 1*, 1992, 185-190 (*furofuranoid lignans, synth*).

Fang, J.M. *et al., Phytochemistry*, 1992, **31**, 3659-3661 (*(+)-Xanthoxylol*).

Arruda, M.S.P. *et al., Phytochemistry*, 1994, **36**, 1303-1306 (*Piperitol 2-hydroxy-3-methyl-3-butenyl ether*).

Ogiku, T. *et al., J. Org. Chem.*, 1995, **60**, 1148-1153 (*furofuranoid lignans, synth, pmr, cmr*).

Pérez, C. *et al., Phytochemistry*, 1995, **40**, 1511-1513 (*Demethylpiperitol*).

Gibbons, S. *et al., Phytochemistry*, 1997, **44**, 1109-1114 (*Neofargesin, (-)-Kobusin*).

Kato, M.J. *et al., Phytochemistry*, 1997, **47**, 583-591 (*biosynth*).

Yoshida, S. *et al., Tetrahedron*, 1997, **53**, 9585-9588 (*Fargesin, synth, abs config*).

Yu, H.-J. *et al., J. Chin. Chem. Soc. (Taipei)*, 1998, **45**, 773-778 (*Magnolia coco constit*).

Kwon, J.C. *et al., Arch. Pharmacal Res.*, 1999, **22**, 417-422 (*Piperitol, isol, activity*).

Brown, R.C.D. *et al., J. Org. Chem.*, 2001, **66**, 6719-6728 (*Fargesin, synth*).

Roy, S.C. *et al., J. Org. Chem.*, 2002, **67**, 3242-3248 (*(±)-Piperitol, (±)-Methylpiperitol, synth*).

Enders, D. *et al., Synthesis*, 2002, 515-522 (*(-)-Methylpiperitol, synth*).

Nakai, M. *et al., J. Agric. Food Chem.*, 2003, **51**, 1666-1670 (*Sesamin metabs*).

Yoda, H. *et al., Heterocycles*, 2005, **65**, 519-522 (*Methylpiperitol, Fargesin, synth*).

Banerjee, B. *et al., Synthesis*, 2005, 2913-2919 (*(-)-Methylpiperitol, synth*).

Nakano, D. *et al., J. Pharmacol. Exp. Ther.*, 2006, **318**, 328-335 (*(+)-Piperitol, vasodilative activity*).

Kim, M.-R. *et al., Arch. Pharmacal Res.*, 2007, **30**, 425-430 (*Styraxjaponoside C*).

Kang, K.W. *et al., J. Toxicol. Public Health*, 2007, **6**, 123-126 (*Geranium thunbergii constit, activity*).

Li, T. *et al., Planta Med.*, 2008, **74**, 1391-1396 (*Lantibesides B,C,D*).

Xu, W. *et al., Chem. Nat. Compd. (Engl. Transl.)*, 2009, **45**, 417-419 (*Horsfieldin, spectra*).

Tian, L.-W. *et al., Helv. Chim. Acta*, 2010, **93**, 2194-2202 (*Demethyl-(-)-piperitol 4′-glucoside*).

Quang, T.H. *et al., Bioorg. Med. Chem. Lett.*, 2012, **22**, 2527-2533 (*(+)-Piperitol, PPARs transcription activator*).

Herrera-Ruiz, M. *et al., Molecules*, 2015, **20**, 13127-13143 (*Tenuifloroside*)

4,7′:4‴,7″-Diepoxy-3′,4″-oxy-3,8′:3‴,8″-diligna-7,7′,7″,7‴-tetraen-4′-ol D103

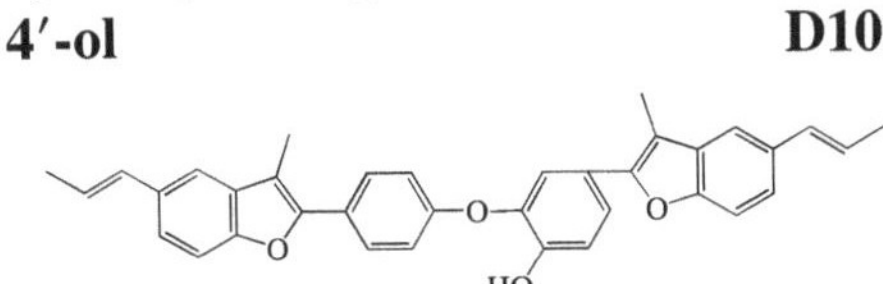

4,7′:4‴,7″-Diepoxy-4′-hydroxy-3′,4″-oxy-3,8′:3‴,8″-diligna-7,7′,7″,7‴-tetraene. ***(4′-O-3′)-Bi(eupomatenoid 6)***

[221666-29-1]

$C_{36}H_{30}O_4$ 526.632

Dimer of 4,7′-Epoxy-3,8′-ligna-7,7′-dien-4′-ol, E61. Constit. of the aerial parts of *Piper aequale*. Viscous oil. λ_{max} 242 (log ε 5.06); 312 (log ε 4.76) (MeOH).

Maxwell, A. *et al., Phytochemistry*, 1999, **50**, 499-504 (*Piper aequale constit*)

7,9′:7′,9-Diepoxy-3,3′,4,4′-tetramethoxy-8,8′-lignan D104

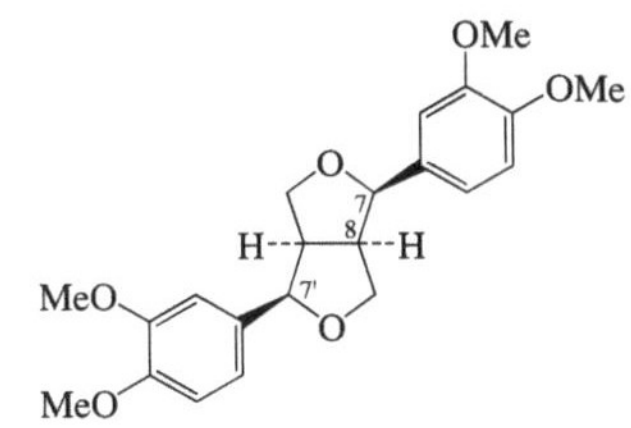

(7R,7′R,8R,8′R)-form

1,4-Bis(3,4-dimethoxyphenyl) tetrahydro-1H,3H-furo[3,4-c]furan, CAS. *2,6-Bis(3,4-dimethoxyphenyl)-3,7-dioxabicyclo[3.3.0]octane. 3,3′,4,4′-Tetramethoxy-7,9′;7′,9-diepoxylignan*

$C_{22}H_{26}O_6$ 386.444

Lignan numbering shown

(7R,7′R,8R,8′R)-form

Diaeudesmin [16499-02-8]

Isol. from *Piper peepuloides*, *Zanthoxylum acanthopodium* and *Zanthoxylum podocarpum*. Mp 160 - 162°. $[\alpha]_D^{22}$ +338.4 (c, 0.6 in $CHCl_3$)

(7R,7′R,8S,8′S)-form

(-)-Eudesmin [526-06-7]

Lignan from the kino of *Eucalyptus hemiphloia*, also from *Liriodendron tulipifera*, *Artemisia gorgonum* and *Orophea enneandra*. Prisms. Sol. MeOH, Et_2O; poorly sol. H_2O. Mp 107°. $[\alpha]_D$ -64.3 (Me_2CO). $[\alpha]_D$ -73.8 (AcOH). λ_{max} 232; 279 (MeOH) (Berdy)

(7R,7′S,8R,8′R)-form

(+)-Epieudesmin [60102-89-8]

Isol. from *Virola sebifera*, *Artemisia absinthium*, *Hernandia nymphaefolia*, *Hernandia peltata* and *Magnolia kobus*. Moderate cancer cell growth inhibitor. Mp 124 - 125°. $[\alpha]_D^{20}$ +117 (c, 0.4 in $CHCl_3$)

(7S,7′R,8S,8′S)-form

(-)-Epieudesmin [4375-03-5]

Constit. of *Asarum* spp., *Carphephorus odoratissimus*, *Magnolia kobus*, *Orophea enneandra* and many other plants. Cryst.. Mp 133 - 134°. $[\alpha]_D^{24}$ -144.8 (c, 2 in $CHCl_3$)

(7S,7′S,8R,8′R)-form

(+)-Eudesmin. Di-*O*-methylpinoresinol. Humbertiol‡ [29106-36-3]

Constit. of the aerial parts of *Haplophyllum acutifolium* and the unripe seeds of *Haplophyllum perforatum* and from *Araucaria angustifolia, Araucaria araucana, Pseuderanthemum carruthersii, Fagara heitzii, Magnolia fargesii, Magnolia kobus* var. *borealis, Magnolia sieboldii* cells and many other plant sources. Inhibits tumour necrosis factor-α prodn. and T-cell proliferation. Moderate inhibitor of HeLa cell growth. Activator of endothelial histamine H_1 receptors causing vascular relaxation in rat. Suppressor of phagocyte oxidative burst. Mp 107 - 108°. $[\alpha]_D^{20}$ +60.5 (c, 2.38 in Me_2CO)

(7RS,7′RS,8SR,8′SR)-form

(±)-Eudesmin [38759-91-0]

Constit. of *Carphephorus odoratissimus* and *Parabenzoin praecox*. Shows insect antifeedant activity. Cryst. ($MeOH/CHCl_3$). Mp 99 - 100°.

Erdtman, H. *Justus Liebigs Ann. Chem.*, 1935, **516**, 162-176 (*Eudesmin, struct*).
Pelter, A. *J. Chem. Soc. C*, 1967, 1376-1380 (*ms*).
Atal, C.K. *et al., J. Chem. Soc. C*, 1967, 2228-2230 (*Diaeudesmin*).
Ayres, D.C. *et al., Tetrahedron*, 1969, **25**, 4093-4098 (*epimer, struct*).
Wahlberg, I. *et al., Acta Chem. Scand.*, 1972, **26**, 1383-1388 (*(+)-Eudesmin, (+)-Epieudesmin, struct*).
Razzakova, D.M. *et al., Chem. Nat. Compd. (Engl. Transl.)*, 1972, **8**, 646-647 (*Haplophyllum acutifolium constit*).
Munakata, K. *ACS Symp. Ser.*, 1977, **62**, 185-196 (*Parabenzoin praecox constit, activity*).
Deshpande, V.H. *et al., Indian J. Chem., Sect. B*, 1977, **15**, 95-96 (*Eudesmin, struct*).
Pelter, A. *et al., Tetrahedron Lett.*, 1977, **18**, 4137-4140 (*Eudesmin, Epieudesmin, cmr*).
Greger, H. *et al., Tetrahedron*, 1980, **36**, 3551-3558 (*(+)-Epieudesmin, abs config*).
Pelter, A. *et al., J. Chem. Soc., Perkin Trans. 1*, 1982, 175-181 (*(±)-Eudesmin, synth*).
Iida, T. *et al., Phytochemistry*, 1982, **21**, 673-675 (*7-epimer, struct*).
Razakova, D.M. *et al., Chem. Nat. Compd. (Engl. Transl.)*, 1985, **21**, 259 (*Haplophyllum perforatum constit*).
Vasquez, M. *et al., Acta Crystallogr., Sect. C*, 1990, **46**, 342-344 (*(+)-Eudesmin, cryst struct*).
Jansen, J.F.G.A. *et al., Tetrahedron Lett.*, 1991, **32**, 3239-3242 (*(-)-Eudesmin, synth*).
Van Oeveren, A. *et al., J. Org. Chem.*, 1994, **59**, 5999-6007 (*(-)-Eudesmin, synth*).
Suginome, H. *et al., J. Org. Chem.*, 1995, **60**, 3052-3064 (*(±)-Eudesmin, synth, pmr*).
Miyazawa, M. *et al., Phytochemistry*, 1997, **39**, 1027-1030 (*(+)-Eudesmin, metab*).
Miyauchi, T. *et al., Phytochemistry*, 1998, **47**, 665-670 (*Magnolia kobus var. borealis constit, enzymatic formation*).
Parmar, V.S. *et al., Phytochemistry*, 1998, **49**, 1069-1078 (*Piper peepuloides constit*).
Cho, J.Y. *et al., Arch. Pharmacal Res.*, 1999, **22**, 348-353 (*Magnolia fargesii constit, activity*).
Bouyssi, D. *et al., Curr. Org. Chem.*, 2001, **12**, 1570-1587 (*(+)-Epidesmin, synth*).
Roy, S.C. *et al., J. Org. Chem.*, 2002, **67**, 3242-3248 (*(±)-Eudesmin, synth*).
Estrada-Reyes, R. *et al., Z. Naturforsch., C*, 2002, **57**, 29-31 (*(+)-Epieudesmin, pmr, cmr*).
Waibel, R. *et al., Phytochemistry*, 2003, **62**, 805-811 (*(±)-Eudesmin, synth*).
Kise, N. *et al., Tetrahedron: Asymmetry*, 2003, **14**, 2495-2497 (*(-)-Eudesmin, synth*).
Pettit, G.R. *et al., J. Nat. Prod.*, 2004, **67**, 214-220 (*(+)-Epieudesmin, cryst struct*).
Swain, N.A. *et al., J. Org. Chem.*, 2004, **69**, 122-129 (*(+)-Epieudesmin, synth*).
Hwang, S.J. *Korean J. Biotechnol. Bioeng.*, 2005, **20**, 34-39 (*(+)-Eudesmin, prodn*).
Céspedes, C.L. *et al., Z. Naturforsch., C*, 2006, **61**, 35-43 (*Araucaria aurucana constit*).
Lee, J. *et al., Chem. Pharm. Bull.*, 2009, **57**, 298-301 (*Epieudesmin*).
Raimundo, J.M. *et al., Eur. J. Pharmacol.*, 2009, **606**, 150-154 (*Eudesmin, vascular relaxation activity*).
Mbaze, L.M. *et al., Phytochemistry*, 2009, **70**, 1442-1447 (*(+)-Eudesmin, oxidative burst suppression*).
Anfimov, A.N. *et al., Russ. Chem. Bull. (Engl. Transl.)*, 2011, **60**, 2336-2342 (*(±)-Epieudesmin, (±)-Diaeudesmin, synth*).
Vo, T.N. *et al., Chem. Pharm. Bull.*, 2012, **60**, 1125-1133 (*Pseuderanthemum carruthersii constit, Hela cell growth inhibitor*)

8,9:8′,9′-Diepoxy-3,3′,5,5′-tetramethoxy-4,4′-lignan-7,7′-diol D105

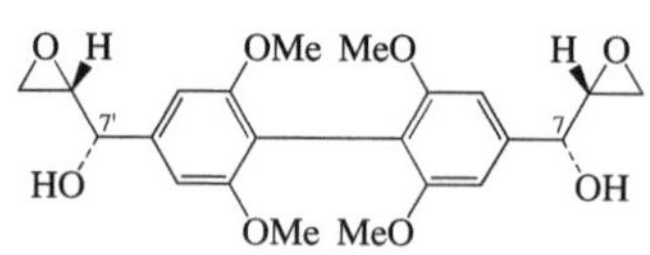

(7R*,7′S*,8R*,8′S*)-form

4,4′-Bis(2,3-epoxy-1-hydroxypropyl)-2,2′,6,6′-tetramethoxybiphenyl.
8,9:8′,9′-Diepoxy-7,7′-dihydroxy-3,3′,5,5′-tetramethoxy-4,4′-neolignan

$C_{22}H_{26}O_8$ 418.442

(7R*,7′S*,8R*,8′S*)-form

[346616-64-6]

Constit. of *Cynanchum atratum, Millettia griffithii* and *Patrinia villosa*. Needles ($CHCl_3$). Mp 175 - 176°. $[\alpha]_D^{25}$ +75 (c, 0.11 in $CHCl_3$)

3,3′-Bis(demethoxy)

8,9:8′,9′-Diepoxy-3,3′-dimethoxy-4,4′-lignan-7,7′-diol.
8,9:8′,9′-Diepoxy-7,7′-dihydroxy-3,3′-dimethoxy-4,4′-lignan.
Griffilignan A
[2007967-88-4]
$C_{20}H_{22}O_6$ 358.39
Constit. of the stems of *Millettia griffithii*. Pale yellow oil. $[\alpha]_D^{25}$ +63.2 (c, 0.1 in $CHCl_3$). λ_{max} 224 (log ε 3.96); 278 (log ε 3.16) (MeOH)

(7R*,7′S*,8S*,8′R*)-form

Constit. of *Patrinia villosa*. Needles.

Day, S.-H. *et al., J. Nat. Prod.*, 2001, **64**, 608-611 (*Cynanchum atratum constit*).
Tang, H. *et al., Bioorg. Med. Chem. Lett.*, 2016, **26**, 4417-4422 (*Griffilignan A*).
Yan, X.-J. *et al., Pharmacogn. Mag.*, 2016, **12**, 1-3 (*Patrinia villosa constits*)

4,7′:4′,7″-Diepoxy-3″,5,5′-trimethoxy-9′-nor-3,8′:3′,8″-sesquilign-7′-ene-4″,9,9″-triol D106

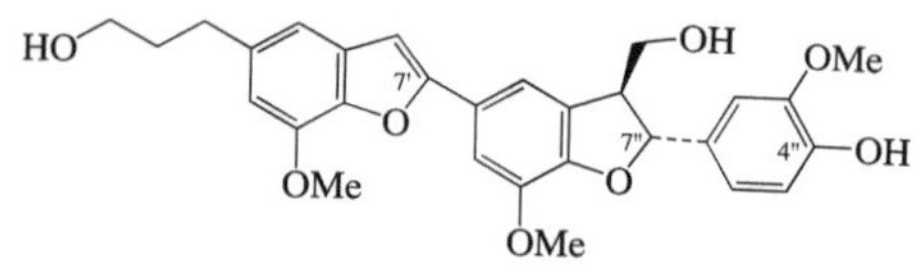

Relative Configuration

2′,3′-Dihydro-2′-(4-hydroxy-3-methoxyphenyl)-3′-(hydroxymethyl)-7,7′-dimethoxy-[2,5′-bibenzofuran]-5-propanol, CAS

[879498-22-3]

$C_{29}H_{30}O_8$ 506.551

Lignan numbering shown. Constit. of the seeds of *Styrax perkinsiae*. Amorph. powder. $[\alpha]_D^{20}$ -90.4 (c, 0.13 in MeOH). λ_{max} 222 (log ε 3.84); 316 (log ε 3.81) (MeOH).

Li, Q.-L. *et al., Planta Med.*, 2005, **71**, 847-851 (*Styrax perkinsiae constit*)

7′,9′:8,9-Diepoxy-3,4′,5-trimethoxy-4,8′-oxylignan-3′,7-diol D107

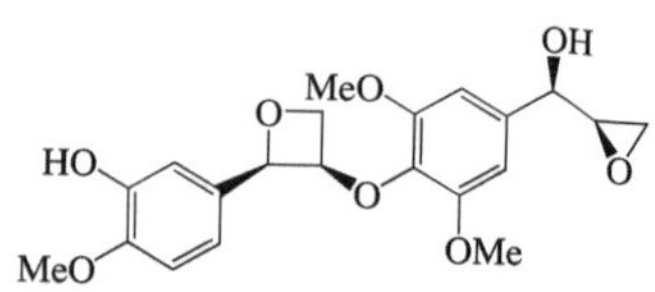

Relative Configuration

7′,9′:8,9-Diepoxy-3′,7-dihydroxy-3,4′,5-trimethoxy-4,8′-oxyneolignan

[1356601-68-7]

$C_{21}H_{24}O_8$ 404.415

Constit. of *Isodon coetsa*. Amorph. powder (MeOH). $[\alpha]_D^{24}$ +34 (c, 0.68 in $MeOH/CHCl_3$).

Zhao, W. *et al., Arch. Pharmacal Res.*, 2011, **34**, 2007-2014 (*Isodon coetsa constit*)

4′,7:4″,7′-Diepoxy-3,5′,5″-trimethoxy-3′,8:3″,8′-sesquilign-7″-ene-4,9,9′,9″-tetrol D108

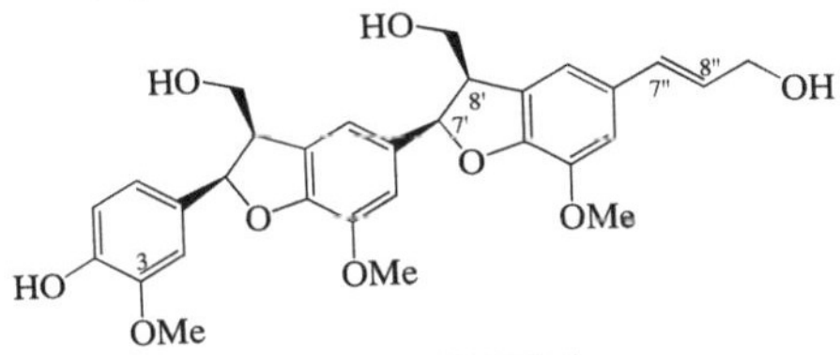

(7R,7′R,7″E,8R,8′R)-form

$C_{30}H_{32}O_9$ 536.577
Stereochem. of this series is confused in the lit.
(7R,7′R,7″E,8R,8′R)-form
7″,8″-Dihydro
Illiciumlignan A
[2226510-55-8]
$C_{30}H_{34}O_9$ 538.593
Constit. of the branches and leaves of *Illicium wardii*. Yellowish oil. $[\alpha]^{20}_{D}$ +3.0 (c, 0.1 in MeOH). λ_{max} 217 (log ε 2.05); 281 (log ε 0.54) (MeOH)
(7R,7′R,7″E,8S,8′S)-form
3-Demethoxy
Sinensiol B
$C_{29}H_{30}O_8$ 506.551
Constit. of *Selaginella sinensis*. Amorph. pale yellow powder. λ_{max} 210 (log ε 3.67); 276 (log ε 2.16) (MeOH)
3-Demethoxy, 7″,8″-dihydro, 9″-carboxylic acid
Sinensiol D
$C_{29}H_{30}O_9$ 522.550
Constit. of *Selaginella sinensis*. Amorph. powder. λ_{max} 208 (log ε 3.16); 283 (log ε 2.07) (MeOH)
(7R*,7′R*,7″E,8S*,8′S*)-form
Herpetotriol [69985-36-0]
Constit. of *Cestrum parqui* and *Herpetospermum caudigerum*. Mp 175 - 176°. [α] D -20 (MeOH)
9″-Aldehyde
Picrasmalignan A
[1314868-59-1]
$C_{30}H_{30}O_9$ 534.561
Constit. of *Cirsium eriophorum* and *Picrasma quassioides*. Amorph. powder. $[\alpha]^{25}_{D}$ -13.2 (c, 0.5 in MeOH). May be enantiomeric with Herpetotriol. λ_{max} 208 (log ε 3.95); 283 (log ε 3.08); 337 (log ε 3.15) (MeOH)
7″,8″-Dihydro
Vitrifol A
[1111080-82-0]
$C_{30}H_{34}O_9$ 538.593
Constit. of the fruit of *Vitex trifolia*. Pale yellow oil. $[\alpha]^{28}_{D}$ +5.9 (c, 0.2 in MeOH). λ_{max} 205 (log ε 4.84); 282 (log ε 3.8); 357 (log ε 2.51) (MeOH)
7″,8″-Dihydro, 4-O-α-L-rhamnopyranoside
Difengpioside A
[2041284-97-1]
$C_{36}H_{44}O_{13}$ 684.735
Constit. of the stem bark of *Illicium difengpi*. Amorph. powder. $[\alpha]^{20}_{D}$ +25 (c, 0.2 in MeOH). λ_{max} 281 (log ε 3.55) (MeOH)
7′,8′-Didehydro
7′,8′-Didehydroherpetotriol
[1161572-47-9]
$C_{30}H_{30}O_9$ 534.561
Constit. of the branches of *Dipteronia dyeriana*. Pale yellow powder. $[\alpha]^{25}_{D}$ +54 (c, 0.5 in MeOH). λ_{max} 234 (log ε 4.21); 279 (log ε 4.15); 300 (log ε 4.11) (MeOH)
3-Demethoxy, 7″,8″-dihydro
Sinensiol A
[1011729-87-5]
$C_{29}H_{32}O_8$ 508.567
Constit. of *Selaginella sinensis*. Solid. Mp 204 - 206°. $[\alpha]^{20}_{D}$ +4 (c, 0.11 in MeOH). λ_{max} 210 (MeOH)
3-Demethoxy, 7″,8″-dihydro, 4-O-β-D-glucopyranoside
Sinensioside A
$C_{35}H_{42}O_{13}$ 670.708
Constit. of *Selaginella sinensis*. Amorph. powder. λ_{max} 207 (log ε 4.35) (MeOH)
(7R*,7′R*,7″Z,8S*,8′S*)-form
[959699-39-9]
Constit. of the leaves of *Cestrum parqui*. Oil. $[\alpha]^{25}_{D}$ -22.3 (c, 0.26 in MeOH).

Favre-Bonvin, J. *et al., Tetrahedron Lett.*, 1978, **19**, 4111-4112 (*Herpetotriol*).
D'Abrosca, B. *et al., Nat. Prod. Res.*, 2006, **20**, 293-298 (*Cestrum parqui constits*).
Wang, Y.Z. *et al., Chin. Chem. Lett.*, 2007, **18**, 1224-1226 (*Sinensiol A*).
Guo, R. *et al., Chin. Chem. Lett.*, 2008, **19**, 1215-1217 (*7′,8′-Didehydroherpetotriol*).
Gu, Q. *et al., J. Asian Nat. Prod. Res.*, 2008, **10**, 499-502 (*Vitrifol A*).
Jiao, W.-H. *et al., Chem. Biodiversity*, 2011, **8**, 1163-1169 (*Picrasmalignan A*).
Chen, H. *et al., Chin. J. Nat. Med.*, 2014, **12**, 148-150 (*Sinensioside A*).
Sólyomváry, A. *et al., Process Biochem.*, 2015, **50**, 853-858 (*Picrasmalignan A*).
Pan, Z.-H. *et al., Molecules*, 2016, **21**, 607:1-8 (*Difengpioside A*).
Ye, F.-M. *et al., Phytochem. Lett.*, 2016, **17**, 263-269 (*Illiciumlignan A*).
Fan, X.-D. *et al., Yaoxue Xuebao*, 2016, **51**, 770-774 (*Herpetotriol, 7′,8′-Didehydroherpetotriol*).
Chen, X. *et al., Fitoterapia*, 2019, **134**, 256-263 (*Sinensiols B,D*)

5,6′-Diethyl-2,3,3′,4′-biphenyltetrol D109

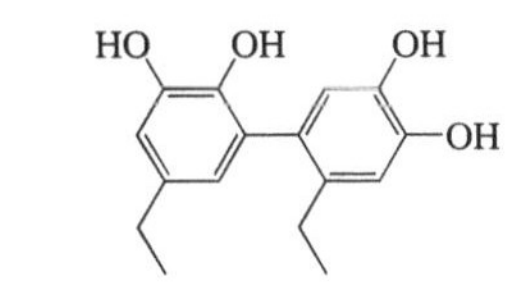

2′,5-Diethyl-2,3,4′,5′-tetrahydroxybiphenyl. 9,9′-Dinor-2,3′-lignan-4,4′,5,5′-tetrol. 4,4′,5,5′-Tetrahydroxy-9,9′-dinor-2,3′-neolignan. ***Glechomol A***
[1439547-44-0]
$C_{16}H_{18}O_4$ 274.316
Isol. from *Glechoma longituba*.
Displays significant protective effects on H_2O_2-induced cardiac cell injury. Brown oil. $[\alpha]^{15}_{D}$ -7.8 (c, 0.087 in MeOH). λ_{max} 217 (log ε 4.20); 285 (log ε 3.57) (MeOH).

Zhu, Q.-F. *et al., J. Asian Nat. Prod. Res.*, 2013, **15**, 258-264 (*Glechomol A, activity*)

6,6′-Diethyl-3,3′,4,4′-biphenyltetrol D110

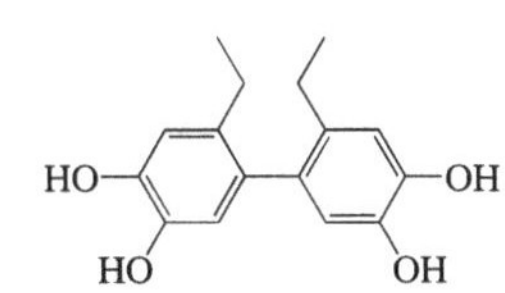

2,2′-Diethyl-4,4′,5,5′-tetrahydroxybiphenyl. 9,9′-Dinor-2,2′-lignan-4,4′,5,5′-tetrol. 4,4′,5,5′-Tetrahydroxy-9,9′-dinor-2,2′-neolignan. ***Glechomol B***
[1439547-52-0]
$C_{16}H_{18}O_4$ 274.316
Isol. from *Glechoma longituba*.
Displays significant protective effects on H_2O_2-induced cardiac cell injury. Brown oil. $[\alpha]^{15}_{D}$ +4.5 (c, 0.086 in MeOH). λ_{max} 215 (log ε 4.13); 288 (log ε 3.51) (MeOH).

Zhu, Q.-F. *et al., J. Asian Nat. Prod. Res.*, 2013, **15**, 258-264 (*Glechomol B, activity*)

Difengpiol B D111

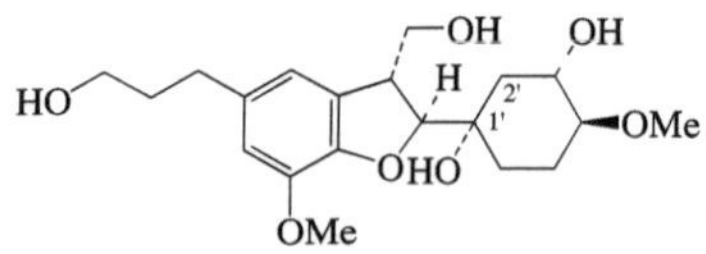
Absolute Configuration

4,7′-Epoxy-1′,2′,3′,4′,5′,6′-hexahydro-4′,5-dimethoxy-3,8′-lignan-1′,3′,9,9′-tetrol.
4,7′-Epoxy-1′,2′,3′,4′,5′,6′-hexahydro-1′,3′,9,9′-tetrahydroxy-3,8′-neolignan
[1227162-59-5]
$C_{20}H_{30}O_7$ 382.453
Constit. of the stem bark of *Illicium difengpi*. Amorph. powder. $[\alpha]_D^{20}$+27 (c, 0.05 in MeOH). λ_{max} 209 (log ε 4.4); 284 (log ε 3.6) (MeOH)

1′-Deoxy, 1′,2′-didehydro, O^4′-de-Me
4,7′-Epoxy-3′,4′,5′,6′-tetrahydro-5-methoxy-3,8′-lignan-3′,4′,9,9′-tetrol.
4,7′-Epoxy-2′,3′,4′,5′-tetrahydro-4′,5′,9,9′-tetrahydroxy-3,8′-lignan. ***Difengpiol A***
[1227162-58-4]
$C_{19}H_{26}O_6$ 350.411
Constit. of the stem bark of *Illicium difengpi*. Amorph. powder. $[\alpha]_D^{20}$ +45 (c, 0.06 in MeOH). λ_{max} 211 (log ε 4.5); 285 (log ε 3.7) (MeOH).

Fang, L. *et al.*, *J. Nat. Prod.*, 2010, **73**, 818-824 *(Difengpiols A,B)*

Dihydrobuddlenol B D112

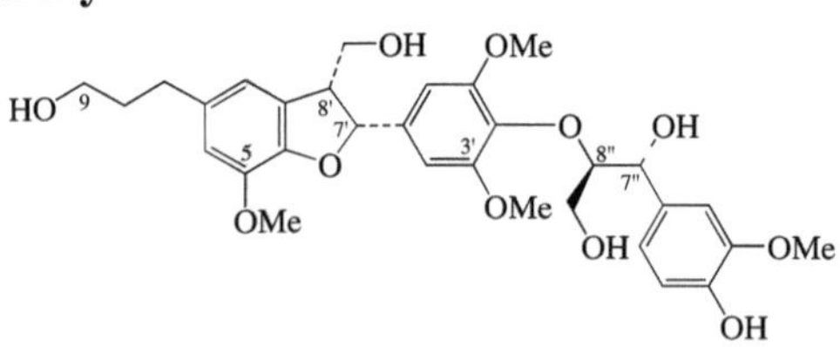
(7′R,7″R,8′R,8″R)-form

[128502-85-2]
$C_{31}H_{38}O_{11}$ 586.634
Abs. config. is uncertain and many isolates are not accompanied by stereochem. data. Compds. more recently isol. have had their abs. and/or rel. configs. assigned and these are listed separately below. Dihydrobuddlenol B usually assigned (7′*R**,8′*S**)-stereochem. with either *threo*-(7″*R**,8″*R**)- or *erythro*-(7″*R**,8″*S**)-stereochem.. Constit. of *Fraxinus rhyncophylla, Illicium verum, Prunus jamasakura, Rhododendron alutaceum* and *Triadica sebifera*. Amorph. powder. $[\alpha]_D^{29}$-28.2 (c, 0.43 in MeOH). λ_{max} 232 (sh); 279 (log ε 3.9) (MeOH)

(7′R,7″R,8′R,8″R)-form

3′-Demethoxy, 9-Ac
Illiciumlignan B
[2226508-77-4]
$C_{32}H_{38}O_{11}$ 598.645
Constit. of the branches and leaves of *Illicium wardii*. Yellowish solid. Mp 105.4 - 106.2°. $[\alpha]_D^{20}$-11.2 (c, 0.13 in MeOH). λ_{max} 214 (log ε 1.95); 230 (log ε 1.51); 282 (log ε 0.55) (MeOH)

(7′R,7″R,8′R,8″S)-form

Constit. of *Phyllanthus glaucus*. $[\alpha]_D^{20}$ +22.4 (c, 0.11 in MeOH). Probably the same as Acernikol. λ_{max} 211 (log ε 4.71); 230 (log ε 4.28); 280 (log ε 3.72) (MeOH)

(7′R,7″R,8′S,8″R)-form

(+)-Phyllanglaucin C [2245722-61-4]
Constit. of *Phyllanthus glaucus*. $[\alpha]_D^{20}$ -11.5 (c, 0.05 in MeOH). $[\alpha]_D^{20}$+7.5 (c, 0.13 in MeOH). There is a conflict of opt. rotation measurements. λ_{max} 210 (log ε 4.70); 230 (log ε 4.27); 280 (log ε 3.69) (MeOH)

9-O-(4-Hydroxy-3-methoxy-E-cinnamoyl)
Sambucasinol B
[1890274-33-5]
$C_{41}H_{46}O_{14}$ 762.805
Constit. of the twigs of *Sambucus williamsii*. Exhibits cytotoxic activity. Yellowish gum. $[\alpha]_D^{25}$+1.2 (c, 0.2 in MeOH). λ_{max} 213 (log ε 4.1); 235 (log ε 2.6); 288 (log ε 1.3); 324 (log ε 1.6) (MeOH)

7R,8R*-Dihydroxy*
Crataeganin A
$C_{31}H_{38}O_{13}$ 618.632
Constit. of the seeds of *Crataegus pinnatifida*. Exhibits significant neuroprotective activity. Light yellow oil. $[\alpha]_D^{20}$-45.2 (c, 0.07 in MeOH). λ_{max} 215 (log ε 2.26); 280 (log ε 0.45) (MeOH)

7R,8S*-Dihydroxy*
Crataeganin C
Constit. of the seeds of *Crataegus pinnatifida*. Exhibits significant neuroprotective activity. Light yellow oil. $[\alpha]_D^{20}$-30.3 (c, 0.23 in MeOH). λ_{max} 213 (log ε 2.54); 280 (log ε 0.71) (MeOH)

7R,8R*-Dihydroxy, 7″-Me ether*
Crataeganin E
$C_{32}H_{40}O_{13}$ 632.659
Constit. of the seeds of *Crataegus pinnatifida*. Exhibits significant neuroprotective activity. Light yellow oil. $[\alpha]_D^{20}$-55.5 (c, 0.15 in MeOH). λ_{max} 220 (log ε 2.91); 280 (log ε 0.84) (MeOH)

7R,8S*-Dihydroxy, 7″-Me ether*
Crataeganin F
$C_{32}H_{40}O_{13}$ 632.659
Constit. of the seeds of *Crataegus pinnatifida*. Exhibits strong neuroprotective activity. Light yellow oil. $[\alpha]_D^{20}$-63.7 (c, 0.12 in MeOH). λ_{max} 215 (log ε 2.51); 280 (log ε 0.46) (MeOH)

(7′R,7″R,8′S,8″S)-form

9-O-(4-Hydroxy-3-methoxy-E-cinnamoyl)
Sambucasinol A
[1890274-32-4]
$C_{41}H_{46}O_{14}$ 762.805
Constit. of the twigs of *Sambucus williamsii*. Exhibits cytotoxic activity. Yellowish gum. $[\alpha]_D^{25}$-6.4 (c, 0.5 in MeOH). λ_{max} 207 (log ε 2.5); 236 (log ε 0.8); 287 (log ε 0.5); 325 (log ε 0.6) (MeOH)

9-O-(4-Hydroxy-3-methoxy-Z-cinnamoyl)
Sambucasinol C
[1890274-34-6]
$C_{41}H_{46}O_{14}$ 762.805
Constit. of the twigs of *Sambucus williamsii*. Exhibits cytotoxic activity. Yellowish gum. $[\alpha]_D^{25}$+1.6 (c, 0.2 in MeOH). λ_{max} 212 (log ε 3.7); 235 (log ε 2.7); 287 (log ε 0.8); 325 (log ε 1.1) (MeOH)

(7′R,7″S,8′R,8″R)-form

Constit. of *Phyllanthus glaucus*. $[\alpha]_D^{20}$ +10.7 (c, 0.11 in MeOH). λ_{max} 211 (log ε 4.72); 230 (log ε 4.33); 281 (log ε 3.74) (MeOH)

3′-Demethoxy
Illiciumlignan E
[2226511-33-5]
$C_{30}H_{36}O_{10}$ 556.608
Constit. of the branches and leaves of *Illicium wardii*. Exhibits moderate inhibitory activity against human tumour cell line SKOV3. Yellowish solid. Mp 90.2 - 91°. $[\alpha]_D^{20}$-0.55 (c, 0.91 in MeOH). λ_{max} 214 (log ε 2.01); 230 (log ε 1.72); 283 (log ε 0.62) (MeOH)

(7′R,7″S,8′R,8″S)-form
(-)-Phyllanglaucin D
Constit. of *Phyllanthus glaucus*. $[\alpha]_D^{20}$ -13 (c, 0.18 in MeOH). λ_{max} 211 (log ε 4.60); 230 (log ε 4.22); 280 (log ε 3.64) (MeOH)
(7′R,7″S,8′S,8″R)-form
4″-Me ether
Berbikonol
[1306741-02-5]
$C_{32}H_{40}O_{11}$ 600.661
Constit. of the trunk of *Berberis koreana*. Exhibits cytotoxic activity. Yellowish gum. $[\alpha]_D^{25}$ +84 (c, 0.8 in MeOH). λ_{max} 232 (log ε 3.8); 281 (log ε 2.1) (MeOH)
7R,8R*-Dihydroxy*
Crataeganin B
$C_{31}H_{38}O_{13}$ 618.632
Constit. of the seeds of *Crataegus pinnatifida*. Exhibits significant neuroprotective activity. Light yellow oil. $[\alpha]_D^{20}$ -23.9 (c, 0.24 in MeOH). λ_{max} 213 (log ε 1.99); 280 (log ε 0.32) (MeOH)
7R,8R*-Dihydroxy, 7″-Me ether*
Crataeganin D
$C_{32}H_{40}O_{13}$ 632.659
Constit. of the seeds of *Crataegus pinnatifida*. Exhibits significant neuroprotective activity. Light yellow oil. $[\alpha]_D^{20}$ -21.7 (c, 0.23 in MeOH). λ_{max} 215 (log ε 2.58); 280 (log ε 0.69) (MeOH)
7R,8S*-Dihydroxy, 7″-Me ether*
Crataeganin G
$C_{32}H_{40}O_{13}$ 632.659
Constit. of the seeds of *Crataegus pinnatifida*. Exhibits significant neuroprotective activity. Light yellow oil. $[\alpha]_D^{20}$ -41.3 (c, 0.24 in MeOH). λ_{max} 215 (log ε 2.36); 280 (log ε 0.35) (MeOH)
3′-Demethoxy
Xanthiumnolic C
[2226004-46-0]
Constit. of the fruit of *Xanthium sibiricum*. NO prodn. inhibitor. Amorph. powder. $[\alpha]_D^{20}$ -37.5 (c, 0.1 in MeOH). λ_{max} 341 (log ε 3.83) (MeOH)
3′-Demethoxy, 4″-O-β-D-glucopyranoside
Xanthiumnolic D
[2226004-47-1]
$C_{36}H_{46}O_{15}$ 718.749
Constit. of the fruit of *Xanthium sibiricum*. NO prodn. inhibitor. Amorph. powder. $[\alpha]_D^{20}$ -46.7 (c, 0.1 in MeOH). λ_{max} 238 (log ε 4.02) (MeOH)
(7′S,7″R,8′R,8″R)-form
(+)-Phyllanglaucin D. Rubuslin B
Constit. of *Phyllanthus glaucus* and *Rubus idaeus*. $[\alpha]_D^{20}$ -40.5 (c, 0.05 in MeOH). $[\alpha]_D^{20}$ +14.4 (c, 0.24 in MeOH). There is a conflict of opt. rotation measurements
5″-Methoxy, 7″-deoxy
Rubuslin A
[2242395-24-8]
$C_{32}H_{40}O_{11}$ 600.661
Constit. of the rhizomes of *Rubus idaeus*. Potent antioxidant. Yellow oil
(7′S,7″R,8′S,8″S)-form
Constit. of *Phyllanthus glaucus*. $[\alpha]_D^{20}$ -10.2 (c, 0.13 in MeOH)
(7′S,7″S,8′R,8″S)-form
(-)-Phyllanglaucin C. Rubuslin C
Constit. of *Phyllanthus glaucus* and *Rubus idaeus*. Yellow oil. $[\alpha]_D^{20}$ -7.4 (c, 0.24 in MeOH). $[\alpha]_D^{20}$ +13.2 (c, 0.05 in MeOH). There is a conflict of opt. rotation measurements
(7′S,7″S,8′S,8″R)-form
Constit. of *Phyllanthus glaucus*. $[\alpha]_D^{20}$ -20 (c, 0.12 in MeOH)
(7′R,7″R,8′R,8″R*)-form*
threo-**Acernikol**
Constit. of *Acer saccharum* syrup. Amorph. yellow powder
(7′R,7″R,8′R,8″S*)-form*
Acernikol [623584-31-6]
Constit. of *Acer nikoense*, *Acer saccharum*, *Illicium simonsii* and *Schima superba*. Powder. $[\alpha]_D^{22}$ +4.7 (c, 0.2 in MeOH). Stereochem. of most isolates named Acernikol is not fully established. λ_{max} 235 (sh); 281 (log ε 3.77) (EtOH)
4″-O-β-D-Glucopyranoside
Acernikol 4″-glucoside
[1350550-88-7]
$C_{37}H_{48}O_{16}$ 748.775
Constit. of the leaves of *Crataegus pinnatifida*. Oil. $[\alpha]_D^{20}$ -15 (c, 0.8 in MeOH)
5-O-De-Me
Alutaceuol
[1416057-02-7]
$C_{30}H_{36}O_{11}$ 572.607
Constit. of the leaves of *Rhododendron alutaceum*. Amorph. powder. $[\alpha]_D^{22}$ -4.6 (c, 0.5 in MeOH). Stereochem. in doubt due to inconsistencies in ref.
(7′R,7″R*,8′S*,8″R*)-form*
9′-O-β-D-Glucopyranoside
Alangisesquin D
[212007-45-9]
$C_{37}H_{48}O_{16}$ 748.775
Constit. of *Alangium premnifolium*. Powder. $[\alpha]_D^{24}$ -3.2 (c, 0.4 in MeOH). λ_{max} 229 (log ε 4.33); 280 (log ε 3.75) (MeOH)
(7′R,7″R*,8′S*,8″S*)-form*
9′-O-β-D-Glucopyranoside
Alangisesquin C
[212007-44-8]
$C_{37}H_{48}O_{16}$ 748.775
Constit. of *Alangium premnifolium*. Powder. $[\alpha]_D^{24}$ -8.7 (c, 0.6 in MeOH). λ_{max} 228 (log ε 4.34); 281 (log ε 3.77) (MeOH)
(7′R,7″ξ,8′R*,8″ξ)-form*
5-O-De-Me, 9-O-(4-hydroxy-3-methoxy-E-cinnamoyl)
Buddlenol G
[873694-47-4]
$C_{40}H_{44}O_{14}$ 748.778
Constit. of the stems of *Sambucus williamsii*. Yellowish oil. $[\alpha]_D^{23}$ -25.2 (c, 0.1 in MeOH). λ_{max} 286 (log ε 3.22); 324 (log ε 3.24) (MeOH)
(7′R,8′R*,8″ξ)-form*
7″-Deoxy
Euryalin A
[1268671-44-8]
$C_{31}H_{38}O_{10}$ 570.635
Constit. of the seeds of *Euryale ferox*. Amorph. solid. $[\alpha]_D^{23}$ -3.1 (c, 0.27 in MeOH). λ_{max} 205 (log ε 4.95); 235 (sh); 281 (log ε 3.82) (MeOH)
3′-Demethoxy, 7″-deoxy
Euryalin B
[1268671-45-9]
$C_{30}H_{36}O_9$ 540.609
Constit. of the seeds of *Euryale ferox*. Amorph. solid. $[\alpha]_D^{23}$ -2.4 (c, 0.22 in MeOH). λ_{max} 205 (log ε 4.89); 228 (sh); 282 (log ε 3.9) (MeOH)
(7′ξ,7″ξ,8′ξ,8″ξ)-form
3′-Demethoxy
Leptolepisol A
[71332-92-8]
$C_{30}H_{36}O_{10}$ 556.608
Similar to Buddlenol B, B82. Constit. of *Illicium simonsii*, *Larix leptolepis*, *Trochodendron*

aralioides and *Vitex rotundifolia*. Shows antioxidant activity. $[\alpha]_D^{27}$ -3.1 (c, 1 in MeOH). λ_{max} 281 (ε 6676) (EtOH).

Miki, K. *et al.*, *Tetrahedron Lett.*, 1979, **20**, 799-802 (*Leptolepisol A*).
Yoshinari, K. *et al.*, *Phytochemistry*, 1990, **29**, 1675-1678 (*Dihydrobuddlenol B*).
Ono, M. *et al.*, *Food Sci. Technol. Int.*, 1998, **4**, 9-13 (*Leptolepisol A, activity*).
Kijima, K. *et al.*, *Phytochemistry*, 1998, **48**, 669-676 (*Alangisesquins C,D*).
Morikawa, T. *et al.*, *Chem. Pharm. Bull.*, 2003, **51**, 62-67 (*Acernikol*).
Yang, X.-J. *et al.*, *J. Asian Nat. Prod. Res.*, 2007, **9**, 583-591 (*Buddlenol G*).
Gao, P.-Y. *et al.*, *Biochem. Syst. Ecol.*, 2010, **38**, 988-992 (*Acernikol 4″-glucoside*).
Kim, K.H. *et al.*, *Bioorg. Med. Chem. Lett.*, 2011, **21**, 2270-2273 (*Berbikonol*).
Song, C.-W. *et al.*, *J. Agric. Food Chem.*, 2011, **59**, 1199-1204 (*Euryalins A,B*).
Li, L. *et al.*, *J. Agric. Food Chem.*, 2011, **59**, 7708-7716 (*Acer saccharum constits*).
Li, H.-Z. *et al.*, *Arch. Pharmacal Res.*, 2012, **35**, 1887-1893 (*Alutaceuol, Dihydrobuddlenol B*).
Suh, W.S. *et al.*, *Bioorg. Med. Chem. Lett.*, 2016, **26**, 1877-1880 (*Sambucasinols A,B,C*).
Ye, F.-M. *et al.*, *Phytochem. Lett.*, 2016, **17**, 263-269 (*Illiciumlignans B,E, cd, activity*).
Wu, Z. *et al.*, *Sci. Rep.*, 2016, **6**, 24809:1-13 (*Acernikols, Phyllanglaucins C,D*).
Jiang, H. *et al.*, *Fitoterapia*, 2017, **117**, 11-15 (*Xanthiumnolics C,D, activity*).
Xu, Y. *et al.*, *J. Funct. Foods*, 2017, **32**, 160-169 (*Rubuslins A,B,C, activity*).
Huang, X.-Y. *et al.*, *Fitoterapia*, 2018, **125**, 6-12 (*Crataeganins A-G, activity*).
Yu, M. *et al.*, *Bioorg. Chem.*, 2020, **103**, 104147:1-6 (*Triadica sebifera constits, abs config*).
Chang, C.-C. *et al.*, *Fitoterapia*, 2020, **144**, 104455:1-7 (*Leptolepisol A*)

7″-Dihydro-5′-demethoxycavaol H D113

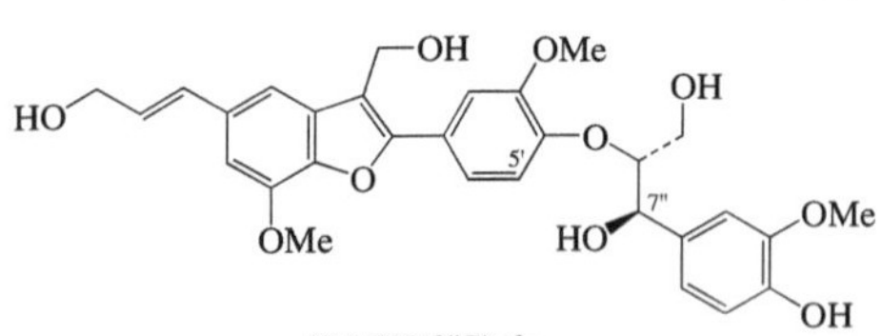

(7E,7″R,8″R)-form

$C_{30}H_{32}O_{10}$ 552.576

Related to Buddlenol B, B82

(7E,7″R,8″R)-form [1245737-46-5]

Constit. of the twigs of *Eurycorymbus cavaleriei*. Yellow gum. $[\alpha]_D^{25}$ -10.7 (c, 0.1 in MeOH)

(7E,7″R,8″S)-form [1245737-47-6]

Constit. of the twigs of *Eurycorymbus cavaleriei*. Yellow gum. $[\alpha]_D^{25}$ -17.7 (c, 0.1 in MeOH)

(7E,7″R*,8″S*)-form

5′-Methoxy

erythro-***7′,8′-Didehydrobuddlenol B***

$C_{31}H_{34}O_{11}$ 582.602

Constit. of the stems of *Dipteronia dyeriana*

(7E,8″ξ)-form

7″-Ketone, 5′-methoxy

Cavaol H

[1245817-30-4]

$C_{31}H_{32}O_{11}$ 580.586

Constit. of the twigs of *Eurycorymbus cavaleriei*. Amorph. powder. $[\alpha]_D^{25}$ -11.3 (c, 0.1 in MeOH).

Ma, Z. *et al.*, *Fitoterapia*, 2009, **80**, 320-326 (*Cavaol H, 7″-Dihydro-5′-demethoxycavaol H*).
Guo, R. *et al.*, *Tianran Chanwu Yanjiu Yu Kaifa*, 2012, **24**, 1007-1013 (*7′,8′-Didehydrobuddlenol B*)

Dihydro-4-(3,4-dihydroxybenzoyl)-2(3*H*)-furanone D114

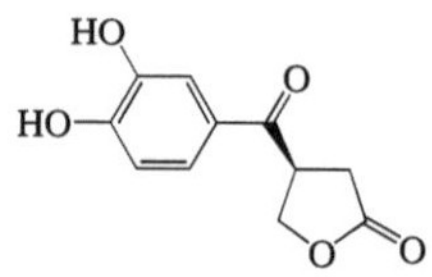

$C_{11}H_{10}O_5$ 222.196

(S)-form

3′-Me ether

Dihydro-4-(4-hydroxy-3-methoxybenzoyl)-2(3H)-furanone

[1584653-02-0]

$C_{12}H_{12}O_5$ 236.223

Constit. of *Croton kongensis*. Yellowish oil. $[\alpha]_D^{20}$ -4 (c, 0.05 in MeOH). λ_{max} 209; 228; 281 (MeOH).

Sun, L. *et al.*, *Nat. Prod. Res.*, 2014, **28**, 563-567 (*Croton kongensis constit*)

3,4-Dihydro-6,7-dihydroxy-2,3-bis(hydroxymethyl)-1(2*H*)-naphthalenone D115

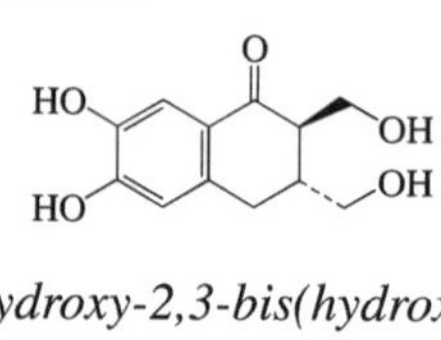

6,7-Dihydroxy-2,3-bis(hydroxymethyl)-1-tetralone

$C_{12}H_{14}O_5$ 238.239

Degraded lignan

(2R,3R)-form

6-Me ether

3,4-Dihydro-7-hydroxy-2,3-bis(hydroxymethyl)-6-methoxy-1(2H)-naphthalenone

[1259948-76-9]

$C_{13}H_{16}O_5$ 252.266

Constit. of the wood of *Taxus yunnanensis*. Amorph. brown solid. $[\alpha]_D^{22}$ -46.6 (c, 0.2 in MeOH).

Tezuka, Y. *et al.*, *J. Nat. Prod.*, 2011, **74**, 102-105 (*Taxus yunnanensis constit*)

3,4-Dihydro-6,7-dihydroxy-2,3-dimethyl-1(2*H*)-naphthalenone D116

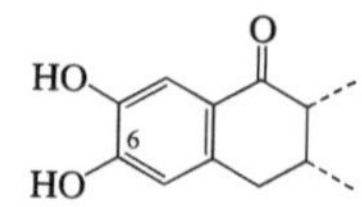

(2R,3R)-form

6,7-Dihydroxy-2,3-dimethyl-1-tetralone. 4,5-Dihydroxy-1′,2′,3′,4′,5′,6′-hexanor-2,7′-cyclolignan-7-one

$C_{12}H_{14}O_3$ 206.241

(2R,3R)-form

6-Me ether

3,4-Dihydro-7-hydroxy-6-methoxy-2,3-dimethyl-1(2H)-naphthalenone

[1310453-59-8]

$C_{13}H_{16}O_3$ 220.268

Constit. of the bark of *Machilus robusta*. Amorph. solid. $[\alpha]_D^{20}$ +90 (c, 0.03 in $CHCl_3$). λ_{max} 207 (log ε 4.72); 233 (log ε 4.62); 274 (log ε 4.55); 315 (log ε 4.17) (MeOH)

(2R,3S)-form

6-Me ether

[1310453-58-7]

Constit. of the bark of *Machilus robusta*. Needles ($CHCl_3$). Mp 159 - 161°. $[\alpha]_D^{20}$ +74.3 (c, 0.01 in $CHCl_3$)

(2ξ,3ξ)-form

6-Me ether

Cinnamophilin D

[1392325-85-7]

$C_{13}H_{16}O_3$ 220.268

Constit. of the leaves of *Cinnamomum philippinense*. Gum. λ_{max} 240 (log ε 3.12); 276 (log ε 2.34) (MeCN).

Li, Y. *et al.*, *J. Nat. Prod.*, 2011, **74**, 1444-1452 (*Machilus robusta constits*).
Cheng, M.-J. *et al.*, *Nat. Prod. Res.*, 2012, **26**, 1433-1435 (*Cinnamophilin D*)

2,3-Dihydro-2,7-dihydroxy-3-(hydroxymethyl)-5-benzofuranpropanol D117

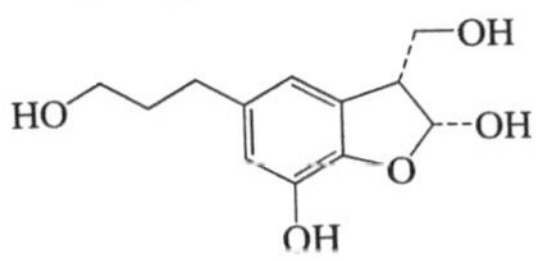

(2R,3R)-form

2,3-Dihydro-3-(hydroxymethyl)-5-(3-hydroxypropyl)benzofuran-2,7-diol
$C_{12}H_{16}O_5$ 240.255
(2R,3R)-form
2-Et, 7-Me ether
[1620489-36-2]
$C_{15}H_{22}O_5$ 282.336
Constit. of the stems of *Rhododendron mariae*. Amorph. powder. $[\alpha]^{20}_{D}$-0.9 (c, 0.05 in MeOH). λ_{max} 285 (MeOH)
(2S,3S)-form
2-Et, 7-Me ether
[1620489-37-3]
Constit. of the stems of *Rhododendron mariae*. Amorph. powder. $[\alpha]^{20}_{D}$+0.06 (c, 0.02 in MeOH). λ_{max} 285 (MeOH).
Guo, Q. *et al.*, *J. Asian Nat. Prod. Res.*, 2014, **16**, 1-10 (*Rhododendron mariae constits*)

2,3-Dihydro-5,7-dimethoxy-3-methyl-2-(3,4-methylenedioxyphenyl)-6-(2-propenyloxy)benzofuran D118

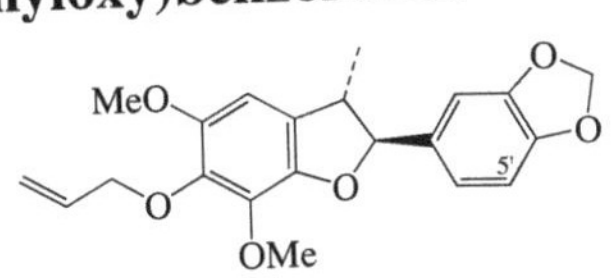

[109194-66-3]
$C_{21}H_{22}O_6$ 370.401
Constit. of *Ocotea catharinensis* and *Ocotea veraguensis*. Plates (EtOAc/petrol). Mp 68 - 70°. $[\alpha]^{21}_{D}$-16 (c, 0.1 in MeOH). λ_{max} 232; 288 (MeOH)
5′-Methoxy
2,3-Dihydro-5,7-dimethoxy-2-(5-methoxy-3,4-methylenedioxyphenyl)-3-methyl-6-(2-propenyloxy)benzofuran
[139906-13-1]
$C_{22}H_{24}O_7$ 400.427
Constit. of *Ocotea catharinensis*.
Khan, M.R. *et al.*, *Phytochemistry*, 1987, **26**, 1155-1158 (*Ocotea veraguensis constit*).
Lordello, A.L.L. *et al.*, *Phytochemistry*, 1997, **46**, 741-744 (*Ocotea catharinensis constits*)

3,4-Dihydro-2,2-dimethyl-5-(2-propenyl)-2*H*-1-benzopyran-3,6,7-triol D119

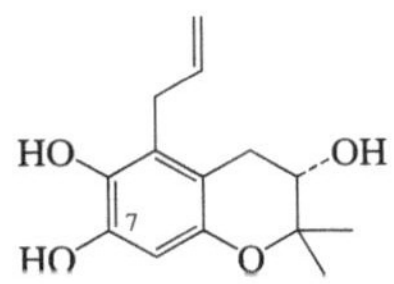

5-Allyl-3,4-dihydro-3,6,7-trihydroxy-2,2-dimethyl-2H-1-benzopyran
$C_{14}H_{18}O_4$ 250.294
Prenylated lignan monomer
(S)-form
7-Me ether
3,4-Dihydro-7-methoxy-2,2-dimethyl-5-(2-propenyl)-2H-1-benzopyran-3,6-diol, CAS. *5-Allyl-3,4-dihydro-3,6-dihydroxy-7-methoxy-2,2-dimethyl-2H-1-benzopyran*. ***Illihenryipyranol A***
[1418206-89-9]
$C_{15}H_{20}O_4$ 264.321
Constit. of roots of *Illicium henryi*. Exhibits strong antioxidant activity. Amorph. powder. $[\alpha]^{20}_{D}$-18.8 (c, 0.53 in MeOH). λ_{max} 207 (log ε 3.04); 289 (log ε 2.31) (MeOH)
7-Et ether
Burmanicumol C
[2052959-93-8]
$C_{16}H_{22}O_4$ 278.348
Constit. of the stem bark of *Illicium burmanicum*. Moderately potent inhibitor of the NF-κβ pathway. Cryst. (MeOH). $[\alpha]^{24}_{D}$-5.4 (c, 2.71 in $CHCl_3$).
Zhuang, P.-Y. *et al.*, *Phytochemistry*, 2013, **86**, 176-183 (*Illihenryipyranol A, activity*).
Huang, D. *et al.*, *Fitoterapia*, 2015, **107**, 22-28 (*Burmanicumol C, activity*)

2,3-Dihydro-2-(4-hydroxy-2,3-dimethoxyphenyl)-9-(5-hydroxy-2-methoxyphenyl)-3-(hydroxymethyl)-6,8-dimethoxy-7*H*-1,4-dioxino[2,3-*h*]chromen-7-one D120

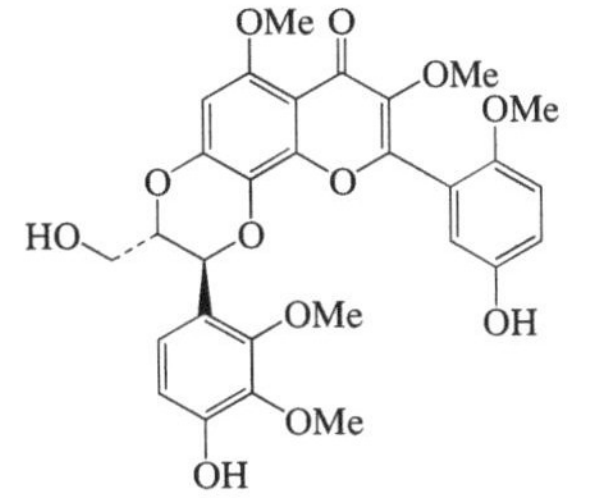

[162663-69-6]
$C_{29}H_{28}O_{12}$ 568.531
Flavonolignan. Constit. of the heartwood of *Distemonanthus benthamianus*. Amorph. powder (as tri-Ac). λ_{max} 274 (ε 14120); 331 (ε 9770) (MeOH) (tri-Ac).
Malan, E. *et al.*, *Phytochemistry*, 1994, **37**, 1771-1772 (*Distemonanthus benthamianus constit*)

Dihydro-3-hydroxy-5-(4-hydroxy-3-methoxyphenyl)-3,4-dimethyl-2(3*H*)-furanone D121

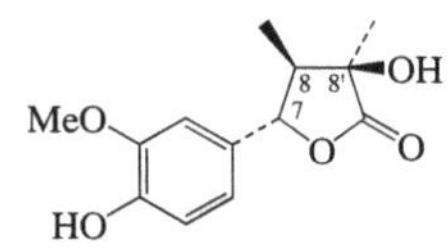

Absolute Configuration

$C_{13}H_{16}O_5$ 252.266
Lignan numbering shown
(7S,8R,8′R)-form [1196997-44-0]
Constit. of the bark of *Machilus wangchiana*. Moderate inhibitor of galactosamine induced cytotoxicity. Amorph. powder. $[\alpha]^{20}_{D}$+14 (c, 0.05 in $CHCl_3$). λ_{max} 203 (log ε 4.2); 231 (log ε 2.9); 280 (log ε 2.46) (MeOH)
4-Me ether
Clinopodphenol B
[1453810-40-6]
$C_{14}H_{18}O_5$ 266.293
Constit. of *Clinopodium urticifolium*. Exhibits modest anti-HIV-1 and anti-tobacco mosaic virus activity
3,4-Methylene ether, 3-O-de-Me
Nicotlactone A
[1384932-96-0]
$C_{13}H_{14}O_5$ 250.25
Isol. from leaves of *Nicotiana tabacum*. Shows modest anti-HIV-1 activity and high anti-tobacco mosaic virus activity. Pale yellow gum.$[\alpha]^{24.8}_{D}$+15.8 (c, 0.25 in MeOH). λ_{max} 205 (log ε 4.35); 235 (log ε 4.54); 282 (log ε 2.22) (MeOH)
3-O-De-Me, 5-methoxy, 3,4-methylene ether
Nicotnorlignan C
[1452306-05-6]
$C_{14}H_{16}O_6$ 280.276
Constit. of *Nicotiana tabacum*. Exhibits modest anti-tobacco mosaic virus activity. Pale yellow gum. $[\alpha]^{24}_{D}$+18.2 (c, 0.22 in MeOH). λ_{max} 210 (log ε 4.41); 238 (log ε 3.16); 280 (log ε 2.86) (MeOH).

Cheng, W. *et al.*, *J. Nat. Prod.*, 2009, **72**, 2145-2152 (*Machilus wangchiana constit*).
Liao, Z. *et al.*, *Asian J. Chem.*, 2012, **24**, 4895-4897 (*Nicotnorlignan C*).
Gao, X. *et al.*, *Heterocycles*, 2012, **85**, 147-153 (*Nicotlactone A, activity*).
Pan, C. *et al.*, *Asian J. Chem.*, 2013, **25**, 1860-1862 (*Clinopodphenol B, activity*).
Krishna, P.R. *et al.*, *Tetrahedron Lett.*, 2013, **54**, 669-671 (*synth*)

2,3-Dihydro-5-hydroxy-2-(4-hydroxy-3-methoxyphenyl)-3-(hydroxymethyl)-7-(2-hydroxyphenyl)-10-methyl-2*H*,6*H*-benzo[1,2-*b*:5,4-*b'*]dipyran-6-one D122

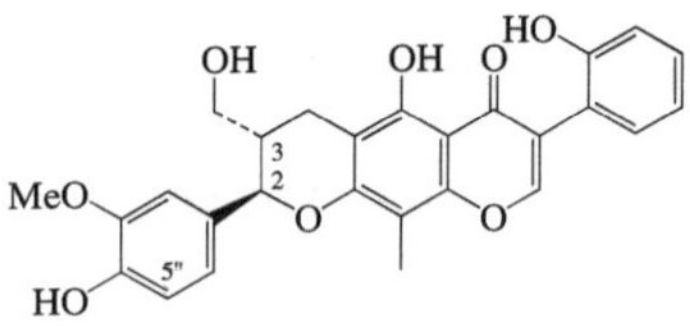

(2*R*,3*S*)-*form*

$C_{27}H_{24}O_8$ 476.481
The nat. prods. were isol. as racemates, resolved by chiral hplc and their diastereoisomers were subsequently named as Pisonones A-D
(2R,3S)-form
Pisonone B
Oil
5″-Methoxy
Pisonone D
Yellow oil
(2S,3R)-form
Pisonone A
Oil
5″-Methoxy
Pisonone C
Yellow oil
(2RS,3SR)-form
Constit. of the stems of *Pisonia umbellifera*. Oil. λ_{max} 237 (log ε 0.36); 271 (log ε 0.12) (MeOH)
5″-Methoxy
$C_{28}H_{26}O_9$ 506.507
Constit. of the stems of *Pisonia umbellifera*. Yellow oil. λ_{max} 239 (log ε 0.47); 270 (log ε 0.11) (MeOH).

Liu, Z. *et al.*, *RSC Adv.*, 2018, **8**, 16383-16391 (*Pisonones A-D, cd*)

2,3-Dihydro-5-hydroxy-2-(4-hydroxyphenyl)-3-benzofuranmethanol D123

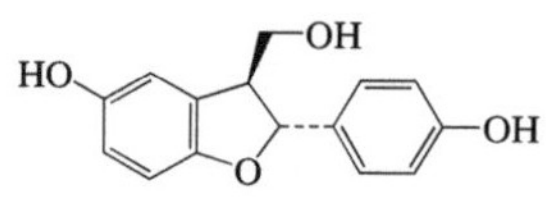

(2*R*,3*S*)-*form*

$C_{15}H_{14}O_4$ 258.273
The nat. prod. Rubussin A was isol. as a partial racemate and resolved by chiral hplc
(2R,3S)-form
(+)-Rubussin A [2247479-95-2]
Oil. $[\alpha]_D^{20}$ +44.8 (c, 0.1 in MeOH)
(2S,3R)-form
(-)-Rubussin A [2247479-97-4]
Exhibits anti $A\beta_{1-42}$ aggregation activity (enantioselective). Oil. $[\alpha]_D^{20}$ -47.6 (c, 0.11 in MeOH)
(2RS,3SR)-form
Rubussin A
Constit. of the fruit of *Rubus ideaus*. Oil. $[\alpha]_D^{20}$ +3.9 (c, 0.25 in MeOH). λ_{max} 207 (log ε 3.80); 227 (log ε 2.49); 286 (log ε 0.65); 302 (log ε 0.61) (MeOH).

Zhou, L. *et al.*, *Bioorg. Chem.*, 2018, **80**, 64-69 (*Rubussin A, cd, activity*)

2-[2,3-Dihydro-3-(hydroxymethyl)-2-phenyl-1,4-benzodioxin-6-yl]-5-hydroxy-3-(4-pentylbenzyloxy)-7-(4-propylbenzyloxy)flavan D124

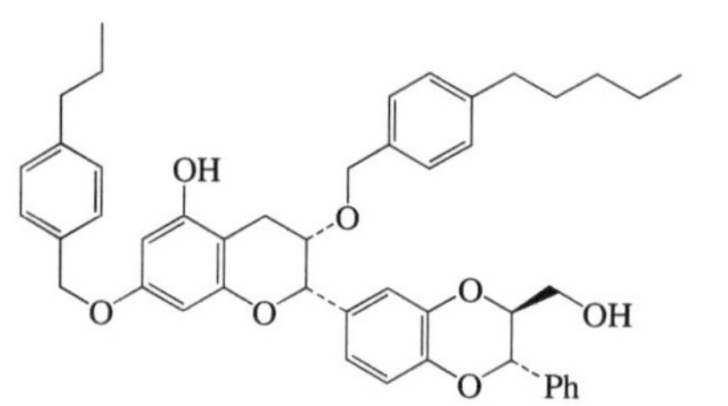

[2095503-90-3]
$C_{46}H_{50}O_7$ 714.899
Flavonolignan. Constit. of the stem bark of *Newtonia griffoniana*. Powder.

Kinyok, M.J. *et al.*, *Nat. Prod. Res.*, 2017, **31**, 2233-2238 (*Newtonia griffoniana constit*)

2,3-Dihydro-2-(4-hydroxyphenyl)-6-[2-(4-hydroxyphenyl)ethenyl]-4(1*H*)-pyridinone D125

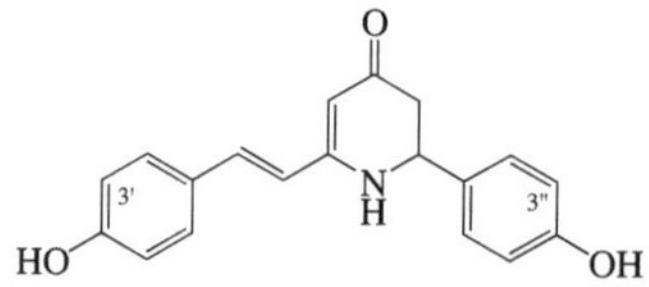

2,3-Dihydro-2-(4-hydroxyphenyl)-6-(4-hydroxystyryl)-4(1H)-pyridinone
$C_{19}H_{17}NO_3$ 307.349
(E,ξ)-form [1221559-37-0]
Alkaloid from the rhizomes of *Curcuma longa*. Yellow powder. Prob. artifact
3′-Methoxy
$C_{20}H_{19}NO_4$ 337.375
Alkaloid from the rhizomes of *Curcuma longa*. Yellow powder. Mp 194 - 196°. Prob. artifact
3′,3″-Dimethoxy
[1221559-35-8]
$C_{21}H_{21}NO_5$ 367.401
Alkaloid from the rhizomes of *Curcuma longa*. Yellow powder. Prob. artifact.

Huo, Y.P. *et al.*, *Chin. Chem. Lett.*, 2009, **20**, 1291-1295 (*synth*).
Bar, F.M.A. *et al.*, *Rec. Nat. Prod.*, 2016, **10**, 582-592 (*Curcuma longa constits*)

2,3-Dihydro-2-(4-hydroxyphenyl)-3-methyl-5-benzofuranol D126

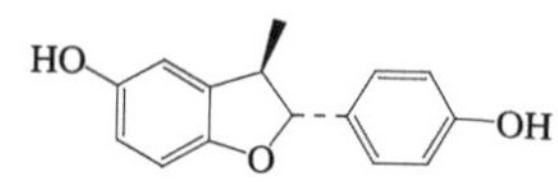

2,3-Dihydro-5-hydroxy-2-(4-hydroxyphenyl)-3-methylbenzofuran
$C_{15}H_{14}O_3$ 242.274
(2R,3R)-form
[119322-05-3] (4′-Ac)
Constit. of *Krameria triandra*. Glassy solid. Mp 50 - 55°. $[\alpha]_D$ +159 (c, 0.1 in $CHCl_3$). λ_{max} 223 (ε 15000); 280 (ε 3500); 298 (ε 4500) (EtOH).

Arnone, A. *et al.*, *Gazz. Chim. Ital.*, 1988, **118**, 675-682 (*Krameria triandra constit*)

4*a*,9*b*-Dihydro-2-[2-hydroxy-5-(2-propenyl)phenyl]-8,9*b*-di-2-propenyl-3(4*H*)-dibenzofuranone, CAS D127

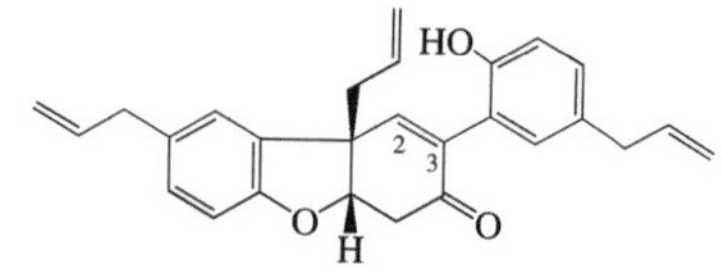

Relative Configuration

[185110-95-6]
$C_{27}H_{26}O_3$ 398.502
Sesquilignan. Related to Fargenone A, F3. Constit. of *Illicium dunnianum*. Gum

2,3α-Dihydro
Simonsol E
[1582261-55-9]
$C_{27}H_{28}O_3$ 400.518
Constit. of *Illicium simonsii*. Isol. as a mixt. with Simonsin A, S68.

Sy, L.-K. *et al.*, *Phytochemistry*, 1996, **43**, 1417-1419 (*Illicium dunnianum constit*).
Dong, C. *et al.*, *Planta Med.*, 2013, **79**, 338-347 (*Simonsol E*)

5,6-Dihydro-3′,4′,5,5′,7′-pentahydroxy-1,8′-lign-8-en-2(1*H*)-one D128

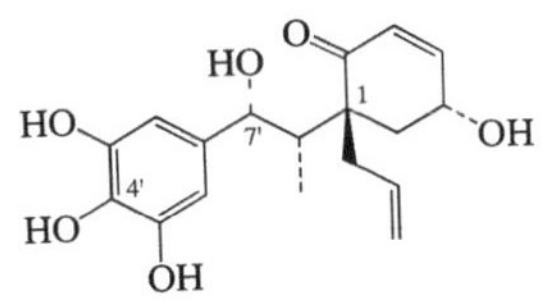

$C_{18}H_{22}O_6$ 334.368
(1R,5R,7′R,8′S)-form
3′,4′,5,5′-Tetra-Me ether
6-[2-Hydroxy-1-methyl-2-(3,4,5-trimethoxyphenyl)ethyl]-4-methoxy-6-(2-propenyl)-2-cyclohexen-1-one, CAS. *5,6-Dihydro-7′-hydroxy-3′,4′,5,5′-tetramethoxy-1,8′-lign-8-en-2(1H)-one.*
Megaphone
[64332-37-2]
[78148-79-5] (7′*RS*,8′*SR*-form)
[106973-35-7] (7′*R**,8′*S**-form)
$C_{22}H_{30}O_6$ 390.476
Constit. of the roots of *Aniba megaphylla*. Shows cytotoxic activity against human KB cancer cells. Cryst. ($CHCl_3/Et_2O$). Mp 151.5 - 152.5°. $[\alpha]^{27}_{D}$ -23 (c, 0.15 in EtOH). λ_{max} 269 (ε 501); 279 (sh) (EtOH) (Derep)
3′,4′,5,5′-Tetra-Me ether, 7′-Ac
Megaphone acetate*.*
Acetylmegaphone
[64332-38-3]
[78148-80-8] (7′*R**,8′*S**-form)
$C_{24}H_{32}O_7$ 432.513
Constit. of *Aniba megaphylla* and *Endlicheria dysodantha*. Shows cytotoxic activity against human KB cancer cells. Oil. $[\alpha]^{22}_{D}$ -2.4 (c, 0.29 in EtOH). λ_{max} 240 (ε 12652); 265 (ε 1957) (MeOH) (Berdy). [neutral] λ_{max} 269 (ε 1056) (EtOH) (Berdy)
3′,4′-Methylene, 5,5′-di-Me ether
5,6-Dihydro-7′-hydroxy-3′,5-dimethoxy-4′,5′-methylenedioxy-1,8′-lign-8-en-2(1H)-one.
Megaphyllone
$C_{21}H_{26}O_6$ 374.433
Not isol.
3′,4′-Methylene, 5,5′-di-Me ether, 7′-Ac
Megaphyllone acetate*.*
Acetylmegaphyllone
[64332-39-4]
[96948-26-4] (rel. config)
$C_{23}H_{28}O_7$ 416.47
Constit. of *Aniba megaphylla* and *Endlicheria dysodantha*. Cytotoxic agent. Oil. λ_{max} 275 (ε 1248) (EtOH) (Berdy).

Kupchan, S.M. *et al.*, *J. Org. Chem.*, 1978, **43**, 586-590 (*Megaphone, Megaphyllone, Acetylmegaphone*).
Büchi, G. *et al.*, *J. Am. Chem. Soc.*, 1981, **103**, 2718-2721 (*Megaphone, synth*).
Zoretic, P.A. *et al.*, *Tetrahedron Lett.*, 1983, **24**, 1125-1128 (*Megaphone, synth*).
Matsumoto, T. *et al.*, *Chem. Lett.*, 1984, 67-70 (*Megaphone, synth*).
Matsumoto, T. *et al.*, *Bull. Chem. Soc. Jpn.*, 1985, **58**, 346-351 (*Megaphyllone acetate, synth*).
Tomioka, K. *et al.*, *Tetrahedron Lett.*, 1985, **26**, 903-906 (*Megaphone, synth*).
Ma, W.-W. *et al.*, *J. Nat. Prod.*, 1991, **54**, 1153-1158 (*Endlicheria dysodantha constits*).
Ma, W.-W. *et al.*, *Heterocycles*, 1992, **34**, 5-11 (*Megaphyllone acetate, activity*)

3,4-Dihydro-2-(4-phenyl-3-butenyl)-2*H*-1-benzopyran-6,7-diol D129

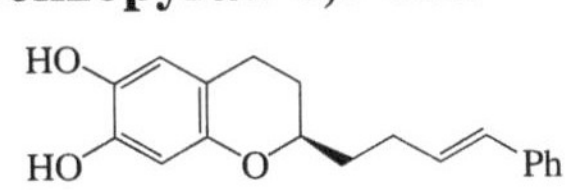

7-(3,4-Dihydroxyphenyl)-5,6′-epoxy-1-phenyl-1-heptene
$C_{19}H_{20}O_3$ 296.366
(R,E)-form
Constit. of the rhizomes of *Curcuma comosa*. Cryst. (EtOAc/hexane). Mp 52 - 53°. $[\alpha]^{29}_{D}$ -43.1 (c, 0.48 in EtOH).

Sornkaew, N. *et al.*, *Nat. Prod. Commun.*, 2015, **10**, 89-93 (*Curcuma comosa constit, cd*)

5,6-Dihydro-3′,4′,5,7′-tetrahydroxy-1,8′-lign-8-en-2(1*H*)-one D130

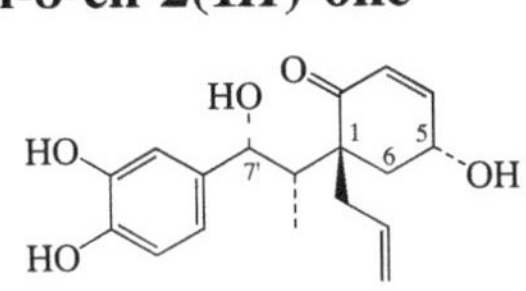

$C_{18}H_{22}O_5$ 318.369
(1R,5R,7′R,8′S)-form
3′,4′,5-Tri-Me ether
5,6-Dihydro-7′-hydroxy-3′,4′,5-trimethoxy-1,8′-lign-8-en-2(1H)-one
[106897-16-9]
$C_{21}H_{28}O_5$ 360.45
Constit. of the wood of *Ocotea porosa*. Cryst. ($CHCl_3/Et_2O$). Mp 149 - 151°
3′,4′-Methylene, 5-Me ether, 7′-Ac
7′-Acetoxy-5,6-dihydro-5-methoxy-3′,4′-methylenedioxy-1,8′-lign-8-en-2(1H)-one. ***Triandrin A***
[119087-41-1]
$C_{22}H_{26}O_6$ 386.444
Constit. of the seeds of *Licaria triandra*. Cryst. (C_6H_6). Mp 115 - 116°. $[\alpha]_D$ +12.2 (c, 3.8 in $CHCl_3$)
7′-Ketone
5,6-Dihydro-3′,4′,5-trihydroxy-1,8′-lign-8-ene-2,7′(1H)-dione
7′-Ketone, 3′,4′,5-tri-Me ether
5,6-Dihydro-3′,4′,5-trimethoxy-1,8′-lign-8-ene-2,7′(1H)-dione
[156665-51-9]
$C_{21}H_{26}O_5$ 358.434
Constit. of the wood of *Ocotea porosa*. Viscous oil. $[\alpha]_D$ -29.6 (c, 1 in $CHCl_3$). λ_{max} 236 (ε 23000); 274 (ε 21900); 302 (ε 12800) (MeOH)
3-Methoxy, 7′-ketone, 3′,4′,5-tri-Me ether
5,6-Dihydro-3,3′,4′,5-tetramethoxy-1,8′-lign-8-ene-2,7′(1H)-dione
[591752-72-6]
$C_{22}H_{28}O_6$ 388.46
Constit. of the leaves of *Ocotea catharinensis*. Viscous oil. $[\alpha]^{21}_{D}$ -17 (c, 0.92 in MeOH).

Dias, D.A. *et al.*, *Phytochemistry*, 1986, **25**, 2613-2616 (*Ocotea porosa tri-Me ether*).
De Carvalho, M.G. *et al.*, *Phytochemistry*, 1988, **27**, 2319-2323 (*Ocotea porosa 7′-ketone*).
Castro-Castillo, O. *et al.*, *Rev. Latinoam. Quim.*, 1988, **19**, 60-62 (*Triandrin A*).
Funasaki, M. *et al.*, *J. Braz. Chem. Soc.*, 2009, **20**, 853-859 (*Ocotea catharinensis constit*)

2′,3′-Dihydro-3′,4,4′,5-tetrahydroxy-2,3′-oxyligna-8,8′-dien-6′(1′*H*)-one D131

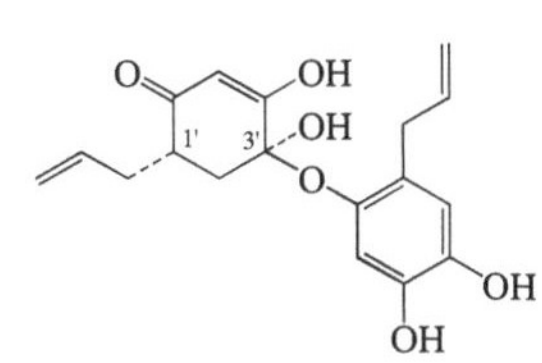

(1′*S*,3′*S*)-form

$C_{18}H_{20}O_6$ 332.352

(1′S,3′S)-form

Bis(methylene) ether

2′,3′-Dihydro-3′,4′:4,5-bis(methylenedioxy)-2,3′-oxyligna-8,8′-dien-6′(1′H)-one. ***Didymochlaenone B***

[877659-98-8]

$C_{20}H_{20}O_6$ 356.374

Constit. of the roots of *Didymochlaena truncatula*. Oil. $[\alpha]_D^{22}$ -27.5 (c, 0.08 in EtOH). λ_{max} 243 (log ε 4.08); 298 (log ε 3.59) (EtOH)

(1′ξ,3′ξ)-form

Tetra-Me ether

2′,3′-Dihydro-3′,4,4′,5-tetramethoxy-2,3′-oxyligna-8,8′-dien-6′(1′H)-one. ***Lancilin***

[74048-73-0]

$C_{22}H_{28}O_6$ 388.46

Constit. of the branches of *Aniba lancifolia*. Oil. λ_{max} 242 (ε 30700); 286 (ε 9050) (MeOH).

Diaz, P.S.P. *et al.*, *Phytochemistry*, 1980, **19**, 285-288 (*Lancilin*).

Cao, S. *et al.*, *J. Nat. Prod.*, 2006, **69**, 284-286 (*Didymochlaenone B*)

2,3-Dihydro-4,5,6-trihydroxy-2-(3,4,5-trihydroxyphenyl)-3-benzofuranmethanol D132

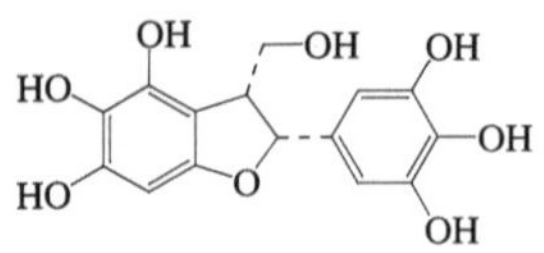

2,3-Dihydro-4,5,6-trihydroxy-3-(hydroxymethyl)-2-(3,4,5-trihydroxyphenyl)benzofuran

$C_{15}H_{14}O_8$ 322.269

(2R,3R*)-form*

3′,4,4′,5′,6-Penta-Me ether

2,3-Dihydro-5-hydroxy-4,6-dimethoxy-2-(3,4,5-trimethoxyphenyl)-3-benzofuranmethanol

[929637-33-2]

$C_{20}H_{24}O_8$ 392.404

Constit. of *Tarenna attenuata*. Shows potent antioxidant activity against H_2O_2-induced impairment in PC 12 cells. Amorph. powder. $[\alpha]_D^{19}$ +7.8 (c, 0.35 in MeOH). λ_{max} 207 (log ε 4.76); 282 (log ε 3.67); 371 (log ε 3.02) (MeOH).

Yang, X.-W. *et al.*, *J. Nat. Prod.*, 2007, **70**, 521-525 (*Tarenna attenuata constit, activity*)

1,2-Dihydro-3′,4′,5′-trimethoxy-3,4-methylenedioxy-3,8′-lign-8-en-6(3*H*)-one D133

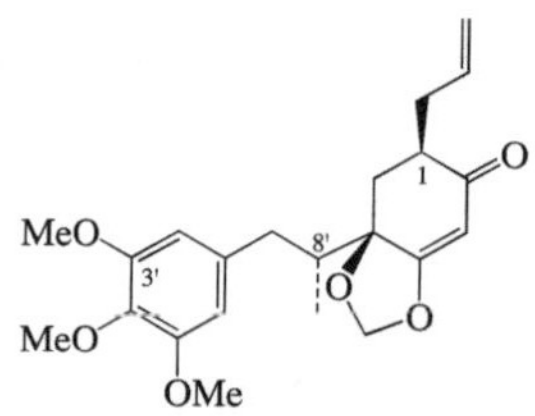

Relative Configuration

7,7a-Dihydro-7a-[1-methyl-2-(3,4,5-trimethoxyphenyl)ethyl]-6-(2-propenyl)-1,3-benzodioxol-5(6H)-one

[165132-58-1]

$C_{22}H_{28}O_6$ 388.46

Constit. of the stem bark of *Ocotea bullata*. Needles. Mp 105°. $[\alpha]_D^{25}$ +175 (c, 0.08 in CH_2Cl_2)

8′-Epimer

1-Epi-5′-methoxybifidenone

[2064109-53-9]

$C_{22}H_{28}O_6$ 388.460

Constit. of the leaves of a *Beilschmiedia* sp.

8′-Epimer, 3′-demethoxy

1-Epibifidenone

[2064109-51-7]

$C_{21}H_{26}O_5$ 358.434

Constit. of the leaves of a *Beilschmiedia* sp.

1,8′-Diepimer

5′-Methoxybifidenone

[2064109-52-8]

$C_{22}H_{28}O_6$ 388.460

Constit. of the leaves of a *Beilschmiedia* sp.

1,8′-Diepimer, 3′-demethoxy

Bifidenone

[2078008-57-6]

$C_{21}H_{26}O_5$ 358.434

Constit. of the leaves of a *Beilschmiedia* sp.. Tubulin polymerisation inhibitor. Induces apoptosis with activation of caspase 3/7.

Drewes, S.E. *et al.*, *Phytochemistry*, 1995, **38**, 1505-1508 (*Ocotea bullata constit*).

Williams, R.B. *et al.*, *J. Nat. Prod.*, 2017, **80**, 616-624 (*Beilschmiedia constits, activity*).

Huang, Z. *et al.*, *J. Org. Chem.*, 2017, **82**, 4235-4241 (*Bifidenone, synth*)

4-(3,4-Dihydroxybenzoyl)-1,2-dihydro-1,2,6,7-naphthalenetetrol D134

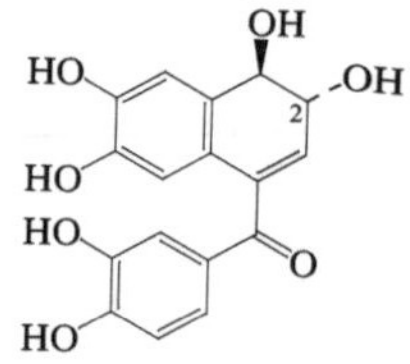

$C_{17}H_{14}O_7$ 330.292

(1R,2R*)-form*

2-O-β-D-Glucopyranoside

Crassifoside C

[910054-01-2]

$C_{23}H_{24}O_{12}$ 492.433

Constit. of the rhizomes of *Curculigo crassifolia*. Powder. $[\alpha]_D^{23}$ -50.5 (c, 0.12 in MeOH). λ_{max} 204 (log ε 3.95); 283 (log ε 4.45) (MeOH).

Li, N. *et al.*, *Z. Naturforsch., B*, 2006, **61**, 611-614 (*isol, pmr, cmr*)

1-(3,4-Dihydroxybenzoyl)-3,6,7-naphthalenetriol D135

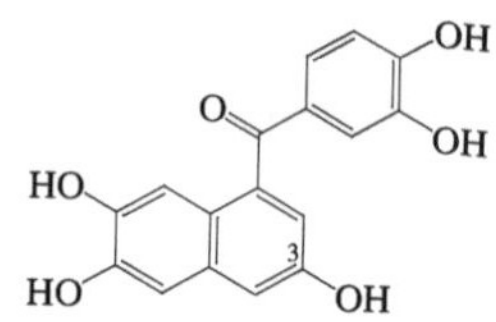

Crassifogenin B

[714231-20-6]

$C_{17}H_{12}O_6$ 312.277

Constit. of the rhizomes of *Curculigo breviscapa* and *Curculigo crassifolia*. Pale yellow powder. λ_{max} 233 (log ε 4.67); 285 (log ε 3.94); 324 (log ε 3.93) (MeOH)

3-O-β-D-Glucopyranoside

Crassifoside A

[714231-21-7]

$C_{23}H_{22}O_{11}$ 474.418

Constit. of the rhizomes of *Curculigo breviscapa* and *Curculigo crassifolia*. Powder. $[\alpha]_D^{29}$ -24.8 (c, 0.1 in MeOH). λ_{max} 234 (log ε 4.61); 285 (log ε 3.93); 324 (log ε 3.88) (MeOH).

Li, N. *et al.*, *Helv. Chim. Acta*, 2004, **87**, 845-850 (*Crassifosides A,B*).

Zhu, C.-C. *et al.*, *Bull. Korean Chem. Soc.*, 2010, **31**, 224 (*Crassifoside A*)

4-(3,4-Dihydroxybenzyl)dihydro-3-(hydroxymethyl)furan-2(3*H*)-one D136

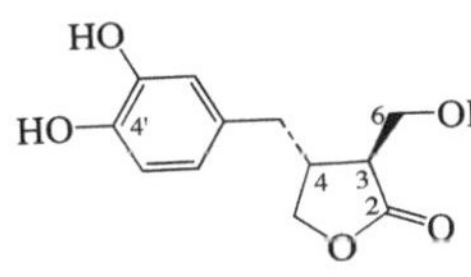

(3*S*,4*R*)-*form*

4-[(3,4-Dihydroxyphenyl)methyl] dihydro-3-(hydroxymethyl)-2(3H)-furanone

$C_{12}H_{14}O_5$ 238.239

(3S,4R)-form

3′,4′-Di-Me ether, 6-O-β-D-glucopyranoside

Arctiiapolignan

[1808085-02-0]

$C_{20}H_{28}O_{10}$ 428.434

Constit. of the fruits of *Arctium lappa*. Amorph. powder. $[\alpha]^{25}_{D}$ +6.97 (c, 0.18 in MeOH). λ_{max} 230 (log ε 3.1); 280 (log ε 2.70) (MeOH)

(3R*,4S*)-form

3′-Me ether

Maninsigin B

[1497273-95-6]

$C_{13}H_{16}O_5$ 252.266

Constit. of *Manglietia insignis*. Yellow oil. $[\alpha]^{25}_{D}$ -0.8 (c, 1.42 in MeOH). λ_{max} 204 (log ε 4.4); 226 (log ε 3.58); 281 (log ε 3.24) (MeOH).

Shang, S.-Z. *et al.*, *Fitoterapia*, 2013, **84**, 58-63 (*Maninsigin B*).
Yang, Y.-N. *et al.*, *J. Agric. Food Chem.*, 2015, **63**, 7958-7966 (*Arctiiapolignan, activity, cd*)

2-(3,4-Dihydroxybenzylidene)-3-(3,4-dihydroxyphenyl)-4-hydroxypentanedioic acid D137

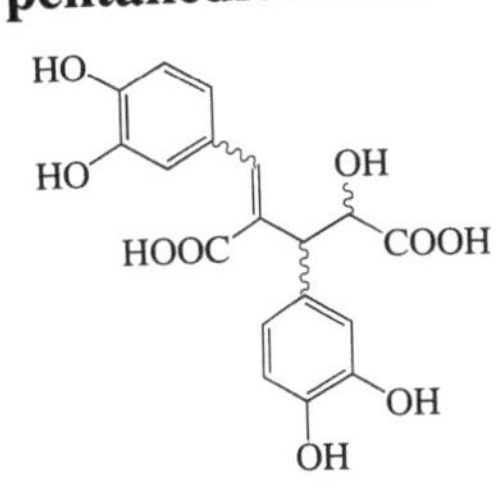

3,3′,4,4′,8-Pentahydroxy-7,8′-lign-7′-ene-9,9′-dioic acid

[1133062-60-8]

$C_{18}H_{16}O_9$ 376.317

Constit. of *Plectranthus amboinicus*. Shows opt. activity.

Chen, Y.-S. *et al.*, *Bioorg. Med. Chem.*, 2014, **22**, 1766-1772 (*Plectranthus amboinicus constit, cd, activity*)

6-*O*-(2,5-Dihydroxycinnamoyl)glucose D138

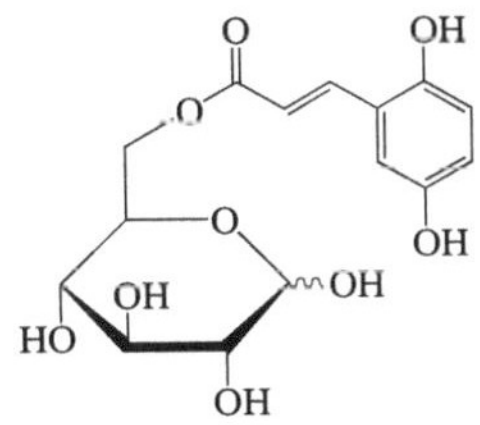

$C_{15}H_{18}O_9$ 342.3

D-***Pyranose-(E)-form*** [1187773-97-2]

Constit. of *Grevillea robusta*. Amorph. powder. $[\alpha]^{28}_{D}$ +28.8 (c, 0.91 in MeOH). λ_{max} 281 (log ε 3.86); 358 (log ε 3.68) (MeOH).

Yamashita-Higuchi, Y. *et al.*, *Chem. Pharm. Bull.*, 2012, **60**, 1347-1350 (*Grevillea robusta constit*)

4,4′-Dihydroxy-3′,5-dimethoxy-3,8′-ligna-7,7′-diene-9,9′-dioic acid D139

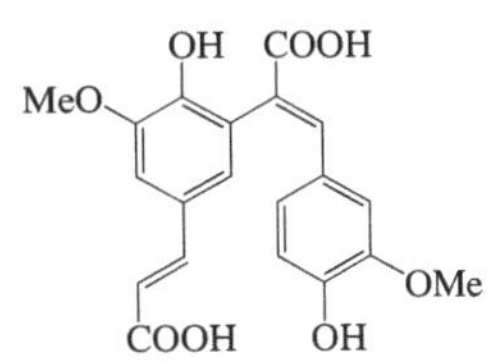

4,4′-Dihydroxy-3,5′-dimethoxy-α,3′-bicinnamic acid. 5,8′-Dehydrodiferulic acid. ***5,8′-Diferulic acid***

$C_{20}H_{18}O_8$ 386.356

Dehydrodimer of 3-(4-Hydroxy-3-methoxyphenyl)-2-propenoic acid, H101

(E,E)-form [160097-30-3]

Constit. of the cell walls of various Gramineae. Isol. from *Beta vulgaris* and *Daucus carota*. Prod. by *Aspergillus unilateralis*. Fine cryst.. Mp 160 - 165° dec.

Bis[2-(4-hydroxyphenyl)ethylamide]

[879223-63-9]

$C_{36}H_{36}N_2O_8$ 624.690

Constit. of the leaves of *Aptenia cordifolia*. Amorph. powder. λ_{max} 246 (log ε 3.2); 315 (log ε 2.6); 336 (log ε 0.9) (MeOH).

Ralph, J. *et al.*, *J. Chem. Soc., Perkin Trans. 1*, 1994, 3485-3498 (*cell wall constit, synth*).
Micard, V. *et al.*, *Phytochemistry*, 1997, **44**, 1365-1368 (*Beta vulgaris constit*).
Ng, A. *et al.*, *J. Agric. Food Chem.*, 1998, **46**, 2933-2939 (*carrot constit*).
Rouau, X. *et al.*, *Phytochemistry*, 2003, **63**, 899-903 (*maize bran constit*).
Capon, R.J. *et al.*, *Org. Biomol. Chem.*, 2005, **3**, 123-129 (*Aspergillus unilateralis metab*).
DellaGreca, M. *et al.*, *Tetrahedron*, 2006, **62**, 2877-2882 (*bis-4-hydroxyphenethylamide*).
Xia, Y. *et al.*, *J. Chem. Res.*, 2015, **39**, 535-538 (*bis-4-hydroxyphenethylamide, synth*)

4,4′-Dihydroxy-5,5′-dimethoxy-3,3′-ligna-7,7′-diene-9,9′-dioic acid D140

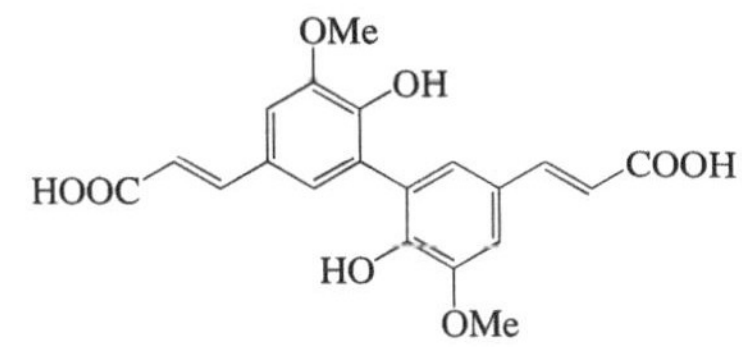

(*E,E*)-*form*

3,3′-(6,6′-Dihydroxy-5,5′-dimethoxy-[1,1′-biphenyl]-3,3′-diyl)bis[2-propenoic acid], CAS. *4,4′-Dihydroxy-5,5′-dimethoxy-3,3′-bicinnamic acid.* ***5,5′-Diferulic acid.*** *5,5′-Dehydrodiferulic acid*

[5649-91-2]

$C_{20}H_{18}O_8$ 386.356

(E,E)-form [60299-54-9]

Isol. from the cell walls of *Lolium multiflorum* and other Gramineae. Prod. by *Aspergillus unilateralis*. Exhibits antimutagenic activity. Amorph. powder. Mp > 280° dec.

Di-L-arabinofuranos-5-yl ester

$C_{30}H_{34}O_{16}$ 650.586

Isol. from the cell walls of maize bran

(β-D-Xylopyranosyl-(1→2)-L-arabinofuranos-5-yl), L-arabinofuranos-5-yl ester

$C_{35}H_{42}O_{20}$ 782.701

Isol. from the cell walls of maize bran

(E,Z)-form [60299-55-0]

Isol. from *Lolium multiflorum*

(Z,Z)-form [60299-56-1]

Isol. from *Lolium multiflorum*.

Hartley, R.D. *et al.*, *Phytochemistry*, 1976, **15**, 1157-1160 (*Lolium multiflorum constits*).
Ralph, J. *et al.*, *J. Chem. Soc., Perkin Trans. 1*, 1994, 3485-3498 (*grass constits*).
Micard, V. *et al.*, *Phytochemistry*, 1997, **44**, 1365-1368 (*sugar beet constit*).
Yamamoto, H. *et al.*, *Biosci., Biotechnol., Biochem.*, 1999, **63**, 390-394 (*synth, anal*).
Saulnier, L. *et al.*, *Carbohydr. Res.*, 1999, **320**, 82-92 (*maize bran esters*).
Ferguson, L.R. *et al.*, *Mutat. Res.*, 2003, **542**, 49-58 (*activity*).
Rouau, X. *et al.*, *Phytochemistry*, 2003, **63**, 899-903 (*maize bran constit*).
Russell, W.R. *et al.*, *Bioorg. Med. Chem.*, 2005, **13**, 2537-2546 (*synth, props*).
Capon, R.J. *et al.*, *Org. Biomol. Chem.*, 2005, **3**, 123-129 (*Aspergillus unilateralis metab*).
Zeller, W.E. *et al.*, *Tetrahedron Lett.*, 2015, **56**, 1076-1079 (*di-Me ether, synth*)

4,4′-Dihydroxy-3,3′-dimethoxy-8,8′-lignan-9,9′-olide D141

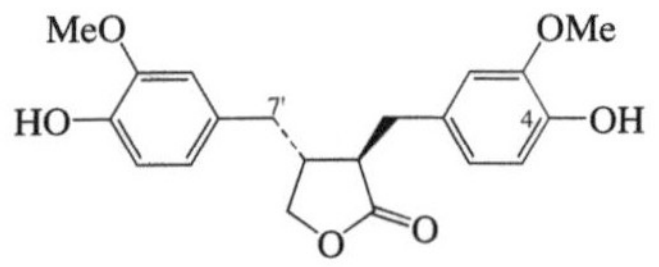

(8R,8′R)-form

Dihydro-3,4-bis[(4-hydroxy-3-methoxyphenyl)methyl]-2(3H)-furanone, CAS
[29764-17-8] [41948-08-7]
[42298-55-5] ((±)-form)
$C_{20}H_{22}O_6$ 358.39

(8R,8′R)-form

Matairesinol [580-72-3]

Constit. of *Abies amabilis*, *Abies pinsapo*, *Arctium lappa*, *Centaurea solstitialis*, *Forsythia intermedia*, *Bupleurum salicifolium*, *Podocarpus spicata*, *Tsuga mertensiana*, *Tsuga heterophylla*, *Wikstroemia viridiflora* and cell cultures of *Linum flavum*. Exhibits significant anti-neuroinflammatory activity. Nematocide. Needles (EtOH or AcOH). Sol. EtOH, $CHCl_3$; fairly sol. C_6H_6, Et_2O, toluene; poorly sol. hexane, H_2O. Mp 119° (anhyd.). $[\alpha]_D$ -40 (c, 0.1 in EtOH). λ_{max} 230; 280 (EtOH) (Berdy)

Dibenzoyl

Cryst. (EtOH). Mp 133°

4-O-β-D-Glucopyranoside

Matairesinoside

[23202-85-9]

$C_{26}H_{32}O_{11}$ 520.531

Constit. of *Arctium lappa*, *Forsythia viridissima*, *Forsythia intermedia*, *Styrax japonica* and *Trachelospermum asiaticum*. Cryst. (EtOAc). Mp 93 - 96°. $[\alpha]_D^{12}$ -46 (EtOH). A glucoside was earlier isol. from safflower *Carthamus tinctorius* but the posn. of glycosylation was not determined. λ_{max} 239 (log ε 4.21); 281 (log ε 3.81) (EtOH)

4,4′-Di-O-β-D-glucopyranoside

[38976-08-8]

$C_{32}H_{42}O_{16}$ 682.672

Constit. of *Arctium lappa* and *Trachelospermum asiaticum*. Powder. Mp 104 - 106°. $[\alpha]_D^{14}$ -24 (c, 0.5 in EtOH)

4′-O-[β-D-Apiofuranosyl-(1→2)-β-D-glucopyranoside]

[211115-82-1]

$C_{31}H_{40}O_{15}$ 652.646

Constit. of safflower oil cake *Carthamus tinctorius*. Mp 115 - 116.5°. $[\alpha]_D^{18}$ -86.3 (c, 0.5 in MeOH). λ_{max} 229 (ε 6900); 280 (ε 2500) (MeOH)

4-O-[β-D-Xylopyranosyl-(1→2)-β-D-glucopyranoside]

[1009838-90-7]

$C_{31}H_{40}O_{15}$ 652.646

Constit. of the roots of *Symplocos caudata*. Amorph. powder. $[\alpha]_D^{25}$ +60 (c, 0.8 in MeOH). λ_{max} 220; 278 (MeOH)

4-O-[β-D-Glucopyranosyl-(1→4)-β-D-glucopyranoside]

Matairesinol 4-O-β-D-cellobioside

$C_{32}H_{42}O_{16}$ 682.672

Constit. of Caulis Trachelospermi. Amorph. powder. $[\alpha]_D^{25}$ -38.1 (c, 0.07 in MeOH)

4-O-[β-D-Glucopyranosyl-(1→6)-β-D-glucopyranoside]

Matairesinol 4-gentiobioside

[106647-14-7]

$C_{32}H_{42}O_{16}$ 682.672

Isol. from *Trachelospermum asiaticum*. Amorph.. $[\alpha]_D^{26}$ -54.7 (c, 0.68 in MeOH)

4-O-(6,7-Dihydroxy-3,7-dimethyl-2-octenyl)

[160195-41-5]

$C_{30}H_{40}O_8$ 528.642

Constit. of *Haplophyllum ptilostylum*. $[\alpha]_D$ -18 (c, 0.4 in $CHCl_3$)

(8S,8′S)-form

(+)-Matairesinol [148409-36-3]

Constit. of *Selaginella doederleinii*. $[\alpha]_D^{20}$ +37 (c, 1 in MeOH). λ_{max} 233 (ε 8710); 283 (ε 5130) (MeOH) (Berdy). [base] λ_{max} 250; 298 (MeOH/NaOH) (Berdy)

4-O-β-D-Glucopyranoside

Styraxlignolide E

[823214-05-7]

$C_{26}H_{32}O_{11}$ 520.531

Constit. of the stem bark of *Styrax japonica*. Plates (MeOH). Mp 102 - 104°. λ_{max} 204 (log ε 4.94); 224 (log ε 4.36); 279 (log ε 3.69) (MeOH)

4′-O-β-D-Glucopyranoside

Styraxlignolide D

[823214-04-6]

$C_{26}H_{32}O_{11}$ 520.531

Constit. of the stem bark of *Styrax japonica*. Plates (MeOH). Mp 100 - 102°. $[\alpha]_D^{23}$ -10 (c, 0.28 in MeOH). λ_{max} 203 (log ε 4.74); 226 (log ε 4.13); 280 (log ε 3.71) (MeOH).

Haworth, R.D. *et al.*, *J. Chem. Soc.*, 1940, 1098-1101 (*synth*).
Haworth, R.D. *et al.*, *J. Chem. Soc.*, 1950, 71-72 (*config*).
Barton, G.M. *et al.*, *J. Org. Chem.*, 1962, **27**, 322-323 (*Tsuga constits, bibl*).
Duffield, A.M. *J. Heterocycl. Chem.*, 1967, **4**, 16-22 (*ms*).
Thieme, H. *et al.*, *Pharmazie*, 1968, **23**, 519 (*Matairesinoside*).
Reisch, J. *et al.*, *Tetrahedron Lett.*, 1969, 3803-3806 (*Maculatin*).
Inagaki, I. *et al.*, *Chem. Pharm. Bull.*, 1972, **20**, 2710-2718 (*Matairesinoside*).
Nishibe, S. *et al.*, *Can. J. Chem.*, 1973, **51**, 1050-1052 (*4,4′-diglucoside*).
Karrer, W. *Konstitution und Vorkommen der Organischen Pflanzenstoffe, 2nd edn., Birkhäuser Verlag*, 1976-1985, no. 1163 (*bibl, occur*).
Tandon, S. *et al.*, *Phytochemistry*, 1976, **15**, 1789-1791 (*Wikstroemia viridiflora constit*).
Mahalanabis, K.K. *et al.*, *Tetrahedron Lett.*, 1982, **23**, 3975-3978 (*synth*).
Nishibe, S. *et al.*, *Chem. Pharm. Bull.*, 1984, **32**, 4653-4657 (*cmr*).
Abe, F. *et al.*, *Chem. Pharm. Bull.*, 1986, **34**, 4340-4345 (*4-gentiobioside*).
Brown, E. *et al.*, *Tetrahedron*, 1989, **45**, 141-154 (*di-Me ether, synth*).
Rahman, M.M.A. *et al.*, *Phytochemistry*, 1990, **29**, 1971-1980 (*Matairesinol 4′-glucoside*).
Tanoguchi, M. *et al.*, *Chem. Pharm. Bull.*, 1991, **39**, 1873-1876 (*cd*).
Lin, R.C. *et al.*, *Planta Med.*, 1994, **60**, 168-170 (*(+)-Matairesinol*).
Ulubelen, A. *et al.*, *Phytochemistry*, 1995, **39**, 417-422 (*dihydroxydimethyloctenyl ether*).
Youssef, D. *et al.*, *Planta Med.*, 1995, **61**, 570-573 (*pmr, cmr*).
Estévez-Braun, A. *et al.*, *Phytochemistry*, 1996, **43**, 885-886 (*cmr*).
Nagatsu, A. *et al.*, *Chem. Pharm. Bull.*, 1998, **46**, 1044-1047 (*4′-apiosylglucoside*).
Eklund, P. *et al.*, *Org. Lett.*, 2003, **5**, 491-493 (*synth*).
Min, B.-S. *et al.*, *J. Nat. Prod.*, 2004, **67**, 1980-1984 (*Styraxlignolides D,E*).
Fischer, J. *et al.*, *Org. Lett.*, 2004, **6**, 1345-1348 (*synth*).
Huo, C. *et al.*, *Phytochemistry*, 2008, **69**, 788-795 (*4-2-xylosylglucoside*).
Ekholm, F.S. *et al.*, *Carbohydr. Res.*, 2010, **345**, 1963-1967 (*4,4′-diglucoside*).
Zhu, C. *et al.*, *Acta Pharm. Sin. B*, 2013, **3**, 109-112 (*4-cellobioside, cd*).
Yang, Y.-N. *et al.*, *J. Agric. Food Chem.*, 2015, **63**, 7958-7966 (*Arctium lappa glucosides*).
Hu, C.-L. *et al.*, *Nat. Prod. Res.*, 2017, **31**, 1358-1364 (*Matairesinol, activity*)

4,4′-Dihydroxy-3,3′-dimethoxy-7′-oxo-7,8′-lignan-9,9′-olide D142

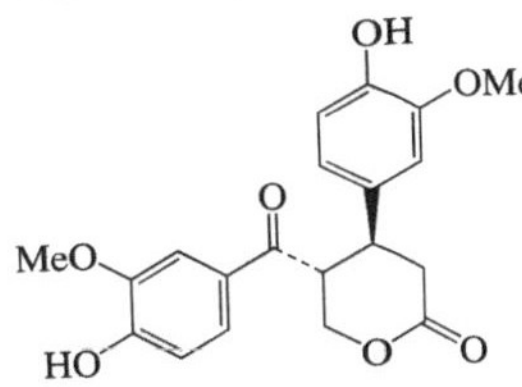

Absolute Configuration

Tetrahydro-5-(4-hydroxy-3-methoxybenzoyl)-4-(4-hydroxy-3-methoxyphenyl)-2H-pyran-2-one
$C_{20}H_{20}O_7$ 372.373
Constit. of the bulbs of *Rhodocodon campanulatus*. Powder (as di-Ac). $[\alpha]_D^{24}$ +5.1 (c, 0.5 in $CHCl_3$) (di-Ac).
Schwikkard, S. *et al., J. Nat. Prod.*, 2017, **80**, 30-37 (*Rhodocodon campanulatus constit, di-Ac, synth, abs config*)

4,4′-Dihydroxy-8,9-dinor-3,8′-lignan-7-al D143

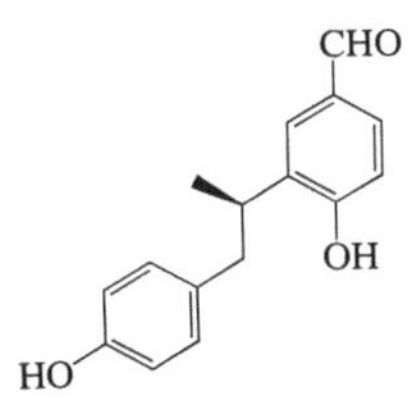

4-Hydroxy-3-[2-(4-hydroxyphenyl)-1-methylethyl]benzaldehyde. 4,4′-Dihydroxy-8,9-dinor-7-oxo-3,8′-neolignan
$C_{16}H_{16}O_3$ 256.301
(R)-form
Constit. of the seeds of *Daphniphyllum macropodum*. Solid. λ_{max} 229; 286 (MeOH).
Ma, X. *et al., RSC Adv.*, 2017, **7**, 52970-52976 (*Daphniphyllum macropodum constit*)

4,4′-Dihydroxy-8′,9′-dinor-3,3′-lignan-9-oic acid D144

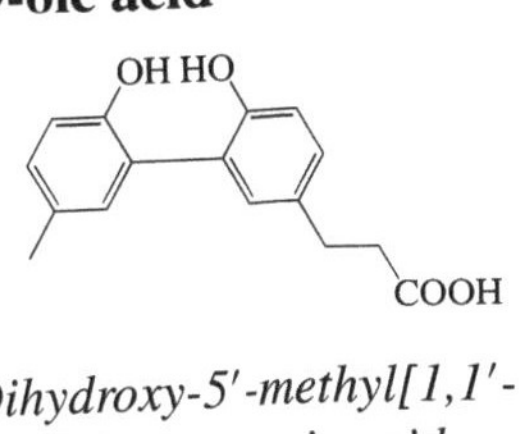

2′,6-Dihydroxy-5′-methyl[1,1′-biphenyl]-3-propanoic acid
[1883800-72-3]
$C_{16}H_{16}O_4$ 272.300
Constit. of the wood of *Betula pendula*. Amorph. powder.
Tsvetkova, D.E. *et al., Russ. Chem. Bull. (Engl. Transl.)*, 2015, **64**, 1413-1418 (*Betula pendula constit*)

4,4′-Dihydroxy-8,9-dinor-3,3′-lign-8′-ene-7-oic acid D145

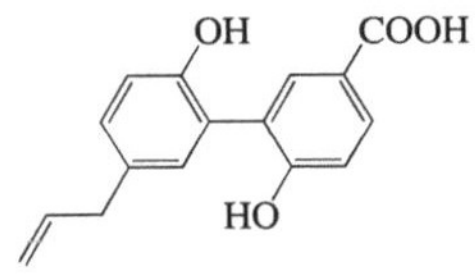

2′,6-Dihydroxy-5′-(2-propenyl)-[1,1′-biphenyl]-3-carboxylic acid, CAS.
Tzumin A
[2055788-34-4]
$C_{16}H_{14}O_4$ 270.284
Constit. of the bark of *Sassafras tzumu*. Exhibits potent acetylcholinesterase inhibitory activity. Needles (MeOH). Mp 175.4°. $[\alpha]_D^{20}$ +53 (c, 0.01 in MeOH). λ_{max} 289; 363; 370 (MeOH).
Lu, H. *et al., Nat. Prod. Res.*, 2017, **31**, 829-834 (*Tzumin A, activity*)

5,7-Dihydroxy-8-[1-(4-hydroxy-3-methoxyphenyl)-2-propenyl]flavanone D146

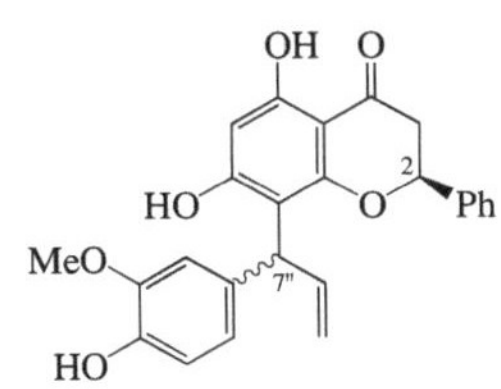

8-[1-(4-Hydroxy-3-methoxyphenyl)-2-propenyl]pinocembrin
$C_{25}H_{22}O_6$ 418.445
Flavonolignan. Similar to 3,5,7-Trihydroxy-8-[1-(4-hydroxy-3-methoxyphenyl)-2-propenyl]flavone, T81
(2S,7″ξ)-form
Isol. from Thai propolis. Yellow-brown solid. $[\alpha]_D^{23}$ +24 (c, 0.15 in $CHCl_3$).
Athikomkulchai, S. *et al., Fitoterapia*, 2013, **88**, 96-100 (*propolis constit*)

2-[4-[1,3-Dihydroxy-2-(4-hydroxy-3-methoxyphenyl)propyl]-2-methoxyphenoxy]-1-(4-hydroxy-3-methoxy phenyl)-1,3-propanediol D147

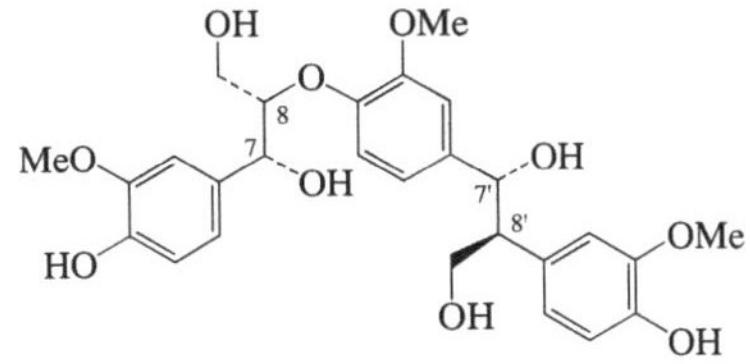

(7R,7′R*,8S,8′S*)-form

[75715-00-3] (unspecified stereochem.)
$C_{27}H_{32}O_{10}$ 516.543
(7R,7′R*,8S,8′S*)-form
7′-Me ether
Crataegusnin C
$C_{28}H_{34}O_{10}$ 530.57
Constit. of the seeds of *Crataegus pinnatifida*. Exhibits potent inhibition of NO prodn.. Yellow oil. $[\alpha]_D^{20}$ +2.6 (c, 0.1 in MeOH). λ_{max} 228 (log ε 0.67); 286 (log ε 0.29) (MeOH)
(7S,7′R*,8R,8′S*)-form
7′-Me ether
Crataegusnin D
$C_{28}H_{34}O_{10}$ 530.57
Constit. of the seeds of *Crataegus pinnatifida*. Exhibits potent inhibition of NO prodn.. Yellow oil. $[\alpha]_D^{20}$ +15 (c, 0.15 in MeOH). λ_{max} 227 (log ε 0.77); 287 (log ε 0.32) (MeOH)
(7S,7′R*,8S,8′R*)-form
7′-Et ether
Crataegusnin F
$C_{29}H_{36}O_{10}$ 544.597
Constit. of the seeds of *Crataegus pinnatifida*. Exhibits potent inhibition of NO prodn.. Yellow oil. $[\alpha]_D^{20}$ +2.2 (c, 0.1 in MeOH). λ_{max} 225 (log ε 0.72); 283 (log ε 0.29) (MeOH)
(7S,7′R*,8S,8′S*)-form
7′-Et ether
Crataegusnin E
$C_{29}H_{36}O_{10}$ 544.597
Constit. of the seeds of *Crataegus pinnatifida*. Exhibits potent inhibition of NO prodn.. Yellow oil. $[\alpha]_D^{20}$ +16.2 (c, 0.1 in MeOH). λ_{max} 226 (log ε 0.63); 285 (log ε 0.26) (MeOH)
7′-Et, 7-Me ether
Crataegusnin G
$C_{30}H_{38}O_{10}$ 558.624
Constit. of the seeds of *Crataegus pinnatifida*. Exhibits potent inhibition of NO prodn.. Yellow oil. $[\alpha]_D^{20}$ +18.5 (c, 0.1 in MeOH). λ_{max} 223 (log ε 0.80); 285 (log ε 0.34) (MeOH)
(7S,7′ξ,8R,8′ξ)-form
Leptolepisol D [13459-21-7]
Constit. of *Acer saccharum*, *Euonymus acanthocarpus*, *Larix leptolepis*,

Quercus acutissima, Picea excelsa and *Picea jezoensis.*

Nimz, H. *et al., Chem. Ber.*, 1966, **99**, 2638-2651 (*Picea excelsa constit*).
Sano, Y. *et al., Mokuzai Gakkaishi*, 1970, **16**, 121-125 (*Picea jezoensis constit*).
Namba, H. *et al., Mokuzai Gakkaishi*, 1980, **26**, 426-431 (*synth*).
Li, L. *et al., J. Agric. Food Chem.*, 2011, **59**, 7708-7716 (*Acer saccharum constit*).
Zhu, J.X. *et al., Arch. Pharmacal Res.*, 2012, **35**, 1739-1747 (*Euonymus acanthocarpus constit*).
Zeng, X.-L. *et al., Nat. Prod. Res.*, 2014, **28**, 1364-1370 (*Quercus acutissima constit*).
Peng, Y. *et al., Bioorg. Med. Chem. Lett.*, 2016, **26**, 5501-5506 (*Crataegusnins C-G, activity*)

4,4′-Dihydroxy-3′-[2-(4-hydroxyphenyl)-1-methylethyl]-7,7′-epoxylignan D148

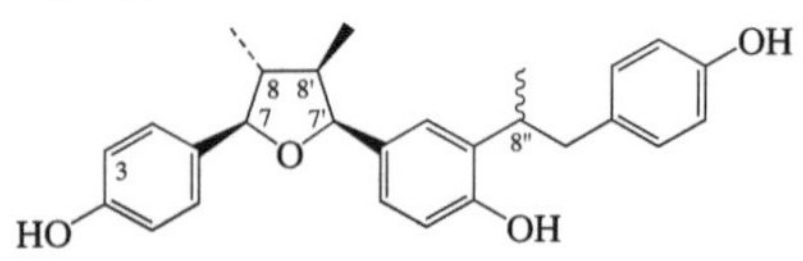

(7*R**,7′*S**,8*R**,8′*R**,8″ξ)-*form*

$C_{27}H_{30}O_4$ 418.533

(7R*,7′S*,8R*,8′R*,8″ξ)-form

3-Hydroxy

Ribesin G

[1469818-68-5]

$C_{27}H_{30}O_5$ 434.532

Isol. from the leaves of *Ribes nigrum*. Displays potent superoxide anion scavenging activity. Solid. $[\alpha]^{25}_{D}$ -31.0 (c, 0.30 in MeOH). λ_{max} 226 (log ε 4.40); 280 (log ε 3.84) (MeOH)

(7R*,7′S*,8R*,8′S*,8″ξ)-form

Ribesin F [1469818-67-4]

Isol. from the leaves of *Ribes nigrum*. Solid. $[\alpha]^{25}_{D}$ -15.7 (c, 0.4 in MeOH). λ_{max} 226 (log ε 2.89); 278 (log ε 2.26) (MeOH)

(7R*,7′S*,8S*,8′R*,8″ξ)-form

Ribesin E [1469818-66-3]

Isol. from the leaves of *Ribes nigrum*. Solid. $[\alpha]^{25}_{D}$ -35.0 (c, 0.56 in MeOH). λ_{max} 226 (log ε 4.50); 279 (log ε 3.87) (MeOH).

Sasaki, T. *et al., Phytochemistry*, 2013, **95**, 333-340 (*Ribesins E,F,G*)

3-[3-[5,7-Dihydroxy-2-(4-hydroxyphenyl)-4-oxo-4*H*-1-benzopyran-8-yl]-4-hydroxyphenyl]-2-propenoic acid D149

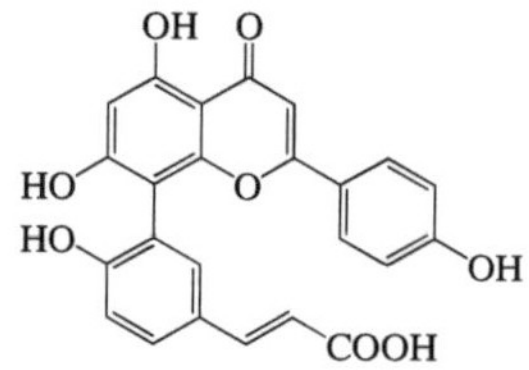

$C_{24}H_{16}O_8$ 432.384

Me ester

Unciflavone B

[1830290-24-8]

$C_{25}H_{18}O_8$ 446.411

Constit. of *Selaginella uncinata*. Exhibits weak cytotoxicity. Yellow powder. λ_{max} 230; 274; 298; 316 (MeOH)

Et ester

Unciflavone A

[1830290-23-7]

$C_{26}H_{20}O_8$ 460.438

Constit. of *Selaginella uncinata*. Exhibits weak cytotoxicity. Yellow powder. λ_{max} 230; 274; 299; 317 (MeOH).

Zou, H. *et al., Fitoterapia*, 2014, **99**, 328-333 (*Unciflavones A,B, activity*)

5,7-Dihydroxy-2-(4-hydroxyphenyl)-4-oxo-β-phenyl-4*H*-1-benzopyran-8-propanoic acid, CAS D150

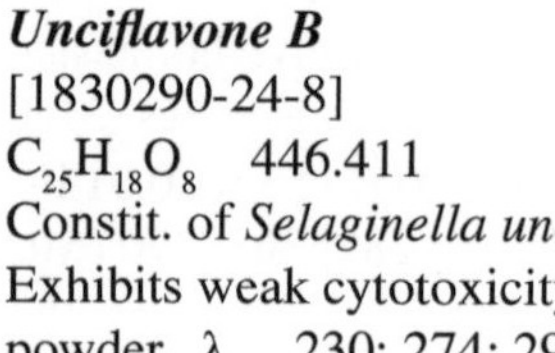

X2

$C_{24}H_{18}O_7$ 418.401

Flavonolignan. Isol. from *Pityrogramma calomelanos* var. *aureoflava* and *Pityrogramma sulphurea*. Yellow powder. Mp 280 - 285°

Me ester

X1

Isol. from *Pityrogramma calomelanos*. Yellow powder. Mp 305 - 307°.

Iinuma, M. *et al., Z. Naturforsch., C*, 1986, **41**, 681-683 (*Pityrogramma calomelanos constits*)

2,3-Dihydroxy-1-(4-hydroxyphenyl)-1-propanone D151

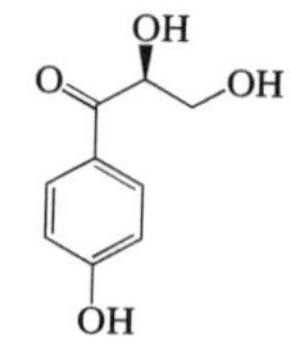

2,3,4′-Trihydroxypropiophenone. 4-(2,3-Dihydroxypropanoyl)phenol

$C_9H_{10}O_4$ 182.175

(S)-form

Schweinfurthinol [343605-25-4]

Constit. of the seeds of *Canarium schweinfurthii*. Powder. Mp 210 - 212°. $[\alpha]_D$ -56.7 (c, 0.11 in MeOH) (natural). $[\alpha]_D$ -25.7 (c, 0.1 in MeOH) (synthetic). λ_{max} 201 (log ε 4.13); 219 (log ε 4.1); 279 (log ε 4.25) (MeOH).

Helene, T. *et al., Bull. Chem. Soc. Ethiop.*, 2000, **14**, 155-159 (*Schweinfurthinol*).
Cui, S. *et al., Chin. J. Chem.*, 2011, **29**, 2759-2762 (*synth, abs config*)

3-[4-[[2,6-Dihydroxy-4-(3-hydroxypropyl)phenyl]amino]-3,5-dihydroxyphenyl]-1,2-propanediol D152

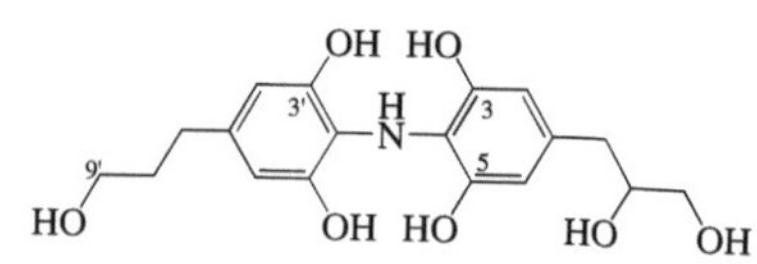

$C_{18}H_{23}NO_7$ 365.382

3,3′,5-Tri-Me ether, 9′-O-α-L-rhamnopyranoside

[403982-57-0]

$C_{27}H_{39}NO_{11}$ 553.605

Constit. of the leaves of *Pieris formosa*. Viscous syrup. $[\alpha]^{23}_{D}$ -16.2 (c, 0.31 in MeOH).

Wang, L. *et al., Fitoterapia*, 2001, **72**, 779-787 (*Pieris formosa constit*)

2,3-Dihydroxy-2-isopropylbutanedioic acid D153

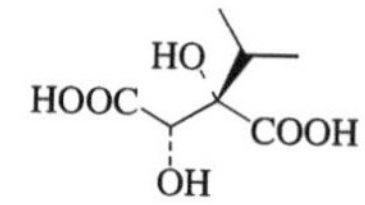

2,3-Dihydroxy-2-(1-methylethyl) butanedioic acid, CAS. *3-Hydroxy-3-isopropylmalic acid*

$C_7H_{12}O_6$ 192.167

(2S,3S)-form

3-O-(3,4-Dihydroxy-E-cinnamoyl)

[2200261-46-5]

$C_{16}H_{18}O_9$ 354.311
Constit. of *Tithonia diversifolia*. Amorph. yellow powder. $[\alpha]_D^{20}$ +38.7 (c, 0.5 in MeOH).
Ccana-Ccapatinta, G.V. *et al.*, *Tetrahedron: Asymmetry*, 2017, **28**, 1823-1828 (*Tithonia diversifolia constit*)

5′-(1,3-Dihydroxyisopropyl)-3,3′,4,4′-tetrahydroxy-8,8′-lignan-9,9′-olide D154

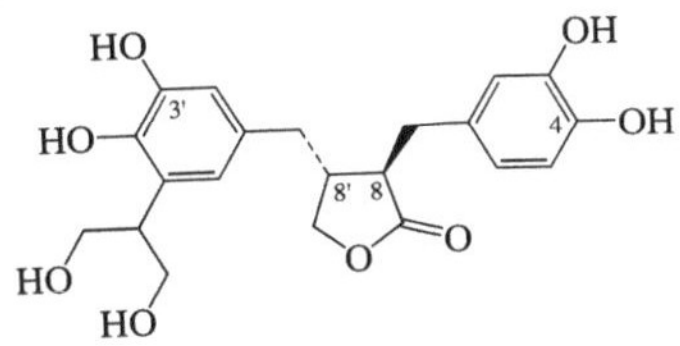

$C_{21}H_{24}O_8$ 404.415
(8R,8′R)-form
3,3′-Di-Me ether, 4-O-β-D-glucopyranoside
5′-Propanediolmatairesinoside
[1807808-86-1]
$C_{29}H_{38}O_{13}$ 594.61
Constit. of the fruit of *Arctium lappa*. Antiinflammatory agent. Amorph. powder. $[\alpha]_D^{25}$ -53.4 (c, 0.11 in MeOH). λ_{max} 230 (log ε 3.48); 280 (log ε 3.10) (MeOH).
Yang, Y.-N. *et al.*, *J. Agric. Food Chem.*, 2015, **63**, 7958-7966 (*5′-Propanediolmatairesinoside*)

7″,8″-Dihydroxyleptolepisol B D155

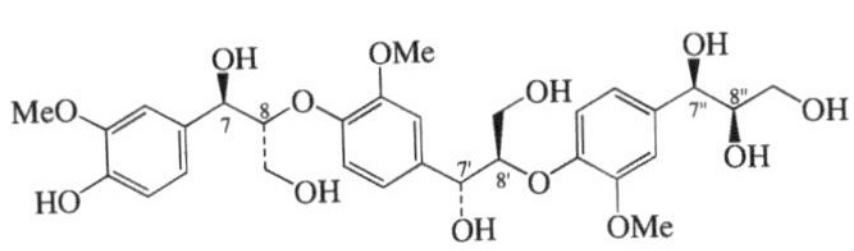

(7R*,8R*)-(7′R*,8′R*)-(7″R*,8″R*)-*form*

$C_{30}H_{38}O_{13}$ 606.621
Similar to Leptolepisol B, L18 and Casearialignan E, C37
(7R*,8R*)-(7′R*,8′R*)-(7″R*,8″R*)-form
Crasesquineolignan E
Constit. of the seeds of *Crataegus pinnatifida*. Exhibits neuroprotective activity. Yellowish oil. $[\alpha]_D^{20}$ +2 (c, 0.2 in MeOH). λ_{max} 231 (log ε 4.21); 279 (log ε 3.77) (MeOH)
(7R*,8S*)-(7′R*,8′S*)-(7″R*,8″R*)-form
Crasesquineolignan G
Constit. of the seeds of *Crataegus pinnatifida*. Exhibits neuroprotective activity. Yellowish oil. $[\alpha]_D$ -1.5 (c, 0.2 in MeOH). λ_{max} 230 (log ε 4.0); 279 (log ε 3.58) (MeOH)
(7R*,8S*)-(7′R*,8′S*)-(7″R*,8″S*)-form
Crasesquineolignan F
Constit. of the seeds of *Crataegus pinnatifida*. Exhibits neuroprotective activity. Yellowish oil. $[\alpha]_D^{20}$ -1.5 (c, 0.2 in MeOH). λ_{max} 230 (log ε 4.22); 279 (log ε 3.81) (MeOH).
Cheng, Z.-Y. *et al.*, *Fitoterapia*, 2019, **133**, 225-230 (*Crasesquineolignans E,F,G, activity*)

4,4′-Dihydroxy-8,9′-ligna-7,7′-dien-9-al D156

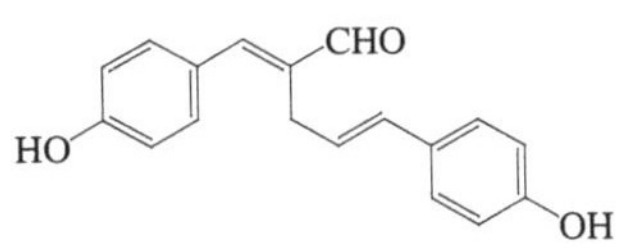

5-(4-Hydroxyphenyl)-2-[(4-hydroxyphenyl)methylene]-4-pentenal. 4,4′-Dihydroxy-9-oxo-8,9′-neoligna-7,7′-diene. ***Galanganal***
[864073-14-3]
$C_{18}H_{16}O_3$ 280.323
Constit. of *Alpinia conchigera* and *Alpinia galanga*. Inhibitor of NO prodn.. Powder. λ_{max} 263 (log ε 4.14); 314 (log ε 3.98) (MeOH).
Morikawa, T. *et al.*, *Chem. Pharm. Bull.*, 2005, **53**, 625-630 (*Galanganal, activity*).
Faustino, H. *et al.*, *Molecules*, 2010, **15**, 9308-9322 (*Galanganal*)

4,6′-Dihydroxy-3,3′-ligna-7,7′-diene-9,9′-dial D157

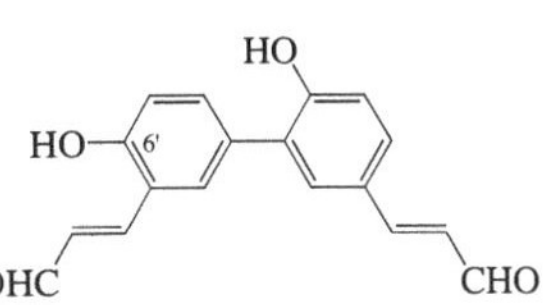

4,6′-Dihydroxy-3,3′-dicinnamaldehyde. 4,6′-Dihydroxy-9,9′-dioxo-3,3′-neoligna-7,7′-diene
$C_{18}H_{14}O_4$ 294.306
(E,E)-form
6′-Me ether
4-Hydroxy-6′-methoxy-3,3′-ligna-7,7′-diene-9,9′-dial. 4-Hydroxy-6′-methoxy-9,9′-dioxo-3,3′-ligna-7,7′-diene. ***4′-Methoxymagndialdehyde***
[957765-05-8]
$C_{19}H_{16}O_4$ 308.333
Constit. of the stem bark of *Magnolia officinalis*. Induces apoptosis in HeLa cells. Yellow powder (MeOH/$CHCl_3$). Mp 168 - 170°. λ_{max} 294 (log ε 3.4); 333 (log ε 3.3) (MeOH).
Youn, U.J. *et al.*, *J. Nat. Prod.*, 2007, **70**, 1687-1689 (*4′-Methoxymagndialdehyde, activity*).
Youn, U.J. *et al.*, *Phytother. Res.*, 2013, **27**, 1419-1422 (*activity*)

4,4′-Dihydroxy-8,8′-ligna-7,7′-diene-9,9′-dioic acid D158

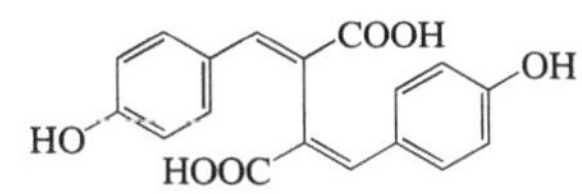

2,3-Bis(4-hydroxybenzylidene) butanedioic acid. 4,4′-Dihydroxy-8,8′-dicinnamic acid. 8,8′-Di-p-coumaric acid
$C_{18}H_{14}O_6$ 326.304
(7E,7′E)-form
Bis(4-methoxybenzyl) ether
Inermidioic acid
[2270884-48-3]
$C_{34}H_{30}O_8$ 566.606
Constit. of *Lawsonia inermis*. Yellow powder. λ_{max} 265 (log ε 4.59); 291 (log ε 4.52) (MeOH).
Yang, C.-S. *et al.*, *Molecules*, 2017, **22**, 936:1-9 (*Inermidioic acid*)

6,6′-Dihydroxy-2,2′-ligna-7,7′-diene-9,9′-dioic acid D159

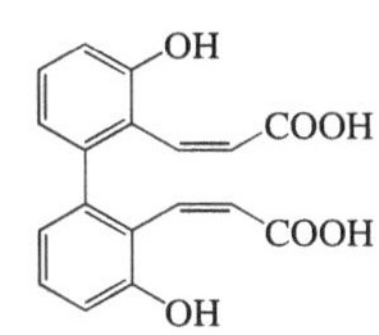

3,3′-(3,3′-Dihydroxy-[1,1′-biphenyl]-2,2′-diyl)bis[2-propenoic acid], CAS
$C_{18}H_{14}O_6$ 326.304
(7Z,7′Z)-form
Di-O-β-D-glucopyranoside
Bis-ortho-cis-melilotoside.
Splendenoside
[2241042-49-7]
$C_{30}H_{34}O_{16}$ 650.586
Constit. of *Artemisia splendens*. Exhibits antioxidant activity. Light brown powder. λ_{max} 253; 290 (sh) (MeOH).
Afshar, F.H. *et al.*, *Rec. Nat. Prod.*, 2017, **11**, 43-50 (*Splendenoside, activity*)

3,3′-Dihydroxy-8,8′-lignan-9,9′-olide D160

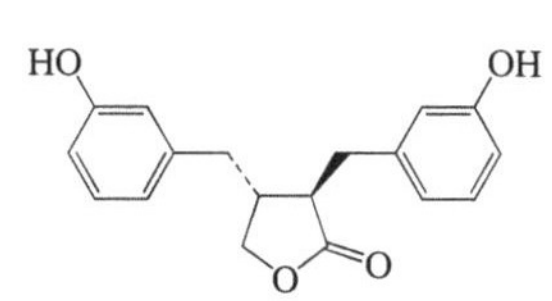

Dihydro-3,4-bis(3-hydroxyphenyl) methyl-2(3H)-furanone, CAS. *2,3-Bis(3-hydroxybenzyl)butyrolactone. HPMF*
[76721-88-5] ((±)-*trans*-form)
$C_{18}H_{18}O_4$ 298.338
(8R,8′R)-form
Enterolactone [78473-71-9]
Isol. from mammalian urine, prod. of intestinal flora acting on lignans present in cereal. Antineoplastic agent. Gum. $[\alpha]_D^{20}$ -40.5 (c, 0.62 in $CHCl_3$).

Setchell, K.D.R. *et al.*, *Nature (London)*, 1980, **287**, 740-742 (*Enterolactone, synth*).
Ganeshpure, P.A. *et al.*, *Chem. Ind. (London)*, 1981, 778 (*synth*).
Pelter, A. *et al.*, *Tetrahedron Lett.*, 1981, **22**, 1549-1550 (*synth*).
Axelson, M. *et al.*, *Nature (London)*, 1982, **298**, 659-660 (*isol*).
Mahalanabis, K.K. *et al.*, *Tetrahedron Lett.*, 1982, **23**, 3975-3978 (*synth*).
Kirk, D.N. *et al.*, *J. Chem. Soc., Perkin Trans. 1*, 1985, 35-37 (*synth*).
Bambagiotti-Alberti, M. *et al.*, *Heterocycles*, 1987, **26**, 1735-1738 (*synth*).
Yoda, H. *et al.*, *Tetrahedron*, 1992, **48**, 3313-3322 (*synth, bibl*).
Van Oeveren, A. *et al.*, *J. Org. Chem.*, 1994, **59**, 5999-6007 (*synth*).
Chenevert, R. *et al.*, *Can. J. Chem.*, 1999, **77**, 223-226 (*synth*).
Raffaelli, B. *et al.*, *J. Chromatogr., B*, 2002, **777**, 29-43 (*rev, biosynth, metab, activity*).
Wang, L.-Q. *J. Chromatogr., B*, 2002, **777**, 289-309 (*rev, chem, activity*).
Sibi, M.P. *et al.*, *J. Org. Chem.*, 2002, **67**, 1738-1745 (*synth*).
Eklund, P. *et al.*, *Org. Lett.*, 2003, **5**, 491-493 (*synth*).
Yan, B. *et al.*, *J. Org. Chem.*, 2004, **69**, 2859-2862 (*synth*).
Danbara, N. *et al.*, *Anticancer Res.*, 2005, **25(3B)**, 2269-2276 (*pharmacol*).
Boccardo, F. *et al.*, *Clin. Chim. Acta*, 2006, **365**, 58-67 (*rev*).
Chen, L.-H. *et al.*, *Mol. Cancer Ther.*, 2007, **6**, 2581-2590 (*pharmacol*).
Hajra, S. *et al.*, *J. Org. Chem.*, 2009, **74**, 7978-7981 (*synth*).
Landete, J.M. *Food Res. Int.*, 2012, **46**, 410-424 (*rev, occur, metab*).
Waulters, Q.T. *et al.*, *ARKIVOC*, 2018, **iv**, 76-93 (*rev, synth, activity*).
Mali, A.V. *et al.*, *Eur. J. Pharmacol.*, 2019, **852**, 107-124 (*rev, activity*)

4,4′-Dihydroxy-8,8′-lign-7′-en-9′,7-olid-9-oic acid D161

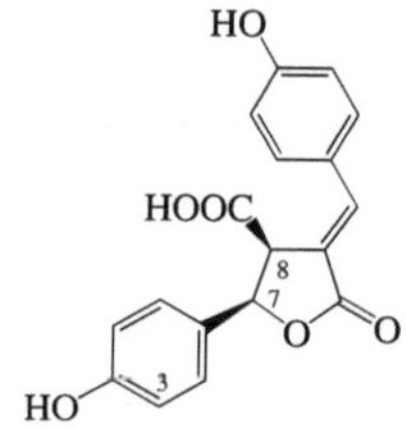

(*7R,7′E,8S*)-*form*

2,3-Dihydro-4-(4-hydroxybenzylidene)-2-(4-hydroxyphenyl)-5-oxo-3-furancarboxylic acid
$C_{18}H_{14}O_6$ 326.304
(7R,7′E,8S)-form
Seric acid D
Constit. of *Oenanthe javanica*. Amorph. pale yellow solid. $[\alpha]_D^{24}$ -164 (c, 0.9 in MeOH). λ_{max} 317 (log ε 4.31) (MeOH)
(7S,7′E,8S)-form
Seric acid C
Constit. of *Oenanthe javanica*. Degranulation inhibitor. Amorph. pale yellow solid. $[\alpha]_D^{27}$ +120 (c, 0.2 in MeOH). λ_{max} 228 (log ε 4.12); 321 (log ε 4.16) (MeOH)
3-Methoxy
4,4′-Dihydroxy-3-methoxy-8,8′-lign-7′-en-9′,7-olid-9-oic acid.
Seric acid E
$C_{19}H_{16}O_7$ 356.330
Constit. of *Oenanthe javanica*. Amorph. pale yellow solid. $[\alpha]_D^{24}$ +188 (c, 0.4 in MeOH). λ_{max} 320 (log ε 4.36) (MeOH)
(7R*,7′E,8R*)-form
Chaerohirsutin A [762242-47-7]
Constit. of the roots of *Chaerophyllum hirsutum*. Pale yellow solid. $[\alpha]_D^{22}$ +68.8 (c, 0.16 in EtOH). λ_{max} 227 (log ε 3.98); 320 (log ε 4.1) (EtOH).

Dall'Acqua, S. *et al.*, *J. Nat. Prod.*, 2004, **67**, 1588-1590 (*Chaerohirsutin A*).
Kim, E. *et al.*, *Synth. Commun.*, 2005, **35**, 1231-1238 (*synth*).
Murata, T. *et al.*, *J. Nat. Prod.*, 2019, **82**, 1518-1526 (*Seric acids C,D,E, activity*)

2,2′-Dihydroxy-5′-methoxy-4,8′-ligna-7,7′-diene-9,9′-dioic acid D162

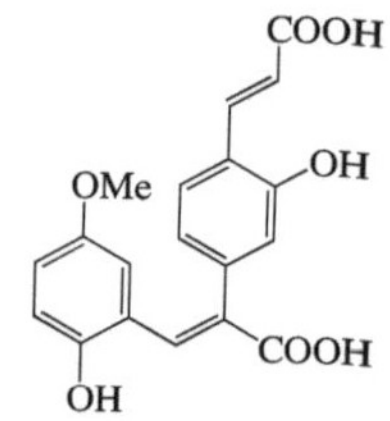

$C_{19}H_{16}O_7$ 356.33
(E,E)-form
Constit. of *Sphallerocarpus gracilis*. Yellow oil.

Shi, M.R. *et al.*, *Biochem. Syst. Ecol.*, 2012, **40**, 1-3 (*Sphallerocarpus gracilis constit*)

2-(3,4-Dihydroxy-5-methoxyphenyl)-2,3-dihydro-3-(hydroxymethyl)-7*H*-pyrano[2,3-*g*]-1,4-benzodioxin-7-one D163

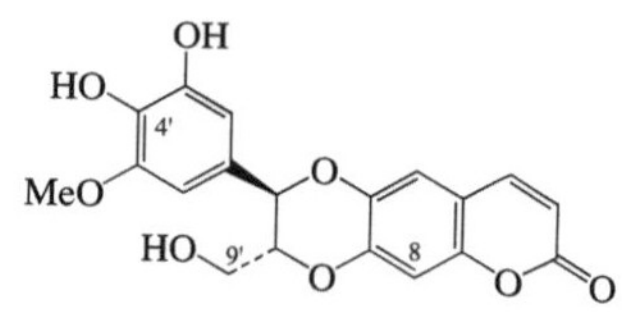

(*7′R,8′R*)-*form*

$C_{19}H_{16}O_8$ 372.329
The nat. prods. Grewin and Sapiumins A and B were isol. as racemates and the latter two were resolved by chiral hplc
(7′R,8′R)-form
Demethoxy
(+)-Sapiumin A
[2265895-20-1]
$C_{18}H_{14}O_7$ 342.303
$[\alpha]_D^{20}$ +41 (c, 0.1 in MeOH)
Demethoxy, 8-methoxy
(+)-Sapiumin B
[2265895-22-3]
$C_{19}H_{16}O_8$ 372.329
$[\alpha]_D^{20}$ +39 (c, 0.1 in MeOH)
(7′S,8′S)-form
Demethoxy
(-)-Sapiumin A
[2265895-19-8]
$[\alpha]_D^{20}$ -42 (c, 0.1 in MeOH)
Demethoxy, 8-methoxy
(-)-Sapiumin B
[2265895-21-2]
$[\alpha]_D^{20}$ -45 (c, 0.1 in MeOH)
(7′RS,8′RS)-form
Grewin [876407-20-4]

Constit. of *Grewia bilamellata*. Exhibits weak antimalarial activity. Amorph. powder

3′,9′-Dideoxy, 5′-O-de-Me, O⁴′-Me

Grewialin

[1418732-27-0]

$C_{19}H_{16}O_6$ 340.331

Constit. of *Grewia optiva*. Amorph. powder. Racemate. λ_{max} 250 (log ε 6.63) (MeOH)

Demethoxy

Maoyancaosu. Sapiumin A

[2265908-41-4]

[72165-29-8] (Maoyancaosu)

$C_{18}H_{14}O_7$ 342.303

Constit. of *Euphorbia lunulata* and *Sapium discolor*. NO prodn. inhibitor. Amorph. powder. Struct. of Maoyancaosu revised in 1984. λ_{max} 230 (log ε 4.36); 261 (log ε 3.79); 289 (log ε 4.09); 344 (log ε 4.06) (MeOH)

Demethoxy, 8-methoxy

Sapiumin B

[2265908-43-6]

$C_{19}H_{16}O_8$ 372.329

Constit. of the leaves of *Sapium discolor*. NO prodn. inhibitor. Amorph. powder. λ_{max} 231 (log ε 4.44); 306 (log ε 4.08) (MeOH)

(7′R*,8′R*)-form

8-Hydroxy, 3′-Me ether

Mucronin A‡

$C_{20}H_{18}O_9$ 402.355

Constit. of *Daphne mucronata*. Shows significant antitubercular activity. Light yellow solid. $[\alpha]^{25}_{D}$ -97.2 (c, 0.31 in MeOH). λ_{max} 230; 235 (sh); 324 (MeOH)

8-Methoxy, 3′-Me ether

Mucronin B‡

$C_{21}H_{20}O_9$ 416.382

Constit. of *Daphne mucronata*. Shows significant antitubercular activity. Light yellow solid. $[\alpha]^{25}_{D}$ -98.2 (c, 0.31 in MeOH). λ_{max} 325 (MeOH).

Wang, D.X. *et al.*, *Yaoxue Xuebao*, 1984, **19**, 261-267 (*Maoyancaosu*).
Ma, C. *et al.*, *J. Nat. Prod.*, 2006, **69**, 346-350 (*Grewin, activity*).
Uddin, G. *et al.*, *Nat. Prod. Res.*, 2013, **27**, 215-220 (*Grewialin*).
Ferheen, S. *et al.*, *Asian J. Chem.*, 2014, **26**, 7262-7264 (*Mucronins A,B, activity*).
Zhang, G.-J. *et al.*, *J. Nat. Prod.*, 2018, **81**, 2251-2258 (*Sapiumins A,B, activity*)

6-(3,4-Dihydroxy-5-methoxyphenyl)-2,3-dihydro-2-[2-(4-hydroxyphenyl)ethyl]-4*H*-pyran-4-one D164

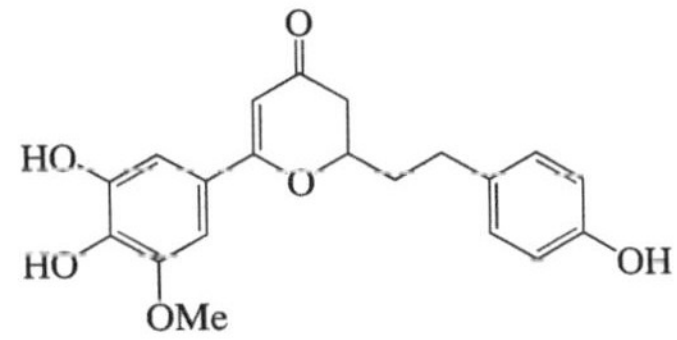

$C_{20}H_{20}O_6$ 356.374

(ξ)-form [2055038-98-5]

Constit. of the fruit of *Amomum tsao-ko*. Yellowish powder.

Zhang, T.-T. *et al.*, *J. Funct. Foods*, 2015, **18**, 423-431 (*Amomum tsao-ko constit*)

1-(3,5-Dihydroxy-4-methylphenyl)-2-propanone D165

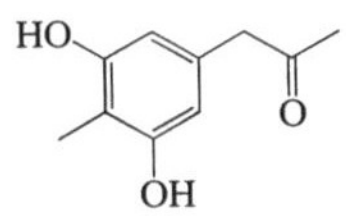

[423171-38-4]

$C_{10}H_{12}O_3$ 180.203

Prod. by a marine-derived *Cladosporium perangustum* FS62. Oil. Bp 205° at 1.4 mm.

Suzuki, T. *et al.*, *J. Heterocycl. Chem.*, 2001, **38**, 1409-1418 (*synth*).
Fan, Z. *et al.*, *J. Asian Nat. Prod. Res.*, 2016, **18**, 1024-1029 (*Cladosporium perangustum constit*)

2,3-Dihydroxy-2-(1-methylpropyl)butanedioic acid D166

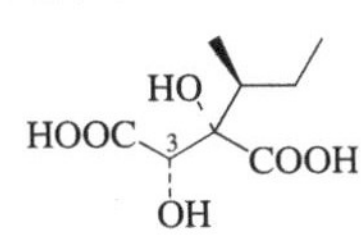

(1′S,2S,3S)-form

2-sec-Butyltartaric acid. 3-Hydroxy-3-sec-butylmalic acid

$C_8H_{14}O_6$ 206.194

(1′S,2S,3S)-form

3-O-(3,4-Dihydroxy-E-cinnamoyl)

[2200278-76-6]

$C_{17}H_{20}O_9$ 368.338

Constit. of *Tithonia diversifolia*. Amorph. yellowish powder. $[\alpha]^{20}_{D}$ +38.5 (c, 0.5 in MeOH)

(1′ξ,2ξ,3ξ)-form

3-O-(3,4-Dihydroxy-E-cinnamoyl)

[1208337-04-5]

Constit. of *Leucosceptrum japonicum*. Yellow oil. $[\alpha]^{22}_{D}$ +20 (c, 0.07 in MeOH). λ_{max} 300 (sh); 331 (log ε 4.3) (MeOH).

Murata, T. *et al.*, *J. Nat. Med.*, 2009, **63**, 402-407 (*Leucosceptrum japonicum constit*).
Ccana-Ccapatinta, G.V. *et al.*, *Tetrahedron: Asymmetry*, 2017, **28**, 1823-1828 (*Tithonia diversifolia constit*)

4,4′-Dihydroxy-9′-nor-7,8′-lign-8-en-7′-one D167

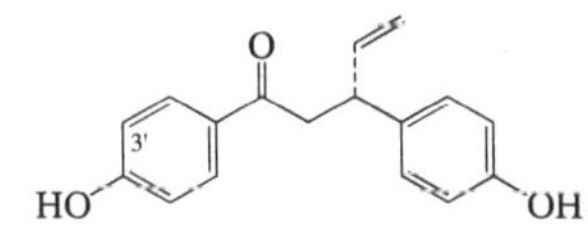

(*R*)-form

1,3-Bis(4-hydroxyphenyl)-4-penten-1-one. 4,4′-Dihydroxy-9′-nor-7′-oxo-7,8′-neolign-8-ene

$C_{17}H_{16}O_3$ 268.312

Provisional norlignan. Biosynth. pathway unclear

(R)-form

3′-Hydroxy

1-(3,4-Dihydroxyphenyl)-3-(4-hydroxyphenyl)-4-penten-1-one. 3′,4,4′-Trihydroxy-9′-nor-7,8′-lign-8-en-7′-one. 3′,4,4′-Trihydroxy-9′-nor-7′-oxo-7,8′-lign-8-ene

[1912455-29-8]

$C_{17}H_{16}O_4$ 284.311

Constit. of the pericarp of *Trapa natans*. Significantly enhances glucose uptake (hypoglycaemic activity) in C2C12 myotubes. Amorph. pale yellow powder. $[\alpha]^{25}_{D}$ -12.4 (c, 0.1 in MeOH). λ_{max} 235 (log ε 4.1); 277 (log ε 3.6); 309 (log ε 3.4) (MeOH)

(ξ)-form [226417-45-4]

Constit. of *Anemarrhena asphodeloides* and *Asparagus cochinchinensis*. Hyaluronidase inhibitor. Amorph. powder. Mp 148 - 150°. $[\alpha]^{26}_{D}$ +3 (c, 0.27 in $CHCl_3$/MeOH). $[\alpha]^{20}_{D}$ -9.8 (c, 0.07 in MeOH). Not clear if the two isolates may be different enantiomers.

Jeong, S.-J. *et al.*, *Planta Med.*, 1999, **65**, 367-368 (*Anemarrhena aspheloides constit, activity*).
Zhang, H-J. *et al.*, *J. Nat. Prod.*, 2004, **67**, 194-200 (*Asparagus cochinchinensis constit*).
Huang, H.-C. *et al.*, *J. Agric. Food Chem.*, 2016, **64**, 3794-3803 (*Trapa natans constit, activity*)

4,4′-Dihydroxy-7′-oxo-8,8′-lign-7-ene-9,9′-dioic acid D168

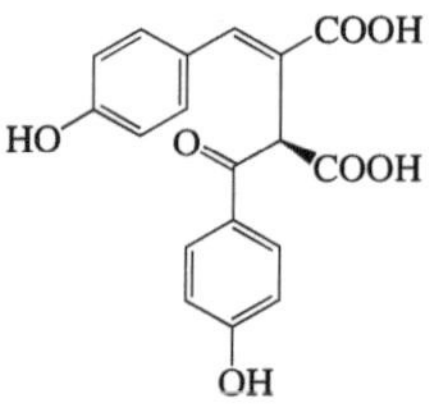

2-(4-Hydroxybenzoyl)-3-(4-hydroxybenzylidene)butanedioic acid
$C_{18}H_{14}O_7$ 342.303
Highly oxidised dibenzylbutane lignan
(R,E)-form
Chaerohirsutin B [762242-48-8]
Constit. of the roots of *Chaerophyllum hirsutum*. Moderately cytotoxic to various human cancer cell lines. Pale yellow solid. $[\alpha]_D^{22}$ +52.1 (c, 0.12 in EtOH). λ_{max} 285 (log ε 4.41) (MeOH).
Dall'Acqua, S. *et al., J. Nat. Prod.*, 2004, **67**, 1588-1590 (*Chaerohirsutin B, activity*)

2-(2,4-Dihydroxyphenyl)-5-benzofurancarboxylic acid D169

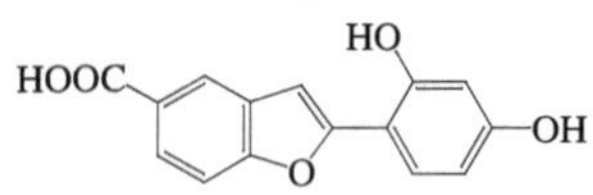

4,7′-Epoxy-2′,4′-dihydroxy-8,9,9′-trinor-3,8′-lign-7′-en-7-oic acid
$C_{15}H_{10}O_5$ 270.24
4′-Me ether, Me ester
Krametosan. *Kramentosan*
[404853-69-6]
$C_{17}H_{14}O_5$ 298.294
Constit. of the roots of *Krameria tomentosa*. Yellow oil. λ_{max} 223 (log ε 4.2); 283 (log ε 3.8); 332 (log ε 3.6) (MeOH).
Silva, S.A. *et al., Nat. Prod. Lett.*, 2001, **15**, 323-329 (*Krametosan*)

10-(2,4-Dihydroxyphenyl)-2,8-bis(3,4-dihydroxyphenyl)-3,4,9,10-tetrahydro-2*H*,8*H*-benzo[1,2-*b*:3,4-*b*′]dipyran-3,9-diol, CAS D170

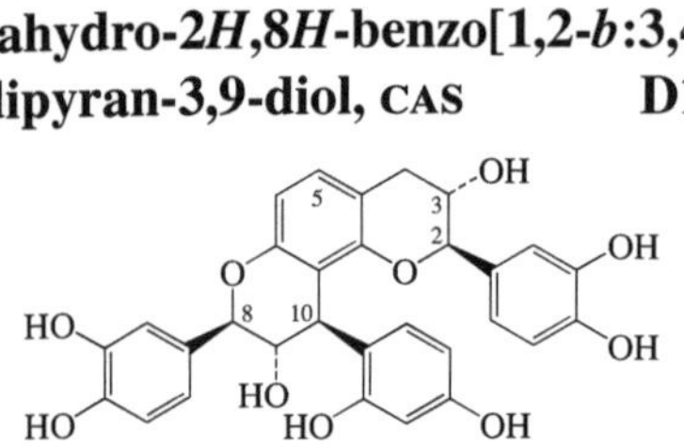

(2*R*,3*S*,8*R*,9*S*,10*R*)-*form*

$C_{30}H_{26}O_{10}$ 546.528
(2R,3S,8R,9S,10R)-form [127644-75-1]
Constit. of *Colophospermum mopane*
(2R,3S,8R,9S,10S)-form [102258-23-1]
Constit. of the heartwood of *Colophospermum mopane*
(2R,3S,8S,9S,10R)-form [127644-77-3]
Constit. of *Colophospermum mopane*
(2S,3S,8R,9S,10S)-form [105880-85-1]
Constit. of *Colophospermum mopane*
(2S,3S,8S,9S,10R)-form [127644-79-5]
Constit. of *Colophospermum mopane*.
Malan, J.C.S. *et al., J. Chem. Soc., Perkin Trans. 1*, 1990, 209-218 (*Colophospermum mopane constit*)

4-(2,4-Dihydroxyphenyl)-2,8-bis(3,4-dihydroxyphenyl)-3,4,9,10-tetrahydro-2*H*,8*H*-benzo[1,2-*b*:3,4-*b*′]dipyran-3,5,9-triol, CAS D171

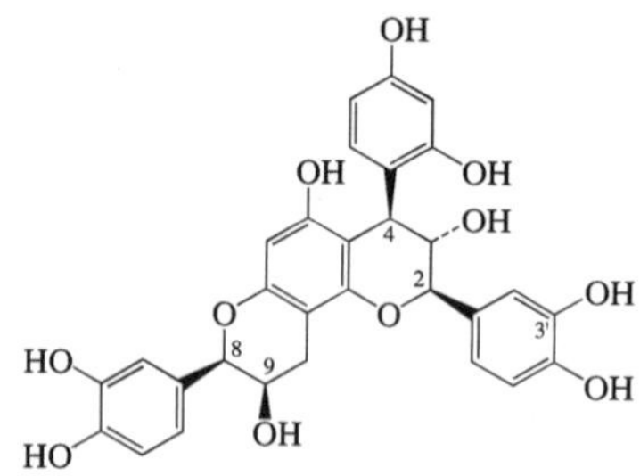

(2*R*,3*S*,4*R*,8*R*,9*R*)-*form*

[105805-43-4]
$C_{30}H_{26}O_{11}$ 562.527
(2R,3S,4R,8R,9R)-form [130932-39-7]
Constit. of *Baikiaea plurijuga* and *Guibourtia coleosperma*
(2R,3S,4R,8R,9S)-form [119065-94-0]
Constit. of *Baikiaea plurijuga* and *Guibourtia coleosperma*
(2R,3S,4S,8R,9R)-form [130932-41-1]
Constit. of *Baikiaea plurijuga*
(2S,3R,4S,8S,9R)-form [130932-51-3]
Constit. of *Baikiaea plurijuga*
(2S,3S,4R,8R,9R)-form [130932-43-3]
Constit. of *Baikiaea plurijuga* and *Guibourtia coleosperma*
(2S,3S,4R,8R,9S)-form [119065-88-2]
Constit. of *Baikiaea plurijuga* and *Guibourtia coleosperma*
3′-Deoxy
[1256087-99-6]
$C_{30}H_{26}O_{10}$ 546.528
Constit. of the roots of *Rosa multiflora*. Amorph. powder. $[\alpha]_D^{25}$ -100 (c, 0.001 in MeOH)
(2S,3S,4R,8S,9R)-form [130932-45-5]
Constit. of *Baikiaea plurijuga*.
Steynberg, J.P. *et al., J. Chem. Soc., Perkin Trans. 1*, 1988, 3323-3329 (*struct, synth*).
Steynberg, J.P. *et al., J. Chem. Soc., Perkin Trans. 1*, 1988, 3331-3338 (*struct, abs config*).
Steynberg, J.P. *et al., J. Chem. Soc., Perkin Trans. 1*, 1989, 1709-1710 (*corrigenda*).
Steynberg, J.P. *et al., Phytochemistry*, 1990, **29**, 2979-2989 (*Baikiaea plurijuga constits, struct, abs config*).
Park, K.H. *et al., Chem. Pharm. Bull.*, 2010, **58**, 1227-1231 (*Rosa multiflora constit*)

8-(2,4-Dihydroxyphenyl)-2,10-bis(3,4-dihydroxyphenyl)-3,4,9,10-tetrahydro-2*H*,8*H*-benzo[1,2-*b*:3,4-*b*′]dipyran-3,5,9-triol, CAS D172

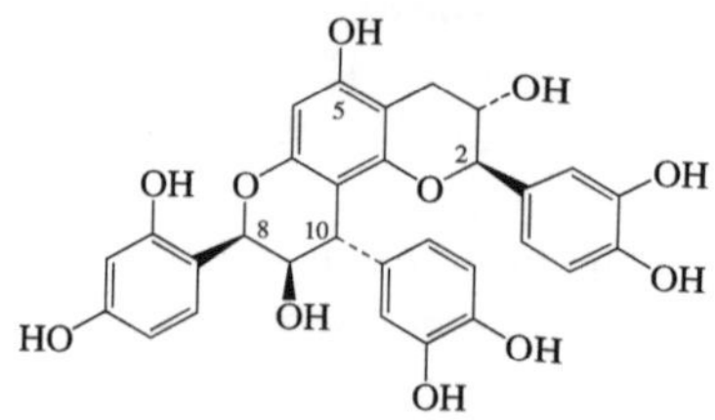

(2*R*,3*S*,8*R*,9*R*,10*S*)-*form*

$C_{30}H_{26}O_{11}$ 562.527
(2R,3S,8R,9R,10S)-form [118964-22-0]
Constit. of the heartwood of *Baikiaea plurijuga*
(2R,3S,8S,9R,10S)-form [119065-84-8]
Constit. of *Baikiaea plurijuga*.
Steynberg, J.P. *et al., J. Chem. Soc., Perkin Trans. 1*, 1988, 3331-3338 (*Baikiaea plurijuga constits, synth*).
Steynberg, J.P. *et al., J. Chem. Soc. Pak.*, 1989, 1709-1710 (*corrigenda*)

9-(3,4-Dihydroxyphenyl)-6,7-dihydro-5*H*-benzocycloheptene-2,3,5,6-tetrol D173

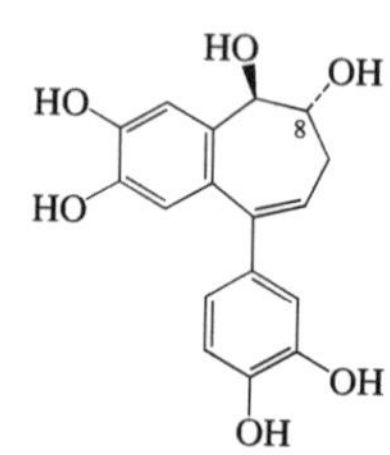

8,9′-Cyclo-9-nor-2,7′-lign-7′-ene-3′,4,4′,5,7,8-hexol. 8′,9-Cyclo-3′,4,4′,5,7,8-hexahydroxy-9′-nor-2,7′-neolign-7′-ene
$C_{17}H_{16}O_6$ 316.309
Lignan numbering shown
(7R*,7R*)-form
8-O-β-D-Glucopyranoside
Crassifoside D
[910054-09-0]
$C_{23}H_{26}O_{11}$ 478.45
Constit. of the rhizomes of *Curculigo crassifolia*. Brown powder. $[\alpha]_D^{29}$ -145.8 (c, 0.27 in MeOH). λ_{max} 208 (log ε 4.49); 363 (log ε 2.58); 375 (log ε 2.39) (MeOH).

Li, N. *et al.*, *Z. Naturforsch., B*, 2006, **61**, 611-614 (*Crassifoside D*)

2-(3,4-Dihydroxyphenyl)-2,3-dihydro-5,6-dihydroxy-3-benzofuranmethanol D174

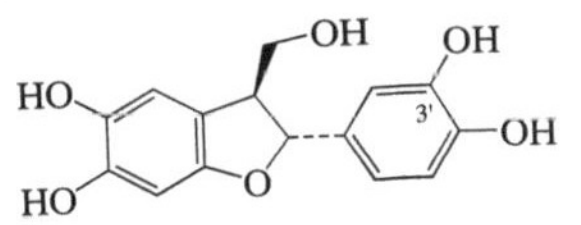

(2*R*,3*S*)-*form*

2-(3,4-Dihydroxyphenyl)-2,3-dihydro-5,6-dihydroxy-3-(hydroxymethyl)benzofuran

$C_{15}H_{14}O_6$ 290.271

(2R,3S)-form

3′,5,6-Tri-Me ether

[2088881-37-0]

Exhibits moderate cytotoxic activity. $[\alpha]^{20}_{D}$ -15 (c, 0.1 in MeOH)

(2S,3R)-form

3′,5,6-Tri-Me ether

[2489663-76-3]

$[\alpha]^{20}_{D}$ +17 (c, 0.1 in MeOH)

(2RS,3SR)-form

3′,5,6-Tri-Me ether

2,3-Dihydro-2-(4-hydroxy-3-methoxyphenyl)-5,6-dimethoxy-3-benzofuranmethanol. ***Crataegifin D***

$C_{18}H_{20}O_6$ 332.352

Constit. of the fruit of *Crataegus pinnatifida*. Yellow oil. $[\alpha]^{20}_{D}$ +1.3 (c, 0.1 in MeOH). Partial racemate resolved by chiral hplc. λ_{max} 288 (log ε 3.44) (MeOH).

Guo, R. *et al.*, *Fitoterapia*, 2019, **137**, 104287:1-8 (*Crataegifins D, activity*)

2-(3,4-Dihydroxyphenyl)-2,3-dihydro-5,7-dihydroxy-3-benzofuranmethanol D175

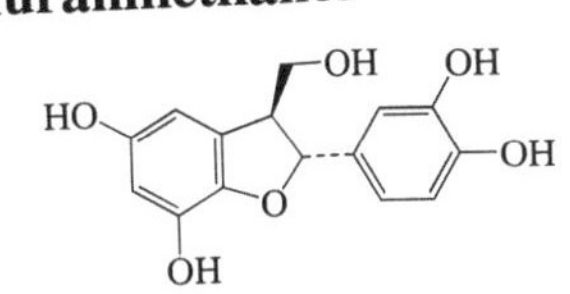

(2*R*,3*S*)-*form*

2-(3,4-Dihydroxyphenyl)-2,3-dihydro-5,7-dihydroxy-3-(hydroxymethyl)benzofuran

$C_{15}H_{14}O_6$ 290.271

(2R,3S)-form

3′,7-Di-Me ether

2,3-Dihydro-5-hydroxy-2-(4-hydroxy-3-methoxyphenyl)-7-methoxy-3-benzofuranmethanol. ***Samwirin***

[1966158-09-7]

$C_{17}H_{18}O_6$ 318.325

Constit. of *Leycesteria formosa* and *Sambucus williamsii*. Promotor of cell proliferation. Yellow powder. $[\alpha]^{26}_{D}$ -6.5 (c, 0.6 in MeOH). λ_{max} 208 (log ε 4.94); 263 (log ε 4.17); 288 (log ε 4.20) (MeOH)

3′,5,7-Tri-Me ether, 1″-O-β-D-xylopyranoside

Difengpioside B

[2041284-98-2]

$C_{23}H_{28}O_{10}$ 464.467

Constit. of the stem bark of *Illicium difengpi*. Amorph. powder. $[\alpha]^{20}_{D}$ +21 (c, 0.2 in MeOH). λ_{max} 206 (log ε 4.87); 281 (log ε 3.58) (MeOH)

(2R*,3R*)-form

3′,5,7-Tri-Me ether

2,3-Dihydro-2-(4-hydroxy-3-methoxyphenyl)-5,7-dimethoxy-3-benzofuranmethanol

[2417849-60-4]

$C_{18}H_{20}O_6$ 332.352

Constit. of stems of *Eleutherococcus senticosus*. Amorph. brown powder. $[\alpha]^{25}_{D}$ -2.1 (c, 0.1 in MeOH). λ_{max} 221; 292 (MeOH).

Xiao, H.-H. *et al.*, *J. Asian Nat. Prod. Res.*, 2015, **17**, 625-632 (*Samwirin, activity*).
Pan, Z.-H. *et al.*, *Molecules*, 2016, **21**, 607:1-8 (*Difengpioside B*).
Zhang, L. *et al.*, *Fitoterapia*, 2017, **121**, 58-63 (*Eleutherococcus senticosus constit*).
Xiang, C.-P. *et al.*, *Tianran Chanwu Yanjiu Yu Kaifa*, 2017, **29**, 58-62 (*Leycesteria formosa constit*)

3-(3,4-Dihydroxyphenyl)-2,3-dihydro-6-hydroxy-1,4-benzodioxin-2-methanol D176

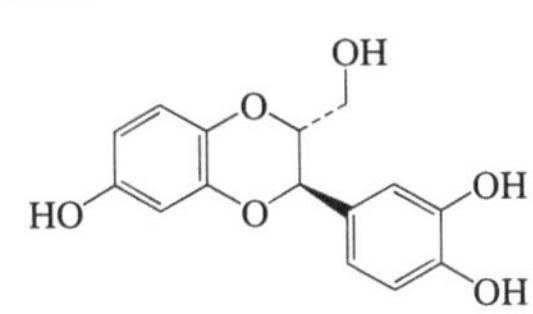

$C_{15}H_{14}O_6$ 290.271

(2R,3R)-form

3′-Me ether

2,3-Dihydro-6-hydroxy-3-(4-hydroxy-3-methoxyphenyl)-1,4-benzodioxin-2-methanol. ***Samwiphenol***

[1966158-10-0]

$C_{16}H_{16}O_6$ 304.298

Constit. of *Sambucus williamsii*. Yellow powder. $[\alpha]^{26}_{D}$ +10.5 (c, 0.6 in MeOH). λ_{max} 206 (log ε 4.74); 231 (log ε 4.16); 286 (log ε 3.88) (MeOH).

Xiao, H.-H. *et al.*, *J. Asian Nat. Prod. Res.*, 2015, **17**, 625-632 (*Samwiphenol, cd*)

2-(3,4-Dihydroxyphenyl)-2,3-dihydro-5-hydroxy-3-benzofuranmethanol D177

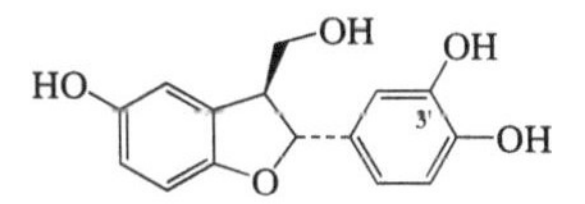

2-(3,4-Dihydroxyphenyl)-2,3-dihydro-5-hydroxy-3-(hydroxymethyl)benzofuran

$C_{15}H_{14}O_5$ 274.272

(2R,3S)-form

3′-Me ether

2,3-Dihydro-5-hydroxy-2-(4-hydroxy-3-methoxyphenyl)-3-benzofuranmethanol. ***Herpetosiol F***

[2374936-69-1]

$C_{16}H_{16}O_5$ 288.299

Constit. of *Herpetospermum pedunculosum*. Amorph. powder. $[\alpha]^{25}_{D}$ -4.9 (c, 0.3 in MeOH). λ_{max} 240 (log ε 3.22); 290 (log ε 2.75) (MeOH).

Ma, Y. *et al.*, *Phytochemistry*, 2019, **164**, 102-110 (*Herpetosiol F*)

2-(3,5-Dihydroxyphenyl)-2,3-dihydro-7-hydroxy-3-(hydroxymethyl)-5-benzofurancarboxylic acid D178

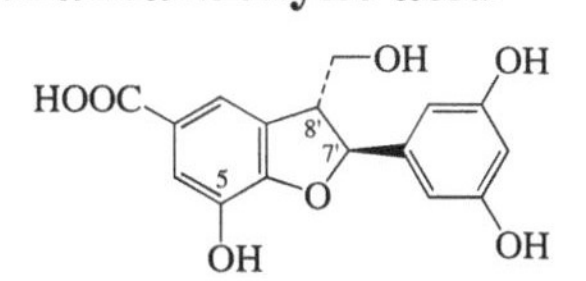

(7′*S*,8′*R*)-*form*

4,7′-Epoxy-3′,5,5′,9′-tetrahydroxy-8,9-dinor-3,8′-lignan-7-oic acid

$C_{16}H_{14}O_7$ 318.281

Structures in this series may need revision. Phenyl 3,5-dioxygenated substn. pattern in doubt as a result of work done by Yang *et al* in 2015

(7′S,8′R)-form

3′,5-Di-Me ether

4,7′-Epoxy-3′,9′-dihydroxy-5,5′-dimethoxy-8,9-dinor-3,8′-lignan-7-oic acid

[1251918-36-1]

$C_{18}H_{18}O_7$ 346.335

Constit. of the fruit of *Livistona chinensis*. Amorph. powder. $[\alpha]^{25}_{D}$ +8.2 (c, 0.2 in MeOH)

(7′R*,8′S*)-form

7-Aldehyde, 3′,5-di-Me ether

4,7′-Epoxy-3′,9′-dihydroxy-5,5′-dimethoxy-8,9-dinor-3,8′-lignan-7-al. ***Curcasinlignan B***

[1427029-44-4]

$C_{18}H_{18}O_6$ 330.336

Constit. of *Acanthopanax senticosus* and *Jatropha curcas*. Pale yellow oil. $[\alpha]_D^{25}$ -13.1 (c, 0.16 in MeOH). λ_{max} 194 (log ε 4.08); 204 (log ε 4.36); 231 (log ε 4.04); 288 (log ε 3.85); 303 (log ε 3.83) (MeOH).

Zeng, X. *et al.*, *Fitoterapia*, 2012, **83**, 104-109 (*Livistona chinensis constit*).

Xu, J.J. *et al.*, *Z. Naturforsch., B*, 2012, **67**, 176-180 (*Curcasinlignan B*).

Wang, Y.-N. *et al.*, *J. Nat. Prod.*, 2015, **78**, 705-711 (*pmr, struct*).

Li, J.-L. *et al.*, *Arch. Pharmacal Res.*, 2016, **39**, (*Curcasinlignan B*)

3-(3,4-Dihydroxyphenyl)-2,3-dihydro-2-(hydroxymethyl)-7*H*-dioxino[2,3-*c*]xanthen-7-one, CAS D179

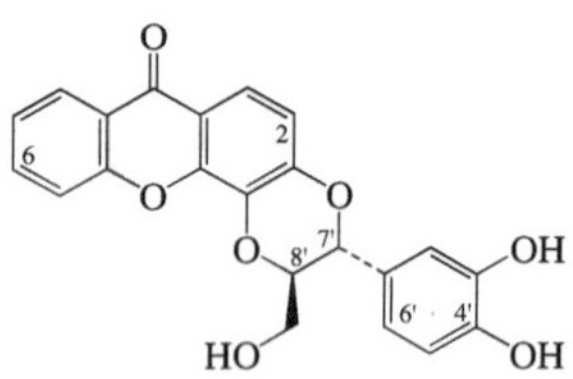

$C_{22}H_{16}O_7$ 392.363

Xanthonolignans. Predominantly isol. as racemates or partial racemates having the same rel. config.

(7′R*,8′R*)-form

2-Hydroxy, 3′-Me ether

2-O-Demethylkielcorin

[1197054-69-5]

$C_{23}H_{18}O_8$ 422.389

Constit. of the stems of *Hypericum chinense*. Yellow powder. $[\alpha]_D$ +15.4 (c, 0.5 in MeOH)

8-Hydroxy, 3′-Me ether

Hyperjaponol A. *Hypejaponol A*

[2374811-38-6]

$C_{23}H_{18}O_8$ 422.389

Constit. of *Hypericum japonicum*. Exhibits moderate cytotoxic activity. Light yellow powder. $[\alpha]_D^{25}$ -7.0 (c, 0.2 in Py). Partial racemate. Two synonyms used. λ_{max} 250 (log ε 3.51); 285 (log ε 2.80); 319 (log ε 2.98) (MeOH)

6,8-Dihydroxy, 3′-Me ether

5′-Demethoxycadensin G

[164365-74-6]

$C_{23}H_{18}O_9$ 438.388

Constit. of *Cratoxylum cochinchinense*, *Cratoxylum glaucum* and *Hypericum chinense*. Exhibits moderate cytotoxic activity. Antioxidant. Gum. Mp 258 - 260°. $[\alpha]_D$ +10.9 (c, 0.14 in MeOH)

2-Methoxy, 3′-Me ether

Kielcorin

[64280-48-4]

$C_{24}H_{20}O_8$ 436.416

Constit. of *Kielmeyera coriacea*, *Psorospermum febrifugum* and several *Hypericum* spp.. Cryst. ($CHCl_3$/EtOH). Mp 250 - 251°. Racemic. λ_{max} 238 (ε 23990); 255 (ε 22387); 277 (ε 6456); 317 (ε 9440) (MeOH) (Berdy). [neutral] λ_{max} 239 (ε 42300); 286 (ε 10800); 318 (ε 10100) (EtOH) (Berdy). [base] λ_{max} 257 (ε 48600); 292 (ε 14120); 318 (ε 17100) (EtOH/NaOH) (Berdy)

2-Methoxy, 4′-Me ether

Hyperielliptone HD

$C_{24}H_{20}O_8$ 436.416

Constit. of the heartwood of *Hypericum geminiflorum*. Powder. Racemic. λ_{max} 210 (log ε 4.2); 235 (log ε 4.1); 312 (log ε 3.7) (MeOH)

2-Methoxy, 4′-Me ether, 9′-Ac

Hypericorin C

$C_{26}H_{22}O_9$ 478.453

Constit. of the roots of *Hypericum oblongifolium*. Amorph. powder. Mp 230 - 232°. $[\alpha]_D^{20}$ +0.3 (c, 0.01 in Me_2CO). λ_{max} 248 (log ε 4.34); 308 (log ε 3.83); 346 (log ε 3.82) (MeOH)

2-Methoxy, 8-hydroxy, 4′-Me ether

Cadensin A

[64280-46-2]

$C_{24}H_{20}O_9$ 452.415

Constit. of *Caraipa densiflora*, *Hypericum henryi* and *Vismia guaramirangae*. Yellow prisms. Mp 260 - 270° dec.. λ_{max} 233 (ε 39000); 256 (ε 43000); 321 (ε 17300); 374 (ε 9600) (EtOH). [base] λ_{max} 251 (ε 47400); 285 (ε 26900); 400 (ε 11800) (EtOH/NaOH aq.)

2-Methoxy, 6,8-dihydroxy, 3′-Me ether

Calophyllumin C

[591753-64-9]

$C_{24}H_{20}O_{10}$ 468.414

Constit. of the roots of *Calophyllum inophyllum*. Mp 240°. $[\alpha]_D^{25}$ -38.2 (c, 0.06 in DMSO). λ_{max} 224 (log ε 3.74); 283 (log ε 3.14); 313 (log ε 3.28) (MeOH)

5′-Methoxy, 2-hydroxy, 3′-Me ether

Subalatin

[119269-69-1]

$C_{24}H_{20}O_9$ 452.415

Constit. of *Hypericum chinense* and *Hypericum subalatum*. Shows moderate cytotoxicity. Pale yellow needles ($CHCl_3$/MeOH). Mp 265° dec.. $[\alpha]_D^{25}$ +600 (c, 0.005 in MeOH)

5′-Methoxy, 1,7-dihydroxy, 3′-Me ether

Esculentin A‡

$C_{24}H_{20}O_{10}$ 468.414

Constit. of *Garcinia esculenta*. Yellow powder. $[\alpha]_D^{25}$ -4 (c, 0.02 in MeOH). λ_{max} 202 (log ε 4.86); 237 (log ε 4.47); 275 (log ε 4.55); 315 (log ε 3.93); 390 (log ε 3.64) (MeOH)

5′-Methoxy, 6,8-dihydroxy, 3′-Me ether

Cadensin G

[124902-10-9]

$C_{24}H_{20}O_{10}$ 468.414

Constit. of *Hypericum chinense*, *Hypericum perforatum* and *Psorospermum febrifugum*. Exhibits moderate cytotoxic activity. Powder (MeOH). Mp 265 - 266°. $[\alpha]_D$ +10.3 (c, 0.08 in MeOH). λ_{max} 218 (ε 31623); 243 (ε 36307); 281 (ε 8130); 320 (ε 16600) (MeOH) (Berdy)

6-Methoxy, 8-hydroxy, 3′-Me ether

5′-Demethoxy-6-O-methylcadensin G

[1260621-01-9]

$C_{24}H_{20}O_9$ 452.415

Constit. of the fruit of *Cratoxylum formosum* var. *pruniflorum*. Yellow powder. $[\alpha]_D^{26}$ +53.4 (c, 0.06 in Me_2CO). λ_{max} 253 (log ε 4.37); 281 (log ε 3.83); 318 (log ε 3.97) ($CHCl_3$)

2,5′-Dimethoxy, 3′-Me ether

Cadensin D. *Hypericorin*

[102349-35-9]

[110901-51-4]

$C_{25}H_{22}O_9$ 466.442

Constit. of *Hypericum canariensis* and *Hypericum mysorense*. Cryst. (MeOH). Mp 243 - 245°. Identity of Cadensin D and Hypericorin not fully established; they both appear to be racemic. λ_{max} 242 (log ε 4.51); 253 (log ε 4.4); 317 (log ε 4.18); 363 (sh) (log ε 3.77) (MeOH) (Cadensin D)

2,5′-Dimethoxy, 2′-hydroxy

Hypericorin D

$C_{24}H_{20}O_{10}$ 468.414

Constit. of the roots of *Hypericum oblongifolium*. Amorph. powder. Mp 250 - 254°. $[\alpha]^{20}_{D}$ +0.6 (c, 0.01 in Me_2CO). λ_{max} 250 (log ε 4.50); 302 (log ε 4.38); 387 (log ε 3.71) (MeOH)

2,5′-Dimethoxy, 8-hydroxy, 3′-Me ether

Cadensin C

[87338-98-5]

$C_{25}H_{22}O_{10}$ 482.441

Constit. of *Vismia guaramirangae*. Needles (EtOH) (as tri-Ac). Mp 210 - 214° (tri-Ac)

2,5′-Dimethoxy, 6,8-dihydroxy, 3′-Me ether

Calophyllumin A

[180031-65-6]

$C_{25}H_{22}O_{11}$ 498.44

Constit. of *Calophyllum inophyllum*. Pale yellow powder. λ_{max} 215; 255; 283; 320; 355 (sh) (MeOH)

2,5′-Dimethoxy, 4′-deoxy

Hypericorin B

[1387006-28-1]

$C_{24}H_{20}O_8$ 436.416

Constit. of the twigs of *Hypericum oblongifolium*. Exhibits antiinflammatory activity. Amorph. powder. Mp 240 - 243°. λ_{max} 236 (log ε 4.38); 286 (log ε 3.81); 315 (log ε 3.71); 359 (log ε 3.39) (MeOH)

2,5′-Dimethoxy, 4′-deoxy, 9′-Ac

Hypericorin A

[1387006-27-0]

$C_{26}H_{22}O_9$ 478.453

Constit. of the twigs of *Hypericum oblongifolium*. Exhibits antiinflammatory activity. Amorph. solid. Mp 235 - 238°. λ_{max} 240 (log ε 4.34); 286 (log ε 3.83); 316 (log ε 3.82); 362 (log ε 3.44) (no solvent reported)

2,6-Dimethoxy, 3′-Me ether

6-Methoxykielcorin

[131189-49-6]

$C_{25}H_{22}O_9$ 466.442

Constit. of *Hypericum reflexum*. λ_{max} 244; 315 (MeOH)

2,6′-Dimethoxy, 8-hydroxy, 4′-Me ether

Cadensin B

[64280-47-3]

$C_{25}H_{22}O_{10}$ 482.441

Constit. of *Caraipa densiflora*. Yellow cryst.. Mp 236 - 238°. λ_{max} 256 (ε 39500); 321 (ε 15000); 371 (ε 7500) (EtOH)

2,8-Dimethoxy, 3′-Me ether

Isocadensin D

[124902-06-3]

$C_{25}H_{22}O_9$ 466.442

Constit. of *Psorospermum febrifugum*. Powder (MeOH). Mp 278 - 279°. λ_{max} 240 (log ε 4.39); 242 (log ε 4.39); 247 (log ε 4.38); 310 (log ε 3.97); 355 (log ε 3.83) (MeOH)

2,8-Dimethoxy, 6-hydroxy, 3′-Me ether

Calophyllumin B

[180031-66-7]

$C_{25}H_{22}O_{10}$ 482.441

Constit. of *Calophyllum austroindicum*. Pale yellow powder. Racemic. λ_{max} 207; 245 (sh); 252; 285; 311; 340 (sh) (MeOH)

5′,6-Dimethoxy, 8-hydroxy, 3′-Me ether

Hyperielliptone HC

[1033905-39-3]

$C_{25}H_{22}O_{10}$ 482.441

Constit. of the heartwood of *Hypericum geminiflorum*. Powder. Racemic. λ_{max} 210 (log ε 4.2); 248 (log ε 4.2); 317 (log ε 3.7) (MeOH)

5′,8-Dimethoxy, 6-hydroxy, 3′-Me ether

Cudraxanthone T

$C_{25}H_{22}O_{10}$ 482.441

Constit. of the roots of *Cudrania fruticosa*. Amorph. powder (MeOH). Mp 268°. $[\alpha]^{20}_{D}$ -8.4 (c, 0.05 in MeOH). λ_{max} 249 (log ε 4.49); 283 (log ε 3.85); 310 (log ε 3.98) (MeOH)

2,5′,8-Trimethoxy, 3′-Me ether

Cadensin F

[124902-08-5]

$C_{26}H_{24}O_{10}$ 496.468

Constit. of *Psorospermum febrifugum*. Powder (MeOH). Mp 237 - 239°. Config. not determined. λ_{max} 244 (log ε 4.88); 249 (log ε 4.88); 310 (log ε 4.4); 353 (log ε 4.25) (MeOH).

Castelão, J.F. *et al.*, *Phytochemistry*, 1977, **16**, 735-740 (*Cadensins A,B, Kielcorin*).

Nielsen, H. *et al.*, *Phytochemistry*, 1978, **17**, 2040-2041 (*Kielcorin, struct*).

Delle Monache, F. *et al.*, *Phytochemistry*, 1983, **22**, 227-232 (*Cadensin C*).

Vishwakarma, R.A. *et al.*, *Indian J. Chem., Sect. B*, 1986, **25**, 1155-1156 (*Hypericorin*).

Cardona, M.L. *et al.*, *J. Nat. Prod.*, 1986, **49**, 95-100 (*Cadensins D,F*).

Pinto, M.N. de M.R.A. *et al.*, *Phytochemistry*, 1987, **26**, 2045-2048 (*synth, bibl*).

Chen, M.-T. *Heterocycles*, 1988, **27**, 2589-2594 (*Subalatin, Kielcorin*).

Tanaka, H. *et al.*, *Chem. Pharm. Bull.*, 1989, **37**, 1916-1918 (*Kielcorin, synth*).

Abou-Shoer, M. *et al.*, *Phytochemistry*, 1989, **28**, 2483-2487 (*Cadensins F,G, Isocadensin D*).

Cardona, M.L. *et al.*, *Phytochemistry*, 1990, **29**, 3003-3006 (*6-Methoxykielcorin*).

Sia, G.L. *et al.*, *Phytochemistry*, 1995, **38**, 1521-1528 (*5′-Demethoxycadensin G*).

Iinuma, M. *et al.*, *Heterocycles*, 1996, **43**, 1521-1527 (*Calophyllumins A,B*).

Venkatraman, G. *et al.*, *Tetrahedron Lett.*, 1996, **37**, 2643-2646 (*5′-Demethoxycadensin G, synth*).

Peres, V. *et al.*, *Phytochemistry*, 1997, **44**, 191-214 (*rev, occur*).

Wu, Q.-L. *et al.*, *Phytochemistry*, 1998, **49**, 1395-1402 (*Kielcorin*).

Pinto, M.M.M. *et al.*, *Curr. Med. Chem.*, 2003, **10**, 1-12 (*rev, chem, biol*).

Wu, Y. *et al.*, *Huaxue Xuebao*, 2003, **61**, 1047-1051 (*Calophyllumin C*).

Sousa, E.P. *et al.*, *Chirality*, 2004, **16**, 279-285 (*resoln*).

Wilairat, R. *et al.*, *Planta Med.*, 2005, **71**, 680-682 (*Kielcorin*).

Wu, C.-C. *et al.*, *J. Nat. Prod.*, 2008, **71**, 1027-1031 (*Hyperielliptones HC,HD*).

Tanaka, N. *et al.*, *Phytochemistry*, 2009, **70**, 1456-1461 (*Hypericum chinense constits*).

Boonnak, N. *et al.*, *Aust. J. Chem.*, 2010, **63**, 1550-1556 (*Demethoxymethylcadensin G*).

Begum, S.A. *et al.*, *Prog. Chem. Org. Nat. Prod.*, 2010, **93**, 1-70 (*rev*).

Sim, W.C. *et al.*, *Asian J. Chem.*, 2011, **23**, 569-572 (*Cratoxylum glaucum constit, activity*).

Ali, M. *et al.*, *Planta Med.*, 2011, **77**, 2013-2018 (*Hypericorins A,B, activity*).

Ali, M. *et al.*, *Fitoterapia*, 2014, **95**, 258-265 (*Hypericorins C,D*).

Zhu, L.-L. *et al.*, *Planta Med.*, 2014, **80**, 1721-1726 (*Esculentin A*).

Liang, B. *et al.*, *Chin. J. Nat. Med.*, 2015, **13**, 781-785 (*Cudraxanthone T*).

Zhu, W. *et al.*, *Phytochemistry*, 2019, **164**, 33-40 (*Hyperjaponol A, cd, activity*)

2-(3,4-Dihydroxyphenyl)-2,3-dihydro-6-(2-propenyl)-2*H*-1-benzopyran-3,8-diol D180

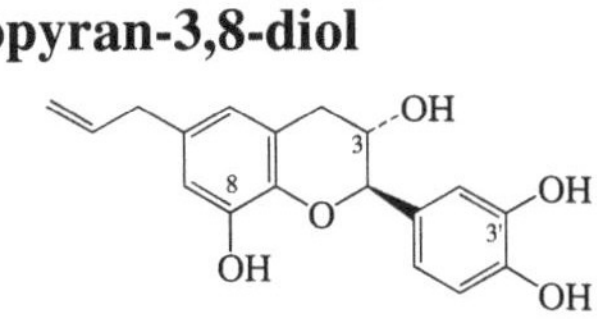

4,7'-Epoxy-3,9'-lign-8-ene-3',4',5,8'-tetrol. 6-Allyl-3,3',4',8-tetrahydroxyflavan
$C_{18}H_{18}O_5$ 314.337
(2R*,3S*)-form
3',8-Di-Me ether
Tulsinol A
[1158305-90-8]
$C_{20}H_{22}O_5$ 342.391
Constit. of *Ocimum sanctum.* Amorph. powder. $[\alpha]_D$ +1.5 (c, 0.04 in MeOH). λ_{max} 238 (ε 10600); 279 (ε 7100) (MeOH)
3-O-[2-Methoxy-4-(2-propenyl)phenyl], 3',8-di-Me ether
Tulsinol B
[1158305-92-0]
$C_{30}H_{32}O_6$ 488.58
Constit. of *Ocimum sanctum.* Pale yellow oil. $[\alpha]_D$ +1.5 (c, 0.02 in MeOH). λ_{max} 222 (sh) (ε 11700); 234 (ε 13300); 279 (ε 6100) (MeOH).

Suzuki, A. *et al., Chem. Pharm. Bull.*, 2009, **57**, 245-251 (*Tulsinols A,B*)

2-(3,4-Dihydroxyphenyl)-2,3-dihydro-4,5,7-trihydroxy-3-benzofuranmethanol D181

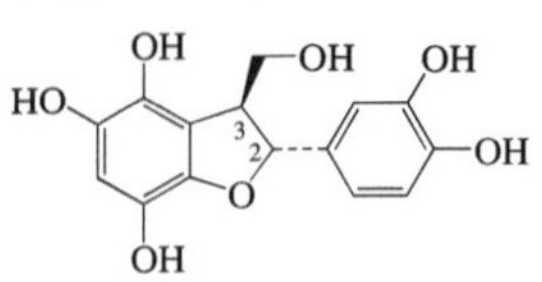

(2*R*,3*S*)-*form*

$C_{15}H_{14}O_7$ 306.270
The nat. prod. Ailanthussin A was isol. as a partial racemate and resolved by chiral hplc
(2R,3S)-form
3',4,5,7-Tetra-Me ether
2,3-Dihydro-2-(4-hydroxy-3-methoxyphenyl)-4,5,7-trimethoxy-3-benzofuranmethanol. (-)-Ailanthussin A
Exhibits significant inhibitory effects against HepG2 cell line. Displays stronger inhibition compared to its enantiomer (+)-Ailanthussin A. $[\alpha]^{20}_D$ -46.4 (c, 0.1 in MeOH)
(2S,3R)-form
3',4,5,7-Tetra-Me ether
(+)-Ailanthussin A
Exhibits inhibitory effects against HepG2 cell line. $[\alpha]^{20}_D$ +43.5 (c, 0.1 in MeOH)
(2RS,3SR)-form
3',4,5,7-Tetra-Me ether
Ailanthussin A
$C_{19}H_{22}O_7$ 362.378
Constit. of the root bark of *Ailanthus altissima.* Exhibits inhibitory effects against HepG2 cell line. Yellow oil. $[\alpha]^{20}_D$ -22.5 (c, 0.1 in MeOH). λ_{max} 210 (log ε 2.44); 285 (log ε 0.39) (MeOH).

Yan, Z.-Y. *et al., Fitoterapia*, 2018, **130**, 234-240 (*Ailanthussin A, cd, activity*)

2-(3,4-Dihydroxyphenyl)-5-[2-(3,4-dihydroxyphenyl)-1,3-dihydroxypropyl]-2,3-dihydro-3-benzofuranmethanol D182

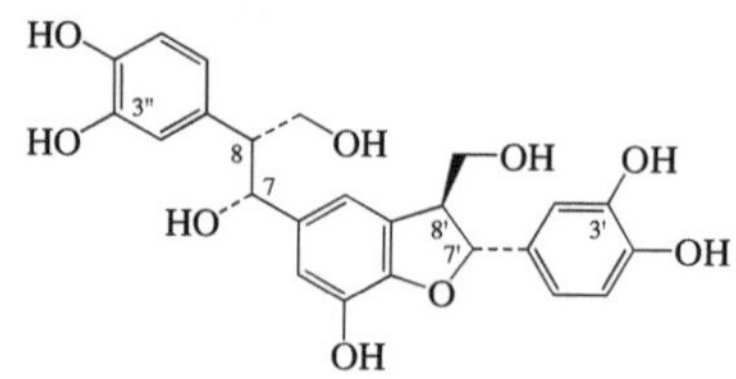

(7*R**,7'*R*,8*R**,8'*S*)-*form*

8-(3,4-Dihydroxyphenyl)-4,7'-epoxy-3,8'-lignan-3',4',5,7,9,9'-hexol. 8-(3,4-Dihydroxyphenyl)-4,7'-epoxy-3',4',5,7,9,9'-hexahydroxy-3,8'-neolignan
$C_{24}H_{24}O_9$ 456.447
(7R*,7'R,8R*,8'S)-form
3',3'',5,7-Tetra-Me ether
4,7'-Epoxy-8-(4-hydroxy-3-methoxyphenyl)-3',5,7-trimethoxy-3,8'-lignan-4',9,9'-triol. 4,7'-Epoxy-4',9,9'-trihydroxy-8-(4-hydroxy-3-methoxyphenyl)-3',5,7-trimethoxy-3,8'-lignan. ***Hawthornnin A***
[1803409-29-1]
$C_{28}H_{32}O_9$ 512.555
Constit. of the seeds of *Crataegus pinnatifida.* Potent inhibitor of NO and TNF-α prodn. Antioxidant. Yellow oil. $[\alpha]^{20}_D$ -7 (c, 0.1 in MeOH). λ_{max} 233 (log ε 0.63); 283 (log ε 0.31) (MeOH)
(7R*,7'R,8S*,8'S)-form
3',3'',5,7-Tetra-Me ether
Hawthornnin C
[1803478-01-4]
$C_{28}H_{32}O_9$ 512.555
Constit. of the seeds of *Crataegus pinnatifida.* Potent inhibitor of NO and TNF-α prodn. Antioxidant. Yellow oil. $[\alpha]^{20}_D$ -4.5 (c, 0.12 in MeOH). λ_{max} 227 (log ε 0.73); 281 (log ε 0.33) (MeOH)
7-Et, 3',3'',5-tri-Me ether
Hawthornnin F
[1803478-04-7]
$C_{29}H_{34}O_9$ 526.582
Constit. of the seeds of *Crataegus pinnatifida.* Potent inhibitor of NO and TNF-α prodn. Antioxidant. Yellow oil. $[\alpha]^{20}_D$ -8.7 (c, 0.13 in MeOH). λ_{max} 226 (log ε 0.73); 281 (log ε 0.38) (MeOH)
(7R*,7'S,8R*,8'R)-form
3',3'',5,7-Tetra-Me ether
Hawthornnin B
[1803409-30-4]
$C_{28}H_{32}O_9$ 512.555
Constit. of the seeds of *Crataegus pinnatifida.* Potent inhibitor of NO and TNF-α prodn. Antioxidant. Yellow oil. $[\alpha]^{20}_D$ +3.5 (c, 0.13 in MeOH). λ_{max} 227 (log ε 0.66); 281 (log ε 0.32) (MeOH)
(7R*,7'S,8S*,8'R)-form
3',3'',5,7-Tetra-Me ether
Hawthornnin D
[1803478-02-5]
$C_{28}H_{32}O_9$ 512.555
Constit. of the seeds of *Crataegus pinnatifida.* Potent inhibitor of NO and TNF-α prodn. Antioxidant. Yellow oil. $[\alpha]^{20}_D$ +7.2 (c, 0.1 in MeOH). λ_{max} 228 (log ε 0.53); 282 (log ε 0.37) (MeOH)
7-Et, 3',3'',5-tri-Me ether
Hawthornnin E
[1803478-03-6]
$C_{29}H_{34}O_9$ 526.582
Constit. of the seeds of *Crataegus pinnatifida.* Potent inhibitor of NO and TNF-α prodn. Antioxidant. Yellow oil. $[\alpha]^{20}_D$ +4.2 (c, 0.16 in MeOH). λ_{max} 228 (log ε 0.63); 285 (log ε 0.32) (MeOH)
(7ξ,7'R*,8ξ,8'S*)-form
3',3'',5-Tri-Me ether
Leptolepisol C
[74498-19-4]
$C_{27}H_{30}O_9$ 498.528
Constit. of *Euonymus acanthocarpus, Larix leptolepis* and *Tripterygium wilfordii.* Amorph. powder. $[\alpha]^{28}_D$ -1 (c, 1.3 in MeOH) (Larix leptolepis constit.). $[\alpha]^{25}_D$ +1 (c, 1.3 in MeOH) (Euonymus acanthocarpus constit.). λ_{max} 280 (log ε 3.78) (EtOH)

3′,3″,5-Tri-Me ether, 4′-O-β-D-glucopyranoside
[1020000-80-9]
$C_{33}H_{40}O_{14}$ 660.669
Constit. of the roots of *Iodes cirrhosa*. Gum. $[\alpha]^{20}_{D}$ -6 (c, 0.1 in MeOH). λ_{max} 208 (log ε 4.8); 279 (log ε 4) (MeOH).

Miki, K. *et al.*, *Phytochemistry*, 1980, **19**, 449-453 (*Leptolepisol C*).
Gan, M. *et al.*, *J. Nat. Prod.*, 2008, **71**, 647-654 (*Iodes cirrhosa glucoside*).
Zhu, J.X. *et al.*, *Arch. Pharmacal Res.*, 2012, **35**, 1739-1747 (*Euonymus acanthocarpus constit*).
Huang, X.-X. *et al.*, *J. Agric. Food Chem.*, 2015, **63**, 7252-7260 (*Hawthornnins A-F, cd, activity*)

2-(3,4-Dihydroxyphenyl)-6-[2-(3,4-dihydroxyphenyl)ethyl]tetrahydro-2*H*-pyran-3-ol D183

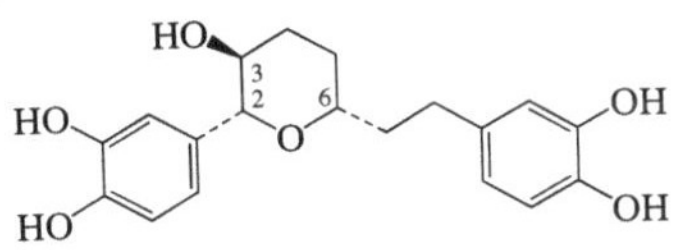

$C_{19}H_{22}O_6$ 346.379
(2R,3S,6S)-form [1869012-34-9]
Constit. of the rhizomes of *Curcuma kwangsiensis*. Yellow oil. $[\alpha]^{25}_{D}$ +9.5 (c, 0.2 in MeOH). λ_{max} 225; 282 (MeOH).

Chen, S.-D. *et al.*, *Fitoterapia*, 2015, **102**, 67-73 (*Curcuma kwangsiensis constit*)

2-(3,4-Dihydroxyphenyl)-6-[2-(3,4-dihydroxyphenyl)ethyl]tetrahydro-2*H*-pyran-4-ol D184

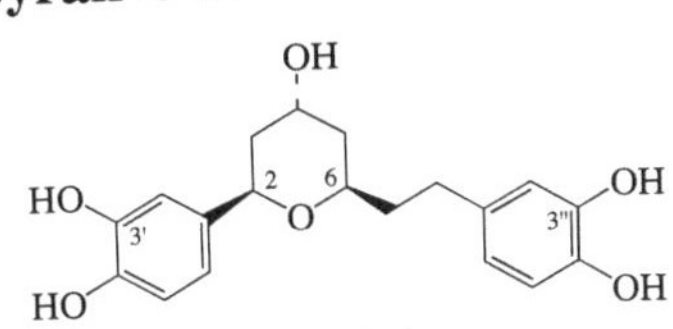

(2*R*,4*R*,6*R*)-form

1,7-Bis(3,4-dihydroxyphenyl)-1,5-epoxy-3-heptanol
$C_{19}H_{22}O_6$ 346.379
(2R,4R,6R)-form
3′,3″-Di-Me ether
Tetrahydro-2-(4-hydroxy-3-methoxyphenyl)-6-[2-(4-hydroxy-3-methoxyphenyl)ethyl]-2H-pyran-4-ol. ***Diosniponol B***
[1443418-29-8]
[1620335-45-6]
$C_{21}H_{26}O_6$ 374.433
Constit. of *Dioscorea nipponica*. Yellowish gum. $[\alpha]^{25}_{D}$ +0.8 (c, 0.005 in MeOH)

(2R,4S,6R)-form
3′,3″-Di-Me ether
Diosniponol A
[1443418-27-6]
[1620335-44-5]
$C_{21}H_{26}O_6$ 374.433
Constit. of *Dioscorea nipponica*. Gum. $[\alpha]^{25}_{D}$ +4.8 (c, 0.06 in MeOH)

(2S,4R,6S)-form
3′,3″-Di-Me ether, 4′-O-β-D-glucopyranoside
Dioscoroside E₁
[2250036-69-0]
$C_{27}H_{36}O_{11}$ 536.574
Constit. of the rhizomes of *Dioscorea septemloba*. Exhibits triglyceride inhibitory effects. Powder. $[\alpha]^{25}_{D}$ +9.5 (c, 0.55 in MeOH). λ_{max} 225 (log ε 4.13); 278 (log ε 3.71) (MeOH).

Woo, K.W. *et al.*, *Bioorg. Med. Chem. Lett.*, 2013, **23**, 3806-3809 (*Diosniponols A,B*).
Yadav, J.S. *et al.*, *Tetrahedron Lett.*, 2014, **55**, 4427-4429 (*Diosniponols A,B, synth, abs config*).
Zhang, Y. *et al.*, *Fitoterapia*, 2017, **117**, 28-33 (*Dioscoroside E_1, activity*)

3-(3,4-Dihydroxyphenyl)-3-[2-(3,4-dihydroxyphenyl)-3,5,7-trihydroxy-4-oxo-4*H*-1-benzopyran-8-yl]propanoic acid D185

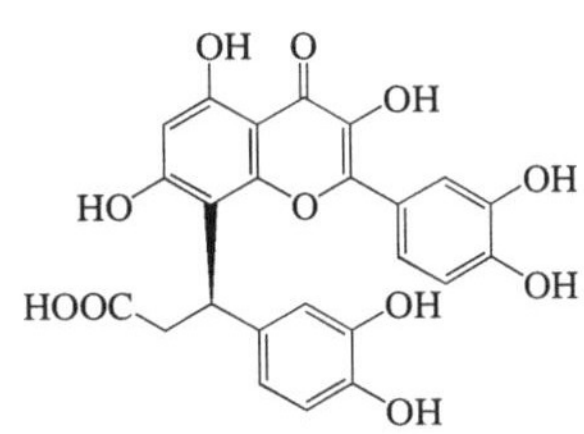

(*R*)-form

$C_{24}H_{18}O_{11}$ 482.397
(R)-form
Me ester
(-)-Uncariol C
$C_{25}H_{20}O_{11}$ 496.424
Constit. of leaves of *Uncaria rhynchophylla*. Potent DPPH radical scavenger (antioxidant). Yellow powder. $[\alpha]^{25}_{D}$ -37.8 (c, 0.1 in MeOH)
Et ester
(-)-Uncariol D
$C_{26}H_{22}O_{11}$ 510.451
Constit. of leaves of *Uncaria rhynchophylla*. Potent DPPH radical scavenger (antioxidant). Yellow powder. $[\alpha]^{25}_{D}$ -68.5 (c, 0.1 in MeOH). Obt. by chiral hplc resoln. of the racemate

(S)-form
Me ester
(+)-Uncariol C
$C_{25}H_{20}O_{11}$ 496.424
Constit. of leaves of *Uncaria rhynchophylla*. Potent DPPH radical scavenger (antioxidant). Yellow powder. $[\alpha]^{25}_{D}$ +39.2 (c, 0.1 in MeOH)
Et ester
(+)-Uncariol D
$C_{26}H_{22}O_{11}$ 510.451
Constit. of leaves of *Uncaria rhynchophylla*. Potent DPPH radical scavenger (antioxidant). Yellow powder. $[\alpha]^{25}_{D}$ +66.2 (c, 0.1 in MeOH)

(±)-form
Me ester
(±)-Uncariol C
$C_{25}H_{20}O_{11}$ 496.424
Constit. of leaves of *Uncaria rhynchophylla*. Potent DPPH radical scavenger (antioxidant). Yellow powder. λ_{max} 259 (log ε 2.60); 274 (log ε 2.56) (MeOH)
Et ester
(±)-Uncariol D
$C_{26}H_{22}O_{11}$ 510.451
Constit. of leaves of *Uncaria rhynchophylla*. Potent DPPH radical scavenger (antioxidant). Yellow powder. λ_{max} 260 (log ε 2.63); 274 (log ε 2.61) (MeOH).

Li, R. *et al.*, *Fitoterapia*, 2017, **116**, 17-23 (*Uncariols C,D, activity*)

3-(3,4-Dihydroxyphenyl)-2,3-dihydroxypropanoic acid D186

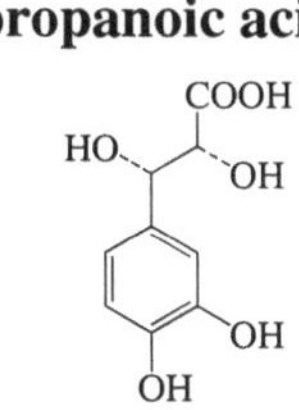

(2*R*,3*S*)-form

3-(3,4-Dihydroxyphenyl)glyceric acid
[879883-04-2] (polymer)
$C_9H_{10}O_6$ 214.173
Isol. as a polymer from *Symphytum asperum* and *Symphytum caucasicum*. Also residue present in Salvianolic acid K, S17. Antioxidant
(2R,3S)-form [1241892-09-0]

Cryst. (EtOH). Mp 157 - 160°. $[\alpha]_D$ +33.3 (c, 0.48 in H_2O)

(2S,3R)-form [1241892-10-3]

Cryst. (EtOH). Mp 155 - 158°. $[\alpha]_D$ -32.5 (c, 0.4 in H_2O)

(2RS,3SR)-form [1241892-08-9]

Cryst. (EtOH). Mp 152 - 158°. λ_{max} 230 (ε 1470); 278 (ε 5170) (H_2O).

Barbakadze, V. *et al.*, *Molecules*, 2005, **10**, 1135-1144 (*Symphytum asperum constit, Symphytum caucasicum constit*).

Merlani, M. *et al.*, *Chirality*, 2010, **22**, 717-725 (*synth, abs config*)

1-(3,4-Dihydroxyphenyl)-2,3-dihydroxy-1-propanone D187

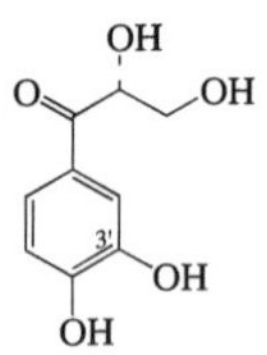

(R)-form

2,3,3′,4′-Tetrahydroxypropiophenone

$C_9H_{10}O_5$ 198.174

(R)-form [1313434-77-3]

Constit. of maple syrup, prod. by boiling sap. Yellowish powder. $[\alpha]^{25}_D$ +267 (c, 0.15 in MeOH)

(±)-form

3′-Me ether, 3-O-(3-methylbutanoyl)

Brachangobinan F

[2359598-26-6] (*R*-form)

$C_{15}H_{20}O_6$ 296.319

Constit. of *Brachanthemum gobicum*. Amorph. yellowish solid. $[\alpha]^{22}_D$+1 (c, 0.3 in MeOH). Partial racemate. λ_{max} 284 (log ε 3.48) (MeOH)

(ξ)-form

3′-Me ether

2,3-Dihydroxy-1-(4-hydroxy-3-methoxyphenyl)-1-propanone. 2,3,4′-Trihydroxy-3′-methoxypropiophenone. ***C-Veratroylglycol***

[168293-10-5]

$C_{10}H_{12}O_5$ 212.201

Constit. of *Cassia laevigata, Endospermum diadenum, Euonymus alatus, Osmanthus fragrans* and *Pisonia umbellifera*. Also obt. from Riesling wine. Gum. λ_{max} 207; 231; 279; 306 (MeOH)

3′-Me ether, 4′-O-[3,4,5-trihydroxybenzoyl-(→6)-β-D-glucopyranoside]

[1259949-54-6]

$C_{23}H_{26}O_{14}$ 526.447

Constit. of the leaves of *Cynara scolymus*. Powder. $[\alpha]^{25}_D$-17 (c, 0.3 in MeOH).

Kijjoa, A. *et al.*, *Phytochemistry*, 1995, **40**, 191-193 (*Endospermum diadenum constit*).

Jones, L. *et al.*, *Fitoterapia*, 2000, **71**, 580-583 (*Cassia laevigata constit*).

Baderschneider, B. *et al.*, *J. Agric. Food Chem.*, 2001, **49**, 2788-2798 (*Riesling constit*).

Liu, R. *et al.*, *Yunnan Zhiwu Yanjiu*, 2009, **31**, 89-92 (*Cynara scolymus constit*).

Li, L. *et al.*, *J. Agric. Food Chem.*, 2011, **59**, 7708-7718 (*maple syrup constit*).

Yan, Z.-H. *et al.*, *Chem. Nat. Compd. (Engl. Transl.)*, 2013, **49**, 340-341 (*Euonymus alatus constit*).

Odonbayar, B. *et al.*, *J. Nat. Prod.*, 2019, **82**, 774-784 (*Brachangobinan F*)

2-(3,4-Dihydroxyphenyl)-5-ethenyl-7-benzofuranol, CAS D188

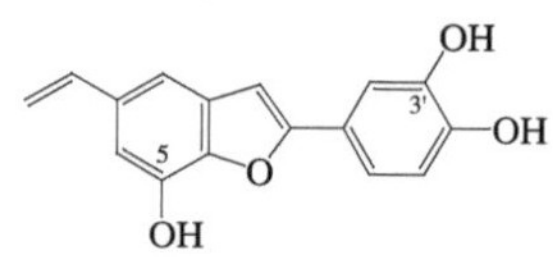

2-(3,4-Dihydroxyphenyl)-7-hydroxy-5-vinylbenzofuran. 4,7′-Epoxy-9,9′-dinor-3,8′-ligna-7,7′-diene-3′,4′,5-triol. 4,7′-Epoxy-3′,4′,5-trihydroxy-9,9′-dinor-3,8′-neoligna-7,7′-diene

$C_{16}H_{12}O_4$ 268.268

Lignan numbering shown

3′,5-Di-Me ether

2-(4-Hydroxy-3-methoxyphenyl)-7-methoxy-5-vinylbenzofuran. 4,7′-Epoxy-3′,5-dimethoxy-9,9′-dinor-3,8′-ligna-7,7′-dien-4′-ol. 4,7′-Epoxy-4′-hydroxy-3′,5-dimethoxy-9,9′-dinor-3,8′-ligna-7,7′-diene. ***Dehydroxymethylailanthoidol***

[129504-59-2]

$C_{18}H_{16}O_4$ 296.322

Constit. of *Litsea acutivena*. Orange prisms (MeOH). Mp 186 - 187°. λ_{max} 231 (log ε 3.68); 272 (log ε 3.64); 313 (log ε 3.67) (EtOH).

Cheng, H.-I. *et al.*, *J. Nat. Prod.*, 2001, **64**, 1502-1505 (*Dehydroxymethylailanthoidol*)

2-[2-(3,4-Dihydroxyphenyl)ethyl]tetrahydro-6-(4-hydroxyphenyl)-2*H*-pyran D189

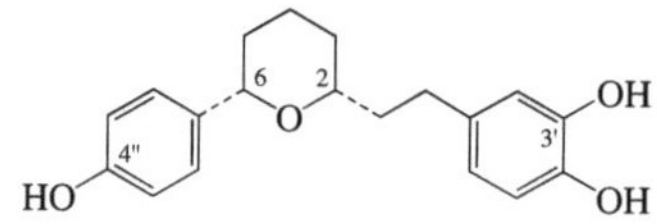

$C_{19}H_{22}O_4$ 314.381

(2R,6S)-form

3′-Me ether

[1869012-33-8]

$C_{20}H_{24}O_4$ 328.408

Constit. of the rhizomes of *Curcuma kwangsiensis*. Yellow oil. $[\alpha]^{25}_D$ -42.2 (c, 0.2 in $CHCl_3$). λ_{max} 228; 283 (MeOH).

Chen, S.-D. *et al.*, *Fitoterapia*, 2015, **102**, 67-73 (*Curcuma kwangsiensis constit*)

2-[2-(3,4-Dihydroxyphenyl)ethyl]tetrahydro-6-(3,4,5-trihydroxyphenyl)-2*H*-pyran-4-ol D190

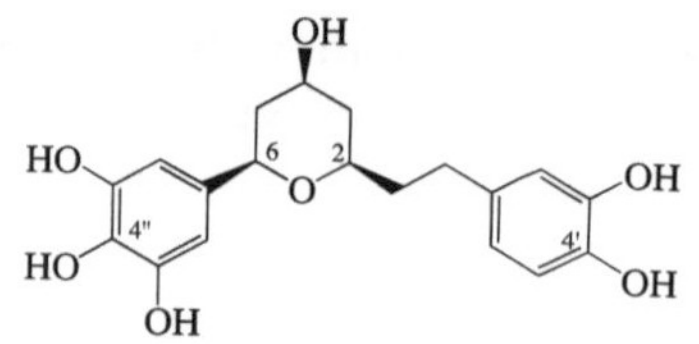

(2R,4S,6R)-form

7-(3,4-Dihydroxyphenyl)-1,5-epoxy-1-(3,4,5-trihydroxyphenyl)-3-heptanol

$C_{19}H_{22}O_7$ 362.378

(2R,4S,6R)-form

3″-Me ether

Alpinin A‡

[2151847-03-7]

$C_{20}H_{24}O_7$ 376.405

Constit. of the rhizomes of *Alpinia officinarum*. Yellowish oil. $[\alpha]^{25}_D$ +51.1 (c, 0.4 in MeOH). λ_{max} 281 (log ε 1.05) (MeOH)

(2S,4R,6S)-form

3′,3″-Di-Me ether

[182227-93-6]

$C_{21}H_{26}O_7$ 390.432

Constit. of *Zingiber officinale*. Amorph.. $[\alpha]^{25}_D$-61.6 (c, 1 in EtOH). λ_{max} 226 (sh); 280 (log ε 3.65) (EtOH)

3′,3″-Di-Me ether, 4-Ac

[182227-92-5]

$C_{23}H_{28}O_8$ 432.469

Constit. of *Zingiber officinale*. Oil. $[\alpha]^{25}_D$-31.2 (c, 1.3 in EtOH). λ_{max} 226 (sh); 280 (log ε 3.75) (EtOH)

3′,3″,5″-Tri-Me ether

[182227-94-7]

$C_{22}H_{28}O_7$ 404.459

Constit. of *Zingiber officinale*. $[\alpha]^{25}_D$-48.6 (c, 0.6 in EtOH). λ_{max} 226 (sh); 280 (log ε 3.59) (EtOH)

(2S,4S,6S)-form

3″-Me ether

[389063-83-6]

$C_{20}H_{24}O_7$ 376.405
Constit. of *Zingiber officinale*. Antioxidant. Amorph. solid. $[\alpha]_D^{25}$ -40.3 (c, 0.05 in MeOH). λ_{max} 214 (log ε 4.29); 225 (sh); 280 (log ε 3.39) (MeOH)

3′,3″-Di-Me ether
[182369-54-6]
Constit. of *Zingiber officinale*. Oil. $[\alpha]_D^{25}$ -41.6 (c, 0.9 in EtOH). λ_{max} 225 (sh); 280 (log ε 3.51) (EtOH)

3′,3″-Di-Me ether, 4-Ac
[182369-53-5]
Constit. of *Zingiber officinale*. Oil. $[\alpha]_D^{25}$ -23.3 (c, 0.5 in EtOH). λ_{max} 226 (sh); 280 (log ε 3.76) (EtOH)

3′,3″,5″-Tri-Me ether
[719270-40-3]
Constit. of *Zingiber officinale*. Oil. $[\alpha]_D^{16}$ -24 (c, 0.19 in EtOH).

Kikuzaki, H. *et al.*, *Phytochemistry*, 1996, **43**, 273-277 (*Zingiber officinale constits*).
He, W.-S. *et al.*, *Zhiwu Xuebao*, 2001, **43**, 757-759 (*2S,4S,6S-3″-Me ether*).
Ma, J. *et al.*, *Phytochemistry*, 2004, **65**, 1137-1143 (*Zingiber officinale constits*).
Fu, G. *et al.*, *J. Agric. Food Chem.*, 2017, **65**, 6608-6614 (*Alpinin A, synth, abs config*)

2-(3,4-Dihydroxyphenyl)-7-hydroxy-4-benzofurancarboxaldehyde D191

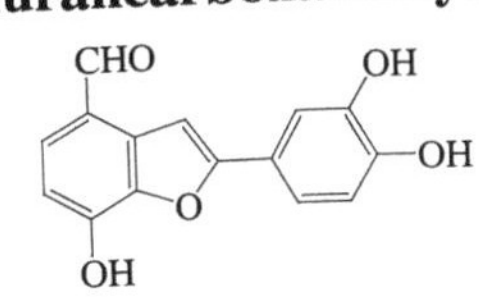

2-(3,4-Dihydroxyphenyl)-4-formyl-7-hydroxybenzofuran. 3,7′-Epoxy-3′,4,4′-trihydroxy-8,9,9′-trinor-2,8′-lign-7′-en-7-al. 3,7′-Epoxy-3′,4,4′-trihydroxy-8,9,9′-trinor-7-oxo-2,8′-neolign-7′-ene. ***Tournefolal***
[434337-15-2]
$C_{15}H_{10}O_5$ 270.24
Constit. of the stems of *Tournefortia sarmentosa*. Anti-lipid peroxidation agent. Inhibits $Cu^{2\oplus}$-induced LDL oxidn. *in vitro*. Pale yellow cryst. (EtOH). Mp 210 - 212°. λ_{max} 276 (log ε 4.01); 350 (log ε 4) (MeOH).

Lin, Y.-L. *et al.*, *J. Nat. Prod.*, 2002, **65**, 745-747 (*Tournefolal, activity*).
Shen, S.-D. *et al.*, *ARKIVOC*, 2012, **vi**, 204-213 (*synth*).
López-Frías, G. *et al.*, *Molecules*, 2015, **20**, 8654-8665 (*synth*)

1-(2,4-Dihydroxyphenyl)-3-hydroxy-2-(4-hydroxyphenyl)-1-propanone D192

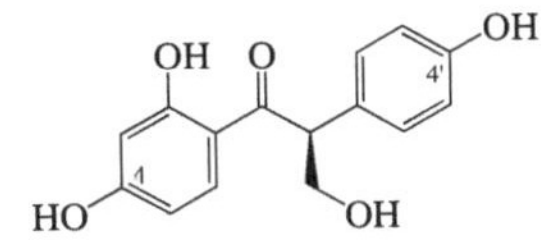

$C_{15}H_{14}O_5$ 274.272
(S)-form
4′-Me ether
1-(2,4-Dihydroxyphenyl)-3-hydroxy-2-(4-methoxyphenyl)-1-propanone
[2226195-55-5]
$C_{16}H_{16}O_5$ 288.299
Constit. of the rhizomes of *Caragana changduensis*. Pale yellow needles. $[\alpha]_D^{20}$ -37.4 (c, 0.01 in MeOH). λ_{max} 215 (log ε 4.6); 280 (log ε 4.35); 320 (log ε 4.15) (MeOH).

Guo, L. *et al.*, *Phytochem. Lett.*, 2017, **20**, 32-35 (*Caragana changduensis constit*)

2-(3,5-Dihydroxyphenyl)-3-hydroxy-1-(4-hydroxyphenyl)-1-propanone D193

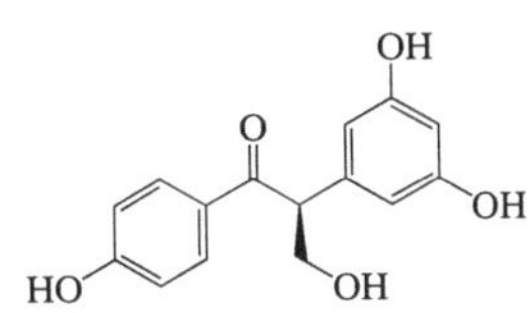

3″,4′,5″-Trihydroxy-α-(hydroxymethyl) deoxybenzoin
$C_{15}H_{14}O_5$ 274.272
Related to Evofolin B, E289
(S)-form
3′-Me ether
3-Hydroxy-2-(3-hydroxy-5-methoxyphenyl)-1-(4-hydroxyphenyl)-1-propanone
$C_{16}H_{16}O_5$ 288.299
Constit. of the roots of *Ficus hirta*. NO prodn. inhibitor. Pale yellow oil. $[\alpha]_D^{24}$ -3.9 (c, 0.84 in MeOH). λ_{max} 217 (log ε 1.62); 277 (log ε 0.62) (MeOH).

Cheng, J. *et al.*, *Fitoterapia*, 2017, **121**, 229-234 (*Ficus hirta constit, activity*)

2-(3,5-Dihydroxyphenyl)-3-(hydroxymethyl)-2,3-dihydro-5,7-benzofurandiol D194

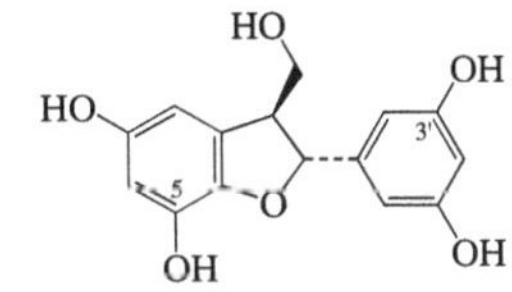

$C_{15}H_{14}O_6$ 290.271
Structures in this series may need revision. Phenyl 3,5-dioxygenated substn. pattern in doubt as a result of work done by Yang *et al* in 2015
(7′R,8′S)-form
3′,5-Di-Me ether
Pinnatifidanin CVIII
[1613724-12-1]
$C_{17}H_{18}O_6$ 318.325
Constit. of the seeds of *Crataegus pinnatifida*. Yellow oil. $[\alpha]_D^{20}$ +6.5 (c, 0.02 in MeOH). λ_{max} 227 (log ε 3.42); 281 (log ε 2.46) (MeOH).

Huang, X.-X. *et al.*, *Fitoterapia*, 2013, **91**, 217-223 (*Pinnatifidanin CVIII*).
Yang, Y.-N. *et al.*, *J. Nat. Prod.*, 2015, **78**, 705-711 (*pmr, struct*)

6-(3,4-Dihydroxyphenyl)-5-(hydroxymethyl)-2-oxo-8-oxabicyclo[3.2.1]oct-3-ene-7-carboxylic acid D195

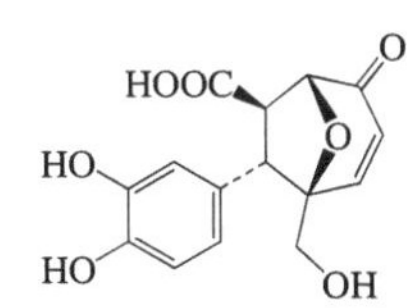

(*1R,5R,6S,7R)-form*

$C_{15}H_{14}O_7$ 306.270
Related to Acorusin C, in H104 and Cartorimin, C36
(1R,5R,6S,7R)-form
Chrysanthemoric acid A
3-Ester with 4-O-(3,4-dihydroxy-E-cinnamoyl)-(-)-quinic acid, 4-O-Caffeoylquinic acid
Chrysanthemorimic acid A
[2089622-91-1]
$C_{31}H_{30}O_{15}$ 642.566
Constit. of the flowers of *Chrysanthemum morifolium*. Exhibits neuroprotectant activity. Amorph. powder. $[\alpha]_D^{20}$ +180 (c, 0.1 in MeOH). λ_{max} 293 (log ε 3.79); 331 (log ε 3.89) (MeOH)
4-Ester with 5-O-Caffeoylquinic acid
Chrysanthemorimic acid C
[2089622-93-3]

$C_{31}H_{30}O_{15}$ 642.566
Constit. of the flowers of *Chrysanthemum morifolium*. Exhibits neuroprotectant activity. Amorph. powder. $[\alpha]_D^{20}$ +162 (c, 0.1 in MeOH). λ_{max} 295 (log ε 3.92); 330 (log ε 4.01) (MeOH)

(1S,5S,6R,7S)-form

Chrysanthemoric acid B

3-Ester with 4-O-Caffeoylquinic acid

Chrysanthemorimic acid B
[2089622-92-2]
$C_{31}H_{30}O_{15}$ 642.566
Constit. of the flowers of *Chrysanthemum morifolium*. Exhibits neuroprotectant activity. Amorph. powder. $[\alpha]_D^{20}$ -285 (c, 0.1 in MeOH). λ_{max} 295 (log ε 3.94); 332 (log ε 4.05) (MeOH).

Yang, P.-F. *et al., J. Nat. Prod.*, 2017, **80**, 1028-1033 (*Chrysanthemorimic acids A,B,C, cd, activity*)

7-(3,4-Dihydroxyphenyl)-1-(4-hydroxyphenyl)-4,6-heptadien-3-one D196

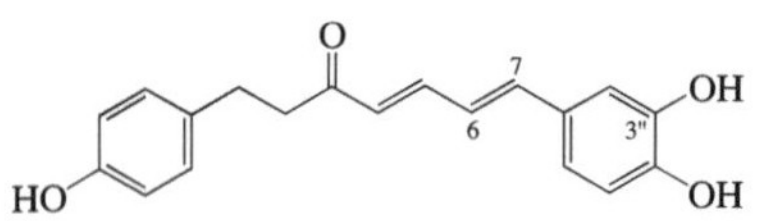

$C_{19}H_{18}O_4$ 310.349

(E,E)-form

Diarylcomosol I [1452487-91-0]
Constit. of *Curcuma comosa* and *Dioscorea collettii*. Displays inhibitory effects on melanogenesis. Pale yellow powder. λ_{max} 365 (log ε 4.10) (MeOH)

3″-Me ether

7-(4-Hydroxy-3-methoxyphenyl)-1-(4-hydroxyphenyl)-4,6-heptadien-3-one. ***Tsaokoarylone***
[811471-20-2]
$C_{20}H_{20}O_4$ 324.376
Constit. of *Amomum tsao-ko*, *Dioscorea collettii* and *Juglans mandshurica*. Yellow solid. Mp 51 - 53°. λ_{max} 272 (log ε 4.01); 361 (log ε 4.21) (MeOH). [neutral] λ_{max} 261 (log ε 3.98); 269 (log ε 3.98); 364 (log ε 4.55) (MeOH)

6,7-Dihydro, 3″-Me ether

7-(4-Hydroxy-3-methoxyphenyl)-1-(4-hydroxyphenyl)-4-hepten-3-one. ***Isogingerenone C***
[475478-41-2]
$C_{20}H_{22}O_4$ 326.392
Constit. of the roots of *Juglans mandshurica* and the rhizomes of *Alpinia officinarum*. Oil. λ_{max} 240 (log ε 3.99); 281 (log ε 3.57) ($CHCl_3$).

Lee, K.-S. *et al., J. Nat. Prod.*, 2002, **65**, 1707-1708 (*Isogingerenone C*).
Li, G. *et al., Bull. Korean Chem. Soc.*, 2004, **25**, 397-399 (*Juglans mandshurica constits*).
Moon, S.-S. *et al., Bull. Korean Chem. Soc.*, 2005, **26**, 447-450 (*Tsaokoarylone*).
Matsumoto, T. *et al., Bioorg. Med. Chem. Lett.*, 2013, **23**, 5178-5181 (*Diarylcomosol I*)

1-(3,4-Dihydroxyphenyl)-7-(4-hydroxyphenyl)-1,5-heptanediol D197

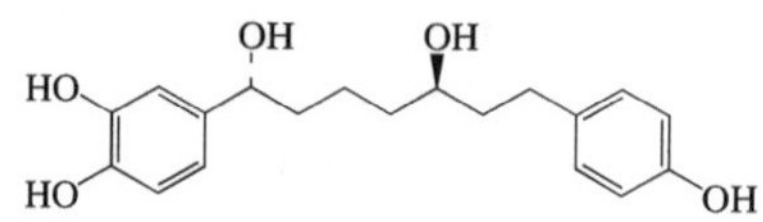

$C_{19}H_{24}O_5$ 332.396

(1R,5R)-form

3′-Me ether

1-(4-Hydroxy-3-methoxyphenyl)-7-(4-hydroxyphenyl)-1,5-heptanediol
[882515-41-5]
$C_{20}H_{26}O_5$ 346.423
Constit. of the roots of *Juglans mandshurica*. Brown solid. $[\alpha]_D^{25}$ -42.2 (c, 0.17 in MeOH). λ_{max} 225 (log ε 4.04); 279 (log ε 3.5) (MeOH).

Li, G. *et al., Bull. Korean Chem. Soc.*, 2005, **26**, 1878-1880 (*Juglans mandshurica constit*)

1-(3,4-Dihydroxyphenyl)-7-(4-hydroxyphenyl)-3,5-heptanediol D198

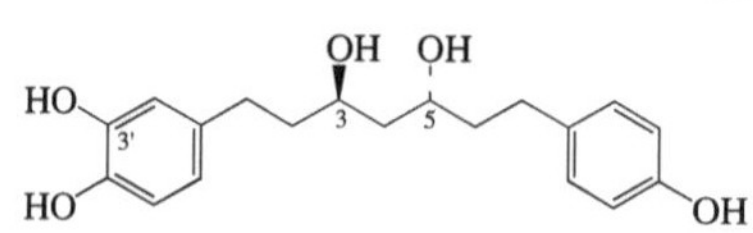

(3R,5R)-form

$C_{19}H_{24}O_5$ 332.396

(3R,5R)-form [408324-00-5]
Constit. of the rhizomes of *Dioscorea zingiberensis* and *Tacca chantrieri*. Exhibits activity preventing pancreatic necrosis. Viscous syrup. $[\alpha]_D^{23}$ +1.7 (c, 0.12 in MeOH). λ_{max} 281 (log ε 3.63) (MeOH)

3-O-β-D-Glucopyranoside
[408324-02-7]
$C_{25}H_{34}O_{10}$ 494.537
Constit. of the rhizomes of *Tacca chantrieri*. Pale yellow solid. $[\alpha]_D^{25}$ -12 (c, 0.1 in MeOH). λ_{max} 282 (log ε 3.55) (MeOH)

5-O-β-D-Glucopyranoside
[408324-03-8]
$C_{25}H_{34}O_{10}$ 494.537
Constit. of the rhizomes of *Tacca chantrieri*. Amorph. solid. $[\alpha]_D^{25}$ -12 (c, 0.1 in MeOH). λ_{max} 281 (log ε 3.21) (MeOH)

3-Ac
[1269839-27-1]
$C_{21}H_{26}O_6$ 374.433
Constit. of the rhizomes of *Curcuma kwangsiensis*

3,5-Di-Ac
[1269839-26-0]
$C_{23}H_{28}O_7$ 416.47
Constit. of the rhizomes of *Curcuma kwangsiensis*. Yellowish oil. $[\alpha]_D^{25}$ +17.2 (c, 0.1 in MeOH). λ_{max} 281 (log ε 3.48) (MeOH)

3′-Me ether, 3-O-β-D-glucopyranoside
[408324-04-9]
$C_{26}H_{36}O_{10}$ 508.564
Constit. of the rhizomes of *Tacca chantrieri*. Amorph. solid. $[\alpha]_D^{25}$ -2 (c, 0.1 in MeOH). λ_{max} 281 (log ε 3.54) (MeOH)

5-Ketone, 3′-Me ether

5-Hydroxy-7-(4-hydroxy-3-methoxyphenyl)-1-(4-hydroxyphenyl)-3-heptanone
[459429-75-5]
$C_{20}H_{24}O_5$ 344.407
Constit. of the dried rhizomes of *Alpinia officinarum*. Pale yellow liq.. $[\alpha]_D^{25}$ +1 (c, 0.8 in EtOH). $[\alpha]_D^{20}$ -13.6 (c, 0.04 in $CHCl_3$). λ_{max} 197 (log ε 6.69); 217 (log ε 6.73); 273 (log ε 6.83) (EtOH)

(3S,5S)-form

3-Ac
[1334172-41-6]
Constit. of the rhizomes of *Curcuma kwangsiensis*. Yellowish oil

3′-Me ether

1-(4-Hydroxy-3-methoxyphenyl)-7-(4-hydroxyphenyl)-3,5-heptanediol
[811471-21-3]
$C_{20}H_{26}O_5$ 346.423
Constit. of the roots of *Juglans mandshurica*. Amorph. solid. $[\alpha]_D^{25}$ +4.3 (c, 0.12 in MeOH). λ_{max} 223 (log ε 3.93); 280 (log ε 3.48) (MeOH)

4″-Me ether, 3-O-β-D-glucopyranoside
[1374348-28-3]
$C_{26}H_{36}O_{10}$ 508.564
Constit. of the leaves of *Pyrostria major*. Amorph. solid. $[\alpha]_D^{25}$ -10 (c, 0.13 in MeOH). λ_{max} 222 (log ε 4.2); 284 (log ε 3.6) (MeOH)

4″-Me ether, 5-O-formyl, 3-O-β-D-glucopyranoside
[1374348-32-9]
$C_{27}H_{36}O_{11}$ 536.574
Constit. of the leaves of *Pyrostria major*. Amorph. solid. $[\alpha]_D^{25}$ -10 (c, 0.03 in MeOH). Stereochem. tentatively assigned. λ_{max} 223 (log ε 4.1); 279 (log ε 3.6) (MeOH)

4′,4″-Di-Me ether, 3-O-β-D-glucopyranoside
[1374348-27-2]
$C_{27}H_{38}O_{10}$ 522.591
Constit. of the leaves of *Pyrostria major*. Amorph. solid. $[\alpha]_D^{25}$ -31 (c, 0.17 in MeOH). λ_{max} 224 (log ε 4.36); 278 (log ε 3.7) (MeOH)

4′,4″-Di-Me ether, 5-O-β-D-glucopyranoside
[1374348-30-7]
$C_{27}H_{38}O_{10}$ 522.591
Constit. of the leaves of *Pyrostria major*. Amorph. solid. $[\alpha]_D^{25}$ -6 (c, 0.05 in MeOH). Stereochem. tentatively assigned. λ_{max} 223 (log ε 4.7); 278 (log ε 3.25) (MeOH)

3-Ketone, 5-O-β-D-xylopyranoside
Alnuside B
[93915-39-0]
$C_{24}H_{30}O_9$ 462.495
Constit. of *Alnus glutinosa*, *Alnus japonica* and *Alnus serrulatoides*. Moderate antioxidant. Viscous liq.. $[\alpha]_D$ -15.5 (c, 0.04 in MeOH)

3-Ketone, 5-O-[2ξ-methylbutanoyl-(→2)-β-D-xylopyranoside]
Alnuside C
[876293-03-7]
$C_{29}H_{38}O_{10}$ 546.613
Constit. of the leaves of *Alnus japonica*. Viscous liq.. $[\alpha]_D$ -15.6 (c, 0.04 in MeOH)

3-Ketone, 5-O-[3,4-dimethoxy-Z-cinnamoyl-(→6)-β-D-glucopyranoside]
[1527469-92-6]
$C_{36}H_{42}O_{13}$ 682.719
Constit. of bark of *Alnus glutinosa*. Brown gum. $[\alpha]_D^{22}$ -10 (c, 1.0 in MeOH). λ_{max} 220 (sh); 288 (log ε 4.23); 310 (log ε 4.36) (MeOH)

3-Ketone, 3′-Me ether
5-Hydroxy-1-(4-hydroxy-3-methoxyphenyl)-7-(4-hydroxyphenyl)-3-heptanone
[1269839-33-9]
$C_{20}H_{24}O_5$ 344.407
Constit. of the rhizomes of *Curcuma kwangsiensis*. Yellowish oil. $[\alpha]_D^{25}$ +12.6 (c, 0.1 in MeOH). λ_{max} 282 (log ε 3.76) (MeOH)

5-Ketone
7-(3,4-Dihydroxyphenyl)-5-hydroxy-1-(4-hydroxyphenyl)-3-heptanone
[1269839-31-7]
$C_{19}H_{22}O_5$ 330.38
Constit. of the rhizomes of *Curcuma kwangsiensis*. Yellowish oil. $[\alpha]_D^{25}$ +11.2 (c, 0.1 in MeOH). λ_{max} 282 (log ε 3.65) (MeOH)

5-Ketone, 3-O-β-D-xylopyranoside
Alnuside A
[93915-38-9]
$C_{24}H_{30}O_9$ 462.495
Constit. of *Alnus glutinosa* and *Alnus japonica*. Moderate antioxidant. Viscous liq.. $[\alpha]_D$ -19.5 (c, 0.03 in MeOH)

5-Ketone, 3-O-β-D-glucopyranoside
[1491148-75-4]
$C_{25}H_{32}O_{10}$ 492.521
Constit. of *Alnus glutinosa*. Exerts a chemoprotective effect on human lymphocytes DNA. Yellow powder. $[\alpha]_D^{22}$ +7.5 (c, 0.67 in MeOH). λ_{max} 222 (log ε 3.62); 281 (log ε 3.16) (MeOH)

5-Ketone, 3-O-[3,4-dimethoxy-E-cinnamoyl-(→6)-β-D-glucopyranoside]
[1527469-91-5]
$C_{36}H_{42}O_{13}$ 682.719
Constit. of bark of *Alnus glutinosa*. $[\alpha]_D^{22}$ -2.5 (c, 2.0 in MeOH). λ_{max} 223 (sh); 286 (log ε 4.21); 323 (log ε 4.19) (MeOH)

5-Ketone, 3-Me ether
7-(3,4-Dihydroxyphenyl)-1-(4-hydroxyphenyl)-5-methoxy-3-heptanone. ***Alnuheptanoid A***
[1622995-01-0]
$C_{20}H_{24}O_5$ 344.407
Constit. of the stem bark of *Alnus japonica*. Brown solid. $[\alpha]_D^{22}$ +19.1 (c, 1 in MeOH). λ_{max} 225 (log ε 4.52); 284 (log ε 4.11) (MeOH)

5-Ketone, 3′-Me ether
[619297-56-2]
$C_{20}H_{24}O_5$ 344.407
Constit. of the roots of *Juglans mandshurica*. Yellow oil. $[\alpha]_D^{25}$ -2.5 (c, 0.09 in MeOH)

(5RS)-form

3-Ketone, 3′-Me ether
5-Hydroxy-1-(4-hydroxy-3-methoxyphenyl)-7-(4-hydroxyphenyl)-3-heptanone
[138870-92-5]
$C_{20}H_{24}O_5$ 344.407
Constit. of rhizomes of *Alpinia officinarum* and *Zingiber officinale*. Oil. $[\alpha]_D^{23}$ 0 (c, 0.2 in EtOH). λ_{max} 222 (log ε 4); 278 (log ε 3.6) (EtOH)

(3R*,5S*)-form [1269839-19-1]
Constit. of the rhizomes of *Curcuma kwangsiensis*. $[\alpha]_D^{25}$ +0.2 (c, 0.1 in MeOH)

3′-Me ether
[1269839-20-4]
$C_{20}H_{26}O_5$ 346.423
Constit. of the rhizomes of *Curcuma kwangsiensis*. $[\alpha]_D^{25}$ +0.4 (c, 0.1 in MeOH)

(3ξ)-form

5-Ketone, 3,3′-di-Me ether
7-(4-Hydroxy-3-methoxyphenyl)-1-(4-hydroxyphenyl)-5-methoxy-3-heptanone
[619314-43-1]
$C_{21}H_{26}O_5$ 358.434
Constit. of the roots of *Juglans mandshurica*. Brown oil. $[\alpha]_D^{21}$ -7.1 (c, 0.42 in MeOH). λ_{max} 224 (log ε 3.99); 281 (log ε 3.48) (MeOH).

Ohta, S. *et al.*, *J. Chem. Soc., Perkin Trans. 1*, 1984, 1635-1642 (*3-ketone 5-xyloside, isol, abs config, pmr, ms, struct*).
Aoki, T. *et al.*, *Phytochemistry*, 1990, **29**, 3611-3614 (*3-ketone 5-xyloside*).
Kikuzaki, H. *et al.*, *Phytochemistry*, 1991, **30**, 3647-3651 (*3-ketone 3′-Me-ether*).
Yokosuka, A. *et al.*, *J. Nat. Prod.*, 2002, **65**, 283-289 (*Tacca chantrieri constits*).
Shin, D. *et al.*, *J. Nat. Prod.*, 2002, **65**, 1315-1318 (*3-ketone 3′-Me ether, 5-ketone 3′-Me ether*).
Li, G. *et al.*, *Chem. Pharm. Bull.*, 2003, **51**, 262-264 (*Juglans mandshurica isolates*).
Li, S.B. *et al.*, *Chin. Chem. Lett.*, 2003, **14**, 251-254 (*3-ketone 3′-Me ether, synth*).
Li, G. *et al.*, *Bull. Korean Chem. Soc.*, 2004, **25**, 397-399 (*3′-Me ether*).
Kuroyanagi, M. *et al.*, *Chem. Pharm. Bull.*, 2005, **53**, 1519-1523 (*Alnusides A,B,C, activity*).
An, N. *et al.*, *Food Chem.*, 2010, **119**, 513-517 (*5-ketone 3′-Me ether*).
Li, J. *et al.*, *Bioorg. Med. Chem. Lett.*, 2011, **21**, 5363-5369 (*Curcuma kwangsiensis constits*).

Beniddir, M.A. *et al.*, *Eur. J. Org. Chem.*, 2012, 1039-1046 (*Pyrostria major glucosides*).
Novakovic, M. *et al.*, *Planta Med.*, 2013, **79**, 499-505 (*Alnus glutinosa constits, activity*).
Ibrahim, S.R.M. *et al.*, *Nat. Prod. Res.*, 2014, **28**, 1765-1771 (*Alnuheptanoid A*).
Novakovic, M. *et al.*, *Phytochemistry*, 2014, **97**, 46-54 (*Alnus glutinosa constits, activity*).
Du, D. *et al.*, *Bioorg. Med. Chem.*, 2017, **27**, 1467-1470 (*Dioscorea zingiberensis constit, activity*)

7-(3,4-Dihydroxyphenyl)-1-(4-hydroxyphenyl)-2,3,5-heptanetriol D199

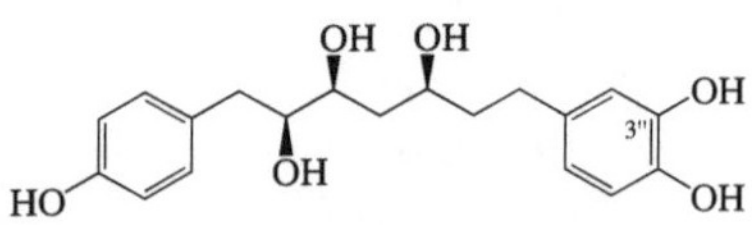

(2*S*,3*S*,5*S*)-form

$C_{19}H_{24}O_6$ 348.395
(2S,3S,5S)-form
3″-Me ether
7-(4-Hydroxy-3-methoxyphenyl)-1-(4-hydroxyphenyl)-2,3,5-heptanetriol.
Rhoiptelol C
[184485-70-9]
$C_{20}H_{26}O_6$ 362.422
Constit. of the fruit of *Rhoiptelea chiliantha* and roots of *Juglans mandshurica*. Amorph. powder + ½H_2O. $[\alpha]_D^{15}$ -24 (c, 0.4 in MeOH)
(2ξ,3ξ,5ξ)-form
3″-Me ether
Juglanol B
$C_{20}H_{26}O_6$ 362.422
Constit. of *Juglans mandshurica* var. *sieboldiana*. $[\alpha]_D^{25}$ -12.1 (c, 0.17 in MeOH). λ_{max} 223 (log ε 4.08); 278 (log ε 3.60) (MeOH).

Jiang, Z.-H. *et al.*, *Phytochemistry*, 1996, **43**, 1049-1054 (*Rhoiptelol C*).
Machida, K. *et al.*, *J. Tohoku Pharm. Univ.*, 2008, **55**, 41-46 (*Juglanol B*).
Reddy, S.P. *et al.*, *Helv. Chim. Acta*, 2014, **97**, 999-1003 (*Rhoiptelol C, synth*)

1-(3,4-Dihydroxyphenyl)-7-(4-hydroxyphenyl)-3-heptanol D200

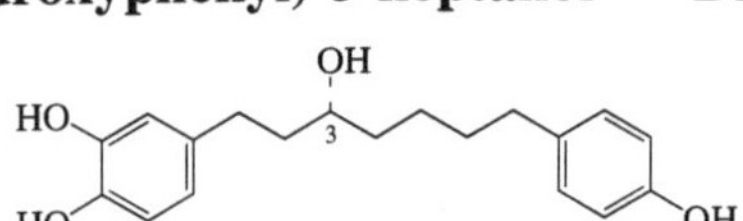

(*R*)-form

$C_{19}H_{24}O_4$ 316.397
(R)-form [1251830-61-1]
Constit. of the rhizomes of *Curcuma kwangsiensis*. Yellowish oil. $[\alpha]_D^{25}$ +2.9 (c, 0.05 in MeOH). λ_{max} 219 (log ε 3.94); 282 (log ε 3.32) (MeOH)
3-Ac
[1251830-63-3]
$C_{21}H_{26}O_5$ 358.434
Constit. of the rhizomes of *Curcuma kwangsiensis*. Yellowish oil. $[\alpha]_D^{25}$ +5.3 (c, 0.05 in MeOH). λ_{max} 209 (log ε 4.17); 262 (log ε 3.12) (MeOH)
3-Ketone
1-(3,4-Dihydroxyphenyl)-7-(4-hydroxyphenyl)-3-heptanone.
Isomuricarpin
[1174389-03-7]
$C_{19}H_{22}O_4$ 314.381
Constit. of *Alnus sibirica* and *Amomum muricarpum*. Also from *nidus vespae*, the nest of *Polistes* sp.. Oil or syrup. λ_{max} 203 (log ε 4.4); 282 (log ε 3.62) (MeOH)
(S)-form [1251830-54-2]
Constit. of the rhizomes of *Curcuma kwangsiensis* and from *nidus vespae*, the nest of *Polistes* sp.. Yellowish oil. $[\alpha]_D^{25}$ -2.9 (c, 0.1 in MeOH)
3-Ac
[1251830-56-4]
Constit. of the rhizomes of *Curcuma kwangsiensis*. Yellowish oil. $[\alpha]_D^{25}$ -5.3 (c, 0.1 in MeOH).

Li, J. *et al.*, *J. Nat. Prod.*, 2010, **73**, 1667-1671 (*Curcuma kwangsiensis constits*).
He, J.-B. *et al.*, *Chem. Biodiversity*, 2011, **8**, 2270-2276 (*nidus vespae constits*).
Giang, P.M. *et al.*, *Nat. Prod. Res.*, 2012, **26**, 1195-1200 (*Isomuricarpin*).
Le, T.T. *et al.*, *Molecules*, 2017, **22**, 1566:1-9 (*Alnus sibirica constit*)

7-(3,4-Dihydroxyphenyl)-1-(4-hydroxyphenyl)-3-heptanol D201

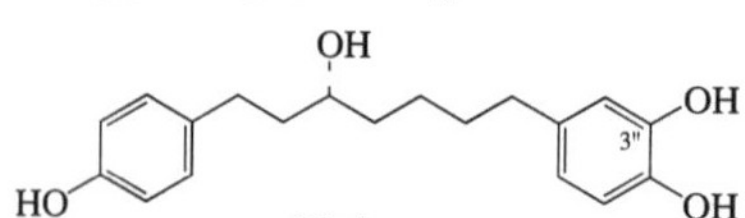

(*R*)-form

$C_{19}H_{24}O_4$ 316.397
(R)-form
3″-Me ether
7-(4-Hydroxy-3-methoxyphenyl)-1-(4-hydroxyphenyl)-3-heptanol.
3′-Deoxy-3″-O-methylrubranol
[811471-19-9]
$C_{20}H_{26}O_4$ 330.424
Constit. of the roots of *Juglans mandshurica*. Yellow solid. $[\alpha]_D^{25}$ -12.3 (c, 0.3 in MeOH). λ_{max} 223 (log ε 4.13); 280 (log ε 3.64); 347 (log ε 2.6) (MeOH)
3-Ketone
7-(3,4-Dihydroxyphenyl)-1-(4-hydroxyphenyl)-3-heptanone.
Muricarpin
[1392299-19-2]
$C_{19}H_{22}O_4$ 314.381
Constit. of *Alnus sibirica* and *Amomum muricarpum*. Syrup
(S)-form [1353746-39-0]
Constit. of the nest of the wasp *Polistes* sp.. Oil. $[\alpha]_D^{16}$ -13.8 (c, 0.13 in MeOH). λ_{max} 222 (log ε 4.06); 281 (log ε 3.53) (MeOH)
(ξ)-form
3-Ac
[1083200-84-3]
$C_{21}H_{26}O_5$ 358.434
Constit. of the rhizomes of *Amomum muricarpum*. $[\alpha]_D^{25}$ -4.4 (c, 0.8 in MeOH).

Li, G. *et al.*, *Bull. Korean Chem. Soc.*, 2004, **25**, 397-399 (*Juglans mandshurica constit*).
He, J.-B. *et al.*, *Chem. Biodiversity*, 2011, **8**, 2270-2276 (*Polistes nest constit*).
Giang, P.M. *et al.*, *Nat. Prod. Res.*, 2012, **26**, 1195-1200 (*Amomum muricarpum constits*).
Le, T.T. *et al.*, *Molecules*, 2017, **22**, 1566:1-9 (*Alnus sibirica constit*)

1-(3,4-Dihydroxyphenyl)-7-(4-hydroxyphenyl)-1,4,6-heptatrien-3-one D202

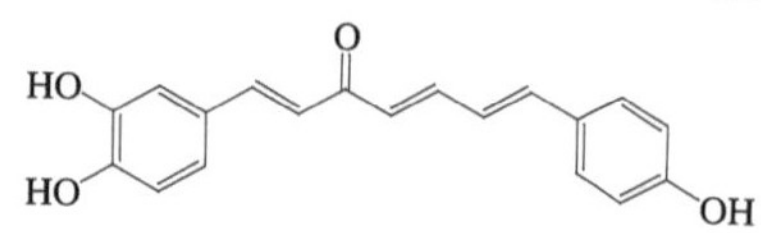

$C_{19}H_{16}O_4$ 308.333
Not yet known as a nat. prod. but analogous compds. are known
1,2,6,7-Tetrahydro
1-(3,4-Dihydroxyphenyl)-7-(4-hydroxyphenyl)-4-hepten-3-one
[886060-78-2]
$C_{19}H_{20}O_4$ 312.365
Constit. of *Alnus japonica*, *Alnus rubra* and *Amomum muricarpum*. Amorph. tan powder
1,2,6,7-Tetrahydro, 3′-Me ether
1-(4-Hydroxy-3-methoxyphenyl)-7-(4-hydroxyphenyl)-4-hepten-3-one.
Gingerenone C
[128701-01-9]
$C_{20}H_{22}O_4$ 326.392
Constit. of *Zingiber officinale*. Oil.

Endo, K. *et al.*, *Phytochemistry*, 1990, **29**, 797-799 (*Gingerenone C*).
Chen, J. *et al.*, *Planta Med.*, 1998, **64**, 74-75 (*tetrahydro*).

Liu, Y.G. *et al.*, *Chin. Chem. Lett.*, 2002, **12**, 1068-1070 (*tetrahydro, synth*).
Li, S.B. *Chin. Chem. Lett.*, 2003, **14**, 251-254 (*Gingerenone C, synth*).
Tung, N.H. *et al.*, *Planta Med.*, 2010, **76**, 626-629 (*tetrahydro*)

1-(3,4-Dihydroxyphenyl)-7-(4-hydroxyphenyl)-1-heptene-3,5-diol D203

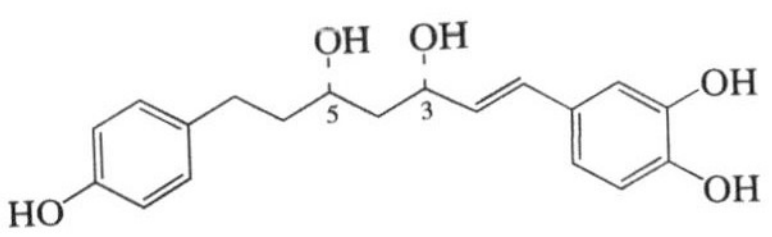

$C_{19}H_{22}O_5$ 330.38

(1E,3S,5S)-form

3-Ketone

1-(3,4-Dihydroxyphenyl)-5-hydroxy-7-(4-hydroxyphenyl)-1-hepten-3-one

[1426399-35-0]

$C_{19}H_{20}O_5$ 328.364

Constit. of *Alnus hirsuta* f. *sibirica*. Yellowish syrup. $[\alpha]^{25}_{D}$ -13.7 (c, 0.2 in MeOH).

Lee, M. *et al.*, *Bioorg. Med. Chem. Lett.*, 2013, **23**, 2069-2073 (*Alnus hirsuta ketone*)

1-(3,4-Dihydroxyphenyl)-7-(4-hydroxyphenyl)-6-hepten-3-ol D204

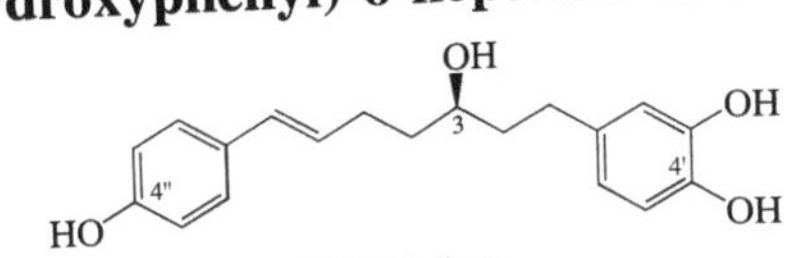

(3*R*,6*E*)-form

$C_{19}H_{22}O_4$ 314.381

(3R,6E)-form [1251830-60-0]

Constit. of the rhizomes of *Curcuma kwangsiensis*. $[\alpha]^{25}_{D}$ +3.7 (c, 0.05 in MeOH). λ_{max} 223 (log ε 3.85); 278 (log ε 3.16) (MeOH)

3-Ac

[1251830-62-2]

$C_{21}H_{24}O_5$ 356.418

Constit. of the rhizomes of *Curcuma kwangsiensis*. Oil. $[\alpha]^{25}_{D}$ +5.7 (c, 0.1 in MeOH). λ_{max} 208 (log ε 4.26); 262 (log ε 3.04) (MeOH)

4″-Deoxy

1-(3,4-Dihydroxyphenyl)-7-phenyl-6-hepten-3-ol

[158697-56-4]

$C_{19}H_{22}O_3$ 298.382

Constit. of the rhizomes of *Curcuma kwangsiensis* and *Curcuma xanthorrhiza*. Cryst. ($CHCl_3/C_6H_6$) or yellowish oil. Mp 100 - 102°. $[\alpha]^{25}_{D}$ +5 (c, 0.05 in MeOH). $[\alpha]^{25}_{D}$ +77.3 (c, 0.19 in MeOH). λ_{max} 201 (log ε 4.15); 283 (log ε 3.64) (MeOH)

(3S,6E)-form [1251830-53-1]

Constit. of the rhizomes of *Curcuma kwangsiensis*. $[\alpha]^{25}_{D}$ -3.7 (c, 0.1 in MeOH)

3-Ac

[1251830-55-3]

Constit. of the rhizomes of *Curcuma kwangsiensis*. Oil. $[\alpha]^{25}_{D}$ -5.7 (c, 0.1 in MeOH)

3′-Me ether

Diarylcomosol II

[1452487-92-1]

$C_{20}H_{24}O_4$ 328.408

Constit. of *Curcuma comosa*. Displays inhibitory effects on melanogenesis. Oil. $[\alpha]^{25}_{D}$ +12.6 (c, 0.50 in MeOH). λ_{max} 261 (log ε 4.11); 288 (log ε 3.45) (MeOH)

4″-Deoxy

[1062082-82-9]

Constit. of the rhizomes of *Curcuma comosa* and *Curcuma kwangsiensis*. Cryst. (CH_2Cl_2/hexane). Mp 100 - 101°. $[\alpha]^{28}_{D}$ -5.2 (c, 0.4 in EtOH). λ_{max} 249 (ε 14100); 282 (ε 4000) (MeOH)

(3ξ,6E)-form [1206788-61-5]

Constit. of the rhizomes of *Curcuma comosa*. Amorph. brown solid. Mp 123.2°. $[\alpha]^{25}_{D}$ -27.1 (c, 0.01 in MeOH). λ_{max} 218 (log ε 3.63); 260 (log ε 3.47) (MeOH).

Suksamrarn, A. *et al.*, *Phytochemistry*, 1994, **36**, 1505-1508 (*Curcuma xanthorrhiza constit*).
Suksamrarn, A. *et al.*, *Bioorg. Med. Chem.*, 2008, **16**, 6891-6902 (*abs config*).
Kaewamatawong, R. *et al.*, *Phytochem. Lett.*, 2009, **2**, 19-21 (*Curcuma comosa constit*).
Li, J. *et al.*, *J. Nat. Prod.*, 2010, **73**, 1667-1671 (*Curcuma kwangsiensis constits*).
Matsumoto, T. *et al.*, *Bioorg. Med. Chem. Lett.*, 2013, **23**, 5178-5181 (*Diarylcomosol II*)

7-(3,4-Dihydroxyphenyl)-1-(3-hydroxyphenyl)-6-hepten-3-ol D205

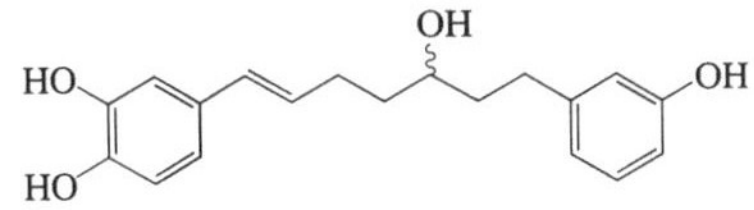

$C_{19}H_{22}O_4$ 314.381

(3ξ,6E)-form

3-Me ether

1-(3,4-Dihydroxyphenyl)-7-(3-hydroxyphenyl)-5-methoxy-1-heptene

[1206788-62-6]

$C_{20}H_{24}O_4$ 328.408

Constit. of the rhizomes of *Curcuma comosa*. Amorph. brown solid. Mp 125.1°. $[\alpha]^{25}_{D}$ -36.7 (c, 0.01 in MeOH). λ_{max} 212 (log ε 3.67); 281 (log ε 3.24) (MeOH).

Kaewamatawong, R. *et al.*, *Phytochem. Lett.*, 2009, **2**, 19-21 (*Curcuma comosa constit*)

7-(3,4-Dihydroxyphenyl)-1-(4-hydroxyphenyl)-4-hepten-3-one D206

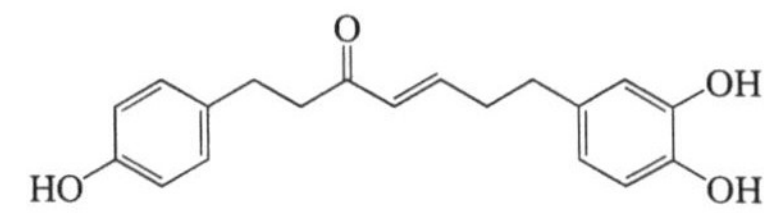

$C_{19}H_{20}O_4$ 312.365

(E)-form

Alusenone [1198360-86-9]

Constit. of *Alnus japonica* and *Renealmia alpinia*. Syrup. λ_{max} 250 (log ε 2.49) (MeOH).

Tung, N.H. *et al.*, *Planta Med.*, 2010, **76**, 626-629 (*Alusenone, activity*)

2-(3,4-Dihydroxyphenyl)-1-(4-hydroxyphenyl)-1,3-propanediol D207

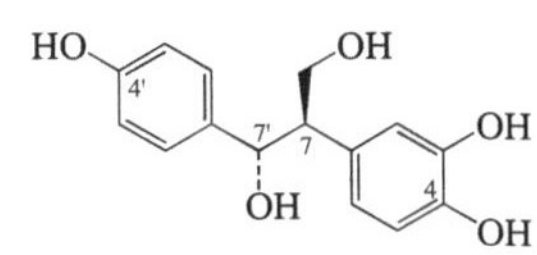

(7*R*,7′*S*)-form

8′,9,9′-Trinor-7,7′-lignan-3,4,4′,7′,8-pentol. 3,4,4′,7′,8-Pentahydroxy-8′,9,9′-trinor-7,7′-neolignan

$C_{15}H_{16}O_5$ 276.288

Lignan numbering shown

(7R,7′S)-form

4,7′-Di-Me ether

Carayensin C

[1365029-69-1]

$C_{17}H_{20}O_5$ 304.342

Constit. of the shells of *Carya cathayensis*. Powder. $[\alpha]^{25}_{D}$ +41 (c, 0.9 in MeOH). λ_{max} 227 (log ε 4.67); 278 (log ε 4.28) (MeOH)

7′-Ketone

2-(3,4-Dihydroxyphenyl)-3-hydroxy-1-(4-hydroxyphenyl)-1-propanone. 3′,4,4′,8′-Tetrahydroxy-8,9,9′-trinor-7,7′-lignan-7-one

7′-Ketone, 3-Me ether

3-Hydroxy-2-(4-hydroxy-3-methoxyphenyl)-1-(4-hydroxyphenyl)-1-propanone. 4,4′,8′-Trihydroxy-3′-methoxy-8,9,9′-trinor-7,7′-lignan-7-one

[2088724-90-5]
$C_{16}H_{16}O_5$ 288.299
Constit. of the fruit of *Juglans mandshurica*. Amorph. powder. $[\alpha]^{20}_{D}$ +21.7 (c, 0.2 in MeOH)

(7S,7'R)-form

4,7'-Di-Me ether

2-(3-Hydroxy-4-methoxyphenyl)-3-(4-hydroxyphenyl)-3-methoxy-1-propanol. 4,7'-dimethoxy-8',9,9'-trinor-7,7'-lignan-3,4',8-triol. 3,4',8-Trihydroxy-4,7'-dimethoxy-8',9,9'-trinor-7,7'-lignan. ***Carayensin B***
[1365029-68-0]
$C_{17}H_{20}O_5$ 304.342
Constit. of the shells of *Carya cathayensis*. Powder. $[\alpha]^{25}_{D}$ -38 (c, 0.9 in MeOH). λ_{max} 224 (log ε 4.83); 277 (log ε 4.4) (MeOH)

8-O-(2-Hydroxyethyl), 4-Me ether

Carayensin A
[1365029-67-9]
$C_{18}H_{22}O_6$ 334.368
Constit. of the shells of *Carya cathayensis*. Powder. $[\alpha]^{25}_{D}$ -33 (c, 0.8 in MeOH). λ_{max} 226 (log ε 3.62); 278 (log ε 3.07) (MeOH)

(7RS,7'RS)-form

3-Me ether

2-(4-Hydroxy-3-methoxyphenyl)-1-(4-hydroxyphenyl)-1,3-propanediol. 3-Methoxy-8',9,9'-trinor-7,7'-lignan-4,4',7',8-tetrol
[2088724-91-6]
$C_{16}H_{18}O_5$ 290.315
Constit. of the fruit of *Juglans mandshurica*. Amorph. powder

3,7'-Di-Me ether

2-(4-Hydroxy-3-methoxyphenyl)-3-(4-hydroxyphenyl)-3-methoxy-1-propanol. 3,7'-dimethoxy-8',9,9'-trinor-7,7'-lignan-4,4',8-triol
[2088724-92-7]
$C_{17}H_{20}O_5$ 304.342
Constit. of the fruit of *Juglans mandshurica*. Amorph. powder.

Wu, W. *et al.*, *Bioorg. Med. Chem. Lett.*, 2012, **22**, 1895-1898 (*Carayensins A,B,C*).
Park, S. *et al.*, *Phytochemistry*, 2017, **137**, 87-93 (*Juglans mandshurica constits*)

1-(2,4-Dihydroxyphenyl)-2-(4-hydroxyphenyl)-1-propanone D208

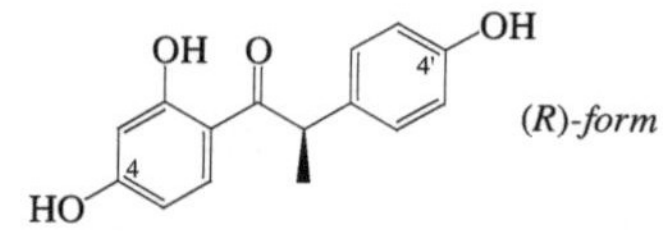

(R)-form

2',4',4''-Trihydroxy-α-methyldeoxybenzoin. O-Demethylangolensin
[21255-69-6]
$C_{15}H_{14}O_4$ 258.273

(R)-form

4'-Me ether

1-(2,4-Dihydroxyphenyl)-2-(4-methoxyphenyl)-1-propanone, CAS. ***Angolensin***
[4842-48-2]
[642-39-7]
$C_{16}H_{16}O_4$ 272.3
Constit. of *Pericopsis elata*, *Pericopsis mooniana*, *Pterocarpus angolensis*, *Pterocarpus erinaceus*, *Pterocarpus indicus* and *Afrormosia elata*. Exhibits inhibition of mushroom tyrosinase. Inhibitor of aflatoxin B_1 metabolism in chicken liver cytosol. Prisms (petrol). Mp 120 - 121°. $[\alpha]^{24}_{D}$ -114 (c, 1 in $CHCl_3$). λ_{max} 277; 317 (MeOH) (Berdy)

2,4'-Di-Me ether

1-(4-Hydroxy-2-methoxyphenyl)-2-(4-methoxyphenyl)-1-propanone. ***2-O-Methylangolensin***
[58822-06-3]
$C_{17}H_{18}O_4$ 286.327
Constit. of *Pericopsis elata*. Oil. $[\alpha]^{24}_{D}$ +5.3 ($CHCl_3$)

4,4'-Di-Me ether

4-O-Methylangolensin
[4842-51-7]
Constit. of the bark of *Cassipourea malosana*. Amorph. powder. Mp 62°. $[\alpha]^{24}_{D}$ -142 (c, 2 in $CHCl_3$)

Tri-Me ether

Mp 49°. $[\alpha]^{24}_{D}$ +26 ($CHCl_3$)

(S)-form

4,4'-Di-Me ether

[75946-85-9]
$C_{17}H_{18}O_4$ 286.327
Isol. from heartwood of *Pterocarpus angolensis*. Needles (MeOH). Mp 28 - 30°

(±)-form

Cryst. (Et_2O/hexane). Mp 103°

4'-Me ether

[21848-33-9]
Constit. of *Pericopsis mooniana*, roots of *Erythrina poeppigiana* and *Pterocarpus indicus*. Cryst. (C_6H_6/petrol). Mp 86 - 88°. λ_{max} 215 (ε 37400); 280 (ε 26800); 319 (ε 17000) (EtOH)

Tri-Me ether

Mp 70°

(ξ)-form

Found in urine of humans, sheep and cattle.

King, F.E. *et al.*, *J. Chem. Soc.*, 1952, 1920-1924 (*Pterocarpus angolensis constit, struct*).
Foxall, C.D. *et al.*, *J. Chem. Soc.*, 1963, 5573-5575 (*Angolensin, Afrormosia elata constit, struct*).
Cooke, R.G. *et al.*, *Aust. J. Chem.*, 1964, **17**, 379-384 (*Pterocarpus indicus constit*).
Ollis, W.D. *et al.*, *Aust. J. Chem.*, 1965, **18**, 1787-1790 (*abs config*).
Clark-Lewis, J.W. *et al.*, *Aust. J. Chem.*, 1965, **18**, 1791-1797 (*abs config*).
Aggarwal, S.K. *et al.*, *Indian J. Chem.*, 1970, **8**, 478-479 (*synth*).
Fitzgerald, M.A. *et al.*, *J. Chem. Soc., Perkin Trans. 1*, 1976, 186-191 (*2-O-Methylangolensin*).
Bezuidenhoudt, B.C.B. *et al.*, *J. Chem. Soc., Perkin Trans. 1*, 1980, 2179-2183 (*S-4-O-Methylangolensin*).
Ingham, J.L. *Prog. Chem. Org. Nat. Prod.*, 1983, **43**, 1-266 (*Angolensin, rev, occur*).
Salakka, A. *et al.*, *J. Chem. Soc., Perkin Trans. 1*, 1999, 2601-2604 (*Demethylangolensin, Angolensin, 4'-O-Methylangolensin*).
Shimizu, K. *et al.*, *Planta Med.*, 2000, **66**, 11-15 (*R-form, activity*).
Lee, S.-E. *et al.*, *J. Agric. Food Chem.*, 2001, **49**, 5171-5177 (*R-form, activity*).
Wähälä, K. *et al.*, *J. Chem. Soc., Perkin Trans. 1*, 2001, 642-644 (*Angolensin, 4,4'-di-Me ether, synth*).
Tanaka, H. *et al.*, *Phytochemistry*, 2002, **60**, 789-794 (*Erythrina poeppigiana constit*).
Nishiyama, Y. *et al.*, *J. Nat. Med.*, 2019, **73**, 289-296 (*R-4-O-Methylangolensin*)

3-(3,4-Dihydroxyphenyl)-2-hydroxypropanoic acid D209

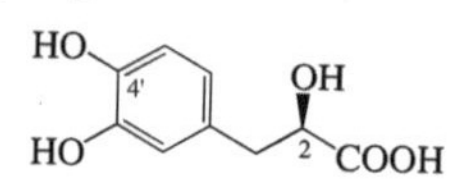

(R)-form

α,3,4-Trihydroxybenzenepropanoic acid, CAS. *3-(3,4-Dihydroxyphenyl)lactic acid. Salvianic acid*
[227782-88-9]
[153381-08-9] (*S*-3',4'-di-Me ether Me ester) [950665-05-1] (*R*-isopropyl ester)

$C_9H_{10}O_5$ 198.174
(R)-form
Danshensu. Danshensuan A
[76822-21-4]
Constit. of *Coptis chinensis*, *Pulmonaria officinalis* and *Salvia miltiorrhiza*. Coronary vasodilator. Prisms ($EtOAc/C_6H_6$). Mp 84 - 87°. $[\alpha]_D^{20}$ +15 (c, 0.02 in MeOH). $[\alpha]_D^{20}$ +25.4 (c, 0.5 in H_2O)
4′-O-β-D-Glucopyranoside
[96552-85-1]
$C_{15}H_{20}O_{10}$ 360.315
Constit. of *Coptis chinensis* root. Hygroscopic powder. $[\alpha]_D^{25}$ -41 (c, 1.00 in MeOH)
2-Sulfate
3-(3,4-Dihydroxyphenyl)-2-(sulfooxy)propanoic acid. ***Tichocarpol B***
[735328-84-4]
$C_9H_{10}O_8S$ 278.231
Isol. from the red alga *Tichocarpus crinitus*. Amorph. solid. $[\alpha]_D^{28}$ +5.9 (c, 0.34 in H_2O)
Me ester, 2-O-β-D-glucopyranoside
[1409971-72-7]
$C_{16}H_{22}O_{10}$ 374.342
Isol. from *Sanguisorba officinalis*. Displays moderate antimicrobial activity. Amorph. powder
Butyl ester, 4′-O-β-D-glucopyranoside
Ternatoside B
[875057-33-3]
$C_{19}H_{28}O_{10}$ 416.423
Constit. of *Ranunculus chinensis* and *Ranunculus ternatus*. Powder. $[\alpha]_D^{20}$ -41.2 (c, 0.5 in MeOH)
2,3-Dihydroxypropyl ester
Glyceryl 3,4-dihydroxyphenyllactate. ***Ranunchinensin A***
[1227366-75-7]
$C_{12}H_{16}O_7$ 272.253
Constit. of *Ranunculus chinensis*. Light yellow gum. $[\alpha]_D^{22}$ +5.6 (c, 0.44 in MeOH)
Amide
2-Hydroxy-3-(3,4-dihydroxyphenyl)propanamide. 3-(3,4-Dihydroxyphenyl)lactamide
[293735-05-4]
$C_9H_{11}NO_4$ 197.190
Alkaloid from the rhizomes of *Salvia miltiorrhiza*. Amorph. yellow powder. $[\alpha]_D^{20}$ +15.8 (c, 0.1 in MeOH). λ_{max} 208 (log ε 4.04); 280 (log ε 3.54) (MeOH)
2-O-(4-Hydroxy-E-cinnamoyl)
Clinopodic acid A. Isorinic acid
[145757-61-5]
Constit. of *Clinopodium chinense* var. *parviflorum* and *Sarcandra glabra*. Amorph. powder. $[\alpha]_D^{23}$ +83.8 (c, 1.6 in MeOH). $[\alpha]_D^{24}$ +114 (c, 0.84 in MeOH). λ_{max} 230 (log ε 4.84); 279 (log ε 4.23); 308 (log ε 4.01); 319 (log ε 4.02) (MeOH)
2-O-(4-Hydroxy-E-cinnamoyl), Me ester
Methyl isorinate
$C_{19}H_{18}O_7$ 358.346
Constit. of *Sarcandra glabra*. Exhibits antiinflammatory activity. $[\alpha]_D^{25}$ +39 (c, 1 in MeOH)
2-O-(4-Hydroxy-E-cinnamoyl), 4′-O-β-D-glucopyranoside
[251926-08-6]
$C_{24}H_{26}O_{12}$ 506.460
Constit. of the fruit of *Helicteres isora*. Amorph. powder. $[\alpha]_D$ -16 (c, 0.3 in H_2O). λ_{max} 219 (log ε 4.26); 286 (log ε 4.16); 319 (log ε 4.07) (MeOH)
2-O-(3,4-Dihydroxy-E-cinnamoyl)
Labiatenic acid. ***Rosmarinic acid.*** *Rosemarinic acid. Rosemary acid*
[20283-92-5]
[537-15-5] (unspecified stereochem.)
$C_{18}H_{16}O_8$ 360.318
Constit. of *Rosmarinus officinalis*, *Melissa officinalis*, *Momordica balsamina*, *Mentha piperita*, *Perilla frutescens*, *Salvia officinalis*, *Teucrium scorodonia*, *Sanicula europaea*, *Coleus blumei*, *Thymus* spp., *Hyptis verticillata*, *Lithospermum erythrorhizon* and other plant spp. Isol. from hornwort *Anthoceros agrestis*. Exhibits antithrombotic and antiplatelet effects. Antiinflammatory agent. Shows anti-HIV, antibacterial and antifungal effects; antiemetic, antigenotoxic activities etc. Plant growth inhibitor. Also shows activity against β-amyloid aggregation. Uv protectant. Antioxidant. Cryst. + $2H_2O$. Mp 204° dec.. $[\alpha]_D^{20}$ +145. Log P 1.01 (calc). λ_{max} 290 (sh); 328 (log ε 3.42) (MeOH)
▶ GD8990000. LD_{50} (mus, ivn) 561 mg/kg
2-O-(3,4-Dihydroxy-E-cinnamoyl), 3′-O-β-D-glucopyranoside
3′-O-β-D-Glucopyranosylrosmarinic acid
[853915-58-9]
$C_{24}H_{26}O_{13}$ 522.459
Constit. of the roots of *Eryngium alpinum* and from hornwort *Anthoceros agrestis*. Amorph. yellow powder. Mp 154 - 155° dec.. $[\alpha]_D^{20}$ +11.2 (c, 0.1 in MeOH)
2-O-(3,4-Dihydroxy-E-cinnamoyl), 4′-O-β-D-glucopyranoside
4′-O-β-D-Glucopyranosylrosmarinic acid
[910028-78-3]
$C_{24}H_{26}O_{13}$ 522.459
Constit. of *Helicteres isora* and *Sanicula lamelligera*. Amorph. powder. $[\alpha]_D$ +9 (c, 0.9 in MeOH). $[\alpha]_D^{25}$ -16.7 (c, 0.6 in H_2O). λ_{max} 204 (log ε 4.52); 287 (log ε 4.22); 319 (log ε 4.11) (MeOH)
2-O-(3,4-Dihydroxy-E-cinnamoyl), Me ester
Methyl rosmarinate
[99353-00-1]
[179462-75-0] (unspecified stereochem.)
$C_{19}H_{18}O_8$ 374.345
Constit. of *Clinopodium laxiflorum*, *Glechoma hederacea* var. *longituba*, *Perilla frutescens* and *Salvia* spp.. Amorph.. $[\alpha]_D^{18}$ +138 (c, 0.6 in MeOH). λ_{max} 287; 326 (MeOH)
2-O-(3,4-Dihydroxy-E-cinnamoyl), Et ester
Ethyl rosmarinate
[174591-47-0]
$C_{20}H_{20}O_8$ 388.372
Constit. of *Glechoma hederacea* var. *longituba* and *Nepeta prattii*
2-O-(3,4-Dihydroxy-E-cinnamoyl), butyl ester
Butyl rosmarinate
[222713-83-9]
$C_{22}H_{24}O_8$ 416.426
Constit. of *Isodon oresbius*. Amorph. yellow powder. $[\alpha]_D$ +56 (c, 0.86 in $CHCl_3$)
2-O-[β-D-Glucopyranosyl-(→3)-3,4-dihydroxy-E-cinnamoyl]
Salviaflaside
[178895-25-5]
$C_{24}H_{26}O_{13}$ 522.459
Constit. of *Salvia flava*, *Salvia miltiorrhiza* and *Perilla frutescens*.

Amorph. yellow powder. [α] D +18.5 (c, 0.065 in MeOH)

2-O-[β-D-Glucopyranosyl-(→3)-3,4-dihydroxy-E-cinnamoyl], Me ester
Salviaflaside methyl ester
[178860-70-3]
$C_{25}H_{28}O_{13}$ 536.486
Constit. of *Salvia flava*. Amorph. yellow powder. [α] D -9.1 (c, 0.12 in MeOH). λ_{max} 205 (log ε 4.55); 287 (log ε 4.07); 320 (log ε 4.11) (no solvent reported)

2-O-[β-D-Glucopyranosyl-(→4)-3,4-dihydroxy-E-cinnamoyl], 4′-O-β-D-glucopyranoside
[251926-07-5]
$C_{30}H_{36}O_{18}$ 684.600
Constit. of the fruit of *Helicteres isora*. Amorph. powder. [α] D -17.9 (c, 1 in H_2O). λ_{max} 216 (log ε 4.34); 284 (log ε 4.24); 317 (log ε 4.12) (MeOH)

2-O-(3,4-Dihydroxy-Z-cinnamoyl)
(Z)-Rosmarinic acid
[845830-82-2]
$C_{18}H_{16}O_8$ 360.318
Constit. of *Clinopodium urticifolium* and *Melissa officinalis*. Pale yellow powder. λ_{max} 219; 290; 328 (MeOH)

2-O-(3-Hydroxy-4-methoxy-E-cinnamoyl)
Shimobashiric acid B
[865999-05-9]
$C_{19}H_{18}O_8$ 374.345
Constit. of *Keiskea japonica*. Amorph. solid. $[\alpha]_D^{21}$ +47.0 (c, 0.2 in MeOH). λ_{max} 204 (log ε 4.60); 288 (log ε 4.13); 324 (log ε 4.09) (MeOH)

2-O-(3-Hydroxy-4-methoxy-E-cinnamoyl), Me ester
Methyl 4″-O-methylrosmarinate
[1312304-53-2]
$C_{20}H_{20}O_8$ 388.372
Constit. of *Glechoma hederacea* var. *longituba*. Amorph. yellow powder. $[\alpha]_D^{25}$ +1.9 (c, 0.1 in MeOH). λ_{max} 252 (log ε 3.2); 292 (log ε 7.1); 327 (log ε 8.4) (MeOH)

2-O-(4-Hydroxy-3-methoxy-E-cinnamoyl)
3″-O-Methylrosmarinic acid
[737003-24-6]
$C_{19}H_{18}O_8$ 374.345
Constit. of *Glechoma hederacea* var. *longituba*

2-O-(3,4-dimethoxy-E-cinnamoyl), Me ester
[854237-30-2]
$C_{21}H_{22}O_8$ 402.399
Constit. of *Ehretia obtusifolia*. Amorph. yellow powder. $[\alpha]_D^{24}$ +75 (c, 0.02 in MeOH). λ_{max} 231 (log ε 4.3); 290 (log ε 3.91); 329 (log ε 3.72) (MeOH)

3′-Me ether, 2-O-(3,4-dihydroxy-E-cinnamoyl)
Clinopodic acid B
[1174920-55-8]
$C_{19}H_{18}O_8$ 374.345
Constit. of *Clinopodium chinense* var. *parviflorum*. Amorph. solid. $[\alpha]_D^{23}$ +79.3 (c, 1.91 in MeOH). λ_{max} 220 (sh); 232 (sh); 288 (log ε 4.07); 326 (log ε 4.17) (MeOH)

3′-Me ether, 2-O-(3,4-dihydroxy-E-cinnamoyl), Me ester
Methyl 3′-O-methylrosmarinate
[1561146-67-5]
$C_{20}H_{20}O_8$ 388.372
Constit. of *Salvia palaestina*. Potent antioxidant. Brown solid. $[\alpha]_D^{25}$ +1.09 (c, 0.0004 in MeOH). λ_{max} 292; 332 (MeOH)

3′,4′-Di-Me ether
2-Hydroxy-3-(3,4-dimethoxyphenyl)propanoic acid
[54844-37-0]
$C_{11}H_{14}O_5$ 226.228
Solid. Mp 102 - 103°. $[\alpha]_D^{20}$ +11 (c, 1 in $CHCl_3$) (99% ee)

3′,4′-Di-Me ether, Me ester
[54814-41-4]
$C_{12}H_{16}O_5$ 240.255
Oil. $[\alpha]_D^{20}$ +18 (c, 1 in $CHCl_3$) (99% ee)

3′-Octadecyl ether, 2-O-(3,4-dihydroxy-E-cinnamoyl)
3′-O-Octadecylrosmarinic acid.
Cordicilin
[1311377-83-9]
$C_{36}H_{52}O_8$ 612.804
Constit. of the fruit of *Cordia latifolia*. Amorph. powder. λ_{max} 290 (log ε 4.42); 331 (log ε 4.51) (MeOH)

(S)-form [42085-50-7]
Yellow solid. $[\alpha]_D^{25}$ -10 (c, 0.5 in MeOH). $[\alpha]_D^{20}$ -24 (c, 0.5 in H_2O)

Me ester
Oresbiusin A
[457893-80-0]
$C_{10}H_{12}O_5$ 212.201
Constit. of *Isodon oresbius* and *Ranunculus chinensis*. Pale yellow solid. $[\alpha]_D^{22}$ +7.9 (c, 0.4 in MeOH). Config. revised in 2012

Isopropyl ester
[1280727-98-1]
No phys. props reported. $[\alpha]_D^{25}$ +9.7 (c, 1.0 in MeOH) (97.8% ee)

2-Ac, Me ester
Lavandunat
$C_{12}H_{14}O_6$ 254.238
Isol. from the remaining material obt. from *Lavandula angustifolia* after distillation of the essential oil. Moderate antioxidant. Amorph. yellow powder (MeOH). $[\alpha]_D^{25}$ -1.0 (c, 0.1 in MeOH). λ_{max} 280 (log ε 3.33) (MeOH)

2-O-(3,4-Dihydroxy-E-cinnamoyl)
(-)-Rosmarininc acid
[179188-11-5]
Amorph. light yellow solid. $[\alpha]_D^{25}$ -86 (c, 0.54 in 95% EtOH)

(±)-form [23028-17-3]

2-O-(4-Hydroxy-3-methoxy-E-cinnamoyl), Me ester
Oresbiusin B
[181701-84-8]
$C_{20}H_{20}O_8$ 388.372
Constit. of *Isodon oresbius*. Yellow solid. Racemate. Both (*R*)-(+)- and (*S*)-(-)-forms have been synth.

3′-Me ether
2-Hydroxy-3-(4-hydroxy-3-methoxyphenyl)propanoic acid.
Vanillactic acid
[2475-56-1]
$C_{10}H_{12}O_5$ 212.201
Cryst. (EtOH). Mp 170°

3′-Me ether, Me ester
[1258863-27-2]
$C_{11}H_{14}O_5$ 226.228
Cryst. (Et_2O). Mp 62°

(ξ)-form

Et ester
[475597-86-5]
$C_{11}H_{14}O_5$ 226.228
Constit. of *Nepeta prattii*. Amorph. yellow solid

Isopropyl ester
[868785-50-6]
$C_{12}H_{16}O_5$ 240.255
Solid

2-O-(3,4-Dihydroxy-E-cinnamoyl), propyl ester
Propyl rosmarinate

[618460-92-7]
$C_{21}H_{22}O_8$ 402.399
Isol. from *Thymus serpyllum*. Displays significant antibacterial activity. Amorph. solid. λ_{max} 290; 330 (MeOH)

4′-O-(4-Hydroxyphenyl), 2-O-(3,4-dihydroxy-E-cinnamoyl), Me ester
[1173885-61-4]
$C_{25}H_{22}O_9$ 466.442
Constit. of *Lindelofia stylosa*. Antioxidant. Gum. $[\alpha]^{25}_{D}$ +28.4 (c, 0.4 in MeOH). λ_{max} 207 (log ε 5.2); 220 (log ε 4.28); 292 (log ε 4.15); 332 (log ε 4.26) (MeOH).

Schopf, C. *et al., Justus Liebigs Ann. Chem.*, 1940, **544**, 30-62 (*(±)-form, synth, (±)-3′-Me ether*).
Kelley, C.J. *et al., J. Org. Chem.*, 1975, **40**, 1804-1815 (*Rosmarinic acid*).
Kelley, C.J. *et al., J. Org. Chem.*, 1976, **41**, 449-459 (*Rosmarinic acid*).
Yahara, S. *et al., Chem. Pharm. Bull.*, 1985, **33**, 527-531 (*Coptis chinensis constits*).
Parnham, M.J. *et al., Drugs of the Future*, 1985, **10**, 756 (*Rosmarinic acid, rev*).
Rao, N.N. *et al., Arch. Pharm. (Weinheim, Ger.)*, 1988, **321**, 179-180 (*enzymic synth*).
Englberger, W. *et al., Int. J. Immunopharmacol.*, 1988, **10**, 729-737 (*Rosmarinic acid, pharmacol*).
Peake, P.W. *et al., Int. J. Immunopharmacol.*, 1991, **13**, 853-857 (*Rosmarinic acid, pharmacol*).
Pabsch, K. *et al., Recl. Trav. Chim. Pays-Bas*, 1991, **110**, 199-205 (*Rosmarinic acid, synth*).
Mahmood, N. *Antiviral Chem. Chemother.*, 1993, **4**, 235 (*Rosmarinic acid, anti-HIV activity, nmr, ms*).
Zou, Z.W. *et al., Yaoxue Xuebao*, 1993, **28**, 241 (*Rosmarinic acid, pharmacol*).
Zhao, L.M. *et al., Chin. Chem. Lett.*, 1996, **7**, 449-452 (*Salviaflaside*).
Abraham, S.K. *Food Chem. Toxicol.*, 1996, **34**, 15-20 (*Rosmarinic acid, activity*).
Huang, H. *et al., J. Nat. Prod.*, 1996, **59**, 1079-1080 (*Oresbiusins A,B*).
Binutu, O.A. *et al., Planta Med.*, 1996, **62**, 352-353 (*Rosmarinic acid, activity*).
Robinson, W.E. *et al., Proc. Natl. Acad. Sci. U.S.A.*, 1996, **93**, 6326-6331 (*Rosmarinic acid, activity*).
Eicher, T. *et al., Synthesis*, 1996, 755-762 (*Rosmarinic acid, synth*).
Bogucki, D.E. *et al., Can. J. Chem.*, 1997, **75**, 1783-1794 (*S-form, S-Rosmarinic acid, synth*).
Reimann, E. *et al., Monatsh. Chem.*, 1997, **128**, 995-1008 (*Rosmarinic acid, synth*).
Kusano, G. *et al., Biol. Pharm. Bull.*, 1998, **21**, 997-999 (*Rosmarinic acid, activity*).
Reimann, E. *et al., Monatsh. Chem.*, 1998, **129**, 187-193 (*Rosmarinic acid, Me ester, synth*).
Satake, T. *et al., Chem. Pharm. Bull.*, 1999, **47**, 1444-1447 (*glucosides, Isorinic acid*).
Huang, H. *et al., Planta Med.*, 1999, **65**, 92-93 (*Butyl rosmarinate*).
Kuo, Y.-H. *et al., J. Chin. Chem. Soc. (Taipei)*, 2000, **47**, 241-246 (*Methyl rosmarinate*).
Choi, J.S. *et al., Fitoterapia*, 2001, **72**, 30-34 (*amide*).
Hou, Z.-F. *et al., J. Chin. Chem. Soc. (Taipei)*, 2002, **49**, 255-258 (*Ethyl rosmarinate*).
Petersen, M. *et al., Phytochemistry*, 2003, **62**, 121-125 (*Rosmarinic acid, rev, biosynth*).
Wei, X.-M. *et al., J. Chin. Chem. Soc. (Taipei)*, 2004, **51**, 1043-1049 (*Z-Rosmarinic acid*).
Ishii, T. *et al., J. Nat. Prod.*, 2004, **67**, 1764-1766 (*Tichocarpol B*).
Iqbal, K. *et al., Chem. Biodiversity*, 2005, **2**, 104-111 (*Ehretia obtusifolia constit*).
Le Claire, E. *et al., J. Agric. Food Chem.*, 2005, **53**, 4367-4372 (*3′-Glucosylrosmarinic acid*).
Tian, J.-K. *et al., J. Asian Nat. Prod. Res.*, 2006, **8**, 35-39 (*Ternatoside B*).
Vogelsang, K. *et al., Planta*, 2006, **223**, 369-373 (*Anthoceros agrestis constits*).
Zhou, L.-Y. *et al., Z. Naturforsch., B*, 2006, **61**, 607-610 (*Sanicula lamelligera constit*).
Choudhary, M.I. *et al., Chem. Biodiversity*, 2008, **5**, 2676-2683 (*Lindelofia stylosa ester*).
Murata, T. *et al., J. Nat. Prod.*, 2009, **72**, 1379-1384 (*Clinopodic acids A,B*).
Petersen, M. *et al., Phytochemistry*, 2009, **70**, 1663-1679 (*Rosmarinic acid, biosynth, rev*).
Nan, Y.-F. *et al., Acta Crystallogr., Sect. E*, 2010, **66**, o1105 (*isopropyl ester, cryst struct*).
Zou, Y. *et al., Chem. Nat. Compd. (Engl. Transl.)*, 2010, **46**, 19-21 (*Ranunchinensin A, Oresbiusin A, Ternatoside B*).
Kim, J.-P. *et al., Bioorg. Med. Chem. Lett.*, 2011, **21**, 3483-3487 (*Glechoma hederacea constits*).
Begum, S. *et al., Chem. Biodiversity*, 2011, **8**, 850-861 (*Cordicilin*).
Chen, M. *et al., Tetrahedron: Asymmetry*, 2011, **22**, 4-7 (*S-isopropyl ester, synth*).
Murata, T. *et al., Chem. Pharm. Bull.*, 2012, **60**, 121-128 (*Shimobashiric acid B*).
Hu, J. *et al., Chin. Chem. Lett.*, 2012, **23**, 1243-1246 (*Me ester 2-glucoside*).
Hwu, J.R. *et al., Eur. J. Org. Chem.*, 2012, 4684-4688 (*Oresbiusins A,B, synth, abs config, activity*).
Petersen, M. *Phytochem. Rev.*, 2013, **12**, 207-227 (*Rosmarinic acid, rev, occur, biosynth*).
Aziz, S. *et al., Chem. Nat. Compd. (Engl. Transl.)*, 2014, **49**, 1023-1027 (*Propyl rosmarinate, activity*).
Bloch, S.E. *et al., ChemBioChem*, 2014, **15**, 2393-2401 (*biosynth*).
Zhou, X.-J. *et al., Food Chem.*, 2014, **164**, 150-157 (*Perilla frutescens constits*).
Al-Qudah, M.A. *et al., Phytochemistry*, 2014, **99**, 115-120 (*Salvia palaestina constit, activity*).
Kim, G.-D. *et al., Appl. Microbiol. Biotechnol.*, 2015, **99**, 2083-2092 (*Rosmarinic acid, rev, prodn, use*).
Wang, B. *et al., Planta*, 2015, **241**, 711-725 (*Rosmarinic acid, biosynth*).
Amoah, S.K.S. *et al., Planta Med.*, 2016, **82**, 388-406 (*Rosmarinic acid, rev, occur, biosynth, activity*).
Zheng, Y. *et al., Tetrahedron*, 2016, **72**, 5047-5050 (*R-di-Me ether, R-di-Me ether Me ester*).
Tsai, Y.-C. *et al., J. Agric. Food Chem.*, 2017, **65**, 6497-6505 (*Isorinic acid, Methyl isorinate, activity*).
Biotechnology of Natural Products, (eds. Schwab, W. et al), Springer, 2018, 25-60 (*Rosmarinic acid, rev, prodn*).
Liu, J. *et al., Chem. Nat. Compd. (Engl. Transl.)*, 2018, **54**, 1076-1080 (*R-form, cd*).
Ngo, Y.L. *et al., Food Chem. Toxicol.*, 2018, **121**, 687-700 (*Rosmarinic acid, rev, isol, activity*).
Sidoryk, K. *et al., J. Mol. Struct.*, 2018, **1153**, 135-148 (*R-form, S-form, synth*).
Yadikar, N. *et al., Phytochem. Lett.*, 2018, **23**, 149-154 (*Lavandunat, activity*).
Li, C. *et al., ACS Sustainable Chem. Eng.*, 2019, **7**, 18247-18253 (*synth*).
Luo, C. *et al., Front. Pharmacol.*, 2020, **11**, 153:1-11 (*Rosmarinic acid, rev, activity*).
Hitl, M. *et al., Planta Med.*, 2021, **87**, 273-282 (*Rosmarinic acid, rev*)

3-(3,4-Dihydroxyphenyl)-3-hydroxypropanoic acid D210

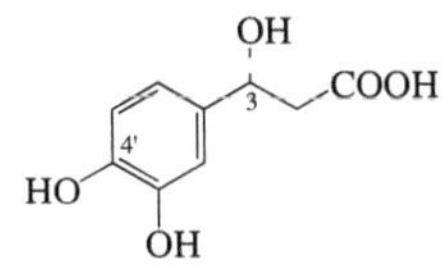

(R)-form

β,3,4-Trihydroxybenzenepropanoic acid, CAS
$C_9H_{10}O_5$ 198.174

(R)-form

3′-Me ether, Me ester

Pisoninol I
[1346752-20-2]
$C_{11}H_{14}O_5$ 226.228
Constit. of the stems of *Pisonia umbellifera*. Oil. $[\alpha]^{25}_{D}$ +37.8 (c, 0.04 in $CHCl_3$). λ_{max} 228 (log ε 3.85); 280 (log ε 3.1) (MeOH). [base] λ_{max} 235 (log ε 3.95); 283 (log ε 3.28) (MeOH/KOH aq.)

(S)-form

3-O-(3,4-Dihydroxy-E-cinnamoyl), Me ester
[345892-22-0]
$C_{19}H_{18}O_8$ 374.345
Constit. of *Isodon excisus*. Amorph. yellow powder. $[\alpha]^{25}_{D}$ +26 (c, 1 in MeOH). λ_{max} 218 (log ε 4.35); 291 (log ε 4.2); 329 (log ε 4.31) (MeOH)

3-Me ether, Me ester

Taraxafolin
[626236-63-3]
$C_{11}H_{14}O_5$ 226.228
Constit. of the aerial parts of *Taraxacum formosanum*. Syrup. $[\alpha]_D$ -32 (c, 0.1 in $CHCl_3$). λ_{max} 234 (sh); 249 (sh) (log ε); 291 (log ε 4.4); 326 (log ε 4.4) (MeOH)

3,3′-Di-Me ether, Me ester

3′-O-Methyltaraxafolin
[1013933-11-3]
$C_{12}H_{16}O_5$ 240.255

Constit. of the roots of *Ligularia dentata*. $[\alpha]_D$ -32.3 ($CHCl_3$).

Lee, C. *et al.*, *J. Nat. Prod.*, 2001, **64**, 659-660 (*Isodon excisus caffeate*).
Leu, Y.-L. *et al.*, *Chem. Pharm. Bull.*, 2003, **51**, 599-601 (*Taraxafolin*).
Baba, H. *et al.*, *J. Nat. Med.*, 2007, **61**, 472-473 (*3′-O-Methyltaraxafolin*).
Kuo, H.-T. *et al.*, *Planta Med.*, 2011, **77**, 736-741 (*Pisoninol I*)

1-(3,4-Dihydroxyphenyl)-2-hydroxy-1-propanone D211

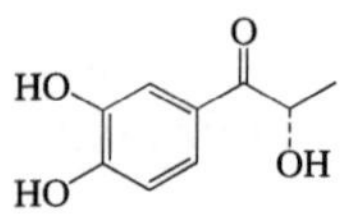

2,3′,4′-Trihydroxypropiophenone
$C_9H_{10}O_4$ 182.175
(S)-form [1644631-96-8]
Constit of *Trollius chinensis*
3′-Me ether
2-Hydroxy-1-(4-hydroxy-3-methoxyphenyl)-1-propanone. 2,4′-Dihydro-3′-methoxypropiophenone. α-Hydroxypropiovanillone. ***Evofolin A***
[168146-21-2]
$C_{10}H_{12}O_4$ 196.202
Isol. from *Betula verrucosa*, *Tetradium glabrifolium* (*Evodia meliaefolia*) and *Sida acuta*. Present in the Lycopodiaceae. Pale yellow oil. $[\alpha]_D$ -28.6 (c, 0.021 in $CHCl_3$). λ_{max} 228 (log ε 4.08); 280 (log ε 3.91); 310 (log ε 3.79) (MeOH)
(±)-form
3′-Me ether
[2034-61-9]
$C_{10}H_{12}O_4$ 196.202
Cryst. (C_6H_6). Mp 109 - 110°
(ξ)-form [90536-60-0]
Hibbert ketone obt. by ethanolysis of lignin
α-Et ether, 3′-Me ether
α-Ethoxypropiovanillone. 2-Ethoxy-1-(4-hydroxy-3-methoxyphenol)-1-propanone, CAS
[2034-62-0]
$C_{12}H_{16}O_4$ 224.256
Hibbert ketone obt. from lignin
5′-Methoxy, 3′-Me ether
2-Hydroxy-1-(4-hydroxy-3,5-dimethoxyphenyl)-1-propanone, CAS. *α-Hydroxypropiosyringone*
[38209-33-5]
$C_{11}H_{14}O_5$ 226.228
Hibbert ketone obt. by ethanolysis of lignin. Mp 126 - 127°.

Cramer, A.B. *et al.*, *J. Am. Chem. Soc.*, 1939, **61**, 2204-2206 (*3′-Me ether, synth*).
Hergert, H.L. *J. Org. Chem.*, 1960, **25**, 405-413 (*ir*).
Towers, G.H.N. *et al.*, *Phytochemistry*, 1965, **4**, 57-66 (*3′-Me ether, occur*).
Skurikhin, I.M. *Chem. Nat. Compd. (Engl. Transl.)*, 1967, **3**, 286-289 (*uv*).
Lundquist, K. *Acta Chem. Scand.*, 1973, **27**, 2597-2606 (*Betula verrucosa constit*).
US Pat., 1979, 4 169 858 (Merck) (*synth*).
Wu, T.-S. *et al.*, *Phytochemistry*, 1995, **40**, 121-124 (*Evofolin A*).
Jang, D.S. *et al.*, *Arch. Pharmacal Res.*, 2003, **26**, 585-590 (*Sida acuta constit*).
Peng, K. *et al.*, *Tetrahedron Lett.*, 2005, **46**, 1217-1220 (*Evofolin A, synth, abs config*).
Li, D.-Y. *et al.*, *J. Asian Nat. Prod. Res.*, 2014, **16**, 1018-1023 (*Trollius chinensis constit*)

1-(3,4-Dihydroxyphenyl)-3-hydroxy-1-propanone D212

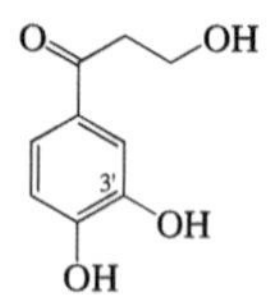

3,3′,4′-Trihydroxypropiophenone. 2-(3,4-Dihydroxybenzoyl)ethanol
$C_9H_{10}O_4$ 182.175
3′-Me ether
3-Hydroxy-1-(4-hydroxy-3-methoxyphenyl)-1-propanone, CAS. *3,4′-Dihydroxy-3′-methoxypropiophenone,* CAS. *ω-Hydroxypropioguaiacone. β-Hydroxypropiovanillone*
[2196-18-1]
$C_{10}H_{12}O_4$ 196.202
Isol. from *Bauhinia manca*, *Selaginella doederleinii*, *Leptactina senegambica*, *Pinus armandii* and *Pinus sylvestris*. Metab. of *Pseudomonas acidovorans* and *Epichloe typhina*. Cryst. (C_6H_6). Mp 109 - 110°
3′-Me ether, 3-O-β-D-glucopyranoside
[114226-07-2]
$C_{16}H_{22}O_9$ 358.343
Isol. from *Pseudomonas sylvestris* and *Picea abies*. $[\alpha]_D^{20}$ -9.5 (c, 1 in MeOH)
3′-Me ether, di-Ac
[107682-24-6]
$C_{14}H_{16}O_6$ 280.276
Cryst. (Et_2O). Mp 62°
3′-Me ether, 3-angeloyl
Smallanactone A
[1192353-92-6]
$C_{15}H_{18}O_5$ 278.304
Constit. of the leaves of *Smallanthus sonchifolius*. Antioxidant. Amorph. solid. λ_{max} 205 (log ε 5.15); 235 (log ε 3.86); 278 (log ε 4.18) (MeOH)
4′-Me ether
3-Hydroxy-1-(3-hydroxy-4-methoxyphenyl)-1-propanone. 3,3′-Dihydroxy-4′-methoxypropiophenone.
Optivanin
[1418732-28-1]
$C_{10}H_{12}O_4$ 196.202
Constit. of the stem bark of *Grewia optiva*. Amorph. powder. λ_{max} 229 (log ε 4.09) (MeOH)
3-Et, 3′-Me ether
3-Ethoxy-1-(4-hydroxy-3-methoxyphenyl)-1-propanone
[91555-69-0]
$C_{12}H_{16}O_4$ 224.256
Constit. of *Diospyros maritima*. Mp 72 - 74°. λ_{max} 230 (log ε 4.17); 277 (log ε 4.11); 303 (log ε 3.94) (MeOH).

Baker, S.B. *et al.*, *J. Am. Chem. Soc.*, 1948, **70**, 60-63 (*synth*).
Kovacik, V. *et al.*, *Chem. Ber.*, 1969, **102**, 1513-1522 (*ms*).
Andersson, R. *et al.*, *Phytochemistry*, 1988, **27**, 559-562 (*glucoside*).
Achenbach, H. *et al.*, *Phytochemistry*, 1988, **27**, 1835-1841 (*3′-Me ether*).
Lin, R.C. *et al.*, *Planta Med.*, 1994, **60**, 168-170 (*3′-Me ether*).
Kuo, Y.-H. *et al.*, *Phytochemistry*, 1998, **49**, 2505-2507 (*3-Et 3′-Me ether*).
Ruan, S. *et al.*, *Asian J. Chem.*, 2010, **22**, 5776-5778 (*Smallanactone A*).
Jiang, Z.-Y. *et al.*, *Carbohydr. Res.*, 2013, **382**, 19-24 (*Ficus tikoua glycoside*).
Uddin, G. *et al.*, *Nat. Prod. Res.*, 2013, **27**, 215-220 (*Optivanin*)

1-(3,4-Dihydroxyphenyl)-3-hydroxy-2-propanone D213

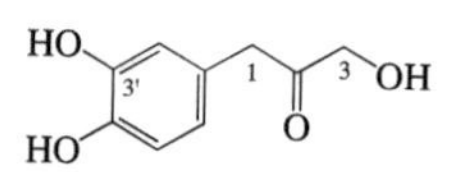

$C_9H_{10}O_4$ 182.175
3-O-(3-Hydroxy-4-methoxy-E-cinnamoyl)
Cimiciphenol. ***Cimiracemate A***
[478294-16-5]
$C_{19}H_{18}O_7$ 358.346
Constit. of the rhizomes of *Cimicifuga racemosa*. Pale yellow or brown powder. λ_{max} 210 (log ε 4.26); 291 (log ε 3.77); 325 (log ε 3.96) (MeOH) (Cimiracemate A)

3-O-(4-Hydroxy-3-methoxy-E-cinnamoyl)
Cimiracemate B
[478294-17-6]
$C_{19}H_{18}O_7$ 358.346
Constit. of the rhizomes of *Cimicifuga racemosa*. Light brown powder. Mp 86 - 88.5°. λ_{max} 208 (log ε 4.28); 327 (log ε 3.92) (MeOH)

3′-Me ether
3-Hydroxy-1-(4-hydroxy-3-methoxyphenyl)-2-propanone
[4899-74-5]

3′-Me ether, 3-benzoyl
[934248-67-6]
$C_{17}H_{16}O_5$ 300.31
Constit. of the resin of *Styrax tonkinensis*. Pale yellow oil

3′-Et ether, 4′-(oxoacetyl)
2-Ethoxy-4-(3-hydroxy-2-oxopropyl)phenyl oxoacetate
[1375747-97-9]
$C_{13}H_{14}O_6$ 266.249
Constit. of the roots of *Paeonia suffruticosa*. IL-1β inhibitor. Yellow oil. λ_{max} 195 (log ε 2.45); 278 (log ε 4.05) (MeOH)

1ξ-Methoxy, 3-O-(3-hydroxy-4-methoxy-E-cinnamoyl)
Cimiracemate C
[488804-01-9]
$C_{20}H_{20}O_8$ 388.372
Constit. of the rhizomes of *Cimicifuga racemosa*. Light brown powder. Mp 88 - 90°. $[\alpha]_D^{20}$ -6.8 (c, 0.15 in MeOH). λ_{max} 212 (log ε 4.35); 293 (log ε 3.87); 326 (log ε 4.09) (MeOH)

1ξ-Methoxy, 3-O-(4-hydroxy-3-methoxy-E-cinnamoyl)
Cimiracemate D
[488804-02-0]
$C_{20}H_{20}O_8$ 388.372
Constit. of the rhizomes of *Cimicifuga racemosa*. Light brown powder. Mp 100 - 102°. $[\alpha]_D^{20}$ -6.2 (c, 0.15 in MeOH). λ_{max} 212 (log ε 4.35); 326 (log ε 4.09) (MeOH).

Chen, S.-N. *et al.*, *Phytochemistry*, 2002, **61**, 409-413 (*Cimiracemates A-D*).
Chen, S.-N. *et al.*, *Nat. Prod. Res.*, 2005, **19**, 287-290 (*Cimiracemate B, synth*).
Stromeier, S. *et al.*, *Planta Med.*, 2005, **71**, 495-500 (*Cimiciphenol*).
Fache, F. *et al.*, *Tetrahedron*, 2005, **61**, 5261-5266 (*Cimiracemate B, synth*).
Wang, F. *et al.*, *J. Asian Nat. Prod. Res.*, 2006, **8**, 137-141 (*Styrax tonkinensis constit*).
Choi, Y.-H. *et al.*, *Arch. Pharmacal Res.*, 2012, **35**, 801-805 (*Paeonia suffruticosa constit*)

1-(3,5-Dihydroxyphenyl)-1-hydroxy-2-propanone D214

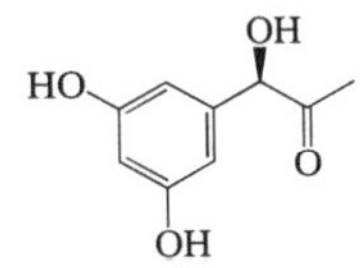

$C_9H_{10}O_4$ 182.175
(R)-form [2219330-71-7]
Prod. by *Conoideocrella tenuis* BCC 44534. Light brown oil. $[\alpha]_D^{23}$ -230 (c, 0.002 in MeOH). λ_{max} 230 (log ε 3.65); 285 (log ε 3.51); 423 (log ε 3.20) (MeOH).

Saepua, S. *et al.*, *Tetrahedron*, 2018, **74**, 859-866 (*Conoideocrella tenuis metab*)

1-(3,5-Dihydroxyphenyl)-2-hydroxy-1-propanone D215

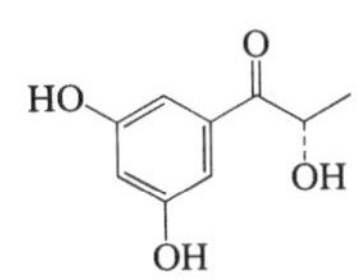

2,3′,5′-Trihydroxypropiophenone
$C_9H_{10}O_4$ 182.175
(S)-form [2222679-10-7]
Prod. by *Conoideocrella tenuis* BCC 44534. Light brown oil. $[\alpha]_D^{25}$ -81.1 (c, 0.01 in MeOH). λ_{max} 231 (log ε 3.76); 259 (log ε 3.73); 417 (log ε 3.34) (MeOH).

Saepua, S. *et al.*, *Tetrahedron*, 2018, **74**, 859-866 (*Conoideocrella tenuis metab*)

1-(3,4-Dihydroxyphenyl)-7-phenyl-4,6-heptadien-3-ol D216

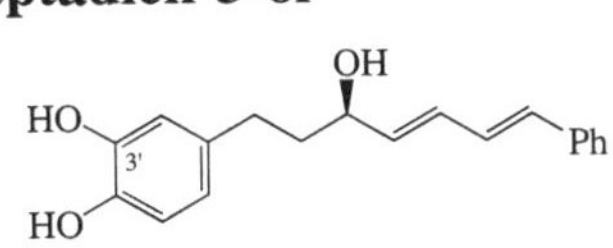

(3R,4E,6E)-form

$C_{19}H_{20}O_3$ 296.366
(3R,4E,6E)-form
3,3′-Di-Me ether
1-(4-Hydroxy-3-methoxyphenyl)-3-methoxy-7-phenyl-4,6-heptadiene
$C_{21}H_{24}O_3$ 324.42
Constit. of the rhizomes of *Curcuma comosa*

(3S,4E,6E)-form
3,3′-Di-Me ether
Constit. of the rhizomes of *Curcuma comosa*.

Sornkaew, N. *et al.*, *Nat. Prod. Commun.*, 2015, **10**, 89-93 (*Curcuma comosa constits*)

1-(3,4-Dihydroxyphenyl)-7-phenyl-3,5-heptanediol D217

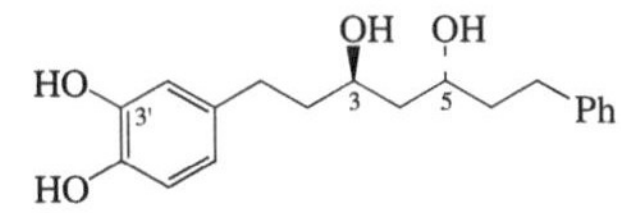

(3R,5R)-form

$C_{19}H_{24}O_4$ 316.397
(3R,5R)-form [1206788-64-8]
Constit. of the rhizomes of *Curcuma comosa*. Mp 120.9°. $[\alpha]_D^{25}$ +16.7 (c, 0.01 in MeOH). λ_{max} 203 (log ε 3.5); 212 (log ε 3.67); 287 (log ε 3.59) (MeOH)

3-Ac
[1869012-36-1]
$C_{21}H_{26}O_5$ 358.434
Constit. of the rhizomes of *Curcuma kwangsiensis*. Yellow oil. $[\alpha]_D^{25}$ +14.7 (c, 0.2 in MeOH). λ_{max} 224; 283 (MeOH)

(3ξ,5ξ)-form
3-Ac
Constit. of the rhizomes of *Curcuma comosa*. Sticky solid. $[\alpha]_D^{29}$ +5.3 (C, 1.3 in EtOH)

3′-Me ether
1-(4-Hydroxy-3-methoxyphenyl)-7-phenyl-3,5-heptanediol
[460052-56-6]
$C_{20}H_{26}O_4$ 330.424
Constit. of the dried rhizomes of *Alpinia officinarum*. Yellow liq.. $[\alpha]_D^{25}$ +6.9 (c, 0.4 in $CHCl_3$). λ_{max} 197 (log ε 6.69); 217 (log ε 6.73); 273 (log ε 6.83) (EtOH)

5-Ketone, 3′-Me ether
5-Hydroxy-7-(4-hydroxy-3-methoxyphenyl)-1-phenyl-3-heptanone
[79559-61-8]
$C_{20}H_{24}O_4$ 328.408
Constit. of the rhizomes of *Alpinia officinarum*. Inhibitor of pancreatic lipase and prostaglandin biosynth.. Pale yellow needles (petrol). Mp 59 - 60°. $[\alpha]_D^{16}$ -10.9 (c, 0.56 in $CHCl_3$)

5-Ketone, 3,3′-di-Me ether
7-(4-Hydroxy-3-methoxyphenyl)-5-methoxy-1-phenyl-3-heptanone, *CAS*
[83161-95-9]
$C_{21}H_{26}O_4$ 342.435
Isol. from *Alpinia officinarum*. Inhibitor of prostaglandin biosynth.. Oil. $[\alpha]_D^{21}$ -5.5 (c, 0.04 in $CHCl_3$). Probably an artifact

5-Ketone, 3′-Et, 3-Me ether
5-Ethoxy-7-(4-hydroxy-3-methoxyphenyl)-1-phenyl-3-heptanone
[1304020-21-0]
$C_{22}H_{28}O_4$ 356.462
Constit. of the rhizomes of *Alpinia officinarum*. Pale yellow oil. $[\alpha]_D^{21}$ +1.9 (c, 0.73 in $CHCl_3$). λ_{max} 242 (log ε 4.18); 281 (log ε 4.08) ($CHCl_3$)
Diketone, 3′-Me ether
1-(4-Hydroxy-3-methoxyphenyl)-7-phenyl-3,5-heptanedione
[83161-94-8]
$C_{20}H_{22}O_4$ 326.392
Isol. from *Alpinia officinarum*. Inhibitor of prostaglandin biosynth.. Oil
(3S)-form
5-Ketone
7-(3,4-Dihydroxyphenyl)-5-hydroxy-1-phenyl-3-heptanone
[1022082-05-8]
$C_{19}H_{22}O_4$ 314.381
Constit. of the rhizomes of *Alpinia officinarum*. Pale yellow oil. $[\alpha]_D^{20}$ +3.1 (c, 0.06 in $CHCl_3$). λ_{max} 204 (log ε 4.45); 283 (log ε 3) (MeOH).

Inoue, T. *et al.*, *Yakugaku Zasshi*, 1978, **98**, 1255-1257 (*5-ketone 3′-Me ether*).
Kiuchi, F. *et al.*, *Chem. Pharm. Bull.*, 1982, **30**, 2279-2282 (*5-ketone 3,3′-di-Me ether, diketone 3′-Me ether*).
Shin, D. *et al.*, *J. Nat. Prod.*, 2002, **65**, 1315-1318 (*3′-Me ether*).
Shin, J.-E. *et al.*, *Biol. Pharm. Bull.*, 2004, **27**, 138-140 (*5-ketone 3′-Me ether*).
An, N. *et al.*, *Fitoterapia*, 2008, **79**, 27-31 (*5-ketone*).
Kaewamatawong, R. *et al.*, *Phytochem. Lett.*, 2009, **2**, 19-21 (*Curcuma comosa constit*).
Zhao, L. *et al.*, *Chin. J. Nat. Med.*, 2010, **8**, 241-243 (*5-ketones*).
Chen, S.-D. *et al.*, *Fitoterapia*, 2015, **102**, 67-73 (*Curcuma kwangsiensis 3-Ac*).
Sornkaew, N. *et al.*, *Nat. Prod. Commun.*, 2015, **10**, 89-93 (*Curcuma comosa 3-Ac*).
Gamre, S. *et al.*, *J. Nat. Prod.*, 2021, **84**, 352-363 (*synth, activity*)

1-(3,4-Dihydroxyphenyl)-7-phenyl-3-heptanol D218

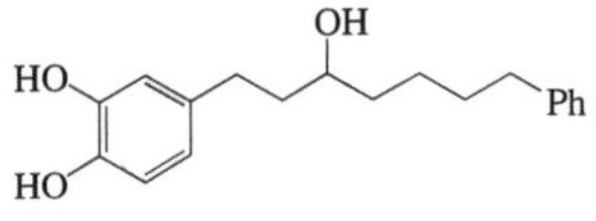

$C_{19}H_{24}O_3$ 300.398
(ξ)-form
3′-Me ether
1-(4-Hydroxy-3-methoxyphenyl)-7-phenyl-3-heptanol. ***Oxyphyllacinol***
[87657-77-0]
$C_{20}H_{26}O_3$ 314.425
Constit. of the fruit of *Alpinia oxyphylla*. Antioxidant. $[\alpha]_D^{20}$ +2.4 (c, 0.17 in EtOH).

Luo, X. *et al.*, *Yaoxue Xuebao*, 2000, **35**, 204-207 (*Oxyphyllacinol*).
Han, J.-T. *et al.*, *Food Sci. Biotechnol.*, 2007, **16**, 1060-1063 (*Oxyphyllacinol, activity*).
Chen, F. *et al.*, *Molecules*, 2014, **19**, 4510-4523 (*Oxyphyllacinol, occur*)

1-(3,4-Dihydroxyphenyl)-7-phenyl-1,4,6-heptatrien-3-one D219

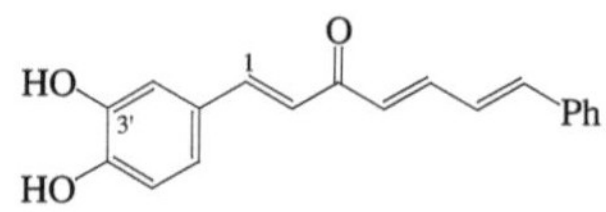

$C_{19}H_{16}O_3$ 292.334
4,5,6,7-Tetrahydro, 3′-Me ether
1-(4-Hydroxy-3-methoxyphenyl)-7-phenyl-1-hepten-3-one. ***Yakuchinone B***
[81840-57-5]
$C_{20}H_{22}O_3$ 310.393
Constit. of the fruit of *Alpinia oxyphylla*. Yellow needles (MeOH). Mp 100.5°. λ_{max} 225 (log ε 4.08); 336 (log ε 4.40) (MeOH). [base] λ_{max} 255; 404 (MeOH/NaOH)
Hexahydro, 3′-Me ether
1-(4-Hydroxy-3-methoxyphenyl)-7-phenyl-3-heptanone. ***Yakuchinone A***
[78954-23-1]
$C_{20}H_{24}O_3$ 312.409
Constit. of the fruit of *Alpinia oxyphylla*. Antioxidant. Pungent yellowish oil. λ_{max} 228 (sh); 282 (log ε 3.48) (MeOH). [base] λ_{max} 244; 292 (MeOH/NaOH)
5′-Hydroxy, hexahydro, 4′-Me ether
1-(3,5-Dihydroxy-4-methoxyphenyl)-7-phenyl-3-heptanone. ***5′-Hydroxyyakuchinone A***
[1606145-76-9]
$C_{20}H_{24}O_4$ 328.408
Constit. of the fruit of *Alpinia oxyphylla*. Antioxidant. Pale yellow oil. λ_{max} 238; 292 (MeOH).

Itokawa, H. *et al.*, *Phytochemistry*, 1981, **20**, 769-771 (*Yakuchinone A*).
Itokawa, H. *et al.*, *Phytochemistry*, 1982, **21**, 241-243 (*Yakuchinone B*).
Han, J.-T. *et al.*, *Food Sci. Biotechnol.*, 2007, **16**, 1060-1063 (*Yakuchinone A, activity*).
Bian, Q.-Y. *et al.*, *J. Asian Nat. Prod. Res.*, 2013, **15**, 1094-1099 (*5′-Hydroxyyakuchinone A*).
Chen, F. *et al.*, *J. Pharm. Biomed. Anal.*, 2014, **97**, 166-177 (*Yakuchinones A,B*)

7-(3,4-Dihydroxyphenyl)-1-phenyl-1-heptene-3,5-dione D220

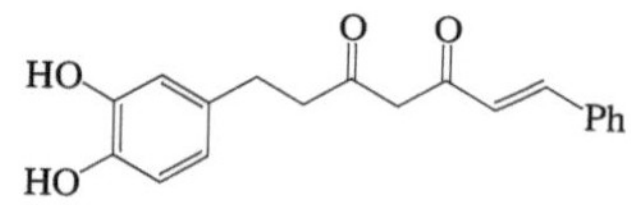

$C_{19}H_{18}O_4$ 310.349
3″-Me ether
7-(4-Hydroxy-3-methoxyphenyl)-1-phenyl-1-heptene-3,5-dione
[952209-86-8]
$C_{20}H_{20}O_4$ 324.376
Constit. of the rhizomes of *Alpinia officinarum*. Cryst. (petrol). Mp 86 - 88°. λ_{max} 228 (sh); 290 (sh); 342 (log ε 4.5) (MeOH).

An, N. *et al.*, *J. Asian Nat. Prod. Res.*, 2006, **8**, 637-641 (*Alpinia officinarum constit*)

1-(2,5-Dihydroxyphenyl)-1,2-propanediol D221

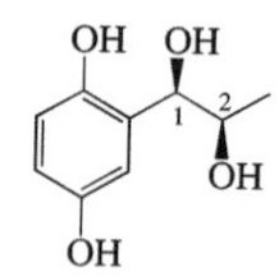

(*1R*,2R**)-form*

$C_9H_{12}O_4$ 184.191
(1R*,2R*)-form
1,5′-Di-Me ether
1-(2-Hydroxy-5-methoxyphenyl)-1-methoxy-2-propanol. ***Thellungianin E***
[187827-76-5]
$C_{11}H_{16}O_4$ 212.245
Constit. of *Pimpinella thellungiana* and *Pimpinella tragium* ssp. *lithophila*. λ_{max} 220; 225; 296 (EtOH)
1-Et, 5′-Me ether
[187827-64-1]
$C_{12}H_{18}O_4$ 226.272
Constit. of the roots of *Pimpinella thellungiana*
(1R*,2S*)-form
1-Et, 5′-Me ether
β-Ethoxy-2-hydroxy-5-methoxy-α-methylbenzeneethanol.
1-Ethoxy-1-(2-hydroxy-5-methoxyphenyl)-2-propanol. ***Thellungianin A.*** *Llungianin A (incorr.)*
[187827-62-9]
$C_{12}H_{18}O_4$ 226.272

Constit. of the roots of *Pimpinella thellungiana*. λ_{max} 215 (sh); 220 (sh); 296 (EtOH)

(1ξ,2ξ)-form

5′-Me ether

1-(2-Hydroxy-5-methoxyphenyl)-1,2-propanediol

[216012-45-2]

$C_{10}H_{14}O_4$ 198.218

Constit. of *Aeschynanthus bracteatus* and *Pimpinella thellungiana*.

Qiao, B. *et al.*, *Yaoxue Xuebao*, 1997, **32**, 56-58 (*Pimpinella thellungiana constits*).
Li, F.-X. *et al.*, *Zhiwu Xuebao*, 1998, **40**, 88-90 (*Thellungianin E*).
Shi, H. *et al.*, *Zhongguo Zhongyao Zazhi*, 1998, **23**, 421-422 (*5′-Me ether*)

1-(3,4-Dihydroxyphenyl)-1,2-propanediol D222

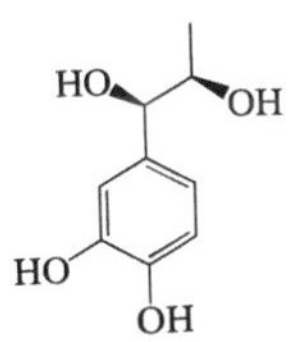

(1R,2R)-form

4-(1,2-Dihydroxypropyl)-1,2-benzenediol

$C_9H_{12}O_4$ 184.191

(1S,2R)-form

1,3′-Di-Me ether

1-(4-Hydroxy-3-methoxyphenyl)-1-methoxy-2-propanol. ***Alatusol E***

[1578237-43-0]

$C_{11}H_{16}O_4$ 212.245

Constit. of *Euonymus alatus*. Displays cytotoxic activity and NO inhibitory activity. Gum. $[\alpha]^{25}_{D}$ +32.4 (c, 0.3 in MeOH). λ_{max} 226 (log ε 4.2); 242 (log ε 2.7); 282 (log ε 1.6) (MeOH)

3-Chloro, 3′-Me ether

3-Chloro-1-(4-hydroxy-3-methoxyphenyl)-1,2-propanediol

[261948-96-3]

$C_{10}H_{13}ClO_4$ 232.66

Constit. of the fruit of *Pimenta dioica*. Antioxidant. Needles ($CHCl_3$). Mp 121°. $[\alpha]^{25}_{D}$ -2 (c, 0.52 in EtOH)

(1R*,2R*)-form

1,2-Cyclic carbonate, 3′,4′-di-Me ether

4-(3,4-Dimethoxyphenyl)-5-methyl-1,3-dioxolan-2-one. ***Saucarbonate***

$C_{12}H_{14}O_5$ 238.239

Constit. of *Saururus chinensis*. Amorph. powder. $[\alpha]^{25}_{D}$ +5.0 (c, 0.5 in $CHCl_3$)

(1ξ,2ξ)-form

3′-Me ether

1-(4-Hydroxy-3-methoxyphenyl)-1,2-propanediol

[24762-60-5]

$C_{10}H_{14}O_4$ 198.218

Constit. of *Zingiber officinale* and *Urtica dioica*. Oil

1,2-O-Isopropylidene, 3′-Me ether

[1885883-31-7]

$C_{13}H_{18}O_4$ 238.283

Constit. of *Viburnum fordiae*. Exhibits moderate antifungal activity against *Aspergillus niger*. Powder. Mp 289 - 291°.

Kikuzaki, H. *et al.*, *Phytochemistry*, 1999, **52**, 1307-1312 (*Pimenta dioica constit*).
Wang, Q. *et al.*, *J. Chem. Res., Synop.*, 2004, 504-505 (*Pimenta dioica constit, synth, abs config*).
Ma, J. *et al.*, *Phytochemistry*, 2004, **65**, 1137-1143 (*3′-Me ether*).
Kim, K.H. *et al.*, *Planta Med.*, 2013, **79**, 361-364 (*Alatusol E*).
Chen, J. *et al.*, *Chem. Nat. Compd. (Engl. Transl.)*, 2016, **52**, 222-223 (*1,2-isopropylidene 3′-Me ether, activity*).
Jin, Q. *et al.*, *J. Nat. Prod.*, 2019, **82**, 3002-3009 (*Saucarbonate*)

1-(3,4-Dihydroxyphenyl)-1,3-propanediol D223

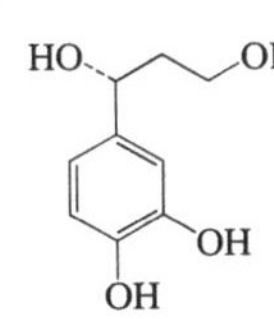

(R)-form

4-(1,3-Dihydroxypropyl)-1,2-benzenediol

$C_9H_{12}O_4$ 184.191

(R)-form [1232138-68-9]

1,3′-Di-Me ether

3-(4-Hydroxy-3-methoxyphenyl)-3-methoxy-1-propanol

[1362738-63-3]

$C_{11}H_{16}O_4$ 212.245

Constit. of the stem bark of *Ginkgo biloba*. $[\alpha]^{25}_{D}$ +9 (c, 0.12 in MeOH)

(ξ)-form [138799-01-6]

3′-Me ether, 4′-O-[β-D-apiofuranosyl-(1→2)-β-D-glucopyranoside]

[371976-20-4]

$C_{21}H_{32}O_{13}$ 492.474

Constit. of the roots of *Astragalus zahlbruckneri*. $[\alpha]^{25}_{D}$ -56 (c, 0.5 in MeOH). λ_{max} 213 (log ε 4.94); 226 (log ε 5) (MeOH).

Calis, I. *et al.*, *J. Nat. Prod.*, 2001, **64**, 1179-1182 (*3′-Me ether 4′-apiosylglucoside*).
Wei, W.-J. *et al.*, *Chirality*, 2010, **22**, 604-611 (*synth*).
Ngan, N.T.T. *et al.*, *J. Agric. Food Chem.*, 2012, **60**, 2815-2824 (*Ginkgo biloba constit*)

2-(2,5-Dihydroxyphenyl)-1,3-propanediol D224

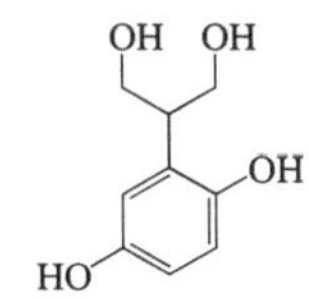

$C_9H_{12}O_4$ 184.191

5′-Me ether

2-(2-Hydroxy-5-methoxyphenyl)-1,3-propanediol. ***Corimen***

[1357928-32-5]

$C_{10}H_{14}O_4$ 198.218

Constit. of *Cotula coronopifolia*. Yellow oil.

Mahjoub, M.A. *et al.*, *Chem. Nat. Compd. (Engl. Transl.)*, 2011, **47**, 955-958 (*Corimen*)

2-(3,4-Dihydroxyphenyl)-1,3-propanediol D225

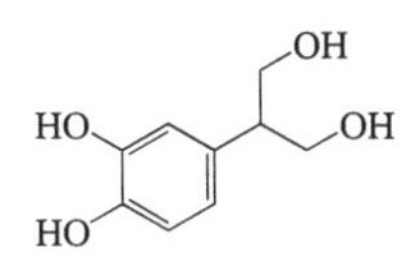

$C_9H_{12}O_4$ 184.191

3′-Me ether

2-(4-Hydroxy-3-methoxyphenyl)-1,3-propanediol. ***Junipediol A***

[86548-91-6]

$C_{10}H_{14}O_4$ 198.218

Constit. of *Apollonias barbujana*, *Cephalotaxus harringtonia* var. *harringtonia*, *Juniperus phoenicea*, *Peucedanum japonicum* and *Saussurea medusa*. Needles. Mp 117 - 120°. λ_{max} 227; 277 (MeOH)

3′-Me ether, 1-O-β-D-glucopyranoside

Junipediol A 8-glucoside

[188894-18-0]

$C_{16}H_{24}O_9$ 360.359

Constit. of *Apium graveolens*, *Juniperus phoenicea* and *Peucedanum japonicum*. Angiotensin-converting enzyme inhibitor. λ_{max} 227; 277 (MeOH)

3′-Me ether, 1-O-[3,4,5-trihydroxybenzoyl-(→6)-β-D-glucopyranoside]

Stroside C

[2211126-76-8]

$C_{23}H_{28}O_{13}$ 512.464
Constit. of the stems of *Lagerstroemia indica*. Brown gum. $[\alpha]^{25}_{D}$ +9.3 (c, 0.03 in MeOH). λ_{max} 206 (log ε 1.5); 228 (log ε 0.8); 275 (log ε 0.5) (MeOH).

Fang, J.-M. *et al.*, *Phytochemistry*, 1992, **31**, 3659-3661 (*Junipediol B*).
Perez, C. *et al.*, *Planta Med.*, 1995, **61**, 535-536 (*3′-Me ether*).
Comte, G. *et al.*, *Phytochemistry*, 1997, **44**, 1169-1173 (*Junipediol 8-glucosides*).
Simaratanamongkol, A. *et al.*, *Food Chem.*, 2014, **165**, 92-97 (*Junipediol A 8-glucoside, activity*).
Woo, K.W. *et al.*, *Heterocycles*, 2015, **91**, 2355-2366 (*Stroside C*)

3-(3,4-Dihydroxyphenyl)-1,2-propanediol D226

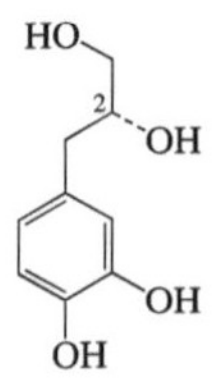

(*R*)-*form*

4-(2,3-Dihydroxypropyl)-1,2-benzenediol
$C_9H_{12}O_4$ 184.191
(R)-form
3′-Me ether
3-(4-Hydroxy-3-methoxyphenyl)-1,2-propanediol
[640293-43-2]
$C_{10}H_{14}O_4$ 198.218
Constit. of *Ficus microcarpa* and *Ginkgo biloba*. Antioxidant. Powder. $[\alpha]^{21}_{D}$ +4.2 (c, 0.9 in MeOH). $[\alpha]^{25}_{D}$ +29.1 (c, 0.15 in MeOH)
3′-Me ether, 2-O-β-D-glucopyranoside
Carthamoside B₅
[1042959-09-0]
$C_{16}H_{24}O_9$ 360.359
Constit. of *Carthamus tinctorius*. Yellow oil. $[\alpha]^{25}_{D}$ -8.2 (c, 0.02 in MeOH). λ_{max} 205 (log ε 4.09); 226 (log ε 3.8); 281 (log ε 3.35) (MeOH)
3′-Me ether, 2-O-[α-L-rhamnopyranosyl-(1→6)-β-D-glucopyranoside]
Carthamoside B₄
[1042959-03-4]
$C_{22}H_{34}O_{13}$ 506.501
Constit. of *Carthamus tinctorius*. Yellow oil. $[\alpha]^{25}_{D}$ -16.8 (c, 0.02 in MeOH). λ_{max} 206 (log ε 4.14); 227 (log ε 3.86); 281 (log ε 3.51) (MeOH)

(S)-form
3′-O-β-D-Galactopyranoside
Obovatoside B
[1819381-01-5]
$C_{15}H_{22}O_9$ 346.332
Constit. of the fruit of *Magnolia obovata*. Pale yellow viscous oil. $[\alpha]_{D}$ -51.8 (c, 0.12 in MeOH)
3′-Me ether, 1-O-[3,4,5-trihydroxybenzoyl-(→6)-β-D-glucopyranoside]
[1378532-50-3]
$C_{23}H_{28}O_{13}$ 512.464
Constit. of the fruit of *Pimenta dioica* and the bark of *Macropteranthes leichhardtii*. Pale yellow powder. $[\alpha]^{23}_{D}$ -3.8 (c, 1 in EtOH)
3′-Me ether, 1-Ac
[2024584-02-7]
$C_{12}H_{16}O_5$ 240.255
Constit. of the stem bark of *Cananga latifolia*. Oil. $[\alpha]^{23}_{D}$ -24.5 (c, 0.1 in $CHCl_3$). λ_{max} 203 (log ε 3.95); 282 (log ε 2.93) ($CHCl_3$)
(ξ)-form [54850-10-1]
3′-Me ether
[27391-18-0]
[220006-74-6] ((-)-form)
$C_{10}H_{14}O_4$ 198.218
Constit. of *Pimenta dioica* and *Zantedeschia aethiopica*. $[\alpha]_{D}$ -6. $[\alpha]_{D}$ -12 (c, 1 in EtOH). Both isolates were partial racemates
3′-Me ether, 1-O-β-D-glucopyranoside
Stroside B
[2095244-18-9]
$C_{16}H_{24}O_9$ 360.359
Constit. of the stems of *Lagerstroemia indica*. Brown gum. $[\alpha]^{25}_{D}$ +8 (c, 0.06 in MeOH). λ_{max} 208 (log ε 1.9); 230 (log ε 1.2); 281 (log ε 0.5) (MeOH)
3′-Me ether, 2-O-[3,4,5-trihydroxybenzoyl-(→6)-β-D-glucopyranoside]
[275363-68-3]
$C_{23}H_{28}O_{13}$ 512.464
Constit. of the fruit of *Pimenta dioica*. Viscous liq.. $[\alpha]^{20}_{D}$ -27.7 (c, 0.15 in MeOH).

Rosenmund, K.W. *et al.*, *Ber. Dtsch. Chem. Ges.*, 1923, **56**, 1262-1269 (*Safrolglycol, synth*).
Kikuzaki, H. *et al.*, *Phytochemistry*, 1999, **52**, 1307-1312 (*3′-Me ether, config*).
Kikuzaki, H. *et al.*, *J. Nat. Prod.*, 2000, **63**, 749-752 (*Pimenta dioica constit*).
Ouyang, M.-A. *et al.*, *Nat. Prod. Res.*, 2007, **21**, 769-774 (*3′-Me ether*).
Zhou, Y.-Z. *et al.*, *Helv. Chim. Acta*, 2008, **91**, 1277-1285 (*Carthamosides B₄,B₅*).
Yoshimura, M. *et al.*, *Biosci., Biotechnol., Biochem.*, 2011, **75**, 2207-2212 (*Pimenta dioica constit*).
Seo, K.-H. *et al.*, *Carbohydr. Res.*, 2015, **416**, 70-74 (*Obovatoside B*).
Woo, K.W. *et al.*, *Heterocycles*, 2015, **91**, 2355-2366 (*Stroside B*).
Patchana, R. *et al.*, *Phytochem. Lett.*, 2015, **13**, 147-151 (*Cananga latifolia acetate*)

1-(3,4-Dihydroxyphenyl)-1,2,3-propanetriol D227

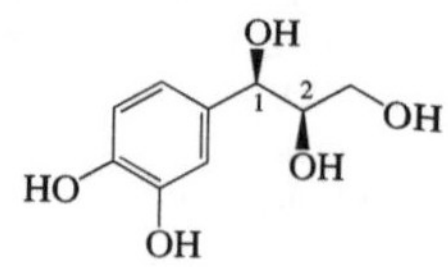

(1*R*,2*R*)-*form*

α-(3,4-Dihydroxyphenyl)glycerol
$C_9H_{12}O_5$ 200.19
(1R,2R)-form

3′-Me ether
1-(4-Hydroxy-3-methoxyphenyl)-1,2,3-propanetriol. Guaiacylglycerol
[84799-27-9]
$C_{10}H_{14}O_5$ 214.217
Constit. of *Zantedeschia aethiopica* and *Vitex rotundifolia*. Cryst. (Me_2CO/CH_2Cl_2). Mp 133 - 134°. $[\alpha]^{23}_{D}$ -26.1 (c, 1.4 in EtOH)
3′-Me ether, 1-O-β-D-glucopyranoside
Guaiacylglycerol 7-glucoside
[142632-41-5]
$C_{16}H_{24}O_{10}$ 376.358
Constit. of the leaves of *Osmanthus asiaticus*. Amorph. powder. $[\alpha]_{D}$ -81.2 (c, 1 in MeOH). λ_{max} 208 (log ε 3.93); 229 (log ε 3.77); 278 (log ε 3.38) (MeOH)
3′-Me ether, 2-O-[3,4,5-trihydroxybenzoyl-(→6)-β-D-glucopyranoside]
[109280-38-8]
$C_{23}H_{28}O_{14}$ 528.463
Constit. of the fresh acorns of *Quercus mongolica*. Needles + 1½H_2O. Mp 143 - 145°. $[\alpha]^{17}_{D}$ -47.5 (c, 0.5 in MeOH)
3′-Me ether, 3-O-β-D-glucopyranoside
Guaiacylglycerol 9-glucoside
[499135-12-5]
$C_{16}H_{24}O_{10}$ 376.358
Constit. of *Pimpinella anisum* and *Saccharum officinarum*. Amorph. powder. $[\alpha]^{22}_{D}$ -13 (c, 1 in MeOH)

1,3′-Di-Me ether
3-(4-Hydroxy-3-methoxy-phenyl)-3-methoxy-1,2-propanediol. 7-O-Methylguaiacylglycerol.
(-)-Alatusol D
[1578237-42-9]
$C_{11}H_{16}O_5$ 228.244
Constit. of *Euonymus alatus* and *Euonymus ulmoides*. Displays cytotoxic activity. Gum. $[\alpha]^{25}_{D}$ -42.1 (c, 0.25 in MeOH). λ_{max} 226 (log ε 4.2); 243 (log ε 2.7); 280 (log ε 1.5) (MeOH)
1,3′-Di-Me ether, 3-Ac
(-)-Alatusol A
[1578237-39-4]
$C_{13}H_{18}O_6$ 270.281
Constit. of *Ailanthus altissima* and *Euonymus alatus*. Displays cytotoxic activity. Gum. $[\alpha]^{25}_{D}$ -45.2 (c, 0.3 in MeOH). λ_{max} 215 (log ε 4.2); 235 (log ε 3.7); 278 (log ε 1.4) (MeOH)
(1R,2S)-form
3′-Me ether, 2-O-β-D-glucopyranoside
[1018987-09-1]
$C_{16}H_{24}O_{10}$ 376.358
Constit. of the roots of *Iodes cirrhosa*. Amorph. powder. $[\alpha]^{20}_{D}$ -21.5 (c, 0.55 in MeOH). λ_{max} 230 (log ε 3.8); 278 (log ε 3.2) (MeOH)
1-Et, 3′-Me ether
[2097562-25-7]
$C_{12}H_{18}O_5$ 242.271
Constit. of the bark of *Eucommia ulmoides*. Powder. $[\alpha]^{20}_{D}$ -19.7 (c, 0.2 in MeOH)
(1S,2R)-form
3′-Me ether
[84822-99-1]
Cryst. (Me_2CO/CH_2Cl_2). Mp 82 - 84°. $[\alpha]^{22}_{D}$ +11 (c, 0.2 in MeOH)
3′-Me ether, 3-O-β-D-glucopyranoside
[499135-13-6]
Constit. of the fruit of *Pimpinella anisum*, and *Saccharum officinarum*. Amorph. powder. $[\alpha]^{22}_{D}$ -20 (c, 0.8 in MeOH)
1,3′-Di-Me ether, 3-Ac
(+)-Alatusol B
[1578237-40-7]
$C_{13}H_{18}O_6$ 270.281
Constit. of *Euonymus alatus*. Displays cytotoxic activity. Gum. $[\alpha]^{25}_{D}$ +28.4 (c, 0.2 in MeOH). λ_{max} 215 (log ε 4.2); 235 (log ε 3.7); 278 (log ε 1.4) (MeOH)
1-Et, 3′-Me ether
[2097562-24-6]
$C_{12}H_{18}O_5$ 242.271
Constit. of the bark of *Eucommia ulmoides*. Powder. $[\alpha]^{20}_{D}$ +19.7 (c, 0.2 in MeOH)
(1S,2S)-form
3′ Me ether
[7095-33-2]
$[\alpha]^{23}_{D}$ +25 (c, 0.11 in MeOH)
3′-Me ether, 2-O-[3,4,5-trihydroxybenzoyl-(→6)-β-D-glucopyranoside]
[109194-54-9]
Isol. from acorns of *Quercus mongolica*. Amorph. powder + 1½H_2O. Mp 143 - 145°. $[\alpha]^{17}_{D}$ -21.5 (c, 0.7 in MeOH)
3′-Me ether, 3-O-β-D-glucopyranoside
[109280-40-2]
Constit. of *Saccharum officinarum* and *Hypericum erectum*. Amorph. powder. $[\alpha]^{25}_{D}$ -12 (c, 0.2 in MeOH)
1,3′-Di-Me ether
(+)-Alatusol D. Forsythiayanoside D
[2231227-08-8]
$C_{11}H_{16}O_5$ 228.244
Constit. of *Euonymus ulmoides* and *Forsythia suspensa*. Antioxidant. Needles (MeOH). $[\alpha]^{25}_{D}$ +19.2 (c, 0.1 in MeOH). λ_{max} 208; 229; 280 (MeOH)
(1R*,2R*)-form
3′-Me ether
[27391-16-8]
$C_{10}H_{14}O_5$ 214.217
Constit. of *Eucommia ulmoides*, *Picea* spp., *Pinus* spp. and *Scorodocarpus borneensis*. A degradn. prod. of lignin. Intermed. in lignin biosynth.. Syrup
3′-Me ether, 1-O-β-D-glucopyranoside
[7213-02-7]
$C_{16}H_{24}O_{10}$ 376.358
Isol. from *Pinus sylvestris* needles. Cryst.. Mp 162 - 166°. $[\alpha]^{22}_{D}$ +45
3′-Me ether, 2-O-β-D-glucopyranoside
[109280-39-9]
$C_{16}H_{24}O_{10}$ 376.358
Isol. from *Pinus sylvestris* and *Piper aduncum*. Inhibitor of tyrosinase. Amorph.. $[\alpha]^{22}_{D}$ -16
3′-Me ether, 3-O-[4-hydroxybenzoyl-(→6)-β-D-glucopyranoside]
[1010686-03-9]
$C_{23}H_{28}O_{12}$ 496.465
Constit. of *Dracocephalum forrestii*. Amorph. powder. $[\alpha]^{19}_{D}$ -2.9 (c, 0.4 in MeOH). λ_{max} 207 (log ε 3.42); 237 (log ε 3.22); 288 (log ε 3.68) (MeOH)
3′-Me ether, 3-O-[4-hydroxy-3-methoxybenzoyl-(→6)-β-D-glucopyranoside]
Brachyanin F
[2113695-62-6]
$C_{24}H_{30}O_{13}$ 526.491
Constit. of stems and roots of *Stauntonia brachyanthera*. Antioxidant. Powder (MeOH). $[\alpha]^{25}_{D}$ +23.5 (c, 0.56 in MeOH)
3′-Me ether, 3-O-[4-hydroxy-E-cinnamoyl-(→6)-β-D-glucopyranoside]
[1010686-05-1]
$C_{25}H_{30}O_{12}$ 522.503
Constit. of *Dracocephalum forrestii*. Amorph. solid. λ_{max} 225 (log ε 4.42); 315 (log ε 4.23) (MeOH)
3′-Me ether, 3-O-[4-hydroxy-Z-cinnamoyl-(→6)-β-D-glucopyranoside]
[1010686-06-2]
$C_{25}H_{30}O_{12}$ 522.503
Constit. of *Dracocephalum forrestii*. Amorph. solid. λ_{max} 212 (log ε 3.67); 315 (log ε 4.11) (MeOH)
3′-Me ether, 3-O-[3,4-dihydroxy-E-cinnamoyl-(→6)-β-D-glucopyranoside]
Akeqintoside C
[1620439-23-7]
$C_{25}H_{30}O_{13}$ 538.502
Constit. of stems of *Akebia quinata*. Exhibits moderate inhibitory activity against IL-6 prodn. in TNF-α stimulated MG-63 cells. Amorph. solid. $[\alpha]^{28}_{D}$ -26.7 (c, 0.21 in MeOH). λ_{max} 218 (log ε 4.25); 289 (log ε 3.60); 328 (log ε 4.05) (MeOH)
3′-Me ether, tetra-Ac
[14065-28-2]
Mp 113 - 114°
4′-Me ether
1-(3-Hydroxy-4-methoxyphenyl)-1,2,3-propanetriol.
Irisdichotin D. *Irisdichototin D (incorr.)*
[1384433-84-4]
$C_{10}H_{14}O_5$ 214.217
Constit. of the stems of *Iris dichotoma*. Powder. $[\alpha]^{25}_{D}$ +9 (c, 0.05 in MeOH). λ_{max} 228 (MeOH)

1-Et ether
Tazettone G
[2237944-41-9]
$C_{11}H_{16}O_5$ 228.244
Constit. of the bulbs of *Narcissus tazetta*. Amorph. powder. $[\alpha]^{20}_{D}$ -3.9 (c, 0.08 in MeOH). Possible artifact

(1R*,2S*)-form
3'-Me ether
[27431-21-6]
Constit. of *Eucommia ulmoides*, *Picea* spp., *Pinus* spp. and *Scorodocarpus borneensis*. Mp 83 - 84°

3'-Me ether, 3-O-(6-O-benzoyl-β-D-glucopyranoside) (1)
Casearicoside C
[1271737-26-8]
$C_{23}H_{28}O_{11}$ 480.466
Constit. of the stems of *Casearia velutina*. Amorph. powder. $[\alpha]^{25}_{D}$ -2.5 (c, 0.02 in MeOH)

3'-Me ether, 3-O-(6-O-benzoyl-β-D-glucopyranoside) (2)
Casearicoside B
[1271737-25-7]
$C_{23}H_{28}O_{11}$ 480.466
Constit. of the stems of *Casearia velutina*. Amorph. powder. $[\alpha]^{25}_{D}$ +3 (c, 0.01 in MeOH). λ_{max} 286 (log ε 3.95) (MeOH)

(1ξ,2ξ)-form [117722-92-6]
3-O-(3,4-Dihydroxy-E-cinnamoyl)
[771531-56-7]
$C_{18}H_{18}O_8$ 362.334
Constit. of *Hypericum sampsonii*. $[\alpha]^{25}_{D}$ +35 (c, 1 in MeOH). λ_{max} 235 (log ε 3.95); 261 (log ε 3.9); 285 (log ε 3.84); 313 (log ε 3.8) (MeOH)

3'-Me ether
[1208-42-0]
Isol. from the leaves of *Portulaca oleracea* induced with $CuCl_2$

3'-Me ether, 2-O-β-D-glucopyranoside
[7162-98-3]
[109280-39-9] ((*R*,*R*)-form)
Constit. of *Pinus resinosa*, *Pinus sylvestris*, *Alangium premnifolium* and *Saccharum officinarum*. Amorph. powder. $[\alpha]^{25}_{D}$ +17 (c, 0.04 in MeOH). $[\alpha]^{29}_{D}$ -16.5 (c, 0.37 in MeOH)

3'-Me ether, 2-O-[4-hydroxybenzoyl-(→6)-β-D-glucopyranoside]
Taxumaiglucoside C
[252636-47-8]
$C_{23}H_{28}O_{12}$ 496.465
Constit. of *Taxus mairei*. Oil (as hepta-Ac). $[\alpha]^{28}_{D}$ +3.4 (c, 1 in $CHCl_3$)

3'-Me ether, 3-O-β-D-glucopyranoside
[7094-73-7]
Constit. of *Pinus resinosa* and *Pinus sylvestris*

3'-Me ether, 3-O-[6S-hydroxy-2,6-dimethyl-2E,7-octadienoyl-(→6)-β-D-glucopyranoside]
[1370004-83-3]
$C_{26}H_{38}O_{12}$ 542.578
Isol. from the leaves of *Portulaca oleracea* induced with $CuCl_2$. Powder. Mp 111 - 113°. $[\alpha]^{20}_{D}$ +46 (c, 0.001 in MeOH). λ_{max} 210 (log ε 4.56); 254 (log ε 3.88); 280 (log ε 3.91) (MeOH)

4'-Me ether, 3-O-[3,4,5-trihydroxybenzoyl-(→6)-β-D-glucopyranoside]
$C_{23}H_{28}O_{14}$ 528.463
Constit. of the leaves of *Baseonema acuminatum*. Yellow powder. Mp 174° dec.. $[\alpha]^{25}_{D}$ +22 (c, 0.1 in MeOH). λ_{max} 272 (log ε 3.89); 375 (sh) (MeOH)

1,3'-Di-Me ether
[1251855-59-0]
$C_{11}H_{16}O_5$ 228.244
Constit. of the roots of *Incarvillea mairei*

1,3'-Di-Me ether, 3-tetracosanoyl
[1370508-11-4]
$C_{35}H_{62}O_6$ 578.875
Constit. of the roots of *Saposhnikovia divaricata*. Amorph. powder

1,3'-Di-Me ether, 3-O-(4-hydroxy-3-methoxy-E-cinnamoyl)
Angeliferulate
[879496-41-0]
$C_{21}H_{24}O_8$ 404.415
Constit. of the roots of *Angelica sinensis*. Pale yellow oil. $[\alpha]^{20}_{D}$ +2 (c, 0.2 in CH_2Cl_2)

1-Et, 3'-Me ether
3-Ethoxy-3-(4-hydroxy-3-methoxyphenyl)-1,2-propanediol.
7-O-Ethylguaiacylglycerol
[929688-54-0]
$C_{12}H_{18}O_5$ 242.271
Constit. of *Tarenna attenuata*. Amorph. powder. $[\alpha]^{27}_{D}$ +11.2 (c, 0.27 in MeOH). Possible artifact. λ_{max} 204 (log ε 4.17); 229 (log ε 3.49); 280 (log ε 3.06) (MeOH).

Adler, E. *et al., Acta Chem. Scand.*, 1963, **17**, 27-36 (*synth, config, bibl*).
Theander, O. *Acta Chem. Scand.*, 1965, **19**, 1792-1793 (*Guaiacylglycerol glucosides*).
Rudloff, E. *Chem. Ind. (London)*, 1965, 180-181 (*Guaiacylglycerol*).
Brezny, R. *et al., Chem. Zvesti*, 1978, **32**, 684-690 (*synth*).
Lundgren, L.N. *et al., Acta Chem. Scand., Ser. B*, 1982, **36**, 695-699 (*Guaiacylglycerol, synth*).
Deyama, T. *et al., Chem. Pharm. Bull.*, 1986, **34**, 523-527 (*Eucommia ulmoides constits*).
Ishimaru, K. *et al., Phytochemistry*, 1987, **26**, 1147-1152 (*galloylglucosides*).
Kijima, H. *et al., Phytochemistry*, 1987, **44**, 1551-1557 (*Alangium premnifolium constit*).
Sugiyama, M. *et al., Chem. Pharm. Bull.*, 1992, **40**, 325-326 (*Guaiacylglycerol 7-glucoside*).
Fuzzati, N. *et al., Phytochemistry*, 1995, **39**, 409-412 (*Bis-2-methylpropanoylguaiacylglycerol*).
Okuyama, E. *et al., Chem. Pharm. Bull.*, 1998, **46**, 655-662 (*Vitex rotundifolia constit*).
Della Greca, M. *et al., Phytochemistry*, 1998, **49**, 1299-1304 (*Zantedeschia aethiopica constit*).
Ishikawa, T. *et al., Chem. Pharm. Bull.*, 2002, **50**, 1460-1466 (*Pimpinella anisum glucosides*).
Lee, S.W. *et al., Arch. Pharmacal Res.*, 2003, **26**, 591-593 (*Taxumaiglucoside C*).
Don, M.-J. *et al., Chem. Pharm. Bull.*, 2004, **52**, 866-869 (*Hypericum sampsonii constit*).
De Leo, M. *et al., Planta Med.*, 2004, **70**, 841-846 (*Baseonema acuminatum constit*).
Deng, S. *et al., J. Nat. Prod.*, 2006, **69**, 536-541 (*Angeliferulate*).
Takara, K. *et al., Biosci., Biotechnol., Biochem.*, 2007, **71**, 183-191 (*Saccharum officinarum glucosides*).
Li, G.-P. *et al., J. Asian Nat. Prod. Res.*, 2007, **9**, 457-461 (*Dracocephalum forrestii benzoylglucosides*).
Yang, X.-W. *et al., J. Nat. Prod.*, 2007, **70**, 521-525 (*7-O-Ethylguaiacylglycerol*).
Gan, M. *et al., J. Nat. Prod.*, 2008, **71**, 647-654 (*Iodes cirrhosa glucoside*).
Huang, Z.-S. *et al., Zhongguo Zhongyao Zazhi*, 2009, **34**, 1672-1675 (*7-O-Methylguaiacylglycerol*).
Chai, X.-Y. *et al., Planta Med.*, 2010, **76**, 91-93 (*Casearicosides B,C*).
Chin, Y.-W. *et al., Bull. Korean Chem. Soc.*, 2011, **32**, 2132-2134 (*Saposhnikovia ester*).
Wu, B. *et al., Carbohydr. Res.*, 2012, **351**, 68-73 (*Portulaca oleracea constits*).
Huang, L. *et al., Helv. Chim. Acta*, 2012, **95**, 1033-1036 (*Irisdichotin D*).
Kim, K.H. *et al., Planta Med.*, 2013, **79**, 361-364 (*Alatusols A,B,D, activity*).
Jin, H.-G. *et al., Chem. Pharm. Bull.*, 2014, **62**, 288-293 (*Akeqintoside C, activity*).
Fu, K.-L. *et al., Fitoterapia*, 2016, **113**, 110-116 (*Tazettone G*).
Chen, J. *et al., Chem. Nat. Compd. (Engl. Transl.)*, 2017, **53**, 254-256 (*Eucommia ulmoides constits*).
Yan, X.-J. *et al., J. Asian Nat. Prod. Res.*, 2017, **19**, 254-259 (*Forsythiayanoside D, activity*)

1-(3,5-Dihydroxyphenyl)-1,2,3-propanetriol D228

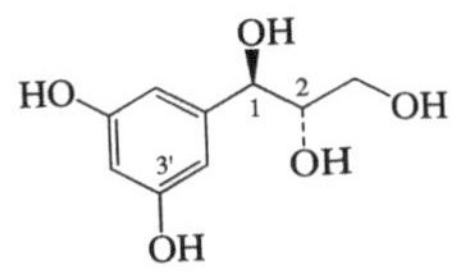

(1*R*,2*S*)-*form*

α-(3,5-Dihydroxyphenyl)glycerol
$C_9H_{12}O_5$ 200.19
Structs. in this series may need revision. Phenyl 3,5-dioxygenated substn. pattern in doubt as a result of work done by Yang *et al* in 2015

(1R,2S)-form

1,3′-Di-Me ether
3-(3-Hydroxy-5-methoxyphenyl)-3-methoxy-1,2-propanediol
[1338576-75-2]
$C_{11}H_{16}O_5$ 228.244
Isol. from the leaves and stems of *Toona ciliata* var. *pubescens*. Shows mild antiproliferative activity against tumour cell lines. Brown oil. $[\alpha]_D^{25}$ -13.8 (c, 1.0 in MeOH)

(1R*,2R*)-form

3′-Me ether, 2-O-β-D-glucopyranoside
[2097336-00-8]
$C_{16}H_{24}O_{10}$ 376.358
Constit. of the stems of *Zanthoxylum armatum*. Amorph. powder. $[\alpha]_D^{20}$ -22.5 (c, 0.1 in H_2O)

3′-Me ether, 3-O-β-D-glucopyranoside
Meliadanoside A
[286369-46-8]
$C_{16}H_{24}O_{10}$ 376.358
Constit. of the fruit of *Melia toosendan*. Amorph. powder. $[\alpha]_D$ -22.6 (c, 1.3 in MeOH). λ_{max} 229; 280 (MeOH)

(1ξ,2ξ)-form

3′-Me ether
1-(3-Hydroxy-5-methoxyphenyl)-1,2,3-propanetriol
[1374605-49-8]
$C_{10}H_{14}O_5$ 214.217
Constit. of *Metasequoia glyptostroboides*. Yellowish gum. $[\alpha]_D^{25}$ +106.7 (c, 0.08 in MeOH). λ_{max} 300 (sh) (MeOH)

3′-Et, 5′-Me ether
1-(3-Ethoxy-5-methoxyphenyl)-1,2,3-propanetriol
[1374605-48-7]
$C_{12}H_{18}O_5$ 242.271
Constit. of *Metasequoia glyptostroboides*. Yellowish gum. $[\alpha]_D^{25}$ +40.4 (c, 0.05 in MeOH). λ_{max} 245 (log ε 0.62); 310 (log ε 1.52) (MeOH).

Chang, J. *et al.*, *Zhiwu Xuebao*, 1999, **41**, 1245-1248 (*Meliadanoside A*).
Liu, Y.-B. *et al.*, *Helv. Chim. Acta*, 2011, **94**, 1685-1691 (*1,3′-di-Me ether*).
Zeng, Q. *et al.*, *Helv. Chim. Acta*, 2012, **95**, 606-612 (*Metasequoia ethers*).
Yang, Y.-N. *et al.*, *J. Nat. Prod.*, 2015, **78**, 705-711 (*pmr, struct*).
Guo, T. *et al.*, *Nat. Prod. Res.*, 2017, **31**, 2335-2340 (*Zanthoxylum armatum constit*)

3-(2,4-Dihydroxyphenyl)propanoic acid D229

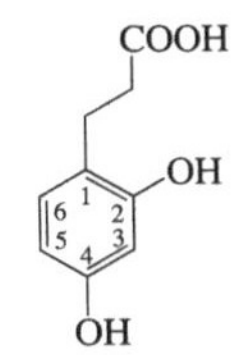

2,4-Dihydroxybenzenepropanoic acid, CAS. *2,4-Dihydroxyhydrocinnamic acid*, CAS. *Hydroumbellic acid*
[5631-68-5]
$C_9H_{10}O_4$ 182.175
Prod. by a marine-derived *Engyodontium album*. Cryst. (C_6H_6 or H_2O). Mp 165°

2-O-β-D-Glucopyranoside
[663627-25-6]
$C_{15}H_{20}O_9$ 344.316
Constit. of the leaves of *Peucedanum japonicum*. Powder. $[\alpha]_D^{21}$ -22.3 (c, 0.1 in EtOH). λ_{max} 227 (log ε 2.94); 280 (log ε 2.62) (EtOH)

Me ester, 2-O-β-D-glucopyranoside
[663627-26-7]
$C_{16}H_{22}O_9$ 358.343
Constit. of the leaves of *Peucedanum japonicum*. Powder. $[\alpha]_D^{25}$ -44.2 (c, 0.1 in EtOH). λ_{max} 212 (log ε 3.25); 292 (log ε 2.77) (EtOH)

4-Me ether
2-Hydroxy-4-methoxybenzenepropanoic acid
[21144-17-2]
$C_{10}H_{12}O_4$ 196.202
Needles (Et_2O/hexane). Mp 91 - 92°

4-Me ether, 2-O-β-D-glucopyranoside
[477873-63-5]
$C_{16}H_{22}O_9$ 358.343
Constit. of *Gnidia polycephala* and *Tagetes lucida*. Amorph. powder. Mp 195 - 197° (168-169°). $[\alpha]_D^{20}$ -38.2 (c, 0.5 in MeOH). $[\alpha]_D^{22}$ -80 (c, 1.6 in MeOH). λ_{max} 224 (log ε 3.15); 278 (log ε 2.75); 321 (log ε 2.08) (MeOH)

4-Me ether, Me ester
[69471-28-9]
$C_{11}H_{14}O_4$ 210.229
Isol. from *Machaerium kuhlmannii*. λ_{max} 211 (ε 9000); 222 (sh); 281 (ε 2200) (EtOH)

4-Me ether, Me ester, 2-O-β-D-glucopyranoside
[477873-64-6]
$C_{17}H_{24}O_9$ 372.37
Constit. of *Gnidia polycephala* and *Tagetes lucida*. Amorph. powder. Mp 88 - 90°. $[\alpha]_D^{20}$ -37.1 (c, 0.2 in MeOH). $[\alpha]_D^{22}$ -85 (c, 0.94 in MeOH). λ_{max} 223 (log ε 3.21); 277 (log ε 2.68); 315 (log ε 2.04) (MeOH)

Di-Me ether
2,4-Dimethoxybenzenepropanoic acid
[22174-29-4]
$C_{11}H_{14}O_4$ 210.229
Cryst. (EtOH). Mp 105°.

Hurd, C.D. *et al.*, *J. Am. Chem. Soc.*, 1939, **61**, 249-254 (*synth*).
Amakasa, T. *et al.*, *J. Org. Chem.*, 1966, **31**, 1433-1436 (*ir*).
Kametani, T. *et al.*, *J. Chem. Soc. C*, 1969, 4-9 (*synth, 4-Me ether*).
Ollis, W.D. *et al.*, *Phytochemistry*, 1978, **17**, 1383-1388 (*4-Me ether Me ester*).
Aquino, R. *et al.*, *J. Nat. Prod.*, 2002, **65**, 1773-1776 (*Tagetes lucida constits*).
Munkombwe, N.M. *et al.*, *Phytochemistry*, 2003, **64**, 1401-1404 (*Gnidia polycephala constits*).
Hisamoto, M. *et al.*, *J. Agric. Food Chem.*, 2004, **52**, 445-450 (*Peucedanum 2-glucosides*).
Wang, W. *et al.*, *Chem. Nat. Compd. (Engl. Transl.)*, 2017, **53**, 224-226 (*Engyodontium album constit*)

3-(3,4-Dihydroxyphenyl)propanoic acid D230

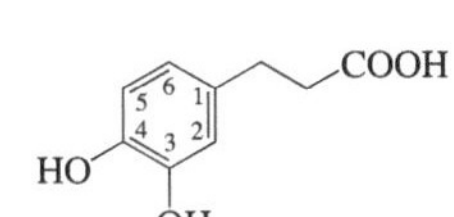

3,4-Dihydroxybenzenepropanoic acid, CAS. *3,4-Dihydroxyhydrocinnamic acid*, CAS. *Hydrocaffeic acid*
[1078-61-1]
$C_9H_{10}O_4$ 182.175
Occurs in spores of *Lycopodium clavatum* and also in higher plants

incl. *Trigonella foenum-graecum*. Some occurrence in complex natural products e.g. Tarchonanthuslactone and Podospermic acid, in T72. Exhibits antioxidant activity. Leaflets (H_2O). Mp 139°. pK_{a1} 4.56; pK_{a2} 9.36; pK_{a3} 11.6 (30°,0.1*M* $NaClO_4$)

▶ MW5143500

4-O-[3,4-Dihydroxy-E-cinnamoyl-(→6)-β-D-glucopyranoside]
Dracunculifoside C
[388077-46-1]
$C_{24}H_{26}O_{12}$ 506.46
Constit. of *Baccharis dracunculifolia*. Amorph. powder. $[\alpha]^{25}_{D}$ -58.9 (c, 0.59 in MeOH). λ_{max} 203 (log ε 4.82); 218 (log ε 4.29); 247 (sh); 287 (sh); 301 (log ε 4.11); 329 (log ε 4.2) (MeOH)

Me ester, 4-O-[β-D-glucopyranosyl-(1→6)-β-D-glucopyranoside]
Citrusin F
[134860-04-1]
$C_{22}H_{32}O_{14}$ 520.484
Isol. from *Citrus limon*. Oil. $[\alpha]^{22}_{D}$ -4.68 (c, 0.1 in MeOH)

Et ester
[3967-57-5]
$C_{11}H_{14}O_4$ 210.229
Mp 50 - 52° (46-47°)

4-Methylpentyl ester
[160661-36-9]
$C_{15}H_{22}O_4$ 266.337
Constit. of *Cheirolophus mauritanicus*. Gum

3-Me ether
3-(4-Hydroxy-3-methoxyphenyl) propanoic acid. Dihydroferulic acid. Hydroferulic acid. ***Shorbic acid***
[1135-23-5]
$C_{10}H_{12}O_4$ 196.202
Constit. of *Aptenia cordifolia*, *Selaginella stautoniana* and *Shorea robusta*. Plates (H_2O). Mp 89 - 90°

3-Me ether, Me ester
[56024-44-3]
$C_{11}H_{14}O_4$ 210.229
Liq.

3-Me ether, Me ester, 4-O-β-D-glucopyranoside
Citrusin E
[134860-03-0]
$C_{17}H_{24}O_9$ 372.37
Isol. from oil of *Citrus limon*. Hypotensive agent in rats. Mp 108°. $[\alpha]^{17}_{D}$ -35.68 (c, 0.1 in MeOH)

3-Me ether, Et ester, 4-O-β-D-glucopyranoside
Carthamoside B$_3$
[1173197-29-9]
$C_{18}H_{26}O_9$ 386.397
Constit. of the petals of *Carthamus tinctorius*. Amorph. powder

3-Me ether, propyl ester
Propyl dihydroferulate
[934749-54-9]
$C_{13}H_{18}O_4$ 238.283
Constit. of the stems of *Cinnamomum reticulatum*

3-Me ether, butyl ester, 4-O-β-D-glucopyranoside
Dichotomoside E
[845673-21-4]
$C_{20}H_{30}O_9$ 414.451
Constit. of the roots of *Stellaria dichotoma* var. *lanceolata*. Powder. $[\alpha]^{27}_{D}$ -29.5 (c, 1.6 in MeOH). λ_{max} 275 (log ε 3.83) (MeOH)

3-Me ether, pentyl ester
Pentyl dihydroferulate
[1064639-63-9]
$C_{15}H_{22}O_4$ 266.337
Constit. of the stems of *Cinnamomum reticulatum*

3-Me ether, tetradecyl ester
Tetradecyl dihydroferulate
[1239038-87-9]
$C_{24}H_{40}O_4$ 392.58
Constit. of the stems of *Cinnamomum reticulatum*

3-Me ether, heptadecyl ester
Heptadecyl dihydroferulate
[934749-56-1]
$C_{27}H_{46}O_4$ 434.661
Constit. of the stems of *Cinnamomum reticulatum*

3-Me ether, eicosyl ester
Eicosyl dihydroferulate
[1239038-90-4]
$C_{30}H_{52}O_4$ 476.742
Constit. of the stems of *Cinnamomum reticulatum*

4-Me ether
3-(3-Hydroxy-4-methoxyphenyl) propanoic acid. Hydrohesperetic acid. Hydroisoferulic acid
[1135-15-5]
$C_{10}H_{12}O_4$ 196.202
Needles (H_2O). Mp 146°

4-Me ether, 3-O-β-D-glucopyranoside
Carthamoside B$_6$
[1042959-14-7]
$C_{16}H_{22}O_9$ 358.343
Isol. from *Carthamus tinctorius*. Powder. $[\alpha]^{25}_{D}$ -40 (c, 0.002 in MeOH). λ_{max} 206 (log ε 4.15); 227 (log ε 3.95); 280 (log ε 3.3) (MeOH)

4-Me ether, Me ester
[129150-61-4]
$C_{11}H_{14}O_4$ 210.229
Cryst. (EtOH). Mp 94 - 95°

Di-Me ether
3-(3,4-Dimethoxyphenyl)propanoic acid. 3,4-Dimethoxyhydrocinnamic acid
[2107-70-2]
$C_{11}H_{14}O_4$ 210.229
Constit. of *Aptenia cordifolia*. Exhibits phytotoxic activity. Needles. Mp 98 - 99°

Di-Me ether, Me ester
[27798-73-8]
$C_{12}H_{16}O_4$ 224.256
Constit. of *Aptenia cordifolia*. Solid. Mp 37°. Bp 175 - 180° at 12 mm

Di-Me ether, Et ester
[5462-13-5]
$C_{13}H_{18}O_4$ 238.283
Constit. of *Aptenia cordifolia*. Oil. Mp 13°. Bp 193° at 20 mm

Di-Me ether, amide
3-(3,4-Dimethoxyphenyl) propanamide
[14773-41-2]
$C_{11}H_{15}NO_3$ 209.245
Alkaloid from *Piper arboricola*. Needles (C_6H_6). Mp 120 - 121°

Di-Me ether, 2-methylpropylamide
3-(3,4-Dimethoxyphenyl)propanoic acid 2-methylpropylamide. 3,4-Dimethoxyhydrocinnamic isobutylamide
[210689-26-2]
$C_{15}H_{23}NO_3$ 265.353
Alkaloid from *Melicope melanophloia*

4-O-[2-Hydroxy-1-(hydroxymethyl) ethyl], 3-Me ether
Guizhi acid A
$C_{13}H_{18}O_6$ 270.281
Constit. of *Cinnamomum cassia*. Amorph. powder.

Aldrich Library of FT-IR Spectra, 1st edn., 1985, **2**, 168C , 171A (*ir*).
Aldrich Library of [13]*C and* [1]*H FT-NMR Spectra*, 1992, **2**, 1034C , 1038A, 1235B (*nmr*).
Lapworth, A. *et al.*, *J. Chem. Soc., Trans.*, 1917, **111**, 790-798 (*synth*).
Zetzsche, F. *et al.*, *Helv. Chim. Acta*, 1927, **10**, 472-474 (*isol, Lycopodium*).

Robinson, R. *et al.*, *J. Chem. Soc.*, 1931, 3163-3172 (*4-Me ether, 4-Me ether Me ester, synth*).
Harwood, H.J. *et al.*, *J. Am. Chem. Soc.*, 1933, **55**, 2555-2559 (*di-Me ether, di-Me ether amide, synth*).
Tamura, S. *et al.*, *Nippon Nogei Kagaku Kaishi*, 1952, **26**, 410-412 (*Et ester, synth*).
Neish, A.C. *Can. J. Biochem.*, 1959, **37**, 1431-1438 (*synth*).
Saha, P.K. *et al.*, *Fitoterapia*, 1979, **50**, 7-9 (*Shorbic acid*).
He, Q. *et al.*, *Zhongcaoyao*, 1981, **12**, 433-435 (*di-Me ether amide*).
Matsubara, Y. *et al.*, *Agric. Biol. Chem.*, 1991, **55**, 647-650 (*Citrusins*).
Marco, J.A. *et al.*, *Phytochemistry*, 1994, **37**, 1101-1107 (*4-methylpentyl ester*).
Blase, F.R. *et al.*, *Synth. Commun.*, 1995, **25**, 3187-3197 (*4-Me ether Me ester, synth, pmr, cmr*).
Latif, Z. *et al.*, *Biochem. Syst. Ecol.*, 1998, **26**, 467-468 (*di-Me ether isobutylamide*).
Nagatani, Y. *et al.*, *Chem. Pharm. Bull.*, 2001, **49**, 1388-1394 (*Dracunculifoside C*).
Haadsma-Svensson, S.R. *et al.*, *J. Med. Chem.*, 2001, **44**, 4716-4732 (*di-Me ether, synth, pmr, ir, ms*).
Doyle, M.P. *et al.*, *J. Org. Chem.*, 2002, **67**, 2954-2959 (*3-Me ether Me ester, synth, pmr, cmr*).
Morikawa, T. *et al.*, *Chem. Pharm. Bull.*, 2004, **52**, 1194-1199 (*Dichotomoside E*).
Cousin, D. *et al.*, *Org. Biomol. Chem.*, 2006, **4**, 54-62 (*4-Me ether Me ester*).
Della Greca, M. *et al.*, *Chem. Biodiversity*, 2007, **4**, 118-128 (*Aptenia cordifolia constits*).
Zhou, Y.-Z. *et al.*, *Helv. Chim. Acta*, 2008, **91**, 1277-1285 (*Carthamoside* B_6).
Zhou, Y.-Z. *et al.*, *J. Asian Nat. Prod. Res.*, 2008, **10**, 817-821 (*Carthamoside* B_3).
Lin, I.-J. *et al.*, *Nat. Prod. Res.*, 2010, **24**, 775-780 (*Cinnamomum dihydroferulates*).
Audubert, C. *et al.*, *Org. Lett.*, 2017, **19**, 4407-4410 (*di-Me ether Me ester*).
Wang, D. *et al.*, *Zhongcaoyao*, 2019, **50**, 8-12 (*Guizhi acid A*)

1-(2,3-Dihydroxyphenyl)-2-propanol D231

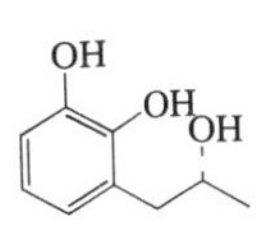

(*R*)*-form*

3-(2-Hydroxypropyl)-1,2-benzenediol
$C_9H_{12}O_3$ 168.192
(R)-form [1210043-00-7]
Prod. by *Lecythophora* sp. 30.5 obt, from *Alyxia reinwardtii*. No phys. props. reported
(ξ)-form [161679-00-1]
Prod. by an unidentified fungus, PSU-N24.

Sommart, U. *et al.*, *Chem. Pharm. Bull.*, 2008, **56**, 1687-1690 (*PSU-N24 metab*).
Sugijanto, N.E. *et al.*, *Nat. Prod. Commun.*, 2009, **4**, 1485-1488 (*Lecythophora metab*)

1-(2,4-Dihydroxyphenyl)-2-propanol D232

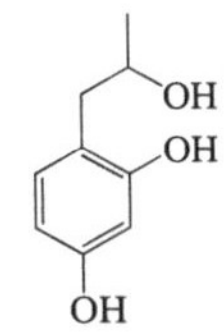

4-(2-Hydroxypropyl)-1,3-benzenediol
$C_9H_{12}O_3$ 168.192
(ξ)-form [101383-05-5]
4′-Me ether
1-(2-Hydroxy-4-methoxyphenyl)-2-propanol. ***Verimol J***
[212516-43-3]
$C_{10}H_{14}O_3$ 182.219
Constit. of *Illicium verum*. Oil.

Sy, L.K. *et al.*, *J. Nat. Prod.*, 1998, **61**, 987-992 (*Verimol J*)

1-(3,5-Dihydroxyphenyl)-2-propanol D233

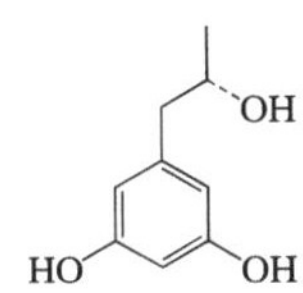

5-(2-Hydroxypropyl)-1,3-benzenediol. ***Orcinotriol***
$C_9H_{12}O_3$ 168.192
Structures in this series may need revision. Phenyl 3,5-dioxygenated substn. pattern in doubt as a result of work done by Yang *et al* in 2015
(S)-form [205648-56-2]
Prod. by marine-derived *Aureobasidium pullulans* and *Daldinia escscholzii*. $[\alpha]_D^{25}$ +6 (c, 1.1 in MeOH). λ_{max} 208 (ε 17000); 276 (ε 5900) (MeOH).

Shigemori, H. *et al.*, *J. Nat. Prod.*, 1998, **61**, 696-698 (*S-Orcinotriol*).
Hu, Z.-X. *et al.*, *Mar. Drugs*, 2014, **12**, 5563-5575 (*Daldinia escscholzii metab*).
Yang, Y.-N. *et al.*, *J. Nat. Prod.*, 2015, **78**, 705-711 (*pmr, struct*)

3-(3,4-Dihydroxyphenyl)-1-propanol D234

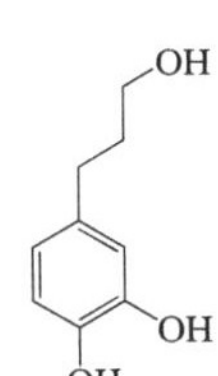

4-(3-Hydroxypropyl)-1,2-benzenediol, CAS. *4-(3-Hydroxypropyl)catechol.* ***Dihydrocaffeyl alcohol***
[46118-02-9]
$C_9H_{12}O_3$ 168.192
Constit. of *Onopordon corymbosum*. Oil
3′-O-β-D-Glucopyranoside
[359875-81-3]
$C_{15}H_{22}O_8$ 330.333
Constit. of the fruit of *Carum ajowan*. Amorph. powder. $[\alpha]_D^{23}$ -41 (c, 0.2 in MeOH)
4′-O-[2-Methylbutanoyl-(→6)-β-D-glucopyranoside]
Cylindrin B
[943326-41-8]
$C_{20}H_{30}O_9$ 414.451
Constit. of *Viburnum cylindricum*
1-O-(4-Methylpentanoyl)
[160661-34-7]
$C_{15}H_{22}O_4$ 266.337
Constit. of *Cheirolophus mauritanicus*. Gum
1-Tetradecanoyl
3-(3,4-Dihydroxyphenyl)propyl myristate
[196883-87-1]
$C_{23}H_{38}O_4$ 378.553
Constit. of *Phonus arborescens* (*Carthamus arborescens*)
1-Hexadecanoyl
3-(3,4-Dihydroxyphenyl)propyl palmitate
[196883-88-2]
$C_{25}H_{42}O_4$ 406.607
Constit. of *Phonus arborescens* (*Carthamus arborescens*)
1-Octadecanoyl
3-(3,4-Dihydroxyphenyl)propyl stearate
[80454-06-4]
$C_{27}H_{46}O_4$ 434.661
Constit. of *Symphyopappus compressus*, *Symphyopappus reticulatus* and *Phonus arborescens* (*Carthamus arborescens*)
1-Eicosanoyl
3-(3,4-Dihydroxyphenyl)propyl arachidate
[80454-07-5]
$C_{29}H_{50}O_4$ 462.715
Constit. of *Symphyopappus compressus*, *Taxus baccata* and *Phonus arborescens* (*Carthamus arborescens*)
1-Heneicosanoyl
$C_{30}H_{52}O_4$ 476.742
Constit. of the needles of *Taxus baccata*
1-Docosanoyl
[499794-91-1]
$C_{31}H_{54}O_4$ 490.769

Constit. of the needles of *Taxus baccata*

1-Tricosanoyl

$C_{32}H_{56}O_4$ 504.796

Constit. of the needles of *Taxus baccata*

1-Tetracosanoyl

[499794-93-3]

$C_{33}H_{58}O_4$ 518.823

Constit. of the needles of *Taxus baccata*

1-Pentacosanoyl

$C_{34}H_{60}O_4$ 532.85

Constit. of the needles of *Taxus baccata*

1-Hexacosanoyl

$C_{35}H_{62}O_4$ 546.877

Constit. of the needles of *Taxus baccata*

1-Heptacosanoyl

$C_{36}H_{64}O_4$ 560.904

Constit. of the needles of *Taxus baccata*

1-Octacosanoyl

$C_{37}H_{66}O_4$ 574.931

Constit. of the needles of *Taxus baccata*

1-Nonacosanoyl

$C_{38}H_{68}O_4$ 588.958

Constit. of the needles of *Taxus baccata*

1-Triacontanoyl

$C_{39}H_{70}O_4$ 602.985

Constit. of the needles of *Taxus baccata*

1-Hentriacontanoyl

$C_{40}H_{72}O_4$ 617.012

Constit. of the needles of *Taxus baccata*

1-Dotriacontanoyl

$C_{41}H_{74}O_4$ 631.039

Constit. of the needles of *Taxus baccata*

1-O-(4-Hydroxy-3-methoxy-E-cinnamoyl)

[110917-94-7]

$C_{19}H_{20}O_6$ 344.363

Constit. of *Baccharis punctulata*

3′-Me ether

4-Hydroxy-3-methoxybenzenepropanol. 3-(4-Hydroxy-3-methoxyphenyl)-1-propanol. ***Dihydroconiferyl alcohol***

[2305-13-7]

$C_{10}H_{14}O_3$ 182.219

Constit. of *Pinus* spp. and other plant spp. Isol. from *Pholiota lubrica*. Mp 65° (63°). Bp 197° at 15 mm

3′-Me ether, 1-O-β-D-glucopyranoside

[64181-82-4]

$C_{16}H_{24}O_8$ 344.36

Constit. of *Pinus contorta* and *Pinus ponderosa*. Amorph. powder. $[\alpha]^{20}_{D}$ -21.6 (c, 0.5 in MeOH)

3′-Me ether, 1-Ac

[14574-06-2]

$C_{12}H_{16}O_4$ 224.256

Constit. of *Pinus ponderosa*. Bp 128 - 130° at 0.01 mm

3′-Me ether, 1-O-hexacosanoyl

3-(4-Hydroxy-3-methoxyphenyl) propyl hexacosanoate

[98770-69-5]

$C_{36}H_{64}O_4$ 560.904

Constit. of the bark of *Tsuga chinensis* var. *formosana*

3′-Me ether, 1-O-hexatetracontanoyl

3-(4-Hydroxy-3-methoxyphenyl) propyl hexatetracontanoate

[845786-99-4]

$C_{56}H_{104}O_4$ 841.444

Constit. of the roots of *Vitex negundo*. Greenish oil. λ_{max} 203; 227; 282 (MeOH)

3′-Me ether, 1-O-[3-(4-hydroxyphenyl)propanoyl]

Dihydroconiferyl dihydro-p-coumarate

[152543-09-4]

$C_{19}H_{22}O_5$ 330.38

Constit. of *Ephemerantha fimbriata*. Amorph. solid

3′-Me ether, 1-benzoyl

Dihydroconiferyl benzoate

[147318-30-7]

$C_{17}H_{18}O_4$ 286.327

Constit. of *Croton hutchinsonianus* and *Gardenia taitensis*. Pale yellow oil. λ_{max} 229 (log ε 4.35); 280 (log ε 3.66) (MeOH)

3′-Me ether, 1-O-(4-hydroxy-E-cinnamoyl)

3-(4-Hydroxy-3-methoxyphenyl) propyl 4-hydroxycinnamate. ***Dihydroconiferyl p-coumarate***

[252205-23-5]

$C_{19}H_{20}O_5$ 328.364

Constit. of Brazilian propolis. Amorph. powder. λ_{max} 224 (log ε 4.36); 290 (log ε 4.24); 311 (log ε 4.26) (MeOH)

3′-Me ether, 1-O-(3,4-dihydroxy-E-cinnamoyl)

Dihydroconiferyl trans-caffeate

[1044565-77-6]

$C_{19}H_{20}O_6$ 344.363

Constit. of *Ligularia duciformis*

3′-Me ether, 1-O-(4-hydroxy-3-methoxy-E-cinnamoyl)

Dihydroconiferyl ferulate

[277749-31-2]

$C_{20}H_{22}O_6$ 358.39

Constit. of *Peganum nigellastrum*. Pale yellow semi-solid. λ_{max} 203 (log ε 4.7); 216 (sh); 229 (log ε 4.34); 289 (log ε 4.25); 325 (log ε 4.39) (MeOH)

1-O-(2-Hydroxy-1-hydroxymethyl) ethyl, 3′-Me ether

Lawsorosemarinol. *Dihydroconiferyl 2-glycerol ether*

[1499023-87-8]

$C_{13}H_{20}O_5$ 256.298

Constit. of *Kadsura longipedunculata* and *Lawsonia alba*. Gummy solid. λ_{max} 226; 297 (MeOH).

Aldrich Library of ^{13}*C and* ^{1}*H FT-NMR Spectra*, 1992, **2**, 413A (*nmr*).
Higuchi, R. *et al.*, *Phytochemistry*, 1977, **16**, 1007-1011 (*Dihydroconiferin*).
Cram, D.J. *et al.*, *J. Org. Chem.*, 1978, **43**, 2758-2772 (*synth*).
Bohlmann, F. *et al.*, *Phytochemistry*, 1981, **20**, 1657-1663 (*isol, esters*).
Manners, G.D. *et al.*, *J. Agric. Food Chem.*, 1982, **30**, 401-404 (*Pinus pondera constits*).
Fang, J.M. *et al.*, *J. Chin. Chem. Soc. (Taipei)*, 1985, **32**, 75-80 , 477 - 480 (*isol*).
Jakupovic, J. *et al.*, *Rev. Latinoam. Quim.*, 1987, **18**, 75-76 (*ferulate*).
Bianco, A. *et al.*, *Synth. Commun.*, 1988, **18**, 1765-1771 (*synth, deriv*).
Morgan, K.R. *et al.*, *J. Am. Chem. Soc.*, 1990, **112**, 4-7 (*Dihydroconiferyl alcohol, cmr*).
Cardona, M.L. *et al.*, *Phytochemistry*, 1990, **29**, 629-631 (*Onopordon corymbosum constit*).
Tezuka, Y. *et al.*, *Chem. Pharm. Bull.*, 1993, **41**, 1346-1349 (*Ephemerantha fimbriata constit*).
Marco, J.A. *et al.*, *Phytochemistry*, 1994, **37**, 1101-1107 (*Cheirolophus mauritanicus constit*).
Barrero, A.F. *et al.*, *J. Nat. Prod.*, 1997, **60**, 1026-1030 (*isol, esters*).
Tazawa, S. *et al.*, *Chem. Pharm. Bull.*, 1999, **47**, 1388-1392 (*Dihydroconiferyl p-coumarate*).
Ma, Z.-Z. *et al.*, *Phytochemistry*, 2000, **53**, 1075-1078 (*Dihydroconiferyl ferulate*).
Ishikawa, T. *et al.*, *Chem. Pharm. Bull.*, 2001, **49**, 840-844 (*3′-glucoside*).
Jetter, R. *et al.*, *Phytochemistry*, 2002, **61**, 579-587 (*Taxus baccata esters*).
Athikomkulchai, S. *et al.*, *Chem. Pharm. Bull.*, 2006, **54**, 262-264 (*Dihydroconiferyl benzoate*).
Zhu, X. *et al.*, *Yunnan Zhiwu Yanjiu*, 2006, **28**, 91-94 (*Cylindrin B*).
Fu, K. *et al.*, *Yaoxue Xuebao*, 2007, **42**, 621-624 (*Dihydroconiferyl caffeate*).
Uddin, N. *et al.*, *Fitoterapia*, 2013, **84**, 202-207 (*Lawsorosemarinol*).
Chen, J.-B. *et al.*, *Zhongcaoyao*, 2015, **46**, 178-184 (*Lawsorosemarinol*)

1-(2,4-Dihydroxyphenyl)-1-propanone, CAS D235

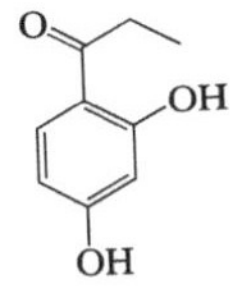

2′,4′-Dihydroxypropiophenone, CAS. *4-Propionylresorcinol. Respropiophenone. Resopropiophenone*
[5792-36-9]
$C_9H_{10}O_3$ 166.176
Used as 1% soln. in EtOH as metallochromic indicator in titrimetric detn. of Fe; spot test for Fe. Yellow needles (EtOH). Mp 98 - 100°

Oxime
[22919-58-0]
$C_9H_{11}NO_3$ 181.191
Used as soln. in aq. EtOH for photometric detn. of Fe(*III*) (λ_{max} 510 nm, ε 2300); metallochromic indicator in titrimetric detn. of Fe(*III*). Yellow needles. Mp 186 - 187°

4-Me ether
1-(2-Hydroxy-4-methoxyphenyl)-1-propanone. 2′-Hydroxy-4′-methoxypropiophenone
[6270-44-6]
$C_{10}H_{12}O_3$ 180.203
Constit. of the defensive secretion of *Tribolium brevicornis*. Plates (EtOH). Mp 58°

Di-Me ether
1-(2,4-Dimethoxyphenyl)-1-propanone. 2′,4′-Dimethoxypropiophenone
[831-00-5]
$C_{11}H_{14}O_3$ 194.23
Cryst. (EtOH aq.). Mp 67°.

Aldrich Library of Infrared Spectra, 3rd edn., 1981, 872C (*ir*).
Aldrich Library of ^{13}C *and* ^{1}H *FT-NMR Spectra*, 1992, **2**, 858A (*nmr*).
Gandhi, M.H. *et al.*, *Anal. Chem.*, 1967, **39**, 1643 (*detn, Fe*).
Széll, T. *J. Chem. Soc. C*, 1967, 2041 (*synth, pmr*).
Desai, M.N. *et al.*, *Microchem. J.*, 1968, **18**, 500 (*spot test*).
Mezheritskii, V.V. *et al.*, *Zh. Org. Khim.*, 1969, **5**, 515 (*synth*).
Desai, M.N. *et al.*, *Fresenius' Z. Anal. Chem.*, 1972, **258**, 127 (*detn, Fe*).
Desai, M.N. *et al.*, *CA*, 1973, **79**, 13100p (*detn, Fe*).
Desai, M.N. *et al.*, *J. Indian Chem. Soc.*, 1973, **50**, 369 (*detn, Fe*).
Howard, R.W. *et al.*, *J. Chem. Ecol.*, 1987, **13**, 1707-1722 (*Tribolium brevicornis constit*)

1-(3,4-Dihydroxyphenyl)-1-propanone, CAS D236

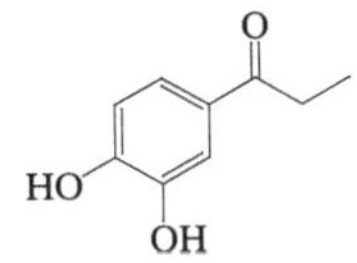

3′,4′-Dihydroxypropiophenone, CAS. *4-Propionylcatechol*
[7451-98-1]
[829-76-5]
$C_9H_{10}O_3$ 166.176
Yellow needles (C_6H_6). Mp 146°
▶ UH0450000. LD_{50} (mus, orl) 760 mg/kg

3-Me ether
1-(4-Hydroxy-3-methoxyphenyl)-1-propanone. 4′-Hydroxy-3′-methoxypropiophenone. ***Propiovanillone****. Propioguaiacone*
[1835-14-9]
$C_{10}H_{12}O_3$ 180.203
Degradn. prod. of lignin, a minor constit. of woods. Constit. of various plant spp.. Cryst. (hexane). Mp 61 - 62° (54°). Bp 182° at 12 mm.

Buu-Hoï, N.P. *et al.*, *J. Org. Chem.*, 1961, **26**, 2401-2402 (*3-Me ether*).
La Manna, A. *et al.*, *Farmaco, Ed. Sci.*, 1967, **22**, 667 (*synth*)

1-(3,5-Dihydroxyphenyl)-2-propanone D237

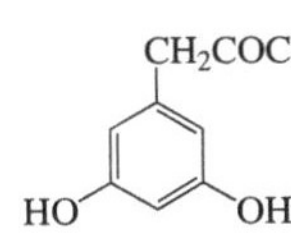

*3,5-Dihydroxyphenylacetone. α-****Acetylorcinol****. 5-Acetonylresorcinol*
[15580-07-1]
$C_9H_{10}O_3$ 166.176
Structures in this series may need revision. Phenyl 3,5-dioxygenated substn. pattern in doubt as a result of work done by Yang *et al* in 2015. Prod. by *Alternaria brassicola, Alternaria tenuissima, Cochliobolus lunata, Stagonospora apocyni* and *Verticillium* sp. Also from the marine-derived *Aureobasidium pullulans* and *Capnodium* sp. SZ-F22. Oil. Bp 185 - 187° at 0.05 mm. λ_{max} 280 (ε 2300); 285 (ε 2300) (EtOH). [neutral] λ_{max} 282 (ε 2400) (MeOH)

O-β-D-Glucopyranoside
Grammatophylloside D
[1425047-29-5]
$C_{15}H_{20}O_8$ 328.317
Constit. of *Grammatophyllum speciosum*. Amorph. powder. $[\alpha]_D^{26}$ -32.9 (c, 0.5 in MeOH).

Stetter, H. *et al.*, *Monatsh. Chem.*, 1967, **98**, 755-762 (*synth*).
Nukina, M. *et al.*, *Agric. Biol. Chem.*, 1977, **41**, 717 (*α-Acetylorcinol*).
Venkatasubbaiah, P. *et al.*, *J. Phytopathol.*, 1992, **135**, 309-316 (*Stagonospora apocyni constit*).
Schulze, A. *Dissertation*, Univ. of Göttingen, 2003, (*occur*).
Sahakitpichan, P. *et al.*, *Tetrahedron*, 2013, **69**, 1031-1037 (*Grammatophylloside D*).
Yang, Y.-N. *et al.*, *J. Nat. Prod.*, 2015, **78**, 705-711 (*pmr, struct*)

3-(3,4-Dihydroxyphenyl)-2-propenal D238

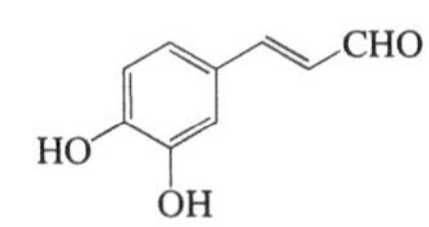

3,4-Dihydroxycinnamaldehyde. Caffealdehyde
[68149-78-0]
$C_9H_8O_3$ 164.16
(E)-form [141632-15-7]
Cryst.. Mp 208 - 210°

4′-O-[3-O-Acetyl-α-L-rhamnopyranosyl-(1→6)-β-D-glucopyranoside]
[1413935-34-8]
$C_{23}H_{30}O_{13}$ 514.48
Isol. from *Conyza japonica*. Amorph. powder. $[\alpha]_D^{20}$ -159 (c, 0.003 in MeOH). λ_{max} 205 (log ε 2.78) (MeOH)

3′-Me ether
3-(4-Hydroxy-3-methoxyphenyl)-2-propenal. 4-Hydroxy-3-methoxycinnamaldehyde. ***Coniferaldehyde****. Hadromal. Coniferylaldehyde. Ferulaaldehyde. Ferulaldehyde*
[20649-42-7]
[458-36-6]
$C_{10}H_{10}O_3$ 178.187
Occurs freely in wood and as a component of lignin; isol. from *Kalopanax* sp., *Daphne feddei, Gleditsia sinensis, Clausena lansium, Maytenus salicifolia* and others, and propolis. Light yellow needles (petrol or MeOH). Mp 84° (75-76°)

3′-Me ether, 2,4-dinitrophenylhydrazone
Mp 265 - 266°

4′-Me ether
3-(3-Hydroxy-4-methoxyphenyl)-2-propenal.

3-Hydroxy-4-methoxycinnamaldehyde. ***Isoferulaldehyde.*** *Isoconiferaldehyde*
[110037-35-9]
[192120-46-0] (unspecified stereochem)
$C_{10}H_{10}O_3$ 178.187
Constit. of *Euonymus acanthocarpus*, *Jatropha curcas* and *Tamarix nilotica*. Pale yellow oil.

Tiemann, F. *Ber. Dtsch. Chem. Ges.*, 1885, **18**, 3481-3493 (*Coniferaldehyde*).
Adler, E. *et al.*, *Acta Chem. Scand.*, 1948, **2**, 839-840 (*Coniferaldehyde, occur, woods*).
Black, R.A. *et al.*, *J. Am. Chem. Soc.*, 1953, **75**, 5344-5346 (*Coniferaldehyde, isol, uv*).
Liptaj, T. *et al.*, *Coll. Czech. Chem. Comm.*, 1980, **45**, 330-334 (*Coniferaldehyde, pmr*).
Barakat, H.H. *et al.*, *Phytochemistry*, 1987, **26**, 1837-1838 (*Isoferulaldehyde, isol*).
Gangloff, A.R. *et al.*, *J. Org. Chem.*, 1990, **55**, 3679-3682 (*synth, ir, pmr, cmr*).
Kulkarni, M.G. *et al.*, *Tetrahedron Lett.*, 1990, **31**, 4497-4500 (*synth*).
Duran, E. *et al.*, *Holzforschung*, 1992, **46**, 21-24 (*Isoferulaldehyde*).
Daubresse, N. *et al.*, *Synthesis*, 1994, 369-371 (*Coniferaldehyde, synth*).
Demin, P. *et al.*, *Bioorg. Med. Chem.*, 2004, **12**, 3019-3026 (*synth*).
Liang, S. *et al.*, *J. Nat. Prod.*, 2008, **71**, 1902-1905 (*Daphne feddei constit*).
Zhu, J.X. *et al.*, *Arch. Pharmacal Res.*, 2012, **35**, 1739-1747 (*Isoferulaldehyde*).
Hu, J. *et al.*, *Chem. Nat. Compd. (Engl. Transl.)*, 2012, **48**, 782-784 (*rhamnosylglucosides*)

3-(2,3-Dihydroxyphenyl)-2-propenoic acid, CAS D239

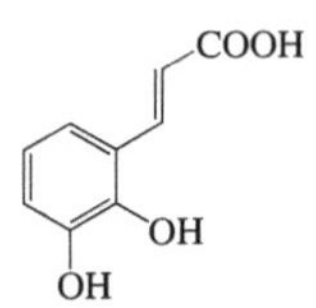

2,3-Dihydroxycinnamic acid
[31082-90-3]
$C_9H_8O_4$ 180.159

(E)-form

Anthenobilic acid [38481-04-8]
Constit. of *Anthemis nobilis*. Mp 218° dec.

2-Me ether
3-(3-Hydroxy-2-methoxyphenyl)-2-propenoic acid. 3-Hydroxy-2-methoxycinnamic acid
[1373444-63-3]
$C_{10}H_{10}O_4$ 194.186
Solid (EtOH aq.). Mp 217 - 220°

2-Me ether, nitrile
3-(2-Cyanoethenyl)-2-methoxyphenol
$C_{10}H_9NO_2$ 175.187
Solid. Mp 149 - 151°

3-Me ether
3-(2-Hydroxy-3-methoxyphenyl)-2-propenoic acid, CAS. 2-Hydroxy-3-methoxycinnamic acid. o-Ferulic acid. Orthoferulic acid
[3626-94-6]
$C_{10}H_{10}O_4$ 194.186
Present in wheat shoots. Mp 183 - 184°

3-Me ether, Me ester
$C_{11}H_{12}O_4$ 208.213
Solid. Mp 114°

Di-Me ether
3-(2,3-Dimethoxyphenyl)-2-propenoic acid, CAS. 2,3-Dimethoxycinnamic acid, CAS
[7345-82-6]
[7461-60-1] (unspecified stereochem.)
$C_{11}H_{12}O_4$ 208.213
Cryst. (EtOH aq.). Mp 182 - 184°

Di-Me ether, Me ester
[15854-60-1]
$C_{12}H_{14}O_4$ 222.24
Cryst.. Mp 53 - 54°

Di-Me ether, decyl ester
Decyl 2,3-dimethoxycinnamate
[1246077-70-2]
$C_{21}H_{32}O_4$ 348.483
Constit. of *Daphne oleoides*

Methylene ether
3-(2,3-Methylenedioxyphenyl) propenoic acid. 3-(1,3-Benzodioxol-4-yl)-2-propenoic acid, CAS. 2,3-Methylenedioxycinnamic acid, CAS
[38489-70-2]
$C_{10}H_8O_4$ 192.17
Prisms (MeOH aq. or EtOAc). Mp 194°

(Z)-form

2-Me ether, nitrile
Solid. Mp 57 - 60°.

Perkin, W.H. *et al.*, *J. Chem. Soc.*, 1926, 2925-2932 (*E-methylene ether, synth*).
El-Basyouni, S. *et al.*, *Can. J. Biochem.*, 1964, **42**, 493-497 (*o-Ferulic acid*).
Herisset, A. *et al.*, *Plant. Med. Phytother.*, 1970, **4**, 189-200 (*Anthemis nobilis constit*).
Santavy, *et al.*, *Coll. Czech. Chem. Comm.*, 1972, **37**, 1825-1850 (*E-methylene ether*).
Mathison, I.W. *et al.*, *J. Org. Chem.*, 1974, **39**, 2852-2855 (*synth*).
Dupin, J.F.E. *et al.*, *Synth. Commun.*, 1985, **15**, 581-586 (*synth*).
Peterson, J.R. *et al.*, *J. Chem. Eng. Data*, 1988, **33**, 534-537 (*E-di-Me ether, synth*).
Ullah, N. *Asian J. Chem.*, 2010, **22**, 2456-2458 (*Decyl dimethoxycinnamate*).
Boeck, F. *et al.*, *Beilstein J. Org. Chem.*, 2012, **8**, 1630-1636 (*E-3-Me ether Me ester*).
Levasseur, J.L. *et al.*, *Synth. Commun.*, 2012, **42**, 292-298 (*E-2-Me ether, E-2-Me ether nitrile, Z-2-Me ether nitrile*)

3-(2,4-Dihydroxyphenyl)-2-propenoic acid D240

2,4-Dihydroxycinnamic acid, CAS. 3-(2,4-Dihydroxyphenyl)acrylic acid. Umbellic acid
[614-86-8]
$C_9H_8O_4$ 180.159
pK_{a1} 4.1 (30°,1.0*M* LiCl)

(E)-form

Cryst. (EtOH); needles (H_2O). Mp 146 - 147°

4-Me ether
3-(2-Hydroxy-4-methoxyphenyl)-2-propenoic acid
[38489-77-9]
$C_{10}H_{10}O_4$ 194.186
Needles (H_2O) or solid. Mp 195 - 198° (180-185°) dec.

4-Me ether, 2-O-β-D-glucopyranoside
*trans-**4-Methoxymelilotoside***
[31564-49-5]
$C_{16}H_{20}O_9$ 356.327
Constit. of *Artemisia herba-alba*, *Artemisia splendens* and *Lavandula angustifolia*. Precursor of coumarins in lavender (via the *Z*-isomer). $[\alpha]^{25}_{D}$ -70.6 (c, 0.26 in MeOH). λ_{max} 203 (log ε 0.8); 279 (log ε 0.3); 309 (log ε 0.26) (MeOH)

4-Me ether, Me ester
[93198-68-6]
$C_{11}H_{12}O_4$ 208.213
Constit. of *Clausena harmandiana*. Solid. Mp 144.3 - 145.3°

4-Me ether, Me ester, 2-O-β-D-glucopyranoside
[1386936-79-3]
$C_{17}H_{22}O_9$ 370.354
Constit. of the leaves of *Calotropis procera*

4-Me ether, hexadecyl ester
[1370450-28-4]
$C_{26}H_{42}O_4$ 418.618
Constit. of *Raputia praetermissa*. Amorph. solid

Di-Me ether
3-(2,4-Dimethoxyphenyl)-2-propenoic acid. 2,4-Dimethoxycinnamic acid, CAS
[16909-09-4]
[6972-61-8] (unspecified config.)
$C_{11}H_{12}O_4$ 208.213
Needles. Mp 192 - 194° (186-187°)

Di-Me ether, Me ester
[66417-42-3]
[15854-55-4] (unspecified config.)
$C_{12}H_{14}O_4$ 222.24
Solid. Mp 86.5 - 87°
Di-Me ether, Et ester
$C_{13}H_{16}O_4$ 236.267
Needles (petrol). Mp 61°. Bp 208 - 212° at 13 mm
Di-Me ether, nitrile
[404938-06-3]
$C_{11}H_{11}NO_2$ 189.214
Oil
Di-Et ether
3-(2,4-Diethoxyphenyl)-2-propenoic acid
$C_{13}H_{16}O_4$ 236.267
Plates (EtOH aq.). Mp 106.5°
(Z)-form
Di-Me ether
[38489-71-3]
Cryst. (MeOH aq.). Mp 139 - 140°
Di-Me ether, Me ester
[84385-10-4]
No phys. props. reported
Di-Me ether, nitrile
Oil.

Aldrich Library of Infrared Spectra, 3rd edn., 1981, 954D (*ir*).
Aldrich Library of [13]*C and* [1]*H FT-NMR Spectra*, 1992, **2**, 1056A (*nmr*).
Will, W. *et al.*, *Ber. Dtsch. Chem. Ges.*, 1886, **19**, 1777-1786 (*di-Et ether*).
Perkin, W.H. *et al.*, *J. Chem. Soc., Trans.*, 1904, **85**, 159-165 (*E-di-Me ether, di-Me ether Et ester*).
Reimer, M. *et al.*, *J. Am. Chem. Soc.*, 1930, **52**, 341-347 (*di-Me ether, di-Me ether Et ester*).
Böhme, H. *Arch. Pharm. Ber. Dtsch. Pharm. Ges.*, 1957, **290**, 405-412 (*E-di-Me ether, Z-di-Me ether*).
Halpem, O. *et al.*, *Helv. Chim. Acta*, 1957, **40**, 758-778 (*E-4-Me ether, E-4-Me ether Me ester, E-di-Me ether Me ester*).
Jones, B. *et al.*, *J. Chem. Soc.*, 1958, 4064-4069 (*di-Me ether, dii-Me ether Et ester*).
Nasipuri, D. *et al.*, *J. Chem. Soc.*, 1962, 3105-3108 (*di-Me ether*).
Brown, S.A. *Phytochemistry*, 1963, **2**, 137-144 (*4-Me ether glucoside*).
Šantavý, F. *et al.*, *Coll. Czech. Chem. Comm.*, 1972, **37**, 1825-1850 (*uv*).
Kaul, S.K. *et al.*, *Indian J. Chem., Sect. B*, 1982, **21**, 472-474 (*Z-di-Me ether, Z-di-Me ether Me ester*).
Hofer, O. *et al.*, *Monatsh. Chem.*, 1986, **117**, 1219-1222 (*E-di-Me ether Me ester, E-4-Me ether, isol*).
Potapov, V.M. *et al.*, *J. Org. Chem. USSR (Engl. Transl.)*, 1989, **25**, 1694-1699 (*di-Me ether, di-Me ether Et ester*).
Mahajan, R.P. *et al.*, *Org. Prep. Proced. Int.*, 2005, **37**, 286-290 (*E-di-Me ether Me ester*).
De Simone, F. *et al.*, *Org. Lett.*, 2009, **11**, 1023-1026 (*E-di-Me ether*).
Rosas, L.V. *et al.*, *J. Braz. Chem. Soc.*, 2011, **22**, 1346-1353 (*Raputia praetermissa ester*).
Mohamed, M.A. *et al.*, *Z. Naturforsch., C*, 2011, **66**, 547-554 (*Calotropis glucoside*).
Dahmani-Hamzaoui, N. *et al.*, *Helv. Chim. Acta*, 2012, **95**, 564-576 (*2-Glucosyloxy-4-methoxycinnamic acid*).
Maneerat, W. *et al.*, *Phytochem. Lett.*, 2013, **6**, 18-20 (*Clausena harmandiana constit*).
Lanari, D. *et al.*, *Org. Lett.*, 2016, **18**, 2680-2683 (*E-di-Me ether nitrile, Z-di-Me ether nitrile*).
Ortega, V. *et al.*, *Org. Lett.*, 2017, **19**, 6236-6239 (*E-di-Me ether Me ester*).
Afshar, F.H. *et al.*, *Rec. Nat. Prod.*, 2017, **11**, 43-50 (*cis-, trans-4-Methoxymelilotosides*)

3-(2,5-Dihydroxyphenyl)-2-propenoic acid, CAS D241

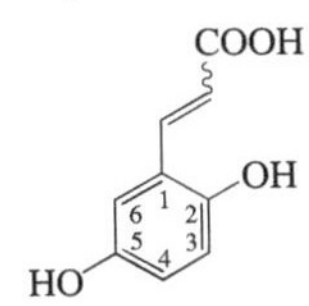

2,5-Dihydroxycinnamic acid, CAS. ***Grevillic acid***. *Yinhua acid*
[636-01-1]
[38489-67-7] ((E)-form)
$C_9H_8O_4$ 180.159
Constit. of *Grevillea robusta*, *Heliciopsis lobata* and *Incarvillea younghusbandii*. Photographic intermed.. Needles (H_2O). Mp 207° dec. pK_{a1} 3.69 (30°,0.1*M* LiCl)
Me ester
[63177-57-1]
[123064-80-2] ((*E*)-form)
$C_{10}H_{10}O_4$ 194.186
Constit. of *Azadirachta indica* and *Murraya paniculata*. Cryst. (hexane). Mp 173 - 174°. λ_{max} 242; 291; 320 (MeOH)
2-Me ether
5-Hydroxy-2-methoxycinnamic acid
[105233-53-2]
$C_{10}H_{10}O_4$ 194.186
Yellow cryst. (H_2O). Mp 179 - 180°
2-Me ether, Me ester
[949462-55-9]
[188426-71-3] ((*E*)-form)
$C_{11}H_{12}O_4$ 208.213
Constit. of *Murraya paniculata*. λ_{max} 218 (log ε 3.46); 296 (log ε 3.54); 324 (log ε 3.63) (MeOH)
Di-Me ether
2,5-Dimethoxycinnamic acid
[10538-51-9]
$C_{11}H_{12}O_4$ 208.213
Yellowish needles (H_2O). Mp 147°
5-Et ether, Me ester
[1383835-68-4]
$C_{12}H_{14}O_4$ 222.24
Constit. of the leaves of *Grevillea robusta*. Amorph. yellow powder. Mp 98 - 100°. λ_{max} 249 (log ε 3.65); 278 (log ε 3.77); 351 (log ε 3.41) (MeOH).

Aldrich Library of FT-IR Spectra, 1st edn., 1985, **2**, 183B (*ir*).
Aldrich Library of [13]*C and* [1]*H FT-NMR Spectra*, 1992, **2**, 1056B (*nmr*).
Updegraff, I.H. *et al.*, *J. Am. Chem. Soc.*, 1949, **71**, 407-410 (*synth*).
De Graw, J.I. *et al.*, *J. Chem. Eng. Data*, 1969, **14**, 509-510 (*synth*).
Šantavý, F. *et al.*, *Coll. Czech. Chem. Comm.*, 1972, **37**, 1825-1850 (*uv*).
Cannon, J.R. *et al.*, *Aust. J. Chem.*, 1973, **26**, 2257-2275 (*Grevillic acid*).
Dupin, J.F.E. *et al.*, *Synth. Commun.*, 1985, **15**, 581-586 (*synth*).
Atta-ur-Rahman, *et al.*, *Phytochemistry*, 1997, **44**, 683-685 (*Murraya paniculata esters*).
Chuang, T.-H. *et al.*, *Molecules*, 2011, **16**, 9331-9339 (*5-Et ether Me ester*)

3-(2,6-Dihydroxyphenyl)-2-propenoic acid, CAS D242

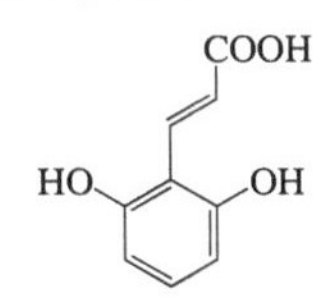

(*E*)-*form*

2,6-Dihydroxycinnamic acid
[38489-75-7] (Z-di-Me ether)
$C_9H_8O_4$ 180.159
(E)-form [1012326-86-1]
Di-Me ether
[76197-69-8]
$C_{11}H_{12}O_4$ 208.213
Constit. of the fruits of *Narthecium ossifragum*
(ξ)-form [408536-05-0]
Cryst.. Mp 205° dec.
Di-Me ether
3-(2,6-Dimethoxyphenyl)-2-propenoic acid, CAS. *2,6-Dimethoxycinnamic acid*, CAS
[59131-19-0]
Cryst. (EtOH aq.). Mp 151 - 153°.

Shah, H.A. *et al.*, *J. Chem. Soc.*, 1938, 1832-1833 (*synth*).
Adams, R. *et al.*, *J. Am. Chem. Soc.*, 1952, **74**, 5346-5348 (*synth*).
Šantavý, F. *et al.*, *Coll. Czech. Chem. Comm.*, 1972, **37**, 1825-1850 (*uv*).
Snook, M.E. *et al.*, *J. Chromatogr.*, 1985, **324**, 141-151 (*synth*).
Saeed, A. *et al.*, *Eur. J. Med. Chem.*, 2014, **78**, 43-53 (*synth*).

Vu, M. *et al.*, *Phytochemistry*, 2016, **132**, 76-85 (*Narthecium ossifragum constit*)

3-(3,4-Dihydroxyphenyl)-2-propenoic acid, CAS D243

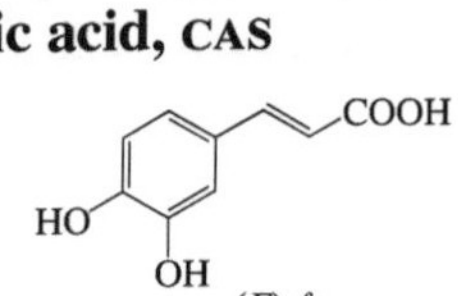

(*E*)-*form*

3,4-Dihydroxycinnamic acid, CAS.
Caffeic acid
[331-39-5]
[110993-57-2] (Na salt, unspecified stereochem.)
$C_9H_8O_4$ 180.159
Log P 0.82 (calc)
▶ GD8950000.
▶ Possible human carcinogen (IARC 2B)
(E)-form [501-16-6]
Widespread in plants, free and as glycosides. Found by Bate-Smith in 66% of investigated dicotyledonous and 50% of investigated monocotyledonous plant spp.. Used for spot detn. of Fe. Enzyme inhibitor. Inhibitor of aflatoxin biosynth. Strongly inhibits neutrophil elastase. Antineoplastic agent. Shows anti-HIV, choleretic and hepatotropic activity. Antioxidant. Yellow cryst.. Sol. MeOH, EtOAc; fairly sol. H_2O. Mp 223 - 225° dec. pK_{a1} 4.62; pK_{a2} 9.07 (25°). Log P 0.82 (calc). Forms a monohydrate. Mod. allergen

3′-O-β-D-Glucopyranoside
[143729-78-6]
$C_{15}H_{18}O_9$ 342.3
Constit. of the fruit of *Ugni molinae*

3′-O-[β-D-Glucopyranosyl-(1→6)-β-D-glucopyranosyl-(1→2)-β-D-glucopyranoside]
Parviside B
$C_{27}H_{38}O_{19}$ 666.582
Constit. of *Galinsoga parviflora*. Gummy solid

4′-O-β-D-Glucopyranoside
Glucocaffeic acid. *Linocaffein*
[17093-82-2]
$C_{15}H_{18}O_9$ 342.3
Isol. from *Linum usitatissimum*, *Polypodium vulgare*, *Ribes rubrum* and other plants. Needles (MeOH). Mp 133 - 135°. Mp 208°. $[\alpha]_D^{25}$ -87.3 (MeOH). Some doubt about the identity of Linocaffein and Glucocaffeic acid in view of the widely differing Mp's reported

4′-O-[3-Hydroxy-2ξ-methylpropanoyl-(→6)-β-D-glucopyranoside]
[1967838-62-5]
$C_{19}H_{24}O_{11}$ 428.39
Constit. of the flowers of *Telekia speciosa*. Amorph. solid. $[\alpha]_D^{26}$ -16 (c, 0.25 in MeOH)

4′-O-[2ξ-Hydroxy-3ξ-methylbutanoyl-(→6)-β-D-glucopyranoside]
[1967838-61-4]
$C_{20}H_{26}O_{11}$ 442.417
Constit. of the flowers of *Telekia speciosa*. Amorph. yellow solid. $[\alpha]_D^{26}$ -55.4 (c, 1 in MeOH)

4′-O-[α-L-Rhamnopyranosyl-(1→4)-β-D-glucopyranoside]
[77690-82-5]
$C_{21}H_{28}O_{13}$ 488.442
Constit. of *Leonurus cardiaca*. Mp 151 - 154°. $[\alpha]_D^{21}$ -5.4 (c, 0.4 in MeOH)

4′-O-[β-D-Glucopyranosyl-(1→4)-3,4-dihydroxy-E-cinnamoyl-(→6)-β-D-glucopyranoside]
Scrocaffeside B
[1034143-62-8]
$C_{30}H_{34}O_{17}$ 666.585
Constit. of the roots of *Picrorhiza scrophulariiflora*. Amorph. powder. $[\alpha]_D$ -23.5 (c, 0.2 in MeOH). λ_{max} 245 (log ε 5.5); 332 (log ε 2.35) (MeOH)

β-D-Glucopyranosyl ester, 3′-O-[4-hydroxy-E-cinnamoyl-(→6)-β-D-glucopyranoside]
Clematiside F
[1800599-16-9]
$C_{30}H_{34}O_{16}$ 650.586
Constit. of the aerial parts of *Clematis tashiroi*. Yellowish powder. $[\alpha]_D^{25}$ -141.1 (c, 0.2 in MeOH). λ_{max} 230 (log ε 4.45); 303 (sh); 315 (log ε 4.55) (MeOH)

β-D-Glucopyranosyl ester, 3′-O-[3,4-dihydroxy-E-cinnamoyl-(→6)-β-D-glucopyranoside]
Clematiside E
[1800599-15-8]
$C_{30}H_{34}O_{17}$ 666.585
Constit. of the aerial parts of *Clematis tashiroi*. Yellowish powder. $[\alpha]_D^{25}$ -95.4 (c, 0.2 in MeOH). λ_{max} 217 (log ε 4.40); 234 (sh); 298 (sh); 322 (log ε 4.41) (MeOH)

β-D-Glucopyranosyl ester, 4′-O-β-D-glucopyranoside
Glucosyl glucocaffeate
[851232-93-4]
$C_{21}H_{28}O_{14}$ 504.441
Constit. of the flowers of *Moricandia arvensis*. Oil. $[\alpha]_D^{25}$ -38.5 (c, 0.2 in MeOH). λ_{max} 217 (log ε 4.1); 233 (log ε 3.8); 285 (log ε 4.15) (MeOH)

β-D-Glucopyranosyl ester, 4′-O-[4-hydroxy-E-cinnamoyl-(→6)-β-D-glucopyranoside]
Clematiside C
[1800599-03-4]
$C_{30}H_{34}O_{16}$ 650.586
Constit. of the aerial pats of *Clematis tashiroi*. Yellowish powder. $[\alpha]_D^{25}$ -91.3 (c, 0.2 in MeOH). λ_{max} 219 (log ε 4.38); 229 (log ε 4.34); 297 (log ε 4.49); 310 (log ε 4.47) (MeOH)

β-D-Glucopyranosyl ester, 4′-O-[3,4-dihydroxy-E-cinnamoyl-(→6)-β-D-glucopyranoside]
Clematiside A
[1800598-97-3]
$C_{30}H_{34}O_{17}$ 666.585
Constit. of the aerial parts of *Clematis tashiroi*. Antioxidant. Yellowish powder. $[\alpha]_D^{25}$ -272.6 (c, 0.2 in MeOH). λ_{max} 217 (log ε 4.86); 243 (log ε 4.74); 294 (log ε 4.84); 324 (log ε 4.83) (MeOH)

β-D-Glucopyranosyl ester, 4′-O-[4-hydroxy-3-methoxy-E-cinnamoyl-(→6)-β-D-glucopyranoside]
Clematiside B
[1800599-02-3]
$C_{31}H_{36}O_{17}$ 680.612
Constit. of the aerial parts of *Clematis tashiroi*. Amorph. yellow powder. $[\alpha]_D^{25}$ -71.9 (c, 0.2 in MeOH). λ_{max} 218 (log ε 4.50); 238 (log ε 4.41); 293 (log ε 4.48); 320 (log ε 4.49) (MeOH)

[β-D-Apiofuranosyl-(1→6)-β-D-glucopyranosyl] ester
[223261-30-1]
$C_{20}H_{26}O_{13}$ 474.415
Constit. of *Salvia officinalis*. Cryst. (MeOH). Mp 69 - 71°. $[\alpha]_D^{23}$ -21.8 (c, 0.15 in MeOH)

[α-L-Xylopyranosyl-(4→2)-β-D-glucopyranosyl] ester, 3′-O-β-D-glucopyranoside
Hydrangeifolin II
[868633-35-6]

$C_{26}H_{36}O_{18}$ 636.556
Constit. of the leaves of *Coussarea hydrangeifolia*. Amorph. powder. $[\alpha]_D^{20}$ +6.5 (c, 0.004 in MeOH). λ_{max} 209; 282 (MeOH)

[β-D-Glucopyranosyl-(1→2)-β-D-glucopyranosyl] ester
1-O-Caffeoyllaminaribiose
[73343-13-2]
$C_{21}H_{28}O_{14}$ 504.441
Isol. from fern *Asplenium adiantum-nigrum*. λ_{max} 300; 314 (sh) (MeOH)

Me ester
[67667-67-8]
[3843-74-1] (unspecified stereochem.)
$C_{10}H_{10}O_4$ 194.186
Isol. from plants, e.g. *Morus alba* and *Gaillardia pulchella*. Light yellow needles (EtOH aq.). Sol. MeOH, C_6H_6; poorly sol. H_2O, hexane. Mp 158 - 160°. λ_{max} 217 (ε 15500); 242 (ε 11900); 326 (ε 18900) (MeOH) (Berdy)

Et ester
[66648-50-8]
$C_{11}H_{12}O_4$ 208.213
Isol. from *Solanum melongena*, *Solanum integrifolium*, *Aristotelia chilensis* and *Salvia guarantica*. Benzodiazepine antagonist. Cryst. (MeOH aq.). Mp 149 - 150°

Et ester, 3′-O-β-D-glucopyranoside
Lycibarbarphenylpropanoid I
[1988787-11-6]
$C_{17}H_{22}O_9$ 370.354
Constit. of the fruit of *Lycium barbarum*. Amorph. yellowish powder. $[\alpha]_D^{25}$ -49.6 (c, 0.5 in MeOH). λ_{max} 205 (log ε 4.08); 214 (log ε 4.10); 233 (log ε 4.08); 316 (log ε 4.27) (MeOH)

Et ester, 4′-O-β-D-galactopyranoside
[1812868-73-7]
$C_{17}H_{22}O_9$ 370.354
Constit. of *Clematis lasiandra*. Exhibits cytotoxicity against tumour cell lines HL-60, Hep-G2 and SGC-7901. Pale yellow solid. Mp 206 - 207°. $[\alpha]_D^{23}$ +49.6 (c, 0.1 in MeOH)

2-Methylpropyl ester
Isobutyl caffeate
[119644-12-1]
$C_{13}H_{16}O_4$ 236.267
Constit. of *Plazia daphnoides*

Heptyl ester
Heptyl caffeate
[888029-59-2]
$C_{16}H_{22}O_4$ 278.348
Constit. of the stem bark of *Flacourtia indica*. Cryst.. Mp 230 - 235°

Decyl ester
Decyl caffeate
[618455-80-4]
$C_{19}H_{28}O_4$ 320.429
Constit. of the seeds of *Phleum pratense*. Antioxidant, shows antitumour props.. λ_{max} 303 (sh); 323 (MeOH)

Undecyl ester
Daphnecinnamate B
[325148-07-0]
$C_{20}H_{30}O_4$ 334.456
Constit. of *Daphne oleoides*. Inhibitor of LPS-induced NO prodn.. Amorph. powder (Et_2O). Mp 109 - 110.5°. Mp 170 - 172°. λ_{max} 229 (log ε 3.12); 244 (log ε 3.21); 278 (log ε 4.86) (CH_2Cl_2)

Tridecyl ester
Daphnecinnamate A
[325148-06-9]
$C_{22}H_{34}O_4$ 362.51
Constit. of *Daphne oleoides*. Amorph. powder. Mp 156 - 158°. λ_{max} 224 (log ε 3.1); 243 (log ε 3.2); 280 (log ε 4.88) (CH_2Cl_2)

Pentadecyl ester
Pentadecyl caffeate
[873928-47-3]
$C_{24}H_{38}O_4$ 390.564
Constit. of *Ceratoides latens*

Hexadecyl ester
Hexadecyl caffeate
[69573-61-1]
$C_{25}H_{40}O_4$ 404.591
Constit. of *Anabasis aphylla* and Egyptian propolis

Octadecyl ester
Octadecyl caffeate
[69573-60-0]
$C_{27}H_{44}O_4$ 432.645
Constit. of *Merremia tuberosa*, *Ormosia henryi*, *Sophora prostrata* and *Artemisia argyi*. Powder. Mp 90 - 92°

Nonadecyl ester
Nonadecyl caffeate
[69573-63-3]
$C_{28}H_{46}O_4$ 446.672
Constit. of *Ceratoides latens*, *Sophora tomentosa*, *Heteroplexis microcephala* and *Sophora tomentosa*

Eicosyl ester
Eicosyl caffeate
[28593-90-0]
$C_{29}H_{48}O_4$ 460.699
Constit. of *Echinosophora koreensis*, *Sophora tomentosa*, *Heteroplexis microcephala* and *Glycyrrhiza glabra*. Elastase inhibitor. Waxy cryst.. Mp 113°

Docosyl ester
Docosyl caffeate
[28593-92-2]
[50432-89-8] (unspecified config.)
$C_{31}H_{52}O_4$ 488.753
Constit. of *Acer triflorum*, *Echinosophora koreensis*, *Glycyrrhiza glabra*, *Halocnemum strobilaceum*, *Lithospermum erythrorhizon*, *Wikstroemia scytophylla*, *Caragana conferta* and others. Elastase inhibitor. Waxy cryst.. Mp 115° (110°)

Tetracosyl ester
Tetracosyl caffeate
[156186-67-3]
$C_{33}H_{56}O_4$ 516.807
Constit. of *Anabasis aphylla*, *Gypsophila oldhamiana*, *Caesalpinia mimosoides* and *Hypericum oblongifolium*. Urease inhibitor

Hexacosyl ester
Hexacosyl caffeate
[18642-11-0]
$C_{35}H_{60}O_4$ 544.861
Constit. of *Pongamia glabra* and *Sophora subprostrata*

Octacosyl ester
Octacosyl caffeate
[329281-30-3]
$C_{37}H_{64}O_4$ 572.915
Constit. of the heartwood of *Gliricidia sepium*. Prisms ($CHCl_3$/hexane). Mp 98 - 99°

Triacontyl ester
Triacontyl caffeate
[138590-92-8]
$C_{39}H_{68}O_4$ 600.969
Constit. of *Pongamia glabra*

Hentriacontyl ester
Hentriacontyl caffeate
[654663-59-9]
$C_{40}H_{70}O_4$ 614.996
Constit. of *Hypericum laricifolium*. λ_{max} 220; 244; 299; 330 (MeOH)

Tetratriacontyl ester
Tetratriacontyl caffeate. ***Galinosoate C***

[1427086-83-6]
$C_{43}H_{76}O_4$ 657.077
Constit. of *Galinsoga parviflora*. Gummy solid. λ_{max} 209 (log ε 3.7) (MeOH)

Heptatriacontyl ester
Heptatriacontyl caffeate
[1428438-06-5]
$C_{46}H_{82}O_4$ 699.158
Constit. of *Artemisia argyi*. Amorph. powder

Vinyl ester
Vinyl caffeate
[179694-77-0]
$C_{11}H_{10}O_4$ 206.197
Constit. of *Perilla frutescens* var. *crispa*. Antioxidant. Cryst.. Mp 133 - 140°. λ_{max} 221 (ε 17000); 247 (ε 13000); 339 (ε 22000) (EtOH)

Benzyl ester
Benzyl caffeate
[130734-47-3]
$C_{16}H_{14}O_4$ 270.284
Constit. of propolis and *Populus* sp.

4-Hydroxyphenyl ester
p-Hydroxyphenyl caffeate
[1777786-74-9]
$C_{15}H_{12}O_5$ 272.256
Constit. of *Wedelia trilobata*. Exhibits *in vitro* α-glucosidase and tyrosinase inhibitory activities. Amorph. yellow powder. λ_{max} 333 (log ε 4.04) (MeOH)

[3-(4-Hydroxyphenyl)-1-methylpropyl] ester
[709002-69-7]
$C_{19}H_{20}O_5$ 328.364
Constit. of *Zuccagnia punctata*. [α] D -27 (c, 0.39 in MeOH)

[3-(3,4-Dihydroxyphenyl)-1-methylpropyl] ester
[709002-70-0]
$C_{19}H_{20}O_6$ 344.363
Constit. of *Zuccagnia punctata*. [α] D -3.6 (c, 0.25 in MeOH)

[3-(4-Hydroxy-3-methoxyphenyl)-1-methylpropyl] ester
[149492-46-6]
$C_{20}H_{22}O_6$ 358.39
Constit. of *Onosma heterophylla*. Yellow viscous solid

[3-(3,4-Dihydroxyphenyl)-2-oxopropyl] ester
Petasiphenol
[145904-50-3]
$C_{18}H_{16}O_7$ 344.319
Isol. from the scapes of *Petasites japonicum*. Bio-antimutagen

Amide, 4′-O-β-D-glucopyranoside
Fortunamide
[1365253-25-3]
$C_{15}H_{19}NO_8$ 341.316
Constit. of the rhizomes of *Drynaria fortunei*. Powder

Anhydride
Caffeic anhydride
[854237-32-4]
$C_{18}H_{14}O_7$ 342.303
Constit. of *Ehretia obtusifolia*. Amorph. yellow powder. λ_{max} 230 (log ε 4.2); 288 (log ε 3.88); 328 (log ε 3.7) (MeOH)

3′-Ac, 4′-O-β-D-glucopyranoside
[1430201-50-5]
$C_{17}H_{20}O_{10}$ 384.337
Constit. of the stems of *Alsophila spinulosa*

Di-Ac
$C_{13}H_{12}O_6$ 264.233
Cryst.. Mp 198°

3′-O-(4-Hydroxy-E-cinnamoyl), 4′-O-β-D-glucopyranoside
Ligusticumacid D
[2573174-20-4]
$C_{24}H_{24}O_{11}$ 488.445
Constit. of the rhizomes of *Ligusticum chuanxiong*. Amorph. yellow powder. $[\alpha]^{20}_{D}$ -13 (c, 0.1 in MeOH). λ_{max} 210 (log ε 4.31); 221 (log ε 4.30); 293 (log ε 4.53) (MeOH)

3′-Me ether
3-(4-Hydroxy-3-methoxyphenyl)-2-propenoic acid, H101

3′-Me ether, benzyl ester
Benzyl ferulate
[132335-97-8]
$C_{17}H_{16}O_4$ 284.311
Constit. of Korean propolis. Moderate antioxidant

4′-Me ether
3-(3-Hydroxy-4-methoxyphenyl)-2-propenoic acid, CAS. *3-Hydroxy-4-methoxycinnamic acid.* ***Isoferulic acid.*** *Hesperetic acid. Hesperetinic acid*
[25522-33-2]
$C_{10}H_{10}O_4$ 194.186
Isol. from *Cimicifuga racemosa*, *Sibiraea angustata*, *Catalpa ovata*, *Ixora javanica* and other plants. Plates; cryst. (EtOAc). Mp 227 - 230° dec.. λ_{max} 241 (sh); 291 (ε 26900); 322 (ε 27500) (MeOH) (Derep)
▶ UD3365400

4′-Me ether, 3′-O-β-D-glucopyranoside
Aphyllin. Pervoside B
[1173087-62-1]
$C_{16}H_{20}O_9$ 356.327
Constit. of *Perovskia atriplicifolia* and flowers of *Tamarix aphylla*. Amorph. powder. Mp 240 - 242°. $[\alpha]^{25}_{D}$ -46.5 (c, 0.2 in MeOH). λ_{max} 205; 218; 230; 285; 315 (MeOH)

4′-Me ether, 3′-sulfate
[342431-60-1] (K salt)
$C_{10}H_{10}O_7S$ 274.243
Constit. of the flowers of *Tamarix amplexicaulis*. Isol. as K salt

4′-Me ether, Me ester
[97966-29-5]
$C_{11}H_{12}O_4$ 208.213
Isol. from fruits of *Pseudopanax arboreum*. Prisms (hexane). Mp 78 - 79°

4′-Me ether, hexacosyl ester
Hexacosyl (E)-isoferulate. ***Erythrinasinate D.*** *Erythrinassinate D*
[132282-80-5]
$C_{36}H_{62}O_4$ 558.888
Constit. of *Erythrina eriotricha* and bark of *Abies firma*. Amorph. powder. Mp 68 - 70°

4′-Me ether, octacosyl ester
Octacosyl (E)-isoferulate. ***Erythrinasinate A.*** *Erythrinassinate A*
[102607-46-5]
$C_{38}H_{66}O_4$ 586.942
Constit. of *Erythrina glauca*, *Erythrina milbraedii*, *Erythrina senegalensis* and *Jatropha podagrica*. Cryst. (CH_2Cl_2/hexane). Mp 75 - 76°

4′-Me ether, 1-methyl-2-phenylethyl ester
[149492-45-5]
$C_{19}H_{20}O_4$ 312.365
Constit. of *Onosma heterophylla*. Yellow viscous gum

Di-Me ether
3-(3,4-Dimethoxyphenyl)-2-propenoic acid, CAS. *3,4-Dimethoxycinnamic acid,* CAS. *Dimethylcaffeic acid*
[14737-89-4]
[2316-26-9] (unspecified stereochem)
Isol. from *Piper methysticum*, *Veronica virginica*, *Cardamine komarovii*, *Hosta ensata*, *Sibiraea angusta*, *Verbesina gigantea* and

other plants. Cryst. (EtOH aq.). Mp 180 - 181.5°

Di-Me ether, α-L-rhamnopyranosyl ester

[1417327-44-6]

$C_{17}H_{22}O_8$ 354.355

Constit. of *Microlepia pilosissima*. Amorph. powder. $[\alpha]_D^{23}$ +6.2 (c, 0.33 in MeOH). λ_{max} 208 (log ε 3.73); 225 (log ε 3.36); 311 (log ε 4.00) (MeOH)

Di-Me ether, [β-D-glucopyranosyl-(1→2)-α-L-rhamnopyranosyl] ester

[1417327-42-4]

$C_{23}H_{32}O_{13}$ 516.496

Constit. of *Microlepia pilosissima*. Antioxidant. Amorph. powder. $[\alpha]_D^{23}$ +61.3 (c, 0.2 in MeOH). λ_{max} 212 (log ε 3.60); 228 (log ε 3.45); 314 (log ε 4.45) (MeOH)

Di-Me ether, Me ester

[30461-77-9]

$C_{12}H_{14}O_4$ 222.24

Isol. from *Pseudopanax arboreum*, *Sideritis* spp. and *Securidaca longipedunculata*. Needles or prisms (hexane). Mp 63 - 64° (needles). Mp 68 - 70° (prisms)

Di-Me ether, 3-phenyl-2-propenyl ester

[391682-57-8]

$C_{20}H_{20}O_4$ 324.376

Constit. of Korean propolis

Di-Me ether, amide

[130973-10-3]

$C_{11}H_{13}NO_3$ 207.229

Solid. Mp 162 - 164°

Di-Me ether, nitrile

[37629-85-9]

[6443-72-7] (unspecified stereochem.)

$C_{11}H_{11}NO_2$ 189.214

Solid. Mp 97 - 99°

Di-Me ether, anhydride

[88909-07-3]

$C_{22}H_{22}O_7$ 398.411

Solid. Mp 178.5 - 179.8°

(Z)-form

Isocaffeic acid [4361-87-9]

Found in various plants

Eicosyl ester

[1389315-92-7]

$C_{29}H_{48}O_4$ 460.699

Constit. of *Anabasis aphylla*. Amorph. powder

Benzyl ester

[115610-24-7]

Constit. of propolis and *Populus* sp.

Di-Me ether

[14737-88-3]

$C_{11}H_{12}O_4$ 208.213

Cryst. (Et_2O). Mp 104°

Di-Me ether, Me ester

[30461-78-0]

$C_{12}H_{14}O_4$ 222.24

Isol. from uredospores of the stripe rust fungus *Puccinia striiformis* and oat crown fungus *Puccinia coronata avenae* and other *Puccinia* spp. Constit. of *Aragoa lucidula*, *Sideritis* spp. and roots of *Securidaca longipedunculata*. Germination self-inhibitor in *Puccinia*. Cryst. (Et_2O). Mp 92 - 93°

(ξ)-form

Di-Me ether, 2-methylpropylamide

3,4-Dimethoxycinnamic acid isobutylamide. *N-(3,4-Dimethoxycinnamoyl)isobutylamide*

[102934-28-1]

$C_{15}H_{21}NO_3$ 263.337

Constit. of *Piper amalago*. Config. not determined.

Aldrich Library of FT-IR Spectra, 1st edn., 1985, **2**, 184B , 184D (*ir*).

Aldrich Library of ^{13}*C and* ^{1}*H FT-NMR Spectra*, 1992, **2**, 1057C , 1058B (*nmr*).

Hermann, K. *Pharmazie*, 1956, **11**, 433-449 (*rev*).

Bate-Smith, E.C. *Sci. Proc. R. Dublin Soc.*, 1956, **27**, 165-176 (*occur*).

Klosterman, H.J. *et al.*, *J. Am. Chem. Soc.*, 1959, **81**, 2188-2191 (*Linocaffein*).

Alder, E. *et al.*, *Acta Chem. Scand.*, 1963, **17**, 27-36 (*Z-di-Me ether, Z-di-Me ether Me ester*).

Jizba, J. *et al.*, *Coll. Czech. Chem. Comm.*, 1967, **32**, 2867-2874 (*Glucocaffeic acid*).

Dewick, P.M. *et al.*, *Chem. Commun. (London)*, 1968, 673-675 (*biosynth*).

Swain, T. *et al.*, *Phytochemistry*, 1970, **9**, 2115-2122 (*biosynth*).

Achenbach, H. *et al.*, *Chem. Ber.*, 1971, **104**, 1468-1477 (*di-Me ether, Piper methysticum constit*).

Bowden, B.F. *et al.*, *Aust. J. Chem.*, 1975, **28**, 91-107 (*Pseudopanax arboreum constits*).

Kelley, C.J. *et al.*, *J. Org. Chem.*, 1976, **41**, 449-455 (*cmr*).

Karrer, W. *Konstitution und Vorkommen der Organischen Pflanzenstoffe, 2nd edn., Birkhäuser Verlag*, 1976-1985, no. 957.

Macko, V. *et al.*, *Phytopathology*, 1977, **67**, 1473-1474 (*di-Me ether Me ester, isol, activity*).

De Silva, S.O. *et al.*, *Can. J. Chem.*, 1979, **57**, 1598-1605 (*E-di-Me ether, synth*).

Imperato, F. *Chem. Ind. (London)*, 1979, 553-554 (*1-O-Caffeoyllaminaribiose*).

Tschesche, R. *et al.*, *Phytochemistry*, 1980, **19**, 2783 (*4-rhamnosylglucoside*).

Baba, K. *et al.*, *Chem. Pharm. Bull.*, 1981, **29**, 2182-2187 (*Hesperetic acid*).

Tsurushima, T. *et al.*, *Agric. Biol. Chem.*, 1984, **48**, 2147-2148 (*di-Me ether Me ester, isol, activity*).

Fomum, Z.T. *et al.*, *Phytochemistry*, 1986, **25**, 757-759 (*Erythrinasinate A*).

Achenbach, H. *et al.*, *Planta Med.*, 1986, **52**, 12-18 (*ξ-di-Me ether isobutylamide*).

García-Granda, S. *et al.*, *Acta Crystallogr., Sect. C*, 1987, **43**, 683-685 (*cryst struct*).

Kreis, W. *Antiviral Res.*, 1990, **14**, 323-337 (*anti-HIV activity*).

Bankova, V. *J. Nat. Prod.*, 1990, **53**, 821-824 (*caffeates*).

Saha, M.M. *et al.*, *Phytochemistry*, 1991, **30**, 3834-3836 (*Pongamia glabra esters*).

Iriye, R. *et al.*, *Biosci., Biotechnol., Biochem.*, 1992, **56**, 1773-1775 (*Petasiphenol*).

IARC Monogr., 1993, **56**, 115 (*Caffeic acid, rev, tox*).

Mellidis, A.S. *et al.*, *J. Nat. Prod.*, 1993, **56**, 949-952 (*Onosma heterophylla esters*).

Nyandat, E. *et al.*, *Phytochemistry*, 1993, **33**, 1493-1496 (*Glucocaffeic acid*).

Yaoita, Y. *et al.*, *Phytochemistry*, 1994, **37**, 1773-1774 (*Petasiphenol*).

Blase, F.R. *et al.*, *Synth. Commun.*, 1995, **25**, 3187-3197 (*synth, Isoferulic acid*).

Tada, M. *et al.*, *Biosci., Biotechnol., Biochem.*, 1996, **60**, 1093-1095 (*vinyl ester, isol*).

Nkengfack, A.E. *et al.*, *Phytochemistry*, 1997, **46**, 573-578 (*Erythrinasinate D*).

Kavitha, J. *et al.*, *Indian J. Chem., Sect. B*, 1999, **38**, 1280-1281 (*vinyl ester, synth*).

Wang, M. *et al.*, *J. Nat. Prod.*, 1999, **62**, 454-456 (*apiosylglucosyl ester*).

Gibbons, S. *et al.*, *Phytochemistry*, 1999, **51**, 465-467 (*Docosyl caffeate*).

Garcia-Argaez, A.N. *et al.*, *Planta Med.*, 1999, **65**, 678-679 (*octadecyl ester*).

Herath, H.M.T.B. *et al.*, *Fitoterapia*, 2000, **71**, 722-724 (*octacosyl ester*).

El-Moussallami, A.M.D. *et al.*, *Nat. Prod. Sci.*, 2000, **6**, 193-198 (*Isoferulic acid 3'-sulfate*).

Löser, B. *et al.*, *Planta Med.*, 2000, **66**, 751-753 (*activity*).

Ullah, N. *et al.*, *Z. Naturforsch., B*, 2000, **55**, 982-984 (*Daphnecinnamates*).

Haadsma-Svensson, S.R. *et al.*, *J. Med. Chem.*, 2001, **44**, 4716-4732 (*di-Me ether, synth, pmr, ir, ms*).

El-Seedi, H.R. *et al.*, *Chem. Pharm. Bull.*, 2003, **51**, 1439-1440 (*hentriacontyl ester*).

Hoeneisen, M. *et al.*, *Z. Naturforsch., C*, 2003, **58**, 39-41 (*isobutyl ester, prenyl ester*).

Svetaz, L. *et al.*, *J. Agric. Food Chem.*, 2004, **52**, 3297-3300 (*hydroxyphenyl-methylpropyl esters*).

Li, S. *et al.*, *Xibei Zhiwu Xuebao*, 2004, **24**, 872-873 (*pentadecyl ester, nonadecyl ester*).

Iqbal, K. *et al.*, *Chem. Biodiversity*, 2005, **2**, 104-111 (*anhydride*).

Sarker, S.D. *et al.*, *Chem. Nat. Compd. (Engl. Transl.)*, 2005, **41**, 293-296 (*decyl ester*).

Braham, H. *et al.*, *J. Nat. Prod.*, 2005, **68**, 517-522 (*Glucosyl glucocaffeate*).

Hamerski, L. *et al.*, *Phytochemistry*, 2005, **66**, 1927-1932 (*Hydrangeifolin II*).

Cousin, D. *et al.*, *Org. Biomol. Chem.*, 2006, **4**, 54-62 (*4'-Me ether Me ester*).

Schwarz, B. *et al.*, *J. Agric. Food Chem.*, 2007, **55**, 1394-1404 (*Glucocaffeic acid, Ribes rubrum constit*).

Uwai, K. *et al.*, *Bioorg. Med. Chem.*, 2008, **16**, 7795-7803 (*Daphnecinnamate B, activity*).

Perveen, S. *et al.*, *J. Asian Nat. Prod. Res.*, 2008, **10**, 1105-1108 (*Pervoside B*).
Zhu, T.F. *et al.*, *Molecules*, 2008, **13**, 729-735 (*Scrocaffeside B*).
Linder, J. *et al.*, *Org. Biomol. Chem.*, 2008, **6**, 3908-3916 (*di-Me ether amide, di-Me ether nitrile*).
Madan, S. *et al.*, *Nat. Prod. Commun.*, 2009, **4**, 381-384 (*heptyl ester*).
Dey, S. *et al.*, *Nat. Prod. Res.*, 2009, **23**, 1657-1663 (*eicosyl ester, docosyl ester*).
Nawwar, M.A.M. *et al.*, *Pharmazie*, 2009, **64**, 342-347 (*Aphyllin*).
Wang, L. *et al.*, *Acta Crystallogr., Sect. E*, 2011, **67**, o354 (*Me ester, cryst struct*).
Trinh, P.T.N. *et al.*, *Coll. Czech. Chem. Comm.*, 2011, **76**, 1133-1139 (*Fortunamide*).
Li, W. *et al.*, *Zhongcaoyao*, 2011, **42**, 837-840 (*Anabasis aphylla esters*).
Hu, J. *et al.*, *Arch. Pharmacal Res.*, 2012, **35**, 2127-2133 (*Microlepia pilosissima di-Me ether constits*).
Afza, N. *et al.*, *J. Asian Nat. Prod. Res.*, 2012, **14**, 424-428 (*Galinosoate C*).
Paz, C. *et al.*, *Acta Crystallogr., Sect. C*, 2013, **69**, 765-769 (*Et ester, isol, cryst struct*).
Wu, S.-L. *et al.*, *Asian J. Chem.*, 2013, **25**, 2317-2318 (*Alsophila spinulosa glucoside*).
Lv, J.-L. *et al.*, *Chem. Nat. Compd. (Engl. Transl.)*, 2013, **49**, 8-11 (*heptatriacontyl ester*).
Lee, I.-K. *et al.*, *Bioorg. Med. Chem. Lett.*, 2014, **24**, 3503-3505 (*Korean propolis constits, activity*).
Afza, N. *et al.*, *Nat. Prod. Commun.*, 2014, **9**, 1171-1172 (*Parviside B*).
Tian, X.-R. *et al.*, *Phytochem. Lett.*, 2014, **10**, 168-172 (*Et ester 4'-galactoside, activity*).
Zhang, L.-J. *et al.*, *J. Nat. Prod.*, 2015, **78**, 1586-1592 (*Clematisides A,B,C,E,F, activity*).
Junqueira-Goncalves, M.P. *et al.*, *Molecules*, 2015, **20**, 5698-5713 (*3'-glucoside*).
Stojakowska, A. *et al.*, *Phytochem. Lett.*, 2015, **12**, 257-261 (*Telekia speciosa glucosides*).
Damasceno, S.S. *et al.*, *Curr. Pharm. Des.*, 2017, **23**, 3015-3023 (*rev, biol, activity*).
Zhou, Z.-Q. *et al.*, *Food Chem.*, 2017, **214**, 644-654 (*Lycibarbarphenylpropanoid I*).
Ren, H. *et al.*, *Phytochem. Lett.*, 2017, **19**, 18-22 (*4-hydroxyphenyl ester, activity*).
Raja, G.C.E. *et al.*, *Synth. Commun.*, 2017, **47**, 2449-2455 (*E-di-Me ether anhydride*).
Aguilar-Hernández, I. *et al.*, *Vib. Spectrosc.*, 2017, **89**, 113-122 (*Raman*).
Zhang, X. *et al.*, *Fitoterapia*, 2018, **125**, 147-154 (*Ligusticumacid D*).
Lewis, R.J. *Sax's Dangerous Properties of Industrial Materials, 10th edn., J. Wiley*, 2000, CAK375

3-(3,5-Dihydroxyphenyl)-2-propenoic acid, CAS D244

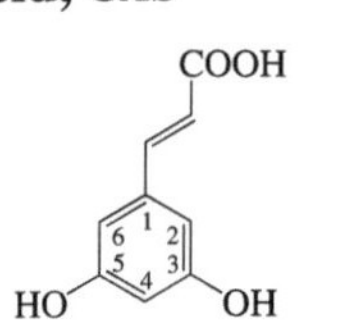

3,5-Dihydroxycinnamic acid, CAS
[28374-93-8]
$C_9H_8O_4$ 180.159
Structs. in this series may need revision. Phenyl 3,5-dioxygenated substn. pattern in doubt as a result of work done by Yang *et al* in 2015

(E)-form [127791-54-2]
Constit. of *Acanthopanax sessiliflorus* and *Prunus persica*. Plant growth inhibitor. Needles (H_2O). Mp 245 - 246°

O-β-D-Glucopyranoside
Grevilloside F
[1094004-50-8]
$C_{15}H_{18}O_9$ 342.3
Constit. of the leaves of *Grevillea robusta*. Amorph. powder. $[\alpha]_D^{27}$ -58.9 (c, 0.52 in MeOH). λ_{max} 222 (log ε 4.2); 277 (log ε 4) (MeOH)

3-O-[6-O-Acetyl-β-D-glucopyranoside]
[1496508-22-5]
$C_{17}H_{20}O_{10}$ 384.337
Constit. of *Saussurea graminea*. Amorph. powder. λ_{max} 225 (log ε 3.44); 310 (log ε 3.78) (MeOH)

Me ester
[351343-42-5]
[63631-37-8] (unspecified stereochem.)
$C_{10}H_{10}O_4$ 194.186
Isol. from *Cuscuta lehmanniana*. Mp 162 - 163°

Me ester, O-β-D-glucopyranoside
[851232-94-5]
$C_{16}H_{20}O_9$ 356.327
Constit. of the flowers of *Moricandia arvensis*. Powder. Mp 195 - 197°. $[\alpha]_D^{25}$ +100 (c, 0.2 in MeOH). λ_{max} 215 (log ε 3.9); 230 (log ε 3.1); 290 (log ε 4) (MeOH)

Octacosyl ester
[1092975-37-5]
$C_{37}H_{64}O_4$ 572.915
Constit. of the stems of *Caesalpinia decapetala*

3-Me ether
3-(3-Hydroxy-5-methoxyphenyl)-2-propenoic acid. 3-Hydroxy-5-methoxycinnamic acid

3-Me ether, 5-O-(4-O-methyl-β-D-glucopyranoside)
[934008-84-1]
$C_{17}H_{22}O_9$ 370.354
Isol. from *Cordyceps cicadae* and *Cordyceps ninchukispora*. Amorph. powder. λ_{max} 216 (log ε 3.86); 225 (log ε 3.8); 281 (log ε 3.85); 308 (log ε 3.76) (MeOH)

3-Me ether, Me ester, 5-O-β-D-glucopyranoside
Linusitamarin
[152574-10-2]
$C_{17}H_{22}O_9$ 370.354
Constit. of *Linum usitatissimum*. Amorph. powder

Di-Me ether
3,5-Dimethoxycinnamic acid
[20767-04-8]
[16909-11-8] (unspecified stereochem.) [149243-19-6] (*Z*-form)
$C_{11}H_{12}O_4$ 208.213
Needles (H_2O). Mp 175 - 176°. pK_{a1} 4.36 (25°)

Di-Me ether, pyrrolidide
N-(3,5-Dimethoxycinnamoyl)pyrrolidine. *3,5-Dimethoxycinnamic acid pyrrolidide*
[807372-38-9]
[102934-29-2] (unspecified stereochem.)
$C_{15}H_{19}NO_3$ 261.321
Alkaloid from *Piper amalago*.

Aldrich Library of FT-IR Spectra, 1st edn., 1985, **2**, 184A (*ir*).
Aldrich Library of ^{13}C and ^{1}H FT-NMR Spectra, 1992, **2**, 1057B (*nmr*).
Mauthner, F. *J. Prakt. Chem.*, 1925, **110**, 126-128 (*synth*).
Altree-Williams, S. *et al.*, *Aust. J. Plant Physiol.*, 1975, **2**, 105-109 (*Prunus persica constit*).
Mukherjee, J. *et al.*, *Indian J. Chem.*, 1975, **13**, 859-860 (*di-Me ether, synth*).
Kamilov, Kh.M. *et al.*, *Chem. Nat. Compd. (Engl. Transl.)*, 1977, 99-100 (*E-Me ester, isol*).
Achenbach, H. *et al.*, *Planta Med.*, 1986, **52**, 12-18 (*di-Me ether pyrrolidide*).
Luyengi, L. *et al.*, *J. Nat. Prod.*, 1993, **56**, 2012-2015 (*Linusitamarin*).
Braham, H. *et al.*, *J. Nat. Prod.*, 2005, **68**, 517-522 (*Me ester glucoside*).
Zhang, S.-W. *et al.*, *Helv. Chim. Acta*, 2007, **90**, 404-410 (*3-Me ether 4-methylglucoside*).
Zhang, Q. *et al.*, *Chin. J. Nat. Med.*, 2008, **6**, 168-171 (*octacosyl ester*).
Yamashita, Y. *et al.*, *Phytochemistry*, 2008, **69**, 2749-2752 (*Grevilloside F*).
In, S.-J. *et al.*, *J. Appl. Biol. Chem.*, 2012, **55**, 217-220 (*Acanthopanax sessiliflorus constit*).
Hu, J. *et al.*, *Chem. Nat. Compd. (Engl. Transl.)*, 2013, **49**, 819-821 (*3-6-acetylglucoside*).
Yang, Y.-N. *et al.*, *J. Nat. Prod.*, 2015, **78**, 705-711 (*pmr, struct*)

1-(3,4-Dihydroxyphenyl)-2-propen-1-ol D245

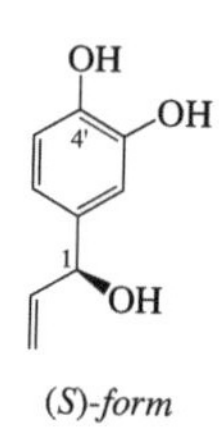

(S)-form

1′,3-Dihydroxychavicol
$C_9H_{10}O_3$ 166.176
(S)-form
3′-Me ether
1-(4-Hydroxy-3-methoxyphenyl)-2-propen-1-ol. ***1′-Hydroxyeugenol***
$C_{10}H_{12}O_3$ 180.203
Found in Shenqi Fuzheng Injection, a traditional Chinese herb formula. Amorph. powder. $[\alpha]^{25}_{D}$ -10.8 (c, 0.81 in MeOH). λ_{max} 204 (log ε 4.20); 230 (log ε 3.76); 277 (log ε 3.20); 302 (log ε 3.08) (MeOH)
3′-Me ether, 1,4′-di-Ac
1′-Acetoxyeugenol acetate
[52946-23-3]
[53890-24-7] (unspecified stereochem.) [108093-85-2] (unspecified stereochem.)
$C_{14}H_{16}O_5$ 264.277
Constit. of *Alpinia galanga*. Antiinflammatory and cytotoxic agent. Melanogenesis inhibitor. Oil. $[\alpha]^{20}_{D}$ -17.8 (c, 0.44 in EtOH)
(ξ)-form
3′-Me ether, 4′-Ac
$C_{12}H_{14}O_4$ 222.24
Constit. of *Alpinia galanga*
3′-Me ether, 4′-O-(2-methylpropanoyl), 1-Ac
1′-Acetoxy-4-O-isobutyryleugenol
[142824-99-5]
[20665-84-3] [142824-48-4]
$C_{16}H_{20}O_5$ 292.331
Constit. of the roots of *Cosmos caudatus*, *Coreopsis lanceolata* and *Coreopsis tinctoria*. Antifungal agent. Dark yellow oil. Bp 130° at 0.005 mm. $[\alpha]^{25}_{D}$ -81 (c, 0.7 in $CHCl_3$). λ_{max} 219 (ε 7080); 273 (ε 2344) (MeOH) (Berdy)
3′-Me ether, 1,4′-bis-O-(2-methylpropanoyl)
[55249-41-7]
$C_{18}H_{24}O_5$ 320.385
Constit. of *Bidens* sp., *Coreopsis* spp. and *Cosmos pringlei*. Oil. Bp 130° at 0.005 mm
3′-Me ether, 4′-O-(2-methylpropanoyl), 1-O-(2-methylbutanoyl)
[120692-49-1]
$C_{19}H_{26}O_5$ 334.412
Constit. of *Coreopsis* spp.
3′-Me ether, 4′-O-(2-methylpropanoyl), 1-O-(3-methylbutanoyl)
[120692-50-4]
$C_{19}H_{26}O_5$ 334.412
Constit. of *Coreopsis* spp. and *Cosmos pringlei*. Oil. $[\alpha]^{25}_{D}$ -150 (c, 1 in $CHCl_3$). λ_{max} 218 (log ε 3.9); 273 (log ε 3.4) (MeOH)
3′,4′-Di-Me ether
1-(3,4-Dimethoxyphenyl)-2-propen-1-ol
$C_{11}H_{14}O_3$ 194.23
Constit. of *Anthriscus sylvestris*. Pale yellow oil. $[\alpha]^{25}_{D}$ +1.46 (c, 1 in MeOH)
Tri-Me ether
1,2-dimethoxy-4-(1-methoxy-2-propenyl)benzene. 1,2-dimethoxy-4-(1-methoxyallyl)benzene. 3-(3,4-Dimethoxyphenyl)-3-methoxy-1-propene. ***Morinin G***
[84744-55-8]
$C_{12}H_{16}O_3$ 208.257
Constit. of the leaves of *Bupleurum fruticosum* and of *Morina chinensis*. Insect antifeedant. Yellowish oil. $[\alpha]^{25}_{D}$ +6.5. $[\alpha]^{25}_{D}$ -1.2 (c, 1.38 in $CHCl_3$). λ_{max} 243 (log ε 3.59); 280 (log ε 3.35) ($CHCl_3$).

Bohlmann, F. *et al., Chem. Ber.*, 1968, **101**, 3243-3254 (*Coreopsis nuecensis constit, synth*).
Noro, T. *et al., Chem. Pharm. Bull.*, 1988, **36**, 244-248 (*1′-Acetoxyeugenol acetate, isol, activity*).
Thron, U. *et al., Z. Naturforsch., C*, 1989, **44**, 7, 573 - 577 (*Coreopsis tinctoria constit, pmr, cmr, ms*).
Derigner, H.P. *et al., Arzneim.-Forsch.*, 1994, **44**, 956-961 (*3′-Me ether 4′-(2-methylpropanoyl) 1-Ac, synth*).
Pistelli, L. *et al., J. Nat. Prod.*, 1995, **58**, 112-116 (*tri-Me ether, isol*).
Fuzzati, N. *et al., Phytochemistry*, 1995, **39**, 409-412 (*Cosmos caudata constit, struct*).
Ikeda, R. *et al., Chem. Pharm. Bull.*, 1998, **46**, 875-878 (*3′,4′-di-Me ether*).
Su, B.N. *et al., J. Nat. Prod.*, 1999, **62**, 1363-1366 (*Morinin G, isol*).
Mata, R. *et al., J. Nat. Prod.*, 2002, **65**, 1030-1032 (*Cosmos pringlei esters, isol*).
Masuda, H. *et al., Bioorg. Med. Chem. Lett.*, 2005, **15**, 1949-1954 (*3′-Me 1,4-di-Ac, activity*).
Morikawa, T. *et al., Chem. Pharm. Bull.*, 2005, **53**, 625-630 (*3′-Me 1,4-di-Ac, activity*).
Oonmetta-aree, J. *et al., LWT - Food Sci. Technol.*, 2006, **39**, 1214-1220 (*Alpinia galanga constits, activity*).
Zhu, X. *et al., Zhongguo Xiandai Zhongyau*, 2008, **10**, 13-15 (*Alpinia galanga constits, props*).
Zhang, S.-Y. *et al., J. Asian Nat. Prod. Res.*, 2015, **17**, 274-279 (*3′-Me ether*).
Manse, Y. *et al., Bioorg. Med. Chem.*, 2016, **24**, 6215-6224 (*1′-Acetoxyeugenol acetate, synth, activity*)

3-(3,4-Dihydroxyphenyl)-2-propen-1-ol D246

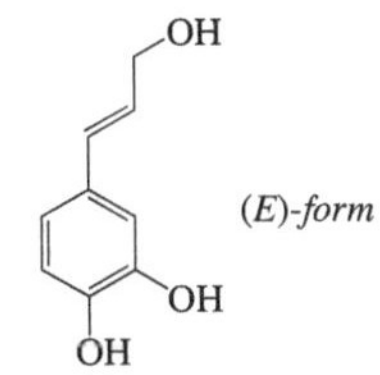

4-(3-Hydroxy-1-propenyl)-1,2-benzenediol, CAS. ***Caffeyl alcohol.*** *3,4-Dihydroxycinnamyl alcohol*
[3598-26-3]
$C_9H_{10}O_3$ 166.176
(*E*)-Config. assumed for derivs. except where (*Z*)-config. is certain
(E)-form
Mp 144 - 145°
1-O-[β-D-Glucopyranosyl-(1→4)-α-L-rhamnopyranosyl-(1→3)-[3,4-dihydroxy-E-cinnamoyl-(→4)]-β-D-glucopyranoside]
Rossicaside E
[869810-52-6]
$C_{36}H_{46}O_{20}$ 798.744
Constit. of *Boschniakia rossica*. Amorph. powder. $[\alpha]^{25}_{D}$ -100 (c, 0.21 in MeOH). λ_{max} 250 (log ε 4.1); 306 (log ε 4.2); 331 (log ε 4.2) (MeOH)
1-O-[β-D-Glucopyranosyl-(1→4)-α-L-rhamnopyranosyl-(1→3)-[3,4-dihydroxy-E-cinnamoyl-(→6)]-β-D-glucopyranoside]
Rossicaside K
[1528759-68-3]
$C_{36}H_{46}O_{20}$ 798.744
Constit. of *Boschniakia rossica*. Brown powder. $[\alpha]^{25}_{D}$ -41 (c, 0.40 in MeOH). λ_{max} 288 (log ε 3.84); 324 (log ε 3.83) (MeOH)
3′-O-β-D-Glucopyranoside
[1303610-03-8]
$C_{15}H_{20}O_8$ 328.317
Constit. of the corms of *Remusatia vivipara*. Amorph. powder. $[\alpha]^{25}_{D}$ -35.2 (c, 0.12 in MeOH). λ_{max} 263 (log ε 3.54) (MeOH)
1-Me ether
4-(3-Methoxy-1-propenyl)-1,2-benzenediol. 1-(3,4-Dihydroxyphenyl)-3-methoxy-1-propene
$C_{10}H_{12}O_3$ 180.203
Constit. of stems of *Acanthopanax divaricatus*. Pale yellow solid

1-Me ether, 3′-Ac
Taiwanensol A
[1610427-01-4]
$C_{12}H_{14}O_4$ 222.24
Constit. of the roots of *Piper taiwanense*. Mixt. of Taiwanensols A and B show potent antitubercular activity (against *Mycobacterium tuberculosis* H37Rv) and moderate inhibitory activity against anti-platelet aggregation induced by collagen. Yellow oil. Obt. as a mixt. with Taiwanensol B, to which the props. refer. λ_{max} 265 (log ε 3.57); 305 (log ε 3.19) (MeOH)

1-Me ether, 4′-Ac
Taiwanensol B
[1610427-02-5]
$C_{12}H_{14}O_4$ 222.24
Constit. of the roots of *Piper taiwanense*. Obt. as a mixt. with Taiwanensol A

3′-Me ether
3-(4-Hydroxy-3-methoxyphenyl)-2-propen-1-ol. 4-(3-Hydroxy-1-propenyl)-2-methoxyphenol, CAS.
Coniferyl alcohol. *Lubanol*
[32811-40-8]
[458-35-5] (unspecified stereochem.)
$C_{10}H_{12}O_3$ 180.203
Lignin degradn. substance found in sulfite liquor and in Siam benzoin gum. Predominant monomer of Lignin in softwood together with *p*-Coumaryl alcohol, in H128 and Sinapyl alcohol, in H81 and in hardwood with Sinapyl alcohol. Sex pheromone of the bee *Apis mellifera* and the fruit fly *Batrocera dorsalis*. Constit. of stems of *Acanthopanax divaricatus*. Prisms. Mp 78° (74-75°). Bp 163 - 165° at 3 mm. pK_{a1} 9.54 (25°). Resinified by acids

3′-Me ether, 1-O-β-D-glucopyranoside
Citrusin D. *Isoconiferin*
[65995-51-9]
$C_{16}H_{22}O_8$ 342.344
Isol. from *Citrus limon*, *Citrus unshiu*, *Fortunella japonica* and *Pinus sylvestris*. Shows antihypertensive props.. $[\alpha]^{20}_{D}$ -16.9 (c, 4.1 in MeOH). λ_{max} 275 (ε 5000) (MeOH)

3′-Me ether, 1-O-[β-D-apiofuranosyl-(1→6)-β-D-glucopyranoside]
[825651-69-2]
$C_{21}H_{30}O_{12}$ 474.459
Constit. of the seeds of *Punica granatum*. Antioxidant. Powder. Mp 276 - 278°. $[\alpha]^{20}_{D}$+8.7 (c, 0.5 in H_2O). λ_{max} 265 (log ε 2.29) (H_2O)

3′-Me ether, 1-O-[α-L-arabinofuranosyl-(1→6)-β-D-glucopyranoside]
Junipercomnoside E
[828921-62-6]
$C_{21}H_{30}O_{12}$ 474.459
Constit. of *Juniperus communis* var. *depressa*. Powder. $[\alpha]_D$ -45.3 (c, 0.47 in MeOH)

3′-Me ether, 1-O-[β-D-glucopyranosyl-(1→3)-β-D-glucopyranoside]
Lycibarbarphenylpropanoid E
[1988786-77-1]
$C_{22}H_{32}O_{13}$ 504.485
Constit. of the fruit of *Lycium barbarum*. Amorph. yellowish powder. $[\alpha]^{25}_{D}$-43.3 (c, 0.5 in MeOH). λ_{max} 209 (log ε 4.40); 268 (log ε 4.16); 302 (log ε 3.93); 335 (log ε 3.59) (MeOH)

3′-Me ether, 1-O-[β-D-glucopyranosyl-(1→6)-β-D-glucopyranoside]
Isoconiferoside
[1347015-00-2]
$C_{22}H_{32}O_{13}$ 504.485
Constit. of the seeds of *Panax ginseng*. Powder. $[\alpha]^{16}_{D}$-37.5 (c, 0.29 in MeOH)

3′-Me ether, 1-O-(3-methylbutanoyl)
Coniferyl isovalerate
[62706-32-5]
[77355-78-3] (unspecified stereochem.)
$C_{15}H_{20}O_4$ 264.321
Constit. of *Coreopsis parvifolia*, *Eremanthus incanus* and *Lychnophora* sp.. Oil

3′-Me ether, 1-angeloyl
[95925-04-5]
$C_{15}H_{18}O_4$ 262.305
Constit. of *Diplostephium floribundum* and *Solidago multiradiata*

3′-Me ether, 1-hexadecanoyl
Coniferyl palmitate
[156332-23-9]
$C_{26}H_{42}O_4$ 418.618
Constit. of the peel of *Cydonia oblonga* and the roots of *Platycodon grandiflorum*. Antioxidant. λ_{max} 262 (log ε 1.3); 300 (sh) (MeOH)

3′-Me ether, 1-O-(8-methylnonanoyl)
Coniferyl 8-methylnonanoate.
Dihydrocapsiconiate
[946572-74-3]
Constit. of the fruit of *Capsicum baccatum* var. *praetermissum*. Oil. λ_{max} 268 (log ε 3.9); 297 (log ε 3.6) (MeOH)

3′-Me ether, 1-O-(8-methyl-6E-nonenoyl)
Coniferyl 8-methyl-6E-nonenoate.
Capsiconiate
[946572-73-2]
$C_{20}H_{28}O_4$ 332.44
Constit. of the fruit of *Capsicum baccatum* var. *praetermissum*. Oil. λ_{max} 268 (log ε 3.9); 297 (log ε 3.6) (MeOH)

3′-Me ether, 1-octadecanoyl
Coniferyl stearate
[684284-26-2]
$C_{28}H_{46}O_4$ 446.672
Constit. of *Cydonia oblonga* and *Opopanax chironium*. Powder. Mp 89°

3′-Me ether, 1-O-(9Z-octadecenoyl)
Coniferyl oleate
[813466-10-3]
$C_{28}H_{44}O_4$ 444.656
Constit. of the roots of *Platycodon grandiflorum*. Antioxidant. λ_{max} 262 (log ε 1.3); 300 (sh) (MeOH)

3′-Me ether, 1-O-(4Z-decenoyl)
Coniferyl obtusilate
[1595287-61-8]
$C_{20}H_{28}O_4$ 332.440
Constit. of a *Saussurea* sp.

3′-Me ether, 1-benzoyl
Lubanol benzoate
[91599-11-0]
[4159-29-9] (unspecified stereochem.)
$C_{17}H_{16}O_4$ 284.311
Isol. from Siam benzoin. Cryst. (Et_2O/petrol). Mp 72°

3′-Me ether, 1-cinnamoyl
Coniferyl cinnamate
[89926-68-1]
$C_{19}H_{18}O_4$ 310.349
Constit. of *Coreopsis nodosa*. Component of An Xi Xiang

3′-Me ether, 1-O-(4-hydroxy-3-methoxy-E-cinnamoyl)
Coniferyl ferulate
[62706-35-8]
[63644-62-2] [123821-69-2]
$C_{20}H_{20}O_6$ 356.374
Constit. of *Coreopsis longula*, *Cnidium officinale* and *Ligusticum wallichii*. Oil

4'-Me ether, 1-O-[5-O-E-cinnamoyl-β-D-apiofuranosyl-(1→3)-β-D-glucopyranoside]
Visartiside F
[1212005-22-5]
$C_{30}H_{36}O_{13}$ 604.605
Constit. of *Viscum articulatum*. Yellow powder. $[\alpha]^{25}_{D}$ -34.8 (c, 1.35 in MeOH). λ_{max} 275 (log ε 2.5) (McOH)

1-Et, 3'-Me ether
4-(3-Ethoxy-1-propenyl)-2-methoxyphenol
[94930-75-3]
$C_{12}H_{16}O_3$ 208.257
Constit. of the resin of *Styrax tonkinensis*. Pale yellow oil

1-O-[2-(4-β-D-Glucopyranosyloxy-3-methoxyphenyl)ethyl], 3'-Me ether
[412029-05-1]
$C_{25}H_{32}O_{10}$ 492.521
Constit. of cane sugar. Amorph. powder. λ_{max} 270; 300 (sh) (MeOH)

1-O-[2-(4-β-D-Glucopyranosyloxy-3,5-dimethoxyphenyl)ethyl], 3'-Me ether
[412029-04-0]
$C_{26}H_{34}O_{11}$ 522.547
Constit. of cane sugar. Amorph. powder. λ_{max} 268; 300 (sh) (MeOH)

(Z)-form

3'-Me ether
(Z)-Coniferyl alcohol
[69056-21-9]
Constit. of various plant spp.. Pale yellow cryst. (CH_2Cl_2/petrol). Mp 72 - 74°. Mp 106.6 - 107.3°

3'-Me ether, 1-O-β-D-glucopyranoside
Faguside. *cis-Isoconiferin*
[66791-74-0]
$C_{16}H_{22}O_8$ 342.344
Isol. from bark of *Fagus silvatica* and *Fagus grandifolia*. Antifeedant against insect pests. $[\alpha]^{33.5}_{D}$ -27.6 (c, 0.13 in MeOH)

3'-Me ether, 1-O-(9Z,12Z-octadecadienoyl)
[1019641-78-1]
$C_{28}H_{42}O_4$ 442.64
Constit. of the peel of *Cydonia oblonga*.

Aldrich Library of FT-IR Spectra, 1st edn., 1985, **1**, 1124B (*ir*).
Aldrich Library of [13]*C and* [1]*H FT-NMR Spectra*, 1992, **2**, 334A (*nmr*).
Tiemann, F. *et al.*, *Ber. Dtsch. Chem. Ges.*, 1878, **11**, 659-675 (*synth*).
Zincke, A. *et al.*, *Monatsh. Chem.*, 1921, **41**, 423-441 (*Coniferyl alcohol*).
Allen, C.F.H. *et al.*, *J. Am. Chem. Soc.*, 1949, **71**, 2683-2684 (*synth*).
Freudenberg, K. *et al.*, *Chem. Ber.*, 1950, **83**, 600-604 (*Coniferyl alcohol, props*).
Hermann, K. *Pharmazie*, 1953, **8**, 303-305 (*Caffeyl alcohol*).
Freudenberg, K. *et al.*, *Chem. Ber.*, 1955, **88**, 10-16 (*Coniferyl alcohol, props*).
Brunow, G. *Acta Chem. Scand.*, 1972, **26**, 1123-1129 (*Coniferyl alcohol*).
Harmatha, J. *et al.*, *Coll. Czech. Chem. Comm.*, 1978, **43**, 774-780 (*Faguside*).
Campbell, W.E. *et al.*, *Phytochemistry*, 1982, **21**, 1455-1456 (*Me ethers, isol, synth*).
Bohlmann, F. *et al.*, *Phytochemistry*, 1983, **22**, 2858-2859 (*cinnamate*).
Metwally, M.A. *et al.*, *Phytochemistry*, 1985, **24**, 182-183 (*Coniferyl ferulate*).
Ishimaru, K. *et al.*, *Phytochemistry*, 1987, **26**, 1147-1152 (*cis-Coniferyl alcohol galloylglucoside*).
Nishida, R. *et al.*, *Chem. Express*, 1988, **3**, 207-210 (*Batrocera Me ethers*).
Sawabe, A. *et al.*, *Nippon Nogei Kagaku Kaishi*, 1988, **62**, 1067-1071 (*Citrusin D*).
Lewis, N.-G. *et al.*, *Phytochemistry*, 1988, **27**, 2119-2121 (*Faguside*).
Quideau, S. *et al.*, *J. Agric. Food Chem.*, 1992, **40**, 1108-1110 (*Coniferyl alcohol, synth*).
Daubresse, N. *et al.*, *Synthesis*, 1994, 369-371 (*Coniferyl alcohol, synth*).
Gao, K. *et al.*, *Phytochemistry*, 1998, **47**, 269-272 (*hydroxydimethyloctadienyl derivs*).
Greca, M.D. *et al.*, *Phytochemistry*, 1998, **49**, 1299-1304 (*Coniferyl alcohol, Citrusin D*).
Panossian, A. *et al.*, *Phytomedicine*, 1998, **5**, 11-17 (*Coniferin, activity*).
Daubresse, N. *et al.*, *Synthesis*, 1998, 157-161 (*Coniferin, synth*).
Ralph, J. *et al.*, *Tetrahedron*, 1998, **54**, 1349-1354 (*Z-3'-Me ether, synth*).
Rivero, I.A. *et al.*, *Synth. Commun.*, 2000, **30**, 1351-1357 (*cis-Coniferyl alcohol*).
Takara, K. *et al.*, *Biosci., Biotechnol., Biochem.*, 2002, **66**, 29-35 (*cane sugar glycosides*).
Lin, L.-C. *et al.*, *Chin. Pharm. J. (Taipei)*, 2004, **56**, 77-85 (*Rossicaside E*).
Appendino, G. *et al.*, *J. Nat. Prod.*, 2004, **67**, 532-536 (*Coniferyl stearate*).
Wang, R.-F. *et al.*, *J. Nat. Prod.*, 2004, **67**, 2096-2098 (*pomegranate diglycoside*).
Lee, J.-Y. *et al.*, *Phytochemistry*, 2004, **65**, 3033-3039 (*Coniferyl palmitate, Coniferyl oleate*).
Wang, F. *et al.*, *J. Asian Nat. Prod. Res.*, 2006, **8**, 137-141 (*E-1-Et 3'-Me ether*).
Fiorentino, A. *et al.*, *J. Agric. Food Chem.*, 2008, **56**, 2660-2667 (*quince esters*).
Kobata, K. *et al.*, *Phytochemistry*, 2008, **69**, 1179-1184 (*Capsiconiate, Dihydrocapsiconiate*).
Kuo, Y.-J. *et al.*, *J. Nat. Prod.*, 2010, **73**, 109-114 (*Visartiside F*).
Tang, G. *et al.*, *Yunnan Zhiwu Yanjiu*, 2010, **32**, 87-90 (*E-3'-glucoside*).
Kim, J.A. *et al.*, *Molecules*, 2011, **16**, 6577-6581 (*Isoconiferoside*).
Zhang, Y. *et al.*, *Fitoterapia*, 2013, **85**, 69-75 (*Rossicaside K*).
Saito, Y. *et al.*, *Nat. Prod. Commun.*, 2013, **8**, 631-634 (*Coniferyl obtusilate*).
Sun, Y.N. *et al.*, *Biochem. Syst. Ecol.*, 2014, **57**, 164-168 (*Acanthopanax divaricatus constits*).
Chen, S. *et al.*, *Chem. Biodiversity*, 2014, **11**, 792-799 (*Taiwanensols A,B, activity*).
Zhou, Z.-Q. *et al.*, *Food Chem.*, 2017, **214**, 644-654 (*Lycibarbarphenylpropanoid E*)

2-(3,4-Dihydroxyphenyl)-5-(2-propenyl)-7-benzofuranol D247

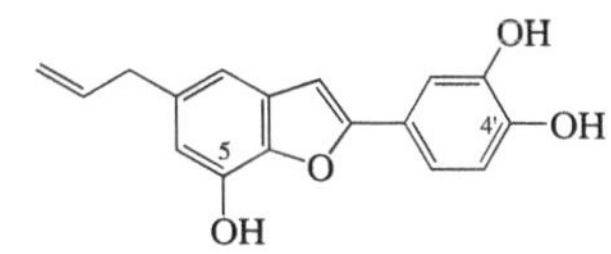

5-Allyl-2-(3,4-dihydroxyphenyl)-7-hydroxybenzofuran. 4,7'-Epoxy-9'-nor-3,8'-ligna-7',8-diene-3',4',5-triol. 4,7'-Epoxy-3',4',5-trihydroxy-9'-nor-3,8'-neoligna-7',8-diene
$C_{17}H_{14}O_4$ 282.295
Lignan numbering shown

3',4'-Methylene, 5-Me ether
7-Methoxy-2-(3,4-methylenedioxyphenyl)-5-(2-propenyl)benzofuran. 4,7'-Epoxy-5-methoxy-3',4'-methylenedioxy-9'-nor-3,8'-ligna-7',8-diene. ***9-Nor-7,8-dehydroisolicarin B***. *7,8-Dehydro-9-norisolicarin B*
[23343-24-0]
$C_{19}H_{16}O_4$ 308.333
Constit. of *Anaxagorea clavata* and *Nectandra lineata*. Potent COX-2 inhibitor and PAF-antagonist. Cryst. (toluene). Mp 69 - 70°. λ_{max} 218 (log ε 4.78); 248 (sh); 300 (sh); 315 (log ε 4.7); 330 (sh) (EtOH).

Ritchie, E. *et al.*, *Aust. J. Chem.*, 1969, **22**, 1329-1330 (*3',4'-methylene 5-Me ether, synth*).
Puentes de Diaz, A.M. *et al.*, *Phytochemistry*, 1997, **44**, 345-346 (*Anaxagorea clavata constit*).
Mali, R.S. *et al.*, *J. Chem. Res., Synop.*, 1998, 230-231 (*3',4'-methylene 5-Me ether, synth*).
Pal, G. *et al.*, *J. Chem. Res., Synop.*, 2003, 142-143 (*3',4'-methylene 5-Me ether, synth*).
Zhang, J. *et al.*, *Tetrahedron*, 2003, **59**, 5609-5616 (*3',4'-methylene 5-Me ether, synth*).
Coy, E.D. *et al.*, *Bioorg. Med. Chem. Lett.*, 2009, **19**, 6922-6925 (*9-Nor-7,8-dehydroisolicarin B, activity*)

2′-(3,4-Dihydroxyphenyl)-2,2′,3,3′-tetrahydro-7,7′-dihydroxy-3,3′-bis(hydroxymethyl)-2,5′-dibenzofuran-5-carboxaldehyde **D248**

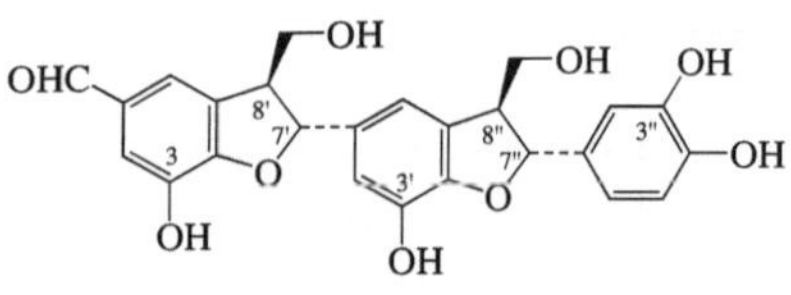

(7′R,7″R,8′S,8″S)-form

$C_{25}H_{22}O_9$ 466.442
Lignan numbering shown
(7′R,7″R,8′S,8″S)-form
5-Carboxylic acid, 3″-deoxy, 3,3′-di-Me ether
Sinensiol C
$C_{27}H_{26}O_9$ 494.496
Constit. of *Selaginella sinensis*. Amorph. powder. λ_{max} 209 (log ε 3.40); 267 (log ε 2.04) (MeOH)
(7′R*,7″R*,8′R*,8″S*)-form
3,3′,3″-Tri-Me ether
[1954754-18-7]
$C_{28}H_{28}O_9$ 508.523
Constit. of the fruit of *Melia toosendan*. Exhibits weak agonistic activity on human melatonin receptor MT_1 on HEK293 cell line. Amorph. pale yellow powder. Mp 67 - 69°. $[\alpha]^{20}_{D}$ -10.38 (c, 0.07 in MeOH). λ_{max} 235 (log ε 4.33); 288 (log ε 4.10); 378 (log ε 2.36) (MeOH).

Wang, H. *et al.*, *Planta Med.*, 2015, **81**, 847-854 (*3,3′,3″-tri-Me ether, activity*).
Chen, X. *et al.*, *Fitoterapia*, 2019, **134**, 256-263 (*Sinensiol C*)

[2-(3,4-Dihydroxyphenyl)tetrahydro-4-hydroxy-3-furanyl]methyl 3,4-dihydroxybenzoate **D249**

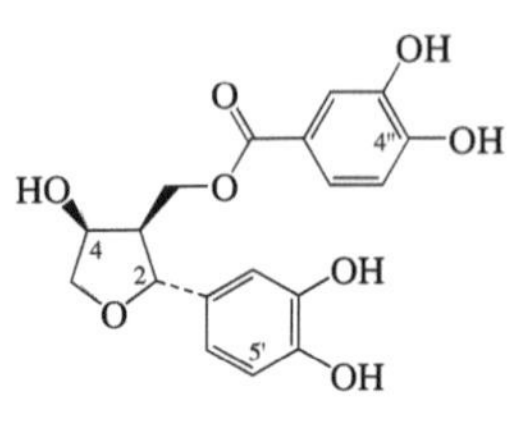

(2R,3R,4S)-form

$C_{18}H_{18}O_8$ 362.334
Abeolignan
(2R,3R,4S)-form
3′,4′-Methylene, 3″,4″-di-Me ether
Zanthplanispin
[2009354-02-1]
$C_{21}H_{22}O_8$ 402.399
Constit. of the roots of *Zanthoxylum planispinum*. NO prodn. inhibitor. Oil. $[\alpha]^{20}_{D}$ +32.5 (c, 0.1 in $CHCl_3$). λ_{max} 220; 264; 288 (MeOH)
(2R,3S,4R)-form
3′,3″,4′,4″-Tetra-Me ether
Magnostellin B
[87562-71-8]
$C_{22}H_{26}O_8$ 418.442
Constit. of *Magnolia stellata* and *Magnolia praecocissima*. Oil. $[\alpha]^{24}_{D}$ +33.9 (c, 0.41 in $CHCl_3$)
(2S,3S,4S)-form
3′,4′:3″,4″-Bis(methylene) ether
Linderin A
[1456695-30-9]
$C_{20}H_{18}O_8$ 386.356
Constit. of *Lindera obtusiloba*. Possesses antiinflammatory activity. Yellow wax. $[\alpha]^{20}_{D}$ +53.3 (c, 0.23 in MeOH)
5′-Methoxy, 3′,3″,4′,4″-tetra-Me ether
[874448-66-5]
$C_{23}H_{28}O_9$ 448.468
Constit. of the flower buds of *Magnolia fargesii*. Oil. $[\alpha]^{20}_{D}$ +34 (c, 0.5 in $CHCl_3$). λ_{max} 265 (log ε 4.06); 290 (log ε 3.79) ($CHCl_3$).

Iida, T. *et al.*, *Phytochemistry*, 1983, **22**, 211-213 (*Magnostellin B*).
Takahashi, H. *et al.*, *Chem. Pharm. Bull.*, 2002, **50**, 541-543 (*Magnostellin B, abs config*).
Lee, J. *et al.*, *Bull. Korean Chem. Soc.*, 2005, **26**, 913-915 (*5′-methoxy deriv*).
Lee, J. *et al.*, *Chem. Pharm. Bull.*, 2009, **57**, 298-301 (*5′-methoxy deriv, abs config*).
Choi, H.G. *et al.*, *Arch. Pharmacal Res.*, 2014, **37**, 467-472 (*Linderin A*).
Su, G.-Y. *et al.*, *Chem. Biodiversity*, 2017, **14**, e1600214 (*Zanthplanispin, activity*)

2-(3,4-Dihydroxyphenyl)tetrahydro-6-[2-(4-hydroxyphenyl)ethyl]-2*H*-pyran **D250**

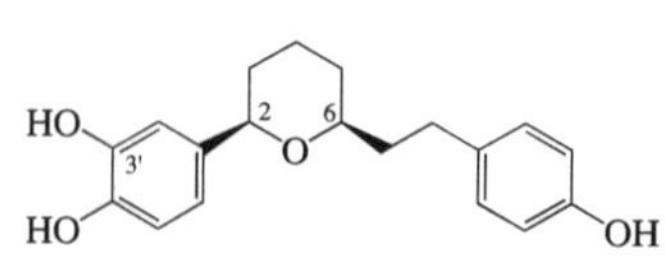

1-(3,4-Dihydroxyphenyl)-1,5-epoxy-7-(4-hydroxyphenyl)heptane
$C_{19}H_{22}O_4$ 314.381
(2R*,6S*)-form [1353086-26-6]
Constit. of the leaves of *Alnus nepalensis*. Needles. Mp 210 - 212°. $[\alpha]^{24}_{D}$ -216.8 (c, 0.05 in MeOH)
3′-Me ether
Engelheptanoxide A
[1397702-19-0]
$C_{20}H_{24}O_4$ 328.408
Constit. of *Engelhardtia roxburghiana*, *Juglans hopeiensis* and *Juglans regia*. Yellowish oil. $[\alpha]^{25}_{D}$ -66.9 (c, 0.12 in $CHCl_3$). λ_{max} 223 (log ε 4.13); 273 (log ε 3.63) (MeOH). [base] λ_{max} 242 (log ε 4.18); 291 (log ε 3.69) (MeOH/KOH).

Phan, M.G. *et al.*, *Chem. Nat. Compd. (Engl. Transl.)*, 2011, **47**, 735-737 (*Alnus nepalensis constit*).
Wu, H.-C. *et al.*, *Phytochemistry*, 2012, **82**, 118-127 (*Engelheptanoxide A*)

2-(3,4-Dihydroxyphenyl)tetrahydro-6-[2-(4-hydroxyphenyl)ethyl]-2*H*-pyran-4-ol **D251**

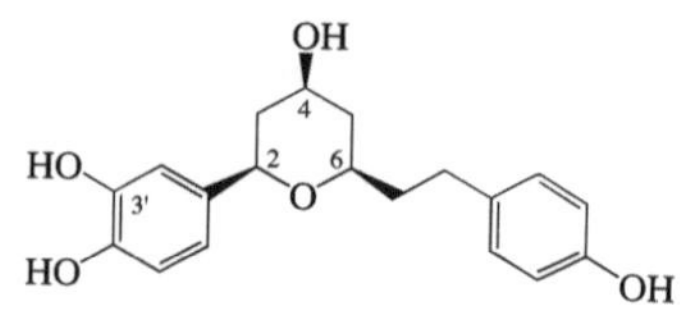

$C_{19}H_{22}O_5$ 330.38
(2R*,4S*,6R*)-form
Phaeoheptanoxide [1801534-15-5]
Constit. of the rhizomes of *Curcuma phaeocaulis*. Yellow powder (MeOH). $[\alpha]^{25}_{D}$ -64.8 (c, 0.13 in MeOH). λ_{max} 224 (log ε 3.82) (MeOH)
3′-Me ether
Engelheptanoxide C
[1397702-23-6]
$C_{20}H_{24}O_5$ 344.407
Constit. of the stems of *Engelhardtia roxburghiana*. Yellowish oil. $[\alpha]^{25}_{D}$ -17.4 (c, 0.14 in MeOH). λ_{max} 206 (log ε 4.27); 225 (log ε 4.05); 279 (log ε 3.56) (MeOH). [base] λ_{max} 210 (log ε 4.6); 244 (log ε 4.04); 288 (log ε 3.6) (MeOH/KOH).

Wu, H.-C. *et al.*, *Phytochemistry*, 2012, **82**, 118-127 (*Engelheptanoxide C*).
Ma, J.-H. *et al.*, *Org. Biomol. Chem.*, 2015, **13**, 8349-8358 (*Phaeoheptanoxide*)

1-(3,4-Dihydroxyphenyl)-7-(3,4,5-trihydroxyphenyl)-4,6-heptadien-3-one **D252**

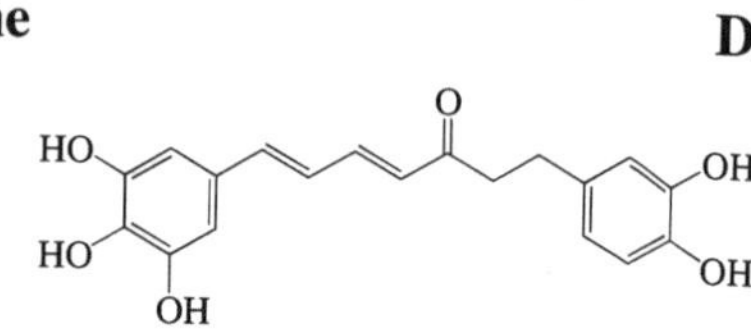

$C_{19}H_{18}O_6$ 342.347

(E,E)-form
3′,3″-Di-Me ether
7-(3,4-Dihydroxy-5-methoxyphenyl)-1-(4-hydroxy-3-methoxyphenyl)-4,6-heptadien-3-one
[1426153-03-8]
$C_{21}H_{22}O_6$ 370.401
Constit. of *Zingiber officinale*. Yellow powder. $[\alpha]^{21}_{D}$ -10.7 (c, 0.23 in $CHCl_3$). λ_{max} 239 (log ε 3.64); 265 (log ε 3.65); 358 (log ε 3.96) (MeOH).

Li, N. *et al.*, *Chin. J. Chem.*, 2012, **30**, 1351-1355 (*Zingiber officinale constit*)

1-(3,4-Dihydroxyphenyl)-7-(3,4,5-trihydroxyphenyl)-3,5-heptanediol, CAS D253

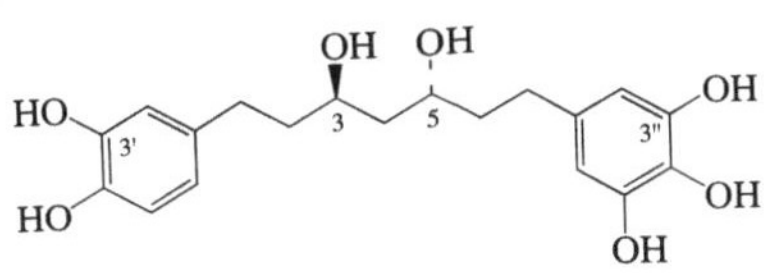

(3*R*,5*R*)-*form*

$C_{19}H_{24}O_7$ 364.394
(3R,5R)-form
3″-Me ether
1-(3,4-Dihydroxy-5-methoxyphenyl)-7-(3,4-dihydroxyphenyl)-3,5-heptanediol. ***Alpinin B‡***
[2125947-85-3]
$C_{20}H_{26}O_7$ 378.421
Constit. of the rhizomes of *Alpinia officinarum*. Yellowish oil. $[\alpha]^{25}_{D}$ +14.6 (c, 0.2 in MeOH). λ_{max} 281 (log ε 2.11) (MeOH)
(3S,5S)-form
3′,3″,5″-Tri-Me ether
1-(4-Hydroxy-3,5-dimethoxyphenyl)-7-(4-hydroxy-3-methoxyphenyl)-3,5-heptanediol
[145888-84-2]
$C_{22}H_{30}O_7$ 406.475
Constit. of the dried rhizomes of *Zingiber officinale*
3-Ketone, 3′,3″-di-Me ether
7-(3,4-Dihydroxy-5-methoxyphenyl)-5-hydroxy-1-(4-hydroxy-3-methoxyphenyl)-3-heptanone
[718639-10-2]
[1246280-52-3] (5*R*-form)
$C_{21}H_{26}O_7$ 390.432
Constit. of the rhizomes of *Alpinia officinarum* and *Zingiber officinale*. Oil. *Alpinia* Constit. has (5*R*)-config.
5-Ketone, 3′,3″-di-Me ether
1-(3,4-Dihydroxy-5-methoxyphenyl)-5-hydroxy-7-(4-hydroxy-3-methoxyphenyl)-3-heptanone
[718639-05-5]
$C_{21}H_{26}O_7$ 390.432
Constit. of the rhizomes of *Zingiber officinale*
(3R*,5S*)-form
Quasi-racemic stereoisomer
3′,3″,5′-Tri-Me ether
[718638-77-8]
$C_{22}H_{30}O_7$ 406.475
Constit. of the rhizomes of *Zingiber officinale*. Oil
3′,3″,5″-Tri-Me ether, 3,5-di-Ac
[153322-46-4]
$C_{26}H_{34}O_9$ 490.549
Constit. of the rhizomes of *Zingiber officinalis*. Oil. $[\alpha]^{20}_{D}$ -1.4 (c, 0.7 in $CHCl_3$)
(3ξ,5ξ)-form
3′,3″-Di-Me ether, 3,5-di-Ac
[1012312-65-0]
$C_{25}H_{32}O_9$ 476.522
Constit. of the rhizomes of *Zingiber officinale*. Oil. $[\alpha]^{20}_{D}$ +5 (c, 0.52 in $CHCl_3$)
3-Ketone, 3′,3″,5″-tri-Me ether
5-Hydroxy-7-(4-hydroxy-3,5-dimethoxyphenyl)-1-(4-hydroxy-3-methoxyphenyl)-3-heptanone
[138870-95-8]
$C_{22}H_{28}O_7$ 404.459
Constit. of the rhizomes of *Zingiber officinale*. Not obt. pure
5-Ketone, 3′,3″,5″-tri-Me ether
5-Hydroxy-1-(4-hydroxy-3,5-dimethoxyphenyl)-7-(4-hydroxy-3-methoxyphenyl)-3-heptanone
[138870-96-9]
$C_{22}H_{28}O_7$ 404.459
Constit. of the rhizomes of *Zingiber officinale*. Not obt. pure.

Kikuzaki, H. *et al.*, *Phytochemistry*, 1991, **30**, 3647-3651 (*ketone tri-Me ethers*).
Yamahara, J. *et al.*, *Yakugaku Zasshi*, 1992, **112**, 645-655 (*3S,5S-tri-Me ether, abs config*).
Ma, J. *et al.*, *Phytochemistry*, 2004, **65**, 1137-1143 (*Zingiber officinale constits*).
Zhou, C.X. *et al.*, *Chin. Chem. Lett.*, 2007, **18**, 1243-1246 (*3′,3″-di-Me ether 3,5-di-Ac*).
An, N. *et al.*, *Food Chem.*, 2010, **119**, 513-517 (*Alpinia officinarum constit*).
Fu, G. *et al.*, *J. Agric. Food Chem.*, 2017, **65**, 6608-6614 (*Alpinin B, cd*)

1-(3,4-Dihydroxyphenyl)-7-(3,4,5-trihydroxyphenyl)-2,3,5-heptanetriol D254

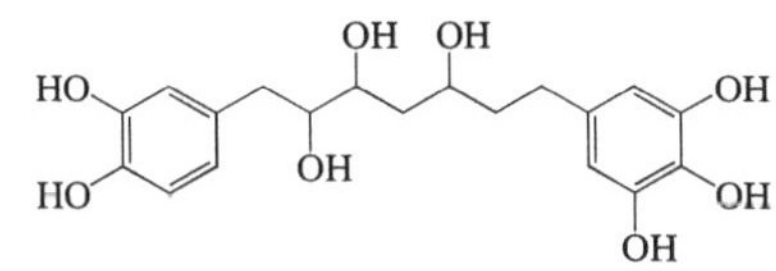

$C_{19}H_{24}O_8$ 380.393
3′,3″,5″-Tri-Me ether
7-(4-Hydroxy-3,5-dimethoxyphenyl)-1-(4-hydroxy-3-methoxyphenyl)-2,3,5-heptanetriol
[1269839-29-3]
$C_{22}H_{30}O_8$ 422.474
Isol. from the rhizomes of *Curcuma kwangsiensis*. Shows strong NO inhibitory activity. Yellow oil.

Jun, L.I. *et al.*, *Chin. J. Nat. Med.*, 2011, **9**, 329-333 (*Curcuma kwangsiensis constit*)

7-(3,4-Dihydroxyphenyl)-1-(3,4,5-trihydroxyphenyl)-3-heptanone D255

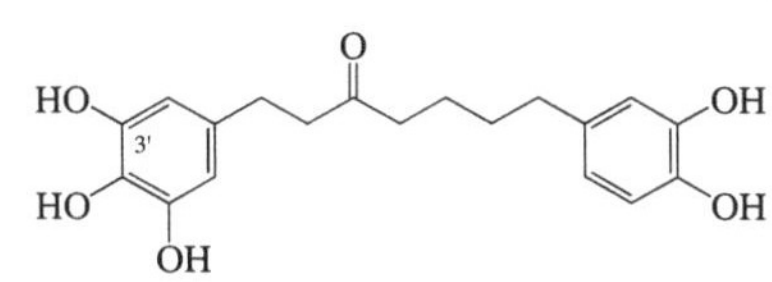

$C_{19}H_{22}O_6$ 346.379
3′-Me ether
1-(3,4-Dihydroxy-5-methoxyphenyl)-7-(3,4-dihydroxyphenyl)-3-heptanone. ***Dihydrogingerenone A***
[1376267-62-7]
$C_{20}H_{24}O_6$ 360.406
Constit. of the seeds of *Aframomum melegueta*. Antioxidant. Exhibits cytotoxic activity. Shows estrogen receptor binding ability. Yellow-brown oil. Misleading synonym, not the dihydro deriv. of Gingerenone A. λ_{max} 210 (log ε 4.55); 285 (log ε 3.73) (MeOH)
3′,5′-Di-Me ether
7-(3,4-Dihydroxyphenyl)-1-(4-hydroxy-3,5-dimethoxyphenyl)-3-heptanone. ***Dihydrogingerenone B***
[1376267-63-8]
$C_{21}H_{26}O_6$ 374.433
Constit. of the seeds of *Aframomum melegueta*. Antioxidant. Exhibits cytotoxic activity. Shows estrogen receptor binding ability. Yellow-brown oil. Misleading synonym, not the dihydro deriv.

of Gingerenone B. λ_{max} 209 (log ε 4.55); 280 (log ε 3.57) (MeOH).

El-Halawany, A.M. *et al., Food Chem.*, 2012, **134**, 219-226 (*Dihydrogingerenones A,B, activity*).
Al-Abbasi, F.A. *et al., Molecules*, 2016, **21**, 886:1-18 (*activity*).
Abdel-Naim, A.B. *et al., Molecules*, 2017, **22**, 1467:1-14 (*activity*)

1-(3,4-Dihydroxyphenyl)-2-(3,4,5-trihydroxyphenyl)-1,3-propanediol D256

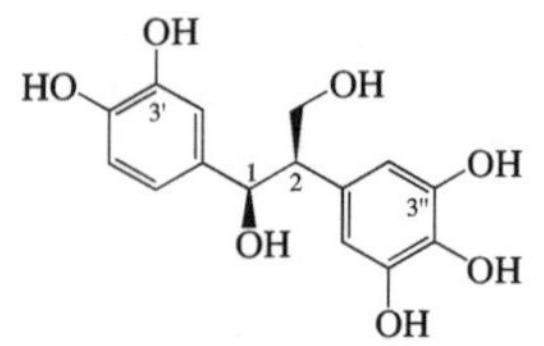

$C_{15}H_{16}O_7$ 308.286
(1R*,2R*)-form
1-Et, 3′,3″,5″-tri-Me ether
3-Ethoxy-3-(4-hydroxy-3-methoxyphenyl)-2-(4-hydroxy-3,5-dimethoxyphenyl)-1-propanol
[1624572-71-9]
$C_{20}H_{26}O_7$ 378.421
Constit. of *Clerodendranthus spicatus* and *Crataegus pinnatifida*. Yellow oil. $[\alpha]^{20}_{D}$ +38.1 (c, 0.7 in MeOH).

Zhou, C.-C. *et al., J. Asian Nat. Prod. Res.*, 2014, **16**, 169-174 (*Crataegus pinnatifida constit*).
Li, X.-Z. *et al., Tianran Chanwu Yanjiu Yu Kaifa*, 2017, **29**, 183-189 (*Clerodendranthus spicatus constit*)

1-(3,5-Dihydroxyphenyl)-2-(3,4,5-trihydroxyphenyl)-1,3-propanediol D257

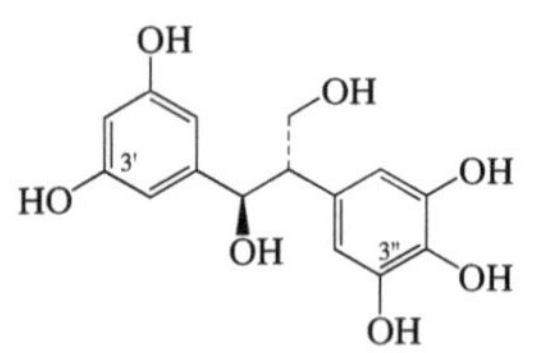

$C_{15}H_{16}O_7$ 308.286
Structures in this series may need revision. Phenyl 3,5-dioxygenated substn. pattern in doubt as a result of work done by Yang *et al* in 2015
(1R*,2S*)-form
3′,3″,5″-Tri-Me ether
2-(4-Hydroxy-3,5-dimethoxyphenyl)-1-(3-hydroxy-5-methoxyphenyl)-1,3-propanediol. ***Sapnol A***
[1665292-24-9]
$C_{18}H_{22}O_7$ 350.367
Isol. from the sap of *Acer saccharum*. Amorph. powder. $[\alpha]^{22}_{D}$ -12.2 (c, 1.6 in MeOH).

Yoshikawa, K. *et al., Molecules*, 2013, **18**, 9641-9649 (*Sapnol A*).
Yang, Y.-N. *et al., J. Nat. Prod.*, 2015, **78**, 705-711 (*pmr, struct*)

2,4-Dihydroxy-3-prenylcinnamic acid D258

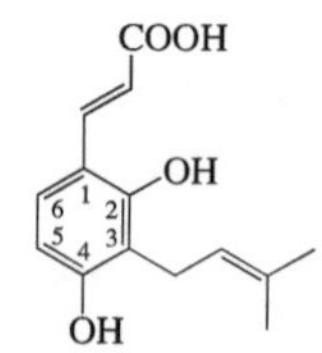

3-[2,4-Dihydroxy-3-(3-methyl-2-butenyl)phenyl]-2-propenoic acid
$C_{14}H_{16}O_4$ 248.278
(E)-form [155377-77-8]
Constit. of *Baccharis linearis*. Oil.

Brown, G.D. *Phytochemistry*, 1994, **35**, 1037-1042 (*Baccharis linearis constit*)

3,4-Dihydroxy-5-prenylcinnamic acid D259

3-[3,4-Dihydroxy-5-(3-methyl-2-butenyl)phenyl]-2-propenoic acid. 5-Prenylcaffeic acid
[252553-55-2]
$C_{14}H_{16}O_4$ 248.278
(E)-form [255054-19-4]
Constit. of Brazilian propolis. Antioxidant. Pale yellow powder. Mp 178 - 180°. λ_{max} 223 (ε 17600); 322 (ε 16200) (MeOH)
3-Me ether
3-(4-Hydroxy-3-methoxy-5-prenylphenyl)-2-propenoic acid. 4-Hydroxy-3-methoxy-5-prenylcinnamic acid. ***5-Prenylferulic acid***
[155377-76-7]
$C_{15}H_{18}O_4$ 262.305
Constit. of *Baccharis linearis*. Oil
Aldehyde, 3-Me ether
4-Hydroxy-3-methoxy-5-prenylcinnamaldehyde. ***Parvifloral***
[21973-34-2]
Constit. of the wood of *Zanthoxylum parviflorum*. Pale yellow needles (MeOH). Mp 124 - 125°
Aldehyde, 3-Me ether, 2′ξ,3′-epoxide
3-(2,3-Epoxy-3-methylbutyl)-4-hydroxy-5-methoxycinnamaldehyde. ***Glomeral***
[1443298-68-7]
$C_{15}H_{18}O_4$ 262.305
Constit. of *Vepris glomerata*. Red-brown solid. Mp 58 - 60°. $[\alpha]^{20}_{D}$ +27.8 (c, 0.02 in $CHCl_3$). λ_{max} 253 (log ε 1.92) ($CHCl_3$).

Diment, J.A. *et al., Aust. J. Chem.*, 1967, **20**, 565-573 (*Parvifloral*).
Brown, G.D. *Phytochemistry*, 1994, **35**, 1037-1042 (*5-Prenylferulic acid*).
Hayashi, K. *et al., Chem. Pharm. Bull.*, 1999, **47**, 1521-1524 (*propolis constit*).
Kiplimo, J.J. *et al., Phytochem. Lett.*, 2012, **5**, 438-442 (*Glomeral*)

2,3-Dihydroxy-1-(3,4,5-trihydroxyphenyl)-1-propanone D260

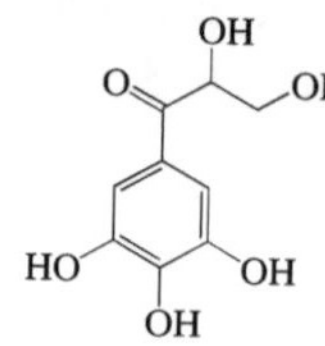

2,3,3′,4′,5′-Pentahydroxypropiophenone
$C_9H_{10}O_6$ 214.173
(±)-form
3′,5′-Di-Me ether
2,3-Dihydroxy-1-(4-hydroxy-3,5-dimethoxyphenyl)-1-propanone. 2,3,4′-Trihydroxy-3′,5′-dimethoxypropiophenone
[33900-74-2]
$C_{11}H_{14}O_6$ 242.227
Constit. of *Ficus beecheyana* and *Phragmites communis*. Metab. of *Ganoderma applanatum*. Amorph. powder. λ_{max} 286 (log ε 3.61) (MeOH).

Lee, T.-H. *et al., J. Nat. Prod.*, 2002, **65**, 1497-1500 (*Ficus beecheyana constit*).
Bu, Q.-H. *et al., Tianran Chanwu Yanjiu Yu Kaifa*, 2016, **28**, 650-654 (*Ganoderma applanatum metab*)

Diinnovanoside A D261

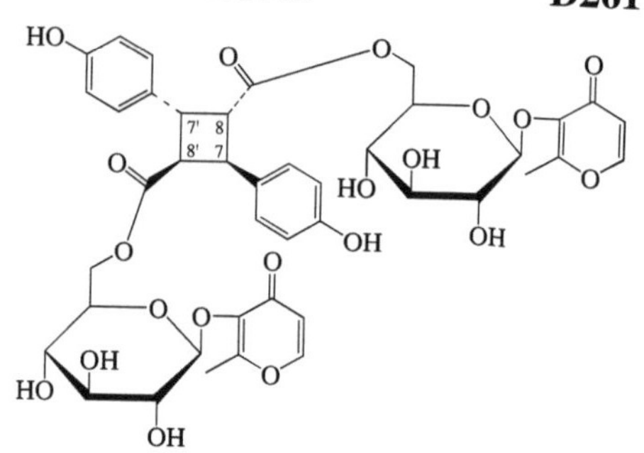

Relative Configuration

[1448439-13-1]
$C_{42}H_{44}O_{20}$ 868.794
Dimer of Innovanoside. Lignan numbering shown. Constit. of the stem

bark of *Daphne aurantiaca*. Oil. $[\alpha]_D^{20}$ -47 (c, 0.12 in MeOH). λ_{max} 231 (log ε 2.82); 259 (log ε 2.41) (MeOH)

7,8-Diepimer

Diinnovanoside B

[1448439-14-2]

$C_{42}H_{44}O_{20}$ 868.794

Constit. of the stem bark of *Daphne aurantiaca*. NO prodn. inhibitor. Oil. $[\alpha]_D^{20}$ -48 (c, 0.08 in MeOH). λ_{max} 230 (log ε 3.34); 256 (log ε 2.58) (MeOH).

Liang, S. *et al.*, *Chem. Commun.*, 2013, **49**, 6968-6970 (*Diinnovanosides A,B*)

Dilatioside A D262

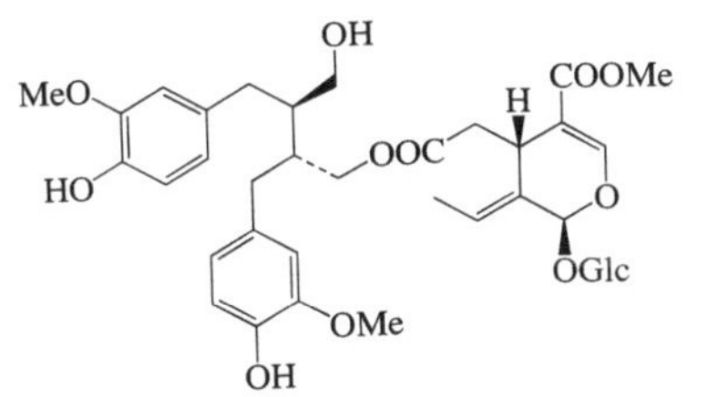

[2301086-65-5]

$C_{37}H_{48}O_{16}$ 748.775

Constit. of *Syringa oblata* var. *dilatata*. Gum. $[\alpha]_D^{25}$ -182 (c, 0.56 in MeOH). λ_{max} 231 (log ε 0.7); 282 (log ε 0.3) (MeOH).

Park, K.J. *et al.*, *Chem. Pharm. Bull.*, 2017, **65**, 359-364 (*Dilatioside A*)

3,3′-Dimethoxy-8,8′-lignan-4,4′-diol D263

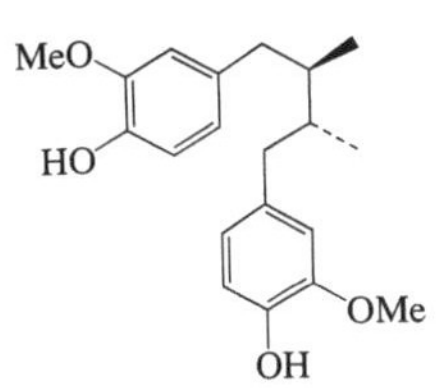

(8R,8′R)-form

4,4′-(2,3-Dimethyl-1,4-butanediyl) bis[2-methoxyphenol], CAS.

4,4′-Dihydroxy-3,3′-dimethoxylignan.

Dihydroguaiaretic acid

[36469-60-0]

$C_{20}H_{26}O_4$ 330.424

(8R,8′R)-form [124649-78-1]

Isol. from *Endiandra anthropophagorum*, *Larrea divaricata*, *Pycnanthus angolensis* and *Saururus cernuus*. Lipoxygenase inhibitor. Antioxidant. Cryst. (Et_2O/petrol). Mp 87 - 88°. $[\alpha]_D$ -46 (c, 1.0 in $CHCl_3$). $[\alpha]_D^{20}$ -27 (c, 0.9 in $CHCl_3$)

(8R,8′S)-form

[66322-34-7]

Constit. of *Porlieria chilensis*, *Machilus thunbergii*, *Myristica argentea*, *Myristica fragrans*, *Larrea divaricata*, *Schisandra glaucescens* and *Saururus chinensis*. Exhibits cytotoxic activity. Insecticide. Cryst. (MeOH aq.). Poorly sol. hexane. Mp 90° (87-88°)

(8S,8′S)-form [174291-55-5]

Isol. from *Sargentodoxa cuneata* and from root of *Cinnamomum philippinense*. Lipooxygenase inhibitor. Exhibits antibacterial activity. Mp 82 - 82.5°. $[\alpha]_D^{23}$ +4.87 (c, 0.914 in $CHCl_3$). $[\alpha]_D^{20}$ +25 (c, 1.0 in $CHCl_3$).

Schrecker, A.W. *J. Am. Chem. Soc.*, 1957, **79**, 3823-3827 (*8R,8′S-form*).
Gisvold, O. *et al.*, *J. Pharm. Sci.*, 1974, **63**, 1905-1907 (*8R,8′S-form*).
Han, G.-Q. *et al.*, *Yaoxue Xuebao*, 1986, **21**, 68-70 (*8S,8′S-form*).
Torres, R. *et al.*, *J. Nat. Prod.*, 1989, **52**, 402-403 (*8R,8′S-form*).
Rao, K.V. *et al.*, *J. Org. Chem.*, 1990, **55**, 1427-1429 (*Saururus cernuus constit, synth*).
Wu, T.-S. *et al.*, *Phytochemistry*, 1994, **36**, 785-788 (*Cinnamomum philippinense constit*).
Njoku, C.J. *et al.*, *Planta Med.*, 1997, **63**, 580-581 (*8R,8′R-form, activity*).
Filleur, F. *et al.*, *Planta Med.*, 2001, **67**, 700-704 (*8R,8′S-form*).
Yamauchi, S. *et al.*, *Biosci., Biotechnol., Biochem.*, 2008, **72**, 2981-2986 (*8S,8′S-form, activity*).
Kawaguchi, Y. *et al.*, *Biosci., Biotechnol., Biochem.*, 2009, **73**, 1806-1810 (*8S,8′S-form, activity*).
Lim, C.J. *et al.*, *Bull. Korean Chem. Soc.*, 2012, **33**, 3079-3082 (*Saururus chinensis constit, 8R,8′S-form*).
Wukirsari, T. *et al.*, *J. Agric. Food Chem.*, 2013, **6**, 4318-4325 (*8R,8′S-form, activity*).
Yu, H.-Y. *et al.*, *J. Nat. Prod.*, 2014, **77**, 1311-1320 (*Schisandra glaucescens constit*).
Le, T.V.T. *et al.*, *Nat. Prod. Sci.*, 2017, **23**, 21-28 (*8R,8′S-form, activity*)

3,3′-Dimethoxy-8,8′-lignan-4,4′,9,9′-tetrol D264

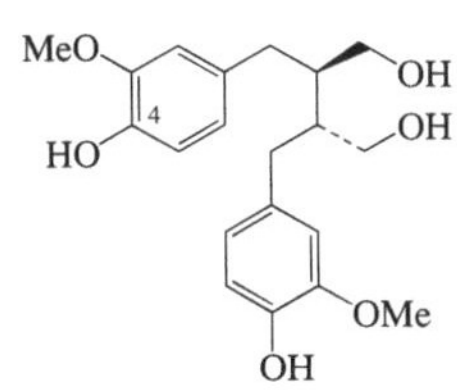

(8R,8′R)-form

2,3-Bis[(4-hydroxy-3-methoxyphenyl) methyl]-1,4-butanediol, CAS.

4,4′,9,9′-Tetrahydroxy-3,3′-dimethoxylignan. ***Secoisolariciresinol.***

seco-Isolariciresinol

[145265-02-7]

$C_{20}H_{26}O_6$ 362.422

Lignan numbering shown

(8R,8′R)-form [29388-59-8]

Constit. of *Podocarpus spicata* and *Larix* spp. heartwood, *Taxus baccata*, *Araucaria angustifolia*, *Picea abies* attacked by *Fomes annosus*, *Ainsliaea macrocephala* and *Clematis armandii*. Also from *Simaba cuneata* as an unseparated mixt. of 9- and 9′-esters with long-chain acids. Shows potent cytotoxic activity. Cryst. (C_6H_6/Me_2CO). Mp 112 - 114°. $[\alpha]_D^{25}$ -32 (c, 1 in Me_2CO). λ_{max} 229; 282 (MeOH) (Berdy)

4,4′-Di-O-α-L-rhamnopyranoside

[123564-48-7]

$C_{32}H_{46}O_{14}$ 654.706

Constit. of *Parsonsia laevigata*. Prisms (MeOH). Mp 112 - 116°. $[\alpha]_D^{27}$ -0.16 (c, 1.04 in MeOH)

4-O-β-D-Glucopyranoside

[129102-89-2]

$C_{26}H_{36}O_{11}$ 524.563

Isol. from *Antidesma membranaceum*, *Glehnia littoralis* and *Urtica dioica*. Amorph. solid. $[\alpha]_D$ -182.1 (c, 0.05 in MeOH). λ_{max} 230; 270 (MeOH)

4-O-[4-Hydroxy-3-methoxybenzoyl-(→6)-β-D-glucopyranoside]

Glehlinoside A

[402740-12-9]

$C_{34}H_{42}O_{14}$ 674.696

Constit. of *Glehnia littoralis*. Amorph. solid. $[\alpha]_D^{25}$ -158.3 (c, 0.08 in MeOH). λ_{max} 230; 275 (MeOH)

4-O-[3,4-Dimethoxybenzoyl-(→6)-β-D-glucopyranoside]

4‴-O-Methylglehlinoside A

[809230-40-8]

$C_{35}H_{44}O_{14}$ 688.723

Constit. of the bark of *Maytenus laevis*. Amorph. yellow powder. $[\alpha]_D$ -31.2 (c, 0.5 in MeOH)

4-O-[3,4-Dihydroxy-E-cinnamoyl-(→6)-β-D-glucopyranoside]

Glehlinoside F

[1181368-51-3]

$C_{35}H_{42}O_{14}$ 686.707

Constit. of the roots of *Glehnia littoralis*. Amorph. solid. $[\alpha]_D$ -38.4 (c, 0.25 in MeOH)

4-O-[4-Hydroxy-3-methoxy-E-cinnamoyl-(→6)-β-D-glucopyranoside]

Glehlinoside E

[1181368-49-9]

$C_{36}H_{44}O_{14}$ 700.734

Constit. of the roots of *Glehnia littoralis*. Amorph. solid. [α] D -55 (c, 0.18 in MeOH)

9-O-α-L-Arabinopyranoside

$C_{25}H_{34}O_{10}$ 494.537

Constit. of the pericarp of *Litchi chinensis*. Amorph. powder. $[\alpha]^{20}_{D}$ -13.4 (c, 0.75 in MeOH). λ_{max} 231 (log ε 3.59); 283 (log ε 3.46) (MeOH)

9-O-β-D-Xylopyranoside

[97065-26-4]

$C_{25}H_{34}O_{10}$ 494.537

Constit of *Pinus massoniana*. Amorph.. $[\alpha]^{22}_{D}$ -28 (c, 0.8 in MeOH)

9-O-α-L-Rhamnopyranoside

[312930-64-6]

$C_{26}H_{36}O_{10}$ 508.564

Constit. of *Rubus amabilis*. Gum. $[\alpha]^{20}_{D}$ -49.5 (c, 0.3 in Me_2CO)

9,9′-Di-O-α-L-rhamnopyranoside

Difengpioside D

[2041285-00-9]

$C_{32}H_{46}O_{14}$ 654.706

Constit. of the stem bark of *Illicium difengpi*. Amorph. powder. $[\alpha]^{20}_{D}$ -32 (c, 0.2 in MeOH). λ_{max} 281 (log ε 3.64) (MeOH)

9-O-β-D-Glucopyranoside

[63320-67-2]

$C_{26}H_{36}O_{11}$ 524.563

Constit. of *Berchemia racemosa, Picea abies, Pinus massoniana, Pinus sylvestris* and *Urtica dioica*. Amorph. powder. [α] D -22.9 (c, 0.5 in EtOH). $[\alpha]^{20}_{D}$ -20.5 (c, 3.2 in EtOH). λ_{max} 223 (sh); 282 (log ε 3.88) (MeOH). [neutral] λ_{max} 210; 282 (EtOH)

9,9′-Di-O-β-D-glucopyranoside

SDG

[158932-33-3]

$C_{32}H_{46}O_{16}$ 686.704

Constit. of the seeds of *Linum usitatissimum*. Precursor of the mammalian lignan 3,3′-Dihydroxy-8,8′-lignan-9,9′-olide, D160. Anticancer agent. Antioxidant. Powder. [α] D -4.4 (c, 0.2 in MeOH)

9-O-[4-Hydroxy-E-cinnamoyl-(→6)-β-D-glucopyranoside]

Hypenol

[1447292-06-9]

$C_{35}H_{42}O_{13}$ 670.708

Constit. of *Hypenia salzmannii*. Dark brown oil. $[\alpha]^{20}_{D}$ -1.2 (c, 0.001 in MeOH)

9-O-[4-Carboxy-3-hydroxy-3-methylbutanoyl-(→6)-β-D-glucopyranoside], 9′-O-β-D-glucopyranoside

6′-HMG SDG

[382156-76-5]

$C_{38}H_{54}O_{20}$ 830.83

Constit. of the seeds of *Linum usitatissimum*. Powder. Forms dimers and trimers. λ_{max} 280 (log ε 3.59) (no solvent reported)

9,9′-Bis-O-[4-carboxy-3-hydroxy-3-methylbutanoyl-(→6)-β-D-glucopyranoside]

[382156-77-6]

$C_{44}H_{62}O_{24}$ 974.956

Constit. of the seeds of *Linum usitatissimum*

Tetra-Ac

[41025-80-3]

Viscous oil. $[\alpha]^{25}_{D}$ -8 (c, 1.0 in $CHCl_3$)

9,9′-Dibenzoyl

[1159990-78-9]

$C_{34}H_{34}O_8$ 570.638

Constit. of *Maytenus apurimacensis*. Amorph. orange solid. $[\alpha]^{20}_{D}$ -20.2 (c, 0.5 in $CHCl_3$). Probable abs. config.. λ_{max} 229 (log ε 4.3); 281 (log ε 3.7) (EtOH)

9-O-(4-Hydroxy-E-cinnamoyl)

9-O-trans-p-Coumaroylsecoisolariciresinol. ***Artopithecin I***

[2387296-67-3]

$C_{29}H_{32}O_8$ 508.567

Constit. of *Artocarpus pithecogallus*. Exhibits moderate mushroom tyrosinase inhibitory activity. Amorph. pale yellow powder. $[\alpha]^{20}_{D}$ -25 (c, 0.2 in MeOH). Obt. as a 1:1 mixt. of *E*- and *Z*-isomers. λ_{max} 226 (log ε 4.82); 287 (log ε 4.40); 311 (log ε 4.36) (MeOH)

9,9′-Bis-O-(4-hydroxy-E-cinnamoyl)

9,9′-Di-O-trans-p-coumaroylsecoisolariciresinol

[2250279-58-2]

$C_{38}H_{38}O_{10}$ 654.712

Constit. of the twigs of *Sageretia theezans*. Antioxidant. Gum. $[\alpha]^{20}_{D}$ -11.9 (c, 0.1 in MeOH). λ_{max} 228 (log ε 2.49); 291 (log ε 2.49); 313 (log ε 2.56) (MeOH)

9-O-(4-Hydroxy-Z-cinnamoyl)

9-O-cis-p-Coumaroylsecoisolariciresinol. ***Artopithecin J***

[2387296-68-4]

$C_{29}H_{32}O_8$ 508.567

Constit. of *Artocarpus pithecogallus*. Exhibits moderate mushroom tyrosinase inhibitory activity. Amorph. pale yellow solid. $[\alpha]^{20}_{D}$ -20 (c, 0.2 in MeOH). Obt. as a 1:1 mixt. of *E*- and *Z*-isomers. λ_{max} 227 (log ε 4.82); 289 (log ε 4.40); 314 (log ε 4.34) (MeOH)

9,9′-Bis-O-(4-hydroxy-Z-cinnamoyl)

9,9′-Di-O-cis-p-coumaroylsecoisolariciresinol

[2250279-60-6]

$C_{38}H_{38}O_{10}$ 654.712

Constit. of the twigs of *Sageretia theezans*. Antioxidant. Gum. $[\alpha]^{20}_{D}$ +7.3 (c, 0.1 in MeOH). λ_{max} 230 (log ε 2.42); 290 (log ε 2.31); 312 (log ε 2.35) (MeOH)

9-O-(4-Hydroxy-E-cinnamoyl), 9′-O-(4-hydroxy-Z-cinnamoyl)

9-O-cis-p-Coumaroyl-9′-O-trans-p-coumaroylsecoisolariciresinol

[2250279-59-3]

$C_{38}H_{38}O_{10}$ 654.712

Constit. of the twigs of *Sageretia theezans*. Antioxidant. Gum. $[\alpha]^{20}_{D}$ -44.9 (c, 0.1 in MeOH). λ_{max} 223 (log ε 2.39); 291 (log ε 2.34); 312 (log ε 2.40) (MeOH)

9-O-(4-Hydroxy-3-methoxy-E-cinnamoyl)

9-O-trans-Feruloylsecoisolariciresinol. ***Hanultarin***

[1067247-32-8]

$C_{30}H_{34}O_9$ 538.593

Constit. of the seeds of *Trichosanthes kirilowii*. Cytotoxic. Light brown oil. $[\alpha]^{20}_{D}$ -19.7 (c, 0.36 in MeOH). λ_{max} 204 (log ε 5); 288 (log ε 4.35); 326 (log ε 4.4) (MeOH)

9-O-(4-Hydroxy-3-methoxy-E-cinnamoyl), 4′-O-β-D-glucopyranoside

Tinoposide D

$C_{36}H_{44}O_{14}$ 700.734

Constit. of the caulis of *Tinospora sinensis*. $[\alpha]^{20}_{D}$ -37 (c, 0.1 in MeOH). λ_{max} 230 (log ε 4.51); 286 (log ε 4.32); 326 (log ε 4.35) (MeOH)

9,9′-Bis-O-(4-hydroxy-3-methoxy-E-cinnamoyl)

9,9′-Di-O-trans-feruloylsecoisolariciresinol

[56973-66-1]

$C_{40}H_{42}O_{12}$ 714.764
Constit. of *Betula ermanii, Alnus japonica, Hypericum androsaemum* and *Microtropis japonica*. Cytotoxic. Pale yellow powder or needles (EtOAc). Mp 204 - 205° (*Alnus* isolate). $[\alpha]_D$ -41.2 (c, 0.7 in MeOH). $[\alpha]_D^{16}$ -52.5 (c, 0.6 in Py) (*Alnus* isolate). λ_{max} 217 (ε 33880); 230 (ε 28840); 289 (ε 21380); 300 (ε 20420); 326 (ε 27540) (MeOH)

9,9′-Bis-O-(4-hydroxy-3-methoxy-Z-cinnamoyl)
9,9′-Di-O-cis-feruloylsecoisolariciresonol
[1030358-30-5]
$C_{40}H_{42}O_{12}$ 714.764
Constit. of the stems of *Microtropis japonica*. Pale yellow powder. $[\alpha]_D^{25}$ -42.6 (c, 0.11 in MeOH). λ_{max} 231 (log ε 4.46); 290 (log ε 4.36); 326 (log ε 4.48) (MeOH)

9-O-(4-β-D-Glucopyranosyloxy-3-methoxy-E-cinnamoyl)
Tinoposide C
$C_{36}H_{44}O_{14}$ 700.734
Constit. of the caulis of *Tinospora sinensis*. $[\alpha]_D^{20}$ -50 (c, 0.1 in MeOH). λ_{max} 230 (log ε 4.49); 286 (log ε 4.28); 326 (log ε 4.30) (MeOH)

(8S,8′S)-form
Cryst.. Mp 114 - 115°. $[\alpha]_D^{20}$ +31 (c, 0.3 in Me_2CO)

9-O-β-D-Xylopyranoside
[145213-58-7]
Constit. of *Laurus nobilis*

9,9′-Dibenzoyl
[1391976-08-1]
Isol. from *Pseudolarix kaempferi*. Amorph. powder. $[\alpha]_D^{17}$ +25.7 (c, 0.7 in $CHCl_3$). λ_{max} 202 (log ε 4.91); 228 (log ε 4.6); 281 (log ε 3.9) (MeOH)

(8RS,8′RS)-form

9,9′-Di-Ac
[226543-61-9]
$C_{24}H_{30}O_8$ 446.496
Constit. of *Virola sebifera*. Cryst. ($CHCl_3$). Mp 112°. λ_{max} 233 (log ε 4.27); 285 (log ε 3.74) (MeOH)

9,9′-Bis-O-(4-hydroxy-3-methoxy-E-cinnamoyl)
[133097-13-9]
$C_{40}H_{42}O_{12}$ 714.764
Constit. of *Machilus thunbergi* and *Ochroma lagopus*

(8RS,8′SR)-form
[57759-55-4]
Constit. of *Cedrus deodara*. Viscous liq.

9,9′-Bis-O-(4-hydroxy-3-methoxy-E-cinnamoyl)
[1243310-59-9]
Constit. of the stems of *Lindera obtusiloba*

(8ξ,8′ξ)-form

4-O-[4-Methoxybenzoyl-(→6)-β-D-glucopyranoside]
[761458-04-2]
$C_{34}H_{42}O_{13}$ 658.697
Constit. of the bark of *Tabebuia impetiginosa*. Amorph. powder. $[\alpha]_D^{22}$ -35 (c, 0.63 in MeOH). λ_{max} 203 (log ε 5.01); 226 (log ε 4.16); 257 (log ε 4.18) (MeOH)

4-O-[3,4-Dimethoxybenzoyl-(→6)-β-D-glucopyranoside]
[761458-06-4]
$C_{35}H_{44}O_{14}$ 688.723
Constit. of the bark of *Tabebuia impetiginosa*. Amorph. powder. $[\alpha]_D^{24}$ -43 (c, 0.5 in MeOH). λ_{max} 203 (log ε 4.99); 219 (log ε 4.56); 262 (log ε 4.15); 280 (sh) (MeOH).

Briggs, L.H. *et al.*, *Tetrahedron Lett.*, no. 4, 1959, **1**, 14-15 (*struct*).
Freudenberg, K. *et al.*, *Tetrahedron Lett.*, no. 17, 1959, **1**, 19-22 (*Larix constit*).
Traveso, G. *Gazz. Chim. Ital.*, 1962, **90**, 792-807 (*abs config*).
Majumdar, R.B. *et al.*, *Indian J. Chem.*, 1972, **10**, 677-680 (*Taxus baccata constit*).
Fonseca, S.F. *et al.*, *Phytochemistry*, 1978, **17**, 499-502 (*Araucaria angustifolia constit, cmr*).
Nomura, M. *et al.*, *Phytochemistry*, 1981, **20**, 1097-1104 (*diferulate*).
Agrawal, P.K. *et al.*, *Phytochemistry*, 1982, **21**, 1459-1461 (*Cedrus deodara constit*).
Mahalanabis, K.K. *et al.*, *Tetrahedron Lett.*, 1982, **23**, 3975-3978 (*synth*).
Lundgren, L.N. *et al.*, *Acta Chem. Scand., Ser. B*, 1985, **39**, 241-248 (*Pinus massoniana constit*).
Inoshiri, S. *et al.*, *Phytochemistry*, 1987, **26**, 2811-2814 (*9-glucoside*).
Abe, F. *et al.*, *Phytochemistry*, 1989, **28**, 1737-1741 (*dirhamnoside*).
Rodrigues Filho, E. *et al.*, *Phytochemistry*, 1992, **31**, 2115-2116 (*Simaba cuneata esters*).
Marinos, V.A. *et al.*, *Phytochemistry*, 1992, **31**, 4307-4312 (*9-glucoside, wine constit*).
Yahara, S. *et al.*, *Shoyakugaku Zasshi*, 1992, **46**, 184-186 (*9-xyloside*).
Shen, Y.-C. *et al.*, *Chin. Pharm. J. (Taipei)*, 1997, **49**, 285-296 (*Taxus constit, activity*).
Milanesio, M. *et al.*, *Croat. Chem. Acta*, 1999, **72**, 351-363 (*cryst struct*).
Chimichi, S. *et al.*, *Magn. Reson. Chem.*, 1999, **37**, 860-863 (*Linum usitatissimum constits*).
Martinez V., J.C. *et al.*, *Phytochemistry*, 1999, **50**, 883-886 (*Virola sebifera di-Ac*).
Chen, X.C. *et al.*, *Chin. Chem. Lett.*, 2000, **11**, 897-900 (*9-rhamnoside*).
Buske, A. *et al.*, *Eur. J. Org. Chem.*, 2001, 3537-3543 (*4-glucoside*).
Baderschneider, B. *et al.*, *J. Agric. Food Chem.*, 2001, **49**, 2788-2798 (*9-glucoside*).
Ford, J.D. *et al.*, *J. Nat. Prod.*, 2001, **64**, 1388-1397 (*9,9′-diglucoside, bibl*).
Chen, X. *et al.*, *Planta Med.*, 2001, **67**, 270-273 (*9-rhamnoside*).
Yuan, Z. *et al.*, *Chem. Pharm. Bull.*, 2002, **50**, 73-77 (*4-glucoside, Glehlinoside A*).
Nakagawa, H. *et al.*, *J. Nat. Prod.*, 2004, **67**, 1919-1924 (*4‴-O-Methylglehlinoside A*).
Warashina, T. *et al.*, *Phytochemistry*, 2004, **65**, 2003-2011 (*Tabebuia impetiginosa glucosides*).
Sugahara, T. *et al.*, *Biosci., Biotechnol., Biochem.*, 2007, **71**, 2962-2968 (*meso-form, synth, pmr, cmr*).
Moon, S.-S. *et al.*, *Bioorg. Med. Chem.*, 2008, **16**, 7264-7269 (*Hanultarin*).
Kong, W.X. *et al.*, *Chin. Chem. Lett.*, 2008, **19**, 1459-1461 (*Glehlinosides E,F*).
Chen, J.-J. *et al.*, *J. Nat. Prod.*, 2008, **71**, 1016-1021 (*di-cis-feruloyl*).
Vazdekis, N.E.J. *et al.*, *J. Nat. Prod.*, 2009, **72**, 1045-1048 (*Maytenus apurimacensis constit*).
Stasevich, O.V. *et al.*, *Chem. Nat. Compd. (Engl. Transl.)*, 2010, **46**, 704-709 (*activity*).
Lee, K.Y. *et al.*, *Planta Med.*, 2010, **76**, 294-297 (*Lindera obtusiloba constit*).
Kim, K.H. *et al.*, *Bioorg. Med. Chem. Lett.*, 2011, **21**, 2270-2273 (*activity*).
Lee, E. *et al.*, *Bioorg. Med. Chem. Lett.*, 2011, **21**, 6245-6248 (*Hanultarin, synth, activity*).
He, W.-J. *et al.*, *Planta Med.*, 2011, **77**, 1924-1931 (*8S,8′S-9,9′-dibenzoyl*).
Mishra, O.P *Bioorg. Med. Chem. Lett.*, 2013, **23**, 5325-5328 (*9,9′-diglycosides, synth*).
Wu, Z-J. *et al.*, *Chem. Nat. Compd. (Engl. Transl.)*, 2013, **49**, 167-179 (*Ainsliaea macrocephala constit*).
Sousa de Lucena, H.B. *et al.*, *Helv. Chim. Acta*, 2013, **96**, 1121-1125 (*Hypenol*).
Xiong, J. *et al.*, *J. Ethnopharmacol.*, 2014, **153**, 737-743 (*Clematis armandii constit*).
Ma, Q. *et al.*, *J. Funct. Foods*, 2014, **9**, 156-161 (*Litchi 9-arabinoside*).
Pan, Z.-H. *et al.*, *Molecules*, 2016, **21**, 607:1-8 (*Difengpioside D*).
Kang, K.B. *et al.*, *J. Nat. Prod.*, 2018, **81**, 1819-1828 (*Sageretia theezans constits, activity*).
Jiang, H. *et al.*, *Zhongcaoyao*, 2018, **49**, 2336-2344 (*Tinoposides C,D, cd*).
Xu, L. *et al.*, *Magn. Reson. Chem.*, 2019, **57**, 506-511 (*Artopithecins I,J, activity*)

3,4-Dimethoxy-3′,4′-methylenedioxy-8,8′-lignan-9,9′-olide D265

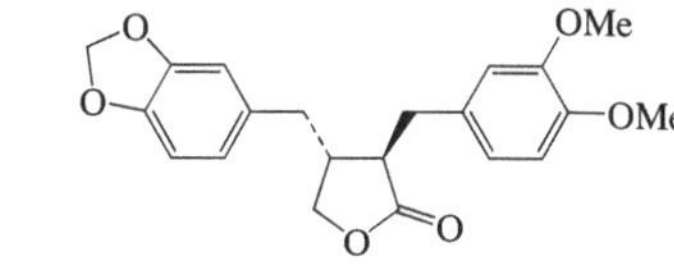

(8R,8′R)-form

$C_{21}H_{22}O_6$ 370.401

(8R,8′R)-form
Bursehernin. Methyl pluviatolide. 5′-Demethoxyyatein [40456-51-7]

Isol. from twigs and bark of *Hernandia ovigera, Hernandia guianensis, Phyllanthus niruri, Bupleurum salicifolium* and *Bursera schlechtendalii*. Antifeedant. Shows antineoplastic activity. Oil. $[\alpha]^{23}_{D}$ -28.7 (c, 2 in $CHCl_3$). $[\alpha]^{28}_{D}$ -151 ($CHCl_3$). The much higher opt. rotn. is from the *P. niruri* isolate (Satyanarayana *et al*). λ_{max} 232 (ε 13000); 246; 283 (ε 9200); 286 (MeOH) (Berdy)

(8S,8′S)-form

Dextrobursehernin [610767-19-6]

Constit. of *Phyllanthus urinaria*. Liq.. $[\alpha]^{25}_{D}$ +36 (c, 1 in $CHCl_3$). λ_{max} 225 (sh) (log ε 4.22); 283 (log ε 3.9) (MeCN).

McDoniel, P.B. *et al., J. Pharm. Sci.*, 1972, **61**, 1992-1994 (*Bursera schlechtendalii constit, struct, activity*).
Yamaguchi, H. *et al., Yakugaku Zasshi*, 1979, **99**, 674-677 (*Hernandia ovigera constit*).
Richomme, P. *et al., Planta Med.*, 1984, **50**, 20-22 (*ms*).
Satyanarayana, P. *et al., J. Nat. Prod.*, 1988, **51**, 44-49 (*Phyllanthus niruri constit, uv, ir, ms*).
Gonzalez, A.G. *et al., Phytochemistry*, 1990, **29**, 1981-1983 (*Bupleurum salicifolium constit*).
Tanoguchi, M. *et al., Chem. Pharm. Bull.*, 1991, **39**, 1873-1876 (*cd*).
Estevez-Braun, A. *et al., Phytochemistry*, 1996, **43**, 885-886 (*cmr*).
Chang, C.-C. *et al., Phytochemistry*, 2003, **63**, 825-833 (*(+)-form*).
Velázquez-Jiménez, R. *et al., Phytochemistry*, 2011, **72**, 2237-2243 (*vcd, abs config*)

1,6-Dimethoxy-3′,4′-methylenedioxy-3,8′-lign-8-en-4(1*H*)-one D266

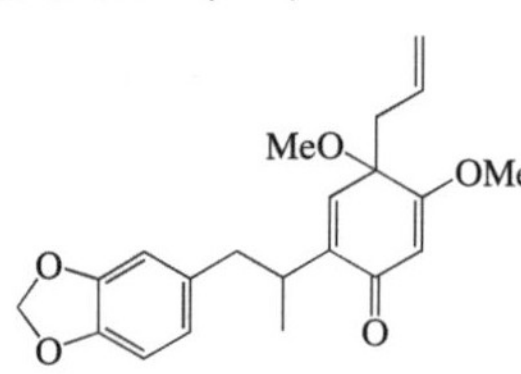

2-[1-(1,3-Benzodioxol-5-yl)-1-methylethyl]-4,5-dimethoxy-4-(2-propenyl)-2,5-cyclohexadien-1-one, CAS. *4-Allyl-4,5-dimethoxy-2-[2-(3,4-methylenedioxyphenyl)-1-methylethyl]-2,5-cyclohexadien-1-one*

[158506-59-3]

$C_{21}H_{24}O_5$ 356.418

Constit. of the stems of *Piper wightii*. Viscous yellow oil. $[\alpha]^{24}_{D}$ -22.7 (c, 0.2 in $CHCl_3$). λ_{max} 248; 286 ($CHCl_3$).

Prasad, A.K. *et al., Tetrahedron*, 1994, **50**, 10579-10586 (*Piper wightii constit*)

3,3′-Dimethoxy-4,4′-oxylign-7′-ene-7,8,9,9′-tetrol D267

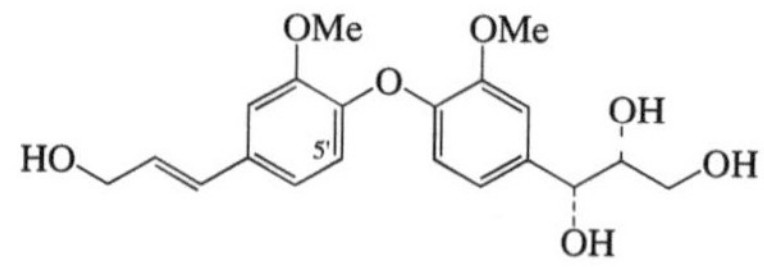

(7R,7′E,8R)-form

1-[4-[4-(3-Hydroxy-1-propenyl)-2-methoxyphenoxy]-3-methoxyphenyl]-1,2,3-propanetriol. 7,8,9,9′-Tetrahydroxy-3,3′-dimethoxy-4,4′-oxyneolign-7′-ene

$C_{20}H_{24}O_7$ 376.405

(7R,7′E,8R)-form [515813-61-3]

Constit. of *Eurycoma longifolia* and *Saussurea pulchella*. Yellow needles. Mp 60 - 62° (57-60°). $[\alpha]_D$ -8 (c, 0.12 in MeOH)

5,5′-Dimethoxy, 7-Me ether

3,3′,5,5′,7-Pentamethoxy-4,4′-oxylign-7′-ene-8,9,9′-triol. 8,9,9′-Trihydroxy-3,3′,5,5′,7-pentamethoxy-4,4′-oxylign-7′-ene. ***Alismaine B***

[2413671-65-3]

$C_{23}H_{30}O_9$ 450.484

Constit. of the rhizome of *Alisma orientale*. Amorph. brown powder. $[\alpha]^{20}_{D}$ +32.0 (c, 0.08 in MeOH). λ_{max} 270 (log ε 4.11) (MeOH)

(7S,7′E,8R)-form

5,5′-Dimethoxy, 7-Me ether

Alismaine A

[2413671-64-2]

$C_{23}H_{30}O_9$ 450.484

Constit. of the rhizomes of *Alisma orientale*. Amorph. brown powder. $[\alpha]^{20}_{D}$ +33.6 (c, 0.06 in MeOH). λ_{max} 270 (log ε 4.20) (MeOH)

(7S,7′E,8S)-form [515813-60-2]

Constit. of *Eurycoma longifolia*. Yellow needles. Mp 56 - 58° (53-56°). $[\alpha]_D$ +1.3 (c, 1.6 in MeOH)

(7R*,7′E,8S*)-form

5′-Methoxy

3,3′,5′-Trimethoxy-4,4′-oxylign-7′-ene-7,8,9,9′-tetrol. 7,8,9,9′-Tetrahydroxy-3,3′,5′-trimethoxy-4,4′-oxylign-7′-ene. ***Utilin B‡***

[1616592-16-5]

$C_{21}H_{26}O_8$ 406.431

Constit. of *Zanthoxylum utile*. Oil. $[\alpha]_D$ +6.6 (c, 0.33 in MeOH).

Morita, H. *et al., Phytochemistry*, 1992, **31**, 3993-3995 (*Eurycoma longifolia constits*).
Yang, Y. *et al., Tetrahedron: Asymmetry*, 2002, **13**, 2689-2692 (*synth*).
Yang, M.C. *et al., Arch. Pharmacal Res.*, 2007, **30**, 1067-1074 (*Saussurea pulchella constit*).
Derong, Z. *et al., Youji Huaxue*, 2013, **33**, 1345-1348 (*Utilin B*).
Liu, S.-S. *et al., Bioorg. Chem.*, 2019, **92**, 103226:1-9 (*Alismaines A,B, cd*)

7-(3,4-Dimethoxyphenyl)-2,3-dimethoxy-8-methyl-6-(2-propenyl)tricyclo[4.2.0.0^{2,8}]oct-3-en-5-one D268

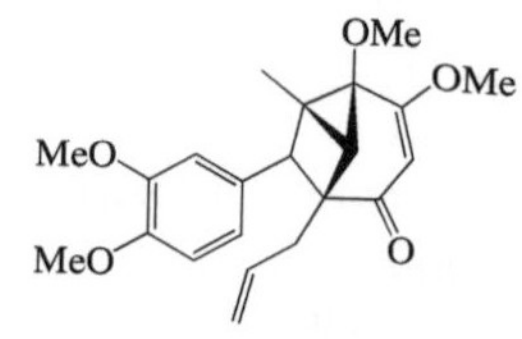

Relative Configuration

6-Allyl-7-(3,4-dimethoxyphenyl)-2,3-dimethoxy-8-methyltricyclo[4.2.0.0^{2,8}]oct-3-en-5-one. 2,8′:3,8′-Bicyclo-2,3-dihydro-3,3′,4,4′-tetramethoxy-1,7′-lign-8-en-6(1H)-one

[853074-90-5]

$C_{22}H_{26}O_5$ 370.445

Constit. of the flower buds of *Magnolia denudata*. Amorph. powder. $[\alpha]_D$ +67.2 (c, 1.7 in $CHCl_3$).

Li, J. *et al., Chem. Pharm. Bull.*, 2005, **53**, 235-237 (*Magnolia denudata constit*)

8-[2,6-Dimethoxy-4-(1,2,3-trihydroxypropyl)phenyl]-7-hydroxy-2*H*-1-benzopyran-2-one, CAS D269

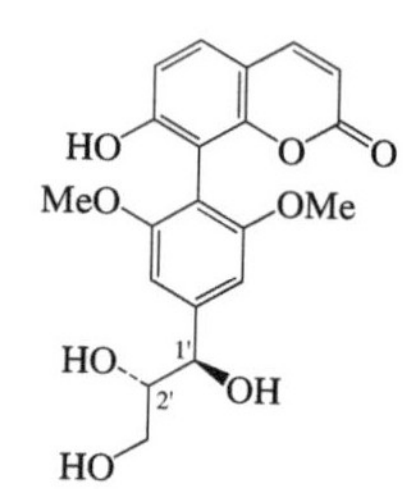

7-Hydroxy-8-(4-syringylglyceryl) coumarin

$C_{20}H_{20}O_8$ 388.372

Coumarinolignan

(1′R*,2′S*)-form [1450817-93-2]

Constit. of the aerial parts of *Rhododendron lepidotum*. Amorph. powder. Mp 206 - 210°. λ_{max} 219; 261; 309 (MeOH).

Shakeel-u-Rahman, *et al., Nat. Prod. Res.*, 2013, **27**, 2033-2038 (*Rhododendron lepidotum constit*)

1-[2-(3,7-Dimethyl-2,6-octadienyl)-6-[2-(4-hydroxy-3-methoxyphenyl) ethyl]-3-methyl-3-cyclohexenyl]-3-phenyl-1-propanone, CAS D270

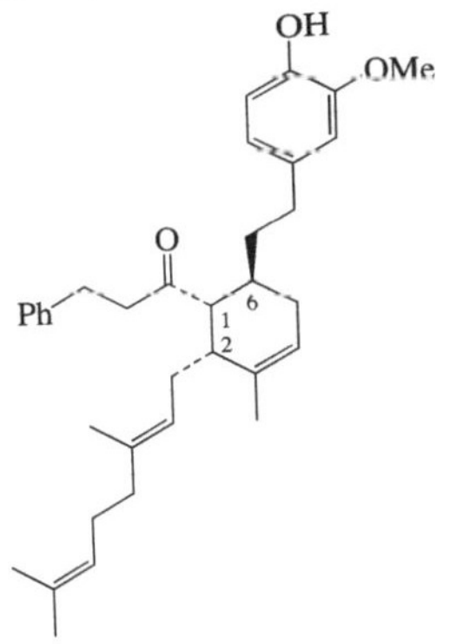

(1*R*,2*R*,6*S*)-*form*

$C_{35}H_{46}O_3$ 514.750
The nat. prods. Alpininoids C, D and E were isol. as racemates and resolved by chiral hplc
(1R,2R,6S)-form
(-)-Alpininoid D [2216000-46-1]
Amorph. powder. $[\alpha]_D^{25}$ -18 (c, 0.2 in MeOH)
Demethoxy
(-)-Alpininoid E
[2210238-38-1]
Amorph. powder. $[\alpha]_D^{25}$ -14 (c, 0.2 in MeOH)
(1R,2S,6S)-form
(-)-Alpininoid C [2216000-45-0]
Amorph. powder. $[\alpha]_D^{25}$ -24 (c, 0.2 in MeOH)
(1S,2R,6R)-form
(+)-Alpininoid C [2216000-43-8]
Amorph. powder. $[\alpha]_D^{25}$ +26 (c, 0.2 in MeOH)
(1S,2S,6R)-form
(+)-Alpininoid D [2216000-44-9]
Amorph. powder. $[\alpha]_D^{25}$ +19 (c, 0.2 in MeOH)
Demethoxy
(+)-Alpininoid E
[2210238-35-8]
Amorph. powder. $[\alpha]_D^{25}$ +14 (c, 0.2 in MeOH)
(1RS,2RS,6SR)-form
Alpininoid D [2216000-50-7]
Constit. of the rhizomes of *Alpinia officinarum*. Amorph. powder. λ_{max} 205 (log ε 3.42); 282 (log ε 2.25) (MeOH)
Demethoxy
Alpininoid E
$C_{34}H_{44}O_2$ 484.724
Constit. of the rhizomes of *Alpinia officinarum*. Amorph. powder. λ_{max} 206 (log ε 3.39); 280 (log ε 2.33) (MeOH)
(1RS,2SR,6SR)-form
Alpininoid C [2216000-49-4]
Constit. of the rhizomes of *Alpinia officinarum*. Amorph. powder. λ_{max} 205 (log ε 3.33); 283 (log ε 2.04) (MeOH).
Liu, H. *et al.*, *J. Nat. Prod.*, 2018, **81**, 162-170 (*Alpininoids C,D,E, cd*)

4-(3,7-Dimethyl-2,6-octadienyl)-7-(3-hydroxy-4-methoxyphenyl)-5-(2-methyl-2-butenyl)-1-phenyl-3-heptanone D271

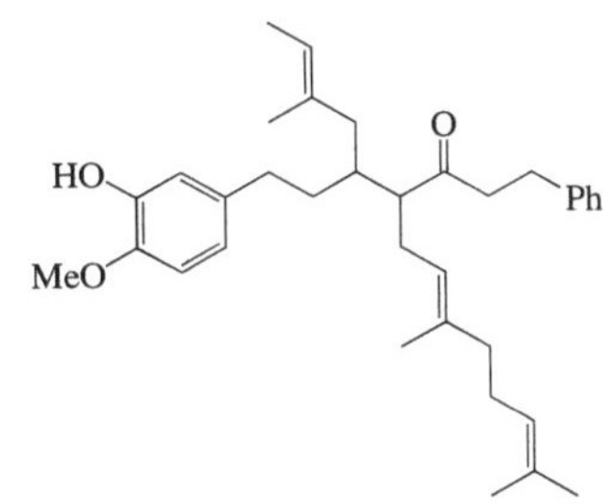

$C_{35}H_{48}O_3$ 516.766
Constit. of the rhizomes of *Alpinia officinarum*. Needles.
Xin, M. *et al.*, *Molecules*, 2017, **22**, 647:1-12 (*Alpinia officinarum constit*)

3-(1,1-Dimethyl-2-propenyl)-5-(2-propenyl)-1,2-benzenediol D272

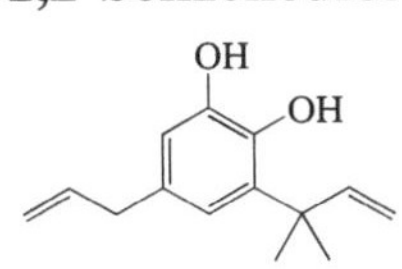

5-Allyl-3-(1,1-dimethylallyl)-1,2-benzenediol
$C_{14}H_{18}O_2$ 218.296
1-Me ether
2-(1,1-Dimethyl-2-propenyl)-6-methoxy-4-(2-propenyl)phenol
[1369516-16-4]
$C_{15}H_{20}O_2$ 232.323
Constit. of the twigs of *Illicium anisatum*. Oil. λ_{max} 283 (ε 1000) (EtOH).
Kubo, M. *et al.*, *Chem. Pharm. Bull.*, 2011, **59**, 898-901 (*Illicium anisatum constit*)

2-(1,1-Dimethyl-2-propenyl)-4-(2-propenyl)phenol D273

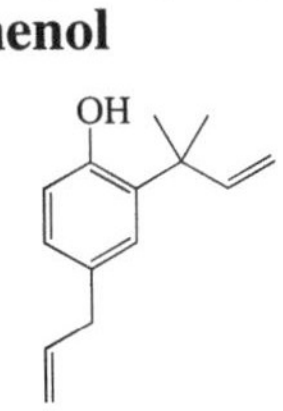

4-Allyl-2-(1,1-dimethylallyl)phenol
[1083166-66-8]
$C_{14}H_{18}O$ 202.297
Constit. of the dried pericarps of *Illicium fargesii*. Oil. λ_{max} 280 (ε 2100) (EtOH).
Moriyama, M. *et al.*, *Chem. Pharm. Bull.*, 2008, **56**, 1201-1204 (*Illicium fargesii constit*)

9,9′-Dinor-2,7′-lignan-3′,4,4′,5-tetrol D274

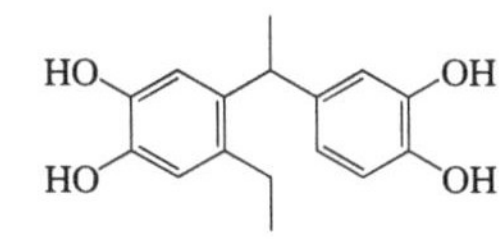

4-Ethyl-5-[1-(3,4-dihydroxyphenyl) ethyl]-1,2-benzenediol.
3′,4,4′,5-Tetrahydroxy-9,9′-dinor-2,7′-neolignan
$C_{16}H_{18}O_4$ 274.316
(ξ)-form
Glechomol C [1439547-60-0]
Isol. from *Glechoma longituba*. Displays significant protective effects on H_2O_2-induced cardiac cell injury. Brown oil. $[\alpha]_D^{15}$ +10.6 (c, 0.083 in MeOH). λ_{max} 214 (log ε 4.03); 286 (log ε 3.45) (MeOH).
Zhu, Q.-F. *et al.*, *J. Asian Nat. Prod. Res.*, 2013, **15**, 258-264 (*Glechomol C, activity*)

8,9-Dinor-4,8′-oxylignan-3,3′,4′,5,7,7′,9′-heptol D275

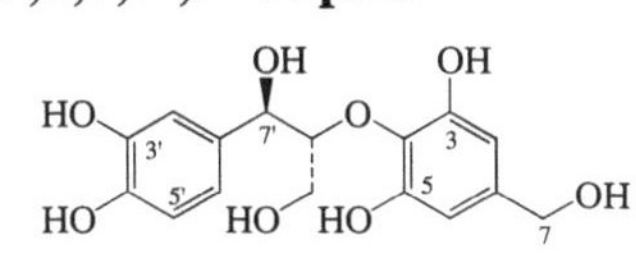

(7′*R*,8′*R*)-*form*

3,3′,4′,5,7,7′,9′-Heptahydroxy-8,9-dinor-4,8′-oxyneolignan
$C_{16}H_{18}O_8$ 338.312
(7′R,8′R)-form
3,3′,5-Tri-Me ether
3,3′,5-Trimethoxy-8,9-dinor-4,8′-oxylignan-4′,7,7′,9′-tetrol.
4′,7,7′,9′-Tetrahydroxy-3,3′,5-trimethoxy-8,9-dinor-4,8′-oxylignan.
Fordianole B
[2230065-39-9]
$C_{19}H_{24}O_8$ 380.393
Constit. of *Viburnum fordiae*. Antioxidant. Amorph. powder. $[\alpha]_D^{23}$ +8.3 (c, 0.1 in MeOH). λ_{max} 205 (log ε 4.95); 279 (log ε 3.68) (MeOH)
7-Aldehyde
3,3′,4′,5,7′,9′-Hexahydroxy-8,9-dinor-4,8′-oxylignan-7-al.

3,3′,4′,5,7′,9′-Hexahydroxy-8,9-dinor-7-oxo-4,8′-oxylignan
7-Aldehyde, 3,3′,5-tri-Me ether
4′,7′,9′-Trihydroxy-3,3′,5-trimethoxy-8,9-dinor-4,8′-oxylignan-7-al. 4′,7′,9′-Trihydroxy-3,3′,5-trimethoxy-8,9-dinor-7-oxo-4,8′-oxylignan
[1447940-07-9]
[1620217-09-5] (7′*S*,8′*S*-form)
$C_{19}H_{22}O_8$ 378.377
Constit. of *Crataegus pinnatifida* and *Euonymus oblongifolius*. Exhibits cytotoxic activity. Oil. $[\alpha]^{20}_{D}$ +7.7 (c, 0.1 in MeOH) (*E. oblongifolius*). $[\alpha]^{20}_{D}$ +12.7 (c, 0.38 in MeOH) (*C. pinnatifida*). λ_{max} 207 (log ε 3.27); 285 (log ε 2.79) (MeOH)

(7′R,8′S)-form
3,3′,5-Tri-Me ether
Wilfordiol B
[1415556-24-9]
$C_{19}H_{24}O_8$ 380.393
Constit. of the leaves of *Tripterygium wilfordii*. Light yellow oil. $[\alpha]^{20}_{D}$ +1.9 (c, 0.1 in MeOH). λ_{max} 230 (log ε 3.91); 279 (log ε 3.34) (MeOH)
7-Aldehyde, 3,3′,5-tri-Me ether
4′,7′,9′-Trihydroxy-3,3′,5-trimethoxy-8,9-dinor-4,8′-oxylignan-7-al. ***Coriarianeolignan A***
$C_{19}H_{22}O_8$ 378.377
Constit. of the roots of *Coriaria nepalensis*. Light yellow solid. $[\alpha]^{20}_{D}$ +0.04 (c, 0.08 in MeOH)
7-Carboxylic acid, 3,3′,5-tri-Me ether
4′,7′,9′-Trihydroxy-3,3′,5-trimethoxy-8,9-dinor-4,8′-oxylignan-7-oic acid. ***Asprenol G***
$C_{19}H_{22}O_9$ 394.376
Constit. of *Ilex asprella*. Amorph. light yellow powder. $[\alpha]^{25}_{D}$ +10 (c, 0.1 in MeOH). λ_{max} 208 (log ε 3.15); 266 (log ε 2.55) (MeOH)

(7′S,8′R)-form
3,3′,5-Tri-Me ether
Fordianole A
[2230065-38-8]
$C_{19}H_{24}O_8$ 380.393
Constit. of *Viburnum fordiae*. Antioxidant. Amorph. powder. $[\alpha]^{23}_{D}$ -13.1 (c, 0.1 in MeOH). λ_{max} 209 (log ε 5.12); 226 (log ε 4.67); 279 (log ε 4.05) (MeOH)
7-Aldehyde, 3,3′,5-tri-Me ether
[1447940-08-0]
$C_{19}H_{22}O_8$ 378.377
Constit. of the stems of *Euonymus oblongifolius*. Oil. $[\alpha]^{20}_{D}$ -2.9 (c, 0.1 in MeOH). λ_{max} 207 (log ε 3.1); 285 (log ε 2.62) (MeOH)

(7′S,8′S)-form
3,3′,5-Tri-Me ether
Wilfordiol A
[1415556-23-8]
$C_{19}H_{24}O_8$ 380.393
Constit. of *Acer saccharum* and *Tripterygium wilfordii*. Exhibits moderate antioxidant activity. Light yellow oil. $[\alpha]^{20}_{D}$ -2.9 (c, 0.1 in MeOH). λ_{max} 233 (log ε 3.98); 279 (log ε 3.5) (MeOH)
7-Carboxylic acid, 3,3′,5-tri-Me ether
[1290633-24-7]
$C_{19}H_{22}O_9$ 394.376
Constit. of the stems of *Sinocalamus affinis*. Amorph. powder. Mp 159 - 161° (synthetic). $[\alpha]^{20}_{D}$ +13.4 (c, 0.13 in MeOH). $[\alpha]^{20}_{D}$ +48 (c, 0.1 in MeOH) (synthetic). λ_{max} 256 (log ε 3.37); 284 (log ε 3.08) (MeOH)

(7′R*,8′R*)-form
3,3′,5-Tri-Me ether, 4′-O-β-D-glucopyranoside
Hovetrichoside F
[210050-41-2]
$C_{25}H_{34}O_{13}$ 542.534
Constit. of the bark of *Hovenia trichocarea*. Amorph. powder. $[\alpha]^{25}_{D}$ -28.2 (c, 1 in MeOH). λ_{max} 230 (log ε 4.1); 280 (log ε 3.6) (MeOH)

(7′R*,8′S*)-form
7-Aldehyde, 3,3′,5-tri-Me ether, di-Me acetal
Fabagoin B
$C_{21}H_{28}O_9$ 424.446
Constit. of *Zygophyllum fabago*. Oil. $[\alpha]^{21.9}_{D}$ -6.4 (c, 0.25 in MeOH).

Yoshikawa, K. *et al., J. Nat. Prod.*, 1998, **61**, 786-790 (*Hovetrichoside F*).
Xiong, L. *et al., J. Nat. Prod.*, 2011, **74**, 1188-1200 (*Sinocalamus affinis constit*).
Cao, X. *et al., Fitoterapia*, 2012, **83**, 343-347 (*Wilfordiols A,B*).
Li, F. *et al., J. Asian Nat. Prod. Res.*, 2012, **14**, 755-758 (*Euonymus oblongifolius constits*).
Li, L.-Z. *et al., Chin. J. Nat. Med.*, 2013, **11**, 411-414 (*Crataegus pinnatifida constit*).
Yoshikawa, K. *et al., Molecules*, 2013, **18**, 9641-9649 (*Wilfordiol A, activity*).
He, J. *et al., Phytochem. Lett.*, 2016, **15**, 116-120 (*Fabagoin B*).
Chen, J. *et al., Arch. Pharmacal Res.*, 2018, **41**, 625-632 (*Fordianoles A,B, activity*).
Zhao, F. *et al., Chin. Chem. Lett.*, 2018, **29**, 467-470 (*Coriarianeolignan A*).
Du, B. *et al., Fitoterapia*, 2018, **129**, 220-225 (*Asprenol G*)

8,9-Dinor-2,8′-oxylignan-3,3′,4′,5,7′,9′-hexol D276

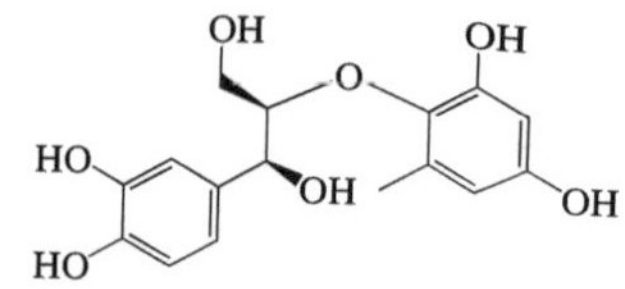

3,3′,4′,5,7′,9′-Hexahydroxy-8,9-dinor-2,8′-oxyneolignan
$C_{16}H_{18}O_7$ 322.313

(7′S,8′R)-form
3,3′,5-Tri-Me ether
8,9-Dinor-3,3′,5-trimethoxy-2,8′-oxylignan-4′,7′,9′-triol.
Neoasarinin A
[2211144-23-7]
$C_{19}H_{24}O_7$ 364.394
Constit. of the roots and rhizomes of *Asarum heterotropoides*. Exhibits antiinflammatory activity. Amorph. powder. $[\alpha]^{20}_{D}$ +10 (c, 0.6 in MeOH). λ_{max} 231 (log ε 4.05); 280 (log ε 3.70); 308 (log ε 3.01) (MeOH).

Jing, Y. *et al., Molecules*, 2017, **22**, 125:1-20 (*Neoasarinin A, activity*)

8,9-Dinor-4,8′-oxylignan-3,3′,4′,5′,7,7′-hexol D277

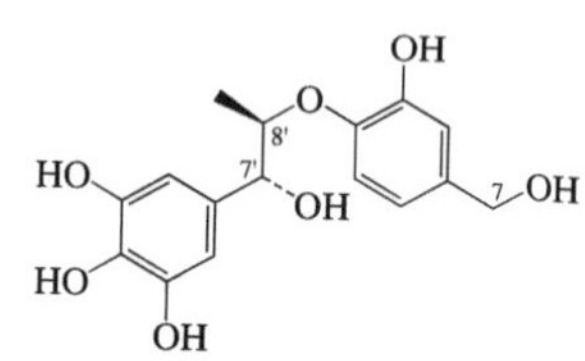

(7′*R*,8′*R*)-*form*

3,3′,4′,5′,7,7′-Hexahydroxy-8,9-dinor-4,8′-oxyneolignan
$C_{16}H_{18}O_7$ 322.313

(7′R,8′R)-form
7-Carboxylic acid, 3,3′,4′,5′-tetra-Me ether
7′-Hydroxy-3,3′,4′,5′-tetramethoxy-8,9-dinor-4,8′-oxylignan-7-oic acid
7-Carboxylic acid, 3,3′,4′,5′-tetra-Me ether, Me ester
(-)-Acortatarinowin C
[1807543-31-2]
$C_{21}H_{26}O_8$ 406.431
Constit. of *Acorus tatarinowii*. Light yellow gum. $[\alpha]^{20}_{D}$ -15 (c, 0.01 in MeOH). Resolved by chiral hplc

(7′S,8′S)-form
7-Carboxylic acid, 3,3′,4′,5′-tetra-Me ether, Me ester
(+)-Acortatarinowin C
[1807543-30-1]
$C_{21}H_{26}O_8$ 406.431
Constit. of *Acorus tatarinowii*. Exhibits weak inhibitory activity against NO prodn. in activated macrophages. Light yellow gum. $[\alpha]^{20}_{D}$+15 (c, 0.03 in MeOH). Resolved by chiral hplc

(7′RS,8′RS)-form
7-Carboxylic acid, 3,3′,4′,5′-tetra-Me ether, Me ester
Acortatarinowin C
[1807543-40-3]
$C_{21}H_{26}O_8$ 406.431
Constit. of *Acorus tartarinowii*. Light yellow gum. $[\alpha]^{20}_{D}$+13 (c, 0.03 in MeOH). Resolved by chiral hplc. λ_{max} 206 (log ε 4.60); 262 (log ε 3.99); 292 (log ε 3.71) (MeOH).

Lu, Y. *et al.*, *J. Nat. Prod.*, 2015, **78**, 2205-2214 (*Acortatarinowin C, activity*)

8,9-Dinor-4,8′-oxylignan-3,3′,4′,7,7′,9′-hexol D278

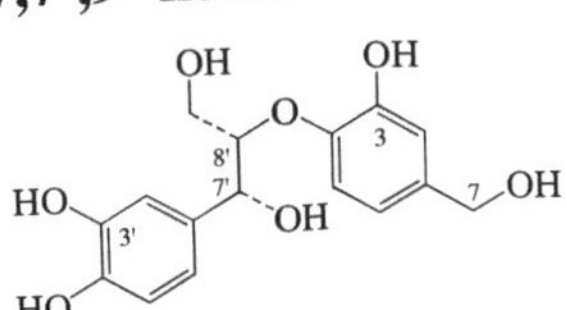

(7′R,8′S)-form

3,3′,4′,7,7′,9′-Hexahydroxy-8,9-dinor-4,8′-oxyneolignan
$C_{16}H_{18}O_7$ 322.313

(7′R,8′S)-form
3,3′-Di-Me ether
3,3′-dimethoxy-8,9-dinor-4,8′-oxylignan-4′,7,7′,9′-tetrol. 4′,7,7′,9′-Tetrahydroxy-3,3′-dimethoxy-8,9-dinor-4,8′-oxylignan
$[\alpha]^{20}_{D}$-18 (c, 0.1 in MeOH)

(7′S,8′R)-form
3,3′-Di-Me ether
Exhibits moderate neuroprotective activity. $[\alpha]^{20}_{D}$+22 (c, 0.1 in MeOH)
7-Carboxylic acid, 3′,4′-methylene, 3-Me ether, 7-β-D-glucopyranosyl ester
Neoasarininoside A
[2211144-26-0]
$C_{24}H_{28}O_{13}$ 524.475
Constit. of the roots and rhizomes of *Asarum heterotropoides*. Exhibits strong antiinflammatory activity. Amorph. powder. $[\alpha]^{20}_{D}$ +30.6 (c, 0.52 in MeOH). λ_{max} 220 (log ε 4.01); 263 (log ε 3.81); 289 (log ε 3.71) (MeOH)

(7′R*,8′R*)-form
7-Aldehyde
3,3′,4′,7′,9′-Pentahydroxy-8,9-dinor-4,8′-oxylignan-7-al
7-Aldehyde, 3,3′-di-Me ether
2-(4-Formyl-2-methoxyphenoxy)-3-(4-hydroxy-3-methoxyphenyl)-1,3-propanediol. 4-[2-Hydroxy-2-(4-hydroxy-3-methoxyphenyl)-1-(hydroxymethyl) ethoxy]-3-methoxybenzaldehyde, CAS. 4′,7′,9′-Trihydroxy-3,3′-dimethoxy-8,9-dinor-4,8′-oxylignan-7-al
[631922-79-7]
[17614-17-4] (unspecified stereochem.) [79171-61-2] [79389-53-0] [79389-56-3]
$C_{18}H_{20}O_7$ 348.351
Constit. of *Crataegus pinnatifida* and *Larix leptolepis*. Oil. $[\alpha]^{20}_{D}$+2.1 (c, 1.1 in MeOH). Partial racemate resolved by chiral hplc. λ_{max} 230 (log ε 4.24); 280 (log ε 4.03); 310 (log ε 3.92) (EtOH)
7-Aldehyde, 7′-Et, 3,3′-di-Me ether
7′-Ethoxy-4′,9′-dihydroxy-3,3′-dimethoxy-8,9-dinor-4,8′-oxylignan-7-al
[1954754-13-2]
$C_{20}H_{24}O_7$ 376.405
Constit. of the fruit of *Melia toosendan*. Pale yellow oil. $[\alpha]^{20}_{D}$ -4.5 (c, 0.09 in MeOH). λ_{max} 228 (log ε 3.85); 279 (log ε 3.67); 309 (log ε 3.50) (MeOH)
7-Carboxylic acid
3,3′,4′,7′,9′-Pentahydroxy-8,9-dinor-4,8′-oxylignan-7-oic acid
7-Carboxylic acid, 3,3′-di-Me ether
4-[2-Hydroxy-2-(4-hydroxy-3-methoxyphenyl)-1-(hydroxymethyl) ethoxy]-3-methoxybenzoic acid. Guaiacylglycerol-β-vanillic acid ether. 4′,7′,9′-Trihydroxy-3,3′-dimethoxy-8,9-dinor-4,8′-oxylignan-7-oic acid
[79317-44-5]
$C_{18}H_{20}O_8$ 364.35
Constit. of the fruit of *Boreava orientalis*. Amorph. powder. Mp 85 - 90°

(7′R*,8′S*)-form
7-Aldehyde, 3,3′-di-Me ether
[547714-32-9]
$C_{18}H_{20}O_7$ 348.351
Constit. of *Crataegus pinnatifida* and *Larix leptolepis*. Oil
7-Aldehyde, 7′-Et, 3,3′-di-Me ether
[1954754-15-4]
$C_{20}H_{24}O_7$ 376.405
Constit. of the fruit of *Melia toosendan*. Pale yellow oil. $[\alpha]^{20}_{D}$ -9.1 (c, 0.09 in MeOH). λ_{max} 228 (log ε 3.88); 279 (log ε 3.71); 309 (log ε 3.57) (MeOH)

(7′RS,8′SR)-form
3,3′-Di-Me ether
$C_{18}H_{22}O_7$ 350.367
Constit. of the stem bark of *Ailanthus altissima*. Light yellow gum. $[\alpha]^{20}_{D}$+1.6 (c, 0.1 in MeOH). Partial racemate resolved by chiral hplc.

Miki, K. *et al.*, *Phytochemistry*, 1980, **19**, 449-453 (*Larix leptolepis constits*).
Sakushima, A. *et al.*, *Phytochemistry*, 1995, **40**, 257-261 (*Boreava orientalis constit*).
Chen, X. *et al.*, *Synth. Commun.*, 2002, **32**, 557-564 (*synth*).
Huang, X.-X. *et al.*, *Bioorg. Med. Chem. Lett.*, 2013, **23**, 5599-5604 (*Crataegus pinnatifida constits*).
Wang, H. *et al.*, *Planta Med.*, 2015, **81**, 847-854 (*Melia toosendan Et ethers*).
Jing, Y. *et al.*, *Molecules*, 2017, **22**, 125:1-20 (*Neoasarininoside A, activity*).
Zhao, P. *et al.*, *Fitoterapia*, 2019, **136**, 104164:1-7 (*Crataegus pinnatifida constit*).
Du, Y.-Q. *et al.*, *Fitoterapia*, 2019, **139**, 104403:1-9 (*Ailanthus altissima constit, cd, activity*)

8,9-Dinor-4,8′-oxylignan-3,3′,4′,5,5′,7,7′,9′-octol D279

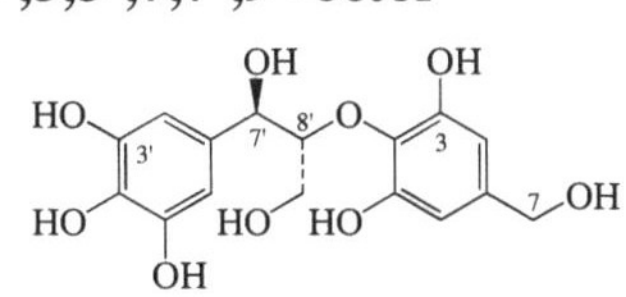

(7′R,8′R)-form

3,3′,4′,5,5′,7,7′,9′-Octahydroxy-8,9-dinor-4,8′-oxyneolignan
$C_{16}H_{18}O_9$ 354.311

(7′R,8′R)-form
7-Carboxylic acid, 3,3′,5,5′-tetra-Me ether
4′,7′,9′-Trihydroxy-3,3′,5,5′-tetramethoxy-8,9-dinor-4,8′-oxylignan-7-oic acid. ***Asprenol D***
$C_{20}H_{24}O_{10}$ 424.402
Constit. of *Ilex asprella*. Amorph. light yellow powder. $[\alpha]^{25}_{D}$-4.0 (c,

0.1 in MeOH). λ_{max} 209 (log ε 3.15); 263 (log ε 2.43) (MeOH)

(7′R,8′S)-form

3,3′,5,5′-Tetra-Me ether

3,3′,5,5′-Tetramethoxy-8,9-dinor-4,8′-oxylignan-4′,7,7′,9′-tetrol. 4′,7,7′,9′-Tetrahydroxy-3,3′,5,5′-tetramethoxy-8,9-dinor-4,8′-oxylignan. ***Wilfordiol D***

[51149-34-9]

$C_{20}H_{26}O_9$ 410.419

Constit. of the leaves of *Tripterygium wilfordii*. Light yellow oil. $[\alpha]^{20}_{D}$ +3.7 (c, 0.04 in MeOH). λ_{max} 206 (log ε 4.5); 228 (log ε 3.93); 241 (log ε 3.66); 275 (log ε 3.13) (MeOH)

(7′S,8′R)-form

7-Carboxylic acid, 3,3′,5,5′-tetra-Me ether

Asprenol C

$C_{20}H_{24}O_{10}$ 424.402

Constit. of the stems of *Ilex asprella*. Amorph. light yellow powder. $[\alpha]^{25}_{D}$ -4.0 (c, 0.1 in MeOH). λ_{max} 209 (log ε 3.21); 263 (log ε 2.47) (MeOH)

(7′S,8′S)-form

3,3′,5,5′-Tetra-Me ether

Wilfordiol C

[51149-33-8]

$C_{20}H_{26}O_9$ 410.419

Constit. of the leaves of *Tripterygium wilfordii*. Light yellow oil. $[\alpha]^{20}_{D}$ +15.6 (c, 0.1 in MeOH). λ_{max} 207 (log ε 4.42); 229 (log ε 3.87); 242 (log ε 3.65); 273 (log ε 3.13) (MeOH)

7-Aldehyde, 3,3′,5,5′-tetra-Me ether

4′,7′,9′-Trihydroxy-3,3′,5,5′-tetramethoxy-8,9-dinor-4,8′-oxylignan-7-al

7-Aldehyde, 3,3′,5,5′-tetra-Me ether, 4′-O-β-D-glucopyranoside

[2056028-46-5]

$C_{26}H_{34}O_{14}$ 570.544

Constit. of *Dendrobium aurantiacum* var. *denneanum*. Gum. $[\alpha]^{20}_{D}$ -31.2 (c, 0.04 in MeOH).

Cao, X. *et al., Fitoterapia*, 2012, **83**, 343-347 (*Wilfordiols C,D*).
Li, X.-H. *et al., Rec. Nat. Prod.*, 2016, **10**, 113-116 (*Dendrobium aurantiacum constit*).
Du, B. *et al., Fitoterapia*, 2018, **129**, 220-225 (*Asprenols C,D*)

8,9-Dinor-4,8′-oxylignan-3,3′,4′,7,7′-pentol D280

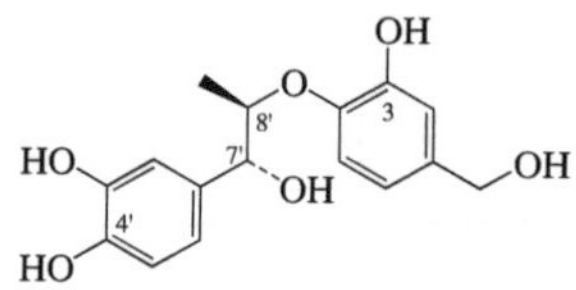

(7′R,8′R)-form

3,3′,4′,7,7′-Pentahydroxy-8,9-dinor-4,8′-oxyneolignan

$C_{16}H_{18}O_6$ 306.314

(7′R,8′R)-form

7-Carboxylic acid, 3,3′,4′-tri-Me ether

7′-Hydroxy-3,3′,4′-trimethoxy-8,9-dinor-4,8′-oxylignan-7-oic acid

7-Carboxylic acid, 3,3′,4′-tri-Me ether, Me ester

(-)-Acortatarinowin A

[1807543-26-5]

$C_{20}H_{24}O_7$ 376.405

Constit. of *Acorus tatarinowii*. Light yellow gum. $[\alpha]^{20}_{D}$ -18 (c, 0.03 in MeOH). Resolved by chiral hplc

(7′R,8′S)-form

7-Carboxylic acid, 3,3′,4′-tri-Me ether, Me ester

(-)-Acortatarinowin B

[1807543-28-7]

$C_{20}H_{24}O_7$ 376.405

Constit. of *Acorus tatarinowii*. Light yellow gum. $[\alpha]^{20}_{D}$ -10 (c, 0.01 in MeOH). Resolved by chiral hplc

(7′S,8′R)-form

7-Carboxylic acid, 3,3′,4′-tri-Me ether, Me ester

(+)-Acortatarinowin B

[1807543-29-8]

$C_{20}H_{24}O_7$ 376.405

Constit. of *Acorus tatarinowii*. Light yellow gum. $[\alpha]^{20}_{D}$ +8 (c, 0.01 in MeOH). Resolved by chiral hplc

(7′S,8′S)-form

7-Carboxylic acid, 3,3′,4′-tri-Me ether, Me ester

(+)-Acortatarinowin A

[1807543-25-4]

$C_{20}H_{24}O_7$ 376.405

Constit. of *Acorus tatarinowii*. Exhibits weak inhibitory activity against NO prodn. in activated macrophages. Light yellow gum. $[\alpha]^{20}_{D}$ +18 (c, 0.05 in MeOH). Resolved by chiral hplc

(7′RS,8′RS)-form

7-Carboxylic acid, 3,3′,4′-tri-Me ether, Me ester

Acortatarinowin A

[1807543-38-9]

$C_{20}H_{24}O_7$ 376.405

Constit. of *Acorus tatarinowii*. Light yellow gum. $[\alpha]^{20}_{D}$ +13 (c, 0.1 in MeOH). Partial racemate. λ_{max} 208 (log ε 4.61); 263 (log ε 4.23); 295 (log ε 3.92) (MeOH)

(7′RS,8′SR)-form

7-Carboxylic acid, 3,3′,4′-tri-Me ether, Me ester

Acortatarinowin B

[1807543-39-0]

$C_{20}H_{24}O_7$ 376.405

Constit. of *Acorus tatarinowii*. Light yellow gum. $[\alpha]^{20}_{D}$ -8 (c, 0.04 in MeOH). Partial racemate. λ_{max} 205 (log ε 4.5); 235 (log ε 4.23); 263 (log ε 4.04) (MeOH).

Lu, Y. *et al., J. Nat. Prod.*, 2015, **78**, 2205-2214 (*Acortatarinowins A,B, activity*)

8,9-Dinor-4,4′-oxylign-7′-ene-7,9′-dioic acid D281

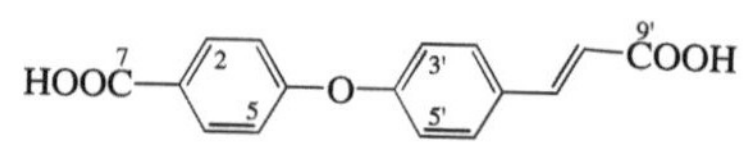

4-[4-(2-Carboxyethenyl)phenoxy] benzoic acid. 3-[4-(4-Carboxyphenoxy) phenyl]-2-propenoic acid

$C_{16}H_{12}O_5$ 284.267

(E)-form [1150187-90-8]

Constit. of *Oxalis pes-caprae*. Powder. λ_{max} 280 (log ε 2.88); 310 (log ε 2.86) (MeOH)

9′-Aldehyde, 7-alcohol, 2,6-dimethoxy, 3′,5′-dihydroxy

3,5,7′-Trihydroxy-2′,6′-dimethoxy-8′,9′-dinor-4,4′-oxylign-7-en-9-al. 3,5,7′-Trihydroxy-2′,6′-dimethoxy-8′,9′-dinor-9-oxo-4,4′-oxylign-7-ene

[1987879-22-0]

$C_{18}H_{18}O_7$ 346.335

Constit. of the stems of *Acanthopanax senticosus*. Inhibitor of PTP1B. Yellow powder. Structure requires revision following synthetic studies in 2018. λ_{max} 295 (log ε 4.21); 340 (log ε 4.32) (MeOH)

Dialdehyde, 3,3′,5-trimethoxy

3,3′,5-Trimethoxy-8,9-dinor-4,4′-oxylign-7′-ene-7,9′-dial. 3,3′,5-Trimethoxy-8,9-dinor-7,9′-dioxo-4,4′-oxylign-7′-ene

[1984790-55-7]
$C_{19}H_{18}O_6$ 342.347
Constit. of the stems of *Acanthopanax senticosus*. Inhibitor of PTP1B. Yellow powder. Structure requires revision following synthetic studies in 2018. λ_{max} 287 (log ε 4.27); 335 (log ε 4.34) (MeOH).

DellaGreca, M. *et al.*, *Chem. Biodiversity*, 2009, **6**, 459-465 (*Oxalis pes-caprae constit*).
Li, N. *et al.*, *Phytochem. Lett.*, 2015, **13**, 286-289 (*Acanthopanax senticosus constits, activity*).
Noshita, T. *et al.*, *Phytochem. Lett.*, 2018, **27**, 214-218 (*Acanthopanax senticosus constits, synth, struct*)

Sinocalamus affinis Dinorsesquilignan D282

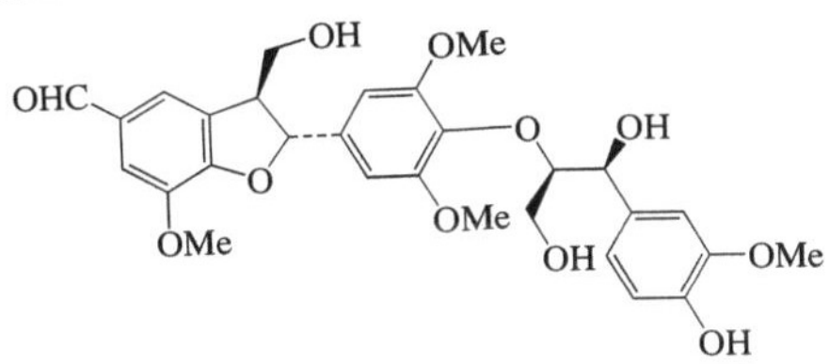

Absolute Configuration

4,7′-Epoxy-4″,7″,9′,9″-tetrahydroxy-3′,3″,5,5′-tetramethoxy-8,9-dinor-4′,8″-oxy-3,8′-sesquilignan-7-al
[1290633-25-8]
$C_{29}H_{32}O_{11}$ 556.564
Constit. of the stems of *Sinocalamus affinis*. Amorph. powder. $[\alpha]^{20}_{D}$ -3 (c, 0.07 in CH_2Cl_2). λ_{max} 205 (log ε 4.22); 235 (log ε 3.77); 286 (log ε 3.44); 310 (log ε 3.35) (MeOH).

Xiong, L. *et al.*, *J. Nat. Prod.*, 2011, **74**, 1188-1200 (*Sinocalamus affinis constit*)

Diobulbinone D283

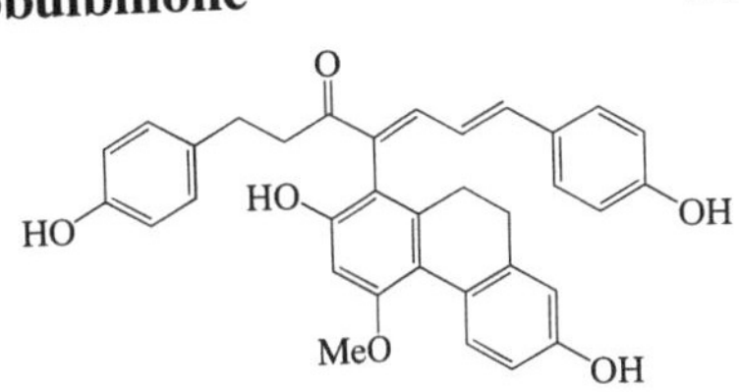

[1351956-49-4]
$C_{34}H_{30}O_6$ 534.608
Constit. of the rhizomes of *Dioscorea bulbifera*. Amorph. yellow powder. $[\alpha]^{25}_{D}$ -42.6 (c, 0.12 in MeOH). λ_{max} 207 (log ε 0.91); 278 (log ε 0.43); 359 (log ε 0.5) (MeOH).

Liu, H. *et al.*, *Chem. Biodiversity*, 2011, **8**, 2110-2116 (*Diobulbinone*)

Diocollettine A D284

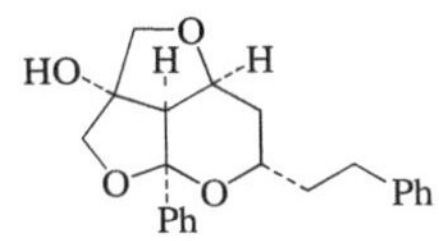

Diocollettines A
[1964492-21-4]
$C_{22}H_{24}O_4$ 352.43
Fused tricyclic diarylheptanoid. Constit. of the rhizomes of *Dioscorea collettii*. Exhibits moderate cytotoxic activity. Cryst.. $[\alpha]^{20}_{D}$ -126.8 (c, 0.1 in MeOH). λ_{max} 195; 205; 256 (MeOH).

Jing, S.-S. *et al.*, *Tetrahedron Lett.*, 2016, **57**, 3215-3217 (*Diocollettine A, cd, cryst struct, activity*).
Kawamoto, Y. *et al.*, *Org. Lett.*, 2019, **21**, 5813-5816 (*synth*).
Kawamoto, Y. *et al.*, *Org. Lett.*, 2019, **21**, 10099-10101 (*synth*)

Dioscorol A D285

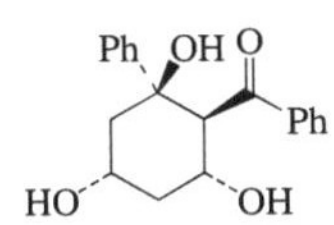

Relative Configuration

2-Benzoyl-1-phenyl-1,3,5-cyclohexanetriol. 2,4,6-Trihydroxy-2-phenylcyclohexyl phenyl ketone
[2250036-68-9]
$C_{19}H_{20}O_4$ 312.365
Constit. of the rhizomes of *Dioscorea septemloba*. Powder. $[\alpha]^{25}_{D}$ +2.5 (c, 0.24 in MeOH). λ_{max} 247 (log ε 3.95); 285 (sh); 343 (sh) (MeOH).

Zhang, Y. *et al.*, *Fitoterapia*, 2017, **117**, 28-33 (*Dioscorol A*)

1,7-Diphenyl-4,6-heptadien-3-ol D286

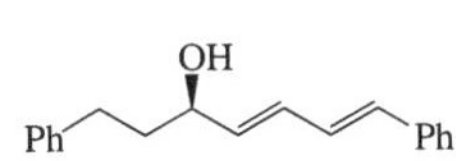

(3*R*,4*E*,6*E*)-*form*

$C_{19}H_{20}O$ 264.368
(3R,4E,6E)-form [152323-34-7]
Constit. of *Curcuma comosa*, *Curcuma xanthorrhiza* and the floral nectar of *Anigozanthos flavidus*. Phytoestrogen. Accelerates human osteoblast proliferation and differentiation. Powder (CH_2Cl_2/hexane). Mp 79 - 80°. $[\alpha]^{27}_{D}$ -64.2 (c, 0.3 in EtOH).

Claeson, P. *et al.*, *Planta Med.*, 1993, **59**, 451-454 (*Curcuma xanthorrhiza constit*).
Suksamrarn, A. *et al.*, *Bioorg. Med. Chem.*, 2008, **16**, 6891-6902 (*Curcuma comosa constit, abs config*).
Hölscher, D. *et al.*, *J. Nat. Prod.*, 2008, **71**, 251-257 (*Anigozanthos flavidus constit, synth*).
Intapad, S. *et al.*, *J. Agric. Food Chem.*, 2012, **60**, 758-764 (*Curcuma comosa constit, activity*).
Tantikanlayaporn, D. *et al.*, *PLoS One*, 2013, **8**, e78739 (*activity*)

1,7-Diphenyl-4,6-heptadien-3-one, CAS D287

$PhCH_2CH_2COCH{=}CHCH{=}CHPh$
Alnustone
[38940-11-3]
[58506-34-6]
$C_{19}H_{18}O$ 262.352
(E,E)-form [33457-62-4]
Constit. of *Alnus pendula*, *Alpinia katsumadai*, and *Curcuma xanthorrhiza*. Pale yellow needles (EtOH). Mp 63 - 63.5° (61-62.5°)
2,4-Dinitrophenylhydrazone
[38940-16-8]
Red cryst. (EtOAc). Mp 188 - 189°
4,5-Dihydro
1,7-Diphenyl-6-hepten-3-one
[87030-31-7]
$C_{19}H_{20}O$ 264.368
Constit. of *Curcuma comosa*. Oil.

Sakakibaba, M. *et al.*, *Agric. Biol. Chem.*, 1972, **36**, 1825-1827 (*synth*).
Suga, T. *et al.*, *Bull. Chem. Soc. Jpn.*, 1972, **45**, 2058-2060 (*Alnustone*).
Vig, O.P. *et al.*, *Indian J. Chem.*, 1975, **13**, 1129-1130 (*synth*).
Jurgens, T.M. *et al.*, *J. Nat. Prod.*, 1994, **57**, 230-235 (*1,7-Diphenyl-6-hepten-3-one*).
Göksu, S. *et al.*, *Turk. J. Chem.*, 2003, **27**, 21-30 (*synth*).
Hölscher, D. *et al.*, *J. Nat. Prod.*, 2008, **16**, 6891-6902 (*synth*).
Xiao, X. *et al.*, *Sep. Purif. Technol.*, 2011, **81**, 265-269 (*Alpinia katsumadai constit*)

1,7-Diphenyl-3,5-heptanediol D288

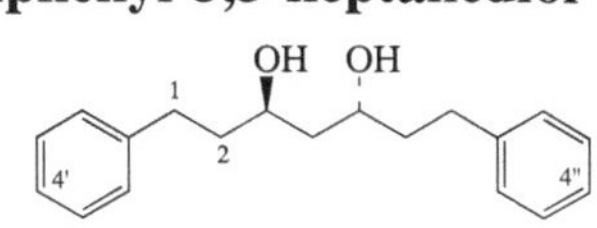

(3*R*,5*R*)-*form*

[52393-54-1] [55836-43-6]
[87095-80-5] [93603-99-7]
[99781-95-0]
$C_{19}H_{24}O_2$ 284.399
(3R,5R)-form
Yashabushidiol B [103729-38-0]
Constit. of male flowers of *Alnus sieboldiana*. Antiproliferative agent. Cryst.. Mp 92 - 93°. $[\alpha]_D$ +7.2 ($CHCl_3$)
3-Ketone
5-Hydroxy-1,7-diphenyl-3-heptanone.
5-Epidihydroyashabushiketol

[100761-20-4]
$C_{19}H_{22}O_2$ 282.383
Constit. of the rhizomes of *Alpinia officinarum*. Needles. Mp 48.5 - 49.5°. $[\alpha]_D$ -16.9 (c, 1.2 in $CHCl_3$). λ_{max} 212 (ε 2020); 248 (ε 430); 253 (ε 490); 259 (ε 520); 268 (ε 390); 279 (sh) (ε 140); 298 (sh) (ε 50) (EtOH)

Diketone
1,7-Diphenyl-3,5-heptanedione
[38572-30-4]
$C_{19}H_{20}O_2$ 280.367
Constit. of the rhizomes of *Alpinia conchigera* and *Alpinia officinarum*. Pale yellow oil

4'-Hydroxy, 1,2-didehydro(E-), 3-ketone
5-Hydroxy-1-(4-hydroxyphenyl)-7-phenyl-1-hepten-3-one. ***4'-Hydroxyyashabushiketol***
[1335038-20-4]
$C_{19}H_{20}O_3$ 296.366
Constit. of the seeds of *Alpinia katsumadai*. Yellow powder. $[\alpha]_D^{25}$ -18.5 (c, 0.1 in MeOH). λ_{max} 233 (log ε 4.6); 325 (log ε 4.7) (MeOH)

4''-Hydroxy, 1,2-didehydro(E-), 5-ketone
5-Hydroxy-1-(4-hydroxyphenyl)-7-phenyl-6-hepten-3-one
[208462-98-0]
$C_{19}H_{20}O_3$ 296.366
Constit. of *Alpinia katsumadai*. Oil. $[\alpha]_D$ +10.7 (c, 0.1 in $CHCl_3$)

(3S,5S)-form [103729-39-1]
Constit. of the seeds of *Alpinia katsumadai*. Cryst.. Mp 91 - 92°. $[\alpha]_D$ -7.3 ($CHCl_3$)

3-Ketone
Dihydroyashabushiketol
[24192-01-6]
$C_{19}H_{22}O_2$ 282.383
Constit. of male flowers of *Alnus sieboldiana* and *Alnus firma*. Cryst.. Mp 72.5 - 73°. $[\alpha]_D$ +14.6 (c, 0.75 in $CHCl_3$)

1,2-Didehydro(E-)
1,7-Diphenyl-1-heptene-3,5-diol, CAS
[87095-75-8]
$C_{19}H_{22}O_2$ 282.383
Constit. of the seeds of *Alpinia katsumadai*. Needles (C_6H_6/hexane). Mp 75 - 77°. $[\alpha]_D^{23}$ +25.19 (c, 0.63 in EtOH). Has *S,S*-config

1,2-Didehydro(E-), 3-ketone
5-Hydroxy-1,7-diphenyl-1-hepten-3-one. ***Yashabushiketol***
[24192-00-5]
$C_{19}H_{20}O_2$ 280.367
Constit. of male flowers of *Alnus sieboldiana* and *Alnus firma*. Needles (hexane). Mp 59.5 - 60.5°. $[\alpha]_D^{20}$ +29 (c, 1.05 in $CHCl_3$)

1,2-Didehydro(E-), 5-ketone
5-Hydroxy-1,7-diphenyl-6-hepten-3-one
[87095-74-7]
$C_{19}H_{20}O_2$ 280.367
Constit. of *Alpinia katsumadai* and *Curcuma comosa*. Yellow needles (C_6H_6/hexane). Mp 59.5 - 60.5°. $[\alpha]_D^{23}$ +17.8 (c, 0.67 in EtOH). Has *R*-config.. λ_{max} 251 (log ε 4.31); 283 (log ε 3.61); 292 (log ε 3.43) (MeOH)

1,2-Didehydro(E), diketone
1,7-Diphenyl-1-heptene-3,5-dione. 5-Hydroxy-1,7-diphenyl-4,6-heptadien-3-one
[99742-15-1]
[87095-77-0] (enol, *E,E*-form)
[93559-31-0] (enol, unspecified stereochem.) [478313-96-1] (enol, 4*Z*,6*E*-form)
$C_{19}H_{18}O_2$ 278.351
Constit. of *Alpinia blepharocalyx*, *Alpinia katsumadai* and *Dioscorea villosa*. Orange-yellow needles (hexane). Mp 74 - 75°. Mixt. of tautomers

3-Deoxy
see 1,7-Diphenyl-6-hepten-3-ol, D292

(3RS,5SR)-form
Yashabushidiol A [52393-53-0]
Constit. of *Alnus sieboldiana* and *Alpinia katsumadai*. Cryst.. Mp 80 - 81° (77-79°)

(5ξ)-form
3-Ketone, Me ether
5-Methoxy-1,7-diphenyl-3-heptanone
[100667-54-7]
$C_{20}H_{24}O_2$ 296.41
Constit. of the rhizomes of *Alpinia officinarum*. Oil. λ_{max} 210 (ε 12600); 260 (ε 500) (EtOH).

Asakawa, Y. *Bull. Chem. Soc. Jpn.*, 1970, **43**, 2223-2229 (*Alnus firma constits*).
Asakawa, Y. *et al., Bull. Chem. Soc. Jpn.*, 1975, **45**, 1794-1797 (*Alnus sieboldiana constits*).
Suga, T. *et al., Bull. Chem. Soc. Jpn.*, 1983, **56**, 3353-3357 (*Dihydroyashabushiketol, abs config*).
Kuroyanagi, M. *et al., Chem. Pharm. Bull.*, 1983, **31**, 1544-1550 (*Alpinia katsumadai constits*).
Itokawa, H. *et al., Chem. Pharm. Bull.*, 1985, **33**, 4889-4893 (*Alpinia officinarum constits*).
Hashimoto, T. *et al., Chem. Pharm. Bull.*, 1986, **34**, 1846-1849 (*Alnus sieboldiana constits*).
Rychnovsky, S.D. *et al., J. Org. Chem.*, 1991, **56**, 5161-5169 (*synth, ir, pmr, cmr, ms*).
Solladie, G. *et al., Tetrahedron Lett.*, 1992, **33**, 931-934 (*Yashabushiketol, synth*).
Zhao, H. *et al., Chin. Chem. Lett.*, 1994, **5**, 367-370 (*Yashabushiketol, synth*).
Jurgens, T.M. *et al., J. Nat. Prod.*, 1994, **57**, 230-235 (*5-Hydroxy-1,7-diphenyl-6-hepten-3-one*).
Athamaprasangsa, S. *et al., Phytochemistry*, 1994, **37**, 871-873 (*1,7-Diphenyl-3,5-heptanedione*).
Sirat, H.M. *et al., Planta Med.*, 1996, **62**, 188-189 (*Hydroxydiphenylheptenone*).
Ngo, K.S. *et al., Phytochemistry*, 1998, **47**, 1117-1123 (*Alpinia katsumadai constits*).
Narasimhulu, M. *et al., Bioorg. Med. Chem. Lett.*, 2009, **19**, 3125-3127 (*Yashabushidiol, synth, activity*).
Nam, J.-W. *et al., J. Nat. Prod.*, 2011, **74**, 2109-2115 (*4'-Hydroxyyashabushiketol*).
Pawar, V.U. *et al., Tetrahedron: Asymmetry*, 2011, **22**, 8-11 (*3S,5S-form, synth*).
Romanski, J. *et al., Tetrahedron: Asymmetry*, 2011, **22**, 787-790 (*Dihydroyashibushiketol, synth*).
Ho, T.-L. *et al., J. Chin. Chem. Soc. (Taipei)*, 2012, **58**, 455-458 (*synth*).
Halperin, S.D. *et al., Org. Biomol. Chem.*, 2013, **11**, 1702-1705 (*Dihydroyashibushiketol, synth*).
Liu, D. *et al., Chin. J. Nat. Med.*, 2014, **12**, 139-141 (*1,7-Diphenyl-3,5-heptanedione*).
Merad, J. *et al., Org. Lett.*, 2015, **17**, 2118-2121 (*3RS,5SR-form, synth, pmr, cmr*).
Gadakh, S.K. *et al., Tetrahedron: Asymmetry*, 2015, **26**, 118-123 (*3S,5S-form, synth*)

1,7-Diphenyl-1,3,5,7-heptanetetrol
D289

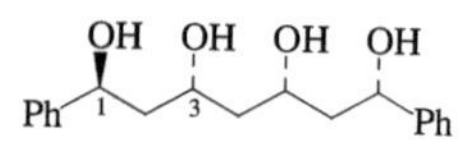

$C_{19}H_{24}O_4$ 316.397
(1S,3R,5S,7S)-form
Diospongin C [791104-65-9]
Constit. of the rhizomes of *Dioscorea spongiosa*. Exhibits antiosteoporotic activity. Light yellow oil. $[\alpha]_D^{25}$ -45.5 (c, 0.5 in $CHCl_3$).

Yin, J. *et al., Planta Med.*, 2004, **70**, 54-58 (*Diospongin C, activity*)

1,7-Diphenyl-1,3,5-heptanetriol
D290

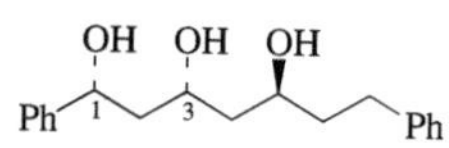

(1R,3R,5S)-form

$C_{19}H_{24}O_3$ 300.398

(1R,3R,5S)-form
Yashabushitriol [103654-25-7]
Constit. of male flowers of *Alnus sieboldiana*. Cryst.. Mp 89 - 90°. [α]D +30.3 ($CHCl_3$)
3-Ketone
1,5-Dihydroxy-1,7-diphenyl-3-heptanone. ***Yashabushiketodiol A*** [103654-23-5]
$C_{19}H_{22}O_3$ 298.382
Constit. of male flowers of *Alnus sieboldiana*. Cryst.. Mp 62 - 63°. [α]D +21.4 (MeOH)
(1S,5S)-form
3-Ketone
Yashabushiketodiol B
[103654-24-6]
$C_{19}H_{22}O_3$ 298.382
Constit. of *Alnus sieboldiana*. Cryst.. Mp 60 - 61°. [α]D -28.6 ($CHCl_3$). [α]D -16.3 (MeOH).

Hashimoto, T. *et al.*, *Chem. Pharm. Bull.*, 1986, **34**, 1846-1849 (*Yashabushitriol, Yashabushiketodiols A,B*).
Feldman, K.S. *et al.*, *Tetrahedron Lett.*, 1989, **30**, 6985-6988 (*Yashabushitriol, synth*).
Miyashita, M. *et al.*, *Chem. Lett.*, 1990, **19**, 791-794 (*Yashabushitriol, Yashabushiketodiols A,B, synth*).
Clarkson, G.J. *et al.*, *Tetrahedron Lett.*, 2013, **50**, 6834-6837 (*Yashabushitriol, synth*).
Sperandio, C. *et al.*, *Eur. J. Org. Chem.*, 2020, 2493-2496 (*Yashabushitriol, synth*)

1,7-Diphenyl-1,3,5-heptatriene D291

PhCH=CHCH=CHCH=$CHCH_2$Ph
1,1′-(1,3,5-Heptatriene-1,7-diyl) bisbenzene, CAS
$C_{19}H_{18}$ 246.353
(1E,3E,5E)-form [130727-57-0]
Isol. from rhizomes of *Curcuma comosa*. Shows moderate estrogenic activity. Cryst. (hexane). Mp 82°.

Tolbert, L.M. *et al.*, *J. Am. Chem. Soc.*, 1990, **112**, 9519-9527 (*synth*).
Thiele, G. *et al.*, *J. Am. Chem. Soc.*, 1994, **116**, 446-454 (*synth*).
Suksamrarn, A. *et al.*, *Bioorg. Med. Chem.*, 2008, **16**, 6891-6902 (*Curcuma comosa constit*)

1,7-Diphenyl-6-hepten-3-ol, CAS D292

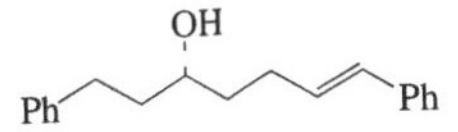

(R,E)-form

α-(4-Phenyl-3-butenyl)benzenepropanol, CAS
[169383-33-9]
$C_{19}H_{22}O$ 266.384
(R,E)-form [152242-59-6]
Constit. of *Curcuma xanthorrhiza*. Mp 47 - 49°. $[\alpha]^{23}_D$ +7 (c, 1.2 in $CHCl_3$)
(S,E)-form [1065306-33-3]
Constit. of the rhizomes of *Curcuma comosa*. Needles (CH_2Cl_2/hexane). Mp 46 - 48°. [α]D -2.6 (c, 0.3 in EtOH)
(±)-(E)-form [87095-76-9]
Constit. of *Alpinia katsumadai* and *Curcuma comosa*. Plates (Me_2CO/hexane). Mp 44 - 46°. λ_{max} 252 (log ε 4.20); 283 (log ε 3.49); 292 (log ε 3.33) (MeOH)
Ac
[155167-05-8]
$C_{21}H_{24}O_2$ 308.421
Constit. of *Curcuma comosa*. Oil.

Kuroyanagi, M. *et al.*, *Chem. Pharm. Bull.*, 1983, **31**, 1544-1550 (*Alpinia katsumadai constit*).
Claeson, P. *et al.*, *Planta Med.*, 1993, **59**, 451-454 (*Curcuma xanthorrhiza constit*).
Jurgens, T.M. *et al.*, *J. Nat. Prod.*, 1994, **57**, 230-235 (*Curcuma comosa constits*).
Suksamrarn, A. *et al.*, *Bioorg. Med. Chem.*, 2008, **16**, 6891-6902 (*abs config*)

1,7-Diphenyl-4-hepten-3-one D293

$PhCH_2CH_2CH$=$CHCOCH_2CH_2Ph$
[79559-59-4]
$C_{19}H_{20}O$ 264.368
Constit. of *Alpinia officinarum*, *Alnus maximowiczii* and *Dioscorea opposita*. PAF receptor binding antagonist. Oil.

Itokawa, H. *et al.*, *Chem. Pharm. Bull.*, 1981, **29**, 2383-2385 (*1,7-Diphenyl-4-hepten-3-one*).
Kato, N. *et al.*, *Chem. Pharm. Bull.*, 1984, **32**, 3323-3326 (*synth*).
Tori, M. *et al.*, *Phytochemistry*, 1995, **40**, 1263-1264 (*Alnus maximowiczii constit*).
Fan, G. *et al.*, *Bioorg. Med. Chem. Lett.*, 2007, **17**, 6720-6722 (*activity*).
Xin, M. *et al.*, *Molecules*, 2017, **22**, 647:1-12 (*Alpinia officinarum constit*)

1,7-Diphenyl-5-hepten-3-one D294

$PhCH_2CH_2COCH_2CH$=$CHCH_2Ph$
$C_{19}H_{20}O$ 264.368
(E)-form [1309048-53-0]
Constit. of the rhizomes of *Alpinia officinarum*. Antibacterial agent. Oil. $[\alpha]^{20}_D$ +2.6 (c, 0.45 in MeOH). Optical rotation unaccounted for. λ_{max} 250; 282 (MeOH).

Zhang, B.-B. *et al.*, *Fitoterapia*, 2010, **81**, 948-952 (*Alpinia officinarum constit*)

1,7-Diphenyl-4-(2-phenylethyl)-1-heptene-3,5-dione D295

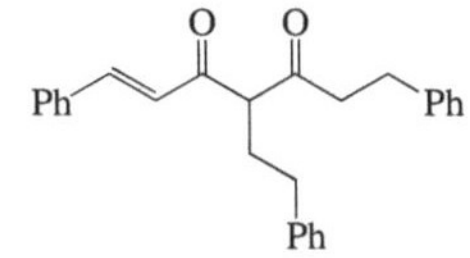

$C_{27}H_{26}O_2$ 382.503
(ξ)-form [1309048-54-1]
Constit. of the rhizomes of *Alpinia officinarum*. Antibacterial agent. Oil.

Zhang, B.-B. *et al.*, *Fitoterapia*, 2010, **81**, 948-952 (*Alpinia officinarum constit*)

Diphyllin‡ D296

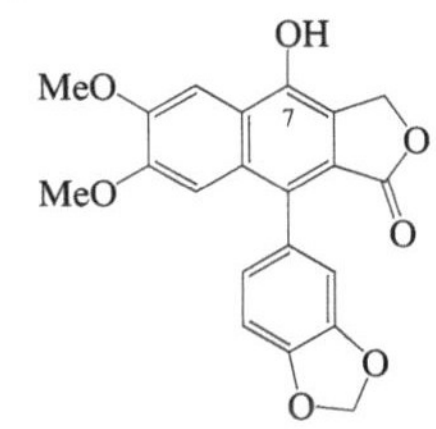

9-(1,3-Benzodioxol-5-yl)-4-hydroxy-6,7-dimethoxynaphtho[2,3-c]-furan-1(3H)-one, CAS. *2,7′-Cyclo-7-hydroxy-4,5-dimethoxy-3′,4′-methylenedioxy-8,8′-ligna-7,7′-dien-9′,9-olide*. *7-Hydroxy-4,5-dimethoxy-3′,4′-methylenedioxy-2,7′-cycloligna-7,7′-dien-9′,9-olide*
[22055-22-7]
$C_{21}H_{16}O_7$ 380.352
Constit. of *Diphylleia grayi*, *Diphylleia cymosa*, *Haplophyllum tuberculatum*, *Cleistanthus collinus*, *Justicia procumbens*, *Justicia patentiflora* and *Haplophyllum hispanicum*. Cytotoxin. Potent V-ATPase inhibitor on influenza viruses. Pale yellow needles (EtOH). Mp 291°. λ_{max} 230; 267; 358 (EtOH) (Berdy)
O-β-D-Apiofuranoside
Tuberculatin. *Diphyllin apioside*
[90706-10-8]
$C_{26}H_{24}O_{11}$ 512.467
Constit. of *Haplophyllum buxbaumii*, *Haplophyllum tuberculatum* and *Justicia procumbens*. Plates (EtOAc). Mp 245 - 248°. $[\alpha]^{20}_D$ -107.6 (c, 1.1 in 20% MeOH/$CHCl_3$). λ_{max} 263 (log ε 4.66); 294 (log ε 3.99); 310 (log ε 3.99); 352 (log ε 3.66) ($CHCl_3$)
O-(5-O-Acetyl-β-D-apiofuranoside)
Diphyllin 5-acetylapioside. *5″-O-Acetyltuberculatin*
[111647-42-8]
$C_{28}H_{26}O_{12}$ 554.504

Isol. from *Haplophyllum buxbaumii*. Amorph.. [α] D -56.2 (c, 0.98 in MeOH). λ_{max} 226 (log ε 4.39); 262 (log ε 4.73); 292 (log ε 4.04); 312 (log ε 4.06); 350 (log ε 3.72) (MeOH)

O-(3,5-Isopropylidene-β-D-apiofuranoside)
Procumbenoside J
[1624617-95-3]
$C_{29}H_{28}O_{11}$ 552.532
Constit. of *Justicia procumbens*. Displays cytotoxic activity. Amorph. yellow powder. $[\alpha]_D^{20}$ -14.7 (c, 0.075 in MeOH). λ_{max} 241 (log ε 3.17); 361 (log ε 3.42) (MeOH)

O-α-L-Arabinopyranoside
Diphyllin arabinoside. **Phyllanthuspermin C. Phyllanthusmin C**
[914639-26-2]
$C_{26}H_{24}O_{11}$ 512.467
Constit. of *Phyllanthus glaucus* and *Phyllanthus oligosperma*. Exhibits cytotoxic activity. Prisms (MeOH). Mp 247 - 249°. [α] D -23.3 (c, 0.015 in MeOH). λ_{max} 261 (log ε 4.27); 292 (log ε 3.89); 310 (log ε 3.97); 353 (log ε 3.67) (MeOH)

O-(3-O-Acetyl-α-L-arabinopyranoside)
Phyllanthusmin E‡
[1613505-43-3]
$C_{28}H_{26}O_{12}$ 554.504
Constit. of *Phyllanthus poilanei*. Exhibits moderate cytotoxicity against HT-29 human colon cancer cells. Amorph. powder. $[\alpha]_D^{20}$ -4 (c, 0.1 in $CHCl_3$). Not to confused with Phyllanthusmin E, also in D296. λ_{max} 260 (log ε 4.58) (no solvent reported)

O-(4-O-Acetyl-α-L-arabinopyranoside)
Phyllanthuspermin B. Phyllanthusmin B
[914639-25-1]
$C_{28}H_{26}O_{12}$ 554.504
Constit. of stems and roots of *Phyllanthus oligosperma*. Prisms (MeOH). Mp 253 - 255°. [α] D -21 (c, 0.03 in MeOH). λ_{max} 261 (log ε 3.74); 298 (log ε 3.17); 315 (log ε 3.02); 347 (log ε 2.64) (MeOH)

O-(3,4-Di-O-acetyl-α-L-arabinopyranoside)
Phyllanthusmin D‡
[1613505-42-2]
$C_{30}H_{28}O_{13}$ 596.541
Constit. of *Flueggea virosa* and *Phyllanthus poilanei*. Exhibits strong cytotoxicity against HT-29 human colon cancer cells. Needles (Me_2CO/hexane). Mp 210 - 211°. $[\alpha]_D^{20}$ -3 (c, 0.1 in $CHCl_3$). Not to confused with Phyllanthusmin D, also in D296. λ_{max} 260 (log ε 4.54) (MeOH)

O-(2-O-Methyl-α-L-arabinopyranoside)
Acutissimalignan A
[1005390-46-4]
$C_{27}H_{26}O_{11}$ 526.494
Constit. of *Phyllanthus acutissima*. Powder. Mp 177 - 178°. $[\alpha]_D^{24}$ +12.1 (c, 0.54 in $CHCl_3$). λ_{max} 220 (log ε 3.61); 257 (log ε 3.94); 293 (log ε 3.26); 314 (log ε 3.26); 350 (log ε 2.92) (EtOH)

O-(2,3,5-Tri-O-methyl-D-xylofuranoside)
Cleistanthin D
[81417-78-9]
$C_{29}H_{30}O_{11}$ 554.548
Isol. from heartwood of *Cleistanthus collinus*. Plates (MeOH). Mp 199 - 200°. [α] D +25. λ_{max} 264 (log ε 4.78); 294 (log ε 3.98); 315 (log ε 4.01); 335 (log ε 3.66) (EtOH)

O-D-Xylopyranoside
Diphyllinin. *Diphyllin xyloside*
[52847-81-1]
$C_{26}H_{24}O_{11}$ 512.467
Constit. of *Haplophyllum hispanicum*. λ_{max} 261; 295; 317 (EtOH)

O-(Acetyl-D-xylopyranoside)
Diphyllinin monoacetate
[52948-65-9]
$C_{28}H_{26}O_{12}$ 554.504
Constit. of *Haplophyllum hispanicum*. Cryst.. Position of Ac group not determined. λ_{max} 261; 295; 317 (EtOH)

O-(4-O-Methyl-β-D-xylopyranoside)
Cleistanthoside B
[109145-66-6]
$C_{27}H_{26}O_{11}$ 526.494
Constit. of *Cleistanthus patulus* and *Cleistanthus collinus*. Feathery needles (MeOH). Mp 178 - 179°. [α] D -62.2 (c, 1 in dioxan). λ_{max} 214 (ε 4365); 274 (ε 12020); 293 (ε 13180); 314 (ε 28800); 348 (ε 6300) (MeOH) (Berdy)

O-(3,4-Di-O-methyl-D-xylopyranoside)
Cleistanthin A
[25047-48-7]
$C_{28}H_{28}O_{11}$ 540.521
Isol. from leaves of *Cleistanthus collinus*. Exhibits antiproliferative activity. Mp 135 - 136°. $[\alpha]_D^{25}$ -67.2 (c, 2 in $CHCl_3$). λ_{max} 262 (ε 63000); 292 (ε 9770); 315 (ε 10200); 335 (ε 4786) (MeOH) (Berdy)
► GF1694000. Toxic

O-(2,3,4-Tri-O-methyl-β-D-xylopyranoside)
Cleistanthin A methyl ether
[23953-54-0]
$C_{29}H_{30}O_{11}$ 554.548
Constit. of *Phyllanthus taxodiifolius*. Powder (EtOH). Mp 211 - 215°. $[\alpha]_D^{27}$ -50 (c, 1 in $CHCl_3$)

O-[2-Butenoyl-(→?)-D-xylopyranoside]
Diphyllinin crotonate
[52828-27-0]
$C_{30}H_{28}O_{12}$ 580.542
Isol. from *Haplophyllum hispanicum*. Cryst.. [α] D -68 ($CHCl_3$). Posn. of acyl group not determined. λ_{max} 261; 295; 317 (EtOH)

O-β-L-Fucopyranoside
Patentiflorin B. *Diphyllin fucoside*
[857050-64-7]
$C_{27}H_{26}O_{11}$ 526.494
Constit. of the leaves of *Justicia patentiflora*. Cytotoxic against KB, MCF-7 and HCTH6 cell lines. Powder. $[\alpha]_D^{25}$ -28.3 (c, 1.2 in CH_2Cl_2)

O-(4-O-Acetyl-β-L-fucopyranoside)
4″-O-Acetylpatentiflorin B
[856676-47-6]
$C_{29}H_{28}O_{12}$ 568.531
Constit. of the leaves of *Justicia patentiflora*. Cytotoxic against MCF-7 cell line. Powder. $[\alpha]_D^{25}$ -62.5 (c, 1.1 in CH_2Cl_2). λ_{max} 203 (log ε 4.56); 261 (log ε 4.49); 306 (log ε 3.43) (MeOH)

O-(6-Deoxy-β-D-glucopyranoside)
Patentiflorin A
[857050-63-6]
$C_{27}H_{26}O_{11}$ 526.494
Constit. of the leaves of *Justicia patentiflora*. Cytotoxic against KB, MCF-7, HCT116 and H2-60 cell lines. Anti-HIV agent. Powder. $[\alpha]_D^{25}$ -54.3 (c, 1 in CH_2Cl_2). λ_{max}

204 (log ε 4.69); 224 (log ε 4.51); 261 (log ε 4.8) (MeOH)

O-α-L-Rhamnopyranoside

Mananthoside A. *Diphyllin rhamnoside*

[493037-14-2]

$C_{27}H_{26}O_{11}$ 526.494

Constit. of *Mananthes patentiflora* (*Justicia patentiflora*, preferred name). Cytotoxic. Amorph. powder. Mp 216 - 218°. λ_{max} 261 (log ε 4.75) (MeOH)

O-β-D-Glucopyranoside

Diphyllin glucoside. ***Cleistanthin B***

[30021-77-3]

$C_{27}H_{26}O_{12}$ 542.493

Constit. of *Cleistanthus collinus*, *Dysosma versipellis* and *Haplophyllum buxbaumii*. Exhibits cytotoxic activity. Displays atropisomerism

O-(3-O-Methyl-β-D-glucopyranoside)

3″-O-Methylcleistanthin B

[213891-69-1]

$C_{28}H_{28}O_{12}$ 556.52

Constit. of *Cleistanthus collinus*. Needles ($CHCl_3$/MeOH). Mp 152 - 153°. $[\alpha]^{20}_{D}$ -20.2. λ_{max} 234 (log ε 4.01); 268 (log ε 4.45); 292 (log ε 3.43); 310 (log ε 3.73); 364 (log ε 3.17) (MeOH)

O-[β-D-Apiofuranosyl-(1→6)-β-D-glucopyranoside]

[1356021-67-4]

$C_{32}H_{34}O_{16}$ 674.608

Constit. of *Justicia prostrata*

O-[α-L-Arabinofuranosyl-(1→6)-β-D-glucopyranoside]

Reticulatuside A

[1479021-21-0]

$C_{32}H_{34}O_{16}$ 674.608

Constit. of *Phyllanthus reticulatus*. Amorph. yellow powder. $[\alpha]^{25}_{D}$ -82 (c, 0.06 in MeOH). λ_{max} 263 (log ε 4.69) (MeOH)

O-[α-L-Arabinopyranosyl-(1→2)-β-D-apiofuranoside]

Procumbenoside A

[408358-51-0]

$C_{31}H_{32}O_{15}$ 644.582

Constit. of *Justicia procumbens*. Powder (MeOH). $[\alpha]^{25}_{D}$ -16 (c, 0.1 in MeOH). λ_{max} 205 (log ε 4.3); 225 (log ε 4.24); 257 (log ε 4.53); 292 (log ε 3.89); 307 (log ε 3.87); 350 (log ε 3.59) (MeOH)

O-[α-L-Arabinopyranosyl-(1→3)-α-L-arabinopyranoside]

[1698916-29-8]

$C_{31}H_{32}O_{15}$ 644.582

Constit. of *Phyllanthus glaucus*. Amorph. powder. $[\alpha]^{23}_{D}$ -1.4 (c, 0.2 in MeOH). λ_{max} 260 (log ε 3.79); 296 (log ε 3.07); 314 (log ε 3.09) (MeOH)

O-[α-L-Arabinopyranosyl-(1→3)-4-O-acetyl-α-L-arabinopyranoside]

Phyllanthusmin D‡

$C_{33}H_{34}O_{16}$ 686.619

Constit. of roots of *Phyllanthus flexuosus*. Amorph. powder. $[\alpha]^{16}_{D}$ +10.5 (c, 0.07 in MeOH). Not to confused with Phyllanthusmin D (1613505-42-2). λ_{max} 260 (log ε 4.94) (MeOH)

O-[α-L-Arabinopyranosyl-(1→6)-β-D-galactopyranoside]

Mananthoside I

[1069137-63-8]

$C_{32}H_{34}O_{16}$ 674.608

Constit. of *Mananthes patentiflora*. Amorph. solid (MeOH). $[\alpha]^{25}_{D}$ -12.4 (c, 0.38 in MeOH). λ_{max} 225 (log ε 4.36); 262 (log ε 4.58); 309 (log ε 3.89); 350 (log ε 3.54) (MeOH)

O-[α-D-Arabinopyranosyl-(1→6)-3-O-acetyl-β-D-galactopyranoside]

Mananthoside B

[493037-13-1]

$C_{34}H_{36}O_{17}$ 716.645

Constit. of *Mananthes patentiflora* (*Justicia patentiflora*, preferred name). Cryst.. Mp 177 - 179°. λ_{max} 210 (log ε 4.82); 261 (log ε 4.85) (MeOH)

O-[α-D-Arabinopyranosyl-(1→6)-3,4-di-O-acetyl-β-D-galactopyranoside]

4″-O-Acetylmananthoside B

[856676-48-7]

$C_{36}H_{38}O_{18}$ 758.682

Constit. of the leaves of *Justicia patentiflora*. Cytotoxic. Powder. $[\alpha]^{25}_{D}$ -74.3 (c, 0.95 in CH_2Cl_2). λ_{max} 262 (log ε 4.73); 312 (log ε 4.04); 340 (log ε 3.72) (MeOH)

O-[α-L-Arabinopyranosyl-(1→6)-β-D-glucopyranoside]

Arabelline

[133084-20-5]

$C_{32}H_{34}O_{16}$ 674.608

Isol. from *Haplophyllum buxbaumii*. $[\alpha]^{24}_{D}$ -30.8 (c, 0.26 in MeOH)

O-[β-D-Xylopyranosyl-(1→2)-β-D-apiofuranoside]

Majidine

[133084-18-1]

$C_{31}H_{32}O_{15}$ 644.582

Isol. from *Haplophyllum buxbaumii*. $[\alpha]^{24}_{D}$ -32.3 (c, 0.62 in MeOH)

O-[β-D-Xylopyranosyl-(1→5)-β-D-apiofuranoside]

Procumbenoside H

[1821497-29-3]

$C_{31}H_{32}O_{15}$ 644.582

Constit. of *Justicia procumbens*. Yellowish powder. $[\alpha]^{20}_{D}$ -49 (c, 0.05 in MeOH)

O-[2,3-Di-O-methyl-β-D-xylopyranosyl-(1→4)-β-D-glucopyranoside]

Cleistanthin C

[57576-58-6]

$C_{34}H_{38}O_{16}$ 702.662

From *Cleistanthus collinus*. Mp 220 - 221°. $[\alpha]_{D}$ -33 (c, 1 in dioxan)

O-[β-D-Glucopyranosyl-(1→2)-β-D-apiofuranoside]

Procumbenoside B

[775342-86-4]

$C_{32}H_{34}O_{16}$ 674.608

Constit. of *Justicia procumbens*. Amorph. solid. $[\alpha]^{25}_{D}$ -40 (c, 0.08 in MeOH). λ_{max} 213 (log ε 3.78); 260 (log ε 3.28); 313 (sh) (log ε 2.73) (MeOH)

O-[β-D-Glucopyranosyl-(1→2)-β-D-xylopyranoside]

Neesiinoside A

[346420-59-5]

$C_{32}H_{34}O_{16}$ 674.608

Constit. of *Justicia neesii*. Cryst. ($CHCl_3$/MeOH). Mp 274 - 276°. $[\alpha]^{20}_{D}$ +13.6 (c, 2.1 in DMSO). λ_{max} 260 (log ε 4.19) (MeOH)

O-[β-D-Glucopyranosyl-(1→4)-2,3-di-O-methyl-β-D-xylopyranoside]

Ramontoside

[133882-75-4]

$C_{34}H_{38}O_{16}$ 702.662

Isol. from *Flacourtia ramontchii*. Cryst.. Mp 140 - 142°. λ_{max} 265 (ε 126000); 320 (ε 26400); 355 (ε 11220) (MeOH) (Berdy)

O-[β-D-Glucopyranosyl-(1→2)-3,4-di-O-methyl-D-xylopyranoside]
Cleistanthoside A
[86402-39-3]
$C_{34}H_{38}O_{16}$ 702.662
Constit. of *Cleistanthus patulus* and *Cleistanthus collinus*. Exhibits antiproliferative activity. Cryst.. Mp 238 - 241°. $[\alpha]_D$ +7.3 (c, 1.04 in dioxan)

O-[6-O-Acetyl-β-D-glucopyranosyl-(1→2)-3,4-di-O-methyl-β-D-xylopyranoside]
Taxodiifoloside
[904326-49-4]
$C_{36}H_{40}O_{17}$ 744.699
Constit. of *Phyllanthus taxodiifolius*. Cytotoxic. Powder (EtOAc/hexane). Mp 142.3 - 143.1°. $[\alpha]_D^{27}$ -34 (c, 0.5 in $CHCl_3$). λ_{max} 223 (log ε 3.92); 258 (log ε 4.31); 294 (log ε 3.67); 314 (log ε 3.7); 349 (log ε 3.46) (EtOH)

O-[β-D-Glucopyranosyl-(1→4)-6-deoxy-β-D-glucopyranoside]
Mananthoside C
[881421-98-3]
$C_{33}H_{36}O_{16}$ 688.635
Constit. of *Mananthes patentiflora*. Amorph. powder. Mp 181 - 183°. $[\alpha]_D^{26}$ -13.7 (c, 0.3 in MeOH). λ_{max} 205 (log ε 4.53); 261 (log ε 4.5); 290 (log ε 3.86); 314 (log ε 3.84); 384 (log ε 2.62) (MeOH)

O-[β-D-Glucopyranosyl-(1→2)-β-D-glucopyranoside]
Diphyllin sophoroside. ***Neesiinoside B***
[346420-60-8]
$C_{33}H_{36}O_{17}$ 704.634
Constit. of *Justicia neesii*. Cryst. ($CHCl_3$/MeOH). Mp 276 - 278°. $[\alpha]_D^{20}$ +27.8 (c, 0.9 in DMSO). λ_{max} 259 (log ε 4.53) (MeOH)

O-[β-D-Glucopyranosyl-(1→6)-β-D-glucopyranoside]
Procumbenoside G
$C_{33}H_{36}O_{17}$ 704.634
Constit. of the rhizomes of *Diphylleia sinensis*. λ_{max} 254; 365 (no solvent reported)

O-[β-D-Apiofuranosyl-(1→4)-β-D-xylopyranosyl-(1→2)-β-D-apiofuranoside]
Patavine
[464195-59-3]
$C_{36}H_{40}O_{19}$ 776.697
Constit. of shoot cultures of *Haplophyllum patavinum*. Amorph. yellowish powder. $[\alpha]_D^{24}$ -18 (c, 1.1 in MeOH). λ_{max} 205; 224; 262; 290; 313; 350 (EtOH)

O-[α-L-Arabinopyranosyl-(1→2)-[β-D-xylopyranosyl-(1→5)]-β-D-apiofuranoside]
Ciliatoside A
[303084-57-3]
$C_{36}H_{40}O_{19}$ 776.697
Constit. of *Justicia ciliata*. Antiinflammatory agent. Oil. $[\alpha]_D^{25}$ -272 (c, 0.15 in MeOH). λ_{max} 262 (log ε 4.78); 288 (log ε 4.22); 313 (sh); 350 (sh) (MeOH)

O-[α-L-Arabinopyranosyl-(1→6)-[β-D-glucopyranosyl-(1→3)]-4-O-acetyl-β-D-galactopyranoside]
Mananthoside E
[881422-00-0]
$C_{40}H_{46}O_{22}$ 878.786
Constit. of *Mananthes patentiflora*. Amorph. powder. Mp 186 - 187°. $[\alpha]_D^{26}$ -2.7 (c, 0.5 in MeOH). λ_{max} 205 (log ε 4.69); 261 (log ε 4.72); 315 (log ε 4.05) (MeOH)

O-[α-L-Arabinopyranosyl-(1→6)-[β-D-glucopyranosyl-(1→3)]-2,4-di-O-acetyl-β-D-galactopyranoside]
Mananthoside D
[881421-99-4]
$C_{42}H_{48}O_{23}$ 920.823
Constit. of *Mananthes patentiflora*. Amorph. powder. Mp 185 - 187°. λ_{max} 205 (log ε 4.42); 224 (log ε 4.4); 261 (log ε 4.63); 294 (log ε 3.98); 315 (log ε 3.99); 346 (log ε 3.66) (MeOH)

O-[α-L-Xylopyranosyl-(1→2)-[α-L-xylopyranosyl-(1→5)]-β-D-apiofuranoside]
Azizin. Procumbenoside E
[220182-12-7]
$C_{36}H_{40}O_{19}$ 776.697
Constit. of *Haplophyllum buxbaumii* and *Justicia procumbens*. $[\alpha]_D^{24}$ -47.4 (c, 0.2 in MeOH) (Azizin). Procumbenoside E may not be identical with Azizin. It may possess all-β-D-sugars. λ_{max} 262 (log ε 4.21) (MeOH)

O-[β-D-Galactopyranosyl-(1→6)-β-D-glucopyranosyl-(1→3)-α-L-arabinopyranoside]
Phyllanthusmin F
$C_{38}H_{44}O_{21}$ 836.749
Constit. of roots of *Phyllanthus flexuosus*. Amorph. powder. $[\alpha]_D^{22}$ -40.6 (c, 0.3 in MeOH). λ_{max} 203 (log ε 4.41); 262 (log ε 4.56) (MeOH)

O-[β-D-Glucopyranosyl-(1→3)-α-L-arabinopyranosyl-(1→5)-2-O-acetyl-β-D-apiofuranoside]
Qudsine
[133084-19-2]
$C_{39}H_{44}O_{21}$ 848.76
Isol. from *Haplophyllum buxbaumii*. $[\alpha]_D^{24}$ -60.5 (c, 0.38 in MeOH)

O-[β-D-Apiofuranosyl-(1→3)-α-L-arabinopyranosyl-(1→2)-[β-D-xylopyranosyl-(1→5)]-β-D-apiofuranoside]
Ciliatoside B
[303085-28-1]
$C_{41}H_{48}O_{23}$ 908.812
Constit. of *Justicia ciliata*. Antiinflammatory agent. Oil. $[\alpha]_D^{25}$ -40 (c, 0.1 in MeOH). λ_{max} 262 (log ε 4.87); 294 (log ε 4.23); 319 (sh); 352 (log ε 3.98) (MeOH)

O-[β-D-Apiofuranosyl-(1→3)-β-D-xylopyranosyl-(1→2)-[β-D-xylopyranosyl-(1→5)]-β-D-apiofuranoside]
Procumbenoside F
[1627958-97-7]
$C_{41}H_{48}O_{23}$ 908.812
Constit. of *Justicia procumbens*

O-[α-L-Arabinopyranosyl-(1→6)-β-D-glucopyranosyl-(1→3)-[α-L-arabinopyranosyl-(1→6)]-β-D-glucopyranoside]
Phyllanthuoside C
[1819380-93-2]
$C_{43}H_{52}O_{25}$ 968.864
Constit. of *Phyllanthus cochinchinensis*. Amorph. powder. $[\alpha]_D^{22}$ -14.9 (c, 0.2 in MeOH). λ_{max} 221 (log ε 4.16); 259 (log ε 4.46); 294 (log ε 3.78); 313 (log ε 3.79) (MeOH)

O-[α-L-Arabinopyranosyl-(1→6)-β-D-glucopyranosyl-(1→3)-[α-L-arabinopyranosyl-(1→6)]-4-O-acetyl-β-D-galactopyranoside]
Mananthoside J
[1069137-64-9]
$C_{45}H_{54}O_{26}$ 1010.901
Constit. of *Mananthes patentiflora*. Amorph. solid (MeOH). $[\alpha]_D^{25}$ +7.9 (c, 0.82 in H_2O). λ_{max} 199 (log ε

4.58); 225 (log ε 4.28); 261 (log ε 4.61); 314 (log ε 3.95); 390 (log ε 2.54) (MeOH)

Ac

Acetyldiphyllin

[25001-54-1]

$C_{23}H_{18}O_8$ 422.389

Constit. of *Haplophyllum telephioides*. Cryst. (C_6H_6). Mp 236 - 240°

Benzoyl

Benzoyldiphyllin

[163982-55-6]

$C_{28}H_{20}O_8$ 484.46

Constit. of *Sauropus quadrangularis*. Cryst.. Mp 232 - 233°

O-(4-Nitrobenzoyl)

(4-Nitrobenzoyl)diphyllin

[163982-56-7]

$C_{28}H_{19}NO_{10}$ 529.457

Constit. of *Sauropus quadrangularis*. Cryst.. Mp 245 - 246°

O-(3,4,5-Trimethoxybenzoyl)

(3,4,5-Trimethoxybenzoyl) diphyllin

[163982-61-4]

$C_{31}H_{26}O_{11}$ 574.538

Constit. of *Sauropus quadrangularis*. Cryst.. Mp 273 - 275°

O-(3,4-Methylenedioxybenzoyl)

(3,4-Methylenedioxybenzoyl) diphyllin

[163982-58-9]

$C_{29}H_{20}O_{10}$ 528.469

Constit. of *Sauropus quadrangularis*. Cryst.. Mp 244 - 245°

O-(6-Bromo-3,4-dimethoxybenzoyl)

(6-Bromo-3,4-dimethoxybenzoyl) diphyllin

[163982-59-0]

$C_{30}H_{23}BrO_{10}$ 623.408

Constit. of *Sauropus quadrangularis*. Cryst.. Mp 261 - 262°

O-(6-Bromo-3,4-methylenedioxybenzoyl)

(6-Bromo-3,4-methylenedioxybenzoyl)diphyllin

[163982-60-3]

$C_{29}H_{19}BrO_{10}$ 607.365

Constit. of *Sauropus quadrangularis*. Cryst.. Mp 255 - 257°

Cinnamoyl

Cinnamoyldiphyllin

[163982-57-8]

$C_{30}H_{22}O_8$ 510.498

Constit. of *Sauropus quadrangularis*. Cryst.. Mp 195 - 196°

5-O-De-Me, 5-O-β-D-glucopyranoside, 7-O-α-L-arabinopyranoside

Phyllanthusmin E‡

$C_{31}H_{32}O_{16}$ 660.581

Constit. of roots of *Phyllanthus flexuosus*. Amorph. powder. $[\alpha]^{27}_{D}$ -48.7 (c, 0.28 in MeOH). Not to confused with Phyllanthusmin E (1613505-43-3). λ_{max} 203 (log ε 4.24); 259 (log ε 4.55) (MeOH).

Murakami, T. *et al.*, *Yakugaku Zasshi*, 1961, **81**, 1596-1600 (*Diphyllin*).

Horii, Z. *et al.*, *Chem. Commun. (London)*, 1968, 653-655 (*struct*).

Govindachari, T.R. *et al.*, *Tetrahedron*, 1969, **25**, 2815-2821 (*Diphyllin, struct*).

Lakshmi, T.G. *et al.*, *Curr. Sci.*, 1970, **39**, 395-396 (*Cleistanthin B*).

Okigawa, M. *et al.*, *Tetrahedron*, 1970, **26**, 4301-4305 (*Justicia procumbens constit*).

Stevenson, R. *et al.*, *J. Org. Chem.*, 1971, **36**, 3450-3453 (*synth*).

Arnold, B.J. *et al.*, *J. Chem. Soc., Perkin Trans. 1*, 1973, 1266-1270 (*synth*).

Anjaneyulu, A.S.R. *et al.*, *Indian J. Chem., Sect. B*, 1977, **15**, 10-11 (*Cleistanthus collinus constit*).

Momose, T. *et al.*, *Chem. Pharm. Bull.*, 1978, **26**, 3195-3198 (*synth*).

Gonzalez, A.G. *et al.*, *Planta Med.*, 1979, **36**, 200-203 (*Diphyllinins*).

Anjaneyulu, A.S.R. *et al.*, *Tetrahedron*, 1981, **37**, 3641-3652 (*Cleistanthins*).

Sastry, K.V. *et al.*, *Planta Med.*, 1983, **47**, 227-229 (*Cleistanthoside, Cleistanthin A, ms*).

Sheriha, G.M. *et al.*, *Phytochemistry*, 1984, **23**, 151-153 (*Tuberculatin*).

Gözler, T. *et al.*, *Tetrahedron*, 1984, **40**, 1145-1150 (*pmr, cmr*).

Nukul, G.S. *et al.*, *J. Nat. Prod.*, 1987, **50**, 748-750 (*5-acetylapioside*).

Sastry, K.V. *et al.*, *Phytochemistry*, 1987, **26**, 1153-1154 (*Cleistanthoside B*).

Al-Abed, Y. *et al.*, *J. Nat. Prod.*, 1990, **53**, 1152-1161 (*Majidine, Arabelline*).

Satyanarayana, V. *et al.*, *Phytochemistry*, 1991, **30**, 1026-1029 (*Ramontoside*).

Ulubelen, A. *et al.*, *Phytochemistry*, 1994, **35**, 1600-1601 (*Acetyldiphyllin*).

Satyanarayana, P. *et al.*, *Indian J. Chem., Sect. B*, 1995, **34**, 571-572 (*Saururus quadrangularis constits*).

Kumar, C.P.P. *et al.*, *Mutagenesis*, 1996, **11**, 553-557 (*Cleistanthin B, activity*).

Prieto, J.M. *et al.*, *Z. Naturforsch., C*, 1996, **51**, 618-622 (*5"-acetylapioside*).

Satyanarayana, P. *et al.*, *Indian J. Chem., Sect. B*, 1998, **37**, 717-719 (*3"-Methylcleistanthin B*).

Al-Abed, Y. *et al.*, *Phytochemistry*, 1998, **49**, 1779-1781 (*Azizin*).

Day, S.-H. *et al.*, *J. Nat. Prod.*, 2000, **63**, 1560-1562 (*Ciliatosides, Procumbenoside A*).

Subbaraju, G.V. *et al.*, *Indian J. Chem., Sect. B*, 2001, **40**, 313-319 (*Neesiinosides*).

Ivanova, A. *Z. Naturforsch., C*, 2001, **56**, 329-333 (*Arabelline*).

Innocenti, G. *et al.*, *Chem. Pharm. Bull.*, 2002, **50**, 844-846 (*Patavine*).

Chen, B. *et al.*, *Chin. Chem. Lett.*, 2002, **13**, 959-962 (*Mananthoside A,B*).

Ramesh, C. *et al.*, *Chem. Pharm. Bull.*, 2003, **51**, 1299-1300 (*Cleistanthins A,D*).

Weng, J.-R. *et al.*, *Arch. Pharm. (Weinheim, Ger.)*, 2004, **337**, 207-212 (*Procumbenoside B*).

Susplugas, S. *et al.*, *J. Nat. Prod.*, 2005, **68**, 734-738 (*Patentiflorins A,B, 4"-Acetylmananthoside B, Diphyllin, activity*).

Wu, S.-J. *et al.*, *Chem. Pharm. Bull.*, 2006, **54**, 1223-1225 (*Phyllanthuspermins B,C*).

Tian, J. *et al.*, *Helv. Chim. Acta*, 2006, **89**, 291-298 (*Mananthosides C,D,E*).

Tuchinda, P. *et al.*, *Planta Med.*, 2006, **72**, 60-62 (*Taxodiifoloside, Cleistanthin A methyl ether*).

Tian, J.-M. *et al.*, *J. Asian Nat. Prod. Res.*, 2008, **10**, 228-232 (*Mananthosides I,J*).

Yu, Z. *et al.*, *J. Carbohydr. Chem.*, 2008, **27**, 113-119 (*Patentiflorins A,B, activity*).

Tuchinda, P. *et al.*, *J. Nat. Prod.*, 2008, **71**, 655-663 (*Acutissimalignan A*).

Liu, G. *et al.*, *Magn. Reson. Chem.*, 2008, **46**, 283-286 (*Tuberculatin*).

Pendey, D.P. *et al.*, *Asian J. Chem.*, 2011, **23**, 2125-2127 (*6-apiosylglucoside*).

Shi, D-K. *et al.*, *Eur. J. Med. Chem.*, 2012, **47**, 424-431 (*Patentiflorin A, activity*).

Ma, J.-X. *et al.*, *J. Asian Nat. Prod. Res.*, 2012, **14**, 1073-1077 (*Reticulatuside A*).

Wu, W. *et al.*, *Zhongguo Yiyao Gongye Zazhi*, 2012, **43**, 669-672 (*Procumbenoside E*).

Chen, H.-W. *et al.*, *Antiviral Res.*, 2013, **99**, 371-382 (*Diphyllin, activity*).

Wu, W.-W. *et al.*, *Zhongchengyao*, 2013, **35**, 985-988 (*Procumbenoside F*).

Zhang, Z. *et al.*, *Eur. J. Med. Chem.*, 2014, **82**, 466-471 (*Cleistanthoside A, Cleistanthin A, synth*).

Jin, H. *et al.*, *Fitoterapia*, 2014, **94**, 70-76 (*Procumbenoside J*).

Zhao, J.-Q. *et al.*, *J. Braz. Chem. Soc.*, 2014, **25**, 1446-1454 (*Phyllanthuoside C*).

Ren, Y. *et al.*, *J. Nat. Prod.*, 2014, **77**, 1494-1504 (*Phyllanthus poilanei constits, cryst struct, activity*).

Zhao, J.-Q. *et al.*, *Nat. Prod. Bioprospect.*, 2014, **4**, 233-242 (*Phyllanthusmins D,E,F, Phyllanthus flexuosus constits*).

Jin, H. *et al.*, *J. Asian Nat. Prod. Res.*, 2015, **17**, 33-39 (*Procumbenoside H*).

Zhou, P. *et al.*, *Molecules*, 2015, **20**, 7048-7058 (*Justicidin A, isol, purifn*).

Naresh, G. *et al.*, *Org. Lett.*, 2015, **17**, 3446-3449 (*synth*).

Yu, S. *et al.*, *Nat. Prod. Res.*, 2016, **30**, 419-425 (*Phyllanthus glaucus constits, activity*).

Kim, T. *et al.*, *Eur. J. Org. Chem.*, 2017, 1704-1712 (*synth*).

Liu, L. *et al.*, *Eur. J. Org. Chem.*, 2017, 3674-3680 (*Phyllsnthusmin D, 4"-O-Acetylmananthoside B, synth*).

Zhang, H.-J. *et al.*, *J. Nat. Prod.*, 2017, **80**, 1798-1807 (*Patentiflorin A, activity*).

Thornburg, C.C. *et al.*, *ACS Chem. Biol.*, 2018, **13**, 2484-2497 (*Phyllanthusmin D*).

Zhou, S.-W. *et al.*, *Nat. Prod. Commun.*, 2018, **13**, 735-736 (*Cleistanthin B, atropisomerism*).

Bai, W. *et al.*, *Zhongcaoyao*, 2018, **49**, 1504-1507 (*Procumbenoside G*).

Cai, R. *et al.*, *Fitoterapia*, 2019, **137**, 104198:1-5 (*synth*).

Lewis, R.J. *Sax's Dangerous Properties of Industrial Materials, 8th edn., Van Nostrand Reinhold*, 1992, CMV375

Dipiperitylmagnolol D297

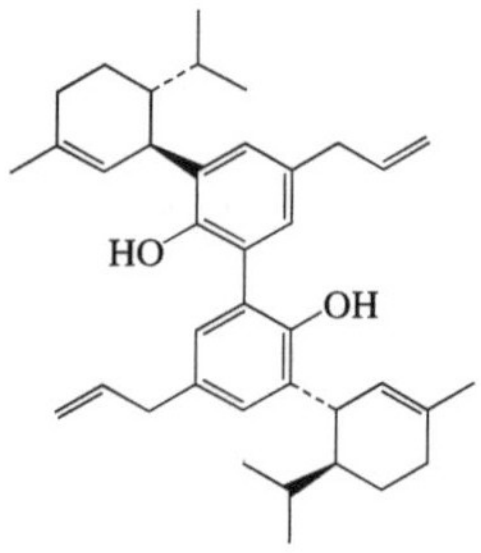

Absolute Configuration

[138591-05-6]

$C_{38}H_{50}O_2$ 538.816

Constit. of *Magnolia officinalis*. Powder. $[\alpha]_D$ -140 (c, 0.31 in $CHCl_3$). λ_{max} 290 (ε 10200) (MeOH).

Yahara, S. *et al., Chem. Pharm. Bull.*, 1991, **39**, 2024-2036 (*Dipiperitylmagnolol*)

α-Diplicatin B D298

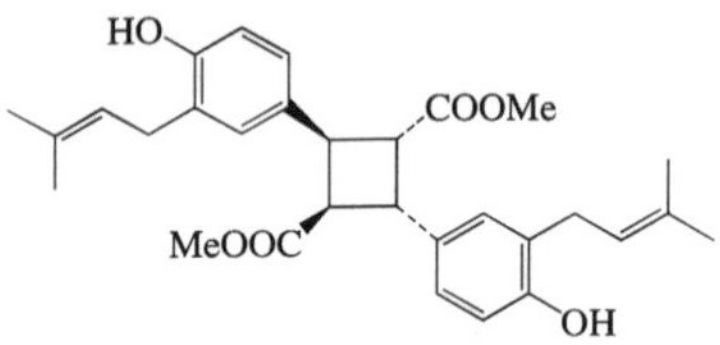

Relative Configuration

[194204-63-2]

$C_{30}H_{36}O_6$ 492.612

Dimer of Plicatin A in D304. Constit. of *Psoralea plicata*.

Hamed, A.I. *et al., Phytochemistry*, 1997, **45**, 1257-1261 (*α-Diplicatin B*)

Diplochalcolin A D299

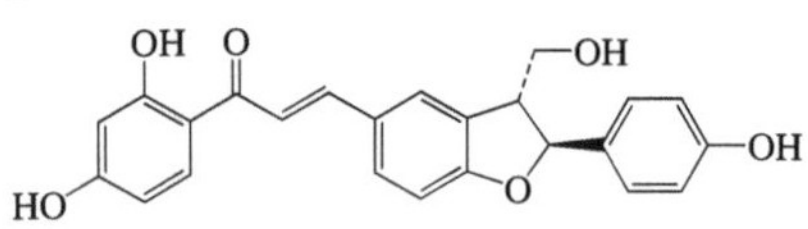

Absolute Configuration

3-[2,3-Dihydro-3-(hydroxymethyl)-2-(4-hydroxyphenyl)-5-benzofuranyl]-1-(2,4-dihydroxyphenyl)-2-propen-1-one, CAS. *Diploflavolin A*

[2034142-38-4]

$C_{24}H_{20}O_6$ 404.418

Constit. of the roots of *Mimosa diplotricha*. $[\alpha]_D^{26}$0 (c, 0.2 in MeOH). λ_{max} 371 (log ε 3.87) (MeOH).

Chiou, C.-T. *et al., J. Nat. Prod.*, 2016, **79**, 2439-2445 (*Diplochalcolin A*)

Diplochalcolin B D300

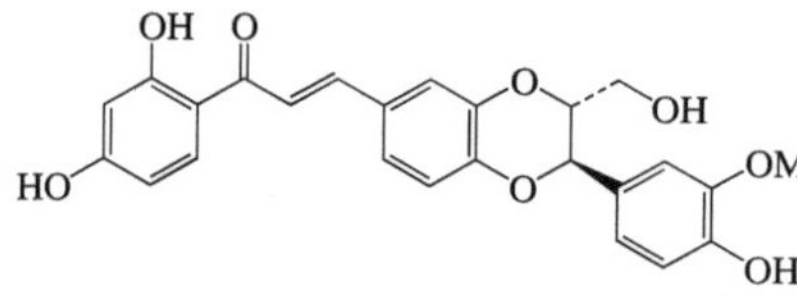

Relative Configuration

Diploflavolin B

[2034142-39-5]

$C_{25}H_{22}O_8$ 450.443

Constit. of the roots of *Mimosa diplotricha*. No phys. props. reported.

Chiou, C.-T. *et al., J. Nat. Prod.*, 2016, **79**, 2439-2445 (*Diplochalcolin B*)

Disaminyl ether D301

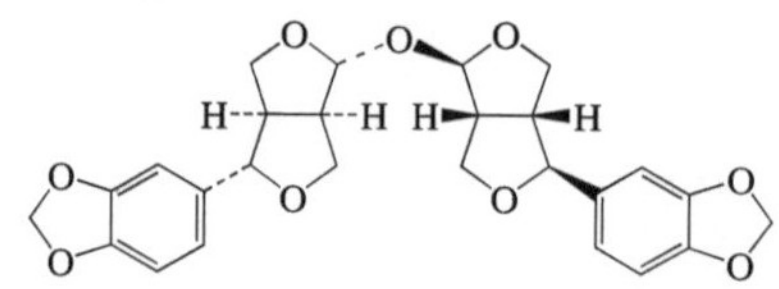

Absolute Configuration

[30186-93-7]

$C_{26}H_{26}O_9$ 482.485

Dimer of Samin, S23. Constit. of the seeds of *Sesamum indicum*. Cryst. (MeOH). Mp 191 - 192° (188-189°). $[\alpha]_D$ +160.1 (c, 1.4 in $CHCl_3$). λ_{max} 243 (log ε 3.8); 288 (log ε 3.9); 337 (log ε 1.32) ($CHCl_3$).

Marchand, P.A. *et al., J. Nat. Prod.*, 1997, **60**, 1189-1192 (*synth*).

Grougnet, R. *et al., J. Agric. Food Chem.*, 2012, **60**, 108-111 (*Disaminyl ether*)

Divaccinoside B D302

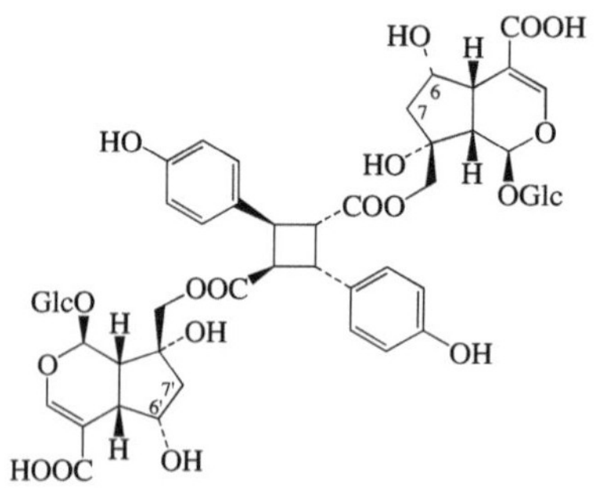

[2106839-36-3]

$C_{50}H_{60}O_{28}$ 1109.002

Constit. of *Vaccinium bracteatum*. Amorph. yellow powder. $[\alpha]_D^{20}$-56.1 (c, 0.43 in MeOH). λ_{max} 228 (log ε 4.41) (MeOH)

6-Deoxy, 6,7-didehydro

Divaccinoside C

[2106839-37-4]

$C_{50}H_{58}O_{27}$ 1090.987

Constit. of *Vaccinium bracteatum*. Amorph. yellow powder. $[\alpha]_D^{20}$ -66.5 (c, 0.17 in MeOH). λ_{max} 228 (log ε 4.28) (MeOH)

6'-Deoxy, 6',7'-didehydro

Divaccinoside D

[2106839-52-3]

$C_{50}H_{58}O_{27}$ 1090.987

Constit. of *Vaccinium bracteatum*. Amorph. yellow powder. $[\alpha]_D^{20}$-46 (c, 0.52 in MeOH). λ_{max} 228 (log ε 4.46) (MeOH)

6,6'-Dideoxy, 6,6',7,7'-tetradehydro

Divaccinoside A

[2106839-35-2]

$C_{50}H_{56}O_{26}$ 1072.972

Constit. of *Vaccinium bracteatum*. Amorph. yellow powder. $[\alpha]_D^{20}$ -32.0 (c, 0.4 in MeOH). λ_{max} 229 (log ε 4.48) (MeOH).

Ren, Y.-M. *et al., Tetrahedron Lett.*, 2017, **58**, 2385-2388 (*Divaccinosides A-D*)

Divanillin D303

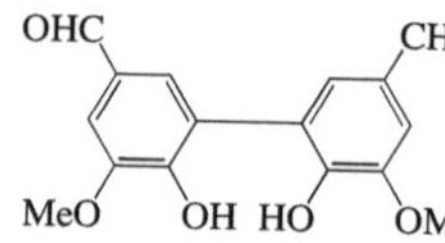

6,6'-Dihydroxy-5,5'-dimethoxy-[1,1'-biphenyl]-3,3'-dicarboxaldehyde, CAS. *5,5'-Diformyl-2,2'-dihydroxy-3,3'-dimethoxybiphenyl. Dehydrodivanillin. FEMA 4107*

[2092-49-1]

$C_{16}H_{14}O_6$ 302.282

Isol. from chemical or enzymatic degradn. of plant lignins incl. *Pinus* sp. Constit. of cured or fermented vanilla beans (*Vanilla* spp.). Antioxidant. Flavouring agent and taste enhancer. Cryst. (AcOH). Mp 312 - 315° (306-307°).

Bobbit, J.H. *et al., Heterocycles*, 1992, **33**, 641-648 (*synth*).

Katahira, R. *et al., J. Wood Sci.*, 2001, **47**, 378-382 (*lignin degrad prod*).

Gatfield, I. *et al., Perfum. Flavor.*, 2006, **31**, 18-20 (*Vanilla constit*).

Delomenede, M. *et al., J. Med. Chem.*, 2008, **51**, 3171-3181 (*synth*).

Gaur, M. *et al., Bull. Korean Chem. Soc.*, 2009, **30**, 2895-2898 (*synth*).

Schwarz, B. *et al., J. Agric. Food Chem.*, 2009, **57**, 3729-3737 (*Vanilla constit*).

Lahtinen, M. *et al., J. Agric. Food Chem.*, 2009, **57**, 8357-8365 (*synth*).

Fenaroli's Handbook of Flavor Ingredients, 6th edn., (ed. Burdock, G.A.), CRC Press, 2010, 526-527 (*use, props*).

Nishimura, R.T. *J. Chem. Educ.*, 2010, **87**, 526-527 (*synth*).

Krings, U. *et al.*, *Flavour Fragrance J.*, 2015, **30**, 362-365 (*rev, occur, synth*).
Imer, M.R. *et al.*, *Acta Crystallogr., Sect. C*, 2018, **74**, 1768-1773 (*cryst struct*)

Drupanin D304

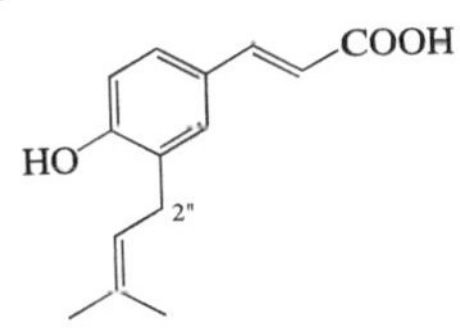

3-[4-Hydroxy-3-(3-methyl-2-butenyl) phenyl]-2-propenoic acid, CAS.
4-Hydroxy-3-prenylcinnamic acid
[53755-58-1]
[78834-99-8] (unspecified stereochem.)
$C_{14}H_{16}O_3$ 232.279
Constit. of the fruits of *Psoralea drupacea*. Also from *Baccharis* spp. Isol. from Brazilian propolis. Acicular cryst.. Mp 147 - 148°

Me ester
Plicatin B‡
[72704-01-9]
$C_{15}H_{18}O_3$ 246.306
Constit. of *Psoralea juncea*, *Psoralea plicata*, *Senecio mathewsii* and *Werneria stuebelii*. Antimutagenic agent. Cryst. (Et_2O/hexane). Mp 70 - 72° (69-71°). [base] λ_{max} 253 (ε 8710); 310 (sh); 369 (ε 26900) (MeOH/NaOH) (Derep). [neutral] λ_{max} 215 (ε 13800); 234 (ε 13800); 297 (sh); 316 (ε 22900) (MeOH) (Derep)

Aldehyde
3-[4-Hydroxy-3-(3-methyl-2-butenyl)phenyl]-2-propenal. 4-Hydroxy-3-prenylcinnamaldehyde
[151455-12-8]
$C_{14}H_{16}O_2$ 216.280
Prod. by injured peel of *Citrus paradisi* and *Citrus sinensis*. Antifungal agent. Mp 144.5 - 146°

2″-Oxo
Capillaric acid
[123452-65-3]
$C_{14}H_{14}O_4$ 246.262
Root growth promotor

2″-Oxo, Me ester
Methyl 4-hydroxy-3-(3-methyl-2-butenoyl)cinnamate. ***Capillarol***
[107140-61-4]
$C_{15}H_{16}O_4$ 260.289
Constit. of *Artemisia capillaris*, *Senecio mathewsii* and *Werneria stuebelii*. Root growth promotor. Yellow cryst. (Et_2O/petrol). Mp 116 - 118° (115.5-116°). λ_{max} 281 (log ε 4.41); 296 (sh); 350 (log ε 3.53) (MeOH)

2″R-Hydroxy, Me ester
Plicatin A‡
[131889-82-2]
$C_{15}H_{18}O_4$ 262.305
Constit. of *Psoralea plicata*. Cryst. (Me_2CO). Mp 182°. $[\alpha]_D$ -59.7 (c, 0.067 in $CHCl_3$). [base] λ_{max} 253 (ε 8710); 310 (sh); 369 (ε 26900) (MeOH/NaOH) (Derep). [neutral] λ_{max} 215 (ε 13800); 234 (ε 13800); 297 (sh); 316 (ε 22900) (MeOH) (Derep)

(Z)-Isomer
(Z)-Drupanin
[640271-16-5]
$C_{14}H_{16}O_3$ 232.279
Constit. of the leaves of *Petasites formosanus*. Cryst.. Mp 122 - 123°.

Golovina, L.A. *et al.*, *Chem. Nat. Compd. (Engl. Transl.)*, 1973, **9**, 672-675 (*Drupanin*).
Ueda, J. *et al.*, *Agric. Biol. Chem.*, 1986, **50**, 3083-3086 (*Capillarol*).
Rasool, N. *et al.*, *Phytochemistry*, 1990, **29**, 3979-3981 (*Plicatins A,B*).
Mitscher, L.A. *et al.*, *Phytochemistry*, 1991, **30**, 3569-3570 (*Plicatin B, Drupanin*).
Stange, R.R. *et al.*, *J. Nat. Prod.*, 1993, **56**, 1627-1629 (*Citrus, aldehyde*).
Bates, R.W. *et al.*, *Tetrahedron*, 1995, **51**, 8199-8212 (*synth*).
Okada, K. *et al.*, *Biosci., Biotechnol., Biochem.*, 1999, **63**, 257-260 (*Capillarol, synth*).
Menon, S.R. *et al.*, *J. Nat. Prod.*, 1999, **62**, 102-106 (*Plicatin B, synth, activity*)

Dumosaol D305

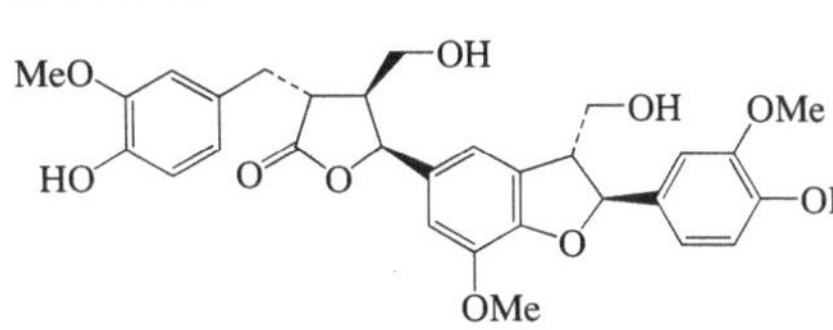

Relative Configuration

[897930-20-0]
$C_{30}H_{32}O_{10}$ 552.576
Sesquineolignan. Constit. of the heartwood of *Tsuga dumosa*. Mp 70 - 72°. $[\alpha]^{27}_D$ +30.4 (c, 0.77 in MeOH).

Zhao, Y. *et al.*, *Yunnan Zhiwu Yanjiu*, 2005, **27**, 217-222 (*Dumosaol*)

Dunalianoside F D306

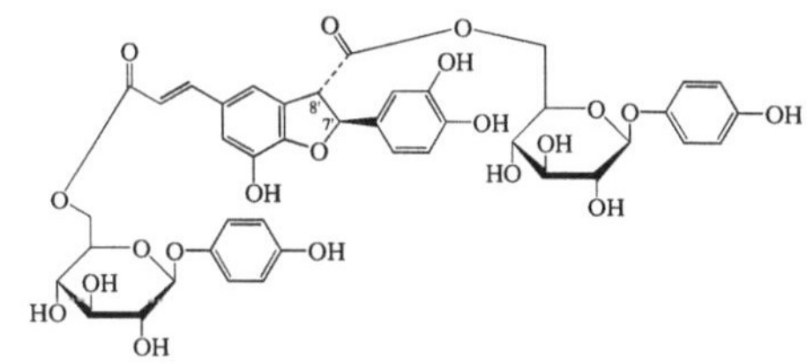

Absolute Configuration

[1042905-48-5]
$C_{42}H_{42}O_{20}$ 866.778
Constit. of the buds of *Vaccinium dunalianum*. Off-white powder. $[\alpha]^{26}_D$ -32.6 (c, 0.1 in MeOH). λ_{max} 223 (log ε 4.59); 289 (log ε 4.26); 330 (log ε 4.31) (MeOH)

7′,8′-Diepimer
Dunalianoside G
[1042905-50-9]
$C_{42}H_{42}O_{20}$ 866.778
Constit. of the buds of *Vaccinium dunalianum*. Off-white powder. $[\alpha]^{26}_D$ -36.1 (c, 0.1 in MeOH). λ_{max} 223 (log ε 4.6); 290 (log ε 4.26); 328 (log ε 4.32) (MeOH).

Zhao, P. *et al.*, *Phytochemistry*, 2008, **69**, 3087-3094 (*Dunalianosides F,G*)

Dunalianoside H D307

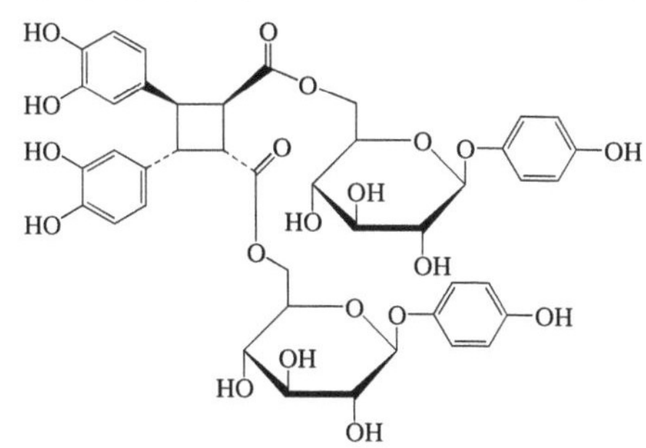

Relative Configuration

[1042905-52-1]
$C_{42}H_{44}O_{20}$ 868.794
Constit. of the buds of *Vaccinium dunalianum*. Off-white powder. $[\alpha]^{26}_D$ -37.2 (c, 0.1 in MeOH). λ_{max} 285 (log ε 4.07); 325 (sh) (MeOH).

Zhao, P. *et al.*, *Phytochemistry*, 2008, **69**, 3087-3094 (*Dunalianoside H*)

Dunnianol D308

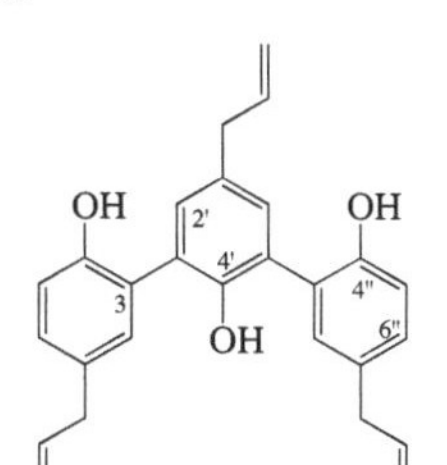

3,3′:3″,5′-Sesquiligna-8,8′,8″-triene-4,4′,4″-triol. 4,4′,4″-Trihydroxy-3,3′:3″,5′-sesquineoligna-8,8′,8″-triene
[139726-29-7]
$C_{27}H_{26}O_3$ 398.502
Constit. of *Illicium dunnianum* and *Illicium simonsii*. Exhibits antibacterial activity. Cryst. (EtOAc/hexane). Mp 132 - 133°

6″-Hydroxy
3,3′:3″,5′-Sesquiligna-8,8′,8″-triene-4,4′,4″,6″-tetrol. 4,4′,4″,6″-Tetrahydroxy-3,3′:3″,5′-sesquiligna-8,8′,8″-triene. ***Simonsienol B***
[1426534-53-3]
$C_{27}H_{26}O_4$ 414.501
Constit. of the stem bark of *Illicium simonsii*. Amorph. solid. $[\alpha]_D^{26}$ -2.0 (c, 0.004 in MeOH). λ_{max} 217 (log ε 3.91); 303 (log ε 3.21) ($CHCl_3$)

4′-Deoxy, 2′-hydroxy
3,3′:3″,5′-Sesquiligna-8,8′,8″-triene-2′,4,4″-triol. 2′,4,4″-Trihydroxy-3,3′:3″,5′-sesquiligna-8,8′,8″-triene. ***Macranthol***
[123941-73-1]
$C_{27}H_{26}O_3$ 398.502
Constit. of *Garcinia cantleyana, Illicium fargesii, Illicium macranthum* and *Illicium simonsii*. Neuroprotectant. Exhibits antibacterial activity. Cytotoxic. Needles (hexane/EtOAc). Mp 140 - 141°. $[\alpha]_D^{18}$ +2 (c, 0.3 in EtOH)

4″-Deoxy, 6″-hydroxy
3,3′:3″,5′-Sesquiligna-8,8′,8″-triene-4,4′,6″-triol. 4,4′,6″-Trihydroxy-3,3′:3″,5′-sesquiligna-8,8′,8″-triene. ***Simonsinol***
[155709-40-3]
$C_{27}H_{26}O_3$ 398.502
Constit. of the bark of *Illicium simonsii*. Oil.

Kouno, I. *et al.*, *Chem. Pharm. Bull.*, 1989, **37**, 1291-1292 (*Macranthol*).
Kouno, I. *et al.*, *Chem. Pharm. Bull.*, 1991, **39**, 2606-2608 (*Dunnianol*).
Kouno, I. *et al.*, *Chem. Pharm. Bull.*, 1994, **42**, 112-114 (*Simonsinol*).
Shadid, K.A. *et al.*, *Phytochemistry*, 2007, **68**, 2537-2544 (*Garcinia cantleyana constit, activity*).
Moriyama, M. *et al.*, *Chem. Pharm. Bull.*, 2008, **56**, 1201-1204 (*Macranthol, activity*).
Liu, J.-F. *et al.*, *Planta Med.*, 2010, **76**, 1464-1467 (*Dunnianol, Macranthol, activity*).
Denton, R.M. *et al.*, *Synlett*, 2010, 633-634 (*synth*).
Dong, C.-F. *et al.*, *Nat. Prod. Bioprospect.*, 2012, **2**, 133-137 (*Simonsienol B*)

Dysosmanorlignan A D309

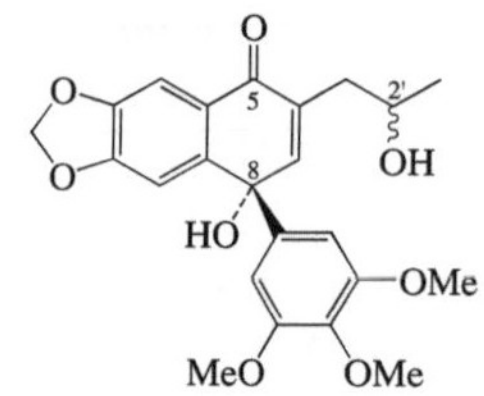

Absolute Configuration

8-Hydroxy-6-(2-hydroxypropyl)-8-(3,4,5-trimethoxyphenyl)naphtho[2,3-d][1,3]dioxol-5(8H)-one
$C_{23}H_{24}O_8$ 428.437
Constit. of roots of *Dysosma versipellis*. Amorph. yellow powder. $[\alpha]_D^{25}$ +80 (c, 0.1 in MeOH). λ_{max} 218 (log ε 2.45); 246 (log ε 2.13) (MeOH)

8-Epimer
Dysosmanorlignan B
$C_{23}H_{24}O_8$ 428.437
Constit. of roots of *Dysosma versipellis*. Amorph. yellow powder. $[\alpha]_D^{25}$ -50 (c, 0.1 in MeOH). λ_{max} 210 (log ε 1.24); 246 (log ε 2.13) (MeOH).

Zheng, Y. *et al.*, *Phytochem. Lett.*, 2016, **16**, 75-81 (*Dysosmanorlignans A,B*)

Edgeworic acid E1

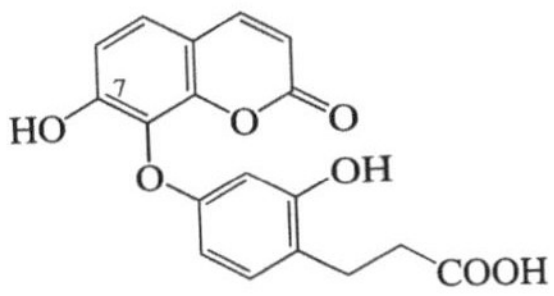

3-[2-Hydroxy-4-[(7-hydroxy-2-oxo-2H-1-benzopyran-8-yl)oxy]phenyl] propanoic acid
[1698044-25-5]
$C_{18}H_{14}O_7$ 342.303
Constit. of the flower buds of *Edgeworthia chrysantha*. Powder. λ_{max} 322 (log ε 4.20) (MeOH)

7-O-β-D-Glucopyranoside
Gardnerol B
[1809303-27-2]
$C_{24}H_{24}O_{12}$ 504.444
Constit. of the flowers of *Edgeworthia gardneri*. Powder (MeOH/$CHCl_3$). $[\alpha]_D^{25}$ -10.5 (c, 0.38 in MeOH). λ_{max} 214 (log ε 4.26); 252 (log ε 3.62); 286 (log ε 3.93); 308 (log ε 3.98) (MeOH)

Me ester
Gardnerol A
[1809274-63-2]
$C_{19}H_{16}O_7$ 356.33
Constit. of the flowers of *Edgeworthia gardneri*. Cryst. (MeOH/$CHCl_3$). Mp 191 - 192°. λ_{max} 208 (log ε 4.40); 212 (log ε 4.41); 285 (log ε 3.90); 322 (log ε 4.12) (MeOH)

Me ester, 7-O-β-D-glucopyranoside
Edgeworoside D
[1245551-16-9]
$C_{25}H_{26}O_{12}$ 518.471
Constit. of *Edgeworthia chrysantha*. Powder. Mp 199 - 200°.

Hu, X.-J. *et al., Nat. Prod. Res.*, 2009, **23**, 1259-1264 (*Edgeworoside D*).
Li, X,-N. *et al., Molecules*, 2014, **19**, 2042-2048 (*Edgeworic acid*).
Ma, Y.Y. *et al., J. Agric. Food Chem.*, 2015, **63**, 8162-8169 (*Gardnerols A,B, cryst struct*)

Egonol E2

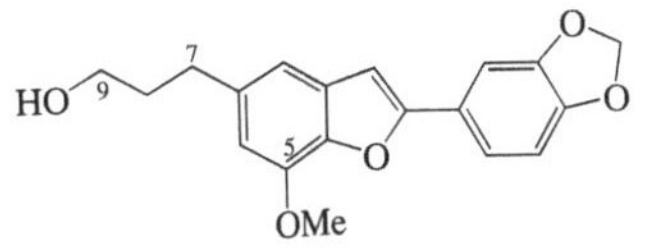

2-(1,3-Benzodioxol-5-yl)-7-methoxy-5-benzofuranpropanol, CAS. *4,7′-Epoxy-5-methoxy-3′,4′-methylenedioxy-9′-nor-3,8′-lign-7′-en-9-ol. 4,7′-Epoxy-9-hydroxy-5-methoxy-3′,4′-methylenedioxy-9′-nor-3,8′-neolign-7′-ene*
[530-22-3]
$C_{19}H_{18}O_5$ 326.348
Lignan numbering shown. Constit. of *Garcinia mangostana, Laetiporus sulphureus* var. *miniatus, Styrax agrestis, Styrax americana, Styrax ferrugineus, Styrax formosanus, Styrax japonica, Styrax obassia* and *Styrax pohlii*. Antiulcer agent. Exhibits antibacterial and antifungal activity against *Staphylococcus aureus, Cladosporium sphaerospermum* and *Candida albicans*. Shows anticomplement activity. Inhibitor of amyloid-β (Aβ) aggregation. Inhibitor of COX-1 and COX-2. Plates (butanol). Mp 117.5 - 118°. Bp 228 - 230° at 15 mm

O-β-D-Glucopyranoside
Egonol glucoside
[77690-83-6]
$C_{25}H_{28}O_{10}$ 488.489
Constit. of *Laetiporus sulphureus* var. *miniatus, Styrax obassia, Styrax officinalis* and *Styrax ferrugineus*. Shows antimicrobial and antifungal activity against *Staphylococcus aureus, Cladosporium sphaerospermum* and *Candida albicans*. Brown needles (EtOH). Mp 169 - 170°. $[\alpha]_D^{22}$ -15.2 (c, 0.6 in MeOH)

O-[β-D-Xylopyranosyl-(1→6)-β-D-glucopyranoside]
Egonol primeveroside. ***Masutakeside I***
[337307-54-7]
$C_{30}H_{36}O_{14}$ 620.604
Constit. of *Laetiporus sulphureus* var. *miniatus* and *Styrax japonica*. Shows moderate anticomplement activity. Amorph. powder. $[\alpha]_D^{25}$ -19.9 (c, 2.6 in MeOH). λ_{max} 210 (log ε 4.49); 233 (log ε 4.46); 278 (log ε 4.17) (MeOH)

O-[β-D-Glucopyranosyl-(1→6)-β-D-glucopyranoside]
Egonol gentiobioside
[77701-99-6]
$C_{31}H_{38}O_{15}$ 650.63
Isol. from *Laetiporus sulphureus* var. *miniatus, Styrax officinalis* and *Styrax obassia*. Modest inhibitor of COX-2. Cryst. (EtOAc/petrol) (as hepta-Ac). Mp 147 - 149° (as hepta-Ac). $[\alpha]_D^{20}$ -27.6 (c, 0.1 in MeOH). Poss. identical with an egonol glycoside, Mp 143°, $[\alpha]^{21}_D$ -29.8°, earlier isol. from *Styrax formosanus*

O-[β-D-Glucopyranosyl-(1→6)-β-D-glucopyranosyl-(1→6)-β-D-glucopyranoside]
Egonol gentiotrioside
[77690-87-0]
$C_{37}H_{48}O_{20}$ 812.771
Constit. of the seeds of *Styrax officinalis* and *Styrax perkinsiae*. Powder (MeOH). Mp 196 - 199°. $[\alpha]_D^{20}$ -35.2 (c, 0.1 in MeOH)

Ac
[15434-00-1]
$C_{21}H_{20}O_6$ 368.385
Constit. of *Nectandra lineata, Styrax obassia, Styrax agrestis* and *Styrax officinalis*. Shows moderate antitrypanosomal activity. Inhibitor of amyloid-β (Aβ) aggregation. Cryst. (EtOH). Mp 127 - 130° (107°)

Propanoyl
Propanoylegonol. Egonol propanoate
[1007348-79-9]
$C_{22}H_{22}O_6$ 382.412
Constit. of *Styrax agrestis* and *Styrax obassia*. Cryst.. Mp 86 - 87°. $[\alpha]_D^{20}$ +4.7 (c, 0.22 in $CHCl_3$). λ_{max} 242 (log ε 3.9); 318 (log ε 4.3) ($CHCl_3$)

2-Methylpropanoyl
[1204762-79-7]
$C_{23}H_{24}O_6$ 396.439
Constit. of *Styrax agrestis* and *Styrax japonica*. Pale yellow needles (EtOH). Mp 61 - 62°

Butanoyl
Butanoylegonol. Egonol butyrate
[879495-13-3]
$C_{23}H_{24}O_6$ 396.439
Constit. of the seeds of *Styrax perkinsiae*. Pale yellow solid. λ_{max} 220 (log ε 3.97); 300 (log ε 3.86); 315 (log ε 4.01) (MeOH)

2S-Methylbutanoyl
2-Methylbutanoylegonol. Egonol 2-methylbutanoate
[115334-06-0]
$C_{24}H_{26}O_6$ 410.466
Constit. of *Riccardia multifida, Styrax agrestis, Styrax obassia* and *Styrax officinalis*. Inhibitor of acetylcholinesterase (AChE) and AChE-induced amyloid-β (Aβ)

aggregation. Pale yellow oil. Mp 40 - 41°. $[\alpha]_D^{25}$ +5.2 (c, 1 in CH_2Cl_2). λ_{max} 211 (log ε 4.05); 228 (log ε 4.09); 318 (log ε 4.22) (CH_2Cl_2)

9Z-Octadecenoyl

Egonol oleate

[1027369-38-5]

$C_{37}H_{50}O_6$ 590.801

Constit. of thc fruit of *Styrax agrestis*. Inhibitor of acetylcholinesterase (AChE). Oil. λ_{max} 230; 318 (MeOH)

9Z,12Z-Octadecadienoyl

Egonol linoleate

[1336884-95-7]

$C_{37}H_{48}O_6$ 588.785

Constit. of the fruit of *Styrax agrestis*. Inhibitor of acetylcholinesterase (AChE) and AChE-induced amyloid-β (Aβ) aggregation. Pale yellow oil. λ_{max} 211 (log ε 4.03); 228 (log ε 4.21); 318 (log ε 4.23) (CH_2Cl_2)

Benzoyl

Benzoylegonol. Egonol benzoate

[610784-71-9]

$C_{26}H_{22}O_6$ 430.456

Constit. of the seeds of *Styrax officinalis*. Cryst.. Mp 71.9 - 74.5°. λ_{max} 205 (sh); 230 (sh); 300; 316 (EtOH)

9-Me ether

4,7′-Epoxy-5,9-dimethoxy-3′,4′-methylenedioxy-9′-nor-3,8′-lign-7′-ene. ***9-O-Methylegonol***

[1639286-97-7]

$C_{20}H_{20}O_5$ 340.375

Constit. of the roots of *Heliopsis helianthoides*. Amorph. solid

O-De-Me

4,7′-Epoxy-3′,4′-methylenedioxy-9′-nor-3,8′-lign-7′-ene-5,9-diol. 4,7′-Epoxy-5,9-dihydroxy-3′,4′-methylenedioxy-9′-nor-3,8′-lign-7′-ene. ***De-O-methylegonol***

[879495-14-4]

$C_{18}H_{16}O_5$ 312.321

Constit. of the seeds of *Styrax perkinsiae*. Shows significant cytotoxic activity against human breast cancer cell lines. Amorph. powder. λ_{max} 204 (log ε 4.69); 302 (log ε 4.47); 317 (log ε 4.54) (MeOH)

9-Deoxy, 8,9-didehydro, 7-oxo

4,7′-Epoxy-5-methoxy-3′,4′-methylenedioxy-9′-nor-3,8′-ligna-7′,8-dien-7-one. 4,7′-Epoxy-5-methoxy-3′,4′-methylenedioxy-9′-nor-7-oxo-3,8′-ligna-7′,8-diene. ***2″,3″-Anhydromachicenonol***

[1426955-26-1]

$C_{19}H_{14}O_5$ 322.316

Constit. of the bark of *Styrax perkinsiae*. Powder. λ_{max} 263 (log ε 4.48); 327 (log ε 4.22) ($CHCl_3$)

Demethoxy

2-(1,3-Benzodioxol-5-yl)-5-benzofuranpropanol, CAS. *4,7′-Epoxy-3′,4′-methylenedioxy-9′-nor-3,8′-lign-7′-en-9-ol. 4,7′-Epoxy-9-hydroxy-3′,4′-methylenedioxy-9′-nor-3,8′-lign-7′-ene.* ***Demethoxyegonol***

[53279-35-9]

$C_{18}H_{16}O_4$ 296.322

Constit. of *Styrax obassia* and *Styrax officinalis*. Isol. from *Laetiporus sulphureus* var. *miniatus*. Needles (MeOH aq.). Mp 124 - 125° (118-119°). λ_{max} 214 (ε 35200); 317 (ε 32300); 331 (ε 26200) (EtOH)

Demethoxy, O-[β-D-glucopyranosyl-(1→6)-β-D-glucopyranoside]

Demethoxyegonol gentiobioside

[909801-17-8]

$C_{30}H_{36}O_{14}$ 620.604

Constit. of the seeds of *Styrax macranthus* and *Styrax perkinsiae*. Amorph. off-white powder. $[\alpha]_D^{20}$ -39 (c, 0.2 in DMSO). λ_{max} 215 (log ε 4.45); 303 (log ε 4.3); 318 (log ε 4.42) (MeOH)

Demethoxy, Ac

[53279-36-0]

Constit. of *Styrax agrestis*, *Styrax japonica* and *Styrax perkinsiae*. Cryst. (EtOH). Mp 98° (90-91°)

Demethoxy, 2S-methylbutanoyl

Demethoxyegonol 2-methylbutanoate

[115334-07-1]

$C_{23}H_{24}O_5$ 380.44

Constit. of *Styrax obassia* and *Styrax agrestis*. Needles (MeOH). Mp 55.5 - 56°. $[\alpha]_D^{25}$ +8.5 (c, 0.5 in CH_2Cl_2)

Demethoxy, 9Z-octadecenoyl

7-Demethoxyegonol oleate

[1336884-97-9]

$C_{36}H_{48}O_5$ 560.775

Constit. of the fruit of *Styrax agrestis*. Inhibitor of acetylcholinesterase (AChE). Oil. λ_{max} 210 (log ε 4.14); 229 (log ε 4.05); 319 (log ε 4.23) (MeOH)

Demethoxy, 9Z,12Z-octadecadienoyl

7-Demethoxyegonol linoleate

[1336884-96-8]

$C_{36}H_{46}O_5$ 558.759

Constit. of the fruit of *Styrax agrestis*. Inhibitor of acetylcholinesterase (AChE). Oil. λ_{max} 228 (log ε 4.04); 320 (log ε 4.23) (MeOH).

Kawai, S. *et al.*, *Ber. Dtsch. Chem. Ges.*, 1938, **71**, 2438-2443 (*Egonol*).

Takanashi, M. *et al.*, *Chem. Lett.*, 1974, 869-871 (*Demethoxyegonol*).

Schreiber, F.G. *et al.*, *J. Chem. Soc., Perkin Trans. 1*, 1976, 1514-1518 (*Egonol, Demethoxyegonol, Ac, synth, bibl*).

Anil, H. *Phytochemistry*, 1980, **19**, 2784-2786 (*glucoside, gentiobioside, gentiotrioside*).

Takanashi, M. *et al.*, *Phytochemistry*, 1988, **27**, 1224-1226 (*Egonol glucoside, Egonol 2-methylbutanoate, Demethoxyegonol 2-methylbutanoate*).

Aoyagi, Y. *et al.*, *Heterocycles*, 1995, **41**, 1077-1084 (*Egonol, synth*).

Mali, R.S. *et al.*, *J. Chem. Res., Synop.*, 1998, 230-231 (*Egonol, synth*).

Pauletti, P.M. *et al.*, *Phytochemistry*, 2000, **55**, 597-602 (*Styrax ferrugineus constits, activity*).

Yoshikawa, K. *et al.*, *Chem. Pharm. Bull.*, 2001, **49**, 327-329 (*Masutakeside I*).

Takanashi, M. *et al.*, *J. Oleo Sci.*, 2002, **51**, 151-155 (*Styrax japonica constits*).

Pal, G. *et al.*, *J. Chem. Res., Synop.*, 2003, 142-143 (*Egonol, synth*).

Akgul, Y.Y. *et al.*, *Phytochemistry*, 2003, **63**, 939-943 (*Egonol, Ac, 2-methylbutanoate, benzoyl, bibl*).

Zhang, J. *et al.*, *Tetrahedron*, 2003, **59**, 5609-5616 (*Egonol, synth*).

Min, B.-S. *et al.*, *Planta Med.*, 2004, **70**, 1210-1215 (*Egonol, Masutakeside I, activity*).

Kinoshita, T. *et al.*, *Heterocycles*, 2005, **65**, 1471-1480 (*Egonol gentiobioside*).

Chérigo, L. *et al.*, *Nat. Prod. Res.*, 2005, **19**, 373-377 (*Ac, activity*).

Li, Q.-L. *et al.*, *Planta Med.*, 2005, **71**, 847-851 (*O-de-Me, butanoyl, activity*).

Park, S.-Y. *et al.*, *Bull. Korean Chem. Soc.*, 2007, **28**, 1874-1876 (*Propanoylegonol*).

Luo, Y. *et al.*, *Fitoterapia*, 2007, **78**, 211-214 (*Styrax macranthus constit, struct*).

Emirdag-Oztürk, S. *et al.*, *Bioorg. Med. Chem.*, 2008, **16**, 4431-4437 (*Egonol, Egonol benzoate, Egonol oleate, synth, activity*).

Choi, D.H. *et al.*, *Bull. Korean Chem. Soc.*, 2008, **29**, 1594-1596 (*Egonol, synth*).

Lee, H.-J. *et al.*, *Chem. Nat. Compd. (Engl. Transl.)*, 2008, **44**, 435-439 (*Egonol, Ac, methylbutanoate, demethoxy methylbutanoate*).

Duan, X.-F. *et al.*, *Synthesis*, 2010, 1181-1187 (*Egonol, Demethoxyegonol, acetates, synth*).

Emirdag-Oztürk, S. *et al.*, *Bioorg. Med. Chem.*, 2011, **19**, 1179-1188 (*Egonol derivs, activity*).

Liu, J. *et al.*, *J. Nat. Prod.*, 2011, **74**, 2081-2088 (*Styrax agrestis constits, AChE inhibitor, Aβ aggregation inhibitor*).

Chen, Q.-F. *et al.*, *Chin. J. Nat. Med.*, 2012, **10**, 92-97 (*2″,3″-Anhydromachicenonol, Egonol, gentiotrioside, gentiobioside, glucoside, Masutakeside I*).

Bertanha, C.S. *et al.*, *Nat. Prod. Res.*, 2012, **26**, 2323-2329 (*Egonol, Egonol gentiobioside, activity*).
Bertanha, C.S. *et al.*, *J. Herbs, Spices Med. Plants*, 2014, **20**, 394-401 (*activity*).
Hajdu, Z. *et al.*, *J. Nat. Prod.*, 2014, **77**, 2641-2650 (*9-O-Methylegonol*).
More, K.R. *et al.*, *Tetrahedron*, 2016, **72**, 7496-7504 (*Egonol, synth*).
Sivaraman, A. *et al.*, *J. Nat. Prod.*, 2020, **83**, 3354-3362 (*Egonol, synth*)

Ehletianol C E3

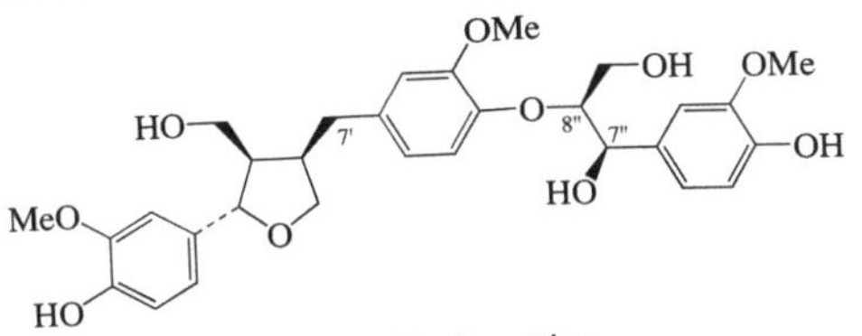

Relative Configuration

[166547-59-7]
$C_{30}H_{36}O_{10}$ 556.608
Only the configs. of the chiral centres in each half of the molecule are known, not the relationship between the halves. Similar to Cerberalignan F, C49 and Chushizisin H, C75. Constit. of *Ehretia ovalifolia*, *Larix olgensis* var. *koreana* and *Rhododendron alutaceum*. Also from *Litchi chinensis*. Amorph. powder. $[\alpha]_D^{20}$ -10.6 (c, 0.7 in MeOH)

Stereoisomer(1)
Sesquimarocanol A
[173522-91-3]
$C_{30}H_{36}O_{10}$ 556.608
Constit. of *Abies marocana* and *Pinus massoniana*. Oil (as penta-Ac). $[\alpha]_D^{25}$ +2 (c, 0.8 in $CHCl_3$) (penta-Ac). Has diastereomeric (7″*R**,8″*R**)-config

Stereoisomer(2), 7′-hydroxy
[845890-67-7]
$C_{30}H_{36}O_{11}$ 572.607
Isol. from *Picea abies*. No stereochem. determined.

Yoshikawa, K. *et al.*, *Phytochemistry*, 1995, **39**, 659-664 (*Ehletianol C*).
Barrero, A.F. *et al.*, *Phytochemistry*, 1996, **41**, 605-609 (*Sesquimarocanol A*).
Willfoer, S. *et al.*, *Holzforschung*, 2004, **58**, 345-354 (*Picea abies lignan*).
Wen, L. *et al.*, *J. Funct. Foods*, 2014, **8**, 26-34 (*Litchi constit*)

Elshrugulosain E4

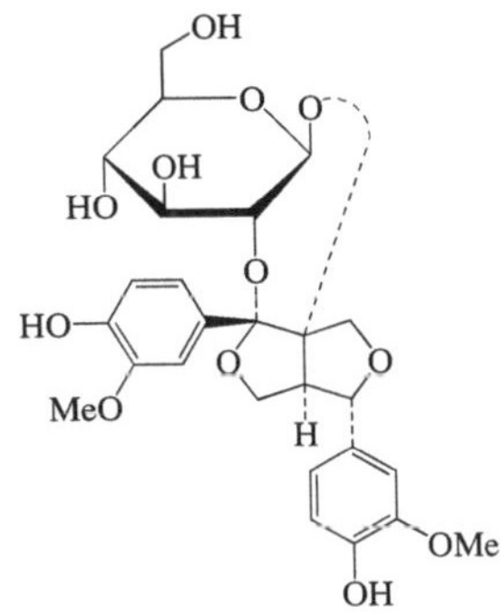

[1416911-35-7]
$C_{26}H_{30}O_{12}$ 534.514
Constit. of *Elsholtzia rugulosa*. Amorph. powder. $[\alpha]_D^{25}$ +36 (c, 0.11 in MeOH). λ_{max} 232 (log ε 3.86); 276 (log ε 3.72) (MeOH).

Liu, B. *et al.*, *J. Asian Nat. Prod. Res.*, 2012, **14**, 89-96 (*Elshrugulosain*)

7′,8′-Epidioxy-3′,5,5′-trimethoxy-4,4′-lign-8-ene-3,9′-diol E5

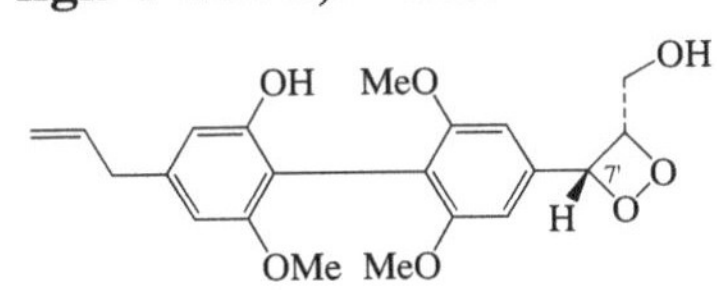

(7′*R*,8′*S*)-*form*

7′,8′-Epidioxy-3,9′-dihydroxy-3′,5,5′-trimethoxy-4,4′-neolign-8-ene
$C_{21}H_{24}O_7$ 388.416

(7′R,8′S)-form
Cinncassin I. Cinnacassin I
[2232873-48-0]
Constit. of the twigs of *Cinnamomum cassia*. Exhibits neuroprotective activity. Amorph. yellow solid. $[\alpha]_D^{15}$ -15 (c, 0.3 in MeOH). λ_{max} 205 (log ε 3.81); 245 (log ε 3.54); 275 (log ε 3.18) (MeOH)

(7′S,8′R)-form
Cinncassin J. Cinnacassin J
[2232873-49-1]
Constit. of the twigs of *Cinnamomum cassia*. Exhibits neuroprotective activity. Amorph. yellow solid. $[\alpha]_D^{25}$ +9 (c, 0.2 in MeOH). λ_{max} 210 (log ε 3.75); 245 (log ε 3.34); 275 (log ε 3.07) (MeOH).

Liu, X. *et al.*, *J. Nat. Prod.*, 2018, **81**, 1333-1342 (*Cinncassins I,J, cd, activity*)

Epiphyllic acid iridoid diester E6

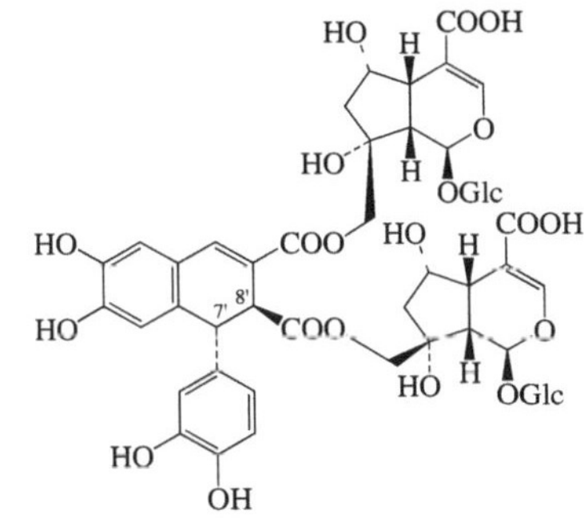

[2101458-52-8]
$C_{50}H_{58}O_{30}$ 1138.984
Constit. of *Vaccinium bracteatum*. Amorph. yellow powder. $[\alpha]_D^{20}$ -72.5 (c, 0.13 in MeOH). λ_{max} 230 (log ε 3.25); 290 (log ε 2.66); 342 (log ε 2.80) (MeOH)

7′,8′-Diepimer
[2101458-53-9]
$C_{50}H_{58}O_{30}$ 1138.984
Constit. of *Vaccinium bracteatum*. Amorph. yellow powder. $[\alpha]_D^{20}$ +11.4 (c, 0.098 in MeOH). λ_{max} 230 (log ε 3.07); 292 (log ε 2.57); 312 (log ε 2.62); 344 (log ε 2.64) (MeOH).

Ren, Y.-M. *et al.*, *Tetrahedron*, 2017, **73**, 3213-3219 (*Vaccinium bracteatum constits*)

4,7′-Epoxy-9′*a*,9′*a*-bishomo-3′,4′,5-trihydroxy-3,8′-lign-7-ene-9-carboxylic acid E7

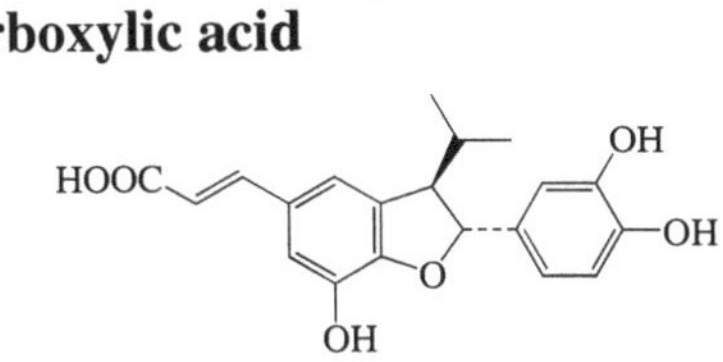

$C_{20}H_{20}O_6$ 356.374

(7E,7′R*,8′R*)-form

3′,5-Di-Me ether
4,7′-Epoxy-9′a,9′a-bishomo-4′-hydroxy-3′,5-dimethoxy-3,8′-lign-7-ene-9-carboxylic acid

3′,5-Di-Me ether, Me ester
Apiculin B
$C_{23}H_{26}O_6$ 398.455
Constit. of the bark of *Helietta apiculata*.

Fernandes, T.S. *et al.*, *Phytochemistry*, 2017, **141**, 131-139 (*Apiculin B, cryst struct*)

9,9′-Epoxy-3,4:3′,4′-bis(methylenedioxy)-8,8′-lignan-9-ol E8

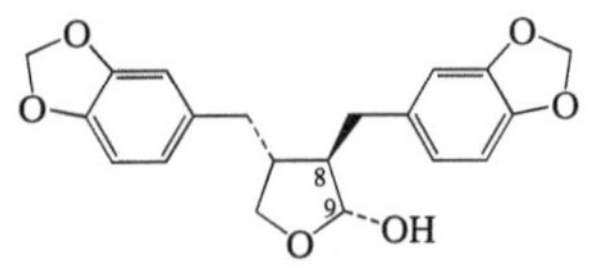

(8*R*,8′*R*,9*R*)-form

3,4-Bis(1,3-benzodioxol-5-ylmethyl)tetrahydro-2-furanol, CAS. *9,9′-Epoxy-9-hydroxy-3,4:3′,4′-bis(methylenedioxy)-8,8′-lignan. 9-Hydroxy-3,4:3′,4′-bis(methylenedioxy)-9,9′-epoxylignan.*
Cubebin
[112448-61-0] [112448-62-1]
$C_{20}H_{20}O_6$ 356.374
(8R,8′R,9R)-form
α-Cubebin [112458-74-9]
Constit. of *Aristolochia birostris*
Me ether
9,9′-Epoxy-9-methoxy-3,4:3′,4′-bis(methylenedioxy)-8,8′-lignan. 9-Methoxy-3,4:3′,4′-bis(methylenedioxy)-9,9′-epoxylignan. ***O-Methyl-α-cubebin***
Constit. of *Syringa oblata* and *Virola surinamensis*. Oil. [α]$_D$ -17.9 (c, 0.2 in $CHCl_3$)
(8R,8′R,9S)-form
β-Cubebin [18423-69-3]
Isol. from unripe fruit of *Piper cubeba*, from *Piper nigrum*, *Aristolochia* spp., *Virola* spp. and other plants. Shows urinary stimulant props., trypanocidal and analgesic activity. Melanogenesis stimulant. Cryst. (EtOH or C_6H_6). Mp 131 - 132°. [α]$^{14}_{D}$ -17.1 (Me_2CO). [α]$^{14}_{D}$ -45.7 ($CHCl_3$). Readily epimerises at C-9
Me ether
O-Methyl-β-cubebin
[173450-55-0]
$C_{21}H_{22}O_6$ 370.401
Constit. of *Syringa oblata* and *Virola surinamensis*. Shows analgesic activity. Amorph. solid. [α]$_D$ -2.2 (c, 0.2 in $CHCl_3$)
Et ether
9,9′-Epoxy-9-ethoxy-3,4:3′,4′-bis(methylenedioxy)-8,8′-lignan. 9-Ethoxy-3,4:3′,4′-bis(methylenedioxy)-9,9′-epoxylignan. ***O-Ethylcubebin***
[146830-09-3]
$C_{22}H_{24}O_6$ 384.428
Constit. of *Piper cubeba* and *Piper clusii*. Viscous solid
(8R,8′R,9ξ)-form
9-(15-Kauren-17-yl) ether
9-(15-Kauren-17-yl)cubebin
[1039033-59-4]
$C_{40}H_{50}O_6$ 626.834
Constit. of *Aristolochia constricta*. Gum. [α]$^{24}_{D}$ -10.9 (c, 0.11 in $CHCl_3$)
(8S,8′R,9S)-form
cis-Cubebin
Structure revised to that of the dimeric Bicubebin B in 2017.

Chatterjee, A. *et al., J. Indian Chem. Soc.*, 1968, **45**, 723-725 (*Piper cubeba constits*).
Burden, R.S. *et al., J. Chem. Soc. C*, 1969, 693-701 (*ord*).
Batterbee, J.E. *et al., J. Chem. Soc. C*, 1969, 2470-2477 (*synth*).
Chen, W.M. *et al., Arch. Pharm. (Weinheim, Ger.)*, 1987, **320**, 374-376 (*struct, stereochem*).
Badheka, L. *et al., Phytochemistry*, 1987, **26**, 2033-2036 (*O-Ethylcubebin*).
Rehnberg, N. *et al., J. Org. Chem.*, 1990, **55**, 4340-4349 (*synth*).
Conserva, L.M. *et al., Phytochemistry*, 1990, **29**, 257-260 (*Aristolochia birostris constit*).
De Almeida Blumenthal, E.E. *et al., Phytochemistry*, 1997, **46**, 745-749 (*O-Methylcubebins*).
Bastos, J.K. *et al., Planta Med.*, 1999, **65**, 541-544 (*activity*).
Borsato, M.L.C. *et al., Phytochemistry*, 2000, **55**, 809-813 (*activity*).
Matsuda, H. *et al., Biol. Pharm. Bull.*, 2004, **27**, 1611-1616 (*Piper nigrum constit*).
Bruno-Colmenárez, J. *et al., Acta Crystallogr., Sect. E*, 2007, **63**, o2046-o2047 (*cryst struct*).
Hirata, N. *et al., Bioorg. Med. Chem.*, 2007, **15**, 4897-4902 (*activity*).
Medola, J.F. *et al., Food Chem. Toxicol.*, 2007, **45**, 638-642 (*Piper cubeba constit*).
Zhang, G. *et al., J. Nat. Prod.*, 2008, **71**, 1167-1172 (*9-(15-Kauren-17-yl)cubebin*).
Zhao, M. *et al., Zhongcaoyao*, 2012, **43**, 251-254 (*Syringa O-Methylcubebins*).
Davidson, S.J. *et al., Org. Lett.*, 2017, **19**, 5368-5371 (*cis-Cubebin, synth*)

Epoxyconiferyl alcohol E9

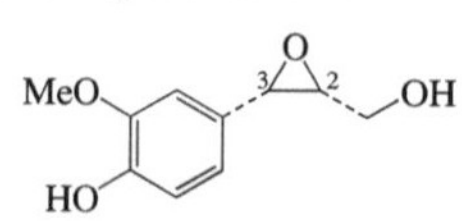

3-(4-Hydroxy-3-methoxyphenyl)oxiranemethanol, CAS
[162614-25-7]
$C_{10}H_{12}O_4$ 196.202
The struct. of the nat. prod. has been shown to be incorrect; however no reference to the derivs. below was made (Phytochemistry, 2000). Many references to isolations have also been made since this paper
(2R*,3S*)-form
[1821386-78-0]
[1243251-11-7] (2*R*,3*S*-form)
Constit. of *Antiaris toxicaria*, *Cirsium leucopsis*, *Fraxinus oxycarpa*, *Hyoscyamus niger*, *Iris suaveolens* and *Nardostachys chinensis*. Powder. [α]$^{20}_{D}$ +51.6 (c, 0.4 in $CHCl_3$). λ_{max} 206; 229; 280 (MeOH)
1-O-(2-Methylpropanoyl)
[2106918-01-6]
$C_{14}H_{18}O_5$ 266.293
Constit. of roots of *Bidens sulphurea*. Oil. [α]$^{20}_{D}$ -5.1 (c, 0.2 in $CHCl_3$). λ_{max} 239 (log ε 3.07); 275 (log ε 2.76) ($CHCl_3$).

Kostova, I. *et al., Phytochemistry*, 1995, **38**, 801-802 (*Fraxinus oxycarpa constit*).
Guz, N.R. *et al., Phytochemistry*, 2000, **54**, 897-899 (*synth, struct*).
Hacibekiroglu, I. *Phytother. Res.*, 2011, **25**, 522-529 (*Iris suaveolens constit, pmr, cmr*).
Da Silva, B.P. *et al., J. Agric. Food Chem.*, 2017, **65**, 5161-5172 (*Bidens sulphurea constit*)

4,7′-Epoxy-2,3-dihydro-3,3′,4′,5,5′-pentahydroxy-3,8′-lign-8-en-6(1*H*)-one E10

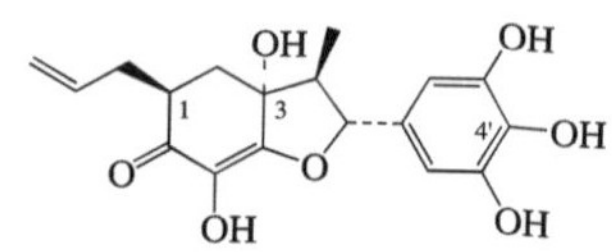

(1*S*,3*R*,7′*R*,8′*S*)-form

3,3a,4,5-Tetrahydro-3a,7-dihydroxy-3-methyl-5-(2-propenyl)-2-(3,4,5-trihydroxyphenyl)benzofuran-6(2H)-one
$C_{18}H_{20}O_7$ 348.351
(1S,3R,7′R,8′S)-form
Penta-Me ether
Heterochromin A
[2244685-39-8]
$C_{23}H_{30}O_7$ 418.486
Constit. of *Ocotea heterochroma*. Potent PAF (platelet activating factor) antagonist. Light yellow liq.. [α]$^{25}_{D}$ +35 (c, 0.06 in $CHCl_3$). λ_{max} 207 (log ε 4.65); 265 (log ε 4.28) (MeOH)
3′,4′-Methylene, 3,5,5′-tri-Me ether
Heterochromin B
[2244685-40-1]
$C_{22}H_{26}O_7$ 402.443
Constit. of *Ocotea heterochroma*. Moderate PAF (platelet activating factor) antagonist. Light yellow liq.. [α]$^{25}_{D}$ +41 (c, 0.03 in $CHCl_3$). λ_{max} 209 (log ε 4.89); 262 (log ε 4.33) (MeOH)

(1S,3R,7′S,8′S)-form
3′,4′-Methylene, 3,5,5′-tri-Me ether
Heterochromin C
[2244685-41-2]
$C_{22}H_{26}O_7$ 402.443
Constit. of *Ocotea heterochroma*. Light yellow liq.. $[\alpha]_D^{25}$ +12 (c, 0.01 in $CHCl_3$). λ_{max} 210 (log ε 4.25); 275 (log ε 3.78) (MeOH).
Rozo-Lugo, C. *et al.*, *J. Nat. Prod.*, 2018, **81**, 1968-1975 (*Heterochromins A,B,C, activity*)

2,7′-Epoxy-2,3-dihydro-3,3′,4,4′-tetrahydroxy-3,8′-lign-8-en-6(1*H*)-one E11

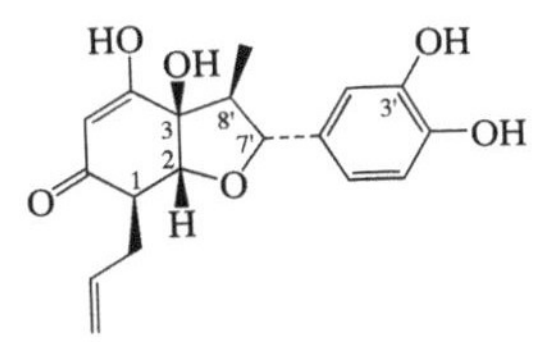
(1*R*,2*R*,3*S*,7′*R*,8′*S*)-*form*

$C_{18}H_{20}O_6$ 332.352
The stereochem. of this series is somewhat confused in the lit. Therefore assignment of stereochemical descriptors is tentative
(1R,2R,3S,7′R,8′S)-form
Tetra-Me ether
2,7′-Epoxy-2,3-dihydro-3,3′,4,4′-tetramethoxy-3,8′-lign-8-en-6(1H)-one. ***Fargesone D***
[1629056-70-7]
$C_{22}H_{28}O_6$ 388.46
Constit. of the flower buds of *Magnolia sprengeri*. Powder. λ_{max} 239 (MeOH)
(1S,2R,3S,7′R,8′S)-form
3′,4′-Methylene, 3,4-di-Me ether
2,7′-Epoxy-2,3-dihydro-3,4-dimethoxy-3′,4′-methylenedioxy-3,8′-lign-8-en-6(1H)-one.
Fargesone B
[116424-70-5]
$C_{21}H_{24}O_6$ 372.417
Constit. of *Magnolia fargesii*, *Magnolia denudata*, *Magnolia sprengeri* and *Piper wightii*. Possesses calcium antagonistic activity. Viscous oil. $[\alpha]_D$ -196 (c, 1 in $CHCl_3$). Isol. of Fargesone A by Du *et al* in 2001 may in fact be Fargesone B
(1S,2S,3R,7′S,8′R)-form
3′,4′-Methylene, 3,4-di-Me ether
Fargesone A
[116424-69-2]
$C_{21}H_{24}O_6$ 372.417
Constit. of *Magnolia fargesii*, *Magnolia denudata*, *Magnolia sprengeri* and *Piper wightii*. Possesses calcium antagonistic activity. Inhibitor of NO prodn.. Viscous oil. $[\alpha]_D$ -150 (c, 1 in $CHCl_3$). Isol. by Du *et al* in 2001 may in fact be Fargesone B not A.
Chen, C.C. *et al.*, *Chem. Pharm. Bull.*, 1988, **36**, 1791-1795 (*Fargesones A,B*).
Prasad, A.K. *et al.*, *Tetrahedron*, 1994, **50**, 10579-10586 (*Piper wightii constits*).
Du, J. *et al.*, *J. Asian Nat. Prod. Res.*, 2001, **3**, 313-319 (*abs config*).
Noshita, T. *et al.*, *Biosci., Biotechnol., Biochem.*, 2008, **72**, 2775-2778 (*Fargesone A, activity*).
Sun, Y. *et al.*, *J. Chromatogr., B*, 2011, **879**, 3775-3779 (*Fargesones A,B*).
Sun, Y. *et al.*, *Chem. Nat. Compd. (Engl. Transl.)*, 2014, **50**, 611-612 (*Fargesone D, cd*)

2,7′-Epoxy-1,6-dihydro-2,3′,4′-trihydroxy-1,8′-lign-8-en-3(2*H*)-one E12

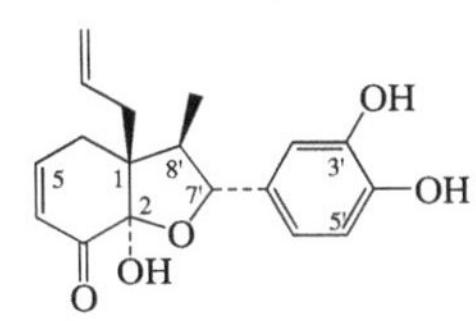
(1*R**,2*R**,7′*R**,8′*R**)-*form*

$C_{18}H_{20}O_5$ 316.353
The rel. configs. of this series have been revised on several occasions
(1R*,2R*,7′R*,8′R*)-form
3′,4′-Methylene ether
2,7′-Epoxy-1,6-dihydro-2-hydroxy-3′,4′-methylenedioxy-1,8′-lign-8-en-3(2H)-one. ***Ferrearin C***
[93528-36-0]
$C_{19}H_{20}O_5$ 328.364
Constit. of *Ocotea aciphylla* and *Ocotea catharinensis*. Exhibits antitumour activity. Cryst. (MeOH). Mp 135 - 136°
(1R*,2R*,7′S*,8′R*)-form
3′,4′-Methylene ether
Ferrearin B
[75666-78-3]
$C_{19}H_{20}O_5$ 328.364
Constit. of *Aniba ferrea*, *Ocotea aciphylla* and *Ocotea catharinensis*. Oil. $[\alpha]_D^{21}$ -22 (c, 1.4 in MeOH)
5′-Methoxy, 3′,4′-methylene ether
2,7′-Epoxy-1,6-dihydro-2-hydroxy-3′-methoxy-4′,5′-methylenedioxy-1,8′-lign-8-en-3(2H)-one. ***Ferrearin A***
[75666-77-2]
$C_{20}H_{22}O_6$ 358.39
Constit. of *Aniba ferrea*. Oil
(1R*,2S*,7′R*,8′S*)-form
3′-Me ether
2,7′-Epoxy-1,6-dihydro-2,4′-dihydroxy-3′-methoxy-1,8′-lign-8-en-3(2H)-one. ***Ferrearin F***
[139955-53-6]
$C_{19}H_{22}O_5$ 330.38
Constit. of *Aniba ferrea*. Amorph. solid. $[\alpha]_D$ -6.7 (c, 1.9 in MeOH)
3′,4′-Di-Me ether
2,7′-Epoxy-1,6-dihydro-2-hydroxy-3′,4′-dimethoxy-1,8′-lign-8-en-3(2H)-one. ***Ferrearin E***
[200052-92-2]
$C_{20}H_{24}O_5$ 344.407
Constit. of *Ocotea catharinensis*. Amorph. solid. λ_{max} 231 (ε 10500); 278 (ε 3590); 308 (ε 1120) (EtOH)
3′,4′-Methylene ether
[200052-91-1]
$C_{19}H_{20}O_5$ 328.364
Constit. of *Ocotea catharinensis*. Oil. $[\alpha]_D^{21}$ -22 (c, 1.4 in MeOH)
3′,4′-Methylene, 2-Me ether
2,7′-Epoxy-1,6-dihydro-2-methoxy-3′,4′-methylenedioxy-1,8′-lign-8-en-3(2H)-one
[200052-94-4]
$C_{20}H_{22}O_5$ 342.391
Constit. of *Ocotea catharinensis*. Oil. $[\alpha]_D^{21}$ -93 (c, 0.5 in MeOH)
5-Methoxy, 3′,4′-di-Me ether
2,7′-Epoxy-1,6-dihydro-2-hydroxy-3′,4′,5-trimethoxy-1,8′-lign-8-en-3(2H)-one. ***5-Methoxyferrearin E***
[1181279-96-8]
$C_{21}H_{26}O_6$ 374.433
Constit. of the leaves of *Ocotea catharinensis*. Viscous oil. $[\alpha]_D^{21}$ -130 (c, 0.2 in MeOH)
5′-Methoxy, 3′,4′-di-Me ether
2,7′-Epoxy-1,6-dihydro-2-hydroxy-3′,4′,5′-trimethoxy-1,8′-lign-8-en-3(2H)-one. ***5′-Methoxyferrearin E***
$C_{21}H_{26}O_6$ 374.433
Constit. of *Ocotea catharinensis*. Viscous oil. $[\alpha]_D^{21}$ -51 (c, 0.2 in MeOH)
(1R*,2S*,7′S*,8′S*)-form
3′-Me ether
Ferrearin H
[140147-72-4]
$C_{19}H_{22}O_5$ 330.38
Constit. of *Aniba ferrea*. Amorph. solid. $[\alpha]_D$ -78.4 (c, 2.3 in MeOH)
3′,4′-Di-Me ether
[200052-90-0]

$C_{20}H_{24}O_5$ 344.407
Constit. of *Ocotea catharinensis*. Oil

5′-Methoxy, 3′,4′-methylene ether

Ferrearin G
[140147-71-3]
$C_{20}H_{22}O_6$ 358.39
Constit. of *Aniba ferrea*. Exhibits antitumour activity. Amorph. solid. $[\alpha]_D$ -150.1 (c, 0.18 in MeOH).

Andrade, C.H.S. *et al.*, *Phytochemistry*, 1980, **19**, 1191-1194 (*Ferrearins A,B*).
Romoff, P. *et al.*, *Phytochemistry*, 1984, **23**, 2101-2104 (*Ferrearins B,C*).
Ishige, M. *et al.*, *Phytochemistry*, 1991, **30**, 4121-4128 (*Ferrearins C,E, cryst struct*).
Rodrigues, D.C. *et al.*, *Phytochemistry*, 1992, **31**, 271-273 (*Ferrearins F,G,H*).
Lordello, A.L.L. *et al.*, *Phytochemistry*, 1997, **46**, 741-744 (*Ocotea catharinensis constits*).
Eur. Pat., 2008, 2 009 007 (*Ferrearins C,G, activity*).
Funasaki, M. *et al.*, *J. Braz. Chem. Soc.*, 2009, **20**, 853-859 (*Ocotea catharinensis constit*)

2,7′-Epoxy-5,6-dihydro-3′,4′,5-trihydroxy-1,8′-lign-8-en-4(1*H*)-one E13

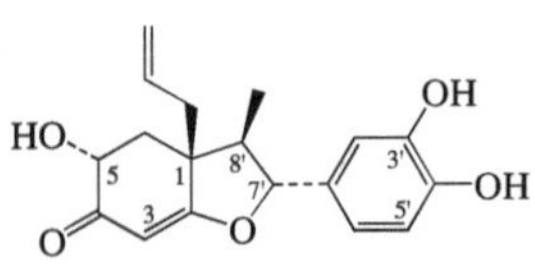

(1*R*,5*R*,7′*R*,8′*R*)-*form*

$C_{18}H_{20}O_5$ 316.353

(1R,5R,7′R,8′R)-form

3′,4′-Methylene, 5-Me ether

2,7′-Epoxy-5,6-dihydro-5-methoxy-3′,4′-methylenedioxy-1,8′-lign-8-en-4(1H)-one. ***Canellin F***
[107078-91-1]
$C_{20}H_{22}O_5$ 342.391
Constit. of *Ocotea porosa*. Oil. λ_{max} 260 (ε 11650) (MeOH)

5′-Methoxy, 3′,4′,5-tri-Me ether

2,7′-Epoxy-5,6-dihydro-3′,4′,5,5′-tetramethoxy-1,8′-lign-8-en-4(1H)-one. ***Canellin D***
[90898-97-8]
$C_{22}H_{28}O_6$ 388.46
Constit. of an *Aniba* sp.. Cryst. (Me_2CO). Mp 141 - 142°. λ_{max} 260 (ε 18250) (MeOH)

3-Methoxy, 3′,4′-methylene, 5-Me ether

2,7′-Epoxy-5,6-dihydro-3,5-dimethoxy-3′,4′-methylenedioxy-1,8′-lign-8-en-4(1H)-one. ***Canellin B***
[54835-73-3]
$C_{21}H_{24}O_6$ 372.417
Constit. of *Licaria canella* and *Ocotea catharinensis*. Oil. $[\alpha]_D^{21}$ +178 (c, 0.45 in MeOH). Struct. revised in 1978

3,5′-Dimethoxy, 3′,4′-methylene, 5-Me ether

2,7′-Epoxy-5,6-dihydro-3,3′,5-trimethoxy-4′,5′-methylenedioxy-1,8′-lign-8-en-4(1H)-one. ***Canellin E***
[90898-96-7]
$C_{22}H_{26}O_7$ 402.443
Constit. of *Aniba* sp.. Oil. λ_{max} 262 (ε 12600) (MeOH)

(1R,5R,7′R,8′S)-form

5′-Methoxy, 3′,4′,5-tri-Me ether

Dysodanthin B
[137491-87-3]
$C_{22}H_{28}O_6$ 388.46
Constit. of *Endlicheria dysodantha*. Cytotoxic agent. Oil

3-Methoxy, 3′,4′-methylene, 5-Me ether

7′-Epiarmenin B
[199916-60-4]
$C_{21}H_{24}O_6$ 372.417
Constit. of *Ocotea catharinensis*. Viscous oil. $[\alpha]_D^{21}$ +179 (c, 0.5 in MeOH)

5′-Methoxy, 3′,4′-methylene, 5-Me ether

2,7′-Epoxy-5,6-dihydro-3′,5-dimethoxy-4′,5′-methylenedioxy-1,8′-lign-8-en-4(1H)-one. ***Dysodanthin A***
[137491-86-2]
$C_{21}H_{24}O_6$ 372.417
Constit. of *Endlicheria dysodantha*. Cytotoxic agent. Needles (CH_2Cl_2/Et_2O). Mp 172°

(1R,5R,7′S,8′S)-form

3′,4′-Methylene, 5-Me ether

Armenin A‡
[70522-39-3]
$C_{20}H_{22}O_5$ 342.391
Constit. of *Licaria armeniaca*. Oil. λ_{max} 255 (ε 12500); 285 (ε 3800) (MeOH)

3-Methoxy, 3′,4′-methylene, 5-Me ether

Armenin B
[70561-33-0]
$C_{21}H_{24}O_6$ 372.417
Constit. of *Licaria armeniaca*, *Ocotea aciphylla*, *Ocotea catharinensis* and *Endlicheria* spp.. Inhibitor of DNA polymerase β-lyase. Cryst. (Me_2CO). Mp 122 - 125°. λ_{max} 239 (sh); 268 (ε 14000); 285 (sh) (MeOH)

5′-Methoxy, 3′,4′-methylene, 5-Me ether

Armenin C
[90851-23-3]
$C_{21}H_{24}O_6$ 372.417
Constit. of *Aniba* sp.. Oil. λ_{max} 258 (ε 6250) (MeOH)

3,5′-Dimethoxy, 3′,4′,5-tri-Me ether

2,7′-Epoxy-5,6-dihydro-3,3′,4′,5,5′-pentamethoxy-1,8′-lign-8-en-4(1H)-one
[1178974-86-1]
$C_{23}H_{30}O_7$ 418.486
Constit. of the leaves of *Ocotea catharinensis*. Viscous oil

(1R,5S,7′R,8′S)-form

3′,4′-Methylene, 5-Me ether
[107078-90-0]
$C_{20}H_{22}O_5$ 342.391
Constit. of the wood of *Ocotea porosa*. Cryst. (MeOH). Mp 112 - 113°. λ_{max} 260 (ε 14000); 290 (sh) (MeOH)

(1S,5S,7′S,8′R)-form

3′,4′,5-Tri-Me ether

2,7′-Epoxy-5,6-dihydro-3′,4′,5-trimethoxy-1,8′-lign-8-en-4(1H)-one. ***Porosin***
[40951-68-6]
$C_{21}H_{26}O_5$ 358.434
Constit. of *Aniba ferrea*, *Ocotea porosa* and *Urbanodendron verrucosum*. Cryst. (C_6H_6/petrol). Mp 133 - 135°. Struct. revised in 1976. λ_{max} 237 (sh); 258 (ε 21800); 285 (sh) (EtOH)

3′,4′-Methylene, 5-Me ether

Porosin B
[83144-65-4]
$C_{20}H_{22}O_5$ 342.391
Constit. of *Ocotea porosa* and *Urbanodendron verrucosum*. Cryst. (MeOH aq.). Mp 114 - 116°. λ_{max} 260 (ε 14000); 290 (sh) (MeOH)

3-Methoxy, 3′,4′,5-tri-Me ether

2,7′-Epoxy-5,6-dihydro-3,3′,4′,5-tetramethoxy-1,8′-lign-8-en-4(1H)-one. ***3-Methoxyporosin***
[199916-61-5]
$C_{22}H_{28}O_6$ 388.46
Constit. of *Ocotea catharinensis*. Prisms (CCl_4). Mp 159 - 161°. λ_{max} 234 (ε 8350); 276 (ε 17100) (EtOH).

Aiba, C.J. *et al.*, *Phytochemistry*, 1976, **15**, 1031-1032 (*Porosin*).

Wenkert, E. *et al.*, *Phytochemistry*, 1976, **15**, 1547-1551 (*Porosin, cmr*).
Tomita, K. *et al.*, *Acta Crystallogr., Sect. B*, 1978, **34**, 3441-3444 (*Porosin, cryst struct*).
Cavalcante, S. de H. *et al.*, *Phytochemistry*, 1978, **17**, 983-985 (*Canellin B*).
Aiba, C.J. *et al.*, *Phytochemistry*, 1978, **17**, 2038-2039 (*Armenin A*).
Dias, A. de F. *et al.*, *Phytochemistry*, 1982, **21**, 1137-1139 (*Porosin, Porosin B*).
Trevisan, L.M.V. *et al.*, *Phytochemistry*, 1984, **23**, 661-665 (*Canellins, Armenins, Porosins, cd, abs config*).
Dias, D.A. *et al.*, *Phytochemistry*, 1986, **25**, 2613-2616 (*Ocotea porosa constits*).
Ma, W.W. *et al.*, *J. Nat. Prod.*, 1991, **54**, 1153-1158 (*Dysodanthins A,B, activity*).
Ishige, M. *et al.*, *Phytochemistry*, 1991, **30**, 4121-4128 (*Ocotea catharinensis constits*).
Matsumoto, T. *et al.*, *Chem. Pharm. Bull.*, 1995, **43**, 2099-2104 (*Porosin, synth*).
Lordello, A.L.L. *et al.*, *Phytochemistry*, 1997, **46**, 741-744 (*Ocotea catharinensis constits*).
Chaturvedula, V.S.P. *et al.*, *J. Nat. Prod.*, 2004, **67**, 964-967 (*Endlicheria constit, activity*).
Funasaki, M. *et al.*, *J. Braz. Chem. Soc.*, 2009, **20**, 853-859 (*Ocotea catharinensis constit*)

4,7′-Epoxy-4′,9′-dihydroxy-3′,5-dimethoxy-9-nor-3,8′-lignan-7-one E14

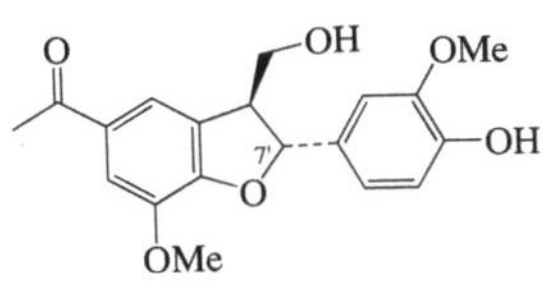

(7′R,8′S)-form

5-Acetyl-2,3-dihydro-2-(4-hydroxy-3-methoxyphenyl)-3-(hydroxymethyl)-7-methoxybenzofuran.
4,7′-Epoxy-4′,9′-dihydroxy-3′,5-dimethoxy-9-nor-7-oxo-3,8′-neolignan
$C_{19}H_{20}O_6$ 344.363

(7′R,8′S)-form

Bombasin [1004985-19-6]
Constit. of the flowers of *Bombax malabaricum*. Gum. $[\alpha]_D^{20}$ -61.8 (c, 7.6 in MeOH). λ_{max} 231 (log ε 4.4); 288 (log ε 4.18) (MeOH)

4′-O-β-D-Glucopyranoside
[1004321-11-2]
$C_{25}H_{30}O_{11}$ 506.504
Constit. of the flowers of *Bombax malabaricum*. Amorph. powder. $[\alpha]_D^{20}$ -74 (c, 1.3 in MeOH). λ_{max} 231 (log ε 4.23); 283 (log ε 4.17) (MeOH)

(7′S,8′R)-form

4′-O-β-D-Glucopyranoside
Lycocernuaside C
[2301994-33-0]
$C_{25}H_{30}O_{11}$ 506.504
Constit. of *Palhinhaea cernua*. Moderate xanthine oxidase inhibitor. Amorph. powder. $[\alpha]_D^{25}$ -66.9 (c, 0.1 in MeOH). λ_{max} 235 (log ε 4.44); 287 (log ε 4.35) (MeOH).

Wu, J. *et al.*, *Helv. Chim. Acta*, 2008, **91**, 136-143 (*Bombax malabaricum constits*).
Li, J. *et al.*, *Fitoterapia*, 2017, **119**, 45-50 (*Lycocernuaside C, activity*)

4,7′-Epoxy-3′,4′-dihydroxy-8,9-dinor-3,8′-lignan-7-al E15

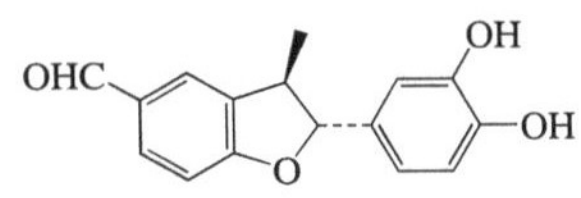

(7′R,8′R)-form

2-(3,4-Dihydroxyphenyl)-2,3-dihydro-3-methyl-5-benzofurancarboxaldehyde
$C_{16}H_{14}O_4$ 270.284

(7′R,8′R)-form

3′-Me ether
4,7′-Epoxy-4′-hydroxy-3′-methoxy-8,9-dinor-3,8′-lignan-7-al
[1669432-19-2]
$C_{17}H_{16}O_4$ 284.311
Isol. from the faeces of *Naupactus bipes* feeding on *Piper regnellii*

(7′S,8′S)-form

Decurrenal A
Constit. of *Mitrephora wangii*. Viscous oil. $[\alpha]_D^{25}$ +27 (c, 0.2 in MeOH). Not to be confused with Decurrenal, in E48. λ_{max} 224 (log ε 2.65); 291 (log ε 2.5) (MeOH)

Di-Me ether
4,7′-Epoxy-3′,4′-dimethoxy-8,9-dinor-3,8′-lignan-7-al
$C_{18}H_{18}O_4$ 298.338
Constit. of the roots of *Dasymaschalon sootepensis*. Yellowish gum. $[\alpha]_D^{28}$ +88.5 (c, 0.15 in MeOH). λ_{max} 230 (log ε 3.48); 287 (log ε 3.31) (MeOH).

Ramos, C.S. *et al.*, *Chemoecology*, 2013, **23**, 143-148 (*Naupactus bipes 3′-Me ether*).
Hongthong, S. *et al.*, *Nat. Prod. Commun.*, 2016, **11**, 809-813 (*Dasymaschalon sootepensis di-Me ether*).
Jaidee, W. *et al.*, *Fitoterapia*, 2018, **130**, 219-224 (*Decurrenal A*)

4,7′-Epoxy-3′,4′-dihydroxy-8,9-dinor-3,8′-lignan-7-oic acid E16

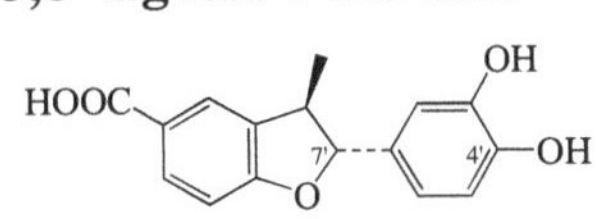

(7′R,8′R)-form

2-(3,4-Dihydroxyphenyl)-2,3-dihydro-3-methyl-5-benzofurancarboxylic acid
$C_{16}H_{14}O_5$ 286.283

(7′R,8′R)-form

7-Aldehyde, 3′,4′-methylene ether
4,7′-Epoxy-3′,4′-methylenedioxy-8,9-dinor-3,8′-lignan-7-al
[1609472-94-7]
$C_{17}H_{14}O_4$ 282.295
Constit. of the aerial parts of *Aristolochia fordiana*. Oil. $[\alpha]_D^{25}$ +37 (c, 0.324 in $CHCl_3$). There are inconsistencies regarding stereochem. in ref.. λ_{max} 204 (log ε 4.48); 229 (log ε 3.99); 291 (log ε 4.03) (no solvent reported)

(7′R,8′S)-form

7-Aldehyde, 3′,4′-methylene ether
[1609472-91-4]
$C_{17}H_{14}O_4$ 282.295
Constit. of the aerial parts of *Aristolochia fordiana*. Oil. $[\alpha]_D^{25}$ -62 (c, 1.18 in $CHCl_3$). There are inconsistencies regarding stereochem. in ref.. λ_{max} 203 (log ε 4.35); 230 (log ε 3.84); 290 (log ε 3.85) (no solvent reported)

7-Amide, 3′,4′-methylene ether
2,3-Dihydro-3-methyl-2-(3,4-methylenedioxyphenyl)-5-benzofurancarboxamide.
4,7′-Epoxy-3′,4′-methylenedioxy-8,9-dinor-3,8′-lignan-7-amide
[1802051-87-1]
$C_{17}H_{15}NO_4$ 297.31
Constit. of the stems of *Aristolochia fordiana*. Oil. $[\alpha]_D^{20}$ -3 (c, 0.5 in MeOH). λ_{max} 211 (log ε 4.38); 259 (log ε 4.07); 286 (log ε 3.88) ($CHCl_3$).

Zhou, Z.-B. *et al.*, *Planta Med.*, 2013, **79**, 1730-1735 (*Aristolochia fordiana constits*).
Tang, G.-H. *et al.*, *J. Nat. Prod.*, 2015, **78**, 1894-1903 (*Aristolochia fordiana amide*)

4,7′-Epoxy-4′,9′-dihydroxy-8,9-dinor-3,8′-lignan-7-oic acid E17

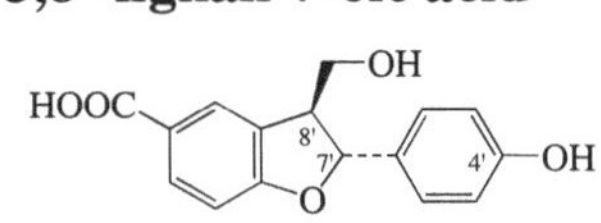

(7′R,8′S)-form

2,3-Dihydro-3-(hydroxymethyl)-2-(4-hydroxyphenyl)-5-benzofurancarboxylic acid
$C_{16}H_{14}O_5$ 286.283
Lignan numbering shown

(7′R,8′S)-form

7-Alcohol

2,3-Dihydro-2-(4-hydroxyphenyl)-3,5-benzofurandimethanol. 4,7′-Epoxy-8,9-dinor-3,8′-lignan-4′,7,9′-triol. (+)-Rubussin B

[2247479-96-3]

$C_{16}H_{16}O_4$ 272.300

Exhibits anti $A\beta_{1-42}$ aggregation activity (enantioselective). Oil. $[\alpha]_D^{20}$ +49.5 (c, 0.1 in MeOH)

7-Alcohol, 7-Me ether

(+)-Idaeusin B

$C_{17}H_{18}O_4$ 286.327

Oil. $[\alpha]_D^{20}$ +51.2 (c, 0.1 in MeOH)

7-Aldehyde

(+)-Idaeusin A

$C_{16}H_{14}O_4$ 270.284

Exhibits high ABTS and moderate DPPH radical scavenging activity. Oil. $[\alpha]_D^{20}$ +26.8 (c, 0.1 in MeOH)

(7′S,8′R)-form [1264177-66-3]

Constit. of *Arundina gramnifolia* and *Selaginella moellendorffii*. Exhibits modest anti-tobacco mosaic virus activity. Amorph. solid. $[\alpha]_D^{26}$ +96.3 (c, 0.52 in MeOH). λ_{max} 258 (log ε 3.82); 373 (log ε 2.96) (MeOH)

7-Alcohol

(-)-Rubussin B

[2247479-98-5]

$C_{16}H_{16}O_4$ 272.300

Oil. $[\alpha]_D^{20}$ -56.9 (c, 0.12 in MeOH)

7-Alcohol, 7-Me ether

(-)-Idaeusin B

$C_{17}H_{18}O_4$ 286.327

Exhibits moderate anti $A\beta_{1-42}$ aggregation activity. Antioxidant. $[\alpha]_D^{20}$ -46.3 (c, 0.1 in MeOH)

7-Aldehyde

(-)-Idaeusin A

$C_{16}H_{14}O_4$ 270.284

Antioxidant. Oil. $[\alpha]_D^{20}$ -34.7 (c, 0.1 in MeOH)

(7′RS,8′SR)-form

7-Alcohol

Rubussin B

$C_{16}H_{16}O_4$ 272.300

Constit. of the fruit of *Rubus ideaus*. Oil. $[\alpha]_D^{20}$ +2.8 (c, 0.2 in MeOH). The nat. prod. Rubussin B was isol. as a partial racemate and resolved by chiral hplc. λ_{max} 230 (log ε 1.84); 283 (log ε 0.55) (MeOH)

7-Alcohol, 7-Me ether

4,7′-Epoxy-7-methoxy-8,9-dinor-3,8′-lignan-4′,9′-diol. ***Idaeusin B***

$C_{17}H_{18}O_4$ 286.327

Constit. of the fruit of *Rubus idaeus*. Oil. $[\alpha]_D^{20}$ +6.6 (c, 0.2 in MeOH). The nat. prods Idaeusins A and B were isol. as racemates and resolved by hplc. λ_{max} 207 (log ε 3.55); 226 (log ε 2.56); 283 (log ε 0.82) (MeOH)

7-Aldehyde

4,7′-Epoxy-4′,9′-dihydroxy-8,9-dinor-3,8′-lignan-7-al. ***Idaeusin A***

$C_{16}H_{14}O_4$ 270.284

Constit. of the fruit of *Rubus idaeus*. Oil. $[\alpha]_D^{20}$ +2.3 (c, 0.1 in MeOH). The nat. prods Idaeusins A and B were isol. as racemates and resolved by hplc. λ_{max} 226 (log ε 3.13); 286 (log ε 0.99) (MeOH).

Wang, Y.-H. *et al.*, *Helv. Chim. Acta*, 2010, **93**, 2467-2477 (*Selaginella moellendorffii constit*).
Gao, X. *et al.*, *Bull. Korean Chem. Soc.*, 2012, **33**, 2447-2449 (*Arundina gramnifolia constit*).
Zhou, L. *et al.*, *Bioorg. Chem.*, 2018, **80**, 64-69 (*Rubussins B, activity*).
Zhou, L. *et al.*, *Fitoterapia*, 2018, **127**, 56-61 (*Idaeusins A,B, activity*)

4,7′-Epoxy-3′,4′-dihydroxy-3,8′-ligna-7,7′-dien-9-al E18

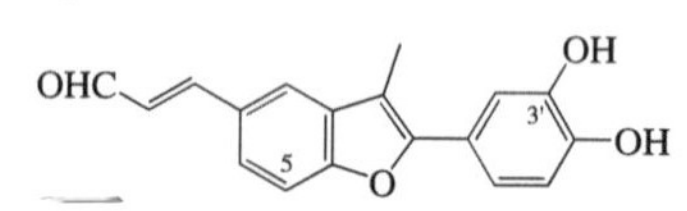

3-[2-(3,4-Dihydroxyphenyl)-3-methyl-5-benzofuranyl]-2-propenal. 2-(3,4-Dihydroxyphenyl)-3-methyl-5-(3-oxo-1-propenyl)benzofuran. 4,7′-Epoxy-3′,4′-dihydroxy-9-oxo-3,8′-neoligna-7,7′-diene

$C_{18}H_{14}O_4$ 294.306

(E)-form

3′,4′-Methylene ether

3-Methyl-2-(3,4-methylenedioxyphenyl)-5-(3-oxo-1-propenyl)benzofuran. 4,7′-Epoxy-3′,4′-methylenedioxy-3,8′-ligna-7,7′-dien-9-al. 4,7′-Epoxy-3′,4′-methylenedioxy-9-oxo-3,8′-ligna-7,7′-diene

[198136-83-3]

$C_{19}H_{14}O_4$ 306.317

Constit. of *Caryodaphnopsis baviensis*. Amorph.

5-Methoxy, 3′-Me ether

4,7′-Epoxy-4′-hydroxy-3′,5-dimethoxy-3,8′-ligna-7,7′-dien-9-al. 4,7′-Epoxy-4′-hydroxy-3′,5-dimethoxy-9-oxo-3,8′-ligna-7,7′-diene. ***Papillarin A***

[220570-30-9]

$C_{20}H_{18}O_5$ 338.359

Constit. of the roots of *Aristolochia papillaris*.

Anh, N.H. *et al.*, *Phytochemistry*, 1997, **46**, 569-571 (*Caryodaphnopsis baviensis constit*).
Franca, V.C. *et al.*, *Quim. Bras.*, 2010, **4**, 7-14 (*Papillarin A*)

4,7′-Epoxy-4′,6-dihydroxy-3,8′-ligna-7,7′-diene-9′-oic acid E19

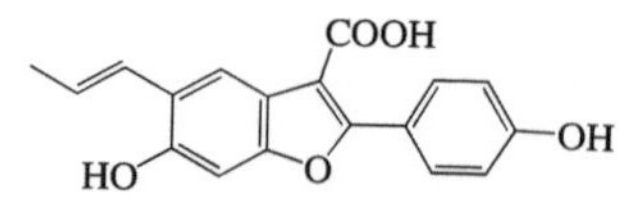

6-Hydroxy-2-(4-hydroxyphenyl)-5-(1-propenyl)-3-benzofurancarboxylic acid

$C_{18}H_{14}O_5$ 310.305

(E)-form

Me ester

$C_{19}H_{16}O_5$ 324.332

Constit. of the seeds of *Daphniphyllum macropodum*. Exhibits apoptosis in A549 cells. Solid. λ_{max} 211; 237; 345 (MeOH).

Ma, X. *et al.*, *RSC Adv.*, 2017, **7**, 52970-52976 (*Me ester, activity*)

4,7′-Epoxy-4′,5-dihydroxy-3,8′-lign-7′-en-8-one E20

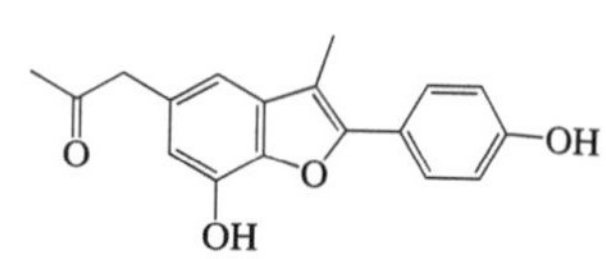

1-[7-Hydroxy-2-(4-hydroxyphenyl)-3-methyl-5-benzofuranyl]-2-propanone. 4,7′-Epoxy-4′,5-dihydroxy-8-oxo-3,8′-neolign-7′-ene

$C_{18}H_{16}O_4$ 296.322

Di-Me ether

4,7′-Epoxy-4′,5-dimethoxy-3,8′-lign-7′-en-8-one

$C_{20}H_{20}O_4$ 324.376

Constit. of the seeds of *Arctium lappa*. Exhibits anti-tobacco mosaic virus activity. Pale yellow gum. λ_{max} 210 (log ε 4.26); 280 (log ε 3.48); 323 (log ε 3.65) (MeOH)

5-Deoxy, 6-hydroxy, 4′-Me ether

4,7′-Epoxy-6-hydroxy-4′-methoxy-3,8′-lign-7′-en-8-one

$C_{19}H_{18}O_4$ 310.349

Constit. of the seeds of *Arctium lappa*. Exhibits anti-tobacco mosaic virus activity. Pale yellow gum.

λ_{max} 210 (log ε 4.20); 275 (log ε 3.47); 316 (log ε 3.68) (MeOH)

6-Methoxy, 5-deoxy, 4′-Me ether

4,7′-Epoxy-4′,6-dimethoxy-3,8′-lign-7′-en-8-one

$C_{20}H_{20}O_4$ 324.376

Constit. of the seeds of *Arctium lappa*. Exhibits anti-tobacco mosaic virus activity. Pale yellow gum. λ_{max} 210 (log ε 4.15); 278 (log ε 3.27); 320 (log ε 3.62) (MeOH).

Zeng, W.-L. *et al.*, *Heterocycles*, 2019, **98**, 437-443 (*Arctium lappa constits, activity*)

2′,4-Epoxy-4′,6-dihydroxy-3′-methoxy-3,7′-lign-7-en-8′-one E21

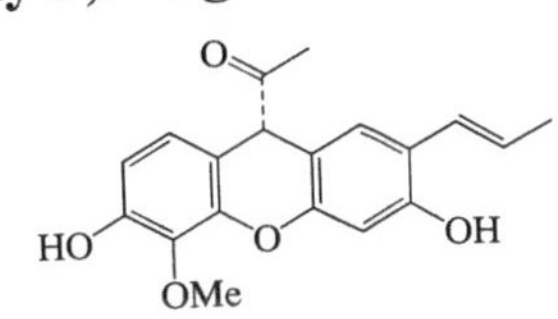

(R,E)-form

1-(3,6-Dihydroxy-5-methoxy-2-propenyl-9H-xanthen-9-yl)ethanone. 2′,4-Epoxy-4′,6-dihydroxy-3′-methoxy-8′-oxo-3,7′-neolign-7-ene

$C_{19}H_{18}O_5$ 326.348

(R,E)-form

Penchinone C [2043024-07-1]

Constit. of *Penthorum chinense*. Gum. $[\alpha]^{20}_{D}$+22.7 (c, 0.12 in MeOH). λ_{max} 224 (log ε 4.33); 268 (log ε 3.96); 308 (log ε 3.53) (MeOH)

(R,Z)-form

Penchinone D [2043024-08-2]

Constit. of *Penthorum chinense*. Gum. $[\alpha]^{20}_{D}$+14.9 (c, 0.04 in MeOH). λ_{max} 222 (log ε 4.32); 259 (log ε 3.98); 301 (log ε 3.55) (MeOH).

He, Y.-C. *et al.*, *RSC Adv.*, 2015, **5**, 76788-76794 (*Penchinones C,D, abs config*)

3,7′-Epoxy-4′,6-dihydroxy-9-nor-2,8′-lignan-7-one E22

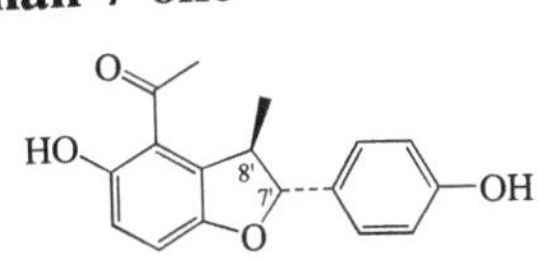

(7′R,8′R)-form

1-[2,3-Dihydro-5-hydroxy-3-methyl-2-(4-hydroxyphenyl)benzofuran-4-yl] ethanone. 4-Acetyl-2,3-dihydro-5-hydroxy-2-(4-hydroxyphenyl)-3-methylbenzofuran

$C_{17}H_{16}O_4$ 284.311

The nat. prods. Cynwilforones E and F were isol. as racemates and resolved by chiral hplc

(7′R,8′R)-form

(-)-Cynwilforone E

Exhibits neuroprotectant activity. $[\alpha]^{17}_{D}$-62 (c, 0.05 in MeOH)

(7′R,8′S)-form

(+)-Cynwilforone F

Exhibits neuroprotectant activity. $[\alpha]^{17}_{D}$+42 (c, 0.04 in MeOH)

(7′S,8′R)-form

(-)-Cynwilforone F

Exhibits neuroprotectant activity. $[\alpha]^{17}_{D}$-36 (c, 0.04 in MeOH)

(7′S,8′S)-form

(+)-Cynwilforone E

Exhibits neuroprotectant activity. $[\alpha]^{17}_{D}$+70 (c, 0.06 in MeOH)

(7′RS,8′RS)-form

Cynwilforone E

Constit. of the root bark of *Cynanchum wilfordii*. Amorph. yellow powder

(7′RS,8′SR)-form

Cynwilforone F

Constit. of the root bark of *Cynanchum wilfordii*. Amorph. yellow powder.

Jiang, H.-W. *et al.*, *Phytochem. Lett.*, 2018, **24**, 179-183 (*Cynwilforones E,F, cd, activity*)

4,7′-Epoxy-3′,4′-dihydroxy-9′-nor-3,8′-lign-7′-en-9-oic acid E23

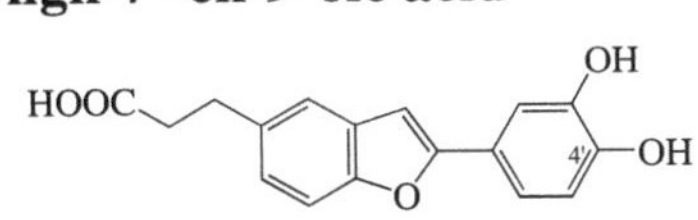

2-(3,4-Dihydroxyphenyl)-5-benzofuranpropanoic acid

$C_{17}H_{14}O_5$ 298.294

3′,4′-Methylene ether

4,7′-Epoxy-3′,4′-methylenedioxy-9′-nor-3,8′-lign-7′-en-9-oic acid

3′,4′-Methylene ether, Me ester

[58335-54-9]

$C_{19}H_{16}O_5$ 324.332

Constit. of the seeds of *Styrax obassia*. Powder. Mp 126 - 128°. $[\alpha]^{20}_{D}$+4 (c, 0.1 in $CHCl_3$). Opt. rotation unaccounted for. λ_{max} 238 (log ε 3.6); 318 (log ε 4) ($CHCl_3$).

Lee, H.-J. *et al.*, *Chem. Nat. Compd. (Engl. Transl.)*, 2008, **44**, 435-439 (*Styrax obassia constit*)

3,7′-Epoxy-4′,6-dihydroxy-9-nor-2,8′-lign-7′-en-7-one E24

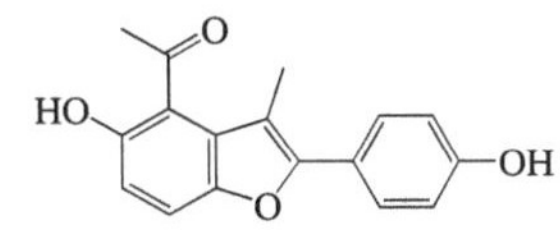

4-Acetyl-5-hydroxy-2-(4-hydroxyphenyl)-3-methylbenzofuran.

Cynwilforone G

$C_{17}H_{14}O_4$ 282.295

Constit. of the root bark of *Cynanchum wilfordii*. Brownish-yellow powder.

Jiang, H.-W. *et al.*, *Phytochem. Lett.*, 2018, **24**, 179-183 (*Cynwilforone G*)

7,7′-Epoxy-3,3′-dimethoxy-8,8′-lignan-4,4′-diol E25

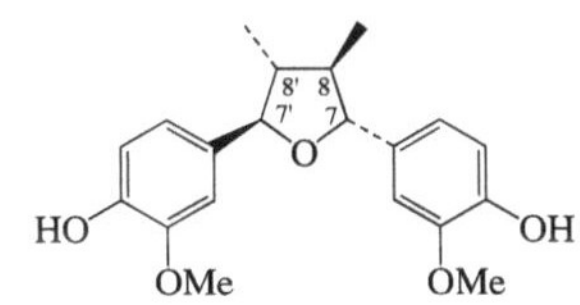

(7R,7′R,8R,8′R)-form

4,4′-(Tetrahydro-3,4-dimethyl-2,5-furandiyl)bis[2-methoxyphenol], CAS. Tetrahydro-2,5-bis(4-hydroxy-3-methoxyphenyl)-3,4-dimethylfuran. 4,4′-Dihydroxy-3,3′-dimethoxy-7,7′-epoxylignan

[128524-98-1]

$C_{20}H_{24}O_5$ 344.407

A common lignan various stereoisomers of which have been repeatedly isol. under different names. The abs. configs. require clarification however

(7R,7′R,8R,8′R)-form

Fragransin A$_2$ [112652-46-7]

Constit. of *Acorus tatarinowii*, *Knema furfuracea*, *Myristica fragrans* and *Myristica dactyloides*. Cryst.. Mp 207 - 209°. $[\alpha]^{20}_{D}$-61 (c, 0.05 in MeOH) (+79)

(7R,7′R,8R,8′S)-form

4-O-β-D-Glucopyranoside

[1566606-30-1]

$C_{26}H_{34}O_{10}$ 506.548

Constit. of bark of *Machilus robusta*. Yellowish solid. $[\alpha]^{20}_{D}$ -54.5 (c, 0.46 in MeOH). Diastereoisomer of the 4′-glucoside. Descr. in the lit. as (7R,7′R,8S,8′R)-4′-glucoside. λ_{max} 231 (log ε 0.33); 281 (log ε 0.12) (MeOH)

4′-O-β-D-Glucopyranoside

[1566606-27-6]

$C_{26}H_{34}O_{10}$ 506.548

Constit. of bark of *Machilus robusta*. Yellowish solid. $[\alpha]_D^{20}$ -25.3 (c, 0.29 in MeOH). Descr. in the lit. as (7*R*,7′*R*,8*S*,8′*R*)-4-glucoside but the stereodescriptor given here should take priority. λ_{max} 230 (log ε 0.38); 278 (log ε 0.16) (MeOH)

(7R,7′S,8R,8′R)-form

Odoratisol C. *ent*-Verrucosin

[891182-95-9]

Constit. of the bark of *Machilus odoratissima* and *Myristica dactyloides* and arils of *Virola oleifera*. Antifungal agent. Oil. $[\alpha]_D^{25}$ -26 (c, 2.8 in $CHCl_3$). $[\alpha]_D^{20}$ -48 (c, 0.3 in $CHCl_3$). λ_{max} 232 (log ε 4.05); 280 (log ε 3.71) (MeOH)

(7R,7′S,8R,8′S)-form

Nectandrin B. Tetrahydrofuroguaiacin A. Calophyllin. Malabaricanol.

[74683-16-2]

Isol. from *Aristolochia ponticum*, *Endiandra anthropophagorum*, *Guaiacum officinale*, *Machilus thunbergii*, *Myristica* spp., *Nectandra rigida*, *Pottsia laxiflora* and *Virola calophylla*. Neuroprotetctive agent. Inhibits platelet aggregation. Oil. Opt. inactive (*meso-*)

4′-O-β-D-Glucopyranoside

Nectandrin B 4-O-β-D-glucopyranoside

[1566606-22-1]

$C_{26}H_{34}O_{10}$ 506.548

Constit. of bark of *Machilus robusta*. Yellowish solid. $[\alpha]_D^{20}$ -22.5 (c, 0.1 in MeOH). Nectandrin B is *meso-* but Glc substn. gives potentially 2 diastereomeric compds. as 7,8-*trans*-7′,8′-*trans*-config. The full abs. config. was not established. λ_{max} 229 (log ε 0.18); 279 (log ε 0.07) (MeOH)

(7R,7′S,8S,8′R)-form

Isonectandrin B.

Tetrahydrofuroguaiacin B

[131829-51-1]

Isol. from *Guaiacum officinale*, *Myristica fragrans* and *Piper polysyphorum*. Neuroprotective agent. Opt. inactive (*meso-*)

(7R,7′S,8S,8′S)-form

Verrucosin. Epimachilin A [83198-63-4]

Isol. from *Urbanodendron verrucosum* and roots of *Jatropha grossidentata*. Oil. $[\alpha]_D^{22}$ +38 (c, 0.25 in $CHCl_3$). Props. refer to Verrucosin.

Epimachilin A incompletely characterised. λ_{max} 280 (ε 17100) (McOH)

(7S,7′S,8S,8′S)-form

(-)-Fragransin A₂ [1538628-90-8]

Constit. of *Acorus tatarinowii*. Amorph. powder. $[\alpha]_D$ +67 (c, 0.2 in MeOH)

(7R*,7′R*,8R*,8′S*)-form

Machilin I. 3,3′-Dimethoxylarreatricin

[114422-21-8]

Isol. from *Machilus thunbergii* and *Larrea tridentata*. Gum. $[\alpha]_D^{25}$ -93 (c, 0.13 in $CHCl_3$). λ_{max} 210 (log ε 4.24); 280 (log ε 3.85) (EtOH).

Blears, J.G. *et al.*, *J. Chem. Soc.*, 1958, 1985-1987 (*synth*).
Birch, A.J. *et al.*, *J. Chem. Soc.*, 1958, 4471-4476 (*stereochem*).
King, F.E. *J. Chem. Soc.*, 1964, 4011-4024 (*Tetrahydrofuroguaiacins*).
Sarkanen, K.V. *et al.*, *J. Chem. Soc., Perkin Trans. 1*, 1973, 1869-1878 (*synth*).
Le Quesne, P.W. *et al.*, *J. Nat. Prod.*, 1980, **43**, 353-359 (*Nectandrin B*).
Dias, A. de F. *et al.*, *Phytochemistry*, 1982, **21**, 1137-1139 (*Verrucosin*).
Purushothaman, K.K. *et al.*, *Indian J. Chem., Sect. B*, 1984, **23**, 46-48 (*Malabaricanol*).
Hattori, M. *et al.*, *Chem. Pharm. Bull.*, 1987, **35**, 3315-3322 (*Fragransin A₂*).
Alvarez, V. *et al.*, *CA*, 1988, **108**, 183594 (*Calophyllin*).
Shimomura, H. *et al.*, *Phytochemistry*, 1988, **27**, 634-636 (*Machilin I*).
Konno, C. *et al.*, *J. Nat. Prod.*, 1990, **53**, 396-406 (*3,3′-Dimethoxylarreatricin*).
Schmeda-Hirschmann, G. *et al.*, *Phytochemistry*, 1992, **31**, 1731-1735 (*Epimachilin A*).
Herath, H.M.T.B. *et al.*, *Phytochemistry*, 1996, **42**, 1439-1442 (*Myristica dactyloides constit*).
Herath, H.M.T.B. *et al.*, *Phytochemistry*, 1997, **44**, 699-703 (*Malabaricanol, Fragransin A₂*).
Sartorelli, P. *et al.*, *Phytochemistry*, 1998, **47**, 1003-1006 (*(-)-Verrucosin*).
Wang, B.-G. *et al.*, *Planta Med.*, 2000, **66**, 511-515 (*Nectandrin B, activity*).
Zhai, H. *et al.*, *Biol. Pharm. Bull.*, 2005, **28**, 289-293 (*Nectandrin B, Isonectandrin B, activity*).
Giang, P.M. *et al.*, *Chem. Pharm. Bull.*, 2006, **54**, 380-383 (*Odoratisol C*).
Akiyama, K. *et al.*, *Biosci., Biotechnol., Biochem.*, 2007, **71**, 1028-1035 (*(-)-Verrucosin*).
Kim, H. *et al.*, *Org. Lett.*, 2007, **9**, 3965-3968 (*synth*).
Rangkaew, N. *et al.*, *Fitoterapia*, 2009, **80**, 377-379 (*Knema furfuracea constit*).
Nguyen, P.H. *et al.*, *Bioorg. Med. Chem. Lett.*, 2010, **20**, 1004-1007 (*Isonectandrin B*).
Mu, L-H. *et al.*, *Chem. Nat. Compd. (Engl. Transl.)*, 2013, **48**, 1004-1007 (*Pottsia laxiflora constit*).
Bu, P.-B. *et al.*, *J. Asian Nat. Prod. Res.*, 2013, **15**, 482-491 (*Machilus robusta constits*).
Nishiwaki, H. *et al.*, *J. Agric. Food Chem.*, 2014, **62**, 651-659 (*synth*).
Lu, Y. *et al.*, *J. Nat. Prod.*, 2015, **78**, 2205-2214 (*Fragransin A₂*)

4,9′-Epoxy-3′,5-dimethoxy-3,8′-lign-7-ene-4′,9-diol E26

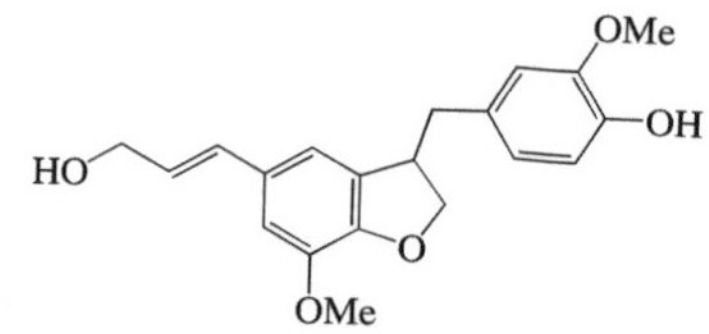

4,9′-Epoxy-4′,9-dihydroxy-3′,5-dimethoxy-3,8′-neolign-7-ene

$C_{20}H_{22}O_5$ 342.391

(7E,8′ξ)-form

4′-O-β-D-Glucopyranoside

Cordirazanol

[2102581-67-7]

$C_{26}H_{32}O_{10}$ 504.532

Constit. of the fruit of *Cordia latifolia*. Amorph. powder. $[\alpha]_D^{24}$ +13.4 (c, 0.3 in MeOH). λ_{max} 258 (log ε 4.9) (MeOH).

Fatima, M. *et al.*, *Chem. Nat. Compd. (Engl. Transl.)*, 2017, **53**, 432-435 (*Cordirazanol*)

8′,9′-Epoxy-3′,5-dimethoxy-3,4′-lign-7-ene-4,7′,9-triol E27

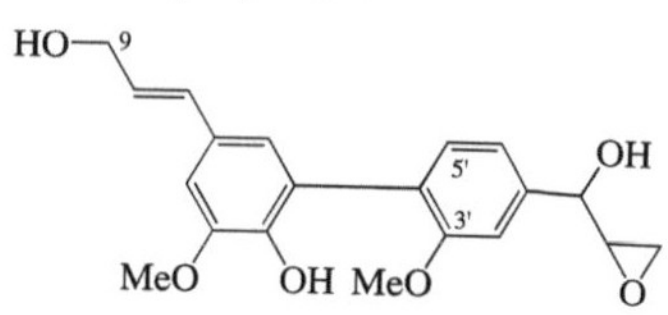

4′-(2,3-Epoxy-1-hydroxypropyl)-2-hydroxy-5-(3-hydroxy-1-propenyl)-2′,3-dimethoxybiphenyl. 8′,9′-Epoxy-4,7′,9-trihydroxy-3′,5-dimethoxy-3,4′-neolign-7-ene

[145224-78-8]

$C_{20}H_{22}O_6$ 358.39

Lignan numbering shown. Constit. of *Eurycoma longifolia*. Yellow needles. Mp 65 - 66°. $[\alpha]_D$ -9.7 (c, 0.7 in MeOH)

5′-Methoxy

4′-(2,3-Epoxy-1-hydroxypropyl)-2-hydroxy-5-(3-hydroxy-1-propenyl)-2′,3,6′-trimethoxybiphenyl. 8′,9′-Epoxy-3′,5,5′-trimethoxy-3,4′-lign-7-ene-4,7′,9-triol. 8′,9′-Epoxy-4,7′,9-trihydroxy-3′,5,5′-trimethoxy-3,4′-lign-7-ene

[145224-77-7]

$C_{21}H_{24}O_7$ 388.416

Constit. of *Eurycoma longifolia*. Yellow needles. Mp 68 - 70°. $[\alpha]_D$ -2.6 (c, 0.8 in MeOH)

5′-Methoxy, 9-aldehyde

8′,9′-Epoxy-4,7′-dihydroxy-3′,5,5′-trimethoxy-3,4′-lign-7-en-9-al.

8′,9′-Epoxy-4,7′-dihydroxy-3′,5,5′-trimethoxy-9-oxo-3,4′-lign-7-ene
[185681-69-0]
$C_{21}H_{22}O_7$ 386.4
Constit. of the wood of *Irvingia malayana*. Amorph. yellow powder. $[\alpha]_D$ +3.1 (c, 0.16 in MeOH). λ_{max} 242 (sh); 341 (log ε 4.9) (MeOH).

Morita, H. *et al., Phytochemistry*, 1992, **31**, 3993-3995 (*Eurycoma longifolia constits*).
Mitsunaga, K. *et al., Nat. Med. (Tokyo)*, 1996, **50**, 325-327 (*Irvingia malayana constit*)

9,9′-Epoxy-3,5-dimethoxy-7,9′-oxylign-7′-ene-4,8-diol E28

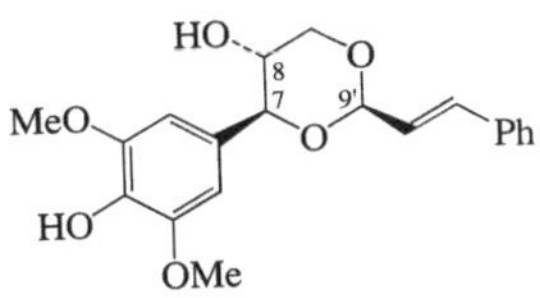

4-(4-Hydroxy-3,5-dimethoxyphenyl)-2-(2-phenylethenyl)-1,3-dioxan-5-ol. 5-Hydroxy-4-(4-hydroxy-3,5-dimethoxyphenyl)-2-styryl-1,3-dioxane. Cinnamic aldehyde cyclic syringylglycerol 1,3-acetal. 9,9′-Epoxy-4,8-dihydroxy-3,5-dimethoxy-7,9′-oxyneolign-7′-ene
$C_{20}H_{22}O_6$ 358.390
Lignan numbering shown
(7R*,8R*,9′R*)-form [1103512-91-9]
Constit. of the bark of *Cinnamomum burmani*. Powder ($CHCl_3$). $[\alpha]_D^{22}$ +36 (c, 0.05 in $CHCl_3$).

Subehan, S. *et al., Planta Med.*, 2008, **74**, 1474-1480 (*Cinnamomum burmani constit*)

7,7′-Epoxy-9,9′-dinor-8,8′-lignan-3,3′-diol E29

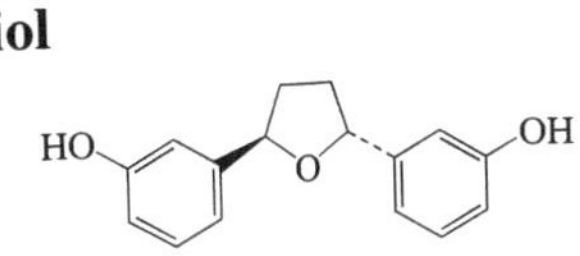

Tetrahydro-2,5-bis(3-hydroxyphenyl)furan. 3,3′-(Tetrahydro-2,5-furandiyl)bisphenol, CAS. *3,3′-Dihydroxy-9,9-dinor-7,7′-epoxylignan.* ***Ammaniol***
$C_{16}H_{16}O_3$ 256.301
(2R*,5R*)-form
[1393082-64-8]
Constit. of *Ammannia multiflora*. Moderately active against *M. tuberculosis* H37Rv. Cryst.. Mp 209°.

Upadhyay, H.C. *et al., Planta Med.*, 2012, **78**, 79-81 (*Ammaniol*)

4,7′-Epoxy-8,9-dinor-3,8′-lignan-4′,5,7,9′-tetrol E30

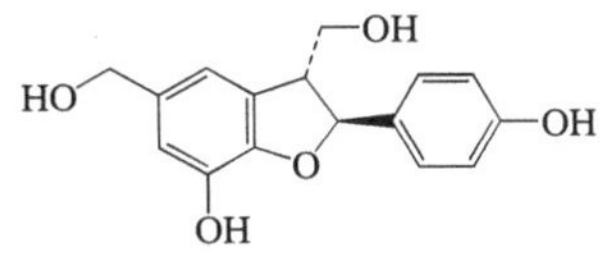

2,3-Dihydro-7-hydroxy-2-(4-hydroxyphenyl)-3,5-benzofurandimethanol. 4,7′-Epoxy-4′,5,7,9′-tetrahydroxy-8,9-dinor-3,8′-neolignan
$C_{16}H_{16}O_5$ 288.299
(7′S,8′R)-form
5-Me ether
4,7′-Epoxy-5-methoxy-8,9-dinor-3,8′-lignan-4′,7,9′-triol. 4,7′-Epoxy-4′,7,9′-trihydroxy-5-methoxy-8,9-dinor-3,8′-lignan. ***Mubiesin D***
[2088724-93-8]
$C_{17}H_{18}O_5$ 302.326
Constit. of *Juglans mandshurica* and *Momordica cochinchinensis*. Oil. $[\alpha]_D^{20}$ +12.8 (c, 0.2 in MeOH). λ_{max} 226 (log ε 3.60); 278 (log ε 2.44) (MeOH).

Park, S. *et al., Phytochemistry*, 2017, **137**, 87-93 (*Juglans mandshurica constit, cd*).
Wang, M. *et al., Fitoterapia*, 2019, **139**, 104360:1-6 (*Mubiesin D, cd*)

4,7′-Epoxy-8,9-dinor-3,8′-lign-7′-ene-4′,5-diol E31

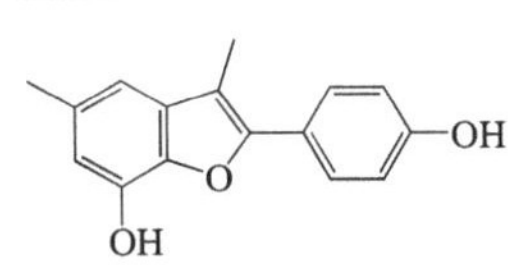

2-(4-Hydroxyphenyl)-3,5-dimethyl-7-benzofuranol. 7-Hydroxy-2-(4-hydroxyphenyl)-3,5-dimethylbenzofuran. 4,7′-Epoxy-4′,5-dihydroxy-8,9-dinor-3,8′-neolign-7′-ene
$C_{16}H_{14}O_3$ 254.285
Di-Me ether
7-Methoxy-2-(4-methoxyphenyl)-3,5-dimethylbenzofuran. 4,7′-Epoxy-4′,5-dimethoxy-8,9-dinor-3,8′-lign-7′-ene
$C_{18}H_{18}O_3$ 282.339
Constit. of the twigs of *Cassia alata*. Exhibits cytotoxic activity. λ_{max} 210 (log ε 4.05); 270 (log ε 3.08); 312 (log ε 3.38) (MeOH).

Ma, H.-Y. *et al., Zhongcaoyao*, 2016, **47**, 2975-2978 (*Cassia alata constit, activity*)

4,7′-Epoxy-8,9-dinor-3,8′-lign-7′-ene-4′,7-diol E32

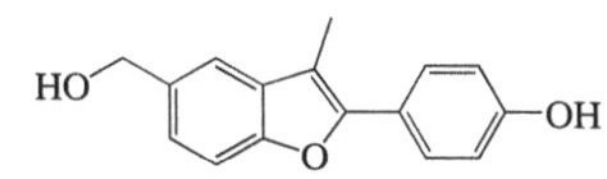

2-(4-Hydroxyphenyl)-3-methyl-5-benzofuranmethanol, CAS. *5-(Hydroxymethyl)-2-(4-hydroxyphenyl)-3-methylbenzofuran. 4,7′-Epoxy-4′,7-dihydroxy-8,9-dinor-3,8′-neolign-7′-ene*
[119322-03-1]
$C_{16}H_{14}O_3$ 254.285
Constit. of *Krameria triandra*. Powder. Mp 187 - 192°
Aldehyde
2-(4-Hydroxyphenyl)-3-methyl-5-benzofurancarboxaldehyde. 5-Formyl-2-(4-hydroxyphenyl)-3-methylbenzofuran. 4,7′-Epoxy-4′-hydroxy-8,9-dinor-3,8′-lign-7′-en-7-al. 4,7′-Epoxy-4′-hydroxy-8,9-dinor-7-oxo-3,8′-lign-7′-ene
[178560-43-5]
$C_{16}H_{12}O_3$ 252.269
Constit. of *Caryodaphnopsis tonkinensis*. Amorph.
Carboxylic acid
2-(4-Hydroxyphenyl)-3-methyl-5-benzofurancarboxylic acid. 4,7′-Epoxy-4′-hydroxy-8,9-dinor-3,8′-lign-7′-en-7-oic acid
$C_{16}H_{12}O_4$ 268.268
Isol. from faeces of *Naupactus bipes* feeding on *Piper regnellii*.

Arnone, A. *et al., Gazz. Chim. Ital.*, 1988, **118**, 675-682 (*Krameria triandra constit*).
Anh, N.H. *et al., Phytochemistry*, 1996, **42**, 1167-1169 (*aldehyde*).
Ramos, C.S. *et al., Chemoecology*, 2013, **23**, 143-148 (*carboxylic acid*)

3,7′-Epoxy-8,9-dinor-4,8′-oxylignan-3′,4′,5,5′,7,9′-hexol E33

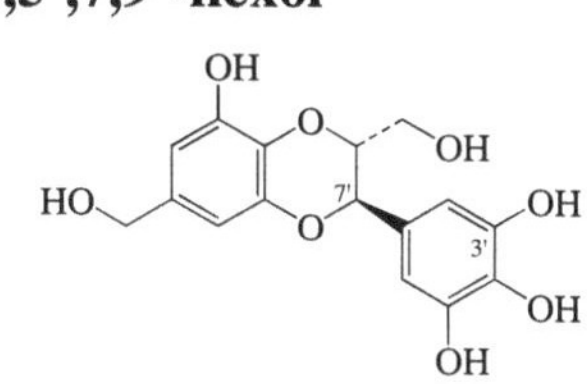

(7′R,8′R)-form

2,3-Dihydro-8-hydroxy-3-(3,4,5-trihydroxyphenyl)-1,4-benzodioxin-2,6-dimethanol, CAS
$C_{16}H_{16}O_8$ 336.296
The nat. prod. Pithecellobiumin A was isol. as a racemate and resolved by hplc

(7′R,8′R)-form

7-Carboxylic acid, 3′,5′-di-Me ether

3,7′-Epoxy-4′,5,9′-trihydroxy-3′,5′-dimethoxy-8,9-dinor-4,8′-oxylignan-7-oic acid

7-Carboxylic acid, 3′,5′-di-Me ether, Me ester

(-)-Pithecellobiumin A

[2225884-60-4]

Inhibitor of β-amyloid aggregation. Pale yellow solid. $[\alpha]_D^{20}$ -15 (c, 0.1 in MeOH)

(7′S,8′S)-form

7-Carboxylic acid, 3′,5′-di-Me ether, Me ester

(+)-Pithecellobiumin A

[2225884-58-0]

Inhibitor of β-amyloid aggregation. Pale yellow solid. $[\alpha]_D^{20}$ +22 (c, 0.1 in MeOH)

(7′RS,8′RS)-form

7-Carboxylic acid, 3′,5′-di-Me ether, Me ester

Pithecellobiumin A

$C_{19}H_{20}O_9$ 392.360

Constit. of the twigs and leaves of *Pithecellobium clypearia*. Pale yellow solid. $[\alpha]_D^{20}$ -1 (c, 0.1 in MeOH). λ_{max} 210 (log ε 6.21); 276 (log ε 3.34) (MeOH).

Wang, Y.-X. *et al.*, *Bioorg. Chem.*, 2018, **77**, 579-585 (*Pithecellobiumin A, cd, activity*)

3,7′-Epoxy-8,9-dinor-4,8′-oxylignan-3′,4′,5,5′,7-pentol E34

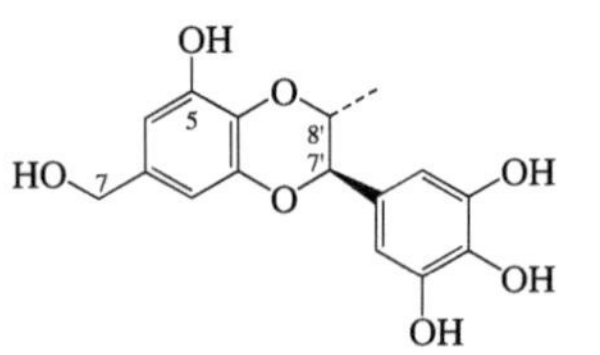

3,7′-Epoxy-3′,4′,5,5′,7-pentahydroxy-8,9-dinor-4,8′-oxyneolignan

$C_{16}H_{16}O_7$ 320.297

(7′R,8′R)-form

7-Aldehyde, tetra-Me ether

3,7′-Epoxy-3′,4′,5,5′-tetramethoxy-8,9-dinor-4,8′-oxylignan-7-al. 3,7′-Epoxy-3′,4′,5,5′-tetramethoxy-8,9-dinor-7-oxo-4,8′-oxylignan.

Eusiderin L

[126176-82-7]

$C_{20}H_{22}O_7$ 374.389

Constit. of *Licaria chrysophylla* and *Nicotiana tabacum*. Shows modest anti-HIV-1 activity. Oil. $[\alpha]_D$ -4.8 (c, 1.7 in MeOH) (synthetic). Abs. config. of natural Eusiderin L unconfirmed, since no chiroptical data was reported. λ_{max} 226 (ε 28650); 308 (ε 5800) (EtOH).

Da Silva, M.S. *et al.*, *Phytochemistry*, 1989, **28**, 3477-3482 (*Eusiderin L*).
Gao, X. *et al.*, *Heterocycles*, 2012, **85**, 147-153 (*Nicotiana tabacum constit, activity*).
Pilkington, L.I. *et al.*, *J. Org. Chem.*, 2012, **77**, 8156-8166 (*Eusiderin L, synth, cd*)

3,7′-Epoxy-8,9-dinor-4,8′-oxylignan-3′,4′,7,9′-tetrol E35

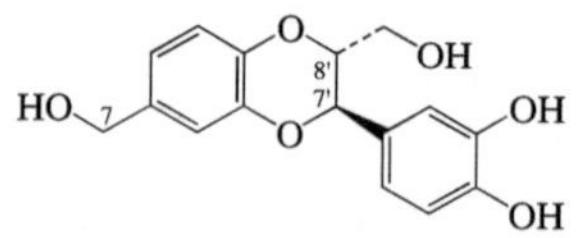

3-(3,4-Dihydroxyphenyl)-2,3-dihydro-1,4-benzodioxin-2,6-dimethanol, CAS. *3,7′-Epoxy-3′,4′,7,9′-tetrahydroxy-8,9-dinor-4,8′-oxyneolignan*

$C_{16}H_{16}O_6$ 304.298

(7′RS,8′RS)-form

7-Carboxylic acid

3-(3,4-Dihydroxyphenyl)-2,3-dihydro-2-(hydroxymethyl)-1,4-benzodioxin-6-carboxylic acid, CAS. *3,7′-Epoxy-3′,4′,9′-trihydroxy-8,9-dinor-4,8′-oxylignan-7-oic acid.*

Isoamericanoic acid A‡

[696603-13-1] (Me ester)

$C_{16}H_{14}O_7$ 318.281

Constit. of the seeds of *Phytolacca americana*. Amorph. solid (as Me ester). Not to be confused with Isoamericanoic acid A, in E218. λ_{max} 204 (ε 40000); 261 (ε 6900); 288 (ε 5000) (EtOH) (Me ester)

7-Carboxylic acid, 3′-Me ether

3,7′-Epoxy-4′,9′-dihydroxy-3′-methoxy-8,9-dinor-4,8′-oxylignan-7-oic acid. ***3-O-Methylisoamericanoic acid A***

[1191906-20-3]

$C_{17}H_{16}O_7$ 332.308

Constit. of the bark of *Picea jezoensis* var. *jezoensis*. Amorph. powder. λ_{max} 229 (log ε 3.89); 255 (log ε 3.76); 285 (log ε 3.62) (EtOH).

Takahashi, H. *et al.*, *Chem. Pharm. Bull.*, 2003, **51**, 1377-1381 (*Isoamericanoic acid A*).
Wada, S. *et al.*, *Bioorg. Med. Chem.*, 2009, **17**, 6414-6421 (*3-O-Methylisoamericanoic acid A*)

4,7′-Epoxy-8,9-dinor-3,8′-oxylignan-3′,4′,7,9′-tetrol E36

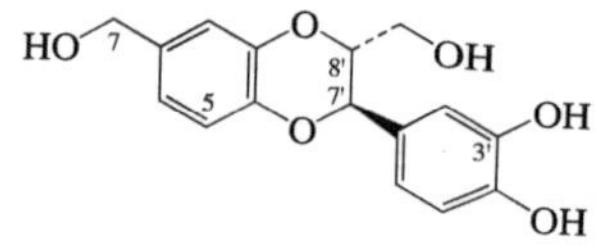

(7′R,8′R)-form

3-(3,4-Dihydroxyphenyl)-2,3-dihydro-1,4-benzodioxin-2,7-dimethanol, CAS. *4,7′-Epoxy-3′,4′,7,9′-tetrahydroxy-8,9-dinor-3,8′-oxyneolignan*

$C_{16}H_{16}O_6$ 304.298

(7′R,8′R)-form

7-Aldehyde

4,7′-Epoxy-3′,4′,9′-trihydroxy-8,9-dinor-3,8′-oxylignan-7-al

[1489173-37-6]

$C_{16}H_{14}O_6$ 302.282

Constit. of *Morinda citrifolia*. Amorph. yellow powder. $[\alpha]_D^{25}$ +2.5 (c, 0.2 in MeOH). λ_{max} 224 (log ε 4.26); 252 (log ε 4.11); 300 (log ε 4.02) (MeOH)

7-Carboxylic acid, 3′-Me ether

4,7′-Epoxy-4′,9′-dihydroxy-3′-methoxy-8,9-dinor-3,8′-oxylignan-7-oic acid

[1489173-24-1]

$C_{17}H_{16}O_7$ 332.308

Constit. of *Morinda citrifolia*. Amorph. brown powder. $[\alpha]_D^{25}$ +4.7 (c, 0.2 in MeOH). λ_{max} 224 (log ε 4.30); 253 (log ε 4.12); 299 (log ε 4.01) (MeOH)

(7′S,8′S)-form

5-Hydroxy, 7-carboxylic acid, 3′-Me ether

Inovonoid A

$C_{17}H_{16}O_8$ 348.307

Constit. of *Balanophora involucrate*. Moderate α-glucosidase inhibitor. Amorph. powder. $[\alpha]_D^{25}$ -18.4 (c, 0.1 in MeOH). λ_{max} 232 (log ε 3.15); 290 (log ε 3.08) (MeOH)

(7′RS,8′RS)-form

7-Carboxylic acid

2-(3,4-Dihydroxyphenyl)-2,3-dihydro-3-(hydroxymethyl)-1,4-benzodioxin-6-carboxylic acid. 4,7′-Epoxy-3′,4′,9′-trihydroxy-8,9-dinor-3,8′-oxylignan-7-oic acid.

Americanoic acid A‡

[696603-06-2] (Me ester)

$C_{16}H_{14}O_7$ 318.281

Constit. of the seeds of *Phytolacca americana*. Neuritogenic agent.

Amorph. (as Me ester). Not to be confused with Americanoic acid A in E220. λ_{max} 206 (ε 30200); 261 (ε 4500); 290 (ε 3000) (EtOH) (Me ester)

(7′R*,8′R*)-form

7,9′-Dicarboxylic acid

4,7′-Epoxy-3′,4′-dihydroxy-8,9-dinor-3,8′-oxylignan-7,9′-dioic acid

7,9′-Dicarboxylic acid, 7-(4-β-D-glucopyranosyloxybenzyl), 9′-Me ester

Origalignanol

[700367-71-1]

$C_{30}H_{30}O_{14}$ 614.556

Constit. of the aerial parts of *Origanum vulgare*. Amorph. powder. $[\alpha]_D^{25}$ -14 (c, 0.5 in MeOH). Possesses (7′*R**,8′*S**)-config. due to change in Cahn-Ingold-Prelog priorities. λ_{max} 258 (log ε 4.33); 289 (log ε 4.13); 295 (sh) (MeOH).

Takahasi, H. *et al.*, *Chem. Pharm. Bull.*, 2003, **51**, 1377-1381 (*Americanoic acid A*).
Lin, Y.-L. *et al.*, *J. Chin. Chem. Soc. (Taipei)*, 2003, **50**, 1079-1083 (*Origalignanol*).
Nguyen, P.H. *et al.*, *J. Nat. Prod.*, 2013, **76**, 2080-2087 (*Morinda citrifolia constits*).
Wei, J. *et al.*, *Fitoterapia*, 2017, **121**, 129-135 (*Inovonoid A, activity*)

3,7′-Epoxy-8,9-dinor-4,8′-oxylignan-3′,4′,7-triol E37

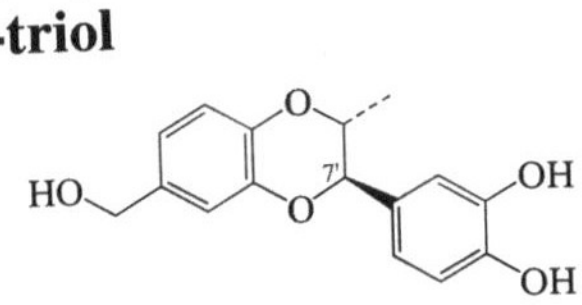

3,7′-Epoxy-3′,4′,7-trihydroxy-8,9-dinor-4,8′-oxyneolignan

$C_{16}H_{16}O_5$ 288.299

(7′R*,8′R*)-form

7-Aldehyde, 3′,4′-methylene ether

3,7′-Epoxy-3′,4′-methylenedioxy-8,9-dinor-4,8′-oxylignan-7-al.

Melaleucin B

$C_{17}H_{14}O_5$ 298.294

Constit. of the leaves of *Melaleuca bracteata*. Oil. λ_{max} 239 (log ε 4.98); 285 (log ε 4.85) ($CHCl_3$).

Li, C. *et al.*, *Fitoterapia*, 2017, **120**, 171-176 (*Melaleucin B*)

7,9′-Epoxy-2′,3,3′,4,4′,5,5′-heptahydroxy-8,8′-lignan-9,7′-olide E38

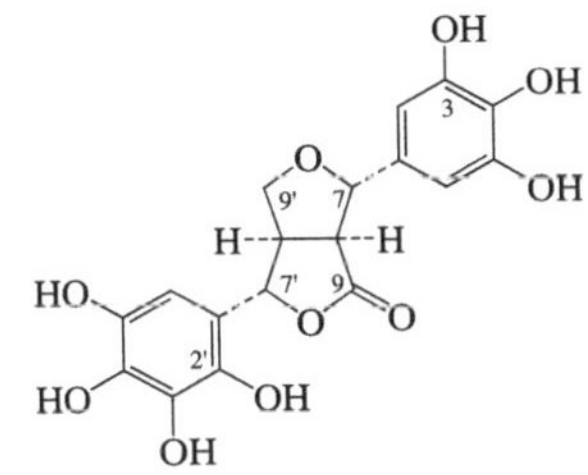

Absolute Configuration

2′,3,3′,4,4′,5,5′-Heptahydroxy-7,9′-epoxylignan-9,7′-olide

$C_{18}H_{16}O_{10}$ 392.316

(7S,7′S,8S,8′R)-form

4,5-Methylene, 3,3′,4′-tri-Me ether, 2′-O-β-D-glucopyranoside

Terminaloside N

$C_{28}H_{32}O_{15}$ 608.549

Constit. of the leaves of *Terminalia citrina*. Amorph. pale yellow powder. $[\alpha]_D^{25}$ +26.7 (c, 0.2 in MeOH). λ_{max} 216 (log ε 4.41); 281 (log ε 3.66) (MeOH)

4,5-Methylene, 2′,3,3′,4′-tetra-Me ether, 5′-O-β-D-glucopyranoside

Terminaloside P

$C_{29}H_{34}O_{15}$ 622.576

Constit. of the leaves of *Terminalia citrina*. Exhibits weak antiestrogenic activity. Amorph. pale yellow powder. $[\alpha]_D^{25}$ +12.6 (c, 0.2 in MeOH). λ_{max} 225 (log ε 4.11); 278 (log ε 3.53) (MeOH)

4,5-Methylene, 2′,3,4′,5′-tetra-Me ether, 3′-O-β-D-glucopyranoside

Terminaloside M

$C_{29}H_{34}O_{15}$ 622.576

Constit. of the leaves of *Terminalia citrina*. Exhibits strong antiestrogenic activity against MCF-7 cells. Amorph. powder. $[\alpha]_D^{25}$ +52.4 (c, 0.1 in MeOH). λ_{max} 216 (log ε 4.48); 280 (log ε 3.65) (MeOH)

4,5-Methylene, 3,3′,4′,5′-tetra-Me ether, 2′-O-β-D-glucopyranoside

Terminaloside O

$C_{29}H_{34}O_{15}$ 622.576

Constit. of the leaves of *Terminalia citrina*. Exhibits weak antiestrogenic activity. Amorph. pale yellow powder. $[\alpha]_D^{25}$ +70.2 (c, 0.1 in MeOH). λ_{max} 216 (log ε 4.52); 280 (log ε 3.66) (MeOH).

Muhit, M.A. *et al.*, *Fitoterapia*, 2016, **113**, 74-79 (*Terminalosides M-P, activity*)

2,7′-Epoxy-1,2,3,4,5,6-hexahydro-1,8′-lign-8-ene-2,3′,4,4′,5-pentol E39

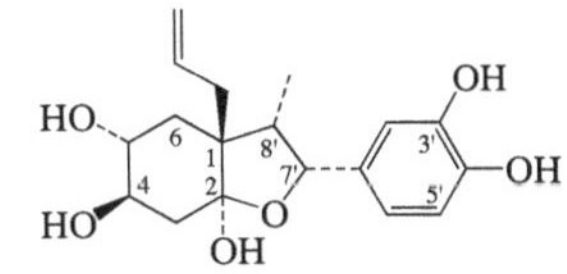

(1*R**,2*S**,4*R**,5*R**,7′*R**,8′*S**)-*form*

2,7′-Epoxy-1,2,3,4,5,6-hexahydro-2,3′,4,4′,5-pentahydroxy-1,8′-neolign-8-ene

$C_{18}H_{24}O_6$ 336.384

(1R*,2S*,4R*,5R*,7′R*,8′S*)-form

3′,4′,5-Tri-Me ether

2-(3,4-Dimethoxyphenyl) hexahydro-5-methoxy-3-methyl-3a-(2-propenyl)-6,7a(2H)-benzofurandiol.

2,7′-Epoxy-1,2,3,4,5,6-hexahydro-3′,4′,5-trimethoxy-1,8′-lign-8-ene-2,4-diol.

2,7′-Epoxy-1,2,3,4,5,6-hexahydro-2,4-dihydroxy-3′,4′,5-trimethoxy-1,8′-lign-8-ene

[106897-17-0]

$C_{21}H_{30}O_6$ 378.465

Constit. of the wood of *Ocotea porosa*. Cryst. (EtOH). Mp 68 - 71°. λ_{max} 230 (ε 80200); 275 (ε 12000) (MeOH)

5′-Methoxy, 3′,4′,5-tri-Me ether

2,7′-Epoxy-1,2,3,4,5,6-hexahydro-3′,4′,5,5′-tetramethoxy-1,8′-lign-8-ene-2,4-diol.

2,7′-Epoxy-1,2,3,4,5,6-hexahydro-2,4-dihydroxy-3′,4′,5,5′-tetramethoxy-1,8′-lign-8-ene.

Dysodanthin D

[142795-94-6]

$C_{22}H_{32}O_7$ 408.491

Constit. of the roots of *Endlicheria dysodantha*. Exhibits antitumour activity. Oil. $[\alpha]_D$ -29 (c, 0.1 in MeOH)

(1R*,2S*,4S*,5R*,7′R*,8′S*)-form

3′,4′,5-Tri-Me ether

[106897-18-1]

$C_{21}H_{30}O_6$ 378.465

Constit. of the wood of *Ocotea porosa*. Cryst. (EtOH). Mp 55 - 56°. λ_{max} 220 (ε 27000); 287 (ε 9000) (MeOH)

5′-Methoxy, 3′,4′,5-tri-Me ether

Dysodanthin E

[142795-95-7]

$C_{22}H_{32}O_7$ 408.491

Constit. of the roots of *Endlicheria dysodantha*. Exhibits antitumour activity. Oil. [α] D +11 (c, 0.1 in MeOH)

5′-Methoxy, 2,3′,4′,5-tetra-Me ether

2,7′-Epoxy-1,2,3,4,5,6-hexahydro-2,3′,4′,5,5′-pentamethoxy-1,8′-lign-8-en-4-ol. 2,7′-Epoxy-1,2,3,4,5,6-hexahydro-4-hydroxy-2,3′,4′,5,5′-pentamethoxy-1,8′-lign-8-ene. ***Dysodanthin F***

[142768-54-5]

$C_{23}H_{34}O_7$ 422.518

Constit. of the roots of *Endlicheria dysodantha*. Exhibits antitumour activity. Oil. [α] D -21 (c, 0.15 in MeOH).

Dias, D.A. *et al.*, *Phytochemistry*, 1986, **25**, 2613-2616 (*Ocotea porosa constits*).
Ma, W.-W. *et al.*, *Heterocycles*, 1992, **34**, 5-11 (*Dysodanthins D,E,F, activity*)

7,9′-Epoxy-3,3′,4,4′,5,5′-hexahydroxy-8,8′-lignan-9,7′-olide E40

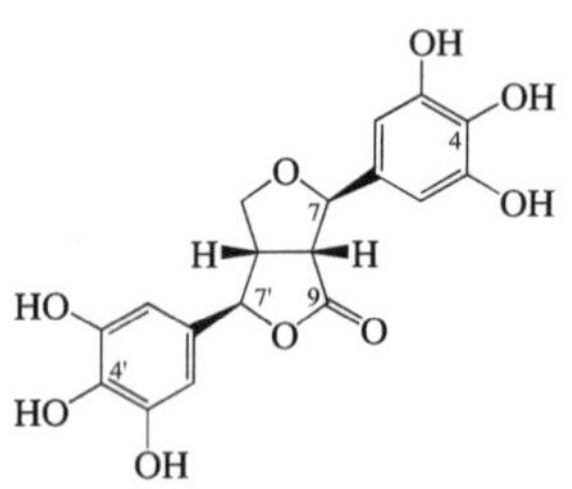

(7R,7′R,8R,8′S)-form

3,3′,4,4′,5,5′-Hexahydroxy-7,9′-epoxylignan-9,7′-olide

$C_{18}H_{16}O_9$ 376.317

(7R,7′R,8R,8′S)-form

3,3′,5,5′-Tetra-Me ether

7,9′-Epoxy-4,4′-dihydroxy-3,3′,5,5′-tetramethoxy-8,8′-lignan-9,7′-olide. ***Dendroresinol***

$C_{22}H_{24}O_9$ 432.425

Constit. of *Dendrobium aurantiacum* var. *denneanum*. Exhibits cytotoxic activity. $[\alpha]^{20}_{D}$ -5.2 (c, 0.26 in MeOH)

(7S,7′S,8S,8′R)-form

4,5-Methylene, 3,3′,4′-tri-Me ether

7,9′-Epoxy-3′-hydroxy-3,4′,5′-trimethoxy-4,5-methylenedioxy-8,8′-lignan-9,7′-olide. 3′-Hydroxy-3,4′,5′-trimethoxy-4,5-methylenedioxy-7,9′-epoxylignan-9,7′-olide

4,5-Methylene, 3,3′,4′-tri-Me ether, 5′-O-β-D-glucopyranoside

Terminaloside L

$C_{28}H_{32}O_{14}$ 592.55

Constit. of the leaves of *Terminalia citrina*. Suppresses E_2-enhanced T47D estrogen responsive breast cancer cell line proliferation. Amorph. pale yellow powder. $[\alpha]^{25}_{D}$ +42 (c, 0.3 in MeOH). λ_{max} 216 (log ε 4.46); 272 (log ε 3.47) (MeOH).

Yang, L. *et al.*, *Zhongcaoyao*, 2015, **46**, 2193-2197 (*Dendroresinol, activity*).
Muhit, M.A. *et al.*, *Fitoterapia*, 2016, **113**, 74-79 (*Terminaloside L, activity*)

4,7′-Epoxy-3′,5,5′,8,9,9′-hexahydroxy-3,8′-lignan-7-one E41

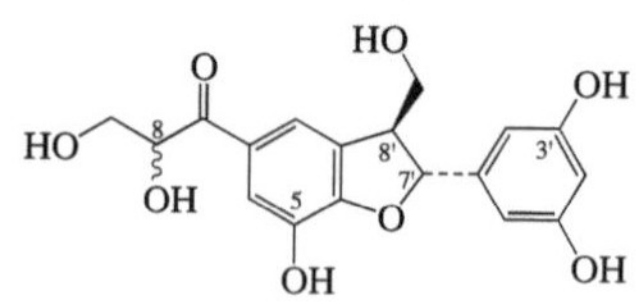

(7′R,8ξ,8′S)-form

4,7′-Epoxy-3′,5,5′,8,9,9′-hexahydroxy-7-oxo-3,8′-neolignan

$C_{18}H_{18}O_8$ 362.334

Structures in this series may need revision. Phenyl 3,5-dioxygenated substn. pattern in doubt as a result of work done by Yang *et al* in 2015

(7′R,8ξ,8′S)-form

3′,5-Di-Me ether

4,7′-Epoxy-3′,8,9,9′-tetrahydroxy-5,5′-dimethoxy-3,8′-lignan-7-one. 4,7′-Epoxy-3′,8,9,9′-tetrahydroxy-5,5′-dimethoxy-7-oxo-3,8′-lignan. ***Pinnatifidanin CII***

[1616701-72-4]

$C_{20}H_{22}O_8$ 390.388

Constit. of the seeds of *Crataegus pinnatifida*. Yellow oil. $[\alpha]^{20}_{D}$ +11.2 (c, 0.04 in MeOH). λ_{max} 229 (log ε 3.60); 279 (log ε 2.16) (MeOH)

(7′S,8′R)-form

8-Deoxy

4,7′-Epoxy-3′,5,5′,9,9′-pentahydroxy-3,8′-lignan-7-one. 4,7′-Epoxy-3′,5,5′,9,9′-pentahydroxy-7-oxo-3,8′-lignan

8-Deoxy, 3′,5-di-Me ether

4,7′-Epoxy-3′,9,9′-trihydroxy-5,5′-dimethoxy-3,8′-lignan-7-one. 4,7′-Epoxy-3′,9,9′-trihydroxy-5,5′-dimethoxy-7-oxo-3,8′-lignan. ***Pinnatifidanin CI***

[1613723-95-7]

$C_{20}H_{22}O_7$ 374.389

Constit. of the seeds of *Crataegus pinnatifida*. Yellow oil. $[\alpha]^{20}_{D}$ -6.3 (c, 0.03 in MeOH). λ_{max} 230 (log ε 3.73); 278 (log ε 2.32) (MeOH).

Huang, X.-X. *et al.*, *Fitoterapia*, 2013, **91**, 217-223 (*Pinnatifidanins CI,CII*).
Yang, Y.-N. *et al.*, *J. Nat. Prod.*, 2015, **78**, 705-711 (*pmr, struct*)

7,9′-Epoxy-3,3′,4,4′,5′,9-hexahydroxy-8,8′-lignan-7′-one E42

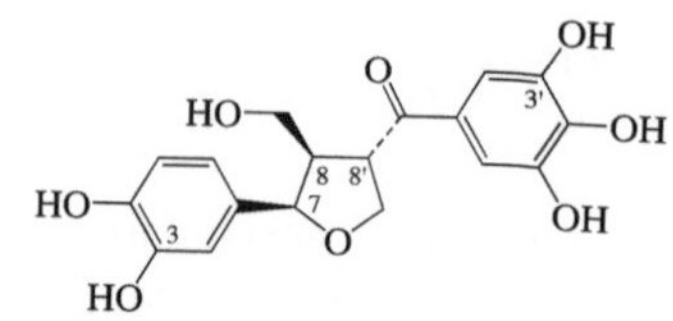

(7R,8R,8′S)-form

3,3′,4,4′,5′,9-Hexahydroxy-7′-oxo-7,9′-epoxylignan

$C_{18}H_{18}O_8$ 362.334

(7R,8R,8′S)-form

3,3′,4,4′,5′-Penta-Me ether

7,9′-Epoxy-9-hydroxy-3,3′,4,4′,5′-pentamethoxy-8,8′-lignan-7′-one. 9-Hydroxy-3,3′,4,4′,5′-pentamethoxy-7′-oxo-7,9′-epoxylignan. ***Sylvone***

[95263-43-7]

$C_{23}H_{28}O_8$ 432.469

Isol. from seeds of *Piper sylvaticum*. Mp 138 - 139°. $[\alpha]^{21}_{D}$ +9.6 ($CHCl_3$)

(7S,8R,8′R)-form

3,3′,4,4′,5′-Penta-Me ether

Hernone

[194026-35-2]

Constit. of *Hernandia nymphaeifolia*. Cytotoxic agent. Shows anti-platelet aggregation activity. Prisms (CH_2Cl_2/Me_2CO). Mp 126 - 128°. $[\alpha]^{25}_{D}$ -29.7 (c, 0.2 in $CHCl_3$). Config. incorr. assigned in ref.. λ_{max} 204 (log ε 4.73); 218 (sh); 282 (log ε 4.09) (EtOH)

3,4-Methylene, 3′,4′,5′-tri-Me ether

7,9′-Epoxy-9-hydroxy-3′,4′,5′-trimethoxy-3,4-methylenedioxy-8,8′-lignan-7′-one. 9-Hydroxy-3′,4′,5′-trimethoxy-3,4-methylenedioxy-7′-oxo-7,9′-epoxylignan. ***Nymphone***

[194026-36-3]

$C_{22}H_{24}O_8$ 416.426

Constit. of *Hernandia nymphaeifolia*. Cytotoxic agent. Prisms (CH_2Cl_2/Me_2CO). Mp 123 - 125°.

$[\alpha]_D^{25}$ -47.4 (c, 0.08 in $CHCl_3$). Config. incorr. assigned in ref.. λ_{max} 204 (log ε 4.63); 216 (sh); 287 (log ε 4.11) (EtOH)

(7S,8R,8′S)-form

3,3′,5′-Tri-Me ether

7,9′-Epoxy-4,4′,9-trihydroxy-3,3′,5′-trimethoxy-8,8′-lignan-7′-one. 4,4′,9-Trihydroxy-3,3′,5′-trimethoxy-7′-oxo-7,9′-epoxylignan. ***Wikstrone***

[878007-01-3]

$C_{21}H_{24}O_8$ 404.415

Constit. of *Wikstroemia hainanensis*. Amorph. powder. $[\alpha]_D^{21}$ -22 (c, 0.09 in $CHCl_3$). λ_{max} 230 (log ε 4.22); 285 (log ε 3.95); 308 (log ε 3.96) (MeOH).

Banerji, A. *et al.*, *Tetrahedron*, 1984, **40**, 5047-5052 (*Sylvone*).
Banerji, A. *et al.*, *J. Indian Chem. Soc.*, 1992, **69**, 321-325 (*Sylvone*).
Chen, I.-S. *et al.*, *Phytochemistry*, 1997, **45**, 991-996 (*Hernone, Nymphone*).
Chen, J.-J. *et al.*, *Planta Med.*, 2000, **66**, 251-256 (*Hernone, activity*).
Liao, S.G. *et al.*, *Helv. Chim. Acta*, 2006, **89**, 73-80 (*Wikstrone*).
Naveschuk, C.G. *et al.*, *Synlett*, 2008, 126-128 (*Sylvone, synth*).
Lee, E. *et al.*, *J. Org. Chem.*, 2015, **80**, 10359-10363 (*Sylvone, synth*)

7,9′-Epoxy-3,3′,4,4′,8′,9-hexahydroxy-8,8′-lignan-7′-one E43

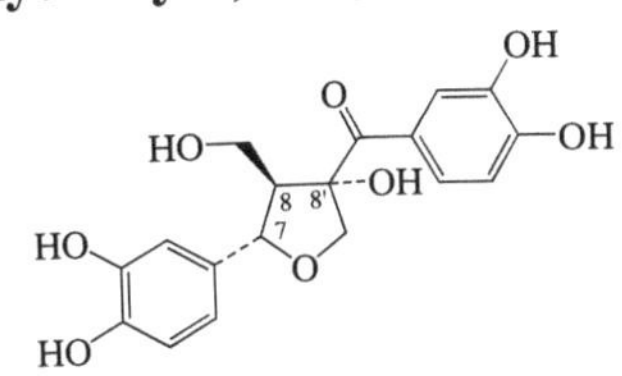

3,3′,4,4′,8′,9-Hexahydroxy-7′-oxo-7,9′-epoxylignan

$C_{18}H_{18}O_8$ 362.334

(7S,8R,8′S)-form

3,4:3′,4′-Bis(methylene) ether

7,9′-Epoxy-8′,9-dihydroxy-3,4:3′,4′-bis(methylenedioxy)-8,8′-lignan-7′-one. 8′,9-Dihydroxy-3,4:3′,4′-bis(methylenedioxy)-7′-oxo-7,9′-epoxylignan. ***Arborone‡***

[108069-03-0]

$C_{20}H_{18}O_8$ 386.356

Isol. from heartwood of *Gmelina arborea*. Mp 216 - 218°. $[\alpha]_D^{28}$ +81.5 ($CHCl_3$).

Satyanarayana, P. *et al.*, *J. Nat. Prod.*, 1986, **49**, 1061-1064 (*Arborone*)

7,9′-Epoxy-3,3′,4,4′,8,9-hexahydroxy-8,8′-lignan-7′-one E44

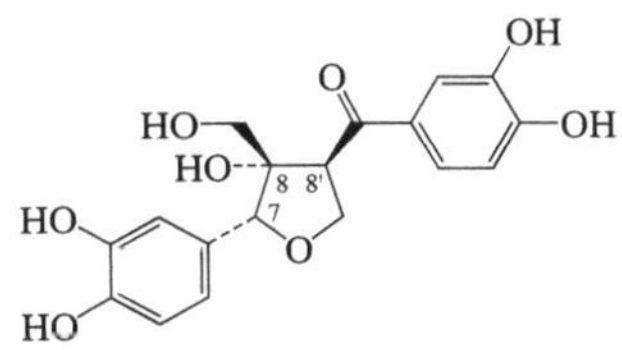

3,3′,4,4′,8,9-Hexahydroxy-7′-oxo-7,9′-epoxylignan

$C_{18}H_{18}O_8$ 362.334

(7R,8S,8′S)-form

3,3′,4,4′-Tetra-Me ether

7,9′-Epoxy-8,9-dihydroxy-3,3′,4,4′-tetramethoxy-8,8′-lignan-7′-one. 8,9-Dihydroxy-3,3′,4,4′-tetramethoxy-7′-oxo-7,9′-epoxylignan.

7′-Oxodihydrogmelinol

[108044-10-6]

$C_{22}H_{26}O_8$ 418.442

Isol. from *Gmelina arborea*. Needles (C_6H_6/petrol). Mp 70°. $[\alpha]_D^{28}$ +63.1 (c, 1.134 in $CHCl_3$). λ_{max} 231; 276; 310 (MeOH).

Satyanarayana, P. *et al.*, *J. Nat. Prod.*, 1986, **49**, 1061-1064 (*7′-Oxodihydrogmelinol*)

7,7′-Epoxy-3,3′,4,4′,5,5′-hexahydroxy-7,8-seco-8,8′-lignan-7,8-dione E45

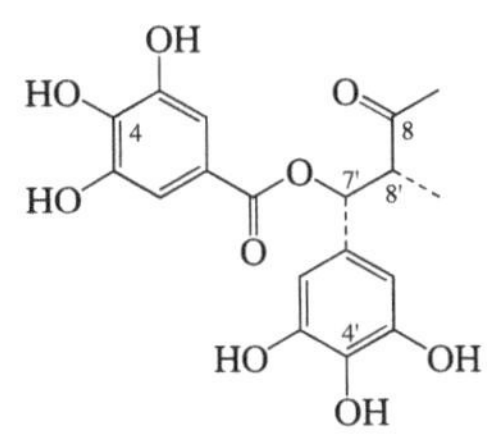

(7′R,8′R)-form

2-Methyl-3-oxo-1-(3,4,5-trihydroxyphenyl)butyl 3,4,5-trihydroxybenzoate, CAS

$C_{18}H_{18}O_9$ 378.333

(7′R,8′R)-form

3′,4′-Methylene, tetra-Me ether

7,7′-Epoxy-3,3′,4,5-tetramethoxy-4′,5′-methylenedioxy-7,8-seco-8,8′-lignan-7,8-dione. ***Tsangin B***

[916656-03-6]

$C_{23}H_{26}O_9$ 446.452

Constit. of the stems of *Beilschmiedia tsangii*. Exhibits cytotoxic activity against mouse lymphocytic leukaemia cells. Oil. $[\alpha]_D^{13}$ -21.8 (c, 0.1 in $CHCl_3$). λ_{max} 217 (log ε 4.68); 268 (log ε 4.18) (MeOH)

(7′R,8′S)-form

3,3′,4,4′-Tetra-Me ether

7,7′-Epoxy-3,3′-dihydroxy-4,4′,5,5′-tetramethoxy-7,8-seco-8,8′-lignan-7,8-dione. ***Manglisin F***

[1621958-66-4]

$C_{22}H_{26}O_9$ 434.441

Constit. of the carpels of *Manglietiastrum sinicum*. Oil. $[\alpha]_D^{26}$ +24 (c, 0.2 in MeOH). λ_{max} 207 (log ε 4.60); 265 (log ε 3.84) (MeOH)

(7′S,8′R)-form

3,4-Methylene, tetra-Me ether

7,7′-Epoxy-3,3′,4′,5′-tetramethoxy-4,5-methylenedioxy-7,8-seco-8,8′-lignan-7,8-dione. ***Tsangin A***

[916656-02-5]

$C_{23}H_{26}O_9$ 446.452

Constit. of the stems of *Beilschmiedia tsangii*. Exhibits cytotoxic activity against mouse lymphocytic leukaemia cells. Oil. $[\alpha]_D^{24}$ +21.1 (c, 0.12 in $CHCl_3$). λ_{max} 217 (log ε 4.63); 281 (log ε 4.15) (MeOH)

(7′ξ,8′ξ)-form

Hexa-Me ether

7,7′-Epoxy-3,3′,4,4′,5,5′-hexamethoxy-7,8-seco-8,8′-lignan-7,8-dione. ***Meiocarpin***

[865154-37-6]

$C_{24}H_{30}O_9$ 462.495

Constit. of the stem bark of *Meiocarpidium lepidotum*. Amorph. powder.

Ngouela, S. *et al.*, *Bull. Chem. Soc. Ethiop.*, 2004, **18**, 221-224 (*Meiocarpin*).
Chen, J.J. *et al.*, *Planta Med.*, 2006, **72**, 351-357 (*Tsangins A,B, activity*).
Ding, J.-Y. *et al.*, *J. Nat. Prod.*, 2014, **77**, 1800-1805 (*Manglisin F*)

7,9′-Epoxy-3,3′,3″,5,5′,5″-hexamethoxy-4,7″:8,8′-sesquilignan-4′,4″,8″,9,9″-pentol E46

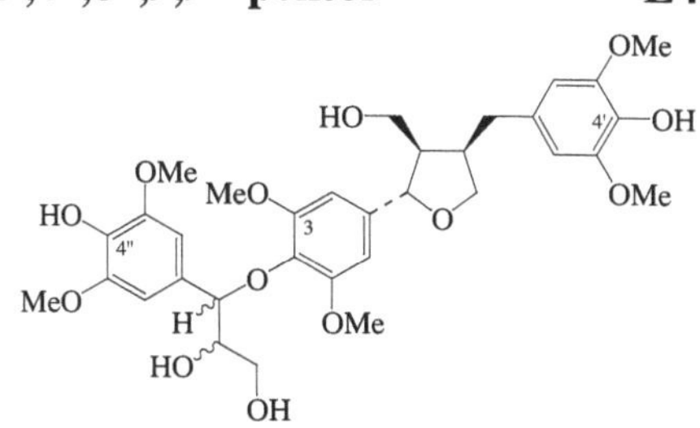

4-O-[α-(1,2-Dihydroxyethyl)syringyl]-5,5′-dimethoxylariciresinol

$C_{33}H_{42}O_{13}$ 646.686

4′-O-[4,6-Bis-O-(4-hydroxy-3-methoxy-E-cinnamoyl)-β-D-glucopyranoside]
[947494-36-2]
$C_{59}H_{68}O_{24}$ 1161.169
Constit. of the roots of *Rhus javanica* var. *roxburghiana*. Powder. $[\alpha]_D^{27}$ +5.9 (c, 0.9 in Me_2CO). λ_{max} 227 (log ε 4.44); 232 (log ε 4.01); 278 (log ε 4.12); 302 (log ε 4.05); 331 (log ε 4.11) (MeOH).

Ouyang, M.-A. *et al.*, *Helv. Chim. Acta*, 2007, **90**, 1099-1106 (*Rhus javanica constit*)

4,7′-Epoxy-9′*a*-homo-3′,4′,5-trihydroxy-3,8′-lign-7-ene-9-carboxylic acid E47

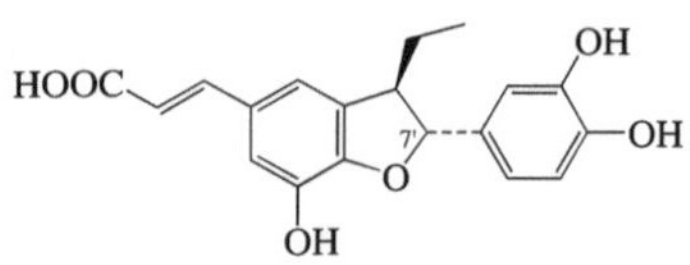

$C_{19}H_{18}O_6$ 342.347
(7E,7′R*,8′R*)-form
3′,5-Di-Me ether
4,7′-Epoxy-9′a-homo-4′-hydroxy-3′,5-dimethoxy-3,8′-lign-7-ene-9-carboxylic acid
3′,5-Di-Me ether, Me ester
Apiculin A
$C_{22}H_{24}O_6$ 384.428
Constit. of the bark of *Helietta apiculata*.

Fernandes, T.S. *et al.*, *Phytochemistry*, 2017, **141**, 131-139 (*Apiculin A, cryst struct*)

4,7′-Epoxy-4′-hydroxy-8,9-dinor-3,8′-lignan-7-al E48

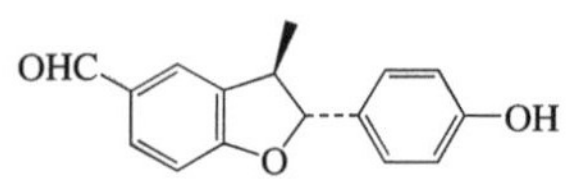

(7′R,8′R)-form

2,3-Dihydro-2-(4-hydroxyphenyl)-3-methyl-5-benzofurancarboxaldehyde. 4,7′-Epoxy-4′-hydroxy-8,9-dinor-7-oxo-3,8′-neolignan
$C_{16}H_{14}O_3$ 254.285
(7′R,8′R)-form
Decurrenal. [172617-91-3]
Constit. of *Aristolochia fordiana*, *Krameria tomentosa*, *Miliusa mollis*, *Piper aequale*, *Piper decurrens* and *Piper nigrum*. Exhibits weak cytotoxic activity. Oil. $[\alpha]_D^{25}$ +32 (c, 0.25 in MeOH). λ_{max} 230 (ε 2400); 290 (ε 1590) (MeOH)
4′-Me ether
4,7′-Epoxy-4′-methoxy-8,9-dinor-3,8′-lignan-7-al. 4,7′-Epoxy-4′-methoxy-8,9-dinor-7-oxo-3,8′-lignan. ***4′-O-Methyldecurrenal***
[1802051-88-2]
$C_{17}H_{16}O_3$ 268.312
Constit. of the stems of *Aristolochia fordiana*. Oil. $[\alpha]_D^{20}$ +111 (c, 0.1 in MeOH). λ_{max} 208 (log ε 4.03); 229 (log ε 4.05); 297 (log ε 3.84) (MeOH)
Carboxylic acid
2,3-Dihydro-2-(4-hydroxyphenyl)-3-methyl-5-benzofurancarboxylic acid. 4,7′-Epoxy-4′-hydroxy-8,9-dinor-3,8′-lignan-7-oic acid
Carboxylic acid, Me ester
[249744-79-4]
$C_{17}H_{16}O_4$ 284.311
Constit. of *Piper regnellii*. Yellow oil. $[\alpha]_D^{21}$ +30.1 (c, 0.08 in MeOH)
(7′S,8′R)-form
[249744-81-8]
Constit. of *Piper regnellii*. Yellow oil. $[\alpha]_D^{21}$ -29.1 (c, 0.1 in MeOH). λ_{max} 230 (ε 5940); 285 (ε 3330) (MeOH)
(7′S,8′S)-form
4′-Me ether
[1802051-89-3]
$C_{17}H_{16}O_3$ 268.312
Constit. of the stems of *Aristolochia fordiana*. Oil. $[\alpha]_D^{20}$ -108 (c, 0.1 in MeOH). λ_{max} 208 (log ε 3.99); 229 (log ε 3.99); 295 (log ε 3.78) (MeOH).

Chauret, D.C. *et al.*, *J. Nat. Prod.*, 1996, **59**, 152-155 (*Decurrenal*).
Maxwell, A. *et al.*, *Phytochemistry*, 1999, **50**, 499-504 (*Piper aequale constit*).
Benevides, P.J.C. *et al.*, *Phytochemistry*, 1999, **52**, 339-343 (*Piper regnellii constits, abs config*).
Silva, S.A.S. *et al.*, *Nat. Prod. Lett.*, 2001, **15**, 323-329 (*Krameria tomentosa constit*).
Sawasdee, K. *et al.*, *Fitoterapia*, 2013, **85**, 49-56 (*cd, activity*).
Tang, G.H *et al.*, *J. Nat. Prod.*, 2015, **78**, 1894-1903 (*4′-O-Methyldecurrenal*)

4,7′-Epoxy-4′-hydroxy-9-nor-3,8′-oxylignan-7-one E49

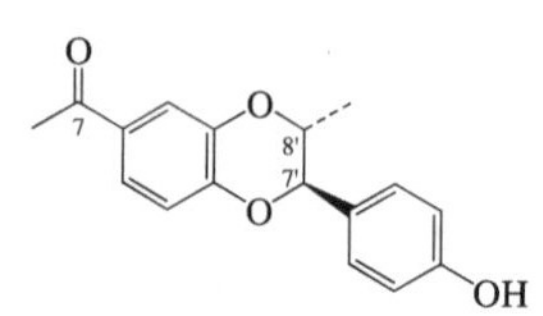

(7′R,8′R)-form

1-[2,3-Dihydro-2-(4-hydroxyphenyl)-3-methyl-1,4-benzodioxin-6-yl] ethanone. 6-Acetyl-2,3-dihydro-2-(4-hydroxyphenyl)-3-methyl-1,4-benzodioxin
$C_{17}H_{16}O_4$ 284.311
The nat. prod. Cynwilforone D was isol. as a racemate and resolved by chiral hplc
(7′R,8′R)-form
(+)-Cynwilforone D
Exhibits neuroprotectant activity. $[\alpha]_D^{17}$ +43 (c, 0.03 in MeOH)
(7′S,8′S)-form
(-)-Cynwilforone D
Exhibits neuroprotectant activity. $[\alpha]_D^{17}$ -38 (c, 0.03 in MeOH)
(7′RS,8′RS)-form
Cynwilforone D
Constit. of the root bark of *Cynanchum wilfordii*. Amorph. yellow powder.

Jiang, H.-W. *et al.*, *Phytochem. Lett.*, 2018, **24**, 179-183 (*Cynwilforone D, cd, cryst struct, activity*)

4,7′-Epoxy-4′-hydroxy-3′,3″,5-trimethoxy-4″,8-oxy-3,8′-sesquiligna-7,7″-diene-9,9′,9″-trioic acid E50

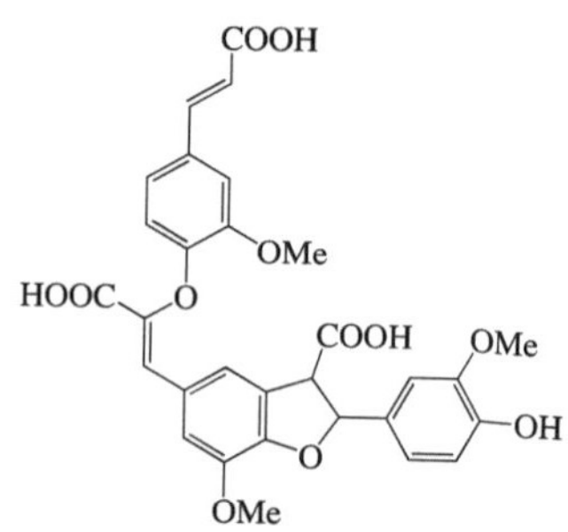

8-O-4/8-5(Cyclic)-triferulic acid
[1995908-20-7]
$C_{30}H_{26}O_{12}$ 578.526
Identified in alkaline hydrolysates of insol. fibres from maize, wheat and sugar beet.

Waterstraat, M. *et al.*, *J. Agric. Food Chem.*, 2016, **64**, 7244-7250 (*8-O-4/8-5(Cyclic)-triferulic acid, synth*)

4,7′-Epoxy-3,8′-ligna-7,7′-diene-3′,4′-diol E51

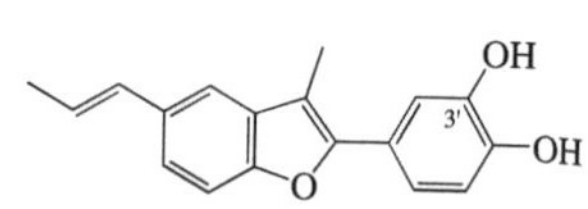

2-(3,4-Dihydroxyphenyl)-3-methyl-5-(1-propenyl)benzofuran. 4,7′-Epoxy-3′,4′-dihydroxy-3,8′-neoligna-7,7′-diene

$C_{18}H_{16}O_3$ 280.323
(E)-form
3′-Me ether
4,7′-Epoxy-3′-methoxy-3,8′-ligna-7,7′-dien-4′-ol. 4,7′-Epoxy-4′-hydroxy-3′-methoxy-3,8′-ligna-7,7′-diene. ***Eupomatenoid 5***
[41744-28-9]
$C_{19}H_{18}O_3$ 294.35
Constit. of *Aristolochia fordiana*, *Eupomatia* spp., *Piper decurrens* and *Piper solmsianum*. Antifungal and antitrypanosomal agent. Cytotoxic. Active against MRSA. Needles (C_6H_6/petrol). Mp 114 - 115°
Di-Me ether
4,7′-Epoxy-3′,4′-dimethoxy-3,8′-ligna-7,7′-diene. ***Eupomatenoid 4***
[41744-29-0]
$C_{20}H_{20}O_3$ 308.377
Constit. of *Caryodaphnopsis tonkinensis* and *Eupomatia* spp.. Prisms (MeOH). Mp 96°
Methylene ether
5-[3-Methyl-5-(1-propenyl)-2-benzofuranyl]-1,3-benzodioxole. 3-Methyl-2-(3,4-methylenedioxyphenyl)-5-(1-propenyl)benzofuran. 4,7′-Epoxy-3′,4′-methylenedioxy-3,8′-ligna-7,7′-diene. ***Eupomatenoid 3***
[41744-27-8]
$C_{19}H_{16}O_3$ 292.334
Constit. of *Caryodaphnopsis baviensis*, *Eupomatia laurina*, *Piper aequale* and *Piper solmsianum*. Prisms (C_6H_6/petrol). Mp 111°. λ_{max} 214 (log ε 4.26); 216 (log ε 4.26); 240 (log ε 4.44); 256 (log ε 4.38); 300 (log ε 4.31); 304 (log ε 4.32); 320 (log ε 4.38) (MeOH).

Bowden, B.F. *et al.*, *Aust. J. Chem.*, 1972, **25**, 2659-2669 (*Eupomatenoids 3,4,5*).
McKittrick, B.A. *et al.*, *J. Chem. Soc., Perkin Trans. 1*, 1983, 475-482 (*Eupomatenoids 4,5, synth*).
Himmelreich, U. *et al.*, *Magn. Reson. Chem.*, 1995, **33**, 280-282 (*Eupomatenoids 3,4,5*).
Chauret, D.C. *et al.*, *J. Nat. Prod.*, 1996, **59**, 152-155 (*Piper decurrens constit*).
Anh, H. *et al.*, *Phytochemistry*, 1997, **46**, 569-571 (*Caryodaphnopsis baviensis constit*).
Bach, T. *et al.*, *Synthesis*, 2003, 925-939 (*Eupomatenoids 4,5, synth*).
De Campos, M.P. *et al.*, *Biol. Pharm. Bull.*, 2005, **28**, 1527-1530 (*Piper solmsianum constits*).
Murakami, K. *et al.*, *Angew. Chem., Int. Ed.*, 2014, **53**, 7510-7513 (*Eupomatenoid 3, synth*)

4,7′-Epoxy-3,8′-ligna-7,7′-diene-4′,5-diol E52

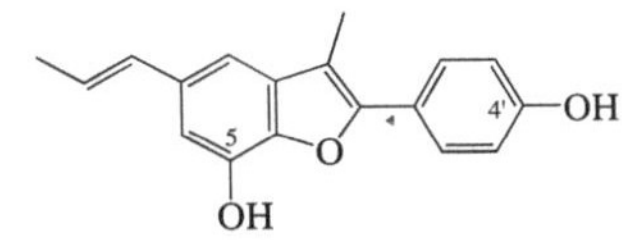

2-(4-Hydroxyphenyl)-3-methyl-5-(1-propenyl)-7-benzofuranol. 4,7′-Epoxy-4′,5-dihydroxy-3,8′-neoligna-7,7′-diene
$C_{18}H_{16}O_3$ 280.323
(E)-form
5-Me ether
4,7′-Epoxy-5-methoxy-3,8′-ligna-7,7′-dien-4′-ol. 4,7′-Epoxy-4′-hydroxy-5-methoxy-3,8′-ligna-7,7′-diene. ***Eupomatenoid 13.*** *Kachirachirol A*
[73027-11-9]
$C_{19}H_{18}O_3$ 294.35
Constit. of *Aristolochia fordiana*, *Caryodaphnopsis tonkinensis*, *Eupomatia laurina*, *Magnolia kachirachirai* and *Krameria cystisoides*. Cytotoxic. Prisms ($CHCl_3$). Mp 193 - 196°
Di-Me ether
4,7′-Epoxy-4′,5-dimethoxy-3,8′-ligna-7,7′-diene. ***Eupomatenoid 16***
[139220-09-0]
$C_{20}H_{20}O_3$ 308.377
Constit. of *Eupomatia bennettii*. Needles (MeOH). Mp 123 - 124°. λ_{max} 229 (ε 16500); 267 (ε 33170); 299 (ε 26800); 306 (sh) (EtOH).

Read, R.W. *et al.*, *Aust. J. Chem.*, 1979, **32**, 2317-2321 (*Eupomatenoid 13*).
McKittrick, B.A. *et al.*, *J. Chem. Soc., Perkin Trans. 1*, 1983, 475-482 (*Eupomatenoid 13, synth*).
Ito, K. *et al.*, *Phytochemistry*, 1984, **23**, 2643-2645 (*Kachirachirol A*).
Achenbach, H. *et al.*, *Phytochemistry*, 1987, **26**, 1159-1166 (*Eupomatenoid 13*).
Carroll, A.R. *et al.*, *Aust. J. Chem.*, 1991, **44**, 1627-1633 (*Eupomatenoid 16*)

4,7′-Epoxy-3,8′-ligna-7′,8-diene-4′,5-diol E53

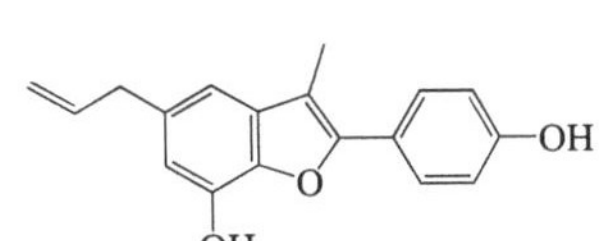

2-(4-Hydroxyphenyl)-3-methyl-5-(2-propenyl)-7-benzofuranol. 4,7′-Epoxy-4′,5-dihydroxy-3,8′-neoligna-7′,8-diene
$C_{18}H_{16}O_3$ 280.323
Di-Me ether
4,7′-Epoxy-4′,5-dimethoxy-3,8′-ligna-7′,8-diene. ***Eupomatenoid 17***
[139220-10-3]
$C_{20}H_{20}O_3$ 308.377
Constit. of *Eupomatia bennettii*. Gum. λ_{max} 213 (ε 20590); 250 (ε 9740); 296 (ε 20350); 306 (sh); 322 (sh) (EtOH).

Carroll, A.R. *et al.*, *Aust. J. Chem.*, 1991, **44**, 1627-1633 (*Eupomatenoid 17*).
Murphy, S.K. *et al.*, *Angew. Chem., Int. Ed.*, 2014, **53**, 2455-2459 (*Eupomatenoid 17, synth*)

4,7′-Epoxy-3,8′-ligna-7,7′-diene-3′,4′,5,5′,9,9′-hexol E54

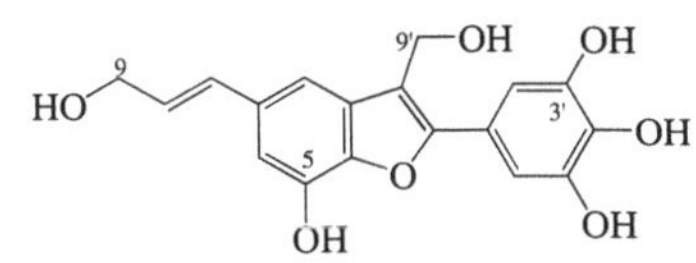

7-Hydroxy-5-(3-hydroxy-1-propenyl)-2-(3,4,5-trihydroxyphenyl)-3-benzofuranmethanol. 4,7′-Epoxy-3′,4′,5,5′,9,9′-hexahydroxy-3,8′-neoligna-7,7′-diene
$C_{18}H_{16}O_7$ 344.319
(E)-form
3′,5,5′-Tri-Me ether
4,7′-Epoxy-3′,5,5′-trimethoxy-3,8′-ligna-7,7′-diene-4′,9,9′-triol.
Herpetol B
$C_{21}H_{22}O_7$ 386.400
Constit. of the roots of *Datura metel*. Exhibits weak cytotoxic activity. Amorph. powder. Mp 119 - 120°. $[\alpha]^{22}_{D}$ -4 (c, 0.1 in MeOH)
3′,5,5′-Tri-Me ether, 9′-O-β-D-glucopyranoside
Herpetol C
$C_{27}H_{32}O_{12}$ 548.541
Constit. of the roots of *Datura metel*. Exhibits weak cytotoxic activity. Amorph. powder. Mp 94 - 95°. $[\alpha]^{22}_{D}$ -4 (c, 0.1 in MeOH)
9-Carboxylic acid
see 4,7′-Epoxy-3′,4′,5,5′,9′-pentahydroxy-3,8′-ligna-7,7′-dien-9-oic acid, E229.

Yang, B.-Y. *et al.*, *Phytochem. Lett.*, 2018, **28**, 8-12 (*Herpetols B,C, activity*)

4,7′-Epoxy-3,8′-ligna-7,7′-diene-3′,4′,5,9,9′-pentol E55

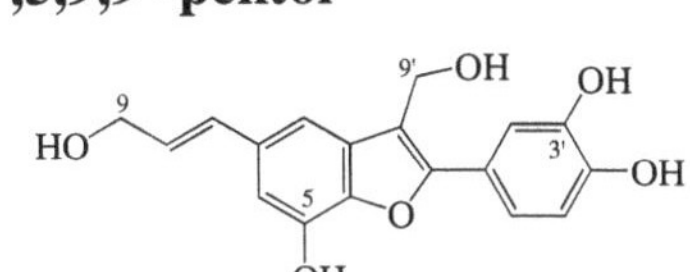

2-(3,4-Dihydroxyphenyl)-7-hydroxy-5-(3-hydroxy-1-propenyl)-3-benzofuranmethanol.

4,7′-Epoxy-3′,4′,5,9,9′-pentahydroxy-3,8′-neoligna-7,7′-diene
$C_{18}H_{16}O_6$ 328.32
(E)-form
3′-Me ether
4,7′-Epoxy-3′-methoxy-3,8′-ligna-7,7′-diene-4′,5,9,9′-tetrol. 4,7′-Epoxy-4′,5,9,9′-tetrahydroxy-3′-methoxy-3,8′-ligna-7,7′-diene. ***Vibsanol***
[179234-02-7]
$C_{19}H_{18}O_6$ 342.347
Constit. of *Solidago canadensis*, *Viburnum awabuki* and *Viburnum erosum*. Shows moderate antioxidant and hypolipidaemic activities. Oil. λ_{max} 228 (ε 17000); 271 (ε 20000); 305 (ε 17600) (EtOH)
3′,4′-Di-Me ether
4,7′-Epoxy-3′,4′-dimethoxy-3,8′-ligna-7,7′-diene-5,9,9′-triol. 4,7′-Epoxy-5,9,9′-trihydroxy-3′,4′-dimethoxy-3,8′-ligna-7,7′-diene. ***4′-O-Methylvibsanol****. 4′-Methoxyvibsanol (incorr.)*
[1627677-79-5]
$C_{20}H_{20}O_6$ 356.374
Constit. of *Viburnum ternatum*. Oil. λ_{max} 274 (log ε 4.0); 314 (log ε 3.95) (MeOH)
3′,5-Di-Me ether
4,7′-Epoxy-3′,5-dimethoxy-3,8′-ligna-7,7′-diene-4′,9,9′-triol. 4,7′-Epoxy-4′,9,9′-trihydroxy-3′,5-dimethoxy-3,8′-ligna-7,7′-diene. ***Herpetol***
[90902-38-8]
$C_{20}H_{20}O_6$ 356.374
Constit. of *Herpetospermum caudigerum* and *Viburnum erosum*. Cytotoxic against a range of human cancer cells. Amorph. solid. Mp 167 - 168° (synthetic). λ_{max} 210; 230 (sh); 270; 305 (sh) (MeOH). [base] λ_{max} 210; 240; 285; 332 (MeOH/NaOH)
3′,9′-Di-Me ether
4,7′-Epoxy-3′,9′-dimethoxy-3,8′-ligna-7,7′-diene-4′,5,9-triol. 4,7′-Epoxy-4′,5,9-trihydroxy-3′,9′-dimethoxy-3,8′-ligna-7,7′-diene. ***9′-O-Methylvibsanol***
[179234-03-8]
$C_{20}H_{20}O_6$ 356.374
Constit. of *Viburnum awabuki* and *Viburnum sargentii*. Exhibits potent cytotoxic activity against MCF-7 and A549 cells. Oil. λ_{max} 224 (ε 15000); 278 (ε 13300); 306 (ε 14600) (EtOH)
3′,4′,5-Tri-Me ether
4,7′-Epoxy-3′,4′,5-trimethoxy-3,8′-ligna-7,7′-diene-9,9′-diol. 4,7′-Epoxy-9,9′-dihydroxy-3′,4′,5-trimethoxy-3,8′-ligna-7,7′-diene. ***4′,5-Di-O-methylvibsanol***
$C_{21}H_{22}O_6$ 370.401
Constit. of branches and leaves of *Viburnum sargentii*
9-Aldehyde, 3′-Me ether
4,7′-Epoxy-4′,5,9′-trihydroxy-3′-methoxy-3,8′-ligna-7,7′-dien-9-al. 4,7′-Epoxy-4′,5,9′-trihydroxy-3′-methoxy-9-oxo-3,8′-ligna-7,7′-diene. ***Vibsanol-9-al***
[1583276-05-4]
$C_{19}H_{16}O_6$ 340.331
Constit. of *Viburnum betulifolium* and *Viburnum ternatum*. Oil. λ_{max} 275 (log ε 4.10); 314 (log ε 4.05) (MeOH)
9-Aldehyde, 3′,5-di-Me ether
4,7′-Epoxy-4′,9′-dihydroxy-3′,5-dimethoxy-3,8′-ligna-7,7′-dien-9-al. ***Betulifoliumin A***
[2197123-99-0]
$C_{20}H_{18}O_6$ 354.358
Constit. of the aerial parts of *Viburnum betulifolium*. Oil. λ_{max} 271 (log ε 4.08); 305 (log ε 3.88) (MeOH)
9-Aldehyde, 3′,4′-methylene ether
4,7′-Epoxy-5,9′-dihydroxy-3′,4′-methylenedioxy-3,8′-ligna-7,7′-dien-9-al. ***Betulifoliumin B***
[2197124-00-6]
$C_{19}H_{14}O_6$ 338.315
Constit. of the aerial parts of *Viburnum betulifolium*. Oil. λ_{max} 278 (log ε 4.10); 306 (log ε 3.98) (MeOH)
9′-Aldehyde, 3′,5-di-Me ether
4,7′-Epoxy-4′,9-dihydroxy-3′,5-dimethoxy-3,8′-ligna-7,7′-dien-9′-al. 4,7′-Epoxy-4′,9-dihydroxy-3′,5-dimethoxy-9′-oxo-3,8′-ligna-7,7′-diene. ***Herpetal***
[70610-20-7]
$C_{20}H_{18}O_6$ 354.358
Constit. of *Herpetospermum caudigerum*. Cryst.. Mp 160 - 161°. λ_{max} 220; 250; 290 (sh); 340 (MeOH).

Kaouadji, M. *et al.*, *Phytochemistry*, 1978, **17**, 2134-2135 (*Herpetal*).
Kaouadji, M. *et al.*, *Z. Naturforsch., C*, 1984, **39**, 307-308 (*Herpetol*).
Fukuyama, Y. *et al.*, *Chem. Pharm. Bull.*, 1996, **44**, 1418-1420 (*Viburnum awabuki constits, activity*).
Sakai, A. *et al.*, *Heterocycles*, 2000, **52**, 643-659 (*Vibsanol, synth*).
Fan, H.-F. *et al.*, *J. Chem. Res.*, 2010, **34**, 233-235 (*Herpetol, synth, activity*).
Bae, K.-E. *et al.*, *Molecules*, 2010, **15**, 4599-4609 (*Viburnum sargentii constit, activity*).
Huang, Y. *et al.*, *J. Asian Nat. Prod. Res.*, 2013, **15**, 319-324 (*Solidago canadensis constit, activity*).
Hu, J. *et al.*, *J. Asian Nat. Prod. Res.*, 2014, **16**, 703-708 (*Viburnum ternatum constits*).
In, S.-J. *et al.*, *Arch. Pharmacal Res.*, 2015, **38**, 26-34 (*Herpetol, Vibsanol*).
Hu, J. *et al.*, *J. Asian Nat. Prod. Res.*, 2016, **18**, 831-836 (*Betulifoliumins A,B*).
Zhu, Y. *et al.*, *Zhongcaoyao*, 2017, **48**, 2177-2181 (*4′,5-Di-O-methylvibsanol*)

4,7′-Epoxy-3,8′-ligna-7,7′-diene-3′,4′,5,5′-tetrol E56

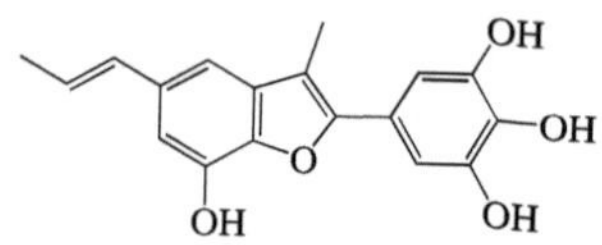

3-Methyl-5-(2-propenyl)-2-(3,4,5-trihydroxyphenyl)-7-benzofuranol. 4,7′-Epoxy-3′,4′,5,5′-tetrahydroxy-3,8′-neoligna-7,7′-diene
$C_{18}H_{16}O_5$ 312.321
(E)-form
3′,5-Di-Me ether
4,7′-Epoxy-5,5′-dimethoxy-3,8′-ligna-7,7′-diene-3′,4′-diol. 4,7′-Epoxy-3′,4′-dihydroxy-5,5′-dimethoxy-3,8′-ligna-7,7′-diene. ***Obovaten***
[214197-47-4]
$C_{20}H_{20}O_5$ 340.375
Constit. of the leaves of *Persea obovatifolia*. Cytotoxic agent. Needles (C_6H_6). Mp 159 - 161°. λ_{max} 238 (log ε 4.67); 270 (log ε 4.61); 306 (log ε 4.6) (EtOH).

Tsai, I.-L. *et al.*, *Phytochemistry*, 1998, **48**, 1371-1375 (*Obovaten, activity*).
Kao, C.-L. *et al.*, *J. Org. Chem.*, 2002, **67**, 6772-6787 (*Obovaten, synth*)

4,7′-Epoxy-3,8′-ligna-7,7′-diene-3′,4′,5-triol E57

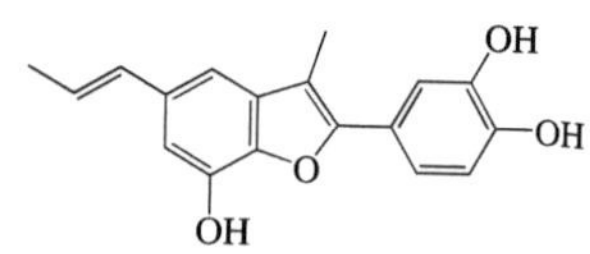

(E)-form

2-(3,4-Dihydroxyphenyl)-3-methyl-5-(1-propenyl)-7-benzofuranol. 4,7′-Epoxy-

3′,4′,5-trihydroxy-3,8′-neoligna-7,7′-diene
$C_{18}H_{16}O_4$ 296.322
(E)-form
3′,5-Di-Me ether
4,7′-Epoxy-3′,5-dimethoxy-3,8′-ligna-7,7′-dien-4′-ol. 4,7′-Epoxy-4′-hydroxy-3′,5-dimethoxy-3,8′-ligna-7,7′-diene. ***Eupomatenoid 7***
[41744-30-3]
$C_{20}H_{20}O_4$ 324.376
Constit. of *Aristolochia elegans*, *Eupomatia* spp., *Piper ribesioides* and *Senna lindheimeriana*. Antitrypanocidal agent. Needles (petrol). Mp 105 - 106°
Tri-Me ether
4,7′-Epoxy-3′,4′,5-trimethoxy-3,8′-ligna-7,7′-diene. ***Eupomatenoid 12***
[41365-37-1]
$C_{21}H_{22}O_4$ 338.403
Constit. of *Eupomatia* spp.. Needles (EtOH). Mp 115 - 116°
3′,4′-Methylene, 5-Me ether
4,7′-Epoxy-5-methoxy-3′,4′-methylenedioxy-3,8′-ligna-7,7′-diene. ***Eupomatene****. Eupomatenoid 1*
[23357-64-4]
$C_{20}H_{18}O_4$ 322.36
Constit. of *Aristolochia elegans*, *Eupomatia laurina* and *Caryodaphnopsis baviensis*. Cryst. (Et_2O). Mp 154 - 156°
(Z)-form
3′,5-Di-Me ether
cis-Eupomatenoid 7
$C_{20}H_{20}O_4$ 324.376
Constit. of *Aristolochia cordigera*. Amorph. yellow solid. λ_{max} 210; 231; 260; 300 (MeOH).

McCredie, R.S. *et al.*, *Aust. J. Chem.*, 1969, **22**, 1011-1032 (*Eupomatene*).
Bowden, B.F. *et al.*, *Aust. J. Chem.*, 1972, **25**, 2659-2669 (*Eupomatenoid 7*).
Picker, K. *et al.*, *Aust. J. Chem.*, 1973, **26**, 1111-1119 (*Eupomatenoid 12*).
Ahmed, R. *et al.*, *Phytochemistry*, 1975, **14**, 2710-2712 (*Eupomatenoids 7,12, synth*).
Watanabe, M. *et al.*, *Chem. Pharm. Bull.*, 1991, **39**, 41-48 (*Eupomatenoid 1, synth*).
Engler, T.A. *et al.*, *Tetrahedron Lett.*, 1996, **37**, 6969-6970 (*Eupomatenoid 12, synth*).
Murphy, S.K. *et al.*, *Angew. Chem., Int. Ed.*, 2014, **53**, 2455-2459 (*Eupomatenoid 12, synth*).
Pereira, M.D.P. *et al.*, *Planta Med.*, 2017, **83**, 912-920 (*cis-Eupomatenoid 7*)

4,7′-Epoxy-3,8′-ligna-7,7′-diene-4′,5,9′-triol E58

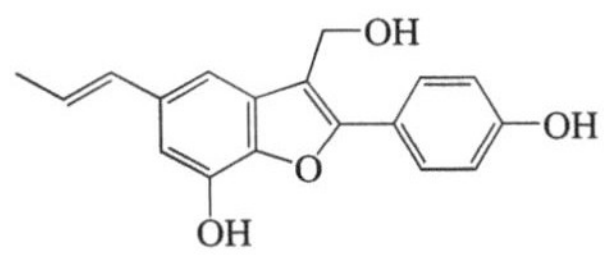

7-Hydroxy-3-(hydroxymethyl)-2-(4-hydroxyphenyl)-5-(1-propenyl) benzofuran. 4,7′-Epoxy-4′,5,9′-trihydroxy-3,8′-neoligna-7,7′-diene
$C_{18}H_{16}O_4$ 296.322
(E)-form
9′-Aldehyde
7-Hydroxy-2-(4-hydroxyphenyl)-5-(1-propenyl)-3-benzofurancarboxaldehyde, CAS. *4,7′-Epoxy-4′,5-dihydroxy-3,8′-ligna-7,7′-dien-9′-al. 4,7′-Epoxy-4′,5-dihydroxy-9′-oxo-3,8′-ligna-7,7′-diene*
9′-Aldehyde, 5-Me ether
2-(4-Hydroxyphenyl)-7-methoxy-5-(1-propenyl)-3-benzofurancarboxaldehyde, CAS. *4,7′-Epoxy-4′-hydroxy-5-methoxy-3,8′-ligna-7,7′-dien-9′-al. 4,7′-Epoxy-4′-hydroxy-5-methoxy-9′-oxo-3,8′-ligna-7,7′-diene*
[109194-68-5]
$C_{19}H_{16}O_4$ 308.333
Constit. of *Krameria cytisoides*. Yellow solid. Mp 258 - 261°. λ_{max} 250 (log ε 4.43); 292 (log ε 4.06); 332 (log ε 4.11) (MeOH). [base] λ_{max} 260 (log ε 4.42); 380 (log ε 4.28) (MeOH/NaOH).

Achenbach, H. *et al.*, *Phytochemistry*, 1987, **26**, 1159-1166 (*Krameria cytisoides constit*)

4,7′-Epoxy-3,8′-ligna-7,7′-diene-4′,6,9′-triol E59

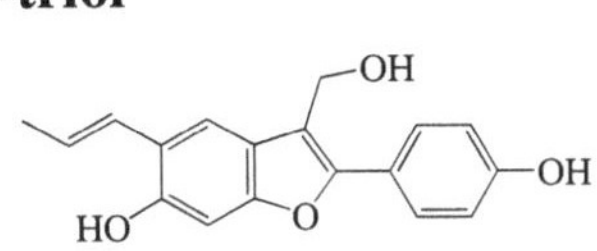

6-Hydroxy-2-(4-hydroxyphenyl)-5-(1-propenyl)-3-benzofuranmethanol. 4,7′-Epoxy-4′,6,9′-trihydroxy-3,8′-neoligna-7,7′-diene
$C_{18}H_{16}O_4$ 296.322
(E)-form
Constit. of the seeds of *Daphniphyllum macropodum*. Solid. λ_{max} 207; 251; 334 (MeOH).

Ma, X. *et al.*, *RSC Adv.*, 2017, **7**, 52970-52976 (*Daphniphyllum macropodum constit*)

4,7′-Epoxy-3,8′-ligna-7′,8-diene-3′,4′,5-triol E60

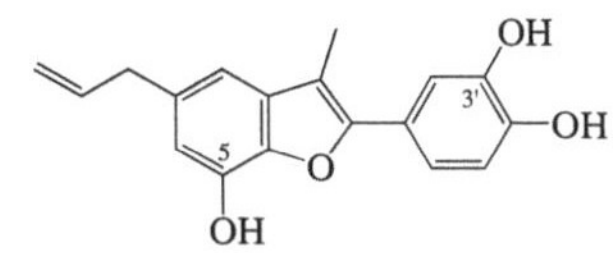

2-(3,4-Dihydroxyphenyl)-3-methyl-5-(2-propenyl)-7-benzofuranol. 4,7′-Epoxy-3′,4′,5-trihydroxy-3,8′-neoligna-7′,8-diene
$C_{18}H_{16}O_4$ 296.322
3′,5-Di-Me ether
4,7′-Epoxy-3′,5-dimethoxy-3,8′-ligna-7′,8-dien-4′-ol. 4,7′-Epoxy-4′-hydroxy-3′,5-dimethoxy-3,8′-ligna-7′,8-diene. ***Carinatidin***
[88640-84-0]
$C_{20}H_{20}O_4$ 324.376
Constit. of *Virola carinata*. Needles. Mp 104 - 105°
Tri-Me ether
4,7′-Epoxy-3′,4′,5-trimethoxy-3,8′-ligna-7′,8-diene. ***Carinatin. Eupomatenoid 18***
[82843-80-9]
$C_{21}H_{22}O_4$ 338.403
Constit. of *Eupomatia bennettii*, *Virola carinata* and *Virola pavonis*. Cryst. (EtOH). Mp 88 - 91°. λ_{max} 307 (log ε 4.43) ($CHCl_3$).

Kawanishi, K. *et al.*, *Phytochemistry*, 1983, **21**, 929-931 (*Carinatin*).
Kawanishi, K. *et al.*, *Phytochemistry*, 1983, **21**, 2277-2280 (*Carinatidin*).
Carroll, A.R. *et al.*, *Aust. J. Chem.*, 1991, **44**, 1627-1633 (*Eupomatenoid 18*).
Watanabe, M. *et al.*, *Chem. Pharm. Bull.*, 1991, **39**, 41-48 (*Carinatin, synth*).
Marques, M.O.M. *et al.*, *Phytochemistry*, 1992, **31**, 4380-4381 (*Carinatin*).
Murphy, S.K. *et al.*, *Angew. Chem., Int. Ed.*, 2014, **53**, 2455-2459 (*Eupomatenoid 18, synth*)

4,7′-Epoxy-3,8′-ligna-7,7′-dien-4′-ol E61

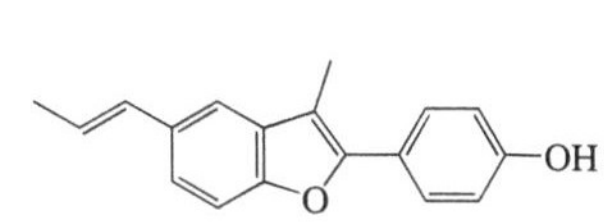

4-[3-Methyl-5-(1-propenyl)-2-benzofuranyl]phenol, CAS. *2-(4-Hydroxyphenyl)-3-methyl-5-(1-propenyl)benzofuran. 4,7′-Epoxy-4′-hydroxy-3,8′-neoligna-7,7′-diene*
$C_{18}H_{16}O_2$ 264.324
(E)-form
Eupomatenoid 6. Ratanhiaphenol II
[41744-26-7]
Constit. of *Eupomatia* spp., *Krameria* spp., *Piper decurrens* and *Piper*

solmsianum. Shows antimicrobial, antitrypanosomal, antioxidant, cytotoxic and insecticidal activities. Uv protectant. Needles (C_6H_6/petrol). Mp 148 - 150°

α-L-Rhamnopyranoside

Eupomatenoid 2

[41744-25-6]

$C_{24}H_{26}O_6$ 410.466

Constit. of *Eupomatia* spp.. Heat shock protein-90 modulator. Cryst. (EtOH). Mp 154°. $[\alpha]_D$ -140 (c, 1 in $CHCl_3$)

Me ether

2-(4-Methoxyphenyl)-3-methyl-5-(1-propenyl)benzofuran. 4,7′-Epoxy-4′-methoxy-3,8′-ligna-7,7′-diene. ***Eupomatenoid 15***

[40341-58-0]

$C_{19}H_{18}O_2$ 278.351

Constit. of *Eupomatia bennettii*. Needles (MeOH). Mp 118°. λ_{max} 225 (ε 18200); 253 (ε 26300); 300 (ε 26300) (EtOH).

Bowden, B.F. *et al.*, *Aust. J. Chem.*, 1972, **25**, 2659-2669 (*Eupomatenoids 2,6*).
Stahl, E. *et al.*, *Planta Med.*, 1981, **42**, 144-154 (*Ratanhiaphenol II*).
McKittrick, B.A. *et al.*, *J. Chem. Soc., Perkin Trans. 1*, 1983, 475-482 (*Eupomatenoid 6, synth*).
Achenbach, H. *et al.*, *Phytochemistry*, 1987, **26**, 1159-1166 (*Eupomatenoid 6*).
Carroll, A.R. *et al.*, *Aust. J. Chem.*, 1991, **44**, 1627-1633 (*Eupomatenoid 15*).
Himmelreich, U. *et al.*, *Magn. Reson. Chem.*, 1995, **33**, 280-282 (*Eupomatenoid 6*).
Chauret, D.C. *et al.*, *J. Nat. Prod.*, 1996, **59**, 152-155 (*Piper decurrens constit*).
Pat. Coop. Treaty (WIPO), 1998, 98 17 294 (*Eupomatenoid 6, activity*).
Bach, T. *et al.*, *Synthesis*, 2003, 925-939 (*Eupomatenoids 6,15, synth*).
Miyata, O. *et al.*, *Org. Lett.*, 2004, **6**, 1761-1763 (*Eupomatenoid 6, synth*).
Baumgartner, L. *et al.*, *J. Nat. Prod.*, 2011, **74**, 1779-1786 (*Ratanhiaphenol II, activity*).
Damodar, K. *et al.*, *Bull. Korean Chem. Soc.*, 2014, **35**, 3618-3622 (*Eupomatenoid 6, synth*).
Macedo, A.L. *et al.*, *Nat. Prod. Res.*, 2018, **32**, 2701-2704 (*Eupomatenoid 6, activity*)

6,8-Epoxy-3,3′-ligna-7,8′-dien-4′-ol E62

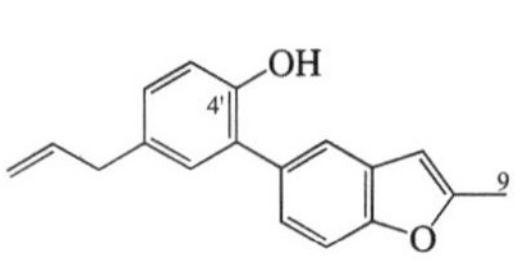

5-[2-Hydroxy-5-(2-propenyl)phenyl]-2-methylbenzofuran. 2-(2-Methyl-5-benzofuranyl)-4-(2-propenyl)phenol. 6,8-Epoxy-4′-hydroxy-3,3′-neoligna-7,8′-diene

[1416567-14-0]

$C_{18}H_{16}O_2$ 264.324

Lignan numbering shown. Constit. of the stem bark *Magnolia officinalis*. Displays potent antiinflammatory activity. Inhibits fMLP/CB-induced elastase release and superoxide anion generation in human neutrophils. Amorph. powder. λ_{max} 238 (log ε 4.31); 293 (log ε 4.00) (MeOH)

9-Hydroxy

5-[2-Hydroxy-5-(2-propenyl) phenyl]-2-benzofuranmethanol. 6,8-Epoxy-3,3′-ligna-7,8′-diene-4′,9-diol. 6,8-Epoxy-4′,9-dihydroxy-3,3′-ligna-7,8′-diene

[1586815-86-2]

$C_{18}H_{16}O_3$ 280.323

Constit. of the twigs of *Magnolia denudata*. Inhibits fMLP/CB-induced elastase release. Amorph. yellow powder. λ_{max} 238 (log ε 4.32); 292 (log ε 4.01) (MeOH).

Kuo, W.L. *et al.*, *Phytochemistry*, 2013, **85**, 153-160 (*Magnolia officinalis constit*).
Chung, C.-Y. *et al.*, *Chem. Biodiversity*, 2015, **12**, 1263-1270 (*Magnolia denudata constit, activity*).
Lin, D. *et al.*, *J. Asian Nat. Prod. Res.*, 2019, **21**, 1052-1067 (*synth*)

4,7′-Epoxy-3,8′-lignan-4′,8-diol E63

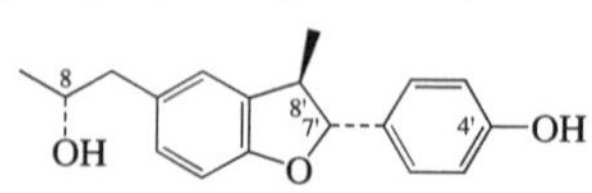

2,3-Dihydro-2-(4-hydroxyphenyl)-5-(2-hydroxypropyl)-3-methylbenzofuran. 4,7′-Epoxy-4′,8-dihydroxy-3,8′-neolignan

$C_{18}H_{20}O_3$ 284.355

(7′R,8R,8′R)-form [184587-70-0]

Constit. of *Krameria parvifolia*. Amorph.. $[\alpha]^{20}_D$ +21.2 (c, 0.6 in MeOH). λ_{max} 228 (log ε 4.06); 275 (sh); 283 (log ε 3.46); 295 (sh) (MeOH)

8-Ketone

4,7′-Epoxy-4′-hydroxy-3,8′-lignan-8-one. 4,7′-Epoxy-4′-hydroxy-8-oxo-3,8′-lignan. ***Miliumollinone***

[1528762-17-5]

$C_{18}H_{18}O_3$ 282.339

Constit. of the leaves of *Miliusa mollis*. Exhibits weak activity against herpes simplex virus. Weakly cytotoxic. Amorph. powder. $[\alpha]^{20}_D$ +23 (c, 0.14 in $CHCl_3$). λ_{max} 207 (log ε 4.35); 229 (log ε 4.34); 284 (log ε 3.9) (MeOH).

Achenbach, H. *et al.*, *Phytochemistry*, 1996, **43**, 1093-1095 (*Krameria parvifolia constit*).
Sawasdee, K. *et al.*, *Fitoterapia*, 2013, **85**, 49-56 (*Miliumollinone, activity*)

4,7′-Epoxy-3,8′-lignan-4′,9-diol E64

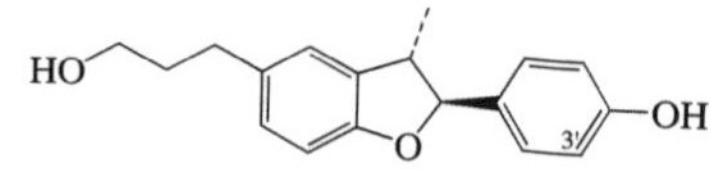

2,3-Dihydro-2-(4-hydroxyphenyl)-3-methyl-5-benzofuranpropanol, CAS. *4,7′-Epoxy-4′,9-dihydroxy-3,8′-neolignan*

$C_{18}H_{20}O_3$ 284.355

(7′S,8′S)-form [109194-67-4]

Constit. of *Krameria cytisoides*. Cryst.. Mp 123 - 125°. $[\alpha]^{21}_D$ +4.1 (c, 0.56 in MeOH). Stereochem. revised in 2000. λ_{max} 229 (ε 14454); 285 (ε 4677); 296 (sh) (MeOH).

Achenbach, H. *et al.*, *Phytochemistry*, 1987, **26**, 1159-1166 (*Krameria cytisoides constit, cd*).
Kurtán, T. *et al.*, *J. Chem. Soc., Perkin Trans. 1*, 2000, 453-461 (*abs config*)

7,7′-Epoxy-8,8′-lignan-4,4′-diol E65

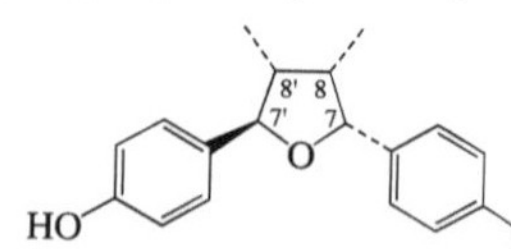

(7R,7′R,8S,8′R)-form

4,4′-(Tetrahydro-3,4-dimethyl-2,5-furandiyl)bisphenol, CAS. *Tetrahydro-2,5-bis(4-hydroxyphenyl)-3,4-dimethylfuran. 4,4′-Dihydroxy-7,7′-epoxylignan*

$C_{18}H_{20}O_3$ 284.355

(7R,7′R,8S,8′R)-form

(+)-Larreatricin [109194-74-3]

Isol. from *Krameria cystisoides*, *Krameria interior* and *Larrea tridentata*. Mp 158 - 162°. $[\alpha]^{21}_D$ +78 (c, 0.15 in MeOH)

(7S,7′R,8R,8′R)-form

4-Me ether

Odoratisol E

$C_{19}H_{22}O_3$ 298.382

Constit. of *Mitrephora wangii*. Oil. $[\alpha]^{22}_D$ -49.7 (c, 0.6 in MeOH). λ_{max} 214 (log ε 2.95); 229 (log ε 2.79); 275 (log ε 2.33) (MeOH)

(7S,7′S,8R,8′S)-form

(-)-Larreatricin [130062-03-2]

Constit. of *Larrea tridentata*, *Krameria ixina*, *Krameria triandria* and *Ribes nigrum*. Needles (C_6H_6/Et_2O). Mp 161 - 163°. $[\alpha]_D$ -48. $[\alpha]^{20}_D$ -19.3

(c, 0.15 in MeOH). Identity of samples not established. Mp refers to Larreatricin from *Larrea tridentata*,. λ_{max} 206 (ε 168500); 225 (ε 18600); 274 (ε 3080) (EtOH) (Berdy)

(7S,7′S,8S,8′S)-form

8′-Epilarreatricin [119365-44-5]

Constit. of *Larrea tridentata* and *Krameria triandria*. Needles (Me_2CO). Mp 230 - 232°. $[\alpha]^{20}_{D}$ -39 (c, 0.01 in MeOH). λ_{max} 223 (ε 17800); 272 (ε 3000) (EtOH) (Berdy)

(7RS,7′SR,8RS,8′RS)-form

3,3′-Didemethoxyverrucosin [618359-59-4]

Constit. of *Larrea tridentata*. Oil. λ_{max} 228 (log ε 4.67); 275 (log ε 3.76) (MeOH). [base] λ_{max} 228 (log ε 4.72); 246 (log ε 4.85); 276 (log ε 4.1) (MeOH/NaOH)

(7RS,7′SR,8RS,8′SR)-form

3,3′-Didemethoxynectandrin B [130325-40-5]

Constit. of *Krameria interior* and *Larrea tridentata*. Oil. One of two possible *meso*-stereoisomers. λ_{max} 224 (log ε 4.46); 280 (log ε 3.69) (MeOH)

(7RS,7′SR,8SR,8′RS)-form

Anolignan C [158189-15-2]

Constit. of *Anogeissus acuminata*. HIV reverse transcriptase inhibitor. Cryst.. Mp 158°. *meso*-. λ_{max} 209 (ε 8511); 228 (ε 16218); 277 (ε 3715) (MeOH) (Berdy)

4-Me ether

7,7′-Epoxy-4′-methoxy-8,8′-lignan-4-ol. 4-Hydroxy-4′-methoxy-7,7′-epoxylignan

[948909-96-4]

$C_{19}H_{22}O_3$ 298.382

Constit. of the stem bark of *Terminalia superba*. Powder. Mp 110 - 113°. $[\alpha]^{25}_{D}$+35 (c, 0.007 in MeOH). Chiral Me ether of the *meso*-stereoisomer having (7*S*,7′*R*,8*R*,8′*S*)-abs. config. where 7- is the aryl ring bearing the methoxy substit.. λ_{max} 227 (log ε 3.6); 240 (log ε 2.64); 250 (log ε 2.5); 277 (log ε 3.87) (MeOH)

Di-Me ether

7,7′-Epoxy-4,4′-dimethoxylignan. 4,4′-Dimethoxy-7,7′-epoxylignan

[948909-97-5]

$C_{20}H_{24}O_3$ 312.409

Constit. of the stem bark of *Terminalia superba*. Powder. Mp 98 - 100°. Opt. inactive (*meso*-). λ_{max} 200 (log ε 3.6); 240 (log ε 2.64); 250 (log ε 2.5); 275 (log ε 3.87) (MeOH).

Achenbach, H. *et al., Phytochemistry*, 1987, **26**, 1159-1166 (*Krameria cystisoides constit*).

Arnone, A. *et al., Gazz. Chim. Ital.*, 1988, **118**, 675-682 (*Krameria triandria constit*).

Konno, C. *et al., J. Nat. Prod.*, 1990, **53**, 396-406 (*Larrea tridentata constits*).

Dominguez, X.A. *et al., Phytochemistry*, 1990, **29**, 2651-2653 (*Krameria interior constits*).

Achenbach, H. *et al., Phytochemistry*, 1991, **30**, 3753-3757 (*Krameria ixina constit*).

Rimando, A.M. *et al., J. Nat. Prod.*, 1994, **57**, 896-904 (*Anolignan C*).

Moinuddin, S.G.A *et al., Org. Biomol. Chem.*, 2003, **1**, 2307-2313 (*Didemethoxyverrucosin, Larreatricin, synth*).

Wansi, J.D. *et al., Phytochemistry*, 2007, **68**, 2096-2100 (*Terminalia superba constits*).

Favela-Hernández, J.M.de J. *et al., Acta Crystallogr., Sect. E*, 2012, **68**, o3019-o3020 (*8′-Epilarreatricin, cryst struct*).

Schmidt, T.J. *et al., Phytochem. Lett.*, 2012, **5**, 632-638 (*Larreatricin, abs config*).

Sasaki, T. *et al., Phytochemistry*, 2013, **95**, 333-340 (*Ribes nigrum constit*).

Jaidee, W. *et al., Fitoterapia*, 2018, **130**, 219-224 (*Odoratisol E*)

9,9′-Epoxy-8,8′-lignan-3,3′-diol E66

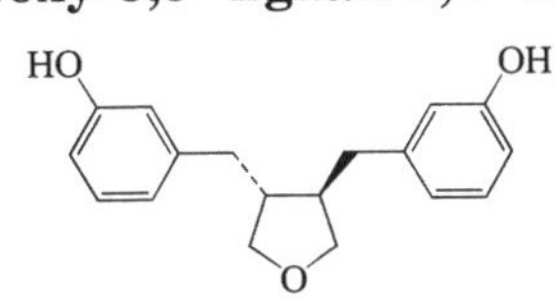

Tetrahydro-3,4-bis[(3-hydroxyphenyl)methyl]furan. 3,4-Bis(3-hydroxybenzyl)tetrahydrofuran. 3,3′-Dihydroxy-9,9′-epoxylignan

$C_{18}H_{20}O_3$ 284.355

(8R*,8′R*)-form

Enterofuran [201532-62-9]

Mammalian lignan metab.. Phytoestrogen. Shows *in vitro* binding to human sex hormone binding globulin (SHBG). Cryst. (MeOH). Mp 130 - 133° (112-114°).

Schoettner, M. *et al., Planta Med.*, 1997, **63**, 529-532 (*activity*).

Liggins, J. *et al., Anal. Biochem.*, 2000, **287**, 102-109 (*occur, ms*).

Ward, R.S. *et al., Tetrahedron*, 2001, **57**, 2057-2064 (*synth, pmr, cmr*).

Raffaelli, B. *et al., J. Chromatogr., B*, 2002, **777**, 29-43 (*rev, biosynth, activity*)

4,7′-Epoxy-3,8′-lignan-2′,3′,5,5′,7,9,9′-heptol E67

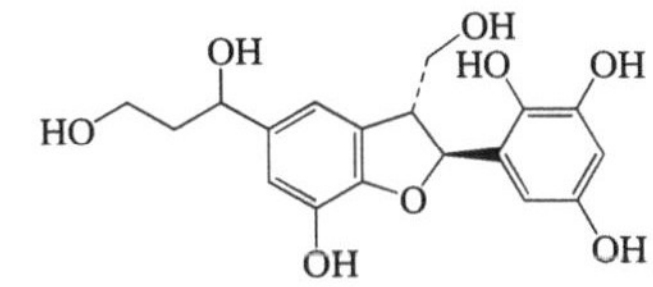

$C_{18}H_{20}O_8$ 364.35

(7′S,8′R)-form

7-Ketone, 2′,3′,5-tri-Me ether

4,7′-Epoxy-5′,9,9′-trihydroxy-2′,3′,5-trimethoxy-3,8′-lignan-7-one. ***Tobdihydrofuran A****. Tobadihydrofuran A*

$C_{21}H_{24}O_8$ 404.415

Constit. of the stems of flue-cured *Nicotiana tabacum*. Exhibits anti-tobacco mosaic virus activity. Pale yellow gum. $[\alpha]^{25}_{D}$-8.5 (c, 0.05 in MeOH). Two different spellings of synonym. λ_{max} 210 (log ε 4.12); 230 (log ε 3.64); 290 (log ε 3.58) (MeOH).

Zhang, T. *et al., Asian J. Chem.*, 2015, **27**, 2964-2966 (*Tobdihydrofuran A, cd, activity*)

4,7′-Epoxy-3,8′-lignan-3′,4′,5,7,8,9,9′-heptol E68

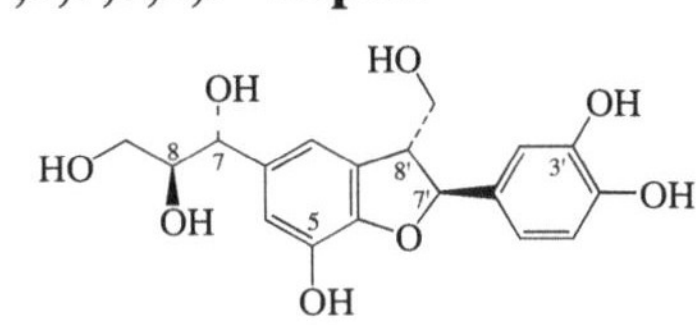

(7*R*,7′*S*,8*S*,8′*R*)-*form*

4,7′-Epoxy-3′,4′,5,7,8,9,9′-heptahydroxy-3,8′-neolignan

$C_{18}H_{20}O_8$ 364.35

(7R,7′S,8S,8′R)-form

3′,5-Di-Me ether

4,7′-Epoxy-3′,5-dimethoxy-3,8′-lignan-4′,7,8,9,9′-pentol. 4,7′-Epoxy-4′,7,8,9,9′-pentahydroxy-3′,5-dimethoxy-3,8′-lignan. ***Meliasendanin B***

[1820034-03-4]

$C_{20}H_{24}O_8$ 392.404

Constit. of the fruit of *Melia toosendan*. λ_{max} 256 (log ε 3.18) (MeOH)

(7S,7′R,8S,8′S)-form

3′,5,7-Tri-Me ether

4,7′-Epoxy-3′,5,7-trimethoxy-3,8′-lignan-4′,8,9,9′-tetrol. 4,7′-Epoxy-4′,8,9,9′-tetrahydroxy-3′,5,7-trimethoxy-3,8′-lignan. ***Pinnatifidanin CIV***

[1613723-75-3]

$C_{21}H_{26}O_8$ 406.431
Constit. of the seeds of *Crataegus pinnatifida*. Yellow oil. $[\alpha]_D^{20}$ +10.2 (c, 0.03 in MeOH). λ_{max} 230 (log ε 3.12); 278 (log ε 2.45) (MeOH)

7-Et, 3′,5-di-Me ether
4,7′-Epoxy-7-ethoxy-3′,5-dimethoxy-3,8′-lignan-4′,8,9,9′-tetrol. 4,7′-Epoxy-7-ethoxy-4′,8,9,9′-tetrahydroxy-3′,5-dimethoxy-3,8′-lignan. ***Pinnatifidanin CVI***
[1616701-73-5]
$C_{22}H_{28}O_8$ 420.458
Constit. of the seeds of *Crataegus pinnatifida*. Yellow oil. $[\alpha]_D^{20}$ +7.0 (c, 0.02 in MeOH). λ_{max} 228 (log ε 2.40); 280 (log ε 1.66) (MeOH)

(7S,7′S,8R,8′R)-form

3′,5-Di-Me ether
Meliasendanin C
[1820034-04-5]
$C_{20}H_{24}O_8$ 392.404
Constit. of the fruit of *Melia toosendan*. Gum. $[\alpha]_D$ +3.3 (c, 0.16 in MeOH). λ_{max} 256 (log ε 3.18) (MeOH)

7-Ketone, 3′,5-di-Me ether
4,7′-Epoxy-4′,8,9,9′-tetrahydroxy-3′,5-dimethoxy-3,8′-lignan-7-one.
(+)-7-Dehydrosisymbrifolin
[1800538-71-9]
$C_{20}H_{22}O_8$ 390.388
Constit. of the fruit of *Xanthium sibiricum*. Amorph. powder. $[\alpha]_D^{25}$ +30 (c, 0.04 in MeOH). λ_{max} 289 (log ε 2.61) (MeOH)

(7S,7′S,8S,8′R)-form

3′,5-Di-Me ether
Meliasendanin D
[1820034-05-6]
$C_{20}H_{24}O_8$ 392.404
Constit. of the fruit of *Melia toosendan*. Gum. $[\alpha]_D^{20}$ -1 (c, 0.51 in MeOH). λ_{max} 256 (log ε 3.18) (MeOH)

3′,5,7-Tri-Me ether
Pinnatifidanin CIII
[1613724-50-7]
$C_{21}H_{26}O_8$ 406.431
Constit. of the seeds of *Crataegus pinnatifida*. Yellow oil. $[\alpha]_D^{20}$ -21.2 (c, 0.08 in MeOH). λ_{max} 228 (log ε 3.32); 280 (log ε 2.01) (MeOH)

7-Et, 3′,5-di-Me ether
Pinnatifidanin CV
[1613724-24-5]
$C_{22}H_{28}O_8$ 420.458
Constit. of the seeds of *Crataegus pinnatifida*. Yellow oil. $[\alpha]_D^{20}$ -8.5 (c, 0.02 in MeOH). λ_{max} 229 (log ε 3.71); 280 (log ε 2.81) (MeOH)

7-Ketone, 3′,5-di-Me ether
(-)-7-Dehydrosisymbrifolin
[1800538-70-8]
$C_{20}H_{22}O_8$ 390.388
Constit. of the fruit of *Xanthium sibiricum*. Amorph. powder. $[\alpha]_D^{25}$ -33 (c, 0.04 in MeOH). λ_{max} 289 (log ε 2.62) (MeOH)

(7R*,7′R,8R*,8′S)-form

3′,5-Di-Me ether
[2260837-72-5]
$C_{20}H_{24}O_8$ 392.404
Constit. of the seeds of *Pharbitis nil*. NO prodn. inhibitor. Gum. $[\alpha]_D$ +18.5 (c, 0.05 in MeOH). λ_{max} 234 (log ε 2.65) (MeOH)

3′,5-Di-Me ether, 9′-O-β-D-glucopyranoside
[1818336-44-5]
$C_{26}H_{34}O_{13}$ 554.545
Constit. of the fruit of *Euterpe oleracea*. Antioxidant. Inhibits generation of ROS in PMA-induced HL-60 cells. Amorph. solid. $[\alpha]_D^{20}$ -9.6 (c, 0.6 in MeOH). λ_{max} 236 (log ε 4.24); 280 (log ε 2.36) (MeOH)

(7R*,7′S,8R*,8′R)-form

3′,5-Di-Me ether, 9′-O-β-D-glucopyranoside
[1818336-45-6]
Constit. of the fruit of *Euterpe oleracea*. Antioxidant. Inhibits generation of ROS in PMA-induced HL-60 cells. Amorph. solid. $[\alpha]_D^{20}$ +6.22 (c, 0.5 in MeOH). λ_{max} 236 (log ε 3.65); 280 (log ε 2.13) (MeOH)

(7R*,7′S,8S*,8′R)-form

3′,5-Di-Me ether
7′,8′-Dihydroxydihydrodehydrodiconiferyl alcohol. ***Sisymbrifolin***
[173936-61-3]
$C_{20}H_{24}O_8$ 392.404
Constit. of *Cucumis sativus*, *Litchi chinensis*, *Salix alba* and *Solanum sisymbriifolium*. Allelopathic agent. Amorph. solid. $[\alpha]_D$ +10.7 (c, 2 in $CHCl_3$) (as penta-Ac). λ_{max} 258 (ε 660); 279 (ε 1500) (MeOH) (penta-Ac)

3′,5-Di-Me ether, 4′-O-α-L-rhamnopyranoside
Sisymbrifolin 4′-rhamnoside
[1135616-11-3]
$C_{26}H_{34}O_{12}$ 538.546
Constit. of the stems of *Jacaranda caucana*. Orange-yellow solid. $[\alpha]_D^{25}$ -45 (c, 1 in MeOH). λ_{max} 229 (log ε 4.03); 281 (log ε 3.6); 328 (log ε 2.83) (MeOH)

3′,5-Di-Me ether, 9′-Ac
$C_{22}H_{26}O_9$ 434.441
Constit. of *Mikania micrantha*. Antioxidant. Oil. $[\alpha]_D^{20}$ +45.71 (c, 0.07 in MeOH). λ_{max} 282 (log ε 3.70) (MeOH)

(7R*,7′R*,8R*,8′S*)-form

3′,5-Di-Me ether, 8-O-β-D-glucopyranoside
Flacoside A
[1204224-34-9]
$C_{26}H_{34}O_{13}$ 554.545
Constit. of *Flacourtia ramontchi*. Syrup. $[\alpha]_D^{20}$ -3.2 (c, 0.11 in MeOH). λ_{max} 207 (log ε 4.83); 281 (log ε 3.78) (MeOH)

(7ξ,7′S,8ξ,8′R)-form

3′,5-Di-Me ether, 9-O-β-D-glucopyranoside
Radulignan
[1165742-63-1]
$C_{26}H_{34}O_{13}$ 554.545
Constit. of *Lantana radula* and *Solanum coagulans*. Amorph. pale yellow solid. $[\alpha]_D^{20}$ -135.9 (c, 0.1 in MeOH)

(7ξ,7′R*,8ξ,8′S*)-form

7-Butyl, 3′-Me ether
7-Butoxy-4,7′-epoxy-3′-methoxy-3,8′-lignan-4′,5,8,9,9′-pentol. 7-Butoxy-4,7′-epoxy-4′,5,8,9,9′-pentahydroxy-3′-methoxy-3,8′-lignan
[1384866-25-4]
$C_{23}H_{30}O_8$ 434.485
Constit. of the leaves of *Platycladus orientalis*. Pale yellow powder. $[\alpha]_D^{20}$ +11.7 (c, 0.12 in MeOH)

(7′R,8R,8′S)-form

7-Ketone, 3′,5-di-Me ether
Prunustosanan AII
[1620571-53-0]
$C_{20}H_{22}O_8$ 390.388
Constit. of the seeds of *Prunus tomentosa*. Oil. $[\alpha]_D^{20}$ -19.7 (c, 0.17 in MeOH). λ_{max} 232 (log ε 0.23); 287 (log ε 0.14) (MeOH)

(7′R,8S,8′R)-form

7-Ketone, 3′,5-di-Me ether
Prunustosanan AIII
[1620571-54-1]

$C_{20}H_{22}O_8$ 390.388
Constit. of the seeds of *Prunus tomentosa*. Oil. $[\alpha]^{20}_{D}$ -6.3 (c, 0.09 in MeOH). λ_{max} 227 (log ε 0.32); 285 (log ε 0.18) (MeOH).

Chakravarty, A.K. *et al.*, *Phytochemistry*, 1996, **41**, 935-939 (*Sisymbrifolin*).
Du, Q. *et al.*, *Nat. Prod. Res.*, 2007, **21**, 451-454 (*Sisymbrifolin*).
Martin, F. *et al.*, *J. Nat. Prod.*, 2009, **72**, 852-856 (*Sisymbrifolin 4′-rhamnoside*).
Filho, J.G.S. *et al.*, *J. Nat. Prod.*, 2009, **72**, 1344-1347 (*Radulignan*).
Chai, X.-Y. *et al.*, *Planta Med.*, 2009, **75**, 1246-1252 (*Flacoside A*).
Fan, S.-Y. *et al.*, *Planta Med.*, 2011, **77**, 1623-1630 (*Platycladus orientalis constit*).
Kato-Noguchi, H. *et al.*, *Acta Physiol. Plant.*, 2012, **34**, 2045-2049 (*Cucumis sativus constit, activity*).
Huang, X-X. *et al.*, *Fitoterapia*, 2013, **91**, 217-223 (*Pinnatifidanins CIII-CVI*).
Wang, L. *et al.*, *Fitoterapia*, 2014, **99**, 92-98 (*Meliasendanins B,C,D*).
Hu, J. *et al.*, *Fitoterapia*, 2014, **99**, 178-183 (*Euterpe oleracea 9′-glucosides*).
Liu, Q.-B. *et al.*, *J. Agric. Food Chem.*, 2014, **62**, 7796-7803 (*Prunustosanans AII,AIII*).
Ma, Q. *et al.*, *J. Funct. Foods*, 2014, **9**, 156-161 (*Litchi chinensis constit*).
Shi, Y.-S. *et al.*, *J. Nat. Prod.*, 2015, **78**, 1526-1535 (*7-Dehydrosisymbrifolins*).
Dong, L.-M. *et al.*, *Molecules*, 2017, **22**, 1140:1-10 (*Mikania micrantha constit, activity*).
Lee, S.R. *et al.*, *Phytochem. Lett.*, 2017, **20**, 98-101 (*Pharbitis nil constit, activity*)

7,7′-Epoxy-8,8′-lignan-3,3′,4,4′,5,5′,9-heptol E69

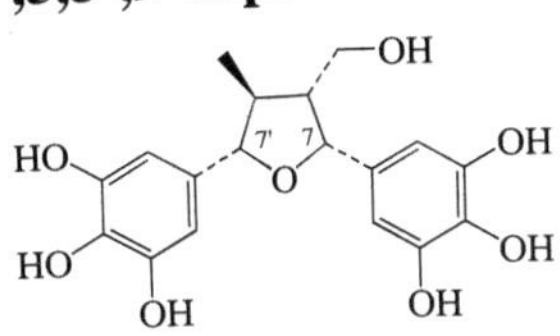

3,3′,4,4′,5,5′,9-Heptahydroxy-7,7′-epoxylignan
$C_{18}H_{20}O_8$ 364.35
(7R*,7′S*,8R*,8′S*)-form
3′,4′-Methylene, 3,4,5,5′-tetra-Me ether
7,7′-Epoxy-3,3′,4,5-tetramethoxy-4′,5′-methylenedioxy-8,8′-lignan-9-ol. 9-Hydroxy-3,3′,4,5-tetramethoxy-4′,5′-methylenedioxy-7,7′-epoxylignan
[1009642-59-4]
$C_{23}H_{28}O_8$ 432.469
Constit. of the aerial parts of *Peperomia blanda*. Pale yellow oil. $[\alpha]^{21}_{D}$ -29.1 (c, 0.2 in MeOH). Incorrect C-8 config. assigned in ref.. λ_{max} 254 (log ε 4.1); 273 (log ε 3.7) (MeOH).

Felippe, L.G. *et al.*, *Phytochemistry*, 2008, **69**, 445-450 (*Peperomia blanda constit*)

7,7′-Epoxy-8,8′-lignan-3,3′,4,4′,5,7′,8′-heptol E70

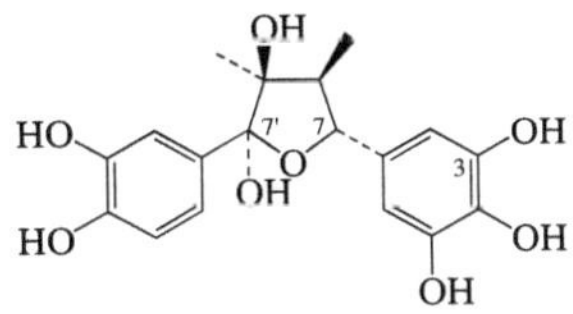

(7R,7′S,8S,8′S)-form

3,3′,4,4′,5,7′,8′-Heptahydroxy-7,7′-epoxylignan
$C_{18}H_{20}O_8$ 364.35
(7R,7′S,8S,8′S)-form
3′,4′-Methylene, 3,4,7′-tri-Me ether
7,7′-Epoxy-4,5,7′-trimethoxy-3′,4′-methylenedioxy-8,8′-lignan-3,8′-diol. 3,8′-Dihydroxy-4,5,7′-trimethoxy-3′,4′-methylenedioxy-7,7′-epoxylignan. **Schisphenlignan H**
[1542234-12-7]
$C_{22}H_{26}O_8$ 418.442
Isol. from the leaves and stems of *Schisandra sphenanthera*. Amorph. powder. $[\alpha]^{27}_{D}$ +51.2 (c, 0.1 in $CHCl_3$). λ_{max} 240 (log ε 3.22) ($CHCl_3$)
3′,4′-Methylene, 3,4,5,7′-tetra-Me ether
7,7′-Epoxy-3,4,5,7′-tetramethoxy-3′,4′-methylenedioxy-8,8′-lignan-8′-ol. 8′-Hydroxy-3,4,5,7′-tetramethoxy-3′,4′-methylenedioxy-7,7′-epoxylignan
[1886973-77-8]
$C_{23}H_{28}O_8$ 432.469
Constit. of *Schisandra sphenanthera*. Yellow oil. $[\alpha]^{25}_{D}$ +50 (c, 0.1 in MeOH). λ_{max} 232 (log ε 0.42) (EtOH)
3,4:3′,4′-Bis(methylene), 7′-Me ether
7,7′-Epoxy-7′-methoxy-3′,4′:4,5-bis(methylenedioxy)-8,8′-lignan-3,8′-diol. 3,8′-Dihydroxy-7′-methoxy-3′,4′:4,5-bis(methylenedioxy)-7,7′-epoxylignan
[1886973-78-9]
$C_{21}H_{22}O_8$ 402.399
Constit. of *Schisandra sphenanthera*. Yellow oil. $[\alpha]^{25}_{D}$ +50 (c, 0.1 in MeOH). λ_{max} 226 (log ε 0.25) (EtOH)
(7S,7′R,8R,8′S)-form
3′,4′-Methylene, 3,4,5,7′-tetra-Me ether
Kadlongirin C
$C_{23}H_{28}O_8$ 432.469
Constit. of the stems of *Schisandra nigra*. Amorph. powder. Mp 132.1 - 133.5°. $[\alpha]^{26}_{D}$ +49.9 (c, 0.73 in MeOH). λ_{max} 206 (log ε 4.92); 234 (log ε 4.15); 282 (log ε 3.72) (MeOH)
(7R*,7′S*,8S*,8′R*)-form
3,3′,4,4′,5,7′-Hexa-Me ether
7,7′-Epoxy-3,3′,4,4′,5,7′-hexamethoxy-8,8′-lignan-8′-ol. 8′-Hydroxy-3,3′,4,4′,5,7′-hexamethoxy-7,7′-epoxylignan.
Kadlongirin A
[1063682-42-7]
$C_{24}H_{32}O_8$ 448.512
Constit. of *Kadsura longipedunculata*. Yellowish oil. $[\alpha]^{28}_{D}$ +46 (c, 0.73 in MeOH). λ_{max} 206 (log ε 5.07); 276 (log ε 3.77); 314 (log ε 2.87); 360 (log ε 2.91) (MeOH).

Pu, J.-X. *et al.*, *Chem. Pharm. Bull.*, 2008, **56**, 1143-1146 (*Kadlongirin A*).
Liang, C.-Q. *et al.*, *Fitoterapia*, 2013, **86**, 171-177 (*Schisphenlignan H*).
Jiang, K. *et al.*, *Fitoterapia*, 2015, **103**, 63-70 (*Schisandra sphenanthera constits, activity, abs config*).
Narukawa, Y. *et al.*, *J. Nat. Med.*, 2016, **70**, 460-466 (*Kadlongirin C*)

7,9′-Epoxy-8,8′-lignan-2′,3,3′,4,4′,7′,9-heptol E71

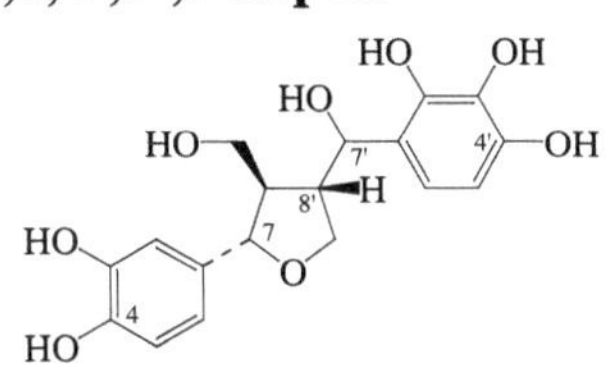

2′,3,3′,4,4′,7′,9-Heptahydroxy-7,9′-epoxylignan
$C_{18}H_{20}O_8$ 364.35
(7S,8R,8′S)-form
7′-Ketone, 3,4:3′,4′-bis(methylene) ether
7,9′-Epoxy-2′,9-dihydroxy-3,4:3′,4′-bis(methylenedioxy)-8,8′-lignan-7′-one. 2′,9-Dihydroxy-3,4:3′,4′-bis(methylenedioxy)-7′-oxo-7,9′-epoxylignan.
2′-Hydroxysesaminone
[1198618-00-6]
$C_{20}H_{18}O_8$ 386.356

Constit. of *Mananthes patentiflora*. Amorph. powder. Mp 123 - 125°. $[\alpha]_D^{21}$ -51.2 (c, 0.9 in $CHCl_3$). λ_{max} 230 (log ε 3.77); 243 (log ε 4.19); 295 (log ε 4.18) (MeOH)

7′-Ketone, 3,4:3′,4′-bis(methylene) ether, 9-O-[β-D-apiofuranosyl-(1→3)-β-D-glucopyranoside]
Mananthoside G
[1198617-98-9]
$C_{31}H_{36}O_{17}$ 680.612
Constit. of *Mananthes patentiflora*. Amorph. powder. Mp 109 - 111°. $[\alpha]_D^{21}$ -53.1 (c, 0.57 in MeOH). λ_{max} 204 (log ε 4.59); 230 (log ε 4.19); 240 (log ε 4.24); 293 (log ε 4.2); 395 (log ε 2.67) (MeOH)

7′-Ketone, 3,4:3′,4′-bis(methylene), 2′-Me ether
7,9′-Epoxy-9-hydroxy-2′-methoxy-3,4:3′,4′-bis(methylenedioxy)-8,8′-lignan-7′-one.
9-Hydroxy-2′-methoxy-3,4:3′,4′-bis(methylenedioxy)-7′-oxo-7,9′-epoxylignan. ***Altissinone***.
2′-Methoxysesaminone
[865147-59-7]
$C_{21}H_{20}O_8$ 400.383
Constit. of the leaves of *Vitex altissima*. Pale green flakes. Mp 151 - 152°. $[\alpha]_D^{25}$ -40.3 (c, 0.5 in $CHCl_3$). λ_{max} 226 (log ε 7) ($CHCl_3$).

Sridhar, C. *et al.*, *Phytochemistry*, 2005, **66**, 1707-1712 (*Altissinone*).
Tian, J. *et al.*, *Planta Med.*, 2009, **75**, 534-537 (*2′-Hydroxysesaminone, Mananthoside G*)

7,9′-Epoxy-8,8′-lignan-3,3′,4,4′,5,5′,9-heptol E72

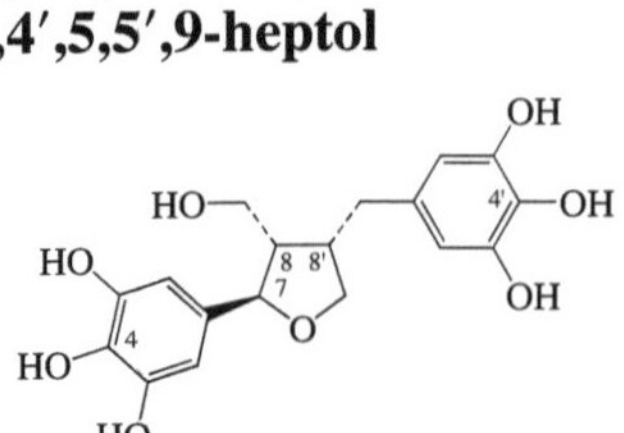

(7*R*,8*S*,8′*S*)-*form*

3,3′,4,4′,5,5′,9-Heptahydroxy-7,9′-epoxylignan
$C_{18}H_{20}O_8$ 364.35

(7R,8S,8′S)-form

3,3′,5,5′-Tetra-Me ether, 9-O-β-D-glucopyranoside
Ligalbumoside A
[833486-02-5]
$C_{28}H_{38}O_{13}$ 582.599
Constit. of *Viscum album*. Powder. $[\alpha]_D^{25}$ -20.2 (c, 0.3 in MeOH)

(7S,8R,8′R)-form

3,3′,5,5′-Tetra-Me ether
7,9′-Epoxy-3,3′,5,5′-tetramethoxy-8,8′-lignan-4,4′,9-triol.
4,4′,9-Trihydroxy-3,3′,5,5′-tetramethoxy-7,9′-epoxylignan.
5,5′-Dimethoxylariciresinol
[116498-58-9]
$C_{22}H_{28}O_8$ 420.458
Isol. from *Bauhinia manca*, *Lindera glauca*, *Campylotropis hirtella* and *Virola elongata*. Cryst.. Mp 124 - 126°. $[\alpha]_D^{21}$ +5 (c, 0.27 in MeOH)

3,3′,5,5′-Tetra-Me ether, 4′-O-β-D-glucopyranoside
Tortoside B. *Manglieside E*
[154418-16-3]
[190655-17-5]
$C_{28}H_{38}O_{13}$ 582.599
Constit. of *Manglietia phuthoensis*, *Phellodendron amurense*, *Helichrysum arenarium*, *Caragana brachyantha* and *Pedicularis torta*. Shows antiosteoporotic and moderate antioxidant activity. Needles (H_2O). Mp 201 - 203°. $[\alpha]_D$ -18.9 (c, 1.1 in Py) (Tortoside B). $[\alpha]_D^{25}$ -92 (c, 1 in MeOH) (Manglieside E)

3,3′,5,5′-Tetra-Me ether, 4,4′-di-O-β-D-glucopyranoside
Salvadoraside
[143522-30-9]
$C_{34}H_{48}O_{18}$ 744.74
Isol. from stems of *Salvadora persica*. Powder. $[\alpha]_D^{23}$ -82 (c, 0.025 in DMSO)

3,3′,5,5′-Tetra-Me ether, 4′-O-[4,6-bis-O-(4-hydroxy-3-methoxy-E-cinnamoyl)-β-D-glucopyranoside]
[947494-35-1]
$C_{48}H_{54}O_{19}$ 934.941
Constit. of the roots of *Rhus javanica* var. *roxburghiana*. Powder. $[\alpha]_D^{27}$ -11.6 (c, 2.1 in Me_2CO). λ_{max} 224 (log ε 4.4); 232 (sh); 288 (log ε 4.1); 301 (log ε 4.03); 330 (log ε 4.1) (MeOH)

3,3′,5,5′-Tetra-Me ether, 4′-O-[β-D-apiofuranosyl-(1→2)-β-D-glucopyranoside]
$C_{33}H_{46}O_{17}$ 714.714
Constit. of *Albizzia julibrissin*. Amorph. powder. $[\alpha]_D^{25}$ -2.7 (c, 0.4 in MeOH)

3,3′,5,5′-Tetra-Me ether, 4,4′-di-O-α-L-rhamnopyranoside
5,5′-Dimethoxylariciresinol 4,4′-dirhamnoside
[123564-47-6]
$C_{34}H_{48}O_{16}$ 712.742
Constit. of the leaves of *Parsonsia laevigata*. Prisms (MeOH). Mp 225 - 230°. $[\alpha]_D^{28}$ -120.7 (c, 1.4 in MeOH)

3,3′,5,5′-Tetra-Me ether, 9-O-β-D-xylopyranoside
Prupaside
[126882-52-8]
$C_{27}H_{36}O_{12}$ 552.573
Isol. from *Prunus padus* and *Asclepias syriaca*. Powder. $[\alpha]_D^{24}$ -38 (c, 0.87 in MeOH)

3,3′,5,5′-Tetra-Me ether, 9-O-β-D-glucopyranoside
Alangilignoside C
[191481-00-2]
$C_{28}H_{38}O_{13}$ 582.599
Constit. of *Alangium premnifolium*, *Tabernaemontana cymosa*, *Strobilanthes cusia*, *Capparis flavicans* and *Viscum album*. Oil or needles (MeOH). $[\alpha]_D^{25}$ +21.5 (c, 0.3 in MeOH). λ_{max} 208 (log ε 4.95); 237 (log ε 4.24); 271 (log ε 3.52) (MeOH)

3,3′,5,5′-Tetra-Me ether, 9-O-[4-hydroxy-E-cinnamoyl-(→6)-β-D-glucopyranoside]
[1062606-18-1]
$C_{37}H_{44}O_{15}$ 728.744
Constit. of *Asclepias curassavica*. Amorph. powder. $[\alpha]_D^{21}$ +18 (c, 0.26 in MeOH). λ_{max} 207 (log ε 4.9); 222 (sh); 312 (log ε 4.26) (MeOH)

3,3′,5,5′-Tetra-Me ether, 9-O-[4-hydroxy-3-methoxy-E-cinnamoyl-(→6)-β-D-glucopyranoside]
[1062606-10-3]
$C_{38}H_{46}O_{16}$ 758.77
Constit. of *Asclepias curassavica*. Amorph. powder. $[\alpha]_D^{22}$ +15.6 (c, 1.2 in MeOH). λ_{max} 209 (log ε 4.71); 233 (log ε 4.26); 284 (sh); 299 (sh); 326 (log ε 4.12) (MeOH)

3,3′,5,5′-Tetra-Me ether, 9-O-[4-hydroxy-3-methoxy-Z-cinnamoyl-(→6)-β-D-glucopyranoside]
[1062606-14-7]
$C_{38}H_{46}O_{16}$ 758.77
Constit. of *Asclepias curassavica*. Amorph. powder. $[\alpha]_D^{22}$ +6.2 (c, 0.6 in MeOH). λ_{max} 210 (log ε 4.67);

230 (sh); 282 (log ε 3.87); 324 (log ε 4.04) (MeOH)

3,3′,5,5′-Tetra-Me ether, 9-O-[4-hydroxy-3,5-dimethoxy-E-cinnamoyl-(→6)-β-D-glucopyranoside]
[1062606-23-8]
$C_{39}H_{48}O_{17}$ 788.796
Constit. of *Asclepias curassavica*. Amorph. powder. $[\alpha]^{21}_{D}$ +17 (c, 0.32 in MeOH). λ_{max} 212 (log ε 5.25); 233 (sh); 283 (sh); 327 (log ε 4.26) (MeOH)

3,3′,5,5′-Tetra-Me ether, 9-O-(4-hydroxy-E-cinnamoyl)
9-O-trans-p-Coumaroyl-5,5′-dimethoxylariciresinol
[1383686-29-0]
$C_{31}H_{34}O_{10}$ 566.603
Constit. of the stems of *Cipadessa baccifera*. Amorph. powder. $[\alpha]^{25}_{D}$ +5.3 (c, 0.4 in MeOH)

3,3′,5,5′-Tetra-Me ether, 9-O-(4-hydroxy-3-methoxy-E-cinnamoyl)
9-O-trans-Feruloyl-5,5′-dimethoxylariciresinol
[166322-14-1]
$C_{32}H_{36}O_{11}$ 596.629
Constit. of *Lindera obtusiloba*, *Zanthoxylum ailanthoides*, *Zanthoxylum avicennae* and *Lindera glauca*. Shows moderate cytotoxic activity against several human cancer cell lines. Inhibitor of NO prodn. in LPS-activated cells. Amorph. powder. $[\alpha]^{25}_{D}$ -27.7 (c, 0.01 in MeOH). $[\alpha]^{25}_{D}$ +20.8 (c, 0.1 in MeOH). One isolation incorrectly reported as from *Aralia bipinnata*, corrected to *Z. ailanthoides* in an erratum. Opposite opt. rotns. reported (from two different *Zanthoxylum* spp.). May be an error or 2 different enantiomers. λ_{max} 286 (log ε 3.85); 298 (log ε 3.89); 324 (log ε 4.03) (MeOH)

3,3′,5,5′-Tetra-Me ether, 9-O-(4-hydroxy-3-methoxy-Z-cinnamoyl)
9-O-cis-Feruloyl-5,5′-dimethoxylariciresinol
[1002753-45-8]
$C_{32}H_{36}O_{11}$ 596.629
Constit. of the stems of *Zanthoxylum avicennae* and *Lindera glauca*. Shows moderate cytotoxic activity against several human cancer cell lines. Inhibitor of NO prodn. in LPS-activated cells. Amorph. powder. $[\alpha]^{25}_{D}$ +23.4 (c, 0.12 in MeOH). λ_{max} 287 (log ε 3.82); 299 (log ε 3.89); 325 (log ε 4.04) (MeOH)

3,3′,4′,5,5′-Penta-Me ether
7,9′-Epoxy-3,3′,4′,5,5′-pentamethoxy-8,8′-lignan-4,9-diol. 4,9-Dihydroxy-3,3′,4′,5,5′-pentamethoxy-7,9′-epoxylignan

3,3′,4′,5,5′-Penta-Me ether, 9-O-(4-hydroxy-3-methoxy-E-cinnamoyl)
Linderuca A
[1829588-29-5]
$C_{33}H_{38}O_{11}$ 610.656
Constit. of the twigs of *Lindera glauca*. Exhibits antiproliferative activity against A549, SK-OV-3, SK-MEL-2 and HCT-15 tumour cell lines; inhibits NO prodn.. Light yellow gum. $[\alpha]^{25}_{D}$ +42.3 (c, 0.5 in MeOH). λ_{max} 230 (log ε 3.7); 286 (log ε 4.0); 298 (log ε 3.3); 324 (log ε 4.5) (MeOH)

3,3′,4′,5,5′-Penta-Me ether, 9-O-(4-hydroxy-3-methoxy-Z-cinnamoyl)
Linderuca B
[1829587-95-2]
$C_{33}H_{38}O_{11}$ 610.656
Constit. of the twigs of *Lindera glauca*. Exhibits antiproliferative activity against A549, SK-OV-3, SK-MEL-2 and HCT-15 tumour cell lines; inhibits NO prodn.. Light yellow gum. $[\alpha]^{25}_{D}$ +42.3 (c, 0.5 in MeOH). λ_{max} 230 (log ε 3.8); 287 (log ε 3.8); 299 (log ε 3.6); 325 (log ε 4.1) (MeOH)

3,3′,4,4′,5,5′-Hexa-Me ether
7,9′-Epoxy-3,3′,4,4′,5,5′-hexamethoxy-8,8′-lignan-9-ol. 9-Hydroxy-3,3′,4,4′,5,5′-hexamethoxy-7,9′-epoxylignan

3,3′,4,4′,5,5′-Hexa-Me ether, 9-angeloyl
5,5′-Dimethoxyleoligin
$C_{29}H_{38}O_{9}$ 530.614
Constit. of the roots of *Leontopodium alpinum*. Gum. $[\alpha]^{20}_{D}$ +21.4 (c, 0.5 in MeOH)

3,4-Methylene, 3′,4′,5,5′-tetra-Me ether
7,9′-Epoxy-3,3′,4′,5′-tetramethoxy-4,5-methylenedioxy-8,8′-lignan-9-ol. 9-Hydroxy-3,3′,4′,5′-tetramethoxy-4,5-methylenedioxy-7,9′-epoxylignan. ***Dihydrosesartemin***
[94608-21-6]
$C_{23}H_{28}O_{8}$ 432.469
Isol. from *Virola elongata*. Oil. $[\alpha]^{25}_{D}$ +11.8 ($CHCl_3$)

3,4:3′,4′-Bis(methylene), 5,5′-di-Me ether
7,9′-Epoxy-3,3′-dimethoxy-4,5:4′,5′-bis(methylenedioxy)-8,8′-lignan-9-ol. 9-Hydroxy-3,3′-dimethoxy-4,5:4′,5′-bis(methylenedioxy)-7,9′-epoxylignan
[929692-67-1]
$C_{22}H_{24}O_{8}$ 416.426
Constit. of *Peperomia duclouxii*. Gum. $[\alpha]^{25}_{D}$ +1.6 (c, 0.25 in $CHCl_3$). λ_{max} 244 (log ε 2.9); 276 (log ε 3.9) ($CHCl_3$).

MacRae, W.D. *et al.*, *Phytochemistry*, 1985, **24**, 561-566 (*Dihydrosesartemin*).
Achenbach, H. *et al.*, *Phytochemistry*, 1988, **27**, 1835-1841 (*5,5′-Dimethoxylariciresinol*).
Yoshinari, K. *et al.*, *Chem. Pharm. Bull.*, 1989, **37**, 3301-3303 (*Prupaside*).
Abe, F. *et al.*, *Phytochemistry*, 1989, **28**, 1737-1741 (*5,5′-Dimethoxylariciresinol 4,4′-dirhamnoside*).
Kinjo, J. *et al.*, *Chem. Pharm. Bull.*, 1991, **39**, 2952-2955 (*3,3′,5,5′-tetra-Me ether 4′-apiosylglucoside*).
Kamel, M.S. *et al.*, *Phytochemistry*, 1992, **31**, 2469-2471 (*Salvadoraside*).
Ida, Y. *et al.*, *Phytochemistry*, 1994, **35**, 209-215 (*3,3′,5,5′-tetra Me ether 4′-glucoside*).
Hsiao, J.-J. *et al.*, *Phytochemistry*, 1995, **39**, 899-902 (*Feruloyldimethoxylariciresinol*).
Changzeng, W. *et al.*, *Phytochemistry*, 1997, **45**, 159-166 (*Tortoside B*).
Achenbach, H. *et al.*, *Phytochemistry*, 1997, **45**, 325-335 (*Alangilignoside C*).
Yuasa, K. *et al.*, *Phytochemistry*, 1997, **45**, 611-615 (*Alangilignoside C*).
Kwon, H.C. *et al.*, *Arch. Pharmacal Res.*, 1999, **22**, 417-422 (*Feruloyldimethoxylariciresinol, activity*).
Schwaiger, S. *et al.*, *Planta Med.*, 2004, **70**, 978-985 (*5,5′-Dimethoxyleoligin*).
Ouyang, M.-A. *et al.*, *Helv. Chim. Acta*, 2007, **90**, 1099-1106 (*Rhus diferuloylglucoside*).
Li, N. *et al.*, *J. Nat. Prod.*, 2007, **70**, 544-548 (*bismethylene 5,5′-di-Me ether*).
Warashina, T. *et al.*, *Chem. Pharm. Bull.*, 2008, **56**, 1159-1163 (*Asclepias curassavica constits*).
Kiem, P.V. *et al.*, *Chem. Pharm. Bull.*, 2008, **56**, 1270-1275 (*Manglieside E*).
Chen, J.-J. *et al.*, *J. Nat. Prod.*, 2008, **71**, 212-217 (*9-Feruloyldimethoxylariciresinols*).
Ning, J. *et al.*, *Yunnan Zhiwu Yanjiu*, 2010, **32**, 466-468 (*Cipadessa baccifera constit*).
Nhiem, N.X. *et al.*, *Magn. Reson. Chem.*, 2012, **50**, 772-777 (*Ligalbumoside A, Alangilignoside C*).
Kim, K.H. *et al.*, *Chem. Pharm. Bull.*, 2014, **62**, 1136-1140 (*Linderucas A,B, cd, activity*).
Perveen , S. *et al.*, *J. Chem. Res.*, 2014, **38**, 46-49 (*Tortoside B, activity*)

7,9′-Epoxy-8,8′-lignan-3,3′,4,4′,5,7′,9-heptol E73

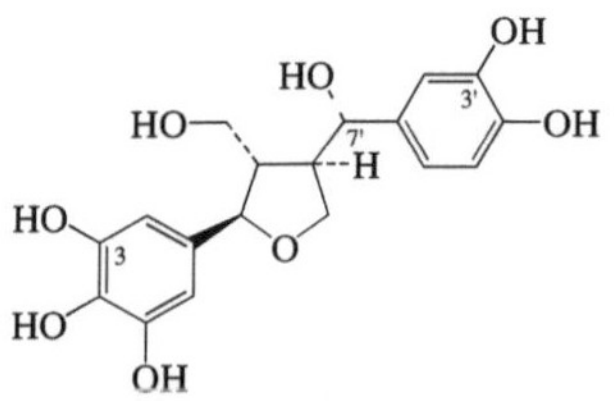

(7R,7′S,8S,8′R)-form

3,3′,4,4′,5,7′,9-Heptahydroxy-7,9′-epoxylignan
$C_{18}H_{20}O_8$ 364.35
(7R,7′S,8S,8′R)-form
3,3′,4,4′,5-Penta-Me ether
7,9′-Epoxy-3,3′,4,4′,5-pentamethoxy-8,8′-lignan-7′,9-diol. 7′,9-Dihydroxy-3,3′,4,4′,5-pentamethoxy-7,9′-epoxylignan
[1158256-43-9]
$C_{23}H_{30}O_8$ 434.485
Constit. of the buds of *Magnolia fargesii*. Superoxide radical scavenger. Oil. $[\alpha]^{20}_D$ -33.6 (c, 0.4 in $CHCl_3$). λ_{max} 228 (log ε 4.06); 278 (log ε 3.4) (MeOH)
(7R,7′ξ,8S,8′S)-form
3′,4′-Methylene, 3,4,5-tri-Me ether
7,9′-Epoxy-3,4,5-trimethoxy-3′,4′-methylenedioxy-8,8′-lignan-7′,9-diol. 7′,9-Dihydroxy-3,4,5-trimethoxy-3′,4′-methylenedioxy-7,9′-epoxylignan. ***Sieversol***
[215037-80-2]
$C_{22}H_{26}O_8$ 418.442
Constit. of *Artemisia sieversiana*. Gum
(7S,7′R,8R,8′R)-form
7′-Ketone
7,9′-Epoxy-3,3′,4,4′,5,9-hexahydroxy-8,8′-lignan-7′-one. 3,3′,4,4′,5,9-Hexahydroxy-7′-oxo-7,9′-epoxylignan
7′-Ketone, 3,3′,4,4′,5-penta-Me ether
7,9′-Epoxy-9-hydroxy-3,3′,4,4′,5-pentamethoxy-8,8′-lignan-7′-one. 9-Hydroxy-3,3′,4,4′,5-pentamethoxy-7′-oxo-7,9′-epoxylignan. ***Magnone B***
[209904-70-1]
$C_{23}H_{28}O_8$ 432.469
Constit. of the flower buds of *Magnolia fargesii*. Platelet activating factor antagonist. Shows superoxide radical scavenging activity. Aldose reductase inhibitor. Shows antiinflammatory activity. Oil. $[\alpha]^{25}_D$ -2.8 (c, 1.2 in MeOH). λ_{max} 226 (log ε 4.47); 274 (log ε 4.16); 303 (log ε 3.98) (MeOH)
(7S,7′R,8R,8′S)-form
3,3′,5-Tri-Me ether
7,9′-Epoxy-3,3′,5-trimethoxy-8,8′-lignan-4,4′,7′,9-tetrol. 4,4′,7′,9-Tetrahydroxy-3,3′,5-trimethoxy-7,9′-epoxylignan
3,3′,5-Tri-Me-ether, 7′-O-β-D-glucopyranoside
Tanegoside C
[131653-22-0]
$C_{27}H_{36}O_{13}$ 568.572
Isol. from whole plant of *Trachelospermum liukiuense*. $[\alpha]^{29}_D$ -32.6 (c, 1.5 in MeOH)
7′-Ketone, 3′,4′-methylene, 3,4,5-tri-Me ether, 9-O-β-D-glucopyranoside
Terminaloside Q
[2222336-15-2]
$C_{28}H_{34}O_{13}$ 578.567
Constit. of *Terminalia citrina*. Amorph. powder. $[\alpha]^{25}_D$ +16 (c, 0.2 in MeOH). λ_{max} 212 (log ε 4.37); 227 (log ε 4.38); 272 (log ε 3.94); 306 (log ε 3.95) (MeOH).

Abe, F. *et al., Chem. Pharm. Bull.*, 1990, **38**, 2143-2145 (*Tanegoside C*).
Jung, K.Y. *et al., J. Nat. Prod.*, 1998, **61**, 808-811 (*Magnone B*).
Tan, R.X. *et al., Phytochemistry*, 1998, **49**, 157-161 (*Sieversol*).
Ahn, K-S. *et al., Biol. Pharm. Bull.*, 2001, **24**, 1085-1087 (*Magnone B, antiinflammatory activity*).
Lee, J. *et al., Chem. Pharm. Bull.*, 2009, **57**, 298-301 (*Magnolia fargesii constits, radical scavenging activity, aldose reductase inhibitory activity*).
Muhit, M.A. *et al., Phytochemistry*, 2018, **145**, 161-167 (*Terminaloside Q*)

7,9′-Epoxy-8,8′-lignan-3,3′,4,4′,5,8,9-heptol E74

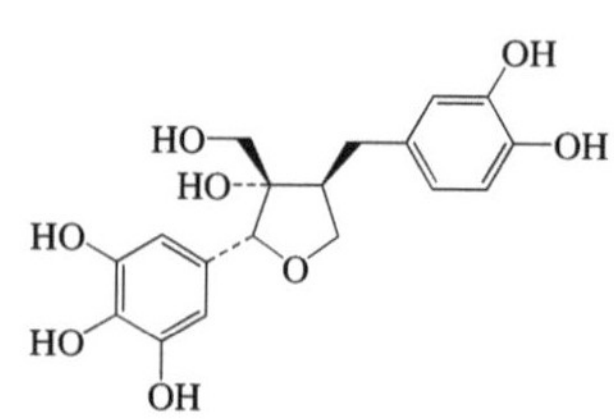

3,3′,4,4′,5,8,9-Heptahydroxy-7,9′-epoxylignan
$C_{18}H_{20}O_8$ 364.35
(7R,8S,8′S)-form
3,3′,5-Tri-Me ether
7,9′-Epoxy-3,3′,5-trimethoxy-8,8′-lignan-4,4′,8,9-tetrol. 4,4′,8,9-Tetrahydroxy-3,3′,5-trimethoxy-7,9′-epoxylignan.
Gentioluteol
[1002101-84-9]
$C_{21}H_{26}O_8$ 406.431
Constit. of the roots of *Gentiana lutea*.

Ando, H. *et al., J. Nat. Med.*, 2007, **61**, 269-279 (*Gentioluteol*)

7,9′-Epoxy-8,8′-lignan-3,3′,4,4′,7,8′,9-heptol E75

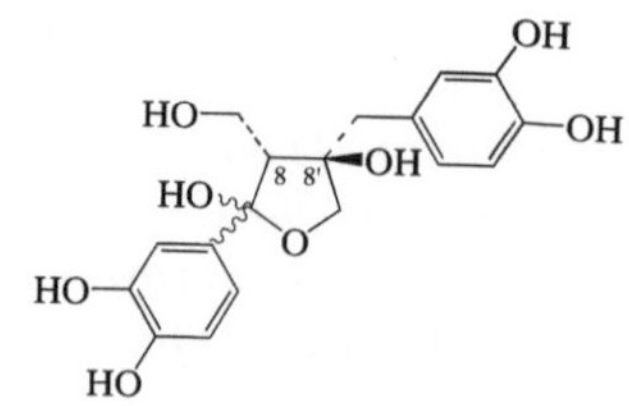

3,3′,4,4′,7,8′,9-Heptahydroxy-7,9′-epoxylignan
$C_{18}H_{20}O_8$ 364.35
(7ξ,8R*,8′R*)-form
3,3′-Di-Me ether
7,9′-Epoxy-3,3′-dimethoxy-8,8′-lignan-4,4′,7,8′,9-pentol. 4,4′,7,8′,9-Pentahydroxy-3,3′-dimethoxy-7,9′-epoxylignan.
7-Hydroxyolivil
[133561-40-7]
$C_{20}H_{24}O_8$ 392.404
Isol. from *Carduus assoi*. Noncryst..

Fernandez, I. *et al., Phytochemistry*, 1991, **30**, 1030-1032 (*7-Hydroxyolivil*)

7,9′-Epoxy-8,8′-lignan-3,3′,4,4′,7′,8′,9-heptol E76

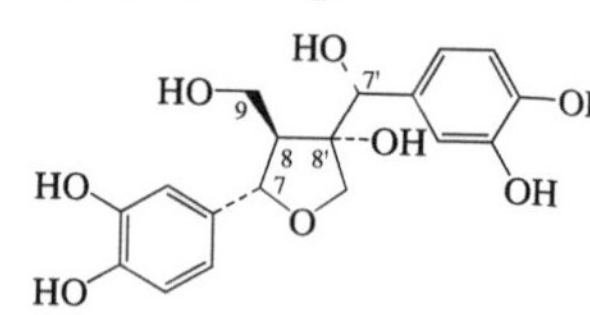

(7S,7′R,8R,8′S)-form

3,3′,4,4′,7′,8′,9-Heptahydroxy-7,9′-epoxylignan
$C_{18}H_{20}O_8$ 364.350
(7S,7′R,8R,8′S)-form
7′,9-Isopropylidene, 3,4:3′,4′-bis(methylene) ether
Premnan C
$C_{23}H_{24}O_8$ 428.437
Constit. of the wood of *Premna serratifolia*. Exhibits melanogenesis-enhancing activity. Amorph. pale yellow solid. $[\alpha]^{25}_D$ +4.7 (c, 0.1 in MeOH). Possible artifact. λ_{max}

245 (log ε 3.67); 285 (log ε 3.74) (MeOH)

(7R*,7′R*,8S*,8′R*)-form

3,4:3′,4′-Bis(methylene) ether

7,9′-Epoxy-3,4:3′,4′-bis(methylenedioxy)-8,8′-lignan-7′,8′,9-triol. 7′,8′,9-Trihydroxy-3,4:3′,4′-bis(methylenedioxy)-7,9′-epoxylignan. ***Taungtangyiol A***

$C_{20}H_{20}O_8$ 388.372

Constit. of the wood of *Premna integrifolia*. Melanogenesis inhibitor. Plates ($CHCl_3$). Mp 165 - 170°. $[\alpha]^{25}_{D}$ -139 (c, 0.1 in MeOH). λ_{max} 236 (log ε 3.97); 286 (log ε 4.02) (MeOH).

Win, N.N. *et al., Fitoterapia*, 2018, **127**, 308-313 (*Taungtangyiol A, cryst struct, activity*).
Woo, S.-Y. *et al., Fitoterapia*, 2019, **133**, 35-42 (*Premnan C, activity*)

7,9′-Epoxy-8,8′-lignan-3,3′,4,4′,7′,8,9-heptol E77

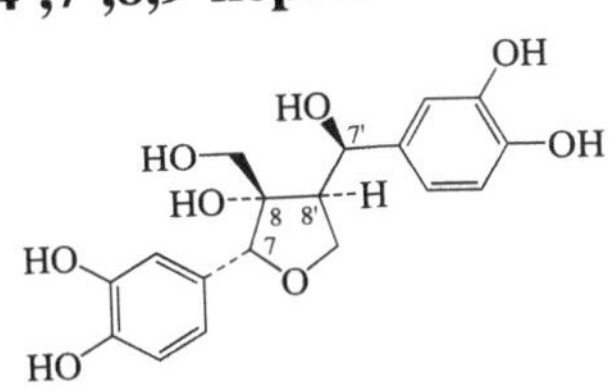

3,3′,4,4′,7′,8,9-Heptahydroxy-7,9′-epoxylignan

$C_{18}H_{20}O_8$ 364.350

(7R*,7′R*,8S*,8′R*)-form

3,4:3′,4′-Bis(methylene) ether

7,9′-Epoxy-3,4:3′,4′-bis(methylenedioxy)-8,8′-lignan-7′,8,9-triol. 7′,8,9-Trihydroxy-3,4:3′,4′-bis(methylenedioxy)-7,9′-epoxylignan. ***Taungtangyiol B***

$C_{20}H_{20}O_8$ 388.372

Constit. of the wood of *Premna integrifolia*. Melanogenesis inhibitor. Oil. $[\alpha]^{25}_{D}$ -68 (c, 0.1 in MeOH). λ_{max} 236 (log ε 3.97); 286 (log ε 3.96) (MeOH).

Win, N.N. *et al., Fitoterapia*, 2018, **127**, 308-313 (*Taungtangyiol B, activity*)

7,9′-Epoxy-8,8′-lignan-3,3′,4,4′,8,8′,9-heptol E78

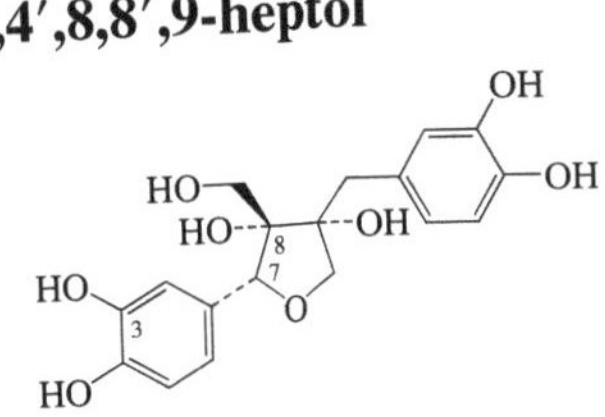

3,3′,4,4′,8,8′,9-Heptahydroxy-7,9′-epoxylignan

$C_{18}H_{20}O_8$ 364.35

(7R,8S,8′R)-form

3,3′-Di-Me ether

7,9′-Epoxy-3,3′-dimethoxy-8,8′-lignan-4,4′,8,8′,9-pentol. 4,4′,8,8′,9-Pentahydroxy-3,3′-dimethoxy-7,9′-epoxylignan. ***Massoniresinol****. Vladinol A*

[96087-10-4]

$C_{20}H_{24}O_8$ 392.404

Constit. of *Pinus massoniana*, *Vladimiria souliei*, *Vladimiria muliensis* and *Abies spectabilis*. Shows moderate free radical scavenging activity. Amorph.. $[\alpha]^{22}_{D}$ -31.4 (c, 0.79 in MeOH)

3,3′-Di-Me ether, 4-O-β-D-glucopyranoside

[448939-03-5]

$C_{26}H_{34}O_{13}$ 554.545

Constit. of the leaves of *Sambucus sieboldiana*. Amorph. powder. $[\alpha]^{26}_{D}$ -63.8 (c, 0.19 in MeOH). λ_{max} 206 (log ε 4.35); 227 (log ε 4.12); 277 (log ε 3.66) (MeOH)

3,3′-Di-Me ether, 4′-O-β-D-glucopyranoside

[148245-84-5]

$C_{26}H_{34}O_{13}$ 554.545

Constit. of *Saussurea lappa*, *Sambucus sieboldiana* and *Valeriana officinalis*. Powder. $[\alpha]_{D}$ -67 (c, 1 in MeOH).

Shen, Z. *et al., Phytochemistry*, 1985, **24**, 364-365 (*Massoniresinol*).
Tan, R.X. *et al., Planta Med.*, 1990, **56**, 475-477 (*Vladinol A*).
Matsuda, H. *et al., Tetrahedron*, 2000, **56**, 7763-7777 (*Massoniresinol 4′-glucoside*).
Machida, K. *et al., Chem. Pharm. Bull.*, 2002, **50**, 669-671 (*Sambucus sieboldiana constits*).
Chen, J-J. *et al., Planta Med.*, 2013, **79**, 1470-1473 (*Vladimiria muliensis constit, activity*)

9,9′-Epoxy-8,8′-lignan-3,3′,4,4′,5,5′,7-heptol E79

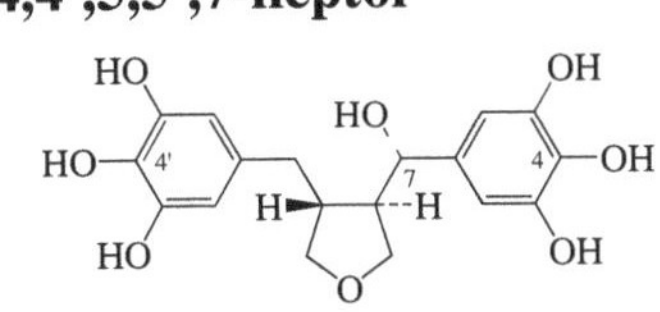

(7S,8R,8′R)-form

3,3′,4,4′,5,5′,7-Heptahydroxy-9,9′-epoxylignan

$C_{18}H_{20}O_8$ 364.35

(7S,8R,8′R)-form

3,3′,5,5′-Tetra-Me ether

9,9′-Epoxy-3,3′,5,5′-tetramethoxy-8,8′-lignan-4,4′,7-triol. 4,4′,7-Trihydroxy-3,3′,5,5′-tetramethoxy-9,9′-epoxylignan

[2126807-05-2]

$C_{22}H_{28}O_8$ 420.458

Constit. of *Phaeanthus vietnamensis*. Amorph. powder. $[\alpha]^{25}_{D}$ -10 (c, 0.01 in $CHCl_3$)

(7ξ,8R*,8′R*)-form

3,3′,5,5′-Tetra-Me ether

Tripterygiol

[940909-08-0]

$C_{22}H_{28}O_8$ 420.458

Constit. of *Euonymus acanthocarpus* and *Tripterygium wilfordii*. Suppressor of proinflammatory gene expression. Amorph. yellow powder. $[\alpha]^{25}_{D}$ +48.3 (c, 0.001 in MeOH). λ_{max} 208; 238; 272 (MeCN).

Ma, J. *et al., Phytochemistry*, 2007, **68**, 1172-1178 (*Tripterygiol*).
Zhu, J.X. *et al., Arch. Pharmacal Res.*, 2012, **35**, 1739-1747 (*Tripterygiol*).
Nhiem, N.X. *et al., Chem. Biodiversity*, 2017, **14**, e1700013:1-7 (*Phaeanthus vietnamensis constit*)

4,7′-Epoxy-3,8′-lignan-3′,4′,5,5′,9,9′-hexol E80

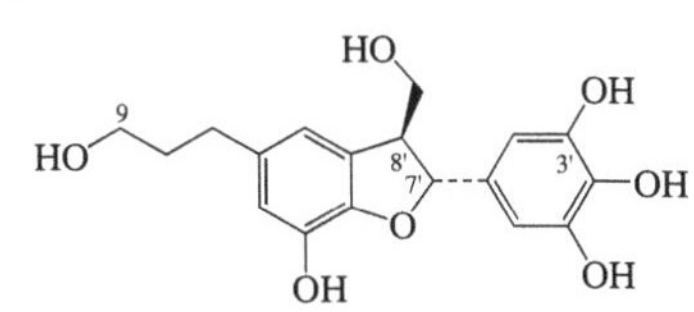

(7′R,8′S)-form

2,3-Dihydro-7-hydroxy-3-(hydroxymethyl)-2-(3,4,5-trihydroxyphenyl)-5-benzofuranpropanol. 4,7′-Epoxy-3′,4′,5,5′,9,9′-hexahydroxy-3,8′-neolignan

$C_{18}H_{20}O_7$ 348.351

Numerous derivs. have been isol. with unspecified stereochem. - some as mixts. of stereoisomers

(7′R,8′S)-form

3′,5-Di-Me ether

4,7′-Epoxy-5,5′-dimethoxy-3,8′-lignan-3′,4′,9,9′-tetrol. 4,7′-Epoxy-3′,4′,9,9′-tetrahydroxy-5,5′-dimethoxy-3,8′-lignan

$C_{20}H_{24}O_7$ 376.405

Isol. from Pu-erh ripe tea prod. from *Camellia sinensis*. Amorph.

powder. $[\alpha]_D^{20}$ -0.8 (c, 0.1 in MeOH). λ_{max} 208 (log ε 2.73); 254 (log ε 2.34) (MeOH)

3′,5′-Di-Me ether

4,7′-Epoxy-3′,5′-dimethoxy-3,8′-lignan-4′,5,9,9′-tetrol. 4,7′-Epoxy-4′,5,9,9′-tetrahydroxy-3′,5′-dimethoxy-3,8′-lignan. ***(+)-Phyllanglaucin A***

$C_{20}H_{24}O_7$ 376.405

Constit. of *Phyllanthus glaucus*. Gum. $[\alpha]_D^{20}$ +12.8 (c, 0.06 in MeOH). λ_{max} 211 (log ε 4.65); 230 (log ε 4.24); 282 (log ε 3.62) (MeOH)

3′,5,5′-Tri-Me ether, 4′-O-[4-hydroxy-3,5-dimethoxybenzoyl-(→6)-β-D-glucopyranoside]

Pittogoside A

[1426258-71-0]

$C_{36}H_{44}O_{16}$ 732.732

Constit. of the roots of *Pittosporum glabratum*. Amorph. powder. $[\alpha]_D^{20}$ -9 (c, 0.9 in MeOH aq.). λ_{max} 206 (log ε 4.28); 243 (sh); 279 (log ε 3.44) (MeOH aq.)

3′,5,5′-Tri-Me ether, 4′-O-[β-D-glucopyranosyl-(1→4)-β-D-glucopyranoside], 9-O-β-D-glucopyranoside

$C_{39}H_{56}O_{22}$ 876.855

Constit. of *Securidaca inappendiculata*. Yellow solid (MeOH). $[\alpha]_D^{25}$ -16.5 (c, 0.1 in MeOH). λ_{max} 279 (log ε 3.44) (MeOH)

3′,5,5′-Tri-Me ether, 9-O-β-D-glucopyranoside

Tortoside E

[176903-29-0]

$C_{27}H_{36}O_{12}$ 552.573

Constit. of *Eurya japonica* and *Pedicularis torta*. $[\alpha]_D^{14}$ +3.3 (c, 0.6 in MeOH)

3′,5,5′-Tri-Me ether, 9′-O-β-D-xylopyranoside

Linderanoside C

[1809032-80-1]

$C_{26}H_{34}O_{11}$ 522.547

Constit. of the trunk of *Lindera glauca*. Exhibits moderate to weak cytotoxicity against A498 cells. Gum. $[\alpha]_D^{25}$ -6.0 (c, 0.30 in MeOH). λ_{max} 210 (log ε 4.3); 288 (log ε 3.2) (MeOH)

3′,5,5′-Tri-Me ether, 9,9′-bis-O-(4-hydroxy-3-methoxy-E-cinnamoyl)

Boehmenan D

[171674-62-7]

$C_{41}H_{42}O_{13}$ 742.774

Constit. of *Drypetes littoralis, Hibiscus cannabinus* and *Ochroma lagopus*. Yellowish oil

4′-O-[2-Hydroxy-1-(hydroxymethyl)ethyl], 3′,5,5′-tri-Me ether

4,7′-Epoxy-4′-(2-glycyryloxy)-3′,5,5′-trimethoxy-3,8′-lignan-9,9′-diol. 4,7′-Epoxy-4′-(2-glycyryloxy)-9,9′-dihydroxy-3′,5,5′-trimethoxy-3,8′-lignan. ***(+)-Sakuraresinol***

$C_{24}H_{32}O_9$ 464.511

Constit. of the sap of *Acer saccharum*

4′-O-[2-Hydroxy-1-(hydroxymethyl)ethyl], 3′,5,5′-tri-Me ether, 9′-O-β-D-glucopyranoside

Prunustosananside AII

[1620571-57-4]

$C_{30}H_{42}O_{14}$ 626.652

Constit. of the seeds of *Prunus tomentosa*. Oil. $[\alpha]_D^{20}$ -19.3 (c, 0.21 in MeOH). λ_{max} 231 (log ε 0.48); 279 (log ε 0.18) (MeOH)

4′-O-[2-Hydroxy-1ξ-(methoxycarbonyl)ethyl], 3′,5,5′-tri-Me ether

Dunnianeolignan A

[1453807-97-0]

$C_{25}H_{32}O_{10}$ 492.521

Constit. of the roots of *Illicium dunnianum*. Syrup. $[\alpha]_D^{20}$ +31.8 (c, 0.004 in MeOH). λ_{max} 237 (log ε 4.67); 281 (log ε 4.15) (MeOH)

(7′S,8′R)-form

3′,5′-Di-Me ether

(-)-Phyllanglaucin A. Rubuslin D. Salicifoneoliganol

[2089652-54-8]

$C_{20}H_{24}O_7$ 376.405

Constit. of *Phyllanthus glaucus, Rubus idaeus* and *Spiraea salicifolia*. Antioxidant. Exhibits weak neuroprotective activity. Yellow oil. $[\alpha]_D^{20}$ -14.1 (c, 0.19 in MeOH). λ_{max} 211 (log ε 4.65); 230 (log ε 4.24); 282 (log ε 3.62) (MeOH)

3′,5,5′-Tri-Me ether

4,7′-Epoxy-3′,5,5′-trimethoxy-3,8′-lignan-4′,9,9′-triol. 4,7′-Epoxy-4′,9,9′-trihydroxy-3′,5,5′-trimethoxy-3,8′-lignan

[873077-50-0]

$C_{21}H_{26}O_7$ 390.432

Constit. of *Selaginella moellendorffii*. Amorph. solid. $[\alpha]_D^{16}$ -3.9 (c, 0.8 in MeOH). λ_{max} 208 (log ε 4.39); 284 (log ε 3.44) (no solvent reported)

3′,5,5′-Tri-Me ether, 4′-O-β-D-xylopyranoside

[1428904-02-2]

$C_{26}H_{34}O_{11}$ 522.547

Constit. of *Eurya japonica*. Powder. Mp 218°. $[\alpha]_D^{25}$ -19.1 (c, 0.2 in MeOH). λ_{max} 230 (sh); 280 (log ε 2.98) (MeOH)

3′,5,5′-Tri-Me ether, 4′-O-β-D-glucopyranoside

[1159495-72-3]

$C_{27}H_{36}O_{12}$ 552.573

Constit. of *Selaginella moellendorffii*. Amorph. solid. $[\alpha]_D^{26}$ -19.3 (c, 0.2 in MeOH). λ_{max} 208 (log ε 4.26); 291 (log ε 3.52); 377 (log ε 3) (MeOH)

3′,5,5′-Tri-Me ether, 9′-O-β-D-xylopyranoside

Litseasin A

[2274645-14-4]

$C_{26}H_{34}O_{11}$ 522.547

Constit. of the root bark of *Litsea glutinosa*. Amorph. powder (MeOH). $[\alpha]_D^{25}$ -29 (c, 0.1 in MeOH). λ_{max} 280 (log ε 3.85) (MeOH)

3′,5,5′-Tri-Me ether, 9′-O-β-D-glucopyranoside

Prunustosananside AI

[1620571-56-3]

$C_{27}H_{36}O_{12}$ 552.573

Constit. of the seeds of *Prunus tomentosa*. Oil. $[\alpha]_D^{20}$ -7.5 (c, 0.19 in MeOH). λ_{max} 230 (log ε 1.54); 281 (log ε 0.57) (MeOH)

4′-O-[2-Hydroxy-1-(hydroxymethyl)ethyl], 3′,5,5′-tri-Me ether

(-)-Sakuraresinol

Powder. $[\alpha]_D^{26}$ -14.1 (c, 0.2 in MeOH). λ_{max} 240 (sh); 281 (log ε 3.2) (EtOH)

4′-O-[2-Hydroxy-1-(hydroxymethyl)ethyl], 3′,5,5′-tri-Me ether, 9′-O-β-D-glucopyranoside

Sakuraresinoside

[623584-29-2]

$C_{30}H_{42}O_{14}$ 626.652

Constit. of the stem bark of *Acer nikoense*. Powder. $[\alpha]_D^{22}$ -8.8 (c, 0.2 in MeOH). λ_{max} 235 (sh); 281 (log ε 3.58) (EtOH)

9-Carboxylic acid, 3′,5,5′-tri-Me ether
4,7′-Epoxy-4′,9′-dihydroxy-3′,5,5′-trimethoxy-3,8′-lignan-9-oic acid.
Sinensiol E
$C_{21}H_{24}O_8$ 404.415
Constit. of *Selaginella sinensis*. Oil. λ_{max} 207 (log ε 4.50); 294 (log ε 2.37) (MeOH)

9-Carboxylic acid, 3′,5,5′-tri-Me ether, Me ester
[1264177-63-0]
$C_{22}H_{26}O_8$ 418.442
Constit. of *Selaginella moellendorffii*. Amorph. solid. $[\alpha]^{26}_{D}$ -4.4 (c, 0.36 in MeOH). λ_{max} 209 (log ε 4.45); 290 (log ε 3.52); 379 (log ε 3.07) (MeOH)

3′,5,5′-Tri-Me ether, 4′-O-[β-D-apiofuranosyl-(1→2)-β-D-glucopyranoside]
$C_{32}H_{44}O_{16}$ 684.688
Constit. of the branches of *Alangium kurzii*. Pale yellow oil. $[\alpha]^{20}_{D}$ -4.2 (c, 0.39 in MeOH)

(7′R*,8′R*)-form
3′,5,5′-Tri-Me ether, 4′-O-β-D-glucopyranoside
[545436-58-6]
$C_{27}H_{36}O_{12}$ 552.573
Isol. from non-centrifuged cane sugar (*Saccharum officinarum*). Oil

(7′R*,8′S*)-form
3′,5,5′-Tri-Me ether, 9-O-β-D-glucopyranoside
[1264177-70-9]
$C_{27}H_{36}O_{12}$ 552.573
Constit. of *Selaginella moellendorffii* and *Smilax trinervula*. Exhibits cytotoxic activity. Amorph. solid. $[\alpha]^{16}_{D}$ -33 (c, 0.67 in MeOH). λ_{max} 208 (log ε 4.6); 281 (log ε 3.39) (MeOH)

(7′RS,8′SR)-form
4′-O-[2-Hydroxy-1-(hydroxymethyl)ethyl], 3′,5,5′-tri-Me ether
Sakuraresinol
[128502-87-4]
$C_{24}H_{32}O_9$ 464.511
Constit. of *Crataegus pinnatifida*, *Illicium henryi* and *Prunus jamasakura*. Amorph. powder. λ_{max} 236 (sh); 330 (log ε 3.70) (MeOH)

(7′ξ,8′ξ)-form
4′-O-[1-(Butoxycarbonyl)-2-hydroxyethyl], 3′,5,5′-tri-Me ether
[149415-65-6]
$C_{28}H_{38}O_{10}$ 534.602
Constit. of the bark of *Illicium difengpi*. Syrup. $[\alpha]^{15}_{D}$ -14.2 (c, 0.88 in MeOH). Possible artifact.

Yoshinari, K. *et al.*, *Phytochemistry*, 1990, **29**, 1675-1678 (*Sakuraresinol*).
Kouno, I. *et al.*, *Phytochemistry*, 1993, **32**, 1573-1577 (*Illicium difengpi constit*).
Paula, V.F. *et al.*, *Tetrahedron*, 1995, **51**, 12453-12462 (*Boehmenan D*).
Changzeng, W. *et al.*, *Phytochemistry*, 1997, **45**, 159-166 (*Tortoside E*).
Lin, M.-S. *et al.*, *J. Nat. Prod.*, 2001, **64**, 707-709 (*Boehmenan D*).
Seca, A.M.L. *et al.*, *Phytochemistry*, 2001, **56**, 759-767 (*Boehmenan D*).
Takara, K. *et al.*, *Biosci., Biotechnol., Biochem.*, 2003, **67**, 376-379 (*Saccharum officinarum constit*).
Morikawa, T. *et al.*, *Chem. Pharm. Bull.*, 2003, **51**, 62-67 (*Sakuraresinoside, Sakuraresinol*).
Wang, Y.-H. *et al.*, *Helv. Chim. Acta*, 2010, **93**, 2467-2477 (*Selaginella moellendorffii constits*).
Bai, J. *et al.*, *J. Asian Nat. Prod. Res.*, 2012, **14**, 940-949 (*Dunnianeolignan A*).
Zhao, H. *et al.*, *Phytochem. Lett.*, 2012, **5**, 240-243 (*Pittogoside A*).
Kuo, L.-M.Y. *et al.*, *J. Nat. Prod.*, 2013, **76**, 580-587 (*Eurya japonica xyloside*).
Yoshikawa, K. *et al.*, *Molecules*, 2013, **18**, 9641-9649 (*Acer saccharum constit*).
Liu, Q.-B. *et al.*, *J. Agric. Food Chem.*, 2014, **62**, 7796-7803 (*Prunustosananside AI,AII*).
Suh, W.S. *et al.*, *Helv. Chim. Acta*, 2015, **98**, 1087-1094 (*Linderanoside C, activity*).
Matsuo, Y. *et al.*, *Chem. Pharm. Bull.*, 2016, **64**, 918-923 (*Pu-erh tea constit*).
Wu, Z. *et al.*, *Sci. Rep.*, 2016, **6**, 24809 (*Phyllanglaucin A*).
Yan, S.-D. *et al.*, *Zhongcaoyao*, 2016, **47**, 2806-2811 (*Salicifoneoligaonol, cd*).
Ren, J. *et al.*, *Fitoterapia*, 2017, **121**, 152-158 (*Alangium kurzii constit*).
Xu, Y. *et al.*, *J. Funct. Foods*, 2017, **32**, 160-169 (*Rubuslin D, activity*).
Wu, Y. *et al.*, *Phytochem. Lett.*, 2017, **20**, 259-262 (*Litseasin A*).
Chen, X. *et al.*, *Fitoterapia*, 2019, **134**, 256-263 (*Sinensiol E*).
Ji, J. *et al.*, *Fitoterapia*, 2019, **137**, 104271:1-7 (*Securidaca inappendiculata constit, cd*)

4,7′-Epoxy-3,8′-lignan-3′,4′,5,7,9,9′-hexol E81

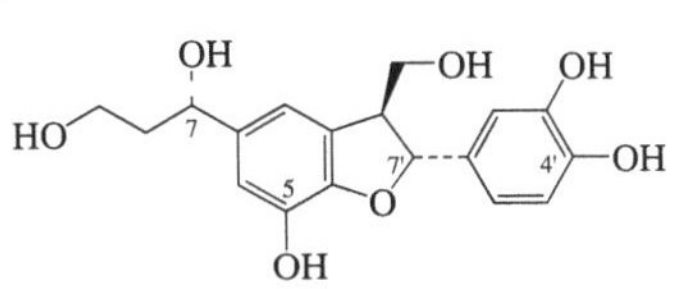

(7S,7′R,8′S)-form

2-(3,4-Dihydroxyphenyl)-5-(1,3-dihydroxypropyl)-2,3-dihydro-7-hydroxy-3-(hydroxymethyl)benzofuran. 4,7′-Epoxy-3′,4′,5,7,9,9′-hexahydroxy-3,8′-neolignan
$C_{18}H_{20}O_7$ 348.351

(7S,7′R,8′S)-form
3′,5-Di-Me ether
4,7′-Epoxy-3′,5-dimethoxy-3,8′-lignan-4′,7,9,9′-tetrol. 4,7′-Epoxy-4′,7,9,9′-tetrahydroxy-3′,5-dimethoxy-3,8′-lignan

3′,5-Di-Me ether, 4′-O-β-D-glucopyranoside
[1627676-85-0]
$C_{26}H_{34}O_{12}$ 538.546
Constit. of the fruit of *Arctium lappa*. Exhibits strong hepatoprotective activity. Amorph. powder. $[\alpha]^{20}_{D}$ -40.3 (c, 0.07 in MeOH)

7-Ketone, 3′,5-di-Me ether
4,7′-Epoxy-4′,9,9′-trihydroxy-3′,5-dimethoxy-3,8′-lignan-7-one. 4,7′-Epoxy-4′,9,9′-trihydroxy-3′,5-dimethoxy-7-oxo-3,8′-lignan.
Prunustosanan AI
[1620571-52-9]
Constit. of the seeds of *Prunus tomentosa*. Oil. $[\alpha]^{20}_{D}$ -16.5 (c, 0.13 in MeOH). λ_{max} 228 (log ε 0.53); 287 (log ε 0.3) (MeOH)

7-Ketone, 3′,5-di-Me ether, 4′-O-β-D-glucopyranoside
$C_{26}H_{32}O_{12}$ 536.53
Constit. of the fruit of *Arctium lappa*. Amorph. powder. $[\alpha]^{25}_{D}$ -17.5 (c, 0.4 in MeOH). λ_{max} 235 (log ε 2.98); 285 (log ε 2.91) (MeOH)

(7S,7′S,8′R)-form
3′,5-Di-Me ether, 4′-O-β-D-glucopyranoside
[1627676-84-9]
Constit. of the fruit of *Arctium lappa*. Exhibits strong hepatoprotective activity. Amorph. powder. $[\alpha]^{20}_{D}$ -32.8 (c, 0.07 in MeOH)

3′,4′-Methylene, 5-Me ether
4,7′-Epoxy-5-methoxy-3′,4′-methylenedioxy-3,8′-lignan-7,9,9′-triol. 4,7′-Epoxy-7,9,9′-trihydroxy-5-methoxy-3′,4′-methylenedioxy-3,8′-lignan
[1571800-60-6]
$C_{20}H_{22}O_7$ 374.389
Isol. from *Breynia fruticosa*. Yellow oil. $[\alpha]^{26}_{D}$ -18.8 (c, 0.5 in MeOH). λ_{max} 207 (log ε 1.3); 286 (log ε 0.4) (MeOH)

3′,4′-Methylene, 5,7-di-Me ether
4,7′-Epoxy-5,7-dimethoxy-3′,4′-methylenedioxy-3,8′-lignan-9,9′-diol. 4,7′-Epoxy-9,9′-dihydroxy-5,7-

dimethoxy-3′,4′-methylenedioxy-3,8′-lignan
[1571800-62-8]
$C_{21}H_{24}O_7$ 388.416
Isol. from *Breynia fruticosa*. Yellow oil. $[\alpha]^{26}_{D}$ -21.8 (c, 0.5 in MeOH). λ_{max} 207 (log ε 1.4); 238 (log ε 0.8); 287 (log ε 0.5) (MeOH)
(7ξ,7′R,8′S)-form
7-Et, 3′,5-di-Me ether
4,7′-Epoxy-7-ethoxy-3′,5-dimethoxy-3,8′-lignan-4′,9,9′-triol. 4,7′-Epoxy-7-ethoxy-4′,9,9′-trihydroxy-3′,5-dimethoxy-3,8′-lignan. ***Pinnatifidanin CVII***
[1616701-74-6]
$C_{22}H_{28}O_7$ 404.459
Constit. of the seeds of *Crataegus pinnatifida*. Yellow oil. $[\alpha]^{20}_{D}$ +12.5 (c, 0.06 in MeOH). λ_{max} 228 (log ε 2.63); 278 (log ε 1.75) (MeOH).

Huang, X.-X. *et al.*, *Fitoterapia*, 2013, **91**, 217-223 (*Pinnatifidanin CVII*).
Li, Y.-P. *et al.*, *Phytochem. Lett.*, 2013, **6**, 281-285 (*Breynia fruticosa constits*).
Liu, Q.-B. *et al.*, *J. Agric. Food Chem.*, 2014, **62**, 7796-7803 (*Prunustosanan AI*).
Yang, Y.-N. *et al.*, *J. Agric. Food Chem.*, 2014, **62**, 9095-9102 (*Arctium lappa 4′-glucosides*).
Huang, X.-Y. *et al.*, *J. Asian Nat. Prod. Res.*, 2015, **17**, 504-511 (*7-ketone 3′,5-di-Me ether 4′-glucoside*)

4,7′-Epoxy-3,8′-lignan-3′,4′,5,8,9,9′-hexol E82

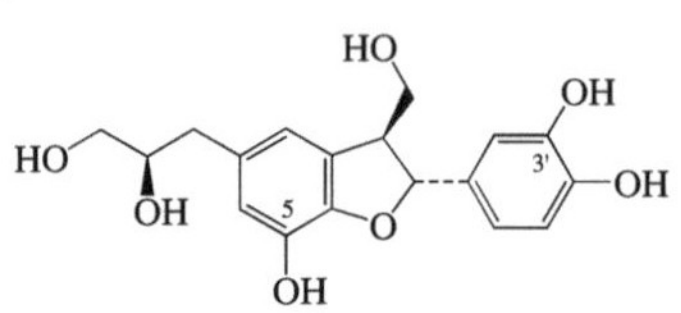

4,7′-Epoxy-3′,4′,5,8,9,9′-hexahydroxy-3,8′-neolignan
$C_{18}H_{20}O_7$ 348.351
(7′R,8R,8′S)-form
3′,5-Di-Me ether
4,7′-Epoxy-3′,5-dimethoxy-3,8′-lignan-4′,8,9,9′-tetrol. 4,7′-Epoxy-4′,8,9,9′-tetrahydroxy-3′,5-dimethoxy-3,8′-lignan.
Prunustosanan AIV
[1620571-55-2]
$C_{20}H_{24}O_7$ 376.405
Constit. of the seeds of *Prunus tomentosa*. Oil. $[\alpha]^{20}_{D}$ -14.1 (c, 0.11 in MeOH). λ_{max} 231 (log ε 1.24); 280 (log ε 0.34) (MeOH).

Liu, Q.-B. *et al.*, *J. Agric. Food Chem.*, 2014, **62**, 7796-7803 (*Prunustosanan AIV*)

7,7′-Epoxy-8,8′-lignan-3,3′,4,4′,5,5′-hexol E83

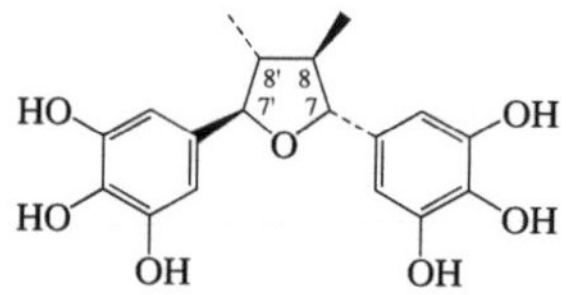

(7R,7′R,8R,8′R)-form

Tetrahydro-3,4-dimethyl-2,5-bis(3,4,5-trihydroxyphenyl)furan. 3,3′,4,4′,5,5′-Hexahydroxy-7,7′-epoxylignan
[50393-98-1] [63284-54-8]
$C_{18}H_{20}O_7$ 348.351
(7R,7′R,8R,8′R)-form
3,3′,5,5′-Tetra-Me ether
7,7′-Epoxy-3,3′,5,5′-tetramethoxy-8,8′-lignan-4,4′-diol. 4,4′-Dihydroxy-3,3′,5,5′-tetramethoxy-7,7′-epoxylignan.
Fragransin B_2
[112572-55-1]
$C_{22}H_{28}O_7$ 404.459
Isol. from *Myristica fragrans*. Oil. $[\alpha]_D$ 0 ($CHCl_3$)
3,3′,4,5,5′-Penta-Me ether
7,7′-Epoxy-3,3′,4′,5,5′-pentamethoxy-8,8′-lignan-4-ol. 4-Hydroxy-3,3′,4′,5,5′-pentamethoxy-7,7′-epoxylignan.
De-O-methylgrandisin
[124078-30-4]
$C_{23}H_{30}O_7$ 418.486
Constit. of *Licaria aurea*. Mp 135 - 137°
Hexa-Me ether
7,7′-Epoxy-3,3′,4,4′,5,5′-hexamethoxylignan. 3,3′,4,4′,5,5′-Hexamethoxy-7,7′-epoxylignan. ***(+)-Grandisin***
[137255-18-6]
$C_{24}H_{32}O_7$ 432.513
Constit. of *Piper polysyphorum*
(7R,7′R,8R,8′S)-form
Hexa-Me ether
Epigrandisin
[271586-50-6]
$C_{24}H_{32}O_7$ 432.513
Constit. of *Aglaia leptantha*
(7R,7′S,8R,8′S)-form
3,4-Methylene, 3′,4′,5,5′-tetra-Me ether
7,7′-Epoxy-3,3′,4′,5′-tetramethoxy-4,5-methylenedioxylignan. 3,3′,4′,5′-Tetramethoxy-4,5-methylenedioxy-7,7′-epoxylignan.
Saurucinol I
[1521111-76-1]
$C_{23}H_{28}O_7$ 416.47
Constit. of *Saururus chinensis*. Oil. $[\alpha]^{20}_{D}$ -1.8 (c, 0.06 in $CHCl_3$). λ_{max} 217 (log ε 4.42); 273 (log ε 3.31) (MeOH)
3,3′,4,5,5′-Penta-Me ether
Saurucinol J
[1521111-79-4]
$C_{23}H_{30}O_7$ 418.486
Constit. of *Saururus chinensis*. Amorph. solid. $[\alpha]^{20}_{D}$ -10 (c, 0.05 in $CHCl_3$). λ_{max} 220 (log ε 4.09); 271 (log ε 3.29) (MeOH)
(7R,7′S,8S,8′S)-form
3,3′,5,5′-Tetra-Me ether
Fragransin B_3
[112572-56-2]
$C_{22}H_{28}O_7$ 404.459
Isol. from *Myristica fragrans*. Oil. $[\alpha]_D$ +12.5 (c, 1.02 in $CHCl_3$)
3,3′,4,4′,5-Penta-Me ether
Eupobennettin
[139220-11-4]
$C_{23}H_{30}O_7$ 418.486
Constit. of *Eupomatia bennettii*. Gum. $[\alpha]^{20}_{D}$ +10.4 (c, 0.5 in $CHCl_3$)
Hexa-Me ether
Beilschmin B
[916656-00-3]
$C_{24}H_{32}O_7$ 432.513
Constit. of the stems of *Beilschmiedia tsangii*. Oil. $[\alpha]^{25}_{D}$ +13.6 (c, 0.12 in $CHCl_3$). λ_{max} 216 (log ε 4.96); 272 (log ε 4.13) (MeOH)
3,4-Methylene, 3′,4′,5,5′-tetra-Me ether
Beilschmin A
[916655-99-7]
$C_{23}H_{28}O_7$ 416.47
Constit. of the stems of *Beilschmiedia tsangii*. Oil. $[\alpha]^{25}_{D}$ +15.4 (c, 0.15 in $CHCl_3$). λ_{max} 217 (log ε 4.83); 276 (log ε 4.16) (MeOH)
(7S,7′R,8R,8′R)-form
3,4-Methylene, 3′,4′,5,5′-tetra-Me ether
Constit. of *Piper solmsianum*. Pale yellow oil. $[\alpha]^{21}_{D}$ -12.1. λ_{max} 207 (ε 2370); 246 (ε 14950); 270 (ε 92300) (MeOH)
(7S,7′S,8S,8′S)-form
3,3′,5,5′-Tetra-Me ether
Di-de-O-methylgrandisin
$C_{22}H_{28}O_7$ 404.459
Constit. of leaves of *Virola pavonis*. Cryst. (Et_2O). Mp 157 - 159°. $[\alpha]^{25}_{D}$

-2.8 (c, 1.6 in $CHCl_3$). λ_{max} 239 (log ε 4.2); 271 (log ε 3.4) (MeOH)

Hexa-Me ether

(-)-Grandisin

[53250-50-3]

$C_{24}H_{32}O_7$ 432.513

Constit. of *Litsea grandis*, *Cryptocarya crassinervia*, *Piper solmsianum* and *Virola surinamensis*. Shows trypanosomicidal activity. Toxic to brine shrimp. Platelet aggregating factor antagonist. Cryst. (Et_2O). Mp 118 - 120°. $[\alpha]_D$ -57.1 (c, 0.35 in $CHCl_3$)

(7R*,7′R*,8R*,8′R*)-form

3,3′,4,4′,5-Penta-Me ether

Beilschminol B

[1467073-21-7]

$C_{23}H_{30}O_7$ 418.486

Constit. of the roots of *Beilschmiedia tsangii*. Light yellow oil. $[\alpha]_D^{25}$ -18.6 (c, 0.18 in MeOH). λ_{max} 209 (log ε 4.63); 240 (sh); 270 (log ε 3.12) (MeOH)

3,4-Methylene, 3′,4′,5,5′-tetra-Me ether

[623552-83-0]

Constit. of the inflorescences of *Piper solmsianum*. Pale yellow oil. $[\alpha]_D^{21}$ -4.5 (c, 0.01 in MeOH). λ_{max} 254 (ε 12650); 273 (ε 5440) (MeOH)

3,4:3′,4′-Bis(methylene), di-Me ether

7,7′-Epoxy-3,3′-dimethoxy-4,5:4′,5′-bis(methylenedioxy)lignan. 3,3′-Dimethoxy-4,5:4′,5′-bis(methylenedioxy)-7,7′-epoxylignan

[623552-85-2]

$C_{22}H_{24}O_7$ 400.427

Constit. of the inflorescences of *Piper solmsianum*. Pale yellow oil. $[\alpha]_D^{21}$ -10.1 (c, 0.01 in MeOH). λ_{max} 254 (ε 15745); 277 (ε 4270) (MeOH)

(7R*,7′S*,8R*,8′R*)-form

3,3′,4,5′,5-Penta-Me ether

[1009642-56-1]

$C_{23}H_{30}O_7$ 418.486

Constit. of *Peperomia blanda*. Pale yellow oil. $[\alpha]_D^{21}$ -31.2 (c, 0.12 in MeOH). λ_{max} 254 (log ε 4.1); 273 (log ε 3.7) (MeOH)

3,4-Methylene, 3′,5,5′-tri-Me ether

7,7′-Epoxy-3,3′,5-trimethoxy-4′,5′-methylenedioxy-8,8′-lignan-4-ol. 4-Hydroxy-3,3′,5-trimethoxy-4′,5′-methylenedioxy-7,7′-epoxylignan

[1009642-54-9]

$C_{22}H_{26}O_7$ 402.443

Constit. of *Peperomia blanda*. Pale yellow oil. $[\alpha]_D^{21}$ -52.8 (c, 0.12 in MeOH). λ_{max} 254 (log ε 4.1); 273 (log ε 3.7) (MeOH)

3,4-Methylene, 3′,4′,5,5′-tetra-Me ether

[1009642-55-0]

Constit. of *Peperomia blanda*. Pale yellow oil. $[\alpha]_D^{21}$ -30.7 (c, 0.22 in MeOH). λ_{max} 254 (log ε 4.1); 273 (log ε 3.7) (MeOH)

3,4:3′,4′-Bis(methylene), di-Me ether

[1009642-58-3]

Constit. of *Peperomia blanda*. Pale yellow oil. $[\alpha]_D^{21}$ -25.6 (c, 0.12 in MeOH). λ_{max} 254 (log ε 4.2); 277 (log ε 3.6) (MeOH)

(7R*,7′S*,8R*,8′S*)-form

3,3′,5,5′-Tetra-Me ether

Fragransin B_1

[112516-03-7]

$C_{22}H_{28}O_7$ 404.459

Isol. from *Myristica fragrans*. Cryst.. Mp 100 - 102°. Opt. inactive (*meso-*)

3,3′,4,4′,5-Penta-Me ether

Schinlignin A

[1352185-19-3]

$C_{23}H_{30}O_7$ 418.486

Constit. of the fruit of *Schisandra chinensis* and roots of *Saururus chinensis*. Yellow oil. $[\alpha]_D^{26}$ -6.9 (c, 0.18 in MeOH). λ_{max} 207 (log ε 4.62); 270 (log ε 3.31) (MeOH)

Hexa-Me ether

4,4′-Di-O-methylfragransin B_1

[50394-05-3]

$C_{24}H_{32}O_7$ 432.513

Isol. from *Aristolochia birostris*. Opt. inactive (*meso-*)

3′,4′-Methylene, 3,4,5,5′-tetra-Me ether

Saurufurin C

[1532522-57-8]

$C_{23}H_{28}O_7$ 416.47

Constit. of *Saururus chinensis*. Light yellow oil. $[\alpha]_D^{26}$ +2 (c, 1.0 in MeOH). λ_{max} 214 (sh); 240 (sh); 276 (log ε 3.47) (MeOH)

(7RS,7′SR,8SR,8′RS)-form

Hexa-Me ether

$C_{24}H_{32}O_7$ 432.513

Constit. of *Piper solmsianum*. Powder. Opt. inactive (*meso-*). λ_{max} 225 (MeOH)

3′,4′-Methylene, 3,4,5,5′-tetra-Me ether

$C_{23}H_{28}O_7$ 416.470

Constit. of *Piper solmsianum*. Pale yellow oil. $[\alpha]_D^{21}$ +80.7 (c, 0.004 in CH_2Cl_2). λ_{max} 225 (MeOH).

Sarkanen, K.V. *et al.*, *J. Heterocycl. Chem.*, 1973, **10**, 1025-1027 (*hexa-Me ethers, pmr, struct*).
Holloway, D. *et al.*, *Phytochemistry*, 1974, **13**, 1233-1236 (*Grandisin*).
Stevenson, R. *et al.*, *Tetrahedron*, 1977, **33**, 285-288 (*Grandisin, synth*).
Hattori, M. *et al.*, *Chem. Pharm. Bull.*, 1987, **35**, 3315-3322 (*Fragransins B_1,B_2,B_3*).
Barbosa-Filho, J.M. *et al.*, *Phytochemistry*, 1989, **28**, 2209-2211 (*De-O-methylgrandisin*).
Conserva, L.M. *et al.*, *Phytochemistry*, 1990, **29**, 257-260 (*4,4′-Di-O-methylfragransin B_1*).
Carroll, A.R. *et al.*, *Aust. J. Chem.*, 1991, **44**, 1627-1633 (*Eupobennettin*).
Ferri, P.H. *et al.*, *Phytochemistry*, 1991, **30**, 4204-4205 (*Dide-O-methylgrandisin*).
Ma, Y. *et al.*, *Yaoxue Xuebao*, 1991, **26**, 345-350 (*(+)-Grandisin*).
Fun, H.-K. *et al.*, *Acta Crystallogr., Sect. C*, 1996, **52**, 414-416 (*hexa-Me ether, cryst struct*).
Lopes, N.P. *et al.*, *Planta Med.*, 1998, **64**, 667-669 (*Grandisin, activity*).
Greger, H. *et al.*, *J. Nat. Prod.*, 2000, **63**, 616-620 (*Grandisin, Epigrandisin*).
Martins, R.C.C. *et al.*, *Phytochemistry*, 2000, **55**, 843-846 (*Piper solmsianum derivs*).
Martins, R.C.C. *et al.*, *Phytochemistry*, 2003, **64**, 667-670 (*Piper solmsianum derivs*).
Chen, J.-J. *et al.*, *Planta Med.*, 2006, **72**, 351-357 (*Beilschmins A,B*).
Felippe, L.G. *et al.*, *Phytochemistry*, 2008, **69**, 445-450 (*Peperomia blanda lignans*).
Xue, Y.-B. *et al.*, *Chem. Pharm. Bull.*, 2010, **58**, 1606-1612 (*Schinlignin A*).
Huang, Y.-T. *et al.*, *Int. J. Mol. Sci.*, 2012, **13**, 16430-16443 (*Beilschminol B*).
Cui, H. *et al.*, *J. Nat. Prod.*, 2014, **77**, 100-110 (*Saurucinols I,J*).
Tsai, W.-J. *et al.*, *J. Nat. Prod.*, 2014, **77**, 125-131 (*Saurufurin C*).
Ramos, C.S. *et al.*, *RSC Adv.*, 2017, **7**, 46932-46937 (*Piper solmsianum constits*)

7,7′-Epoxy-8,8′-lignan-3,3′,4,4′,5,7-hexol E84

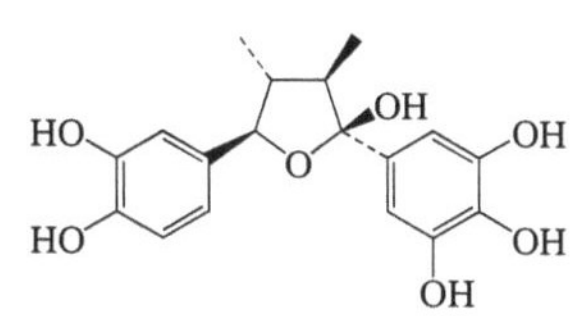

3,3′,4,4′,5,7-Hexahydroxy-7,7′-epoxylignan

$C_{18}H_{20}O_7$ 348.351

(7R,7′R,8R,8′R)-form

3′,4′-Methylene, tetra-Me ether

7,7′-Epoxy-3,4,5,7-tetramethoxy-3′,4′-methylenedioxylignan. 3,4,5,7-Tetramethoxy-3′,4′-methylenedioxy-7,7′-epoxylignan

[1886973-81-4]

$C_{23}H_{28}O_7$ 416.47
Constit. of the roots of *Schisandra sphenanthera*. Yellow oil. $[\alpha]_D^{25}$ +40 (c, 0.1 in MeOH). λ_{max} 227 (log ε 0.24) (EtOH).

Jiang, K. *et al.*, *Fitoterapia*, 2015, **103**, 63-70 (*Schisandra sphenanthera constit*)

7,7′-Epoxy-8,8′-lignan-3,3′,4,4′,7,8-hexol E85

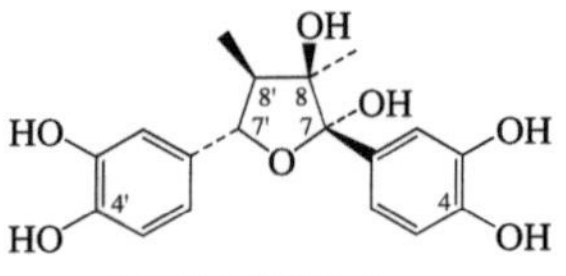

(7R,7′S,8R,8′R)-form

3,3′,4,4′,7,8-Hexahydroxy-7,7′-epoxylignan
$C_{18}H_{20}O_7$ 348.351
(7R,7′S,8R,8′R)-form
3,3′,4,4′,7-Penta-Me ether
7,7′-Epoxy-3,3′,4,4′,7-pentamethoxy-8,8′-lignan-8-ol. 8-Hydroxy-3,3′,4,4′,7-pentamethoxy-7,7′-epoxylignan. ***Henricine A***
[1158301-82-6]
$C_{23}H_{30}O_7$ 418.486
Constit. of the stems of *Schisandra henryi*. Flakes. $[\alpha]_D^{20}$ +77.3 (c, 0.07 in $CHCl_3$). λ_{max} 208 (log ε 4.48); 232 (log ε 4.29); 279 (log ε 3.78) (MeOH)
3,4-Methylene, 3′,4′,7-tri-Me ether
7,7′-Epoxy-3′,4′,7-trimethoxy-3,4-methylenedioxy-8,8′-lignan-8-ol. 8-Hydroxy-3′,4′,7-trimethoxy-3,4-methylenedioxy-7,7′-epoxylignan. ***Schiglaucin A***
[1514662-83-9]
$C_{22}H_{26}O_7$ 402.443
Constit. of the stems of *Schisandra glaucescens*. Oil. $[\alpha]_D^{20}$ +53.56 (c, 12.51 in $CHCl_3$). λ_{max} 237 (log ε 3.84); 282 (log ε 3.56) (CH_2Cl_2)
3′,4′-Methylene, 3,4,7-tri-Me ether
7,7′-Epoxy-3,4,7-trimethoxy-3′,4′-methylenedioxy-8,8′-lignan-8-ol. 8-Hydroxy-3,4,7-trimethoxy-3′,4′-methylenedioxy-7,7′-epoxylignan. ***Schiglaucin B***
[1514662-84-0]
$C_{22}H_{26}O_7$ 402.443
Constit. of the stems of *Schisandra glaucescens*. Oil. $[\alpha]_D^{20}$ +58.63 (c, 9.95 in $CHCl_3$). λ_{max} 233 (log ε 3.90); 285 (log ε 3.60) (CH_2Cl_2)
3,4:3′,4′-Bis(methylene), 7-Me ether
7,7′-Epoxy-7-methoxy-3,4:3′,4′-bis(methylenedioxy)-8,8′-lignan-8-ol. 8-Hydroxy-7-methoxy-3,4:3′,4′-bis(methylenedioxy)-7,7′-epoxylignan
[1571072-48-4]
$C_{21}H_{22}O_7$ 386.4
Constit. of the stems of *Schisandra glaucescens*. Pale yellow oil. $[\alpha]_D^{20}$ +69.1 (c, 0.06 in $CHCl_3$). λ_{max} 205 (log ε 4.25); 230 (log ε 3.68) (MeOH)
(7S,7′R,8R,8′S)-form
3,4-Methylene, 3′,4′-di-Me ether
7,7′-Epoxy-3′,4′-dimethoxy-3,4-methylenedioxy-8,8′-lignan-7,8-diol. 7,8-Dihydroxy-3′,4′-dimethoxy-3,4-methylenedioxy-7,7′-epoxylignan. ***Schpropinrin D***
[1565857-30-8]
$C_{21}H_{24}O_7$ 388.416
Constit. of the leaves and stems of *Schisandra propinqua*. Yellow oil. $[\alpha]_D^{24}$ +1.4 (c, 0.23 in MeOH). λ_{max} 231 (log ε 3.45); 282 (log ε 3.17) (MeOH)
3,4-Methylene, 3′,7-di-Me ether
7,7′-Epoxy-3′,7-dimethoxy-3,4-methylenedioxy-8,8′-lignan-4′,8-diol. 4′,8-Dihydroxy-3′,7-dimethoxy-3,4-methylenedioxy-7,7′-epoxylignan. ***Schpropinrin C***
[1565857-29-5]
$C_{21}H_{24}O_7$ 388.416
Constit. of the leaves and stems of *Schisandra propinqua*. Amorph. yellow powder. $[\alpha]_D^{25}$ +41.3 (c, 0.3 in MeOH). λ_{max} 202 (log ε 3.90); 229 (log ε 3.28); 283 (log ε 2.99) (MeOH)
3,4-Methylene, 3′,4′,7-tri-Me ether
Schpropinrin A
[1565857-27-3]
$C_{22}H_{26}O_7$ 402.443
Constit. of the leaves and stems of *Schisandra propinqua*. Yellow oil. $[\alpha]_D^{24}$ +66 (c, 0.14 in MeOH). λ_{max} 231 (log ε 3.43); 282 (log ε 3.11) (MeOH)
3′,4′-Methylene, 3,4,7-tri-Me ether
Schpropinrin B
[1565857-28-4]
$C_{22}H_{26}O_7$ 402.443
Constit. of the leaves and stems of *Schisandra propinqua*. Yellow oil. $[\alpha]_D^{24}$ +76.8 (c, 0.2 in MeOH). λ_{max} 205 (log ε 3.98); 231 (log ε 3.41); 282 (log ε 3.11) (MeOH)
3,4:3′,4′-Bis(methylene) ether
7,7′-Epoxy-3,4:3′,4′-bis(methylenedioxy)-8,8′-lignan-7,8-diol. 7,8-Dihydroxy-3,4:3′,4′-bis(methylenedioxy)-7,7′-epoxylignan. ***Zuihonin F****. Zuonin F*
[1453168-48-3]
$C_{20}H_{20}O_7$ 372.373
Constit. of the stems of *Schisandra bicolor*. Yellowish oil. $[\alpha]_D^{20}$ +11.3 (c, 0.15 in MeOH). λ_{max} 234 (log ε 3.63); 276 (log ε 3.31) (MeOH)
3,4:3′,4′-Bis(methylene), 7-Me ether
Zuihonin E*. Zuonin E*
[1453168-47-2]
$C_{21}H_{22}O_7$ 386.4
Constit. of the stems of *Schisandra bicolor*. Yellowish oil. $[\alpha]_D^{20}$ +21.3 (c, 0.45 in MeOH). λ_{max} 231 (log ε 3.75); 281 (log ε 3.47) (MeOH).

Liu, H.-T. *et al.*, *Chem. Pharm. Bull.*, 2009, **57**, 405-407 (*Henricine A*).
Huang, R.-M. *et al.*, *J. Asian Nat. Prod. Res.*, 2012, **14**, 1116-1121 (*Zuihonins E,F*).
Yu, H.-Y. *et al.*, *Planta Med.*, 2012, **78**, 1962-1966 (*Schiglaucins A,B*).
Zhang, P.-P. *et al.*, *J. Asian Nat. Prod. Res.*, 2013, **15**, 466-472 (*Schisandra glaucescens bis(methylene) 7-Me ether*).
Shang, S.-Z. *et al.*, *Nat. Prod. Bioprospect.*, 2013, **3**, 56-60 (*Schpropinrins A-D, abs config*)

7,7′-Epoxy-8,8′-lignan-3,3′,4,4′,9,9′-hexol E86

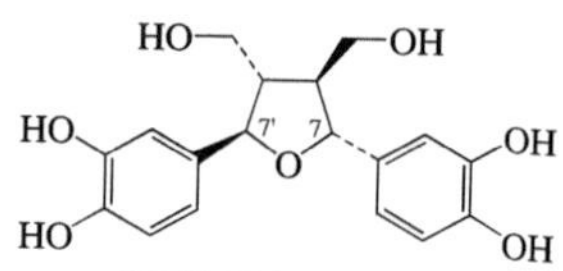

(7R,7′R,8S,8′S)-form

2,5-Bis(3,4-dihydroxyphenyl)tetrahydro-3,4-bis(hydroxymethyl)furan. 3,3′,4,4′,9,9′-Hexahydroxy-7,7′-epoxylignan
$C_{18}H_{20}O_7$ 348.351
(7R,7′R,8S,8′S)-form
3,3′-Di-Me ether
7,7′-Epoxy-3,3′-dimethoxy-8,8′-lignan-4,4′,9,9′-tetrol. 4,4′,9,9′-Tetrahydroxy-3,3′-dimethoxy-7,7′-epoxylignan. ***Neoolivil****. Biconiferyl*
[77790-55-7]
[486-02-2] [129065-76-5]
[129387-92-4] [129387-93-5]
$C_{20}H_{24}O_7$ 376.405

Constit. of *Thymus longiflorus*, *Urtica dioica* and *Daphne feddei*. Powder. $[\alpha]_D^{23}$ +49 (c, 0.16 in EtOH). Abs. config. revised in 1997. λ_{max} 204 (log ε 4.5); 231 (log ε 4.0); 280 (log ε 3.6) (MeOH)

3,3′-Di-Me ether, 4-O-β-D-glucopyranoside

[104402-97-3]

$C_{26}H_{34}O_{12}$ 538.546

Constit. of *Swertia japonica* and *Urtica dioica*. Amorph. powder. $[\alpha]_D^{24}$ +6.4 (c, 0.11 in MeOH). λ_{max} 228 (log ε 4.2); 277 (log ε 3.7) (MeOH)

3,3′-Di-Me ether, 4-O-[β-D-apiofuranosyl-(1→2)-β-D-glucopyranoside]

[371968-12-6]

$C_{31}H_{42}O_{16}$ 670.661

Constit. of the roots of *Astragalus zahlbruckneri*. $[\alpha]_D^{25}$ -69.9 (c, 0.5 in MeOH). λ_{max} 218 (log ε 5) (MeOH)

3,3′-Di-Me ether, 9-O-β-D-xylopyranoside

[1620489-33-9]

$C_{25}H_{32}O_{11}$ 508.52

Constit. of the stems of *Rhododendron mariae*. Amorph. powder. $[\alpha]_D^{20}$ -4.4 (c, 0.06 in MeOH). λ_{max} 231; 280 (MeOH)

3,3′-Di-Me ether, 9-O-β-D-glucopyranoside

[129170-78-1]

$C_{26}H_{34}O_{12}$ 538.546

Constit. of *Osmanthus fragrans* var. *aurantiacus* and *Urtica dioica*. Amorph. powder. $[\alpha]_D^{25}$ -18.2 (c, 0.12 in MeOH). Two stereoisomers characterised from *Urtica dioica* but abs. configs. are not clear following the revision of abs. config. for Neoolivil. λ_{max} 205 (log ε 4.5); 231 (log ε 4.1); 279 (log ε 3.7) (MeOH)

3,3′-Di-Me ether, 9-Ac, 4-O-β-D-glucopyranoside

[104425-71-0]

$C_{28}H_{36}O_{13}$ 580.583

Constit. of *Urtica dioica*

3,3′-Di-Me ether, 9,9′-di-Ac, 4-O-β-D-glucopyranoside

[104380-18-9]

$C_{30}H_{38}O_{14}$ 622.62

Constit. of *Urtica dioica*

3,3′-Di-Me ether, tetra-Ac

Amorph. solid. $[\alpha]_D^{23}$ +17 (EtOH)

9-Et, 3,3′-di-Me ether

7,7′-Epoxy-9′-ethoxy-3,3′-methoxy-8,8′-lignan-4,4′,9-triol. 9-Ethoxy-4,4′,9′-trihydroxy-3,3′-methoxy-7,7′-epoxylignan. ***Taxumairin***

[200929-70-0]

$C_{22}H_{28}O_7$ 404.459

Constit. of *Taxus mairei* roots. Amorph. solid. $[\alpha]_D^{25}$ +2 (c, 0.2 in MeOH). λ_{max} 232 (log ε 3.56); 280 (log ε 3.17) (MeOH)

(7S,7′R,8S,8′S)-form

3,4-Methylene, 3′,4′,9,9′-tetra-Me ether

7,7′-Epoxy-3,4,9,9′-tetramethoxy-3′,4′-methylenedioxy-8,8′-lignan. 3,4,9,9′-Tetramethoxy-3′,4′-methylenedioxy-7,7′-epoxylignan. ***Virgatusin***

[176050-44-5]

$C_{23}H_{28}O_7$ 416.47

Constit. of *Phyllanthus urinaria* and *Phyllanthus virgatus*. Viscous oil. $[\alpha]_D^{25}$ -12.7 (c, 0.5 in CH_2Cl_2). λ_{max} 239 (log ε 4.99); 241 (log ε 5); 283 (log ε 5.07) (CH_2Cl_2)

3,4:3′,4′-Bis(methylene), 9,9′-di-Me ether

7,7′-Epoxy-9,9′-dimethoxy-3,4;3′,4′-bis(methylenedioxy)-8,8′-lignan. 9,9′-Dimethoxy-3,4:3′,4′-bis(methylenedioxy)-7,7′-epoxylignan. ***Urinalignan.*** *Urinaligran*

[610313-87-6]

$C_{22}H_{24}O_7$ 400.427

Constit. of *Phyllanthus urinaria*. Shows antifungal activity against *Colletotrichium lagenarium*. Liq.. $[\alpha]_D^{25}$ +19 (c, 1 in $CHCl_3$). λ_{max} 235 (log ε 4.01); 286 (log ε 3.89) (MeCN)

(7R*,7′S*,8S*,8′R*)-form

3,3′-Di-Me ether, 4-O-β-D-glucopyranoside

Sinolignan C

[1426550-87-9]

$C_{26}H_{34}O_{12}$ 538.546

Constit. of *Sinopodophyllum emodi*. Amorph. powder. $[\alpha]_D^{25}$ -10 (c, 0.3 in MeOH). λ_{max} 208 (log ε 1.82); 228 (log ε 0.5); 278 (log ε 0.16) (MeOH)

3,3′-Di-Me ether, 9-O-β-D-glucopyranoside

[1260685-64-0]

$C_{26}H_{34}O_{12}$ 538.546

Constit. of the leaves of *Osmanthus fragrans* var. *aurantiacus*. Amorph. powder. $[\alpha]_D^{25}$ -20 (c, 0.15 in MeOH). λ_{max} 205 (log ε 4.4); 229 (log ε 4); 278 (log ε 3.7) (MeOH)

(7RS,7′RS,8SR,8′SR)-form

5-Methoxy, 3,3′-di-Me ether

7,7′-Epoxy-3,3′,5-trimethoxy-8,8′-lignan-4,4′,9,9′-tetrol

[2170866-06-3]

$C_{21}H_{26}O_8$ 406.431

Constit. of the rhizomes of *Iris tectorum*. Pale yellow oil. $[\alpha]_D^{20}$ +17.6 (c, 0.15 in MeOH). Partial racemate. λ_{max} 205 (log ε 4.97) (MeOH)

(7RS,7′SR,8SR,8′SR)-form

3,3′-Di-Me ether

[1536262-88-0]

[1538608-41-1] (enantiomer)

$C_{20}H_{24}O_7$ 376.405

Constit. of the rhizomes of *Iris tectorum*. Pale yellow oil. $[\alpha]_D^{20}$ -9.8 (c, 0.15 in MeOH). Partial racemate. λ_{max} 204 (log ε 4.98) (MeOH).

Hernandez, A. *et al.*, *Phytochemistry*, 1981, **20**, 181-182 (*Neoolivil*).

Kraus, R. *et al.*, *Liebigs Ann. Chem.*, 1990, 1205-1213 (*Neoolivil glucosides*).

Huang, Y.-L. *et al.*, *J. Nat. Prod.*, 1996, **59**, 520-521 (*Phyllanthus virgatus constit*).

Schöttner, M. *et al.*, *Phytochemistry*, 1997, **46**, 1107-1109 (*Neoolivil, abs config*).

Shen, Y.-C. *et al.*, *Phytochemistry*, 1997, **46**, 1111-1113 (*Taxumairin*).

Yoda, H. *et al.*, *Tetrahedron Lett.*, 1999, **40**, 4701-4702 (*Virgatusin, synth*).

Calis, I. *et al.*, *J. Nat. Prod.*, 2001, **64**, 1179-1182 (*Neoolivil 4-apiosylglucoside*).

Chang, C.-C. *et al.*, *Phytochemistry*, 2003, **63**, 825-833 (*Urinalignan, abs config*).

Kikuchi, M. *et al.*, *Chem. Pharm. Bull.*, 2005, **53**, 48-51 (*Neoolivil 4-glucoside*).

Yamauchi, S. *et al.*, *Org. Biomol. Chem.*, 2005, **3**, 1670-1675 (*Virgatusin, synth*).

Akindele, T. *et al.*, *Org. Lett.*, 2005, **7**, 3685-3688 (*Virgatusin, synth*).

Akiyama, K. *et al.*, *Biosci., Biotechnol., Biochem.*, 2007, **71**, 1028-1035 (*Urinalignan, activity*).

Yamauchi, S. *et al.*, *Biosci., Biotechnol., Biochem.*, 2007, **71**, 2248-2255 (*Virgatusin, synth*).

Liang, S. *et al.*, *J. Nat. Prod.*, 2008, **71**, 1902-1905 (*Daphne feddei constit*).

Matcha, K. *et al.*, *Tetrahedron Lett.*, 2008, **49**, 3433-3436 (*Virgatusin, synth*).

Sanders, S.D. *et al.*, *Chem. Commun.*, 2009, 5135-5137 (*(+)-Virgatusin, synth*).

Machida, K. *et al.*, *Helv. Chim. Acta*, 2010, **93**, 2164-2175 (*Osmanthus fragrans glucosides*).

Sun, Y.-J. *et al.*, *Planta Med.*, 2012, **78**, 480-484 (*Sinolignan C*).

Guo, Q. *et al.*, *J. Asian Nat. Prod. Res.*, 2014, **16**, 1-10 (*Neoolivil 9-xyloside*).

Zhang, C.-L. *et al.*, *Fitoterapia*, 2016, **108**, 93-97 (*Iris tectorum constits*).

Shen, Y. *et al.*, *J. Org. Chem.*, 2018, **83**, 12549-12558 (*Urinalignan, synth*)

7,7′-Epoxy-8,8′-lignan-3,3′,4,5′,7,8-hexol E87

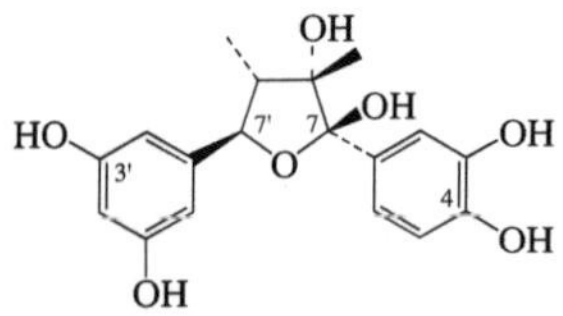

3,3′,4,5′,7,8-Hexahydroxy-7,7′-epoxylignan

$C_{18}H_{20}O_7$ 348.351

Structures in this series may need revision. Phenyl 3,5-dioxygenated substn. pattern in doubt as a result of work done by Yang *et al* in 2015

(7S,7′R,8S,8′S)-form

3,4-Methylene, 3′,7-di-Me ether

7,7′-Epoxy-5′,7-dimethoxy-3,4-methylenedioxy-8,8′-lignan-3′,8-diol.

3′,8-Dihydroxy-5′,7-dimethoxy-3,4-methylenedioxy-7,7′-epoxylignan

[1886973-76-7]

$C_{21}H_{24}O_7$ 388.416

Constit. of the roots of *Schisandra sphenanthera*. Antioxidant. Yellow oil. $[\alpha]^{25}_{D}$ +60 (c, 0.1 in MeOH). λ_{max} 228 (log ε 0.15) (EtOH).

Jiang, K. *et al.*, *Fitoterapia*, 2015, **103**, 63-70 (*Schisandra sphenanthera constit*).
Yang, Y.-N. *et al.*, *J. Nat. Prod.*, 2015, **78**, 705-711 (*pmr, struct*)

7,7′-Epoxy-8,8′-lignan-3,3′,4,5′,9,9′-hexol E88

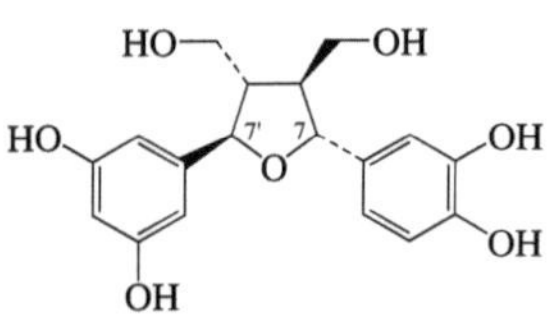

(7R*,7′R*,8S*,8′S*)-form

3,3′,4,5′,9,9′-Hexahydroxy-7,7′-epoxylignan

$C_{18}H_{20}O_7$ 348.351

Structures in this series may need revision. Phenyl 3,5-dioxygenated substn. pattern in doubt as a result of work done by Yang *et al* in 2015

(7R*,7′R*,8S*,8′S*)-form

3,3′-Di-Me ether

7,7′-Epoxy-3′,5-dimethoxy-8,8′-lignan-3,4′,9,9′-tetrol.

3′,4,9,9′-Tetrahydroxy-3,5′-dimethoxy-7,7′-epoxylignan.

Laxanol

[1059576-70-3]

$C_{20}H_{24}O_7$ 376.405

Constit. of *Justicia laxa*

(7R*,7′S*,8S*,8′S*)-form

3′,4-Di-Me ether, 9,9′-di-Ac

Curcasinlignan D

[1426542-86-0]

$C_{24}H_{28}O_9$ 460.479

Constit. of *Jatropha curcas*. $[\alpha]^{18.5}_{D}$ -4.47 (c, 0.2 in $CHCl_3$). λ_{max} 204 (log ε 4.76); 231 (log ε 4.19); 281 (log ε 3.87) (MeOH).

Hu, J.-F. *et al.*, *J. Chem. Res.*, 2008, 20-21 (*Laxanol*).
Xu, J.J. *et al.*, *Z. Naturforsch., B*, 2012, **67**, 176-180 (*Curcasinlignan D*).
Yang, Y.-N. *et al.*, *J. Nat. Prod.*, 2015, **78**, 705-711 (*pmr, struct*)

7′,8-Epoxy-7,8′-lignan-2,2′,4,4′,5,5′-hexol E89

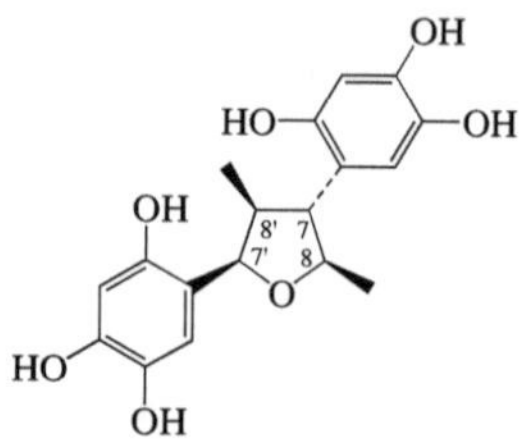

(7RS,7′RS,8RS,8′SR)-form

7′,8-Epoxy-2,2′,4,4′,5,5′-hexahydroxy-7,8′-neolignan

$C_{18}H_{20}O_7$ 348.351

4 Stereoisomers of hexa-Me ether characterised, all racemic

(7RS,7′RS,8RS,8′SR)-form

Hexa-Me ether

7′,8-Epoxy-2,2′,4,4′,5,5′-hexamethoxy-7,8′-lignan

[194594-57-5]

Constit. of *Aniba puchury-minor*

(7RS,7′SR,8RS,8′RS)-form

Hexa-Me ether

Acortatarinowin E

[1807543-46-9]

[1807543-34-5] (7*S*,7′*R*,8*S*,8′*S*-form) [1807543-35-6] (7*S*,7′*R*,8*R*,8′*R*-form)

$C_{24}H_{32}O_7$ 432.513

Constit. of the rhizomes of *Acorus tatarinowii*. Cryst. (MeOH). Mp 104 - 105°. Partial racemate. Resolved by chiral hplc. λ_{max} 204 (log ε 4.78); 233 (log ε 4.24); 291 (log ε 3.94) (MeOH)

(7RS,7′SR,8RS,8′SR)-form

Hexa-Me ether

[104596-94-3]

Constit. of *Aniba puchury-minor*

(7RS,7′SR,8SR,8′SR)-form

Hexa-Me ether

Magnosalicin

[93376-03-5]

[104596-93-2]

Constit. of the buds of *Magnolia salicifolia* and of *Aniba puchury-minor*. Shows antiallergic activity. Cryst. (MeOH). Mp 134.5 - 135°.

Tsuruga, T. *et al.*, *Tetrahedron Lett.*, 1984, **25**, 4129-4132 (*Magnosalicin, cryst struct*).
Mori, K. *et al.*, *Tetrahedron*, 1986, **42**, 523-528 (*synth, pmr, cmr*).
Muraoka, O. *et al.*, *Chem. Pharm. Bull.*, 1993, **41**, 772-774 (*synth*).
Da Silva, M.D.F. *et al.*, *Phytochemistry*, 1997, **45**, 1527-1528 (*Aniba puchury-minor constits*).
Lu, Y. *et al.*, *J. Nat. Prod.*, 2015, **78**, 2205-2214 (*Acortatarinowin E*)

7,9′-Epoxy-8,8′-lignan-3,3′,4,4′,5,9-hexol E90

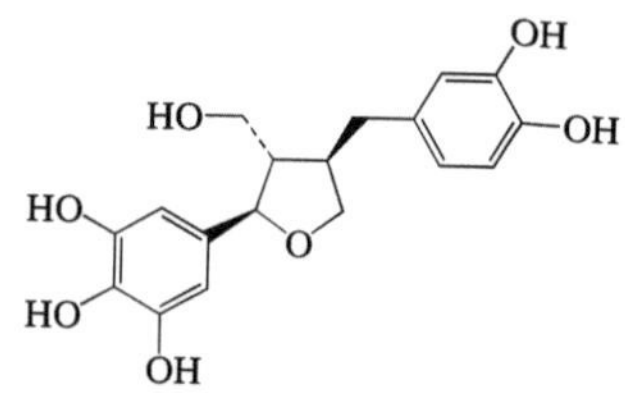

(7R,8S,8′R)-form

3,3′,4,4′,5,9-Hexahydroxy-7,9′-epoxylignan

$C_{18}H_{20}O_7$ 348.351

(7R,8S,8′R)-form

3,3′,5-Tri-Me ether, 9-O-β-D-xylopyranoside

Litseasin B

[2274645-18-8]

$C_{26}H_{34}O_{11}$ 522.547

Constit. of the root bark of *Litsea glutinosa*. Amorph. powder (MeOH). $[\alpha]^{25}_{D}$ +38 (c, 0.1 in MeOH). λ_{max} 239 (log ε 4.03) (MeOH)

(7R,8S,8′S)-form

3,3′,5-Tri-Me ether, 4′-O-β-D-glucopyranoside

Armandiside

[724422-23-5]

$C_{27}H_{36}O_{12}$ 552.573

Constit. of *Clematis armandii* and *Syringa reticulata*. Superoxide radical scavenger. Mp 144 - 145°. $[\alpha]^{22}_{D}$ -17.5 (c, 0.3 in MeOH)

3,3′,5-Tri-Me ether, 4,4′-di-O-β-D-glucopyranoside

[2410993-13-2]

$C_{33}H_{46}O_{17}$ 714.714

Constit. of the flowers of *Urtica fissa*. Amorph. powder. $[\alpha]_D^{25}$ -18.1 (c, 0.31 in MeOH). λ_{max} 230; 280 (MeOH)

3,3′,5-Tri-Me ether, 9-benzoyl

$C_{28}H_{30}O_8$ 494.540

Constit. of *Tripterygium wilfordii*. NO prodn. inhibitor. Yellow oil. $[\alpha]_D^{25}$ +17.7 (c, 0.1 in MeOH). λ_{max} 229 (log ε 2.89); 279 (log ε 2.32) (MeOH)

(7S,8R,8′R)-form

3,3′,5-Tri-Me ether, 4-O-α-D-galactopyranoside

[332875-90-8]

$C_{27}H_{36}O_{12}$ 552.573

Constit. of *Daphne oleoides* and *Syringa reticulata*. Superoxide radical scavenger. Amorph. powder

3,3′,5-Tri-Me ether, 4-O-β-D-glucopyranoside

Conicaoside

[916888-05-6]

$C_{27}H_{36}O_{12}$ 552.573

Constit. of *Acanthopanax senticosus* and *Saussurea conica*. Amorph. powder. $[\alpha]_D^{20}$ -17.5 (c, 0.95 in Py)

3,3′,5-Tri-Me ether, 4,4′-di-O-β-D-glucopyranoside

[192877-39-7]

$C_{33}H_{46}O_{17}$ 714.714

Constit. of *Galium sinaicum*. Amorph. powder. Mp 139 - 140°. $[\alpha]_D^{26}$ -40 (c, 0.3 in Py). Abs. config. not certain. λ_{max} 230; 280 (MeOH)

3,3′,5-Tri-Me ether, 9-O-β-D-glucopyranoside

Alangilignoside D

[193153-53-6]

$C_{27}H_{36}O_{12}$ 552.573

Constit. of *Alangium premnifolium* and *Capparis flavicans*. Amorph. powder. $[\alpha]_D^{24}$ +10.3 (c, 0.9 in MeOH). λ_{max} 207 (log ε 4.69); 228 (log ε 4.14); 281 (log ε 3.61) (MeOH)

3,3′,5-Tri-Me ether, 9-O-[4-hydroxy-3-methoxy-E-cinnamoyl-(→6)-β-D-glucopyranoside]

[1062606-27-2]

$C_{37}H_{44}O_{15}$ 728.744

Constit. of *Asclepias curassavica*. Amorph. powder. $[\alpha]_D^{21}$ +15.3 (c, 1.08 in MeOH). λ_{max} 209 (log ε 4.55); 229 (log ε 4.28); 288 (log ε 4.02); 326 (log ε 4.15) (MeOH)

3,3′,5-Tri-Me ether, 9-O-[4-hydroxy-3-methoxy-Z-cinnamoyl-(→6)-β-D-glucopyranoside]

[1062606-31-8]

$C_{37}H_{44}O_{15}$ 728.744

Constit. of *Asclepias curassavica*

3,3′,5-Tri-Me ether, 9-O-(4-hydroxy-3-methoxycinnamoyl)

9-O-Feruloyl-5-methoxylariciresinol

[166322-15-2]

$C_{31}H_{34}O_{10}$ 566.603

Constit. of the wood of *Aralia bipinnata*. Amorph. powder. $[\alpha]_D^{25}$ +20.1 (c, 0.005 in MeOH)

3,4′,5-Tri-Me ether

7,9′-Epoxy-3,4′,5-trimethoxy-8,8′-lignan-3′,4,9-triol. 3′,4,9-Trihydroxy-3,4′,5-trimethoxy-7,9′-epoxylignan

$C_{21}H_{26}O_7$ 390.432

Constit. of the stems of *Eleutherococcus senticosus*. Exhibits weak inhibitory selectivity on DGAT2 with respect to DGAT1 (diacylglycerol acyltransferase). Amorph. brown powder. $[\alpha]_D^{25}$ -12.1 (c, 0.1 in MeOH). λ_{max} 229 (log ε 3.56); 320 (log ε 3.37) (MeOH)

3,3′,4,4′,5-Penta-Me ether, 9-angeloyl

5-Methoxyleoligin

[846060-80-8]

$C_{28}H_{36}O_8$ 500.588

Constit. of the roots of *Leontopodium alpinum*. Stimulates CYP26B1-dependent angiogenesis *in vitro* and induces arteriogenesis *in vivo* (rat). Gum. $[\alpha]_D^{20}$ +20.9 (c, 0.3 in MeOH). λ_{max} 208; 227; 280 (no solvent reported)

3,3′,5-Tri-Me ether, 9-O-[β-D-apiofuranosyl-(1→6)-β-D-glucopyranoside]

$C_{32}H_{44}O_{16}$ 684.688

Constit. of the branches of *Alangium kurzii*. Brown oil. $[\alpha]_D^{20}$ -2.7 (c, 0.17 in MeOH)

(7S,8R,8′S)-form

3,3′,5-Tri-Me ether, 9-O-β-D-xylopyranoside

Litseasin C

[2274645-20-2]

$C_{26}H_{34}O_{11}$ 522.547

Constit. of the root bark of *Litsea glutinosa*. Amorph. powder (MeOH). $[\alpha]_D^{25}$ -42 (c, 0.1 in MeOH). λ_{max} 236 (log ε 4.05) (MeOH)

(7RS,8SR,8′SR)-form

3,3′,5-Tri-Me ether

7,9′-Epoxy-3,3′,5-trimethoxy-8,8′-lignan-4,4′,9-triol. 4,4′,9-Trihydroxy-3,3′,5-trimethoxy-7,9′-epoxylignan.

5-Methoxylariciresinol

[105256-12-0]

$C_{21}H_{26}O_7$ 390.432

Constit. of *Wikstroemia elliptica* and *Rubia yunnanensis*. Cytotoxic agent. Powder. Sol. MeOH, $CHCl_3$; poorly sol. H_2O. Called 5′-Methoxylariciresinol in the lit. (Duh, 1986). The numbering of the primed and unprimed rings is inverted according to the usual scheme. λ_{max} 230 (ε 51290); 281 (ε 15490) (MeOH)

3,3′,4,5-Tetra-Me ether

7,9′-Epoxy-3,3′,4,5-tetramethoxy-8,8′-lignan-4′,9-diol. 4′,9-Dihydroxy-3,3′,4,5-tetramethoxy-7,9′-epoxylignan

[129684-05-5]

$C_{22}H_{28}O_7$ 404.459

Constit. of *Virola elongata*. Oil

3,3′,4,4′,5-Penta-Me ether

7,9′-Epoxy-3,3′,4,4′,5-pentamethoxy-8,8′-lignan-9-ol. 9-Hydroxy-3,3′,4,4′,5-pentamethoxy-7,9′-epoxylignan

[129684-04-4]

$C_{23}H_{30}O_7$ 418.486

Constit. of *Virola elongata*.

Duh, C.Y. *et al.*, *J. Nat. Prod.*, 1986, **49**, 706-709 (*5-Methoxylariciresinol*).

Kato, M.J. *et al.*, *Phytochemistry*, 1990, **29**, 1799-1810 (*Virola elongata constits*).

Hsiao, J.J. *et al.*, *Phytochemistry*, 1995, **39**, 899-902 (*Aralia bipinnata constit*).

El-Gamal, A.A. *et al.*, *Phytochemistry*, 1997, **45**, 597-600 (*tri-Me ether diglucoside*).

Yuasa, K. *et al.*, *Phytochemistry*, 1997, **45**, 611-615 (*Alangilignoside D*).

Riaz, M. *et al.*, *Z. Naturforsch., B*, 2000, **55**, 1216-1220 (*Daphne oleoides galactoside*).

Huang, W.-W. *et al.*, *Zhongguo Tianran Yaowu*, 2003, **1**, 199-203 (*Armandiside*).

Schwaiger, S. *et al.*, *Planta Med.*, 2004, **70**, 978-985 (*5-Methoxyleoligin*).

Fan, C.-Q. *et al.*, *Planta Med.*, 2006, **72**, 590-595 (*Conicaoside*).

Warashina, T. *et al.*, *Chem. Pharm. Bull.*, 2008, **56**, 1159-1163 (*Asclepias curassavica constits*).

Bi, X. *et al.*, *Bioorg. Med. Chem. Lett.*, 2011, **21**, 6426-6429 (*Armandiside, activity*).

Messner , B. *et al.*, *PLoS One*, 2013, **8**, e58342 (*5-Methoxyleoligin, activity*).

Li, J.-L. *et al.*, *Phytochem. Lett.*, 2016, **15**, 147-151 (*3,4′,5-tri-Me ether, activity*).

Ren, J. *et al.*, *Fitoterapia*, 2017, **121**, 152-158 (*Alangium kurzii constit*).

Wu, Y. *et al.*, *Phytochem. Lett.*, 2017, **20**, 259-262 (*Litseasins B,C*).
Chen, F. *et al.*, *Fitoterapia*, 2018, **129**, 133-137 (*Tripterygium wilfordii constit, activity*).
Wang, M. *et al.*, *Chem. Nat. Compd. (Engl. Transl.)*, 2019, **55**, 839-842 (*Urtica fissa constit*)

7,9′-Epoxy-8,8′-lignan-3,3′,4,4′,5′,9-hexol E91

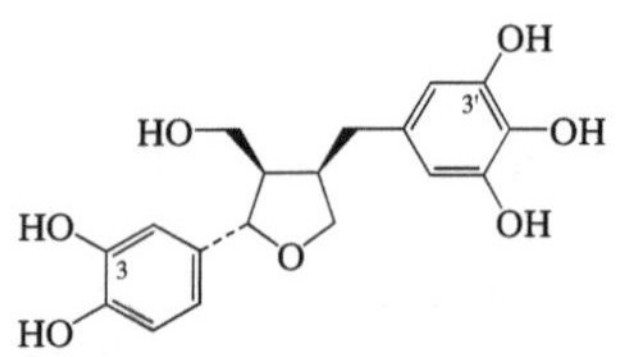

3,3′,4,4′,5′,9-Hexahydroxy-7,9′-epoxylignan
$C_{18}H_{20}O_7$ 348.351
(7S,8R,8′R)-form
3,3′,5′-Tri-Me ether
7,9′-Epoxy-3,3′,5′-trimethoxy-8,8′-lignan-4,4′,9-triol. 4,4′,9-Trihydroxy-3,3′,5′-trimethoxy-7,9′-epoxylignan. ***Justiciresinol****. 3-Methoxylariciresinol. Jusglaucinol*
[136051-41-7]
$C_{21}H_{26}O_7$ 390.432
Isol. from *Justicia glauca* and *Cestrum aurantiacum*. Cryst.. Mp 127°. Mp 153 - 155°. [α] D +21.95 (c, 0.22 in $CHCl_3$). [α] D +24.3 (c, 3.5 in MeOH). λ_{max} 207; 229; 281 (MeOH) (Derep)
3,3′,4′,5′-Tetra-Me ether
7,9′-Epoxy-3,3′,4′,5′-tetramethoxy-8,8′-lignan-4,9-diol. 4,9-Dihydroxy-3,3′,4′,5′-tetramethoxy-7,9′-epoxylignan
3,3′,4′,5′-Tetra-Me ether, 9-O-(4-hydroxy-3-methoxy-E-cinnamoyl)
Linderuca C
[1829594-30-0]
$C_{32}H_{36}O_{10}$ 580.63
Constit. of the twigs of *Lindera glauca*. Exhibits antiproliferative activity against A549, SK-OV-3, SK-MEL-2 and HCT-15 tumour cell lines; inhibits NO prodn.. Light yellow gum. $[\alpha]^{25}_{D}$ +23.4 (c, 0.5 in MeOH). λ_{max} 230 (log ε 3.8); 286 (log ε 3.9); 298 (log ε 3.2); 324 (log ε 4.0) (MeOH).

Rajendiran, C. *et al.*, *Indian J. Chem., Sect. B*, 1991, **30**, 681-683 (*Jusglaucinol*).
Subbaraju, G.V. *et al.*, *J. Nat. Prod.*, 1991, **54**, 1639-1641 (*Justiciresinol*).
Kim, K.H. *et al.*, *Chem. Pharm. Bull.*, 2014, **62**, 1136-1140 (*Linderuca C, cd, activity*)

7,9′-Epoxy-8,8′-lignan-3,3′,4,4′,7′,9-hexol E92

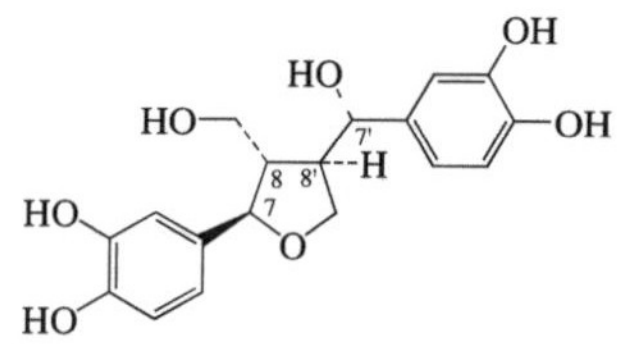

(7R,7′S,8S,8′R)-form

3,3′,4,4′,7′,9-Hexahydroxy-7,9′-epoxylignan
$C_{18}H_{20}O_7$ 348.351
Various errors and uncertainties in assignment of *R,S*-configs. noted
(7R,7′S,8S,8′R)-form
3,3′-Di-Me ether, 4-O-β-D-glucopyranoside
[1627677-51-3]
$C_{26}H_{34}O_{12}$ 538.546
Constit. of fruit of *Arctium lappa*. Exhibits strong hepatoprotective activity. Amorph. powder. $[\alpha]^{20}_{D}$ -85.1 (c, 0.07 in MeOH)
3,3′-Di-Me ether, 4′-O-β-D-glucopyranoside
[1627677-57-9]
$C_{26}H_{34}O_{12}$ 538.546
Constit. of fruit of *Arctium lappa*. Exhibits strong hepatoprotective activity. Amorph. powder. $[\alpha]^{20}_{D}$ -84.2 (c, 0.08 in MeOH)
3,3′-Di-Me ether, 9-O-β-D-glucopyranoside
[1260540-41-7]
$C_{26}H_{34}O_{12}$ 538.546
Constit. of the leaves of *Osmanthus fragrans* var. *aurantiacus*. Amorph. powder. $[\alpha]^{25}_{D}$ -12 (c, 0.13 in MeOH). λ_{max} 229 (log ε 4); 278 (log ε 3.6) (MeOH)
3,3′,7′-Tri-Me ether, 9-O-β-D-glucopyranoside
Lippianoside A
$C_{27}H_{36}O_{12}$ 552.573
Constit. of the aerial parts of *Lippia triphylla*. Powder. $[\alpha]^{25}_{D}$ +7.9 (c, 0.61 in MeOH). λ_{max} 229 (log ε 4.07); 279 (log ε 3.70) (MeOH)
3,3′,4,4′-Tetra-Me ether
7,9′-Epoxy-3,3′,4,4′-tetramethoxy-8,8′-lignan-7′,9-diol. 7′,9-Dihydroxy-3,3′,4,4′-tetramethoxy-7,9′-epoxylignan
[929894-96-2]
Constit. of the flower buds of *Magnolia fargesii*. Viscous oil. $[\alpha]^{20}_{D}$ -5 (c, 0.5 in $CHCl_3$). λ_{max} 231 (log ε 4.06); 278 (log ε 3.59) (MeOH)
(7S,7′R,8R,8′R)-form
3′-Me ether
7,9′-Epoxy-3′-methoxy-8,8′-lignan-3,4,4′,7′,9-pentol. 3,4,4′,7′,9-Pentahydroxy-3′-methoxy-7,9′-epoxylignan.
7R-Hydroxytaxiresinol
[1259948-77-0]
$C_{19}H_{22}O_7$ 362.378
Constit. of the wood of *Taxus yunnanensis*. Amorph. yellow solid. $[\alpha]^{22}_{D}$ -34.6 (c, 0.4 in MeOH)
3,3′-Di-Me ether
7,9′-Epoxy-3,3′-dimethoxy-8,8′-lignan-4,4′,7′,9-tetrol. 4,4′,7′,9-Tetrahydroxy-3,3′-dimethoxy-7,9′-epoxylignan.
7′R-Hydroxylariciresinol
[1259948-78-1]
$C_{20}H_{24}O_7$ 376.405
Constit. of *Abies pinsapo* and *Taxus yunnanensis*. Shows hepatoprotectant activity
7′-Ketone, 3,3′,4-tri-Me ether
7,9′-Epoxy-4′,9-dihydroxy-3,3′,4-trimethoxy-8,8′-lignan-7′-one.
Forsythialan E
$C_{21}H_{24}O_7$ 388.416
Constit. of the fruit of *Forsythia suspensa*. Powder. $[\alpha]^{20}_{D}$ +50 (c, 0.05 in MeOH). λ_{max} 227 (log ε 3.88); 275 (log ε 3.56); 310 (log ε 3.28) (MeOH)
(7S,7′R,8R,8′S)-form
3,3′-Di-Me ether, 4-O-β-D-glucopyranoside
[1627677-53-5]
$C_{26}H_{34}O_{12}$ 538.546
Constit. of fruit of *Arctium lappa*. Exhibits strong hepatoprotective activity. Amorph. powder. $[\alpha]^{20}_{D}$ +74.6 (c, 0.06 in MeOH)
3,3′-Di-Me ether, 4′-O-β-D-glucopyranoside
[1627677-59-1]
$C_{26}H_{34}O_{12}$ 538.546
Constit. of fruit of *Arctium lappa*. Exhibits strong hepatoprotective activity. Amorph. powder. $[\alpha]^{20}_{D}$ +74.5 (c, 0.06 in MeOH)
3,3′-Di-Me ether, 9-O-β-D-glucopyranoside
[1260540-40-6]
Constit. of the leaves of *Osmanthus fragrans* var. *aurantiacus*. Amorph. powder. $[\alpha]^{25}_{D}$ -16.7 (c, 0.13 in

MeOH). λ_{max} 205 (log ε 4.4); 229 (log ε 4.1); 278 (log ε 3.7) (MeOH)

7′-Et, 3,3′-di-Me ether

Asarinin A

[2211144-28-2]

$C_{22}H_{28}O_7$ 404.459

Constit. of the roots and rhizomes of *Asarum heterotropoides*. Amorph. powder. Stereochem. not certain

7′-Butyl, 3,3′-di-Me ether, 9-O-β-D-glucopyranoside

[1443052-67-2]

$C_{30}H_{42}O_{12}$ 594.654

Isol. from the whole plant of *Anemone rivularis* var. *flore-minore*. Amorph. powder. $[\alpha]_D^{23}$ +17.4 (c, 0.10 in MeOH)

7′-Ketone, 3,3′-di-Me ether, 4′-O-β-D-glucopyranoside

$C_{26}H_{32}O_{12}$ 536.53

Constit. of the fruit of *Arctium lappa*. Amorph. powder. $[\alpha]_D^{25}$ -53.4 (c, 0.22 in MeOH). λ_{max} 228 (log ε 3.52); 271 (log ε 3.36); 300 (log ε 3.10) (MeOH)

7′-Ketone, 3,3′,4′-tri-Me ether, 4-O-β-D-glucopyranoside

Forsythiayanoside A

$C_{27}H_{34}O_{12}$ 550.557

Constit. of fruit of *Forsythia suspense*. Solid (MeOH). $[\alpha]_D^{25}$ -12.4 (c, 0.1 in MeOH). λ_{max} 207; 228; 276; 302 (MeOH)

(7S,7′R,8S,8′R)-form

3,3′,4,4′-Tetra-Me ether

Kobusinol B

[177313-81-4]

$C_{22}H_{28}O_7$ 404.459

Constit. of *Magnolia kobus*

(7S,7′S,8R,8′R)-form

3,3′-Di-Me ether

8′-Epitanegool. *7S-Hydroxylariciresinol*

[158249-99-1]

$C_{20}H_{24}O_7$ 376.405

Constit. of *Abies pinsapo*, *Brassica fruiticulosa*, *Helianthus annuus* and *Taxus mairei*. Gum. $[\alpha]_D^{25}$ -10 (c, 2.5 in MeOH). Some confusion over names and stereochem. Yang (1999) reports (-)-8′-Epitanegool from *T. mairei* with the opposite 7′-config. Also reported (Cutillo, 2003) as (-)-Tanegol (sic.)

3,3′,4,4′-Tetra-Me ether

[929894-95-1]

Constit. of the flower buds of *Magnolia fargesii*. Viscous oil. $[\alpha]_D^{20}$ +25.4 (c, 0.5 in $CHCl_3$). λ_{max} 230 (log ε 3.84); 277 (log ε 3.37) (MeOH)

3,4:3′,4′-Bis(methylene) ether

7,9′-Epoxy-3,4:3′,4′-bis(methylenedioxy)-8,8′-lignan-7′,9-diol. 7′,9-Dihydroxy-3,4:3′,4′-bis(methylenedioxy)-7,9′-epoxylignan.

7′-Hydroxydihydrosesamin

[139955-56-9] (di-Ac)

$C_{20}H_{20}O_7$ 372.373

Isol. from leaves of *Juniperus thurifera*. Oil (as di-Ac). λ_{max} 235 (ε 23800); 284 (ε 15300) (MeOH) (Berdy)

7′-Ketone, 3,3′-di-Me ether

7,9′-Epoxy-4,4′,9-trihydroxy-3,3′-dimethoxy-8,8′-lignan-7′-one.

Vladinol D

[133084-47-6]

$C_{20}H_{22}O_7$ 374.389

Isol. from underground parts of *Vladimiria souliei*. Also from *Vladimiria muliensis* and *Rubia yunnanensis*. Oil

7′-Ketone, 3,3′-di-Me ether, 4-O-β-D-glucopyranoside

Vladinol D 4-glucoside

[1593092-62-6]

$C_{26}H_{32}O_{12}$ 536.53

Constit. of *Vladimiria muliensis*. Amorph. powder. $[\alpha]_D^{20}$ -31.8 (c, 2.2 in MeOH). λ_{max} 204 (log ε 3.82); 224 (log ε 3.58); 275 (log ε 3.68); 301 (log ε 3.09) (MeOH)

7′-Ketone, 3,3′,4,4′-tetra-Me ether

7,9′-Epoxy-9-hydroxy-3,3′,4,4′-tetramethoxy-8,8′-lignan-7′-one.

Magnone A

[209904-68-7]

$C_{22}H_{26}O_7$ 402.443

Constit. of flower buds of *Magnolia fargesii* and *Zanthoxylum podocarpum*. Platelet activating factor antagonist. Shows antiinflammatory activity. Cryst. (MeOH). Mp 149 - 150°. $[\alpha]_D^{25}$ +24.7 (c, 0.5 in MeOH). λ_{max} 230 (log ε 4.37); 277 (log ε 4.11); 304 (log ε 3.91) (MeOH)

7′-Ketone, 3,4:3′,4′-bis(methylene) ether

7,9′-Epoxy-9-hydroxy-3,4:3′,4′-bis(methylenedioxy)-8,8′-lignan-7′-one. ***8′-Episesaminone***

[198014-98-1]

$C_{20}H_{18}O_7$ 370.357

Constit. of the seeds of *Sesamum indicum* and of *Wikstroemia chamaedaphne*. $[\alpha]_D$ +126.2 (c, 0.1 in $CHCl_3$). λ_{max} 230 (log ε 4.76); 280 (log ε 3.75); 312 (log ε 2.87) (MeOH)

7′-Ketone, 3,4:3′,4′-bis(methylene) ether, 9-O-[β-D-glucopyranosyl-(1→2)-β-D-glucopyranoside]

8′-Episesaminone 9-sophoroside

[914107-16-7]

$C_{32}H_{38}O_{17}$ 694.639

Constit. of the perisperm of *Sesamum indicum*. Prisms (Me_2CO). Mp 253°. $[\alpha]_D$ +20 (c, 0.2 in MeOH). Config. incorrectly shown in the lit.

(7S,7′S,8R,8′S)-form

Dide-O-methyltanegool [936478-88-5]

Constit. of the fruits of *Morinda citrifolia*. Potent inhibitor of 15-lipoxygenase. Greenish-brown semisolid. $[\alpha]_D^{25}$ +5.6 (c, 0.16 in MeOH). λ_{max} 227 (log ε 4.08); 279 (log ε 3.53) (MeOH)

3-Me ether

7,9′-Epoxy-3-methoxy-8,8′-lignan-3′,4,4′,7′,9-pentol. 3′,4,4′,7′,9-Pentahydroxy-3-methoxy-7,9′-epoxylignan.

3′-O-Demethyltanegool

[252333-75-8]

$C_{19}H_{22}O_7$ 362.378

Constit. of *Taxus mairei*. Gum. $[\alpha]_D^{24}$ +66.4 (c, 1.1 in MeOH)

3,3′-Di-Me ether

Tanegool

[131723-84-7]

Constit. of *Helianthus annuus*, *Magnolia fargesii* and *Taxus yunnanensis*. Also from the extract of Caulis Trachelospermi. Shows superoxide radical scavenging ability. $[\alpha]_D^{29}$ +64.7 (c, 0.45 in MeOH). Yang (1999) and Macias (2004) give incorrect stereochem. for Tanegool and its homologues. The correct stereochem. is given in more recent refs. (Lee 2009; Albertson 2015)

3,3′-Di-Me ether, 4-O-β-D-glucopyranoside

[1627677-55-7]

$C_{26}H_{34}O_{12}$ 538.546

Constit. of fruits of *Arctium lappa*. Hepatoprotectant. Amorph. powder. $[\alpha]_D^{20}$ +29.9 (c, 0.07 in MeOH)

3,3′-Di-Me ether, 4′-O-β-D-glucopyranoside

[1627677-61-5]

$C_{26}H_{34}O_{12}$ 538.546

Constit. of fruits of *Arctium lappa*. Hepatoprotectant. Amorph. powder. $[\alpha]_D^{20}$ +55.2 (c, 0.06 in MeOH)

3,3′-Di-Me ether, 7′-O-β-D-allopyranoside

Tanegoside B

[131723-82-5]

$C_{26}H_{34}O_{12}$ 538.546

Isol. from whole plant of *Trachelospermum liukiuense*. $[\alpha]_D^{30}$ -24.7 (c, 1.94 in MeOH)

3,3′-Di-Me ether, 7′-O-β-D-glucopyranoside

Tanegoside A

[131653-21-9]

$C_{26}H_{34}O_{12}$ 538.546

Constit. of *Osmanthus fragrans* var. *aurantiacus* and *Trachelospermum liukiuense*. $[\alpha]_D^{31}$ -20 (c, 2.53 in MeOH)

3,3′,4′-Tri-Me ether, 4-O-β-D-glucopyranoside

Tinosposide B

[725743-82-8]

$C_{27}H_{36}O_{12}$ 552.573

Constit. of the stems of *Tinospora sinensis*. Amorph. powder. $[\alpha]_D^{25}$ -10 (c, 0.3 in MeOH). Ref (Li, 2004) quotes 7′*S*- config. but draws as correct 7′*R*-. λ_{max} 228 (log ε 4.3); 277 (log ε 3.8) (MeOH)

3,3′,4,4′-Tetra-Me ether

Fargesol

[128855-64-1]

Isol. from flowerbuds of *Magnolia fargesii* and from *Magnolia praecocissima*. Viscous oil. $[\alpha]_D$ -35.4 (c, 0.10 in MeOH). Errors in *R,S*-assignment in the 2002 paper assigning abs. config.

3,3′,4,4′-Tetra-Me ether, 9-Ac

9-O-Acetylfargesol

[449172-61-6]

$C_{24}H_{30}O_8$ 446.496

Constit. of the seeds of *Magnolia praecocissima*. Oil. $[\alpha]_D^{21}$ +35.2 (c, 1.2 in $CHCl_3$). Called 9′-acetyl in the ref. Errors in *R,S*- assignments in the paper assigning config. (2002)

3′,4′-Methylene, 3-Me ether

7,9′-Epoxy-3-methoxy-3′,4′-methylenedioxy-8,8′-lignan-4,7′,9-triol. 4,7′,9-Trihydroxy-3-methoxy-3′,4′-methylenedioxy-7,9′-epoxylignan

$C_{20}H_{22}O_7$ 374.389

Constit. of roots of *Asiasarum heterotropoides*. Amorph. solid. $[\alpha]_D^{20}$ +39.2 (c, 0.1 in MeOH). λ_{max} 227 (log ε 4.92); 283 (log ε 4.44) (MeOH)

7′-Ketone, 3,3′-di-Me ether

[252333-76-9]

Constit. of *Taxus mairei* and *Melia toosendan*. Gum. $[\alpha]_D^{24}$ -60.4 (c, 1 in Me_2CO)

7′-Ketone, 3,3′-di-Me ether, 4,4′-di-O-β-D-glucopyranoside

Arctiinoside B

[1421638-97-2]

$C_{32}H_{42}O_{17}$ 698.671

Constit. of the dried fruit of *Arctium lappa*. Amorph. powder. $[\alpha]_D^{20}$ -14.7 (c, 0.03 in MeOH). λ_{max} 225; 278; 300 (MeOH)

7′-Ketone, 3,4-methylene, 3′,4′-di-Me ether

7,9′-Epoxy-9-hydroxy-3′,4′-dimethoxy-3,4-methylenedioxy-8,8′-lignan-7′-one. ***Magnolone***

[212703-86-1]

$C_{21}H_{22}O_7$ 386.4

Constit. of *Magnolia coco* and *Zanthoxylum armatum*. Insect antifeedant. Viscous oil. $[\alpha]_D^{20}$ -31 (c, 0.2 in $CHCl_3$). $[\alpha]_D^{20}$ -19 (c, 0.4 in MeOH). λ_{max} 202 (log ε 3.34); 217 (log ε 3.84); 227 (log ε 3.84); 230 (log ε 3.7); 249 (log ε 2.16) (MeOH)

7′-Ketone, 3,4:3′,4′-bis(methylene) ether

Sesaminone

[157207-93-7]

Prod. by *Streptomyces* sp. IT-44. Also constit. of *Zanthoxylum ailanthoides*, *Zanthoxylum nitidum*, *Dioscorea spongiosa* and *Magnolia coco*. Shows antiosteoporotic activity. Syrup. Sol. MeOH, $CHCl_3$, C_6H_6, EtOAc; poorly sol. H_2O. $[\alpha]_D^{28}$ -25 (c, 0.14 in MeOH). λ_{max} 204 (ε 63400); 231 (ε 20700); 278 (ε 7900); 310 (ε 7150) (MeOH) (Berdy)

7′-Ketone, 3,4:3′,4′-bis(methylene) ether, 9-O-[β-D-glucopyranosyl-(1→3)-β-D-glucopyranoside]

Mananthoside H

[1198617-99-0]

$C_{32}H_{38}O_{17}$ 694.639

Constit. of *Mananthes patentiflora*. Shows antioxidant activity against hydrogen peroxide induced impairment of PC12 cells. Amorph. powder. Mp 115 - 117°. $[\alpha]_D^{21}$ +13.7 (c, 0.55 in MeOH). λ_{max} 195 (log ε 4.41); 204 (log ε 4.94); 225 (log ε 4.44); 281 (log ε 4) (MeOH)

(7S,7′ξ,8R,8′R)-form

3,3′,4,4′-Tetra-Me ether

Biondinin B

[160550-16-3]

[172584-53-1] (unspecified stereochem.)

Constit. of the flower buds of *Magnolia biondii*. Oil. $[\alpha]_D$ +80 (c, 0.1 in $CHCl_3$). λ_{max} 204 (log ε 4.04); 229 (log ε 3.94); 278 (log ε 3.64) (MeOH)

3,3′,4,4′,7′-Penta-Me ether

7,9′-Epoxy-3,3′,4,4′,7′-pentamethoxy-8,8′-lignan-9-ol. 9-Hydroxy-3,3′,4,4′,7′-pentamethoxy-7,9′-epoxylignan.

Biondinin E

[160472-98-0]

$C_{23}H_{30}O_7$ 418.486

Constit. of the flower buds of *Magnolia biondii*. Oil. $[\alpha]_D$ +16 (c, 0.1 in $CHCl_3$). Error of assignment of 8,8′-configs. in the text of the ref.. λ_{max} 206 (log ε 4.37); 228 (log ε 4.05); 277 (log ε 3.58) (MeOH)

(7S,7ξ,8R,8′S)-form

3,3′,7′-Tri-Me ether, 4-O-β-D-glucopyranoside

Forsythiayanoside B

$C_{27}H_{36}O_{12}$ 552.573

Constit. of fruits of *Forsythia suspense*. Viscous oil. $[\alpha]_D^{25}$ -10.3 (c, 0.05 in MeOH). λ_{max} 207; 227; 279 (MeOH)

(7R*,7′R*,8S*,8′S*)-form

3,3′,7′-Tri-Me ether, 4-O-β-D-glucopyranoside

[2169952-26-3]

$C_{27}H_{36}O_{12}$ 552.573

Constit. of roots of *Cyclea racemosa*. Amorph. powder

7′-Ketone, 3,3′-di-Me ether

[2225711-75-9]

$C_{20}H_{22}O_7$ 374.389

Constit. of stems of *Syringa pinnatifolia*. Moderate antioxidant. Needles. Mp 201 - 203°. $[\alpha]_D^{25}$ +15.2 (c, 0.1 in MeOH)

(7R*,7′S*,8S*,8′R*)-form

3,3′-Di-Me ether, 4-O-β-D-glucopyranoside

Lanicepside B. Neoarctin A‡

[944746-46-7]

Constit. of *Arctium lappa* and *Saussurea laniceps*. Amorph. powder (MeOH). $[\alpha]_D^{25}$ -61 (c, 0.15 in MeOH) (Lanicepside B). $[\alpha]_D^{20}$ -15.7 (c, 0.1 in MeOH) (Neoarctin A). Possible enantiomers. Not to be confused with Neoarctin A, N4. λ_{max} 229 (log ε 4.2); 279 (log ε 3.7) (MeOH) (Lanicepside B). [neutral] λ_{max} 247; 268; 273 (MeOH) (Neoarctin A)

3,3′-Di-Me ether, 4′-O-β-D-glucopyranoside

Lanicepside A

[944746-27-4]

$C_{26}H_{34}O_{12}$ 538.546

Constit. of *Boschniakia himalaica, Saussurea laniceps* and *Sinopodophyllum emodi*. Powder. $[\alpha]_D^{25}$ -63 (c, 0.16 in MeOH). λ_{max} 229 (log ε 4.2); 278 (log ε 3.8) (MeOH)

3,3′-Di-Me ether, 9-Ac, 4-O-β-D-glucopyranoside

9-O-Acetyllanicepside B

[1403337-50-7]

$C_{28}H_{36}O_{13}$ 580.583

Constit. of *Boschniakia himalaica*. Light yellow powder. $[\alpha]_D^{20}$ -19 (c, 0.1 in MeOH)

(7R*,7′ξ,8R*,8′R*)-form

3,4:3′,4′-Bis(methylene) ether, 9-O-β-D-glucopyranoside

Tibeticoside A

[241482-41-7]

$C_{26}H_{30}O_{12}$ 534.514

Constit. of *Lancea tibetica*. Amorph. powder. $[\alpha]_D^{20}$ -120 (c, 0.5 in MeOH). λ_{max} 207 (log ε 4.52); 235 (log ε 3.34); 286 (log ε 3.32) (MeOH)

(7R*,7′ξ,8S*,8′R*)-form

3,3′,7′-Tri-Me ether

7,9′-Epoxy-3,3′,7-trimethoxy-8,8′-lignan-4,4′,9-triol. 4,4′,9-Trihydroxy-3,3′,7′-trimethoxy-7,9′-epoxylignan. Tanegool 7′-methyl ether.

7′-O-Methyltanegool

[848845-01-2]

$C_{21}H_{26}O_7$ 390.432

Constit. of *Diphylleia sinensis* and *Sinopodophyllum emodi*. Shows potent cytotoxic acivity against HeLa and KB cells. Oil. $[\alpha]_D^{25}$ -19.4 (c, 0.14 in $CHCl_3$). λ_{max} 208 (log ε 2.73); 230 (log ε 1.82); 280 (log ε 0.81) (MeOH)

(7R*,7′ξ,8S*,8′S*)-form

3,3′-Di-Me ether, 9-O-(4-hydroxy-E-cinnamoyl)

[864452-89-1]

$C_{29}H_{30}O_9$ 522.55

Constit. of the bark of *Larix olgensis* var. *koreana*. Powder (MeOH). Mp 106 - 152° (sic.). $[\alpha]_D^{20}$ -23 (c, 2.03 in MeOH)

3,3′,7′-Tri-Me ether

(+)-7′-Methoxylariciresinol

[1426550-86-8]

$C_{21}H_{26}O_7$ 390.432

Constit. of *Firmiana simplex* and *Sinopodophyllum emodi*. Shows moderate cytotoxic activity against HeLa cells. Oil. $[\alpha]_D^{25}$ +9 (c, 0.07 in $CHCl_3$). $[\alpha]_D^{25}$ +25 (c, 0.03 in MeOH). λ_{max} 208 (log ε 2.54); 230 (log ε 1.34); 280 (log ε 0.65) (MeOH)

(7ξ,7′ξ,8ξ,8′ξ)-form

7′-Et, 3,3′-di-Me ether

7,9′-Epoxy-7′-ethoxy-3,3′-dimethoxy-8,8′-lignan-4,4′,9-triol. 7′-Ethoxy-4,4′,9-trihydroxy-3,3′-dimethoxy-7,9′-epoxylignan

[834888-44-7]

$C_{22}H_{28}O_7$ 404.459

Constit. of *Patrinia scabra*. Mp 114 - 116°

(7R,8R,8′S)-form

7′-Ketone, 3,4:3′,4′-bis(methylene) ether

7-Episesaminone

[1616736-26-5]

$C_{20}H_{18}O_7$ 370.357

Constit. of roots of *Asarum heterotropoides*. Exhibits significant antiinflammatory activity. Solid (MeOH). $[\alpha]_D^{21}$ -23.1 (c, 0.55 in MeOH). λ_{max} 209; 230; 280 (MeOH)

(7R,8S,8′S)-form

7′-Ketone, 3,3′-di-Me ether

[163341-69-3]

$C_{20}H_{22}O_7$ 374.389

Constit. of *Abies mariesii*

7′-Ketone, 3,3′-di-Me ether, 4′-O-β-D-glucopyranoside

Forsythialanside B

[1616079-09-4]

$C_{26}H_{32}O_{12}$ 536.53

Constit. of fruits of *Forsythia suspensa*. Exhibits mild antiviral activity against influenza A (H1N1) virus. Yellow gum. $[\alpha]_D^{20}$ -23.2 (c, 0.25 in MeOH). λ_{max} 227 (log ε 3.65); 273 (log ε 3.42); 305 (log ε 3.16) (MeOH)

7′-Ketone, 3,3′,4′-tri-Me ether, 4-O-β-D-glucopyranoside

Forsythialanside A

[1616079-08-3]

$C_{27}H_{34}O_{12}$ 550.557

Constit. of fruits of *Forsythia suspensa*. Inhibitor of respiratory syncytial virus (RSV). Yellow gum. $[\alpha]_D^{20}$ -36.8 (c, 0.25 in MeOH). λ_{max} 229 (log ε 4.18); 276 (log ε 3.92); 304 (log ε 3.72) (MeOH)

(7RS,7′RS,8SR,8′RS)-form

9-Carboxylic acid, 7′-Et, 3,3′-di-Me ether, 2-(4-hydroxyphenyl)ethylamide

[2252244-26-9]

$C_{30}H_{35}NO_8$ 537.609

Constit. of the rhizomes of *Alocasia macrorrhiza*. NO prodn. inhibitor. Amorph. yellow powder. $[\alpha]_D^{27}$ +0.6 (c, 0.5 in MeOH). λ_{max} 224; 280; 318 (MeOH).

Abe, F. *et al., Chem. Pharm. Bull.*, 1990, **38**, 2143-2145 (*Tanegosides A,B*).

Huang, Y.L. *et al., Planta Med.*, 1990, **56**, 237-239 (*Fargesol*).

Tan, R.X. *et al., Planta Med.*, 1990, **56**, 475-478 (*Vladinol D*).

San Feliciano, A. *et al., Phytochemistry*, 1992, **31**, 267-270 (*7′-Hydroxydihydrosesamin*).

Chiung, Y.-M. *et al., J. Antibiot.*, 1994, **47**, 487-491 (*Sesaminone*).

Barrero, A.F. *et al., J. Nat. Prod.*, 1994, **57**, 713-719 (*7-Hydroxylariciresinols*).

Omori, S. *et al., Mokuzai Gakkaishi*, 1994, **40**, 1107-1118 (*Abies mariesii constit*).

Kim, Y.G. *et al., CA*, 1996, **125**, 5468t (*Kobusinol B*).

Ma, Y.-L. *et al., Phytochemistry*, 1996, **41**, 287-288 (*Biondinins B,E*).

Marchand, P.A. *et al., J. Nat. Prod.*, 1997, **60**, 1189-1192 (*8′-Episesaminone*).

Maioli, A.T. *et al., J. Org. Chem.*, 1997, **62**, 7413-7417 (*Sesaminone, synth, abs config*).

Jung, K.Y. *et al., J. Nat. Prod.*, 1998, **61**, 808-811 (*Magnone A*).

Yu, H.-J. *et al., J. Nat. Prod.*, 1998, **61**, 1017-1019 (*Magnolone*).

Yang, S.-J. *et al., J. Chin. Chem. Soc. (Taipei)*, 1999, **46**, 811-818 (*3′-O-Demethyltanegool, 8′-Epitanegool*).

Su, B.N. *et al.*, *Planta Med.*, 1999, **65**, 558-561 (*Tibeticoside A*).
Ahn, K-S. *et al.*, *Biol. Pharm. Bull.*, 2001, **24**, 1085-1087 (*Magnone A, activity*).
Yoda, H. *et al.*, *Synlett*, 2001, 400-402 (*Sesaminone, synth*).
Takahashi, H. *et al.*, *Chem. Pharm. Bull.*, 2002, **50**, 541-543 (*Fargesol, 9-O-Acetylfargesol, abs config*).
Cutillo, F. *et al.*, *J. Agric. Food Chem.*, 2003, **51**, 6165-6172 (*8′-Epitanegool*).
Li, T.-Z. *et al.*, *Yaoxue Xuebao*, 2003, **38**, 520-522 (*Patrinia scabra lignan*).
Li, W. *et al.*, *Chem. Pharm. Bull.*, 2004, **52**, 638-640 (*Tinosposide B*).
Macias, F.A. *et al.*, *J. Agric. Food Chem.*, 2004, **52**, 6443-6447 (*Tanegool*).
Nguyen, N.T. *et al.*, *Planta Med.*, 2004, **70**, 29-33 (*7′R-Hydroxylariciresinol, hepatoprotective activity*).
Yin, J. *et al.*, *Planta Med.*, 2004, **70**, 54-58 (*Sesaminone, activity*).
Yang, B.-H. *et al.*, *J. Nat. Prod.*, 2005, **68**, 1175-1179 (*Larix olgensis coumarate*).
Grougnet, R. *et al.*, *J. Agric. Food Chem.*, 2006, **54**, 7570-7574 (*Episesaminone 9-sophoroside*).
Lee, J. *et al.*, *Chem. Pharm. Bull.*, 2007, **55**, 137-139 (*Magnolia fargesii tetra-Me ethers*).
Zhou, Z.-W. *et al.*, *Helv. Chim. Acta*, 2007, **90**, 951-956 (*Lanicepsides A,B*).
Yong, M. *et al.*, *J. Asian Nat. Prod. Res.*, 2007, **9**, 541-544 (*Neoarctin A*).
Deng, S. *et al.*, *J. Nat. Prod.*, 2007, **70**, 859-862 (*Dide-O-methyltanegool*).
Nakato, T. *et al.*, *J. Nat. Prod.*, 2007, **70**, 1588-1592 (*Magnolone, synth, abs config*).
Pat. Coop. Treaty (WIPO), 2008, 2008 067410 (Tahitian Noni International) (*Dide-O-methyltanegool, activity*).
Lee, J. *et al.*, *Chem. Pharm. Bull.*, 2009, **57**, 298-301 (*Tanegool, activity*).
Tian, J. *et al.*, *Planta Med.*, 2009, **75**, 534-537 (*Mananthoside H*).
Machida, K. *et al.*, *Helv. Chim. Acta*, 2010, **93**, 2164-2175 (*Osmanthus fragrans glucosides*).
Pandey, G. *et al.*, *Tetrahedron Lett.*, 2010, **51**, 2975-2978 (*Magnolone, synth*).
Tezuka, Y. *et al.*, *J. Nat. Prod.*, 2011, **74**, 102-105 (*7R-Hydroxytaxiresinol*).
Yang, X. *et al.*, *Chem. Nat. Compd. (Engl. Transl.)*, 2012, **48**, 555-558 (*9-O-Acetyllanicepside B*).
Yang, Y.-N. *et al.*, *J. Asian Nat. Prod. Res.*, 2012, **14**, 981-985 (*Arctiinoside B*).
Ding, Y. *et al.*, *Phytochem. Lett.*, 2012, **5**, 668-672 (*7′-butyl 3,3′-di-Me 9-glucoside*).
Sun, Y.-J. *et al.*, *Planta Med.*, 2012, **78**, 480-484 (*Sinopodophyllum 3,3′,7′-tri-Me ethers*).
Chen, J.-J. *et al.*, *Planta Med.*, 2013, **79**, 1470-1473 (*Vladinol D 4-glucoside*).
Yang, Y.-N. *et al.*, *J. Agric. Food Chem.*, 2014, **62**, 9095-9102 (*Arctium lappa constits*).
Huang, J. *et al.*, *J. Asian Nat. Prod. Res.*, 2014, **16**, 387-392 (*7-Episesaminone*).
Yan, X.J. *et al.*, *J. Asian Nat. Prod. Res.*, 2014, **16**, 602-610 (*Forsythiayanosides A,B, cd*).
Lee, J. *et al.*, *Molecules*, 2014, **19**, 122-138 (*Asiasarum heterotropoides constit, cd*).
Liu, X.-T. *et al.*, *Molecules*, 2014, **19**, 11560-11571 (*Caulis Trachelospermi extract constit*).
Li, C. *et al.*, *Phytochemistry*, 2014, **104**, 105-113 (*Forsythialansides A,B, cd, activity*).
Albertson, A.K.F. *et al.*, *Angew. Chem., Int. Ed.*, 2015, **54**, 2204-2208 (*Tanegool, synth*).
Huang, X.-Y. *et al.*, *J. Asian Nat. Prod. Res.*, 2015, **17**, 504-511 (*Arctium lappa 7′-ketone 4′-glucoside*).
Zhang, Y. *et al.*, *Molecules*, 2015, **20**, 21946-21959 (*Lippianoside A*).
Woo, K.W. *et al.*, *Bioorg. Med. Chem. Lett.*, 2016, **26**, 730-733 (*(+)-7′-Methoxylariciresinol*).
Shao, J.-H. *et al.*, *Chem. Nat. Compd. (Engl. Transl.)*, 2017, **53**, 1025-1027 (*Cyclea racemosa constit*).
Huang, W. *et al.*, *Fitoterapia*, 2017, **117**, 126-132 (*Alocasia macrorrhiza constit, activity*).
Jing, Y. *et al.*, *Molecules*, 2017, **22**, 125:1-20 (*Asarinin A*).
Wang, Q.-H. *et al.*, *Chem. Nat. Compd. (Engl. Transl.)*, 2018, **54**, 18-21 (*Syringa pinnatifolia constit, activity*).
Zhang, W. *et al.*, *Molecules*, 2018, **23**, 617:1-10 (*Magnone A, Magnolone, activity*).
Bai, W. *et al.*, *Zhongcaoyao*, 2018, **49**, 1504-1507 (*7′-O-Methyltanegool*).
Li, C. *et al.*, *Phytochem. Lett.*, 2019, **32**, 115-118 (*Forsythialan E, cd*)

7,9′-Epoxy-8,8′-lignan-3,3′,4,4′,8,9-hexol E93

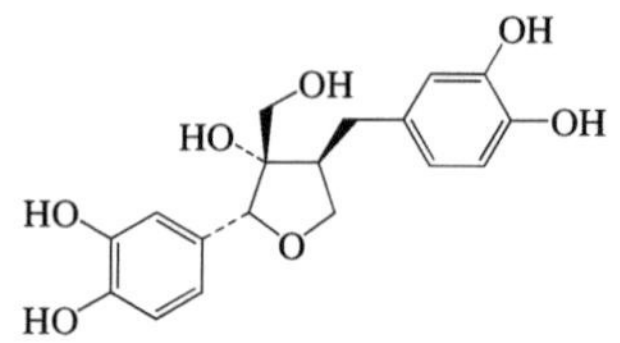

(*7R,8S,8′S*)*-form*

3,3′,4,4′,8,9-Hexahydroxy-7,9′-epoxylignan

$C_{18}H_{20}O_7$ 348.351

(7R,8S,8′S)-form

3,3′-Di-Me ether

7,9′-Epoxy-3,3′-dimethoxy-8,8′-lignan-4,4′,8,9-tetrol. 4,4′,8,9-Tetrahydroxy-3,3′-dimethoxy-7,9′-epoxylignan. ***Berchemol***. *Vladinol B*

[126882-59-5]

$C_{20}H_{24}O_7$ 376.405

Constit. of *Berchemia racemosa, Saussurea medusa, Centaurea montana, Comastoma pedunculatum, Vladimiria muliensis* and *Vladimiria souliei*. Potent free radical scavenger. Needles (MeOH). Mp 188 - 189°. $[\alpha]^{20}_{D}$ -7.9 (c, 0.3 in MeOH). Sakurai *et al* (1989) give wrong name, corrected in 1990 erratum. λ_{max} 231 (log ε 4.1); 281 (log ε 3.7) (MeOH)

3,3′-Di-Me ether, 4-O-β-D-glucopyranoside

Medusaside B

[640756-25-8]

$C_{26}H_{34}O_{12}$ 538.546

Constit. of *Saussurea medusa*. Amorph. powder. $[\alpha]^{20}_{D}$ -34.9 (c, 0.53 in MeOH). λ_{max} 227 (log ε 3.89); 279 (log ε 3.43) (MeOH)

3,3′-Di-Me ether, 4′-O-β-D-glucopyranoside

Berchemol 4′-O-glucoside

[473443-03-7]

$C_{26}H_{34}O_{12}$ 538.546

Constit. of the roots of *Valeriana officinalis* and from *Centaurea montana*. $[\alpha]^{21}_{D}$ -15 (c, 0.01 in DMSO). λ_{max} 225; 278 (MeOH aq.)

3′,4′-Methylene, 3-Me ether

7,9′-Epoxy-3-methoxy-3′,4′-methylenedioxy-8,8′-lignan-4,8,9-triol. 4,8,9-Trihydroxy-3-methoxy-3′,4′-methylenedioxy-7,9′-epoxylignan. ***Mandshuricol B***

[1253121-48-0]

$C_{20}H_{22}O_7$ 374.389

Constit. of the leaves of *Syringa reticulata* var. *mandshurica*. PAF antagonist. Oil. $[\alpha]^{25}_{D}$ +24.8 (c, 0.01 in MeOH). Incorrect name given in ref.. λ_{max} 204 (log ε 4.52); 219 (log ε 3.38); 280 (log ε 3.2) (MeOH)

3,4:3′,4′-Bis(methylene) ether

7,9′-Epoxy-3,4:3′,4′-bis(methylenedioxy)-8,8′-lignan-8,9-diol. 8,9-Dihydroxy-3,4:3′,4′-bis(methylenedioxy)-7,9′-epoxylignan. ***Mandshuricol A***

[172342-92-6]

$C_{20}H_{20}O_7$ 372.373

Constit. of the leaves of *Syringa reticulata* var. *mandshurica*. PAF antagonist. Oil. $[\alpha]^{25}_{D}$ +26.5 (c, 0.02 in MeOH). λ_{max} 205 (log ε 4.55); 218 (log ε 3.36); 279 (log ε 3.25) (MeOH)

(7R*,8R*,8′R*)-form

3,3′-Di-Me ether

[1257096-68-6]

$C_{20}H_{24}O_7$ 376.405

Constit. of *Viburnum propinquum*. Powder. $[\alpha]^{25}_{D}$ -23.2 (c, 0.1 in MeOH). λ_{max} 230 (log ε 1.62); 280 (log ε 0.58) (MeOH).

Sakurai, N. *et al.*, *Chem. Pharm. Bull.*, 1989, **37**, 3311-3315 (*Berchemol, cryst struct, abs config*).
Sakurai, N. *et al.*, *Chem. Pharm. Bull.*, 1990, **38**, 1801 (*Berchemol, erratum*).
Tan, R.X. *et al.*, *Planta Med.*, 1990, **56**, 475-477 (*Vladinol B*).

Schumacher, B. *et al.*, *J. Nat. Prod.*, 2002, **65**, 1479-1485 (*Berchemol 4′-glucoside*).
Fan, C.-Q. *et al.*, *Bioorg. Med. Chem.*, 2003, **11**, 703-708 (*Medusaside B*).
Wang, X.-Y. *et al.*, *Planta Med.*, 2009, **75**, 1262-1265 (*Viburnum 3,3′-di-Me ether*).
Xu, Q.-M. *et al.*, *Chem. Nat. Compd. (Engl. Transl.)*, 2010, **46**, 366-369 (*Mandshuricols A,B*).
Chen, J-W. *et al.*, *Planta Med.*, 2013, **79**, 1470-1473 (*Berchemol, activity*)

7,9′-Epoxy-8,8′-lignan-3,3′,4,4′,8′,9-hexol E94

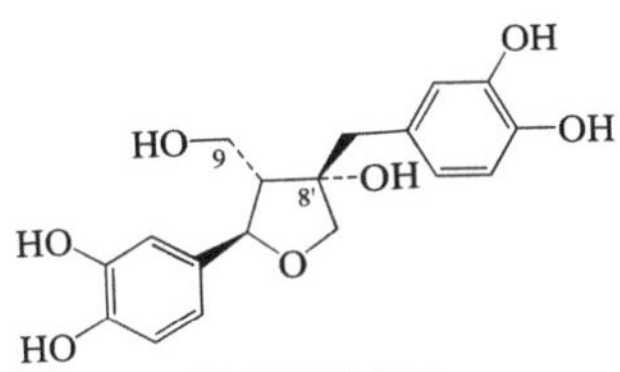

(7R,8S,8'S)-form

3,3′,4,4′,8′,9-Hexahydroxy-7,9′-epoxylignan
$C_{18}H_{20}O_7$ 348.351
(7R,8S,8′S)-form
3,3′-Di-Me ether
7,9′-Epoxy-3,3′-dimethoxy-8,8′-lignan-4,4′,8′,9-tetrol. 4,4′,8′,9-Tetrahydroxy-3,3′-dimethoxy-7,9′-epoxylignan.
Diepiolivil
[1449572-44-4]
$C_{20}H_{24}O_7$ 376.405
Isol. from stems of *Cameraria latifolia*. Amorph. powder. $[\alpha]^{20}_{D}$ -53.5 (c, 0.15 in MeOH). λ_{max} 206 (log ε 4.62); 230 (log ε 4.23); 280 (log ε 3.89) (MeOH)
(7S,8R,8′S)-form
3,3′-Di-Me ether
Olivil. *Vladinol C*
[2955-23-9]
[95701-39-6] (unspecified stereochem.)
$C_{20}H_{24}O_7$ 376.405
Isol. from resin of *Olea europaea*, from *Olea africana. Abies tazaotana, Cerbera manghas* and *Vladimiria souliei*. First isol. in 1816. Antioxidant. Potent feeding stimulant for the olive weevil *Dyscerus perforatus*. Cryst. + $1H_2O$ (H_2O). Mp 127° (105°). Mp 142 - 143° (anhyd.). $[\alpha]^{20}_{D}$-127 (no solvent reported). $[\alpha]^{20}_{D}$-49 (c, 0.007 in $CHCl_3$)
3,3′-Di-Me ether, 4-O-β-D-glucopyranoside
[56440-73-4]
$C_{26}H_{34}O_{12}$ 538.546
Constit. of *Cerbera manghas*, other *Cerbera* spp. and *Eucommia ulmoides*. Amorph. powder. $[\alpha]^{23}_{D}$ -66.9 (c, 0.48 in MeOH) (-33.3). λ_{max} 229; 279 (MeOH)
3,3′-Di-Me ether, 4′-O-β-D-glucopyranoside
[76880-93-8]
$C_{26}H_{34}O_{12}$ 538.546
Constit. of *Cerbera manghas, Cerbera distachya*, other *Cerbera* sp., *Epimedium* sp., *Ligustrum* spp., *Syringa* sp. and other plant spp.. $[\alpha]^{23}_{D}$-76 (c, 0.73 in MeOH) (-22.7)
3,3′-Di-Me ether, 9-O-β-D-glucopyranoside
[1105067-00-2]
$C_{26}H_{34}O_{12}$ 538.546
Constit. of the roots of *Urtica triangularis*. Amorph. powder. $[\alpha]^{20}_{D}$-10.3 (c, 0.1 in MeOH)
3,3′-Di-Me ether, 9-O-[β-D-apiofuranosyl-(1→6)-β-D-glucopyranoside]
Yemuoside YM$_2$
[122127-68-8]
$C_{31}H_{42}O_{16}$ 670.661
Isol. from *Stauntonia chinensis*. Powder. Mp 124 - 126°. $[\alpha]^{23}_{D}$-48.7 (c, 0.08 in MeOH)
3,3′-Di-Me ether, 4,4′-di-O-β-D-glucopyranoside
$C_{32}H_{44}O_{17}$ 700.687
Isol. from *Eucommia ulmoides*. Powder. Mp 157°. $[\alpha]^{25}_{D}$+65 (H_2O)
3,3′-Di-Me ether, 4′-O-β-D-glucopyranoside, 9-O-[α-D-glucopyranosyl-(1→2)-β-D-fructofuranos-6-ide]
$C_{38}H_{54}O_{22}$ 862.828
Constit. of the roots of *Valeriana officinalis*. Adenosine receptor agonist. $[\alpha]^{21}_{D}$-27 (c, 0.05 in DMSO). λ_{max} 227; 278 (MeOH aq.)
3,3′-Di-Me ether, 9-Ac
Icariresinol
[56440-71-2]
$C_{22}H_{26}O_8$ 418.442
Constit. of *Epimedium grandiflorum*. Cryst.. $[\alpha]^{15}_{D}$-72 (c, 0.22 in MeOH). λ_{max} 231 (log ε 4.36); 282 (log ε 3.9) (EtOH)
3,3′-Di-Me ether, 9-butanoyl
Salvicin A‡
[1351793-59-3]
$C_{24}H_{30}O_8$ 446.496
Constit. of *Salvia santolinifolia*. Amorph. solid. $[\alpha]^{25}_{D}$+38.2 (c, 0.03 in MeOH). λ_{max} 235 (log ε 3.91); 272 (log ε 3.45); 280 (sh) (MeOH)
3,3′-Di-Me ether, 9-O-(4-hydroxybutanoyl)
Salvicin B
[1351793-60-6]
$C_{24}H_{30}O_9$ 462.495
Constit. of *Salvia santolinifolia*. Amorph. solid. $[\alpha]^{25}_{D}$+51.1 (c, 0.04 in MeOH). λ_{max} 235 (log ε 3.91); 272 (log ε 3.45); 280 (sh) (MeOH)
3,3′,4-Tri-Me ether
7,9′-Epoxy-3,3′,4-trimethoxy-8,8′-lignan-4′,8′,9-triol. 4′,8′,9-Trihydroxy-3,3′,4-trimethoxy-7,9′-epoxylignan
3,3′,4-Tri-Me ether, 8′-Ac
8′-O-Acetyl-4-O-methylolivil
[1105066-98-5]
$C_{23}H_{28}O_8$ 432.469
Constit. of the roots of *Urtica triangularis*. Needles (Me_2CO). Mp 101 - 102°. $[\alpha]^{20}_{D}$-36.1 (c, 0.4 in MeOH)
3,3′,4-Tri-Me ether, 8′-Ac, 4′-O-[α-arabinopyranosyl-(1→6)-β-D-glucopyranoside]
[1105066-99-6]
$C_{34}H_{46}O_{17}$ 726.725
Constit. of the roots of *Urtica triangularis*. Amorph. powder. $[\alpha]^{20}_{D}$-30.8 (c, 0.1 in MeOH)
3,3′,4,4′-Tetra-Me ether
7,9′-Epoxy-3,3′,4,4′-tetramethoxy-8,8′-lignan-8′,9-diol. 8′,9-Dihydroxy-3,3′,4,4′-tetramethoxy-7,9′-epoxylignan
Cryst.. Mp 156°
(7S,8S,8′S)-form
3′,4′-Di-Me ether
7,9′-Epoxy-3′,4′-dimethoxy-8,8′-lignan-3,4,8′,9-tetrol. 3,4,8′,9-Tetrahydroxy-3′,4′-dimethoxy-7,9′-epoxylignan.
Santolinol
[1017553-68-2]
$C_{20}H_{24}O_7$ 376.405
Constit. of *Salvia santolinifolia*. Inhibitor of soya lipoxygenase 1B
(7R*,8R*,8′S*)-form
9-Carboxylic acid
7,9′-Epoxy-3,3′,4,4′,8′-pentahydroxy-8,8′-lignan-9-oic acid
$C_{18}H_{18}O_8$ 362.334

9-Carboxylic acid, 3,3′-di-Me ether, Me ester
Alyterinate C
[1616247-84-7]
$C_{21}H_{24}O_8$ 404.415
Isol. from *Alyxia schlechteri*. Shows good antifungal activity against *Pythium insidiosum*. Solid. Mp 120 - 122°. $[\alpha]_D^{25}$ -11.3 (c, 0.001 in MeOH). λ_{max} 242 (log ε 4.81); 280 (log ε 3.84) ($CHCl_3$).

Smith, M. *Tetrahedron Lett.*, 1963, **4**, 991-992 (*Olivil*).
Ayres, D.C. *et al.*, *J. Chem. Soc.*, 1965, 3586-3589 (*Olivil*).
Tokuoka, Y. *et al.*, *Yakugaku Zasshi*, 1975, **95**, 557-563 (*Icariresinol*).
Karrer, W. *Konstitution und Vorkommen der Organischen Pflanzenstoffe, 2nd edn., Birkhäuser Verlag*, 1976-1985, no. 1161 (*Olivil, bibl*).
Deyama, T. *et al.*, *Chem. Pharm. Bull.*, 1985, **33**, 3651-3657 (*Olivil 4,4′-diglucoside*).
Abe, F. *et al.*, *Chem. Pharm. Bull.*, 1988, **36**, 795-799 (*Cerbera manghas constits*).
Wang, H.B. *et al.*, *J. Nat. Prod.*, 1989, **52**, 342-345 (*Yemuoside YM_2*).
Schumacher, B. *et al.*, *J. Nat. Prod.*, 2002, **65**, 1479-1485 (*Valerian glycoside*).
Kadowaki, E. *et al.*, *Biosci., Biotechnol., Biochem.*, 2003, **67**, 415-419 (*Olivil, activity*).
Yamauchi, S. *et al.*, *Biosci., Biotechnol., Biochem.*, 2005, **69**, 113-122 (*synth*).
Yan, X.-G. *et al.*, *Chem. Pharm. Bull.*, 2008, **56**, 1463-1465 (*Urtica triangularis constits*).
Mehmood, S. *et al.*, *Pol. J. Chem.*, 2008, **82**, 571-575 (*Santolinol*).
Mehmood, S. *et al.*, *J. Asian Nat. Prod. Res.*, 2011, **13**, 588-591 (*Salvicins A,B*).
Sriphana, U. *et al.*, *Fitoterapia*, 2013, **91**, 39-43 (*Alyterinate C*).
Ren, Y. *et al.*, *Tetrahedron Lett.*, 2013, **54**, 4854-4858 (*Diepiolivil*).
Vakiti, J.R. *et al.*, *Org. Lett.*, 2020, **22**, 3345-3350 (*Olivil, synth, abs config*)

7,9′-Epoxy-8,8′-lignan-3,3′,4,4′,9,9′-hexol E95

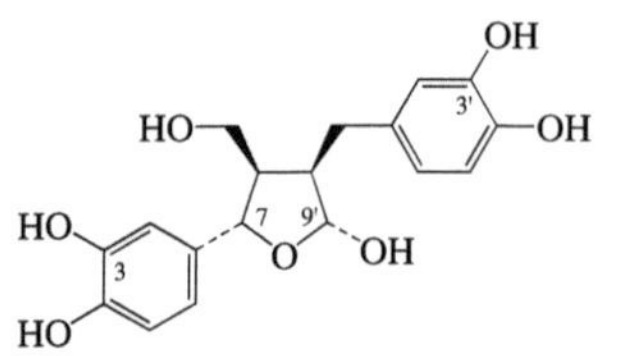

(7S,8R,8′R,9′R)-form

3,3′,4,4′,9,9′-Hexahydroxy-7,9′-epoxylignan
$C_{18}H_{20}O_7$ 348.351

(7S,8R,8′R,9′R)-form

3,3′-Di-Me ether
7,9′-Epoxy-3,3′-dimethoxy-8,8′-lignan-4,4′,9,9′-tetrol. 4,4′,9,9′-Tetrahydroxy-3,3′-dimethoxy-7,9′-epoxylignan. **9′-Hydroxylariciresinol**
[158042-37-6]
$C_{20}H_{24}O_7$ 376.405
Constit. of the wood of *Abies pinsapo*. Oil (as tetra-Ac). $[\alpha]_D^{25}$ +2 (c, 1.7 in $CHCl_3$) (tetra-Ac)

3,4:3′,4′-Bis(methylene) ether
7,9′-Epoxy-3,4:3′,4′-bis(methylenedioxy)-8,8′-lignan-9,9′-diol. 9,9′-Dihydroxy-3,4:3′,4′-bis(methylenedioxy)-7,9′-epoxylignan. **Parabenzoinol**
[169527-41-7]
$C_{20}H_{20}O_7$ 372.373
Constit. of the leaves of *Parabenzoin trilobum* (Lauraceae). Needles. Mp 191 - 192°. $[\alpha]_D$ -33 (c, 1.4 in DMSO-d_2)

9′-Ketone (lactone)
3,3′,4,4′,9-Pentahydroxy-8,8′-lignan-9′,7-olide

9′-Ketone (lactone), 3,3′-di-Me ether
4,4′,9-Trihydroxy-3,3′-dimethoxy-8,8′-lignan-9′,7-olide. **Isohydroxymatairesinol**
[293744-18-0]
$C_{20}H_{22}O_7$ 374.389
Constit. of knots of *Picea abies*. Powder. $[\alpha]_D^{25}$ +51 (c, 0.005 in THF)

(7R,8R,8′R)-form

9′-Ketone (lactone), 3,3′-di-Me ether
Epiisohydroxymatairesinol
[718614-97-2]
$C_{20}H_{22}O_7$ 374.389
Constit. of knots of *Picea abies*. Powder. $[\alpha]_D^{25}$ +59.8 (c, 0.005 in THF).

Barrero, A.F. *et al.*, *J. Nat. Prod.*, 1994, **57**, 713-719 (*9′-Hydroxylariciresinol*).
Takaoka, D. *et al.*, *Chem. Lett.*, 1995, 915-916 (*Parabenzoinol, cryst struct*).
Eklund, P.C. *et al.*, *J. Nat. Prod.*, 2004, **67**, 927-931 (*Isohydroxymatairesinol, Epiisohydroxymatairesinol*).
Eklund , P.C. *et al.*, *J. Mass Spectrom.*, 2008, **43**, 97-107 (*Isohydroxymatairesinol, ms*)

7,9′-Epoxy-8,8′-lignan-3,3′,4,5′,7′,9-hexol E96

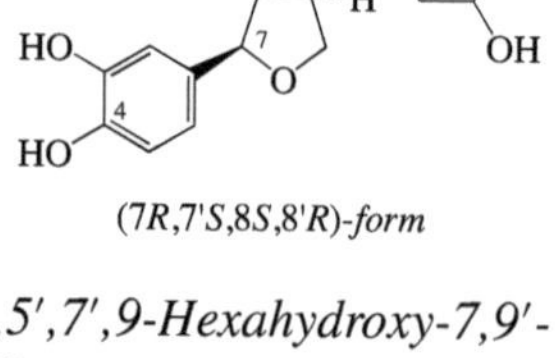

(7R,7′S,8S,8′R)-form

3,3′,4,5′,7′,9-Hexahydroxy-7,9′-epoxylignan
$C_{18}H_{20}O_7$ 348.351

(7R,7′S,8S,8′R)-form

3,3′-Di-Me ether, 4-O-β-D-glucopyranoside
Koreanaside B
[2231630-69-4]
$C_{26}H_{34}O_{12}$ 538.546
Constit. of the flowers of *Forsythia koreana*. Antioxidant. Pale yellow needles. Mp 97 - 98°. $[\alpha]_D$ -5.3 (c, 0.1 in MeOH)

(7R*,7′S*,8S*,8′S*)-form

3,3′-Di-Me ether
7,9′-Epoxy-3,5′-dimethoxy-8,8′-lignan-3′,4,7′,9-tetrol. 3′,4,7′,9-Tetrahydroxy-3,5′-dimethoxy-7,9′-epoxylignan.
Dysosmarol
[926887-99-2]
$C_{20}H_{24}O_7$ 376.405
Constit. of the roots of *Dysosma versipellis*. Amorph. powder. $[\alpha]_D^{20}$ +32 (c, 0.1 in MeOH).

Jiang, R.-W. *et al.*, *J. Nat. Prod.*, 2007, **70**, 283-286 (*Dysosmarol*).
Lee, Y.-G. *et al.*, *Chem. Biodiversity*, 2018, **15**, e1800026:1-8 (*Koreanaside B, activity*)

7,9′-Epoxy-8,8′-lignan-3,3′,4′,5,7′,9-hexol E97

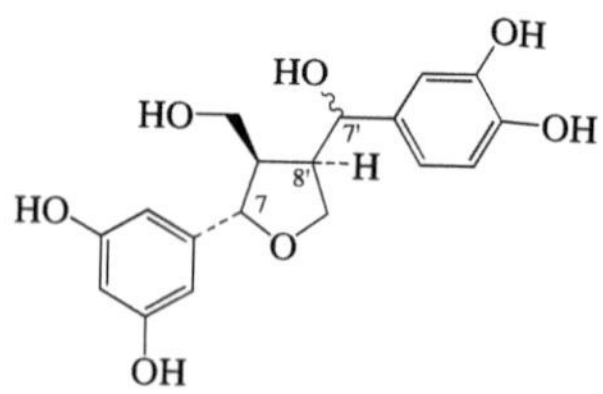

3,3′,4′,5,7′,9-Hexahydroxy-7,9′-epoxylignan
$C_{18}H_{20}O_7$ 348.351
Structures in this series may need revision. Phenyl 3,5-dioxygenated substn. pattern in doubt as a result of work done by Yang *et al* in 2015

(7S,8R,8′R)-form

7′-Ketone
7,9-Epoxy-3,3′,4′,5,9-pentahydroxy-8,8′-lignan-7′-one. 3,3′,4′,5,9-Pentahydroxy-7′-oxo-7,9′-epoxylignan

7′-Ketone, 3,3′-di-Me ether
7,9′-Epoxy-3,4′,9-trihydroxy-3′,5-dimethoxy-8,8′-lignan-7′-one. 3,4′,9-Trihydroxy-3′,5-dimethoxy-7′-oxo-7,9′-epoxylignan. **Forsythialan A**
[1019859-20-1]
$C_{20}H_{22}O_7$ 374.389

Constit. of the fruit of *Forsythia suspensa*. Shows antioxidant activity in renal epithelial cells. Oil. $[\alpha]_D^{25}$ -7.2 (c, 0.56 in $CHCl_3$)

7′-Ketone, 3,3′,4′-tri-Me ether

7,9′-Epoxy-3,9-dihydroxy-3′,4′,5-trimethoxy-8,8′-lignan-7′-one. 3,9-Dihydroxy-3′,4′,5-trimethoxy-7′-oxo-7,9′-epoxylignan. ***Forsythialan B***

[1019859-21-2]

$C_{21}H_{24}O_7$ 388.416

Constit. of the fruit of *Forsythia suspensa*. Shows antioxidant activity in renal epithelial cells. Oil. $[\alpha]_D^{25}$ -21.5 (c, 0.9 in $CHCl_3$).

Piao, X.-L. *et al.*, *Bioorg. Med. Chem. Lett.*, 2008, **18**, 1980-1984 (*Forsythialans A,B*).
Yang, Y.-N. *et al.*, *J. Nat. Prod.*, 2015, **78**, 705-711 (*pmr, struct*)

7,9′-Epoxy-8,8′-lignan-3,3′,4′,5,8,9-hexol E98

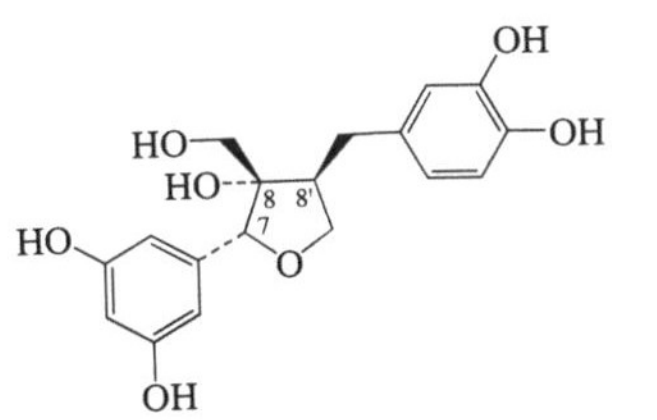

(7R,8S,8′S)-form

3,3′,4′,5,8,9-Hexahydroxy-7,9′-epoxylignan

$C_{18}H_{20}O_7$ 348.351

Structures in this series may need revision. Phenyl 3,5-dioxygenated substn. pattern in doubt as a result of work done by Yang *et al* in 2015

(7R,8S,8′S)-form

3,3′-Di-Me ether

7,9′-Epoxy-3′,5-dimethoxy-8,8′-lignan-3,4′,8,9-tetrol. 3,4′,8,9-Tetrahydroxy-3′,5-dimethoxy-7,9′-epoxylignan

[1227937-39-4]

$C_{20}H_{24}O_7$ 376.405

Constit. of *Saussurea cordifolia* and *Viburnum erosum*. Amorph. powder. $[\alpha]_D^{20}$ -5.5 (c, 0.28 in MeOH). λ_{max} 229 (log ε 3.97); 282 (log ε 3.55) (MeOH)

(7S,8R,8′R)-form

3,3′-Di-Me ether

Deltoignan A

[1427038-08-1]

$C_{20}H_{24}O_7$ 376.405

Constit. of *Saussurea deltoidea*. Oil. $[\alpha]_D^{23}$ +5.65 (c, 0.25 in MeOH).

Li, X.-W. *et al.*, *Phytochemistry*, 2010, **71**, 682-687 (*Saussurea cordifolia constit*).
Xu, J.-J. *et al.*, *Fitoterapia*, 2012, **83**, 1125-1130 (*Deltoignan A*).
In, S.-J. *et al.*, *Arch. Pharmacal Res.*, 2015, **38**, 26-34 (*Viburnum erosum constit*).
Yang, Y.-N. *et al.*, *J. Nat. Prod.*, 2015, **78**, 705-711 (*pmr, struct*)

7,9′-Epoxy-8,8′-lignan-3,3′,4,5′,8′,9-hexol E99

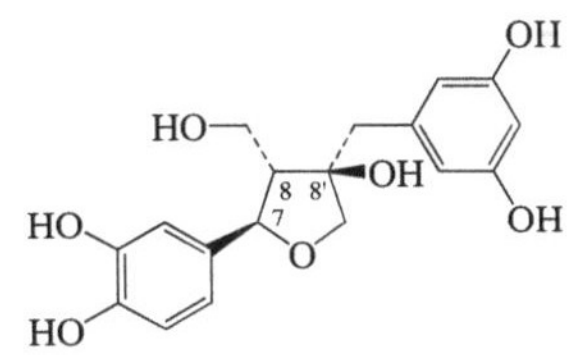

3,3′,4,5′,8′,9-Hexahydroxy-7,9′-epoxylignan

$C_{18}H_{20}O_7$ 348.351

Structures in this series may need revision. Phenyl 3,5-dioxygenated substn. pattern in doubt as a result of work done by Yang *et al* in 2015

(7R*,8S*,8′R*)-form

3,3′-Di-Me ether

7,9′-Epoxy-3,5′-dimethoxy-8,8′-lignan-3′,4,8′,9-tetrol. 3′,4,8′,9-Tetrahydroxy-3,5′-dimethoxy-7,9′-epoxylignan. ***Ginkgool***

[1818205-99-0]

$C_{20}H_{24}O_7$ 376.405

Constit. of *Ginkgo biloba* and *Jasminum tortuosum*. Needles.

Wei, X.-L. *et al.*, *Chem. Nat. Compd. (Engl. Transl.)*, 2015, **51**, 819-821 (*Ginkgool*).
Yang, Y.-N. *et al.*, *J. Nat. Prod.*, 2015, **78**, 705-711 (*pmr, struct*)

7,9′-Epoxy-8,8′-lignan-3,3′,4′,5,8′,9-hexol E100

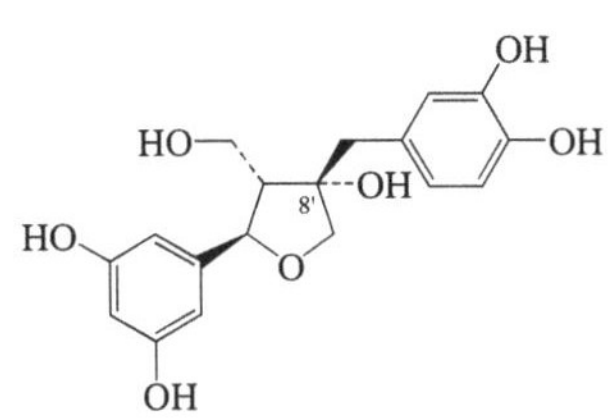

3,3′,4′,5,8′,9-Hexahydroxy-7,9′-epoxylignan

$C_{18}H_{20}O_7$ 348.351

(7R,8S,8′S)-form

3,3′-Di-Me ether

7,9′-Epoxy-3′,5-dimethoxy-8,8′-lignan-3,4′,8′,9-tetrol. 3,4′,8′,9-Tetrahydroxy-3′,5-dimethoxy-7,9′-epoxylignan

3,3′-Di-Me ether, 9-Ac

Melodinin C

$C_{22}H_{26}O_8$ 418.442

Constit. of the stems of *Melodinus suaveolens*. Amorph. powder. $[\alpha]_D^{25}$ +4 (c, 0.05 in MeOH).

Fang, L. *et al.*, *Phytochem. Lett.*, 2018, **26**, 134-137 (*Melodinin C, cd*)

9,9′-Epoxy-8,8′-lignan-3,3′,4,4′,5,5′-hexol E101

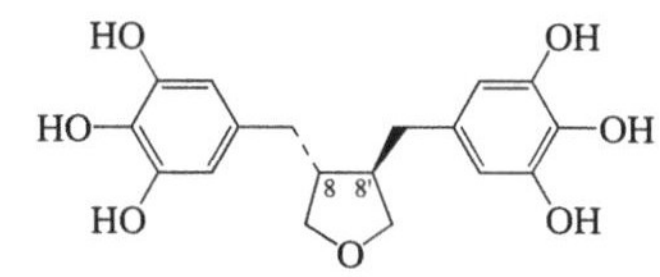

Tetrahydro-3,4-bis[(3,4,5-trihydroxyphenyl)methyl] furan. 3,4-Bis(3,4,5-trihydroxybenzyl)tetrahydrofuran. 3,3′,4,4′,5,5′-Hexahydroxy-9,9′-epoxylignan

$C_{18}H_{20}O_7$ 348.351

(8R*,8′R*)-form

Hexa-Me ether

9,9′-Epoxy-3,3′,4,4′,5,5′-hexamethoxylignan. *3,3′,4,4′,5,5′-Hexamethoxy-9,9′-epoxylignan*

[138306-22-6]

$C_{24}H_{32}O_7$ 432.513

Constit. of the bark of *Aglaia elaeagnoidea*. Dark yellow oil. λ_{max} 232 (log ε 4.15); 280 (log ε 3.75) (MeOH).

Coran, S.A. *et al.*, *Eur. J. Med. Chem.*, 1991, **26**, 643-650 (*synth*).
Fuzzati, N. *et al.*, *Phytochemistry*, 1996, **42**, 1395-1398 (*Aglaia eleagnoidea constit*)

9,9′-Epoxy-8,8′-lignan-3,3′,4,4′,5,7-hexol E102

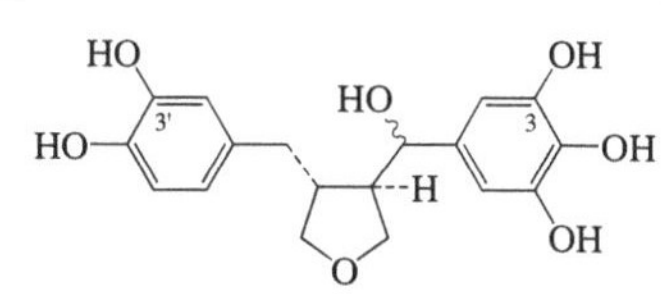

3,3′,4,4′,5,7-Hexahydroxy-9,9′-epoxylignan

$C_{18}H_{20}O_7$ 348.351

(7ξ,8R*,8′R*)-form

3,3′,5-Tri-Me ether

9,9′-Epoxy-3,3′,5-trimethoxy-8,8′-lignan-4,4′,7-triol. 4,4′,7-Trihydroxy-3,3′,5-trimethoxy-9,9′-epoxylignan. ***Busaliol***

[163564-90-7]

$C_{21}H_{26}O_7$ 390.432

Constit. of the leaves of *Bupleurum salicifolium*. Yellow oil. $[\alpha]_D^{20}$ +25 (c, 0.2 in $CHCl_3$).

Estevez-Braun, A. *et al.*, *J. Nat. Prod.*, 1995, **58**, 887-892 (*Busaliol*)

9,9′-Epoxy-8,8′-lignan-3,3′,4,4′,5,9-hexol E103

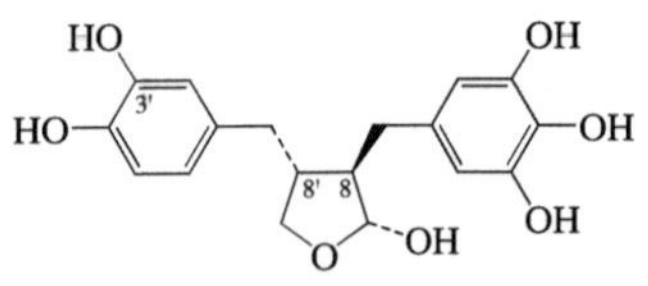

(8*R*,8′*R*,9*R*)-*form*

3,3′,4,4′,5,9-Hexahydroxy-9,9′-epoxylignan

$C_{18}H_{20}O_7$ 348.351

(8R,8′R,9R)-form

3′,4′-Methylene, 3,4,5,9-tetra-Me ether

3,4,5,9-Tetramethoxy-3′,4′-methylenedioxy-9,9′-epoxylignan. O-Methyl-α-clusin

[329371-32-6]

$C_{23}H_{28}O_7$ 416.47

Constit. of *Lychnophora ericoides*. Oil. $[\alpha]_D^{23}$ -19.3 (c, 0.1 in $CHCl_3$)

(8R,8′R,9S)-form

3′,4′-Methylene, 3,4,5,9-tetra-Me ether

O-Methyl-β-clusin

[329371-33-7]

$C_{23}H_{28}O_7$ 416.47

Constit. of *Lychnophora ericoides*. Oil. $[\alpha]_D^{23}$ -3.5 (c, 0.2 in $CHCl_3$)

3′,4′-Methylene, 9-Et, 3,4,5-tri-Me ether

9-Ethoxy-3,4,5-trimethoxy-3′,4′-methylenedioxy-9,9′-epoxylignan. O-Ethyl-α-clusin

[93395-17-6]

$C_{24}H_{30}O_7$ 430.497

Constit. of *Piper clusii*. Gummy solid. $[\alpha]_D^{36}$ -31.6 (MeOH)

(8R,8′R,9ξ)-form

3′,4′-Methylene, 3,4,5-tri-Me ether

9,9′-Epoxy-3,4,5-trimethoxy-3′,4′-methylenedioxy-8,8′-lignan-9-ol. 9-Hydroxy-3,4,5-trimethoxy-3′,4′-methylenedioxy-9,9′-epoxylignan. ***Clusin***

[86992-94-1]

[72627-54-4] (unspecified stereochem.)

$C_{22}H_{26}O_7$ 402.443

Constit. of *Piper clusii* and *Piper cubeba*. Gum. $[\alpha]_D$ -34.4 (c, 1.0 in $CHCl_3$). Stereochem. not confirmed

3,4:3′,4′-Bis(methylene), 5-Me ether

9,9′-Epoxy-3-methoxy-3′,4′:4,5-bis(methylenedioxy)-8,8′-lignan-9-ol. 9-Hydroxy-3-methoxy-3′,4′:4,5-bis(methylenedioxy)-9,9′-epoxylignan. ***Trichostin***

[115713-34-3]

[128051-21-8] (8*R*,8′*R*,9*R*-form)

[128052-63-1] (8*R*,8′*R*,9*S*-form)

$C_{21}H_{22}O_7$ 386.4

Constit. of *Piper trichostachyon*. Gum. $[\alpha]_D^{20}$ -62.25 (c, 0.8 in MeOH). Stereochem. not confirmed

(8R*,8′R*,9ξ)-form

3,3′,4,4′,5-Penta-Me ether

9,9′-Epoxy-3,3′,4,4′,5-pentamethoxy-8,8′-lignan-9-ol. 9-Hydroxy-3,3′,4,4′,5-pentamethoxy-9,9′-epoxylignan.

3′-Desmethoxycubebinin

$C_{23}H_{30}O_7$ 418.486

Constit. of the leaves of *Piper nigrum*. Viscous solid. $[\alpha]_D^{25}$ -27.4 (c, 1 in $CHCl_3$). λ_{max} 228 (log ε 4.02); 281 (log ε 3.58) (MeOH).

Koul, S.K. *et al.*, *Phytochemistry*, 1983, **22**, 999-1000 (*Piper clusii constits*).
Koul, S.K. *et al.*, *Phytochemistry*, 1984, **23**, 2099-2101 (*Ethylclusin*).
Koul, S.K. *et al.*, *Phytochemistry*, 1985, **24**, 329-332 (*Piper cubeba constit*).
Koul, S.K. *et al.*, *Phytochemistry*, 1988, **27**, 1479-1282 (*Trichostin*).
Rehnberg, N. *et al.*, *J. Org. Chem.*, 1990, **55**, 4340-4349 (*Trichostin, synth*).
Borsato, M.L.C. *et al.*, *Phytochemistry*, 2000, **55**, 809-813 (*Lychnophorea ericoides constits*).
Matsuda, H. *et al.*, *Biol. Pharm. Bull.*, 2004, **27**, 1611-1616 (*3′-Demethoxycubebinin*)

9,9′-Epoxy-8,8′-lignan-3,3′,4,4′,7,7′-hexol E104

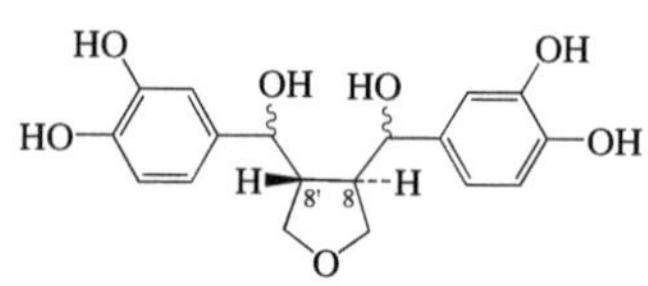

(7ξ,7′ξ,8*R**,8′*R**)-*form*

3,3′,4,4′,7,7′-Hexahydroxy-9,9′-epoxylignan

$C_{18}H_{20}O_7$ 348.351

(7ξ,7′ξ,8R*,8′R*)-form

3,3′-Di-Me ether

9,9′-Epoxy-3,3′-dimethoxy-8,8′-lignan-4,4′,7,7′-tetrol. 4,4′,7,7′-Tetrahydroxy-3,3′-dimethoxy-9,9′-epoxylignan. ***Liovil***

[484-39-9]

$C_{20}H_{24}O_7$ 376.405

Constit. of *Picea excelsa*, *Abies nephrolepis* and *Viburnum foetidum*. Cryst. (MeOH aq.). Mp 173.5 - 174.5° (165-167°). $[\alpha]_D^{25}$ -32.8 (c, 4 in MeOH)

3,3′-Di-Me ether, 4′-O-β-D-glucopyranoside

Liovil 4′-O-β-D-glucopyranoside

[2052271-09-5]

$C_{26}H_{34}O_{12}$ 538.546

Constit. of the fruits of *Narthecium ossifragum*

3,3′-Di-Me ether, 7,7′-di-Ac

Sanjidin B

[186885-67-6]

$C_{24}H_{28}O_9$ 460.479

Constit. of *Pleione bulbocodioides*. Oil. λ_{max} 230 (sh); 270 (log ε 4.09) (MeOH)

7-Et, 3,3′-di-Me ether

9,9′-Epoxy-7′-ethoxy-3,3′-dimethoxy-8,8′-lignan-4,4′,7-triol. 7-Ethoxy-4,4′,7′-trihydroxy-3,3′-dimethoxy-9,9′-epoxylignan.

Busalicifol

[163564-91-8]

$C_{22}H_{28}O_7$ 404.459

Isol. from leaves of *Bupleurum salicifolium*. Pale yellow oil. $[\alpha]_D^{20}$ +45 (c, 0.2 in $CHCl_3$)

(7ξ,7′ξ,8R*,8′S*)-form

3,3′-Di-Me ether, 7,7′-di-Ac

Sanjidin A

[186885-66-5]

$C_{24}H_{28}O_9$ 460.479

Constit. of *Pleione bulbocodioides*. Oil. λ_{max} 236 (log ε 4.25); 286 (log ε 4.15) (MeOH).

Weinges, K. *Chem. Ber.*, 1961, **94**, 3032-3043 (*Liovil*).
Leont'eva, V.G. *et al.*, *Chem. Nat. Compd. (Engl. Transl.)*, 1973, **9**, 252-253 (*Abies nephrolepis constit*).
Estevez-Braun, A. *et al.*, *J. Nat. Prod.*, 1995, **58**, 887-892 (*Busalicifol*).
Bai, L. *et al.*, *Phytochemistry*, 1997, **44**, 341-343 (*Sanjidins*).
Chen, X-Q. *et al.*, *Chem. Pharm. Bull.*, 2009, **57**, 1129-1131 (*Viburnum foetidum constit*).
Pat. Coop. Treaty (WIPO), 2014, 2014 083 171 (*Liovil*).
Vu, M. *et al.*, *Phytochemistry*, 2016, **132**, 76-85 (*Liovil 4-glucoside*)

9,9′-Epoxy-8,8′-lignan-3,3′,4,4′,7,9′-hexol **E105**

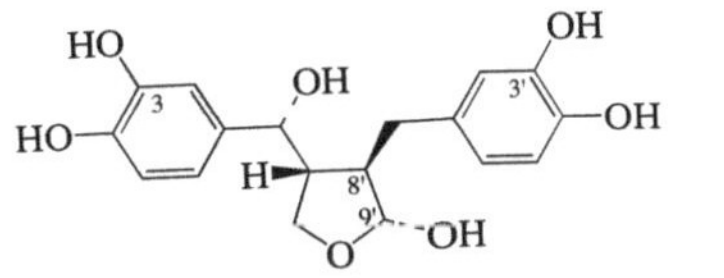

(7R,8R,8′R,9′R)-form

3,3′,4,4′,7,9′-Hexahydroxy-9,9′-epoxylignan
$C_{18}H_{20}O_7$ 348.351
(7R,8R,8′R,9′R)-form
3,3′,7,9′-Tetra-Me ether
9,9′-Epoxy-3,3′,7,9′-tetramethoxy-8,8′-lignan-4,4′-diol. 4,4′-Dihydroxy-3,3′,7,9′-tetramethoxy-9,9′-epoxylignan.
Iso-α-intermedianol
[103019-35-8]
$C_{22}H_{28}O_7$ 404.459
Isol. from wood of *Dacrydium intermedium*. Glass. Mp 65 - 67°. $[\alpha]^{20}_{D}$-53.5 (c, 0.7 in $CHCl_3$)
(7R,8R,8′S,9′R)-form
3,3′,9′-Tri-Me ether
9,9′-Epoxy-3,3′,9′-trimethoxy-8,8′-lignan-4,4′,7-triol. 4,4′,7-Trihydroxy-3,3′,9′-trimethoxy-9,9′-epoxylignan.
Holophyllol C
[1493768-70-9]
$C_{21}H_{26}O_7$ 390.432
Isol. from trunkwood of *Abies holophylla*. Needles. Mp 164 - 166°. $[\alpha]^{25}_{D}$+12.0 (c, 0.05 in MeOH). λ_{max} 229 (log ε 3.4); 281 (log ε 1.6) (MeOH)
(7R,8R,8′S,9′S)-form
3,3′,9′-Tri-Me ether
Holophyllol B
[1493768-67-4]
$C_{21}H_{26}O_7$ 390.432
Isol. from trunkwood of *Abies holophylla*. Needles. Mp 159 - 161°. $[\alpha]^{25}_{D}$-25.0 (c, 0.10 in MeOH). λ_{max} 229 (log ε 3.0); 281 (log ε 1.6) (MeOH)
(7S,8R,8′R,9′R)-form
3,3′-Di-Me ether
9,9′-Epoxy-3,3′-dimethoxy-8,8′-lignan-4,4′,7,9′-tetrol. 4,4′,7,9′-Tetrahydroxy-3,3′-dimethoxy-9,9′-epoxylignan.
Sesamolactol
[914253-79-5]
$C_{20}H_{24}O_7$ 376.405
Constit. of *Picea pungens* and *Sesamum indicum*. Prisms. Mp 168°. $[\alpha]_{D}$ -35 (c, 0.2 in MeOH). $[\alpha]_{D}$ -61.5 (c, 1 in THF)
3,3′,9′-Tri-Me ether
Holophyllol A
[1493768-65-2]
$C_{21}H_{26}O_7$ 390.432
Isol. from trunkwood of *Abies holophylla*. Needles. Mp 175 - 178°. $[\alpha]^{25}_{D}$-50.6 (c, 0.75 in MeOH). λ_{max} 230 (log ε 3.0); 281 (log ε 1.5) (MeOH)
3,3′,7,9′-Tetra-Me ether
α-Intermedianol
[103019-34-7]
$C_{22}H_{28}O_7$ 404.459
Isol. from wood of *Dacrydium intermedium*. Pale cream plates (EtOAc). Mp 141 - 142°. $[\alpha]^{20}_{D}$ -96.3 (c, 1 in $CHCl_3$)
(7S,8R,8′R,9′S)-form
3,3′-Di-Me ether
[917874-31-8]
Constit. of the heartwood of *Picea pungens*
3,3′,9′-Tri-Me ether
$C_{21}H_{26}O_7$ 390.432
Isol. from trunkwood of *Abies holophylla*. Needles. Mp 168 - 170°. $[\alpha]^{25}_{D}$+20.0 (c, 0.2 in MeOH). λ_{max} 228 (log ε 2.9); 281 (log ε 1.6) (MeOH)
3,3′,7,9′-Tetra-Me ether
β-Intermedianol
[103019-33-6]
$C_{22}H_{28}O_7$ 404.459
Isol. from wood of *Dacrydium intermedium*. Prisms (EtOAc/hexane). Mp 141.5 - 142°. $[\alpha]^{20}_{D}$ +23.7 (c, 1 in $CHCl_3$). α- and β-Intermedianols were isol. after chromatog. and prob. equilibrate
(7R*,8R*,8′R*,9′R*)-form
3,4:3′,4′-Bis(methylene), 9′-Me ether
Dolichanthin B
[2222919-90-4]
$C_{21}H_{22}O_7$ 386.400
Constit. of the twigs of *Wikstroemia dolichantha*. Yellow powder. $[\alpha]^{32}_{D}$-40 (c, 0.2 in MeOH). λ_{max} 207 (log ε 2.69); 221 (log ε 1.11); 235 (log ε 1.53) (MeOH)
(7R*,8R*,8′R*,9′S*)-form
3,3′,9′-Tri-Me ether
Pseudolarkaemin A
[1391976-10-5]
$C_{21}H_{26}O_7$ 390.432
Isol. from *Pseudolarix kaempferi*. Powder. $[\alpha]^{17}_{D}$+38.9 (c, 0.15 in MeOH). λ_{max} 205 (log ε 4.78); 222 (log ε 4.77); 281 (log ε 3.9) (MeOH)
(7R*,8S*,8′S*,9′R*)-form
3,4:3′,4′-Bis(methylene), 9′-Me ether
Dolichanthin A
[2222919-89-1]
$C_{21}H_{22}O_7$ 386.400
Constit. of the twigs of *Wikstroemia dolichantha*. Weak acetylcholinesterase inhibitor. Yellow powder. $[\alpha]^{32}_{D}$-90 (c, 0.2 in MeOH). λ_{max} 206 (log ε 2.65); 221 (log ε 1.0); 234 (log ε 1.28) (MeOH)
(7ξ,8R*,8′R*,9′ξ)-form
3,3′-Di-Me ether
Todolactol A
[119716-94-8]
$C_{20}H_{24}O_7$ 376.405
Isol. from wood of *Abies sachalinensis*. Mp 183 - 184°. $[\alpha]^{33}_{D}$-80.6 (c, 1.02 in dioxan). Claimed to be a mixt. of C-7 epimers
3,3′,9′-Tri-Me ether
Koreanol
[158372-30-6]
$C_{21}H_{26}O_7$ 390.432
Constit. of *Abies koreana*
(7ξ,8ξ,8′ξ,9′ξ)-form
3,3′-Di-Me ether
Isoliovil
[82039-13-2]
$C_{20}H_{24}O_7$ 376.405
Constit. of *Taxus wallichiana*. Cryst. (MeOH). Mp 170 - 172°. $[\alpha]^{23}_{D}$-47.9 (c, 0.22 in EtOH).

Miller, R.W. *et al.*, *J. Nat. Prod.*, 1982, **45**, 78-82 (*Isoliovil*).
Cambie, R.C. *et al.*, *Aust. J. Chem.*, 1985, **38**, 1631-1635 (*Intermedianols, cryst struct*).
Ozawa, S. *et al.*, *Mokuzai Gakkaishi*, 1988, **34**, 851-857 (*Todolactol A*).
Kim, Y.-G. *et al.*, *Mokuzai Gakkaishi*, 1994, **40**, 414-418 (*Koreanol*).
Willför, S. *et al.*, *Holzforschung*, 2005, **59**, 413-417 (*Picea pungens constits, cryst struct*).
Grougnet, R. *et al.*, *J. Agric. Food Chem.*, 2006, **54**, 7570-7574 (*Sesamolactol*).
He, W.-J. *et al.*, *Planta Med.*, 2011, **77**, 1924-1931 (*Pseudolarkaemin A*).
Kim, C.S. *et al.*, *J. Nat. Prod.*, 2013, **76**, 2131-2135 (*Abies holophylla constits*).
Ma, Q.-Y. *et al.*, *Chem. Nat. Compd. (Engl. Transl.)*, 2018, **54**, 22-25 (*Dolichanthins A,B, activity*)

7,9′-Epoxy-8,8′-lignan-2′,3,3′,4,4′,5,5′,7′,9-nonol E106

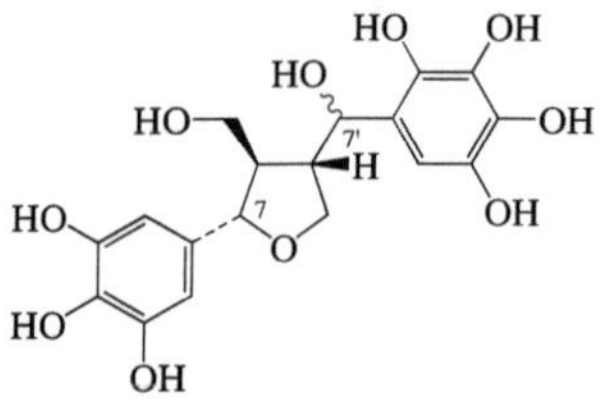

2′,3,3′,4,4′,5,5′,7′,9-Nonahydroxy-7,9′-epoxylignan
$C_{18}H_{20}O_{10}$ 396.348
(7S,8R,8′S)-form
7′-Ketone, 3,3′,4,4′,5,5′-hexa-Me ether
7,9′-Epoxy-2′,9-dihydroxy-3,3′,4,4′,5,5′-hexamethoxy-8,8′-lignan-7-one
7′-Ketone, 3,3′,4,4′,5,5′-hexa-Me ether, 9-O-β-D-glucopyranoside
Terminaloside U
[2222336-23-2]
$C_{30}H_{40}O_{15}$ 640.635
Constit. of the leaves of *Terminalia citrina*. Amorph. powder. $[\alpha]^{25}_{D}$+37 (c, 0.2 in MeOH). λ_{max} 211 (log ε 4.2); 220 (log ε 4.25); 285 (log ε 4.02) (MeOH).

Muhit, M.A. *et al., Phytochemistry*, 2018, **145**, 161-167 (*Terminaloside U*)

4,7′-Epoxy-3,8′-lignan-3′,4′,5,5′,7,8,9,9′-octol E107

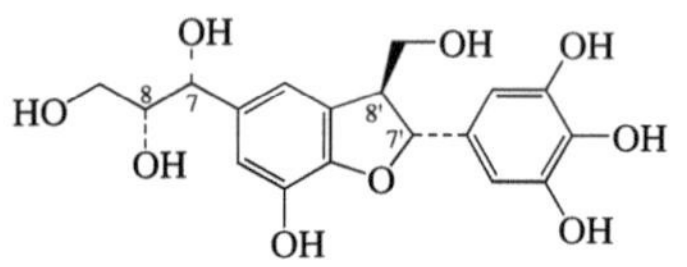

(7R*,7′R,8R*,8′S)-form

4,7′-Epoxy-3′,4′,5,5′,7,8,9,9′-octahydroxy-3,8′-neolignan
$C_{18}H_{20}O_{9}$ 380.349
(7R*,7′R,8R*,8′S)-form
3′,5,5′-Tri-Me ether
4,7′-Epoxy-3′,5,5′-trimethoxy-3,8′-lignan-4′,7,8,9,9′-pentol. 4,7′-Epoxy-4′,7,8,9,9′-pentahydroxy-3′,5,5′-trimethoxy-3,8′-lignan. **Hawthornnin H**
[1803478-05-8]
$C_{21}H_{26}O_{9}$ 422.43
Constit. of the seeds of *Crataegus pinnatifida*. Yellow oil. $[\alpha]^{20}_{D}$-6.5 (c, 0.1 in MeOH). λ_{max} 230 (log ε 0.63); 280 (log ε 0.28) (MeOH)
(7R*,7′R*,8R*,8′S*)-form
3′,5,5′-Tri-Me ether
Bennettin
[1204224-33-8]
$C_{21}H_{26}O_{9}$ 422.43
Constit. of *Bennettiodendron leprosipes*. Syrup. $[\alpha]^{20}_{D}$+10.5 (c, 0.51 in MeOH). λ_{max} 237 (log ε 3.89); 281 (log ε 3.3) (MeOH)
(7ξ,7′S,8ξ,8′R)-form
3′,5,5′-Tri-Me ether, 7-O-β-D-glucopyranoside
Lyciumlignan D
$C_{27}H_{36}O_{14}$ 584.571
Constit. of the root bark of *Lycium chinense*. Amorph. powder. $[\alpha]^{20}_{D}$ +22.9 (c, 0.1 in MeOH). λ_{max} 238 (log ε 4.02); 282 (log ε 3.53) (MeOH).

Chai, X.-Y. *et al., Planta Med.*, 2009, **75**, 1246-1252 (*Bennettin*).
Huang, X.-X. *et al., J. Agric. Food Chem.*, 2015, **63**, 7252-7260 (*Hawthornnin H*).
Yang, Y. *et al., Acta Pharm. Sin. B*, 2017, **7**, 491-495 (*Lyciumlignan D, cd*)

7,7′-Epoxy-8,8′-lignan-3,3′,4,4′,5,5′,9,9′-octol E108

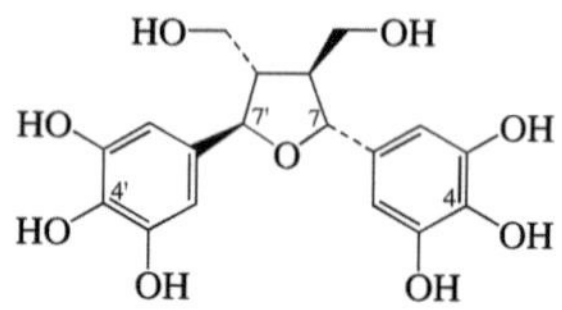

(7R,7′R,8S,8′S)-form

3,3′,4,4′,5,5′,9,9′-Octahydroxy-7,7′-epoxylignan
$C_{18}H_{20}O_{9}$ 380.349
(7R,7′R,8S,8′S)-form
3,3′,5,5′-Tetra-Me ether, 4-O-β-D-glucopyranoside
Foliachinenoside C
[1041180-87-3]
$C_{28}H_{38}O_{14}$ 598.598
Constit. of the leaves of *Salacia chinensis*. $[\alpha]^{25}_{D}$-0.2 (c, 0.9 in MeOH). λ_{max} 213 (log ε 4.62); 238 (log ε 4.06); 271 (log ε 3.28) (MeOH)
3,3′,5,5′-Tetra-Me ether, 9-O-β-D-xylopyranoside
[545436-28-0]
$C_{27}H_{36}O_{13}$ 568.572
Isol. from non-centrifuged cane sugar (*Saccharum officinarum*) and from leaves of *Neoalsomitra integrifoliola*. Oil
3,3′,5,5′-Tetra-Me ether, 9-O-β-D-glucopyranoside
Ligraminol B
[1335126-25-4]
$C_{28}H_{38}O_{14}$ 598.598
Isol. from rhizomes of *Acorus gramineus*. Gum. $[\alpha]^{25}_{D}$+2.4 (c, 0.35 in MeOH). λ_{max} 218 (log ε 4.3); 241 (log ε 2.7); 273 (log ε 1.5) (MeOH)
3,3′,5,5′-Tetra-Me ether, 9,9′-di-O-(4-hydroxy-E-cinnamoyl)
Oryzativol A
[2222876-00-6]
$C_{40}H_{40}O_{13}$ 728.747
Constit. of the roots of *Oryza sativa*. Enables both suppression of adipogenesis and promotion of osteogenesis. Oryzativol A is preferred to Oryzativol B due to its structural stability. Yellowish gum. $[\alpha]^{25}_{D}$-12.3 (c, 0.07 in MeOH). λ_{max} 228 (log ε 3.5); 315 (log ε 1.9) (MeOH)
3,3′,5,5′-Tetra-Me ether, 9-O-(4-hydroxy-Z-cinnamoyl), 9′-O-(4-hydroxy-E-cinnamoyl)
Oryzativol B
[2222876-01-7]
$C_{40}H_{40}O_{13}$ 728.747
Constit. of the roots of *Oryza sativa*. Enables both suppression of adipogenesis and promotion of osteogenesis. Yellowish gum. $[\alpha]^{25}_{D}$ -6.7 (c, 0.08 in MeOH). λ_{max} 228 (log ε 3.5); 314 (log ε 1.9) (MeOH)
3,3′,4′,5,5′-Penta-Me ether
7,7′-Epoxy-3,3′,4′,5,5′-pentamethoxy-8,8′-lignan-4,9,9′-triol. 4,9,9′-Trihydroxy-3,3′,4′,5,5′-pentamethoxy-7,7′-epoxylignan
[1290633-28-1]
$C_{23}H_{30}O_{9}$ 450.484
Constit. of the stems of *Sinocalamus affinis*. Amorph. powder. $[\alpha]^{20}_{D}$ +25.5 (c, 0.04 in MeOH). λ_{max} 207 (log ε 4.48); 241 (log ε 3.72); 277 (log ε 3.03) (MeOH)
(7S,7′S,8R,8′R)-form
3,3′,5,5′-Tetra-Me ether
7,7′-Epoxy-3,3′,5,5′-tetramethoxy-8,8′-lignan-4,4′,9,9′-tetrol. 4,4′,9,9′-Tetrahydroxy-3,3′,5,5′-tetramethoxy-7,7′-epoxylignan.
Icariol A$_2$
[135820-79-0]
$C_{22}H_{28}O_{9}$ 436.457
Constit. of *Aeschynanthus bracteatus*, *Epimedium sagittatum*, *Grewia bilamellata* and *Machilus robusta*. Amorph. powder. $[\alpha]^{23}_{D}$-22 (c, 0.59 in MeOH). λ_{max} 238 (sh); 271 (log ε 3.7) (MeOH)

3,3′,5,5′-Tetra-Me ether, 9-O-α-L-rhamnopyranoside
[1807733-32-9]
$C_{28}H_{38}O_{13}$ 582.599
Constit. of the leaves and twigs of *Osteomeles schwerinae*. Powder. $[\alpha]^{25}_{D}$ -51.4 (c, 0.1 in MeOH). λ_{max} 238 (log ε 4.0); 271 (log ε 3.6) (MeOH)

3,3′,5,5′-Tetra-Me ether, 9-O-β-D-glucopyranoside
Icariol A₂ 9-glucoside
[942911-52-6]
$C_{28}H_{38}O_{14}$ 598.598
Constit. of *Glechoma hederacea*. Amorph. powder. $[\alpha]^{27}_{D}$ +2.6 (c, 0.11 in MeOH). λ_{max} 214 (log ε 4.3); 237 (log ε 4.1); 270 (log ε 3.5) (MeOH)

3,4:3′,4′-Bis(methylene), 5,5′-di-Me ether, 9-Ac
[869804-41-1]
$C_{24}H_{26}O_{10}$ 474.462
Constit. of *Peperomia dindygulensis*. Pale yellow gum. $[\alpha]^{20}_{D}$ +20.7 (c, 0.03 in $CHCl_3$). λ_{max} 214; 245; 280 (MeOH)

(7R*,7′R*,8R*,8′R*)-form

3,3′,5,5′-Tetra-Me ether, 4-O-β-D-glucopyranoside
[441772-84-5]
$C_{28}H_{38}O_{14}$ 598.598
Constit. of the aerial parts of *Phagnalon rupestre*. Amorph. powder. $[\alpha]_{D}$ -13 (c, 0.1 in MeOH). λ_{max} 212; 273 (MeOH)

3,3′,5,5′-Tetra-Me ether, 9-O-β-D-glucopyranoside
[1076221-85-6]
$C_{28}H_{38}O_{14}$ 598.598
Constit. of the stems of *Hydnocarpus hainanensis*. Light yellow powder. $[\alpha]^{25}_{D}$ -26 (c, 0.1 in MeOH). λ_{max} 214 (log ε 2.52); 272 (log ε 0.2) (MeOH)

(7R*,7′R*,8S*,8′S*)-form

3,3′,5,5′-Tetra-Me ether
Huazhongilexin
[161657-71-2]
$C_{22}H_{28}O_9$ 436.457
Constit. of *Ilex centrochinensis*, *Neoalsomitra integrifoliola* and *Tetracentron sinense*. Cryst.. Mp 157 - 158°

3,3′,5,5′-Tetra-Me ether, 9-O-[4-hydroxy-3-methoxy-E-cinnamoyl-(→2)-β-D-xylopyranoside]
$C_{37}H_{44}O_{16}$ 744.743
Constit. of the leaves of *Neoalsomitra integrifoliola*. Pale yellow powder. $[\alpha]^{20}_{D}$ -33.3 (c, 0.2 in MeOH). λ_{max} 208 (log ε 2.21); 238 (log ε 0.69); 325 (log ε 0.47) (MeOH)

3,3′,4,4′,5,5′-Hexa-Me ether, 9-O-α-L-arabinopyranoside
[1022983-49-8]
$C_{29}H_{40}O_{13}$ 596.626
Constit. of the leaves of *Neoalsomitra integrifoliola*. Amorph. powder. $[\alpha]^{20}_{D}$ +1 (c, 0.04 in MeOH). λ_{max} 204 (log ε 1.34); 223 (log ε 0.67) (MeOH)

3,3′,4,4′,5,5′-Hexa-Me ether, 9-O-β-D-xylopyranoside
[1022983-43-2]
$C_{29}H_{40}O_{13}$ 596.626
Constit. of the leaves of *Neoalsomitra integrifoliola*. Amorph. powder. $[\alpha]^{20}_{D}$ -23 (c, 0.11 in MeOH). λ_{max} 207 (log ε 1.57); 225 (log ε 0.16) (MeOH)

(7R*,7′S*,8R*,8′R*)-form

3′,4′-Methylene, 3,5,5′-tri-Me ether, 9,9′-di-Ac
[869804-42-2]
$C_{26}H_{30}O_{11}$ 518.515
Constit. of *Peperomia dindygulensis*. Pale yellow gum. $[\alpha]^{20}_{D}$ -23.4 (c, 0.25 in $CHCl_3$). λ_{max} 212; 244; 280 (MeOH)

3′,4′-Methylene, 3,4,5,5′-tetra-Me ether, 9′-Ac
[869804-43-3]
$C_{25}H_{30}O_{10}$ 490.505
Constit. of *Peperomia dindygulensis*. Pale yellow gum. $[\alpha]^{20}_{D}$ -21.8 (c, 0.47 in $CHCl_3$). λ_{max} 212; 244; 280 (MeOH)

3,4:3′,4′-Bis(methylene), 5,5′-di-Me ether, 9-Ac
[869804-40-0]
$C_{24}H_{26}O_{10}$ 474.462
Constit. of *Peperomia dindygulensis*. Pale yellow gum. $[\alpha]^{20}_{D}$ +17.8 (c, 0.09 in $CHCl_3$). λ_{max} 213; 242; 280 (MeOH)

3,4:3′,4′-Bis(methylene), 5,5′-di-Me ether, 9′-Ac
[869804-39-7]
$C_{24}H_{26}O_{10}$ 474.462
Constit. of *Peperomia dindygulensis*. Pale yellow gum. $[\alpha]^{20}_{D}$ -11 (c, 1.69 in $CHCl_3$). λ_{max} 213; 246; 281 (MeOH)

(7R*,7′S*,8S*,8′R*)-form

3,3′,5,5′-Tetra-Me ether, 9-O-β-D-glucopyranoside
[339175-42-7]
$C_{28}H_{38}O_{14}$ 598.598
Constit. of *Acanthus ilicifolius*. Amorph. powder. $[\alpha]^{22}_{D}$ -25.5 (c, 1 in MeOH)

(7R*,7′S*,8S*,8′S*)-form

3,3′,5,5′-Tetra-Me ether, 9,9′-di-Ac
[878383-48-3]
$C_{26}H_{32}O_{11}$ 520.531
Constit. of *Peperomia pellucida*. Amorph. powder. λ_{max} 239; 271 (MeOH aq.)

3,4-Methylene, 3′,5,5′-tri-Me ether, 9-Ac
[878028-83-2]
$C_{24}H_{28}O_{10}$ 476.478
Constit. of *Peperomia pellucida* and *Peperomia heyneana*. Amorph. powder or gum. $[\alpha]^{20}_{D}$ -8.1 (c, 0.63 in Me_2CO). $[\alpha]^{20}_{D}$ -42.4 (c, 0.16 in MeOH). λ_{max} 239; 276 (MeOH aq.). [neutral] λ_{max} 245; 280 (MeOH aq.).

Matsushita, H. *et al.*, *Phytochemistry*, 1991, **30**, 2025-2027 (*Icariol A₂*).
Lin, L.-D. *et al.*, *Huaxue Xuebao*, 1995, **53**, 98-101 (*Huazhongilexin*).
Kanchanapoom, T. *et al.*, *Phytochemistry*, 2001, **56**, 369-372 (*Acanthus ilicifolius glucoside*).
Gongora, L. *et al.*, *Phytochemistry*, 2002, **59**, 857-860 (*Phagnalon rupestre glucoside*).
Takara, K. *et al.*, *Biosci., Biotechnol., Biochem.*, 2003, **67**, 376-379 (*cane sugar xyloside*).
Wu, J.-L. *et al.*, *J. Nat. Prod.*, 2005, **68**, 1656-1660 (*Peperomia dindygulensis lignans*).
Xu, S. *et al.*, *J. Nat. Prod.*, 2006, **69**, 247-250 (*Peperomia pellucida lignans*).
Yamauchi, H. *et al.*, *Chem. Pharm. Bull.*, 2007, **55**, 346-347 (*Icariol A₂, Icariol A₂ 9-glucoside*).
Zhang, G.-L. *et al.*, *J. Nat. Prod.*, 2007, **70**, 662-664 (*Peperomia heyneana lignan*).
Nakamura, S. *et al.*, *Heterocycles*, 2008, **75**, 1435-1446 (*Foliachinenoside C*).
Su, D. *et al.*, *J. Nat. Prod.*, 2008, **71**, 784-788 (*Neoalsomitra integrifoliola glycosides*).
Shi, H.-M. *et al.*, *Nat. Prod. Res.*, 2008, **22**, 633-637 (*Hydnocarpus hainanensis glucoside*).
Xiong, L. *et al.*, *J. Nat. Prod.*, 2011, **74**, 1188-1200 (*Sinocalamus affinis penta-Me ether*).
Kim, K.H. *et al.*, *J. Nat. Prod.*, 2011, **74**, 2187-2192 (*Ligraminol B*).
Lee, I.-S. *et al.*, *J. Nat. Prod.*, 2015, **78**, 2249-2254 (*3,3′,5,5′-tetra-Me ether 9-rhamnoside*).
Kang, H.R. *et al.*, *J. Agric. Food Chem.*, 2018, **66**, 2677-2684 (*Oryzativols A,B, activity*)

7,9′-Epoxy-8,8′-lignan-3,3′,4,4′,5,5′,7′,9-octol E109

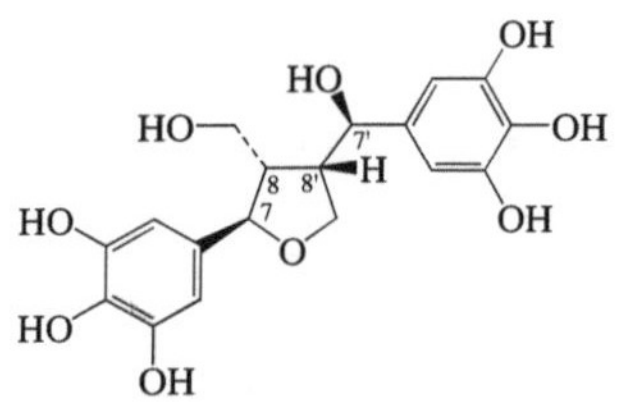

(7R,7′R,8S,8′S)-form

3,3′,4,4′,5,5′,7′,9-Octahydroxy-7,9′-epoxylignan
$C_{18}H_{20}O_9$ 380.349
(7R,7′R,8S,8′S)-form
3,3′,5,5′,7′-Penta-Me ether
7,9′-Epoxy-3,3′,5,5′,7′-pentamethoxy-8,8′-lignan-4,4′,9-triol. 4,4′,9-Trihydroxy-3,3′,5,5′,7′-pentamethoxy-7,9′-epoxylignan
3,3′,5,5′,7′-Penta-Me ether, 9-O-β-D-glucopyranoside
Ligalbumoside E
[1408245-03-3]
$C_{29}H_{40}O_{14}$ 612.625
Constit. of *Viscum album*. Powder. $[\alpha]_D^{25}$ -24.5 (c, 0.3 in MeOH)
3,3′,5,5′,7′-Penta-Me ether, 9-Ac
Cinncassin H. *Cinnacassin H*
[2232873-47-9]
$C_{25}H_{32}O_{10}$ 492.521
Constit. of twigs of *Cinnamomum cassia*. Exhibits neuroprotective activity. Yellow oil. $[\alpha]_D^{25}$ +5 (c, 0.2 in MeOH). λ_{max} 205 (log ε 4.02); 235 (log ε 3.60); 275 (log ε 3.47) (MeOH)
7′-Ketone
7,9′-Epoxy-3,3′,4,4′,5,5′,9-heptahydroxy-8,8′-lignan-7′-one. 3,3′,4,4′,5,5′,9-Heptahydroxy-7′-oxo-7,9′-epoxylignan
7′-Ketone, 3,3′,5,5′-tetra-Me ether
7,9′-Epoxy-4,4′,9-trihydroxy-3,3′,5,5′-tetramethoxy-8,8′-lignan-7′-one. 4,4′,9-Trihydroxy-3,3′,5,5′-tetramethoxy-7′-oxo-7,9′-epoxylignan
7′-Ketone, 3,3′,5,5′-tetra-Me ether, 9-O-β-D-glucopyranoside
Aketrilignoside C
[1020079-73-5]
$C_{28}H_{36}O_{14}$ 596.582
Constit. of the stems of *Akebia trifoliata*. Gum. $[\alpha]_D^{20}$ -77 (c, 0.09 in MeOH). Incorrectly assigned 8*R*-config. in the lit.. λ_{max} 210 (log ε 4.7); 306 (log ε 4.1) (MeOH)

(7R,7′S,8R,8′S)-form
3,3′,5,5′,7′-Penta-Me ether, 9-O-β-D-glucopyranoside
Ligalbumoside C
[1408244-84-7]
$C_{29}H_{40}O_{14}$ 612.625
Constit. of *Viscum album*. Powder. $[\alpha]_D^{25}$ -45 (c, 0.3 in MeOH)
7′-Ketone, 3,3′,5,5′-tetra-Me ether
$C_{22}H_{26}O_9$ 434.441
Constit. of the rhizomes of *Sinomenium acutum*. Gum. $[\alpha]_D$ +1.8 (c, 0.13 in MeCN). $[\alpha]_D$ +14 (c, 0.13 in $CHCl_3$)
(7R,7′S,8S,8′S)-form
3,3′,5,5′,7′-Penta-Me ether, 9-Ac
Cinncassin G. *Cinnacassin G*
[2232873-46-8]
$C_{25}H_{32}O_{10}$ 492.521
Constit. of twigs of *Cinnamomum cassia*. Yellow oil. $[\alpha]_D^{25}$ -1 (c, 0.7 in MeOH). λ_{max} 205 (log ε 4.40); 215 (log ε 4.38); 235 (log ε 4.16); 275 (log ε 3.34) (MeOH)
(7S,7′S,8R,8′R)-form
3,3′,5,5′,7′-Penta-Me ether, 9-O-β-D-glucopyranoside
Ligalbumoside D
[1408244-90-5]
$C_{29}H_{40}O_{14}$ 612.625
Constit. of *Viscum album*. Powder. $[\alpha]_D^{25}$ +28.1 (c, 0.3 in MeOH)
7′-Ketone, 3,3′,5,5′-tetra-Me ether, 4-O-β-D-glucopyranoside
[2247433-01-6]
$C_{28}H_{36}O_{14}$ 596.582
Constit. of the rhizomes of *Smilax trinervula*. Powder (MeOH). $[\alpha]_D^{25}$ -31.41 (c, 0.19 in MeOH). λ_{max} 280 (log ε 4.94) (MeOH)
7′-Ketone, 3,3′,5,5′-tetra-Me ether, 4′-O-β-D-glucopyranoside
[2247433-02-7]
$C_{28}H_{36}O_{14}$ 596.582
Constit. of the rhizomes of *Smilax trinervula*. Powder (MeOH). $[\alpha]_D^{25}$ -61.73 (c, 0.24 in MeOH). λ_{max} 225 (log ε 4.95); 305 (log ε 4.89) (MeOH)
7′-Ketone, 3,3′,5,5′-tetra-Me ether, 9-O-β-D-glucopyranoside
Aketrilignoside B
[1020108-72-8]
$C_{28}H_{36}O_{14}$ 596.582
Constit. of *Akebia trifoliata*, *Erycibe hainanensis* and *Viola tianshanica*. Gum. $[\alpha]_D^{20}$ -6 (c, 0.35 in MeOH). Incorrectly assigned 8*S*-config. in the lit.. λ_{max} 208 (log ε 4.5); 302 (log ε 3.9) (MeOH)
7′-Ketone, 3′,4′-methylene, 3,4,5,5′-tetra-Me ether, 9-O-β-D-glucopyranoside
Terminaloside W
[2222336-25-4]
$C_{29}H_{36}O_{14}$ 608.593
Constit. of *Terminalia citrina*. Amorph. powder. $[\alpha]_D^{25}$ +78 (c, 0.2 in MeOH). λ_{max} 216 (log ε 4.53); 301 (log ε 4.12) (MeOH)
7′-Ketone, 4′,5′-methylene, 3,3′,4-tri-Me ether, 9-O-β-D-glucopyranoside
Terminaloside V
[2222336-24-3]
$C_{28}H_{34}O_{14}$ 594.566
Constit. of *Terminalia citrina*. Amorph. powder. $[\alpha]_D^{25}$ +58 (c, 0.2 in MeOH). λ_{max} 231 (log ε 4.18); 302 (log ε 3.93) (MeOH)
(7S,7′S,8S,8′R)-form
3,3′,5,5′,7′-Penta-Me ether, 9-O-β-D-glucopyranoside
Ligalbumoside B
[1408244-81-4]
$C_{29}H_{40}O_{14}$ 612.625
Constit. of *Viscum album*. Powder. $[\alpha]_D^{25}$ -5 (c, 0.3 in MeOH)
(7R*,7′ξ,8R*,8′R*)-form
3,3′,5,5′-Tetra-Me ether
7,9′-Epoxy-3,3′,5,5′-tetramethoxy-8,8′-lignan-4,4′,7′,9-tetrol. 4,4′,7′,9-Tetrahydroxy-3,3′,5,5′-tetramethoxy-7,9′-epoxylignan
3,3′,5,5′-Tetra-Me ether, 9-O-β-D-glucopyranoside
Pseuderanoside
[1451255-63-2]
$C_{28}H_{38}O_{14}$ 598.598
Constit. of the roots of *Pseuderanthemum carruthersii* var. *atropurpureum*. Shows moderate cytotoxic activity. Light yellow gum. $[\alpha]_D^{20}$ -166 (c, 0.6 in MeOH)
(7R,8S,8′R)-form
7′-Ketone, 3,3′,5,5′-tetra-Me ether, 4′-O-β-D-glucopyranoside
Officinalioside
[1812868-20-4]
$C_{28}H_{36}O_{14}$ 596.582
Constit. of *Borago officinalis*. Antioxidant. Amorph. powder. $[\alpha]_D^{24}$ -8.1 (c, 1 in MeOH). λ_{max} 255 (log ε 4.63); 310 (log ε 4.66); 360 (log ε 4.74); 380 (log ε 4.68) (MeOH)

(7S,8R,8′S)-form

7′-Ketone, 3,3′,5,5′-tetra-Me ether
[1292294-22-4]
$C_{22}H_{26}O_9$ 434.441
Constit. of the stems of *Sinocalamus affinis* and *Cinnamomum cassia*. Inhibits secretion of fibronectin, collagen IV and IL-6 implicated in diabetic nephropathy. Amorph. powder. $[\alpha]^{20}_{D}$ -1.5 (c, 0.2 in MeCN). λ_{max} 208 (log ε 4.18); 233 (log ε 3.73); 303 (log ε 3.47) (MeOH)

7′-Ketone, 3,3′,4,4′,5,5′-hexa-Me ether, 9-O-β-D-glucopyranoside
Terminaloside T
[2222336-22-1]
$C_{30}H_{40}O_{14}$ 624.636
Constit. of *Terminalia citrina*. Inhibits cell growth in T47D cells. Amorph. powder. $[\alpha]^{25}_{D}$ +22 (c, 0.2 in MeOH). λ_{max} 224 (log ε 4.26); 279 (log ε 4.04) (MeOH)

7′-Ketone, 3,4-methylene, 3′,4′,5,5′-tetra-Me ether, 9-O-β-D-glucopyranoside
Terminaloside S
[2222336-21-0]
$C_{29}H_{36}O_{14}$ 608.593
Constit. of *Terminalia citrina*. Amorph. powder. $[\alpha]^{25}_{D}$ +25 (c, 0.2 in MeOH). λ_{max} 224 (log ε 4.16); 280 (log ε 4.05) (MeOH)

7′-Ketone, 3′,4′-methylene, 3,4,5,5′-tetra-Me ether, 9-O-β-D-glucopyranoside
Terminaloside R
[2222336-20-9]
$C_{29}H_{36}O_{14}$ 608.593
Constit. of *Terminalia citrina*. Inhibits cell growth in T47D and MCF-7 cell lines. Amorph. powder. $[\alpha]^{25}_{D}$ +54 (c, 0.2 in MeOH). λ_{max} 219 (log ε 4.36); 298 (log ε 3.99) (MeOH)

(7R,8R*,8′S*)-form*

7′-Ketone, 3,3′,4,4′,5,5′-hexa-Me ether
7,9′-Epoxy-9-hydroxy-3,3′,4,4′,5,5′-hexamethoxy-8,8′-lignan-7′-one. 9-Hydroxy-3,3′,4,4′,5,5′-hexamethoxy-7′-oxo-7,9′-epoxylignan. ***Arborone‡***
[957627-41-7]
$C_{24}H_{30}O_9$ 462.495
Constit. of the stems of *Piper arborescens*. Prisms (MeOH). Mp 141 - 143°. $[\alpha]^{25}_{D}$ +11.5 (c, 0.05 in CHCl$_3$). λ_{max} 216 (log ε 4.43); 282 (log ε 4.03) (MeOH)

(7R,8S*,8′R*)-form*

7′-Ketone, 3,3′,5,5′-tetra-Me ether, 9-O-β-D-xylopyranoside
[1022983-42-1]
$C_{27}H_{34}O_{13}$ 566.556
Constit. of the leaves of *Neoalsomitra integrifoliola*. Amorph. powder. $[\alpha]^{20}_{D}$ -77.3 (c, 0.14 in MeOH). λ_{max} 209 (log ε 2.2); 229 (log ε 1.02); 302 (log ε 0.59) (MeOH)

(7R,8S*,8′S*)-form*

7′-Ketone, 3,3′,4,4′,5,5′-hexa-Me ether
Seselinone
[1009297-75-9]
$C_{24}H_{30}O_9$ 462.495
Constit. of the aerial parts of *Seseli annuum*. Oil. $[\alpha]^{20}_{D}$ -13.3 (c, 1.5 in CH_2Cl_2). λ_{max} 206; 224 (sh); 282 (MeOH)

(7ξ,8ξ,8′ξ)-form

7′-Ketone, 3,3′,5,5′-tetra-Me ether
Ciwujiatone
[218901-26-9]
$C_{22}H_{26}O_9$ 434.441
Isol. from *Acanthopanax senticosus*, *Euonymus alatus* and *Grewia bilamellata*. Cryst. (CHCl$_3$). Mp 140 - 141° (112-114°). In equilib. with cyclic hemiacetal (furanofuranoid) form. Stereochem. of the different isolates need not be the same

7′-Ketone, 3,3′,5,5′-tetra-Me ether, 4-O-[β-D-apiofuranosyl-(1→2)-β-D-glucopyranoside]
[139701-49-8]
$C_{33}H_{44}O_{18}$ 728.697
Constit. of the stem bark of *Albizzia julibrissin*. Amorph. powder. $[\alpha]^{20}_{D}$ -46.4 (c, 0.58 in MeOH). λ_{max} 287 (log ε 4.73); 301 (log ε 3.88) (MeOH)

7′-Ketone, 3,3′,5,5′-tetra-Me ether, 4′-O-β-D-glucopyranoside
Magnolenin C
[66779-67-7]
$C_{28}H_{36}O_{14}$ 596.582
Isol. from *Magnolia grandiflora*, *Acanthus ebracteatus* and *Stellera chamaejasme*.

Rao, K.V. *et al.*, *Lloydia*, 1978, **41**, 56-62 (*Magnolenin C*).
Kinjo, J. *et al.*, *Chem. Pharm. Bull.*, 1991, **39**, 2952-2955 (*Albizzia apiosylglucoside*).
Wu, L.J. *et al.*, *Zhongguo Yaoli Xuebao*, 1999, **34**, 294-296 (*Ciwujiatone*).
Tsai, I.-L. *et al.*, *Planta Med.*, 2005, **71**, 535-542 (*Arborone*).
Takahashi, M. *et al.*, *Heterocycles*, 2007, **71**, 2113-2118 (*Arborone, synth*).
Vučković, I. *et al.*, *Phytother. Res.*, 2007, **21**, 790-792 (*Seselinone*).
Su, D. *et al.*, *J. Nat. Prod.*, 2008, **71**, 784-788 (*Neoalsomitra integrifoliola constit*).
Guan, S.H. *et al.*, *Magn. Reson. Chem.*, 2008, **46**, 186-190 (*Aketrilignosides B,C*).
Xiong, L. *et al.*, *J. Nat. Prod.*, 2011, **74**, 1188-1200 (*Sinocalamus affinis constit*).
Vo, T.N. *et al.*, *Chem. Pharm. Bull.*, 2012, **60**, 1125-1133 (*Pseuderanoside*).
Nhiem, N.X. *et al.*, *Magn. Reson. Chem.*, 2012, **50**, 772-777 (*Ligalbumosides B-E*).
Luo, Q. *et al.*, *Molecules*, 2013, **18**, 10930-10943 (*Cinnamomum cassia constit, activity*).
Qin, Y. *et al.*, *Molecules*, 2013, **18**, 13636-13644 (*Aketrilignoside B*).
Samy, M.N. *et al.*, *Nat. Prod. Res.*, 2016, **30**, 967-972 (*Officinalioside*).
Lee, S.Y. *et al.*, *Phytochem. Lett.*, 2016, **17**, 79-84 (*Sinomenium acutum constit*).
Shu, J. *et al.*, *Phytochem. Lett.*, 2017, **20**, 1-8 (*Smilax trinervula constits*).
Liu, X. *et al.*, *J. Nat. Prod.*, 2018, **81**, 1333-1342 (*Cinncassins G,H, activity*).
Muhit, M.A. *et al.*, *Phytochemistry*, 2018, **145**, 161-167 (*Terminalosides R,S,T,V,W, activity*)

9,9′-Epoxy-8,8′-lignan-3,3′,4,4′,5,5′,7,7′-octol E110

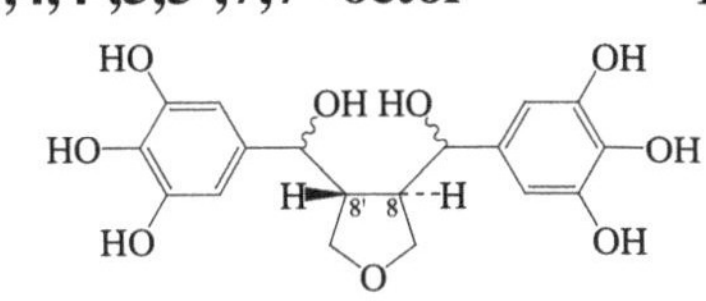

3,3′,4,4′,5,5′,7,7′-Octahydroxy-9,9′-epoxylignan
$C_{18}H_{20}O_9$ 380.349

(7ξ,7′ξ,8R,8′R*)-form*

7-Ketone
9,9′-Epoxy-3,3′,4,4′,5,5′,7′-heptahydroxy-8,8′-lignan-7-one. 3,3′,4,4′,5,5′,7-Heptahydroxy-7′-oxo-9,9′-epoxylignan

7-Ketone, 3,3′,4,4′,5,5′-hexa-Me ether
9,9′-Epoxy-7′-hydroxy-3,3′,4,4′,5,5′-hexamethoxy-8,8′-lignan-7-one. 7-Hydroxy-3,3′,4,4′,5,5′-hexamethoxy-7′-oxo-9,9′-epoxylignan. ***Aglacin K***
[686744-63-8]
$C_{24}H_{30}O_9$ 462.495
Constit. of the stem bark of *Aglaia cordata*. Waxy solid. $[\alpha]^{20}_{D}$ -25.5 (c, 0.62 in CHCl$_3$). λ_{max} 287 (log ε 4.06) (CHCl$_3$).

Wang, B.-G. *et al.*, *J. Nat. Prod.*, 2004, **67**, 682-684 (*Aglacin K*).
Takahashi, M. *et al.*, *Heterocycles*, 2007, **71**, 2113-2118 (*synth*)

3,7′-Epoxy-4,8′-lignan-3′,4′,5,9,9′-pentol E111

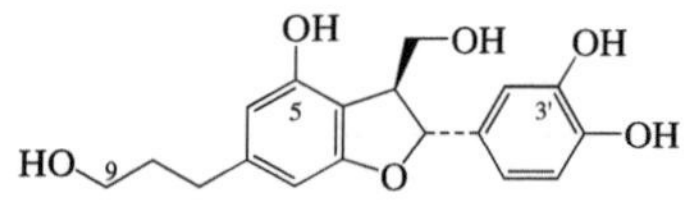

Absolute Configuration

4-[2,3-Dihydro-4-hydroxy-3-hydroxymethyl-6-(3-hydroxypropyl) benzofuran-2-yl]benzene-1,2-diol. 3,7′-Epoxy-3′,4′,5,9,9′-pentahydroxy-4,8′-neolignan

$C_{18}H_{20}O_6$ 332.352

(7′R,8′S)-form

3′,5-Di-Me ether

3,7′-Epoxy-3′,5-dimethoxy-4,8′-lignan-4′,9,9′-triol

3′,5-Di-Me ether, 9-Ac

Rebaneolignan B

[2101546-34-1]

$C_{22}H_{26}O_7$ 402.443

Constit. of the stems of *Mangifera reba*. α-Glucosidase inhibitor. Amorph. yellow solid.

Duong, T.T.T. *et al., Tetrahedron Lett.*, 2017, **58**, 2280-2283 (*Rebaneolignan B, activity*)

4,7′-Epoxy-3,8′-lignan-3′,4′,5,7,8-pentol E112

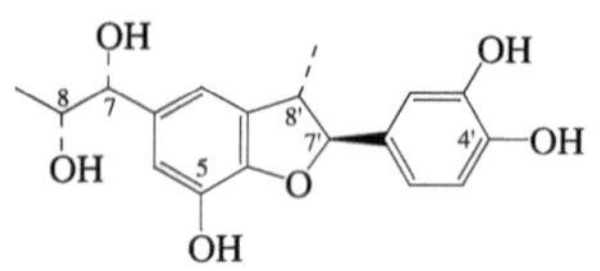

(7R,7′S,8R,8′S)-form

2-(3,4-Dihydroxyphenyl)-5-(1,2-dihydroxypropyl)-2,3-dihydro-7-hydroxy-3-methylbenzofuran. 4,7′-Epoxy-3′,4′,5,7,8-pentahydroxy-3,8′-neolignan

$C_{18}H_{20}O_6$ 332.352

(7R,7′S,8R,8′S)-form

3′,5-Di-Me ether

4,7′-Epoxy-3′,5-dimethoxy-3,8′-lignan-4′,7,8-triol. 4,7′-Epoxy-4′,7,8-trihydroxy-3′,5-dimethoxy-3,8′-lignan. ***Licarinediol B***

[250655-00-6]

$C_{20}H_{24}O_6$ 360.406

Constit. of *Aristolochia pubescens*. Yellow oil. $[\alpha]^{25}_{D}$ +3.6 (c, 1.02 in $CHCl_3$). Data shown is for a mixt. with Licarinediol A. λ_{max} 242 (log ε 3.91); 282 (log ε 3.8); 310 (sh) ($CHCl_3$)

3′,4′,5-Tri-Me ether

4,7′-Epoxy-3′,4′,5-trimethoxy-3,8′-lignan-7,8-diol. 4,7′-Epoxy-7,8-dihydroxy-3′,4′,5-trimethoxy-3,8′-lignan. ***5-O-Methyllicarinediol B***

$C_{21}H_{26}O_6$ 374.433

Constit. of *Aristolochia pubescens*. Yellow oil. $[\alpha]^{25}_{D}$ +3.5 (c, 1.04 in $CHCl_3$). Data shown is for a mixt. with 5-*O*-Methyllicarinediol A. λ_{max} 242 (log ε 3.83); 282 (log ε 3.68); 310 (sh) ($CHCl_3$)

(7S,7′S,8S,8′S)-form

3′,5-Di-Me ether

Licarinediol A

[250654-95-6]

$C_{20}H_{24}O_6$ 360.406

Constit. of *Aristolochia pubescens*. Config. revised in 2000

3′,4′,5-Tri-Me ether

5-O-Methyllicarinediol A

[250600-26-1]

$C_{21}H_{26}O_6$ 374.433

Constit. of *Aristolochia pubescens*

(7ξ,7′R,8ξ,8′R)-form

3′,5-Di-Me ether

Machilusol C

[289040-75-1]

$C_{20}H_{24}O_6$ 360.406

Constit. of *Machilus obovatifolia*. Cytotoxic agent. Oil. $[\alpha]^{25}_{D}$ -15.7 (c, 0.09 in $CHCl_3$). λ_{max} 230 (log ε 2.84); 290 (log ε 2.29) (EtOH)

3′,4′,5-Tri-Me ether

Machilusol D

[289040-76-2]

$C_{21}H_{26}O_6$ 374.433

Constit. of *Machilus obovatifolia*. Cytotoxic agent. Oil. $[\alpha]^{25}_{D}$ -18.2 (c, 0.05 in $CHCl_3$). λ_{max} 230 (sh); 280 (log ε 2.17) (EtOH)

3′,4′-Methylene, 5-Me ether

4,7′-Epoxy-5-methoxy-3′,4′-methylenedioxy-3,8′-lignan-7,8-diol. 4,7′-Epoxy-7,8-dihydroxy-5-methoxy-3′,4′-methylenedioxy-3,8′-lignan. ***Machilusol B***

[289040-74-0]

$C_{20}H_{22}O_6$ 358.39

Constit. of *Machilus obovatifolia*. Oil. $[\alpha]^{25}_{D}$ -21.5 (c, 0.9 in $CHCl_3$). λ_{max} 286 (log ε 3.38) (EtOH)

5′-Hydroxy, 3′,4′,5-tri-Me ether

4,7′-Epoxy-4′,5,5′-trimethoxy-3,8′-lignan-3′,7,8-triol. 4,7′-Epoxy-3′,7,8-trihydroxy-4′,5,5′-trimethoxy-3,8′-lignan. ***Machilusol E***

[289040-77-3]

$C_{21}H_{26}O_7$ 390.432

Constit. of *Machilus obovatifolia*. Cytotoxic agent. Oil. $[\alpha]^{25}_{D}$ -22.3 (c, 0.03 in $CHCl_3$). λ_{max} 215 (log ε 3.98); 285 (log ε 3.03) (EtOH)

(7R*,7′R*,8R*,8′R*)-form

3′,4′,5,7-Tetra-Me ether

4,7′-Epoxy-3′,4′,5,7-tetramethoxy-3,8′-lignan-8-ol

[1801289-81-5]

$C_{22}H_{28}O_6$ 388.46

Constit. of *Machilus obovatifolia*. Oil. $[\alpha]^{25}_{D}$ -8 (c, 0.31 in $CHCl_3$). Obt. as mixt. with stereoisomer. Stereochem. in doubt

(7R*,7′R*,8S*,8′R*)-form

3′,5,7-Tri-Me ether

4,7′-Epoxy-3′,5,7-trimethoxy-3,8′-lignan-4′,8-diol. 4,7′-Epoxy-4′,8-dihydroxy-3′,5,7-trimethoxy-3,8′-lignan. ***Fragransol A***

[114394-19-3]

$C_{21}H_{26}O_6$ 374.433

Constit. of *Machilus obovatifolia* and *Myristica fragrans*. Oil. Conflicting information in the lit. about stereochem.

(7R*,7′S*,8R*,8′S*)-form

3′,5,7-Tri-Me ether

Isofragransol A

[1801690-33-4]

$C_{21}H_{26}O_6$ 374.433

Constit. of *Machilus obovatifolia*. Oil. $[\alpha]^{25}_{D}$ -15.6 (c, 0.1 in $CHCl_3$). Obt. as mixt. with Fragransol A. Stereochem. in doubt. λ_{max} 209 (log ε 4.72); 236 (sh); 281 (log ε 3.81) (MeOH)

3′,4′,5,7-Tetra-Me ether

[1801290-27-6]

$C_{22}H_{28}O_6$ 388.46

Constit. of the roots of *Machilus obovatifolia*. Oil. $[\alpha]^{25}_{D}$ -8 (c, 0.31 in $CHCl_3$). Obt. as mixt. with stereoisomer. Stereochem. in doubt. λ_{max} 209 (log ε 4.65); 234 (log ε 4.18); 280 (log ε 3.69) (MeOH).

Hada, S. *et al., Phytochemistry*, 1988, **27**, 563-568 (*Fragransol A*).
Nascimento, I.R. *et al., Phytochemistry*, 1999, **52**, 345-350 (*Licarinediols*).
Juhasz, L. *et al., J. Nat. Prod.*, 2000, **63**, 866-870 (*Fragransol A, synth*).
Tsai, I.L. *et al., Planta Med.*, 2000, **66**, 403-407 (*Machilusols B-E, activity*).
Nascimento, I.R. *et al., Tetrahedron*, 2000, **56**, 9181-9193 (*Licarinediols A,B, synth, abs config*).

Lin, S.-Y. *et al.*, *Chem. Biodiversity*, 2015, **12**, 1057-1067 (*Machilus obovatifolia constits*)

4,7′-Epoxy-3,8′-lignan-3′,4′,5,9,9′-pentol E113

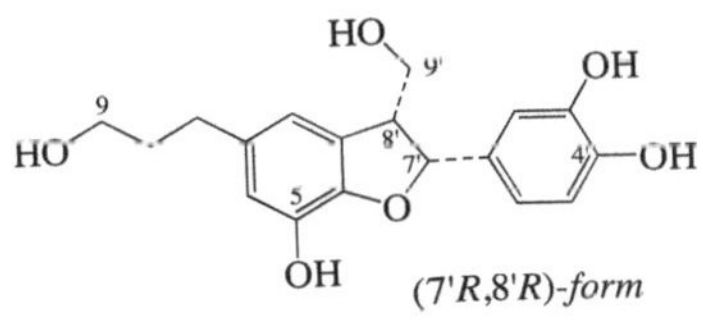

(7′R,8′R)-form

2-(3,4-Dihydroxyphenyl)-2,3-dihydro-7-hydroxy-3-(hydroxymethyl)-5-benzofuranpropanol. 4,7′-Epoxy-3′,4′,5,9,9′-pentahydroxy-3,8′-neolignan
[14049-57-1] (7′ξ,8′ξ)-(3′,5-di-Me ether) [57687-35-1] (7′ξ,8′ξ)-(3′,4-di-*O*-β-D-Glc) [57687-37-3] (7′ξ,8′ξ)-(3′-Me ether) [60536-58-5] (7′*R**,8′*R**)-(3′,5-di-Me ether) [73135-77-0] (*trans*-form)-(3′,5-di-Me ether) [125874-71-7] (3-Me ether, ±-*trans*-form) [125874-72-8] (3′,5-di-Me ether, ±-*trans*-form) [125874-73-9] (3′,5-di-Me tri-Ac, ±-*trans*-form)
$C_{18}H_{20}O_6$ 332.352
Lignan numbering shown. Numerous derivs. have been isol. with unspecified stereochem. - some as mixts. of stereoisomers

(7′R,8′R)-form

3′-Me ether, 4′-O-α-L-rhamnopyranoside
[167819-20-7]
[110770-30-4] (unspecified stereochem.)
$C_{25}H_{32}O_{10}$ 492.521
Constit. of *Picea abies* and *Pinus* spp.

3′-Me ether, 4′-O-(3-O-methyl-α-L-rhamnopyranoside)
[167638-62-2]
$C_{26}H_{34}O_{10}$ 506.548
Constit. of the root bark of *Picea abies*. [α] D -25 (c, 0.4 in MeOH)

3′-Me ether, 9′-O-β-D-glucopyranoside
[351331-39-0]
$C_{25}H_{32}O_{11}$ 508.52
Isol. from Riesling wine. λ_{max} 207; 282 (MeOH)

3′,5-Di-Me ether
Vladinol F
[133318-48-6]
$C_{20}H_{24}O_6$ 360.406
Constit. of *Acer buergerianum* and *Vladimiria souliei*. Shows significant anti-HIV activity. The synonym Vladinol F was originally assigned to the (7′*R*,8′*R*)-stereoisomer but later references also attribute a (7′*R*,8′*S*)-config. Also obt. as a mixt. with Vladinol E (see 4,7′-Epoxy-3,8′-lign-7-ene-3′,4′,5,9,9′-pentol, E163). λ_{max} 216 (ε 17000); 228 (ε 14500); 282 (ε 6170) (MeOH) (Derep)

3′,5-Di-Me ether, 9-O-β-D-glucopyranoside
7′-Epiglochidioboside
[935753-11-0]
$C_{26}H_{34}O_{11}$ 522.547
Constit. of *Acer truncatum* and *Camellia amplexicaulis*. Promotes osteoblast differentiation. Amorph. powder. $[\alpha]^{25}_{D}$ -10 (c, 0.69 in MeOH). λ_{max} 206 (log ε 4.76); 281 (log ε 3.82) (MeOH)

3′,5-Di-Me ether, 9′-O-β-D-arabinofuranoside
Illiciumlignan D
$C_{25}H_{32}O_{10}$ 492.521
Constit. of the leaves of *Illicium wardii*. Exhibits moderate inhibitory activity against human tumour cell line SKOV3. Yellowish solid. Mp 84.4 - 85.3°. λ_{max} 214 (log ε 1.98); 232 (log ε 1.58); 283 (log ε 0.61) (MeOH)

3′,5-Di-Me ether, 9′-O-β-D-glucopyranoside
[351346-10-6]
$C_{26}H_{34}O_{11}$ 522.547
Isol. from Riesling wine. λ_{max} 213; 282 (MeOH)

3′,5-Di-Me ether, 9-Ac, 9′-O-β-D-arabinofuranoside
Illiciumlignan F
$C_{27}H_{34}O_{11}$ 534.558
Constit. of the leaves of *Illicium wardii*. Yellowish cryst.. Mp 99.3 - 100.2°. λ_{max} 208 (log ε 0.69); 236 (log ε 0.21); 283 (log ε 0.08) (MeOH)

(7′R,8′S)-form

3′-Me ether, 4′-O-α-L-rhamnopyranoside
Isomassonianoside B
[889864-63-5]
[57687-36-2] (unspecified stereochem.)
$C_{25}H_{32}O_{10}$ 492.521
Constit. of *Illicium* sp., *Larix* sp., *Picea* sp. and *Pinus* sp.. Solid. $[\alpha]^{27}_{D}$ -120 (c, 0.75 in MeOH). λ_{max} 204 (log ε 4.13); 281 (log ε 5.15) (MeOH)

3′-Me ether, 4′-O-(3-O-methyl-α-L-rhamnopyranoside)
[889864-62-4]
$C_{26}H_{34}O_{10}$ 506.548
Constit. of *Abies holophylla*. Gum. $[\alpha]^{25}_{D}$ -16.0 (c, 0.05 in MeOH). λ_{max} 231 (log ε 4.0); 280 (log ε 1.3) (MeOH)

3′-Me ether, 4′-O-α-L-rhamnopyranoside, 5-O-β-D-glucopyranoside
[1355468-17-5]
$C_{31}H_{42}O_{15}$ 654.662
Constit. of the roots of *Rhodiola crenulata*. Amorph. solid. $[\alpha]^{20}_{D}$ -31.5 (c, 0.01 in MeOH). λ_{max} 238; 280 (MeOH)

3′-Me ether, 5-O-β-D-glucopyranoside
Lophanthoside C
[1256295-75-6]
$C_{25}H_{32}O_{11}$ 508.52
Constit. of *Rabdosia lophanthoides* and *Rhodiola crenulata*. Powder. Mp 214 - 216°. $[\alpha]^{20}_{D}$-9.6 (c, 0.11 in MeOH) (Lophanthoside C). $[\alpha]^{20}_{D}$+46.9 (c, 0.03 in MeOH) (Rhodiola crenulata constit). λ_{max} 214; 233; 282 (MeOH)

3′-Me ether, 9′-O-[4-hydroxy-3,5-dimethoxybenzoyl-(→6)-β-D-glucopyranoside]
Lophanthoside B
[1256295-74-5]
$C_{34}H_{40}O_{15}$ 688.679
Constit. of *Rabdosia lophanthoides*. Amorph. powder. Mp 232 - 233°. $[\alpha]^{20}_{D}$-11.6 (c, 0.16 in MeOH). λ_{max} 280 (MeOH)

9-Carboxylic acid, 3′,5-di-Me ether, Et ester
[1309659-18-4]
$C_{22}H_{26}O_7$ 402.443
Constit. of *Lobelia chinensis*. Oil. $[\alpha]^{25}_{D}$+12.2 (c, 0.25 in MeOH)

3′,4′-Di-Me ether, 5-O-β-D-glucopyranoside
Isodonoside IV
[2095793-39-6]
$C_{26}H_{34}O_{11}$ 522.547
Constit. of the aerial parts of *Isodon japonicus*. Exhibits protection against benzo[*a*]pyrene (BaP) induced cytotoxicity. Powder. $[\alpha]^{25}_{D}$ -24.1 (c, 0.48 in MeOH). λ_{max} 281 (log ε 3.1) (MeOH)

3′,4′-Di-Me ether, 9-O-β-D-glucopyranoside
Isodonoside V
[2095793-40-9]
$C_{26}H_{34}O_{11}$ 522.547
Constit. of the aerial parts of *Isodon japonicus*. Powder. $[\alpha]_D^{25}$ -43.4 (c, 0.25 in MeOH). λ_{max} 282 (log ε 3.2) (MeOH)

3′,4′-Di-Me ether, 9′-O-β-D-glucopyranoside
Isodonoside III
[2095793-38-5]
$C_{26}H_{34}O_{11}$ 522.547
Constit. of the aerial parts of *Isodon japonicus*. Powder. $[\alpha]_D^{25}$ -135.6 (c, 0.1 in MeOH). λ_{max} 281 (log ε 2.3) (MeOH)

3′,5-Di-Me ether
4,7′-Epoxy-3′,5-dimethoxy-3,8′-lignan-4′,9,9′-triol. **Dihydrodehydrodiconiferyl alcohol**
[126253-41-6]
$C_{20}H_{24}O_6$ 360.406
Constit. of *Betula maximowicziana*, *Euterpe oleracea*, *Illicium henryi*, *Juniperus chinensis*, *Kalopanax septemlobus* and *Viburnum awabuki*. Hydroxyl radical scavenger. Exhibits neuritogenic activity. Oil. $[\alpha]_D$ -25 (c, 0.1 in MeOH). Has been named Vladinol F but strictly speaking this is the (7′*R*,8′*R*)-stereoisomer. λ_{max} 204; 231 (sh); 282 (MeOH)

3′,5-Di-Me ether, 4′-O-α-L-rhamnopyranoside
Icariside E_4
[126253-42-7]
$C_{26}H_{34}O_{10}$ 506.548
Constit. of *Epimedium diphyllum*, *Juniperus communis* var. *depressa*, *Quercus macrocarpa* and *Tabebuia roseo-alba*. Exhibits antinociceptive activity. Amorph. powder. $[\alpha]_D$ -59.2 (c, 1 in MeOH)

3′,5-Di-Me ether, 4′-O-(3-O-methyl-α-L-rhamnopyranoside)
Juniperigiside
[1356829-13-4]
$C_{27}H_{36}O_{10}$ 520.575
Constit. of *Juniperus rigida*. Gum. $[\alpha]_D^{25}$ -50 (c, 0.25 in MeOH). λ_{max} 289 (log ε 4.08) (no solvent reported)

3′,5-Di-Me ether, 4′-O-β-D-glucopyranoside
[71144-68-8]
[167819-21-8] ((7′*R*,8′*R*)-form)
$C_{26}H_{34}O_{11}$ 522.547
Constit. of *Epimedium diphyllum*, *Larix leptolepis*, *Viburnum awabuki*, *Picea abies* and *Trachelospermum asiaticum*. $[\alpha]_D^{27}$ -34.6 (c, 1.4 in MeOH)

3′,5-Di-Me ether, 4′-O-[4-hydroxy-3-methoxybenzoyl-(→6)-β-D-glucopyranoside]
Saccharumoside A
[1346171-00-3]
$C_{34}H_{40}O_{14}$ 672.68
Constit. of the bark of *Acer saccharum*. Amorph. solid. $[\alpha]_D^{20}$ -31 (c, 0.5 in MeOH). λ_{max} 223 (log ε 3.91); 267 (log ε 3.73); 293 (log ε 3.45) (MeOH)

3′,5-Di-Me ether, 4′-O-[4-hydroxy-3-methoxybenzoyl-(→6)-β-D-glucopyranoside], 9-O-β-D-glucopyranoside
Viburfordoside F
[2257425-91-3]
$C_{40}H_{50}O_{19}$ 834.821
Constit. of the fruit of *Viburnum fordiae*. Moderate inhibitor of α-glucosidase and LPS-induced NO prodn.. Powder. $[\alpha]_D^{23}$ -35.9 (c, 0.1 in MeOH). λ_{max} 266 (log ε 4.45) (MeOH)

3′,5-Di-Me ether, 4′-O-[3,4,5-trihydroxybenzoyl-(→6)-β-D-glucopyranoside]
[1039751-07-9]
$C_{33}H_{38}O_{15}$ 674.652
Constit. of the stem bark of *Trewia nudiflora*. Amorph. yellow solid. $[\alpha]_D^{20}$ -44.4 (c, 1.2 in MeOH). λ_{max} 205 (log ε 6.2); 280 (log ε 1) (MeOH)

3′,5-Di-Me ether, 4′-O-[β-D-apiofuranosyl-(1→2)-β-D-glucopyranoside]
Alangiplatanoside
[178666-25-6]
$C_{31}H_{42}O_{15}$ 654.662
Constit. of *Alangium platanifolium*. Amorph. powder. $[\alpha]_D^{25}$ -77.1 (c, 1.7 in MeOH)

3′,5-Di-Me ether, 4′-O-[β-D-apiofuranosyl-(1→6)-β-D-glucopyranoside]
Symplolignanoside A
[852945-99-4]
$C_{31}H_{42}O_{15}$ 654.662
Constit. of the roots of *Symplocos caudata*. Powder. Mp 130 - 132°. $[\alpha]_D^{25}$ -20.8 (c, 0.05 in MeOH). λ_{max} 210; 285 (MeOH)

3′,5-Di-Me ether, 4′-O-[β-D-glucopyranosyl-(1→2)-β-D-glucopyranoside]
$C_{32}H_{44}O_{16}$ 684.688
Constit. of the roots of *Codonopsis pilosula*. $[\alpha]_D$ -29.2 (c, 0.04 in MeOH). λ_{max} 205 (log ε 3.86); 231 (log ε 3.25); 282 (log ε 2.71) (MeOH)

3′,5-Di-Me ether, 4′-O-[β-D-glucopyranosyl-(1→4)-β-D-glucopyranoside]
$C_{32}H_{44}O_{16}$ 684.688
Constit. of the stems of *Securidaca inappendiculata*. Exhibits moderate hepatoprotective activity. Yellow solid (MeOH). $[\alpha]_D^{25}$ +65 (c, 0.1 in MeOH)

3′,5-Di-Me ether, 4′-O-[β-D-glucopyranosyl-(1→4)-β-D-glucopyranoside], 9-O-β-D-glucopyranoside
$C_{38}H_{54}O_{21}$ 846.829
Constit. of *Securidaca inappendiculata*. Exhibits moderate hepatoprotective activity. Yellow solid (MeOH). $[\alpha]_D^{25}$ -23 (c, 0.1 in MeOH). λ_{max} 279 (log ε 3.59) (MeOH)

3′,5-Di-Me ether, 9-O-[β-D-apiofuranosyl-(1→6)-β-D-glucopyranoside]
Cinnacassoside B
[1691247-47-8]
$C_{31}H_{42}O_{15}$ 654.662
Constit. of bark of *Cinnamomum cassia*. Yellow solid. $[\alpha]_D^{15.4}$ -38 (c, 0.2 in MeOH). λ_{max} 205 (log ε 4.74); 283 (log ε 3.72) (MeOH)

3′,5-Di-Me ether, 9-O-[α-L-rhamnopyranosyl-(1→6)-β-D-glucopyranoside], 9′-O-β-D-xylopyranoside
[1006034-09-8]
$C_{37}H_{52}O_{19}$ 800.804
Constit. of the fruit of *Illicium oligandrum*. Powder. Mp 167 - 169°. $[\alpha]_D^{25}$ -12 (c, 0.5 in MeOH). λ_{max} 231 (log ε 4.45); 282 (log ε 3.51) (MeOH)

3′,5-Di-Me ether, 9′-O-β-D-glucopyranoside
[421546-95-4]
$C_{26}H_{34}O_{11}$ 522.547
Constit. of *Juniperus communis* var. *depressa*, *Pedicularis torta*, *Phlomis viscosa*, *Prunus domestica* and *Rabdosia lophanthoides*. Antioxidant. Amorph. powder. $[\alpha]_D$ -12.4 (c, 0.18 in MeOH).

$[\alpha]^{20}_{D}$ +38.7 (c, 0.08 in MeOH) (Prunus domestica). λ_{max} 210 (log ε 4.54); 226 (sh) (log ε 4.21); 282 (log ε 3.8) (MeOH)

3′,5-Di-Me ether, 9′-ester with shikimic acid, 4′-O-β-D-glucopyranoside
[1006034-10-1]
$C_{33}H_{42}O_{15}$ 678.684
Constit. of the fruit of *Illicium oligandrum*. Powder. Mp 181 - 183°. $[\alpha]^{25}_{D}$ -12.9 (c, 0.03 in MeOH). λ_{max} 235 (log ε 4.51); 281 (log ε 3.23) (MeOH)

3′,5-Di-Me ether, 9-Ac
[352005-77-7]
$C_{22}H_{26}O_7$ 402.443
Constit. of *Lobelia chinensis* and *Sambucus nigra*. Oil. $[\alpha]^{25}_{D}$ +11.3 (c, 0.18 in MeOH)

3′,5-Di-Me ether, 9-Ac, 9′-O-β-D-glucopyranoside
[1648932-58-4]
$C_{28}H_{36}O_{12}$ 564.584
Constit, of *Arenaria serpyllifolia*. Inhibitor of NO prodn.. Light yellow powder. $[\alpha]^{20}_{D}$ -24 (c, 0.26 in MeOH). λ_{max} 295 (log ε 2.88) (MeOH)

3′,5-Di-Me ether, 9′-Ac
(-)-Crataegifin A
$C_{22}H_{26}O_7$ 402.443
$[\alpha]^{20}_{D}$ -16.5 (c, 0.2 in MeOH)

3′,5-Di-Me ether, 9-O-(3-methylbutanoyl)
Dihydrodehydrodiconiferyl 9-isovalerate
[1803565-12-9]
$C_{25}H_{32}O_7$ 444.524
Constit. of the roots of *Valeriana officinalis* var. *latifolia*. Amorph. solid. $[\alpha]^{23}_{D}$ -1.3 (c, 0.31 in MeOH). λ_{max} 206 (log ε 4.70); 282 (log ε 3.76) (MeOH)

3′,5-Di-Me ether, 9′-O-(4-hydroxy-E-cinnamoyl)
[864378-13-2]
$C_{29}H_{30}O_8$ 506.551
Constit. of the bark of *Larix olgensis* var. *koreana*. Powder (MeOH). Mp 103 - 105°. $[\alpha]^{20}_{D}$ -50 (c, 0.24 in MeOH)

3′,5-Di-Me ether, 9′-O-(3,4-dihydroxy-E-cinnamoyl), 9-O-β-D-glucopyranoside
Xanthiumnolic E
[2226004-48-2]
$C_{35}H_{40}O_{14}$ 684.691
Constit. of the fruit of *Xanthium sibiricum*. Exhibits antiinflammatory activity and strong inhibitory activity against NO prodn.. Amorph. powder. $[\alpha]^{20}_{D}$ -19.7 (c, 0.1 in MeOH). λ_{max} 243 (log ε 4.02) (MeOH)

3′,5-Di-Me ether, 9,9′-bis-O-(4-hydroxy-3-methoxy-E-cinnamoyl)
(-)-Boehmenan
$C_{40}H_{40}O_{12}$ 712.748
Constit. of *Helicteres hirsuta*. Powder. $[\alpha]_{D}$ -14.3 (c, 0.03 in $CHCl_3$)

3′,5-Di-Me ether, 9′-O-(4-hydroxy-3-methoxy-E-cinnamoyl), 9-O-(4-hydroxy-3-methoxy-Z-cinnamoyl)
Boehmenan B
[171869-65-1]
$C_{40}H_{40}O_{12}$ 712.748
Constit. of *Ochroma lagopus*

3′,5-Di-Me ether, 9′-O-(4-hydroxy-3-methoxy-Z-cinnamoyl), 9-O-(4-hydroxy-3-methoxy-E-cinnamoyl)
Boehmenan C
[171868-04-5]
$C_{40}H_{40}O_{12}$ 712.748
Constit. of *Ochroma lagopus*

3′,4′,5-Tri-Me ether
4,7′-Epoxy-3′,4′,5-trimethoxy-3,8′-lignan-9,9′-diol
[166021-14-3]
[127179-41-3] (7′*S*,8′*R*-form)
$C_{21}H_{26}O_6$ 374.433
Constit. of *Croton* spp. and *Magnolia denudata*. Cell proliferation inhibitor and thymidine incorporation inhibitor. $[\alpha]^{23}_{D}$ -8.5 (c, 0.25 in $CHCl_3$). Abs. config. revised in 1998. Formerly assigned as dimethylcedrusin

3′,4′,5-Tri-Me ether, 9-O-β-D-glucopyranoside
[446878-60-0]
$C_{27}H_{36}O_{11}$ 536.574
Constit. of the aerial parts of *Phlomis chimerae* and *Phlomis viscosa*

3′,4′,5-Tri-Me ether, 9′-O-β-D-glucopyranoside
[862677-62-1]
$C_{27}H_{36}O_{11}$ 536.574
Constit. of the aerial parts of *Phlomis viscosa*

3′,4′,5-Tri-Me ether, 9′-Ac
Mitredrusin
[1882124-40-4]
$C_{23}H_{28}O_7$ 416.470
Constit. of the leaves of *Mitrephora teysmannii*. Liq.. $[\alpha]^{25}_{D}$ -22.3 (c, 0.04 in $CHCl_3$). λ_{max} 207 (log ε 3.40); 237 (log ε 3.10); 278 (log ε 3.00) (MeOH)

4′-O-(1-Hydroxymethyl-2-hydroxyethyl), 3′-Me ether
[1334309-87-3]
$C_{22}H_{28}O_8$ 420.458
Constit. of the root bark of *Illicium henryi*. Amorph. powder

4′-O-(2-Hydroxy-1-methoxycarbonylethyl), 3′,5-di-Me ether
Dunnianeolignan B
[1453807-98-1]
$C_{24}H_{30}O_9$ 462.495
Constit. of the roots of *Illicium dunnianum*. Syrup. $[\alpha]^{20}_{D}$ +6.6 (c, 0.08 in MeOH). λ_{max} 207 (log ε 4.16); 229 (log ε 3.72); 280 (log ε 3.25) (MeOH)

9-Carboxylic acid, 3′,5-di-Me ether, 4′-O-β-D-glucopyranoside
Lycocernuaside A
[2012588-10-0]
$C_{26}H_{32}O_{12}$ 536.53
Constit. of *Lycopodiella cernua*. Amorph. powder. $[\alpha]^{25}_{D}$ -36.2 (c, 0.53 in MeOH). λ_{max} 230 (log ε 3.12); 281 (log ε 2.50) (MeOH)

(7′S,8′R)-form

3′-Me ether
Cedrusin
[75775-36-9]
$C_{19}H_{22}O_6$ 346.379
Constit. of *Cedrus deodara*, *Eucommia ulmoides*, *Forsythia suspensa*, *Ginkgo biloba*, *Pinus* spp. and *Tsuga chinensis*. NO prod. inhibitor. Shows moderate inhibition of LDL oxidation. Powder. $[\alpha]_{D}$ +4.39 (c, 0.91 in MeOH)

3′-Me ether, 4′-O-β-D-xylopyranoside
Junipercomnoside A
[675881-22-8]
$C_{24}H_{30}O_{10}$ 478.494
Constit. of *Juniperus communis* var. *depressa* and *Quercus macrocarpa*. Amorph. powder. $[\alpha]_{D}$ -26.3 (c, 0.8 in MeOH)

3′-Me ether, 4′-O-α-L-rhamnopyranoside
Massonianoside B
[188300-19-8]
$C_{25}H_{32}O_{10}$ 492.521
Constit. of *Picea neoveitchii* and *Pinus massoniana*. Exhibits

moderate antifungal activity. Mp 127 - 128°. [α] D -26.4. λ_{max} 231; 281 (MeOH)

3′-Me ether, 4′-O-β-D-glucopyranoside

Massonianoside D

[85115-04-4]

$C_{25}H_{32}O_{11}$ 508.52

Constit. of *Pinus massoniana, Pseudotsuga menziesii* and *Trochodendron aralioides*. [α] D -28.2 (c, 4.9 in MeOH). λ_{max} 231; 281 (MeOH)

3′-Me ether, 4′-O-[6-O-benzoyl-β-D-glucopyranoside]

[1449786-28-0]

$C_{32}H_{36}O_{12}$ 612.628

Constit. of *Abies recurvata*. Yellow powder. $[\alpha]^{20}_{D}$ -22.5 (c, 0.45 in MeOH). λ_{max} 223 (log ε 4.05); 278 (log ε 3.8) (MeOH)

3′-Me ether, 5-O-β-D-glucopyranoside

Clemastanin A

[172670-47-2]

$C_{25}H_{32}O_{11}$ 508.52

Constit. of *Clematis stans* and *Rhodiola crenulata*. Amorph. powder. $[\alpha]^{20}_{D}$ -44.3 (c, 1.2 in MeOH)

3′-Me ether, 9-O-β-D-glucopyranoside

Isodonoside VI

[2091853-58-4]

$C_{25}H_{32}O_{11}$ 508.520

Constit. of the aerial parts of *Isodon japonicus*. Powder. $[\alpha]^{25}_{D}$ -18.2 (c, 0.27 in MeOH). λ_{max} 283 (log ε 3.0) (MeOH)

3′-Me ether, 9′-O-α-L-rhamnopyranoside

Massonianoside A

[623945-11-9]

$C_{25}H_{32}O_{10}$ 492.521

Constit. of *Juniperus communis* var. *depressa* and *Pinus massoniana*. Mp 130 - 131°. [α] D -14.8 (c, 3.3 in MeOH). λ_{max} 231; 281 (MeOH)

3′-Me ether, 9′-ester with shikimic acid

Massonianoid A

[854935-56-1]

$C_{26}H_{30}O_{10}$ 502.516

Constit. of the needles of *Pinus massoniana*. Mp 142 - 144°. $[\alpha]^{25}_{D}$ -60.2 (c, 0.05 in MeOH). λ_{max} 209 (MeOH)

3′,4′-Di-Me ether

4′-O-Methylcedrusin

[149340-29-4]

$C_{20}H_{24}O_6$ 360.406

Constit. of *Clinopodium urticifolium* and *Croton* spp.

3′,5-Di-Me ether

5-O-Methylcedrusin

[28199-69-1]

$C_{20}H_{24}O_6$ 360.406

Constit. of *Berberis koreana, Cornus kousa, Eucommia ulmoides, Euonymus alatus, Illicium henryi* and *Licaria chrysophylla*. NO prodn. inhibitor. Shows moderate cytotoxic activity. Active against hepatitus B virus.. Needles (Me_2CO aq.). Mp 102 - 103°. $[\alpha]^{27}_{D}$ +58 (c, 0.1 in MeOH)

3′,5-Di-Me ether, 4′-O-β-D-xylopyranoside

Cornuskoside A

[1058203-85-2]

$C_{25}H_{32}O_{10}$ 492.521

Constit. of *Cornus kousa* and *Quercus glauca*. Amorph. powder (MeOH) or yellow solid. Mp 115°. $[\alpha]^{26}_{D}$ -24 (c, 0.1 in MeOH). $[\alpha]^{25}_{D}$ -42.3 (c, 2.6 in MeOH). Incorrect stereochem. shown in paper. λ_{max} 243; 282 (MeOH)

3′,5-Di-Me ether, 4′-O-α-L-rhamnopyranoside

Massonianoside C

[188300-17-6]

$C_{26}H_{34}O_{10}$ 506.548

Constit. of the needles of *Pinus massoniana* and bark of *Pseudotsuga menziesii*. Mp 97 - 99°. [α] D -24.5 (c, 4.8 in MeOH). λ_{max} 231; 281 (MeOH)

3′,5-Di-Me ether, 4′,9-di-O-α-L-rhamnopyranoside

Forsythialanside D

[1616079-11-8]

$C_{32}H_{44}O_{14}$ 652.69

Constit. of the fruit of *Forsythia suspensa*. Exhibits mild antiviral activity against influenza A (H1N1) virus. Yellow gum. $[\alpha]^{20}_{D}$ -32.8 (c, 0.4 in MeOH). λ_{max} 206 (log ε 4.57); 233 (log ε 4.08); 279 (log ε 3.71) (MeOH)

3′,5-Di-Me ether, 4′-O-β-D-glucopyranoside

Urolignoside

[131723-83-6]

$C_{26}H_{34}O_{11}$ 522.547

Constit. of *Camellia amplexicaulis, Chloranthus japonicus, Ginkgo biloba, Helichrysum italicum, Jasminum urophyllum* and *Viburnum awabuki*. Shows significant activity against *Pseudomonas aeruginosa*. Osteoblast differentiation promotor. Amorph. powder. [α] D -43.6 (c, 0.1 in MeOH)

3′,5-Di-Me ether, 4′-O-β-D-glucopyranoside, 9-O-α-L-rhamnopyranoside

Forsythialanside C

[1616079-10-7]

$C_{32}H_{44}O_{15}$ 668.689

Constit. of the fruit of *Forsythia suspensa*. Yellow gum. $[\alpha]^{20}_{D}$ -53.5 (c, 0.6 in MeOH). λ_{max} 206 (log ε 4.31); 230 (log ε 3.92); 279 (log ε 3.50) (MeOH)

3′,5-Di-Me ether, 4′-O-[3,4,5-trihydroxybenzoyl-(→6)-β-D-glucopyranoside]

Davidioside A

[1015476-47-7]

$C_{33}H_{38}O_{15}$ 674.652

Constit. of the bark of *Davidia involucrata*. Amorph. yellow powder. Mp 136 - 138°. $[\alpha]^{21}_{D}$ -21 (c, 0.5 in MeOH). λ_{max} 229 (log ε 3.92); 281 (log ε 3.45); 310 (log ε 3.14) (MeOH)

3′,5-Di-Me ether, 9-O-β-D-glucopyranoside

Glochidioboside

[220745-70-0]

$C_{26}H_{34}O_{11}$ 522.547

Constit. of *Chloranthus glabra, Glochidion obovatum, Glochidion zeylanicum* and *Sarcandra glabra*. Amorph. powder. [α] D -14.1 (c, 1 in MeOH). λ_{max} 229 (ε 13530); 282 (ε 5730) (MeOH)

3′,5-Di-Me ether, 9-O-β-D-glucopyranoside, 9′-O-α-L-rhamnopyranoside

Junipercomnoside B

[675881-24-0]

$C_{32}H_{44}O_{15}$ 668.689

Constit. of the aerial parts of *Juniperus communis* var. *depressa*. Amorph. powder. [α] D -14.8 (c, 0.23 in MeOH)

3′,5-Di-Me ether, 9-O-[β-D-apiofuranosyl-(1→6)-β-D-glucopyranoside]

Samwiside

[1621072-56-7]

$C_{31}H_{42}O_{15}$ 654.662

Constit. of *Sambucus williamsii*. Amorph. pale yellow powder. $[\alpha]_D^{26}$ -60.4 (c, 0.6 in MeOH). λ_{max} 206 (log ε 4.92); 236 (log ε 4.30); 283 (log ε 3.90) (MeOH)

3′,5-Di-Me ether, 9′-O-β-D-apiofuranoside

Schisphenlignan K

[1542234-16-1]

$C_{25}H_{32}O_{10}$ 492.521

Isol. from the leaves and stems of *Schisandra sphenanthera*. Amorph. powder. $[\alpha]_D^{15}$ -99.7 (c, 0.13 in MeOH). λ_{max} 204 (log ε 4.53); 221 (log ε 4.68) ($CHCl_3$)

3′,5-Di-Me ether, 9′-O-β-D-xylopyranoside

[1015175-06-0]

$C_{25}H_{32}O_{10}$ 492.521

Constit. of *Davidia involucrata* and *Illicium henryi*

3′,5-Di-Me ether, 9′-O-(3-O-acetyl-β-D-xylopyranoside)

[1309459-87-7]

$C_{27}H_{34}O_{11}$ 534.558

Constit. of *Illicium henryi*. Shows moderate activity against hepatitis B virus. Amorph. powder. $[\alpha]_D^{28}$ -4.6 (c, 0.21 in MeOH). λ_{max} 282 (log ε 3.73) ($CHCl_3$)

3′,5-Di-Me ether, 9′-O-α-L-rhamnopyranoside

[1252572-36-3]

$C_{26}H_{34}O_{10}$ 506.548

Constit. of *Juniperus communis* var. *depressa*. Amorph. powder. $[\alpha]_D$ -4.5 (c, 0.31 in MeOH). λ_{max} 210 (log ε 4.48); 225 (sh); 282 (log ε 3.85) (MeOH)

3′,5-Di-Me ether, 9′-O-β-D-glucopyranoside

[126371-34-4]

Constit. of *Chloranthus japonicus*, *Glochidion zeylanicum*, *Licaria chrysophylla*, *Sarcandra glabra* and *Vitex rotundifolia*

3′,5-Di-Me ether, 9′-O-[β-D-apiofuranosyl-(1→6)-β-D-glucopyranoside]

[1022980-94-4]

$C_{31}H_{42}O_{15}$ 654.662

Constit. of the leaves of *Neoalsomitra integrifoliola*. Amorph. powder. $[\alpha]_D^{20}$ -25.7 (c, 0.1 in MeOH). λ_{max} 207 (log ε 1.84); 233 (log ε 0.48); 281 (log ε 0.18) (MeOH)

3′,5-Di-Me ether, 9′-O-[α-L-rhamnopyranosyl-(1→6)-β-D-glucopyranoside]

[2099167-57-2]

$C_{32}H_{44}O_{15}$ 668.689

Constit. of twigs of *Litsea cubeba*. Exhibits a reduction of acetaminophen-induced HepG2 cell injury. Amorph. powder. $[\alpha]_D^{20}$ -12 (c, 0.2 in MeOH). λ_{max} 234 (sh); 282 (log ε 0.56) (MeOH)

3′,5-Di-Me ether, 9′-O-[β-D-glucopyranosyl-(1→2)-β-D-glucopyranoside]

Saposide B

[1686144-41-1]

$C_{32}H_{44}O_{16}$ 684.688

Constit. of the sap of *Acer saccharum*. Amorph. powder. $[\alpha]_D^{22}$ -6 (c, 0.8 in MeOH). λ_{max} 222 (log ε 4.09); 228 (log ε 4.1); 282 (log ε 3.75) (MeOH)

3′,5-Di-Me ether, 9,9′-di-O-β-D-glucopyranoside

Viburfordoside E

[2257425-90-2]

$C_{32}H_{44}O_{16}$ 684.688

Constit. of the fruit of *Viburnum fordiae*. Powder. $[\alpha]_D^{23}$ +14.8 (c, 0.1 in MeOH). λ_{max} 282 (log ε 3.92) (MeOH)

3′,5-Di-Me ether, 9-O-sulfate

[289708-37-8]

$C_{20}H_{24}O_9S$ 440.463

Constit. of *Glochidion zeylanicum* and *Pseudotsuga menziesii*. Amorph. powder. $[\alpha]_D^{22}$ +5.1 (c, 0.43 in MeOH). λ_{max} 212 (log ε 4.4); 228 (sh); 282 (log ε 3.75) (MeOH)

3′,5-Di-Me ether, 9-Ac

[352005-74-4]

Constit. of *Lobelia chinensis*. Oil. $[\alpha]_D^{25}$ -9.8 (c, 0.2 in MeOH)

3′,5-Di-Me ether, 9′-Ac

(+)-Crataegifin A

$[\alpha]_D^{20}$ +18 (c, 0.2 in MeOH)

3′,5-Di-Me ether, 9,9′-di-Ac

Tiruneesin

[352005-75-5]

$C_{24}H_{28}O_8$ 444.48

Constit. of *Justicia neesii*. Cryst. (EtOAc/petrol). Mp 98 - 100°. $[\alpha]_D$ -35.2 (c, 0.2 in $CHCl_3$). λ_{max} 206; 212; 225; 282 (MeOH)

3′,5-Di-Me ether, 9-O-(4-hydroxy-3-methoxy-E-cinnamoyl), 9′-O-(4-hydroxy-E-cinnamoyl)

Boehmenan X

[1253289-73-4]

$C_{39}H_{38}O_{11}$ 682.722

Constit. of *Durio carinatus* and *Durio oxleyanus*. Amorph. solid. $[\alpha]_D^{28}$ +11.2 (c, 0.41 in $CHCl_3$)

3′,5-Di-Me ether, 9,9′-bis-O-(4-hydroxy-3-methoxy-E-cinnamoyl)

(+)-Boehmenan

[57296-22-7]

$C_{40}H_{40}O_{12}$ 712.748

Constit. of *Boehmeria tricuspis*, *Durio affinis*, *Durio zibethinus*, *Hibiscus ficulneus* and *Ochroma lagopus*. Shows moderate cytotoxic activity. Shows Wnt signal inhibitory activity. Powder. $[\alpha]_D^{25}$ +10.2 (c, 0.5 in MeOH)

3′,4′-Methylene, 5-Me ether

4,7′-Epoxy-5-methoxy-3′,4′-methylenedioxy-3,8′-lignan-9,9′-diol

$C_{20}H_{22}O_6$ 358.39

Constit. of *Taiwania cryptomerioides*. $[\alpha]_D^{25}$ -12 (c, 1.1 in MeOH)

9-Carboxylic acid, 4′-Me ether, Me ester

Trogopterin A

[1656284-94-4]

$C_{20}H_{22}O_7$ 374.389

Constit. of faeces of *Trogopterus xanthipes*. Exhibits moderate cytotoxicity against HL-60 cells. Light brown oil. $[\alpha]_D^{25}$ +33.62 (c, 1.0 in EtOH)

9-Carboxylic acid, 3′,5-di-Me ether

4,7′-Epoxy-4′,9′-dihydroxy-3′,5-dimethoxy-3,8′-lignan-9-oic acid

9-Carboxylic acid, 3′,5-di-Me ether, 4′-O-β-D-glucopyranoside

$C_{26}H_{32}O_{12}$ 536.53

Constit. of the leaves of *Litsea japonica*. Powder. $[\alpha]_D^{25}$ -43 (c, 0.1 in MeOH). λ_{max} 230; 276 (MeOH)

(7′S,8′S)-form

3′,5-Di-Me ether, tri-Ac

[168751-77-7]

$C_{26}H_{30}O_9$ 486.517

Constit. of the leaves of *Cryptomeria japonica*. Gum. $[\alpha]_D^{25}$ -66 (c, 0.8 in $CHCl_3$)

3′,5-Di-Me ether, 9-O-β-D-glucopyranoside

[1526987-82-5]

$C_{26}H_{34}O_{11}$ 522.547

Constit. of *Pulsatilla koreana*. Sortase A inhibitor in *Streptococcus mutans*. Solid. $[\alpha]_D^{25}$ -11.3 (c, 0.5 in MeOH). λ_{max} 282 (log ε 3.63) (MeOH)

3′,5-Di-Me ether, 9′-O-β-D-xylopyranoside

Illiciumlignan C

$C_{25}H_{32}O_{10}$ 492.521

Constit. of the leaves of *Illicium wardii*. Yellowish solid. Mp 95.7 - 96.6°. λ_{max} 213 (log ε 1.74); 230 (log ε 1.72); 284 (log ε 0.43) (MeOH)

3′,5-Di-Me ether, 9′-O-β-D-glucopyranoside

[1527462-15-2]

$C_{26}H_{34}O_{11}$ 522.547

Constit. of *Pulsatilla koreana*. Sortase A inhibitor in *Streptococcus mutans*. Solid. $[\alpha]_D^{25}$ -17.1 (c, 0.25 in MeOH). λ_{max} 286 (log ε 2.21) (MeOH)

(7′R*,8′R*)-form

3′-Me ether, 4′-O-α-L-rhamnopyranoside

[255881-96-0]

$C_{25}H_{32}O_{10}$ 492.521

Constit. of *Betula platyphylla* var. *japonica*

3′,5-Di-Me ether, 9′-O-β-D-glucopyranoside

Artemordolignan glycoside A

$C_{26}H_{34}O_{11}$ 522.547

Constit. of *Artemisia ordosica*. Powder. λ_{max} 258 (log ε 3.12) (MeOH)

(7′R*,8′S*)-form

3′-Me ether, 5-O-α-L-rhamnopyranoside

Juniperoside A‡

[1370732-70-9]

$C_{25}H_{32}O_{10}$ 492.521

Constit. of the fruit of *Juniperus communis*. $[\alpha]_D^{25}$ +34.4

3′,5-Di-Me ether

Lawsonicin

[85165-02-2]

$C_{20}H_{24}O_6$ 360.406

Constit. of the aerial parts of *Lawsonia alba*. Powder. Racemate. Struct. of Lawsonicin revised in 2010. λ_{max} 282 (log ε 2.87) (MeOH)

3′,5-Di-Me ether, 4′-O-β-D-xylopyranoside

Davidioside B

[1015476-48-8]

$C_{25}H_{32}O_{10}$ 492.521

Constit. of the bark of *Davidia involucrata*. Amorph. powder. Mp 85 - 86°. $[\alpha]_D^{21}$ -14 (c, 0.16 in MeOH). λ_{max} 229 (log ε 4.02); 282 (log ε 3.65) (MeOH)

3′,5-Di-Me ether, 4′,9-di-O-β-D-glucopyranoside

[1058130-35-0]

$C_{32}H_{44}O_{16}$ 684.688

Constit. of *Oplopanax elatus*. Powder. $[\alpha]_D$ -42.1 (c, 0.01 in MeOH)

3′,5-Di-Me ether, 4′-O-[β-D-glucopyranosyl-(1→2)-[β-D-glucopyranosyl-(1→6)]-β-D-glucopyranoside]

Lawsonicin 4′-O-neopolitanoside

$C_{38}H_{54}O_{21}$ 846.829

Isol. from the exocarp of *Castanea henryi*. Displays tyrosinase inhibitory activity. Powder. Mp 124 - 126°. $[\alpha]_D^{20}$ -23 (c, 0.001 in MeOH). λ_{max} 210 (log ε 4.01); 254 (log ε 3.56); 280 (log ε 4.08) (MeOH)

3′,5-Di-Me ether, 9-O-β-D-xylopyranoside

Mayselignoside

[1631180-14-7]

$C_{25}H_{32}O_{10}$ 492.521

Constit. of the leaves of *Maytenus senegalensis*. Amorph. light brown powder. $[\alpha]_D^{20}$ -32 (c, 0.02 in MeOH). λ_{max} 236; 282 (PDA)

3′,5-Di-Me ether, 9-O-(4-hydroxy-3-methoxy-E-cinnamoyl)

Boehmenan H

[344944-00-9]

$C_{30}H_{32}O_9$ 536.577

Constit. of *Hibiscus cannabinus*. Yellowish oil

4′-O-[2-Hydroxy-1-(hydroxymethyl)ethyl], 3′,5-di-Me ether

Demethoxysakuraresinol

[149415-63-4]

$C_{23}H_{30}O_8$ 434.485

Constit. of the bark of *Illicium difengpi* and *Illicium henryi*. Amorph. powder. $[\alpha]_D^{25}$ +3.7 (c, 1 in MeOH)

9-Carboxylic acid, 3′-Me ether, Me ester

[1800422-46-1]

$C_{20}H_{22}O_7$ 374.389

Constit. of the aerial parts of *Abies fabri*. Amorph. powder. $[\alpha]_D^{20}$ +1.2 (c, 0.5 in MeOH)

9-Carboxylic acid, 3′,5-di-Me ether

[545436-56-4]

$C_{20}H_{22}O_7$ 374.389

Isol. from non-centrifuged cane sugar (*Saccharum officinarum*). Oil

(7′RS,8′SR)-form

3′,5-Di-Me ether, 9′-Ac

9′-O-Acetyl-4,7′-epoxy-3′,5-dimethoxy-3,8′-lignan-4′,9-diol.

Crataegifin A

$C_{22}H_{26}O_7$ 402.443

Constit. of the fruit of *Crataegus pinnatifida*. Yellow oil. $[\alpha]_D^{20}$ -1.8 (c, 0.1 in MeOH). Partial racemate resolved by chiral hplc. λ_{max} 282 (log ε 3.21) (MeOH)

(7′ξ,8′ξ)-form

3′-Me ether, 4′-O-β-D-glucopyranoside

[85317-74-4]

$C_{25}H_{32}O_{11}$ 508.52

Constit. of *Cedrus deodara*, *Picea abies* and *Pinus* spp.

3′-Me ether, 9-O-α-L-rhamnopyranoside

[67029-29-2]

$C_{25}H_{32}O_{10}$ 492.521

Constit. of *Thuja plicata* and *Thujopsis dolabrata*

3′-Me ether, 9′-O-α-L-rhamnopyranoside

[71144-69-9]

$C_{25}H_{32}O_{10}$ 492.521

Constit. of *Larix leptolepis* and *Pinus massoniana*

3′,5-Di-Me ether, 4′-O-α-L-rhamnopyranoside

[97055-94-2]

$C_{26}H_{34}O_{10}$ 506.548

Constit. of *Chamaecyparis obtusa* and *Pinus massoniana*. Exhibits hypotensive activity. $[\alpha]_D^{22}$ -55 (c, 0.9 in MeOH)

3′,5-Di-Me ether, 9-O-α-L-rhamnopyranoside

[187342-00-3]

$C_{26}H_{34}O_{10}$ 506.548

Constit. of *Illicium dunnianum*, *Pinus massoniana*, *Trachelospermum asiaticum* and *Thujopsis dolabrata*. Amorph. solid. $[\alpha]_D^{25}$ -15.5 (c, 2 in MeOH)

3′,5-Di-Me ether, 9′-O-β-D-xylopyranoside

[149415-64-5]

$C_{25}H_{32}O_{10}$ 492.521

Constit. of *Gaultheria yunnanensis* and *Illicium difengpi*. Powder. $[\alpha]_D^{20}$ +13.4 (c, 0.69 in MeOH)

3′,5-Di-Me ether, 9′-O-α-L-rhamnopyranoside

Massonianoside E

[97055-93-1]

$C_{26}H_{34}O_{10}$ 506.548

Constit. of *Baseonema acuminatum* and *Pinus massoniana*. Amorph.. Mp 94 - 95°. $[\alpha]_D^{25}$ -10.3 (c, 3.1 in MeOH)

3′,5-Di-Me ether, 9′-O-[4-hydroxybenzoyl-(→2)-β-D-glucopyranoside]

$C_{33}H_{38}O_{13}$ 642.654

Constit. of the dried fruits of *Vitex rotundifolia*. Amorph. powder. $[\alpha]_D^{25}$ -30 (c, 0.1 in MeOH). λ_{max} 238 (log ε 4.26); 260 (log ε 4.23); 276 (sh) (EtOH).

Agrawal, P.K. *et al.*, *Phytochemistry*, 1980, **19**, 1260-1261 (*Cedrusin*).
Agarwal, P.K. *Phytochemistry*, 1982, **21**, 1459-1461 (*Cedrusin*).
Agrawal, P.K. *et al.*, *Org. Magn. Reson.*, 1983, **21**, 119-121 (*cmr*).
Lundgren, L.N. *et al.*, *Acta Chem. Scand., Ser. B*, 1985, **39**, 241-248 (*Pinus massoniana constits*).
Abe, F. *et al.*, *Chem. Pharm. Bull.*, 1986, **34**, 4340-4345 (*3′-Me ether glucosides*).
Da Silva, M.S. *et al.*, *Phytochemistry*, 1989, **28**, 3477-3482 (*3′,5-di-Me, 9′-glucoside*).
Miyase, T. *et al.*, *Phytochemistry*, 1989, **28**, 3483-3487 (*Icariside E_4*).
Antus, R.X. *et al.*, *Liebigs Ann. Chem.*, 1990, 495-497 (*Cedrusin, synth*).
Tan, R.X. *et al.*, *Planta Med.*, 1990, **56**, 475-477 (*Vladinol F*).
Pieters, L. *et al.*, *J. Nat. Prod.*, 1993, **56**, 899-906 (*3′,4′,5-tri-Me ether*).
Pieters, L. *et al.*, *Magn. Reson. Chem.*, 1993, **31**, 692-693 (*pmr*).
Kouno, I. *et al.*, *Phytochemistry*, 1993, **32**, 1573-1577 (*Illicium difengpi constits*).
Kizu, H. *et al.*, *Chem. Pharm. Bull.*, 1995, **43**, 2187-2194 (*Clemastanin A*).
Lemiere, G. *et al.*, *J. Chem. Soc., Perkin Trans. 1*, 1995, 1775-1779 (*3′,4′,5-tri-Me ether*).
Pan, H. *et al.*, *Phytochemistry*, 1995, **39**, 1423-1428 (*4′-methylrhamnoside*).
Su, W.-C. *et al.*, *Phytochemistry*, 1995, **40**, 563-566 (*Cryptomeria japonica constit*).
Paula, V.F. *et al.*, *Tetrahedron*, 1995, **51**, 12453-12462 (*Boehmenans*).
Matsuda, N. *et al.*, *Chem. Pharm. Bull.*, 1996, **44**, 1122-1123 (*3′,5-di-Me 4-glucosides*).
Fukuyama, Y. *et al.*, *Chem. Pharm. Bull.*, 1996, **44**, 1418-1420 (*Dihydrodehydrodiconiferyl alcohol*).
Otsuka, H. *et al.*, *Phytochemistry*, 1996, **42**, 1435-1438 (*Alangiplatanoside*).
Wang, C. *et al.*, *Phytochemistry*, 1997, **45**, 159-166 (*Pedicularis torta constit*).
Hanawa, F. *et al.*, *Phytochemistry*, 1997, **45**, 589-595 (*Dihydrodehydrodiconiferyl alcohol*).
Okuyama, E. *et al.*, *Chem. Pharm. Bull.*, 1998, **46**, 655-662 (*Vitex hydroxybenzoylglucoside*).
Shen, Y.C. *et al.*, *Phytochemistry*, 1998, **48**, 719-723 (*Urolignoside*).
Takeda, Y. *et al.*, *Phytochemistry*, 1998, **49**, 2137-2139 (*Glochidioboside*).
Yuen, M.S.M. *et al.*, *Tetrahedron*, 1998, **54**, 12429-12444 (*3′,4-Dimethylcedrusin, synth, abs config*).
Shen, Y. *et al.*, *J. Wood Sci.*, 1999, **45**, 326-331 (*3′-Me 4′-rhamnoside*).
Lin, W.-H. *et al.*, *Phytochemistry*, 1999, **50**, 653-658 (*3′,4′-methylene-5-Me ether*).
Otsuka, H. *et al.*, *Chem. Pharm. Bull.*, 2000, **48**, 1084-1086 (*3′,5-di-Me 9-sulfate*).
Van Dyck, S.M.O. *et al.*, *Molecules*, 2000, **5**, 153-161 (*4-O-Methylcedrusin, synth*).
Gopalaiah, K. *et al.*, *Indian J. Chem., Sect. B*, 2001, **40**, 596-600 (*Tiruneesin*).
Baderschneider, B. *et al.*, *J. Agric. Food Chem.*, 2001, **49**, 2788-2798 (*3′-Me 9′-glucoside, 3′,5-di-Me 9′-glucoside*).
Seca, A.M.L. *et al.*, *Phytochemistry*, 2001, **56**, 759-767 (*Boehmenans*).
Van Dyck, S.M.O. *et al.*, *Tetrahedron: Asymmetry*, 2001, **12**, 785-789 (*Dihydrodehydrodiconiferyl alcohol, resoln*).
Bi, Y.-F. *et al.*, *Yaoxue Xuebao*, 2001, **36**, 832-835 (*Massonianoside B*).
Bi, Y.-F. *et al.*, *Yaoxue Xuebao*, 2002, **37**, 626-629 (*Massonianosides A,C*).
Ersöz, T. *et al.*, *Z. Naturforsch., C*, 2002, **57**, 221-225 (*3′,4′,5-tri-Me 9-glucoside*).
Takara, K. *et al.*, *Biosci., Biotechnol., Biochem.*, 2003, **67**, 376-379 (*cane sugar neolignans*).
Bi, Y.-F. *et al.*, *Yaoxue Xuebao*, 2003, **38**, 199-202 (*Massonianoside D*).
Feng, W.-S. *et al.*, *Yaoxue Xuebao*, 2003, **38**, 927-930 (*Massonianoid A*).
Wu, P.L. *et al.*, *Bioorg. Med. Chem.*, 2004, **12**, 2193-2197 (*Boehmenan, activity*).
Kikuzaki, H. *et al.*, *J. Agric. Food Chem.*, 2004, **52**, 344-349 (*Prunus domestica constit, activity*).
Nakanishi, T. *et al.*, *Phytochemistry*, 2004, **65**, 207-213 (*Junipercomnosides, Icariside E_4, Massonianoside A*).
De Leo, M. *et al.*, *Planta Med.*, 2004, **70**, 841-846 (*Basionema acuminatum constit*).
Shin, J.S. *et al.*, *Arch. Pharmacal Res.*, 2005, **28**, 1337-1340 (*Dihydrodehydrodiconiferyl alcohol, activity*).
Jiang, J.S. *et al.*, *Chem. Pharm. Bull.*, 2005, **53**, 110-113 (*Symplolignanoside A*).
Yang, B. *et al.*, *J. Nat. Prod.*, 2005, **68**, 1175-1179 (*Larix olgensis coumarate*).
Calis, I. *et al.*, *Turk. J. Chem.*, 2005, **29**, 71-81 (*3′,4′,5-tri-Me ether 9′-glucoside*).
Dong, L.-P. *et al.*, *Molecules*, 2006, **11**, 1009-1014 (*7′-Epiglochidioboside*).
Chin, Y.-W. *et al.*, *Phytother. Res.*, 2006, **20**, 62-65 (*(-)-Boehmenan*).
Garcia-Muñoz, S. *et al.*, *Tetrahedron*, 2006, **62**, 12182-12190 (*Dihydrodehydrodiconiferyl alcohol, synth*).
Subbaraju, G.V. *et al.*, *Indian J. Chem., Sect. B*, 2007, **46**, 357-359 (*Tiruneesin, synth*).
Ouyang, M.-A. *et al.*, *J. Asian Nat. Prod. Res.*, 2007, **9**, 487-492 (*Davidiosides A,B*).
Tang, W.-Z. *et al.*, *Planta Med.*, 2007, **73**, 484-490 (*Illicium oligandrum glycosides*).
Lee, D.-Y. *et al.*, *Arch. Pharmacal Res.*, 2008, **31**, 830-833 (*Cornuskoside A*).
Chang, M.-J. *et al.*, *Biosci., Biotechnol., Biochem.*, 2008, **72**, 2750-2755 (*Forsythia suspensum constit, activity*).
Chin, Y.-W. *et al.*, *J. Agric. Food Chem.*, 2008, **56**, 7759-7764 (*Euterpe oleracea constit, activity*).
Su, D. *et al.*, *J. Nat. Prod.*, 2008, **71**, 784-788 (*3′,5-di-Me 9′-apiosylglucoside*).
Kang, Q.-J. *et al.*, *Planta Med.*, 2008, **74**, 445-448 (*Trewia nudiflora 6-galloylglucoside*).
Tung, N.H. *et al.*, *Chem. Pharm. Bull.*, 2009, **57**, 65-68 (*Camellia amplexicaulis constits, activity*).
Dou, D.-Q. *et al.*, *Nat. Prod. Res.*, 2009, **23**, 334-342 (*Oplopanax elatus 4′,9-diglucoside*).
Iida, N. *et al.*, *Chem. Pharm. Bull.*, 2010, **58**, 742-746 (*Juniperus communis glycosides*).
Xiang, W.-J. *et al.*, *Fitoterapia*, 2010, **81**, 1228-1231 (*4′-glyceryl 3′-Me ether*).
Wang, Y.-H. *et al.*, *Helv. Chim. Acta*, 2010, **93**, 2467-2477 (*Selaginella moellendorffii constits*).
Feng, W.-S. *et al.*, *J. Asian Nat. Prod. Res.*, 2010, **12**, 557-561 (*Lophanthosides B,C*).
Rudiyansyah, *et al.*, *J. Nat. Prod.*, 2010, **73**, 1649-1654 (*Boehmenan X*).
Chen, J.-X. *et al.*, *Nat. Prod. Commun.*, 2010, **5**, 1627-1630 (*Lobelia chinensis constits*).
Meng, J. *et al.*, *Org. Biomol. Chem.*, 2010, 107-113 (*Dihydrodehydrodiconiferyl alcohol, Lawsonicin, synth*).
Woo, K.W. *et al.*, *Arch. Pharmacal Res.*, 2011, **34**, 2043-2049 (*Juniperigiside*).
Kim, K.H. *et al.*, *Bioorg. Med. Chem. Lett.*, 2011, **21**, 2270-2273 (*Berberis koreana constit, activity*).
Jeong, E.J. *et al.*, *Bioorg. Med. Chem. Lett.*, 2011, **21**, 2283-2286 (*Euonymus alatus constits, activity*).
Liu, J.-F. *et al.*, *Chem. Biodiversity*, 2011, **8**, 692-698 (*Illicium henryi constits, activity*).
He, Y.-Q. *et al.*, *Chem. Nat. Compd. (Engl. Transl.)*, 2011, **47**, 472-473 (*Quercus macrocarpa constits*).
Hu, Q.-F. *et al.*, *Chem. Pharm. Bull.*, 2011, **59**, 1421-1424 (*Vladinol F, activity*).
Yuan, T. *et al.*, *J. Nat. Prod.*, 2011, **74**, 2472-2476 (*Saccharumoside A*).
De Marino, S. *et al.*, *Planta Med.*, 2011, **77**, 1822-1828 (*Juniperoside A*).
He, W.J. *et al.*, *Z. Naturforsch., B*, 2011, **66**, 733-739 (*Isomassonianoside B*).
Wu, B. *et al.*, *Carbohydr. Res.*, 2012, **355**, 45-49 (*Lawsonicin 4′-neopolitanoside*).
Shen, C.-C. *et al.*, *Chem. Pharm. Bull.*, 2012, **60**, 924-929 (*Quercus glauca constit*).
Yang, Y.-N. *et al.*, *J. Agric. Food Chem.*, 2012, **60**, 964-972 (*Lophanthoside C, Rhodiola crenulata glycosides*).
Bai, J. *et al.*, *J. Asian Nat. Prod. Res.*, 2012, **14**, 940-949 (*Dunnianeolignan B*).
Wu, M. *et al.*, *Molecules*, 2012, **17**, 5212-5218 (*Sarcandra glabra constits*).
Li, Y.-L. *et al.*, *Planta Med.*, 2012, **78**, 1574-1578 (*Cedrusin 6-benzoylglucoside*).
D'Abrosca, B. *et al.*, *Bioorg. Med. Chem.*, 2013, **21**, 7038-7046 (*Helichrysum italicum constits, activity*).
Liang, C.-Q. *et al.*, *Fitoterapia*, 2013, **86**, 171-177 (*Schisphenlignan K*).
Wang, P.-C. *et al.*, *J. Braz. Chem. Soc.*, 2013, **24**, 1544-1548 (*3′,5-di-Me ether 9-isovalerate*).

Kim, C.S. *et al., J. Nat. Prod.*, 2013, **76**, 2131-2135 (*Abies 3-O-methylrhamnoside*).
Yoshikawa, K. *et al., Molecules*, 2013, **18**, 9641-9649 (*Saposide B*).
Luo, Q. *et al., Molecules*, 2013, **18**, 10930-10943 (*Cinnacassoside B*).
Zhou, G.-S. *et al., Planta Med.*, 2013, **79**, 687-692 (*Arenaria serpyllifolia constit, activity*).
Baek, S. *et al., Beilstein J. Org. Chem.*, 2014, **10**, 2955-2962 (*Trogopterin A, activity*).
Lee, S. *et al., Bioorg. Med. Chem. Lett.*, 2014, **24**, 44-48 (*Pulsatilla koreana constits*).
Cho, N. *et al., Bioorg. Med. Chem. Lett.*, 2014, **24**, 5675-5678 (*NO prodn inhibitors*).
Xiao, H.-H. *et al., Fitoterapia*, 2014, **94**, 29-35 (*Samwiside*).
Xia, Y. *et al., J. Chem. Sci. (Berlin, Ger.)*, 2014, **126**, 813-820 (*Boehmenan, synth*).
Li, C. *et al., Phytochemistry*, 2014, **104**, 105-113 (*Forsythialansides C,D, activity*).
Rudiyansyah, L.K. *et al., Rec. Nat. Prod.*, 2014, **8**, 195-198 (*(+)-Boehmenan, abs config*).
Ferreira-Júnior, J.C. *et al., Arch. Pharmacal Res.*, 2015, **38**, 950-956 (*Icariside E_4, activity*).
Shono, T. *et al., Bioorg. Med. Chem. Lett.*, 2015, **25**, 2735-2738 (*Boehmenan, activity*).
Hung, T.M. *et al., Chem. -Biol. Interact.*, 2015, **240**, 74-82 (*Lycocernuaside A*).
Hemelaere, R. *et al., Eur. J. Org. Chem.*, 2015, 2470-2481 (*Dihydrodehydrodiconiferyl alcohol, synth*).
Okoye, F.B.C *et al., Nat. Prod. Res.*, 2015, **29**, 109-115 (*Mayselignoside*).
Li, Y.-L. *et al., Phytochemistry*, 2015, **117**, 135-143 (*Abies fabri constit*).
Rayanil, K.-O. *et al., Nat. Prod. Res.*, 2016, **30**, 2675-2681 (*Mitredrusin*).
Ye, F.-M. *et al., Phytochem. Lett.*, 2016, **17**, 263-269 (*Illiciumlignans C,D,F, activity*).
Jiang, Y.-P. *et al., Yaoxue Xuebao*, 2016, **51**, 616-625 (*Codonopsis pilosula diglucoside*).
Jiang, H. *et al., Fitoterapia*, 2017, **117**, 11-15 (*Xanthiumnolic E, activity*).
Wang, L.-Y. *et al., J. Nat. Prod.*, 2017, **80**, 1808-1818 (*Litsea cubeba constit, activity*).
Matsumoto, T. *et al., Phytochemistry*, 2017, **137**, 101-108 (*Isodonosides III-VI, activity*).
Lee, I.-S. *et al., Planta Med.*, 2017, **83**, 318-325 (*Litsea japonica constit*).
Zhao, C. *et al., J. Agric. Food Chem.*, 2018, **66**, 10421-10430 (*Viburfordosides E,F, activity*).
Wang, Q.-H. *et al., Chem. Nat. Compd. (Engl. Transl.)*, 2019, **55**, 1007-1009 (*Artemordolignan glycoside A*).
Ji, J. *et al., Fitoterapia*, 2019, **137**, 104271:1-7 (*Securidaca inappendiculata constit, cd, activity*).
Guo, R. *et al., Fitoterapia*, 2019, **137**, 104287:1-8 (*Crataegifins A, cd*).
Ji, J. *et al., Phytochem. Lett.*, 2019, **31**, 58-62 (*Securidaca inappendiculata constit, activity*).
Chang, C.-C. *et al., Fitoterapia*, 2020, **144**, 104455:1-7 (*Massonianoside D, cd*)

4,7′-Epoxy-3,8′-lignan-3′,4′,5′,9,9′-pentol E114

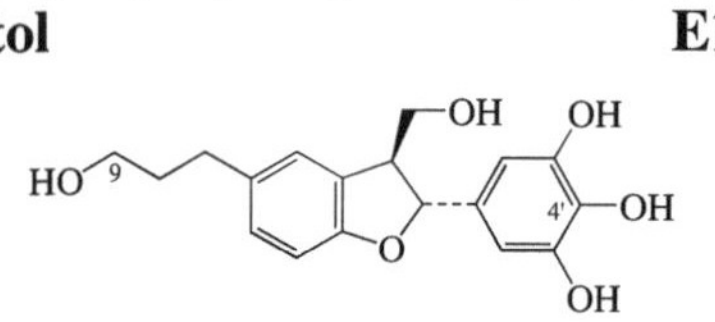

4,7′-Epoxy-3′,4′,5′,9,9′-pentahydroxy-3,8′-neolignan

$C_{18}H_{20}O_6$ 332.352

3′,5′-Di-Me ether

4,7′-Epoxy-3′,5′-dimethoxy-3,8′-lignan-4′,9,9′-triol. 4,7′-Epoxy-4′,9,9′-trihydroxy-3′,5′-dimethoxy-3,8′-lignan. ***Hyaloserin***

[845729-25-1]

$C_{20}H_{24}O_6$ 360.406

Constit. of *Benincasa hispida* and *Hyaloseris andrade-limae*. Gum.

De Trimarco, J.T. *et al., Biochem. Syst. Ecol.*, 2004, **32**, 1063-1067 (*Hyaloserin*)

4,7′-Epoxy-3,8′-lignan-3′,5,5′,9,9′-pentol E115

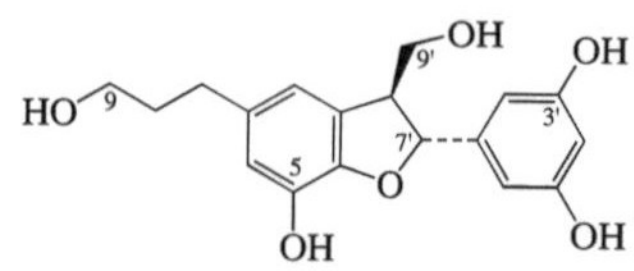

4,7′-Epoxy-3′,5,5′,9,9′-pentahydroxy-3,8′-neolignan

$C_{18}H_{20}O_6$ 332.352

Structures in this series may need revision. Phenyl 3,5-dioxygenated substn. pattern in doubt as a result of work done by Yang *et al* in 2015

(7′R,8′S)-form

3′,5-Di-Me ether

4,7′-Epoxy-5,5′-dimethoxy-3,8′-lignan-3′,9,9′-triol. 4,7′-Epoxy-3′,9,9′-trihydroxy-5,5′-dimethoxy-3,8′-lignan

[1426955-32-9]

$C_{20}H_{24}O_6$ 360.406

Constit. of the aerial parts of *Leontopodium leontopodiodes*. NO prodn. inhibitor. Amorph. yellow solid. $[\alpha]_D^{25}$ +6.8 (c, 0.13 in MeOH). λ_{max} 212 (log ε 4); 229 (log ε 3.73); 282 (log ε 3.35) (MeOH)

3′,5-Di-Me ether, 9,9′-di-Ac

Rebaneolignan A

[2101546-33-0]

$C_{24}H_{28}O_8$ 444.480

Constit. of *Mangifera reba*. α-Glucosidase inhibitor. Amorph. yellowish solid.

Li, X. *et al., Fitoterapia*, 2012, **83**, 883-887 (*Leontopodium ether, activity*).
Yang, Y.-N. *et al., J. Nat. Prod.*, 2015, **78**, 705-711 (*pmr, struct*).
Duong, T.T.T. *et al., Tetrahedron Lett.*, 2017, **58**, 2280-2283 (*Rebaneolignan A, activity*)

7,7′-Epoxy-8,8′-lignan-2,3′,4,4′,5-pentol E116

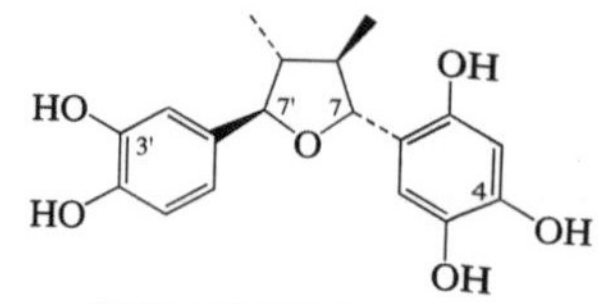

(7R*,7′R*,8R*,8′R*)-form

2,3′,4,4′,5-Pentahydroxy-7,7′-epoxylignan

$C_{18}H_{20}O_6$ 332.352

(7R*,7′R*,8R*,8′R*)-form

3′,4′:4,5-Bis(methylene) ether

7,7′-Epoxy-3′,4′:4,5-bis(methylenedioxy)-8,8′-lignan-2-ol. 2-Hydroxy-3′,4′:4,5-bis(methylenedioxy)-7,7′-epoxylignan. ***Saucerneol K***

[1198210-46-6]

$C_{20}H_{20}O_6$ 356.374

Constit. of the roots of *Saururus chinensis*. Oil. $[\alpha]_D^{22}$ +21.9 (c, 0.08 in $CHCl_3$). λ_{max} 244 (log ε 3.24); 291 (log ε 3.01) ($CHCl_3$)

(7R*,7′S*,8R*,8′R*)-form

3′,4,4′,5-Tetra-Me ether

7,7′-Epoxy-3′,4,4′,5-tetramethoxy-8,8′-lignan-2-ol. 2-Hydroxy-3′,4,4′,5-tetramethoxy-7,7′-epoxylignan. ***Nectandrin C***

[656253-34-8]

$C_{22}H_{28}O_6$ 388.46

Constit. of the leaves of *Nectandra megapotamica*. Exhibits moderate trypanocidal activity. Oil. $[\alpha]_D^{26}$ +1.3 (c, 0.01 in $CHCl_3$)

(7S*,7′S*,8S*,8′S*)-form

3′,4′:4,5-Bis(methylene) ether

Saurufurin D

[1532522-58-9]

$C_{20}H_{20}O_6$ 356.374

Constit. of *Saururus chinensis*. Yellow oil. $[\alpha]_D^{26}$ -29 (c, 0.2 in MeOH). Presumably enantiomeric with Saucerneol K, although not certian as opt. rotn. solvents differ. λ_{max} 237 (log ε 3.32); 294 (log ε 3.25) (MeOH).

Filho, A.A. da S. *et al., J. Nat. Prod.*, 2004, **67**, 42-45 (*Nectandrin C*).
Seo, C.S. *et al., Bull. Korean Chem. Soc.*, 2009, **30**, 716-718 (*Saucerneol K*).
Tsai, W.-J. *et al., J. Nat. Prod.*, 2014, **77**, 125-131 (*Saurufurin D*)

7,7′-Epoxy-8,8′-lignan-2,3′,4′,5,5′-pentol E117

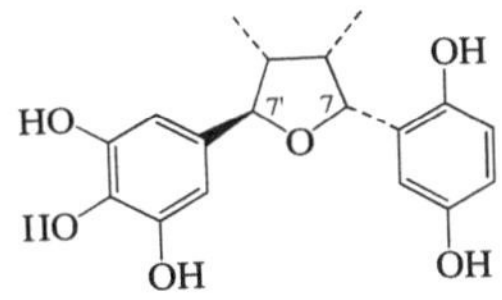

2,3′,4′,5,5′-Pentahydroxy-7,7′-epoxylignan
$C_{18}H_{20}O_6$ 332.352
(7R,7′R,8S,8′R)-form
Penta-Me ether
7,7′-Epoxy-2,3′,4′,5,5′-pentamethoxylignan. 2,3′,4′,5,5′-Pentamethoxy-7,7′-epoxylignan
[165054-90-0]
$C_{23}H_{30}O_6$ 402.487
Constit. of the leaves of *Ocotea foetens*. $[\alpha]^{20}_{D}$-19 (c, 1.6 in $CHCl_3$).

Lopez, H. *et al.*, *J. Nat. Prod.*, 1995, **58**, 782-785 (*Ocotea foetens constit*)

7,7′-Epoxy-8,8′-lignan-3,3′,4,4′,5-pentol E118

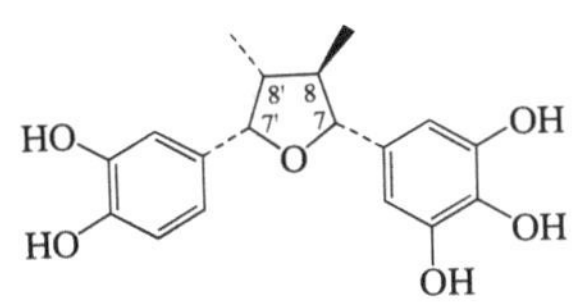

(7R,7′S,8R,8′R)-form

3,3′,4,4′,5-Pentahydroxy-7,7′-epoxylignan
$C_{18}H_{20}O_6$ 332.352
(7R,7′S,8R,8′R)-form
Penta-Me ether
7,7′-Epoxy-3,3′,4,4′,5-pentamethoxylignan. 3,3′,4,4′,5-Pentamethoxy-7,7′-epoxylignan. ***(-)-Acortatarinowin I***
$C_{23}H_{30}O_6$ 402.487
Constit. of the rhizomes of *Acorus tatarinowii*. Light yellow gum. $[\alpha]^{20}_{D}$-32 (c, 0.1 in MeOH). Resolved by chiral hplc
(7R,7′S,8R,8′S)-form
3,3′,5-Tri-Me ether
7,7′-Epoxy-3,3′,5-trimethoxy-8,8′-lignan-4,4′-diol. 4,4′-Dihydroxy-3,3′,5-trimethoxy-7,7′-epoxylignan
[1310453-68-9]
$C_{21}H_{26}O_6$ 374.433
Isol. from the bark of *Machilus robusta*. Amorph. powder. $[\alpha]^{20}_{D}$ -2.2 (c, 0.31 in MeOH). λ_{max} 232 (log ε 3.83); 280 (log ε 3.3) (MeOH)
3,3′,4′,5-Tetra-Me ether
7,7′-Epoxy-3,3′,4′,5-tetramethoxy-8,8′-lignan-4-ol. 4-Hydroxy-3,3′,4′,5-tetramethoxy-7,7′-epoxylignan. ***Saurucinol H***
[1521111-73-8]
$C_{22}H_{28}O_6$ 388.46
Constit. of *Saururus chinensis*. Oil. $[\alpha]^{20}_{D}$-13 (c, 0.02 in $CHCl_3$). λ_{max} 237 (log ε 4.05); 277 (log ε 3.68) (MeOH)
Penta-Me ether
4-O-Methylsaurucinol H
[1512869-86-1]
$C_{23}H_{30}O_6$ 402.487
Constit. of *Saururus chinensis*. Amorph. solid. $[\alpha]^{20}_{D}$-8.9 (c, 0.05 in $CHCl_3$). λ_{max} 215 (log ε 4.32); 227 (log ε 4.27); 277 (log ε 3.68) (MeOH)
(7S,7′R,8S,8′S)-form
3,3′,4,4′-Tetra-Me ether
7,7′-Epoxy-3′,4,4′,5-tetramethoxy-8,8′-lignan-3-ol. 3-Hydroxy-3′,4,4′,5-tetramethoxy-7,7′-epoxylignan. ***3-Hydroxyveraguensin***
[1446341-01-0]
$C_{22}H_{28}O_6$ 388.46
Constit. of *Miliusa fragrans*. Viscous liq.. $[\alpha]^{20}_{D}$+37.7 (c, 0.17 in $CHCl_3$). λ_{max} 216 (log ε 4.30); 233 (log ε 4.25); 279 (log ε 3.67) (MeOH)
Penta-Me ether
(+)-Acortatarinowin I
[184486-88-2]
$C_{23}H_{30}O_6$ 402.487
Constit. of *Virola surinamensis* and the rhizomes of *Acorus tatarinowii*. Yellow oil or light yellow gum. $[\alpha]_D$ +16 (c, 0.1 in $CHCl_3$). $[\alpha]^{20}_{D}$ +32 (c, 0.3 in MeOH). Resolved by chiral hplc
(7S,7′S,8S,8′R)-form
Penta-Me ether
Ligraminol A. *5′-Methoxyganschisandrin*
[1335126-23-2]
$C_{23}H_{30}O_6$ 402.487
Isol. from the rhizomes of *Acorus gramineus*. Shows selective cytotoxicity against tumour and normal cell lines and antineuroinflammatory activity. Oil. $[\alpha]^{25}_{D}$ +17.3 (c, 0.15 in MeOH). λ_{max} 217 (log ε 4.5); 238 (log ε 2.5); 277 (log ε 1.7) (MeOH)
3,4:3′,4′-Bis(methylene), 5-Me ether
7,7′-Epoxy-3-methoxy-3′,4′:4,5-bis(methylenedioxy) lignan. 3-Methoxy-3′,4′:4,5-bis(methylenedioxy)-7,7′-epoxylignan. ***5-Methoxyzuonin A***
$C_{21}H_{22}O_6$ 370.401
Constit. of the stems of *Schisandra nigra*. Oil. $[\alpha]^{26}_{D}$+99 (c, 0.1 in $CHCl_3$). λ_{max} 205 (log ε 4.38); 285 (log ε 3.34) (MeOH)
(7S,7′S,8S,8′S)-form
3′,4′-Methylene, 3,4,5-tri-Me ether
7,7′-Epoxy-3,4,5-trimethoxy-3′,4′-methylenedioxylignan. 3,4,5-Trimethoxy-3′,4′-methylenedioxy-7,7′-epoxylignan
[184486-89-3]
$C_{22}H_{26}O_6$ 386.444
Constit. of *Virola surinamensis*. Yellow oil. $[\alpha]_D$ -10 (c, 0.02 in $CHCl_3$)
Penta-Me ether
Constit. of *Virola surinamensis*. Oil. $[\alpha]_D$ -12.6 (c, 0.1 in $CHCl_3$)
(7R*,7′R*,8R*,8′R*)-form
3,3′,5-Tri-Me ether
Fragransin C_2
[112572-54-0]
$C_{21}H_{26}O_6$ 374.433
Isol. from *Myristica fragrans*. Oil. $[\alpha]_D$ +20.2 (c, 0.11 in $CHCl_3$)
3,3′,4-Tri-Me ether
[2089242-49-7]
$C_{21}H_{26}O_6$ 374.433
Constit. of the fruit of *Schisandra chinensis*. Potent inhibitor of PCSK9 mRNA expression. Yellow oil. $[\alpha]^{20}_{D}$-7.4 (c, 0.61 in $CHCl_3$). λ_{max} 279 (log ε 3.56) (MeOH)
(7R*,7′R*,8R*,8′S*)-form
3′,4′-Methylene, 3,4,5-tri-Me ether
Henricine
[107783-46-0]
$C_{22}H_{26}O_6$ 386.444
Constit. of *Schisandra henryi*. Cryst.. Mp 73 - 75°. $[\alpha]^{21}_{D}$+102 ($CHCl_3$)
(7R*,7′S*,8R*,8′R*)-form
3,3′,4-Tri-Me ether
7,7′-Epoxy-3′,4,5-trimethoxy-8,8′-lignan-3,4′-diol. 3,4′-Dihydroxy-3′,4,5-trimethoxy-7,7′-epoxylignan.
Manglisin E
[1621958-65-3]
$C_{21}H_{26}O_6$ 374.433
Constit. of carpels of *Manglietiastrum sinicum*. Exhibits moderate

antimicrobial activity against *Staphylococcus aureus* and MRSA types 82, 92 and 98. Yellow oil. $[\alpha]_D^{23}$ -6 (c, 0.1 in MeOH). λ_{max} 204 (log ε 3.91); 279 (log ε 2.63) (MeOH)

3,3′,5-Tri-Me ether

Fragransin C_{3a}

[112572-53-9]

$C_{21}H_{26}O_6$ 374.433

Constit. of *Myristica fragrans*. Oil. $[\alpha]_D$ +19.6 (c, 0.29 in $CHCl_3$)

3,4-Methylene, 3′,4′,5-tri-Me ether

7,7′-Epoxy-3,3′,4′-trimethoxy-4,5-methylenedioxylignan. 3,3′,4′-Trimethoxy-4,5-methylenedioxy-7,7′-epoxylignan.

Saurufurin A

[1532522-55-6]

$C_{22}H_{26}O_6$ 386.444

Constit. of *Saururus chinensis*. Amorph. powder. $[\alpha]_D^{26}$ +18 (c, 0.2 in MeOH). Stereochem. not certain. (7*S*,7′*R*,8*S*,8′*S*)-form depicted in the ref. but descr. as (7*S*,7′*S*,8*S*,8′*R*) in the text. Assignment based on comparison with Fragransin B_3 which has same stereochem. as that depicted in struct.. λ_{max} 234 (sh); 279 (log ε 3.68) (MeOH)

(7R*,7′S*,8R*,8′S*)-form

3,3′,4-Tri-Me ether

Schinlignin B

[1352185-22-8]

$C_{21}H_{26}O_6$ 374.433

Constit. of the fruit of *Schisandra chinensis* and mature carpels of *Manglietiastrum sinicum*. Yellow oil. $[\alpha]_D^{27}$ -4.2 (c, 0.24 in MeOH). λ_{max} 205 (log ε 4.56); 280 (log ε 3.52); 375 (log ε 3.14) (MeOH)

3,3′,5-Tri-Me ether

Fragransin C_1*. Machilin H*

[112572-57-3]

$C_{21}H_{26}O_6$ 374.433

Isol. from *Myristica fragrans* and *Machilus thunbergii*. AMPK activator showing potential for treatment of obesity and type 2 diabetes. Oil. $[\alpha]_D$ +3.8 (c, 0.60 in $CHCl_3$). $[\alpha]_D$ +8.8 (c, 0.37 in $CHCl_3$)

3,3′,4,5-Tetra-Me ether

7,7′-Epoxy-3,3′,4′,5′-tetramethoxy-8,8′-lignan-4-ol. 4-Hydroxy-3,3′,4′,5′-tetramethoxy-7,7′-epoxylignan. ***Fragransin D_1***

[114394-21-7]

$C_{22}H_{28}O_6$ 388.46

Constit. of *Myristica fragrans*. Oil. $[\alpha]_D$ +18.38 (c, 0.136 in $CHCl_3$)

3,4-Methylene, 3′,4′,5-tri-Me ether

Saurufurin B

[1532522-56-7]

$C_{22}H_{26}O_6$ 386.444

Constit. of *Saururus chinensis*. Oil. $[\alpha]_D^{26}$ -2 (c, 0.2 in MeOH). λ_{max} 233 (sh); 278 (log ε 3.73) (MeOH)

3′,4′-Methylene, 3,4-di-Me ether

7,7′-Epoxy-4,5-dimethoxy-3′,4′-methylenedioxy-8,8′-lignan-3-ol. 3-Hydroxy-4,5-dimethoxy-3′,4′-methylenedioxy-7,7′-epoxylignan.

Manglisin C

[1621958-63-1]

$C_{21}H_{24}O_6$ 372.417

Constit. of carpels of *Manglietiastrum sinicum*. Exhibits inhibitory activity against *Staphylococcus aureus* and MRSA types 82, 92, 98 and 331. Oil. $[\alpha]_D^{16}$ +5 (c, 0.1 in MeOH). λ_{max} 203 (log ε 4.84); 283 (log ε 3.64) (MeOH)

(7R*,7′S*,8S*,8′R*)-form

3,3′,4,5-Tetra-Me ether

Fragransin D_2

[114422-23-0]

$C_{22}H_{28}O_6$ 388.46

Constit. of *Myristica fragrans*. Oil. $[\alpha]_D$ +30.49 (c, 0.106 in $CHCl_3$)

(7R*,7′S*,8S*,8′S*)-form

3,3′,5-Tri-Me ether

Fragransin C_{3b}

[112503-91-0]

$C_{21}H_{26}O_6$ 374.433

Constit. of *Myristica fragrans*. Oil. $[\alpha]_D$ +7.2 (c, 0.24 in $CHCl_3$)

3,3′,4′,5-Tetra-Me ether

Fragransin D_3

[114422-24-1]

$C_{22}H_{28}O_6$ 388.46

Constit. of *Myristica fragrans*. Oil. $[\alpha]_D$ +11.45 (c, 0.262 in $CHCl_3$)

3′,4′-Methylene, 3,4-di-Me ether

Manglisin D

[1621958-64-2]

$C_{21}H_{24}O_6$ 372.417

Constit. of carpels of *Manglietiastrum sinicum*. Exhibits inhibitory activity against *Staphylococcus aureus* and MRSA types 82, 92, 98 and 331. Oil. $[\alpha]_D^{16}$ +17 (c, 0.1 in MeOH). λ_{max} 203 (log ε 4.64); 285 (log ε 3.49) (MeOH)

(7RS,7′SR,8RS,8′RS)-form

Penta-Me ether

(±)-Acortatarinowin I

$C_{23}H_{30}O_6$ 402.487

Constit. of the rhizomes of *Acorus tatarinowii*. Light yellow gum. $[\alpha]_D^{20}$ +13 (c, 0.4 in MeOH). Racemate. Resolved by chiral hplc. λ_{max} 207 (log ε 4.72); 232 (log ε 4.51); 278 (log ε 4.04) (MeOH).

Lian-niang, L. *et al.*, *Planta Med.*, 1986, 493-494 (*Henricine*).

Hattori, M. *et al.*, *Chem. Pharm. Bull.*, 1987, **35**, 3315-3322 (*Fragransins*).

Hada, S. *et al.*, *Phytochemistry*, 1988, **27**, 563-568 (*Myristica fragrans constits*).

Shimomura, H. *et al.*, *Phytochemistry*, 1988, **27**, 634-636 (*Machilin H*).

Lopes, N.P. *et al.*, *Phytochemistry*, 1996, **43**, 1089-1092 (*Virola surinamensis constits*).

Xue, Y.-B. *et al.*, *Chem. Pharm. Bull.*, 2010, **58**, 1606-1612 (*Schinlignin B*).

Li, Y. *et al.*, *J. Nat. Prod.*, 2011, **74**, 1444-1452 (*Machilus robusta tri-Me ether*).

Kim, K.H. *et al.*, *J. Nat. Prod.*, 2011, **74**, 2187-2192 (*Ligraminol A*).

Sawasdee, K. *et al.*, *Tetrahedron Lett.*, 2013, **54**, 4259-4263 (*3-Hydroxyveraguensin*).

Cui, H. *et al.*, *J. Nat. Prod.*, 2014, **77**, 100-110 (*Saurucinol H, 4-O-Methylsaurucinol H*).

Tsai, W.-J. *et al.*, *J. Nat. Prod.*, 2014, **77**, 125-131 (*Saurufurins A,B*).

Ding, J.-Y. *et al.*, *J. Nat. Prod.*, 2014, **77**, 1800-1805 (*Manglisins C,D,E*).

Narukawa, Y. *et al.*, *J. Nat. Med.*, 2016, **70**, 460-466 (*5-Methoxyzuonin A, activity*).

Lu, Y. *et al.*, *Sci. Rep.*, 2016, **6**, 22909 (*Acortatarinowin I*).

Chaimanee, S. *et al.*, *Org. Biomol. Chem.*, 2017, **15**, 3985-2994 (*Fragransin C_1, synth*).

Pel, P. *et al.*, *Phytochemistry*, 2017, **136**, 119-124 (*Schisandra chinensis constit, activity*)

7,7′-Epoxy-8,8′-lignan-3,3′,4,4′,7-pentol E119

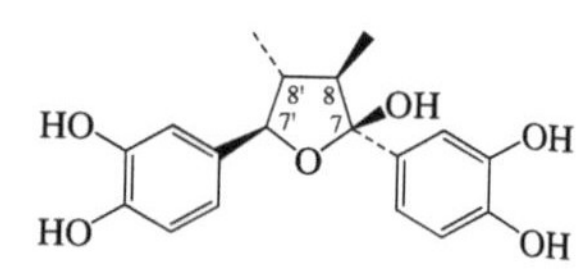

3,3′,4,4′,7-Pentahydroxy-7,7′-epoxylignan

$C_{18}H_{20}O_6$ 332.352

(7R,7′R,8R,8′R)-form

3,4-Methylene, tri-Me ether

7,7′-Epoxy-3′,4′,7-trimethoxy-3,4-methylenedioxylignan. 3′,4′,7-Trimethoxy-3,4-methylenedioxy-7,7′-epoxylignan

[1886973-79-0]

$C_{22}H_{26}O_6$ 386.444

Constit. of the roots of *Schisandra sphenanthera*. Yellow oil. $[\alpha]_D^{25}$ +20 (c, 0.1 in MeOH). λ_{max} 232 (log ε 0.25) (EtOH).

Jiang, K. *et al.*, *Fitoterapia*, 2015, **103**, 63-70 (*Schisandra sphenanthera constit*)

7,7′-Epoxy-8,8′-lignan-3,3′,4,4′,9-pentol E120

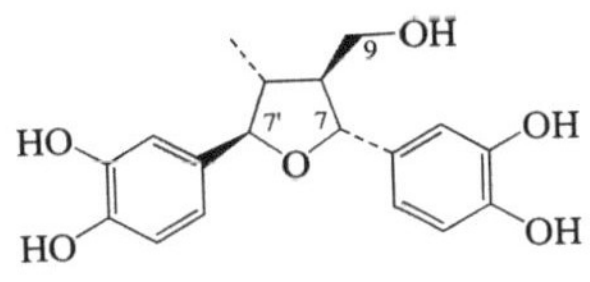

(7*R*,7′*R*,8*S*,8′*R*)-*form*

3,3′,4,4′,9-Pentahydroxy-7,7′-epoxylignan

$C_{18}H_{20}O_6$ 332.352

(7R,7′R,8S,8′R)-form

3,4:3′,4′-Bis(methylene) ether

7,7′-Epoxy-3,4:3′,4′-bis(methylenedioxy)-8,8′-lignan-9-ol. 9-Hydroxy-3,4:3′,4′-bis(methylenedioxy)-7,7′-epoxylignan. ***9-Hydroxygalbacin***

[1194586-66-7]

$C_{20}H_{20}O_6$ 356.374

Constit. of the bark of *Machilus thunbergii*. Amorph. powder. $[\alpha]^{25}_{D}$ +19 (c, 2.4 in $CHCl_3$). Incorrect synonym assigned in ref.

(7R,7′S,8R,8′S)-form

3,3′-Di-Me ether

7,7′-Epoxy-3,3′-dimethoxy-8,8′-lignan-4,4′,9-triol. 4,4′,9-Trihydroxy-3,3′-dimethoxy-7,7′-epoxylignan

$C_{20}H_{24}O_6$ 360.406

Constit. of leaves and stems of *Schisandra chinensis*. Yellow gum. $[\alpha]^{27}_{D}$ +5.3 (c, 0.11 in MeOH). λ_{max} 205 (log ε 4.03); 230 (log ε 3.52); 281 (log ε 3.13) (MeOH)

(7R,7′S,8S,8′R)-form

3,3′-Di-Me ether

$C_{20}H_{24}O_6$ 360.406

Constit. of leaves and stems of *Schisandra chinensis*. Yellow gum. $[\alpha]^{27}_{D}$ +47 (c, 0.1 in MeOH). λ_{max} 205 (log ε 3.95); 213 (log ε 3.44); 281 (log ε 3.02) (MeOH)

(7S,7′R,8S,8′R)-form

3,3′-Di-Me ether, 9-Ac

Schisphenlignan G

[1542234-11-6]

$C_{22}H_{26}O_7$ 402.443

Isol. from the leaves and stems of *Schisandra sphenanthera*. Powder. $[\alpha]^{15}_{D}$ +8.3 (c, 0.34 in $CHCl_3$). λ_{max} 239 (log ε 3.14) ($CHCl_3$)

3′,4′-Methylene, 3-Me ether, 9-Ac

Schibitubin H

[2094994-51-9]

$C_{22}H_{24}O_7$ 400.427

Constit. of the fruit of *Schisandra bicolor* var. *tuberculata*. Exhibits neuroprotective activity. Amorph. solid. $[\alpha]^{20}_{D}$ +18.7 (c, 0.08 in $CHCl_3$). λ_{max} 230 (log ε 3.54); 285 (log ε 2.93); 370 (log ε 2.15) (MeOH).

Ma, C.J. *et al.*, *J. Enzyme Inhib. Med. Chem.*, 2009, **24**, 1117-1121 (*9-Hydroxygalbacin*).

Liang, C.-Q. *et al.*, *Fitoterapia*, 2013, **86**, 171-177 (*Schisphenlignan G*).

Shi, Y. *et al.*, *Chin. J. Chem.*, 2014, **32**, 734-740 (*Schisandra chinensis constits*).

Liu, Y. *et al.*, *J. Nat. Prod.*, 2017, **80**, 1117-1124 (*Schibitubin H, activity*)

7,9′-Epoxy-8,8′-lignan-2′,3,4′,5,9-pentol E121

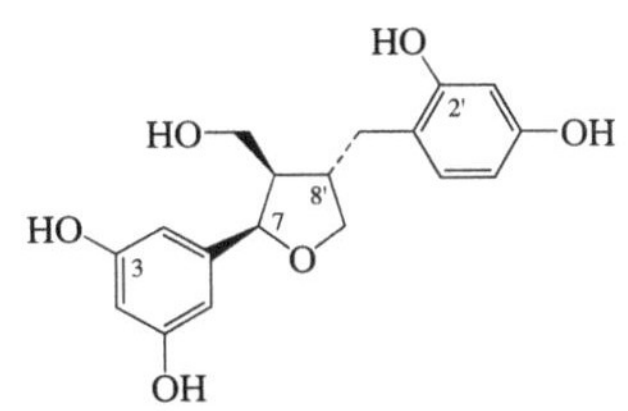

2′,3,4′,5,9-Pentahydroxy-7,9′-epoxylignan

$C_{18}H_{20}O_6$ 332.352

Structures in this series may need revision. Phenyl 3,5-dioxygenated substn. pattern in doubt as a result of work done by Yang *et al* in 2015

(7R*,8R*,8′S*)-form

3,5-Di-Me ether

7,9′-Epoxy-3,5-dimethoxy-8,8′-lignan-2′,4′,9-triol. 2′,4′,9-Trihydroxy-3,5-dimethoxy-7,9′-epoxylignan

[875056-44-3]

$C_{20}H_{24}O_6$ 360.406

Constit. of the arils of *Torreya grandis*. Mp 220 - 223°.

Zhou, D.-Z. *et al.*, *Yaoxue Xuebao*, 2004, **39**, 269-271 (*Torreya grandis constit*).

Yang, Y.-N. *et al.*, *J. Nat. Prod.*, 2015, **78**, 705-711 (*pmr, struct*)

7,9′-Epoxy-8,8′-lignan-3,3′,4,4′,5′-pentol E122

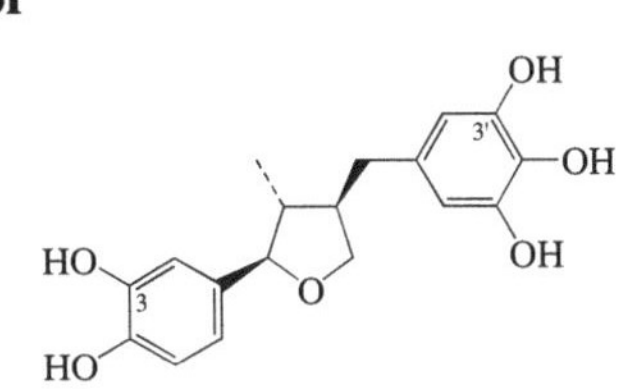

3,3′,4,4′,5′-Pentahydroxy-7,9′-epoxylignan

$C_{18}H_{20}O_6$ 332.352

Strict application of IUPAC principles would exchange the numbering of the rings (3,3′,4,4′,5-pentol) but the 3,3′,4,4′,5′-numbering is retained here for consistency

(7R,8R,8′R)-form

3,3′,5′-Tri-Me ether

7,9′-Epoxy-3,3′,5′-trimethoxy-8,8′-lignan-4,4′-diol. 4,4′-Dihydroxy-3,3′,5′-trimethoxy-7,9′-epoxylignan

[1310453-69-0]

$C_{21}H_{26}O_6$ 374.433

Isol. from the bark of *Machilus robusta*. Amorph. solid. $[\alpha]^{20}_{D}$ -25.6 (c, 0.08 in MeOH). λ_{max} 232 (log ε 3.26); 280 (log ε 2.7) (MeOH).

Li, Y. *et al.*, *J. Nat. Prod.*, 2011, **74**, 1444-1452 (*tri-Me ether*)

7,9′-Epoxy-8,8′-lignan-3,3′,4,4′,5-pentol E123

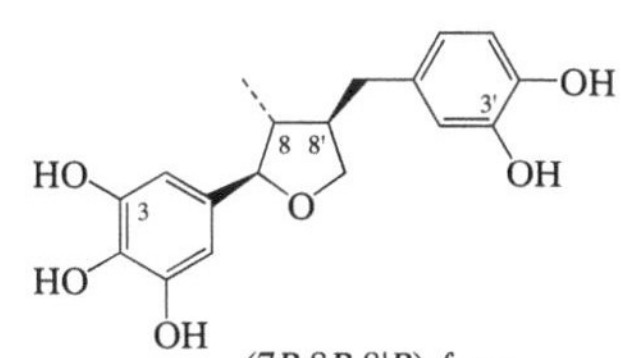

(7*R*,8*R*,8′*R*)-*form*

3,3′,4,4′,5-Pentahydroxy-7,9′-epoxylignan

$C_{18}H_{20}O_6$ 332.352

(7R,8R,8′R)-form

3,3′,5-Tri-Me ether

7,9′-Epoxy-3,3′,5-trimethoxy-8,8′-lignan-4,4′-diol. 4,4′-Dihydroxy-3,3′,5-trimethoxy-7,9′-epoxylignan. ***Schisphenlignan J***

[1542234-15-0]

$C_{21}H_{26}O_6$ 374.433

Isol. from the leaves and stems of *Schisandra sphenanthera*. Amorph. powder. $[\alpha]^{27}_{D}$ +17.4 (c, 0.07 in $CHCl_3$). λ_{max} 240 (log ε 3.49) ($CHCl_3$)

(7R,8R,8′S)-form

3,3′,4,5-Tetra-Me ether

7,9′-Epoxy-3,3′,4,5-tetramethoxy-8,8′-lignan-4′-ol. 4′-Hydroxy-3,3′,4,5-tetramethoxy-7,9′-epoxylignan. ***Acortatarinowin L***

$C_{22}H_{28}O_6$ 388.46

Constit. of the rhizomes of *Acorus tatarinowii*. Exhibits strong DPPH radical scavenging activity. Light yellow gum. $[\alpha]^{20}_{D}$ +18 (c, 0.1 in

MeOH). λ_{max} 205 (log ε 4.81); 230 (log ε 4.15); 280 (log ε 3.53) (MeOH).

Liang, C.-Q. *et al.*, *Fitoterapia*, 2013, **86**, 171-177 (*Schisphenlignan J*).
Lu, Y. *et al.*, *Sci. Rep.*, 2016, **6**, 22909 (*Acortatarinowin L*)

7,9'-Epoxy-8,8'-lignan-3,3',4,4',7'-pentol E124

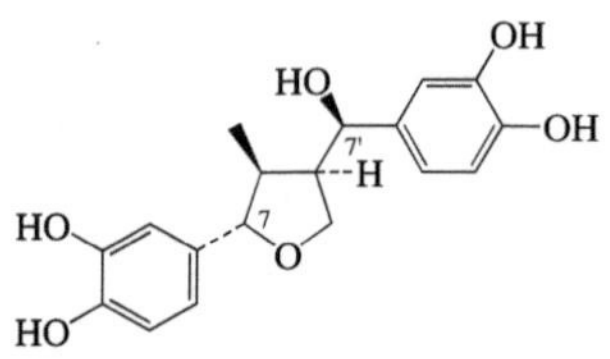

(7S,7'R,8S,8'R)-form

3,3',4,4',7'-Pentahydroxy-7,9'-epoxylignan
$C_{18}H_{20}O_6$ 332.352

(7S,7'R,8S,8'R)-form

3,3',4'-Tri-Me ether
7,9'-Epoxy-3,3',4'-trimethoxy-8,8'-lignan-4,7'-diol. 4,7'-Dihydroxy-3,3',4'-trimethoxy-7,9'-epoxylignan. ***Kobusinol A***
[177080-50-1]
$C_{21}H_{26}O_6$ 374.433
Constit. of *Magnolia kobus* var. *borealis*

3,3',4,4'-Tetra-Me ether
7,9'-Epoxy-3,3',4,4'-tetramethoxy-8,8'-lignan-7'-ol. 7'-Hydroxy-3,3',4,4'-tetramethoxy-7,9'-epoxylignan. ***Magnostellin A***
[87562-70-7]
$C_{22}H_{28}O_6$ 388.46
Isol. from leaves of *Magnolia stellata*, *Magnolia fargesii* and fruits of *Virola elongata*. Oil. $[\alpha]_D$ +75 (c, 0.75 in $CHCl_3$)

3,4-Methylene, 3',4'-di-Me ether
7,9'-Epoxy-3',4'-dimethoxy-3,4-methylenedioxy-8,8'-lignan-7'-ol. 7'-Hydroxy-3',4'-dimethoxy-3,4-methylenedioxy-7,9'-epoxylignan.
Magnostellin C
[101205-03-2]
$C_{21}H_{24}O_6$ 372.417
Isol. from *Virola elongata*. Oil. $[\alpha]_D$ +56.3 (c, 1.67 in $CHCl_3$)

(7R*,8R*,8'R*)-form

7'-Ketone, 3,3'-di-Me ether
7',9-Epoxy-4,4'-dihydroxy-3,3'-dimethoxy-8,8'-lignan-7-one.
(7R*,8R*,8'R*)-Forsythialan C
[2089058-30-8]
$C_{20}H_{22}O_6$ 358.390
Constit. of the fruit of *Forsythia suspensa*. Syrup. $[\alpha]_D^{25}$ +2.2 (c, 0.1 in MeOH). λ_{max} 244 (log ε 2.69); 275 (log ε 3.09); 301 (log ε 2.93) (MeOH)

(7R*,8S*,8'R*)-form

7'-Ketone, 3,3'-di-Me ether
(7R*,8S*,8'R*)-Forsythialan C
[2089054-57-7]
$C_{20}H_{22}O_6$ 358.390
Constit. of the fruit of *Forsythia suspensa*. Syrup. $[\alpha]_D^{25}$ +30 (c, 0.3 in MeOH). λ_{max} 244 (log ε 2.97); 273 (log ε 3.33); 301 (log ε 3.20) (MeOH)

(7RS,8RS,8'RS)-form

7'-Ketone, tetra-Me ether
7,9'-Epoxy-3,3',4,4'-tetramethoxy-8,8'-lignan-7'-one. ***Tatarinoid D***
[2222357-82-4]
$C_{22}H_{26}O_6$ 386.444
Constit. of the rhizomes of *Acorus tatarinowii*. Exhibits antialzheimers disease (AD) props.. Solid. $[\alpha]_D^{27}$ +6.2 (c, 0.5 in MeOH). The nat. prod. Tatarinoid D was isol. as a partial racemate and resolved by chiral hplc. λ_{max} 209 (log ε 4.60); 228 (log ε 4.32); 277 (log ε 4.06); 310 (log ε 3.93) (MeOH).

Iida, T. *et al.*, *Phytochemistry*, 1983, **22**, 211-213 (*Magnostellin A*).
Kato, M.J. *et al.*, *Phytochemistry*, 1986, **25**, 279-280 (*Magnostellins A,C, abs config*).
Kim, Y.-G. *et al.*, *Enshurin Kenkyu Hokoku (Hokkaido Daigaku Nogakubu)*, 1996, **53**, 1-28 (*Kobusinol A*).
Yamauchi, S. *et al.*, *Biosci., Biotechnol., Biochem.*, 2001, **65**, 1559-1567 (*Magnostellin C, synth*).
Kuo, P.-C. *et al.*, *J. Nat. Prod.*, 2017, **80**, 1055-1064 (*Forsythialans C, cd*).
Zhang, W.-Y. *et al.*, *Bioorg. Med. Chem. Lett.*, 2018, **28**, 814-819 (*Tatarinoid D, cd, cryst struct, activity*)

7,9'-Epoxy-8,8'-lignan-3,3',4,4',9-pentol E125

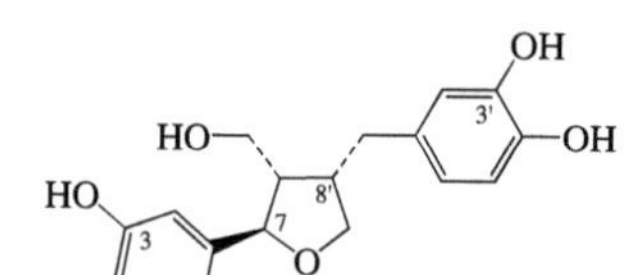

(7R,8S,8'S)-form

Tetrahydro-2-(3,4-dihydroxyphenyl)-4-[(3,4-dihydroxyphenyl) methyl]-3-furanmethanol, CAS.
3,3',4,4',9-Pentahydroxy-7,9'-epoxylignan
$C_{18}H_{20}O_6$ 332.352

(7R,8S,8'S)-form

3,3'-Di-Me ether
7,9'-Epoxy-3,3'-dimethoxy-8,8'-lignan-4,4',9-triol. 4,4',9-Trihydroxy-3,3'-dimethoxy-7,9'-epoxylignan. **(-)-Lariciresinol.** *NSC 329247*
[83327-19-9]
$C_{20}H_{24}O_6$ 360.406
Constit. of *Daphne odora*, *Daphne tangutica*, *Populus tremula*, *Picea abies*, *Viburnum sargentii*, *Diplomorpha canescens* and *Dirca occidentalis*. Cytotoxic agent. Shows antioxidant activity. Plant growth regulator. Sol. MeOH, $CHCl_3$; poorly sol. H_2O. $[\alpha]_D$ -17.8 (Me_2CO). λ_{max} 232 (ε 14800); 283 (ε 10500) (MeOH) (Berdy)

3,3'-Di-Me ether, 4'-sulfate
[1923816-93-6]
$C_{20}H_{24}O_9S$ 440.463
Constit. of the aerial parts of *Zygophyllum aegyptium*. Amorph. powder. $[\alpha]_D^{25}$ +12 (c, 0.1 in MeOH). λ_{max} 204; 277 (MeOH)

3,3'-Di-Me ether, 9-O-(2-methylbutanoyl)
9-O-(2-Methylbutanoyl)-(-)-lariciresinol
[1257982-43-6]
$C_{25}H_{32}O_7$ 444.524
Constit. of the roots of *Aster flaccidus*. Oil. $[\alpha]_D^{20}$ +15 (c, 1.5 in $CHCl_3$). λ_{max} 242; 281 (MeOH)

3,3'-Di-Me ether, 9-benzoyl
$C_{27}H_{28}O_7$ 464.514
Constit. of *Tripterygium wilfordii*. NO prodn. inhibitor. Yellow oil. $[\alpha]_D^{25}$ +36.3 (c, 0.1 in MeOH). λ_{max} 229 (log ε 2.67); 281 (log ε 2.32) (MeOH)

3,3'-Di-Me ether, 9-O-(4-hydroxybenzoyl)
Agastinol
[413615-33-5]
$C_{27}H_{28}O_8$ 480.513
Constit. of *Agastache rugosa*. Amorph. yellow powder. $[\alpha]_D^{25}$ -30 (c, 1 in MeOH). λ_{max} 232 (log ε 4.32); 259 (log ε 4.34); 334 (log ε 3.26) (MeOH)

3,3'-Di-Me ether, 9-O-[3-[2R-hydroxy-2-(methoxycarbonyl)ethyl] benzoyl]
Hyosmin
[926628-62-8]
$C_{31}H_{34}O_{10}$ 566.603

Constit. of the seeds of *Hyoscyamus niger*. Amorph. solid. $[\alpha]_D^{25}$ -17.9 (c, 0.58 in MeOH). λ_{max} 232; 281 (MeOH)

3,3′-Di-Me ether, 9-O-(3,4,5-trihydroxybenzoyl), 4′-O-β-D-glucopyranoside

Balaxiflorin A

[1203964-83-3]

$C_{33}H_{38}O_{15}$ 674.652

Constit. of *Balanophora laxiflora*. Brown powder. $[\alpha]_D^{21}$ -29.5 (c, 0.01 in Me_2CO). Abs. config. not determined

3,4:3′,4′-Bis(methylene) ether

7,9′-Epoxy-3,4:3′,4′-bis(methylenedioxy)-8,8′-lignan-9-ol. 9-Hydroxy-3,4:3′,4′-bis(methylenedioxy)-7,9′-epoxylignan. ***(-)-Dihydrosesamin***

[83708-70-7]

$C_{20}H_{20}O_6$ 356.374

Isol. from *Daphne tangutica* and *Diplomorpha canescens*. Cryst. (MeOH). Mp 98 - 99°. $[\alpha]_D^{25}$ -15.9 (c, 0.67 in Py)

(7S,8R,8′R)-form

3-Me ether

7,9′-Epoxy-3-methoxy-8,8′-lignan-3′,4,4′,9-tetrol. 3′,4,4′,9-Tetrahydroxy-3-methoxy-7,9′-epoxylignan

3-Me ether, 3′-O-β-D-glucopyranoside

Alashinol C

[2238886-74-1]

$C_{25}H_{32}O_{11}$ 508.520

Constit. of the stem bark of *Syringa pinnatifolia*. Powder. $[\alpha]_D^{25}$ -3.1 (c, 0.1 in MeOH). λ_{max} 206 (log ε 4.59); 227 (log ε 4.20); 280 (log ε 3.86) (MeOH)

3′-Me ether

7,9′-Epoxy-3′-methoxy-8,8′-lignan-3,4,4′,9-tetrol. 3,4,4′,9-Tetrahydroxy-3′-methoxy-7,9′-epoxylignan. ***Taxiresinol***

[40951-69-7]

$C_{19}H_{22}O_6$ 346.379

Isol. from the heartwood of *Taxus baccata* and from *Taxus meieri*, *Taxus wallichiana*, *Taxus yunnanensis*, *Daphne oleides* and *Podocarpus spicata*. Shows cytotoxic activity against various human cancer cell lines. Hepatoprotectant. Needles (Me_2CO). Mp 158°. $[\alpha]_D$ +32.2 (EtOH). $[\alpha]_D^{25}$ +68 (c, 1.0 in MeOH). The 7-config. is incorrectly descr. as *R*- by Chattopadhyay *et al.*. λ_{max} 226; 282 (MeOH) (Berdy)

3′-Me ether, 9-Ac

9-O-Acetyltaxiresinol

[252333-74-7]

$C_{21}H_{24}O_7$ 388.416

Constit. of *Taxus mairei*. Gum

3′-Me ether, tetra Ac

[211371-46-9]

Cryst. (MeOH). Mp 90°

3,3′-Di-Me ether

(+)-Lariciresinol

[27003-73-2]

$C_{20}H_{24}O_6$ 360.406

Constit. of the wound resin of *Larix decidua*. Also found in *Araucaria angustifolia*, *Berberis koreana*, *Euterpe oleracea* and *Taxus yunnanensis*. Shows cytotoxic acivity against A549 and XF498 human cancer cell lines. NO prodn. inhibitor. Shows antioxidant and cytoprotectant activity. Potent plant growth inhibitor. Needles (MeOH). Mp 167 - 168° (162-164°). $[\alpha]_D^{25}$ +18 (c, 1.0 in Me_2CO)

3,3′-Di-Me ether, 4-O-β-D-glucopyranoside

Indigoticoside A*. Indigocalignanoside A*

[107110-16-7]

$C_{26}H_{34}O_{11}$ 522.547

Constit. of *Asclepias subulata*, *Clematis stans*, *Syringa amurensis* and *Osmanthus asiaticus*. Powder. $[\alpha]_D^{25}$ -25.3 (c, 1.33 in MeOH) (-19.3)

3,3′-Di-Me ether, 4-O-[β-D-glucopyranosyl-(1→3)-β-D-glucopyranoside]

(+)-Lariciresinol 4-O-β-laminaribioside

[639857-95-7]

$C_{32}H_{44}O_{16}$ 684.688

Constit. of the leaves of *Syringa reticulata*. Amorph. powder. $[\alpha]_D^{25}$ -20 (c, 0.25 in MeOH). λ_{max} 205 (log ε 4.38); 225 (log ε 4.07); 278 (log ε 3.63) (MeOH)

3,3′-Di-Me ether, 4′-O-β-D-glucopyranoside

Lariciresinol 4′-glucoside

[143663-00-7]

$C_{26}H_{34}O_{11}$ 522.547

Constit. of *Arum italicum*, *Osmanthus asiaticus*, *Rhodiola rosea*, *Clematis stans*, *Sedum sarmentosum* and *Syringa vulgaris*. Shows significant hepatoprotectant activity. Amorph. light yellow powder. $[\alpha]_D^{20}$ -25.9 (c, 1.1 in EtOH). $[\alpha]_D^{22}$ -33.1 (c, 1.71 in MeOH) (-19.3). Incorrectly descr. as (-)-Lariciresinol 4′-glucosidein the original ref. (Kurkin *et al* 1991)

3,3′-Di-Me ether, 4,4′-di-O-β-D-glucopyranoside

Clemastanin B

[112747-98-5]

$C_{32}H_{44}O_{16}$ 684.688

Constit. of *Clematis stans*, *Sedum sarmentosum*, *Eucommia* sp. and *Galium sinaicum*. Amorph. powder. Mp 200 - 203°. $[\alpha]_D^{24}$ -50 (c, 0.16 in Py). $[\alpha]_D^{24}$ -42.7 (c, 2 in H_2O). λ_{max} 209 (ε 16220); 226 (ε 15850); 276 (ε 5010) (MeOH)

3,3′-Di-Me ether, 4,4′-di-O-α-L-rhamnopyranoside

(+)-Lariciresinol 4,4′-dirhamnoside

[123592-90-5]

$C_{32}H_{44}O_{14}$ 652.69

Constit. of *Parsonsia laevigata*. Fine prisms (MeOH). Mp 124 - 127°. $[\alpha]_D^{26}$ -100 (c, 1.3 in MeOH)

3,3′-Di-Me ether, 4′-O-[E-cinnamoyl-(→6)-β-D-glucopyranoside]

[1477720-87-8]

$C_{35}H_{40}O_{12}$ 652.693

Constit. of *Jasminum giraldii*. Powder. $[\alpha]_D^{20}$ -10.91 (c, 0.1 in MeOH). λ_{max} 203; 216; 278 (MeOH)

3,3′-Di-Me ether, 4′-O-[4-hydroxy-3-methoxy-E-cinnamoyl-(→6)-β-D-glucopyranoside]

[947494-33-9]

$C_{36}H_{42}O_{14}$ 698.718

Constit. of the roots of *Rhus javanica* var. *roxburghiana*. Powder. $[\alpha]_D^{27}$ -30.8 (c, 0.21 in Me_2CO). λ_{max} 227 (log ε 4.16); 232 (log ε 4.01); 276 (log ε 4); 301 (sh); 330 (log ε 4.08) (MeOH)

3,3′-Di-Me ether, 4′-O-[4,6-bis-O-(4-hydroxy-3-methoxy-E-cinnamoyl)-β-D-glucopyranoside]

[947494-34-0]

$C_{46}H_{50}O_{17}$ 874.889

Constit. of the roots of *Rhus javanica* var. *roxburghiana*. Powder. $[\alpha]_D^{27}$ -21.4 (c, 0.2 in Me_2CO). λ_{max} 227 (log ε 4.34); 233

(log ε 4.07); 278 (sh); 300 (log ε 4.01); 331 (log ε 4.11) (MeOH)

3,3′-Di-Me ether, 9-O-β-D-xylopyranoside
(+)-Lariciresinol 9-xyloside
[1013648-95-7]
$C_{25}H_{32}O_{10}$ 492.521
Constit. of *Humulus japonicus*. Yellowish gum. [α] D +21.3 (c, 0.33 in MeOH). λ_{max} 206 (log ε 4.26); 228 (log ε 3.84); 281 (log ε 3.48) (MeOH)

3,3′-Di-Me ether, 9-O-β-D-glucopyranoside
[67308-35-4]
$C_{26}H_{34}O_{11}$ 522.547
Constit. of *Arum italicum*, *Stewartia koreana* and *Caryopteris incana*

3,3′-Di-Me ether, 9-O-[4-hydroxy-3-methoxy-E-cinnamoyl-(→6)-β-D-glucopyranoside]
[2128271-74-7]
$C_{36}H_{42}O_{14}$ 698.718
Constit. of the stem bark of *Catalpa ovata*. Potent scavenger of peroxynitrite. Amorph. light yellow gum. $[\alpha]^{22}_{D}$ +12 (c, 0.1 in MeOH). λ_{max} 218 (log ε 4.33); 230 (log ε 4.33); 287 (log ε 4.13); 327 (log ε 4.21) (MeOH)

3,3′-Di-Me ether, 9-O-[β-D-apiofuranosyl-(1→6)-β-D-glucopyranoside]
$C_{31}H_{42}O_{15}$ 654.662
Constit. of the branches of *Alangium kurzii*. Brown oil. $[\alpha]^{20}_{D}$ -2.35 (c, 0.53 in MeOH)

3,3′-Di-Me ether, 9-O-[β-D-glucopyranosyl-(1→2)-β-D-glucopyranoside]
Tianshanoside A
[1696402-36-4]
$C_{32}H_{44}O_{16}$ 684.688
Constit. of *Viola tianshanica*. Amorph. powder. $[\alpha]^{25}_{D}$ +15.1 (c, 0.2 in MeOH). λ_{max} 230; 280 (MeOH)

3,3′-Di-Me ether, 9-O-[α-L-rhamnopyranosyl-(1→2)-β-D-glucopyranoside]
$C_{32}H_{44}O_{15}$ 668.689
Constit. of *Pilea cavaleriei*. Amorph. powder. $[\alpha]^{20}_{D}$ -33 (c, 0.20 in MeOH). λ_{max} 206; 279 (MeOH)

3,3′-Di-Me ether, 9-Ac
9-O-Acetyl-(+)-lariciresinol
[79114-77-5]
$C_{22}H_{26}O_7$ 402.443
Constit. of *Aglaia elaeagnoidea* and *Araucaria angustifolia*. [α] D +17 (c, 0.3 in $CHCl_3$). λ_{max} 230 (log ε 4.11); 282 (log ε 3.79) (MeOH)

3,3′-Di-Me ether, tri-Ac
[67560-67-2]
Viscous oil. $[\alpha]^{25}_{D}$ +8 (c, 1.0 in $CHCl_3$)

3,3′-Di-Me ether, 9-O-(3-methylbutanoyl)
9-O-(3-Methylbutanoyl)-(+)-lariciresinol.
9-Isovaleroyl-(+)-lariciresinol
[1210965-78-8]
$C_{25}H_{32}O_7$ 444.524
Constit. of *Aegle marmelos*, *Valeriana jatamansi* and *Viburnum odoratissimum*. Shows significant cytotoxicity against PC-3M and HCT-8 cell lines. Inhibitor of hypoxia-induced HIF-1 activation. Oil. $[\alpha]^{20}_{D}$ +23.2 (c, 0.22 in MeOH)

3,3′-Di-Me ether, 9-O-(3-methyl-2-butenoyl)
9-O-Senecioyl-(+)-lariciresinol
[1952356-78-3]
$C_{25}H_{30}O_7$ 442.508
Constit. of the branches and leaves of *Viburnum odoratissimum*. Exhibits moderate inhibitory activity against human A431 and T47D tumour cell lines. Light brown oil. $[\alpha]^{15}_{D}$ +22 (c, 0.1 in MeOH). λ_{max} 218 (log ε 4.37); 282 (log ε 3.74) (MeOH)

3,3′-Di-Me ether, 9-octadecanoyl
(+)-Lariciresinol 9-stearate
[1132696-65-1]
$C_{38}H_{58}O_7$ 626.875
Constit. of aerial parts of *Tephrosia vogelii*. Yellow powder. $[\alpha]^{20}_{D}$ +8.1 (c, 1.0 in MeOH). λ_{max} 220 (log ε 4.37); 277 (log ε 4) (MeOH)

3,3′-Di-Me ether, 9-pentatriacontanoyl
(+)-Lariciresinol 9-pentatriacontanoate
$C_{55}H_{92}O_7$ 865.334
Constit. of the flowers of *Daphne genkwa*. $[\alpha]^{20}_{D}$ +15 (c, 0.35 in $CHCl_3$)

3,3′-Di-Me ether, 9-angeloyl, 4-O-β-D-glucopyranoside
Leontopodioside C
$C_{31}H_{40}O_{12}$ 604.649
Constit. of *Leontopodium leontopodioides*. Weak α-glucosidase inhibitor. Amorph. powder. $[\alpha]^{20}_{D}$ -10.7 (c, 0.42 in MeOH). λ_{max} 206 (log ε 4.46); 279 (log ε 3.61) (MeOH)

3,3′-Di-Me ether, 9-O-(4-hydroxy-E-cinnamoyl)
9-O-p-Coumaroyl-(+)-lariciresinol
[864452-88-0]
$C_{29}H_{30}O_8$ 506.551
Constit. of the bark of *Larix olgensis* var. *koreana*. Powder (MeOH). Mp 120 - 121°. $[\alpha]^{20}_{D}$ +47.4 (c, 0.51 in MeOH). Abs. config. not determined

3,3′-Di-Me ether, 9-O-(3,4-dihydroxy-E-cinnamoyl)
9-O-Caffeoyl-(+)-lariciresinol
[864452-90-4]
$C_{29}H_{30}O_9$ 522.55
Constit. of the bark of *Larix olgensis* var. *koreana*. Powder (MeOH). $[\alpha]^{20}_{D}$ -23 (c, 2 in MeOH). Abs. config. not determined

3,3′-Di-Me ether, 9-O-(4-hydroxy-3-methoxy-E-cinnamoyl)
9-O-trans-Feruloyl-(+)-lariciresinol
[1353749-60-6]
$C_{30}H_{32}O_9$ 536.577
Constit. of *Abies nephrolepis*. Brown powder. $[\alpha]^{24}_{D}$ +10.4 (c, 0.34 in MeOH)

3,3′,4-Tri-Me ether
7,9′-Epoxy-3,3′,4-trimethoxy-8,8′-lignan-4′,9-diol.
4′,9-Dihydroxy-3,3′,4-trimethoxy-7,9′-epoxylignan. ***Daphneligin***.
4-O-Methyllariciresinol
[73354-13-9]
$C_{21}H_{26}O_6$ 374.433
Constit. of *Turraea nilotica*, *Aphanamixis grandifolia* and *Daphne oleoides*. Cytotoxic agent. Cryst.. Mp 136 - 138° (125-127°). $[\alpha]^{25}_{D}$ +12.9 (c, 0.8 in $CHCl_3$) (+11.5). λ_{max} 202 (log ε 4.15); 229 (log ε 3.45); 278 (log ε 3) (EtOH)

3,3′,4-Tri-Me ether, 4′-O-β-D-glucopyranoside
Daphneligin 4′-glucoside
[334699-27-3]
$C_{27}H_{36}O_{11}$ 536.574
Constit. of *Daphne oleoides*. Amorph. powder. Mp 192 - 193°. $[\alpha]^{20}_{D}$ -30.5 (c, 0.1 in MeOH). λ_{max} 229 (EtOH)

3,3′,4,4′-Tetra-Me ether
7,9′-Epoxy-3,3′,4,4′-tetramethoxy-8,8′-lignan-9-ol. 9-Hydroxy-3,3′,4,4′-tetramethoxy-7,9′-epoxylignan. ***Lariciresinol dimethyl ether***
[67560-68-3]
$C_{22}H_{28}O_6$ 388.46
Constit. of *Monechma ciliatum*, *Berberis koreana* and *Virola michelii*. Cytotoxic against a number of human cancer cell lines. Inhibitor of NO prodn. in LPS-activated BV-Z cells. Needles (Et_2O) or viscous oil. Mp 75 - 76°. $[\alpha]_D^{25}$ +58 (c, 0.9 in Me_2CO)

3,3′,4,4′-Tetra-Me ether, 9-angeloyl
Leoligin
[619282-88-1]
$C_{27}H_{34}O_7$ 470.562
Constit. of the roots of *Leontopodium nivale* ssp. *alpinum*. Activates cholesteryl ester transfer protein and prevents cholesterol efflux from THP-1 macrophages. Inhibitor of leukotriene biosynth. and intimal hyperplasia of venous bypass grafts. Amorph. solid. $[\alpha]_D^{20}$ +25 (c, 0.002 in CH_2Cl_2)

3,4-Methylene, 3′-Me ether
7,9′-Epoxy-3′-methoxy-3,4-methylenedioxy-8,8′-lignan-4′,9-diol. 4′,9-Dihydroxy-3′-methoxy-3,4-methylenedioxy-7,9′-epoxylignan.
Acuminatin‡
[118632-19-2]
$C_{20}H_{22}O_6$ 358.39
Isol. from aerial parts of *Helichrysum acuminatum* and from *Machilus obovatifolia*. Oil. $[\alpha]_D$ +11.1 (c, 0.09 in $CHCl_3$) (semisynthetic by hydrol. of Actifolin)

3,4-Methylene, 3′-Me ether, 9-Ac
Actifolin
[120931-63-7]
$C_{22}H_{24}O_7$ 400.427
Isol. from *Actinodaphne longifolia* and *Lindera obtusiloba*. Cytotoxic agent. Oil. $[\alpha]_D$ +13.3 (c, 0.48 in $CHCl_3$)

3,4-Methylene, 3′,4′-di-Me ether
7,9′-Epoxy-3′,4′-dimethoxy-3,4-methylenedioxy-8,8′-lignan-9-ol. 9-Hydroxy-3′,4′-dimethoxy-3,4-methylenedioxy-7,9′-epoxylignan.
4′-O-Methylacuminatin
[170515-46-5]
$C_{21}H_{24}O_6$ 372.417
Constit. of the kernels of *Virola michelii*. Viscous oil. $[\alpha]_D^{25}$ +4.5 (c, 0.2 in Me_2CO)

3′,4′-Methylene, 3-Me ether
7,9′-Epoxy-3-methoxy-3′,4′-methylenedioxy-8,8′-lignan-4,9-diol. 4,9-Dihydroxy-3-methoxy-3′,4′-methylenedioxy-7,9′-epoxylignan.
Sanshodiol
[54854-91-0]
$C_{20}H_{22}O_6$ 358.39
Constit. of *Syringa pinnatifolia* and *Zanthoxylum piperitum*. Inhibitor of LPS-induced NO prodn.. Needles. Mp 148 - 149°. $[\alpha]_D^{25}$ -14.8 (c, 0.1 in MeOH). λ_{max} 206 (log ε 3.94); 231 (log ε 4.19); 284 (log ε 3.22) (MeOH)

3′,4′-Methylene, 3-Me ether, 4-O-β-D-glucopyranoside
Alashinol B
[2238886-75-2]
$C_{26}H_{32}O_{11}$ 520.531
Constit. of the stem bark of *Syringa pinnatifolia*. Needles. Mp 174.5 - 175°. $[\alpha]_D^{25}$ -19.2 (c, 0.1 in MeOH). λ_{max} 206 (log ε 4.67); 228 (log ε 4.20); 282 (log ε 3.90) (MeOH)

3′,4′-Methylene, 3,4-di-Me ether
7,9′-Epoxy-3,4-dimethoxy-3′,4′-methylenedioxy-8,8′-lignan-9-ol. 9-Hydroxy-3,4-dimethoxy-3′,4′-methylenedioxy-7,9′-epoxylignan
[209908-87-2]
$C_{21}H_{24}O_6$ 372.417
Constit. of *Zanthoxylum culantrillo*. Viscous oil. $[\alpha]_D^{25}$ +14 (c, 1.7 in $CHCl_3$). λ_{max} 229 (log ε 3.06); 281 (log ε 2.98) (MeOH)

3,4:3′,4′-Bis(methylene) ether
(+)-Dihydrosesamin
[133644-85-6]
$C_{20}H_{20}O_6$ 356.374
Isol. from *Chamaecyparis obtusa*. Mp 97 - 98°. $[\alpha]_D^{20}$ +15.9 (c, 0.75 in Py) (synthetic). Props. of nat. prod. not well documented. Acetylation gave (+)-9-*O*-Acetyldihydrosesamin

3,4:3′,4′-Bis(methylene) ether, 9-Ac
9-O-Acetyldihydrosesamin
[133576-09-7]
$C_{22}H_{22}O_7$ 398.411
Constit. of the leaves of *Chamaecyparis obtusa*. Needles (MeOH). Mp 87 - 89°. $[\alpha]_D^{25}$ +20.2 (c, 0.03 in MeOH). λ_{max} 236 (ε 5060); 287 (ε 6050) (MeOH)

(7R*,8R*,8′S*)-form

9-Carboxylic acid, β-D-glucopyranosyl ester
[2409947-30-2]
$C_{24}H_{28}O_{12}$ 508.476
Constit. of the roots of *Hypochaeris radicata*. Yellow gum. $[\alpha]_D^{20}$ +47.1 (c, 2.1 in MeOH)

9-Carboxylic acid, [2-O-(2-methyl-2-propenoyl)-β-D-glucopyranosyl] ester
[2409947-31-3]
$C_{28}H_{32}O_{13}$ 576.551
Constit. of the roots of *Hypochaeris radicata*. Red gum. $[\alpha]_D^{20}$ +67.4 (c, 0.84 in MeOH)

(7R*,8S*,8′S*)-form

3,3′-Di-Me ether, 4-O-[6-O-acetyl-β-D-glucopyranoside]
Yunnanensin B
[1426545-06-3]
$C_{28}H_{36}O_{12}$ 564.584
Constit. of *Parakmeria yunnanensis*. Yellow oil. $[\alpha]_D^{23}$ -15.7 (c, 1.39 in MeOH). λ_{max} 202 (log ε 3.78); 226 (log ε 3.18); 279 (log ε 2.73) (MeOH)

3,3′-Di-Me ether, 4-O-[4-hydroxy-3,5-dimethoxybenzoyl-(→6)-β-D-glucopyranoside]
Trigonoheteran
$C_{35}H_{42}O_{15}$ 702.706
Constit. of the stems of *Trigonostemon heterophyllus*. Powder. $[\alpha]_D^{22}$ -50 (c, 0.02 in MeOH). λ_{max} 219; 228; 278 (MeOH)

3,4-Methylene, 3′-Me ether, 9-O-β-D-glucopyranoside
[942624-12-6]
$C_{26}H_{32}O_{11}$ 520.531
Constit. of *Kadsura longipedunculata*. Pale yellow powder. $[\alpha]_D^{15}$ -22.4 (c, 1 in Py)

3′,4′-Methylene, 3-Me ether, 9-Ac
Isoactifolin
[400716-48-5]
$C_{22}H_{24}O_7$ 400.427
Constit. of *Chamaecyparis obtusa*. Abs. config. not confirmed

9-Carboxylic acid
7,9′-Epoxy-3,3′,4,4′-tetrahydroxy-8,8′-lignan-9-oic acid

9-Carboxylic acid, 3′,3′-di-Me ether, Me ester

Alyterinate A

[1616247-82-5]

$C_{21}H_{24}O_7$ 388.416

Isol. from *Alyxia schlechteri*. Solid. Mp 114-115°. $[\alpha]_D^{26}$ +10.3 (c, 0.001 in $CHCl_3$). Has 8*R**-config. (change of Cahn-Ingold-Prelog priorities). λ_{max} 243 (log ε 3.44); 281 (log ε 3.06) ($CHCl_3$).

Haworth, R.D. *et al., J. Chem. Soc.*, 1937, 1645-1649 (*Lariciresinol, struct*).
Haworth, R.D. *et al., J. Chem. Soc.*, 1939, 1054-1057 (*Lariciresinol, stereochem*).
Majumdar, R.B. *et al., Indian J. Chem.*, 1972, **10**, 677-680 (*Taxiresinol*).
Abe, F. *et al., Chem. Pharm. Bull.*, 1974, **22**, 2650-2655 (*Sanshodiol*).
Fonseca, F.S. *et al., Phytochemistry*, 1978, **17**, 499-502 (*cmr*).
Zhuang, L. *et al., Planta Med.*, 1982, **45**, 172-176 (*(-)-Lariciresinol, (-)-Dihydrosesamin*).
Lin-Gen, Z. *et al., Phytochemistry*, 1983, **22**, 265-267 (*Dihydrosesamin, cryst struct*).
Ayoub, S.M.H. *et al., J. Nat. Prod.*, 1984, **47**, 875-876 (*4-O-Methyllariciresinol*).
Maat, L. *et al., Magn. Reson. Chem.*, 1985, **23**, 385-386 (*Lariciresinol dimethyl ether*).
Jolad, S.D. *et al., Phytochemistry*, 1986, **25**, 2581-2590 (*Lariciresinol 4-glucoside*).
Jakupovic, J. *et al., Phytochemistry*, 1987, **26**, 803-807 (*Acuminatin*).
Tanaka, H. *et al., Phytochemistry*, 1989, **28**, 952-954 (*Actifolin*).
Abe, F. *et al., Phytochemistry*, 1989, **28**, 1737-1741 (*Lariciresinol 4,4′-dirhamnoside*).
Kurkin, V.A. *et al., Chem. Nat. Compd. (Engl. Transl.)*, 1991, **27**, 678-680 (*(+)-Lariciresinol 4′-glucoside*).
Stevens, D.R. *et al., J. Chem. Soc., Perkin Trans. 1*, 1992, 633-637 (*(±)-Dihydrosesamin, synth*).
Sugiyama, M. *et al., Heterocycles*, 1993, **36**, 117-121 (*Osmanthus asiaticus glucosides*).
Della Greca, M. *et al., Heterocycles*, 1993, **36**, 2081-2086 (*Arum italicum glucosides*).
Umezawa, T. *et al., Biosci., Biotechnol., Biochem.*, 1994, **58**, 230-234 (*(±)-Lariciresinol, resoln*).
Kizu, H. *et al., Chem. Pharm. Bull.*, 1995, **43**, 2187-2194 (*Clemastanin B*).
Maiti, G. *et al., J. Chem. Soc., Perkin Trans. 1*, 1995, 927-929 (*Dihydrosesamin, Lariciresinol, synth*).
Vidigal, M.C.S. *et al., Phytochemistry*, 1995, **40**, 1259-1261 (*4′-O-Methylacuminatin*).
El-Gamal, A.A. *et al., Phytochemistry*, 1997, **45**, 597-600 (*Clemastanin B*).
Brader, G. *et al., J. Nat. Prod.*, 1998, **61**, 1482-1490 (*9-O-Acetyl-(+)-lariciresinol*).
Cuca S., L.E. *et al., Phytochemistry*, 1998, **47**, 1437-1439 (*3′,4′-methylene 3,4-di-Me ether*).
Kwon, H.C. *et al., Arch. Pharmacal Res.*, 1999, **22**, 417-422 (*Lindera obtusiloba constit, cytotoxicity*).
Yang, S.-J. *et al., J. Chin. Chem. Soc. (Taipei)*, 1999, **46**, 811-820 (*9-O-Acetyltaxiresinol*).
Ullah, N. *et al., Phytochemistry*, 1999, **50**, 147-149 (*Daphneligin*).
Rajasekhar, D. *et al., J. Asian Nat. Prod. Res.*, 2000, **2**, 153-156 (*9-Acetyllariciresinol, synth*).
Ullah, N. *et al., Helv. Chim. Acta*, 2001, **84**, 157-162 (*Daphneligin 4′-glucoside*).
Yamauchi, S. *et al., J. Chem. Soc., Perkin Trans. 1*, 2001, 2158-2160 (*(+)-Dihydrosesamin, synth*).
Takaku, N. *et al., J. Wood Sci.*, 2001, **47**, 493-496 (*Isoactifolin*).
Lee, C. *et al., J. Nat. Prod.*, 2002, **65**, 414-416 (*Agastinol*).
Roy, S.C. *et al., J. Org. Chem.*, 2002, **67**, 3242-3248 (*synth*).
Chattopadhyay, S.K. *et al., Bioorg. Med. Chem.*, 2003, **11**, 4945-4948 (*Taxiresinol, pmr, cmr, cryst struct, abs config, cytotoxicity*).
Machida, K. *et al., Chem. Pharm. Bull.*, 2003, **51**, 883-884 (*Lariciresinol 4-laminaribioside*).
Dobner, M.J. *et al., Helv. Chim. Acta*, 2003, **86**, 733-738 (*Leoligin*).
Nguyen, N.T. *et al., Planta Med.*, 2004, **70**, 29-33 (*Taxiresinol, hepatoprotective activity*).
Yang, B.-H. *et al., J. Nat. Prod.*, 2005, **68**, 1175-1179 (*Larix olgensis cinnamates*).
Banerjee, B. *et al., Synthesis*, 2005, 2913-2919 (*Dihydrosesamin, synth*).
Begum, A.S. *et al., J. Chem. Res.*, 2006, 675-677 (*Hyosmin*).
Pu, J.-X. *et al., Helv. Chim. Acta*, 2007, **90**, 723-729 (*Kadsura longipedunculata glycoside*).
Ouyang, M.-A. *et al., Helv. Chim. Acta*, 2007, **90**, 1099-1106 (*Rhus 4′-feruloylglucosides*).
Zhang, Y. *et al., Heterocycles*, 2007, **71**, 1565-1576 (*Sedum sarmentosum constits, hepatoprotective avtivity*).
Yu, B.C. *et al., Arch. Pharmacal Res.*, 2008, **30**, 1471-1475 (*Lariciresinol 9-xyloside*).
Kuroyanagi, M. *et al., Chem. Pharm. Bull.*, 2008, **56**, 60-63 (*(+)-Dihydrosesamin, 9-Acetyldihydrosesamin*).
Chin, Y.-W. *et al., J. Agric. Food Chem.*, 2008, **56**, 7759-7764 (*(+)-Lariciresinol, antioxidant activity, cytoprotective activity*).
Reisinger, U. *et al., Cardiovasc. Res.*, 2009, **82**, 542-549 (*Leoligin, activity*).
She, G.-M. *et al., Chem. Biodiversity*, 2009, **6**, 875-880 (*Balaxiflorin A*).
Wei, H.-H. *et al., Helv. Chim. Acta*, 2009, **92**, 370-374 (*Tephrosia vogelii constit*).
Redzynia, I. *et al., Molecules*, 2009, **14**, 4147-4158 (*(-)-Lariciresinol, cryst struct, antioxidant activity*).
Liu, Z.-L. *et al., Fitoterapia*, 2010, **81**, 140-144 (*9-(2-Methylbutanoyl)lariciresinol*).
Lin, S. *et al., J. Nat. Prod.*, 2010, **73**, 632-638 (*9-Isovaleroyllariciresinol*).
Duwensee, K. *et al., Atherosclerosis (Amsterdam)*, 2011, **219**, 109-115 (*Leoligin, activity*).
Kim, K.H. *et al., Bioorg. Med. Chem. Lett.*, 2011, **21**, 2270-2273 (*Berberis koreana constits, cytotoxicity, NO production inhibitor*).
Li, Y.-L. *et al., Chem. Biodiversity*, 2011, **8**, 2299-2309 (*9-Feruloyllariciresinol*).
Nishiwaki, H. *et al., J. Agric. Food Chem.*, 2011, **59**, 13089-13095 (*(-)-Lariciresinol, synth, plant growth inhibitor*).
Lacret, R. *et al., Phytochem. Lett.*, 2012, **5**, 382-386 (*(+)-Lariciresinol, plant growth inhibitor*).
Shang, S.-Z. *et al., Arch. Pharmacal Res.*, 2013, **36**, 1223-1230 (*Yunnanensin B*).
Li, J. *et al., Bioorg. Med. Chem.*, 2013, **21**, 1795-1803 (*Aegle marmelos constit, HIF-1 activation inhibitor*).
Zeng, Q. *et al., Chem. Nat. Compd. (Engl. Transl.)*, 2013, **49**, 486-492 (*Aphanamixis grandifolia constit, spectra*).
Sriphana, U. *et al., Fitoterapia*, 2013, **91**, 39-43 (*Alyterinate A*).
Yue, Z. *et al., Molecules*, 2013, **18**, 4766-4775 (*Jasminum giraldii constit*).
Qin, Y. *et al., Molecules*, 2013, **18**, 13636-13644 (*Tianshanoside A*).
Li, Y.-X. *et al., J. Asian Nat. Prod. Res.*, 2014, **16**, 549-553 (*Trigonoheteran*).
Zhou, Y. *et al., J. Chin. Pharm. Sci.*, 2014, **23**, 425-428 (*Pilea cavaleriei constit*).
Kim, M.H. *et al., Nat. Prod. Res.*, 2014, **28**, 1409-1412 (*Lariciresinol 9-glucoside*).
Li, F.-J. *et al., J. Asian Nat. Prod. Res.*, 2015, **17**, 475-481 (*9-Senecioyl-(+)-lariciresinol, activity*).
Su, G. *et al., Fitoterapia*, 2016, **114**, 63-68 (*Alashinols B,C, Sanshodiol, cryst struct, activity*).
Wang, L. *et al., J. Nat. Prod.*, 2016, **79**, 1651-1657 (*Leoligin, activity*).
Zaki, A.A. *et al., Magn. Reson. Chem.*, 2016, **54**, 771-773 (*Zygophyllum aegyptium sulfate*).
Xiao, Y. *et al., Phytochem. Lett.*, 2016, **17**, 247-250 (*Leontopodioside C, activity*).
Wang, R. *et al., Zhongcaoyao*, 2016, **47**, 2408-2411 (*(+)-Lariciresinol 9-pentatriacontanoate*).
Ren, J. *et al., Fitoterapia*, 2017, **121**, 152-158 (*Alangium kurzii constit*).
Kil, Y.-S. *et al., J. Nat. Prod.*, 2017, **80**, 2240-2251 (*Catalpa ovata constit, activity*).
Chen, F. *et al., Fitoterapia*, 2018, **129**, 133-137 (*Tripterygium wilfordii constit, activity*).
Shulha, O. *et al., Phytochemistry*, 2019, **165**, 112047:1-12 (*Hypochaeris radicata constits*)

7,9′-Epoxy-8,8′-lignan-3,3′,4′,5,9-pentol E126

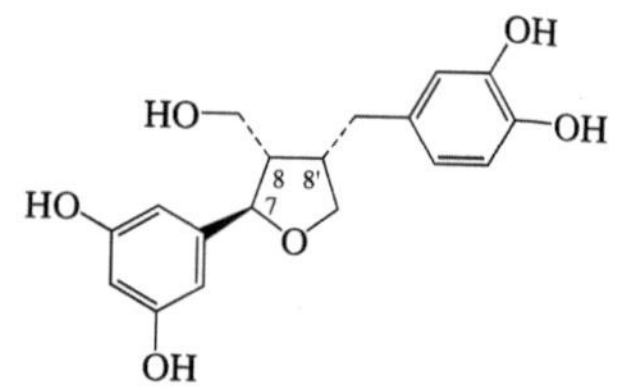

(7*R*,8*S*,8′*S*)-*form*

3,3′,4′,5,9-Pentahydroxy-7,9′-epoxylignan

$C_{18}H_{20}O_6$ 332.352

Structures in this series may need revision. Phenyl 3,5-dioxygenated substn. pattern in doubt as a result of work done by Yang *et al* in 2015

(7R,8S,8′S)-form

3,3′-Di-Me ether

Wikstresinol

$C_{20}H_{24}O_6$ 360.406

Constit. of the roots of *Wikstroemia indica*. Light yellow gum. $[\alpha]_D^{20}$ -16.8 (c, 0.07 in MeOH). λ_{max} 231 (log ε 3.10); 292 (log ε 3.48) (MeOH)

3,3′-Di-Me ether, 9-Ac

Acetylwikstresinol

$C_{22}H_{26}O_7$ 402.443

Constit. of the roots of *Wikstroemia indica*. Light yellow gum. $[\alpha]^{20}_{D}$ -15.5 (c, 0.07 in MeOH). λ_{max} 233 (log ε 3.10); 292 (log ε 3.48) (MeOH)

(7S,8R,8′R)-form

3,3′-Di-Me ether

7,9′-Epoxy-3,3′-dimethoxy-8,8′-lignan-4′,5,9-triol. 4′,5,9-Trihydroxy-3,3′-dimethoxy-7,9′-epoxylignan. **Vibruresinol**

[1591729-37-1]

$C_{20}H_{24}O_6$ 360.406

Constit. of the stems of *Viburnum erosum*. Oil. $[\alpha]^{26}_{D}$+10.9 (c, 0.41 in MeOH)

(7R*,8R*,8′R*)-form

3,3′-Di-Me ether

[1394179-93-1]

Isol. from the flower buds of *Magnolia liliflora*. Purple oil. $[\alpha]^{20}_{D}$ -6.2 (c, 0.07 in Me_2CO). λ_{max} 239 (log ε 2.90); 298 (log ε 3.90) (MeOH).

Wang, W.-S. *et al.*, *Planta Med.*, 2012, **78**, 141-147 (*Magnolia liliflora constit*).
In, S.-J. *et al.*, *Arch. Pharmacal Res.*, 2015, **38**, 26-34 (*Vibruresinol, cd*).
Yang, Y.-N. *et al.*, *J. Nat. Prod.*, 2015, **78**, 705-711 (*pmr, struct*).
Chang, H. *et al.*, *Fitoterapia*, 2017, **121**, 31-37 (*Wilstreisinol, Acetylwikstresinol, cd*)

8′,9′-Epoxy-8,8′-lignan-3,3′,4,4′,9-pentol E127

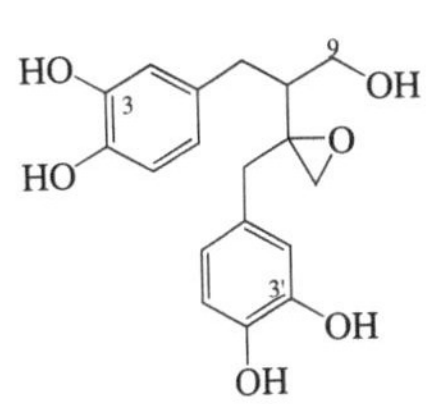

8′,9′-Epoxy-3,3′,4,4′,9-pentahydroxylignan

$C_{18}H_{20}O_6$ 332.352

(8ξ,8′ξ)-form

3,3′-Di-Me ether

8′,9′-Epoxy-3,3′-dimethoxy-8,8′-lignan-4,4′,9-triol. 8′,9′-Epoxy-4,4′,9-trihydroxy-3,3′-dimethoxylignan

[1791438-95-3]

$C_{20}H_{24}O_6$ 360.406

Constit. of *Schisandra chinensis*. Yellow gum. $[\alpha]^{23}_{D}$-49 (c, 0.12 in MeOH). λ_{max} 204 (log ε 3.95); 226 (log ε 3.45); 282 (log ε 3.08) (MeOH).

Shi, Y. *et al.*, *Chin. J. Chem.*, 2014, **32**, 734-740 (*Schisandra chinensis constit*)

9,9′-Epoxy-8,8′-lignan-3,3′,4,4′,5-pentol E128

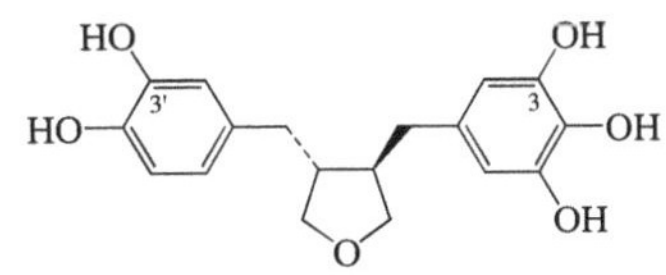

3,3′,4,4′,5-Pentahydroxy-9,9′-epoxylignan

$C_{18}H_{20}O_6$ 332.352

(8R*,8′R*)-form

3′,4′-Methylene, 3,4,5-tri-Me ether

9,9′-Epoxy-3,4,5-trimethoxy-3′,4′-methylenedioxylignan. 3,4,5-Trimethoxy-3′,4′-methylenedioxy-9,9′-epoxylignan.

Burseran

[73465-36-8]

[23284-23-3] ((+)-form)

$C_{22}H_{26}O_6$ 386.444

Constit. of *Bursera microphylla*. Antineoplastic agent. Pale yellow oil. $[\alpha]^{25}_{D}$-45.1 (c, 1.1 in $CHCl_3$) (-34.8). Log P 2.58 (calc). Abs. config. of nat. prod. not determined, no opt. rotn. reported. Props. refer to synthetic (-)-enantiomer. (+)-Enantiomer also known.

Cole, J.H. *et al.*, *J. Pharm. Sci.*, 1969, **58**, 175-176 (*Burseran*).
Tomioka, K. *et al.*, *Chem. Pharm. Bull.*, 1985, **33**, 4333-4337 (*synth*).
Rehnberg, N. *et al.*, *J. Org. Chem.*, 1990, **55**, 4340-4349 (*synth*).
Hanessian, S. *et al.*, *Synlett*, 1992, 402-404 (*synth*).
Gaboury, J.A. *et al.*, *J. Org. Chem.*, 1993, **58**, 2173-2180 (*synth*).
Garçon, S. *et al.*, *J. Org. Chem.*, 2001, **66**, 4069-4073 (*synth*).
Enders, D. *et al.*, *Synthesis*, 2002, 515-522 (*Burseran, synth*).
Pohjoispää, M. *Molecules*, 2013, **18**, 13124-13138 (*Burseran, synth*)

9,9′-Epoxy-8,8′-lignan-3,3′,4,4′,7-pentol E129

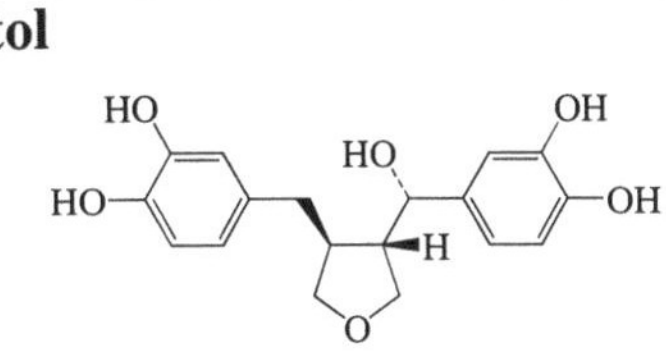

(7S,8S,8′S)-form

3,3′,4,4′,7-Pentahydroxy-9,9′-epoxylignan

$C_{18}H_{20}O_6$ 332.352

(7S,8S,8′S)-form

3,4-Methylene, 3′,4′-di-Me ether

9,9′-Epoxy-3′,4′-dimethoxy-3,4-methylenedioxy-8,8′-lignan-7-ol. 7-Hydroxy-3′,4′-dimethoxy-3,4-methylenedioxy-9,9′-epoxylignan.

Aristelegin C

[477199-88-5]

$C_{21}H_{24}O_6$ 372.417

Constit. of roots and stems of *Aristolochia elegans*. Oil. $[\alpha]^{25}_{D}$-53 (c, 0.02 in $CHCl_3$). λ_{max} 240 (log ε 4.52); 270 (log ε 3.31) (MeOH)

3,3′,4,4′-Tetra-Me ether

9,9′-Epoxy-3,3′,4,4′-tetramethoxy-8,8′-lignan-7-ol. 7-Hydroxy-3,3′,4,4′-tetramethoxy-9,9′-epoxylignan. **Aristelegin B**

[476687-29-3]

$C_{22}H_{28}O_6$ 388.46

Constit. of roots and stems of *Aristolochia elegans*. Oil. $[\alpha]^{25}_{D}$-9.9 (c, 0.07 in $CHCl_3$). λ_{max} 231 (log ε 4.13); 280 (log ε 3.67) (MeOH)

(7R*,8R*,8′R*)-form

3,4:3′,4′-Bis(methylene) ether

9,9′-Epoxy-3,4:3′,4′-bis(methylenedioxy)-8,8′-lignan-7-ol. 7-Hydroxy-3,4:3′,4′-bis(methylenedioxy)-9,9′-epoxylignan. **Isocubebin**

[1073342-10-5]

$C_{20}H_{20}O_6$ 356.374

Constit. of the stem bark and leaves of *Daphne feddei*. Cytotoxic to human HL-60, HepG2 and KB cell lines. Inhibitor of LPS-induced NO prodn.. Pale yellow oil. $[\alpha]^{20}_{D}$-9 (c, 0.2 in $CHCl_3$)

(7ξ,8R*,8′R*)-form

3,3′-Di-Me ether

9,9′-Epoxy-3,3′-dimethoxy-8,8′-lignan-4,4′,7-triol. 4,4′,7-Trihydroxy-3,3′-dimethoxy-9,9′-epoxylignan

[104086-77-3]

$C_{20}H_{24}O_6$ 360.406

Constit. of *Bupleurum salicifolium*, *Tarenna attenuata* and *Tinospora cordifolia*. Flakes. (EtOAc/petrol). Mp 168°. $[\alpha]^{22}_{D}$-30 (c, 0.5 in Me_2CO). λ_{max} 205 (ε 16000); 232 (ε 11000); 285 (MeOH)

3,3′-Di-Me ether, 4-O-β-D-glucopyranoside
$C_{26}H_{34}O_{11}$ 522.547
Constit. of the roots of *Aconitum pendulum*. Powder. $[\alpha]^{20}_{D}$ -2.9 (c, 0.37 in MeOH)

(7ξ,8R*,8′S*)-form

3,3′-Di-Me ether
Constit. of *Chrysolaena verbascifolia* (preferred genus name *Vernonia*)

3,4-Methylene, 3′-Me ether
9,9′-Epoxy-3-methoxy-3′,4′-methylenedioxy-8,8′-lignan-4,7′-diol. 4′,7-Dihydroxy-3′-methoxy-3,4-methylenedioxy-9,9′-epoxylignan
[942624-13-7]
$C_{20}H_{22}O_{6}$ 358.39
Constit. of *Kadsura longipedunculata*. Pale yellow powder. $[\alpha]^{15}_{D}$ +1.6 (c, 0.32 in MeOH). λ_{max} 283 (log ε 4) (MeOH).

Hanuman, J.B. *et al.*, *J. Chem. Soc., Perkin Trans. 1*, 1986, 1181-1185 (*Tinospora cordifolia lignan*).
Bardon, A. *et al.*, *Phytochemistry*, 1993, **34**, 253-259 (*Chrysolaena verbascifolia lignan*).
Estevez-Braun, A. *et al.*, *J. Nat. Prod.*, 1995, **58**, 887-892 (*Bupleurum lignan*).
Wu, T.-S. *et al.*, *J. Nat. Prod.*, 2002, **65**, 1522-1525 (*Aristelegins*).
Pu, J.-X. *et al.*, *Helv. Chim. Acta*, 2007, **90**, 723-729 (*Kadsura longipedunculata lignan*).
Yang, X-W. *et al.*, *J. Nat. Prod.*, 2007, **70**, 521-525 (*Tarenna attenuata constits*).
Liang, S. *et al.*, *J. Nat. Prod.*, 2008, **71**, 1902-1905 (*Isocubebin*).
Hu, Q-F. *et al.*, *Chem. Pharm. Bull.*, 2011, **59**, 1421-1424 (*Isocubebin, activity*).
Wang, J. *et al.*, *Nat. Prod. Bioprospect.*, 2019, **9**, 419-423 (*Aconitum pendulum constit*)

9,9′-Epoxy-8,8′-lignan-3,3′,4,4′,8-pentol E130

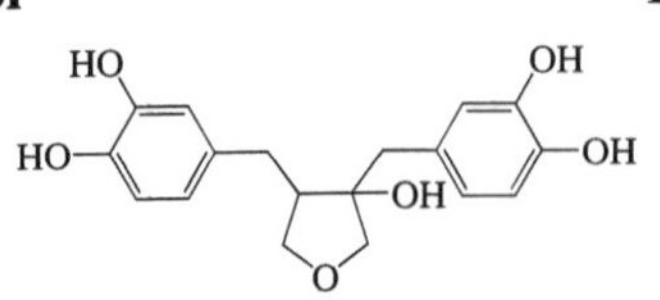

3,4-Bis(3,4-dihydroxybenzyl)-3-furanol. 3,3′,4,4′,8-Pentahydroxy-9,9′-epoxylignan
$C_{18}H_{20}O_{6}$ 332.352

(8ξ,8′ξ)-form

3,3′-Di-Me ether
9,9′-Epoxy-3,3′-dimethoxy-8,8′-lignan-4,4′,8-triol. 4,4′,8-Trihydroxy-3,3′-dimethoxy-9,9′-epoxylignan. ***Syripinnalignan A***
[1402917-80-9]
$C_{20}H_{24}O_{6}$ 360.406
Constit. of the stems of *Syringa pinnatifolia*. Pale yellow gum. $[\alpha]^{25}_{D}$ +12.7 (c, 0.1 in MeOH). λ_{max} 215 (log ε 4.87); 265 (log ε 4.11); 316 (log ε 3.1) (MeOH)

3,4:3′,4′-Bis(methylene) ether
9,9′-Epoxy-3,4:3′,4′-bis(methylenedioxy)-8,8′-lignan-8-ol. 8-Hydroxy-3,4:3′,4′-bis(methylene)-9,9′-epoxylignan. ***Syripinnalignan B***
[1402917-82-1]
$C_{20}H_{20}O_{6}$ 356.374
Constit. of the stems of *Syringa pinnatifolia*. Pale yellow gum. $[\alpha]^{25}_{D}$ +11.5 (c, 0.1 in MeOH). λ_{max} 220 (log ε 4.95) (MeOH).

Ao, W.-L.-J. *et al.*, *J. Asian Nat. Prod. Res.*, 2012, **14**, 396-400 (*Syripinnalignans A,B*)

9,9′-Epoxy-8,8′-lignan-3,3′,4,4′,9-pentol E131

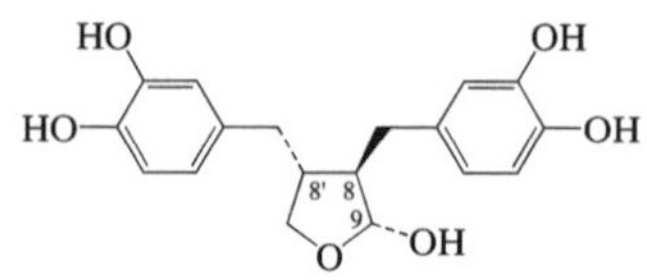

(8R,8′R,9R)-form

3,3′,4,4′,9-Pentahydroxy-9,9′-epoxylignan
$C_{18}H_{20}O_{6}$ 332.352

(8R,8′R,9R)-form

3,3′-Di-Me ether
9,9′-Epoxy-3,3′-dimethoxy-8,8′-lignan-4,4′,9-triol. 4,4′,9-Trihydroxy-3,3′-dimethoxy-9,9′-epoxylignan
[158042-33-2]
$C_{20}H_{24}O_{6}$ 360.406
Constit. of *Abies pinsapo*, *Daphne feddei* and *Arnica chamissonis*. Inhibitor of LPS-induced NO prodn.. Amorph. powder. $[\alpha]^{25}_{D}$ -40.9 (c, 1 in $CHCl_3$). Obt. as a mixt. with C-9 epimer. Data given is for this mixt.

3,3′-Di-Me ether, 9-O-β-D-glucopyranoside
Haplomarin
$C_{26}H_{34}O_{11}$ 522.547
Constit. of *Arnica chamissonis* and *Haplophyllum suaveolens*. Amorph. powder. λ_{max} 208; 230; 280 (MeOH)

3,9-Di-Me ether
9,9′-Epoxy-3,9-dimethoxy-8,8′-lignan-3′,4,4′-triol. 3′,4,4′-Trihydroxy-3,9-dimethoxy-9,9′-epoxylignan
[1448614-93-4]
$C_{20}H_{24}O_{6}$ 360.406
Constit. of *Viburnum foetidum*. Gum. $[\alpha]^{23}_{D}$ -30.1 (c, 0.25 in MeOH). λ_{max} 204 (log ε 3.1); 224 (log ε 3.6); 281 (log ε 4.3) (MeOH)

3,3′,9-Tri-Me ether
9,9′-Epoxy-3,3′,9-trimethoxy-8,8′-lignan-4,4′-diol. 4,4′-Dihydroxy-3,3′,9-trimethoxy-9,9′-epoxylignan
$C_{21}H_{26}O_{6}$ 374.433
Constit. of *Arnica chamissonis* and *Viburnum foetidum* var. *foetidum*. $[\alpha]_{D}$ -96.1 (c, 1.8 in MeOH). λ_{max} 225; 285 (MeOH)

3,4′,9-Tri-Me ether
9,9′-Epoxy-3,4′,9-trimethoxy-8,8′-lignan-3′,4-diol. 3′,4-Dihydroxy-3,4′,9-trimethoxy-9,9′-epoxylignan
[1448614-94-5]
$C_{21}H_{26}O_{6}$ 374.433
Constit. of *Viburnum foetidum*. Gum. $[\alpha]^{23}_{D}$ -21.1 (c, 0.11 in MeOH). λ_{max} 205 (log ε 3.3); 223 (log ε 3.6); 283 (log ε 4.1) (MeOH)

9-Butyl, 3,3′-di-Me ether
9-Butoxy-9,9′-epoxy-3,3′-dimethoxy-8,8′-lignan-4,4′-diol. 9-Butoxy-4,4′-dihydroxy-3,3′-dimethoxy-9,9′-epoxylignan
$C_{24}H_{32}O_{6}$ 416.514
Constit. of *Arnica chamissonis*. $[\alpha]^{20}_{D}$ -33.3 (c, 0.3 in MeOH). λ_{max} 225; 285 (MeOH)

3,4-Methylene, 3′,4′-di-Me ether
9,9′-Epoxy-3′,4′-dimethoxy-3,4-methylenedioxy-8,8′-lignan-9-ol. 9-Hydroxy-3′,4′-dimethoxy-3,4-methylenedioxy-9,9′-epoxylignan. 3′,4′-Dimethoxy-3′,4′-desmethylenecubebin
[67533-86-2]
$C_{21}H_{24}O_{6}$ 372.417
Constit. of *Aristolochia triangularis* and *Piper nigrum*. Cryst. (C_6H_6/hexane). Mp 66°. $[\alpha]^{25}_{D}$ -15.9 (c, 0.2 in $CHCl_3$)

3′,4′-Methylene, 3,4-di-Me ether
9,9′-Epoxy-3,4-dimethoxy-3′,4′-methylenedioxy-8,8′-lignan-9-ol. 9-Hydroxy-3,4-dimethoxy-3′,4′-methylenedioxy-9,9′-epoxylignan.

3,4-Dimethoxy-3,4-desmethylenecubebin
[77533-71-2]
[210161-38-9]
$C_{21}H_{24}O_6$ 372.417
Constit. of *Aristolochia triangularis*, *Piper chaba*, *Piper nigrum* and *Virola michellii*. Needles (C_6H_6/hexane). Mp 89 - 91°. $[\alpha]_D^{25}$ -52.9 (c, 0.3 in $CHCl_3$)

Bis(methylene) ether
see 9,9′-Epoxy-3,4:3′,4′-bis(methylenedioxy)-8,8′-lignan-9-ol, E8

(8R,8′R,9S)-form

3,3′-Di-Me ether
[158042-32-1]
Constit. of *Abies pinsapo* and *Arnica chamissonis*. Obt. as a mixt. with C-9 epimer

3,3′,9-Tri-Me ether
Constit. of *Arnica chamissonis*. $[\alpha]_D^{20}$ -24.5 (c, 2.0 in MeOH). λ_{max} 225; 280 (MeOH)

3′,4′-Methylene, 3,4-di-Me ether
Constit. of *Piper chaba*

(8R,8′R,9ξ)-form

3,3′,4,4′-Tetra-Me ether
9,9′-Epoxy-3,3′,4,4′-tetramethoxy-8,8′-lignan-9-ol. 9-Hydroxy-3,3′,4,4′-tetramethoxy-9,9′-epoxylignan. **Kusunokinol**
[58311-19-6]
$C_{22}H_{28}O_6$ 388.46
Constit. of *Cinnamomum camphora* and *Virola mitchellii*. Needles (CH_2Cl_2/petrol or Et_2O). Mp 147 - 148° (140-141.5°). $[\alpha]_D^{25}$ -87 (c, 1.4 in $CHCl_3$). $[\alpha]_D^{27}$ -23.5 (c, 0.51 in EtOH). A mixt. of C-9 epimers

3,4-Methylene, 3′-Me ether
9,9′-Epoxy-3′-methoxy-3,4-methylenedioxy-8,8′-lignan-4′,9-diol. 4′,9-Dihydroxy-3′-methoxy-3,4-methylenedioxy-9,9′-epoxylignan. ***Haplomyrfolol***
[177857-23-7]
$C_{20}H_{22}O_6$ 358.39
Constit. of *Haplophyllum vulcanicum*. Amorph. solid. $[\alpha]_D$ -48 (c, 0.1 in MeOH). Exists as a mixt. of C-9 epimers. λ_{max} 206 (log ε 4.59); 230 (log ε 4.1); 284 (log ε 3.91) (MeOH)

(8R*,8′R*,9S*)-form

3,3′,9-Tri-Me ether
[1206464-65-4]
$C_{21}H_{26}O_6$ 374.433
Constit. of *Viburnum foetidum* var. *foetidum*. Gum. $[\alpha]_D^{23}$ +76.5 (c, 0.19 in MeOH). λ_{max} 203 (log ε 3); 221 (log ε 3.7); 282 (log ε 4.3) (MeOH)

9-Et, 3,3′-di-Me ether
9,9′-Epoxy-9-ethoxy-3,3′-dimethoxy-8,8′-lignan-4,4′-diol. 9-Ethoxy-4,4′-dihydroxy-3,3′-dimethoxy-9,9′-epoxylignan
[1073173-22-4]
$C_{22}H_{28}O_6$ 388.46
Constit. of the stem bark of *Daphne feddei*. Pale yellow oil. $[\alpha]_D^{20}$ +32 (c, 0.15 in $CHCl_3$). Possible artifact

9-Butyl, 3,3′-di-Me ether
[1073173-21-3]
Constit. of the stem bark of *Daphne feddei*. Pale yellow oil. $[\alpha]_D^{20}$ +50 (c, 0.2 in $CHCl_3$). Possible artifact

(8R*,8′S*,9R*)-form

3′,4′-Methylene, 9-Et, 3,4-di-Me ether
9-Ethoxy-3,4-dimethoxy-3′,4′-methylenedioxy-9,9′-epoxylignan
[235426-83-2]
$C_{23}H_{28}O_6$ 400.471
Constit. of *Aristolochia peltato-deltoidea*

(8R*,8′S*,9S*)-form

3′,4′-Methylene, 9-Et, 3,4-di-Me ether
[235426-82-1]
Constit. of *Aristolochia peltato-deltoidea*.

Inagaki, I. *et al.*, *Chem. Pharm. Bull.*, 1972, **20**, 2710-2718 (*Cinnamomum camphora constit*).
Takaoka, D. *et al.*, *Nippon Kagaku Kaishi*, 1975, 2192-2196 (*Kusunokinol*).
Ruecker, G. *et al.*, *Planta Med.*, 1981, **41**, 143-149 (*Dimethoxydesmethylenecubebins*).
Sumathykutty, M.A. *et al.*, *Indian J. Chem., Sect. B*, 1988, **27**, 388-389 (*Dimethoxydesmethylenecubebins*).
Brown, E. *et al.*, *Tetrahedron*, 1989, **45**, 141-154 (*Kusunokinol, synth, pmr*).
Barrero, A.F. *et al.*, *J. Nat. Prod.*, 1994, **57**, 713-719 (*Abies pinsapo constits*).
Vidigal, M.C.S. *et al.*, *Phytochemistry*, 1995, **40**, 1259-1261 (*Virola mitchellii constits*).
Gözler, B. *et al.*, *Phytochemistry*, 1996, **42**, 695-699 (*Haplomyrfolol*).
Bhandari, S.P.S. *et al.*, *Phytochemistry*, 1998, **47**, 1435-1436 (*Piper chaba constit*).
Da Silva, A.P.F. *et al.*, *J. Braz. Chem. Soc.*, 1999, **10**, 122-126 (*Aristolochia peltato-deltoidea constits*).
Ivanova, A. *et al.*, *Z. Naturforsch., C*, 2001, **56**, 329-333 (*Haplomarin*).
Schmidt, T.J. *et al.*, *Nat. Prod. Res.*, 2006, **20**, 443-453 (*Arnica chamissonis constits*).
Liang, S. *et al.*, *J. Nat. Prod.*, 2008, **71**, 1902-1905 (*Daphne feddei ethers, activity*).
Chen, X.-Q. *et al.*, *Chem. Pharm. Bull.*, 2009, **57**, 1129-1131 (*Viburnum foetidum constits*).
Li, H. *et al.*, *Arch. Pharmacal Res.*, 2013, **36**, 1211-1214 (*Viburnum foetidum constits*).
Shang, S.-Z. *et al.*, *Fitoterapia*, 2013, **84**, 58-63 (*Maninsigin A*)

9,9′-Epoxy-8,8′-lignan-3,3′,4,5′,7′-pentol E132

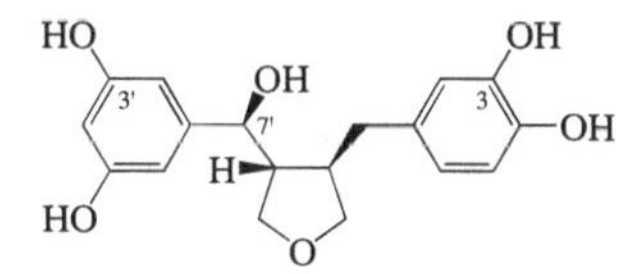

(7′S,8R,8′R)-form

3,3′,4,5′,7′-Pentahydroxy-9,9′-epoxylignan
$C_{18}H_{20}O_6$ 332.352
Structures in this series may need revision. Phenyl 3,5-dioxygenated substn. pattern in doubt as a result of work done by Yang *et al* in 2015

(7′S,8R,8′R)-form

3′,5′-Di-Me ether
9,9′-Epoxy-3′,5′-dimethoxy-8,8′-lignan-3,4,7′-triol. 3,4,7′-Trihydroxy-3′,5′-dimethoxy-9,9′-epoxylignan. ***Acanthosessilin A***
[1467037-30-4]
$C_{20}H_{24}O_6$ 360.406
Constit. of the fruit of *Acanthopanax sessiliflorus*. Cryst. (MeOH). Mp 123 - 125°. $[\alpha]_D^{25}$ -43.5 (c, 0.5 in MeOH). λ_{max} 280 (MeOH)

(7′S,8R*,8′S*)-form

3,3′-Di-Me ether
9,9′-Epoxy-3,5′-dimethoxy-8,8′-lignan-3′,4,7′-triol. 3′,4,7′-Trihydroxy-3,5′-dimethoxy-9,9′-epoxylignan
[929022-92-4]
$C_{20}H_{24}O_6$ 360.406
Constit. of the roots of *Rubia cordifolia* and stems of *Stewartia koreana*. Yellowish cryst.. $[\alpha]_D$ +12 (c, 0.1 in MeOH). Stereochem. not certain. λ_{max} 282 (MeOH).

Son, J.K. *et al.*, *Bull. Korean Chem. Soc.*, 2006, **27**, 1231-1234 (*Rubia cordifolia constit*).
Lee, D.Y. *et al.*, *Molecules*, 2013, **18**, 41-49 (*Acanthosessilin A*).
Kim, M.H. *et al.*, *Nat. Prod. Res.*, 2014, **28**, 1409-1412 (*Stewartia koreana constit*).
Yang, Y.-N. *et al.*, *J. Nat. Prod.*, 2015, **78**, 705-711 (*pmr, struct*)

4,7′-Epoxy-3,8′-lignan-3′,4′,5,8-tetrol E133

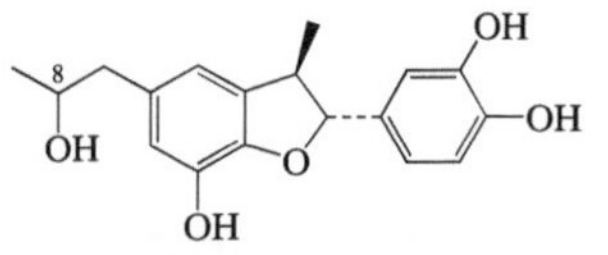

(7′R,8′R)-form

2-(3,4-Dihydroxyphenyl)-2,3-dihydro-7-hydroxy-5-(2-hydroxypropyl)-3-methylbenzofuran. 4,7′-Epoxy-3′,4′,5,8-tetrahydroxy-3,8′-neolignan

$C_{18}H_{20}O_5$ 316.353

(7′R,8′R)-form

8-Ketone

2-(3,4-Dihydroxyphenyl)-2,3-dihydro-7-hydroxy-3-methyl-5-(2-oxopropyl)benzofuran. 4,7′-Epoxy-3′,4′,5-trihydroxy-3,8′-lignan-8-one. 4,7′-Epoxy-3′,4′,5-trihydroxy-8-oxo-3,8′-lignan

8-Ketone, 3′,5-di-Me ether

4,7′-Epoxy-4′-hydroxy-3′,5-dimethoxy-3,8′-lignan-8-one. 4,7′-Epoxy-4′-hydroxy-3′,5-dimethoxy-8-oxo-3,8′-lignan. ***Machilusol F***

[289040-78-4]

$C_{20}H_{22}O_5$ 342.391

Constit. of the stems of *Machilus obovatifolia*. Oil. $[\alpha]^{25}_{D}$ +34.8 (c, 0.08 in $CHCl_3$). λ_{max} 220 (log ε 4.46); 283 (log ε 3.95) (EtOH)

(7′R*,8′R*)-form

8-Ketone, 3′,4′-methylene, 5-Me ether

4,7′-Epoxy-5-methoxy-3′,4′-methylenedioxy-3,8′-lignan-8-one. 4,7′-Epoxy-5-methoxy-3′,4′-methylenedioxy-8-oxo-3,8′-lignan. ***Saurusine B***

$C_{20}H_{20}O_5$ 340.375

Constit. of *Saururus chinensis*. Gum. λ_{max} 206 (log ε 4.59); 288 (log ε 3.79) (MeOH).

Tsai, I.-L. *et al.*, *Planta Med.*, 2000, **66**, 403-407 (*Machilusol F*).

Peng, B. *et al.*, *Zhongcaoyao*, 2016, **47**, 2221-2225 (*Saurusine B*)

4,7′-Epoxy-3,8′-lignan-3′,4′,5,9-tetrol E134

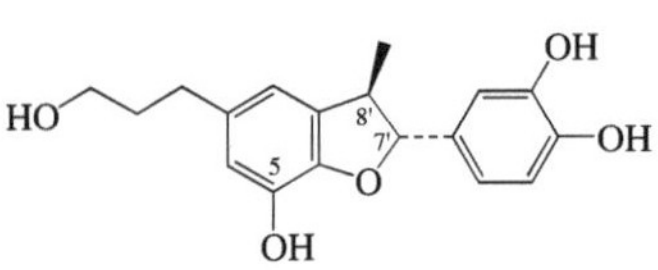

(7′R,8′R)-form

2-(3,4-Dihydroxyphenyl)-2,3-dihydro-7-hydroxy-3-methyl-5-benzofuranpropanol. 4,7′-Epoxy-3′,4′,5,9-tetrahydroxy-3,8′-neolignan

$C_{18}H_{20}O_5$ 316.353

(7′R,8′R)-form

3′,5-Di-Me ether, 4′-O-β-D-glucopyranoside

[1904578-09-1]

$C_{26}H_{34}O_{10}$ 506.548

Constit. of the roots of *Paeonia lactiflora*. Exhibits moderate anti-β-amyloid aggregation activity. Amorph. powder. $[\alpha]^{25}_{D}$ -17.5 (c, 0.1 in MeOH). λ_{max} 208 (log ε 4.31); 280 (log ε 3.32) (MeOH)

(7′S,8′S)-form

3′,5-Di-Me ether, 4′-O-β-D-glucopyranoside

[1904578-11-5]

$C_{26}H_{34}O_{10}$ 506.548

Constit. of the roots of *Paeonia lactiflora*. Exhibits strong anti-β-amyloid aggregation activity. Amorph. powder. $[\alpha]^{25}_{D}$ -11.5 (c, 0.11 in MeOH). λ_{max} 208 (log ε 4.31); 280 (log ε 3.32) (MeOH)

(7′R*,8′R*)-form

3′,5-Di-Me ether

4,7′-Epoxy-3′,5-dimethoxy-3,8′-lignan-4′,9-diol. 4,7′-Epoxy-4′,9-dihydroxy-3′,5-dimethoxy-3,8′-lignan. ***9′-Deoxyvladinol F****. 9′-Dehydroxyvladinol F*

[1383810-01-2]

$C_{20}H_{24}O_5$ 344.407

Constit. of *Arundina graminifolia* and *Daphne feddei*. Stereochem. is uncertain

3′,4′-Methylene, 5-Me ether

4,7′-Epoxy-5-methoxy-3′,4′-methylenedioxy-3,8′-lignan-9-ol. 4,7′-Epoxy-9-hydroxy-5-methoxy-3′,4′-methylenedioxy-3,8′-lignan. ***Phyllnirurin***

[120396-53-4]

$C_{20}H_{22}O_5$ 342.391

Constit. of *Phyllanthus niruri*. Stereochem. is uncertain. λ_{max} 252; 286; 324 (no solvent reported).

Singh, B. *et al.*, *J. Nat. Prod.*, 1989, **52**, 48-51 (*Phyllnirurin*).

Hu, Q.-F. *et al.*, *Chem. Pharm. Bull.*, 2011, **59**, 1421-1424 (*9′-Deoxyvladinol F*).

Liu, X. *et al.*, *Fitoterapia*, 2015, **103**, 136-142 (*Paeonia lactiflora constits, activity*)

4,7′-Epoxy-3,8′-lignan-3′,4′,9,9′-tetrol E135

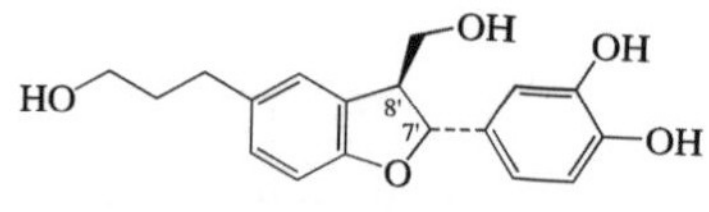

(7′R,8′S)-form

4,7′-Epoxy-3′,4′,9,9′-tetrahydroxy-3,8′-neolignan

$C_{18}H_{20}O_5$ 316.353

(7′R,8′S)-form

3′-Me ether, 4′-O-α-L-rhamopyranoside

[1493598-33-6]

$C_{25}H_{32}O_9$ 476.522

Constit. of the trunk of *Abies holophylla*. Gum. $[\alpha]^{25}_{D}$ +16.0 (c, 0.05 in MeOH). λ_{max} 229 (log ε 4.0); 282 (log ε 1.6) (MeOH)

(7′S,8′R)-form

3′-Me ether

4,7′-Epoxy-3′-methoxy-3,8′-lignan-4′,9,9′-triol. 4,7′-Epoxy-4′,9,9′-trihydroxy-3′-methoxy-3,8′-lignan. ***Cedrusinin***

[83728-85-2]

$C_{19}H_{22}O_5$ 330.38

Constit. of *Abies nephrolepis*, *Cedrus deodara*, *Cryptomeria japonica* and *Pinus sylvestris*. Syrup. $[\alpha]_{D}$ +4.2 (c, 1 in MeOH). λ_{max} 236; 283 (EtOH)

3′-Me ether, 4′-O-α-L-rhamnopyranoside

Cedrusinin 4′-rhamnoside

[1403967-10-1]

$C_{25}H_{32}O_9$ 476.522

Constit. of *Cathaya argyrophylla*. Amorph. solid. $[\alpha]^{23}_{D}$ -50 (c, 0.1 in MeOH). λ_{max} 207 (log ε 3.87); 225 (log ε 4.06); 283 (log ε 3.5) (MeOH)

3′-Me ether, 4′-O-β-D-glucopyranoside

Cedrusinin 4′-glucoside. ***3′-Methoxyisocupressoside B***

[1363328-81-7]

$C_{25}H_{32}O_{10}$ 492.521

Constit. of the twigs of *Keteleeria evelyniana*. Amorph. solid. $[\alpha]^{27}_{D}$ -36.7 (c, 0.3 in MeOH). λ_{max} 205 (log ε 4.63); 226 (log ε 4.08); 283 (log ε 3.6) (MeOH).

Agrawal, P.K. *et al.*, *Phytochemistry*, 1982, **21**, 1459-1461 (*Cedrusinin*).

Su, W.-C. *et al.*, *Phytochemistry*, 1995, **40**, 563-566 (*Cedrusinin tri-Ac*).

Sinkkonen, J. *et al.*, *Magn. Reson. Chem.*, 2006, **44**, 633-636 (*Cedrusinin*).

He, W.J. *et al.*, *Z. Naturforsch., B*, 2011, **66**, 733-739 (*Cedrusinin 4′-glucoside*).
He, W.-J. *et al.*, *Phytochemistry*, 2012, **83**, 63-69 (*Cedrusinin 4′-rhamnoside*).
Kim, C.S. *et al.*, *J. Nat. Prod.*, 2013, **76**, 2131-2135 (*Abies holophylla constit*)

4,7′-Epoxy-3,8′-lignan-4′,5,9,9′-tetrol E136

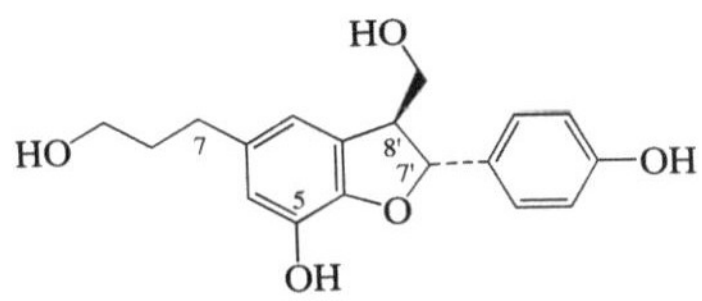

(7′R,8′S)-form

4,7′-Epoxy-4′,5,9,9′-tetrahydroxy-3,8′-neolignan
$C_{18}H_{20}O_5$ 316.353
(7′R,8′S)-form
5-Me ether
4,7′-Epoxy-5-methoxy-3,8′-lignan-4′,9,9′-triol. 4,7′-Epoxy-4′,9,9′-trihydroxy-5-methoxy-3,8′-lignan. ***Chushizisin E***
[1142925-34-5]
$C_{19}H_{22}O_5$ 330.38
Constit. of the fruit of *Broussonetia papyrifera* and from the seeds of *Momordica cochinchinensis*. Exhibits antioxidant activity. Amorph. powder. $[\alpha]^{27}_{D}$ -9.4 (c, 0.1 in MeOH). λ_{max} 206 (log ε 4.61); 225 (log ε 4.32); 282 (log ε 3.63) (MeOH)
7-Oxo, 5-Me ether
(+)-Idaeusin C
$C_{19}H_{20}O_6$ 344.363
Exhibits moderate anti $A\beta_{1-42}$ aggregation activity. Antioxidant. Oil. $[\alpha]^{20}_{D}$ +48.6 (c, 0.1 in MeOH)
(7′S,8′R)-form
7-Oxo, 5-Me ether
(-)-Idaeusin C
$C_{19}H_{20}O_6$ 344.363
Exhibits neuroprotective activity and moderate anti $A\beta_{1-42}$ aggregation activity. Antioxidant. Oil. $[\alpha]^{20}_{D}$ -51.3 (c, 0.1 in MeOH)
(7′R*,8′S*)-form
7-Oxo, 5-Me ether
4,7′-Epoxy-4′,9,9′-trihydroxy-5-methoxy-3,8′-lignan-7-one. 4,7′-Epoxy-4′,9,9′-trihydroxy-5-methoxy-7-oxo-3,8′-lignan. ***Chushizisin F***
[1142925-35-6]
$C_{19}H_{20}O_6$ 344.363
Constit. of the fruit of *Broussonetia papyrifera*. Exhibits antioxidant activity. Amorph. powder. $[\alpha]^{27}_{D}$ -18.9 (c, 0.1 in MeOH). λ_{max} 204 (log ε 4.45); 283 (log ε 3.96) (MeOH)
(7′RS,8′SR)-form
7-Oxo, 5-Me ether
Epoxy-4,9,9′-trihydroxy-3,8′-lignan-7-one. ***Idaeusin C***
$C_{19}H_{20}O_6$ 344.363
Constit. of the fruits of *Rubus idaeus*. Oil. $[\alpha]^{20}_{D}$ -3.4 (c, 0.1 in MeOH). λ_{max} 223 (log ε 3.30); 286 (log ε 1.18) (MeOH).

Mei, R.-Q. *et al.*, *J. Nat. Prod.*, 2009, **72**, 621-625 (*Chushizisins E,F, activity*).
Zhou, L. *et al.*, *Fitoterapia*, 2018, **127**, 56-61 (*Idaeusin C, activity*)

7,7′-Epoxy-8,8′-lignan-2,3′,4′,5-tetrol E137

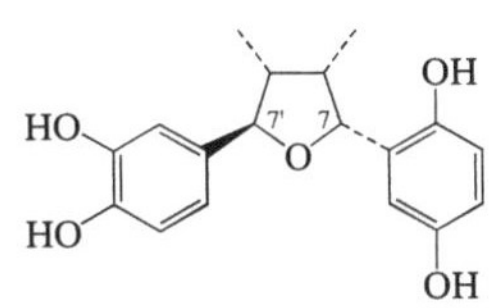

2-(2,5-Dihydroxyphenyl)-5-(3,4-dihydroxyphenyl)tetrahydro-3,4-dimethylfuran. 2,3′,4′,5-Tetrahydroxy-7,7′-epoxylignan
$C_{18}H_{20}O_5$ 316.353
(7R,7′R,8S,8′R)-form
Tetra-Me ether
7,7′-Epoxy-2,3′,4′,5-tetramethoxylignan. 2,3′,4′,5-Tetramethoxy-7,7′-epoxylignan
[165054-91-1]
$C_{22}H_{28}O_5$ 372.461
Constit. of the leaves of *Ocotea foetens*. $[\alpha]^{20}_{D}$ +73.7 (c, 2 in $CHCl_3$).

Lopez, H. *et al.*, *J. Nat. Prod.*, 1995, **58**, 782-785 (*Ocotea foetens constit*)

7,7′-Epoxy-8,8′-lignan-3,3′,4,4′-tetrol E138

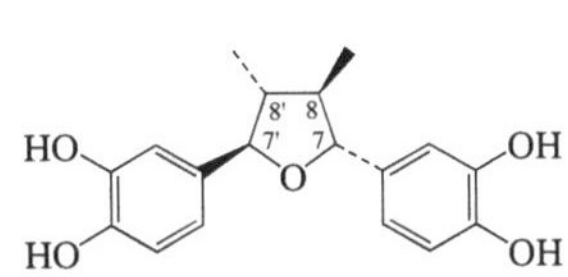

(7R,7′R,8R,8′R)-form

2,5-Bis(3,4-dihydroxyphenyl) tetrahydro-3,4-dimethylfuran. 3,3′,4,4′-Tetrahydroxy-7,7′-epoxylignan
$C_{18}H_{20}O_5$ 316.353
(7R,7′R,8R,8′R)-form
3,4-Methylene ether
7,7′-Epoxy-3′,4′-methylenedioxy-8,8′-lignan-3,4-diol. 3,4-Dihydroxy-3′,4′-methylenedioxy-7,7′-epoxylignan
[1886973-84-7]
$C_{19}H_{20}O_5$ 328.364
Constit. of the roots of *Schisandra sphenanthera*. Antioxidant. Amorph. solid. $[\alpha]^{25}_{D}$ +10 (c, 0.1 in MeOH). λ_{max} 228 (log ε 0.15) (EtOH)
3,4:3′,4′-Bis(methylene) ether
5,5′-(Tetrahydro-3,4-dimethyl-2,5-furandiyl)bis(1,3-benzodioxole). Tetrahydro-2,5-bis(3,4-methylenedioxyphenyl)-3,4-dimethylfuran. 3,4:3′,4′-Bis(methylenedioxy)-7,7′-epoxylignan. ***(+)-Galbacin***
[61891-31-4]
Constit. of *Machilus japonica* leaves. Cryst. (EtOAc). Mp 115.5 - 116°. $[\alpha]_{D}$ +117 ($CHCl_3$)
(7R,7′R,8R,8′S)-form
3,4:3′,4′-Bis(methylene) ether
(-)-Zuonin A*. Zuihonin A*
[79120-58-4]
Constit. of *Aristolochia chilensis*, *Chamaecyparis obtusa* var. *formosana* and *Piper schmidtii*. Mp 119 - 121°. $[\alpha]_{D}$ -139.4 (c, 1 in $CHCl_3$)
(7R,7′S,8R,8′S)-form
meso-Stereoisomer
3,3′,4-Tri-Me ether
7,7′-Epoxy-3,3′,4′-trimethoxy-8,8′-lignan-4-ol. 4-Hydroxy-3,3′,4′-trimethoxy-7,7′-epoxylignan. ***Nectandrin A***
[74683-15-1]
$C_{21}H_{26}O_5$ 358.434
Isol. from bark of *Machilus thunbergii* and from *Aristolochia arcuata*, *Nectandra rigida* and *Myristica fragrans*. Shows weak antiprotozoal activity. Oil. $[\alpha]^{23}_{D}$ -8.9 (c, 0.40 in $CHCl_3$). $[\alpha]^{24}_{D}$ +10 (c, 1.4 in MeOH). Chiral deriv. of *meso*- parent. The original sample of Nectandrin A from *N. rigida* was dextrorotatory (MeOH), the sample from *M. thunbergii* was laevorotatory ($CHCl_3$). They appear to be enantiomeric though it could be a solv. effect

3,3′,4′-Tri-Me ether

Saurucinol K

[1521111-82-9]

$C_{21}H_{26}O_5$ 358.434

Constit. of *Saururus chinensis*. Amorph. solid. $[\alpha]^{20}_{D}$ +63 (c, 0.01 in $CHCl_3$). Chiral tri-Me ether of a *meso*-stereoisomer. λ_{max} 210 (log ε 4.31); 230 (log ε 4.18); 277 (log ε 3.71) (MeOH)

3,4:3′,4′-Bis(methylene) ether

Zuihonin B. ***Zuonin B***

[79120-59-5]

Constit. of *Aristolochia chilensis* and *Myristica dactyloides*. Oil. Opt. inactive (*meso-*)

(7S,7′R,8S,8′S)-form

3,3′,4-Tri-Me ether

Aristolignin

[110268-34-3]

$C_{21}H_{26}O_5$ 358.434

Constit. of *Aristolochia arcuata*, *Aristolochia chilensis* and leaves of *Nectandra megapotamica*. Shows moderate antiprotozoal activity against *Trypanosoma cruzi*. Neuroprotectant. Oil. $[\alpha]^{21}_{D}$ +27.5 (c, 0.31 in $CHCl_3$) (+24.3)

(7S,7′S,8S,8′R)-form

3,4:3′,4′-Bis(methylene) ether

(+)-Zuonin A

[84709-25-1]

Constit. of *Saururus cernuus* and *Schisandra nigra*. Oil. Stated to be the (+)-isomer of Zuonin A but reported with a negative opt. rotn. (1990)

(7S,7′S,8S,8′S)-form [1446317-92-5]

Constit. of *Larrea tridentata*. Resin. Abs. config. revised in 2012. λ_{max} 200 (log ε 4.71); 227 (sh); 282 (log ε 3.63) (MeOH)

4-Me ether

7,7′-Epoxy-4′-methoxy-8,8′-lignan-3,3′,4-triol. 3,3′,4-Trihydroxy-4′-methoxy-7,7′-epoxylignan

[793674-57-4]

$C_{19}H_{22}O_5$ 330.38

Constit. of the leaves of *Larrea tridentata*. Red-brown powder. $[\alpha]^{25}_{D}$ -60 (c, 0.001 in MeOH). λ_{max} 208; 230; 282 (MeOH)

3,3′-Di-Me ether

see 7,7′-Epoxy-3,3′-dimethoxy-8,8′-lignan-4,4′-diol, E25

3,4-Methylene, 3′-Me ether

see 7,7′-Epoxy-3-methoxy-3′,4′-methylenedioxy-8,8′-lignan-4-ol, E184

3,4:3′,4′-Bis(methylene) ether

(-)-Galbacin

[528-64-3]

Isol. from *Himantandra baccata*, *Saururus chinensis* and *Machilus thunbergii*. Moderate inhibitor of topoisomerase I activity. Bipyramids (EtOH). Mp 116°. $[\alpha]^{20}_{D}$ -114 (c, 0.8 in $CHCl_3$)

(7R*,7′R*,8R*,8′S*)-form

3,3′,4-Tri-Me ether

Nectandrin E

[656253-35-9]

$C_{21}H_{26}O_5$ 358.434

Constit. of the leaves of *Nectandra megapotamica*. Oil. $[\alpha]^{26}_{D}$ +1.1 (c, 0.01 in $CHCl_3$)

(7R*,7′R*,8S*,8′R*)-form

3,3′,4-Tri-Me ether

Nectandrin D

[656253-36-0]

$C_{21}H_{26}O_5$ 358.434

Constit. of the leaves of *Nectandra megapotamica*. Oil. $[\alpha]^{26}_{D}$ +1.9 (c, 0.008 in $CHCl_3$)

(7R*,7′S*,8R*,8′R*)-form

3,4:3′,4′-Bis(methylene) ether

[178740-32-4]

Constit. of *Myristica dactyloides*. Oil. $[\alpha]_{D}$ -20 (c, 0.2 in CH_2Cl_2). Only relative config. known

(7R*,7′S*,8S*,8′S*)-form

Futokadsurin A [852459-89-3]

Constit of *Piper futokadsura*. Oil. $[\alpha]_{D}$ +12.3 (c, 1.1 in $CHCl_3$). λ_{max} 233 (log ε 5.19); 280 (log ε 4.76) (EtOH).

Hughes, G.K. *et al., Aust. J. Chem.*, 1954, **7**, 104-112 (*(-)-Galbacin*).
Pelter, A. *et al., J. Heterocycl. Chem.*, 1966, **3**, 191-197 (*ms*).
Tanaoka, D. *et al., Bull. Chem. Soc. Jpn.*, 1976, **49**, 3564-3566 (*(+)-Galbacin*).
Fonseca, S.F. *et al., Can. J. Chem.*, 1979, **57**, 441-443 (*cmr*).
Le Quesne, P.W. *et al., J. Nat. Prod.*, 1980, **43**, 353-359 (*Nectandrin A*).
Urzúa, A. *et al., Phytochemistry*, 1987, **26**, 1509-1511 (*(-)-Zuonin A, Aristolignin*).
Shimomura, H. *et al., Phytochemistry*, 1988, **27**, 634-636 (*Nectandrin A*).
Rao, K.V. *et al., J. Nat. Prod.*, 1990, **53**, 212-215 (*(+)-Zuonin A*).
Fernandes, A. *et al., Phytochemistry*, 1993, **32**, 1567-1572 (*Aristolignin, cd*).
Herath, H.M.T.B. *et al., Phytochemistry*, 1996, **42**, 1439-1442 (*Myristica dactyloides constits*).
Kuo, Y.-H. *et al., Chem. Pharm. Bull.*, 2002, **50**, 978-980 (*Chamaecyparis obtusa constit*).
Li, G. *et al., Biol. Pharm. Bull.*, 2004, **27**, 1147-1150 (*Machilus thunbergii constits, activity*).
Da Silva Filho, A.A. *et al., J. Nat. Prod.*, 2004, **67**, 42-45 (*Nectandrins A,D,E, Aristolignin, activity*).
Abou-Gazar, H. *et al., Phytochemistry*, 2004, **65**, 2499-2505 (*Larrea tridentata constits*).
Zhai, H. *et al., Biol. Pharm. Bull.*, 2005, **28**, 289-293 (*Aristolignin, activity*).
Konishi, T. *et al., Chem. Pharm. Bull.*, 2005, **53**, 121-124 (*Futokadsurin A*).
Nguyen, P.H. *et al., Bioorg. Med. Chem. Lett.*, 2010, **20**, 4128-4131 (*Myristica fragrans constit*).
Schmidt, T.J. *et al., Phytochem. Lett.*, 2012, **5**, 632-638 (*Larrea tridentata constit, abs config*).
Hazra, S. *et al., RSC Adv.*, 2013, **3**, 22834-22836 (*(+)-Galbacin, synth*).
Cui, H. *et al., J. Nat. Prod.*, 2014, **77**, 100-110 (*Saurucinol K*).
Tsai, W-J. *et al., J. Nat. Prod.*, 2014, **77**, 125-131 (*Saururus chinensis constit*).
Jiang, K. *et al., Fitoterapia*, 2015, **103**, 63-70 (*Schisandra sphenanthera constit*).
Narukawa, Y. *et al., J. Nat. Med.*, 2016, **70**, 460-466 (*(+)-Zuonin, activity*).
Henrion, S. *et al., Org. Biomol. Chem.*, 2018, **16**, 1672-1678 (*(-)-Galbacin, synth*)

7,9′-Epoxy-8,8′-lignan-3,3′,4,4′-tetrol E139

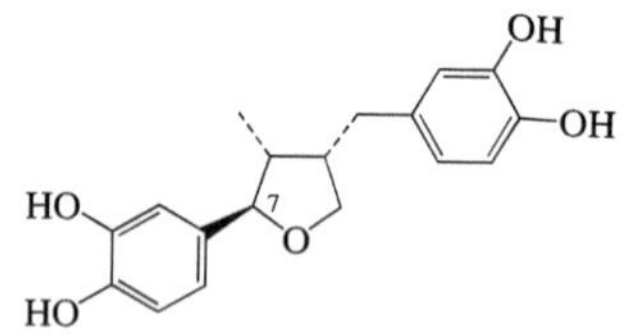

3,3′,4,4′-Tetrahydroxy-7,9′-epoxylignan

$C_{18}H_{20}O_5$ 316.353

(7R*,8R*,8′S*)-form

3,3′-Di-Me ether

7,9′-Epoxy-3,3′-dimethoxy-8,8′-lignan-4,4′-diol. 4,4′-Dihydroxy-3,3′-dimethoxy-7,9′-epoxylignan.

Tobphenol A

[1268614-99-8]

$C_{20}H_{24}O_5$ 344.407

Constit. of the stems of *Nicotiana tabacum*. Exhibits anti-HIV activity. Amorph. solid. $[\alpha]^{24}_{D}$ +3.6 (c, 0.26 in MeOH). λ_{max} 206 (log ε 5.36); 280 (log ε 4.18) (MeOH).

Kong, W.-S. *et al., Asian J. Chem.*, 2011, **23**, 5308-5311 (*Tobphenol A, activity*)

7,9′-Epoxy-8,8′-lignan-4,4′,7′,9-tetrol E140

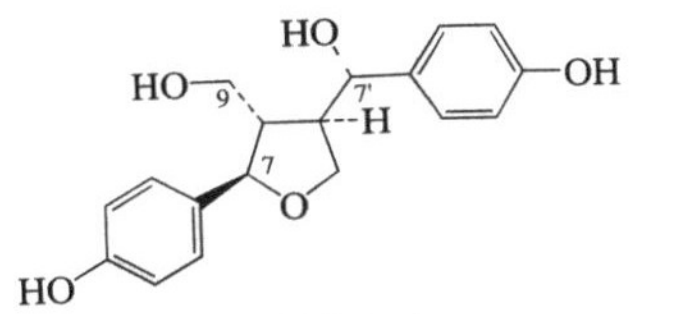

(7R,7′S,8S,8′R)-form

4,4′,7′,9-Tetrahydroxy-7,9′-epoxylignan

$C_{18}H_{20}O_5$ 316.353

(7R,7′S,8S,8′R)-form

4′-Me ether

7,9′-Epoxy-4′-methoxy-8,8′-lignan-4,7′,9-triol. 4,7′,9-Trihydroxy-4′-methoxy-7,9′-epoxylignan

4′-Me ether, 7′,9-di-Ac

Acerifuranoid A

[1256450-30-2]

$C_{23}H_{26}O_7$ 414.454

Constit. of the roots of *Aceriphyllum rossii*. Viscous yellow oil. $[\alpha]_D^{22}$ -6.3 (c, 0.5 in $CHCl_3$). λ_{max} 239; 275 ($CHCl_3$)

(7R*,7′ξ,8S*,8′R*)-form

7′-Me ether

***Chushizisin G**. 7,9′-Epoxy-7′-methoxy-8,8′-lignan-4,4′,9-triol. 4,4′,9-Trihydroxy-7′-methoxy-7,9′-epoxylignan*

[1142925-36-7]

$C_{19}H_{22}O_5$ 330.38

Constit. of the fruit of *Broussonetia papyrifera*. Shows moderate antioxidant activity against peroxide-induced impairment in PC12 cells. Oil. $[\alpha]_D^{27}$ +17.2 (c, 0.1 in MeOH). λ_{max} 202 (log ε 4.18); 226 (log ε 4.19); 277 (log ε 3.48) (MeOH)

(7R,8S,8′S)-form

7′-Ketone

7,9′-Epoxy-4,4′,9′-trihydroxy-8,8′-lignan-7′-one. 4,4′,9-Trihydroxy-7′-oxo-7,9′-epoxylignan. ***Ligballinone***

[1174182-09-2]

Constit. of *Gynostemma pentaphyllum*. Amorph. yellow powder.

Wang, X.W. *et al.*, *Chin. Chem. Lett.*, 2009, **20**, 589-591 (*Ligballinone*).

Mei, R.-Q. *et al.*, *J. Nat. Prod.*, 2009, **72**, 621-625 (*Chushizisin G*).

Van, L.T.K. *et al.*, *Helv. Chim. Acta*, 2010, **93**, 1803-1807 (*Acerifuranoid A*)

9,9′-Epoxy-8,8′-lignan-3,3′,4,4′-tetrol E141

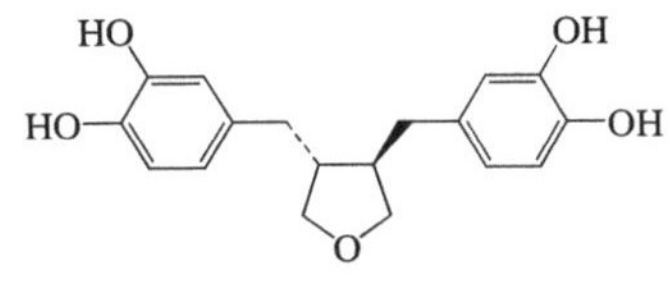

(8R,8′R)-form

3,4-Bis[(3,4-dihydroxyphenyl)methyl]tetrahydrofuran. 3,4-Bis(3,4-dihydroxybenzyl)tetrahydrofuran. 3,3′,4,4′-Tetrahydroxy-9,9′-epoxylignan

$C_{18}H_{20}O_5$ 316.353

(8R,8′R)-form

Tetra-Me ether

3,4-Bis(3,4-dimethoxybenzyl)tetrahydrofuran. 9,9′-Epoxy-3,3′,4,4′-tetramethoxylignan. 3,3′,4,4′-Tetramethoxy-9,9′-epoxylignan. ***Brassilignan***

[62624-76-4]

$C_{22}H_{28}O_5$ 372.461

Constit. of *Flindersia brassi*. Needles (EtOAc/C_6H_6 or MeOH). Mp 120°. $[\alpha]_D^{23}$ -45.5 (c, 0.5 in $CHCl_3$)

(8S,8′S)-form

Tetra-Me ether

[10352-26-8]

Needles (MeOH). Mp 117° (114-116°). $[\alpha]_D^{30}$ +53 ($CHCl_3$)

(8RS,8′RS)-form

Tetra-Me ether

Mp 90 - 90.5°

(8R*,8′R*)-form

3,3′-Di-Me ether

4,4′-[(Tetrahydro-3,4-furandiyl)bis(methylene)]bis[2-methoxyphenol]. 9,9′-Epoxy-3,3′-dimethoxy-8,8′-lignan-4,4′-diol. 4,4′-Dihydroxy-3,3′-dimethoxy-9,9′-epoxylignan. Divanillyltetrahydrofuran. ***Shonanin**. Anhydrosecoisolariciresinol*

[15429-99-9] [29388-32-7] ((±)-form) [29388-33-8] ((-)-form) [34730-78-4] (unspecified stereochem.)

$C_{20}H_{24}O_5$ 344.407

Isol. from *Abies sibirica*, *Calocedrus formosana*, *Larix dahurica*, *Picea abies*, *Picea excelsa*, *Picea koraiensis* and *Picea obovata*. Weak insecticide. Prisms; cryst. (EtOAc/hexane). Mp 136 - 137° (116-117°). $[\alpha]_D^{25}$ -52.2 (c, 1.4 in THF). Shonanin, descr. more recently as a new compd., appears to be identical with previous isolates

3,3′-Di-Me ether, 4-O-(4-hydroxy-3-methoxy-E-cinnamoyl)

Divanillyltetrahydrofuran ferulate

[96917-11-2]

$C_{30}H_{32}O_8$ 520.578

Constit. of *Diospyros kaki*. Needles (EtOAc). Mp 184 - 185°. $[\alpha]_D^{25}$ -58.3 (c, 0.1 in THF). λ_{max} 233 (log ε 4.3); 290 (log ε 4.2); 327 (log ε 4.3) (MeOH)

(8RS,8′SR)-form

Tetra-Me ether

Mp 120 - 120.4°.

Freudenberg, K. *et al.*, *Chem. Ber.*, 1957, **90**, 2857-2869 (*Picea constit*).

Row, L.R. *et al.*, *Tetrahedron*, 1967, **23**, 1915-1918 (*8S,8′S-tetra-Me ether, synth*).

Lapteva, K.I. *et al.*, *Chem. Nat. Compd. (Engl. Transl.)*, 1971, **7**, 802-803 (*Larix dahurica constit*).

Modonova, L.D. *et al.*, *Chem. Nat. Compd. (Engl. Transl.)*, 1972, **8**, 170-174 (*Picea obovata constit*).

Leont'eva, V.G. *et al.*, *Chem. Nat. Compd. (Engl. Transl.)*, 1974, **10**, 399-400 (*Picea koraiensis constit*).

Luedemann, H.D. *et al.*, *Makromol. Chem.*, 1974, **175**, 2393-2407 (*cmr*).

Nimgirawath, S. *et al.*, *Aust. J. Chem.*, 1977, **30**, 451-453 (*Brassilignan*).

Matsuura, S. *et al.*, *Phytochemistry*, 1985, **24**, 626-628 (*Divanillyl tetrahydroferulate*).

Fang, J.-M. *et al.*, *Phytochemistry*, 1989, **28**, 3553-3555 (*Calocedrus formosana constit, cryst struct*).

Brown, E. *et al.*, *Tetrahedron*, 1989, **45**, 141-154 (*8R,8′R-tetra-Me ether, synth, ir, pmr*).

Daugan, A. *et al.*, *J. Nat. Prod.*, 1991, **54**, 110-118 (*(-)-Anhydrosecoisolariciresinol, synth, ir, pmr*).

Gaboury, J.A. *et al.*, *J. Org. Chem.*, 1993, **58**, 2173-2180 (*Brassilignan, synth*).

Pan, H. *et al.*, *Phytochemistry*, 1995, **39**, 1423-1428 (*Picea abies constit*).

Xia, Y-M. *et al.*, *J. Chem. Sci. (Berlin, Ger.)*, 2010, **122**, 433-436 (*Divanillyl tetrahydroferulate, synth*).

Wukirsari, T. *et al.*, *J. Agric. Food Chem.*, 2013, **61**, 4318-4325 (*activity*)

4,7′-Epoxy-3,8′-lignan-3′,4′,9-triol E142

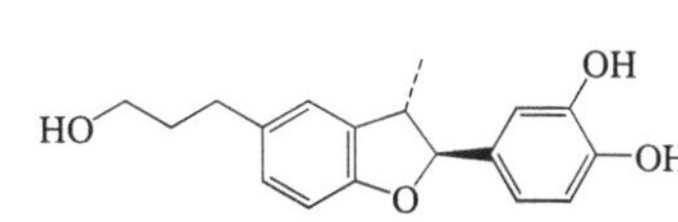

2-(3,4-Dihydroxyphenyl)-2,3-dihydro-3-methyl-5-benzofuranpropanol. 4,7′-Epoxy-3′,4′,9-trihydroxy-3,8′-neolignan

$C_{18}H_{20}O_4$ 300.354

(7′S,8′S)-form

3′-Me ether

4,7′-Epoxy-3′-methoxy-3,8′-lignan-4′,9-diol. 4,7′-Epoxy-4′,9-dihydroxy-3′-methoxy-3,8′-lignan

[109225-42-5]

$C_{19}H_{22}O_4$ 314.381

Constit. of *Krameria cytisoides*. Oil. $[\alpha]_D^{21}$ +37 (c, 0.76 in MeOH). Stereochem. revised in 2000. λ_{max} 231 (log ε 4.21); 285 (log ε 3.84) (MeOH). [base] λ_{max} 236 (log ε 4.11); 252 (log ε 4.11); 291 (log ε 3.93) (MeOH/NaOH).

Achenbach, H. *et al.*, *Phytochemistry*, 1987, **26**, 1159-1166 (*Krameria cytisoides constit*).
Kurtán, T. *et al.*, *J. Chem. Soc., Perkin Trans. 1*, 2000, 453-461 (*abs config*)

4,7′-Epoxy-3,8′-lignan-4′,9,9′-triol E143

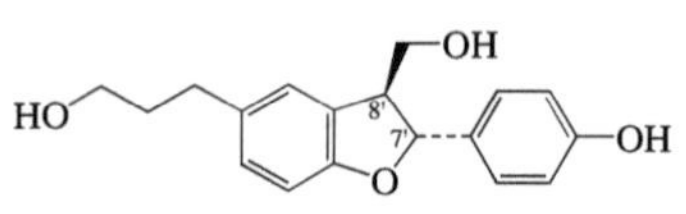

(7′R,8′S)-form

2,3-Dihydro-3-(hydroxymethyl)-2-(4-hydroxyphenyl)-5-benzofuranpropanol. 4,7′-Epoxy-4′,9,9′-trihydroxy-3,8′-neolignan

$C_{18}H_{20}O_4$ 300.354

(7′R,8′S)-form

(+)-Idaeusin D

Exhibits neuroprotective activity. Antioxidant. Oil. $[\alpha]_D^{20}$ +30.9 (c, 0.1 in MeOH)

4′-O-β-D-Glucopyranoside

Cupressoside B

[934371-89-8]

$C_{24}H_{30}O_9$ 462.495

Constit. of *Cupressus duclouxiana*. Amorph. powder. $[\alpha]_D^{23}$ -7.3 (c, 0.41 in MeOH). λ_{max} 202; 223; 228 (no solvent reported)

(7′S,8′R)-form

(-)-Idaeusin D

Antioxidant. Oil. $[\alpha]_D^{20}$ -31.4 (c, 0.12 in MeOH)

4′-O-β-D-Glucopyranoside

Isocupressoside B

[1363328-79-3]

$C_{24}H_{30}O_9$ 462.495

Constit. of the twigs of *Keteleeria evelyniana*. Amorph. solid. λ_{max} 202 (log ε 4.36); 223 (log ε 3.97); 284 (log ε 3.32) (MeOH)

(7′RS,8′SR)-form

Idaeusin D

Constit. of the fruits of *Rubus idaeus*. Oil. $[\alpha]_D^{20}$ -2.6 (c, 0.11 in MeOH). λ_{max} 220 (log ε 4.19); 285 (log ε 1.90) (MeOH).

Xu, J.-F. *et al.*, *J. Asian Nat. Prod. Res.*, 2006, **8**, 181-185 (*Cupressoside B*).
He, W.J. *et al.*, *Z. Naturforsch., B*, 2011, **66**, 733-739 (*Isocupressoside B*).
Zhou, L. *et al.*, *Fitoterapia*, 2018, **127**, 56-61 (*Idaeusin D, activity*)

7,7′-Epoxy-8,8′-lignan-3,4,4′-triol E144

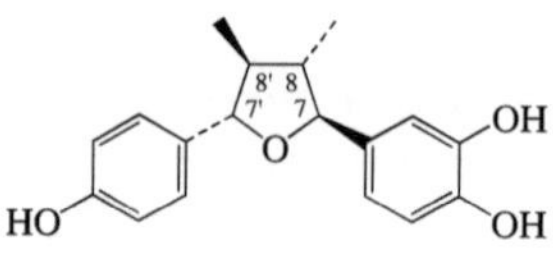

(7S,7′S,8S,8′S)-form

2-(3,4-Dihydroxyphenyl)tetrahydro-5-(4-hydroxyphenyl)-3,4-dimethylfuran. 3,4,4′-Trihydroxy-7,7′-epoxylignan

$C_{18}H_{20}O_4$ 300.354

(7S,7′S,8S,8′S)-form

3-Hydroxy-8′-epilarreatricin

[1446317-90-3]

Constit. of *Larrea tridentata*. Plates (hexane/$CHCl_3$/Me_2CO). Mp 188 - 190°. CD $[\theta]_{207}$ -32100. λ_{max} 225 (log ε 4.36); 281 (log ε 3.72) (MeOH). [base] λ_{max} 225 (log ε 4.62); 246 (log ε 4.39); 285 (log ε 4.04) (MeOH/NaOH)

(7R*,7′R*,8R*,8′S*)-form

3-Me ether

7,7′-Epoxy-3-methoxy-8,8′-lignan-4,4′-diol. 4,4′-Dihydroxy-3-methoxy-7,7′-epoxylignan

[109194-75-4]

$C_{19}H_{22}O_4$ 314.381

Isol. from *Krameria cystisoides*. Cryst. (petrol). $[\alpha]_D^{21}$ +56 (c, 0.4 in MeOH)

(7R*,7′S*,8R*,8′S*)-form

3-Me ether

[109280-45-7]

Isol. from *Krameria cystisoides*. Oil. $[\alpha]_D$ +3.7 (c, 0.54 in MeOH).

Achenbach, H. *et al.*, *Phytochemistry*, 1987, **26**, 1159-1166 (*Krameria cystisoides constits*).
Schmidt, T.J. *et al.*, *Phytochem. Lett.*, 2012, **5**, 632-638 (*3-Hydroxy-8′-epilarreatricin, abs config*)

7,9′-Epoxy-8,8′-lignan-4,4′,9-triol E145

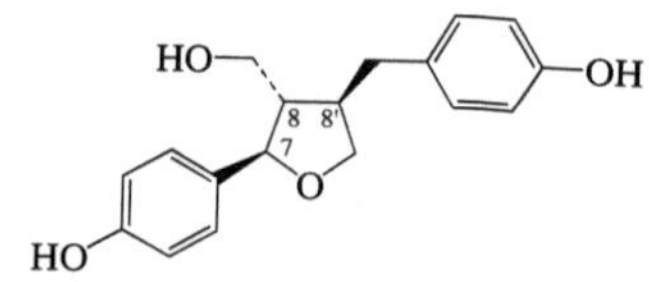

(7R,8S,8′R)-form

4,4′,9-Trihydroxy-7,9′-epoxylignan

$C_{18}H_{20}O_4$ 300.354

(7R,8S,8′R)-form [1365029-66-8]

Constit. of the shells of *Carya cathayensis*. Shows significant cytotoxic acivity against several human cancer cell lines. Powder. $[\alpha]_D^{25}$ +108.3 (c, 0.5 in MeOH). λ_{max} 202 (log ε 3.73); 225 (log ε 3.75); 276 (log ε 2.97) (MeOH)

(7S,8R,8′R)-form

4,4′-Di-Me ether

7,9′-Epoxy-4,4′-dimethoxy-8,8′-lignan-9-ol. 9-Hydroxy-4,4′-dimethoxy-7,9′-epoxylignan

[1992030-86-0]

$C_{20}H_{24}O_4$ 328.408

Constit. of *Oresitrophe rupifraga*. Exhibits antineuroinflammatory effects. Amorph. powder. $[\alpha]_D^{22}$ +11.7 (c, 0.2 in MeOH). λ_{max} 226 (log ε 4.23); 276 (log ε 3.36) (MeOH).

Wu, W. *et al.*, *Bioorg. Med. Chem. Lett.*, 2012, **22**, 1895-1898 (*Carya cathayensis constit*).
Wu, X.-Y. *et al.*, *Chem. Biodiversity*, 2016, **13**, 1030-1037 (*Oresitrophe rupifraga constit, activity*)

2,7′-Epoxy-3,8′-lign-8-ene-4′,5-diol E146

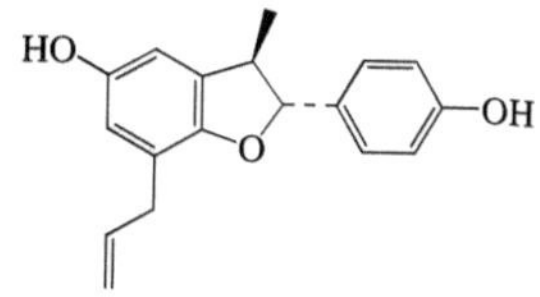

2,3-Dihydro-2-(4-hydroxyphenyl)-3-methyl-7-(2-propenyl)-5-benzofuranol. 7-Allyl-2,3-dihydro-5-hydroxy-2-(4-hydroxyphenyl)-3-methylbenzofuran. 2,7′-Epoxy-4′,5-dihydroxy-3,8′-neolign-8-ene

$C_{18}H_{18}O_3$ 282.339

(7′R,8′R)-form

5-Me ether

7-Allyl-2,3-dihydro-2-(4-hydroxyphenyl)-5-methoxy-3-methylbenzofuran. 2,7′-Epoxy-5-methoxy-3,8′-lign-8-en-4′-ol.

2,7′-Epoxy-4′-hydroxy-5-methoxy-3,8′-lign-8-ene
[221666-28-0]
$C_{19}H_{20}O_3$ 296.366
Constit. of *Piper aequale*. Oil. $[\alpha]^{25}_{D}$ +35 (c, 1.9 in MeOH). λ_{max} 210 (log ε 5.08); 228 (log ε 4.78); 282 (log ε 4.19) (MeOH).

Maxwell, A. *et al.*, *Phytochemistry*, 1999, **50**, 499-504 (*Piper aequale constit*)

4,7′-Epoxy-3,8′-lign-7-ene-3′,4′-diol E147

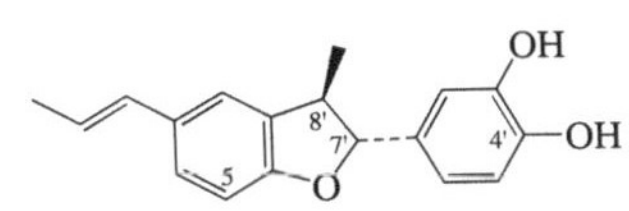

(7E,7′R,8′R)-form

2-(3,4-Dihydroxyphenyl)-2,3-dihydro-3-methyl-5-(1-propenyl)benzofuran. 4,7′-Epoxy-3′,4′-dihydroxy-3,8′-neolign-7-ene
$C_{18}H_{18}O_3$ 282.339
Lignan numbering shown
(7E,7′R,8′R)-form
3′-Me ether
2,3-Dihydro-2-(4-hydroxy-3-methoxyphenyl)-3-methyl-5-(1-propenyl)benzofuran. 4,7′-Epoxy-3′-methoxy-3,8′-lign-7-en-4′-ol. 4,7′-Epoxy-4′-hydroxy-3′-methoxy-3,8′-lign-7-ene
[109225-40-3]
$C_{19}H_{20}O_3$ 296.366
Constit. of *Krameria cystisoides*. Cryst.. Mp 105 - 107°. $[\alpha]^{21}_{D}$ +94 (c, 1.1 in MeOH). λ_{max} 208 (log ε 4.53); 264 (log ε 4.36); 287 (sh); 305 (log ε 3.68) (MeOH). [base] λ_{max} 263 (log ε 4.43); 298 (log ε 4.06) (MeOH/NaOH)
3′,4′-Di-Me ether
4,7′-Epoxy-3′,4′-dimethoxy-3,8′-lign-7-ene
[1802051-90-6]
$C_{20}H_{22}O_3$ 310.393
Constit. of the stems of *Aristolochia fordiana*. Oil. $[\alpha]^{20}_{D}$ -31 (c, 0.1 in MeOH). λ_{max} 210 (log ε 4.43); 228 (log ε 4.28); 273 (log ε 4.00) ($CHCl_3$)
3′,4′-Methylene ether
4,7′-Epoxy-3′,4′-methylenedioxy-3,8′-lign-7-ene. ***Regnellin***
[249902-94-1]
$C_{19}H_{18}O_3$ 294.35
Constit. of *Piper regnellii*. Amorph.. $[\alpha]^{21}_{D}$ +85.5 (c, 0.06 in MeOH). λ_{max} 232 (ε 11250); 265 (ε 8750); 290 (ε 7500) (MeOH)
(7E,7′R,8′S)-form
3′-Me ether
Constit. of the stems of *Aristolochia fordiana*. Oil. $[\alpha]^{25}_{D}$ +12.4 (c, 0.277 in $CHCl_3$). λ_{max} 204 (log ε 4.19); 266 (log ε 3.63) (MeOH)
3′,4′-Di-Me ether
[1802051-86-0]
Constit. of the stems of *Aristolochia fordiana*. Oil. $[\alpha]^{20}_{D}$ -2 (c, 0.2 in MeOH). λ_{max} 208 (log ε 4.38); 260 (log ε 3.86) (MeOH)
3′,4′-Methylene ether
[1609472-90-3]
Constit. of the stems of *Aristolochia fordiana*. Oil. $[\alpha]^{20}_{D}$ -26 (c, 0.2 in MeOH). λ_{max} 212 (log ε 4.44); 263 (log ε 4.28) (MeOH)
(7E,7′S,8′S)-form
3′-Me ether
Parakmerin A
[119555-99-6]
$C_{19}H_{20}O_3$ 296.366
Constit. of *Aristolochia fordiana*, *Myristica fragrans* and *Parakmeria yunnanensis* (preferred genus name *Magnolia*). Moderate inhibitor of NO prodn.. Mp 68 - 70°. $[\alpha]^{21}_{D}$ -101.5 (c, 0.5 in $CHCl_3$)
3′,4′-Di-Me ether
[1802051-91-7]
Constit. of the stems of *Aristolochia fordiana*. Oil. $[\alpha]^{20}_{D}$ +29 (c, 0.2 in MeOH). λ_{max} 210 (log ε 4.40); 228 (log ε 4.21); 272 (log ε 3.99) (MeOH).

Achenbach, H. *et al.*, *Phytochemistry*, 1987, **26**, 1159-1166 (*Krameria cystisoides constit*).
Benevides, P.J.C. *et al.*, *Phytochemistry*, 1999, **52**, 339-343 (*Regnellin*).
Cheng, Y.-X. *et al.*, *Yunnan Zhiwu Yanjiu*, 2001, **23**, 352-356 (*Parakmerin A*).
Zhou, Z.-B. *et al.*, *Planta Med.*, 2013, **79**, 1730-1735 (*Aristolochia fordiana constits*).
Tang, G.-H. *et al.*, *J. Nat. Prod.*, 2015, **78**, 1894-1903 (*Aristolochia fordiana constits*)

4,7′-Epoxy-3,8′-lign-7′-ene-4′,8-diol E148

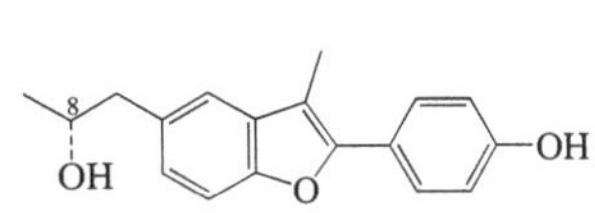

(R)-form

2-(4-Hydroxyphenyl)-α,3-dimethyl-5-benzofuranethanol, CAS. 2-(4-Hydroxyphenyl)-5-(2-hydroxypropyl)-3-methylbenzofuran. 4,7′-Epoxy-4′,8-dihydroxy-3,8′-neolign-7′-ene
[127988-90-3]
$C_{18}H_{18}O_3$ 282.339
(R)-form [152367-38-9]
Constit. of *Krameria grayi* and *Krameria paucifolia*. Cryst.. Mp 161 - 164°. $[\alpha]^{20}_{D}$ -6 (c, 2.1 in MeOH)
(S)-form [125205-70-1]
Constit. of *Krameria lanceolata*. Oil. $[\alpha]^{21}_{D}$ +9 (c, 0.3 in MeOH). λ_{max} 308 (log ε 3.56) (MeOH). [base] λ_{max} 322 (log ε 3.61) (MeOH/NaOH)
8-Ketone
2-(4-Hydroxyphenyl)-3-methyl-5-(2-oxopropyl)benzofuran. 4,7′-Epoxy-4′-hydroxy-3,8′-lign-7′-en-8-one. 4,7′-Epoxy-4′-hydroxy-8-oxo-3,8′-lign-7′-ene
[125205-71-2]
$C_{18}H_{16}O_3$ 280.323
Constit. of *Krameria lanceolata*. Amorph.. λ_{max} 309 (log ε 4.19) (MeOH). [base] λ_{max} 326 (log ε 4.24) (MeOH/NaOH).

Achenbach, H. *et al.*, *Phytochemistry*, 1989, **28**, 1959-1962 (*Krameria lanceolata constits*).
Achenbach, H. *et al.*, *Phytochemistry*, 1993, **34**, 835-837 (*Krameria paucifolia constit*)

4,7′-Epoxy-3,8′-lign-7′-ene-4′,9-diol E149

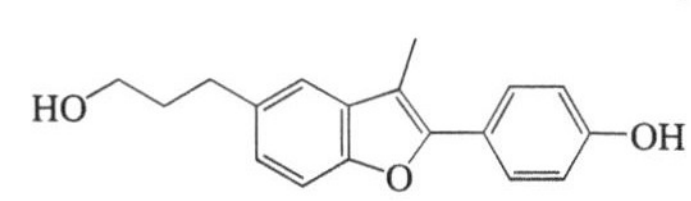

2-(4-Hydroxyphenyl)-3-methyl-5-benzofuranpropanol, CAS. 4,7′-Epoxy-4′,9-dihydroxy-3,8′-neolign-7′-ene.
Olmecol
[109145-65-5]
$C_{18}H_{18}O_3$ 282.339
Constit. of *Krameria cystisoides*. Cryst.. Mp 161 - 162°. λ_{max} 242 (ε 7080); 309 (ε 23440) (MeOH).

Achenbach, H. *et al.*, *Phytochemistry*, 1987, **26**, 1159-1166 (*Olmecol*)

4,7′-Epoxy-3,8′-lign-7-ene-4′,9′-diol E150

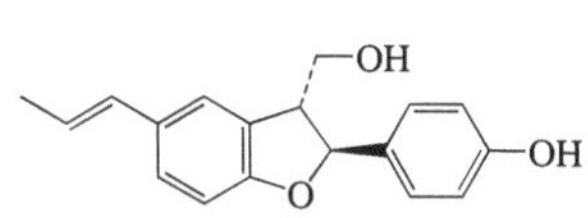

2,3-Dihydro-2-(4-hydroxyphenyl)-5-(1-propenyl)-3-benzofuranmethanol, CAS. 4,7′-Epoxy-4′,9′-dihydroxy-3,8′-neolign-7-ene
$C_{18}H_{18}O_3$ 282.339

(7E,7′S,8′R)-form [119322-04-2]

Constit. of *Krameria lappacea* and *Krameria triandra*. Glassy solid. Mp 56 - 58°. [α] D -19.3 (c, 0.2 in $CHCl_3$).

Arnone, A. *et al.*, *Gazz. Chim. Ital.*, 1988, **118**, 675-682 (*Krameria triandra constit*).

Baumgartner, L. *et al.*, *J. Nat. Prod.*, 2011, **74**, 1779-1786 (*Krameria lappacea constit, activity*)

4,7′-Epoxy-3,8′-lign-7-ene-4′,9-diol E151

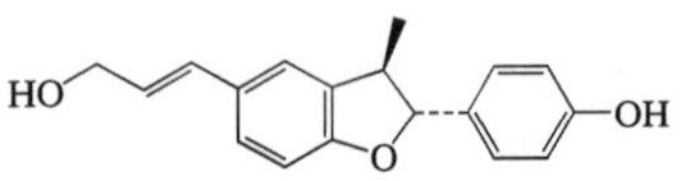

4,7′-Epoxy-4′,9-dihydroxy-3,8′-neolign-7-ene

$C_{18}H_{18}O_3$ 282.339

(7E,7′R,8′R)-form

Mappiodoinin A [2158272-33-2]

Constit. of *Mappianthus iodoies*. Exhibits cytotoxic activity. Amorph. powder. $[\alpha]^{19}_{D}$ -52.6 (c, 0.12 in MeOH). λ_{max} 218 (log ε 4.33); 269 (log ε 2.52); 309 (log ε 3.16) (MeOH)

9-Aldehyde

4,7′-Epoxy-4′-hydroxy-3,8′-lign-7-en-9-al. 4,7′-Epoxy-4′-hydroxy-9-oxo-3,8′-lign-7-ene

[1669432-18-1]

$C_{18}H_{16}O_3$ 280.323

Isol. from the faeces of *Naupactus bipes* feeding on *Piper regnellii*. Pale yellow powder.

Ramos, C.S. *et al.*, *Chemoecology*, 2013, **23**, 143-148 (*9-aldehyde*).

Jiang, Z.-H. *et al.*, *Bioorg. Chem.*, 2017, **75**, 260-264 (*Mappiodoinin A, activity*)

4,7′-Epoxy-3,8′-lign-8-ene-3′,4′-diol E152

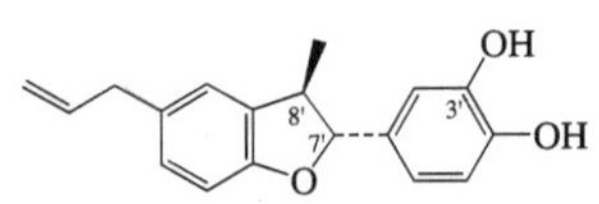

2-(3,4-Dihydroxyphenyl)-2,3-dihydro-3-methyl-5-(2-propenyl)benzofuran. 4,7′-Epoxy-3′,4′-dihydroxy-3,8′-neolign-8-ene

$C_{18}H_{18}O_3$ 282.339

(7′R,8′R)-form

3′-Me ether

4,7′-Epoxy-3′-methoxy-3,8′-lign-8-en-4′-ol. 4,7′-Epoxy-4′-hydroxy-3′-methoxy-3,8′-lign-8-ene.

3′-Methoxymiliumollin

[1528762-15-3]

$C_{19}H_{20}O_3$ 296.366

Constit. of the leaves of *Miliusa mollis*. Exhibits weak cytotoxic activity. Viscous liq.. $[\alpha]^{20}_{D}$ +8.60 (c, 0.18 in $CHCl_3$). λ_{max} 208 (log ε 4.46); 233 (log ε 4.22); 283 (log ε 3.95) (MeOH).

Sawasdee, K. *et al.*, *Fitoterapia*, 2013, **85**, 49-56 (*3′-Methoxymiliumollin, activity*)

4,7′-Epoxy-3,8′-lign-8-ene-4′,5-diol E153

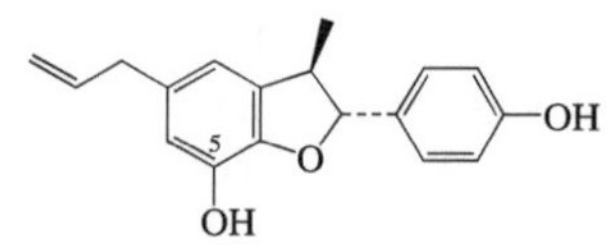

2,3-Dihydro-2-(4-hydroxyphenyl)-3-methyl-5-(2-propenyl)-7-benzofuranol. 4,7′-Epoxy-4′,5-dihydroxy-3,8′-neolign-8-ene

$C_{18}H_{18}O_3$ 282.339

Lignan numbering shown

(7′R,8′R)-form

5-Me ether

4,7′-Epoxy-5-methoxy-3,8′-lign-8-en-4′-ol. 4,7′-Epoxy-4′-hydroxy-5-methoxy-3,8′-lign-8-ene.

5-Methoxymiliumollin

[1528762-16-4]

$C_{19}H_{20}O_3$ 296.366

Constit. of the leaves of *Miliusa mollis*. Viscous liq.. $[\alpha]^{20}_{D}$ +15.7 (c, 0.07 in $CHCl_3$). λ_{max} 211 (log ε 4.46); 228 (log ε 4.30); 280 (log ε 3.82) (MeOH).

Sawasdee, K. *et al.*, *Fitoterapia*, 2013, **85**, 49-56 (*5-Methoxymiliumollin*)

7,7′-Epoxy-8,8′-lign-8-ene-4,4′-diol E154

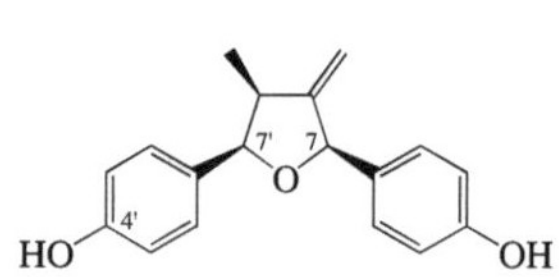

(7S,7′R,8′S)-form

4,4′-Dihydroxy-7,7′-epoxylign-8-ene

$C_{18}H_{18}O_3$ 282.339

(7S,7′R,8′S)-form

4′-Me ether

7,7′-Epoxy-4′-methoxy-8,8′-lign-8-en-4-ol. 4-Hydroxy-4′-methoxy-7,7-epoxylign-8-ene. ***Larreatridenticin***

[128450-68-0]

$C_{19}H_{20}O_3$ 296.366

Constit. of *Larrea tridentata*. Cryst. (Me_2CO). Mp 152 - 153°

(7S,7′S,8′R)-form

Ribesin C [1469818-64-1]

Constit. of the leaves of *Ribes nigrum*. Solid. $[\alpha]^{25}_{D}$ +33 (c, 1.6 in MeOH). λ_{max} 229 (log ε 4.41); 277 (log ε 3.60) (MeOH).

Konno, C. *et al.*, *J. Nat. Prod.*, 1990, **53**, 396-406 (*Larreatridenticin*).

Sasaki, T. *et al.*, *Phytochemistry*, 2013, **95**, 333-340 (*Ribesin C*)

7,7′-Epoxy-8,8′-lign-8(8′)-ene-4,4′-diol E155

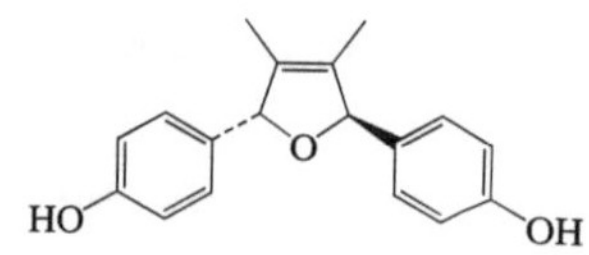

4,4′-Dihydroxy-7,7′-epoxylign-8(8′)-ene

$C_{18}H_{18}O_3$ 282.339

(7R,7′R)-form

Dehydrolarreatricin. Ribesin A

[128450-69-1]

Constit. of *Larrea tridentata* and *Ribes nigrum*. Cryst.. Mp 211 - 213° (Dehydrolarreatricin). $[\alpha]^{25}_{D}$ -88 (c, 0.4 in MeOH) (Ribesin A). λ_{max} 232 (log ε 4.39); 276 (log ε 3.59) (MeOH).

Schmidt, T.J. *et al.*, *Phytochem. Lett.*, 2012, **5**, 632-638 (*Dehydrolarreatricin*).

Sasaki, T. *et al.*, *Phytochemistry*, 2013, **95**, 333-340 (*Ribesin A*)

4,7′-Epoxy-3,8′-lign-7-ene-3′,4′,5,5′,9,9′-hexol E156

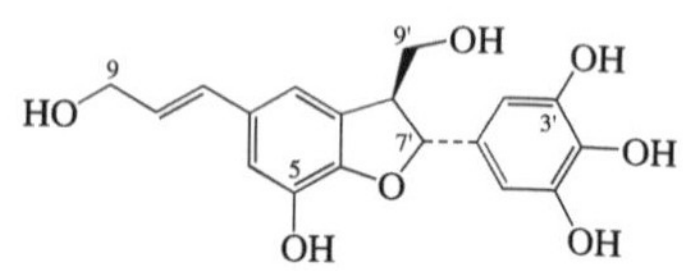

(7E,7′R,8′S)-form

4,7′-Epoxy-3′,4′,5,5′,9,9′-hexahydroxy-3,8′-neolign-7-ene

$C_{18}H_{18}O_7$ 346.335

(7E,7′R,8′S)-form

3′,5,5′-Tri-Me ether

4,7′-Epoxy-3′,5,5′-trimethoxy-3,8′-lign-7-ene-4′,9,9′-triol. 4,7′-Epoxy-4′,9,9′-trihydroxy-3′,5,5′-trimethoxy-3,8′-lign-7-ene.

Simulanol

[500574-38-9]

$C_{21}H_{24}O_7$ 388.416

Constit. of *Euonymus acanthocarpus* and *Zanthoxylum simulans*. Oil. $[\alpha]^{25}_{D}$ -13.1 (c, 0.06 in $CHCl_3$). λ_{max} 221 (log ε 4.48); 267 (log ε 4.41); 303 (log ε 4.11) (EtOH)

3′,5,5′-Tri-Me ether, 4′-O-β-D-glucopyranoside
Alangilignoside B. Scorzonoside
[217650-81-2]
$C_{27}H_{34}O_{12}$ 550.557
Constit. of *Alangium premnifolium* and *Scorzonera hispanica*. Needles. Mp 116 - 118°. $[\alpha]_D^{20}$ -15.2 (c, 0.3 in MeOH). Stereochem. of Alangilignoside B revised in 2009. λ_{max} 208 (log ε 4.64); 224 (log ε 4.44) (MeOH)

3′,5,5′-Tri-Me ether, 9,9′-di-Ac
(-)-Jatrointelignan C
[1620489-34-0]
$C_{25}H_{28}O_9$ 472.49
Constit. of *Jatropha integerrima* and *Rhododendron mariae*. Inhibits NO prodn. in LPS-activated C57BL6/J mouse macrophages. Amorph. powder. $[\alpha]_D^{20}$ -4.0 (c, 0.01 in MeOH). Isol. as a racemate. Resolved by chiral hplc. λ_{max} 273 (MeOH)

3′,5,5′-Tri-Me ether, 4′-O-[4-hydroxy-3-methoxybenzoyl-(→6)-β-D-glucopyranoside], 9-O-β-D-glucopyranoside
Viburfordoside B
[2257425-87-7]
$C_{41}H_{50}O_{20}$ 862.831
Constit. of the fruit of *Viburnum fordiae*. Powder. $[\alpha]_D^{23}$ -50.4 (c, 0.1 in MeOH). λ_{max} 213 (log ε 4.28); 268 (log ε 3.97) (MeOH)

3′,5,5′-Tri-Me ether, 4′-O-[4-hydroxy-3,5-dimethoxybenzoyl-(→6)-β-D-glucopyranoside], 9-O-β-D-glucopyranoside
Viburfordoside C
[2257425-88-8]
$C_{42}H_{52}O_{21}$ 892.857
Constit. of the fruit of *Viburnum fordiae*. Moderate α-glucosidase inhibitor. Powder. $[\alpha]_D^{23}$ -16.2 (c, 0.1 in MeOH). λ_{max} 276 (log ε 4.78) (MeOH)

3′,5,5′-Tri-Me ether, 9-O-(4-hydroxy-3-methoxy-E-cinnamoyl)
Litsecol B
[2170722-67-3]
$C_{31}H_{32}O_{10}$ 564.587
Constit. of roots and stems of *Litsea cubeba*. NO prodn. inhibitor. Amorph. powder. $[\alpha]_D^{25}$ -9.1 (c, 0.08 in MeOH). λ_{max} 211 (log ε 4.71); 285 (log ε 4.13); 324 (log ε 4.19) (MeOH)

3′,4′,5,5′-Tetra-Me ether
4,7′-Epoxy-3′,4′,5,5′-tetramethoxy-3,8′-lign-7-ene-9,9′-diol. 4,7′-Epoxy-9,9′-dihydroxy-3′,4′,5,5′-tetramethoxy-3,8′-lign-7-ene. ***Woorenogenin***
[166990-13-2]
[1290633-26-9] (*ent*-form)
$C_{22}H_{26}O_7$ 402.443
Constit. of the rhizomes of *Coptis japonica* and from stems of *Sinocalamus affinis*. Amorph. powder. $[\alpha]_D^{20}$ -7.8 (c, 0.1 in MeOH). $[\alpha]_D^{20}$ -9 (c, 0.6 in MeOH). Some confusion over abs. configs. Isol. of both enantiomers claimed; both assigned same abs. config. and both have reported -ve opt. rotns. Xiang *et al* state that the earlier -ve rotation for Woorenogenin from *Coptis* was 'ambiguous'. λ_{max} 208 (ε 43650); 275 (ε 16220) (EtOH)

3′,4′,5,5′-Tetra-Me ether, 9-O-β-D-glucopyranoside
Woorenoside I
[166990-14-3]
$C_{28}H_{36}O_{12}$ 564.584
Constit. of the rhizomes of *Coptis japonica*. Shows antiinflammatory effects. Suppresses TNF-α and nitric oxide prodn.. Amorph. powder. $[\alpha]_D^{20}$ -21.3 (c, 6 in MeOH)

3′,4′,5,5′-Tetra-Me ether, 9′-Ac, 9-O-β-D-glucopyranoside
Woorenoside II
[166990-15-4]
$C_{30}H_{38}O_{13}$ 606.621
Constit. of the rhizomes of *Coptis japonica*. Shows antiinflammatory effects. Suppresses TNF-α and nitric oxide prodn.. Amorph. powder. $[\alpha]_D^{20}$ -26.7 (c, 10 in MeOH)

3′,4′,5,5′-Tetra-Me ether, 9-O-[4-hydroxy-2-methylenebutanoyl-(→6)-β-D-glucopyranoside]
Woorenoside III
[166990-16-5]
$C_{33}H_{42}O_{14}$ 662.685
Constit. of the rhizomes of *Coptis japonica*. Shows antiinflammatory effects. Suppresses TNF-α and nitric oxide prodn.. Amorph. powder. $[\alpha]_D^{20}$ -24.2 (c, 10 in MeOH)

3′,4′,5,5′-Tetra-Me ether, 9′-Ac, 9-O-[4-hydroxy-2-methylenebutanoyl-(→6)-β-D-glucopyranoside]
Woorenoside IV
[166990-17-6]
$C_{35}H_{44}O_{15}$ 704.722
Constit. of the rhizomes of *Coptis japonica*. Shows antiinflammatory effects. Suppresses TNF-α and nitric oxide prodn.. Amorph. powder. $[\alpha]_D^{20}$ -17.6 (c, 3 in MeOH)

3′,5,5′,9-Tetra-Me ether
4,7′-Epoxy-3′,5,5′,9-tetramethoxy-3,8′-lign-7-ene-4′,9′-diol. 4,7′-Epoxy-4′,9′-dihydroxy-3′,5,5′,9-tetramethoxy-3,8′-lign-7-ene. ***Tirpitzin A***
[2170998-22-6]
$C_{22}H_{26}O_7$ 402.443
Constit. of *Solanum erianthum* and *Tirpitzia ovoidea*. Exhibits moderate cytotoxic activity. Gum. $[\alpha]_D^{25}$ -8.5 (c, 0.28 in $CHCl_3$) (Solanum erianthum constit). $[\alpha]_D^{23}$ +34 (c, 0.1 in MeOH) (Tirpitzin A). Both isolates have same config. claimed but have conflicting opt. rotation data. λ_{max} 213 (log ε 4.25); 276 (log ε 3.88) (MeOH)

9-Et, 3′,5,5′-tri-Me ether
4,7′-Epoxy-9-ethoxy-3′,5,5′-trimethoxy-3,8′-lign-7-ene-4′,9′-diol. 4,7′-Epoxy-9-ethoxy-4′,9′-dihydroxy-3′,5,5′-trimethoxy-3,8′-lign-7-ene. ***(-)-Crataegifin C***
[1428538-16-2]
$C_{23}H_{28}O_7$ 416.47
Constit. of seedlings of *Stellera chamaejasme*. Pale yellow oil. $[\alpha]_D^{20}$ -3 (c, 0.1 in MeOH). $[\alpha]_D^{20}$ -25.6 (c, 0.1 in MeOH). λ_{max} 208 (log ε 4.55); 279 (log ε 3.86); 315 (log ε 3.84) (MeOH)

9-Aldehyde, 3′,5-di-Me ether
4,7′-Epoxy-4′,5′,9′-trihydroxy-3′,5-dimethoxy-3,8′-lign-7-en-9-al.
Cassyformin‡
[749917-08-6]
$C_{20}H_{20}O_7$ 372.373
Constit. of *Cassytha filiformis* and *Michelia compressa* var. *formosana*. Pale yellow oil. $[\alpha]_D^{22}$ -50.9 (c, 0.001 in $CHCl_3$). Abs. config. not confirmed. λ_{max} 340 (log ε 4.34) (MeOH)

9-Aldehyde, 3′,5,5′-tri-Me ether
4,7′-Epoxy-4′,9′-dihydroxy-3′,5,5′-trimethoxy-3,8′-lign-7-en-9-al.

4,7′-Epoxy-4′,9′-dihydroxy-3′,5,5′-trimethoxy-9-oxo-3,8′-lign-7-ene.
(-)-5-Methoxybalanophonin
[953393-40-3]
$C_{21}H_{22}O_7$ 386.4
Constit. of the stems of *Firmiana simplex* and *Picrasma quassioides*. NO prodn. inhibitor. Exhibits weak cytotoxic activity. $[\alpha]^{20}_{D}$ -26 (c, 0.1 in MeOH)

9-Carboxylic acid, 3′,5,5′-tri-Me ether, 2-(4-hydroxyphenyl)ethylamide
[2129609-43-2]
$C_{29}H_{31}NO_8$ 521.566
Constit. of the fruit of *Nicandra physaloides*. Yellow oil. $[\alpha]^{25}_{D}$ -86 (c, 0.1 in MeOH). λ_{max} 228 (log ε 2.5) (MeOH)

(7E,7′S,8′R)-form

3′,5,5′-Tri-Me ether, 4′-O-β-D-glucopyranoside
[193153-52-5]
$C_{27}H_{34}O_{12}$ 550.557
Constit. of the leaves of *Osmanthus heterophyllus*. Amorph. powder. Mp 116 - 118°. $[\alpha]^{27}_{D}$ -5.7 (c, 0.18 in MeOH). λ_{max} 208 (log ε 4.6); 220 (sh); 270 (log ε 4.2) (MeOH)

3′,5,5′-Tri-Me ether, 9-O-β-D-glucopyranoside
Aquilignanoside A. *Aquilaroside A*
[1771703-86-6]
$C_{27}H_{34}O_{12}$ 550.557
Constit. of the stems of *Aquilaria sinensis*. Amorph. powder (MeOH). $[\alpha]^{25}_{D}$ +72 (c, 0.02 in MeOH). Authors have called the nat. prod. both Aquilaroside A and Aquilignanoside A in the paper. λ_{max} 206 (log ε 0.38); 274 (log ε 0.12); 338 (log ε 0.04) (MeOH)

3′,5,5′-Tri-Me ether, 4′,9-di-O-β-D-glucopyranoside
Viburfordoside A
[2099167-56-1]
$C_{33}H_{44}O_{17}$ 712.698
Constit. of twigs of *Litsea cubeba* and the fruit of *Viburnum fordiae*. Exhibits a reduction of acetaminophen induced HepG2 cell injury. Amorph. solid or powder. $[\alpha]^{20}_{D}$ -9 (c, 0.1 in MeOH). $[\alpha]^{23}_{D}$ -24 (c, 0.1 in MeOH) (Viburfordoside A). λ_{max} 240 (log ε 1.01); 277 (log ε 3.35) (MeOH). [neutral] λ_{max} 208 (log ε 4.89); 274 (log ε 4.46) (MeOH) (Viburfordoside A)

3′,5,5′-Tri-Me ether, 9,9′-di-Ac
(+)-Jatrointelignan C
[1842427-57-9]
$C_{25}H_{28}O_9$ 472.49
Constit. of the trunk of *Jatropha integerrima*. Gum. $[\alpha]^{20}_{D}$ +102 (c, 0.1 in $CHCl_3$). Isol. as a racemate. Resolved by chiral hplc. λ_{max} 213 (log ε 4.46); 275 (log ε 4.15) (MeOH)

3′,5,5′,9-Tetra-Me ether
[2170998-21-5]
Constit. of the stems of *Solanum erianthum*. Gum. $[\alpha]^{25}_{D}$ +10.5 (c, 0.21 in $CHCl_3$). λ_{max} 213 (log ε 4.25); 276 (log ε 3.88) (MeOH)

9-Et, 3′,5,5′-tri-Me ether
(+)-Crataegifin C
Exhibits low cytotoxicity. $[\alpha]^{20}_{D}$ +28 (c, 0.1 in MeOH)

9-Aldehyde, 3′,5,5′-tri-Me ether
(+)-5-Methoxybalanophonin
[158531-55-6]
$C_{21}H_{22}O_7$ 386.4
Constit. of *Chosenia arbutifolia* and *Picrasma quassioides*. Exhibits weak cytotoxic activity. $[\alpha]^{20}_{D}$ +22 (c, 0.1 in MeOH)

9-Aldehyde, 3′,5,5′-tri-Me ether, 9′-O-β-D-glucopyranoside
Plucheoside D$_2$
[133733-96-7]
$C_{27}H_{32}O_{12}$ 548.541
Isol. from *Pluchea indica*. Powder. $[\alpha]^{21}_{D}$ -67.5 (c, 0.4 in MeOH). λ_{max} 341 (log ε 4.33) (MeOH)

9-Aldehyde, 3′,5,5′-tri-Me ether, 9′-O-(4-hydroxybenzoyl)
[158531-56-7]
$C_{28}H_{26}O_9$ 506.507
Constit. of the wood of *Chosenia arbutifolia* (preferred genus name *Salix*)

(7E,7′R*,8′R*)-form

9-Carboxylic acid, 3′,5,5′-tri-Me ether, 9′-Ac, 2-(4-hydroxyphenyl)ethylamide
Tatarine E
[1844922-69-5]
$C_{31}H_{33}NO_9$ 563.603
Constit. of the rhizomes of *Acorus tatarinowii*. Yellow powder. Racemate

(7E,7′R*,8′S*)-form

3′,5,5′-Tri-Me ether, 9,9′-bis-O-(4-hydroxybenzoyl)
Quiquesetinerviusin B
[1250376-63-6]
$C_{35}H_{32}O_{11}$ 628.63
Constit. of the stems of *Calamus quiquesetinervius*. Amorph. powder. Mp 175°. $[\alpha]^{25}_{D}$ +4.8 (c, 0.63 in MeOH). λ_{max} 210 (log ε 4.06); 267 (log ε 3.71); 305 (sh) (MeOH)

9-Carboxylic acid, 3′,5,5′-tri-Me ether
4,7′-Epoxy-4′,9′-dihydroxy-3′,5,5′-trimethoxy-3,8′-lign-7-en-9-oic acid
[1264177-64-1]
$C_{21}H_{22}O_8$ 402.399
Constit. of *Selaginella moellendorffii*. Amorph. solid. $[\alpha]^{25}_{D}$ -48 (c, 0.08 in MeOH)

(7E,7′RS,8′SR)-form

3′,5,5′,9-Tetra-Me ether
Constit. of the stems of *Solanum erianthum*. Gum

9-Et, 3′,5,5′-tri-Me ether
Crataegifin C
$C_{23}H_{28}O_7$ 416.470
Constit. of the fruit of *Crataegus pinnatifida*. Yellow oil. $[\alpha]^{20}_{D}$ +1.8 (c, 0.1 in MeOH). Partial racemate resolved by chiral hplc. λ_{max} 209 (log ε 4.45); 272 (log ε 3.73) (MeOH)

9-Aldehyde, 3′,5,5′-tri-Me ether
(±)-5-Methoxybalanophonin
[61424-15-5]
$C_{21}H_{22}O_7$ 386.400
Constit. of the stems of *Picrasma quassioides*. Exhibits weak cytotoxic activity. Yellow oil. Resolved by chiral hplc

9-Carboxylic acid, 3′,5,5′-tri-Me ether, 2-(4-hydroxyphenyl)ethylamide
[2252244-23-6]
$C_{29}H_{31}NO_8$ 521.566
Constit. of the rhizomes of *Alocasia macrorrhiza*. NO prodn. inhibitor. Gum. $[\alpha]^{27}_{D}$ +0.8 (c, 0.5 in MeOH). λ_{max} 223; 319 (MeOH)

(7Z,7′RS,8′SR)-form

9-Carboxylic acid, 3′,5,5′-tri-Me ether, 2-(4-hydroxyphenyl)ethylamide
[2252244-25-8]
$C_{29}H_{31}NO_8$ 521.566
Constit. of the rhizomes of *Alocasia macrorrhiza*. NO prodn. inhibitor. Gum. $[\alpha]^{27}_{D}$ +1.1 (c, 0.5 in MeOH). λ_{max} 283 (MeOH).

Uchiyama, T. *et al.*, *Phytochemistry*, 1991, **30**, 655-657 (*Plucheoside D_2*).
Hashimoto, Y. *et al.*, *Mokuzai Gakkaishi*, 1994, **40**, 549-553 (*Chosenia arbutifolia constits*).
Yoshikawa, K. *et al.*, *Chem. Pharm. Bull.*, 1995, **43**, 578-581 (*Woorenosides, Woorenogenin*).
Yuasa, K. *et al.*, *Phytochemistry*, 1997, **45**, 611-615 (*Alangilignoside B*).
Tolstikhina, V.V. *et al.*, *Rastit. Resur.*, 1998, **34**, 77-80 (*Scorzonoside*).
Cho, J.Y. *et al.*, *J. Nat. Prod.*, 2000, **63**, 1205-1209 (*Woorenosides, activity*).
Yang, Y.-P. *et al.*, *Phytochemistry*, 2002, **61**, 567-572 (*Simulanol*).
Ho, J.-C. *et al.*, *J. Chin. Chem. Soc. (Taipei)*, 2004, **51**, 221-223 (*Cassyformin*).
Machida, K. *et al.*, *J. Nat. Med.*, 2009, **63**, 227-231 (*Alangilignoside B*).
Wang, Y.-H. *et al.*, *Helv. Chim. Acta*, 2010, **93**, 2467-2477 (*Selaginella moellendorffii constit*).
Chang, C.-L. *et al.*, *J. Nat. Prod.*, 2010, **73**, 1482-1488 (*Quiquesetinerviusin B*).
Xiong, L. *et al.*, *J. Nat. Prod.*, 2011, **74**, 1188-1200 (*Woorenogenin*).
Zhu, J.X. *et al.*, *Arch. Pharmacal Res.*, 2012, **35**, 1739-1747 (*Simulanol*).
Qiao, L.-R. *et al.*, *Planta Med.*, 2012, **78**, 711-719 (*Stellera chamaejasme Et ether*).
Wu, Y. *et al.*, *Biochem. Syst. Ecol.*, 2014, **55**, 41-45 (*Aquilignanoside A*).
Guo, Q. *et al.*, *J. Asian Nat. Prod. Res.*, 2014, **16**, 1-10 (*Rhododendron mariae constit*).
Zhu, J.-Y. *et al.*, *RSC Adv.*, 2015, **5**, 12202-12208 (*Jatrointelignan C*).
Woo, K.W. *et al.*, *Bioorg. Med. Chem. Lett.*, 2016, **26**, 730-733 (*(-)-5-Methoxybalanophonin*).
Li, H. *et al.*, *Chirality*, 2016, **28**, 259-263 (*Solanum erianthum constits*).
Guo, Q. *et al.*, *J. Asian Nat. Prod. Res.*, 2016, **18**, 51-58 (*Litsecol B, activity*).
Feng, X.-L. *et al.*, *Magn. Reson. Chem.*, 2016, **54**, 396-399 (*Tatarine E*).
Yang, X.-Y. *et al.*, *Chin. J. Nat. Med.*, 2017, **15**, 938-943 (*Tirpitzin A, activity*).
Huang, W. *et al.*, *Fitoterapia*, 2017, **117**, 126-132 (*Alocasia macrorrhiza constits, activity*).
Wang, L.-Q. *et al.*, *J. Funct. Foods*, 2017, **31**, 229-236 (*Nicandra physaloides amide*).
Wang, L.-Y. *et al.*, *J. Nat. Prod.*, 2017, **80**, 1808-1818 (*Litsea cubeba constit, activity*).
Lou, L.-L. *et al.*, *Bioorg. Med. Chem. Lett.*, 2018, **28**, 1263-1268 (*5-Methoxybalanophonin, cd, resoln, activity*).
Zhao, C. *et al.*, *J. Agric. Food Chem.*, 2018, **66**, 10421-10430 (*Viburfordosides A,B,C, activity*).
Guo, R. *et al.*, *Fitoterapia*, 2019, **137**, 104287:1-8 (*Crataegifins C, cd, activity*)

4,7′-Epoxy-3,8′-lign-7′-ene-3′,4′,5,5′,9,9′-hexol E157

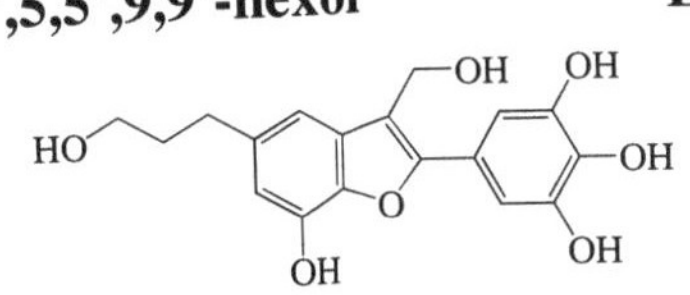

7-Hydroxy-3-(hydroxymethyl)-5-(3-hydroxypropyl)-2-(3,4,5-trihydroxyphenyl)benzofuran.
4,7′-Epoxy-3′,4′,5,5′,9,9′-hexahydroxy-3,8′-neolign-7′-ene

$C_{18}H_{18}O_7$ 346.335

3′,5,5′-Tri-Me ether

4,7′-Epoxy-3′,5,5′-trimethoxy-3,8′-lign-7′-ene-4′,9,9′-triol. 4,7′-Epoxy-4′,9,9′-trihydroxy-3′,5,5′-trimethoxy-3,8′-lign-7′-ene. ***Selamoellenin B***

$C_{21}H_{24}O_7$ 388.416

Constit. of *Selaginella moellendorffii*. Yellowish powder. λ_{max} 215; 258; 300 (MeOH)

3′,5,5′-Tri-Me ether, 9-O-β-D-glucopyranoside

[1264177-72-1]

$C_{27}H_{34}O_{12}$ 550.557

Constit. of *Selaginella moellendorffii*. Amorph. solid. $[\alpha]_D^{28}$ -46.6 (c, 0.27 in MeOH). λ_{max} 219 (log ε 4.21); 307 (log ε 4.04) (MeOH)

9′-Aldehyde, 3′,5-di-Me ether

4,7′-Epoxy-3′,4′,9-trihydroxy-5,5′-dimethoxy-3,8′-lign-7′-en-9′-al. 4,7′-Epoxy-3′,4′,9-trihydroxy-5,5′-dimethoxy-9′-oxo-3,8′-lign-7′-ene. 5′-Hydroxysalvinal. ***Mururin C***

[497846-77-2]

[350221-57-7] (allene compd.)

$C_{20}H_{20}O_7$ 372.373

Constit. of the bark of *Brosimum acutifolium*. Amorph. yellow solid. $[\alpha]_D^{22}$ -36 (c, 0.07 in MeOH). In 2000 an isolate from *Brosimum acutifolium* was assigned an unusual allene struct. but this was shown to be incorrect in 2008. λ_{max} 255 (sh); 329 (MeOH)

9′-Aldehyde, 3′,5,5′-tri-Me ether

4,7′-Epoxy-4′,9-dihydroxy-3′,5,5′-trimethoxy-3,8′-lign-7′-en-9′-al. 4,7′-Epoxy-4′9-dihydroxy-3′,5,5′-trimethoxy-9′-oxo-3,8′-lign-7′-ene. ***Selamoellenin A***

[2097275-71-1]

$C_{21}H_{22}O_7$ 386.400

Constit. of *Selaginella moellendorffii*. Amorph. light yellow powder. λ_{max} 254 (log ε 6.54); 338 (log ε 6.25) (MeOH).

Takashima, J. *et al.*, *Planta Med.*, 2002, **68**, 621-625 (*Mururin C*).
Hu, G. *et al.*, *Org. Lett.*, 2008, **10**, 5493-5496 (*Brosimum acutifolium constit*).
Wang, Y.-H. *et al.*, *Helv. Chim. Acta*, 2010, **93**, 2467-2477 (*Selaginella moellendorffii glucoside*).
Zeng, W. *et al.*, *Nat. Prod. Res.*, 2017, **31**, 2223-2227 (*Selamoellenin A, activity*).
Zhu, Y. *et al.*, *Fitoterapia*, 2018, **130**, 281-289 (*Selamoellenin B*)

7,7′-Epoxy-8,8′-lign-7-ene-3,3′,4,4′,5,5′-hexol E158

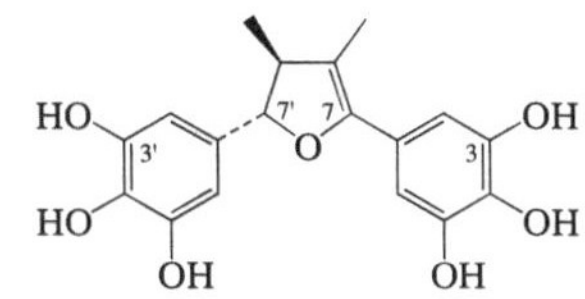

3,3′,4,4′,5,5′-Hexahydroxy-7,7′-epoxylign-7-ene

$C_{18}H_{18}O_7$ 346.335

(7′S,8′S)-form

3,4-Methylene, 3′,4′,5,5′-tetra-Me ether

7,7′-Epoxy-3,3′,4′,5′-tetramethoxy-4,5-methylenedioxy-8,8′-lign-7-ene. 3,3′,4′,5′-Tetramethoxy-4,5-methylenedioxy-7,7′-epoxylign-7-ene. ***Beilschmin C***

[916656-01-4]

$C_{23}H_{26}O_7$ 414.454

Constit. of the stems of *Beilschmiedia tsangii*. Cytotoxic to human colon carcinoma and mouse lymphocyte leukaemia P-388 cells. Oil. $[\alpha]_D^{24}$ +22.6 (c, 0.12 in $CHCl_3$). λ_{max} 216 (log ε 2.72); 275 (log ε 4.16) (MeOH).

Chen, J.-J. *et al.*, *Planta Med.*, 2006, **72**, 351-357 (*Beilschmin C*)

7,9′-Epoxy-8,8′-lign-7′-ene-3,3′,4,4′,5,9-hexol E159

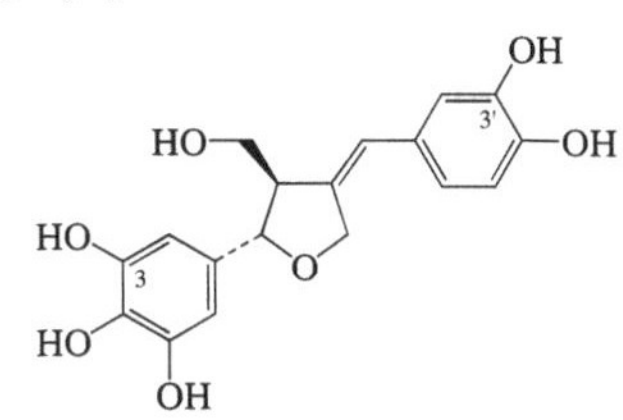

3,3′,4,4′,5,9-Hexahydroxy-7,9′-epoxylign-7′-ene

$C_{18}H_{18}O_7$ 346.335

(7S,7′Z,8R)-form

3,3′,4,4′,5-Penta-Me ether

7,9′-Epoxy-3,3′,4,4′,5-pentamethoxy-8,8′-lign-7′-en-9-ol. 9-Hydroxy-3,3′,4,4′,5-pentamethoxy-7,9′-epoxylign-7′-ene. ***Magnofargesin***

[156027-84-8]

$C_{23}H_{28}O_7$ 416.47

Constit. of the flower buds of *Magnolia fargesii*. PAF activating factor. Shows bronchodilatory activity. Cryst.. [α]D -33.3 (c, 3 in $CHCl_3$).

Miyazawa, M. *et al.*, *Phytochemistry*, 1996, **42**, 531-533 (*Magnofargesin*).

Japan. Pat., 2000, 6 065 224 (Nissan Chemical Industries) (*Magnofargesin, activity*).

Wardrop, D.J. *et al.*, *Org. Lett.*, 2006, **8**, 3659-3662 (*synth*).

Chakraborty, P. *et al.*, *Tetrahedron Lett.*, 2012, **53**, 6584-6587 (*synth*)

2,7′-Epoxy-3,8′-lign-8-ene-3′,4,4′,5,5′-pentol E160

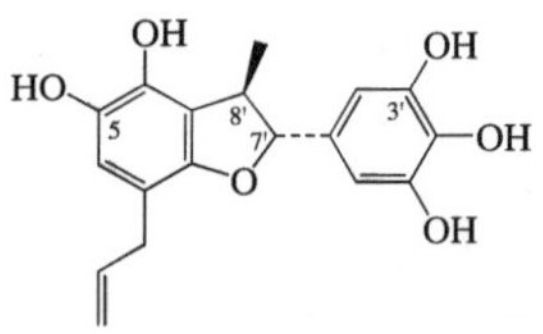

2,3-Dihydro-3-methyl-7-(2-propenyl)-2-(3,4,5-trihydroxyphenyl)-4,5-benzofurandiol. 2,7′-Epoxy-3′,4,4′,5,5′-pentahydroxy-3,8′-neolign-8-ene

$C_{18}H_{18}O_6$ 330.336

(7′R*,8′R*)-form

4,5-Methylene, 3′,4′-di-Me ether

2,7′-Epoxy-4′,5′-dimethoxy-4,5-methylenedioxy-3,8′-lign-8-en-3′-ol. 2,7′-Epoxy-3′-hydroxy-4′,5′-dimethoxy-4,5-methylenedioxy-3,8′-lign-8-ene. ***Ococymosin***

[1651840-80-0]

$C_{21}H_{22}O_6$ 370.401

Constit. of the stems of *Ocotea cymosa*. Exhibits activity against *Aedes aegypti* and *Plasmodium falciparum*. Oil. $[\alpha]_D^{25}$ +42 (c, 0.9 in MeOH). λ_{max} 240; 300 (MeOH aq.).

Rakotondraibe, L.H. *et al.*, *J. Nat. Prod.*, 2015, **78**, 431-440 (*Ococymosin, activity*)

3,7′-Epoxy-4,8′-lign-7-ene-3′,4′,6,9,9′-pentol E161

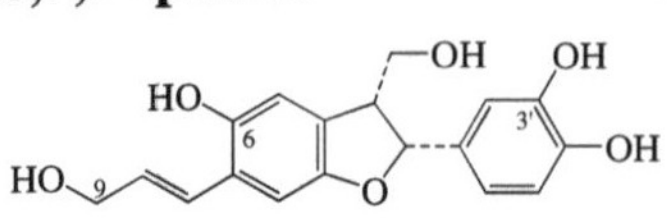

(7′R,8′R)-form

4-[5-Hydroxy-3-(hydroxymethyl)-6-(3-hydroxy-1-propenyl)-2,3-dihydrobenzofuran-2-yl]benzene-1,2-diol. 3,7′-Epoxy-3′,4′,6,9,9′-pentahydroxy-4,8′-neolign-7-ene

$C_{18}H_{18}O_6$ 330.336

(7′R,8′R)-form

3′,9-Di-Me ether, 6-O-β-D-glucopyranoside

Akeqintoside B

[1620439-22-6]

$C_{26}H_{32}O_{11}$ 520.531

Constit. of the stems of *Akebia quinata*. Exhibits moderate inhibitory activity against IL-6 prodn. in TNF-α stimulated MG-63 cells. Amorph. powder (MeOH). $[\alpha]_D^{27}$ -50.4 (c, 0.77 in MeOH). Abs. config. determined. However, while text and cd data indicate 7′*R*,8′*R*-form, the struct. indicates 7′*S*,8′*S*-form. λ_{max} 218 (log ε 4.22); 289 (log ε 3.63); 328 (log ε 4.0) (MeOH)

(7′S,8′S)-form

7,8-Dihydro, 3′-Me ether, 6-O-β-D-glucopyranoside

Akeqintoside A

[1620439-21-5]

$C_{25}H_{32}O_{11}$ 508.52

Constit. of the stems of *Akebia quinata*. Amorph. powder (MeOH). $[\alpha]_D^{20}$ -35.1 (c, 0.12 in MeOH). Abs. config. determined. However, while text and cd data indicate 7′*S*,8′*S*-form, the struct. indicates 7′*R*,8′*R*-form. λ_{max} 217 (log ε 4.23); 289 (log ε 3.61); 328 (log ε 4.07) (MeOH).

Jin, H.-G. *et al.*, *Chem. Pharm. Bull.*, 2014, **62**, 288-293 (*Akeqintosides A,B, activity*)

4,7′-Epoxy-3,8′-lign-7-ene-3′,4′,5,5′,9-pentol E162

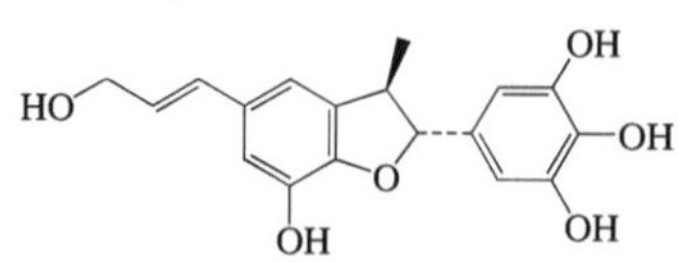

4,7′-Epoxy-3′,4′,5,5′,9-pentahydroxy-3,8′-neolign-7-ene

$C_{18}H_{18}O_6$ 330.336

(7E,7′R,8′R)-form

3′,4′-Methylene, 5,5′-di-Me ether

4,7′-Epoxy-3′,5-dimethoxy-4′,5′-methylenedioxy-3,8′-lign-7-en-9-ol. 4,7′-Epoxy-9-hydroxy-3′,5-dimethoxy-4′,5′-methylenedioxy-3,8′-lign-7-ene. ***Fragransol D***

[114892-43-2]

$C_{21}H_{22}O_6$ 370.401

Constit. of the arils of *Myristica fragrans*. Oil. [α]D +47.6 (c, 1 in $CHCl_3$)

3′,4′-Methylene, 5,5′-di-Me ether, 9-Ac

Maceneolignan C

[1967042-50-7]

$C_{23}H_{24}O_7$ 412.438

Constit. of the arils of *Myristica fragrans*. Oil. $[\alpha]_D^{25}$ +25 (c, 0.2 in $CHCl_3$). λ_{max} 213 (log ε 4.45); 278 (log ε 3.89); 306 (sh) (MeOH).

Hattori, M. *et al.*, *Chem. Pharm. Bull.*, 1988, **36**, 648-653 (*Fragransol D*).

Morikawa, T. *et al.*, *J. Nat. Prod.*, 2016, **79**, 2005-2013 (*Maceneolignan C*)

4,7′-Epoxy-3,8′-lign-7-ene-3′,4′,5,9,9′-pentol E163

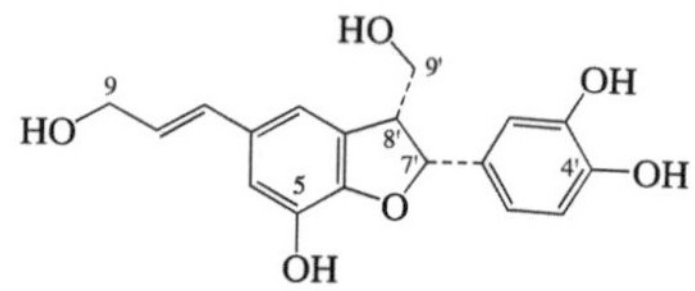

(7E,7′R,8′R)-form

2-(3,4-Dihydroxyphenyl)-2,3-dihydro-7-hydroxy-5-(3-hydroxy-1-propenyl)-5-benzofuranmethanol, CAS. *4,7′-Epoxy-3′,4′,5,9,9′-pentahydroxy-3,8′-neolign-7-ene*

[28199-68-0] [86363-02-2] [112835-77-5] [150407-42-4]

$C_{18}H_{18}O_6$ 330.336

Numerous derivs. have been isol. with unspecified stereochem. - some as mixts. of stereoisomers

(7E,7′R,8′R)-form

3′,5-Di-Me ether

4,7′-Epoxy-3′,5-dimethoxy-3,8′-lign-7-ene-4′,9,9′-triol. 4,7′-Epoxy-4′,9,9′-trihydroxy-3′,5-dimethoxy-3,8′-lign-7-ene. ***Vladinol E***

[133318-47-5]

$C_{20}H_{22}O_6$ 358.39

Constit. of *Vladimiria souliei* and *Vetiveria zizanioides*. Stereochem. in doubt; may have *trans*-config. Obt. as a mixt. with Vladinol F in E113

3′,5-Di-Me ether, 4′-O-β-D-glucopyranoside

Alaschanioside C

[157543-21-0]

[749858-83-1] (unspecified stereochem.)

$C_{26}H_{32}O_{11}$ 520.531

Constit. of *Aeschynanthus bracteatus*, *Euphrasia rostkoviana*, *Notopterygium incisum* and *Pedicularis alaschanica*. Amorph.

powder. $[\alpha]^{23}_{D}$ -100 (c, 0.1 in MeOH)

3′,5-Di-Me ether, 4′-O-[6-O-(4-hydroxy-3,5-dimethoxybenzoyl)-β-D-glucopyranoside], 9-O-β-D-glucopyranoside

Viburfordoside D

[2257425-89-9]

$C_{41}H_{50}O_{20}$ 862.831

Constit. of the fruit of *Viburnum fordiae*. Moderate inhibitor of α-glucosidase and of LPS-induced NO prodn.. Powder. $[\alpha]^{23}_{D}$ -13.3 (c, 0.1 in MeOH). λ_{max} 277 (log ε 4.64) (MeOH)

9-Carboxylic acid, 3′,5-di-Me ether, Me ester

Xylobuxin. *Gardenifolin H*

[165306-73-0]

[68340-37-4] [215319-52-1]

$C_{21}H_{22}O_7$ 386.400

Constit. of the stem bark of *Xylopia buxifolia* and *Gardenia ternifolia*. Amorph. powder. $[\alpha]_D$ +7 (c, 0.3 in MeOH) (Xylobuxin). $[\alpha]^{23}_{D}$ -20 (c, 0.04 in Me_2CO) (Gardenifolin H). λ_{max} 226 (log ε 2.3); 287 (log ε 1.7); 327 (log ε 4.3) (Me_2CO) (Gardenifolin H)

(7E,7′R,8′S)-form

3′-Me ether

4,7′-Epoxy-3′-methoxy-3,8′-lign-7-ene-4′,5,9,9′-tetrol. 4,7′-Epoxy-4′,5,9,9′-tetrahydroxy-3′-methoxy-3,8′-lign-7-ene. **Hierochin D**

[155759-04-9]

$C_{19}H_{20}O_6$ 344.363

Constit. of *Anastatica hierochuntica* and *Rehmannia glutinosa*. $[\alpha]^{25}_{D}$ -46.5 (c, 0.35 in MeOH)

3′-Me ether, 9-O-β-D-glucopyranoside

Glehlinoside I

[1335048-33-3]

$C_{25}H_{30}O_{11}$ 506.504

Constit. of the roots of *Glehnia littoralis*. Amorph. solid. $[\alpha]^{20}_{D}$ -16.1 (c, 0.3 in MeOH). λ_{max} 214 (log ε 5); 246 (log ε 4.5); 270 (log ε 4.7) (MeOH)

3′,5-Di-Me ether

(-)-Dehydrodiconiferyl alcohol

[155836-29-6]

[4263-87-0] (unspecified stereochem)

$C_{20}H_{22}O_6$ 358.39

Constit. of *Clematis armandii*, *Euonymus alatus*, *Illicium dunnianum*, *Picrasma quassioides*, *Sambucus williamsii* and *Tribulus terrestris*. Exhibits moderate cytotoxic activity. NO prodn. inhibitor. Mp 141 - 142°. $[\alpha]^{23}_{D}$ -11 (c, 0.8 in $CHCl_3$). $[\alpha]^{20}_{D}$ -25 (c, 0.1 in MeOH)

3′,5-Di-Me ether, 4′-O-β-D-glucopyranoside

[107870-87-1]

$C_{26}H_{32}O_{11}$ 520.531

Constit. of *Citrus* spp., *Pedicularis torta* and *Vinca rosea*. $[\alpha]^{20}_{D}$ -71.2 (c, 0.56 in MeOH)

3′,5-Di-Me ether, 4′,9′-di-O-β-D-glucopyranoside

[142674-79-1]

$C_{32}H_{42}O_{16}$ 682.672

Constit. of *Angelica dahurica* and *Verbascum thapsus*. Amorph. powder. Mp 203 - 204°. $[\alpha]^{19}_{D}$ -58.1 (c, 0.17 in MeOH aq.)

3′,5-Di-Me ether, 4′-O-[β-D-glucopyranosyl-(1→2)-[β-D-glucopyranosyl-(1→3)]-β-D-glucopyranoside]

Hovetrichoside E

[210050-40-1]

$C_{38}H_{52}O_{21}$ 844.813

Constit. of the bark of *Hovenia trichocarea*. Needles. Mp 227 - 229°. $[\alpha]^{25}_{D}$ -18.3 (c, 0.9 in Py). λ_{max} 210 (log ε 4.49); 233 (log ε 4.46); 278 (log ε 4.17) (MeOH)

3′,5-Di-Me ether, 9-O-β-D-glucopyranoside

[130756-26-2]

$C_{26}H_{32}O_{11}$ 520.531

Constit. of *Cistanche tubulosa*. Powder. $[\alpha]^{25}_{D}$ -7.4 (c, 1.6 in MeOH)

3′,5-Di-Me ether, 9-O-[4-hydroxy-2-methylenebutanoyl-(→6)-β-D-glucopyranoside]

Woorenoside V

[166990-18-7]

$C_{31}H_{38}O_{13}$ 618.632

Isol. from rhizomes of *Coptis japonica*. Exhibits antiinflammatory activity. Amorph. powder. $[\alpha]^{20}_{D}$ -11.8 (c, 2.4 in MeOH)

3′,5-Di-Me ether, 9-O-[β-D-apiofuranosyl-(1→2)-β-D-glucopyranoside]

Barlericin

[1307910-13-9]

$C_{31}H_{40}O_{15}$ 652.646

Constit. of *Barleria acanthoides*. Viscous solid. $[\alpha]^{20}_{D}$ -8 (c, 0.13 in MeOH). λ_{max} 219 (sh); 285; 300 (sh) (no solvent reported)

3′,5-Di-Me ether, 9-O-[β-D-apiofuranosyl-(1→6)-β-D-glucopyranoside]

Zanthoxyloside A

[1773515-06-2]

$C_{31}H_{40}O_{15}$ 652.646

Constit. of the stems of *Zanthoxylum schinifolium*. Significant inhibitor of NF-κB activity. Oil. $[\alpha]^{25}_{D}$ -19.2 (c, 0.3 in MeOH)

3′,5-Di-Me ether, 9′-O-β-D-glucopyranoside

[107911-09-1]

$C_{26}H_{32}O_{11}$ 520.531

Constit. of *Vinca rosea* and *Verbascum thapsus*

3′,5-Di-Me ether, 9′-O-(4-O-methyl-β-D-glucopyranoside)

Tortoside D

[176903-28-9]

$C_{27}H_{34}O_{11}$ 534.558

Constit. of *Pedicularis torta*. Amorph. powder. $[\alpha]^{14}_{D}$ +25.2 (c, 1.31 in MeOH)

3′,5-Di-Me ether, 9′-O-[β-D-glucopyranosyl-(1→2)-β-D-glucopyranoside]

[828264-66-0]

$C_{32}H_{42}O_{16}$ 682.672

Constit. of *Globularia cordifolia*. Amorph. powder. λ_{max} 222; 278; 340 (MeOH)

3′,5-Di-Me ether, 9′-Ac

(-)-Schisphenlignan I

[1542234-14-9]

$C_{22}H_{24}O_7$ 400.427

Constit. of *Clematis armandii*, *Jatropha integerrima* and *Schisandra sphenanthera*. Inhibitor of NO and TNF-α in LPS-activated BV-2 cells. Amorph. powder. $[\alpha]^{22}_{D}$ -13.5 (c, 0.17 in MeOH). λ_{max} 204 (log ε 2.73) ($CHCl_3$)

3′,5-Di-Me ether, 9,9′-di-Ac

[1260636-11-0]

$C_{24}H_{26}O_8$ 442.464

Constit. of *Limonia acidissima*, *Jatropha integerrima* and *Petradoria pumila*. Gum. $[\alpha]^{25}_{D}$ -60.5 (c, 0.4 in MeOH). λ_{max} 278 (log ε 4.1); 303 (log ε 3.5); 345 (log ε 3) (MeOH)

3′,9-Di-Me ether

4,7′-Epoxy-3′,9-dimethoxy-3,8′-lign-7-ene-4′,5,9′-triol. 4,7′-Epoxy-4′,5,9′-trihydroxy-3′,9-dimethoxy-3,8′-lign-7-ene. **Hierochin A**

[650610-26-7]
$C_{20}H_{22}O_6$ 358.39
Constit. of *Anastatica hierochuntica*. Melanogenesis inhibitor. Pale yellow oil. $[\alpha]_D^{24}$ -32.2 (c, 0.53 in MeOH). λ_{max} 279 (log ε 4.2) (MeOH)

3′,9-Di-Me ether, 4′-O-β-D-glucopyranoside
Glehlinoside J
[1335048-34-4]
$C_{26}H_{32}O_{11}$ 520.531
Constit. of the roots of *Glehnia littoralis*. Amorph. solid. $[\alpha]_D^{20}$ -30.9 (c, 0.15 in MeOH). λ_{max} 216 (log ε 5); 246 (log ε 4.5); 270 (log ε 4.6) (MeOH)

3′,9-Di-Me ether, 5-O-β-D-glucopyranoside
Coixide A
$C_{26}H_{32}O_{11}$ 520.531
Constit. of *Coix lachryma-jobi* var. *ma-yuen*. Yellow gum. $[\alpha]_D$ -55.1 (c, 0.08 in MeOH). λ_{max} 224; 274 (MeOH)

3′,4′,5-Tri-Me ether, 9-O-β-D-glucopyranoside
Longifloroside A
[175556-08-8]
$C_{27}H_{34}O_{11}$ 534.558
Constit. of *Pedicularis longiflora* and *Phlomis chimerae*. Amorph. powder. $[\alpha]_D^{14}$ +49 (c, 2.3 in MeOH)

3′,5,9-Tri-Me ether
4,7′-Epoxy-3′,5,9-trimethoxy-3,8′-lign-7-ene-4′,9′-diol. 4,7′-Epoxy-4′,9′-dihydroxy-3′,5,9-trimethoxy-3,8′-lign-7-ene. **Hawthornnin G.** *(-)-Jatrointelignan D*
[1621072-64-7]
$C_{21}H_{24}O_6$ 372.417
Constit. of *Crataegus pinnatifida*, *Jatropha integerrima*, *Melia azedarach* and *Sambucus williamsii*. Potent inhibitor of NO and TNF-α prodn.. Pale gum or yellow oil. $[\alpha]_D^{20}$ -63 (c, 0.15 in $CHCl_3$) ((-)-Jatrointelignan D). $[\alpha]_D^{20}$ -5.8 (c, 0.11 in MeOH) (Hawthornnin G). λ_{max} 215 (log ε 4.14); 276 (log ε 3.76) (MeOH) ((-)-Jatrointelignan D). [neutral] λ_{max} 230 (log ε 0.72); 280 (log ε 0.41) (MeOH) (Hawthornnin G)

9-Et, 3′,5-di-Me ether
4,7′-Epoxy-9-ethoxy-3′,5-dimethoxy-3,8′-lign-7-ene-4′,9′-diol. 4,7′-Epoxy-9-ethoxy-4′,9′-dihydroxy-3′,5-dimethoxy-3,8′-lign-7-ene
[1428538-15-1]
$C_{22}H_{26}O_6$ 386.444
Constit. of seedlings of *Stellera chamaejasme*. Pale yellow oil. $[\alpha]_D^{20}$ -18 (c, 0.1 in MeOH). λ_{max} 206 (log ε 4.52); 279 (log ε 4.15) (MeOH)

9-Et, 3′,5-di-Me ether, 4′-O-β-D-glucopyranoside
Acutoside A
[1643674-91-2]
$C_{28}H_{36}O_{11}$ 548.585
Constit. of *Gentianella acute*. Exhibits cardioprotective effect against H_2O_2-induced apoptosis in H9c2 cardiomyoblasts. Amorph. pale yellow solid. $[\alpha]_D^{25}$ -1.4 (c, 0.9 in MeOH). λ_{max} 221; 278 (MeOH)

9′-Et, 3′,5-di-Me ether, 9-O-β-D-glucopyranoside
Acutoside B
[1643674-92-3]
$C_{28}H_{36}O_{11}$ 548.585
Constit. of *Gentianella acute*. Amorph. pale yellow solid. $[\alpha]_D^{25}$ -8.26 (c, 0.5 in MeOH). λ_{max} 222; 278 (MeOH)

9-Aldehyde, 3′,5-di-Me ether
4,7′-Epoxy-4′,9′-dihydroxy-3′,5-dimethoxy-3,8′-lign-7-en-9-al. 4,7′-Epoxy-4′,9′-dihydroxy-3′,5-dimethoxy-9-oxo-3,8′-lign-7-ene.
(-)-Balanophonin‡
[80286-36-8]
$C_{20}H_{20}O_6$ 356.374
Constit. of *Artemisia annua*, *Balanophora japonica*, *Cirsium eriophorum*, *Chrysolaena verbascifora*, *Jatropha integerrima* and *Picrasma quassioides*. Exhibits cytotoxic activity. Pale yellow oil. $[\alpha]_D^{20}$ -15 (c, 0.1 in MeOH). Abs. config. revised in 1998. Not to be confused with Balanophonin A, B1

9-Aldehyde, 3′,5-di-Me ether, 4′-O-β-D-glucopyranoside
Balanophonin 4′-glucoside
[215319-51-0]
$C_{26}H_{30}O_{11}$ 518.515
Constit. of the whole plants of *Aeginetia indica*. Pale yellow powder. Mp 254 - 255°. $[\alpha]_D^{22}$ -92.3 (c, 1 in MeOH)

9-Aldehyde, 3′,5-di-Me ether, 9′-O-β-D-glucopyranoside
Tortoside F
[190655-19-7]
$C_{26}H_{30}O_{11}$ 518.515
Constit. of *Pedicularis torta*. Amorph. powder. $[\alpha]_D^{14}$ +35.9 (c, 0.32 in MeOH)

9-Aldehyde, 3′,5-di-Me ether, 9′-O-[4-hydroxy-3-methoxy-E-cinnamoyl-(→6)-β-D-glucopyranoside]
Ovatoside A
[2129610-85-9]
$C_{36}H_{38}O_{14}$ 694.686
Constit. of the stem bark of *Catalpa ovata*. Potent scavenger of peroxynitrite. Amorph. gum. $[\alpha]_D^{22}$ -102 (c, 0.1 in MeOH). λ_{max} 230 (log ε 4.45); 290 (log ε 4.26); 329 (log ε 4.49) (MeOH)

9-Aldehyde, 3′,5-di-Me ether, 9′-Ac
(-)-Crataegifin B
Exhibits selective cytotoxic activity. $[\alpha]_D^{20}$ -45.6 (c, 0.2 in MeOH)

9-Aldehyde, 3′,5-di-Me ether, 9′-O-(3-methylbutanoyl)
9′-O-(3-Methylbutanoyl)balanophonin.
9′-O-Isovaleryloxybalanophonin
[1470306-05-8]
$C_{25}H_{28}O_7$ 440.492
Isol. from the roots of *Otanthus maritimus*. Yellow oil. $[\alpha]_D^{25}$ +9.3 (c, 0.08 in CH_2Cl_2). λ_{max} 207 (log ε 1.0); 229 (log ε 0.2); 341 (log ε 0.4) (CH_2Cl_2)

9-Aldehyde, 3′,5-di-Me ether, 9′-O-(4-hydroxybenzoyl)
9′-O-(4-Hydroxybenzoyl) balanophonin
[158531-57-8]
$C_{27}H_{24}O_8$ 476.481
Constit. of *Calamus quiquesetinervius*, *Chosenia arbutifolia* and *Posidonia oceanica*

9-Aldehyde, 3′,4′,5-tri-Me ether
4,7′-Epoxy-9′-hydroxy-3′,4′,5-trimethoxy-3,8′-lign-7-en-9-al. 4,7′-Epoxy-9′-hydroxy-3′,4′,5-trimethoxy-9-oxo-3,8′-lign-7-ene.
4-O-Methylbalanophonin
[215228-47-0]
$C_{21}H_{22}O_6$ 370.401
Constit. of the whole plant of *Cassytha filiformis*. Pale yellow oil. $[\alpha]_D^{22}$ -110.4 (c, 0.01 in $CHCl_3$). Abs. config. probably stated incorr. in ref.. λ_{max} 342 (log ε 4.28) (EtOH)

9-Aldehyde, 3′,4′-methylene, 5-Me ether
4,7′-Epoxy-9′-hydroxy-5-methoxy-3′,4′-methylenedioxy-3,8′-lign-7-en-9-al. 4,7′-Epoxy-9′-hydroxy-5-methoxy-3′,4′-methylenedioxy-9-oxo-3,8′-lign-7-ene. **Hyoscyamal**
[1199225-19-8]
$C_{20}H_{18}O_6$ 354.358
Constit. of the seeds of *Hyoscyamus niger*. Yellow oil. [α] D -13.8 (c, 0.04 in MeOH). λ_{max} 225; 332 (MeOH)

9-Carboxylic acid, 3′,4′-di-Me ether
4,7′-Epoxy-5,9′-dihydroxy-3′,4′-dimethoxy-3,8′-lign-7-en-9-oic acid. **Avellanedae A**
[1476060-00-0]
$C_{20}H_{20}O_7$ 372.373
Isol. from *Tabebuia avellanedae*. Displays antiinflammatory activity. Amorph. light yellow solid. $[\alpha]_D^{30}$ +7.0 (c, 1.0 in MeOH). λ_{max} 212 (log ε 4.40); 323 (log ε 3.99) (MeOH)

9-Carboxylic acid, 3′,5-di-Me ether
4,7′-Epoxy-4′,9′-dihydroxy-3′,5-dimethoxy-3,8′-lign-7-en-9-oic acid

9-Carboxylic acid, 3′,5-di-Me ether, 4′-O-α-L-rhamnopyranoside
Gelsemiunoside A
[1170710-46-9]
$C_{26}H_{30}O_{11}$ 518.515
Constit. of the whole plant of *Gelsemium elegans*. Pale yellow powder. Mp 150 - 152°. $[\alpha]_D^{22}$ -108.3 (c, 0.33 in MeOH). λ_{max} 222 (log ε 4.01); 322 (log ε 3.89) (MeOH)

9-Carboxylic acid, 3′,5-di-Me ether, 4′-O-β-D-glucopyranoside
Aegineoside
[752209-48-6]
$C_{26}H_{30}O_{12}$ 534.514
Constit. of the whole plant of *Aeginetia indica*. Pale yellow powder. Mp 262 - 264°. $[\alpha]_D^{22}$-56.7 (c, 1 in MeOH). λ_{max} 290 (log ε 3.17); 325 (sh) (EtOH)

9-Carboxylic acid, 3′,5-di-Me ether, 9′-O-β-D-glucopyranoside
Cimicifugaside F
$C_{26}H_{30}O_{12}$ 534.514
Constit. of the rhizomes of *Cimicifuga dahurica*. Yellow solid

9-Carboxylic acid, 3′,5-di-Me ether, Me ester
Gardenifolin D
[215319-52-1]
$C_{21}H_{22}O_7$ 386.400
Constit. of the stem bark of *Gardenia ternifolia*. Exhibits high cytotoxicity against human cancer HeLa cell lines. Amorph. powder. $[\alpha]_D^{23}$-21 (c, 0.03 in Me_2CO). Obt. by chiral hplc. λ_{max} 226 (log ε 2.3); 287 (log ε 1.7); 327 (log ε 4.3) (Me_2CO)

9-Carboxylic acid, 3′,5-di-Me ether, 1ξ,3-bis(methoxycarbonyl)propylamide
[185311-15-3]
$C_{27}H_{31}NO_{10}$ 529.542
Constit. of *Boreava orientalis*. Pale yellow powder. Mp 82 - 84.5°. λ_{max} 288 (log ε 3.3); 320 (log ε 3.32) (MeOH)

(7E,7′S,8′R)-form

3′-Me ether
[155759-02-7]
$C_{19}H_{20}O_6$ 344.363
Constit. of *Sambucus williamsii*. [α] D -4 (c, 1.2 in EtOH). Isol. as partial racemate

3′-Me ether, 5-O-β-D-glucopyranoside
Yemuoside YM*₁*
[140850-36-8]
$C_{25}H_{30}O_{11}$ 506.504
Constit. of *Cichorium intybus* and *Stauntonia chinensis*. Powder. Mp 123 - 127°. $[\alpha]_D^{23}$-30.7 (c, 0.101 in MeOH)

5-Me ether
4,7′-Epoxy-5-methoxy-3,8′-lign-7-ene-3′,4′,9,9′-tetrol. 4,7′-Epoxy-3′,4′,9,9′-tetrahydroxy-5-methoxy-3,8′-lign-7-ene.
3-De-O-methyldehydrodiconiferyl alcohol
$C_{19}H_{20}O_6$ 344.363
Constit. of *Taxus mairei*. $[\alpha]_D^{24}$-10 (c, 1 in MeOH)

3′,5-Di-Me ether
(+)-Dehydrodiconiferyl alcohol
[97465-82-2]
Constit. of *Anastatica hierochuntica*, *Clematis delavayi*, *Euonymus alatus* and *Picrasma quassioides*. NO prodn. inhibitor. Melanogenesis inhibitor. Mp 141 - 142°. $[\alpha]_D^{23}$ +11.6 (c, 0.13 in $CHCl_3$). $[\alpha]_D^{20}$+20 (c, 0.1 in MeOH)

3′,5-Di-Me ether, 4′-O-β-D-glucopyranoside
[107870-88-2]
Constit. of *Capparis flavicans*, *Cistanche tubulosa*, *Urtica dioica*, *Sinomenium acutum* and *Vinca rosea*. Powder. $[\alpha]_D^{25}$-56.4 (c, 1.1 in MeOH)

3′,5-Di-Me ether, 9′-O-β-D-glucopyranoside
[107870-89-3]
Constit. of *Boreava orientalis*, *Cynomorium songaricum*, *Verbascum thapsus* and *Vinca rosea*. Pale yellow powder. $[\alpha]_D^{15}$-19.5 (c, 0.4 in MeOH)

3′,5-Di-Me ether, 4′,9-di-O-β-D-glucopyranoside
[109792-90-7]
$C_{32}H_{42}O_{16}$ 682.672
Constit. of *Eucommia ulmoides*. Powder (EtOH). Mp 134.7°. $[\alpha]_D^{28}$ -62.3 (c, 0.14 in MeOH)

3′,5-Di-Me ether, 9′-Ac
(+)-Schisphenlignan I
[1842427-58-0]
$C_{22}H_{24}O_7$ 400.427
Constit. of *Jatropha integerrima*. Yellow oil. $[\alpha]_D^{20}$+72 (c, 0.17 in $CHCl_3$). Resolved by chiral hplc. λ_{max} 215 (log ε 3.94); 276 (log ε 3.73) (MeOH)

3′,5-Di-Me ether, 9,9′-di-Ac
[184046-40-0]
[238098-72-1]
$C_{24}H_{26}O_8$ 442.464
Constit. of *Baccharis* spp., *Euphorbia* spp., *Jatropha integerrima*, *Lasiolaena morii* and propolis. Prob. a partial racemate

3′,4′,5-Tri-Me ether
4,7′-Epoxy-3′,4′,5-trimethoxy-3,8′-lign-7-ene-9,9′-diol. 4,7′-Epoxy-9,9′-dihydroxy-3′,4′,5-trimethoxy-3,8′-lign-7-ene. **Hierochin B**
[215228-14-1]
$C_{21}H_{24}O_6$ 372.417
Constit. of *Anastatica hierochuntica*. Melanogenesis inhibitor. Oil. $[\alpha]_D^{24}$+29.7 (c, 0.4 in MeOH). λ_{max} 279 (log ε 4.2) (MeOH)

3′,5,9-Tri-Me ether
(+)-Jatrointelignan D
[125564-66-1]
$C_{21}H_{24}O_6$ 372.417
Constit. of *Jatropha integerrima*. Pale gum. $[\alpha]_D^{20}$+74 (c, 0.14 in $CHCl_3$). Resolved by chiral hplc.

λ_{max} 215 (log ε 4.14); 276 (log ε 3.76) (MeOH)

3′,5,9-Tri-Me ether, 4′-O-β-D-glucopyranoside
[336103-99-2]
$C_{27}H_{34}O_{11}$ 534.558
Constit. of the bulbs of *Urginea maritima*. Amorph. powder. $[\alpha]_D^{28}$ -54.8 (c, 0.67 in MeOH). λ_{max} 204 (log ε 4.84); 219 (log ε 4.51); 278 (log ε 4.26); 292 (sh) (MeOH)

3′,5,9-Tri-Me ether, 9′-O-β-D-glucopyranoside
Plucheoside D$_3$
[133733-97-8]
$C_{27}H_{34}O_{11}$ 534.558
Isol. from *Pluchea indica*. Powder. $[\alpha]_D^{21}$ -56.8 (c, 0.22 in MeOH)

3′,5,9-Tri-Me ether, 9′-O-[4-hydroxy-3-methoxy-E-cinnamoyl-(→6)-β-D-glucopyranoside]
Ovatoside B
[2129610-87-1]
$C_{37}H_{42}O_{14}$ 710.729
Constit. of the stem bark of *Catalpa ovata*. Potent scavenger of peroxynitrite. Amorph. gum. $[\alpha]_D^{22}$ -33 (c, 0.1 in MeOH). λ_{max} 218 (log ε 4.15); 281 (log ε 3.91); 320 (log ε 3.79) (MeOH)

3′,5,9,9′-Tetra-Me ether
4,7′-Epoxy-3′,5,9,9′-tetramethoxy-3,8′-lign-7-en-4′-ol. 4,7′-Epoxy-4′-hydroxy-3′,5,9,9′-tetramethoxy-3,8′-lign-7-ene
[2244938-41-6]
$C_{22}H_{26}O_6$ 386.444
Constit. of *Cnidium officinale*. Yellowish oil. $[\alpha]_D^{24}$ -3.7 (c, 0.1 in MeOH). λ_{max} 220 (log ε 4.25) (MeOH)

9-Et, 3′,5-di-Me ether
[1290633-27-0]
$C_{22}H_{26}O_6$ 386.444
Constit. of the stems of *Sinocalamus affinis*. Amorph. powder. $[\alpha]_D^{20}$ +11.8 (c, 0.1 in MeOH). λ_{max} 204 (log ε 4.16); 278 (log ε 3.67) (MeOH)

9-Butyl, 3′,5-di-Me ether, 4′-O-β-D-glucopyranoside
[1831849-01-4]
$C_{30}H_{40}O_{11}$ 576.639
Constit. of the roots of *Heracleum dissectum* and the aerial parts of *Prunella vulgaris*. Exhibits significant cytotoxicity against HepG2 and moderate cytotoxicity against HCT-8 and HeLa cells. Amorph. powder. $[\alpha]_D^{20}$ -6.5 (c, 0.05 in MeOH). $[\alpha]_D^{25}$ -42.5 (c, 1.0 in MeOH). λ_{max} 232 (log ε 3.54) (MeOH). [neutral] λ_{max} 277 (log ε 3.54) (MeOH)

9-Aldehyde, 3′,5-di-Me ether
(+)-Balanophonin‡
[215319-47-4]
$C_{20}H_{20}O_6$ 356.374
Constit. of *Anastatica hierochuntica, Jatropha integerrima, Picrasma quassioides* and *Schisandra lancifolia*. Melanogenesis inhibitor. Exhibits antiviral and cytotoxic activities. $[\alpha]_D^{20}$ +10 (c, 0.1 in MeOH). Not to be confused with Balanophonin A, B1

9-Aldehyde, 3′,5-di-Me ether, 4′-O-β-D-glucopyranoside
(+)-Balanophonin 4′-glucoside
[850089-58-6]
$C_{26}H_{30}O_{11}$ 518.515
Constit. of *Balanophora japonica* and *Tabebuia impetiginosa*. Amorph. powder. $[\alpha]_D^{24}$ +7.5 (c, 0.94 in MeOH). λ_{max} 227 (log ε 4.3); 240 (sh); 286 (sh); 340 (log ε 4.32) (MeOH)

9-Aldehyde, 3′,5-di-Me ether, 9′-O-β-D-glucopyranoside
Plucheoside D$_1$
[133733-95-6]
$C_{26}H_{30}O_{11}$ 518.515
Isol. from *Pluchea indica*. Powder. $[\alpha]_D^{21}$ -116.4 (c, 0.64 in MeOH)

9-Aldehyde, 3′,5-di-Me ether, 9′-Ac
(+)-Crataegifin B
Exhibits selective cytotoxic activity. $[\alpha]_D^{20}$ +47.3 (c, 0.2 in MeOH)

9-Aldehyde, 3′,5-di-Me ether, Me hemiacetal
4,7′-Epoxy-3′,5,9-trimethoxy-3,8′-lign-7-ene-4′,9,9′-triol. 4,7′-Epoxy-4′,9,9′-trihydroxy-3′,5,9-trimethoxy-3,8′-lign-7-ene.
Viterolignan A
[1469986-06-8]
$C_{21}H_{24}O_7$ 388.416
Constit. of the fruit of *Vitex rotundifolia*. Inhibits LPS-induced NO prodn.. Yellow oil. $[\alpha]_D^{25}$ +14.7 (c, 0.12 in MeOH). λ_{max} 221 (log ε 3.73); 280 (log ε 3.56) (MeOH)

9-Aldehyde, 4′,5-di-Me ether
4,7′-Epoxy-3′,9′-dihydroxy-4′,5-dimethoxy-3,8′-lign-7-en-9-al. 4,7′-Epoxy-3′,9′-dihydroxy-4′,5-dimethoxy-9-oxo-3,8′-lign-7-ene
[1810717-78-2]
$C_{20}H_{20}O_6$ 356.374
Constit. of *Acanthopanax senticosus*. Protein tyrosine phosphatase 1B inhibitor. Amorph. powder. $[\alpha]_D^{25}$ -2.6 (c, 0.1 in MeOH). λ_{max} 205 (log ε 4.45); 225 (log ε 4.21); 290 (log ε 3.83); 340 (log ε 4.21) (MeOH)

9-Carboxylic acid
4,7′-Epoxy-3′,4′,5,9′-tetrahydroxy-3,8′-lign-7-en-9-oic acid.
Morindolin
[777061-46-8]
$C_{18}H_{16}O_7$ 344.319
Constit. of the fruit of *Morinda citrifolia*. Potent inhibitor of low-density lipoprotein oxidn.. Pale yellow powder. $[\alpha]_D^{24}$ +36 (c, 0.3 in MeOH). λ_{max} 206 (log ε 4.47); 209 (log ε 4.48); 224 (log ε 4.34); 324 (log ε 4.21) (MeOH)

9-Carboxylic acid, 3′-Me ether
4,7′-Epoxy-4′,5,9′-trihydroxy-3′-methoxy-3,8′-lign-7-en-9-oic acid.
Inovonoid B
$C_{19}H_{18}O_7$ 358.346
Constit. of *Balanophora involucrate*. Moderate α-glucosidase inhibitor. Amorph. powder. $[\alpha]_D^{25}$ -8.5 (c, 0.1 in MeOH). λ_{max} 232 (log ε 3.15); 290 (log ε 3.08) (MeOH)

9-Carboxylic acid, 3′,5-di-Me ether, Me ester
Gardenifolin C
[215319-48-5]
$C_{21}H_{22}O_7$ 386.400
Constit. of the stem bark of *Gardenia ternifolia*. Amorph. powder. $[\alpha]_D^{23}$ +21 (c, 0.04 in Me_2CO). Obt. by chiral hplc. λ_{max} 226 (log ε 4.3); 287 (log ε 1.9); 327 (log ε 5.2) (Me_2CO)

9-Carboxylic acid, 3′,5-di-Me ether, glycine amide
$C_{22}H_{23}NO_8$ 429.425
Constit. of the root bark of *Cordia alliodora*. Brown powder. $[\alpha]_D^{20}$ +6.9 (c, 0.05 in MeOH)

(7E,7′S,8′S)-form
3′,5-Di-Me ether, 4′-O-β-D-glucopyranoside
[96738-84-0]
Constit. of *Chamaecyparis obtusa*, *Citrus* spp., *Fortunella japonica* and *Plagiorhegma dubium*. Mp 125°
3′,5-Di-Me ether, 9′-O-β-D-glucopyranoside
[96738-86-2]
$C_{26}H_{32}O_{11}$ 520.531
Isol. from *Plagiorhegma dubium* cell cultures. Mp 115°
9-Carboxylic acid, 3′,5-di-Me ether, Me ester
Gardenifolin G
[2098350-35-5]
$C_{21}H_{22}O_7$ 386.400
Constit. of the stem bark of *Gardenia ternifolia*. Amorph. powder. $[\alpha]_D^{23}$ +26 (c, 0.05 in Me_2CO). Obt. by chiral hplc. λ_{max} 226 (log ε 4.3); 287 (log ε 1.9); 327 (log ε 5.2) (Me_2CO)
9,9′-Dicarboxylic acid, di-Et ester
Pharbilignan C. *Pharbilignanol C*
[1620558-56-6]
$C_{22}H_{22}O_8$ 414.41
Constit. of the seeds of *Pharbitis nil*. Exhibits potent cytotoxicity against human tumour cell lines AS549, SK-OV-3, SK-MEL-2 and HCT-15. Amorph. powder. $[\alpha]_D^{25}$ -26.3 (c, 0.65 in MeOH). Possesses (7′S,8′R)-config. due to change in Cahn-Ingold-Prelog priorities. λ_{max} 229 (log ε 4.23); 256 (log ε 3.98); 322 (log ε 3.75) (MeOH)
(7E,7′R,8′S*)-form*
9′-O-(4-Hydroxy-3-methoxy-E-cinnamoyl), 9-O-(4-hydroxy-E-cinnamoyl)
Boehmenan K
[344944-03-2]
$C_{39}H_{36}O_{11}$ 680.706
Constit. of *Hibiscus cannabinus*. Yellowish oil
3′,5-Di-Me ether, 9′-O-(4-hydroxybenzoyl)
[1456786-56-3]
$C_{27}H_{26}O_8$ 478.497
Constit. of *Posidonia oceanica*. Oil. $[\alpha]_D$ -8.2 (c, 0.08 in MeOH)
3′,5-Di-Me ether, 9,9′-bis-O-(4-hydroxybenzoyl)
Quiquesetinerviusin A
[1250376-62-5]
$C_{34}H_{30}O_{10}$ 598.604
Constit. of *Calamus quiquesetinervius* and *Posidonia oceanica*. Amorph. powder. Mp 110°. $[\alpha]_D^{25}$ -4.3 (c, 0.47 in MeOH). λ_{max} 262 (log ε 4.52); 308 (sh) (MeOH)
3′,4′,9-Tri-Me ether, 9′-Ac
[468081-54-1]
$C_{23}H_{26}O_7$ 414.454
Constit. of the stem wood of *Euphorbia quinquecostata*. Yellowish oil. $[\alpha]_D^{23}$ -73.3 (c, 0.1 in $CHCl_3$). λ_{max} 209 (log ε 3.68); 279 (log ε 3.38); 339 (log ε 3.12) (MeOH)
9-Aldehyde
4,7′-Epoxy-3′,4′,5,9′-tetrahydroxy-3,8′-lign-7-en-9-al. 4,7′-Epoxy-3′,4′,5,9′-tetrahydroxy-9-oxo-3,8′-lign-7-ene. ***Americanin D***
[77053-45-3]
[106622-79-1] ((±)-form)
$C_{18}H_{16}O_6$ 328.32
Constit. of *Morinda citrifolia* and *Phytolacca americana*. Cryst.. Mp 197 - 200°
9-Aldehyde, 3′,5-di-Me ether, 9′-Ac
9′-O-Acetylbalanophonin
[210161-63-0]
$C_{22}H_{22}O_7$ 398.411
Constit. of Chilean propolis and *Feroniella lucida*. Exhibits cytotoxic activity against a range of of human cancer cell lines. Oil. $[\alpha]_D^{25}$ +13.1 (c, 1.1 in CH_2Cl_2). λ_{max} 287 (log ε 3.75); 338 (log ε 4.07) (MeOH)
9-Carboxylic acid, 3′,5-di-Me ether, 9′-O-β-D-glucopyranoside
[1860824-39-0]
$C_{26}H_{30}O_{12}$ 534.514
Constit. of the fruit of *Solanum torvum*
9-Carboxylic acid, 3′,5-di-Me ether, Me ester
[107296-48-0]
$C_{21}H_{22}O_7$ 386.4
Constit. of *Brassica napus* and *Zizyphus jujuba*. Prostaglandin I_2 inducer
9-Carboxylic acid, 3′,5-di-Me ether, 2-(4-hydroxyphenyl)ethylamide
Grossamide K
[403647-07-4]
$C_{28}H_{29}NO_7$ 491.54
Constit. of the bark of *Hibiscus cannabinus*. Yellowish oil
9,9′-Dicarboxylic acid
4,7′-Epoxy-3′,4′,5-trihydroxy-3,8′-lign-7-ene-9,9′-dioic acid
C-8′ config. changes due to Cahn-Ingold-Prelog priorities
9,9′-Dicarboxylic acid, 3′,5-di-Me ether
3-[3-Carboxy-2,3-dihydro-2-(4-hydroxy-3-methoxyphenyl)-7-methoxy-5-benzofuranyl]-2-propenoic acid, CAS. 4,7′-Epoxy-4′-hydroxy-3′,5-dimethoxy-3,8′-lign-7-ene-9,9′-dioic acid. 4-O-7′:5,8′-Diferulic acid. 5,8′-Diferulic acid (benzofuran-form)
[160169-55-1]
[151998-30-0] [180579-78-6]
$C_{20}H_{18}O_8$ 386.356
Isol. by saponification from wheat bran, cocksfoot, switch grass and cultured corn cells. Present in plant cell walls as cross-linked polysaccharide esters. Also found in *Beta vulgaris* and *Daucus carota*. Antioxidant.Exhibits anticancer activity. Fine cryst.. Mp 169 - 174° dec.. Possesses (7′R*,8′R*)-config. due to change in Cahn-Ingold-Prelog priorities
9,9′-Dicarboxylic acid, 3′,5-di-Me ether, di-Et ester
[211190-63-5]
[160169-54-0]
Needles (Me_2CO/petrol). Mp 152.8 - 153.1°
9,9′-Dicarboxylic acid, 3′,5-di-Me ether, bis[(4-hydroxyphenyl)ethylamide]
see Grossamide, G29
7R,8R*-Epoxide, 3′,5-di-Me ether*
4,7′:7,8-Diepoxy-3′,5-dimethoxy-3,8′-lignan-4′,9,9′-triol. 4,7′:7,8-Diepoxy-4′,9,9′-trihydroxy-3′,5-dimethoxy-3,8′-lignan. ***Lyciumnan***
[1940189-88-7]
$C_{20}H_{22}O_7$ 374.389
Constit. of the leaves of *Lycium barbarum*. Exhibits cytotoxicity against A549, HeLa and PC-3 cell lines. Oil. $[\alpha]_D^{20}$ -12 (c, 0.05 in MeOH)
(7E,7′RS,8′RS)-form
3′,5-Di-Me ether
[60536-58-5]
Mp 160 - 161°

(7E,7′RS,8′SR)-form

3′,5-Di-Me ether

(±)-Dehydrodiconiferyl alcohol

[86363-02-2]

$C_{20}H_{22}O_6$ 358.390

Constit. of *Picrasma quassioides* and *Schisandra lancifolia*. Exhibits moderate cytotoxic activity. Yellow oil. Resolved by chiral hplc

3′,5-Di-Me ether, 9,9′-dibenzoyl

(±)-Dehydrodiconiferyl 9,9′-dibenzoate. ***Japonin A***

[478178-35-7]

$C_{34}H_{30}O_8$ 566.606

Constit. of the roots of *Euterpe precatoria* and *Stemona japonica*. Shows moderate antiplasmodial activity. Oil. λ_{max} 250 (log ε 1.35) (MeCN aq.)

9-Aldehyde, 3′,5-di-Me ether

(±)-Balanophonin

[118916-57-7]

$C_{20}H_{20}O_6$ 356.374

Constit. of the stems of *Picrasma quassioides*. Pale yellow solid. Resolved by chiral hplc

9-Aldehyde, 3′,5-di-Me ether, 9′-Ac

Crataegifin B

Constit. of the fruit of *Crataegus pinnatifida*. Yellow oil. $[\alpha]^{20}_{D}$ +2.2 (c, 0.1 in MeOH). λ_{max} 228 (log ε 3.85); 338 (log ε 3.76) (MeOH)

(7E,7′ξ,8′ξ)-form

3′,5-Di-Me ether

[28199-68-0]

$C_{20}H_{22}O_6$ 358.39

Isol. from *Silybum marianum*. Isolate was dextrorotatory, no stereochem. assigned

3′,5-Di-Me ether, 9-O-β-D-glucopyranoside

$C_{26}H_{32}O_{11}$ 520.531

Constit. of cane sugar. Oil. λ_{max} 218 (sh); 278; 300 (sh) (MeOH)

9′-Carboxylic acid, 3′,5-di-Me ether

4,7′-Epoxy-4′,9-dihydroxy-3′,5-dimethoxy-3,8′-lign-7-en-9′-oic acid. Glycosmisic acid. ***Spicatolignan B***

[443908-19-8]

$C_{20}H_{20}O_7$ 372.373

Constit. of *Cinnamomum cassia*, *Mentha spicata* and *Glycosmis citrifolia*. Amorph. powder (MeOH). Mp 150 - 152°

(7Z,7′R,8′R)-form

9-Carboxylic acid, 3′,5-di-Me ether, Me ester

Gardenifolin F

[2098350-34-4]

$C_{21}H_{22}O_7$ 386.400

Constit. of the stem bark of *Gardenia ternifolia*. Amorph. powder. $[\alpha]^{23}_{D}$ -23 (c, 0.03 in Me_2CO). Obt. by chiral hplc. λ_{max} 207 (log ε 3.1); 226 (log ε 4.2); 288 (log ε 1.7); 321 (log ε 5.2) (Me_2CO)

(7Z,7′R,8′S)-form

3′,5-Di-Me ether, 4′-O-β-D-glucopyranoside

Tenuiside E

[1473421-74-7]

$C_{26}H_{32}O_{11}$ 520.531

Constit. of *Polygala tenuifolia*. Powder. $[\alpha]^{20}_{D}$ -20.0 (c, 0.3 in MeOH). λ_{max} 308 (log ε 3.89) (MeOH)

9-Carboxylic acid, 3′,5-di-Me ether, Me ester

Gardenifolin B

[2098350-32-2]

$C_{21}H_{22}O_7$ 386.400

Constit. of the stem bark of *Gardenia ternifolia*. Amorph. powder. $[\alpha]^{23}_{D}$ -15 (c, 0.02 in Me_2CO). Obt. by chiral hplc. λ_{max} 207 (log ε 5.1); 226 (log ε 4.2); 288 (log ε 1.9); 321 (log ε 5.2) (Me_2CO)

9-Carboxylic acid, 3′,5-di-Me ether, 1ξ,3-bis(methoxycarbonyl)propylamide

[185311-14-2]

Constit. of *Boreava orientalis*. Pale yellow powder. Mp 95 - 99°. λ_{max} 287 (log ε 3.07); 318 (log ε 1.83) (MeOH)

(7Z,7′S,8′R)-form

9-Aldehyde, 3′,5-di-Me ether, dimethyl acetal

4,7′-Epoxy-3′,5,9,9-tetramethoxy-3,8′-lign-7-ene-4′,9′-diol. 4,7′-Epoxy-4′,9′-dihydroxy-3′,5,9,9-tetramethoxy-3,8′-lign-7-ene. ***Viterolignan B***

[1469986-07-9]

$C_{22}H_{26}O_7$ 402.443

Constit. of the fruit of *Vitex rotundifolia*. Yellow oil. $[\alpha]^{25}_{D}$ +26.3 (c, 0.25 in MeOH). λ_{max} 220 (log ε 3.71); 274 (log ε 3.50) (MeOH)

9-Carboxylic acid, 3′,5-di-Me ether, Me ester

Gardenifolin A

[2098350-31-1]

$C_{21}H_{22}O_7$ 386.400

Constit. of the stem bark of *Gardenia ternifolia*. Amorph. powder. $[\alpha]^{23}_{D}$ +20 (c, 0.02 in Me_2CO). Obt. by chiral hplc. λ_{max} 207 (log ε 4.9); 226 (log ε 4.4); 288 (log ε 3.5); 321 (log ε 4.1) (Me_2CO)

(7Z,7′S,8′S)-form

9-Carboxylic acid, 3′,5-di-Me ether, Me ester

Gardenifolin E

[2098350-33-3]

$C_{21}H_{22}O_7$ 386.400

Constit. of the stem bark of *Gardenia ternifolia*. Exhibits high cytotoxicity against human cancer HeLa cell lines. Amorph. powder. $[\alpha]^{23}_{D}$ +27 (c, 0.03 in Me_2CO). Obt. by chiral hplc. λ_{max} 207 (log ε 5.3); 226 (log ε 4.3); 288 (log ε 3.6); 321 (log ε 4.5) (Me_2CO)

(7Z,7′R*,8′S*)-form

9-Carboxylic acid, 3′,5-di-Me ether, 2-(4-hydroxyphenyl)ethylamide

cis-***Grossamide K***

[1224161-94-7]

$C_{28}H_{29}NO_7$ 491.54

Constit. of the bark of *Colocasia antiquorum* var. *esculenta*. Gum. $[\alpha]^{25}_{D}$ +91.5 (c, 0.3 in $CHCl_3$). λ_{max} 225 (log ε 3.2); 284 (log ε 1.8); 310 (log ε 1.7) (MeOH).

Freudenberg, K. *et al.*, *Chem. Ber.*, 1960, **93**, 1354-1366 (*Dehydrodiconiferyl alcohol, reactions*).

Freudenberg, K. *et al.*, *Chem. Ber.*, 1963, **96**, 1265-1270 (*Dehydrodiconiferyl alcohol, reactions*).

Kovacik, V. *et al.*, *Chem. Ber.*, 1969, **102**, 3623-3631 (*ms*).

Weinges, K. *et al.*, *Justus Liebigs Ann. Chem.*, 1970, **736**, 170-172 (*Silybum marianum constit*).

Sih, C.J. *et al.*, *J. Am. Chem. Soc.*, 1976, **98**, 5412-5413 (*7′RS,8′RS 3′,5-di-Me ether, synth*).

Woo, W.S. *et al.*, *Tetrahedron Lett.*, 1980, **21**, 4255-4258 (*Americanin D*).

Haruna, M. *et al.*, *Chem. Pharm. Bull.*, 1982, **30**, 1525-1527 (*Balanophonin*).

Arens, H. *et al.*, *Planta Med.*, 1985, **51**, 52-55 (*Plagiorhegma dubium glucoside*).

Fukuyama, Y. *et al.*, *Planta Med.*, 1986, 501-502 (*carboxylic acid derivs*).

Deyama, T. *et al.*, *Chem. Pharm. Bull.*, 1987, **35**, 1803-1807 (*Eucommia ulmoides diglucoside*).

Antus, S. *et al.*, *Liebigs Ann. Chem.*, 1987, 357-360 (*Americanin D, synth*).

Binns, A.N. *et al.*, *Proc. Natl. Acad. Sci. U.S.A.*, 1987, **84**, 980-984 (*Vinca rosea glucosides*).
Yoshizawa, F. *et al.*, *Chem. Pharm. Bull.*, 1990, **38**, 1927-1930 (*Cistanche tubulosa glucosides*).
Tan, R.X. *et al.*, *Planta Med.*, 1990, **56**, 475-477 (*Vladinol E*).
Uchiyama, T. *et al.*, *Phytochemistry*, 1991, **30**, 655-657 (*Plucheosides*).
Wang, H. *et al.*, *J. Nat. Prod.*, 1992, **55**, 214-216 (*Yemuoside* YM_1).
Warashina, T. *et al.*, *Phytochemistry*, 1992, **31**, 961-965 (*Verbascum thapsus constits*).
Bardon, A. *et al.*, *Phytochemistry*, 1993, **34**, 253-259 (*Chrysolaena verbascifolia constit*).
Hirai, N. *et al.*, *Biosci., Biotechnol., Biochem.*, 1994, **58**, 1679-1684 (*Dehydrodiconiferyl alcohol, abs config, resoln*).
Della Greca, M. *et al.*, *Heterocycles*, 1994, **38**, 1099-1102 (*5-O-de-Me*).
Ralph, J. *et al.*, *J. Chem. Soc., Perkin Trans. 1*, 1994, 3485-3498 (*diacid-di-Me ether, isol, synth, diacid-di-Me ether-di-Et ester, nmr*).
Hashimoto, Y. *et al.*, *Mokuzai Gakkaishi*, 1994, **40**, 549-553 (*9′-4-Hydroxybenzoylbalanophonin*).
Yoshikawa, K. *et al.*, *Chem. Pharm. Bull.*, 1995, **43**, 578-581 (*Woorenoside V*).
Gao, J. *et al.*, *Indian J. Chem., Sect. B*, 1995, **34**, 466-468 (*Alaschanioside C*).
Wahl, A. *et al.*, *J. Nat. Prod.*, 1995, **58**, 786-789 (*Xylobuxin*).
Sakushima, A. *et al.*, *Phytochemistry*, 1996, **43**, 1349-1354 (*Boreava orientalis amides*).
Bankova, V. *et al.*, *Z. Naturforsch., C*, 1996, **51**, 735-737 (*3′,5-di-Me 9,9′-di-Ac*).
Changzeng, W. *et al.*, *Phytochemistry*, 1997, **45**, 159-166 (*Tortosides*).
Li, S. *et al.*, *Phytochemistry*, 1997, **46**, 929-934 (*Vladinol E, config*).
Wang, C.Z. *et al.*, *Planta Med.*, 1997, **63**, 241-244 (*Longifloroside A*).
Garcia-Conesa, M.T. *et al.*, *Biochem. Soc. Trans.*, 1998, **26**, 5166 , 5168 (*diacid di-Me ether, isol, props*).
Ralph, J. *et al.*, *J. Agric. Food Chem.*, 1998, **46**, 2531-2532 (***synth***, *diacid di-Me ether di-Et ester*).
Valcic, S. *et al.*, *J. Nat. Prod.*, 1998, **61**, 771-775 (*9′-Acetylbalanophonin, 3′,5-di-Me 9,9′-di-Ac*).
Yoshikawa, K. *et al.*, *J. Nat. Prod.*, 1998, **61**, 786-790 (*Hovetrichoside E*).
Banskota, A.H. *et al.*, *J. Nat. Prod.*, 1998, **61**, 896-900 (*3′,5-di-Me 9,9′-di-Ac*).
Yuen, M.S.M. *et al.*, *Tetrahedron*, 1998, **54**, 12429-12444 (*Dehydrodiconiferyl alcohol, Balanophonin, abs config*).
Bolzacchini, E. *et al.*, *Tetrahedron Lett.*, 1998, **39**, 3291-3294 (*diacid di-Me ether, synth*).
Yang, S.-J. *et al.*, *J. Chin. Chem. Soc. (Taipei)*, 1999, **46**, 811-820 (*3-O-Demethyldehydrodiconiferyl alcohol*).
Sy, L.K. *et al.*, *Phytochemistry*, 1999, **50**, 781-785 (*Balanophonin*).
Cho, J.-Y. *et al.*, *J. Nat. Prod.*, 2000, **63**, 1205-1209 (*Woorenoside V, activity*).
Juhasz, L. *et al.*, *Tetrahedron Lett.*, 2000, **41**, 2491-2494 (*9-carboxylic acid 3′,5 di-Me ether Me ester, synth*).
Okazaki, M. *et al.*, *Biosci., Biotechnol., Biochem.*, 2001, **65**, 1134-1140 (*(+)-Dehydrodiconiferyl alcohol, synth*).
Iizuka, M. *et al.*, *Chem. Pharm. Bull.*, 2001, **49**, 282-286 (*3′,5,9-tri-Me 4′-glucoside*).
Jiang, Z.-H. *et al.*, *Chem. Pharm. Bull.*, 2001, **49**, 1036-1038 (*7′S,8′R-3′,5-di-Me 9′-glucoside*).
Larsen, E. *et al.*, *J. Agric. Food Chem.*, 2001, **49**, 3471-3475 (*9,9′-dicarboxylic acid 3′,5-di-Me ether, synth*).
Seca, A.M.L. *et al.*, *Phytochemistry*, 2001, **56**, 759-767 (*Boehmenan K*).
Seca, A.M.L. *et al.*, *Phytochemistry*, 2001, **58**, 1219-1223 (*Grossamide K*)
Orlandi, M. *et al.*, *Tetrahedron*, 2001, **57**, 371-378 (*Dehydrodiconiferyl alcohol, synth*).
Takara, K. *et al.*, *Biosci., Biotechnol., Biochem.*, 2002, **66**, 29-35 (*3′,5-di-Me 9-glucoside*).
Su, B.-N. *et al.*, *J. Nat. Prod.*, 2002, **65**, 1278-1282 (*Euphorbia quinquecostata neolignan*).
Jensen, J.F. *et al.*, *J. Nat. Prod.*, 2002, **65**, 1915-1917 (*Japonin A*).
Shen, X. *et al.*, *Zhongguo Yaoxue Zazhi*, 2002, **37**, 14-17 (*Glycosmisic acid*).
Yoshikawa, M. *et al.*, *Heterocycles*, 2003, **60**, 1787-1792 (*Hierochins*).
Ho, J.-C. *et al.*, *J. Chin. Chem. Soc. (Taipei)*, 2003, **50**, 1271-1274 (*Aegineoside, Balanophonin 4′-glucoside*).
Kamiya, K. *et al.*, *J. Agric. Food Chem.*, 2004, **52**, 5843-5848 (*Morindolin, activity*).
Ho, J.-C. *et al.*, *J. Chin. Chem. Soc. (Taipei)*, 2004, **51**, 221-223 (*4-O-Methylbalanophonin*).
Kirmizibekmez, H. *et al.*, *Turk. J. Chem.*, 2004, **28**, 455-460 (*Globularia cordifolia glucoside*).
Warashina, T. *et al.*, *Phytochemistry*, 2005, **66**, 589-597 (*(+)-Balanophonin 4′-glucoside*).
Bruschi, M. *et al.*, *J. Phys. Org. Chem.*, 2006, **19**, 592-596 (*Dehydrodiconiferyl alcohol, synth*).
Rakotondramanana, D.L.A. *et al.*, *Bioorg. Med. Chem.*, 2007, **15**, 6018-6026 (*9-carboxylic acid 3′,5-di-Me ether Me ester, synth*).
Zhao, X.Z. *et al.*, *Chin. Chem. Lett.*, 2007, **18**, 168-170 (*Angelica dahurica diglucoside*).
Yang, X.-Z. *et al.*, *Helv. Chim. Acta*, 2007, **90**, 318-325 (*Japonin A*).
Zheng, J. *et al.*, *J. Asian Nat. Prod. Res.*, 2007, **9**, 431-435 (*Spicatolignan B*).
Mitsuhashi, S. *et al.*, *Bioorg. Med. Chem.*, 2008, **16**, 2645-2650 (*Dehydrodiconiferyl alcohol, synth, activity*).
Hua, W. *et al.*, *Chin. Chem. Lett.*, 2008, **19**, 1327-1329 (*Gelsemiunoside A*).
Begum, A.S. *et al.*, *Nat. Prod. Res.*, 2009, **23**, 595-600 (*Hyoscyamal*).
Nakashima, S. *et al.*, *Bioorg. Med. Chem.*, 2010, **18**, 2337-2345 (*Anastatica hierochuntica constits, activity*).
Xiao, W.-L. *et al.*, *Chem. Pharm. Bull.*, 2010, **58**, 852-855 (*(+)-Balanophonin, activity*).
Kim, K.-H. *et al.*, *J. Agric. Food Chem.*, 2010, **58**, 4779-4785 (*cis-Grossamide K*).
Karim, A. *et al.*, *J. Asian Nat. Prod. Res.*, 2010, **12**, 714-718 (*Barlericin*).
Manna, S.K. *et al.*, *J. Biol. Chem.*, 2010, **285**, 22318-22327 (*Benfur, activity*).
Kim, K.H. *et al.*, *J. Enzyme Inhib. Med. Chem.*, 2010, **25**, 887-892 (*Limonia acidissima 9,9′-di-Ac*).
Chang, C.-L. *et al.*, *J. Nat. Prod.*, 2010, **73**, 1482-1488 (*Quiquesetinerviusin A*).
Xu, Y. *et al.*, *Planta Med.*, 2010, **76**, 1706-1709 (*Glehlinosides I,J*).
Jeong, E.J. *et al.*, *Bioorg. Med. Chem. Lett.*, 2011, **21**, 2283-2286 (*Euonymus alatus constits*).
Xiong, L. *et al.*, *J. Nat. Prod.*, 2011, **74**, 1188-1200 (*9-Et 3′,5-di-Me ether*).
Suo, M. *et al.*, *Fitoterapia*, 2012, **83**, 1484-1488 (*Avellanedae A*).
Constantin, M.-A. *et al.*, *Green Chem.*, 2012, **14**, 2375-2379 (*9,9′-dicarboxylic acid 3′,5-di-Me ether di-Et ester, synth*).
Fang, J. *et al.*, *J. Agric. Food Chem.*, 2012, **60**, 10523-10529 (*Brassica napus esters*).
Bitam, F. *et al.*, *Phytochem. Lett.*, 2012, **5**, 696-699 (*Posidonia oceanica constits*).
Qiao, L.-R. *et al.*, *Planta Med.*, 2012, **78**, 711-719 (*Stellera chamaejasme 9-Et ether*).
Matsumoto, S. *et al.*, *Synlett*, 2012, **23**, 1082-1084 (*(-)-Dehydrodiconiferyl alcohol, synth*).
Ruiu, S. *et al.*, *Bioorg. Med. Chem.*, 2013, **21**, 7074-7082 (*9′-O-(3-Methylbutanoyl)balanophonin*).
Shi, T.-X. *et al.*, *Bioorg. Med. Chem. Lett.*, 2013, **23**, 5904-5908 (*Tenuiside E*).
Lee, C. *et al.*, *Bioorg. Med. Chem. Lett.*, 2013, **23**, 6010-6014 (*Viterolignans A,B, activity*).
Liang, C.-Q. *et al.*, *Fitoterapia*, 2013, **86**, 171-177 (*Schisphenlignan I*).
Malarz, J. *et al.*, *Phytochem. Lett.*, 2013, **6**, 59-61 (*Yemuoside* YM_1).
Hong, S.S. *et al.*, *Tetrahedron Lett.*, 2013, **54**, 3967-3970 (*Tribulus terrestris constit, activity*).
Li, W. *et al.*, *Chem. Pharm. Bull.*, 2014, **62**, 196-202 (*Zanthoxyloside A, activity*).
Xiao, H.-H. *et al.*, *Fitoterapia*, 2014, **94**, 29-35 (*Sambucus williamsii constit*).
Kim, K.H. *et al.*, *J. Agric. Food Chem.*, 2014, **62**, 7746-7752 (*Pharbilignan C, activity*).
Xiong, J. *et al.*, *J. Ethnopharmacol.*, 2014, **153**, 737-743 (*Clematis armandii constits, activity*).
Gao, Y. *et al.*, *Phytochem. Lett.*, 2014, **10**, 276-280 (*9-butyl 3′,5-di-Me ether 4′-glucoside*).
Yu, Y. *et al.*, *Rec. Nat. Prod.*, 2014, **8**, 234-241 (*Acutosides A,B, activity*).
Sriyatep, T. *et al.*, *Tetrahedron*, 2014, **70**, 1773-1779 (*Feroniella lucida constit, activity*).
Li, J.S. *et al.*, *Zhongguo Zhongyao Zazhi*, 2014, **39**, 2670-2673 (*Solanum torvum glycoside*).
Huang, X.-X. *et al.*, *J. Agric. Food Chem.*, 2015, **63**, 7252-7260 (*Hawthornnin G, cd, activity*).
Sólyomváry, A.. *et al.*, *Process Biochem.*, 2015, **50**, 853-858 (*Balanophonin*).
Zhu, J.-Y. *et al.*, *RSC Adv.*, 2015, **5**, 12202-12208 (*Jatropha integerrima constits, activity*).
Wang, Y. *et al.*, *Chem. Nat. Compd. (Engl. Transl.)*, 2016, **52**, 396-398 (*Lyciumnan, activity*).
Fouseki, M.M. *et al.*, *Fitoterapia*, 2016, **115**, 9-14 (*Cordia alliodora amide*).
Madeiros, T.C.T. *et al.*, *J. Braz. Chem. Soc.*, 2016, **27**, 136-143 (*diacid, di-Me ester, pmr, cmr*).
Kim, S.Y. *et al.*, *Nat. Prod. Commun.*, 2016, **11**, 229-231 (*Coixide A*).
Li, J.-L. *et al.*, *Arch. Pharmacal Res.*, 2017, **40**, 1265-1270 (*Acanthopanax senticosus constit*).
Lee, C. *et al.*, *Chem. Nat. Compd. (Engl. Transl.)*, 2017, **53**, 24-27 (*Prunella vulgaris constit*).
Wei, J. *et al.*, *Fitoterapia*, 2017, **121**, 129-135 (*Inovonoid B, activity*).
Tshitenge, D.T. *et al.*, *J. Nat. Prod.*, 2017, **80**, 1604-1614 (*Gardenifolins A-H, activity*).
Kil, Y.-S. *et al.*, *J. Nat. Prod.*, 2017, **80**, 2240-2251 (*Ovatosides A,B, activity*).
Tran, H.N.K. *et al.*, *Arch. Pharmacal Res.*, 2018, **41**, 977-985 (*Cnidium officinale tetra-Me ether*).

Lou, L.-L. *et al.*, *Bioorg. Med. Chem. Lett.*, 2018, **28**, 1263-1268 (*Dehydrodiconiferyl alcohol, Balanophonin, resoln, cd, activity*).
Zhao, C. *et al.*, *J. Agric. Food Chem.*, 2018, **66**, 10421-10430 (*Viburfordoside D, activity*).
Guo, R. *et al.*, *Fitoterapia*, 2019, **137**, 104287:1-8 (*Crataegifins B, cd, activity*).
Lu, Q. *et al.*, *Zhongcaoyao*, 2019, **50**, 3261-3268 (*Cimicifugaside F, cd*)

4,7′-Epoxy-3,8′-lign-7′-ene-3′,4′,5,9,9′-pentol E164

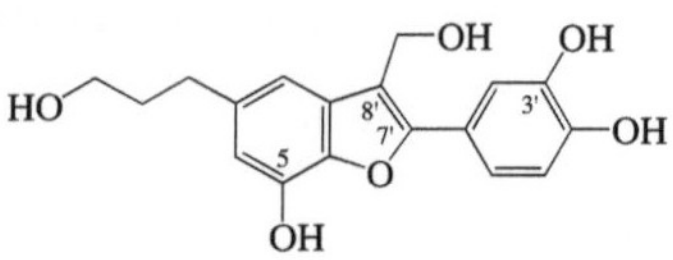

2-(3,4-Dihydroxyphenyl)-7-hydroxy-3-(hydroxymethyl)-5-benzofuranpropanol. 4,7′-Epoxy-3′,4′,5,9,9′-pentahydroxy-3,8′-neolign-7′-ene
$C_{18}H_{18}O_6$ 330.336
3′,5-Di-Me ether
4,7′-Epoxy-3′,5-dimethoxy-3,8′-lign-7′-ene-4′,9,9′-triol. 4,7′-Epoxy-4′,9,9′-trihydroxy-3′,5-dimethoxy-3,8′-lign-7′-ene
[70610-22-9]
$C_{20}H_{22}O_6$ 358.39
Constit. of the bark of *Abies sachalinensis*. Pale yellow powder. Mp 122 - 123° (synthetic). λ_{max} 232 (log ε 3.83); 286 (log ε 3.78); 294 (log ε 3.78) (EtOH)
3′,5-Di-Me ether, 9′-O-β-D-glucopyranoside
Pittogoside B
[1426258-72-1]
$C_{26}H_{32}O_{11}$ 520.531
Constit. of the roots of *Pittosporum glabratum*. Amorph. yellow powder. $[\alpha]^{20}_{D}$ -22.1 (c, 0.5 in MeOH). λ_{max} 216 (log ε 4.21); 306 (log ε 4.04) (MeOH)
9′-Aldehyde, 3′,5-di-Me ether
2-(4-Hydroxy-3-methoxyphenyl)-5-(3-hydroxypropyl)-7-methoxy-3-benzofurancarboxaldehyde, CAS. *4,7′-Epoxy-4′,9-dihydroxy-3′,5-dimethoxy-3,8′-lign-7′-en-9′-al. 4,7′-Epoxy-4′,9-dihydroxy-3′,5-dimethoxy-9′-oxo-3,8′-lign-7′-ene.* ***Salvinal***. *Tanshinone 278A. XH 14*
[135040-83-4]
$C_{20}H_{20}O_6$ 356.374
Constit. of *Liriodendron tulipifera* and *Salvia miltiorrhiza*. Microtubule inhibitor. Active against various cancer cells. Adenosine A_1 receptor ligand. Also shows antifibrotic activity. Cryst.. Mp 78 - 80° (synthetic)
7′,8′-Dihydro
4,7′-Epoxy-3,8′-lignan-3′,4′,5,9,9′-pentol, see E113.

Yang, Z. *et al.*, *Tetrahedron Lett.*, 1991, **32**, 2061-2064 (*Salvinal*).
Kuo, Y.-H. *et al.*, *J. Nat. Prod.*, 1996, **59**, 625-628 (*Salvinal, synth*).
Hutchinson, S.A. *et al.*, *Bioorg. Med. Chem. Lett.*, 1997, **7**, 3081-3084 (*Salvinal, synth*).
Luetjens, H. *et al.*, *Tetrahedron Lett.*, 1998, **39**, 6581-6584 (*Salvinal, synth*).
Kao, C.-L. *et al.*, *J. Org. Chem.*, 2002, **67**, 6772-6787 (*Salvinal, synth*).
Lin, S.-Y. *et al.*, *J. Org. Chem.*, 2003, **68**, 2968-2971 (*Salvinal, synth*).
Chang, J.-Y. *et al.*, *Mol. Pharmacol.*, 2004, **65**, 77-84 (*Salvinal, synth, activity*).
Miert, S.V. *et al.*, *Bioorg. Med. Chem.*, 2005, **13**, 661-669 (*3′,5-di-Me ether, synth*).
Wang, E.-C. *et al.*, *Tetrahedron Lett.*, 2006, **47**, 9195-9197 (*Salvinal, synth*).
Wada, S. *et al.*, *Helv. Chim. Acta*, 2009, **92**, 1610-1620 (*3′,5-di-Me ether*).
Bang, H.B. *et al.*, *Synth. Commun.*, 2009, **39**, 506-515 (*Salvinal, synth*).
Fan, H.-F. *et al.*, *J. Chem. Res.*, 2010, **34**, 233-235 (*3′,5-di-Me ether, synth*).
Zhao, H. *et al.*, *Phytochem. Lett.*, 2012, **5**, 240-243 (*Pittogoside B*).
Jeong, E.J. *et al.*, *Biol. Pharm. Bull.*, 2015, **38**, 228-234 (*Salvinal, isol, activity*)

4,7′-Epoxy-3,8′-lign-7-ene-3′,5,5′,9,9′-pentol E165

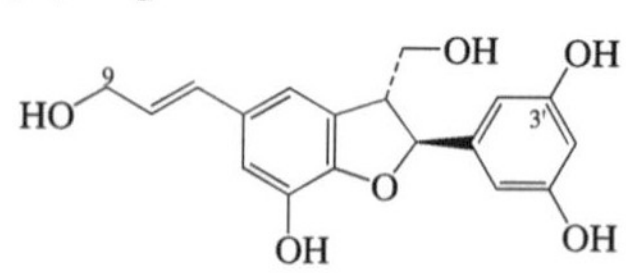

(7E,7′S,8′R)-form

4,7′-Epoxy-3′,5,5′,9,9′-pentahydroxy-3,8′-neolign-7-ene
$C_{18}H_{18}O_6$ 330.336
Structs. in this series may need revision. Phenyl 3,5-dioxygenated substn. pattern in doubt as a result of work done by Yang *et al* in 2015
(7E,7′S,8′R)-form
9-Et, 3′,5-di-Me ether
4,7′-Epoxy-9-ethoxy-5,5′-dimethoxy-3,8′-lign-7-ene-3′,9′-diol. 4,7′-Epoxy-9-ethoxy-3′,9′-dihydroxy-5,5′-dimethoxy-3,8′-lign-7-ene
$C_{22}H_{26}O_6$ 386.444
Constit. of the roots of *Cynanchum stauntonii*. Brown gum. $[\alpha]^{20}_{D}$ +1.4 (c, 0.1 in MeOH)
9-Carboxylic acid, 3′,5-di-Me ether
4,7′-Epoxy-3′,9′-dihydroxy-5,5′-dimethoxy-3,8′-lign-7-en-9-oic acid. ***Cimicifugic acid‡***
[616216-51-4]
$C_{20}H_{20}O_7$ 372.373
Constit. of the rhizomes of *Cimicifuga foetida*. Cryst. (MeOH). Mp 233 - 235°. $[\alpha]^{20}_{D}$ +5.2 (c, 0.48 in MeOH). λ_{max} 220; 316 (MeOH)
(7E,7′R*,8′R*)-form
3′,5-Di-Me ether
4,7′-Epoxy-5,5′-dimethoxy-3,8′-lign-7-ene-3′,9,9′-triol
$C_{20}H_{22}O_6$ 358.390
Constit. of the stems of *Eleutherococcus senticosus*. Exhibits selective inhibitory activity on protein tyrosine phosphatase 1B. Amorph. brown powder. $[\alpha]^{25}_{D}$ -17.2 (c, 0.1 in MeOH). λ_{max} 213; 226; 271 (MeOH)
3′,5,9-Tri-Me ether
4,7′-Epoxy-5,5′,9-trimethoxy-3,8′-lign-7-ene-3′,9′-diol
$C_{21}H_{24}O_6$ 372.417
Constit. of the stems of *Eleutherococcus senticosus*. Exhibits selective inhibitory activity on protein tyrosine phosphatase 1B. Amorph. brown powder. $[\alpha]^{25}_{D}$ -12.2 (c, 0.1 in MeOH). λ_{max} 222; 290 (MeOH)
9,9′-Dicarboxylic acid, 3′,5-di-Me ether, dioctadecyl ester
Brugnanin
[1085337-88-7]
$C_{56}H_{90}O_8$ 891.328
Constit. of the stem bark of the mangrove plant *Bruguiera gymnorrhiza*. Cryst. (hexane). Mp 114 - 115°. Possesses (7′*R**,8′*S**)-config. due to change in Cahn-Ingold-Prelog priorities
(7E,7′R*,8′S*)-form
9-Aldehyde, 3′,5-di-Me ether
4,7′-Epoxy-3′,9′-dihydroxy-5,5′-dimethoxy-3,8′-lign-7-en-9-al. 4,7′-Epoxy-3′,9′-dihydroxy-5,5′-dimethoxy-9-oxo-3,8′-lign-7-ene.
Curcasinlignan A
[1426543-22-7]
$C_{20}H_{20}O_6$ 356.374
Constit. of *Acanthopanax senticosus*, *Jatropha curcas* and *Trigonostemon chinensis*. Pale yellow oil. $[\alpha]^{25}_{D}$ -4.3 (c, 0.38 in MeOH). λ_{max}

227 (log ε 4.26); 289 (log ε 3.89); 340 (log ε 4.21) (MeOH)

(7E,7′RS,8′RS)-form

9,9′-Dicarboxylic acid, 3′,5-di-Me ether

4,7′-Epoxy-3′-hydroxy-5,5′-dimethoxy-3,8′-lign-7-ene-9,9′-dioic acid

9,9′-Dicarboxylic acid, 3′,5-di-Me ether, di-Et ester

Catalpafurxin

[948846-86-4]

$C_{24}H_{26}O_8$ 442.464

Constit. of the fruit of *Catalpa ovata*. Mp 145 - 147°. Possesses (7′*RS*,8′*SR*)-config. due to change in Cahn-Ingold-Prelog priorities. λ_{max} 237; 325 (MeOH)

(7E,7′RS,8′SR)-form

9-Carboxylic acid, 3′,5-di-Me ether, 2-(4-hydroxyphenyl)ethylamide

[2252244-22-5]

$C_{28}H_{29}NO_7$ 491.540

Constit. of the rhizomes of *Alocasia macrorrhiza*. NO prodn. inhibitor. Gum. $[\alpha]^{27}_{D}$ +1.2 (c, 0.5 in MeOH). λ_{max} 225; 319 (MeOH)

(7Z,7′RS,8′SR)-form

9-Carboxylic acid, 3′,5-di-Me ether, 2-(4-hydroxyphenyl)ethylamide

[2252244-24-7]

Constit. of the rhizomes of *Alocasia macrorrhiza*. NO prodn. inhibitor. Gum. $[\alpha]^{27}_{D}$ +0.6 (c, 0.5 in MeOH). λ_{max} 209; 284 (MeOH).

Zhao, X.-H. *et al.*, *Yaoxue Xuebao*, 2002, **37**, 535-538 (*Cimicifugic acid*).
Wang, Q.-Z. *et al.*, *Zhongcaoyao*, 2005, **36**, 164-166 (*Catalpafurxin*).
Shang, S. *et al.*, *Chem. Nat. Compd. (Engl. Transl.)*, 2008, **44**, 186-189 (*Brugnanin*).
Xu, J.J. *et al.*, *Z. Naturforsch., B*, 2012, **67**, 176-180 (*Curcasinlignan A*).
Yang, Y.-N. *et al.*, *J. Nat. Prod.*, 2015, **78**, 705-711 (*pmr, struct*).
Yu, J.-Q. *et al.*, *Phytochem. Lett.*, 2015, **13**, 355-359 (*Cynanchum stauntonii constit*).
Li, J.-L. *et al.*, *Arch. Pharmacal Res.*, 2016, **39**, (*Curcasinlignan A*).
Huang, W. *et al.*, *Fitoterapia*, 2017, **117**, 126-132 (*Alocasia macrorrhiza constits, activity*).
Zhang, L. *et al.*, *Fitoterapia*, 2017, **121**, 58-63 (*Eleutherococcus senticosus constits, activity*)

7,9′-Epoxy-8,8′-lign-7′-ene-3,3′,4,4′,9-pentol E166

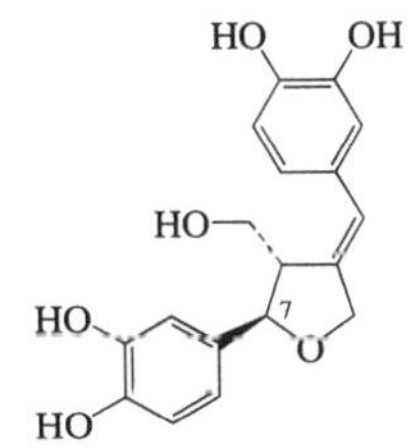

3,3′,4,4′,9-Pentahydroxy-7,9′-epoxylign-7′-ene

$C_{18}H_{18}O_6$ 330.336

(7R,7′E,8S)-form

3,3′-Di-Me ether, 9-O-(4-hydroxybenzoyl)

Agastenol

[413615-34-6]

$C_{27}H_{26}O_8$ 478.497

Constit. of *Agastache rugosa*. Potent inhibitor of Etoposide-induced apoptosis in U937 cells. Amorph. yellow powder. $[\alpha]^{25}_{D}$ +45 (c, 1 in MeOH). λ_{max} 264 (log ε 4.5); 300 (log ε 3.82) (MeOH).

Lee, C. *et al.*, *J. Nat. Prod.*, 2002, **65**, 414-416 (*Agastenol*)

2,7′-Epoxy-3,8′-lign-7-ene-3′,4,4′,5-tetrol E167

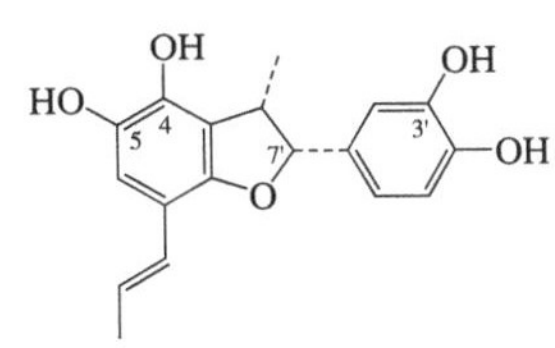

2-(3,4-Dihydroxyphenyl)-2,3-dihydro-7-(1-propenyl)-4,5-benzofurandiol. 2,7′-Epoxy-3′,4,4′,5-tetrahydroxy-3,8′-neolign-7-ene

$C_{18}H_{18}O_5$ 314.337

(7E,7′R,8′S)-form

4,5-Methylene, 3′-Me ether

2,7′-Epoxy-3′-methoxy-4,5-methylenedioxy-3,8′-lign-7-en-4′-ol.

2,7′-Epoxy-4′-hydroxy-3′-methoxy-4,5-methylenedioxy-3,8′-lign-7-ene

[217082-65-0]

$C_{20}H_{20}O_5$ 340.375

Constit. of *Marrubium alysson*.

Hosny, M. *et al.*, *Al-Azhar J. Pharm. Sci.*, 1997, **19**, 165-178 (*Marrubium alysson constit*)

2,7′-Epoxy-3,8′-lign-8-ene-3′,4,4′,5-tetrol E168

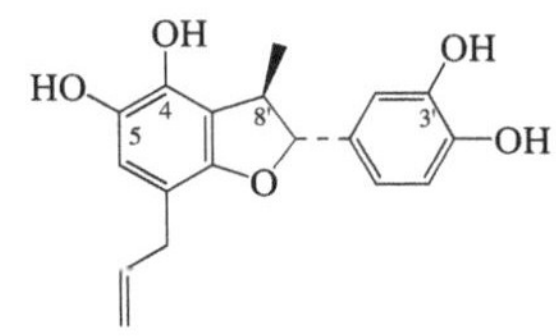

2-(3,4-Dihydroxyphenyl)-2,3-dihydro-3-methyl-7-(2-propenyl)-4,5-benzofurandiol. 7-Allyl-2-(3,4-dihydroxyphenyl)-2,3-dihydro-4,5-dihydroxy-3-methylbenzofuran. 2,7′-Epoxy-3′,4,4′,5-tetrahydroxy-3,8′-neolign-8-ene

$C_{18}H_{18}O_5$ 314.337

(7′R*,8′R*)-form

4,5-Methylene, 3′-Me ether

2,7′-Epoxy-3′-methoxy-4,5-methylenedioxy-3,8′-lign-8-en-4′-ol.

2,7′-Epoxy-4′-hydroxy-3′-methoxy-4,5-methylenedioxy-3,8′-lign-8-ene

[138195-42-3]

$C_{20}H_{20}O_5$ 340.375

Constit. of the roots of *Piper capense*. $[\alpha]^{22}_{D}$ +3.4 (c, 0.02 in EtOAc)

3′,4′:4,5-Bis(methylene) ether

2,7′-Epoxy-3′,4′:4,5-bis(methylenedioxy)-3,8′-lign-8-ene

[138195-43-4]

$C_{20}H_{18}O_5$ 338.359

Constit. of the roots of *Piper capense*. $[\alpha]^{22}_{D}$ +5.7 (c, 0.02 in EtOAc).

Green, T.P. *et al.*, *Phytochemistry*, 1991, **30**, 1649-1652 (*Piper capense constits, cryst struct*)

4,7′-Epoxy-3,8′-lign-7-ene-3′,4′,5,5′-tetrol E169

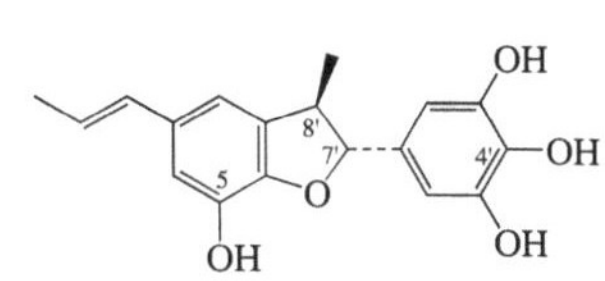

(7E,7′R,8′R)-form

2,3-Dihydro-7-hydroxy-3-methyl-5-(1-propenyl)-2-(3,4,5-trihydroxyphenyl) benzofuran. 4,7′-Epoxy-3′,4′,5,5′-tetrahydroxy-3,8′-neolign-7-ene

$C_{18}H_{18}O_5$ 314.337

(7E,7′R,8′R)-form

3′,5,5′-Tri-Me ether

4,7′-Epoxy-3′,5,5′-trimethoxy-3,8′-lign-7-en-4′-ol.

4,7′-Epoxy-4′-hydroxy-3′,5,5′-trimethoxy-3,8′-lign-7-ene. ***Maceneolignan A***
[1967042-42-7]
$C_{21}H_{24}O_5$ 356.418
Constit. of the arils of *Myristica fragrans*. Inhibits CCR3 mediated chemotaxis. Potential candidate as CCR3 antagonist. Oil. $[\alpha]^{26}_{D}$ +34 (c, 1.6 in $CHCl_3$). λ_{max} 217 (sh); 223 (log ε 4.40); 270 (log ε 4.18) (MeOH)

Tetra-Me ether
4,7′-Epoxy-3′,4′,5,5′-tetramethoxy-3,8′-lign-7-ene. ***Licarin C***
[60297-83-8]
[56353-18-5] (unspecified stereochem.)
$C_{22}H_{26}O_5$ 370.445
Constit. of *Myristica fragrans*, *Nectandra miranda* and *Licaria* sp.. Cryst. (MeOH). Mp 100 - 101°. λ_{max} 222 (ε 36300); 268 (ε 21500) (MeOH)

3′,4′-Methylene, 5,5′-di-Me ether
4,7′-Epoxy-3′,5-dimethoxy-4′,5′-methylenedioxy-3,8′-lign-7-ene. 5′-Methoxyeupomatenoid 8. ***Maceneolignan B***
[1613723-68-4]
[50354-06-8] (unspecified stereochem.) [1246343-71-4]
$C_{21}H_{22}O_5$ 354.402
Constit. of the arils and seeds of *Myristica fragrans*. Also from *Embelia ribes*. Oil. $[\alpha]^{23}_{D}$ +43.5 (c, 1.65 in $CHCl_3$). λ_{max} 214 (log ε 4.58); 270 (log ε 4.11); 306 (sh) (MeOH)

(7E,7′S,8′S)-form
3′,5-Di-Me ether
4,7′-Epoxy-5,5′-dimethoxy-3,8′-lign-7-ene-3′,4′-diol. 4,7′-Epoxy-3′,4′-dihydroxy-5,5′-dimethoxy-3,8′-lign-7-ene. ***Obovatifol***
[81264-50-8]
$C_{20}H_{22}O_5$ 342.391
Constit. of *Persea obovatifolia* and *Machilus odoratissima*. Exhibits potent cytotoxic activity against KB16 cancer cell line. Amorph.. $[\alpha]^{25}_{D}$ -50 (c, 0.3 in $CHCl_3$). λ_{max} 210 (log ε 4.68); 271 (log ε 4.24) (EtOH)

3′,4′,5-Tri-Me ether
4,7′-Epoxy-4′,5,5′-trimethoxy-3,8′-lign-7-en-3′-ol. 4,7′-Epoxy-3′-hydroxy-4′,5,5′-trimethoxy-3,8′-lign-7-ene. ***Machilusol A***
[289040-73-9]
$C_{21}H_{24}O_5$ 356.418
Constit. of *Machilus obovatifolia*. Cytotoxic agent. Oil. $[\alpha]^{25}_{D}$ -31 (c, 0.22 in $CHCl_3$). λ_{max} 220 (log ε 4.43); 271 (log ε 4.05) (EtOH)

3′,5,5′-Tri-Me ether
Odoratisol A
[891182-93-7]
$C_{21}H_{24}O_5$ 356.418
Constit. of the bark of *Machilus odoratissima*. Oil. $[\alpha]^{25}_{D}$ -35.1 (c, 1.2 in $CHCl_3$). λ_{max} 219 (log ε 4.43); 270 (log ε 4.18) (MeOH).

Isogai, A. *et al.*, *Agric. Biol. Chem.*, 1973, **37**, 889-895 (*3′,4′-methylene 5,5′-di-Me ether*).
Isogai, A. *et al.*, *Agric. Biol. Chem.*, 1973, **37**, 1479-1486 (*3′,4′-methylene 5,5′-di-Me ether*).
Braz Filho, R. *et al.*, *Tetrahedron Lett.*, 1976, **17**, 1157-1160 (*Licarin C*).
Aiba, C.J. *et al.*, *Phytochemistry*, 1977, **16**, 745-748 (*Licarin C*).
Davis, D.V. *et al.*, *J. Agric. Food Chem.*, 1982, **30**, 495-504 (*Licarin C, ms*).
Tsai, I.-L. *et al.*, *Phytochemistry*, 1998, **48**, 1371-1375 (*Obovatifol, activity*).
Tsai, I.-L. *et al.*, *Planta Med.*, 2000, **66**, 403-407 (*Machilusol A*).
Giang, P.M. *et al.*, *Chem. Pharm. Bull.*, 2006, **54**, 380-383 (*Odoratisol A*).
Francis, K.S. *et al.*, *Nat. Prod. Res.*, 2014, **28**, 1664-1668 (*Licarin C, cryst struct*).
Morikawa, T. *et al.*, *J. Nat. Prod.*, 2016, **79**, 2005-2013 (*Maceneolignans A,B, activity*)

4,7′-Epoxy-3,8′-lign-7-ene-3′,4′,5,9-tetrol E170

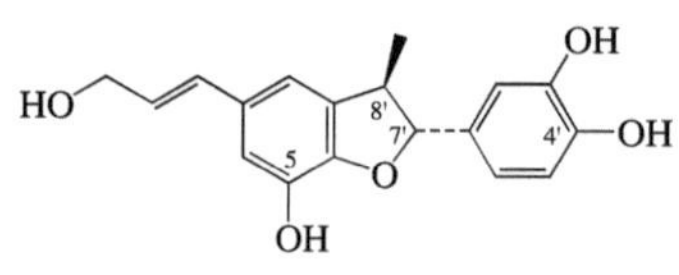

(7E,7′R,8′R)-form

4,7′-Epoxy-3′,4′,5,9-tetrahydroxy-3,8′-neolign-7-ene
$C_{18}H_{18}O_5$ 314.337

(7E,7′R,8′R)-form
9-Aldehyde
4,7′-Epoxy-3′,4′,5-trihydroxy-3,8′-lign-7-en-9-al. 4,7′-Epoxy-3′,4′,5-trihydroxy-9-oxo-3,8′-lign-7-ene

9-Aldehyde, 3′,4′-methylene, 5-Me ether
4,7′-Epoxy-5-methoxy-3′,4′-methylenedioxy-3,8′-lign-7-en-9-al. 4,7′-Epoxy-5-methoxy-3′,4′-methylenedioxy-9-oxo-3,8′-lign-7-ene. ***Papillarin B***
[1418183-14-8]
$C_{20}H_{18}O_5$ 338.359
Constit. of *Aristolochia papillaris*

(7E,7′S,8′S)-form
3′,4′,5-Tri-Me ether
4,7′-Epoxy-3′,4′,5-trimethoxy-3,8′-lign-7-en-9-ol. 4,7′-Epoxy-9-hydroxy-3′,4′,5-trimethoxy-3,8′-lign-7-ene. ***Fragransol C***
[114926-96-4]
$C_{21}H_{24}O_5$ 356.418
Constit. of the arils of *Myristica fragrans*. Oil. $[\alpha]_{D}$ -44 (c, 0.1 in $CHCl_3$)

3′,4′-Methylene, 5-Me ether
4,7′-Epoxy-5-methoxy-3′,4′-methylenedioxy-3,8′-lign-7-en-9-ol. 4,7′-Epoxy-9-hydroxy-5-methoxy-3′,4′-methylenedioxy-3,8′-lign-7-ene. ***Machilin B***
[110209-90-0]
$C_{20}H_{20}O_5$ 340.375
Constit. of the bark of *Machilus thunbergii*. Oil. $[\alpha]^{25}_{D}$ -40.1 (c, 0.11 in $CHCl_3$)

9-Aldehyde, 3′,5-di-Me ether
4,7′-Epoxy-4′-hydroxy-3′,5-dimethoxy-3,8-lign-7-en-9-al. 4,7′-Epoxy-4′-hydroxy-3′,5-dimethoxy-9-oxo-3,8′-lign-7-ene. ***Maceneolignan D***
[1967042-58-5]
$C_{20}H_{20}O_5$ 340.375
Constit. of the arils of *Myristica fragrans*. Moderate inhibitor of CCR3 mediated chemotaxis. Oil. $[\alpha]^{22}_{D}$ -51 (c, 0.1 in $CHCl_3$). λ_{max} 230 (log ε 4.21); 289 (log ε 3.80); 342 (log ε 4.18) (MeOH).

Shimomura, H. *et al.*, *Phytochemistry*, 1987, **26**, 1513-1515 (*Machilin B*).
Hattori, M. *et al.*, *Chem. Pharm. Bull.*, 1988, **36**, 648-653 (*Fragransol C*).
Juhasz, L. *et al.*, *J. Nat. Prod.*, 2000, **63**, 866-870 (*Fragransol C, synth*).
Franca, V.C. *et al.*, *Quim. Bras.*, 2010, **4**, 7-14 (*Papillarin B*).
Morikawa, T. *et al.*, *J. Nat. Prod.*, 2016, **79**, 2005-2013 (*Maceneolignan D*)

4,7′-Epoxy-3,8′-lign-7′-ene-3′,4′,7,8-tetrol E171

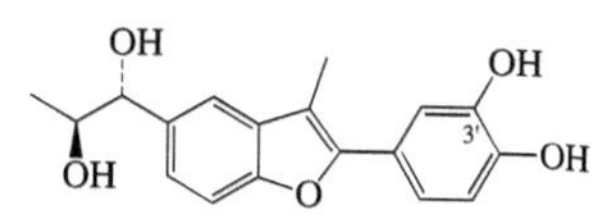

2-(3,4-Dihydroxyphenyl)-5-(1,2-dihydroxypropyl)-3-methylbenzofuran. 4,7′-Epoxy-3′,4′,7,8-tetrahydroxy-3,8′-neolign-7′-ene
$C_{18}H_{18}O_5$ 314.337

(7R*,8S*)-form

3′-Me ether

5-(1,2-Dihydroxypropyl)-2-(4-hydroxy-3-methoxyphenyl)-3-methylbenzofuran. 4,7′-Epoxy-3′-methoxy-3,8′-lign-7′-ene-4′,7,8-triol. 4,7′-Epoxy-4′,7,8-trihydroxy-3′-methoxy-3,8′-lign-7′-ene. **Eupomatenoid 9**

[41365-34-8]

$C_{19}H_{20}O_5$ 328.364

Constit. of *Caryodaphnopsis tonkinensis* and *Eupomatia laurina*. Cryst. ($CHCl_3$/MeOH). Mp 165 - 168° (natural). Mp 169 - 171° (synthetic). $[\alpha]_D^{25}$ +11.2 (c, 0.6 in MeOH). Identity of the two isolates not certain

3′,4′-Methylene ether

4,7′-Epoxy-3′,4′-methylenedioxy-3,8′-lign-7′-ene-7,8-diol. 4,7′-Epoxy-7,8-dihydroxy-3′,4′-methylenedioxy-3,8′-lign-7′-ene

[164229-82-7]

$C_{19}H_{18}O_5$ 326.348

Constit. of the bark of *Caryodaphnopsis tonkinensis*. Amorph.. $[\alpha]_D^{25}$ +12 (c, 0.7 in $CHCl_3$)

3′-Deoxy

5-(1,2-Dihydroxypropyl)-2-(4-hydroxyphenyl)-3-methylbenzofuran. 4,7′-Epoxy-3,8′-lign-7′-ene-4′,7,8-triol. 4,7′-Epoxy-4′,7,8-trihydroxy-3,8′-lign-7′-ene

[164229-81-6]

$C_{18}H_{18}O_4$ 298.338

Constit. of the bark of *Caryodaphnopsis tonkinensis*. Cryst. ($CHCl_3$/MeOH). Mp 162 - 166°. $[\alpha]_D^{25}$ +15.8 (c, 0.7 in MeOH).

Picker, K. *et al.*, *Aust. J. Chem.*, 1973, **26**, 1111-1119 (*Eupomatenoid 9, synth*).
Ripperger, H. *et al.*, *Phytochemistry*, 1995, **39**, 213-215 (*Caryodaphnopsis tonkinensis constits, synth*).
Bach, T. *et al.*, *Tetrahedron Lett.*, 2002, **43**, 9125-9127 ((±)-*Eupomatenoid 9, synth*)

4,7′-Epoxy-3,8′-lign-7-ene-4′,5,9,9′-tetrol E172

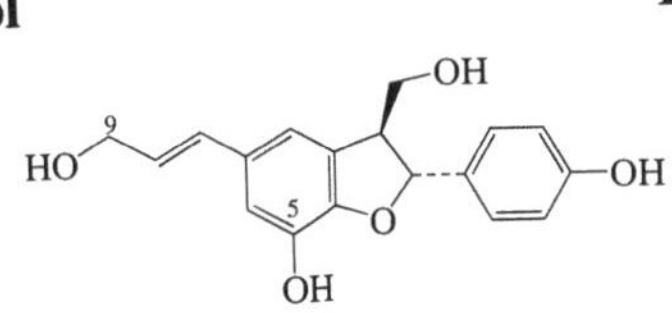

(7E,7′R,8′S)-form

2,3-Dihydro-7-hydroxy-2-(4-hydroxyphenyl)-5-(3-hydroxy-1-propenyl)-3-benzofuranmethanol. 4,7′-Epoxy-4′,5,9,9′-tetrahydroxy-3,8′-neolign-7-ene

$C_{18}H_{18}O_5$ 314.337

(7E,7′R,8′S)-form

9-Aldehyde

4,7′-Epoxy-4′,5,9′-trihydroxy-3,8′-lign-7-en-9-al

9-Aldehyde, 5-Me ether

4,7′-Epoxy-4′,9′-dihydroxy-5-methoxy-3,8′-lign-7-en-9-al. **Mappiodoinin C**

[2158272-35-4]

$C_{19}H_{18}O_5$ 326.348

Constit. of *Mappianthus iodoies*. Exhibits cytotoxic activity. Amorph. powder. $[\alpha]_D^{19}$ -98.2 (c, 0.11 in MeOH). λ_{max} 228 (log ε 4.56); 282 (log ε 2.82); 326 (log ε 3.58) (MeOH)

(7E,7′S,8′R)-form

9,9′-Dicarboxylic acid, 5-Me ether

4,7′-Epoxy-4′-hydroxy-5-methoxy-3,8′-lign-7-ene-9,9′-dioic acid

9,9′-Dicarboxylic acid, 5-Me ether, 4′-O-β-D-glucopyranoside

Ligusticumacid C

[2573174-09-9]

$C_{25}H_{26}O_{12}$ 518.471

Constit. of the rhizomes of *Ligusticum chuanxiong*. Amorph. powder. $[\alpha]_D^{20}$ -62 (c, 0.1 in MeOH). Possesses (7′*S*,8′*S*)-config. due to change in Cahn-Ingold-Prelog priorities. λ_{max} 224 (log ε 4.46); 322 (log ε 4.30) (MeOH).

Jiang, Z.-H. *et al.*, *Bioorg. Chem.*, 2017, **75**, 260-264 (*Mappiodoinin C, activity*).
Zhang, X. *et al.*, *Fitoterapia*, 2018, **125**, 147-154 (*Ligusticumacid C*)

4,7′-Epoxy-3,8′-lign-8-ene-3′,4′,5,9′-tetrol E173

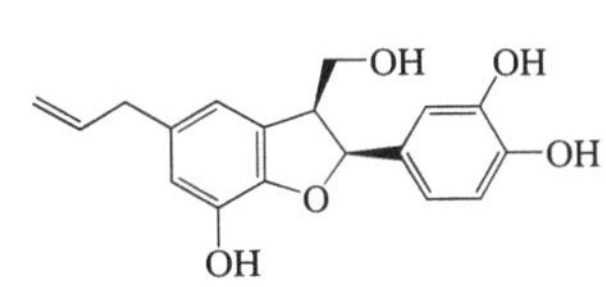

4,7′-Epoxy-3′,4′,5,9′-tetrahydroxy-3,8′-neolign-8-ene

$C_{18}H_{18}O_5$ 314.337

(7′S,8′S)-form

3′,4′,5-Tri-Me ether

4,7′-Epoxy-3′,4′,5-trimethoxy-3,8′-lign-8-en-9′-ol. 4,7′-Epoxy-9′-hydroxy-3′,4′,5-trimethoxy-3,8′-lign-8-ene. **Dihydrocarinatinol**

[85769-57-9]

$C_{21}H_{24}O_5$ 356.418

Constit. of the bark of *Virola carinata*. Resin. $[\alpha]_D^{25}$ -12.3 ($CHCl_3$). λ_{max} 249 (log ε 3.83); 283 (log ε 3.80); 286 (log ε 3.79) ($CHCl_3$).

Kawanishi, K. *et al.*, *Phytochemistry*, 1982, **21**, 2725-2728 (*Dihydrocarinatinol*).
Pieters, L. *et al.*, *Magn. Reson. Chem.*, 1993, **31**, 692 693 (*pmr, cmr*)

7,7′-Epoxy-8,8′-lign-8-ene-3,3′,4,4′-tetrol E174

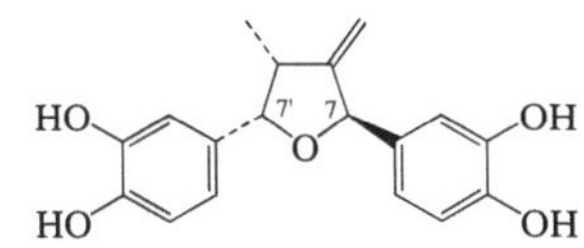

3,3′,4,4′-Tetrahydroxy-7,7′-epoxylign-8-ene

$C_{18}H_{18}O_5$ 314.337

(7S,7′S,8′R)-form

3,3′-Di-Me ether

7,7′-Epoxy-3,3′-dimethoxy-8,8′-lign-8-ene-4,4′-diol. 4,4′-Dihydroxy-3,3′-dimethoxy-7,7′-epoxylign-8-ene. **Ribesin D**

[1469818-65-2]

$C_{20}H_{22}O_5$ 342.391

Constit. of leaves of *Ribes nigrum*. Displays potent superoxide anion scavenging activity. Solid. $[\alpha]_D^{20}$ +20.8 (c, 0.05 in MeOH). λ_{max} 229 (log ε 4.13); 280 (log ε 3.83) (MeOH).

Sasaki, T. *et al.*, *Phytochemistry*, 2013, **95**, 333-340 (*Ribesin D*)

7,7′-Epoxy-8,8′-lign-8(8′)-ene-3,3′,4,4′-tetrol E175

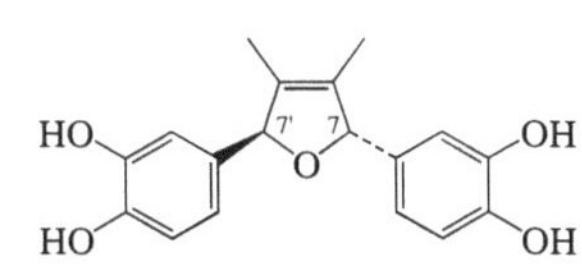

3,3′,4,4′-Tetrahydroxy-7,7′-epoxylign-8(8′)-ene

$C_{18}H_{18}O_5$ 314.337

(7R,7′R)-form

3,3′-Di-Me ether

7,7′-Epoxy-3,3′-dimethoxy-8,8′-lign-8(8′)-ene-4,4′-diol. 4,4′-Dihydroxy-3,3′-dimethoxy-7,7′-epoxylign-8(8′)-ene. **Ribesin B**

[1469818-63-0]

$C_{20}H_{22}O_5$ 342.391

Constit. of leaves of *Ribes nigrum*. $[\alpha]_D^{25}$ -45.2 (c, 0.25 in MeOH). λ_{max} 232 (log ε 4.21); 282 (log ε 3.84) (MeOH).

Sasaki, T. *et al.*, *Phytochemistry*, 2013, **95**, 333-340 (*Ribesin B*)

7,9′-Epoxy-8,8′-lign-7′-ene-3,3′,4′,5-tetrol E176

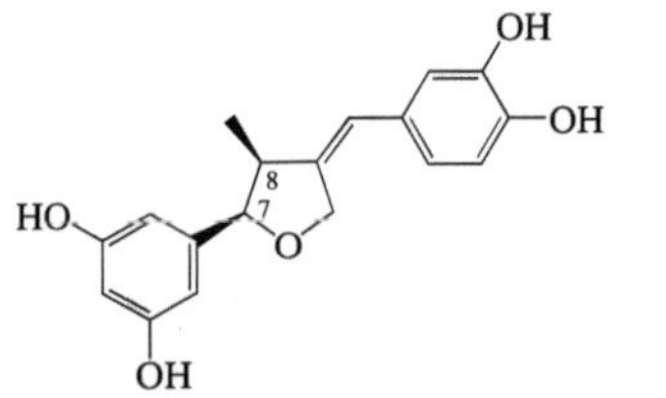

3,3′,4′,5-Tetrahydroxy-7,9′-epoxylign-7′-ene

$C_{18}H_{18}O_5$ 314.337

Structures in this series may need revision. Phenyl 3,5-dioxygenated substn. pattern in doubt as a result of work done by Yang *et al* in 2015

(7R*,7′Z,8S*)-form

Tetra-Me ether

7,9′-Epoxy-3,3′,4′,5-tetramethoxy-8,8′-lign-7′-ene. 3,3′,4′,5-Tetramethoxy-7,9′-epoxylign-7′-ene

[1394179-94-2]

$C_{22}H_{26}O_5$ 370.445

Isol. from the flower buds of *Magnolia liliflora*. Oil. $[\alpha]^{20}_{D}$ -26 (c, 0.02 in MeOH). λ_{max} 239 (log ε 2.96); 299 (log ε 4.03) (MeOH).

Wang, W.-S. *et al.*, *Planta Med.*, 2012, **78**, 141-147 (*Magnolia liliflora constit*).
Yang, Y.-N. *et al.*, *J. Nat. Prod.*, 2015, **78**, 705-711 (*pmr, struct*)

7,9′-Epoxy-8,8′-lign-8-ene-3,3′,4,4′-tetrol E177

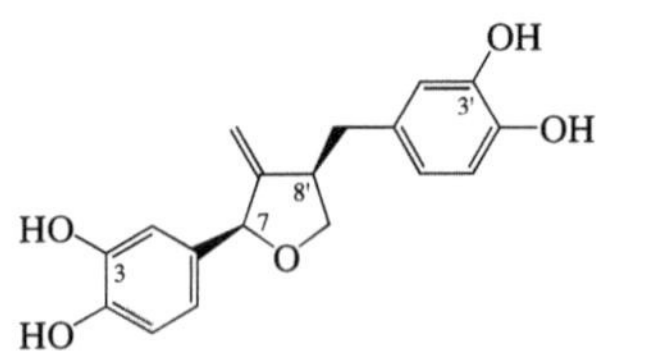

(7R*,8′R*)-form

3,3′,4,4′-Tetrahydroxy-7,9′-epoxylign-8-ene

$C_{18}H_{18}O_5$ 314.337

(7R*,8′R*)-form

3,3′-Di-Me ether

7,9′-Epoxy-3,3′-dimethoxy-8,8′-lign-8-ene-4,4′-diol. 4,4′-Dihydroxy-3,3′-dimethoxy-7,9′-epoxylign-8-ene. ***Pinnatifolin***

$C_{20}H_{22}O_5$ 342.391

Constit. of the roots of *Syringa pinnatifolia*. Moderate antioxidant. Oil. $[\alpha]_D$ -26.8 (c, 0.1 in MeOH). λ_{max} 207; 237; 281 (MeOH)

(7R*,8′S*)-form

3,3′-Di-Me ether

Isopinnatifolin

[1821386-92-8]

$C_{20}H_{22}O_5$ 342.391

Constit. of the roots of *Syringa pinnatifolia*. Moderate antioxidant. Oil. $[\alpha]^{25}_{D}$ -15.3 (c, 0.1 in MeOH). λ_{max} 207; 237; 281 (MeOH).

Zhang, Z.-M. *et al.*, *Chem. Pharm. Bull.*, 2014, **62**, 1009-1012 (*Pinnatifolin, Isopinnatifolin, activity*)

4,7′-Epoxy-3,8′-lign-7-ene-3′,4′,5-triol E178

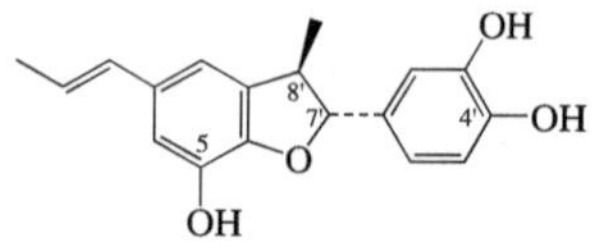

(7E,7′R,8′R)-form

2-(3,4-Dihydroxyphenyl)-2,3-dihydro-3-methyl-5-(1-propenyl)-7-benzofuranol, CAS. *4,7′-Epoxy-3′,4′,5-trihydroxy-3,8′-neolign-7-ene*

[23342-98-5]

$C_{18}H_{18}O_4$ 298.338

(7E,7′R,8′R)-form

3′-Me ether

4,7′-Epoxy-3′-methoxy-3,8′-lign-7-ene-4′,5-diol. 4,7′-Epoxy-4′,5-dihydroxy-3′-methoxy-3,8′-lign-7-ene. ***Callislignan A***

[1416369-71-5]

$C_{19}H_{20}O_4$ 312.365

Constit. of the leaves of *Callistemon lanceolatus*. Active against two strains of *Staphylococcus aureus*. Yellowish gum. $[\alpha]^{29}_{D}$ +56.7 (c, 0.8 in MeOH)

5-Me ether

4,7′-Epoxy-5-methoxy-3,8′-lign-7-ene-3′,4′-diol. 4,7′-Epoxy-3′,4′-dihydroxy-5-methoxy-3,8′-lign-7-ene

$C_{19}H_{20}O_4$ 312.365

Constit. of the fruit of *Myristica fragrans*. Moderately cytotoxic. Oil. $[\alpha]^{20}_{D}$ +46.3 (c, 0.04 in $CHCl_3$)

3′,5-Di-Me ether

4,7′-Epoxy-3′,5-dimethoxy-3,8′-lign-7-en-4′-ol. 4,7′-Epoxy-4′-hydroxy-3′,5-dimethoxy-3,8′-lign-7-ene. ***Dehydrodiisoeugenol.*** *(+)-Licarin A*

[83377-50-8]

[23518-30-1] (rel. config.)

[69744-09-8] ((±)-form)

[88034-58-6] ((±)-form)

[127000-67-3] (*trans*-form)

$C_{20}H_{22}O_4$ 326.392

Constit. of *Aristolochia pubescens, Krameria grayi, Myristica fragrans, Nectandra amazonum* and *Piper kadsura*. 5-Lipoxygenase inhibitor. Antiosteoporotic and neuroprotectant. Cryst. (MeOH or hexane). Mp 118 - 119° (106-108°). $[\alpha]^{25}_{D}$ +43.5 (c, 1.4 in $CHCl_3$). $[\alpha]_D$ +66 (c, 0.37 in $CHCl_3$). λ_{max} 218 (log ε 4.06); 274 (log ε 3.82) (MeOH)

Tri-Me ether

4,7′-Epoxy-3′,4′,5-trimethoxy-3,8′-lign-7-ene. ***(+)-Acuminatin‡.*** *(+)-Licarin D*

[41744-39-2]

[6544-71-4] (unspecified stereochem.) [58456-59-0] ((±)-form) [61277-74-5] (7ξ,7′*R*,8′*S*-form) [72881-08-4] (7*E*,7′*R**,8′*R**-form) [73679-50-2] (7ξ,7′*S*,8′*S*-form) [83144-66-5] (Licarin D incorr.)

$C_{21}H_{24}O_4$ 340.419

Constit. of *Magnolia acuminata, Nectandra miranda, Piper futokadsura* and *Urbanodendron verrucosum*. Moderate inhibitor of cyclooxygenase-2. Moderate inhibitor of NO prodn. Inhibitor of DNA topoisomerase. Cryst. (diisopropyl ether). Mp 77.5°. $[\alpha]^{27}_{D}$ +4.3 (c, 0.22 in MeOH). $[\alpha]_D$ +44 (c, 0.3 in MeOH) (synthetic). Struct. of Acuminatin revised in 1982

3′,4′-Methylene, 5-Me ether

4,7′-Epoxy-5-methoxy-3′,4′-methylenedioxy-3,8′-lign-7-ene. ***Eupomatenoid 8.*** *(+)-Licarin B. Licarin E*

[42016-15-9]

[124041-75-4] (Licarin E) [183508-57-8] (*Z*-form) [1609472-88-9] ((±)-form)

$C_{20}H_{20}O_4$ 324.376

Constit. of *Aristolochia elegans, Embelia ribes, Eupomatia* sp., *Nectandra glabrescens, Ocotea macrophylla* and *Virola oleifera*. Cryst. (MeOH). Mp 90 - 91°. $[\alpha]_D$ +43 (c, 0.84 in $CHCl_3$). Synonym Licarin B originally assigned solely to (7′*S*,8′*S*)-isomer

(7E,7′R,8′S)-form

Tri-Me ether

(+)-cis-Acuminatin

[2001069-77-6]

$C_{21}H_{24}O_4$ 340.419

Constit. of the aerial parts of *Piper bonii*

3′,4′-Methylene, 5-Me ether

[1609472-88-9]

$C_{20}H_{20}O_4$ 324.376

Constit. of *Aristolochia fordiana*. Oil. $[\alpha]_D^{20}$ +69 (c, 0.1 in MeOH). $[\alpha]_D^{25}$ +45.2 (c, 0.69 in $CHCl_3$). λ_{max} 213 (log ε 4.35); 272 (log ε 3.98) (MeOH)

(7E,7′S,8′R)-form

3′,5-Di-Me ether

Isolicarin A

[955094-15-2]

[72274-58-9] (rel. config.)

$C_{20}H_{22}O_4$ 326.392

Constit. of the arils of *Myristica fragrans*. Oil. $[\alpha]_D^{20}$ -24.2 (c, 1.1 in $CHCl_3$). λ_{max} 221 (log ε 3.94); 272 (log ε 3.67) (MeOH)

(7E,7′S,8′S)-form

5-Me ether

Kachirachirol B

[94513-61-8]

$C_{19}H_{20}O_4$ 312.365

Constit. of *Magnolia kachirachirai*. Needles (C_6H_6). Mp 64 - 66°. $[\alpha]_D$ -60 (c, 0.65 in $CHCl_3$)

3′,5-Di-Me ether

(-)-Licarin A

[51020-86-1]

[2680-81-1] (unspecified stereochem.) [64870-40-2] (7′*S*-*trans*-form)

$C_{20}H_{22}O_4$ 326.392

Constit. of *Aristolochia fordiana*, *Licaria aritu*, *Machilus thunbergii*, *Saururus cernuus*, *Virola koschnyi*, *Urbanodendron verrucosum* and *Krameria cytisoides*. Also isol. from the liverwort *Jackiella javanica*. Moderate inhibitor of ACAT, COX-2 and 5-LOX. Exhibits antiproliferative, schistosomicidal and trypanocidal activities. Cytotoxic. Induces apoptosis via caspase-3 activation. Shows neuroprotective activity. Exhibits antitubercular activity. Cryst. (MeOH). Mp 114 - 116° (106-108°). $[\alpha]_D$ -62 (c, 0.002 in $CHCl_3$). Species given as Krameria cystisoides in ref.

Tri-Me ether

(-)-Acuminatin. *(-)-Licarin D*

[150134-24-0]

[83144-66-5] (Licarin D)

[155734-04-6]

$C_{21}H_{24}O_4$ 340.419

Constit. of *Machilus japonica*, *Machilus thunbergii*, *Magnolia denudata* and *Piper betle*. Shows antiproliferative props. Moderate ACAT inhibitor. Hepatoprotectant. Insect growth inhibitor. Exhibits synergistic antibacterial activity with Norfloxacin. Oil. $[\alpha]_D$ -61 (c, 0.0025 in MeOH)

3′,4′-Methylene, 5-Me ether

(-)-Licarin B

[51020-87-2]

$C_{20}H_{20}O_4$ 324.376

Constit. of *Aristolochia fordiana*, *Machilus thunbergii*, *Magnolia denudata*, *Myristica fragrans* and *Saururus chinensis*. Moderate inhibitor of cyclooxygenase-2. Shows antiosteoporotic activity. Specific inhibitor of cytochrome P450 3A4. Needles (hexane). Mp 86.5 - 88°. $[\alpha]_D$ -50.8 (c, 0.25 in $CHCl_3$). Synonym Licarin B originally assigned solely to (7′*S*,8′*S*)-isomer. λ_{max} 273 (log ε 2.56) ($CHCl_3$)

(7E,7′RS,8′RS)-form

3′,4′-Methylene, 5-Me ether

(±)-Licarin B

[55515-49-6]

Constit. of *Eupomatia* spp.. Oil.

Bowden, B.F. *et al.*, *Aust. J. Chem.*, 1972, **25**, 2659-2669 (*Eupomatenoid 8*).

Aiba, C.J. *et al.*, *Phytochemistry*, 1973, **12**, 1163-1164 (*Licarin A, Dehydrodiisoeugenol*).

Aiba, C.J. *et al.*, *Phytochemistry*, 1975, **14**, 253-255 (*Licarin B, synth*).

Takaoka, D. *et al.*, *Bull. Chem. Soc. Jpn.*, 1976, **49**, 3564-3566 (*Licarin B*).

Wenkert, E. *et al.*, *Phytochemistry*, 1976, **15**, 1547-1551 (*pmr*).

Aiba, C.J. *et al.*, *Phytochemistry*, 1977, **16**, 745-748 (*Nectandra miranda constit, abs config*).

Gottlieb, O.R. *et al.*, *Phytochemistry*, 1977, **16**, 1003-1006 (*(-)-Licarins A,B, (+)-Acuminatin, abs config*).

El-Feraly, F.S. *et al.*, *Phytochemistry*, 1982, **21**, 1133-1135 (*Acuminatin, Dehydrodiisoeugenol*).

De F. Dias, A. *et al.*, *Phytochemistry*, 1982, **21**, 1137-1139 (*Licarin D*).

Ito, K. *et al.*, *Phytochemistry*, 1984, **23**, 2643-2645 (*Kachirachirol B*).

Achenbach, H. *et al.*, *Phytochemistry*, 1987, **26**, 1159-1166 (*Licarin A*).

Barbosa-Filho, J.M. *et al.*, *Phytochemistry*, 1989, **28**, 1991 (*Licarin E*).

Nagashima, F. *et al.*, *Phytochemistry*, 1990, **29**, 2169-2174 (*Licarin A, Jackiella constit*).

Kim, Y.B. *et al.*, *Arch. Pharmacal Res.*, 1991, **14**, 1-6 (*Licarin B, cryst struct*).

Fernandes, A.M.A.P. *et al.*, *Phytochemistry*, 1993, **32**, 1567-1572 (*Eupomatenoid 8*).

Gonzalez-Coloma, A. *et al.*, *Phytochemistry*, 1994, **35**, 607-610 (*(-)-Licarin A, (-)-Acuminatin, activity*).

Engler, T.A. *et al.*, *Tetrahedron Lett.*, 1996, **37**, 6969-6970 (*Licarin B, synth*).

Nascimento, I.R. *et al.*, *Phytochemistry*, 1999, **52**, 345-350 (*Licarin A*).

Kwon, B.M. *et al.*, *Planta Med.*, 1999, **65**, 74-76 (*Licarin A, activity*).

Nascimento, I.R. *et al.*, *Tetrahedron*, 2000, **56**, 9181-9193 (*Licarin A, synth*).

Lee, J.S. *et al.*, *Arch. Pharmacal Res.*, 2004, **27**, 1043-1047 (*Acuminatin, activity*).

Park, B.-Y. *et al.*, *Biol. Pharm. Bull.*, 2004, **27**, 1305-1307 (*Licarins A,B, activity*).

Choong, J.M. *et al.*, *Br. J. Pharmacol.*, 2005, **146**, 752-759 (*Licarin A, activity*).

Li, F. *et al.*, *Helv. Chim. Acta*, 2007, **90**, 1491-1496 (*Dehydrodiisoeugenol, Isolicarin A*).

Coy, E.D. *et al.*, *Bioorg. Med. Chem. Lett.*, 2009, **19**, 6922-6925 (*Licarins A,B, Acuminatin, activity*).

Duan, L. *et al.*, *Planta Med.*, 2009, **75**, 1241-1245 (*Myristica 5-Me ether*).

Pereira, A.C. *et al.*, *J. Chromatogr., A*, 2011, **1218**, 7051-7054 (*Licarin A, Dehydrodiisoeugenol, synth, resoln*).

Pereira, A.C. *et al.*, *Phytochemistry*, 2011, **72**, 1424-1430 (*Licarin A, resoln, activity*).

Rattanaburi, S. *et al.*, *Phytochem. Lett.*, 2012, **5**, 18-21 (*Callislignan A, activity*).

León-Diaz, R. *et al.*, *Arch. Med. Res.*, 2013, **44**, 99-104 (*(-)-Licarin A, activity*).

Zhou, Z.-B. *et al.*, *Planta Med.*, 2013, **79**, 1730-1735 (*Aristolochia fordiana constits, activity*).

Lopez-Sanchez, C. *et al.*, *Tetrahedron*, 2013, **69**, 5511-5516 (*(+)-Acuminatin, synth*).

Tang, G.-H. *et al.*, *J. Nat. Prod.*, 2015, **78**, 1894-1903 (*Aristolochia fordiana constit*).

Mardhatillah, A. *et al.*, *Int. J. Chem.*, 2016, **8**, 1-7 (*Dehydrodiisoeugenol, synth*).

Ding, D.-D. *et al.*, *Phytochemistry*, 2016, **129**, 36-44 (*(+)-cis-Acuminatin*).

Sun, Z.-L. *et al.*, *RSC Adv.*, 2016, **6**, 43518-43525 (*(-)-Acuminatin, cd, activity*)

4,7′-Epoxy-3,8′-lign-7-ene-4′,9,9′-triol E179

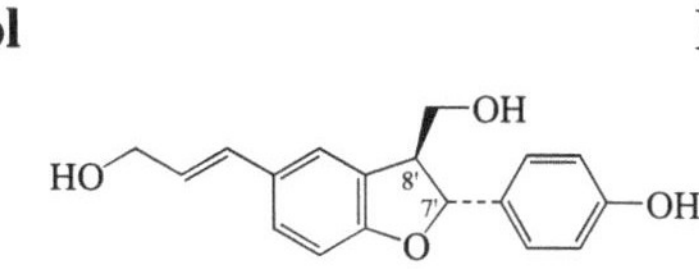

(7E,7′R,8′S)-form

2,3-Dihydro-2-(4-hydroxyphenyl)-5-(3-hydroxy-1-propenyl)-3-benzofuranmethanol. 4,7′-Epoxy-4′,9,9′-trihydroxy-3,8′-neolign-7-ene

$C_{18}H_{18}O_4$ 298.338

(7E,7′R,8′S)-form

Trichobenzolignan [1651866-59-9]

Constit. of the roots of *Trichosanthes kirilowii*. Exhibits moderate cytotoxic activity. Amorph. powder. $[\alpha]^{25}_{D}$ -38 (c, 0.15 in MeOH)

9-Me ether

4,7′-Epoxy-9-methoxy-3,8′-lign-7-ene-4′,9′-diol. ***Mappiodoinin B***

[2158272-34-3]

$C_{19}H_{20}O_4$ 312.365

Constit. of *Mappianthus iodoies*. Exhibits cytotoxic activity. Amorph. powder. $[\alpha]^{19}_{D}$ -42.8 (c, 0.13 in MeOH). λ_{max} 222 (log ε 4.49); 276 (log ε 2.78); 312 (log ε 3.86) (MeOH)

(7E,7′S,8′R)-form

9-Aldehyde

4,7′-Epoxy-4′,9′-dihydroxy-3,8′-lign-7-en-9-al. 4,7′-Epoxy-4′,9′-dihydroxy-9-oxo-3,8′-lign-7-ene. ***Morrisonicolanin***

[906089-01-8]

$C_{18}H_{16}O_4$ 296.322

Constit. of the heartwood of *Picea morrisonicola*. Amorph. solid. $[\alpha]^{24}_{D}$ +6.7 (c, 1.4 in $CHCl_3$). λ_{max} 230 (log ε 4); 330 (log ε 3.6) (MeOH).

Lee, T.-H. *et al.*, *Chem. Pharm. Bull.*, 2006, **54**, 693-695 (*Morrisonicolanin*).
Van Minh, C. *et al.*, *Arch. Pharmacal Res.*, 2015, **38**, 1443-1448 (*Trichobenzolignan, activity*).
Jiang, Z.-H. *et al.*, *Bioorg. Chem.*, 2017, **75**, 260-264 (*Mappiodoinin B, activity*)

4,7′-Epoxy-3,8′-lign-8-ene-3′,4′,5-triol E180

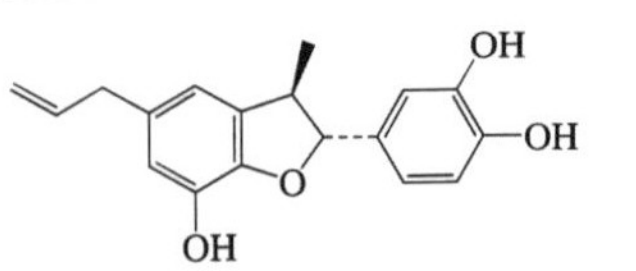

(7′R,8′R)-form

4,7′-Epoxy-3′,4′,5-trihydroxy-3,8′-neolign-8-ene

$C_{18}H_{18}O_4$ 298.338

(7′R,8′R)-form

3′,5-Di-Me ether

4,7′-Epoxy-3′,5-dimethoxy-3,8′-lign-8-en-4′-ol. 4,7′-Epoxy-4′-hydroxy-3′,5-dimethoxy-3,8′-lign-8-ene. ***Isodihydrocarinatidin***

[955094-13-0]

$C_{20}H_{22}O_4$ 326.392

Constit. of the arils of *Myristica fragrans*. Oil. $[\alpha]^{20}_{D}$ +15 (c, 0.6 in $CHCl_3$). λ_{max} 238 (log ε 4.13); 282 (log ε 3.81) (MeOH)

(7′S,8′R)-form

3′,5-Di-Me ether

Maceneolignan E

[1967042-64-3]

$C_{20}H_{22}O_4$ 326.392

Constit. of the arils of *Myristica fragrans*. Oil. $[\alpha]^{25}_{D}$ -81 (c, 0.8 in $CHCl_3$). λ_{max} 227 (log ε 4.38); 282 (sh) (MeOH)

(7′S,8′S)-form

3′,5-Di-Me ether

Dihydrocarinatidin

[88640-85-1]

$C_{20}H_{22}O_4$ 326.392

Constit. of *Krameria tomentosa*, *Virola carinata* and *Virola pavonis*. $[\alpha]_{D}$ -12.7 ($CHCl_3$). λ_{max} 244 (log ε 3.93); 282 (log ε 3.83); 285 (log ε 3.82) ($CHCl_3$)

Tri-Me ether

4,7′-Epoxy-3′,4′,5-trimethoxy-3,8′-lign-8-ene. ***Dihydrocarinatin***

[146340-01-4]

$C_{21}H_{24}O_4$ 340.419

Constit. of *Virola pavonis*. λ_{max} 246 (log ε 4.08); 282 (log ε 3.87); 286 (log ε 3.86) ($CHCl_3$).

Kawanishi, K. *et al.*, *Phytochemistry*, 1982, **21**, 929-931 (*Dihydrocarinatin*).
Kawanishi, K. *et al.*, *Phytochemistry*, 1983, **22**, 2277-2280 (*Dihydrocarinatidin*).
Marques, M.O.M. *et al.*, *Phytochemistry*, 1992, **31**, 4380-4381 (*Dihydrocarinatin*).
Li, F. *et al.*, *Helv. Chim. Acta*, 2007, **90**, 1491-1496 (*Isodihydrocarinatidin*).
Morikawa, T. *et al.*, *J. Nat. Prod.*, 2016, **79**, 2005-2013 (*Maceneolignan E*)

4,7′-Epoxy-3,8′-lign-8-ene-3′,4′,6-triol E181

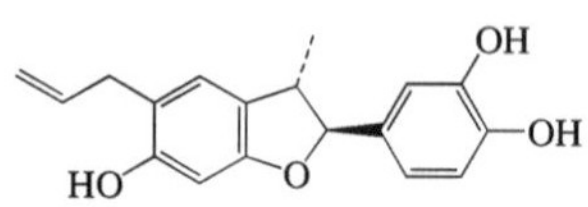

2-(3,4-Dihydroxyphenyl)-2,3-dihydro-5-(2-propenyl)-6-benzofuranol. 4,7′-Epoxy-3′,4′,6-trihydroxy-3,8′-neolign-8-ene

$C_{18}H_{18}O_4$ 298.338

(7′S,8′S)-form

3′,4′-Di-Me ether

4,7′-Epoxy-3′,4′-dimethoxy-3,8′-lign-8-en-6-ol. 4,7′-Epoxy-6-hydroxy-3′,4′-dimethoxy-3,8′-lign-8-ene. ***Liliflol B***

[87402-79-7]

$C_{20}H_{22}O_4$ 326.392

Constit. of *Magnolia liliflora*. Viscous oil

3′,4′-Methylene ether

4,7′-Epoxy-3′,4′-methylenedioxy-3,8′-lign-8-en-6-ol. 4,7′-Epoxy-6-hydroxy-3′,4′-methylenedioxy-3,8′-lign-8-ene. ***Liliflol A***

[87402-78-6]

$C_{19}H_{18}O_4$ 310.349

Constit. of *Magnolia liliflora* and *Nectandra amazonum*. Oil.

Iida, T. *et al.*, *Phytochemistry*, 1983, **22**, 763-766 (*Liliflols A,B*).
Horne, D.A. *et al.*, *Tetrahedron Lett.*, 1999, **40**, 5443-5447 (*Liliflol B, synth*)

4,7′-Epoxy-3,8′-lign-7-en-4′-ol E182

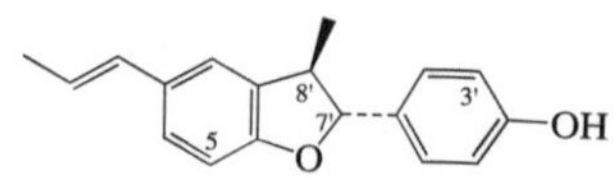

(7E,7′R,8′R)-form

2,3-Dihydro-2-(4-hydroxyphenyl)-3-methyl-5-(1-propenyl)benzofuran. 4,7′-Epoxy-4′-hydroxy-3,8′-neolign-7-ene

[118201-52-8] (unspecified stereochem.)

$C_{18}H_{18}O_2$ 266.34

Lignan numbering shown. Abs. configs. finalised in 2008-2009

(7E,7′R,8′R)-form

(-)-Conocarpan [56319-02-9]

Constit. of *Conocarpus erectus*, *Piper aequale*, *Piper rivinoides* and *Krameria* spp.. Shows antitrypanosomal activity. Antifungal agent. Cryst.. Mp 128 - 130°. $[\alpha]^{25}_{D}$ -103 (c, 3.2 in MeOH). λ_{max} 242 (ε 5010); 270 (ε 10700); 303 (ε 2820) (MeOH)

Ac

Cryst. (petrol). Mp 94 - 95°

5-Methoxy

4,7′-Epoxy-5-methoxy-3,8′-lign-7-en-4′-ol. 4,7′-Epoxy-4′-hydroxy-5-methoxy-3,8′-lign-7-ene. ***5-Methoxyconocarpan***

[1802051-92-8]

$C_{19}H_{20}O_3$ 296.366

Constit. of the stems of *Aristolochia fordiana*. Oil. $[\alpha]^{20}_{D}$ -15 (c, 0.6 in MeOH). λ_{max} 222 (log ε 4.55); 269 (log ε 4.28) (MeOH)

(7E,7′R,8′S)-form

(-)-Epiconocarpan [249744-77-2]

Constit. of the roots of *Piper regnellii*. Amorph.. $[\alpha]^{21}_{D}$ -33.3 (c, 0.03 in MeOH). λ_{max} 220 (ε 12750); 275 (ε 6380) (MeOH)

(7E,7′S,8′S)-form

(+)-Conocarpan [221666-27-9]

Constit. of *Aristolochia fordiana*, *Krameria cytisoides* and *Piper decurrens*. Shows antiinflammatory and insecticidal activities. Cryst.. Mp

133 - 135°. $[\alpha]^{21}_{D}$+122 (c, 1.03 in MeOH)

Me ether

4,7′-Epoxy-4′-methoxy-3,8′-lign-7-ene. ***O-Methyl-(+)-conocarpan*** [139573-13-0]

$C_{19}H_{20}O_2$ 280.367

Constit. of *Krameria grayi* and *Krameria ixina*. Mp 68 - 73°. $[\alpha]^{20}_{D}$ +98 (c, 2.3 in MeOH). λ_{max} 226 (sh); 263 (log ε 4.09); 270 (sh); 302 (log ε 3.35) (no solvent reported)

3′-Hydroxy

see 4,7′-Epoxy-3,8′-lign-7-ene-3′,4′-diol, E147

5-Methoxy

5-Methoxy-(+)-conocarpan [109225-41-4]

$C_{19}H_{20}O_3$ 296.366

Constit. of *Aristolochia fordiana* and *Krameria cytisoides*. NO prodn. inhibitor. Cryst.. Mp 100 - 103°. $[\alpha]^{21}_{D}$+57 (c, 0.67 in MeOH).

Hayashi, T. *Phytochemistry*, 1975, **14**, 1085-1087 (*Ac, struct*).
Achenbach, H. *et al.*, *Phytochemistry*, 1987, **26**, 1159-1166 (*Conocarpan*).
Arnone, A. *et al.*, *Gazz. Chim. Ital.*, 1988, **118**, 675-682 (*(-)-form, isol, activity*).
Achenbach, H. *et al.*, *Phytochemistry*, 1991, **30**, 3753-3757 (*Krameria ixina constit*).
Achenbach, H. *et al.*, *Phytochemistry*, 1995, **39**, 413-415 (*Krameria grayi constit*).
Chauret, D.C. *et al.*, *J. Nat. Prod.*, 1996, **59**, 152-155 (*isol, pmr, cmr*).
Snider, B.B. *et al.*, *J. Org. Chem.*, 1997, **62**, 6978-6984 (*Conocarpan, synth*).
Maxwell, A. *et al.*, *Phytochemistry*, 1999, **50**, 499-504 (*(-)-form*).
Benevides, P.J.C. *et al.*, *Phytochemistry*, 1999, **52**, 339-343 (*8-Epiconocarpan*).
Kurtan, T. *et al.*, *J. Chem. Soc., Perkin Trans. 1*, 2000, 453-461 (*abs config*).
Sartorelli, P. *et al.*, *Plant Sci. (Shannon, Irel.)*, 2001, **161**, 1083-1088 (*Conocarpan, synth, biosynth*).
Zheng, S.-L. *et al.*, *Tetrahedron Lett.*, 2003, **44**, 1445-1447 (*8-Epiconocarpan, synth*).
Luize, P.S. *et al.*, *Biol. Pharm. Bull.*, 2006, **29**, 2126-2130 (*Conocarpan, activity*).
Clive, D.L.J. *et al.*, *Org. Biomol. Chem.*, 2008, **6**, 1831-1842 (*synth, abs config*).
Natori, Y. *et al.*, *J. Org. Chem.*, 2009, **74**, 4418-4421 (*Conocarpan, 8-Epiconocarpan, synth, abs config*).
Baumgartner, L. *et al.*, *J. Nat. Prod.*, 2011, **74**, 1779-1786 (*(+)-Conocarpan, activity*).
Zhou, Z.-B. *et al.*, *Planta Med.*, 2013, **79**, 1730-1735 (*Aristolochia fordiana constits*).
Chen, C.-Y. *et al.*, *Synlett*, 2013, **24**, 189-192 (*(+)-Conocarpan, synth*).
Chen, P.Y. *et al.*, *Tetrahedron*, 2013, **69**, 653-657 (*Conocarpan, synth*).
Tang, G.-H. *et al.*, *J. Nat. Prod.*, 2015, **78**, 1894-1903 (*5-Methoxyconocarpan*).
Moreira, D.de L. *et al.*, *Rec. Nat. Prod.*, 2016, **10**, 472-484 (*Conocarpan, activity*)

4,7′-Epoxy-3,8′-lign-8-en-4′-ol E183

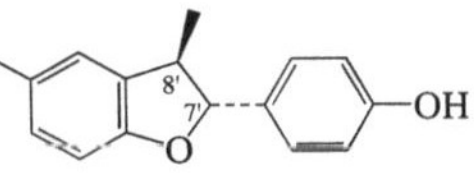

(7′R,8′R)-form

2,3-Dihydro-2-(4-hydroxyphenyl)-3-methyl-5-(2-propenyl)benzofuran. 4,7′-Epoxy-4′-hydroxy-3,8′-neolign-8-ene

$C_{18}H_{18}O_2$ 266.34

(7′R,8′R)-form

Miliumollin [1529771-47-8]

Constit. of the leaves of *Miliusa mollis*. Exhibits weak cytotoxic activity. Viscous liq.. $[\alpha]^{20}_{D}$+32.4 (c, 0.17 in $CHCl_3$). λ_{max} 209 (log ε 4.30); 230 (log ε 4.31); 285 (log ε 3.82) (MeOH)

(7′S,8′S)-form

ent-Miliumollin

Me ether

4,7′-Epoxy-4′-methoxy-3,8′-lign-8-ene. ***4′-O-Methyl-ent-miliumollin*** [1529771-46-7]

$C_{19}H_{20}O_2$ 280.367

Constit. of the leaves of *Miliusa mollis*. Viscous liq.. $[\alpha]^{20}_{D}$-25.0 (c, 0.12 in $CHCl_3$). λ_{max} 212 (log ε 3.95); 229 (log ε 4.09); 283 (log ε 3.63) (MeOH).

Sawasdee, K. *et al.*, *Fitoterapia*, 2013, **85**, 49-56 (*Miliumollin, 4′-O-Methyl-ent-miliumollin*)

7,7′-Epoxy-3-methoxy-3′,4′-methylenedioxy-8,8′-lignan-4-ol E184

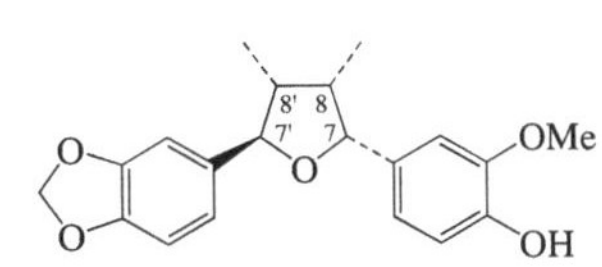

(7R,7′R,8S,8′R)-form

4-[5-(1,3-Benzodioxol-5-yl)tetrahydro-3,4-dimethyl-2-furanyl]-2-methoxyphenol, CAS. *Tetrahydro-2-(4-hydroxy-3-methoxyphenyl)-5-(3,4-methylenedioxyphenyl)-3,4-dimethylfuran. 4-Hydroxy-3-methoxy-3′,4′-methylenedioxy-7,7′-epoxylignan*

$C_{20}H_{22}O_5$ 342.391

(7R,7′R,8S,8′R)-form

(+)-Chicanine [78919-28-5]

Constit. of *Schisandra* spp.. Prisms ($CHCl_3$/hexane). Mp 122 - 124°. $[\alpha]^{31}_{D}$+118.8 (c, 4.6 in $CHCl_3$). Abs. config. revised in 2011. λ_{max} 233 (ε 12589); 284 (ε 7413) (MeOH)

(7R,7′S,8R,8′R)-form

Odoratisol D [891182-96-0]

Constit. of the bark of *Machilus odoratissima*. $[\alpha]^{25}_{D}$-12.8 (c, 4 in $CHCl_3$). λ_{max} 234 (ε 9332); 282 (ε 6309) (MeOH)

Me ether

7,7′-Epoxy-3,4-dimethoxy-3′,4′-methylenedioxylignan. 3,4-Dimethoxy-3′,4′-methylenedioxy-7,7′-epoxylignan. ***Futokadsurin B*** [852459-90-6]

$C_{21}H_{24}O_5$ 356.418

Constit. of *Piper futokadsura*. Needles. Mp 102°. $[\alpha]^{23}_{D}$+33.7 (c, 1.2 in $CHCl_3$)

(7R,7′S,8S,8′R)-form

Me ether

[155835-53-3]

Constit. of *Machilus japonica*. Oil. $[\alpha]_{D}$ -39 (c, 0.002 in MeOH)

(7R,7′S,8S,8′S)-form

Austrobailignan 7 [55890-25-0]

Constit. of *Austrobaileya scandens*, *Aristolochia taliscana* and from mace (*Myristica fragrans*). Pale yellow gum. Bp 130° at 0.02 mm. $[\alpha]^{25}_{D}$+12 (c, 2.9 in $CHCl_3$)

Ac

[55730-99-9]

Mp 81 - 82°

Me ether

Calopeptin. ***Calopiptin*** [19950-67-5]

$C_{21}H_{24}O_5$ 356.418

Constit. of *Piptocalyx moorei*, *Saruma henryi*, *Piper schmidtii*, *Pycnanthus angolensis* and *Virola surinamensis*. Cryst. (C_6H_6 or petrol). Mp 95.5° (79-83°). $[\alpha]^{14}_{D}$ +30 (c, 0.1 in $CHCl_3$) (+28). λ_{max} 233; 283 (EtOH)

(7S,7′R,8S,8′R)-form

Me ether

Futokadsurin C [852459-91-7]

$C_{21}H_{24}O_5$ 356.418

Constit. of *Piper futokadsura*. Inhibitor of LPS-induced NO prodn.. Oil. $[\alpha]^{23}_{D}$-11.7 (c, 3.3 in $CHCl_3$). λ_{max} 230 (log ε 5.01); 279 (log ε 4.83) (EtOH)

(7S,7′S,8R,8′S)-form

(-)-Chicanine [627875-49-4]

Isol. from *Leucas aspera*. Amorph. solid. [α] D -134.3 ($CHCl_3$). Abs. config. revised in 2011

Me ether

Machilusin. *Zuonin C. Zuihonin C*

[61989-57-9]

[79120-60-8] (Zuonin C)

$C_{21}H_{24}O_5$ 356.418

Constit. of *Leucas aspera*, *Machilus japonica*, *Piper futokadsura* and *Schisandra propinqua* var. *sinensis*. Plates. Mp 121 - 122°. [α] D -130.6 ($CHCl_3$)

(7S,7′S,8S,8′S)-form

Talaumidin [199994-33-7]

Constit. of *Aristolochia arcuata*, *Talauma hodgsonii* and *Virola surinamensis*. Neurotrophic agent. Viscous gum or oil. $[\alpha]^{20}_D$ -90.2 (c, 0.01 in $CHCl_3$). [α] D -17.9 (c, 0.1 in $CHCl_3$). λ_{max} 233 (log ε 4); 283 (log ε 3.7) (MeOH)

(7R*,7′R*,8R*,8′R*)-form

Me ether

[158569-73-4]

Constit. of *Piper wightii*. Visc. oil. $[\alpha]^{24}_D$ -41.7 (c, 0.15 in $CHCl_3$). Relative config. only

(7R*,7′S*,8R*,8′S*)-form

Machilin F. Fragransin E_1. 3-Epiaustrobailignan 7 [114488-90-3]

Constit. of mace (*Myristica fragrans*), *Machilus thunbergii* and *Aristolochia chilensis*. Oil. Abs. config. unknown (quasi-*meso*)

Me ether

Machilin G

[114422-19-4]

$C_{21}H_{24}O_5$ 356.418

Constit. of *Machilus thunbergii* and *Piper schmidtii*. Oil. $[\alpha]^{24}_D$ +3.9 (c, 0.3 in $CHCl_3$). Abs. config. unknown (quasi-*meso*). The *Piper schmidtii* isolate has negative opt. rotn.. λ_{max} 217 (ε 28184); 282 (ε 15488) (EtOH).

McAlpine, J.B. *et al.*, *Aust. J. Chem.*, 1968, **21**, 2095-2106 (*Calopiptin*).
Murphy, S.T. *et al.*, *Aust. J. Chem.*, 1975, **28**, 81-90 (*Austrobailignan 7*).
Liu, S.-J. *et al.*, *Can. J. Chem.*, 1981, **59**, 1680-1684 (*Chicanine*).
Urzua, A. *et al.*, *Phytochemistry*, 1987, **26**, 1509-1511 (*Zuonin C*).
Hada, S. *et al.*, *Phytochemistry*, 1988, **27**, 563-568 (*Fragransin E_1*).
Shimomura, H. *et al.*, *Phytochemistry*, 1988, **27**, 634-636 (*Machilin F*).
Tyagi, O.D. *et al.*, *Phytochemistry*, 1993, **32**, 445-448 (*Piper schmidtii constits*).
Urzua, A. *et al.*, *Phytochemistry*, 1993, **34**, 874-875 (*3-Epiaustrobailignan 7*).
Gonzalez-Coloma, A. *et al.*, *Phytochemistry*, 1994, **35**, 607-610 (*Machilusin*).
Prasad, A.K. *et al.*, *Tetrahedron*, 1994, **50**, 10579-10586 (*Piper wightii constit*).
Lopes, N.P. *et al.*, *Phytochemistry*, 1996, **43**, 1089-1092 (*Calopiptin*).
De Almeida Blumenthal, E.E. *et al.*, *Phytochemistry*, 1997, **46**, 745-749 (*Talaumidin*).
Vieira, L.M. *et al.*, *Phytochemistry*, 1998, **48**, 1079-1081 (*Talaumidin*).
Uck, Y. *et al.*, *J. Pharm. Pharmacol.*, 2000, **52**, 1163-1169 (*Machilin G*).
Sadhu, S.K. *et al.*, *Chem. Pharm. Bull.*, 2003, **51**, 595-598 (*(-)-Chicanine*).
Zhai, H. *et al.*, *Planta Med.*, 2004, **70**, 598-602 (*Talaumidin, activity*).
Konishi, T. *et al.*, *Chem. Pharm. Bull.*, 2005, **53**, 121-124 (*Futokadsurins B,C*).
Giang, P.M. *et al.*, *Chem. Pharm. Bull.*, 2006, **54**, 380-383 (*Odoratisol D*).
Kim, H. *et al.*, *Org. Lett.*, 2007, **9**, 3965-3968 (*synth*).
Wabo, H.K. *et al.*, *Planta Med.*, 2007, **73**, 187-189 (*Pycnanthus angolensis constit*).
Esumi, T. *et al.*, *Heterocycles*, 2008, **76**, 551-567 (*Talaumidin, synth*).
Matcha, K. *et al.*, *Tetrahedron Lett.*, 2008, **49**, 3433-3436 (*Talaumidin, synth*).
Xue, P. *et al.*, *J. Asian Nat. Prod. Res.*, 2009, **11**, 281-287 (*Talaumidin, synth*).
Kim, K.H. *et al.*, *Bioorg. Med. Chem. Lett.*, 2010, **20**, 409-412 (*Futokadsurin C, actyivity*).
Harada, K. *et al.*, *Heterocycles*, 2011, **82**, 1127-1132 (*Machidin F, synth*).
Harada, K. *et al.*, *Tetrahedron Lett.*, 2011, **52**, 3005-3008 (*Chicanine, synth, abs config*).
Harada, K. *et al.*, *J. Org. Chem.*, 2015, **80**, 7076-7088 (*Talaumidin, synth*).
Harada, K. *et al.*, *Front. Chem.*, 2020, **8**, 301:1-14 (*Talaumidin, rev, synth, activity*)

1,2-Epoxy-1-(3,4-methylenedioxyphenyl)-6-methoxybenzocyclobutene-4-propanol E185

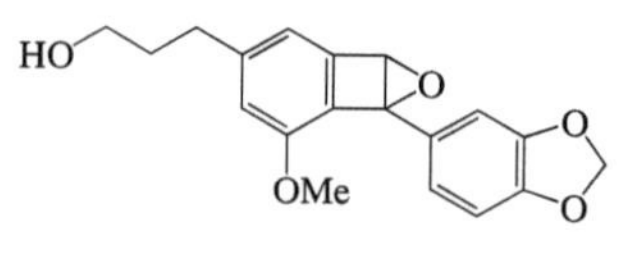

3-[7-(1,3-Benzodioxol-5-yl)-5-methoxy-8-oxatricyclo[4.3.0.0^{7,9}]nona-1,3,5-trienyl]-1-propanol. 4,7′-Cyclo-7′,8′-epoxy-5-methoxy-3′,4′-methylenedioxy-9′-nor-3,8′-lignan-9-ol. 4,7′-Cyclo-7′,8′-epoxy-9-hydroxy-5-methoxy-3′,4′-methylenedioxy-9′-nor-3,8′-neolignan

$C_{19}H_{18}O_5$ 326.348

2-Methylbutanoyl

[612058-47-6]

$C_{24}H_{26}O_6$ 410.466

Constit. of the seeds of *Styrax officinalis*. Viscous yellow oil. λ_{max} 206 (sh); 289; 316 (MeOH).

Akgul, Y.Y. *et al.*, *Phytochemistry*, 2003, **63**, 939-943 (*Styrax officinalis constit*)

4,7′-Epoxy-9′-nor-3,8′-ligna-7,7′-diene E186

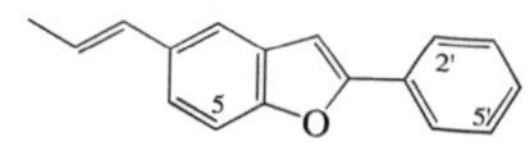

2-Phenyl-5-(1-propenyl)benzofuran. 4,7′-Epoxy-9′-nor-3,8′-neoligna-7,7′-diene

$C_{17}H_{14}O$ 234.298

Parent not isol.

(E)-form

4′-Hydroxy

4-[5-(1-Propenyl)-2-benzofuranyl]phenol. 2-(4-Hydroxyphenyl)-5-(1-propenyl)benzofuran. 4,7′-Epoxy-9′-nor-3,8′-ligna-7,7′-dien-4′-ol. 4,7′-Epoxy-4′-hydroxy-9′-nor-3,8′-ligna-7,7′-diene

[109194-69-6]

$C_{17}H_{14}O_2$ 250.297

Constit. of *Krameria cytisoides* and *Krameria lappacea*. Cryst.. Mp 208 - 211°. λ_{max} 225 (log ε 4.23); 254 (log ε 4.31); 290 (log ε 4.3); 305 (log ε 4.31); 319 (log ε 4.27); 332 (sh) (MeOH)

2′,4′-Dihydroxy

2-(2,4-Dihydroxyphenyl)-5-(1-propenyl)benzofuran. 4,7′-Epoxy-9′-nor-3,8′-ligna-7,7′-diene-2′,4′-diol. 4,7′-Epoxy-2′,4′-dihydroxy-9′-nor-3,8′-ligna-7,7′-diene

[109194-71-0]

$C_{17}H_{14}O_3$ 266.296

Constit. of *Krameria cytisoides*, *Krameria lappacea* and *Krameria triandra*. Shows antifungal and antibacterial activities. Cryst.. Mp 181 - 184° (180-182°). λ_{max} 230 (ε 24500); 250 (ε 22900); 285 (ε 18200); 315 (ε 25000); 325 (ε 23300) (EtOH) (Berdy)

2′-Methoxy, 4′-hydroxy

2-(4-Hydroxy-2-methoxyphenyl)-5-(1-propenyl)benzofuran. 4,7′-Epoxy-2′-methoxy-9′-nor-3,8′-ligna-7,7′-dien-4′-ol. 4,7′-Epoxy-4′-hydroxy-2′-methoxy-9′-nor-3,8′-ligna-7,7′-diene. ***Ratanhiaphenol III***

[91432-06-3]

$C_{18}H_{16}O_3$ 280.323
Constit. of *Krameria lanceolata*, *Krameria lappacea*, *Krameria parvifolia* and *Krameria triandra*. Exhibits antibacterial and antifungal activities. Cryst.. Sol. MeOH, C_6H_6; fairly sol. hexane; poorly sol. H_2O. Mp 134 - 135°. Mp 145 - 150°

4′-Methoxy
2-(4-Methoxyphenyl)-5-(1-propenyl)benzofuran. 4,7′-Epoxy-4′-methoxy-9′-nor-3,8′-ligna-7,7′-diene
[166660-91-9]
$C_{18}H_{16}O_2$ 264.324
Constit. of *Krameria grayi*. Amorph.. λ_{max} 224 (log ε 4.26); 256 (log ε 4.33); 272 (sh); 291 (log ε 4.31); 304 (log ε 4.29); 318 (log ε 4.22); 325 (sh); 333 (sh) (MeOH)

4′-Methoxy, 2′-hydroxy
2-(2-Hydroxy-4-methoxyphenyl)-5-(1-propenyl)benzofuran. 4,7′-Epoxy-4′-methoxy-9′-nor-3,8′-ligna-7,7′-dien-2′-ol. 4,7′-Epoxy-2′-hydroxy-4′-methoxy-9′-nor-3,8′-ligna-7,7′-diene. ***Ratanhiaphenol I***
[79214-54-3]
$C_{18}H_{16}O_3$ 280.323
Constit. of *Krameria triandra*, *Krameria cytisoides*, *Krameria lappacea* and *Krameria tomentosa*. Cryst.. Mp 182 - 184° (180-182°). λ_{max} 233 (log ε 4.36); 255 (log ε 4.34); 275 (log ε 4.26); 288 (log ε 4.25); 313 (sh); 320 (log ε 4.44); 327 (log ε 4.37); 335 (log ε 4.39) (MeOH)

4′-Methoxy, 2′,6′-dihydroxy
2-(2,6-Dihydroxy-4-methoxyphenyl)-5-(1-propenyl)benzofuran. 4,7′-Epoxy-4′-methoxy-9′-nor-3,8′-ligna-7,7′-diene-2′,6′-diol. 4,7′-Epoxy-2′,6′-dihydroxy-4′-methoxy-9′-nor-3,8′-ligna-7,7′-diene
[110784-21-9]
$C_{18}H_{16}O_4$ 296.322
Constit. of *Krameria ramosissima*. Cryst.. Mp 167 - 168°. λ_{max} 236 (log ε 4.56); 247 (sh); 258 (sh); 293 (log ε 4.47); 305 (log ε 4.47); 318 (log ε 4.45); 326 (sh); 332 (sh) (MeOH)

5-Methoxy, 4′-hydroxy
2-(4-Hydroxyphenyl)-7-methoxy-5-(1-propenyl)benzofuran. 4,7′-Epoxy-5-methoxy-9′-nor-3,8′-ligna-7,7′-dien-4′-ol. 4,7′-Epoxy-4′-hydroxy-5-methoxy-9′-nor-3,8′-ligna-7,7′-diene
[109194-70-9]
$C_{18}H_{16}O_3$ 280.323
Constit. of *Krameria cytisoides*. Cryst.. Mp 177 - 179°

5-Methoxy, 2′,4′-dihydroxy
2-(2,4-Dihydroxyphenyl)-7-methoxy-5-(1-propenyl)benzofuran. 4,7′-Epoxy-5-methoxy-9′-nor-3,8′-ligna-7,7′-diene-2′,4′-diol. 4,7′-Epoxy-2′,4′-dihydroxy-5-methoxy-9′-nor-3,8′-ligna-7,7′-diene
[109194-72-1]
$C_{18}H_{16}O_4$ 296.322
Constit. of *Krameria cytisoides*. Cryst.. Mp 172 - 174°

2′,4′-Dimethoxy
2-(2,4-Dimethoxyphenyl)-5-(1-propenyl)benzofuran. 4,7′-Epoxy-2′,4′-dimethoxy-9′-nor-3,8′-ligna-7,7′-diene
[110784-22-0]
$C_{19}H_{18}O_3$ 294.35
Constit. of *Krameria grayi*, *Krameria ramosissima* and *Krameria triandra*. Cryst.. Mp 146 - 147° (140-145°). λ_{max} 228 (ε 6300); 252 (ε 5650); 271 (ε 4500); 284 (ε 4450); 309 (ε 5860); 317 (ε 6300); 324 (ε 6130); 333 (ε 6500) (EtOH)

2′,4′-Dimethoxy, 6′-hydroxy
2-(2-Hydroxy-4,6-dimethoxyphenyl)-5-(1-propenyl)benzofuran. 4,7′-Epoxy-4′,6′-dimethoxy-9′-nor-3,8′-ligna-7,7′-dien-2′-ol. 4,7′-Epoxy-2′-hydroxy-4′,6′-dimethoxy-9′-nor-3,8′-ligna-7,7′-diene
[110784-20-8]
$C_{19}H_{18}O_4$ 310.349
Constit. of *Krameria ramosissima* and *Krameria tomentosa*. Cryst.. Mp 119 - 120°. λ_{max} 233 (log ε 4.68); 246 (sh); 256 (sh); 294 (log ε 4.49); 307 (sh); 320 (sh); 332 (sh) (MeOH)

4′,5-Dimethoxy, 2′-hydroxy
2-(2-Hydroxy-4-methoxyphenyl)-7-methoxy-5-(1-propenyl)benzofuran. 4,7′-Epoxy-4′,5-dimethoxy-9′-nor-3,8′-ligna-7,7′-dien-2′-ol. 4,7′-Epoxy-2′-hydroxy-4′,5-dimethoxy-9′-nor-3,8′-ligna-7,7′-diene. ***Toltecol***
[109145-64-4]
$C_{19}H_{18}O_4$ 310.349
Constit. of *Krameria cytisoides*. Cryst.. Mp 101 - 102°. λ_{max} 232 (log ε 4.29); 267 (log ε 4.38); 289 (log ε 4.26); 316 (log ε 4.34); 324 (sh); 331 (log ε 4.26) (MeOH)

2′,4′,5′-Trimethoxy
5-(1-Propenyl)-2-(2,4,5-trimethoxyphenyl)benzofuran. 4,7′-Epoxy-2′,4′,5′-trimethoxy-9′-nor-3,8′-ligna-7,7′-diene
[138916-48-0]
$C_{20}H_{20}O_4$ 324.376
Constit. of *Krameria ixina*. Cryst.. Mp 131 - 134°. λ_{max} 219 (sh); 239 (log ε 4.21); 256 (sh); 278 (log ε 4.00); 291 (log ε 3.99); 315 (sh); 332 (log ε 4.22); 345 (log ε 4.18) (MeOH)

2′,4′,6′-Trimethoxy
5-(1-Propenyl)-2-(2,4,6-trimethoxyphenyl)benzofuran. 4,7′-Epoxy-2′,4′,6′-trimethoxy-9′-nor-3,8′-ligna-7,7′-diene
[110801-34-8]
$C_{20}H_{20}O_4$ 324.376
Constit. of *Krameria ramosissima*. Cryst.. Mp 139 - 140°. λ_{max} 232 (log ε 4.55); 255 (sh); 320 (sh) (MeOH)

2′,3′,4′,6′-Tetramethoxy
5-(1-Propenyl)-2-(2,3,4,6-tetramethoxyphenyl)benzofuran. 4,7′-Epoxy-2′,3′,4′,6′-tetramethoxy-9′-nor-3,8′-ligna-7,7′-diene
[110801-33-7]
$C_{21}H_{22}O_5$ 354.402
Constit. of *Krameria grayi* and *Krameria ramosissima*. Cryst.. Mp 125 - 127°. λ_{max} 234 (log ε 4.64); 258 (sh); 292 (sh); 320 (sh) (MeOH)

2′,3′,4′,5′,6′-Pentamethoxy
2-(2,3,4,5,6-Pentamethoxyphenyl)-5-(1-propenyl)benzofuran. 4,7′-Epoxy-2′,3′,4′,5′,6′-pentamethoxy-9′-nor-3,8′-ligna-7,7′-diene. ***Ramosissin***
[110784-16-2]
$C_{22}H_{24}O_6$ 384.428
Constit. of the roots of *Krameria ramosissima*. Cryst.. Mp 92 - 94°. λ_{max} 233 (log ε 4.55); 255 (sh); 270 (sh); 285 (sh); 317 (sh) (MeOH).

Stahl, E. *et al., Planta Med.*, 1981, **42**, 144-154 (*Krameria triandra constits, Ratanhiaphenol I*).
Achenbach, H. *et al., Phytochemistry*, 1987, **26**, 1159-1166 (*Krameria cytisoides constits*).
Achenbach, H. *et al., Phytochemistry*, 1987, **26**, 2041-2043 (*Krameria ramosissima constits*).
Arnone, A. *et al., Gazz. Chim. Ital.*, 1988, **118**, 675-682 (*Krameria triandra constits, activity*).
Achenbach, H. *et al., Phytochemistry*, 1989, **28**, 1959-1962 (*Ratanhiaphenol III*).
US Pat., 1989, 4 886 667 (*Ratanhiaphenol III, activity*).
Achenbach, H. *et al., Phytochemistry*, 1991, **30**, 3753-3757 (*Krameria ixina constit*).
Donnelly, D.M.X. *et al., Tetrahedron Lett.*, 1991, **32**, 3835-3836 (*synth*).
Achenbach, H. *et al., Phytochemistry*, 1995, **39**, 413-415 (*Krameria grayi constits*).
Pat. Coop. Treaty (WIPO), 1998, 98 17 294 (*activity*).
Silva, S.A.S. *et al., Nat. Prod. Lett.*, 2001, **15**, 323-329 (*Ratanhiaphenol I*).
Seo, P.-J. *et al., Arch. Pharmacal Res.*, 2004, **27**, 1189-1193 (*synth*).
Duan, X.-F. *et al., Synthesis*, 2010, 2547-2552 (*synth*).
Baumgartner, L. *et al., J. Nat. Prod.*, 2011, **74**, 1779-1786 (*Krameria lappacea constits*)

4,7′-Epoxy-9′-nor-3,8′-ligna-7,7′-diene-3′,4′,5,9-tetrol E187

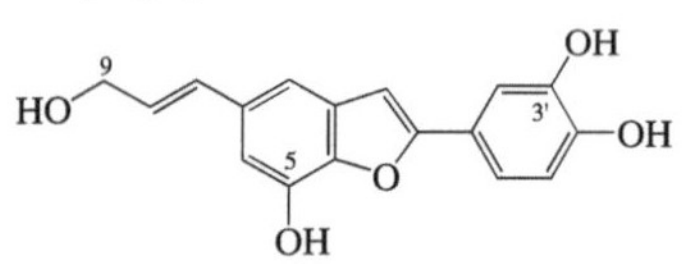

3-[2-(3,4-Dihydroxyphenyl)-7-hydroxy-5-benzofuranyl]-2-propen-1-ol. 4,7′-Epoxy-3′,4′,5,9-tetrahydroxy-9′-nor-3,8′-neoligna-7,7′-diene

$C_{17}H_{14}O_5$ 298.294

(E)-form

3′-Me ether

4,7′-Epoxy-3′-methoxy-9′-nor-3,8′-ligna-7,7′-diene-4′,5,9-triol. 4,7′-Epoxy-4′,5,9-trihydroxy-3′-methoxy-9′-nor-3,8′-ligna-7,7′-diene. 8-Dehydroxymethylvibsanol. ***De-O-methylailanthoidol***

[1583276-03-2]

$C_{18}H_{16}O_5$ 312.321

Constit. of *Solidago canadensis*. Displays hypolipidaemic effects in HFD-fed hamsters. Oil. $[\alpha]_D^{23}$ -7.03 (c, 0.12 in MeOH). λ_{max} 275 (log ε 4.11); 315 (log ε 4.00) (MeOH)

3′,5-Di-Me ether

4,7′-Epoxy-3′,5-dimethoxy-9′-nor-3,8′-ligna-7,7′-diene-4′,9-diol. 4,7′-Epoxy-4′,9-dihydroxy-3′,5-dimethoxy-9′-nor-3,8′-ligna-7,7′-diene. ***Ailanthoidol***

[156398-61-7]

$C_{19}H_{18}O_5$ 326.348

Constit. of *Sarcomelicope megistophylla* and *Zanthoxylum ailanthoides*. Shows potential antiinflammatory activity via inhibition of nitric oxide (NO) formn. in stimulated RAW 264.7 macrophages. Prisms (MeOH). Mp 199 - 201°. λ_{max} 215 (sh); 233 (log ε 4.38); 275 (log ε 4.45); 298 (sh); 314 (log ε 4.49); 335 (sh) (EtOH)

3′,4′-Methylene, 5-Me ether

4,7′-Epoxy-5-methoxy-3′,4′-methylenedioxy-9′-nor-3,8′-ligna-7,7′-dien-9-ol. 4,7′-Epoxy-9-hydroxy-5-methoxy-3′,4′-methylenedioxy-9′-nor-3,8′-ligna-7,7′-diene. ***Zanthocapensol***

[1252931-57-9]

$C_{19}H_{16}O_5$ 324.332

Constit. of the roots of *Zanthoxylum capense*. Amorph. solid

3′,4′-Methylene, 5,9-di-Me ether

4,7′-Epoxy-5,9-dimethoxy-3′,4′-methylenedioxy-9′-nor-3,8′-ligna-7,7′-diene

[1639286-96-6]

$C_{20}H_{18}O_5$ 338.359

Constit. of the roots of *Heliopsis helianthoides*. Amorph. solid

9-Aldehyde, 3′-Me ether

4,7′-Epoxy-4′,5-dihydroxy-3′-methoxy-9′-nor-3,8′-ligna-7,7′-dien-9-al. ***Betulifoliumin C***

[2197124-01-7]

$C_{18}H_{14}O_5$ 310.305

Constit. of the aerial parts of *Viburnum betulifolium*. Oil. λ_{max} 275 (log ε 4.13); 315 (log ε 4.02) (MeOH)

9-Aldehyde, 3′,4′-di-Me ether

4,7′-Epoxy-5-hydroxy-3′,4′-dimethoxy-9′-nor-3,8′-ligna-7,7′-dien-9-al. ***Betulifoliumin D***

[2197124-02-8]

$C_{19}H_{16}O_5$ 324.332

Constit. of the aerial parts of *Viburnum betulifolium*. Yellow oil. $[\alpha]_D^{20}$ -5.1 (c, 0.14 in MeOH). λ_{max} 276 (log ε 4.12); 310 (log ε 3.95) (MeOH)

9-Aldehyde, 3′,5-di-Me ether

4,7′-Epoxy-4′-hydroxy-3′,5-dimethoxy-9′-nor-3,8′-ligna-7,7′-dien-9-al. 4,7′-Epoxy-4′-hydroxy-3′,5-dimethoxy-9′-nor-9-oxo-3,8′-ligna-7,7′-diene. ***Sarcomeginal***

[623930-35-8]

$C_{19}H_{16}O_5$ 324.332

Constit. of *Sarcomelicope megistophylla* and *Viburnum betulifolium*. Amorph. solid. λ_{max} 286 (sh); 297 (log ε 3.74); 316 (log ε 3.8); 327 (sh) (MeOH)

9-Aldehyde, 3′,4′-methylene, 5-Me ether

3-[7-Methoxy-2-(3,4-methylenedioxyphenyl)-5-benzofuranyl]-2-propenal. 4,7′-Epoxy-5-methoxy-3′,4′-methylenedioxy-9′-nor-3,8′-ligna-7,7′-dien-9-al. 4,7′-Epoxy-5-methoxy-3′,4′-methylenedioxy-9′-nor-9-oxo-3,8′-ligna-7,7′-diene

[879495-12-2]

$C_{19}H_{14}O_5$ 322.316

Constit. of the seeds of *Styrax perkinsiae*. Needles (Me_2CO). Mp 187 - 189°. λ_{max} 216 (log ε 4.08); 299 (log ε 4.3); 319 (log ε 4.4) (MeOH)

9-Carboxylic acid, 3′,4′-methylene, 5-Me ether

4,7′-Epoxy-5-methoxy-3′,4′-methylenedioxy-9′-nor-3,8′-ligna-7,7′-dien-9-oic acid

9-Carboxylic acid, 3′,4′-methylene, 5-Me ether, Me ester

Zanthocapensate

[1396033-28-5]

$C_{20}H_{16}O_6$ 352.342

Constit. of the roots of *Zanthoxylum capense*. Exhibits apoptosis inducing activity. Amorph. solid.

Sheen, W.-S. *et al., Phytochemistry*, 1994, **36**, 213-215 (*Ailanthoidol*).
Bates, R.W. *et al., Synlett*, 1995, 1151-1152 (*Ailanthoidol, synth*).
Kao, C.-L. *et al., J. Org. Chem.*, 2002, **67**, 6772-6787 (*Ailanthoidol, synth*).
Lin, S.-Y. *et al., J. Org. Chem.*, 2003, **68**, 2968-2971 (*Ailanthoidol, synth*).
Fokialakis, N. *et al., Planta Med.*, 2003, **69**, 566-568 (*Sarcomelicope megistophylla constits*).
Li, Q.-L. *et al., Planta Med.*, 2005, **71**, 847-851 (*Styrax perkinsiae constit*).
Hwang, J.W. *et al., Bull. Korean Chem. Soc.*, 2010, **31**, 965-970 (*Ailanthoidol, synth, activity*).
Duan, X.-F. *et al., Synthesis*, 2010, 1181-1187 (*Ailanthoidol, synth*).
Rao, M.L.N. *et al., Tetrahedron Lett.*, 2010, **51**, 1979-1981 (*Ailanthoidol, synth*).
Lee, N.L. *et al., Bull. Korean Chem. Soc.*, 2012, **33**, 1907-1912 (*Ailanthoidol derivs, activity*).
Luo, X. *et al., Planta Med.*, 2012, **78**, 148-153 (*Zanthocapensol, Zanthocapensate*).
Huang, Y. *et al., J. Asian Nat. Prod. Res.*, 2013, **15**, 319-324 (*Solidago canadensis constit*).
Hajdu, Z. *et al., J. Nat. Prod.*, 2014, **77**, 2641-2650 (*Heliopsis helianthoides constit*).
Hu, J. *et al., J. Asian Nat. Prod. Res.*, 2016, **18**, 831-836 (*Betulifoliumins C,D*).

Sivaraman, A. *et al.*, *J. Nat. Prod.*, 2020, **83**, 3354-3362 (*Ailanthoidol, synth*)

4,7′-Epoxy-9′-nor-3,8′-ligna-7,7′-diene-2′,4′,9-triol E188

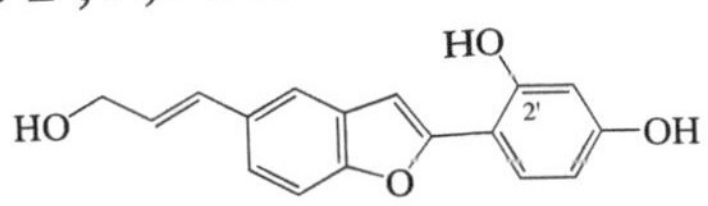

3-[2-(2,4-Dihydroxyphenyl)-5-benzofuranyl]-2-propen-1-ol.
4,7′-Epoxy-2′,4′,9-trihydroxy-9′-nor-3,8′-neoligna-7,7′-diene
$C_{17}H_{14}O_4$ 282.295
(E)-form
2′-Me ether
3-[2-(4-Hydroxy-2-methoxyphenyl)-5-benzofuranyl]-2-propen-1-ol. 4,7′-Epoxy-2′-methoxy-9′-nor-3,8′-ligna-7,7′-diene-4′,9-diol. 4,7′-Epoxy-4′,9-dihydroxy-2′-methoxy-9′-nor-3,8′-ligna-7,7′-diene
[143001-80-3]
$C_{18}H_{16}O_4$ 296.322
Constit. of the roots of *Krameria interior*. Amorph. solid. λ_{max} 233 (log ε 4.38); 258 (log ε 4.37); 266 (sh); 275 (sh); 289 (log ε 4.26); 313 (sh); 320 (log ε 4.40); 327 (sh); 335 (log ε 4.37) (MeOH)
4′-Me ether
3-[2-(2-Hydroxy-4-methoxyphenyl)-5-benzofuranyl]-2-propen-1-ol. 4,7′-Epoxy-4′-methoxy-9′-nor-3,8′-ligna-7,7′-diene-2′,9-diol. 4,7′-Epoxy-2′,9-dihydroxy-4′-methoxy-9′-nor-3,8′-ligna-7,7′-diene
[152367-37-8]
$C_{18}H_{16}O_4$ 296.322
Constit. of *Krameria grayi* and *Krameria paucifolia*. Amorph.. λ_{max} 233 (log ε 3.95); 257 (log ε 3.96); 275 (sh); 287 (log ε 3.76); 313 (sh); 320 (log ε 3.99); 325 (sh); 330 (log ε 3.94); 335 (log ε 3.95) (MeOH).

Dominguez, X.A. *et al.*, *Phytochemistry*, 1990, **29**, 2651-2653 (*Krameria interior constit*).
Achenbach, H. *et al.*, *Phytochemistry*, 1993, **34**, 835-837 (*Krameria paucifolia constit*)

4,7′-Epoxy-9-nor-3,8′-lignan-3′,4′,5,5′,7,9′-hexol E189

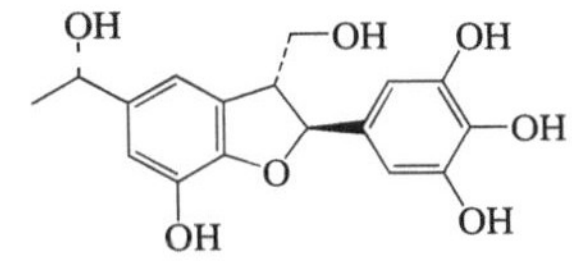

(7S,7′S,8′R)-form

4,7′-Epoxy-3′,4′,5,5′,7,9′-hexahydroxy-9-nor-3,8′-neolignan
$C_{17}H_{18}O_7$ 334.324
(7S,7′S,8′R)-form
3′,5,5′-Tri-Me ether
4,7′-Epoxy-3′,5,5′-trimethoxy-9-nor-3,8′-lignan-4′,7,9′-triol. 4,7′-Epoxy-4′,7,9′-trihydroxy-3′,5,5′-trimethoxy-9-nor-3,8′-lignan. ***Tamariscinol T***
$C_{20}H_{24}O_7$ 376.405
Constit. of *Selaginella tamariscina*. Gum. $[\alpha]^{22}_{D}$ -16.8 (c, 0.03 in MeOH). λ_{max} 212 (log ε 3.79); 280 (log ε 0.93) (MeOH)
3′,5,5′-Tri-Me ether, 4′-O-β-D-glucopyranoside
Tamariscinoside C
[884310-83-2]
$C_{26}H_{34}O_{12}$ 538.546
Constit. of *Selaginella tamariscina*. Powder. Mp 238 - 240°
(7RS,7′R,8′S)-form
3′,5,5′,7-Tetra-Me ether
4,7′-Epoxy-3′,5,5′,7-tetramethoxy-9-nor-3,8′-lignan-4′,9′-diol. 4,7′-Epoxy-4′,9′-dihydroxy-3′,5,5′,7-tetramethoxy-9-nor-3,8′-lignan. ***Tamariscinol U***
$C_{21}H_{26}O_7$ 390.432
Constit. of *Selaginella tamariscina*. Strong inhibitor of mushroom tyrosinase. Yellow gum. Mixt. of two inseparable C-7 epimers (1:1)
7-Et, 3′,5,5′-tri-Me ether
4,7′-Epoxy-7-ethoxy-3′,5,5′-trimethoxy-9-nor-3,8′-lignan-4′,9′-diol. 4,7′-Epoxy-4′,9′-dihydroxy-7-ethoxy-3′,5,5′-trimethoxy-9-nor-3,8′-lignan. ***Tamariscinol V***
$C_{22}H_{28}O_7$ 404.459
Constit. of *Selaginella tamariscina*. Inhibitor of mushroom tyrosinase. Yellow gum. Mixt. of two inseparable C-7 epimers
7-Butyl, 3′,5,5′-tri-Me ether
7-Butoxy-4,7′-epoxy-3′,5,5′-trimethoxy-9-nor-3,8′-lignan-4′,9′-diol. 7-Butoxy-4,7′-epoxy-4′,9′-dihydroxy-3′,5,5′-trimethoxy-9-nor-3,8′-lignan. ***Tamariscinol W***
$C_{24}H_{32}O_7$ 432.513
Constit. of *Selaginella tamariscina*. Moderate inhibitor of mushroom tyrosinase. Yellow gum. Mixt. of two inseparable C-7 epimers.

Zheng, X.K. *et al.*, *Yaoxue Xuebao*, 2004, **39**, 719-721 (*Tamariscinoside C*).
Kim, J.H. *et al.*, *Bull. Korean Chem. Soc.*, 2015, **36**, 300-304 (*Tamariscinol T, cd*).
He, X.-R. *et al.*, *Phytochem. Lett.*, 2019, **34**, 79-83 (*Tamariscinols U,V,W, cd, activity*)

7,7′-Epoxy-9-nor-8,8′-lignan-3,3′,4,5′,8,9′-hexol E190

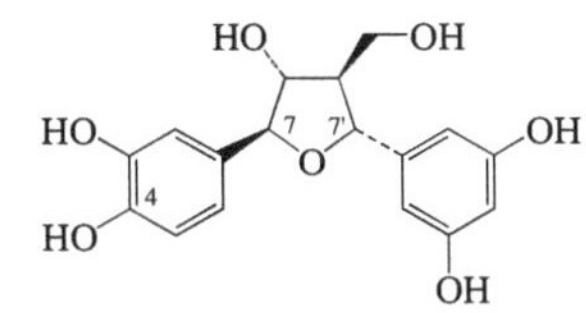

5-(3,4-Dihydroxyphenyl)-2-(3,5-dihydroxyphenyl)tetrahydro-4-hydroxy-3-furanmethanol.
3,3′,4,5′,8,9′-Hexahydroxy-9-nor-7,7′-epoxylignan
$C_{17}H_{18}O_7$ 334.324
Structures in this series may need revision. Phenyl 3,5-dioxygenated substn. pattern in doubt as a result of work done by Yang *et al* in 2015. Lignan numbering shown. Numbering priority between the two lignan moeties varies according to nomenclature used
(7S,7′S,8R,8′R)-form
4-O-β-D-Glucopyranoside
Cestrumoside
[943348-26-3]
$C_{23}H_{28}O_{12}$ 496.465
Constit. of the leaves of *Cestrum diurnum*. Amorph. powder. $[\alpha]^{23}_{D}$ -24.3 (c, 0.23 in MeOH). λ_{max} 227 (log ε 4.2); 275 (log ε 4) (MeOH).

Mohamed, K.M. *et al.*, *ARKIVOC*, 2007, **xiii**, 63-70 (*Cestrumoside*).
Yang, Y.-N. *et al.*, *J. Nat. Prod.*, 2015, **78**, 705-711 (*pmr, struct*)

4,7′-Epoxy-9′-nor-3,8′-lign-7′-ene-2′,4′,6′,7,9-pentol E191

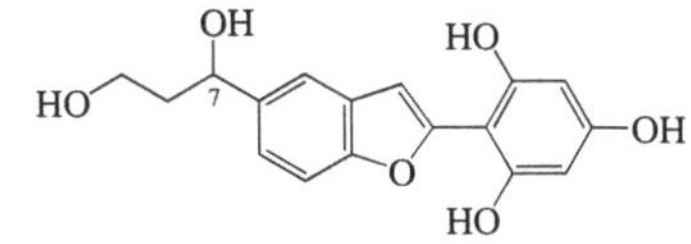

1-[2-(2,4,6-Trihydroxyphenyl)-5-benzofuranyl]-1,3-propanediol. 5-(1,3-Dihydroxypropyl)-2-(2,4,6-trihydroxyphenyl)benzofuran.

4,7′-Epoxy-2′,4′,6′,7,9-pentahydroxy-9′-nor-3,8′-neolign-7′-ene
$C_{17}H_{16}O_6$ 316.309
7-Ketone, 2′,4′-di-Me ether
3-Hydroxy-1-[2-(2-hydroxy-4,6-dimethoxyphenyl)-5-benzofuranyl]-1-propanone, CAS. 4,7′-Epoxy-2′,9-dihydroxy-4′,6′-dimethoxy-9′-nor-3,8′-lign-7′-en-7-one. 4,7′-Epoxy-2′,9-dihydroxy-4′,6′-dimethoxy-9′-nor-7-oxo-3,8′-lign-7′-ene. ***Fargesilignan A***
[1686087-40-0]
$C_{19}H_{18}O_6$ 342.347
Constit. of the flower buds of *Magnolia fargesii*. Powder. λ_{max} 210 (log ε 4.48); 256 (log ε 3.72); 308 (log ε 3.65); 352 (log ε 4.02) (MeOH).

Yang, Y.-K. *et al.*, *Asian J. Chem.*, 2013, **25**, 9610-9612 (*Fargesilignan A*)

4,7′-Epoxy-9′-nor-3,8′-lign-7′-ene-3′,4′,5,7,9-pentol E192

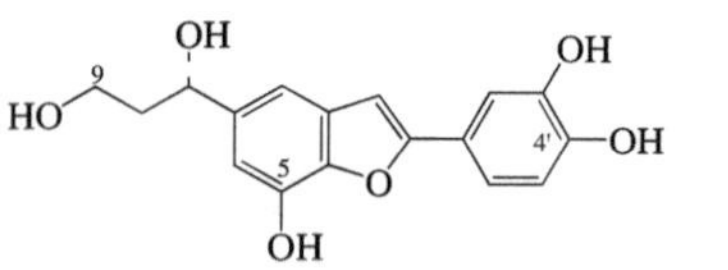

(*S*)-*form*

2-(3,4-Dihydroxyphenyl)-5-(1,3-dihydroxypropyl)-7-hydroxybenzofuran. 4,7′-Epoxy-3′,4′,5,7,9-pentahydroxy-9′-nor-3,8′-neolign-7′-ene
$C_{17}H_{16}O_6$ 316.309
(S)-form
3′,4′,5-Tri-Me ether
4,7′-Epoxy-3′,4′,5-trimethoxy-9′-nor-3,8′-lign-7′-ene-7,9-diol. 4,7′-Epoxy-7,9-dihydroxy-3′,4′,5-trimethoxy-9′-nor-3,8′-lign-7′-ene
[1820833-37-1]
$C_{20}H_{22}O_6$ 358.39
Constit. of the stem bark of *Styrax obassia*. NO prodn. inhibitor. Suppresses expression of LPS-induced COX-2 protein. Amorph. powder. $[\alpha]_D^{25}$ -18.6 (c, 0.01 in CH_2Cl_2). λ_{max} 230 (log ε 1.16); 315 (log ε 1.55) (MeOH)
3′,4′-Methylene, 5-Me ether
4,7′-Epoxy-5-methoxy-3′,4′-methylenedioxy-9′-nor-3,8′-lign-7′-ene-7,9-diol. 4,7′-Epoxy-7,9-dihydroxy-5-methoxy-3′,4′-methylenedioxy-9′-nor-3,8′-lign-7′-ene. ***Machicendiol***
[62306-72-3]
[62306-76-7] (unspecified stereochem.)
$C_{19}H_{18}O_6$ 342.347
Constit. of *Breynia fruticosa*, *Machilus glaucescens*, *Stewartia koreana* and *Styrax obassia*. Cryst. (EtOAc/petrol). Mp 127°. Mp 148 - 149° (synthetic). $[\alpha]_D^{26}$ -11.4 (c, 0.22 in MeOH)
3′,4′-Methylene, 5-Me ether, 9-O-β-D-glucopyranoside
Obassioside B
[860800-68-6]
$C_{25}H_{28}O_{11}$ 504.488
Constit. of the bark of *Styrax obassia* and *Styrax perkinsiae*. Amorph. solid. $[\alpha]_D^{25}$ -13 (c, 0.23 in MeOH). λ_{max} 216 (log ε 4.39); 301 (log ε 4.26); 316 (log ε 4.33) (MeOH)
3′,4′-Methylene, 5-Me ether, 9-O-[β-D-glucopyranosyl-(1→6)-β-D-glucopyranoside]
Obassioside C
[860800-69-7]
$C_{31}H_{38}O_{16}$ 666.629
Constit. of the bark of *Styrax obassia*. Amorph. solid. $[\alpha]_D^{25}$ -17.3 (c, 0.23 in MeOH). λ_{max} 217 (log ε 4.08); 301 (log ε 3.93) (MeOH)
3′,4′-Methylene, 5-Me ether, di-Ac
[62306-73-4]
[62357-40-8] (unspecified stereochem.)
Cryst. (MeOH/petrol). Mp 109°
7-Ketone, 3′,4′-methylene, 5-Me ether
4,7′-Epoxy-9-hydroxy-5-methoxy-3′,4′-methylenedioxy-9′-nor-3,8′-lign-7′-en-7-one. 4,7′-Epoxy-9-hydroxy-5-methoxy-3′,4′-methylenedioxy-9′-nor-7-oxo-3,8′-lign-7′-ene. ***Machicenonol***
[72430-54-7]
$C_{19}H_{16}O_6$ 340.331
Constit. of *Helicteres angustifolia*, *Machilus glaucescens* and *Styrax obassia*. Needles ($CHCl_3$). Mp 175°
9-O-(1-Hydroxy-1-methylethyl), 3′,4′-methylene, 5-Me ether
[1820833-38-2]
$C_{22}H_{24}O_7$ 400.427
Constit. of the stem bark of *Styrax obassia*. Amorph. powder. $[\alpha]_D^{25}$ -45.8 (c, 0.01 in CH_2Cl_2). λ_{max} 231 (log ε 1.41); 319 (log ε 1.76) (MeOH)
(ξ)-form
3′,4′,5-Tri-Me ether, 9-O-β-D-glucopyranoside
Obassioside A
[860800-67-5]
$C_{26}H_{32}O_{11}$ 520.531
Constit. of the bark of *Styrax obassia*. Amorph. solid. $[\alpha]_D^{25}$ -14.7 (c, 0.24 in MeOH). λ_{max} 216 (log ε 4.36); 312 (log ε 4.33) (MeOH).

Talapatra, S.K. *et al.*, *Indian J. Chem., Sect. B*, 1976, **14**, 613-614 (*Machicendiol, struct, synth*).
Talapatra, B. *et al.*, *J. Indian Chem. Soc.*, 1978, **55**, 1204-1208 (*Machicendiol, Machicenonol, struct, synth*).
Schneiders, G.E. *et al.*, *J. Org. Chem.*, 1979, **44**, 4710-4711 (*Machicendiol, di-Ac, synth*).
Aoyagi, Y. *et al.*, *Heterocycles*, 1995, **41**, 1077-1084 (*(±)-Machicendiol, synth*).
Kinoshita, T. *et al.*, *Heterocycles*, 2005, **65**, 1471-1480 (*Obassiosides A,B,C*).
Li, Y.-P. *et al.*, *Phytochem. Lett.*, 2013, **6**, 281-285 (*Machicendiol, cd*).
Cao, T.Q. *et al.*, *Bioorg. Med. Chem. Lett.*, 2015, **25**, 5087-5091 (*Styrax obassia constits, activity*).
More, K.R. *et al.*, *Synth. Commun.*, 2017, **47**, (*Machicendiol, synth*)

4,7′-Epoxy-9′-nor-3,8′-lign-7′-ene-2′,4′,9-triol E193

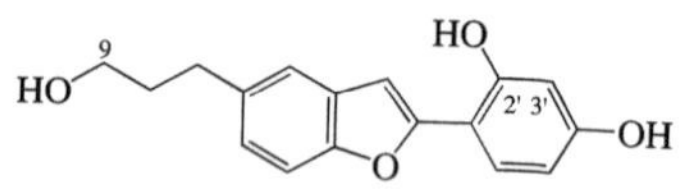

2-(2,4-Dihydroxyphenyl)-5-benzofuranpropanol. 2-(2,4-Dihydroxyphenyl)-5-(3-hydroxypropyl)benzofuran. 4,7′-Epoxy-2′,4′,9-trihydroxy-9′-nor-3,8′-neolign-7′-ene
[119321-97-0]
$C_{17}H_{16}O_4$ 284.311
Constit. of the roots of *Krameria triandra*. Yellow cryst.. Mp 112 - 115°. λ_{max} 207 (ε 28300); 278 (ε 14200); 284 (ε 13300); 315 (ε 23950); 338 (ε 21100) (EtOH) (Berdy)
2′-Me ether
2-(4-Hydroxy-2-methoxyphenyl)-5-benzofuranpropanol, CAS. 4,7′-Epoxy-2′-methoxy-9′-nor-3,8′-lign-7′-ene-4′,9-diol. 4,7′-Epoxy-4′,9-dihydroxy-2′-methoxy-9′-nor-3,8′-lign-7′-ene
[119321-96-9]
$C_{18}H_{18}O_4$ 298.338
Constit. of *Krameria triandra*. Pale yellow cryst.. Mp 143 - 145°. λ_{max} 205 (ε 38150); 277 (ε 18160); 287

(ε 19500); 312 (ε 40800); 328 (ε 34750) (no solvent reported)

2′-Me ether, 9-Ac

[119321-99-2]

$C_{20}H_{20}O_5$ 340.375

Constit. of *Krameria triandra*. Yellow solid. Mp 115 - 118°. λ_{max} 205 (ε 40000); 277 (ε 17150); 288 (ε 18200); 312 (ε 26800); 328 (ε 29300) (EtOH) (Berdy)

4′-Me ether

2-(2-Hydroxy-4-methoxyphenyl)-5-benzofuranpropanol, CAS. *4,7′-Epoxy-4′-methoxy-9′-nor-3,8′-lign-7′-ene-2′,9-diol. 4,7′-Epoxy-2′,9-dihydroxy-4′-methoxy-9′-nor-3,8′-lign-7′-ene*

[119321-98-1]

$C_{18}H_{18}O_4$ 298.338

Constit. of *Krameria lappacea* and *Krameria triandra*. Exhibits antiinflammatory activity. Yellow solid. Mp 80 - 82°. λ_{max} 207 (ε 33000); 278 (ε 17100); 287 (ε 32900); 313 (ε 32950); 329 (ε 28900) (EtOH) (Berdy)

3′-Hydroxy, 2′-Me ether

5-(3-Hydroxypropyl)-2-(2-methoxy-3,4-dihydroxyphenyl)benzofuran. 4,7′-Epoxy-2′-methoxy-9′-nor-3,8′-lign-7′-ene-3′,4′,9-triol. 4,7′-Epoxy-3′,4′,9-trihydroxy-2′-methoxy-9′-nor-3,8′-lign-7′-ene

[119322-00-8]

$C_{18}H_{18}O_5$ 314.337

Constit. of *Krameria triandra*. Yellow cryst.. Mp 140 - 143°. λ_{max} 210 (ε 14350); 307 (ε 14650); 320 (ε 12500) (EtOH) (Berdy)

3′-Methoxy, 2′-Me ether

2-(4-Hydroxy-2,3-dimethoxyphenyl)-5-benzofuranpropanol, CAS. *4,7′-Epoxy-2′,3′-dimethoxy-9′-nor-3,8′-lign-7′-ene-4′,9-diol. 4,7′-Epoxy-4′,9-dihydroxy-2′,3′-dimethoxy-9′-nor-3,8′-lign-7′-ene*

[119322-01-9]

$C_{19}H_{20}O_5$ 328.364

Constit. of *Krameria triandra*. Yellow glass. λ_{max} 310 (ε 8350); 325 (ε 7130) (EtOH).

Arnone, A. *et al.*, *Gazz. Chim. Ital.*, 1988, **118**, 675-682 (*Krameria triandra constits*).
Baumgartner, L. *et al.*, *J. Nat. Prod.*, 2011, **74**, 1779-1786 (*Krameria lappacea constit, activity*)

4,7′-Epoxy-9′-nor-3,8′-lign-7′-ene-3′,4′,9-triol E194

2-(3,4-Dihydroxyphenyl)-5-benzofuranpropanol, CAS. *2-(3,4-Dihydroxyphenyl)-5-(3-hydroxypropyl)benzofuran. 4,7′-Epoxy-3′,4′,9-trihydroxy-9′-nor-3,8′-neolign-7′-ene*

$C_{17}H_{16}O_4$ 284.311

3′,4′-Di-Me ether

4,7′-Epoxy-3′,4′-dimethoxy-9′-nor-3,8′-lign-7′-en-9-ol

3′,4′-Di-Me ether, 3″-O-[β-D-glucopyranosyl-(1→6)-β-D-glucopyranoside]

[1204762-80-0]

$C_{31}H_{40}O_{14}$ 636.647

Constit. of the seeds of *Styrax japonica*. Pale yellow cryst. (EtOH aq.). Mp 230.5°. $[\alpha]^{23}_{D}$ -16.9 (c, 0.3 in $CHCl_3$/MeOH).

Takanashi, M. *et al.*, *J. Oleo Sci.*, 2002, **51**, 151-155 (*Styrax japonica constit*)

4,7′-Epoxy-9-nor-3,8′-lign-7′-ene-4′,5,8-triol E195

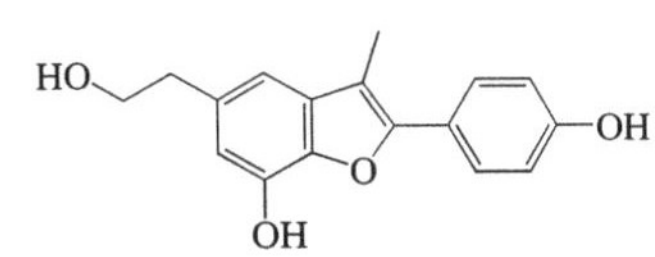

7-Hydroxy-2-(4-hydroxyphenyl)-3-methyl-5-benzofuranethanol. 4,7′-Epoxy-4′,5,8-trihydroxy-9-nor-3,8′-neolign-7′-ene

$C_{17}H_{16}O_4$ 284.311

5-Me ether

4,7′-Epoxy-5-methoxy-9-nor-3,8′-lign-7′-ene-4′,8-diol

[2218430-99-8]

$C_{18}H_{18}O_4$ 298.338

Constit. of *Lavandula angustifolia*. Exhibits anti-tobacco mosaic virus activity. Pale yellow gum. λ_{max} 210 (log ε 4.22); 270 (log ε 3.38); 309 (log ε 3.64) (MeOH)

4′,5-Di-Me ether

7-Methoxy-2-(4-methoxyphenyl)-3-methyl-5-benzofuranethanol. 4,7′-Epoxy-4′,5-dimethoxy-9-nor-3,8′-lign-7′-en-8-ol

[2218430-96-5]

$C_{19}H_{20}O_4$ 312.365

Constit. of *Lavandula angustifolia*. Exhibits anti-tobacco mosaic virus activity. Pale yellow gum. λ_{max} 210 (log ε 4.22); 272 (log ε 3.35); 312 (log ε 3.58) (MeOH).

Tang, S. *et al.*, *Phytochem. Lett.*, 2017, **19**, 60-63 (*Lavandula angustifolia constits, activity*).
Sivaraman, A. *et al.*, *J. Nat. Prod.*, 2020, **83**, 3354-3362 (*synth*)

4,7′-Epoxy-9-nor-3,8′-lign-7′-ene-4′,6,8-triol E196

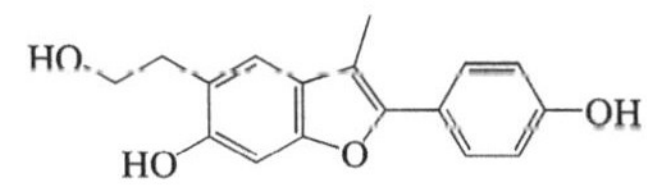

6-Hydroxy-2-(4-hydroxyphenyl)-3-methyl-5-benzofuranethanol. 4,7′-Epoxy-4′,6,8-trihydroxy-9-nor-3,8′-neolign-7′-ene

$C_{17}H_{16}O_4$ 284.311

4′,6-Di-Me ether

4,7′-Epoxy-4′,6-dimethoxy-9-nor-3,8′-lign-7′-en-8-ol

[2218431-00-4]

$C_{19}H_{20}O_4$ 312.365

Constit. of *Arctium lappa* and *Lavandula angustifolia*. Exhibits anti-tobacco mosaic virus activity. Pale yellow gum. λ_{max} 210 (log ε 4.13); 275 (log ε 3.42); 310 (log ε 3.72) (MeOH).

Tang, S. *et al.*, *Phytochem. Lett.*, 2017, **19**, 60-63 (*Lavandula angustifolia constit, activity*).
Zeng, W.-L. *et al.*, *Heterocycles*, 2019, **98**, 437-443 (*Arctium lappa constit*).
Sivaraman, A. *et al.*, *J. Nat. Prod.*, 2020, **83**, 3354-3362 (*synth*)

3,7′-Epoxy-9-nor-4,8′-oxylignan-3′,4′,5′,8,9′-pentol E197

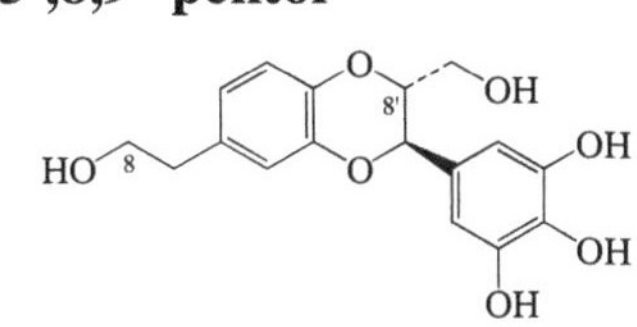

Relative Configuration

3,7′-Epoxy-9-nor-3′,4′,5′,8,9′-pentahydroxy-4,8′-oxyneolignan

$C_{17}H_{18}O_7$ 334.324

(7′RS,8′RS)-form

3′,5′-Di-Me ether

3,7′-Epoxy-3′,5′-dimethoxy-9-nor-4,8′-oxylignan-4′,8,9′-triol. ***Tatarinoid H***

[2222357-86-8]

$C_{19}H_{22}O_7$ 362.378

Constit. of the rhizomes of *Acorus tatarinowii*. λ_{max} 207 (log ε 5.19); 243 (log ε 4.37); 282 (log ε 4.14) (MeOH).

Zhang, W.-Y. *et al.*, *Bioorg. Med. Chem. Lett.*, 2018, **28**, 814-819 (*Tatarinoid H*)

3,7′-Epoxy-9′-nor-4,8′-oxylignan-3′,4′,7′,8′,9-pentol E198

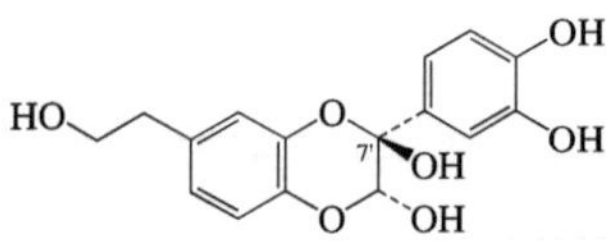

3-(3,4-Dihydroxyphenyl)-2,3-dihydro-2,3-dihydroxy-1,4-benzodioxin-6-ethanol. 3,7′-Epoxy-3′,4′,7′,8′,9-pentahydroxy-9′-nor-4,8′-oxyneolignan
$C_{16}H_{16}O_7$ 320.297
(2RS,3RS)-form
7′-Me ether
3,7′-Epoxy-7′-methoxy-9′-nor-4,8′-oxylignan-3′,4′,8′,9-tetrol. ***Blapsol D***
[1788033-62-4]
[1788033-63-5] ((+)-form)
[1788033-64-6] ((-)-form)
$C_{17}H_{18}O_7$ 334.324
Isol. from the stink beetle *Blaps japanensis*. COX-2 inhibitor. Enantiomers have been sepd.
7′-Et ether
Blapsol B
[1788033-55-5]
[1788033-56-6] ((+)-form)
[1788033-57-7] ((-)-form)
$C_{18}H_{20}O_7$ 348.351
Isol. from the stink beetle *Blaps japanensis*. COX-2 inhibitor. Enantiomers have been sepd..
Yan, Y.-M. *et al.*, *Bioorg. Med. Chem. Lett.*, 2015, **25**, 2469-2472 (*Blapsols B,D*)

4,7′-Epoxy-9′-nor-3,8′-oxylignan-3′,4′,7′,8′,9-pentol E199

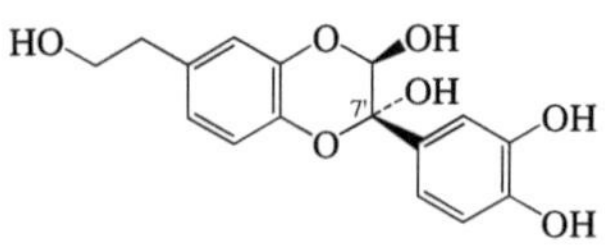

2-(3,4-Dihydroxyphenyl)-2,3-dihydro-2,3-dihydroxy-1,4-benzodioxin-6-ethanol. 4,7′-Epoxy-3′,4′,7′,8′,9-pentahydroxy-9′-nor-3,8′-oxyneolignan
$C_{16}H_{16}O_7$ 320.297
(2RS,3RS)-form
7′-Me ether
4,7′-Epoxy-7′-methoxy-9′-nor-3,8′-oxylignan-3′,4′,8′,9-tetrol. ***Blapsol C***
[1788033-58-8]
[1788033-59-9] ((+)-form)
[1788033-60-2] ((-)-form)
$C_{17}H_{18}O_7$ 334.324
Isol. from the stink beetle *Blaps japanensis*. COX-2 inhibitor. Enantiomers have been sepd.
7′-Et ether
Blapsol A
[1788033-52-2]
[1788033-53-3] ((+)-form)
[1788033-54-4] ((-)-form)
$C_{18}H_{20}O_7$ 348.351
Isol. from the stink beetle *Blaps japanensis*. COX-2 inhibitor. Enantiomers have been sepd..
Yan, Y.-M. *et al.*, *Bioorg. Med. Chem. Lett.*, 2015, **25**, 2469-2472 (*Blapsols A,C, cryst struct*)

4,7′-Epoxy-9-nor-3,8′-oxylignan-3′,4′,8,9′-tetrol E200

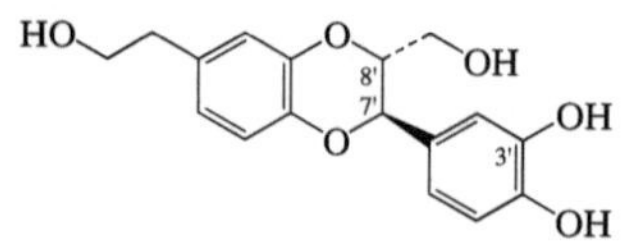

2-(3,4-Dihydroxyphenyl)-2,3-dihydro-3-(hydroxymethyl)-1,4-benzodioxin-6-ethanol. 4,7′-Epoxy-3′,4′,8,9′-tetrahydroxy-9-nor-3,8′-oxyneolignan
$C_{17}H_{18}O_6$ 318.325
(7′R*,8′R*)-form
3′-Me ether
4,7′-Epoxy-3′-methoxy-9-nor-3,8′-oxylignan-4′,8,9′-triol. 4,7′-Epoxy-4′,8,9′-trihydroxy-3′-methoxy-9-nor-3,8′-oxylignan
[1623008-47-8]
$C_{18}H_{20}O_6$ 332.352
Constit. of *Quercus pannosa*. $[\alpha]_D^{24}$ +7 (c, 0.5 in MeOH). λ_{max} 221; 225; 281 (MeOH).
Ou, Y.-B. *et al.*, *Zhongcaoyao*, 2013, **44**, 1872-1876 (*Quercus pannosa constit*)

4,4′-Epoxy-3,3′-oxyligna-7,7′-diene-9,9′-dioic acid E201

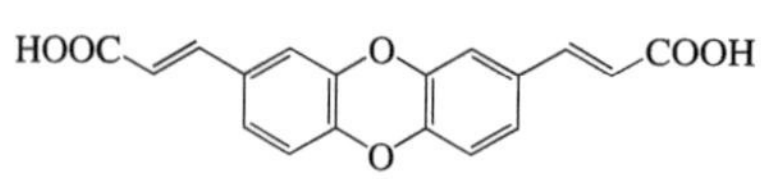

3,3′-(Dibenzo[b,e]dioxin-2,8-diyl) bis[2-propenoic acid]. 4,4′-Epoxy-3,3′-oxyneoligna-7,7′-diene-9,9′-dioic acid
[2485841-18-5]
$C_{18}H_{12}O_6$ 324.288
(E,E)-form
Sacric acid A
Constit. of *Artemisia sacrorum*. Needles. Mp 228 - 230°. λ_{max} 257 (log ε 4.87) (MeOH).
Wang, Q. *et al.*, *Nat. Prod. Commun.*, 2016, **11**, 489-490 (*Sacric acid A*)

4,7′-Epoxy-3,8′-oxylignan-3′,4′,5,5′,9,9′-hexol E202

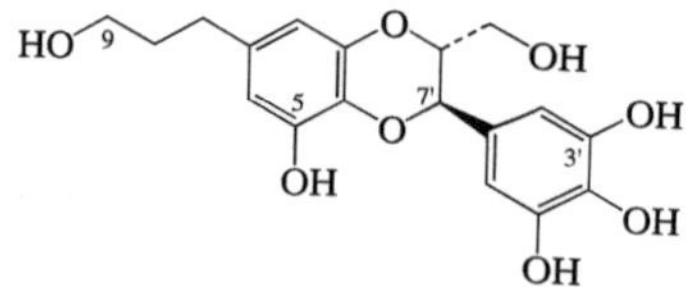
Relative Configuration

2,3-Dihydro-8-hydroxy-3-(hydroxymethyl)-2-(3,4,5-trihydroxyphenyl)-1,4-benzodioxin-6-propanol, CAS. *4,7′-Epoxy-3′,4′,5,5′,9,9′-hexahydroxy-3,8′-oxyneolignan. 3′,4′,5,5′,9,9′-Hexahydroxy-3,8′:4,7′-diepoxylignan*
$C_{18}H_{20}O_8$ 364.350
(7′R*,8′R*)-form
3′,5′-Di-Me ether
4,7′-Epoxy-3′,5′-dimethoxy-3,8′-oxylignan-4′,5,9,9′-tetrol. 4,7′-Epoxy-4′,5,9,9′-tetrahydroxy-3′,5′-dimethoxy-3,8′-oxylignan. 4′,5,9,9′-Tetrahydroxy-3′,5′-dimethoxy-3,8′:4,7′-diepoxylignan
3′,5′-Di-Me ether, 5-O-β-D-glucopyranoside
[2247433-04-9]
$C_{26}H_{34}O_{13}$ 554.545
Constit. of the rhizomes of *Smilax trinervula*. Powder (MeOH). $[\alpha]_D^{25}$ -34.3 (c, 0.18 in MeOH). λ_{max} 225 (log ε 4.90); 275 (log ε 3.85) (MeOH).
Shu, J. *et al.*, *Phytochem. Lett.*, 2017, **20**, 1-8 (*Smilax trinervula constit*)

3,7′-Epoxy-4,8′-oxylignan-3′,4′,5,5′,9-pentol E203

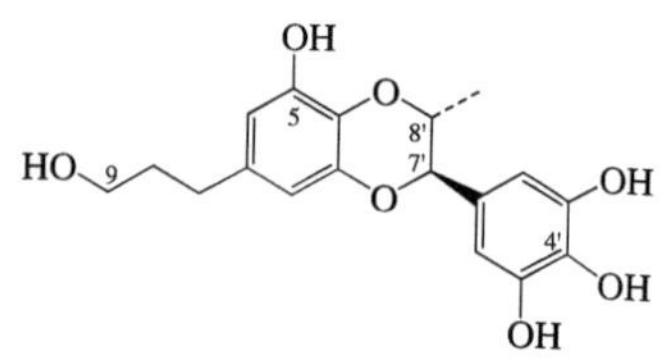

3,7′-Epoxy-3′,4′,5,5′,9-pentahydroxy-4,8′-oxyneolignan
$C_{18}H_{20}O_7$ 348.351
(7′R,8′R)-form
3′,5′-Di-Me ether
3,7′-Epoxy-3′,5′-dimethoxy-4,8′-oxylignan-4′,5,9-triol. 3,7′-Epoxy-4′,5,9-trihydroxy-3′,5′-dimethoxy-4,8′-oxylignan
3′,5′-Di-Me ether, 5-O-β-D-glucopyranoside
Eusiderin N

[1428946-69-3]
$C_{26}H_{34}O_{12}$ 538.546
Constit. of the stems of *Eurya japonica*. Powder. Mp 195°. $[\alpha]^{25}_{D}$ -5.0 (c, 0.2 in MeOH). λ_{max} 230 (sh); 273 (log ε 3.48) (MeOH).

Kuo, L.M.Y. *et al., J. Nat. Prod.*, 2013, **76**, 580-587 (*Eusiderin N*)

3,7′-Epoxy-4,8′-oxylignan-3′,4′,5,9,9′-pentol E204

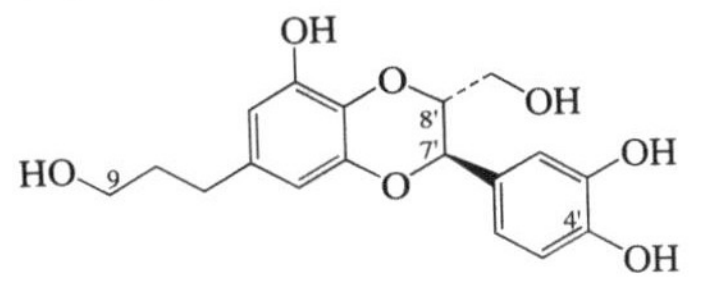

(7′R,8′R)-form

3,7′-Epoxy-3′,4′,5,9,9′-pentahydroxy-4,8′-oxyneolignan
$C_{18}H_{20}O_7$ 348.351
(7′R,8′R)-form
3′,5-Di-Me ether
3,7′-Epoxy-3′,5-dimethoxy-4,8′-oxylignan-4′,9,9′-triol. 3,7′-Epoxy-4′,9,9′-trihydroxy-3′,5-dimethoxy-4,8′-oxylignan
3′,5-Di-Me ether, 9′-Ac
$C_{22}H_{26}O_8$ 418.442
Constit. of *Pinus thunbergii*. Oil. $[\alpha]^{24}_{D}$ -34 (c, 1 in EtOH)
(7′S,8′S)-form
3′-Me ether
3,7′-Epoxy-3′-methoxy-4,8′-oxylignan-4′,5,9,9′-tetrol. 3,7′-Epoxy-4′,5,9,9′-tetrahydroxy-3′-methoxy-4,8′-oxylignan
3′-Me ether, 5-O-β-D-glucopyranoside
[1616981-18-0]
$C_{25}H_{32}O_{12}$ 524.519
Constit. of the leaves of *Liquidambar formosana*. Amorph. powder. $[\alpha]^{20}_{D}$ -28.6 (c, 0.22 in MeOH). λ_{max} 278 (MeOH)
3′,5-Di-Me ether, 9-O-α-L-rhamnopyranoside
5-Methoxycupressoside A. ***Cupressoside C***
[1232236-62-2]
$C_{26}H_{34}O_{11}$ 522.547
Constit. of *Epimedium koreanum*. $[\alpha]^{20}_{D}$ -8.9 (c, 1 in MeOH). Stereochem. not confirmed. λ_{max} 280; 304; 346 (MeOH).

Jiang, F. *et al., Zhongcaoyao*, 2008, **39**, 1281-1285 (*Cupressoside C*).
Zhong, R.-J. *et al., J. Asian Nat. Prod. Res.*, 2013, **15**, 1249-1255 (*3′-Me ether 5-glucoside*).
Sono, M. *et al., Nat. Prod. Commun.*, 2017, **12**, 1181-1184 (*3′,5-di-Me ether 9′-Ac*)

3,7′-Epoxy-4,8′-oxylignan-3′,4′,5′,9,9′-pentol E205

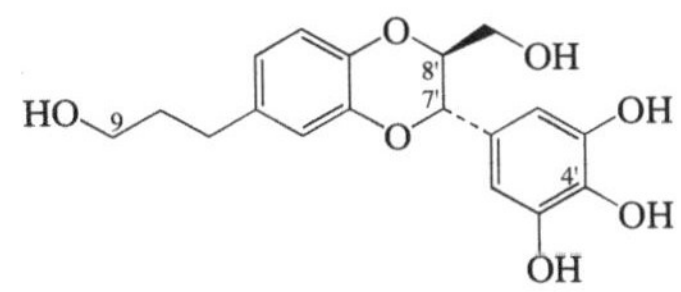

3,7′-Epoxy-3′,4′,5′,9,9′-pentahydroxy-4,8′-oxyneolignan
$C_{18}H_{20}O_7$ 348.351
(7′S,8′S)-form
3′,5′-Di-Me ether
3,7′-Epoxy-3′,5′-dimethoxy-4,8′-oxylignan-4′,9,9′-triol. 3,7′-Epoxy-4′,9,9′-trihydroxy-3′,5′-dimethoxy-4,8′-oxylignan
3′,5′-Di-Me ether, 9-O-α-L-rhamnopyranoside
5′-Methoxycupressoside A
[1392407-94-1]
$C_{26}H_{34}O_{11}$ 522.547
Constit. of *Geum japonicum* var. *chinense*. Gum. $[\alpha]^{25}_{D}$ -19.6 (c, 0.1 in MeOH). λ_{max} 228 (sh); 280 (log ε 3.8) (MeOH).

Cheng, X. *et al., Planta Med.*, 2011, **77**, 2061-2065 (*5′-Methoxycupressoside A*)

3,7′-Epoxy-4,8′-oxylignan-3′,4′,9,9′-tetrol E206

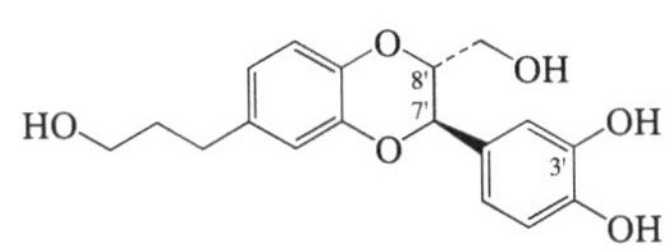

(7′R,8′R)-form

3-(3,4-Dihydroxyphenyl)-2,3-dihydro-2-(hydroxymethyl)-1,4-benzodioxin-6-propanol, CAS. *3,7′-Epoxy-3′,4′,9,9′-tetrahydroxy-4,8′-oxyneolignan*
$C_{18}H_{20}O_6$ 332.352
(7′R,8′R)-form
3′-Me ether, 9-O-β-rhamnopyranoside
[245084-93-9]
[245084-97-3] (penta-Ac)
$C_{25}H_{32}O_{10}$ 492.521
Constit. of *Chamaecyparis formosensis*. Gum (as penta-Ac). $[\alpha]^{25}_{D}$ -25 (c, 4 in $CHCl_3$) (penta-Ac)
3′-Me ether, 9′-O-β-D-glucopyranoside
Junipercomnoside D
[828921-61-5]
$C_{25}H_{32}O_{11}$ 508.52
Constit. of *Juniperus communis* var. *depressa*. Powder. $[\alpha]_D$ +16.9 (c, 0.5 in MeOH)
(7′R,8′S)-form
3′-Me ether
3,7′-Epoxy-3′-methoxy-4,8′-oxylignan-4′,9,9′-triol. 3,7′-Epoxy-4′,9,9′-trihydroxy-3′-methoxy-4,8′-oxylignan
[144939-27-5]
$C_{19}H_{22}O_6$ 346.379
Constit. of *Juniperus chinensis* and *Pinus thunbergii*. Oil. $[\alpha]^{25}_{D}$ -2.6 (c, 1.7 in MeOH). λ_{max} 227 (ε 15700); 280 (ε 7320) (MeOH)
3′-Me ether, 9-O-β-rhamnopyranoside
[403655-46-9]
Constit. of *Juniperus chinensis* var. *kaizuka*. Yellowish oil. $[\alpha]^{25}_{D}$ -32 (c, 0.4 in $CHCl_3$)
(7′S,8′S)-form
3′-Me ether
[144881-21-0]
$C_{19}H_{22}O_6$ 346.379
Constit. of *Pinus thunbergii*, *Juniperus chinensis* and *Santalum album*. Oil. $[\alpha]^{25}_{D}$ +0.34 (c, 1.4 in MeOH). λ_{max} 227 (ε 33750); 277 (ε 19170) (MeOH)
3′-Me ether, 9-O-α-L-rhamnopyranoside
Cupressoside A
[934371-82-1]
$C_{25}H_{32}O_{10}$ 492.521
Constit. of *Cupressus duclouxiana*, *Epimedium koreanum*, *Pinus densiflora* and *Juniperus chinensis* var. *kaizuka*. Amorph. powder. $[\alpha]^{23}_{D}$ -22.8 (c, 0.37 in MeOH)
3′-Me ether, 9-O-β-D-glucopyranoside
Junipercomnoside C
[828921-37-5]
$C_{25}H_{32}O_{11}$ 508.52
Constit. of *Juniperus communis* var. *depressa*. Powder. $[\alpha]_D$ -5 (c, 0.72 in MeOH)
(7R*,8R*)-form
9,9′-Dicarboxylic acid, 9,9′-bis[2-(4-hydroxyphenyl)ethylamide]
Lyciumamide E
$C_{34}H_{34}N_2O_8$ 598.652
Constit. of the root bark of *Lycium chinense*. Exhibits potent antihyperlipidaemic activity. Amorph. powder. λ_{max} 206 (log ε 4.39); 224 (log ε 3.93); 279 (log ε 3.41) (MeOH).

Fang, J.-M. *et al., Phytochemistry*, 1992, **31**, 3659-3661 (*3′-Me ethers*).

Lin, T.-C. *et al.*, *Phytochemistry*, 1999, **51**, 793-801 (*Chamaecyparis rhamnoside*).
Gu, W. *et al.*, *Tetrahedron Lett.*, 2000, **41**, 6079-6082 (*3′-Me ether, synth*).
Lee, C.-K. *et al.*, *J. Chin. Chem. Soc. (Taipei)*, 2001, **48**, 1077-1080 (*Juniperus chinensis rhamnosides*).
Nakanishi, T. *et al.*, *Heterocycles*, 2004, **63**, 2573-2580 (*Junipercomnosides C,D*).
Xu, J.-F. *et al.*, *J. Asian Nat. Prod. Res.*, 2006, **8**, 181-185 (*Cupressoside A*).
Chen, H. *et al.*, *Fitoterapia*, 2017, **122**, 119-125 (*Lyciumamide E, activity*).
Sono, M. *et al.*, *Nat. Prod. Commun.*, 2017, **12**, 1181-1184 (*Pinus thunbergii constits*)

4,7′-Epoxy-3,8′-oxylignan-3′,4′,9,9′-tetrol E207

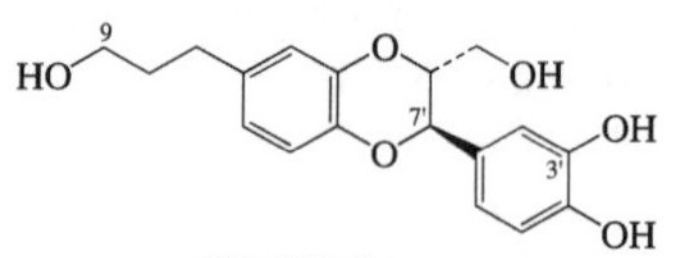

(7′R,8′R)-form

2-(3,4-Dihydroxyphenyl)-2,3-dihydro-3-(hydroxymethyl)-1,4-benzodioxin-6-propanol, CAS. *4,7′-Epoxy-3′,4′,9,9′-tetrahydroxy-3,8′-oxyneolignan*
$C_{18}H_{20}O_6$ 332.352
(7′R,8′R)-form
3′-Me ether, 9′-O-β-D-xylopyranoside
[1334770-29-4]
$C_{24}H_{30}O_{10}$ 478.494
Constit. of the root bark of *Illicium henryi*. Amorph. powder. $[\alpha]^{20}_{D}$ +13 (c, 0.11 in MeOH)
(7′R*,8′R*)-form
3′-Me ether
4,7′-Epoxy-3′-methoxy-3,8′-oxylignan-4′,9,9′-triol. 4,7′-Epoxy-4′,9,9′-trihydroxy-3′-methoxy-3,8′-oxylignan
[1309075-56-6]
$C_{19}H_{22}O_6$ 346.379
Constit. of the leaves of *Juniperus brevifolia*. Yellowish oil
9,9′-Dicarboxylic acid, 9,9′-bis[2-(4-hydroxyphenyl)ethylamide]
Lyciumamide D
$C_{34}H_{34}N_2O_8$ 598.652
Constit. of the root bark of *Lycium chinense*. Exhibits potent antihyperlipidaemic activity. Amorph. powder. λ_{max} 224 (log ε 3.92); 279 (log ε 3.39) (MeOH)
(7′ξ,8′ξ)-form
3′-Me ether, 9-O-α-L-rhamnopyranoside
Massonianoside F. *Massonianoside E‡*
[872037-79-1]
$C_{25}H_{32}O_{10}$ 492.521
Constit. of the needles of *Pinus massoniana*. Mp 146 - 148°. $[\alpha]^{25}_{D}$ -41.9 (c, 0.18 in MeOH). It is probable that the synonym was incorrectly given the wrong code letter E, and has been renamed with code letter F. λ_{max} 210 (log ε 2.15); 226 (log ε 1.25); 284 (log ε 0.24) (no solvent reported).

Feng, W.-S. *et al.*, *Yaoxue Xuebao*, 2004, **39**, 190-193 (*Massonianoside F*).
Xiang, W.-J. *et al.*, *Fitoterapia*, 2010, **81**, 1228-1231 (*Illicium henryi xyloside*).
Seca, A.M.L. *et al.*, *Phytochem. Lett.*, 2010, **3**, 126-128 (*3′-Me ether*).
Chen, H. *et al.*, *Fitoterapia*, 2017, **122**, 119-125 (*Lyciumamide D, activity*)

3,7′-Epoxy-4,8′-oxylign-7-ene-2′,4′-diol E208

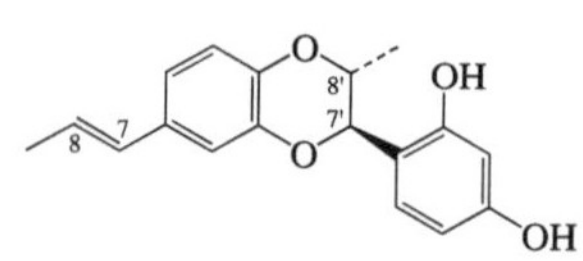

(7E,7′R,8′R)-form

4-[2,3-Dihydro-3-methyl-7-(1-propenyl)-1,4-benzodioxin-2-yl]-1,3-benzenediol. 3-(2,4-Dihydroxyphenyl)-2,3-dihydro-2-methyl-6-(1-propenyl)-1,4-benzodioxin. 3,7′-Epoxy-2′,4′-dihydroxy-4,8′-oxyneolign-7-ene
$C_{18}H_{18}O_4$ 298.338
Absolute configuration revised in 2015
(7E,7′R,8′R)-form
trans-**Rodgersinine A** [1662687-98-0]
[666250-52-8] (7′S,8′S-form)
Constit. of the rhizomes of *Rodgersia podophylla*. Amorph. powder. $[\alpha]^{20}_{D}$ +24.8 (c, 0.07 in MeOH). λ_{max} 267 (log ε 3.97) (MeOH)
7,8-Didehydro
*3-(2,4-Dihydroxyphenyl)-2,3-dihydro-2-methyl-6-(1-propynyl)-1,4-benzodioxin. 3,7′-Epoxy-4,8′-oxylign-7-yne-2′,4′-diol. 3,7′-Epoxy-2′,4′-dihydroxy-4,8′-oxylign-7-yne. trans-**Rodgersinine B***
[1662687-91-3]
[666250-54-0] (7′S,8′S-form)
$C_{18}H_{16}O_4$ 296.322
Constit. of the rhizomes of *Rodgersia podophylla*. Amorph. powder
(7E,7′S,8′R)-form
cis-**Rodgersinine A** [666250-51-7]
Constit. of the rhizomes of *Rodgersia podophylla*. Amorph. powder. $[\alpha]^{20}_{D}$ +79.9 (c, 0.18 in MeOH). λ_{max} 266 (log ε 3.88) (MeOH)
7,8-Didehydro
cis-Rodgersinine B
[666250-53-9]
$C_{18}H_{16}O_4$ 296.322
Constit. of the rhizomes of *Rodgersia podophylla*. Amorph. powder. $[\alpha]^{20}_{D}$ +62 (c, 0.03 in MeOH). λ_{max} 256 (log ε 4.24); 285 (log ε 3.84); 301 (log ε 3.76) (MeOH).

Chin, Y.-W. *et al.*, *Tetrahedron Lett.*, 2004, **45**, 339-341 (*Rodgersia podophylla constits*).
Chin, Y.-W. *et al.*, *Tetrahedron Lett.*, 2004, **45**, 2867 (*corringendum*).
Miao, Q. *et al.*, *J. Chin. Pharm. Sci.*, 2007, **16**, 41-42 (*synth*).
Pilkington, L.I. *et al.*, *Org. Lett.*, 2015, **17**, 1046-1049 (*synth, cd, abs config*)

3,7′-Epoxy-4,8′-oxylign-8-ene-3′,4′-diol E209

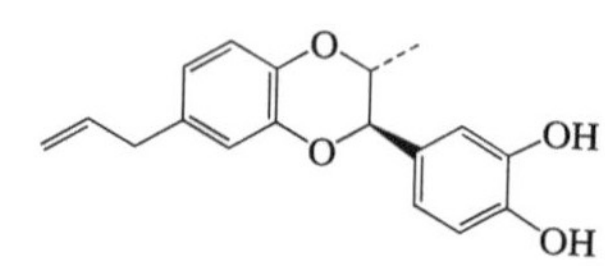

3-(3,4-Dihydroxyphenyl)-2,3-dihydro-2-methyl-6-(2-propenyl)-1,4-benzodioxin. 3,7′-Epoxy-3′,4′-dihydroxy-4,8′-oxyneolign-8-ene
$C_{18}H_{18}O_4$ 298.338
(7′R*,8′R*)-form
3′-Me ether
3,7′-Epoxy-3′-methoxy-4,8′-oxylign-8-en-4′-ol. ***Melaleucin A***
$C_{19}H_{20}O_4$ 312.365
Constit. of *Melaleuca bracteata*. Exhibits antibacterial activity. Oil. λ_{max} 239 (log ε 4.98); 285 (log ε 4.85) ($CHCl_3$).

Li, C. *et al.*, *Fitoterapia*, 2017, **120**, 171-176 (*Melaleucin A, activity*)

3,7′-Epoxy-4,8′-oxylign-7-ene-3′,4′,5,5′,9,9′-hexol E210

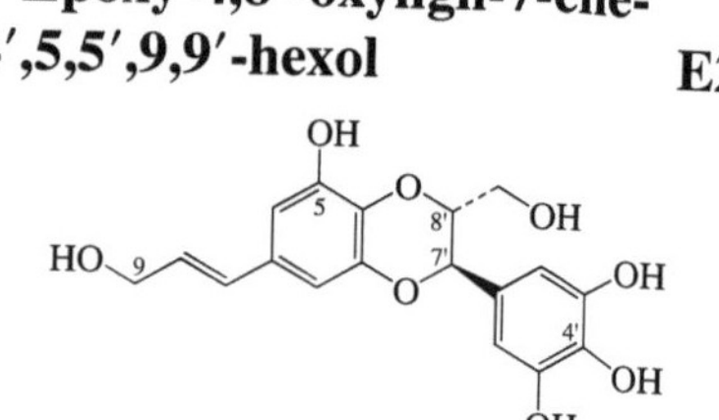

(7E,7′R,8′R)-form

2,3-Dihydro-8-hydroxy-6-(3-hydroxy-1-propenyl)-3-(3,4,5-trihydroxyphenyl)-1,4-benzodioxin-2-methanol, CAS. *3,7′-Epoxy-3′,4′,5,5′,9,9′-hexahydroxy-4,8′-oxyneolign-7-ene*

$C_{18}H_{18}O_8$ 362.334
(7E,7′R,8′R)-form
9-Aldehyde, 3′,5-di-Me ether
3,7′-Epoxy-3′,4′,9′-trihydroxy-5,5′-dimethoxy-4,8′-oxylign-7-en-9-al. 3,7′-Epoxy-3′,4′,9′-trihydroxy-5,5′-dimethoxy-9-oxo-4,8′-oxylign-7-ene
[1248482-39-4]
$C_{20}H_{20}O_8$ 388.372
Constit. of the heartwood of *Santalum album*. Pale yellow solid. $[\alpha]^{23}_{D}$ +2.4 (c, 0.8 in MeOH). λ_{max} 225 (log ε 3.36); 275 (log ε 2.66) (MeOH)
(7E,7′S,8′S)-form
3′,5-Di-Me ether
3,7′-Epoxy-5,5′-dimethoxy-4,8′-oxylign-7-ene-3′,4′,9,9′-tetrol. 3,7′-Epoxy-3′,4′,9,9′-tetrahydroxy-5,5′-dimethoxy-4,8′-oxylign-7-ene
[864079-22-1]
$C_{20}H_{22}O_8$ 390.388
Constit. of *Excoecaria acerifolia* and *Santalum album*. Oil. $[\alpha]^{20}_{D}$ +8 (c, 1 in MeOH). λ_{max} 223 (log ε 2.43); 273 (log ε 2) (MeOH)
3′,5,5′-Tri-Me ether
3,7′-Epoxy-3′,5,5′-trimethoxy-4,8′-oxylign-7-ene-4′,9,9′-triol. 3,7′-Epoxy-4′,9,9′-trihydroxy-3′,5,5′-trimethoxy-4,8′-oxylign-7-ene. (-)-Nitidanin
[864168-11-6]
$C_{21}H_{24}O_8$ 404.415
Constit. of *Excoecaria acerifolia* and *Santalum album*. Oil. $[\alpha]^{20}_{D}$ -16 (c, 0.5 in MeOH). λ_{max} 224 (log ε 2.45); 273 (log ε 1.98) (MeOH)
3′,5,9-Tri-Me ether
3,7′-Epoxy-5,5′,9-trimethoxy-4,8′-oxylign-7-ene-3′,4′,9′-triol. 3,7′-Epoxy-3′,4′,9′-trihydroxy-5,5′,9-trimethoxy-4,8′-oxylign-7-ene
[1416732-04-1]
$C_{21}H_{24}O_8$ 404.415
Constit. of *Excoecaria acerifolia*. $[\alpha]_D$ -2.6 (c, 0.003 in MeOH). λ_{max} 194; 209; 274 (MeOH)
3′,5,5′,9-Tetra-Me ether
3,7′-Epoxy-3′,5,5′,9-tetramethoxy-4,8′-oxylign-7-ene-4′,9′-diol. 3,7′-Epoxy-4′,9′-dihydroxy-3′,5,5′,9-tetramethoxy-4,8′-oxylign-7-ene
[1864843-78-6]
$C_{22}H_{26}O_8$ 418.442
Constit. of the aerial parts of *Euphorbia antiquorum*. Oil. $[\alpha]^{23.3}_{D}$ -16.6 (c, 0.53 in MeOH). λ_{max} 224 (log ε 2.40); 275 (log ε 2.05) (MeOH)
9-Aldehyde, 3′,4′,5′-tri-Me ether
3,7′-Epoxy-5,9′-dihydroxy-3′,4′,5′-trimethoxy-4,8′-oxylign-7-en-9-al. 3,7′-Epoxy-5,9′-dihydroxy-3′,4′,5′-trimethoxy-9-oxo-4,8′-oxylign-7-ene
[1002753-43-6]
$C_{21}H_{22}O_8$ 402.399
Constit. of the stem wood of *Zanthoxylum avicennae*. Pale yellow oil. $[\alpha]^{25}_{D}$ -21.5 (c, 0.1 in MeOH). λ_{max} 277 (log ε 4.13); 331 (log ε 4.22) (MeOH)
9-Aldehyde, 3′,5,5′-tri-Me ether
3,7′-Epoxy-4′,9′-dihydroxy-3′,5,5′-trimethoxy-4,8′-oxylign-7-en-9-al. 3,7′-Epoxy-4′,9′-dihydroxy-3′,5,5′-trimethoxy-9-oxo-4,8′-oxylign-7-ene
[1003863-33-9]
$C_{21}H_{22}O_8$ 402.399
Constit. of *Santalum album* and *Zanthoxylum avicennae*. Pale yellow powder. $[\alpha]^{25}_{D}$ -20.2 (c, 0.12 in MeOH). λ_{max} 276 (log ε 4.15); 332 (log ε 4.31) (MeOH)
9-Carboxylic acid, 3′,5-di-Me ether
3,7′-Epoxy-3′,4′,9′-trihydroxy-5,5′-dimethoxy-4,8′-oxylign-7-en-9-oic acid. ***Reevesic acid***
$C_{20}H_{20}O_9$ 404.371
Constit. of *Reevesia formosana*. Oil. $[\alpha]^{25}_{D}$ -8.1 (c, 0.14 in MeOH). λ_{max} 210 (ε 4.09); 229 (sh); 299 (log ε 3.63) (MeOH)
(7E,7′RS,8′RS)-form
3′,5-Di-Me ether
Simplidin
[855418-64-3]
$C_{20}H_{22}O_8$ 390.388
Constit. of *Excoecaria acerifolia* and *Firmiana simplex*. Mp 194°. λ_{max} 224; 272 (MeOH)
3′,5,5′-Tri-Me ether
(±)-Nitidanin
[171674-89-8]
$C_{21}H_{24}O_8$ 404.415
Constit. of *Euonymus acanthocarpus*, *Firmiana simplex*, *Grewia bilamellata* and *Zanthoxylum nitidum*. Powder. Mp 240° (232°). λ_{max} 224; 272 (MeOH)
3′,5,5′-Tri-Me ether, 9,9′-bis-O-(3-methylbutanoyl)
Nitidanin 9,9′-diisovalerate
[1326706-47-1]
$C_{31}H_{40}O_{10}$ 572.651
Constit. of *Echinacea purpurea* and *Onopordum acanthium*. Amorph. solid. λ_{max} 208 (log ε 3.31); 226 (log ε 3.17); 275 (log ε 2.82) (MeOH)
(7E,7′R*,8′R*)-form
9-Aldehyde, 3′,5,5′-tri-Me ether
Bilagrewin
[876407-21-5]
$C_{21}H_{22}O_8$ 402.399
Constit. of *Grewia bilamellata*. Amorph. yellow powder. $[\alpha]^{23}_{D}$ +3.3 (c, 0.18 in MeOH).

Ishikawa, T. *et al.*, *Chem. Pharm. Bull.*, 1995, **43**, 2014-2018 (*Nitidanin*).
Son, Y.K. *et al.*, *Arch. Pharmacal Res.*, 2005, **28**, 34-38 (*Simplidin, Nitidanin*).
Kim, T.H. *et al.*, *Chem. Pharm. Bull.*, 2005, **53**, 641-644 (*Santalum album constits*).
Ma, C. *et al.*, *J. Nat. Prod.*, 2006, **69**, 346-350 (*Bilagrewin*).
Kuboki, A. *et al.*, *Tetrahedron Lett.*, 2007, **48**, 771-774 (*Nitidanin, synth*).
Chen, J.-J. *et al.*, *J. Nat. Prod.*, 2008, **71**, 212-217 (*Zanthoxylum avicennae constits*).
Matsuo, Y. *et al.*, *Chem. Pharm. Bull.*, 2010, **58**, 587-590 (*Santalum album constits*).
Hohmann, J. *et al.*, *Phytochemistry*, 2011, **72**, 1848-1853 (*Nitidanin diisovalerate*).
Hu, J. *et al.*, *Zhongguo Zhongyao Zazhi*, 2011, **36**, 1969-1974 (*Excoecaria acerifolia constits*).
Zhu, J.X. *et al.*, *Arch. Pharmacal Res.*, 2012, **35**, 1739-1747 (*Nitidanin*).
Hu, J. *et al.*, *Chem. Nat. Compd. (Engl. Transl.)*, 2016, **52**, 48-51 (*Euphorbia tetra-Me ether*).
Chang, H.-S. *et al.*, *RSC Adv.*, 2017, **7**, 27040-27047 (*Reevesic acid*)

3,7′-Epoxy-4,8′-oxylign-7-ene-3′,4′,5,5′,9-pentol E211

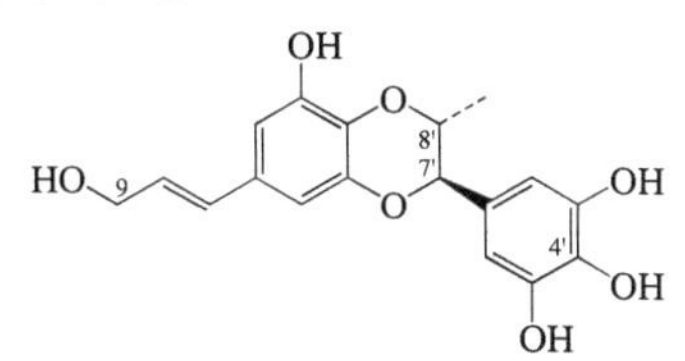

3,7′-Epoxy-3′,4′,5,5′,9-pentahydroxy-4,8′-oxyneolign-7-ene
$C_{18}H_{18}O_7$ 346.335
(7E,7′R,8′R)-form
3′,4′,5,5′-Tetra-Me ether
3,7′-Epoxy-3′,4′,5,5′-tetramethoxy-4,8′-oxylign-7-en-9-ol. 3,7′-Epoxy-9-hydroxy-3′,4′,5,5′-tetramethoxy-4,8′-oxylign-7-ene. ***Eusiderin M***
[1400817-40-4]
[126176-83-8] (natural)

$C_{22}H_{26}O_7$ 402.443
Constit. of *Licaria chrysophylla*. Oil. [α] D -16.9 (c, 0.77 in MeOH) (synthetic). Abs. config. of natural isolate could not be confirmed since no chiroptical data had been reported. λ_{max} 245 (ε 16400); 293 (ε 5750) (EtOH)

9-Aldehyde, 3′,4′,5,5′-tetra-Me ether
3,7′-Epoxy-3′,4′,5,5′-tetramethoxy-4,8′-oxylign-7-en-9-al. ***Eusiderin G***
[1400817-53-9]
[101508-18-3] (natural)
$C_{22}H_{24}O_7$ 400.427
Constit. of an *Aniba* sp.. Oil. [α] D -19.4 (c, 0.54 in MeOH) (synthetic). Abs. config. of natural isolate could not be confirmed since no chiroptical data had been reported. λ_{max} 230 (ε 8850); 280 (ε 1700); 328 (ε 3400) (MeOH).

Dias, S.M.C. *et al.*, *Phytochemistry*, 1986, **25**, 213-217 (*Eusiderin G*).
Da Silva, M.S. *et al.*, *Phytochemistry*, 1989, **28**, 3477-3482 (*Eusiderin M*).
Jing, X.-B. *et al.*, *J. Chin. Chem. Soc. (Taipei)*, 2001, **48**, 59-63 (*Eusiderins G,M, synth*).
Jing, X.-B. *et al.*, *J. Chin. Chem. Soc. (Taipei)*, 2004, **51**, 1001-1004 (*Eusiderins G,M, synth*).
Pilkington, L.I. *et al.*, *J. Org. Chem.*, 2012, **77**, 8156-8166 (*Eusiderins G,M, synth, abs config*)

3,7′-Epoxy-4,8′-oxylign-7-ene-3′,4′,5,9,9′-pentol E212

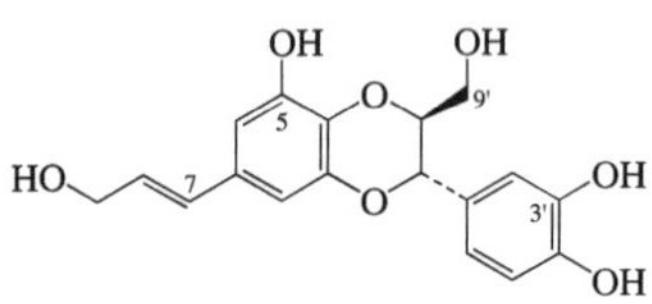

(7E,7′S,8′S)-form

3-(3,4-Dihydroxyphenyl)-2,3-dihydro-8-hydroxy-6-(3-hydroxy-1-propenyl)-1,4-benzodioxin-2-methanol, CAS.
3,7′-Epoxy-3′,4′,5,9,9′-pentahydroxy-4,8′-oxyneolign-7-ene
$C_{18}H_{18}O_7$ 346.335

(7E,7′S,8′S)-form

3′,5,9-Tri-Me ether, 4′,9′-di-Ac
[1864843-79-7]
$C_{25}H_{28}O_9$ 472.49
Constit. of the aerial parts of *Euphorbia antiquorum*. Oil.$[\alpha]^{23.3}_{D}$ -24 (c, 0.15 in MeOH). λ_{max} 274 (log ε 4.17); 334 (log ε 4.10) (MeOH)

9-Aldehyde, 3′,5-di-Me ether
3,7′-Epoxy-4′,9′-dihydroxy-3′,5-dimethoxy-4,8′-oxylign-7-en-9-al.
3,7′-Epoxy-4′,9′-dihydroxy-3′,5-dimethoxy-9-oxo-4,8′-oxylign-7-ene
[1004264-58-7]
$C_{20}H_{20}O_7$ 372.373
Constit. of the stem wood of *Zanthoxylum avicennae*. Pale yellow oil. $[\alpha]^{25}_{D}$-23.8 (c, 0.14 in MeOH). λ_{max} 275 (log ε 4.18); 334 (log ε 4.17) (MeOH)

(7E,7′R*,8′R*)-form

5-Me ether
3,7′-Epoxy-5-methoxy-4,8′-oxylign-7-ene-3′,4′,9,9′-tetrol. 3,7′-Epoxy-3′,4′,9,9′-tetrahydroxy-5-methoxy-4,8′-oxylign-7-ene. ***Jatrophasin D***
[1612832-91-3]
$C_{19}H_{20}O_7$ 360.362
Constit. of the seeds of *Jatropha curcas*. Brown powder. Racemate

3′,5-Di-Me ether
3,7′-Epoxy-3′,5-dimethoxy-4,8′-oxylign-7-ene-4′,9,9′-triol. 3,7′-Epoxy-4′,9,9′-trihydroxy-3′,5-dimethoxy-4,8′-oxylign-7-ene. ***trans-Nocomtol***
[844503-10-2]
$C_{20}H_{22}O_7$ 374.389
Constit. of *Excoecaria acerifolia* and *Populus* spp.

3′,5-Di-Me ether, 4′-O-β-D-glucopyranoside
Verticillatoside A
[193615-01-9]
[74708-36-4] [133906-61-3]
$C_{26}H_{32}O_{12}$ 536.53
Constit. of *Pedicularis verticillata*. Amorph. powder. $[\alpha]^{20}_{D}$-8 (c, 0.5 in MeOH). Related to 3,7′-Epoxy-4,8′-oxylign-7-ene-3′,4′,9,9′-tetrol, E218. λ_{max} 201; 266 (MeOH)

9-Aldehyde, 3′,5-di-Me ether
trans-Nocomtal
[844503-12-4]
$C_{20}H_{20}O_7$ 372.373
Constit. of *Populus* spp.

(7E,7′R*,8′S*)-form

3′,5-Di-Me ether
cis-Nocomtol
[844503-11-3]
$C_{20}H_{22}O_7$ 374.389
Constit. of *Excoecaria acerifolia* and *Populus* spp.

3′,5-Di-Me ether, 4′-O-β-D-glucopyranoside
Verticillatoside B
[193699-62-6]
$C_{26}H_{32}O_{12}$ 536.53
Constit. of *Pedicularis verticillata*. Amorph. powder. $[\alpha]^{20}_{D}$-32 (c, 0.5 in MeOH). λ_{max} 200; 266 (MeOH)

9-Aldehyde, 3′,5-di-Me ether
cis-Nocomtal
[844503-13-5]
$C_{20}H_{20}O_7$ 372.373
Constit. of *Populus* spp..

Bao-Ning, S. *et al.*, *Phytochemistry*, 1997, **45**, 1271-1273 (*Verticillatosides A,B*).
Morreel, K. *et al.*, *Plant Physiol.*, 2004, **136**, 4023-4026 (*Nocomtol, Noncomtal, synth*).
Chen, J.-J. *et al.*, *J. Nat. Prod.*, 2008, **71**, 212-217 (*7′S,8′S-Nocomtal*).
Li, X. *et al.*, *Nat. Prod. Res.*, 2014, **28**, 1985-1991 (*Jatrophasin D*).
Hu, J. *et al.*, *Chem. Nat. Compd. (Engl. Transl.)*, 2016, **52**, 48-51 (*Euphorbia tri-Me ether di-Ac*)

3,7′-Epoxy-4,8′-oxylign-7-ene-3′,4′,5′,9,9′-pentol E213

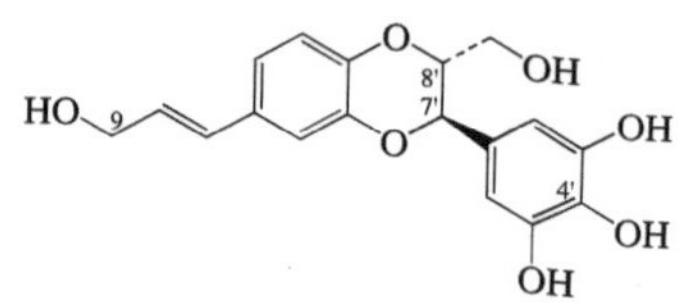

(7E,7′R,8′R)-form

2,3-Dihydro-6-(3-hydroxy-1-propenyl)-3-(3,4,5-trihydroxyphenyl)-1,4-benzodioxin-2-methanol, CAS.
3,7′-Epoxy-3′,4′,5′,9,9′-pentahydroxy-4,8′-oxyneolign-7-ene
$C_{18}H_{18}O_7$ 346.335

(7E,7′R,8′R)-form

9-Carboxylic acid, 3′,5′-di-Me ether
3,7′-Epoxy-4′,9′-dihydroxy-3′,5′-dimethoxy-4,8′-oxylign-7-en-9-oic acid

9-Carboxylic acid, 3′,5′-di-Me ether, 9′-Ac, hexacosyl ester
Lucidanin
[1564313-96-7]
$C_{48}H_{74}O_9$ 795.111
Constit. of the twigs of *Feroniella lucida*. Yellow solid. Mp 78 - 79°. $[\alpha]^{24}_{D}$-22.7 (c, 0.83 in MeOH). λ_{max} 244 (log ε 2.56); 290 (log ε 2.61); 319 (log ε 2.7) ($CHCl_3$)

(7E,7′S,8′S)-form

9′-Aldehyde, 3′,5′-di-Me ether
3,7′-Epoxy-4′,9-dihydroxy-3′,5′-dimethoxy-4,8′-oxylign-7-en-9′-al
[1002753-42-5]
$C_{20}H_{20}O_7$ 372.373
Constit. of the stem wood of *Zanthoxylum avicennae*. Oil. $[\alpha]^{25}_{D}$ -19.7 (c, 0.13 in MeOH). λ_{max} 305 (log ε 4.27); 332 (log ε 4.33) (MeOH)

(7E,7′RS,8′RS)-form

3′-Me ether

3,7′-Epoxy-5′-methoxy-4,8′-oxylign-7-ene-3′,4′,9,9′-tetrol. 3,7′-Epoxy-3′,4′,9,9′-tetrahydroxy-5′-methoxy-4,8′-oxylign-7-ene. ***Jatrophasin C***

[1612832-90-2]

$C_{19}H_{20}O_7$ 360.362

Constit. of the seeds of *Jatropha curcas*. Brown powder.

Chen, J.-J. *et al., J. Nat. Prod.*, 2008, **71**, 212-217 (*Zanthoxylum avicennae constit*).

Li, X. *et al., Nat. Prod. Res.*, 2014, **28**, 1985-1991 (*Jatrophasin C*).

Sriyatep, T. *et al., Tetrahedron*, 2014, **70**, 1773-1779 (*Lucidanin*)

4,7′-Epoxy-3,8′-oxylign-7-ene-3′,4′,5′,9,9′-pentol E214

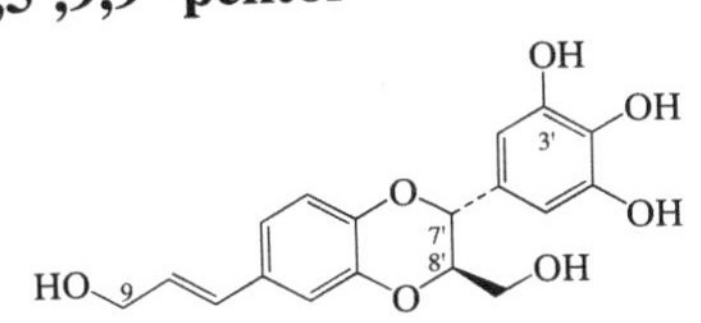

(7E,7′RS,8′RS)-form

4,7′-Epoxy-3′,4′,5′,9,9′-pentahydroxy-3,8′-oxyneolign-7-ene

$C_{18}H_{18}O_7$ 346.335

(7E,7′RS,8′RS)-form

9-Carboxylic acid, 3′,5′-di-Me ether

4,7′-Epoxy-4′,9′-dihydroxy-3′,5′-dimethoxy-3,8′-oxylign-7-en-9-oic acid

9-Carboxylic acid, 3′,5′-di-Me ether, 2-(4-hydroxyphenyl)ethylamide

Thoreliamide B

[1052115-06-6]

$C_{28}H_{29}NO_8$ 507.539

Constit. of *Lycium chinense* and *Mitrephora thorelii*. Amorph. powder. λ_{max} 293 (log ε 4.23); 318 (log ε 4.22) (MeOH)

(7Z,7′RS,8′RS)-form

9-Carboxylic acid, 3′,5′-di-Me ether, 2-(4-hydroxyphenyl)ethylamide

Z-Thoreliamide B

$C_{28}H_{29}NO_8$ 507.539

Constit. of the stems of *Lycium barbarum*. Pale yellow powder. $[\alpha]^{21}_{D}$ -11.1 (c, 0.1 in MeOH). Partial racemate. λ_{max} 207 (log ε 4.7); 240 (log ε 4.3); 278 (log ε 4.0) (MeOH).

Ge, F. *et al., Helv. Chim. Acta*, 2008, **91**, 1023-1030 (*Thoreliamide B*).

Zhang, J.-X. *et al., J. Nat. Prod.*, 2013, **76**, 51-58 (*Lycium chinense amide*).

Zhu, P.-F. *et al., Nat. Prod. Bioprospect.*, 2017, **7**, 421-431 (*Z-Thoreliamide B*)

3,7′-Epoxy-4,8′-oxylign-7-ene-2′,4′,5,9-tetrol E215

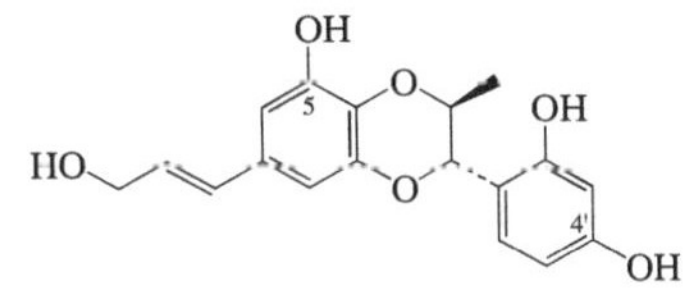

3-[3-(2,4-Dihydroxyphenyl)-2,3-dihydro-8-hydroxy-2-methyl-1,4-benzodioxin-6-yl]-2-propen-1-ol. 3,7′-Epoxy-2′,4′,5,9-tetrahydroxy-4,8′-oxyneolign-7-ene

$C_{18}H_{18}O_6$ 330.336

(7E,7′S,8′S)-form

5-Me ether

3,7′-Epoxy-5-methoxy-4,8′-oxylign-7-ene-2′,4′,9-triol. 3,7′-Epoxy-2′,4′,9-trihydroxy-5-methoxy-4,8′-oxylign-7-ene

5-Me ether, 4′-O-[β-D-apiofuranosyl-(1→6)-β-D-glucopyranoside]

Multifidoside A‡

[1071607-61-8]

$C_{30}H_{38}O_{15}$ 638.619

Constit. of the roots of *Pteris multifida*. Amorph. powder. Mp 216 - 218°. $[\alpha]^{20}_{D}$ -11.2 (c, 0.45 in MeOH). λ_{max} 208; 266 (MeOH).

Zheng, X.-D. *et al., Bull. Korean Chem. Soc.*, 2008, **29**, 1250-1252 (*Multifidoside A*).

Zheng, X.-D. *et al., Indian J. Chem., Sect. B*, 2008, **47**, 773-777 (*Multifidoside A*)

3,7′-Epoxy-4,8′-oxylign-7-ene-2′,4′,9,9′-tetrol E216

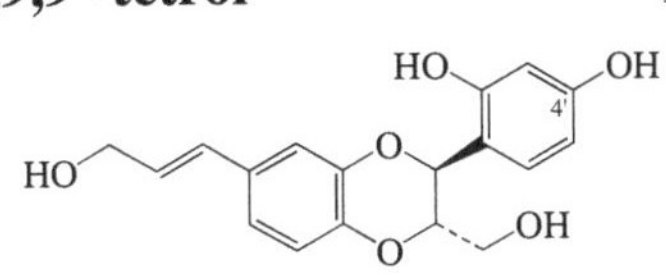

3-[3-(2,4-Dihydroxyphenyl)-2,3-dihydro-2-(hydroxymethyl)-1,4-benzodioxin-6-yl]-2-propen-1-ol. 3,7′-Epoxy-2′,4′,9,9′-tetrahydroxy-4,8′-oxyneolign-7-ene

$C_{18}H_{18}O_6$ 330.336

(7E,7′S,8′S)-form

4′-O-[β-D-Apiofuranosyl-(1→6)-β-D-glucopyranoside]

Brachyposide A‡. Multifidoside B‡

[1071608-00-8]

$C_{29}H_{36}O_{15}$ 624.592

Constit. of *Acanthopanax brachypus* and *Pteris multifida*. Amorph. powder (MeOH). Mp 212 - 215°. $[\alpha]^{20}_{D}$ -10.8 (c, 0.45 in MeOH). λ_{max} 209; 266 (MeOH).

Zheng, X.-D. *et al., Indian J. Chem., Sect. B*, 2008, **47**, 773-777 (*Multifidoside B*).

Hu, H.-B. *et al., Acta Chim. Slov.*, 2009, **56**, 470-474 (*Brachyposide A*)

3,7′-Epoxy-4,8′-oxylign-7-ene-3′,4′,5,5′-tetrol E217

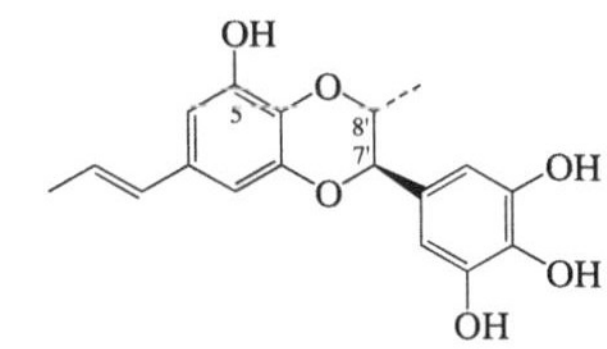

3,7′-Epoxy-3′,4′,5,5′-tetrahydroxy-4,8′-oxyneolign-7-ene

$C_{18}H_{18}O_6$ 330.336

(7E,7′R*,8′R*)-form

3′,5,5′-Tri-Me ether

3,7′-Epoxy-3′,5,5′-trimethoxy-4,8′-oxylign-7-en-4′-ol. 3,7′-Epoxy-4′-hydroxy-3′,5,5′-trimethoxy-4,8′-oxylign-7-ene. ***Eusiderin E***

[97730-86-4]

$C_{21}H_{24}O_6$ 372.417

Constit. of *Virola carinata* and *Virola pavonis*. Oil. λ_{max} 210 (ε 11100); 275 (ε 2150) (MeOH).

Cavalcante, S.H. *et al., Phytochemistry*, 1985, **24**, 1051-1055 (*Eusiderin E*).

Jing, X.-B. *et al., J. Chin. Chem. Soc. (Taipei)*, 2004, **51**, 1001-1004 (*Eusiderin E, synth*).

Pilkington, L.I. *et al., J. Org. Chem.*, 2012, **77**, 8156-8166 (*Eusiderin E*)

3,7′-Epoxy-4,8′-oxylign-7-ene-3′,4′,9,9′-tetrol E218

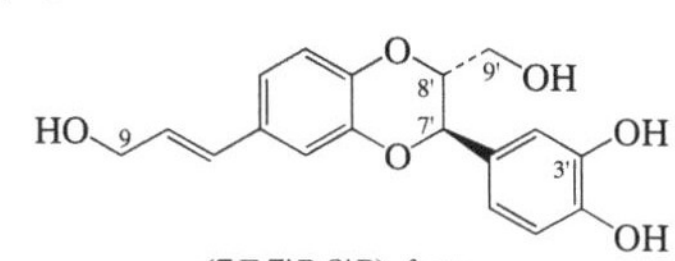

(7E,7′R,8′R)-form

3,7′-Epoxy-3′,4′,9,9′-tetrahydroxy-4,8′-oxyneolign-7-ene

$C_{18}H_{18}O_6$ 330.336

(7E,7′R,8′R)-form

9,9′-Dicarboxylic acid

3,7′-Epoxy-3′,4′-dihydroxy-4,8′-oxylign-7-ene-9,9′-dioic acid

9,9′-Dicarboxylic acid, 9-(3,4-dihydroxyphenyl)ethenyl ester(Z-)

[1588874-29-6]

$C_{26}H_{20}O_{10}$ 492.436

Constit. of *Hyptis rhomboides*. Xanthine oxidase inhibitor. Possesses (7′*R*,8′*S*)-config. due to change in Cahn-Ingold-Prelog priorities. λ_{max} 289 (log ε 3.91); 333 (log ε 3.95) (MeOH)

(7E,7′S,8′R)-form

9,9′-Dicarboxylic acid, 9-(3,4-dihydroxyphenyl)ethenyl ester(E-)
[1588874-26-3]
$C_{26}H_{20}O_{10}$ 492.436
Constit. of *Hyptis rhomboides*. Xanthine oxidase inhibitor. Possesses (7′*S*,8′*S*)-config. due to change in Cahn-Ingold-Prelog priorities. λ_{max} 286 (log ε 4.02); 334 (log ε 4.13) (MeOH)

9,9′-Dicarboxylic acid, 9-(3,4-dihydroxyphenyl)ethenyl ester(Z-)
[1588874-27-4]
$C_{26}H_{20}O_{10}$ 492.436
Constit. of *Hyptis rhomboides*. Xanthine oxidase inhibitor. Possesses (7′*S*,8′*S*)-config. due to change in Cahn-Ingold-Prelog priorities. λ_{max} 287 (log ε 4.08); 336 (log ε 4.21) (MeOH)

9,9′-Dicarboxylic acid, 9-(3,4-dihydroxyphenyl)ethenyl ester(Z-), 9′-Me ester
[1588874-30-9]
$C_{27}H_{22}O_{10}$ 506.463
Constit. of *Hyptis rhomboides*. Xanthine oxidase inhibitor. Possesses (7′*S*,8′*S*)-config. due to change in Cahn-Ingold-Prelog priorities. λ_{max} 288 (log ε 4.03); 331 (log ε 4.03) (MeOH)

(7E,7′R*,8′R*)-form

Isoamericanol A [133838-66-1]
Constit. of *Capparis flavicans*, *Colocasia antiquorum* var. *esculenta*, *Hevea brasiliensis*, *Jatropha curcas*, *Joannesia princeps* and *Phytolacca americana*. Shows neurotrophic activity. Cell proliferation inhibitor. Prisms (Me_2CO/EtOAc). Mp 157 - 159°. $[\alpha]_D$ +6.25 (c, 0.08 in MeOH). Isoamericanol A and many of its derivs. have been isol. as racemates or partial racemates of the (7′,8′-*trans*)-isomer. λ_{max} 218 (ε 26800); 224 (ε 24440); 234 (ε 21100); 250 (ε 16700); 290 (ε 14600); 310 (ε 19500) (EtOH)

9-Me ether
3,7′-Epoxy-9-methoxy-4,8′-oxylign-7-ene-3′,4′,9′-triol. 3,7′-Epoxy-3′,4′,9′-trihydroxy-9-methoxy-4,8′-oxylign-7-ene. ***9-O-Methylisoamericanol A***
[844697-27-4]
$C_{19}H_{20}O_6$ 344.363
Constit. of the seeds of *Opuntia ficus-indica* var. *saboten* and *Trewia nudiflora*. Exhibits potent cytoprotective activity. Powder. $[\alpha]^{23}_D$ -40 (c, 0.05 in MeOH). λ_{max} 206; 270; 380 (MeOH)

9-Aldehyde
3,7′-Epoxy-3′,4′,9′-trihydroxy-4,8′-oxylign-7-en-9-al. 3,7′-Epoxy-3′,4′,9′-trihydroxy-9-oxo-4,8′-oxylign-7-ene. ***Isoamericanin A***
[109063-85-6]
$C_{18}H_{16}O_6$ 328.32
Constit. of *Artabotrys hexapetalus*, *Jatropha curcas*, *Joannesia princeps* and *Phytolacca americana*. Prostaglandin I_2 inducer. Mp 177 - 178°

9-Carboxylic acid
3,7′-Epoxy-3′,4′,9′-trihydroxy-4,8′-oxylign-7-en-9-oic acid. ***Isoamericanoic acid A‡***
[1005382-37-5]
$C_{18}H_{16}O_7$ 344.319
Constit. of the fruit of *Morinda citrifolia*. Isol. from *Ophioceras dolichostomum* YMF1.00988. Exhibits moderate antifungal activity. Mp 210 - 212°. $[\alpha]^{25}_D$ -3.7 (c, 2.7 in MeOH). Not to be confused with Isoamericanoic acid A in E35. λ_{max} 203 (log ε 3.31); 219 (log ε 3.04); 286 (log ε 2.91) (MeOH)

9-Carboxylic acid, 3′-Me ether
3,7′-Epoxy-4′,9′-dihydroxy-3′-methoxy-4,8′-oxylign-7-en-9-oic acid. ***3′-O-Methylisoamericanoic acid***
[1966942-31-3]
$C_{19}H_{18}O_7$ 358.346
Constit. of the roots of *Rhodiola crenulata*. Moderate stimulator of IFN-γ expression. Yellow viscous powder. $[\alpha]^{25}_D$ -9 (c, 0.2 in MeOH). λ_{max} 202; 224; 288; 316 (MeOH)

9-Carboxylic acid, 3′-Me ether, 9′-docosanoyl, docosyl ester
[944257-62-9]
$C_{63}H_{104}O_8$ 989.517
Constit. of the roots of *Glycyrrhiza glabra*

9-Carboxylic acid, 3′-Me ether, 9′-docosanoyl, eicosyl ester
[944257-61-8]
$C_{61}H_{100}O_8$ 961.463
Constit. of the roots of *Glycyrrhiza glabra*

9-Carboxylic acid, 3′-Me ether, 9′-tetracosanoyl, docosyl ester
[944257-63-0]
$C_{65}H_{108}O_8$ 1017.571
Constit. of the roots of *Glycyrrhiza glabra*

9-Carboxylic acid, 3′-Me ether, 9′-Ac, 12-oleanen-3-yl ester
Uragogin
[1308252-38-1]
$C_{51}H_{68}O_8$ 809.097
Constit. of *Crossopetalum uragoga*. Glass. $[\alpha]^{20}_D$ +25.8 (c, 0.12 in $CHCl_3$). λ_{max} 217 (log ε 4.08); 235 (log ε 4.01); 288 (log ε 3.89); 297 (log ε 3.89); 320 (log ε 3.88) (EtOH)

9,9′-Dicarboxylic acid, bis[(4-hydroxyphenyl)ethyl]amide
Cannabisin Q
$C_{34}H_{32}N_2O_8$ 596.636
Constit. of the seeds of *Cannabis sativa*. Exhibits neuroantiinflammatory activity. Amorph. powder. $[\alpha]^{20}_D$ -2.2 (c, 1.0 in MeOH). λ_{max} 223; 285; 315 (MeOH)

(7Z,7′ξ,8′ξ)-form

9′-Carboxylic acid, 3′-Me ether
3,7′-Epoxy-4′,9-dihydroxy-3′-methoxy-4,8′-oxylign-7-en-9′-oic acid. ***Smallanlignan B***
[1392836-32-6]
$C_{19}H_{18}O_7$ 358.346
Constit. of the leaves of *Smallanthus sonchifolius*. Amorph. solid. $[\alpha]^{25}_D$ -26.3 (c, 0.2 in $CHCl_3$). λ_{max} 205 (log ε 4.45); 261 (log ε 3.92); 285 (log ε 3.65) ($CHCl_3$).

Hasegawa, T. *et al.*, *Chem. Lett.*, 1987, 329-332 (*Isoamericanin A*).
Fukuyama, Y. *et al.*, *Chem. Pharm. Bull.*, 1992, **40**, 252-254 (*Isoamericanol A*).
She, X. *et al.*, *Synth. Commun.*, 1999, **29**, 2625-2628 (*synth, pmr*).
Gu, W.X. *et al.*, *Chin. Chem. Lett.*, 2001, **12**, 485-486 (*synth*).
Takahashi, H. *et al.*, *Heterocycles*, 2002, **56**, 245-256 (*synth*).
Waibel, R. *et al.*, *Phytochemistry*, 2003, **62**, 805-811 (*Isoamericanol A, Isoamericanin A*).
Zhao, P.J. *et al.*, *Chin. Chem. Lett.*, 2004, **15**, 921-924 (*9′-O-Methylisoamericanol A*).
Jing, X. *et al.*, *Synth. Commun.*, 2004, **34**, 1723-1727 (*synth*).
Chin, Y.-W. *et al.*, *J. Agric. Food Chem.*, 2007, **55**, 4691-4697 (*Glycyrriza glabra constits*).
Lin, C.F. *et al.*, *Nat. Prod. Res.*, 2007, **21**, 1199-1204 (*Isoamericanoic acid A*).
Dong, J.-Y. *et al.*, *Nat. Prod. Res.*, 2010, **24**, 1004-1012 (*Isoamericanoic acid A, activity*).

Núñez, M.J. *et al.*, *Tetrahedron*, 2011, **67**, 3030-3033 (*Uragogin*).
Lv, J. *et al.*, *Asian J. Chem.*, 2012, **24**, 312-314 (*Smallanlignan B*).
Pilkington, L.I. *et al.*, *Eur. J. Org. Chem.*, 2014, 1037-1046 (*synth*).
Tsai, S.F. *et al.*, *Phytochemistry*, 2014, **101**, 121-127 (*Hyptis rhomboides esters*).
Zhou, J.-T. *et al.*, *Molecules*, 2015, **20**, 13725-13739 (*3′-O-Methylisoamericanoic acid, activity*).
Zhou, Y. *et al.*, *Phytochem. Lett.*, 2018, **23**, 57-61 (*Cannabisin Q, activity*)

3,7′-Epoxy-4,8′-oxylign-8-ene-3′,4′,5,5′-tetrol E219

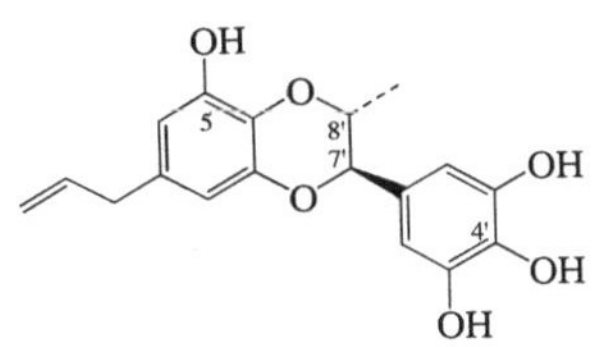

(7′R,8′R)-form

2,3-Dihydro-5-hydroxy-3-methyl-7-(2-propenyl)-2-(3,4,5-trihydroxyphenyl)-1,4-benzodioxin. 3,7′-Epoxy-3′,4′,5,5′-tetrahydroxy-4,8′-oxyneolign-8-ene
$C_{18}H_{18}O_6$ 330.336
(7′R,8′R)-form
3′,5,5′-Tri-Me ether
3,7′-Epoxy-3′,5,5′-trimethoxy-4,8′-oxylign-8-en-4′-ol. 3,7′-Epoxy-4′-hydroxy-3′,5,5′-trimethoxy-4,8′-oxylign-8-ene. ***Eusiderin K***
[126176-81-6]
$C_{21}H_{24}O_6$ 372.417
Constit. of *Licaria chrysophylla*. Cryst.. Mp 85 - 87°. λ_{max} 230 (ε 7050); 280 (ε 1850) (MeOH)
Tetra-Me ether
2,3-Dihydro-5-methoxy-3-methyl-7-(2-propenyl)-2-(3,4,5-trimethoxyphenyl)-1,4-benzodioxin. 3,7′-Epoxy-3′,4′,5,5′-tetramethoxy-4,8′-oxylign-8-ene. ***(-)-Eusiderin A***
[59332-00-2]
$C_{22}H_{26}O_6$ 386.444
Constit. of *Aniba* sp., *Eusideroxylon zwageri*, *Licaria aurea*, *Licaria chrysophylla* and *Virola pavonis*. Needles (MeOH). Mp 94°. $[\alpha]_D^{19}$ -25.4 (c, 1.8 in MeOH). Named solely as Eusiderin in early lit.. λ_{max} 272 (log ε 0.60) (MeOH)
(7′R,8′S)-form
Tetra-Me ether
(-)-Eusiderin C
[76333-70-5]
$C_{22}H_{26}O_6$ 386.444
Constit. of *Virola pavonis*. λ_{max} 219 (log ε 4.47); 270 (log ε 3.23) (MeOH)
(7′S,8′R)-form
3′,4′,5-Tri-Me ether
3,7′-Epoxy-4′,5,5′-trimethoxy-4,8′-oxylign-8-en-3′-ol. 3,7′-Epoxy-3′-hydroxy-4′,5,5′-trimethoxy-4,8′-oxylign-8-ene.
3-O-Demethyl-(+)-eusiderin C
[1446340-98-2]
$C_{21}H_{24}O_6$ 372.417
Constit. of *Miliusa fragrans*. Cytotoxic. Viscous liq.. $[\alpha]_D^{20}$ +65.5 (c, 0.46 in $CHCl_3$). λ_{max} 220 (log ε 4.51); 275 (log ε 3.63) (MeOH)
3′,5,5′-Tri-Me ether
4′-O-Demethyl-(+)-eusiderin C
[1446741-16-7]
$C_{21}H_{24}O_6$ 372.417
Constit. of *Miliusa fragrans*. Shows antiherpetic activity. Cytotoxic. Viscous liq.. $[\alpha]_D$ +96.9 (0.16 in $CHCl_3$). λ_{max} 215 (log ε 4.59); 276 (log ε 4.43) (MeOH)
Tetra-Me ether
(+)-Eusiderin C
[1400817-50-6]
$C_{22}H_{26}O_6$ 386.444
Constit. of *Licaria chrysophylla* and *Miliusa fragrans*. Oil. $[\alpha]_D$ +31.1 (c, 0.28 in MeOH) (synthetic). λ_{max} 274 (log ε 0.19) (MeOH)
(7′S,8′S)-form
Tetra-Me ether
(+)-Eusiderin A
[1446341-00-9]
$C_{22}H_{26}O_6$ 386.444
Constit. of *Miliusa fragrans*. Amorph. solid. $[\alpha]_D^{20}$ +13.6 (c, 0.33 in $CHCl_3$). λ_{max} 215 (log ε 4.77); 273 (log ε 3.60) (MeOH)
(7′R*,8′R*)-form
3′,4′,5-Tri-Me ether
Eusiderin H
$C_{21}H_{24}O_6$ 372.417
Constit. of *Virola pavonis*. Stereochem. unconfirmed
3′,4′-Methylene, 5,5′-di-Me ether
3,7′-Epoxy-3′,5-dimethoxy-4′,5′-methylenedioxy-4,8′-oxylign-8-ene.
Eusiderin I
[127420-50-2]
$C_{21}H_{22}O_6$ 370.401
Constit. of *Eusideroxylon zwageri* and *Licaria chrysophylla*. Oil
(7′R*,8′S*)-form
3′,4′-Methylene, 5,5′-di-Me ether
Eusiderin J
[127420-51-3]
$C_{21}H_{22}O_6$ 370.401
Constit. of *Licaria chrysophylla*. Oil.

Hobbs, J.J. *et al.*, *J. Chem. Soc.*, 1960, 4732-4738 (*Eusiderin*).
Merlini, L. *et al.*, *Tetrahedron Lett.*, 1975, **16**, 3621-3622 (*synth*).
Wenkert, E. *et al.*, *Phytochemistry*, 1976, **15**, 1547-1551 (*cmr*).
Braz Filho, R. *et al.*, *Tetrahedron Lett.*, 1976, **17**, 1157-1160 (*Eusiderin, struct*).
Fernandes, J.B. *et al.*, *Phytochemistry*, 1980, **19**, 1523-1525 (*Eusiderins A,C*).
Carneiro, M.O. *et al.*, *Cienc. Cult. (Sao Paulo)*, (Suppl.), 1983, **35**, 457 (*Eusiderin H*).
Rodrigues, M. de M. *et al.*, *Phytochemistry*, 1984, **23**, 667-669 (*cryst struct*).
Arnoldi, A. *et al.*, *J. Chem. Soc., Perkin Trans. 1*, 1985, 2555-2557 (*Eusiderins A,C, synth, cd*).
Da Silva, M.S. *et al.*, *Phytochemistry*, 1989, **28**, 3477-3482 (*Eusiderins A,K*).
Jing, X.-B. *et al.*, *J. Chin. Chem. Soc. (Taipei)*, 2004, **51**, 1001-1004 (*(±)-Eusiderins J,K, synth*).
Pilkington, L.I. *et al.*, *J. Org. Chem.*, 2012, **77**, 8156-8166 (*Eusiderins A,C, synth, abs config*).
Sawasdee, K. *et al.*, *Tetrahedron Lett.*, 2013, **54**, 4259-4263 (*Miliusa fragrans constits, activity*)

4,7′-Epoxy-3,8′-oxylign-7-ene-3′,4′,9,9′-tetrol E220

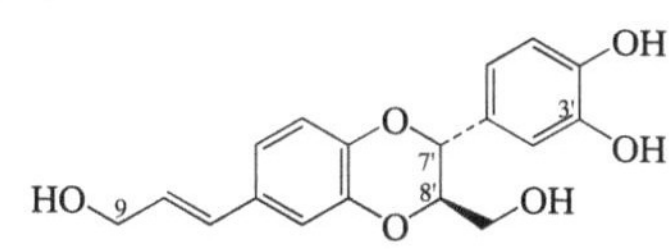

(7E,7′R,8′R)-form

3-(3,4-Dihydroxyphenyl)-2,3-dihydro-7-(3-hydroxy-1-propenyl)-1,4-benzodioxin-2-methanol, CAS. *4,7′-Epoxy-3′,4′,9,9′-tetrahydroxy-3,8′-oxyneolign-7-ene*
$C_{18}H_{18}O_6$ 330.336
(7E,7′R,8′R)-form
9-Carboxylic acid, 3′-Me ether
4,7′-Epoxy-4′,9′-dihydroxy-3′-methoxy-3,8′-oxylign-7-en-9-oic acid. ***Arteminorin D***
[1158294-48-4]
$C_{19}H_{18}O_7$ 358.346
Constit. of *Artemisia minor*, *Morinda citrifolia* and *Rhodiola crenulata*. Moderate stimulator of IFN-γ expression. Amorph. powder. $[\alpha]_D^{25}$ -4.7 (c, 0.2 in MeOH). λ_{max} 202; 222; 288; 319 (MeOH)

(7E,7′R,8′S)-form

9,9′-Dicarboxylic acid

4,7′-Epoxy-3′,4′-dihydroxy-3,8′-oxylign-7-ene-9,9′-dioic acid

[134453-07-9]

9,9′-Dicarboxylic acid, 9′-Et ester

Pharbilignan B. Pharbilignanol B

[1620558-55-5]

$C_{20}H_{18}O_8$ 386.356

Constit. of the seeds of *Pharbitis nil*. Exhibits moderate cytotoxicity against SK-OV-3, SK-MEL-2 and HCT-15 human tumour cell lines. Amorph. powder. $[\alpha]^{25}_{D}$ -15.6 (c, 0.35 in MeOH). Possesses (7′*R*,8′*R*)-config. due to change in Cahn-Ingold-Prelog priorities. λ_{max} 259 (log ε 4.05); 306 (log ε 3.63) (MeOH)

(7E,7′S,8′R)-form

9-Carboxylic acid

4,7′-Epoxy-3′,4′,9′-trihydroxy-3,8′-oxylign-7-en-9-oic acid. **8′-Epiamericanoic acid**

[1489173-46-7]

$C_{18}H_{16}O_7$ 344.319

Constit. of *Morinda citrifolia*. Amorph. yellow powder. $[\alpha]^{25}_{D}$ -26.8 (c, 0.1 in MeOH). λ_{max} 224 (log ε 4.25); 253 (log ε 4.13); 310 (log ε 4.10) (MeOH)

9,9′-Dicarboxylic acid, 9-(3,4-dihydroxyphenyl)ethenyl ester(Z-)

[1588874-28-5]

$C_{26}H_{20}O_{10}$ 492.436

Constit. of *Hyptis rhomboides*. Xanthine oxidase inhibitor. Possesses (7′*S*,8′*S*)-config. due to change in Cahn-Ingold-Prelog priorities. λ_{max} 288 (log ε 4.11); 332 (log ε 4.15) (MeOH)

(7E,7′S,8′S)-form

9,9′-Dicarboxylic acid, 9′-Et ester

Pharbilignan A. Pharbilignanol A

[1620558-54-4]

$C_{20}H_{18}O_8$ 386.356

Constit. of the seeds of *Pharbitis nil*. Exhibits moderate cytotoxicity against SK-OV-3, SK-MEL-2 and HCT-15 human tumour cell lines. Amorph. powder. $[\alpha]^{25}_{D}$ -9.1 (c, 0.80 in MeOH). Possesses (7′*S*,8′*R*)-config. due to change in Cahn-Ingold-Prelog priorities. λ_{max} 259 (log ε 4.05); 306 (log ε 3.62) (MeOH)

(7E,7′RS,8′RS)-form

9-Carboxylic acid, 3′-Me ether, 2-(4-hydroxyphenyl)ethylamide

[1416474-58-2]

$C_{27}H_{27}NO_7$ 477.513

Constit. of *Lycium chinense*. Powder. λ_{max} 285 (log ε 4.07); 319 (log ε 3.97) (MeOH)

(7E,7′R,8′R*)-form*

Americanol A [133838-65-0]

Constit. of *Capparis flavicans*, *Colocasia antiquorum* var. *esculenta*, *Hevea brasiliensis*, *Joannesia princeps*, *Morinda citrifolia* and *Phytolacca americana*. Shows neurotrophic activity. Antioxidant. Prisms (Me_2CO/EtOAc). Mp 125 - 128°. $[\alpha]^{22}_{D}$ +7.2 (c, 1.4 in MeOH). Americanol A and many of its derivs. have been isol. as racemates or partial racemates of the (7′,8′-*trans*)-isomer. λ_{max} 260 (ε 18800); 268 (ε 18800); 300 (ε 6400); 312 (ε 4100) (EtOH)

9-Me ether

4,7′-Epoxy-9-methoxy-3,8′-oxylign-7-ene-3′,4′,9′-triol. 4,7′-Epoxy-3′,4′,9′-trihydroxy-9-methoxy-3,8′-oxylign-7-ene. **9-O-Methylamericanol A**

[696603-21-1]

$C_{19}H_{20}O_6$ 344.363

Constit. of the seeds of *Phytolacca americana* and *Trewia nudiflora*. Powder. $[\alpha]^{23}_{D}$ -6.7. The *P. americana* isolate was racemic. λ_{max} 203 (ε 64300); 263 (ε 16700) (EtOH). [neutral] λ_{max} 204 (log ε 4.7); 269 (log ε 3.32); 341 (log ε 2.73); 376 (log ε 2.48) (MeOH)

9-Et ether

4,7′-Epoxy-9-ethoxy-3,8′-oxylign-7-ene-3′,4′,9′-triol. 4,7′-Epoxy-9-ethoxy-3′,4′,9′-trihydroxy-3,8′-oxylign-7-ene. **9-O-Ethylamericanol A**

[844697-28-5]

$C_{20}H_{22}O_6$ 358.39

Constit. of the seeds of *Trewia nudiflora*. Yellow powder. $[\alpha]^{23}_{D}$ -1.7 (c, 0.6 in MeOH). λ_{max} 206; 270; 380 (MeOH)

9-Butyl ether

9-Butoxy-4,7′-epoxy-3,8′-oxylign-7-ene-3′,4′,9′-triol. 9-Butoxy-4,7′-epoxy-3′,4′,9′-trihydroxy-3,8′-oxylign-7-ene. **9-O-Butylamericanol A**

[844697-29-6]

$C_{22}H_{26}O_6$ 386.444

Constit. of the seeds of *Trewia nudiflora*. Yellow powder. $[\alpha]^{23}_{D}$ -7.9 (c, 1.4 in MeOH). λ_{max} 206; 270; 380 (MeOH)

9-Aldehyde

4,7′-Epoxy-3′,4′,9′-trihydroxy-3,8′-oxylign-7-en-9-al. 4,7′-Epoxy-3′,4′,9′-trihydroxy-9-oxo-3,8′-oxylign-7-ene. **Americanin A**

[69506-79-2]

$C_{18}H_{16}O_6$ 328.32

Constit. of *Excoecaria acerifolia*, *Morinda citrifolia*, *Phytolacca americana* and cured vanilla beans. Yellow plates (MeOH). Mp 246 - 247°. $[\alpha]^{17}_{D}$ +23.7. λ_{max} 245; 333 (MeCN)

9-Carboxylic acid

Americanoic acid A‡

[777061-45-7]

$C_{18}H_{16}O_7$ 344.319

Constit. of the fruit of *Morinda citrifolia*. Pale yellow powder. $[\alpha]^{24}_{D}$ +26.2 (c, 0.7 in MeOH). Not to be confused with Americanoic acid A in E36. λ_{max} 218 (log ε 4.33); 234 (log ε 4.24); 287 (log ε 4.18); 318 (log ε 4.1) (MeOH)

9,9′-Dicarboxylic acid, bis[2-(4-hydroxyphenyl)ethylamide]

Cannabisin M

[1831134-13-4]

$C_{34}H_{32}N_2O_8$ 596.636

Constit. of the seeds of *Cannabis sativa*. Antioxidant. Amorph. powder. $[\alpha]^{20}_{D}$ -5 (c, 0.1 in MeOH). λ_{max} 243 (log ε 3.49); 291 (log ε 3.30); 318 (log ε 3.0) (MeOH)

(7Z,7′RS,8′RS)-form

9-Carboxylic acid, 3′-Me ether, 2-(4-hydroxyphenyl)ethylamide

[1416474-59-3]

$C_{27}H_{27}NO_7$ 477.513

Constit. of *Lycium chinense*. Powder. λ_{max} 277 (log ε 3.69); 305 (log ε 3.54) (MeOH).

Antus, S. *et al.*, *Liebigs Ann. Chem.*, 1986, 647-654 (*Americanin A*).

Fukuyama, Y. *et al.*, *Chem. Pharm. Bull.*, 1992, **40**, 252-254 (*Americanol A*).

She, X.G. *et al.*, *J. Chem. Res., Synop.*, 1998, 436-437 (*Americanin A, synth*).

Matsumoto, K. *et al.*, *Tetrahedron Lett.*, 1999, **40**, 3185-3186 (*Americanol A, synth*).

Takahashi, H. *et al.*, *Heterocycles*, 2002, **56**, 245-256 (*Americanol A, synth*).

Takahashi, H. *et al.*, *Chem. Pharm. Bull.*, 2003, **51**, 1377-1381 (*9′-O-Methylamericanol A*).

Waibel, R. *et al.*, *Phytochemistry*, 2003, **62**, 805-811 (*Americanol A*).
Zhao, P.J. *et al.*, *Chin. Chem. Lett.*, 2004, **15**, 921-924 (*Trewia nudiflora constits*).
Kamiya, K. *et al.*, *J. Agric. Food Chem.*, 2004, **52**, 5843-5848 (*Americanoic acid A, activity*).
Schwarz, B. *et al.*, *J. Agric. Food Chem.*, 2009, **57**, 3729-3737 (*Americanin A, isol, vanilla bean, props*).
He, Z.-Z. *et al.*, *J. Nat. Prod.*, 2009, **72**, 1198-1201 (*Arteminorin D*).
Zhang, J.-X. *et al.*, *J. Nat. Prod.*, 2013, **76**, 51-58 (*Lycium chinense amides*).
Nguyen, P.-H. *et al.*, *J. Nat. Prod.*, 2013, **76**, 2080-2087 (*8-Epiamericanoic acid*).
Kim, K.H. *et al.*, *J. Agric. Food Chem.*, 2014, **62**, 7746-7752 (*Pharbilignans A,B, activity*).
Tsai, S.-F. *et al.*, *Phytochemistry*, 2014, **101**, 121-127 (*Hyptis rhomboides constit*).
Yan, X. *et al.*, *J. Agric. Food Chem.*, 2015, **63**, 10611-10619 (*Cannabisin M, activity*).
Zhou, J.-T. *et al.*, *Molecules*, 2015, **20**, 13725-13739 (*Arteminorin D, cd, activity*)

4,7′-Epoxy-3,8′-oxylign-7-ene-3′,5′,9,9′-tetrol E221

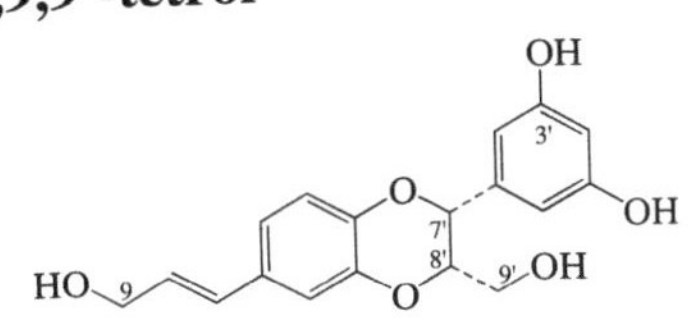

(7E,7′R*,8′S*)-form

4,7′-Epoxy-3′,5′,9,9′-tetrahydroxy-3,8′-oxyneolign-7-ene

$C_{18}H_{18}O_6$ 330.336

Structures in this series may need revision. Phenyl 3,5-dioxygenated substn. pattern in doubt as a result of work done by Yang *et al* in 2015

(7E,7′R*,8′S*)-form

9,9′-Dicarboxylic acid, 3′-Me ether, bis[2-(4-hydroxyphenyl)ethylamide]

Cannabisin N

[1831134-14-5]

$C_{35}H_{34}N_2O_8$ 610.663

Constit. of the seeds of *Cannabis sativa*. Amorph. powder. $[\alpha]_D^{20}$ -9.8 (c, 0.1 in MeOH). Has (7′*R**,8′*R**)-config. due to change in Cahn-Ingold-Prelog priorities. λ_{max} 240 (log ε 3.78); 285 (log ε 3.63); 315 (log ε 3.50) (MeOH)

(7E,7′RS,8′RS)-form

9,9′-Dicarboxylic acid, 3′-Me ether

4,7′-Epoxy-3′-hydroxy-5′-methoxy-3,8′-oxylign-7-ene-9,9′-dioic acid

9,9′-Dicarboxylic acid, 3′-Me ether, bis[2-(4-hydroxyphenyl)ethylamide]

[1416474-60-6]

$C_{35}H_{34}N_2O_8$ 610.663

Constit. of the root bark of *Lycium chinense*. Powder. Possesses (7′*RS*,8′*SR*)-config. due to change in Cahn-Ingold-Prelog priorities. λ_{max} 285 (log ε 3.86); 318 (log ε 3.73) (MeOH)

(7Z,7′RS,8′RS)-form

9,9′-Dicarboxylic acid, 3′-Me ether, bis[2-(4-hydroxyphenyl)ethylamide]

[1416474-61-7]

Constit. of the root bark of *Lycium chinense*. Powder. Possesses (7′*RS*,8′*SR*)-config. due to change in Cahn-Ingold-Prelog priorities. λ_{max} 278 (log ε 3.85); 306 (log ε 3.67) (MeOH).

Zhang, J.-X. *et al.*, *J. Nat. Prod.*, 2013, **76**, 51-58 (*Lycium chinense constits*).
Yan, X. *et al.*, *J. Agric. Food Chem.*, 2015, **63**, 10611-10619 (*Cannabisin N*).
Yang, Y.-N. *et al.*, *J. Nat. Prod.*, 2015, **78**, 705-711 (*pmr, struct*)

3,7′-Epoxy-4,8′-oxylign-7-ene-2′,3′,4′-triol E222

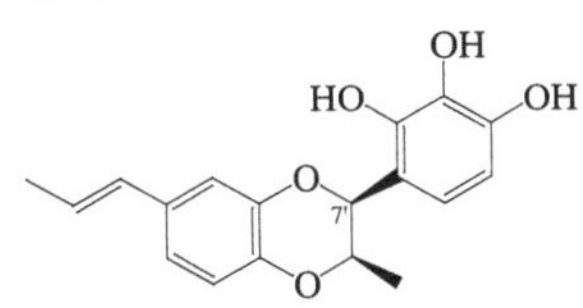

2,3-Dihydro-2-methyl-6-(1-propenyl)-3-(2,3,4-trihydroxyphenyl)-1,4-benzodioxin. 3,7′-Epoxy-2′,3′,4′-trihydroxy-4,8′-oxyneolign-7-ene

$C_{18}H_{18}O_5$ 314.337

(7E,7′S,8′R)-form

2′-Me ether

3,7′-Epoxy-2′-methoxy-4,8′-oxylign-7-ene-3′,4′-diol. 3,7′-Epoxy-3′,4′-dihydroxy-2′-methoxy-4,8′-oxylign-7-ene

[852836-21-6]

$C_{19}H_{20}O_5$ 328.364

Constit. of the aerial parts of *Rodgersia podophylla*. $[\alpha]_D^{20}$ +24.2 (c, 0.06 in MeOH). λ_{max} 258 (log ε 4.04); 302 (log ε 3.62) (MeOH)

7,8-Didehydro, 2′-Me ether

3,7′-Epoxy-2′-methoxy-4,8′-oxylign-7-yne-3′,4′-diol. 3,7′-Epoxy-3′,4′-dihydroxy-2′-methoxy-4,8′-oxylign-7-yne

[852836-22-7]

$C_{19}H_{18}O_5$ 326.348

Constit. of the aerial parts of *Rodgersia podophylla*. λ_{max} 256 (ε 10715); 295 (ε 3390) (MeOH).

Kim, Y.L. *et al.*, *Chem. Pharm. Bull.*, 2005, **53**, 103-104 (*Rodgersia podophylla constits*)

3,7′-Epoxy-4,8′-oxylign-7-ene-3′,4′,5-triol E223

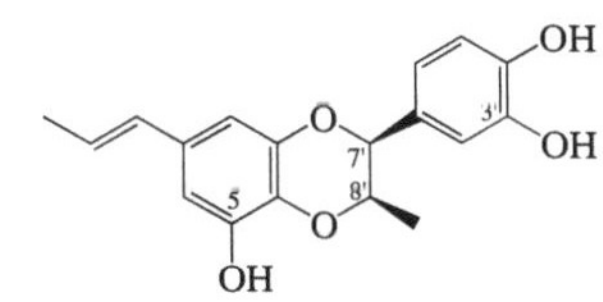

2-(3,4-Dihydroxyphenyl)-2,3-dihydro-5-hydroxy-3-methyl-7-(1-propenyl)-1,4-benzodioxin. 3,7′-Epoxy-3′,4′,5-trihydroxy-4,8′-oxyneolign-7-ene

$C_{18}H_{18}O_5$ 314.337

(7E,7′S,8′R)-form

3′,5-Di-Me ether

3,7′-Epoxy-3′,5-dimethoxy-4,8′-oxylign-7-en-4′-ol. 3,7′-Epoxy-4′-hydroxy-3′,5-dimethoxy-4,8′-oxylign-7-ene

[182172-29-8]

$C_{20}H_{22}O_5$ 342.391

Constit. of *Iryanthera paraensis*. Viscous oil. $[\alpha]_D^{25}$ -8 (c, 0.2 in $CHCl_3$).

Motter Magri, F.M. *et al.*, *Phytochemistry*, 1996, **43**, 669-671 (*Iryanthera paraensis constit*)

3,7′-Epoxy-4,8′-oxylign-7-ene-3′,4′,9-triol E224

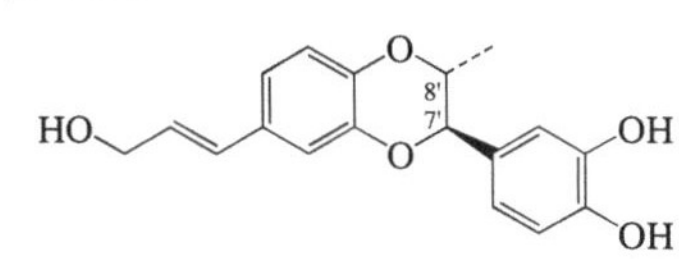

(7E,7′R,8′R)-form

3,7′-Epoxy-3′,4′,9-trihydroxy-4,8′-oxyneolign-7-ene

$C_{18}H_{18}O_5$ 314.337

(7E,7′R,8′R)-form

9-Aldehyde, 3′,4′-methylene ether

3,7′-Epoxy-3′,4′-methylenedioxy-4,8′-oxylign-7-en-9-al. ***Melaleucin C***

$C_{19}H_{16}O_5$ 324.332

Constit. of *Melaleuca bracteata*. Oil. $[\alpha]_D^{20}$ +14.2 (c, 0.3 in $CHCl_3$). Stereochem. uncertain due to errors in lit.. λ_{max} 249 (log ε 5.06); 334 (log ε 5.16) ($CHCl_3$)

(7Z,7′R*,8′R*)-form

3′-Me ether

3,7′-Epoxy-3′-methoxy-4,8′-oxylign-7-ene-4′,9-diol. 3,7′-Epoxy-4′,9-dihydroxy-3′-methoxy-4,8′-oxylign-7-ene. ***Isoamericanol B‡***

[1383714-32-6]
$C_{19}H_{20}O_5$ 328.364
Constit. of the leaves of *Elaeagnus lanceolata* and *Elaeagnus oldhamii*. Exhibits antiinflammatory activity. Amorph. solid or yellowish oil. $[\alpha]_D^{24}$ +98.1 (c, 0.4 in MeOH) (*Elaeagnus lanceolata* isolate). $[\alpha]_D^{23}$ +32.1 (c, 0.2 in MeOH) (*Elaeagnus oldhamii* isolate). Not to be confused with Isoamericanol B, in I28. λ_{max} 216 (log ε 4.44); 259 (log ε 4.05); 286 (log ε 3.76) (MeOH) (*Elaeagnus lanceolata* isolate).

Song, W.W. *et al.*, *Yunnan Zhiwu Yanjiu*, 2010, **32**, 455-462 (*Elaeagnus lanceolata constit*).
Liao, C.-R. *et al.*, *Molecules*, 2013, **18**, 13218-13227 (*Elaeagnus oldhamii constit, activity*).
Li, C. *et al.*, *Fitoterapia*, 2017, **120**, 171-176 (*Melaleucin C*)

3,7′-Epoxy-4,8′-oxylign-8-ene-3′,4′,5-triol E225

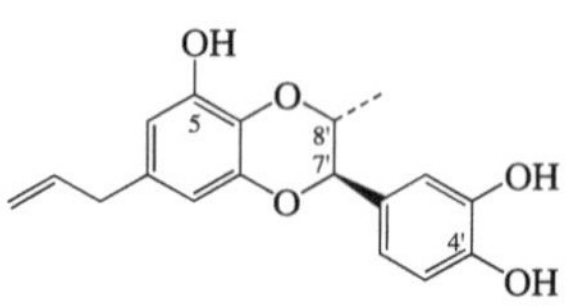

(7′R,8′R)-form

2-(3,4-Dihydroxyphenyl)-2,3-dihydro-5-hydroxy-3-methyl-7-(2-propenyl)-1,4-benzodioxin. 3,7′-Epoxy-3′,4′,5-trihydroxy-4,8′-oxyneolign-8-ene
$C_{18}H_{18}O_5$ 314.337
(7′R,8′R)-form
3′,5-Di-Me ether
3,7′-Epoxy-3′,5-dimethoxy-4,8′-oxylign-8-en-4′-ol. 3,7′-Epoxy-4′-hydroxy-3′,5-dimethoxy-4,8′-oxylign-8-ene. ***Miliusfragrin***
[1446741-17-8]
$C_{20}H_{22}O_5$ 342.391
Constit. of *Miliusa fragrans*. Shows appreciable cytotoxic activity against KB, MCF7 and NCI-H187 cancer cell lines. Solid. $[\alpha]_D^{20}$ -23.6 (c, 0.33 in $CHCl_3$). λ_{max} 215 (log ε 4.42); 281 (log ε 3.73) (MeOH)
Tri-Me ether
3,7′-Epoxy-3′,4′,5-trimethoxy-4,8′-oxylign-8-ene. ***4′-O-Methylmiliusfragrin***
[1446340-99-3]
$C_{21}H_{24}O_5$ 356.418
Constit. of *Miliusa fragrans*. Viscous liq.. $[\alpha]_D^{20}$ -15.3 (c, 0.28 in $CHCl_3$). λ_{max} 215 (log ε 4.97); 278 (log ε 4.09) (MeOH)
3′,4′-Methylene, 5-Me ether
3,7′-Epoxy-5-methoxy-3′,4′-methylenedioxy-4,8′-oxylign-8-ene. ***(-)-Eusiderin B***
[60297-82-7]
$C_{20}H_{20}O_5$ 340.375
Constit. of *Licaria chrysophylla* and *Licaria rigida*. Oil. $[\alpha]_D$ -2.3 (c, 0.8 in MeOH) (synthetic). Abs. config. of nat. Eusiderin B could not be confirmed since no opt. rotn. data was taken. λ_{max} 285 (log ε 1.07) (MeOH)
(7′R,8′S)-form
Tri-Me ether
(-)-Eusiderin D
[75680-33-0]
$C_{21}H_{24}O_5$ 356.418
Constit. of *Virola pavonis*. Oil. Abs. config. not confirmed. λ_{max} 219 (log ε 4.53); 232 (sh); 276 (log ε 3.73) (MeOH).

Fernandes, J.B. *et al.*, *Phytochemistry*, 1980, **19**, 1523-1525 (*Eusiderin D*).
Braz Filho, R. *et al.*, *Phytochemistry*, 1981, **20**, 2049-2050 (*Eusiderin B*).
Pilkington, L.I. *et al.*, *J. Org. Chem.*, 2012, **77**, 8156-8166 (*Eusiderin B, synth, abs config*).
Sawasdee, K. *et al.*, *Tetrahedron Lett.*, 2013, **54**, 4259-4263 (*Miliusa fragrans constits, activity*)

7,7′-Epoxy-4″,8′-oxy-2′,8-sesquilign-7″-ene-3,3″,4,4′,5′,9,9′,9″-octol E226

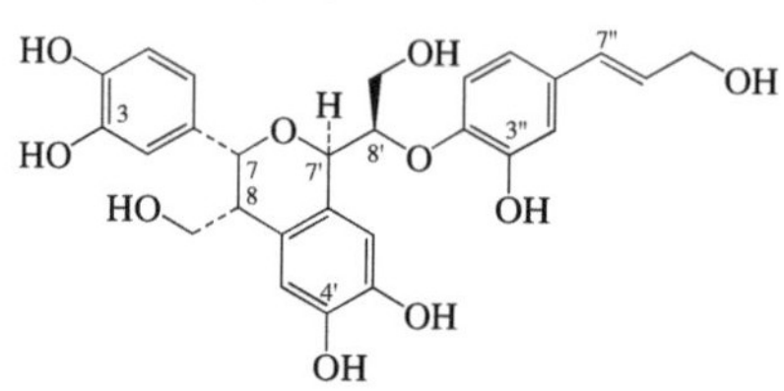

(7R*,7′R*,7″E,8R*,8′R*)-form

7,7′-Epoxy-3,3″,4,4′,5′,9,9′,9″-octahydroxy-4″,8′-oxy-2′,8-sesquineolign-7″-ene
$C_{27}H_{28}O_{10}$ 512.511
(7R*,7′R*,7″E,8R*,8′R*)-form
3,3″,4′-Tri-Me ether, penta-Ac
[221302-68-7]
$C_{40}H_{44}O_{15}$ 764.777
Constit. of degraded wood of *Pinus taeda*. Tentative stereochemistry assigned
(7R*,7′R*,7″E,8S*,8′S*)-form
3,3″,4′-Tri-Me ether, penta-Ac
[221302-67-6]
Constit. of degraded wood of *Pinus taeda*. Tentative stereochemistry assigned.

Peng, J. *et al.*, *Phytochemistry*, 1999, **50**, 659-666 (*Pinus taeda constits*)

4,7′-Epoxy-3′,4′,5,5′,9′-pentahydroxy-8,9-dinor-3,8′-lignan-7-al E227

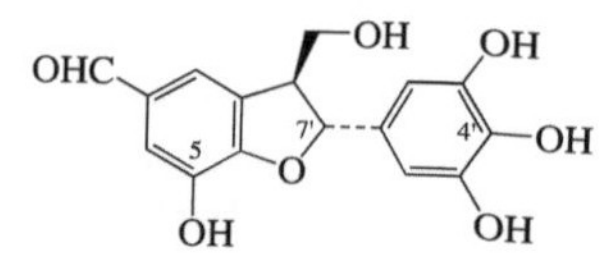

(7′R,8′S)-form

2,3-Dihydro-7-hydroxy-3-(hydroxymethyl)-2-(3,4,5-trihydroxyphenyl)-5-benzofurancarboxaldehyde.
4,7′-Epoxy-3′,4′,5,5′,9′-pentahydroxy-8,9-dinor-7-oxo-3,8′-neolignan
$C_{16}H_{14}O_7$ 318.281
(7′R,8′S)-form
3′,5,5′-Tri-Me ether
4,7′-Epoxy-4′,9′-dihydroxy-3′,5,5′-trimethoxy-8,9-dinor-3,8′-lignan-7-al. 4,7′-Epoxy-4′,9′-dihydroxy-3′,5,5′-trimethoxy-8,9-dinor-7-oxo-3,8′-lignan. ***7′R,8′S-Tomentosanan A***
$C_{19}H_{20}O_7$ 360.362
Constit. of *Hedyotis uncinella*. Amorph. red powder. $[\alpha]_D^{20}$ -1 (c, 0.1 in MeOH)
3′,5,5′-Tri-Me ether, 4′-O-β-D-glucopyranoside
Moellenoside A
[1029007-27-9]
$C_{25}H_{30}O_{12}$ 522.503
Constit. of *Selaginella moellendorffii*. Powder. Mp 232 - 234°. $[\alpha]_D^{25}$ -9.2 (c, 0.11 in MeOH). λ_{max} 208 (no solvent reported)
(7′S,8′R)-form
3′,5,5′-Tri-Me ether
7′S,8′R-Tomentosanan A
[1771744-25-2]
$C_{19}H_{20}O_7$ 360.362
Constit. of *Prunus tomentosa*. Oil. $[\alpha]_D^{20}$ -16.5 (c, 0.05 in MeOH).

Zheng, X.K. *et al.*, *Chin. Chem. Lett.*, 2008, **19**, 79-81 (*Moellenoside A*).
Liu, Q.-B. *et al.*, *Biochem. Syst. Ecol.*, 2014, **55**, 236-240 (*7′S,8′R-Tomentosanan A*).
Pan, Y. *et al.*, *Nat. Prod. Commun.*, 2016, **11**, 1115-1116 (*7′R,8′S-Tomentosanan A*)

4,7′-Epoxy-3′,4′,5,5′,9′-pentahydroxy-8,9-dinor-3,8′-lignan-7-oic acid E228

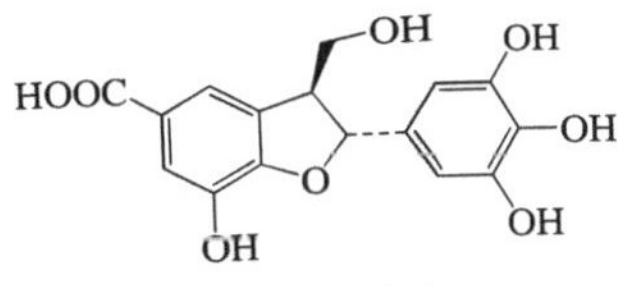

(7′R,8′S)-form

2,3-Dihydro-7-hydroxy-3-(hydroxymethyl)-2-(3,4,5-trihydroxyphenyl)-5-benzofurancarboxylic acid

$C_{16}H_{14}O_8$ 334.28

(7′R,8′S)-form

3′,5,5′-Tri-Me ether, Me ester, 9′-O-β-D-glucopyranoside

[1021524-77-5]

$C_{26}H_{32}O_{13}$ 552.529

Constit. of *Chione venosa* var. *venosa*. [α]D -33.3 (c, 0.06 in MeOH). λ_{max} 228 (sh); 278; 300 (sh) (MeOH aq.)

(7′S,8′R)-form

3′,5,5′-Tri-Me ether

4,7′-Epoxy-4′,9′-dihydroxy-3′,5,5′-trimethoxy-8,9-dinor-3,8′-lignan-7-oic acid. ***5′-Methoxyceplignan***

$C_{19}H_{20}O_8$ 376.361

Constit. of the active fraction of Re-Du-Ning injection (RDNI).

Werner, I. *et al.*, *Nat. Prod. Commun.*, 2008, **3**, 179-184 (*Chione venosa constit*).

Li, H.-B. *et al.*, *Zhongcaoyao*, 2016, **47**, 1643-1649 (*5′-Methoxyceplignan*)

4,7′-Epoxy-3′,4′,5,5′,9′-pentahydroxy-3,8′-ligna-7,7′-dien-9-oic acid E229

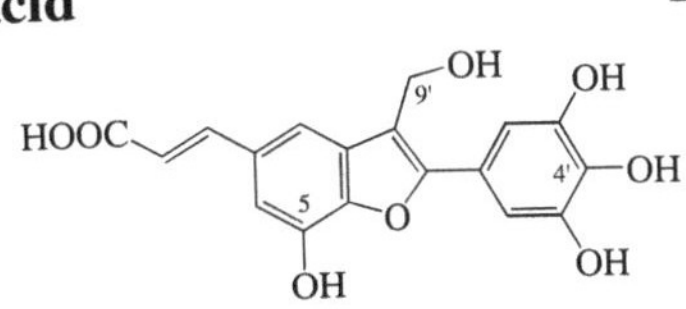

3-[7-Hydroxy-3-(hydroxymethyl)-2-(3,4,5-trihydroxyphenyl)-5-benzofuranyl]-2-propenoic acid

$C_{18}H_{14}O_8$ 358.302

(E)-form

3′,5,5′-Tri-Me ether

4,7′-Epoxy-4′,9′-dihydroxy-3′,5,5′-trimethoxy-3,8′-ligna-7,7′-dien-9-oic acid

3′,5,5′-Tri-Me ether, 9′-O-β-D-glucopyranoside

Spicatolignan A

[1010821-33-6]

$C_{27}H_{30}O_{13}$ 562.524

Constit. of *Mentha spicata*. Needles (MeOH). Mp 175 - 177°.

Zheng, J. *et al.*, *J. Asian Nat. Prod. Res.*, 2007, **9**, 431-435 (*Spicatolignan A*)

7,9′-Epoxy-3,3′,4,4′,5′-pentahydroxy-8,8′-lignan-9,7′-olide E230

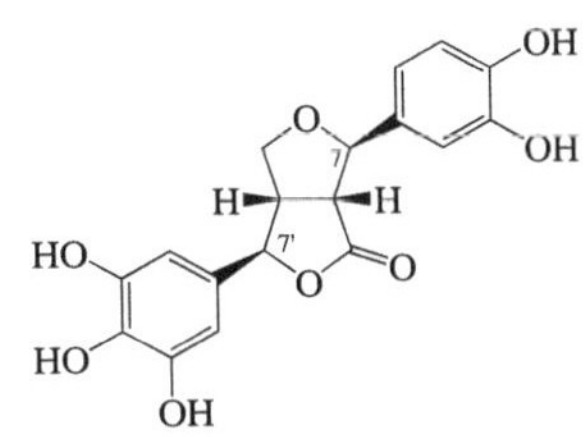

(7R,7′R*,8R*,8′S*)-form*

3,3′,4,4′,5′-Pentahydroxy-7,9′-epoxylignan-9,7′-olide

$C_{18}H_{16}O_8$ 360.318

(7R*,7′R*,8R*,8′S*)-form

Syzygiresinol B [1151670-59-5]

Constit. of the stem bark of *Syzygium cumini*. Pale yellow cryst. ($CHCl_3$/petrol). Mp 250 - 251°. λ_{max} 240 (log ε 6.5); 370 (log ε 6.3) (MeOH)

3′,5′-Di-Me ether

7,9′-Epoxy-3,4,4′-trihydroxy-3′,5′-dimethoxy-8,8′-lignan-9,7′-olide. 3,4,4′-Trihydroxy-3′,5-dimethoxy-7,9′-epoxylignan-9,7′-olide

[408513-60-0]

$C_{20}H_{20}O_8$ 388.372

Constit. of *Saussurea medusa*. Amorph. powder. $[\alpha]^{25}_{D}$ -16 (c, 0.7 in MeOH). λ_{max} 232 (log ε 4.08); 280 (log ε 3.73) (MeOH)

3,3′,5′-Tri-Me ether

7,9′-Epoxy-4,4′-dihydroxy-3,3′,5′-trimethoxy-8,8′-lignan-9,7′-olide. 4,4′-Dihydroxy-3,3′,5′-trimethoxy-7,9′-epoxylignan-9,7′-olide. ***Mayuenolide***

[474511-43-8]

$C_{21}H_{22}O_8$ 402.399

Constit. of *Coix lachryma-jobi* var. *ma-yuen*. Light yellow cryst.. Mp 215 - 217°. $[\alpha]^{25}_{D}$ -25.5 (no solvent reported). λ_{max} 280 (log ε 3.1) (MeOH)

(7R*,7′R*,8S*,8′R*)-form

3′,4,5′-Tri-Me ether

7,9′-Epoxy-3,4′-dihydroxy-3′,4,5′-trimethoxy-8,8′-lignan-9,7′-olide. 3,4′-Dihydroxy-3′,4,5′-trimethoxy-7,9′-epoxylignan-9,7′-olide. ***Amislignol***

[1707310-81-3]

$C_{21}H_{22}O_8$ 402.399

Constit. of *Amischotolype hispida*. Solid. $[\alpha]^{25}_{D}$ -5.2 (c, 0.04 in $CHCl_3$). λ_{max} 210 (log ε 4.62); 232 (sh); 280 (log ε 3.68) (MeOH).

Kuo, C.-C. *et al.*, *J. Agric. Food Chem.*, 2002, **50**, 5850-5855 (*Mayuenolide*).

Duan, H. *et al.*, *Phytochemistry*, 2002, **59**, 85-90 (*Saussurea medusa constit*).

Mir, Q.-Y. *et al.*, *Nat. Prod. Res.*, 2009, **23**, 422-430 (*Syzygiresinol B*).

Chang, C.-W. *et al.*, *Helv. Chim. Acta*, 2015, **98**, 347-358 (*Amislignol*)

7,9′-Epoxy-3,3′,4′,5,5′-pentahydroxy-8,8′-lignan-9,7′-olide E231

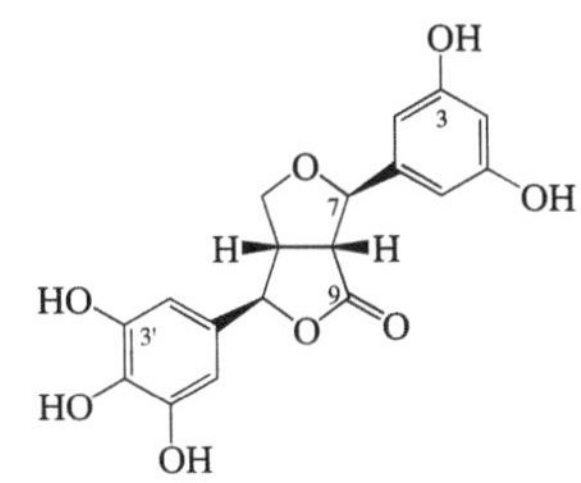

$C_{18}H_{16}O_8$ 360.318

Structures in this series may need revision. Phenyl 3,5-dioxygenated substn. pattern in doubt as a result of work done by Yang *et al* in 2015

(7R,7′R,8R,8′S)-form

3,3′,4′-Tri-Me ether

7,9′-Epoxy-3,3′-dihydroxy-4′,5,5′-trimethoxy-8,8′-lignan-9,7′-olide. ***Graminone B***

[161407-73-4]

$C_{21}H_{22}O_8$ 402.399

Constit. of the rhizomes of *Imperata cylindrica* and roots of *Fagopyrum tataricum*. Vasodilator. Oil. $[\alpha]^{25}_{D}$ -4 (c, 0.1 in $CHCl_3$).

Matsunaga, K. *et al.*, *J. Nat. Prod.*, 1994, **57**, 1734-1736 (*Graminone B*).

Hu, C-L. *et al.*, *Chem. Nat. Compd. (Engl. Transl.)*, 2012, **48**, 303-304 (*Fagopyrum tataricum constit*).

Yang, Y.-N. *et al.*, *J. Nat. Prod.*, 2015, **78**, 705-711 (*pmr, struct*)

7,9′-Epoxy-3,3′,4,4′,5-pentahydroxy-8,8′-lignan-7′-one E232

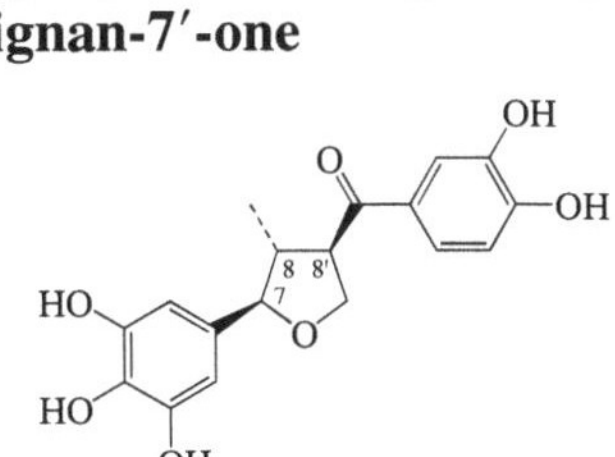

(7R,8R,8′R)-form

3,3′,4,4′,5-Pentahydroxy-7′-oxo-7,9′-epoxylignan
$C_{18}H_{18}O_7$ 346.335
(7R,8R,8′R)-form
Penta-Me ether
7,9′-Epoxy-3,3′,4,4′,5-pentamethoxy-8,8′-lignan-7′-one. ***Acortatarinowin J***
$C_{23}H_{28}O_7$ 416.47
Constit. of the rhizomes of *Acorus tatarinowii*. Block cryst.. Mp 158 - 159°. $[\alpha]_D^{20}$ -1.4 (c, 0.4 in CH_2Cl_2). λ_{max} 207 (log ε 5.19); 231 (log ε 4.60); 278 (log ε 4.11); 303 (log ε 3.95) (MeOH)
(7RS,8RS,8′RS)-form
Penta-Me ether
Tatarinoid E
[2222357-83-5]
$C_{23}H_{28}O_7$ 416.470
Constit. of the rhizomes of *Acorus tatarinowii*. $[\alpha]_D^{27}$ +2.2 (c, 0.5 in MeOH). Partial racemate. λ_{max} 209 (log ε 4.63); 228 (log ε 4.35); 277 (log ε 4.09); 310 (log ε 3.96) (MeOH).

Lu, Y. *et al.*, *Sci. Rep.*, 2016, **6**, 22909 (*Acortatarinowin J, cryst struct*).
Zhang, W.-Y. *et al.*, *Bioorg. Med. Chem. Lett.*, 2018, **28**, 814-819 (*Tatarinoid E*)

2,8′-Epoxy-3,3′,4′,5,5′-pentahydroxy-1,7′-lign-8-en-4(1*H*)-one E233

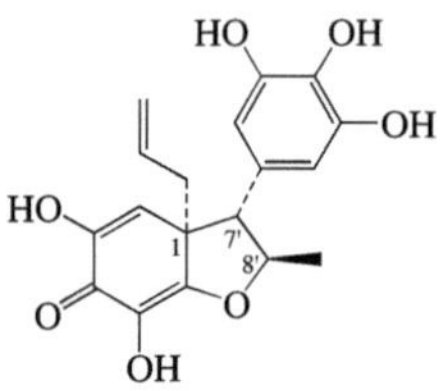

$C_{18}H_{18}O_7$ 346.335
(1R*,7′S*,8′R*)-form
Penta-Me ether
2,8′-Epoxy-3,3′,4′,5,5′-pentamethoxy-1,7′-lign-8-en-4(1H)-one. ***Chrysophyllon IB***
[106897-10-3]
$C_{23}H_{28}O_7$ 416.47
Constit. of *Licaria chrysophylla*. Oil
3′,4′-Methylene, 3,5,5′-tri-Me ether
2,8′-Epoxy-3,3′,5-trimethoxy-4′,5′-methylenedioxy-1,7′-lign-8-en-4(1H)-one. ***Chrysophyllon IA***
[106897-09-0]
$C_{22}H_{24}O_7$ 400.427
Constit. of *Licaria chrysophylla*. Oil. λ_{max} 269 (ε 8700) (MeOH)
7′,8′-Didehydro, penta-Me ether
2,8′-Epoxy-3,3′,4′,5,5′-pentamethoxy-1,7′-ligna-7′,8-dien-4(1H)-one. ***Chrysophyllon IVB***
[1454880-24-0]
$C_{23}H_{26}O_7$ 414.454
Constit. of *Licaria chrysophylla*.

Lopes, M.N. *et al.*, *Phytochemistry*, 1986, **25**, 2609-2612 (*Chrysophyllons IA,IB*).
Bezerra, A. de M. *et al.*, *Quim. Nova*, 2012, **35**, 2226-2228 (*Chrysophyllon IVB*)

4,8′-Epoxy-1,3′,4′,5,5′-pentahydroxy-3,7′-lign-8-en-6(1*H*)-one E234

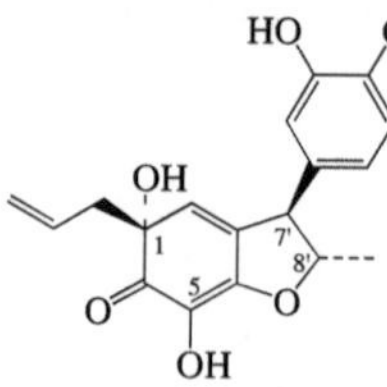

$C_{18}H_{18}O_7$ 346.335
(1R*,7′R*,8′S*)-form
Penta-Me ether
4,8′-Epoxy-1,3′,4′,5,5′-pentamethoxy-3,7′-lign-8-en-6(1H)-one. ***Chrysophyllon IIB***
[106897-12-5]
$C_{23}H_{28}O_7$ 416.47
Constit. of *Licaria chrysophylla*. Cryst. (C_6H_6). Mp 138 - 140°. λ_{max} 334 (ε 7200) (MeOH)
3′,4′-Methylene, 1,5,5′-tri-Me ether
4,8′-Epoxy-1,3′,5-trimethoxy-4′,5′-methylenedioxy-3,7′-lign-8-en-6(1H)-one. ***Chrysophyllon IIA***
[106897-11-4]
$C_{22}H_{24}O_7$ 400.427
Constit. of *Licaria chrysophylla*. Oil.

Lopes, M.N. *et al.*, *Phytochemistry*, 1986, **25**, 2609-2612 (*Chrysophyllons IIA,IIB*)

4,7′-Epoxy-3′,4′,5,5′,9′-pentahydroxy-9-nor-3,8′-lignan-7-one E235

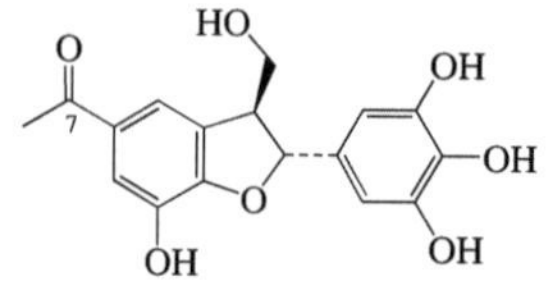

5-Acetyl-2,3-dihydro-7-hydroxy-3-(hydroxymethyl)-2-(3,4,5-trihydroxyphenyl)benzofuran.
4,7′-Epoxy-3′,4′,5,5′,9′-pentahydroxy-9-nor-7-oxo-3,8′-neolignan
$C_{17}H_{16}O_7$ 332.308
(7′R,8′S)-form
3′,5,5′-Tri-Me ether
5-Acetyl-2,3-dihydro-3-(hydroxymethyl)-2-(4-hydroxy-3,5-dimethoxyphenyl)-7-methoxybenzofuran.
4,7′-Epoxy-4′,9′-dihydroxy-3′,5,5′-trimethoxy-9-nor-3,8′-lignan-7-one.
4,7′-Epoxy-4′,9′-dihydroxy-3′,5,5′-trimethoxy-9-nor-7-oxo-3,8′-lignan
[849431-11-4]
$C_{20}H_{22}O_7$ 374.389
Constit. of *Selaginella tamariscina*. $[\alpha]_D^{25}$ -9.6 (c, 1.8 in MeOH). λ_{max} 210; 280 (no solvent reported)
3′,5,5′-Tri-Me ether, 4′-O-β-D-glucopyranoside
Tamariscinoside E
$C_{26}H_{32}O_{12}$ 536.53
Constit. of *Selaginella tamariscina*. Inhibitor of NO prodn. in LPS-stimulated RAW 264.7 cells. Amorph. powder
3′,5,5′-Tri-Me ether, 4′-O-[β-D-glucopyranosyl-(1→6)-β-D-glucopyranoside]
Tamariscinoside F
$C_{32}H_{42}O_{17}$ 698.671
Constit. of *Selaginella tamariscina*. Liq.
3′,5,5′-Tri-Me ether, 9′-O-β-D-glucopyranoside
Tamariscinoside D
$C_{26}H_{32}O_{12}$ 536.53
Constit. of *Selaginella tamariscina*. Inhibitor of NO prodn. in LPS-stimulated RAW 246.7 cells. Amorph. powder.

Bi, Y.-F. *et al.*, *Yaoxue Xuebao*, 2004, **39**, 41-45 (*3′,5,5′-tri-Me ether*).
Dat, L.D. *et al.*, *Bioorg. Med. Chem. Lett.*, 2017, **27**, 524-529 (*Tamariscinosides D,E,F, activity*)

7,7′-Epoxy-3,3′,4,4′,5′-pentahydroxy-7,8-seco-8,8′-lignan-7,8-dione E236

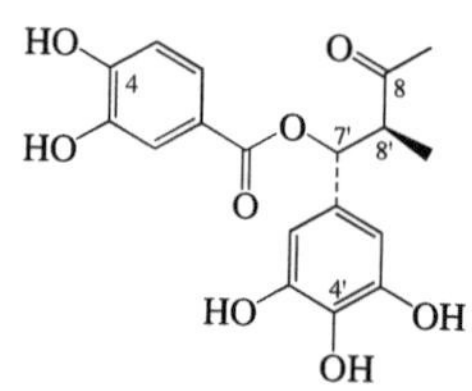

(7′R,8′S)-form

2-Methyl-3-oxo-1-(3,4,5-trihydroxyphenyl)butyl 3,4-dihydroxybenzoate, CAS
$C_{18}H_{18}O_8$ 362.334
(7′R,8′S)-form
3,3′,4′-Tri-Me ether
7,7′-Epoxy-3′,4-dihydroxy-3,4′,5′-trimethoxy-7,8-seco-8,8′-lignan 7,8-dione. ***Manglisin G***
[1621958-67-5]
$C_{21}H_{24}O_8$ 404.415
Constit. of carpels of *Manglietiastrum sinicum*. Oil. $[\alpha]^{26}_{D}$ +43 (c, 0.2 in MeOH). λ_{max} 264 (log ε 4.10); 291 (log ε 3.87) (MeOH)
(7′R*,8′S*)-form
3′,4′-Di-Me ether
Neglectahenol E
$C_{20}H_{22}O_8$ 390.388
Constit. of the fruit of *Schisandra neglecta*. Amorph. powder. $[\alpha]^{25}_{D}$ +28.6 (c, 0.2 in MeOH). λ_{max} 272 (log ε 3.48); 356 (log ε 1.24) (MeOH)
3,3′,4,4′,5′-Penta-Me ether
7,7′-Epoxy-3,3′,4,4′,5′-pentamethoxy-7,8-seco-8,8′-lignan-7,8-dione
[2089124-42-3]
$C_{23}H_{28}O_8$ 432.469
Constit. of *Saururus chinensis*. Amorph. powder. $[\alpha]^{23}_{D}$ -105 (c, 0.01 in MeOH). λ_{max} 206 (log ε 4.38); 263 (log ε 3.77); 293 (log ε 3.50) (MeOH)
3,4-Methylene, 3′,4′-di-Me ether
7,7′-Epoxy-3′-hydroxy-4′,5′-dimethoxy-3,4-methylenedioxy-7,8-seco-8,8′-lignan-7,8-dione. ***Marphenol B***
[1260657-87-1]
$C_{21}H_{22}O_8$ 402.399
Constit. of *Schisandra neglecta* and *Schisandra wilsoniana*. Amorph. powder. $[\alpha]^{26}_{D}$ +18.4 (c, 0.28 in MeOH). λ_{max} 205 (log ε 5.08); 275 (log ε 3.24); 350 (log ε 1.68) (MeOH)
3,4-Methylene, tri-Me ether
Neglectahenol F. Turbacenlignan A
$C_{22}H_{24}O_8$ 416.426
Constit. of *Nectandra turbacensis* and *Schisandra neglecta*. Amorph. powder. $[\alpha]^{25}_{D}$ +25.6 (c, 0.2 in MeOH) (Neglectahenol F). $[\alpha]^{24}_{D}$ +33 (c, 0.5 in $CHCl_3$) (Turbacenlignan A). λ_{max} 273 (log ε 3.42); 355 (log ε 1.08) (MeOH)
8ξ-Alcohol, 3,3′,4,4′-tetra-Me ether
7,7′-Epoxy-3′,8-dihydroxy-3,4,4′,5′-tetramethoxy-7,8-seco-8,8′-lignan-7-one. ***Neglectahenol D***
[1465784-56-8]
$C_{22}H_{28}O_8$ 420.458
Constit. of *Schisandra neglecta*. Exhibits modest anti-HIV-1 activity. Amorph. powder. $[\alpha]^{26}_{D}$ +16.9 (c, 0.2 in MeOH). Possesses (7′*R**,8′*R**)-config. due to change in Cahn-Ingold-Prelog priorities. λ_{max} 210 (log ε 4.15); 275 (log ε 3.87); 340 (log ε 2.69) (MeOH)
8ξ-Alcohol, 3,3′,4,4′,5′-penta-Me ether
7,7′-Epoxy-8-hydroxy-3,3′,4,4′,5′-pentamethoxy-7,8-seco-8,8′-lignan-7-one. ***Neglectahenol C***
[1465784-55-7]
$C_{23}H_{30}O_8$ 434.485
Constit. of *Schisandra neglecta*. Exhibits modest anti-HIV-1 activity. Amorph. powder. $[\alpha]^{26}_{D}$ +18.9 (c, 0.21 in MeOH). Possesses (7′*R**,8′*R**)-config. due to change in Cahn-Ingold-Prelog priorities. λ_{max} 210 (log ε 4.21); 275 (log ε 3.93); 340 (log ε 2.76) (MeOH)
8ξ-Alcohol, 3,4-methylene, 3′,4′-di-Me ether
7,7′-Epoxy-3′,8-dihydroxy-4′,5′-dimethoxy-3,4-methylenedioxy-7,8-seco-8,8′-lignan-7-one. ***Neglectahenol A***
[1465784-51-3]
$C_{21}H_{24}O_8$ 404.415
Constit. of *Schisandra neglecta*. Amorph. powder. $[\alpha]^{26}_{D}$ +22.6 (c, 0.22 in MeOH). Possesses (7′*R**,8′*R**)-config. due to change in Cahn-Ingold-Prelog priorities. λ_{max} 210 (log ε 4.26); 276 (log ε 3.88); 340 (log ε 2.78) (MeOH).

Zhang, X.-J. *et al.*, *Chem. Biodiversity*, 2010, **7**, 2692-2701 (*Marphenol B*).
Gao, X. *et al.*, *J. Braz. Chem. Soc.*, 2012, **23**, 1853-1857 (*Neglectahenols A,C,D, activity*).
Ding, J.-Y. *et al.*, *J. Nat. Prod.*, 2014, **77**, 1800-1805 (*Manglisin G*).
Wang, H. *et al.*, *J. Asian Nat. Prod. Res.*, 2015, **17**, 27-32 (*Neglectahenols E,F*).
Macias-Villamizar, V. *et al.*, *Rec. Nat. Prod.*, 2016, **10**, 654-658 (*Turbacenlignan A*).
Li, C. *et al.*, *Nat. Prod. Res.*, 2017, **31**, 1598-1603 (*Saururus chinensis constit*)

2,7′-Epoxy-1,2,5,6-tetrahydro-3,5-dimethoxy-3′,4′-methylenedioxy-1,8′-lign-8-en-2-ol E237

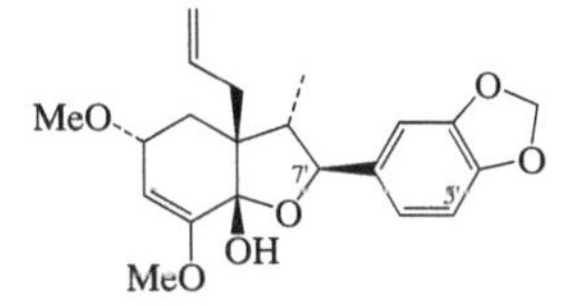

Relative Configuration

3,3a,4,5-Tetrahydro-7a-hydroxy-5,7-dimethoxy-3-methyl-2-(3,4-methylenedioxyphenyl)-3a-(2-propenyl)-2H-benzofuran. 2,7′-Epoxy-1,2,5,6-tetrahydro-2-hydroxy-3,5-dimethoxy-3′,4′-methylenedioxy-1,8′-neolign-8-ene
[1181279-98-0]
$C_{21}H_{26}O_6$ 374.433
Lignan numbering shown. Constit. of the leaves of *Ocotea catharinensis*. Viscous oil
7′-Epimer, 5′-methoxy
2,7′-Epoxy-1,2,5,6-tetrahydro-3,5,5′-trimethoxy-3′,4′-methylenedioxy-1,8′-lign-8-en-2-ol. 2,7′-Epoxy-1,2,5,6-tetrahydro-2-hydroxy-3,5,5′-trimethoxy-3′,4′-methylenedioxy-1,8′-lign-8-ene
[1181280-00-1]
$C_{22}H_{28}O_7$ 404.459
Constit. of the leaves of *Ocotea catharinensis*. Viscous oil.

Funasaki, M. *et al.*, *J. Braz. Chem. Soc.*, 2009, **20**, 853-859 (*Ocotea catharinensis constits*)

2,4′-Epoxy-1,2,3,4-tetrahydro-1,3′-ligna-8,8′-dien-4-ol E238

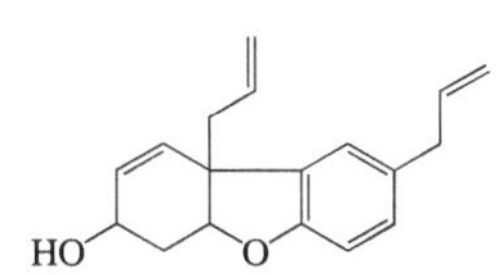

3,4,4a,9b-Tetrahydro-8,9b-di-2-propenyl-3-dibenzofuranol, CAS. *2,4′-Epoxy-1,2,3,4-tetrahydro-4-hydroxy-1,3′-neoligna-8,8′-diene*
[2139312-29-9]
$C_{18}H_{20}O_2$ 268.356
Constit. of *Magnolia grandiflora*. Exhibits weak antiplasmodial activity. Oil. $[\alpha]^{21}_{D}$ -33.8 (c, 0.02 in MeOH). λ_{max} 225; 285 (MeCN aq.).

Latif, A. *et al.*, *Chem. Biodiversity*, 2017, **14**, e1700209:1-7 (*Magnolia grandiflora constit, activity*)

4,7′-Epoxy-3,4,5,6-tetrahydro-3,8′-lign-8-ene-3,3′,4,4′,6-pentol E239

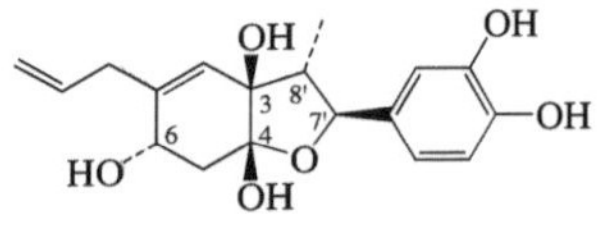

(3S,4R,6S,7′S,8′R)-form

4,7′-Epoxy-3,4,5,6-tetrahydro-3,3′,4,4′,6-pentahydroxy-3,8′-neolign-8-ene

$C_{18}H_{22}O_6$ 334.368

(3S,4R,6S,7′S,8′R)-form

3,3′,4,4′-Tetra-Me ether

4,7′-Epoxy-3,4,5,6-tetrahydro-3,3′,4,4′-tetramethoxy-3,8′-lign-8-en-6-ol. 4,7′-Epoxy-3,4,5,6-tetrahydro-6-hydroxy-3,3′,4,4′-tetramethoxy-3,8′-lign-8-ene

[61307-70-8]

$C_{22}H_{30}O_6$ 390.476

Constit. of *Piper wightii*. Mp 72.5 - 73.5° (semisynthetic). $[\alpha]^{22}_{D}$ -228 (c, 0.3 in MeOH)

3′,4′-Methylene, 3,4-di-Me ether

4,7′-Epoxy-3,4,5,6-tetrahydro-3,4-dimethoxy-3′,4′-methylenedioxy-3,8′-lign-8-en-6-ol. 4,7′-Epoxy-3,4,5,6-tetrahydro-6-hydroxy-3,4-dimethoxy-3′,4′-methylenedioxy-3,8′-lign-8-ene. ***Kadsurin B****. Schmiditin*

[99257-52-0]

$C_{21}H_{26}O_6$ 374.433

Constit. of *Piper kadsura*, *Piper schmidtii*, *Piper attenuatum* and *Piper argyrophylum*. Cryst.. Mp 101 - 102°. $[\alpha]^{22}_{D}$ -18.5. $[\alpha]^{22}_{D}$ -217.1 (c, 0.47 in MeOH). Struct. of Schmiditin revised in 1995. λ_{max} 213; 234; 285 (MeOH)

6-Ketone, 3,3′,4-tri-Me ether

4,7′-Epoxy-3,4-dihydro-4′-hydroxy-3,3′,4-trimethoxy-3,8′-lign-8-en-6(5H)-one. 4′-O-Demethylpiperenone. ***Liliflone***

[87402-80-0]

$C_{21}H_{26}O_6$ 374.433

Constit. of *Magnolia liliflora*. Viscous oil

6-Ketone, 3,3′,4,4′-tetra-Me ether

2-(3,4-Dimethoxyphenyl)-3,3a,7,7a-tetrahydro-3a,7a-dimethoxy-3-methyl-5-(2-propenyl)-6(2H)-benzofuranone. 4,7′-Epoxy-3,4-dihydro-3,3′,4,4′-tetramethoxy-3,8′-lign-8-en-6(5H)-one. ***Piperenone***

[57625-31-7]

$C_{22}H_{28}O_6$ 388.46

Constit. of *Piper kadsura*, *Piper cubeb* and *Piper cuneifolium*. Insect antifeedant. Mp 86 - 88°. $[\alpha]_D$ -129 (c, 1.16 in MeOH). λ_{max} 231; 280; 285 (MeOH)

6-Ketone, 3′,4′-methylene, 3,4-di-Me ether

4,7′-Epoxy-3,4-dihydro-3,4-dimethoxy-3′,4′-methylenedioxy-3,8′-lign-8-en-6(5H)-one. ***Kadsurin A‡***

[99340-07-5]

$C_{21}H_{24}O_6$ 372.417

Constit. of *Piper cubeb*, *Piper futokadsura*, *Piper schmidtii*, *Piper wightii* and *Magnolia denudata*. Oil. $[\alpha]^{22}_{D}$ -104.4 (c, 0.46 in $CHCl_3$). Struct. of *Piper futokadsura* isolate in doubt. λ_{max} 287 (log ε 4.91) (MeOH)

(3R*,4R*,7′S*,8′S*)-form

6-Ketone, 3′,4′-methylene, 3,4-di-Me ether

[155418-00-1]

$C_{21}H_{24}O_6$ 372.417

Constit. of *Piper wightii*. Viscous yellow oil. $[\alpha]^{24}_{D}$ +60.7 (c, 0.2 in $CHCl_3$). Relative config. only known

(3R*,4S*,7′S*,8′R*)-form

1S,2-Dihydro, 6-ketone, 3,4-methylene, 3′,4′-di-Me ether*

4,7′-Epoxy-1,2,4,5-tetrahydro-3′,4′-dimethoxy-3,4-methylenedioxy-3,8′-lign-8-en-6(3H)-one. ***7,4′-Furanobifidenone***

[2064109-54-0]

$C_{21}H_{26}O_6$ 374.433

Constit. of the leaves of *Beilschmiedia* sp..

Matsui, K. *et al.*, *Agric. Biol. Chem.*, 1976, **40**, 1113-1118 (*Piperenone, synth*).

Matsui, K. *et al.*, *Bull. Chem. Soc. Jpn.*, 1976, **49**, 62-65 (*Piperenone, cryst struct*).

Iida, T. *et al.*, *Phytochemistry*, 1983, **22**, 763-766 (*Liliflone, Piperenone*).

Chang, M.N. *et al.*, *Phytochemistry*, 1985, **24**, 2079-2082 (*Kadsurins A,B*).

Tyagi, O.D. *et al.*, *Phytochemistry*, 1993, **32**, 445-448 (*Kadsurin A*).

Tyagi, O.D. *et al.*, *Acta Chem. Scand.*, 1994, **48**, 1007-1011 (*Piper wightii constit*).

Tyagi, O.D. *et al.*, *Acta Chem. Scand.*, 1995, **49**, 142-148 (*Kadsurin B, Schmiditin*).

Koul, J.L. *et al.*, *Phytochemistry*, 1996, **41**, 1097-1099 (*Kadsurin A, Piperenone*).

Boll, P.M. *et al.*, *Recl. Trav. Chim. Pays-Bas*, 1996, **115**, 9-12 (*Piper wightii constit*).

Williams, R.B. *et al.*, *J. Nat. Prod.*, 2017, **80**, 616-624 (*7,4′-Furanobifidenone*)

4,7′-Epoxy-3′,4′,5,5′-tetrahydroxy-8,9-dinor-3,8′-lignan-7-al E240

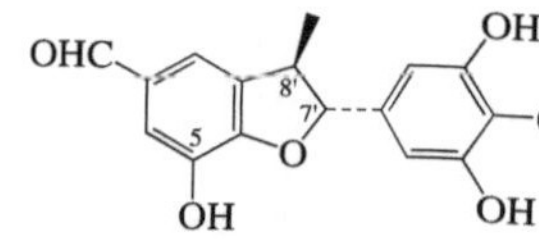

(7′R,8′R)-form

2,3-Dihydro-7-hydroxy-3-methyl-2-(3,4,5-trihydroxyphenyl)-5-benzofurancarboxaldehyde. 4,7′-Epoxy-3′,4′,5,5′-tetrahydroxy-8,9-dinor-7-oxo-3,8′-neolignan

$C_{16}H_{14}O_6$ 302.282

(7′R,8′R)-form

3′,4′-Methylene, 5,5′-di-Me ether

4,7′-Epoxy-3′,5-dimethoxy-4′,5′-methylenedioxy-8,9-dinor-3,8′-lignan-7-al. 4,7′-Epoxy-3′,5-dimethoxy-4′,5′-methylenedioxy-8,9-dinor-7-oxo-3,8′-lignan. ***Myrisfrageal B***

$C_{19}H_{18}O_6$ 342.347

Constit. of the seeds of *Myristica fragrans*. NO prodn. inhibitor. Yellowish oil. $[\alpha]^{20}_{D}$ +54 (c, 1 in CH_3Cl). λ_{max} 212 (log ε 4.14); 236 (log ε 3.72); 300 (log ε 3.57) (MeOH)

(7′S,8′S)-form

3′,5-Di-Me ether

4,7′-Epoxy-3′,4′-dihydroxy-5,5′-dimethoxy-8,9-dinor-3,8′-lignan-7-al. 4,7′-Epoxy-3′,4′-dihydroxy-5,5′-dimethoxy-8,9-dinor-7-oxo-3,8′-lignan. ***Perseal E***

[269739-06-2]

$C_{18}H_{18}O_6$ 330.336

Constit. of the leaves of *Persea obovatifolia*

3′,4′,5-Tri-Me ether

4,7′-Epoxy-3′-hydroxy-4′,5,5′-trimethoxy-8,9-dinor-3,8′-lignan-7-al. 4,7′-Epoxy-3′-hydroxy-4′,5,5′-trimethoxy-8,9-dinor-7-oxo-3,8′-lignan. ***Obovatinal***

[186033-35-2]

$C_{19}H_{20}O_6$ 344.363

Constit. of the leaves of *Persea obovatifolia*. Exhibits cytotoxic activity. Oil. $[\alpha]_D$ -37 (c, 0.1 in $CHCl_3$). λ_{max} 230 (log ε 4.44); 280

(log ε 4.22); 310 (log ε 4.12) (EtOH).

Tsai, I.-L. *et al.*, *Phytochemistry*, 1996, **43**, 1261-1263 (*Obovatinal*).
Tsai, I.-L. *et al.*, *Chin. Pharm. J. (Taipei)*, 1999, **51**, 335-345 (*Perseal E*).
Cao, G.-Y. *et al.*, *Food Chem. Toxicol.*, 2013, **62**, 167-171 (*Myrisfrageal B, activity*)

4,7′-Epoxy-3′,4′,5,9′-tetrahydroxy-8,9-dinor-3,8′-lignan-7-oic acid E241

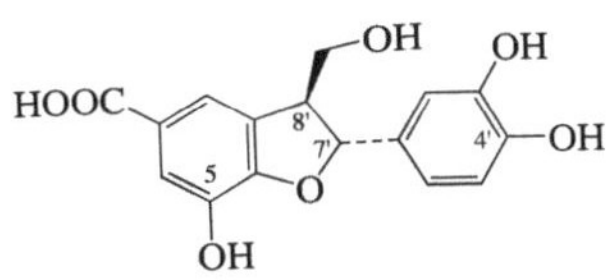

(7′R,8′S)-form

2-(3,4-Dihydroxyphenyl)-2,3-dihydro-7-hydroxy-3-(hydroxymethyl)-5-benzofurancarboxylic acid
$C_{16}H_{14}O_7$ 318.281
(7′R,8′S)-form
3′,5-Di-Me ether
4,7′-Epoxy-4′,9′-dihydroxy-3′,5-dimethoxy-8,9-dinor-3,8′-lignan-7-oic acid. **(-)-Ceplignan**
[185244-77-3]
$C_{18}H_{18}O_7$ 346.335
Constit. of *Acorus tatarinowii*, *Boreava orientalis* and *Sambucus williamsii*. Pale yellow powder. Mp 82.5 - 85°. $[\alpha]^{20}_{D}$ -5.4 (c, 0.16 in MeOH). λ_{max} 280 (log ε 2.14) (MeOH)
3′,5-Di-Me ether, 4′-O-β-D-glucopyranoside
[1309791-02-3]
$C_{24}H_{28}O_{12}$ 508.476
Constit. of the aerial parts of *Urena lobata*. Pale yellow powder. $[\alpha]^{26}_{D}$ -49.2 (c, 0.4 in H_2O). λ_{max} 264 (log ε 3.14) (MeOH)
3′,5-Di-Me ether, 4′-O-α-L-rhamnopyranoside
Gelsemiunoside B
[1170710-53-8]
$C_{24}H_{28}O_{11}$ 492.477
Constit. of *Gelsemium elegans*. Pale yellow powder. Mp 138 - 140°. $[\alpha]^{22}_{D}$ -32.8 (c, 0.14 in MeOH). λ_{max} 223 (log ε 4.13); 324 (log ε 3.64) (MeOH)
3′,5-Di-Me ether, Me ester
(-)-Curlignan
$[\alpha]^{20}_{D}$ -74.2 (c, 0.1 in MeOH)
3′,5-Di-Me ether, Me ester, 9′-O-β-D-glucopyranoside
[1021524-76-4]
$C_{25}H_{30}O_{12}$ 522.503
Constit. of *Chione venosa* var. *venosa*. $[\alpha]_{D}$ -34 (c, 0.05 in MeOH). λ_{max} 228 (sh); 278; 300 (sh) (MeOH aq.)
3′,4′-Methylene, 5-Me ether, Me ester
Toonin C
[1481547-09-4]
$C_{19}H_{18}O_7$ 358.346
Constit. of the roots of *Toona sinensis*. Powder (MeOH). $[\alpha]^{25}_{D}$ -22.6 (c, 0.1 in MeOH)
7-Aldehyde
4,7′-Epoxy-3′,4′,5,9′-tetrahydroxy-8,9-dinor-3,8′-lignan-7-al. 4,7′-Epoxy-3′,4′,5,9′-tetrahydroxy-8,9-dinor-7-oxo-3,8′-lignan
7-Aldehyde, 3′-Me ether
4,7′-Epoxy-4′,5,9′-trihydroxy-3′-methoxy-8,9-dinor-3,8′-lignan-7-al. 4,7′-Epoxy-4′,5,9′-trihydroxy-3′-methoxy-8,9-dinor-7-oxo-3,8′-lignan. **Hierochin C**
[650600-31-0]
$C_{17}H_{16}O_6$ 316.309
Constit. of *Anastatica hierochuntica*. Pale yellow oil. $[\alpha]^{27}_{D}$ -60.1 (c, 1 in MeOH). λ_{max} 235 (log ε 4.28); 289 (log ε 4.1) (MeOH)
7-Aldehyde, 3′,5-di-Me ether
4,7′-Epoxy-4′,9′-dihydroxy-3′,5-dimethoxy-8,9-dinor-3,8′-lignan-7-al. 4,7′-Epoxy-4′,9′-dihydroxy-3′,5-dimethoxy-8,9-dinor-7-oxo-3,8′-lignan. **(+)-Ficusal**
[321991-55-3]
[70379-30-5] (unspecified stereochem.) [73022-33-0] (*trans*-form)
$C_{18}H_{18}O_6$ 330.336
Constit. of *Artemisia anomala*, *Catunaregam spinosa*, *Coix lachryma-jobi*, *Ficus microcarpa* and *Sambucus williamsii*. Significant inhibitor of MDA-MB-231 cells. Pale yellow oil. $[\alpha]^{25}_{D}$ +3 (c, 0.18 in $CHCl_3$). λ_{max} 235 (log ε 4.6); 284 (log ε 4.15) (MeOH)
7-Aldehyde, 3′,5-di-Me ether, 4′-O-β-D-glucopyranoside
Ligusticumaldehyde A
[2573174-19-1]
$C_{24}H_{28}O_{11}$ 492.477
Constit. of the rhizomes of *Ligusticum chuanxiong*. Amorph. powder. λ_{max} 205 (log ε 4.69); 231 (log ε 4.47); 307 (log ε 4.41) (MeOH)
7-Aldehyde, 3′,5-di-Me ether, 9′-O-(3-methylbutanoyl)
(-)-Brachangobinan I
[2357185-56-7]
Amorph. solid. $[\alpha]^{28}_{D}$ -51 (c, 0.14 in MeOH)
(7′S,8′R)-form
3′-Me ether
4,7′-Epoxy-4′,5,9′-trihydroxy-3′-methoxy-8,9-dinor-3,8′-lignan-7-oic acid
3′-Me ether, Me ester
Norcurlignan
[1432056-53-5]
$C_{18}H_{18}O_7$ 346.335
Constit. of the roots of *Liriope muscari*. Amorph. solid. $[\alpha]^{25}_{D}$ +52.3 (c, 0.016 in MeOH). λ_{max} 274 (log ε 4.3) (MeOH)
3′,5-Di-Me ether
(+)-Ceplignan
[185244-78-4]
$C_{18}H_{18}O_7$ 346.335
Constit. of *Acorus tatarinowii*, *Cephalomappa sinensis*, *Gardenia jasminoides* and *Inula cappa*. $[\alpha]^{20}_{D}$ +29 (c, 0.2 in MeOH). λ_{max} 223 (log ε 4.46); 273 (log ε 4.13) (MeOH). [neutral] λ_{max} 275 (MeOH)
3′,5-Di-Me ether, β-D-glucopyranosyl ester
Pomegralignan
[1562492-32-3]
$C_{24}H_{28}O_{12}$ 508.476
Constit. of the arils of *Punica granatum*. Off-white powder. $[\alpha]^{23}_{D}$ +44.9 (c, 0.5 in MeOH). λ_{max} 233 (log ε 4.29); 280 (log ε 4.15) (MeOH)
3′,5-Di-Me ether, [3,4,5-trihydroxybenzoyl-(→1)-β-D-glucopyranos-6-yl] ester
Punicatannin C
[1620439-20-4]
$C_{31}H_{32}O_{16}$ 660.581
Constit. of the flowers of *Punica granatum*. Amorph. solid. $[\alpha]^{20}_{D}$ +66 (c, 0.03 in MeOH). λ_{max} 218 (log ε 3.92); 278 (log ε 4.1) (MeOH)
3′,5-Di-Me ether, Me ester
(+)-Curlignan
[220736-54-9]
$C_{19}H_{20}O_7$ 360.362
Amorph. powder. $[\alpha]^{20}_{D}$ +70.6 (c, 0.1 in MeOH). λ_{max} 228 (sh); 276 (log ε 4.19); 298 (sh) (MeOH).

[base] λ_{max} 231 (sh); 273 (log ε 4.26); 303 (sh) (MeOH/NaOH)

3′,4′-Methylene, 5-Me ether
4,7′-Epoxy-9′-hydroxy-5-methoxy-3′,4′-methylenedioxy-8,9-dinor-3,8′-lignan-7-oic acid. ***Cedralin A***
[1220883-78-2]
$C_{18}H_{16}O_7$ 344.319
Constit. of the leaves of *Cedrela sinensis*. Amorph. powder. $[\alpha]^{25}_{D}$ -32 (c, 0.2 in $CHCl_3$). λ_{max} 278 (log ε 3.42) (MeOH)

3′,4′-Methylene, 5-Me ether, Me ester, 9′-O-β-D-glucopyranoside
Cedralin B
[1220883-79-3]
$C_{25}H_{28}O_{12}$ 520.487
Constit. of the leaves of *Cedrela sinensis*. Amorph. powder. $[\alpha]^{25}_{D}$ -6.7 (c, 0.2 in MeOH). λ_{max} 280 (log ε 3.22) (MeOH)

7-Alcohol
4,7′-Epoxy-8,9-dinor-3,8′-lignan-3′,4′,5,7,9′-pentol. 4,7′-Epoxy-3′,4′,5,7,9′-pentahydroxy-8,9-dinor-3,8′-lignan

7-Alcohol, 3′,5-di-Me ether
4,7′-Epoxy-3′,5-dimethoxy-8,9-dinor-3,8′-lignan-4′,7,9′-triol. 4,7′-Epoxy-4′,7,9′-trihydroxy-3′,5-dimethoxy-8,9-dinor-3,8′-lignan.
Tomentosanan B
[1771744-26-3]
$C_{18}H_{20}O_6$ 332.352
Constit. of the seeds of *Prunus tomentosa*. Oil. $[\alpha]^{20}_{D}$ -9.8 (c, 0.044 in MeOH)

7-Aldehyde, 3′,5-di-Me ether
(-)-Ficusal
[125638-35-9]
$C_{18}H_{18}O_6$ 330.336
Constit. of *Crataegus pinnatifida* and *Prunus tomentosa*. Potent inhibitor of NO prodn. and TNF-α prodn.

7-Aldehyde, 3′,5-di-Me ether, 4′-O-β-D-glucopyranoside
[2056257-68-0]
$C_{24}H_{28}O_{11}$ 492.477
Constit. of *Dendrobium fimbriatum*. Amorph. powder. Stereochem. uncertain but may have (7′*S*,8′*R*)-config.

7-Aldehyde, 3′,5-di-Me ether, 9′-O-(3-methylbutanoyl)
(+)-Brachangobinan I
[2357185-57-8]
Amorph. solid. $[\alpha]^{28}_{D}$ +64 (c, 0.09 in MeOH)

7-Aldehyde, 3′,4′-methylene, 5-Me ether
4,7′-Epoxy-9′-hydroxy-5-methoxy-3′,4′-methylenedioxy-8,9-dinor-3,8′-lignan-7-al. 4,7′-Epoxy-9′-hydroxy-5-methoxy-3′,4′-methylenedioxy-8,9-dinor-7-oxo-3,8′-lignan. ***Perseal C***
[214197-59-8]
$C_{18}H_{16}O_6$ 328.32
Constit. of the leaves of *Persea obovatifolia*. Oil. $[\alpha]^{25}_{D}$ -20 (c, 0.03 in $CHCl_3$). λ_{max} 211 (log ε 4.66); 240 (sh); 305 (log ε 4.15) (EtOH)

(7′RS,8′SR)-form

3′,5-Di-Me ether, Me ester
Curlignan
$C_{19}H_{20}O_7$ 360.362
Constit. *Ailanthus altissima* and *Curculigo capitulata*. $[\alpha]^{20}_{D}$ -54.7 (c, 0.1 in MeOH) (*Ailanthus* constit.). $[\alpha]^{23}_{D}$ +40 (c, 0.5 in MeOH) (*Curculigo* constit.). The nat. prod. Curlignan has been isol. as a partial racemate and resolved by chiral hplc. λ_{max} 207 (log ε 1.81); 274 (log ε 0.74) (MeOH)

3′,5-Di-Me ether, 9′-O-(3-methylbutanoyl)
Brachangobinan J
[2359598-30-2]
$C_{23}H_{26}O_8$ 430.453
Constit. of *Brachanthemum gobicum*. Amorph. yellowish solid. $[\alpha]^{22}_{D}$ -5 (c, 0.14 in MeOH). Partial racemate. λ_{max} 205 (log ε 4.35); 263 (log ε 3.77) (MeOH)

7-Aldehyde, 3′,5-di-Me ether, 9′-O-(3-methylbutanoyl)
Brachangobinan I
[2359598-29-9]
$C_{23}H_{26}O_7$ 414.454
Constit. of *Brachanthemum gobicum*. Exhibits moderate inhibitory activity against *Trypanosma congolense*. Amorph. solid. $[\alpha]^{22}_{D}$ -4 (c, 0.6 in MeOH). The nat. prod. Brachangobinan I was isol. as a partial racemate and resolved by chiral hplc. λ_{max} 233 (log ε 4.11); 288 (log ε 3.88) (MeOH)

(7′ξ,8′ξ)-form

7-Aldehyde, 3′-Me ether, 4′-O-β-D-glucopyranoside
Rosea A
[1453193-13-9]
$C_{23}H_{26}O_{11}$ 478.45
Constit. of *Alcea rosea*. Amorph. powder.

Sakushima, A. *et al.*, *Phytochemistry*, 1996, **43**, 1349-1354 (*(-)-Ceplignan*).
Tsai, I.-L. *et al.*, *Phytochemistry*, 1998, **48**, 1371-1375 (*Perseal C*).
Chang, W.-L. *et al.*, *Phytochemistry*, 1998, **49**, 2133-2136 (*Curlignan*).
Li, Y.-C. *et al.*, *Chem. Pharm. Bull.*, 2000, **48**, 1862-1865 (*(+)-Ficusal*).
Yoshikawa, M. *et al.*, *Heterocycles*, 2003, **60**, 1787-1792 (*Hierochin C*).
Hua, W. *et al.*, *Chin. Chem. Lett.*, 2008, **19**, 1327-1329 (*Gelsemiunoside B*).
Werner, I. *et al.*, *Nat. Prod. Commun.*, 2008, **3**, 179-184 (*Chione venosa constit*).
Nakashima, S. *et al.*, *Bioorg. Med. Chem.*, 2010, **18**, 2337-2345 (*Hierochin C, activity*).
Lee, I.-S. *et al.*, *Helv. Chim. Acta*, 2010, **93**, 272-276 (*Cedralins A,B*).
Jia, L. *et al.*, *J. Asian Nat. Prod. Res.*, 2010, **12**, 962-967 (*Urena lobata glucoside*).
Zan, K. *et al.*, *J. Chin. Pharm. Sci.*, 2010, **19**, 95-99 (*(+)-Ficusal*).
Yu, Y. *et al.*, *Zhongcaoyao*, 2010, **41**, 509-514 (*Ceplignan, Ficusal*).
Chung, C.-P. *et al.*, *J. Agric. Food Chem.*, 2011, **59**, 1185-1194 (*Ficusal, activity*).
Li, W.-J. *et al.*, *Molecules*, 2012, **17**, 8773-8781 (*Norcurlignan*).
Cheng, W. *et al.*, *Asian J. Chem.*, 2013, **25**, 2243-2245 (*Rosea A*).
Dong, X.-J. *et al.*, *Molecules*, 2013, **18**, 2840-2850 (*Toonin C*).
Yuan, T. *et al.*, *Planta Med.*, 2013, **79**, 1674-1679 (*Punicatannin C*).
Liu, Q.-B. *et al.*, *Biochem. Syst. Ecol.*, 2014, **55**, 236-240 (*Tomentosanan B*).
Ito, H. *et al.*, *Food Chem.*, 2014, **152**, 323-330 (*Pomegralignan*).
Liang, S. *et al.*, *Bioorg. Med. Chem. Lett.*, 2015, **25**, 4214-4218 (*(+)-Ceplignan, (-)-Ceplignan*).
Huang, X.-X. *et al.*, *J. Agric. Food Chem.*, 2015, **63**, 7252-7260 (*(-)-Ficusal, activity*).
Xu, F.-Q. *et al.*, *Nat. Prod. Res.*, 2017, **31**, 797-801 (*Dendrobium fimbriatum constit*).
Zhang, X. *et al.*, *Fitoterapia*, 2018, **125**, 147-154 (*Ligusticumaldehyde A*).
Yan, Z.-Y. *et al.*, *Fitoterapia*, 2018, **130**, 234-240 (*Curlignan, resoln, cd*).
Odonbayar, B. *et al.*, *J. Nat. Prod.*, 2019, **82**, 774-784 (*Brachangobinans I,J, cd, activity*)

4,7′-Epoxy-3′,4′,5,5′-tetrahydroxy-8,9-dinor-3,8′-lign-7′-en-7-al E242

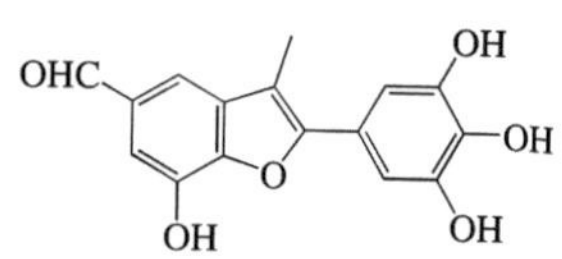

7-Hydroxy-3-methyl-2-(3,4,5-trihydroxyphenyl)-5-benzofurancarboxaldehyde. 4,7′-Epoxy-3′,4′,5,5′-tetrahydroxy-8,9-dinor-7-oxo-3,8′-neoligan-7′-ene
$C_{16}H_{12}O_6$ 300.266

3′,5-Di-Me ether
4,7′-Epoxy-3′,4′-dihydroxy-5,5′-dimethoxy-8,9-dinor-3,8′-lign-7′-en-7-al. 4,7′-Epoxy-3′,4′-dihydroxy-5,5′-dimethoxy-8,9-dinor-7-oxo-3,8′-lign-7′-ene. ***Perseal D***
[214197-63-4]
$C_{18}H_{16}O_6$ 328.32
Constit. of *Persea obovatifolia*. Exhibits cytotoxic activity. Oil. λ_{max} 205 (log ε 4.46); 237 (log ε 4.47); 295 (log ε 4.52) (EtOH)
Tetra-Me ether
4,7′-Epoxy-3′,4′,5,5′-tetramethoxy-8,9-dinor-3,8′-lign-7′-en-7-al. 4,7′-Epoxy-3′,4′,5,5′-tetramethoxy-8,9-dinor-7-oxo-3,8′-lign-7′-ene.
Ocophyllal B
[1189119-89-8]
$C_{20}H_{20}O_6$ 356.374
Constit. of the leaves of *Ocotea macrophylla*. Moderate inhibitor of COX-2. Orange oil.

Tsai, I.-L. *et al.*, *Phytochemistry*, 1998, **48**, 1371-1375 (*Perseal D*).
Coy, E.D. *et al.*, *Bioorg. Med. Chem. Lett.*, 2009, **19**, 6922-6925 (*Ocophyllal B, activity*).
Coy-Barrera, E.D. *et al.*, *Phytochemistry*, 2009, **70**, 1309-1314 (*Ocophyllal B*)

7,7′-Epoxy-3,3′,4,4′-tetrahydroxy-8,8′-lignan-9,9′-dioic acid E243

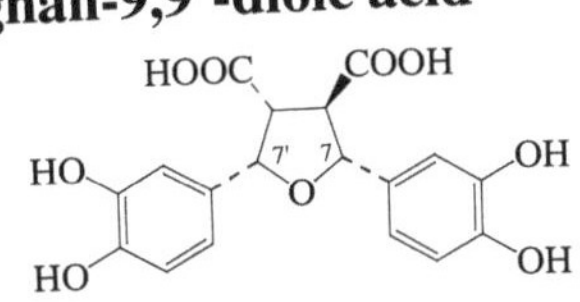

2,5-Bis(3,4-dihydroxyphenyl)tetrahydro-3,4-furandicarboxylic acid. 3,3′,4,4′-Tetrahydroxy-7,7′-epoxylignan-9,9′-dioic acid
$C_{18}H_{16}O_9$ 376.317
(7R*,7′S*,8R*,8′R*)-form
3,3′-Di-Me ether
7,7′-Epoxy-4,4′-dihydroxy-3,3′-dimethoxy-8,8′-lignan-9,9′-dioic acid. 4,4′-Dihydroxy-3,3′-dimethoxy-7,7′-epoxylignan-9,9′-dioic acid
[908298-65-7]
$C_{20}H_{20}O_9$ 404.371
Obt. from hydrol. of cereal plant cell walls. Prisms (MeOH) (as di-Ac). Mp 137 - 140° (di-Ac).

Schatz, P.F. *et al.*, *Org. Biomol. Chem.*, 2006, **4**, 2801-2806 (*3,3′-Di-Me ether, synth*)

7,7′-Epoxy-3,3′,4,4′-tetrahydroxy-8,8′-lignan-9,9′-olide E244

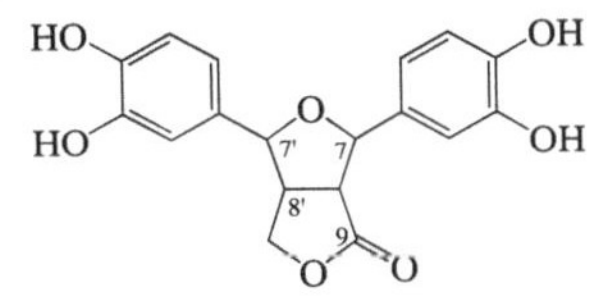

4,6-Bis(3,4-dihydroxyphenyl)tetrahydro-1H,3H-furo[3,4-c]furan-1-one. 3,3′,4,4′-Tetrahydroxy-7,7′-epoxylignan-9,9′-olide
$C_{18}H_{16}O_7$ 344.319
(7ξ,7′ξ,8ξ,8′ξ)-form
3,3′-Di-Me ether
7,7′-Epoxy-4,4′-dihydroxy-3,3′-dimethoxy-8,8′-lignan-9,9′-olide. 4,4′-Dihydroxy-3,3′-dimethoxy-7,7′-epoxylignan-9,9′-olide.
Neoolivilic acid
[1605308-40-4]
$C_{20}H_{20}O_7$ 372.373
Constit. of the stems of *Zea mays*. Exhibits antiinflammatory and neuroprotective activities. Pale brown powder (MeOH). $[\alpha]_D^{24}$ -5.4 (c, 0.5 in MeOH).

Eklund, P.C. *et al.*, *Tetrahedron*, 2003, **59**, 4515-4523 (*synth*).
Jung, Y.-J. *et al.*, *Arch. Pharmacal Res.*, 2015, **38**, 178-185 (*Neolivilic acid*)

7,9′-Epoxy-3,3′,4,4′-tetrahydroxy-8,8′-lignan-9,7′-olide E245

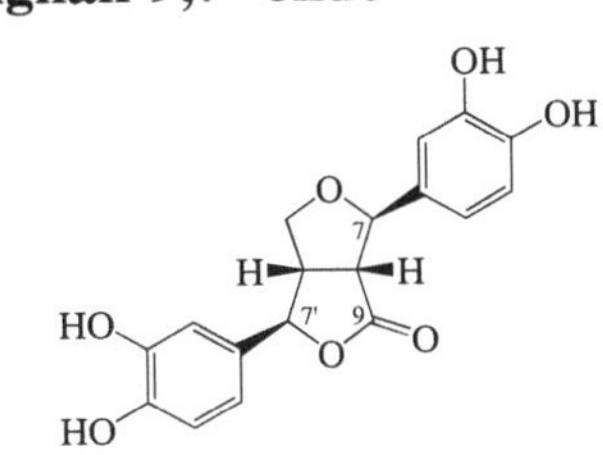

(7R,7′R,8R,8′S)-form

3,3′,4,4′-Tetrahydroxy-7,9′-epoxylignan-9,7′-olide. Bis(demethylene)aptosimone
$C_{18}H_{16}O_7$ 344.319
(7R,7′R,8R,8′S)-form [1492980-16-1]
Constit. of *Morinda citrifolia*. Amorph. brown powder. $[\alpha]_D^{25}$ -5.8 (c, 0.1 in MeOH). λ_{max} 231 (log ε 4.41); 282 (log ε 3.71) (MeOH)
3′,4′-Methylene, 3-Me ether
7,9′-Epoxy-4-hydroxy-3-methoxy-3′,4′-methylenedioxy-8,8′-lignan-9,7′-olide. 4-Hydroxy-3-methoxy-3′,4′-methylenedioxy-7,9′-epoxylignan-9,7′-olide. ***Styraxin***
[69742-32-1]
$C_{20}H_{18}O_7$ 370.357
Isol. from *Styrax officinalis*. Shows antitumour activity. Mp 146 - 147°
3′,4′-Methylene, 3-Me ether, 4-O-β-D-glucopyranoside
Styraxlignolide B
[823214-02-4]
$C_{26}H_{28}O_{12}$ 532.498
Constit. of the stem bark of *Styrax japonica*. Brown plates (MeOH). Mp 111 - 113°. $[\alpha]_D^{23}$ -96.8 (c, 0.31 in MeOH). λ_{max} 205 (log ε 4.74); 228 (log ε 4.14); 282 (log ε 3.8) (MeOH)
3′,4′-Methylene, 3,4-di-Me ether
7,9′-Epoxy-3,4-dimethoxy-3′,4′-methylenedioxy-8,8′-lignan-9,7′-olide. 3,4-Dimethoxy-3′,4′-methylenedioxy-7,9′-epoxylignan-9,7′-olide
Mp 115 - 117°
3,4:3′,4′-Bis(methylene) ether
7,9′-Epoxy-3,4:3′,4′-bis(methylenedioxy)-8,8′-lignan-9,7′-olide. ***(-)-Aptosimone***. *Genkdaphin*
[139402-21-4]
$C_{20}H_{16}O_7$ 368.341
Constit. of *Daphne genkwa*, *Sesamum indicum* and *Wikstroemia lanceolata*. Exhibits antiinflammatory activity. Cryst.. Mp 118 - 119.5°. $[\alpha]_D^{26}$ -101.5 (c, 2 in $CHCl_3$). λ_{max} 239 (log ε 4.37); 287 (log ε 4.34) (MeOH)
(7S,7′S,8S,8′R)-form [936478-87-4]
Constit. of the fruit of *Morinda citrifolia*. Light brown semi-solid. $[\alpha]_D^{25}$ +7.5 (c, 0.15 in MeOH). λ_{max} 230 (log ε 4.11); 282 (log ε 3.71) (MeOH)
3,3′-Di-Me ether
7,9′-Epoxy-4,4′-dihydroxy-3,3′-dimethoxy-8,8′-lignan-9,7′-olide. 4,4′-Dihydroxy-3,3′-dimethoxy-7,9′-epoxylignan-9,7′-olide. ***Balanophonin B***. *4-Ketopinoresinol*
[66288-89-9]
$C_{20}H_{20}O_7$ 372.373
Constit. of *Amischotolype hispida*, *Coix lachryma-jobi* var. *ma-yuen*, *Cucurbita moschata*, *Eragrostis ferruginea*, *Lithocarpus pachylepis* and *Phyllanthus emblica*. ARE activator. Amorph. powder. $[\alpha]_D^{20}$ +54.7 (c, 0.71 in MeOH). λ_{max} 230 (log ε 4.36); 279 (log ε 3.25) (MeOH)

3,4:3′,4′-Bis(methylene) ether

(+)-Aptosimone

[61254-17-9]

$C_{20}H_{16}O_7$ 368.341

Constit. of *Aptosimum spinescens* and *Cuscuta chinensis*. Needles. Mp 123°. $[\alpha]_D^{20}$ +70 (c, 2 in $CHCl_3$)

(7R*,7′R*,8R*,8′S*)-form

3′-Me ether

7,9′-Epoxy-3,4,4′-trihydroxy-3′-methoxy-8,8′-lignan-9,7′-olide. 3,4,4′-Trihydroxy-3′-methoxy-7,9′-epoxylignan-9,7′-olide. ***Syzygiresinol A***

[1151670-51-7]

$C_{19}H_{18}O_7$ 358.346

Constit. of *Alyxia schlechteri*, *Rhodiola crenulata* and *Syzygium cumini*. Cream cryst. (Me_2CO). Mp 210 - 211°. $[\alpha]_D^{25}$ +5 (c, 0.2 in MeOH). λ_{max} 250 (log ε 6.5); 361 (log ε 3.8) (MeOH). [neutral] λ_{max} 203; 282 (MeOH)

3,3′-Di-Me ether

Monoepoxylignanolide*. 9-Oxopinoresinol. MEL*

[125638-31-5]

$C_{20}H_{20}O_7$ 372.373

Constit. of *Aegilops ovata* and *Senecio scandens*. Germination inhibitor. Antioxidant. Mp 190 - 192°

(7R*,7′R*,8S*,8′R*)-form

3,3′-Di-Me ether

[1374694-16-2]

$C_{20}H_{20}O_7$ 372.373

Constit. of the tubers of *Polygonum perfoliatum*. Powder. Mp 109 - 111°. $[\alpha]_D^{25}$ -3.5 (c, 0.2 in MeOH).

Lavie, D. *et al., Nature (London)*, 1974, **249**, 388 (*Monoepoxylignanolide*).
Brieskorn, C.H. *et al., Tetrahedron Lett.*, 1976, **17**, 2221-2224 (*Aptomium spinescens consit*).
Ulubelen, A. *Planta Med.*, 1978, **34**, 403-407 (*Styraxin, ms, cryst struct*).
Velde, V.V. *et al., J. Chem. Soc., Perkin Trans. 1*, 1984, 1159-1163 (*Monoepoxylignanolide, synth*).
Stevens, D.R. *et al., Tetrahedron Lett.*, 1986, **27**, 4629-4632 (*(±)-Styraxin, (±)-Aptosimone, synth*).
Otsuka, H. *et al., Phytochemistry*, 1989, **28**, 883-886 (*4-Ketopinoresinol*).
Wang, M. *et al., Yaoxue Xuebao*, 1990, **25**, 866-868 (*Genkdaphin*).
Stevens, D.R. *et al., J. Chem. Soc., Perkin Trans. 1*, 1992, 185-190 (*Monoepoxylignanolide, synth*).
Yoshida, S.-I. *et al., Synthesis*, 1997, 1475-1480 (*Styraxin, synth*).
Yamauchi, S. *et al., Biosci., Biotechnol., Biochem.*, 2003, **67**, 838-846 (*Aptosimone, synth*).
Min, B.-S. *et al., J. Nat. Prod.*, 2004, **67**, 1980-1984 (*Styraxlignolide B*).
Lin, R.-W. *et al., Planta Med.*, 2004, **70**, 234-238 (*(-)-Aptosimone*).
Deng, S. *et al., J. Nat. Prod.*, 2007, **70**, 859-862 (*Morinda citrifolia constit*).
Moon, S.-S. *et al., Bioorg. Med. Chem.*, 2008, **16**, 7264-7269 (*4-Ketopinoresinol*).
Lee, M.-Y. *et al., Int. Immunopharmacol.*, 2009, **9**, 878-885 (*Aptosimone, activity*).
Mir, Q.Y. *et al., Nat. Prod. Res.*, 2009, **23**, 422-430 (*Syzygiresinol A*).
Nguyen, P. H. *et al., J. Nat. Prod.*, 2013, **76**, 2080-2087 (*Morinda citrifolia constit*).
Wang, K.-W. *et al., Nat. Prod. Res.*, 2013, **27**, 568-573 (*Polygonum perforatum constit*).
Ma, G. *et al., Phytochem. Lett.*, 2013, **6**, 152-155 (*Balanophonin B*).
Chang, C.-I. *et al., J. Funct. Foods*, 2014, **10**, 260-273 (*4-Ketopinoresinol*).
Zhou, J.-T. *et al., Molecules*, 2015, **20**, 13725-13739 (*(+)-Syzygiresinol A*).
Sribuhom, T. *et al., Phytochem. Lett.*, 2015, **11**, 80-84 (*Syzygiresinol A*)

7,9′-Epoxy-3,3′,5,5′-tetrahydroxy-8,8′-lignan-9,7′-olide E246

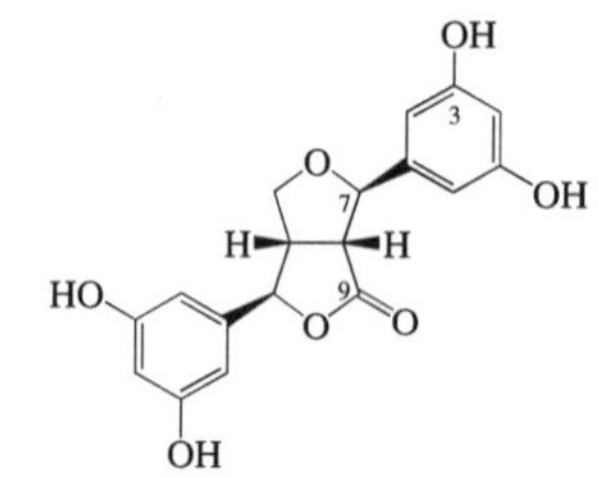

$C_{18}H_{16}O_7$ 344.319

Structures in this series may need revision. Phenyl 3,5-dioxygenated substn. pattern in doubt as a result of work done by Yang *et al* in 2015

(7R,7′R,8R,8′S)-form

3,3′-Di-Me ether

7,9′-Epoxy-3,3′-dihydroxy-5,5′-dimethoxy-8,8′-lignan-9,7′-olide.

Graminone A

[161407-72-3]

$C_{20}H_{20}O_7$ 372.373

Constit. of the rhizomes of *Imperata cylindrica*. Oil. $[\alpha]_D^{25}$ -1 (c, 0.1 in $CHCl_3$).

Matsunaga, K. *et al., J. Nat. Prod.*, 1994, **57**, 1734-1736 (*Graminone A*).
Yang, Y.-N. *et al., J. Nat. Prod.*, 2015, **78**, 705-711 (*pmr, struct*)

7,9′-Epoxy-3,3′,4′,5-tetrahydroxy-8,8′-lignan-7′-one E247

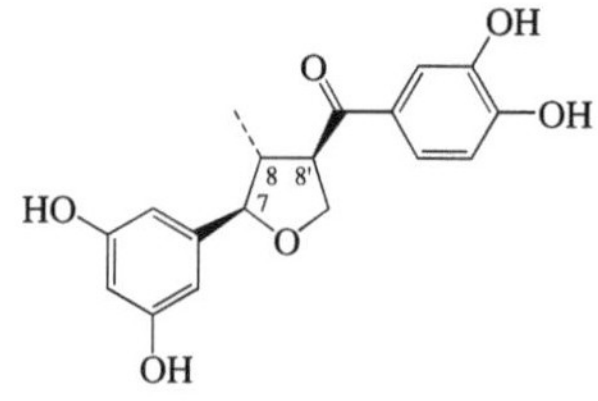

3,3′,4′,5-Tetrahydroxy-7′-oxo-7,9′-epoxylignan

$C_{18}H_{18}O_6$ 330.336

Structures in this series may need revision. Phenyl 3,5-dioxygenated substn. pattern in doubt as a result of work done by Yang *et al* in 2015

(7R,8R,8′R)-form

Tetra-Me ether

7,9′-Epoxy-3,3′,4′,5-tetramethoxy-8,8′-lignan-7′-one. ***Acortatarinowin K***

$C_{22}H_{26}O_6$ 386.444

Constit. of the rhizomes of *Acorus tatarinowii*. Amorph. powder. $[\alpha]_D^{20}$ -1.2 (c, 0.1 in MeOH). λ_{max} 205 (log ε 4.45); 231 (log ε 4.24); 278 (log ε 3.93); 308 (log ε 3.77) (MeOH).

Yang, Y.-N. *et al., J. Nat. Prod.*, 2015, **78**, 705-711 (*pmr, struct*).
Lu, Y. *et al., Sci. Rep.*, 2016, **6**, 22909 (*Acortatarinowin K*)

4,7′-Epoxy-3′,4′,6,9-tetrahydroxy-3,8′-lign-7′-en-9′-al E248

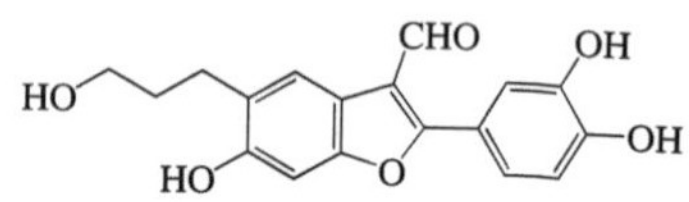

2-(3,4-Dihydroxyphenyl)-6-hydroxy-5-(3-hydroxypropyl)-3-benzofurancarboxaldehyde, CAS. *4,7′-Epoxy-3′,4′,6,9-tetrahydroxy-9′-oxo-3,8′-neolign-7′-ene*

$C_{18}H_{16}O_6$ 328.32

3′,6-Di-Me ether

4,7′-Epoxy-4′,9-dihydroxy-3′,6-dimethoxy-3,8′-lign-7′-en-9′-al. 4,7′-Epoxy-4′,9-dihydroxy-3′,6-dimethoxy-9′-oxo-3,8′-lign-7′-ene

[949161-74-4]

$C_{20}H_{20}O_6$ 356.374

Constit. of the stem bark of *Ginkgo biloba*. Exhibits antiplatelet activity.

US Pat., 2006, 7 341 750 (Candleflower) (*activity*).
Ngan, N.T.T. *et al., J. Agric. Food Chem.*, 2012, **60**, 2815-2824 (*Ginkgo biloba constit*)

3,7′-Epoxy-3′,4′,5,9′-tetrahydroxy-2,8′-lign-7-en-9-oic acid E249

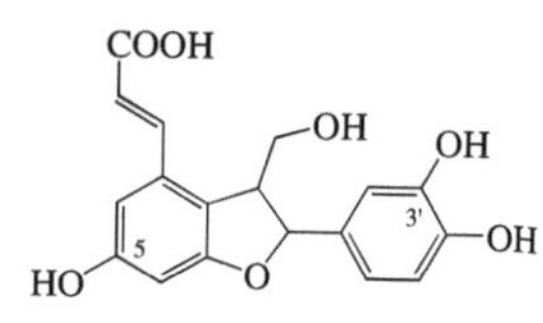

4-(2-Carboxyethenyl)-2,3-dihydro-6-hydroxy-2-(3,4-dihydroxyphenyl)-3-benzofuranmethanol
$C_{18}H_{16}O_7$ 344.319
3′,5-Di-Me ether
3,7′-Epoxy-4′,9′-dihydroxy-3′,5-dimethoxy-2,8′-lign-7-en-9-oic acid. ***Lepidiumlignan B***
$C_{20}H_{20}O_7$ 372.373
Constit. of the seeds of *Lepidium apetalum*. $[\alpha]_D^{20}$ +6.6 (c, 0.1 in MeOD). λ_{max} 205 (log ε 2.7); 290 (log ε 0.64); 325 (log ε 0.81) (MeOH).
Meng, L. *et al., Zhongcaoyao*, 2018, **49**, 3981-3986 (*Lepidiumlignan B*)

7,9′-Epoxy-3,3′,4,4′-tetrahydroxy-8,8′-lign-7′-en-9-oic acid E250

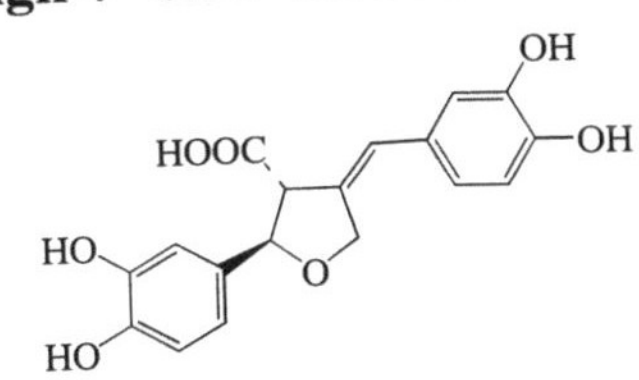

$C_{18}H_{16}O_7$ 344.319
(7R*,7′Z,8R*)-form
3,3′-Di-Me ether, Me ester
Alyterinate B
[1616247-83-6]
$C_{21}H_{22}O_7$ 386.4
Isol. from *Alyxia schlechteri*. Pale yellow oil. $[\alpha]_D^{23}$ +14.7 (c, 0.001 in $CHCl_3$). λ_{max} 239 (log ε 4.26); 274 (log ε 4.26) ($CHCl_3$).
Sriphana, U. *et al., Fitoterapia*, 2013, **91**, 39-43 (*Alyterinate B*)

4,7′-Epoxy-3′,4′,5,9′-tetrahydroxy-9-nor-3,8′-lignan-8-oic acid E251

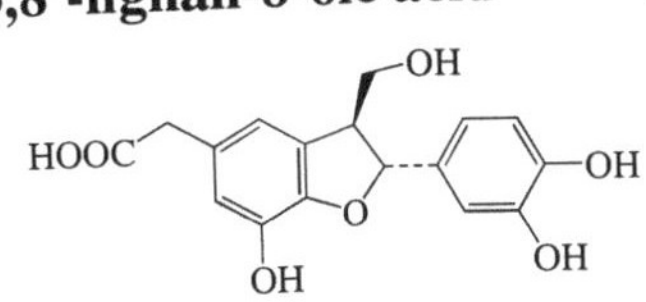

(7′R,8′S)-form

2-(3,4-Dihydroxyphenyl)-2,3-dihydro-7-hydroxy-3-(hydroxymethyl)-5-benzofuranacetic acid
$C_{17}H_{16}O_7$ 332.308
(7′R,8′S)-form
3′,5-Di-Me ether, Et ester
(+)-Crataegusin B
[2357252-58-3]
$[\alpha]_D^{20}$ +24 (c, 0.1 in MeOH)
(7′S,8′R)-form
3′,5-Di-Me ether, Et ester
(-)-Crataegusin B
[2357252-59-4]
$[\alpha]_D^{20}$ -20 (c, 0.1 in MeOH)
(7′RS,8′SR)-form
3′,5-Di-Me ether
4,7′-Epoxy-4′,9′-dihydroxy-3′,5-dimethoxy-9-nor-3,8′-lignan-8-oic acid
3′,5-Di Me ether, Et ester
Crataegusin B
$C_{21}H_{24}O_7$ 388.416
Constit. of the fruit of *Crataegus pinnatifida*. Yellow oil. $[\alpha]_D^{20}$ +1.3 (c, 0.1 in MeOH). Resolved by chiral hplc. λ_{max} 230 (log ε 3.59); 280 (log ε 2.92) (MeOH).
Guo, R. *et al., Bioorg. Chem.*, 2019, **85**, 469-474 (*Crataegusin B*)

3,7′-Epoxy-3′,4,4′,5′-tetrahydroxy-9-nor-2,8′-lignan-7-one E252

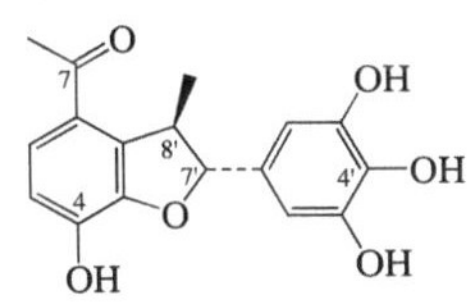

1-[2,3-Dihydro-7-hydroxy-3-methyl-2-(3,4,5-trihydroxyphenyl)benzofuran-4-yl]ethanone
(7′R*,8′R*)-form
3′,5′-Di-Me ether
3,7′-Epoxy-4,4′-dihydroxy-3′,5′-dimethoxy-9-nor-2,8′-lignan-7-one
$C_{19}H_{20}O_6$ 344.363
Constit. of the roots of *Adelostemma gracillimum*. Light yellow gum. $[\alpha]_D^{20}$ +8.7 (c, 0.31 in MeOH). λ_{max} 228; 335 (MeOH)
5′-Deoxy
3,7′-Epoxy-3′,4,4′-trihydroxy-9-nor-2,8′-lignan-7-one
5′-Deoxy, 3′-Me ether
3,7′-Epoxy-4,4′-dihydroxy-3′-methoxy-9-nor-2,8′-lignan-7-one
$C_{18}H_{18}O_5$ 314.337
Constit. of the roots of *Adelostemma gracillimum*. Exhibits significant neuroprotection against NMDA-induced cell death. Pale yellow gum. $[\alpha]_D^{20}$ +2.3 (c, 0.94 in MeOH). λ_{max} 231; 336 (MeOH).
Zhang, J. *et al., Phytochem. Lett.*, 2016, **16**, 1-7 (*Adelostemma gracillimum constits, activity*)

4,7′-Epoxy-3′,5,5′,9′-tetrahydroxy-9-nor-3,8′-lignan-7-one E253

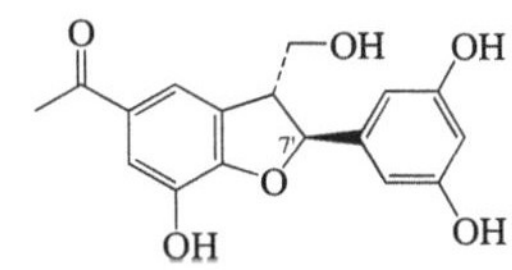

5-Acetyl-2-(3,5-dihydroxyphenyl)-2,3-dihydro-7-hydroxy-3-(hydroxymethyl) benzofuran. 4,7′-Epoxy-3′,5,5′,9′-tetrahydroxy-9-nor-7-oxo-3,8′-neolignan
$C_{17}H_{16}O_6$ 316.309
Structures in this series may need revision. Phenyl 3,5-dioxygenated substn. pattern in doubt as a result of work done by Yang *et al* in 2015
(7′S,8′R)-form
3′,5-Di-Me ether
4,7′-Epoxy-3′,9′-dihydroxy-5,5′-dimethoxy-9-nor-3,8′-lignan-7-one.
Lycocernuaside B
$C_{19}H_{20}O_6$ 344.363
Constit. of *Palhinhaea cernua*. Moderate inhibitor of xanthine oxidase. Amorph. yellow powder. $[\alpha]_D^{25}$ -66 (c, 0.1 in MeOH).
Yang, Y.-N. *et al., J. Nat. Prod.*, 2015, **78**, 705-711 (*pmr, struct*).
Li., J. *et al., Fitoterapia*, 2017, **119**, 45-50 (*Lycocernuaside B, activity*)

7,7′-Epoxy-3,3′,4,4′-tetrahydroxy-7,8-seco-8,8′-lignan-7,8-dione E254

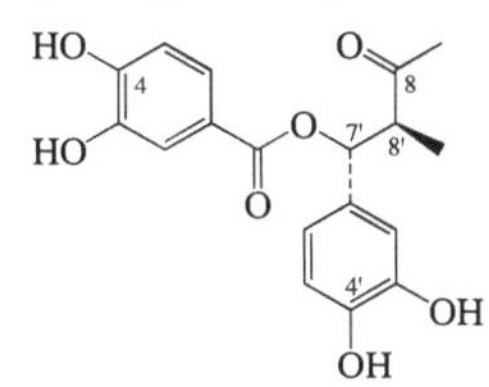

(7′R,8′S)-form

1-(3,4-Dihydroxyphenyl)-2-methyl-3-oxobutyl 3,4-dihydroxybenzoate, CAS
$C_{18}H_{18}O_7$ 346.335
(7′R,8′S)-form
Tetra-Me ether
7,7′-Epoxy-3,3′,4,4′-tetramethoxy-7,8-seco-8,8′-lignan-7,8-dione
[165054-92-2]
$C_{22}H_{26}O_7$ 402.443
Constit. of the leaves of *Ocotea foetens* and *Schisandra sphenanthera*. Inhibitor of HSV-2 and adenovirus. Oil
3,4-Methylene, 3′,4′-di-Me ether
Iso-7,8-secoholostylone B
[1259294-30-8]

$C_{21}H_{22}O_7$ 386.400
Constit. of the fruit of *Schisandra sphenanthera* and *Schisandra glaucescens*. Exhibits neuroprotective activity. $[\alpha]^{26}_{D}$ +65.7 (c, 0.18 in $CHCl_3$). λ_{max} 221 (log ε 3.91); 265 (log ε 3.4); 288 (log ε 3.35) (MeOH)

3′,4′-Methylene, 3,4-di-Me ether
7,8-Secoholostylone B
[694493-86-2]
$C_{21}H_{22}O_7$ 386.400
Constit. of *Holostylis reniformis*, *Schisandra glaucescens*, *Schisandra neglecta* and *Schisandra wilsoniana*. Exhibits neuroprotective activity. Amorph. yellow solid. $[\alpha]^{25}_{D}$ +37 (c, 0.13 in $CHCl_3$). λ_{max} 221 (log ε 3.8); 264 (log ε 3.4); 289 (log ε 3.3) (MeOH)

(7′S,8′R)-form
Tetra-Me ether
Schibitubin I
[1609628-72-9]
$C_{22}H_{26}O_7$ 402.443
Constit. of the fruit of *Schisandra bicolor* var. *tuberculata* and *Schisandra sphenanthera*. Exhibits neuroprotective activity. Oil. $[\alpha]^{20}_{D}$ +35.5 (c, 3.66 in $CHCl_3$) (Schibitubin I). $[\alpha]_D$ -50.3 (c, 0.85 in $CHCl_3$). λ_{max} 205 (log ε 4.45); 265 (log ε 3.78); 370 (log ε 2.88) (MeOH) (Schibitubin I). [neutral] λ_{max} 219 (log ε 4.41); 263 (log ε 4.09); 285 (log ε 3.91); 294 (log ε 3.85) (MeOH)

3′,4′-Methylene, 3-Me ether
7,7′-Epoxy-4-hydroxy-3-methoxy-3′,4′-methylenedioxy-7,8-seco-8,8′-lignan-7,8-dione. ***Marphenol A***
[1260657-83-7]
$C_{20}H_{20}O_7$ 372.373
Constit. of *Schisandra neglecta* and *Schisandra wilsoniana*. Exhibits anti-HIV-1 activity. Amorph. powder. $[\alpha]^{26}_{D}$ +30.6 (c, 0.36 in MeOH). λ_{max} 203 (log ε 5.6); 272 (log ε 3.57); 356 (log ε 1.68) (MeOH)

3,4-Methylene, 3′,4′-di-Me ether
7,7′-Epoxy-3′,4′-dimethoxy-3,4-methylenedioxy-7,8-seco-8,8′-lignan-7,8-dione. ***Schisandlignan A***
[1610338-25-4]
$C_{21}H_{22}O_7$ 386.400
Constit. of *Schisandra sphenanthera*. Oil. $[\alpha]^{20}_{D}$ -73 (c, 1.5 in $CHCl_3$). λ_{max} 220 (log ε 4.48); 267 (log ε 3.99); 298 (log ε 3.94) (MeOH)

3′,4′-Methylene, 3,4-di-Me ether
7,7′-Epoxy-3,4-dimethoxy-3′,4′-methylenedioxy-7,8-seco-8,8′-lignan-7,8-dione. ***Schisandlignan B***
[1610338-26-5]
$C_{21}H_{22}O_7$ 386.400
Constit. of *Schisandra sphenanthera*. Oil. $[\alpha]_D$ -51 (c, 0.1 in $CHCl_3$). λ_{max} 220 (log ε 4.36); 263 (log ε 4.07); 289 (log ε 4.01) (MeOH)

(7′S,8′S)-form
Tetra-Me ether
Schisandlignan C
[1610338-27-6]
$C_{22}H_{26}O_7$ 402.443
Constit. of *Schisandra sphenanthera*. Amorph. powder. $[\alpha]_D$ -100 (c, 0.015 in $CHCl_3$). λ_{max} 250 (log ε 4.15); 289 (log ε 4.02) (MeOH)

(7′RS,8′RS)-form
3′,4′-Methylene, 3-Me ether
Machilolin B
[1639932-16-3]
$C_{20}H_{20}O_7$ 372.373
Constit. of the leaves of *Machilus zuihoensis*. Pale yellow oil. λ_{max} 232 (log ε 3.05); 277 (log ε 1.99) (MeCN)

(7′R*,8′S*)-form
8ξ-Alcohol, 3,4-methylene, 3′-Me ether
7,7′-Epoxy-4′,8-dihydroxy-3′-methoxy-3,4-methylenedioxy-7,8-seco-8,8′-lignan-7-one. ***Neglectahenol B***
[1465784-53-5]
$C_{20}H_{22}O_7$ 374.389
Constit. of the leaves and stems of *Schisandra neglecta*. Exhibits modest anti-HIV-1 activity. Amorph. powder. $[\alpha]^{26}_{D}$ +18.4 (c, 0.2 in MeOH). Possesses (7′*R**,8′*R**)-config. due to change in Cahn-Ingold-Prelog priorities. λ_{max} 210 (log ε 4.19); 275 (log ε 3.86); 340 (log ε 2.82) (MeOH)

(7′ξ,8′ξ)-form
Bis(methylene) ether
7,7′-Epoxy-3,4:3′,4′-bis(methylenedioxy)-7,8-seco-8,8′-lignan-7,8-dione. ***Machilolin A***
[1174181-71-5]
$C_{20}H_{18}O_7$ 370.357
Constit. of *Machilus zuihoensis* and *Schisandra sphenanthera*. Pale yellow oil. Racemate. λ_{max} 232 (log ε 3.06); 277 (log ε 1.98) (MeCN).

Lopez, H. *et al.*, *J. Nat. Prod.*, 1995, **58**, 782-785 (*Ocotea foetens constit*).
Da Silva, T. *et al.*, *Phytochemistry*, 2004, **65**, 751-759 (*7,8-Secoholostylone B*).
Wang, C.R. *et al.*, *Nat. Prod. Commun.*, 2009, **4**, 1571-1574 (*Iso-7,8-secoholostylone B*).
Chen, C.-Y. *et al.*, *Nat. Prod. Res.*, 2009, **23**, 871-875 (*Machilolin A*).
Zhang, X.-J. *et al.*, *Chem. Biodiversity*, 2010, **7**, 2692-2701 (*Marphenol A*).
Gao, X. *et al.*, *J. Braz. Chem. Soc.*, 2012, **23**, 1853-1857 (*Neglectahenol B, activity*).
Li, Y.-F. *et al.*, *J. Asian Nat. Prod. Res.*, 2013, **15**, 934-940 (*Schisandra sphenanthera constits*).
Song, Q.-Y. *et al.*, *Phytochem. Lett.*, 2013, **6**, 174-178 (*tetra-Me ether, activity*).
Lin, R.-J. *et al.*, *Chem. Nat. Compd. (Engl. Transl.)*, 2014, **50**, 978-981 (*Machilolin B*).
Yu, H.-Y. *et al.*, *J. Nat. Prod.*, 2014, **77**, 1311-1320 (*Schisandra glaucescens constits, abs config, activity*).
Liu, Y. *et al.*, *J. Nat. Prod.*, 2017, **80**, 1117-1124 (*Schibitubin I, activity*)

7,7′-Epoxy-3,3′,4,4′-tetramethoxy-8,8′-lignan E255

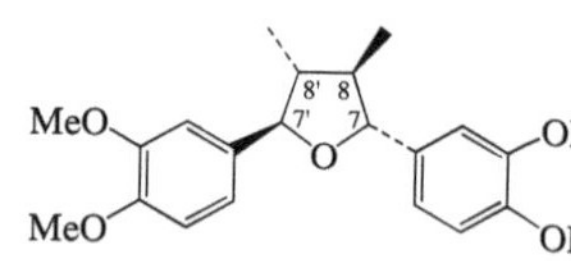

(7R,7′R,8R,8′R)-form

2,5-Bis(3,4-dimethoxyphenyl)tetrahydro-3,4-dimethylfuran, CAS. *3,3′,4,4′-Tetramethoxy-7,7′-epoxylignan*
$C_{22}H_{28}O_5$ 372.461

(7R,7′R,8R,8′R)-form
(+)-Galbelgin [61949-24-4]
Constit. of *Machilus japonica* leaves and *Machilus thunbergii*. Cryst. (EtOAc). Mp 142 - 142.5°. $[\alpha]_D$ +129 ($CHCl_3$) (+80.7)

(7R,7′S,8R,8′S)-form
Galgravin [528-63-2]
Constit. of *Galbulimima (Himantandra) belgraveana* bark and of *Magnolia acuminata*. Cryst. (MeOH or petrol). Mp 131°. Opt. inactive (*meso-*)

(7R,7′S,8S,8′R)-form
Di-O-methyltetrahydrofuriguaiacin B
[27686-82-4]
Constit. of *Illicium floridanum* and *Saururus chinensis*. Cryst. (MeOH). Mp 132 - 133°. $[\alpha]^{20}_{D}$ +43 (c, 0.5 in $CHCl_3$) (Di-*O*-methyltetrahydrofuriguaiacin B). Opt. inactive (*meso-*).

The stereochem. illustrated for Di-*O*-methyltetrahydrofuriguaiacin B in the lit. must be erroneous, as this is a *meso*-compd.. λ_{max} 235; 280 (MeOH)

(7R,7′S,8S,8′S)-form

Veraguensin [19950-55-1]

Isol. from *Ocotea veraguensis* and *Virola surinamensis*. Shows trypanosomicidal activity. Neuroprotectant. Rhombs (Et_2O). Mp 128 - 129°. $[\alpha]_D^{25}$ +34.2 (c, 1.1 in $CHCl_3$). $[\alpha]_D$ +11.5 (c, 0.28 in MeOH)

(7S,7′R,8R,8′R)-form

(-)-Veraguensin

Solid. $[\alpha]_D^{30}$ -34.9 (c, 1.14 in $CHCl_3$)

(7S,7′S,8S,8′R)-form

Ganschisandrin [122872-00-8]

Isol. from *Piper clarkii* and *Schisandra sphenanthera*. Flakes (Et_2O/hexane). Mp 114 - 115°. $[\alpha]_D^{20}$ +173.1 (c, 0.26 in $CHCl_3$). $[\alpha]_D^{22}$ +48.6 (c, 0.1 in MeOH)

(7S,7′S,8S,8′S)-form

(-)-Galbelgin [10569-12-7]

Constit. of *Himantandra belgraveana* bark and from *Piper thomsonii*. Mp 138°. $[\alpha]_D$ -102 (c, 0.04 in $CHCl_3$)

(7R*,7′R*,8S*,8′S*)-form

Saucernetin [83377-13-3]

Constit. of *Saururus cernuus*. Cryst.. Mp 80 - 81°. $[\alpha]_D$ +48 ($CHCl_3$)

(7RS,7′RS,8RS,8′RS)-form

[50393-84-5]

Prisms (MeOH). Mp 124 - 127°.

Hughes, G.K. *et al.*, *Aust. J. Chem.*, 1954, **7**, 104-112 (*Galgravin*).
Birch, A.J. *et al.*, *J. Chem. Soc.*, 1958, 4471-4476 (*(-)-Galbelgin*).
Freudenberg, K. *et al.*, *Tetrahedron*, 1961, **15**, 115-128 (*abs config*).
Crossley, N.S. *et al.*, *J. Chem. Soc.*, 1962, 1459-1462 (*Veraguensin*).
Pelter, A. *et al.*, *J. Heterocycl. Chem.*, 1966, **3**, 191-197 (*ms*).
Doskotch, R.W. *et al.*, *Tetrahedron*, 1972, **28**, 4711-4717 (*Galgravin, Veraguensin*).
Takaoka, D. *et al.*, *Bull. Chem. Soc. Jpn.*, 1976, **49**, 3564-3566 (*Machilus japonica constit*).
Ahmed, R. *et al.*, *Tetrahedron*, 1976, **32**, 1339-1344 (*Veraguensin, synth*).
Stevenson, R. *et al.*, *Tetrahedron*, 1977, **33**, 285-288 (*(±)-Galbelgin, synth*).
Fonseca, S.F. *et al.*, *Can. J. Chem.*, 1979, **57**, 441-443 (*cmr*).
Rao, K.V. *et al.*, *J. Nat. Prod.*, 1982, **45**, 393-397 (*Saucernetin*).
Yue, J.M. *et al.*, *Phytochemistry*, 1989, **28**, 1774-1776 (*Ganschisandrine*).
Prasad, A.K. *et al.*, *Phytochemistry*, 1995, **39**, 655-658 (*Piper clarkii constit*).
Lopes, N.P. *et al.*, *Planta Med.*, 1998, **64**, 667-669 (*Veraguensin, activity*).
Uck, Y. *et al.*, *J. Pharm. Pharmacol.*, 2000, **52**, 1163-1169 (*Machilus thunbergii constit*).
Schmidt, T.J. *et al.*, *Planta Med.*, 2000, **66**, 749-751 (*Illicium floridanum constit*).
Sung, S.H. *et al.*, *Chem. Pharm. Bull.*, 2001, **49**, 1192-1194 (*Dimethyltetrahydrofuriguaiacin B*).
Zhai, H. *et al.*, *Biol. Pharm. Bull.*, 2005, **28**, 289-293 (*Veraguensin, activity*).
Kim, H. *et al.*, *Org. Lett.*, 2007, **9**, 3965-3968 (*(+) Galbelgin, (-)-Veraguensin, synth, ir, pmr, cmr*).
Xue, P. *et al.*, *J. Asian Nat. Prod. Res.*, 2009, **11**, 281-287 (*Galbelgin, synth*).
Matcha, K. *et al.*, *Tetrahedron Lett.*, 2010, **51**, 6924-6927 (*Veraguensin, synth*).
Rye, C.E. *et al.*, *J. Org. Chem.*, 2011, **76**, 6636-6648 (*(+)-Galbelgin, synth, pmr, cmr*).
Hazra, S. *et al.*, *RSC Adv.*, 2013, **3**, 22834-22836 (*(+)-Galbeigin, synth, pmr, cmr*)

4,7′-Epoxy-3′,3″,4″,5-tetramethoxy-3,8′:4′,8″-sesquilignan-9,9′,9″-triol E256

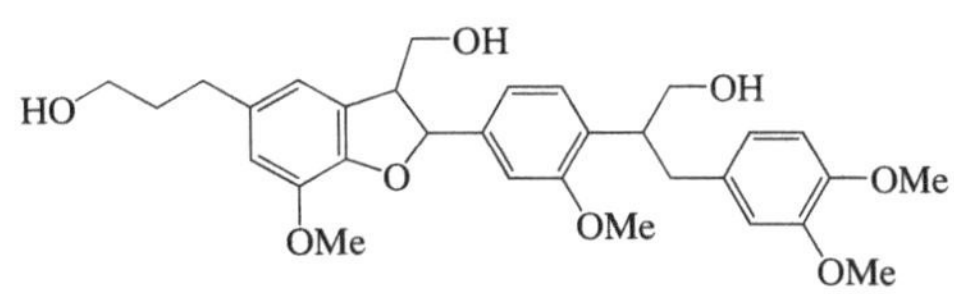

4,7′-Epoxy-9,9′,9″-trihydroxy-3′,3″,4″,5-tetramethoxy-3,8′:4′,8″-sesquineolignan

[248259-09-8]

$C_{31}H_{38}O_8$ 538.637

Related to Schinifolisatin A, S31. Constit. of the flowers of *Onopordum illyricum*. Pale yellow oil. $[\alpha]_D^{25}$ +36 (c, 0.05 in MeOH). λ_{max} 239 (sh); 282 (log ε 3.46) (MeOH).

Braca, A. *et al.*, *J. Nat. Prod.*, 1999, **62**, 1371-1375 (*Onopordum illyricum constit*)

2′,4-Epoxy-4′,6,7-trihydroxy-3′,5′-dimethylflavan E257

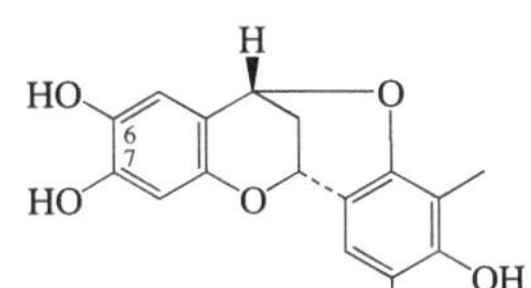

2,4-Dimethyl-6,12-methano-6H,12H-dibenz[b,f][1,5]dioxocin-3,8,9-triol. 4′,6,7-Trihydroxy-3′,5′-dimethyl-4-O-2′-cycloflavan

$C_{17}H_{16}O_5$ 300.310

(2R,4R)-form

6,7-Methylene ether

2′,4-Epoxy-4′-hydroxy-6,7-methylenedioxy-3′,5′-dimethylflavan

[1942089-03-3]

$C_{18}H_{16}O_5$ 312.321

Constit. of *Horsfieldia glabra*. Amorph. brown powder. $[\alpha]_D^{20}$ -20.6 (c, 0.001 in MeOH).

Peng, W. *et al.*, *Nat. Prod. Res.*, 2016, **30**, 2350-2355 (*Horsfieldia glabra constit*)

4,7′-Epoxy-3′,4′,5-trihydroxy-8,9-dinor-3,8′-lignan-7-al E258

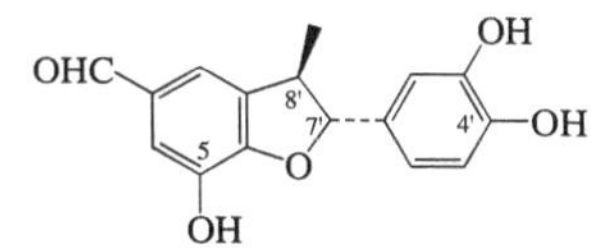

(7′R,8′R)-form

2-(3,4-Dihydroxyphenyl)-2,3-dihydro-7-hydroxy-3-methyl-5-benzofurancarboxaldehyde. 4,7′-Epoxy-3′,4′,5-trihydroxy-8,9-dinor-7-oxo-3,8′-neolignan

$C_{16}H_{14}O_5$ 286.283

(7′R,8′R)-form

3′,5-Di-Me ether

4,7′-Epoxy-4′-hydroxy-3′,5-dimethoxy-8,9-dinor-3,8′-lignan-7-al. 4,7′-Epoxy-4′-hydroxy-3′,5-dimethoxy-8,9-dinor-7-oxo-3,8′-lignan

[250600-23-8]

$C_{18}H_{18}O_5$ 314.337

Constit. of *Aristolochia pubescens*. $[\alpha]_D$ +28.6 ($CHCl_3$)

Tri-Me ether

4,7′-Epoxy-3′,4′,5-trimethoxy-8,9-dinor-3,8′-lignan-7-al. 4,7′-Epoxy-3′,4′,5-trimethoxy-8,9-dinor-7-oxo-3,8′-lignan. (+)-Kadsurenin M

[250600-28-3]

Yellow oil. $[\alpha]_D^{25}$ +25.1 (c, 0.38 in $CHCl_3$)

3′,4′-Methylene, 5-Me ether

4,7′-Epoxy-5-methoxy-3′,4′-methylenedioxy-8,9-dinor-3,8′-lignan-7-al. 4,7′-Epoxy-5-methoxy-3′,4′-methylenedioxy-8,9-dinor-7-oxo-3,8′-lignan. ***Myrisfrageal A***

[1613724-21-2]

$C_{18}H_{16}O_5$ 312.321

Constit. of the seeds of *Myristica fragrans*. NO prodn. inhibitor. Yellowish oil. $[\alpha]_D^{20}$ +55.6 (c, 1 in CH_3Cl). λ_{max} 203 (log ε 4.14); 236 (log ε 3.75); 294 (log ε 3.68) (MeOH)

Carboxylic acid, 3′,5-di-Me ether

2,3-Dihydro-2-(4-hydroxy-3-methoxyphenyl)-7-methoxy-3-methyl-5-benzofurancarboxylic acid. 4,7′-Epoxy-4′-hydroxy-3′,5-

dimethoxy-8,9-dinor-3,8′-lignan-7-oic acid
[250600-25-0]
$C_{18}H_{18}O_6$ 330.336
Constit. of *Aristolochia pubescens*. Yellow oil. $[\alpha]_D^{25}$ +14.7 (c, 0.57 in $CHCl_3$). λ_{max} 244 (log ε 3.88); 278 (log ε 3.74); 304 (sh) (log ε 3.42) ($CHCl_3$)

(7′R,8′S)-form

3′,4′-Methylene, 5-Me ether
[1609472-92-5]
$C_{18}H_{16}O_5$ 312.321
Constit. of *Aristolochia fordiana*. Oil. $[\alpha]_D^{25}$ +112 (c, 0.425 in $CHCl_3$). There are inconsistencies regarding stereochem. in ref.. λ_{max} 192 (log ε 3.83); 204 (log ε 4.26); 236 (log ε 3.80); 290 (log ε 2.70) (no solvent reported)

(7′S,8′S)-form

3′,5-Di-Me ether
[100663-36-3]
Constit. of *Machilus thunbergii*, *Magnolia ovata* and *Ocotea porosa*. Moderate inhibitor of topoisomerase I. Oil

4′,5-Di-Me ether
4,7′-Epoxy-3′-hydroxy-4′,5-dimethoxy-8,9-dinor-3,8′-lignan-7-al. 4,7′-Epoxy-3′-hydroxy-4′,5-dimethoxy-8,9-dinor-7-oxo-3,8′-lignan. ***Perseal F***
[371196-59-7]
$C_{18}H_{18}O_5$ 314.337
Constit. of *Machilus obovatifolia*. Cytotoxic. Oil. $[\alpha]_D^{25}$ -75 (c, 0.13 in $CHCl_3$). λ_{max} 236 (log ε 3.89); 291 (log ε 3.7) (EtOH)

Tri-Me ether
(-)-Kadsurenin M
[150133-00-9]
$C_{19}H_{20}O_5$ 328.364
Constit. of *Piper kadsura*. Platelet activating factor antagonist. Yellow oil. $[\alpha]_D^{25}$ -25.3 (c, 0.5 in $CHCl_3$)

3′,4′-Methylene, 5-Me ether
Perseal G
[371196-95-1]
[156665-53-1]
$C_{18}H_{16}O_5$ 312.321
Constit. of *Machilus obovatifolia* and *Ocotea porosa*. Oil.

Ma, Y. *et al.*, *Yaoxue Xuebao*, 1993, **28**, 370-373 (*(-)-Kadsurenin M*).
David, J.M. *et al.*, *Phytochemistry*, 1994, **36**, 491-499 (*Ocotea porosa constit*).
Nascimento, I.R. *et al.*, *Phytochemistry*, 1999, **52**, 345-350 (*Aristolochia pubescens constits*).
Tsai, I.-L. *et al.*, *Planta Med.*, 2001, **67**, 559-560 (*Perseal F*).
Li, G. *et al.*, *Biol. Pharm. Bull.*, 2004, **27**, 1147-1150 (*Machilus thunbergii constit, activity*).
Cao, G.-Y. *et al.*, *Food Chem. Toxicol.*, 2013, **62**, 167-171 (*Myrisfrageal A*).
Zhou, Z.-B. *et al.*, *Planta Med.*, 2013, **79**, 1730-1735 (*Aristolochia fordiana constit*)

4,7′-Epoxy-3′,4′,5′-trihydroxy-8,9-dinor-3,8′-lignan-7-al E259

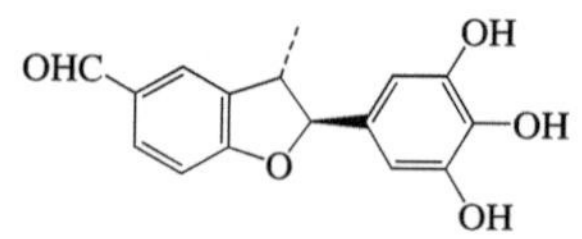

2,3-Dihydro-3-methyl-2-(3,4,5-trihydroxyphenyl)-5-benzofurancarboxaldehyde.
4,7′-Epoxy-3′,4′,5′-trihydroxy-8,9-dinor-7-oxo-3,8′-neolignan
$C_{16}H_{14}O_5$ 286.283

(7′S,8′S)-form

3′,4′-Methylene, 5′-Me ether
4,7′-Epoxy-3′-methoxy-4′,5′-methylenedioxy-8,9-dinor-3,8′-lignan-7-al. ***Myticaganal B***
$C_{18}H_{16}O_5$ 312.321
Constit. of the seeds of *Myristica fragrans*. Exhibits moderate cytotoxic activity. Gum. $[\alpha]_D^{28}$ +44.9 (c, 0.1 in MeOH). λ_{max} 231 (log ε 3.71); 288 (log ε 1.66) (MeOH).

Chumkaew, P. *et al.*, *J. Nat. Med.*, 2019, **73**, 273-277 (*Myticaganal B, activity*)

4,7′-Epoxy-3′,5,5′-trihydroxy-8,9-dinor-3,8′-lignan-7-oic acid E260

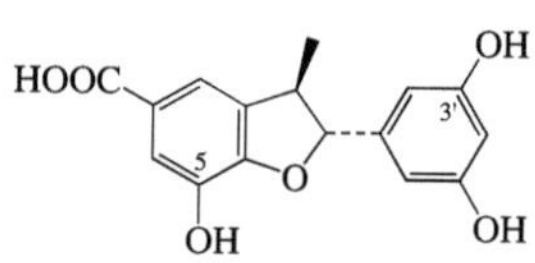

2-(3,5-Dihydroxyphenyl)-2,3-dihydro-7-hydroxy-3-methyl-5-benzofurancarboxylic acid
$C_{16}H_{14}O_6$ 302.282
Structures in this series may need revision. Phenyl 3,5-dioxygenated substn. pattern in doubt as a result of work done by Yang *et al* in 2015

(7′R*,8′R*)-form

7-Aldehyde, 3′,5-di-Me ether
4,7′-Epoxy-3′-hydroxy-5,5′-dimethoxy-8,9-dinor-3,8′-lignan-7-al. ***Curcasinlignan C***
[1426544-10-6]
$C_{18}H_{18}O_5$ 314.337
Constit. of the aerial parts of *Jatropha curcas*. Pale yellow oil. $[\alpha]_D^{18}$ -4.7 (c, 0.2 in $CHCl_3$). λ_{max} 207 (log ε 4.33); 234 (log ε 4.13); 289 (log ε 3.93); 301 (log ε 3.90) (MeOH).

Xu, J.J. *et al.*, *Z. Naturforsch., B*, 2012, **67**, 176-180 (*Curcasinlignan C*).
Yang, Y.-N. *et al.*, *J. Nat. Prod.*, 2015, **78**, 705-711 (*pmr, struct*).
Li, J.-L. *et al.*, *Arch. Pharmacal Res.*, 2017, **40**, 1265-1270 (*Acanthopanax senticosus constit*)

4,7′-Epoxy-3′,4′,5-trihydroxy-8,9-dinor-3,8′-lign-7′-en-7-al E261

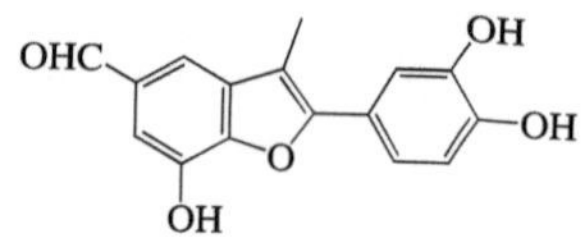

2-(3,4-Dihydroxyphenyl)-7-hydroxy-3-methyl-5-benzofurancarboxaldehyde.
4,7′-Epoxy-3′,4′,5-trihydroxy-8,9-dinor-7-oxo-3,8′-neolign-7′-ene
$C_{16}H_{12}O_5$ 284.267

3′,4′-Methylene, 5-Me ether
4,7′-Epoxy-5-methoxy-3′,4′-methylenedioxy-8,9-dinor-3,8′-lign-7′-en-7-al. 4,7′-Epoxy-5-methoxy-3′,4′-methylenedioxy-8,9-dinor-7-oxo-3,8′-lign-7′-ene.
Ocophyllal A
[1189119-88-7]
$C_{18}H_{14}O_5$ 310.305
Constit. of the leaves of *Ocotea macrophylla*. Exhibits antiproliferative activity. Moderate inhibitor of COX-2. Orange oil.

Coy, E.D. *et al.*, *Bioorg. Med. Chem. Lett.*, 2009, **19**, 6922-6925 (*Ocophyllal A, activity*).
Coy-Barrera, E.D. *et al.*, *Phytochemistry*, 2009, **70**, 1309-1314 (*Ocophyllal A*)

4,7′-Epoxy-2,3′,4′-trihydroxy-3,8′-ligna-7,7′-dien-9′-al E262

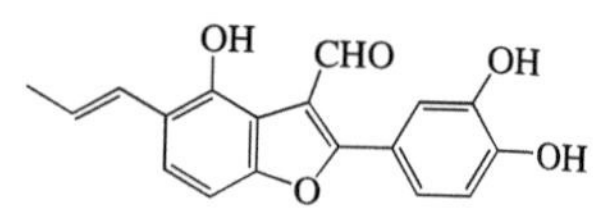

2-(3,4-Dihydroxyphenyl)-4-hydroxy-5-(1-propenyl)-3-benzofurancarboxaldehyde.
4,7′-Epoxy-2,3′,4′-trihydroxy-9′-oxo-3,8′-ligna-7,7′-diene
$C_{18}H_{14}O_5$ 310.305

(E)-form

Iteafuranal A [1651825-57-8]
Constit. of *Itea ilicifolia*. Prod. by *Phomopsis* sp. YK-7 on *Tetrastigma hemsleyanum*. Antioxidant. Yellow

needles (MeOH). Mp 199°. λ_{max} 226 (log ε 4.1); 286 (log ε 3.8); 344 (log ε 3.8) (MeOH)

4′-Me ether

4,7′-Epoxy-2,3′-dihydroxy-4′-methoxy-3,8′-ligna-7,7′-dien-9′-al. 4,7′-Epoxy-2,3′-dihydroxy-4′-methoxy-9′-oxo-3,8′-ligna-7,7′-diene. ***Iteafuranal B***

[1651825-59-0]

$C_{19}H_{16}O_5$ 324.332

Constit. of *Arctium lappa* and *Itea ilicifolia*. Antioxidant. Yellow powder. λ_{max} 230 (log ε 4.2); 298 (log ε 3.9); 353 (log ε 3.8) (MeOH).

Luo, G. *et al.*, *Phytochem. Lett.*, 2014, **10**, 19-22 (*Iteafuranals A,B, cryst struct, activity*).
Huang, Y.-P. *et al.*, *Zhongcaoyao*, 2017, **48**, 3032-3036 (*Phomopsis YK-7 metab*).
Zeng, W.-L. *et al.*, *Heterocycles*, 2019, **98**, 437-443 (*Arctium lappa constit*)

4,7′-Epoxy-3′,4′,5-trihydroxy-3,8′-ligna-7,7′-diene-9,9′-dioic acid E263

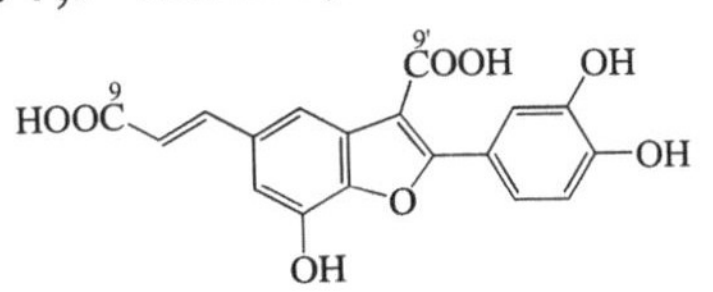

5-(2-Carboxyethenyl)-2-(3,4-dihydroxyphenyl)-7-hydroxy-3-benzofurancarboxylic acid

$C_{18}H_{12}O_8$ 356.286

(E)-form

Schizotenuin D [162439-97-6]

Constit. of *Schizonepeta tenuifolia*. Moderate inhibitor of lipid peroxidation. Powder. Mp 300°

9-[1R-Carboxy-2-(3,4-dihydroxyphenyl)ethyl] ester

Schizotenuin C₂

[181364-83-0]

$C_{27}H_{20}O_{12}$ 536.445

Constit. of *Schizonepeta tenuifolia*

9′-[1R-Carboxy-2-(3,4-dihydroxyphenyl)ethyl] ester

Schizotenuin C₁

[181492-66-0]

$C_{27}H_{20}O_{12}$ 536.445

Constit. of *Schizonepeta tenuifolia*

9,9′-Bis[1R-carboxy-2-(3,4-dihydroxyphenyl)ethyl] ester

Schizotenuin A

[144608-09-3]

$C_{36}H_{28}O_{16}$ 716.604

Isol. from *Lycopus lucidus* and *Schizonepeta tenuifolia*. Inhibitor of lipid peroxidation. Hyaluronidase inhibitor

9-[1R-Methoxycarbonyl-2-(3,4-dihydroxyphenyl)ethyl], 9′-[1R-carboxy-2-(3,4-dihydroxyphenyl) ethyl] ester

9″-O-Methylschizotenuin A. ***Schizotenuin B₂***

[181364-73-8]

$C_{37}H_{30}O_{16}$ 730.631

Constit. of *Schizonepeta tenuifolia*.

Maeda, S. *et al.*, *Chem. Pharm. Bull.*, 1994, **42**, 2500-2505 (*Schizotenuin D, synth, activity*).
Maeda, S. *et al.*, *Chem. Pharm. Bull.*, 1994, **42**, 2536-2545 (*synth*).
Maeda, S. *et al.*, *Chem. Pharm. Bull.*, 1995, **43**, 84-90 (*Schizotenuin A, activity*).
Matsuta, M. *et al.*, *Nat. Med. (Tokyo)*, 1996, **50**, 204-211 (*Schizotenuins A-D*).
Murata, T. *et al.*, *Chem. Pharm. Bull.*, 2010, **58**, 394-397 (*Schizotenuin A, activity*)

4,7′-Epoxy-4′,9,9′-trihydroxy-3,8′-lignan-7-one E264

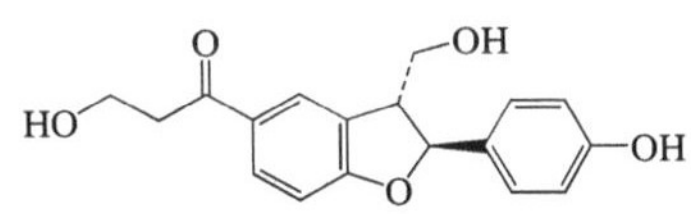

1-[2,3-Dihydro-3-(hydroxymethyl)-2-(4-hydroxyphenyl)-5-benzofuranyl]-3-hydroxy-1-propanone. 4,7′-Epoxy-4′,9,9′-trihydroxy-7-oxo-3,8′-neolignan

$C_{18}H_{18}O_5$ 314.337

(7′S,8′R)-form

Mubiesin C

Constit. of the seeds of *Momordica cochinchinensis*. Amorph. powder. $[\alpha]_D^{25}$ -18.9 (c, 0.1 in MeOH). λ_{max} 225 (log ε 3.36); 284 (log ε 3.27) (MeOH).

Wang, M. *et al.*, *Fitoterapia*, 2019, **139**, 104360:1-6 (*Mubiesin C*)

3,7′-Epoxy-3′,4,4′-trihydroxy-2,8′-lign-7-ene-9,9′-dioic acid E265

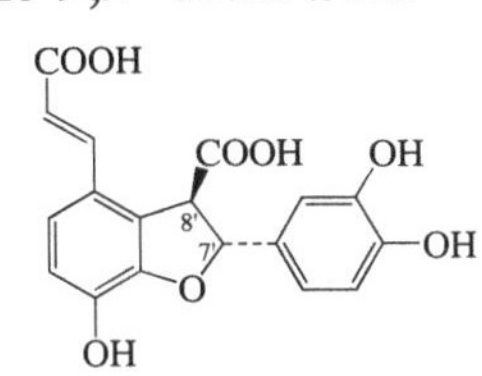

(7E,7′R,8′R)-form

4-(2-Carboxyethenyl)-2-(3,4-dihydroxyphenyl)-2,3-dihydro-7-hydroxy-3-benzofurancarboxylic acid, CAS

[172285-71-1]

$C_{18}H_{14}O_8$ 358.302

(7E,7′R,8′R)-form

7-Epiblechnic acid [145554-85-4]

Constit. of *Blechnum orientale*, *Brainea insignis*, *Meehania fargesii*, *Salvia miltiorrhiza* and *Woodwardia* spp.. Powder. Mp 231.5 - 232.5°. $[\alpha]_D^{23}$ -145 (c, 1 in MeOH). λ_{max} 254 (log ε 4.15); 295 (log ε 4.09); 307 (log ε 4.1); 327 (log ε 4.03) (MeOH)

(7E,7′S,8′R)-form

Blechnic acid [146805-53-0]

Constit. of *Blechnum orientale*, *Blechnum spicant*, *Pteridium aquilinum* var. *latiusculum* and *Woodwardia* spp.. Needles. Mp 198 - 198.5°. $[\alpha]_D^{23}$ -28 (c, 1 in MeOH). λ_{max} 253 (log ε 4.13); 292 (log ε 4.11); 305 (log ε 4.1); 327 (log ε 4.02) (MeOH)

9-(3-Carboxy-5,6-dihydroxy-3-cyclohexen-1-yl)(1R,5R,6R-) ester

Brainic acid

[146713-96-4]

$C_{25}H_{22}O_{12}$ 514.439

Constit. of *Blechnum orientale*, *Blechnum spicant*, *Brainea insignis*, *Pteridium aquilinum* var. *latiusculum* and *Woodwardia* spp.. Amorph. powder. $[\alpha]_D^{24}$ -42 (c, 1 in MeOH). λ_{max} 250 (log ε 4.12); 292 (log ε 4.15); 327 (log ε 4.02) (MeOH)

(7E,7′S,8′S)-form

8-Epiblechnic acid. Prolithospermic acid. ***Przewalskinic acid*** [145554-86-5] [136112-75-9]

Constit. of *Blechnum orientale*, *Clinopodium gracile*, *Melissa officinalis*, *Salvia miltiorrhiza* and *Salvia przewalskii*. Vasodilator. $[\alpha]_D^{23}$ +142 (c, 1.1 in MeOH)

9′-Et ester

9′-Ethyl prolithospermate

[2098346-16-6]

$C_{20}H_{18}O_8$ 386.356

Constit. of *Salvia yunnanensis*. Yellow powder. $[\alpha]_D^{25}$ +85 (c, 0.8 in MeOH). λ_{max} 205 (log ε 1.27); 290 (log ε 0.37) (MeOH)

9′-Butyl ester

9′-Butyl prolithospermate

[2098346-17-7]

$C_{22}H_{22}O_8$ 414.410

Constit. of *Salvia yunnanensis*. Yellow powder. $[\alpha]_D^{25}$ +40 (c, 0.6 in MeOH). λ_{max} 246 (log ε 0.28); 291 (log ε 0.24) (MeOH)

(7Z,7′S,8′R)-form [333459-72-6]

Constit. of *Blechnum spicant*. $[\alpha]_D^{20}$ -128 (c, 0.58 in MeOH). λ_{max} 238; 280 (MeOH).

Lu, X.Z. *et al.*, *Chin. Chem. Lett.*, 1991, **2**, 301-302 (*Przewalskinic acid*).

Wada, H. *et al.*, *Chem. Pharm. Bull.*, 1992, **40**, 2099-2101 (*Blechnum orientale constits*).
Nagai, M. *et al.*, *Biol. Pharm. Bull.*, 1996, **19**, 228-232 (*Salvia miltiorrhiza constit, activity*).
Ai, C.B. *et al.*, *Chin. Chem. Lett.*, 1996, **7**, 427-430 (*synth*).
Davin, L.B. *et al.*, *Phytochemistry*, 2003, **62**, 501-511 (*Blechnum spicant constits, biosyth*).
Tian, S.-M. *et al.*, *Zhongguo Yaoxue Zazhi*, 2011, **46**, 1238-1241 (*Pteridium aquilinum constits*).
Li, C. *et al.*, *J. Ind. Microbiol. Biotechnol.*, 2014, **41**, 887-892 (*Przewalskinic acid, synth*).
Ito, M. *et al.*, *Synlett*, 2014, **25**, 288-292 (*synth*).
Yu, Y. *et al.*, *Nat. Prod. Res.*, 2017, **31**, 2505-2512 (*9′-Ethyl prolithospermate, 9′-Butyl prolithospermate*)

2,7′-Epoxy-3′,4′,5-trihydroxy-1,8′-lign-8-en-4(1*H*)-one E266

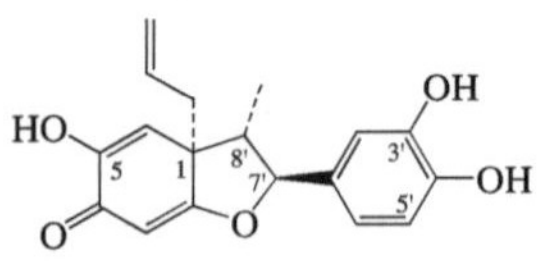

(1*R*,7′*S*,8′*S*)-*form*

$C_{18}H_{18}O_5$ 314.337

(1R,7′S,8′S)-form

Tri-Me ether

2-(3,4-Dimethoxyphenyl)-3,3a-dihydro-5-methoxy-3-methyl-3a-(2-propenyl)-6(2H)-benzofuranone, CAS. *2,7′-Epoxy-3′,4′,5-trimethoxy-1,8′-lign-8-en-4(1H)-one*

[61240-28-6]

$C_{21}H_{24}O_5$ 356.418

Constit. of *Aniba* spp. and *Nectandra* sp.. Viscous oil

3′,4′-Methylene, 5-Me ether

2-(1,3-Benzodioxol-5-yl)-3,3a-dihydro-5-methoxy-3-methyl-3a-(2-propenyl)-6(2H)-benzofuranone, CAS. *2,7′-Epoxy-5-methoxy-3′,4′-methylenedioxy-1,8′-lign-8-en-4(1H)-one*. **Burchellin**

[38276-59-4]

[69175-80-0] (relative config.) [69175-83-3] (unspecified stereochem.)

$C_{20}H_{20}O_5$ 340.375

Constit. of *Aniba burchellii*, *Magnolia biondii*, *Ocotea cymbarum*, *Ocotea porosa* and *Piper wightii*. Larvicidal agent against *Aedes aegypti*. Cryst. (MeOH). Mp 154 - 156°. $[\alpha]_D^{22}$ +42 (c, 0.06 in $CHCl_3$). λ_{max} 213 (ε 20250); 261 (ε 16000); 287 (sh) (EtOH)

3-Methoxy, 3′,4′-methylene, 5-Me ether

2,7′-Epoxy-3,5-dimethoxy-3′,4′-methylenedioxy-1,8′-lign-8-en-4(1H)-one. ***3-Methoxyburchellin***

[104701-07-7]

$C_{21}H_{22}O_6$ 370.401

Constit. of *Licaria puchury-major* and *Ocotea aciphylla*. Viscous oil

(1S,7′R,8′S)-form

Tri-Me ether

[162239-54-5]

Constit. of *Piper schmidtii* and *Piper wightii*. Viscous oil. $[\alpha]_D^{22}$ -20.8 (c, 0.06 in MeOH)

3′,4′-Methylene, 5-Me ether

2-Epi-3a-epiburchellin. ***1,7′-Diepiburchellin.*** *cis-Burchellin*

[57457-99-5]

$C_{20}H_{20}O_5$ 340.375

Constit. of *Aniba terminalis*, *Magnolia biondii*, *Ocotea porosa*, *Piper argyrophyllum*, *Piper clarkii* and *Piper wightii*. Cryst. (MeOH). Mp 121 - 122°. $[\alpha]_D^{22}$-67.4 (c, 0.5 in MeOH). λ_{max} 257 (log ε 4.24); 285 (log ε 3.96) (EtOH)

3-Methoxy, 3′,4′-methylene, 5-Me ether

1,7′-Diepi-3-methoxyburchellin

[87592-88-9]

$C_{21}H_{22}O_6$ 370.401

Constit. of *Ocotea aciphylla* and *Ocotea catharinensis*. Cryst. (CCl_4). Mp 137 - 139°. λ_{max} 267 (ε 17400); 285 (ε 9200) (MeOH)

5′-Methoxy, 3′,4′-methylene, 5-Me ether

2,7′-Epoxy-3′,5-dimethoxy-4′,5′-methylenedioxy-1,8′-lign-8-en-4(1H)-one. **Dysodanthin C**

[62211-15-8]

$C_{21}H_{22}O_6$ 370.401

Constit. of the roots of *Endlicheria dysodantha*. Plates (CH_2Cl_2/Et_2O). Mp 182°. Synonym not assigned in ref..

Lima, O.A. *et al.*, *Phytochemistry*, 1972, **11**, 2031-2037 (*Burchellin*).
Gottlieb, O.R. *et al.*, *Phytochemistry*, 1975, **14**, 1825-1827 (*Diepiburchellin*).
Fernandes, J.B. *et al.*, *Phytochemistry*, 1976, **15**, 1033-1036 (*Aniba constit*).
Wenkert, E. *et al.*, *Phytochemistry*, 1976, **15**, 1547-1551 (*Burchellin, cmr*).
Büchi, G. *et al.*, *J. Am. Chem. Soc.*, 1977, **99**, 8073-8075 (*synth*).
Cavalcante, S. de H. *et al.*, *Phytochemistry*, 1978, **17**, 983-985 (*Aniba terminalis constit*).
Haraguchi, M. *et al.*, *Phytochemistry*, 1983, **22**, 561-563 (*1,7′-Diepi-3-methoxyburchellin*).
Felicio, J.D'Arc. *et al.*, *Phytochemistry*, 1986, **25**, 1707-1710 (*3-Methoxyburchellin*).
Tyagi, O.D. *et al.*, *Acta Chem. Scand.*, 1994, **48**, 1007-1011 (*Burchellin, Diepiburchellin*).
Engler, T.A. *et al.*, *J. Org. Chem.*, 1994, **59**, 6588-6599 (*Burchellin, synth*).
Jensen, S. *et al.*, *Phytochemistry*, 1994, **36**, 789-792 (*Piper clarkii constit*).
Tyagi, O.D. *et al.*, *Acta Chem. Scand.*, 1995, **49**, 142-147 (*Piper schmidtii constit, abs config*).
Singh, S.K. *et al.*, *Phytochemistry*, 1996, **43**, 1355-1360 (*Piper argyrophyllum constit*).
Narciso, J.O.A. *et al.*, *Parasites Vectors*, 2014, **7**, 172 (*Burchellin, activity*).
Wang, R.-B. *et al.*, *Org. Biomol. Chem.*, 2020, **18**, 9081-9087 (*Burchellin, synth*)

4,7′-Epoxy-1,3′,4′-trihydroxy-3,8′-lign-8-en-6(1*H*)-one E267

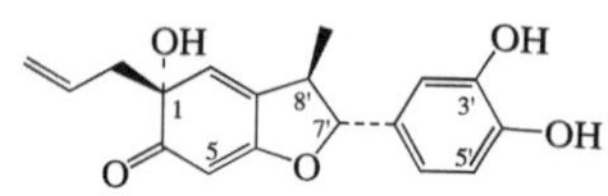

(1*R*,7′*R*,8′*R*)-*form*

2-(3,4-Dihydroxyphenyl)-3,5-dihydro-5-hydroxy-3-methyl-5-(2-propenyl)-6(2H)-benzofuranone, CAS

$C_{18}H_{18}O_5$ 314.337

(1R,7′R,8′R)-form

3′,4′-Methylene, 1-Me ether

4,7′-Epoxy-1-methoxy-3′,4′-methylenedioxy-3,8′-lign-8-en-6(1H)-one

[155551-60-3]

[62163-21-7] (unspecified stereochem.) [69175-82-2] (rel. config.)

$C_{20}H_{20}O_5$ 340.375

Constit. of *Piper kadsura*

(1R,7′R,8′S)-form

3′,4′-Methylene, 1-Me ether

[67920-49-4]

$C_{20}H_{20}O_5$ 340.375

Constit. of *Mezilaurus itauba*

(1R,7′S,8′S)-form

Tri-Me ether

4,7′-Epoxy-1,3′,4′-trimethoxy-3,8′-lign-8-en-6(1H)-one

[112456-79-8]

$C_{21}H_{24}O_5$ 356.418

Constit. of *Aniba riparia* and *Piper kadsura*

3′,4′-Methylene, 1-Me ether

[67920-48-3]

Constit. of *Aniba riparia*, *Licaria armeniaca* and *Ocotea veraguensis*. Oil. $[\alpha]_D^{21}$ +60 (c, 0.05 in MeOH)

5′-Methoxy, tri-Me ether

4,7′-Epoxy-1,3′,4′,5′-tetramethoxy-3,8′-lign-8-en-6(1H)-one

[57462-51-8]

$C_{22}H_{26}O_6$ 386.444

Constit. of *Aniba riparia* and other *Aniba* spp.. Viscous oil

5-Methoxy, 3′,4′-methylene, 1-Me ether

4,7′-Epoxy-1,5-dimethoxy-3′,4′-methylenedioxy-3,8′-lign-8-en-6(1H)-one

[87592-89-0]

$C_{21}H_{22}O_6$ 370.401

Constit. of *Ocotea catharinensis* and *Ocotea veraguensis*. Oil. $[\alpha]_D^{21}$ +40 (c, 0.06 in MeOH). λ_{max} 238 (ε 16450); 285 (ε 11100); 325 (ε 9200) (MeOH)

(1S,7′R,8′R)-form

Tri-Me ether

[156558-82-6]

Constit. of *Piper clarkii*. Oil. $[\alpha]_D^{22}$ +61.4 (c, 0.2 in MeOH)

3′,4′-Methylene, 1-Me ether

[155551-59-0]

Constit. of *Piper argyrophylum*, *Piper clarkii* and *Piper kadsura*. Oil. $[\alpha]_D^{22}$ -84.3 (c, 1.5 in MeOH)

(1S,7′S,8′S)-form

3′,4′-Methylene, 1-Me ether

[57430-03-2]

Constit. of *Aniba affinis*, *Aniba terminalis* and *Aniba burchellii*. Oil

5-Methoxy, 3′,4′-methylene, 1-Me ether

[112464-24-1]

$C_{21}H_{22}O_6$ 370.401

Constit. of *Ocotea veraguensis*.

Gottlieb, O.R. *et al.*, *Phytochemistry*, 1975, **14**, 1825-1827 (*Aniba terminalis constit*).

De Alvarenga, M.A. *et al.*, *Phytochemistry*, 1977, **16**, 1797-1799 (*Aniba burchellii constit*).

Cavalcante, S. de H. *et al.*, *Phytochemistry*, 1978, **17**, 983-985 (*pmr*).

Yanez, X. *et al.*, *Phytochemistry*, 1986, **25**, 1953-1956 (*Mezilaurus itauba constit*).

Khan, M.R. *et al.*, *Phytochemistry*, 1987, **26**, 1155-1158 (*Ocotea veraguensis constit*).

Dodson, C.D. *et al.*, *Phytochemistry*, 1987, **26**, 2037-2040 (*Ocotea veraguensis constit*).

Barbosa-Filho, J.M. *et al.*, *Phytochemistry*, 1987, **26**, 2615-2617 (*Aniba riparia constits*).

Ma, Y. *et al.*, *Zhiwu Xuebao*, 1993, **35**, 687-692 (*Piper kadsura constits*).

Jensen, S. *et al.*, *Phytochemistry*, 1994, **36**, 789-792 (*Piper clarkii constits*).

Prasad, A.K. *et al.*, *Phytochemistry*, 1995, **39**, 655-658 (*Piper clarkii constit*).

Singh, S.K. *et al.*, *Phytochemistry*, 1996, **43**, 1355-1360 (*Piper argyrophylum constit*)

4,7′-Epoxy-3,3′,4′-trihydroxy-3,8′-lign-8-en-6(3*H*)-one E268

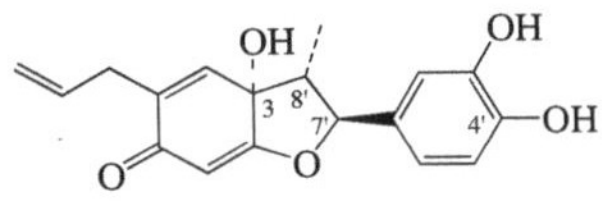

(3R,7′S,8′R)-form

2-(3,4-Dihydroxyphenyl)-3,3a dihydro-3a-hydroxy-3-methyl-5-(2-propenyl)-6(2H)-benzofuranone

$C_{18}H_{18}O_5$ 314.337

(3R,7′S,8′R)-form

Tri-Me ether

4,7′-Epoxy-3,3′,4′-trimethoxy-3,8′-lign-8-en-6(3H)-one. ***(+)-Denudatin B***

[87402-88-8]

$C_{21}H_{24}O_5$ 356.418

Constit. of *Magnolia denudata*, *Magnolia fargesii*, *Nectandra amazonum*, *Piper betle* and *Piper clarkii*. NO prodn. inhibitor. Exhibits synergistic antibacterial activity combined with Norfloxacin. Oil. $[\alpha]_D$ +82.7 (c, 2.67 in MeOH). λ_{max} 208 (ε 18010); 235 (ε 17820); 285 (ε 6900); 300 (ε 6350) (MeOH)

3′,4′-Methylene, 3-Me ether

4,7′-Epoxy-3-methoxy-3′,4′-methylenedioxy-3,8′-lign-8-en-6(3H)-one. ***Denudatin A***

[87402-87-7]

$C_{20}H_{20}O_5$ 340.375

Constit. of *Magnolia denudata*, *Magnolia soulangeana* and *Piper kadsura*. NO prodn. inhibitor. Moderate PAF inhibitor. Cryst. (Et_2O). Mp 106 - 107°. $[\alpha]_D$ +122.3 (c, 2.5 in MeOH)

(3S,7′R,8′R)-form

3′,4′-Methylene, 3-Me ether

[160821-05-6]

$C_{20}H_{20}O_5$ 340.375

Constit. of *Magnolia denudata* and *Piper wightii*. Oil. $[\alpha]_D^{22}$ -38.8 (c, 0.4 in MeOH)

(3S,7′R,8′S)-form

Tri-Me ether

(-)-Denudatin B

[150134-23-9]

$C_{21}H_{24}O_5$ 356.418

Isol. from *Piper kadsura*. Potent PAF antagonist. $[\alpha]_D^{15}$ -76.4 (c, 0.11 in $CHCl_3$)

(3S,7′S,8′R)-form

Tri-Me ether

Kadsurenone

[95851-37-9]

$C_{21}H_{24}O_5$ 356.418

Isol. from *Piper kadsura* and *Piper bonii*. PAF binding inhibitor. COX-1 and 5-LOX inhibitor. Cryst.. Mp 62 - 65°. $[\alpha]_D^{22}$ +3.2

3′,4′-Methylene, 3-Me ether

Hancinone

[104013-61-8]

$C_{20}H_{20}O_5$ 340.375

Constit. of *Magnolia denudata*, *Piper hancei* and *Piper wightii*. $[\alpha]_D^{22}$ +18 (c, 0.15 in $CHCl_3$).

Iida, T. *et al.*, *Phytochemistry*, 1982, **21**, 2939-2941 (*Denudatins A,B*).

Chang, M.N. *et al.*, *Phytochemistry*, 1985, **24**, 2079-2082 (*Kadsurenone*).

Han, G. *et al.*, *Yaoxue Xuebao*, 1986, **21**, 361-365 (*Hancinone*).

Ponpipom, M.M. *et al.*, *J. Med. Chem.*, 1987, **30**, 136-142 (*Kadsurenone, activity*).

Chen, C.-C. *et al.*, *Chem. Pharm. Bull.*, 1988, **36**, 1791-1795 (*Denudatin B*).

Ma, Y. *et al.*, *Yaoxue Xuebao*, 1993, **28**, 370-373 (*(-)-Denudatin B, activity*).

Tyagi, O.D. *et al.*, *Acta Chem. Scand.*, 1994, **48**, 1007-1011 (*Piper wightii constits*).

Engler, T.A. *et al.*, *J. Org. Chem.*, 1994, **59**, 6567-6587 (*Denudatin B, Kadsurenone, synth*).

Horne, D.A. *et al.*, *Tetrahedron Lett.*, 1999, **40**, 5443-5447 (*Denudatin B, synth*).

Kuroyanagi, M. *et al.*, *Chem. Pharm. Bull.*, 2000, **48**, 832-837 (*Denudatin A, activity*).

Song, Q. *et al.*, *Acta Crystallogr., Sect. E*, 2001, **57**, 1105-1106 (*Denudatin A, cryst struct*).

Wang, M. *et al.*, *J. Org. Chem.*, 2002, **67**, 5405-5407 (*Denudatin B, Kadsurenone, synth*).

Noshita, T. *et al.*, *Biosci., Biotechnol., Biochem.*, 2008, **72**, 2775-2778 (*Denudatins A,B, activity*).

Coy, E.D. *et al.*, *Bioorg. Med. Chem. Lett.*, 2009, **19**, 6922-6925 (*Kadsurenone, activity*).

Sun, Z.-L. *et al.*, *RSC Adv.*, 2016, **6**, 43518-43525 (*Denudatin B, cd, activity*)

2′,4-Epoxy-4′,6,7-trihydroxy-3′-methylflavan E269

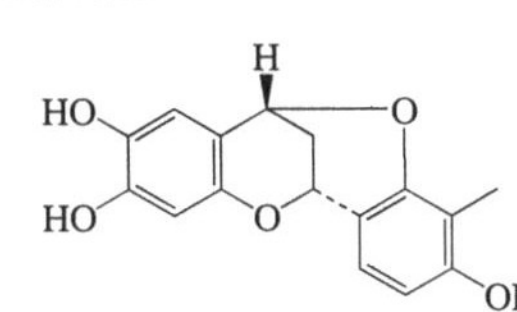

4-Methyl-6,12-methano-6H,12H-dibenz[b,f][1,5]dioxocin-3,8,9-triol.

4′,6,7-Trihydroxy-3′-methyl-4-O-2′-cycloflavan

$C_{16}H_{14}O_5$ 286.283

(2R,4R)-form
6,7-Methylene ether
2′,4-Epoxy-4′-hydroxy-6,7-methylenedioxy-3′-methylflavan
[1942089-04-4]
$C_{17}H_{14}O_5$ 298.294
Constit. of *Horsfieldia glabra*. Amorph. brown powder. $[\alpha]^{20}_{D}$ -17.6 (c, 0.001 in MeOH).

Peng, W. *et al.*, *Nat. Prod. Res.*, 2016, **30**, 2350-2355 (*Horsfieldia glabra constit*)

4,7′-Epoxy-3′,4′,5-trihydroxy-9′-nor-3,8′-lign-7′-en-9-oic acid E270

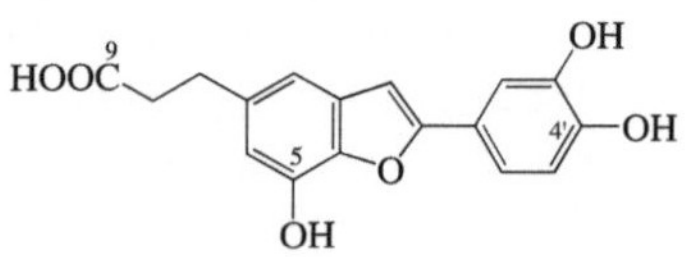

2-(3,4-Dihydroxyphenyl)-7-hydroxy-5-benzofuranpropanoic acid, CAS
$C_{17}H_{14}O_6$ 314.293
3′,4′-Methylene, 5-Me ether
7-Methoxy-2-(3,4-methylenedioxyphenyl)-5-benzofuranpropanoic acid. 4,7′-Epoxy-5-methoxy-3′,4′-methylenedioxy-9′-nor-3,8′-lign-7′-en-9-oic acid. ***Egonoic acid***
[15433-97-3]
$C_{19}H_{16}O_6$ 340.331
Constit. of *Anaxagorea clavata*, *Nectandra purpurascens* and *Styrax agrestis*. Cryst. (EtOH). Mp 113 - 114°. Mp 148.2 - 149°. Mp 179 - 180°. λ_{max} 216 (log ε 4.15); 247 (sh); 300 (sh); 316 (log ε 4.05); 332 (sh) (log ε 3.91) (EtOH)
3′,4′-Methylene, 5-Me ether, Me ester
Methyl egonoate
[61222-91-1]
$C_{20}H_{18}O_6$ 354.358
Constit. of the seeds of *Styrax obassia*. Powder. Mp 154 - 155° (105-107°). $[\alpha]^{20}_{D}$ +2.6 (c, 0.1 in $CHCl_3$). Opt. rotation unaccounted for. λ_{max} 238 (log ε 3.4); 316 (log ε 3.9) ($CHCl_3$)
3′,4′-Methylene, 5-Me ether, ester with Egonol, E2
Egonyl egonoate
[955997-33-8]
$C_{38}H_{32}O_{10}$ 648.664
Constit. of the seeds of *Styrax macranthus*. Amorph. powder. λ_{max} 242 (log ε 4.44); 302 (log ε 4.63); 318 (log ε 4.68) (MeOH).

Hopkins, C.Y. *et al.*, *Can. J. Chem.*, 1967, **45**, 1425-1428 (*Egonoic acid, Me ester, synth*).
Puentes de Diaz, A.M. *et al.*, *Phytochemistry*, 1997, **44**, 345-346 (*Anaxagorea clavata constit*).
Pal, G. *et al.*, *J. Chem. Res., Synop.*, 2003, 142-143 (*3′,4′-methylene 5-Me ether, synth*).
Luo, Y. *et al.*, *Fitoterapia*, 2007, **78**, 211-214 (*Egonyl egonate*).
Lee, H.-J. *et al.*, *Chem. Nat. Compd. (Engl. Transl.)*, 2008, **44**, 435-439 (*Methyl egonoate*).
Emirdag-Oztürk, S. *et al.*, *Bioorg. Med. Chem.*, 2011, **19**, 1179-1188 (*Egonoic acid, synth*).
Liu, J. *et al.*, *J. Nat. Prod.*, 2011, **74**, 2081-2088 (*Egonoic acid, synth*)

3,7′-Epoxy-3′,4′,8′-trihydroxy-9′-nor-4,8′-oxylign-7-en-9-oic acid E271

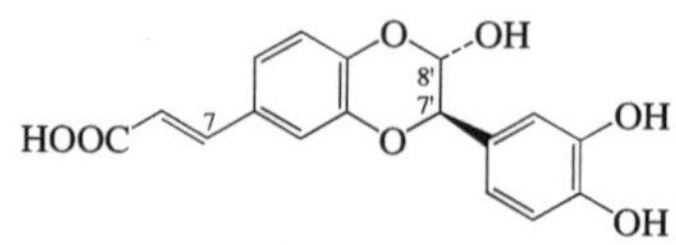

3-[3-(3,4-Dihydroxyphenyl)-2,3-dihydro-2-hydroxy-1,4-benzodioxin-6-yl]-2-propenoic acid
$C_{17}H_{14}O_7$ 330.292
(7E,7′R,8′S)-form
8′-O-(3,4-Dihydroxy-E-cinnamoyl)
[1588874-32-1]
$C_{26}H_{20}O_{10}$ 492.436
Constit. of *Hyptis rhomboides*. Xanthine oxidase inhibitor. λ_{max} 286 (log ε 4.1); 332 (log ε 4.04) (MeOH).

Tsai, S.-F. *et al.*, *Phytochemistry*, 2014, **101**, 121-127 (*Hyptis rhomboides constit*)

4,7′-Epoxy-3′,4′,8′-trihydroxy-9′-nor-3,8′-oxylign-7-en-9-oic acid E272

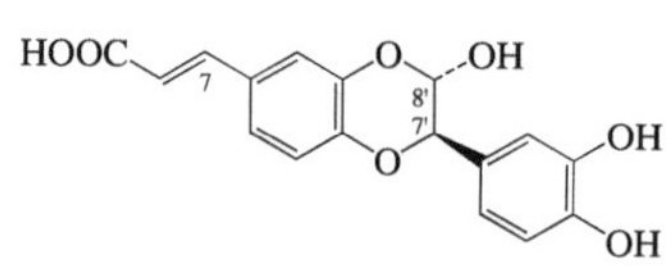

3-[2-(3,4-Dihydroxyphenyl)-2,3-dihydro-3-hydroxy-1,4-benzodioxin-6-yl]-2-propenoic acid
$C_{17}H_{14}O_7$ 330.292
(7E,7′R,8′S)-form
8′-O-(3,4-Dihydroxy-E-cinnamoyl)
[1588874-31-0]
$C_{26}H_{20}O_{10}$ 492.436
Constit. of *Hyptis rhomboides*. Xanthine oxidase inhibitor. λ_{max} 289 (log ε 4.3); 331 (log ε 4.27) (MeOH).

Tsai, S.-F. *et al.*, *Phytochemistry*, 2014, **101**, 121-127 (*Hyptis rhomboides constit*)

4,7′-Epoxy-3′,4′,5-trihydroxy-8,9,9′-trinor-3,8′-lignan-7-oic acid E273

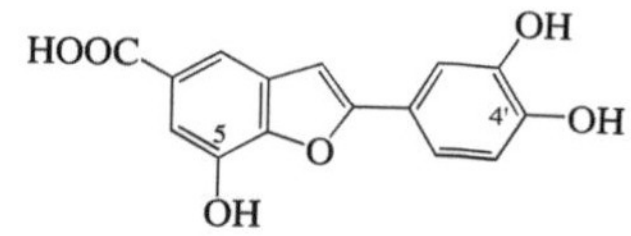

2-(3,4-Dihydroxyphenyl)-7-hydroxy-5-benzofurancarboxylic acid
$C_{15}H_{10}O_6$ 286.239
Lignan numbering shown
3′,4′-Methylene, 5-Me ether
4,7′-Epoxy-5-methoxy-3′,4′-methylenedioxy-8,9,9′-trinor-3,8′-lignan-7-oic acid
3′,4′-Methylene, 5-Me ether, Me ester
[1310351-00-8]
$C_{18}H_{14}O_6$ 326.304
Constit. of the bark of *Nectandra purpurascens*. Amorph. powder. Mp 183 - 184°.

Rios-Motta, J. *et al.*, *Nat. Prod. Commun.*, 2010, **5**, 1063-1066 (*Nectandra purpurascens constit*)

2′,8′-Epoxy-3,5,5′-trimethoxy-4,7′-oxylign-8-ene E274

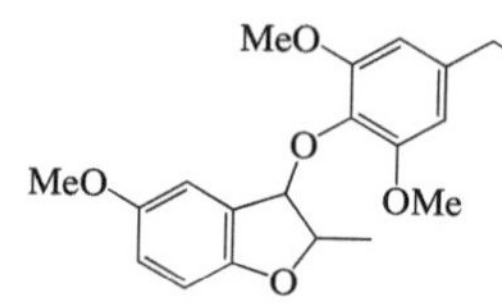

3-[2,6-Dimethoxy-4-(2-propenyl) phenoxy]-2,3-dihydro-5-methoxy-2-methylbenzofuran. 2′,8′-Epoxy-3,5,5′-trimethoxy-4,7′-oxyneolign-8-ene
$C_{21}H_{24}O_5$ 356.418
Constit. of the kernel of *Myristica fragrans*.

Chiu, S. *et al.*, *Nat. Prod. Commun.*, 2016, **11**, 483-488 (*Myristica fragrans constit*)

7′,7″-Epoxy-3″,5,5′-trimethoxy-4,4′,4″-trihydroxy-3,3′:8′,8″-sesquilign-7-ene-9,9′,9″-trioic acid E275

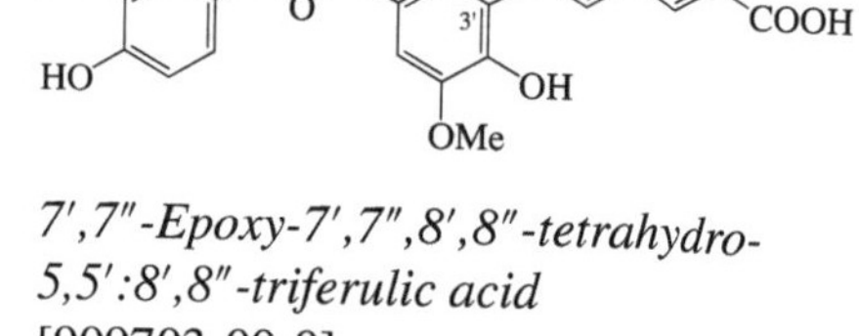

7′,7″-Epoxy-7′,7″,8′,8″-tetrahydro-5,5′:8′,8″-triferulic acid
[909703-00-0]
$C_{30}H_{28}O_{13}$ 596.541

Constit. of insoluble maize bran fibre (*Zea mays*). λ_{max} 288; 324 (no solvent reported).

Bunzel, M. *et al., J. Agric. Food Chem.*, 2006, **54**, 6409-6418 (*Zea mays constit*)

Erathrin A E276

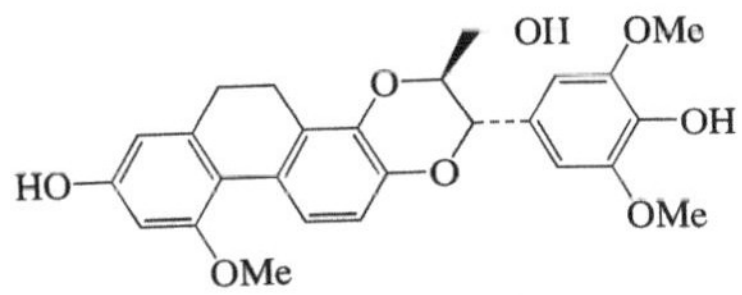

Relative Configuration

$C_{26}H_{26}O_8$ 466.486

Phenanthrenolignan. Constit. of *Eria bambusifolia*. Exhibits cytotoxic activity. Oil. $[\alpha]_D^{21}$ -6.5 (c, 0.24 in MeOH). λ_{max} 208 (log ε 4.02); 281 (log ε 3.53) (MeOH).

Zhan, R. *et al., Chin. J. Nat. Med.*, 2016, **14**, 621-625 (*Erathrin A, activity*)

Erimopyrone E277

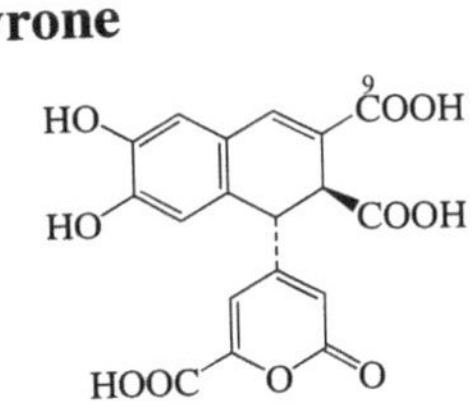

Absolute Configuration

[248255-41-6]

$C_{18}H_{12}O_{10}$ 388.284

Seconaphthalenoid lignan. Constit. of the liverwort *Moerckia erimona*. Light yellow powder. $[\alpha]_D$ -69.8 (c, 0.09 in MeOH). λ_{max} 250 (log ε 3.73); 306 (log ε 3.57); 332 (sh) (MeOH)

9-Me ester

[247923-72-4]

$C_{19}H_{14}O_{10}$ 402.311

Constit. of *Moerckia erimona*. Light yellow powder. $[\alpha]_D$ -68.8 (c, 0.74 in MeOH). λ_{max} 252 (log ε 4.29); 308 (log ε 4.15); 331 (sh) (MeOH).

Tazaki, H. *et al., Biosci., Biotechnol., Biochem.*, 1999, **63**, 1238-1241 (*Moerckia erimona constits*)

Erythraddison B E278

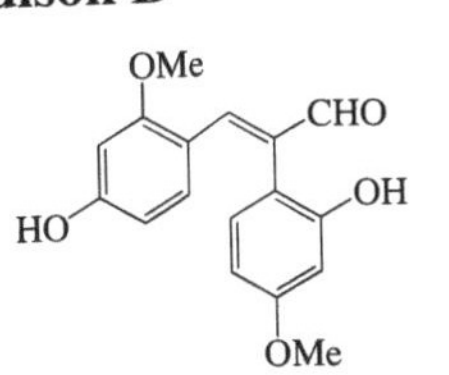

2-(2-Hydroxy-4-methoxyphenyl)-3-(4-hydroxy-2-methoxyphenyl)-2-propenal

$C_{17}H_{16}O_5$ 300.31

(E)-form [1257862-73-9]

Constit. of the root bark of *Erythrina addisoniae*. Inhibitor of viral neuraminidases. Yellow-brown powder. λ_{max} 209; 229; 266; 365; 383 (MeOH).

Nguyen, P.H. *et al., Bioorg. Med. Chem. Lett.*, 2010, **20**, 6430-6434 (*Erythraddison B*)

Eudeshonokiol A E279

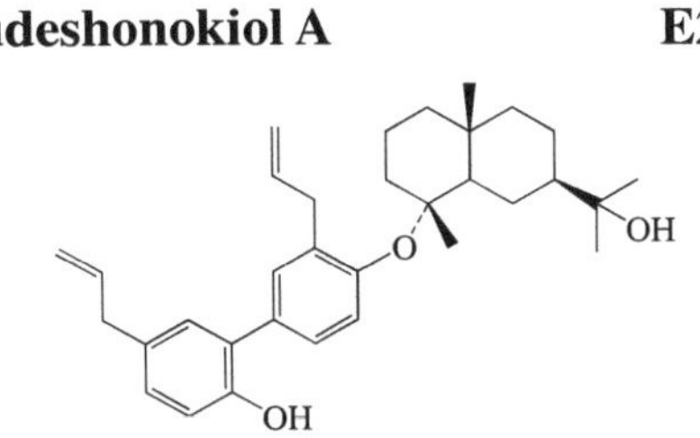

Relative Configuration

[126654-55-5]

$C_{33}H_{44}O_3$ 488.712

Constit. of bark of *Magnolia obovata*. Oil. $[\alpha]_D$ -48.6 (c, 0.45 in EtOH).

Fukuyama, Y. *et al., Tetrahedron*, 1992, **48**, 377-392 (*Eudeshonokiol A*)

Eudeshonokiol B E280

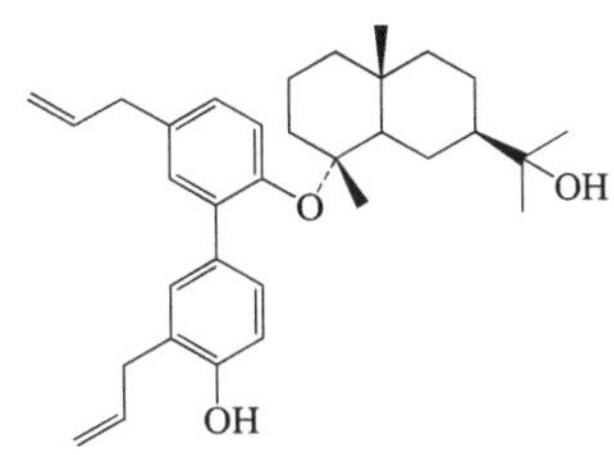

[133056-10-7]

$C_{33}H_{44}O_3$ 488.712

Constit. of *Magnolia obovata*. Oil. $[\alpha]_D^{22}$ -71.7 (c, 1.08 in $CHCl_3$).

Fukuyama, Y. *et al., Tetrahedron*, 1992, **48**, 377-392 (*Eudeshonokiol B*)

Eudesmagnolol E281

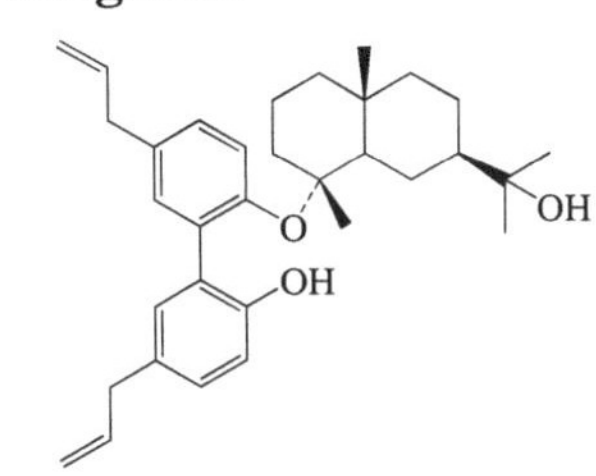

Relative Configuration

[126654-54-4]

$C_{33}H_{44}O_3$ 488.712

Constit. of bark of *Magnolia obovata*. Oil. $[\alpha]_D$ -74.8 (c, 9.2 in $CHCl_3$).

Fukuyama, Y. *et al., Tetrahedron*, 1992, **48**, 377-382 (*Eudesmagnolol*)

Eudesobovatol A E282

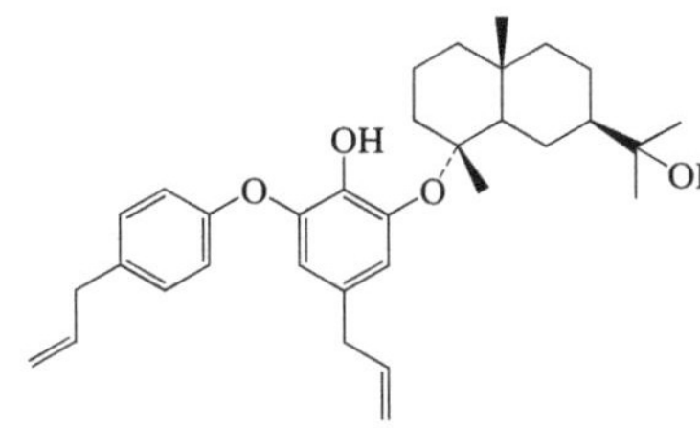

[125196-77-2]

$C_{33}H_{44}O_4$ 504.711

Constit. of *Magnolia obovata* and *Manglietia hookeri*. Neurotrophic agent. Oil. $[\alpha]_D$ -46.1 (c, 2.5 in $CHCl_3$). λ_{max} 208 (ε 58000); 274 (ε 7200); 281 (ε 6700) (EtOH) (Derep).

Fukuyama, Y. *et al., Tetrahedron*, 1992, **48**, 377-392 (*Eudesobovatol A*).

Hu, X. *et al., Nat. Prod. Res.*, 2016, **30**, 1477-1483 (*Manglietia hookeri constit*)

Eudesobovatol B E283

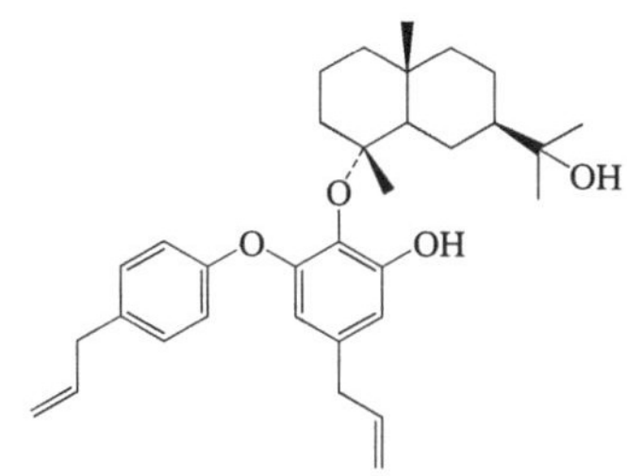

[125196-78-3]

$C_{33}H_{44}O_4$ 504.711

Constit. of *Magnolia obovata* and *Manglietia hookeri*. Neurotrophic agent. Oil. $[\alpha]_D$ -26.1 (c, 1.15 in $CHCl_3$). λ_{max} 208 (ε 58000); 274 (ε 7200); 281 (ε 6700) (EtOH) (Derep).

Fukuyama, Y. *et al., Tetrahedron*, 1992, **48**, 377-392 (*Eudesobovatol B*).

Hu, X. *et al., Nat. Prod. Res.*, 2016, **30**, 1477-1483 (*Manglietia hookeri constit*)

Eupodienone 8 E284

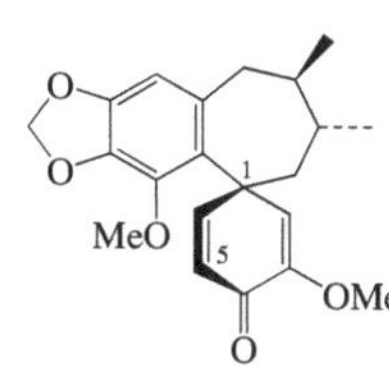

Absolute Configuration

1,2′-Cyclo-3,3′-dimethoxy-4′,5′-methylenedioxy-8,8′-lignan-4(1H)-one

[133084-49-8]

$C_{21}H_{24}O_5$ 356.418

Constit. of the flowers of *Eupomatia laurina*. Prisms (Et_2O). Mp 180 - 181°. $[\alpha]_D$ -143.2 (c, 0.5 in $CHCl_3$). λ_{max} 219 (ε 42300); 246 (ε 13700); 276 (ε 5380) (EtOH)

5-Methoxy

1,2′-Cyclo-3,3′,5-trimethoxy-4′,5′-methylenedioxy-8,8′-lignan-4(1H)-one. ***Manglisin A***

[1621958-61-9]

$C_{22}H_{26}O_6$ 386.444

Constit. of the carpels of *Manglietiastrum sinicum*. Exhibits moderate antibacterial activity, incl. MRSA. Cryst. (MeOH). Mp 118 - 119°. $[\alpha]^{16}_{D}$ -64 (c, 0.2 in MeOH). λ_{max} 216 (log ε 4.59); 245 (log ε 3.86); 280 (log ε 4.06) (MeOH)

1-Epimer

Eupodienone 9

[133162-96-6]

$C_{21}H_{24}O_5$ 356.418

Constit. of the flowers of *Eupomatia laurina*. Needles (Et_2O/petrol). Mp 136 - 137°. $[\alpha]_D$ -61.7 (c, 0.3 in $CHCl_3$). λ_{max} 218 (ε 39390); 249 (ε 15250); 289 (sh); 356 (ε 1370) (EtOH).

Carroll, A.R. *et al., Aust. J. Chem.*, 1990, **43**, 1871-1879 (*Eupodienones 8,9*).
Ding, J.-Y. *et al., J. Nat. Prod.*, 2014, **77**, 1800-1805 (*Manglisin A, cd, cryst struct*)

Eupomatenoid 10 E285

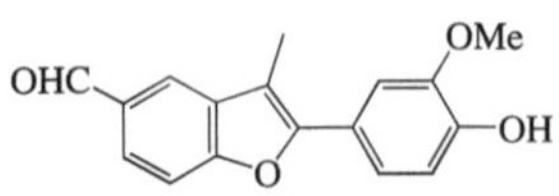

2-(4-Hydroxy-3-methoxyphenyl)-3-methyl-5-benzofurancarboxaldehyde, CAS. *4,7′-Epoxy-4′-hydroxy-3′-methoxy-8,9-dinor-3,8′-lign-7′-en-7-al. 4,7′-Epoxy-4′-hydroxy-3′-methoxy-8,9-dinor-7-oxo-3,8′-neolign-7′-ene*

[41365-35-9]

$C_{17}H_{14}O_4$ 282.295

Constit. of the bark and leaves of *Eupomatia laurina*. Plates (EtOH). Mp 145 - 147°. May be an artifact formed from Eupomatenoid 5 during isolation.

Picker, K. *et al., Aust. J. Chem.*, 1973, **26**, 1111-1119 (*Eupomatenoid 10*)

Eupomatenoid 11 E286

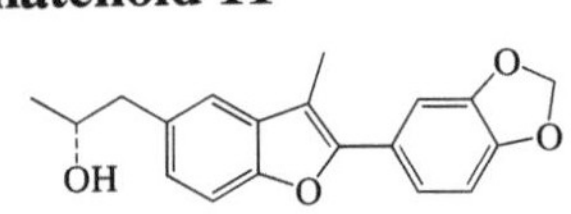

2-(1,3-Benzodioxol-5-yl)-α,3-dimethyl-5-benzofuranethanol, CAS. *5-(2-Hydroxypropyl)-3-methyl-2-(3,4-methylenedioxyphenyl)benzofuran. 4,7′-Epoxy-3′,4′-methylenedioxy-3,8′-lign-7′-en-8-ol. 4,7′-Epoxy-8-hydroxy-3′,4′-methylenedioxy-3,8′-lign-7′-ene*

[41365-36-0]

$C_{19}H_{18}O_4$ 310.349

Constit. of the bark of *Eupomatia laurina*. Needles (petrol). Mp 80 - 81°. $[\alpha]^{20}_{D}$ -12 (c, 1 in $CHCl_3$)

8-Ketone

2-(3,4-Methylenedioxyphenyl)-3-methyl-5-(2-oxopropyl)benzofuran. 4,7′-Epoxy-3′,4′-methylenedioxy-3,8′-lign-7′-en-8-one. 4,7′-Epoxy-3′,4′-methylenedioxy-8-oxo-3,8′-lign-7′-ene

[41206-98-8]

$C_{19}H_{16}O_4$ 308.333

Constit. of *Piper aequale*. Plates (C_6H_6/petrol). Mp 73 - 74°. λ_{max} 216 (log ε 4.25); 316 (log ε 4.18) (MeOH).

Picker, K. *et al., Aust. J. Chem.*, 1973, **26**, 1111-1119 (*Eupomatenoid 11, 8-ketone, synth*).
Maxwell, A. *et al., Phytochemistry*, 1999, **50**, 499-504 (*Piper aequale constit*)

Euryalin C E287

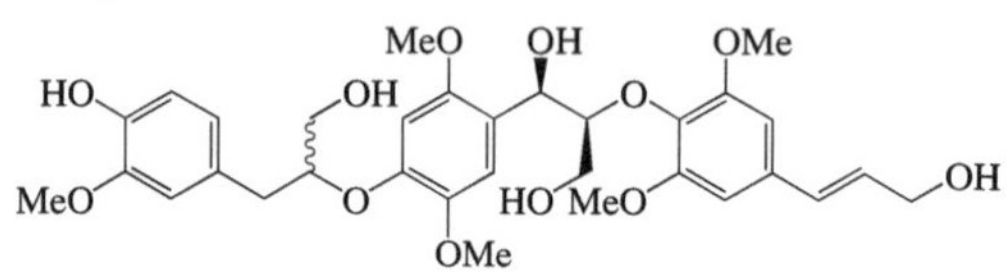

Relative Configuration

[1268671-46-0]

$C_{32}H_{40}O_{12}$ 616.66

Constit. of the seeds of *Euryale ferox*. Gum. $[\alpha]^{25}_{D}$ -9.3 (c, 0.22 in MeOH). λ_{max} 206 (log ε 4.81); 228 (sh); 271 (log ε 4.2) (MeOH).

Song, C.-W. *et al., J. Agric. Food Chem.*, 2011, **59**, 1199-1204 (*Euryalin C*)

Eusiderin F E288

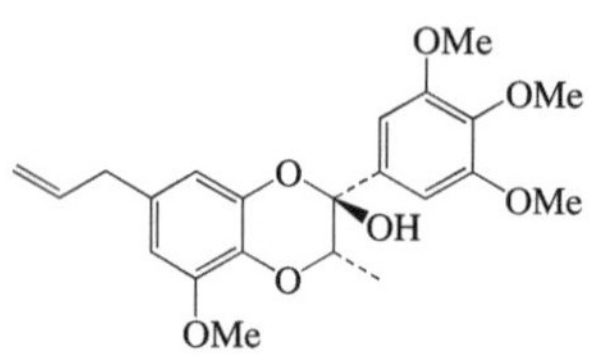

Relative Configuration

3,7′-Epoxy-3′,4′,5,5′-tetramethoxy-4,8′-oxylign-8-en-7′-ol. 3,7′-Epoxy-7′-hydroxy-3′,4′,5,5′-tetramethoxy-4,8′-oxyneolign-8-ene

[101508-19-4]

$C_{22}H_{26}O_7$ 402.443

Constit. of an *Aniba* sp.. Oil. λ_{max} 235 (sh); 255 (ε 1500) (MeOH).

Dias, S.M.C. *et al., Phytochemistry*, 1986, **25**, 213-217 (*Eusiderin F*).
Jing, X.-B. *et al., J. Chin. Chem. Soc. (Taipei)*, 2004, **51**, 1001-1004 (*synth*)

Evofolin B E289

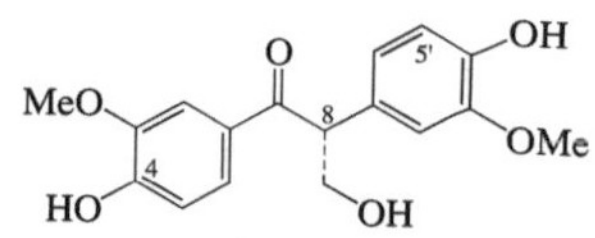

(*R*)-*form*

3-Hydroxy-1,2-bis(4-hydroxy-3-methoxyphenyl)-1-propanone, CAS. *4′,4″-Dihydroxy-3′,3″-dimethoxy-α-(hydroxymethyl)deoxybenzoin*

$C_{17}H_{18}O_6$ 318.325

(R)-form [1961305-60-1]

Constit. of *Dysoxylum hainanense* and *Sambucus williamsii*. Yellow oil. $[\alpha]^{26}_{D}$ +16.7 (c, 0.3 in MeOH). λ_{max} 231; 279 (MeOH)

9-Me ether, 4-O-β-D-glucopyranoside

Phyllanthuoside B

[1819380-90-9]

$C_{24}H_{30}O_{11}$ 494.493

Constit. of *Phyllanthus cochinchinensis*. Amorph. powder. $[\alpha]^{20}_{D}$ -19.7 (c, 0.3 in MeOH). λ_{max} 227 (log ε 4.29); 280 (log ε 4.02); 307 (log ε 3.93) (MeOH)

3′-O-De-Me, 4′-Me ether

3-Hydroxy-1-(4-hydroxy-3-methoxyphenyl)-2-(3-hydroxy-4-methoxyphenyl)-1-propanone

[2260774-73-8]

$C_{17}H_{18}O_6$ 318.325

Constit. of *Eleutherococcus senticosus*. Exhibits moderate inhibitory activity against diacylglycerol acyltransferase. Pale yellow powder. $[\alpha]^{25}_{D}$ -7.6 (c, 0.1 in MeOH). λ_{max} 230 (log ε 4.25); 276 (log ε 4.55); 308 (log ε 4.83) (MeOH)

(S)-form [168254-96-4]

Constit. of *Anastatica hierochuntica, Juglans mandshurica, Lagerstroemia indica, Tetradium glabrifolium* (*Evodia meliaefolia*) and *Schisandra verruculosa*. Exhibits quinone reductase inducing activity. Pale yellow oil. $[\alpha]_D$ -14.3 (c, 0.02 in MeOH)

4-O-β-D-Glucopyranoside

Stroside A

[2211126-75-7]

$C_{23}H_{28}O_{11}$ 480.466

Constit. of the stems of *Lagerstroemia indica*. Brown gum. [α] D +3 (c, 0.05 in $CHCl_3$). λ_{max} 207 (log ε 1.8); 227 (log ε 1.3); 272 (log ε 0.7) (MeOH)

5′-Methoxy

3-Hydroxy-2-(4-hydroxy-3,5-dimethoxyphenyl)-1-(4-hydroxy 3-methoxyphenyl)-1-propanone. ***5′-Methoxyevofolin B***

[1033546-79-0]

$C_{18}H_{20}O_7$ 348.351

Constit. of the stems of *Microtropis japonica*. Pale yellow oil. $[\alpha]_D^{25}$ -16.2 (c, 0.11 in $CHCl_3$). λ_{max} 229 (log ε 4.61); 278 (log ε 4.31); 307 (log ε 4.12) (MeOH)

9-Me ether, 4′-O-β-D-glucopyranoside

Phyllanthuoside A

[1819380-89-6]

$C_{24}H_{30}O_{11}$ 494.493

Constit. of *Phyllanthus cochinchinensis*. Amorph. powder. $[\alpha]_D^{20}$ -38.2 (c, 0.2 in MeOH). λ_{max} 224 (log ε 4.14); 272 (log ε 3.87) (MeOH)

(±)-form [138171-64-9]

Constit. of *Saussurea deltoidea*

(ξ)-form

3-Ac

3-O-Acetylevofolin B. ***Lagerstroemiate A***

$C_{19}H_{20}O_7$ 360.362

Constit. of *Lagerstroemia indica*. $[\alpha]_D^{20}$ +53 (c, 0.02 in MeOH). λ_{max} 208 (MeOH).

Wu, T.-S. *et al.*, *Phytochemistry*, 1995, **40**, 121-124 (*Tetradium glabrifolium constit*).
Luo, X.-D. *et al.*, *Yunnan Zhiwu Yanjiu*, 2001, **23**, 368-372 (*(+)-form*).
Jang, D.S. *et al.*, *Arch. Pharmacal Res.*, 2004, **27**, 169-172 (*isol, activity*).
Wilairat, R. *et al.*, *Pharm. Biol.*, 2006, **44**, 411-415 (*Schisandra verruculosa constit*).
Chen, J.-J. *et al.*, *J. Nat. Prod.*, 2008, **71**, 1016-1021 (*5′-Methoxyevofolin B*).
Huang, H.-Q. *et al.*, *Zhongguo Xinyao Zazhi* , 2011, **20**, 1569-1572 (*(±)-form*).
Zhao, J.-Q. *et al.*, *J. Braz. Chem. Soc.*, 2014, **25**, 1446-1454 (*Phyllanthuosides A,B, cd*).
Woo, K.W. *et al.*, *Heterocycles*, 2015, **91**, 2355-2366 (*Stroside A*).
Zhang, D. *et al.*, *Zhongcaoyao*, 2015, **46**, 2209-2211 (*Lagerstroemiate A*).
Li, B.-B. *et al.*, *Phytochem. Lett.*, 2017, **20**, 123-127 (*Eleutherococcus senticosus constit, activity*)

Excoecanol A E290

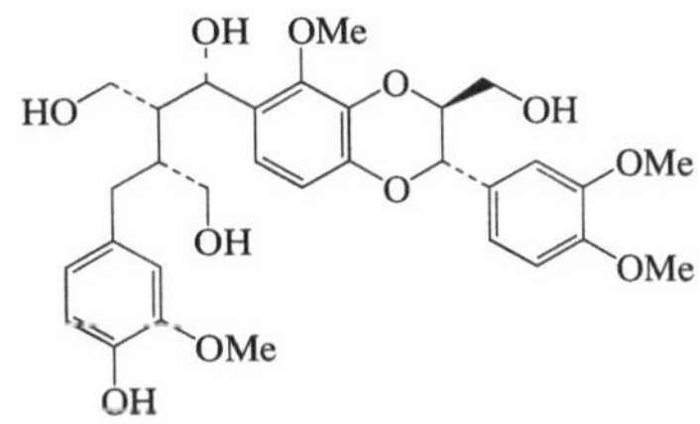

Absolute Configuration

[1632401-80-9]

$C_{31}H_{38}O_{11}$ 586.634

Constit. of the stems of *Excoecaria acerifolia*. Oil. $[\alpha]_D^{17}$ +0.74 (c, 0.28 in MeOH). λ_{max} 280 (log ε 3.79) (MeOH).

Huang, S.-Z. *et al.*, *Chem. Biodiversity*, 2014, **11**, 1406-1416 (*Excoecanol A*)

Falloconvolin B F1

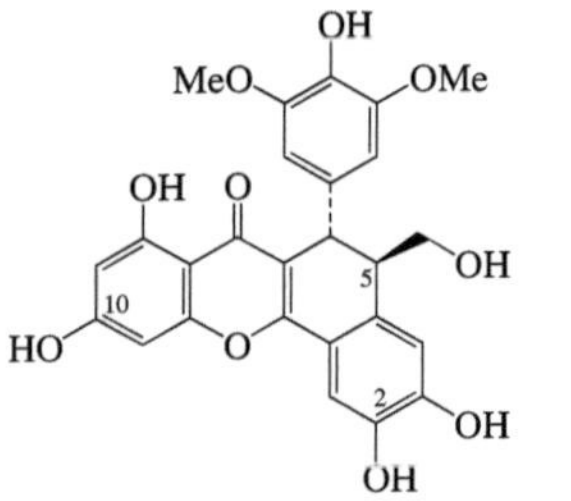

Relative Configuration

2,3,8,10-Tetrahydroxy-6-(4-hydroxy-3,5-dimethoxyphenyl)-5-(hydroxymethyl)-5H-benzo[c]xanthen-7(6H)-one

[1340493-18-6]

$C_{26}H_{22}O_{10}$ 494.452

Similar to Neohydnocarpin, N9. Constit. of aerial parts of *Fallopia convolvulus*. Yellow powder (MeOH). Mp 203 - 204°. $[\alpha]^{25}_{D}$ -20.5 (c, 0.1 in MeOH). λ_{max} 226 (log ε 4.73); 260 (sh); 374 (log ε 4.55) (MeOH).

Zhang, C.-F. *et al.*, *J. Asian Nat. Prod. Res.*, 2011, **13**, 136-142 (*Falloconvolin B*)

Fargenin F2

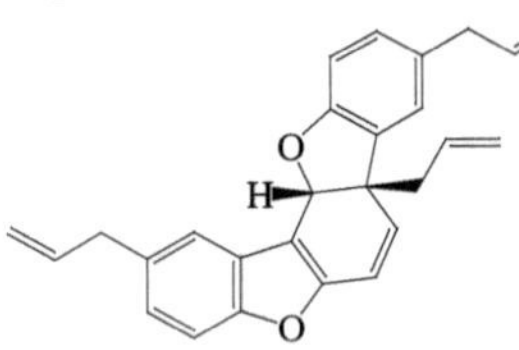

Relative Configuration

[937785-18-7]

$C_{27}H_{24}O_2$ 380.487

Constit. of the pericarps of *Illicium fargesii*. Amorph. solid. $[\alpha]^{12}_{D}$ -12.7 (c, 0.8 in $CHCl_3$). λ_{max} 293 (ε 3300); 306 (ε 3200) (EtOH).

Moriyama, M. *et al.*, *Tetrahedron*, 2007, **63**, 4243-4249 (*Fargenin*)

Fargenone A F3

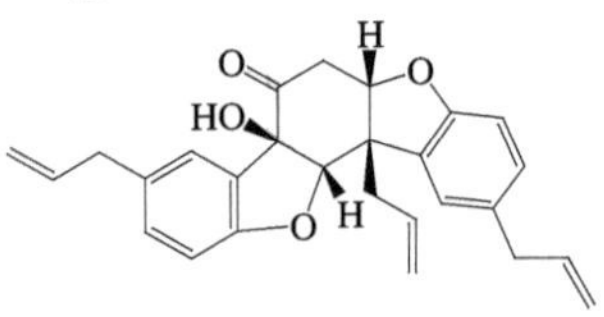

Relative Configuration

[937785-19-8]

$C_{27}H_{26}O_4$ 414.501

Related to Simonin A, S63. Constit. of the pericarps of *Illicium fargesii*. Amorph. powder. $[\alpha]^{21}_{D}$ +14.2 (c, 1.53 in $CHCl_3$). λ_{max} 292 (ε 3100) (EtOH).

Moriyama, M. *et al.*, *Tetrahedron*, 2007, **63**, 4243-4249 (*Fargenone A*)

Fargenone B F4

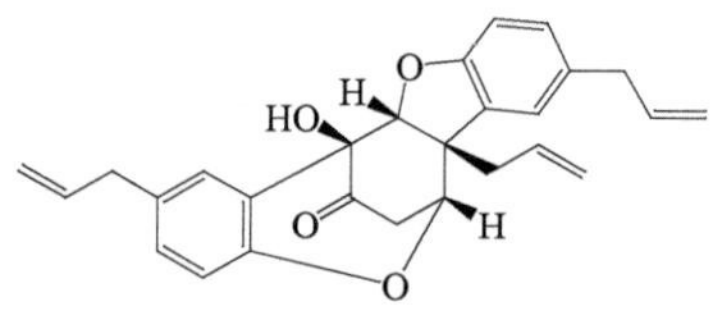

Relative Configuration

[937785-20-1]

$C_{27}H_{26}O_4$ 414.501

Constit. of the pericarps of *Illicium fargesii*. Amorph. powder. $[\alpha]^{22}_{D}$ -19.2 (c, 1.54 in $CHCl_3$). MF incorrect in the ref.. λ_{max} 286 (ε 2200) (EtOH).

Moriyama, M. *et al.*, *Tetrahedron*, 2007, **63**, 4243-4249 (*Fargenone B*)

Feddeiticin F5

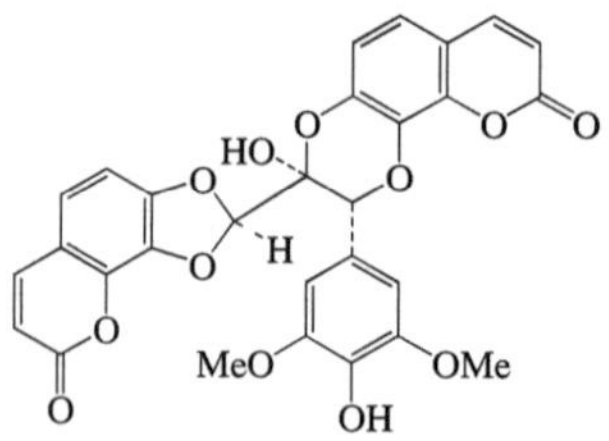

Relative Configuration

[1140273-36-4]

$C_{29}H_{20}O_{12}$ 560.467

Constit. of the stem bark of *Daphne feddei*. Powder (MeOH). Mp 188 - 189°. Racemic. λ_{max} 248 (log ε 4.11); 312 (log ε 4.21) (MeOH).

Liang, S. *et al.*, *Helv. Chim. Acta*, 2009, **92**, 133-138 (*Feddeiticin*)

Ferrearin D F6

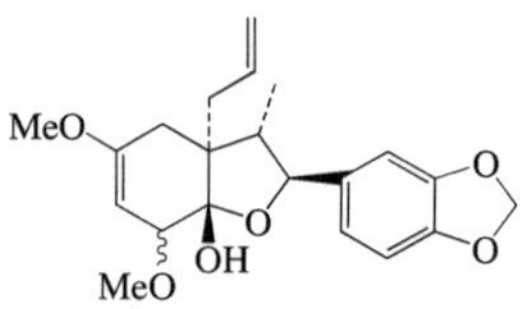

Relative Configuration

2,7′-Epoxy-1,2,3,6-tetrahydro-3,5-dimethoxy-3′,4′-methylenedioxy-1,8′-lign-8-en-2-ol. 2,7′-Epoxy-1,2,3,6-tetrahydro-2-hydroxy-3,5-dimethoxy-3′,4′-methylenedioxy-1,8′-neolign-8-ene

[104701-05-5]

$C_{21}H_{26}O_6$ 374.433

Constit. of the wood of *Ocotea aciphylla*. Oil. λ_{max} 236; 288 (MeOH).

Felício, J.D.'.A. *et al.*, *Phytochemistry*, 1986, **25**, 1707-1710 (*Ferrearin D*)

2-*O*-Feruloylglycerol F7

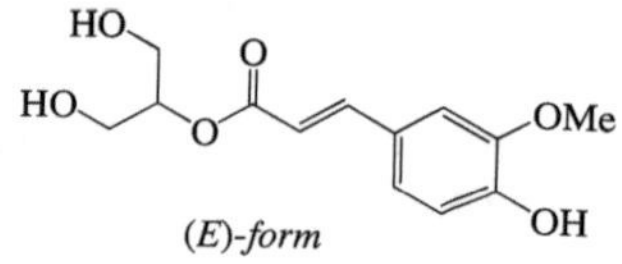

(E)-form

2-O-(4-Hydroxy-3-methoxycinnamoyl) glycerol. 2-Glyceryl ferulate

$C_{13}H_{16}O_6$ 268.265

(E)-form [1699543-75-3]

Constit. of *Oplopanax horridus*. Light yellow gum. $[\alpha]^{20}_{D}$ +52.1 (c, 0.5 in MeOH). Opt. rotn. unexplained. λ_{max} 210 (log ε 0.91); 226 (log ε 1.84); 264 (log ε 2.75) (MeOH)

(ξ)-form

Di-O-docosanoyl

Glycerol 1,3-didocosanoate 2-ferulate. 1,3-Dibehenoyl-2-feruloylglycerol. NSC 332563

[80324-81-8]

$C_{57}H_{100}O_8$ 913.419

Constit. of *Aquilaria malaccensis*. Antitumour agent. Needles (CH_2Cl_2/hexane). Mp 84 - 85°. Config. of double bond not determined. λ_{max} 218 (log ε 4.06); 238 (log ε 3.99); 298 (sh); 327 (log ε 4.17) (MeOH).

Gunasekera, S.P. *et al.*, *J. Nat. Prod.*, 1981, **44**, 569-572 (*Aquilaria malaccensis constit*).

Sun, S.-D. *et al.*, *J. Agric. Food Chem.*, 2008, **56**, 442-447 (*synth*).

Zeng, Y. *et al.*, *J. Mol. Catal. B: Enzym.*, 2014, **110**, 140-146 (*synth*).

Huang, W.-H. *et al.*, *Helv. Chim. Acta*, 2015, **98**, 201-209 (*2-Glyceryl ferulate*)

Ficuscarpanoside B F8

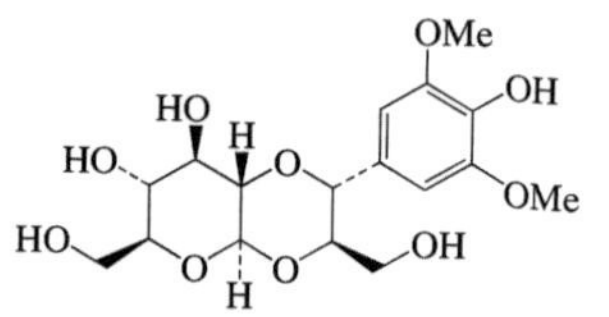

Cremanthodioside

[848347-67-1]

$C_{17}H_{24}O_{10}$ 388.369

Regioisomer of Fissistigmoside, F10. Constit. of *Cremanthodium ellisii* and *Ficus microcarpa*. Needles (MeOH). Mp 197 - 201° (189-190°). $[\alpha]^{20}_{D}$ +75 (c, 0.2 in MeOH). λ_{max} 209 (log ε 3.92); 232 (log ε 3.45); 284 (log ε 3.24) (MeOH)

3′-Demethoxy

Ficuscarpanoside A

[1005764-37-3]

$C_{16}H_{22}O_9$ 358.343

Constit. of the aerial roots of *Ficus microcarpa*. Amorph. powder. Mp 171 - 174°. $[\alpha]_D^{21}$ +26 (c, 0.7 in MeOH). λ_{max} 206 (log ε 4.02); 221 (log ε 3.75); 277 (log ε 3.34) (MeOH).

Wang, A.-X. *et al.*, *Pharmazie*, 2004, **59**, 889-892 (*Cremanthodioside*).
Ouyang, M.-A. *et al.*, *J. Asian Nat. Prod. Res.*, 2006, **8**, 625-630 (*Ficuscarpanoside B*).
Ouyang, M.-A. *et al.*, *Nat. Prod. Res.*, 2007, **21**, 769-774 (*Ficuscarpanoside A*)

Ficusesquilignan B F9

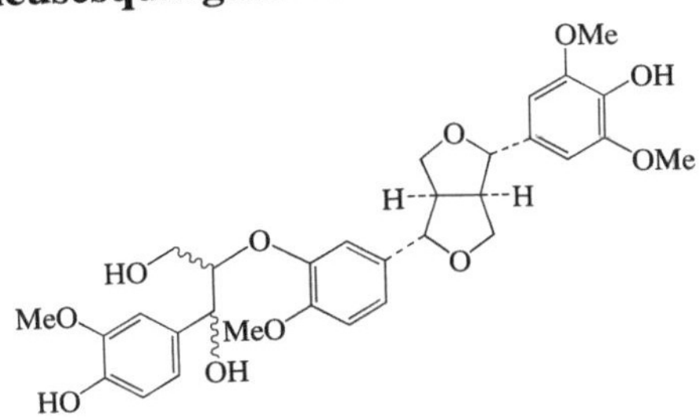

[321991-57-5]
$C_{31}H_{36}O_{11}$ 584.618
Constit. of *Endospermum chinense*, *Ficus microcarpa* and *Toona sinensis*. Pale yellow oil. $[\alpha]_D^{27}$ -8 (c, 0.24 in $CHCl_3$). λ_{max} 240 (log ε 3.86); 279 (log ε 3.47) (MeOH).

Li, Y.-C. *et al.*, *Chem. Pharm. Bull.*, 2000, **48**, 1862-1865 (*Ficusesquilignan B*).
Zou, T. *et al.*, *Tianran Chanwu Yanjiu Yu Kaifa*, 2014, **26**, 1610-1617 (*Endospermum chinense constit*)

Fissistigmoside F10

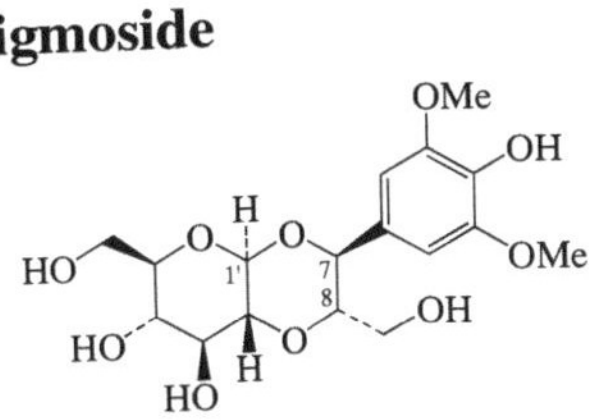

Absolute Configuration

[237764-36-2]
$C_{17}H_{24}O_{10}$ 388.369
Regioisomer of Ficuscarpanoside B, F8. Constit. of *Fissistigma polyanthum*. Mp 188 - 190°. λ_{max} 210; 268 (MeOH)

1′,7,8-Triepimer

Ericarboside
[1221446-69-0]
$C_{17}H_{24}O_{10}$ 388.369
Constit. of *Erica arborea*. Gum. $[\alpha]_D^{25}$ +20 (c, 0.18 in MeOH). λ_{max} 208 (log ε 3.75); 232 (log ε 3.45); 272 (log ε 2.49); 292 (sh) (MeOH).

Liao, Y.-H. *et al.*, *Yaoxue Xuebao*, 1999, **34**, 207-209 (*Fissistigmoside*).
Demirkiran, O. *et al.*, *Helv. Chim. Acta*, 2010, **93**, 77-83 (*Ericarboside*)

Flavipesolide C F11

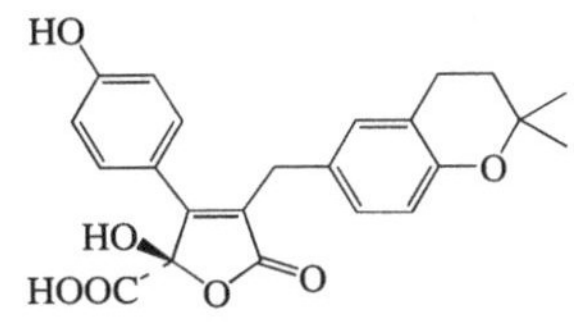

Absolute Configuration

[2062613-48-1]
$C_{23}H_{22}O_7$ 410.422
Related to Asperteretal B, A65. Prod. by the marine-derived *Aspergillus flavipes* HN4-13 and *Aspergillus terreus* OUCMDZ-2739. α-Glucosidase inhibitor. Light brown solid. $[\alpha]_D^{25}$ -22 (c, 0.1 in CH_2Cl_2). λ_{max} 204 (log ε 3.11); 226 (log ε 1.21); 306 (log ε 0.81) (MeOH)

Me ester

Flavipesolide B
[2062613-47-0]
$C_{24}H_{24}O_7$ 424.449
Prod. by the marine-derived *Aspergillus flavipes* HN4-13 and *Aspergillus terreus* OUCM-DZ-2739. α-Glucosidase inhibitor. Light brown solid. $[\alpha]_D^{25}$ -34 (c, 0.1 in CH_2Cl_2). λ_{max} 202 (log ε 2.61); 228 (log ε 2.39); 308 (log ε 3.98) (MeOH)

Et ester

Flavipesolide A
[2062613-46-9]
$C_{25}H_{26}O_7$ 438.476
Prod. by the marine-derived *Aspergillus flavipes* HN4-13 and *Aspergillus terreus* OUCM-DZ-2739. α-Glucosidase inhibitor. Light brown solid. $[\alpha]_D^{25}$ -28 (c, 0.1 in CH_2Cl_2). λ_{max} 210 (log ε 2.60); 230 (log ε 2.43); 310 (log ε 2.31) (MeOH).

Wang, C. *et al.*, *J. Nat. Prod.*, 2016, **79**, 2977-2981 (*Flavipesolides A,B,C, activity*)

Flaxseed lignan-HMGA oligomer F12

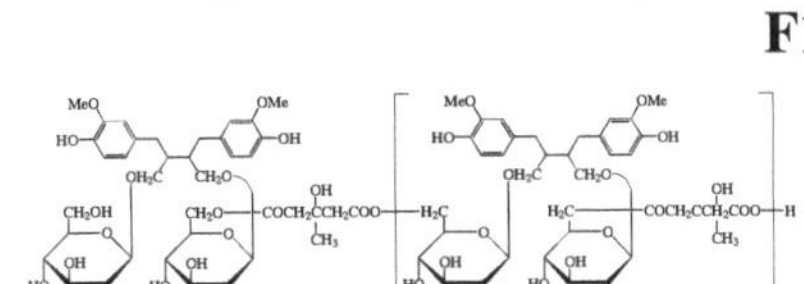

$C_{190}H_{256}O_{100}$ 4140.038
Oligomer consisting of secoisolariciresinol 9,9′-diglucosides and 3-hydroxy-3-methylglutaric acid residues. Major lignan has $n = 4$ to which MF refers. Constit. of the seeds of *Linum usitatissimum*.

Kamal-Eldin, A. *et al.*, *Phytochemistry*, 2001, **58**, 587-590 (*Linum usitatissimum constit*)

3-Formyl-2,2′-dihydroxy-5,5′-di-2-propenylbiphenyl F13

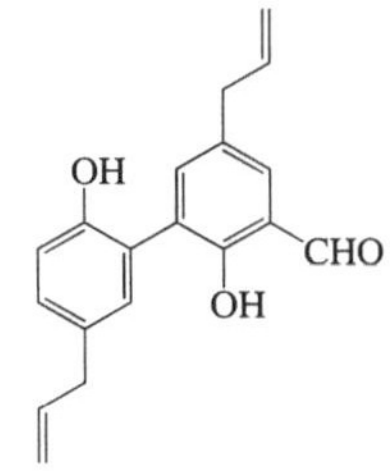

2,2′-Dihydroxy-5,5′-di-2-propenyl[1,1′-biphenyl]-3-carboxaldehyde, CAS. *5-Formyl-3,3′-ligna-8,8′-diene-4,4′-diol. 5-Formyl-4,4′-dihydroxy-3,3′-neoligna-8,8′-diene. 5-Formylmagnolol.*

Houpulin E
[1448425-85-1]
$C_{19}H_{18}O_3$ 294.35
Homolignan. Constit. of the roots of *Magnolia officinalis*. Syrup. λ_{max} 283 (log ε 3.88); 345 (log ε 3.72) (MeOH).

Shih, H.-C. *et al.*, *Bioorg. Med. Chem.*, 2016, **24**, 1439-1445 (*Houpulin E*)

Fragransol B F14

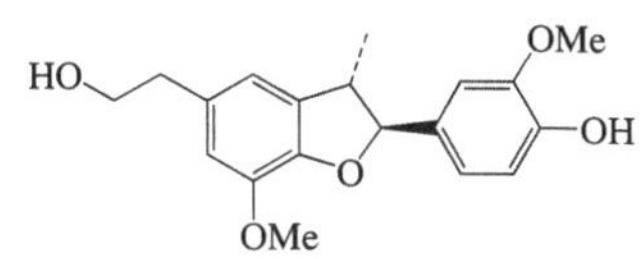

Relative Configuration

4,7′-Epoxy-3′,5-dimethoxy-9-nor-3,8′-lignan-4′,8-diol. 4,7′-Epoxy-4′,8-dihydroxy-3′,5-dimethoxy-9-nor-3,8′-neolignan
[114394-20-6]
$C_{19}H_{22}O_5$ 330.38
Constit. of the arils of *Myristica fragrans*. Oil.

Hada, S. *et al.*, *Phytochemistry*, 1988, **27**, 563-568 (*Fragransol B*).
Juhasz, L. *et al.*, *J. Nat. Prod.*, 2000, **63**, 866-870 (*(±)-form, synth*)

Furoguaiacin F15

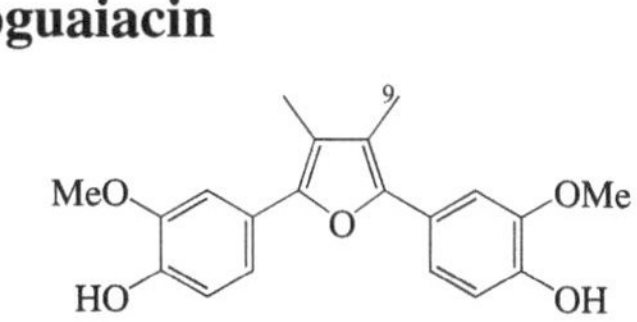

7,7′-Epoxy-3,3′-dimethoxy-8,8′-ligna-7,7′-diene-4,4′-diol. 4,4′-Dihydroxy-3,3′-dimethoxy-7,7′-epoxyligna-7,7′-diene. 2,5-Bis(4-

hydroxy-3-methoxyphenyl)-3,4-dimethylfuran. α-Guaiaconic acid
[10035-27-5]
$C_{20}H_{20}O_5$ 340.375
Constit. of *Guaiacum officinale*. Inhibitor of arachidonic metab.. Cryst. ($Me_2CO/AcOH/H_2O$ or CH_2Cl_2/Et_2O). Mp 149° (142-144°). λ_{max} 228 (ε 24500); 252 (ε 22900); 285 (ε 18200); 315 (ε 25000); 335 (ε 23300) (EtOH) (Berdy)

Mono-Me ether
Methylfuroguaiacin
$C_{21}H_{22}O_5$ 354.402
Constit. of *Guaiacum officinale*

Di-Me ether
$C_{22}H_{24}O_5$ 368.429
Cryst. (MeOH). Mp 170 - 171°

9-Methoxy
7,7′-Epoxy-3,3′,9-trimethoxy-8,8′-ligna-7,7′-diene-4,4′-diol. 4,4′-Dihydroxy-3,3′,9-trimethoxy-7,7′-epoxyligna-7,7′-diene.
Furoguaiacidin
[52465-57-3]
$C_{21}H_{22}O_6$ 370.401
Constit. of heartwood of *Guaiacum officinale*

9-Methoxy, di-Et ether
[52465-56-2]
$C_{25}H_{30}O_6$ 426.509
Mp 135°. λ_{max} 258 (log ε 4.16); 324 (log ε 4.5) (EtOH).

King, F.E. *et al., J. Chem. Soc.*, 1964, 4011-4024 (*Furoguaiacin, Methylfuroguaiacin*).
Kratochvil, J.F. *et al., Phytochemistry*, 1971, **10**, 2529-2531 (*α-Guaiaconic acid*).
Schreiber, F.G. *et al., J. Org. Chem.*, 1975, **40**, 386-387 (*Furoguaiacidin diethyl ether, synth*).
Stevenson, R. *et al., Org. Prep. Proced. Int.*, 1976, **8**, 179-183 (*synth*).
Fonseca, S.F. *et al., Can. J. Chem.*, 1979, **57**, 441-443 (*Furoguaiacin di-Me ether, cmr*).
Ito, K. *et al., Int. J. Cancer*, 1993, **55**, 660-666 (*Furoguaiacin, activity*).
Majumder, P.L. *et al., J. Indian Chem. Soc.*, 1998, **75**, 776-783 (*Furoguaiacidin*)

Furoguaiaoxidin F16

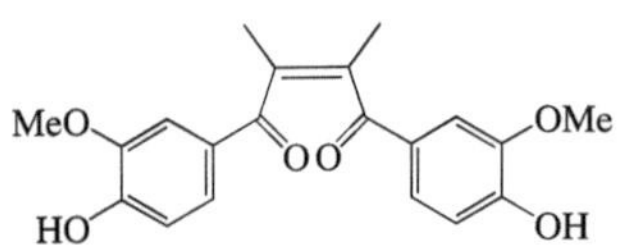

1,4-Bis(4-hydroxy-3-methoxyphenyl)-2,3-dimethyl-2-butene-1,4-dione, CAS. *4,4′-Dihydroxy-3,3′-dimethoxy-8,8′-lign-8(8′)-ene-7,7′-dione*
[58096-91-6]
$C_{20}H_{20}O_6$ 356.374
Constit. of *Guaiacum officinale*. Isol. as diethyl ether

Di-Me ether
1,4-Bis(3,4-dimethoxyphenyl)-2,3-dimethyl-2-butene-1,4-dione. 3,3′,4,4′-Tetramethoxy-8,8′-lign-8(8′)-ene-7,7′-dione
[2089124-41-2]
$C_{22}H_{24}O_6$ 384.428
Constit. of *Saururus chinensis*. Amorph. powder. $[\alpha]^{24}_{D}$ -6.5 (c, 0.3 in MeOH). λ_{max} 228 (log ε 3.93); 279 (log ε 3.73); 315 (log ε 3.67) (MeOH)

Di-Et ether
[58096-85-8]
$C_{24}H_{28}O_6$ 412.482
Mp 105°. Mp 120° (double Mp). λ_{max} 230 (log ε 4.52); 280 (log ε 4.21); 315 (log ε 4.17) (EtOH)

(E)-Isomer, bis(demethoxy)
1,4-Bis(4-hydroxyphenyl)-2,3-dimethyl-2-butene-1,4-dione, CAS. *4,4′-Dihydroxy-8,8′-lign-8(8′)-ene-7,7′-dione*
[793674-59-6]
$C_{18}H_{16}O_4$ 296.322
Constit. of the leaves of *Larrea tridentata*. Off-white powder.

Majumder, P.L. *et al., J. Chem. Soc., Chem. Commun.*, 1975, 702-703 (*Furoguaiaoxidin*).
Majumder, P.L. *et al., J. Indian Chem. Soc.*, 1998, **75**, 776-783 (*Furoguaiaoxidin*).
Abou-Gazar, H. *et al., Phytochemistry*, 2004, **65**, 2499-2505 (*Larrea tridentata constit*).
Li, C. *et al., Nat. Prod. Res.*, 2017, **31**, 1598-1603 (*Saururus chinensis constit*)

Futoenone F17

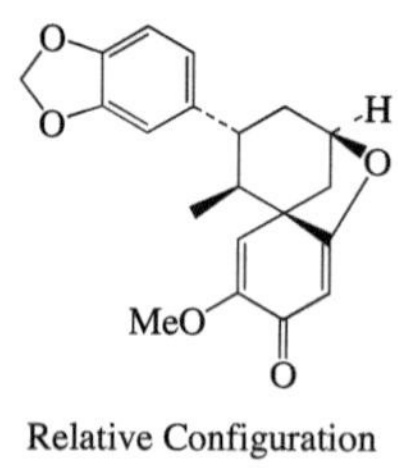

Relative Configuration

4-(1,3-Benzodioxol-5-yl)-2,3,4,5-tetrahydro-7-methoxy-5-methyl-8H-2,5a-methano-1-benzoxepin-8-one, CAS. *7′,9-Cyclo-2,8-epoxy-5-methoxy-3′,4′-methylenedioxy-1,8′-lignan-4(1H)-one*
[19913-01-0]
[20156-97-2] (unspecified stereochem.)
[150652-81-6] ((±)-form)
$C_{20}H_{20}O_5$ 340.375
Similar to Denudatone, D20. Constit. of *Magnolia denudata, Magnolia soulangeana* and *Piper futokadsura*. Matrix metalloproteinase (MMP) inhibitor. Mp 197°. $[\alpha]_D$ -58 ($CHCl_3$).

Ogiso, A. *et al., Tetrahedron Lett.*, 1968, **9**, 2003-2008 (*Futoenone*).
Woods, M.C. *et al., Tetrahedron Lett.*, 1968, **9**, 2009-2014 (*pmr*).
Ogiso, A. *et al., Chem. Pharm. Bull.*, 1970, **18**, 105-114 (*synth, struct*).
Büchi, G. *et al., J. Am. Chem. Soc.*, 1977, **99**, 8073-8075 (*synth*).
Angle, S.R. *et al., J. Org. Chem.*, 1993, **58**, 5360-5369 (*synth*).
Yeh, L.A. *et al., Bioorg. Med. Chem. Lett.*, 1995, **5**, 1637-1642 (*activity*).
Song, Q. *et al., Acta Crystallogr., Sect. E*, 2001, **57**, 1094-1095 (*cryst struct*)

Futoquinol F18

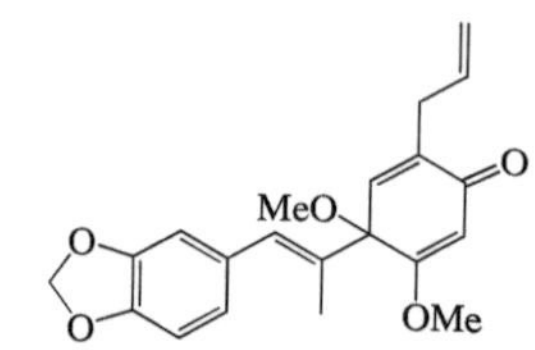

3,4-Dimethoxy-3′,4′-methylenedioxy-3,8′-ligna-7′,8-dien-6(3H)-one.
Hancinone D
[28178-92-9]
$C_{21}H_{22}O_5$ 354.402
Constit. of *Clinopodium urticifolium, Piper futokadsura, Piper hancei* and *Piper schmidtii*. Exhibits anti-neuroinflammatory activity. Cryst. (MeOH). Mp 97 - 98°. Racemate. λ_{max} 260 (ε 16700); 295 (ε 10760) (MeOH)

(Z)-Isomer
Isofutoquinol B
[62532-61-0]
$C_{21}H_{22}O_5$ 354.402
Isol. from leaves of *Piper futokadsura*. Mp 81 - 83°. Racemate. λ_{max} 237 (ε 12300); 288 (ε 6270) (EtOH).

Takahashi, S. *et al., Chem. Pharm. Bull.*, 1970, **18**, 100-104 (*Futoquinol*).
Matsui, K. *et al., Tetrahedron Lett.*, 1976, **17**, 4371-4374 (*Isofutoquinol B*).
Shizuri, Y. *et al., Tetrahedron Lett.*, 1986, **27**, 727-730 (*synth*).
Han, G.Q. *et al., Yaoxue Xuebao*, 1989, **24**, 438 (*Hancinone D*).
Tyagi, O.D. *et al., Phytochemistry*, 1993, **32**, 445-448 (*Futoquinol*).
Lin, L.-C. *et al., J. Nat. Prod.*, 2006, **69**, 842-844 (*Futoquinol, activity*).
Kim, K.-H. *et al., Bioorg. Med. Chem. Lett.*, 2010, **20**, 409-412 (*Futoquinol, activity*)

Galanganol C **G1**

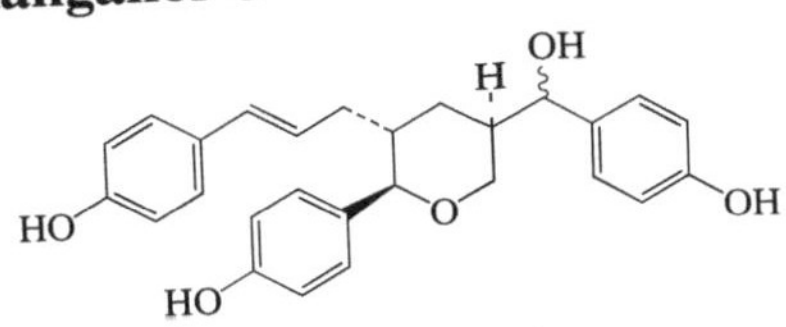

Relative Configuration

$C_{27}H_{28}O_5$ 432.516
Constit. of the rhizomes of *Alpinia galanga*. NO prodn. inhibitor. Powder. Racemic. λ_{max} 263 (log ε 4.31) (MeOH).
Morikawa, T. *et al.*, *Chem. Pharm. Bull.*, 2005, **53**, 625-630 (*Galanganol C, activity*)

Galeon **G2**

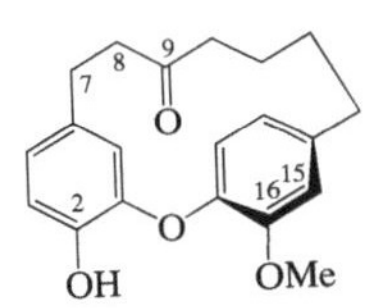

(*R*)-*form*

4-Hydroxy-17-methoxy-2-oxatricyclo[13.2.2.1^{3,7}]eicosa-3,5,7(20),15,17,18-hexaen-10-one, CAS
[61576-08-7]
$C_{20}H_{22}O_4$ 326.392
Related to Acerogenin A, A4 and Acerogenin B, A5. Exhibits potent cytotoxic activity and shows antitubercular props. Moderate topoisomerase inhibitor

(R)-form [191999-64-1]
Constit. of *Myrica gale*. Plates (hexane/EtOAc). Mp 178 - 180°. $[\alpha]_D$ +24.9 (c, 1.4 in $CHCl_3$)

O-De-Me
Engelhardione. Pterocarine
[862261-09-4]
[862702-70-3] (Engelhardione)
$C_{19}H_{20}O_4$ 312.365
Constit. of *Engelhardtia roxburghiana*, *Juglans hopeiensis*, *Juglans sigillata* and *Pterocarya tonkinensis*. Cell cycle inhibitor. Amorph. solid. $[\alpha]_D^{27}$ +60.5 (c, 1 in $CHCl_3$). Struct. of Engelhardione revised and shown to be identical with Pterocarine in 2011. λ_{max} 204 (log ε 4.37); 222 (sh); 281 (log ε 3.42) (no solvent reported)

8R-(β-D-Glucopyranosyloxy)
Juglaside A
[207564-47-4]
$C_{26}H_{32}O_{10}$ 504.532
Constit. of *Juglans mandshurica*. Powder. $[\alpha]_D^{25}$ +4.1 (c, 0.5 in MeOH). λ_{max} 205 (log ε 4.54); 279 (log ε 3.71) (MeOH)

8R-(β-D-Glucopyranosyloxy), 9R-alcohol
Jugcathayenoside
[1441641-10-6]
$C_{26}H_{34}O_{10}$ 506.548
Constit. of the root bark of *Juglans cathayensis*. Needles (MeOH). Mp 127 - 128°. $[\alpha]_D$ -18 (no solvent reported)

(S)-form
Constit. of *Myrica gale*. Mp 178 - 181°. $[\alpha]_D^{20}$ -16 (c, 2.03 in $CHCl_3$)

9R-Alcohol, 2-Me ether
Platycarynol
[206202-74-6]
$C_{21}H_{26}O_4$ 342.435
Constit. of *Platycarya strobilacea*. Needles (hexane/EtOAc). Mp 137 - 138°. $[\alpha]_D^{20}$ -17.1 (c, 0.7 in $CHCl_3$)

15-Hydroxy
Myricatomentogenin
[191999-66-3]
$C_{20}H_{22}O_5$ 342.391
Constit. of *Alnus hirsuta*, *Juglans hopeiensis*, *Juglans mandshurica* and *Pterocarya tonkinensis*. Powder. $[\alpha]_D$ -50 (c, 0.1 in $CHCl_3$)

15-Hydroxy, 2-Me ether
Juglanin A. Jugcathanin
[943989-90-0]
$C_{21}H_{24}O_5$ 356.418
Constit. of the pericarps of *Juglans regia* and *Juglans cathayensis*. Needles. Mp 148 - 150°. $[\alpha]_D^{20}$ -64 (c, 0.5 in MeOH)

15-Hydroxy, 9R-alcohol, 2-Me ether
16-Hydroxyplatycarynol
[355850-97-4]
$C_{21}H_{26}O_5$ 358.434
Constit. of *Alnus hirsuta*, *Juglans regia* and *Rhoiptelea chiliantha*. Amorph. powder. $[\alpha]_D^{15}$ -58.5 (c, 0.3 in $CHCl_3$)

15-β-D-Glucopyranosyloxy
Myricatomentoside I
[191999-61-8]
$C_{26}H_{32}O_{10}$ 504.532
Constit. of *Myrica gale* var. *tomentosa*. Powder. $[\alpha]_D$ +8.6 (c, 0.5 in MeOH). λ_{max} 281 (log ε 3.58) (MeOH)

(±)-form
8R-Hydroxy, 9S*-alcohol*
Engelhardiol A
[1397702-27-0]
$C_{20}H_{24}O_5$ 344.407
Constit. of *Engelhardtia roxburghiana* and *Juglans regia* (walnut). Amorph. solid. λ_{max} 208 (log ε 4.4); 230 (log ε 3.98); 281 (log ε 3.5) (MeOH). [base] λ_{max} 210 (log ε 4.59); 252 (log ε 3.71); 288 (log ε 3.5) (MeOH/KOH)

8R-Hydroxy, 9R*-alcohol, O-de-Me*
Engelhardiol B
[1397702-29-2]
$C_{19}H_{22}O_5$ 330.38
Constit. of the stems of *Engelhardtia roxburghiana*. Needles (MeOH). Mp 122 - 124°. $[\alpha]_D^{20}$ +186.9 (c, 0.6 in MeOH). λ_{max} 208 (log ε 4.4); 224 (sh); 282 (log ε 3.52) (MeOH). [base] λ_{max} 211 (log ε 4.58); 241 (sh); 293 (log ε 3.52) (MeOH/KOH)

(ξ)-form
9R-Alcohol
16-Methoxyacerogenin B
[619297-55-1]
$C_{20}H_{24}O_4$ 328.408
Constit. of *Juglans mandshurica* and *Juglans regia*. Exhibits moderate cytotoxic activity. Brown oil. $[\alpha]_D^{25}$ -39.1 (c, 0.2 in MeOH). λ_{max} 279 (log ε 3.6) (MeOH)

9ξ-Alcohol, 9-O-[β-D-apiofuranosyl-(1→6)-β-D-glucopyranoside]
[1276043-55-0]
$C_{31}H_{42}O_{13}$ 622.664
Constit. of the bark of *Myrica rubra*. Amorph. solid. λ_{max} 221 (log ε 3.81); 283 (log ε 3.33) (EtOH)

7ξ-Hydroxy
Hydroxygaleon
[61576-09-8]
$C_{20}H_{22}O_5$ 342.391
Constit. of *Myrica gale*. Mp 183 - 184°

8R-Hydroxy
[619297-54-0]
$C_{20}H_{22}O_5$ 342.391
Constit. of the roots of *Juglans mandshurica*. Brown solid. $[\alpha]_D^{25}$ -81.1 (c, 0.03 in MeOH). λ_{max} 282 (log ε 3.53) (MeOH)

8R-Hydroxy, 9R*-alcohol*
[1663474-99-4]
$C_{20}H_{24}O_5$ 344.407
Constit. of the roots of *Juglans mandshurica*. Light yellow powder. $[\alpha]_D^{20}$ -75.6 (c, 0.17 in MeOH). λ_{max} 226 (log ε 2.75); 268 (log ε 1.15) (MeOH)

10ξ-Hydroxy, 9ξ-alcohol

Myricananadiol

[957200-77-0]

$C_{20}H_{24}O_5$ 344.407

Constit. of the roots of *Myrica nana*. Powder. $[\alpha]^{24}_{D}$ -71.4 (c, 0.07 in Me_2CO). λ_{max} 279 (MeOH)

15-Hydroxy, 9R-alcohol

[1607448-30-5]

$C_{20}H_{24}O_5$ 344.407

Constit. of the leaves of *Juglans mandshurica*. Brown powder. $[\alpha]^{20}_{D}$ -97.7 (c, 0.07 in MeOH). λ_{max} 209; 280 (MeOH).

Malterud, K.E. *et al., Tetrahedron Lett.*, 1976, **17**, 3069-3072 (*Galeon, Hydroxygaleon, cryst struct*).

Morihara, M. *et al., Chem. Pharm. Bull.*, 1997, **45**, 820-823 (*R-form, abs config*).

Kim, S.-H. *et al., J. Nat. Prod.*, 1998, **61**, 643-645 (*Juglaside A*).

Tanaka, T. *et al., Phytochemistry*, 1998, **47**, 851-854 (*Platycarynol*).

Jiang, Z.-H. *et al., Chem. Pharm. Bull.*, 2001, **49**, 737-740 (*16-Hydroxyplatycarynol, cd*).

Li, G. *et al., Chem. Pharm. Bull.*, 2003, **51**, 262-264 (*9R-alcohol, 8R-hydroxy*).

Liu, H.B. *et al., Chin. Chem. Lett.*, 2005, **16**, 215-218 (*Pterocarine*).

Jin, W.Y. *et al., Arch. Pharmacal Res.*, 2007, **30**, 412-418 (*Alnus hirsuta constits*).

Yu, Y.-F. *et al., Chem. Biodiversity*, 2007, **90**, 1691-1696 (*Myricananadiol*).

Wang, Q. *et al., Synth. Commun.*, 2007, **37**, 675-681 (*synth, Pterocarine*).

Machida, K. *et al., J. Tohoku Pharm. Univ.*, 2008, **55**, 41-46 (*Juglaside A*).

Liu, J.X. *et al., Planta Med.*, 2008, **74**, 754-759 (*Juglanin A, 16-Hydroxyplatycarynol*).

Akazawa, H. *et al., J. Oleo Sci.*, 2011, **59**, 213-221 (*9-apiosylglucoside*).

Shen, L. *et al., Tetrahedron Lett.*, 2011, **52**, 4570-4574 (*Pterocarine, Engelhardione, synth*).

Salih, M.Q. *et al., Org. Lett.*, 2012, **14**, 4026-4029 (*synth*).

Wu, H.-C. *et al., Phytochemistry*, 2012, **82**, 118-127 (*Engelhardiols A,B*).

Li, J. *et al., Chin. Chem. Lett.*, 2013, **24**, 521-523 (*Jugcathayenoside*).

Salih, M.Q. *et al., Org. Lett.*, 2013, **15**, 4540-4543 (*synth*).

Pattawong, O. *et al., Org. Biomol. Chem.*, 2014, **12**, 3303-3309 (*conformn, chirality*).

Yao, D.L. *et al., Arch. Pharmacal Res.*, 2015, **38**, 480-484 (*15-hydroxy 9R-alcohol*).

Jin, M. *et al., Nat. Prod. Res.*, 2015, **29**, 1839-1844 (*Juglans mandshurica constit*).

Ding, Q. *et al., Org. Lett.*, 2017, **19**, 1804-1807 (*synth*).

Salih, M.Q. *et al., Tetrahedron Lett.*, 2017, **58**, 2023-2025 (*synth*).

Rahman, A.F.M.M. *et al., Molecules*, 2020, **25**, 5903:1-26 (*Galeon, ms, metab, activity*)

Garrettilignan A G3

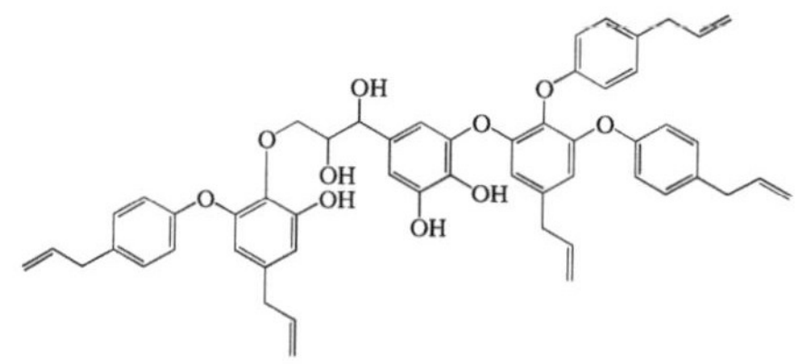

[1240907-40-7]

$C_{54}H_{52}O_{10}$ 861.000

Constit. of the leaves of *Magnolia garrettii*. Pale brown solid. $[\alpha]^{21}_{D}$ +6.4 (c, 2.84 in $CHCl_3$). λ_{max} 207 (log ε 5); 274 (log ε 4.2) (MeOH).

Schuehly, W. *et al., J. Nat. Prod.*, 2010, **73**, 1381-1384 (*Garrettilignan A*)

Garrettilignan B G4

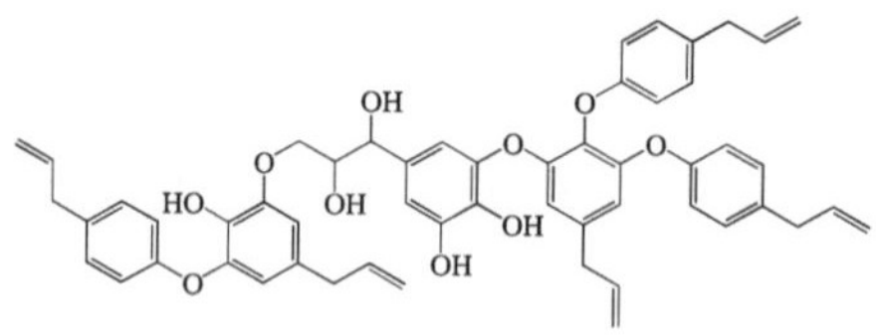

[1240907-41-8]

$C_{54}H_{52}O_{10}$ 861.000

Constit. of the leaves of *Magnolia garrettii*. Dark brown solid. $[\alpha]^{21}_{D}$ +1.4 (c, 10.9 in $CHCl_3$). λ_{max} 207 (log ε 5.1); 275 (log ε 4.4) (MeOH).

Schuehly, W. *et al., J. Nat. Prod.*, 2010, **73**, 1381-1384 (*Garrettilignan B*)

Garugamblin I G5

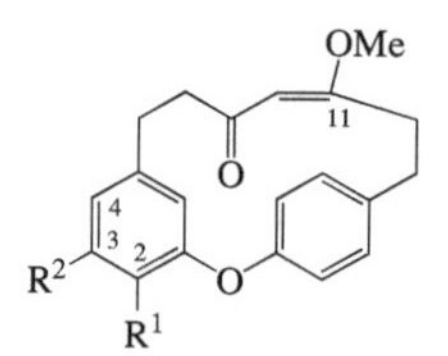

R^1 = OMe, R^2 = H

Garugamblin 1. ***Garuganin IV***

[118382-42-6]

[152726-64-2] (Garuganin IV)

$C_{21}H_{22}O_4$ 338.403

Shown to be achiral molecules in 2013 despite their reported opt. rotations. The struct. of Garuganin IV was revised to that of Garugamblin I in 2013. Related to Acerogenin A, A4, Acerogenin B, A5, and Galeon, G2. Isol. from the bark of *Garuga gamblei* and *Garuga pinnata*. Solid (C_6H_6). Mp 205 - 206°. $[\alpha]^{30}_{D}$ +48 (c, 0.2 in $CHCl_3$)

11-O-De-Me

11-O-Demethylgarugamblin I

[920317-43-7]

$C_{20}H_{20}O_4$ 324.376

Constit. of the stem bark of *Garuga pinnata*. Gum. $[\alpha]^{20}_{D}$ +6.4 (c, 0.28 in $CHCl_3$). Enolised β-diketone

Di-O-de-Me

Garugamblin III. *Garugamblin 3*

[752236-21-8]

$C_{19}H_{18}O_4$ 310.349

Constit. of the fruit of *Alnus japonica*. Cryst.. Mp 190 - 192°. λ_{max} 279 (log ε 4.01) (MeOH)

3-Hydroxy, 11-O-de-Me

2,11-Di-O-demethylgaruganin III (incorr.). ***Garuganin VII***

[920317-44-8]

$C_{20}H_{20}O_5$ 340.375

Constit. of the stem bark of *Garuga pinnata*. Solid. $[\alpha]^{20}_{D}$ -1 (c, 0.5 in $CHCl_3$). Struct. of Didemethylgaruganin III was revised; and renamed as Garuganin VII in 2013

3-Methoxy, 2-O-de-Me

2-O-Demethylgaruganin III. *1-Desmethylgaruganin III*

[1622997-27-6]

$C_{21}H_{22}O_5$ 354.402

Constit. of the stem bark of *Garuga pinnata*. Yellowish solid. Struct. may be erroneous as the authors do not seem to have taken into account the struct. revision of Garuganin III in 1993

2,4-Dimethoxy

Garuganin I

[90906-53-9]

$C_{22}H_{24}O_5$ 368.429

Constit. of leaves and bark of *Garuga pinnata*. Cryst. (MeOH). Mp 165 - 166°. λ_{max} 207; 263; 298 (MeOH) (Berdy).

Pattabhi, V. *et al., Acta Crystallogr., Sect. C*, 1984, **40**, 832-834 (*Garuganin I, cryst struct*).

Haribal, M.M. *et al., Tetrahedron*, 1985, **41**, 4949-4951 (*Garuganin I*).

Kalchhauser, H. *et al., Monatsh. Chem.*, 1988, **119**, 1047-1050 (*Garugamblins I,II*).

Kalchhauser, H. *et al., Magn. Reson. Chem.*, 1989, **27**, 635-639 (*pmr, cmr*).

Nethaji, M. *et al., Acta Crystallogr., Sect. C*, 1990, **46**, 307-310 (*Garugamblin I, cryst struct*).

Venkatraman, G. *et al., Phytochemistry*, 1993, **33**, 1221-1225 (*Garuganin IV*).

Vermes, B. *et al., Tetrahedron*, 1993, **49**, 4893-4900 (*Garugamblin I, synth*).

Kang, H.-M. *et al., Nat. Prod. Res.*, 2004, **18**, 295-299 (*Garugamblin 3*).

Ara, K. *et al., Phytochemistry*, 2006, **67**, 2659-2662 (*11-O-Demethylgarugamblin 1, 2,11-Di-O-demethylgaruganin III*).

Ara, K. *et al., Dhaka Univ. J. Pharm. Sci.*, 2012, **11**, 165-167 (*2-O-Demethylgaruganin III*).

Zhu, Z.-Q. *et al., J. Org. Chem.*, 2013, **78**, 2881-2896 (*synth, struct revision*).

Zhu, Z.-Q. *et al.*, *J. Org. Chem.*, 2013, **78**, 3336-3341 (*Garuganins IV,VII, Garugamblin I, synth*).
Pattawong, O. *et al.*, *Org. Biomol. Chem.*, 2014, **12**, 3303-3309 (*conformn, chirality*)

Garugamblin II G6

As Garugamblin I, G5 with
R^1R^2 = –OCH_2O–
Garugamblin 2
[118382-41-5]
$C_{21}H_{20}O_5$ 352.386
Shown to be an achiral molecule despite reported opt. rotation. Isol. from bark of *Garuga gamblei*. Solid (C_6H_6). Mp 195 - 197°. $[\alpha]^{30}_{D}$+51 (c, 0.2 in $CHCl_3$).
Kalchauser, H. *et al.*, *Monatsh. Chem.*, 1988, **119**, 1047 (*Garugamblin II*).
Kalchauser, H. *et al.*, *Magn. Reson. Chem.*, 1989, **27**, 635 (*pmr, cmr*).
Nethaji, M. *et al.*, *Acta Crystallogr., Sect. C*, 1990, **46**, 307-310 (*cryst struct*).
Keseru, G.M. *et al.*, *Liebigs Ann. Chem.*, 1994, 361 (*synth, pmr*).
Zhu, Z.-Q. *et al.*, *J. Org. Chem.*, 2013, **78**, 2881-2896 (*synth, struct*)

Garuganin II G7

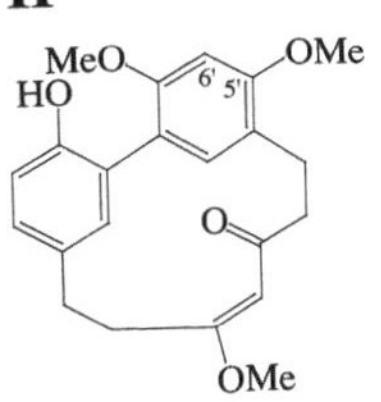

[111456-81-6]
$C_{22}H_{24}O_5$ 368.429
Achiral molecules despite reported opt. rotations. Isol. from *Garuga pinnata*. Shows antibiotic props.

5'-Demethoxy
Garuganin V
[152726-65-3]
$C_{21}H_{22}O_4$ 338.403
Constit. of the stem bark of *Garuga pinnata*. Cryst. (MeOH). Mp 180 - 181°. $[\alpha]^{20}_{D}$+102 (c, 0.18 in MeOH)

5'-Demethoxy, 6'-hydroxy
6'-Hydroxygaruganin V
[920743-30-2]
$C_{21}H_{22}O_5$ 354.402
Constit. of the stem bark of *Garuga pinnata*. Cryst.. $[\alpha]^{20}_{D}$-10.9 (c, 1 in $CHCl_3$).

Krishnaswamy, S. *et al.*, *Acta Crystallogr., Sect. C*, 1987, **43**, 527-530 (*cryst struct*).
Venkatraman, G. *et al.*, *Phytochemistry*, 1993, **33**, 1221-1225 (*pmr, cmr, Garuganin V*).
Naveen, C. *et al.*, *Z. Kristallogr.*, 1995, **210**, 365-366 (*Garuganin V, cryst struct*).
Ara, K. *et al.*, *Phytochemistry*, 2006, **67**, 2659-2662 (*6'-Hydroxygaruganin V*).
Zhu, Z.-Q. *et al.*, *J. Org. Chem.*, 2013, **78**, 2881-2896 (*synth, struct*)

Garuganin VI G8

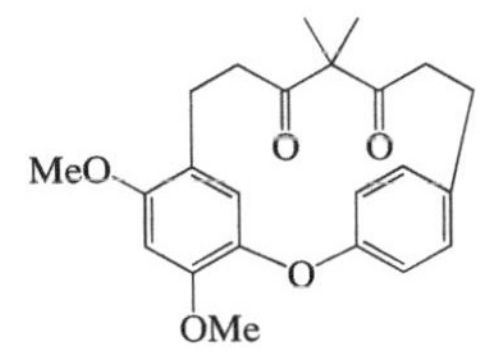

4,6-Dimethoxy-11,11-dimethyl-2-oxatricyclo[13.2.2.1^{3,7}]eicosa-3,5,7(20),15,17,18-hexaene-10,12-dione, CAS
[152686-84-5]
$C_{23}H_{26}O_5$ 382.456
Achiral molecule. Related to Acerogenin A, A4 and Galeon, G2. Constit. of the stem bark of *Garuga pinnata*. Cryst. (MeOH). Mp 205 - 206°.
Venkatraman, G. *et al.*, *Phytochemistry*, 1993, **33**, 1221-1225 (*Garuganin VI*).
Zhu, Z.-Q. *et al.*, *J. Org. Chem.*, 2013, **78**, 2881-2896 (*synth, struct*)

Gastradefurphenol G9

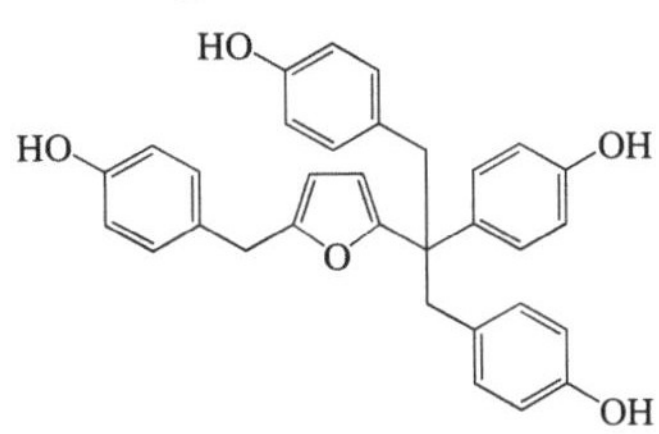

8,8'-Epoxy-7,7-bis(4-hydroxybenzyl)-9,9'-ligna-8,8'-diene-4,4'-diol
[2097633-06-0]
$C_{32}H_{28}O_5$ 492.571
Constit. of the rhizomes of *Gastrodia elata*. Amorph. yellow powder. λ_{max} 222 (log ε 2.61); 279 (log ε 1.94) (MeOH).
Zhou, X. *et al.*, *Chin. Chem. Lett.*, 2017, **28**, 1185-1189 (*Gastradefurphenol*)

Gemixanthone A G10

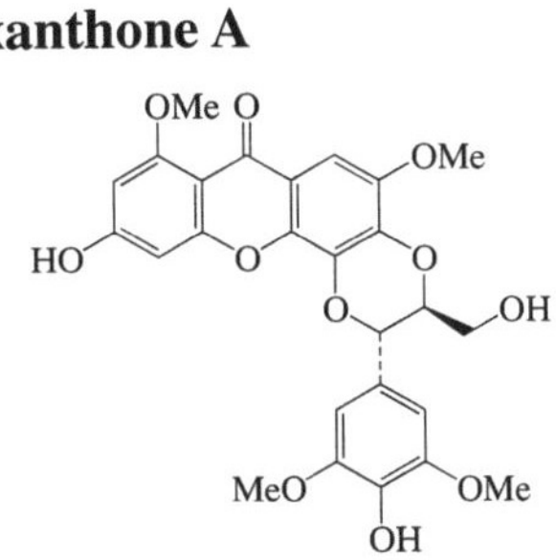

Relative Configuration

[238750-51-1]
$C_{26}H_{24}O_{11}$ 512.467
Constit. of *Hypericum geminiflorum*. Cryst. ($CHCl_3$) (as tri-Ac). Mp 241 - 242° (tri-Ac). $[\alpha]^{26}_{D}$-21 (c, 0.07 in $CHCl_3$) (tri-Ac). λ_{max} 250 (log ε 4.26); 286 (log ε 3.61); 319 (log ε 3.78); 402 (sh) (MeOH) (tri-Ac).
Chung, M.-I. *et al.*, *J. Nat. Prod.*, 1999, **62**, 1033-1035 (*Gemixanthone A*)

Geniculatin G11

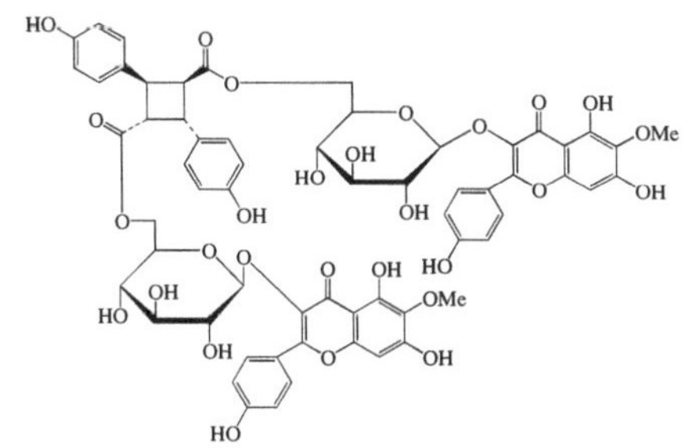

[1372408-20-2]
$C_{62}H_{56}O_{28}$ 1249.102
Constit. of the flowers of *Paepalanthus geniculatis*. Amorph. powder. $[\alpha]^{25}_{D}$-7.4 (c, 0.1 in MeOH). λ_{max} 265 (log ε 3.8); 282 (sh); 355 (log ε 4.1) (MeOH).
Amaral, F.P.D. *et al.*, *J. Nat. Prod.*, 2012, **75**, 547-556 (*Geniculatin*)

Giffonin F G12

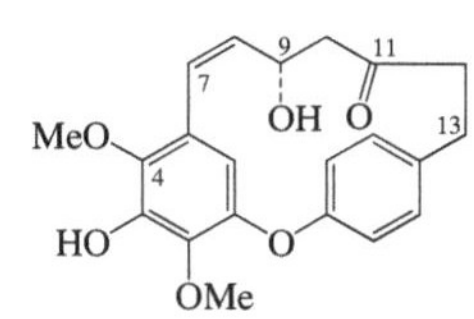

Absolute Configuration

[1638542-51-4]
$C_{21}H_{22}O_6$ 370.401
Constit. of the leaves of *Corylus avellana*. Amorph. solid. $[\alpha]^{25}_{D}$-24 (c, 0.03 in MeOH)

9-Deoxy
Giffonin D
[1638542-49-0]
$C_{21}H_{22}O_5$ 354.402
Constit. of the leaves of *Corylus avellana*. Amorph. solid

9-Deoxy, 11R-alcohol
Giffonin C
[1638542-48-9]
$C_{21}H_{24}O_5$ 356.418
Constit. of the leaves of *Corylus avellana*. Amorph. solid. $[\alpha]^{25}_{D}$-28 (c, 0.03 in MeOH)

9-Deoxy, 9,10-didehydro(E-), 11R-alcohol
Giffonin G
[1638542-52-5]

$C_{21}H_{22}O_5$ 354.402
Constit. of the leaves of *Corylus avellana*. Amorph. solid. $[\alpha]^{25}_{D}$ -71 (c, 0.1 in MeOH)

9-Deoxy, 13R-hydroxy, 11R*-alcohol*
Giffonin J
[1830306-87-0]
$C_{21}H_{24}O_6$ 372.417
Constit. of the leaves of *Corylus avellana*. Amorph. solid. $[\alpha]^{25}_{D}$ -15 (c, 0.1 in MeOH)

9-Deoxy, 13S-hydroxy, 9,10-didehydro(E-), 11R*-alcohol*
Giffonin K
[1830306-88-1]
$C_{21}H_{22}O_6$ 370.401
Constit. of the leaves of *Corylus avellana*. Amorph. solid. $[\alpha]^{25}_{D}$ -5 (c, 0.1 in MeOH)

11-Deoxo
Giffonin E
[1638542-50-3]
$C_{21}H_{24}O_5$ 356.418
Constit. of the leaves of *Corylus avellana*. Amorph. solid. $[\alpha]^{25}_{D}$ -28 (c, 0.03 in MeOH)

4-Demethoxy, 9-deoxy, 9,10-didehydro(E-), 11R-alcohol
Giffonin H
[1638542-53-6]
$C_{20}H_{20}O_4$ 324.376
Constit. of the leaves of *Corylus avellana*. Amorph. solid. $[\alpha]^{25}_{D}$ -22 (c, 0.03 in MeOH)

7E-Isomer, 9-deoxy
Giffonin A
[1638542-46-7]
$C_{21}H_{22}O_5$ 354.402
Constit. of the leaves of *Corylus avellana*. Amorph. solid

7E-Isomer, 9-deoxy, 13S-hydroxy
Giffonin B
[1638542-47-8]
$C_{21}H_{22}O_6$ 370.401
Constit. of the leaves of *Corylus avellana*. Amorph. solid. $[\alpha]^{25}_{D}$ -14 (c, 0.1 in MeOH).

Masullo, M. *et al.*, *J. Nat. Prod.*, 2015, **78**, 17-25 (*Giffonins A-H*).
Masullo, M. *et al.*, *J. Nat. Prod.*, 2015, **78**, 2975-2982 (*Giffonins J,K*).
Park, S. *et al.*, *Org. Chem. Front.*, 2019, **6**, 704-708 (*synth*)

Giffonin L G13

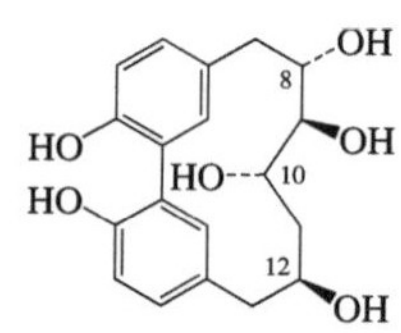

Relative Configuration

Tricyclo[12.3.1.1^2,6]nonadeca-1(18),2,4,6(19),14,16-hexaene-3,8,9,10,12,17-hexol, CAS
[1830306-89-2]
$C_{19}H_{22}O_6$ 346.379
Constit. of the leaves of *Corylus avellana*. Amorph. solid. $[\alpha]^{25}_{D}$ +10 (c, 0.1 in MeOH)

10-Ketone
Giffonin O
[1830306-92-7]
$C_{19}H_{20}O_6$ 344.363
Constit. of the leaves of *Corylus avellana*. Amorph. solid. $[\alpha]^{25}_{D}$ +27 (c, 0.1 in MeOH). Rel. config. determined. May be the same as Asadanin

8-Epimer, 11R-hydroxy, 12-ketone
Giffonin U
[2098225-05-7]
$C_{19}H_{20}O_7$ 360.362
Constit. of the green leafy covers of *Corylus avellana*. Amorph. solid. $[\alpha]^{25}_{D}$ -21 (c, 0.1 in MeOH). Possesses ($8R,9R,10R,11R,S_{axial}$)-config.

10-Epimer, 8-deoxy, 9-ketone
Giffonin M
[1830306-90-5]
$C_{19}H_{20}O_5$ 328.364
Constit. of the leaves of *Corylus avellana*. Amorph. solid. $[\alpha]^{25}_{D}$ -14 (c, 0.1 in MeOH). Rel. config. determined

10-Epimer, 8-deoxy, 9-ketone, 12-O-β-D-glucopyranoside
Giffonin N
[1830306-91-6]
$C_{25}H_{30}O_{10}$ 490.505
Constit. of the leaves of *Corylus avellana*. Amorph. solid. $[\alpha]^{25}_{D}$ +6 (c, 0.03 in MeOH). Rel. config. determined

12-Epimer, 11S-hydroxy
Giffonin P
[1830306-93-8]
$C_{19}H_{22}O_7$ 362.378
Constit. of the kernelsand leaves of *Corylus avellana*. Amorph. solid. $[\alpha]^{25}_{D}$ -37 (c, 0.1 in MeOH). Rel. config. determined

Stereoisomer
Asadanol. Epiasadanol. Isoasadanol
[32479-45-1]
$C_{19}H_{22}O_6$ 346.379
Constit. of *Ostrya japonica*. Three stereoisomers were isol. from the same source but there stereochem. was undetermined

Stereoisomer, 10-ketone
Asadanin
[22756-44-1]
$C_{19}H_{20}O_6$ 344.363
Constit. of *Corylus avellana* and *Ostrya japonica*. Mp 236 - 239°. $[\alpha]^{17}_{D}$ +84. Stereochem. undetermined. May be the same as Giffonin O

Stereoisomer, 10-deoxy
Deoxoasadanin
[22756-47-4]
$C_{19}H_{22}O_5$ 330.38
Constit. of *Ostrya japonica*. Mp 279 - 283°. Stereochem. undetermined.

Yasue, M. *et al.*, *Mokuzai Gakkaishi*, 1965, **11**, 146 (*Asadanins*).
Yasue, M. *et al.*, *Mokuzai Gakkaishi*, 1965, **11**, 152 (*Asadanins*).
Yasue, M. *Ringyo Shikenjo Kenkyu Hokoku*, 1968, **209**, 77-168 (*Asadanol, Epiasadanol, Isoasadanol*).
Masullo, M. *et al.*, *J. Nat. Prod.*, 2015, **78**, 2975-2982 (*Giffonins L-P*).
Singldinger, B. *et al.*, *J. Agric. Food Chem.*, 2017, **65**, 1677-1683 (*Asadanin*).
Cerulli, A. *et al.*, *J. Nat. Prod.*, 2017, **80**, 1703-1713 (*Giffonin U, cd, abs config*).
Singldinger, B. *et al.*, *J. Agric. Food Chem.*, 2018, **66**, 4660-4673 (*Giffonin P, Asadanin, props*).
Jahng, Y. *et al.*, *Molecules*, 2018, **23**, 3107:1-42 (*rev, isol, synth, activity*)

Globoidnan A G14

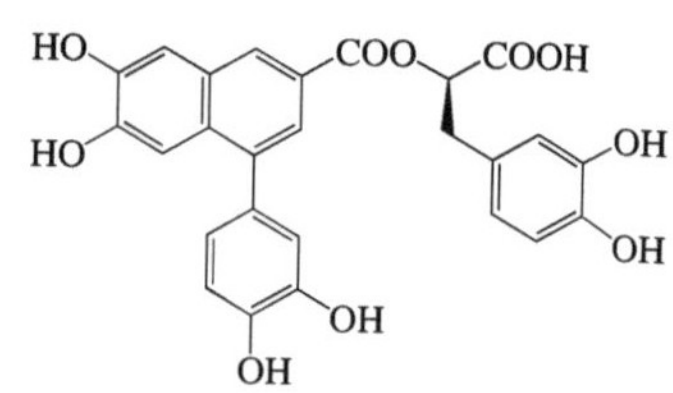

Absolute Configuration

Eritrichin
[827322-48-5]
$C_{26}H_{20}O_{10}$ 492.436
Constit. of *Eritrichium sericeum*, *Eucalyptus globoidea*, *Origanum glandulosum*, *Paris verticillata*, *Symphytum officinale* and *Thymus*

praecox ssp. *grossheimii*. HIV integrase inhibitor. Active against β-amyloid aggregation. Weakly cytotoxic against SK-OV-3 ovary ascites tumour cells. Brown powder. $[\alpha]_D$ +14.4 (c, 0.12 in MeOH) (Globoidnan A). $[\alpha]_D^{20}$ +39 (c, 0.4 in MeOH) (Eritrichin). λ_{max} 224; 260; 320 (MeCN aq.) (Globoidnan A). [neutral] λ_{max} 253 (log ε 4.38); 285 (log ε 4.25); 344 (log ε 4.23) (MeOH) (Eritrichin).

Ovenden, S.P.B. *et al., Phytochemistry*, 2004, **65**, 3255-3259 (*Globoidnan A*).
Fedoreyev, S.A. *et al., Planta Med.*, 2005, **71**, 446-451 (*Eritrichin*).
Lee, K.H. *et al., Molecules*, 2008, **13**, 41-45 (*Globoidnan A, activity*).
Basli, A. *et al., Rec. Nat. Prod.*, 2014, **8**, 208-216 (*Globoidnan A, activity*).
Erenler, R. *et al., Rec. Nat. Prod.*, 2016, **10**, 766-770 (*Globoidnan A, activity*).
Trifan, A. *et al., Phytochem. Anal.*, 2020, **31**, (*Symphytum officinale constit*)

Glochinin A G15

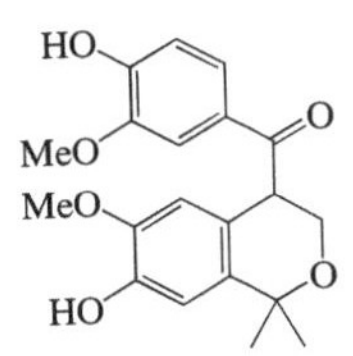

[1118571-77-9]
$C_{20}H_{22}O_6$ 358.39
Constit. of *Glochidion puberum*. Powder. $[\alpha]_D^{24}$ -13.5 (c, 0.37 in MeOH). λ_{max} 208 (log ε 4.25); 228 (log ε 3.97); 281 (log ε 3.75) (MeOH).

Liu, M. *et al., Chem. Nat. Compd. (Engl. Transl.)*, 2008, **44**, 588-590 (*Glochinin A*)

Glomexanthone A G16

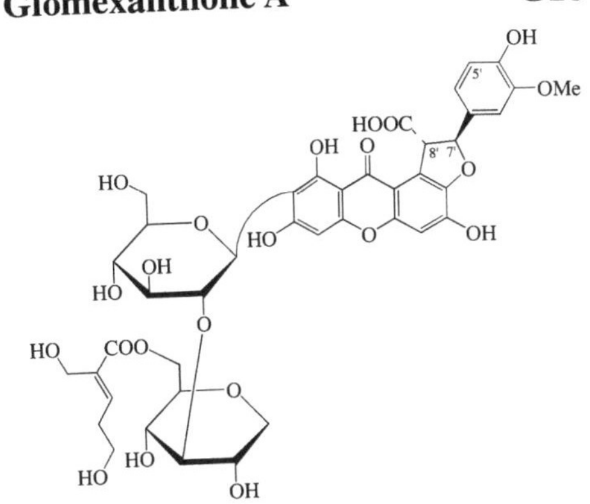

Absolute Configuration

Glomerxanthone A
[1621720-11-3]
$C_{41}H_{44}O_{22}$ 888.781
Xanthonolignan *C*-glycosides. Isol. from *Polygala glomerata*. Displays moderate neuroprotective effects in neuroblastoma cells. Yellow amorph. powder. $[\alpha]_D^{20}$ +11 (c, 0.10 in MeOH). λ_{max} 204; 262; 319; 372 (MeOH)

5′-Methoxy

Glomexanthone C. *Glomerxanthone C*
[1621720-14-6]
$C_{42}H_{46}O_{23}$ 918.807
Isol. from *Polygala glomerata*. Displays moderate neuroprotective effects in neuroblastoma cells. $[\alpha]_D^{20}$ +24 (c, 0.08 in MeOH). λ_{max} 204; 262; 319; 372 (MeOH)

7′,8′-Diepimer

Glomexanthone B. *Glomerxanthone B*
[1621720-13-5]
$C_{41}H_{44}O_{22}$ 888.781
Isol. from *Polygala glomerata*. Displays moderate neuroprotective effects in neuroblastoma cells. Yellow amorph. powder. $[\alpha]_D^{20}$ +19 (c, 0.08 in MeOH). λ_{max} 204; 262; 319; 372 (MeOH).

Li, C.-J. *et al., Fitoterapia*, 2014, **93**, 175-181 (*Glomexanthones A,B,C, activity*)

Glycerol tris(4-hydroxy-3,5-dimethoxyphenyl-2-propenoate) G17

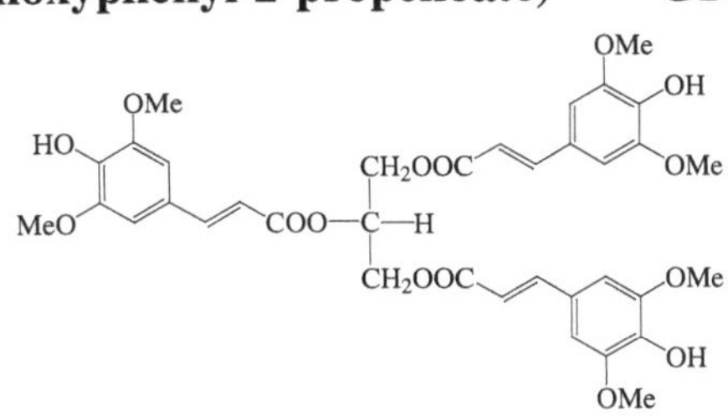

Glycerol trisinapate. Tri-O-sinapoylglycerol
[959577-09-4]
$C_{36}H_{38}O_{15}$ 710.685
Constit. of *Raphanus sativus*. Component of Lai Fu Zi.

Traditional Chinese Medicines, (ed. Milne, G.A.), Ashgate , 1999, 2694

Glypentoside C G18

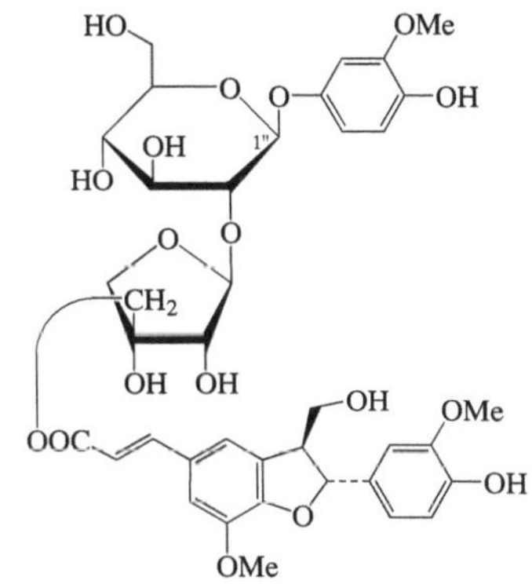

Absolute Configuration

Glycopentoside C
[879215-60-8]
$C_{38}H_{44}O_{18}$ 788.752
Constit. of *Clerodendranthus spicatus* and *Glycosmis pentaphylla*. Amorph. powder. $[\alpha]_D^{25}$ -49.7 (c, 0.84 in MeOH). Two similar synonyms assigned in ref.. λ_{max} 205 (log ε 4.7); 224 (log ε 4.44); 290 (log ε 4.14); 329 (log ε 4.34) (MeOH)

1″-O-[2-(4-Hydroxyphenyl)ethyl] analogue

Bunginoside A
[1612285-11-6]
$C_{39}H_{46}O_{17}$ 786.78
Constit. of the roots of *Clerodendron bungei*. $[\alpha]_D^{20}$ -7.7 (c, 0.05 in MeOH). λ_{max} 330 (log ε 4.24) (MeOH).

Wang, J. *et al., Phytochemistry*, 2006, **67**, 486-491 (*Glypentoside C*).
Chen, Y.-L. *et al., Zhongcaoyao*, 2009, **40**, 689-693 (*Clerodendranthus spicatus constit*).
Liu, Q. *et al., Phytochemistry*, 2014, **103**, 196-202 (*Bunginoside A*)

Gmelanone G19

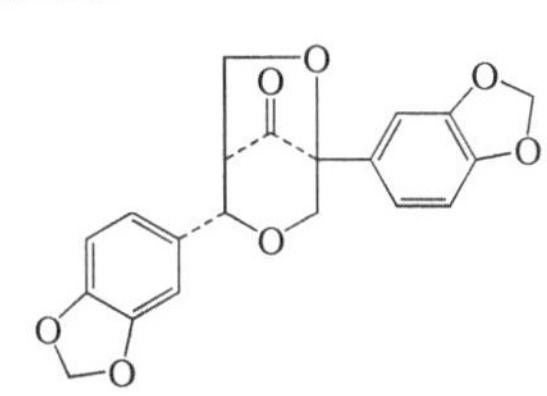

2,5-Bis(1,3-benzodioxol-5-yl)-3,6-dioxabicyclo[3.2.1]octan-8-one, CAS
[54826-95-8]
$C_{20}H_{16}O_7$ 368.341
Acid-cat. rearr. prod. of Arboreol, in D73. Constit. of *Gmelina arborea*. Cryst. (C_6H_6). Mp 201 - 202° (190°). $[\alpha]_D^{25}$ -72 (c, 0.5 in $CHCl_3$).

Row, L.R. *et al., J. Chem. Soc., Chem. Commun.*, 1974, 476-477 (*Gmelanone*).

Anjaneyulu, A.S.R. *et al., Tetrahedron*, 1975, **31**, 1277-1285 (*Gmelina arborea constit*).
Row, R.L. *et al., Tetrahedron Lett.*, 1980, **21**, 2919-2922 (*synth*).
Venkateswarlu, R. *et al., Tetrahedron*, 2005, **61**, 8956-8961 (*synth*)

Gnetifolin F G20

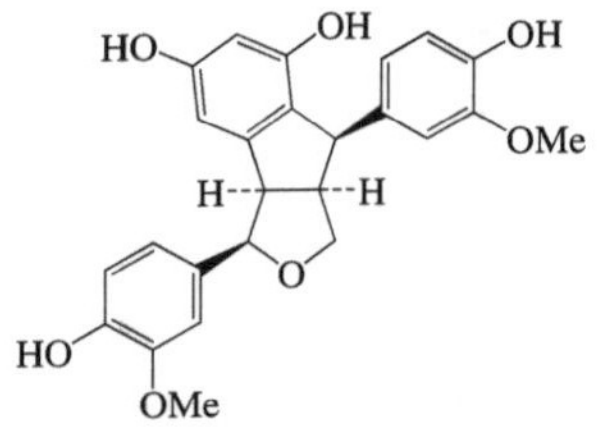

Relative Configuration

3,3a,8,8a-Tetrahydro-3,8-bis(4-hydroxy-3-methoxyphenyl)-1H-indeno[1,2-c]furan-5,7-diol
[140208-77-1]
$C_{25}H_{24}O_7$ 436.46
Stilbenolignan. Similar to Maackolin, M1. Constit. of *Gnetum klossii* and *Gnetum parvifolium*. Shows moderate antiinflammatory activity. Off-white prisms (C_6H_6/EtOH). Mp 144 - 148°. Racemic. λ_{max} 235 (log ε 4.37); 285 (log ε 4.1) (MeOH).

Lin, M. *et al., Phytochemistry*, 1991, **30**, 4201-4203 (*Gnetifolin F, cryst struct*).
Ali, Z. *et al., J. Nat. Prod.*, 2003, **66**, 558-560 (*Gnetum klossii constit*).
Yao, C.S. *et al., Chem. Pharm. Bull.*, 2006, **54**, 1053-1057 (*synth, activity*)

Gnetucleistol F G21

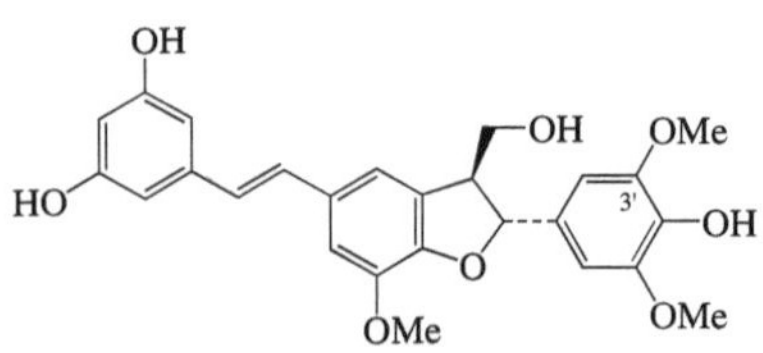

Relative Configuration

[913529-99-4]
$C_{26}H_{26}O_8$ 466.486
Stilbenolignan. Structures in this series may need revision. Phenyl 3,5-dioxygenated substn. pattern in doubt as a result of work done by Yang *et al* in 2015. Constit. of the lianas of *Gnetum cleistostachyum*. Amorph. powder. $[\alpha]^{20}_D$+4.7 (c, 0.1 in MeOH). λ_{max} 235 (sh); 310 (sh); 327 (log ε 4.29) (MeOH)

Enantiomer
$C_{26}H_{26}O_8$ 466.486
Constit. of the caulis of *Gnetum montanum*. Amorph. brown powder. λ_{max} 210 (log ε 4.80); 329 (log ε 4.43) (MeOH)

Enantiomer, 3'-demethoxy
Gnetofuran A
[526214-78-8]
$C_{25}H_{24}O_7$ 436.460
Constit. of the stems of *Gnetum klossii*. Amorph. powder. $[\alpha]^{26}_D$-7 (c, 0.2 in MeOH). λ_{max} 220 (sh); 288 (log ε 4.25); 308 (log ε 4.33) (MeOH).

Ali, Z. *et al., J. Nat. Prod.*, 2003, **66**, 558-560 (*Gnetofuran A*).
Yao, C.-S. *et al., Chem. Pharm. Bull.*, 2006, **54**, 1053-1057 (*(+)-Gnetucleistol F*).
Yang, Y.-N. *et al., J. Nat. Prod.*, 2015, **78**, 705-711 (*pmr, struct*).
Ma, Y.-Q. *et al., Phytochem. Lett.*, 2017, **21**, 42-45 (*(-)-Gnetucleistol F*)

Gnetumonin A G22

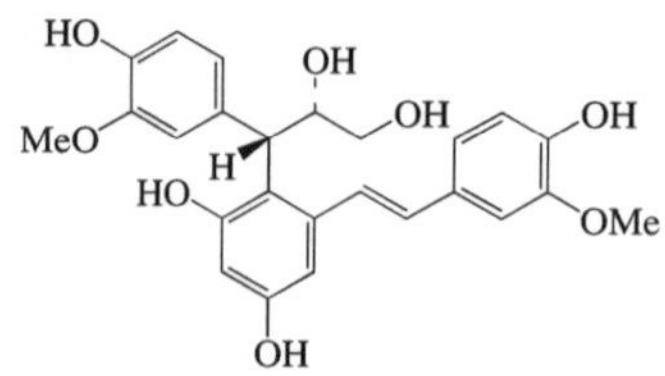

Relative Configuration

$C_{25}H_{26}O_8$ 454.475
Stilbenolignan. Constit. of the caulis of *Gnetum montanum*. Amorph. yellow powder. $[\alpha]^{25}_D$-59.3 (c, 0.05 in MeOH). λ_{max} 288 (log ε 4.19); 316 (log ε 4.18) (MeOH).

Ma, Y.-Q. *et al., Phytochem. Lett.*, 2017, **21**, 42-45 (*Gnetumonin A*)

Gnetumonin B G23

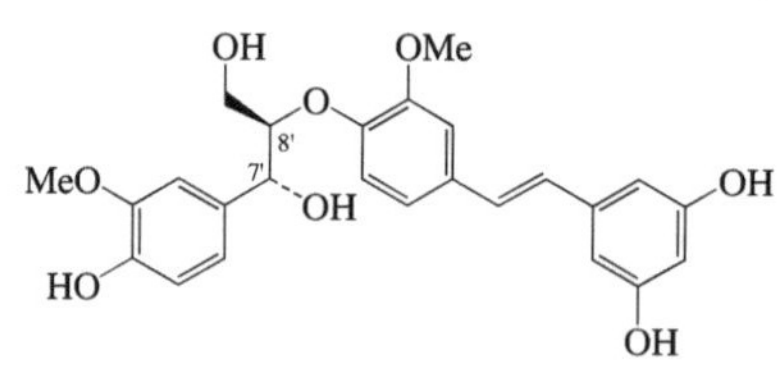

Relative Configuration

$C_{25}H_{26}O_8$ 454.475
Structures in this series may need revision. Phenyl 3,5-dioxygenated substn. pattern in doubt as a result of work done by Yang *et al* in 2015. Related to Dendrocandin M, D17. Constit. of the caulis of *Gnetum montanum*. Amorph. yellowish powder. $[\alpha]^{25}_D$-3.9 (c, 0.05 in MeOH). λ_{max} 323 (log ε 3.62) (MeOH)

7'- or 8'-Epimer
Gnetumonin C
$C_{25}H_{26}O_8$ 454.475
Constit. of the caulis of *Gnetum montanum*. Amorph. yellowish powder. $[\alpha]^{25}_D$-6.7 (c, 0.05 in MeOH). λ_{max} 322 (log ε 4.35) (MeOH).

Yang, Y.-N. *et al., J. Nat. Prod.*, 2015, **78**, 705-711 (*pmr, struct*).
Ma, Y.-Q. *et al., Phytochem. Lett.*, 2017, **21**, 42-45 (*Gnetumonins B,C*)

Gnetumontanin C G24

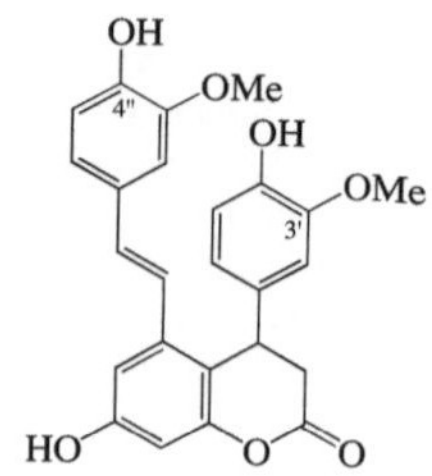

[809237-88-5]
$C_{25}H_{22}O_7$ 434.444
Stilbenolignan. Constit. of the lianas of *Gnetum montanum* f. *megalocarpum* and *Gnetum cleistostachyum*. Exhibits moderate inhibition on TNF-α prodn. by murine peritoneal macrophages. Amorph. yellow powder. Mp 72 - 73°. Racemic. λ_{max} 210 (log ε 4.56); 230 (sh); 290 (log ε 4.14); 324 (log ε 4.24) (MeOH)

4''-O-β-D-Glucopyranoside
Gnetumontanin D
[809237-89-6]
$C_{31}H_{32}O_{12}$ 596.585
Constit. of the lianas of *Gnetum montanum* f. *megalocarpum*. Amorph. yellow powder. $[\alpha]^{20}_D$-24 (c, 0.07 in MeOH). λ_{max} 207 (log ε 4.76); 289 (log ε 4.22); 319 (log ε 4.25) (MeOH)

Di-O-de-Me
Gnetumontanin F
[1300029-27-9]
$C_{23}H_{18}O_7$ 406.39
Constit. of the rhizomes of *Smilax corbularia*. Brown powder. $[\alpha]^{25}_D$ +0.2 (c, 0.1 in MeOH). λ_{max} 210 (log ε 4.65); 230 (log ε 4.44); 288 (log ε 4.15); 327 (log ε 4.13) (MeOH)

3''-Demethoxy
Gnetumontanin G
[1300029-28-0]
$C_{24}H_{20}O_6$ 404.418
Constit. of the rhizomes of *Smilax corbularia*. Pale brown powder. $[\alpha]^{25}_D$+11 (c, 0.1 in MeOH). λ_{max} 208 (log ε 4.64); 230 (log ε 4.45);

290 (log ε 4.35); 305 (log ε 4.39); 319 (log ε 4.4) (MeOH)

3''-Demethoxy, 3'-O-de-Me

Gnetumontanin E

[1300029-26-8]

$C_{23}H_{18}O_6$ 390.391

Constit. of the rhizomes of *Smilax corbularia*. Brown powder. $[\alpha]^{25}_{D}$ +7.4 (c, 0.1 in MeOH). λ_{max} 209 (log ε 4.65); 296 (log ε 4.44); 309 (log ε 4.45); 318 (log ε 4.46) (MeOH).

Li, X.-M. *et al.*, *Planta Med.*, 2004, **70**, 160-165 (*Gnetumontanins C,D*).
Yao, C.-S. *et al.*, *Chem. Pharm. Bull.*, 2006, **54**, 1053-1057 (*Gnetum cleistostachyum constit, activity*).
Wungsintaweekul, B. *et al.*, *Phytochemistry*, 2011, **72**, 495-502 (*Gnetumontanins E,F,G*)

Gobicusin A G25

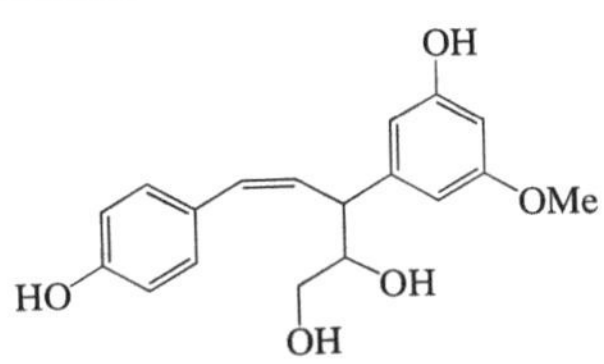

3-(3-Hydroxy-5-methoxyphenyl)-5-(4-hydroxyphenyl)-4-pentene-1,2-diol. 5-Methoxy-9'-nor-7,8'-lign-7'-ene-3,4',8,9-tetrol. 3,4',8,9-Tetrahydroxy-5-methoxy-9'-nor-7,8'-neolign-7'-ene

[835613-07-5]

$C_{18}H_{20}O_5$ 316.353

Structure may need revision. Phenyl 3,5-dioxygenated substn. pattern in doubt as a result of work done by Yang *et al* in 2015. Provisional norlignan. Biosynth. pathway unclear. Constit. of the roots of *Asparagus gobicus* and *Asparagus filicinus*. Gum. $[\alpha]^{21}_{D}$ +51.6 (c, 5.5 in Me_2CO).

Yang, C.-X. *et al.*, *Planta Med.*, 2004, **70**, 446-451 (*Gobicusin A*).
Tao, H.-M. *et al.*, *Zhongcaoyao*, 2011, **42**, 2181-2185 (*Asparagus filicinus constit*).
Yang, Y.-N. *et al.*, *J. Nat. Prod.*, 2015, **78**, 705-711 (*pmr, struct*)

Gomisin D G26

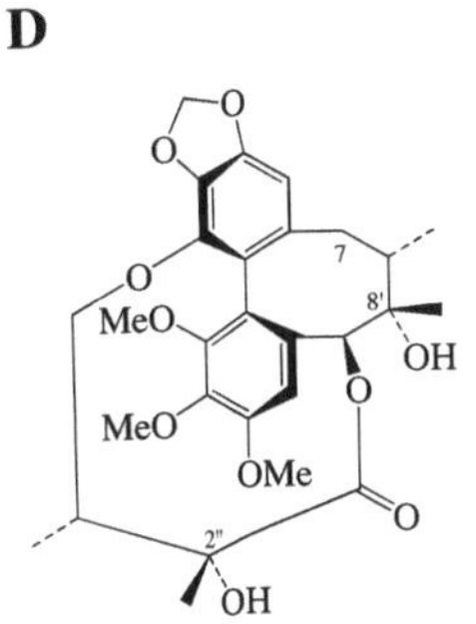

Absolute Configuration

[60546-10-3]

$C_{28}H_{34}O_{10}$ 530.57

Isol. from fruit of *Schizandra chinensis* and from *Kadsura coccinea*. Prisms (hexane/Et_2O). Mp 194°. $[\alpha]^{25}_{D}$ -58.8 (c, 0.265 in $CHCl_3$)

2''-Me ether

2''-O-Methylgomisin D, ***Marlignan M‡***

$C_{29}H_{36}O_{10}$ 544.597

Constit. of *Schisandra wilsoniana*. Amorph. powder. $[\alpha]^{24}_{D}$ -46.2 (c, 0.2 in MeOH). Not to be confused with Marlignan M, in C225, also isol. by the same workers. λ_{max} 215 (log ε 5.02); 254 (log ε 3.87); 290 (log ε 1.56) (MeOH)

7-Acetoxy

Kadsuphilin N

[1594808-29-3]

$C_{30}H_{36}O_{12}$ 588.606

Constit. of the leaves and stems of *Kadsura philippinensis*. $[\alpha]^{26}_{D}$ -2.4 (c, 1.3 in CH_2Cl_2). λ_{max} 212; 259; 292 (MeOH)

2''-Epimer

Marlignan N‡

$C_{28}H_{34}O_{10}$ 530.570

Constit. of *Schisandra wilsoniana*. Amorph. pale yellow solid. $[\alpha]^{25}_{D}$ -35.3 (c, 0.2 in MeOH). Not to be confused with Marlignan N, in C225, also isol. by the same workers. λ_{max} 210 (log ε 5.24); 255 (log ε 3.55); 290 (log ε 1.65) (MeOH)

2''-Epimer, 3''S-hydroxy

Marlignan O‡

$C_{28}H_{34}O_{11}$ 546.569

Constit. of *Schisandra wilsoniana*. Exhibits antiviral activity. Yellow gum. $[\alpha]^{25}_{D}$ -32.7 (c, 0.2 in MeOH). Not to be confused with Marlignan O, in C225, also isol. by the same workers. λ_{max} 210 (log ε 5.40); 251 (log ε 3.84); 292 (log ε 1.62) (MeOH)

8'-Epimer, 8'-deoxy

Gomisin E‡

[72960-21-5]

$C_{28}H_{34}O_9$ 514.571

Isol. from *Schizandra chinensis* and *Kadsura coccinea*. Amorph. powder. $[\alpha]^{24}_{D}$ +28.1 (c, 0.28 in $CHCl_3$).

Ikeya, Y. *et al.*, *Chem. Pharm. Bull.*, 1979, **27**, 1395-1401 (*Gomisin D*).
Ikeya, Y. *et al.*, *Chem. Pharm. Bull.*, 1979, **27**, 2695-2709 (*Gomisin E*).
Lin, Y.-C. *et al.*, *Molecules*, 2013, **18**, 6573-6583 (*Kadsuphilin N, cd*).
Yang, G.-Y. *et al.*, *Nat. Prod. Commun.*, 2013, **8**, 467-470 (*Marlignans M,N,O, cd, activity*)

Grahamine‡ G27

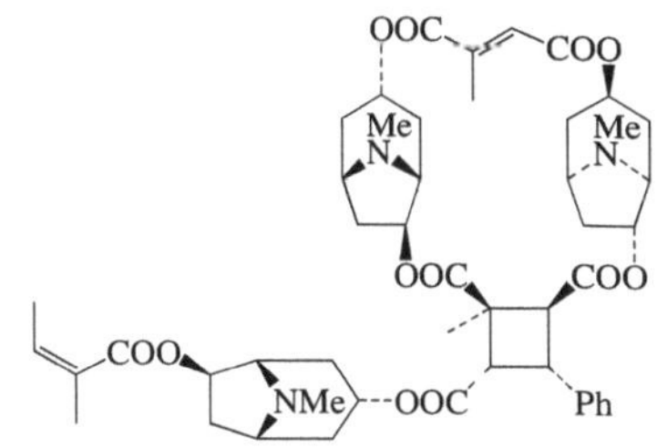

[125685-27-0]

$C_{48}H_{61}N_3O_{12}$ 872.025

Alkaloid from *Schizanthus grahamii*. Oil. $[\alpha]^{20}_{D}$ +6.21 ($CHCl_3$).

Hartmann, R. *et al.*, *Angew. Chem., Int. Ed.*, 1990, **29**, 385-386 (*Grahamine*)

Griffinone E G28

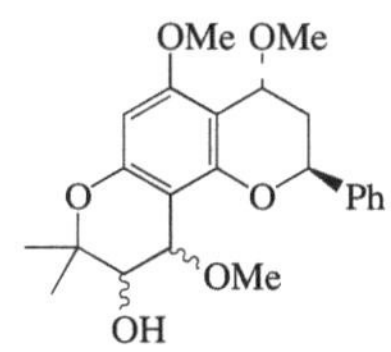

Absolute Configuration

3,4,9,10-Tetrahydro-4,5,10-trimethoxy-8,8-dimethyl-2-phenyl-2H,8H-benzo[1,2-b:3,4-b']dipyran-9-ol, CAS

[2007967-87-3]

$C_{23}H_{28}O_6$ 400.471

Constit. of the stems of *Millettia griffithii*. Yellow powder. $[\alpha]^{25}_{D}$ -67.5 (c, 0.2 in $CHCl_3$). λ_{max} 215 (log ε 4.22); 247 (log ε 3.76) (MeOH).

Tang, H. *et al.*, *Bioorg. Med. Chem. Lett.*, 2016, **26**, 4417-4422 (*Griffinone E*)

Grossamide G29

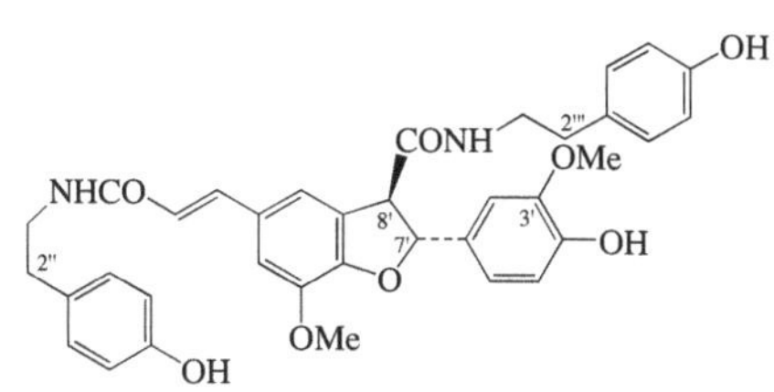

Relative Configuration

[80510-06-1]

[870472-08-5] (7'-Epimer)

$C_{36}H_{36}N_2O_8$ 624.69

Deriv. of 4,7'-Epoxy-3,8'-lign-7-ene-3',4',5,9,9'-pentol, E163. Lignan numbering shown. Isol. from *Annona crassiflora*, *Cannabis sativa*, *Capsicum annuum* var. *grossum*, *Gastrodia elata*, *Hibiscus cannabinus*, *Hyoscyamus niger*, *Lycium barbarum*, *Solanum tuberosum*, *Tetrapterys mucronata* and *Withania somnifera*. Protease inhibitor. Shows moderate activity against human prostate cancer cells. Germination inhibitor. Cryst. (MeOH/$CHCl_3$/cyclohexane). Mp 174 - 175° (133-135°)

4′-Me ether

4-O-Methylgrossamide

$C_{37}H_{38}N_2O_8$ 638.717

Constit. of the stems of *Lycium barbarum*. Powder. $[\alpha]_D$ -16.8 (c, 0.11 in DMSO). Possesses 7′*R*,8′*R*-config.. λ_{max} 249 (log ε 4.2); 289 (log ε 4.4); 320 (log ε 4.4) (MeOH)

3′-O-De-Me

Demethylgrossamide

[170171-94-5]

$C_{35}H_{34}N_2O_8$ 610.663

Isol. from seeds of *Xylopia aethiopica*. Amorph. powder

2″ξ-Hydroxy

Melongenamide B

[1676050-23-9]

$C_{36}H_{36}N_2O_9$ 640.689

Constit. of the roots of *Solanum melongena*. NO prodn. inhibitor. Powder. $[\alpha]_D^{21}$ -24 (c, 0.1 in MeOH). λ_{max} 223 (log ε 4.48); 287 (log ε 4.32); 320 (log ε 4.39) (MeOH)

4′-Deoxy, 5′-hydroxy

Thoreliamide A

[1052115-05-5]

$C_{36}H_{36}N_2O_8$ 624.69

Constit. of the stems of *Mitrephora thorelii*. Amorph. powder. Racemate. λ_{max} 225 (log ε 4.67); 286 (log ε 4.39); 321 (log ε 4.41) (MeOH)

7′-Epimer

Tribulusamide A

[218622-84-5]

$C_{36}H_{36}N_2O_8$ 624.69

Alkaloid from *Humulus scandens*, *Solanum surattense* and *Tribulus terrestris*. Hepatoprotective agent. Amorph. solid. $[\alpha]_D^{25}$ -10.9 (c, 0.1 in MeOH). λ_{max} 220 (log ε 4.22); 287 (log ε 4.01); 340 (log ε 4.07) (MeOH)

7′-Epimer, 5-O-de-Me, 4′-Me ether

Dimorphamide A

[1297298-16-8]

$C_{36}H_{36}N_2O_8$ 624.69

Alkaloid from *Atriplex dimorphostagia*. Amorph. solid. Racemic. λ_{max} 223 (log ε 4.27); 287 (log ε 3.97); 319 (log ε 3.96) ($CHCl_3$)

7′-Epimer, 2‴-oxo

Tribulusamide B

[218622-86-7]

$C_{36}H_{34}N_2O_9$ 638.673

Alkaloid from the fruit of *Tribulus terrestris*. Hepatoprotective agent. Amorph. solid. $[\alpha]_D^{25}$ -4.9 (c, 0.2 in MeOH). λ_{max} 220 (log ε 4.38); 285 (log ε 4.32); 320 (log ε 4.18) (MeOH)

Z-Isomer

Tataramide B. *(Z)-Grossamide*

[187655-56-7]

[870472-09-6] (Z-Grossamide)

$C_{36}H_{36}N_2O_8$ 624.69

Isol. from *Acorus tatarinowii*, *Solanum tuberosum* and *Withania somnifera*. Germination inhibitor.

Yoshihara, T. *et al.*, *Agric. Biol. Chem.*, 1981, **45**, 2593-2598 (*Grossamide, Dihydrogrossamide, synth*).
Yoshihara, T. *et al.*, *Agric. Biol. Chem.*, 1983, **47**, 217-220 (*Grossamide, config*).
Sakakibara, I. *et al.*, *Phytochemistry*, 1991, **30**, 3013-3016 (*Cannabis sativa constit*).
Lajide, L. *et al.*, *Phytochemistry*, 1995, **40**, 1105-1112 (*Demethylgrossamide*).
Wang, M.F. *et al.*, *Chin. Chem. Lett.*, 1997, **8**, 35-36 (*Tataramide B*).
Li, J.-X. *et al.*, *Planta Med.*, 1998, **64**, 628-631 (*Tribulusamides A,B, activity*).
Seca, A.M.L. *et al.*, *Phytochemistry*, 2001, **58**, 1219-1223 (*Hibiscus cannabinus constit*).
King, R.R. *et al.*, *Phytochemistry*, 2005, **66**, 2468-2473 (*E,Z-Grossamides, potato scab constit*).
Baxendale, I.R. *et al.*, *Synlett*, 2006, 427-430 (*synth*).
Ge, F. *et al.*, *Helv. Chim. Acta*, 2008, **91**, 1023-1030 (*Thoreliamide A*).
King, R.R. *et al.*, *Phytochemistry*, 2010, **71**, 2187-2189 (*Grossamide, potato constit*).
Karim, A. *et al.*, *Helv. Chim. Acta*, 2011, **94**, 528-533 (*Dimorphamide A*).
Zhang, W.-N. *et al.*, *J. Agric. Food Chem.*, 2012, **60**, 1682-1687 (*E,Z-Grossamides, activity*).
Bolleddula, J. *et al.*, *Rapid Commun. Mass Spectrom.*, 2012, **26**, 1277-1290 (*Withania somnifera constit*).
Sun, J. *et al.*, *Fitoterapia*, 2014, **98**, 110-116 (*Melongenamide B, activity*).
Feng, J. *et al.*, *Asian J. Chem.*, 2016, **28**, 1820-1822 (*Tribulusamide A*).
Zhu, P.-F. *et al.*, *Nat. Prod. Bioprospect.*, 2017, **7**, 421-431 (*4-O-Methylgrossamide*)

4″-*O*-(8-Guaiacylglycerol)buddlenol A

G30

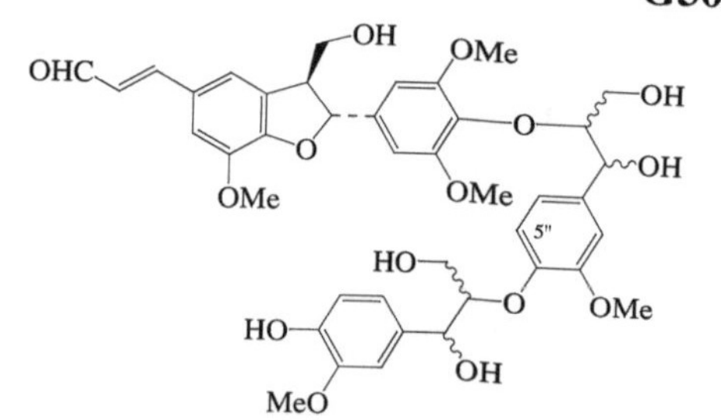

[929688-52-8]

$C_{41}H_{46}O_{15}$ 778.804

Constit. of *Tarenna attenuata*. Amorph. powder. $[\alpha]_D^{27}$ +5.4 (c, 0.34 in MeOH). λ_{max} 204 (log ε 4.84); 281 (log ε 3.88); 335 (log ε 3.9) (MeOH)

5″-Methoxy

4″-O-(8-Guaiacylglycerol)-5″-methoxybuddlenol A

[929688-53-9]

$C_{42}H_{48}O_{16}$ 808.83

Constit. of *Tarenna attenuata*. Amorph. powder. $[\alpha]_D^{27}$ +5.9 (c, 0.38 in MeOH). λ_{max} 205 (log ε 4.78); 279 (log ε 3.83); 334 (log ε 3.72) (MeOH).

Yang, X.-W. *et al.*, *J. Nat. Prod.*, 2007, **70**, 521-525 (*Tarenna attenuata constits*)

7-*O*-(8-Guaiacylglycerol)diphyllin

G31

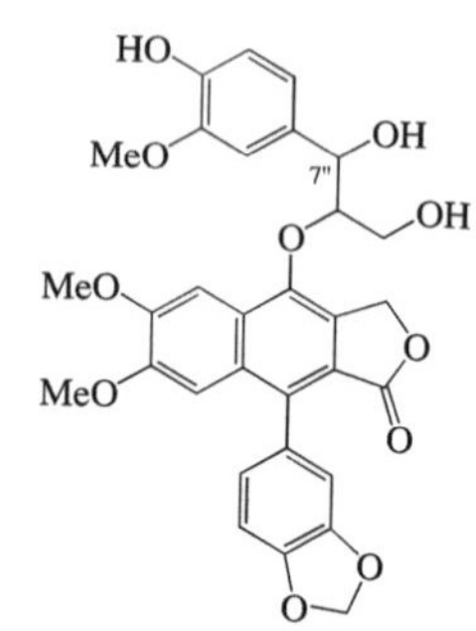

$C_{31}H_{28}O_{11}$ 576.554

(7″ξ,8″ξ)-form

7″-O-β-D-Glucopyranoside

Procumbenoside M

[2243471-13-6]

$C_{37}H_{38}O_{16}$ 738.695

Constit. of *Justicia procumbens*. Amorph. powder (MeOH).

Jin, H. *et al.*, *J. Asian Nat. Prod. Res.*, 2017, **19**, 1-8 (*Procumbenoside M*)

4-*O*-(8-Guaiacylglycerol)evofolin B G32

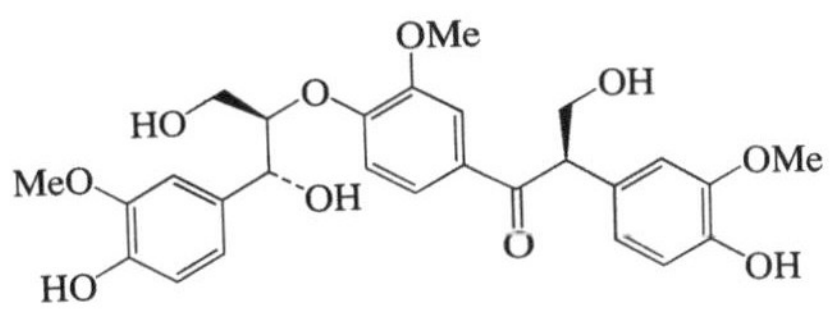

[1621072-58-9]

$C_{27}H_{30}O_{10}$ 514.527

Constit. of *Sambucus williamsii*. Amorph. pale yellow solid. $[\alpha]_D^{26}$ -6.7 (c, 0.60 in MeOH). λ_{max} 206 (log ε 4.95); 229 (log ε 4.57); 280 (4.32); 314 (sh) (MeOH).

Xiao, H.-H. *et al.*, *Fitoterapia*, 2014, **94**, 29-35 (*4-8-Guaiacylglycerolevofolin B*)

Gymnothebutyllignan A G33

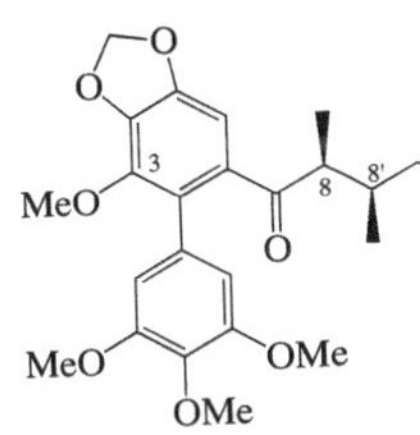

Absolute Configuration

[2095349-94-1]

$C_{23}H_{28}O_8$ 432.469

Constit. of *Gymnotheca involucrata*. Cryst. (MeOH). $[\alpha]_D^{20}$ +37 (c, 0.11 in MeOH). λ_{max} 297 (log ε 3.77) (MeOH)

8′-Epimer

Gymnothebutyllignan B

[2095349-96-3]

$C_{23}H_{28}O_8$ 432.469

Constit. of *Gymnotheca involucrata*. Powder (MeOH). $[\alpha]_D^{20}$ -44 (c, 0.32 in MeOH). λ_{max} 297 (log ε 3.77) (MeOH).

Xiao, S.-J. *et al.*, *Chin. Chem. Lett.*, 2017, **28**, 1049-1051 (*Gymnothebutyllignans A,B, cryst struct*)

Gymnothedelignan B G34

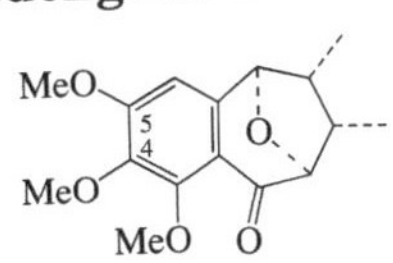

Absolute Configuration

[1607438-65-2]

$C_{16}H_{20}O_5$ 292.331

Degraded lignan. Constit. of *Gymnotheca chinensis*. Cryst.. $[\alpha]_D^{20}$ -180 (c, 0.01 in MeOH). λ_{max} 226 (log ε 3.88) (MeOH)

4,5-Di-O-de-Me, 4,5-methylene ether

Gymnothedelignan A

[1607438-64-1]

$C_{15}H_{16}O_5$ 276.288

Constit. of *Gymnotheca chinensis*. Cryst.. $[\alpha]_D^{20}$ -214 (c, 0.01 in MeOH). λ_{max} 227 (log ε 4.06) (MeOH).

Xiao, S.-J. *et al.*, *Tetrahedron Lett.*, 2014, **55**, 2869-2871 (*Gymnothedelignans A,B, cryst struct*)

Gymnothelignan A G35

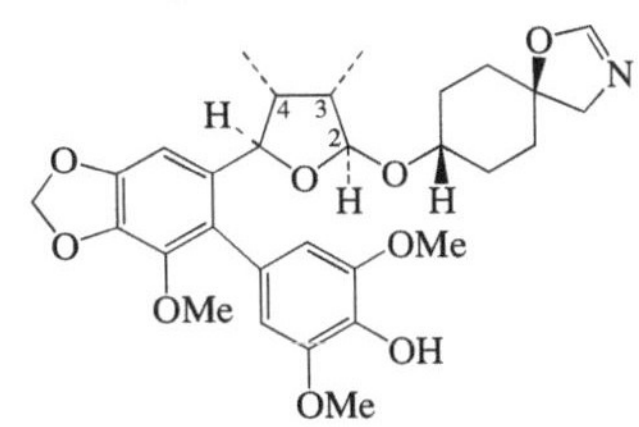

Absolute Configuration

[1401068-78-7]

$C_{30}H_{37}NO_9$ 555.624

Constit. of *Gymnotheca chinensis*. Cryst.. Mp 171 - 172°. $[\alpha]_D^{20}$ -9 (c, 0.09 in MeOH). λ_{max} 284 (MeOH)

2,3-Diepimer

Gymnothelignan B

[1401068-79-8]

$C_{30}H_{37}NO_9$ 555.624

Constit. of *Gymnotheca chinensis*. Powder. $[\alpha]_D^{20}$ -65 (c, 0.04 in MeOH). λ_{max} 281 (MeOH)

2,4-Diepimer

Gymnothelignan C

[1401068-80-1]

$C_{30}H_{37}NO_9$ 555.624

Constit. of *Gymnotheca chinensis*. Powder. $[\alpha]_D^{20}$ -30 (c, 0.03 in MeOH). λ_{max} 280 (MeOH).

He, D. *et al.*, *J. Org. Chem.*, 2012, **77**, 8435-8443 (*Gymnothelignans A,B,C*)

Gymnothelignan K G36

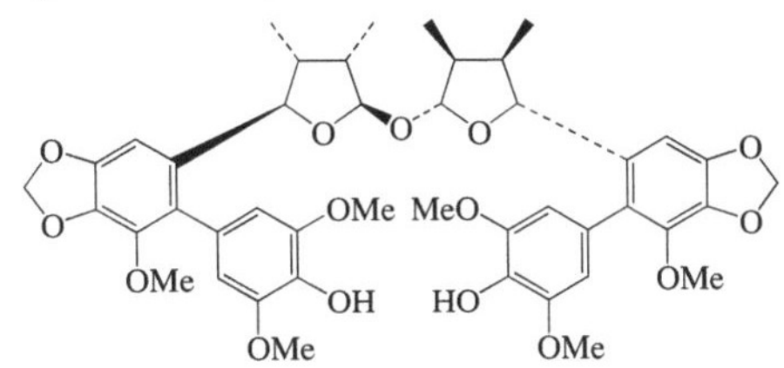

Absolute Configuration

[1401068-87-8]

$C_{44}H_{50}O_{15}$ 818.869

Constit. of *Gymnotheca chinensis*. Powder. $[\alpha]_D^{20}$ -10 (c, 0.09 in MeOH). Struct. illus. in ref. is incorrect. λ_{max} 280 (MeOH).

He, D. *et al.*, *J. Org. Chem.*, 2012, **77**, 8435-8443 (*Gymnothelignan K*)

Gymnothelignan N G37

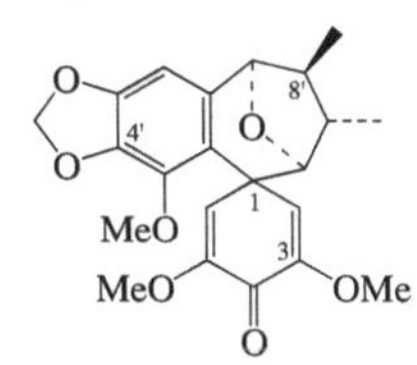

Absolute Configuration

1,2′-Cyclo-7,7′-epoxy-3,3′,5-trimethoxy-4′,5′-methylenedioxy-8,8′-lignan-4(1H)-one

[1401068-90-3]

$C_{22}H_{24}O_7$ 400.427

Constit. of *Gymnotheca chinensis*. Cryst.. Mp 201 - 202°. $[\alpha]_D^{20}$ +13 (c, 0.04 in MeOH). λ_{max} 290 (MeOH)

3-Demethoxy

1,2′-Cyclo-7,7′-epoxy-3′,5-dimethoxy-4′,5′-methylenedioxy-8,8′-lignan-4(1H)-one. ***Gymnothelignan U***

[1562241-81-9]

$C_{21}H_{22}O_6$ 370.401

Constit. of *Gymnotheca chinensis*. Powder. $[\alpha]_D^{20}$ +17 (c, 0.30 in MeOH). Has 1*R*-config.

8′-Epimer

Gymnothelignan O

[1401068-91-4]

$C_{22}H_{24}O_7$ 400.427

Constit. of *Gymnotheca chinensis*. Powder. $[\alpha]_D^{20}$ +15 (c, 0.04 in MeOH). λ_{max} 283 (MeOH)

8′-Epimer, 4′,5′-demethylene, 4′,5′-di-Me ether

1,2′-Cyclo-7,7′-epoxy-3,3′,4′,5,5′-pentamethoxy-8,8′-lignan-4(1H)-one. ***Gymnothelignan T***

[1562241-78-4]

$C_{23}H_{28}O_7$ 416.47

Constit. of *Gymnotheca chinensis*. Yellow gum. $[\alpha]_D^{20}$ +20 (c, 0.35 in MeOH).

He, D. *et al.*, *J. Org. Chem.*, 2012, **77**, 8435-8443 (*Gymnothelignans N,O*).

Xiao, S.J. *et al.*, *Chin. Chem. Lett.*, 2014, **25**, 463-464 (*Gymnothelignans T,U*).

Li, H. *et al.*, *Org. Lett.*, 2014, **16**, 4440-4443 (*synth*)

Gymnothespirolignan A G38

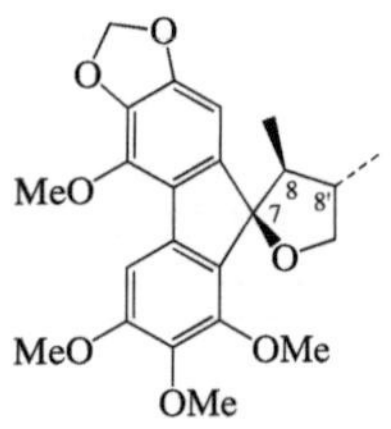

[1631711-16-4]
$C_{23}H_{26}O_7$ 414.454
Lignan numbering shown. Constit. of *Gymnotheca involucrata*. Exhibits antiviral activity against RSV. Cryst. (MeOH). $[\alpha]_D^{20}$ +5.4 (c, 0.54 in MeOH). λ_{max} 223 (log ε 4.41); 245 (log ε 4.42); 307 (log ε 3.81) (MeOH)

7,8-Diepimer

Gymnothespirolignan B
[1631711-17-5]
$C_{23}H_{26}O_7$ 414.454
Constit. of *Gymnotheca involucrata*. Exhibits antiviral activity against RSV. Cryst. (MeOH). $[\alpha]_D^{20}$ -69 (c, 0.16 in MeOH). λ_{max} 242 (log ε 4.41); 310 (log ε 3.82) (MeOH).

Xiao, S.-J. *et al., Tetrahedron Lett.*, 2014, **55**, 5949-5951 (*Gymnothespirolignans A,B, cryst struct, activity*).
Ali, G. *et al., J. Org. Chem.*, 2021, **86**, 10517-10525 (*Gymnothespirolignan B, synth*)

Gymnothespirolignan E G39

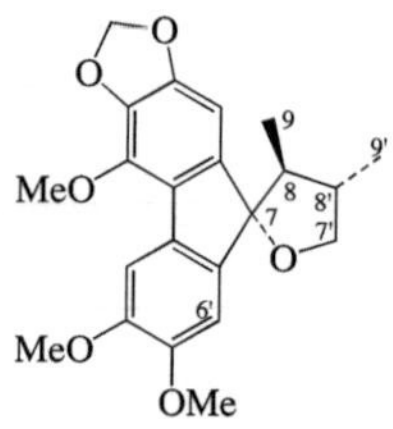

[2172644-44-7]
$C_{22}H_{24}O_6$ 384.428
Constit. of *Gymnotheca involucrata*. Cryst.. $[\alpha]_D^{20}$ +16 (c, 0.08 in MeOH). λ_{max} 243 (log ε 4.35); 309 (log ε 3.81) (MeOH)

6′-Methoxy

Gymnothespirolignan C
[2172644-46-9]
$C_{23}H_{26}O_7$ 414.454
Constit. of *Gymnotheca involucrata*. Exhibits moderate insecticidal activity against *Diabrotica balteata*. Powder. $[\alpha]_D^{20}$ -3 (c, 0.1 in MeOH). λ_{max} 242 (log ε 4.41); 309 (log ε 3.82) (MeOH)

8-Epimer, 6′-methoxy, 8α-hydroxy

Gymnothespirolignan F
[2172644-47-0]
$C_{23}H_{26}O_8$ 430.453
Constit. of *Gymnotheca involucrata*. Powder. $[\alpha]_D^{20}$ -3 (c, 0.1 in MeOH). λ_{max} 242 (log ε 4.36); 311 (log ε 3.83) (MeOH)

8,8′-Diepimer

Gymnothespirolignan D
[2172644-45-8]
$C_{22}H_{24}O_6$ 384.428
Constit. of *Gymnotheca involucrata*. Exhibits weak antifungal activity against *Uromyces viciae-fabae*. Powder. $[\alpha]_D^{20}$ -44 (c, 0.22 in MeOH). Relative config.. λ_{max} 243 (log ε 4.36); 308 (log ε 3.81) (MeOH).

Xiao, S.-J. *et al., Planta Med.*, 2016, **82**, 723-728 (*Gymnothespirolignans C-F, cryst struct, activity*).
Ali, G. *et al., J. Org. Chem.*, 2021, **86**, 10517-10525 (*Gymnothespirolignan C, synth*)

Haedoxan I H1

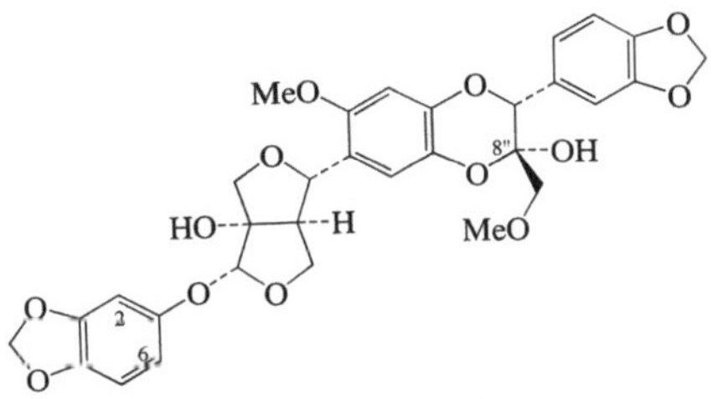

Absolute Configuration

[123522-86-1]
$C_{31}H_{30}O_{13}$ 610.568
Constit. of *Phryma leptostachya*. Insecticide

8''-Deoxy

Haedoxan E
[124019-43-8]
$C_{31}H_{30}O_{12}$ 594.569
Constit. of *Phryma leptostachya*. Insecticide

6-Methoxy

Haedoxan H
[123522-87-2]
$C_{32}H_{32}O_{14}$ 640.594
Constit. of *Phryma leptostachya*. Insecticide

6-Methoxy, 8''-deoxy

Haedoxan D
[123619-27-2]
$C_{32}H_{32}O_{13}$ 624.595
Constit. of *Phryma leptostachya*. Insecticide

2,6-Dimethoxy

Haedoxan C
[123522-88-3]
$C_{33}H_{34}O_{15}$ 670.62
Constit. of *Phryma leptostachya*. Insecticide

2,6-Dimethoxy, 8''-deoxy

Haedoxan A
[123619-26-1]
$C_{33}H_{34}O_{14}$ 654.621
Constit. of *Phryma leptostachya*. Insecticide. Rods (Et_2O). Mp 158 - 159°. $[\alpha]_D^{18}$ +125 (c, 0.32 in $EtOH/CH_2Cl_2$). λ_{max} 234 (ε 11750); 292 (ε 11480) (EtOH).

Taniguchi, E. *et al., Tennen Yuki Kagobutsu Toronkai Koen Yoshishu*, 1988, **30**, 1-8 (*Haedoxans C,H,I*).
Taniguchi, E. *et al., Agric. Biol. Chem.*, 1989, **53**, 631-643 (*Haedoxan A*).
Ishibashi, F. *et al., Agric. Biol. Chem.*, 1989, **53**, 1565-1573 (*Haedoxans A,D,E, synth*).
Ishibashi, F. *et al., Phytochemistry*, 1998, **49**, 613-622 (*Haedoxan A, synth, abs config*).
Chen, Y. *et al., Front. Chem.*, 2020, **8**, 460:1-6 (*rev, synth, activity*)

Haedoxan B H2

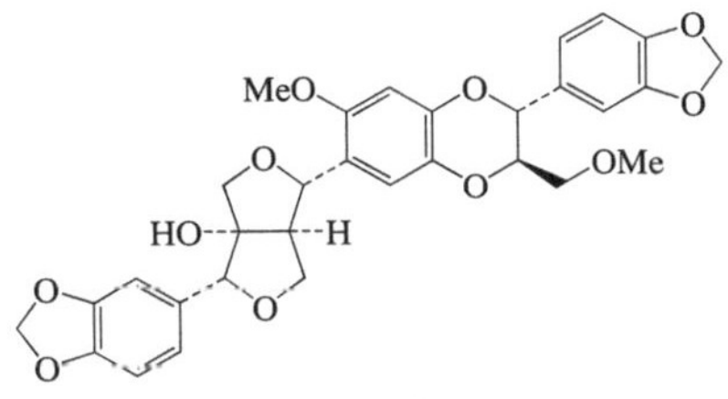

Absolute Configuration

[124019-44-9]
$C_{31}H_{30}O_{11}$ 578.57
Sesquilignan. Related to Princepin, P71. Constit. of *Phryma leptostachya*. Insecticide.

Taniguchi, E. *et al., Tennen Yuki Kagobutsu Toronkai Koen Yoshishu*, 1988, **30**, 1-8 (*Haedoxan B*)

Haedoxan J H3

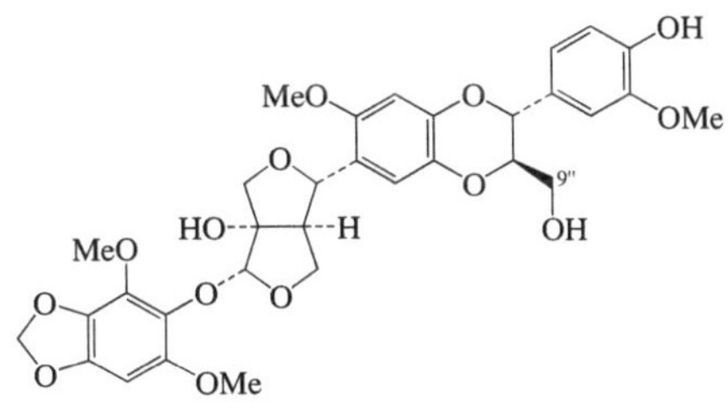

Relative Configuration

[1363160-26-2]
$C_{32}H_{34}O_{14}$ 642.61
Constit. of *Phryma leptostachya*. Prisms (MeOH). Mp 158 - 159°. $[\alpha]_D^{25}$ +125 (c, 0.32 in $EtOH/CH_2Cl_2$). λ_{max} 234 (log ε 4.07); 292 (log ε 4.06) (MeOH)

9''-O-β-D-Glucopyranoside

Haedoxancoside A
[1449376-83-3]
$C_{38}H_{44}O_{19}$ 804.751
Constit. of *Phryma leptostachya*. Powder. Mp 181 - 183°. $[\alpha]_D^{20}$ +1.4 (c, 1.6 in MeOH). λ_{max} 234 (log ε 4.07); 292 (log ε 4.06) (MeOH).

Chen, C. *et al., Helv. Chim. Acta*, 2012, **95**, 333-338 (*Haedoxan J*).
Chen, C. *et al., Helv. Chim. Acta*, 2013, **96**, 1392-1396 (*Haedoxancoside A*)

Hedycaffeoylglucoside A H4

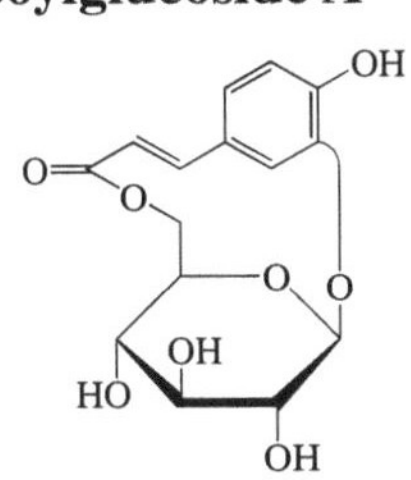

$C_{15}H_{16}O_8$ 324.285

Constit. of *Hedyotis caudatifolia*. Exhibits cytotoxic activity. Amorph. powder.

Chen, M.-L. *et al., Zhongcaoyao*, 2019, **50**, 25-29 (*Hedycaffeoylglucoside A, activity*)

Hedyotisol A H5

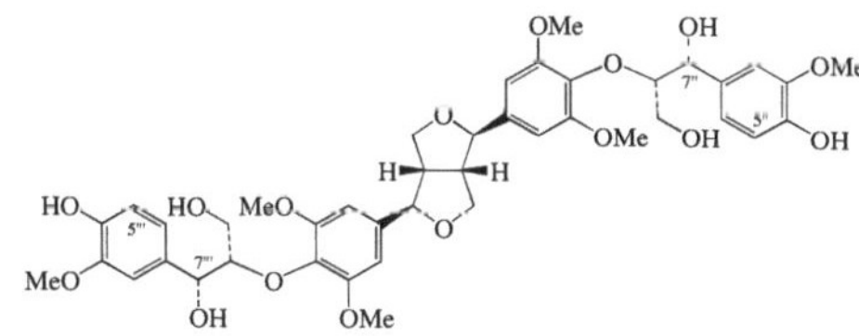

Absolute Configuration

[95732-59-5]
$C_{42}H_{50}O_{16}$ 810.846
Constit. of *Brassica fruticulosa*, *Campylotropis hirtella*, *Cassia occidentalis*, *Hedyotis lawsoniae*, *Secale cereale* and *Sinocalamus affinis*. Amorph. powder. $[\alpha]_D^{20}$ +1.2 (c, 0.25 in MeCN). λ_{max} 273 (log ε 3.92); 279 (log ε 3.9) (no solvent reported) (hexa-Ac). [neutral] λ_{max} 204 (log ε 4.44); 232 (log ε 3.81); 278 (log ε 3.20) (MeOH)

5''-Methoxy

5''-Methoxyhedyotisol A
[929688-51-7]
$C_{43}H_{52}O_{17}$ 840.872
Constit. of *Secale cereale* and *Tarenna attenuata*. Amorph. powder. $[\alpha]_D^{27}$ +6.5 (c, 0.21 in MeOH). λ_{max} 205 (log ε 4.92); 280 (log ε 3.69); 357 (log ε 3.21) (MeOH)

7''-Epimer

Hedyotisol B
[95839-45-5]
[1292294-30-4]
$C_{42}H_{50}O_{16}$ 810.846
Constit. of *Campylotropis hirtella*, *Euonymus alatus*, *Hedyotis lawsoniae* and *Sinocalamus affinis*. Amorph. powder. $[\alpha]_D^{20}$ +2.1 (c, 0.1 in MeCN). λ_{max} 233 (log ε 3.79); 278 (log ε 3.18) (MeOH)

7'',7'''-Diepimer

Hedyotisol C
[95839-46-6]
[1292294-28-0]
$C_{42}H_{50}O_{16}$ 810.846
Constit. of *Campylotropis hirtella*, *Euonymus alatus*, *Hedyotis lawsoniae* and *Sinocalamus affinis*. Amorph. powder. $[\alpha]_D^{20}$ +1.2 (c, 0.1 in MeCN). λ_{max} 205 (log ε 4.39); 234 (log ε 3.77); 278 (log ε 3.20) (MeOH)

7″,7‴-Diepimer, 5″,5‴-dimethoxy

5″,5‴-Dimethoxyhedyotisol C

[942047-39-4]

$C_{44}H_{54}O_{18}$ 870.898

Constit. of the leaves of *Aptenia cordifolia*. $[\alpha]_D^{25}$ +3.1 (c, 0.05 in CH_2Cl_2). Rel. stereochem..

Matsuda, S. *et al., Chem. Pharm. Bull.*, 1984, **32**, 5066-5069 (*Hedyotisols A,B,C*).
Cutillo, F. *et al., J. Agric. Food Chem.*, 2003, **51**, 6165-6172 (*Hedyotisol A, Brassica constit*).
Greca, M.D. *et al., Chem. Biodiversity*, 2007, **4**, 118-128 (*Aptenia cordifolia constit*).
Yang, X.-W. *et al., J. Nat. Prod.*, 2007, **70**, 521-525 (*5″-Methoxyhedyotisol A*).
Han, H.-Y. *et al., Nat. Prod. Res.*, 2008, **22**, 984-989 (*Campylotropis hirtella constits*).
Jeong, E.J. *et al., Bioorg. Med. Chem. Lett.*, 2011, **21**, 2283-2286 (*Euonymus alatus constits, activity*).
Xiong, L. *et al., J. Nat. Prod.*, 2011, **74**, 1188-1200 (*Hedyotisols A,B,C, abs config*).
Hanhineva, K. *et al., Metabolomics*, 2012, **8**, 399-409 (*rye bran constits*)

Hedyotol A H6

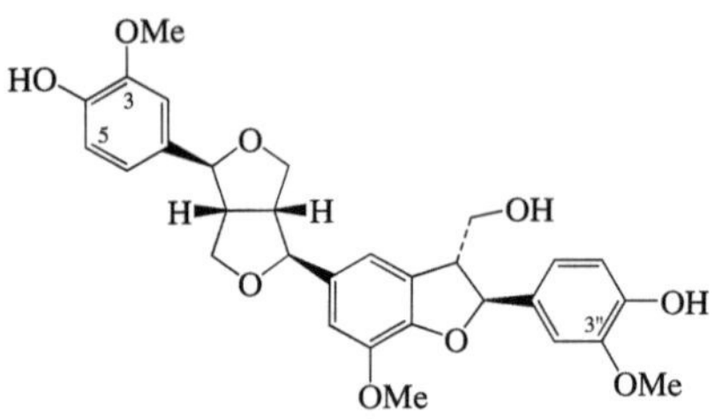

Absolute Configuration

[97400-01-6]

$C_{30}H_{32}O_9$ 536.577

Constit. of leaves of *Hedyotis lawsoniae*. $[\alpha]_D^{20}$ +51.6 (c. 0.72 in $CHCl_3$) (synthetic)

Tri-Ac

[97400-02-7]

Amorph. powder. $[\alpha]_D^{21}$ +29 (c, 1.0 in $CHCl_3$)

5-Methoxy

Hedyotol B

[97400-03-8]

$C_{31}H_{34}O_{10}$ 566.603

Constit. of *Hedyotis lawsoniae*

5-Methoxy, tri-Ac

[97400-04-9]

Amorph. powder

3,3″-Didemethoxy

Chushizisin I

[1142925-39-0]

$C_{28}H_{28}O_7$ 476.525

Constit. of the fruit of *Broussonetia papyrifera*. Antioxidant. Oil. $[\alpha]_D^{27}$ +20.7 (c, 0.1 in MeOH). λ_{max} 208 (log ε 4.7); 225 (log ε 4.45); 278 (log ε 3.8) (MeOH).

Kikuchi, T. *et al., Chem. Pharm. Bull.*, 1985, **33**, 1444-1451 (*Hedyotols A,B*).
Mei, R.-Q. *et al., J. Nat. Prod.*, 2009, **72**, 621-625 (*Chushizisin I, activity*).
Kawabe, Y. *et al., Org. Lett.*, 2014, **16**, 1976-1979 (*synth*)

Helicterin parent acid H7

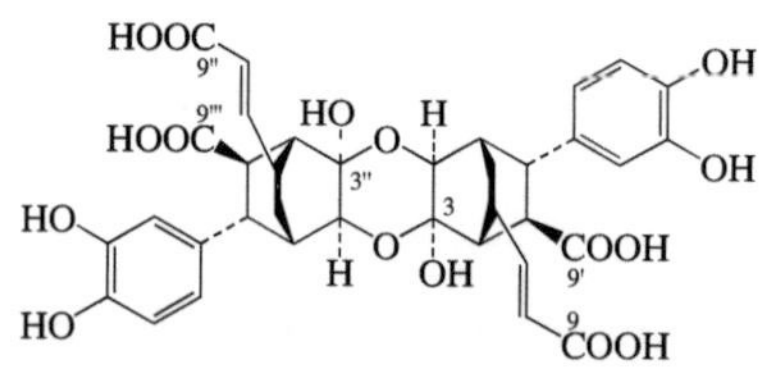

Relative Configuration

$C_{36}H_{32}O_{16}$ 720.636

Lignan numbering shown

3-Me ether, 9,9″,9‴-tris[2-(3,4-dihydroxyphenyl)-1R-(methoxycarbonyl)ethyl], 9′-Me ester

Helicterin E

[317821-21-9]

$C_{68}H_{66}O_{28}$ 1331.248

Constit. of the fruit of *Helicteres isora*. Amorph. solid. $[\alpha]_D^{26}$ +17 (c, 0.74 in MeOH). λ_{max} 278 (log ε 4.71) (MeOH)

3-Me ether, 9,9′,9″,9‴-tetrakis[2-(3,4-dihydroxyphenyl)-1R-(methoxycarbonyl)ethyl] ester

Helicterin B

[317821-18-4]

$C_{77}H_{74}O_{32}$ 1511.407

Constit. of the fruit of *Helicteres isora*. Amorph. solid. $[\alpha]_D^{26}$ +4.2 (c, 0.4 in MeOH). λ_{max} 280 (log ε 4.59) (MeOH)

3″-Me ether, 9′,9″,9‴-tris[2-(3,4-dihydroxyphenyl)-1R-(methoxycarbonyl)ethyl], 9-Me ester

Helicterin F

[317821-22-0]

$C_{68}H_{66}O_{28}$ 1331.248

Constit. of the fruit of *Helicteres isora*. Amorph. solid. $[\alpha]_D^{26}$ +7.4 (c, 0.86 in MeOH). λ_{max} 280 (log ε 4.5) (MeOH)

3,3″-Di-Me ether, 9,9″,9‴-tris[2-(3,4-dihydroxyphenyl)-1R-(methoxycarbonyl)ethyl], 9′-Me ester

Helicterin C

[317821-19-5]

$C_{69}H_{68}O_{28}$ 1345.275

Constit. of the fruit of *Helicteres isora*. Amorph. solid. $[\alpha]_D^{26}$ -10 (c, 0.32 in MeOH). λ_{max} 280 (log ε 4.65) (MeOH)

3,3″-Di-Me ether, 9′,9″,9‴-tris[2-(3,4-dihydroxyphenyl)-1R-(methoxycarbonyl)ethyl], 9-Me ester

Helicterin D

[317821-20-8]

$C_{69}H_{68}O_{28}$ 1345.275

Constit. of the fruit of *Helicteres isora*. Amorph. solid. $[\alpha]_D^{26}$ +0.3 (c, 0.87 in MeOH). λ_{max} 279 (log ε 4.66) (MeOH)

3,3″-Di-Me ether, 9,9′,9″,9‴-tetrakis[2-(3,4-dihydroxyphenyl)-1R-(methoxycarbonyl)ethyl] ester

Helicterin A

[317821-17-3]

$C_{78}H_{76}O_{32}$ 1525.434

Constit. of the fruit of *Helicteres isora*. Amorph. solid. $[\alpha]_D^{26}$ +2.1 (c, 0.4 in MeOH). λ_{max} 281 (log ε 4.29) (MeOH).

Tezuka, Y. *et al., Helv. Chim. Acta*, 2000, **83**, 2908-2919 (*Helicterins A-F*).
Snyder, S.A. *et al., J. Am. Chem. Soc.*, 2009, **131**, 1745-1752 (*Helicterin B, synth*)

Heliotropamide H8

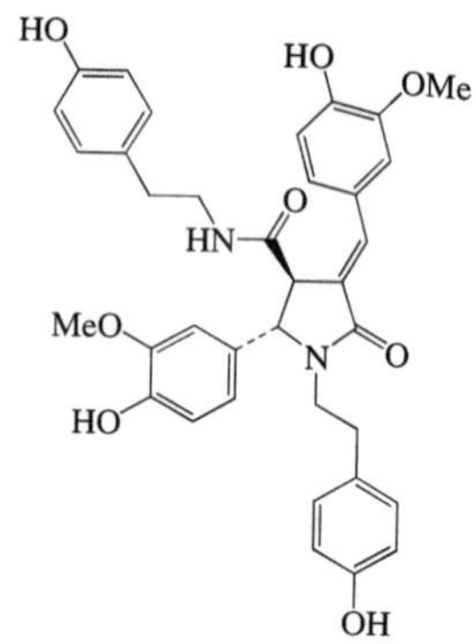

Relative Configuration

[646990-38-7]

$C_{36}H_{36}N_2O_8$ 624.69

Alkaloid from *Heliotropium ovalifolium* and *Hyoscyamus niger*. Amorph. powder. $[\alpha]_D^{21}$ +14 (c, 0.01 in MeOH). λ_{max} 223 (log ε 4.16); 287 (log ε 3.86); 327 (log ε 3.94) (MeOH)

Epimer, di-O-de-Me

Di-O-demethyl-cis-heliotropamide

[1829539-27-6]

$C_{34}H_{32}N_2O_8$ 596.636

Constit. of the seeds of *Cannabis sativa*. Antioxidant. Acetylcholinesterase inhibitor. Amorph. powder. $[\alpha]_D^{20}$ +2.5 (c, 0.1 in MeOH). λ_{max}

251 (log ε 3.73); 301 (log ε 3.69); 338 (log ε 4.17) (MeOH).

Guntern, A. *et al.*, *J. Nat. Prod.*, 2003, **66**, 1550-1553 (*Heliotropamide*).
Younai, A. *et al.*, *J. Org. Chem.*, 2010, **75**, 8333-8336 (*synth*).
Younai, A. *et al.*, *J. Org. Chem.*, 2011, **76**, 1184 (*synth, corrigendum*).
Yan, X. *et al.*, *J. Agric. Food Chem.*, 2015, **63**, 10611-10619 (*Di-O-demethyl-cis-heliotropamide*)

Heliotropiumide A H9

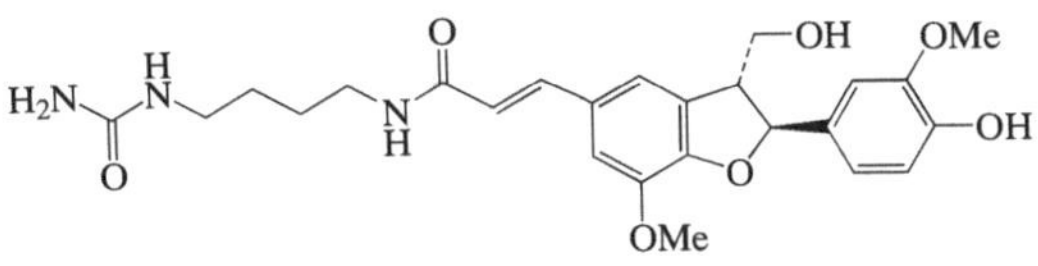

Absolute Configuration

[2143090-76-8]
$C_{25}H_{31}N_3O_7$ 485.537
Alkaloid from the seeds of *Heliotropium foertherianum*. Solid. [α] D +14 (c, 0.12 in MeOH). λ_{max} 288; 318 (MeOH).

Cai, Y.-S. *et al.*, *Bioorg. Med. Chem. Lett.*, 2017, **27**, 4630-4634 (*Heliotropiumide A*)

Helisorin H10

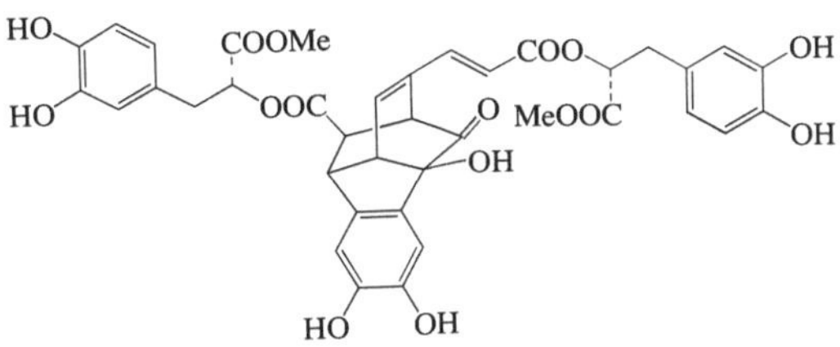

[1115210-85-9]
[223436-40-6] (deleted registry number)
$C_{38}H_{34}O_{16}$ 746.674
Constit. of the fruit of *Helicteres isora*. Amorph. yellow solid. $[\alpha]^{26}_{D}$ +28.4 (c, 0.94 in MeOH). λ_{max} 279 (log ε 4.03); 317 (sh) (MeOH).

Tezuka, Y. *et al.*, *Helv. Chim. Acta*, 1999, **82**, 408-417 (*Helisorin*).
Snyder, S.A. *et al.*, *J. Am. Chem. Soc.*, 2009, **131**, 1745-1752 (*synth*)

Hemidesminine H11

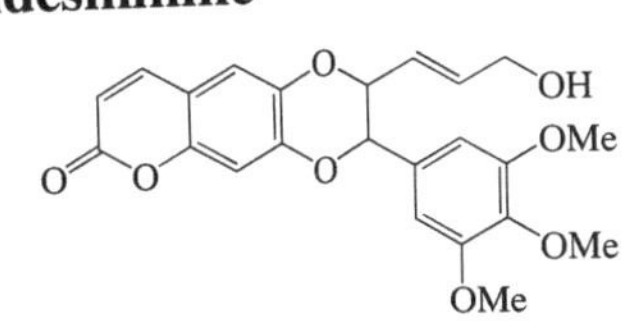

[136051-40-6]
$C_{23}H_{22}O_8$ 426.421
Constit. of *Hemidesmus indicus*. Amorph. pale yellow solid.

Mandal, S. *et al.*, *Indian J. Chem., Sect. B*, 1991, **30**, 712-713 (*Hemidesminine*)

3,3′,4,4′,5,5′,7-Heptahydroxy-8,8′-lignan-9,9′-olide H12

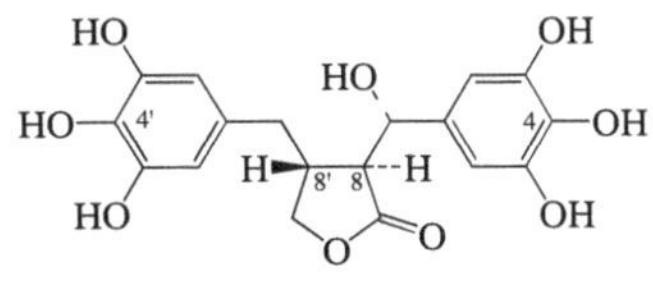

(7*S*,8*S*,8′*R*)-*form*

$C_{18}H_{18}O_9$ 378.333
(7S,8S,8′R)-form
3′,4′-Methylene, 3,5,5′-tri-Me ether
4,7-Dihydroxy-3,3′,5-trimethoxy-4′,5′-methylenedioxy-8,8′-lignan-9,9′-olide
[878142-52-0]
$C_{22}H_{24}O_9$ 432.425
Constit. of *Peperomia duclouxii*. Gum. $[\alpha]^{25}_{D}$ -38.9 (c, 0.38 in $CHCl_3$). λ_{max} 242; 275 ($CHCl_3$)
3′,4′-Methylene, 3,4,5,5′-tetra-Me ether
7-Hydroxy-3,3′,4,5-tetramethoxy-4′,5′-methylenedioxy-8,8′-lignan-9,9′-olide. ***5′-Methoxypodorhizol***
[94359-21-4]
$C_{23}H_{26}O_9$ 446.452
Constit. of *Hernandia cordigera*, *Hernandia ovigera* and *Hernandia nymphaeifolia*. Cryst.. Mp 112 - 113°. [α] D -20.5 (c, 1 in $CHCl_3$)
3′,4′-Methylene, 3,4,5,5′-tetra-Me ether, 7-Ac
5′-Methoxypodorhizol acetate
[225669-53-4]
$C_{25}H_{28}O_{10}$ 488.489
Constit. of the seeds of *Hernandia sonora*. Amorph. solid. [α] D -35.9 (c, 3.4 in $CHCl_3$)
(8R,8′S)-form
7-Ketone, 3′,4′-methylene, 3,5,5′-tri-Me ether
4-Hydroxy-3,3′,5-trimethoxy-4′,5′-methylenedioxy-7-oxo-8,8′-lignan-9,9′-olide
[878142-53-1]
$C_{22}H_{22}O_9$ 430.409
Constit. of *Peperomia duclouxii*. Gum. $[\alpha]^{25}_{D}$ -58.2 (c, 0.07 in $CHCl_3$).

Richomme, P. *et al.*, *Heterocycles*, 1985, **23**, 309-312 (*Hernandia cordigera constit*).
Tanoguchi, M. *et al.*, *Chem. Pharm. Bull.*, 1991, **39**, 1873-1876 (*Hernandia ovigera constit*).
Udino, L. *et al.*, *Planta Med.*, 1999, **65**, 279-281 (*5′-Methoxypodorhizol acetate*).
Chen, J.-J. *et al.*, *Planta Med.*, 2000, **66**, 251-256 (*Hernandia nymphaeifolia constit*).
Ito, C. *et al.*, *Planta Med.*, 2001, **67**, 166-168 (*5-Methoxypodorhizol, activity*).
Li, N. *et al.*, *J. Nat. Prod.*, 2006, **69**, 234-239 (*Peperomia duclouxii constits*)

3,3′,4,4′,5,8,8′-Heptahydroxy-8,8′-lignan-9,9′-olide H13

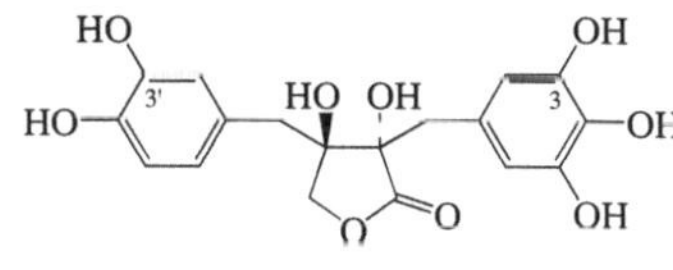

(8*S*,8′*R*)-*form*

$C_{18}H_{18}O_9$ 378.333
(8S,8′R)-form
3,3′,5-Tri-Me ether
4,4′,8,8′-Tetrahydroxy-3,3′,5-trimethoxy-8,8′-lignan-9,9′-olide. ***Dihydroxythujaplicatin methyl ether***
[13594-98-4]
$C_{21}H_{24}O_9$ 420.414
From *Thuja plicata*. Cryst. (MeOH). Mp 98 - 100°. $[\alpha]^{24}_{D}$ -97.2 (c, 0.68 in $CHCl_3$)
(8R*,8′S*)-form
3,3′-Di-Me ether
3,4,4′,8,8′-Pentahydroxy-3′,5-dimethoxy-8,8′-lignan-9,9′-olide.
Dihydroxythujaplicatin
[15797-37-2]
$C_{20}H_{22}O_9$ 406.387
Isol from *Thuja plicata*. Amorph.. Could not be crystallised.

Maclean, H. *et al.*, *Can. J. Chem.*, 1967, **45**, 305-309 (*Dihydroxythujaplicatin methyl ether*).
Maclean, W. *et al.*, *Can. J. Chem.*, 1967, **45**, 739-740 (*Dihydroxythujaplicatin*).
Nishibe, S. *et al.*, *Chem. Ind. (London)*, 1973, 328 (*Dihydroxythujaplicatin methyl ether, abs config*)

3,3′,4,4′,5,5′,9′-Heptahydroxy-8,8′-lign-8-ene-7,7′-dione H14

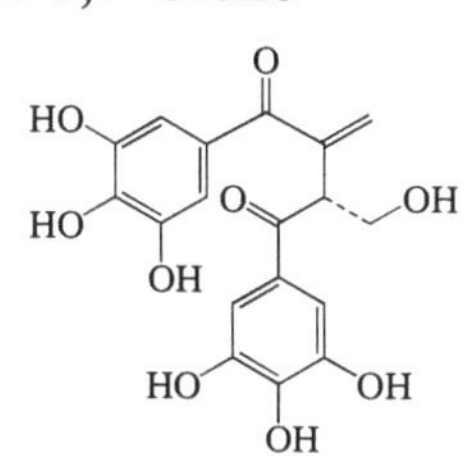

Absolute Configuration

$C_{18}H_{16}O_9$ 376.317
3,3′,4,4′,5,5′-Hexa-Me ether
9′-Hydroxy-3,3′,4,4′,5,5′-hexamethoxy-8,8′-lign-8-ene-7,7′-dione.
Rumphiin
[1803405-73-3]
$C_{24}H_{28}O_9$ 460.479

Constit. of the stems of *Polyalthia rumphii*. Exhibits moderate antitumour props.. Pale yellow solid. $[\alpha]_D^{21}$ +4.4 (c, 0.4 in MeOH).

Wang, T.-S. *et al.*, *Nat. Prod. Commun.*, 2013, **8**, 1427-1429 (*Rumphiin, abs config, activity*)

Herpepentol **H15**

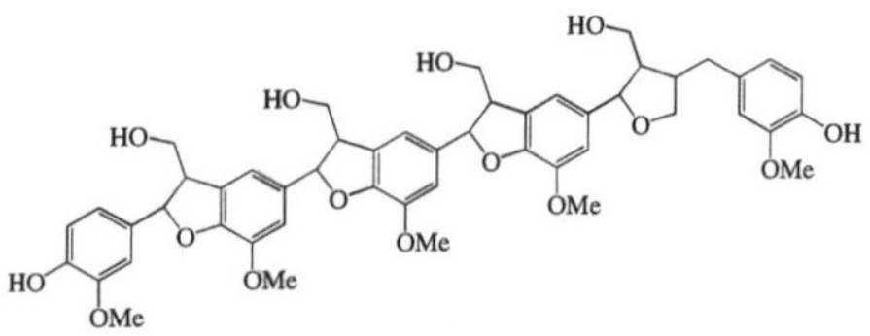

[94818-13-0]

$C_{50}H_{54}O_{15}$ 894.967

Constit. of the seeds of *Herpetospermum caudigerum*. Amorph..

Kaouadji, M. *et al.*, *Tetrahedron Lett.*, 1984, **25**, 5137-5138 (*Herpepentol*)

Herpepropenal **H16**

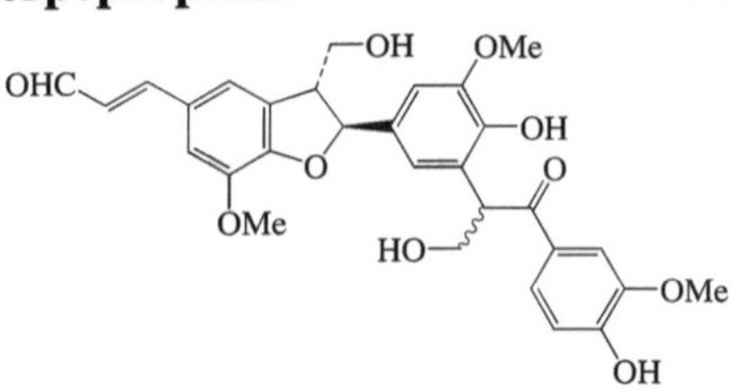

Relative Configuration

[1258011-61-8]

$C_{30}H_{30}O_{10}$ 550.56

Constit. of the seeds of *Herpetospermum caudigerum*. Inhibitor of hepatitis B virus. Amorph. yellow solid (MeOH). Mp 119 - 120°. $[\alpha]_D^{26}$ -30.3 (c, 1 in MeOH).

Yang, F. *et al.*, *Chem. Pharm. Bull.*, 2010, **58**, 402-404 (*Herpepropenal, activity*)

Herpetetradione **H17**

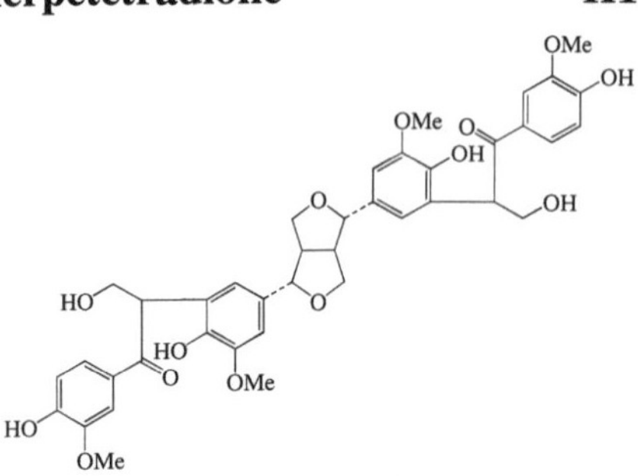

Relative Configuration

[94818-14-1]

$C_{40}H_{42}O_{14}$ 746.762

Constit. of seeds of *Herpetospermum caudigerum*. Amorph..

Kaouadji, M. *et al.*, *Tetrahedron Lett.*, 1984, **25**, 5135-5136 (*Herpetetradione*)

Herpetetrol **H18**

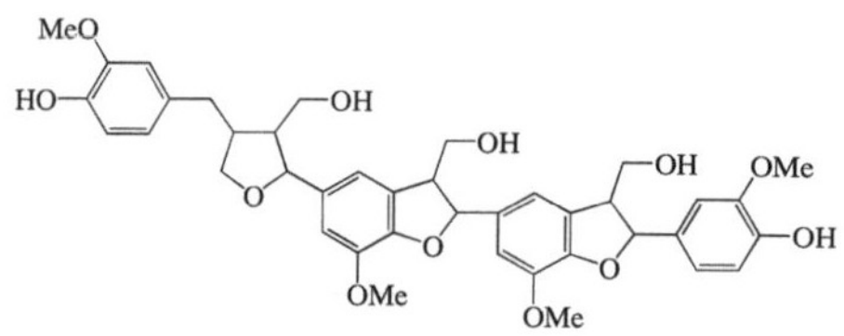

[73140-42-8]

$C_{40}H_{44}O_{12}$ 716.78

Constit. of the seeds of *Herpetospermum caudigerum*. $[\alpha]_D^{20}$ +56 (c, 0.6 in MeOH).

Kaouadji, M. *et al.*, *Z. Naturforsch., C*, 1979, **34**, 1129-1132 (*Herpetetrol*)

Herpetetrone **H19**

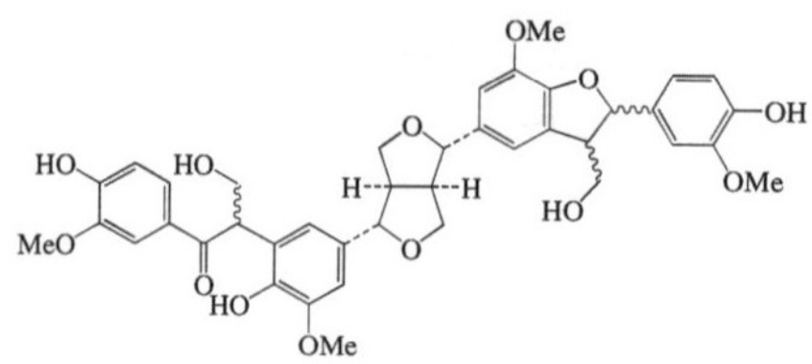

Relative Configuration

[112899-36-2]

$C_{40}H_{42}O_{13}$ 730.763

Constit. of the seeds of *Herpetospermum caudigerum*. Active against hepatitis B virus. Amorph. powder.

Kaouadji, M. *et al.*, *J. Nat. Prod.*, 1987, **50**, 1089-1094 (*Herpetetrone*).

Cong, L.-B. *et al.*, *Biomed. Chromatogr.*, 2008, **22**, 1084-1090 (*detn, hplc*).

Yu, J.-Q. *et al.*, *Phytochem. Lett.*, 2014, **10**, 230-234 (*Herpetospermum caudigerum constit, activity*)

Herpetin **H20**

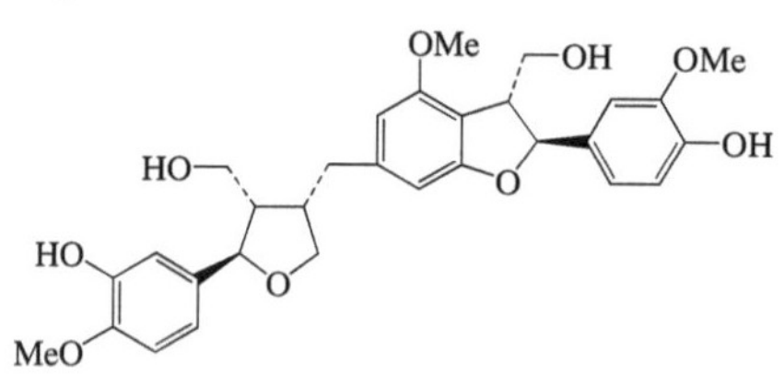

Relative Configuration

[911052-87-4]

$C_{30}H_{34}O_9$ 538.593

Constit. of *Aquilaria sinensis* and *Herpetospermum caudigerum*. Reduces replication and expression of HBsAg and HBeAg and significant inhibitor of HBV-DNA. Exhibits antiviral activity. Amorph. solid. Mp 163°. $[\alpha]_D^{20}$ +19.7 (c, 0.61 in MeOH). λ_{max} 230 (MeOH).

Yuan, H.-L. *et al.*, *Chem. Pharm. Bull.*, 2006, **54**, 1592-1594 (*Herpetin, activity*).

Gong, P.-Y. *et al.*, *Molecules*, 2017, **22**, 14:1-10 (*isol, activity*)

Herpetone **H21**

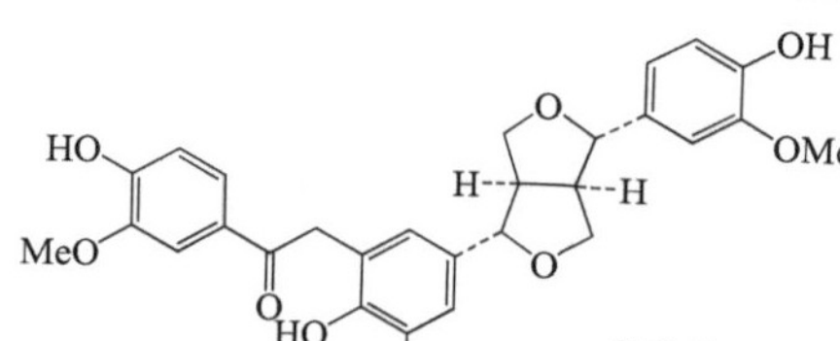

Relative Configuration

[951677-22-8]

$C_{29}H_{30}O_9$ 522.55

Constit. of *Herpetospermum caudigerum* and *Herpetospermum pedunculosum*. Exhibits protective effect on CCl_4 induced hepatocyte injury. Exhibits antiviral activity. Cryst. (EtOAc/petrol).

Zhang, M. *et al.*, *Yaoxue Xuebao*, 2006, **41**, 659-661 (*Herpetone*).

Zhang, M. *et al.*, *Chem. Pharm. Bull.*, 2008, **56**, 192-193 (*activity*).

Gong, P.-Y. *et al.*, *Molecules*, 2017, **22**, 14:1-10 (*isol, activity*)

Herpetosiol D **H22**

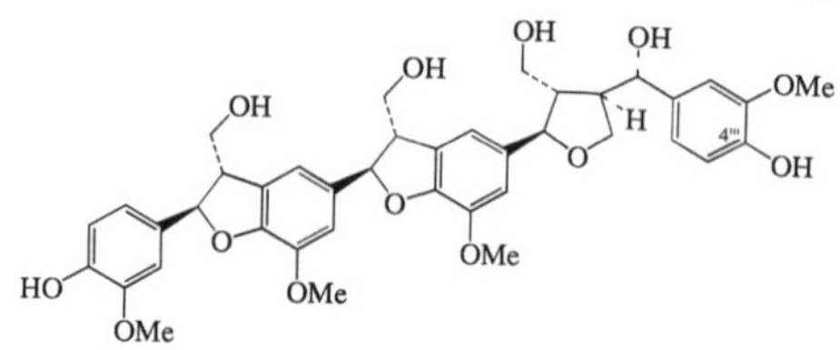

Absolute Configuration

[2374936-67-9]

$C_{40}H_{44}O_{13}$ 732.779

Related to Herpetetrol, H18. Constit. of *Herpetospermum pedunculosum*. Amorph. powder. $[\alpha]_D^{25}$ -57 (c, 0.6 in MeOH). λ_{max} 230 (log ε 3.1); 280 (log ε 2.7) (MeOH)

4'''-Deoxy, 5'''-hydroxy

Herpetosiol E

[2374936-68-0]

$C_{40}H_{44}O_{13}$ 732.779

Constit. of *Herpetospermum pedunculosum*. Amorph. powder. $[\alpha]_D^{25}$ -32.9 (c, 0.5 in MeOH). λ_{max} 230 (log ε 3.01); 280 (log ε 3.0) (MeOH).

Ma, Y. *et al.*, *Phytochemistry*, 2019, **164**, 102-110 (*Herpetosiols D,E*)

Herpetrione H23

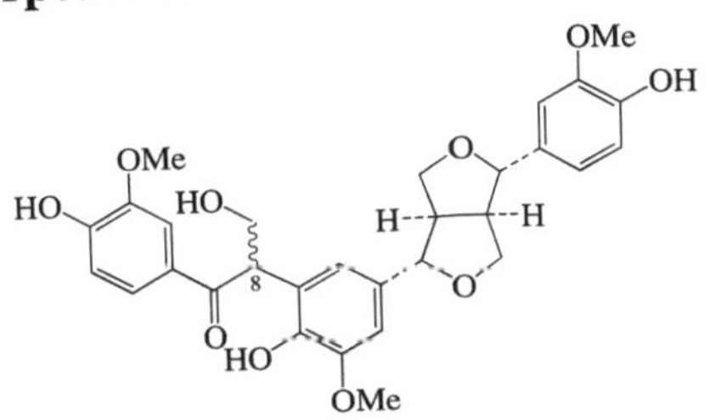

Relative Configuration

[89020-08-6]

$C_{30}H_{32}O_{10}$ 552.576

Constit. of seeds of *Herpetospermum caudigerum*. Potent antiviral agent. Shows hepatoprotectant activity. Amorph.. Mp 141 - 143°. $[\alpha]_D$ +140 (c, 0.2 in MeOH)

9-Deoxy, 8,9-didehydro

8,9-Anhydroherpetrione

[1903721-97-0]

$C_{30}H_{30}O_9$ 534.561

Constit. of the seeds of *Herpetospermum caudigerum*. Inhibits HBsAg and HBeAg secretion; cytotoxic against HepG2 cells. Amorph. powder. $[\alpha]_D^{20}$ +60.8 (c, 0.63 in MeOH).

Kaouadji, M. *et al.*, *Tetrahedron Lett.*, 1983, **24**, 5881-5884 (*Herpetrione*).
Guo, J.-J. *et al.*, *Int. J. Pharm. (Amsterdam, Neth.)*, 2013, **441**, 227-233 (*activity*).
Yu, J.-Q. *et al.*, *Phytochem. Lett.*, 2014, **10**, 230-234 (*8,9-Anhydroherpetrione*).
Shen, B. *et al.*, *J. Ethnopharmacol.*, 2015, **164**, 46-52 (*activity*)

Heteroclitin I‡ H24

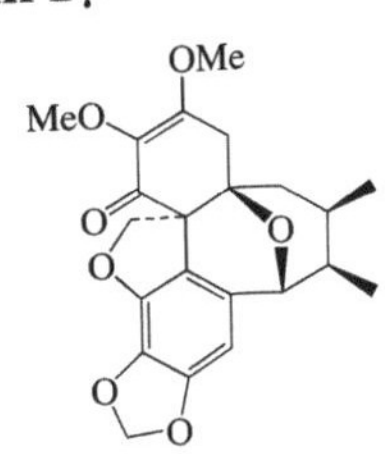

Absolute Configuration

[1011762-96-1]

$C_{22}H_{24}O_7$ 400.427

A 2′-homo-2,2′-cyclolignan. Constit. of the stems of *Kadsura heteroclita*. Powder. $[\alpha]_D^{15}$ -241.6 (c, 0.69 in Py). λ_{max} 218 (log ε 4.8); 271 (log ε 4.4) (MeOH).

Pu, J.-X. *et al.*, *Phytochemistry*, 2008, **69**, 1266-1272 (*Heteroclitin I*)

Heteroclitin J‡ H25

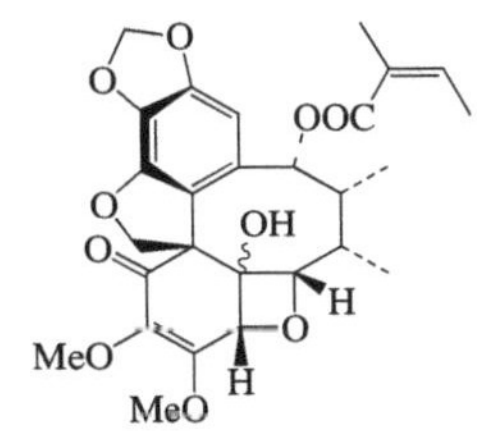

Absolute Configuration

[1011762-99-4]

$C_{27}H_{30}O_{10}$ 514.527

Unique diepoxy-2,2′-cyclolignan struct.. Constit. of the stems of *Kadsura heteroclita*. Powder. $[\alpha]_D^{17}$ +51.3 (c, 0.8 in Py). λ_{max} 217 (log ε 4.91); 255 (log ε 4.35); 268 (log ε 4.44) (MeOH)

Deangeloyl, 7-Ac

Longipedlignan P

[2408243-29-6]

$C_{24}H_{26}O_{10}$ 474.462

Constit. of roots of *Kadsura longipedunculata*. Amorph. yellow powder. $[\alpha]_D^{20}$ +35 (c, 0.1 in MeOH). λ_{max} 216 (log ε 1.74); 263 (log ε 1.36) (MeOH)

Deangeloyl, 7-benzoyl

Longipedlignan O

[2408242-35-1]

$C_{29}H_{28}O_{10}$ 536.533

Constit. of roots of *Kadsura longipedunculata*. Amorph. yellow powder. $[\alpha]_D^{20}$ +20 (c, 0.1 in MeOH). λ_{max} 216 (log ε 1.76); 268 (log ε 1.25) (MeOH).

Pu, J.-X. *et al.*, *Phytochemistry*, 2008, **69**, 1266-1272 (*Heteroclitin J*).
Lu, J. *et al.*, *J. Nat. Prod.*, 2019, **82**, 2842-2851 (*Longipedlignans O,P*)

Heterotropanone H26

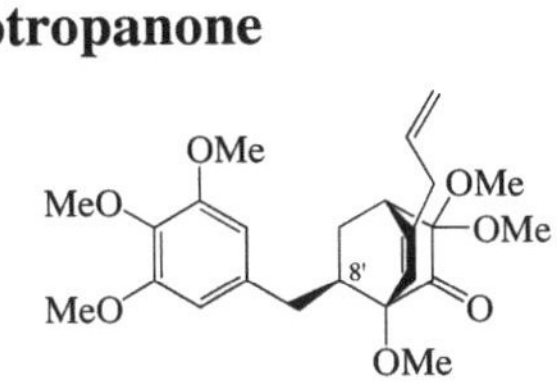

Relative Configuration

1,3,3-Trimethoxy-5-(2-propenyl)-7-[(3,4,5-trimethoxyphenyl)methyl]bicyclo[2.2.2]oct-5-en-2-one, CAS. *5-Allyl-1,3,3-trimethoxy-7-(3,4,5-trimethoxybenzyl)bicyclo[2.2.2]oct-5-en-2-one. 5,8′-Cyclo-2,3-dihydro-3,3,3′,4′,5,5′-hexamethoxy-2,9′-lign-8-en-4(5H)-one*

[70280-34-1]

$C_{24}H_{32}O_7$ 432.513

Constit. of *Heterotropa takaoi* (preferred genus name *Asarum*). Viscous liq.. Racemate

8′-Epimer

Isoheterotropanone

[70287-69-3]

$C_{24}H_{32}O_7$ 432.513

Constit. of *Heterotropa takaoi*. Viscous liq.. Racemate.

Yamamura, S. *et al.*, *Tetrahedron Lett.*, 1978, **19**, 4891-4894 (*Heterotropa takaoi constits*).
Yamamura, S. *et al.*, *Bull. Chem. Soc. Jpn.*, 1982, **55**, 3573-3579 (*Heterotropanone, Isoheterotropanone, synth*)

Heterotropatrione H27

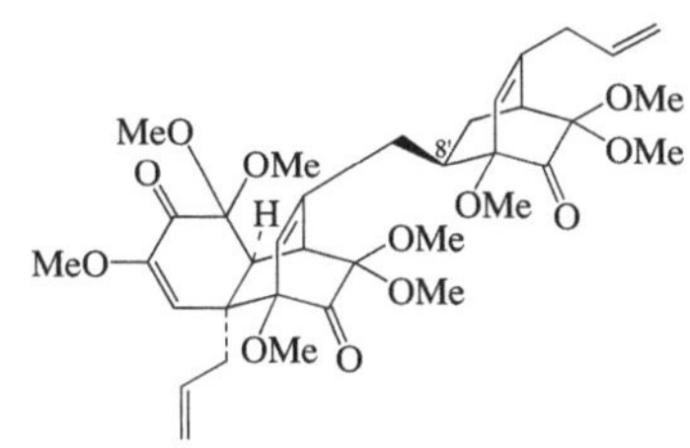

Relative Configuration

[71332-97-3]

$C_{36}H_{48}O_{12}$ 672.768

Constit. of *Heterotropa takaoi* (preferred genus name *Asarum*). Viscous liq.. Racemate

8′-Epimer

6-Epiheterotropatrione. Isoheterotropatrione

[71358-32-2]

$C_{36}H_{48}O_{12}$ 672.768

Constit. of *Heterotropa takaoi*. Viscous liq.. Racemate.

Niwa, M. *et al.*, *Bull. Chem. Soc. Jpn.*, 1982, **55**, 3573-3579 (*Heterotropatrione, Isoheterotropatrione*).
Nishiyama, A. *et al.*, *Agric. Biol. Chem.*, 1985, **49**, 1197-1198 (*synth*)

3,3′,4′,5,5′,7′-Hexahydroxy-8,9-dinor-4,8′-oxylignan-7-oic acid H28

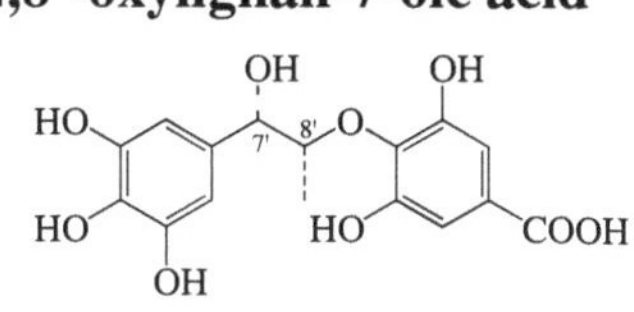

$C_{16}H_{16}O_9$ 352.295

(7′S,8′R)-form

3,3′,4′,5,5′-Penta-Me ether

7′-Hydroxy-3,3′,4′,5,5′-pentamethoxy-8,9-dinor-4,8′-oxylignan-7-oic acid

3,3′,4′,5,5′-Penta-Me ether, Me ester

Acortatarinowin M

$C_{22}H_{28}O_9$ 436.457

Constit. of the rhizomes of *Acorus tatarinowii*. Light yellow gum. $[\alpha]_D^{20}$ +12 (c, 0.1 in MeOH). λ_{max} 208 (log ε 4.73); 272 (log ε 4.01) (MeOH).

Lu, Y. *et al.*, *Sci. Rep.*, 2016, **6**, 22909 (*Acortatarinowin M*)

3,3′,4,4′,5,5′-Hexahydroxy-8,8′-ligna-7,7′-diene-9,9′-dioic acid H29

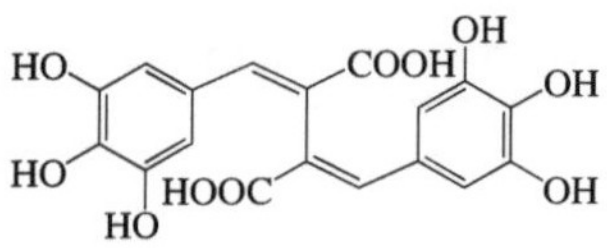

2,3-Bis(3,4,5-trihydroxybenzylidene)butanedioic acid
$C_{18}H_{14}O_{10}$ 390.3
(E,E)-form
3,3′,5,5′-Tetra-Me ether
2,3-Bis(4-hydroxy-3,5-dimethoxybenzylidene)butanedioic acid. 4,4′-Dihydroxy-3,3′,5,5′-tetramethoxy-8,8′-ligna-7,7′-diene-9,9′-dioic acid. ***8,8′-Disinapic acid.*** *8,8′-Dehydrodisinapic acid*
[202407-94-1]
[1424177-94-5] (unspecified stereochem.)
$C_{22}H_{22}O_{10}$ 446.408
Found in wheat and rice. Also from *Brassica rapa* var. *chinensis*. Dehydrodimer of Sinapic acid, in T106
3,3′,5,5′-Tetra-Me ether, di-Me ester
[1402923-23-2]
$C_{24}H_{26}O_{10}$ 474.462
Constit. of the seeds of *Brassica napus*
3,3′,5,5′-Tetra-Me ether, di-Et ester
Diethyl 8,8′-disinapate. **Phebalarin**
[56136-37-9]
$C_{26}H_{30}O_{10}$ 502.516
Constit. of the bark of *Phebalium nudum*. Antioxidant. Yellow needles (EtOH or Et_2O/toluene). Mp 125 - 126° (90°). λ_{max} 225; 236; 245; 332 (EtOH)
3,3′,5,5′-Tetra-Me ether, di-Ac, di-Et ester
[56136-39-1]
Needles (MeOH). Mp 119°
3,3′,5,5′-Tetra-Me ether, mono-[2-(trimethylamino)ethyl] ester
[1402923-17-4]
$C_{27}H_{34}NO_{10}^{\oplus}$ 532.566 (ion)
Constit. of the seeds of *Brassica napus*. λ_{max} 241; 331 (MeOH aq.)
3,3′,5,5′-Tetra-Me ether, 9-[2-(trimethylamino)ethyl] ester, 9′-Me ester
[1402923-21-0]
$C_{28}H_{36}NO_{10}^{\oplus}$ 546.593 (ion)
Constit. of the seeds of *Brassica napus*. λ_{max} 242; 338 (MeOH aq.)
3,3′,5,5′-Tetra-Me ether, bis[2-(trimethylamino)ethyl] ester
Dicholinyl 8,8′-disinapate
[1402923-16-3]
$C_{32}H_{46}N_2O_{10}^{2\oplus}$ 618.724 (ion)
Constit. of the seeds of *Brassica napus*. Counterion not specified. λ_{max} 243; 339 (MeOH aq.).

Briggs, L.H. *et al.*, *Tetrahedron*, 1958, **2**, 256-270 (*Phebalarin*).
Brown, K.L. *et al.*, *Aust. J. Chem.*, 1975, **28**, 1327-1337 (*Phebalarin, synth*).
Bunzel, M. *et al.*, *J. Agric. Food Chem.*, 2003, **51**, 1427-1434 (*synth, biosynth*).
Neudörffer, A. *et al.*, *J. Agric. Food Chem.*, 2004, **52**, 2084-2091 (*Phebalarin, activity*).
Fang, J. *et al.*, *J. Agric. Food Chem.*, 2012, **60**, 10523-10529 (*Brassica napus esters*).
Grúz, J. *et al.*, *Food Chem.*, 2015, **171**, 280-286 (*synth, detn*)

3,3′,4,4′,5,5′-Hexahydroxy-2,2′-lignan-8,8′-dione H30

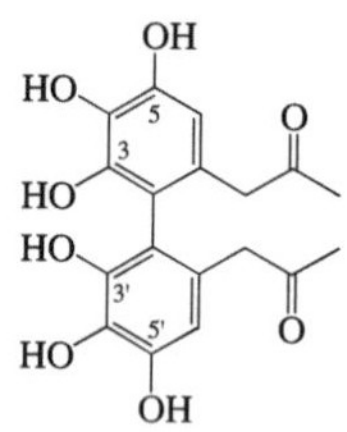

4,4′,5,5′,6,6′-Hexahydroxy-[1,1′-biphenyl]-2,2′-diylbis[2-propanone]. 2,2′,3,3′,4,4′-Hexahydroxy-6,6′-bis(2-oxopropyl)biphenyl. 3,3′,4,4′,5,5′-Hexahydroxy-8,8′-dioxo-2,2′-neolignan
$C_{18}H_{18}O_8$ 362.334
8,8′-Seco-2,2′-cyclolignans. Lignan numbering shown
4,4′,5,5′-Tetra-Me ether
3,3′-Dihydroxy-4,4′,5,5′-tetramethoxy-2,2′-lignan-8,8′-dione. 3,3′-Dihydroxy-4,4′,5,5′-tetramethoxy-8,8′-dioxo-2,2′-lignan. ***Neglignan D***
[1441710-21-9]
$C_{22}H_{26}O_8$ 418.442
Isol. from the stems of *Schisandra neglecta*. Displays cytotoxic activity against NB4 and SHSY5Y cancer cell lines. Yellow gum. λ_{max} 210 (log ε 4.65); 276 (log ε 3.68); 320 (log ε 1.42) (MeOH)
4,5:4′,5′-Bis(methylene), 3,3′-di-Me ether
3,3′-Dimethoxy-4,5:4′,5′-bis(methylenedioxy)-2,2′-lignan-8,8′-dione. 3,3′-Dimethoxy-4,5:4′,5′-bis(methylenedioxy)-8,8′-dioxo-2,2′-lignan. ***Neglignan C***
[1441710-14-0]
$C_{22}H_{22}O_8$ 414.41
Isol. from the stems of *Schisandra neglecta*. Yellow gum. λ_{max} 210 (log ε 4.68); 276 (log ε 3.56); 320 (log ε 1.55) (MeOH).

Gao, X.-M. *et al.*, *J. Nat. Prod.*, 2013, **76**, 1052-1057 (*Neglignans C,D, activity*)

3,3′,4,4′,5,5′-Hexahydroxy-8,8′-lignan-7,7′-dione H31

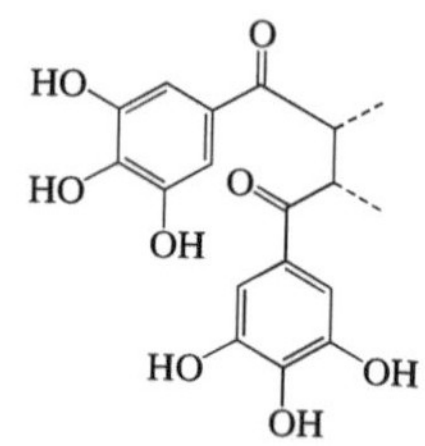

2,3-Dimethyl-1,4-bis(3,4,5-trihydroxyphenyl)-1,4-butanedione. 2,3-Bis(3,4,5-trihydroxybenzoyl)butane
$C_{18}H_{18}O_8$ 362.334
(2RS,3SR)-form
Hexa-Me ether
3,3′,4,4′,5,5′-Hexamethoxy-8,8′-lignan-7,7′-dione. 2,3-Dimethyl-1,4-bis(3,4,5-trimethoxyphenyl)-1,4-butanedione, CAS
[72730-18-8]
$C_{24}H_{30}O_8$ 446.496
Constit. of *Saururus chinensis*. Exhibits significant inhibitory activity on ADP-induced platelet aggregation. Needles. Opt. inactive (*meso-*).

Qu, W. *et al.*, *Chem. Nat. Compd. (Engl. Transl.)*, 2014, **50**, 814-818 (*Saururus chinensis constit, activity*)

3,3′,4,4′,8,9′-Hexahydroxy-8,8′-lignan-9-oic acid H32

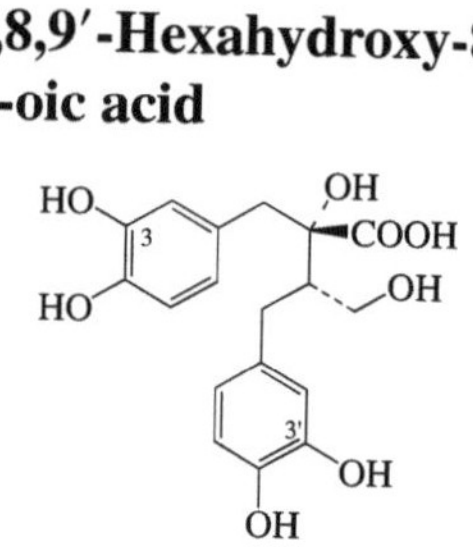

2,3-Bis(3,4-dihydroxybenzyl)-2,4-dihydroxybutanoic acid
$C_{18}H_{20}O_8$ 364.35

(8S,8′S)-form

3,3′,4′-Tri-Me ether, amide

Trachelogenin amide

[132472-34-5]

$C_{21}H_{27}NO_7$ 405.447

Alkaloid from *Trachelospermum jasminoides*. Amorph. solid. $[\alpha]_D^{25}$ -9 (c, 0.55 in MeOH).

Tan, X.-Q. *et al.*, *Planta Med.*, 2005, **71**, 93-95 (*Trachelogenin amide*)

2′,3,4,4′,5,5′-Hexahydroxy-8,8′-lignan-9,9′-olide H33

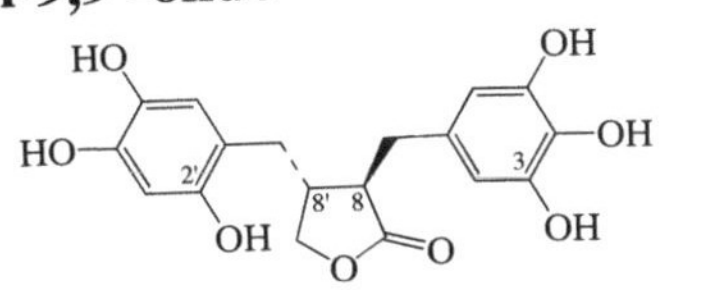

Dihydro-4-(2,4,5-trihydroxybenzyl)-3-(3,4,5-trihydroxybenzyl)-2(3H)-furanone

$C_{18}H_{18}O_8$ 362.334

(8R,8′R)-form

4′,5′-Methylene, 3,4,5-tri-Me ether

2′-Hydroxy-3,4,5-trimethoxy-4′,5′-methylenedioxy-8,8′-lignan-9,9′-olide.

6′-Hydroxyyatein

[193978-43-7]

$C_{22}H_{24}O_8$ 416.426

Constit. of the bark of *Hernandia nymphaeifolia*. Cytotoxic agent against P-388, KB16, A 549 and HT-29 cell lines. Amorph. powder. $[\alpha]_D^{25}$ -29.3 (c, 0.08 in $CHCl_3$). λ_{max} 206 (log ε 4.42); 230 (sh); 303 (log ε 3.48) (EtOH).

Chen, I.-S. *et al.*, *Phytochemistry*, 1997, **45**, 991-996 (*6′-Hydroxyyatein*)

2′,3,4,4′,5′,7-Hexahydroxy-8,8′-lignan-9,9′-olide H34

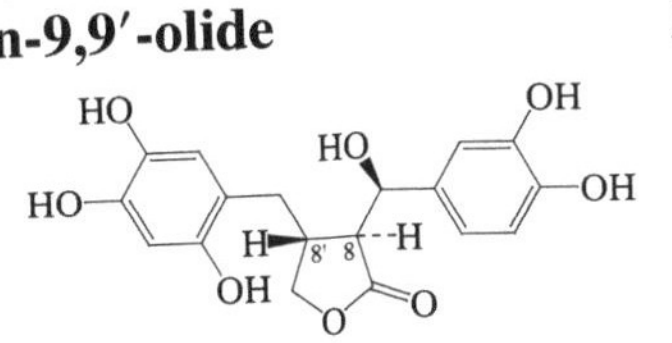

$C_{18}H_{18}O_8$ 362.334

(7R,8S,8′R)-form

4′,5′-Methylene, 3,4-di-Me ether

2′,7-Dihydroxy-3,4-dimethoxy-4′,5′-methylenedioxy-8,8′-lignan-9,9′-olide.

6′,7-Dihydroxybursehernin

[511541-39-2]

$C_{21}H_{22}O_8$ 402.399

Constit. of *Linum flavum* var. *compactum*.

Mikame, K. *et al.*, *J. Wood Sci.*, 2002, **48**, 440-445 (*6′,7-Dihydroxybursehernin*)

3,3′,4,4′,5,5′-Hexahydroxy-8,8′-lignan-9,9′-olide H35

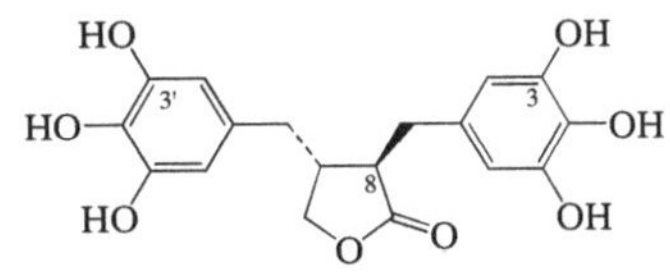

(8R,8′R)-form

Dihydro-3,4-bis(3,4,5-trihydroxybenzyl)-2(3H)-furanone

$C_{18}H_{18}O_8$ 362.334

Lignan numbering shown

(8R,8′R)-form

3,3′,4′,5,5′-Penta-Me ether

4-Hydroxy-3,3′,4′,5,5′-pentamethoxy-8,8′-lignan-9,9′-olide.

Hernolactone

[112667-54-6]

$C_{23}H_{28}O_8$ 432.469

Isol. from seeds of *Hernandia ovigera*. Pale yellow solid (C_6H_6/Et_2O). Mp 116 - 118°. $[\alpha]_D$ -30 (c, 0.36 in $CHCl_3$) (-24.9). λ_{max} 225 (ε 15490); 272 (ε 1950) (EtOH) (Berdy)

Hexa-Me ether

3,3′,4,4′,5,5′-Hexamethoxy-8,8′-lignan-9,9′-olide. **Cubebininolide**. *Cordigerin*

[93395-16-5]

$C_{24}H_{30}O_8$ 446.496

Constit. of *Hernandia cordigera, Piper clusii* and *Piper cubeba*. $[\alpha]_D$ -42.4 (MeOH). $[\alpha]_D^{25}$ -17.6 (c, 0.23 in $CHCl_3$). λ_{max} 226 (ε 16982); 272 (ε 1995) (MeOH)

3,4-Methylene, 3′,5,5′-tri-Me ether

4′-Hydroxy-3,3′,5′-trimethoxy-4,5-methylenedioxy-8,8′-lignan-9,9′-olide.

4′-O-Demethylcubebinolide

[845897-02-1]

$C_{22}H_{24}O_8$ 416.426

Constit. of the fruit of *Piper cubeba*. Pale yellow oil. $[\alpha]_D^{25}$ -30 (c, 0.08 in $CHCl_3$)

3,4-Methylene, 3′,4′,5,5′-tetra-Me ether

3,3′,4′,5′-Tetramethoxy-4,5-methylenedioxy-8,8′-lignan-9,9′-olide.

Cubebinone

[101751-71-7]

$C_{23}H_{26}O_8$ 430.453

Constit. of *Piper cubeba*. Cryst. (EtOH). Mp 108 - 110°. $[\alpha]_D^{25}$ -36.1 (c, 0.083 in $CHCl_3$). λ_{max} 275 (ε 5370) (MeOH)

3′,4′-Methylene, 3,4,5,5′-tetra-Me ether

3,3′,4,5-Tetramethoxy-4′,5′-methylenedioxy-8,8′-lignan-9,9′-olide.

5′-Methoxyyatein

[94359-20-3]

$C_{23}H_{26}O_8$ 430.453

Constit. of *Hernandia cordigera*. Amorph. powder. $[\alpha]_D$ -21 (c, 1 in $CHCl_3$)

9-Alcohol(lactol)

9,9′-Epoxy-8,8′-lignan-3,3′,4,4′,5,5′,9-heptol. 9,9′-Epoxy-3,3′,4,4′,5,5′,9′-heptahydroxylignan. Tetrahydro-3,4-bis[(3,4,5-trihydroxyphenyl)methyl]-2-furanol

9ξ-Alcohol(lactol), 3,3′,5,5′-tetra-Me ether

9,9′-Epoxy-3,3′,5,5′-tetramethoxy-8,8′-lignan-4,4′,9-triol. 9,9′-Epoxy-4,4′,9-trihydroxy-3,3′,5,5′-tetramethoxylignan

$C_{22}H_{28}O_8$ 420.458

9S-Alcohol(lactol), 3′,4′-methylene, 3,4,5,5′-tetra-Me ether

9,9′-Epoxy-3,3′,4,5-tetramethoxy-4′,5′-methylenedioxy-8,8′-lignan-9-ol. 9,9′-Epoxy-9-hydroxy-3,3′,4,5-tetramethoxy-4′,5′-methylenedioxylignan. ***5′-Methoxyclusin***

[845897-03-2]

$C_{23}H_{28}O_8$ 432.469

Constit. of the fruit of *Piper cubeba*. Potent inhibitor of CYP3A4. Pale yellow oil. $[\alpha]_D^{25}$ -53.4 (c, 0.3 in $CHCl_3$)

9ξ-Alcohol(lactol), 3,3′,5,5′-tetra-Me ether, 4,4′-di-O-β-D-glucopyranoside

[155913-98-7]

$C_{34}H_{48}O_{18}$ 744.74

Constit. of the roots of *Pyracantha coccinea*. Viscous oil. $[\alpha]_D^{25}$ -43 (c, 0.8 in MeOH). λ_{max} 240; 268; 320 (MeOH)

9ξ-Alcohol(lactol), 3,3′,4,4′,5,5′-hexa-Me ether

9,9′-Epoxy-3,3′,4,4′,5,5′-hexamethoxy-8,8′-lignan-9-ol. 9,9′-Epoxy-9-hydroxy-3,3′,4,4′,5,5′-hexamethoxylignan. Tetrahydro-3,4-bis[(3,4,5-trimethoxyphenyl)methyl]-2-furanol, CAS.

Cubebinin

[96238-92-5]

$C_{24}H_{32}O_8$ 448.512

Constit. of *Piper cubeba*. Gum. $[\alpha]_D^{25}$ -23.3 (c, 0.3 in $CHCl_3$). λ_{max} 226 (ε 17378); 270 (ε 1585) (MeOH)

(8S,8′S)-form

3′,4′-Methylene, 3,5′-di-Me ether

3,4-Dihydroxy-3′,5-dimethoxy-4′,5′-methylenedioxy-8,8′-lignan-9,9′-olide

[878142-50-8]

$C_{21}H_{22}O_8$ 402.399

Constit. of *Peperomia duclouxii*. Gum. $[\alpha]_D^{25}$ +19.2 (c, 0.07 in $CHCl_3$). λ_{max} 242; 277 ($CHCl_3$)

3′,4′-Methylene, 3,5,5′-tri-Me ether

4-Hydroxy-3,3′,5-trimethoxy-4′,5′-methylenedioxy-8,8′-lignan-9,9′-olide

[635752-18-0]

$C_{22}H_{24}O_8$ 416.426

Constit. of *Peperomia duclouxii*. Pale yellow gum. $[\alpha]_D^{25}$ +30.5 (c, 0.38 in $CHCl_3$). λ_{max} 242; 276 ($CHCl_3$)

3′,4′-Methylene, 3,5,5′-tri-Me ether, 4-O-β-D-glucopyranoside

[1115239-79-6]

$C_{28}H_{34}O_{13}$ 578.567

Constit. of *Peperomia duclouxii*. Amorph. powder. $[\alpha]_D^{22}$ +7 (c, 0.09 in MeOH)

3′,4′-Methylene, 3,4,5,5′-tetra-Me ether

[635752-16-8]

Constit. of *Peperomia duclouxii*. Pale yellow gum. $[\alpha]_D^{25}$ +24.8 (c, 0.39 in $CHCl_3$). λ_{max} 241; 275 ($CHCl_3$)

3,4:3′,4′-Bis(methylene), 5,5′-di-Me ether

3,3′-Dimethoxy-4,5:4′,5′-bis(methylenedioxy)-8,8′-lignan-9,9′-olide

[635752-15-7]

$C_{22}H_{22}O_8$ 414.41

Constit. of *Peperomia duclouxii*. Pale yellow gum. $[\alpha]_D^{25}$ +29 (c, 0.55 in $CHCl_3$). λ_{max} 243; 278 ($CHCl_3$).

Richomme, P. *et al.*, *J. Nat. Prod.*, 1984, **47**, 879-881 (*Cordigerin*).

Koul, S.K. *et al.*, *Phytochemistry*, 1984, **23**, 2099-2101 (*Cubebininolide*).

Richomme, P. *et al.*, *Heterocycles*, 1985, **23**, 309-312 (*5′-Methoxyyatein*).

Prabhu, B.R. *et al.*, *Phytochemistry*, 1985, **24**, 329-332 (*Cubebinin*).

Badheka, L.P. *et al.*, *Phytochemistry*, 1986, **25**, 487 (*Cubebinone*).

Lalami, K. *et al.*, *Heterocycles*, 1988, **27**, 1131-1134 (*Cubebinin, Cubebinone, synth*).

Tanoguchi, M. *et al.*, *Chem. Pharm. Bull.*, 1989, **37**, 68-72 (*Hernolactone*).

Bilia, A.R. *et al.*, *Tetrahedron*, 1994, **50**, 5181-5188 (*Pyracantha diglucoside*).

Li, N. *et al.*, *J. Nat. Prod.*, 2003, **66**, 1421-1426 (*Peperomia duclouxii lignans*).

Jang, D.S. *et al.*, *Phytochemistry*, 2004, **70**, 893-896 (*Hernolactone*).

Usia, T. *et al.*, *J. Nat. Prod.*, 2005, **68**, 64-68 (*4′-O-Demethylcubebinolide, 5′-Methoxyclusin, Cordigerin*).

Li, N. *et al.*, *J. Nat. Prod.*, 2006, **69**, 234-239 (*Peperomia dulouxii lignans*).

Zhu, Y.H. *et al.*, *Nat. Prod. Res.*, 2008, **22**, 1483-1486 (*Peperomia duclouxii glucoside*).

Amancha, P.K. *et al.*, *Eur. J. Org. Chem.*, 2010, 3473-3480 (*(±)-5′-Methoxyyatein, (±)-5′-Methoxyyclusin, synth*)

3,3′,4,4′,5,7-Hexahydroxy-8,8′-lignan-9,9′-olide H36

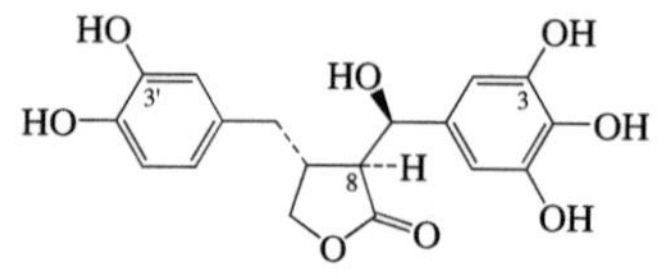

(7R,8S,8′R)-form

$C_{18}H_{18}O_8$ 362.334

(7R,8S,8′R)-form

3′,4′-Methylene, 3,4,5-tri-Me ether

7-Hydroxy-3,4,5-trimethoxy-3′,4′-methylenedioxy-8,8′-lignan-9,9′-olide.

Epipodorhizol

[17187-82-5]

[59366-92-6] ((±)-form)

$C_{22}H_{24}O_8$ 416.426

Constit. of *Juniperus thurifera*. Cryst. (CH_2Cl_2/hexane). Mp 129 - 130°. $[\alpha]_D$ -20.8 ($CHCl_3$)

(7S,8S,8′R)-form

3′,4′-Methylene, 3,4,5-tri-Me ether

Podorhizol

[17187-78-9]

[59366-91-5] ((±)-form)

$C_{22}H_{24}O_8$ 416.426

Isol. from *Podophyllum emodi, Hernandia nymphaeifolia, Hernandia peltata* and *Hernandia ovigera*. Shows antimicrobial and cytotoxic props.. Cryst. (Me_2CO/Et_2O/pentane). Mp 125 - 126°. $[\alpha]_D^{21}$ -51.8 (c, 1.042 in $CHCl_3$)

3′,4′-Methylene, 3,4,5-tri-Me ether, 7-O-β-D-glucopyranoside

Podorhizol glucoside. *Lignan F*

[17187-73-4]

$C_{28}H_{34}O_{13}$ 578.567

Constit. of *Podophyllum* spp. such as *Podophyllum peltanum*. Amorph. powder + ½H_2O. Poorly sol. hexane. Mp 103 - 104°. $[\alpha]_D^{21}$ -55.5 (c, 0.6 in $CHCl_3$). λ_{max} 285 (ε 3630) (EtOH) (Berdy)

3′,4′-Methylene, 3,4,5-tri-Me ether, 7-Ac

Podorhizol acetate

$C_{24}H_{26}O_9$ 458.463

Constit. of *Juniperus sabina*. Mp 47 - 49°. $[\alpha]_D^{21}$ -45.6 (c, 0.614 in $CHCl_3$)

7-Ketone, 3′,4′-methylene, 3,4,5-tri-Me ether

3,4,5-Trimethoxy-3′,4′-methylenedioxy-7-oxo-8,8′-lignan-9,9′-olide. Podorhizone

[17187-75-6]

$C_{22}H_{22}O_8$ 414.41

Semisynthetic. Fine needles. Mp 129 - 130°. $[\alpha]_D^{23}$ +79.7 (c, 0.5 in $CHCl_3$).

Kuhn, M. *et al.*, *Helv. Chim. Acta*, 1967, **50**, 1546-1565 (*Podorhizol, Podorhizol glucoside*).

Ayres, D.C. *Tetrahedron Lett.*, 1969, **10**, 883-886 (*Podorhizol, biosynth*).

Brown, E. *et al.*, *J. Chem. Soc., Chem. Commun.*, 1978, 556-557 (*Podorhizone, synth*).

Ziegler, F.E. *et al.*, *J. Org. Chem.*, 1978, **43**, 985-991 (*Podorhizone, synth*).

Yamaguchi, H. *et al.*, *Yakugaku Zasshi*, 1979, **99**, 674-677 (*Podorhizol, Epipodorhizol*).

Tomioka, K. *et al.*, *Chem. Pharm. Bull.*, 1982, **30**, 4304-4313 (*Podorhizone, synth*).

Robin, J.-P. *et al.*, *Tetrahedron*, 1982, **38**, 3667-3671 (*Podorhizol, Podorhizone, synth*).

Posner, G.H. *et al.*, *Tetrahedron Lett.*, 1984, **25**, 2627-2630 (*Podorhizone, synth*).

Kosugi, H. *et al.*, *J. Chem. Soc., Perkin Trans. 1*, 1989, 935-943 (*Podorhizone, synth*).

San Feliciano, A. *et al.*, *Phytochemistry*, 1989, **28**, 2863-2866 (*Juniperus thurifera constit*).

Brown, E. *et al.*, *Tetrahedron*, 1989, **45**, 141-154 (*Podorhizone, synth*).

San Felliciano, A. *et al.*, *Phytochemistry*, 1990, **29**, 1335-1338 (*Juniperus sabina constit*).

Tanoguchi, M. *et al.*, *Chem. Pharm. Bull.*, 1991, **39**, 1873-1876 (*Hernandia ovigera constit, cd*).

Morimoto, T. *Tetrahedron*, 1993, **49**, 1793-1806 (*Podorhizone, synth*).

Yamauchi, S. *et al.*, *Biosci., Biotechnol., Biochem.*, 1999, **63**, 1453-1462 (*(-)-Podorhizol, synth*).

Chen, J.-J. *et al.*, *Planta Med.*, 2000, **66**, 251-256 (*Hernandia nymphaeifolia constit*).

Ito, C. *et al.*, *Planta Med.*, 2001, **67**, 166-168 (*Hernandia ovigera constit, activity*).

Pettit, G.R. *et al.*, *J. Nat. Prod.*, 2004, **67**, 214-220 (*Hernandia nymphaeifolia constit, Hernandia peltata constit, activity*)

3,3′,4,4′,5′,7′-Hexahydroxy-8,8′-lignan-9,9′-olide H37

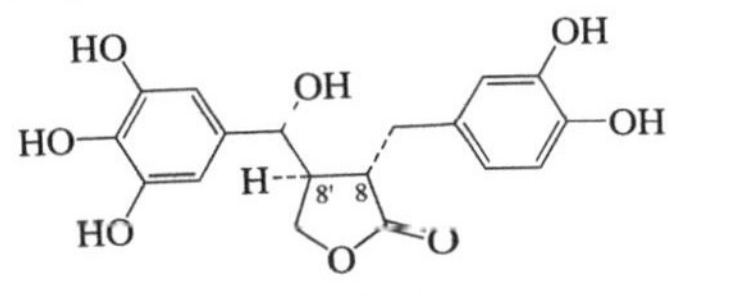

(7′R,8S,8′S)-form

$C_{18}H_{18}O_8$ 362.334

(7′R,8S,8′S)-form

3,3′,5′-Tri-Me ether

4,4′,7′-Trihydroxy-3,3′,5′-trimethoxy-8,8′-lignan-9,9′-olide

3,3′,5′-Tri-Me ether, 4′-O-β-D-glucopyranoside

Wasabiside C

[2022212-54-8]

$C_{27}H_{34}O_{13}$ 566.556

Constit. of the roots of *Wasabia japonica*. Gum. $[\alpha]_D^{25}$ +10.3 (c, 0.3 in MeOH). λ_{max} 234 (log ε 2.20); 280 (log ε 1.15) (MeOH)

(7′S,8R,8′R)-form

3,3′,5′-Tri-Me ether, 7′-O-β-D-glucopyranoside

Patrinian B

[2226140-32-3]

$C_{27}H_{34}O_{13}$ 566.556

Constit. of *Patrinia villosa*. Potent antioxidant. Exhibits neuroprotective effects. Yellow oil. $[\alpha]_D^{20}$ +26.3 (c, 1.52 in MeOH). λ_{max} 232 (log ε 3.88); 281 (log ε 3.90) (MeOH)

(8R*,8′R*)-form

7′-Ketone, 3,3′,5′-tri-Me ether

4,4′-Dihydroxy-3,3′,5′-trimethoxy-7′-oxo-8,8′-lignan-9,9′-olide.

5′-Methoxy-7′-oxomatairesinol

[1807319-95-4]

$C_{21}H_{22}O_8$ 402.399

Constit. of *Chonemorpha griffithii*. Amorph. powder. $[\alpha]_D^{23}$ -109 (c, 0.05 in MeOH).

Bai, J.-Q. *et al.*, *Biochem. Syst. Ecol.*, 2013, **51**, 171-174 (*5′-Methoxy-7′-oxomatairesinol*).

Kim, C.S. *et al.*, *J. Nat. Prod.*, 2016, **79**, 2652-2657 (*Wasabiside C*).

Bai, M. *et al.*, *J. Funct. Foods*, 2017, **28**, 106-113 (*Patrinian B, activity*)

3,3′,4,4′,5,8-Hexahydroxy-8,8′-lignan-9,9′-olide H38

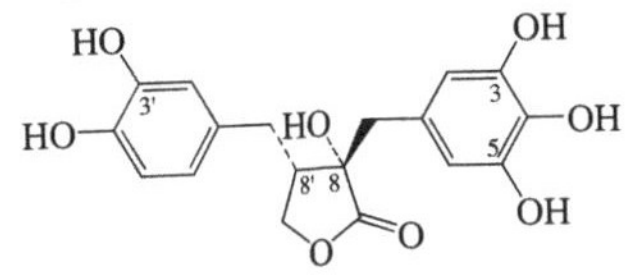

(8S,8′S)-form

$C_{18}H_{18}O_8$ 362.334

(8S,8′S)-form

3,3′,4′,5-Tetra-Me ether

4,8-Dihydroxy-3,3′,4′,5-tetramethoxy-8,8′-lignan-9,9′-olide.

5-Methoxytrachelogenin

[957467-21-9]

$C_{22}H_{26}O_8$ 418.442

Constit. of the heartwood of *Cunninghamia lanceolata*. $[\alpha]_D^{26}$ -84.5 (c, 0.3 in $CHCl_3$) (as di-Ac)

3′,4′-Methylene, 3-Me ether

3,4,8-Trihydroxy-5-methoxy-3′,4′-methylenedioxy-8,8′-lignan-9,9′-olide.

5,8-Dihydroxypluviatolide

[222849-86-7]

$C_{20}H_{20}O_8$ 388.372

Constit. of *Taiwania cryptomerioides*. $[\alpha]_D^{25}$ -37 (c, 0.9 in $CHCl_3$)

(8R*,8′R*)-form

3,3′,5-Tri-Me ether

4,4′,8-Trihydroxy-3,3′,5-trimethoxy-8,8′-lignan-9,9′-olide. ***Hydroxythujaplicatin methyl ether.***

5-Methoxywikstromol

[6744-84-9]

$C_{21}H_{24}O_8$ 404.415

Constit. of *Cunninghamia lanceolata, Thuja occidentalis* and *Thuja plicata*. Cryst. (Et_2O). Mp 128.5 - 129.5°. $[\alpha]_D^{26}$ -85.2 (c, 0.5 in $CHCl_3$) (as tri-Ac).

Maclean, H. *et al.*, *Can. J. Chem.*, 1966, **44**, 1827-1830 (*Hydroxythujaplicatin methyl ether*).

Khamlach, K. *et al.*, *Heterocycles*, 1990, **31**, 2195-2199 (*synth*).

Lin, W-H. *et al.*, *Phytochemistry*, 1999, **50**, 653-658 (*5,8-Dihydroxypluviatolide*).

Lee, T.-H. *et al.*, *Biosci., Biotechnol., Biochem.*, 2007, **71**, 2075-2078 (*Cunninghamia lanceolata constits*)

3,3′,4,4′,5′,8-Hexahydroxy-8,8′-lignan-9,9′-olide H39

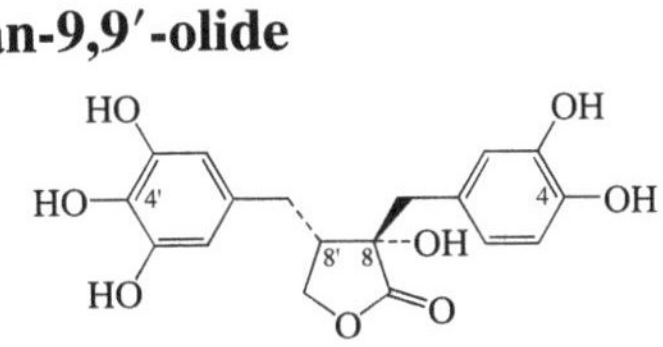

$C_{18}H_{18}O_8$ 362.334

(8S,8′S)-form

3,3′,4′,5′-Tetra-Me ether

4,8-Dihydroxy-3,3′,4′,5′-tetramethoxy-8,8′-ligan-9,9′-olide.

5′-Methoxytrachelogenin

[1321810-66-5]

$C_{22}H_{26}O_8$ 418.442

Constit. of the stems and leaves of *Trachelospermum jasminoides*. Amorph. powder. $[\alpha]_D^{26}$ -48.2 (c, 0.08 in MeOH). λ_{max} 230 (log ε 4.18); 280 (log ε 3.61) (MeOH)

3,3′,4′,5′-Tetra-Me ether, 4-O-β-D-glucopyranoside

5′-Methoxytracheloside

[1321810-65-4]

$C_{28}H_{36}O_{13}$ 580.583

Constit. of the stems and leaves of *Trachelospermum jasminoides*. Amorph. powder. $[\alpha]_D^{26}$ -45.6 (c, 0.09 in MeOH). λ_{max} 230 (log ε 4.15); 278 (log ε 3.6) (MeOH).

Jing, L. *et al.*, *Chin. Chem. Lett.*, 2011, **22**, 1075-1077 (*5′-Methoxytrachelogenin, 5′-Methoxytracheloside*)

3,3′,4,4′,7,7′-Hexahydroxy-8,8′-lignan-9,9′-olide H40

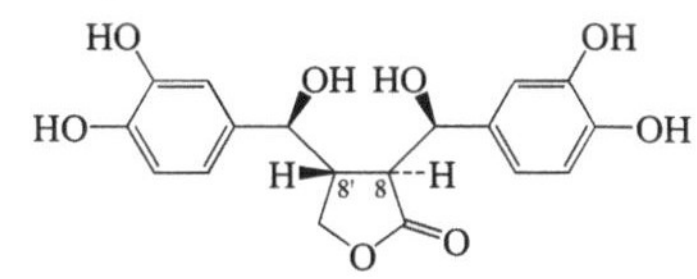

(7R,7′S,8S,8′R)-form

Dihydro-3,4-bis[hydroxy(3,4-dihydroxyphenyl)methyl]-2(3H)-furanone, CAS

$C_{18}H_{18}O_8$ 362.334

(7R,7′S,8S,8′R)-form

7-Ketone, 3,4-methylene, 3′,4′-di-Me ether

7′-Hydroxy-3′,4′-dimethoxy-3,4-methylenedioxy-7-oxo-8,8′-lignan-9,9′-olide

7-Ketone, 3,4-methylene, 3′,4′-di-Me ether, 7′-Ac

Sanguinolignan C

[1255645-44-3]

$C_{23}H_{22}O_9$ 442.42

Constit. of the leaves of *Piper sanguineispicum*. Powder. $[\alpha]_D^{20}$ -1.2 (c, 0.85 in CH_2Cl_2). λ_{max} 282; 319 (CH_2Cl_2)

7-Ketone, 3,4:3′,4′-bis(methylene) ether

7′-Hydroxy-3,4:3′,4′-bis(methylenedioxy)-7-oxo-8,8′-lignan-9,9′-olide. ***Sanguinolignan A***

[1255645-42-1]

$C_{20}H_{16}O_8$ 384.34

Constit. of the leaves of *Piper sanguineispicum*. Powder. $[\alpha]_D^{20}$ -24 (c, 1 in Me_2CO). λ_{max} 283; 317 (MeOH)

7-Ketone, 3,4:3′,4′-bis(methylene) ether, 7′-Ac
Sanguinolignan D
[1255645-45-4]
$C_{22}H_{18}O_9$ 426.377
Constit. of the leaves of *Piper sanguineispicum*. Powder. $[\alpha]_D^{20}$ -47.4 (c, 2.15 in CH_2Cl_2). λ_{max} 283; 317 (CH_2Cl_2)

(7R,7′S,8S,8′S)-form
7-Ketone, 3,4:3′,4′-bis(methylene) ether
Sanguinolignan B
[1255645-43-2]
$C_{20}H_{16}O_8$ 384.34
Constit. of the leaves of *Piper sanguineispicum*. Powder. $[\alpha]_D^{20}$ -34.3 (c, 1.2 in CH_2Cl_2). λ_{max} 284; 318 (CH_2Cl_2)

(7ξ,7′ξ,8S,8′R)-form
3,3′-Di-Me ether
4,4′,7,7′-Tetrahydroxy-3,3′-dimethoxy-8,8′-lignan-9,9′-olide.
7,7′-Dihydroxymatairesinol
[108656-97-9]
$C_{20}H_{22}O_8$ 390.388
Constit. of *Bombax ceiba*, *Larix leptolepis* and *Lindera obtusiloba*. Cytotoxic agent

3′,4′-Di-Me ether
[1683551-46-3]
$C_{20}H_{22}O_8$ 390.388
Constit. of *Zea mays*. Yellow-brown powder. λ_{max} 236; 279 (MeOH)

3′,4′-Methylene, 3-Me ether
4,7,7′-Trihydroxy-3-methoxy-3′,4′-methylenedioxy-8,8′-lignan-9,9′-olide

3′,4′-Methylene, 3-Me ether, 4-O-β-D-glucopyranoside
Styraxjaponoside A
[848757-11-9]
$C_{26}H_{30}O_{13}$ 550.513
Constit. of the stem bark of *Styrax japonica*. Cryst.. Mp 156.8 - 156.9° dec.. $[\alpha]_D^{25}$ +0.25 (c, 0.5 in MeOH). λ_{max} 234 (log ε 1.93); 281 (log ε 1.15) (MeOH)

3′,4′-Methylene, 3,4-di-Me ether
7,7′-Dihydroxy-3,4-dimethoxy-3′,4′-methylenedioxy-8,8′-lignan-9,9′-olide.
7,7′-Dihydroxybursehernin
[445389-48-0]
$C_{21}H_{22}O_8$ 402.399
Constit. of *Geranium thunbergii*. Oil. $[\alpha]_D^{24}$ +30 (c, 0.5 in MeOH). λ_{max} 234 (MeOH).

Sakakibara, A. *et al.*, *Holzforschung*, 1987, **41**, 1-11 (*7,7′-Dihydroxymatairesinol*).
Kwon, H.C. *et al.*, *Arch. Pharmacal Res.*, 1999, **22**, 417-422 (*7,7′-Dihydroxymatairesinol*).
Kim, M.-R. *et al.*, *Chem. Pharm. Bull.*, 2004, **52**, 1466-1469 (*Styraxjaponoside A*).
Liu, Q.-H. *et al.*, *Arch. Pharmacal Res.*, 2006, **29**, 1109-1113 (*7,7′-Dihydroxybursehernin*).
Cabanillas, B.J. *et al.*, *J. Nat. Prod.*, 2010, **73**, 1884-1890 (*Sanguinolignans A-D*).
Ashour, A. *et al.*, *Molecules*, 2013, **18**, 13823-13830 (*Zea mays 3′,4′-di-Me ether*)

3,3′,4,4′,7′,8-Hexahydroxy-8,8′-lignan-9,9′-olide H41

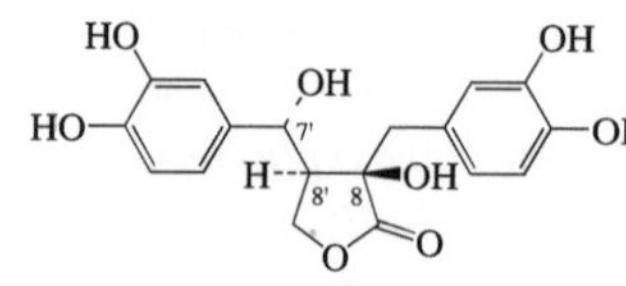

(7′R,8R,8′S)-form

$C_{18}H_{18}O_8$ 362.334

(7′R,8R,8′S)-form
3,3′,7′-Tri-Me ether
4,4′,8-Trihydroxy-3,3′,7′-trimethoxy-8,8′-lignan-9,9′-olide.
Wikstronin B
Constit. of the roots of *Wikstroemia indica*. Gum. $[\alpha]_D^{20}$ +63.3 (c, 0.1 in MeOH). λ_{max} 228 (log ε 3.12); 284 (log ε 2.69) (MeOH)

(7′R,8S,8′R)-form
3,3′-Di-Me ether
4,4′,7′,8-Tetrahydroxy-3,3′-dimethoxy-8,8′-lignan-9,9′-olide.
7′-Hydroxynortrachelogenin
[252333-68-9]
$C_{20}H_{22}O_8$ 390.388
Constit. of *Taxus mairei*. Oil. $[\alpha]_D^{29}$ +12 (c, 0.9 in $CHCl_3$)

3,3′,7′-Tri-Me ether
Wikstronin A
$C_{21}H_{24}O_8$ 404.415
Constit. of the roots of *Wikstroemia indica*. Gum. $[\alpha]_D^{20}$ -58.1 (c, 0.08 in MeOH). λ_{max} 228 (log ε 2.87); 284 (log ε 2.54) (MeOH)

7-Ketone, 3,3′-di-Me ether
4,4′,8-Trihydroxy-3,3′-dimethoxy-7-oxo-8,8′-lignan-9,9′-olide.
8-Hydroxyoxomatairesinol
[293744-17-9]
$C_{20}H_{20}O_8$ 388.372
Constit. of *Tsuga heterophylla*. Pale yellow powder. $[\alpha]_D^{25}$ +97.3 (c, 0.38 in MeOH). λ_{max} 201 (log ε 4.62); 232 (log ε 4.29); 281 (log ε 4.08); 309 (log ε 3.94) (MeOH).

Yang, S.-J. *et al.*, *J. Chin. Chem. Soc. (Taipei)*, 1999, **46**, 811-818 (*7′-Hydroxynortrachelogenin*).
Kawamura, F. *et al.*, *Phytochemistry*, 2000, **54**, 439-444 (*8-Hydroxyoxomatairesinol*).
Chang, H. *et al.*, *Fitoterapia*, 2017, **121**, 31-37 (*Wikstronins A,B*)

3,3′,4,4′,8,8′-Hexahydroxy-8,8′-lignan-9,9′-olide H42

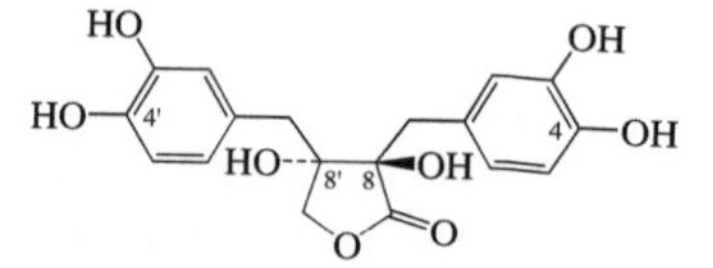

(8R,8′S)-form

$C_{18}H_{18}O_8$ 362.334

(8R,8′S)-form
3′,4′-Methylene, 3,4-di-Me ether
8,8′-Dihydroxy-3,4-dimethoxy-3′,4′-methylenedioxy-8,8′-lignan-9,9′-olide
[1574555-99-9]
$C_{21}H_{22}O_8$ 402.399
Constit. of wood of *Cunninghamia konishii*. Yellowish oil. $[\alpha]_D^{26}$ -124.7 (c, 0.4 in $CHCl_3$). λ_{max} 207 (log ε 4.40); 230 (log ε 4.05); 283 (log ε 3.81) (MeOH)

(8S,8′R)-form
3,3′-Di-Me ether, 4-O-β-D-glucopyranoside
$C_{26}H_{32}O_{13}$ 552.529
Constit. of *Chamaecyparis obtusa* var. *breviramea*. Oil. $[\alpha]_D^{10}$ -106.9 (c, 0.45 in MeOH)

3′,4′-Methylene, 3-Me ether
4,8,8′-Trihydroxy-3-methoxy-3′,4′-methylenedioxy-8,8′-lignan-9,9′-olide.
Calomacrol A.
8,8′-Dihydroxypluviatolide
[1441421-91-5]
$C_{20}H_{20}O_8$ 388.372
Constit. of *Calocedrus macrolepis* and roots of *Cunninghamia lanceolata*. Oil or amorph. powder. $[\alpha]_D^{21}$ -202.4 (c, 0.09 in MeOH) (8,8′-Dihydroxypluviatolide). $[\alpha]_D^{25}$ -73.1 (c, 0.6 in $CHCl_3$) (Calomacrol A). Rel. config. only determined for Calomacrol A from *C. macrolepis* but the opt. rotns. of the two isolates are both strongly negative. λ_{max} 231 (log ε 4.21); 284 (log ε 4.0) (MeOH) (8,8′-Dihydroxypluviatolide). [neutral] λ_{max} 231 (ε 3321); 283 (ε 5696) (MeOH) (Calomacrol A)

(8R,8′S*)-form*

3,3′-Di-Me ether

4,4′,8,8′-Tetrahydroxy-3,3′-dimethoxy-8,8′-lignan-9,9′-olide.

Thujastandin

[19473-16-6]

$C_{20}H_{22}O_8$ 390.388

Constit. of *Juniperus rigida* and *Thuja standishii*. Gum. $[\alpha]_D^{25}$ -11.6 (c, 0.3 in MeOH). Config. not confirmed.

Woo, K.W. *et al., Arch. Pharmacal Res.*, 2011, **34**, 2043-2049 (*Thujastandin*).
Li, Z *et al., Chem. Nat. Compd. (Engl. Transl.)*, 2013, **49**, 229-231 (*Cunninghamia lanceolata constit*).
Wu, X.-D. *et al., Fitoterapia*, 2013, **85**, 154-160 (*Calomacrol A*).
Chang, C.-I. *et al., Nat. Prod. Commun.*, 2013, **8**, 805-806 (*Cunninghamia konishii constit*).
Xu, J. *et al., Nat. Prod. Commun.*, 2014, **9**, 215-216 (*Chamaecyparis glucoside*)

3,3′,4,4′,8′,9-Hexahydroxy-8,8′-lignan-9′,7-olide H43

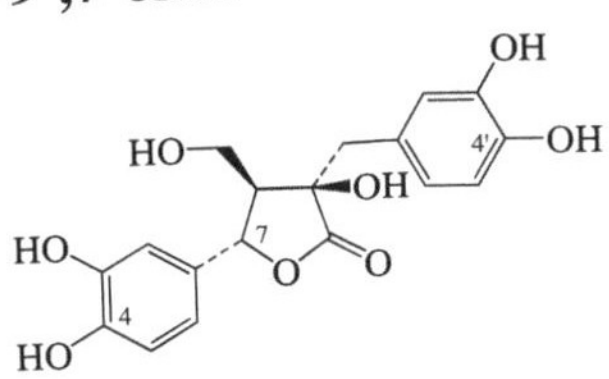

(7S,8R,8′R)-form

$C_{18}H_{18}O_8$ 362.334

(7S,8R,8′R)-form

3,3′-Di-Me ether

4,4′,8′,9-Tetrahydroxy-3,3′-dimethoxy-8,8′-lignan-9′,7-olide.

Cephafortin A

$C_{20}H_{22}O_8$ 390.388

Constit. of the twigs and leaves of *Cephalotaxus fortunei*. Gum. $[\alpha]_D^{27}$ +15 (c, 0.2 in MeOH). Abs. config. revised in 2020. λ_{max} 232 (log ε 4.01); 282 (log ε 3.63) (MeOH)

3,3′,4-Tri-Me ether

4′,8′,9-Trihydroxy-3,3′,4-trimethoxy-8,8′-lignan-9′,7-olide

[1313372-83-6]

$C_{21}H_{24}O_8$ 404.415

Constit. of *Acer* sp.. Amorph. powder. $[\alpha]_D^{25}$+17 (c, 1.5 in MeOH). Abs. config. revised in 2020

3,3′,4-Tri-Me ether, 4′-O-α-L-rhamnopyranoside

[1312614-75-7]

$C_{27}H_{34}O_{12}$ 550.557

Constit. of the wood of *Acer saccharum*. Amorph. solid. $[\alpha]_D$ -40.1 (c, 1.7 in MeOH)

3,3′,4,4′-Tetra-Me ether

8′,9-Dihydroxy-3,3′,4,4′-tetramethoxy-8,8′-ligan-9′,7-olide.

4,4′-Di-O-methylcephafortin A

$C_{22}H_{26}O_8$ 418.442

Constit. of the twigs and leaves of *Cephalotaxus fortunei*. Gum. $[\alpha]_D^{27}$ +16 (c, 0.1 in MeOH). Abs. config. revised in 2020. λ_{max} 233 (log ε 4.08); 279 (log ε 3.70) (MeOH)

(7S,8R,8′S)-form

3,3′,4-Tri-Me ether, 4′-O-(3-O-methyl-α-L-rhamnopyranoside)

[960399-02-4]

$C_{28}H_{36}O_{12}$ 564.584

Constit. of *Cephalotaxus koreana*. Amorph. solid. $[\alpha]_D^{24}$ -4.8 (c, 0.1 in MeOH). λ_{max} 227 (log ε 2.12); 277 (log ε 1.68) (MeOH).

Yoon, K.D. *et al., J. Nat. Prod.*, 2007, **70**, 2029-2032 (*Cephalotaxus koreana constit*).
Li, L. *et al., J. Agric. Food Chem.*, 2011, **59**, 7708-7716 (*Canadian maple syrup constit*).
Yoshikawa, K. *et al., J. Nat. Med.*, 2011, **65**, 191-193 (*Acer saccharum constit*).
Zhao, J.-X. *et al., J. Nat. Prod.*, 2017, **80**, 356-362 (*Cephafortin A, Dimethylcephafortin A*).
Vakiti, J.R. *et al., Org. Lett.*, 2020, **22**, 3345-3350 (*synth, abs config*)

3′,4,4′,5′,7,7′-Hexahydroxy-8,8′-lignan-9,9′-olide H44

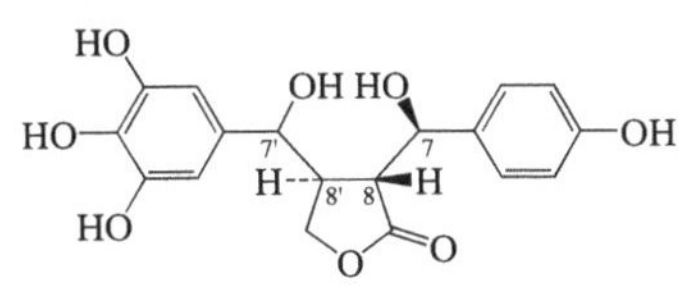

(7R,8R,8′S)-form

$C_{18}H_{18}O_8$ 362.334

(7R,8R,8′S)-form

7′-Ketone

3′,4,4′,5′,7-Pentahydroxy-7′-oxo-8,8′-lignan-9,9′-olide

7′-Ketone, 3′,5′-di-Me ether

4,4′,7-Trihydroxy-3′,5′-dimethoxy-7′-oxo-8,8′-lignan-9,9′-olide.

Tadehaginosin

[1650566-28-1]

$C_{20}H_{20}O_8$ 388.372

Constit. of the aerial parts of *Tadehagi triquetrum*. Increases glucose consumption by HepG2 cells. Oil. $[\alpha]_D^{20}$ -19 (c, 0.01 in MeOH). λ_{max} 307 (log ε 4.08) (MeOH)

(7R,7′ξ,8R*,8′S)-form*

3′,5′-Di-Me ether

4,4′,7,7′-Tetrahydroxy-3′,5′-dimethoxy-8,8′-lignan-9,9′-olide

3′,5′-Di-Me ether, 7′-O-β-D-glucopyranoside

Bambulignan A

[1432625-95-0]

$C_{26}H_{32}O_{13}$ 552.529

Constit. of *Bambusa tuldoides*. Amorph. powder. $[\alpha]_D$ +3.78 (c, 0.6 in MeOH). λ_{max} 231; 275 (MeOH).

Sun, J. *et al., J. Agric. Food Chem.*, 2013, **61**, 4556-4562 (*Bambulignan A*).
Wu, J. *et al., Nat. Prod. Res.*, 2015, **29**, 1723-1727 (*Tadehaginosin, activity*)

2,3,3′,4,4′,6-Hexahydroxy-8,8′-lign-7-ene-9,9′-dioic acid H45

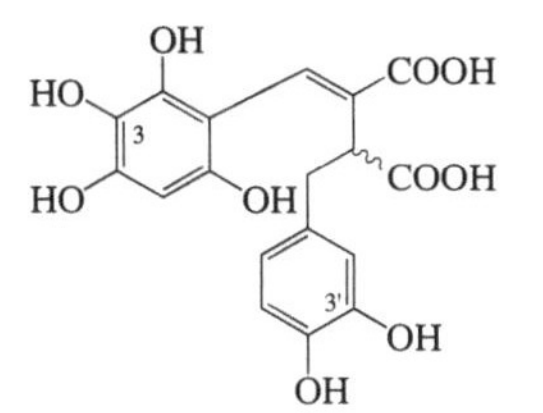

$C_{18}H_{16}O_{10}$ 392.316

(7E,8′ξ)-form

3,4:3′,4′-Bis(methylene), 2,6-di-Me ether

2,6-Dimethoxy-3,4:3′,4′-bis(methylenedioxy)-8,8′-lign-7-ene-9,9′-dioic acid

3,4:3′,4′-Bis(methylene), 2,6-di-Me ether, di-Me ester

2,6-Dimethoxydehydroheliobuphthalmin

$C_{24}H_{24}O_{10}$ 472.446

Constit. of *Hypoestes verticillaris*. Exhibits mild antiplasmodial activity. Cryst.. Mp 141 - 143°. $[\alpha]_D^{23}$ -184 (c, 0.1 in CH_2Cl_2).

Omole, R.A. *et al., Phytochem. Lett.*, 2019, **30**, 194-200 (*Hypoestes verticillaris constit, activity*)

2,3,3′,4,4′,6-Hexahydroxy-8,8′-lign-7-en-9,9′-olide H46

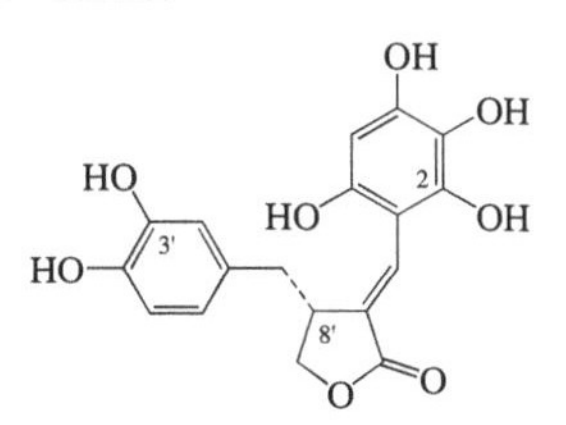

4-(3,4-Dihydroxybenzyl)dihydro-3-(2,3,4,6-tetrahydroxybenzylidene)-2(3H)-furanone

$C_{18}H_{16}O_8$ 360.318

(7E,8′R)-form

3,4:3′,4′-Bis(methylene), 2,6-di-Me ether

2,6-Dimethoxy-3,4:3′,4′-bis(methylenedioxy)-8,8′-lign-7-en-9,9′-olide. ***2,6-Dimethoxysavinin***

$C_{22}H_{20}O_8$ 412.394

Constit. of *Hypoestes verticillaris*. Exhibits mild antiplasmodial activity. Light yellow solid. $[\alpha]_D^{24}$ -499 (c, 2.0 in CH_2Cl_2). λ_{max} 258 (log ε 0.7); 293 (log ε 0.76); 340 (log ε 0.6) (no solvent reported).

Omole, R.A. *et al.*, *Phytochem. Lett.*, 2019, **30**, 194-200 (*2,6-Dimethoxysavinin, activity*)

3,3′,4,4′,5,5′-Hexahydroxy-8,8′-lign-7-en-9,9′-olide H47

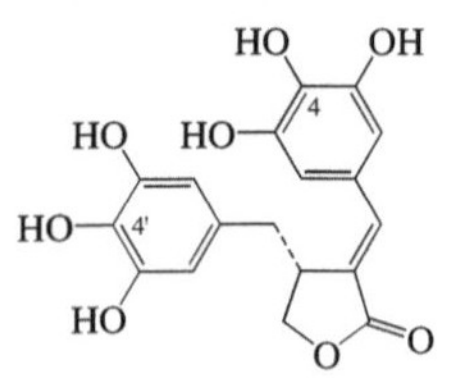

(7E,8′R)-form

$C_{18}H_{16}O_8$ 360.318

(7E,8′R)-form

3,3′,4′,5,5′-Penta-Me ether

4-Hydroxy-3,3′,4′,5,5′-pentamethoxy-8,8′-lign-7-en-9,9′-olide. ***Piperphilippinin III***

[949102-28-7]

$C_{23}H_{26}O_8$ 430.453

Constit. of the stems of *Piper philippinum*. Shows anticoagulant props.. Prisms (hexane/EtOAc). Mp 90 - 91°. $[\alpha]_D^{22}$ -52 (c, 0.15 in $CHCl_3$). λ_{max} 250 (sh); 265 (sh); 336 (log ε 4.22) (MeOH). [base] λ_{max} 265 (log ε 4.22); 406 (log ε 4.34) (MeOH/KOH)

(7E,8′S)-form

3,3′,4′,5,5′-Penta-Me ether

[171565-72-3]

Synthetic. Prisms (EtOAc/hexane). Mp 105.5 - 107°. $[\alpha]_D^{24}$ +78.3 (c, 1.13 in $CHCl_3$)

3′,4′-Methylene, 3,5,5′-tri-Me ether

4-Hydroxy-3,3′,5-trimethoxy-4′,5′-methylenedioxy-8,8′-lign-7-en-9,9′-olide

[878142-51-9]

$C_{22}H_{22}O_8$ 414.41

Constit. of *Peperomia duclouxii*. Gum. $[\alpha]_D^{25}$ +65.7 (c, 0.41 in $CHCl_3$). λ_{max} 244; 329 ($CHCl_3$).

Tanaka, M. *et al.*, *Tetrahedron*, 1995, **51**, 11703-11724 (*7E,8′S-penta-Me ether, synth*).

Li, N. *et al.*, *J. Nat. Prod.*, 2006, **69**, 234-239 (*Peperomia duclouxii constit*).

Chen, Y.-C. *et al.*, *Phytochemistry*, 2007, **68**, 2101-2111 (*Piperphilippinin III*)

3,3′,4,4′,5,5′-Hexahydroxy-9-nor-8,8′-lign-7′-en-9′,7-olide H48

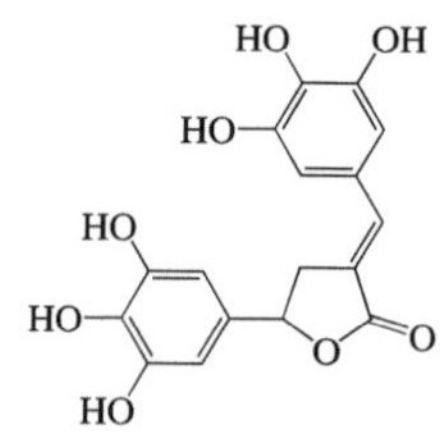

Dihydro-3-(3,4,5-trihydroxybenzylidene)-5-(3,4,5-trihydroxyphenyl)-2(3H)-furanone

$C_{17}H_{14}O_8$ 346.291

(7ξ,7′E)-form

3,3′,5,5′-Tetra-Me ether

4,4′-Dihydroxy-3,3′,5,5′-tetramethoxy-9-nor-8,8′-lign-7′-en-9′,7-olide. ***Descurainolide B***

[848138-65-8]

$C_{21}H_{22}O_8$ 402.399

Constit. of the seeds of *Descurainia sophia*. Needles. Mp 201 - 203°. $[\alpha]_D^{20}$ +2.3 (c, 0.37 in MeOH).

Sun, K. *et al.*, *Chem. Pharm. Bull.*, 2004, **52**, 1483-1486 (*Descurainolide B*)

Hexakis(demethyl)polemannone C H49

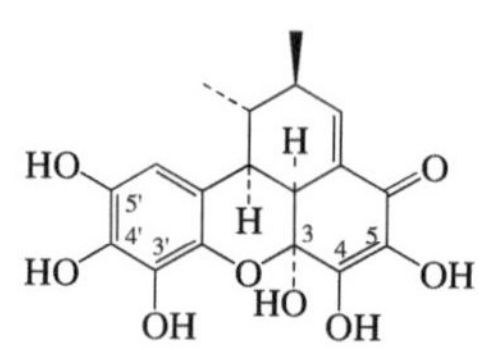

Relative Configuration

2,7′-Cyclo-2′,3-epoxy-2,3-dihydro-3,3′,4,4′,5,5′-hexahydroxy-8,8′-lign-1(7)-en-6(1H)-one

$C_{18}H_{18}O_8$ 362.334

Lignan numbering shown. Related to Bis(demethylene)sauchinone, B27. All Polemannones were racemates

Hexa-Me ether

2,7′-Cyclo-2′,3-epoxy-2,3-dihydro-3,3′,4,4′,5,5′-hexamethoxy-8,8′-lign-1(7)-en-6(1H)-one. ***Polemannone C.*** *4,4′,5,5′-Tetramethoxypolemannone*

[111010-29-8]

$C_{24}H_{30}O_8$ 446.496

Constit. of the roots of *Polemannia montana*. Cryst.. Mp 140°

3,4-Methylene, 3′,4′,5,5′-tetra-Me ether

2,7′-Cyclo-2′,3-epoxy-2,3-dihydro-3′,4′,5,5′-tetramethoxy-3,4-methylenedioxy-8,8′-lign-1(7)-en-6(1H)-one. ***Polemannone A.*** *4,5-Dimethoxy-4′,5′-methylenedioxypolemannone*

[110979-07-2]

$C_{23}H_{26}O_8$ 430.453

Constit. of the roots of *Polemannia montana*. Cryst.. Mp 197°

3,4:4′,5′-Bis(methylene), 3′,5-di-Me ether

2,7′-Cyclo-2′,3-epoxy-2,3-dihydro-3′,5-dimethoxy-4,5:4′,5′-bis(methylenedioxy)-8,8′-lign-1(7)-en-6(1H)-one. ***Polemannone B.*** *4,5;4′,5′-Bismethylenedioxypolemannone*

[111010-28-7]

$C_{22}H_{22}O_8$ 414.41

Constit. of the roots of *Polemannia montana*. Cryst.. Mp 170°.

Jakupovic, J. *et al.*, *Phytochemistry*, 1987, **26**, 2427-2429 (*Polemannia montana constits*).

Fadeyi, O.O. *et al.*, *Tetrahedron Lett.*, 2009, **50**, 3084-3087 (*Polemannones B,C, synth*).

Aka, E.C. *et al.*, *J. Org. Chem.*, 2019, **84**, 14101-14112 (*Polemannone B, synth*)

2,2′,4,4′,5,5′-Hexamethoxy-7,7′-cyclolignan H50

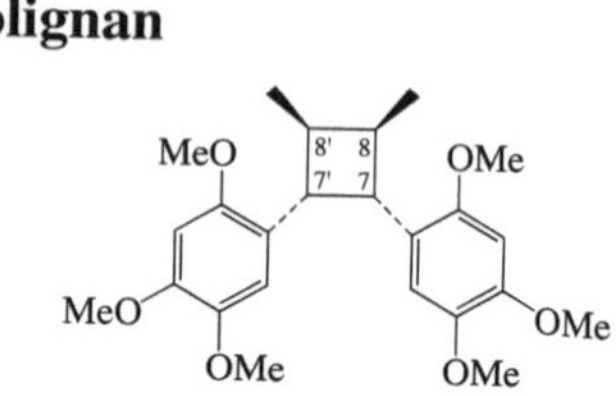

(7α,7′α,8β,8′β)-form

1,2-Dimethyl-3,4-bis(2,4,5-trimethoxyphenyl)cyclobutane. *1,1′-(3,4-Dimethyl-1,2-cyclobutanediyl) bis[2,4,5-trimethoxybenzene],* CAS

$C_{24}H_{32}O_6$ 416.514

(7α,7′α,8β,8′β)-form

Heterotropan [70280-35-2]

Constit. of *Heterotropa takaoi* (preferred genus name *Asarum*). Also from *Piper cubeba*. Viscous liq.. Opt. inactive (*meso-*)

(7β,7′α,8α,8′β)-form

Andamanicin [130323-08-9]

Constit. of *Perilla frutescens* and *Piper sumatranum* var. *andamanica*. Inhibitor of NO synthase and tumour necrosis factor α. Powder ($CHCl_3$ or diisopropyl ether). Mp 110 - 112° (96-98°). $[\alpha]_D^{20}$ +2.8 (c, 0.5 in $CHCl_3$). Chiral molecule, may be racemic. Config. revised in 1993. λ_{max} 204 (ε 69183); 231 (ε 18620); 291 (ε 10715) (EtOH)

(7β,7′α,8β,8′α)-form

Magnosalin [81861-74-7]

Constit. of *Magnolia salicifolia*, *Piper cubeba* and *Perilla frutescens*. Angiogenesis inhibitor. Cryst. (Et_2O/hexane). Mp 98 - 99°. $[\alpha]_D^{23}$ -4.6 (c, 0.25 in $CHCl_3$). Chiral molecule, may be racemic. Config. revised in 1993. λ_{max} 208 (ε 40738); 231 (ε 16980); 298 (ε 10715) (MeOH) (Berdy)

(7ξ,7′ξ,8ξ,8′ξ)-form

Acoradin. Bisasaricin [73036-51-8]

Constit. of *Acorus calamus*, *Acorus gramineus*, *Cupressus dupreziana* and *Cupressus sempervirens*. Prisms (Et_2O/petrol). Mp 101°. $[\alpha]_D$ 0 ($CHCl_3$).

Patra, A. *et al.*, *Indian J. Chem., Sect. B*, 1979, **17**, 412-414 (*Acoradin*).
Yamamura, S. *et al.*, *Bull. Chem. Soc. Jpn.*, 1982, **55**, 3573-3579 (*Heterotropan*).
Kikuchi, T. *et al.*, *Chem. Pharm. Bull.*, 1983, **31**, 1112-1114 (*Magnosalin*).
Badheka, L.P. *et al.*, *Phytochemistry*, 1987, **26**, 2033-2036 (*Piper cubeba constits*).
Kadota, S. *et al.*, *Tetrahedron Lett.*, 1987, **28**, 2857-2860 (*synth*).
Kimura, M. *et al.*, *Int. Arch. Allergy Appl. Immunol.*, 1990, **93**, 365-370 (*activity*).
Malhotra, S. *et al.*, *Phytochemistry*, 1990, **29**, 2733-2734 (*Andamanicin*).
Mahindru, R.N. *et al.*, *Phytochemistry*, 1993, **32**, 1073-1075 (*cryst struct*).
Takahashi, K. *et al.*, *Phytother. Res.*, 1996, **10**, 42-48 (*Magnosalin, activity*).
Ryu, J.-H. *et al.*, *Bioorg. Med. Chem. Lett.*, 2002, **12**, 649-651 (*Perilla frutescens constits, activity*)

2,2′,4,4′,5,5′-Hexamethoxy-9,9′-dinor-7,7′-cyclolignan H51

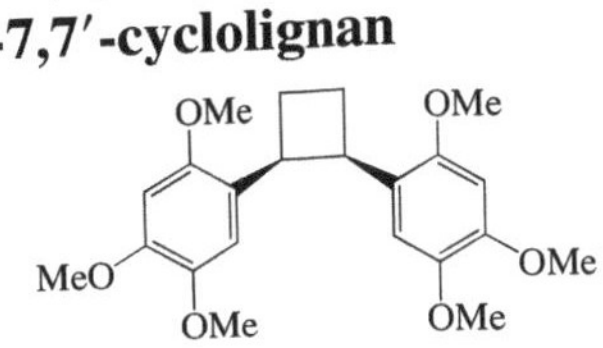

(1*RS*,2*SR*)-*form*

1,1′-(1,2-Cyclobutanediyl)bis[2,4,5-trimethoxybenzene], CAS. *1,2-Bis(2,4,5-trimethoxyphenyl)cyclobutane*

$C_{22}H_{28}O_6$ 388.46

Norlignan

(7RS,7′SR)-form

Pellucidin A. Pachypophyllin [114542-60-8]

Constit. of *Peperomia pellucida* and *Pachypodanthium staudtii*. Amorph. solid (hexane/EtOAc) or cryst. (Et_2O/petrol). Mp 109.6 - 110.5°. Mp 158°. Higher Mp and uv refer to Pachpophyllin (crystalline). λ_{max} 225 (log ε 4.11); 285 (log ε 3.85) (MeOH).

Ngadjui, B.T. *et al.*, *Phytochemistry*, 1989, **28**, 231-234 (*Pachypophyllin*).
Bayma, J. de C. *et al.*, *Phytochemistry*, 2000, **55**, 779-782 (*Pellucidin A*)

Hinokiresinol H52

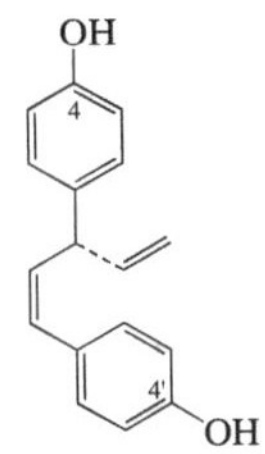

(*R*)-(*Z*)-*form*

4,4′-(3-Ethenyl-1-propene-1,3-diyl)bisphenol, CAS. *1,3-Bis(4-hydroxyphenyl)-1,4-pentadiene. 9′-Nor-7,8′-ligna-7′,8-diene-4,4′-diol. 4,4′-Dihydroxy-9′-nor-7,8′-neoligna-7′,8-diene*

[4825-69-8] [16453-39-7] [79056-21-6] [79056-22-7] [175520-27-1] (ξ,*E*-form) [230292-85-0] (ξ,*Z*-form) [1208867-86-0] (*R*,*E*-form)

$C_{17}H_{16}O_2$ 252.313

Lignan numbering shown. Stereochem. of the various isomers has been confused. This appears to have been clarified (2005)

(R,Z)-form

cis-Hinokiresinol. (-)-Nyasol

[96895-25-9]

Constit. of *Anemarrhena asphodeloides*, *Hypoxis nyasica* and *Sonneratia ovata*. Estrogen agonist. Synergistic with azole antifungals on *Candida albicans*. Broad spectrum inhibitor of eicosanoid and NO prodn.. Oil. $[\alpha]_D^{22}$ -198 (c, 0.13 in MeOH). Nat. prod. is a partial racemate. λ_{max} 202 (ε 37400); 231 (ε 10700); 259 (ε 15700) (EtOH)

4-O-β-D-Glucopyranoside

Mononyasin A

[126431-35-4]

$C_{23}H_{26}O_7$ 414.454

Constit. of *Hypoxis nyasica*. Cryst. (EtOAc). Mp 169 - 170°. $[\alpha]_D^{20}$ -166 (c, 0.6 in MeOH)

4-O-[β-D-Apiofuranosyl-(1→6)-β-D-glucopyranoside], 4′-O-β-D-glucopyranoside

Nyaside

[126595-87-7]

$C_{34}H_{44}O_{16}$ 708.71

Constit. of *Hypoxis angustifolia* and *Hypoxis nyasica*. Cryst. (EtOH/EtOAc). Mp 108 - 110°. $[\alpha]_D^{20}$ -156 (MeOH). λ_{max} 208 (log ε 4.32); 257 (log ε 4.18); 300 (sh) (MeOH)

4′-O-β-D-Glucopyranoside

Mononyasin B

[126394-53-4]

$C_{23}H_{26}O_7$ 414.454

Constit. of *Hypoxis nyasica*. Cryst. (EtOAc/hexane). Mp 147 - 149°. $[\alpha]_D^{20}$ -198 (c, 0.6 in MeOH)

Di-O-β-D-glucopyranoside

Nyasoside

[96848-98-5]

$C_{29}H_{36}O_{12}$ 576.595

Constit. of rhizome of *Hypoxis nyasica*. Cryst. (MeOH/Me_2CO). Mp 110 - 112°. $[\alpha]_D^{20}$ -87 (c, 0.8 in MeOH)

4′-Me ether

4′-Methoxy-9′-nor-7,8′-ligna-7′,8-dien-4-ol. 4-Hydroxy-4′-methoxy-9′-nor-7,8′-ligna-7′,8-diene. 4′-O-Methylhinokiresinol. 4′-O-Methylnyasol. ***Ellisinin A***

[79004-25-4]

$C_{18}H_{18}O_2$ 266.34

Constit. of *Anemarrhena asphodeloides*, *Cremanthodium ellisii* and *Sonneratia ovata*. Amorph. powder or gum. Mp 77 - 79°. $[\alpha]_D^{25}$ -42.2 (c, 0.64 in $CHCl_3$). λ_{max} 244 (log ε 4.78); 285 (log ε 4.23) ($CHCl_3$)

3′-Hydroxy, 4′-Me ether

4′-Methoxy-9′-nor-7,8′-ligna-7′,8-diene-3′,4-diol. 3′,4-Dihydroxy-4′-methoxy-9′-nor-7,8′-ligna-7′,8-diene. 3′-Hydroxy-4′-O-methylnyasol. 3′-Hydroxy-4′-O-methylhinokiresinol. ***Sonnerphenolic B***

[1627516-10-2]

$C_{18}H_{18}O_3$ 282.339

Constit. of the mangrove-derived *Sonneratia ovata*. Yellowish oil. $[\alpha]_D^{25}$ -282.8 (c, 0.18 in MeOH)

(S,E)-form

trans-Hinokiresinol [17676-24-3]

Constit. of *Chamaecyparis obtusa*. Estrogen agonist. Mp 102 - 103°. [α] D -3 (c, 1 in Me_2CO). Partial racemate. λ_{max} 191 (ε 2990); 204 (ε 32500); 265 (ε 22000) (EtOH)

4-O-β-D-Glucopyranoside
[2166163-80-8]
$C_{23}H_{26}O_7$ 414.454
Constit. of the twigs of *Cephalotaxus oliveri*. Amorph. powder. $[\alpha]_D^{20}$ -16.6 (c, 0.05 in MeOH). λ_{max} 206 (log ε 4.30) (MeOH)

4′-O-β-D-Glucopyranoside
[2166163-81-9]
$C_{23}H_{26}O_7$ 414.454
Constit. of the twigs of *Cephalotaxus oliveri*. Amorph. powder. $[\alpha]_D^{20}$ -38.3 (c, 0.13 in MeOH). λ_{max} 208 (log ε 4.04); 261 (log ε 4.01) (MeOH)

(S,Z)-form

(+)-Nyasol [185020-38-6]
Constit. of *Anemarrhena asphodeloides*, *Asparagus cochinchinensis*, *Chamaecyparis obtusa*, *Cryptomeria japonica*, and *Hypoxis* spp.. Shows strong estrogen receptor binding activity. Leukotriene B_4 binding inhibitor. Exhibits cytotoxic activities. Mp 102 - 103°. $[\alpha]_D^{22}$ +170 (c, 0.06 in MeOH)

4-O-β-D-Glucopyranoside
[2166163-78-4]
Constit. of the twigs of *Cephalotaxus oliveri*. Amorph. powder. $[\alpha]_D^{20}$ +119 (c, 0.03 in MeOH). λ_{max} 207 (log ε 4.02); 257 (log ε 3.76) (MeOH)

4′-O-β-D-Glucopyranoside
[2166163-79-5]
Constit. of the twigs of *Cephalotaxus oliveri*. $[\alpha]_D^{20}$ +95 (c, 0.1 in MeOH). λ_{max} 206 (log ε 3.93); 252 (log ε 3.72) (MeOH)

4′-Me ether
[851884-74-7]
$C_{18}H_{18}O_2$ 266.34
Constit. of *Asparagus filicinus*. Brown oil. $[\alpha]_D^{20}$ +162 (c, 0.1 in MeOH). λ_{max} 241 (log ε 4.57); 286 (log ε 4.12) (MeOH)

Di-Me ether
[4825-70-1]
Mp 64°. [α] D +8.4

3-Methoxy
3-(4-Hydroxy-3-methoxyphenyl)-1-(4-hydroxyphenyl)-1,4-pentadiene. 3-Methoxy-9′-nor-7,8′-ligna-7′,8-diene-4,4′-diol. 4,4′-Dihydroxy-3-methoxy-9′-nor-7,8′-ligna-7′,8-diene. ***3″-Methoxynyasol***
[835613-05-3]
[90332-37-9]
$C_{18}H_{18}O_3$ 282.339
Constit. of the roots of *Asparagus gobicus*, *Agathis* sp. and *Chamaecyparis obtusa*. Gum. $[\alpha]_D^{21}$ +55.8 (c, 4.8 in $CHCl_3$)

(ξ,Z)-form

3-Hydroxy, 4-Me ether
4-Methoxy-9′-nor-7,8′-ligna-7′,8-diene-3,4′-diol. 3,4′-Dihydroxy-4-methoxy-9′-nor-7,8′-ligna-7′,8-diene. 3″-Hydroxy-4″-O-methylhinokiresinol. ***3″-Hydroxy-4″-O-methylnyasol***
[671240-65-6]
$C_{18}H_{18}O_3$ 282.339
Constit. of the roots of *Asparagus cochinchinensis*. Powder. $[\alpha]_D^{20}$ +85.8 (c, 0.09 in MeOH). λ_{max} 198 (log ε 3.67); 255 (log ε 0.89) (no solvent reported)

3-Methoxy
3″-Methoxyhinokiresinol
[671774-39-3]
$C_{18}H_{18}O_3$ 282.339
Constit. of *Asparagus cochinchinensis* and *Agathis* sp.. Stereochem. of the different isolates of 3″-Methoxynyasol is not clear (see also above).

Hirose, Y. *et al., Tetrahedron Lett.*, 1965, **6**, 3665-3668 (*Hinokiresinol, S,E-form*).
Beracierta, A.P. *et al., Tetrahedron Lett.*, 1976, **17**, 2367-2370 (*di-Me ether, synth*).
Fujita, N. *et al., Mokuzai Gakkaishi*, 1984, **30**, 264-268 (*3″-Methoxynyasol*).
Marini-Bettolo, G.B. *et al., Tetrahedron*, 1985, **41**, 665-670 (*Nyasin, Nyasol*).
Galeffi, C. *et al., Tetrahedron*, 1987, **43**, 3519-3522 (*Nyasol*).
Galeffi, C. *et al., Gazz. Chim. Ital.*, 1989, **119**, 565-568 (*Nyaside*).
Messana, I. *et al., Phytochemistry*, 1989, **28**, 2807 (*Mononyasins*).
Galeffi, C. *et al., Gazz. Chim. Ital.*, 1990, **120**, 215 (*Nyaside, erratum*).
Muraoka, O. *et al., J. Chem. Soc., Perkin Trans. 1*, 1996, 405-411 (*(+)-di-Me ether, synth*).
Tsui, W.-Y. *et al., Phytochemistry*, 1996, **43**, 1413-1415 (*(+)-Nyasol*).
Jeong, S.-J. *et al., Planta Med.*, 1999, **65**, 367-368 (*4′-O-Methylhinokiresinol*).
Lee, H.J. *et al., Planta Med.*, 1999, **65**, 391 (*Anemarrhena asphodeloides constit, activity*).
Minami, E. *et al., Chem. Pharm. Bull.*, 2000, **48**, 389-392 (*abs config, bibl*).
Su, B.-N. *et al., Phytochemistry*, 2000, **53**, 1103-1108 (*4′-O-Methylhinokiresinol*).
Iida, Y. *et al., Planta Med.*, 2000, **66**, 435-438 (*activity*).
Park, H.J. *et al., Phytochemistry*, 2003, **64**, 997-1001 (*Nyasol, isol, activity*).
Zhang, H.-J. *et al., J. Nat. Prod.*, 2004, **67**, 194-200 (*3″-Hydroxy-4-O-methylnyasol*).
Yang, C.-X. *et al., Planta Med.*, 2004, **70**, 446-451 (*3″-Methoxynyasol, Nyasol, activity*).
Li, Y.F. *et al., J. Asian Nat. Prod. Res.*, 2005, **7**, 43-47 (*4′-Me ethers*).
Lassen, P.R. *et al., J. Nat. Prod.*, 2005, **68**, 1603-1609 (*Nyasol, synth, conformn, abs config, bibl*).
Lim, H. *et al., Arch. Pharmacal Res.*, 2009, **32**, 1509-1514 (*(-)-Nyasol, activity*).
Kwon, J. *et al., Bull. Korean Chem. Soc.*, 2013, **34**, 1247-1249 (*(±)-form, synth*).
Hamilton, J.Y. *et al., J. Am. Chem. Soc.*, 2013, **135**, 994-997 (*(-)-form, synth*).
Nguyen, T.-H.-T. *et al., Bioorg. Med. Chem. Lett.*, 2015, **25**, 2366-2371 (*Sonneratia ovata constits*).
Fang, Y. *et al., Synth. Commun.*, 2015, **45**, 137-142 (*(±)-form, synth*).
Meng, F.-C. *et al., Tetrahedron: Asymmetry*, 2017, **28**, 1686-1689 (*Cephalotaxus oliveri constits*)

Homoegonol H53

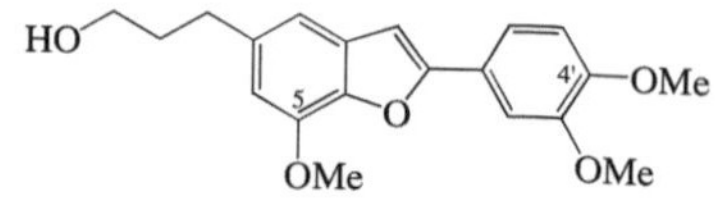

2-(3,4-Dimethoxyphenyl)-7-methoxy-5-benzofuranpropanol, CAS. *4,7′-Epoxy-3′,4′,5-trimethoxy-9′-nor-3,8′-lignan-9-ol. 4,7′-Epoxy-9-hydroxy-3′,4′,5-trimethoxy-9′-nor-3,8′-neolignan*
[17375-66-5]
$C_{20}H_{22}O_5$ 342.391
Constit. of *Styrax officinalis*, *Styrax pohlii* and *Machilus glaucescens*. Weak inhibitor of COX-1 and COX-2. Cryst. (C_6H_6/MeOH). Mp 120 - 122°

O-β-D-Glucopyranoside
Homoegonol glucoside
[325791-19-3]
$C_{26}H_{32}O_{10}$ 504.532
Constit. of *Styrax ferrugineus* and *Styrax pohlii*. Moderate inhibitor of COX-1. Powder. $[\alpha]_D^{25}$ -22.2 (c, 0.8 in MeOH)

O-[β-D-Xylopyranosyl-(1→6)-β-D-glucopyranoside]
Homoegonol primeveroside. ***Styraxlignolide A***
[857289-36-2]
$C_{31}H_{40}O_{14}$ 636.647
Constit. of the stem bark of *Styrax japonica*. Amorph. powder. $[\alpha]_D^{22}$ -53.7 (c, 0.8 in MeOH). λ_{max} 217

(log ε 4.39); 312 (log ε 4.38) (MeOH)

O-[β-D-Glucopyranosyl-(1→6)-β-D-glucopyranoside]

Homoegonol gentiobioside

[77702-00-2]

$C_{32}H_{42}O_{15}$ 666.673

Isol. from *Styrax officinalis* and *Styrax pohlii*. Weak inhibitor of COX-2. Cryst. (EtOAc/petrol) (as hepta-Ac). Mp 131 - 133° (hepta-Ac). $[\alpha]^{22}_{D}$ -14.2 (c, 1.1 in $CHCl_3$) (hepta-Ac)

Ac

Cryst. (C_6H_6/MeOH). Mp 90 - 91°

2-Methylbutanoyl

Homoegonol 2-methylbutanoate

[612058-46-5]

$C_{25}H_{30}O_6$ 426.509

Constit. of the seeds of *Styrax officinalis*. Cryst. (EtOH). Mp 62 - 65°. λ_{max} 215 (sh); 304; 313 (MeOH). [neutral] λ_{max} 216; 304; 314; 328 (EtOH)

Hexadecanoyl

Homoegonol palmitate. ***Machicenin***

[72430-55-8]

$C_{36}H_{52}O_6$ 580.806

Constit. of *Machilus glaucescens*. Cryst.. Mp 94°

4′,5-Di-O-de-Me, 5-O-β-D-glucopyranoside

Paeonibenzofuran

$C_{24}H_{28}O_{10}$ 476.478

Constit. of the roots of Edulis superba (horticultural cultivar of *Paeonia lactiflora*). Exhibits antiallergic activity. Amorph. powder. $[\alpha]^{20}_{D}$ -44.8 (c, 0.11 in MeOH).

Segal, R. *et al., J. Chem. Soc. C*, 1967, 2402-2404 (*Homoegonol*).
Schreiber, F.G. *et al., J. Chem. Soc., Perkin Trans. 1*, 1976, 1514-1518 (*Homoegonol, synth*).
Talapatra, B. *et al., J. Indian Chem. Soc.*, 1978, **55**, 1204-1208 (*Machicenin, synth*).
Anil, H. *Phytochemistry*, 1980, **19**, 2784-2786 (*gentiobioside*).
Mali, R.S. *et al., J. Chem. Res., Synop.*, 1998, 230-231 (*Homoegonol, synth*).
Pauletti, P.M. *et al., Phytochemistry*, 2000, **55**, 597-601 (*Homoegonol*).
Akgul, Y.Y. *et al., Phytochemistry*, 2003, **63**, 939-943 (*2-methylbutanoate*).
Min, B.-S. *et al., Planta Med.*, 2004, **70**, 1210-1215 (*Styraxlignolide A*).
Duan, X.-F. *et al., Synthesis*, 2010, 515-519 (*Homoegonol, synth*).
Bertanha, C.S. *et al., Nat. Prod. Res.*, 2012, **26**, 2323-2329 (*gentiobioside, activity*).
Bertanha, C.S. *et al., J. Herbs, Spices Med. Plants*, 2014, **20**, 394-401 (*activity*).
Shi, Y.-N. *et al., J. Nat. Med.*, 2016, **70**, 234-240 (*Paeonibenzofuran, activity*).
More, K.R. *et al., Tetrahedron*, 2016, **72**, 7496-7504 (*Homoegonol, synth*).
Sivaraman, A. *et al., J. Nat. Prod.*, 2020, **83**, 3354-3362 (*Homoegonol, synth*)

9a-Homo-4,8′-oxylignan-3,3′,4′,5,5′,7′,9,9′-octol H54

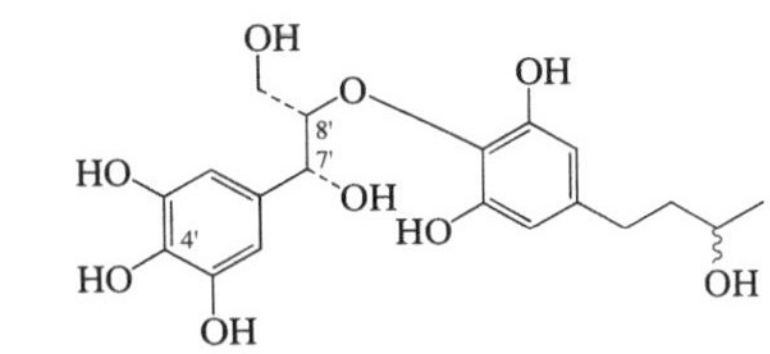

9a-Homo-3,3′,4′,5,5′,7′,9,9′-octahydroxy-4,8′-oxyneolignan

$C_{19}H_{24}O_9$ 396.392

(7′R,8′S,9ξ)-form

3,3′,5,5′-Tetra-Me ether

9a-Homo-3,3′,5,5′-tetramethoxy-4,8′-oxylignan-4′,7′,9,9′-tetrol

3,3′,5,5′-Tetra-Me ether, 4′-O-β-D-glucopyranoside

[2101222-64-2]

$C_{29}H_{42}O_{14}$ 614.641

Constit. of the twigs of *Litsea cubeba*. Exhibits a redn. of acetaminophen-induced HepG2 cell injury. Amorph. solid. $[\alpha]^{20}_{D}$ -10 (c, 0.1 in MeOH). λ_{max} 207 (log ε 4.12); 275 (log ε 0.14) (MeOH).

Wang, L.-Y. *et al., J. Nat. Prod.*, 2017, **80**, 1808-1818 (*Litsea cubeba constit, activity*)

Hordatine A H55

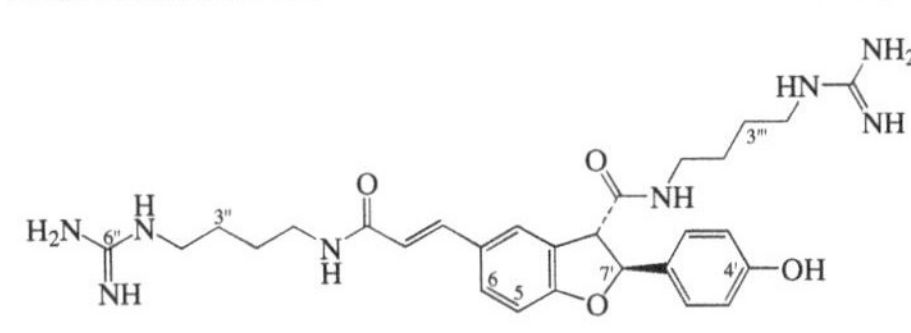

Absolute Configuration

[909118-83-8]

[7073-64-5] (*R,R*-form) [16995-91-8] ((±)-form) [47818-76-8] [96270-91-6] (unspecified stereochem.) [909118-82-7] (*2S,3R*-form)

$C_{28}H_{38}N_8O_4$ 550.664

Neolignan alkaloid. Found in beer. Alkaloid from *Hordeum vulgare*, *Hordeum bulbosum*, *Hordeum distichon*, *Hordeum jubatum* and *Hordeum murinum*. Antifungal agent. Muscarinic M3 receptor antagonist. α1A-Adrenoceptor antagonist. Gastric motility stimulant. Sol. H_2O. Mp 127 - 128° (as dipicrate). $[\alpha]^{23}_{D}$ +69 (c, 1 in H_2O). $[\alpha]^{20}_{D}$ +83.8 (c, 0.2 in MeOH) (hydrochloride salt). λ_{max} 229 (ε 20700); 298 (ε 18400); 307 (ε 18400) (MeOH) (Berdy). [base] λ_{max} 235 (ε 17000); 297 (ε 19400); 305 (ε 18700) (MeOH/NaOH) (Berdy)

4′-O-β-D-Glucopyranoside

Hordatine A β-D-glucoside. ***4′-O-β-D-Glucopyranosylhordatine A***

[887471-08-1]

$C_{34}H_{48}N_8O_9$ 712.805

Isol. from ungerminated grains of barley (*Hordeum vulgare*) and barley malt

4′-O-[α-D-Glucopyranosyl-(1→4)-β-D-glucopyranoside]

Hordatine A β-D-maltoside. ***4′-O-β-D-Maltosylhordatine A***

[1395383-30-8]

$C_{40}H_{58}N_8O_{14}$ 874.946

Isol. from acrospires in barley malt (*Hordeum vulgare*)

$N^{6''}$-Me

$N^{6''}$-Methylhordatine A

$C_{29}H_{40}N_8O_4$ 564.691

Isol. from the seeds of *Hordeum vulgare*

3″ξ-Hydroxy

3″-Hydroxyhordatine A. ***Hordatine A_1***

$C_{28}H_{38}N_8O_5$ 566.663

Isol. from the seeds of *Hordeum vulgare*

5-Hydroxy, 4′-O-α-D-glucopyranoside

$C_{34}H_{48}N_8O_{10}$ 728.804

Component of Hordatine M from barley (see above)

3″ξ,3‴ξ-Dihydroxy

Hordatine A_2

$C_{28}H_{38}N_8O_6$ 582.662

Isol. from the seeds of *Hordeum vulgare*

5-Methoxy

Hordatine B

[10502-21-3]

$C_{29}H_{40}N_8O_5$ 580.69

Found in beer. Alkaloid from *Hordeum vulgare*, *Hordeum bulbosum*, *Hordeum jubatum*, *Hordeum murinum* and *Hordeum spontaneum*. Antifungal agent. Phytotoxin. Sol. H_2O. Mp 132 - 135° (as dipicrate). $[\alpha]^{23}_{D}$ +54 (c, 1 in H_2O). Deriv. of 4,7′-Epoxy-3,8′-lign-7-ene-3′,4′,5,9,9′-pentol,

E163. λ_{max} 223 (ε 22000); 301 (ε 15500); 316 (ε 16200) (EtOH) (Berdy)

5-Methoxy, 4′-O-α-D-glucopyranoside
$C_{35}H_{50}N_8O_{10}$ 742.831
Present in barley (*Hordeum vulgare*) as an inseparable mixt. with Hordatine A glucoside, known as Hordatine M. $[\alpha]^{23}_{D}$ +60 (c, 1 in H_2O)

5-Methoxy, 4′-O-β-D-glucopyranoside
Hordatine B β-D-glucoside. ***4′-O-β-D-Glucopyranosylhordatine B***
[1395383-27-3]
$C_{35}H_{50}N_8O_{10}$ 742.831
Isol. from acrospires in barley malt (*Hordeum vulgare*)

5-Methoxy, N^6″-Me
N^6″-Methylhordatine B
$C_{30}H_{42}N_8O_5$ 594.717
Isol. from the seeds of *Hordeum vulgare*

5-Methoxy, 3″ξ-hydroxy
3″-Hydroxyhordatine B. ***Hordatine B₁***
$C_{29}H_{40}N_8O_6$ 596.689
Isol. from the seeds of *Hordeum vulgare*

5-Methoxy, 3″ξ,3‴ξ-dihydroxy
Hordatine B_2
$C_{29}H_{40}N_8O_7$ 612.688
Isol. from the seeds of *Hordeum vulgare*

3′,5-Dimethoxy
Hordatine C
[1700622-71-4]
$C_{30}H_{42}N_8O_6$ 610.716
Isol. from the seeds of *Hordeum vulgare*

3′,5-Dimethoxy, N^6″-Me
N^6″-Methylhordatine C
$C_{31}H_{44}N_8O_6$ 624.743
Isol. from the seeds of *Hordeum vulgare*

3′,5-Dimethoxy, 3″ξ-hydroxy
3″-Hydroxyhordatine C. ***Hordatine C_1***
[1704729-67-8]
$C_{30}H_{42}N_8O_7$ 626.715
Isol. from the seeds of *Hordeum vulgare*

3′,5-Dimethoxy, 3″ξ,3‴ξ-dihydroxy
Hordatine C_2
$C_{30}H_{42}N_8O_8$ 642.714
Isol. from the seeds of *Hordeum vulgare*

3′,5,6-Trimethoxy
Hordatine D
[1695572-75-8]
$C_{31}H_{44}N_8O_7$ 640.742
Isol. from the seeds of *Hordeum vulgare*

3′,5,6-Trimethoxy, 3″ξ-hydroxy
3″-Hydroxyhordatine D. ***Hordatine D_1***
[1704729-68-9]
$C_{31}H_{44}N_8O_8$ 656.741
Isol. from the seeds of *Hordeum vulgare*

(Z)-Isomer, 4′-O-β-D-glucopyranoside
4′-O-β-D-Glucopyranosyl-cis-hordatine A
[1443149-19-6]
$C_{34}H_{48}N_8O_9$ 712.805
Isol. from the seeds of *Hordeum vulgare*

(Z)-Isomer, 4′-O-[α-D-glucopyranosyl-(1→4)-β-D-glucopyranoside]
4′-O-β-D-Maltosyl-cis-hordatine A
[1443149-20-9]
$C_{40}H_{58}N_8O_{14}$ 874.946
Isol. from the seeds of *Hordeum vulgare*

(Z)-Isomer, 5-methoxy, 4′-O-β-D-glucopyranoside
4′-O-β-D-Glucopyranosyl-cis-hordatine B
[1443149-21-0]
$C_{35}H_{50}N_8O_{10}$ 742.831
Isol. from the seeds of *Hordeum vulgare*

7′-Epimer
Aperidine
[639519-71-4]
$C_{28}H_{38}N_8O_4$ 550.664
Found in beer from *Hordeum vulgare*. Muscarinic M_3 receptor antagonist. $[\alpha]_D$ -54 (c, 0.14 in MeOH).

Koshimizu, K. *et al.*, *Can. J. Bot.*, 1963, **41**, 744-746 (*activity*).
Stoessl, A. *Can. J. Chem.*, 1967, **45**, 1745-1760 (*Hordatines A,B, synth*).
US Pat., 1969, 3 475 459 (*Hordatines A,B, synth, activity*).
Stoessi, A. *et al.*, *Can. J. Bot.*, 1970, **48**, 465-470 (*Hordatines A,B, activity*).
Smith, T.A. *et al.*, *Phytochemistry*, 1978, **17**, 1093-1098 (*Hordatines A,B,M, occur*).
Bird, C.R. *et al.*, *Phytochemistry*, 1981, **20**, 2345-2346 (*biosynth*).
Yamaji, N. *et al.*, *Alcohol.: Clin. Exp. Res.*, 2007, **31**, S9-S14 (*Aperidine, Hordatine A, activity*).
Nomura, T. *et al.*, *Genes Genet. Syst.*, 2007, **82**, 455-464 (*biosynth*).
Wakimoto, T. *et al.*, *Bioorg. Med. Chem. Lett.*, 2009, **19**, 5905-5908 (*Hordatine A, Aperidine, abs config, activity*).
Wakimoto, T. *et al.*, *Org. Lett.*, 2010, **13**, 2789-2791 (*Aperidine, synth*).
Kageyama, N. *et al.*, *J. Am. Soc. Brew. Chem.*, 2011, **69**, 255-259 (*Hordatines A,B β-D-glycosides*).
Kageyama, N. *et al.*, *J. Am. Soc. Brew. Chem.*, 2012, **70**, 133-136 (*Hordatines A,B β-D-glycosides*).
Kohyama, N. *et al.*, *J. Agric. Food Chem.*, 2013, **61**, 1112-1116 (*Hordatine A β-D-glucoside*).
Pihlava, J.-M. *J. Cereal Sci.*, 2014, **60**, 645-652 (*Hordatines A_1-C_2, N-Methylhordatines*).
Gorzolka, K. *et al.*, *Planta*, 2014, **239**, 1321-1335 (*Hordatines A-D, 3″-Hydroxyhordatines A-D*)

Houpulin I H56

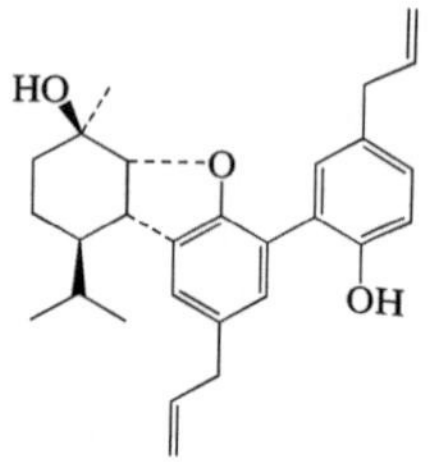

[1884599-71-6]
$C_{28}H_{34}O_3$ 418.577
Constit. of the roots of *Magnolia officinalis*. Significant inhibitor of superoxide anion generation and elastase release. Syrup. $[\alpha]^{25}_{D}$ +15 (c, 1.3 in MeOH). λ_{max} 297 (log ε 3.91) (MeOH).

Shih, H.-C. *et al.*, *Bioorg. Med. Chem.*, 2016, **24**, 1439-1445 (*Houpulin I, activity*)

Houpulin A H57

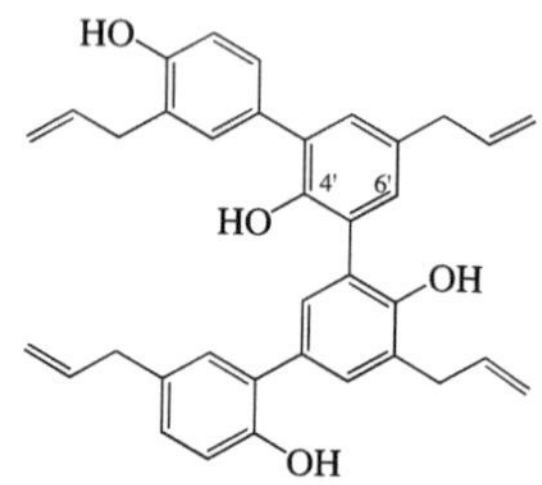

[1578247-77-4]
$C_{36}H_{34}O_4$ 530.664
Dimer of 3,3′-Ligna-8,8′-diene-4,6′-diol, L21. Constit. of the roots of *Magnolia officinalis*. Brown syrup. $[\alpha]^{25}_{D}$ -7.9 (c, 0.83 in MeOH)

4′-Deoxy, 6′-hydroxy
Houpulin B
[1578247-79-6]
$C_{36}H_{34}O_4$ 530.664
Constit. of the roots of *Magnolia officinalis*. Antiinflammatory agent. Brown syrup. $[\alpha]_D$ -8 (c, 0.7 in MeOH).

Shih, H.-C. *et al.*, *PLoS One*, 2013, **8**, e59502 (*Houpulins A,B*)

Houpulin C H58

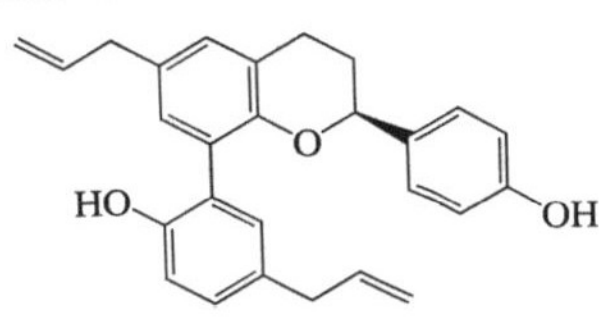

Absolute Configuration

[1578247-82-1]
$C_{27}H_{26}O_3$ 398.502
Constit. of *Magnolia officinalis*. Brown gum. $[\alpha]_D^{25}$ -65.7 (c, 0.11 in MeOH).
Shih, H.-C. *et al.*, *PLoS One*, 2013, **8**, e59502 (*Houpulin C*)

Houpulin D H59

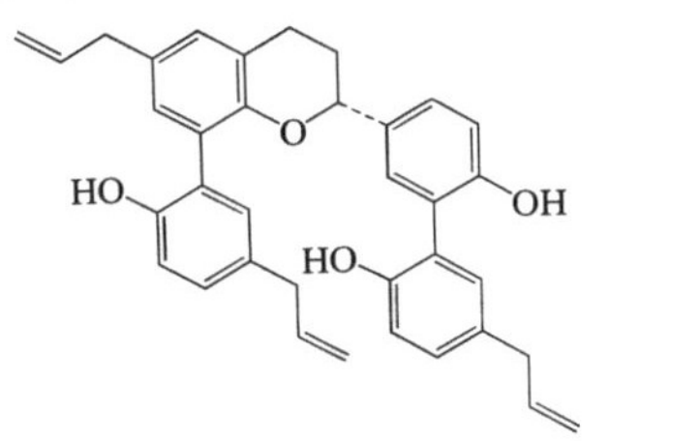

Absolute Configuration

[1578247-85-4]
$C_{36}H_{34}O_4$ 530.664
Constit. of the roots of *Magnolia officinalis*. Brown syrup. $[\alpha]_D^{25}$ -57.6 (c, 0.23 in MeOH).
Shih, H.-C. *et al.*, *PLoS One*, 2013, **8**, e59502 (*Houpulin D*)

Houpulin F H60

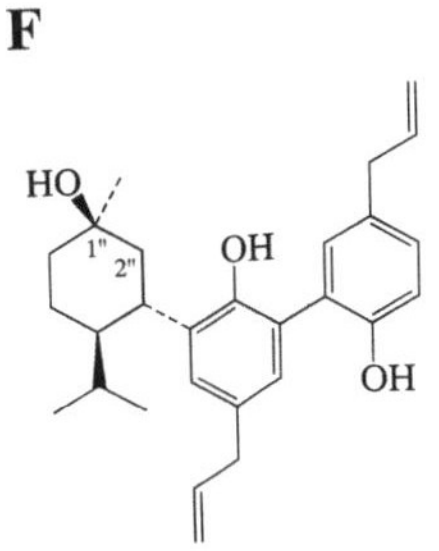

[1884599-67-0]
$C_{28}H_{36}O_3$ 420.593
Related to Piperitylmagnolol, P60. Constit. of the roots of *Magnolia officinalis*. Syrup. $[\alpha]_D^{25}$ -3 (c, 4.8 in MeOH). λ_{max} 290 (log ε 2.81) (MeOH)

2″α-Hydroxy
Houpulin H
[1884599-70-5]
$C_{28}H_{36}O_4$ 436.592
Constit. of the roots of *Magnolia officinalis*. Syrup. $[\alpha]_D^{25}$ -47 (c, 0.2 in MeOH). λ_{max} 290 (log ε 3.82) (MeOH)

1″-Epimer
Houpulin G
[1884599-69-2]
$C_{28}H_{36}O_3$ 420.593
Constit. of the roots of *Magnolia officinalis*. Significant inhibitor of superoxide anion generation and elastase release. Syrup. $[\alpha]_D^{25}$ -66 (c, 0.1 in MeOH).
Shih, H.-C. *et al.*, *Bioorg. Med. Chem.*, 2016, **24**, 1439-1445 (*Houpulins F,G,H, activity*)

Houpulin J H61

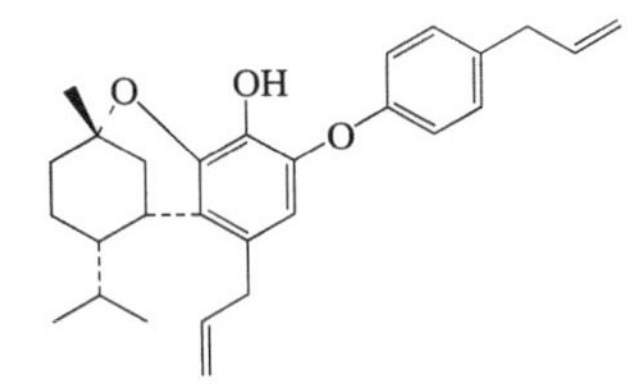

[1884599-72-7]
$C_{28}H_{34}O_3$ 418.577
Constit. of the roots of *Magnolia officinalis*. Significantly inhibits superoxide anion generation and elastase release. Syrup. $[\alpha]_D^{25}$ -2 (c, 2.7 in MeOH). λ_{max} 291 (log ε 4.35) (MeOH).
Shih, H.-C. *et al.*, *Bioorg. Med. Chem.*, 2016, **24**, 1439-1445 (*Houpulin J, activity*)

Houpulin K H62

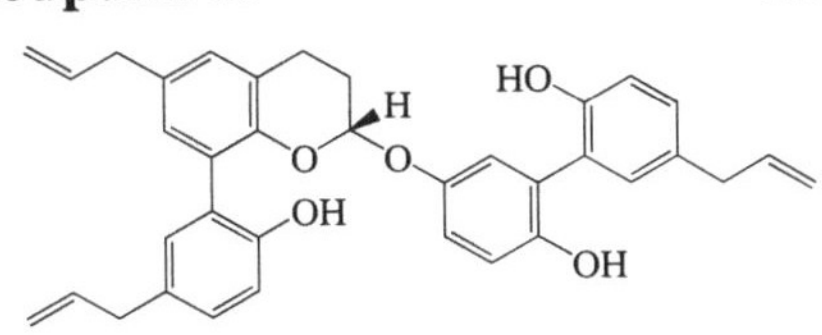

Absolute Configuration

[1884599-73-8]
$C_{36}H_{34}O_5$ 546.663
Constit. of the roots of *Magnolia officinalis*. Syrup. $[\alpha]_D^{25}$ -8 (c, 1.3 in MeOH). λ_{max} 290 (log ε 4.04) (MeOH).
Shih, H.-C. *et al.*, *Bioorg. Med. Chem.*, 2016, **24**, 1439-1445 (*Houpulin K*)

Houpulin L H63

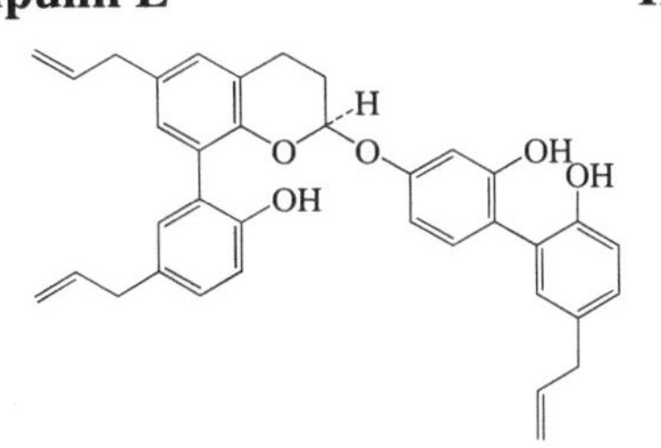

Absolute Configuration

[1884599-74-9]
$C_{36}H_{34}O_5$ 546.663
Constit. of the roots of *Magnolia officinalis*. Syrup. $[\alpha]_D^{25}$ -72 (c, 0.1 in MeOH). λ_{max} 290 (log ε 3.80) (MeOH).
Shih, H.-C. *et al.*, *Bioorg. Med. Chem.*, 2016, **24**, 1439-1445 (*Houpulin L*)

Hydnocarpin H64

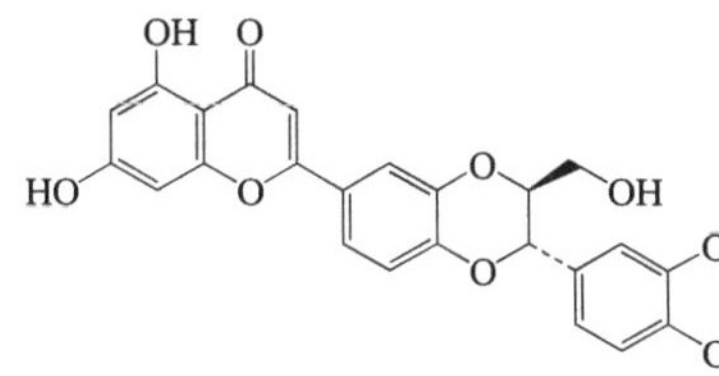

Absolute Configuration

2-[2,3-Dihydro-2-(4-hydroxy-3-methoxyphenyl)-3-(hydroxymethyl)-1,4-benzodioxin-6-yl]-5,7-dihydroxy-4H-1-benzopyran-4-one, CAS
[51419-48-8]
[298185-77-0] ((±)-form)
[298704-61-7] (7″*R**,8″*R**-form)
[1147503-23-8] (7″*S*,8″*S*-form)
[1803194-27-5] (7″*R*,8″*R*-form)
$C_{25}H_{20}O_9$ 464.426
Flavonolignan. Regioisomer of Hydnocarpin D, H65. Constit. of *Hydnocarpus wightiana*, *Brucea javanica*, *Cassia absus*, *Inula wissmanniana*, *Lonicera japonica*, *Onopordon corymbosum* and *Verbascum sinaiticum*. Cytotoxic. Antiproliferative agent. Bacterial multidrug resistant (MDR) efflux pump inhibitor. Synergistic with Berberine on *Staphylococcus aureus*. Exhibits bacteriostatic props.. Cryst. (MeOH aq.). Mp 262 - 264°. Mp 278 - 280°. $[\alpha]_D^{23}$ -32 (c, 0.5 in DMSO). Registry number refers to original isolate with unspecified stereochem.. λ_{max} 271; 285; 340 (MeOH)

7-O-β-D-Glucopyranoside
Hydnocarpin 7-glucoside
[1309133-19-4]
$C_{31}H_{30}O_{14}$ 626.567
Constit. of *Inula cappa*. Yellow powder. Mp 192 - 194°. $[\alpha]_D^{20}$ -20 (c, 0.2 in MeOH). λ_{max} 215 (log ε 4.36); 271 (log ε 4.09); 340 (log ε 4.17) (MeOH)

7-O-[4-Hydroxy-E-cinnamoyl-(→4)-β-D-glucopyranoside]
[1141888-82-5]
$C_{40}H_{36}O_{16}$ 772.712
Constit. of *Mallotus metcalfianus*. Obt. as a mixt. with the analogous Hydnocarpin D deriv. (see in H65)

Tetra-Ac

Mp 196 - 198°. $[\alpha]_D^{23}$ -21.2 (c, 0.05 in MeOH)

3″-Demethoxy

3″-Demethoxyhydnocarpin. ***Sinaiticin***

[152340-74-4]

$C_{24}H_{18}O_8$ 434.4

Constit. of *Hydnocarpus anthelminthica* and *Verbascum sinaiticum*. Cytotoxic agent. Inhibitor of NO prodn. in LPS-stimulated RAW 264.7 macrophage cell lines. Mp 192 - 193°. $[\alpha]_D$ -15.2 (c, 0.003 in MeOH). λ_{max} 270 (ε 52500); 341 (ε 29600) (MeOH) (Berdy). [base] λ_{max} 278; 368 (MeOH/NaOH) (Berdy)

5″-Methoxy

5″-Methoxyhydnocarpin

[126794-19-2]

$C_{26}H_{22}O_{10}$ 494.452

Constit. of the aerial parts of *Onopordon corymbosum*. Cytotoxic agent. Bacterial multidrug resistant (MDR) efflux pump inhibitor. Synergistic with Berberine on *Staphylococcus aurens*. Yellow cryst.. Mp 260°.

Ranganathan, K.R. *et al., Tetrahedron Lett.*, 1973, **14**, 3481-3482 (*Hydnocarpus wightiana constit*).
Kostova, I. *et al., Indian J. Chem., Sect. B*, 1977, **15**, 764-765 (*Cassia absus constit*).
Antus, S. *et al., Liebigs Ann. Chem.*, 1986, 647-654 (*struct*).
Cardona, M.L. *et al., Phytochemistry*, 1990, **29**, 629-631 (*Hydnocarpin, 5″-Methoxyhydnocarpin*).
Afifi, M.S.A. *et al., Phytochemistry*, 1993, **34**, 839-841 (*Sinaiticin, Hydnocarpin*).
She, X. *et al., Tetrahedron Lett.*, 1999, **40**, 4567-4570 (*Sinaiticin, synth*).
Guz, N.R. *et al., J. Nat. Prod.*, 2000, **63**, 1140-1145 (*synth*).
Guz, N.R. *et al., J. Med. Chem.*, 2001, **44**, 261-268 (*synth, activity*).
Pan, L. *et al., Bioorg. Med. Chem.*, 2009, **17**, 2219-2224 (*Hydnocarpin, abs config*).
Rivière, C. *et al., Phytochemistry*, 2009, **70**, 91-99 (*7-4-coumaroylglucoside*).
Wang, Y.-L. *et al., J. Asian Nat. Prod. Res.*, 2010, **12**, 765-769 (*Hydnocarpin 7-glucoside, Sinaiticin, activity*).
Wang, C. *et al., Arch. Pharmacal Res.*, 2013, **36**, 1516-1524 (*Inula wissmanniana constit*).
Lee, M.-A. *et al., Bioorg. Med. Chem. Lett.*, 2013, **23**, 5511-5514 (*Hydnocarpin, activity*).
Pérez, L.B. *et al., Phytother. Res.*, 2013, **27**, 1735-1738 (*Hydnocarpin, activity*).
Sahoo, M.R. *et al., J. Ethnopharmacol.*, 2014, **154**, 17-25 (*rev, props*).
Vimberg, V. *et al., J. Nat. Prod.*, 2015, **78**, 2095-2103 (*synth*).
Huang, G. *et al., Beilstein J. Org. Chem.*, 2016, **12**, 662-669 (*synth*)

Hydnocarpin D H65

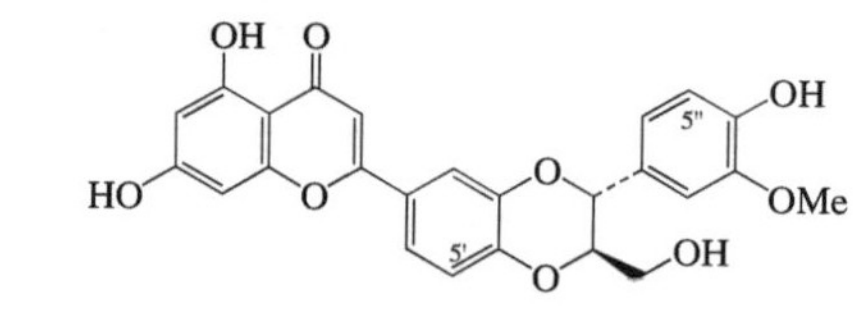

Relative Configuration

2-[2,3-Dihydro-3-(4-hydroxy-3-methoxyphenyl)-2-(hydroxymethyl)-1,4-benzodioxin-6-yl]-5,7-dihydroxy-4H-1-benzopyran-4-one, CAS

[298185-77-0]

[942611-66-7] (unspecified stereochem.) [1803192-76-8] (7″*R*,8″*R*-form) [1803192-77-9] (7″*S*,8″*S*-form)

$C_{25}H_{20}O_9$ 464.426

Flavonolignan. Regioisomer of Hydnocarpin, H64. Constit. of *Hydnocarpus anthelminthica*, *Hymenaea palustris* and *Epimedium sagittatum*. Cytotoxic. Amorph. pale yellow powder (MeOH). Mp 245° dec.. D in synonym denotes "D-ring down" in structure

7-O-[4-Hydroxy-E-cinnamoyl-(→4)-β-D-glucopyranoside]

[1141888-83-6]

$C_{40}H_{36}O_{16}$ 772.712

Constit. of *Mallotus metcalfianus*. Obt. as a mixt. with the analogous Hydnocarpin deriv. (see Hydnocarpin, H64)

5′-Methoxy

5′-Methoxyhydnocarpin D

[54431-77-5]

$C_{26}H_{22}O_{10}$ 494.452

Constit. of *Hydnocarpus wightiana*, *Hymenaea palustris*, *Berberis fremontii* and *Berberis repens*. Pale yellow powder (MeOH). Mp 224 - 226°. Racemic. Originally assigned the regioisomeric struct. of 5′-Methoxyhydnocarpin

5′,5″-Dimethoxy

5′,5″-Dimethoxyhydnocarpin D. ***Palstatin***

[503068-80-2]

$C_{27}H_{24}O_{11}$ 524.478

Constit. of *Epimedium sagittatum* and *Hymenaea palustris*. Cytotoxic. Amorph. pale yellow powder (MeOH). Mp 238 - 239°. Racemic. λ_{max} 269 (log ε 4); 341 (log ε 4.1) (MeOH).

Guz, N.R. *et al., J. Nat. Prod.*, 2000, **63**, 1140-1145 (*synth, struct*).
Stermitz, F.R. *et al., J. Nat. Prod.*, 2000, **63**, 1146-1149 (*Berberis fremontii constit*).
Pettit, G.R. *et al., J. Nat. Prod.*, 2003, **66**, 259-262 (*Palstatin, Hydnocarpin D*).
Chan, K.-F. *et al., Tetrahedron*, 2005, **61**, 4149-4156 (*5′-Methoxyhydnocarpin D, synth*).
Feng, J.-P. *et al., Chin. J. Chem.*, 2006, **24**, 215-218 (*Palstatin, synth*).
Wang, G.-J. *et al., Phytochemistry*, 2007, **68**, 2455-2464 (*Epimedium sagittatum constit*).
Rivière, C. *et al., Phytochemistry*, 2009, **70**, 91-99 (*7-4-coumaroylglucoside*).
Sahoo, M.R. *et al., J. Ethnopharmacol.*, 2014, **154**, 17-25 (*rev, props*).
Vimburg, V. *et al., J. Nat. Prod.*, 2015, **78**, 2095-2103 (*synth*).
Huang, G. *et al., Beilstein J. Org. Chem.*, 2016, **12**, 662-669 (*synth*)

Hydnowightin H66

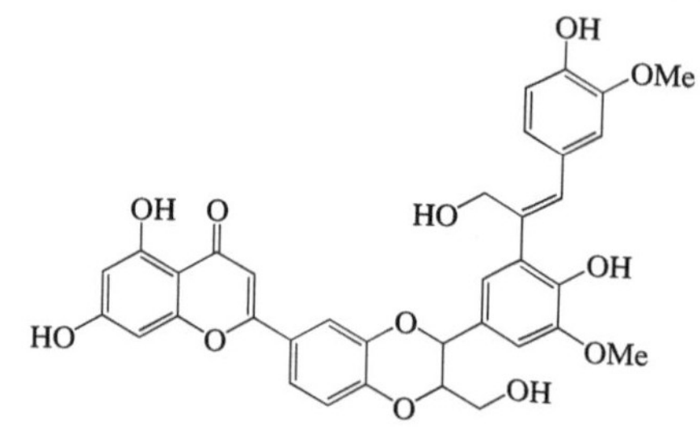

[71392-06-8]

$C_{35}H_{30}O_{12}$ 642.613

Constit. of the seeds of *Hydnocarpus wightiana*. Exhibits hypolipidaemic, antiinflammatory and antineoplastic activities. Shows competitive inhibitory activity against milefosine. Cryst.. Mp 239 - 241°. $[\alpha]_D$ +40 (MeOH).

Sharma, D.K. *et al., Planta Med.*, 1979, **37**, 79-83 (*Hydnowightin*).
Sharma, D.K. *J. Nat. Prod.*, 1991, **54**, 1298-1302 (*activity*).
Shinde, S. *et al., Gene*, 2014, **550**, 214-222 (*activity*)

Hydrangeside C H67

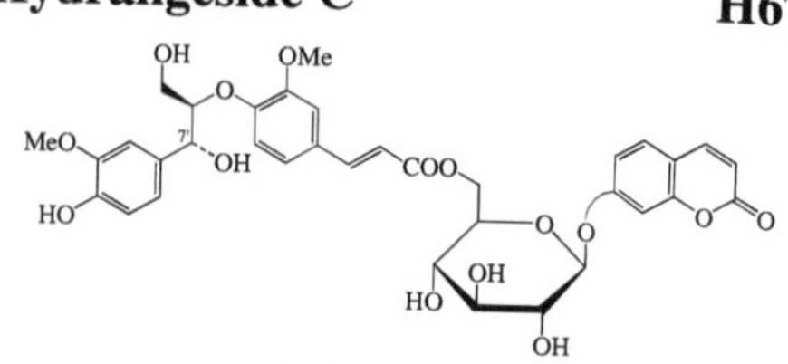

Absolute Configuration

[1645267-44-2]

$C_{35}H_{36}O_{15}$ 696.658

Constit. of the stems of *Hydrangea paniculata*. Amorph. powder. $[\alpha]_D^{20}$ -80.4 (c, 0.08 in MeOH). λ_{max} 203 (log ε 4.56); 330 (log ε 4.08) (MeOH)

7′-Epimer

Hydrangeside D

[1645267-45-3]
$C_{35}H_{36}O_{15}$ 696.658
Constit. of the stems of *Hydrangea paniculata*. Amorph. powder. $[\alpha]_D^{20}$ -102.5 (c, 0.07 in MeOH). λ_{max} 203 (log ε 4.50); 330 (log ε 4.03) (MeOH).
Shi, J. *et al.*, *Fitoterapia*, 2014, **96**, 138-145 (*Hydrangesides C,D*)

3-(4-Hydroperoxyphenyl)-2-propenoic acid H68

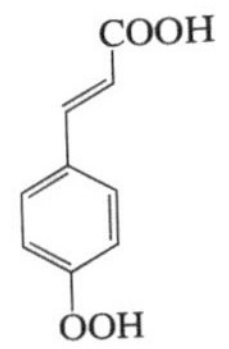

4-Hydroperoxycinnamic acid
$C_9H_8O_4$ 180.159
(E)-form
Nonadecyl ester
Nonadecyl trans-p-hydroperoxycinnamate
[1629058-30-5]
$C_{28}H_{46}O_4$ 446.672
Constit. of the bark of *Erythrina excelsa*. Exhibits activity against cancer cell lines. Gum.
Kwamou, G.M.N. *et al.*, *Nat. Prod. Res.*, 2015, **29**, 921-925 (*Erythrina excelsa constit, activity*)

2-(4-Hydroxybenzyl)-6-(4-hydroxy-3,5-dimethoxybenzylidene)-2*H*-pyran-3,5(4*H*,6*H*)-dione H69

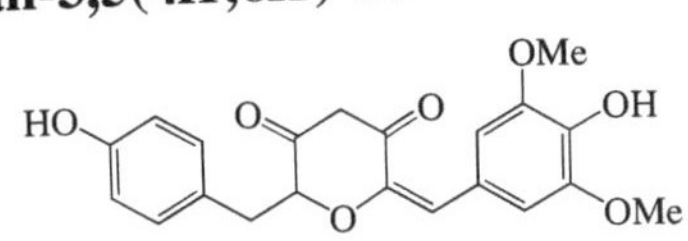

$C_{21}H_{20}O_7$ 384.384
(ξ)-form [2055038-97-4]
Constit. of the fruit of *Amomum tsao-ko*. Antioxidant. Active against tumours.
Zhang, T.-T. *et al.*, *J. Funct. Foods*, 2015, **18**, 423-431 (*Amomum tsao-ko constit*)

3-(4-Hydroxybenzyl)-6-[2-(4-hydroxyphenyl)ethyl]-2,2-dimethyl-4-piperidinone H70

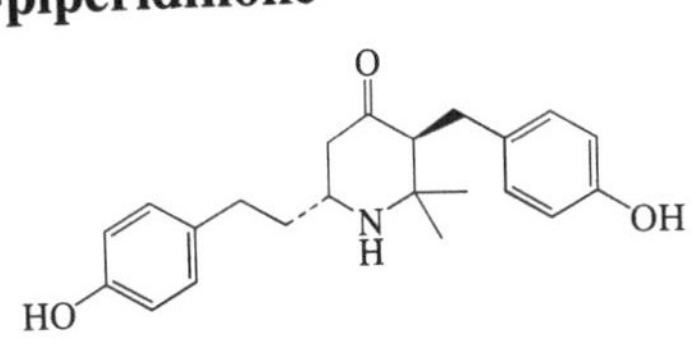

$C_{22}H_{27}NO_3$ 353.462
(3S,6R)-form [193805-10-6]
Alkaloid from the leaves of *Pellacalyx saccardianus*. COX-2 inhibitor. Cryst.. Mp 109 - 111°.
Abed, S.A. *et al.*, *Phytochem. Lett.*, 2015, **13**, 20-29 (*Pellacalyx saccardianus constit, cryst struct*)

7-Hydroxy-1,7-bis(4-hydroxy-3-methoxyphenyl)-1-heptene-3,5-dione, CAS H71

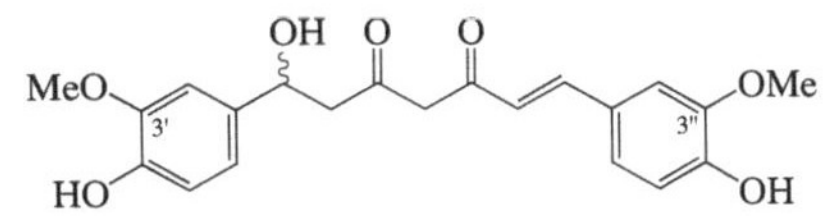

1,5-Dihydroxy-1,7-bis(4-hydroxy-3-methoxyphenyl)-4,6-heptadien-3-one.
1,2-Dihydro-1-hydroxycurcumin.
1-Hydroxydihydrocurcumin
[112494-43-6]
[1192453-07-8]
$C_{21}H_{22}O_7$ 386.4
Closely related to Curcumin, C130. Constit. of *Curcuma longa* and *Curcuma xanthorrhiza*. Exhibits neuroprotectant activity. Yellow powder. Mp 92 - 96° (92-94°). $[\alpha]_D$ +12.2 (c, 0.06 in EtOH). λ_{max} 230 (sh); 260 (ε 21000); 283 (ε 18000); 370 (ε 56000) (EtOH)
3′-Demethoxy
7-Hydroxy-1-(4-hydroxy-3-methoxyphenyl)-7-(4-hydroxyphenyl)-1-heptene-3,5-dione.
1,5-Dihydroxy-7-(4-hydroxy-3-methoxyphenyl)-1-(4-hydroxyphenyl)-4,6-heptadien-3-one
[1192557-85-9]
$C_{20}H_{20}O_6$ 356.374
Constit. of the rhizomes of *Curcuma longa*. Yellow powder. Mp 113 - 115°. $[\alpha]_D^{20}$ +0.3 (c, 0.71 in MeOH). λ_{max} 222; 260; 368 (MeCN aq.)
3″-Demethoxy
7-Hydroxy-7-(4-hydroxy-3-methoxyphenyl)-1-(4-hydroxyphenyl)-1-heptene-3,5-dione.
1,5-Dihydroxy-1-(4-hydroxy-3-methoxyphenyl)-7-(4-hydroxyphenyl)-4,6-heptadien-3-one
[1192557-84-8]
$C_{20}H_{20}O_6$ 356.374
Constit. of the rhizomes of *Curcuma longa*. Yellow powder. Mp 107 - 109°. $[\alpha]_D^{20}$ -0.2 (c, 0.43 in MeOH). λ_{max} 234; 288; 358 (MeCN aq.)
Bis(demethoxy)
7-Hydroxy-1,7-bis(4-hydroxyphenyl)-1-heptene-3,5-dione.
1,5-Dihydroxy-1,7-bis(4-hydroxyphenyl)-4,6-heptadien-3-one
[1192557-83-7]
$C_{19}H_{18}O_5$ 326.348
Constit. of the rhizomes of *Curcuma longa*. Yellow powder. Mp 116 - 118°. $[\alpha]_D^{20}$ +0.2 (c, 0.6 in MeOH). λ_{max} 228; 350 (MeCN aq.).
Uehara, S. *et al.*, *Chem. Pharm. Bull.*, 1987, **35**, 3298-3304 (*1-Hydroxydihydrocurcumin*).
Masuda, T. *et al.*, *Phytochemistry*, 1992, **31**, 3645-3647 (*Curcuma xanthorrhiza constit*).
Li, W. *et al.*, *Magn. Reson. Chem.*, 2009, **47**, 902-908 (*Curcuma longa constits*)

3-Hydroxy-1,7-bis(4-hydroxyphenyl)-6-heptene-1,5-dione H72

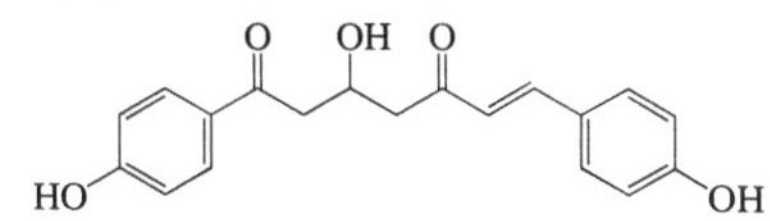

[1192557-86-0]
$C_{19}H_{18}O_5$ 326.348
Constit. of the rhizomes of *Curcuma longa*. Orange powder. Mp 88 - 91°. $[\alpha]_D^{20}$ -2 (c, 0.03 in MeOH). λ_{max} 224; 286; 324 (MeCN aq.).
Li, W. *et al.*, *Magn. Reson. Chem.*, 2009, **47**, 902-908 (*Curcuma longa constit*)

7-Hydroxy-1,7-bis(4-hydroxyphenyl)-5-heptene-3-sulfonic acid H73

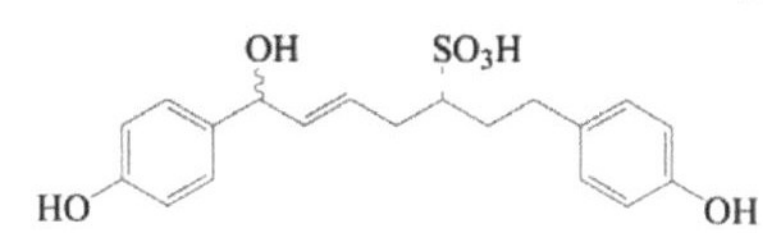

(3*R*,5*E*,7ξ)*-form*

$C_{19}H_{22}O_6S$ 378.439
(3R,5E,7ξ)-form [1008165-48-7]
Constit. of the rhizomes of *Zingiber officinale*. Amorph. yellow powder. $[\alpha]_D^{25}$ -16.4 (c, 0.8 in MeOH). Isol. as a mixt. with (3*S*,5*E*,7ξ)-isomer to which data refers
(3S,5E,7ξ)-form [1008165-47-6]
Constit. of the rhizomes of *Zingiber officinale*. Isol. as a mixt. with (3*R*,5*E*,7ξ)-isomer.
Tao, Q.-F. *et al.*, *J. Nat. Prod.*, 2008, **71**, 12-17 (*isol, pmr, cmr*)

9′-Hydroxy-3,4:3′,4′-bis(3,4-methylenedioxy)-8,8′-lign-7-en-9,9′-olide H74

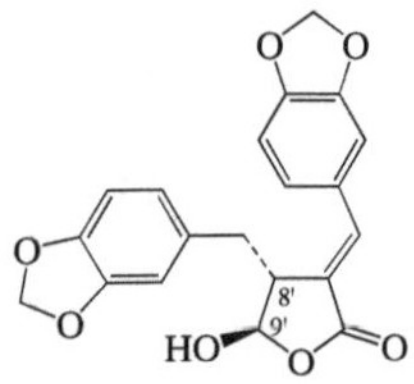

(7E,8′R,9′R)-form

$C_{20}H_{16}O_7$ 368.341

(7E,8′R,9′R)-form

Calocedrin [98891-33-9]

Constit. of the wood of *Calocedrus formosana* and *Pterocarpus santalinus*. Moderate inhibitor of tumour necrosis factor prodn. in LPS-stimulated cells. Cryst. (EtOH). Mp 187 - 188°. $[\alpha]^{25}_{D}$ +6 (c, 0.9 in Me_2CO). $[\alpha]_D$ -87 (c, 0.95 in $CHCl_3$). λ_{max} 237 (ε 10400); 294 (ε 9560); 330 (ε 11600) (EtOH)

(7E,8′R,9′S)-form

9′-Epicalocedrin [1268462-05-0]

Constit. of the roots of *Juniperus chinensis*. Light yellow solid. Mp 184 - 186°. $[\alpha]^{25}_{D}$ +4.6 (c, 0.13 in MeOH). Called 9-Epicalocedrin in the lit.. λ_{max} 238 (log ε 4.02); 294 (log ε 3.98); 330 (log ε 4.23) (EtOH).

Fang, J.-M. *et al.*, *Phytochemistry*, 1985, **24**, 1863-1864 (*Calocedrin*).

Wang, Y. *et al.*, *Acta Crystallogr., Sect. C*, 1987, **43**, 1006-1008 (*Calocedrin, cryst struct*).

Cho, J.Y. *et al.*, *Biol. Pharm. Bull.*, 2001, **24**, 167-171 (*Pterocarpus santalinus constit, activity*).

Chang, C.-I. *et al.*, *Nat. Prod. Commun.*, 2010, **5**, 55-57 (*9′-Epicalocedrin*)

3-*O*-(2-Hydroxycinnamoyl)quinic acid H75

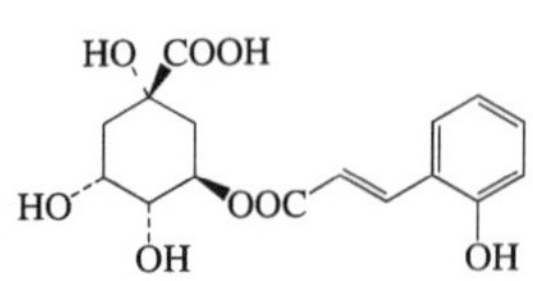

3-O-o-Coumaroylquinic acid

$C_{16}H_{18}O_8$ 338.312

(E)-form

Me ester

[1686087-95-5]

$C_{17}H_{20}O_8$ 352.339

Constit. of *Ageratina adenophora*. Exhibits antimicrobial activities. Yellowish gum. $[\alpha]_D$ -22.2 (c, 0.09 in MeOH). λ_{max} 212 (log ε 3.86); 276 (log ε 3.92) (MeOH).

Zhang, M. *et al.*, *Molecules*, 2013, **18**, 14096-14104 (*Me ester*)

4-Hydroxy-3,5-dimethoxycinnamaldehyde H76

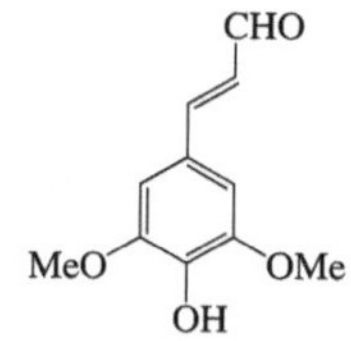

3-(4-Hydroxy-3,5-dimethoxyphenyl)-2-propen-1-al. ***Sinapaldehyde.*** *Sinapinaldehyde. Sinapyl aldehyde* [87345-53-7]

$C_{11}H_{12}O_4$ 208.213

(E)-form [4206-58-0]

Present in *Quercus* spp., *Juglans* spp., *Populus tremuloides*, *Eucalyptus sieberi*, *Betula pubescens*, *Drypetes hoaensis*, *Diplomorpha canescens*, *Michelia lanuginosa* and many other woods, especially hardwoods. Also in other plants such as grasses. Light yellow needles (C_6H_6 or EtOH). Mp 108.5 - 109.5° (105-106°)

2,4-Dinitrophenylhydrazone

Mp 273° dec..

Pauly, H. *et al.*, *Ber. Dtsch. Chem. Ges.*, 1929, **62**, 2277-2284 (*synth*).

Black, R.A. *et al.*, *J. Am. Chem. Soc.*, 1953, **75**, 5344-5346 (*Quercus constit*).

Aulin-Erdtman, G. *et al.*, *Acta Chem. Scand.*, 1968, **22**, 1187-1209 (*synth*).

Karrer, W. *Konstitution und Vorkommen der Organischen Pflanzenstoffe, 2nd edn., Birkhäuser Verlag*, 1976-1985, no. 407 (*occur*).

Talapatra, S.K. *et al.*, *J. Indian Chem. Soc.*, 1978, **55**, 1152-1160 (*Sinapaldehyde*).

Olstein, R. *et al.*, *Aust. J. Chem.*, 1979, **32**, 681-686 (*synth*).

Dunan, E. *et al.*, *Tetrahedron Lett.*, 1984, **25**, 2755-2758 (*synth*)

7′-Hydroxy-1,4-dimethoxy-3′,4′-methylenedioxy-3,8′-lign-8-en-6(1*H*)-one H77

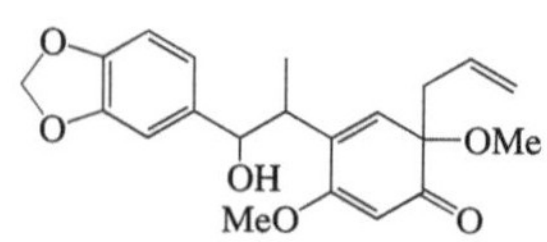

4-[2-(1,3-Benzodioxol-5-yl)-2-hydroxy-1-methylethyl]-3,6-dimethoxy-6-(2-propenyl)-2,4-cyclohexadien-1-one, CAS. *6-Allyl-4-[2-hydroxy-1-methyl-2-(3,4-methylenedioxyphenyl)]-3,6-dimethoxy-2,4-cyclohexadien-1-one*

[184954-90-3]

$C_{21}H_{24}O_6$ 372.417

Stereochem. not assigned. Constit. of the stems of *Piper argyrophylum*. Oil.

Singh, S.K. *et al.*, *Phytochemistry*, 1996, **43**, 1355-1360 (*Piper argyrophylum constit*)

4′-Hydroxy-3,3′-dimethoxy-4,8′-oxyligna-7,7′-diene-9,9′-dioic acid H78

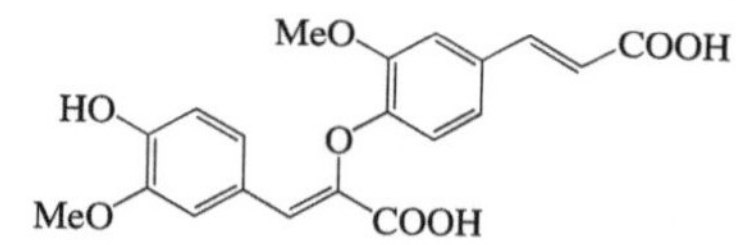

α-[4-(2-Carboxyvinyl)-2-methoxyphenoxy]-4-hydroxy-3-methoxycinnamic acid. 4-O-8′-Dehydrodiferulic acid. ***4-O-8′-Diferulic acid***

$C_{20}H_{18}O_8$ 386.356

(7E,7′Z)-form [180387-79-5]

Isol. from sugar beet and maize bran. Constit. of various Gramineae cell walls. Also from a marine-derived *Aspergillus* sp.. Pale yellow solid (Me_2CO/CCl_4). Mp 208 - 210°

Di-L-arabinofuranos-5-yl ester

[847805-77-0]

$C_{30}H_{34}O_{16}$ 650.586

Isol. from maize bran insoluble fibre

O-β-D-Glucopyranoside

Ligusticumacid A

[2573174-07-7]

$C_{26}H_{28}O_{13}$ 548.497

Constit. of the rhizomes of *Ligusticum chuanxiong*. Amorph. yellow powder. $[\alpha]^{20}_{D}$ -41 (c, 0.1 in MeOH). λ_{max} 216 (log ε 4.30); 230 (log ε 4.25); 290 (log ε 4.34); 323 (log ε 4.39) (MeOH).

Ralph, J. *et al.*, *J. Chem. Soc., Perkin Trans. 1*, 1994, 3485-3498 (*cell wall constit, synth*).

Micard, V. *et al.*, *Phytochemistry*, 1997, **44**, 1365-1368 (*Beta vulgaris constit*).

Bunzel, M. *et al.*, *J. Sep. Sci.*, 2004, **27**, 1080-1086 (*maize bran constit*).

Allerdings, E. *et al.*, *Phytochemistry*, 2005, **66**, 113-124 (*diarabinosyl ester*).

Funk, C. *et al.*, *Phytochemistry*, 2005, **66**, 363-371 (*maize bran constit*).

Zhou, Y. *et al.*, *Tetrahedron Lett.*, 2014, **55**, 2789-2792 (*Aspergillus metab*).

Grúz, J. *et al.*, *Food Chem.*, 2015, **171**, 280-286 (*detn, biosynth*).

Zhang, X. *et al.*, *Fitoterapia*, 2018, **125**, 147-154 (*Ligusticumacid A*)

4-Hydroxy-3′,5-dimethoxy-3,4′-oxyligna-7,7′-diene-9,9′-dioic acid H79

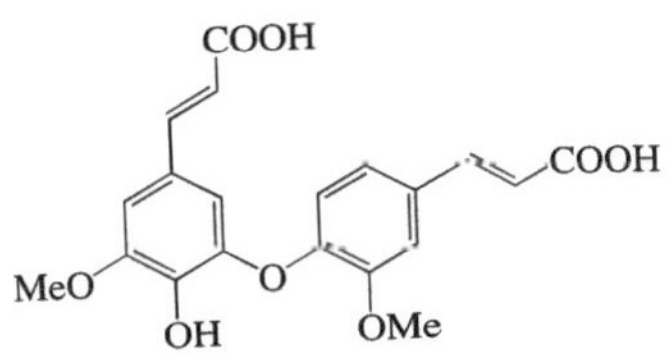

(7*E*,7′*E*)-form

3-[4-(2-Carboxyvinyl)-2-methoxyphenoxy]-4-hydroxy-5-methoxycinnamic acid. 4-O-5′-Dehydrodiferulic acid. ***4-O-5′-Diferulic acid***

$C_{20}H_{18}O_8$ 386.356

Dimer of 3-(4-Hydroxy-3-methoxyphenyl)-2-propenoic acid, H101

(7E,7′E)-form [160097-29-0]

Constit. of *Oryza sativa, Triticum aestivum, Zea mays* and the cell walls of other Gramineae. Cryst. (Me_2CO/toluene). Mp 212 - 214.5° dec.

(7Z,7′E)-form

N,N′-Bis[2-(4-hydroxyphenyl)ethylamide]

[1810715-25-3]

$C_{36}H_{36}N_2O_8$ 624.69

Constit. of the fruit of *Lycium barbarum*. Antioxidant.

Ralph, J. *et al., J. Chem. Soc., Perkin Trans. 1*, 1994, 3485-3498 (*Gramineae cell wall constits, synth*).

Bunzel, M. *et al., J. Agric. Food Chem.*, 2000, **48**, 3166-3169 (*occur*).

Lu, F. *et al., J. Agric. Food Chem.*, 2012, **60**, 8272-8277 (*synth*).

Forino, M. *et al., Food Chem.*, 2016, **194**, 1254-1259 (*bis(4-hydroxyphenylethylamide)*)

1-(4-Hydroxy-3,5-dimethoxyphenyl)-1,2,3-propanetriol, CAS H80

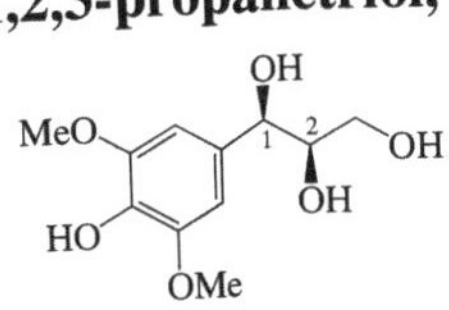

(1*R*,2*R*)-form

α-(4-Hydroxy-3,5-dimethoxyphenyl)glycerol. Syringoylglycerol. ***Syringylglycerol***

$C_{11}H_{16}O_6$ 244.243

(1R,2R)-form [2174104-98-2]

2-O-β-D-Glucopyranoside

[189456-52-8]

$C_{17}H_{26}O_{11}$ 406.384

Constit. of *Alangium premnifolium, Hypericum erectum* and *Machilus robusta*. Needles (MeOH). Mp 204 - 205°. $[\alpha]_D^{20}$ -48 (c, 0.53 in MeOH). λ_{max} 212 (log ε 4.18); 231 (log ε 3.8); 271 (log ε 3.11) (MeOH)

3-O-β-D-Glucopyranoside

[934588-98-4]

$C_{17}H_{26}O_{11}$ 406.384

Constit. of *Saccharum officinarum*. Amorph. powder. $[\alpha]_D^{25}$ -7 (c, 0.4 in MeOH)

1-Me ether

3-(4-Hydroxy-3,5-dimethoxyphenyl)-3-methoxy-1,2-propanediol. *threo*-***7-O-Methylsyringylglycerol***

[1220531-16-7]

$C_{12}H_{18}O_6$ 258.27

Constit. of *Laguncularia racemosa*. Oil. $[\alpha]_D^{20}$ +48.4 (c, 0.1 in MeOH). λ_{max} 270 (MeOH)

1-Me ether, 3-Ac

Alatusol C

[1578237-41-8]

$C_{14}H_{20}O_7$ 300.307

Constit. of *Euonymus alatus*. Displays cytotoxic activity. Gum. $[\alpha]_D^{25}$ -33.5 (c, 0.2 in MeOH). λ_{max} 215 (log ε 4.2); 235 (log ε 3.7); 278 (log ε 1.5) (MeOH)

(1R,2S)-form

2-O-β-D-Glucopyranoside

[1019851-19-4]

Constit. of the roots of *Iodes cirrhosa*. Amorph. powder. $[\alpha]_D^{20}$ -14.3 (c, 0.3 in MeOH). λ_{max} 211 (log ε 4.2); 232 (log ε 3.7); 271 (log ε 3.4) (MeOH)

1-Me ether

erythro-***7-Methoxysyringylglycerol***

[1714994-90-7]

$C_{12}H_{18}O_6$ 258.27

Constit. of the seeds of *Crataegus pinnatifida*

(1S,2R)-form [951627-70-6]

3-O-β-D-Glucopyranoside

[934588-97-3]

Constit. of *Saccharum officinarum*. Amorph. powder. $[\alpha]_D^{25}$ -10 (c, 0.2 in MeOH)

(1S,2S)-form [674359-50-3]

2-O-β-D-Glucopyranoside

[189456-46-0]

Constit. of *Alangium premnifolium* and *Pieris japonica*. Needles (MeOH). Mp 204 - 206°. $[\alpha]_D^{29}$ +7.9 (c, 0.63 in MeOH). λ_{max} 211 (log ε 4.11); 230 (log ε 3.72); 271 (log ε 3.03) (MeOH)

2-O-[4-Hydroxy-3,5-dimethoxybenzoyl-(→6)-β-D-glucopyranoside]

Brachyanin E

$C_{26}H_{34}O_{15}$ 586.543

Constit. of *Stauntonia brachyanthera*. Antioxidant. Oil. $[\alpha]_D^{25}$ +7.0 (c, 0.63 in MeOH)

3-O-β-D-Glucopyranoside

[189456-38-0]

Constit. of *Hyssopus officinalis* and *Saccharum officinarum*. α-Glucosidase inhibitor. Oil. $[\alpha]_D^{25}$ -18 (c, 0.6 in MeOH)

3-O-[E-Cinnamoyl-(→6)-β-D-glucopyranoside]

[674359-48-9]

$C_{26}H_{32}O_{12}$ 536.530

Constit. of *Hyssopus officinalis*. α-Glucosidase inhibitor. Oil. $[\alpha]_D^{23}$ -31.5 (c, 1 in MeOH)

3-Benzoyl

[1411560-49-0]

$C_{18}H_{20}O_7$ 348.351

Constit. of *Euonymus acanthocarpus*. Yellow oil. $[\alpha]_D^{25}$ +7.6 (c, 0.1 in MeOH)

(1R*,2R*)-form [121748-11-6]

Constit. of the roots of *Coix lachryma-jobi*. λ_{max} 230 (sh); 270; 278 (sh) (MeOH)

3-O-β-D-Glucopyranoside

Constit. of *Alangium premnifolium*. Amorph. powder. $[\alpha]_D$ -11.9 (c, 1.4 in MeOH). λ_{max} 210 (log ε 4.18); 231 (log ε 3.75); 271 (log ε 2.99) (MeOH)

1-Et ether

3-Ethoxy-3-(4-hydroxy-3,5-dimethoxyphenyl)-1,2-propanediol

[468094-67-9]

$C_{13}H_{20}O_6$ 272.297

Constit. of the roots of *Ficus beecheyana*. Amorph. powder. $[\alpha]_D^{25}$ +28 (c, 0.1 in $CHCl_3$). λ_{max} 271 (log ε 3.59) (MeOH)

(1R*,2S*)-form [38789-97-8]

Constit. of the roots of *Coix lachryma-jobi*. λ_{max} 225 (sh); 275 (MeOH)

3-O-(4-Hydroxy-E-cinnamoyl), 1-O-β-D-glucopyranoside

[2170740-71-1]

$C_{26}H_{32}O_{13}$ 552.529

Constit. of leaves of *Indocalamus latifolius*. Amorph. powder

(MeOH). $[\alpha]^{20}_{D}$ -43.8 (c, 0.1 in MeOH). λ_{max} 314 (MeOH aq.).

Otsuka, H. *et al.*, *Phytochemistry*, 1989, **28**, 883-886 (*Coix lachryma-jobi constit, synth*).
Kijima, H. *et al.*, *Phytochemistry*, 1997, **44**, 1551-1557 (*Alangium premnifolium constits*).
Lee, T.-H. *et al.*, *J. Nat. Prod.*, 2002, **65**, 1497-1500 (*1-Et ether*).
Matsuura, H. *et al.*, *Phytochemistry*, 2004, **65**, 91-97 (*Hyssopus officinalis glycosides*).
Matsuoka, E. *et al.*, *J. Tohoku Pharm. Univ.*, 2006, **53**, 37-41 (*Hypericum erectum glucoside*).
Takara, K. *et al.*, *Biosci., Biotechnol., Biochem.*, 2007, **71**, 183-191 (*Saccharum officinarum glucosides*).
Gan, M. *et al.*, *J. Nat. Prod.*, 2008, **71**, 647-654 (*Iodes cirrhosa 2-glucoside*).
Yao, G. *et al.*, *Yaoxue Xuebao*, 2008, **43**, 284-290 (*Pieris japonica glucosides*).
Shi, C. *et al.*, *Phytochemistry*, 2010, **71**, 435-442 (*threo-1-Me ether*).
Zhu, J.X. *et al.*, *Arch. Pharmacal Res.*, 2012, **35**, 1739-1747 (*Euonymus acanthocarpus constit*).
Bu, P.-B. *et al.*, *J. Asian Nat. Prod. Res.*, 2013, **15**, 482-491 (*Machilus robusta glucoside*).
Kim, K.H. *et al.*, *Planta Med.*, 2013, **79**, 361-364 (*Alatusol C*).
Huang, X.-X. *et al.*, *Biochem. Syst. Ecol.*, 2014, **54**, 208-212 (*erythro-1-Me ether*).
Zhao, J. *et al.*, *J. Funct. Foods*, 2015, **14**, 374-383 (*Brachyanin E, activity*).
Sun, J. *et al.*, *J. Asian Nat. Prod. Res.*, 2016, **18**, 360-365 (*3-coumaroyl 1-glucoside*)

3-(4-Hydroxy-3,5-dimethoxyphenyl)-2-propen-1-ol, CAS H81

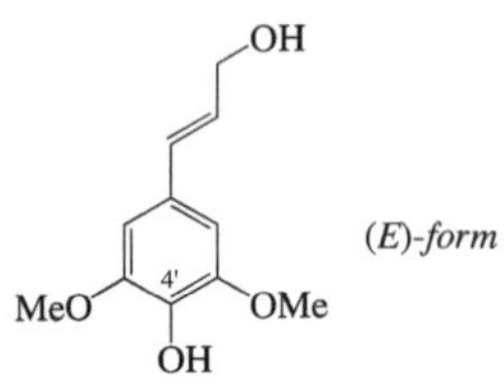

4-(3-Hydroxy-1-propenyl)-2,6-dimethoxyphenol, CAS. **Sinapyl alcohol.** *Syringenin*
[537-33-7]
$C_{11}H_{14}O_4$ 210.229
(E)-form [20675-96-1]
Widespread in higher plants. Principal building block of angiosperm lignin together with *p*-Coumaryl alcohol (in H129)/Coniferyl alcohol (in D24) (softwoods) and Coniferyl alcohol (hardwoods). Needles (petrol). Mp 63 - 65°
1-O-β-D-Glucopyranoside
[220006-75-7]
$C_{17}H_{24}O_9$ 372.37
Constit. of *Pinus stratiotes* and *Zantedeschia aethiopica*
1-O-[β-D-Apiofuranosyl-(1→6)-β-D-glucopyranoside]
[825651-70-5]
$C_{22}H_{32}O_{13}$ 504.485
Constit. of the seeds of *Punica granatum*. Antioxidant. Powder. Mp 284 - 286°. $[\alpha]^{20}_{D}$ +13.2 (c, 0.5 in H_2O). λ_{max} 275 (log ε 2.42) (H_2O)
1-O-(3-Methylbutanoyl)
Sinapyl isovalerate
[78869-17-7]
$C_{16}H_{22}O_5$ 294.347
Constit. of *Odontocline glabra*. Oil
1-O-(3-Methyl-2-butenoyl)
Sinapyl senecioate
[78869-18-8]
$C_{16}H_{20}O_5$ 292.331
Constit. of *Odontocline laciniata* and roots of *Ligularia duciformis*. Oil
1-Angeloyl
Sinapyl angelate
[78869-19-9]
$C_{16}H_{20}O_5$ 292.331
Constit. of *Odontocline laciniata* and *Solidago multiradiata*. Oil.

Freudenberg, K. *et al.*, *Chem. Ber.*, 1951, **84**, 67-70 (*Sinapyl alcohol, struct, synth*).
Aulin-Erdtman, G. *et al.*, *Acta Chem. Scand.*, 1968, **22**, 1187-1209 (*uv*).
Bohlmann, F. *et al.*, *Phytochemistry*, 1981, **20**, 831-832 (*Odontocline constits*).
Zanarotti, A. *Tetrahedron Lett.*, 1982, **23**, 3815-3818 (*synth*).
Quideau, S. *et al.*, *J. Agric. Food Chem.*, 1992, **40**, 1108-1111 (*synth*).
Daubresse, N. *et al.*, *Synthesis*, 1994, 369-371 (*synth*).
Greca, M.D. *et al.*, *Phytochemistry*, 1998, **49**, 1299-1304 (*1-glucoside*).
Wang, R.-F. *et al.*, *J. Nat. Prod.*, 2004, **67**, 2096-2098 (*pomegranate diglycoside*).
Schmidt, B. *et al.*, *J. Org. Chem.*, 2017, **83**, 4386-4395 (*synth*)

6-Hydroxy-1,7-diphenyl-4-hepten-3-one H82

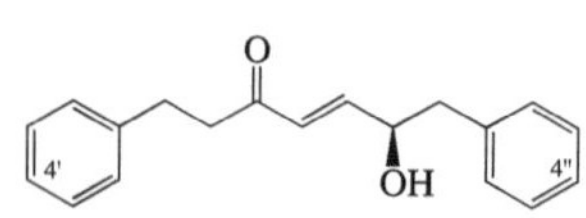

Alpinoid C
[1000878-38-5] (unspecified stereochem.)
$C_{19}H_{20}O_2$ 280.367
(4E,6R)-form [1005405-97-9]
Constit. of the rhizomes of *Alpinia officinarum*. PAF receptor binding antagonist. Oil. $[\alpha]^{25}_{D}$ -3.9 (c, 0.02 in $CHCl_3$). λ_{max} 238 (log ε 3.4); 279 (log ε 2.8) (MeOH)
3''-Methoxy, 4''-hydroxy
6-Hydroxy-7-(4-hydroxy-3-methoxyphenyl)-1-phenyl-4-hepten-3-one. ***Alpinoid B***
[1005405-98-0]
$C_{20}H_{22}O_4$ 326.392
Constit. of the rhizomes of *Alpinia officinarum*. Oil. $[\alpha]^{25}_{D}$ -3.9 (c, 0.05 in $CHCl_3$). λ_{max} 214 (log ε 3.9); 241 (log ε 3.5); 281 (log ε 2.3) (MeOH).

Fan, G.-J. *et al.*, *Bioorg. Med. Chem. Lett.*, 2007, **17**, 6720-6722 (*Alpinia officinarum constit, activity*).
Sun, Y. *et al.*, *Helv. Chim. Acta*, 2008, **91**, 118-123 (*Alpinoids B,C*).
Reddy, S.P. *et al.*, *Bioorg. Med. Chem. Lett.*, 2012, **22**, 4182-4184 (*synth*)

9-*O*-(11-Hydroxyeudesman-4-yl) dihydrosesamin H83

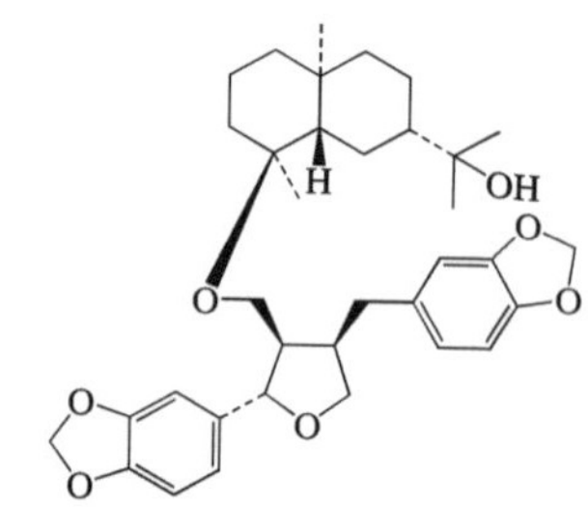

[1039089-73-0]
$C_{35}H_{46}O_7$ 578.746
Ether of *ent*-Dihydrosesamin (see under 7,9′-Epoxy-8,8′-lignan-3,3′,4,4′,9-pentol, E125) and *ent*-Cryptomeridiol. Constit. of the leaves of *Chamaecyparis obtusa*. Promotes neurite outgrowth in neuronal PC12 cells. Amorph. solid. $[\alpha]^{25}_{D}$ -8.3 (c, 0.02 in MeOH). λ_{max} 237 (ε 5440); 287 (ε 6750) (no solvent reported).

Kuroyanagi, M. *et al.*, *Chem. Pharm. Bull.*, 2008, **56**, 60-63 (*Chamaecyparis obtusa constit*)

6-Hydroxyhexanoic acid, CAS H84

$HOCH_2(CH_2)_4COOH$
6-Hydroxycaproic acid
[1191-25-9]
$C_6H_{12}O_3$ 132.159
Hygroscopic solid. Mp 41 - 43°
K salt
[21810-34-4]
Solid. Mp 203 - 205°
Me ester
[4547-43-7]
$C_7H_{14}O_3$ 146.186
Oil

Et ester

[5299-60-5]

$C_8H_{16}O_3$ 160.213

d_4^{20} 0.98. Bp 134° at 15 mm.

n_D^{20} +1.4390

Hydrazide

[1694-83-3]

$C_6H_{14}N_2O_2$ 146.19

Powder. Mp 115 - 116°

Nitrile

6-Hydroxyhexanenitrile. 5-Cyano-1-pentanol. Pentamethylene cyanohydrin

[2453-48-7]

$C_6H_{11}NO$ 113.16

Liq.. Bp 97 - 98° at 3 mm

O-(4-Hydroxy-3-methoxy-E-cinnamoyl), Me ester

Rhodiolate‡

$C_{17}H_{22}O_6$ 322.357

Constit. of *Rhodiola crenulata* and *Tinospora sinensis*. Light yellow solid. λ_{max} 197; 324 (MeOH).

Aldrich Library of [13]*C and* [1]*H FT-NMR Spectra*, 1992, **1**, 1040C (*nmr*).
Robinson, R. *et al., J. Chem. Soc.*, 1937, 371-374 (*synth, Et ester*).
Heilbron, I. *et al., J. Chem. Soc.*, 1947, 1588-1589 (*synth, hydrazide*).
Fore, S.P. *et al., J. Am. Oil Chem. Soc.*, 1963, **40**, 30-33 (*synth*).
Nohira, H. *et al., Bull. Chem. Soc. Jpn.*, 1965, **38**, 897-901 (*nitrile*).
UK Pat., 1968, 1 129 292 (Courtaulds) (*synth*).
Ger. Pat., 1970, 2 013 524 (Teijin) (*synth*).
Srinivasan, K.G. *et al., J. Am. Chem. Soc.*, 1978, **100**, 2789 (*nitrile*).
Hasegawa, A. *et al., J. Carbohydr. Chem.*, 1986, **5**, 371-385 (*Me ester*).
Binns, F. *et al., J. Chem. Soc., Perkin Trans. 1*, 1993, 899-904 (*synth, Me ester*).
Qi, Y. *et al., Macromolecules*, 2003, **36**, 3146-3151 (*hydrazide*).
Stephan, M.M.S. *et al., Org. Process Res. Dev.*, 2006, **10**, 481-483 (*synth*).
Rajabi, M. *et al., Synth. Commun.*, 2014, **44**, 1149-1154 (*synth*).
Zhou, J.-T. *et al., Molecules*, 2015, **20**, 13725-13739 (*Rhodiolate*)

3-Hydroxy-5-(4-hydroxybenzyl)-4-(4-hydroxyphenyl)-2(5*H*)-furanone H85

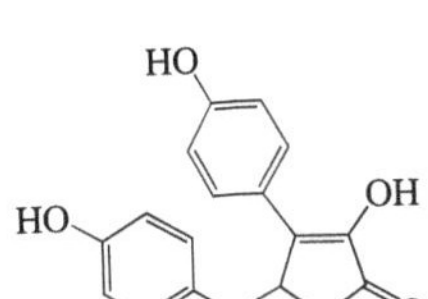

4,4′,8-Trihydroxy-9′-nor-7,8′-lign-7-en-9,8′-olide. Demethoxyisobutyrolactone II

[87414-46-8]

$C_{17}H_{14}O_5$ 298.294

Provisional norlignan. Biosynth. pathway unclear. Prod. by *Aspergillus* sp. CPCC 400735 and the marine-derived *Aspergillus terreus* SCSGAF0162. Also from a co-culture of *Aspergillus terreus* and *Bacillus subtilis*. Antioxidant. Mp 248 - 249°. $[\alpha]_D^{25}$ -1.9 (c, 0.05 in MeOH). Prob. racemic or a partial racemate. λ_{max} 220; 302 (H_2O). [base] λ_{max} 218; 247; 339 (NaOH aq.)

3-Me ether

5-(4-Hydroxybenzyl)-4-(4-hydroxyphenyl)-3-methoxy-2(5H)-furanone

[2170646-05-4]

$C_{18}H_{16}O_5$ 312.321

Prod. by *Aspergillus* sp. CPCC 400735.

Cotelle, P. *et al., Bioorg. Med. Chem.*, 2003, **11**, 1087-1093 (*synth, activity*).
He, F. *et al., J. Nat. Prod.*, 2013, **76**, 1182-1186 (*Aspergillus terreus constit*).
Chen, H. *et al., Phytochem. Lett.*, 2015, **12**, 35-41 (*Aspergillus terreus/Bacillus subtilis constit*).
Pang, X. *et al., J. Nat. Prod.*, 2017, **80**, 2595-2601 (*Aspergillus CPCC 400735 constits*)

7-Hydroxy-3-[2-hydroxy-4′-[4-(-hydroxymethyl)-1,2-dioxetan-3-yl]-2′,6,6′-trimethoxy-[1,1′-biphenyl]-4-yl]-4*H*-1-benzopyran-4-one H86

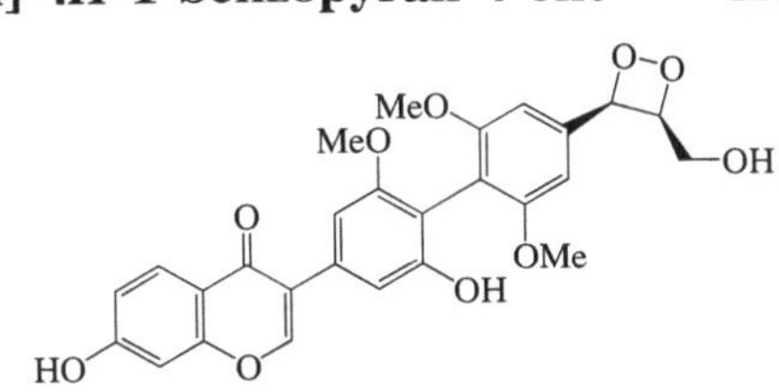

Absolute Configuration

[2226033-45-8]

$C_{27}H_{24}O_{10}$ 508.479

Constit. of the roots of *Pongamia pinnata*. NO prodn. inhibitor. Amorph. light yellow powder. $[\alpha]_D^{25}$ +4 (c, 0.1 in MeOH). λ_{max} 208 (log ε 4.61); 265 (log ε 4.09); 302 (log ε 3.76) (MeOH).

Wen, R. *et al., Bioorg. Med. Chem. Lett.*, 2018, **28**, 1050-1055 (*Pongamia pinnata constit, activity*)

3-Hydroxy-1-(2-hydroxy-5-methoxyphenyl)-2-(4-hydroxy-3-methoxyphenyl)-1-propanone H87

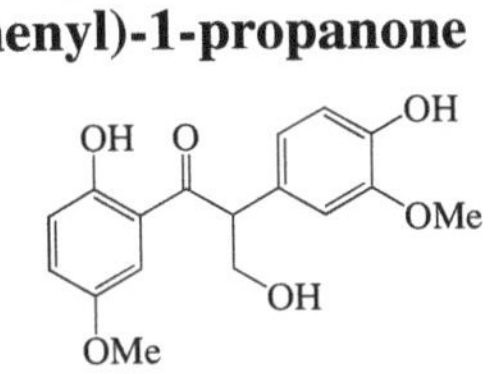

2′,4″-Dihydroxy-3″,5′-dimethoxy-α-(hydroxymethyl)deoxybenzoin

$C_{17}H_{18}O_6$ 318.325

(ξ)-form [1008993-47-2]

Constit. of the seeds of *Trichosanthes kirilowii*. Light brown powder. Mp 47 - 53°. λ_{max} 204 (log ε 4.83); 229 (log ε 4.57); 279 (log ε 4.35) (MeOH).

Rahman, M.A.A. *et al., Bull. Korean Chem. Soc.*, 2007, **28**, 1261-1264 (*isol, pmr, cmr*)

7-Hydroxy-3-(3-hydroxy-4-methoxyphenyl)-8-[4-[4-(hydroxymethyl)-1,2-dioxetan-3-yl]-2,6-dimethoxyphenyl]-4*H*-1-benzopyran-4-one H88

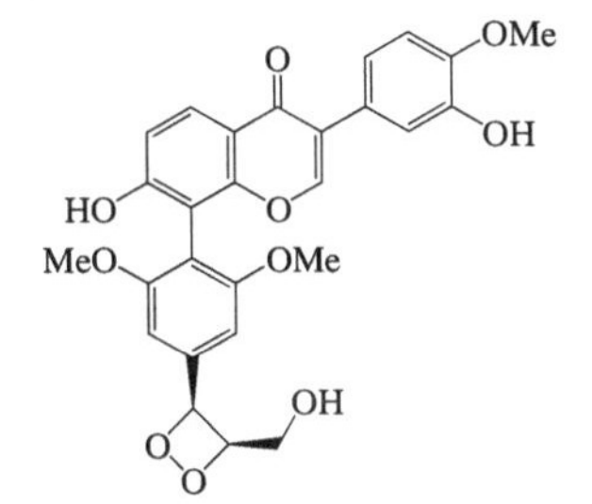

Absolute Configuration

[2226033-44-7]

$C_{27}H_{24}O_{10}$ 508.479

Constit. of the roots of *Pongamia pinnata*. NO prodn. inhibitor. Amorph. light yellow powder. $[\alpha]_D^{25}$ +4 (c, 0.1 in MeOH). λ_{max} 205 (log ε 4.86); 260 (log ε 4.52) (MeOH).

Wen, R. *et al., Bioorg. Med. Chem. Lett.*, 2018, **28**, 1050-1055 (*Pongamia pinnata constit, activity*)

4-Hydroxy-4-(4-hydroxy-3-methoxyphenyl)-3-methyl-2-butanone H89

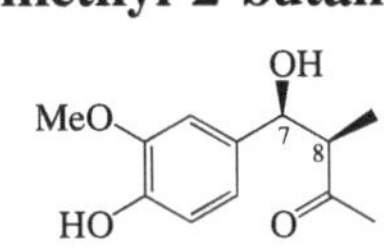

$C_{12}H_{16}O_4$ 224.256

Norlignan

(7S,8R)-form [1196997-43-9]

Constit. of the bark of *Machilus wangchiana*. Moderate inhibitor of galactosamine induced cytotoxicity. Amorph. powder. $[\alpha]_D^{20}$ +15 (c, 0.05 in $CHCl_3$). λ_{max} 230 (log ε 4.29); 280 (log ε 3.92) (MeOH).

Cheng, W. *et al., J. Nat. Prod.*, 2009, **72**, 2145-2152 (*Machilus wangchiana constit, activity*)

5-[3-Hydroxy-2-[(4-hydroxy-3-methoxyphenyl)methyl]-1-(hydroxymethyl)-2-(4-hydroxyphenyl) propyl]-1,3-benzenediol H90

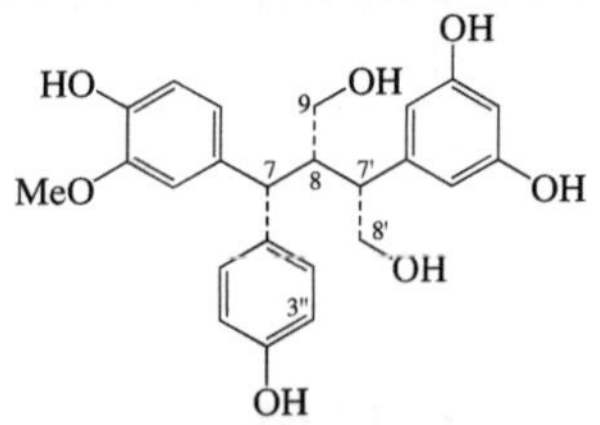

(7*R**,7'*R**,8*S**)-form

$C_{24}H_{26}O_7$ 426.465
Structures in this series may need revision. Phenyl 3,5-dioxygenated substn. pattern in doubt as a result of work done by Yang *et al* in 2015

(7R*,7'R*,8S*)-form

3"-Methoxy

Pouzolignan H
[1966155-20-3]
$C_{25}H_{28}O_8$ 456.491
Constit. of the aerial parts of *Pouzolzia zeylanica*. Inhibits LPS-induced NO prodn. in mouse peritoneal macrophages. Amorph. brown powder. $[\alpha]_D^{25}$ +34.2 (c, 0.1 in MeOH). C-7 is achiral. λ_{max} 281 (log ε 3.74) (MeOH)

3"-Methoxy, 9-Ac

Pouzolignan F
[1966155-18-9]
$C_{27}H_{30}O_9$ 498.528
Constit. of the aerial parts of *Pouzolzia zeylanica*. Amorph. yellow powder. C-7 is achiral. λ_{max} 281 (log ε 3.16) (MeOH)

3",5"-Dimethoxy

Pouzolignan I
[1966155-21-4]
$C_{26}H_{30}O_9$ 486.517
Constit. of the aerial parts of *Pouzolzia zeylanica*. Inhibits LPS-induced NO prodn. in mouse peritoneal macrophages. Amorph. brown powder. $[\alpha]_D^{25}$ +36.4 (c, 0.1 in MeOH). λ_{max} 281 (log ε 3.74) (MeOH)

3",5"-Dimethoxy, 9-Ac

Pouzolignan G
[1966155-19-0]
$C_{28}H_{32}O_{10}$ 528.554
Constit. of the aerial parts of *Pouzolzia zeylanica*. Amorph. yellow powder. $[\alpha]_D^{25}$ +51.7 (c, 0.1 in MeOH). λ_{max} 281 (log ε 3.24) (MeOH)

3",5"-Dimethoxy, 4"-Me ether

Pouzolignan J
[1966155-22-5]
$C_{27}H_{32}O_9$ 500.544
Constit. of the aerial parts of *Pouzolzia zeylanica*. Brown gum. $[\alpha]_D^{25}$ +42.2 (c, 0.1 in MeOH). λ_{max} 281 (log ε 2.95) (MeOH)

(7R*,7'S*,8R*)-form

Pouzolignan B [1299486-06-8]
Constit. of *Pouzolzia occidentalis*. Brown gum. $[\alpha]_D^{25}$ -29.6 (c, 0.1 in MeOH). λ_{max} 270 (log ε 3.94) (MeOH)

(7ξ,7'ξ,8ξ)-form

3"-Methoxy, 8'-Ac

Pouzolignan L
$C_{27}H_{30}O_9$ 498.528
Constit, of the aerial parts of *Pouzolzia zeylanica*. Powder. $[\alpha]_D^{20}$ +29.7 (c, 0.1 in MeOH). λ_{max} 275 (log ε 3.25) (MeOH)

3",5"-Dimethoxy, 8'-Ac

Pouzolignan M
$C_{28}H_{32}O_{10}$ 528.554
Constit. of aerial parts of *Pouzolzia zeylanica*. Powder. $[\alpha]_D^{20}$ +36.1 (c, 0.1 in MeOH). λ_{max} 275 (log ε 3.34) (MeOH).

Mohammed, M. *et al., Phytochem. Lett.*, 2010, **3**, 29-32 (*Pouzolignan B*).
Mohammed, M. *et al., Phytochem. Lett.*, 2011, **4**, 56 (*erratum*).
Zhong, C.-Q. *et al., Heterocycles*, 2015, **91**, 1926-1936 (*Pouzolignans L,M*).
Chen, Z.-H. *et al., J. Asian Nat. Prod. Res.*, 2015, **17**, 959-966 (*Pouzolignans F-J, activity*).
Yang, Y.-N. *et al., J. Nat. Prod.*, 2015, **78**, 705-711 (*pmr, struct*)

5-Hydroxy-2-(2-hydroxyphenyl)-4-benzofurancarboxylic acid H91

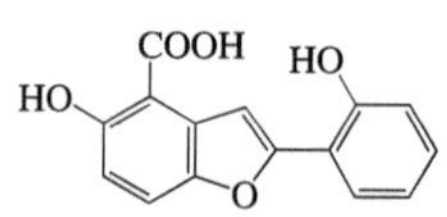

3,7'-Epoxy-2',6-dihydroxy-8,9,9'-trinor-2,8'-lign-7'-en-7-oic acid
$C_{15}H_{10}O_5$ 270.240

Me ester
[1994336-50-3]
$C_{16}H_{12}O_5$ 284.267
Constit. of the tubers of *Sparganium stoloniferum*. Yellow needles. Mp 206.3 - 206.7°. λ_{max} 244; 297; 321; 357 (MeOH).

Hu, X. *et al., J. Asian Nat. Prod. Res.*, 2016, **18**, 643-647 (*Me ester*).
Wu, D.-W. *et al., Nat. Prod. Res.*, 2017, **31**, 131-137 (*Me ester*)

3-Hydroxy-4-(4-hydroxyphenyl)-5-(4-hydroxy-3-prenylbenzyl)-2(5*H*)-furanone H92

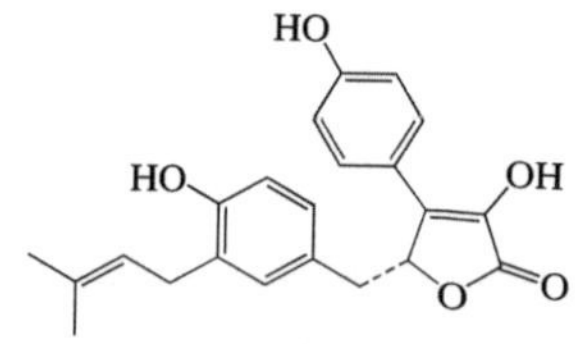

3-Hydroxy-5-[[4-hydroxy-3-(3-methyl-2-butenyl)phenyl]methyl]-4-(4-hydroxyphenyl)-2(5H)-furanone, CAS. *4,4',8-Trihydroxy-9'-nor-3'-prenyl-7,8'-lign-7-en-9,8'-olide*
$C_{22}H_{22}O_5$ 366.413

(R)-form [156003-02-0]
Prod. by *Aspergillus terreus* strains PR-P-2 and and marine-derived *Aspergillus terreus* PT06-2. Oil. $[\alpha]_D^{20}$ -14.3 (c, 0.29 in MeOH).

Wang, Y. *et al., Mar. Drugs*, 2011, **9**, 1368-1378 (*Aspergillus terreus metab*).
Guo, F. *et al., Fitoterapia*, 2016, **113**, 44-50 (*Aspergillus terreus metab*)

2-Hydroxy-3-(2-hydroxyphenyl) propanoic acid H93

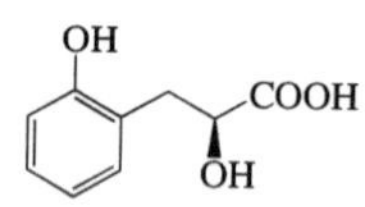

(*S*)*-form*

α,2-Dihydroxybenzenepropanoic acid, CAS. *3-(2-Hydroxyphenyl)lactic acid*
$C_9H_{10}O_4$ 182.175

(S)-form [775574-92-0]
Constit. of the pollen of *Crocus sativus*. Powder (as Na salt). $[\alpha]_D$ +59 (c, 0.08 in MeOH). $[\alpha]_D$ +77 (c, 0.07 in MeOH) (Na salt). λ_{max} 253; 258; 264 (MeOH) (Na salt)

(±)-form [6803-09-4]
[112714-96-2] (Na salt)
Constit. of the fruit of *Prunus cerasus*. Powder. Dec. at 241° (as Na salt). λ_{max} 218 (log ε 3.04); 253 (log ε 3.42); 289 (log ε 3.66) (MeOH).

Shaw, K.N.F. *et al., J. Org. Chem.*, 1956, **21**, 601-604 (*synth*).
Wang, H. *et al., J. Nat. Prod.*, 1999, **62**, 86-88 (*Prunus cerasus constit*).
Li, C.-Y. *et al., Chem. Pharm. Bull.*, 2002, **50**, 1305-1309 (*Crocus sativus constit*)

2-Hydroxy-3-(4-hydroxyphenyl) propanoic acid H94

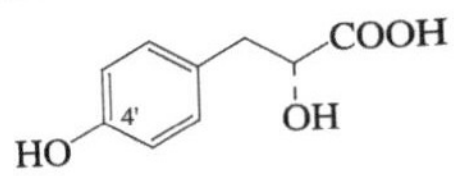

(R)-form

α,4-Dihydroxybenzenepropanoic acid, CAS. *3-(4-Hydroxyphenyl)lactic acid*, CAS
$C_9H_{10}O_4$ 182.175
Mp 142 - 143°
▶ OD5500000
▶ Tumourigenic

(R)-form [89919-57-3]
Isol. from the heartwood of *Pterocarpus marsupium* and *Aristolochia kaempferi*. Solid. Mp 169 - 170° (164-165°). $[\alpha]_D^{26}$ -19.6 (c, 1.3 in H_2O). $[\alpha]_D$ -13 (c, 0.15 in MeOH)

Na salt
[217188-19-7]
Mp > 300°. $[\alpha]_D$ -33 (c, 0.01 in MeOH)

2-O-Sulfate
3-(4-Hydroxyphenyl)-2-(sulfooxy)propanoic acid. **Tichocarpol A**
[735328-83-3] (unspecified stereochem.)
$C_9H_{10}O_7S$ 262.232
Isol. from the red alga *Tichocarpus crinitus*. Amorph. solid. $[\alpha]_D^{26}$ +2.7 (c, 0.4 in H_2O)

2-O-(3,4-Dihydroxy-E-cinnamoyl), Me ester
Perovskoate
[909094-34-4]
$C_{19}H_{18}O_7$ 358.346
Constit. of *Perovskia atriplicifolia*. Pale yellow gum. $[\alpha]_D^{20}$ +42 (c, 0.02 in MeOH). λ_{max} 220; 247; 330 (MeOH)

2-O-[β-D-Glucopyranosyl-(→3)-3,4-dihydroxy-E-cinnamoyl]
[1619935-11-3]
$C_{24}H_{26}O_{12}$ 506.46
Constit. of the seed flour of *Perilla frutescens*. Antioxidant. Yellow powder. $[\alpha]_D$ -11.9 (c, 0.3 in H_2O). λ_{max} 230; 320 (MeOH)

(S)-form [23508-35-2]
Metab. of *Ceratocystis* spp., a *Candida* sp. and *Lactobacillus plantarum*. Oxidase inhibitor. Antioxidant. Mp 173° (167-169°). $[\alpha]_D$ +8.8 (c, 0.5 in MeOH). $[\alpha]_D^{30}$ +20 (c, 2.5 in EtOH)

Me ester
Crypticin B. Latifolicinin C
[123359-33-1]
$C_{10}H_{12}O_4$ 196.202
Constit. of the fruit of *Cordia latifolia*. Prod. by *Diaporthella cryptica* CBS140348. Amorph. powder. $[\alpha]_D^{25}$ -4.9 (c, 0.6 in MeOH). λ_{max} 200 (log ε 6.57); 226 (log ε 6.41); 278 (log ε 5.79) (MeOH)

Et ester
Latifolicinin B
[267228-41-1]
$C_{11}H_{14}O_4$ 210.229
Constit. of the fruit of *Cordia latifolia*. Amorph. powder. λ_{max} 223 (log ε 3.8); 278 (log ε 3.23) (MeOH)

Butyl ester
Latifolicinin A
[884315-17-7]
$C_{13}H_{18}O_4$ 238.283
Constit. of the fruit of *Cordia latifolia*. Amorph. powder. λ_{max} 225 (log ε 4.08); 278 (log ε 3.98) (no solvent reported)

(±)-form
Cryst.. Mp 147.5 - 149° (139-140°)

(ξ)-form [306-23-0]
Constit. of human urine, wheat grains and various other plant spp. Metab. of tyrosine in humans.

Bubl, E.C. *et al., J. Am. Chem. Soc.*, 1951, **73**, 4972 (*synth*).
Holden, J.T. *et al., J. Biol. Chem.*, 1951, **191**, 559-576 (*synth*).
Buckle, A.L.J. *et al., J. Chem. Soc.*, 1954, 3981-3986 (*synth*).
Matsuura, T. *et al., J. Am. Chem. Soc.*, 1959, **81**, 871-878 (*synth*).
Stoessl, A. *Biochem. Biophys. Res. Commun.*, 1969, **35**, 186-191 (*Ceratocystis metab, synth*).
Rauschenbach, M.O. *et al., Cancer Res.*, 1975, **35**, 577-585 (*tox*).
Narayanan, T.K. *et al., Can. J. Microbiol.*, 1976, **22**, 364-389 (*Candida metab*).
Danifshefsky, S. *et al., J. Am. Chem. Soc.*, 1979, **101**, 7013-7018 (*(±)-form, synth*).
Lohaus, E. *et al., Z. Naturforsch., C*, 1983, **38**, 524-530 (*occur*).
Maurya, R. *et al., J. Nat. Prod.*, 1984, **47**, 179-181 (*Pterocarpus marsupium constit*).
Ayer, W.A. *et al., Can. J. Chem.*, 1986, **64**, 904-909 (*Ceratocystis metab*).
Wu, T.-S. *et al., Chem. Pharm. Bull.*, 1998, **46**, 1624-1626 (*Aristolochia kaempferi constit*).
Ishii, T. *et al., J. Nat. Prod.*, 2004, **67**, 1764-1766 (*Tichocarpol A*).
Linderberg, M.T. *et al., Org. Process Res. Dev.*, 2004, **8**, 838-845 (*2-Et ether*).
Aikins, J.A. *et al., J. Org. Chem.*, 2005, **70**, 4695-4705 (*S-form, !synth*).
Siddiqui, B.S. *et al., Nat. Prod. Res.*, 2006, **20**, 131-137 (*Latifolicinins A-D*).
Brenna, E. *et al., Tetrahedron: Asymmetry*, 2009, **20**, 2594-2599 (*S-2-Me ether Et ester, Latifolicinin D, S-2-Et ether Et ester, (±)-4′-Me ether, synth*).
Brenna, E. *et al., Tetrahedron: Asymmetry*, 2009, **20**, 2694-2698 (*S-2-Me ether Et ester, S-di-Me ether, synth*).
Suzuki, Y. *et al., Biosci., Biotechnol., Biochem.*, 2013, **77**, 1299-1302 (*Lactobacillus plantarum constit*).
Zhou, X.-J. *et al., Food Chem.*, 2014, **164**, 150-157 (*Perilla frutescens constit*).
Lewis, R.J. *Sax's Dangerous Properties of Industrial Materials, 12th edn., J. Wiley*, 2012, HNG700

3-Hydroxy-1-(4-hydroxyphenyl)-1-propanone, CAS H95

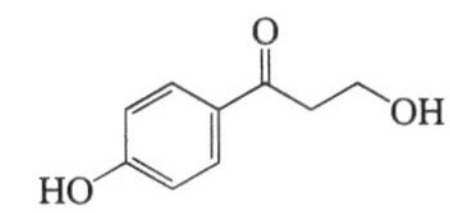

3,4′-Dihydroxypropiophenone. 4-(3-Hydroxypropionyl)phenol. 2-(p-Hydroxybenzoyl)ethanol. 2-(Hydroxyethyl) 4-hydroxyphenyl ketone
[53170-93-7]
$C_9H_{10}O_3$ 166.176
Constit. of *Betula platyphylla*, *Carissa edulis*, *Pinus armandii* and *Trichosanthes kirilowii*. Pale yellow prisms (Me_2CO). Mp 142 - 143°

3-O-β-D-Glucopyranoside
[53170-92-6]
$C_{15}H_{20}O_8$ 328.317
Constit. of *Betula alba* and *Betula platyphylla*. Needles (MeOH). Mp 158 - 163°. $[\alpha]_D^{23}$ -20.4 (c, 1 in MeOH)

3-O-[3,4,5-Trihydroxybenzoyl-(→6)-β-D-glucopyranoside]
[2101478-54-8]
$C_{22}H_{24}O_{12}$ 480.422
Constit. of the stem bark of *Schinopsis brasiliensis*. Exhibits free radical scavenging activity. Amorph. solid

Dibenzoyl
Cryst. (Me_2CO aq.). Mp 157°

4′-Me ether, 3-hexadecanoyl
3-(4-Methoxyphenyl)-3-oxopropyl hexadecanoate
[877995-13-6]
$C_{26}H_{42}O_4$ 418.618
Constit. of the stems of *Carissa spinarum*. Pale yellow solid. Mp 78 - 80°.

Zigeuner, G. *et al., Monatsh. Chem.*, 1949, **80**, 801-811 (*synth*).

Tschesche, R. *et al.*, *Phytochemistry*, 1974, **13**, 518-519 (*Betula alba glucoside*).
Achenbach, H. *et al.*, *Phytochemistry*, 1983, **22**, 749-753 (*Carissa edulis constit*).
Fang, J.M. *et al.*, *J. Chin. Chem. Soc. (Taipei)*, 1991, **38**, 61-64 (*Pinus armandii constit*).
Mori, K. *et al.*, *Liebigs Ann. Chem.*, 1992, 485-487 (*3-glucoside, synth*).
Rao, R.J. *et al.*, *Nat. Prod. Res.*, 2005, **19**, 763-769 (*Carissa spinarum constit*).
Jacubert, M. *et al.*, *Tetrahedron*, 2010, **66**, 3775-3787 (*synth*).
Duan, W. *et al.*, *Asian J. Chem.*, 2015, **27**, 2756-2758 (*Trichosanthes kirilowii constit*).
Santos, C.C.S. *et al.*, *Phytochemistry*, 2017, **140**, 45-51 (*3-6′-galloylglucoside, activity*)

3-[2-Hydroxy-4-methoxy-5-(7-methoxy-2-oxo-2*H*-1-benzopyran-6-yl) phenyl]propanoic acid H96

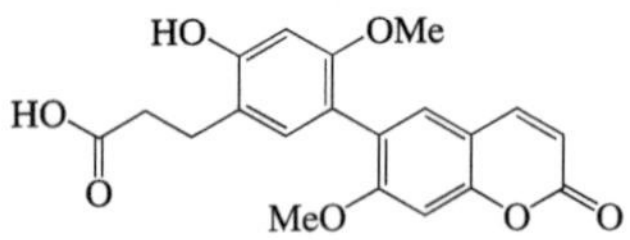

[1621459-75-3]
$C_{20}H_{18}O_7$ 370.357
Constit. of the stem bark of *Hypericum riparium*. Yellowish-white cryst.. Mp 162 - 164°. $[\alpha]^{25}_{D}$ +5.1 (c, 2.4 in MeOH). λ_{max} 286 (log ε 1.31); 324 (log ε 1.21) (MeOH).

Tanemossu, S.A.F. *et al.*, *Phytochemistry*, 2014, **105**, 171-177 (*Hypericum riparium constit*)

7-Hydroxy-6-methoxy-8-[2-methoxy-5-(1,2,3-trihydroxypropyl)phenoxy]-2*H*-1-benzopyran-2-one H97

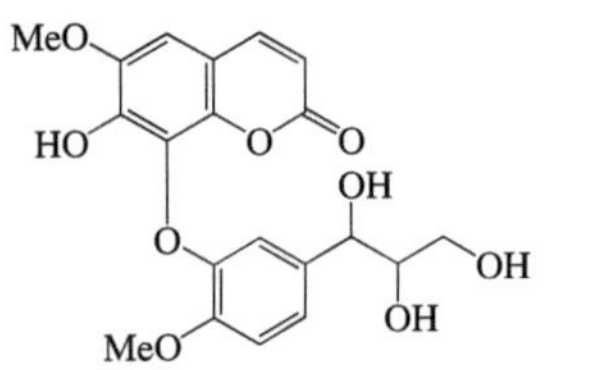

[1392435-24-3]
$C_{20}H_{20}O_9$ 404.371
Constit. of the leaves and twigs of *Aglaia odorata*. Needles. $[\alpha]^{20}_{D}$ +23.5 (c, 0.02 in MeOH).

Zhang, H. *et al.*, *Biochem. Syst. Ecol.*, 2012, **41**, 35-40 (*Aglaia odorata constit*)

3-(4-Hydroxy-3-methoxyphenyl)-2,3-bis[4-(3-hydroxy-1-propenyl)-2-methoxyphenoxy]-1-propanol H98

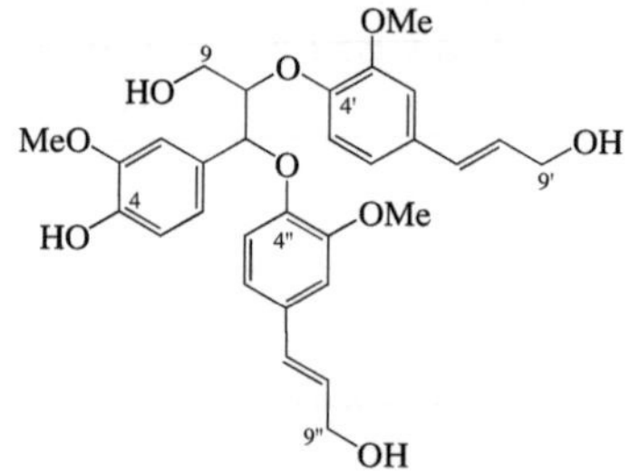

$C_{30}H_{34}O_9$ 538.593
Lignan numbering shown
9,9′,9″-Tri-Ac
[210161-52-7]
$C_{36}H_{40}O_{12}$ 664.704
Constit. of Chilean propolis. Oil. $[\alpha]^{25}_{D}$ -0.6 (c, 1.5 in CH_2Cl_2). Stereochem. not determined. λ_{max} 266 (log ε 4.45); 300 (log ε 4.14) (MeOH).

Valcic, S. *et al.*, *J. Nat. Prod.*, 1998, **61**, 771-775 (*propolis constit*)

1-[6-[2-(4-Hydroxy-3-methoxyphenyl)ethyl]-3-(4-methyl-3-pentenyl)-3-cyclohexenyl]-3-phenyl-1-propanone, CAS H99

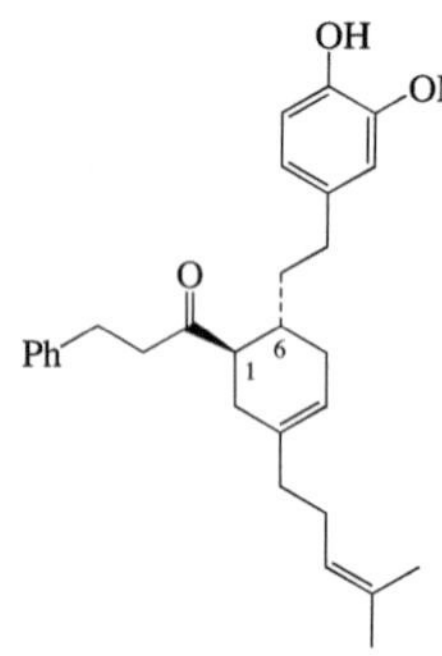

(1*R*,6*R*)-form

$C_{30}H_{38}O_3$ 446.631
The nat. prod. Alpininoid A was isol. as a racemate and resolved by chiral hplc
(1R,6R)-form
(-)-Alpininoid A [2210238-36-9]
Yellow oil. $[\alpha]^{25}_{D}$ -25 (c, 0.1 in MeOH)
(1S,6S)-form
(+)-Alpininoid A [2210238-33-6]
Exhibits neuroprotectant activity. Yellow oil. $[\alpha]^{25}_{D}$ +25 (c, 0.1 in MeOH)
(1RS,6RS)-form
Alpininoid A [2216000-47-2]
Constit. of the rhizomes of *Alpinia officinarum*. Yellow oil. λ_{max} 282 (log ε 1.89) (MeOH).

Liu, H. *et al.*, *J. Nat. Prod.*, 2018, **81**, 162-170 (*Alpininoid A, cd, activity*)

1-[6-[2-(4-Hydroxy-3-methoxyphenyl)ethyl]-4-(4-methyl-3-pentenyl)-3-cyclohexenyl]-3-phenyl-1-propanone, CAS H100

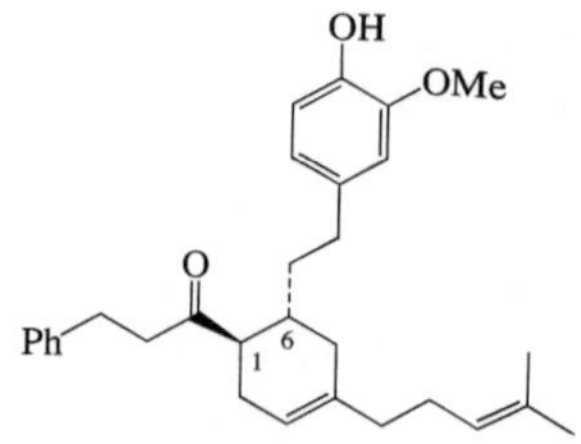

(1*R*,6*R*)-form

$C_{30}H_{38}O_3$ 446.631
The nat. prod. Alpininoid B was isol. as a racemate and resolved by chiral hplc. Closely related to Officinaruminane B, O9
(1R,6R)-form
(-)-Alpininoid B [2210238-37-0]
Yellow oil. $[\alpha]^{25}_{D}$ -20 (c, 0.2 in MeOH)
(1S,6S)-form
(+)-Alpininoid B [2210238-34-7]
Yellow oil. $[\alpha]^{25}_{D}$ +21 (c, 0.2 in MeOH)
(1RS,6RS)-form
Alpininoid B [2216000-48-3]
Constit. of the rhizomes of *Alpinia officinarum*. Yellow oil. λ_{max} 205 (log ε 3.37); 282 (log ε 2.28) (MeOH).

Liu, H. *et al.*, *J. Nat. Prod.*, 2018, **81**, 162-170 (*Alpininoid B, cd*)

3-(4-Hydroxy-3-methoxyphenyl)-2-propenoic acid, CAS H101

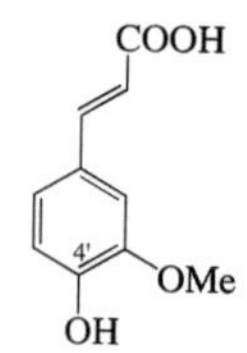

(*E*)-form

4-Hydroxy-3-methoxycinnamic acid, CAS. ***Ferulic acid***. *3-O-Methylcaffeic acid*
[1135-24-6]
[115374-26-0]
$C_{10}H_{10}O_4$ 194.186
Log P 1.27 (calc)
▶ UD3365500
(E)-form [537-98-4]
Obtainable by microbial degradn. of agricultural waste, e.g wheat bran.

Widely distributed in plants, first isol. from *Ferula foetida*. Metab. of a *Penicillium* sp. Found. by Bate-Smith in 33% of investigated dicotyledonous and 67% of monocotyledonous spp.. Antioxidant used in the food industry to inhibit oxidn. of fats. Investigated as starting material for microbial conversion to vanillin. Shows antineoplastic props. and cytostatic activity. Choleretic agent. Antifungal agent used to prevent fruit spoilage. Orthorhombic needles (H_2O). Sol. MeOH, EtOAc; fairly sol. H_2O, Et_2O; poorly sol. C_6H_6, hexane. Mp 174°. Undergoes $SOCl_2$-activated polycondensation. λ_{max} 236; 322 (EtOH) (Berdy)

▶ GD9275000. Exp. reprod. effects

Na salt

Sodium ferulate. FEMA 3812

[24276-84-4]

Flavouring ingredient with a sweet clovery, phenolic odour. Light yellow powder

[β-D-Xylopyranosyl-(1→6)-β-D-glucopyranosyl] ester

[143205-76-9]

$C_{21}H_{28}O_{13}$ 488.442

Constit. of the roots of *Asiasarum sieboldii*. Powder. $[\alpha]^{24}_{D}$ -22 (c, 0.21 in MeOH)

[β-D-Glucopyranosyl-(1→2)-β-D-glucopyranosyl] ester

Tuberonoid A

[1709795-29-8]

$C_{22}H_{30}O_{14}$ 518.468

Constit. of aerial parts of *Allium tuberosum*. Gum. λ_{max} 297 (log ε 3.85); 325 (log ε 3.87) (MeOH)

[β-D-Glucopyranosyl-(1→3)-β-D-glucopyranosyl] ester

Lycibarbarphenylpropanoid C

[1988786-73-7]

$C_{22}H_{30}O_{14}$ 518.468

Constit. of the fruit of *Lycium barbarum*. Amorph. yellowish powder. $[\alpha]^{25}_{D}$ -43.7 (c, 0.5 in MeOH). λ_{max} 204 (log ε 3.79); 219 (log ε 3.88); 237 (log ε 3.85); 328 (log ε 4.14) (MeOH)

[β-D-Glucopyranosyl-(1→4)-β-D-glucopyranosyl] ester

Lycibarbarphenylpropanoid D

[1988786-75-9]

$C_{22}H_{30}O_{14}$ 518.468

Constit. of the fruit of *Lycium barbarum*. Amorph. yellowish powder. $[\alpha]^{25}_{D}$ -39.2 (c, 0.5 in MeOH). λ_{max} 204 (log ε 3.95); 216 (log ε 3.92); 239 (log ε 3.82); 330 (log ε 4.14) (MeOH)

[α-L-Arabinofuranosyl-(1→3)-[β-D-glucopyranosyl-(1→6)]-β-D-glucopyranosyl] ester

[234761-03-6]

$C_{27}H_{38}O_{18}$ 650.583

Constit. of soybean molasses (*Glycine* sp.). Amorph. powder. $[\alpha]^{25}_{D}$ -23.8 (c, 0.52 in MeOH). λ_{max} 227 (log ε 3.9); 275 (log ε 4.15) (MeOH)

Me ester

[22329-76-6]

[2309-07-1]

$C_{11}H_{12}O_4$ 208.213

Isol. from leaves of *Buddleja globosa* and *Tetragonia tetragonoides*, from *Lycopodium* sp., *Meum athamanticum*, *Coreopsis* sp. and many other plants. Germination self-inhibitor of wheat stem rust uredospores. Mp 63 - 64°

Et ester

[28028-62-8]

[4046-02-0]

$C_{12}H_{14}O_4$ 222.24

Isol. from *Solanum melongena*, *Solanum integrifolium* and *Coptis japonica*. Mp 63 - 65°. Bp 164 - 166° at 0.5 mm

Octyl ester

Octyl ferulate

[1132010-84-4]

$C_{18}H_{26}O_4$ 306.402

Constit. of *Arnebia euchroma*. Oil

Tetradecyl ester

Tetradecyl (E)-ferulate. ***Erythrinasinate C****. Erythrinassinate C*

[169694-26-2]

$C_{24}H_{38}O_4$ 390.564

Constit. of *Erythrina sigmoidea*. Amorph. powder. Mp 76°

Pentadecyl ester

Pentadecyl (E)-ferulate

[791603-37-7]

$C_{25}H_{40}O_4$ 404.591

Constit. of *Salicornia herbacea*. Antioxidant. Exhibits antiproliferative props.. Amorph. powder

Hexadecyl ester

Hexadecyl (E)-ferulate

[158306-36-6]

[64190-80-3]

$C_{26}H_{42}O_4$ 418.618

Isol. from *Astragalus membranaceus*, *Euchresta horsfieldii* and *Ipomoea batatas*

Octadecyl ester

Octadecyl (E)-ferulate

[133882-77-6]

[64190-81-4]

$C_{28}H_{46}O_4$ 446.672

Constit. of *Euphorbia pekinensis*, *Lupinus luteus*, *Pavetta owariensis*, *Rubia tinctorum*, *Withania somnifera* and other plant spp.. Needles. Mp 110 - 111°. λ_{max} 204 (log ε 4.24); 237 (log ε 3.95); 324 (log ε 4.25) (MeOH)

9Z-Octadecenyl ester

9-Octadecenyl ferulate

[197250-17-2]

$C_{28}H_{44}O_4$ 444.656

Constit. of *Lupinus luteus*

11Z-Octadecenyl ester

11-Octadecenyl ferulate

[197250-18-3]

$C_{28}H_{44}O_4$ 444.656

Constit. of *Lupinus luteus*

13Z-Octadecenyl ester

[197250-19-4]

$C_{28}H_{44}O_4$ 444.656

Constit. of *Lupinus luteus*

14Z-Octadecenyl ester

[197250-20-7]

$C_{28}H_{44}O_4$ 444.656

Constit. of *Lupinus luteus*

9Z,12Z-Octadecadienyl ester

Linoleyl ferulate

[2119632-78-7]

$C_{28}H_{42}O_4$ 442.640

Constit. of the roots of *Astragalus membranaceus*. α-Glucosidase inhibitor

Nonadecyl ester

Nonadecyl ferulate

[126192-42-5]

$C_{29}H_{48}O_4$ 460.699

Constit. of *Lupinus luteus*, *Pavetta owariensis* and *Spiranthes sinensis*. *E*-Config. not confirmed

Eicosyl ester

Eicosyl (E)-ferulate. Icosyl ferulate

[133882-79-8]

[64190-82-5]

$C_{30}H_{50}O_4$ 474.726

Constit. of *Baccharis genistelloides*, *Larix laricina Lupinus luteus* and *Withania somnifera*. Cryst. (cyclohexane). Mp 61 - 62°

11Z-Eicosenyl ester
11-Eicosenyl ferulate. 11-Icosenyl ferulate
[197250-21-8]
$C_{30}H_{48}O_4$ 472.71
Constit. of *Lupinus luteus*. Geometry not certain

Heneicosyl ester
Heneicosyl ferulate
[145205-02-3]
[126192-41-4]
Constit. of *Pavetta owariensis* and *Spiranthes sinensis*

Docosyl ester
Docosyl (E)-ferulate. Docosyl trans-ferulate
[101927-24-6]
[62267-81-6]
$C_{32}H_{54}O_4$ 502.78
Isol. from *Bauhinia manca, Lupinus luteus, Platycladus orientalis, Pavetta owariensis, Ligularia nelumbifolia, Teucrium divaricatum* ssp. *villosum* and *Withania somnifera*. Needles. Mp 57 - 58° (52-53°). λ_{max} 219 (log ε 3.6); 223 (log ε 3.6); 293 (sh); 325 (log ε 4) (MeOH)

Tricosyl ester
Tricosyl (E)-ferulate
[101959-30-2]
[126192-40-3]
$C_{33}H_{56}O_4$ 516.807
Isol. from the stems of *Bauhinia manca*

Tetracosyl ester
Tetracosyl (E)-ferulate. Lignoceryl ferulate
[101927-25-7]
[29018-62-0]
$C_{34}H_{58}O_4$ 530.834
Constit. of *Bauhinia manca, Croton megalocarpus, Hedysarum polybotrys, Lupinus luteus, Pygeum africanum*, roots of *Vanda roxburghii* and many other plants. λ_{max} 217; 246; 296; 327 (EtOH) (Berdy). [base] λ_{max} 216; 250; 308; 362 (EtOH/NaOH) (Berdy)

Pentacosyl ester
Pentacosyl (E)-ferulate
[101959-33-5]
[126192-39-0]
$C_{35}H_{60}O_4$ 544.861
Isol. from the stems of *Bauhinia manca*

Hexacosyl ester
Hexacosyl (E)-ferulate
[63034-29-7]
[18594-58-6]
$C_{36}H_{62}O_4$ 558.888
Isol. from the stems of *Bauhinia manca* and from *Pinus roxburghii*. Mp 66.5°

Heptacosyl ester
Heptacosyl (E)-ferulate
[101959-35-7]
[126192-38-9]
$C_{37}H_{64}O_4$ 572.915
Isol. from the stems of *Bauhinia manca*

Octacosyl ester
Octacosyl (E)-ferulate. ***Erythrinasinate B***. *Erythrinassinate B*
[101959-37-9]
[35321-71-2]
$C_{38}H_{66}O_4$ 586.942
Constit. of *Bauhinia manca, Bombax malabaricum, Clusia nemorosa, Croton megalocarpus, Erythrina excelsa, Erythrina indica, Rhodiola yunnanensis* and *Tecomella undulata*. Cryst. (hexane/EtOAc). Mp 80°. λ_{max} 230 (log ε 4.03); 320 (log ε 4.05) (EtOH)

Triacontyl ester
Triacontyl ferulate
[146555-74-0]
[115374-26-0]
$C_{40}H_{70}O_4$ 614.996
Constit. of *Bombax malabaricum, Bridelia ovata, Kalanchoe daigremontiana* and *Ailanthus integrifolia*. Amorph. powder. *E*-Config. of *K.* sp. isolate not confirmed

Dotriacontyl ester
Dotriacontyl (E)-ferulate
[1372653-48-9]
$C_{42}H_{74}O_4$ 643.05
Constit. of the stem bark of *Callistemon lanceolatus*

Pentatriacontyl ester
Pentatriacontyl ferulate
[148454-84-6]
$C_{45}H_{80}O_4$ 685.131
Constit. of *Jatropha elliptica*

20-Hydroxyeicosyl ester
20-Hydroxyeicosyl (E)-ferulate. ***Lyciumol B***
[1228555-13-2]
$C_{30}H_{50}O_5$ 490.725
Constit. of the root bark of *Lycium chinense*. Amorph. powder. λ_{max} 236 (log ε 3.92); 327 (log ε 4.07) ($CHCl_3$)

22-Hydroxydocosyl ester
22-Hydroxydocosyl (E)-ferulate. ***Lyciumol A***
[1228555-12-1]
$C_{32}H_{54}O_5$ 518.779
Constit. of the root bark of *Lycium chinense*. Amorph. powder. λ_{max} 241 (log ε 3.88); 321 (log ε 4.05) ($CHCl_3$)

32-Hydroxydotriacontyl ester
32-Hydroxydotriacontyl (E)-ferulate
[1372653-49-0]
$C_{42}H_{74}O_5$ 659.049
Constit. of the stem bark of *Callistemon lanceolatus* and the leaves of *Colebrookea oppositifolia*. Brownish solid. Mp 88°

4-Hydroxy-3-methylbutyl ester
[144978-91-6]
$C_{15}H_{20}O_5$ 280.32
Constit. of *Eupatorium buniifolium*. *E*-Config. not confirmed

Amide
Ferulamide.
4-Hydroxy-3-methoxycinnamide
[61012-31-5]
[19272-90-3]
$C_{10}H_{11}NO_3$ 193.202
Alkaloid from *Glaucium grandiflorum* var. *grandiflorum*. Prisms (EtOH). Mp 151 - 153°

Anhydride
[1631068-68-2]
$C_{20}H_{18}O_7$ 370.357
Constit. of leaves of *Taxus cuspidata*. Powder

(Z)-form [1014-83-1]
Mp 108 - 111°. λ_{max} 316 (EtOH)

[β-D-Glucopyranosyl-(1→2)-β-D-glucopyranosyl] ester
Tuberonoid B
[1709795-30-1]
$C_{22}H_{30}O_{14}$ 518.468
Constit. of the aerial parts of *Allium tuberosum*. Gum. λ_{max} 270 (log ε 3.83); 293 (log ε 3.97) (MeOH)

Me ester
[34298-89-0]
$C_{11}H_{12}O_4$ 208.213
Isol. from *Murraya paniculata* and *Tetragonia tetragonoides*. Germination self-inhibitor of wheat stem rust uredospores. Oil

Octadecyl ester
[133882-78-7]
Constit. of *Pavetta owariensis*

Docosyl ester

Docosyl (Z)-ferulate. Docosyl cis-ferulate

[133882-81-2]

Constit. of *Pavetta owariensis*, *Platycladus orientalis* and *Teucrium divaricatum* ssp. *villosum*. Mp 57 - 58°. λ_{max} 218 (log ε 3.8); 232 (log ε 3.6); 300 (sh); 326 (log ε 4) (MeOH)

Octacosyl ester

Octacosyl (Z)-ferulate

[945481-23-2]

$C_{38}H_{66}O_4$ 586.942

Constit. of the roots of *Euphorbia hylonoma*. Cryst.. Mp 101 - 103°.

Aldrich Library of FT-IR Spectra, 1st edn., 1985, **2**, 184C (*ir*).

Aldrich Library of ^{13}C *and* ^{1}H *FT-NMR Spectra*, 1992, **2**, 1058A , 1235A (*nmr*).

Pearl, I.A. *et al.*, *J. Org. Chem.*, 1951, **16**, 216-220 (*synth*).

Bate-Smith, E.C. *Sci. Proc. R. Dublin Soc.*, 1956, **27**, 165-176 (*occur*).

Comte, P. *et al.*, *C. R. Hebd. Seances Acad. Sci.*, 1957, **245**, 1144-1145 (*Z-form, synth*).

Herrmann, K. *Naturwissenschaften*, 1958, **45**, 111-112 (*occur*).

Nair, G.V. *et al.*, *Can. J. Chem.*, 1959, **37**, 1608-1613 (*eicosyl ester*).

Challice, J.S. *et al.*, *J. Chromatogr.*, 1966, **21**, 357-362 (*isom, uv*).

Macko, V. *et al.*, *Science (Washington, D.C.)*, 1971, **173**, 835-836 (*E-form Me ester, Z-form Me ester, isol, synth*).

Von Massow, F. *et al.*, *J. Chem. Soc., Perkin Trans. 2*, 1976, 977-980 (*cmr*).

Kelley, C.J. *et al.*, *J. Org. Chem.*, 1976, **41**, 449-455 (*cmr*).

Karrer, W. *Konstitution und Vorkommen der Organischen Pflanzenstoffe, 2nd edn., Birkhäuser Verlag*, 1976-1985, no. 959 (*occur*).

Bohlmann, F. *et al.*, *Phytochemistry*, 1980, **19**, 2471-2472 (*E-form Me ester, isol*).

Kuo, Y.H. *et al.*, *CA*, 1983, **99**, 53253 (*Z-form Me ester, synth*).

Chenault, J. *et al.*, *Synth. Commun.*, 1984, **14**, 1059-1065 (*synth*).

Elias, H.G. *et al.*, *Makromol. Chem.*, 1985, **186**, 1027-1045 (*polym*).

Achenbach, H. *et al.*, *Z. Naturforsch., C*, 1986, **41**, 164-168 (*esters*).

Nethaji, M. *et al.*, *Acta Crystallogr., Sect. C*, 1988, **44**, 275-277 (*cryst struct*).

Yoshihara, T. *et al.*, *CA*, 1989, **110**, 151334 (*E-form Et ester, isol*).

Tillett, R. *et al.*, *Arch. Mikrobiol.*, 1990, **154**, 206-208 (*Penicillium metab*).

Wandji, J. *et al.*, *J. Nat. Prod.*, 1990, **53**, 1425-1429 (*Hexacosyl ferulate, Octacosyl ferulate*).

Mizuno, M. *et al.*, *Phytochemistry*, 1990, **29**, 2663-2665 (*Hexadecyl ferulate*).

Sakagami, H. *et al.*, *Chem. Pharm. Bull.*, 1991, **39**, 950-955 (*polym*).

Balde, A.M. *et al.*, *Phytochemistry*, 1991, **30**, 1024-1026 (*Pavetta owariensis esters*).

Naim, M. *et al.*, *ACS Symp. Ser.*, 1992, **506**, 180-191 (*occur*).

Booyaratavej, S. *et al.*, *J. Nat. Prod.*, 1992, **55**, 1761-1763 (*triacontyl ester*).

Hashimoto, K. *et al.*, *Phytochemistry*, 1992, **31**, 2477-2480 (*xylosylglucosyl ester*).

Kosuge, K. *et al.*, *Chem. Pharm. Bull.*, 1994, **42**, 1669-1671 (*triacontyl ester*).

Li, J.K. *et al.*, *J. Chin. Chem. Soc. (Taipei)*, 1994, **41**, 213-216 (*E-form Me ester, Z-form Me ester, Murraya paniculata constits*).

Zhao, Y. *et al.*, *Phytochemistry*, 1994, **37**, 1149-1152 (*docosyl ester*).

Ulubelen, A. *et al.*, *Phytochemistry*, 1994, **37**, 1371-1375 (*docosyl esters*).

Katagiri, Y. *et al.*, *Phytochemistry*, 1997, **46**, 347-352 (*Lupinus luteus esters*).

Nkengfack, A.E. *et al.*, *Phytochemistry*, 1997, **46**, 573-578 (*Erythrinasinate C*).

Kavitha, J. *et al.*, *J. Asian Nat. Prod. Res.*, 1999, **2**, 51-54 (*Tetradecyl ferrulate, synth*).

Kuo, Y.H. *et al.*, *J. Chin. Chem. Soc. (Taipei)*, 1999, **46**, 819-825 (*docosyl esters*).

Faulds, C. *et al.*, *Chem. Br.*, May, 2000, 48-50 (*rev, use*).

Sari, A. *Acta Pharm. Turc.*, 2001, **43**, 89-92 (*Ferulamide, Glaucium grandiflorum constit*).

Nomura, E. *et al.*, *Bioorg. Med. Chem.*, 2003, **11**, 3807-3813 (*amide, synth, ir, pmr, cmr*).

Ruan, H.-L. *et al.*, *Fitoterapia*, 2007, **78**, 72-73 (*Octacosyl Z-ferulate*).

Bazin, M.-A. *et al.*, *Synth. Commun.*, 2008, **38**, 3861-3874 (*Me ester*).

Pan, S. *et al.*, *J. Asian Nat. Prod. Res.*, 2009, **11**, 681-685 (*Lyciumols A,B*).

Zhu, L.-C. *et al.*, *Acta Crystallogr., Sect. E*, 2010, **66**, o2953 (*cryst struct*).

Fenaroli's Handbook of Flavor Ingredients, 6th edn., (ed. Burdock, G.A.), CRC Press, 2010, 1846 (*Na salt*).

Liu, H. *et al.*, *J. Asian Nat. Prod. Res.*, 2010, **12**, 286-292 (*octyl ester*).

Salum, M.L. *et al.*, *Org. Lett.*, 2010, **12**, 4808-4811 (*Z-form, synth, uv, pmr, cmr, ms*).

Hennig, L. *et al.*, *ARKIVOC*, 2011, **vi**, 74-81 (*eicosyl ester*).

Jain, R. *et al.*, *J. Indian Chem. Soc.*, 2011, **88**, 735-737 (*dotriacontyl ester, 32-hydroxydotriacontyl ester*).

Verma, S.K. *et al.*, *Indian J. Chem., Sect. B*, 2012, **51**, 1502-1503 (*32-hydroxydotriacontyl ester*).

Wang, X. *et al.*, *Food Chem.*, 2013, **141**, 2066-2074 (*pentadecyl ester*).

Wu, Y. *et al.*, *Chem. Nat. Compd. (Engl. Transl.)*, 2014, **50**, 603-605 (*anhydride*).

Han, S.H. *et al.*, *Arch. Pharmacal Res.*, 2015, **38**, 1312-1316 (*Tuberonoids A,B*).

De Oliveiria, D.M. *et al.*, *Plant Biotechnol. J.*, 2015, **13**, 1224-1232 (*rev, biosynth*).

Damasceno, S.S. *et al.*, *Curr. Pharm. Des.*, 2017, **23**, 3015-3023 (*rev, biol, activity*).

Zhou, Z.-Q. *et al.*, *Food Chem.*, 2017, **214**, 644-654 (*Lycibarbarphenylpropanoids C,D,H*).

Liu, Y. *et al.*, *Molecules*, 2017, **22**, 411:1-12 (*Linoleyl ferulate, activity*).

Aguilar-Hernández, I. *et al.*, *Vib. Spectrosc.*, 2017, **89**, 113-122 (*Raman*).

Lewis, R.J. *Sax's Dangerous Properties of Industrial Materials, 10th edn., J. Wiley*, 2000, FBP200

3-[4-Hydroxy-3-(3-methyl-2-butenyl)phenyl]propanol H102

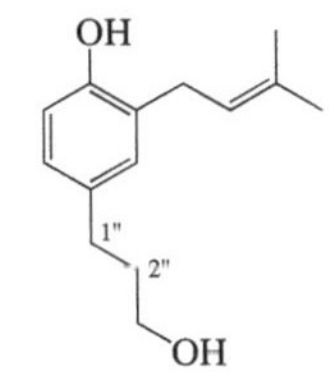

4-Hydroxy-3-(3-methyl-2-butenyl)benzenepropanol, CAS. *4-(3-Hydroxypropyl)-2-prenylphenol.* ***Zanthoxylol***

[13515-57-6]

$C_{14}H_{20}O_2$ 220.312

Constit. of the heartwood and roots of *Zanthoxylum zanthoxyloides*. Antisickling agent. Oil. λ_{max} 280 (no solvent reported)

1″,2″-Didehydro(E-)

3-[4-Hydroxy-3-(3-methyl-2-butenyl)phenyl]-2-propen-1-ol. 3-(4-Hydroxy-3-prenylphenyl)-2-propen-1-ol. 4-Hydroxy-3-(3-methyl-2-butenyl)cinnamyl alcohol. 4-Hydroxy-3-prenylcinnamyl alcohol

[104358-48-7]

$C_{14}H_{18}O_2$ 218.296

Constit. of the stem wood of *Zanthoxylum integrifoliolum*. Yellowish needles ($CHCl_3$/MeOH). Mp 88 - 90°. λ_{max} 219 (log ε 4.4); 263 (log ε 4.31) (MeOH).

Eshiet, I.T. *et al.*, *J. Chem. Soc. C*, 1968, 481-484 (*Zanthoxylol*).

Elujoba, A.A. *et al.*, *J. Pharm. Biomed. Anal.*, 1985, **3**, 447-451 (*Zanthoxylol*).

Menon, S.R. *et al.*, *J. Nat. Prod.*, 1999, **62**, 102-106 (*synth*).

Chen, J.-J. *et al.*, *J. Nat. Prod.*, 2007, **70**, 1444-1448 (*didehydro*)

2-Hydroxy-4,5-methylenedioxycinnamic acid H103

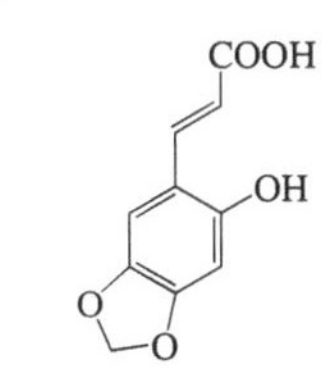

(*E*)-*form*

$C_{10}H_8O_5$ 208.169

(E)-form

4-Aminobutylamide

[1971064-92-2]

$C_{14}H_{18}N_2O_4$ 278.308

Alkaloid from *Drypetes staudtii*. Light brown solid. λ_{max} 230; 318 (MeOH)

(4-Aminobutyl)methylamide

[1971064-93-3]

$C_{15}H_{20}N_2O_4$ 292.335

Alkaloid from *Drypetes staudtii*. Light brown solid. λ_{max} 232; 317 (MeOH)

Me ether

2-Methoxy-4,5-methylenedioxycinnamic acid

Me ether, 2-methylpropylamide

2-Methoxy-4,5-methylenedioxycinnamoyl isobutylamide. *5′-Methoxyfagaramide*

[259258-85-0]

$C_{15}H_{19}NO_4$ 277.320

Alkaloid from the roots of *Piper amalago* var. *nigrinodum*. Cryst.. Mp 150 - 152°

Me ether, pyrrolidide

2-Methoxy-4,5-methylenedioxycinnamoyl pyrrolidide

[82682-47-1]

$C_{15}H_{17}NO_4$ 275.304

Alkaloid from the roots of *Piper amalago*. Yellowish cryst.. Mp 178°

Me ether, piperidide

2-Methoxy-4,5-methylenedioxycinnamoyl piperidide

[67879-91-8]

$C_{16}H_{19}NO_4$ 289.331

Alkaloid from the roots of *Piper amalago*. Yellowish plates (hexane/Me_2CO). Mp 120°

(Z)-form

Me ether, pyrrolidide

[105705-79-1]

Alkaloid from the leaves of *Piper peepuloides* and from *Piper hispidum*. Needles (EtOAc/petrol). Mp 118°

Me ether, piperidide

[73885-65-1]

Alkaloid from *Piper peepuloides*. Cryst. (EtOAc/petrol). Mp 99°. λ_{max} 240; 275; 333 (MeOH).

Sehgal, C.K. *et al.*, *Phytochemistry*, 1979, **18**, 1865-1867 (*cis-piperidide*).

Domínguez, X.A. *et al.*, *Phytochemistry*, 1986, **25**, 239-240 (*trans-piperidide, trans-pyrrolidide*).

Shah, S. *et al.*, *Phytochemistry*, 1986, **25**, 1997-1998 (*cis-pyrrolidide*).

Jacobs, H. *et al.*, *J. Indian Chem. Soc.*, 1999, **76**, 713-717 (*isobutylamide*).

Grace, D. *et al.*, *Chem. Biodiversity*, 2016, **13**, 913-917 (*Drypetes staudtii amides*)

5-(Hydroxymethyl)-7-methyl-6-(2,4,5-trimethoxyphenyl)-8-oxabicyclo[3.2.1]oct-3-en-2-one H104

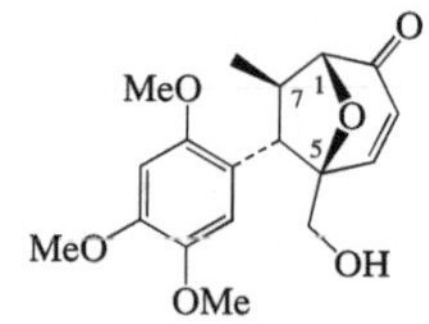

(1*R*,5*R*,6*S*,7*R*)-*form*

$C_{18}H_{22}O_6$ 334.368

(1R,5R,6S,7R)-form

(+)-Acortatarone A. ***Acorusin C***

Related to Cartorimin, C36. The nat. prod. Acortatarone A was isol. as a racemate and resolved by chiral hplc. Constit. of the rhizomes of *Acorus tatarinowii*. Exhibits moderate cytotoxic activity. Solid. $[\alpha]^{20}_{D}$ +75.8 (c, 0.06 in MeOH) (Acorusin C). $[\alpha]_D$ +70 (c, 0.5 in MeOH) ((+)-Acortatarone A). λ_{max} 228 (log ε 4.56); 291 (log ε 3.98) (MeOH)

(1S,5S,6R,7S)-form

(-)-Acortatarone A

$[\alpha]^{32}_{D}$ -79.2 (c, 0.5 in MeOH)

(1RS,5RS,6SR,7RS)-form

Acortatarone A

Constit. of the rhizomes of *Acorus tatarinowii*. α-Glucosidase inhibitor. λ_{max} 205 (log ε 3.88); 230 (log ε 3.40); 288 (log ε 3.20) (MeOH).

Ni, G. *et al.*, *Planta Med.*, 2016, **82**, 632-638 (*Acorusin C, cd, activity*).

Gao, E. *et al.*, *Fitoterapia*, 2017, **122**, 7-10 (*Acortatarone A, cd, activity*)

1-(5-Hydroxy-2-methylphenyl)-1,2-propanediol H105

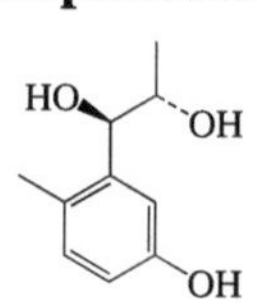

$C_{10}H_{14}O_3$ 182.219

(1R*,2S*)-form [2145115-88-2]

Prod. by *Microsphaeropsis arundinis* PH30472. Exhibits antifungal activity. Amorph. solid. $[\alpha]^{25}_{D}$ -16.7 (c, 0.2 in MeOH). λ_{max} 209; 283 (MeOH).

Qin, S.-H. *et al.*, *Chem. Nat. Compd. (Engl. Transl.)*, 2017, **53**, 877-879 (*Microsphaeropsis arundinis metab, activity*)

4-Hydroxy-3,4′-oxylign-7-ene-9,9′-dioic acid H106

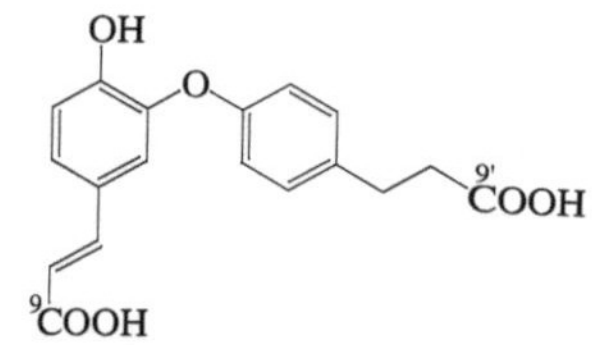

3-[3-[4-(2-Carboxyethyl)phenoxy]-4-hydroxyphenyl]-2-propenoic acid

$C_{18}H_{16}O_6$ 328.32

O-β-D-Glucopyranoside

Spiraformin D

[833485-21-5]

$C_{24}H_{26}O_{11}$ 490.461

Constit. of the stems of *Spiraea formosana*. Syrup. $[\alpha]^{25}_{D}$ -2.1 (c, 0.9 in MeOH). λ_{max} 222; 289 (MeOH)

Di-Me ester

Spiraformin A

[834867-95-7]

$C_{20}H_{20}O_6$ 356.374

Constit. of the stems of *Spiraea formosana*. Syrup. λ_{max} 223; 313 (MeOH)

9′-Et, 9-Me diester

Spiraformin B

[834867-96-8]

$C_{21}H_{22}O_6$ 370.401

Constit. of the stems of *Spiraea formosana*. Syrup. λ_{max} 286 (MeOH)

9′-Butyl, 9-Me diester

Spiraformin C

[834867-97-9]

$C_{23}H_{26}O_6$ 398.455

Constit. of the stems of *Spiraea formosana*. Syrup. λ_{max} 222; 312 (MeOH).

Wu, T.-S. *et al.*, *Chem. Pharm. Bull.*, 2004, **52**, 1227-1230 (*Spiraformins A-D*).

Jung, N. *et al.*, *Eur. J. Org. Chem.*, 2009, 4494-4502 (*Spiraformin A, synth*)

3-(4-Hydroxyphenyl)-2-oxopropanoic acid H107

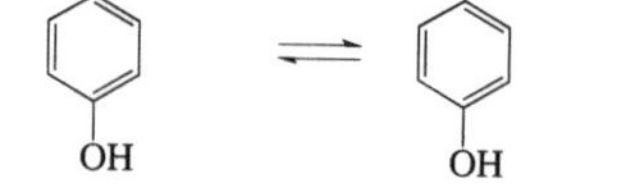

(Z)-*enol-form*

2-Hydroxy-3-(4-hydroxyphenyl)-2-propenoic acid. 4-Hydroxy-α-oxobenzenepropanoic acid. 4-Hydroxyphenylpyruvic acid

[156-39-8]

[10589-28-3] (enol-form)
$C_9H_8O_4$ 180.159
Exhibits keto-enol tautom. Enol form predominates. Constit. of *Erythraea centaurium* (preferred genus name *Centaurium*), *Euphrasia rostkoviana*, *Lysimachia nummularia*, *Phlox* spp. and *Thuja plicata*. Prod. by *Aerobacter aerogenes*. Isol. from the red alga *Ceratodictyon spongiosum* and its sponge symbiont *Haliclona cymaeformis*. Biosynthetic intermed. for aromatic compds. Enzyme substrate. Cryst. (H_2O). Mp 220° dec. (212-214°). Rapidly oxidised in alkaline soln.

4-Sulfate
4-Sulfooxyphenylpyruvic acid
[457955-46-3]
$C_9H_8O_7S$ 260.216
Constit. of the alga *Ceratodictyon spongiosum* and its sponge symbiont *Haliclona cymaeformis*. Amorph. solid. λ_{max} 208 (log ε 3.6); 220 (log ε 3.6); 298 (log ε 3.8) (MeOH)

Oxime
4-Hydroxy-α-(hydroxyimino) benzenepropanoic acid, CAS. ***α-Oxyiminotyrosine***
[56401-28-6]
$C_9H_9NO_4$ 195.174
Dehydroaminoacid from the sponge *Hymeniacidon sanguinea*. Cryst.. Mp 168 - 169°

Phenylhydrazone
Mp 161 - 162°

Me ester
[214534-53-9] (Z-form)
[214534-54-0] (*E*-form)
$C_{10}H_{10}O_4$ 194.186
Constit. of the green alga *Caulerpa taxifolia*. Isol. as a mixt. of *E*/*Z*-enol isomers (Z-predominant)

Me ester, oxime
Cryst. ($CHCl_3$). Mp 135.5 - 137°

4-Me ether
4-Methoxy-α-oxobenzenepropanoic acid. 4-Methoxyphenylpyruvic acid
[28030-16-2]
$C_{10}H_{10}O_4$ 194.186
Mp 194° dec. (190° dec.)

Z-enol-form
α-Me ether, amide
Botryllamide L
[1238387-74-0]
$C_{10}H_{11}NO_3$ 193.202
Isol. from *Aplidium altarium*. Powder. λ_{max} 220 (log ε 3.88); 297 (log ε 4.12) (MeOH).

Gaudruy, R. *et al., Can. J. Chem.*, 1951, **29**, 427-438 (*synth*).
Org. Synth., 1963, **43**, 49.
Brandner, G. *et al., Acta Chem. Scand.*, 1964, **18**, 574-575 (*Phlox decussata constit*).
Org. Synth., Coll. Vol. 5, 1973, 627-632 (*synth, bibl*).
Cimino, G. *et al., Experientia*, 1975, **31**, 756-757 (*Hymeniacidon sanguinea constit*).
Forrester, A.R. *et al., J. Chem. Soc., Perkin Trans. 1*, 1975, 2340-2348 (*synth*).
Moeller, B.L. *et al., Acta Chem. Scand., Ser. B*, 1977, **31**, 343-344 (*nitrile, synth*).
Moeller, B.L. *et al., Carlsberg Res. Commun.*, 1979, **44**, 367-379 (*ms*).
Klein, C. *et al., Liebigs Ann. Chem.*, 1983, 1638-1639 (*synth*).
Luczak, S. *et al., Acta Pol. Pharm.*, 1989, **46**, 381-385 (*Lysimachia nummularia constit*).
Hanai, K. *et al., J. Phys. Chem.*, 1989, **93**, 6013-6016 (*ir, Raman, tautom*).
Simchen, G. *et al., Synthesis*, 1989, 945-946 (*synth*).
Mancini, I. *et al., Helv. Chim. Acta*, 1998, **81**, 1681-1691 (*Me ester*).
Bugni, T.S. *et al., Phytochemistry*, 2002, **60**, 361-363 (*4-Sulfooxyphenylpyruvic acid*).
Yin, S. *et al., Tetrahedron Lett.*, 2010, **51**, 3403-3405 (*Botryllamide L*).
Maeda, H. *et al., Annu. Rev. Plant Biol.*, 2012, **63**, 73-105 (*rev, biosynth*)

3-(4-Hydroxyphenyl)-3-oxopropanoic acid H108

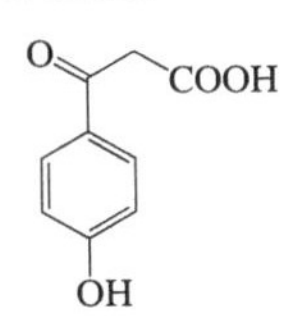

4-Hydroxy-β-oxobenzenepropanoic acid, CAS. *4-Hydroxybenzoylacetic acid*
[62024-30-0]
$C_9H_8O_4$ 180.159

Me ester
[32066-29-8]
$C_{10}H_{10}O_4$ 194.186
Isol. from flowers of *Bellendena montana*. Cryst.. Mp 83 - 84°

4-Ac
[1878-81-5]
$C_{11}H_{10}O_5$ 222.196
Mp 175 - 177°

Me ether
4-Methoxybenzoylacetic acid. Anisoylacetic acid. 3-Oxo-3-(4-methoxyphenyl)propanoic acid
[13422-77-0]
$C_{10}H_{10}O_4$ 194.186
Isol. from *Bellendena montana*. Mp 180° dec. (83-84°)

Me ether, Me ester
[22027-50-5]
$C_{11}H_{12}O_4$ 208.213
Mp 27 - 28°. Bp 190 - 192° at 10 mm

Me ether, Et ester
[2881-83-6]
$C_{12}H_{14}O_4$ 222.24
Bp 180 - 190° at 11 mm (dec.) (lit. gives a pressure range)

Me ether, nitrile
[3672-47-7]
$C_{10}H_9NO_2$ 175.187
Pale yellow solid. Mp 129 - 130°.

Wahl, A. *et al., Bull. Soc. Chim. Fr.*, 1912, **11**, 25-34 (*synth*).
Sonn, A. *Ber. Dtsch. Chem. Ges.*, 1918, **51**, 821-827 (*synth*).
Bick, I.C. *et al., Phytochemistry*, 1971, **10**, 475-477 (*Bellendena montana constits*).
Wierenga, W. *et al., J. Org. Chem.*, 1979, **44**, 310-311 (*synth*).
Herbert, R.B. *et al., J. Chem. Soc., Perkin Trans. 1*, 1984, 825-931 (*synth*).
Nowill, R.W. *et al., Tetrahedron Lett.*, 2011, **52**, 2440-2442 (*Me ether nitrile*).
Merad, J. *et al., Monatsh. Chem.*, 2017, **148**, 57-62 (*Me ether*).
Nagata, T. *et al., Org. Lett.*, 2017, **19**, 4672-4675 (*Me ether nitrile*)

1-(4-Hydroxyphenyl)-7-phenyl-4,6-heptadien-3-ol H109

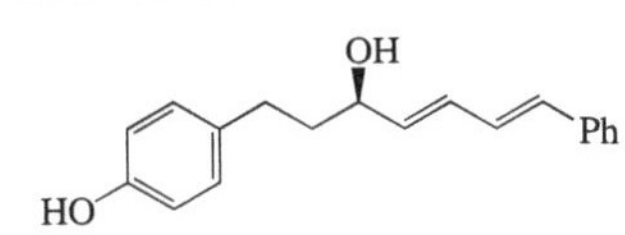

(3*R*,4*E*,6*E*)-*form*

$C_{19}H_{20}O_2$ 280.367

(3R,4E,6E)-form
Constit. of the rhizomes of *Curcuma comosa*

3-Me ether
1-(4-Hydroxyphenyl)-3-methoxy-7-phenyl-4,6-heptadiene
$C_{20}H_{22}O_2$ 294.394
Constit. of the rhizomes of *Curcuma comosa*

(3S,4E,6E)-form
Constit. of the rhizomes of *Curcuma comosa*

3-Me ether
Constit. of the rhizomes of *Curcuma comosa*.

Sornkaew, N. *et al., Nat. Prod. Commun.*, 2015, **10**, 89-93 (*Curcuma comosa constits*)

1-(4-Hydroxyphenyl)-7-phenyl-4,6-heptadien-3-one H110

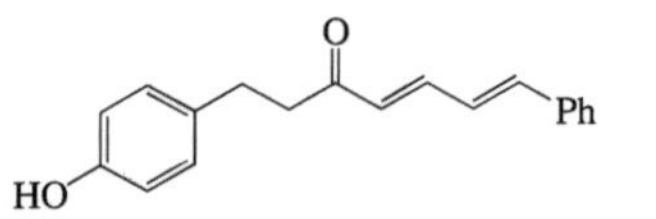

$C_{19}H_{18}O_2$ 278.351

(4E,6E)-form [562810-28-0]

Isol. from rhizomes of *Curcuma comosa*. Prisms. Mp 129 - 130°

4,5-Dihydro

1-(4-Hydroxyphenyl)-7-phenyl-6-hepten-3-one

[1062082-72-7]

$C_{19}H_{20}O_2$ 280.367

Constit. of the rhizomes of *Curcuma comosa* and *Curcuma kwangsiensis*. Viscous oil.

Baranovsky, A. *et al.*, *Synth. Commun.*, 2003, **33**, 1019-1045 (*synth*).
Suksamrarn, A. *et al.*, *Bioorg. Med. Chem.*, 2008, **16**, 6891-6902 (*Curcuma comosa constits*).
Li, J. *et al.*, *J. Nat. Prod.*, 2010, **73**, 1667-1671 (*Curcuma kwangsiensis constit*)

1-(4-Hydroxyphenyl)-7-phenyl-3,5-heptanediol H111

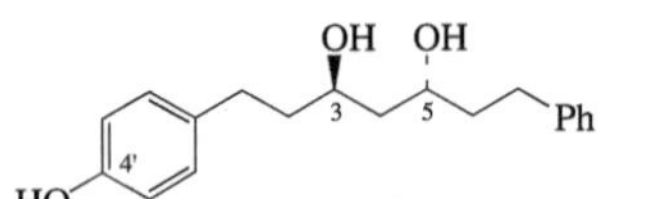

(3R,5R)-form

$C_{19}H_{24}O_3$ 300.398

(3R,5R)-form [112494-44-7]

Constit. of *Alpinia officinarum*. Needles. Mp 109 - 111°. $[\alpha]_D$ +8.3 (c, 0.4 in $CHCl_3$). λ_{max} 215 (ε 8000); 225 (ε 8500); 275 (ε 1500) (EtOH)

(3S)-form

5-Ketone, 3-Ac

5-Acetoxy-7-(4-hydroxyphenyl)-1-phenyl-3-heptanone

[1246280-51-2]

$C_{21}H_{24}O_4$ 340.419

Constit. of the rhizomes of *Alpinia officinarum*. Oil. $[\alpha]^{20}_D$ +5.4 (c, 0.05 in $CHCl_3$). λ_{max} 210; 270 (MeOH)

(5S)-form

3-Ketone

5-Hydroxy-1-(4-hydroxyphenyl)-7-phenyl-3-heptanone

[1269839-32-8]

$C_{19}H_{22}O_3$ 298.382

Constit. of the rhizomes of *Curcuma kwangsiensis*. Yellowish oil. $[\alpha]^{25}_D$ +15.3 (c, 0.1 in MeOH). λ_{max} 278 (log ε 3.03) (MeOH)

(3ξ)-form

5-Ketone

5-Hydroxy-7-(4-hydroxyphenyl)-1-phenyl-3-heptanone

[83161-96-0]

$C_{19}H_{22}O_3$ 298.382

Constit. of *Alpinia officinarum*. Inhibitor of prostaglandin biosynth.. Oil. $[\alpha]^{30}_D$ -13.3 (c, 1.04 in $CHCl_3$)

5-Ketone, 3-Me ether

7-(4-Hydroxyphenyl)-5-methoxy-1-phenyl-3-heptanone

[100667-53-6]

$C_{20}H_{24}O_3$ 312.409

Constit. of *Alpinia officinarum*. Needles. Mp 68 - 70°. λ_{max} 209 (ε 24800); 224 (sh); 279 (ε 3600) (EtOH).

Kiuchi, F. *et al.*, *Chem. Pharm. Bull.*, 1982, **30**, 2279-2282 (*5-ketone*).
Itokawa, H. *et al.*, *Chem. Pharm. Bull.*, 1985, **33**, 4889-4893 (*5-ketone 3-Me ether*).
Uehara, S. *et al.*, *Chem. Pharm. Bull.*, 1987, **35**, 3298-3304 (*Alpinia officinarum constit*).
An, N. *et al.*, *Food Chem.*, 2010, **119**, 513-517 (*5-ketone 3-Ac*).
Li, J. *et al.*, *Bioorg. Med. Chem. Lett.*, 2011, **21**, 5363-5369 (*Curcuma kwangsiensis ketone*)

7-(4-Hydroxyphenyl)-1-phenyl-1-heptene-3,5-diol H112

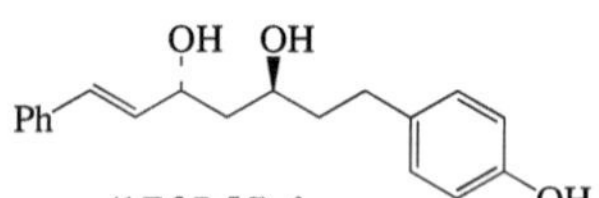

(1E,3R,5S)-form

Alpinikatin

$C_{19}H_{22}O_3$ 298.382

(1E,3R,5S)-form

$[\alpha]^{24}_D$ +19 (c, 0.9 in $CHCl_3$)

3-Me ether

1-(4-Hydroxyphenyl)-5-methoxy-7-phenyl-6-hepten-3-ol

[332371-78-5]

$C_{20}H_{24}O_3$ 312.409

Constit. of the seeds of *Alpinia blepharocalyx*. Amorph. light brown powder. $[\alpha]^{25}_D$ +21 (c, 0.08 in MeOH)

3-Et ether

5-Ethoxy-1-(4-hydroxyphenyl)-7-phenyl-6-hepten-3-ol

[332371-80-9]

$C_{21}H_{26}O_3$ 326.436

Constit. of the seeds of *Alpinia blepharocalyx*. Amorph. light yellow solid. $[\alpha]^{25}_D$ +73.9 (c, 0.04 in MeOH)

(1E,3S,5S)-form [1337550-58-9]

Constit. of the seeds of *Alpinia katsumadai*. Yellow powder. Mp 92°. $[\alpha]^{20}_D$ +23 (c, 0.47 in MeOH). λ_{max} 227 (log ε 4.2); 252 (log ε 4.4) (MeOH)

3-Me ether

[332371-77-4]

$C_{20}H_{24}O_3$ 312.409

Constit. of the seeds of *Alpinia blepharocalyx*. Amorph. light brown solid. $[\alpha]_D$ +21.3 (c, 0.13 in MeOH)

3-Et ether

[332371-79-6]

$C_{21}H_{26}O_3$ 326.436

Constit. of the seeds of *Alpinia blepharocalyx*. Amorph. light yellow solid. $[\alpha]^{25}_D$ +49.5 (c, 0.1 in MeOH)

(1E,3RS,5SR)-form

3-Ac

[1486465-16-0]

$C_{21}H_{24}O_4$ 340.419

Constit. of *Alpinia katsumadai*. Gum

5-Ac

[1486465-15-9]

$C_{21}H_{24}O_4$ 340.419

Constit. of *Alpinia katsumadai*. Gum.

Ali, M.S. *et al.*, *J. Nat. Prod.*, 2001, **64**, 289-293 (*Alpinia blepharocalyx constits*).
Nam, J.-W. *et al.*, *J. Nat. Prod.*, 2011, **74**, 2109-2115 (*Alpinikatin*).
Nam, J.-W. *et al.*, *Helv. Chim. Acta*, 2013, **96**, 1670-1680 (*Alpinikatin acetates*).
Cornil, J. *et al.*, *Eur. J. Org. Chem.*, 2014, **23**, 4958-4962 (*Alpinikatin, synth*).
Venkatesham, K. *et al.*, *Helv. Chim. Acta*, 2015, **98**, 1307-1314 (*Alpinikatin, synth*)

1-(4-Hydroxyphenyl)-7-phenyl-6-hepten-3-ol H113

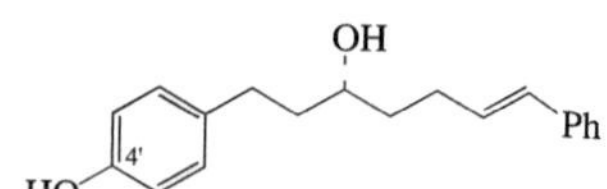

(R)-form

[158697-54-2]

$C_{19}H_{22}O_2$ 282.383

Information on abs. configs. is fragmentary. Isol. from rhizomes of *Curcuma xanthorrhiza* and *Curcuma comosa*. Cryst. ($CHCl_3/C_6H_6$) or oil. Mp 94 - 95°. $[\alpha]^{25}_D$ -66.8 (c, 0.17 in MeOH). Opt. rotn. refers to *C. xanthorrhiza* isolate of undetermined abs. config. No opt. rotn. reported for *C. comosa* isolate. λ_{max} 249 (ε 17400); 278 (ε 3900) (MeOH)

4′-Me ether
1-(4-Methoxyphenyl)-7-phenyl-6-hepten-3-ol
[1062082-65-8]
$C_{20}H_{24}O_2$ 296.41
Constit. of the rhizomes of *Curcuma comosa*. Isol. as a partial racemate *R*:S 2:1 (33.3% e.e.).

Suksamrarn, A. *et al.*, *Phytochemistry*, 1994, **36**, 1505-1508 (*Curcuma xanthorrhiza constit*).
Suksamrarn, A. *et al.*, *Bioorg. Med. Chem.*, 2008, **16**, 6891-6902 (*Curcuma comosa constit*)

7-(4-Hydroxyphenyl)-1-phenyl-4-hepten-3-one H114

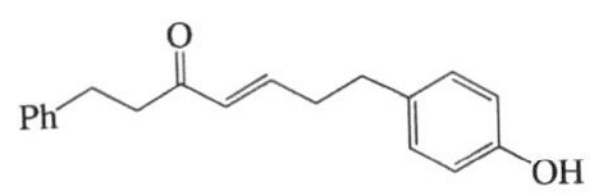

$C_{19}H_{20}O_2$ 280.367
(E)-form [100667-52-5]
Constit. of rhizomes of *Alpinia officinarum*. Oil. λ_{max} 216 (sh); 223 (ε 8100); 276 (ε 1400) (EtOH). [base] λ_{max} 214; 236; 295 (EtOH/NaOH aq.)
3″-Methoxy
7-(4-Hydroxy-3-methoxyphenyl)-1-phenyl-4-hepten-3-one
[478314-07-7]
$C_{20}H_{22}O_3$ 310.393
Constit. of rhizomes of *Alpinia officinarum*. Prostaglandin biosynth. inhibitor. Oil.

Kiuchi, F. *et al.*, *Chem. Pharm. Bull.*, 1982, **30**, 2279-2282 (*3″-methoxy, activity*).
Itokawa, H. *et al.*, *Chem. Pharm. Bull.*, 1985, **33**, 4889-4893 (*Alpinia officinarum constit*).
An, N. *et al.*, *Food Chem.*, 2010, **119**, 513-517 (*Alpinia officinarum constits*).
Gamre, S. *et al.*, *J. Nat. Prod.*, 2021, **84**, 352-363 (*synth, activity*)

1-(4-Hydroxyphenyl)-1,2-propanediol, CAS H115

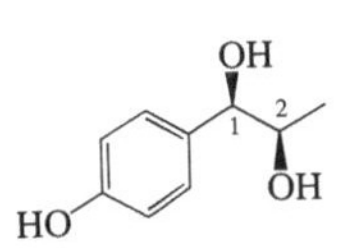

(1*R*,2*R*)-form

1-(p-Hydroxyphenyl)propylene glycol
[22805-45-4]
$C_9H_{12}O_3$ 168.192
(1R,2R)-form
[178405-44-2]
Constit. of *Narvalina domingensis*. Cryst. (MeOH). Mp 109 - 111°. $[\alpha]^{25}_{D}$ -42.5 (c, 0.7 in MeOH)
2-O-β-D-Glucopyranoside
[499135-10-3]
$C_{15}H_{22}O_8$ 330.333
Constit. of the fruit of *Pimpinella anisum*. Amorph. powder. $[\alpha]^{22}_{D}$ -51 (c, 0.8 in MeOH)
(1R,2S)-form
2-O-β-D-Glucopyranoside
[499135-08-9]
Constit. of the fruit *Pimpinella anisum*. Amorph. powder. $[\alpha]^{21}_{D}$ -33 (c, 0.3 in MeOH)
(1S,2R)-form
2-O-β-D-Glucopyranoside
[499135-09-0]
Constit. of the fruit of *Pimpinella anisum*. Amorph. powder. $[\alpha]^{21}_{D}$ -16 (c, 0.2 in MeOH)
(1S,2S)-form [499201-61-5]
Amorph. powder. $[\alpha]^{22}_{D}$ +19 (c, 0.2 in MeOH)
2-O-β-D-Glucopyranoside
[499135-11-4]
Constit. of the fruit of *Pimpinella anisum*. Amorph. powder. $[\alpha]^{22}_{D}$ +21 (c, 1 in MeOH)
(1R*,2R*)-form [28282-82-8]
1-Me ether
1-(4-Hydroxyphenyl)-1-methoxy-2-propanol
[178405-45-3]
$C_{10}H_{14}O_3$ 182.219
Constit. of *Narvalina domingensis*. Cryst. (MeOH). Mp 118 - 120°. $[\alpha]^{28}_{D}$ -81.2 (c, 0.4 in MeOH)
1-Et ether
1-Ethoxy-1-(4-hydroxyphenyl)-2-propanol
[178405-47-5]
$C_{11}H_{16}O_3$ 196.246
Constit. of *Narvalina domingensis*. Cryst. (MeOH). Mp 114 - 115°. $[\alpha]^{23}_{D}$ -90.6 (c, 0.3 in MeOH)
(1R*,2S*)-form
[178405-43-1]
[28282-81-7]
Constit. of *Narvalina domingensis*. Cryst. (MeOH). Mp 153 - 154°. $[\alpha]^{25}_{D}$ +23.1 (c, 0.6 in MeOH)
1-Et ether
[178405-46-4]
Constit. of *Croton sellowii* and *Narvalina domingensis*. Oil. $[\alpha]^{28}_{D}$ +39.5 (c, 0.2 in MeOH).

Zuman, P. *et al.*, *Coll. Czech. Chem. Comm.*, 1958, **23**, 1237-1251 (*config*).
Yamaguchi, A. *et al.*, *CA*, 1969, **71**, 3092h ; 1970, 72, 132209w (*synth*).
Balboul, B.A.A.A. *et al.*, *Phytochemistry*, 1996, **42**, 1191-1193 (*Narvalina domingensis derivs*).
Ishikawa, T. *et al.*, *Chem. Pharm. Bull.*, 2002, **50**, 1460-1466 (*Pimpinella anisum constits*)

3-(4-Hydroxyphenyl)-1,2-propanediol H116

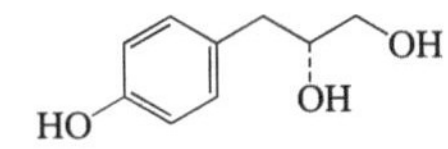

(*R*)-form

[28188-87-6]
$C_9H_{12}O_3$ 168.192
(R)-form
Lathyroxin B [1932463-97-2]
Prod. by *Ascochyta lentis* var. *lathyri*. Exhibits phytotoxic activity. Amorph. solid. λ_{max} 224 (log ε 3.56); 278 (log ε 3.07) (MeOH)
(S)-form
4′-O-β-D-Galactopyranoside
Obovatoside A
[1819381-00-4]
$C_{15}H_{22}O_8$ 330.333
Constit. of the fruit of *Magnolia obovata*. Potential antidiabetic agent. Orange needles. Mp 98 - 100°. $[\alpha]_{D}$ -62 (c, 0.11 in MeOH)
4′-Me ether
[127852-64-6]
$C_{10}H_{14}O_3$ 182.219
Brownish oil. $[\alpha]^{20}_{D}$ -12.7 (c, 0.93 in $CHCl_3$)
2,4′-Di-Me ether
β,4-Dimethoxybenzenepropanol
[1214260-84-0]
$C_{11}H_{16}O_3$ 196.246
Yellow oil. $[\alpha]^{20}_{D}$ -3.4 (c, 1.16 in EtOH) (95% ee)
(±)-form [78472-12-5]
Metab. of the fungus *Nidula niveo-tomentosa*. Mp 90 - 92° (81-82°)
4′-O-β-D-Glucopyranoside
[217973-59-6]
$C_{15}H_{22}O_8$ 330.333
Constit. of *Foeniculum vulgare*. Amorph. powder. $[\alpha]^{23}_{D}$ -57 (c, 0.2 in MeOH)
(ξ)-form [22805-47-6]
Prod. by *Aspergillus flavus*
4′-O-[β-D-Allopyranosyl-(1→4)-4-hydroxy-3-methoxybenzoyl-(→6)-β-D-allopyranoside]
Magnoloside Y
[1946004-98-3]
$C_{29}H_{38}O_{16}$ 642.607
Constit. of the stem bark of *Magnolia officinalis*. Amorph. pale yellow solid

4′-O-[α-L-Rhamnopyranosyl-(1→4)-4-hydroxy-3-methoxybenzoyl-(→6)-β-D-allopyranoside]
Magnoloside Z
[1946004-99-4]
$C_{29}H_{38}O_{15}$ 626.608
Constit. of the stem bark of *Magnolia officinalis*. Amorph. pale yellow solid. $[\alpha]_D^{20}$ -89 (c, 0.15 in MeOH). λ_{max} 256 (log ε 4.08); 295 (log ε 3.78) (H_2O)

4′-O-[α-L-Rhamnopyranosyl-(1→4)-4-hydroxy-3,5-dimethoxybenzoyl-(→6)-β-D-allopyranoside]
Magnoloside X
[1946004-97-2]
$C_{30}H_{40}O_{16}$ 656.634
Constit. of the stem bark of *Magnolia officinalis*. Moderate α-glucosidase inhibitor. Amorph. pale yellow solid. $[\alpha]_D^{20}$ -86.7 (c, 0.15 in MeOH). λ_{max} 266 (log ε 3.88) (H_2O)

4′-O-[β-D-Glucopyranosyl-(1→4)-α-L-rhamnopyranosyl-(1→4)-4-hydroxy-3-methoxybenzoyl-(→6)-β-D-allopyranoside]
Magnoloside W
[1946004-96-1]
$C_{35}H_{48}O_{20}$ 788.749
Constit. of the stem bark of *Magnolia officinalis*. Amorph. solid. $[\alpha]_D^{20}$ -20.8 (c, 0.1 in MeOH). λ_{max} 256 (log ε 4.48); 293 (log ε 4.18) (H_2O)

4′-O-[β-D-Glucopyranosyl-(1→4)-α-L-rhamnopyranosyl-(1→4)-4-hydroxy-3,5-dimethoxybenzoyl-(→6)-β-D-allopyranoside]
Magnoloside V
[1946004-95-0]
$C_{36}H_{50}O_{21}$ 818.775
Constit. of the stem bark of *Magnolia officinalis*. Amorph. solid

4′-Me ether
3-(4-Methoxyphenyl)-1,2-propanediol. ***Prestragol***
[17131-20-3]
$C_{10}H_{14}O_3$ 182.219
Constit. of *Artemisia dracunculus*. Powder. Mp 49°. λ_{max} 225; 275; 285 (MeOH).

Yamaguchi, A. *et al.*, *CA*, 1970, **72**, 132209w (*synth*).
Ayer, W.A. *et al.*, *Phytochemistry*, 1980, **19**, 2717-2721 (*Nidula niveo-tomentosa constit*).
Kitajima, J. *et al.*, *Chem. Pharm. Bull.*, 1998, **46**, 1591-1594 (*glucoside*).
Brenna, E. *et al.*, *Tetrahedron: Asymmetry*, 2009, **20**, 2694-2698 (*S-4′-Me ether, S-2,4′-di-Me ether*).
Forseth, R.R. *et al.*, *Angew. Chem., Int. Ed.*, 2013, **52**, 1590-1594 (*Aspergillus flavus constit*).
Seo, K.-H. *et al.*, *Carbohydr. Res.*, 2015, **416**, 70-74 (*Obovatoside A, activity*).
Talbi, M. *et al.*, *J. Asian Nat. Prod. Res.*, 2016, **18**, 724-729 (*Prestragol*).
Xue, Z. *et al.*, *Phytochemistry*, 2016, **127**, 50-62 (*Magnolosides V-Z, activity*).
Masi, M. *et al.*, *J. Nat. Prod.*, 2018, **81**, 1093-1097 (*Lathyroxin B, activity*)

1-(4-Hydroxyphenyl)-1,2-propanedione H117

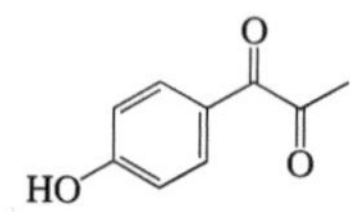

4-Hydroxyphenyl methyl diketone
[10087-36-2]
$C_9H_8O_3$ 164.16
Constit. of *Cynanchum auriculatum* and *Gastrodia elata*. Cryst.. Mp 92°.

Higuchi, T. *et al.*, *Holzforschung*, 1966, **20**, 16-21 (*synth*).
Deng, Y. *et al.*, *Tianran Chanwu Yanjiu Yu Kaifa*, 2013, **25**, 729-732 (*Cynanchum auriculatum constit*).
Guo, Q. *et al.*, *Acta Pharm. Sin. B*, 2015, **5**, 350-357 (*Gastrodia elata constit*)

1-(4-Hydroxyphenyl)-1,2,3-propanetriol, CAS H118

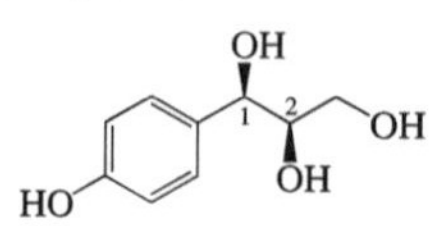

(1*R*,2*R*)-*form*

1-(4-Hydroxyphenyl)glycerol
[22805-48-7]
$C_9H_{12}O_4$ 184.191

(1R,2R)-form
[84799-28-0]
Mp 144 - 146°. $[\alpha]_D^{23}$ -33.8 (c, 1.2 in EtOH)

4′-O-(2-Methyl-2Z-butenoyl), 3-Ac
Charuol B
[2097437-94-8]
$C_{16}H_{20}O_6$ 308.330
Constit. of the leaves of *Ophryosporus charua*. Germination inhibitor. Oil. $[\alpha]_D^{22}$ -0.2 (c, 0.2 in MeOH). λ_{max} 224 (MeCN)

1-Me ether
3-(4-Hydroxyphenyl)-3-methoxy-1,2-propanediol
[1621347-17-8]
$C_{10}H_{14}O_4$ 198.218
Constit. of *Abies delavayi*. Amorph. powder. $[\alpha]_D^{20}$ -8.2 (c, 0.5 in MeOH). λ_{max} 224 (log ε 3.90); 276 (log ε 3.13) (MeOH)

1,2-O-Isopropylidene
[1621347-18-9]
$C_{12}H_{16}O_4$ 224.256
Constit. of *Abies delavayi*. Amorph. powder. $[\alpha]_D^{23}$ +1.0 (c, 0.25 in MeOH)

(1R,2S)-form
1-Et ether
[1621513-43-6]
$C_{11}H_{16}O_4$ 212.245
Constit. of *Abies nukiangensis*. Amorph. powder. $[\alpha]_D^{20}$ -3.0 (c, 0.15 in MeOH). Possible artifact

(1S,2R)-form
[84822-96-8]
Mp 140 - 141°. $[\alpha]_D^{23}$ +21.3 (c, 0.6 in EtOH)

3-O-(6-O-Benzoyl-β-D-glucopyranoside)
Prupersin A
[1452403-63-2]
$C_{22}H_{26}O_{10}$ 450.44
Constit. of *Prunus davidiana*. Antioxidant. $[\alpha]_D^{20}$ -25.3 (c, 2.35 in MeOH). λ_{max} 205 (log ε 3.77) (MeOH)

4′-O-(2-Methyl-2Z-butenoyl), 3-Ac
Charuol A
[2097437-92-6]
$C_{16}H_{20}O_6$ 308.330
Constit. of the leaves of *Ophryosporus charua*. Germination inhibitor. Oil. $[\alpha]_D^{22}$ -0.1 (c, 0.18 in MeOH). λ_{max} 220 (log ε 1.72) (MeCN)

(1RS,2RS)-form
[25743-67-3]
Constit. of *Pinus massoniana* and *Pinus sylvestris*

(1RS,2SR)-form
[38916-91-5]
Constit. of *Pinus sylvestris*. Mp 149 - 151°

(1R*,2R*)-form
2-O-[4-Hydroxybenzoyl-(→6)-β-D-glucopyranoside]
Taxumaiglucoside A
[252636-45-6]
$C_{22}H_{26}O_{11}$ 466.439
Constit. of *Taxus mairei*. Oil (as hepta-Ac). $[\alpha]_D^{28}$ +9 (c, 0.5 in $CHCl_3$) (hepta-Ac)

2-O-[4-Hydroxy-3-methoxybenzoyl-(→6)-β-D-glucopyranoside]
Taxumaiglucoside B
[252636-46-7]
$C_{23}H_{28}O_{12}$ 496.465
Constit. of *Taxus mairei*. Oil (as hepta-Ac). $[\alpha]_D^{28}$ +4 (c, 0.5 in $CHCl_3$)

(1R*,2S*)-form
1-Me ether
[1800422-47-2]
$C_{10}H_{14}O_4$ 198.218
Constit. of aerial parts of *Abies fabri*. Amorph. powder. $[\alpha]_D^{20}$ -1.0 (c, 0.15 in MeOH)

(1ξ,2ξ)-form
1-O-β-D-Glucopyranoside
[67151-72-8]
$C_{15}H_{22}O_9$ 346.332
Constit. of *Picea glehnii*
2-O-β-D-Glucopyranoside
[84744-60-5]
$C_{15}H_{22}O_9$ 346.332
Constit. of *Pinus sylvestris*
3-O-β-D-Glucopyranoside
[67151-73-9]
$C_{15}H_{22}O_9$ 346.332
Constit. of *Picea glehnii* and *Pinus sylvestris*
4′-Me ether
1-(4-Methoxyphenyl)-1,2,3-propanetriol. 1-(4-Methoxyphenyl)glycerol
[108334-70-9]
$C_{10}H_{14}O_4$ 198.218
Constit. of the fruit of *Illicium verum*. Oil. $[\alpha]_D^{25}$ +11.9 (c, 0.21 in MeOH)
4′-Butyl ether
1-(4-Butoxyphenyl)-1,2,3-propanetriol. 1-(4-Butoxyphenyl)glycerol
[1492016-12-2]
$C_{13}H_{20}O_4$ 240.299
Constit. of *Araucaria cunninghamii*. Amorph. powder. $[\alpha]_D^{20}$ -27 (c, 0.1 in MeOH). λ_{max} 229; 279 (MeOH).

Lundgren, L.N. *et al.*, *Acta Chem. Scand., Ser. B*, 1982, **36**, 695-699 (*Pinus sylvestris constits, synth*).
Lundgren, L.N. *et al.*, *Acta Chem. Scand., Ser. B*, 1985, **39**, 241-248 (*Pinus massoniana constit*).
Yang, S.-J. *et al.*, *J. Chin. Chem. Soc. (Taipei)*, 1999, **46**, 811-818 (*Taxumaiglucosides A,B*).
Lee, S.W. *et al.*, *Arch. Pharmacal Res.*, 2003, **26**, 591-593 (*(Methoxyphenyl)glycerol*).
Chen, X.-Y. *et al.*, *J. Nat. Prod.*, 2013, **76**, 1528-1534 (*Prupersin A*).
Chen, J. *et al.*, *Phytochem. Lett.*, 2013, **6**, 41-45 (*(Butoxyphenyl)glycerol*).
Yang, X.-W. *et al.*, *Phytochemistry*, 2014, **105**, 164-170 (*Abies delavayi constits*).
Li, Y.-L. *et al.*, *Phytochemistry*, 2014, **106**, 116-123 (*1-Et ether*).
Li, Y.-L. *et al.*, *Phytochemistry*, 2015, **117**, 135-143 (*Abies fabri constit*).
Bravetti, M.M. del M. *et al.*, *Phytochemistry*, 2017, **138**, 145-151 (*Charuols A,B, activity*).
Chakraborty, I. *et al.*, *Tetrahedron Lett.*, 2019, **60**, 1597-1599 (*synth, abs config*)

3-(2-Hydroxyphenyl)propanoic acid H119

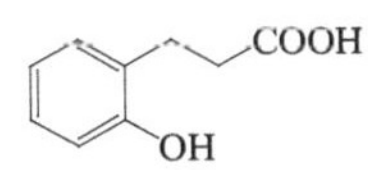

2-Hydroxybenzenepropanoic acid, CAS. 3-(o-Hydroxyphenyl)propionic acid, CAS. Hydrocoumaric acid. o-Hydroxyhydrocinnamic acid. ***Melilotic acid***
[495-78-3]
$C_9H_{10}O_3$ 166.176
Constit. of *Angraecum fragrans, Melilotus alba, Dipteryx odorata* and other plants. Cryst. (H_2O) or needles (C_6H_6). Mp 88 - 89°. pK_{a1} 4.75 (25°)
β-D-Glucopyranosyl ester
Dihydrocinnacasside
[1160707-56-1]
$C_{15}H_{20}O_8$ 328.317
Constit. of the twigs of *Cinnamomum cassia*. Amorph. powder. Mp 174 - 175°. $[\alpha]_D^{25}$ -29.1 (c, 0.4 in MeOH). λ_{max} 214 (log ε 2.55) (MeOH)
Me ester
[20349-89-7]
$C_{10}H_{12}O_3$ 180.203
Solid (hexane/Et_2O). Mp 40 - 41°
2-Ethylhexyl ester
Meliloester
[1384180-98-6]
$C_{17}H_{26}O_3$ 278.392
Constit. of *Melilotus alba*. Gum. $[\alpha]_D^{25}$ -117.6 (c, 0.17 in MeOH). Synthetic studies suggest struct. is incorrect. λ_{max} 282 (MeOH)
Piperidide
1-[3-(2-Hydroxyphenyl)propanoyl]piperidine
[1455247-73-0]
$C_{14}H_{19}NO_2$ 233.311
Alkaloid from the leaves of *Ficus rumphii*. Cryst. (MeOH/$CHCl_3$). Mp 167 - 168°
3-(2,6-Dihydroxyphenyl)propanoyl
Rhinacanthinic acid B
[1803120-00-4]
$C_{18}H_{18}O_6$ 330.336
Constit. of the stems of *Rhinacanthus nasutus*. Exhibits moderate activity against herpes simplex virus type-2 (HSV-2). Gum. λ_{max} 230 (log ε 3.31); 265 (log ε 2.81) (MeOH)
2-Hydroxy-E-cinnamoyl
Rhinacanthinic acid A
[1803119-99-4]
$C_{18}H_{16}O_5$ 312.321
Constit. of the stems of *Rhinacanthus nasutus*. Exhibits moderate activity against herpes simplex virus type-2 (HSV-2). Pale yellow gum. λ_{max} 225 (log ε 3.24); 268 (log ε 2.83) (MeOH).

Aldrich Library of FT-IR Spectra, 1st edn., 1985, **2**, 152A (*ir*).
Aldrich Library of ¹³C and ¹H FT-NMR Spectra, 1992, **2**, 1004C (*nmr*).
v. Auwers, *Justus Liebigs Ann. Chem.*, 1918, **415**, 98-168 (*Me ether, Me ether Me ester, synth*).
Claisen, L. *et al.*, *Ber. Dtsch. Chem. Ges.*, 1921, **54**, 200-203 (*Et ester*).
Matsuura, T. *et al.*, *J. Org. Chem.*, 1964, **29**, 3168-3170 (*synth*).
Amsberry, K.L. *et al.*, *J. Org. Chem.*, 1990, **55**, 5867-5877 (*synth*).
Begum, N.S. *et al.*, *Acta Crystallogr., Sect. C*, 1992, **48**, 1076-1078 (*cryst struct*).
Malakov, P.Y. *et al.*, *Fitoterapia*, 1998, **69**, 552 (*cmr*).
Sugimoto, H. *et al.*, *Adv. Synth. Catal.*, 2007, **349**, 669-679 (*Me ester*).
Ngoc, T.M. *et al.*, *J. Nat. Prod.*, 2009, **72**, 1205-1208 (*Dihydrocinnacasside*).
Khatoon, R. *et al.*, *Nat. Prod. Commun.*, 2012, **7**, 61-62 (*Meliloester*).
Ding, L. *et al.*, *Synthesis*, 2012, **44**, 3296-3300 (*Meliloester, synth*).
Parveen, M. *et al.*, *Nat. Prod. Res.*, 2014, **28**, 646-652 (*piperidide*).
Thongchuai, B. *et al.*, *Tetrahedron Lett.*, 2015, **56**, 5161-5163 (*Rhinacanthinic acids A,B, activity*)

3-(4-Hydroxyphenyl)propanoic acid H120

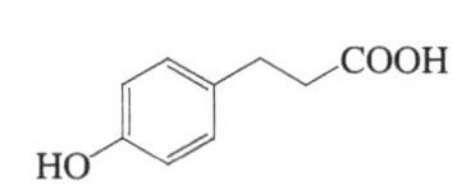

4-Hydroxybenzenepropanoic acid, CAS. 3-(p-Hydroxyphenyl)propionic acid, CAS. Dihydro-p-coumaric acid. p-Hydroxyhydrocinnamic acid. Phloretic acid. Phloretinic acid. HPPA
[501-97-3]
$C_9H_{10}O_3$ 166.176
Constit. of urine. Isol. from sulfite liquors, prod. by *Clostridium butyricum*. Isol. from heartwood of *Haplormosia monophylla*. A prod. of tyrosine metab.; conc. in urine increases in patients with gastrointestinal diseases. Fluorescent

substrate for the detn. of peroxidase. Prisms (Et_2O). Mp 129 - 130°. pK_a 4.76 (25°)

▶ MW5342000

Me ester

[5597-50-2]

$C_{10}H_{12}O_3$ 180.203

Cryst. (Et_2O/petrol). Mp 40 - 41°. Bp 186 - 187° at 17 mm

Et ester

[23795-02-0]

$C_{11}H_{14}O_3$ 194.23

Mp 45°. Bp 193° at 18 mm

Amide

3-(4-Hydroxyphenyl)propanamide. ***Phloretamide***

[23838-70-2]

$C_9H_{11}NO_2$ 165.192

Constit. of the young fruit and xylem sap of *Malus domestica*. Phytohormone. Exhibits chemopreventative anticancer props.. Prisms. Mp 127 - 128°

Methylamide

3-(4-Hydroxyphenyl)-N-methylpropanamide

[876302-34-0]

$C_{10}H_{13}NO_2$ 179.219

Prod. by *Micromonospora* sp. P1068. Pale yellow solid. Mp 86°.

Aldrich Library of FT-IR Spectra, 1st edn., 1985, **2**, 161A , 161D (*ir*).
Aldrich Library of [13]*C and* [1]*H FT-NMR Spectra*, 1992, **2**, 1018A , 1019B, 1215A (*nmr*).
Bowden, E. *et al.*, *J. Am. Chem. Soc.*, 1940, **62**, 2422-2423 (*synth*).
Grimmer, G. *et al.*, *Biochem. Z.*, 1966, **346**, 186-190 (*Clostridium butyricum metab*).
Wittstruck, T.A. *et al.*, *J. Am. Chem. Soc.*, 1967, **89**, 3803-3809 (*pmr*).
Ballantine, J.A. *et al.*, *Org. Mass Spectrom.*, 1970, **3**, 1215-1217 (*ms*).
Hayashi, Y. *et al.*, *Mokuzai Gakkaishi*, 1974, **20**, 595-599 (*Haplormosia monophylla constit*).
Rybicka, H. *Biochem. Physiol. Pflanz.*, 1986, **181**, 131-134 (*Phloretamide*).
Okabe, N. *et al.*, *Acta Crystallogr., Sect. C*, 1995, **51**, 1324-1325 (*cryst struct*).
Gutierrez-Lugo, M.-T. *et al.*, *Nat. Prod. Res.*, 2005, **19**, 645-652 (*methylamide*).
Narsaiah, A.V. *et al.*, *Synth. Commun.*, 2011, **41**, 1603-1608 (*Me ester*).
Krajka-Kuzniak, V. *et al.*, *Food Chem. Toxicol.*, 2013, **51**, 202-209 (*Phloretamide, activity*).
Wawryzniak, D. *et al.*, *Biotechnologia*, 2016, **97**, 51-54 (*Phloretamide, synth*)

3-(2-Hydroxyphenyl)-1-propanol H121

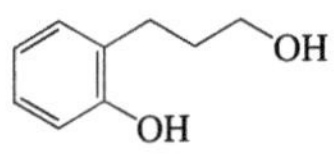

2-Hydroxybenzenepropanol, CAS. *2-γ-Hydroxypropylphenol*

[1481-92-1]

$C_9H_{12}O_2$ 152.193

Yellow oil. $d_4^{15.1}$ 1.13. Bp 177 - 179° at 12 mm.

1-O-(Methylmalonyl)

3-(2-Hydroxyphenyl)propyl methylmalonate

[1494654-26-0]

$C_{13}H_{16}O_5$ 252.266

Constit. of roots of *Ageratina adenophora*. Exhibits inhibitory activity against *Arabidopsis thaliana* seed germination. Oil. λ_{max} 214 (log ε 3.83); 274 (log ε 3.32) (no solvent reported)

2′-Me ether

2-Methoxybenzenepropanol, CAS

[10493-37-5]

$C_{10}H_{14}O_2$ 166.22

Oil. Bp 145 - 146° at 10 mm. Bp 84 - 87° at 0.2 mm

2′-Me ether, 1-O-β-D-glucopyranoside

Tazettoside E

[2077176-31-7]

$C_{16}H_{24}O_7$ 328.361

Constit. of the flowers of *Narcissus tazetta*. Powder. $[\alpha]_D^{25}$ -10.4 (c, 0.34 in MeOH).

Lockett, J. *et al.*, *J. Chem. Soc.*, 1939, 787-790 (*synth, 2′-Me ether*).
Thakar, G.P. *et al.*, *Indian J. Chem.*, 1965, **3**, 74-77 (*synth*).
Billman, J.H. *et al.*, *J. Pharm. Sci.*, 1971, **60**, 1188-1192 (*synth, 2′-Me ether*).
Botta, M. *et al.*, *Org. Biomol. Chem.*, 2004, **2**, 570-577 (*synth*).
Liang, H. *et al.*, *Org. Lett.*, 2010, **12**, 1760-1763 (*2′-Me ether*).
Zhou, Z.-Y. *et al.*, *J. Agric. Food Chem.*, 2013, **61**, 11792-11799 (*methylmalonyl*).
Morikawa, T. *et al.*, *J. Nat. Med.*, 2016, **70**, 89-101 (*Tazettoside E*)

3-(4-Hydroxyphenyl)-1-propanol H122

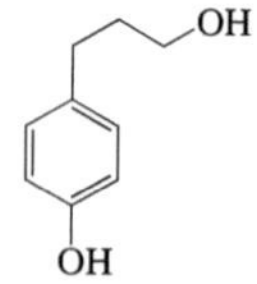

4-Hydroxybenzenepropanol, CAS. *4-γ-Hydroxypropylphenol. 4-(3-Hydroxypropyl)phenol. Dihydro-p-coumaroyl alcohol*

[10210-17-0]

[125092-37-7]

$C_9H_{12}O_2$ 152.193

Constit. of *Fucus vesiculosus*, *Pinus sylvestris*, *Taxus baccata* and *Zanthoxylum cuspidatum*. Prod. by a marine-derived *Engyodontium album* 3A00174. Cryst. (Et_2O/petrol). Mp 55°

1-O-[3-(3,4-Dihydroxyphenyl)propanoyl]

[536740-20-2]

$C_{18}H_{20}O_5$ 316.353

Constit. of the stems of the mangrove *Lumnitzera racemosa*. Needles (MeOH). Mp 185°

1-Nonanoyl

Nepetanoate

[1174171-52-8]

$C_{18}H_{28}O_3$ 292.419

Constit. of *Nepeta juncea*. Cryst. ($CHCl_3$). $[\alpha]_D^{23}$ +2.8 ($CHCl_3$). Opt. rotation unaccounted for. λ_{max} 276 (MeOH)

1-Eicosanoyl

$C_{29}H_{50}O_3$ 446.716

Constit. of *Taxus baccata*

1-Heneicosanoyl

$C_{30}H_{52}O_3$ 460.743

Constit. of *Taxus baccata*

1-Docosanoyl

[289893-78-3]

$C_{31}H_{54}O_3$ 474.77

Constit. of *Ozothamnus hookeri* and *Taxus baccata*. Oil

1-Tricosanoyl

$C_{32}H_{56}O_3$ 488.797

Constit. of *Taxus baccata*

1-Tetracosanoyl

3-(4-Hydroxyphenyl)propyl tetracosanoate

[141361-42-4]

$C_{33}H_{58}O_3$ 502.824

Constit. of *Piper clarkii*, *Taxus baccata* and *Taxus canadensis*. Cryst. (petrol). Mp 73 - 74° (synthetic)

1-Pentacosanoyl

3-(4-Hydroxyphenyl)propyl pentacosanoate

[219591-42-1]

$C_{34}H_{60}O_3$ 516.851

Constit. of *Taxus baccata* and *Taxus canadensis*

1-Hexacosanoyl

3-(4-Hydroxyphenyl)propyl hexacosanoate

[219591-43-2]

$C_{35}H_{62}O_3$ 530.878

Constit. of *Taxus baccata* and *Taxus canadensis*

1-Heptacosanoyl
$C_{36}H_{64}O_3$ 544.905
Constit. of *Taxus baccata*
1-Octacosanoyl
$C_{37}H_{66}O_3$ 558.932
Constit. of *Taxus baccata*
1-Nonacosanoyl
$C_{38}H_{68}O_3$ 572.959
Constit. of *Taxus baccata*
1-Triacontanoyl
$C_{39}H_{70}O_3$ 586.986
Constit. of *Taxus baccata*
1-Hentriacontanoyl
$C_{40}H_{72}O_3$ 601.013
Constit. of *Taxus baccata*
1-Dotriacontanoyl
3-(4-Hydroxyphenyl)propyl dotriacontanoate
[219591-44-3]
$C_{41}H_{74}O_3$ 615.04
Constit. of *Taxus baccata* and *Taxus canadensis*
1-Tritriacontanoyl
$C_{42}H_{76}O_3$ 629.067
Constit. of *Taxus baccata*
1-Tetratriacontanoyl
3-(4-Hydroxyphenyl)propyl tetratriacontanoate
[219591-45-4]
$C_{43}H_{78}O_3$ 643.094
Constit. of *Taxus baccata* and *Taxus canadensis*
1-Pentatriacontanoyl
$C_{44}H_{80}O_3$ 657.121
Constit. of *Taxus baccata*
1-Hexatriacontanoyl
$C_{45}H_{82}O_3$ 671.148
Constit. of *Taxus baccata*
1-Benzoyl
3-(4-Hydroxyphenyl)propyl benzoate
[887470-46-4]
$C_{16}H_{16}O_3$ 256.301
Constit. of *Croton hutchinsonianus* and *Stemona parviflora*. Pale yellow oil. λ_{max} 228 (log ε 4.08); 279 (log ε 3.26) (MeOH)
1-Me ether
4-(3-Methoxypropyl)phenol
[116087-10-6]
$C_{10}H_{14}O_2$ 166.22
Mp 25 - 26°. Bp 149° at 10 mm.

Aldrich Library of FT-IR Spectra, 1st edn., 1985, **1**, 1167A , 1167B (*ir*).
Aldrich Library of FT-IR Spectra: Vapor Phase, 1989, **3**, 1086B , 1086C (*ir*).
Aldrich Library of [13]*C and* [1]*H FT-NMR Spectra*, 1992, **2**, 409B , 409C (*nmr*).
v. Braun, J. *et al.*, *Ber. Dtsch. Chem. Ges.*, 1912, **45**, 2504-2522 (*synth*).
Perold, G.W. *et al.*, *J. Chem. Soc., Perkin Trans. 1*, 1972, 2450-2457 (*synth*).
Henley-Smith, P. *et al.*, *J. Chem. Soc., Perkin Trans. 1*, 1980, 614-622 (*synth*).
Boll, P.M. *et al.*, *Phytochemistry*, 1992, **31**, 1035-1037 (*tetracosanoyl*).
Olsen, C.E. *et al.*, *Indian J. Chem., Sect. B*, 1998, **37**, 828-831 (*Taxus canadensis esters*).
Hori, M. *et al.*, *J. Org. Chem.*, 1998, **63**, 889-894 (*synth, pmr*).
Rumbero, A. *et al.*, *Z. Naturforsch., C*, 2000, **55**, 318-322 (*docosanoyl*).
Jetter, R. *et al.*, *Phytochemistry*, 2002, **61**, 579-587 (*Taxus baccata esters*).
Anjaneyulu, A.S.R. *et al.*, *ARKIVOC*, 2003, **iii**, 25-30 (*3,4-dihydroxyphenylpropanoate*).
Athikomkulchai, S. *et al.*, *Chem. Pharm. Bull.*, 2006, **54**, 262-264 (*1-benzoate*).
Hussain, J. *et al.*, *Magn. Reson. Chem.*, 2009, **47**, 625-627 (*Nepetanoate*)

3-(4-Hydroxyphenyl)-2-propenal, CAS H123

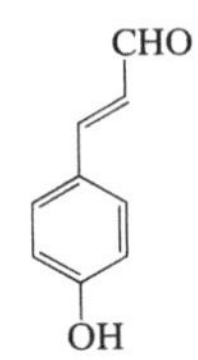

p-Hydroxycinnamaldehyde, CAS. *p-Coumaraldehyde. trans-p-Coumaryl aldehyde*
[2538-87-6]
$C_9H_8O_2$ 148.161
Some derivs. not specifically assigned as (*E*)- in the lit. but this can be inferred
(E)-form [20711-53-9]
Constit. of *Alpinia galanga* rhizomes, *Cucurbita maxima* and the insect *Sarcophyte sanguinea*. Antifungal and antibacterial agent. Cryst. (EtOAc). Mp 140° (136-138°)
Semicarbazone
Mp 224°
Ac
[157096-53-2]
[58505-83-2]
$C_{11}H_{10}O_3$ 190.198
Yellowish cryst.. Mp 77 - 78°
Me ether
3-(4-Methoxyphenyl)-2-propenal. p-Methoxycinnamaldehyde. FEMA 3567
[24680-50-0]
[1963-36-6]
$C_{10}H_{10}O_2$ 162.188
Isol. from oils of *Artemisia dracunculus*, *Orthodon methylchavicoliferum*, *Agastache rugosa* and *Sphaeranthus indicus*. Antitrichomonas agent. Shows weak antibacterial activity. Yellow needles (EtOH aq.). Mp 58 - 59°. Bp 277°. Bp 171° at 15 mm.

Pauly, H. *et al.*, *Ber. Dtsch. Chem. Ges.*, 1923, **56**, 603-610 (*synth*).
Friedrich, K. *et al.*, *Chem. Ber.*, 1961, **94**, 838-839 (*Me ether*).
Aulin-Erdtman, G. *et al.*, *Acta Chem. Scand.*, 1968, **22**, 1187-1209 (*uv*).
Barik, B.R. *et al.*, *Phytochemistry*, 1987, **26**, 2126-2127 (*Alpinia galanga constit*).
Naidoo, L.A.C. *et al.*, *Phytochemistry*, 1992, **31**, 3929-3931 (*Sarcophyte sanguinea constit, pmr, cmr*).
Bellassoued, M. *et al.*, *J. Org. Chem.*, 1993, **58**, 2517-2522 (*Me ether, synth, pmr, ir*).
Francesch, C. *et al.*, *Synthesis*, 1994, 369-371 (*Ac*).
Encyclopedia of Food and Color Additives, (ed. Burdock, G.A.), CRC Press , 1997, 1694-1695 (*Me ether*).
Iliefski, T. *et al.*, *Acta Chem. Scand.*, 1998, **52**, 1177-1182 (*Me ether*).
Stange, R.R. *et al.*, *Phytochemistry*, 1999, **52**, 41-43 (*Cucurbita maxima constit, pmr, cmr*).
Fenaroli's Handbook of Flavor Ingredients, 6th edn., (ed. Burdock, G.A.), CRC Press, 2010, 1192 (*Me ether, use, occur*)

3-(2-Hydroxyphenyl)-2-propenoic acid, CAS H124

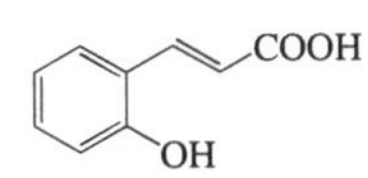

(*E*)-*form*

o-Hydroxycinnamic acid, CAS. *o-Coumaric acid. o-Cumaric acid*
[583-17-5]
$C_9H_8O_3$ 164.16
Used as a 1% soln. in EtOH as a fluorescence acid-base indicator (pH range 7.2-9.0; colour change: non-fluorescent → green). pK_{a1} 3.7 (30°,1.0*M* LiCl)
(E)-form
FEMA 4700 [614-60-8]
Occurs in many plants, e.g. *Melilotus officinalis* and *Angraecum fragrans*. Intermed. in biosynth. of coumarins. Antioxidant. Bitter taste inhibitor. Needles (H_2O). Sol. EtOH, hot H_2O, Et_2O; insol. CS_2, $CHCl_3$. Mp 207 - 208° dec.. pK_a 4.61. Nonvolatile in steam. Dec. on distiln.
▶ GD9090000. LD_{50} (mus, ivn) 180 mg/kg
(Z)-form
Coumarinic acid [495-79-4]
Obt. by photoisomerisation. Readily loses H_2O giving coumarin.

Aldrich Library of FT-IR Spectra, 1st edn., 1985, **2**, 176C , 176D (*ir*).

Aldrich Library of ¹³*C and* ¹*H FT-NMR Spectra*, 1992, **2**, 1046A , 1046B (*nmr*).
Tiemann, F. *Ber. Dtsch. Chem. Ges.*, 1877, **10**, 283-288 (*synth*).
Ebert, G. *et al.*, *Justus Liebigs Ann. Chem.*, 1883, **216**, 139-161 (*synth*).
Ebert, G. *et al.*, *Justus Liebigs Ann. Chem.*, 1884, **226**, 347-355 (*synth*).
Roth, W.A. *et al.*, *Ber. Dtsch. Chem. Ges.*, 1913, **46**, 260-280 (*synth*).
Sen, R.N. *et al.*, *J. Indian Chem. Soc.*, 1930, **7**, 247-250 (*synth*).
Jensen, K.A. *Z. Anorg. Chem.*, 1933, **94**, 177-179 (*use, indicator*).
Dippy, F.J.F. *et al.*, *J. Chem. Soc.*, 1938, 357-363 (*synth*).
Baddar, F.G. *J. Chem. Soc.*, 1947, 224-227 (*synth*).
Narayana, M. *et al.*, *J. Org. Chem.*, 1962, **27**, 4704-4706 (*synth, uv*).
Wheeler, O.H. *et al.*, *J. Org. Chem.*, 1963, **28**, 2015-2018 (*uv*).
Hadjoudis, E. *et al.*, *J. Chem. Soc., Perkin Trans. 2*, 1972, 1060-1062 (*synth, ir*).
Eur. Pat., 2011, 2 340 719 (*E-form, use*).
Kowczyk-Sadowy, M. *et al.*, *Molecules*, 2015, **20**, 3146-3169 (*pmr, cmr, ft-ir, ft-Raman, uv*).
Tabatabaei Resaei, S.J. *et al.*, *Appl. Organomet. Chem.*, 2017, **31**, e3707:1-10 (*synth*).
Lewis, R.J. *Sax's Dangerous Properties of Industrial Materials, 12th edn., J. Wiley*, 2012, MEJ775

3-(3-Hydroxyphenyl)-2-propenoic acid, CAS H125

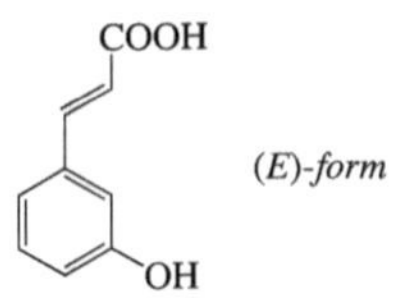

(*E*)-*form*

m-Hydroxycinnamic acid. m-Coumaric acid. m-Cumaric acid
[588-30-7]
[25429-38-3]
$C_9H_8O_3$ 164.16
(E)-form [14755-02-3]
Constit. of the roots of *Citrus sinensis, Citrus limon* and *Citrus paradisi.* Prisms (H_2O). Sol. EtOH, Et_2O, C_6H_6, hot H_2O. Mp 191° (194°). pK_a 4.44
Me ester
[66417-46-7]
$C_{10}H_{10}O_3$ 178.187
Leaflets. Mp 85°.

Aldrich Library of FT-IR Spectra, 1st edn., 1985, **2**, 178B , 178C (*ir*).
Aldrich Library of ¹³*C and* ¹*H FT-NMR Spectra*, 1992, **2**, 1048C , 1049A (*nmr*).
Jones, J.I. *et al.*, *J. Chem. Soc.*, 1935, 1600-1604 (*synth*).
Dippy, F.J.F. *et al.*, *J. Chem. Soc.*, 1938, 357-363 (*synth*).
Katritzky, A.R. *et al.*, *Spectrochim. Acta*, 1960, **16**, 954-963 (*ir*).
Feldman, A.W. *et al.*, *Nature (London)*, 1965, **207**, 985-986 (*Citrus constit*).
Kobuke, Y. *et al.*, *J. Polym. Sci., Part A: Polym. Chem.*, 1970, **8**, 3155-3162 (*synth*).
Santavy, F. *et al.*, *Coll. Czech. Chem. Comm.*, 1972, **37**, 1825-1850 (*uv*)

3-(4-Hydroxyphenyl)-2-propenoic acid, CAS H126

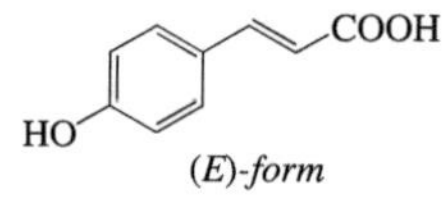

(*E*)-*form*

p-Hydroxycinnamic acid, CAS. *p-Hydroxyphenylacrylic acid.* ***p-Coumaric acid***. *p-Cumaric acid. Naringenic acid. Naringeninic acid*
[7400-08-0]
$C_9H_8O_3$ 164.16
▶ GD9094000
▶ LD_{50} (mus, ipr) 657 mg/kg. Exp. reprod. effects
(E)-form [501-98-4]
Widespread in plants, e.g. *Prunus serotina, Trifolium pratense* and *Daviesia latifolia*. Isol. from *Mycena chlorophos*. Occurs as many glycosides. Found by Bate-Smith in 48% of investigated dicotyledonous and 55% of monocotyledonous spp.. Shows antifungal, antiinflammatory and cytostatic activities. Immunoactive agent, inhibitor of stilbene oxidase. Acts as a bioluminescence activator in fungi. Cryst. + $1H_2O$ (cold H_2O); anhyd. cryst. (hot H_2O). Mp 210 - 213°. pK_{a1} 4.64; pK_{a2} 9.45 (25°). $SOCl_2$-activated polycondensation gives the homopolyester with an all-*trans* struct. Polyesters with other hydroxy acids are also known. λ_{max} 225; 290; 310 (MeOH) (Berdy). [neutral] λ_{max} 223 (ε 14450); 286 (ε 19000) (EtOH) (Berdy)
▶ GD9095000
α-L-Rhamnopyranosyl ester
1-O-p-Coumaroylrhamnose
[102719-86-8]
$C_{15}H_{18}O_7$ 310.302
Constit. of the fruit of *Lappula echinata*. Component of Dong Bei He Shi. Cryst. ($CHCl_3$/MeOH/Me_2CO). Mp 81 - 83°. Mp 188 - 190° (dimorph.)
6-O-Acetyl-β-D-glucopyranosyl ester
[1409971-71-6]
$C_{17}H_{20}O_9$ 368.338
Isol. from *Sanguisorba officinalis*. Displays moderate antimicrobial activity. Amorph. solid
[β-D-Xylopyranosyl-(1→6)-β-D-glucopyranosyl] ester
[143205-77-0]
$C_{20}H_{26}O_{12}$ 458.416
Constit. of the roots of *Asiasarum sieboldii*. Powder. $[\alpha]^{24}_{D}$ -23 (c, 0.42 in MeOH)
[β-D-Glucopyranosyl-(1→3)-β-D-glucopyranosyl] ester
Lycibarbarphenylpropanoid A
[1988786-71-5]
$C_{21}H_{28}O_{13}$ 488.442
Constit. of the fruit of *Lycium barbarum*. Exhibits prominent oxygen radical absorbance capacity. Amorph. yellowish powder. $[\alpha]^{25}_{D}$ -38.2 (c, 0.5 in MeOH). λ_{max} 212 (log ε 3.80); 229 (log ε 3.86); 316 (log ε 4.23) (MeOH)
[β-D-Glucopyranosyl-(1→4)-β-D-glucopyranosyl] ester
Lycibarbarphenylpropanoid B
[1988786-72-6]
$C_{21}H_{28}O_{13}$ 488.442
Constit. of the fruit of *Lycium barbarum*. Exhibits prominent oxygen radical absorbance capacity. Amorph. yellowish powder. $[\alpha]^{25}_{D}$ -34.9 (c, 0.5 in MeOH). λ_{max} 213 (log ε 3.77); 230 (log ε 3.93); 315 (log ε 4.36) (MeOH)
[β-D-Glucopyranosyl-(1→6)-β-D-glucuronopyranosyl] ester
Cranioside B
[430433-46-8]
$C_{21}H_{26}O_{14}$ 502.425
Constit. of *Craniotome furcata*. Amorph. powder
[α-L-Rhamnopyranosyl-(1→4)-α-L-rhamnopyranosyl-(1→3)-[4-hydroxy-E-cinnamoyl-(→4)]-α-D-glucopyranosyl-(1→2)-β-D-fructofuranos-3-yl] ester
Quillajaside A
[1817003-59-0]
$C_{42}H_{54}O_{23}$ 926.871
Constit. of inner bark of *Quillaja saponaria*. Amorph. powder
[β-D-Apiofuranosyl-(1→4)-α-L-rhamnopyranosyl-(1→4)-α-L-rhamnopyranosyl-(1→3)-[4-hydroxy-E-cinnamoyl-(→4)]-α-D-glucopyranosyl-(1→2)-β-D-fructofuranos-3-yl] ester
Quillajaside B
[1817003-60-3]
$C_{47}H_{62}O_{27}$ 1058.986

Constit. of inner bark of *Quillaja saponaria*. Amorph. powder

Me ester

[19367-38-5]

[3943-97-3] (unspecified stereochem.)

$C_{10}H_{10}O_3$ 178.187

Constit. of *Pseudosindora palustris*, *Juglans mandshurica*, *Microtropis fokiensis* and *Pueraria lobata* callus tissue. Exhibits antifungal props.. Needles (H_2O); tablets (EtOH). Mp 137°. $[\alpha]_D^{23}$ +12.06 (c, 0.12 in MeOH)

Et ester

[7362-39-2]

[2979-06-8] (unspecified stereochem.)

$C_{11}H_{12}O_3$ 192.214

Prod. by *Pharbitis nil*, *Solanum melongena*, *Solanum integrifolium* and *Ziziphora clinopioides*. NO prodn. inhibitor. Cytotoxic. Antihypertensive. Mp 83°. Bp 175 - 184° at 0.08 mm

Tetradecyl ester

Tetradecyl p-coumarate

[948827-73-4]

$C_{23}H_{36}O_3$ 360.538

Constit. of the bark of *Acacia* sp.. *E*-Config. not specifically determined but probable

Hexadecyl ester

Hexadecyl (E)-p-coumarate. Hexadecyl 4-hydroxycinnamate

[150113-05-6]

[208927-14-4] (unspecified stereochem.)

$C_{25}H_{40}O_3$ 388.592

Constit. of *Agropyron repens*, *Argyreia speciosa* and *Taraxacum linearisquameum*. Cryst.. Mp 78° (63-65°). λ_{max} 203; 225; 309 (MeOH)

16-Hydroxyhexadecyl ester

16-Hydroxyhexadecyl (E)-p-coumarate

[150113-07-8]

$C_{25}H_{40}O_4$ 404.591

Constit. of *Agropyron repens*. Waxy solid. Mp 107 - 111°

Octadecyl ester

Octadecyl (E)-p-coumarate. Octadecyl 4-hydroxycinnamate. Stearyl p-coumarate

[72943-88-5]

[42933-21-1] (unspecified stereochem.)

$C_{27}H_{44}O_3$ 416.646

Constit. of *Argyreia populifera*, *Artemisia monosperma*, *Chromolaena* spp., *Dorstenia psilurus*, *Araucaria angustifolia* and *Ipomoea batatas*. AChE and BChE inhibitors. Also inhibitor of DNA polymerases α and β. Cryst. ($CHCl_3$). Mp 99 - 100°. λ_{max} 203 (log ε 3.97); 228 (log ε 3.85); 311 (log ε 4.12); 424 (log ε 2.68) (MeOH)

Nonadecyl ester

Nonadecyl (E)-p-coumarate

[605663-40-9]

$C_{28}H_{46}O_3$ 430.673

Constit. of the bark of *Tapirira guianensis*. Oil

Eicosyl ester

Eicosyl (E)-p-coumarate

[112561-76-9]

$C_{29}H_{48}O_3$ 444.7

Constit. of *Aerva javanica*, *Crepis taraxacifolia* and *Psiadia punctulata*

Docosyl ester

Docosyl (E)-p-coumarate

[101959-29-9]

$C_{31}H_{52}O_3$ 472.754

Isol. from the stems of *Bauhinia manca* and *Psiadia punctulata*. Flakes (hexane/EtOAc). Mp 84 - 86°

Tricosyl ester

Tricosyl (E)-p-coumarate

[101959-31-3]

$C_{32}H_{54}O_3$ 486.781

Isol. from the stems of *Bauhinia manca*

Tetracosyl ester

Tetracosyl (E)-p-coumarate

[101959-32-4]

$C_{33}H_{56}O_3$ 500.808

Isol. from *Bauhinia manca*, *Dendrobium falconeri* and *Artemisia igniana*. Mp 95 - 96°

Pentacosyl ester

Pentacosyl (E)-p-coumarate

[101959-34-6]

$C_{34}H_{58}O_3$ 514.835

Isol. from the stems of *Bauhinia manca*

Hexacosyl ester

Hexacosyl (E)-p-coumarate

[74602-24-7]

$C_{35}H_{60}O_3$ 528.862

Isol. from the stems of *Bauhinia manca* and the resin of *Gardenia lucida*. Constit. of *Agrimonia* sp., *Artemisia* sp., *Gardenia* sp. and *Tamarix* sp.. Cryst. (C_6H_6). Mp 95 - 96°. λ_{max} 211 (ε 6918); 227 (ε 7413); 312 (ε 14126) (MeOH)

Heptacosyl ester

Heptacosyl (E)-p-coumarate

[101959-36-8]

$C_{36}H_{62}O_3$ 542.889

Isol. from the stems of *Bauhinia manca*

Octacosyl ester

Octacosyl (E)-p-coumarate

[101959-38-0]

$C_{37}H_{64}O_3$ 556.916

Constit. of *Bauhinia manca*, *Cyrtopodium macrobulbon* and *Eriosema chinensis*

Triacontyl ester

Triacontyl (E)-p-coumarate. ***Defuscin***

[120727-03-9]

$C_{39}H_{68}O_3$ 584.97

Constit. of *Agrimonia pilosa*, *Bombacopsis glabra*, *Dendrobium fuscescens* and *Erythrina* sp.. Inhibitor of snake venom metalloproteinases. Needles ($CHCl_3$/petrol). Mp 101° (94°)

Tetratriacontyl ester

Tetratriacontyl (E)-p-coumarate

[126223-27-6]

$C_{43}H_{76}O_3$ 641.078

Constit. of *Baccharis myrsinites*. Powder. Incorrect name given in the paper. λ_{max} 228 (log ε 4.04); 312 (log ε 2.01) (MeOH)

Benzyl ester

Benzyl (E)-p-coumarate

[61844-62-0]

[27727-38-4] (unspecified stereochem.)

$C_{16}H_{14}O_3$ 254.285

Constit. of *Eriodictyon angustifolium*, *Populus* spp. and propolis. Cryst.. Mp 90 - 92°

1-Phenylethyl ester

[133367-29-0]

$C_{17}H_{16}O_3$ 268.312

Constit. of *Populus trichocarpa*

2-Phenylethyl ester

Phenethyl p-coumarate

[115610-25-8]

[115610-76-9] (unspecified stereochem.)

$C_{17}H_{16}O_3$ 268.312

Constit. of *Populus* spp. and propolis. Potent inhibitor of melanogenesis. Mp 90 - 92°

3-Phenyl-2E-propenyl ester
[115610-30-5]
$C_{18}H_{16}O_3$ 280.323
Constit. of Korean propolis
4′-O-(3,4-Dihydroxy-E-cinnamoyl), Me ester
[1291071-87-8]
$C_{19}H_{16}O_6$ 340.331
Constit. of the rhizomes of *Ligularia vellerea*. Pale yellow powder

(Z)-form [4501-31-9]
Found in *Larix sibirica* bark and *Miscanthus floridulus*. Cryst. (C_6H_6 or H_2O). Mp 134 - 134.5° dec. (subl. from 124°)

▶ GD9095100

4′-O-[4-Hydroxy-E-cinnamoyl-(→6)-β-D-glucopyranoside]
Pulsatilloside B‡
$C_{24}H_{24}O_{10}$ 472.446
Isol. from the aerial parts of *Pulsatilla koreana*. Displays antiadipogenic activity. Pale yellow gum. $[\alpha]_D^{25}$ +225.9 (c, 0.13 in MeOH). λ_{max} 228 (log ε 2.89); 311 (log ε 3.90) (MeOH)

Hexadecyl ester
Hexadecyl (Z)-p-coumarate
[150113-06-7]
$C_{25}H_{40}O_3$ 388.592
Constit. of *Agropyron repens* and *Taraxacum linearisquameum*. Waxy solid. Mp 42 - 47°

16-Hydroxyhexadecyl ester
16-Hydroxyhexadecyl (Z)-p-coumarate
[150113-08-9]
$C_{25}H_{40}O_4$ 404.591
Constit. of *Agropyron repens*. Waxy solid. Mp 105 - 108°

Octadecyl ester
Octadecyl (Z)-p-coumarate
[72943-89-6]
$C_{27}H_{44}O_3$ 416.646
Constit. of *Artemisia igniaria*, *Chromolaena turnariensis*, *Araucaria angustifolia* and *Ipomoea batatas*. Inhibitor of DNA polymerases α and β

Eicosyl ester
Eicosyl (Z)-p-coumarate
[158306-37-7]
$C_{29}H_{48}O_3$ 444.7
Constit. of *Artemisia igniaria* and *Psiadia punctulata*

Docosyl ester
Docosyl (Z)-p-coumarate
[145204-87-1]
$C_{31}H_{52}O_3$ 472.754
Constit. of *Rosa rugosa* and *Psiadia punctulata*. Flakes (hexane/EtOAc). Mp 61 - 63°. λ_{max} 214; 230; 300; 360 (MeOH)

Tetracosyl ester
Tetracosyl (Z)-p-coumarate
[143200-81-1]
$C_{33}H_{56}O_3$ 500.808
Constit. of *Bauhinia manca*, *Dendrobium falconeri* and *Artemisia igniaria*. Mp 98 - 99°

Hexacosyl ester
Hexacosyl (Z)-p-coumarate
[145204-89-3]
$C_{35}H_{60}O_3$ 528.862
Constit. of *Bauhinia manca*, *Broussonetia papyrifera* and *Rosa rugosa*. Possible artifact

Octacosyl ester
Octacosyl (Z)-p-coumarate
[143200-84-4]
$C_{37}H_{64}O_3$ 556.916
Isol. from the stems of *Brassica manca* and leaves of *Rosa rugosa*. Possible artifact

Triacontyl ester
Triacontyl (Z)-p-coumarate
[143200-86-6]
$C_{39}H_{68}O_3$ 584.97
Constit. of *Baccharis myrsinites*. Powder. λ_{max} 228 (log ε 4.2); 284 (log ε 2.21); 312 (sh) (log ε 2.01) (MeOH)

Benzyl ester
Benzyl (Z)-p-coumarate
[115610-22-5]
Constit. of *Populus* spp. and propolis.

Aldrich Library of FT-IR Spectra, 1st edn., 1985, **2**, 180C , 181A, 440D (*ir*).
Aldrich Library of ^{13}*C and* ^{1}*H FT-NMR Spectra*, 1992, **2**, 1051C , 1506A (*nmr*).
Zincke, T. *et al.*, *Justus Liebigs Ann. Chem.*, 1902, **322**, 220-238 (*synth*).
Dippy, F.J.F. *et al.*, *J. Chem. Soc.*, 1938, 357-363 (*synth*).
Bate-Smith, E.C. *Sci. Proc. R. Dublin Soc.*, 1956, **27**, 165-176 (*occur*).
Utsumi, H. *et al.*, *Bull. Chem. Soc. Jpn.*, 1967, **40**, 426 (*cryst struct*).
Aulin-Erdtman, G. *et al.*, *Acta Chem. Scand.*, 1968, **22**, 1187-1209 (*Z-form, synth, uv*).
Karrer, W. *Konstitution und Vorkommen der Organischen Pflanzenstoffe, 2nd edn., Birkhäuser Verlag*, 1976-1985, nos. 951 , 953, 955 (*occur*).
Coutrot, P. *et al.*, *Synthesis*, 1978, 128-130 (*synth*).
Bohlmann, F. *et al.*, *Phytochemistry*, 1979, **18**, 1177-1179 (*octadecyl esters*).
Tachibana, S. *et al.*, *Holzforschung*, 1980, **34**, 131 (*E-Me ester, occur*).
Chatterjee, A. *et al.*, *Indian J. Chem., Sect. B*, 1980, **19**, 421-422 (*hexacosyl p-coumarate*).
Elias, H.C. *et al.*, *Makromol. Chem.*, 1985, **186**, 893-899 (*polycondensation*).
Wang, X. *et al.*, *Yaoxue Xuebao*, 1986, **21**, 183-186 (*rhamnosyl ester*).
Achenbach, H. *et al.*, *Z. Naturforsch., C*, 1986, **41**, 164-168 (*Bauhinia manca constits*).
El-Negoumy, S.I. *et al.*, *Rev. Latinoam. Quim.*, 1987, **107**, 88 (*isol, lentils*).
Wandji, J. *et al.*, *J. Nat. Prod.*, 1990, **53**, 1425-1429 (*Defuscin*).
Sashida, Y. *et al.*, *Chem. Pharm. Bull.*, 1991, **39**, 709-711 (*coumaroylglucoside, acetylcoumaroylglucoside*).
Balde, A.M. *et al.*, *Phytochemistry*, 1991, **30**, 1024-1026 (*octadecyl esters*).
Koetter, U. *et al.*, *Planta Med.*, 1993, **59**, 279-280 (*E-hexadecyl ester, E-16-hydroxyhexadecyl ester*).
Parmar, V.S. *et al.*, *Phytochemistry*, 1994, **36**, 507-511 (*Hexacosyl p-coumarate*).
Lemmich, J. *Phytochemistry*, 1995, **38**, 427-432 (*glucosides*).
Kisiel, W. *et al.*, *Planta Med.*, 1995, **61**, 87-88 (*eicosyl ester*).
Keriko, J.M. *et al.*, *Biosci., Biotechnol., Biochem.*, 1997, **61**, 2127-2128 (*eicosyl esters*).
Srivastava, A. *et al.*, *Indian J. Chem., Sect. B*, 1998, **37**, 192-194 (*hexadecyl ester*).
Garcia, M.D. *et al.*, *Z. Naturforsch., C*, 1998, **53**, 1093-1095 (*activity*).
Fonseca, F.N. *et al.*, *Phytochemistry*, 2000, **55**, 575-580 (*octadecyl esters*).
Juma, B.F. *et al.*, *Phytochemistry*, 2001, **57**, 571-574 (*docosyl coumarates*).
Ye, W. *et al.*, *Planta Med.*, 2002, **68**, 183-186 (*Pulsatillosides B,C*).
Freire, C.S.R. *et al.*, *Phytochem. Anal.*, 2007, **18**, 151-156 (*tetradecyl ester*).
Nishimura, K. *et al.*, *Chem. Pharm. Bull.*, 2009, **57**, 476-480 (*tetracosyl esters, synth*).
Pichette, A. *et al.*, *Can. J. Chem.*, 2010, **88**, 104-110 (*phenethyl coumarate, pmr, cmr*).
Salum, M.L. *et al.*, *Org. Lett.*, 2010, **12**, 4808-4811 (*Z-form, synth, uv, pmr, cmr, ms*).
Hattori, H. *et al.*, *Bioorg. Med. Chem.*, 2011, **19**, 5392-5401 (*3-phenylpropanoyl, propolis constit*).
Nidiry, E.S.J. *et al.*, *Nat. Prod. Commun.*, 2011, **6**, 1889-1892 (*octadecyl esters*).
Hu, J. *et al.*, *Arch. Pharmacal Res.*, 2012, **35**, 2127-2133 (*Microlepia pilosissima constits*).
Avó, J. *et al.*, *Org. Lett.*, 2014, **16**, 2582-2585 (*E-Me ester*).
Zhang, L.-J. *et al.*, *J. Nat. Prod.*, 2015, **78**, 1586-1592 (*Clematiside D*).
Lewis, R.J. *Sax's Dangerous Properties of Industrial Materials, 8th edn., Van Nostrand Reinhold*, 1992, CNU825

1-(4-Hydroxyphenyl)-2-propen-1-ol H127

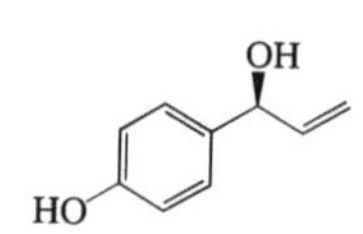

(S)-form

α-Ethenyl-4-hydroxybenzenemethanol, CAS. 4-(1-Hydroxy-2-propenyl)phenol. 4-Hydroxy-α-vinylbenzyl alcohol. 1′-Hydroxychavicol
[53580-61-3]
[51410-44-7] (4′-Me ether)
[53890-21-4] (di-Ac, unspecified stereochem.) [108093-86-3] (4′-Ac)
$C_9H_{10}O_2$ 150.177
(±)-form
1-Ac
1′-Acetoxychavicol
[151986-17-3]
$C_{11}H_{12}O_3$ 192.214
Constit. of *Acorus calamus*. Superoxide anion generation inhibitor. Inhibits tumour promotion by co-carcinogen A1 on ICR mouse skin.

Mitsui, S. *et al., Chem. Pharm. Bull.*, 1976, **24**, 2377-2382 (*isol, synth, activity*).
Rajashekkar, B. *et al., J. Biol. Chem.*, 1984, **259**, 6925-6930 (*synth*).
Chowdhury, A.K.A. *et al., Pharmazie*, 1993, **48**, 786-787 (*synth*).
Murakami, A. *et al., Oncology*, 1996, **53**, 386-391 (*use*).
Matsuda, H. *et al., Bioorg. Med. Chem. Lett.*, 2003, **13**, 3197-3202 (*activity, sar*).
Yang, X. *et al., J. Agric. Food Chem.*, 2009, **57**, 3286-3290 (*synth*)

3-(2-Hydroxyphenyl)-2-propen-1-ol H128

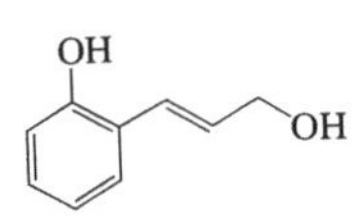

(E)-form

2-(3-Hydroxy-1-propenyl)phenol, CAS. 2-Hydroxycinnamyl alcohol. o-Coumaryl alcohol
[13523-27-8]
$C_9H_{10}O_2$ 150.177
(E)-form [51764-86-4]
Constit. of the twigs of *Cinnamomum cassia*. Mp 86 - 87°. Mp 110°
2′-O-[β-D-Apiofuranosyl-(1→6)-β-D-glucopyranoside]
Cinnacasolide B
[1393363-33-1]
$C_{20}H_{28}O_{11}$ 444.433
Constit. of the twigs of *Cinnamomum cassia*. Amorph. solid
2′-O-[β-D-Apiofuranosyl-(1→6)-1-O-benzyl-β-D-glucopyranosyl-(2→2)-[α-L-rhamnopyranosyl-(1→6)]-β-D-glucopyranoside]
Cinnacassiol*. Cinnacassinol*
[1160707-55-0]
$C_{39}H_{54}O_{20}$ 842.841
Constit. of the twigs of *Cinnamomum cassia*. Amorph. powder
1-Ac
Cryptamygin A
[475632-38-3]
$C_{11}H_{12}O_3$ 192.214
Constit. of the stem bark of *Cryptocarya amygdalina*. Oil. λ_{max} 212; 251; 303 (MeOH)
2′-Me ether
3-(2-Methoxyphenyl)-2-propen-1-ol. 2-Methoxycinnamyl alcohol
[1504-61-6]
$C_{10}H_{12}O_2$ 164.204
Bp 95 - 100° at 0.4 mm
(Z)-form [14864-24-5]
Long rectangular prisms (C_6H_6/EtOAc). Mp 110 - 111°
2′-Me ether, 3,5-dinitrobenzoyl
[13523-28-9]
Pale yellow needles (C_6H_6). Mp 165 - 166°.

Karrer, P. *et al., Helv. Chim. Acta*, 1949, **32**, 1692-1693 (*synth*).
Jorgenson, M.J. *Tetrahedron Lett.*, 1962, **3**, 559-562 (*synth*).
Kumari, D. *et al., Tetrahedron*, 1966, **22**, 3491-3499 (*Z-form*).
Chan, Y.-Y. *et al., J. Chin. Chem. Soc. (Taipei)*, 2002, **49**, 263-268 (*Cryptamygin A*).
Ngoc, T.M. *et al., J. Nat. Prod.*, 2009, **72**, 1205-1208 (*Cinnacassiol*).
Ngoc, T.M. *et al., Bioorg. Med. Chem. Lett.*, 2012, **22**, 4625-4628 (*Cinnacasolide B*).
Zhu, J.-B. *et al., Org. Lett.*, 2013, **15**, 3054-3057 (*synth*).
Meng, J. *et al., Asian J. Chem.*, 2017, **6**, 1724-1727 (*synth*)

3-(4-Hydroxyphenyl)-2-propen-1-ol H129

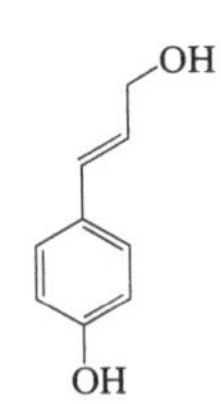

4-(3-Hydroxy-1-propenyl)phenol, CAS. 4-Hydroxycinnamyl alcohol. 4′-Hydroxystyrone. p-Coumaryl alcohol
[3690-05-9]
$C_9H_{10}O_2$ 150.177
Props. and derivs. all appear to belong to the *E*-form although this is not always explicit in the lit.
(E)-form [20649-40-5]
Minor lignan monomer found in *Alpinia galanga* and *Malus domestica* wax. Present in softwoods together with coniferyl alcohol, in D246 and sinapyl alcohol, in H81. Also present in grass and forage type lignins. Moderate NO inhibitor. Prisms (Et_2O/petrol or H_2O). Mp 124° (116-118°,89.3-90.5°). Acid-sensitive
1-O-β-D-Glucopyranoside
p-Coumaryl β-D-glucopyranoside. ***Sachaliside 1****. Triandrin*
[132294-76-9]
[19764-35-3] (unspecified stereochem.)
$C_{15}H_{20}O_7$ 312.318
Constit. of *Boschniakia rossica*, *Salix triandra*, *Salix sachalinesis* and many other *Salix* spp.. Needles (EtOAc). Mp 177 - 179° (152-152.5°). $[\alpha]^{20}_{D}$ -60.5 (c, 1.85 in H_2O). Stereochem. of Triandrin not detd. but prob. *E*-. Stereochem. of Sachaliside 1 detd. as *E*-
1-O-[β-D-Apiofuranosyl-(1→6)-β-D-glucopyranoside]
[2247995-06-6]
$C_{20}H_{28}O_{11}$ 444.433
Constit. of the root bark of *Lycium chinensis*. $[\alpha]^{20}_{D}$ -10.4 (c, 0.1 in MeOH). λ_{max} 206 (log ε 4.25); 263 (log ε 4.05); 297 (log ε 3.46) (MeOH)
1-O-[α-L-Arabinopyranosyl-(1→6)-β-D-glucopyranoside]
Rossicasin B
[886465-83-4]
$C_{20}H_{28}O_{11}$ 444.433
Constit. of *Boschniakia rossica*. Brown syrup. $[\alpha]^{23}_{D}$ -51.7 (c, 0.3 in H_2O). λ_{max} 263 (log ε 4.02) (MeOH)
1-O-[β-D-Xylopyranosyl-(1→6)-β-D-glucopyranoside]
Rossicasin A
[886465-81-2]
$C_{20}H_{28}O_{11}$ 444.433
Constit. of *Boschniakia rossica*. Needles (MeOH). Mp 168 - 170°. $[\alpha]^{23}_{D}$ -92.5 (c, 0.4 in H_2O). λ_{max} 263 (log ε 4.36) (MeOH)
1-O-[α-L-Rhamnopyranosyl-(1→3)-[4-hydroxy-3-methoxy-E-cinnamoyl-(→4)]-β-D-glucopyranoside]
[1000872-46-7]
$C_{31}H_{38}O_{14}$ 634.631
Constit. of *Pedicularis rex*
1-O-[β-D-Glucopyranosyl-(1→2)-β-D-glucopyranoside]
[1834500-20-7]

$C_{21}H_{30}O_{12}$ 474.459
Constit. of *Boschniakia rossica*. Powder. $[\alpha]_D^{20}$ -16.5 (MeOH). λ_{max} 265 (log ε 4.38) (MeOH)

1-O-[β-D-Glucopyranosyl-(1→2)-[4-hydroxy-E-cinnamoyl-(→4)]-β-D-glucopyranoside]
Rossicaside D
[80458-57-7]
$C_{30}H_{36}O_{14}$ 620.604
Isol. from the parasitic plant *Boschniakia rossica*. The dried stem has been used as a tonic in Japan. Needles + 1½H_2O (MeOH aq.). Mp 198 - 201.5° dec.. $[\alpha]_D^{20}$ -58.3 (c, 0.5 in MeOH). λ_{max} 230 (ε 26915); 272 (ε 32359); 300 (ε 33884); 312 (ε 36308) (EtOH)

1-O-[β-D-Glucopyranosyl-(1→2)-[4-hydroxy-E-cinnamoyl-(→6)]-β-D-glucopyranoside]
Rossicaside C
[80458-56-6]
$C_{30}H_{36}O_{14}$ 620.604
Isol. from the parasitic plant *Boschniakia rossica*. The dried stem has been used as a tonic in Japan. Needles + ½H_2O (MeOH). Mp 236 - 237.5° dec.. $[\alpha]_D^{19}$ -18.3 (c, 1.1 in Py). λ_{max} 232 (ε 28184); 269 (ε 60256); 300 (ε 48978); 310 (ε 53703) (EtOH)

1-O-[β-D-Xylopyranosyl-(1→4)-α-L-rhamnopyranosyl-(1→3)-[3,4-dihydroxy-E-cinnamoyl-(→4)]-β-D-glucopyranoside]
Rossicaside J
[1528759-67-2]
$C_{35}H_{44}O_{18}$ 752.719
Constit. of *Boschniakia rossica*. Brown powder. $[\alpha]_D^{25}$ -61.5 (c, 0.27 in MeOH). λ_{max} 258 (log ε 4.12); 295 (log ε 3.86); 327 (log ε 3.85) (MeOH)

1-O-[β-D-Glucopyranosyl-(1→4)-α-L-rhamnopyranosyl-(1→3)-β-D-glucopyranoside]
[885702-23-8]
$C_{27}H_{40}O_{16}$ 620.601
Constit. of *Boschniakia rossica*. Brown syrup. λ_{max} 263 (log ε 4.12) (MeOH)

1-O-[β-D-Glucopyranosyl-(1→4)-α-L-rhamnopyranosyl-(1→3)-[4-hydroxy-E-cinnamoyl-(→4)]-β-D-glucopyranoside]
Rossicaside G
[1589079-41-3]
$C_{36}H_{46}O_{18}$ 766.746
Constit. of *Boschniakia rossica*. Significantly inhibited TG accumulation in HepG2 cells. Brown powder. $[\alpha]_D^{25}$ -65.9 (c, 0.34 in MeOH). λ_{max} 265 (log ε 4.40); 307 (log ε 4.33) (MeOH)

1-O-[β-D-Glucopyranosyl-(1→4)-α-L-rhamnopyranosyl-(1→3)-[4-hydroxy-E-cinnamoyl-(→6)]-β-D-glucopyranoside]
Rossicaside H
[1589079-54-8]
$C_{36}H_{46}O_{18}$ 766.746
Constit. of *Boschniakia rossica*. Significantly inhibited TG accumulation in HepG2 cells. Brown powder. $[\alpha]_D^{25}$ -45.7 (c, 0.11 in MeOH). λ_{max} 265 (log ε 4.21); 307 (log ε 4.15) (MeOH)

1-O-[β-D-Glucopyranosyl-(1→4)-α-L-rhamnopyranosyl-(1→3)-[3,4-dihydroxy-E-cinnamoyl-(→4)]-β-D-glucopyranoside]
Rossicaside B
[80458-55-5]
$C_{36}H_{46}O_{19}$ 782.745
Isol. from the parasitic plant *Boschniakia rossica*. The dried stem has been used as a tonic in Japan. Needles (H_2O or MeOH aq.). Mp 177 - 178° dec.. $[\alpha]_D^{17}$ -66.6 (c, 1.1 in MeOH). λ_{max} 263 (ε 24547); 281 (ε 18621); 299 (ε 11749); 340 (ε 13490) (EtOH)

1-O-[β-D-Glucopyranosyl-(1→4)-α-L-rhamnopyranosyl-(1→3)-[3,4-dihydroxy-E-cinnamoyl-(→6)]-β-D-glucopyranoside]
Rossicaside I
[1528759-66-1]
$C_{36}H_{46}O_{19}$ 782.745
Constit. of *Boschniakia rossica*. Brown powder. $[\alpha]_D^{25}$ -41.8 (c, 0.44 in MeOH). λ_{max} 259 (log ε 4.21); 295 (log ε 4.00); 322 (log ε 3.95) (MeOH)

1-Hexadecanoyl
[1019641-76-9]
$C_{25}H_{40}O_3$ 388.592
Constit. of the peel of *Cydonia oblonga*

1-Octadecanoyl
[852547-56-9]
$C_{27}H_{44}O_3$ 416.646
Constit. of Annurca apple fruit (*Malus* sp.) and the peel of *Cydonia oblonga*

1-O-(9Z-Octadecenoyl)
[852547-54-7]
$C_{27}H_{42}O_3$ 414.63
Constit. of Annurca apple fruit (*Malus* sp.) and the peel of *Cydonia oblonga*

1-O-(9Z,12Z-Octadecadienoyl)
[852547-52-5]
$C_{27}H_{40}O_3$ 412.614
Constit. of Annurca apple fruit (*Malus* sp.) and the peel of *Cydonia oblonga*

1-Eicosanoyl
[852547-60-5]
$C_{29}H_{48}O_3$ 444.7
Constit. of Annurca apple fruit (*Malus* sp.)

1-Docosanoyl
[125003-14-7]
$C_{31}H_{52}O_3$ 472.754
Constit. of Annurca apple fruit (*Malus* sp.) and the peel of *Cydonia oblonga*. Also isol. from the root bark of *Buddleia globosa*. Cryst. (Me_2CO). Mp 60 - 61°. λ_{max} 232 (log ε 3.6); 275 (log ε 3.1); 295 (sh) (MeOH). [base] λ_{max} 248 (log ε 3.5); 345 (log ε 3.7) (MeOH/NaOH)

1-O-(13Z-Docosenoyl)
[1019641-80-5]
$C_{31}H_{50}O_3$ 470.738
Constit. of the peel of *Cydonia oblonga*

1-Tetracosanoyl
[125003-13-6]
$C_{33}H_{56}O_3$ 500.808
Constit. of the root bark of *Buddleia globosa*. Cryst. (Me_2CO). Mp 59 - 60°. λ_{max} 232 (log ε 3.6); 275 (log ε 3.1); 295 (sh) (MeOH). [base] λ_{max} 248 (log ε 3.5); 345 (log ε 3.7) (MeOH/NaOH)

1-O-(4-Hydroxy-E-cinnamoyl)
[122135-45-9]
$C_{18}H_{16}O_4$ 296.322
Constit. of *Populus candicans*. Solid

4′-Me ether, 1-O-(4-hydroxy-E-cinnamoyl)
[885703-08-2]
$C_{19}H_{18}O_4$ 310.349
Constit. of *Etlingera pavieana*

(Z)-form [124076-60-4]

Constit. of *Angelica keikei*

1-Hexadecanoyl

[1019641-77-0]

Constit. of the peel of *Cydonia oblonga*

1-Octadecanoyl

[852547-55-8]

Constit. of Annurca apple fruit (*Malus* sp.)

1-O-(9Z-Octadecenoyl)

[852547-53-6]

Constit. of Annurca apple fruit (*Malus* sp.)

1-O-(9Z,12Z-Octadecadienoyl)

[852547-51-4]

Constit. of Annurca apple fruit (*Malus* sp.)

1-O-(9Z,12Z,15Z-Octadecatrienoyl)

[877880-88-1]

$C_{27}H_{38}O_3$ 410.598

Constit. of Annurca apple fruit (*Malus* sp.)

1-Eicosanoyl

[852547-57-0]

Constit. of Annurca apple fruit (*Malus* sp.)

1-Docosanoyl

[852547-58-1]

Constit. of Annurca apple fruit (*Malus* sp.)

1-O-(13Z-Docosenoyl)

[1019641-81-6]

Constit. of the peel of *Cydonia oblonga*

1-Tetracosanoyl

[852547-59-2]

Constit. of Annurca apple fruit (*Malus* sp.) and the peel of *Cydonia oblonga*.

Karrer, P. *Helv. Chim. Acta*, 1928, **11**, 1209-1213 (*Me ether*).
Freudenberg, K. *et al.*, *Chem. Ber.*, 1951, **84**, 443-451 (*synth*).
Marshall, D. *et al.*, *J. Chem. Soc.*, 1956, 4082-4088 (*synth, uv*).
Steglich, W. *et al.*, *Chem. Ber.*, 1978, **111**, 3939-3948 (*synth, pmr*).
Konishi, T. *et al.*, *Chem. Pharm. Bull.*, 1981, **29**, 2807-2815 (*Rossicasides*).
Houghton, P.J. *et al.*, *Phytochemistry*, 1989, **28**, 2693-2695 (*docosanoyl, tetracosanoyl*).
Whatley, F.R. *et al.*, *Z. Naturforsch., C*, 1989, **44**, 353-356 (*Populus candicans constit*).
Quideau, S. *et al.*, *J. Agric. Food Chem.*, 1992, **40**, 1108-1110 (*synth*).
Naidoo, L.A.C. *et al.*, *Phytochemistry*, 1992, **31**, 3929-3931 (*Triandrin*).
Farrell, R. *et al.*, *J. Nat. Prod.*, 1996, **59**, 446-447 (*synth, pmr, cmr*).
Whitaker, B.D. *et al.*, *J. Agric. Food Chem.*, 2001, **49**, 3787-3792 (*Malus domestica constits*).
Fiorentino, A. *et al.*, *J. Agric. Food Chem.*, 2008, **56**, 2660-2667 (*Cydonia oblonga esters*).
Wu, C.H. *et al.*, *Chin. Herb. Med.*, 2013, **5**, 5-8 (*Boschniakia rossica glycosides*).
Zhang, Y. *et al.*, *Fitoterapia*, 2013, **85**, 69-75 (*Rossicasides B-D,G-J*).
Tachai, S. *et al.*, *Nat. Prod. Res.*, 2016, **30**, 2215-2219 (*Etlingera pavieana constit*).
Yang, Y.-N. *et al.*, *RSC Adv.*, 2017, **7**, 805-812 (*Lycium chinensis 6-apiosylglucoside*)

1-(4-Hydroxyphenyl)-7-(3,4,5-trihydroxyphenyl)-3,5-heptanediol H130

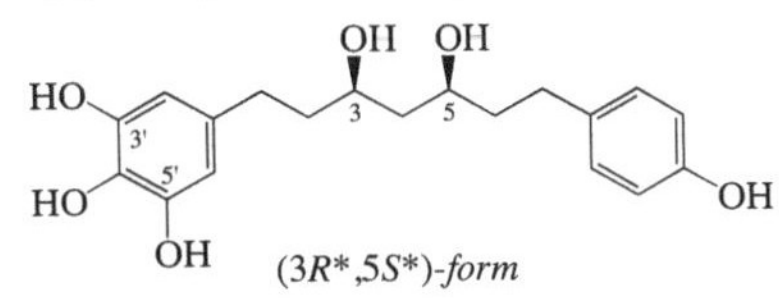

(3*R**,5*S**)-form

$C_{19}H_{24}O_6$ 348.395

(3R*,5S*)-form

3′-Me ether

1-(3,4-Dihydroxy-5-methoxyphenyl)-7-(4-hydroxyphenyl)-3,5-heptanediol

[1334172-40-5]

$C_{20}H_{26}O_6$ 362.422

Constit. of the rhizomes of *Curcuma kwangsiensis*. Yellowish oil. λ_{max} 279 (log ε 3.71) (MeOH)

(3RS,5SR)-form

3′,5′-Di-Me ether

1-(4-Hydroxy-3,5-dimethoxyphenyl)-7-(4-hydroxyphenyl)-3,5-heptanediol. ***Diarylcomosol III***

[1452487-93-2]

$C_{21}H_{28}O_6$ 376.449

Constit. of *Curcuma comosa*. Oil. λ_{max} 279 (log ε 3.39) (MeOH).

Li, J. *et al.*, *Bioorg. Med. Chem. Lett.*, 2011, **21**, 5363-5369 (*Curcuma kwangsiensis constit*).
Matsumoto, T. *et al.*, *Bioorg. Med. Chem. Lett.*, 2013, **23**, 5178-5181 (*Diarylcomosol III*)

1-(4-Hydroxyphenyl)-7-(3,4,5-trihydroxyphenyl)-2,3,5-heptanetriol H131

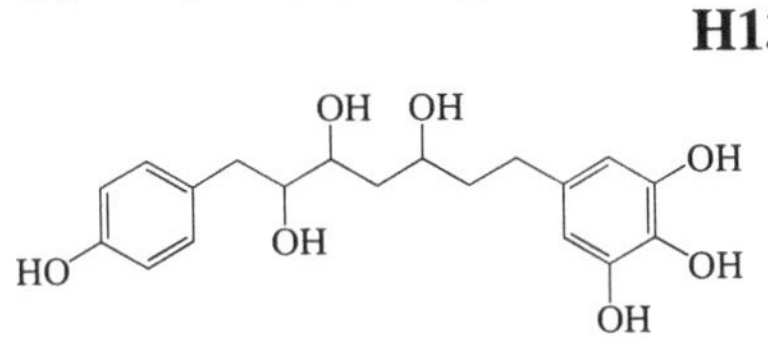

$C_{19}H_{24}O_7$ 364.394

3″,5″-Di-Me ether

7-(4-Hydroxy-3,5-dimethoxyphenyl)-1-(4-hydroxyphenyl)-2,3,5-heptanetriol

[1269839-30-6]

$C_{21}H_{28}O_7$ 392.448

Isol. from the rhizomes of *Curcuma kwangsiensis*. Shows strong NO inhibitory activity. Yellow oil.

Jun, L.I. *et al.*, *Chin. J. Nat. Med.*, 2011, **9**, 329-333 (*Curcuma kwangsiensis constit*)

5-Hydroxy-1-phenyl-7-(3,4,5-trihydroxyphenyl)-3-heptanone H132

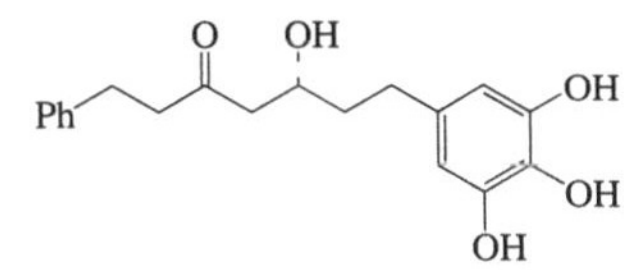

$C_{19}H_{22}O_5$ 330.38

(R)-form

3′-Me ether

7-(3,4-Dihydroxy-5-methoxyphenyl)-5-hydroxy-1-phenyl-3-heptanone

[1022082-06-9]

$C_{20}H_{24}O_5$ 344.407

Constit. of the rhizomes of *Alpinia officinarum*. Yellow oil. $[\alpha]^{20}_{D}$ -10.7 (c, 0.04 in $CHCl_3$). λ_{max} 204 (log ε 4.39); 278 (log ε 3.13) (MeOH).

An, N. *et al.*, *Fitoterapia*, 2008, **79**, 27-31 (*Alpinia officinarum constit*)

7-(4-Hydroxyphenyl)-1-(2,3,4-trihydroxyphenyl)-3-heptanone H133

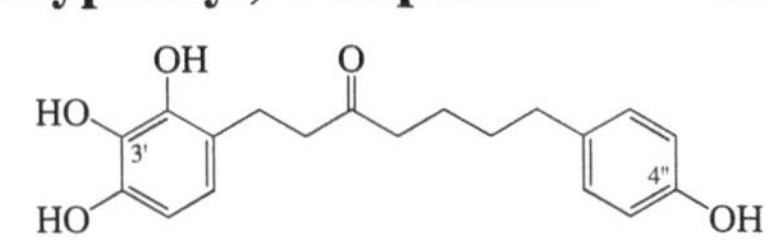

$C_{19}H_{22}O_5$ 330.38

3′,4″-Di-Me ether

1-(2,4-Dihydroxy-3-methoxyphenyl)-7-(4-methoxyphenyl)-3-heptanone

[1606145-77-0]

$C_{21}H_{26}O_5$ 358.434

Constit. of the fruit of *Alpinia oxyphylla*. Antioxidant. Light yellow oil. λ_{max} 242; 296 (MeOH).

Bian, Q.-Y. *et al.*, *J. Asian Nat. Prod. Res.*, 2013, **15**, 1094-1099 (*Alpinia oxyphylla constit*)

2-(4-Hydroxyphenyl)-1-(2,4,6-trihydroxyphenyl)-1-propanone, CAS H134

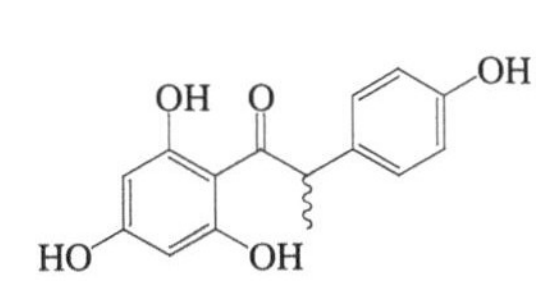

2′,4′,4″,6′-Tetrahydroxy-α-methyldeoxybenzoin. 6′-Hydroxy-O-demethylangolensin

$C_{15}H_{14}O_5$ 274.272

(ξ)-form [153409-52-0]
Tentatively identified in human urine. Antioxidant.

Ingham, J.L. *Prog. Chem. Org. Nat. Prod.*, 1983, **43**, 1-266 (*occur, rev*).
Ruefer, C.E. *et al.*, *J. Agric. Food Chem.*, 2006, **54**, 2926-2931 (*activity*)

6-Hydroxy-1-pyrrolizidinecarboxylic acid **H135**

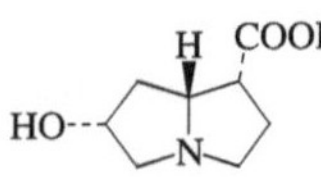

(1R,6R,7aR)-form

Hexahydro-6-hydroxy-1H-pyrrolizine-1-carboxylic acid
[215932-88-0]
$C_8H_{13}NO_3$ 171.196
(1R,6R,7aR)-form
Cryst. (Me_2CO/MeOH)(as hydrochloride). Mp 147 - 148° (hydrochloride). $[\alpha]_D$ +32 (c, 1 in MeOH)
Me ester
Amphorogynine D
[544711-77-5]
$C_9H_{15}NO_3$ 185.223
Alkaloid from *Amphorogyne spicata*. Amorph. gum. $[\alpha]_D$ +17 (c, 1 in $CHCl_3$)
Et ester
[73354-03-7]
$C_{10}H_{17}NO_3$ 199.25
Needles (EtOAc). Mp 110°. $[\alpha]_D^{18}$ +73.4 (c, 4 in $CHCl_3$)
6-O-[3-(4-Hydroxy-3-methoxyphenyl)propanoyl], Me ester
Amphorogynine A
[215932-85-7]
$C_{19}H_{25}NO_6$ 363.410
Alkaloid from *Amphorogyne spicata*. Cryst. (MeOH/heptane). Mp 108°. $[\alpha]_D$ +53 (c, 1 in $CHCl_3$)
(1R,6S,7aR)-form
6-O-[3-(4-Hydroxy-3-methoxyphenyl)propanoyl], Me ester
Amphorogynine C
[215932-87-9]
$C_{19}H_{25}NO_6$ 363.410
Alkaloid from *Amphorogyne spicata*. Cryst. (MeOH). Mp 130°. $[\alpha]_D$ -2 (c, 1 in $CHCl_3$)
(1S,6S,7aR)-form
6-O-[3-(4-Hydroxy-3-methoxyphenyl)propanoyl], Me ester
Amphorogynine B
[215932-86-8]
Alkaloid from *Amphorogyne spicata*. Gum. $[\alpha]_D$ -7 (c, 1 in $CHCl_3$)
(1S,6S,7aS)-form
Et ester
[78175-02-7]
Needles (EtOAc). Mp 110°. $[\alpha]_D^{18}$ -72.1 (c, 4 in $CHCl_3$).

Robins, D.J. *et al.*, *J. Chem. Soc., Perkin Trans. 1*, 1989, 1339-1342 (*synth, ester*).
Huong, D.T.T. *et al.*, *J. Nat. Prod.*, 1998, **61**, 1444-1446 (*Amphorogynines A-D*).
Yoda, H. *et al.*, *Tetrahedron Lett.*, 2003, **44**, 1643-1646 (*Amphorogynine A, synth*).
Roche, C. *et al.*, *J. Org. Chem.*, 2005, **70**, 8352-8363 (*Amphorogynine A, D, synth*).
de Miguel, I. *et al.*, *Eur. J. Org. Chem.*, 2012, 4347-4353 (*Amphorogynine C, synth*)

8-Hydroxytrachelosiaside **H136**

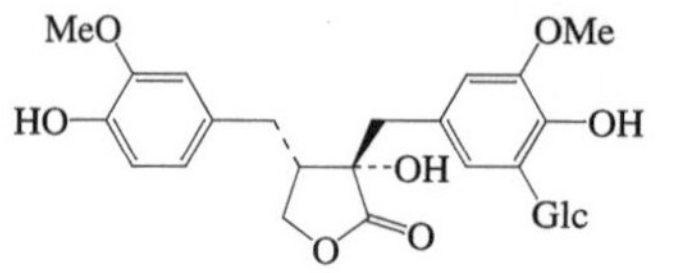

Relative Configuration

3-C-β-D-Glucopyranosyl-4,4′,8-trihydroxy-3′,5-dimethoxy-8,8′-lignan-9,9′-olide. 5-Glucosylnortrachelogenin
[858127-39-6]
$C_{26}H_{32}O_{12}$ 536.53
Constit. of *Trachelospermum jasminoides*. Amorph. powder. $[\alpha]_D^{25}$ -38 (c, 0.62 in MeOH).

Tan, X.-Q. *et al.*, *Planta Med.*, 2005, **71**, 93-95 (*8-Hydroxytrachelosiaside*)

2-Hydroxy-3-(3,4,5-trihydroxyphenyl)propanoic acid **H137**

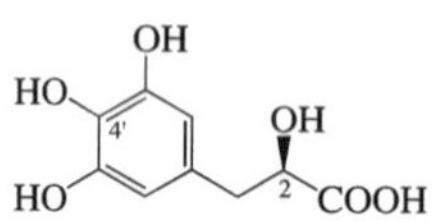

α,3,4,5-Tetrahydroxybenzenepropanoic acid. 3-(3,4,5-Trihydroxyphenyl)lactic acid
$C_9H_{10}O_6$ 214.173
(R)-form
3′,4′,5′-Tri-Me ether, 2-O-(4-hydroxy-3-methoxy-E-cinnamoyl)
Clinopodphenol A
[1380396-22-4]
$C_{22}H_{24}O_9$ 432.425
Constit. of *Clinopodium urticifolium*. Cytotoxic. Pale yellow gum. $[\alpha]_D^{16}$ +11.5 (c, 0.2 in MeOH). Abs. config. not confirmed. λ_{max} 230 (log ε 4.25); 279 (log ε 4.08); 305 (log ε 3.22) (MeOH).

Hu, Q. *et al.*, *Chem. Nat. Compd. (Engl. Transl.)*, 2012, **48**, 198-201 (*Clinopodphenol A*)

3-Hydroxy-3-(3,4,5-trihydroxyphenyl)propanoic acid **H138**

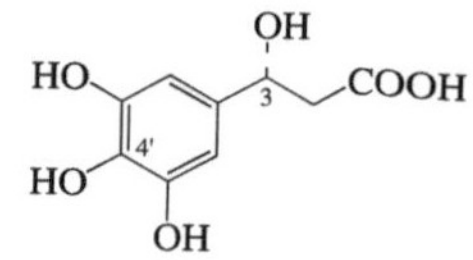

(R)-form

β,3,4,5-Tetrahydroxybenzenepropanoic acid
[2353656-19-4]
$C_9H_{10}O_6$ 214.173
(R)-form [2349684-34-8]
3′,5′-Di-Me ether
3-Hydroxy-3-(4-hydroxy-3,5-dimethoxyphenyl)propanoic acid
3′,5′-Di-Me ether, Me ester
[2170113-26-3]
$C_{12}H_{16}O_6$ 256.254
Constit. of the seeds of *Raphanus sativus*. Gum. $[\alpha]_D^{25}$ +17.3 (c, 0.4 in MeOH). λ_{max} 216 (log ε 3.9); 238 (log ε 2.8); 270 (log ε 1.2) (MeOH)
3′,4′-Methylene, 5′-Me ether, 3-Ac, Me ester
[1393661-34-1]
$C_{14}H_{16}O_7$ 296.275
Constit. of *Peperomia tetraphylla*. Gum. $[\alpha]_D^{20}$ +54.2 (c, 0.02 in $CHCl_3$). λ_{max} 250 ($CHCl_3$)
(S)-form [2348356-39-6]
3′,5′-Di-Me ether, Me ester
[1187551-31-0]
Constit. of *Ruta graveolens*. Amorph. yellow powder. $[\alpha]_D^{25}$ -1 (c, 0.5 in MeOH). λ_{max} 207; 238; 269; 282 (MeOH).

Rollinger, J.M. *et al.*, *Planta Med.*, 2009, **75**, 195-204 (*Ruta graveolens constit*).
Yu, Y. *et al.*, *Chin. J. Chem.*, 2013, **31**, 1336-1340 (*abs config*).
Li, Y.-Z. *et al.*, *Nat. Prod. Res.*, 2013, **27**, 837-840 (*Peperomia tetraphylla constit*).
Kim, K.H. *et al.*, *J. Braz. Chem. Soc.*, 2015, **26**, 2307-2312 (*Raphanus sativus constit*)

3-Hydroxy-1-(2,4,5-trihydroxyphenyl)-1-propanone **H139**

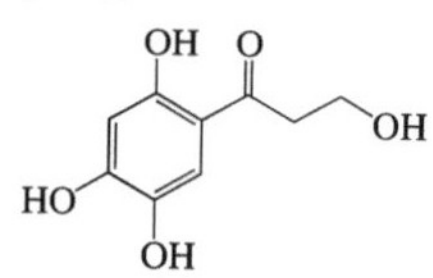

$C_9H_{10}O_5$ 198.174
4′,5′-Di-Me ether
3-Hydroxy-1-(2-hydroxy-4,5-dimethoxyphenyl)-1-propanone.
Cordifoliketone A
[1393715-39-3]

$C_{11}H_{14}O_5$ 226.228
Constit. of the roots of *Codonopsis cordifolioidea*. Cytotoxic. Pale yellow gum. λ_{max} 210 (log ε 4.89); 248 (log ε 3.12); 288 (log ε 4.22); 325 (log ε 2.42) (MeOH)

4′,5′-Di-Me ether, 3-Ac
Cordifoliketone B
[1393715-40-6]
$C_{13}H_{16}O_6$ 268.265
Constit. of the roots of *Codonopsis cordifolioidea*. Cytotoxic. Pale yellow gum. λ_{max} 210 (log ε 4.92); 248 (log ε 3.22); 288 (log ε 4.12); 324 (log ε 2.46) (MeOH).

Hu, Q. *et al.*, *Bull. Korean Chem. Soc.*, 2012, **33**, 278-280 (*Cordifoliketones A,B*)

3-Hydroxy-1-(3,4,5-trihydroxyphenyl)-1-propanone H140

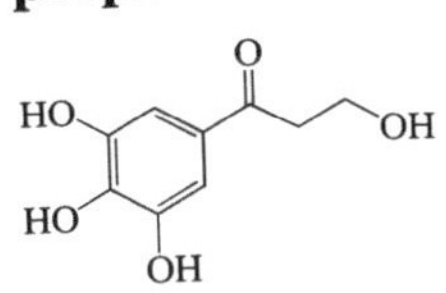

3,3′,4′,5′-Tetrahydroxypropiophenone
$C_9H_{10}O_5$ 198.174

3′,4′-Di-Me ether
3-Hydroxy-1-(3-hydroxy-4,5-dimethoxyphenyl)-1-propanone. 3,3′-Dihydroxy-4′,5′-dimethoxypropiophenone. ***Feddeiketone B***
[1416313-52-4]
$C_{11}H_{14}O_5$ 226.228
Constit. of *Daphne feddei* and *Nicotiana tabacum*. Pale yellow gum. λ_{max} 210 (log ε 4.89); 250 (log ε 3.18); 288 (log ε 4.16); 325 (log ε 2.44) (MeOH)

3′,4′-Di-Me ether, 3-Ac
Feddeiketone A
[1352448-98-6]
$C_{13}H_{16}O_6$ 268.265
Constit. of *Daphne feddei*. Pale yellow gum. λ_{max} 210 (log ε 4.82); 250 (log ε 3.26); 289 (log ε 4.18); 324 (log ε 2.47) (MeOH)

3′,5′-Di-Me ether
3-Hydroxy-1-(4-hydroxy-3,5-dimethoxyphenyl)-1-propanone, CAS. *3,4′-Dihydroxy-3′,5′-dimethoxypropiophenone*
[136196-47-9]
$C_{11}H_{14}O_5$ 226.228
Constit. of *Cassia laevigata, Ficus beecheyana, Nicotiana tabacum, Selaginella doederleinii* and *Tectona grandis*

3′,5′-Di-Me ether, 4′-O-β-D-glucopyranoside
$C_{17}H_{24}O_{10}$ 388.369
Constit. of the leaves of *Bellis sylvestris*. Exhibits antimicrobial action against *Proteus mirabilis* and *Pseudomonas aeruginosa* associated with urinary tract infections. Amorph. powder

3′,5′-Di-Me ether, 3-O-(4-hydroxy-E-cinnamoyl)
[1202573-44-1]
$C_{20}H_{20}O_7$ 372.373
Constit. of the leaves of *Sasa qualpaertensis*. Yellow solid. λ_{max} 297 (MeOH)

3′,4′-Methylene, 5′-Me ether
3-Hydroxy-3′-methoxy-4′,5′-methylenedioxypropiophenone. ***Feddeiketone C***
[1416313-53-5]
$C_{11}H_{12}O_5$ 224.212
Constit. of *Daphne feddei*. Pale yellow gum. λ_{max} 210 (log ε 4.82); 250 (log ε 3.26); 287 (log ε 4.22); 325 (log ε 2.57) (MeOH).

Lin, R.C. *et al.*, *Planta Med.*, 1994, **60**, 168-170 (*Selaginella doederleinii constit*).
Jones, L. *et al.*, *Fitoterapia*, 2000, **71**, 580-583 (*3′,5′-di-Me ether*).
Lee, T.-H. *et al.*, *J. Nat. Prod.*, 2002, **65**, 1497-1500 (*Ficus beecheyana constit*).
Sultana, N. *et al.*, *Bull. Korean Chem. Soc.*, 2009, **30**, 1729-1732 (*3′,5′-di-Me ether 3-coumaroyl*).
Lu, Y. *et al.*, *J. Braz. Chem. Soc.*, 2012, **23**, 656-660 (*Feddeiketones A,B,C*).
Scognamiglio, M. *et al.*, *Phytochem. Lett.*, 2016, **17**, 6-13 (*Bellis sylvestris constit, activity*)

4-Hydroxy-3,3′,4′-trimethoxy-8,8′-lignan-9,9′-olide H141

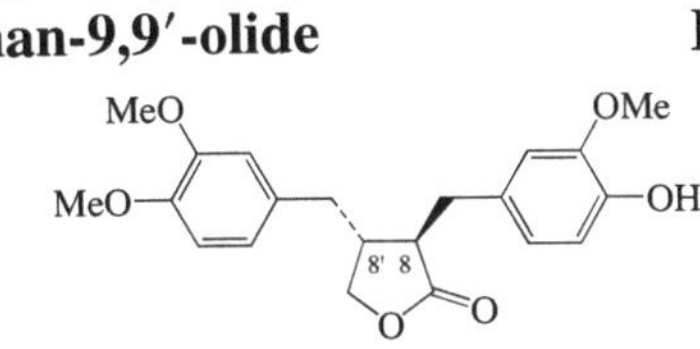

(*8R,8′R*)*-form*

4-[(3,4-Dimethoxyphenyl)methyl]dihydro-3-[(4-hydroxy-3-methoxyphenyl)methyl]-2(3H)-furanone, CAS
[51447-12-2] [81446-30-2] [87678-74-8] [87678-75-9] [144901-91-7]
$C_{21}H_{24}O_6$ 372.417
Log P 1.71 (calc)

(8R,8′R)-form
4′-*O*-Methylmatairesinol. ***(-)-Arctigenin***
[7770-78-7]
Constit. of *Arctium lappa, Cnicus benedictus, Forsythia viridissima, Centaurea pannionica, Centaurea sphareocephala, Ipomoea cairica, Jurinea mollis, Trachelospermum asiaticum* and others. Shows antitumour promoting effect. Antiinflammatory. Topoisomerase-Ii and HIV reverse transcriptase inhibitor, shows anti-HIV and antileukaemic activity. Differentiation inducer. Immunomodulator. Plates ($MeOH/Et_2O$). Mp 102°. $[\alpha]_D$ -37.7 (EtOH) (-28.7). λ_{max} 231 (ε 8800); 281 (ε 4000) (MeOH)

O-β-D-Glucopyranoside
Arctiin. *Arctin*
[20362-31-6]
$C_{27}H_{34}O_{11}$ 534.558
Constit. of the seeds of *Arctium lappa*, stems of *Trachelospermum asiaticum* and from *Forsythia intermedia, Centaurea spaerocephala* and *Centaurea imperialis*. Widespread in the Compositae. Shows anticancer props. in animal models and growth inhibition in human prostate cancer PC-3 cells. Down-regulator of cyclin D1 protein expression. Mp 110 - 112°. $[\alpha]^{27}_D$ -51.5 (c, 2 in EtOH). Arctin was originally descr. as a different compd. and given an erroneous struct.. λ_{max} 229 (ε 5012); 279 (ε 1995) (MeOH) (Berdy)

O-(2-O-Acetyl-β-D-glucopyranoside)
2″-Acetylarctiin
[1310548-47-0]
$C_{29}H_{36}O_{12}$ 576.595
Constit. of the seeds of *Saussurea involucrata*. Light yellow oil. $[\alpha]^{20}_D$ -27.2 (c, 0.33 in MeOH). λ_{max} 203 (MeOH)

O-(3-O-Acetyl-β-D-glucopyranoside)
3″-Acetylarctiin
[1310548-48-1]
$C_{29}H_{36}O_{12}$ 576.595
Constit. of the seeds of *Saussurea involucrata*. Light yellow oil. $[\alpha]^{20}_D$ -32.7 (c, 0.25 in MeOH). λ_{max} 202 (MeOH)

O-(6-O-Acetyl-β-D-glucopyranoside)
6″-Acetylarctiin
[1310548-46-9]
$C_{29}H_{36}O_{12}$ 576.595
Constit. of the seeds of *Saussurea involucrata*. Yellow gum. $[\alpha]^{20}_D$ -45.9 (c, 0.93 in MeOH). λ_{max} 201 (MeOH)

O-[3,4-Dihydroxycinnamoyl-(→3)-β-D-glucopyranoside]
3″-O-Caffeoylarctiin. ***Americanin‡***
$C_{36}H_{40}O_{14}$ 696.702
Constit. of the seeds of *Centaurea americana*. Gum. $[\alpha]_D^{23}$ -34.5 (c, 0.001 in MeOH). λ_{max} 225; 282; 310 (MeOH)

O-[β-D-Apiofuranosyl-(1→6)-β-D-glucopyranoside]
[1807808-85-0]
$C_{32}H_{42}O_{15}$ 666.673
Constit. of fruits of *Arctium lappa*. Amorph. powder. $[\alpha]_D^{25}$ -52.6 (c, 0.12 in MeOH). λ_{max} 229 (log ε 3.58); 278 (log ε 3.19) (MeOH)

O-[α-D-Galactopyranosyl-(1→6)-β-D-glucopyranoside]
[1807808-84-9]
$C_{33}H_{44}O_{16}$ 696.699
Constit. of fruits of *Arctium lappa*. Amorph. powder. $[\alpha]_D^{25}$ +6.3 (c, 0.07 in MeOH). λ_{max} 230 (log ε 3.12); 278 (log ε 2.69) (MeOH)

O-[β-D-Glucopyranosyl-(1→6)-β-D-glucopyranoside]
Arctigenin gentiobioside
[41682-24-0]
$C_{33}H_{44}O_{16}$ 696.699
Constit. of *Arctium lappa* and *Trachelospermum asiaticum*. Mp 179 - 180°. $[\alpha]_D^{19}$ -52 (c, 0.4 in H_2O)

Ac
Needles (EtOAc). Mp 52 - 60°

(8S,8′S)-form
(+)-Arctigenin [84413-77-4]
Constit. of *Wikstroemia indica* and *Wikstroemia viridiflora*. $[\alpha]_D^{23}$ +28.05 (c, 1.23 in EtOH).

Ozawa, T. *Yakugaku Zasshi*, 1952, **72**, 551-553 (*Arctigenin, synth, abs config*).
Hänsel, R. *et al.*, *Z. Naturforsch., B*, 1964, **19**, 727-734 (*Arctiin, occur*).
Nishibe, S. *et al.*, *Chem. Pharm. Bull.*, 1972, **20**, 2710-2719 (*Arctiin, Arctigenin gentiobioside*).
Ingaki, I. *et al.*, *Chem. Pharm. Bull.*, 1973, **21**, 2778-2779 (*Arctigenin gentiobioside, synth*).
Nishibe, S. *et al.*, *Experientia*, 1973, **29**, 17-18 (*Arctigenin gentiobioside*).
Omar, A.A. *Lloydia*, 1978, **41**, 638-639 (*Arctiin*).
Suzuki, H. *et al.*, *Phytochemistry*, 1982, **21**, 1824-1825 (*(+)-Arctigenin*).
Brown, E. *et al.*, *Tetrahedron*, 1989, **45**, 141-154 (*synth*).
Rahman, M.M.A. *et al.*, *Phytochemistry*, 1990, **29**, 1971-1980 (*Arctigenin, Arctiin*).
Eich, E. *et al.*, *Planta Med.*, 1990, **56**, 506 (*activity*).
Imehara, K. *et al.*, *Chem. Pharm. Bull.*, 1993, **41**, 1774-1779 (*Arctigenin, activity*).
Ozawa, S. *et al.*, *Phytochemistry*, 1993, **32**, 643-652 (*biosynth*).
Mitra, J. *et al.*, *Indian J. Chem., Sect. B*, 1994, **33**, 953-956 (*synth*).
Hirano, T. *et al.*, *Life Sci.*, 1994, **55**, 1061-1069 (*activity*).
Youssef, D. *et al.*, *Planta Med.*, 1995, **61**, 570-573 (*pmr, cmr*).
Eich, E. *et al.*, *J. Med. Chem.*, 1996, **39**, 86-89 (*anti-HIV activity*).
Sibi, M.P. *et al.*, *J. Org. Chem.*, 2002, **67**, 1738-1745 (*synth*).
Fischer, J. *et al.*, *Org. Lett.*, 2004, **6**, 1345-1348 (*synth*).
Shoeb, M. *et al.*, *Phytochemistry*, 2006, **67**, 2370-2375 (*Americanin*).
Gao, H. *et al.*, *Acta Crystallogr., Sect. E*, 2008, **64**, o1538 (*cryst struct*).
Matsuzaki, Y. *et al.*, *Oncol. Rep.*, 2008, **19**, 721-727 (*Arctiin, activity*).
Liu, Y.-D. *et al.*, *J. Asian Nat. Prod. Res.*, 2010, **12**, 828-833 (*Acetylarctiins*).
Sun, C. *et al.*, *Chem. Nat. Compd. (Engl. Transl.)*, 2012, **48**, 493-497 (*Wikstroemia indica constit*).
Chen, G-R. *et al.*, *Nat. Prod. Res.*, 2012, **26**, 177-181 (*activity*).
Ifantis, T.M. *et al.*, *Phytochemistry*, 2013, **94**, 159-170 (*Centaurea pannonica constit*).
Liu, F. *et al.*, *J. Sep. Sci.*, 2014, **37**, 376-381 (*Arctigenin, isol, purifn*).
Awale, S. *et al.*, *Nat. Prod. Commun.*, 2014, **9**, 79-82 (*Arctigenin, config, activity*).
Yang, Y.-N. *et al.*, *J. Agric. Food Chem.*, 2015, **63**, 7958-7966 (*Arctium lappa diglycosides*).
Guan, Y.-X. *et al.*, *Zhongcaoyao*, 2015, **46**, 1620-1624 (*Arctigenin, prodn*).
Lü, H. *et al.*, *J. Chromatogr. Sci.*, 2016, **54**, 472-478 (*Arctigenin, isol, purfn*).
Könye, R. *et al.*, *Nat. Prod. Commun.*, 2016, **11**, 1459-1462 (*Jurinea mollis constit, conformn*).
Ping, W. *et al.*, *Youji Huaxue*, 2016, **36**, 1111-1117 (*Arctigenin, synth*)

Hyperione A H142

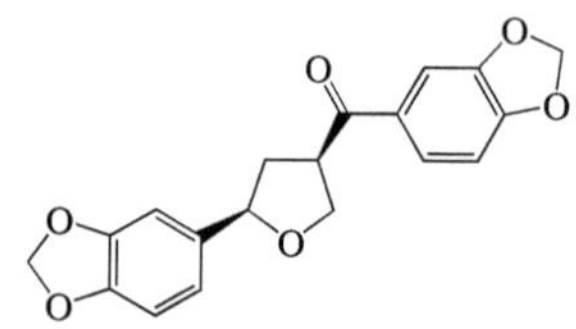

Relative Configuration

Tetrahydro-4-(3,4-methylenedioxybenzoyl)-2-(3,4-methylenedioxyphenyl)furan. 7,9′-Epoxy-3,4:3′,4′-bis(methylenedioxy)-9-nor-8,8′-lignan-7′-one. 3,4:3′,4′-Bis(methylenedioxy)-9-nor-7′-oxo-7,9′-epoxylignan
[1257216-78-6]
$C_{19}H_{16}O_6$ 340.331
Constit. of the leaves of *Hypericum chinense*. Oil. $[\alpha]_D^{20}$ +195.6 (c, 0.04 in $CHCl_3$). λ_{max} 207 (log ε 2.08); 229 (log ε 2.54); 276 (log ε 1.15); 307 (log ε 1.07) (MeOH)

Diastereoisomer
Hyperione B
[1257216-79-7]
$C_{19}H_{16}O_6$ 340.331
Constit. of the leaves of *Hypericum chinense*. Amorph. powder. $[\alpha]_D^{20}$ +48 (c, 0.05 in Me_2CO). λ_{max} 207 (log ε 2.04); 230 (log ε 2.56); 276 (log ε 1.2); 308 (log ε 1.06) (MeOH).

Wang, W. *et al.*, *J. Nat. Prod.*, 2010, **73**, 1815-1820 (*Hyperiones A,B*).
Duhamel, N. *et al.*, *Asian J. Org. Chem.*, 2013, **2**, 491-493 (*synth*).
Sapu, C.M. *et al.*, *Org. Biomol. Chem.*, 2013, **11**, 1376-1382 (*synth*)

Hyperjaponol B H143

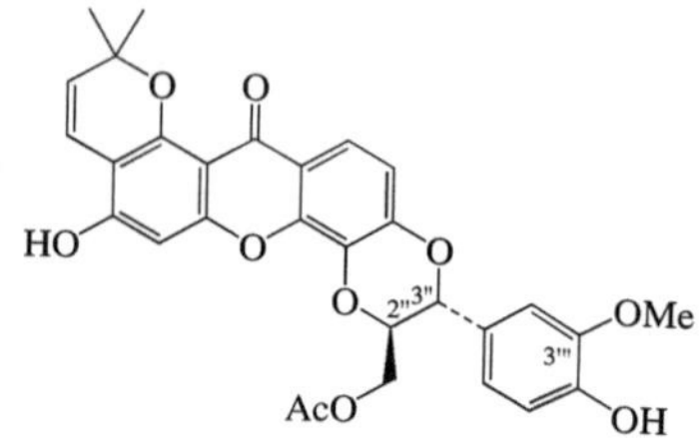

Relative Configuration

Hypejaponol B
[2374811-40-0]
$C_{30}H_{26}O_{10}$ 546.528
Constit. of *Hypericum japonicum*. $[\alpha]_D^{25}$ -8.6 (c, 0.07 in MeOH). Prob. a partial racemate. Two synonyms used. λ_{max} 255 (log ε 3.22); 326 (log ε 2.63) (MeOH).

Zhu, W. *et al.*, *Phytochemistry*, 2019, **164**, 33-40 (*Hyperjaponol B*)

Ilexin C I1

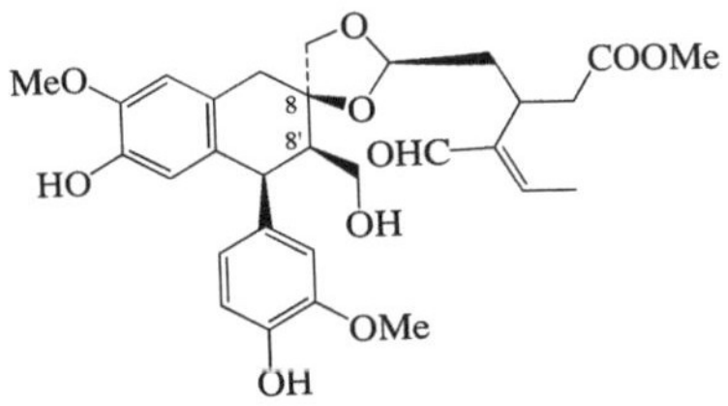

Absolute Configuration

[1042419-14-6]
$C_{30}H_{36}O_{10}$ 556.608
Constit. of the roots of *Ilex pubescens*. Amorph. solid. $[\alpha]^{20}_{D}$ +3.6 (c, 0.008 in MeOH). λ_{max} 231 (log ε 4.23); 284 (log ε 3.68) (MeOH)

8,8'-Diepimer

Ilexin L₁

[1499177-76-2]
$C_{30}H_{36}O_{10}$ 556.608
Constit. of *Ilex pubescens*. Solid.

Zhou, Y.-B. *et al.*, *Helv. Chim. Acta*, 2008, **91**, 1244-1250 (*Ilexin C*).
Chen, J. *et al.*, *Chem. Nat. Compd. (Engl. Transl.)*, 2013, **49**, 848-851 (*Ilexin L₁, cryst struct*)

Ilexlignan A I2

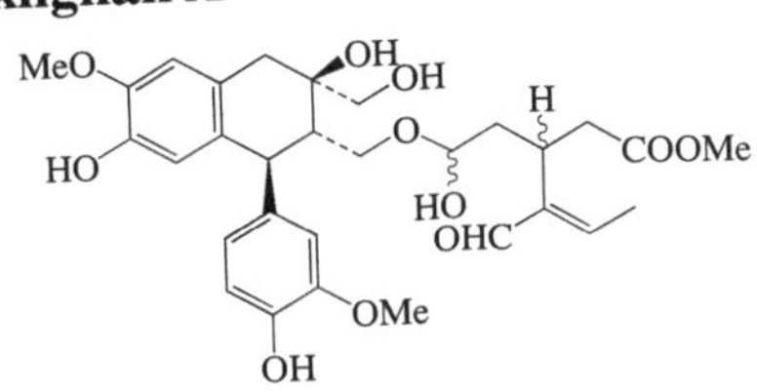

[1120356-40-2]
$C_{30}H_{38}O_{11}$ 574.623
Constit. of the roots of *Ilex pubescens*. Amorph. solid.

Zhou, Y.B. *et al.*, *Chin. Chem. Lett.*, 2008, **19**, 550-552 (*Ilexlignan A*)

Illicidione A I3

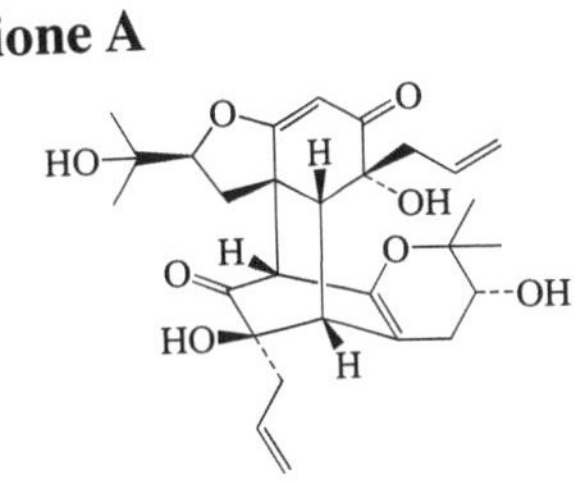

Absolute Configuration

[1304653-71-1]
$C_{28}H_{36}O_{8}$ 500.588
Prenylated neolignan. Constit. of *Illicium henryi* and *Illicium oligandrum*. Exhibits weak antiinflammatory activity. Prisms. Mp 213 - 215°. $[\alpha]^{20}_{D}$ +36.5 (c, 0.5 in MeOH). λ_{max} 265 (log ε 4.26) (MeOH).

Tang, W.-Z. *et al.*, *J. Nat. Prod.*, 2011, **74**, 1268-1271 (*Illicidione A, cryst struct, activity*).
Zhuang, P.-Y. *et al.*, *Phytochem. Lett.*, 2013, **6**, 444-448 (*isol, activity*).
Ren, X.-D. *et al.*, *Tetrahedron*, 2015, **71**, 4821-4829 (*synth*)

Illicidione C I4

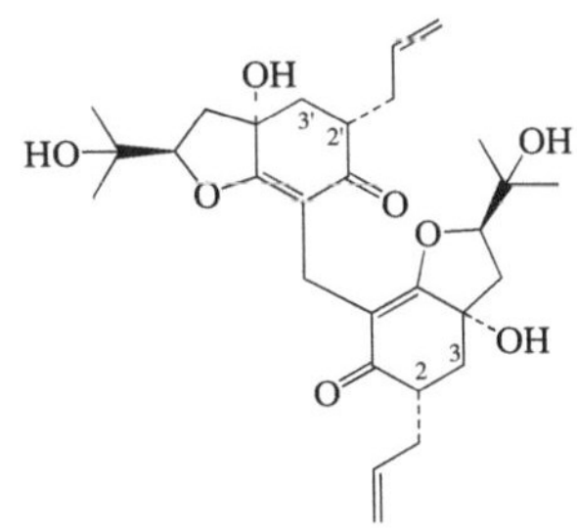

Absolute Configuration

[1304653-73-3]
$C_{29}H_{40}O_{8}$ 516.631
Prenylated neolignan. Constit. of the stem bark of *Illicium oligandrum*. Exhibits weak antiinflammatory activity. Powder. $[\alpha]^{20}_{D}$ +9.4 (c, 0.15 in MeOH). λ_{max} 265 (log ε 4.43) (MeOH)

2,2',3,3'-Tetradehydro

Illicidione B

[1304653-72-2]
$C_{29}H_{36}O_{8}$ 512.599
Constit. of the stem bark of *Illicium oligandrum*. Exhibits weak antiinflammatory activity. Powder. $[\alpha]^{20}_{D}$ -24.4 (c, 0.4 in MeOH). λ_{max} 242 (log ε 4.26); 315 (log ε 3.28) (MeOH).

Tang, W.-Z. *et al.*, *J. Nat. Prod.*, 2011, **74**, 1268-1271 (*Illicidiones B,C, cryst struct, activity*)

Illihendione A I5

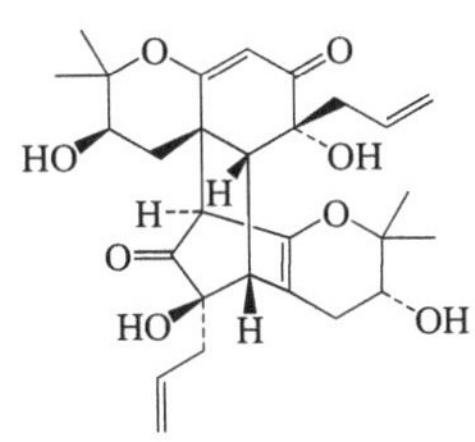

Absolute Configuration

[1607434-76-3]
$C_{28}H_{36}O_{8}$ 500.588
Prenylated sesquineolignan. Isol. from the roots of *Illicium henryi*. Exhibits a weak inhibitory ratio for β-glucuronidase release induced by platelet activating factor in rat polymorphonuclear leukocytes. Amorph. powder. $[\alpha]^{20}_{D}$ +50.6 (c, 0.5 in MeOH). λ_{max} 260 (log ε 3.42) (MeOH).

Zhuang, P.-Y. *et al.*, *Phytochem. Lett.*, 2013, **6**, 444-448 (*Illihendione A*).
Ren, X.-D. *et al.*, *Tetrahedron*, 2015, **71**, 4821-4829 (*synth*)

Illihenryifunol A I6

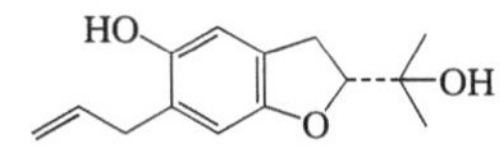

2,3-Dihydro-5-hydroxy-α,α-dimethyl-6-(2-propen-1-yl)-2-benzofuranmethanol, *CAS. 6-Allyl-2,3-dihydro-5-hydroxy-2-(1-hydroxy-1-methylethyl)benzofuran*
$C_{14}H_{18}O_{3}$ 234.295
Prenylated lignan monomer

(R)-form [1418206-88-8]
Constit. of roots of *Illicium henryi*. Exhibits moderate antioxidant activity. Amorph. powder. $[\alpha]^{20}_{D}$ -28.8 (c, 0.55 in MeOH). λ_{max} 202 (log ε 3.17); 236 (log ε 2.33); 300 (log ε 2.51) (MeOH).

Zhuang, P.-Y. *et al.*, *Phytochemistry*, 2013, **86**, 176-183 (*Illihenryifunol A*)

Illihenryifunone A I7

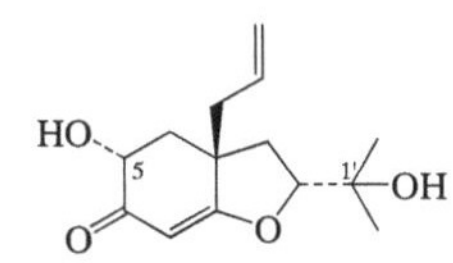

Absolute Configuration

3,3a,4,5-Tetrahydro-5-hydroxy-2-(1-hydroxy-1-methylethyl)-3a-(2-propen-1-yl)-6(2H)-benzofuranone
[1418206-86-6]
$C_{14}H_{20}O_{4}$ 252.31
Prenylated lignan monomer. Constit. of roots of *Illicium henryi*. Oil. $[\alpha]^{20}_{D}$ +131.4 (c, 0.51 in MeOH). λ_{max} 249 (log ε 3.19) (MeOH)

1'-O-β-D-Glucopyranoside

Illihenryifunone B

[1418206-87-7]
$C_{20}H_{30}O_{9}$ 414.451
Constit. of roots of *Illicium henryi*. Amorph. powder. $[\alpha]^{20}_{D}$ +58.8 (c, 0.52 in MeOH). λ_{max} 260 (log ε 3.42) (MeOH).

Zhuang, P.-Y. *et al.*, *Phytochemistry*, 2013, **86**, 176-183 (*Illihenryifunones A,B*)

Illioliganfunone A I8

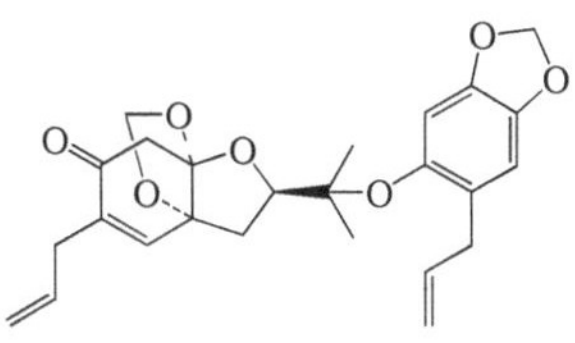

[1268466-43-8]
$C_{25}H_{28}O_{7}$ 440.492

Prenylated lignan. Constit. of roots of *Illicium oligandrum*. Oil. $[\alpha]^{20}_{D}$ -90.8 (c, 0.051 in MeOH). λ_{max} 205 (log ε 4.55); 235 (log ε 4.01); 298 (log ε 3.58) (MeOH).

Ma, S.-G. *et al., Phytochemistry*, 2011, **72**, 115-125 (*Illioliganfunone A*)

Illioliganfuranone D I9

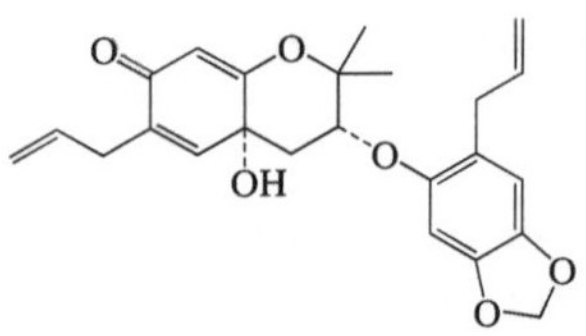

Absolute Configuration

[1924673-34-6]

$C_{24}H_{26}O_6$ 410.466

Prenylated dilignan. Constit. of the roots of *Illicium oligandrum*. Exhibits potent antiviral activity against influenza virus A H3N2. Pale yellow powder. $[\alpha]^{20}_{D}$ +76.7 (c, 0.11 in MeOH). λ_{max} 238 (log ε 4.23); 299 (log ε 3.85) (MeOH).

Ma, S.-G. *et al., Tetrahedron*, 2016, **72**, 3003-3013 (*Illioliganfuranone D, cd, activity*)

Illioliganfuranone E I10

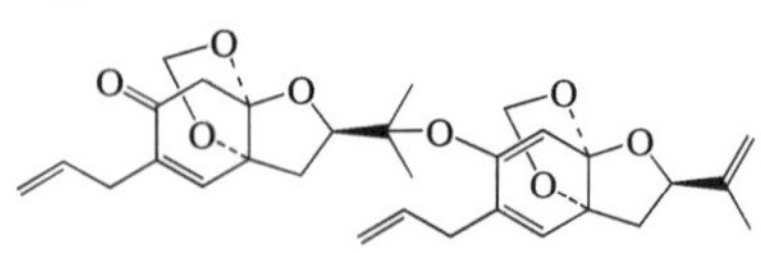

Absolute Configuration

[1924673-30-2]

$C_{30}H_{36}O_8$ 524.61

Constit. of the roots of *Illicium oligandrum*. Exhibits potent antiviral activity against CVB3. Oil. $[\alpha]^{20}_{D}$ -6.9 (c, 0.17 in MeOH). λ_{max} 250 (sh) (MeOH).

Ma, S.-G. *et al., Tetrahedron*, 2016, **72**, 3003-3013 (*Illioliganfuranone E, activity*)

Illioliganfuranone F I11

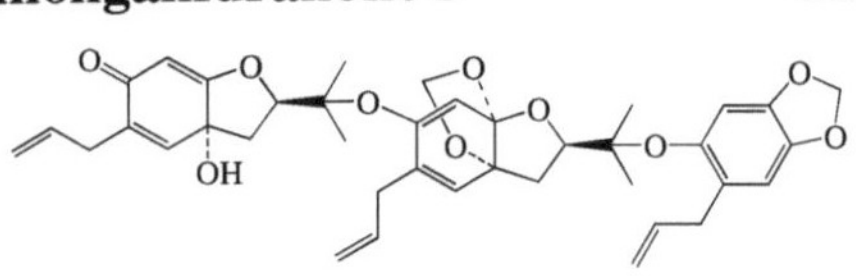

Absolute Configuration

[1924673-32-4]

$C_{39}H_{44}O_{10}$ 672.771

Prenylated trilignan. Constit. of the roots of *Illicium oligandrum*. Exhibits potent antiviral activity against CVB3. Pale yellow powder. $[\alpha]^{20}_{D}$ -81.1 (c, 0.14 in MeOH). λ_{max} 237 (log ε 3.83); 296 (log ε 3.52) (MeOH).

Ma, S.-G. *et al., Tetrahedron*, 2016, **72**, 3003-3013 (*Illioliganfuranone F, activity*)

Illioliganpyranone B I12

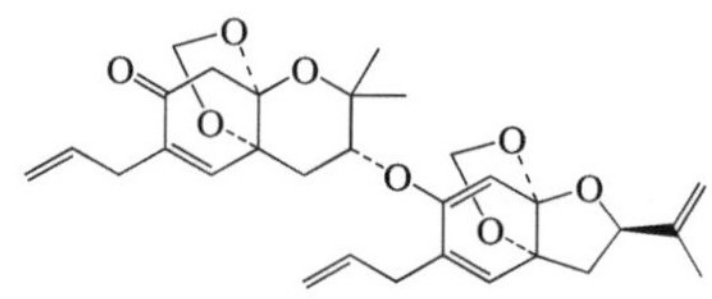

Absolute Configuration

[1924673-28-8]

$C_{30}H_{36}O_8$ 524.61

Prenylated dilignan. Constit. of the roots of *Illicium oligandrum*. Exhibits potent antiviral activity against CVB3. Cryst. (Me_2CO aq.). Mp 118 - 120°. $[\alpha]^{20}_{D}$ +11.9 (c, 0.06 in MeOH). λ_{max} 250 (sh) (MeOH).

Ma, S.-G. *et al., Tetrahedron*, 2016, **72**, 3003-3013 (*Illioliganpyranone B, cryst struct, activity*)

Illioliganpyranone C I13

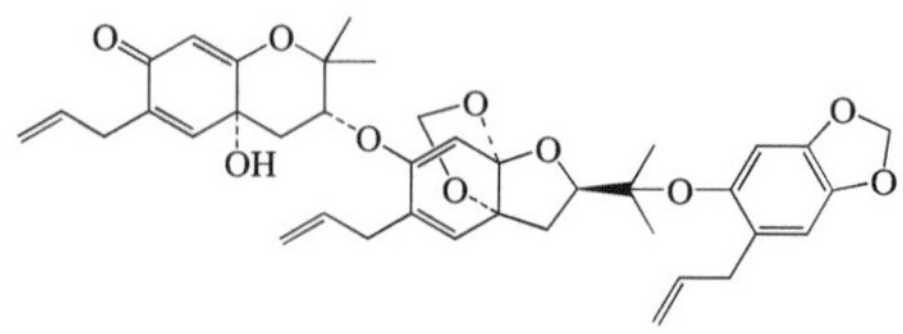

Absolute Configuration

[1924673-31-3]

$C_{39}H_{44}O_{10}$ 672.771

Prenylated trilignan. Constit. of the roots of *Illicium oligandrum*. Exhibits potent antiviral activity against CVB3. Pale yellow powder. $[\alpha]^{20}_{D}$ -33.5 (c, 0.31 in MeOH). λ_{max} 236 (log ε 3.43); 292 (log ε 3.02) (MeOH).

Ma, S.-G. *et al., Tetrahedron*, 2016, **72**, 3003-3013 (*Illioliganpyranone C, activity*)

Impecyloside I14

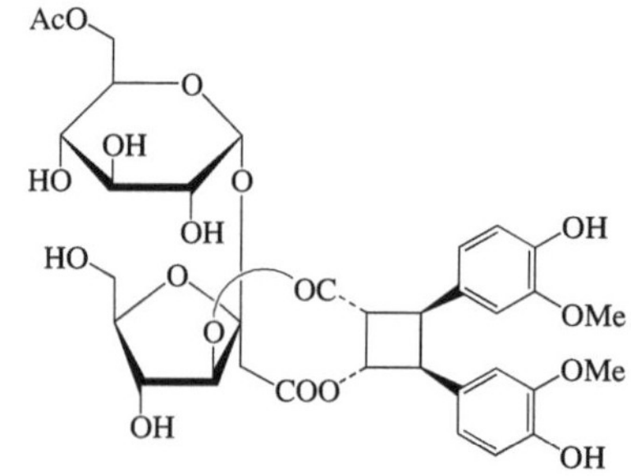

Sucrose 6-acetate 1',3'-(4,4'-dihydroxy-3,3'-dimethoxy-β-truxinate)

[1070967-17-7]

$C_{34}H_{40}O_{18}$ 736.676

Constit. of the rhizomes of *Imperata cylindrica*. Amorph. yellow powder (MeOH). $[\alpha]^{25}_{D}$ +12 (c, 0.01 in MeOH). λ_{max} 212 (MeOH)

O-De-Ac

Deacetylimpecyloside

[1622990-04-8]

$C_{32}H_{38}O_{17}$ 694.639

Isol. from *Imperata cylindrica*. Yellow syrup. $[\alpha]^{20}_{D}$ +22.4 (c, 0.07 in MeOH). λ_{max} 256 (MeOH).

Lee, D.-Y. *et al., J. Asian Nat. Prod. Res.*, 2008, **10**, 299-302 (*Impecyloside*).

Liu, X. *et al., J. Nat. Med.*, 2014, **68**, 295-301 (*Deacetylimpecyloside*)

Incarvillateine I15

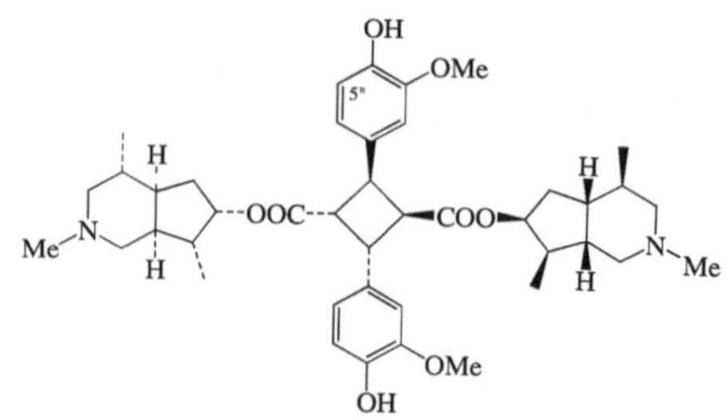

Relative Configuration

[129748-10-3]

$C_{42}H_{58}N_2O_8$ 718.932

Alkaloid from the aerial parts of *Incarvillea sinensis*. Antinociceptive agent, shows analgesic and sedative activity. Cryst. (MeOH). Mp 217.2 - 217.7°. $[\alpha]_D$ -10.8 (MeOH). $[\alpha]^{23}_{D}$ -14.8 (c, 1.39 in $CHCl_3$)

N-Oxide

Incarvillateine N-oxide

[185054-71-1]

$C_{42}H_{58}N_2O_9$ 734.931

Alkaloid from *Incarvillea sinensis*. Powder. $[\alpha]^{18}_{D}$ -19.2 (c, 0.5 in MeOH). Obt. as an inseparable mixt. of the two possible *N*-oxides

5''-Methoxy

Methoxyincarvillateine

[185054-70-0]

$C_{43}H_{60}N_2O_9$ 748.958

Alkaloid from *Incarvillea sinensis*. Powder. $[\alpha]^{20}_{D}$ -4 (c, 0.6 in $CHCl_3$)

Mono(demethoxy)

Incarvillateine D

[244259-23-2]

$C_{41}H_{56}N_2O_7$ 688.906

Alkaloid from *Incarvillea sinensis*. Powder. $[\alpha]^{16}_{D}$ -5.1 (c, 0.3 in $CHCl_3$)

Bis(demethoxy)

Incarvillateine C

[244259-22-1]

$C_{40}H_{54}N_2O_6$ 658.88

Alkaloid from *Incarvillea sinensis*. Powder. $[\alpha]^{17}_{D}$ -3.2 (c, 0.3 in $CHCl_3$).

Chi, Y.-M. *et al.*, *Phytochemistry*, 1990, **29**, 2376-2378 (*Incarvillateine, cryst struct*).
Chi, Y.-M. *et al.*, *Phytochemistry*, 1997, **46**, 763-769 (*Incarvillateine N-oxide, Methoxyincarvillateine*).
Nakamura, M. *et al.*, *J. Nat. Prod.*, 1999, **62**, 1293-1294 (*activity*).
Nakamura, M. *et al.*, *Phytochemistry*, 1999, **51**, 595-597 (*Incarvillaeines C,D*).
Nakamura, M. *et al.*, *Planta Med.*, 2001, **67**, 114-117 (*activity*).
Ichikawa, M. *et al.*, *J. Am. Chem. Soc.*, 2004, **126**, 16553-16558 (*synth*).
Chi, Y-M. *et al.*, *Biol. Pharm. Bull.*, 2005, **28**, 1989-1991 (*activity*).
Chi, Y.-M. *et al.*, *Chem. Pharm. Bull.*, 2005, **53**, 1472-1474 (*hplc, pmr, cmr, cryst struct, bibl*).
Tsai, A.S. *et al.*, *J. Am. Chem. Soc.*, 2008, **130**, 6316-7317 (*synth*).
Zhang, F. *et al.*, *Tetrahedron*, 2009, **65**, 6840-6843 (*synth*)

Incarvilline **I16**

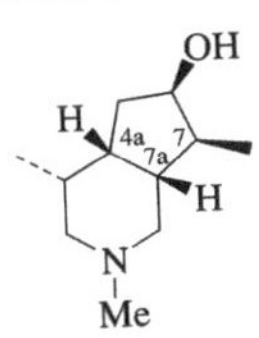

Absolute Configuration

6-Hydroxyskytanthine
[145307-22-8]
$C_{11}H_{21}NO$ 183.295
Numbering systems vary. Alkaloid from aerial parts of *Incarvillea sinensis*. Mp 93.4 - 93.8°. $[\alpha]^{24}_{D}$ -8 (c, 1.24 in $CHCl_3$)

O-(4-Hydroxy-3-methoxy-E-cinnamoyl)
Incarvine C. *Incarvilline 7-O-ferulate*
[168009-89-0]
$C_{21}H_{29}NO_4$ 359.466
Alkaloid from the aerial parts of *Incarvillea sinensis*. Possible precursor for alkaloidal lignans. Powder. $[\alpha]_D$ -20.8 (c, 0.46 in $CHCl_3$).

Chi, Y.-M. *et al.*, *Phytochemistry*, 1992, **31**, 2930-2932 (*Incarvilline*).
Chi, Y.-M. *et al.*, *Phytochemistry*, 1995, **39**, 1485-1487 (*Incarvines B,C*).
Chi, Y.-M. *et al.*, *Chem. Pharm. Bull.*, 1997, **45**, 495-498 (*Incarvilline, cryst struct, abs config, 4a-Hydroxyincarvilline*).
Ichikawa, M. *et al.*, *J. Am. Chem. Soc.*, 2004, **126**, 16553-16558 (*synth*).
Honda, T. *et al.*, *J. Org. Chem.*, 2007, **72**, 6541-6547 (*synth*).
Zhang, F. *et al.*, *Tetrahedron*, 2009, **65**, 6840-6843 (*synth*).
Seo, H. *et al.*, *Org. Lett.*, 2013, **15**, 531-533 (*synth*)

Indicumine C **I17**

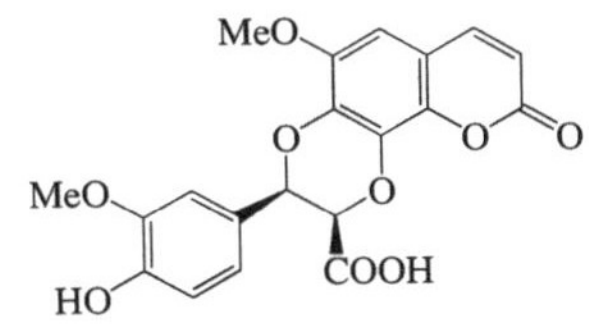

[1497251-70-3]
$C_{20}H_{16}O_9$ 400.339
Related to Cleomiscosin A, C82. Constit. of the seeds of *Solanum indicum*. Amorph. powder. $[\alpha]^{20}_{D}$ -15.7 (c, 0.12 in Py). λ_{max} 286 (log ε 3.90); 319 (log ε 4.09); 325 (log ε 4.10) (MeOH)

Et ester
Indicumine D
[1497251-73-6]
$C_{22}H_{20}O_9$ 428.393
Constit. of the seeds of *Solanum indicum*. Amorph. powder. $[\alpha]^{20}_{D}$ +4 (c, 0.1 in MeOH). λ_{max} 287 (log ε 3.91); 317 (log ε 3.92) (MeOH)

(4-Hydroxyphenylethyl)amide
Indicumine A
[1497251-67-8]
$C_{28}H_{25}NO_9$ 519.506
Constit. of the seeds of *Solanum indicum*. Exhibits moderate anti-HBV activity. Amorph. powder. $[\alpha]^{20}_{D}$ -2.8 (c, 0.11 in MeOH). λ_{max} 288 (log ε 3.83); 320 (log ε 3.93) (MeOH)

2-(1H-Indol-3-yl)-2-oxoethylamide
Indicumine B
[1497251-69-0]
$C_{30}H_{24}N_2O_9$ 556.527
Constit. of the seeds of *Solanum indicum*. Exhibits weak anti-HBV activity. Amorph. red powder. $[\alpha]^{20}_{D}$ +60.9 (c, 0.11 in Py). λ_{max} 299 (log ε 4.37) (MeOH).

Yin, H.-L. *et al.*, *Fitoterapia*, 2013, **84**, 360-365 (*Indicumines A-D*)

Indicumine E **I18**

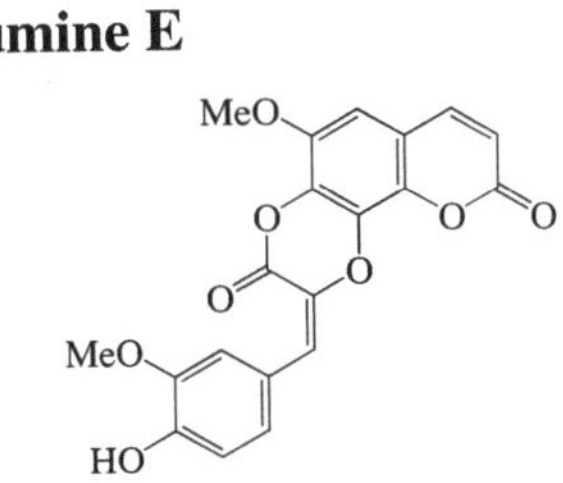

[1615695-92-5]
$C_{20}H_{14}O_8$ 382.324
Coumarinolignan. Constit. of the seeds of *Solanum indicum*. Amorph. yellow powder. $[\alpha]^{20}_{D}$ +14.3 (c, 0.11 in Py). Opt. rotn. unaccounted for. λ_{max} 327 (log ε 4.12) (MeOH).

Yin, H.-L. *et al.*, *J. Asian Nat. Prod. Res.*, 2014, **16**, 153-157 (*Indicumine E*)

Ipobscurine B **I19**

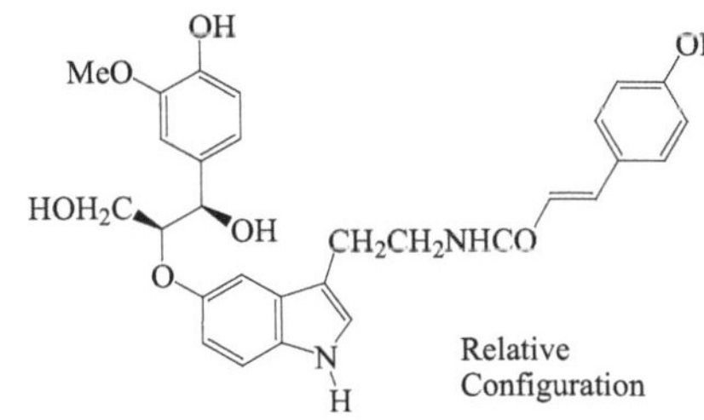

Relative Configuration

[571144-95-1]
$C_{29}H_{30}N_2O_7$ 518.566
Alkaloid from the seeds of *Ipomoea obscura*. Yellow solid. $[\alpha]^{20}_{D}$ -35 (c, 0.28 in MeOH).

Jenett-Siems, K. *et al.*, *Phytochemistry*, 2003, **62**, 1257-1263 (*Ipobscurine B*)

Ipobscurine C **I20**

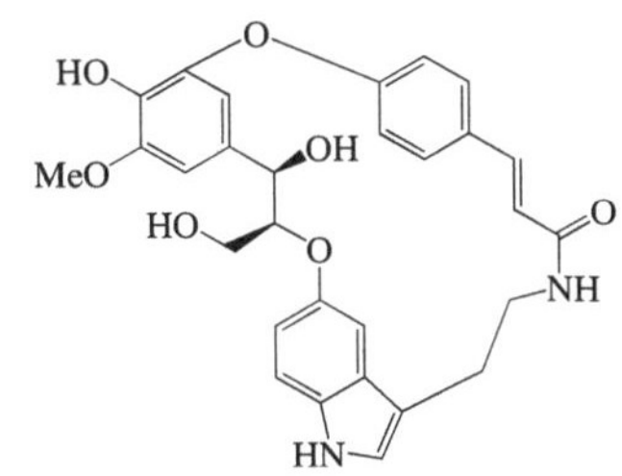

Relative Configuration

[571144-96-2]
$C_{29}H_{28}N_2O_7$ 516.55
Alkaloid from seeds of *Ipomoea obscura*. Cryst.. Mp 193 - 195°. $[\alpha]^{20}_{D}$ -44 (c, 0.16 in $CHCl_3$)

Z-Isomer
Ipobscurine D
[571144-97-3]
$C_{29}H_{28}N_2O_7$ 516.55
Alkaloid from the seeds of *Ipomoea obscura*. Yellow solid. Mp 245 - 247°.

Jenett-Siems, K. *et al.*, *Phytochemistry*, 2003, **62**, 1257-1263 (*Iposcurines C,D, synth*)

Irisjaponoside A **I21**

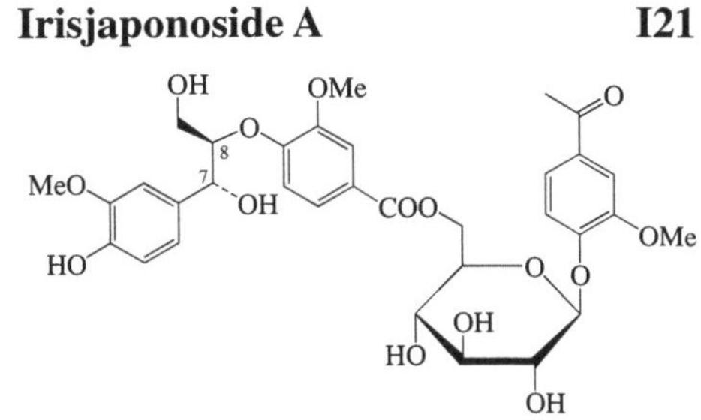

(7R,8R)-form

$C_{33}H_{38}O_{15}$ 674.652

(7R,8R)-form

Irisjaponoside A$_3$ [2185046-41-5]

Constit. of *Iris japonica*. Amorph. powder. $[\alpha]_D^{20}$ -9.4 (c, 0.08 in MeOH). λ_{max} 222 (log ε 4.81); 265 (log ε 4.54) (MeOH)

(7R,8S)-form

Irisjaponoside A$_1$ [2185046-39-1]

Constit. of *Iris japonica*. Amorph. powder. $[\alpha]_D^{20}$ -36.3 (c, 0.08 in MeOH). λ_{max} 222 (log ε 4.86); 265 (log ε 4.62) (MeOH)

(7S,8R)-form

Irisjaponoside A$_2$ [2185046-40-4]

Constit. of *Iris japonica*. Amorph. powder. $[\alpha]_D^{20}$ -27.9 (c, 0.08 in MeOH). λ_{max} 222 (log ε 4.81); 265 (log ε 4.54) (MeOH)

(7S,8S)-form

Irisjaponoside A$_4$ [2185046-42-6]

Constit. of *Iris japonica*. Amorph. powder. $[\alpha]_D^{20}$ -6.5 (c, 0.08 in MeOH). λ_{max} 222 (log ε 4.77); 265 (log ε 4.46) (MeOH).

Shi, G.-R. *et al., J. Asian Nat. Prod. Res.*, 2016, **18**, 921-927 (*Irisjaponosides A$_1$-A$_4$*)

Iryantherin I I122

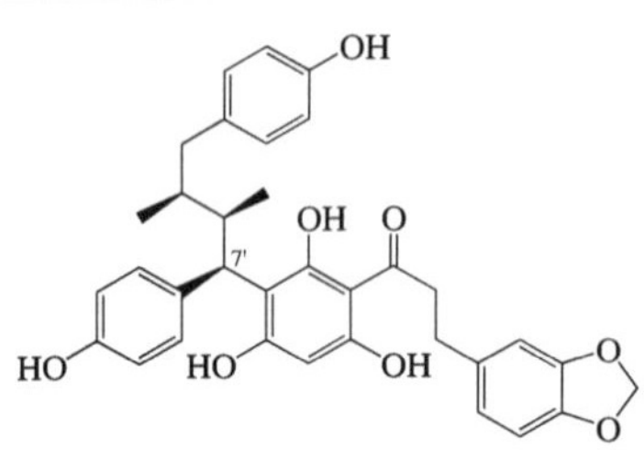

Relative Configuration

[163860-19-3]

$C_{34}H_{34}O_8$ 570.638

Flavonolignoid. Constit. of the fruit of *Iryanthera grandis*. Isol. as penta-Ac

7′-Epimer

Iryantherin J

[163959-60-2]

$C_{34}H_{34}O_8$ 570.638

Constit. of the fruit of *Iryanthera grandis*. Isol. as penta-Ac.

Silva, D.H.S. *et al., Phytochemistry*, 1995, **38**, 1013-1016 (*Iryantherins I,J*)

Iryantherin A I123

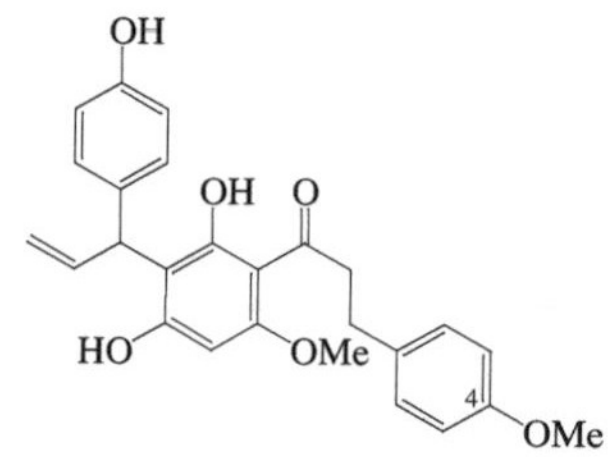

1-[2,4-Dihydroxy-3-[1-(4-hydroxyphenyl)-2-propenyl]-6-methoxyphenyl]-3-(4-methoxyphenyl)-1-propanone, CAS. *3′-(α-Ethenyl-4-hydroxybenzyl)-2′,4′-dihydroxy-4,6′-dimethoxydihydrochalcone*

[113122-18-2]

$C_{26}H_{26}O_6$ 434.488

Flavonolignan. Constit. of the fruit of *Iryanthera laevis* and other *Iryanthera* spp.. Cryst. ($CHCl_3$). Mp 129 - 130°. λ_{max} 225 (ε 38450); 290 (ε 23500) (MeOH)

4-Demethoxy

4-Demethoxyiryantherin A

[883222-14-8]

$C_{25}H_{24}O_5$ 404.462

Constit. of *Sarcandra glabra*. Yellowish oil. $[\alpha]_D^{20}$ +4.7 (c, 0.17 in $CHCl_3$). λ_{max} 215; 241; 286 (no solvent reported).

Garzon, N.L. *et al., Phytochemistry*, 1987, **26**, 2835-2837 (*Iryanthera laevis constit, struct*).
Conserva, L.M. *et al., Phytochemistry*, 1990, **29**, 3911-3918 (*Iryanthera constit*).
Li, Y. *et al., Chin. Chem. Lett.*, 2006, **17**, 207-210 (*4-Demethoxyiryantherin A*)

Iryantherin B I124

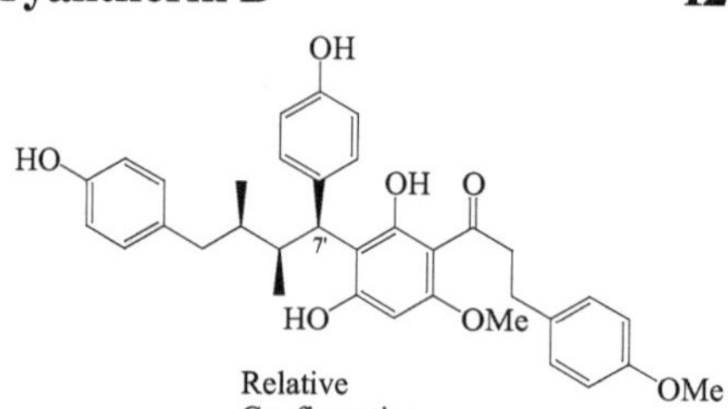

Relative Configuration

Iryantherin L. 1-[3-[1,4-Bis(4-hydroxyphenyl)-2,3-dimethylbutyl]-2,4-dihydroxy-6-methoxyphenyl]-3-(4-methoxyphenyl)-1-propanone, CAS

[250346-09-9]

[121781-35-9]

$C_{35}H_{38}O_7$ 570.682

Lignoflavonoid. Iryantherins B and L probably have the same relative configs.; work on the abs. configs. has not been completed (1999). Constit. of the bark of *Iryanthera laevis*, *Iryanthera lancifolia*, *Iryanthera megistophylla* and *Iryanthera ulei*. Active against Gram-positive bacteria. Moderate inhibitor of potato virus X. $[\alpha]_D^{25}$ +45.6 (c, 1.9 in MeOH) (Iryantherin L). λ_{max} 248 (log ε 4.22); 285 (log ε 4.17) (MeOH)

7′-Epimer

Iryantherin K

[250346-08-8]

$C_{35}H_{38}O_7$ 570.682

Constit. of *Iryanthera lancifolia* and *Iryanthera megistophylla*. $[\alpha]_D^{25}$ -36 (c, 1.6 in MeOH). λ_{max} 245 (log ε 4.81); 288 (log ε 4.31) (MeOH).

Villamil, Z. *et al., Spectroscopy (Ottawa)*, 1988, **6**, 157-165 (*Iryanthera laevis constit*).
Conserva, L.M. *et al., Phytochemistry*, 1990, **29**, 3911-3918 (*Iryantherins B,K*).
Silva, D.H.S. *et al., J. Nat. Prod.*, 1999, **62**, 1475-1478 (*Iryanthera lancifolia constits*).
Ming, D.S. *et al., J. Nat. Prod.*, 2002, **65**, 1412-1416 (*Iryanthera megistophylla constits, activity*)

Iryantherin C I125

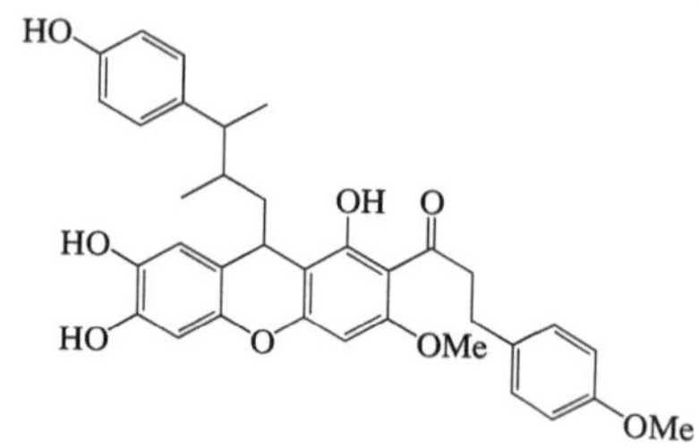

[132396-73-7]

$C_{35}H_{36}O_8$ 584.665

Flavonolignan. Constit. of *Iryanthera laevis*.

Conserva, L.M. *et al., Phytochemistry*, 1990, **29**, 3911-3918 (*Iryantherin C*)

Iryantherin G I126

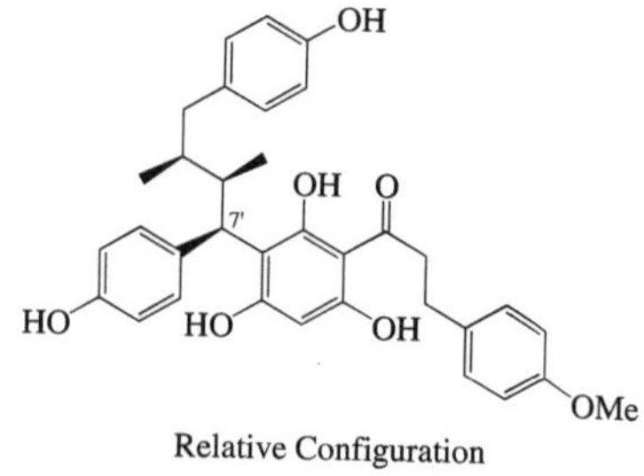

Relative Configuration

[163860-18-2]

$C_{34}H_{36}O_7$ 556.655

Flavonolignoid. Constit. of the fruit of *Iryanthera grandis*. Isol. as penta-Ac

7′-Epimer

Iryantherin H

[163959-59-9]

$C_{34}H_{36}O_7$ 556.655

Constit. of the fruit of *Iryanthera grandis*. Isol. as penta-Ac.

Silva, D.H.S. *et al.*, *Phytochemistry*, 1995, **38**, 1013-1016 (*Iryantherins G,H*)

Isoamericanol B_1 I127

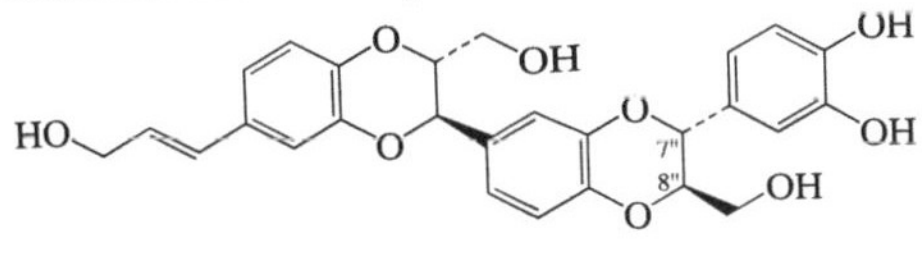

Relative Configuration

[569673-53-6]

$C_{27}H_{26}O_9$ 494.496

Constit. of the seeds of *Joannesia princeps*. Amorph. solid. λ_{max} 206 (log ε 4.72); 229 (sh); 269 (log ε 4); 287 (sh); 304 (sh); 314 (sh) (MeOH)

7″,8″-Diepimer

Isoamericanol B_2

[569673-52-5]

$C_{27}H_{26}O_9$ 494.496

Constit. of the seeds of *Joannesia princeps*. Amorph. solid. λ_{max} 206 (log ε 4.72); 229 (sh); 269 (log ε 4.00); 287 (sh); 304 (sh); 314 (sh) (MeOH).

Waibel, R. *et al.*, *Phytochemistry*, 2003, **62**, 805-811 (*Isoamericanols B_1,B_2*)

Isoamericanol C_1 I128

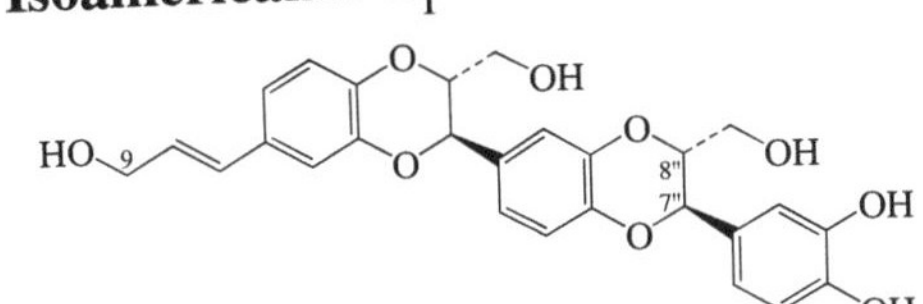

Relative Configuration

[571166-71-7]

$C_{27}H_{26}O_9$ 494.496

Sesquilignan. Similar to Americanin B and may have the same config.. Constit. of the seeds of *Joannesia princeps*. Amorph. solid. λ_{max} 206 (log ε 4.72); 229 (sh); 269 (log ε 4); 287 (sh); 304 (sh); 314 (sh) (MeOH)

7″,8″-Diepimer

Isoamericanol C_2

[571166-72-8]

$C_{27}H_{26}O_9$ 494.496

Constit. of the seeds of *Joannesia princeps*. Amorph. solid. λ_{max} 206 (log ε 4.72); 229 (sh); 269 (log ε 4); 287 (sh); 304 (sh); 314 (sh) (MeOH)

Stereoisomer

***Artabotrycinol**. Isoamericanol B‡*

[362606-64-2]

$C_{27}H_{26}O_9$ 494.496

Constit. of the seeds of *Artabotrys hexapetalus*. Cryst. (MeOH). Mp 194 - 196°. [α]$^{20}_{D}$-7 (c, 0.03 in MeOH). May be identical with Isoamericanol C_1. Not to be confused with Isoamericanol B, in E224. λ_{max} 236 (sh); 282 (MeOH)

Stereoisomer, 9-aldehyde

Americanin B

[77053-44-2]

$C_{27}H_{24}O_9$ 492.48

Constit. of *Artabotrys hexapetalus* and *Phytolacca americana*. Shows antioxidant activity. Mp 258 - 260°. [α]$^{17}_{D}$+1.7 (c, 0.3 in Py). May have same config. as Isoamericanol C_1.

Woo, W.S. *et al.*, *Tetrahedron Lett.*, 1980, **21**, 4255-4258 (*Americanin B*).

Yu, J.-G. *et al.*, *Yaoxue Xuebao*, 2001, **36**, 281-286 (*Americanin B, Artabotrycinol*).

Waibel, R. *et al.*, *Phytochemistry*, 2003, **62**, 805-811 (*Isoamericanols C_1,C_2*)

Isoasatone‡ I129

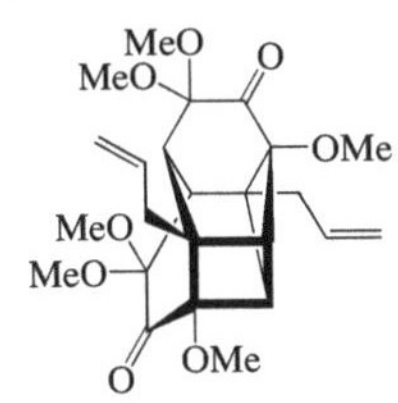

Octahydro-2,4,4,6,6,7a-hexamethoxy-7b,8-di-2-propenyl-1,2,5-metheno-1H-cyclobuta[d,e]naphthalene-3,7-dione, CAS

[52665-70-0]

$C_{24}H_{32}O_8$ 448.512

Highly modified cyclolignan. Not to be confused with Isoasatone A, in A57. Constit. of *Asarum taitonense*. Cryst. (hexane). Mp 156.5 - 158°. Racemate. λ_{max} 221 (ε 2010) (MeOH).

Sasaki, K. *et al.*, *Tetrahedron Lett.*, 1973, **14**, 4881-4884 (*Isoasatone, cryst struct*).

Yamamura, S. *et al.*, *Bull. Chem. Soc. Jpn.*, 1976, **49**, 1940-1948 (*Isoasatone, cryst struct*)

Isobutyrolactone II I130

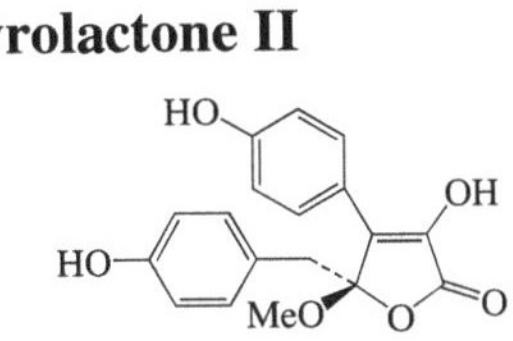

Absolute Configuration

3-Hydroxy-5-(4-hydroxybenzyl)-4-(4-hydroxyphenyl)-5-methoxy-2(5H)-furanone

[1884490-96-3]

$C_{18}H_{16}O_6$ 328.32

Related to, but not an isomer of, Butyrolactone II, B90. Prod. by a marine-derived *Aspergillus terreus* SCSGAF0162 and a co-culture of *Aspergillus terreus* and *Bacillus subtilis*. [α]$^{25}_{D}$-3 (c, 1.4 in MeOH). [α]$^{20}_{D}$+8.7 (c, 0.2 in MeOII). λ_{max} 224 (log ε 3.7); 313 (log ε 3.6) (MeOH)

O-De-Me

3,5-Dihydroxy-5-(4-hydroxybenzyl)-4-(4-hydroxyphenyl)-2(5H)-furanone. ***4-O-Demethylisobutyrolactone II***

[1884463-50-6]

$C_{17}H_{14}O_6$ 314.293

Prod. by co-cultures of *Aspergillus terreus* and *Bacillus subtilis*. [α]$^{20}_{D}$ +13 (c, 0.2 in MeOH). λ_{max} 224 (log ε 4.0); 313 (log ε 3.9) (MeOH).

Nong, X.-H. *et al.*, *Mar. Drugs*, 2014, **12**, 6113-6124 (*Isobutyrolactone II*).

Chen, H. *et al.*, *Phytochem. Lett.*, 2015, **12**, 35-41 (*Isobutyrolactone II, 4-O-Demethylisobutyrolactone II*)

Isobutyrolactone V I131

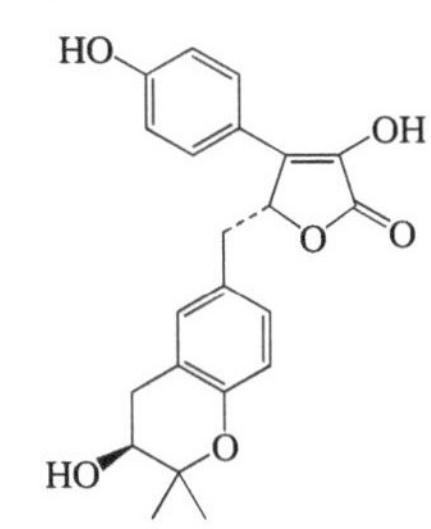

Absolute Configuration

[1851279-13-4]

$C_{22}H_{22}O_6$ 382.412

Related to, but not an isomer of, Butyrolactone V, B92. Prod. by a marine-derived *Aspergillus terreus* SCSGAF0162. Amorph. yellow solid. [α]$^{25}_{D}$-2 (c, 0.4 in MeOH). λ_{max} 205 (log ε 4.27); 220 (log ε 4.02); 302 (log ε 4.10) (MeOH).

Nong, X.-H. *et al.*, *Mar. Drugs*, 2014, **12**, 6113-6124 (*Isobutyrolactone V, cd*)

Isocorniculatolide A I132

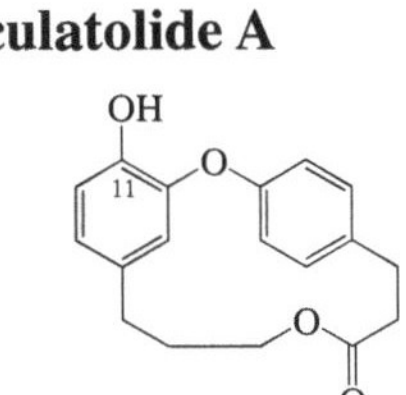

4-Hydroxy-3,4′-oxylignan-9′,9-olide. ***Isocombretastatin D_4***

[1360058-54-3]

$C_{18}H_{18}O_4$ 298.338
Closely related to Combretastatin D_2, C101. Constit. of the bark of *Aegiceras corniculatum* and the stems of *Celastrus hindsii*. Exhibits moderate to weak cytotoxicity against human small cell lung (NCI-H187) and colon (HCT116) cancer cells. Needles (hexane/Me_2CO). Mp 202 - 203°. λ_{max} 279 (ε 2740) (MeOH)

Me ether

4-Methoxy-3,4'-oxylignan-9',9-olide. ***11-O-Methylisocorniculatolide A***
[1358550-62-5]
$C_{19}H_{20}O_4$ 312.365
Constit. of the bark of *Aegiceras corniculatum*. Needles (Me_2CO). Mp 134 - 136°. λ_{max} 278 (ε 1860) (MeOH).

Ponnapalli, M.G. *et al.*, *J. Nat. Prod.*, 2012, **75**, 275-279 (*Isocorniculatolide A, Me ether, cryst struct*).
Raut, G.N. *et al.*, *Tetrahedron Lett.*, 2012, **53**, 6343-6346 (*synth*).
Hu, X.-Q. *et al.*, *Phytochem. Lett.*, 2014, **7**, 169-172 (*Celastrus hindsii constit, activity*)

Isocymodiene I33

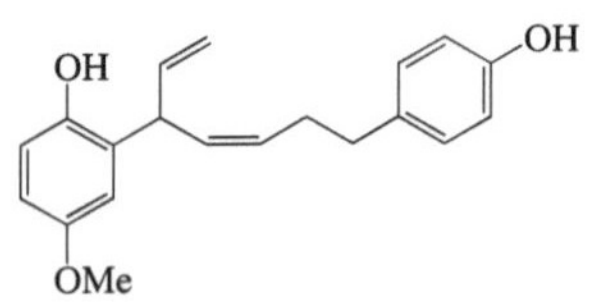

[1012852-19-5]
$C_{20}H_{22}O_3$ 310.393
Isol. from the sea grass *Cymodocea nodosa*. Oil. $[\alpha]_D^{25}$ +0.35 (c, 0.06 in CH_2Cl_2). λ_{max} 229 (log ε 0.61); 239 (log ε 0.97); 285 (log ε 0.84) ($CHCl_3$).

Kontiza, I. *et al.*, *Tetrahedron*, 2008, **64**, 1696-1702 (*Isocymodiene*)

Isodunnianol I34

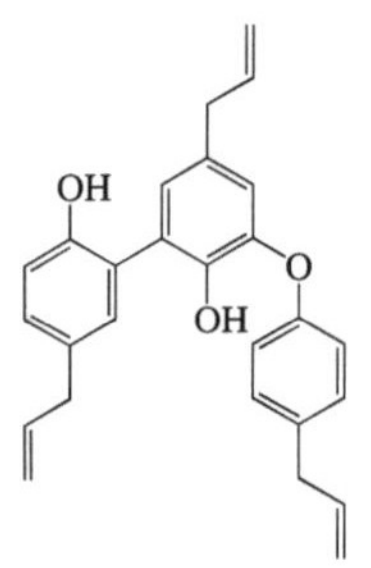

4'',5'-Oxy-3,3'-sesquiligna-8,8',8''-triene-4,4'-diol. 4,4'-Dihydroxy-4'',5'-oxy-3,3'-sesquineoligna-8,8',8''-triene
[139726-30-0]
$C_{27}H_{26}O_3$ 398.502
Constit. of *Illicium dunnianum*, *Illicium fargesii* and *Illicium simonsii*. Neurite outgrowth promotor. Shows antibacterial activity. Oil.

Kouno, I. *et al.*, *Chem. Pharm. Bull.*, 1991, **39**, 2606-2608 (*Isodunnianol*).
Moriyama, M. *et al.*, *Tetrahedron*, 2007, **63**, 4243-4249 (*Illicium fargesii constit, activity*).
Liu, J.-F. *et al.*, *Planta Med.*, 2010, **76**, 1464-1467 (*Illicium simonsii constit, activity*)

Isoeudesobovatol A I35

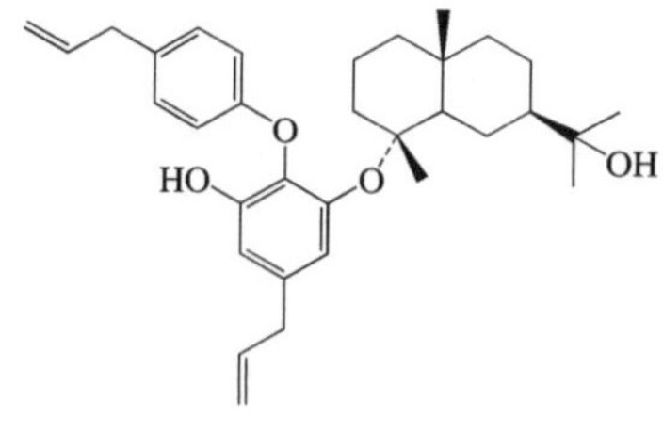

[1929551-32-5]
$C_{33}H_{44}O_4$ 504.711
Regioisomer of Eudesobovatol A, E282. Constit. of *Manglietia hookeri*. Oil. $[\alpha]_D^{25}$ -54 (c, 0.1 in Me_2CO). λ_{max} 208 (log ε 2.67); 275 (log ε 2.67); 281 (log ε 1.83) (EtOH).

Hu, X. *et al.*, *Nat. Prod. Res.*, 2016, **30**, 1477-1483 (*Isoeudesobovatol A*)

Isoferulic isovanillic anhydride I36

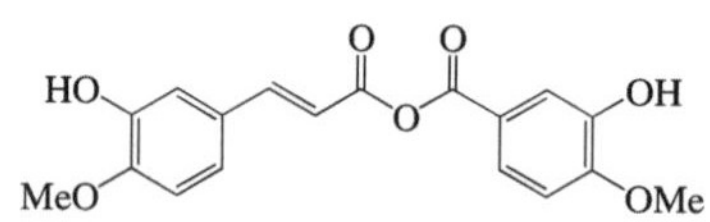

3-Hydroxy-4-methoxybenzoic 3-(3-hydroxy-4-methoxyphenyl)-2-propenoic anhydride
$C_{18}H_{16}O_7$ 344.319
(E)-form [2226515-60-0]
Constit. of the rhizomes of *Cimicifuga dahurica*. Exhibits neuroprotective activity. Amorph. powder. λ_{max} 272; 319 (MeOH).

Qin, R. *et al.*, *Fitoterapia*, 2016, **115**, 52-56 (*Cimicifuga dahurica constit, activity*)

Isofutoquinol A I37

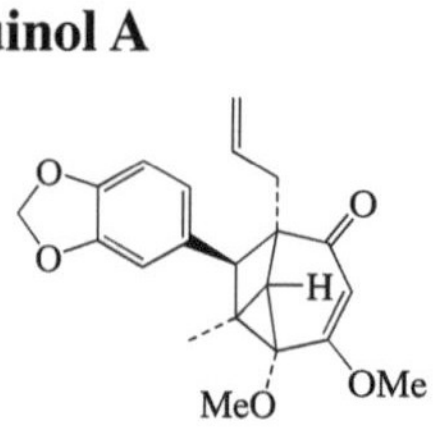

Relative Configuration

7-(1,3-Benzodioxol-5-yl)-2,3-dimethoxy-8-methyl-6-(2-propenyl)tricyclo[4.2.0.0^{2,8}]oct-3-en-5-one, CAS. *2,8':3,8'-Bicyclo-2,3-dihydro-3,4-dimethoxy-3',4'-methylenedioxy-1,7'-lign-8-en-6(1H)-one*
[62499-70-1]
$C_{21}H_{22}O_5$ 354.402
Constit. of *Piper futokadsura* and *Piper wallichii*. Cryst. (hexane). Mp 118 - 119°. λ_{max} 233 (sh); 277 (ε 9520) (EtOH).

Matsui, K. *et al.*, *Tetrahedron Lett.*, 1976, **17**, 4371-4374 (*Isofutoquinol A*).
Shizuri, Y. *et al.*, *Tetrahedron Lett.*, 1986, **27**, 727-730 (*synth*).
Ohba, S. *et al.*, *Acta Crystallogr., Sect. C*, 1987, **43**, 172-173 (*cryst struct*)

Isohydnocarpin I38

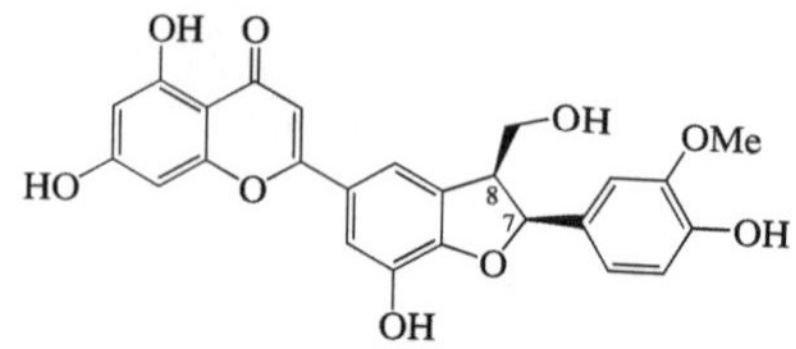

Relative Configuration

2-[2,3-Dihydro-7-hydroxy-2-(4-hydroxy-3-methoxyphenyl)-3-hydroxymethyl-5-benzofuranyl]-5,7-dihydroxy-4H-1-benzopyran-4-one, CAS
[55607-52-8]
$C_{25}H_{20}O_9$ 464.426
Flavonolignan. Closely related to Anthelminthicol A, A33 and HDW84-Z. Regioisomer of Hydnocarpin, H64. Constit. of *Hydnocarpus wightiana*, *Hydnocarpus anthelminthica*, *Inula wissmanniana*, *Nelumbo nucifera* and *Cassia absus*. Inhibitor of NO prodn.. Cryst. (MeOH). Mp 238 - 240°

Penta-Ac

Mp 163 - 165°

Stereoisomer

Isohydnocarpin A
[1639343-44-4]
$C_{25}H_{20}O_9$ 464.426
Constit. of *Nelumbo nucifera*. Amorph. brown powder. $[\alpha]_D^{25}$ +26.4 (c, 0.15 in MeOH). Possesses (7*R**,8*S**)-config. (*trans*-). λ_{max} 268; 345 (MeOH).

Ranganathan, K.R. *et al.*, *Indian J. Chem.*, 1974, **12**, 888-889 (*Isohydnocarpin*).
Kostova, I. *et al.*, *Indian J. Chem., Sect. B*, 1977, **15**, 764-765 (*Cassia absus constit, struct*).
Parthasarathy, M.R. *et al.*, *Phytochemistry*, 1979, **18**, 506-508 (*cmr*).
Wang, J.-F. *et al.*, *J. Asian Nat. Prod. Res.*, 2011, **13**, 80-83 (*Hydnocarpus anthelminthica constit, activity*).
Wang, C. *et al.*, *Arch. Pharmacal Res.*, 2013, **36**, 1516-1524 (*Inula wissmanniana constit*).

Ahn, J.H. *et al.*, *Chem. Nat. Compd. (Engl. Transl.)*, 2014, **50**, 998-1000 (*Isohydnocarpin A*).
Sahoo, M.R. *et al.*, *J. Ethnopharmacol.*, 2014, **154**, 17-25 (*rev, props*).
Vimberg, V. *et al.*, *J. Nat. Prod.*, 2015, **78**, 2095-2103 (*synth*).
Huang, G. *et al.*, *Beilstein J. Org. Chem.*, 2016, **12**, 662-669 (*synth*)

Isolariciresinol guaiacylglyceryl ether I139

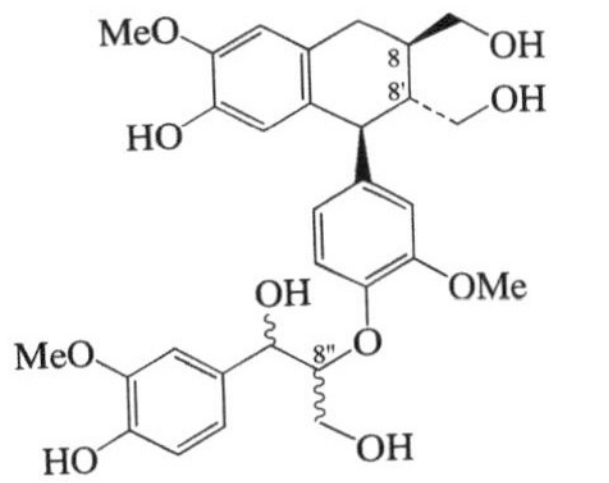

2,7′-Cyclo-3′,3″,5-trimethoxy-4′,8″-oxy-8,8′-sesquilignan-4,4″,7″,9,9′,9″-hexol. 2,7′-Cyclo-4,4″,7″,9,9′,9″-hexahydroxy-3′,3″,5-trimethoxy-4′,8″-oxysesquineolignan
$C_{30}H_{36}O_{10}$ 556.608
(7′S,7″ξ,8R,8′R,8″ξ)-form
[845890-66-6]
Constit. of *Picea abies*.
Willfoer, S. *et al.*, *Holzforschung*, 2004, **58**, 345-354 (*Picea abies constit*)

Isomagnaldehyde I140

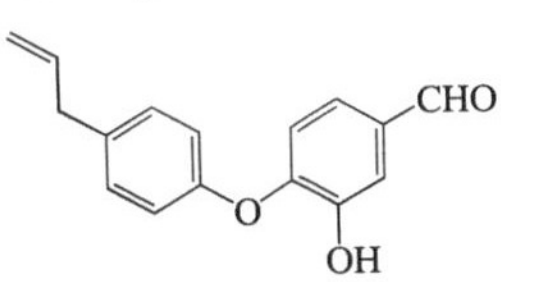

3-Hydroxy-4-[4-(2-propenyl)phenoxy]benzaldehyde. 3-Hydroxy-8,9-dinor-4,4′-oxylign-8′-en-7-al. 3-Hydroxy-8,9-dinor-7-oxo-4,4′-oxyneolign-8′-ene
[1427279-41-1]
$C_{16}H_{14}O_3$ 254.285
Isol. from the roots of *Streblus asper*. Shows antiviral activity against hepatitis B. Purple-black oil. λ_{max} 208 (log ε 4.54); 254 (log ε 2.65); 289 (log ε 2.33) (MeOH)
Me ether
3-Methoxy-8,9-dinor-4,4′-oxylign-8′-en-7-al. 3-Methoxy-8,9-dinor-7-oxo-4,4′-oxylign-8′-ene.
O-Methylisomagnaldehyde
[2196258-67-8]
$C_{17}H_{16}O_3$ 268.312
Constit. of the roots of *Streblus asper*. Exhibits antimicrobial activity against *Saccharomyces cerevisiae, Bacillus subtilis, Pseudomonas aeruginosa, Escherichia coli* and *Staphylococcus aureus*. Purple oil. λ_{max} 209 (log ε 4.50); 254 (log ε 2.60); 287 (log ε 2.32) (MeOH).
Li, J. *et al.*, *Bioorg. Med. Chem. Lett.*, 2013, **23**, 2238-2244 (*Isomagnaldehyde, activity*).
Li, C. *et al.*, *Rapid Commun. Mass Spectrom.*, 2014, **28**, 2363-2370 (*ms*).
Nie, H. *et al.*, *Phytochem. Lett.*, 2016, **18**, 226-231 (*O-Methylisomagnaldehyde, activity*)

Isoprincepin I41

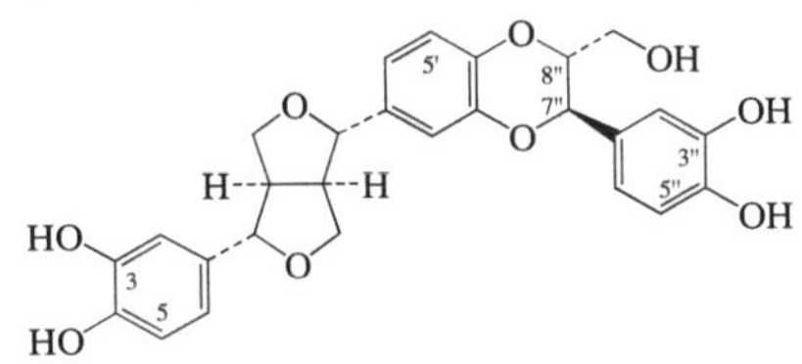

Absolute Configuration

[571166-70-6]
$C_{27}H_{26}O_9$ 494.496
Sesquilignan. Constit. of *Morinda citrifolia* and *Joannesia princeps*. Antioxidant. Pale yellow powder. $[\alpha]^{26}_{D}$ +27.8 (c, 0.4 in MeOH). λ_{max} 208 (log ε 5.01); 225 (sh); 282 (log ε 4.15) (MeOH)
5′-Methoxy, 3,3″-di-Me ether, 4,4″-di-O-β-D-glucopyranoside
Galiveroside A
[2131050-36-5]
$C_{42}H_{52}O_{20}$ 876.858
Constit. of *Galium verum*. Exhibits insecticidal activity towards cotton bollworm. Amorph. solid. $[\alpha]^{20}_{D}$ +10.3 (c, 0.2 in MeOH). Rel. config.
Stereoisomer, 5,5′,5″-trimethoxy
Reevesilignan
$C_{30}H_{32}O_{12}$ 584.574
Constit. of *Reevesia formosana*. Oil. $[\alpha]^{25}_{D}$ -10.5 (c, 0.06 in MeOH). Rel. config. determined. Possibly the enantiomer of Isoprincepin. λ_{max} 214 (log ε 4.75); 239 (sh); 277 (log ε 3.61) (MeOH).
Waibel, R. *et al.*, *Phytochemistry*, 2003, **62**, 805-811 (*Isoprincepin*).
Kamiya, K. *et al.*, *J. Agric. Food Chem.*, 2004, **52**, 5843-5848 (*Morinda citrifolia constit, abs config*).
Chen, J. *et al.*, *Chem. Nat. Compd. (Engl. Transl.)*, 2017, **53**, 626-628 (*Galiveroside A, activity*).
Chang, H.-S. *et al.*, *RSC Adv.*, 2017, **7**, 27040-27047 (*Reevesilignan*).
Kobayashi, M. *et al.*, *J. Org. Chem.*, 2019, **84**, 14227-14240 (*synth*)

Isosalvianolic acid C I42

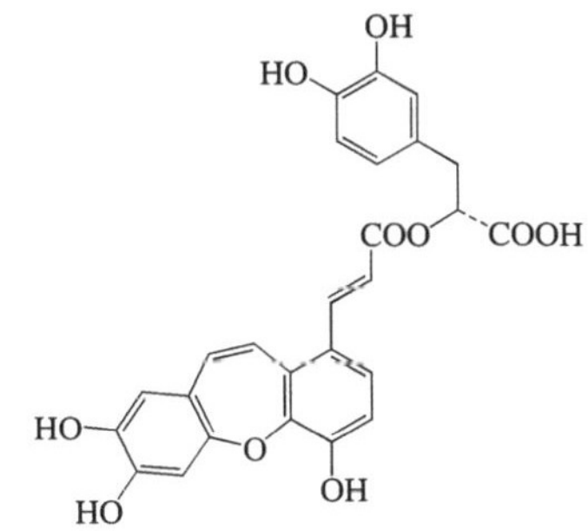

Absolute Configuration

[142115-17-1]
$C_{26}H_{20}O_{10}$ 492.436
Constit. of *Lavandula angustifolia, Salvia chinensis, Salvia miltiorrhiza, Salvia prionitis* and *Tournefortia sarmentosa*. Amorph. yellow powder. $[\alpha]^{12}_{D}$+39 (c, 0.06 in EtOH). λ_{max} 222 (sh); 288 (log ε 4.21); 326 (log ε 4.3); 340 (log ε 4.24) (EtOH)
Me ester
[1347669-70-8]
$C_{27}H_{22}O_{10}$ 506.463
Constit. of the roots of *Salvia miltiorrhiza*. Yellow powder (MeOH). $[\alpha]^{20}_{D}$+21.5 (c, 0.11 in MeOH). λ_{max} 224 (sh); 290 (log ε 4.21); 329 (log ε 4.3) (MeOH).
Qian, T.-X. *et al.*, *Phytochemistry*, 1992, **31**, 1068-1070 (*Isosalvianolic acid C*).
Ma, H.-Y. *et al.*, *J. Nat. Med.*, 2011, **65**, 37-42 (*Me ester*).
Yadikar, N. *et al.*, *Phytochem. Lett.*, 2018, **23**, 149-154 (*Lavandula angustifolia constit*)

Isosilybin I43

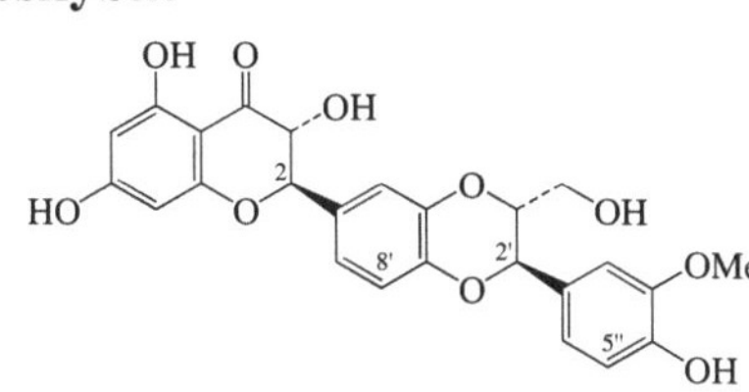

(*2R,2′R,3R,3′R*)*-form*

[72581-71-6]
$C_{25}H_{22}O_{10}$ 482.441
Flavonolignan. Isol. as a mixt. of two diastereoisomeric forms first characterised individually in 2003. Constit. of *Silybum marianum* and one of its fungal endophytes *Aspergillus iizukae*. Antihepatotoxic principle. Cryst.. Mp 239 - 241°. $[\alpha]_D$ +16.9 (Me_2CO). Component of Silymarin, crude milk thistle extract. λ_{max} 288; 322 (MeOH)
(2R,2′R,3R,3′R)-form
Isosilybin A [142796-21-2]

Constit. of *Silybum marianum*, *Euonymus acanthocarpus* and *Inula wissmanniana*. Isol. from *Aspergillus iizukae*. Needles (MeOH aq.). Mp 158 - 160°. Mp 201 - 203°. $[\alpha]_D$ +30.4 (c, 0.3 in MeOH). $[\alpha]_D$ +48.1 (c, 0.27 in Me_2CO). λ_{max} 212 (log ε 5.4); 230 (log ε 5.3); 288 (log ε 5.1) (MeOH)

3-Deoxy

Silandrin A

[70815-32-6]

$C_{25}H_{22}O_9$ 466.442

Constit. of *Silybum marianum*. Shows hepatotoxic activity. Bacterial multi-drug resistant (MDR) efflux pump inhibitor. Synergistic with berberine on *Staphylococcus aureus*. Cryst.. Mp 234 - 236°. $[\alpha]_D$ -42.7

4″-Deoxy, 5″-hydroxy

Isosilybin C

[1289644-25-2]

$C_{25}H_{22}O_{10}$ 482.441

Minor constit. of *Silybum marianum*. Yellow film. $[\alpha]_D$ -13 (c, 0.003 in MeOH). λ_{max} 212 (log ε 1.6); 230 (log ε 1.33); 288 (log ε 1.14) (MeOH)

5″-Methoxy, 8′-hydroxy

Hovenin C

[1423247-46-4]

$C_{26}H_{24}O_{12}$ 528.466

Constit. of the seeds of *Hovenia acerba*. Amorph. powder (MeOH). Mp 172 - 174°. $[\alpha]^{25}_D$ +7.5 (c, 0.04 in MeOH). λ_{max} 210 (log ε 4.62); 248 (log ε 3.99); 371 (log ε 3.78) (MeOH)

(2R,2′S,3R,3′S)-form

Isosilybin B [142796-22-3]

Constit. of *Silybum marianum*. Needles (MeOH aq.). Mp 194 - 196°. Mp 236 - 238°. $[\alpha]_D$ -17.6 (c, 0.3 in MeOH). $[\alpha]_D$ -23.5 (c, 0.3 in Me_2CO). λ_{max} 212 (log ε 5.4); 230 (log ε 5.3); 288 (log ε 5.1) (MeOH)

3-Deoxy

Silandrin B

[1021297-33-5]

$C_{25}H_{22}O_9$ 466.442

Constit. of *Silybum marianum*. Has 2*S*,2′*S*,3′*S*-config.

4″-Deoxy, 5″-hydroxy

Isosilybin D

[1289644-27-4]

$C_{25}H_{22}O_{10}$ 482.441

Minor constit. of *Silybum marianum*. Yellow film. $[\alpha]_D$ +11 (c, 0.02 in MeOH). λ_{max} 212 (log ε 1.24); 230 (log ε 0.95); 288 (log ε 0.87) (MeOH)

5″-Methoxy, 8′-hydroxy

Hovenin D

[1423247-47-5]

$C_{26}H_{24}O_{12}$ 528.466

Constit. of the seeds of *Hovenia acerba*. Amorph. powder (MeOH). Mp 171 - 173°. $[\alpha]^{25}_D$ -7.2 (c, 0.04 in MeOH). λ_{max} 210 (log ε 4.72); 247 (log ε 4.07); 371 (log ε 3.81) (MeOH)

(2S,2′S,3′R)-form

3-Deoxy

Cisilandrin. *cis-Silandrin*

[1021297-36-8]

$C_{25}H_{22}O_9$ 466.442

Constit. of *Silybum marianum*. Has 2*S*,2′*S*,3′*R*-config..

Arnone, A. *et al.*, *J. Chem. Soc., Chem. Commun.*, 1979, 696-697 (*Isosilybin*).
Merlini, L. *et al.*, *J. Chem. Soc., Perkin Trans. 1*, 1980, 775-778 (*synth*).
Szilági, I. *et al.*, *Planta Med.*, no. 10, 1981, **43**, 121-127 (*Silandrin*).
Antus, S. *et al.*, *Liebigs Ann. Chem.*, 1991, 633-641 (*Silandrin, abs config*).
Antus, S. *et al.*, *Liebigs Ann. Chem.*, 1993, 105-109 (*Silandrin, synth*).
Guz, N.R. *et al.*, *J. Med. Chem.*, 2001, **44**, 261-268 (*Silandrin, activity*).
Lee, D.Y.-W. *et al.*, *J. Nat. Prod.*, 2003, **66**, 1171-1174 (*Isosilybins A,B, abs config*).
Kim, N.-C. *et al.*, *Org. Biomol. Chem.*, 2003, **1**, 1684-1689 (*Isosilybins A,B*).
Kim, N.-C. *et al.*, *Org. Biomol. Chem.*, 2003, **1**, 3470 (*erratum*).
Nyiredy, S. *et al.*, *J. Chromatogr. Sci.*, 2008, **46**, 93-96 (*Silybum marianum constits, biosynth*).
Nakashima, S. *et al.*, *Bioorg. Med. Chem. Lett.*, 2010, **18**, 2337-2345 (*Isosilybins A,B, activity*).
Sy-Cordero, A. *et al.*, *Planta Med.*, 2010, **76**, 644-647 (*Isosilybins C,D*).
Sy-Cordero, A.A. *et al.*, *J. Nat. Prod.*, 2012, **75**, 1879-1881 (*cryst struct, abs config*).
Zhang, X.-Q. *et al.*, *Phytochem. Lett.*, 2012, **5**, 292-296 (*Hovenins C,D*).
Wang, C. *et al.*, *Arch. Pharmacal Res.*, 2013, **36**, 1516-1524 (*Inula wissmanniana constit*).
El-Elimat, T. *et al.*, *J. Nat. Prod.*, 2014, **77**, 193-199 (*Isosilybin A*).
McDonald, B.R. *et al.*, *Org. Lett.*, 2015, **17**, 98-101 (*Isosilybin A, synth*)

Isosilychristin I44

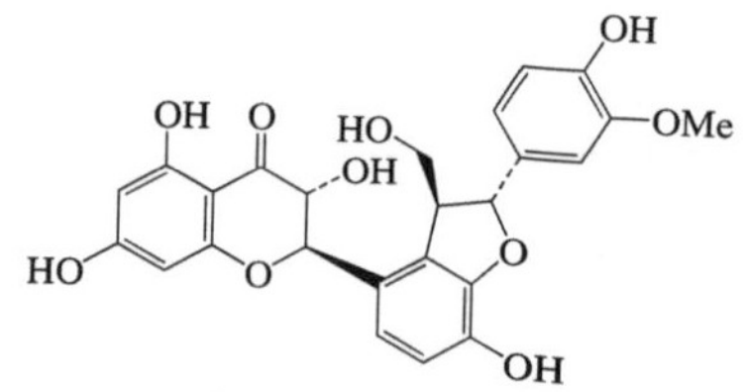

Absolute Configuration

[77182-66-2]

$C_{25}H_{22}O_{10}$ 482.441

Constit. of *Silybum marianum* and *Inula wissmanniana*. Inhibitor of T-cell cytokine prodn. in human. Cryst.. Mp 155 - 157°. $[\alpha]_D$ +176.9 (c, 0.06 in MeOH). $[\alpha]_D$ +245 (Py). $[\alpha]^{25}_D$ +256.4 (c, 0.01 in MeOH). λ_{max} 231 (log ε 5.3); 289 (log ε 5.2) (MeOH).

Kalonga, M. *Z. Naturforsch., B*, 1981, **36**, 262-265 (*Isosilychristin*).
Kim, N.-C. *et al.*, *Org. Biomol. Chem.*, 2003, **1**, 1684-1689 (*Isosilychristin, cd*).
Smith, W.A. *et al.*, *Planta Med.*, 2005, **71**, 877-880 (*Silybum marianum constit*).
Polyak, S.J. *et al.*, *Proc. Natl. Acad. Sci. U.S.A.*, 2010, **107**, 5995-5999 (*activity*).
Wang, C. *et al.*, *Arch. Pharmacal Res.*, 2013, **36**, 1516-1524 (*Inula wissmanniana constit*)

Isostrebluslignaldehyde I45

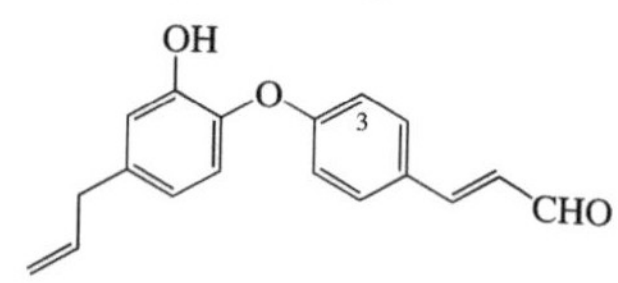

3-[4-[2-Hydroxy-4-(2-propenyl)phenoxy]phenyl]-2-propenal, CAS. *3′-Hydroxy-4,4′-oxyligna-7,8′-dien-9-al. 3′-Hydroxy-9-oxo-4,4′-oxyligna-7,8′-diene. Isostreblusligananaldehyde*

[1427279-42-2]

$C_{18}H_{16}O_3$ 280.323

Isol. from the roots of *Streblus asper*. Shows antiviral activity against hepatitis B. Red oil. λ_{max} 208 (log ε 4.54); 254 (log ε 2.65); 289 (log ε 2.33) (MeOH)

3-Hydroxy

3,3′-Dihydroxy-4,4′-oxyligna-7,8′-dien-9-al. 3,3′-Dihydroxy-9-oxo-4,4′-oxyligna-7,8′-diene.

3-Hydroxyisostrebluslignaldehyde

[2196258-65-6]

$C_{18}H_{16}O_4$ 296.322

Constit. of the roots of *Streblus asper*. Exhibits antimicrobial activity against *Saccharomyces cerevisiae*, *Bacillus subtilis*, *Pseudomonas aeruginosa*, *Escherichia coli* and *Staphylococcus aureus*. Red oil.

λ_{max} 210 (log ε 4.50); 253 (log ε 2.59); 287 (log ε 2.31) (MeOH).

Li, J. *et al.*, *Bioorg. Med. Chem. Lett.*, 2013, **23**, 2238-2244 (*Isostreblusligaldehyde*).

Nie, H. *et al.*, *Phytochem. Lett.*, 2016, **18**, 226-231 (*3-Hydroxyisostreblusligaldehyde, activity*)

Itoside N I46

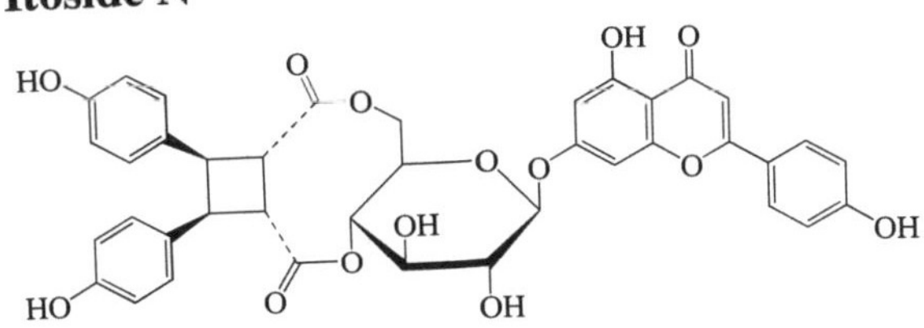

[1016275-83-4]

$C_{39}H_{32}O_{14}$ 724.671

Flavone glycoside-lignan ester. Constit. of *Itoa orientalis*. Cyclooxygenase-2 inhibitor. Yellow-brown powder. $[\alpha]^{25}_{D}$ +10 (c, 0.06 in MeOH). λ_{max} 226 (log ε 4.36); 268 (log ε 4.14); 331 (log ε 4.2) (MeOH).

Chai, X.Y. *et al.*, *J. Nat. Prod.*, 2008, **71**, 814-819 (*Itoside N*)

Jacpaniculine J1

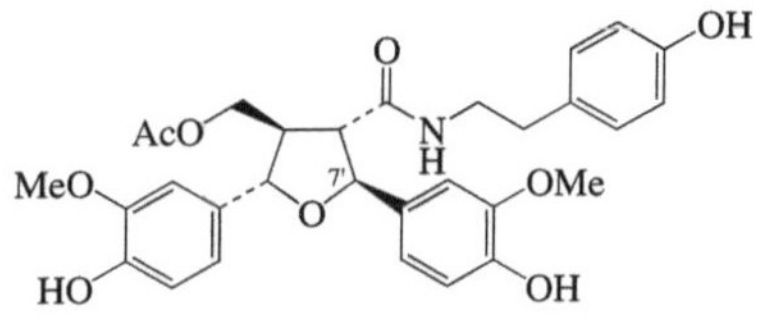

Relative Configuration

[162857-96-7]

$C_{30}H_{33}NO_9$ 551.592

Lignanamide alkaloid from fruits of *Jacquemontia paniculata* var. *paniculata*. Cryst.. Mp not reported. λ_{max} 203; 225; 279 (MeOH)

7′-Epimer

Isojacpaniculine

[162990-76-3]

$C_{30}H_{33}NO_9$ 551.592

Alkaloid from fruits of *Jacquemontia paniculata* var. *paniculata*. Cryst.. Mp not reported.

Henrici, A. *et al.*, *Phytochemistry*, 1994, **37**, 1637-1640 (*Jacpaniculine, Isojacpaniculine*)

Jamesopyrone J2

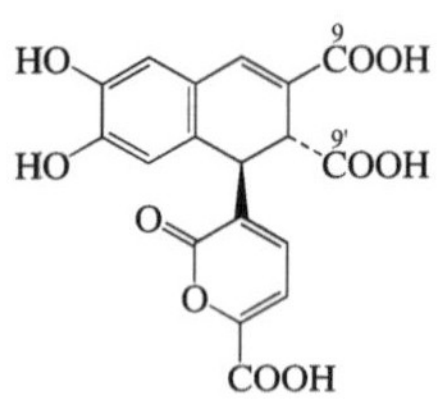

$C_{18}H_{12}O_{10}$ 388.284

Lignan numbering shown. Related to Scapaniapyrone A, S30

(7′R*,8′R*)-form

[171901-67-0]

Isol. from the liverwort *Jamesoniella autumnalis*. Light yellow powder. $[\alpha]^{20}_{Hg}$+54.9 (c, 0.3 in MeOH). λ_{max} 225 (log ε 4.01); 249 (log ε 4.23); 308 (log ε 4.22); 322 (sh) (MeOH)

9-Me ester

[171901-68-1]

$C_{19}H_{14}O_{10}$ 402.311

Isol. from *Jamesoniella autumnalis*. Light yellow powder. $[\alpha]^{20}_{Hg}$ +33.5 (c, 1.7 in MeOH). λ_{max} 225 (log ε 4.03); 250 (log ε 4.21); 308 (log ε 4.18); 321 (sh) (MeOH)

9′-Me ester

[171901-69-2]

$C_{19}H_{14}O_{10}$ 402.311

Isol. from *Jamesoniella autumnalis*. Light yellow powder. $[\alpha]^{20}_{Hg}$ +50.5 (c, 0.95 in MeOH). λ_{max} 221 (log ε 4.14); 250 (log ε 4.24); 309 (log ε 4.20); 319 (sh) (MeOH).

Tazaki, H. *et al.*, *Phytochemistry*, 1995, **40**, 1671-1675 (*Jamesoniella autumnalis constits*).
Tazaki, H. *et al.*, *Tetrahedron Lett.*, 1999, **40**, 101-104 (*biosynth*)

Jasurolignoside J3

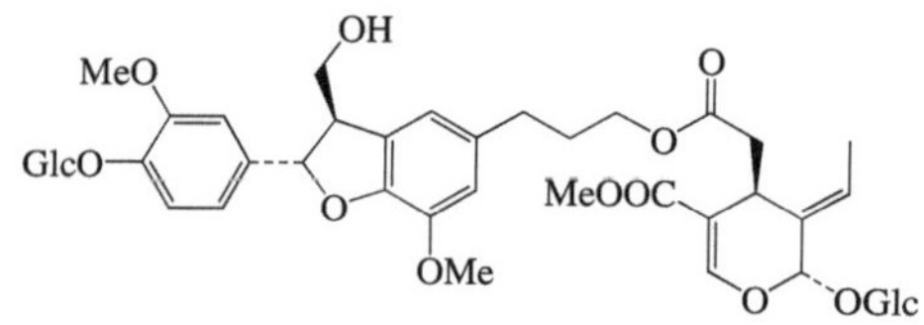

[212065-15-1]

$C_{43}H_{56}O_{21}$ 908.9

Neolignan-secoiridoid glucoside. Constit. of *Jasminum urophyllum*. Amorph. pale yellow solid. $[\alpha]^{25}_{D}$-140 (c, 0.1 in MeOH). λ_{max} 233 (log ε 4.02); 278 (log ε 3.27) (MeOH).

Shen, Y.-C *et al.*, *Phytochemistry*, 1998, **48**, 719-723 (*Jasurolignoside, cd*)

Jezonolide J4

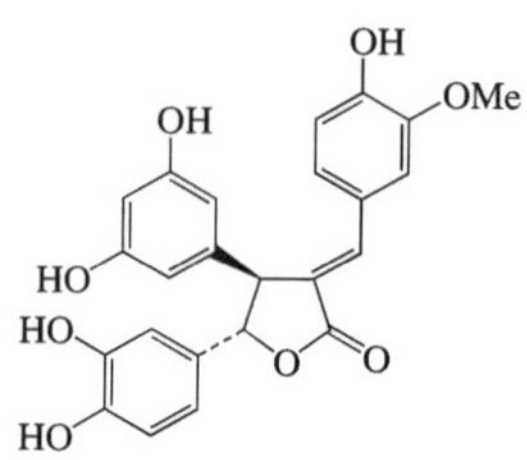

Relative Configuration

[1192159-98-0]

$C_{24}H_{20}O_8$ 436.416

Constit. of the bark of *Picea jezoensis* var. *jezoensis*. Inhibitor of NOR-1 activation. Amorph. brown powder. Racemate. λ_{max} 212 (log ε 4.13); 286 (log ε 3.70); 336 (log ε 3.82) (EtOH).

Wada, S. *et al.*, *Bioorg. Med. Chem.*, 2009, **17**, 6414-6421 (*Jezonolide*)

Junipetrioloside A J5

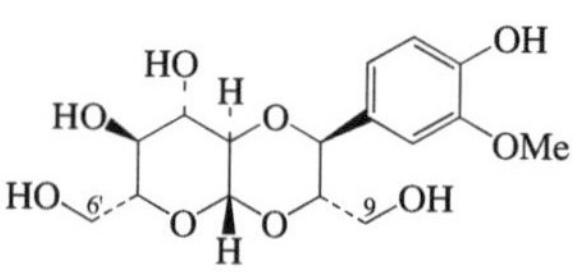

Absolute Configuration

[195056-67-8]

$C_{16}H_{22}O_9$ 358.343

Cyclic glucoside of 1-(3,4-Dihydroxyphenyl)-1,2,3-propanetriol, D227. Isol. from *Juniperus phoenicea*. λ_{max} 229; 277 (MeOH)

6′-Benzoyl

Xylongoside A

[1383435-50-4]

$C_{23}H_{26}O_{10}$ 462.451

Constit. of the stem bark of *Xylosma longifolia*. Cryst.. Mp 137 - 138°. $[\alpha]^{27}_{D}$+57 (c, 1.5 in MeOH). λ_{max} 207 (log ε 4.22); 230 (log ε 4.25); 280 (log ε 3.5) (MeOH)

6′-O-(4-Hydroxy-3-methoxybenzoyl)

Brachyanin B

$C_{24}H_{28}O_{12}$ 508.476

Constit. of *Stauntonia brachyanthera*. Antioxidant. Oil (MeOH). $[\alpha]^{25}_{D}$+44 (c, 0.9 in MeOH)

6′-O-(3,4,5-Trihydroxybenzoyl)

Mallophenol A

[752252-70-3]

$C_{23}H_{26}O_{13}$ 510.448

Constit. of the leaves of *Mallotus furetianus* and the bark of *Macropteranthes leichhardtii*. Amorph. powder. $[\alpha]^{25}_{D}$+19.4 (c, 1 in MeOH). λ_{max} 217 (log ε 4.51); 276 (log ε 4.11) (MeOH)

9-O-(3,4,5-Trihydroxybenzoyl)

Macropteranthol

[1800599-25-0]

$C_{23}H_{26}O_{13}$ 510.448

Constit. of the bark of *Macropteranthes leichhardtii*. Amorph. powder. $[\alpha]_D$ +2.7 (c, 0.04 in MeOH). λ_{max} 219 (log ε 4.12); 278 (log ε 3.76) (MeOH)

9-Deoxy

Hemidesmoside A‡

[1467586-33-9]

$C_{16}H_{22}O_8$ 342.344

Constit. of the roots of *Hemidesmus indicus*. Amorph. powder. $[\alpha]^{20}_{D}$+69.7 (c, 0.3 in MeOH). λ_{max} 230 (log ε 3.83); 282 (log ε 3.53) (MeOH)

9-Deoxy, 6′-O-(3S-hydroxy-3-methylglutaroyl)

Asprenol A

$C_{22}H_{30}O_{12}$ 486.470

Constit. of *Ilex asprella*. Amorph. light yellow powder. $[\alpha]^{25}_{D}$+56 (c, 0.1 in MeOH). λ_{max} 206 (log ε 3.18); 230 (log ε 2.77); 279 (log ε 2.35) (MeOH).

Comte, G. *et al.*, *Phytochemistry*, 1997, **45**, 1679-1682 (*Junipetrolioside A*).
Wei, K. *et al.*, *Chem. Pharm. Bull.*, 2004, **52**, 776-779 (*Mallophenol A*).
Truong, B.N. *et al.*, *Phytochem. Lett.*, 2011, **4**, 250-253 (*Xylongoside A*).
Zhao, Z. *et al.*, *J. Nat. Med.*, 2013, **67**, 137-142 (*Hemidesmoside A*).
Zhao, J. *et al.*, *J. Funct. Foods*, 2015, **14**, 374-383 (*Brachyanin B, activity*).
Tian, L.-W. *et al.*, *J. Nat. Prod.*, 2015, **78**, 1756-1760 (*Macropteranthol*).

Du, B. *et al.*, *Fitoterapia*, 2018, **129**, 220-225 (*Asprenol A*)

Justicidone J6

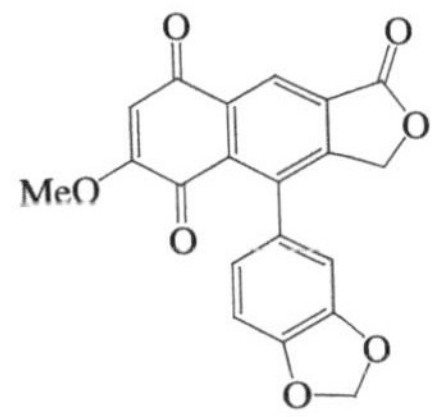

[723302-84-9]
$C_{20}H_{12}O_7$ 364.309
Constit. of the aerial parts of *Justicia hyssopifolia*. Red cryst. (EtOAc/hexane). Mp 114 - 115°. λ_{max} 252 (log ε 4.92); 276 (log ε 4.69); 322 (log ε 4.29) (MeOH).

Pérez, J.A. *et al.*, *Chem. Pharm. Bull.*, 2004, **52**, 130-131 (*Justicidone*).
Boluda, C.J. *et al.*, *Chem. Pharm. Bull.*, 2005, **53**, 930-933 (*synth*)

Justiflorinol J7

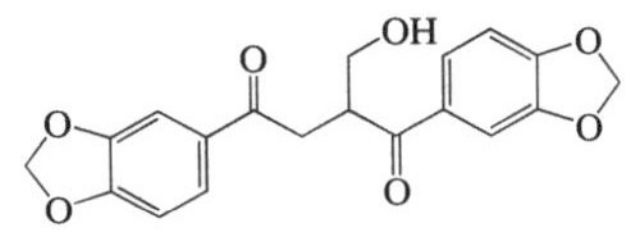

1,4-Bis(1,3-benzodioxol-5-yl)-2-hydroxymethyl-1,4-butanedione. 2,3-Bis(3,4-methylenedioxybenzoyl)-1-propanol. 9-Hydroxy-3,4:3′,4′-bis(methylenedioxy)-9′-nor-8,8′-lignan-7,7′-dione
[857050-62-5]
$C_{19}H_{16}O_7$ 356.33
Constit. of the leaves of *Justicia patentiflora* and *Piper sanguineispicum*. Exhibits modest cytotoxic activity. Powder. $[\alpha]_D^{25}$ -75.6 (c, 0.17 in CH_2Cl_2). λ_{max} 225 (log ε 3.22); 269 (log ε 2.85); 305 (log ε 2.94) (MeOH).

Susplugas, S. *et al.*, *J. Nat. Prod.*, 2005, **68**, 734-738 (*Justiflorinol, activity*).
Cabanillas, B.J. *et al.*, *J. Nat. Prod.*, 2010, **73**, 1884-1890 (*Justiflorinol*)

Kadlongirin B K1

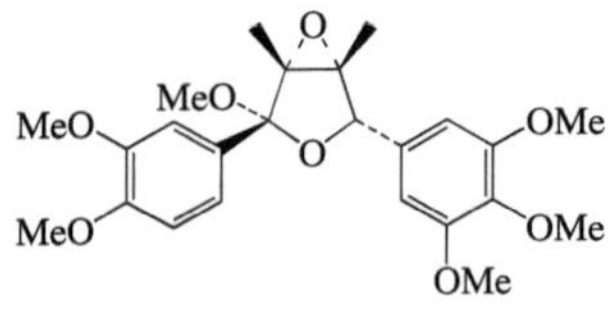

Relative Configuration

7,7′:8,8′-Diepoxy-3,3′,4,4′,5,7′-hexamethoxylignan.
3,3′,4,4′,5,7′-Hexamethoxy-7,7′:8,8′-diepoxylignan
[1063682-44-9]
$C_{24}H_{30}O_8$ 446.496
Constit. of *Kadsura longipedunculata*. Yellowish oil. $[\alpha]^{28}_{D}$ +85 (c, 0.5 in MeOH). λ_{max} 206 (log ε 5.16); 275 (log ε 3.93); 363 (log ε 2.94); 374 (log ε 2.8) (MeOH).

Pu, J.-X. *et al.*, *Chem. Pharm. Bull.*, 2008, **56**, 1143-1146 (*Kadlongirin B*)

Kadsulignan A K2

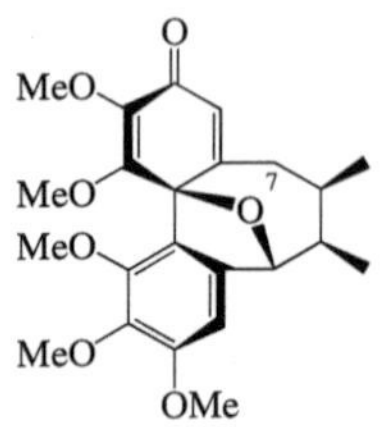

Absolute Configuration

2,2′-Cyclo-2,7′-epoxy-3,3′,4,4′,5′-pentamethoxy-8,8′-lignan-5(2H)-one
[122350-74-7]
$C_{23}H_{28}O_7$ 416.47
2,7′-Epoxy-2,2′-cyclolignan. Similar to Kadsuralignan D, K7. Constit. of *Kadsura coccinea* seeds. Yellow gum. $[\alpha]_D$ +100.89 (c, 0.045 in Me_2CO). λ_{max} 212 (ε 47863); 240 (ε 19055); 283 (ε 5129); 301 (ε 3890) (MeOH)

7α-Acetoxy
Kadsulignan B
[122350-75-8]
$C_{25}H_{30}O_9$ 474.506
Constit. of *Kadsura coccinea* seeds. Yellow gum. $[\alpha]_D$ +97.03 (c, 0.0133 in Me_2CO). λ_{max} 214 (ε 56234); 240 (ε 20893); 284 (ε 5248); 305 (ε 4266) (MeOH).

Liu, J.-S. *et al.*, *Can. J. Chem.*, 1989, **67**, 682-684 (*Kadsulignans A,B*)

Kadsuphilin L K3

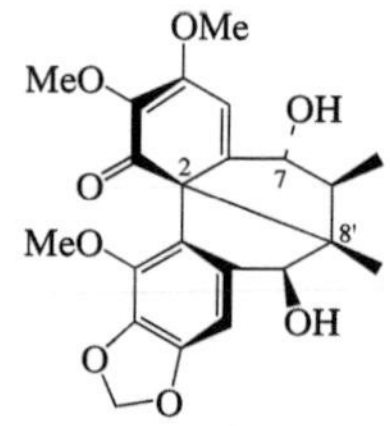

Absolute Configuration

2,2′:2,8′-Bicyclo-7,7′-dihydroxy-3′,4,5-trimethoxy-4′,5′-methylenedioxy-8,8′-lignan-3(2H)-one.
7,7′-Dihydroxy-3′,4,5-trimethoxy-4′,5′-methylenedioxy-3-oxo-2,2′:2,8′-bicyclolignan
[1017242-95-3]
$C_{22}H_{24}O_8$ 416.426
Unusual 2,8′-cyclolignan. Constit. of *Kadsura philippinensis*. Amorph. powder. $[\alpha]^{25}_{D}$ +17 (c, 0.1 in CH_2Cl_2). λ_{max} 225 (log ε 4.38); 285 (log ε 3.37) (MeOH)

7-Epimer
Kadsuphilin M
[1017242-96-4]
$C_{22}H_{24}O_8$ 416.426
Constit. of *Kadsura philippinensis*. Amorph. powder. $[\alpha]^{25}_{D}$ -15 (c, 0.1 in CH_2Cl_2). λ_{max} 231 (log ε 4.3); 287 (log ε 3.25) (MeOH)

7-Deoxy
2,2′:2,8′-Bicyclo-7′-hydroxy-3′,4,5-trimethoxy-4′,5′-methylenedioxy-8,8′-lignan-3(2H)-one.
7-Hydroxy-3,4′,5′-trimethoxy-4,5-methylenedioxy-3′-oxo--2,2′:2′,8-bicyclolignan. ***Heteroclitin G***. *Kadsulignan K*
[144027-74-7]
$C_{22}H_{24}O_7$ 400.427
Constit. of the stems of *Kadsura heteroclita* and the seeds of a *Kadsura* sp.. Yellow cryst. (Me_2CO/petrol). Mp 214 - 215° (205-206°). $[\alpha]^{25}_{D}$ +21.5 (c, 0.91 in MeOH).

Yang, X.-W. *et al.*, *Chem. Pharm. Bull.*, 1992, **40**, 1510-1576 (*Heteroclitin G*).
Liu, J.-S. *et al.*, *Phytochemistry*, 1992, **31**, 1379-1382 (*Kadsulignan K*).
Shen, Y.-C. *et al.*, *Helv. Chim. Acta*, 2008, **91**, 483-494 (*Kadsuphilins L,M*)

Kadsuphilin O K4

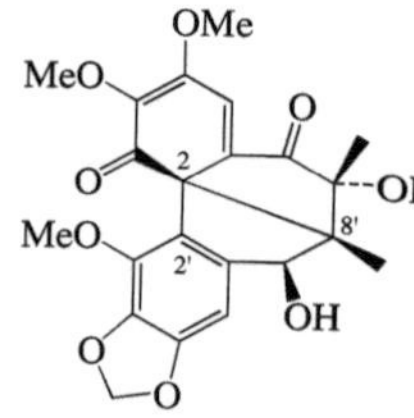

Absolute Configuration

[1594808-34-0]
$C_{22}H_{22}O_9$ 430.409
2,2′:2,8′-Bicyclo-8,8′-lignan. Constit. of the leaves and stems of *Kadsura philippinensis*. Mp 167°. $[\alpha]^{25}_{D}$ -8.0 (c, 0.6 in CH_2Cl_2). λ_{max} 215; 246; 283 (MeOH).

Lin, Y.-C. *et al.*, *Molecules*, 2013, **18**, 6573-6583 (*Kadsuphilin O, cryst struct*)

Kadsuphilol G K5

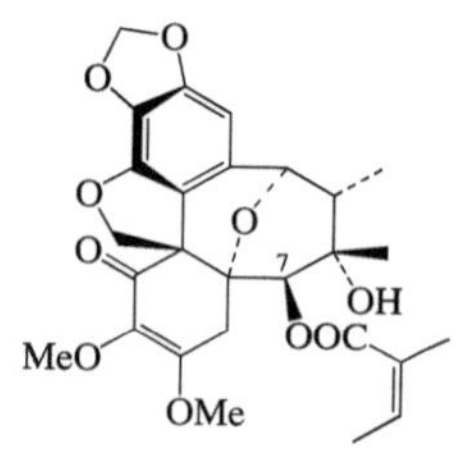

Absolute Configuration

[947256-80-6]
$C_{27}H_{30}O_{10}$ 514.527
2′-Homo-2,2′-cyclolignan. Constit. of *Kadsura philippinensis*. Amorph. yellow powder. $[\alpha]^{25}_{D}$ +150 (c, 0.01 in CH_2Cl_2). λ_{max} 216 (log ε 3.66); 274 (log ε 3.18) (MeOH)

Deangeloyl, 7-benzoyl
Taiwankadsurin F
[1594808-32-8]
$C_{29}H_{28}O_{10}$ 536.533
Constit. of the leaves and stems of *Kadsura philippinensis*. $[\alpha]^{25}_{D}$ -13.2 (c, 0.6 in CH_2Cl_2). λ_{max} 220; 255 (MeOH).

Shen, Y.-C. *et al.*, *J. Nat. Prod.*, 2007, **70**, 1139-1145 (*Kadsuphilol G*).
Lin, Y.-C. *et al.*, *Molecules*, 2013, **18**, 6573-6583 (*Taiwankadsurin F*)

Kadsuphilol H — K6

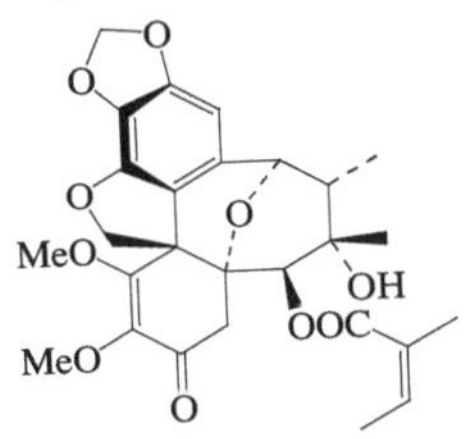

Absolute Configuration

[947256-81-7]
$C_{27}H_{30}O_{10}$ 514.527
A 2′-homo-2,2′-cyclolignan. Constit. of *Kadsura philippinensis*. Amorph. yellow powder. $[\alpha]_D^{25}$ +98 (c, 0.01 in CH_2Cl_2). λ_{max} 219 (log ε 3.86) (MeOH).
Shen, Y.-C. *et al., J. Nat. Prod.*, 2007, **70**, 1139-1145 (*Kadsuphilol H*)

Kadsuralignan D — K7

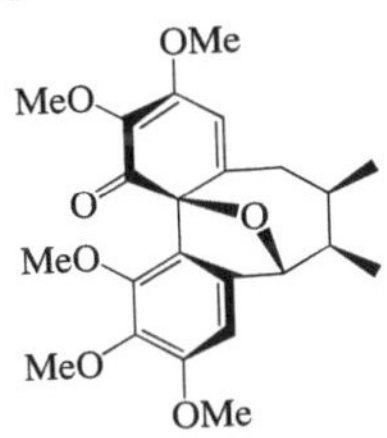

Absolute Configuration

2,2′-Cyclo-2,7′-epoxy-3′,4,4′,5,5′-pentamethoxy-8,8′-lignan-3(2H)-one. 2,7′-Epoxy-3′,4,4′,5,5′-pentamethoxy-3-oxo-2,2′-cyclolignan
[906081-85-4]
$C_{23}H_{28}O_7$ 416.47
2,7′-Epoxy-2,2′-cyclolignan. Similar to Kadsulignan A, K2. Constit. of the rhizomes of *Kadsura coccinea*. Yellow gum. λ_{max} 210 (log ε 4.61); 242 (log ε 3.77); 300 (log ε 3.51) (MeOH).
Li, H. *et al., Heterocycles*, 2006, **68**, 1259-1265 (*Kadsuralignan D, cd*)

Kaempsulfonic acid A — K8

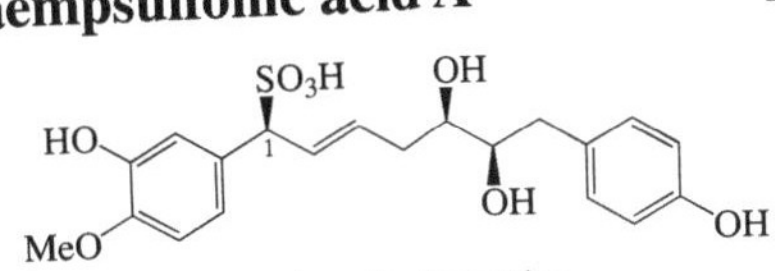

Absolute Configuration

[1611494-41-7]
$C_{20}H_{24}O_8S$ 424.464
Isol. from *Kaempferia galanga*. Yellow oil. $[\alpha]_D^{23}$ +11.1 (c, 0.11 in MeOH). λ_{max} 206 (log ε 4.08); 224 (log ε 3.76); 379 (log ε 3.22) (MeOH)

1-Epimer

Kaempsulfonic acid B
[1611494-42-8]
$C_{20}H_{24}O_8S$ 424.464
Isol. from *Kaempferia galanga*. Yellow oil. $[\alpha]_D^{23}$ +15.2 (c, 0.16 in MeOH). λ_{max} 202 (log ε 3.56); 223 (log ε 3.35); 275 (log ε 2.74) (MeOH).
Wang, F.-L. *et al., Chin. J. Nat. Med.*, 2013, **11**, 171-176 (*Kaempsulfonic acids A,B*)

Katsumadain A‡ — K9

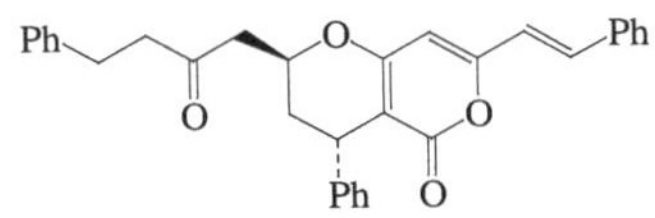

Relative Configuration

[252239-83-1]
$C_{32}H_{28}O_4$ 476.572
Not to be confused with the terpenoid Katsumadain. Constit. of the seeds of *Alpinia katsumadai*. Antiemetic agent. Inhibitor of neuroaminidase. Amorph. light yellow solid ($CHCl_3$). Mp 68 - 70°. $[\alpha]_D^{25}$ +3.7 (c, 0.4 in EtOH). λ_{max} 258 (log ε 3.3); 353 (log ε 3.36) (hexane).
Yang, Y. *et al., J. Nat. Prod.*, 1999, **62**, 1672-1674 (*Katsumadain A*).
Wang, Y. *et al., Beilstein J. Org. Chem.*, 2013, **9**, 1601-1606 (*synth*)

Katsumadain B — K10

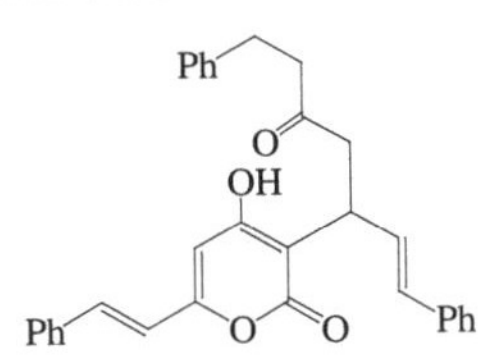

[252239-85-3]
$C_{32}H_{28}O_4$ 476.572
Constit. of the seeds of *Alpinia katsumadai*. Antiemetic agent. Amorph. light yellow solid ($CHCl_3$). Mp 118 - 121°. $[\alpha]_D^{25}$ +10.5 (c, 0.25 in EtOH). λ_{max} 254 (log ε 4.23); 355 (log ε 4.04) (MeOH).
Yang, Y. *et al., J. Nat. Prod.*, 1999, **62**, 1672-1674 (*Katsumadain B*).
Wang, Y. *et al., Beilstein J. Org. Chem.*, 2013, **9**, 1601-1606 (*synth*)

Kielcorin B — K11

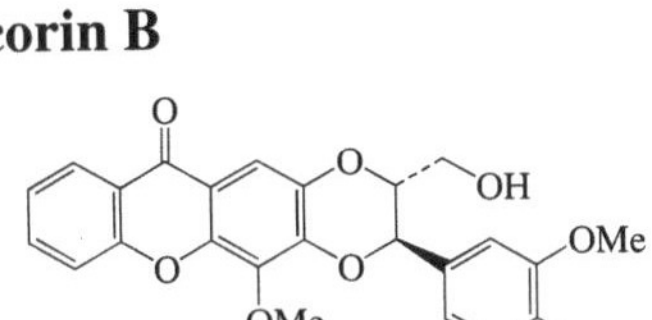

[110784-15-1]
$C_{24}H_{20}O_8$ 436.416
Constit. of *Kielmeyera coriacea*. Mp 232 - 235° (synthetic). Racemate.
Pinto, M.M. de M. *et al., Phytochemistry*, 1987, **26**, 2045-2048 (*Kielcorin B*).
Fernandes, E.G.R. *et al., Heterocycles*, 1999, **51**, 821-828 (*synth*)

Kuhlmanniquinol — K12

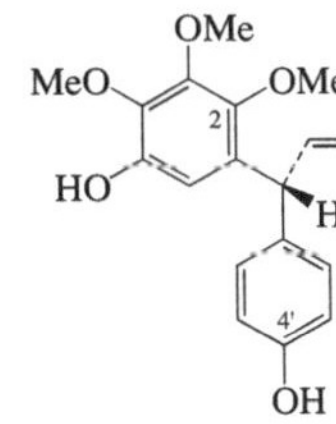

5-[1-(4-Hydroxyphenyl)-2-propenyl]-2,3,4-trimethoxyphenol, CAS. *3-(4-Hydroxyphenyl)-3-(5-hydroxy-2,3,4-trimethoxyphenyl)-1-propene. 4′,5-Dihydroxy-2,3,4′-trimethoxydalbergiquinol*
[28418-55-5] (unspecified stereochem.)
$C_{18}H_{20}O_5$ 316.353

(R)-form [21044-87-1]
Constit. of *Machaerium kuhlmannii* and *Machaerium nictitans*. Yellow oil. Bp 130 - 140° at 1 mm (bath). $[\alpha]_D^{20}$ +15 (c, 1.07 in $CHCl_3$). λ_{max} 227 (sh); 283 (ε 6000) (EtOH).
Ollis, W.D. *et al., Phytochemistry*, 1978, **17**, 1383-1388 (*Kuhlmanniquinol, abs config*)

Lanceolatanin A L1

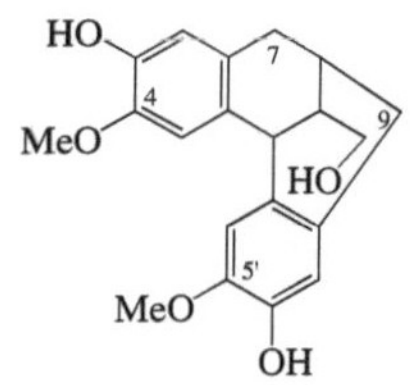

2,7′:2′,9-Bicyclo-4,5′-dimethoxy-8,8′-lignan-4′,5,9′-triol. 4′,5,9′-Trihydroxy-4,5′-dimethoxy-2,7′:2′,9-bicyclolignan
[1194545-58-8]
$C_{20}H_{22}O_5$ 342.391
2′,9-Cyclonaphthalenoid lignans. Lignan numbering shown. Related to Lirionol, L113. Constit. of the heartwood of *Cunninghamia lanceolata*. Oil (as tri-Ac). $[\alpha]^{25}_{D}$ +22.5 (c, 0.6 in $CHCl_3$) (tri-Ac)

O^4-De-Me
2,7′:2′,9-Bicyclo-5′-methoxy-8,8′-lignan-4,4′,5,9′-tetrol. 4,4′,5,9′-Tetrahydroxy-5′-methoxy-2,7′:2′,9-bicyclolignan. ***Lanceolatanin B***
[1194545-60-2]
$C_{19}H_{20}O_5$ 328.364
Constit. of the heartwood of *Cunninghamia lanceolata*. Oil (as tetra-Ac). $[\alpha]^{25}_{D}$ +23.2 (c. 0.6 in $CHCl_3$) (tetra-Ac)

7-Oxo
2,7′:2′,9-Bicyclo-4′,5,9′-trihydroxy-4,5′-dimethoxy-8,8′-lignan-7-one. 4′,5,9′-Trihydroxy-4,5′-dimethoxy-7-oxo-2,7′:2′,9-bicyclolignan. ***Lanceolatanin C***
[1194545-62-4]
$C_{20}H_{20}O_6$ 356.374
Constit. of *Cunninghamia lanceolata* and *Taxus wallichiana*. Oil (as tri-Ac). $[\alpha]^{25}_{D}$ +25.5 (c, 0.5 in $CHCl_3$) (tri-Ac). λ_{max} 235 (log ε 4.21); 279 (log ε 4) (MeOH) (tri-Ac)

7-Oxo, O^4-de-Me
2,7′:2′,9-Bicyclo-4,4′,5,9′-tetrahydroxy-5′-methoxy-8,8′-lignan-7-one. 4,4′,5,9′-Tetrahydroxy-5′-methoxy-7-oxo-2,7′:2′,9-bicyclolignan. ***Lanceolatanin D***
[1194545-64-6]
$C_{19}H_{18}O_6$ 342.347
Constit. of *Cunninghamia lanceolata* and *Taxus wallichiana*. Oil (as tetra-Ac). $[\alpha]^{25}_{D}$ +24.6 (c, 0.7 in $CHCl_3$) (tetra-Ac). λ_{max} 235 (log ε 4.18); 260 (log ε 4); 280 (log ε 3.69) (MeOH) (tetra-Ac).

Lee, C.-K. *et al.*, *Helv. Chim. Acta*, 2009, **92**, 1983-1989 (*Lanceolatanins A-D*)

Lancilignanside A L2

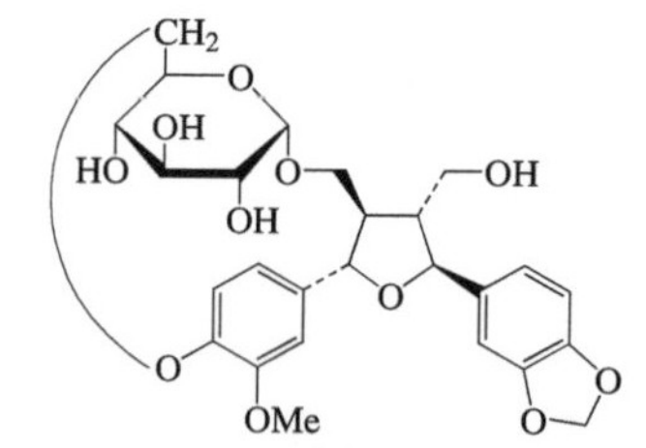

Relative Configuration

$C_{26}H_{30}O_{11}$ 518.515
Constit. of *Schisandra lancifolia*. Anti-HIV agent. Amorph. powder. $[\alpha]^{25}_{D}$ +12.5 (c, 0.11 in MeOH). λ_{max} 210 (log ε 3.87); 240 (log ε 3.65); 272 (log ε 3.1) (MeOH).

Xiao, W.-L. *et al.*, *Chem. Pharm. Bull.*, 2010, **58**, 852-855 (*Lancilignanside A*)

Lappaol A L3

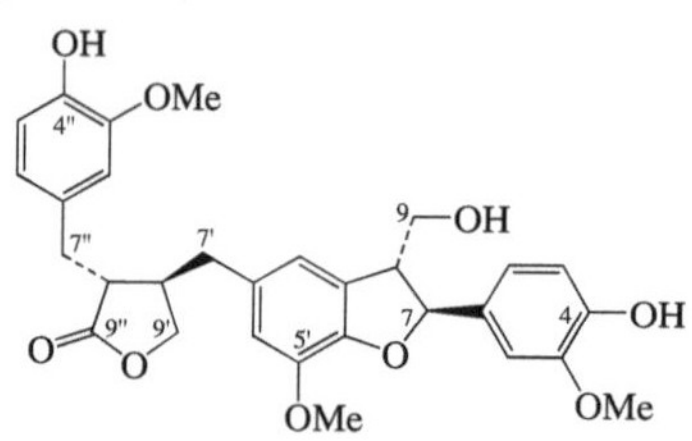

Absolute Configuration

[62333-08-8]
$C_{30}H_{32}O_9$ 536.577
Constit. of *Arctium lappa*, *Cnicus benedictus* and *Centaurea americana*. Amorph.. $[\alpha]^{20}_{D}$ -17.4 (c, 1 in MeOH). λ_{max} 283 (ε 7800) (MeOH) (Berdy)

9″-Deoxo, 9′-oxo
Isolappaol A
[131400-96-9]
$C_{30}H_{32}O_9$ 536.577
Constit. of *Arctium lappa* and *Cnicus benedictus*

9″-Deoxo, 9′-oxo, 4″-Me ether
Lappaol B
[62359-60-8]
[136832-05-8]
$C_{31}H_{34}O_9$ 550.604
Constit. of *Arctium lappa* and *Polygonum orientale*. $[\alpha]^{20}_{D}$ -19.4 (c, 1.0 in MeOH). λ_{max} 282 (ε 7460) (MeOH)

9″-Deoxo, 9′-oxo, 7″R-hydroxy
7′-Hydroxyisolappaol A
[849371-65-9]
$C_{30}H_{32}O_{10}$ 552.576
Constit. of *Saussurea macrota*. Pale yellow gum. $[\alpha]^{20}_{D}$ +12 (c, 2.5 in MeOH)

9″-Deoxo, 9′-oxo, 7-hydroxy, 4″-Me ether
7-Hydroxylappaol B
[144209-74-5]
$C_{31}H_{34}O_{10}$ 566.603
Constit. of *Saussurea gossypiphora*

7,8-Diepimer, 4,4″-di-O-β-D-glucopyranoside
Arctiisesquineolignan A
[1808085-04-2]
$C_{42}H_{52}O_{19}$ 860.859
Constit. of the fruit of *Arctium lappa*. Amorph. powder. $[\alpha]^{25}_{D}$ +18.5 (c, 0.15 in MeOH). λ_{max} 228 (log ε 3.87); 279 (log ε 3.42) (MeOH)

8′,8″-Diepimer, 9″-deoxo, 9′-oxo, 5′-demethoxy, 9-Me ether
Effususin E
[2292181-77-0]
$C_{30}H_{32}O_8$ 520.578
Constit. of the medullae of *Juncus effusus*. NO prodn. inhibitor. Yellow gum. $[\alpha]^{24}_{D}$ +24.9 (c, 1.9 in MeOH). λ_{max} 226 (log ε 4.11); 282 (log ε 3.71) (MeOH)

Stereoisomer, 9-Ac
Furfuraceol A
[1010398-37-4]
$C_{32}H_{34}O_{10}$ 578.614
Constit. of the flowers of *Centaurea furfuracea*. Amorph. solid. λ_{max} 210; 282 (EtOH)

Stereoisomer, 5′-O-de-Me
Deltoignan B
[1427038-09-2]
$C_{29}H_{30}O_9$ 522.55
Constit. of *Saussurea deltoidea*. Yellow gum. $[\alpha]^{21}_{D}$ +26.9 (c, 0.2 in MeOH). Relative config. has been determined. λ_{max} 203 (log ε 4.78) (MeOH)

Stereoisomer, 9″-deoxo, 9′-oxo, 9-Ac
Furfuraceol B
[1010398-38-5]
$C_{32}H_{34}O_{10}$ 578.614
Constit. of the flowers of *Centaurea furfuracea*. Amorph. solid. λ_{max} 210; 282 (EtOH)

Stereoisomer, 9″-deoxo, 9′-oxo, 7′ξ-hydroxy
Saussol
[128718-09-2]
$C_{30}H_{32}O_{10}$ 552.576

Constit. of *Saussurea hieracioides* and *Tsuga dumosa*. Amorph. powder. Mp 141 - 143°. Stereochem. not determined. May be identical with 7′-Hydroxyisolappaol A (above)

Stereoisomer, 9″-deoxo, 9′-oxo, 7′,8′-didehydro(E-), 3″-O-de-Me, 3″,4″-methylene ether

Orientalin‡

[1313196-78-9]

$C_{30}H_{28}O_9$ 532.545

Constit. of *Polygonum orientale*. Mp 126 - 128°. $[\alpha]^{25}_{D}$ +21.4 (c, 2.1 in $CHCl_3$). λ_{max} 220 (ε 9540); 285 (ε 7850) (MeOH).

Ichihara, A. *et al., Tetrahedron Lett.*, 1976, **17**, 3961-3964 (*Lappaols A,B*).
Ichihara, A. *et al., Tetrahedron Lett.*, 1979, **20**, 3735-3736 (*abs config*).
Ichihara, A. *et al., Agric. Biol. Chem.*, 1981, **45**, 2163-2164 (*Lappaol B, synth*).
Liu, Z. *et al., Gaodeng Xuexiao Huaxue Xuebao*, 1989, **10**, 1090-1094 (*Saussol*).
Zheng, S. *et al., Chin. Chem. Lett.*, 1991, **2**, 373-374 (*7-Hydroxylappaol B*).
Umehara, K. *et al., Chem. Pharm. Bull.*, 1993, **41**, 1774-1779 (*Isolappaol A*).
Zheng, S.-Z. *et al., Zhiwu Xuebao*, 1998, **40**, 466-469 (*Orientalin*).
Yang, M. *et al., Pharmazie*, 2004, **59**, 972-976 (*7′-Hydroxyisolappaol A*).
Zhao, Y.-X. *et al., Yunnan Zhiwu Yanjiu*, 2005, **27**, 217-222 (*Saussol*).
Fakhfakh, J.A. *et al., Nat. Prod. Res.*, 2007, **21**, 1037-1041 (*Furfuraceols A,B*).
Xu, J.-J. *et al., Fitoterapia*, 2012, **83**, 1125-1130 (*Deltoignan B*).
Yang, Y.-N. *et al., J. Agric. Food Chem.*, 2015, **63**, 7958-7966 (*Arctiisesquineolignan A*).
Ma, W. *et al., J. Mex. Chem. Soc.*, 2016, **60**, 83-88 (*Effususin E, activity*)

Lappaol C L4

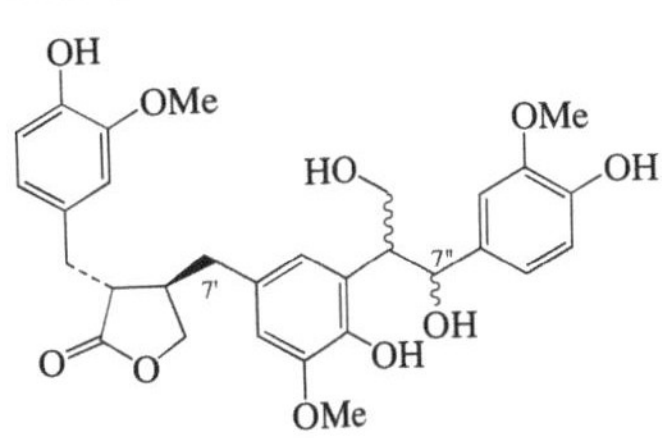

Relative Configuration

[64855-00-1]

$C_{30}H_{34}O_{10}$ 554.592

Isol. from seeds of *Arctium lappa*. Amorph. powder. $[\alpha]^{20}_{D}$ -55 (c, 1 in MeOH)

7″-Ketone

Arctignan B

[155661-09-9]

$C_{30}H_{32}O_{10}$ 552.576

Constit. of *Arctium lappa*. Amorph. powder. $[\alpha]_D$ +82.3 (c, 0.84 in MeOH). Not to be confused with Arctignan B, A45

7′ξ-Hydroxy

Arctignan C‡

$C_{30}H_{34}O_{11}$ 570.591

Constit. of seeds of *Arctium lappa*. Oil. $[\alpha]^{26}_{D}$ -32.1 (c, 0.27 in MeOH). Not to be confused with Arctignan C in L5. λ_{max} 282 (log ε 3.91) (MeOH).

Ichihara, A. *et al., Agric. Biol. Chem.*, 1977, **41**, 1813-1814 (*Lappaol C*).
Umehara, K. *et al., Chem. Pharm. Bull.*, 1993, **41**, 1774-1779 (*Arctignan B*).
Qin, Z.-B. *et al., Phytochem. Lett.*, 2019, **34**, 43-49 (*Arctignan C, cd*)

Lappaol D L5

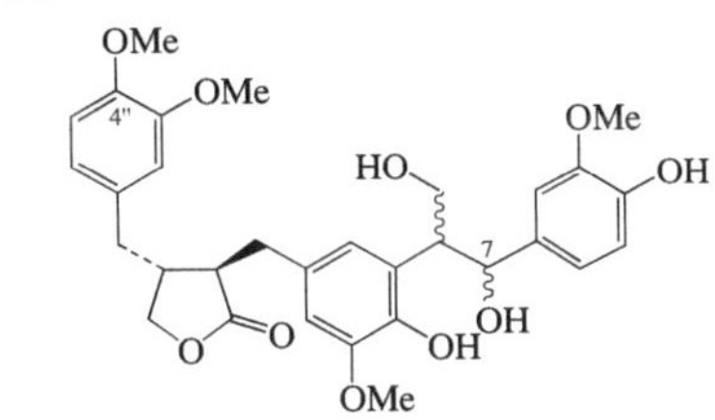

Relative Configuration

[64855-01-2]

$C_{31}H_{36}O_{10}$ 568.619

Constit. of the seeds of *Arctium lappa*. Amorph. powder. $[\alpha]^{25}_{D}$ -50.4 (c, 0.8 in MeOH). λ_{max} 230 (log ε 4.3); 282 (log ε 4) (MeOH)

4″-O-De-Me

Isolappaol C

[929905-15-7]

$C_{30}H_{34}O_{10}$ 554.592

Constit. of the seeds of *Arctium lappa*. Amorph. powder. $[\alpha]^{25}_{D}$ -43.6 (c, 0.1 in MeOH). λ_{max} 230 (log ε 4.4); 281 (log ε 4) (MeOH)

7-Ketone, 4″-O-de-Me

Arctignan C

[155661-10-2]

$C_{30}H_{32}O_{10}$ 552.576

Constit. of the fruit of *Arctium lappa*. Amorph. powder. $[\alpha]_D$ +63.8 (c, 0.91 in MeOH).

Ichihara, A. *et al., Agric. Biol. Chem.*, 1977, **41**, 1813-1814 (*Lappaol D*).
Umehara, K. *et al., Chem. Pharm. Bull.*, 1993, **41**, 1774-1779 (*Arctignan C*).
Park, S.Y. *et al., Chem. Pharm. Bull.*, 2007, **55**, 150-152 (*Isolappaol C*)

Lappaol E L6

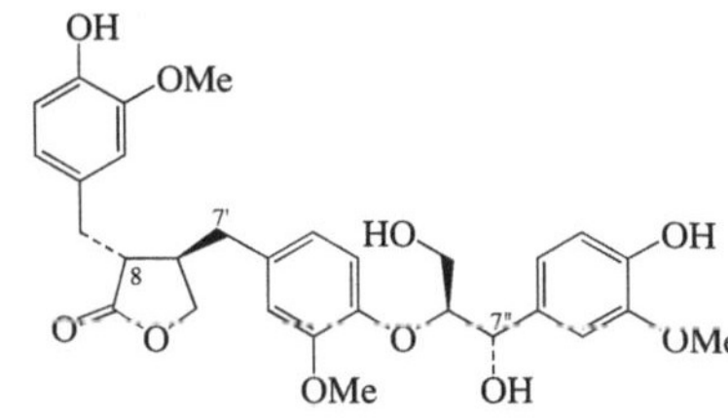

Absolute Configuration

[64855-02-3]

$C_{30}H_{34}O_{10}$ 554.592

Isol. from seeds of *Arctium lappa*. $Ca^{2\oplus}$ antagonist. $[\alpha]^{20}_{D}$ -26.56 (c, 1.0 in MeOH)

7′R-Hydroxy

7′R-Hydroxylappaol E

[189194-13-6]

$C_{30}H_{34}O_{11}$ 570.591

Constit. of *Tsuga heterophylla*. Amorph. powder. $[\alpha]^{25}_{D}$ -3.7 (c, 0.3 in MeOH). λ_{max} 208 (log ε 4.63); 230 (log ε 4.4); 281 (log ε 4.01) (MeOH)

8S-Hydroxy

8-Hydroxylappaol E

[845890-70-2]

$C_{30}H_{34}O_{11}$ 570.591

Constit. of *Picea abies*. Stereochem. not fully established

7″-Epimer, 7′R-hydroxy

7″-Epi-7′R-hydroxylappaol E

[189194-14-7]

$C_{30}H_{34}O_{11}$ 570.591

Constit. of *Tsuga heterophylla*. Amorph. powder (as penta-Ac). $[\alpha]^{25}_{D}$ +48 (c, 0.2 in MeOH) (penta-Ac)

7″-Epimer, 7′S-hydroxy

7″-Epi-7′S-hydroxylappaol E

[845890-82-6]

$C_{30}H_{34}O_{11}$ 570.591

Constit. of *Picea abies*.

Ichihara, A. *et al., Agric. Biol. Chem.*, 1977, **41**, 1813-1814 (*Lappaol E*).
Ichikawa, K. *et al., Chem. Pharm. Bull.*, 1986, **34**, 3514-3517 (*activity*).
Kawamura, F. *et al., Phytochemistry*, 1997, **44**, 1351-1357 (*Tsuga heterophylla constits*).
Kawamura, F. *et al., Phytochemistry*, 2000, **54**, 439-444 (*Tsuga heterophylla constits*).
Willfoer, S. *et al., Holzforschung*, 2004, **58**, 345-354 (*Picea abies constits*)

Lappaol F L7

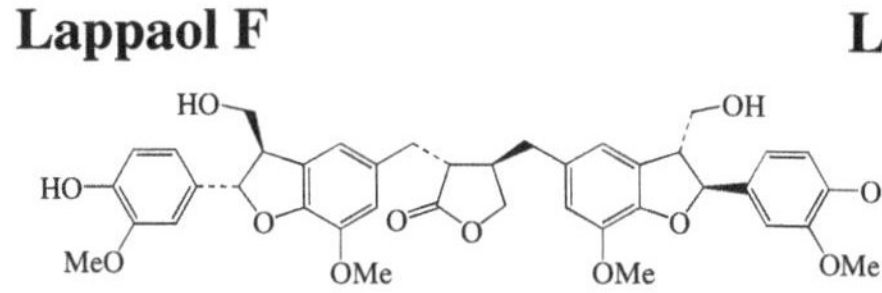

Absolute Configuration

[69394-17-8]

$C_{40}H_{42}O_{12}$ 714.764

Related to Arctignan D, A46 and Arctignan E, A47. Constit. of *Arctium lappa*. Isol from *Amauroderma rude*. Anticancer agent. NO prodn. inhibitor. Amorph. powder. $[\alpha]_D^{20}$ +13.5 (c, 1.33 in MeOH). λ_{max} 229 (log ε 4.2); 282 (log ε 4.0) (MeOH).

Ichihara, A. *et al.*, *Tetrahedron Lett.*, 1978, **19**, 3035-3038 (*Lappaol F*).
Ichihara, A. *et al.*, *Tetrahedron Lett.*, 1979, **20**, 3735-3736 (*Lappaol F, abs config*).
Park, S.Y. *et al.*, *Chem. Pharm. Bull.*, 2007, **55**, 150-152 (*Lappaol F, activity*).
Sun, Q. *et al.*, *Mol. Cancer Ther.*, 2014, **13**, 49-59 (*Lappaol F, activity*).
Dong, M. *et al.*, *Rec. Nat. Prod.*, 2019, **13**, 296-300 (*Amauroderma rude constit*)

Lappaol H L8

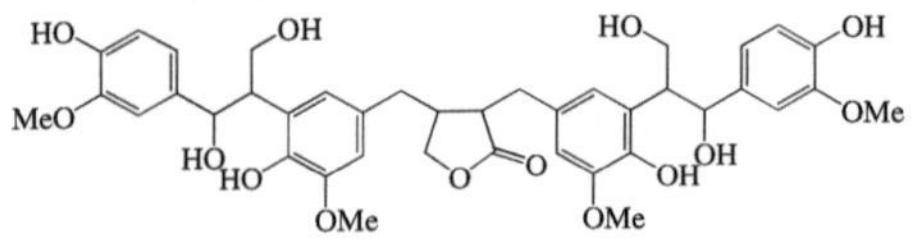

[69394-18-9]

$C_{40}H_{46}O_{14}$ 750.794

Constit. of *Arctium lappa*. Powder. $[\alpha]_D^{20}$ -46.6 (c, 1.33 in MeOH).

Ichihara, A. *et al.*, *Tetrahedron Lett.*, 1978, **19**, 3035-3038 (*Lappaol H*)

Larixnaphthaone L9

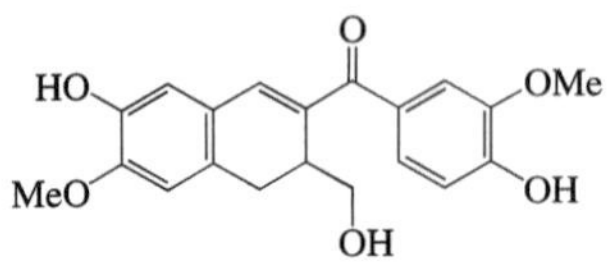

2′,9-Cyclo-4,4′,9′-trihydroxy-3,5′-dimethoxy-8,8′-lign-8-en-7-one. 4,4′,9′-Trihydroxy-3,5′-dimethoxy-7-oxo-2′,9-cyclolign-8-ene

[864452-91-5]

$C_{20}H_{20}O_6$ 356.374

Constit. of the bark of *Larix olgensis* var. *koreana*. Amorph. powder (MeOH). Mp 114 - 115°. $[\alpha]_D^{20}$ -56.1 (c, 0.66 in MeOH).

Yang, B. *et al.*, *J. Nat. Prod.*, 2005, **68**, 1175-1179 (*Larixnaphthaone*)

Larixsin L10

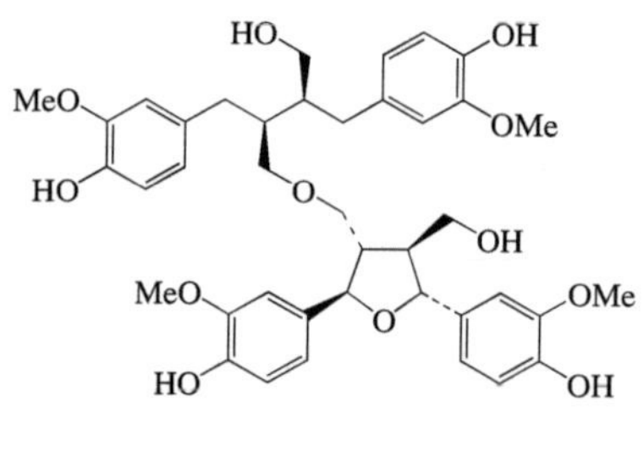

Relative Configuration

[864452-92-6]

$C_{40}H_{48}O_{12}$ 720.812

Heterodimer of 3,3′-Dimethoxy-8,8′-lignan-4,4′,9,9′-tetrol, D264 and Neoolivil in E86. Constit. of the bark of *Larix olgensis* var. *koreana*. Powder (MeOH). $[\alpha]_D^{20}$ -19.1 (c, 0.68 in MeOH).

Yang, B. *et al.*, *J. Nat. Prod.*, 2005, **68**, 1175-1179 (*Larixsin*)

Launeugenol L11

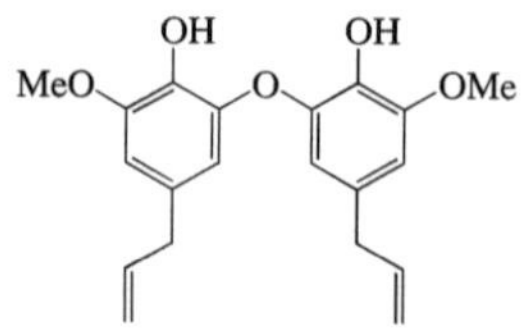

2,2′-Oxybis[6-methoxy-4-(2-propenyl)phenol], CAS. *2,2′-Dihydroxy-3,3′-dimethoxy-5,5′-di-(2-propenyl)diphenyl ether. 5,5′-Dimethoxy-3,3′-oxyligna-8,8′-diene-4,4′-diol. 4,4′-Dihydroxy-5,5′-dimethoxy-3,3′-oxyneoligna-8,8′-diene*

[1632539-74-2]

$C_{20}H_{22}O_5$ 342.391

Constit. of *Launaea intybacea*. Oil. λ_{max} 206 (log ε 4.58); 272 (log ε 4.46) (MeOH).

Saleem, M. *et al.*, *J. Chem. Soc. Pak.*, 2013, **35**, 912-916 (*Launeugenol*)

Lavandudiphenyl A L12

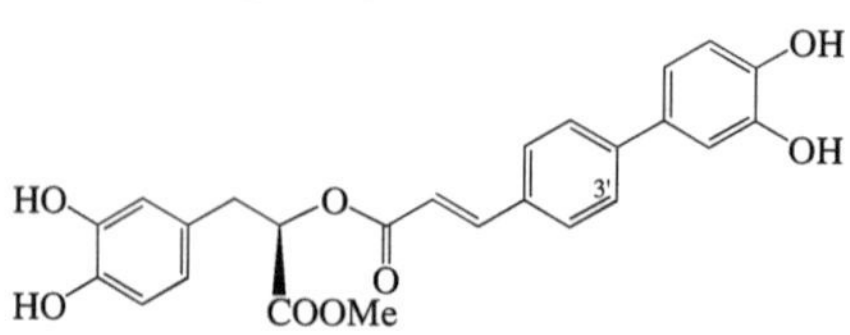

Absolute Configuration

$C_{25}H_{22}O_8$ 450.443

Constit. of *Lavandula angustifolia*. Moderate antioxidant. Amorph. yellow powder (MeOH). $[\alpha]_D^{25}$ +18 (c, 0.1 in MeOH). λ_{max} 283 (log ε 3.89); 334 (log ε 4.05) (MeOH)

3′-(3,4-Dihydroxyphenyl) analogue
Lavandudiphenyl B

$C_{25}H_{22}O_8$ 450.443

Constit. of *Lavandula angustifolia*. Moderate antioxidant. Amorph. yellow powder (MeOH). $[\alpha]_D^{25}$ +14 (c, 0.1 in MeOH). λ_{max} 276 (log ε 4.12) (MeOH).

Yadikar, N. *et al.*, *Phytochem. Lett.*, 2018, **23**, 149-154 (*Lavandudiphenyls A,B, activity*)

Lavandufluoren L13

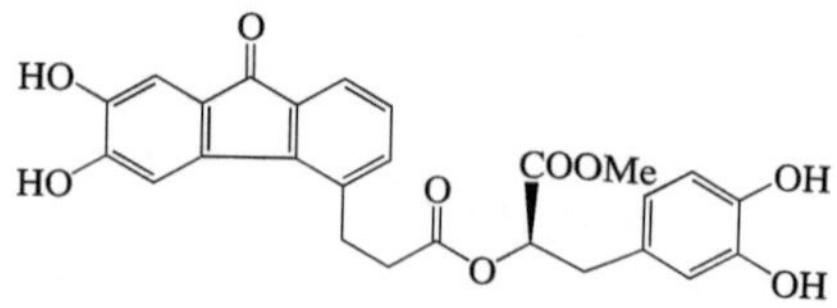

Absolute Configuration

$C_{26}H_{22}O_9$ 478.453

Constit. of *Lavandula angustifolia*. Strong antioxidant. Amorph. yellow powder (MeOH). $[\alpha]_D^{25}$ -7.0 (c, 0.1 in MeOH). λ_{max} 280 (log ε 4.28) (MeOH).

Yadikar, N. *et al.*, *Phytochem. Lett.*, 2018, **23**, 149-154 (*Lavandufluoren, activity*)

Lavandufurandiol L14

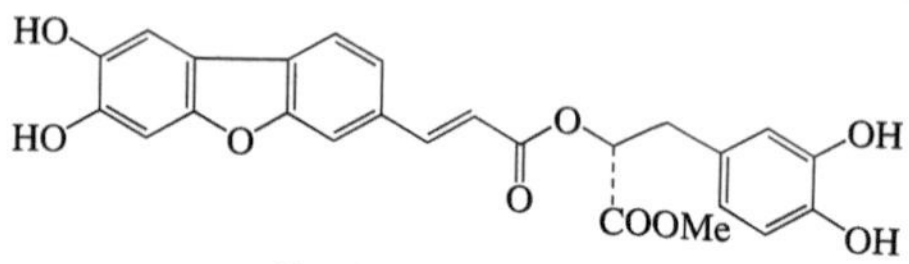

Absolute Configuration

$C_{25}H_{20}O_9$ 464.426

Constit. of *Lavandula angustifolia*. Strong antioxidant. Amorph. yellow powder (Me_2CO). $[\alpha]_D^{25}$ +49.4 (c, 0.09 in MeOH). λ_{max} 354 (log ε 4.02) (MeOH).

Yadikar, N. *et al.*, *Phytochem. Lett.*, 2018, **23**, 149-154 (*Lavandufurandiol, activity*)

Lavandupyrone A L15

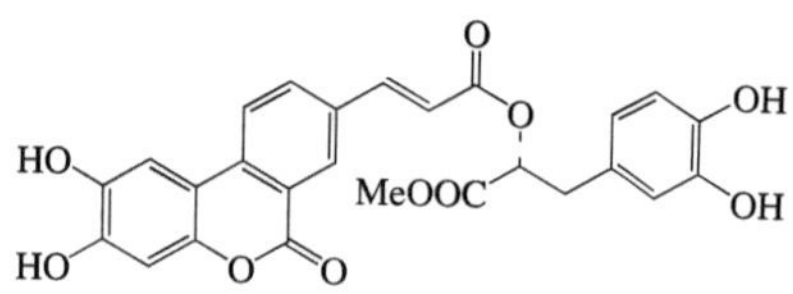

Absolute Configuration

$C_{26}H_{20}O_{10}$ 492.436

Constit. of *Lavandula angustifolia*. Amorph. yellow powder (MeOH). $[\alpha]_D^{25}$ +33 (c, 0.1 in MeOH). λ_{max} 287 (log ε 4.08) (MeOH).

Yadikar, N. *et al.*, *Phytochem. Lett.*, 2018, **23**, 149-154 (*Lavandupyrone A, activity*)

Lehmbachol D L16

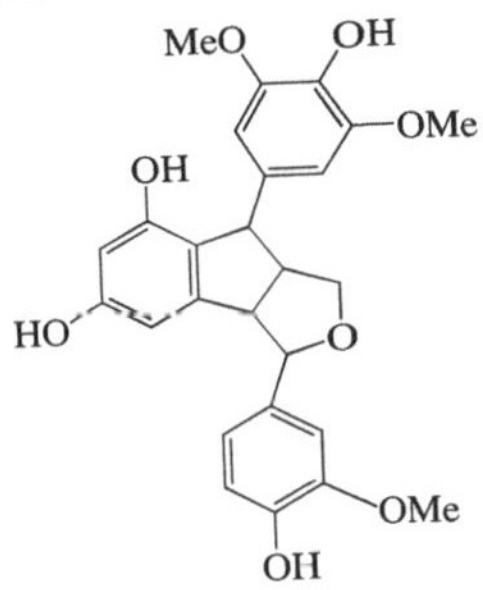

[189295-09-8]

$C_{26}H_{26}O_8$ 466.486

Stilbenolignan. Related to Gnetifolin F, G20 and Maackolin, M1. Constit. of the bark of *Salacia lehmbachii*. Amorph. yellow powder. $[\alpha]_D^{25}$ +25.1 (c, 0.9 in MeOH). λ_{max} 281 (log ε 3.8); 315 (log ε 3.2) (no solvent reported).

Kawazoe, K. *et al.*, *Phytochemistry*, 1997, **44**, 1569-1573 (*Lehmbachol D*)

Lepidovitroside L17

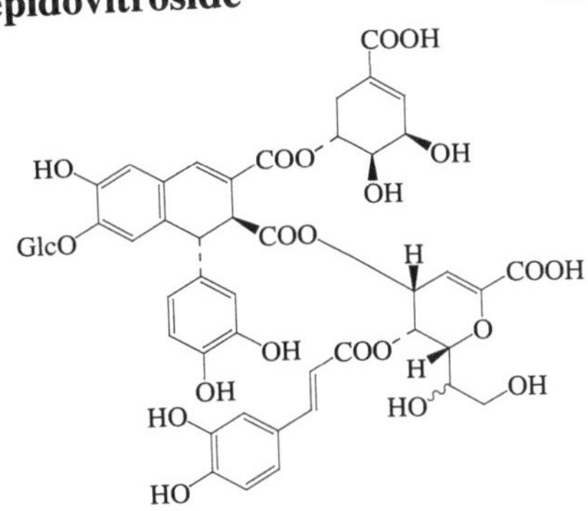

[927207-12-3]

$C_{48}H_{48}O_{26}$ 1040.886

Constit. of *Lepidozia vitrea*. Powder. $[\alpha]_D^{25}$ -0.7 (c, 0.8 in MeOH).

Ma, B. *et al.*, *Helv. Chim. Acta*, 2007, **90**, 58-62 (*Lepidovitroside*)

Leptolepisol B L18

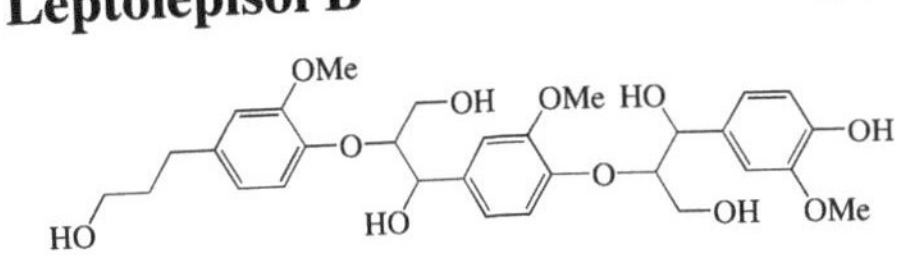

[71332-94-0]

$C_{30}H_{38}O_{11}$ 574.623

Leptolepisol B may be used as a synonym for all compds. with this planar struct. since isolations are not accompanied by stereochem. data. Similar to Casearialignan E, C37. Constit. of *Kadsura longipedunculata* and *Larix leptolepis*. Amorph.. $[\alpha]_D^{27}$ -1.6 (c, 0.6 in MeOH). λ_{max} 280 (ε 6056) (EtOH).

Miki, K. *et al.*, *Tetrahedron Lett.*, 1979, **20**, 799-802 (*Leptolepisol B*).
Chen, J.-B. *et al.*, *Zhongcaoyao*, 2015, **46**, 178-184 (*Kadsura constit*)

Leptolepisol E L19

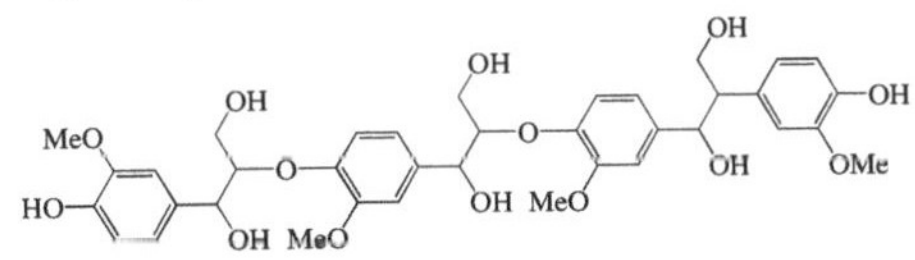

[72468-38-3]

$C_{37}H_{44}O_{14}$ 712.745

Constit. of the wood of *Larix leptolepis*.

Miki, K. *et al.*, *Mokuzai Gakkaishi*, 1979, **25**, 678-679 (*Leptolepisol E*).
Miki, K. *et al.*, *Mokuzai Gakkaishi*, 1980, **26**, 633-636 (*struct*)

3,3′-Ligna-8,8′-diene-4,4′-diol L20

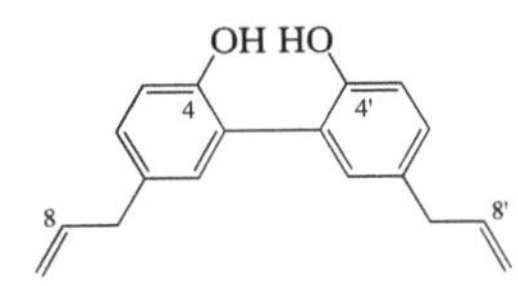

2,2′-Dihydroxy-5,5′-di-2-propenylbiphenyl. 5,5′-Di-2-propenyl-[1,1′-biphenyl]-2,2′-diol, CAS. *5,5′-Diallyl-2,2′-dihydroxybiphenyl. 4,4′-Dihydroxy-3,3′-neoligna-8,8′-diene.* ***Magnolol.*** *2,2′-Bichavicol. FEMA 4559. NSC 293099*

[528-43-8]

$C_{18}H_{18}O_2$ 266.34

Constit. of *Beilschmiedia volckii*, *Cercidiphyllum japonicum*, *Magnolia* spp., *Manglietia hookeri*, *Sassafras randaiense* and *Streblus asper*. Flavouring and cosmetic ingredient. Exhibits antibacterial, antiviral and antifungal activities. Antiplatelet agent. CNS depressant with serotonergic activity. NO prodn. inhibitor. Shows antiinflammatory and analgesic effects. Phytoalexin. Inhibits bone-resorbing activity of osteoclasts. Cannabinoid receptor activator. Exhibits cytotoxic activity. Potent acetylcholinesterase inhibitor. Cryst. (Et_2O/hexane). Poorly sol. hexane. Mp 101.5 - 102°. Log P 4.84 (calc). λ_{max} 293 (ε 7950) (MeOH) (Berdy). [base] λ_{max} 223 (ε 29300); 279 (ε 6420); 320 (ε 10700) (MeOH/NaOH) (Berdy). [neutral] λ_{max} 204 (ε 49500); 275 (ε 5440) (EtOH) (Berdy)

4-O-(Ethoxycarbonyl)

4-[(Ethoxycarbonyl)oxy]magnolol. *Magnolol 4-ethylcarbonate*

[2250441-64-4]

$C_{21}H_{22}O_4$ 338.403

Constit. of the branches of *Illicium burmanicum*. Yellow oil

4-Et ether

4′-Ethoxy-3,3′-ligna-8,8′-dien-4-ol. 4′-Ethoxy-4-hydroxy-3,3′-ligna-8,8′-diene. ***4-O-Ethylmagnolol***

[1356708-23-0]

$C_{20}H_{22}O_2$ 294.394

Constit. of twigs of *Magnolia denudata*. Inhibits fMLP/CB-induced elastase release. Yellowish oil. λ_{max} 211 (log ε 4.66); 288 (log ε 3.91) (MeOH)

4-O-(2-Propenyl) ether

5,5′-Diallyl-2′-(allyloxy)biphenyl-2-ol

[1416567-12-8]

$C_{21}H_{22}O_2$ 306.405

Isol. from *Magnolia officinalis*. Exhibits potent antiinflammatory activity. Yellow oil. λ_{max} 205 (log ε 4.72); 286 (log ε 3.91) (MeOH)

4-O-(3-Methyl-2-butenyl) ether

5,5′-Diallyl-2′-(3-methyl-2-butenyloxy)biphenyl-2-ol

[1416567-13-9]

$C_{23}H_{26}O_2$ 334.459

Isol. from *Magnolia officinalis*. Exhibits potent antiinflammatory activity. Oil. λ_{max} 206 (log ε 4.73); 287 (log ε 3.89) (MeOH)

4-Bornyl ether

Bornylmagnolol

[92911-80-3]

$C_{28}H_{34}O_2$ 402.578

Constit. of *Magnolia officinalis*. Viscous oil. $[\alpha]_D^{26}$ -7.6 (c, 1.67 in $CHCl_3$)

8,8′,9,9′-Tetrahydro

5,5′-Dipropyl-[1,1′-biphenyl]-2,2′-diol. 3,3′-Lignan-4,4′-diol. 4,4′-Dihydroxy-3,3′-lignan. Magnolignan®

[20601-85-8]

$C_{18}H_{22}O_2$ 270.372

Down-regulator of melanin synth. Skin-lightening agent. Powder. Mp 144°.

Erdtman, H. *et al.*, *Acta Chem. Scand.*, 1957, **11**, 1060-1061 (*Magnolol, synth*).
Runeberg, J. *Acta Chem. Scand.*, 1958, **12**, 188-192 (*synth*).
Clark, A.M. *et al.*, *J. Pharm. Sci.*, 1981, **70**, 951-952 (*Magnolol, activity*).
Wang, Y. *et al.*, *J. Chin. Chem. Soc. (Peking)*, 1982, **30**, 215-221 (*cryst struct*).
Namba, T. *et al.*, *Planta Med.*, 1982, **44**, 100-106 (*Magnolol, activity*).
El-Feraly, F.S. *et al.*, *J. Nat. Prod.*, 1983, **46**, 493-498 (*Sassafras randaiense constit, activity*).

Watanabe, K. *et al.*, *Planta Med.*, 1983, **49**, 103-108 (*Magnolol, activity*).
Hattori, M. *et al.*, *Chem. Pharm. Bull.*, 1986, **34**, 158-167 (*metab*).
Takasugi, M. *et al.*, *Phytochemistry*, 1986, **25**, 2751-2752 (*activity*).
Teng, C.-M. *et al.*, *Thromb. Res.*, 1988, **50**, 757-765 (*pharmacol*).
Yahara, S. *et al.*, *Chem. Pharm. Bull.*, 1991, **39**, 2024-2036 (*Bornylmagnolol*).
Nitao, J.K. *et al.*, *Phytochemistry*, 1991, **30**, 2193-2195 (*isol, activity*).
Wang, J.-P. *et al.*, *Naunyn-Schmiedeberg's Arch. Pharmacol.*, 1992, **346**, 707-712 (*pharmacol*).
Tsai, T.-H. *et al.*, *Drug Metab. Dispos.*, 1994, **22**, 518-521 (*hplc, pharmacokinet*).
Agharahimi, M.R. *et al.*, *J. Org. Chem.*, 1995, **60**, 1856-1863 (*struct, synth*).
Tsai, T.-H. *et al.*, *Neurosci. Lett.*, 1995, **186**, 49-52 (*pharmacol*).
Chang, B.S. *et al.*, *Planta Med.*, 1998, **64**, 367-369 (*activity*).
Gu, W. *et al.*, *Tetrahedron: Asymmetry*, 1998, **9**, 1377-1380 (*synth*).
Kim, Y.-K. *et al.*, *Planta Med.*, 1999, **65**, 291-292 (*activity*).
Bang, K.H. *et al.*, *Arch. Pharmacal Res.*, 2000, **23**, 46-49 (*activity*).
Ho, K.-Y. *et al.*, *Phytother. Res.*, 2001, **15**, 139-141 (*activity*).
Wang, X. *et al.*, *J. Chromatogr., A*, 2004, **1036**, 171-175 (*Magnolol, sepn*).
Kong, Z.-L. *et al.*, *Bioorg. Med. Chem. Lett.*, 2005, **15**, 163-166 (*synth*).
Kotani, A. *et al.*, *Chem. Pharm. Bull.*, 2005, **53**, 319-322 (*detn, hplc*).
Pat. Coop. Treaty (WIPO), 2011, 131 445 (*activity, use*).
Rempel, V. *et al.*, *ACS Med. Chem. Lett.*, 2013, **4**, 41-45 (*activity*).
Li, J. *et al.*, *Bioorg. Med. Chem. Lett.*, 2013, **23**, 2238-2244 (*Streblus asper constit, activity*).
Narita, S. *et al.*, *J. Chem.*, 2013, 686242:1-5 (*Magnolignan, synth*).
Kuo, W.-L. *et al.*, *Phytochemistry*, 2013, **85**, 153-160 (*2-propenyl ether, 3-methyl-2-butenyl ether, activity*).
Seo, K.-H. *et al.*, *Planta Med.*, 2013, **79**, 1335-1340 (*activity*).
Chung, C.-Y. *et al.*, *Chem. Biodiversity*, 2015, **12**, 1263-1270 (*4-O-Ethylmagnolol, activity*).
Lu, S.-H. *et al.*, *J. Nat. Prod.*, 2015, **78**, 61-68 (*activity*).
Yang, C. *et al.*, *Mini-Rev. Med. Chem.*, 2016, **16**, 404-426 (*rev, synth, activity*).
Lu, H. *et al.*, *Nat. Prod. Res.*, 2017, **31**, 829-834 (*activity*).
Tian, X. *et al.*, *RSC Adv.*, 2017, **7**, 10849-10854 (*4-ethylcarbonate*).
Cassiano, C. *et al.*, *Front. Chem.*, 2019, **7**, 53:1-8 (*activity, props*)

3,3′-Ligna-8,8′-diene-4,6′-diol L21

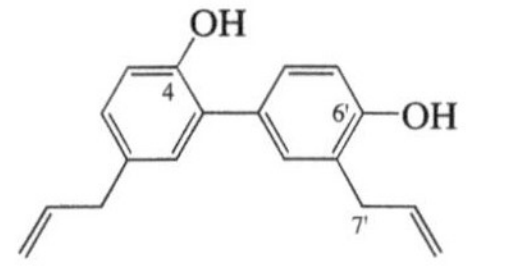

3′,5-Di-2-propenyl-2,4′-biphenyldiol. 2,4′-Dihydroxy-3′,5-di-2-propenylbiphenyl. 3′,5-Diallyl-2,4′-dihydroxybiphenyl. 4,6′-Dihydroxy-3,3′-neoligna-8,8′-diene. ***Honokiol***
[35354-74-6]
$C_{18}H_{18}O_2$ 266.34
Constit. of *Artemisia argyri*, *Magnolia officinalis* and other *Magnolia* spp.. Shows antimicrobial and muscle-relaxant activity. Antineoplastic agent. Induces cell-cycle arrest and apoptosis in myeloid leukaemia. Activates 15-LOX and thereby inhibits peroxisome proliferator-activated receptor-γ (PPAR-γ) and COX-2-dependent signals. Shows antiinflammatory props.. Poorly sol. hexane. Mp 87.5°. Log P 4.84 (calc). λ_{max} 295 (ε 7080) (MeOH). [base] λ_{max} 305 (ε 12590) (MeOH/NaOH)

6′-Me ether
3′,5-Diallyl-2-hydroxy-4′-methoxybiphenyl. 6′-Methoxy-3,3′-ligna-8,8′-dien-4-ol. ***4-O-Methylhonokiol***
[68592-15-4]
$C_{19}H_{20}O_2$ 280.367
Constit. of *Magnolia grandiflora*, *Magnolia officinalis* and *Magnolia virginiana*. Shows antibacterial and antiinflammatory activities. Inhibitor of COX-1 and COX-2. Also inhibits NO and PGF_1 prodn. in macrophages. Pale yellow oil. Poorly sol. hexane. λ_{max} 259 (ε 19900); 293 (ε 15800) (MeOH). [base] λ_{max} 262 (ε 19900) (MeOH/NaOH)

8′,9′-Dihydro, 8′ξ,9′-dihydroxy
3,3′-Lign-8′-ene-4′,6,8,9-tetrol. ***8′,9′-Dihydroxyhonokiol***
[1472626-93-9]
$C_{18}H_{20}O_4$ 300.354
Constit. of *Magnolia officinalis*. Yellow powder. Possible artifact

Δ^7′-Isomer(E-)
*3′-(1-Propenyl)-5-(2-propenyl)-2,4′-biphenyldiol. 3,3′-Ligna-7′,8-diene-4,6′-diol. Δ^7′-****Honokiol***
[1178919-05-5]
$C_{18}H_{18}O_2$ 266.34
Constit. of *Magnolia officinalis*. Exhibits potent antiinflammatory activity. Amorph. powder. λ_{max} 205 (log ε 4.47); 223 (log ε 4.48); 245 (log ε 4.37); 298 (log ε 3.97) (MeOH).

Fujita, M. *et al.*, *Chem. Pharm. Bull.*, 1972, **20**, 212-213 (*Honokiol*).
El-Feraly, F.S. *et al.*, *Lloydia*, 1978, **41**, 442-498 (*4′-Me ether*).
Clark, A.M. *et al.*, *J. Pharm. Sci.*, 1981, **70**, 951-952 (*activity*).
Watanabe, K. *et al.*, *Planta Med.*, 1983, **49**, 103-108 (*activity*).
Nitao, J.K. *et al.*, *Phytochemistry*, 1991, **30**, 2193-2195 (*4-O-Methylhonokiol*).
Fullmer, M.J. *et al.*, *Acta Crystallogr., Sect. C*, 1994, **50**, 1966-1967 (*Honokiol, cryst struct*).
Esumi, T. *et al.*, *Bioorg. Med. Chem. Lett.*, 2004, **14**, 2621-2625 (*synth, activity*).
Wang, X. *et al.*, *J. Chromatogr., A*, 2004, **1036**, 171-175 (*Honokiol, sepn*).
Kotani, A. *et al.*, *Chem. Pharm. Bull.*, 2005, **53**, 319-322 (*detn, hplc*).
Schuehly, W. *et al.*, *Bioorg. Med. Chem.*, 2009, **17**, 4459-4465 (*4-O-Methylhonokiol, synth, activity*).
Chen, C.-M. *et al.*, *Tetrahedron Lett.*, 2009, **50**, 1151-1152 (*synth*).
Liu, S.H. *et al.*, *Br. J. Pharmacol.*, 2010, **160**, 1963-1972 (*activity*).
Chao, L.K. *et al.*, *J. Agric. Food Chem.*, 2010, **58**, 3472-3478 (*activity*).
Kotani, H. *et al.*, *J. Nat. Prod.*, 2010, **73**, 1332-1336 (*isol*).
Schuehly, W. *et al.*, *J. Nat. Prod.*, 2010, **73**, 1381-1384 (*occur*).
Denton, R.M. *et al.*, *Tetrahedron*, 2010, **66**, 8029-8035 (*synth*).
Kwak, J.-H. *et al.*, *Tetrahedron*, 2011, **67**, 9401-9404 (*4-O-Methylhonokiol, synth*).
Lee, B. *et al.*, *Bioorg. Med. Chem.*, 2012, **20**, 2860-2868 (*4-O-Methylhonokiol, activity*).
Tripathi, S. *et al.*, *Bioorg. Med. Chem. Lett.*, 2012, **22**, 216-221 (*synth*).
Tian, W. *et al.*, *Pharmazie*, 2012, **67**, 811-816 (*rev, activity*).
Zhang, B. *et al.*, *J. Chin. Pharm. Sci.*, 2013, **22**, 420-426 (*8′,9′-Dihydroxyhonokiol*).
Kuo, W.-L. *et al.*, *Phytochemistry*, 2013, **85**, 153-160 (*Δ^7′-Honokiol, activity*).
Bernaskova, M. *et al.*, *Molecules*, 2014, **19**, 1223-1237 (*antitumour activity, bibl*).
Reddy, B.V.S. *et al.*, *Tetrahedron Lett.*, 2014, **55**, 1049-1051 (*synth*).
Srinivas, J. *et al.*, *Tetrahedron Lett.*, 2014, **55**, 4295-4297 (*synth*).
Harada, K. *et al.*, *Tetrahedron Lett.*, 2014, **55**, 6001-6003 (*synth*).
Li, H.-Y. *et al.*, *J. Cell Biochem.*, 2015, **116**, 287-298 (*antileukaemic activity*).
Yang, C. *et al.*, *Mini-Rev. Med. Chem.*, 2016, **16**, 404-426 (*rev, synth, activity*).
Wright, A.M. *et al.*, *Tetrahedron Lett.*, 2016, **57**, 3441-3443 (*synth*).
Rauf, A. *et al.*, *Biomed. Pharmacother.*, 2018, **107**, 555-562 (*rev, activity*).
Sarrica, A. *et al.*, *Planta Med.*, 2018, **84**, 1151-1164 (*rev, activity, tox*).
Ong, C.P. *et al.*, *Cancers*, 2020, **12**, 48:1-44 (*rev, activity*).
Khan, P.R. *et al.*, *Tetrahedron Lett.*, 2020, **61**, 152229:1-2 (*synth*).
Chen, C. *et al.*, *Chin. J. Nat. Med.*, 2021, **19**, 481-490 (*rev*)

3,8′-Ligna-7,8′-diene-4,4′-diol L22

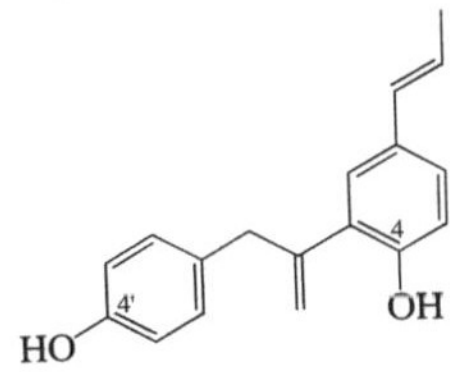

4,4′-Dihydroxy-3,8′-neoligna-7,8′-diene
$C_{18}H_{18}O_2$ 266.34
4-Me ether
4-[2-[2-Methoxy-5-(1-propenyl)phenyl]-2-propenyl]phenol, CAS. *4-Methoxy-3,8′-ligna-7,8′-dien-4′-ol. 4′-Hydroxy-4-methoxy-3,8′-ligna-7,8′-diene.* ***Hermosillol***
[110241-11-7]
$C_{19}H_{20}O_2$ 280.367
Constit. of *Krameria parvifolia* and *Krameria sonorae*. Oil
Di-Me ether
4,4′-Dimethoxy-3,8′-ligna-7,8′-diene. ***O-Methylhermosillol***
[138916-50-4]
$C_{20}H_{22}O_2$ 294.394
Constit. of *Krameria ixina*. Oil.
Dominguez, X.A. *et al.*, *Phytochemistry*, 1987, **26**, 1821-1824 (*Hermosillol*).
Achenbach, H. *et al.*, *Phytochemistry*, 1991, **30**, 3753-3757 (*O-Methylhermosillol*)

4,4′-Ligna-8,8′-diene-3,3′-diol L23

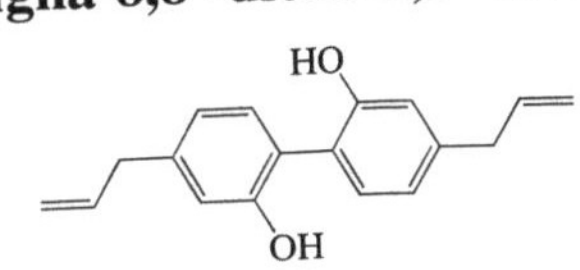

4,4′-Di(2-propenyl)-[1,1′-biphenyl]-2,2′-diol. 2,2′-Dihydroxy-4,4′-di-2-propenylbiphenyl. 3,3′-Dihydroxy-4,4′-neoligna-8,8′-diene. ***Tzumin B***
[2055788-35-5]
$C_{18}H_{18}O_2$ 266.340
Constit. of the bark of *Sassafras tzumu*. Exhibits potent acetylcholinesterase inhibitory activity. Cryst. (EtOAc/petrol). Mp 81.2°. $[\alpha]_D^{20}$ +26 (c, 0.01 in MeOH). λ_{max} 289; 360; 373 (MeOH).
Lu, H. *et al.*, *Nat. Prod. Res.*, 2017, **31**, 829-834 (*Tzumin B, activity*)

8,8′-Ligna-8,8′-diene-4,4′-diol L24

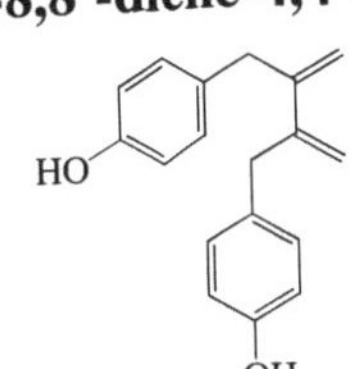

4,4′-[2,3-Bis(methylene)-1,4-butanediyl]bisphenol, CAS. *2,3-Bis(4-hydroxybenzyl)-1,3-butadiene. 4,4′-Dihydroxy-8,8′-lignadiene.*
Anolignan B
[158081-98-2]
$C_{18}H_{18}O_2$ 266.34
Constit. of *Anogeissus acuminata* and fruit of *Terminalia bellerica*. Inhibitor of HIV-1 reverse transcriptase. Cryst.. Mp 147°. Mp 160 - 165°. λ_{max} 230 (ε 17782); 280 (ε 3715) (MeOH) (Berdy).
Rimando, A.M. *et al.*, *J. Nat. Prod.*, 1994, **57**, 896-904 (*Anolignan B*).
Mori, M. *et al.*, *J. Org. Chem.*, 2002, **67**, 224-226 (*synth, struct*)

2,9′-Ligna-7,7′-diene-3,3′,4,4′,5,5′,9-heptol L25

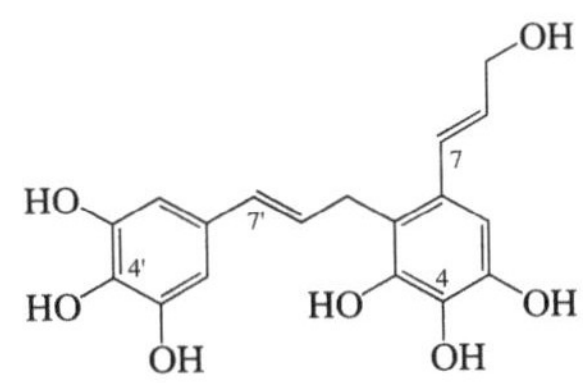

3,3′,4,4′,5,5′,9-Heptahydroxy-2,9′-neoligna-7,7′-diene
$C_{18}H_{18}O_7$ 346.335
(E,E)-form
3,3′,5,5′-Tetra-Me ether
3,3′,5,5′-Tetramethoxy-2,9′-ligna-7,7′-diene-4,4′,9-triol. 4,4′,9-Trihydroxy-3,3′,5,5′-tetramethoxy-2,9′-ligna-7,7′-diene
3,3′,5,5′-Tetra-Me ether, 4,4′-di-O-β-D-glucopyranoside
Tangshenoside III
[129277-39-0]
$C_{34}H_{46}O_{17}$ 726.725
Constit. of *Codonopsis lanceolata*, *Codonopsis tangshen* and *Saussurea involucrata*. Powder + $2H_2O$. $[\alpha]_D^{24}$ -22.1 (c, 1.69 in H_2O). λ_{max} 221 (log ε 4.6); 262 (log ε 4.4) (H_2O)
3,3′,5,5′-Tetra-Me ether, 9-O-[4-carboxy-3S-(β-D-glucopyranosyloxy)-3-methylbutanoyl], 4,4′-di-O-β-D-glucopyranoside
Tangshenoside IV
[129277-40-3]
$C_{46}H_{64}O_{26}$ 1032.992
Constit. of *Codonopsis lanceolata* and *Codonopsis tangshen*. Powder + 1½H_2O. $[\alpha]_D^{20}$ -32.2 (c, 0.8 in H_2O). λ_{max} 227 (log ε 4.4); 263 (log ε 4.4) (H_2O).
Yuda, M. *et al.*, *Phytochemistry*, 1990, **29**, 1989-1993 (*Tangshenosides III,IV*)

2,8′-Ligna-7,7′-diene-3′,4,4′,5,9,9′-hexol L26

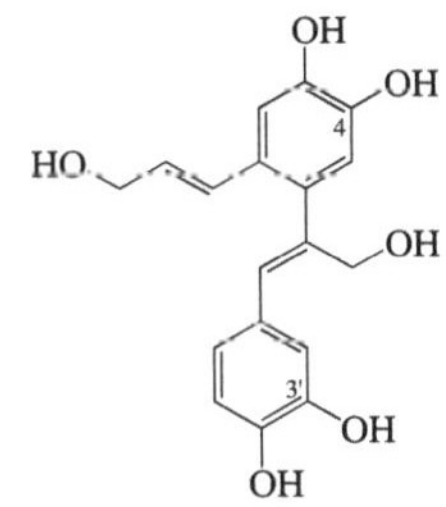

3′,4,4′,5,9,9′-Hexahydroxy-2,8′-neoligna-7,7′-diene
$C_{18}H_{18}O_6$ 330.336
(7E,7′Z)-form
3′,4-Di-Me ether
3′,4-Dimethoxy-2,8′-ligna-7,7′-diene-4′,5,9,9′-tetrol. 4′,5,9,9′-Tetrahydroxy-3′,4-dimethoxy-2,8′-ligna-7,7′-diene
3′,4-Di-Me ether, tetra-Ac
[221302-69-8]
$C_{28}H_{30}O_{10}$ 526.538
Constit. of degraded wood of *Pinus taeda*.
Peng, J. *et al.*, *Phytochemistry*, 1999, **50**, 659-666 (*Pinus taeda constit*)

2,9′-Ligna-7,7′-diene-3,3′,4,4′,5,9-hexol L27

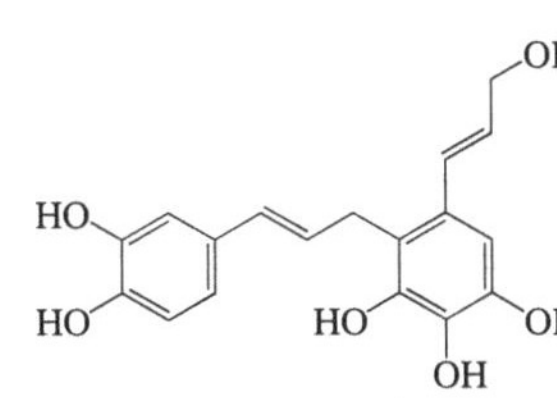

3,3′,4,4′,5,9-Hexahydroxy-2,9′-neoligna-7,7′-diene
$C_{18}H_{18}O_6$ 330.336
(7E,7′E)-form
3,3′,5-Tri-Me ether
3,3′,5-Trimethoxy-2,9′-ligna-7,7′-diene-4,4′,9-triol. 4,4′,9-Trihydroxy-3,3′,5-trimethoxy-2,9′-ligna-7,7′-diene
3,3′,5-Tri-Me ether, 4,4′-di-O-β-D-glucopyranoside
Aralialignan
[2077212-65-6]
$C_{33}H_{44}O_{16}$ 696.699
Constit. of the roots of *Aralia echinocaulis*. Powder. $[\alpha]_D^{20}$ -58.1 (c, 0.12 in MeOH).
Li, Y. *et al.*, *Nat. Prod. Res.*, 2017, **31**, 1047-1051 (*Aralialignan*)

2,8′-Ligna-7,7′-diene-3′,4,4′,5,9-pentol L28

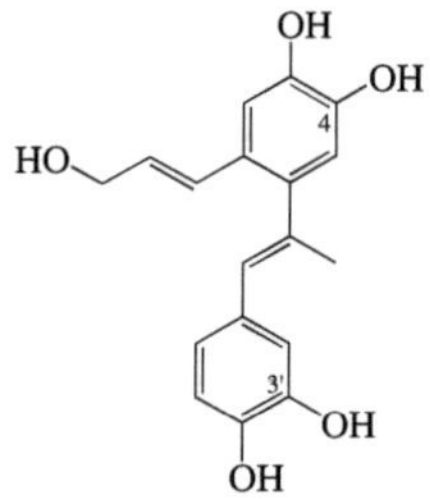

3′,4,4′,5,9-Pentahydroxy-2,8′-neoligna-7,7′-diene

$C_{18}H_{18}O_5$ 314.337

(7E,7′E)-form

3′,4-Di-Me ether

3′,4-Dimethoxy-2,8′-ligna-7,7′-diene-4′,5,9-triol. 4′,5,9-Trihydroxy-3′,4-dimethoxy-2,8′-ligna-7,7′-diene

3′,4-Di-Me ether, tri-Ac

[221302-70-1]

$C_{26}H_{28}O_8$ 468.502

Constit. of degraded wood of *Pinus taeda*.

Peng, J. *et al.*, *Phytochemistry*, 1999, **50**, 659-666 (*Pinus taeda constit*)

2,8′-Ligna-7,8′-diene-3′,4,4′,5,9-pentol L29

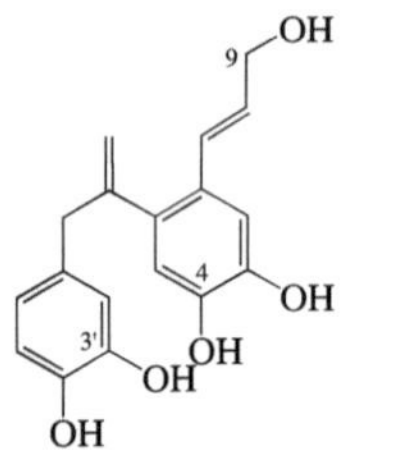

3′,4,4′,5,9-Pentahydroxy-2,8′-neoligna-7,8′-diene

$C_{18}H_{18}O_5$ 314.337

Lignan numbering shown

(E)-form

3′,4-Di-Me ether

3′,4-Dimethoxy-2,8′-ligna-7,8′-diene-4′,5,9-triol. 4′,5,9-Trihydroxy-3′,4-dimethoxy-2,8′-ligna-7,8′-diene

3′,4-Di-Me ether, tri-Ac

[221302-71-2]

$C_{26}H_{28}O_8$ 468.502

Constit. of degraded wood of *Pinus taeda*.

Peng, J. *et al.*, *Phytochemistry*, 1999, **50**, 659-666 (*Pinus taeda constit*)

2,9′-Ligna-7,7′-diene-3′,4,4′,5,9-pentol L30

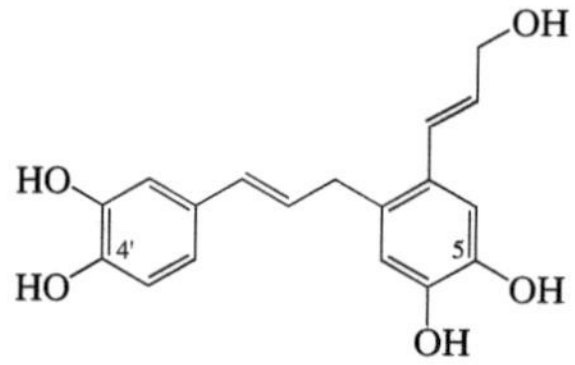

3′,4,4′,5,9-Pentahydroxy-2,9′-neoligna-7,7′-diene

$C_{18}H_{18}O_5$ 314.337

(7E,7′E)-form

3′,5-Di-Me ether

3′,5-Dimethoxy-2,9′-ligna-7,7′-diene-4,4′,9-triol. 4,4′,9-Trihydroxy-3′,5-dimethoxy-2,9′-ligna-7,7′-diene

3′,5-Di-Me ether, 4,4′-di-O-β-D-glucopyranoside

Balajaponin C

[1362086-47-2]

$C_{32}H_{42}O_{15}$ 666.673

Constit. of *Balanophora japonica*. Amorph. powder. $[\alpha]_D^{20}$ -168 (c, 0.2 in H_2O).

Zhou, T. *et al.*, *Planta Med.*, 2011, **77**, 477-481 (*Balajaponin C*)

2,2′-Ligna-8,8′-diene-4,4′,5,5′-tetrol L31

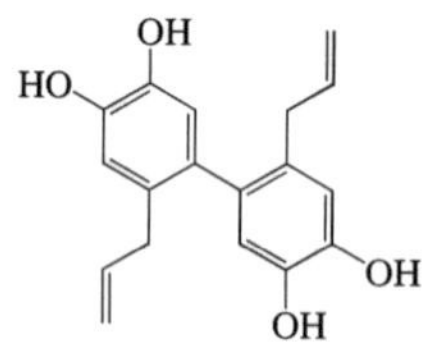

6,6′-Di-2-propenyl-3,3′,4,4′-biphenyltetrol. 6,6′-Diallyl-3,3′,4,4′-tetrahydroxybiphenyl. 4,4′,5,5′-Tetrahydroxy-2,2′-neoligna-8,8′-diene. 5,5′-Bis(4-allylcatechol).

Diallylcatechol

[1449376-76-4]

$C_{18}H_{18}O_4$ 298.338

Constit. of the roots of *Piper taiwanense*. Brownish oil. λ_{max} 284 (log ε 3.40) (MeOH).

Chen, S. *et al.*, *Phytochemistry*, 2013, **93**, 203-209 (*Diallylcatechol*)

2,3′-Ligna-8,8′-diene-4,4′,5,5′-tetrol L32

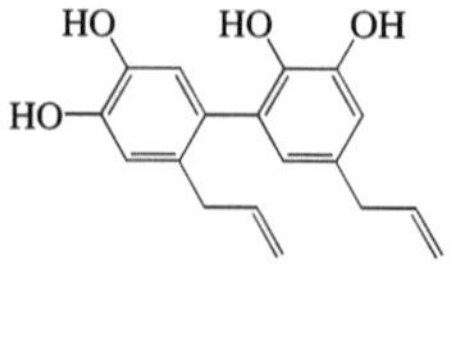

4,4′,5,5′-Tetrahydroxy-2,3′-neoligna-8,8′-diene. 5,6′-Di-2-propenyl-[1,1′-biphenyl]-2,3,3′,4′-tetrol, CAS. *5,6′-Diallyl-2,3,3′,4′-tetrahydroxybiphenyl. 2,3,4′,5′-Tetrahydroxy-2′,5-di-2-propenylbiphenyl.* ***Nudibaccatumol A****. Nudibaccatumin A‡. 2,3′-Bichavicol*

$C_{18}H_{18}O_4$ 298.338

Assigned 2 synonyms by the authors.

Constit. of *Piper nudibaccatum*. Moderate xanthine oxidase inhibitor. Yellowish oil. λ_{max} 207 (log ε 3.81); 289 (log ε 3.10); 319 (log ε 2.71) (MeOH).

Liu, H.-X. *et al.*, *Phytochem. Lett.*, 2015, **12**, 133-137 (*Nudibaccatumol A, Nudibaccatumin A, activity*)

2,7′-Ligna-8,8′-diene-3′,4,4′,5-tetrol L33

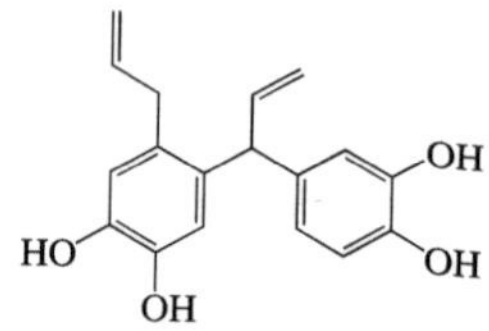

3′,4,4′,5-Tetrahydroxy-2,7′-neoligna-8,8′-diene

$C_{18}H_{18}O_4$ 298.338

(+)-form

Neotaiwanensol B [1449387-72-7]

Constit. of *Piper nudibaccatum* and *Piper taiwanense*. Xanthine oxidase inhibitor. Brown oil. $[\alpha]_D^{25}$ +9.3 (c, 0.08 in MeOH). λ_{max} 236 (sh); 284 (log ε 3.59) (MeOH).

Chen, S. *et al.*, *Phytochemistry*, 2013, **93**, 203-209 (*Neotaiwanensol B*).

Liu, H.X. *et al.*, *Phytochem. Lett.*, 2015, **12**, 133-137 (*activity*)

2,9′-Ligna-7′,8-diene-3′,4,4′,5-tetrol L34

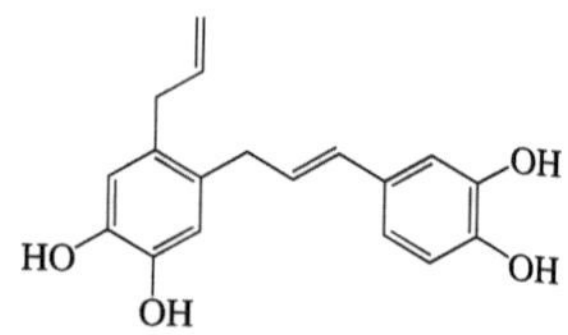

1-(3,4-Dihydroxyphenyl)-3-[4,5-dihydroxy-2-(2-propenyl)phenyl]-1-propene. 3′,4,4′,5-Tetrahydroxy-2,9′-neoligna-7′,8-diene

$C_{18}H_{18}O_4$ 298.338

Dimer of 4-(2-Propenyl)-1,2-benzenediol, P75

(E)-form

Neotaiwanensol A [1449387-71-6]

Constit. of the roots of *Piper betle* and *Piper taiwanense*. Exhibits inhibitory effects on the generation of superoxide anion and the release of elastase by human neutrophils. Brown solid or brown oil. Mp 73 - 75°. λ_{max} 260 (sh); 266 (log ε 3.76); 281 (log ε 3.69); 290 (sh) (MeOH).

Lin, C.-F. *et al.*, *Molecules*, 2013, **18**, 2563-2570 (*Piper betle constit, activity*).
Chen, S. *et al.*, *Phytochemistry*, 2013, **93**, 203-209 (*Neotaiwanensol A*)

3,3′-Ligna-8,8′-diene-4,4′,5,5′-tetrol L35

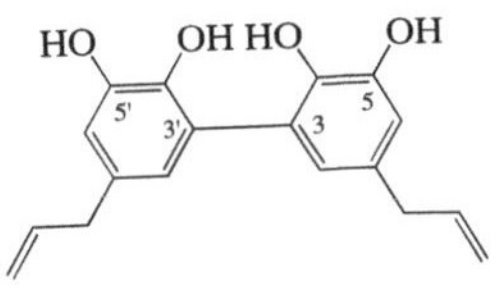

2,2′,3,3′-Tetrahydroxy-5,5′-di-2-propenylbiphenyl. 4,4′,5,5′-Tetrahydroxy-3,3′-neoligna-8,8′-diene
$C_{18}H_{18}O_4$ 298.338

5,5′-Di-Me ether
3,3′-Dimethoxy-5,5′-di-2-propenyl[1,1′-biphenyl]-2,2′-diol, CAS. *5,5′-Diallyl-2,2′-dihydroxy-3,3′-dimethoxybiphenyl. 5,5′-Dimethoxy-3,3′-ligna-8,8′-diene-4,4′-diol. 4,4′-Dihydroxy-5,5′-dimethoxy-3,3′-ligna-8,8′-diene.* ***Dehydrodieugenol.*** *Bieugenol. Dieugenol*
[4433-08-3]
$C_{20}H_{22}O_4$ 326.392
Isol. from Java citronella oil, from *Coix lachryma-jobi* var. *ma-yuen*, *Litsea turfosa*, *Ocimum sanctum*, *Ocotea cymbarum*, *Virola carinata*, *Piper pedicellosum* and *Nectandra polita* and *Syzygium aromaticum*. Potent activator of peroxisome proliferator-activated receptor γ (PPAR γ). Toxic to *Artemia salina*. Shows antimutagenic activity. Plates (EtOH). Mp 106°

4,5,5′-Tri-Me ether
2-Hydroxy-2′,3,3′-trimethoxy-5,5′-di-2-propenylbiphenyl. 4′,5,5′-Trimethoxy-3,3′-ligna-8,8′-dien-4-ol. 4-Hydroxy-4′,5,5′-trimethoxy-3,3′-ligna-8,8′-diene. ***4-O-Methyldehydrodieugenol***
[75225-84-2]
$C_{21}H_{24}O_4$ 340.419
Isol. from *Ocotea cymbarum*, *Virola carinata* and *Nectandra polita*. Oil

Tetra-Me ether
2,2′,3,3′-Tetramethoxy-5,5′-di-2-propenylbiphenyl. 4,4′,5,5′-Tetramethoxy-3,3′-ligna-8,8′-diene. ***Di-O-methyldehydrodieugenol***
[13417-56-6]
$C_{22}H_{26}O_4$ 354.446
Isol. from *Nectandra polita*. Oil.

Jones, H.A. *et al.*, *J. Am. Chem. Soc.*, 1940, **62**, 2558-2559 (*Dehydrodieugenol*).
Holloway, D.M. *et al.*, *Phytochemistry*, 1973, **12**, 1503-1505 (*Litsea turfosa constit*).
de Diaz, A.M.P. *et al.*, *Phytochemistry*, 1980, **19**, 681-682 (*Ocotea cymbarum constit*).
Suarez, M. *et al.*, *Phytochemistry*, 1983, **22**, 609-610 (*Nectandra polita constit*).
De Farias Dias, A. *et al.*, *Phytochemistry*, 1988, **27**, 3008-3009 (*Dehydrodieugenol, synth*).
Farediah, A. *et al.*, *ACGC Chem. Res. Commun.*, 2000, **10**, 63-66 (*activity*).
Miyazawa, M. *et al.*, *J. Agric. Food Chem.*, 2003, **51**, 6413-6422 (*Dehydrodieugenol, activity*).
Delogu, G. *et al.*, *Tetrahedron: Asymmetry*, 2004, **15**, 275-282 (*synth*).
Fakhrudin, N. *et al.*, *Mol. Pharmacol.*, 2010, **77**, 559-566 (*Dehydrodieugenol, activity*)

8,8′-Ligna-8,8′-diene-2,3′,4,4′-tetrol L36

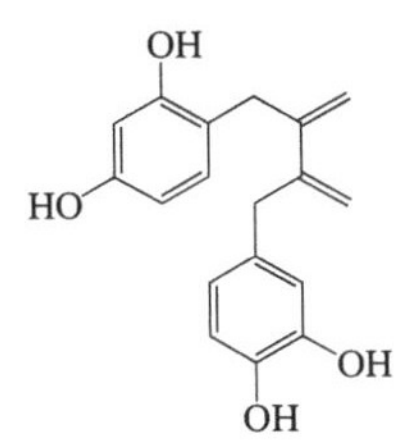

1-(2,4-Dihydroxyphenyl)-4-(3,4-dihydroxyphenyl)-2,3-dimethylenebutane. 2-(2,4-Dihydroxybenzyl)-3-(3,4-dihydroxybenzyl)-1,3-butadiene. 2,3′,4,4′-Tetrahydroxy-8,8′-lignadiene
$C_{18}H_{18}O_4$ 298.338

3′,4′-Methylene ether
4-[3-(1,3-Benzodioxol-5-ylmethyl)-2-methylene-3-butenyl]-1,3-benzenediol, CAS. *2-(2,4-Dihydroxybenzyl)-3-(3,4-methylenedioxybenzyl)-1,3-butadiene. 3′,4′-Methylenedioxy-8,8′-ligna-8,8′-diene-2,4-diol. 2,4-Dihydroxy-3′,4′-methylenedioxy-8,8′-lignadiene.* ***Anolignan A***
[158081-97-1]
$C_{19}H_{18}O_4$ 310.349
Constit. of *Anogeissus acuminata*. Inhibitor of HIV-1 reverse transcriptase. Off-white powder or cryst. (CH_2Cl_2). Mp 97 - 98°. λ_{max} 226 (ε 21878); 284 (ε 8710) (MeOH) (Berdy).

Rimando, A.M. *et al.*, *J. Nat. Prod.*, 1994, **57**, 896-904 (*Anolignan A, isol, activity*).
Luo, M. *et al.*, *Tetrahedron Lett.*, 2000, **41**, 4401-4402 (*Anolignan A, synth*).
Mori, M. *et al.*, *J. Org. Chem.*, 2002, **67**, 224-226 (*Anolignan A, synth, struct*)

2,9′-Ligna-8,8′-diene-3′,4,5-triol L37

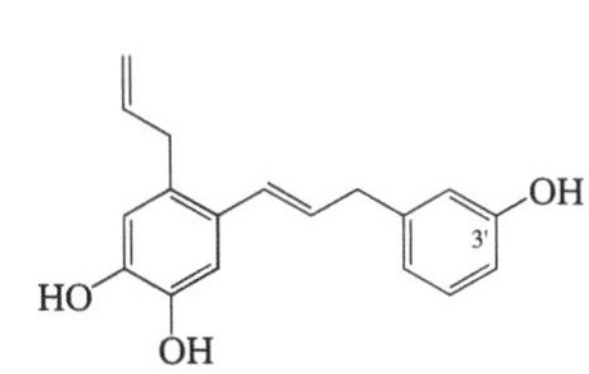

1-[4,5-Dihydroxy-2-(2-propenyl)phenyl]-3-(3-hydroxyphenyl)-1-propene. 3′,4,5-Trihydroxy-2,9′-neoligna-8,8′-diene
$C_{18}H_{18}O_3$ 282.339

(E)-form
3′-Dodecanoyl
[2095554-26-8]
$C_{30}H_{40}O_4$ 464.646
Constit. of *Piper betle* var. *birkoli*. Antioxidant. Exhibits cytotoxic activity. Brown solid. Mp 62 - 68°.

Atiya, A. *et al.*, *Nat. Prod. Res.*, 2017, **31**, 2604-2611 (*Piper betle constit, activity*)

3,3′-Ligna-8,8′-diene-4,4′,5-triol L38

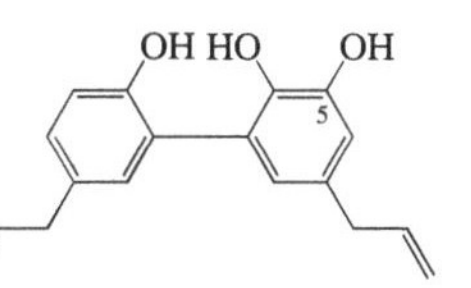

2,2′,3-Trihydroxy-5,5′-di-2-propenylbiphenyl. 5,5′-Di-2-propenyl-[1,1′-biphenyl]-2,2′,3-triol. 5,5′-Diallyl-2,2′,3-trihydroxybiphenyl. 4,4′,5-Trihydroxy-3,3′-neoligna-8,8′-diene. 3-Hydroxymagnolol
$C_{18}H_{18}O_3$ 282.339

5-Me ether
5,5′-Diallyl-2,2′-dihydroxy-3-methoxybiphenyl. 5-Methoxy-3,3′-ligna-8,8′-diene-4,4′-diol. 4,4′-Dihydroxy-5-methoxy-3,3′-ligna-8,8′-diene. ***3-Methoxymagnolol***
[122738-74-3]
$C_{19}H_{20}O_3$ 296.366
Constit. of *Illicium dunnianum*, *Magnolia henryi*, *Magnolia officinalis* and *Manglietia phuthoensis*. Significant inhibitor of superoxide

anion generation and elastase release. Gum.

Kijjoa, A. *et al.*, *Phytochemistry*, 1989, **28**, 1284-1286 (*3-Methoxymagnolol*).
Sy, L.-K. *et al.*, *Phytochemistry*, 1997, **44**, 1099-1108 (*Illicium dunnianum constit*).
van Kiem, P. *et al.*, *Chem. Pharm. Bull.*, 2008, **56**, 1270-1275 (*Manglietia phuthoensis constit*).
Shih, H.-C. *et al.*, *Bioorg. Med. Chem.*, 2016, **24**, 1439-1445 (*Magnolia officinalis constit, activity*)

8,8′-Ligna-8,8′-diene-2,4,4′-triol L39

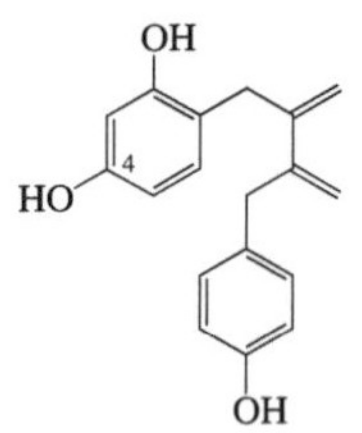

2-(2,4-Dihydroxybenzyl)-3-(4-hydroxybenzyl)-1,3-butadiene. 2,4,4′-Trihydroxy-8,8′-lignadiene
$C_{18}H_{18}O_3$ 282.339
4-Me ether
2-[3-[(4-Hydroxyphenyl)methyl]-2-methylene-3-butenyl]-5-methoxyphenol, CAS. *2-(4-Hydroxybenzyl)-3-(2-hydroxy-4-methoxybenzyl)-1,3-butadiene. 4-Methoxy-8,8′-ligna-8,8′-diene-2,4′-diol. 2,4′-Dihydroxy-4-methoxy-8,8′-lignadiene.* ***Termilignan***
[192220-05-6]
$C_{19}H_{20}O_3$ 296.366
Constit. of the fruit rind of *Terminalia bellerica*. Exhibits weak anti-HIV-1, antimalarial and antifungal activities. Cryst. powder. Mp 106 - 107°.

Valsaraj, R. *et al.*, *J. Nat. Prod.*, 1997, **60**, 739-742 (*Termilignan*)

8,8′-Lignan-4,4′-diol L40

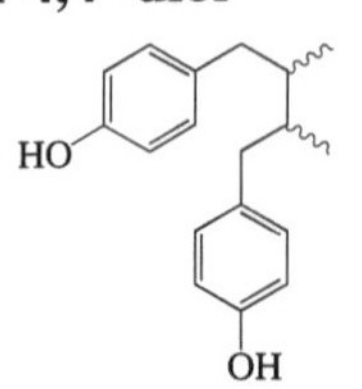

4,4′-(2,3-Dimethyl-1,4-butanediyl)bisphenol. 1,4-Bis(4-hydroxyphenyl)-2,3-dimethylbutane. 2,3-Bis(4-hydroxybenzyl)butane. 4,4′-Dihydroxylignan. ***Conocarpol‡***
[56319-00-7]
$C_{18}H_{22}O_2$ 270.372
Constit. of wood of *Conocarpus erectus*. Opt. inactive
Di-Me ether
1,4-Bis(4-methoxyphenyl)-2,3-dimethylbutane. 4,4′-Dimethoxylignan
[56319-03-0]
$C_{20}H_{26}O_2$ 298.426
Leaflets (EtOH). Mp 90 - 90.5°.

Hayashi, T. *et al.*, *Phytochemistry*, 1975, **14**, 1085-1087 (*Conocarpol*)

3,8′-Lignan-3′,4,4′,5,5′,9,9′-heptol L41

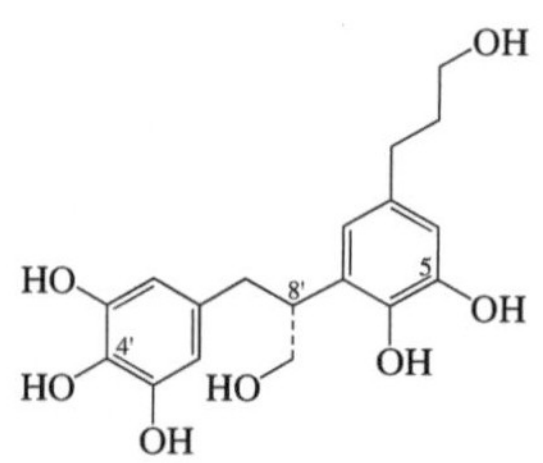

3′,4,4′,5,5′,9,9′-Heptahydroxy-3,8′-neolignan
$C_{18}H_{22}O_7$ 350.367
(R)-form
3′,5,5′-Tri-Me ether
3′,5,5′-Trimethoxy-3,8′-lignan-4,4′,9,9′-tetrol. 4,4′,9,9′-Tetrahydroxy-3′,5,5′-trimethoxy-3,8′-lignan. ***Selanginellol***
$C_{21}H_{28}O_7$ 392.448
Constit. of *Selaginella moellendorffii*. Pale yellow oil. $[\alpha]_D^{24}$ -50.4 (c, 0.4 in MeOH). λ_{max} 228 (log ε 3.98); 280 (log ε 3.31) (MeOH)
3′,5,5′-Tri-Me ether, 4′-O-β-D-glucopyranoside
$C_{27}H_{38}O_{12}$ 554.589
Constit. of *Selaginella moellendorffii*. Pale yellow solid (MeOH). $[\alpha]_D^{22}$ -57.8 (c, 0.26 in MeOH). λ_{max} 274 (log ε 3.80) (MeOH).

Zhuo, J.-X. *et al.*, *Nat. Prod. Bioprospect.*, 2016, **6**, 161-166 (*Selaginellol, Selaginellol 4′-glucoside*)

3,8′-Lignan-3′,4,4′,5,7′,9,9′-heptol L42

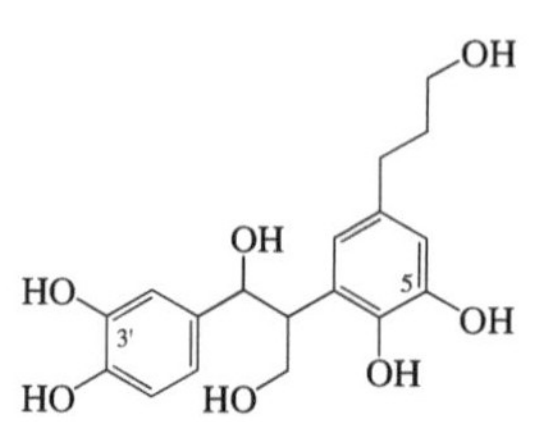

(7′ξ,8′ξ)-*form*

3′,4,4′,5,7′,9,9′-Heptahydroxy-3,8′-neolignan
$C_{18}H_{22}O_7$ 350.367
(7′ξ,8′ξ)-form
3′-Me ether
3′-Methoxy-3,8′-lignan-4,4′,5,7′,9,9′-hexol. 4,4′,5,7′,9,9′-Hexahydroxy-3′-methoxy-3,8′-lignan. ***Pinyunlignan***
[1253190-42-9]
$C_{19}H_{24}O_7$ 364.394
Constit. of the bark of *Pinus yunnanensis*. Potent inhibitor of cyclooxygenase-2 (COX-2). Oil. $[\alpha]_D^{16}$ +6.4 (c, 0.14 in MeOH). There are errors in the ref. which cast doubt on the assigned struct.. λ_{max} 211 (log ε 4.32); 282 (log ε 3.50) (MeOH)
3′,5-Di-Me ether
3′,5-Dimethoxy-3,8′-lignan-4,4′,7′,9,9′-pentol. 4,4′,7′,9,9′-Pentahydroxy-3′,5-dimethoxy-3,8′-lignan. ***Taxyunlignan***
[927868-84-6]
$C_{20}H_{26}O_7$ 378.421
Constit. of the wood of *Taxus yunnanensis*. Amorph. solid. $[\alpha]_D$ +12.1 (c, 0.21 in $CHCl_3$)
7′-Ketone, 3′,5-di-Me ether
4,4′,9,9′-Tetrahydroxy-3′,5-dimethoxy-3,8′-lignan-7′-one. 4,4′,9,9′-Tetrahydroxy-3′,5-dimethoxy-7′-oxo-3,8′-lignan
7′-Ketone, 3′,5-di-Me ether, 4-O-β-D-glucopyranoside
$C_{26}H_{34}O_{12}$ 538.546
Constit. of roots of *Vetiveria zizanioides*. Amorph. solid. $[\alpha]_D^{25}$ +54.1 (c, 0.1 in MeOH). λ_{max} 225 (log ε 3.91); 280 (log ε 3.61); 306 (log ε 3.52) (MeOH)
(8′S)-form
7′-Ketone, 3′,5-di-Me ether, 4-O-β-D-glucopyranoside
Paniculignan
$C_{26}H_{34}O_{12}$ 538.546
Constit. of *Panicum turgidum*. Yellow residue. $[\alpha]_D^{25}$ -40 (c, 0.1 in MeOH). λ_{max} 232; 280 (MeOH).

Koyama, J. *et al.*, *Biol. Pharm. Bull.*, 2006, **29**, 2310-2312 (*Taxyunlignan*).
Feng, T. *et al.*, *Z. Naturforsch., B*, 2010, **65**, 765-769 (*Pinyunlignan, activity*).
Matsuo, Y. *et al.*, *Nat. Prod. Commun.*, 2016, **11**, 983-985 (*Vetiveria zizanioides constit*).
Zaki, A.A. *et al.*, *Nat. Prod. Commun.*, 2016, **11**, 987-988 (*Paniculignan, cd*)

7,8′-Lignan-2,2′,4,4′,5,5′,7′-heptol L43

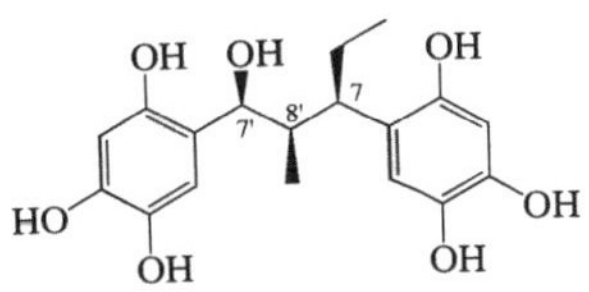

2,2′,4,4′,5,5′,7′-Heptahydroxy-7,8′-neolignan

$C_{18}H_{22}O_7$ 350.367

(7RS,7′SR,8′RS)-form

Hepta-Me ether

2,2′,4,4′,5,5′,7′-Heptamethoxy-7,8′-lignan. **Asarolignan D**

$C_{25}H_{36}O_7$ 448.556

Constit. of the rhizomes of *Acorus tatarinowii*. Cryst.. λ_{max} 205 (log ε 5.08); 230 (log ε 4.45); 291 (log ε 4.22) (MeOH).

Qin, D.-P. *et al., RSC Adv.*, 2017, **7**, 8512-8520 (*Asarolignan D, cryst struct*)

8,8′-Lignan-2,3′,4,4′,5,5′,7-heptol L44

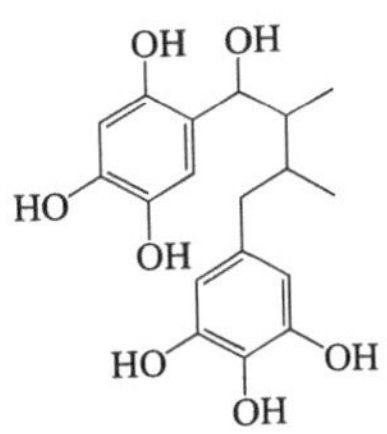

2,3′,4,4′,5,5′,7-Heptahydroxylignan

$C_{18}H_{22}O_7$ 350.367

(7ξ,8ξ,8′ξ)-form

3′,4,4′,5-Tetra-Me ether

4,4′,5,5′-Tetramethoxy-8,8′-lignan-2,3′,7-triol. 2,3′,7-Trihydroxy-4,4′,5,5′-tetramethoxylignan.

Kadcoccilignan

[1415922-03-0]

$C_{22}H_{30}O_7$ 406.475

Constit. of the stems of *Kadsura coccinea*. Pale yellow solid. $[\alpha]^{17}_{D}$ -4.2 (c, 0.076 in MeOH). λ_{max} 206 (log ε 5.1); 277 (log ε 3.64) (MeOH).

Gao, X.-M. *et al., J. Asian Nat. Prod. Res.*, 2012, **14**, 129-134 (*Kadcoccilignan*)

8,8′-Lignan-3,3′,4,4′,5,5′,7-heptol L45

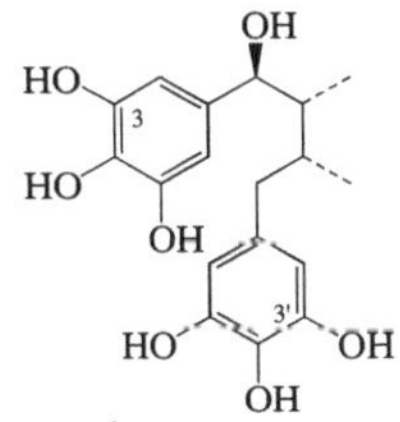

(7S,8R,8′R)-form

2,3-Dimethyl-1,4-bis(3,4,5-trihydroxyphenyl)-1-butanol. 3,3′,4,4′,5,5′,7-Heptahydroxylignan

$C_{18}H_{22}O_7$ 350.367

(7S,8R,8′R)-form

3,3′,4,4′-Tetra-Me ether

4,4′,5,5′-Tetramethoxy-8,8′-lignan-3,3′,7-triol. 3,3′,7-Trihydroxy-4,4′,5,5′-tetramethoxylignan.

Longipedlignan Q

[2408244-34-6]

$C_{22}H_{30}O_7$ 406.475

Constit. of the roots of *Kadsura longipedunculata*. Amorph. powder. $[\alpha]^{20}_{D}$ -12 (c, 0.2 in MeOH). λ_{max} 205 (log ε 2.36); 277 (log ε 1.29) (MeOH)

(7R*,8S*,8′R*)-form

3,3′,4,4′-Tetra-Me ether

Kadsuphilin J

[1017242-94-2]

$C_{22}H_{30}O_7$ 406.475

Constit. of *Kadsura philippinensis*. Amorph. powder. $[\alpha]^{25}_{D}$ -8.8 (c, 0.5 in CH_2Cl_2). λ_{max} 215 (log ε 4.44); 274 (log ε 3.52) (MeOH)

(7ξ,8ξ,8′ξ)-form

3′,4′-Methylene, 3,4,5′-tri-Me ether

3′,4,5-Trimethoxy-4′,5′-methylenedioxy-8,8′-lignan-3,7-diol. 3,7-Dihydroxy-3′,4,5-trimethoxy-4′,5′-methylenedioxylignan.

Kadangustin I

[1013633-27-6]

$C_{22}H_{28}O_7$ 404.459

Constit. of the stems of *Kadsura angustifolia*. Pale yellow solid. $[\alpha]^{17}_{D}$ -3.8 (c, 0.09 in MeOH). λ_{max} 207 (log ε 5.12); 276 (log ε 3.58) (MeOH)

7-Ketone

3,3′,4,4′,5,5′-Hexahydroxy-8,8′-lignan-7-one

7-Ketone, 3,3′,4,4′-tetra-Me ether

3,3′-Dihydroxy-4,4′,5,5′-tetramethoxy-8,8′-lignan-7-one. **Coccilignan A**

[1670237-96-3]

$C_{22}H_{28}O_7$ 404.459

Constit. of roots of *Kadsura coccinea*. Yellow oil. $[\alpha]^{20}_{D}$+45.1 (c, 0.24 in MeOH).

Shen, Y.-C. *et al., Helv. Chim. Acta*, 2008, **91**, 483-494 (*Kadsuphilin J*).
Gao, X.-M. *et al., J. Nat. Prod.*, 2008, **71**, 558-563 (*Kadangustin I*).
Fang, L. *et al., Phytochem. Lett.*, 2014, **9**, 158-162 (*Coccilignan A*).
Liu, J. *et al., J. Nat. Prod.*, 2019, **82**, 2842-2851 (*Longipedlignan Q, cd*)

8,8′-Lignan-3,3′,4,4′,5,9,9′-heptol L46

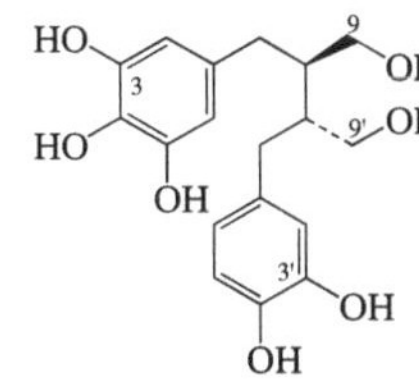

(8R,8′R)-form

3,3′,4,4′,5,9,9′-Heptahydroxylignan

$C_{18}H_{22}O_7$ 350.367

(8R,8′R)-form

3,3′,5-Tri-Me ether

3,3′,5-Trimethoxy-8,8′-lignan-4,4′,9,9′-tetrol. 4,4′,9,9′-Tetrahydroxy-3,3′,5-trimethoxylignan

3,3′,5-Tri-Me ether, 4-O-β-D-glucopyranoside

Glehlinoside G

[1335048-30-0]

$C_{27}H_{38}O_{12}$ 554.589

Constit. of the roots of *Glehnia littoralis*. Amorph. solid. $[\alpha]^{20}_{D}$ -25.3 (c, 0.2 in MeOH). λ_{max} 252 (log ε 4); 272 (log ε 4.3) (MeOH)

3,3′,4′,5-Tetra-Me ether

3,3′,4′,5-Tetramethoxy-8,8′-lignan-4,9,9′-triol. 4,9,9′-Trihydroxy-3,3′,4′,5-tetramethoxylignan

3,3′,5-Tri-Me ether, 4-O-[4-hydroxy-3-methoxybenzoyl-(→6)-β-D-glucopyranoside]

Glehlinoside B

[402740-13-0]

$C_{35}H_{44}O_{15}$ 704.722

Constit. of *Glehnia littoralis*. Amorph. solid. $[\alpha]^{25}_{D}$ -54.5 (c, 0.11 in MeOH). λ_{max} 230; 275 (MeOH)

3,3′,4′,5-Tetra-Me ether, 9,9′-di-Ac

Justin C

[202655-28-5]

$C_{26}H_{34}O_9$ 490.549

Constit. of *Justicia procumbens*. Oil. $[\alpha]_D^{28}$ -37.6 (c, 1.1 in $CHCl_3$). λ_{max} 235; 285 (MeOH)

3,3′,4,4′,5-Penta-Me ether

3,3′,4,4′,5-Pentamethoxy-8,8′-lignan-9,9′-diol. 9,9′-Dihydroxy-3,3′,4,4′,5-pentamethoxylignan

[79288-75-8]

$C_{23}H_{32}O_7$ 420.502

Constit. of *Virola elongata* and *Justicia procumbens*. Amorph.. $[\alpha]_D^{25}$ -22 (c, 0.17 in $CHCl_3$). λ_{max} 227 (ε 15849); 278 (ε 3236) (EtOH)

3,3′,4,4′,5-Penta-Me ether, 9,9′-di-Ac

[202644-38-0]

$C_{27}H_{36}O_9$ 504.576

Constit. of *Justicia procumbens*. Oil. $[\alpha]_D^{28}$ -23.4 (c, 8.5 in $CHCl_3$). λ_{max} 235; 287 (MeOH)

Hepta-Me ether

3,3′,4,4′,5,9,9′-Heptamethoxylignan. ***Linnanthin***. *3-Methoxyphyllanthin*

[139501-42-1]

$C_{25}H_{36}O_7$ 448.556

Isol. from leaves of *Phyllanthus niruri*. Oil. $[\alpha]_D^{28}$ +25 (c, 0.1.6 in $CHCl_3$) (+13.8). λ_{max} 210 (ε 48978); 280 (ε 4074) (MeOH)

3′,4′-Methylene, 3,5-di-Me ether

3,5-Dimethoxy-3′,4′-methylenedioxy-8,8′-lignan-4,9,9′-triol. 4,9,9′-Trihydroxy-3,5-dimethoxy-3′,4′-methylenedioxylignan

3′,4′-Methylene, 3,5-di-Me ether, 9,9′-di-Ac

Justin B

[202655-27-4]

$C_{25}H_{30}O_9$ 474.506

Constit. of *Justicia procumbens* and *Rostellularia procumbens*. Antibacterial agent. Oil. $[\alpha]_D^{28}$ -23.9 (c, 1.6 in $CHCl_3$). λ_{max} 235; 287 (MeOH)

3′,4′-Methylene, 3,4,5-tri-Me ether

3,4,5-Trimethoxy-3′,4′-methylenedioxy-8,8′-lignan-9,9′-diol. 9,9′-Dihydroxy-3,4,5-trimethoxy-3′,4′-methylenedioxylignan. ***Dihydroclusin***

[73149-51-6]

$C_{22}H_{28}O_7$ 404.459

Constit. of *Piper cubeba* and *Piper clusii*. Cryst.. Mp 97 - 98°. $[\alpha]_D^{25}$ -27.13 (c, 0.24 in $CHCl_3$)

3′,4′-Methylene, 3,4,5-tri-Me ether, 9-Ac

[1392820-60-8]

$C_{24}H_{30}O_8$ 446.496

Isol. from *Bursera fagaroides*. $[\alpha]_D^{25}$ -14.5 (c, 0.2 in $CHCl_3$)

3′,4′-Methylene, 3,4,5-tri-Me ether, 9,9′-di-Ac

9,9′-Diacetoxy-3,4,5-trimethoxy-3′,4′-methylenedioxylignan. ***Dihydroclusin diacetate***

[202644-37-9]

$C_{26}H_{32}O_9$ 488.533

Constit. of *Justicia procumbens* and *Bursera fagaroides*. Oil. $[\alpha]_D^{28}$ -17.4 (c, 1.4 in $CHCl_3$). λ_{max} 236; 287 (MeOH)

3′,4′-Methylene, 3,4,5-tri-Me ether, 9′-hexadecanoyl, 9-Ac

[1392820-49-3]

$C_{40}H_{60}O_9$ 684.911

Isol. from *Bursera fagaroides*. $[\alpha]_D^{25}$ -13.5 (c, 0.2 in $CHCl_3$). Incorrecly named as 9-pentadecanoyl in the ref.

(8S,8′S)-form

3,3′,4′,9,9′-Penta-Me ether

3′,4′,5,9,9′-Pentamethoxy-8,8′-lignan-3,4-diol. 3,4-Dihydroxy-3′,4′,5,9,9′-pentamethoxylignan. ***Demethyleneniranthin***

[139501-43-2]

$C_{23}H_{32}O_7$ 420.502

Isol. from *Phyllanthus niruri*. Oil. $[\alpha]_D$ +18 (c, 3.4 in $CHCl_3$)

3,4-Methylene, 3′,4′,5,9,9′-penta-Me ether

3,3′,4′,9,9′-Pentamethoxy-4,5-methylenedioxylignan. ***Niranthin***

[50656-77-4]

$C_{24}H_{32}O_7$ 432.513

Constit. of *Phyllanthus niruri*, *Phyllanthus amarus* and *Phyllanthus urinaria*. Antiinflammatory and cytotoxic agent. Cryst. (hexane). Mp 67 - 69°. $[\alpha]_D^{30}$ +28 (c, 1.29 in $CHCl_3$). Revised abs. config.. λ_{max} 201 (log ε 4.83); 227 (sh) (log ε 4.16); 278 (log ε 3.65) (MeCN)

3,4:3′,4′-Bis(methylene), 5-Me ether

3,4:3′,4′-Bis(methylenedioxy)-5-methoxy-8,8′-lignan-9,9′-diol. 9,9′-Dihydroxy-3,4:3′,4′-bis(methylenedioxy)-5-methoxylignan. ***Dihydrotrichostin***

[115713-37-6]

$C_{21}H_{24}O_7$ 388.416

Constit. of *Peperomia duclouxii* and *Piper trichostachyon*. Gum. $[\alpha]_D^{25}$ +19 (c, 0.13 in $CHCl_3$). $[\alpha]_D^{20}$ -13.3 (c, 0.6 in MeOH). λ_{max} 242; 286 ($CHCl_3$).

Anjaneyulu, A.S.R. *et al.*, *Tetrahedron*, 1973, **29**, 1291-1298 (*Niranthin*).
Prabhu, B.R. *et al.*, *Phytochemistry*, 1985, **24**, 329-331 (*Dihydroclusin*).
Koul, S.K. *et al.*, *Phytochemistry*, 1988, **27**, 1479-1482 (*Dihydrotrichostin*).
Kato, M.J. *et al.*, *Phytochemistry*, 1990, **29**, 1799-1810 (*Virola elongata lignan*).
Satyanarayana, P. *et al.*, *Tetrahedron*, 1991, **47**, 8931-8940 (*Demethyleneniranthin, Linnanthin*).
Chen, C.-C. *et al.*, *J. Nat. Prod.*, 1998, **61**, 227-229 (*Justicia procumbens lignans*).
Yuan, Z. *et al.*, *Chem. Pharm. Bull.*, 2002, **50**, 73-77 (*Glehlinoside B*).
Li, N. *et al.*, *J. Nat. Prod.*, 2003, **66**, 1421-1426 (*Peperomia duclouxii lignan*).
Chang, C.-C. *et al.*, *Phytochemistry*, 2003, **63**, 825-833 (*Niranthin*).
Kassuya, C.A.L. *et al.*, *Planta Med.*, 2005, **71**, 721-726 (*Niranthin, activity*).
Leite, D.F.P. *et al.*, *Planta Med.*, 2006, **72**, 1353-1358 (*Niranthin, activity*).
Zhang, Y. *et al.*, *Planta Med.*, 2007, **73**, 1596-1599 (*Justin B, activity*).
Xu, Y. *et al.*, *Planta Med.*, 2010, **76**, 1706-1709 (*Glehlinoside G*).
Morales-Serna, J.A. *et al.*, *Nat. Prod. Res.*, 2013, **27**, 824-829 (*Bursera fagaroides constits*)

8,8′-Lignan-3,3′,4,4′,7,9,9′-heptol

L47

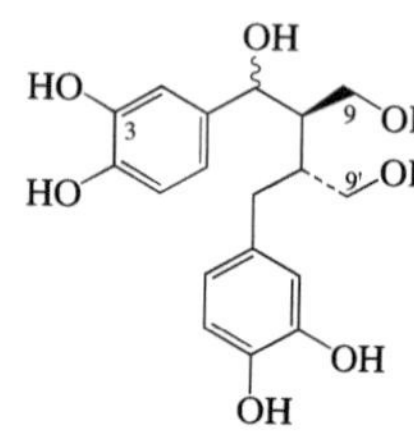

(7ξ,8R*,8′R*)-form

3,3′,4,4′,7,9,9′-Heptahydroxylignan

$C_{18}H_{22}O_7$ 350.367

(7ξ,8R*,8′R*)-form

3,3′-Di-Me ether, 9′-O-β-D-glucopyranoside

[1388857-89-3]

$C_{26}H_{36}O_{12}$ 540.562

Isol. from the exocarp of *Castanea henryi*. Yellow oil. $[\alpha]_D^{20}$ +30 (c, 0.001 in MeOH). λ_{max} 210 (log ε 4.23); 235 (log ε 3.63); 278 (log ε 4.42) (MeOH)

(7ξ,8ξ,8′ξ)-form

3,3′-Di-Me ether, 9-O-β-D-glucopyranoside

Cinnacassoside A

[1691247-46-7]
$C_{26}H_{36}O_{12}$ 540.562
Constit. of bark of *Cinnamomum cassia*. Solid.[α]$^{15.4}_{D}$-31.9 (c, 0.2 in MeOH). $λ_{max}$ 203 (log ε 4.65); 226 (log ε 4.06); 281 (log ε 3.64) (MeOH)
7-Ketone
3,3′,4,4′,9,9′-Hexahydroxy-8,8′-lignan-7-one
7-Ketone, 3,3′-di-Me ether, 9′-O-β-D-glucopyranoside
Medusaside A
[640293-42-1]
$C_{26}H_{34}O_{12}$ 538.546
Constit. of *Saussurea medusa*. Amorph. powder. [α]$^{20}_{D}$-60.9 (c, 0.75 in MeOH). $λ_{max}$ 229 (log ε 4.08); 279 (log ε 3.85); 306 (log ε 3.68) (MeOH)
7-Ketone, 3,4:3′,4′-bis(methylene), 9,9′-di-Me ether
1,4-Bis(1,3-benzodioxol-5-yl)-2,3-bis(methoxymethyl)-1-butanone. 9,9′-Dimethoxy-3,4:3′,4′-bis(methylenedioxy)-8,8′-lignan-7-one.
Di-O-methyl-7-oxodihydrocubebin
[1919044-38-4]
$C_{22}H_{24}O_7$ 400.427
Constit. of the roots of *Phyllanthus amarus*. Exhibits strong cytotoxic activity. Amorph. solid. [α]$^{25}_{D}$-40 (c, 0.2 in MeOH). $λ_{max}$ 230; 279 (MeOH).

Fan, C.-Q. *et al., Bioorg. Med. Chem.*, 2003, **11**, 703-708 (*Medusaside A*).
Wu, B. *et al., Carbohydr. Res.*, 2012, **355**, 45-49 (*Castanea henryi constit*).
Luo, Q. *et al., Molecules*, 2013, **18**, 10930-10943 (*Cinnacassoside A*).
Sparzak, B. *et al., Molecules*, 2015, **20**, 7915-7924 (*Dimethyl-7-oxodihydrocubebin*)

8,8′-Lignan-3,3′,4,4′,8,9,9′-heptol L48

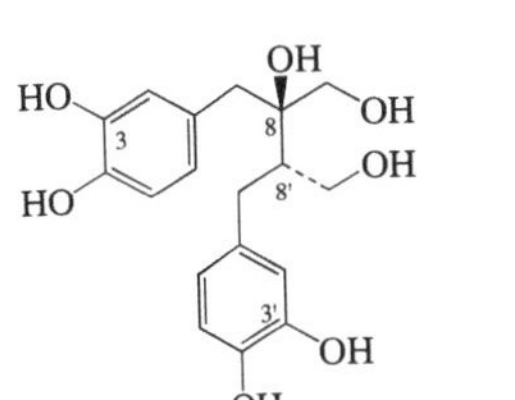
(8R,8′S)-form

2,3-Bis[(3,4-dihydroxyphenyl)methyl]-1,2,4-butanetriol. 2,3-Bis(3,4-dihydroxybenzyl)-1,2,4-butanetriol. 3,3′,4,4′,8,9,9′-Heptahydroxylignan
$C_{18}H_{22}O_7$ 350.367
(8R,8′S)-form
3,3′-Di-Me ether
3,3′-Dimethoxy-8,8′-lignan-4,4′,8,9,9′-pentol. 2,3-Bis(4-hydroxy-3-methoxyphenyl)methyl-1,2,4-butanetriol, CAS. *4,4′,8,9,9′-Pentahydroxy-3,3′-dimethoxylignan.* ***Alashinol F***
[2095103-02-7]
$C_{20}H_{26}O_7$ 378.421
Constit. of the stem bark of *Syringa pinnatifolia*. Exhibits weak antiinflammatory activity and increases cell viability in anti-hypoxia study. Powder. [α]$^{25}_{D}$-2.2 (c, 0.1 in MeOH). $λ_{max}$ 207 (log ε 4.36); 228 (log ε 4.03); 281 (log ε 3.60) (MeOH)
(8S,8′S)-form
3,3′-Di-Me ether
Alashinol G. Carinol
[58139-12-1]
$C_{20}H_{26}O_7$ 378.421
Constit. of *Carissa carandas, Carissa spinarum, Carissa edulis, Cathaya argyrophylla, Kerria japonica, Syringa pinnatifolia* and *Urtica dioica*. Exhibits cytotoxic activity and increases cell viability in anti-hypoxia study. Amorph. powder. [α] D -22.4 (c, 1 in EtOH). $λ_{max}$ 205 (log ε 4.46); 228 (log ε 3.97); 280 (log ε 3.60) (MeOH)
(8ξ,8′S)-form
3,3′,4-Tri-Me ether
3,3′,4′-Trimethoxy-8,8′-lignan-4,8′,9,9′-tetrol. 4′,8,9,9′-Tetrahydroxy-3,3′,4-trimethoxylignan.
Syripinnalignan C
[1582797-11-2]
$C_{21}H_{28}O_7$ 392.448
Constit. of roots and stems of *Syringa pinnatifolia*. [α]$^{20}_{D}$+12.5 (c, 0.1 in MeOH). $λ_{max}$ 216; 268 (MeOH)
3,3′,4,9-Tetra-Me ether
3,3′,4′,9′-Tetramethoxy-8,8′-lignan-4,8′,9-triol. 4′,8,9′-Trihydroxy-3,3′,4,9-tetramethoxylignan.
Syripinnalignan E
[1582797-13-4]
$C_{22}H_{30}O_7$ 406.475
Constit. of roots and stems of *Syringa pinnatifolia*. [α]$^{20}_{D}$+11.2 (c, 0.1 in MeOH). $λ_{max}$ 213; 258 (MeOH)
9-Et, 3,3′,4-tri-Me ether
9′-Ethoxy-3,3′,4′-trimethoxy-8,8′-lignan-4,8′,9-triol. 9-Ethoxy-4′,8,9′-trihydroxy-3,3′,4-trimethoxylignan.
Syripinnalignan D
[1582797-12-3]
$C_{23}H_{32}O_7$ 420.502
Constit. of roots and stems of *Syringa pinnatifolia*. [α]$^{20}_{D}$+11.7 (c, 0.1 in MeOH). $λ_{max}$ 213; 257 (MeOH)
(8ξ,8′ξ)-form
3′,4′-Methylene, 3-Me ether
3-Methoxy-3′,4′-methylenedioxy-8,8′-lignan-4,8,9,9′-tetrol. 4,8,9,9′-Tetrahydroxy-3-methoxy-3′,4′-methylenedioxylignan
[1360831-78-2]
$C_{20}H_{24}O_7$ 376.405
Constit. of *Zanthoxylum monophyllum*. Amorph. powder.

Pal, R. *et al., Phytochemistry*, 1975, **14**, 2302-2303 (*Carinol*).
Achenbach, H. *et al., Phytochemistry*, 1983, **22**, 749-754 (*Carissa edulis constit*).
Khamlach, K. *et al., Heterocycles*, 1990, **31**, 2195-2199 (*Carinol, synth*).
Kraus, R. *et al., Phytochemistry*, 1990, **29**, 1653-1659 (*Urtica dioica constit*).
Sefkow, M. *J. Org. Chem.*, 2001, **66**, 2343-2349 (*Carinol, synth*).
Rodriguez-Guzman, R. *et al., Planta Med.*, 2011, **77**, 1542-1544 (*Zanthoxylum monophyllum constit*).
He, W-J. *et al., Phytochemistry*, 2012, **83**, 63-69 (*Cathaya argyrophylla constit*).
Wangteeraprasert, R. *et al., Phytother. Res.*, 2012, **26**, 1496-1499 (*Carissa spinarum constit, activity*).
Ao, W. *et al., Zhongcaoyao*, 2013, **44**, 790-793 (*Syripinnalignans C,D,E*).
Feng, X. *et al., Nat. Prod. Res.*, 2017, **31**, 1555-1560 (*Alashinols F,G, activity*).
Vakiti, J.R. *et al., Org. Lett.*, 2020, **22**, 3345-3350 (*Alashinols F,G, synth, abs config*)

2,7′-Lignan-3,3′,4,4′,5,5′-hexol L49

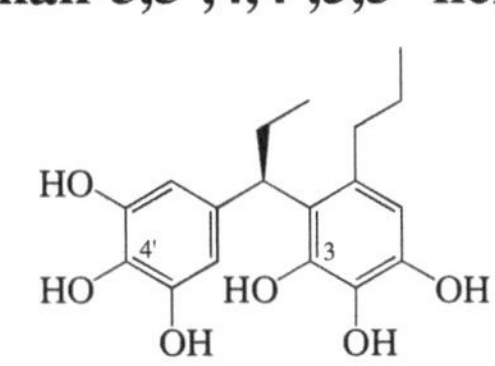

1-(3,4,5-Trihydroxyphenyl)-1-(2,3,4-trihydroxy-6-propylphenyl)propane. 3,3′,4,4′,5,5′-Hexahydroxy-2,7′-neolignan
$C_{18}H_{22}O_6$ 334.368
(R)-form
3,3′,5,5′-Tetra-Me ether
3,3′,5,5′-Tetramethoxy-2,7′-lignan-4,4′-diol.

4,4′-Dihydroxy-3,3′,5,5′-tetramethoxy-2,7′-lignan
3,3′,5,5′-Tetra-Me ether, 4-O-β-D-glucopyranoside
Feligoside. *Pheligoside*
[58497-06-6]
$C_{28}H_{40}O_{11}$ 552.617
Constit. of *Ferula cummosa* and *Ferula kopetdaghensis*. Mp 204 - 205°. $[\alpha]^{20}_{D}$ -37.5 (c, 0.5 in MeOH). λ_{max} 265 (log ε 4.3) (no solvent reported).

Kamilov, Kh.M. *et al., Chem. Nat. Compd. (Engl. Transl.)*, 1975, **11**, 609-612 (*Feligoside*).
Abdullaev, Kh. *et al., CA*, 1979, **91**, 49295t (*isol, activity*)

2,7′-Lignan-3,3′,4′,5,9,9′-hexol L50

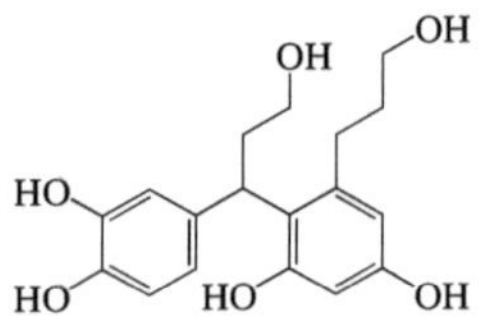

3,3′,4′,5,9,9′-Hexahydroxy-2,7′-neolignan
$C_{18}H_{22}O_6$ 334.368
(ξ)-form
3′,5-Di-Me ether
3′,5-Dimethoxy-2,7′-lignan-3,4′,9,9′-tetrol. 3,4′,9,9′-Tetrahydroxy-3′,5-dimethoxy-2,7′-lignan
3′,5-Di-Me ether, 3-O-β-D-glucopyranoside
Quercussioside
$C_{26}H_{36}O_{11}$ 524.563
Constit. of *Quercus gilva*. $[\alpha]_{D}$ +8 (c, 0.005 in MeOH).

Youn, S.H. *et al., Molecules*, 2017, **22**, 1121:1-13 (*Quercussioside*)

3,8′-Lignan-3′,4,4′,5,9,9′-hexol L51

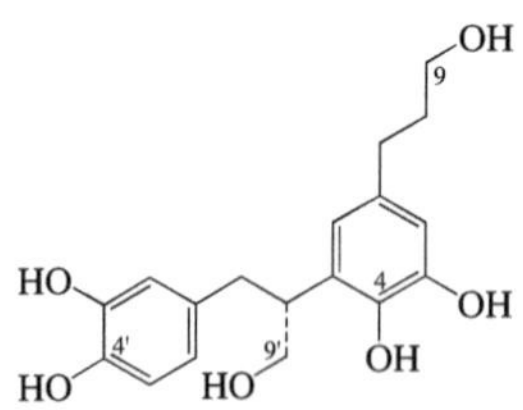

(*R*)-*form*

3′,4,4′,5,9,9′-Hexahydroxy-3,8′-neolignan
$C_{18}H_{22}O_6$ 334.368
(R)-form
3′,5-Di-Me ether
3′,5-Dimethoxy-3,8′-lignan-4,4′,9,9′-tetrol. 4,4′,9,9′-Tetrahydroxy-3′,5-dimethoxy-3,8′-lignan.
Secodihydrodehydrodiconiferyl alcohol
[137822-23-2] (unspecified stereochem.)
$C_{20}H_{26}O_6$ 362.422
Constit. of *Selaginella moellendorfii*. Inhibitor of matrix metalloproteinase 9. Cytotoxic agent
3′,5-Di-Me ether, 9-O-β-D-glucopyranoside
[1812868-75-9]
$C_{26}H_{36}O_{11}$ 524.563
Constit. of *Clematis lasiandra*. Exhibits weak cytotoxicity against the tumour cell lines HL-60, Hep-G2 and SGC-7901. Amorph. powder (Me_2CO). Mp 242 - 243°. $[\alpha]^{25}_{D}$ -89.1 (c, 0.35 in MeOH)
(S)-form
3′,5-Di-Me ether
[117677-67-5]
Amorph. powder. $[\alpha]^{22}_{D}$ -80.6 (c, 0.4 in MeOH)
3′,5-Di-Me ether, 4-O-β-D-glucopyranoside
Icariside E₃
[117613-74-8]
$C_{26}H_{36}O_{11}$ 524.563
Constit. of *Brainea insignis*, *Epimedium grandiflorum* var. *thunbergianum*, *Illicium majus*, *Saraca asoca* and *Ulmus davidiana* var. *japonica*. Moderate antioxidant. Amorph. powder. $[\alpha]^{22}_{D}$ -61.3 (c, 0.8 in MeOH). λ_{max} 278 (ε 4570) (MeOH)
3′,5-Di-Me ether, tetra-Ac
[169534-37-6]
[5260-13-9] (unspecified stereochem.) [117677-66-4]
$C_{28}H_{34}O_{10}$ 530.57
Constit. of the leaves of *Cryptomeria japonica*. Gum. $[\alpha]^{25}_{D}$ -2.5 (c, 2 in $CHCl_3$). Config. incorr. drawn in paper. λ_{max} 274 (ε 7700) (MeOH)
(ξ)-form
3′-Me ether
3′-Methoxy-3,8′-lignan-4,4′,5,9,9′-pentol. 4,4′,5,9,9′-Pentahydroxy-3′-methoxy-3,8′-lignan
3′-Me ether, 4-O-β-D-glucopyranoside
5-O-Demethylicariside E₃
[1363328-77-1]
$C_{25}H_{34}O_{11}$ 510.536
Constit. of the twigs of *Keteleeria evelyniana*. Amorph. solid. $[\alpha]^{25}_{D}$ -13.3 (c, 0.2 in MeOH). λ_{max} 205 (log ε 4.57); 281 (log ε 3.47) (MeOH).

Miyase, T. *et al., Chem. Pharm. Bull.*, 1988, **36**, 2475-2484 (*Icariside E₃*).
Kouno, I. *et al., Chem. Pharm. Bull.*, 1991, **39**, 1773-1778 (*Icariside E₃*).
Su, W.-C. *et al., Phytochemistry*, 1995, **40**, 563-566 (*Cryptomeria japonica tetra-Ac*).
Sadhu, S.K. *J. Nat. Med.*, 2007, **61**, 480-482 (*Icariside E₃, activity*).
He, W.J. *et al., Z. Naturforsch., B*, 2011, **66**, 733-739 (*5-Demethylicariside E₃*).
Tian, X.-R. *et al., Phytochem. Lett.*, 2014, **10**, 168-172 (*Clematis 9-glucoside*).
Zhu, Y. *et al., Fitoterapia*, 2018, **130**, 281-289 (*Selaginella moellendorfii constit, activity*)

8,8′-Lignan-2,2′,4,4′,5,5′-hexol L52

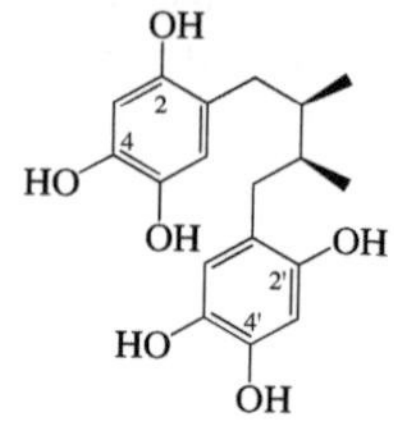

(8*R**,8′*S**)-*form*

2,3-Dimethyl-1,4-bis(2,4,5-trihydroxyphenyl)butane. 2,3-Bis(2,4,5-trihydroxybenzyl)butane. 2,2′,4,4′,5,5′-Hexahydroxylignan
$C_{18}H_{22}O_6$ 334.368
Parent compd. is *meso-*; unsymmetrically substd. derivs. are chiral
(8R*,8′S*)-form
4,5-Methylene, 5′-Me ether
5-Methoxy-4′,5′-methylenedioxy-8,8′-lignan-2,2′,4-triol. 2,2′,4-Trihydroxy-5-methoxy-4′,5′-methylenedioxylignan.
Saurulignan B
[1532522-51-2]
$C_{20}H_{24}O_6$ 360.406
Constit. of *Saururus chinensis*. Amorph. brown powder. $[\alpha]^{25}_{D}$ -3 (c, 0.5 in MeOH). λ_{max} 297 (log ε 2.88) (MeOH)
4,5-Methylene, 2′,4′-di-Me ether
4′,6′-Dimethoxy-4,5-methylenedioxy-8,8′-lignan-2,3′-diol. 2,3′-Dihydroxy-4′,6′-dimethoxy-4,5-methylenedioxylignan. ***Saurulignan D***
[1532522-53-4]
$C_{21}H_{26}O_6$ 374.433
Constit. of *Saururus chinensis*. Brown gum. $[\alpha]^{26}_{D}$ -29 (c, 1.0 in MeOH). λ_{max} 231 (log ε 3.88); 298 (log ε 3.84) (MeOH)

4,5-Methylene, 4′,5′-di-Me ether
4,5-Dimethoxy-4′,5′-methylenedioxy-8,8′-lignan-2,2′-diol. 2,2′-Dihydroxy-4,5-dimethoxy-4′,5′-methylenedioxylignan. **Saurulignan C**
[1532522-52-3]
$C_{21}H_{26}O_6$ 374.433
Constit. of *Saururus chinensis*. Amorph. light yellow powder. $[\alpha]^{25}_{D}$ -2 (c, 0.5 in MeOH). λ_{max} 233 (log ε 3.02); 276 (log ε 2.85); 297 (log ε 2.83) (MeOH)

(8RS,8′SR)-form
4,5:4′,5′-Bis(methylene) ether
4,5:4′,5′-Bis(methylenedioxy)-8,8′-lignan-2,2′-diol. 2,2′-Dihydroxy-4,5:4′,5′-bis(methylenedioxy)lignan. **Saururin A**
[389573-12-0]
$C_{20}H_{22}O_6$ 358.39
Constit. of *Saururus chinensis*. Antioxidant. Pale brown solid. λ_{max} 208 (log ε 3.15); 243 (log ε 2.9); 296 (log ε 2.87) (MeOH).

Ahn, B.-T. *et al.*, *J. Nat. Prod.*, 2001, **64**, 1562-1564 (*Saururin A, activity*).
Tsai, W.-J. *et al.*, *J. Nat. Prod.*, 2014, **77**, 125-131 (*Saurulignans B,C,D*)

8,8′-Lignan-2,3′,4,4′,5,7′-hexol L53

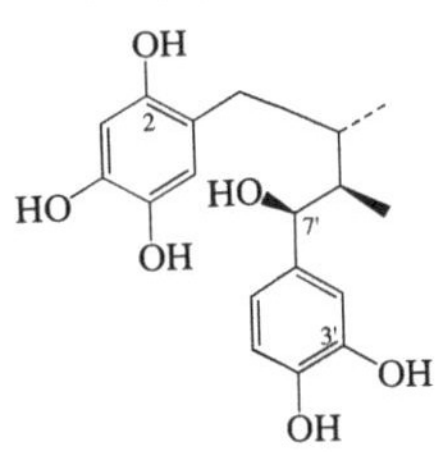

(7′S,8R,8′S)-form

2,3′,4,4′,5,7′-Hexahydroxylignan
$C_{18}H_{22}O_6$ 334.368

(7′S,8S,8′R)-form
3′,4′:4,5-Bis(methylene) ether
3′,4′:4,5-Bis(methylenedioxy)-8,8′-lignan-2,7′-diol. 2,7′-Dihydroxy-3′,4′:4,5-bis(methylenedioxy)lignan. **Saucerneol H**
[1061273-60-6]
$C_{20}H_{22}O_6$ 358.39
Constit. of the roots of *Saururus chinensis*. Exhibits cytotoxic activity. Sticky solid (MeOH aq.). $[\alpha]^{22}_{D}$ -51.9 (c, 0.35 in $CHCl_3$). λ_{max} 231 (log ε 4.1); 290 (log ε 3.93) (MeOH)

(8ξ,8′ξ)-form
7′-Ketone
2′,3,4,4′,5′-Pentahydroxy-8,8′-lignan-7-one

7′-Ketone, 3′,4′:4,5-bis(methylene) ether
2′-Hydroxy-3,4:4′,5′-bis(methylenedioxy)-8,8′-lignan-7-one. **Saucerneol G**
[1061273-58-2]
$C_{20}H_{20}O_6$ 356.374
Constit. of the roots of *Saururus chinensis*. Exhibits cytotoxic activity. Inhibitor of matrix metalloproteinase-9 induction. Brown powder (MeOH aq.). Mp 41 - 43°. $[\alpha]^{22}_{D}$ +13 (c, 0.59 in $CHCl_3$). λ_{max} 229 (log ε 4.14); 276 (log ε 3.69); 306 (log ε 3.88) (MeOH).

Seo, C.-S. *et al.*, *J. Nat. Prod.*, 2008, **71**, 1771-1774 (*Saucerneols G,H, activity*).
Lu, Y. *et al.*, *Biol. Pharm. Bull.*, 2010, **33**, 1944-1948 (*Saucerneol G, activity*)

8,8′-Lignan-3,3′,4,4′,5,5′-hexol L54

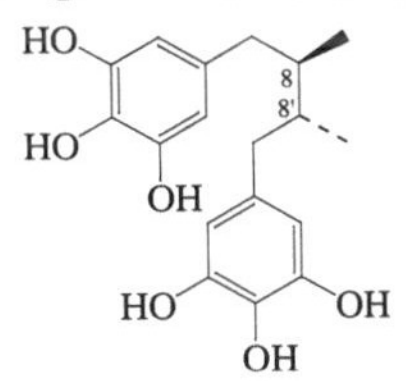

(8R,8′R)-form

2,3-Dimethyl-1,4-bis(3,4,5-trihydroxyphenyl)butane.
2,3-Bis(3,4,5-trihydroxybenzyl)butane.
3,3′,4,4′,5,5′-Hexahydroxylignan
$C_{18}H_{22}O_6$ 334.368

(8R,8′R)-form
3,3′-Di-Me ether
5,5′-Dimethoxy-8,8′-lignan-3,3′,4,4′-tetrol. 3,3′,4,4′-Tetrahydroxy-5,5′-dimethoxylignan. **Sauriol A**
[289497-19-4]
$C_{20}H_{26}O_6$ 362.422
Constit. of *Saururus cernuus*. Shows antifeedant activity. Amorph. solid. $[\alpha]_{D}$ -240 (c, 0.03 in $CHCl_3$). λ_{max} 242 (log ε 3.7); 275 (log ε 3.4) ($CHCl_3$)

3,3′,5-Tri-Me ether
3′,5,5′-Trimethoxy-8,8′-lignan-3,4,4′-triol. 3,4,4′-Trihydroxy-3′,5,5′-trimethoxylignan. **Sauriol B**
[289497-20-7]
$C_{21}H_{28}O_6$ 376.449
Constit. of *Nectandra turbacensis* and *Saururus cernuus*. Shows antifeedant activity. Amorph. solid. $[\alpha]^{27}_{D}$ -36 (c, 1.1 in $CHCl_3$). $[\alpha]^{20}_{D}$ -92 (c, 0.13 in $CHCl_3$). λ_{max} 241 (log ε 3.8); 272 (log ε 3.4) ($CHCl_3$)

(8RS,8′SR)-form
3,3′,4,4′-Tetra-Me ether
4,4′,5,5′-Tetramethoxy-8,8′-lignan-3,3′-diol. 3,3′-Dihydroxy-4,4′,5,5′-tetramethoxylignan. ***Pregomisin***
[66280-26-0]
$C_{22}H_{30}O_6$ 390.476
Isol. from fruits of *Schisandra chinensis* and seeds of *Kadsura ananosma*. Cryst. (Et_2O/hexane). Mp 130 - 131.5°

3,3′,5,5′-Tetra-Me ether
3,3′,5,5′-Tetramethoxy-8,8′-lignan-4,4′-diol. 4,4′-Dihydroxy-3,3′,5,5′-tetramethoxylignan. ***Isopregomisin***
[122222-00-8]
$C_{22}H_{30}O_6$ 390.476
Constit. of *Porlieria chilensis*. Cryst.. Mp 110 - 112°

(8ξ,8′ξ)-form
3,3′,4,5′-Tetra-Me ether
3′,4,5,5′-Tetramethoxy-8,8′-lignan-3,4′-diol. 3,4′-Dihydroxy-3′,4,5,5′-tetramethoxylignan. ***Schisphenlignan F***
[1542234-10-5]
$C_{22}H_{30}O_6$ 390.476
Isol. from the leaves and stems of *Schisandra sphenanthera*. Powder. $[\alpha]^{15}_{D}$ -1.9 (c, 0.17 in $CHCl_3$). λ_{max} 240 (log ε 3.08) ($CHCl_3$).

Ikeya, Y. *et al.*, *Chem. Pharm. Bull.*, 1978, **26**, 682-684 (*Pregomisin*).
Ikeya, Y. *et al.*, *Chem. Pharm. Bull.*, 1979, **27**, 1583-1588 (*Pregomisin, struct*).
Torres, R. *et al.*, *J. Nat. Prod.*, 1989, **52**, 402-403 (*Porlieria chilensis constits*).
Mingyi, W. *et al.*, *J. Chem. Res., Synop.*, 1999, 158-159 (*Isopregomisin, synth*).
Kubanek, J. *et al.*, *Phytochemistry*, 2000, **54**, 281-287 (*Sauriols A,B, activity*).
Mori, N. *et al.*, *Biosci., Biotechnol., Biochem.*, 2006, **70**, 1750-1753 (*Sauriols A,B, synth*).
Huang, X. *et al.*, *J. Mass Spectrom.*, 2007, **42**, 1148-1161 (*Pregomisin, ms*).
Lee, I.S. *et al.*, *Chem. Pharm. Bull.*, 2010, **58**, 1606-1611 (*Schisandra chinensis constit*).
Yang, J-H. *et al.*, *Tetrahedron*, 2011, **67**, 4498-4504 (*Kadsura ananosma constit*).
Chen, Y-C. *et al.*, *Bioorg. Med. Chem. Lett.*, 2013, **23**, 880-885 (*Schisandra arisanensis constit*).
Liang, C.-Q. *et al.*, *Fitoterapia*, 2013, **86**, 171-177 (*Schisphenlignan F*).
Macias-Villamizar, V. *et al.*, *Rev. Colomb. Quim.*, 2014, **43**, 12-16 (*Sauriol B*)

8,8′-Lignan-3,3′,4,4′,5,7-hexol L55

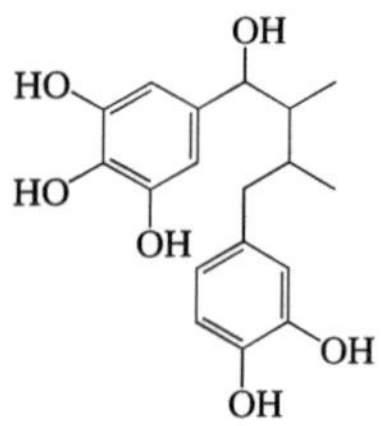

4-(3,4-Dihydroxyphenyl)-2,3-dimethyl-1-(3,4,5-trihydroxyphenyl)-1-butanol. 3,3′,4,4′,5,7-Hexahydroxylignan

$C_{18}H_{22}O_6$ 334.368

(7ξ,8ξ,8′ξ)-form

3,3′,4,4′-Tetra-Me ether

3′,4,4′,5-Tetramethoxy-8,8′-lignan-3,7-diol. 3,7-Dihydroxy-3′,4,4′,5-tetramethoxylignan. ***Kadangustin H***

[1013633-26-5]

$C_{22}H_{30}O_6$ 390.476

Constit. of the stems of *Kadsura angustifolia*. $[\alpha]_D^{17.4}$-4.8 (c, 0.1 in MeOH). λ_{max} 205 (ε 114815); 278 (ε 6760); 349 (ε 537); 370 (ε 240) (MeOH).

Gao, X.-M. *et al., J. Nat. Prod.*, 2008, **71**, 558-563 (*Kadangustin H*)

8,8′-Lignan-3,3′,4,4′,5,9-hexol L56

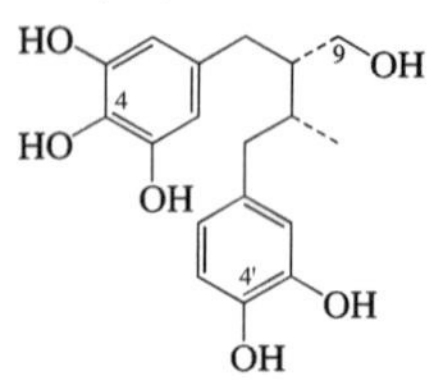

3,3′,4,4′,5,9-Hexahydroxylignan

$C_{18}H_{22}O_6$ 334.368

(8S,8′R)-form

3,3′,5-Tri-Me ether

3,3′,5-Trimethoxy-8,8′-lignan-4,4′,9-triol. 4,4′,9-Trihydroxy-3,3′,5-trimethoxylignan

3,3′,5-Tri-Me ether, 9-Ac

[1310453-67-8]

$C_{23}H_{30}O_7$ 418.486

Constit. of the bark of *Machilus robusta*. Amorph. powder. $[\alpha]_D^{20}$ +13.5 (c, 0.14 in $CHCl_3$). λ_{max} 203 (log ε 4.86); 232 (log ε 4.05); 281 (log ε 3.37) (MeOH).

Li, Y. *et al., J. Nat. Prod.*, 2011, **74**, 1444-1452 (*Machilus robusta constit*)

8,8′-Lignan-3,3′,4,4′,5,9′-hexol L57

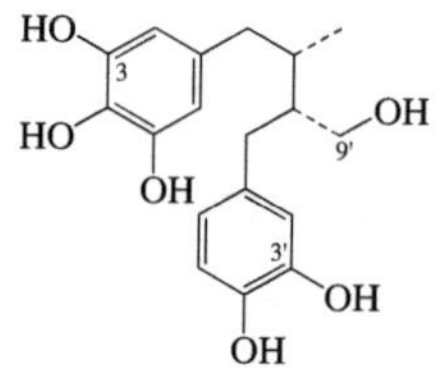

3,3′,4,4′,5,9′-Hexahydroxylignan

$C_{18}H_{22}O_6$ 334.368

(8S,8′R)-form

3,3′,5-Tri-Me ether

3,3′,5′-Trimethoxy-8,8′-lignan-4,4′,9-triol. 4,4′,9-Trihydroxy-3,3′,5′-trimethoxylignan

3,3′,5-Tri-Me ether, 9′-Ac

[1310453-66-7]

$C_{23}H_{30}O_7$ 418.486

Constit. of the bark of *Machilus robusta*. Amorph. powder. $[\alpha]_D^{20}$ +7.3 (c, 0.15 in $CHCl_3$). λ_{max} 203 (log ε 5); 232 (log ε 4.34); 281 (log ε 3.9) (MeOH).

Li, Y. *et al., J. Nat. Prod.*, 2011, **74**, 1444-1452 (*Machilus robusta constit*)

8,8′-Lignan-3,3′,4,4′,7,7′-hexol L58

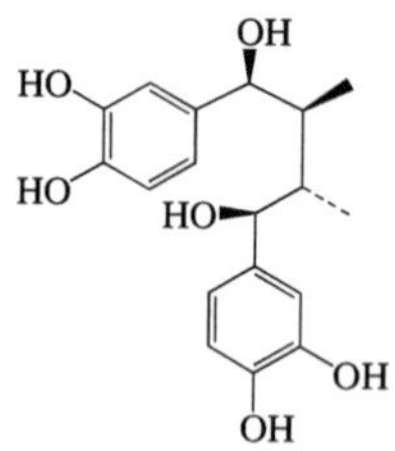

(7S,7′S,8S,8′S)-form

1,4-Bis(3,4-dihydroxyphenyl)-2,3-dimethyl-1,4-butanediol. 3,3′,4,4′,7,7′-Hexahydroxylignan

$C_{18}H_{22}O_6$ 334.368

(7S,7′S,8S,8′S)-form

3,4:3′,4′-Bis(methylene) ether

7,7′-Dihydroxy-3,4:3′,4′-bis(methylenedioxy)lignan. ***Saucerneol I***

[1061273-62-8]

$C_{20}H_{22}O_6$ 358.39

Constit. of the roots of *Saururus chinensis*. Shows strong cytotoxicity against human colon adenocarcinoma HT-29 and moderate activity against liver hepatoblastoma and breast adenocarcinoma cell lines. Powder (MeOH aq.). Mp 138 - 142°. $[\alpha]_D^{25}$-70.3 (c, 0.2 in $CHCl_3$). λ_{max} 235 (log ε 4.07); 287 (log ε 4) (MeOH)

Diketone

3,3′,4,4′-Tetrahydroxy-8,8′-lignan-7,7′-dione

Diketone, bis(methylene) ether

2,3-Dimethyl-1,4-bis(3,4-methylenedioxyphenyl)-1,4-butanedione. 3,4:3′,4′-Bis(methylenedioxy)-8,8′-lignan-7,7′-dione

[488735-13-3]

$C_{20}H_{18}O_6$ 354.358

Constit. of the fruit of *Virola sebifera*. Cryst. (Et_2O/hexane). Mp 212°. $[\alpha]_D^{24}$-149 (c, 0.1 in $CHCl_3$)

(7RS,7′SR,8RS,8′RS)-form

3,3′,4,4′-Tetra-Me ether

1,4-Bis(3,4-dimethoxyphenyl)-2,3-dimethyl-1,4-butanediol. 7,7′-Dihydroxy-3,3′,4,4′-tetramethoxylignan

[103239-13-0]

$C_{22}H_{30}O_6$ 390.476

Constit. of *Schisandra rubriflora*. Bp 183 - 185° at 1 mm. *meso*-Stereoisomer

(7RS,7′RS,8SR,8′SR)-form

3,3′,4,4′-Tetra-Me ether

[68930-19-8]

Flakes ($CHCl_3$/petrol). Mp 125 - 126°.

Traverso, G. *et al., Gazz. Chim. Ital.*, 1959, **89**, 1818-1828 (*tetra-Me ether, synth*).
Biftu, T. *et al., J. Chem. Soc., Perkin Trans. 1*, 1978, 1147-1150 (*synth*).
Wang, H. *et al., Chin. Chem. Lett.*, 1993, **4**, 31-32 (*Schisandra rubriflora constit*).
Rezende, K.R. *et al., Phytochemistry*, 2002, **61**, 427-432 (*Virola sebifera constit*).
Enders, D. *et al., Eur. J. Org. Chem.*, 2005, 1984-1990 (*Virola sebifera constit, abs config*).
Seo, C.-S. *et al., J. Nat. Prod.*, 2008, **71**, 1771-1774 (*Saucerneol I*)

8,8′-Lignan-3,3′,4,4′,9,9′-hexol L59

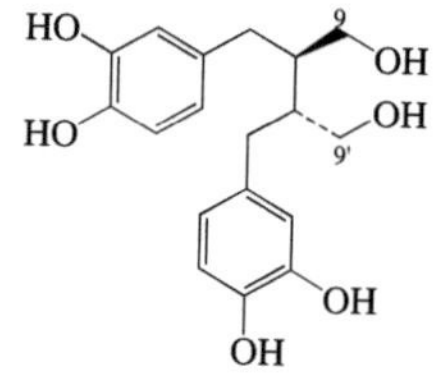

(8R,8′R)-form

2,3-Bis(3,4-dihydroxybenzyl)-1,4-butanediol. 3,3′,4,4′,9,9′-Hexahydroxylignan

[40516-27-6]

$C_{18}H_{22}O_6$ 334.368

(8R,8′R)-form

3-Me ether

3′-Methoxy-8,8′-lignan-3,4,4′,9,9′-pentol. 3,4,4′,9,9′-Pen-

tahydroxy-3′-methoxylignan. ***3-De-O-methylsecoisolariciresinol***
[151453-70-2]
$C_{19}H_{24}O_6$ 348.395
Constit. of the needles of *Taxus baccata*. Cryst. ($CHCl_3$/MeOH). Mp 125 - 126°. $[\alpha]^{25}_{D}$-29.7 (c, 0.65 in MeOH)

3,3′-Di-Me ether
3,3′-Dimethoxy-8,8′-lignan-4,4′,9,9′-tetrol, D264

3,4-Di-Me ether
3′,4′-Dimethoxy-8,8′-lignan-3,4,9,9′-tetrol. 3,4,9,9′-Tetrahydroxy-3′,4′-dimethoxylignan
[157072-28-1]
$C_{20}H_{26}O_6$ 362.422
Constit. of the needles of *Taxus baccata*. Cryst. ($CHCl_3$/MeOH). Mp 137 - 138°. $[\alpha]^{25}_{D}$-27.2 (c, 0.4 in MeOH)

3,4-Di-Me ether, 4′-O-β-D-glucopyranoside
Sargentodoside D
[1868129-15-0]
$C_{26}H_{36}O_{11}$ 524.563
Constit. of the stems of *Sargentodoxa cuneata*. Amorph. solid. $[\alpha]^{20}_{D}$ -168.7 (c, 0.05 in MeOH)

3′,4-Di-Me ether
3′,4-Dimethoxy-8,8′-lignan-3,4′,9,9′-tetrol. 3,4′,9,9′-Tetrahydroxy-3′,4-dimethoxylignan

3′,4-Di-Me ether, 4′-O-β-D-glucopyranoside
Koreanaside C
[2378524-17-3]
$C_{26}H_{36}O_{11}$ 524.563
Constit. of the flowers of *Forsythia koreana*. Amorph. powder (MeOH). Mp 83 - 84°. $[\alpha]_{D}$ -47.8 (c, 0.5 in MeOH)

3,3′,4-Tri-Me ether, 4′-O-β-D-glucopyranoside
Tinosposide A
[725743-81-7]
$C_{27}H_{38}O_{11}$ 538.59
Constit. of the stems of *Tinospora sinensis*. Amorph. powder. $[\alpha]^{25}_{D}$ -39 (c, 1 in MeOH). λ_{max} 227 (log ε 4.3); 278 (log ε 3.8) (MeOH)

3,3′,4,4′-Tetra-Me ether
3,3′,4,4′-Tetramethoxy-8,8′-lignan-9,9′-diol. 9,9′-Dihydroxy-3,3′,4,4′-tetramethoxylignan. Dimethylsecoisolariciresinol
[58311-18-5]
$C_{22}H_{30}O_6$ 390.476
Semisynthetic, from dimethylmatairesinol (see 4,4′-Dihydroxy-3,3′-dimethoxy-8,8′-lignan-9,9′-olide, D141). Constit. of *Justicia procumbens*. Mp 123 - 124°. $[\alpha]_{D}$ -33.5 (c, 1.0 in $CHCl_3$)

3,3′,4,4′-Tetra-Me ether, 9,9′-di-Ac
Dimethylsecoisolariciresinol diacetate
[119030-75-0]
$C_{26}H_{34}O_8$ 474.55
Constit. of *Justicia procumbens* and *Juniperus thurifera*. Oil. $[\alpha]^{28}_{D}$-16.6 (c, 6.7 in $CHCl_3$). λ_{max} 228; 280 (MeOH)

3,3′,4,4′,9-Penta-Me ether
3,3′,4,4′,9′-Pentamethoxy-8,8′-lignan-9-ol. 9-Hydroxy-3,3′,4,4′,9′-pentamethoxylignan. ***Secoisolariciresinol trimethyl ether***
[113851-65-3]
$C_{23}H_{32}O_6$ 404.503
Isol. from *Phyllanthus niruri*. Oil. $[\alpha]^{28}_{D}$-2.3 ($CHCl_3$)

3,4-Methylene, 3′-Me ether
3-Methoxy-3′,4′-methylenedioxy-8,8′-lignan-4,9,9′-triol. 4,9,9′-Trihydroxy-3-methoxy-3′,4′-methylenedioxylignan.
Piperphilippinin VI
[949102-30-1]
$C_{20}H_{24}O_6$ 360.406
Constit. of the stems of *Piper philippinum*. Oil. $[\alpha]^{24}_{D}$-17 (c, 0.07 in $CHCl_3$). λ_{max} 229 (sh); 283 (log ε 3.66) (MeOH). [base] λ_{max} 232 (sh); 287 (log ε 3.66) (MeOH/KOH)

3,4-Methylene, 3′-Me ether, 9′-O-β-D-glucopyranoside
[1613273-68-9]
$C_{26}H_{34}O_{11}$ 522.547
Constit. of the stems of *Sargentodoxa cuneata*. Antioxidant. Amorph. powder. $[\alpha]^{20}_{D}$-18 (c, 0.2 in MeOH). Obt. as an isomeric mixt. with the 9-*O*-β-D-glucopyranoside. Phys. props. assigned to mixt.. λ_{max} 231 (log ε 4.10); 284 (log ε 3.90) (MeOH)

3,4-Methylene, 3′-Me ether, 9,9′-di-Ac
Justin A
[202655-26-3]
$C_{24}H_{28}O_8$ 444.48
Constit. of *Justicia procumbens*. Oil. $[\alpha]^{28}_{D}$-20 (c, 0.1 in $CHCl_3$). λ_{max} 235; 285 (MeOH)

3,4-Methylene, 3′,4′-di-Me ether
3,4-Dimethoxy-3′,4′-methylenedioxy-8,8′-lignan-9,9′-diol. 9,9′-Dihydroxy-3,4-dimethoxy-3′,4′-methylenedioxylignan.
9,9′-Demethylsecolintetralin
[139501-40-9]
$C_{21}H_{26}O_6$ 374.433
Constit. of *Phyllanthus niruri*, *Virola venosa* and *Justicia procumbens*. Cryst. (hexane/EtOAc). Mp 69.5 - 71°. $[\alpha]_{D}$-9.9 (c, 0.36 in $CHCl_3$)

3,4-Methylene, 3′,4′-di-Me ether, 9′-Ac
[1392820-57-3]
$C_{23}H_{28}O_7$ 416.47
Constit. of *Bursera fagaroides*. $[\alpha]^{25}_{D}$-10.0 (c, 0.2 in $CHCl_3$)

3,4-Methylene, 3′,4′-di-Me ether, 9,9′-di-Ac
[139501-41-0]
$C_{25}H_{30}O_8$ 458.507
Constit. of *Bursera fagaroides*, *Justicia procumbens* and *Phyllanthus niruri*. Prisms. Mp 52 - 54°. $[\alpha]^{28}_{D}$-9.6 (c, 2.2 in $CHCl_3$)

3,4:3′,4′-Bis(methylene) ether
2,3-Bis(1,3-benzodioxol-5-ylmethyl)-1,4-butanediol, CAS. *3,4:3′,4′-Bis(methylenedioxy)-8,8′-lignan-9,9′-diol. 9,9′-Dihydroxy-3,4:3′,4′-bis(methylenedioxy)lignan.* **Dihydrocubebin**
[24563-03-9]
$C_{20}H_{22}O_6$ 358.39
Isol. from *Cleistanthus collinus*, *Piper guineense* and *Piper cubeba*. Cryst.. Mp 112°. $[\alpha]_{D}$ -42

3,4:3′,4′-Bis(methylene) ether, 9-O-β-D-glucopyranoside
Tetracentronside B*. 9-O-β-D-Glucopyranosyldihydrocubebin*
[288248-82-8]
$C_{26}H_{32}O_{11}$ 520.531
Constit. of *Tetracentron sinense*. Plates (EtOAc). Mp 156 - 157°. $[\alpha]^{25}_{D}$-12.1 (c, 0.28 in MeOH). λ_{max} 234; 287 (MeOH)

3,4:3′,4′-Bis(methylene) ether, 9′-O-[α-L-arabinofuranosyl-(1→6)-β-D-glucopyranoside]
[1613273-67-8]
$C_{31}H_{40}O_{15}$ 652.646

Constit. of the fruit of *Schisandra glaucescens*. Amorph. powder. $[\alpha]_D^{20}$ -44 (c, 0.2 in MeOH). λ_{max} 236 (log ε 3.97); 288 (log ε 3.96) (MeOH)

3,4:3′,4′-Bis(methylene) ether, 9-Ac
Hemiariensin
[112448-60-9]
$C_{22}H_{24}O_7$ 400.427
Constit. of *Piper cubeba*. Viscous gum. λ_{max} 232 (log ε 4.42); 286 (log ε 4.39) (MeOH)

3,4:3′,4′-Bis(methylene) ether, 9,9′-di-Ac
Ariensin
[81410-43-7]
$C_{24}H_{26}O_8$ 442.464
Constit. of *Bursera ariensis*

3,4:3′,4′-Bis(methylene) ether, 9-dodecanoyl
Dihydrocubebin monolaurate
[199994-29-1]
$C_{32}H_{44}O_7$ 540.697
Constit. of *Virola surinamensis*. Oil. $[\alpha]_D$ -13.3 (c, 0.07 in $CHCl_3$)

(8S,8′S)-form

Hexa-Me ether
3,3′,4,4′,9,9′-Hexamethoxylignan. **Phyllanthin‡**. *Tetra-O-methylsecoisolariciresinol*
[10351-88-9]
$C_{24}H_{34}O_6$ 418.53
Isol. from *Phyllanthus niruri*, *Phyllanthus amarus* and *Phyllanthus urinaria*. Cytotoxicity enhancer. Cryst. (MeOH). Mp 96°. Mp 118°. $[\alpha]_D^{30}$ +12.42 (c, 1.45 in $CHCl_3$). λ_{max} 230 (ε 19950); 280 (ε 141) (EtOH) (Berdy)

3,4-Methylene, 3′,4′,9,9′-tetra-Me ether
3,4,9,9′-Tetramethoxy-3′,4′-methylenedioxylignan.
5-Demethoxyniranthin
[610310-54-8]
$C_{23}H_{30}O_6$ 402.487
Constit. of *Phyllanthus urinaria*. Amorph. solid. $[\alpha]_D^{25}$ +15.4 (c, 0.19 in $CHCl_3$)

(8R*,8′R*)-form

9,9′-Bis-O-(4-hydroxycinnamoyl)
$C_{36}H_{34}O_{10}$ 626.658
Present in the trunkwood of *Myrianthus arboreus*, isol. in the form of Myrianthiphyllin after permethylation

3,3′,4,4′-Tetra-Me ether, 9,9′-bis-O-(4-methoxycinnamoyl)
Myrianthiphyllin
[129272-58-8]
$C_{42}H_{46}O_{10}$ 710.82
Isol. from *Myrianthus arboreus* after permethylation. Mp 157 - 160°. $[\alpha]_D^{20}$ -70 ($CHCl_3$)

3,4:3′,4′-Bis(methylene), 9,9′-di-Me ether
9,9′-Dimethoxy-3,4:3′,4′-bis(methylenedioxy)lignan. ***Di-O-methyldihydrocubebin***. *Dihydrocubebin dimethyl ether*
[874015-53-9]
$C_{22}H_{26}O_6$ 386.444
Constit. of *Phyllanthus niruri*. Amorph. solid

(8R*,8′S*)-form

3,4-Methylene, 3′-Me ether
Neriilignan
[2009362-66-5]
$C_{20}H_{24}O_6$ 360.406
Constit. of the leaves of *Podocarpus neriifolius*. Yellow oil. $[\alpha]_D^{23}$ -53.1 (c, 0.001 in Py)

(8RS,8′RS)-form

3,4-Methylene, 3′-Me ether, 9,9′-di-Ac
[226543-62-0]
$C_{24}H_{28}O_8$ 444.48
Constit. of *Virola sebifera*. Viscous oil. λ_{max} 232 (log ε 4.2); 286 (log ε 3.63) (MeOH)

(8RS,8′SR)-form

3,4:3′,4′-Bis(methylene) ether
meso-Dihydrocubebin
[24562-96-7]
$C_{20}H_{22}O_6$ 358.39
Constit. of *Virola flexuosa*. Cryst. (C_6H_6/petrol). Mp 92°

(8ξ,8′ξ)-form

3,4-Methylene, 3′,4′-di-Me ether, 9,9′-di-Ac
Dihydroprasanthalin
[126380-09-4]
$C_{25}H_{30}O_8$ 458.507
Constit. of *Jatropha gossypifolia*

3,4;3′,4′-Bis(methylene) ether, 9-O-β-D-glucopyranoside
Syringanide
[124846-49-7]
$C_{26}H_{32}O_{11}$ 520.531
Constit. of the leaves of *Syringa vulgaris*.

Row, L.R. *et al.*, *Tetrahedron*, 1966, **22**, 2899-2908 (*Phyllanthin*).
Batterbee, J.E. *et al.*, *J. Chem. Soc. C*, 1969, 2470-2477 (*meso-Dihydrocubebin, synth*).
Dwuma-Badu, D. *et al.*, *Lloydia*, 1975, **38**, 343-345 (*Dihydrocubebin*).
Anjaneyulu, A.S.R. *et al.*, *Tetrahedron Lett.*, 1975, **16**, 2961-2964 (*Dihydrocubebin*).
Mahalanabis, K.K. *et al.*, *Tetrahedron Lett.*, 1982, **23**, 3975-3978 (*synth*).
Hernandez, J.D. *et al.*, *Planta Med.*, 1983, **47**, 215-217 (*Ariensin*).
Burke, J.M. *et al.*, *Planta Med.*, 1985, **51**, 450-452 (*Ariensin, synth*).
Badheka, L.P. *et al.*, *Phytochemistry*, 1987, **26**, 2033-2036 (*Hemiariensin*).
Kikuchi, M. *et al.*, *Annu. Rep. Tohoku Coll. Pharm.*, 1988, 105 (*Syringanide*).
Satyanarayana, P. *et al.*, *J. Nat. Prod.*, 1988, **51**, 44-49 (*Secoisolariciresinol trimethyl ether*).
Banerjee, J. *et al.*, *Indian J. Chem., Sect. B*, 1989, **28**, 711-713 (*Dihydroprasanthalin*).
Brown, E. *et al.*, *Tetrahedron*, 1989, **45**, 141-154 (*synth*).
Ngounou, F.N. *et al.*, *Planta Med.*, 1990, **56**, 227-229 (*Myrianthiphyllin*).
Satyanarayana, P. *et al.*, *Tetrahedron*, 1991, **47**, 8931-8940 (*Phyllanthus niruri constits*).
Kato, M.J. *et al.*, *Phytochemistry*, 1992, **31**, 283-287 (*3,4-methylene 3′,4′-di-Me ether*).
Somanabandhu, A. *et al.*, *J. Nat. Prod.*, 1993, **56**, 233-239 (*Phyllanthin, pmr, cmr*).
Das, B. *et al.*, *Phytochemistry*, 1993, **33**, 1489-1491 (*3-Demethylsecoisolariciresinol*).
Das, B. *et al.*, *Phytochemistry*, 1994, **36**, 1031 (*3′,4′-di-Me ether*).
De Almeida Blumenthal, E.E. *et al.*, *Phytochemistry*, 1997, **46**, 745-749 (*Dihydrocubebin laurate*).
Chen, C.-C. *et al.*, *J. Nat. Prod.*, 1998, **61**, 227-229 (*Justicia procumbens lignans*).
Martinez V., J.C. *et al.*, *Phytochemistry*, 1999, **50**, 883-886 (*Virola sebifera diacetate*).
Yi, J. *et al.*, *Phytochemistry*, 2000, **53**, 1001-1003 (*Tetracentronside B*).
Chang, C.-C. *et al.*, *Phytochemistry*, 2003, **63**, 825-833 (*Phyllanthin, 5-Demethoxyniranthin*).
Li, W. *et al.*, *Chem. Pharm. Bull.*, 2004, **52**, 638-640 (*Tinosposide A*).
Morimoto, T. *et al.*, *Synth. Commun.*, 2005, **35**, 857-865 (*resoln*).
Elfahmi, *et al.*, *J. Nat. Prod.*, 2006, **69**, 55-58 (*Di-O-methyldihydrocubebin*).
Chen, Y.-C. *et al.*, *Phytochemistry*, 2007, **68**, 2101-2111 (*Piperphilippinin VI*).
Enders, D. *et al.*, *Z. Naturforsch., B*, 2007, **62**, 117-120 (*Dihydrocubebin, synth*).
Morales-Serna, J.A. *et al.*, *Nat. Prod. Res.*, 2013, **27**, 824-829 (*Bursera fagaroides constit*).
Yu, H.-Y. *et al.*, *J. Nat. Prod.*, 2014, **77**, 1311-1320 (*Schisandra glaucescens constits, cd, activity*).
Zeng, X. *et al.*, *Fitoterapia*, 2015, **101**, 153-161 (*Sargentodoside D*).
Wu, J.-J. *et al.*, *Nat. Prod. Res.*, 2017, **31**, 239-244 (*Neriilignan*).
Lee, Y.-G. *et al.*, *Chem. Nat. Compd. (Engl. Transl.)*, 2019, **55**, 432-434 (*Koreanaside C, cd*)

8,8′-Lignan-3,3′,5,5′,7,7′-hexol L60

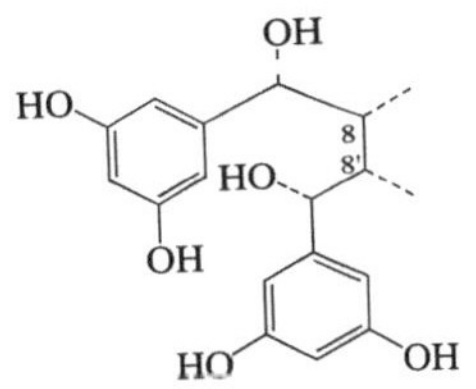

1,4-Bis(3,5-dihydroxyphenyl)-2,3-dimethyl-1,4-butanediol
$C_{18}H_{22}O_6$ 334.368
Structures in this series may need revision. Phenyl 3,5-dioxygenated substn. pattern in doubt as a result of work done by Yang *et al* in 2015
(7R*,7′R*,8R*,8′S*)-form
3,3′-Di-Me ether
5,5′-Dimethoxy-8,8′-lignan-3,3′,7,7′-tetrol
[2252456-24-7]
$C_{20}H_{26}O_6$ 362.422
Constit. of the leaves of *Myristica fatua*. Exhibits *in vitro* cytotoxicity against breast cancer cell line MCF-7. Brown-yellow oil. $[\alpha]^{25}_{D}$ +19.5 (c, 0.33 in MeOH). λ_{max} 231; 236; 282 (MeOH).

Yang, Y.-N. *et al.*, *J. Nat. Prod.*, 2015, **78**, 705-711 (*pmr, struct*).
Fajriah, S. *et al.*, *Phytochem. Lett.*, 2017, **20**, 36-39 (*Myristica fatua constit, activity*)

3,8′-Lignan-3′,4,4′,5,5′,7′,9,9′-octol L61

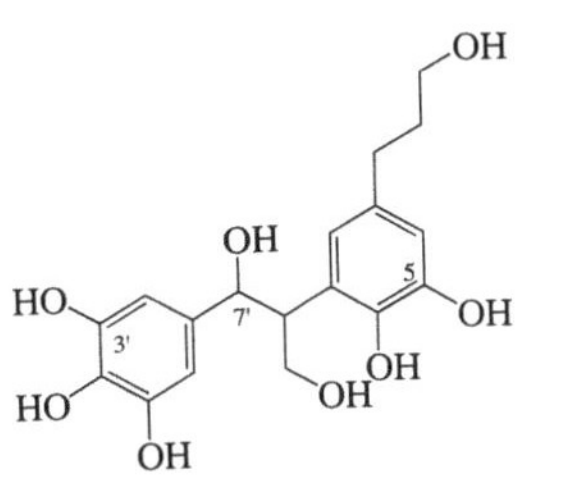

3′,4,4′,5,5′,7′,9,9′-Octahydroxy-3,8′-neolignan
$C_{18}H_{22}O_8$ 366.366
(±)-form
7′-Ketone, 3′,5,5′-tri-Me ether
4,4′,9,9′-Tetrahydroxy-3′,5,5′-trimethoxy-3,8′-lignan-7′-one.
Selamoellenin D
$C_{21}H_{26}O_8$ 406.431
Constit. of *Selaginella moellendorffii*. Pale yellow oil. $[\alpha]^{20}_{D}$+7.0 (c, 0.06 in MeOH). Partial racemate which has been resolved by chiral hplc. λ_{max} 206; 258; 283; 475; 485; 518; 558; 581; 713 (MeOH).

Zhu, Y. *et al.*, *Fitoterapia*, 2018, **130**, 281-289 (*Selamoellenin D*)

8,8′-Lignan-3,3′,4,4′,5,5′,9,9′-octol L62

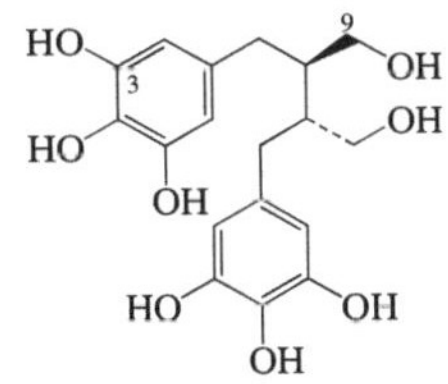

(8R,8′R)-form

2,3-Bis(3,4,5-trihydroxybenzyl)-1,4-butanediol.
3,3′,4,4′,5,5′,9,9′-Octahydroxylignan
$C_{18}H_{22}O_8$ 366.366
(8R,8′R)-form
3,3′,5,5′-Tetra-Me ether, 9-O-(4-hydroxy-3-methoxy-E-cinnamoyl)
9-O-trans-Feruloyl-8,8′-bisdihydrosiringenin
[1207887-19-1]
$C_{32}H_{38}O_{11}$ 598.645
Constit. of *Hypericum petiolulatum*
3,3′,5′,9-Tetra-Me ether, 9′-O-β-D-glucopyranoside
Bonafousioside
[115900-10-2]
$C_{28}H_{40}O_{13}$ 584.615
Isol. from *Bonafousia macrocalyx* (preferred genus name *Tabernaemontana*)
3,4-Methylene, 3′,4′,5,5′-tetra-Me ether
3,3′,4,5-Tetramethoxy-4′,5′-methylenedioxy-8,8′-lignan-9,9′-diol. 9,9′-Dihydroxy-3,3′,4,5-tetramethoxy-4′,5′-methylenedioxylignan
[93395-18-7]
$C_{23}H_{30}O_8$ 434.485
Constit. of *Piper clusii*. Cryst. (EtOH). Mp 58 - 60°. $[\alpha]^{36}_{D}$-28 (MeOH)
3,4-Methylene, 3′,5,5′,9,9′-penta-Me ether
3,3′,5,9,9′-Pentamethoxy-4′,5′-methylenedioxy-8,8′-lignan-4-ol. 4-Hydroxy-3,3′,5,9,9′-pentamethoxy-4′,5′-methylenedioxylignan.
Nirphyllin
[120396-54-5]
$C_{24}H_{32}O_8$ 448.512
Isol. from *Phyllanthus niruri*. Potential antihepatotoxic agent
3,4:3′,4′-Bis(methylene), 5,5′-di-Me ether
3,3′-Dimethoxy-4,5:4′,5′-bis(methylenedioxy)-8,8′-lignan-9,9′-diol. 9,9′-Dihydroxy-3,3′-dimethoxy-4,5:4′,5′-bis(methylenedioxy) lignan
[878142-55-3]
$C_{22}H_{26}O_8$ 418.442
Constit. of *Peperomia duclouxii*. Gum. $[\alpha]^{25}_{D}$-8.5 (c, 0.5 in $CHCl_3$). λ_{max} 242; 278 ($CHCl_3$)
(8S,8′S)-form
3,3′,5,5′-Tetra-Me ether
5,5′-Dimethoxysecoisolariciresinol
[147852-15-1]
$C_{22}H_{30}O_8$ 422.474
Constit. of *Apollonias barbujana* and *Quercus petraea*. Amorph. powder. $[\alpha]^{20}_{D}$+27.3 (c, 1.6 in MeOH)
3,3′,5,5′-Tetra-Me ether, 9-O-xylopyranoside
Ssioriside
[126882-53-9]
$C_{27}H_{38}O_{12}$ 554.589
Isol. from *Prunus ssiori*, *Prunus padus*, *Quercus petraea*, *Cinnamomum osmophloeum* and *Litsea glutinosa*. Powder. Mp 140 - 142°. $[\alpha]^{20}_{D}$-9.2 (c, 0.9 in MeOH). $[\alpha]^{24}_{D}$ +0.7 (c, 1.0 in MeOH). Prob. isol. as partial racemate. λ_{max} 211 (ε 44668); 232 (ε 14125); 320 (ε 2951) (MeOH)
3,3′,5,5′-Tetra-Me ether, 9-O-[4-hydroxy-3,5-dimethoxycinnamoyl-(→6)-β-D-glucopyranoside]
Hazaleanin A
[147742-22-1]
$C_{39}H_{50}O_{17}$ 790.812
Constit. of the bark of *Fagara rhetza*. Amorph. solid. $[\alpha]^{26}_{D}$-3.6 (c, 0.92 in MeOH)
3,3′,5,5′-Tetra-Me ether, 9,9′-bis-O-(4-hydroxy-3-methoxy-E-cinnamoyl)
[1290544-75-0]
$C_{42}H_{46}O_{14}$ 774.816
Constit. of the roots of *Cinnamomum osmophloeum* and *Lindera umbellata*. Powder or amorph. solid. Mp 214 - 216°. $[\alpha]^{24}_{D}$+56.8 (c, 0.5 in DMSO). $[\alpha]^{25}_{D}$+111.5 (c, 0.1 in MeOH/$CHCl_3$). Config. tentatively assigned. λ_{max} 192 (log ε 5.25); 208 (log ε 5.47); 234 (log ε 5); 297 (log ε 4.93); 326 (log ε

5.03) (MeOH). [neutral] λ_{max} 206 (log ε 4.79); 233 (log ε 4.33); 326 (log ε 4.32) (MeOH)

3,4-Methylene, 3′,5,5′-tri-Me ether

3,3′,5-Trimethoxy-4′,5′-methylenedioxy-8,8′-lignan-4,9,9′-triol. 4,9,9′-Trihydroxy-3,3′,5-trimethoxy-4′,5′-methylenedioxylignan

3,4-Methylene, 3′,5,5′-tri-Me ether, 9,9′-di-Ac

[635752-22-6]

$C_{26}H_{32}O_{10}$ 504.532

Constit. of *Peperomia duclouxii*. Gum. $[\alpha]_D^{25}$ +23.6 (c, 0.27 in $CHCl_3$). λ_{max} 242; 277 ($CHCl_3$)

3,4-Methylene, 3′,4′,5,5′-tetra-Me ether, 9,9′-di-Ac

[635752-20-4]

$C_{27}H_{34}O_{10}$ 518.559

Constit. of *Peperomia duclouxii*. Gum. $[\alpha]_D^{25}$ +19.8 (c, 0.31 in $CHCl_3$). λ_{max} 241; 275 ($CHCl_3$)

3,4:3′,4′-Bis(methylene), 5,5′-di-Me ether, 9-Ac

[635752-21-5]

$C_{24}H_{28}O_9$ 460.479

Constit. of *Peperomia duclouxii*. Gum. $[\alpha]_D^{25}$ +9.5 (c, 0.15 in $CHCl_3$)

(8R*,8′R*)-form

3,3′,4,4′,5,5′-Hexa-Me ether, 9,9′-di-Ac

1,4-Diacetoxy-2,3-bis(3,4,5-trimethoxybenzyl)butane

[178631-16-8]

$C_{28}H_{38}O_{10}$ 534.602

Constit. of the bark of *Aglaia elaeagnoidea*. Oil. $[\alpha]_D^{25}$ +22 (c, 0.2 in $CHCl_3$). Abs. config. not determined. λ_{max} 230 (log ε 4.25); 278 (log ε 3.69) (MeOH)

(8RS,8′RS)-form

3,3′,5,5′-Tetra-Me ether

2,3-Bis(4-hydroxy-3,5-dimethoxybenzyl)-1,4-butanediol. 3,3′,5,5′-Tetramethoxy-8,8′-lignan-4,4′,9,9′-tetrol. 4,4′,9,9′-Tetrahydroxy-3,3′,5,5′-tetramethoxylignan. ***8,8′-Bisdihydrosiringenin.*** *8,8′-Bisdihydrosyringenin*

[95172-23-9]

$C_{22}H_{30}O_8$ 422.474

Constit. of *Annona senegalensis*. Amorph. powder. Mp 180 - 181°

(8RS,8′SR)-form

3,3′,5,5′-Tetra-Me ether, 9,9′-bis-O-(4-hydroxy-3-methoxy-E-cinnamoyl)

[1243192-19-9]

$C_{42}H_{46}O_{14}$ 774.816

Constit. of the branches of *Lindera obtusiloba*. Amorph. powder. Mp 180°. λ_{max} 241 (log ε 4.45); 300 (sh); 325 (log ε 4.47) (MeOH)

3,3′,5,5′-Tetra-Me ether, 9,9′-bis-O-(4-hydroxy-3,5-dimethoxy-E-cinnamoyl)

[1243192-21-3]

$C_{44}H_{50}O_{16}$ 834.868

Constit. of the branches of *Lindera obtusiloba*. Powder. Mp 175°. λ_{max} 240 (log ε 4.46); 300 (sh); 326 (log ε 4.48) (MeOH)

3,3′,4,4′,5,5′-Hexa-Me ether

2,3-Bis(3,4,5-trimethoxybenzyl)-1,4-butanediol. 9,9′-Dihydroxy-3,3′,4,4′,5,5′-hexamethoxylignan

[115495-67-5]

$C_{24}H_{34}O_8$ 450.528

Constit. of the bark of *Zanthoxylum heitzii*. Cryst.. Mp 142 - 146°

3,3′,4,4′,5,5′-Hexa-Me ether, 9-Ac

4-Acetoxy-2,3-bis(3,4,5-trimethoxybenzyl)-1-butanol. 9-Acetoxy-9′-hydroxy-3,3′,4,4′,5,5′-hexamethoxylignan

[159598-97-7]

$C_{26}H_{36}O_9$ 492.565

Constit. of the bark of *Zanthoxylum heitzii*. Cryst.. Mp 121 - 122°. Racemic

(8ξ,8′ξ)-form

3,3′,5,5′-Tetra-Me ether, 9-O-α-L-rhamnopyranoside

Ufaside

[1637283-11-4]

$C_{28}H_{40}O_{12}$ 568.616

Constit. of the aerial parts of *Uvaria rufa*. Amorph. powder. λ_{max} 207; 236 (sh); 273 (MeOH).

Koul, S.K. *et al.*, *Phytochemistry*, 1984, **23**, 2099-2101 (*3,4-methylene tetra-Me ether*).

Garnier, J. *J. Nat. Prod.*, 1988, **51**, 484-487 (*Bonafousioside*).

Yoshinari, K. *et al.*, *Chem. Pharm. Bull.*, 1989, **37**, 3301-3303 (*Ssioriside*).

Singh, B. *et al.*, *J. Nat. Prod.*, 1989, **52**, 48-51 (*Nirphyllin*).

Shibuya, H. *et al.*, *Chem. Pharm. Bull.*, 1992, **40**, 2639-2646 (*Hazaleanin A*).

Ngouela, S. *et al.*, *Phytochemistry*, 1994, **37**, 867-869 (*hexa-Me ether*).

You, M. *et al.*, *J. Nat. Prod.*, 1995, **58**, 598-604 (*Bisdihydrosiringenin*).

Perez, C. *et al.*, *Phytochemistry*, 1995, **40**, 1511-1513 (*3,3′,5,5′-tetra-Me ether*).

Fuzzati, N. *et al.*, *Phytochemistry*, 1996, **42**, 1395-1398 (*hexa-Me ether di-Ac*).

Li, N. *et al.*, *J. Nat. Prod.*, 2003, **66**, 1421-1426 (*Peperomia duclouxii lignans*).

Li, N. *et al.*, *J. Nat. Prod.*, 2006, **69**, 234-239 (*Peperomia duclouxii constit*).

Zhao, Q. *et al.*, *Zhongguo Zhongyao Zazhi*, 2009, **34**, 1373-1376 (*9-Feruloylbisdihydrosiringenin*).

Li, X. *et al.*, *Chem. Nat. Compd. (Engl. Transl.)*, 2010, **46**, 106-108 (*Rhododendron primulaeflorum constit*).

Pan, J-Y. *et al.*, *Helv. Chim. Acta*, 2010, **93**, 951-957 (*Litsea glutinosa constit*).

Lee, K.Y. *et al.*, *Planta Med.*, 2010, **76**, 294-297 (*Lindera obtusiloba dicinnamates*).

Chen, T.-H. *et al.*, *Planta Med.*, 2010, **76**, 613-619 (*Cinnamomum osmophloeum constits*).

Kuroda, M. *et al.*, *J. Nat. Med.*, 2011, **65**, 198-201 (*Lindera umbellata diferulate*).

Nguyen, T.H. *et al.*, *Nat. Prod. Res.*, 2015, **29**, 247-252 (*Ufaside*).

Marchal, A. *et al.*, *Tetrahedron*, 2015, **71**, 3148-3156 (*Ssioriside, 5,5′-Dimethoxysecoisolariciresinol*)

8,8′-Lignan-3,3′,4,4′,5,7′,9,9′-octol

L63

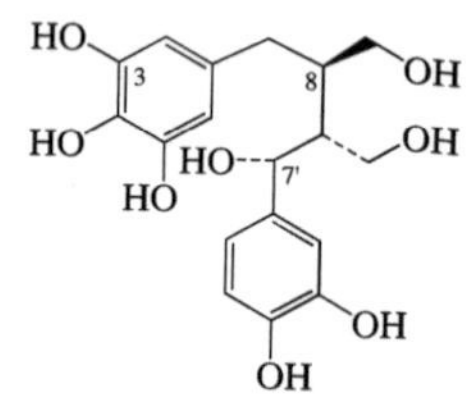

3,3′,4,4′,5,7′,9,9′-Octahydroxylignan

$C_{18}H_{22}O_8$ 366.366

(7′R,8R,8′R)-form

3,4-Methylene, 3′,4′,5,9,9′-penta-Me ether

3,3′,4,9,9′-Pentamethoxy-4′,5′-methylenedioxy-8,8′-lignan-7-ol.

Hydroxyniranthin

[113851-66-4]

$C_{24}H_{32}O_8$ 448.512

Isol. from *Phyllanthus niruri*. Oil. $[\alpha]_D^{28}$ -50 ($CHCl_3$). Isomeric with 3,3′,4′,9,9′-Pentamethoxy-4,5-methylenedioxy-8,8′-lignan-7-ol, in L64 and could be called 7′-Hydroxyniranthin.

Satyanarayana, P. *J. Nat. Prod.*, 1988, **51**, 44-49 (*Hydroxyniranthin*)

8,8′-Lignan-3,3′,4,4′,5,7,9,9′-octol

L64

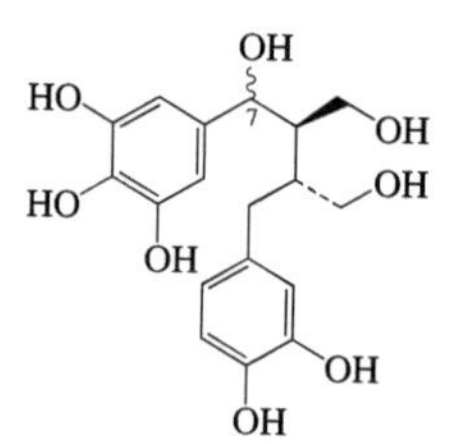

3,3′,4,4′,5,7,9,9′-Octahydroxylignan

$C_{18}H_{22}O_8$ 366.366

(7ξ,8R*,8′R*)-form

3,4-Methylene, 3′,4′,5,9,9′-penta-Me ether

3,3′,4′,9,9′-Pentamethoxy-4,5-methylenedioxy-8,8′-lignan-7-ol. 7-Hydroxy-3,3′,4′,9,9′-pentamethoxy-4,5-methylenedioxylignan. ***7-Hydroxyniranthin***

[1204520-92-2]

$C_{24}H_{32}O_8$ 448.512

Constit. of the leaves of *Phyllanthus amarus*. Viscous solid. $[\alpha]_D^{28}$ -19.6 (c, 0.12 in $CHCl_3$). λ_{max} 226 (log ε 3.75); 276 (log ε 3.25) (MeOH).

Singh, M. *et al., J. Asian Nat. Prod. Res.*, 2009, **11**, 562-568 (*7-Hydroxyniranthin*)

8,8′-Lignan-3,3′,4,4′,8,8′,9,9′-octol L65

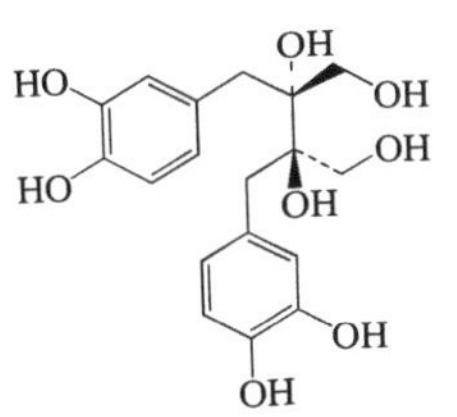

2,3-Bis[(3,4-dihydroxyphenyl)methyl]-1,2,3,4-butanetetrol. 2,3-Bis(3,4-dihydroxybenzyl)-1,2,3,4-butanetetrol. 3,3′,4,4′,8,8′,9,9′-Octahydroxylignan

$C_{18}H_{22}O_8$ 366.366

(8R*,8′R*)-form

3,3′-Di-Me ether

3,3′-Dimethoxy-8,8′-lignan-4,4′,8,8′,9,9′-hexol. 4,4′,8,8′,9,9′-Hexahydroxy-3,3′-dimethoxylignan. ***Kerinol***

[1114891-22-3]

$C_{20}H_{26}O_8$ 394.42

Constit. of the twigs of *Kerria japonica*. Amorph. powder. $[\alpha]_D^{22}$ -30.6 (c, 0.32 in MeOH).

Wu, J. *et al., J. Asian Nat. Prod. Res.*, 2008, **10**, 435-438 (*Kerinol*)

8,8′-Lignan-2,3′,4,4′,5-pentol L66

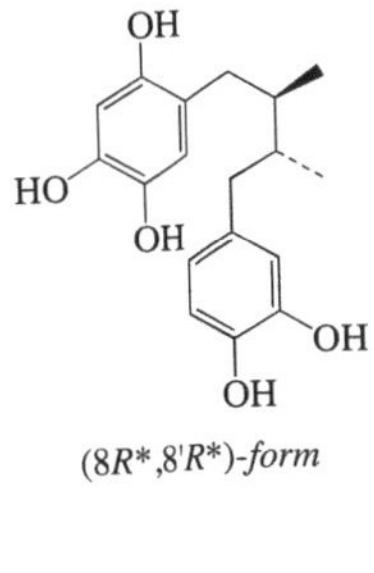

(8R*,8′R*)-form

1-(3,4-Dihydroxyphenyl)-2,3-dimethyl-4-(2,4,5-trihydroxyphenyl)butane. 2,3′,4,4′,5-Pentahydroxylignan

$C_{18}H_{22}O_5$ 318.369

(8R*,8′R*)-form

3′,4′:4,5-Bis(methylene) ether

3′,4′:4,5-Bis(methylenedioxy)-8,8′-lignan-2-ol. 2-Hydroxy-3′,4′:4,5-bis(methylenedioxy)lignan. ***Saururin C***

$C_{20}H_{22}O_5$ 342.391

Constit. of *Saururus chinensis*. Amorph. powder. $[\alpha]_D^{25}$ -15 (c, 0.2 in $CHCl_3$). λ_{max} 236 (log ε 3.1); 293 (log ε 2.84) (MeOH)

(8R*,8′S*)-form

4′,5-Di-Me ether

4′,5-Dimethoxy-8,8′-lignan-2,3′,4-triol. 2,3′,4-Trihydroxy-4′,5-dimethoxylignan. ***2′-Hydroxydihydroguaiaretic acid***

[805230-74-4]

$C_{20}H_{26}O_5$ 346.423

Constit. of *Saururus chinensis*. Antioxidant. Oil. $[\alpha]_D^{25}$ -13.8 (c, 0.29 in $CHCl_3$). λ_{max} 289 (ε 3500) (MeOH)

4,5-Methylene, 3′-Me ether

3′-Methoxy-4,5-methylenedioxy-8,8′-lignan-2,4′-diol. 2,4′-Dihydroxy-3′-methoxy-4,5-methylenedioxylignan. ***Saurulignan A***

[1532522-50-1]

$C_{20}H_{24}O_5$ 344.407

Constit. of *Saururus chinensis*. Amorph. brown powder. $[\alpha]_D^{26}$ -7 (c, 0.2 in MeOH). λ_{max} 231 (sh); 284 (log ε 3.56); 307 (sh) (MeOH).

Lee, W.S. *et al., Bioorg. Med. Chem. Lett.*, 2004, **14**, 5623-5628 (*2′-Hydroxydihydroguaiaretic acid*).
Tsai, W.-J. *et al., J. Nat. Prod.*, 2014, **77**, 125-131 (*Saurulignan A*).
Jin, Q. *et al., J. Nat. Prod.*, 2019, **82**, 3002-3009 (*Saururin C*)

8,8′-Lignan-3,3′,4,4′,5-pentol L67

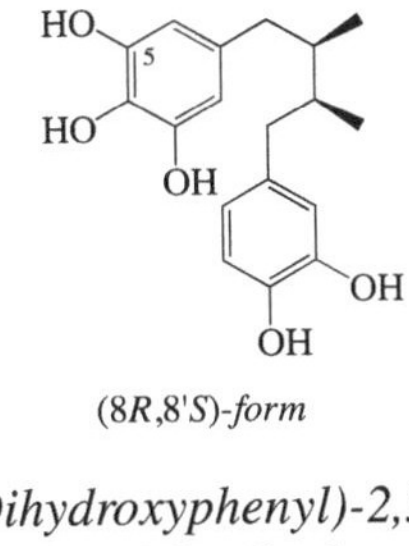

(8R,8′S)-form

1-(3,4-Dihydroxyphenyl)-2,3-dimethyl-4-(3,4,5-trihydroxyphenyl)butane. 2-(3,4-Dihydroxybenzyl)-3-(3,4,5-trihydroxybenzyl)butane. 3,3′,4,4′,5-Pentahydroxylignan

$C_{18}H_{22}O_5$ 318.369

(8R,8′S)-form

3,3′,5-Tri-Me ether, 4-O-β-D-glucopyranoside

[1566606-32-3]

$C_{27}H_{38}O_{10}$ 522.591

Constit. of bark of *Machilus robusta*. Amorph. powder. $[\alpha]_D^{20}$ -18.8 (c, 0.12 in MeOH). λ_{max} 230 (log ε 0.32); 278 (log ε 0.10) (MeOH)

(8S,8′R)-form

3,3′-Di-Me ether

3′,5-Dimethoxy-8,8′-lignan-3,4,4′-triol. 3,4,4′-Trihydroxy-3′,5-dimethoxylignan

[1310453-63-4]

$C_{20}H_{26}O_5$ 346.423

Isol. from the bark of *Machilus robusta*. Reduces galactosamine-induced hepatocyte damage. Amorph. powder. $[\alpha]_D^{20}$ +22.5 (c, 0.08 in $CHCl_3$). λ_{max} 232 (log ε 4.14); 281 (log ε 3.43) (MeOH)

3,3′,5-Tri-Me ether

3,3′,5-Trimethoxy-8,8′-lignan-4,4′-diol. 4,4′-Dihydroxy-3,3′,5-trimethoxylignan

[1310453-64-5]

$C_{21}H_{28}O_5$ 360.45

Isol. from the bark of *Machilus robusta*. Amorph. powder. $[\alpha]_D^{20}$ +6.01 (c, 0.98 in $CHCl_3$). $[\alpha]_D^{20}$ +2.5 (c, 0.56 in $CHCl_3$). λ_{max} 205 (log ε 4.94); 232 (log ε 3.71); 283 (log ε 3.23) (MeOH)

3,3′,4,5-Tetra-Me ether

3,3′,4′,5′-Tetramethoxy-8,8′-lignan-4-ol. 4′-Hydroxy-3,3′,4,5-tetramethoxylignan

[1310453-65-6]

$C_{22}H_{30}O_5$ 374.477

Isol. from the bark of *Machilus robusta*. Amorph. powder. $[\alpha]_D^{20}$ +4.4 (c, 0.09 in $CHCl_3$). λ_{max} 230 (log ε 3.99); 280 (log ε 3.35) (MeOH)

(8R*,8′R*)-form

3,4,4′-Tri-Me ether

4,4′,5-Trimethoxy-8,8′-lignan-3,3′-diol. 3,3′-Dihydroxy-4,4′,5-trimethoxylignan. ***Arisanschinin E***

[1333378-33-8]

$C_{21}H_{28}O_5$ 360.45

Constit. of *Schisandra arisanensis*. Amorph. solid. λ_{max} 229 (log ε 4.01); 279 (log ε 3.55) (MeOH)

3,4:3′,4′-Bis(methylene), 5-Me ether

3-Methoxy-3′,4′:4,5-bis(methylenedioxy)lignan. 5-Methoxyaustrobailignan 5. ***Saururin***

[128364-33-0]

$C_{21}H_{24}O_5$ 356.418

Isol. from *Saururus cernuus*. Glass. $[\alpha]_D$ -24.7 (c, 1 in $CHCl_3$)

(8R*,8′S*)-form

3′,4′-Methylene ether

3′,4′-Methylenedioxy-8,8′-lignan-3,4,5-triol. 3,4,5-Trihydroxy-3′,4′-methylenedioxylignan. ***Schibicolignan B***

[2255378-54-0]

$C_{19}H_{22}O_5$ 330.380

Constit. of the stems of *Schisandra bicolor*. Amorph. yellowish powder. $[\alpha]_D^{20}$ +6.3 (no solvent reported)

(8ξ,8′ξ)-form

3,3′,4-Tri-Me ether

3′,4,5-Trimethoxy-8,8′-lignan-3,4′-diol. 3,4′-Dihydroxy-3′,4,5-trimethoxylignan. ***Schineolignin C***

[1352185-28-4]

$C_{21}H_{28}O_5$ 360.45

Constit. of the fruit of *Schisandra chinensis*. Amorph. solid. $[\alpha]_D^{23}$ -2.9 (c, 0.31 in MeOH). λ_{max} 205 (log ε 4.44); 281 (log ε 3.26) (MeOH)

3,3′,5-Tri-Me ether

Guayacasin

[132194-64-0]

$C_{21}H_{28}O_5$ 360.45

Constit. of *Porlieria chilensis*

3′,4,4′-Tri-Me ether

3′,4,4′-Trimethoxy-8,8′-lignan-3,5-diol. 3,5-Dihydroxy-3′,4,4′-trimethoxylignan. ***Styrlignan A***

[1345413-44-6]

$C_{21}H_{28}O_5$ 360.45

Constit. of the stems of *Styrax japonica*. Exhibits moderate anti-HIV-1 activity. λ_{max} 193 (log ε 3.29); 206 (log ε 5.21); 230 (log ε 2.16); 242 (log ε 3.78); 276 (log ε 3.47) ($CHCl_3$)

3,3′,4,4′-Tetra-Me ether

3′,4,4′,5-Tetramethoxy-8,8′-lignan-3-ol. 3-Hydroxy-3′,4,4′,5-tetramethoxylignan. ***Schineolignin B***

[1352185-26-2]

$C_{22}H_{30}O_5$ 374.477

Constit. of the fruit of *Schisandra chinensis*. Amorph. solid. $[\alpha]_D^{23}$ +14.9 (c, 0.34 in MeOH). λ_{max} 205 (log ε 4.36); 279 (log ε 3.2) (MeOH)

Penta-Me ether

3,3′,4,4′,5-Pentamethoxylignan. ***Tiegusanin N***

[1156503-65-9]

$C_{23}H_{32}O_5$ 388.504

Constit. of *Schisandra propinqua* var. *sinensis*. Amorph. solid. λ_{max} 193 (log ε 3.29); 206 (log ε 3.24); 209 (log ε 3.22); 224 (log ε 3.26); 228 (log ε 3.25); 232 (log ε 3.26); 241 (log ε 3.7); 279 (log ε 3.22) ($CHCl_3$)

3′,4′-Methylene, 3-Me ether

5-Methoxy-3′,4′-methylenedioxy-8,8′-lignan-3,4-diol. 3,4-Dihydroxy-5-methoxy-3′,4′-methylenedioxylignan. ***Kadsurindutin E***

[1670237-95-2]

$C_{20}H_{24}O_5$ 344.407

Constit. of roots of *Kadsura coccinea*. Exhibits inhibitory effect on LPS-induced NO prodn. in BV-2 cells. Yellow oil. $[\alpha]_D^{20}$ +37 (c, 0.39 in MeOH)

3′,4′-Methylene, 3,4-di-Me ether

4,5-Dimethoxy-3′,4′-methylenedioxy-8,8′-lignan-3-ol. 3-Hydroxy-4,5-dimethoxy-3′,4′-methylenedioxylignan. ***Sphenanlignan***

[866347-36-6]

$C_{21}H_{26}O_5$ 358.434

Constit. of the seeds of *Schisandra sphenanthera* and of leaves and stems of *Schisandra propinqua*. $[\alpha]_D$ +17.4 (c, 1.1 in $CHCl_3$).

Rao, K.V. *et al.*, *J. Nat. Prod.*, 1990, **53**, 212-215 (*Saururin*).
Fauré, M. *et al.*, *Phytochemistry*, 1990, **29**, 3773-3775 (*Guayacasin*).
Jiang, S.-J. *et al.*, *Zhongguo Tianran Yaowu*, 2005, **3**, 78-82 (*Sphenanlignan*).
Lei, C. *et al.*, *Chem. Pharm. Bull.*, 2007, **55**, 1281-1283 (*Schisandra propinqua constit*).
Li, X.-N. *et al.*, *J. Nat. Prod.*, 2009, **72**, 1133-1141 (*Tiegusanin N*).
Xue, Y.-B. *et al.*, *Chem. Pharm. Bull.*, 2010, **58**, 1606-1612 (*Schineolignins B,C*).
Liu, C.-C. *et al.*, *Planta Med.*, 2010, **76**, 1605-1610 (*Arisanschinin E*).
Ren, R. *et al.*, *Z. Naturforsch., B*, 2010, **65**, 211-218 (*Schisandra sphenanthera constit*).
Wu, N. *et al.*, *Asian J. Chem.*, 2011, **23**, 931-932 (*Styrlignan A, activity*).
Li, Y. *et al.*, *J. Nat. Prod.*, 2011, **74**, 1444-1452 (*Machilus robusta constits, activity*).
Bu, P.-B. *et al.*, *J. Asian Nat. Prod. Res.*, 2013, **15**, 482-491 (*Machilus robusta glycoside*).
Fang, L. *et al.*, *Phytochem. Lett.*, 2014, **9**, 158-162 (*Kadsurindutin E, isol, activity*).
Mo, X. *et al.*, *Chem. Nat. Compd. (Engl. Transl.)*, 2018, **54**, 872-874 (*Schibicolignan B*)

8,8′-Lignan-3,3′,4,4′,7-pentol L68

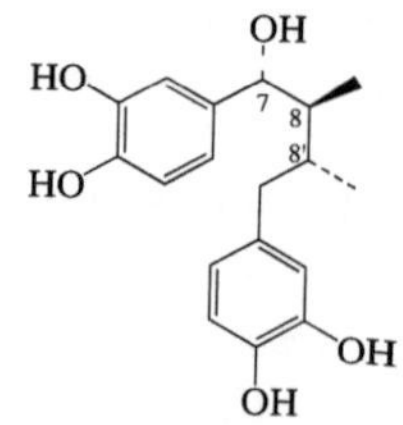

(7R,8S,8′R)-form

3,3′,4,4′,7-Pentahydroxylignan

$C_{18}H_{22}O_5$ 318.369

Lignan numbering shown

(7R,8S,8′R)-form

3,3′,4-Tri-Me ether

3,3′,4′-Trimethoxy-8,8′-lignan-4,7′-diol. 4,7′-Dihydroxy-3,3′,4′-trimethoxylignan. ***Oleiferin G***

[207388-14-5]

$C_{21}H_{28}O_5$ 360.45

Constit. of *Virola oleifera*. Yellow oil. $[\alpha]_D^{21}$ +13.3 (c, 0.8 in $CHCl_3$)

3,3′,4-Tri-Me ether, 7-Ac

7-O-Acetyloleiferin G

[467216-33-7]

$C_{23}H_{30}O_6$ 402.487

Constit. of the bark of *Talauma ovata*. Cryst. (Me_2CO). Mp 85°. $[\alpha]_D^{20}$ +69 (c, 0.02 in $CHCl_3$). λ_{max} 284 (log ε 3.93) (MeOH)

3,4-Methylene, 3′-Me ether

3-Methoxy-3′,4′-methylenedioxy-8,8′-lignan-4,7′-diol. 4,7′-Dihydroxy-3-methoxy-3′,4′-methylenedioxylignan. ***Oleiferin F***

[207388-13-4]

$C_{20}H_{24}O_5$ 344.407

Constit. of *Virola oleifera* and *Talauma ovata*. Yellow oil. $[\alpha]_D^{21}$ +12.6 (c, 0.2 in $CHCl_3$)

3,4-Methylene, 3′-Me ether, 7-Ac

7-O-Acetyloleiferin F

[467216-32-6]

$C_{22}H_{26}O_6$ 386.444

Constit. of the bark of *Talauma ovata*. Cryst. (Me_2CO). Mp 84 - 85°. $[\alpha]_D^{20}$ +80.9 (c, 0.02 in $CHCl_3$)

3,4-Methylene, 3′,4′-di-Me ether

3′,4′-Dimethoxy-3,4-methylenedioxy-8,8′-lignan-7-ol.

7-Hydroxy-3′,4′-dimethoxy-3,4-methylenedioxylignan. ***Oleiferin A***
[149415-70-3]
$C_{21}H_{26}O_5$ 358.434
Constit. of the leaves of *Virola oleifera*. Oil. $[\alpha]_D^{21}$ +41.4 (c, 0.83 in $CHCl_3$)

3′,4′-Methylene, 3,4-di-Me ether
3,4-Dimethoxy-3′,4′-methylenedioxy-8,8′-lignan-7-ol. 7-Hydroxy-3,4-dimethoxy-3′,4′-methylenedioxylignan. ***Oleiferin B***
[149496-15-1]
$C_{21}H_{26}O_5$ 358.434
Constit. of *Virola oleifera*. Oil. $[\alpha]_D^{21}$ +54.4 (c, 0.47 in $CHCl_3$)

3,4:3′,4′-Bis(methylene) ether
3,4:3′,4′-Bis(methylenedioxy)-8,8′-lignan-7-ol. 7-Hydroxy-3,4:3′,4′-bis(methylenedioxy)lignan. ***Oleiferin C***
[111786-59-5]
$C_{20}H_{22}O_5$ 342.391
Constit. of the leaves of *Virola oleifera* and from *Magnolia ovata*, *Iryanthera lancifolia* and *Machilus thunbergi*. Cytotoxic against human leukaemia HL-60 cells. Oil. $[\alpha]_D^{25}$ +44 (c, 0.77 in $CHCl_3$). λ_{max} 251 (ε 1023); 285 (ε 1230) ($CHCl_3$)

3,4:3′,4′-Bis(methylene) ether, 7-Ac
Acetyloleiferin C
[111786-56-2]
$C_{22}H_{24}O_6$ 384.428
Constit. of the bark of *Talauma ovata*. Cryst. (MeOH). Mp 82 - 83°. $[\alpha]_D$ +108.2 (c, 0.03 in $CHCl_3$). λ_{max} 232 (log ε 4.04); 284 (log ε 3.93) (MeOH)

3,4:3′,4′-Bis(methylene), 7-Et ether
7-Ethoxy-3,4:3′,4′-bis(methylenedioxy)lignan. ***O-Ethyloleiferin C***
[1028810-10-7]
$C_{22}H_{26}O_5$ 370.445
Constit. of the stem bark of *Iryanthera lancifolia*. Amorph. solid. $[\alpha]_D^{20}$ +43.2 (c, 0.9 in $CHCl_3$). λ_{max} 235 (log ε 4); 287 (log ε 3.9) (EtOH)

7-Ketone, tetra-Me ether
3,3′,4,4′-Tetramethoxy-7-lignanone
[106874-23-1]
$C_{22}H_{28}O_5$ 372.461
Constit. of *Nectandra puberula*. Viscous oil. $[\alpha]_D^{22}$ +83 (c, 0.9 in MeOH). Config. revised in 1994

7-Ketone, 3,4:3′,4′-bis(methylene) ether
7-Oxoaustrobailignan 5.
Saururinone
[106874-21-9]
$C_{20}H_{20}O_5$ 340.375
Constit. of *Saururus cernuus*. $[\alpha]_D$ +139.6 (c, 1 in $CHCl_3$). Incorrectly named in CA

(7S,8R,8′R)-form
3,4-Methylene, 3′,4′-di-Me ether
Magnovatin B
[1181210-42-3]
$C_{21}H_{26}O_5$ 358.434
Constit. of the leaves of *Magnolia ovata*. Oil. $[\alpha]_D^{25}$ +10.6 (c, 0.5 in $CHCl_3$). λ_{max} 208 (log ε 4.7); 234 (log ε 4.1); 287 (log ε 3.8) (MeOH)

3,4:3′,4′-Bis(methylene) ether
Magnovatin A
[1181210-41-2]
$C_{20}H_{22}O_5$ 342.391
Constit. of the leaves of *Magnolia ovata*. Oil. $[\alpha]_D^{25}$ +27.3 (c, 0.5 in $CHCl_3$). λ_{max} 208 (log ε 4.6); 239 (log ε 3.9); 291 (log ε 3.8) (MeOH)

(7S,8R,8′S)-form
3,4:3′,4′-Bis(methylene) ether
Saucerneol J
[1198210-45-5]
$C_{20}H_{22}O_5$ 342.391
Constit. of the roots of *Saururus chinensis*. Yellow oil. $[\alpha]_D^{20}$ -65.7 (c, 1.81 in $CHCl_3$). λ_{max} 234 (log ε 3.92); 287 (log ε 3.79) (MeOH)

7-Ketone, 3,3′-di-Me ether
4,4′-Dihydroxy-3,3′-dimethoxy-8,8′-lignan-7-one. ***Cinnamophilin‡***
[154677-96-0]
$C_{20}H_{24}O_5$ 344.407
Constit. of *Cinnamomum philippinense*, *Machilus litseifolia* and *Machilus philippinensis*. Shows antioxidant, antiplatelet aggregation and vasorelaxant props. Thromboxane A_2 receptor antagonist. Possesses cytoprotective and neuroprotective activities. Plates (Et_2O). Mp 89 - 91°. $[\alpha]_D$ +60.1 (c, 1.1 in $CHCl_3$). Not to be confused with Cinnamophilin A, in C221

(7S,8S,8′R)-form
3,4:3′,4′-Bis(methylene) ether
Saururinol. *7-Hydroxyaustrobailignan 5*
[169873-76-1]
[1029089-89-1]
$C_{20}H_{22}O_5$ 342.391
Constit. of *Osteophloeum sulcatum*, *Iryanthera lancifolia* and *Myristica fragrans*. Amorph. solid. $[\alpha]_D$ -24.7 (c, 0.47 in $CHCl_3$). $[\alpha]_D^{20}$ -75.2 (c, 0.2 in $CHCl_3$). Abs. configs. of isolates not certain. Abs. config. determined for the *I. lancifolia* constit., which has the higher opt. rotn.. λ_{max} 234 (3.4); 286 (3.3) (EtOH)

(7S,8S,8′S)-form
3,3′,4,4′-Tetra-Me ether
3,3′,4,4′-Tetramethoxy-8,8′-lignan-7-ol. 7-Hydroxy-3,3′,4,4′-tetramethoxylignan.
7-Hydroxy-4,4′-di-O-methyldihydroguaiaretic acid
[94595-78-5]
$C_{22}H_{30}O_5$ 374.477
Isol. from trunkwood of *Nectandra puberula* and fruits of *Virola sebifera*. Oil. Corrected struct.

(7R*,8R*,8′R*)-form
3′,4′-Methylene, 3,4-di-Me ether
[129684-07-7]
$C_{21}H_{26}O_5$ 358.434
Constit. of *Virola elongata*. Oil

(7R*,8R*,8′S*)-form
3,3′,4-Tri-Me ether
Oleiferin H
[207388-15-6]
$C_{21}H_{28}O_5$ 360.45
Constit. of *Virola oleifera*. Yellow oil. $[\alpha]_D^{21}$ -15.2 (c, 0.4 in $CHCl_3$)

7-Ketone, 3′,4′-methylene, 3-Me ether
4-Hydroxy-3-methoxy-3′,4′-methylenedioxy-8,8′-lignan-7-one.
Oleiferin D
[149415-71-4]
$C_{20}H_{22}O_5$ 342.391
Constit. of the leaves of *Virola oleifera*. Oil. $[\alpha]_D^{25}$ -52.5 (c, 0.16 in $CHCl_3$). λ_{max} 228 (ε 6310); 280 (ε 5012) (MeOH)

7-Ketone, 3′,4′-methylene, 3,4-di-Me ether
3,4-Dimethoxy-3′,4′-methylenedioxy-8,8′-lignan-7-one
[98649-95-7]
$C_{21}H_{24}O_5$ 356.418
Constit. of *Virola elongata*. Yellow oil

(7R*,8S*,8′S*)-form
3,3′-Di-Me ether
3,3′-Dimethoxy-8,8′-lignan-4,4′,7-triol. 4,4′,7-Trihydroxy-3,3′-dimethoxylignan.
Myristargenol B

[119802-37-8]
$C_{20}H_{26}O_5$ 346.423
Constit. of *Myristica argentea* and *Leucas aspera*. Cryst.. Mp 93°. $[\alpha]_D$ +14.2 (c, 0.79 in $CHCl_3$). λ_{max} 203 (ε 19055); 228 (ε 4365); 281 (ε 1862) (MeOH)

3′,4′-Methylene, 3-Me ether
Myristargenol A
[118169-28-1]
$C_{20}H_{24}O_5$ 344.407
Constit. of *Myristica argentea*. Shows neuroprotective activity. Needles. Mp 133°. $[\alpha]_D^{24}$ +9.9 (c, 0.81 in EtOH). λ_{max} 203 (ε 27542); 230 (ε 5370); 283 (ε 3090) (MeOH)

(8R,8′R)-form

7-Ketone
3,3′,4,4′-Tetrahydroxy-8,8′-lignan-7-one

7-Ketone, 3,4-methylene, 3′-Me ether
4′-Hydroxy-3′-methoxy-3,4-methylenedioxy-8,8′-lignan-7-one. ***Oxoaustrobailignan 6***
[106874-20-8]
$C_{20}H_{22}O_5$ 342.391
Constit. of *Nectandra puberula*. Viscous oil. $[\alpha]_D^{22}$ +111 (c, 0.79 in MeOH). Config. revised in 1994. Can also be called 7-oxoaustrobailignan 6: in the parent tetrol the 7 and 7′ posns. are equivalent

7-Ketone, 3,4-methylene, 3′,4′-di-Me ether
3′,4′-Dimethoxy-3,4-methylenedioxy-8,8′-lignan-7-one. ***4′-O-Methyl-7-oxoaustrobailignan 6***
[106874-22-0]
$C_{21}H_{24}O_5$ 356.418
Constit. of *Nectandra puberula*. $[\alpha]_D^{22}$ +118 (c, 0.53 in MeOH). Config. revised in 1994

7-Ketone, 3,4:3′,4′-bis(methylene) ether
3,4:3′,4′-Bis(methylenedioxy)-8,8′-lignan-7-one
$C_{20}H_{20}O_5$ 340.375
Constit. of *Nectandra puberula*. Viscous oil. $[\alpha]_D^{22}$ +88 (c, 0.82 in MeOH). Config. revised in 1994

7′-Oxo, 7-ketone, 3,4-methylene, 3′,4′-di-Me ether
4-O-Methyl-7,7′-dioxoaustrobailignan 6
[94513-79-8]
$C_{21}H_{22}O_6$ 370.401

Isol. from trunkwood of *Virola sebifera*. Amorph.. $[\alpha]_D^{25}$ -30.5 ($CHCl_3$). Abs. config. not certain.

Lopes, L.M.X. *et al., Phytochemistry*, 1984, **23**, 2647-2652 (*Virola sebifera constits*).
Martinez, J.C. *et al., Phytochemistry*, 1985, **24**, 1612-1614 (*Virola elongata constit*).
Moro, J.C. *et al., Phytochemistry*, 1987, **26**, 269-272 (*Nectandra puberula derivs, 7-Oxoaustrobailignan 5*).
Hattori, M. *et al., Chem. Pharm. Bull.*, 1988, **36**, 648-653 (*7-Hydroxyaustrobailignan 5*).
Nakatani, N. *et al., Phytochemistry*, 1988, **27**, 3127-3129 (*Myristargenols*).
Rao, K.V. *et al., J. Nat. Prod.*, 1990, **53**, 212-215 (*Saururinone*).
Kato, M.J. *et al., Phytochemistry*, 1990, **29**, 1799-1810 (*Virola elongata constit*).
Fernandes, A.M.A. *et al., Phytochemistry*, 1993, **32**, 1567-1572 (*Oleiferins A-D*).
Yu, S.-M. *et al., Eur. J. Pharmacol.*, 1994, **256**, 85-91 (*Cinnamophilin, activity*).
Fernandes, A.M.A. *et al., Phytochemistry*, 1994, **36**, 533-534 (*Oleiferins A-C, abs config*).
Wu, T.-S. *et al., Phytochemistry*, 1994, **36**, 785-788 (*Cinnamophilin, stereochem*).
Martinez, V. *et al., Rev. Colomb. Quim.*, 1994, **23**, 9-16 (*Saururinol*).
Sartorelli, P. *et al., Phytochemistry*, 1998, **47**, 1003-1006 (*Oleiferins F-H*).
Stefanello, M.E. *et al., Fitoterapia*, 2002, **73**, 135-139 (*Oleiferins C,F, Acetyloleiferins C,F,G*).
Sadhu, S.K. *et al., Chem. Pharm. Bull.*, 2003, **51**, 595-598 (*Leucas aspera constit*).
Park, B-Y. *et al., Biol. Pharm. Bull.*, 2004, **27**, 1305-1307 (*Oleiferin C, activity*).
Mesa-Siverio, D. *et al., Bioorg. Med. Chem.*, 2008, **16**, 3387-3394 (*Iryanthera lancifolia constits*).
Seo, C.S. *et al., Bull. Korean Chem. Soc.*, 2009, **30**, 716-718 (*Saucerneol J*).
Barros, L.F.L. *et al., J. Nat. Prod.*, 2009, **72**, 1529-1532 (*Magnovatins A,B, Oleiferin C*).
Tai, S.-H. *et al., RSC Adv.*, 2019, **9**, 21616-21625 (*Cinnamophilin, activity*).
Park, J.-S. *et al., RSC Adv.*, 2019, **9**, 31247-31254 (*Myristargenol A, activity*)

8,8′-Lignan-3,3′,4,4′,9-pentol L69

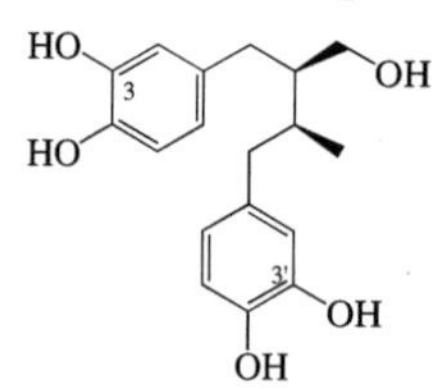

(8*R*,8′*S*)-*form*

2,3-Bis(3,4-dihydroxybenzyl)-1-butanol.
3,3′,4,4′,9-Pentahydroxylignan
$C_{18}H_{22}O_5$ 318.369

(8R,8′S)-form

3,3′-Di-Me ether, 9-Ac
Schibitubin F
[2094994-49-5]
$C_{22}H_{28}O_6$ 388.46
Constit. of the fruit of *Schisandra bicolor* var. *tuberculata*. Exhibits neuroprotective activity. Yellow oil. $[\alpha]_D^{20}$ -12.2 (c, 0.12 in $CHCl_3$). λ_{max} 210 (log ε 4.12); 285 (log ε 3.58); 370 (log ε 2.83) (MeOH)

3,4-Methylene, 3′-Me ether
3-Methoxy-3′,4′-methylenedioxy-8,8′-lignan-4,9′-diol. 4′,9-Dihydroxy-3′-methoxy-3,4-methylenedioxylignan. ***Schibitubin C***
[2094994-46-2]
$C_{20}H_{24}O_5$ 344.407
Constit. of the fruit of *Schisandra bicolor* var. *tuberculata*. Exhibits neuroprotective activity. Yellow oil. $[\alpha]_D^{20}$ -8.4 (c, 0.52 in MeOH). λ_{max} 205 (log ε 4.26); 230 (log ε 3.97); 285 (log ε 3.66); 370 (log ε 2.86) (MeOH)

3,4-Methylene, 3′-Me ether, 9-Ac
Schibitubin D
[2094994-47-3]
$C_{22}H_{26}O_6$ 386.444
Constit. of the fruit of *Schisandra bicolor* var. *tuberculata*. Amorph. solid. $[\alpha]_D^{20}$ -8.4 (c, 0.11 in $CHCl_3$). λ_{max} 205 (log ε 3.70); 285 (log ε 2.91); 370 (log ε 3.29) (MeOH)

3′,4′-Methylene, 3-Me ether, 9-Ac
Schibitubin E
[2094994-48-4]
$C_{22}H_{26}O_6$ 386.444
Constit. of the fruit of *Schisandra bicolor* var. *tuberculata*. Amorph. solid. $[\alpha]_D^{20}$ -7.7 (c, 0.17 in $CHCl_3$). λ_{max} 230 (sh); 285 (log ε 2.96); 370 (log ε 2.24) (MeOH)

(8S,8′R)-form

3,3′-Di-Me ether
3,3′-Dimethoxy-8,8′-lignan-4,4′,9-triol. ***Schibitubin B.*** *4,4′,9-Trihydroxy-3,3′-dimethoxylignan*
[2094994-45-1]
$C_{20}H_{26}O_5$ 346.423
Constit. of the fruit of *Schisandra bicolor* var. *tuberculata*. Exhibits neuroprotective activity. Yellow oil. $[\alpha]_D^{20}$ -2.2 (c, 0.31 in $CDCl_3$). λ_{max} 230 (sh); 285 (log ε 3.62) (MeOH)

3′,4′-Methylene, 3-Me ether
3-Methoxy-3′,4′-methylenedioxy-8,8′-lignan-4,9-diol. ***Schibitubin A.*** *4,9-Dihydroxy-3-methoxy-3′,4′-methylenedioxylignan*
[2094994-44-0]
$C_{20}H_{24}O_5$ 344.407

Constit. of the fruit of *Schisandra bicolor* var. *tuberculata*. Cryst.. $[\alpha]_D^{20}$ +2.9 (c, 0.35 in $CDCl_3$). λ_{max} 230 (sh); 285 (log ε 3.69); 370 (log ε 3.03) (MeOH).

Liu, Y. *et al.*, *J. Nat. Prod.*, 2017, **80**, 1117-1124 (*Schibitubins A-F, cd, cryst struct, activity*)

8,8′-Lignan-3,3′,4′,5,7-pentol L70

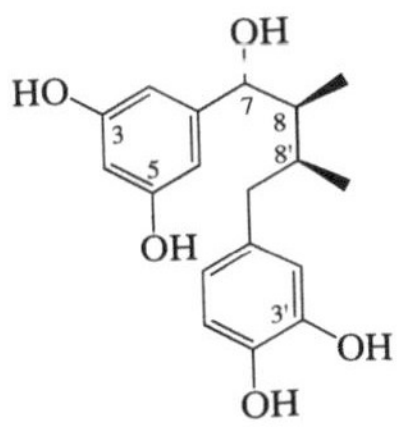

3,3′,4′,5,7-Pentahydroxylignan
$C_{18}H_{22}O_5$ 318.369
Structures in this series may need revision. Phenyl 3,5-dioxygenated substn. pattern in doubt as a result of work done by Yang *et al* in 2015
(7R*,8S*,8′S*)-form
3′,4′-Methylene, 3-Me ether
5-Methoxy-3′,4′-methylenedioxy-8,8′-lignan-3,7-diol. 3,7-Dihydroxy-5-methoxy-3′,4′-methylenedioxylignan
[1886973-85-8]
$C_{20}H_{24}O_5$ 344.407
Constit. of the roots of *Schisandra sphenanthera*. Antioxidant. Amorph. solid.

Jiang, K. *et al.*, *Fitoterapia*, 2015, **103**, 63-70 (*Schisandra sphenanthera constit*).
Yang, Y.-N. *et al.*, *J. Nat. Prod.*, 2015, **78**, 705-711 (*pmr, struct*)

3,8′-Lignan-3′,5,9,9′-tetrol L71

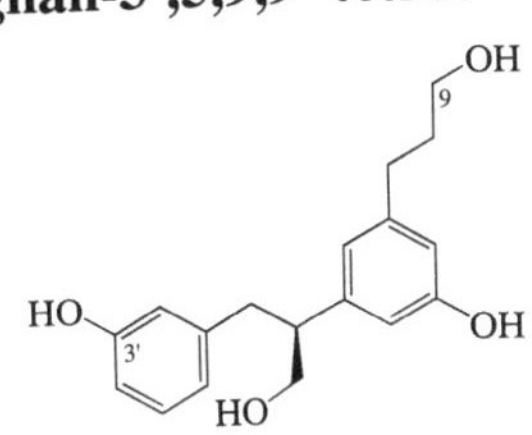

3′,5,9,9′-Tetrahydroxy-3,8′-neolignan
$C_{18}H_{22}O_4$ 302.37
(S)-form
9-Carboxylic acid
3′,5,9′-Trihydroxy-3,8′-lignan-9-oic acid
9-Carboxylic acid, 9-Me ester
Trogopterin B
[1656285-97-0]
$C_{19}H_{22}O_5$ 330.38
Isol. from faeces of *Trogopterus xanthipes*. Exhibits moderate cytotoxicity against HL-60 cells. Light brown oil. $[\alpha]_D^{25}$ +20.2 (c, 1.0 in EtOH).

Baek, S. *et al.*, *Beilstein J. Org. Chem.*, 2014, **10**, 2955-2962 (*Trogopterin B, activity*)

8,8′-Lignan-3,3′,4,4′-tetrol L72

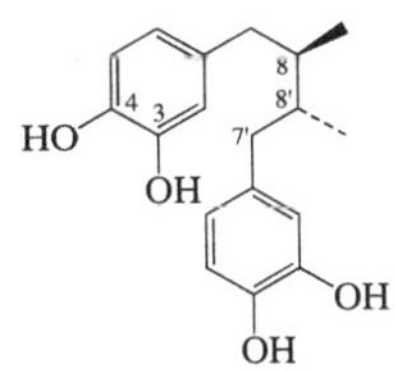

(8*R*,8′*R*)-*form*

1,4-Bis(3,4-dihydroxyphenyl)-2,3-dimethylbutane. 2,3-Bis(3,4-dihydroxybenzyl)butane. 3,3′,4,4′-Tetrahydroxylignan. Bisdemethyldihydroguaiaretic acid. ***Nordihydroguaiaretic acid.*** *Norhydroguaiaretic acid. NDGA*
[500-38-9]
[68964-82-9] [90365-59-6]
[103185-28-0] [119182-23-9]
[124649-80-5]
$C_{18}H_{22}O_4$ 302.37
Lignan numbering shown. Lipoxygenase inhibitor. Antineoplastic agent. Antioxidant. Antibacterial agent. Formerly approved as a food additive; now banned by FDA (USA). Log P 3.92 (calc)
▶ UX1750000
(8R,8′R)-form [119584-40-6]
Isol. from *Guaiacum officinale* and *Larrea* sp.. Inhibitor of human leukocyte elastase
3,3′-Di-Me ether
see 3,3′-Dimethoxy-8,8′-lignan-4,4′-diol, D263
3,4-Methylene, 3′-Me ether
3-Methoxy-3′,4′-methylenedioxy-8,8′-lignan-4-ol. 4-Hydroxy-3-methoxy-3′,4′-methylenedioxylignan. ***Austrobailignan 6***
[55890-24-9]
$C_{20}H_{24}O_4$ 328.408
Constit. of *Austrobaileya scandens*, *Saururus cernuus* and *Kadsura longipedunculata*. Exhibits insecticidal and larvicidal props.. Oil. Bp 100 - 120° at 0.02 mm. $[\alpha]_D^{25}$ -32 (c, 1.3 in $CHCl_3$). $[\alpha]_D^{25}$ -25 (c, 0.9 in $CHCl_3$)
3,4-Methylene, 3′,4′-di-Me ether
3,4-Dimethoxy-3′,4′-methylenedioxylignan. ***Saururenin***
[128594-97-8]
$C_{21}H_{26}O_4$ 342.435
Constit. of *Saururus cernuus* and *Kadsura longipedunculata*. Glass. $[\alpha]_D$ -34 (c, 1 in $CHCl_3$)
Bis(methylene) ether
see 3,4:3′,4′-Bis(methylenedioxy)lignan, B61
(8R,8′S)-form
Masoprocol, INN, USAN. Actinex. CHX 100. [27686-84-6]
Investigated for treatment of actinic keratoses, however potent skin sensitising effects observed. Launched 1992 (US). Cryst.. Mp 185 - 186°. Subl. 175 - 180° at 0.3 mm
3,3′,4,4′-Tetra-O-β-D-glucopyranoside
Larrealignan A
[1393342-98-7]
$C_{42}H_{62}O_{24}$ 950.934
Constit. of *Larrea tridentata*. Amorph. solid. $[\alpha]_D^{25}$ +21.1 (c, 0.1 in MeOH). λ_{max} 273 (log ε 3.48) (MeOH)
Tetra-Me ether
3,3′,4,4′-Tetramethoxylignan. ***Terameprocol,*** *USAN. EM 1421. M4N*
[24150-24-1]
$C_{22}H_{30}O_4$ 358.478
CDC2/CDK1 transcriptional inhibitor. Antineoplastic agent. Cryst. (MeOH). Mp 100 - 101°
3,4-Methylene, 3′-Me ether
Anwulignan. Calophyn
[114127-24-1]
$C_{20}H_{24}O_4$ 328.408
Constit. of *Kadsura longipedunculata*, *Schisandra sephenanthera* and *Virola calophylla*. COX-2 inhibitor. Mp 66 - 68°. $[\alpha]_D$ 0. λ_{max} 234 (log ε 4.04); 286 (log ε 3.86) (MeOH)
3′,4′-Methylene, 3-Me ether
Macelignan
[107534-93-0]
$C_{20}H_{24}O_4$ 328.408
Constit. of *Myristica fragrans*, *Myristica argentea*, bark of *Kadsura longipedunculata* and *Schisandra henryi*. Active against P388 leukaemia and hepatitis B. Cryst. (hexane). Mp 70 - 72°. $[\alpha]_D^{20}$ +5.28 (c, 1.8 in $CHCl_3$). Information on abs. config. and identity of the various samples not fully clear
(8S,8′S)-form [119584-39-3]

Constit. of *Larrea tridentata*
3,3′,4-Tri-Me ether
3,3′,4′-Trimethoxy-8,8′-lignan-4-ol. 4-Hydroxy-3,3′,4′-trimethoxylignan
[174291-51-1]
$C_{21}H_{28}O_4$ 344.451
Constit. of *Larrea tridentata*
3′,4′-Methylene, 3-Me ether
Isoanwulignan
[143601-12-1]
$C_{20}H_{24}O_4$ 328.408
Constit. of *Schisandra viridis* and *Kadsura longipedunculata*
(8R*,8′S*)-form
3-Me ether
3′-Methoxy-8,8′-lignan-3,4,4′-triol. 3,4,4′-Trihydroxy-3′-methoxylignan. Demethyldihydroguaiaretic acid
[71113-15-0]
$C_{19}H_{24}O_4$ 316.397
Constit. of *Larrea tridentata*. Exhibits anti-HIV activities
3-Me ether, 3′,4,4′-tri-O-β-D-glucopyranoside
Larrealignan B
[1393342-99-8]
$C_{37}H_{54}O_{19}$ 802.82
Constit. of *Larrea tridentata*. Amorph. solid. $[\alpha]_D^{26}$ -26.8 (c, 0.1 in MeOH). λ_{max} 275 (log ε 3.55) (MeOH)
4-Me ether
4′-Methoxy-8,8′-lignan-3,3′,4-triol. 3,3′,4-Trihydroxy-4′-methoxylignan
[171204-42-5]
$C_{19}H_{24}O_4$ 316.397
Constit. of *Larrea tridentata*. $[\alpha]_D^{25}$ +76 (c, 0.2 in MeOH)
3,4,4′-Tri-Me ether
3′,4,4′-Trimethoxy-8,8′-lignan-3-ol. 3-Hydroxy-3′,4,4′-trimethoxylignan
[171204-39-0]
$C_{21}H_{28}O_4$ 344.451
Constit. of *Larrea tridentata*
3,4-Methylene, 3′,4′-di-Me ether
[299931-30-9]
Constit. of *Manglietiastrum sinicum* (preferred genus name *Magnolia*). Oil. $[\alpha]_D^{26}$ +0.8 (c, 0.3 in $CHCl_3$).

Aldrich Library of FT-IR Spectra, 1st edn., 1985, **1**, 1119D (*ir*).
Aldrich Library of [13]*C and* [1]*H FT-NMR Spectra*, 1992, **2**, 325A (*nmr*).
Perry, C.W. *et al.*, *J. Org. Chem.*, 1972, **37**, 4371-4376 (*bibl, uv, ir, pmr, synth*).
Gisvold, O. *et al.*, *J. Pharm. Sci.*, 1974, **63**, 1905-1907 (*isol, synth*).
Murphy, S.T. *et al.*, *Aust. J. Chem.*, 1975, **28**, 81-90 (*Austrobailignan 6*).
Biftu, T. *et al.*, *J. Chem. Soc., Perkin Trans. 1*, 1978, 1147-1150 (*synth*).
Liu, J.S. *et al.*, *Huaxue Xuebao*, 1984, **42**, 264-270 (*Anwulignan*).
Woo, W.S. *et al.*, *Phytochemistry*, 1987, **26**, 1542-1543 (*Myristica fragrans constit, cryst struct, Macelignan*).
Alvarez, V. *et al.*, *CA*, 1988, **108**, 183594 (*Calophyn*).
Liu, J.S. *et al.*, *Youji Huaxue*, 1988, **8**, 227 (*Anwulignan*).
Steele, V.E. *et al.*, *Cancer Res.*, 1990, **50**, 2068-2074 (*masoprocol, pharmacol*).
Rao, K.V. *et al.*, *J. Nat. Prod.*, 1990, **53**, 212-215 (*Saururenin*).
Rao, K.V. *et al.*, *J. Org. Chem.*, 1990, **55**, 1427-1429 (*Austrobailignan 6, (R,R)-3,4-Dimethoxy-3′,4′-methylenedioxylignan, synth*).
Olsen, E.A. *et al.*, *J. Am. Acad. Dermatol.*, 1991, **24**, 738-743 (*masoprocol, use*).
Luo, G. *et al.*, *Huaxue Xuebao*, 1992, **50**, 515-520 (*Isoanwulignan*).
Epstein, E. *J. Am. Acad. Dermatol.*, 1994, **31**, 295-297 (*masoprocol, tox*).
Gnabre, J. *et al.*, *Tetrahedron*, 1995, **51**, 12203-12210 (*Larrea tridentata constits, activity*).
Fun, H.-K. *et al.*, *Acta Crystallogr., Sect. C*, 1997, **53**, 1857-1859 (*terameprocol, cryst struct*).
Hwu, J.R. *et al.*, *J. Med. Chem.*, 1998, **41**, 2994-3000 (*terameprocol, synth, pmr, pharmacol*).
Wang, B.-G. *et al.*, *Planta Med.*, 2000, **66**, 511-515 (*Manglietiastrum sinicum constit*).
Heller, J.D. *et al.*, *Cancer Res.*, 2001, **61**, 5499-5504 (*terameprocol, pharmacol*).
Melzig, M.F. *et al.*, *Pharmazie*, 2001, **56**, 967-970 (*8R,8′R-form, activity*).
Gezginci, M.H. *et al.*, *Tetrahedron Lett.*, 2001, **42**, 6083-6085 (*Nordihydroguaiaretic acid, synth*).
Xia, Y.M. *et al.*, *Chin. Chem. Lett.*, 2003, **14**, 359-360 (*Nordihydroguaiaretic acid, synth*).
Lambert, J.D. *et al.*, *Pharm. Biol.*, 2004, **42**, 149-158 (*rev, activity*).
Chang, C.C. *et al.*, *Proc. Natl. Acad. Sci. U.S.A.*, 2004, **101**, 13239-13244 (*terameprocol, pharmacol*).
Hansel, D.E. *et al.*, *Am. J. Surg. Pathol.*, 2005, **29**, 390-399 (*terameprocol, pharmacol*).
Lopez, R.A. *et al.*, *Anti-Cancer Drugs*, 2007, **18**, 933-939 (*terameprocol, pharmacol*).
Martindale, The Complete Drug Reference, 35th edn., Pharmaceutical Press, 2007, 1490 (*NDGA, props*).
Pu, J.-X. *et al.*, *Chem. Pharm. Bull.*, 2008, **56**, 1143-1146 (*Kadsura longipedunculata constit*).
Smolewski, P. *IDrugs*, 2008, **11**, 204-214 (*terameprocol, pharmacol*).
Nishiwaki, H. *et al.*, *Biosci., Biotechnol., Biochem.*, 2011, **75**, 1735-1739 (*Austrobailignan 6, activity*).
Yokosuka, A. *et al.*, *Chem. Pharm. Bull.*, 2011, **59**, 1467-1470 (*Larrealignans A,B*).
Zaugg, J. *et al.*, *Phytochemistry*, 2011, **72**, 2385-2395 (*Saururenin*).
Wukirsari, T. *et al.*, *J. Agric. Food Chem.*, 2013, **61**, 4318-4325 (*Austrobailignan 6, activity*).
Tibes, R. *et al.*, *Invest. New Drugs*, 2015, **33**, 389-396 (*terameprocol, pharmacol*).
Chen, J.-B. *et al.*, *Zhongcaoyao*, 2015, **46**, 178-184 (*Anwulignan*).
Le, T.V.T. *et al.*, *Nat. Prod. Sci.*, 2017, **23**, 21-28 (*Macelignan*).
Zhang, X.-Y. *et al.*, *Fitoterapia*, 2021, **151**, 104872:1-7 (*Anwulignan, activity*).
Lewis, R.J. *Sax's Dangerous Properties of Industrial Materials, 12th edn., J. Wiley*, 2012, NBR000 (*NDGA*)

8,8′-Lignan-3,3′,9,9′-tetrol L73

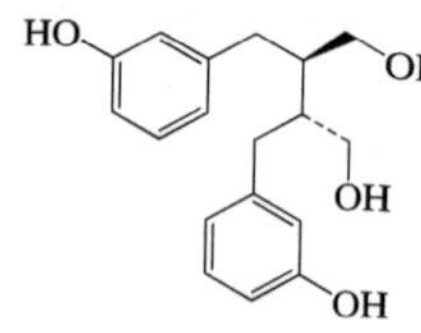

(8*R**,8′*R**)-*form*

2,3-Bis[(3-hydroxyphenyl)methyl]-1,4-butanediol, CAS. *2,3-Bis(3-hydroxybenzyl)-1,4-butanediol. 3,3′,9,9′-Tetrahydroxylignan*
[80226-00-2]
$C_{18}H_{22}O_4$ 302.37
Log P 1.14 (calc)
(8R*,8″R*)-form
Enterodiol
Compd. of lignan-type isol. from mammalian urine. Formed by the action of gut flora on 4,4′-Dihydroxy-3,3′-dimethoxy-8,8′-lignan-9,9′-olide, D141 present in grain
(8RS,8′RS)-form
[76543-16-3]
Reduces binding of dihydrotestosterone to human sex hormone binding globulin. Mp 148 - 150°. Mp 165 - 167°. λ_{max} 276; 283 (no solvent reported).

Setchell, K.D.R. *et al.*, *Nature (London)*, 1980, **287**, 740-742 (*isol*).
Setchell, K.D.R. *et al.*, *Biochem. J.*, 1981, **197**, 447-458 (*urine constit, spectra*).
Cooley, G. *et al.*, *Tetrahedron Lett.*, 1981, **22**, 349-350 (*synth*).
Axelson, M. *et al.*, *Nature (London)*, 1982, **298**, 659-660 (*occur*).
Bannwart, C. *et al.*, *Finn. Chem. Lett.*, 1984, 120-125 (*biosynth*).
Borriello, S.P. *et al.*, *J. Appl. Bacteriol.*, 1985, **58**, 37-43 (*metab*).
Kirk, D.N. *et al.*, *J. Chem. Soc., Perkin Trans. 1*, 1985, 35-37 (*synth*).
Belletire, J.L. *et al.*, *Tetrahedron Lett.*, 1986, **27**, 127-130 (*synth*).
Van Oeveren, A. *et al.*, *J. Org. Chem.*, 1994, **59**, 5999-6007 (*synth*).
Schottner, M. *et al.*, *J. Nat. Prod.*, 1998, **61**, 119-121 (*activity*).
Ward, R.S. *et al.*, *Tetrahedron*, 2001, **57**, 2057-2064 (*synth*).
Raffaelli, B. *et al.*, *J. Chromatogr., B*, 2002, **777**, 29-43 (*rev*).
Wang, L.-Q. *J. Chromatogr., B*, 2002, **777**, 289-309 (*rev, chem, activity*).
Eklund, P. *et al.*, *Org. Lett.*, 2003, **5**, 491-493 (*synth*).

Landete, J.M. *Food Res. Int.*, 2012, **46**, 410-424 (*rev, occur, metab*).
Shoba, V.M. *et al., J. Am. Chem. Soc.*, 2017, **139**, 5740-5743 (*synth*)

8,8′-Lignan-3,4,4′,5-tetrol L74

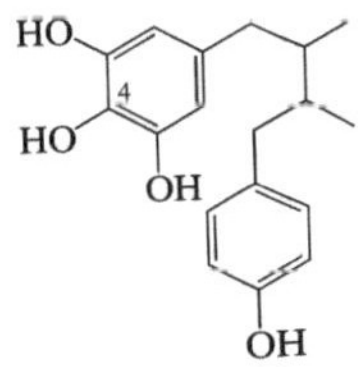

1-(4-Hydroxyphenyl)-2,3-dimethyl-4-(3,4,5-trihydroxyphenyl)butane. 2-(4-Hydroxybenzyl)-3-(3,4,5-trihydroxybenzyl)butane. 3,4,4′,5-Tetrahydroxylignan
$C_{18}H_{22}O_4$ 302.37
(8ξ,8′ξ)-form
3,4-Di-Me ether
3,4′-Dihydroxy-4,5-dimethoxylignan. ***Schineolignin A***
[1352185-24-0]
$C_{20}H_{26}O_4$ 330.424
Constit. of the fruit of *Schisandra chinensis*. Amorph. solid. $[\alpha]_D^{19}$ -18.2 (c, 0.26 in MeOH). λ_{max} 205 (log ε 4.34); 278 (log ε 3.28) (MeOH).
Xue, Y.-B. *et al., Chem. Pharm. Bull.*, 2010, **58**, 1606-1612 (*Schineolignin A*)

8,8′-Lignan-2,4,4′-triol L75

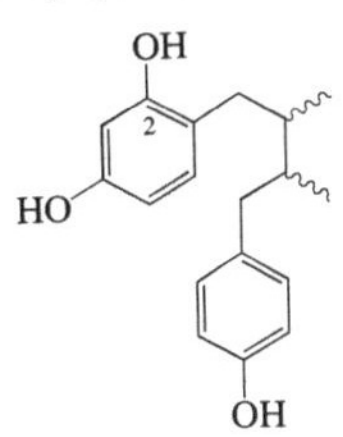

1-(2,4-Dihydroxyphenyl)-4-(4-hydroxyphenyl)-2,3-dimethylbutane. 2-(2,4-Dihydroxybenzyl)-3-(4-hydroxybenzyl)butane. 2,4,4′-Trihydroxylignan
$C_{18}H_{22}O_3$ 286.371
2-Me ether
4-[4-(4-Hydroxyphenyl)-2,3-dimethylbutyl]-3-methoxyphenol, CAS. *1-(4-Hydroxy-2-methoxyphenyl)-4-(4-hydroxyphenyl)-2,3-dimethylbutane. 2-(4-Hydroxybenzyl)-3-(4-hydroxy-2-methoxybenzyl)butane. 2-Methoxy-8,8′-lignan-4,4′-diol. 4,4′-Dihydroxy-2-methoxylignan.* ***2-Methoxyconocarpol***
[56319-01-8]
$C_{19}H_{24}O_3$ 300.398
Isol. from *Conocarpus erectus*. Cryst. (C_6H_6). Mp 129 - 130°. Shows weak opt. activity.
Hayashi, T. *et al., Phytochemistry*, 1975, **14**, 1085-1087 (*2-Methoxyconocarpol*)

8,8′-Lignan-3,4,4′-triol L76

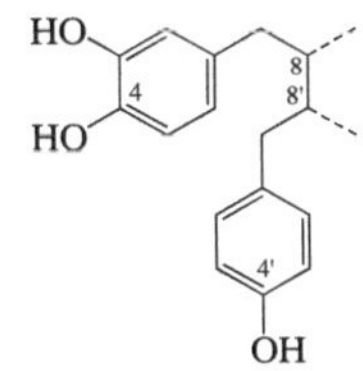

(8*S*,8′*R*)-*form*

1-(3,4-Dihydroxyphenyl)-4-(4-hydroxyphenyl)-2,3-dimethylbutane. 2-(3,4-Dihydroxybenzyl)-3-(4-hydroxybenzyl)butane. 3,4,4′-Trihydroxylignan
$C_{18}H_{22}O_3$ 286.371
Lignan numbering shown
(8S,8′R)-form
3-Me ether
4,4′-Dihydroxy-3-methoxylignan
[1310804-33-1]
$C_{19}H_{24}O_3$ 300.398
Isol. from the bark of *Machilus robusta*. Amorph. powder. $[\alpha]_D^{20}$ +8.7 (c, 0.08 in MeOH). λ_{max} 202 (log ε 4.08); 224 (log ε 3.84); 280 (log ε 3.19) (MeOH)
(8R*,8′R*)-form
3,4-Methylene ether
4′-Hydroxy-3,4-methylenedioxylignan
[1028810-07-2]
$C_{19}H_{22}O_3$ 298.382
Constit. of the stem bark of *Iryanthera lancifolia*. Amorph. solid. $[\alpha]_D^{20}$ -22.9 (c, 0.5 in $CHCl_3$). λ_{max} 226 (log ε 4); 286 (log ε 3.7) (EtOH)
(8R*,8′S*)-form [171204-40-3]
Constit. of the leaves of *Larrea tridentata*. Cryst.. Mp 125 - 127°
3-Me ether
Pycnantolol
[1158791-46-8]
Constit. of the stem bark of *Pycnanthus angolensis*. Oil. $[\alpha]_D$ -13.5 (c, 0.12 in $CHCl_3$).
Gnabre, J. *et al., Tetrahedron*, 1995, **51**, 12203-12210 (*Larrea tridentata constit*).
Mesa-Siverio, D. *et al., Bioorg. Med. Chem.*, 2008, **16**, 3387-3394 (*3,4-methylene ether*).
Abrantes, M. *et al., Planta Med.*, 2008, **74**, 1408-1412 (*Pycnantolol*).
Li, Y. *et al., J. Nat. Prod.*, 2011, **74**, 1444-1452 (*Machilus robusta constit*)

8,8′-Lign-7-ene-4,4′-diol L77

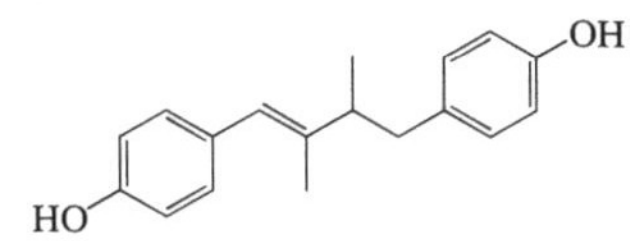

1,4-Bis(4-hydroxyphenyl)-2,3-dimethyl-1-butene. 4,4′-Dihydroxylign-7-ene
$C_{18}H_{20}O_2$ 268.356
(7E,8′ξ)-form [1028810-08-3]
Constit. of the stem bark of *Iryanthera lancifolia*. Shows estrogenic activity. Amorph. solid. $[\alpha]_D^{20}$ -46.2 (c, 0.3 in $CHCl_3$). λ_{max} 226 (log ε 4); 286 (log ε 3.7) (EtOH).
Mesa-Siverio, D. *et al., Bioorg. Med. Chem.*, 2008, **16**, 3387-3394 (*Iryanthera lancifolia constit*)

2,8′-Lign-7-ene-3′,4,4′,5,7′,9,9′-heptol L78

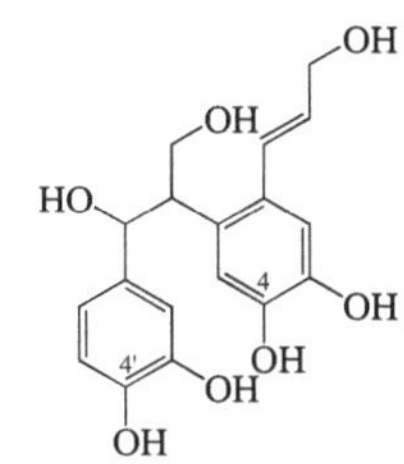

3′,4,4′,5,7′,9,9′-Heptahydroxy-2,8′-neolign-7-ene
$C_{18}H_{20}O_7$ 348.351
(7E,7′ξ,8′ξ)-form
4,4′-Di-Me ether
2-[4-Hydroxy-2-(3-hydroxy-1-propenyl)-5-methoxyphenyl]-1-(3-hydroxy-4-methoxyphenyl)-1,3-propanediol. 4,4′-Dimethoxy-2,8′-lign-7-ene-3′,5,7′,9,9′-pentol. 3′,5,7′,9,9′-Pentahydroxy-4,4′-dimethoxy-2,8′-lign-7-ene
[938054-61-6]
$C_{20}H_{24}O_7$ 376.405
Constit. of *Erythrina lysistemon*. Brown solid. Mp 68 - 71°.
Juma, B.F. *et al., Nat. Prod. Commun.*, 2006, **1**, 101-107 (*Erythrina lysistemon constit*)

3,8′-Lign-7-ene-3′,4,4′,5,7′,9,9′-heptol L79

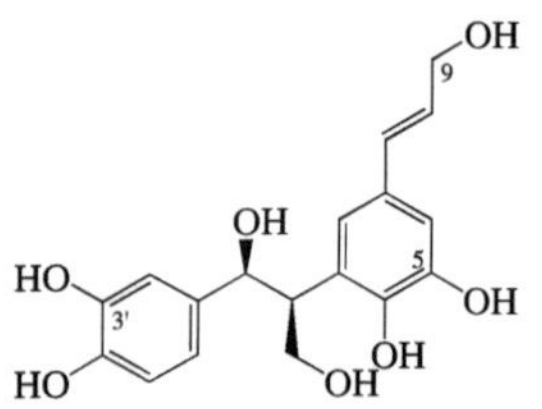

(7E,7′S,8′S)-form

3′,4,4′,5,7′,9,9′-Heptahydroxy-3,8′-neolign-7-ene

$C_{18}H_{20}O_7$ 348.351

(7E,7′S,8′S)-form

9-Aldehyde

3′,4,4′,5,7′,9′-Hexahydroxy-3,8′-lign-7-en-9-al. 3′,4,4′,5,7′,9′-Hexahydroxy-9-oxo-3,8′-lign-7-ene

9-Aldehyde, 3′,5-di-Me ether

4,4′,7′,9′-Tetrahydroxy-3′,5-dimethoxy-3,8′-lign-7-en-9-al. 4,4′,7′,9′-Tetrahydroxy-3′,5-dimethoxy-9-oxo-3,8′-lign-7-ene.

Prebalanophonin

[1887236-26-1]

$C_{20}H_{22}O_7$ 374.389

Constit. of the fruit of *Cirsium eriophorum*

(7E,8′ξ)-form

7′-Ketone, 3′,5-di-Me ether

4,4′,9,9′-Tetrahydroxy-3′,5-dimethoxy-3,8′-lign-7-en-7′-one. 4,4′,9,9′-Tetrahydroxy-3′,5-dimethoxy-7′-oxo-3,8′-lign-7-ene

7′-Ketone, 3′,5-di-Me ether, 4-O-β-D-glucopyranoside

Echinolignan A

[606138-65-2]

$C_{26}H_{32}O_{12}$ 536.53

Constit. of the roots of *Echinops grijsii*. Amorph. yellow powder. Mp 150 - 152°. $[\alpha]^{25}_{D}$ +25.3 (c, 0.1 in MeOH). λ_{max} 220 (log ε 4.52); 260 (log ε 4.08); 306 (log ε 3.95) (MeOH).

Koike, K. *et al.*, *Nat. Med. (Tokyo)*, 2002, **56**, 255-257 (*Echinolignan A*).
Sólyomváry, A. *et al.*, *Process Biochem.*, 2015, **50**, 853-858 (*Prebalanophonin*).
Könye, R. *et al.*, *Fitoterapia*, 2018, **127**, 413-419 (*Prebalanophorin*)

3,9′-Lign-7-ene-3′,4,4′,5,7′,9,9′-heptol L80

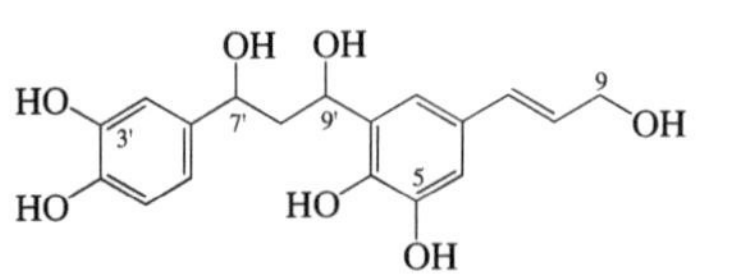

1-[2,3-Dihydroxy-5-(3-hydroxy-1-propenyl)phenyl]-3-(3,4-dihydroxyphenyl)-1,3-propanediol. 3′,4,4′,5,7′,9,9′-Heptahydroxy-3,9′-neolign-7-ene

$C_{18}H_{20}O_7$ 348.351

Lignan numbering shown

(7E,7′ξ,9′ξ)-form

3′,5-Di-Me ether

3′,5-Dimethoxy-3,9′-lign-7-ene-4,4′,7′,9,9′-pentol. 4,4′,7′,9,9′-Pentahydroxy-3′,5-dimethoxy-3,9′-lign-7-ene

3′,5-Di-Me ether, 9-O-β-D-glucopyranoside

Bidenlignaside B

[914616-01-6]

$C_{26}H_{34}O_{12}$ 538.546

Constit. of *Bidens parviflora*. Inhibitor of histamine release. Powder (MeOH). $[\alpha]^{25}_{D}$ -18.2 (c, 0.31 in MeOH). λ_{max} 207 (log ε 4.39); 246 (log ε 4.45); 268 (log ε 2.09) (MeOH)

7′-Ketone, 3′,5-di-Me ether

4,4′,9,9′-Tetrahydroxy-3′,5-dimethoxy-3,9′-lign-7-en-7′-one. 4,4′,9,9′-Tetrahydroxy-3′,5-dimethoxy-7′-oxo-3,9′-lign-7-ene

7′-Ketone, 3′,5-di-Me ether, 4-O-β-D-glucopyranoside

Bidenlignaside A

[501121-33-1]

$C_{26}H_{32}O_{12}$ 536.53

Constit. of *Bidens parviflora*. Inhibitor of histamine release. Powder (MeOH). $[\alpha]^{25}_{D}$ -16.5 (c, 0.14 in MeOH). λ_{max} 222 (log ε 4.34); 248 (log ε 3.95); 267 (log ε 4.04) (MeOH).

Wang, N.L. *et al.*, *Chem. Pharm. Bull.*, 2006, **54**, 1190-1192 (*Bidenlignasides A,B, activity*)

8,9′-Lign-7′-ene-2,2′,4,4′,5,5′,7-heptol L81

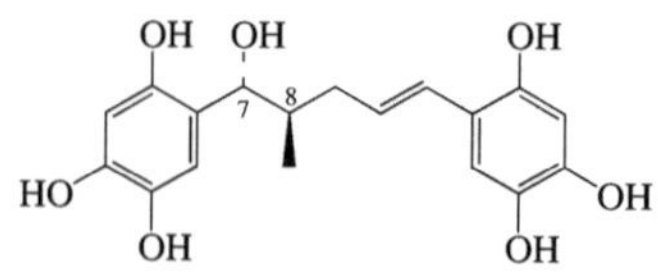

(7RS,7′E,8RS)-form

2,2′,4,4′,5,5′,7-Heptahydroxy-8,9′-neolign-7′-ene

$C_{18}H_{20}O_7$ 348.351

(7RS,7′E,8RS)-form

Hepta-Me ether

2,2′,4,4′,5,5′,7-Heptamethoxy-8,9′-lign-7′-ene. ***Asarolignan E***

$C_{25}H_{34}O_7$ 446.54

Constit. of the rhizomes of *Acorus tatarinowii*. Oil. λ_{max} 206 (log ε 5.0); 262 (log ε 4.58); 310 (log ε 4.30) (MeOH)

(7RS,7′E,8SR)-form

Hepta-Me ether

Asarolignan F

$C_{25}H_{34}O_7$ 446.54

Constit. of the rhizomes of *Acorus tatarinowii*. Oil. λ_{max} 209 (log ε 4.97); 261 (log ε 4.59); 309 (log ε 4.33) (MeOH).

Qin, D.-P. *et al.*, *RSC Adv.*, 2017, **7**, 8512-8520 (*Asarolignans E,F*)

2,9′-Lign-8-ene-3′,4,4′,5,7′,9′-hexol L82

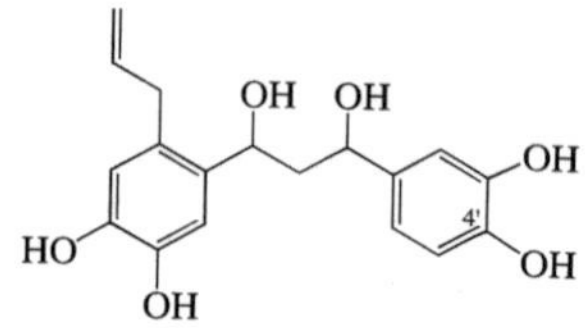

1-(3,4-Dihydroxyphenyl)-3-[4,5-dihydroxy-2-(2-propenyl)phenyl]-1,3-propanediol. 3′,4,4′,5,7′,9′-Hexahydroxy-2,9′-neolign-8-ene

$C_{18}H_{20}O_6$ 332.352

(7′ξ,9′ξ)-form

4′-Dodecanoyl

[2095554-27-9]

$C_{30}H_{42}O_7$ 514.659

Constit. of *Piper betle*. Antioxidant. Exhibits cytotoxic activity. Brown solid. Mp 75 - 78°.

Atiya, A. *et al.*, *Nat. Prod. Res.*, 2017, **31**, 2604-2611 (*Piper betle constit, activity*)

3,8′-Lign-7-ene-3′,4,4′,5,9,9′-hexol L83

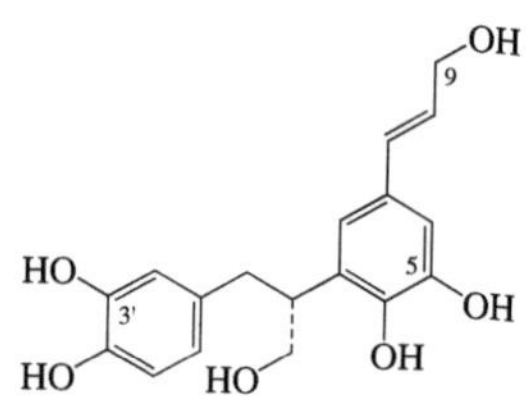

(R,E)-form

3′,4,4′,5,9,9′-Hexahydroxy-3,8′-neolign-7-ene

$C_{18}H_{20}O_6$ 332.352

(R,E)-form

3′,5-Di-Me ether

3′,5-Dimethoxy-3,8′-lign-7-ene-4,4′,9,9′-tetrol.

4,4′,9,9′-Tetrahydroxy-3′,5-dimethoxy-3,8′-lign-7-ene. Isodehydrodiconiferyl alcohol
$[\alpha]^{22}_{D}$ -28 (c, 0.1 in MeOH). λ_{max} 224 (MeOH)

3′,5-Di-Me ether, 4-O-β-D-xylopyranoside
Acutumoside
[1813572-36-9]
$C_{25}H_{32}O_{10}$ 492.521
Constit. of the rhizomes of *Sinomenium acutum*. Exhibits weak antitumour activity. Gum. $[\alpha]^{25}_{D}$ -83.1 (c, 1.2 in MeOH). λ_{max} 217 (log ε 4.3); 267 (log ε 3.7); 281 (log ε 3.3) (MeOH)

3′,5-Di-Me ether, 4-O-β-D-glucopyranoside
Icariside E$_5$
[126176-79-2]
$C_{26}H_{34}O_{11}$ 522.547
Constit. of *Albizzia julibrissin, Brainea insignis, Capsicum annuum* var. *acuminatum, Epimedium diphyllum, Fraxinus sieboldiana* and *Zanthoxylum simulans*. Antioxidant. Amorph. powder. $[\alpha]^{22}_{D}$ -118.8 (c, 1.04 in MeOH). λ_{max} 219 (log ε 4.56); 259 (log ε 4.22); 267 (sh); 288 (sh); 306 (sh) (MeOH)

3′,5-Di-Me ether, 4-O-[4-hydroxy-3-methoxybenzoyl-(→6)-β-D-glucopyranoside]
6″-O-Vanilloylicariside E$_5$
[1203605-97-3]
$C_{34}H_{40}O_{14}$ 672.68
Constit. of the fruit of *Capsicum annuum*. Amorph. powder (MeOH). $[\alpha]^{22}_{D}$ -85 (c, 0.1 in MeOH). λ_{max} 267 (MeOH)

9-Aldehyde, 3′,5-di-Me ether, 4-O-β-D-glucopyranoside
Ehletianol D
[166604-05-3]
$C_{26}H_{32}O_{11}$ 520.531
Constit. of the bark of *Ehretia ovalifolia*. Amorph. powder. $[\alpha]^{20}_{D}$ -165 (c, 2.5 in MeOH). λ_{max} 234 (log ε 4.29); 303 (log ε 4.22) (MeOH)

9-Aldehyde, 4′,5-di-Me ether, 4-O-β-D-glucopyranoside
Dioscoroside H$_1$
[2291325-06-7]
$C_{26}H_{32}O_{11}$ 520.531
Constit. of the rhizomes of *Dioscorea septemloba*. Powder. $[\alpha]^{25}_{D}$ -105.5 (c, 0.11 in MeOH). λ_{max} 304 (log ε 3.30) (MeOH)

(S,E)-form

3′,5-Di-Me ether, 4-O-β-D-glucopyranoside
Saikolignanoside A‡
[807630-72-4]
$C_{26}H_{34}O_{11}$ 522.547
Constit. of *Bupleurum scorzonerifolium, Lycium chinense* and *Magnolia officinalis*. Pale yellow powder. Mp 110 - 112°. $[\alpha]^{23}_{D}$ +32.8 (c, 0.13 in MeOH)

9-Aldehyde, 3′,5-di-Me ether
4,4′,9′-Trihydroxy-3′,5-dimethoxy-3,8′-lign-7-en-9-al. ***Ecdysanol A***
$C_{20}H_{22}O_6$ 358.390
Constit. of the caulis of *Urceola rosea*. Exhibits antiinflammatory activity. Amorph. yellow powder. $[\alpha]^{25}_{D}$ +146.2 (c, 0.11 in MeOH). λ_{max} 344 (log ε 5.34) (MeOH)

(ξ,E)-form

3′,5-Di-Me ether, 4′-O-β-D-glucopyranoside
[1391996-47-6]
$C_{26}H_{34}O_{11}$ 522.547
Isol. from the root bark of *Ailanthus altissima*. Amorph. powder. $[\alpha]^{27}_{D}$ +4.0 (c, 0.27 in MeOH).

Miyase, T. *et al.*, *Phytochemistry*, 1989, **28**, 3483-3485 (*Icariside E$_5$*).
Higuchi, H. *et al.*, *Chem. Pharm. Bull.*, 1992, **40**, 534-535 (*Icariside E$_5$*).
Yoshikawa, K. *et al.*, *Phytochemistry*, 1995, **39**, 659-664 (*Ehletianol D*).
Iorizzi, M. *et al.*, *J. Agric. Food Chem.*, 2001, **49**, 2022-2029 (*Icariside E$_5$*).
Tan, L. *et al.*, *Chin. Chem. Lett.*, 2004, **15**, 1053-1056 (*Saikolignanoside A*).
Lee, D.-Y. *et al.*, *Arch. Pharmacal Res.*, 2009, **32**, 1345-1349 (*6-Vanilloylicariside E$_5$*).
Tan, Q.-W. *et al.*, *Nat. Prod. Res.*, 2012, **26**, 1375-1380 (*Ailanthus altissima constit).*
Kim, K.H. *et al.*, *Biosci., Biotechnol., Biochem.*, 2013, **77**, 2144-2147 (*Acutumoside, activity*).
Zhang, Y. *et al.*, *Fitoterapia*, 2016, **115**, 165-172 (*Dioscoroside H$_1$*).
Dong, D.-D. *et al.*, *Fitoterapia*, 2019, **134**, 96-100 (*Ecdysanol A, activity*)

3,8′-Lign-7-ene-3′,4,5,5′,9,9′-hexol L84

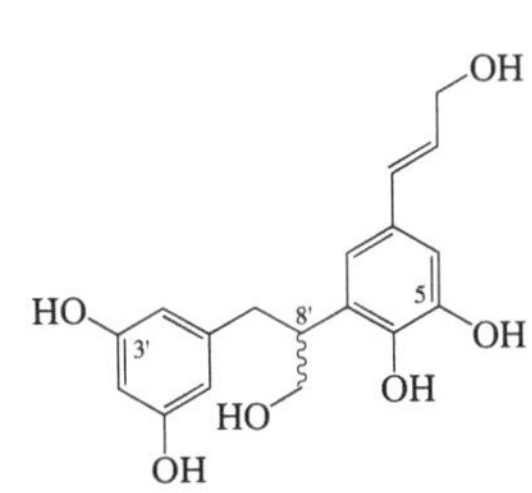

3′,4,5,5′,9,9′-Hexahydroxy-3,8′-neolign-7-ene
$C_{18}H_{20}O_6$ 332.352
Structures in this series may need revision. Phenyl 3,5-dioxygenated substn. pattern in doubt as a result of work done by Yang *et al* in 2015

(7E,8′ξ)-form

5-Me ether
5-Methoxy-3,8′-lign-7-ene-3′,4,5′,9,9′-pentol. 3′,4,5′,9,9′-Pentahydroxy-5-methoxy-3,8′-lign-7-ene

5-Me ether, 4-O-β-D-glucopyranoside
Ecdysantherol B
[1830287-97-2]
$C_{25}H_{32}O_{11}$ 508.52
Constit. of *Ecdysanthera rosea*. Amorph. yellow powder. $[\alpha]^{21}_{D}$ +28.3 (c, 0.1 in MeOH). λ_{max} 203 (log ε 4.65); 246 (log ε 4.12) (MeOH)

3′,5-Di-Me ether
5,5′-Dimethoxy-3,8′-lign-7-ene-3′,4,9,9′-tetrol. 3′,4,9,9′-Tetrahydroxy-5,5′-dimethoxy-3,8′-lign-7-ene

3′,5-Di-Me ether, 4-O-β-D-glucopyranoside
Manglieside D
[1093091-44-1]
$C_{26}H_{34}O_{11}$ 522.547
Constit. of *Ecdysanthera rosea* and *Manglietia phuthoensis*. Amorph. powder. $[\alpha]^{25}_{D}$ -125 (c, 1 in MeOH).

Kiem, P.V. *et al.*, *Chem. Pharm. Bull.*, 2008, **56**, 1270-1275 (*Manglieside D*).
Song, C.-W. *et al.*, *Nat. Prod. Bioprospect.*, 2014, **4**, 319-322 (*Ecdysantherol B*).
Yang, Y.-N. *et al.*, *J. Nat. Prod.*, 2015, **78**, 705-711 (*pmr, struct*)

3,8′-Lign-7-ene-3′,4,5′,6,9,9′-hexol L85

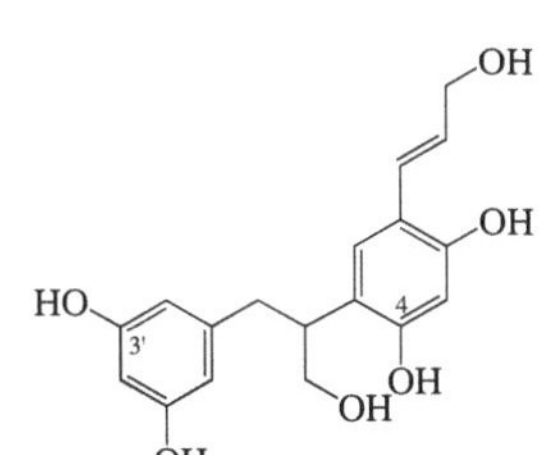

3′,4,5′,6,9,9′-Hexahydroxy-3,8′-neolign-7-ene
$C_{18}H_{20}O_6$ 332.352
Structures in this series may need revision. Phenyl 3,5-dioxygenated substn. pattern in doubt as a result of work done by Yang *et al* in 2015

(7E,8′ξ)-form
3′-Me ether, 4-O-β-D-glucopyranoside
Loropetaliside B
[1680201-16-4]
$C_{25}H_{32}O_{11}$ 508.52
Constit. of *Loropetalum chinense*. Amorph. powder. $[\alpha]^{25}_{D}$ +0.65 (c, 1.2 in MeOH). λ_{max} 203 (log ε 3.72); 258 (log ε 4.27) (MeOH).

Zhang, Q. *et al.*, *Fitoterapia*, 2013, **90**, 228-232 (*Loropetaliside B*).
Yang, Y.-N. *et al.*, *J. Nat. Prod.*, 2015, **78**, 705-711 (*pmr, struct*)

3,8′-Lign-8-ene-3′,4,4′,5,7′,9′-hexol L86

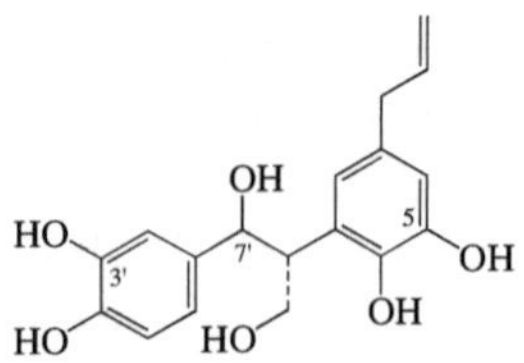

3′,4,4′,5,7′,9′-Hexahydroxy-3,8′-neolign-8-ene
$C_{18}H_{20}O_6$ 332.352
(8′R)-form
7′-Ketone
3′,4,4′,5,9′-Pentahydroxy-3,8′-lign-8-en-7′-one. 3′,4,4′,5,9′-Pentahydroxy-7′-oxo-3,8′-lign-8-ene
7′-Ketone, 3′,4′,5-tri-Me ether
4,9′-Dihydroxy-3′,4′,5-trimethoxy-3,8′-lign-8-en-7′-one. 4,9′-Dihydroxy-3′,4′,5-trimethoxy-7′-oxo-3,8′-lign-8-ene. ***Carinatonol***
[85769-58-0]
$C_{21}H_{24}O_6$ 372.417
Constit. of the bark of *Virola carinata*. Cryst.. Mp 128 - 130°. $[\alpha]^{25}_{D}$ -171.4 ($CHCl_3$). λ_{max} 242 (log ε 4.11); 281 (log ε 4.18); 308 (log ε 4.06) ($CHCl_3$).

Kawanishi, K. *et al.*, *Phytochemistry*, 1982, **21**, 2725-2728 (*Carinatonol*)

8,8′-Lign-7-ene-3,3′,4,4′,9,9′-hexol L87

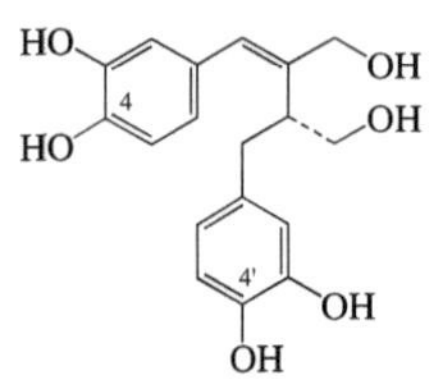

(7E,8′R)-form

2-(3,4-Dihydroxybenzyl)-3-(3,4-dihydroxybenzylidene)-1,4-butanediol. 3,3′,4,4′,9,9′-Hexahydroxylign-7-ene
$C_{18}H_{20}O_6$ 332.352
(7E,8′R)-form
3′,3′-Di-Me ether
3,3′-Dimethoxy-8,8′-lign-7-ene-4,4′,9,9′-tetrol. 4,4′,9,9′-Tetrahydroxy-3,3′-dimethoxylign-7-ene.
Syripinin E
[2241365-99-9]
$C_{20}H_{24}O_6$ 360.406
Constit. of the stems of *Syringa pinnatifolia*. Needles. Stereochem. uncertain
3,3′-Di-Me ether, 4-O-β-D-glucopyranoside
Sagitiside A
[1402932-12-0]
$C_{26}H_{34}O_{11}$ 522.547
Constit. of *Tinospora sagittata* var. *yunnanensis*. Shows antioxidant activity. Amorph. powder. $[\alpha]^{20}_{D}$ -42 (c, 0.1 in MeOH). Config. not confirmed
3,4-Methylene, 3′,4′-di-Me ether
3′,4′-Dimethoxy-3,4-methylenedioxy-8,8′-lign-7-ene-9,9′-diol. 9,9′-Dihydroxy-3′,4′-dimethoxy-3,4-methylenedioxylign-7-ene
[94285-09-3]
$C_{21}H_{24}O_6$ 372.417
Constit. of *Taxus baccata*. Cryst. (C_6H_6). Mp 120 - 121°. $[\alpha]^{25}_{D}$ -35.7 (c, 0.4 in $CHCl_3$). λ_{max} 258 (log ε 3.83) (EtOH)
3′,4′-Methylene, 3,4-di-Me ether
3,4-Dimethoxy-3′,4′-methylenedioxy-8,8′-lign-7-ene-9,9′-diol. 9,9′-Dihydroxy-3,4-dimethoxy-3′,4′-methylenedioxylign-7-ene
[173938-15-3]
$C_{21}H_{24}O_6$ 372.417
Constit. of *Taxus baccata*
3,4-Methylene, 3′,4′-di-Me ether, 9,9′-di-Ac
Prasanthalin
[118040-42-9]
$C_{25}H_{28}O_8$ 456.491
Constit. of *Jatropha gossypifolia*. Oil. $[\alpha]^{25}_{D}$ -32 ($CHCl_3$)
3,4:3′,4′-Bis(methylene) ether
3,4:3′,4′-Bis(methylenedioxy)-8,8′-lign-7-ene-9,9′-diol. 9,9′-Dihydroxy-3,4:3′,4′-bis(methylenedioxy)lign-7-ene
[126313-83-5]
[24562-95-6] (7ξ,8′*R*-form)
$C_{20}H_{20}O_6$ 356.374
Constit. of *Taxus baccata*
(7E,8′S)-form
3,4:3′,4′-Bis(methylene) ether, 9,9′-di-Ac
Gossypilin
[223428-36-2]
$C_{24}H_{24}O_8$ 440.448
Constit. of *Jatropha gossypifolia*. Viscous solid. $[\alpha]^{25}_{D}$ +31 (c, 0.1 in $CHCl_3$).

Banerji, J. *et al.*, *Phytochemistry*, 1984, **23**, 2323-2328 (*3,4-methylene 3′,4′-di-Me ether, synth*).
Chatterjee, A. *et al.*, *Indian J. Chem., Sect. B*, 1988, **27**, 740-741 (*Prasanthalin*).
Das, B. *et al.*, *Phytochemistry*, 1995, **38**, 715-717 (*Taxus baccata constit*).
Das, B. *et al.*, *Nat. Prod. Sci.*, 1998, **4**, 238-240 (*Gossypilin*).
Huang, X.-Z. *et al.*, *Nat. Prod. Res.*, 2012, **26**, 1876-1880 (*Sagitiside A*).
Wang, Q.-H. *et al.*, *Chem. Nat. Compd. (Engl. Transl.)*, 2018, **54**, 435-438 (*Syripinin E*)

8,9′-Lign-7′-ene-2,2′,6,6′,7,9-hexol L88

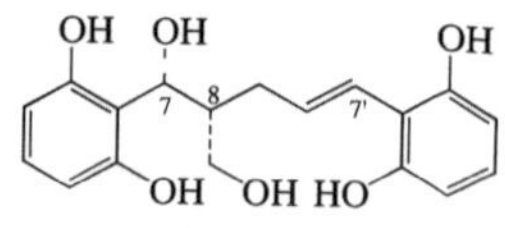

(7R,7′E,8R)-form

2,2′,6,6′,7,9-Hexahydroxy-8,9′-neolign-7′-ene
$C_{18}H_{20}O_6$ 332.352
(7RS,7′E,8RS)-form
2,2′,6,6′-Tetra-Me ether
2,2′,6,6′-Tetramethoxy-8,9′-lign-7′-ene-7,9-diol. 7,9-Dihydroxy-2,2′,6,6′-tetramethoxy-8,9′-lign-7′-ene. ***Torreyayunan A***
[1708116-28-2]
[1778668-64-6] (7*S*,7′*E*,8*S*-form)
[1778669-08-1] (7*R*,7′*E*,8*R*-form)
$C_{22}H_{28}O_6$ 388.46
Constit. of the leaves and twigs of *Torreya yunnanensis*. Light yellow oil. Has been sepd. into the two diastereoisomers. λ_{max} 289 (log ε 3.34) (MeOH)
(7RS,7′E,8SR)-form
2,2′,6,6′-Tetra-Me ether
Torreyayunan B
[1708116-29-3]
[1778669-31-0] (7*S*,7′*E*,8*R*-form)
[1778669-53-6] (7*R*,7′*E*,8*S*-form)
$C_{22}H_{28}O_6$ 388.46
Constit. of the leaves and twigs of *Torreya yunnanensis*. Light yellow oil. Has been sepd. into two diastereoisomers. λ_{max} 288 (log ε 3.32) (MeOH).

Bao, J.M. *et al.*, *Chirality*, 2014, **26**, 825-828 (*Torreyayunans A,B, cd, abs config*)

8,9′-Lign-7′-ene-3,3′,4,4′,7,9-hexol L89

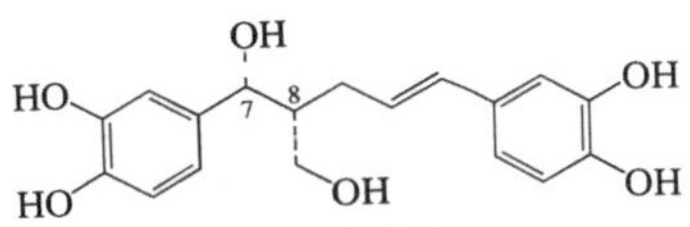

(7*R**,7′*E*,8*R**)-*form*

1,5-Bis(3,4-dihydroxyphenyl)-2-(hydroxymethyl)-4-penten-1-ol. 3,3′,4,4′,7,9-Hexahydroxy-8,9′-neolign-7′-ene

$C_{18}H_{20}O_6$ 332.352

All isolates appear to be partial racemates

(7R*,7′E,8R*)-form

3,3′-Di-Me ether

3,3′-Dimethoxy-8,9′-lign-7′-ene-4,4′,7,9-tetrol. 4,4′,7,9-Tetrahydroxy-3,3′-dimethoxy-8,9′-lign-7′-ene. ***Balajaponin D***

[1362086-49-4]

$C_{20}H_{24}O_6$ 360.406

Constit. of *Balanophora japonica*. Amorph. brown powder. $[\alpha]^{20}_{D}$ -8 (c, 0.25 in MeOH/DMSO)

3,3′-Di-Me ether, 4,4′-di-O-β-D-glucopyranoside

Balajaponin E

[1362086-51-8]

$C_{32}H_{44}O_{16}$ 684.688

Constit. of *Balanophora japonica*. Amorph. brown powder. $[\alpha]^{20}_{D}$ -52 (c, 0.25 in H_2O)

3,3′,4,4′-Tetra-Me ether

3,3′,4,4′-Tetramethoxy-8,9′-lign-7′-ene-7,9-diol. 7,9-Dihydroxy-3,3′,4,4′-tetramethoxy-8,9′-lign-7′-ene. ***Morinol G***

[256380-21-9]

$C_{22}H_{28}O_6$ 388.46

Constit. of *Morina chinensis*. Oil. $[\alpha]^{24}_{D}$ +1.2 (c, 1 in $CHCl_3$). λ_{max} 240 (sh); 272 (end) (log ε 3.94) ($CHCl_3$)

3,3′,4,4′-Tetra-Me ether, 9-O-(2-hydroxymethyl-2Z-butenoyl)

Morinol L

[256380-26-4]

$C_{27}H_{34}O_8$ 486.561

Constit. of *Morina chinensis*. Oil. $[\alpha]^{24}_{D}$ +0.8 (c, 0.9 in $CHCl_3$). λ_{max} 240 (log ε 3.83); 265 (log ε 3.9) ($CHCl_3$)

3,3′,4,4′,9-Penta-Me ether

1,5-Bis(3,4-dimethoxyphenyl)-2-(methoxymethyl)-4-penten-1-ol. 3,3′,4,4′,9-Pentamethoxy-8,9′-lign-7′-en-7-ol. 7-Hydroxy-3,3′,4,4′,9-pentamethoxy-8,9′-lign-7′-ene

[1147563-95-8]

$C_{23}H_{30}O_6$ 402.487

Constit. of *Morina kokonorica*. Oil. $[\alpha]^{20}_{D}$ +1 (c, 0.3 in MeOH). λ_{max} 225 (log ε 2.98); 262 (log ε 2.72) (MeOH)

Hexa-Me ether

1,5-Bis(3,4-dimethoxyphenyl)-5-methoxy-4-(methoxymethyl)-1-pentene. 3,3′,4,4′,7,9-Hexamethoxy-8,9′-lign-7′-ene

[1147563-96-9]

$C_{24}H_{32}O_6$ 416.514

Constit. of *Morina kokonorica*. Oil. $[\alpha]^{20}_{D}$ +2 (c, 1 in MeOH). λ_{max} 231 (log ε 3.38); 267 (log ε 3.36) (MeOH)

7-Deoxy, 7,8-didehydro(Z-), 3,3′,4,4′-tetra-Me ether

3,3′,4,4′-Tetramethoxy-8,9′-ligna-7,7′-dien-9-ol. 9-Hydroxy-3,3′,4,4′-tetramethoxy-8,9′-ligna-7,7′-diene.

Morinol I

[1360820-00-3]

$C_{22}H_{26}O_5$ 370.445

Constit. of *Morina chinensis*. Oil. $[\alpha]^{24}_{D}$ +1.6 (c, 0.21 in $CHCl_3$). Unexplained opt. rotation. Tentative struct. assigned. λ_{max} 241 (sh); 272 (log ε 3.96) ($CHCl_3$)

(7R*,7′E,8S*)-form

3,3′,4,4′-Tetra-Me ether

Morinol H

[256380-22-0]

Constit. of *Morina chinensis*. Oil. $[\alpha]^{24}_{D}$ +2.7 (c, 0.45 in $CHCl_3$). λ_{max} 240 (sh); 272 (log ε 3.94) ($CHCl_3$).

Su, B.-N. *et al.*, *Tetrahedron*, 1999, **55**, 14571-14586 (*Morinols G,H,I,L*).
Zhu, Y. *et al.*, *Helv. Chim. Acta*, 2009, **92**, 536-545 (*penta-Me ether, hexa-Me ether*).
Fleming, F.F. *et al.*, *Eur. J. Org. Chem.*, 2011, 6843-6846 (*Morinol I, synth*).
Zhou, T. *et al.*, *Planta Med.*, 2011, **77**, 477-481 (*Balajaponins D,E*)

3,3′-Lign-8′-ene-4,4′,7,8,9-pentol L90

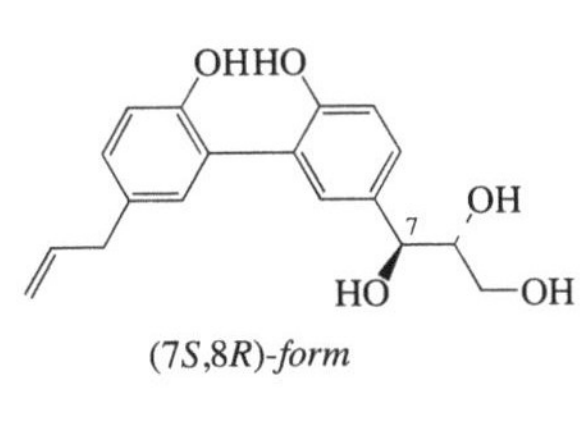

(7*S*,8*R*)-*form*

5-(2-Propenyl)-5′-(1,2,3-trihydroxypropyl)-[1,1′-biphenyl]-2,2′-diol. 4,4′,7,8,9-Pentahydroxy-3,3′-neolign-8′-ene

[138591-07-8] (unspecified stereochem.)

$C_{18}H_{20}O_5$ 316.353

(7S,8R)-form

7-Me ether

7-Methoxy-3,3′-lign-8′-ene-4,4′,8,9-tetrol. 4,4′,8,9-Tetrahydroxy-7-methoxy-3,3′-lign-8′-ene.

Clypearianin B

[2200314-84-5]

$C_{19}H_{22}O_5$ 330.380

Constit. of the leaves and twigs of *Pithecellobium clypearia*. Antioxidant. Yellow oil. $[\alpha]^{23}_{D}$ -30 (c, 0.1 in MeOH). λ_{max} 235 (log ε 0.72); 279 (log ε 0.29) (MeOH)

(7R*,8R*)-form

threo-**Magnolignan B** [1099687-81-6]

Constit. of the bark of *Magnolia officinalis*. Light brown gum. $[\alpha]_{D}$ +1.2 (c, 0.5 in MeOH). λ_{max} 214 (log ε 4.59); 250 (sh); 290 (log ε 3.86) (MeOH)

(7R*,8S*)-form

erythro-**Magnolignan B** [1099687-82-7]

Constit. of the bark of *Magnolia officinalis*. Light brown gum. $[\alpha]_{D}$ +1.3 (c, 1.0 in MeOH). λ_{max} 220 (log ε 4.57); 250 (sh); 289 (log ε 3.97) (MeOH).

Yahara, S. *et al.*, *Chem. Pharm. Bull.*, 1991, **39**, 2024-2036 (*Magnolignan B*).
Shen, C.-C. *et al.*, *J. Nat. Prod.*, 2009, **72**, 168-171 (*Magnolignan B*).
Lou, L.-L. *et al.*, *Fitoterapia*, 2016, **112**, 16-21 (*Clypearianin B, activity*)

3,3′-Lign-8′-ene-4,6′,7,8,9-pentol L91

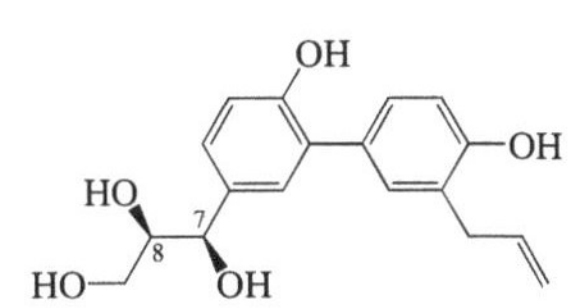

(7*R**,8*R**)-*form*

4,6′,7,8,9-Pentahydroxy-3,3′-neolign-8′-ene

$C_{18}H_{20}O_5$ 316.353

Similar to 3,3′-Lign-8′-ene-4,4′,7,8,9-pentol, L90

(7R*,8R*)-form

threo-**Honokitriol** [1099687-80-5]

Constit. of the stem bark of *Magnolia officinalis* and seeds of *Lithocarpus pachylepis*. Moderate inhibitor of LPS-induced NO prodn.. Amorph. solid. [α] D +9.3 (c, 0.3 in MeOH). λ_{max} 213 (log ε 4.45); 256 (log ε 4.06); 291 (log ε 3.87) (MeOH)

7-Me ether

*7-Methoxy-3,3′-lign-8′-ene-4,6′,8,9-tetrol. 4,6′,8,9-Tetrahydroxy-7-methoxy-3,3′-lign-8′-ene. threo-***Magnolignan D**. *7-O-Methylhonokitriol*

[1788899-44-4]

[138749-67-4] (unspecified stereochem.)

$C_{19}H_{22}O_5$ 330.38

Constit. of *Magnolia officinalis*. Amorph. solid. Magnolignan D is likely to be a racemate of *threo*- and *erythro*-isomers. λ_{max} 259 (ε 14500); 293 (ε 11500) (MeOH)

7-Et ether

7-Ethoxy-3,3′-lign-8′-ene-4,6′,8,9-tetrol. 7-Ethoxy-4,6′,8,9-tetrahydroxy-3,3′-lign-8′-ene. ***7-O-Ethylhonokitriol***

[1791410-56-4]

$C_{20}H_{24}O_5$ 344.407

Constit. of *Magnolia officinalis*. Yellow solid

(7R*,8S*)-form

*erythro-***Honokitriol** [1099687-79-2]

Constit. of the stem bark of *Magnolia officinalis* and seeds of *Lithocarpus pachylepis*. Moderate inhibitor of LPS-induced NO prodn.. Amorph. solid. [α] D -16.7 (c, 1 in MeOH). λ_{max} 213 (log ε 4.77); 256 (log ε 3.87); 291 (log ε 3.67) (MeOH)

7-Me ether

*erythro-***Magnolignan D**

[1791410-55-3]

$C_{19}H_{22}O_5$ 330.38

Constit. of *Magnolia officinalis*. Amorph. solid.

Yahara, S. *et al.*, *Chem. Pharm. Bull.*, 1991, **39**, 2024-2036 (*Magnolignan D*).
Shen, C.-C. *et al.*, *J. Nat. Prod.*, 2009, **72**, 168-171 (*Magnolia officinalis constits*).
Zhang, B. *et al.*, *J. Chin. Pharm. Sci.*, 2013, **22**, 420-426 (*7-O-Ethylhonokitriol, 7-O-Methylhonokitriols*).
Xie, Y. *et al.*, *Molecules*, 2013, **18**, 10397-10403 (*Lithocarpus pachylepis constits, activity*)

3,8′-Lign-7-ene-3′,4′,5,5′,7′-pentol L92

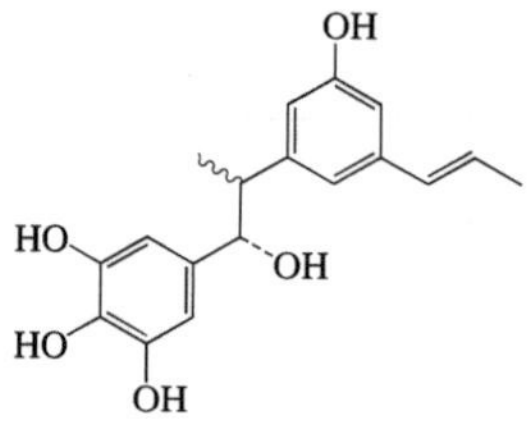

3′,4′,5,5′,7′-Pentahydroxy-3,8′-neolign-7-ene

$C_{18}H_{20}O_5$ 316.353

(7E,7′S,8′ξ)-form

3′,4′,5,5′-Tetra-Me ether

3′,4′,5,5′-Tetramethoxy-3,8′-lign-7-en-7′-ol. 7′-Hydroxy-3′,4′,5,5′-tetramethoxy-3,8′-lign-7-ene

[1879071-56-3]

$C_{22}H_{28}O_5$ 372.461

Constit. of the arils of *Myristica fragrans*. Exhibits moderate PARP-1 and κ-Ras inhibitory activity. Yellow oil. $[\alpha]_D^{20}$ +18 (c, 1.0 in MeOH). λ_{max} 210 (log ε 3.89); 268 (log ε 4.00) (MeOH).

Acuña, U.M. *et al.*, *Food Chem.*, 2016, **202**, 269-275 (*Myristica fragrans constit, activity*)

3,8′-Lign-8-ene-3′,4,4′,5,7′-pentol L93

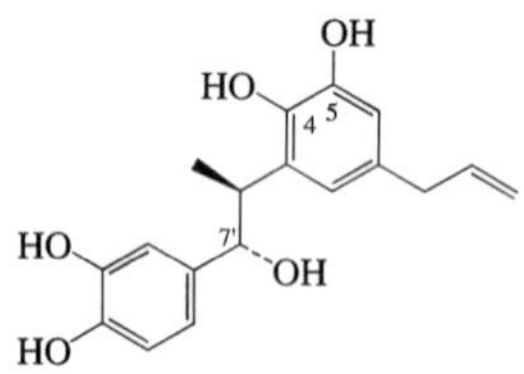

3′,4,4′,5,7′-Pentahydroxy-3,8′-neolign-8-ene

$C_{18}H_{20}O_5$ 316.353

(7′S,8′S)-form

3′,4′,5-Tri-Me ether

3′,4′,5-Trimethoxy-3,8′-lign-8-ene-4,7′-diol. 4,7′-Dihydroxy-3′,4′,5-trimethoxy-3,8′-lign-8-ene. ***Carinatol***

[85769-59-1]

$C_{21}H_{26}O_5$ 358.434

Constit. of *Virola carinata* and *Virola pavonis*. Gummy solid. λ_{max} 229 (ε 19950); 280 (ε 7413) (EtOH aq.)

7′-Ketone

3′,4,4′,5-Tetrahydroxy-3,8′-lign-8-en-7-one. 3′,4,4′,5-Tetrahydroxy-7′-oxo-3,8′-lign-8-ene

7′-Ketone, 3′,4′,5-tri-Me ether

4-Hydroxy-3′,4′,5-trimethoxy-3,8′-lign-8-en-7′-one. 4-Hydroxy-3′,4′,5-trimethoxy-7′-oxo-3,8′-lign-8-ene. ***Carinatone***

[82843-81-0]

$C_{21}H_{24}O_5$ 356.418

Constit. of *Justicia patentiflora*, *Virola carinata* and *Virola pavonis*. Cryst. (C_6H_6/petrol). Mp 103°. $[\alpha]_D^{25}$ -108.7 ($CHCl_3$). λ_{max} 243 (log ε 3.94); 277 (log ε 4.02); 304 (log ε 4.09) ($CHCl_3$).

Kawanishi, K. *et al.*, *Phytochemistry*, 1982, **21**, 929-931 (*Carinatol, Carinatone*).
Watanabe, M. *et al.*, *Chem. Pharm. Bull.*, 1991, **39**, 3123-3131 (*Carinatol, synth*).
Marques, M.O.M. *et al.*, *Phytochemistry*, 1992, **31**, 4380-4381 (*Carinatol, Carinatone*).
Susplugas, S. *et al.*, *J. Nat. Prod.*, 2005, **68**, 734-738 (*Carinatone*)

3,8′-Lign-8-ene-3′,4′,5,6,7′-pentol L94

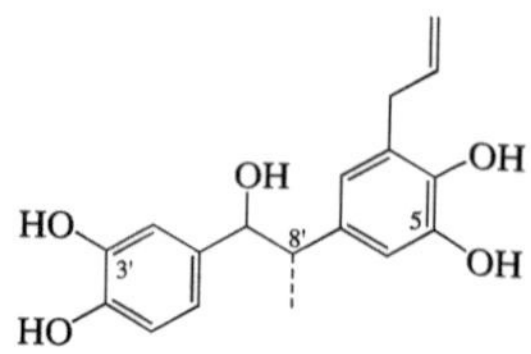

3′,4′,5,6,7′-Pentahydroxy-3,8′-neolign-8-ene

$C_{18}H_{20}O_5$ 316.353

(8′S)-form

7′-Ketone

3′,4′,5,6-Tetrahydroxy-3,8′-lign-8-en-7′-one. 3′,4′,5,6-Tetrahydroxy-7′-oxo-3,8′-lign-8-ene

7′-Ketone, 3′,4′-methylene, 5-Me ether

1-(1,3-Benzodioxol-5-yl)-2-[4-hydroxy-3-methoxy-5-(2-propenyl) phenyl]-1-propanone. 6-Hydroxy-5-methoxy-3′,4′-methylenedioxy-3,8′-lign-8-en-7′-one. 6-Hydroxy-5-methoxy-3′,4′-methylenedioxy-7′-oxo-3,8′-lign-8-ene

[156665-52-0]

$C_{20}H_{20}O_5$ 340.375

Constit. of the wood of *Ocotea porosa*. Cryst. (petrol/Me_2CO). Mp 103 - 105°. λ_{max} 230 (ε 5400); 278 (ε 1900); 312 (ε 1800) (MeOH).

De Carvalho, M.G. *et al.*, *Phytochemistry*, 1988, **27**, 2319-2323 (*Ocotea porosa constit*).
David, J.M. *et al.*, *Phytochemistry*, 1994, **36**, 491-499 (*Ocotea porosa constit*)

8,8′-Lign-8′-ene-2,4,4′,8,9-pentol L95

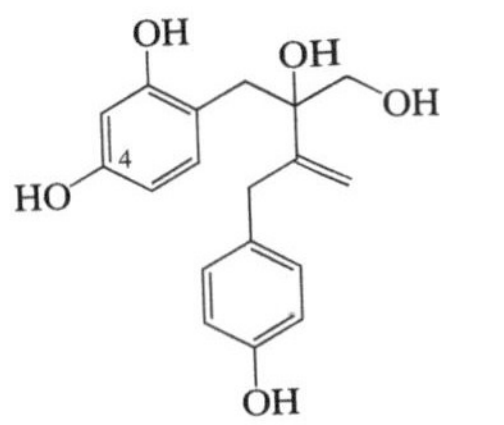

2-(2,4-Dihydroxybenzyl)-3-(4-hydroxybenzyl)-3-butene-1,2-diol. 2,4,4′,8,9-Pentahydroxy-8′-lignene
$C_{18}H_{20}O_5$ 316.353
(ξ)-form
4-Me ether
3-(4-Hydroxybenzyl)-2-(2-hydroxy-4-methoxybenzyl)-3-butene-1,2-diol. 4-Methoxy-8,8′-lign-8′-ene-2,4′,8,9-tetrol. 2,4′,8,9-Tetrahydroxy-4-methoxy-8′-lignene. **Thannilignan**
[192220-06-7]
$C_{19}H_{22}O_5$ 330.38
Constit. of the fruit rind of *Terminalia bellerica*. Exhibits anti-HIV-1, antimalarial and antifungal activities. Cryst. powder. Mp 127 - 129°. $[\alpha]^{24}_{D}$ -63.9 (c, 0.3 in $CHCl_3$).

Valsaraj, R. *et al.*, *J. Nat. Prod.*, 1997, **60**, 739-742 (*Thannilignan, activity*)

8,9′-Lign-7′-ene-3,4,4′,7,9-pentol L96

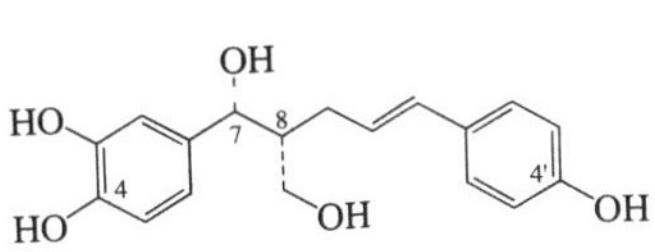

(7R*,7′E,8R*)-form

1-(3,4-Dihydroxyphenyl)-2-(hydroxymethyl)-5-(4-hydroxyphenyl)-4-penten-1-ol. 3,4,4′,7,9-Pentahydroxy-8,9′-neolign-7′-ene
$C_{18}H_{20}O_5$ 316.353
All isolates appear to be partial racemates
(7R*,7′E,8R*)-form
3,4,4′-Tri-Me ether
3,4,4′-Trimethoxy-8,9′-lign-7′-ene-7,9-diol. 7,9-Dihydroxy-3,4,4′-trimethoxy-8,9′-lign-7′-ene. **Morinol C**
[256380-17-3]
$C_{21}H_{26}O_5$ 358.434
Constit. of *Morina chinensis*. Oil. $[\alpha]^{24}_{D}$ -1.7 (c, 1 in $CHCl_3$). λ_{max} 240 (sh); 272 (log ε 3.92) ($CHCl_3$)
3,4,4′-Tri-Me ether, 9-O-(2-hydroxymethyl-2Z-butenoyl)
Morinol J
[256380-24-2]
$C_{26}H_{32}O_7$ 456.535
Constit. of *Morina chinensis*. Oil. $[\alpha]^{24}_{D}$ +0.4 (c, 0.5 in $CHCl_3$). λ_{max} 240 (log ε 3.82); 265 (log ε 3.91) ($CHCl_3$)
(7R*,7′E,8S*)-form
3,4,4′-Tri-Me ether
Morinol D
[256380-18-4]
$C_{21}H_{26}O_5$ 358.434
Constit. of *Morina chinensis*. Oil. $[\alpha]_{D}$ +0.8 (c, 0.25 in $CHCl_3$). λ_{max} 240 (sh); 271 (log ε 3.87) ($CHCl_3$)
9-Carboxylic acid, 3,7-di-Me ether
4,4′-Dihydroxy-3,7-dimethoxy-8,9′-lign-7′-en-9-oic acid
9-Carboxylic acid, 3,7-di-Me ether, Me ester
Conchigeranal A
[1613724-49-4]
$C_{21}H_{24}O_6$ 372.417
Constit. of *Alpinia conchigera*. Oil. $[\alpha]^{19}_{D}$ -5.5 (c, 0.26 in MeOH). λ_{max} 203 (log ε 4.29); 263 (log ε 3.95) (MeOH).

Su, B.-N. *et al.*, *Tetrahedron*, 1999, **55**, 14571-14586 (*Morinols C,D,J*).
Yamauchi, S. *et al.*, *Org. Biomol. Chem.*, 2003, **1**, 1323-1329 (*Morinols C,D, synth*).
Ogura, K. *et al.*, *Biosci., Biotechnol., Biochem.*, 2010, **74**, 1641-1644 (*Morinols C,D, synth*).
Xu, J.-J. *et al.*, *Fitoterapia*, 2013, **91**, 82-86 (*Conchigeranal A*)

8,9′-Lign-7′-ene-3′,4,4′,7,9-pentol L97

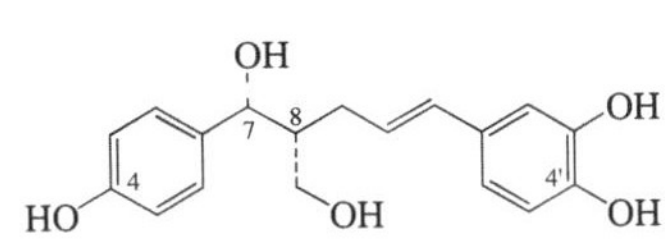

(7R*,7′E,8R*)-form

5-(3,4-Dihydroxyphenyl)-2-(hydroxymethyl)-1-(4-hydroxyphenyl)-4-penten-1-ol. 3′,4,4′,7,9-Pentahydroxy-8,9′-neolign-7′-ene
$C_{18}H_{20}O_5$ 316.353
All isolates appear to be partial racemates
(7R*,7′E,8R*)-form
3′,4,4′-Tri-Me ether
3′,4,4′-Trimethoxy-8,9′-lign-7′-ene-7,9-diol. 7,9-Dihydroxy-3′,4,4′-trimethoxy-8,9′-lign-7′-ene. **Morinol E**
[256380-19-5]
$C_{21}H_{26}O_5$ 358.434
Constit. of *Morina chinensis*. Oil. $[\alpha]^{24}_{D}$ -0.4 (c, 0.9 in $CHCl_3$). λ_{max} 241 (sh); 272 (log ε 3.91) ($CHCl_3$)
3′,4,4′-Tri-Me ether, 9-O-(2-hydroxymethyl-2Z-butenoyl)
Morinol K
[256380-25-3]
$C_{26}H_{32}O_7$ 456.535
Constit. of *Morina chinensis*. Oil. $[\alpha]^{24}_{D}$ +3.2 (c, 0.74 in $CHCl_3$). λ_{max} 240 (log ε 3.83); 265 (log ε 3.92) ($CHCl_3$)
(7R*,7′E,8S*)-form
3′,4,4′-Tri-Me ether
Morinol F
[256380-20-8]
Constit. of *Morina chinensis*. Oil. $[\alpha]^{24}_{D}$ +0.8 (c, 0.48 in $CHCl_3$). λ_{max} 240 (sh); 272 (log ε 3.91) ($CHCl_3$).

Su, B.-N. *et al.*, *Tetrahedron*, 1999, **55**, 14571-14586 (*Morinols E,F,K*)

3,3′-Lign-8′-ene-4,4′,7,8-tetrol L98

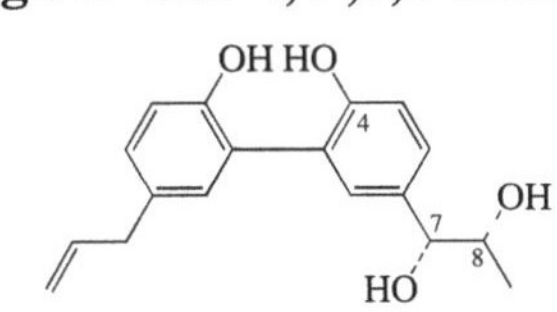

(7R,8R)-form

5-(1,2-Dihydroxypropyl)-5′-(2-propenyl)-[1,1′-biphenyl]-2,2′-diol, CAS. *4,4′,7,8-Tetrahydroxy-3,3′-neolign-8′-ene. 9-Deoxymagnolignan B*
$C_{18}H_{20}O_4$ 300.354
(7R,8R)-form
threo-**Strebluslignanol** [1036038-31-9]
Constit. of the stem bark and heartwood of *Streblus asper*. Exhibits moderate cytotoxic activity. Amorph. powder or yellow oil. Mp 133.5 - 134.3°. $[\alpha]^{25}_{D}$ +28.45 (c, 1.5 in MeOH). λ_{max} 232 (log ε 3.35); 288 (log ε 1.38); 332 (log ε 2.5) (MeOH)
4-O-β-D-Glucopyranoside
[1420837-31-5]
$C_{24}H_{30}O_9$ 462.495
Constit. of the heartwood of *Streblus asper*. Yellow oil. $[\alpha]^{25}_{D}$ -27.2 (c, 0.1 in MeOH). λ_{max} 248 (log ε 2.46); 276 (log ε 1.35) (MeOH)
7-Me ether
7-Methoxy-3,3′-lign-8′-ene-4,4′,8-triol. 4,4′,8-Trihydroxy-7-methoxy-3,3′-lign-8′-ene.
7-O-Methyl-threo-strebluslignanol
[1415341-82-0]

$C_{19}H_{22}O_4$ 314.381
Constit. of the heartwood of *Streblus asper*. Yellowish oil. $[\alpha]_D^{25}$ +27.3 (c, 1.5 in MeOH). λ_{max} 235 (log ε 3.38); 288 (log ε 3.35) (MeOH)

7-Ketone

4,4′,8-Trihydroxy-3,3′-lign-8′-en-7-one. 4,4′,8-Trihydroxy-7-oxo-3,3′-lign-8′-ene. ***Streblusol C***
[1399052-34-6]
$C_{18}H_{18}O_4$ 298.338
Constit. of the stem bark of *Streblus asper*. Displays anti-hepatitis B virus activity. Powder. Mp 102 - 103° (MeOH). $[\alpha]_D^{25}$+41.7 (c, 1.0 in MeOH). λ_{max} 243 (log ε 3.05); 287 (log ε 2.13) (no solvent reported)

9-Chloro

9-Chloro-3,3′-lign-8′-ene-4,4′,7,8-tetrol. 9-Chloro-4,4′,7,8-tetrahydroxy-3,3′-lign-8′-ene. ***Manneoinsigin A***
[1460737-80-7]
$C_{18}H_{19}ClO_4$ 334.796
Constit. of *Manglietia insignis*. Yellow gum.$[\alpha]_D^{23.5}$-11 (c, 0.23 in MeOH). Possesses (7*R*,8*S*)-config. due to change in Cahn-Ingold-Prelog priorities. λ_{max} 214 (log ε 4.01); 289 (log ε 3.27) (MeOH)

(7R,8S)-form

erythro-**Streblusilignanol**
[1415577-62-6]
Constit. of the heartwood of *Streblus asper*. Yellow oil. $[\alpha]_D^{25}$+29.2 (c, 1.5 in MeOH). λ_{max} 240 (log ε 3.36); 290 (log ε 1.39) (MeOH)

7-Me ether

7-O-Methyl-*erythro*-***streblusilignanol***
[1415340-57-6]
$C_{19}H_{22}O_4$ 314.381
Constit. of *Sassafras tzumu* and *Streblus asper*. Yellowish oil. $[\alpha]_D^{25}$ +28.3 (c, 1.5 in MeOH). λ_{max} 243 (log ε 3.4); 288 (log ε 3.35) (MeOH)

4,4′-Di-Me ether

4,4′-Dimethoxy-3,3′-lign-8′-ene-7,8-diol. 7,8-Dihydroxy-4,4′-dimethoxy-3,3′-lign-8′-ene
[2196258-64-5]
$C_{20}H_{24}O_4$ 328.408
Constit. of the roots of *Streblus asper*. Exhibits antimicrobial activity. Yellow oil. $[\alpha]_D^{25}$+29.2 (c, 1.5 in MeOH). λ_{max} 242 (log ε 3.30); 289 (log ε 1.37) (MeOH)

7-Et ether

*7-Ethoxy-3,3′-lign-8′-ene-4,4′,8-triol. 7-Ethoxy-4,4′,8-trihydroxy-3,3′-lign-8′-ene. erythro-****Streblusol B***
[1399052-32-4]
$C_{20}H_{24}O_4$ 328.408
Constit. of the stem bark of *Streblus asper*. Displays anti-hepatitis B virus activity. Yellow oil. $[\alpha]_D^{25}$+25.5 (c, 1.4 in MeOH). λ_{max} 209 (log ε 4.82); 254 (log ε 2.65); 290 (log ε 2.35) (MeOH)

9-Chloro

Manneoinsigin B
[1460737-81-8]
$C_{18}H_{19}ClO_4$ 334.796
Constit. of *Manglietia insignis*. Gum.$[\alpha]_D^{23.5}$-14.6 (c, 0.11 in MeOH). λ_{max} 196 (log ε 3.84); 212 (log ε 4.20); 289 (log ε 3.49) (MeOH)

(7S,8R)-form

Clypearianin F [2200314-88-9]
Constit. of *Pithecellobium clypearia*. Antioxidant. Yellow oil. $[\alpha]_D^{23}$-25.2 (c, 0.1 in MeOH). λ_{max} 236 (log ε 0.81); 279 (log ε 0.31) (MeOH)

7-Me ether

Clypearianin D
[2200314-86-7]
$C_{19}H_{22}O_4$ 314.381
Constit. of *Pithecellobium clypearia*. Shows moderate antioxidant activity. Yellow oil. $[\alpha]_D^{23}$-21.6 (c, 0.1 in MeOH). λ_{max} 230 (log ε 0.78); 279 (log ε 0.32) (MeOH)

(7S,8S)-form

Clypearianin E [2200314-87-8]
Constit. of *Pithecellobium clypearia*. Yellow oil. $[\alpha]_D$ -29.8 (c, 0.1 in MeOH). λ_{max} 234 (log ε 0.79); 279 (log ε 0.28) (MeOH)

7-Et ether

threo-***Streblusol B***
[1399052-33-5]
$C_{20}H_{24}O_4$ 328.408
Constit. of the stem bark of *Streblus asper*. Displays anti-hepatitis B virus activity. Yellow oil. $[\alpha]_D^{25}$-37.3 (c, 2 in MeOH). λ_{max} 209 (log ε 4.80); 254 (log ε 2.64) (no solvent reported)

7-Ketone

Clypearianin C
[2200314-85-6]
$C_{18}H_{18}O_4$ 298.338
Constit. of *Pithecellobium clypearia*. Antioxidant. Yellow oil. $[\alpha]_D^{23}$-20 (c, 0.1 in MeOH). λ_{max} 238 (log ε 0.82); 281 (log ε 0.24) (MeOH)

9-Chloro

Clypearianin G
[2200314-89-0]
$C_{18}H_{19}ClO_4$ 334.796
Constit. of *Pithecellobium clypearia*. Antioxidant. Yellow oil. $[\alpha]_D^{23}$-26 (c, 0.1 in MeOH). λ_{max} 233 (log ε 0.75); 279 (log ε 0.27) (MeOH).

Li, J. *et al.*, *Magn. Reson. Chem.*, 2008, **46**, 497-500 (*Streblusilignanol, activity*).
Li, L.-Q. *et al.*, *Fitoterapia*, 2012, **83**, 303-309 (*Streblus asper constits*).
Shang, S.-Z. *et al.*, *Nat. Prod. Bioprospect.*, 2012, **2**, 227-230 (*Manneoinsigins A,B*).
Li, J. *et al.*, *Nat. Prod. Commun.*, 2012, **7**, 599-602 (*Streblus glucoside*).
Li, J. *et al.*, *Phytochemistry*, 2012, **82**, 100-109 (*Streblus asper constits*).
Lou, L.-L. *et al.*, *Fitoterapia*, 2016, **112**, 16-21 (*Clypearianins C-G, activity*).
Nie, H. *et al.*, *Phytochem. Lett.*, 2016, **18**, 226-231 (*4,4′-di-Me ether, activity*)

3,3′-Lign-8′-ene-4,4′,8,9-tetrol L99

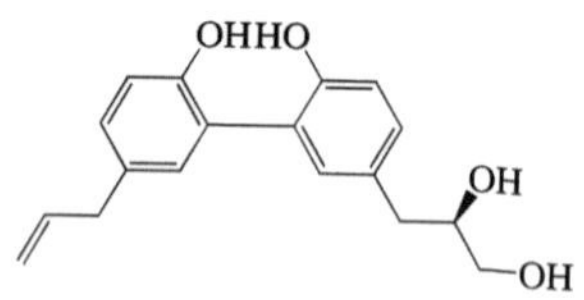

5-(2,3-Dihydroxypropyl)-5′-(2-propenyl)-[1,1′-biphenyl]-2,2′-diol. 4,4′,8,9-Tetrahydroxy-3,3′-neolign-8′-ene. 8,9-Dihydroxydihydromagnolol
$C_{18}H_{20}O_4$ 300.354

(R)-form

Magnolignan A [93673-81-5]
Constit. of *Magnolia officinalis*, *Magnolia rostrata*, *Sassafras randaiense* and *Streblus asper*. Shows antiallergic activity. Powder. Mp 143 - 145°. $[\alpha]_D$ -0.8 (c, 1.5 in MeOH). λ_{max} 256 (ε 18000); 293 (ε 18600) (MeOH)

4′-O-β-D-Glucopyranoside

Magnolignan A 4′-glucoside
[1035846-18-4]
$C_{24}H_{30}O_9$ 462.495
Constit. of *Magnolia officinalis* and *Streblus asper*. Yellow oil. $[\alpha]_D^{25}$-30 (c, 0.08 in MeOH). λ_{max} 248 (ε 19570); 274 (ε 13200) (MeOH).

Yahara, S. *et al.*, *Chem. Pharm. Bull.*, 1991, **39**, 2024-2036 (*Magnolignan A*).
Gu, W. *et al.*, *Tetrahedron: Asymmetry*, 1998, **9**, 1377-1380 (*Magnolignan A, synth*).
Taniguchi, C. *et al.*, *Planta Med.*, 2000, **66**, 607-611 (*activity*).
Li, J. *et al.*, *Magn. Reson. Chem.*, 2008, **46**, 497-500 (*4′-glucoside*)

3,3′-Lign-8′-ene-4,6′,8,9-tetrol L100

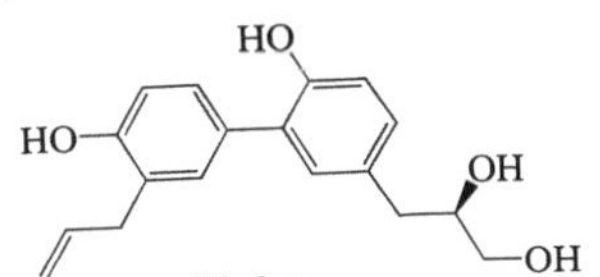

(*R*)-form

5-(2,3-Dihydroxypropyl)-5′-(2-propenyl)-[1,1′-biphenyl]-2,4′-diol. 4,6′,8,9-Tetrahydroxy-3,3′-neolign-8′-ene

$C_{18}H_{20}O_4$ 300.354

(R)-form

Oil. $[\alpha]^{20}_{D}$ +8.9 (c, 0.33 in MeOH). λ_{max} 210 (log ε 4.4); 255 (log ε 3.9); 292 (log ε 3.7) (MeOH)

(S)-form

Magnolignan C [93697-42-8]

Constit. of *Magnolia obovata* and *Magnolia officinalis*. Powder. $[\alpha]^{22}_{D}$ -6.8 (c, 0.9 in MeOH). λ_{max} 259 (ε 19400); 293 (ε 17800) (MeOH)

6′-O-β-D-Glucopyranoside

Magnolignan C 6′-glucoside

[773894-59-0]

$C_{24}H_{30}O_9$ 462.495

Constit. of the bark of *Magnolia rostrata*. $[\alpha]^{25}_{D}$ -17.8 (c, 0.28 in MeOH)

(ξ)-form

4,6′-Di-O-β-D-glucopyranoside

Magnoloside U

[1946004-94-9]

$C_{30}H_{40}O_{14}$ 624.636

Constit. of the stem bark of *Magnolia officinalis*. Amorph. pale yellow solid. $[\alpha]^{20}_{D}$ -44.4 (c, 0.2 in MeOH). λ_{max} 253 (log ε 4.20); 284 (log ε 4.08) (H_2O).

Yahara, S. *et al.*, *Chem. Pharm. Bull.*, 1991, **39**, 2024-2036 (*Magnolignan C*).
Deng, S. *et al.*, *Yunnan Zhiwu Yanjiu*, 2002, **24**, 397-400 (*6′-glucoside*).
Zhang, B. *et al.*, *J. Chin. Pharm. Sci.*, 2013, **22**, 420-426 (*Magnolignan C*).
Wang, Z. *et al.*, *J. Chin. Pharm. Sci.*, 2013, **22**, 531-534 (*R-form, S-form, synth*).
Xue, Z. *et al.*, *Phytochemistry*, 2016, **127**, 50-62 (*Magnoloside U*)

3,8′-Lign-8-ene-3′,4,4′,5-tetrol L101

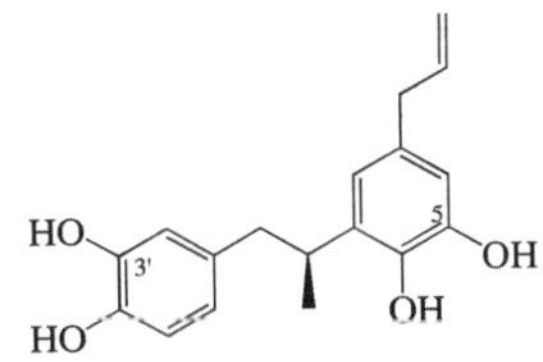

3′,4,4′,5-Tetrahydroxy-3,8′-neolign-8-ene

$C_{18}H_{20}O_4$ 300.354

(S)-form

3′,5-Di-Me ether

3′,5-Dimethoxy-3,8′-lign-8-ene-4,4′-diol. 4,4′-Dihydroxy-3′,5-dimethoxy-3,8′-lign-8-ene. ***1-Deoxycarinatone***

[955084-54-5]

$C_{20}H_{24}O_4$ 328.408

Constit. of the arils of *Myristica fragrans*. Oil. $[\alpha]^{20}_{D}$ +33.3 (c, 0.6 in $CHCl_3$). The synonym 1-Deoxycarinatone is misleading. λ_{max} 230 (log ε 3.94); 280 (log ε 3.49) (MeOH).

Li, F. *et al.*, *Helv. Chim. Acta*, 2007, **90**, 1491-1496 (*1-Deoxycarinatone*)

8,8′-Lign-7-ene-3,3′,4,4′-tetrol L102

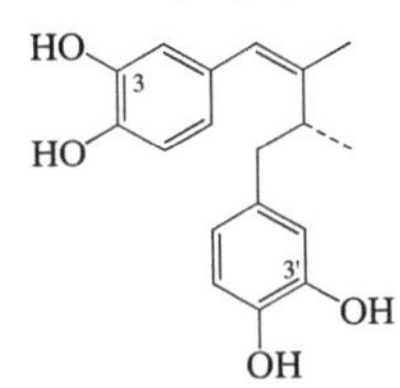

4,4′-(2,3-Dimethyl-1-butene-1,4-diyl)bis-1,2-benzenediol. 1,4-Bis(3,4-dihydroxyphenyl)-2,3-dimethyl-1-butene. 2-(3,4-Dihydroxybenzyl)-3-(3,4-dihydroxybenzylidene)butane. 3,3′,4,4′-Tetrahydroxy-7-lignene

$C_{18}H_{20}O_4$ 300.354

(7E,8′R)-form

3,3′-Di-Me ether

4,4′-(2,3-Dimethyl-1-butene-1,4-diyl)bis[2-methoxyphenol], CAS. *1,4-Bis(4-hydroxy-3-methoxyphenyl)-2,3-dimethyl-1-butene. 3,3′-Dimethoxy-8,8′-lign-7-ene-4,4′-diol. 4,4′-Dihydroxy-3,3′-dimethoxy-7-lignene.* ***Guaiaretic acid***

[500-40-3]

$C_{20}H_{24}O_4$ 328.408

Constit. of resin of *Guaiacum officinale* and *Guaiacum sanctum*. Key compd. in elucidation of lignan abs. configs.. Cryst. (EtOH or Et_2O/hexane). Mp 100.8 - 101.3° (85-87°). $[\alpha]_D$ -94 (EtOH). $[\alpha]^{25}_{D}$ -81 (c, 1.0 in $CHCl_3$). λ_{max} 206; 260; 280 (sh) (EtOH).

Hearon, W.M. *et al.*, *Chem. Rev.*, 1955, **55**, 957-1068 (*Guaiaretic acid, rev*).
Carnmalm, B. *Chem. Ind. (London)*, 1956, 1093 (*abs config*).
Schrecker, A.W. *et al.*, *J. Org. Chem.*, 1956, **21**, 381-382 (*abs config*).
King, F.E. *et al.*, *J. Chem. Soc.*, 1964, 4011-4024 (*Guaiaretic acid, isol*).
Ikeya, Y. *et al.*, *Chem. Pharm. Bull.*, 1979, **27**, 1583-1588 (*Guaiaretic acid, synth*).
Ortega, A.R. *et al.*, *Nat. Prod. Commun.*, 2018, **13**, 981-986 (*cd, abs config, bibl*)

8,8′-Lign-8-ene-3,3′,4,4′-tetrol L103

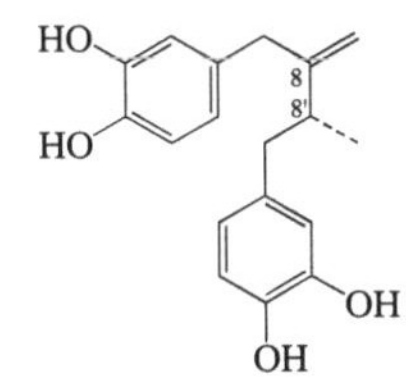

2,3-Bis(3,4-dihydroxybenzyl)-1-butene. 3,3′,4,4′-Tetrahydroxy-8-lignene

$C_{18}H_{20}O_4$ 300.354

(R)-form

3,4-Methylene, 3′-Me ether

3-Methoxy-3′,4′-methylenedioxy-8,8′-lign-8′-en-4-ol. 4′-Hydroxy-3′-methoxy-3,4-methylenedioxy-8-lignene. ***Schibitubin G***

[2094994-50-8]

$C_{20}H_{22}O_4$ 326.392

Constit. of the fruit of *Schisandra bicolor* var. *tuberculata*. Exhibits neuroprotective activity. Amorph. solid. $[\alpha]^{20}_{D}$ +9.2 (c, 0.23 in $CHCl_3$). λ_{max} 230 (log ε 3.54); 285 (log ε 3.29); 370 (log ε 2.41) (MeOH).

Liu, Y. *et al.*, *J. Nat. Prod.*, 2017, **80**, 1117-1124 (*Schibitubin G, activity*)

8,9′-Lign-7′-ene-4,4′,7,9-tetrol L104

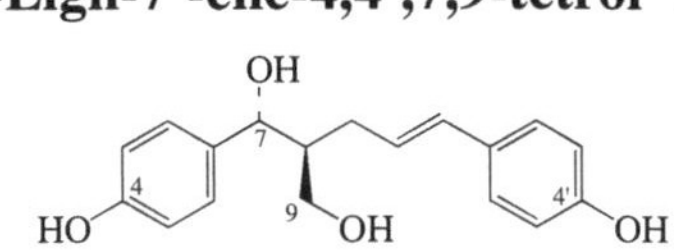

(7*R**,7′*E*,8*S**)-form

2-(Hydroxymethyl)-1,5-bis(4-hydroxyphenyl)-4-penten-1-ol. 4,4′,7,9-Tetrahydroxy-8,9′-neolign-7′-ene

$C_{18}H_{20}O_4$ 300.354

All isolates appear to be racemates or partial racemates

(7R*,7′E,8S*)-form

9-Carboxylic acid, 7-Me ether

4,4′-Dihydroxy-7-methoxy-8,9′-lign-7′-en-9-oic acid

9-Carboxylic acid, 7-Me ether, Me ester

Conchigeranal B

$C_{20}H_{22}O_5$ 342.391

Constit. of *Alpinia conchigera*. Oil. $[\alpha]^{19}_{D}$ -1.6 (c, 0.58 in MeOH). λ_{max} 262 (log ε 4.15) (MeOH)

(7RS,7′E,8RS)-form

Galanganol B [864073-17-6]

Constit. of the rhizomes of *Alpinia galanga* and *Alpinia conchigera*. NO prodn. inhibitor. Powder. λ_{max} 262 (log ε 4.35) (MeOH)

(7RS,7′E,8SR)-form

Galanganol A [864073-15-4]

Constit. of the rhizomes of *Alpinia galanga* and *Alpinia conchigera*. Powder. λ_{max} 262 (log ε 4.33) (MeOH)

(7ξ,7′E,8ξ)-form [596823-75-5]

Constit. of the rhizomes of *Alpinia officinarum*. Antioxidant. Wax. $[\alpha]^{27}_{D}$ -13.2 (c, 0.2 in MeOH). λ_{max} 261 (log ε 4.37) (MeOH)

7-Me ether(1)

7-Methoxy-8,9′-lign-7′-ene-4,4′,9-triol. 4,4′,9-Trihydroxy-7-methoxy-8,9′-lign-7′-ene. ***Conchigeranal D*** [1613724-11-0]

$C_{19}H_{22}O_4$ 314.381

Constit. of *Alpinia conchigera*. Displays cytotoxic activity against A549 cancer cell lines. Oil. $[\alpha]^{19}_{D}$ -6.1 (c, 0.30 in MeOH). λ_{max} 262 (log ε 4.11) (MeOH)

7-Me ether(2)

Conchigeranal E

$C_{19}H_{22}O_4$ 314.381

Constit. of *Alpinia conchigera*. Displays cytotoxic activity against A549 cancer cell lines. Oil. $[\alpha]^{19}_{D}$ -0.6 (c, 0.11 in MeOH). λ_{max} 262 (log ε 4.11) (MeOH)

7-Me ether, 9-Ac

Conchigeranal C

[1613724-43-8]

$C_{21}H_{24}O_5$ 356.418

Constit. of *Alpinia conchigera*. Oil. $[\alpha]^{21}_{D}$ -9.67 (c, 0.17 in MeOH). λ_{max} 263 (log ε 4.13) (MeOH)

9-Me ether

9-Methoxy-8,9′-lign-7′-ene-4,4′,7-triol. 4,4′,7-Trihydroxy-9-methoxy-8,9′-lign-7′-ene

[596823-74-4]

$C_{19}H_{22}O_4$ 314.381

Constit. of the rhizomes of *Alpinia officinarum*. Antioxidant. Wax. $[\alpha]^{27}_{D}$ +4.3 (c, 0.12 in MeOH). λ_{max} 261 (log ε 4.44) (MeOH)

9-Me, 7-O-[3-(4-acetoxyphenyl)-2E-propenyl] ether

[596823-73-3]

$C_{30}H_{32}O_6$ 488.58

Constit. of the rhizomes of *Alpinia officinarum*. Wax. $[\alpha]^{27}_{D}$ +0.6 (c, 0.35 in MeOH). λ_{max} 257 (log ε 4.67) (MeOH)

7,9-Di-Me ether(1)

7,9-Dimethoxy-8,9′-lign-7′-ene-4,4′-diol. 4,4′-Dihydroxy-7,9-dimethoxy-8,9′-lign-7′-ene

[596823-76-6]

$C_{20}H_{24}O_4$ 328.408

Constit. of the rhizomes of *Alpinia officinarum*. Antioxidant. Pale yellow wax. $[\alpha]^{27}_{D}$ -0.6 (c, 0.52 in MeOH). λ_{max} 261 (log ε 4.42) (MeOH)

7,9-Di-Me ether(2)

[596823-71-1]

$C_{20}H_{24}O_4$ 328.408

Constit. of the rhizomes of *Alpinia officinarum*. Antioxidant. $[\alpha]^{27}_{D}$ +1 (c, 0.58 in MeOH). λ_{max} 261 (log ε 4.39) (MeOH)

7-Et, 9-Me ether(1)

7-Ethoxy-9-methoxy-8,9′-lign-7′-ene-4,4′-diol. 7-Ethoxy-4,4′-dihydroxy-9-methoxy-8,9′-lign-7′-ene

[596823-72-2]

$C_{21}H_{26}O_4$ 342.435

Constit. of the rhizomes of *Alpinia officinarum*. Yellow wax. $[\alpha]^{27}_{D}$ +2.9 (c, 0.17 in MeOH). λ_{max} 261 (log ε 4.37) (MeOH)

7-Et, 9-Me ether(2)

[596823-77-7]

$C_{21}H_{26}O_4$ 342.435

Constit. of the rhizomes of *Alpinia officinarum*. Pale yellow wax. $[\alpha]^{27}_{D}$ +3.3 (c, 0.06 in MeOH). λ_{max} 261 (log ε 4.32) (MeOH).

Ly, T.N. *et al.*, *J. Agric. Food Chem.*, 2003, **51**, 4924-4929 (*Alpinia officinarum constits*).

Morikawa, T. *et al.*, *Chem. Pharm. Bull.*, 2005, **53**, 625-630 (*Galanganols A,B*).

Xu, J.-J. *et al.*, *Fitoterapia*, 2013, **91**, 82-86 (*Conchigeranals B-E*)

3,3′-Lign-8′-ene-4,4′,8-triol L105

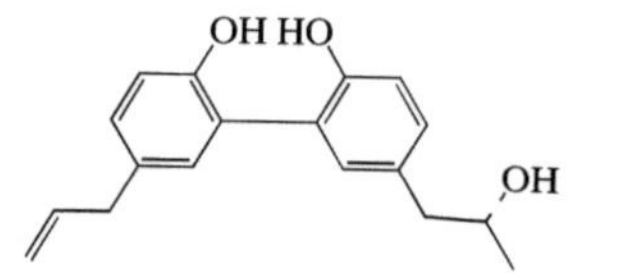

5-(2-Hydroxypropyl)-5′-(2-propenyl)-[1,1′-biphenyl]-2,2′-diol, CAS. *4,4′,8-Trihydroxy-3,3′-neolign-8′-ene*

$C_{18}H_{20}O_3$ 284.355

(R)-form

Clypearianin A [2200314-83-4]

Constit. of the leaves and twigs of *Pithecellobium clypearia*. Moderate antioxidant. Yellow oil. $[\alpha]^{23}_{D}$ -22.4 (c, 0.1 in MeOH). λ_{max} 231 (log ε 0.73); 279 (log ε 0.27) (MeOH).

Lou, L.-L. *et al.*, *Fitoterapia*, 2016, **112**, 16-21 (*Clypearianin A, activity*)

3,8′-Lign-7-ene-4,4′,8′-triol L106

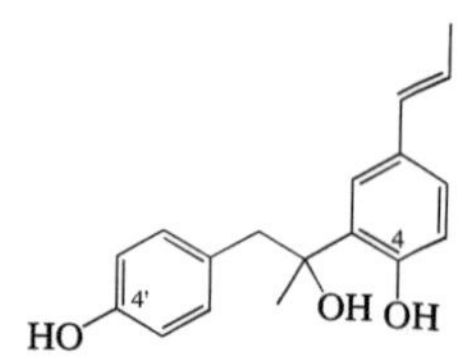

4,4′,8′-Trihydroxy-3,8′-neolign-7-ene

$C_{18}H_{20}O_3$ 284.355

4-Me ether

4-[2-Hydroxy-2-[2-methoxy-5-(2-propenyl)phenyl]propyl]phenol. 1-(4-Hydroxyphenyl)-2-[2-methoxy-5-(2-propenyl)phenyl]-2-propanol. 4-Methoxy-3,8′-lign-7-ene-4′,8′-diol. 4′,8′-Dihydroxy-4-methoxy-3,8′-lign-7-ene

[125205-69-8]

$C_{19}H_{22}O_3$ 298.382

Constit. of *Krameria grayi* and *Krameria lanceolata*. Amorph.. $[\alpha]^{21}_{D}$ +7 (c, 0.2 in MeOH).

Achenbach, H. *et al.*, *Phytochemistry*, 1989, **28**, 1959-1962 (*Krameria lanceolata constit*)

3,8′-Lign-7-ene-4,4′,9-triol L107

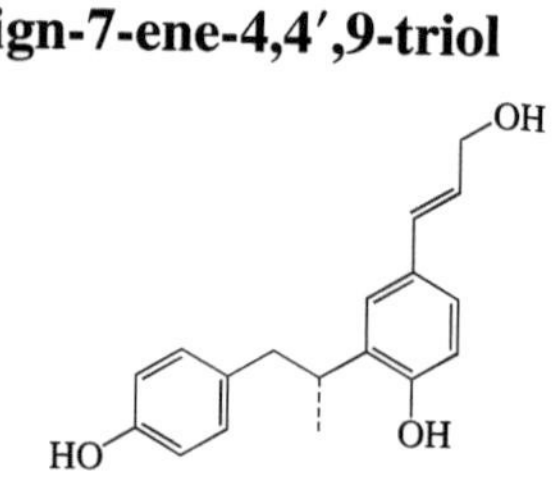

4,4′,9-Trihydroxy-3,8′-neolign-7-ene

$C_{18}H_{20}O_3$ 284.355

(R,E)-form

9-Carboxylic acid

4,4′-Dihydroxy-3,8′-lign-7-en-9-oic acid

9-Carboxylic acid, Me ester

$C_{19}H_{20}O_4$ 312.365

Constit. of the seeds of *Daphniphyllum macropodum*. Solid. $[\alpha]_D^{25}$ -4.5 (c, 0.06 in $CHCl_3$). λ_{max} 205 (sh); 226; 316 (MeOH).

Ma, X. *et al.*, *RSC Adv.*, 2017, **7**, 52970-52976 (*Daphniphyllum macropodum constit*)

3,8′-Lign-8′-ene-4,4′,8-triol L108

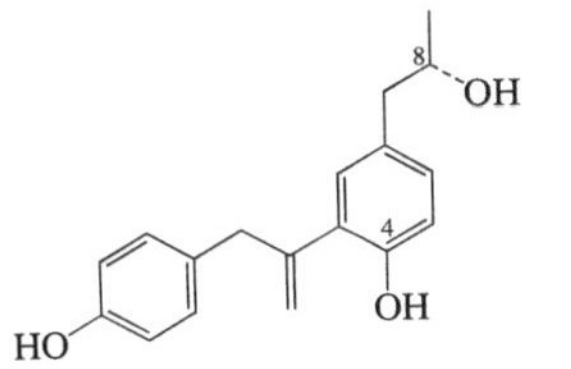

4,4′,8-Trihydroxy-3,8′-neolign-8′-ene

$C_{18}H_{20}O_3$ 284.355

(S)-form

4-Me ether

4-Methoxy-3,8′-lign-8′-ene-4′,8-diol. 4′,8-Dihydroxy-4-methoxy-3,8′-lign-8′-ene. **Zapotecol**

[125205-68-7]

$C_{19}H_{22}O_3$ 298.382

Constit. of *Krameria lanceolata* and *Krameria parvifolia*. Oil. $[\alpha]_D^{21}$ +5 (c, 0.6 in MeOH)

8-Ketone

4,4′-Dihydroxy-3,8′-lign-8′-en-8-one. 4,4′-Dihydroxy-8-oxo-3,8′-lign-8′-ene

8-Ketone, 4-Me ether

4′-Hydroxy-4-methoxy-3,8′-lign-8′-en-8-one. 4′-Hydroxy-4-methoxy-8-oxo-3,8′-lign-8′-ene. **Zapotecone**

[125233-97-8]

$C_{19}H_{20}O_3$ 296.366

Constit. of *Krameria lanceolata*. Oil.

Achenbach, H. *et al.*, *Phytochemistry*, 1989, **28**, 1959-1962 (*Zapotecol, Zapotecone*)

Lindleyanin L109

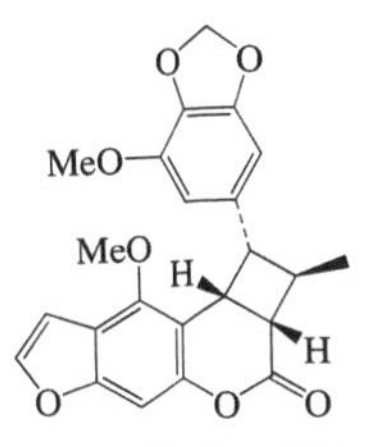

Relative Configuration

[878196-90-8]

$C_{23}H_{20}O_7$ 408.406

Coumarinolignan. Constit. of *Pleurospermum lindleyanum*. Gum. Opt. inactive. λ_{max} 192; 207; 240 ($CHCl_3$).

Tan, J.-J. *et al.*, *Helv. Chim. Acta*, 2006, **89**, 117-121 (*Lindleyanin*)

Linoxepin L110

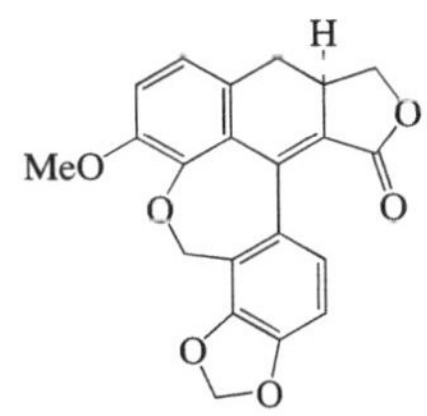

(*R*)-form

$C_{21}H_{16}O_6$ 364.353

(R)-form [1008096-11-4]

Constit. of *Linum perenne*. Amorph. solid. Mp 228° dec.. $[\alpha]_D^{20}$+23 (c, 0.93 in $CHCl_3$). λ_{max} 210 (sh); 228 (log ε 4.06); 258 (sh); 318 (log ε 3.94); 350 (sh) (MeOH)

(S)-form

Cryst.. $[\alpha]_D^{22}$-99.7 (c, 0.87 in $CHCl_3$).

Schmidt, T.J. *et al.*, *Planta Med.*, 2007, **73**, 1574-1580 (*Linoxepin*).

Tietze, L.F. *et al.*, *Angew. Chem., Int. Ed.*, 2013, **52**, 3191-3194 (*synth, cryst struct*).

Weinstabl, H. *et al.*, *Angew. Chem., Int. Ed.*, 2013, **52**, 5305-5308 (*synth*).

Tietze, L.F. *et al.*, *Chem. - Eur. J.*, 2014, **20**, 17119-17124 (*synth*).

Qureshi, Z. *et al.*, *Eur. J. Org. Chem.*, 2014, 4053-4069 (*synth, struct*).

Minami, O. *et al.*, *Heterocycles*, 2017, **95**, 116-126 (*synth*)

Liovil guaiacylglyceryl ether L111

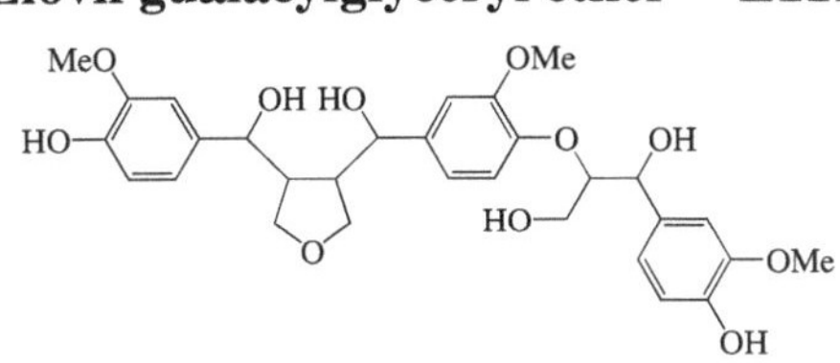

9,9′-Epoxy-3,3′,3″-trimethoxy-4′,8″-oxy-8,8′-sesquilignan-4,4″,7,7′,7″,9″-hexol

[845890-68-8]

$C_{30}H_{36}O_{11}$ 572.607

Constit. of *Picea abies*.

Willfoer, S. *et al.*, *Holzforschung*, 2004, **58**, 345-354 (*Picea abies constit*)

Liriodenol L112

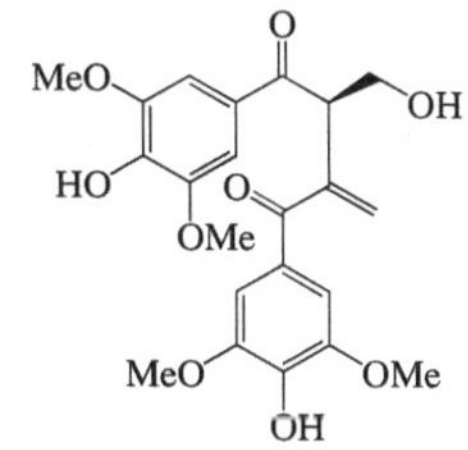

(*R*)-form

1,4-Bis(4-hydroxy-3,5-dimethoxyphenyl)-2-(hydroxymethyl)-3-methylenebutane-1,4-dione. 4,4′,9-Trihydroxy-3,3′,5,5′-tetramethoxy-8,8′-lign-8′-ene-7,7′-dione

$C_{22}H_{24}O_9$ 432.425

(R)-form [1702406-11-8]

Constit. of bark of *Liriodendron hybrid*. Exhibits weak or no cytotoxic effects against four cancer cell lines compared to *S*- and (±)-forms. $[\alpha]_D$ -89.5 (MeOH) (calc.). Resolved by chiral hplc

(S)-form [1702406-08-3]

Constit. of bark of *Liriodendron hybrid*. Exhibits significant cytotoxic effects on four cancer cell lines NIC-H460, MDA-MB-23, BGC-8231 and SGC-7901. Possesses strongest cytotoxicity compared to *R*- and (±)-forms. $[\alpha]_D$ +89.5 (MeOH) (calc.). Resolved by chiral hplc

(±)-form [1702480-54-3]

Constit. of bark of *Liriodendron hybrid*. Exhibits weak/moderate cytotoxic effects on the four cancer cell lines NIC-H460, MDA-MB-231, BGC-823 and SGC-7901. Stronger cytotoxicity compared to *R*- but weaker than *S*-form. Amorph. solid. λ_{max} 208; 235; 303 (MeOH).

Yang, D.-T. *et al.*, *Bioorg. Med. Chem. Lett.*, 2015, **25**, 1976-1978 (*Liriodenol, activity*)

Lirionol L113

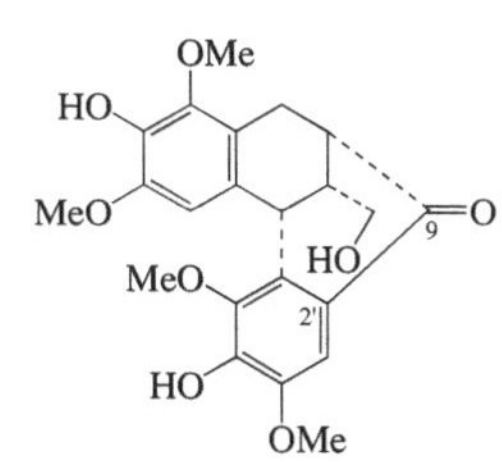

Relative Configuration

2,7′:2′,9-Bicyclo-5,5′,9′-trihydroxy-4,4′,6,6′-tetramethoxy-8,8′-lignan-9-one. 5,5′,9′-Trihydroxy-4,4′,6,6′-

tetramethoxy-9-oxo-2,7′:2′,9-bicyclolignan
[67881-22-5]
$C_{22}H_{24}O_8$ 416.426
2′,9-Cyclonaphthalenoid lignan. Isol. from the bark of *Liriodendron tulipifera*. Cryst. (THF). Mp 216 - 225° dec..
Chen, C.-L. *et al.*, *Phytochemistry*, 1978, **17**, 779-782 (*Lirionol*).
Weeratunga, G. *et al.*, *J. Org. Chem.*, 1985, **50**, 5902-5905 (*synth, sterochem*)

Lithospermic acid L114

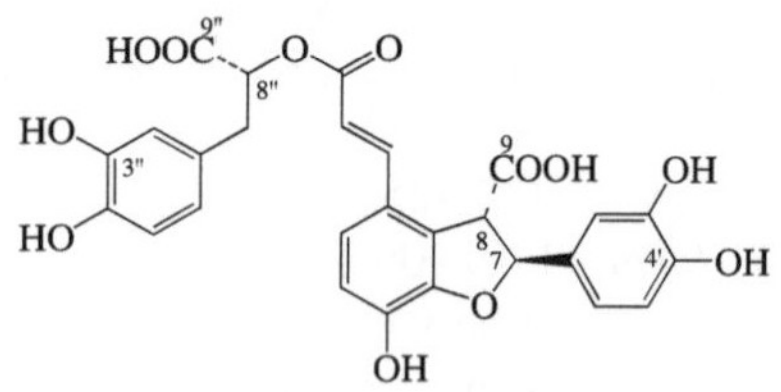

Absolute Configuration

4-[3-[1-Carboxy-2-(3,4-dihydroxyphenyl)ethoxy]-3-oxo-1-propenyl]-2-(3,4-dihydroxyphenyl)-2,3-dihydro-7-hydroxy-3-benzofurancarboxylic acid, CAS. ***Clinopodic acid H***. *Lithospermic acid A*
[28831-65-4]
[1016894-33-9] (Clinopodic acid H)
$C_{27}H_{22}O_{12}$ 538.461
Constit. of *Clinopodium chinense* var. *parviflorum, Celastrus hindsii, Cynoglossum officinale, Cordia spinescens, Lycopus europeus, Pulmonaria officinalis, Salvia miltiorrhiza, Lithospermum ruderale, Lithospermum officinale, Origanum vulgare* and *Tournefortia sarmentosa*. Cardiotonic agent. Shows antigonadotrophic and antithyroidal props. Inhibitor of DNA polymerase and hyaluronidase. Exhibits potent anti-HIV activity. Antioxidant. Amorph. powder. Mp 181 - 186°. $[\alpha]_D^{23}$+131 (c, 2.6 in MeOH). Log P 1.19 (uncertain value) (calc). λ_{max} 252; 286; 304; 332 (MeOH)

3″-O-β-D-Glucopyranoside
Clinopodic acid Q
[1427521-31-0]
$C_{33}H_{32}O_{17}$ 700.602
Constit. of *Clinopodium gracile*. Amorph. solid. $[\alpha]_D^{23}$-87.9 (c, 0.37 in MeOH). λ_{max} 252 (log ε 4.17); 287 (log ε 4.15); 310 (log ε 4.10) (MeOH)

9-Me ester
9-O-Methyllithospermic acid
[933054-33-2]
$C_{28}H_{24}O_{12}$ 552.488
Constit. of the roots of *Lithospermum erythrorhizon*. Stimulates expression of serine palmitoyltransferase in human keratinocyte cells. Yellow-brown gum. $[\alpha]_D^{22}$+67.3 (c, 0.5 in MeOH). λ_{max} 255; 291; 305; 330 (MeOH)

9″-Me ester
9″-O-Methyllithospermic acid
[1151683-53-2]
$C_{28}H_{24}O_{12}$ 552.488
Constit. of the roots of *Lithospermum erythrorhizon*. Stimulates expression of serine palmitoyltransferase in human keratinocyte cells. Yellow-brown gum. $[\alpha]_D^{22}$+92.1 (c, 0.5 in MeOH). λ_{max} 252; 291; 302; 329 (MeOH)

Di-Me ester
9,9″-Di-O-methyllithospermic acid
[54844-34-7]
$C_{29}H_{26}O_{12}$ 566.515
Constit. of *Dracocephalum forrestii, Lindelofia stylosa* and *Salvia miltiorrhiza*. Matrix metalloproteinase inhibitor

9-Et, 9″-Me ester
[875313-63-6]
$C_{30}H_{28}O_{12}$ 580.542
Constit. of *Lindelofia stylosa*. Brown-yellow gum. $[\alpha]_D^{23}$+65.6 (c, 0.25 in MeOH). λ_{max} 224 (log ε 3.56); 254 (log ε 3.46); 290 (log ε 3.46); 309 (log ε 3.46) (MeOH)

9″-Butyl ester
9″-O-Butyllithospermic acid
[1219124-87-4]
$C_{31}H_{30}O_{12}$ 594.569
Constit. of *Dracocephalum forrestii*. Yellow powder. $[\alpha]_D^{20}$ +142.7 (c, 1 in MeOH). λ_{max} 212 (log ε 3.65); 220 (log ε 3.28); 253 (log ε 3.09); 290 (log ε 3.07) (MeOH)

9-[1R-Carboxy-2-(3,4-dihydroxyphenyl)ethyl] ester
Danshensuan B. ***Lithospermic acid B***. *Salvianic acid B*. ***Salvianolic acid B***
[121521-90-2]
[115939-25-8]
$C_{36}H_{30}O_{16}$ 718.62
Constit. of *Lithospermum erythrorhizon, Melissa officinalis, Salvia miltiorrhiza* and other *Salvia* spp. Prod. by *Streptomyces* sp. M4. Lipoxygenase inhibitor. Antihypertensive agent. Exhibits hypotensive effects. Inhibitor of protein–protein interactions mediated by the SH2 domains of Src-family kinases. Antioxidant. Antifungal agent. Yellow powder. $[\alpha]_D^{26}$+146.3 (c, 1.2 in H_2O) (free acid). Isol. as $Mg^{2\oplus}$ or $Mg^{2\oplus}/NH_3^{\oplus}$ salts. Abs. config. revised to that of Lithospermic acid B in 2006. λ_{max} 254 (no solvent reported)

9-[1R-Carboxy-2-(3,4-dihydroxyphenyl)ethyl] ester, 4′-O-β-D-glucopyranoside
Salvinoside
[1148139-61-0]
$C_{42}H_{40}O_{21}$ 880.761
Constit. of the roots of *Salvia miltiorrhiza*. Amorph. yellow powder. $[\alpha]_D^{20}$+86 (c, 0.44 in H_2O). λ_{max} 249; 283; 310 (MeOH)

7-Epimer, 9-[2-(3,4-dihydroxyphenyl)-1R-methoxycarbonylethyl] ester
Orthosiphoic acid C
[1638527-13-5]
$C_{37}H_{32}O_{16}$ 732.647
Constit. of the aerial parts of *Clerodendranthus spicatus*. Exhibits weak anti-HIV-1 protease activity. Amorph. brown powder. $[\alpha]_D^{20}$+21 (c, 0.02 in MeOH). λ_{max} 285 (log ε 0.66) (MeOH)

3″,4′-Bis-O-(24R-stigmast-5-en-3β-yl) ether, 9-[1R-carboxy-2-[4-hydroxy-3-(24R-stigmast-5-en-3β-yloxy)phenyl]ethyl] ester
3″,3‴,4-Tris-O-β-sitosteryl-lithospermic acid B. ***β-Sitosterol lithospermate***
[1469486-14-3]
$C_{123}H_{174}O_{16}$ 1908.729
Constit. of *Salvia columbariae*. Cryst.. $[\alpha]_D^{20}$-17.7 (no solvent reported)

3″-Deoxy
Orthosiphoic acid A
[1638527-26-0]
$C_{27}H_{22}O_{11}$ 522.462
Constit. of the aerial parts of *Clerodendranthus spicatus*. Exhibits weak anti-HIV-1 protease activity. Amorph. brown powder. $[\alpha]_D^{20}$+48.3 (c, 0.03 in MeOH). λ_{max} 284 (log ε 0.63) (MeOH)

7-Epimer

cis-Lithospermic acid A. ***Lithospermic acid C***

[1111082-72-4]

$C_{27}H_{22}O_{12}$ 538.461

Constit. of *Lithospermum erythrorhizon, Salvia miltiorrhiza* and *Salvia yunnanensis*. Brownish powder. $[\alpha]_D^{20}$ +41 (c, 0.08 in MeOH). λ_{max} 253 (log ε 4.28); 288 (log ε 4.22); 309 (log ε 4.38) (MeOH)

7-Epimer, 9-[1R-carboxy-2-(3,4-dihydroxyphenyl)ethyl] ester

7-Episalvianolic acid B. ***Salvianolic acid Y***. *Orthosiphoic acid B*

[1638738-76-7]

$C_{36}H_{30}O_{16}$ 718.62

Constit. of *Clerodendranthus spicatus, Salvia miltiorrhiza* and *Salvia officinalis*. Protector of PC12 cells against H_2O_2-induced injury. Exhibits weak anti-HIV-1 protease activity. Amorph. powder. $[\alpha]_D^{20}$ -47.1 (c, 0.09 in MeOH). $[\alpha]_D^{20}$ +20 (c, 0.01 in MeOH) (Orthosiphoic acid B). λ_{max} 207; 255; 289; 307 (MeOH). [neutral] λ_{max} 288 (log ε 1.05) (MeOH) (Orthosiphoic acid B)

8-Epimer

Clinopodic acid N. ***Monardic acid C***

[1427521-28-5]

$C_{27}H_{22}O_{12}$ 538.461

Constit. of *Clinopodium gracile* and *Monarda fistulosa*. Amorph. solid. $[\alpha]_D^{23}$ -29.0 (c, 1.05 in MeOH) (*C. gracile* constit). $[\alpha]_D^{25}$ +41.4 (c, 0.88 in MeOH) (*M. fistulosa* constit). Same abs. configs. determined for two different isolates but opt. rotation data is conflicting. λ_{max} 276 (log ε 4.21) (MeOH)

8″-Epimer

8″-Epilithospermic acid A

[906482-45-9]

$C_{27}H_{22}O_{12}$ 538.461

Constit. of *Origanum vulgare* ssp. *hirtum*. $[\alpha]_D^{20}$ +99.4 (c, 0.06 in MeOH). λ_{max} 201 (ε 86000); 230 (sh); 253 (ε 17000); 290 (ε 16000); 314 (ε 14000) (MeOH)

7,8-Diepimer

Monardic acid A.

7,8-Diepilithospermic acid A

[1607436-76-9]

$C_{27}H_{22}O_{12}$ 538.461

Constit. of *Monarda fistulosa* and *Pulmonaria officinalis*. Amorph. solid. $[\alpha]_D^{25}$ -86.2 (c, 2.5 in MeOH). λ_{max} 256 (log ε 4.28); 291 (log ε 4.24); 311 (log ε 4.26) (MeOH)

7,8-Diepimer, 9-[1R-carboxy-2-(3,4-dihydroxyphenyl)ethyl] ester

Monardic acid B

[1607436-77-0]

$C_{36}H_{30}O_{16}$ 718.62

Constit. of *Monarda fistulosa*. Amorph. solid. $[\alpha]_D^{24}$ -43.5 (c, 0.46 in MeOH). λ_{max} 254 (log ε 4.25); 288 (log ε 4.26); 307 (log ε 4.19) (MeOH)

Enantiomer, 9-[1S-carboxy-2-(3,4-dihydroxyphenyl)ethyl] ester

ent-Salvianolic acid B

[1838183-14-4]

$C_{36}H_{30}O_{16}$ 718.62

Constit. of *Salvia miltiorrhiza*. DPPH radical scavenger. Brown powder. $[\alpha]_D^{20}$ +17.7 (c, 0.07 in MeOH)

Diastereoisomer, 9-[1ξ-carboxy-2-(3,4-dihydroxyphenyl)ethyl] ester

Epilithospermic acid B

[906561-46-4]

$C_{36}H_{30}O_{16}$ 718.62

Constit. of *Origanum vulgare* ssp. *hirtum* and *Salvia miltiorrhiza*. Epimeric with Lithospermic acid B at either C-8″ or C-8‴. λ_{max} 200 (ε 81000); 228 (sh); 254 (ε 13000); 286 (ε 13000); 321 (ε 12000) (MeOH)

7,8″-Diepimer

8″-Epilithospermic acid C

$C_{27}H_{22}O_{12}$ 538.461

Constit. of the roots of *Salvia miltiorrhiza*. Antioxidant. Amorph. powder.

Kelley, C.J. *et al., J. Org. Chem.*, 1975, **40**, 1804-1815 (*Lithospermic acid*).

Kelley, C.J. *et al., J. Org. Chem.*, 1976, **41**, 449-455 (*Lithospermic acid, cmr*).

Ai, C.-B. *et al., J. Nat. Prod.*, 1988, **51**, 145-149 (*Salvianolic acid B*).

Tanaka, T. *et al., Chem. Pharm. Bull.*, 1989, **37**, 340-344 (*Salvianolic acid B*).

Kamata, K. *et al., Gen. Pharmacol.*, 1994, **25**, 69-73 (*Salvianolic acid B, activity*).

Kang, H.S. *et al., Arch. Pharmacal Res.*, 1997, **20**, 496-500 (*di-Me ester*).

Yuan, J.-P. *et al., J. Agric. Food Chem.*, 1998, **46**, 2651-2654 (*Salvianolic acid B, detn, hplc*).

Lin, Y.-L. *et al., J. Chin. Chem. Soc. (Taipei)*, 2003, **50**, 1079-1083 (*Oreganum vulgare constit*).

Choudhary, M.I. *et al., Chem. Pharm. Bull.*, 2005, **53**, 1469-1471 (*Me Et ester*).

O'Malley, S.J. *et al., J. Am. Chem. Soc.*, 2005, **127**, 13496-13497 (*synth*).

Koukoulitsa, C. *et al., J. Agric. Food Chem.*, 2006, **54**, 5388-5392 (*8″-Epilithospermic acids A,B*).

Watzke, A. *et al., J. Nat. Prod.*, 2006, **69**, 1231-1233 (*Salvianolic acid B, abs config*).

Zhou, B. *et al., Anal. Chim. Acta*, 2008, **628**, 111-116 (*Salvianolic acid B, anal*).

Deng, F. *et al., Asian J. Chem.*, 2008, **20**, 6129-6133 (*Salvinoside*).

Zhao, G.-R. *et al., Food Chem. Toxicol.*, 2008, **46**, 73-81 (*activity*).

Zhang, Z.-F. *et al., J. Asian Nat. Prod. Res.*, 2008, **10**, 391-396 (*cis-Lithospermic acid A*).

Pat. Coop. Treaty (WIPO), 2008, 2008 08806 (*Lithospermic acid, anti-HIV activity*).

Thuong, P.T. *et al., Bioorg. Med. Chem. Lett.*, 2009, **19**, 1815-1817 (*pmr, cmr, mono-Me esters, activity*).

Sperl, B. *et al., Bioorg. Med. Chem. Lett.*, 2009, **19**, 3305-3309 (*Salvianolic acid B, activity*).

Murata, T. *et al., J. Nat. Prod.*, 2009, **72**, 1379-1384 (*Clinopodic acid H*).

Li, S.-M. *et al., Planta Med.*, 2009, **75**, 1591-1596 (*9″-butyl ester*).

Wang, D.-H. *et al., J. Am. Chem. Soc.*, 2011, **133**, 5767-5769 (*synth*).

Fischer, J. *et al., Org. Lett.*, 2011, **13**, 3376-3379 (*synth*).

Wang, L. *et al., Bioorg. Med. Chem.*, 2012, **20**, 4164-4171 (*di-Me ester, activity*).

Aoshima, H. *et al., Chem. Pharm. Bull.*, 2012, **60**, 499-507 (*Clinopodic acids N,Q, activity*).

Adams, J.D. *et al., Open Nat. Prod. J.*, 2012, **5**, 13-15 (*β-Sitosterol lithospermate*).

Varadaraju, T.G. *et al., Org. Biomol. Chem.*, 2012, **10**, 5456-5465 (*synth, activity*).

Ghosh, A.K. *et al., Org. Lett.*, 2012, **14**, 5046-5049 (*synth*).

Murata, T. *et al., Fitoterapia*, 2013, **91**, 51-59 (*Monardic acids A,B,C, Lithospermic acid C*).

Barros, L. *et al., Food Chem.*, 2013, **136**, 1-8 (*Melissa officinalis constit*).

Sun, Z. *et al., Phytochem. Lett.*, 2014, **8**, 16-21 (*Orthosiphoic acids A,B,C, activity*).

Gong, J. *et al., Molecules*, 2015, **20**, 683-692 (*Salvianolic acids B,Y, biosynth*).

Si, Y. *et al., Bioorg. Med. Chem. Lett.*, 2016, **26**, 82-86 (*Salvia miltiorrhiza constits*).

Zhang, J. *et al., Fitoterapia*, 2017, **121**, 46-52 (*8″-Epilithospermic acid C, activity*).

Zheng, Y. *et al., Tetrahedron*, 2018, **20**, 5950-5954 (*synth*).

Sharma, M. *et al., Microbiological Res.*, 2020, **237**, 126478 (*Salvianolic acid B, Streptomyces metab, activity*)

Lobechinenoids L115

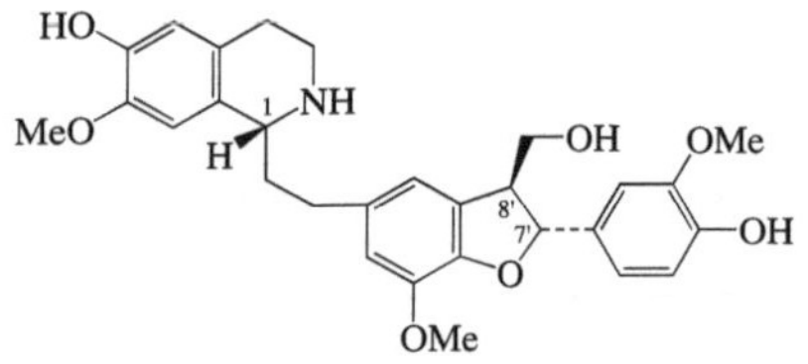

(1*R*,7'*R*,8'*S*)-*form*

[1459155-32-8] (Lobechinenoid A)
[1459155-33-9] (Lobechinenoid B)
[1459155-34-0] (Lobechinenoid C)
[1459155-35-1] (Lobechinenoid D)
$C_{29}H_{33}NO_7$ 507.583
Alkaloid-lignan hybrid. Isol. as a mixt. of two pairs of enantiomers called Lobechinenoids A-D. Config. of C-7'/C-8'- is *trans* in all four enantiomers. Constit. of the aerial parts of *Lobelia chinensis*. Yellowish oil. λ_{max} 235; 285 (MeOH).
Yang, S. *et al.*, *J. Chromatogr., A*, 2013, **1311**, 134-139 (*Lobechinenoids A-D*)

Longifloroside C L116

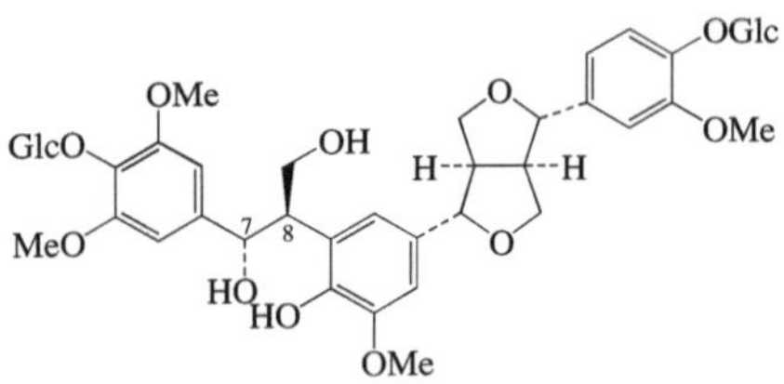

Relative Configuration

[193967-32-7]
$C_{43}H_{56}O_{21}$ 908.9
Constit. of *Pedicularis longiflora*. Obt. as a mixt. with epimer
7 or 8-Epimer
Longifloroside D
[193967-33-8]
$C_{43}H_{56}O_{21}$ 908.9
Constit. of *Pedicularis longiflora*.
Wang, C.Z. *et al.*, *Planta Med.*, 1997, **63**, 241-244 (*Longiflorosides C,D*)

Longipedlignan L L117

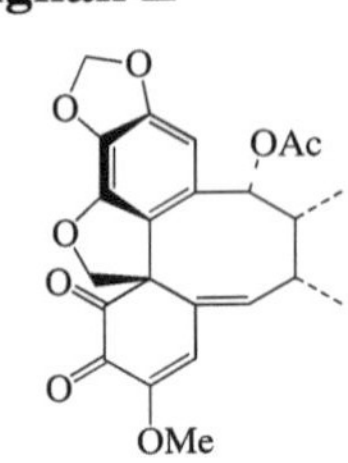

Absolute Configuration

[2377873-85-1]
$C_{23}H_{22}O_8$ 426.421
Constit. of the roots of *Kadsura longipedunculata*. Amorph. yellow powder. $[\alpha]_D^{20}$ +13 (c, 0.1 in MeOH).
Lu, J. *et al.*, *J. Nat. Prod.*, 2019, **82**, 2842-2851 (*Longipedlignan L*)

Lophirone I L118

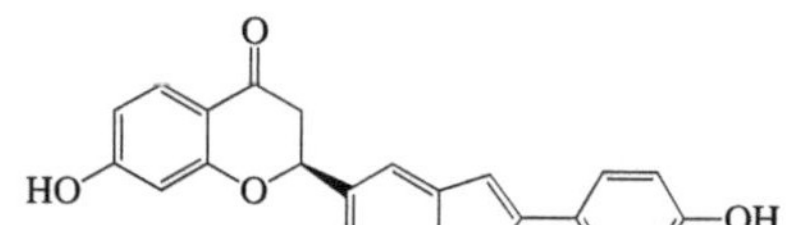

[155548-31-5]
$C_{23}H_{16}O_5$ 372.376
Constit. of the stem bark of *Lophira lanceolata*. Amorph. yellow solid. $[\alpha]_D^{20}$ +2.3 (c, 0.5 in Me_2CO)
Di-Me ether
Lophirone J
[155548-32-6]
$C_{25}H_{20}O_5$ 400.43
Constit. of the stem bark of *Lophira lanceolata*. Amorph. yellow solid. $[\alpha]_D^{20}$ -1.4 (c, 0.4 in Me_2CO).
Tih, R.G. *et al.*, *J. Nat. Prod.*, 1994, **57**, 142-145 (*Lophirones I,J*)

Lyciumamide C L119

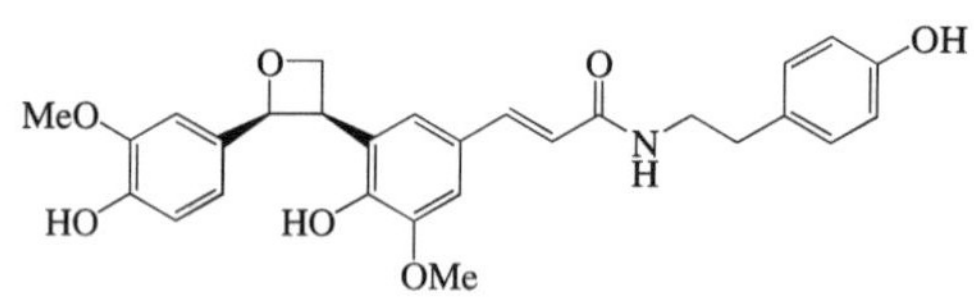

Relative Configuration

[1647111-42-9]
$C_{28}H_{29}NO_7$ 491.54
Alkaloid from the fruit of *Lycium barbarum*. Antioxidant. Yellow powder. $[\alpha]_D^{22}$ -7.9 (c, 0.34 in MeOH)
Z-Isomer
Z-Lyciumamide C
$C_{28}H_{29}NO_7$ 491.540
Constit. of the stems of *Lycium barbarum*. Powder. $[\alpha]_D$ -5 (c, 0.1 in MeOH).
Gao, K. *et al.*, *J. Agric. Food Chem.*, 2015, **63**, 1067-1075 (*Lyciumamide C*).
Zhu, P.-F. *et al.*, *Nat. Prod. Bioprospect.*, 2017, **7**, 421-431 (*Z-Lyciumamide C*)

Lyciumamide F L120

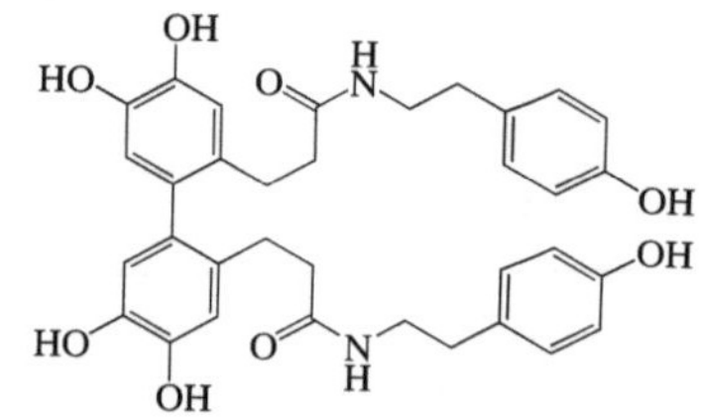

$C_{34}H_{36}N_2O_8$ 600.668
Constit. of the root bark of *Lycium chinense*. Amorph. powder. λ_{max} 226 (log ε 3.87); 285 (log ε 3.39) (MeOH).
Chen, H. *et al.*, *Fitoterapia*, 2017, **122**, 119-125 (*Lyciumamide F*)

Lycopic acid C L121

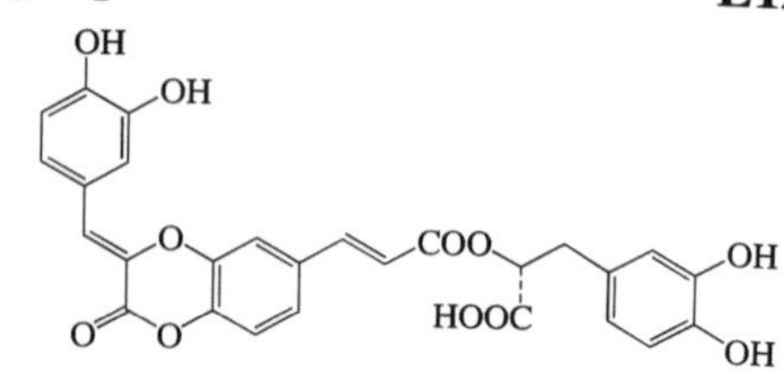

Absolute Configuration

$C_{27}H_{20}O_{11}$ 520.446
Regioisomer of Melitric acid B, M28. Constit. of *Lycopus lucidus* and *Pulmonaria officinalis*. Hyaluronidase inhibitor. Pale brown solid. $[\alpha]_D^{21}$ +98.2 (c, 0.55 in MeOH). λ_{max} 247 (log ε 4.64); 284 (log ε 4.67); 323 (log ε 4.61) (MeOH).
Murata, T. *et al.*, *Chem. Pharm. Bull.*, 2010, **58**, 394-397 (*Lycopic acid C, activity*)

Lycopodone L122

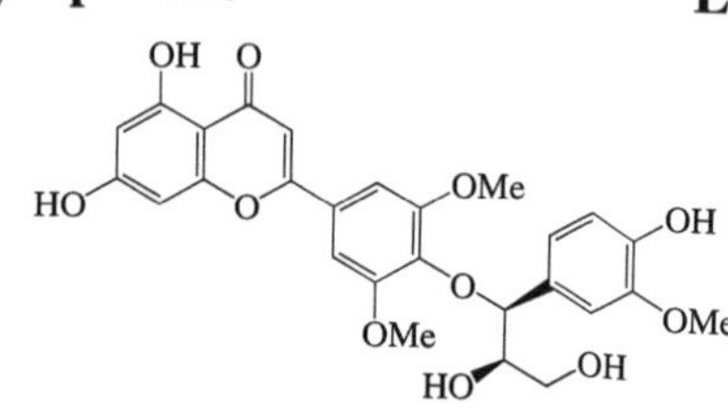

Relative Configuration

[851385-04-1]
$C_{27}H_{26}O_{11}$ 526.494
Flavonolignan. Constit. of *Lycopodium japonicum*. Moderate antitumour agent. Yellow powder. Mp > 300°. $[\alpha]_D^{23}$ -11.8 (c, 1.7 in MeOH). λ_{max} 204 (log ε 4.16); 273 (log ε 3.55); 306 (log ε 3.53); 331 (log ε 3.55) (MeOH).
Yan, J. *et al.*, *Heterocycles*, 2005, **65**, 661-666 (*Lycopodone*)

Lyoniresinol 9′-*O*-(8″-syringylglycerol) ether L123

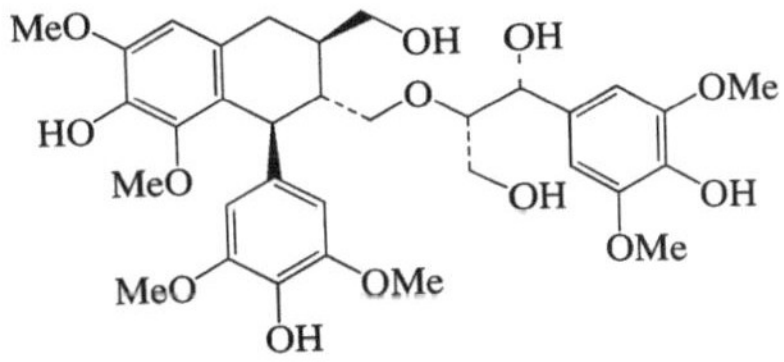

Relative Configuration

[1465820-57-8]
$C_{33}H_{42}O_{13}$ 646.686
Constit. of the rhizomes of *Smilax microphylla*. Amorph. powder. $[\alpha]^{27}_{D}$ -2.8 (c, 1 in MeOH). λ_{max} 300 (MeOH).

Liu, L.-S. *et al.*, *Nat. Prod. Commun.*, 2013, **8**, 113-114 (*Smilax microphylla constit*)

Maackolin M1

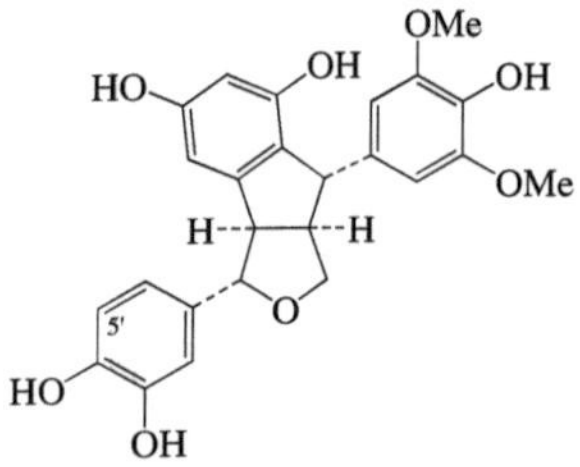

Relative Configuration

Kompasinol A
[170663-51-1]
$C_{25}H_{24}O_8$ 452.459
Stilbenolignan. Related to Gnetifolin F, G20. Constit. of *Koompassia malaccensis*, *Maackia amurensis*, *Smilax glabra* and *Syagrus romanzoffiana*. Antioxidant. Cream powder. Mp 90 - 92° (as penta-Ac). Kompasinol A was definitely racemic. λ_{max} 283 (log ε 3.93); 285; 325 (log ε 3.42) (MeOH)

5'-Hydroxy
13-Hydroxykompasinol A
[1020286-30-9]
$C_{25}H_{24}O_9$ 468.458
Constit. of the seeds of *Syagrus romanzoffiana*. α-Glucosidase type IV inhibitor. Pale yellow solid. Racemic. λ_{max} 277 (log ε 3.73) (MeOH).

Kulesh, N.I. *et al.*, *Phytochemistry*, 1995, **40**, 1001-1003 (*Maackolin*).
Kobayashi, M. *et al.*, *Chem. Pharm. Bull.*, 1996, **44**, 2249-2253 (*Kompasinol A*).
Lam, S.-H. *et al.*, *Phytochemistry*, 2008, **69**, 1173-1178 (*13-Hydroxykompasinol A, activity*).
Utkina, N.K. *et al.*, *Pharm. Chem. J. (Engl. Transl.)*, 2012, **46**, 488-491 (*antioxidant activity*)

Machilutone A M2

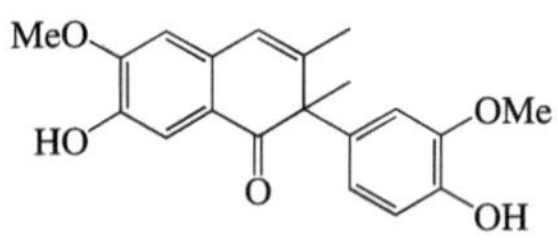

7-Hydroxy-2-(4-hydroxy-3-methoxyphenyl)-6-methoxy-2,3-dimethyl-1(2H)-naphthalenone
$C_{20}H_{20}O_5$ 340.375
Constit. of the stems of *Machilus philippinensis*. Syrup. $[\alpha]^{25}_{D}$ +23 (c, 0.08 in MeOH). λ_{max} 259 (log ε 3.6); 281 (log ε 3.8) (MeOH).

Tai, S.-H. *et al.*, *RSC Adv.*, 2019, **9**, 21616-21625 (*Machilutone A*)

Madangone B M3

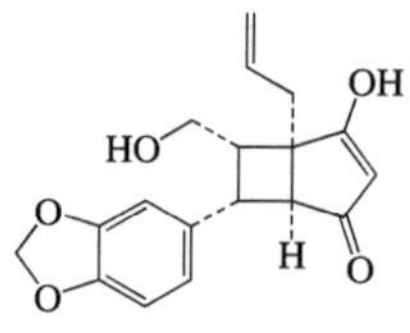

Relative Configuration

$C_{18}H_{18}O_5$ 314.337
Constit. of stem bark of *Beilschmiedia madang*. Exhibits inhibitory activity against COX-2 and acetylcholinesterase. Amorph. solid. $[\alpha]^{20}_{D}$ -0.16 (c, 0.18 in $CHCl_3$).

Salleh, W.M.N.H.W. *et al.*, *Phytochem. Lett.*, 2016, **15**, 168-173 (*Madangone B, activity*)

Magnaldehyde A M4

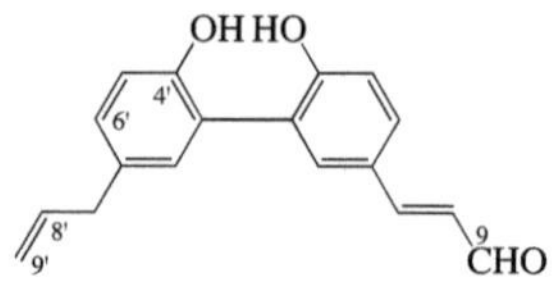

3-[2',6-Dihydroxy-5'-(2-propenyl)[1,1'-biphenyl-3-yl]]-2-propenal, CAS. *4,4'-Dihydroxy-3,3'-ligna-7,8'-dien-9-al. 4,4'-Dihydroxy-9-oxo-3,3'-neoligna-7,8'-diene.* ***Randainal***
[87562-13-8]
[92829-73-7] (unspecified stereochem.)
$C_{18}H_{16}O_3$ 280.323
Lignan numbering shown. Constit. of *Magnolia officinalis* and *Sassafras randaiense*. Inhibitor of induced elastase release by human neutrophils. Cryst. (EtOAc/hexane). Mp 137 - 139°. λ_{max} 330 (ε 28000); 373 (ε 26000) (MeOH)

9-Alcohol
3,3'-Ligna-7,8'-diene-4,4',9-triol. 4,4',9-Trihydroxy-3,3'-ligna-7,8'-diene. ***Randainol***
[93753-25-4]
$C_{18}H_{18}O_3$ 282.339
Constit. of *Sassafras randaiense*. Oil. Dec. rapidly even at -20°

8',9'-Dihydro, 8',9'-dihydroxy
4,4',8',9'-Tetrahydroxy-3,3'-lign-7-en-9-al. 4,4',8',9'-Tetrahydroxy-9-oxo-3,3'-lign-7-ene. ***Magnaldehyde C***
[138591-06-7]
$C_{18}H_{18}O_5$ 314.337
Constit. of *Magnolia officinalis*. Pale yellow powder. $[\alpha]^{29}_{D}$ -15.5 (c, 1.5 in MeOH)

6'-Hydroxy, 4'-deoxy
4,6'-Dihydroxy-3,3'-ligna-7,8'-dien-9-al. 4,6'-Dihydroxy-9-oxo-3,3'-ligna-7,8'-diene. ***Magnaldehyde B***
[92829-72-6]
$C_{18}H_{16}O_3$ 280.323
Constit. of *Magnolia officinalis* and *Magnolia rostrata*. Pale yellow needles ($CHCl_3$); cryst. (C_6H_6/petrol). Mp 161 - 162°. λ_{max} 284 (ε 40000); 324 (ε 33000) (MeOH)

6-Methoxy, 4-deoxy
4'-Hydroxy-6-methoxy-3,3'-ligna-7,8'-dien-9-al. 4'-Hydroxy-6-methoxy-9-oxo-3,3'-ligna-7,8'-diene. ***4-Deoxy-6-methoxymagnaldehyde A****. 4-Methoxymagnaldehyde B (incorr.)*
[1031791-93-1]
$C_{19}H_{18}O_3$ 294.35
Constit. of the stem bark of *Magnolia obovata*. Cytotoxic. Yellowish oil. λ_{max} 265 (log ε 4.3); 300 (log ε 4.3) (no solvent reported)

6'-Methoxy, 4'-deoxy
4-Hydroxy-6'-methoxy-3,3'-ligna-7,8'-dien-9-al. 4-Hydroxy-6'-methoxy-9-oxo-3,3'-ligna-7,8'-diene. ***6'-O-Methylmagnaldehyde B***
[957765-06-9]
$C_{19}H_{18}O_3$ 294.35
Constit. of the stem bark of *Magnolia officinalis*. Yellowish oil. λ_{max} 265 (log ε 4.3); 300 (log ε 4.3) (MeOH).

Chen, F.-C. *et al.*, *Phytochemistry*, 1983, **22**, 616-617 (*Randainal*).
El-Feraly, F.S. *Phytochemistry*, 1984, **23**, 2329-2331 (*Randainol*).
Takeya, T. *et al.*, *Chem. Pharm. Bull.*, 1987, **35**, 1755-1761 (*Magnaldehyde B, synth*).
Yahara, S. *et al.*, *Chem. Pharm. Bull.*, 1991, **39**, 2024-2036 (*Magnaldehydes A,B,C*).
Youn, U.-J. *et al.*, *J. Nat. Prod.*, 2007, **70**, 1687-1689 (*6'-O-Methylmagnaldehyde B*).
Youn, U.J. *et al.*, *Chem. Pharm. Bull.*, 2008, **56**, 115-117 (*4-Deoxy-6-methoxymagnaldehyde A, activity*).
Kuo, W.-L. *et al.*, *Phytochemistry*, 2013, **85**, 153-160 (*Randainal, activity*).
Schmidt, B. *et al.*, *Eur. J. Org. Chem.*, 2015, 3760-3766 (*Magnaldehyde B, synth*).
Xu, T.-T. *et al.*, *Org. Biomol. Chem.*, 2018, **16**, 5350-5258 (*Randainal, synth*)

Magnaldehyde D M5

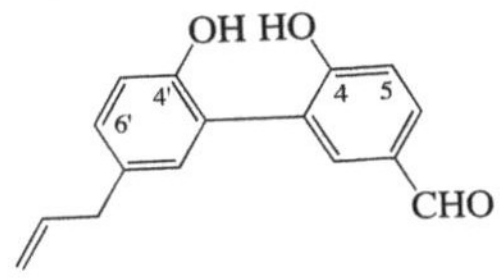

2′,6-Dihydroxy-5′-(2-propenyl)[1,1′-biphenyl]-3-carboxaldehyde, CAS. *5-Allyl-5′-formyl-2,2′-dihydroxybiphenyl. 4,4′-Dihydroxy-8,9-dinor-3,3′-lign-8′-en-7-al. 4,4′-Dihydroxy-8,9-dinor-7-oxo-3,3′-neolign-8′-ene*

[93753-33-4]

$C_{16}H_{14}O_3$ 254.285

Constit. of *Magnolia officinalis* and *Streblus asper*. Inhibitor of LPS-induced NO prodn.. Pale yellow needles ($CHCl_3$/C_6H_6). Mp 140 - 143°

5-Hydroxy, 4′-Me ether

4,5-Dihydroxy-4′-methoxy-8,9-dinor-3,3′-lign-8′-en-7-al. 4,5-Dihydroxy-4′-methoxy-8,9-dinor-7-oxo-3,3′-lign-8′-ene. ***Maninsigin A***

[1434146-09-4]

$C_{17}H_{16}O_4$ 284.311

Constit. of *Manglietia insignis*. Yellow oil. $[\alpha]^{25}_{D}$ +0.9 (c, 1.5 in MeOH). λ_{max} 203 (log ε 4.12); 219 (log ε 4.15); 288 (log ε 3.77) (MeOH)

4′-Deoxy, 6′-hydroxy

4,6′-Dihydroxy-8,9-dinor-3,3′-lign-8′-en-7-al. 4,6′-Dihydroxy-8,9-dinor-7-oxo-3,3′-lign-8′-ene. ***Magnaldehyde E***

[138591-09-0]

$C_{16}H_{14}O_3$ 254.285

Constit. of *Magnolia officinalis* and *Magnolia obovata*. Pale yellow needles ($CHCl_3$/C_6H_6). Mp 160 - 162°

6′-Methoxy, 4′-deoxy

4-Hydroxy-6′-methoxy-8,9-dinor-3,3′-lign-8′-en-7-al. 4-Hydroxy-6′-methoxy-8,9-dinor-7-oxo-3,3′-lign-8′-ene. ***6-O-Methylmagnaldehyde E***

[957765-07-0]

$C_{17}H_{16}O_3$ 268.312

Constit. of the stem bark of *Magnolia officinalis*. Yellowish oil. λ_{max} 271 (log ε 4.3) (MeOH).

Yahara, S. *et al.*, *Chem. Pharm. Bull.*, 1991, **39**, 2024-2036 (*Magnaldehydes D,E*).
Youn, U.-J. *et al.*, *J. Nat. Prod.*, 2007, **70**, 1687-1689 (*6-O-Methylmagnaldehyde E*).
Youn, U.-J. *et al.*, *Chem. Pharm. Bull.*, 2008, **56**, 115-117 (*Magnolia obovata constit*).
Li, J. *et al.*, *Magn. Reson. Chem.*, 2008, **46**, 497-500 (*Streblus asper constit*).
Shen, C.-C. *et al.*, *J. Nat. Prod.*, 2009, **72**, 168-171 (*activity*).
Shang, S.-Z. *et al.*, *Fitoterapia*, 2013, **84**, 58-63 (*Maninsigin A*).
Kuo, W.-L. *et al.*, *Phytochemistry*, 2013, **85**, 153-160 (*activity*).
Schmidt, B. *et al.*, *Eur. J. Org. Chem.*, 2015, 3760-3766 (*Magnaldehyde E, synth*)

Magnobovatol M6

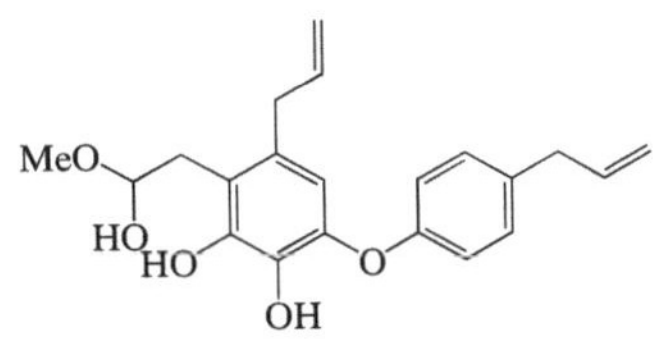

3-(2-Hydroxy-2-methoxyethyl)-4-(2-propenyl)-6-[4-(2-propenyl)phenoxy]-1,2-benzenediol, CAS. *6-(2-Hydroxy-2-methoxyethyl)-3,4′-oxyligna-8,8′-diene-4,5-diol. 4,5-Dihydroxy-6-(2-hydroxy-2-methoxyethyl)-3,4′-oxyneoligna-8,8′-diene*

[1637576-06-7]

$C_{21}H_{24}O_5$ 356.418

Constit. of the fruit of *Magnolia obovata*. Inhibits LPS-induced NO prodn.. Brown oil. $[\alpha]_D$ +7.4 (c, 0.06 in $CHCl_3$).

Seo, K.-H. *et al.*, *Planta Med.*, 2013, **79**, 1335-1340 (*Magnobovatol*)

Magnolianin M7

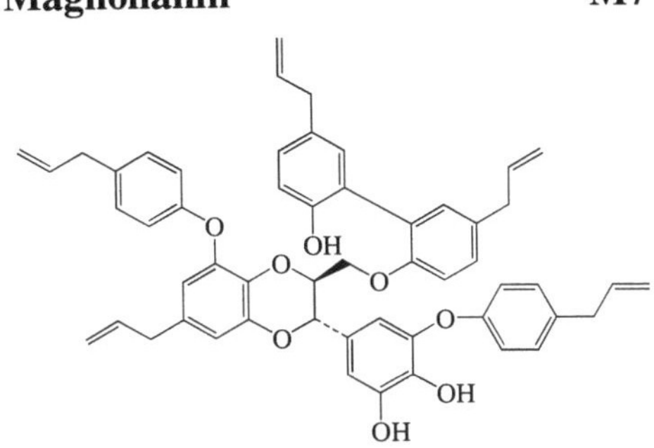

Relative Configuration

[147663-91-0]

$C_{54}H_{50}O_8$ 826.986

Constit. of the bark of *Magnolia obovata*. Potent 5-lipoxygenase (5-LOX) inhibitor. Racemate. λ_{max} 224 (ε 42800); 275 (ε 8300); 280 (ε 8300) (EtOH) (Derep).

Fukuyama, Y. *et al.*, *Tetrahedron Lett.*, 1993, **34**, 1051-1054 (*Magnolianin, activity*)

Magnolignan I M8

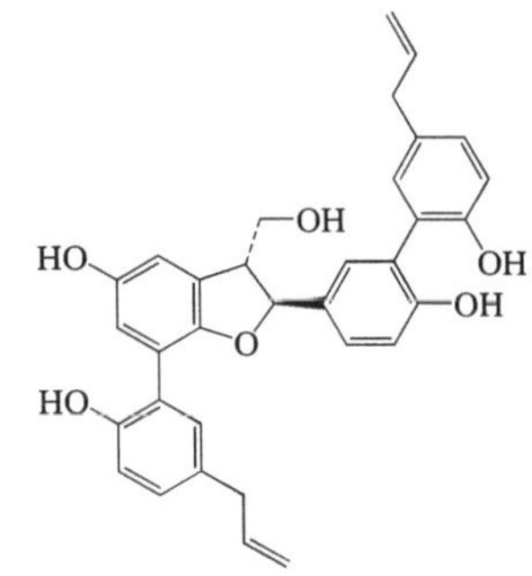

Relative Configuration

[138591-13-6]

$C_{33}H_{30}O_6$ 522.597

Constit. of *Magnolia officinalis*. Pale brown powder. $[\alpha]^{28}_{D}$ +2.1 (c, 1 in MeOH).

Yahara, S. *et al.*, *Chem. Pharm. Bull.*, 1991, **39**, 2024-2036 (*Magnolignan I*)

Magnolignan E M9

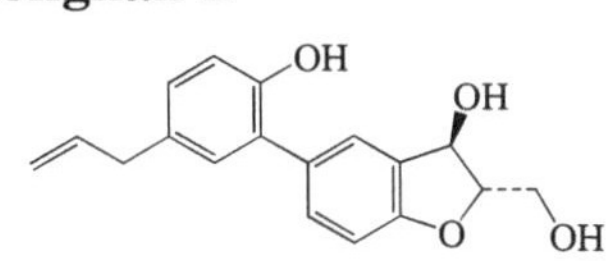

Relative Configuration

6,8-Epoxy-3,3′-lign-8′-ene-4′,7,9-triol. 6,8-Epoxy-4′,7,9-trihydroxy-3,3′-neolign-8′-ene

[138591-08-9]

$C_{18}H_{18}O_4$ 298.338

Constit. of the bark of *Magnolia officinalis*. Powder. $[\alpha]^{29}_{D}$ -2 (c, 2.1 in MeOH).

Yahara, S. *et al.*, *Chem. Pharm. Bull.*, 1991, **39**, 2024-2036 (*Magnolignan E*)

Magnolignan F M10

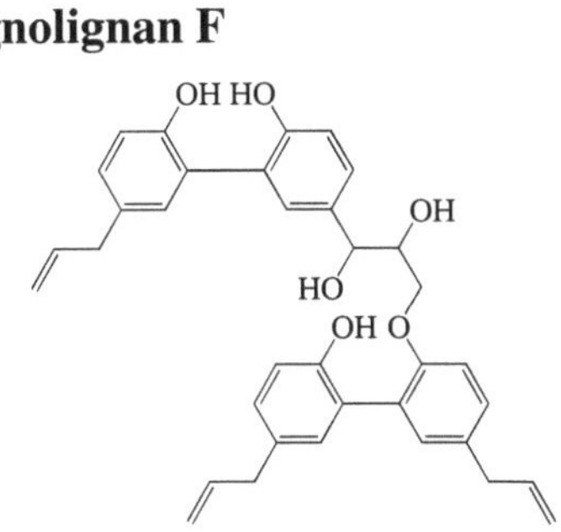

[138591-10-3]

$C_{36}H_{36}O_6$ 564.678

Constit. of *Magnolia officinalis*. Pale brown powder. $[\alpha]^{28}_{D}$ -1.5 (c, 1.04 in MeOH).

Yahara, S. *et al.*, *Chem. Pharm. Bull.*, 1991, **39**, 2024-2036 (*Magnolignan F*)

Magnolignan G — M11

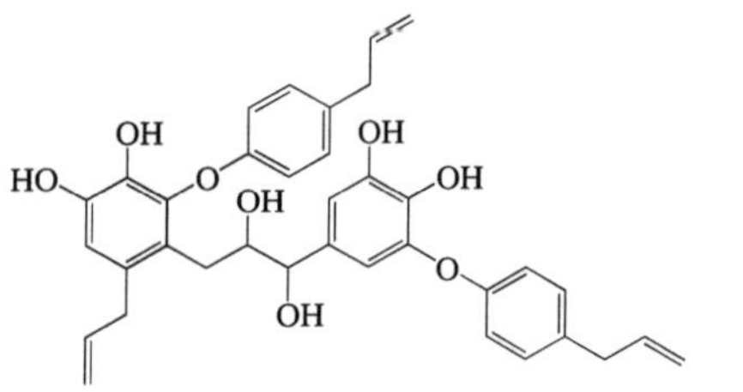

[138591-11-4]

$C_{36}H_{36}O_8$ 596.676

Constit. of *Magnolia officinalis*. Pale brown powder. $[\alpha]_D^{29}$ +0.1 (c, 1.25 in MeOH).

Yahara, S. *et al.*, *Chem. Pharm. Bull.*, 1991, **39**, 2024-2036 (*Magnolignan G*)

Magnolignan H — M12

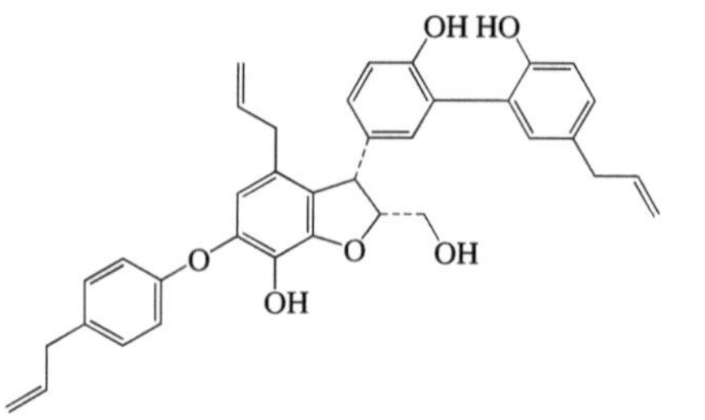

Relative Configuration

[138591-12-5]

$C_{36}H_{34}O_6$ 562.662

Constit. of *Magnolia officinalis*. Pale brown powder. $[\alpha]_D^{28}$ +2 (c, 2.65 in MeOH).

Yahara, S. *et al.*, *Chem. Pharm. Bull.*, 1991, **39**, 2024-2036 (*Magnolignan H*)

Magnoquinone — M13

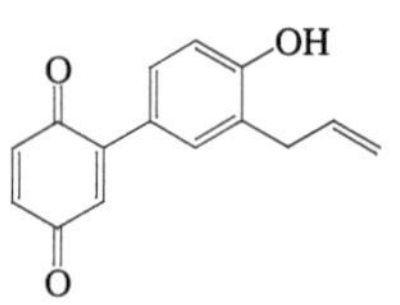

2-[4-Hydroxy-3-(2-propenyl)phenyl]-2,5-cyclohexadiene-1,4-dione, CAS. *2-(3-Allyl-4-hydroxyphenyl)-1,4-benzoquinone*

[349535-92-8]

$C_{15}H_{12}O_3$ 240.258

Constit. of the bark of *Magnolia rostrata*. Mp 73 - 76°. λ_{max} 248 (log ε 4.18); 296 (log ε 3.4); 450 (log ε 3.35) (EtOH).

Deng, S. *et al.*, *Yunnan Zhiwu Yanjiu*, 2001, **23**, 121-125 (*Magnoquinone*)

Majusalignan A — M14

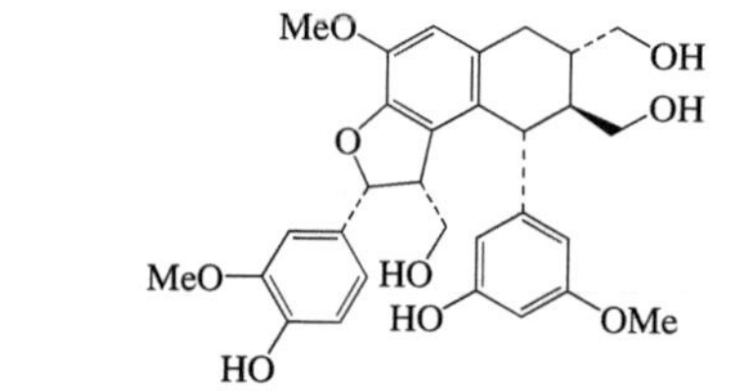

Absolute Configuration

$C_{30}H_{34}O_9$ 538.593

Similar to Spathulatol, S80. Constit. of the leaves and twigs of *Illicium majus*. Inhibitor of β-glucuronidase release in rat leukocytes. Amorph. powder. $[\alpha]_D^{20}$ -39.1 (c, 0.2 in MeOH). λ_{max} 221 (log ε 3.49); 234 (log ε 3.04); 284 (log ε 2.64) (MeOH).

Fang, Z.-F. *et al.*, *Fitoterapia*, 2018, **129**, 42-46 (*Majusalignan A, cd, activity*)

Manassantin A — M15

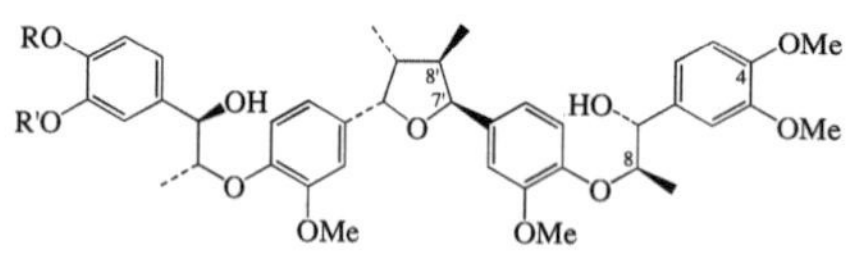

R = R' = Me

Absolute Configuration

[88497-87-4]

$C_{42}H_{52}O_{11}$ 732.867

Constit. of *Saururus cernuus* and *Saururus chinensis*. Has potential neuroleptic activity. Cell adhesion inhibitor. ACAT and HIF1 inhibitor. Immunosuppressant. Also inhibits 5-lipoxygenase-dependent leukotriene C_4 generation in bone-marrow derived mast cells. Powder. Mp 82 - 85°. $[\alpha]_D^{25}$ -102.1 (c, 0.5 in $CHCl_3$). λ_{max} 210 (log ε 4.82); 235 (log ε 4.67); 280 (log ε 4.09) (MeOH)

- ▶ DA4990000
- ▶ Toxic

O^4-De-Me

4-O-Demethylmanassantin A

[943011-47-0]

$C_{41}H_{50}O_{11}$ 718.84

Constit. of the roots of *Saururus chinensis*. Mp 190 - 192°. $[\alpha]_D$ -45 (c, 0.5 in $CHCl_3$). λ_{max} 232; 280 (MeOH)

7′α,8′α-Epoxide

7′,8′-Epoxymanassantin A. ***Saurucinol B***

[1521111-54-5]

$C_{42}H_{50}O_{12}$ 746.85

Constit. of *Saururus chinensis*. Amorph. solid. $[\alpha]_D^{20}$ -80 (c, 0.01 in $CHCl_3$). λ_{max} 214 (log ε 4.28); 233 (log ε 4.37); 279 (log ε 4.00) (MeOH)

Stereoisomer

Saucernetin 8

[350690-48-1]

$C_{42}H_{52}O_{11}$ 732.867

Constit. of the roots of *Saururus chinensis*. Pale brown solid. $[\alpha]_D^{25}$ -15.1 (c, 0.01 in MeOH).

Rao, K.V. *et al.*, *Tetrahedron Lett.*, 1983, **24**, 4947-4950 (*Saururus cernuus constit*).
Min, M. *et al.*, *Zhongcaoyao*, 2001, **32**, 9-11 (*Saucernetin 8*).
Hwang, B.Y. *et al.*, *Phytochemistry*, 2003, **64**, 765-771 (*Manassantin A*).
Park, H.-J. *et al.*, *Planta Med.*, 2003, **69**, 947-950 (*Saucernetin 8*).
Rho, M.-C. *et al.*, *Planta Med.*, 2003, **69**, 1147-1149 (*Saururus chinensis constit, activity*).
Lee, W.S. *et al.*, *Bioorg. Med. Chem. Lett.*, 2004, **14**, 3109-3112 (*Manassantin A, activity*).
Sung, S.H. *et al.*, *Fitoterapia*, 2006, **77**, 487-488 (*4-O-Demethylmanassantin A*).
Hanessian, S. *et al.*, *Org. Lett.*, 2006, **8**, 5477-5480 (*synth, abs config*).
Park, S-Y. *et al.*, *Planta Med.*, 2007, **73**, 674-678 (*Manassantin A, activity*).
Oh, K-S. *et al.*, *Planta Med.*, 2008, **74**, 233-238 (*Saururus chinensis constit*).
Kasper, A.C. *et al.*, *Bioorg. Med. Chem. Lett.*, 2009, **19**, 3783-3786 (*activity*).
Kim, H. *et al.*, *Org. Lett.*, 2009, **11**, 89-92 , 1671 (*synth*).
Kim, S.J. *et al.*, *Biol. Pharm. Bull.*, 2011, **34**, 1769-1772 (*Manassantin A, activity*).
Cui, H. *et al.*, *J. Nat. Prod.*, 2013, **77**, 100-110 (*Saurucinol B, 4-O-Demethylmanassantin A*)

Manassantin B — M16

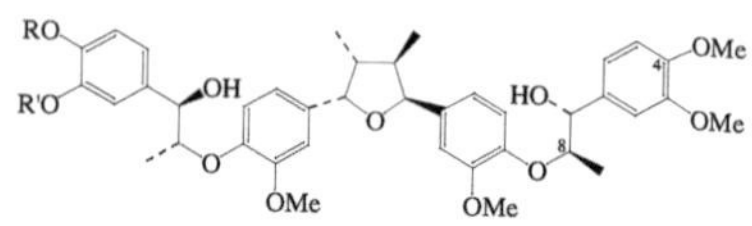

Absolute Configuration

RR′ = –CH₂–

Saucernetin 7

[88497-88-5]

[350690-49-2] (Saucernetin 7)

$C_{41}H_{48}O_{11}$ 716.824

Constit. of *Saururus cernuus* and *Saururus chinensis*. Cell adhesion inhibitor. ACAT inhibitor. Shows antiinflammatory activity. Inhibits COX-2 dependent prostaglandin D_2. Active against Epstein-Barr virus lytic replication. Powder. Mp 83 - 86°. $[\alpha]_D^{25}$ -108.4 (c, 0.22 in $CHCl_3$) (-99.8). λ_{max} 209 (log ε 4.9); 236 (log ε 4.56); 282 (log ε 3.99) (MeOH)

- ▶ Toxic

3-O-De-Me

3-O-Demethylmanassantin B

[1512869-84-9]

$C_{40}H_{46}O_{11}$ 702.797

Constit. of *Saururus chinensis*. Active against Epstein-Barr virus lytic replication. Amorph. solid. $[\alpha]_D^{20}$ -112 (c, 0.02 in $CHCl_3$). λ_{max} 208 (log ε 4.44); 231 (log ε 4.03); 286 (log ε 3.61) (MeOH)

4-O-De-Me

4-O-Demethylmanassantin B

[695151-38-3]

$C_{40}H_{46}O_{11}$ 702.797

Constit. of *Saururus cernuus*. Powder. $[\alpha]_D^{25}$ -45 (c, 0.08 in $CHCl_3$). λ_{max} 208 (log ε 4.78); 234 (log ε 4.35); 282 (log ε 3.36) (MeOH)

8-Epimer

Manassantin B₁

$C_{41}H_{48}O_{11}$ 716.824

Constit. of *Saururus cernuus*. Amorph. solid. [α] D -51.7 (c, 0.33 in MeOH).

Rao, K.V. *et al.*, *Tetrahedron Lett.*, 1983, **24**, 4947-4950 (*Manassantin B*).
Min, M. *et al.*, *Zhongcaoyao*, 2001, **32**, 9-11 (*Saucernetin 7*).
Hwang, B.Y. *et al.*, *Phytochemistry*, 2003, **64**, 765-771 (*Saururus chinensis constit*).
Park, H.-J. *et al.*, *Planta Med.*, 2003, **69**, 947-950 (*Saucernetin 7*).
Rho, M.-C. *et al.*, *Planta Med.*, 2003, **69**, 1147-1149 (*activity*).
Lee, W.S. *et al.*, *Bioorg. Med. Chem. Lett.*, 2004, **14**, 3109-3112 (*Manassantin B, activity*).
Hodges, T.W. *et al.*, *J. Nat. Prod.*, 2004, **67**, 767-771 (*4-O-Demethylmanassantin B, abs config*).
Hossain, C.F. *et al.*, *Biochem. Biophys. Res. Commun.*, 2005, **333**, 1026-1033 (*Manassantin B₁*).
Hanessian, S. *et al.*, *Org. Lett.*, 2006, **8**, 5477-5480 (*synth, abs config*).
Lee, J-H. *et al.*, *Biochem. Pharmacol.*, 2008, **66**, 1925-1933 (*activity*).
Kim, H. *et al.*, *Org. Lett.*, 2009, **11**, 89-92 (*synth*).
Kim, H. *et al.*, *Org. Lett.*, 2009, **11**, 1671 (*erratum*).
Lu, Y. *et al.*, *Biol. Pharm. Bull.*, 2013, **36**, 1370-1374 (*Manassantin B, activity*).
Cui, H. *et al.*, *J. Nat. Prod.*, 2014, **77**, 100-110 (*Manassantin B, activity, 3-O-Demethylmanassantin B*)

Mansoxetane M17

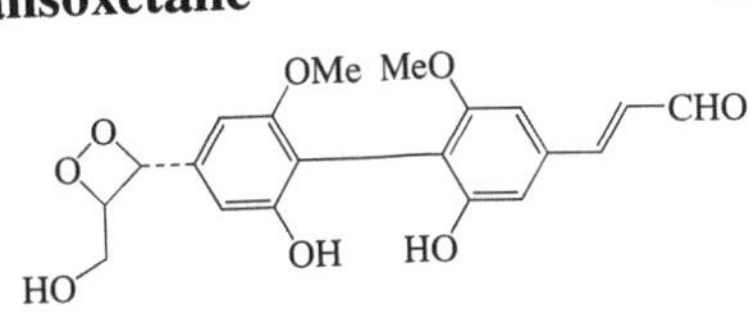

7′,8′-Epidioxy-3,3′,9′-trihydroxy-5,5′-dimethoxy-4,4′-lign-7-en-9-al.
7′,8′-Epidioxy-3,3′,9′-trihydroxy-5,5′-dimethoxy-9-oxo-4,4′-neolign-7-ene

[616880-65-0]

$C_{20}H_{20}O_8$ 388.372

Constit. of the heartwood of *Mansonia gagei*. Pale brown solid. $[\alpha]_D^{15}$ -12.8 (c, 0.27 in $CHCl_3$).

Tiew, P. *et al.*, *Tetrahedron Lett.*, 2003, **44**, 6759-6761 (*Mansoxetane*)

Mariamide A M18

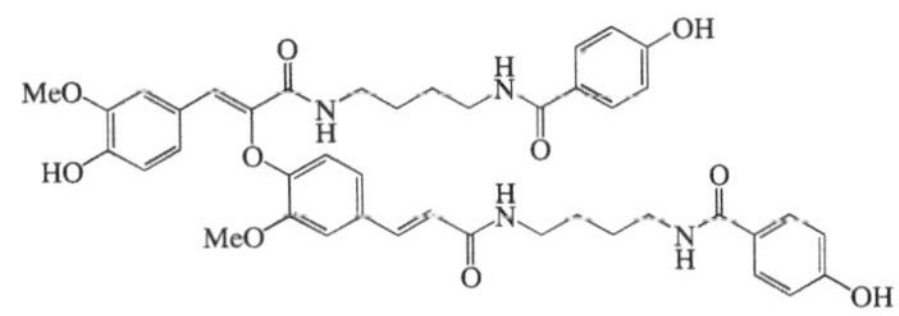

[2263940-18-5]

$C_{42}H_{46}N_4O_{10}$ 766.848

Alkaloid from the seeds of *Silybum marianum*. α-Glucosidase inhibitor. Antioxidant. Amorph. powder. λ_{max} 256 (log ε 4.22) (MeOH).

Qin, N.-B. *et al.*, *Fitoterapia*, 2017, **119**, 83-89 (*Mariamide A, activity*)

Marphenol C M19

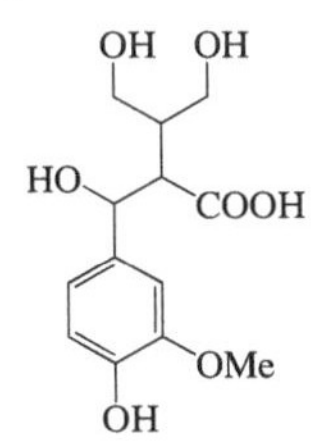

[1420791-97-4]

$C_{13}H_{18}O_7$ 286.28

Norlignan. Constit. of the fruits of *Schisandra wilsoniana*. Shows anti-HIV-1 activity. Amorph. solid. $[\alpha]_D^{26}$ +18.6 (c, 0.2 in MeOH). λ_{max} 205 (log ε 4.87); 278 (log ε 4.12); 326 (log ε 2.26) (MeOH).

Yang, G.-Y. *et al.*, *J. Nat. Prod.*, 2013, **76**, 250-255 (*Marphenol C*)

Marphenol D M20

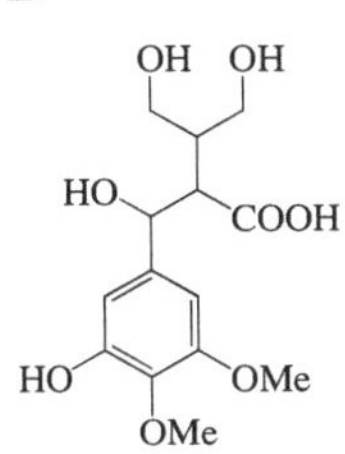

[1420791-98-5]

$C_{14}H_{20}O_8$ 316.306

Norlignan. Constit. of the fruits of *Schisandra wilsoniana*. Amorph. solid. $[\alpha]_D^{25}$ +22.5 (c, 0.21 in MeOH). λ_{max} 210 (log ε 4.82); 276 (log ε 4.15); 325 (log ε 2.30) (MeOH).

Yang, G.-Y. *et al.*, *J. Nat. Prod.*, 2013, **76**, 250-255 (*Marphenol D*)

Maximowicziol A M21

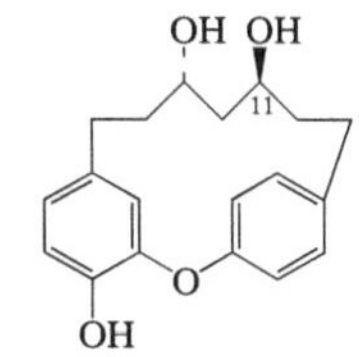

Relative Configuration

[193153-41-2]

$C_{19}H_{22}O_4$ 314.381

Constit. of *Betula maximowicziana*. Needles (MeOH). Mp 220 - 221°. $[\alpha]_D^{23}$ -86.3 (c, 0.7 in MeOH)

9-Ketone

9-Oxoacerogenin A

[1400570-80-0]

$C_{19}H_{20}O_4$ 312.365

Constit. of the bark of *Acer nikoense*. Fine needles (MeOH). Mp 147 - 150°. $[\alpha]_D^{25}$ +0.5 (c, 0.36 in MeOH). C-11 config. not determined. λ_{max} 276 (log ε 3.52) (MeOH)

11-Ketone

Acerogenin D

[150144-87-9]

$C_{19}H_{20}O_4$ 312.365

Constit. of *Acer nikoense*. Needles (MeOH). Mp 171 - 172°. $[\alpha]_D^{17}$ -5.2 (c, 1.6 in $CHCl_3$/MeOH). C-9 config. not determined

11-Ketone, 4-O-β-D-glucopyranoside

Aceroside V

[74390-31-1]

$C_{25}H_{30}O_9$ 474.506

Constit. of *Acer nikoense*. Needles + 1½H_2O (MeOH). Mp 130 - 132°. $[\alpha]_D^{20}$ -57.3 (c, 1 in MeOH).

Nagumo, S. *et al.*, *Chem. Pharm. Bull.*, 1993, **41**, 1255-1257 (*Aceroside V*).
Hanawa, F. *et al.*, *Phytochemistry*, 1997, **45**, 589-595 (*Maximowicziol A, cryst struct*).
Akihisa, T. *et al.*, *Chem. Pharm. Bull.*, 2006, **54**, 735-739 (*Acerogenin D*).
Akihisa, T. *et al.*, *Chem. Biodiversity*, 2012, **9**, 1475-1489 (*9-Oxoacerogenin A*)

Meehanioside E M22

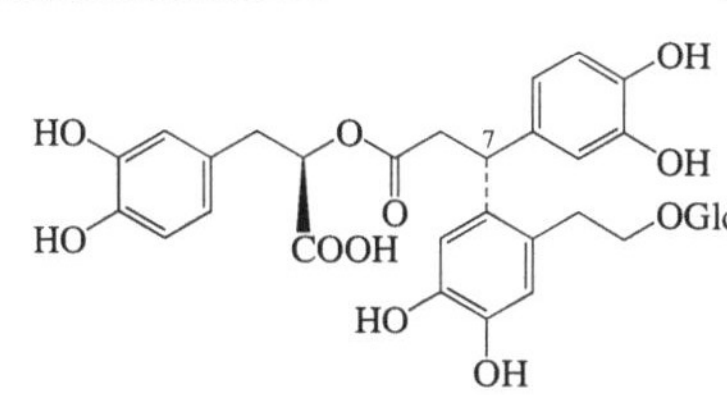

Absolute Configuration

[1387575-23-6]

$C_{32}H_{36}O_{16}$ 676.624

Constit. of *Meehania urticifolia*. Amorph. powder. $[\alpha]_D^{21}$ -26.7 (c, 0.12 in MeOH). λ_{max} 286 (log ε 5); 329 (log ε 4.7) (MeOH)

4‴-O-(3,4-Dihydroxy-E-cinnamoyl)
Meehanioside A
[1387575-18-9]
$C_{41}H_{42}O_{19}$ 838.768
Constit. of *Meehania urticifolia*. Amorph. powder. $[\alpha]_D^{21}$ -13.2 (c, 0.44 in MeOH). λ_{max} 205 (log ε 5.83); 288 (log ε 5.22); 331 (log ε 5.1) (MeOH)

4‴-(3,4-Dihydroxy-E-cinnamoyl), 3‴-O-β-D-xylopyranosyl
Meehanioside C
[1387575-20-3]
$C_{46}H_{50}O_{23}$ 970.883
Constit. of *Meehania urticifolia*. Amorph. powder. $[\alpha]_D^{20}$ -23.3 (c, 0.97 in MeOH). λ_{max} 205 (log ε 5.77); 289 (log ε 5.14); 330 (log ε 5.03) (MeOH)

7-Epimer, 4‴-O-(3,4-dihydroxy-E-cinnamoyl)
Meehanioside B
[1387575-19-0]
$C_{41}H_{42}O_{19}$ 838.768
Constit. of *Meehania urticifolia*. Amorph. powder. $[\alpha]_D^{22}$ -4.9 (c, 0.53 in MeOH). λ_{max} 208 (log ε 5.76); 289 (log ε 5.2); 331 (log ε 5.09) (MeOH)

7-Epimer, 4‴-O-(3,4-dihydroxy-E-cinnamoyl), 3‴-O-β-D-xylopyranosyl
Meehanioside D
[1387575-21-4]
$C_{46}H_{50}O_{23}$ 970.883
Constit. of *Meehania urticifolia*. Amorph. powder. $[\alpha]_D^{21}$ -17.7 (c, 0.79 in MeOH). λ_{max} 203 (log ε 5.91); 289 (log ε 5.21); 330 (log ε 5.11) (MeOH).

Murata, T. *et al.*, *Chem. Pharm. Bull.*, 2011, **59**, 88-95 (*Meehaniosides A-E*)

Megacerotonic acid M23

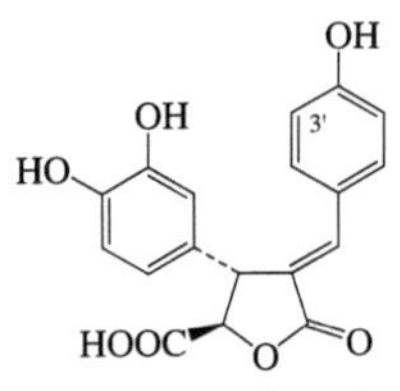

Relative Configuration

3,4,4'-Trihydroxy-7,8'-lign-7'-en-9',8-olid-9-oic acid
[130396-76-8]
$C_{18}H_{14}O_7$ 342.303
Provisional lignan. Biosynth. pathway unclear. An isomeric struct. was briefly proposed before 1990. Isol. from the hornwort *Megaceros flagellaris*. $[\alpha]_D$ +233 (c, 1.65 in 5% AcOH). [acid] λ_{max} 228 (ε 13500); 289 (sh); 319 (ε 18000) (5% AcOH) (Derep)

3'-Methoxy
3,4,4'-Trihydroxy-3'-methoxy-7,8'-lign-7'-en-9',8-olid-9-oic acid.
Shimobashiric acid A
[1393828-21-1]
$C_{19}H_{16}O_8$ 372.329
Constit. of the aerial parts of *Keiskea japonica*. Amorph. solid. $[\alpha]_D^{22}$+205 (c, 0.16 in MeOH). λ_{max} 201 (log ε 5.22); 293 (log ε 4.21); 333 (log ε 4.37) (MeOH).

Takeda, R. *et al.*, *Tetrahedron Lett.*, 1990, **31**, 4159-4162 (*Megacerotonic acid*).
Brown, E. *et al.*, *Tetrahedron*, 1995, **51**, 13061-13072 (*synth, struct*).
Murata, T. *et al.*, *Chem. Pharm. Bull.*, 2012, **60**, 121-128 (*Shimobashiric acid A*).
Krabbe, S.W. *et al.*, *Org. Lett.*, 2015, **17**, 1188-1191 (*synth*)

Meliadanoside B M24

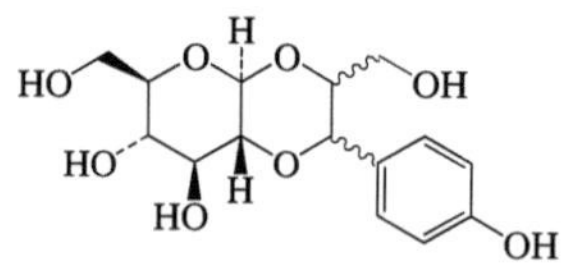

Absolute Configuration

1,2-O-[1-(Hydroxymethyl)-2-(4-hydroxyphenyl)-1,2-ethanediyl]-β-D-glucopyranose, CAS
[286369-47-9]
$C_{15}H_{20}O_8$ 328.317
Constit. of the fruit of *Melia toosendan*. Amorph. powder. λ_{max} 226; 276 (MeOH).

Chang, J. *et al.*, *Zhiwu Xuebao*, 1999, **41**, 1245-1248 (*Meliadanoside B*)

Melianoninol M25

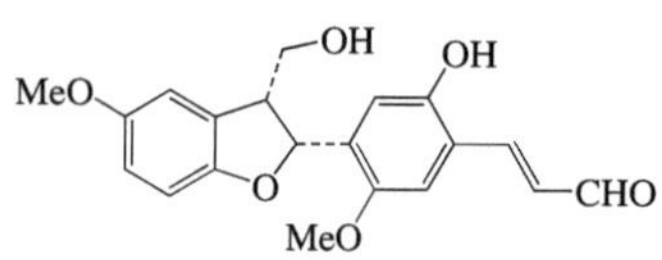

[136880-81-4]
$C_{20}H_{20}O_6$ 356.374
Constit. of the fruit of *Melia azedarach*. Insect antifeedant. Yellow needles. Mp 150 - 152°. $[\alpha]_D$ -4.3 (c, 0.46 in MeOH).

Han, J. *et al.*, *Yaoxue Xuebao*, 1991, **26**, 426-429 (*Melianoninol, activity*).
Xu, R.-S *et al.*, *J. Chin. Pharm. Sci.*, 1992, **1**, 7-10 (*Melianoninol*)

Meliasendanin A M26

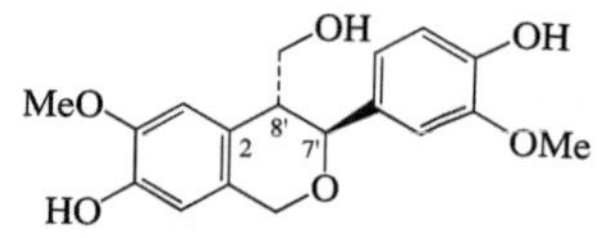

Absolute Configuration

3,4-Dihydro-7-hydroxy-3-(4-hydroxy-3-methoxyphenyl)-6-methoxy-1H-2-benzopyran-4-methanol. 7,7'-Epoxy-3',4-dimethoxy-8,9-dinor-2,8'-lignan-4',5,9'-triol. 7,7'-Epoxy-4',5,9'-trihydroxy-3',4-dimethoxy-8,9-dinor-2,8'-neolignan
[1820034-02-3]
$C_{18}H_{20}O_6$ 332.352
Constit. of the fruits of *Melia toosendan*. Yellow solid. $[\alpha]_D^{20}$+3.8 (c, 0.13 in MeOH). λ_{max} 251 (log ε 3.41) (MeOH).

Wang, L. *et al.*, *Fitoterapia*, 2014, **99**, 92-98 (*Meliasendanin A, cd*)

Melicodin C M27

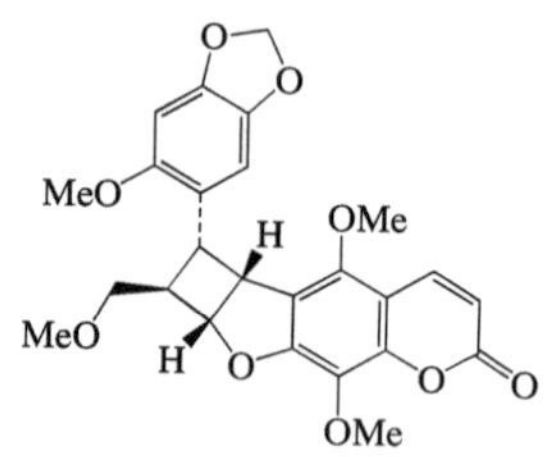

Relative Configuration

[1363039-87-5]
$C_{25}H_{24}O_9$ 468.458
Coumarinolignan. Constit. of the leaves of *Melicope denhamii*. Amorph. yellow solid. Mp 128 - 130°. Racemic. λ_{max} 234 (log ε 4.25); 301 (log ε 3.89); 333 (log ε 3.9) (MeOH).

Nakashima, K.-I. *et al.*, *Tetrahedron*, 2012, **68**, 2421-2428 (*Melicodin C*)

Melitric acid B M28

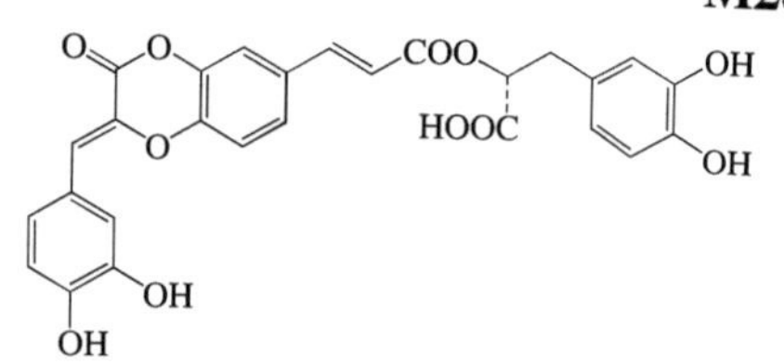

Absolute Configuration

[153765-46-9]
$C_{27}H_{20}O_{11}$ 520.446
Constit. of *Melissa officinalis*. Light brown powder + 1½H_2O. Mp 133 - 135°. $[\alpha]_D^{24}$+119.7 (c, 0.33 in MeOH).

λ_{max} 288 (log ε 4.28); 338 (log ε 4.38) (MeOH).

Agata, I. *et al.*, *Chem. Pharm. Bull.*, 1993, **41**, 1608-1611 (*Melitric acid B*)

p-Menthadunnianol M29

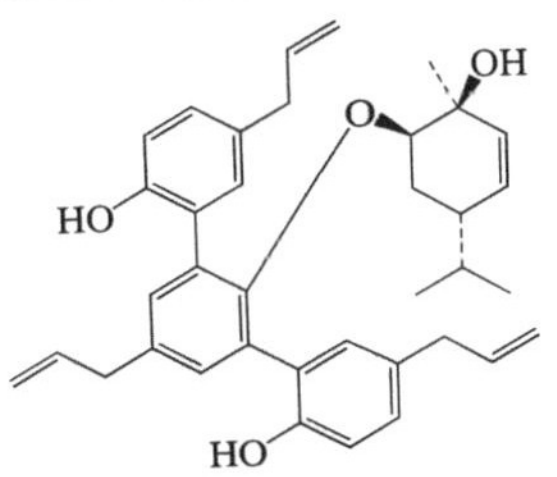

Relative Configuration

Paramenthadunnianol

[1582261-59-3]

$C_{37}H_{42}O_4$ 550.739

Constit. of *Illicium simonsii*. Moderate ACE inhibitor. Gum. $[\alpha]^{22}_{D}$ -13.9 (c, 0.1 in MeOH). λ_{max} 218 (log ε 3.91); 291 (log ε 3.22) (MeOH).

Dong, C. *et al.*, *Planta Med.*, 2013, **79**, 338-347 (*p-Menthadunnianol*)

Merrillianoid M30

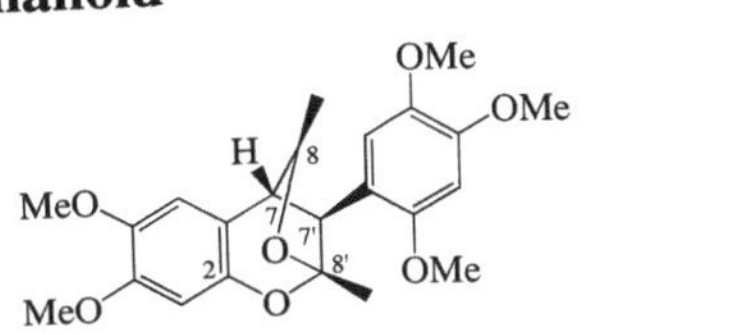

(7R,7'S,8R,8'R)-form

2,8':8,8'-Diepoxy-2',4,4',5,5'-pentamethoxy-7,7'-lignan

$C_{23}H_{28}O_7$ 416.47

Nat. prod. isol. as racemate and sepd. by chiral hplc

(7R,7'S,8R,8'R)-form [1810737-85-9]

$[\alpha]^{25}_{D}$ -48 (c, 0.25 in MeOH)

(7S,7'R,8S,8'S)-form [1810737-84-8]

Exhibits neurotrophic activity. $[\alpha]^{25}_{D}$ +48 (c, 0.25 in MeOH)

(7RS,7'SR,8RS,8'RS)-form

Constit. of *Illicium merrillianum*. Needles (MeOH). Mp 58 - 59°.

Tian, X. *et al.*, *Sci. Rep.*, 2015, **5**, 16982 (*Merrillianoid, cryst struct, cd*)

Merrilliaquinone M31

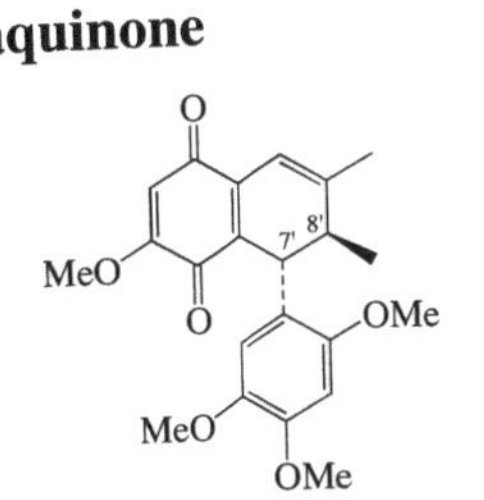

(7'R,8'S)-form

2',4,4',5'-Tetramethoxy-2,7'-cyclolign-7-ene-3,6-dione

$C_{22}H_{24}O_6$ 384.428

Lignan numbering shown

(7'R,8'S)-form [1838627-46-5]

Constit. of the branches and leaves of *Illicium merrillianum*. Exhibits selective cytotoxicity against human hepatoma cell lines SMMC7721 and HuH7. $[\alpha]^{25}_{D}$ +12 (c, 0.3 in MeOH). Resolved by chiral hplc

(7'S,8'R)-form [1838627-45-4]

Constit. of the branches and leaves of *Illicium merrillianum*. Exhibits very weak cytotoxicity on human hepatoma cell lines SMMC7721 and HuH7. $[\alpha]^{25}_{D}$ -11.7 (c, 0.3 in MeOH). Resolved by chiral hplc

(7'RS,8'SR)-form [2003240-22-8]

Constit. of the branches and leaves of *Illicium merrillianum*. Red powder.

Tian, X. *et al.*, *RSC Adv.*, 2015, **5**, 75857-75862 (*Merrilliaquinone, activity*)

Metasequirin I M32

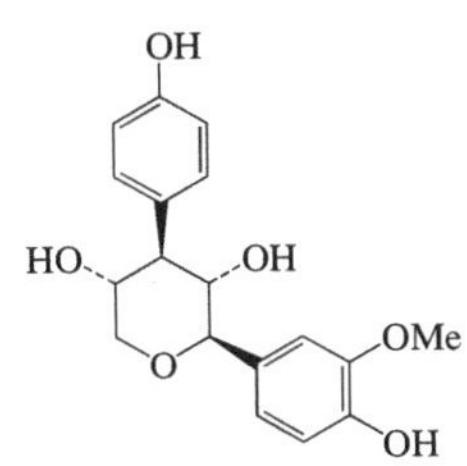

Relative Configuration

7',9-Epoxy-3'-methoxy-9'-nor-7,8'-lignan-4,4',8,8'-tetrol. 7',9-Epoxy-4,4',8,8'-tetrahydroxy-3'-methoxy-9'-nor-7,8'-neolignan

[1374605-47-6]

$C_{18}H_{20}O_6$ 332.352

Provisional norlignan. Biosynth. pathway unclear. Related to Taxodascendin, T7. Constit. of *Metasequoia glyptostroboides*. Brown gum. $[\alpha]^{25}_{D}$ +66.2 (c, 0.06 in MeOH). λ_{max} 200 (log ε 0.63); 220 (log ε 0.48); 250 (log ε 1.04); 270 (log ε 0.96); 280 (log ε 0.89); 305 (sh) (MeOH).

Zeng, Q. *et al.*, *Helv. Chim. Acta*, 2012, **95**, 606-612 (*Metasequirin I*)

Metasequirin A M33

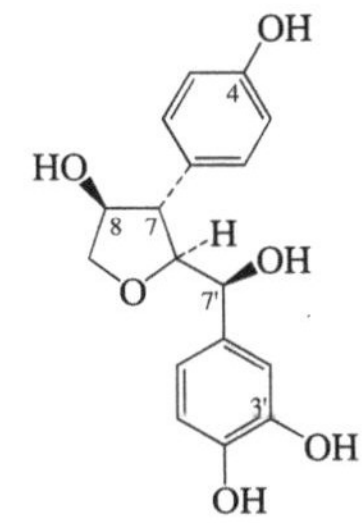

Absolute Configuration

4-[Hydroxy[tetrahydro-4-hydroxy-3-(4-hydroxyphenyl)-2-furanyl]methyl]-1,2-benzenediol, CAS. *8',9-Epoxy-9'-nor-7,8'-lignan-3',4,4',7',8-pentol. 8',9-Epoxy-3',4,4',7',8-pentahydroxy-9'-nor-7,8'-neolignan*

[65559-97-9]

$C_{17}H_{18}O_6$ 318.325

Provisional norlignan. Biosynth. pathway unclear. Constit. of the heartwood of *Metasequoia glyptostroboides*

3-Hydroxy

8',9-Epoxy-9'-nor-7,8'-lignan-3,3',4,4',7',8-hexol. ***Hydroxymetasequirin A***

[65560-06-7]

$C_{17}H_{18}O_7$ 334.324

Constit. of *Metasequoia glyptostroboides*

7,7'-Diepimer, 3'-Me ether

Metasequirin F

[1264695-00-2]

$C_{18}H_{20}O_6$ 332.352

Constit. of *Metasequoia glyptostroboides*. Gum. $[\alpha]^{25}_{D}$ +8.7 (c, 0.32 in MeOH). λ_{max} 226 (log ε 3.71); 279 (log ε 3.18) (MeOH)

7,7'-Diepimer, 3-hydroxy, 3,4-methylene, 3'-Me ether

Nicotnorlignan A

[1384932-95-9]

$C_{19}H_{20}O_7$ 360.362

Constit. of the leaves of *Nicotiana tabacum*. Exhibits modest anti HIV-1 and anti-tobacco mosaic virus activity. Yellow gum. $[\alpha]^{24.8}_{D}$ -25.6 (c, 0.25 in MeOH). λ_{max} 205 (log ε 4.46); 230 (log ε 3.85); 280 (log ε 3.62) (MeOH)

7,7'-Diepimer, 3-methoxy, 3'-Me ether

Metasequirin E

[1264694-98-5]

$C_{19}H_{22}O_7$ 362.378

Constit. of *Metasequoia glyptostroboides*. Gum. $[\alpha]^{25}_{D}$ +22.2 (c,

0.25 in MeOH). λ_{max} 204 (log ε 4.41); 229 (log ε 3.84); 281 (log ε 3.49) (MeOH)

7,8′-Diepimer, 7′-Me ether

Metasequirin G

[1374605-45-4]

$C_{18}H_{20}O_6$ 332.352

Constit. of *Metasequoia glyptostroboides*. Brown gum. $[\alpha]^{25}_{D}$ +32 (c, 0.05 in MeOH). C-7′ config. not determined. λ_{max} 215 (log ε 0.94); 235 (log ε 0.48); 275 (log ε 1.03); 310 (sh) (MeOH)

7,8′-Diepimer, 3′,7′-di-Me ether

Metasequirin H

[1374605-46-5]

$C_{19}H_{22}O_6$ 346.379

Constit. of *Metasequoia glyptostroboides*. Brown gum. $[\alpha]^{25}_{D}$ +46.7 (c, 0.05 in MeOH). C-7′ config. not determined. λ_{max} 275 (log ε 0.94); 305 (sh) (MeOH).

Enoki, A. *et al.*, *Mokuzai Gakkaishi*, 1977, **23**, 579-586 (*Metasequirin A, Hydroxymetasequirin A*).
Dong, L.-B. *et al.*, *J. Nat. Prod.*, 2011, **74**, 234-239 (*Metasequirins E,F*).
Zeng, Q. *et al.*, *Helv. Chim. Acta*, 2012, **95**, 606-612 (*Metasequirins G,H*).
Gao, X. *et al.*, *Heterocycles*, 2012, **85**, 147-153 (*Nicotnorlignan A, activity*).
Wang, P.-S. *et al.*, *Org. Lett.*, 2014, **16**, 976-979 (*synth*)

3-[5-Methoxy-2-[(2-methyl-4-oxo-4*H*-pyran-3-yl)oxy]phenyl]propanoic acid M34

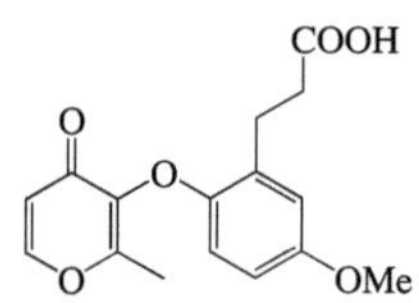

$C_{16}H_{16}O_6$ 304.298

[β-D-Glucopyranosyl-(1→3)-β-D-glucopyranosyl] ester

[490036-67-4]

$C_{28}H_{36}O_{16}$ 628.580

Constit. of *Scleranthus uncinatus*. Amorph. powder. Mp 68 - 70° dec.. $[\alpha]_D$ -15.8 (c, 0.38 in MeOH).

Yayli, N. *et al.*, *Pharm. Biol.*, 2002, **40**, 369-373 (*Scleranthus uncinatus constit*)

3′-Methoxy-2,4′-oxylignan-9,9′-dioic acid M35

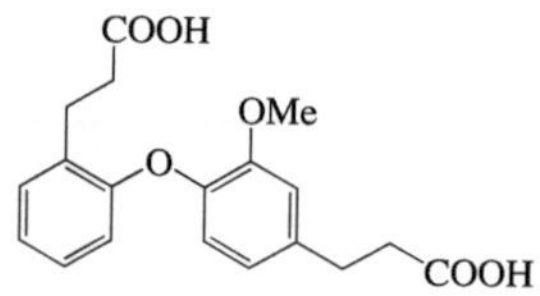

*3′-Methoxy-2,4′-oxybis[3-phenylpropanoic acid]. **Apteniol F***

[871503-87-6]

$C_{19}H_{20}O_6$ 344.363

Constit. of the leaves of *Aptenia cordifolia*. Exhibits phytotoxic effects on germination and root elongation. λ_{max} 280 (log ε 1.4) (MeOH).

DellaGreca, M. *et al.*, *Tetrahedron*, 2005, **61**, 11924-11929 (*Apteniol F, activity*)

3-Methoxy-4,8″-oxy-8,8′-sesquilignan-4′,4″,7″,9″-tetrol M36

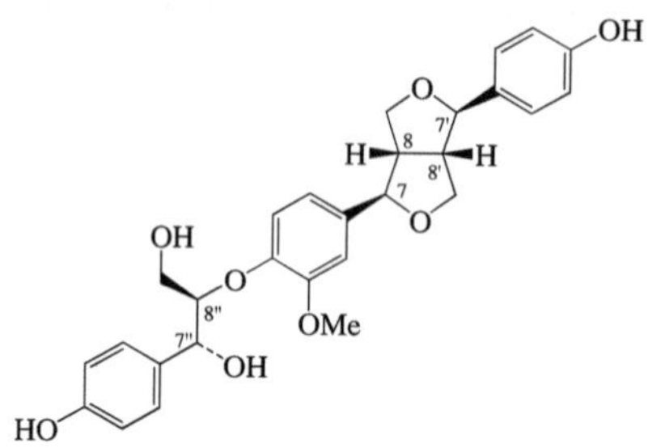

(7R,7′R,7″R,8S,8′S,8″R)-form

4′,4″,7″,9″-Tetrahydroxy-3-methoxy-4,8″-oxy-8,8′-sesquineolignan

$C_{28}H_{30}O_8$ 494.540

Related to Buddlenol E, B83

(7R,7′R,7″R,8S,8′S,8″R)-form

(-)-Idaeusinol D [2376889-73-3]

Light yellow oil. $[\alpha]^{20}_{D}$ -28.3 (c, 0.1 in MeOH)

(7R,7′R,7″S,8S,8′S,8″R)-form

Mubiesin A

Constit. of the seeds of *Momordica cochinchinensis*. Exhibits cytotoxic activity. Amorph. powder. $[\alpha]^{25}_{D}$ -11.8 (c, 0.17 in MeOH). λ_{max} 224 (log ε 3.76); 278 (log ε 2.53) (MeOH)

(7R,7′R,7″S,8S,8′S,8″S)-form

(-)-Idaeusinol C [2368234-79-9]

Light yellow oil. $[\alpha]^{20}_{D}$ -30.6 (c, 0.1 in MeOH)

(7S,7′S,7″R,8R,8′R,8″R)-form

(+)-Idaeusinol C [2376889-71-1]

Exhibits weak neuroprotective activity. Light yellow oil. $[\alpha]^{20}_{D}$ +31.7 (c, 0.1 in MeOH)

(7S,7′S,7″S,8R,8′R,8″S)-form

(+)-Idaeusinol D [2376889-72-2]

Exhibits significant neuroprotective activity. Light yellow oil. $[\alpha]^{20}_{D}$ +29.5 (c, 0.1 in MeOH)

(7RS,7′RS,7″RS,8SR,8′SR,8″RS)-form

Idaeusinol D

Constit. of the fruit of *Rubus idaeus*. Light yellow oil. Resolved by chiral hplc

(7RS,7′RS,7″SR,8SR,8′SR,8″SR)-form

Idaeusinol C

Constit. of the fruit of *Rubus idaeus*. Light yellow oil. Resolved by chiral hplc. λ_{max} 228 (log ε 3.91); 278 (log ε 3.35) (MeOH).

Wang, M. *et al.*, *Fitoterapia*, 2019, **139**, 104360:1-6 (*Mubiesin A, activity*).
Zhou, L. *et al.*, *Phytochemistry*, 2019, **164**, 122-129 (*Idaeusinols C,D, activity*)

2-Methoxy-4-(1-propenyl)phenol, CAS M37

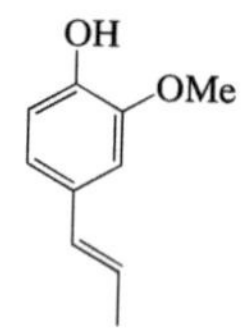

(E)-form

1-(4-Hydroxy-3-methoxyphenyl)-1-propene. 5-(1-Propenyl)guaiacol.

Isoeugenol

$C_{10}H_{12}O_2$ 164.204

Normally isol. as a mixt. of *E*- and *Z*-forms

▶ SL7875000

▶ Skin irritant. LD_{50} (rat, orl) 1560 mg/kg. Fl. p. >74/>100°

(E)-form [5932-68-3]

Constit. of *Petunia hybrida* and numerous fruits eg. *Actinidia chinensis*, *Malus domestica* and *Prunus cerasus* and spices. Cryst.. Mp 33°. Bp 140 - 141° at 12 mm. pK_a 9.88 (25°). n^{20}_{D} +1.5790. λ_{max} 260 (ε 15135) (EtOH)

▶ SL7876000

Ac

Isoeugenol acetate. FEMA 2470

[5912-87-8]

[93-29-8] (unspecified stereochem.)

$C_{12}H_{14}O_3$ 206.241

Flavouring ingredient. Cryst. with a weak rose-carnation, somewhat spicy odour and burning sweet taste. Mp 79°

(Z)-form [5912-86-7]

Oil. Bp 133 - 135° at 13 mm.
λ_{max} 258 (ε 13800); 291 (ε 3890) (EtOH)SL7877000

(ξ)-form

FEMA 2468 [97-54-1]

Occurs in ylang-ylang and other essential oils. Isol. from bark of *Phebalium nudum*. Used in perfumery and flavour industries.

Aldrich Library of FT-IR Spectra, 1st edn., 1985, **1**, 1091B (*ir*).
Aldrich Library of FT-IR Spectra: Vapor Phase, 1989, **3**, 1028B (*ir*).
Aldrich Library of ^{13}*C and* ^{1}*H FT-NMR Spectra*, 1992, **2**, 278C (*nmr*).
Hirao, N. *et al.*, *Nippon Kagaku Kaishi*, 1933, **54**, 505-509 (*synth*).
Morris, W.W. *J. Assoc. Off. Anal. Chem.*, 1973, **56**, 1037-1064 (*ir*).
Opdyke, D.L.J. *Food Cosmet. Toxicol.*, 1975, **13**, 815-817 (*rev, tox*).
Von Massow, F. *et al.*, *J. Chem. Soc., Perkin Trans. 2*, 1976, 977-980 (*cmr*).
Andrieux, J. *et al.*, *J. Chem. Soc., Perkin Trans. 1*, 1977, 359-363 (*synth*).
Purchase, R. *et al.*, *Food Chem. Toxicol.*, 1992, **30**, 475-481 (*tox*).
Fenaroli's Handbook of Flavor Ingredients, 3rd edn., (ed. Burdock, G.A.), CRC Press, 1995, **1**, 399-402.
Encyclopedia of Food and Color Additives, (ed. Burdock, G.A.), CRC Press , 1997, 1464-1471.
Koeduka, T. *et al.*, *Proc. Natl. Acad. Sci. U.S.A.*, 2006, **103**, 10128-10133 (*Isoeugenol, biosynth*).
Chowdhry, B.Z. *et al.*, *J. Phys. Chem. A*, 2015, **119**, 11280-11292 (*ir, Raman, conformn*).
Lewis, R.J. *Sax's Dangerous Properties of Industrial Materials, 8th edn., Van Nostrand Reinhold*, 1992, IKQ000

2-Methoxy-4-(2-propenyl)phenol, M38

CAS

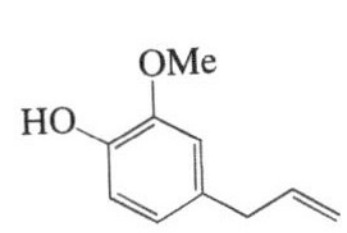

1-Allyl-4-hydroxy-3-methoxybenzene. 4-Allyl-2-methoxyphenol. 5-Allylguaiacol. ***Eugenol,*** *USAN. FEMA 2467*
[97-53-0]
$C_{10}H_{12}O_2$ 164.204

Very widespread occurrence in essential oils. First isol. as a major component of clove oil *Eugenia caryophyllata* (up to 95% of content of clove leaf oil) in 1826. Also found in those of *Cinnamomum* spp., *Cistus* spp., *Camellia* spp., *Pelargonium* spp., *Thymus* spp. and *Citrus* peels. Present in *Malus*, *Prunus*, *Musa* and other fruits. Also in *Eugenia caryophyllata*, *Helichrysum italicum*, *Ocimum syave*, *Carum carvi*, *Ocimum gratissimum*, *Artemisia dracunculus*, *Licaria* sp., *Piper betle*, *Celosia argentea* and many others. Used in dental cement preparations, insect attractants and extensively in the perfumery and flavour industries. Used for the photometric detn. of ozone. Flavouring agent. Antifungal agent, antiseptic, dental analgesic. Shows growth promoting activity on the germination of lettuce. Platelet aggregation inhibitor. Antioxidant. Oil with strong clove-like odour, and spicy pungent taste. d_4^{20} 1.07. Fp -9°. Bp 248°. Bp 128 - 130° at 15 mm. pK_{a1} 10.19 (25°). Log P 2.4 (calc). n_D^{20} +1.5400. Bp also given as 254°

▶ SJ4375000

▶ Skin irritant. LD_{50} (rat, orl) 1930 mg/kg ; LD_{50} (mus, ipr) 365 mg/kg.

Aldrich Library of FT-IR Spectra: Vapor Phase, 1989, **3**, 1028A (*ir*).
Aldrich Library of ^{13}*C and* ^{1}*H FT-NMR Spectra*, 1992, **2**, 278A (*nmr*).
Erdmann, F. *et al.*, *J. Prakt. Chem.*, 1897, **56**, 143-156 (*synth*).
Sachdev, S.L. *et al.*, *Anal. Chim. Acta*, 1972, **58**, 141-147 (*detn,* O_3).
Popa, G. *et al.*, *Rev. Chim. (Bucharest, Rom.)*, 1973, **24**, 635-638 (*spectra*).
Yamaguchi, H. *et al.*, *Chem. Pharm. Bull.*, 1975, **23**, 1169-1170 (*pmr*).
Opdyke, D.L.J. *Food Cosmet. Toxicol.*, 1975, **13**, 545-554 (*rev, tox*).
Klischies, M. *et al.*, *J. Chem. Soc., Chem. Commun.*, 1975, 879-880 (*biosynth*).
Manito, P. *et al.*, *J. Chem. Soc., Perkin Trans. 1*, 1975, 1548-1551 (*biosynth*).
Von Massow, F. *et al.*, *J. Chem. Soc., Perkin Trans. 2*, 1976, 977-980 (*cmr*).
Karrer, W. *Konstitution und Vorkommen der Organischen Pflanzenstoffe, 2nd edn., Birkhäuser Verlag*, 1976-1985, nos. 185-186 (*Eugenol, Gein, occur*).
Wenkert, E. *et al.*, *Synthesis*, 1983, 701-703 (*synth*).
IARC Monogr., 1985, **36**, 75 ; 1987, *Suppl. 7*, 63 (*rev, tox*).
Lewis, R.J. *Food Additives Handbook*, Van Nostrand Reinhold International, New York , 1989, EQR500 , EQS000.
Howes, A.J. *et al.*, *Food Chem. Toxicol.*, 1990, **28**, 537-542 (*cytotoxicity*).
Encyclopedia of Food and Color Additives, (ed. Burdock, G.A.), CRC Press , 1997, 1052-1058 (*occur, props, esters*).
Yuwono, M. *et al.*, *Anal. Profiles Drug Subst.*, 2002, **29**, 149-177 (*rev*).
Koeduka, T. *et al.*, *Proc. Natl. Acad. Sci. U.S.A.*, 2006, **103**, 10129-10133 (*Eugenol, biosynth*).
Martindale, The Complete Drug Reference, 36th edn., Pharmaceutical Press, 2009, 2301.
Fenaroli's Handbook of Flavor Ingredients, 6th edn., (ed. Burdock, G.A.), CRC Press, 2010, 680-683 (*use, esters*).
Pramod, K. *et al.*, *Nat. Prod. Commun.*, 2010, **5**, 1999-2006 (*rev, activity*).
Kamatou, G.P. *et al.*, *Molecules*, 2012, **17**, 6953-6981 (*Eugenol, rev, activity*).
Kaufman, T.S. *J. Braz. Chem. Soc.*, 2015, **26**, 1055-1085 (*rev, use*).
Chowdhry, B.Z. *et al.*, *J. Phys. Chem. A*, 2015, **119**, 11280-11292 (*Eugenol, ir, Raman, conformn*).
Zhang, Q. *et al.*, *Synthesis*, 2015, **47**, 55-64 (*Eugenol, synth*).
Lewis, R.J. *Sax's Dangerous Properties of Industrial Materials, 12th edn., J. Wiley*, 2012, EQR500 , EQS000

2-Methoxy-5-(1-propenyl)phenol, M39

CAS

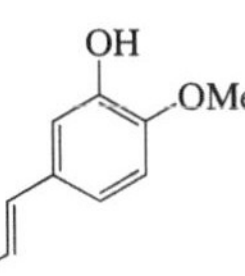

5-(1-Propenyl)guaiacol. ***Isochavibetol***
[501-20-2]
[41243-66-7] ((*Z*)-form)
$C_{10}H_{12}O_2$ 164.204

(E)-form [19784-98-6]

Constit. of *Osmorhiza aristata* and *Pimpinella anisum*. Cryst. (EtOH) with vanilla-carnation odour. Mp 96°. Bp 147° at 19 mm. pK_{a1} 9.9 (25°).

Kovacs, J. *et al.*, *Acta Chim. Acad. Sci. Hung.*, 1958, **16**, 141-147 (*synth*).
Kurihara, T. *et al.*, *Yakugaku Zasshi*, 1979, **99**, 1116-1121 (*Isochavibetol*).
Fujimatu, E. *et al.*, *Phytochemistry*, 2003, **63**, 609-616 (*glucoside*).
Hu, H.-B. *et al.*, *Bull. Korean Chem. Soc.*, 2009, **30**, 703-706 (*glucoside*)

2-Methoxy-5-(2-propenyl)phenol, M40

CAS

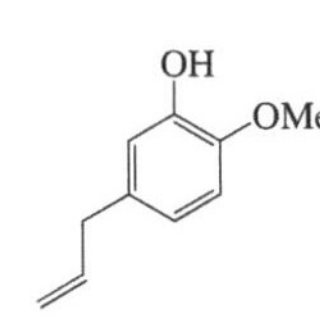

4-Allyl-2-hydroxy-1-methoxybenzene. 4-Allylguaiacol. 3-Hydroxyestragole. ***Chavibetol****. Betelphenol*
[501-19-9]
$C_{10}H_{12}O_2$ 164.204

Constit. of the leaves of *Piper betle*, *Pimenta pseudocaryophyllus* and *Ocimum kilimandscharicum*. Nematocide. Liq.. Mp 8.5°. Bp 254°. Bp 131° at 12 mm. pK_{a1} 10.02 (25°).

Hirao, N. *Bull. Chem. Soc. Jpn.*, 1936, **11**, 179-184 (*synth*).
Schöpf, C. *et al.*, *Justus Liebigs Ann. Chem.*, 1940, **544**, 30-62 (*synth*).
Rimando, A.M. *et al.*, *Arch. Pharmacal Res.*, 1986, **9**, 93-97 (*Piper betle constits*).

Dos Santos, B.C.B. *et al.*, *J. Chromatogr., A*, 2009, **1216**, 4303-4306 (*Pimenta pseudocaryophyllus constit*)

Methyl 2,5-dihydro-α-hydroxy-2-[4-(4-hydroxy-3-methoxyphenyl)-2,3-dimethylbutyl]-5-oxo-2-furanacetate M41

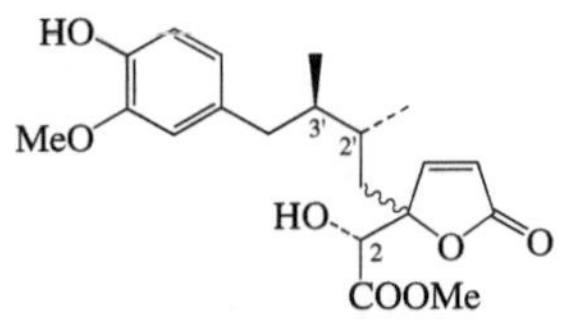

(2*R*,2'*R*,3ξ,3'*R*)-*form*

$C_{20}H_{26}O_7$ 378.421

(2R,2′R,3ξ,3′R)-form [1234858-79-7]

Constit. of the bark of *Machilus thunbergii*. Neuroprotectant. Oil. $[\alpha]_D$ -13.3 (c, 1.7 in $CHCl_3$)

(2R,2′R,3ξ,3′S)-form [1234858-80-0]

Constit. of the bark of *Machilus thunbergii*. Neuroprotectant. Oil. $[\alpha]_D$ -7.2 (c, 0.7 in $CHCl_3$).

Ma, C.J. *et al.*, *Nat. Prod. Res.*, 2010, **24**, 562-568 (*Machilus thunbergii constits*)

2-(3-Methyl-2-butenyl)-4-(2-propenyl)phenol M42

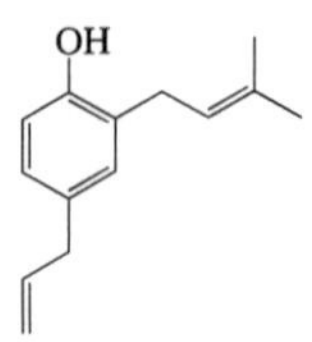

4-Allyl-2-prenylphenol

[1083166-65-7]

$C_{14}H_{18}O$ 202.297

Constit. of the dried pericarps of *Illicium fargesii*. Oil. λ_{max} 280 (ε 1200) (EtOH).

Moriyama, M. *et al.*, *Chem. Pharm. Bull.*, 2008, **56**, 1201-1204 (*Illicium fargesii constit*)

Mirandin A M43

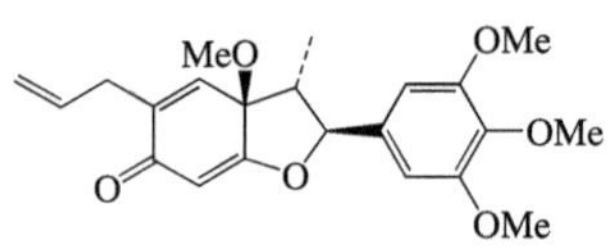

Absolute Configuration

4,7′-Epoxy-3,3′,4′,5′-tetramethoxy-3,8′-lign-8-en-6(3H)-one

[62163-23-9]

$C_{22}H_{26}O_6$ 386.444

Constit. of *Nectandra miranda* and *Pleurothyrium cinereum*. Inhibitor of COX-1, 5-LOX and PAF. Cryst. (C_6H_6/hexane). Mp 89 - 91°

3-Epimer

Mirandin B

[62163-24-0]

Constit. of *Nectandra miranda* and *Magnolia denudata*. Cryst. (C_6H_6/CCl_4). Mp 88 - 92°.

Wenkert, E. *et al.*, *Phytochemistry*, 1976, **15**, 1547-1551 (*Mirandins A,B, cmr*).
Tomita, K. *et al.*, *Acta Crystallogr., Sect. B*, 1977, **33**, 2678-2680 (*Mirandin A, cryst struct*).
Aiba, C.J. *et al.*, *Phytochemistry*, 1977, **16**, 745-748 (*Mirandins A,B, abs config*).
Coy, E.D. *et al.*, *Bioorg. Med. Chem. Lett.*, 2009, **19**, 6922-6925 (*activity*)

Mistletonone M44

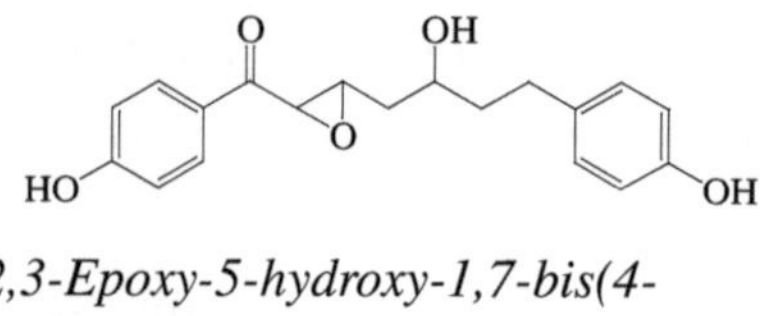

2,3-Epoxy-5-hydroxy-1,7-bis(4-hydroxyphenyl)-1-heptanone

[912537-87-2]

$C_{19}H_{20}O_5$ 328.364

Constit. of *Viscum coloratum*. Antioxidant. Pale yellow powder. λ_{max} 222 (sh); 286 (MeOH).

Yao, H. *et al.*, *Molecules*, 2007, **12**, 312-317 (*Mistletonone, activity*)

Moluccanin M45

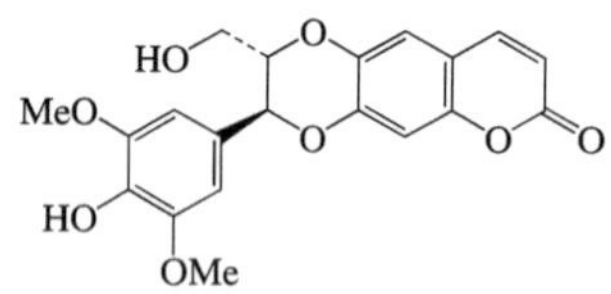

Relative Configuration

[116521-73-4]

$C_{20}H_{18}O_8$ 386.356

Constit. of *Aleurites moluccana* and *Eurycorymbus cavaleriei*. Cryst. (Me_2CO/MeOH). Mp 220°. λ_{max} 260; 295; 340 (MeOH).

Shamsuddin, T. *et al.*, *Phytochemistry*, 1988, **27**, 1908-1909 (*Moluccanin*).
Cheng, L. *et al.*, *J. Chromatogr., A*, 2009, **1216**, 4859-4867 (*Eurycorymbus cavaleriei constit*)

Mongolicumin A M46

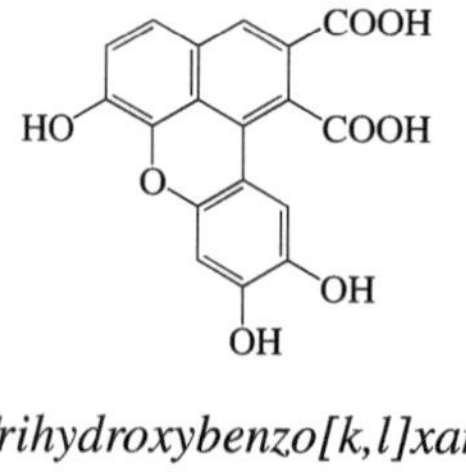

6,9,10-Trihydroxybenzo[k,l]xanthene-1,2-dicarboxylic acid. 2,7′-Cyclo-2′,3-epoxy-4,4′,5′-trihydroxy-8,8′-ligna-7,7′-diene-9,9′-dioic acid

[950169-42-3]

$C_{18}H_{10}O_8$ 354.27

Constit. of the aerial parts of *Taraxacum mongolicum*. Amorph. red powder

Bis[2-hydroxyphenyl-(→1)-β-D-glucopyranos-6-yl] ester

Dodegranoside C

[1198215-17-6]

$C_{42}H_{38}O_{20}$ 862.746

Constit. of the leaves of *Dodecadenia grandiflora*. Antihyperglycaemic agent in traditional medicine. Dark green powder. $[\alpha]_D^{26}$ -14 (c, 0.05 in MeOH). λ_{max} 214; 270 (MeOH)

Di-Me ester

[1206897-88-2]

$C_{20}H_{14}O_8$ 382.324

Constit. of *Balanophora laxiflora* and *Orobanche cernua*. Exhibits cytotoxic and antioxidant props.. Amorph. yellow powder. λ_{max} 270 (ε 24250); 393 (ε 11050) (MeOH)

Bis[1R-carboxy-2-(3,4-dihydroxyphenyl)ethyl] ester

Yunnaneic acid H

[199433-63-1]

$C_{36}H_{26}O_{16}$ 714.588

Constit. of *Salvia yunnanensis*. Amorph. dark yellow powder + 1½H_2O. $[\alpha]_D^{26}$ +55.3 (c, 0.2 in MeOH). Related to Rabdosiin, R1. λ_{max} 278 (ε 24420); 396 (ε 8850) (EtOH)

Anhydride

Rufescidride

[753459-08-4]

$C_{18}H_8O_7$ 336.255

Constit. of *Cordia rufescens* and *Taraxacum mongolicum*. Red powder. Mp 327 - 330°.

Tanaka, T. *et al.*, *Chem. Pharm. Bull.*, 1997, **45**, 1596-1600 (*Yunnaneic acid H*).
Da Silva, S.A.S. *et al.*, *ARKIVOC*, 2004, **vi**, 54-58 (*Rufescidride*).
Shi, S.Y. *et al.*, *Chin. Chem. Lett.*, 2007, **18**, 1367-1370 (*Mongolicumin A*).
Shi, S.Y. *et al.*, *J. Chromatogr., A*, 2008, **1209**, 145-152 (*Mongolicumin A*).
Kumar, M. *et al.*, *Phytochemistry*, 2009, **70**, 1448-1455 (*Dodegranoside C*).
Spatafora, C. *et al.*, *Org. Biomol. Chem.*, 2014, **12**, 2686-2701 (*synth*).
Qu, Z.-Y. *et al.*, *Biochem. Syst. Ecol.*, 2015, **60**, 199-203 (*di-Me ester*).
Quang, D.N. *et al.*, *Nat. Prod. Res.*, 2018, **32**, 767-772 (*di-Me ester, activity*)

Monochaetin‡ M47

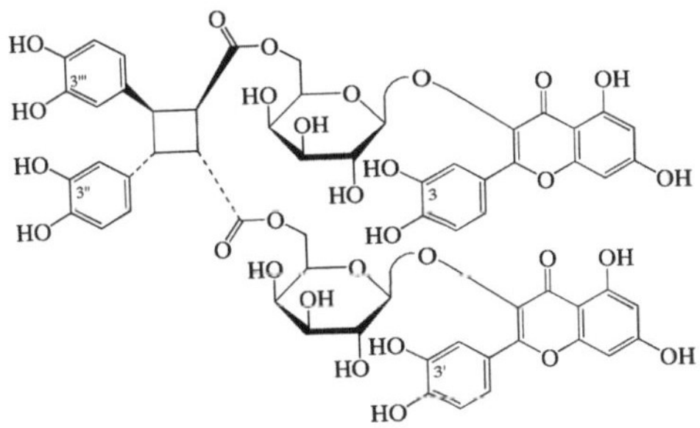

[252199-99-8]

$C_{60}H_{52}O_{30}$ 1253.046

Flavonol glycoside-lignan ester. Related to Stachysetin, S81. Constit. of the leaves of *Monochaetum multiflorum*. Amorph. yellow powder. $[\alpha]_D$ +25 (c, 1 in MeOH). λ_{max} 258 (log ε 4.71); 267 (sh); 286 (sh); 358 (log ε 4.61) (MeOH)

3,3′,3″,3‴-Tetradeoxy

Potentilin A

[1254581-00-4]

$C_{60}H_{52}O_{26}$ 1189.05

Constit. of the rhizomes of *Potentilla anserina*. Amorph. yellow powder. $[\alpha]_D^{25}$ +20.7 (c, 0.05 in MeOH). λ_{max} 224; 270 (sh); 350 (MeOH).

Isaza, M.J.H. *et al., Phytochemistry*, 2001, **58**, 321-327 (*Monochaetin*).

Isaza, M.J.H. *et al., Phytochemistry*, 2002, **59**, 363 (*corrigendum*).

Xu, J.-F. *et al., J. Asian Nat. Prod. Res.*, 2010, **12**, 529-534 (*Potentilin A*)

Mooniine A M48

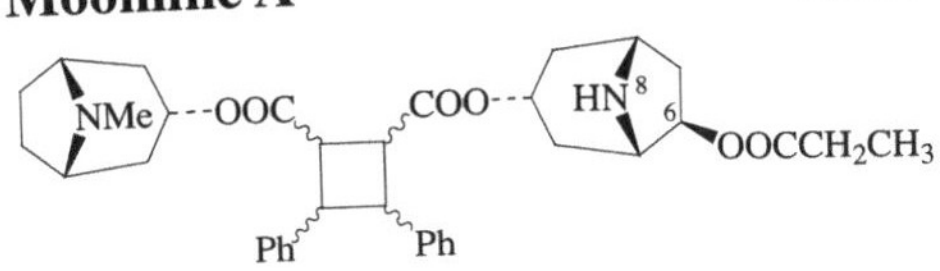

[211444-30-3]

$C_{36}H_{44}N_2O_6$ 600.756

Alkaloid from the leaves of *Erythroxylum moonii*. Solid. λ_{max} 201; 227; 272 ($CHCl_3$)

N^8-Me, O-depropanoyl, Ac

Mooniine B

[211444-31-4]

$C_{36}H_{44}N_2O_6$ 600.756

Alkaloid from the leaves of *Erythroxylum moonii*. Solid. λ_{max} 200; 221; 272 ($CHCl_3$).

Atta-ur-Rahman, *et al., Phytochemistry*, 1998, **48**, 377-383 (*Mooniines A,B*)

Mopanolchin M49

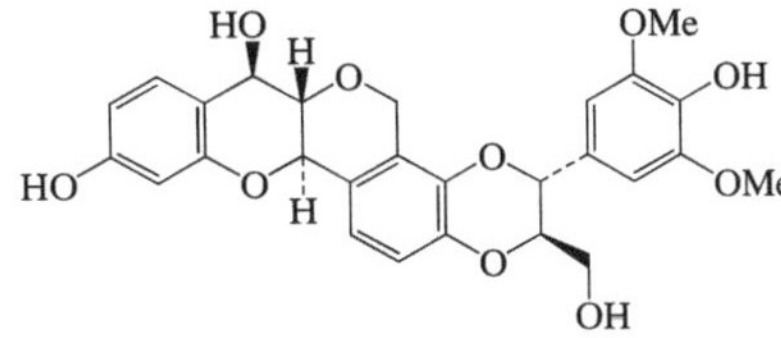

Relative Configuration

[912635-34-8]

$C_{27}H_{26}O_{10}$ 510.495

Flavonolignan. Constit. of the roots of *Lysidice rhodostegia*. Cryst.. Mp 132 - 135°. $[\alpha]_D^{20}$ +23.6 (c, 0.1 in EtOH). λ_{max} 209 (log ε 1.18); 277 (log ε 0.14); 360 (log ε 0.05) (EtOH).

Gao, S. *et al., J. Integ. Plant Biol.*, 2005, **47**, 759-763 (*Mopanolchin*)

Morinol A M50

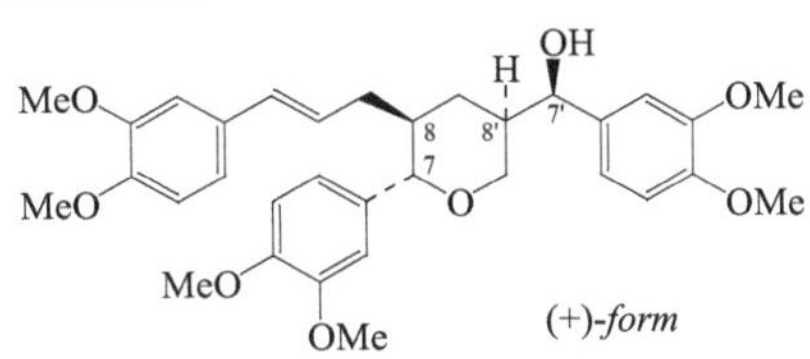

(+)-*form*

[256380-15-1]

$C_{33}H_{40}O_8$ 564.675

Sesquilignan. The nat. prods. were shown to be the 7′,8′-*threo*-isomers illus. with the isolated products almost racemic. The pure enantiomers and also the *erythro*- isomers were synthesised. Constit. of the roots of *Morina chinensis*. Exhibits cytotoxic activity. Oil. $[\alpha]_D^{25}$ +1 (c, 1.02 in $CHCl_3$) (natural). $[\alpha]_D^{20}$ +15 (c, 0.6 in $CHCl_3$) (synthetic). λ_{max} 240 (log ε 3.93); 272 (log ε 3.94) ($CHCl_3$)

7′,8′-Diepimer

Morinol B

[256380-16-2]

$C_{33}H_{40}O_8$ 564.675

Constit. of the roots of *Morina chinensis*. Cytokine prodn. inhibitor. Exhibits cytotoxic activity. Oil. $[\alpha]_D^{25}$ -3.9 (c, 0.51 in $CHCl_3$) (natural). $[\alpha]_D^{20}$ +69 (c, 0.7 in $CHCl_3$) (synthetic). Partial racemate. λ_{max} 241 (log ε 3.89); 272 (log ε 3.95) ($CHCl_3$).

Su, B.-N. *et al., Tetrahedron*, 1999, **55**, 14571-14586 (*Morinols A,B*).

Yamauchi, S. *et al., J. Nat. Prod.*, 2007, **70**, 549-556 (*synth, abs config*).

Akiyama, K. *et al., Biosci., Biotechnol., Biochem.*, 2009, **73**, 129-133 (*synth, activity*).

Yamauchi, S. *et al., Bioorg. Med. Chem. Lett.*, 2013, **23**, 4923-4930 (*Morinols A,B, activity*)

Moslolignan A M51

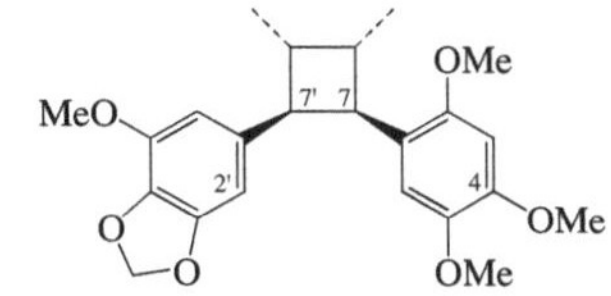

2,3′,4,5-Tetramethoxy-4′,5′-methylenedioxy-7,7′-cyclolignan

[308088-37-1]

$C_{23}H_{28}O_6$ 400.471

Lignan numbering shown. Constit. of *Mosla scabra*. Gum. $[\alpha]_D$ -0.1 (c, 1.6 in MeOH). λ_{max} 207 (log ε 4.93); 233 (sh); 289 (log ε 4.01) (MeOH)

2′-Methoxy

2,2′,4,5,5′-Pentamethoxy-3′,4′-methylenedioxy-7,7′-cyclolignan.

Moslolignan B

[307990-70-1]

$C_{24}H_{30}O_7$ 430.497

Constit. of *Mosla scabra*. Gum. $[\alpha]_D$ -0.4 (c, 1.4 in MeOH). λ_{max} 207 (log ε 4.66); 234 (sh); 292 (log ε 3.74) (MeOH).

Wang, Q. *et al., Phytochemistry*, 2000, **54**, 909-912 (*Moslolignans A,B*)

Mubiesin B M52

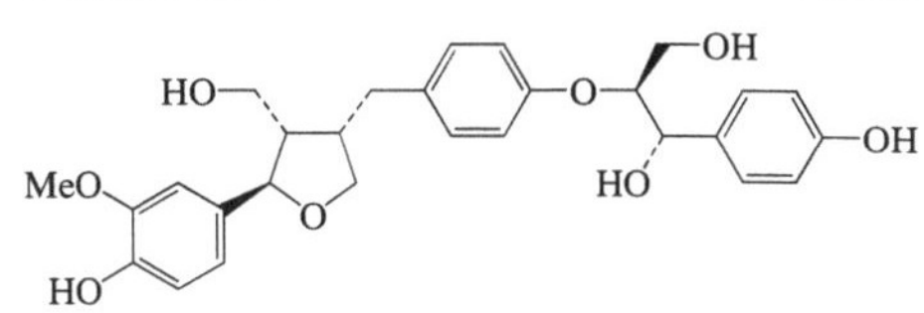

Relative Configuration

2-[4-[[Tetrahydro-5-(4-hydroxy-3-methoxyphenyl)-4-(hydroxymethyl)furan-3-yl]methyl]phenoxy]-1-(4-hydroxyphenyl)-1,3-propanediol

$C_{28}H_{32}O_8$ 496.556

Related to Ehletianol C, E3. Constit. of seeds of *Momordica cochinchinensis*. Exhibits cytotoxic activity. Amorph. powder. $[\alpha]_D^{25}$ -10.6 (c, 0.35 in MeOH). λ_{max} 224 (log ε 3.51); 278 (log ε 2.37) (MeOH).

Wang, M. *et al., Fitoterapia*, 2019, **139**, 104360:1-6 (*Mubiesin B, activity*)

Murrayanine‡ M53

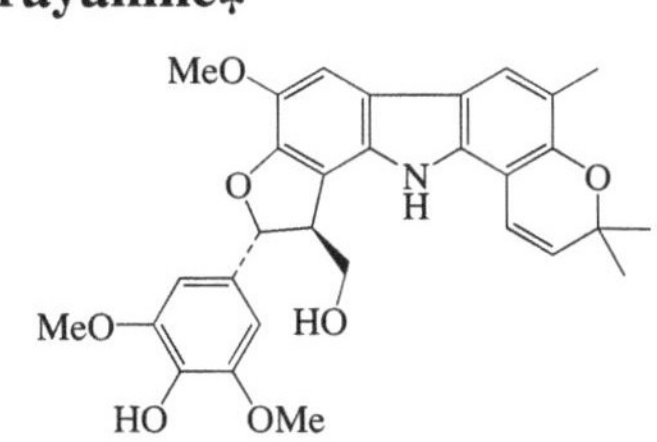

Relative Configuration

[500725-15-5]
$C_{30}H_{31}NO_7$ 517.578
Alkaloid from *Murraya koenigii*. Brown powder. $[\alpha]^{25}_{D}$ +8 (c, 0.74 in MeOH). λ_{max} 206; 225; 238; 300; 342 (MeOH).

Wang, Y.-S. *et al.*, *J. Nat. Prod.*, 2003, **66**, 416-418 (*Murrayanine*)

Mururin A M54

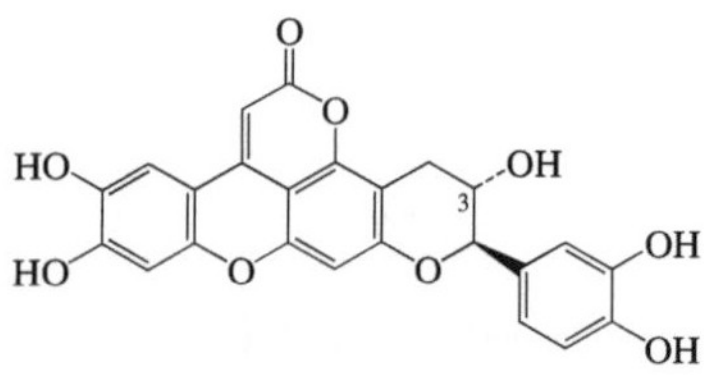

Absolute Configuration

[497845-74-6]
$C_{24}H_{16}O_9$ 448.383
Flavonolignan. Constit. of *Brosimum acutifolium*, *Flacourtia indica* and *Vaccinium ashei*. Protein kinase inhibitor. Shows antiplasmodial props.. Amorph. yellow solid. $[\alpha]^{22}_{D}$ -56.8 (c, 0.12 in MeOH). λ_{max} 220; 235 (sh); 261; 318; 375 (sh); 393 (MeOH)

3-Epimer
Mururin B
[497845-75-7]
$C_{24}H_{16}O_9$ 448.383
Constit. of *Brosimum acutifolium* and *Vaccinium ashei*. Protein kinase inhibitor. Amorph. yellow solid. $[\alpha]^{22}_{D}$ -24 (c, 0.07 in MeOH). λ_{max} 219; 235 (sh); 261; 317; 375 (sh); 392 (MeOH)

2,3-Diepimer
ent-Mururin A
$C_{24}H_{16}O_9$ 448.383
Constit. of *Cecropia obtusifolia*. Exhibits moderate antiplasmodial activity. Amorph. yellow powder.

Takashima, J. *et al.*, *Planta Med.*, 2002, **68**, 621-625 (*Mururins A,B, activity*).
Matsuo, Y. *et al.*, *Food Chem.*, 2010, **121**, 1073-1079 (*Vaccinium ashei constits*).
Sashidhara, K.V. *et al.*, *Eur. J. Med. Chem.*, 2013, **60**, 497-502 (*Flacourtia indica constit, activity*).
Rivera-Mondragon, A. *et al.*, *Phytochem. Lett.*, 2019, **31**, 118-120 (*ent-Mururin A, activity*)

Musaitinerin A M55

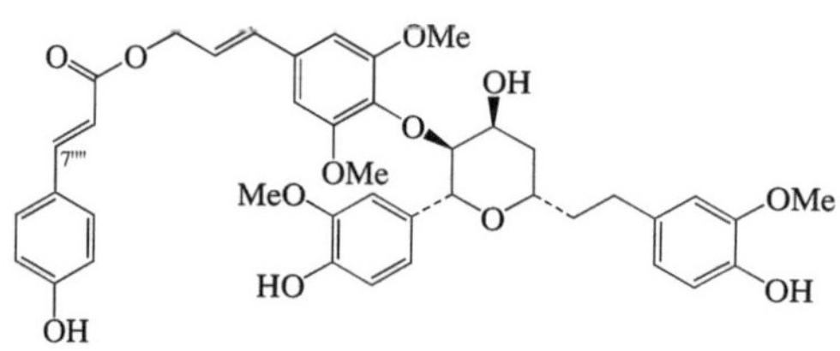

Relative Configuration

[1611534-97-4]
$C_{41}H_{44}O_{12}$ 728.791
Constit. of the fruit of *Musa itinerans*. Amorph. solid. $[\alpha]^{21}_{D}$ -36.3 (c, 0.24 in MeOH). λ_{max} 224 (log ε 3.95); 283 (log ε 3.75); 309 (log ε 3.69) (MeOH)

7‴Z-Isomer
Musaitinerin B
[1611534-98-5]
$C_{41}H_{44}O_{12}$ 728.791
Constit. of the fruit of *Musa itinerans*. Amorph. solid. $[\alpha]^{18}_{D}$ -35.7 (c, 0.29 in MeOH). λ_{max} 223 (log ε 3.96); 280 (log ε 3.72) (MeOH).

Liu, F. *et al.*, *Phytochemistry*, 2014, **103**, 171-177 (*Musaitinerins A,B*)

Mutiflinside M56

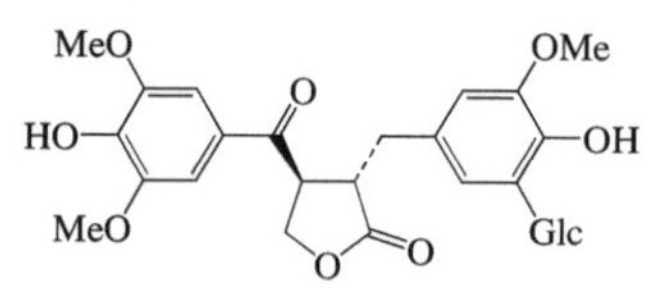

Absolute Configuration

3-C-β-D-Glucopyranosyl-4,4′-dihydroxy-3′,5,5′-trimethoxy-7′-oxolignan-9,9′-olide. Multiflinside
[1008533-27-4]
$C_{27}H_{32}O_{13}$ 564.54
Related to Trachelosiaside, T67. Constit. of *Gardneria multiflora*. λ_{max} 293; 308 (MeOH).

Xie, G.-H. *et al.*, *Zhongguo Tianran Yaowu*, 2007, **5**, 255-258 (*Mutiflinside*)

Myricanol M57

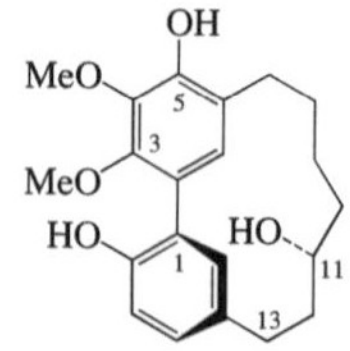

$(11R,S_{axial})$-*form*

16,17-Dimethoxytricyclo[12.3.1.1[2,6]*] nonadeca-1(18),2,4,6(19),14,16-hexaene-3,9,15-triol,* CAS.
3,9,15-Trihydroxy-10,11-dimethoxy[7.0] metacyclophane. Bayberry alcohol
[33606-81-4]
$C_{21}H_{26}O_5$ 358.434
Related to Alnusdiol, A19. Stereochem. information on some of the related compds. included in this entry is not complete

$(11R,S_{axial})$-form [1356848-48-0]
Constit. of *Morella salicifolia*, *Myrica cerifera*, *Myrica nagi*, *Myrica rubra* and *Myrica esculenta*. Promotor of autophagic tau clearance. Needles (Et_2O). Mp 105 - 125°. $[\alpha]^{27.5}_{D}$ -65.5 ($CHCl_3$)

5-O-β-D-Glucopyranoside
Myricanol 5-glucoside
[90052-02-1]
$C_{27}H_{36}O_{10}$ 520.575
Constit. of *Morella salicifolia*, *Myrica arborea* and *Myrica rubra*. Mp 220 - 223°. $[\alpha]_{D}$ -53.7

11-Sulfate
[1403239-92-8]
$C_{21}H_{26}O_8S$ 438.491
Constit. of the bark of *Myrica rubra*. Amorph. off-white powder. $[\alpha]_{D}$ -28.6 (c, 0.5 in MeOH)

4-O-De-Me, 5-Me ether
4-O-Demethyl-5-O-methylmyricanol
[2172890-22-9]
$C_{21}H_{26}O_5$ 358.434
Constit. of *Micromelum integerrimum* and *Myrica rubra*. Amorph. powder. $[\alpha]^{20}_{D}$ -89.2 (c, 2.65 in $CHCl_3$) (Myrica rubra). $[\alpha]^{20}_{D}$ -126.2 (c, 0.06 in $CHCl_3$) (Micromelum integerrimum). λ_{max} 214; 259; 297 (MeOH)

5-Deoxy, 3-O-de-Me, 11-sulfate
[1537119-35-9] (K salt)
$C_{20}H_{24}O_7S$ 408.465
Constit. of *Morella salicifolia* and *Myrica rubra*. Significantly reduces LPS-induced expression of iNOS and COX-2 proteins. Off-white powder. $[\alpha]^{25}_{D}$ -16.9 (c, 0.01 in MeOH)

$(11S,R_{axial})$-form [1356848-45-7]
Constit. of the fruit of *Myrica cerifera*. Reduces tau levels. Potential candidate for treating Alzheimer's disease. $[\alpha]^{20}_{D}$ +48 (c, 1 in $CHCl_3$). λ_{max} 215; 220; 260; 300 (MeOH)

5-Deoxy
[1607448-28-1]

$C_{21}H_{26}O_4$ 342.435
Constit. of the leaves of *Juglans mandshurica*. Brown powder. [α] D +84.2 (c, 0.17 in MeOH). λ_{max} 207; 250; 292 (MeOH)

5-Deoxy, 3-O-de-Me, 11-sulfate
Juglanin B 11-sulfate
[1403239-93-9]
$C_{20}H_{24}O_7S$ 408.465
Constit of the bark of *Myrica rubra*. Amorph. off-white powder. [α] D +23.2 (c, 0.5 in MeOH)

13-Oxo, 5-deoxy
Rhoiptelol A
[184485-68-5]
$C_{21}H_{24}O_5$ 356.418
Constit. of the leaves of *Rhoiptelea chiliantha*. Needles + ⅓H_2O (MeOH aq.). Mp 174 - 175°. $[\alpha]^{22}_D$ -10 (c, 0.1 in $CHCl_3$). Axial config. not determined. λ_{max} 205 (ε 10000); 218 (ε 5128); 294 (ε 3311) (MeOH) (Berdy)

(±)-form
Isol. from *Myrica cerifera*

11-Me ether
11-O-Methylmyricanol. *11-Methoxymyricanol (incorr.)*
[181061-42-7]
$C_{22}H_{28}O_5$ 372.461
Constit. of the bark of *Myrica rubra*. Needles ($CHCl_3$). Mp 190.1 - 192.5°. λ_{max} 236 (log ε 3.30); 348 (log ε 2.25) (MeOH)

(ξ)-form

5-O-[(3,4,5-Trihydroxybenzoyl)-(→6)-β-D-glucopyranoside]
Myricanol 5-(6-galloylglucoside)
[113204-18-5]
$C_{34}H_{40}O_{14}$ 672.68
Constit. of *Morella salicifolia* and *Myrica rubra*. Cryst. (MeOH aq.). Mp 267 - 269°. $[\alpha]^{24}_D$ -78.4 (c, 1 in EtOH). λ_{max} 256 (log ε 4.2); 285 (log ε 4.11) (EtOH)

5-O-[α-L-Arabinofuranosyl-(1→6)-β-D-glucopyranoside]
[150107-09-8]
$C_{32}H_{44}O_{14}$ 652.690
Constit. of *Morella salicifolia* and *Myrica rubra*. Off-white powder. $[\alpha]^{21}_D$ -81.5 (c, 1 in MeOH)

5-O-[β-D-Glucopyranosyl-(1→3)-β-D-glucopyranoside]
Myricanol 5-laminaribioside
[137578-55-3]
$C_{33}H_{46}O_{15}$ 682.716
Constit. of *Myrica rubra*. Off-white powder. $[\alpha]^{21}_D$ -47.7 (c, 1 in MeOH)

5-O-[β-D-Glucopyranosyl-(1→6)-β-D-glucopyranoside]
Myricanol 5-gentiobioside
[116107-17-6]
$C_{33}H_{46}O_{15}$ 682.716
Constit. of *Morella salicifolia* and *Myrica rubra*. Off-white powder. $[\alpha]^{20}_D$ -64.6 (c, 1.1 in Py)

11-O-β-D-Xylopyranoside
Myricanol 11-xyloside
[307313-20-8]
$C_{26}H_{34}O_9$ 490.549
Constit. of *Morella salicifolia* and *Myrica arborea*. Needles (CH_2Cl_2). Mp 229 - 231°. $[\alpha]^{22}_D$ -46 (c, 0.55 in MeOH). λ_{max} 214 (log ε 4.46); 260 (log ε 3.97); 298 (log ε 3.79) (MeOH)

11-O-β-D-Glucopyranoside
Myricanol 11-glucoside
[538321-96-9]
$C_{27}H_{36}O_{10}$ 520.575
Constit. of the bark of *Myrica rubra*. Powder. $[\alpha]^{22}_D$ -3 (c, 0.2 in EtOH). λ_{max} 215 (log ε 4.4); 260 (log ε 3.9); 295 (log ε 3.8) (EtOH)

11-Ketone
Myricanone
[32492-74-3]
$C_{21}H_{24}O_5$ 356.418
Constit. of *Morella salicifolia*, *Myrica cerifera*, *Myrica nagi* and *Myrica rubra*. Exhibits anticancer activities. Cryst. (EtOH). Mp 194 - 196°

11-Ketone, 5-O-β-D-glucopyranoside
Myricanone 5-glucoside
[538321-97-0]
$C_{27}H_{34}O_{10}$ 518.559
Constit. of the bark of *Morella salicifolia* and *Myrica rubra*. Powder. $[\alpha]^{22}_D$ +25.4 (c, 0.1 in EtOH). λ_{max} 217 (log ε 4.6); 254 (log ε 4.1); 295 (log ε 3.8) (EtOH)

11-Ketone, 5-O-[(3,4,5-trihydroxybenzoyl)-(→6)-β-D-glucopyranoside]
Myricanone 5-O-(6′-O-galloyl) glucoside
[1403239-94-0]
$C_{34}H_{38}O_{14}$ 670.664
Constit. of the bark of *Morella salicifolia* and *Myrica rubra*. Amorph. pale yellow powder

11-Ketone, 5-O-[α-L-arabinofuranosyl-(1→6)-β-D-glucopyranoside]
Myricanone 5-(6-arabinosylglucoside)
[1276043-53-8]
$C_{32}H_{42}O_{14}$ 650.674
Constit. of thc bark of *Myrica rubra*. Amorph. solid. λ_{max} 218 (log ε 4.05); 255 (log ε 3.4); 298 (log ε 3.15) (EtOH)

11-Ketone, 17-O-[3,4,5-trihydroxybenzoyl-(→6)-β-D-glucopyranoside]
Myricanone 17-(6-galloylglucoside)
[1276043-54-9]
$C_{34}H_{38}O_{14}$ 670.664
Constit. of the bark of *Myrica rubra*. Needles. Mp 265 - 267°. λ_{max} 215 (log ε 4.01); 256 (log ε 3.5); 298 (log ε 3.16) (EtOH)

11-Ketone, 17-Ac, 5-O-β-D-glucopyranoside
Alnuheptanoid B
$C_{29}H_{36}O_{11}$ 560.596
Constit. of the stem bark of *Alnus japonica*. Amorph. powder. [α] D +43.1 (c, 0.5 in MeOH). λ_{max} 219 (log ε 4.71); 251 (log ε 4.36); 293 (log ε 3.95) (MeOH)

5-Deoxy, 3-O-de-Me
Juglanin B
[958365-42-9]
$C_{20}H_{24}O_4$ 328.408
Constit. of *Juglans regia* and *Juglans sinensis*. Needles (EtOAc). Mp 181 - 183°. $[\alpha]^{20}_D$ +6 (c, 0.5 in MeOH). λ_{max} 256 (log ε 3.96); 286 (log ε 3.85); 298 (log ε 3.81) (MeOH)

5-Deoxy, 3,4-di-O-de-Me, 3,4-methylene ether
Juglanin C
[1093115-72-0]
$C_{20}H_{22}O_4$ 326.392
Constit. of *Juglans mandshurica* and *Juglans sinensis*. Needles. $[\alpha]^{25}_D$ -12.2 (c, 1.13 in MeOH). λ_{max} 298 (log ε 3.74) (MeOH)

5-Deoxy, 11-ketone
5-Deoxymyricanone
[110007-10-8]
$C_{21}H_{24}O_4$ 340.419
Isol. from bacterial galls of *Myrica rubra*. Amorph.

5-Deoxy, 11-ketone, 3-O-de-Me
Myricananin C
[1079941-40-4]

$C_{20}H_{22}O_4$ 326.392
Constit. of *Juglans hopeiensis* and *Myrica nana*. Cryst. (MeOH). Mp 192 - 193°. $[\alpha]^{26}_{D}$+31.7 (c, 0.1 in MeOH). λ_{max} 214 (log ε 4.67); 250 (log ε 4.17); 298 (log ε 4.02) (MeOH)

5-Deoxy, 12-hydroxy, 3-O-de-Me
Myricananin A
[1079941-35-7]
$C_{20}H_{24}O_5$ 344.407
Constit. of the roots of *Myrica nana*. Needles (MeOH). Mp 161 - 162°. $[\alpha]^{20}_{D}$+28.3 (c, 0.1 in MeOH). Possesses (11*R**,12*R**)-config.. λ_{max} 213 (log ε 4.65); 258 (log ε 4.12); 295 (log ε 4.94) (MeOH)

11-Deoxy, 10E,11-didehydro
Myricanene A
[449753-39-3]
$C_{21}H_{24}O_4$ 340.419
Powder

11-Deoxy, 10E,11-didehydro, 5-O-[α-L-arabinofuranosyl-(1→6)-β-D-glucopyranoside]
[449147-51-7]
$C_{32}H_{42}O_{13}$ 634.675
Constit. of the bark of *Myrica rubra*. Powder. $[\alpha]^{22}_{D}$-0.6 (c, 0.1 in EtOH). λ_{max} 215 (log ε 4.2); 257 (log ε 3.9); 300 (log ε 3.6) (EtOH)

11-Deoxy, 11E,12-didehydro
Myricanene B
[449753-40-6]
$C_{21}H_{24}O_4$ 340.419
Powder

11-Deoxy, 11E,12-didehydro, 5-O-[α-L-arabinofuranosyl-(1→6)-β-D-glucopyranoside]
[449147-52-8]
$C_{32}H_{42}O_{13}$ 634.675
Constit. of the stem bark of *Myrica rubra*. Powder. $[\alpha]^{22}_{D}$-10.8 (c, 0.1 in EtOH). λ_{max} 214 (log ε 4.5); 253 (log ε 4.1); 297 (log ε 3.7) (EtOH)

5,11-Dideoxy, 10ξ-hydroxy, 3-O-de-Me
Myricananin F
[1190206-23-5]
$C_{20}H_{24}O_4$ 328.408
Constit. of the roots of *Myrica nana*. Amorph. powder. Mp 169 - 171°. $[\alpha]^{26}_{D}$+58.3 (c, 0.1 in MeOH). λ_{max} 213 (log ε 4.55); 254 (log ε 4.05); 299 (log ε 3.93) (MeOH)

8ξ-Hydroxy, 11-ketone
Myricananone
[957200-76-9]
$C_{21}H_{24}O_6$ 372.417
Constit. of the roots of *Myrica nana*. Powder. Opt. inactive. λ_{max} 258; 295 (MeOH)

12β-Hydroxy
Myricananin H
[1190206-25-7]
$C_{21}H_{26}O_6$ 374.433
Constit. of the roots of *Myrica nana*. Cryst.. Mp 215 - 216°. $[\alpha]^{26}_{D}$ +14.8 (c, 0.14 in MeOH). Possesses (11*R**,12*R**)-config.. λ_{max} 214 (log ε 4.47); 253 (log ε 4); 283 (log ε 3.75); 295 (log ε 3.73) (MeOH)

12ξ-Hydroxy, 11-ketone
12-Hydroxymyricanone
[171438-26-9]
$C_{21}H_{24}O_6$ 372.417
Constit. of *Myrica gale* var. *tomentosa* and *Myrica rubra*. Powder. $[\alpha]_D$ -6.1 (c, 0.5 in $CHCl_3$). λ_{max} 213 (log ε 4.55); 257 (log ε 4.01); 294 (log ε 3.84) (MeOH)

12ξ-Hydroxy, 11-ketone, 5-O-β-D-glucopyranoside
Myricatomentoside II
[191999-63-0]
$C_{27}H_{34}O_{11}$ 534.558
Constit. of *Myrica gale* var. *tomentosa*. Needles (MeOH aq.). Mp 148 - 150°. $[\alpha]_D$ +31.3 (c, 0.5 in MeOH). λ_{max} 213 (log ε 4.48); 249 (log ε 3.99); 295 (log ε 3.67) (MeOH)

12ξ-Hydroxy, 11-ketone, 5-Me ether
Porson
[56222-03-8]
$C_{22}H_{26}O_6$ 386.444
Constit. of *Myrica gale*. Needles (MeOH). Mp 186 - 187°. $[\alpha]_D$ -1.8 (c, 2.3 in $CHCl_3$). Struct. revised in 1995. Abs. config. not confirmed. λ_{max} 213 (ε 38900); 248 (ε 12300); 294 (ε 6025) (EtOH)

16-Hydroxy, 12Z,13-didehydro, 11-ketone
Myricananin D
[1079941-41-5]
$C_{21}H_{22}O_6$ 370.401
Constit. of the roots of *Myrica nana*. Amorph. powder. Mp 107 - 110°. $[\alpha]^{26}_{D}$+61.7 (c, 0.1 in MeOH). λ_{max} 221 (log ε 4.86); 257 (log ε 4.66) (MeOH)

12ξ-Methoxy, 16-hydroxy, 11-ketone
Myricananin E
[1079941-43-7]
$C_{22}H_{26}O_7$ 402.443
Constit. of the roots of *Myrica nana*. Amorph. powder. Mp 129 - 131°. $[\alpha]^{26}_{D}$+45 (c, 0.1 in MeOH). λ_{max} 217 (log ε 4.65); 259 (log ε 4.19) (MeOH)

9-Oxo, 11-deoxy
Neomyricanone
Powder. λ_{max} 217 (log ε 4); 261 (log ε 3.4); 297 (log ε 3.2) (EtOH)

9-Oxo, 11-deoxy, 5-O-β-D-glucopyranoside
Neomyricanone 5-glucoside
[538321-98-1]
$C_{27}H_{34}O_{10}$ 518.559
Constit. of the bark of *Myrica rubra*. Powder. $[\alpha]^{22}_{D}$-1.1 (c, 0.2 in EtOH). λ_{max} 213 (log ε 4.3); 254 (log ε 3.8); 297 (log ε 3.5) (EtOH)

12-Oxo, 11-ketone, 5-Me ether
12-Dehydroporson
[171438-25-8]
$C_{22}H_{24}O_6$ 384.428
Constit. of *Myrica gale*. Pale yellow needles (CH_2Cl_2/MeOH). Mp 191 - 192°. Exists as α-diketo-tautomer. λ_{max} 213 (ε 22910); 249 (ε 6760); 294 (ε 3310) (MeOH)

13-Oxo
13-Oxomyricanol
[74517-63-8]
$C_{21}H_{24}O_6$ 372.417
Constit. of *Myrica nagi*. Cryst. (MeOH aq.). Mp 239.5 - 240°. $[\alpha]^{20}_{D}$-36 (c, 0.3 in $CHCl_3$). λ_{max} 224 (log ε 4.23); 262 (log ε 4.24); 281 (sh) (log ε 4.01) (MeOH)

11-Epimer, 5-O-β-D-glucopyranoside
Epimyricanol 5-glucoside
[449729-89-9]
$C_{27}H_{36}O_{10}$ 520.575
Constit. of the bark of *Myrica rubra*. Powder. $[\alpha]^{22}_{D}$+81.3 (c, 0.2 in EtOH). λ_{max} 216 (log ε 4.5); 256 (log ε 4.2); 299 (log ε 3.7) (EtOH)

11-Epimer, 12α-hydroxy
Myricananin G
[1190206-24-6]
$C_{21}H_{26}O_6$ 374.433
Constit. of the roots of *Myrica nana*. Amorph. powder. Mp 223 - 225°. $[\alpha]^{26}_{D}$-3.3 (c, 0.1 in MeOH). Possesses (11*R**,12*S**)-config.. λ_{max} 214 (log ε 4.53); 258 (log ε 4.01); 295 (log ε 3.8) (MeOH)

11-Epimer, 12β-hydroxy
Myricananin B
[1079941-37-9]
$C_{21}H_{26}O_6$ 374.433
Constit. of the roots of *Myrica nana*. Cryst. (EtOH). Mp 210 - 212°. $[\alpha]_D^{26}$ -41.7 (c, 0.1 in MeOH). Possesses (11*R**,12*R**)-config.. λ_{max} 195 (log ε 4.45); 212 (log ε 4.69); 258 (log ε 4.15); 297 (log ε 3.98) (MeOH).

Campbell, R.V.M. *et al.*, *J. Chem. Soc. D: Chem. Commun.*, 1970, 1206-1207 (*Myricanol, Myricanone*).
Begley, M.J. *et al.*, *J. Chem. Soc. C*, 1971, 3634-3642 (*Myricanol, cryst struct*).
Whiting, D.A. *Tetrahedron Lett.*, 1978, **19**, 2335-2338 (*Myricanol, Myricanone, synth*).
Malterud, K.E. *et al.*, *Phytochemistry*, 1980, **19**, 705-707 (*13-Oxomyricanol*).
Inoue, T. *et al.*, *Yakugaku Zasshi*, 1984, **104**, 37-41 (*5-glucoside*).
Takeda, Y. *et al.*, *Chem. Pharm. Bull.*, 1987, **35**, 2569-2573 (*5-Deoxymyricanone*).
Yaguchi, Y. *et al.*, *Chem. Pharm. Bull.*, 1988, **36**, 1419-1424 (*5-gentiobioside, 5-6-galloylglucoside*).
Sun, D. *et al.*, *Phytochemistry*, 1988, **27**, 579-583 (*Myricanol, Myricanone, cmr*).
Sakurai, N. *et al.*, *Phytochemistry*, 1991, **30**, 3077-3079 (*5-arabinosylglucoside, 5-laminaribioside*).
Nagai, M. *et al.*, *Chem. Pharm. Bull.*, 1995, **43**, 1674-1677 (*Porson, Dehydroporson, 12-Hydroxymyricanone*).
Joshi, B.S. *et al.*, *J. Nat. Prod.*, 1996, **59**, 759-764 (*Myricanol, Myricanone, cryst struct*).
Jiang, Z.-H. *et al.*, *Phytochemistry*, 1996, **43**, 1049-1054 (*Rhoiptelol A*).
Morihara, M. *et al.*, *Chem. Pharm. Bull.*, 1997, **45**, 820-823 (*Myricatomentoside II*).
Tene, M. *et al.*, *Phytochemistry*, 2000, **54**, 975-978 (*11-xyloside*).
Tao, J. *et al.*, *Bioorg. Med. Chem.*, 2002, **10**, 4005-4012 (*11-glucosides, Neomyricanones*).
Matsuda, H. *et al.*, *Chem. Pharm. Bull.*, 2002, **50**, 208-215 (*Epimyricanol, Myricanenes*).
Yu, Y.-F. *et al.*, *Helv. Chim. Acta*, 2007, **90**, 1691-1696 (*Myricananone*).
Wang, J.-F. *et al.*, *Bioorg. Med. Chem.*, 2008, **16**, 8510-8515 (*Myricananins A-E*).
Huang, Y.-D. *et al.*, *Guangdong Yaoxueyuan Xuebao*, 2008, **24**, 141-143 (*11-O-Methylmyricanol*).
Kawai, S. *et al.*, *J. Wood Sci.*, 2008, **54**, 256-260 (*Myricananone, biosynth*).
Liu, J. *et al.*, *Planta Med.*, 2008, **74**, 754-759 (*Juglanin B*).
Wang, J.-F. *et al.*, *Helv. Chim. Acta*, 2009, **92**, 1594-1599 (*Myricananins F,G,H*).
Yang, H. *et al.*, *Eur. J. Org. Chem.*, 2010, 345-351 (*Rhoiptelol B, synth*).
Akazawa, H. *et al.*, *J. Oleo Sci.*, 2010, **59**, 213-221 (*Myricanone 17-galloylglucoside, Myricanone 5-arabinosylglucoside*).
Yadav, J.S. *et al.*, *Synthesis*, 2010, 4300-4306 (*Rhoiptelol B, synth*).
Jones, J.R. *et al.*, *J. Nat. Prod.*, 2011, **74**, 38-44 (*Myricanol, bibl*).
Yang, H. *et al.*, *Planta Med.*, 2011, **77**, 841-845 (*Juglanin C*).
Yoshimura, M. *et al.*, *J. Nat. Prod.*, 2012, **75**, 1798-1802 (*11-sulfates, Myricanone 5-(6-galloyl)glucoside*).
Sun, C. *et al.*, *Plant Foods Hum. Nutr.*, 2013, **68**, 97-106 (*rev, activity*).
Dai, G. *et al.*, *Chemotherapy (Basel, Switz.)*, 2014, **60**, 81-87 (*Myricanone, activity*).
Kim, H.H. *et al.*, *Fitoterapia*, 2014, **92**, 188-193 (*Juglanin B 11R-sulfate*).
Martin, M.D. *et al.*, *ACS Chem. Biol.*, 2015, **10**, 1099-1109 (*Myricanol, synth, cryst struct, activity*).
Yao, D.L. *et al.*, *Arch. Pharmacal Res.*, 2015, **38**, 480-484 (*Juglans mandshurica constit*).
Hu, Y.-B. *et al.*, *Nat. Prod. Commun.*, 2015, **10**, 1711-1714 (*4-O-Demethyl-5-O-methylmyricanol*).
Shen, S.-N. *et al.*, *Yaoxue Xuebao*, 2015, **50**, 746-748 (*4-O-Demethyl-5-O-methylmyricanol*).
Ibrahim, S.R.M. *et al.*, *Rec. Nat. Prod.*, 2016, **10**, 362-368 (*Alnuheptanoid B*).
Makule, E. *et al.*, *Molecules*, 2017, **22**, 2266:1-15 (*Morella salicifolia constis, cd, abs config*).
Jahng, Y. *et al.*, *Molecules*, 2018, **23**, 3107:1-42 (*rev, isol, synth, activity*).
Bochicchio, A. *et al.*, *Org. Biomol. Chem.*, 2018, **16**, 8859-8869 (*Myricanol, synth*)

Myricarborin‡ M58

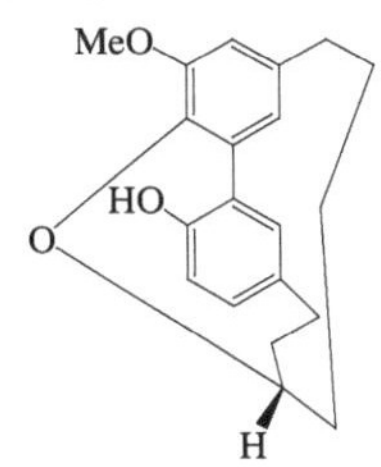

[307554-15-0]
$C_{20}H_{22}O_3$ 310.393
Not to be confused with Myricarborin A, M59. Constit. of *Myrica arborea*. Plates (CH_2Cl_2). Mp 141 - 143°. $[\alpha]_D$ +9.3 (c, 0.41 in MeOH). λ_{max} 214 (log ε 4.32); 256 (log ε 3.82); 328 (log ε 3.71) (MeOH).

Tene, M. *et al.*, *Phytochemistry*, 2000, **54**, 975-978 (*Myricarborin*)

Myricarborin A‡ M59

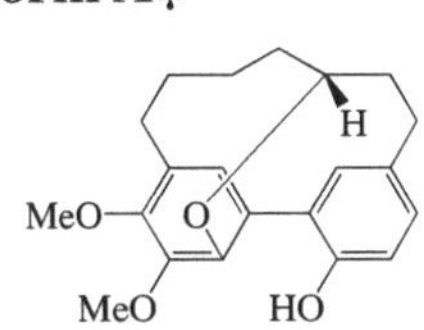

Absolute Configuration

[1190745-45-9]
$C_{21}H_{24}O_4$ 340.419
Not to be confused with Myricarborin, M58 and related to Myricanol, M57. Constit. of the bark of *Myrica rubra*. Oil. $[\alpha]_D^{20}$ -74.3 (c, 0.07 in MeOH).

Liu, Z. *et al.*, *Nat. Prod. Commun.*, 2009, **4**, 513-516 (*Myricarborin A*)

Nectamazin C N1

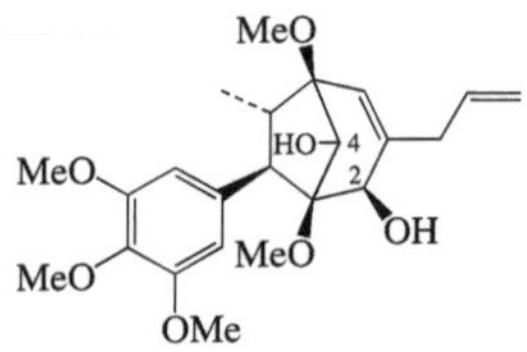

Absolute Configuration

5,8′-Cyclo-2,3,4,5-tetrahydro-3,3′,4′,5,5′-pentamethoxy-3,7′-lign-8-ene-2,4-diol.
5,8′-Cyclo-2,3,4,5-tetrahydro-2,4-dihydroxy-3,3′,4′,5,5′-pentamethoxy-3,7′-neolign-8-ene
[1187947-56-3]
$C_{23}H_{32}O_7$ 420.502
Lignan numbering shown. Related to Cinerin C, C76. Constit. of the leaves of *Nectandra amazonum*. Exhibits dual 5-LOX/COX-2 inhibition. Powder. Mp 177 - 179°. $[\alpha]_D^{25}$+7.5 (c, 0.03 in $CHCl_3$)

2-Ketone
5,8′-Cyclo-4,5-dihydro-4-hydroxy-3,3′,4′,5,5′-pentamethoxy-3,7′-lign-8-en-2(3H)-one. ***Nectamazin B***
[1187947-50-7]
$C_{23}H_{30}O_7$ 418.486
Constit. of the leaves of *Nectandra amazonum*. Oil. $[\alpha]_D^{25}$+9.7 (c, 0.26 in $CHCl_3$)

4-Epimer, 2-ketone
Nectamazin A
[1187947-44-9]
$C_{23}H_{30}O_7$ 418.486
Constit. of the leaves of *Nectandra amazonum*. Yellowish oil. $[\alpha]_D^{25}$ +12.2 (c, 0.32 in $CHCl_3$).

Coy, E.D. *et al., Bioorg. Med. Chem. Lett.*, 2009, **19**, 6922-6925 (*Nectamazin C, activity*).
Barrera, E.D.C. *et al., Chem. Pharm. Bull.*, 2009, **57**, 639-642 (*Nectamazins A,B,C, cd, activity*)

Neglignan H N2

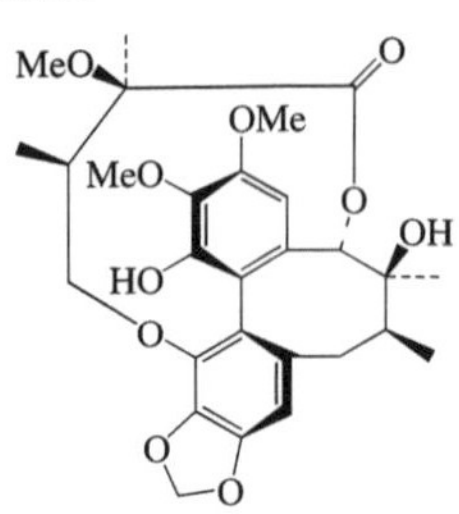

Absolute Configuration

[1706665-47-5]
$C_{28}H_{34}O_{10}$ 530.57
Constit. of the stems of *Schisandra neglecta*. Exhibits modest cytotoxic activity. Amorph. powder. $[\alpha]_D^{25}$-42.5 (c, 0.2 in MeOH). λ_{max} 210 (log ε 4.48); 254 (log ε 3.94); 300 (log ε 1.52) (MeOH).

Ye, Y.-Q. *et al., Asian J. Chem.*, 2014, **26**, 1948-1950 (*Neglignan H, activity*)

Negundin A N3

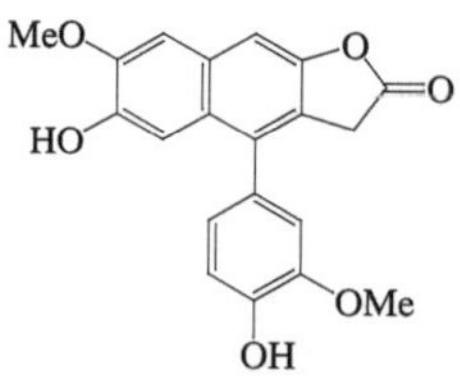

6-Hydroxy-4-(4-hydroxy-3-methoxyphenyl)-7-methoxynaphtho[2,3-b]furan-2(3H)-one
[844692-50-8]
$C_{20}H_{16}O_6$ 352.342
Unusual struct. containing a 9-nor-9′-homolignan. Constit. of the roots of *Vitex negundo*. Amorph. solid. Mp 125°. λ_{max} 256 (log ε 4.64); 289 (log ε 4.07); 314 (log ε 4.1) (MeOH).

Azhar-Ul-Haq, *et al., Chem. Pharm. Bull.*, 2004, **52**, 1269-1272 (*Negundin A*)

Neoarctin A‡ N4

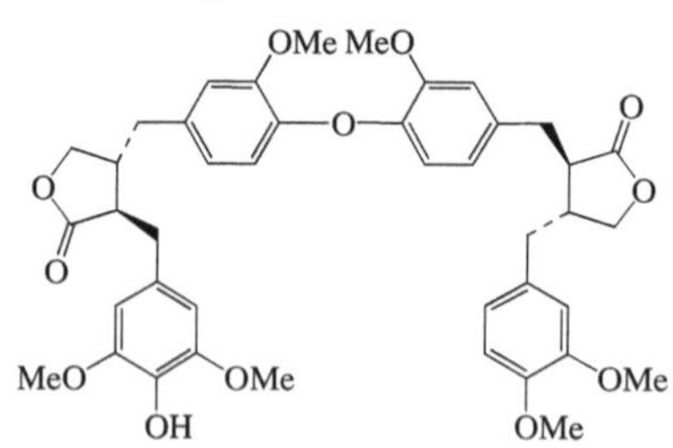

[163318-69-2]
$C_{42}H_{46}O_{12}$ 742.818
Not to be confused with Neoarctin A in E92. Constit. of the seeds of *Arctium lappa*. Amorph. + $2H_2O$. $[\alpha]_D^{24}$-17.5 (c, 0.09 in EtOH). λ_{max} 228 (sh); 280 (log ε 3.66) (EtOH).

Wang, H.Y. *et al., Chin. Chem. Lett.*, 1995, **6**, 217-220 (*Neoarctin A*)

Neoarctin B N5

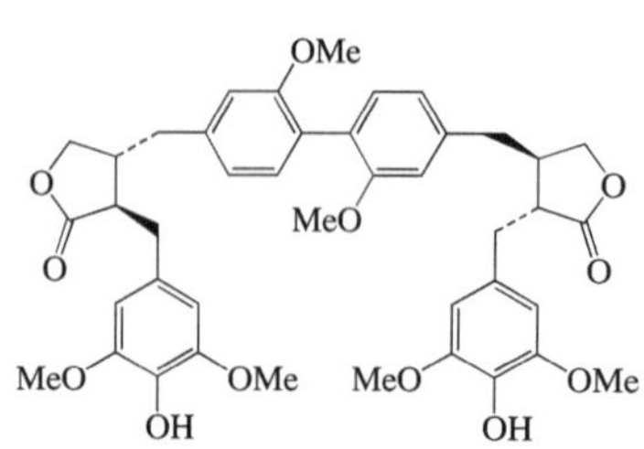

Relative Configuration

[155969-67-8]
$C_{42}H_{46}O_{12}$ 742.818
Constit. of the seeds of *Arctium lappa*. Light yellow powder. Mp 102 - 103.5°. $[\alpha]_D^{15}$-46.9 (c, 0.08 in $CHCl_3$).

Wang, H.Y. *et al., Yaoxue Xuebao*, 1993, **28**, 911-917 (*Neoarctin B*)

Neobonaspectin A N6

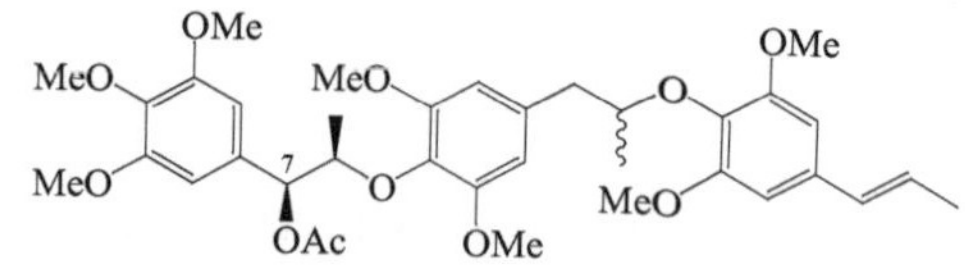

Relative Configuration

[261949-33-1]
$C_{36}H_{46}O_{11}$ 654.753
Sesquilignan. Constit. of *Bonamia spectabilis*. Oil. $[\alpha]_D^{20}$+5.3 (c, 0.23 in $CHCl_3$). λ_{max} 210; 267 (MeOH)

Deacetoxy
Neobonaspectin B
[261949-34-2]
$C_{34}H_{44}O_9$ 596.717
Constit. of *Bonamia spectabilis*. Oil. $[\alpha]_D^{20}$+11.5 (c, 0.2 in $CHCl_3$). λ_{max} 210; 236 (sh); 270 (MeOH).

Tofern, B. *et al., Phytochemistry*, 2000, **53**, 119-128 (*Neobonaspectins A,B*)

Neocassumunarin A N7

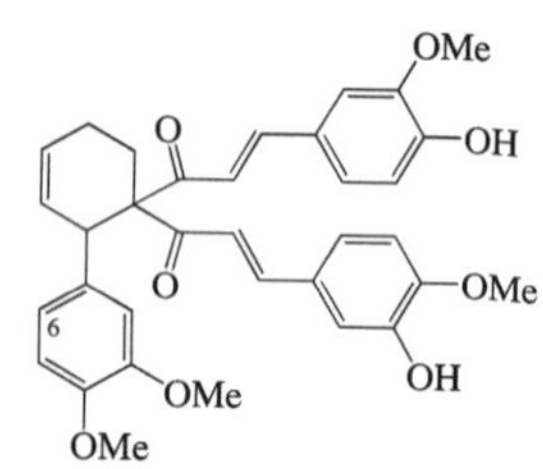

[1672673-96-9]
$C_{34}H_{34}O_8$ 570.638
Constit. of the rhizomes of *Zingiber purpureum*. Neurite outgrowth promotor. Racemate. λ_{max} 224 (log ε 4.23); 260 (log ε 3.92); 359 (log ε 2.65) (MeOH)

6-Methoxy
Neocassumunarin B
[1672673-97-0]
$C_{35}H_{36}O_9$ 600.664
Constit. of the rhizomes of *Zingiber purpureum*. Neurite outgrowth promotor. Racemate. λ_{max} 266 (log ε 3.93); 353 (log ε 3.89) (MeOH).

Kubo, M. *et al., Bioorg. Med. Chem. Lett.*, 2015, **25**, 1586-1591 (*Neocassumunarins A,B*)

Neocleistantoxin N8

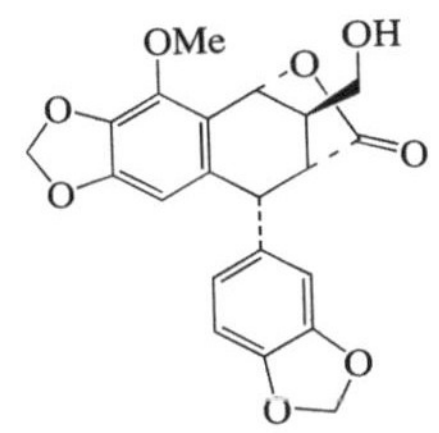

Absolute Configuration

2,7′-Cyclo-9-hydroxy-6-methoxy-3′,4′:4,5-bis(methylenedioxy)-8,8′-lignan-9′,7-olide
[1397199-09-5]
$C_{21}H_{18}O_8$ 398.367
Unusual naphthalenoid 9′→7-olide lignan. Isol. from the fruits of *Cleistanthus indochinensis*. Shows weak to moderate cytotoxicity against human nasopharynx KB cancer cells. Powder. Mp 225 - 226°. $[\alpha]^{30}_{D}$ -44.4 (c, 0.225 in $CHCl_3$). λ_{max} 212 (log ε 3.4); 240 (log ε 4.23); 287 (log ε 3.82) ($CHCl_3$).

Thanh, V.T.T. *et al., J. Nat. Prod.*, 2012, **75**, 1578-1583 (*Neocleistantoxin*)

Neohydnocarpin N9

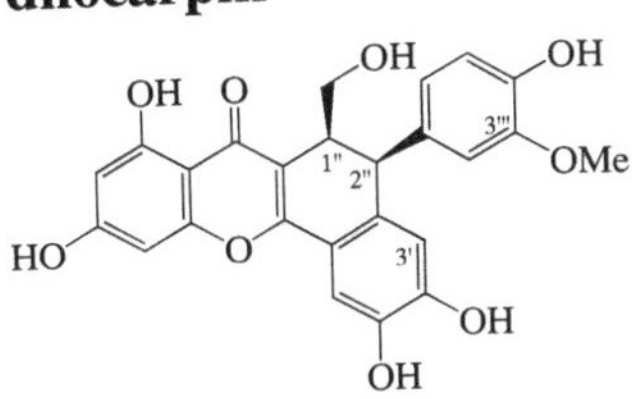

Absolute Configuration

[71417-57-7]
$C_{25}H_{20}O_9$ 464.426
Flavonolignan. Constit. of the seeds of *Hydnocarpus wightiana*. Shows hypolipidaemic and cytotoxic props.. Yellow cryst. (C_6H_6/Me_2CO). Mp 235 - 237°. $[\alpha]_D$ -20.3 (c, 0.59 in MeOH)

Hexa-Ac
Needles (C_6H_6/petrol). Mp 236 - 238°

3′-Methoxy, 5′-Me ether
3′-Methoxy-5′-O-methylneohydnocarpin
[849245-12-1]
$C_{27}H_{24}O_{10}$ 508.479
Constit. of *Avena sativa*. Amorph. yellow powder. $[\alpha]^{23}_{Hg}$ -43.5 (c, 0.1 in MeOH). λ_{max} 271; 365 (MeOH)

3‴-Demethoxy, 3′-methoxy, 5-Me ether
Natansin A
[2025333-57-5]
$C_{26}H_{22}O_9$ 478.453
Constit. of *Elymus natans*. Amorph. yellow powder. $[\alpha]^{25}_{D}$ -20.9 (c, 0.1 in MeOH). λ_{max} 272; 364 (MeOH)

1″-Epimer, 3′-methoxy, 5′-Me ether, 7-O-β-D-glucopyranoside
Indocalatin A
[2170740-72-2]
$C_{33}H_{34}O_{15}$ 670.620
Constit. of leaves of *Indocalamus latifolius*. Amorph. yellow powder. $[\alpha]^{20}_{D}$ +31.5 (c, 0.15 in MeOH). λ_{max} 274; 370 (MeOH aq.).

Parthasarathy, M.R. *et al., Phytochemistry*, 1979, **18**, 506-508 (*cmr*).
Sharma, D.K. *et al., Planta Med.*, 1979, **37**, 79-83 (*Neohydnocarpin*).
Sharma, D.K. *et al., J. Nat. Prod.*, 1991, **54**, 1298-1302 (*props*).
Wenzig, E. *et al., J. Nat. Prod.*, 2005, **68**, 289-292 (*Avena sativa constit*).
Sun, J. *et al., J. Asian Nat. Prod. Res.*, 2016, **18**, 360-365 (*Indocalatin A*).
Liu, Q. *et al., J. Agric. Food Chem.*, 2017, **65**, 1320-1327 (*Natansin A*)

Neoisostegane N10

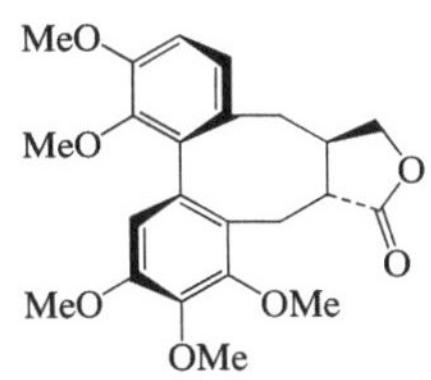

[87084-98-8]
$C_{23}H_{26}O_7$ 414.454
Constit. of *Steganotaenia araliacea*. Cryst. (Et_2O or EtOH). Sol. MeOH, EtOAc; poorly sol. H_2O. Mp 71 - 74°. Mp 156 - 157°. $[\alpha]^{20}_{D}$ +65 (c, 0.35 in $CHCl_3$). $[\alpha]_D$ +107.7 (c, 0.51 in $CHCl_3$). The higher Mp refers to synthetic compd. of higher purity than the nat. isolate. λ_{max} 255 (ε 11200); 287 (ε 5600) (EtOH) (Derep).

Taafrout, M. *et al., Tetrahedron Lett.*, 1983, **24**, 2983-2986 (*Neoisostegane*).
Hicks, R.P. *et al., Tetrahedron Lett.*, 1983, **24**, 2987-2990 (*pmr, cmr*).
Taafrout, M. *et al., J. Nat. Prod.*, 1984, **47**, 600-606 (*cd*).
Taafrout, M. *et al., Tetrahedron Lett.*, 1986, **27**, 1781-1784 (*synth*).
Morimoto, T. *et al., Tetrahedron*, 1993, **49**, 1793-1806 (*synth*).
Hartmann, P. *et al., J. Org. Chem.*, 2019, **84**, 5831-5837 (*synth*)

Neonootkatol N11

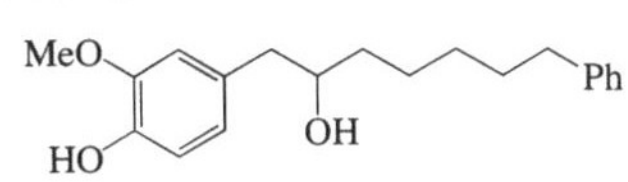

α-[(4-Hydroxy-3-methoxyphenyl)methyl]benzenehexanol, CAS. *1-(4-Hydroxy-3-methoxyphenyl)-7-phenyl-2-heptanol*
[196399-56-1]
$C_{20}H_{26}O_3$ 314.425
Constit. of the fruit of *Alpinia oxyphylla*. λ_{max} 206; 229; 282; 286 (MeOH).

Zhang, O. *et al., Zhongcaoyao*, 1997, **28**, 131-133 (*Neonootkatol*)

Nicotfuran A N12

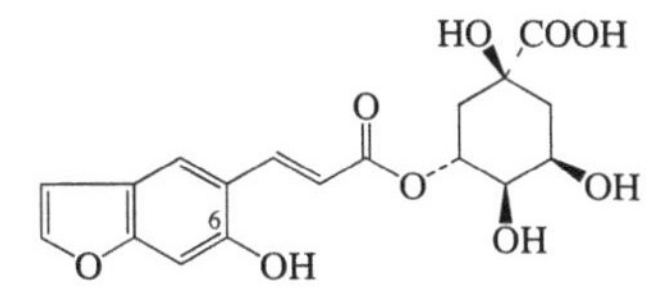

[1361044-15-6]
$C_{18}H_{18}O_9$ 378.333
Isol. from roots and stems of *Nicotiana tabacum*. Shows activity against tobacco mosaic virus. Pale yellow gum. $[\alpha]^{24.2}_{D}$ -42.3 (c, 0.20 in MeOH). λ_{max} 215 (log ε 4.94); 260 (log ε 4.18); 310 (log ε 3.84) (MeOH)

6-Me ether
Nicotfuran B
[1361044-16-7]
$C_{19}H_{20}O_9$ 392.36
Isol. from roots and stems of *Nicotiana tabacum*. Shows activity against tobacco mosaic virus. Pale yellow gum. $[\alpha]^{24.5}_{D}$ -39.4 (c, 0.22 in MeOH). λ_{max} 215 (log ε 4.98); 262 (log ε 4.22); 310 (log ε 3.8) (MeOH).

Tan, J. *et al., Heterocycles*, 2011, **83**, 2381-2385 (*Nicotfurans A,B*)

Nicotlactone B N13

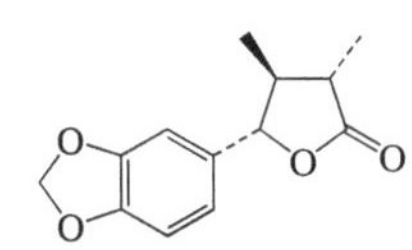

Absolute Configuration

5-(1,3-Benzodioxol-5-yl)dihydro-3,4-dimethyl-2(3H)-furanone, CAS. *Dihydro-3,4-dimethyl-5-(3,4-methylenedioxy)-2(3H)-furanone*
[1935724-09-6]
$C_{13}H_{14}O_4$ 234.251
Constit. of *Austrobaileya scandens*. Weak inhibitor of human prostate cancer PC3 cells. Amorph. powder. $[\alpha]^{25}_{D}$ -18 (c, 0.03 in MeOH). λ_{max} 205 (log ε 4.3); 236 (log ε 3.4); 280 (log ε 2.10) (MeOH).

Hazra, S. *et al., RSC Adv.*, 2013, **3**, 22834-22836 (*Nicotlactone B, synth*).

Tran, T.D. *et al.*, *J. Nat. Prod.*, 2016, **79**, 1514-1523 (*Nicotlactone B, activity*).
Henrion, S. *et al.*, *Org. Biomol. Chem.*, 2018, **16**, 1672-1678 (*synth*)

Nigramide P N14

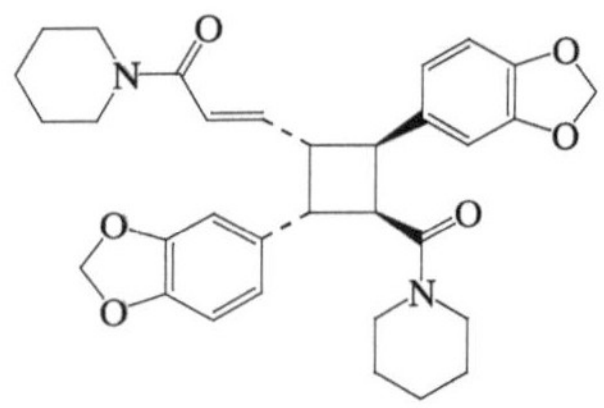

[847552-97-0]
$C_{32}H_{36}N_2O_6$ 544.648
Alkaloid from the roots of *Piper nigrum*. Oil. Racemic. λ_{max} 207 (log ε 4.72); 287 (log ε 4.08) (MeOH).
Wei, K. *et al.*, *J. Org. Chem.*, 2005, **70**, 1164-1176 (*Nigramide P*)

Nigramide Q N15

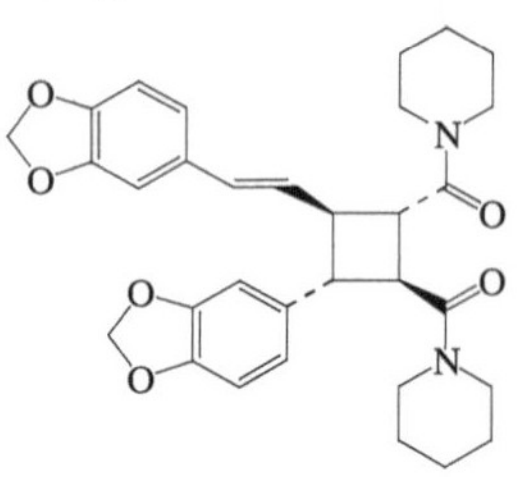

[847552-98-1]
$C_{32}H_{36}N_2O_6$ 544.648
Alkaloid from the roots of *Piper nigrum*. Oil. Racemic. λ_{max} 209 (log ε 4.61); 273 (log ε 4.16) (MeOH).
Wei, K. *et al.*, *J. Org. Chem.*, 2005, **70**, 1164-1176 (*Nigramide Q*)

Nigramide R N16

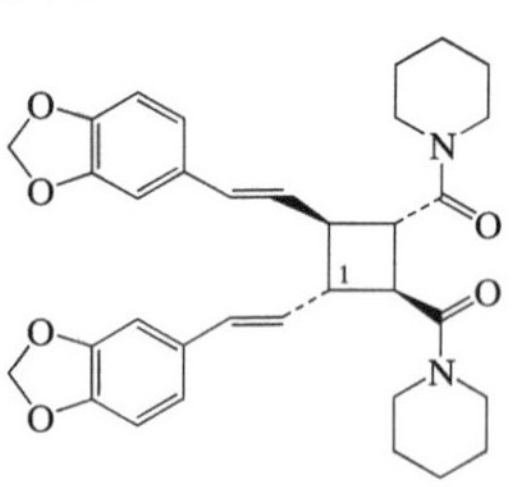

[847552-99-2]
$C_{34}H_{38}N_2O_6$ 570.686
Alkaloid from the roots of *Piper nigrum*. Oil. Racemic. λ_{max} 213 (log ε 4.56); 267 (log ε 4.3) (MeOH)
1-Epimer
Dipiperamide G
[1704700-69-5]
$C_{34}H_{38}N_2O_6$ 570.686
Alkaloid from the fruit of *Piper retrofractum*. Yellow powder. λ_{max} 208; 268; 305 (PDA).
Wei, K. *et al.*, *J. Org. Chem.*, 2005, **70**, 1164-1176 (*Nigramide R*).
Muharini, R. *et al.*, *Tetrahedron Lett.*, 2015, **56**, 2521-2525 (*Dipiperamide G*)

Nigramide S N17

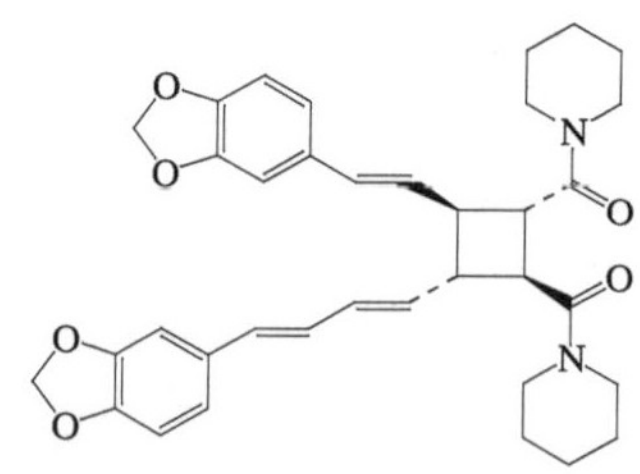

[847553-00-8]
$C_{36}H_{40}N_2O_6$ 596.724
Alkaloid from the roots of *Piper nigrum*. Oil. Racemic. λ_{max} 211 (log ε 4.44); 274 (log ε 4.1) (MeOH).
Wei, K. *et al.*, *J. Org. Chem.*, 2005, **70**, 1164-1176 (*Nigramide S*)

Noralashinol B N18

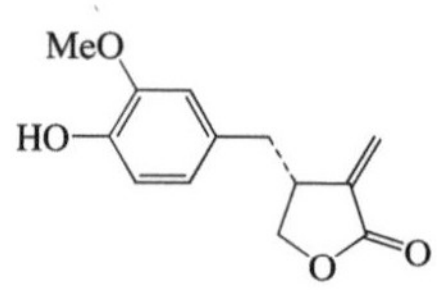

Absolute Configuration

Dihydro-4-(4-hydroxy-3-methoxybenzyl)-3-methylene-2(3H)-furanone
[2301983-62-8]
$C_{13}H_{14}O_4$ 234.251
Constit. of the stem bark of *Syringa pinnatifolia*. Exhibits weak cytotoxic activity. Yellow powder. $[\alpha]^{25}_{D}$ +57.4 (c, 0.1 in MeOH). λ_{max} 211 (log ε 4.17); 282 (log ε 0.97) (MeOH).
Zhang, R.-F. *et al.*, *J. Asian Nat. Prod. Res.*, 2017, **19**, 416-422 (*Noralashinol B, cd, activity*)

9-Nor-3,8′-lignan-3′,4,4′,5,5′,7,9′-heptol N19

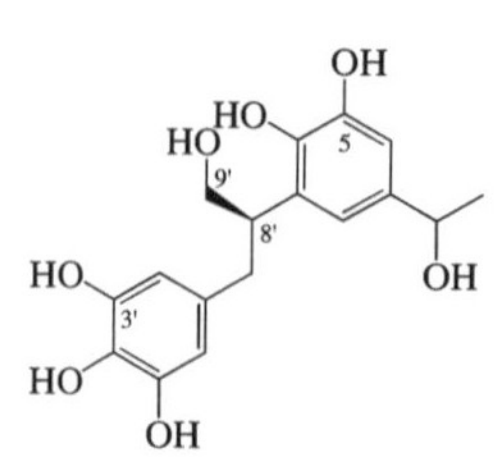

(8′R)-form

2-[2,3-Dihydroxy-5-(1-hydroxyethyl)phenyl]-3-(3,4,5-trihydroxyphenyl)-1-propanol.
3′,4,4′,5,5′,7,9′-Heptahydroxy-9-nor-3,8′-neolignan
$C_{17}H_{20}O_7$ 336.34
(8′R)-form
7-Ketone
3′,4,4′,5,5′,9′-Hexahydroxy-9-nor-3,8′-lignan-7-one
7-Ketone, 3′,5,5′-tri-Me ether, 9′-O-β-D-glucopyranoside
Moellenoside B
[1454889-98-5]
$C_{26}H_{34}O_{12}$ 538.546
Constit. of *Selaginella moellendorffii* and *Selaginella tamariscina*. Amorph. powder. Mp 246 - 248°. $[\alpha]^{20}_{D}$ -36.7 (c, 0.18 in $CHCl_3$). λ_{max} 207 (log ε 3.48); 280 (log ε 4.16) (MeOH)
(7ξ,8′ξ)-form
3′,5,5′-Tri-Me ether
3′,5,5′-Trimethoxy-9-nor-3,8′-lignan-4,4′,7,9′-tetrol
3′,5,5′-Tri-Me ether, 9′-O-β-D-glucopyranoside
Tamariscinoside B
[875056-64-7]
$C_{26}H_{36}O_{12}$ 540.562
Constit. of *Selaginella tamariscina*. Mp 241 - 242°. $[\alpha]^{25}_{D}$ -24 (c, 0.75 in H_2O). λ_{max} 210; 270 (H_2O).
Zheng, X.-K. *et al.*, *Yaoxue Xuebao*, 2004, **39**, 266-268 (*Tamariscinoside B*).
Feng, W.-S. *et al.*, *Acta Pharm. Sin. B*, 2011, **1**, 36-39 (*Moellenoside B*)

9-Nor-8,8′-lignan-3,3′,4,4′,8,9′-hexol N20

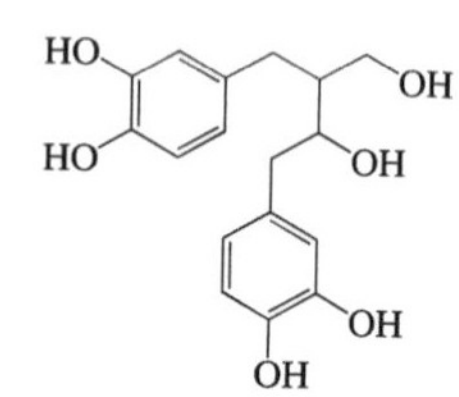

3,3′,4,4′,8,9′-Hexahydroxy-9-norlignan
$C_{17}H_{20}O_6$ 320.341
(8ξ,8′ξ)-form
3,4:3′,4′-Bis(methylene) ether
3,4:3′,4′-Bis(methylenedioxy)-9-nor-8,8′-lignan-8,9′-diol.
8,9′-Dihydroxy-3,4:3′,4′-bis(methylenedioxy)-9-norlignan
3,4:3′,4′-Bis(methylene) ether, 8,9′-di-Ac
Qinghaolignan B
$C_{23}H_{24}O_8$ 428.437
Constit. of *Artemisia annua*. Exhibits antifungal activity. Gum. $[\alpha]^{20}_{D}$ +0.5 (c, 0.86 in MeOH). λ_{max} 234 (log ε 2.78); 287 (log ε 2.73) (MeOH).

Li, K.-M. *et al.*, *Fitoterapia*, 2019, **134**, 323-328 (*Qinghaolignan B, activity*)

9-Nor-7,8′-lignan-3,3′,4,4′,5-pentol N21

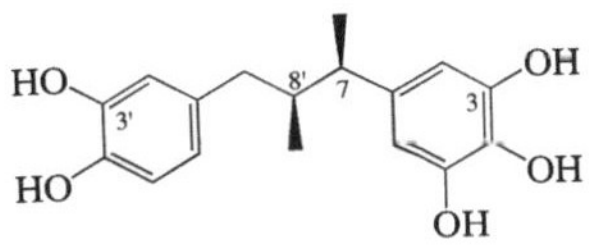

5-[3-(3,4-Dihydroxyphenyl)-1,2-dimethylpropyl]-1,2,3-benzenetriol. 3,3′,4,4′,5-Pentahydroxy-9-nor-7,8′-neolignan

$C_{17}H_{20}O_5$ 304.342

(7R,8′S)-form

3,3′-Di-Me ether

5-[3-(4-Hydroxy-3-methoxyphenyl)-1,2-dimethylpropyl]-3-methoxy-1,2-benzenediol. 3′,5-Dimethoxy-9-nor-7,8′-lignan-3,4,4′-triol. 3,4,4′-Trihydroxy-3′,5-dimethoxy-9-nor-7,8′-lignan

[1345855-24-4]

$C_{19}H_{24}O_5$ 332.396

Constit. of the seeds of *Myristica fragrans*. Moderate inhibitor of LPS-induced NO prodn.. Oil. $[\alpha]_D^{25}$ -8 (c, 0.1 in MeOH). λ_{max} 283 (MeOH).

Cuong, T.D. *et al.*, *Bioorg. Med. Chem. Lett.*, 2011, **21**, 6884-6887 (*Myristica fragrans constit, cd, activity*)

9′-Nor-8,8′-lign-7′-ene-3,3′,4,4′,9-pentol N22

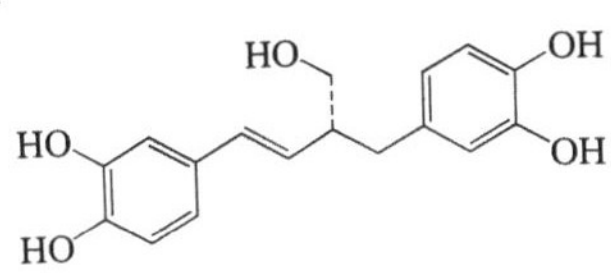

(R,E)-form

β-[2-(3,4-Dihydroxyphenyl)ethenyl]-3,4-dihydroxybenzenepropanol, CAS. 3,3′,4,4′,9-Pentahydroxy-9′-nor-8,8′-lign-7′-ene

$C_{17}H_{18}O_5$ 302.326

(R,E)-form [371759-01-2]

Oil. $[\alpha]_D^{28}$ -104.1 (c, 0.1 in $CHCl_3$) (>99% ee)

(R,Z)-form

Bis(methylene) ether

3,4:3′,4′-Bis(methylenedioxy)-9′-nor-8,8′-lign-7′-en-9-ol. 3,4:3′,4′-Bis(methylenedioxy)-9-hydroxy-9′-norlign-7′-ene. ***Artemisarin***

[1620559-34-3]

$C_{19}H_{18}O_5$ 326.348

Constit. of the leaves of *Artemisia argyi*. Yellowish oil. $[\alpha]_D^{18}$ -60.7 (c, 0.27 in $CHCl_3$). Config. not confirmed. λ_{max} 240 (log ε 3.04); 266 (log ε 3.03); 291 (log ε 3.02) ($CHCl_3$)

(S,E)-form

3,3′-Di-Me ether

3,3′-Dimethoxy-9′-nor-8,8′-lign-7′-ene-4,4′,9-triol. 4,4′,9-Trihydroxy-3,3′-dimethoxy-9′-norlign-7′-ene. ***Imperanene***

[163634-08-0]

$C_{19}H_{22}O_5$ 330.38

Constit. of the rhizomes of *Imperata cylindrica*. Platelet aggregation inhibitor. Oil. $[\alpha]_D^{25}$ +104 (c, 0.01 in $CHCl_3$). λ_{max} 270 (ε 9174) (MeOH).

Matsunaga, K. *et al.*, *J. Nat. Prod.*, 1995, **58**, 138-139 (*Imperanene*).
Shattuck, J.C. *et al.*, *Org. Lett.*, 2001, **3**, 3021-3023 (*synth*).
Doyle, M.P. *et al.*, *J. Org. Chem.*, 2002, **67**, 2954-2959 (*synth*).
Eklund, P.C. *et al.*, *J. Org. Chem.*, 2002, **67**, 7544-7546 (*R-form, synth*).
Davies, H.M.L. *et al.*, *Tetrahedron: Asymmetry*, 2003, **14**, 941-949 (*S-form, synth*).
Carr, J.A. *et al.*, *Org. Lett.*, 2004, **6**, 3297-3300 (*R-form, S-form, synth*).
Takashima, Y. *et al.*, *J. Org. Chem.*, 2009, **74**, 5920-5926 (*S-form, synth*).
Egi, M. *et al.*, *Angew. Chem., Int. Ed.*, 2013, **52**, 3654-3658 (*Imperanene, synth, resoln*).
Song, C. *et al.*, *Chem. Nat. Compd. (Engl. Transl.)*, 2014, **50**, 414-416 (*Artemisarin*)

9′-Nor-8,8′-lign-7′-ene-4,4′,7,9-tetrol N23

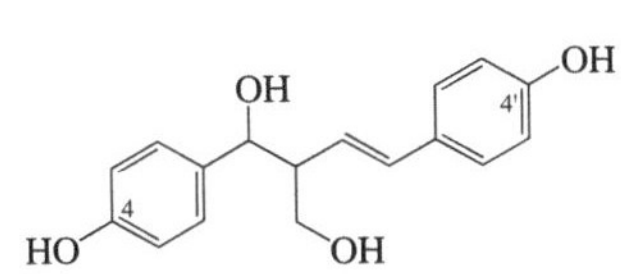

1-(4-Hydroxyphenyl)-2-[2-(4-hydroxyphenyl)ethenyl]-1,3-propanediol, CAS. 4,4′,7,9-Tetrahydroxy-9′-nor-7′-lignene. ***Yateresinol***

$C_{17}H_{18}O_4$ 286.327

(7ξ,7′E,8ξ)-form [73148-04-6]

Isol. from the heartwood of *Cryptomeria japonica* and *Libocedrus yateensis*

4,4′-Di-Me ether

Cryst. (Me_2CO). Mp 156 - 158°.

Erdtman, H. *et al.*, *Phytochemistry*, 1979, **18**, 1495-1500 (*Yateresinol*).
Takahashi, K. *et al.*, *J. Wood Sci.*, 2006, **52**, 134-139 (*Cryptomeria japonica constit*)

Norsalvianolic acid L N24

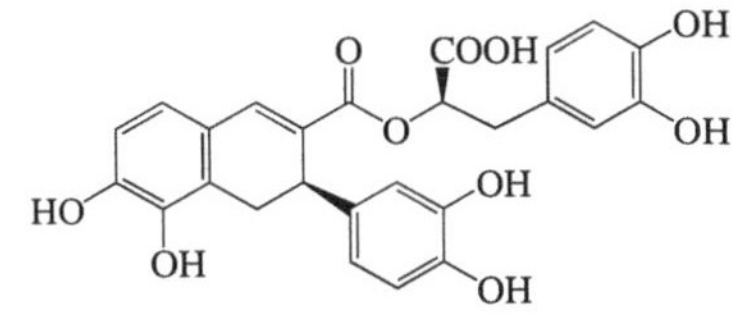

(R,R)-form

$C_{26}H_{22}O_{10}$ 494.452

(R,R)-form [1818285-56-1]

Isol. from Danshen injection, *Salvia miltiorrhizae*. Antioxidant. Yellow powder. $[\alpha]_D^{20}$ -200 (c, 0.21 in MeOH)

(S,S)-form [1818285-57-2]

Isol. from Danshen injection, *Salvia miltiorrhizae*. Antioxidant. Yellow powder. $[\alpha]_D^{20}$ +139.2 (c, 0.38 in MeOH).

Liu, J. *et al.*, *Molecules*, 2014, **19**, 1786-1794 (*Norsalvianolic acid L, cd, activity*)

Nuciferol A N25

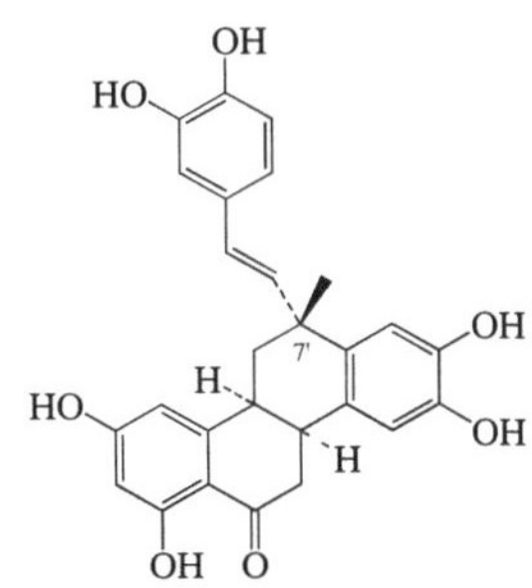

Absolute Configuration

12-[2-(3,4-Dihydroxyphenyl)ethenyl]-4b,10b,11,12-tetrahydro-2,3,7,9-tetrahydroxy-12-methyl-5H-chrysen-6-one

[2379666-76-7]

$C_{27}H_{24}O_7$ 460.482

Constit. of the endocarp of *Cocos nucifera*. Exhibits radical scavenging activity. Amorph. solid. $[\alpha]_D^{24}$ -44.9 (c, 0.29 in EtOH). λ_{max} 262 (no solvent reported)

7′-Epimer

Nuciferol B

[2379666-77-8]

$C_{27}H_{24}O_7$ 460.482

Constit. of the endocarp of *Cocos nucifera*. Exhibits radical scavenging and neuroprotective activities. Amorph. brown solid. $[\alpha]_D^{24}$ -181.5 (c, 0.37 in EtOH).

Elsbaey, M. *et al.*, *Tetrahedron Lett.*, 2019, **60**, 150948:1-4 (*Nuciferols A,B, activity*)

Nudibaccatumin A‡ **N26**

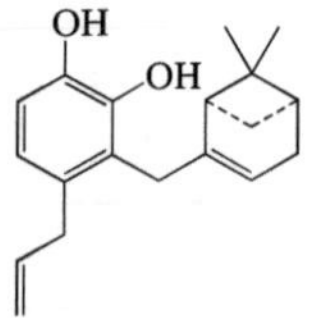

Absolute Configuration

[1685245-70-8]
$C_{19}H_{24}O_2$ 284.399
Assigned duplicated synonym by authors. Constit. of *Piper nudibaccatum*. Yellow oil. $[\alpha]_D^{23}$ -40.8 (c, 0.3 in MeOH). Isol. as a mixt. with Nudibaccatumin B, N27. λ_{max} 205 (log ε 3.18) (MeOH).
Liu, H.-X. *et al.*, *Tetrahedron*, 2015, **71**, 2369-2375 (*Nudibaccatumin A*)

Nudibaccatumin B‡ **N27**

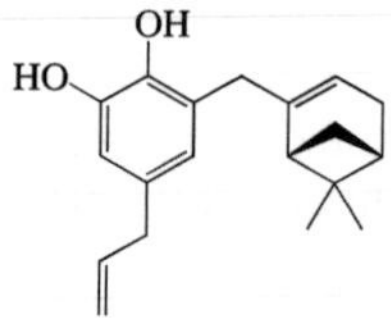

Absolute Configuration

[1685245-71-9]
$C_{19}H_{24}O_2$ 284.399
Assigned duplicated synonym by authors. Constit. of *Piper nudibaccatum*. Yellow oil. $[\alpha]_D^{14}$ -19.1 (c, 0.27 in MeOH). Isol. as a mixt. with Nudibaccatumin A, N26. λ_{max} 205 (log ε 3.18) ($CHCl_3$).
Liu, H.-X. *et al.*, *Tetrahedron*, 2015, **71**, 2369-2375 (*Nudibaccatumin B*)

Obovaaldehyde O1

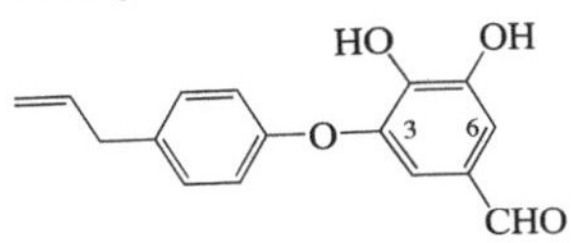

4,5-Dihydroxy-8,9-dinor-3,4′-oxylign-8′-en-7-al
[83864-76-0]
$C_{16}H_{14}O_4$ 270.284
Constit. of leaves of *Magnolia obovata*. Prisms (Me_2CO). Mp 162 - 163°. λ_{max} 203; 232; 335 (MeOH)

6-Hydroxy
4,5,6-Trihydroxy-8,9-dinor-3,4′-oxylign-8′-en-7-al.
2-Hydroxyobovaaldehyde
[1637562-80-1]
$C_{16}H_{14}O_5$ 286.283
Constit. of the fruit of *Magnolia obovata*. Inhibits LPS-induced NO prodn.. Dark brown oil.

Ito, K. *et al.*, *Chem. Pharm. Bull.*, 1982, **30**, 3347-3353 (*Obovaaldehyde*).
Seo, K.-H. *et al.*, *Planta Med.*, 2013, **79**, 1335-1340 (*Obovaaldehyde, 2-Hydroxyobovaaldehyde, activity*)

Obovatalignan I O2

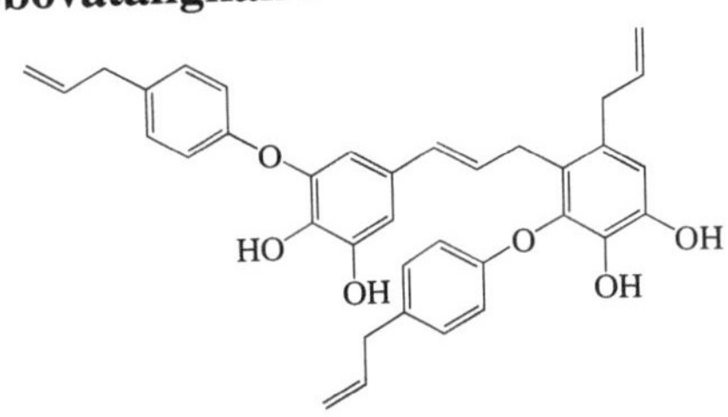

4-[3-[3,4-Dihydroxy-5-[4-(2-propenyl)phenoxy]phenyl]-2-propenyl]-5-(2-propenyl)-3-[4-(2-propenyl)phenoxy]-1,2-benzenediol, CAS
[2095702-48-8]
$C_{36}H_{34}O_6$ 562.662
Constit. of the fruit of *Magnolia obovata*. Amorph. brown powder.

Seo, K.-H. *et al.*, *Phytochemistry*, 2017, **136**, 133-140 (*Obovatalignan I*)

Obovatalignan A O3

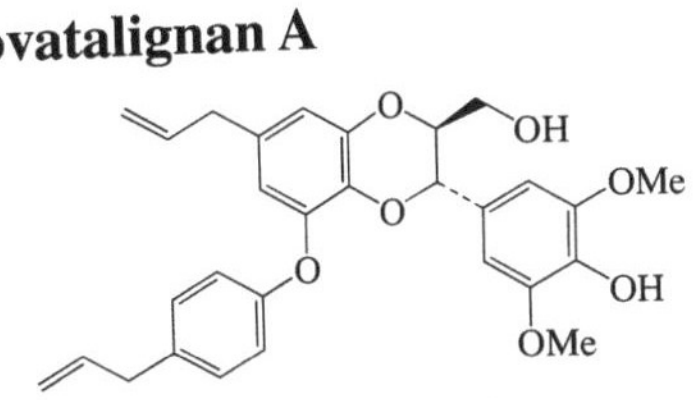

Absolute Configuration

Obovatoside A. *2,3-Dihydro-3-(4-hydroxy-3,5-dimethoxyphenyl)-7-(2-propenyl)-5-[4-(2-propenyl)phenoxy]-1,4-benzodioxin-2-methanol,* CAS
[1942113-39-4]
$C_{29}H_{30}O_7$ 490.552
Constit. of the fruit of *Magnolia obovata*. Exhibits neuroprotective effects. Amorph. powder. $[\alpha]_D^{18}$ +6.5 (c, 0.11 in MeOH). Assigned 2 synonyms. λ_{max} 226 (log ε 2.37); 273 (log ε 1.97) (MeOH).

Seo, K.-H. *et al.*, *Helv. Chim. Acta*, 2016, **99**, 411-415 (*Obovatalignan A, activity*)

Obovatalignan C O4

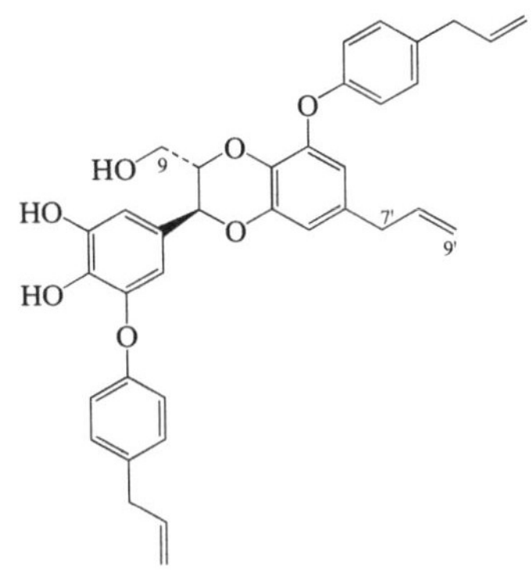

Absolute Configuration

5-[2,3-Dihydro-3-(hydroxymethyl)-7-(2-propenyl)-5-[4-(2-propenyl)phenoxy]-1,4-benzodioxin-2-yl]-3-[4-(2-propenyl)phenoxy]-1,2-benzenediol, CAS
[2095702-43-3]
$C_{36}H_{34}O_7$ 578.661
Constit. of the fruit of *Magnolia obovata*. Amorph. brown powder. $[\alpha]_D^{21}$ -0.7 (c, 0.78 in $CHCl_3$)

9-Me ether
Obovatalignan D
[2095702-44-4]
$C_{37}H_{36}O_7$ 592.688
Constit. of the fruit of *Magnolia obovata*. Amorph. brown powder. $[\alpha]_D^{21}$ -13.7 (c, 0.3 in $CHCl_3$)

$\Delta^{7'}$-Isomer, 9′-methoxy, 9-Me ether
Obovatalignan E
[2095702-45-5]
$C_{38}H_{38}O_8$ 622.714
Constit. of the fruit of *Magnolia obovata*. Amorph. brown powder. $[\alpha]_D^{21}$ +2.0 (c, 0.85 in $CHCl_3$)

$\Delta^{7'}$-Isomer, 9′-oxo, 9-Me ether
Obovatalignan F
[2095702-46-6]
$C_{37}H_{34}O_8$ 606.671
Constit. of the fruit of *Magnolia obovata*. Amorph. brown powder. $[\alpha]_D^{21}$ +2.1 (c, 0.29 in $CHCl_3$)

$\Delta^{7'}$-Isomer, 9′-oxo, 9-methoxy, 9-Me ether
Obovatalignan G
[2095702-47-7]
$C_{38}H_{36}O_9$ 636.697
Constit. of the fruit of *Magnolia obovata*. Amorph. brown powder. $[\alpha]_D^{21}$ +26.7 (c, 0.18 in $CHCl_3$).

Seo, K.-H. *et al.*, *Phytochemistry*, 2017, **136**, 133-140 (*Obovatalignans C-G, abs config*)

Obovatalignan H O5

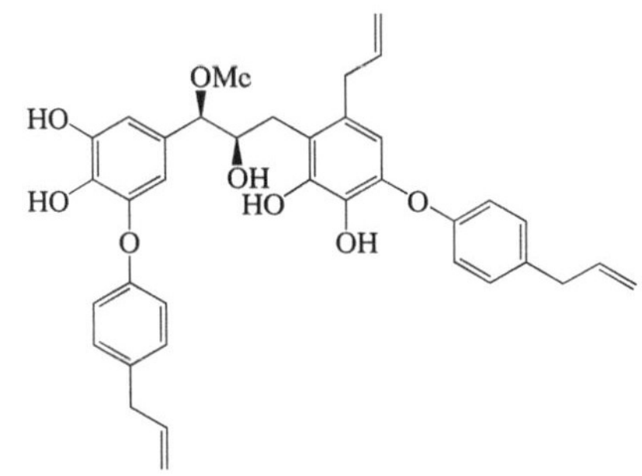

Relative Configuration

[2095702-59-1]
$C_{37}H_{38}O_8$ 610.703
Constit. of the fruit of *Magnolia obovata*. Amorph. brown powder. $[\alpha]_D^{21}$ +0.7 (c, 0.54 in $CHCl_3$).

Seo, K.-H. *et al.*, *Phytochemistry*, 2017, **136**, 133-140 (*Obovatalignan H*)

Obtulignolide O6

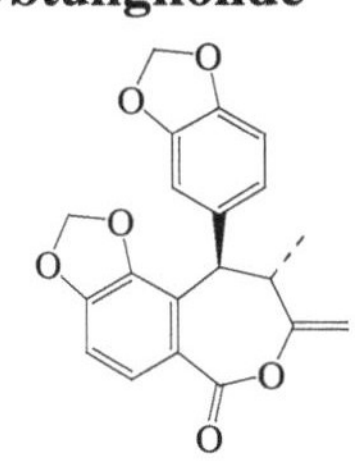

Relative Configuration

[375392-14-6]
$C_{20}H_{16}O_6$ 352.342
Constit. of the heartwood of *Chamaecyparis obtusa* var. *formosana*. Amorph.. $[\alpha]_D^{25}$ -2.8 (c, 0.25 in $CHCl_3$). λ_{max} 226 (log ε 3.91); 291 (log ε 3.57) (MeOH).

Kuo, Y.-H. *et al.*, *Tetrahedron Lett.*, 2001, **42**, 6731-6735 (*Obtulignolide*)

Obtunorlignan A O7

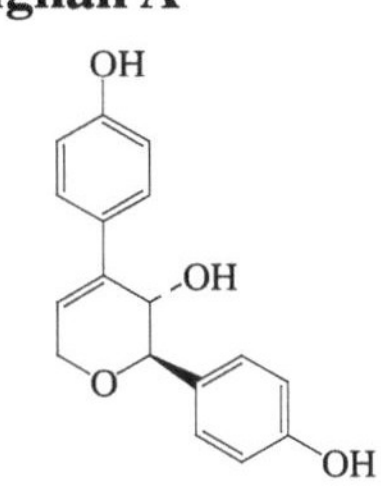

3,6-Dihydro-2,4-bis(4-hydroxyphenyl)-2H-pyran-3-ol. 7′,9-Epoxy-9′-nor-7,8′-lign-7-ene-4,4′,8′-triol.

7′,9-Epoxy-4,4′,8′-trihydroxy-9′-nor-7,8′-neolign-7-ene
[929195-27-7]
$C_{17}H_{16}O_4$ 284.311
Provisional norlignan. Biosynth. pathway unclear. Constit. of the heartwood of *Chamaecyparis obtusa* var. *formosana*. Amorph. solid. $[\alpha]_D^{20}$ -10.3 (c, 0.35 in MeOH). λ_{max} 225 (sh); 255 (log ε 4.01) (MeOH).
Chien, S.-C. *et al.*, *Tetrahedron Lett.*, 2007, **48**, 1567-1569 (*Obtunorlignan A*)

Officinaruminane A O8

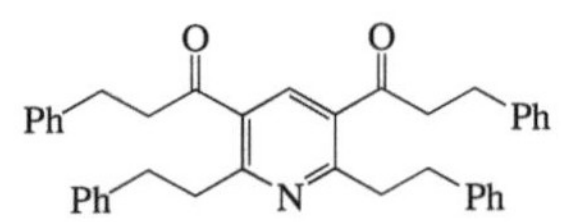

2,6-Bis(2-phenylethyl)-3,5-bis(3-phenylpropanoyl)pyridine. Officinarumane A (incorr.)
[1246282-66-5]
$C_{39}H_{37}NO_2$ 551.73
Alkaloid from the rhizomes of *Alpinia officinarum*. Oil. λ_{max} 202; 250 (sh) (MeOH).
An, N. *et al.*, *Food Chem.*, 2010, **119**, 513-517 (*Officinaruminane A*)

Officinaruminane B O9

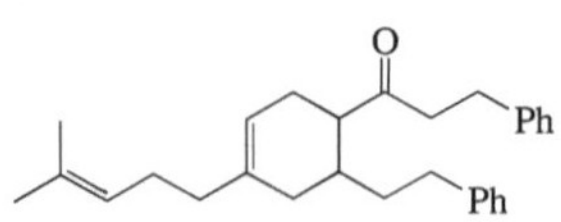

Officinarumane B (incorr.)
[1246282-67-6]
$C_{29}H_{36}O$ 400.606
Constit. of the rhizomes of *Alpinia officinarum*. Oil. λ_{max} 209; 268 (MeOH).
An, N. *et al.*, *Food Chem.*, 2010, **119**, 513-517 (*Officinaruminane B*)

Officinin A O10

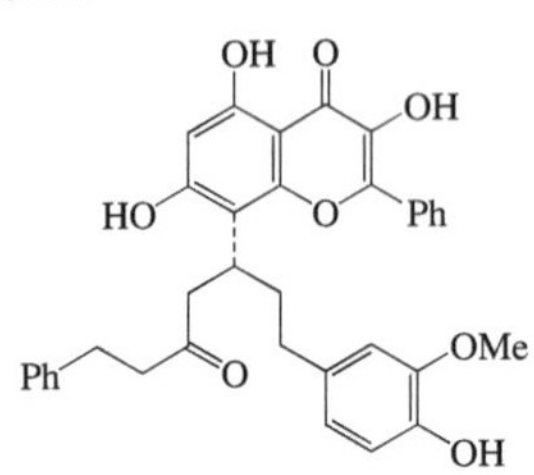

Absolute Configuration

[1258000-65-5]
$C_{35}H_{32}O_8$ 580.633
Constit. of the rhizomes of *Alpinia officinarum*. Amorph. yellow powder. $[\alpha]_D^{25}$ -2.4 (c, 0.16 in MeOH).
Zhao, L. *et al.*, *Chin. Chem. Lett.*, 2010, **21**, 194-196 (*Officinin A*)

Officinine B O11

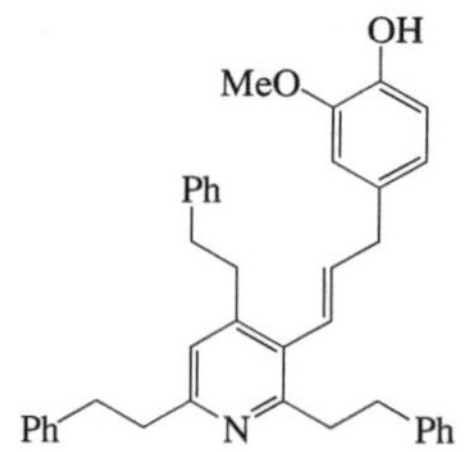

3-[3-(4-Hydroxy-3-methoxyphenyl)-1-propenyl]-2,4,6-tris(2-phenylethyl) pyridine
[1415715-02-4]
$C_{39}H_{39}NO_2$ 553.746
Alkaloid from the rhizomes of *Alpinia officinarum*. Oil.
Zhao, L. *et al.*, *Chem. Nat. Compd. (Engl. Transl.)*, 2012, **48**, 836-838 (*Officinine B*)

Ophiocerol O12

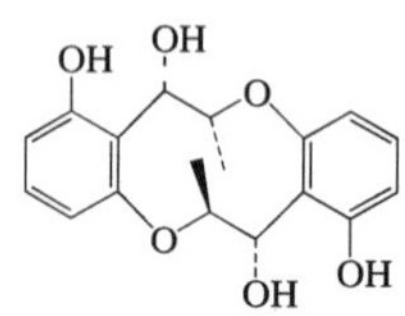

Relative Configuration

2′,8-Epoxy-2,8′-oxylignan-6,6′,7,7′-tetrol. 2′,8-Epoxy-6,6′,7,7′-tetrahydroxy-2,8′-oxyneolignan
[1244030-51-0]
$C_{18}H_{26}O_6$ 338.4
Isol. from *Ophioceras dolichostomum* YMF1.00988. Exhibits moderate antifungal activity. Amorph. yellow solid. $[\alpha]_D^{27}$ -15 (c, 0.8 in Me_2CO). λ_{max} 209 (log ε 4.33); 244 (log ε 3.68); 313 (log ε 3.46) (MeOH).
Dong, J.-Y. *et al.*, *Nat. Prod. Res.*, 2010, **24**, 1004-1012 (*Ophiocerol, activity*)

Origanine B O13

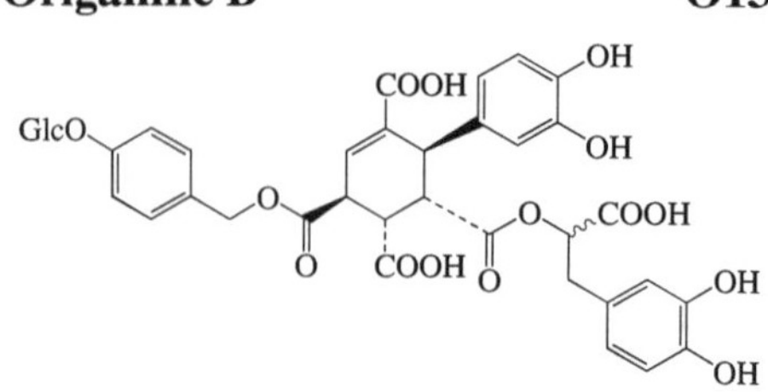

Absolute Configuration

[1353377-72-6]
$C_{38}H_{38}O_{20}$ 814.702
Constit. of the leaves of *Origanum vulgare*. λ_{max} 228; 282 (no solvent reported).
Liu, H. *et al.*, *J. Agric. Food Chem.*, 2012, **60**, 129-135 (*Origanine B*)

Origanine C O14

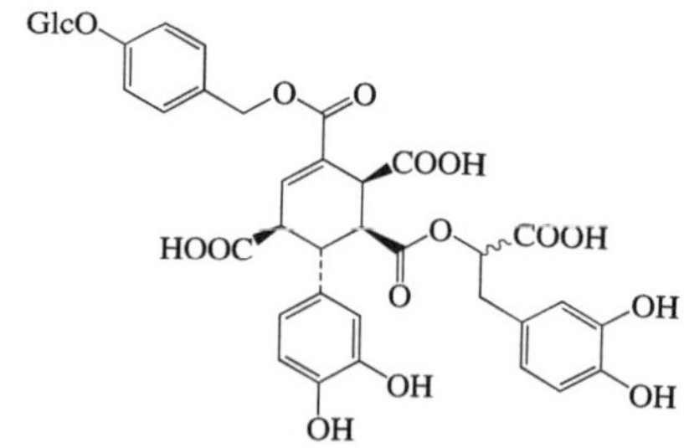

Absolute Configuration

[1353377-98-6]
$C_{38}H_{38}O_{20}$ 814.702
Constit. of the leaves of *Origanum vulgare*. λ_{max} 222; 282; 335 (no solvent reported).
Liu, H. *et al.*, *J. Agric. Food Chem.*, 2012, **60**, 129-135 (*Origanine C*)

Ovafolinin A O15

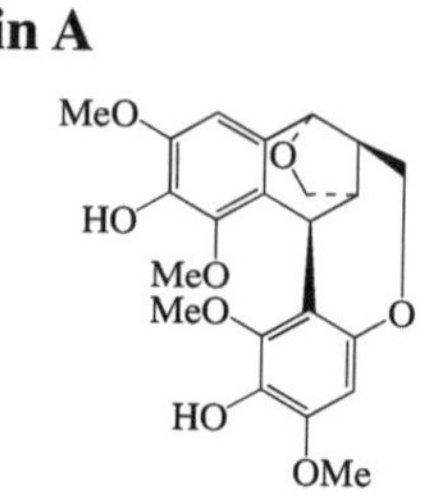

Relative Configuration

2,7′-Cyclo-2′,9:7,9′-diepoxy-3,4′,5,6′-tetramethoxy-8,8′-lignan-4,5′-diol. 2,9:7,9′-Diepoxy-4,5′-dihydroxy-3,4′,5,6′-tetramethoxy-2,7′-cyclolignan
[1252008-40-4]
$C_{22}H_{24}O_8$ 416.426
Constit. of the wood of *Lyonia ovalifolia*. Prisms. $[\alpha]_D^{25}$ -37.3 (c, 0.36 in MeOH). λ_{max} 243 (log ε 3.86); 283 (log ε 3.58) (MeOH).
Kashima, K. *et al.*, *Chem. Pharm. Bull.*, 2010, **58**, 191-194 (*Ovafolinin A, cryst struct*).
Davidson, S.J. *et al.*, *Angew. Chem., Int. Ed.*, 2017, **56**, 9483-9486 (*synth*).
Feng, X. *et al.*, *Chem. Commun.*, 2018, **54**, 7539-7541 (*synth*)

Ovafolinin B O16

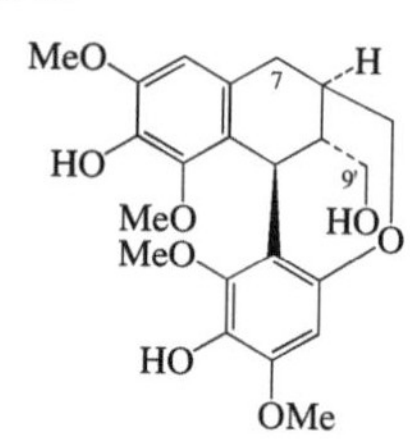

(+)-*form*

2,7′-Cyclo-6′,9-epoxy-2′,3,4′,5-tetramethoxy-8,8′-lignan-3′,4,9′-triol.

6′,9-Epoxy-3′,4,9′-trihydroxy-2′,3,4′,5-tetramethoxy-2,7′-cyclolignan
$C_{22}H_{26}O_8$ 418.442
(+)-form [1252008-41-5]
Constit. of *Lyonia ovalifolia* and *Sinocalamus affinis* (preferred genus name *Dendrocalamus*). Prisms. $[\alpha]^{25}_{D}$ +52 (c, 0.26 in MeOH). λ_{max} 207 (log ε 4.28); 235 (log ε 3.68); 282 (log ε 3.24) (MeOH)
9′-O-β-D-Glucopyranoside
[1428903-98-3]
$C_{28}H_{36}O_{13}$ 580.583
Lignan constit. of *Eurya japonica*. Potent antioxidant. Amorph. yellow solid. $[\alpha]^{25}_{D}$ +193.1 (c, 0.2 in MeOH). λ_{max} 283 (log ε 3.76) (MeOH)
7-Oxo
2,7′-Cyclo-6′,9-epoxy-3′,4,9′-trihydroxy-2′,3,4′,5-tetramethoxy-8,8′-lignan-7-one. 6′,9-Epoxy-3′,4,9′-trihydroxy-2′,3,4′,5-tetramethoxy-2,7′-cyclolignan-7-one. ***Ovafolinin C***
[1252008-43-7]
$C_{22}H_{24}O_9$ 432.425
Constit. of *Lyonia ovalifolia*. Amorph. yellowish solid. $[\alpha]^{25}_{D}$ +105.7 (c, 0.11 in MeOH). λ_{max} 290 (log ε 3.85) (MeOH)
(-)-form
9′-O-β-D-Glucopyranoside
[1428903-99-4]
Constit. of *Eurya japonica*. Potent antioxidant. Amorph. yellow solid. $[\alpha]^{25}_{D}$ -127 (c, 0.2 in MeOH). λ_{max} 283 (log ε 3.92) (MeOH).

Kashima, K. *et al.*, *Chem. Pharm. Bull.*, 2010, **58**, 191-194 (*Ovafolinins B,C*).
Xiong, L. *et al.*, *J. Nat. Prod.*, 2011, **74**, 1188-1200 (*Sinocalamus affinis constit*).
Kuo, L.-M.Y. *et al.*, *J. Nat. Prod.*, 2013, **76**, 580-587 (*Ovafolinin B glucosides*).
Davidson, S.J. *et al.*, *Angew. Chem., Int. Ed.*, 2017, **56**, 9483-9486 (*synth*).
Feng, X. *et al.*, *Chem. Commun.*, 2018, **54**, 7539-7541 (*synth*)

Ovafolinin D O17

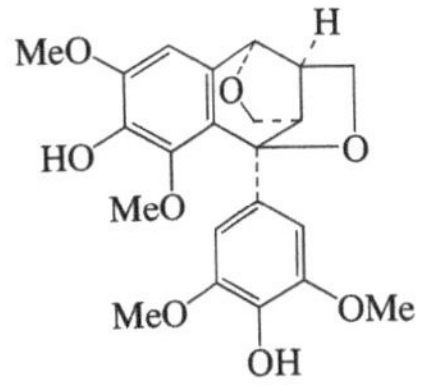

Absolute Configuration

[1252008-44-8]
$C_{22}H_{24}O_8$ 416.426
Constit. of the wood of *Lyonia ovalifolia*. Yellowish amorph. powder. $[\alpha]^{25}_{D}$ -33 (c, 0.09 in MeOH). λ_{max} 276 (log ε 3.07) (MeOH).

Kashima, K. *et al.*, *Chem. Pharm. Bull.*, 2010, **58**, 191-194 (*Ovafolinin D*)

Oxaguianin O18

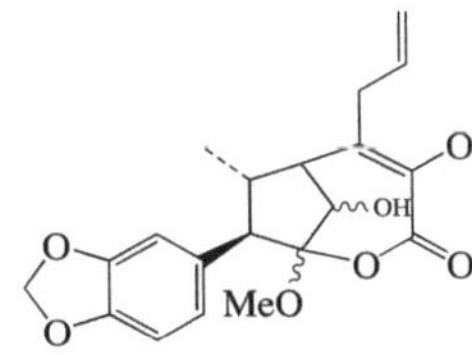

Relative Configuration

[104701-06-6]
$C_{21}H_{24}O_7$ 388.416
Constit. of *Licaria puchury-major* and *Ocotea aciphylla*. Cryst. (MeOH). Mp 174 - 176°. λ_{max} 234; 288 (MeOH).

D'arc Felício, J. *et al.*, *Phytochemistry*, 1986, **25**, 1707-1710 (*Oxaguianin*)

2-Oxatricyclo[13.3.1.1^{3,7}]nonadeca-1(19),3,5,7(20),8,15,17-heptaene-4,12,16-triol O19

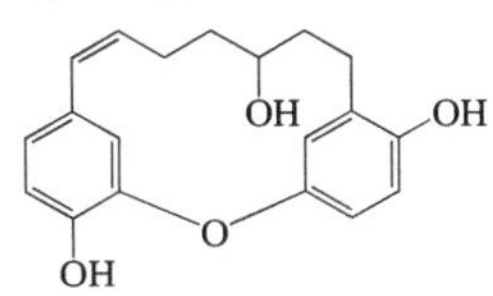

[2227432-50-8]
$C_{19}H_{20}O_4$ 312.365
Constit. of the kernels of *Corylus avellana* infected with 'Cimiciato'. Intense bitter taste.

Singldinger, B. *et al.*, *J. Agric. Food Chem.*, 2018, **66**, 4660-4673 (*Corylus avellana constit, props*)

9′,7-Oxidolanceolatanin C O20

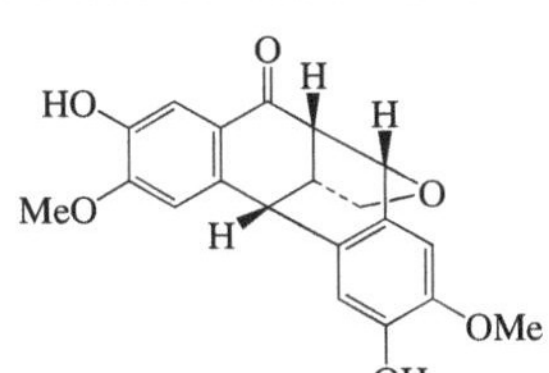

Relative Configuration

2,7′:2′,9-Bicyclo-9,9′-epoxy-5,5′-dihydroxy-4,4′-dimethoxy-8,8′-lignan-7-one. 9,9′-Epoxylanceolatanin C
[2106917-96-6]
$C_{20}H_{18}O_6$ 354.358
Constit. of the roots of *Taxus wallichiana*. Amorph. solid. $[\alpha]^{25}_{D}$ +35.6 (c, 0.01 in MeOH).

Dang, P.H. *et al.*, *J. Nat. Prod.*, 2017, **80**, 1876-1882 (*9′,7-Oxidolanceolatanin C*)

Oxokadsurane O21

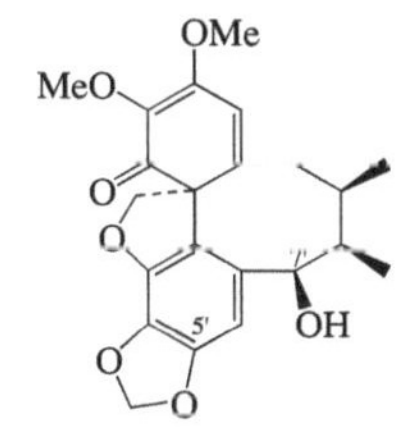

Absolute Configuration

7′-Hydroxy-4,5-dimethoxy-4′,5′-methylenedioxy-2,2′-cyclo-2a-homo,3′-epoxy-8,8′-lignan-3(2H)-one. Deacylkadsulignan H. Deacylschiarisanrin A
$C_{22}H_{24}O_7$ 400.427
The abs configs. of the Oxokadsuranes were assigned only tentatively but appear to be the same as for the Kadsulignans. The opt. rotns. of Kadsulignan I and Propoxyoxokadsurane are similar
Ac
Acetoxyoxokadsurane. *Schiarisanrin B*
[130252-43-6]
[188944-01-6] (deleted)
$C_{24}H_{26}O_8$ 442.464
Isol. from *Kadsura coccinea* and *Schizandra arisanensis*. Amorph. yellow powder or prisms. Mp 98 - 100°. $[\alpha]^{31}_{D}$ -114.7 (c, 0.09 in $CHCl_3$). $[\alpha]_D$ -173 (c, 0.3 in $CHCl_3$) (Schiarisanrin B)
Propanoyl
Kadsulignan I.
Propoxyoxokadsurane
[130252-42-5]
$C_{25}H_{28}O_8$ 456.491
Constit. of the seeds of a *Kadsura* sp. and stems of *Kadsura coccinea*. Pale yellow cryst. (Et_2O/petrol) or amorph. yellow powder. Mp 119 - 120°. $[\alpha]^{23}_{D}$ -100 (c, 0.72 in MeOH) (Kadsulignan I). $[\alpha]^{31}_{D}$ -90 (c, 0.15 in $CHCl_3$) (Propoxyoxokadsurane). λ_{max} 218 (log ε 4.09); 335 (log ε 3.01) (MeOH)
Butanoyl
Kadsulignan H
[144049-91-2]
$C_{26}H_{30}O_8$ 470.518
Constit. of the seeds of a *Kadsura* sp.. Shows weak activity against leukaemia P-388. Pale yellow cryst.

(Et$_2$O/petrol). Mp 137 - 138°. $[\alpha]^{19}_{D}$ -94 (c, 0.23 in CHCl$_3$). λ_{max} 220 (log ε 4.44); 250 (sh); 332 (log ε 3.35) (MeOH)

2S-Methylbutanoyl

Kadsulignan J. *Schiarisanrin A*
[130252-41-4]
[188944-00-5] (deleted)
$C_{27}H_{32}O_8$ 484.545
Constit. of *Schizandra arisanensis* and the seeds of a *Kadsura* sp.. Pale yellow cryst. (Et$_2$O/petrol) or prisms. Mp 175 - 177° (172-173°). $[\alpha]^{15}_{D}$ -28 (c, 1.49 in MeOH). $[\alpha]_D$ -133 (c, 0.3 in CHCl$_3$) (Schiarisanrin A). λ_{max} 220 (log ε 4.63); 250 (sh); 332 (log ε 3.54) (MeOH)

Angeloyl

Heteroclitin D
[140369-76-2]
$C_{27}H_{30}O_8$ 482.529
Constit. of the stems of *Kadsura heteroclita*, roots of *Kadsura coccinea* and from *Schisandra rubriflora*. Yellowish powder. $[\alpha]^{25}_{D}$ -93.9 (c, 1.02 in CHCl$_3$). λ_{max} 221 (ε 32700); 276 (ε 2200); 332 (ε 4000) (MeOH)

Tigloyl

Heteroclitin H
[952287-15-9]
$C_{27}H_{30}O_8$ 482.529
Constit. of the stems of *Kadsura heteroclita*. Needles (petrol/EtOAc). Mp 171 - 173°. $[\alpha]^{22}_{D}$ -30 (c, 0.06 in MeOH). λ_{max} 217 (log ε 4.83); 335 (log ε 3.73) (MeOH)

Benzoyl

Benzoyloxokadsurane. *Schiarisanrin C*
[130252-44-7]
[188944-02-7] (deleted)
$C_{29}H_{28}O_8$ 504.535
Isol. from *Kadsura coccinea* and *Schizandra arisanensis*. Cytotoxic agent active against human nasopharynx KB, colon COLO-205, hepatoma HEPA-2B and cervical HeLa cancer cell lines. Amorph. yellow powder. Mp 130 - 133°. $[\alpha]^{31}_{D}$ -103.3 (c, 0.06 in CHCl$_3$). $[\alpha]_D$ -146 (c, 0.3 in CHCl$_3$) (Schiarisanrin C)

E-Cinnamoyl

Schiarisanrin D
[188943-88-6]
$C_{31}H_{30}O_8$ 530.573
Constit. of *Schizandra arisanensis*. Amorph. yellow powder. Mp 133 - 135°. $[\alpha]_D$ -126 (c, 0.3 in CHCl$_3$)

4′,5′-O-Demethylene, 7′-angeloyl, 4′-O-β-D-glucopyranoside

Longipedunculatin B
[2408236-50-8]
$C_{32}H_{40}O_{13}$ 632.659
Constit. of the roots of *Kadsura longipedunculata*. Strong inhibitor of NO prodn.. Amorph. yellow powder. $[\alpha]^{20}_{D}$ -11 (c, 0.1 in MeOH). λ_{max} 215 (log ε 2.57); 332 (log ε 1.48) (MeOH)

4′,5′-O-Demethylene, 7′-angeloyl, 5′-O-β-D-glucopyranoside

Longipedunculatin A
[2408236-46-2]
$C_{32}H_{40}O_{13}$ 632.659
Constit. of the roots of *Kadsura longipedunculata*. Amorph. yellow powder. $[\alpha]^{20}_{D}$ -12 (c, 0.3 in MeOH). λ_{max} 216 (log ε 2.58); 330 (log ε 1.47) (MeOH)

4′,5′-O-Demethylene, 7′-O-(2ξ-methylbutanoyl), 5′-O-β-D-glucopyranoside

Longipedunculatin C
[2408236-51-9]
$C_{32}H_{42}O_{13}$ 634.675
Constit. of the roots of *Kadsura longipedunculata*. Amorph. yellow powder. $[\alpha]^{20}_{D}$ -8 (c, 0.2 in MeOH). λ_{max} 215 (log ε 2.25); 329 (log ε 1.18) (MeOH)

Stereoisomer, angeloyl

Schiarisanrin E
[697228-90-3]
$C_{27}H_{30}O_8$ 482.529
Constit. of the stems of *Schizandra arisanensis*. Pale yellow solid.

Li, L.-N. *et al., Phytochemistry*, 1990, **29**, 2730 (*Oxokadsuranes*).
Chen, D.-F. *et al., Phytochemistry*, 1992, **31**, 629-632 (*Heteroclitin D*).
Liu, J.-S. *et al., Phytochemistry*, 1992, **31**, 1379-1382 (*Kadsulignans*).
Kuo, Y.-H. *et al., J. Org. Chem.*, 1997, **62**, 3242-3245 (*Schiarisanrins A-D*).
Wu, M.-D. *et al., Chem. Pharm. Bull.*, 2003, **51**, 1233-1236 (*Schiarisanrin E*).
Chen, M. *et al., J. Asian Nat. Prod. Res.*, 2006, **8**, 643-648 (*Heteroclitin H*).
Liu, J. *et al., J. Nat. Prod.*, 2019, **82**, 2842-2851 (*Longipedunculatins A,B,C, cd, activity*)

Oxokadsuranol O22

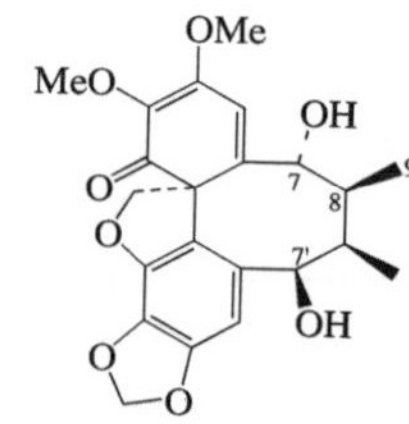

Absolute Configuration

7,7′-Dihydroxy-4,5-dimethoxy-4′,5′-methylenedioxy-2,2′-cyclo-2a-homo,3′-epoxy-8,8′-lignan-3(2H)-one
$C_{22}H_{24}O_8$ 416.426

7′-O-(2ξ-Methylbutanoyl)

Isovaleroyloxokadsuranol
[130252-45-8]
$C_{27}H_{32}O_9$ 500.544
Isol. from stems of *Kadsura coccinea*. Cryst.. Mp 200 - 202°. $[\alpha]^{31}_{D}$ +40.3 (c, 0.05 in CHCl$_3$)

7′-Angeloyl

Heteroclitin E
[140369-77-3]
$C_{27}H_{30}O_9$ 498.528
Constit. of the stems of *Kadsura heteroclita*. Amorph. powder. $[\alpha]^{25}_{D}$ -38 (c, 0.79 in MeOH). λ_{max} 219 (ε 42100); 279 (ε 2600); 329 (ε 3400) (MeOH)

7′-Benzoyl

Benzoyloxokadsuranol
[130252-47-0]
$C_{29}H_{28}O_9$ 520.534
Constit. of the stems of *Kadsura longipedunculata*. Amorph. yellow powder. $[\alpha]^{22}_{D}$ -6.9 (c, 0.15 in MeOH). Has S_{axial}-Config.. λ_{max} 221 (log ε 1.91) (no solvent reported)

7-Ketone, 7′-tigloyl

Heteroclitin R
$C_{27}H_{28}O_9$ 496.512
Constit. of stems of *Kadsura heteroclita*. Amorph. yellow powder. $[\alpha]^{25}_{D}$ +7.9 (c, 0.29 in MeOH). λ_{max} 200 (4.78) (MeOH)

8,9-Didehydro, 8α,9-epoxide, 7-benzoyl, 7′-angeloyl

Heteroclitin O
[1007566-05-3]
$C_{34}H_{32}O_{11}$ 616.619
Constit. of the stems of *Kadsura heteroclita*. Yellow needles. Mp 190 - 191°. $[\alpha]^{20}_{D}$ -134 (c, 0.68 in MeOH). λ_{max} 221 (log ε 3.77); 278 (log ε 2.89); 328 (log ε 3.13) (MeOH).

Li, L.-N. *et al.*, *Phytochemistry*, 1990, **29**, 2730-2732 (*Isovaleroyloxokadsuranol*).
Chen, D.-F. *et al.*, *Phytochemistry*, 1992, **31**, 629-632 (*Heteroclitin E*).
Li, L. *et al.*, *Helv. Chim. Acta*, 2004, **87**, 2943-2947 (*Schisandra rubiflora constit*).
Xu, L.-J. *et al.*, *Helv. Chim. Acta*, 2008, **91**, 220-226 (*Heteroclitin O*).
Chen, M. *et al.*, *Chin. J. Nat. Med.*, 2014, **12**, 689-692 (*Heteroclitin R*).
Fang, L. *et al.*, *Phytochem. Lett.*, 2014, **9**, 158-162 (*Kadsura coccinea constit*).
Zhang, J. *et al.*, *Chem. Nat. Compd. (Engl. Transl.)*, 2018, **54**, 837-840 (*benzoyl oxokadsuranol*)

3,3′-Oxyligna-7,7′-diene-9,9′-dioic acid O23

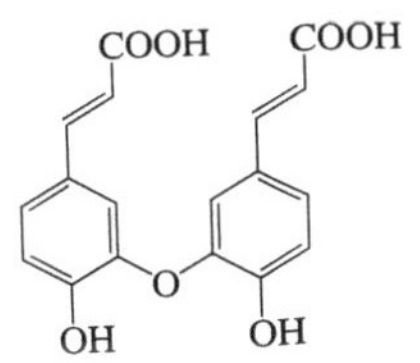

(*E,E*)-*form*

3,3′-Oxybis[4-hydroxycinnamic acid]. 3-O-3′-Dicaffeic acid
$C_{18}H_{14}O_7$ 342.303
(E,E)-form [1345675-33-3]
Constit. of the leaves of *Salvia lanigera*. $[\alpha]^{25}_{D}$ -10.7 (c, 1 in $CHCl_3$). Light yellow powder. λ_{max} 291; 325 (MeOH)
Di-Me ether, dioctyl ester
Dioctyl 4,4′-di-O-methyldicaffeate
[175991-52-3]
$C_{36}H_{50}O_7$ 594.789
Constit. of the roots of *Salvia forskahlei*. λ_{max} 234 (ε 15850); 325 (ε 1585) (MeOH) (Berdy)
(Z,Z)-form
Di-Me ether, dioctyl ester
[175991-51-2]
Constit. of the roots of *Salvia forskhalei*. λ_{max} 234 (log ε 4.2); 293 (sh); 325 (log ε 3.2) (MeOH).

Ulubelen, A. *et al.*, *Phytochemistry*, 1996, **42**, 145-147 (*Salvia forskahlei esters*).
Shaheen, U.Y. *et al.*, *Pharmacogn. J.*, 2011, **3**, 36-48 (*Salvia lanigera constit*)

4,4′-Oxyligna-7,7′-diene-9,9′-dioic acid O24

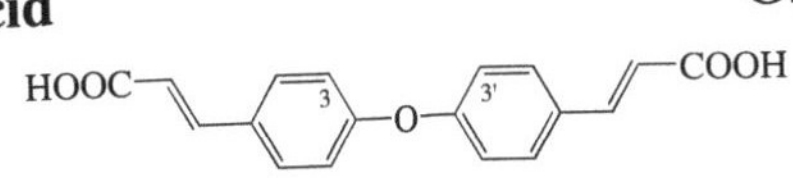

(*E,E*)-*form*

4,4′-Oxybis[3-phenyl-2-propenoic acid]. 4,4′-Oxybiscinnamic acid. 4,4′-Bis(2-carboxyvinyl)diphenyl ether
$C_{18}H_{14}O_5$ 310.305
(E,E)-form [135791-17-2]
Constit. of the rhizomes of *Curcuma chuanyujin*. Yellow powder. λ_{max} 413 (MeOH)
3-Methoxy
3-Methoxy-4,4′-oxyligna-7,7′-diene-9,9′-dioic acid
[300849-02-9]
$C_{19}H_{16}O_6$ 340.331
Constit. of the rhizomes of *Curcuma chuanyujin*. Yellow powder. λ_{max} 417 (MeOH)
3-Methoxy, di-Me ester
[943895-26-9]
$C_{21}H_{20}O_6$ 368.385
Constit. of a mangrove-derived *Penicillium* sp.
3,3′-Dimethoxy
4,4′-Oxybis[3-(3-methoxyphenyl)-2-propenoic acid]. 4,4′-Oxybis[3-methoxycinnamic acid]. 3,3′-Dimethoxy-4,4′-oxyligna-7,7′-diene-9,9′-dioic acid
[300849-01-8]
$C_{20}H_{18}O_7$ 370.357
Constit. of *Curcuma chuanyujin* and *Brassica fruticulosa*. Weak plant growth inhibitor. Yellow powder. λ_{max} 424 (MeOH)
3,3′-Dimethoxy, 7,8-dihydro
3,3′-Dimethoxy-4,4′-oxylign-7-ene-9,9′-dioic acid
3,3′-Dimethoxy, 7,8-dihydro, di-Me ester
[887917-81-9]
$C_{22}H_{24}O_7$ 400.427
Constit. of the leaves of *Cestrum parqui*. Oil
(E,Z)-form [2102516-98-1]
Constit. of fruit of *Cordia latifolia*. Amorph. powder (MeOH). Mp 107 - 108°. λ_{max} 291 (log ε 3.8); 332 (log ε 3.0) (MeOH).

Huang, J. *et al.*, *Chem. Pharm. Bull.*, 2000, **48**, 1228-1229 (*Curcuma chuanyujin constits*).
Cutillo, F. *et al.*, *J. Agric. Food Chem.*, 2003, **51**, 6165-6172 (*Brassica fruticulosa constit*).
D'Abrosca, B. *et al.*, *Nat. Prod. Res.*, 2006, **20**, 293-298 (*Cestrum parqui constit*).
Li, X. *et al.*, *Tianran Chanwu Yanjiu Yu Kaifa*, 2007, **19**, 4-6 (*3-methoxy di-Me ester*).
Fatima, M. *et al.*, *Chem. Nat. Compd. (Engl. Transl.)*, 2017, **53**, 432-435 (*E,Z-form*)

3,3′-Oxyligna-8,8′-diene-4,4′-diol O25

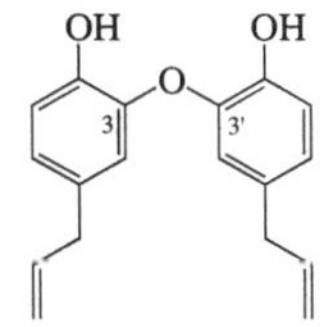

5,5′-Diallyl-2,2′-dihydroxybiphenyl ether. 2,2′-Oxybis[4-(2-propenyl) phenol]. 4,4′-Dihydroxy-3,3′-oxyneoligna-8,8′-diene. ***Isoobovatol***
[1637576-07-8]
$C_{18}H_{18}O_3$ 282.339
Constit. of the fruit of *Magnolia obovata*. Oil.

Baek, Y.-S. *et al.*, *Chem. Nat. Compd. (Engl. Transl.)*, 2016, **52**, 986-988 (*Isoobovatol*)

3,4′-Oxyligna-8,8′-diene-4,5-diol O26

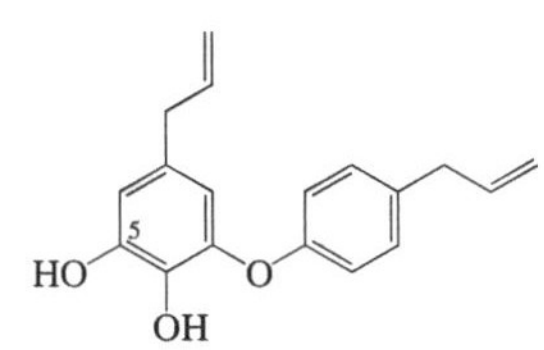

5-(2-Propenyl)-3-[4-(2-propenyl) phenoxy]-1,2-benzenediol. 4′,5-Diallyl-2,3-dihydroxydiphenyl ether. 4,5-Dihydroxy-3,4′-oxyneoligna-8,8′-diene. ***Obovatol***
[83864-78-2]
$C_{18}H_{18}O_3$ 282.339
Constit. of *Magnolia obovata*, *Manglietia hookeri* and *Streblus asper*. Exhibits antibacterial activity. NO prodn. inhibitor. ACAT inhibitor. Oil. λ_{max} 206; 272 (MeOH)
5-Me ether
4′,5-Diallyl-2-hydroxy-3-methoxydiphenyl ether. 5-Methoxy-3,4′-oxyligna-8,8′-dien-4-ol. 4-Hydroxy-5-methoxy-3,4′-oxyligna-8,8′-diene.
5-O-Methylobovatol
[122738-75-4]
$C_{19}H_{20}O_3$ 296.366
Constit. of the bark of *Magnolia henryi*. Gum.

Ito, K. *et al.*, *Chem. Pharm. Bull.*, 1982, **30**, 3347-3353 (*Obovatol*).
Kijjoa, A. *et al.*, *Phytochemistry*, 1989, **28**, 1284-1286 (*5-O-Methylobovatol*).
Kwon, B.-M. *et al.*, *Planta Med.*, 1997, **63**, 550-551 (*Obovatol, activity*).
Choi, M.-S. *et al.*, *Eur. J. Pharmacol.*, 2007, **556**, 181-189 (*Obovatol, activity*).
Choi, W.-S. *et al.*, *J. Antibiot.*, 2017, **70**, 1065-1069 (*Obovatol, activity*)

4,4′-Oxyligna-8,8′-diene-2,3′-diol O27

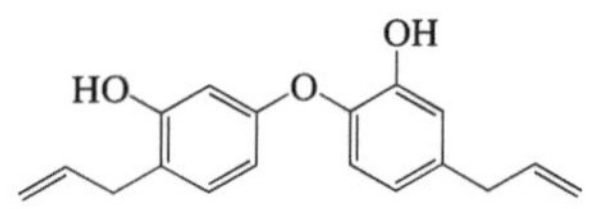

2-[3-Hydroxy-4-(2-propenyl) phenoxy]-5-(2-propenyl)phenol, CAS. *2,3′-Dihydroxy-4,4′-di-2-propenyldiphenyl ether. 4,4′-Diallyl-2,3′-dihydroxydiphenyl ether. 2,3′-Dihydroxy-4,4′-oxyneoligna-8,8′-diene*

[135547-25-0]

$C_{18}H_{18}O_3$ 282.339

Constit. of *Magnolia virginiana* and *Magnolia obovata*. Antifungal and cytotoxic agent. Pale yellow oil. λ_{max} 270 (ε 29900); 290 (ε 6210) (MeOH). [base] λ_{max} 206 (ε 24300); 226 (ε 28200); 319 (ε 7780) (MeOH/NaOH) (Berdy).

Nitao, J.K. *et al.*, *Phytochemistry*, 1991, **30**, 2193-2195 (*Magnolia virginiana constit*).
Chandra, A. *et al.*, *Planta Med.*, 1995, **61**, 192-195 (*Magnolia virginiana constit, activity*).
Johnson, K.S. *et al.*, *J. Chem. Ecol.*, 1996, **22**, 1955-1969 (*activity*).
Kim, Y.-K. *et al.*, *Planta Med.*, 1999, **65**, 291-292 (*Magnolia obovata constit*)

7,9′-Oxyligna-7′,8-diene-4,4′-diol O28

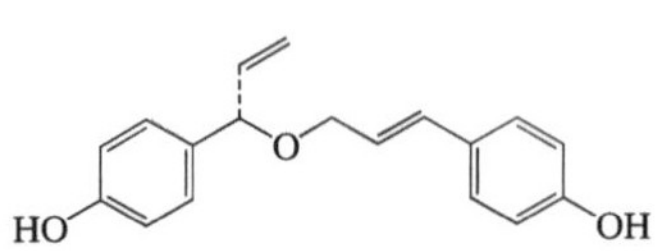

4,4′-Dihydroxy-7,9′-oxyneoligna-7′,8-diene

$C_{18}H_{18}O_3$ 282.339

(S,E)-form

Galanganol D [2055586-04-2]

Oil. $[\alpha]^{25}_{D}$ -17.4 (c, 1.5 in $CHCl_3$). λ_{max} 231 (log ε 4.99); 261 (log ε 4.26) (MeOH)

Di-Ac

4,4′-Diacetoxy-7,9′-oxyligna-7′,8-diene. ***Diacetylgalanganol D***

[2055636-88-7]

$C_{22}H_{22}O_5$ 366.413

Constit. of the fruit of *Alpinia galanga*. Melanogenesis inhibitor. Pale yellow oil. $[\alpha]^{24}_{D}$ -6.2 (c, 0.55 in $CHCl_3$). λ_{max} 225 (log ε 3.62); 271 (log ε 3.19) (MeOH).

Manse, Y. *et al.*, *Bioorg. Med. Chem.*, 2016, **24**, 6215-6224 (*Galanganol D, Diacetylgalanganol D, synth, activity*)

3,4′-Oxyligna-8,8′-diene-3′,4,5,7-tetrol O29

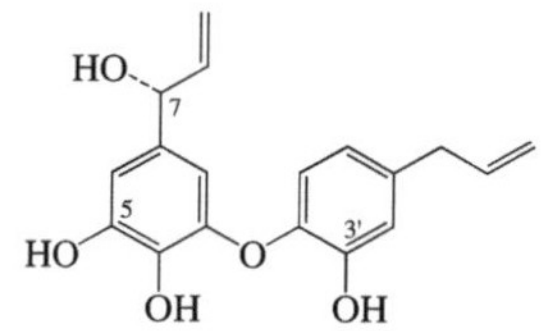

3′,4,5,7-Tetrahydroxy-3,4′-oxyneoligna-8,8′-diene

$C_{18}H_{18}O_5$ 314.337

(R)-form

3′,5-Di-Me ether

3′,5-Dimethoxy-3,4′-oxyligna-8,8′-diene-4,7-diol. 4,7-Dihydroxy-3′,5-dimethoxy-3,4′-oxyligna-8,8′-diene

[1695562-42-5]

$C_{20}H_{22}O_5$ 342.391

Constit. of the twigs of *Nectandra leucantha*. Brownish oil. $[\alpha]^{25}_{D}$ -0.01 (c, 0.01 in MeOH). λ_{max} 275 (log ε 4.0); 285 (log ε 3.9); 325 (log ε 3.7) (MeOH)

3′,4,5-Tri-Me ether

3′,4,5-Trimethoxy-3,4′-oxyligna-8,8′-dien-7-ol. 7-Hydroxy-3′,4,5-trimethoxy-3,4′-oxyligna-8,8′-diene

[2108673-67-0]

$C_{21}H_{24}O_5$ 356.418

Constit. of twigs of *Nectandra leucantha*. Exhibits moderate cytotoxic activity. Brownish oil. $[\alpha]^{25}_{D}$ +1.6 (c, 0.15 in MeOH). λ_{max} 222 (log ε 3.7); 278 (log ε 2.9) (MeOH)

7-Ketone

3′,4,5-Trihydroxy-3,4′-oxyligna-8,8′-dien-7-one

7-Ketone, 3′,5-di-Me ether

4-Hydroxy-3′,5-dimethoxy-3,4′-oxyligna-8,8′-dien-7-one

[2108673-68-1]

$C_{20}H_{20}O_5$ 340.375

Constit. of twigs of *Nectandra leucantha*. Exhibits strong cytotoxic activity. Brownish oil. λ_{max} 284 (log ε 2.8); 321 (log ε 3.0) (MeOH)

7-Ketone, 3′,4,5-tri-Me ether

3′,4,5-Trimethoxy-3,4′-oxyligna-8,8′-dien-7-one

[2108673-69-2]

$C_{21}H_{22}O_5$ 354.402

Constit. of twigs of *Nectandra leucantha*. Brownish oil. λ_{max} 214 (log ε 3.7); 284 (log ε 3.2) (MeOH).

Da Costa-Silva, T.A. *et al.*, *J. Nat. Prod.*, 2015, **78**, 653-657 (*Nectandra leucantha constit*).
De Sousa, F.S. *et al.*, *Phytochemistry*, 2017, **140**, 108-117 (*Nectandra leucantha constits, activity*)

8,8′-Oxyligna-7,7′-diene-3,3′,4,4′-tetrol O30

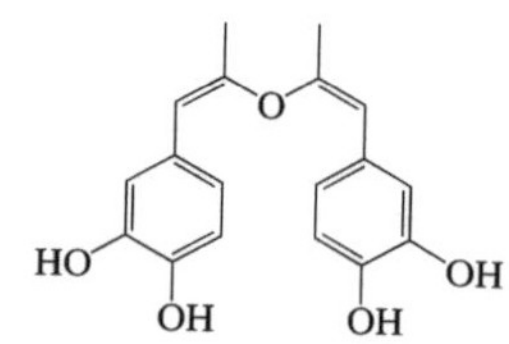

3,3′,4,4′-Tetrahydroxy-8,8′-oxyneoligna-7,7′-diene

$C_{18}H_{18}O_5$ 314.337

(Z,Z)-form

3,3′-Di-Me ether

4,4′-[Oxybis(2-methyl-2,1-ethenediyl)] bis[2-methoxyphenol], CAS. *3,3′-Dimethoxy-8,8′-oxyligna-7,7′-diene-4,4′-diol. 4,4′-Dihydroxy-3,3′-dimethoxy-8,8′-oxyligna-7,7′-diene.* ***Leucosykol***

[149471-06-7]

$C_{20}H_{22}O_5$ 342.391

Constit. of the stem of *Leucosyke quadrinervia*. Prisms ($CHCl_3$/MeOH). Mp 168 - 170°. λ_{max} 237 (log ε 4.22); 280 (log ε 3.83) (EtOH). [base] λ_{max} 253 (log ε 4.37); 292 (log ε 4.00) (EtOH/KOH).

Tsai, I.-L. *et al.*, *Phytochemistry*, 1993, **32**, 1065-1067 (*Leucosykol*)

9,9′-Oxyligna-7,7′-diene-3,3′,4,4′-tetrol O31

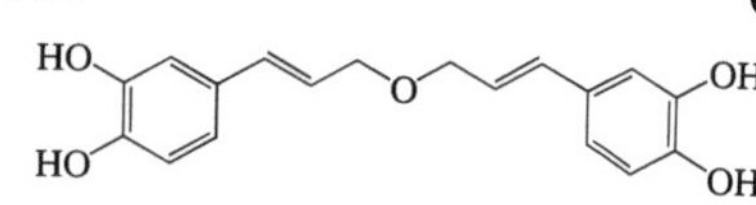

Bis(3,4-dihydroxycinnamyl) ether. 4,4′-Oxybis(1-propene-3,3′-diyl)bis[1,2-benzenediol]. 3,3′,4,4′-Tetrahydroxy-9,9′-oxyneoliga-7,7′diene

$C_{18}H_{18}O_5$ 314.337

(E,E)-form

3,3′-Di-Me ether

3,3′-Dimethoxy-9,9′-oxyligna-7,7′-diene-4,4′-diol. 4,4′-Dihydroxy-3,3′-dimethoxy-9,9′-oxyligna-7,7′-diene

3,3′-Di-Me ether, 4,4′-di-O-β-D-glucopyranoside

Balajaponin B

[1362086-45-0]

$C_{32}H_{42}O_{15}$ 666.673

Constit. of *Balanophora japonica*. Amorph. yellow powder. $[\alpha]^{20}_{D}$ -71 (c, 0.15 in H_2O)

5,5′-Dimethoxy, 3,3′-di-Me ether, 4,4′-di-O-β-D-glucopyranoside

Disyringin ether

[1421957-61-0]

$C_{34}H_{46}O_{17}$ 726.725

Constit. of the bark of *Ilex rotunda*. Mp 223 - 224°. λ_{max} 225; 242; 362 (MeOH).

Zhou, T. *et al.*, *Planta Med.*, 2011, **77**, 477-481 (*Balajaponin B*).

Xu, R. *et al.*, *Zhongcaoyao*, 2011, **42**, 2389-2393 (*Disyringin ether*)

2,3′-Oxyligna-8,8′-diene-4,4′,5-triol O32

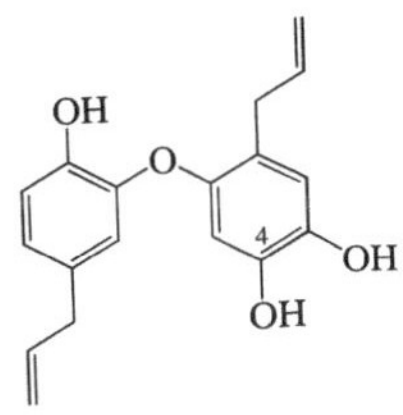

4-[2-Hydroxy-5-(2-propenyl)phenoxy]-5-(2-propenyl)-1,2-benzenediol. 4,4′,5-Trihydroxy-2,3′-oxyneoligna-8,8′-diene

$C_{18}H_{18}O_4$ 298.338

Dimer of 4-(2-Propenyl)-1,2-benzenediol, P75

4-Me ether

4-Methoxy-2,3′-oxyligna-8,8′-diene-4′,5-diol. 4′,5-Dihydroxy-4-methoxy-2,3′-oxyligna-8,8′-diene

[1414838-47-3]

$C_{19}H_{20}O_4$ 312.365

Constit. of the leaves of *Syzygium polyanthum*. Pancreatic lipase inhibitor.

Kato, E. *et al.*, *Food Chem.*, 2013, **136**, 1239-1242 (*Syzygium polyanthum constit, activity*)

2,4′-Oxyligna-8,8′-diene-3′,4,5-triol O33

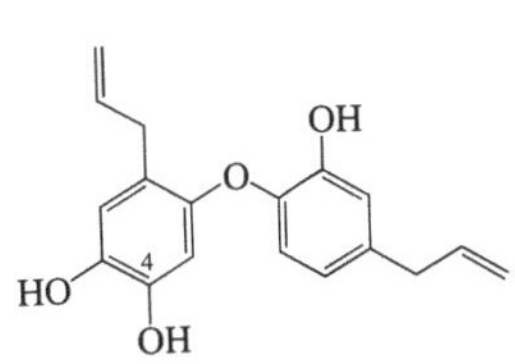

4-[2-Hydroxy-4-(2-propenyl)phenoxy]-5-(2-propenyl)-1,2-benzenediol. 3′,4,5-Trihydroxy-2,4′-oxyneoligna-8,8′-diene

$C_{18}H_{18}O_4$ 298.338

Dimer of 4-(2-Propenyl)-1,2-benzenediol, P75

4-Me ether

4-Methoxy-2,4′-oxyligna-8,8′-diene-3′,5-diol. 3′,5-Dihydroxy-4-methoxy-2,4′-oxyligna-8,8′-diene

[1414838-48-4]

$C_{19}H_{20}O_4$ 312.365

Constit. of the leaves of *Syzygium polyanthum*. Pancreatic lipase inhibitor.

Kato, E. *et al.*, *Food Chem.*, 2013, **136**, 1239-1242 (*Syzygium polyanthum constit, activity*)

3,3′-Oxyligna-8,8′-diene-4,4′,5-triol O34

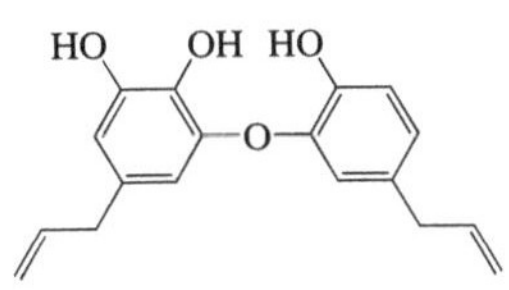

2,2′,3-Trihydroxy-5,5′-di-2-propenyldiphenyl ether. 5,5′-Diallyl-2,2′,3-trihydroxydiphenyl ether. 4,4′,5-Trihydroxy-3,3′-oxyneoligna-8,8′-diene

$C_{18}H_{18}O_4$ 298.338

Tri-Me ether

2,2′,3-Trimethoxy-5,5′-di-2-propenyldiphenyl ether. 4,4′,5-Trimethoxy-3,3′-oxyligna-8,8′-diene

$C_{21}H_{24}O_4$ 340.419

Constit. of leaves and stems of *Illicium lanceolatum*. Exhibits inhibitory activity on LPS-stimulated NO prodn. in RAW 264.7 macrophages. Oil. λ_{max} 319 (log ε 4.32) ($CHCl_3$).

Gui, X. *et al.*, *Fitoterapia*, 2014, **95**, 51-57 (*Illicium lanceolatum constit, activity*)

3,4′-Oxyligna-7,8′-diene-4,5,9-triol O35

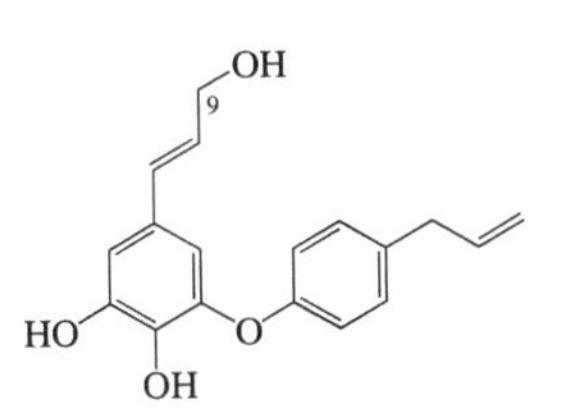

4,5,9-Trihydroxy-3,4′-oxyneoligna-7,8′-diene

$C_{18}H_{18}O_4$ 298.338

(E)-form

9-Me ether

9-Methoxy-3,4′-oxyligna-7,8′-diene-4,5-diol. 4,5-Dihydroxy-9-methoxy-3,4′-oxyligna-7,8′-diene. ***9-Methoxyobovatol***

[1637562-81-2]

$C_{19}H_{20}O_4$ 312.365

Constit. of the fruit of *Magnolia obovata*. NO prodn. inhibitor. Dark brown oil. Sl. misleading synonym

9-Aldehyde

3-[3,4-Dihydroxy-5-[4-(2-propenyl)phenoxy]phenyl]-2-propenal. 4,5-Dihydroxy-3,4′-oxyligna-7,8′-dien-9-al. 4,5-Dihydroxy-9-oxo-3,4′-oxyligna-7,8′-diene. ***Obovatal***

[83864-77-1]

$C_{18}H_{16}O_4$ 296.322

Constit. of leaves of *Magnolia obovata*. Inhibitor of MMP-2. Metastatic agent. Cryst. (Me_2CO). Mp 160.5 - 162°. λ_{max} 210; 235 (sh); 336 (MeOH).

Ito, K. *et al.*, *Chem. Pharm. Bull.*, 1982, **30**, 3347-3353 (*Obovatal*).

Lee, S.K. *et al.*, *Bioorg. Med. Chem.*, 2007, **15**, 4085-4090 (*Obovatal, activity*).

Seo, K.H. *et al.*, *Planta Med.*, 2013, **79**, 1335-1340 (*9-Methoxyobovatol*)

3,4′-Oxyligna-8,8′-diene-3′,4,5-triol O36

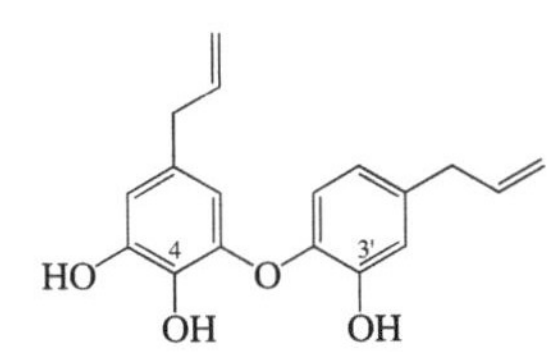

3′,4,5-Trihydroxy-3,4′-oxyneoligna-8,8′-diene

$C_{18}H_{18}O_4$ 298.338

3′,5-Di-Me ether

2-Methoxy-6-[2-methoxy-4-(2-propenyl)phenoxy]-4-(2-propenyl)phenol. 4′,5-Diallyl-2-hydroxy-2′,3-dimethoxydiphenyl ether. 3′,5-Dimethoxy-3,4′-oxyligna-8,8′-dien-4-ol. 4-Hydroxy-3′,5-dimethoxy-3,4′-oxyligna-8,8′-diene. ***Dehydrodieugenol B***

[75225-33-1]

$C_{20}H_{22}O_4$ 326.392

Constit. of *Illicium simonsii*, *Nectandra leucantha*, *Ocimum sanctum*, *Ocotea cymbarum* and *Virola carinata*. Exhibits antileishmanial and immunomodulatory activities. Oil. λ_{max} 275 (sh); 285 (ε 9200); 325 (sh) (MeOH). [neutral] λ_{max} 215 (log ε 3.2); 280 (log ε 2.3) (MeOH)

Tri-Me ether

4′,5-Diallyl-2,2′,3-trimethoxydiphenyl ether. 3′,4,5-Trimethoxy-3,4′-oxyligna-8,8′-diene

[1695562-41-4]
$C_{21}H_{24}O_4$ 340.419
Constit. of *Illicium burmanicum* and *Nectandra leucantha*. Yellowish oil or cryst.. λ_{max} 220 (log ε 3.2); 280 (log ε 2.5) (MeOH).

De Diaz, A.M.P. *et al.*, *Phytochemistry*, 1980, **19**, 681-682 (*Dehydrodieugenol B*).
Kawanishi, K. *et al.*, *Phytochemistry*, 1983, **22**, 2277-2280 (*Virola carinata constit*).
Suzuki, A. *et al.*, *Chem. Pharm. Bull.*, 2009, **57**, 245-251 (*Ocimum sanctum constit, activity*).
Da Costa-Silva, T.A. *et al.*, *J. Nat. Prod.*, 2015, **78**, 653-658 (*Nectandra leucantha constits, activity*).
Tian, X. *et al.*, *RSC Adv.*, 2017, **7**, 10849-10854 (*Illicium burmanicum constit*).
Grecco, S.S. *et al.*, *Acta Crystallogr., Sect. E*, 2018, **74**, 518-521 (*tri-Me ether, cryst struct*)

3,4′-Oxyligna-7,7′-dien-4-ol O37

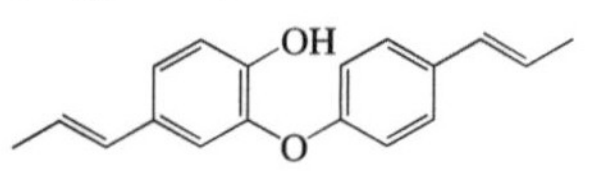

4-(1-Propenyl)-2-[4-(1-propenyl)phenoxy]phenol. 4-Hydroxy-3,4′-oxyneoligna-7,7′-diene
$C_{18}H_{18}O_2$ 266.34
(E,E)-form
Me ether
2-Methoxy-4′,5-di-1-propenyldiphenyl ether. 4-Methoxy-3,4′-oxyligna-7,7′-diene. ***Ottomentosa***
[1469981-57-4]
$C_{19}H_{20}O_2$ 280.367
Constit. of the roots of *Krameria tomentosa*. Oil.

Madeiro, S.A.L. *et al.*, *J. Braz. Chem. Soc.*, 2012, **23**, 2021-2026 (*Ottomentosa*)

3,4′-Oxyligna-8,8′-dien-4-ol O38

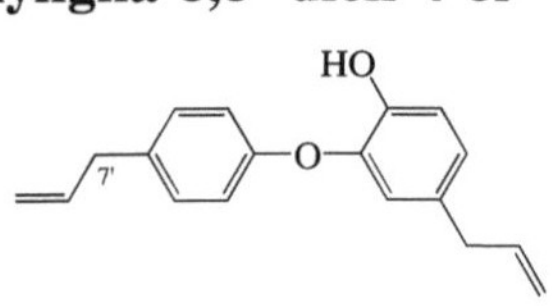

4-Hydroxy-3,4′-oxyneoligna-8,8′-diene. ***Isomagnolol‡***
[87688-90-2]
$C_{18}H_{18}O_2$ 266.34
Constit. of *Magnolia obovata*, *Sassafras randaiense* and *Streblus asper*. Exhibits antimicrobial and antiviral activities. NO prodn. inhibitor. Cytotoxic. Oil. Mp 143.5°. λ_{max} 293 (MeOH) (Berdy)
8′,9′-Dihydro, 7′-oxo
4-Hydroxy-3,4′-oxylign-8-en-7′-one. 4-Hydroxy-7′-oxo-3,4′-oxylign-8-ene. ***Isomagnolone***
[155709-41-4]
$C_{18}H_{18}O_3$ 282.339
Constit. of the bark of *Illicium simsonii*. Oil.

El-Feraly, F.S. *et al.*, *J. Nat. Prod.*, 1983, **46**, 493-498 (*Isomagnolol, activity*).
Kouno, I. *et al.*, *Chem. Pharm. Bull.*, 1994, **42**, 112-114 (*Isomagnolone*).
Panda, A.K. *Indian J. Chem., Sect. B*, 2002, **41**, 181-183 (*Isomagnolone, synth*).
Li, J. *et al.*, *Bioorg. Med. Chem. Lett.*, 2013, **23**, 2238-2244 (*activity*).
Seo, K.-H. *et al.*, *Planta Med.*, 2013, **79**, 1335-1340 (*Magnolia obovata constit, activity*)

4,8′-Oxylignan-3,3′,4′,5,5′,7,7′,8,9,9′-decol O39

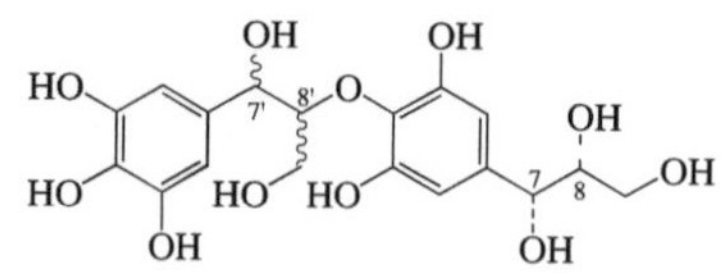

3,3′,4′,5,5′,7,7′,8,9,9′-Decahydroxy-4,8′-oxyneolignan
$C_{18}H_{22}O_{11}$ 414.363
(7R*,7′ξ,8R*,8′ξ)-form
7-Ketone, 3,3′,5,5′-tetra-Me ether
4′,7′,8,9,9′-Pentahydroxy-3,3′,5,5′-tetramethoxy-4,8′-oxylignan-7-one
7-Ketone, 3,3′,5,5′-tetra-Me ether, 9,9′-bis-O-(4-hydroxybenzoyl)
Quiquelignan E
[1215818-68-0]
$C_{36}H_{36}O_{15}$ 708.669
Constit. of the stems of *Calamus quiquesetinervius*. Pale yellow powder. Mp 214°. $[\alpha]^{25}_{D}$ -42.4 (c, 0.3 in MeOH). λ_{max} 235; 258; 275 (sh); 310 (MeOH).

Chang, C.-L. *et al.*, *Bioorg. Med. Chem.*, 2010, **18**, 518-525 (*Quiquelignan E*)

4,4′-Oxylignan-9,9′-dioic acid O40

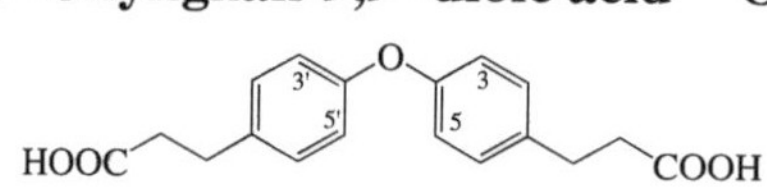

4,4′-Oxybis[benzenepropanoic acid], CAS. *4,4′-Oxybis[3-phenylpropanoic acid]. 4,4′-Bis(2-carboxyethyl)diphenyl ether. Diphenyl ether-4,4′-dipropanoic acid. 4,4′-Oxybishydrocinnamic acid.*
Apteniol A
[13655-10-2]
$C_{18}H_{18}O_5$ 314.337
Synthetic studies in 2014/2015/2017 have put structures of this series in doubt. Constit. of the leaves of *Aptenia cordifolia*. Prod. by the marine-derived *Bacillus licheniformis* SAB1. Antibacterial and antifungal agent. Pale yellow cryst. (C_6H_6/MeOH). Mp 195°. λ_{max} 278 (log ε 3.2) (MeOH)
3-Methoxy
3-Methoxy-4,4′-oxylignan-9,9′-dioic acid. ***Apteniol B***
[871503-82-1]
$C_{19}H_{20}O_6$ 344.363
Constit. of the leaves of *Aptenia cordifolia*. Exhibits phytotoxic activity. λ_{max} 280 (log ε 2.2) (MeOH)
3-Methoxy, di-Me ester
Apteniol C
[871503-84-3]
$C_{21}H_{24}O_6$ 372.417
Constit. of the leaves of *Aptenia cordifolia*. Exhibits phytotoxic activity. λ_{max} 280 (log ε 3.1) (MeOH)
3,3′,5-Trimethoxy
3,3′,5-Trimethoxy-4,4′-oxylignan-9,9′-dioic acid. ***Apteniol D***
[871503-85-4]
$C_{21}H_{24}O_8$ 404.415
Constit. of the leaves of *Aptenia cordifolia*. Exhibits phytotoxic activity. λ_{max} 242 (log ε 2.5); 280 (log ε 2.6); 321 (log ε 0.9) (MeOH)
3,3′,5,5′-Tetramethoxy
4,4′-Oxybis[3-(3,5-dimethoxyphenyl)propanoic acid]. 3,3′,5,5′-Tetramethoxy-4,4′-oxylignan-9,9′-dioic acid. ***Apteniol E***
[871503-86-5]
$C_{22}H_{26}O_9$ 434.441
Constit. of the leaves of *Aptenia cordifolia*. Exhibits phytotoxic activity. λ_{max} 244 (log ε 2.8); 279 (log ε 3); 322 (log ε 1.5) (MeOH).

Cummings, W.A.W. *et al.*, *J. Appl. Chem.*, 1962, **12**, 86-90 (*synth*).
Tortai, J.P. *et al.*, *Bull. Soc. Chim. Fr.*, 1971, 1000-1007 (*synth*).
DellaGreca, M. *et al.*, *Tetrahedron*, 2005, **61**, 11924-11929 (*Apteniols A-E, activity*).
Devi, P. *et al.*, *Mar. Drugs*, 2010, **8**, 1203-1212 (*marine, isol, activity*).
Nishikawa, H. *et al.*, *Biosci., Biotechnol., Biochem.*, 2014, **78**, 1485-1489 (*synth, struct*).
Noshita, T. *et al.*, *Biosci., Biotechnol., Biochem.*, 2015, **79**, 1743-1749 (*Apteniols B,C, synth, struct*).
Noshita, T. *et al.*, *Nat. Prod. Res.*, 2017, **31**, 163-168 (*Apteniol D, synth, struct*)

3,7′-Oxylignan-3′,4,4′,5,8′,9,9′-heptol O41

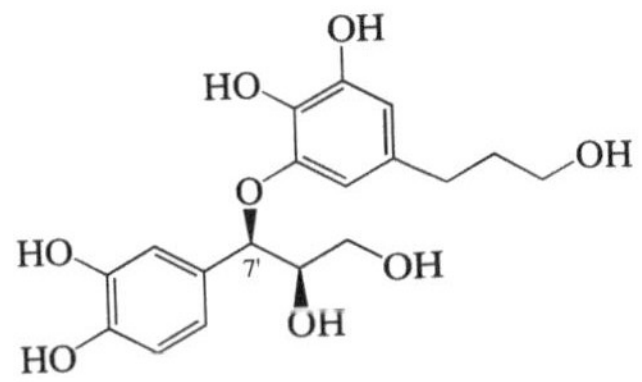

3′,4,4′,5,8′,9,9′-Heptahydroxy-3,7′-oxyneolignan
$C_{18}H_{22}O_8$ 366.366
(7′R,8′R)-form
3′,5-Di-Me ether
3′,5-Dimethoxy-3,7′-oxylignan-4,4′,8′,9,9′-pentol. 4,4′,8′,9,9′-Pentahydroxy-3′,5-dimethoxy-3,7′-oxylignan
3′,5-Di-Me ether, 4′-O-β-D-glucopyranoside
Baekmaoside A
[2552840-00-1]
$C_{26}H_{36}O_{13}$ 556.561
Constit. of the stems of *Chrysanthemum morifolium*. Amorph. yellow powder (MeOH). $[\alpha]_D^{23}$ +43 (c, 0.1 in MeOH). Text and diag. abs. config. do not agree in ref..

Kim, H.-G. *et al., Chem. Nat. Compd. (Engl. Transl.)*, 2019, **55**, 610-613 (*Baekmaoside A*)

4,8′-Oxylignan-3,3′,4′,5,5′,9,9′-heptol O42

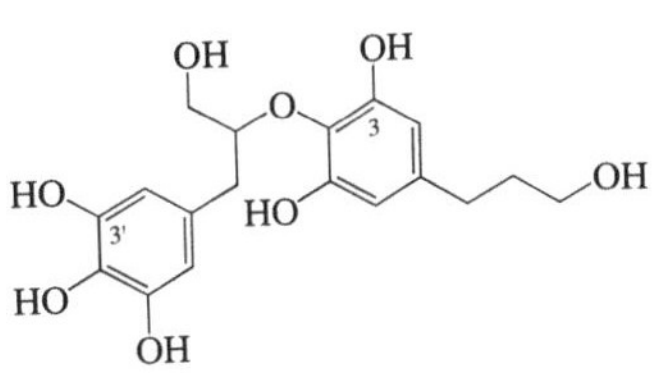

3,3′,4′,5,5′,9,9′-Heptahydroxy-4,8′-oxyneolignan
$C_{18}H_{22}O_8$ 366.366
(ξ)-form
3,3′,5,5′-Tetra-Me ether
3,3′,5,5′-Tetramethoxy-4,8′-oxylignan-4′,9,9′-triol. 4′,9,9′-Trihydroxy-3,3′,5,5′-tetramethoxy-4,8′-oxylignan
3,3′,5,5′-Tetra-Me ether, 4′-O-β-D-glucopyranoside
[1264177-74-3]
$C_{28}H_{40}O_{13}$ 584.615
Constit. of *Selaginella moellendorffii*. Amorph. solid. $[\alpha]_D^{26}$ -25 (c, 0.35 in MeOH). λ_{max} 207 (log ε 4.57); 269 (log ε 3.56) (MeOH).

Wang, Y.-H. *et al., Helv. Chim. Acta*, 2010, **93**, 2467-2477 (*Selaginella moellendorffii constit*)

4,8′-Oxylignan-3,3′,4′,5′,7′,9,9′-heptol O43

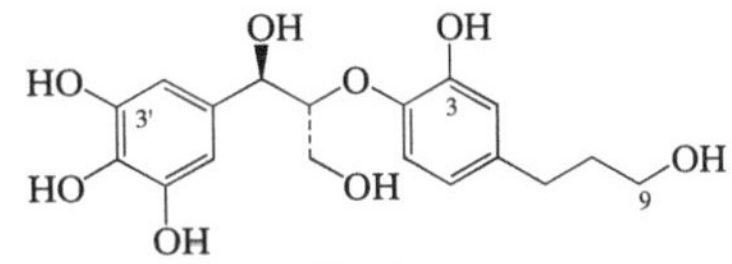

(7′R,8′R)-form

3,3′,4′,5′,7′,9,9′-Heptahydroxy-4,8′-oxyneolignan
$C_{18}H_{22}O_8$ 366.366
(7′R,8′R)-form
3,3′,5′-Tri-Me ether
3,3′,5′-Trimethoxy-4,8′-oxylignan-4′,7′,9,9′-tetrol. 4′,7′,9,9′-Tetrahydroxy-3,3′,5′-trimethoxy-4,8′-oxylignan
$C_{21}H_{28}O_8$ 408.447
Constit. of the roots of *Alangium chinense*. Antioxidant. Amorph. powder. $[\alpha]_D^{20}$ +55.2 (c, 0.3 in MeOH). λ_{max} 207 (log ε 4.73); 233 (log ε 4.08); 280 (log ε 3.57) (MeOH)
4′-O-[2-Hydroxy-1-(hydroxymethyl)ethyl], 3,3′,5′-tri-Me ether
4′-(2-Glyceryloxy)-3,3′,5′-trimethoxy-4,8′-oxylignan-7′,9,9′-triol. 4′-(2-Glyceryloxy)-7′,9,9′-trihydroxy-3,3′,5′-trimethoxy-4,8′-oxylignan
[1227064-00-7]
$C_{24}H_{34}O_{10}$ 482.526
Constit. of the stem bark of *Illicium difengpi*. Amorph. powder. $[\alpha]_D^{20}$ -12.5 (c, 0.07 in MeOH). λ_{max} 208 (log ε 4.4); 277 (log ε 3.2) (MeOH)
(7′S,8′R)-form
3,3′,5′-Tri-Me ether, 4′-O-β-D-glucopyranoside
[2247433-05-0]
$C_{27}H_{38}O_{13}$ 570.588
Constit. of the rhizomes of *Smilax trinervula*. Exhibits cytotoxic activity. Powder (MeOH). $[\alpha]_D^{25}$ -30.6 (c, 0.32 in MeOH). λ_{max} 220 (log ε 4.25); 280 (log ε 3.55) (no solvent reported)
4′-O-[2-Hydroxy-1-(hydroxymethyl)ethyl], 3,3′,5′-tri-Me ether
[1227064-01-8]
$C_{24}H_{34}O_{10}$ 482.526
Constit. of the stem bark of *Illicium difengpi*. Amorph. powder. $[\alpha]_D^{20}$ -3 (c, 0.07 in MeOH). λ_{max} 209 (log ε 4.4); 278 (log ε 3.3) (MeOH)
(7′R*,8′S*)-form
3,3′,5′-Tri-Me ether, 9-O-α-L-rhamnopyranoside
Icariside E₆
[135774-83-3]
$C_{27}H_{38}O_{12}$ 554.589
Constit. of *Epimedium sagittatum*. Amorph. powder. $[\alpha]_D^{23}$ -12.5 (c, 0.3 in MeOH). λ_{max} 230 (sh); 279 (log ε 3.5) (MeOH)
(8′RS)-form
7′-Ketone
3,3′,4′,5′,9,9′-Hexahydroxy-4,8′-oxylignan-7′-one. 3,3′,4′,5′,9,9′-Hexahydroxy-7′-oxo-4,8′-oxylignan
7′-Ketone, 3,3′,5′-tri-Me ether
4′,9,9′-Trihydroxy-3,3′,5′-trimethoxy-4,8′-oxylignan-7′-one.
Icariol A₁
[135743-06-5]
$C_{21}H_{26}O_8$ 406.431
Constit. of the aerial parts of *Epimedium sagittatum*. Amorph. powder. λ_{max} 228 (sh); 286 (sh); 311 (log ε 3.95) (MeOH).

Matsushita, H. *et al., Phytochemistry*, 1991, **30**, 2025-2027 (*Icariside E₆, Icariol A₁*).
Kuono, I. *et al., Phytochemistry*, 1993, **32**, 1573-1577 (*Illicium difengpi constits*).
Fang, L. *et al., J. Nat. Prod.*, 2010, **73**, 818-824 (*Illicium difengpi constits*).
Zhang, Y. *et al., Chin. Chem. Lett.*, 2017, **28**, 32-36 (*Alangium chinense constit, activity*).
Shu, J. *et al., Phytochem. Lett.*, 2017, **20**, 1-8 (*Smilax trinervula constit, activity*)

4,8′-Oxylignan-3,3′,4′,5,7′,9,9′-heptol O44

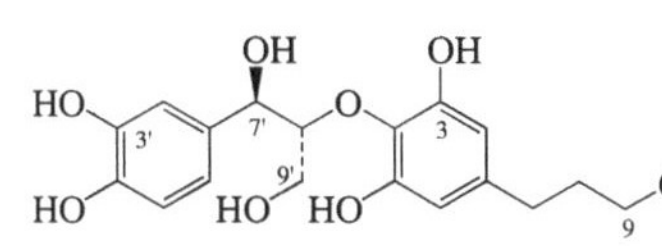

(7′R,8′R)-form

3,3′,4′,5,7′,9,9′-Heptahydroxy-4,8′-oxyneolignan
$C_{18}H_{22}O_8$ 366.366
Lignan numbering shown
(7′R,8′R)-form
3,3′,5-Tri-Me ether
3,3′,5-Trimethoxy-4,8′-oxylignan-4′,7′,9,9′-tetrol. 4′,7′,9,9′-Tetrahydroxy-3,3′,5-trimethoxy-4,8′-oxylignan
$[\alpha]_D^{20}$ -20 (c, 0.1 in MeOH)
3,3′,5-Tri-Me ether, 7′-O-β-D-glucopyranoside
Morineoliganoside C
[2209122-28-9]

$C_{27}H_{38}O_{13}$ 570.588
Constit. of the stems of *Chrysanthemum morifolium*. Exhibits inhibitory effects on xanthine oxidase. Powder. $[\alpha]^{25}_{D}$ -37 (c, 1.01 in MeOH). λ_{max} 226 (log ε 4.11); 279 (log ε 3.52) (MeOH)

3,3′,5-Tri-Me ether, 9-O-α-L-rhamnopyranoside
[1392407-92-9]
$C_{27}H_{38}O_{12}$ 554.589
Constit. of *Geum japonicum* var. *chinense*. Gum. $[\alpha]^{25}_{D}$ -7 (c, 0.1 in MeOH). λ_{max} 230 (sh); 280 (log ε 3.17) (MeOH)

3,3′,5,7′-Tetra-Me ether
$C_{22}H_{30}O_{8}$ 422.474
Exhibits moderate neuroprotective activity. $[\alpha]^{20}_{D}$ -20 (c, 0.1 in MeOH)

(7′R,8′S)-form

3′-Me ether, 3-O-β-D-glucopyranoside
Rubuslin E
[2242395-25-9]
$C_{25}H_{34}O_{13}$ 542.534
Constit. of the rhizomes of *Rubus idaeus*. Weak antioxidant. Exhibits neuroprotective activity. Yellow oil

3,3′-Di-Me ether
3′,5-Dimethoxy-4,8′-oxylignan-3,4′,7′,9,9′-pentol. 3,4′,7′,9,9′-Pentahydroxy-3′,5-dimethoxy-4,8′-oxylignan. ***Pinnatifidanin B VII***
[1462882-66-1]
$C_{20}H_{26}O_{8}$ 394.42
Constit. of *Crataegus pinnatifida*. Yellow oil. $[\alpha]^{20}_{D}$ -2.4 (c, 0.01 in MeOH). λ_{max} 238 (log ε 2.52) (MeOH)

3,3′,5-Tri-Me ether
$C_{21}H_{28}O_{8}$ 408.447
Exhibits neuroprotective activity. $[\alpha]^{20}_{D}$ +20.5 (c, 0.1 in MeOH)

3,3′,5-Tri-Me ether, 9-O-α-L-rhamnopyranoside
[1392407-93-0]
$C_{27}H_{38}O_{12}$ 554.589
Constit. of *Geum japonicum* var. *chinense*. Gum. $[\alpha]^{25}_{D}$ +20 (c, 0.05 in MeOH). λ_{max} 230 (sh); 280 (log ε 3.05) (MeOH)

(7′S,8′R)-form

3,3′-Di-Me ether, 7′-O-α-L-rhamnopyranoside
[1416620-27-3]
$C_{26}H_{36}O_{12}$ 540.562
Constit. of *Sanguisorba officinalis*. Amorph. powder. Mp 188 - 190°. $[\alpha]^{22}_{D}$ -66.5 (c, 0.26 in MeOH). λ_{max} 279 (log ε 3.66) (MeOH)

3,3′,5-Tri-Me ether
$C_{21}H_{28}O_{8}$ 408.447
Exhibits neuroprotective activity. $[\alpha]^{20}_{D}$ -22 (c, 0.1 in MeOH)

3,3′,5-Tri-Me ether, 4′-O-β-D-glucopyranoside
[1006034-11-2]
$C_{27}H_{38}O_{13}$ 570.588
Constit. of *Dendrobium longicornu*, *Helichrysum italicum* and *Illicium oligandrum*. Amorph. yellow powder (Me_2CO). Mp 153 - 155°. Mp 205 - 207°. $[\alpha]^{25}_{D}$ -7.5 (c, 0.04 in MeOH). $[\alpha]^{22}_{D}$ +3.2 (c, 0.6 in MeOH). λ_{max} 231 (log ε 4.72); 277 (log ε 3.13) (MeOH). [neutral] λ_{max} 240 (log ε 3.2); 274 (log ε 2.9) ($CHCl_3$)

3,3′,5-Tri-Me ether, 7′-O-α-L-rhamnopyranoside
[1416620-28-4]
$C_{27}H_{38}O_{12}$ 554.589
Constit. of *Sanguisorba officinalis*. Amorph. powder. Mp 197 - 199°. $[\alpha]^{22}_{D}$ -73.5 (c, 0.41 in MeOH). λ_{max} 205 (log ε 5.01); 280 (log ε 4.23) (MeOH)

3,3′,5-Tri-Me ether, 9′-O-β-D-glucopyranoside
Isodonoside I
[2095793-36-3]
$C_{27}H_{38}O_{13}$ 570.588
Constit. of *Isodon japonicus*. Powder. $[\alpha]^{25}_{D}$ -46.1 (c, 0.4 in MeOH). λ_{max} 230 (log ε 3.3) (MeOH)

4′-O-[2-Hydroxy-1-(hydroxymethyl)ethyl], 3,3′,5-tri-Me ether
4′-O-(2-Glyceryl)-3,3′,5-trimethoxy-4,8′-oxylignan-7′,9,9′-triol
[1227063-99-1]
$C_{24}H_{34}O_{10}$ 482.526
Constit. of the stem bark of *Illicium difengpi*. Shows antiinflammatory activity. Amorph. powder. $[\alpha]^{20}_{D}$ -12 (c, 0.07 in MeOH). λ_{max} 209 (log ε 4.5); 276 (log ε 3) (MeOH)

(7′S,8′S)-form

3,3′-Di-Me ether
[874448-65-4]
$C_{20}H_{26}O_{8}$ 394.42
Constit. of the flower buds of *Magnolia fargesii*. Oil. λ_{max} 282 (log ε 3.84) ($CHCl_3$)

3,3′,5-Tri-Me ether
Asprenol H
$C_{21}H_{28}O_{8}$ 408.447
Constit. of *Ilex asprella*. Amorph. powder. $[\alpha]^{25}_{D}$ +20 (c, 0.1 in MeOH). λ_{max} 208 (log ε 3.20); 279 (log ε 2.06) (MeOH)

3,3′,5-Tri-Me ether, 7′-O-β-D-glucopyranoside
Rourinoside
[906670-57-3]
$C_{27}H_{38}O_{13}$ 570.588
Constit. of *Clethra barbinervis* and *Rourea minor*. Shows moderate antiplasmodial activity. Amorph. powder. $[\alpha]^{20}_{D}$ -21.3 (c, 0.08 in MeOH). λ_{max} 225 (log ε 2.17); 278 (log ε 3.54) (MeOH)

3,3′,5,7′-Tetra-Me ether
$C_{22}H_{30}O_{8}$ 422.474
Exhibits moderate neuroprotective activity. $[\alpha]^{20}_{D}$ +21 (c, 0.1 in MeOH)

(7′R*,8′R*)-form

3,3′-Di-Me ether, 9-O-α-L-rhamnopyranoside
Icariside E_2
[112295-61-1]
$C_{26}H_{36}O_{12}$ 540.562
Constit. of *Epimedium grandiflorum* var. *thunbergianum*. Amorph. powder. $[\alpha]^{23}_{D}$ -12.7 (c, 0.5 in MeOH)

3,3′,5-Tri-Me ether, 9,9′-bis-O-(4-hydroxy-3-methoxy-E-cinnamoyl)
threo-***Carolignan F***
[171261-41-9] (unspecified stereochem.)
$C_{41}H_{44}O_{14}$ 760.789
Constit. of *Ochroma lagopus*

(7′R*,8′S*)-form

3′-Me ether, 3-O-β-D-glucopyranoside
[1179397-87-5]
$C_{25}H_{34}O_{13}$ 542.534
Constit. of the stem bark of *Tetracentron sinense*. Mp 112 - 115°. $[\alpha]_{D}$ -23.2 (c, 0.6 in MeOH)

3,3′-Di-Me ether, 9-O-α-L-rhamnopyranoside
Icariside E_1
[112232-54-9]
$C_{26}H_{36}O_{12}$ 540.562
Constit. of *Epimedium grandiflorum* var. *thunbergianum*. Amorph. powder + ½H_2O. $[\alpha]^{23}_{D}$ -24.3 (c, 0.4 in MeOH). λ_{max} 228 (sh); 279 (ε 4170) (MeOH)

3,3′,5-Tri-Me ether, 4′-O-α-L-rhamnopyranoside
Nymphaeoside A
[524938-02-1]
$C_{27}H_{38}O_{12}$ 554.589
Constit. of the leaves of *Nymphaea odorata*. Amorph. powder. $[\alpha]_D^{22}$ -23.1 (c, 0.5 in MeOH). λ_{max} 214 (log ε 4.05); 230 (sh); 276 (log ε 3.17) (MeOH)

3,3′,5-Tri-Me ether, 4′-O-β-D-glucopyranoside
[1294515-90-4]
$C_{27}H_{38}O_{13}$ 570.588
Constit. of the roots of *Symplocos caudata*. Amorph. powder. $[\alpha]_D^{25}$ -32 (c, 1 in MeOH). Diastereoisomeric mixt. (1:1). λ_{max} 225; 274 (MeOH)

3,3′,5-Tri-Me ether, 9-O-β-D-glucopyranoside
[1047663-56-8]
$C_{27}H_{38}O_{13}$ 570.588
Constit. of the leaves of *Pieris japonica*. Amorph. powder. $[\alpha]_D^{20}$ -15 (c, 0.001 in MeOH). λ_{max} 222 (log ε 4.98); 286 (log ε 3.35) (MeOH)

3,3′,5-Tri-Me ether, 9,9′-bis-O-(4-hydroxy-3-methoxy-E-cinnamoyl)
erythro-**Carolignan F**
[171261-41-9] (unspecified stereochem.)
$C_{41}H_{44}O_{14}$ 760.789
Constit. of *Hibiscus cannabinus* and *Ochroma lagopus*

(7′RS,8′RS)-form

3,3′,5-Tri-Me ether
$C_{21}H_{28}O_8$ 408.447
Constit. of the fruit of *Crataegus pinnatifida*. Gum. $[\alpha]_D^{20}$ -1.5 (c, 0.1 in MeOH). Partial racemate resolved by chiral hplc. λ_{max} 205 (log ε 4.0); 251 (log ε 0.18); 278 (log ε 0.31) (MeOH)

3,3′,5,7′-Tetra-Me ether
3,3′,5,7′-Tetramethoxy-4,8′-oxylignan-4′,9,9′-triol. 4′,9,9′-Trihydroxy-3,3′,5,7′-tetramethoxy-4,8′-oxylignan
$C_{22}H_{30}O_8$ 422.474
Constit. of the root bark of *Ailanthus altissima*. Light yellow gum. $[\alpha]_D^{20}$ +1.2 (c, 0.1 in MeOH). Resolved by chiral hplc. λ_{max} 227 (log ε 4.65); 250 (log ε 4.18) (MeOH)

(7′RS,8′SR)-form

3,3′,5-Tri-Me ether
$C_{21}H_{28}O_8$ 408.447
Constit. of the fruit of *Crataegus pinnatifida*. Gum. $[\alpha]_D^{20}$ -0.8 (c, 0.1 in MeOH). Partial racemate resolved by chiral hplc. λ_{max} 254 (log ε 0.13); 279 (log ε 0.4) (MeOH).

Miyase, T. *et al.*, *Chem. Pharm. Bull.*, 1987, **35**, 3713-3719 (*Icarisides*).
Kuono, I. *et al.*, *Phytochemistry*, 1993, **32**, 1573-1577 (*Illicium difengpi constit*).
Paula, V.F. *et al.*, *Tetrahedron*, 1995, **51**, 12453-12462 (*Carolignan F*).
Zhang, Z. *et al.*, *J. Nat. Prod.*, 2003, **66**, 548-550 (*Nymphaeoside A*).
Lee, J. *et al.*, *Bull. Korean Chem. Soc.*, 2005, **26**, 913-915 (*Magnolia fargesii di-Me ether*).
He, Z.-D. *et al.*, *Phytochemistry*, 2006, **67**, 1378-1384 (*Rourinoside, activity*).
Tang, W.-Z. *et al.*, *Planta Med.*, 2007, **73**, 484-490 (*Illicium oligandrum glucoside*).
Wang, Y.F. *et al.*, *Fitoterapia*, 2008, **79**, 574-576 (*Tetracentron sinense glucoside*).
Hu, J.-M. *et al.*, *Planta Med.*, 2008, **74**, 535-539 (*Dendrobium longicornu glucoside*).
Yao, G. *et al.*, *Yaoxue Xuebao*, 2008, **43**, 284-290 (*Pieris japonica glucoside*).
Huo, C. *et al.*, *Chem. Nat. Compd. (Engl. Transl.)*, 2010, **46**, 343-347 (*Symplocos caudata glucoside*).
Fang, L. *et al.*, *J. Nat. Prod.*, 2010, **73**, 818-824 (*Illicium difengpi constit, activity*).
Cheng, X. *et al.*, *Planta Med.*, 2011, **77**, 2061-2065 (*Geum japonicum rhamnosides*).
Hu, J. *et al.*, *J. Asian Nat. Prod. Res.*, 2012, **14**, 171-175 (*Sanguisorba officinalis constits*).
Huang, X.-X. *et al.*, *Bioorg. Med. Chem. Lett.*, 2013, **23**, 5599-5604 (*Pinnatifidanin B VII*).
Xu, Y. *et al.*, *J. Funct. Foods*, 2017, **32**, 160-169 (*Rubuslin E, activity*).
Qu, L. *et al.*, *Phytochem. Lett.*, 2017, **19**, 39-45 (*Morineoliganoside C, activity*).
Matsumoto, T. *et al.*, *Phytochemistry*, 2017, **137**, 101-108 (*Isodonoside I*).
Du, B. *et al.*, *Fitoterapia*, 2018, **129**, 220-225 (*Asprenol H*).
Zhao, P. *et al.*, *Fitoterapia*, 2019, **136**, 104164:1-7 (*Crataegus pinnatifida constits, cd, activity*).
Du, Y.-Q. *et al.*, *Fitoterapia*, 2019, **139**, 104403:1-9 (*Ailanthus altissima constit, cd, activity*)

4,8′-Oxylignan-3,3′,4′,7,7′,9,9′-heptol O45

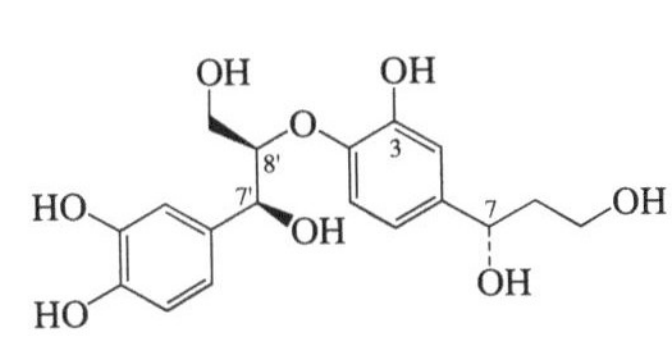

(7S,7′S,8′R)-form

3,3′,4′,7,7′,9,9′-Heptahydroxy-4,8′-oxyneolignan
$C_{18}H_{22}O_8$ 366.366

(7S,8′R)-form

7′-Ketone, 3,3′-di-Me ether
4′,7,9,9′-Tetrahydroxy-3,3′-dimethoxy-4,8′-oxylignan-7′-one. 4′,7,9,9′-Tetrahydroxy-3,3′-dimethoxy-7′-oxo-4,8′-oxylignan

7′-Ketone, 3,3′-di-Me ether, 4′-O-β-D-glucopyranoside
[1627676-86-1]
$C_{26}H_{34}O_{13}$ 554.545
Constit. of the fruits of *Arctium lappa*. Amorph. powder. $[\alpha]_D^{20}$ -31.4 (c, 0.07 in MeOH)

(7S,8′S)-form

7′-Ketone, 3,3′-di-Me ether, 4′-O-β-D-glucopyranoside
[1627676-87-2]
Constit. of the fruits of *Arctium lappa*. Amorph. powder. $[\alpha]_D^{20}$ -44.8 (c, 0.07 in MeOH)

(7′S,8′R)-form

7-Ketone, 3,3′-di-Me ether
4′,7′,9,9′-Tetrahydroxy-3,3′-dimethoxy-4,8′-oxylignan-7-one. 4′,7′,9,9′-Tetrahydroxy-3,3′-dimethoxy-7-oxo-4,8′-oxylignan

7-Ketone, 3,3′-di-Me ether, 4′-O-β-D-glucopyranoside
Arctiinoside A
[1421638-96-1]
$C_{26}H_{34}O_{13}$ 554.545
Constit. of the dried fruit of *Arctium lappa*. Amorph. powder. $[\alpha]_D^{20}$ -6.4 (c, 0.04 in MeOH). λ_{max} 225; 280; 305 (MeOH)

(7′S,8′S)-form

7-Ketone, 3,3′-di-Me ether
Nicotphenol A. *Tobphenol C*
[1356958-65-0]
$C_{20}H_{24}O_8$ 392.404
Constit. of tobacco leaves, *Nicotiana tabacum*. Cytotoxic. Shows modest anti-HIV activity. Amorph. solid. $[\alpha]_D^{24}$ +32.6 (c, 0.22 in MeOH). λ_{max} 210 (log ε 4.91); 287 (log ε 4.02); 328 (log ε 3.12) (MeOH)

7-Ketone, 3,3′-di-Me ether, 9′-Ac
Difenneolignan A
[1870003-59-0]
$C_{22}H_{26}O_9$ 434.441
Constit. of the stem bark of *Illicium difengpi*. Exhibits anti-HIV-1 activity. Amorph. solid. $[\alpha]_D^{24.8}$ +35.3 (c, 0.25 in MeOH). λ_{max} 210 (log ε 4.04); 282 (log ε 3.73); 328 (log ε 3.14) (MeOH)

7-Ketone, 3,3′,4′-tri-Me ether

7′,9,9′-Trihydroxy-3,3′,4′-trimethoxy-4,8′-oxylignan-7-one. 7′,9,9′-Trihydroxy-3,3′,4′-trimethoxy-7-oxo-4,8′-oxylignan.

Nicotphenol B

[1430237-66-3]

$C_{21}H_{26}O_8$ 406.431

Constit. of tobacco leaves, *Nicotiana tabacum*. Cytotoxic and shows modest anti-HIV activity. Amorph. solid. $[\alpha]_D^{24}$ +35.4 (c, 0.24 in MeOH). λ_{max} 210 (log ε 4.86); 285 (log ε 4.06); 325 (log ε 3.08) (MeOH).

Chen, Y.-K. *et al., J. Asian Nat. Prod. Res.*, 2012, **14**, 450-456 (*Nicotphenols A,B, activity*).
Yang, Y.-N. *et al., J. Asian Nat. Prod. Res.*, 2012, **14**, 981-985 (*Arctiinoside A*).
Yang, Y.-N. *et al., J. Agric. Food Chem.*, 2014, **62**, 9095-9102 (*Arctium lappa glucosides*).
Yang, Y. *et al., Chem. Nat. Compd. (Engl. Transl.)*, 2016, **52**, 43-47 (*Difenneolignan A, activity*)

4,8′-Oxylignan-3,3′,4′,7′,8,9,9′-heptol O46

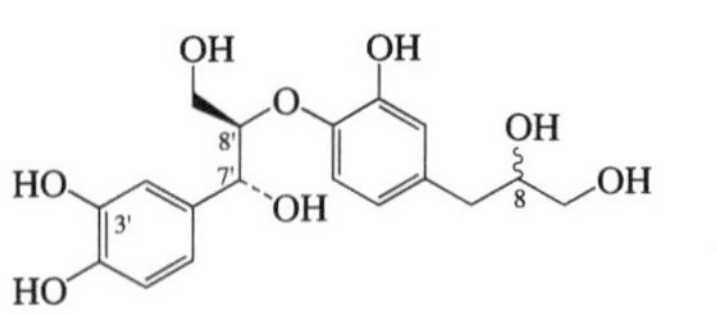

3,3′,4′,7′,8,9,9′-Heptahydroxy-4,8′-oxyneolignan

$C_{18}H_{22}O_8$ 366.366

(7′R*,8ξ,8′R*)-form

3,3′-Di-Me ether

3,3′-Dimethoxy-4,8′-oxylignan-4′,7′,8,9,9′-pentol. 4′,7′,8,9,9′-Pentahydroxy-3,3′-dimethoxy-4,8′-oxylignan

[220028-32-0]

$C_{20}H_{26}O_8$ 394.42

Constit. of *Zantedeschia aethiopica*. $[\alpha]_D$ +24.4 (c, 0.4 in EtOH).

Della Greca, M. *et al., Phytochemistry*, 1998, **49**, 1299-1304 (*Zantedeschia aethiopica constit*)

3,8′-Oxylignan-3′,4,4′,7′,9,9′-hexol O47

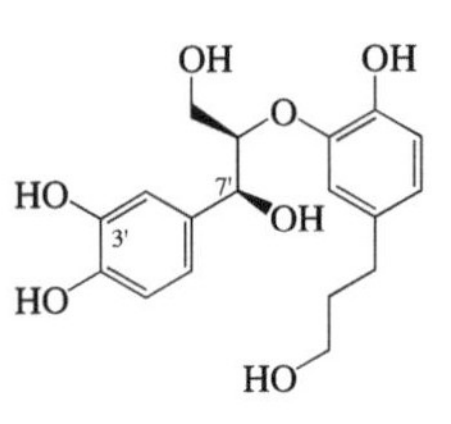

3′,4,4′,7′,9,9′-Hexahydroxy-3,8′-oxyneolignan

$C_{18}H_{22}O_7$ 350.367

(7′S,8′R)-form

3′-Me ether

3′-Methoxy-3,8′-oxylignan-4,4′,7′,9,9′-pentol. 4,4′,7′,9,9′-Pentahydroxy-3′-methoxy-3,8′-oxylignan

3′-Me ether, 4-O-β-D-glucopyranoside

Morineoliganoside B

[2209122-27-8]

$C_{25}H_{34}O_{12}$ 526.535

Constit. of the stems of *Chrysanthemum morifolium*. Powder. $[\alpha]_D^{25}$ -60.4 (c, 1.0 in MeOH). λ_{max} 227 (log ε 4.08); 279 (log ε 4.01) (MeOH).

Qu, L. *et al., Phytochem. Lett.*, 2017, **19**, 39-45 (*Morineoliganoside B, cd*)

4,4′-Oxylignan-3,3′,7,7′,9,9′-hexol O48

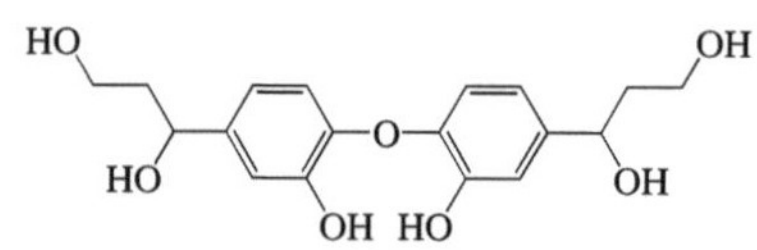

3,3′,7,7′,9,9′-Hexahydroxy-4,4′-oxyneolignan

$C_{18}H_{22}O_7$ 350.367

7,7′-Diketone

3,3′,9,9′-Tetrahydroxy-4,4′-oxylignan-7,7′-dione

7,7′-Diketone, 3,3′-di-Me ether

9,9′-Dihydroxy-3,3′-dimethoxy-4,4′-oxylignan-7,7′-dione.

Claulignan

[1662674-51-2]

$C_{20}H_{22}O_7$ 374.389

Constit. of the roots and stems of *Clausena lansium*. Exhibits cytotoxic activity. Brown powder. $[\alpha]_D^{21}$ -12.9 (c, 0.1 in MeOH). Unexplained opt. rotn.. λ_{max} 205 (log ε 4.48); 228 (log ε 4.48); 276 (log ε 4.30); 303 (log ε 4.23) (MeOH).

Song, W.-W. *et al., Zhiwu Fenlei Yu Ziyuan Xuebao*, 2014, **36**, 545-550 (*Claulignan, activity*)

4,8′-Oxylignan-3,3′,4′,5′,7′,9-hexol O49

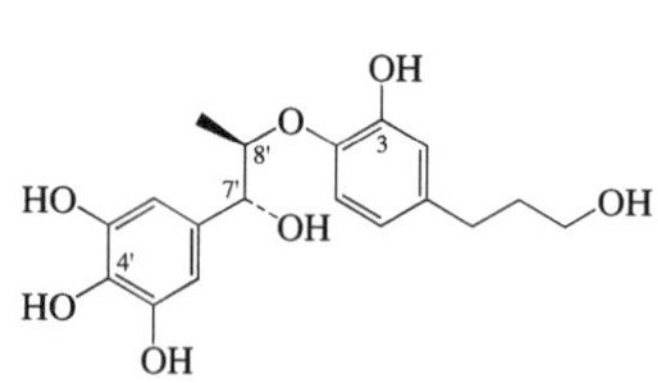

(7′*R*,8′*R*)-*form*

3,3′,4′,5′,7′,9-Hexahydroxy-4,8′-oxyneolignan

$C_{18}H_{22}O_7$ 350.367

(7′R,8′R)-form

3,3′,4′,5′-Tetra-Me ether

3,3′,4′,5′-Tetramethoxy-4,8′-oxylignan-7′,9-diol. 7′,9-Dihydroxy-3,3′,4′,5′-tetramethoxy-4,8′-oxylignan.

Surinamensinol A

[1402544-68-6]

$C_{22}H_{30}O_7$ 406.475

Constit. of the rhizomes of *Acorus gramineus*. Antitumour and antiinflammatory agent. Exhibits antiproliferative activity. NO prodn. inhibitor. Oil. $[\alpha]_D^{25}$ -62.1 (c, 0.3 in MeOH). λ_{max} 206 (log ε 3.8); 231 (log ε 2.6); 282 (log ε 1.9) (MeOH)

(7′S,8′R)-form

3,3′,4′,5′-Tetra-Me ether

Surinamensinol B

[1402544-69-7]

$C_{22}H_{30}O_7$ 406.475

Constit. of the rhizomes of *Acorus gramineus*. Antitumour and antiinflammatory agent. Exhibits antiproliferative activity. NO prodn. inhibitor. Oil. $[\alpha]_D^{25}$ -11.7 (c, 0.2 in MeOH). λ_{max} 206 (log ε 3.9); 232 (log ε 2.7); 281 (log ε 1.8) (MeOH).

Kim, K.H. *et al., Bioorg. Med. Chem. Lett.*, 2012, **22**, 6155-6159 (*Surinamensinols A,B, activity*).
Reddy, P.R. *et al., RSC Adv.*, 2014, **4**, 7432-7434 (*Surinamensinol A, synth*).
Lalwani, K.G. *et al., Eur. J. Org. Chem.*, 2015, 7344-7351 (*Surinamensinol B, synth*)

4,8′-Oxylignan-3,3′,4′,5′,9,9′-hexol O50

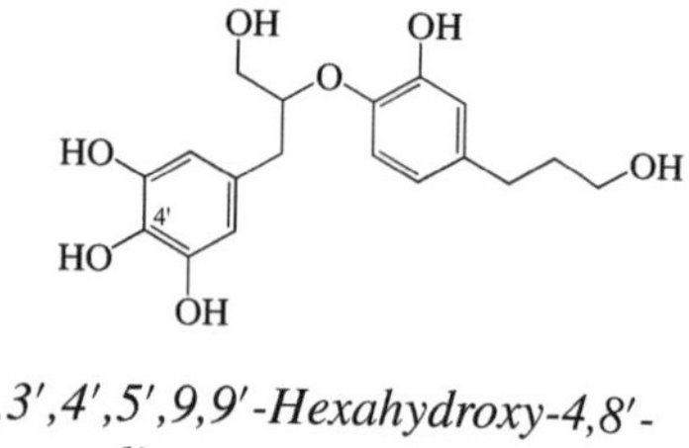

3,3′,4′,5′,9,9′-Hexahydroxy-4,8′-oxyneolignan

$C_{18}H_{22}O_7$ 350.367

(ξ)-form

3′,5′-Di-Me ether

3′,5′-Dimethoxy-4,8′-oxylignan-3,4′,9,9′-tetrol. 3,4′,9,9′-Tetrahydroxy-3′,5′-dimethoxy-4,8′-oxylignan

3′,5′-Di-Me ether, 4′-O-β-D-glucopyranoside
$C_{26}H_{36}O_{12}$ 540.562
Constit. of rhizomes of *Smilax trinervula*. Powder (MeOH). $[\alpha]^{25}_{D}$ -18 (c, 0.003 in MeOH). λ_{max} 231 (log ε 0.27) (MeOH).
Shu, J. *et al.*, *Fitoterapia*, 2015, **104**, 64-68 (*3′,5′-di-Me ether 4′-glucoside*)

4,8′-Oxylignan-3,3′,4′,7′,9,9′-hexol O51

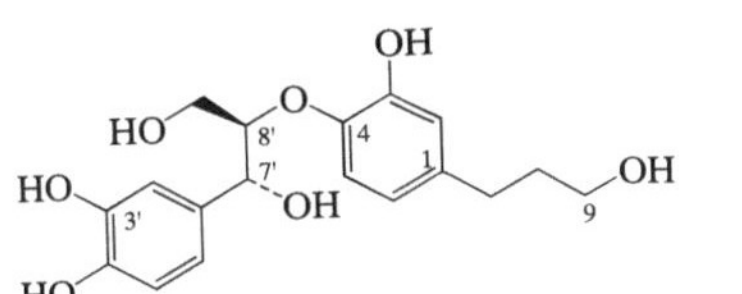

(7′R,8′R)-form

3,3′,4′,7′,9,9′-Hexahydroxy-4,8′-oxyneolignan
[57701-87-8] [171261-40-8]
$C_{18}H_{22}O_7$ 350.367

(7′R,8′R)-form

3′-Me ether
3′-Methoxy-4,8′-oxylignan-3,4′,7′,9,9′-pentol. 3,4′,7′,9,9′-Pentahydroxy-3′-methoxy-4,8′-oxylignan
[182211-72-9]
$C_{19}H_{24}O_7$ 364.394
Constit. of *Crataegus pinnatifida*

3′-Me ether, 3-O-β-D-glucopyranoside
[182057-03-0]
$C_{25}H_{34}O_{12}$ 526.535
Constit. of *Lonicera gracilipes* var. *glandulosa* and *Rhodiola crenulata*. Amorph. powder. λ_{max} 228 (log ε 4.16); 278 (log ε 3.73) (MeOH)

3′-Me ether, 4′-O-β-D-glucopyranoside
[1356023-61-4]
$C_{25}H_{34}O_{12}$ 526.535
Constit. of the roots of *Rhodiola crenulata*. Amorph. powder. $[\alpha]^{20}_{D}$ -35.3 (c, 0.01 in MeOH). λ_{max} 237; 276 (MeOH)

3′-Me ether, 9-O-β-D-glucopyranoside
Camellioside A
[1179525-23-5]
$C_{25}H_{34}O_{12}$ 526.535
Constit. of *Camellia amplexicaulis*. Amorph. powder. $[\alpha]_{D}$ +14.8 (c, 0.3 in MeOH). λ_{max} 227 (log ε 4.2); 277 (log ε 3.8) (MeOH)

3,3′-Di-Me ether
3,3′-Dimethoxy-4,8′-oxylignan-4′,7′,9,9′-tetrol. 4′,7′,9,9′-Tetrahydroxy-3,3′-dimethoxy-4,8′-oxylignan
[97133-59-0]
$C_{20}H_{26}O_7$ 378.421
Constit. of *Crataegus pinnatifida*

3,3′-Di-Me ether, 4′-O-β-D-glucopyranoside
[1009838-88-3]
$C_{26}H_{36}O_{12}$ 540.562
Constit. of *Arctium lappa* and *Symplocos caudata*. Amorph. powder. $[\alpha]^{25}_{D}$ -72 (c, 0.5 in MeOH). λ_{max} 225; 275 (MeOH)

3,3′-Di-Me ether, 7′-O-β-D-glucopyranoside
[1206731-40-9]
$C_{26}H_{36}O_{12}$ 540.562
Constit. of the leaves of *Glochidion rubrum*. Amorph. powder. $[\alpha]^{24}_{D}$ -17.4 (c, 0.35 in MeOH). λ_{max} 206 (log ε 4.05); 228 (log ε 3.71); 280 (log ε 3.26) (MeOH)

3,3′-Di-Me ether, 9-O-β-D-glucopyranoside
[182211-70-7]
Constit. of *Arctium lappa* and *Lonicera gracilipes* var. *glandulosa*. Amorph. powder. $[\alpha]_{D}$ -21.9 (c, 0.3 in MeOH). λ_{max} 227 (log ε 4.11); 277 (log ε 3.68) (MeOH)

3,3′-Di-Me ether, 9,9′-bis-O-(4-hydroxy-3-methoxy-E-cinnamoyl)
(+)-threo-Carolignan E
[1851318-73-4]
$C_{40}H_{42}O_{13}$ 730.763
Constit. of the aerial parts of *Euphorbia sikkimensis*. Amorph. solid. $[\alpha]^{20}_{D}$ +30.4 (c, 0.1 in $CHCl_3$). Obt. from racemate by chiral hplc. λ_{max} 230 (log ε 4.51); 284 (log ε 2.75); 320 (log ε 4.83) (MeOH)

3,3′,7′-Tri-Me ether
3,3′,7′-Trimethoxy-4,8′-oxylignan-4′,9,9′-triol. 4′,9,9′-Trihydroxy-3,3′,7′-trimethoxy-4,8′-oxylignan. ***Pinnatifidanin BV***
[1462882-60-5]
$C_{21}H_{28}O_7$ 392.448
Constit. of *Clematis lasiandra* and *Crataegus pinnatifida*. Antibacterial agent. Yellow oil. $[\alpha]^{20}_{D}$ -34 (c, 0.1 in MeOH). λ_{max} 232 (log ε 2.76) (MeOH)

3,3′,7′-Tri-Me ether, 9,9′-bis-O-(4-hydroxy-3-methoxy-E-cinnamoyl)
7′-O-Methyl-(+)-threo-carolignan E
[1851315-52-0]
$C_{41}H_{44}O_{13}$ 744.79
Constit. of the aerial parts of *Euphorbia sikkimensis*. Amorph. solid. $[\alpha]^{20}_{D}$ +32.6 (c, 0.2 in $CHCl_3$). λ_{max} 232 (log ε 4.36); 288 (log ε 2.92); 326 (log ε 4.64) (MeOH)

4′-O-[2-Hydroxy-1-(hydroxymethyl)ethyl], 3,3′-di-Me ether
4′-O-(2-Glyceryl)-3,3′-dimethoxy-4,8′-oxylignan-7′,9,9′-triol
[1227064-02-9]
$C_{23}H_{32}O_9$ 452.5
Constit. of the stem bark of *Illicium difengpi*. Amorph. powder. $[\alpha]^{20}_{D}$ -9 (c, 0.06 in MeOH). λ_{max} 206 (log ε 4.5); 279 (log ε 3.3) (MeOH)

7′-Ketone, 3,3′-di-Me ether
4′,9,9′-Trihydroxy-3,3′-dimethoxy-4,8′-oxylignan-7-one. 4′,9,9′-Trihydroxy-3,3′-dimethoxy-7-oxo-4,8′-oxylignan

7′-Ketone, 3,3′-di-Me ether, 4′-O-β-D-glucopyranoside
$C_{26}H_{34}O_{12}$ 538.546
Constit. of the fruit of *Arctium lappa*. Amorph. powder. $[\alpha]^{25}_{D}$ -58.3 (c, 0.31 in MeOH). λ_{max} 227 (log ε 3.20); 275 (log ε 2.94); 308 (log ε 2.67) (MeOH)

9-Carboxylic acid, 3,3′-di-Me ether
4′,7′,9′-Trihydroxy-3,3′-dimethoxy-4,8′-oxylignan-9-oic acid.
Asprenol F
$C_{20}H_{24}O_8$ 392.404
Constit. of *Ilex asprella*. Amorph. light yellow powder. $[\alpha]^{25}_{D}$ -8.0 (c, 0.1 in MeOH). λ_{max} 208 (log ε 3.20); 229 (log ε 2.90); 279 (log ε 2.49) (MeOH)

(7′R,8′S)-form

3′-Me ether
[144993-46-4]
Constit. of *Juniperus chinensis*. $[\alpha]^{25}_{D}$ +11.7 (c, 3 in Me_2CO)

3′-Me ether, 3-O-β-D-glucopyranoside
[1018987-05-7]
Constit. of *Iodes cirrhosa* and *Rhodiola crenulata*. Amorph. powder. $[\alpha]^{20}_{D}$ -6.2 (c, 0.08 in MeOH). λ_{max} 205 (log ε 4.3); 261 (log ε 3.9) (MeOH)

3′-Me ether, 9-O-β-D-glucopyranoside
[1018987-07-9]
$C_{25}H_{34}O_{12}$ 526.535
Constit. of the roots of *Iodes cirrhosa*. Gum. $[\alpha]^{20}_{D}$ -12.5 (c, 0.08

in MeOH). λ_{max} 207 (log ε 4.4); 280 (log ε 3.7) (MeOH)

3,3′-Di-Me ether, 4′-O-β-D-glucopyranoside
[1009838-89-4]
Constit. of *Arctium lappa*, *Rhodiola crenulata* and *Symplocos caudata*. Amorph. powder. $[\alpha]_D^{25}$ -28.6 (c, 0.14 in MeOH). λ_{max} 225; 275 (MeOH)

3,3′-Di-Me ether, 4′-O-[β-D-glucopyranosyl-(1→4)-β-D-glucopyranoside]
$C_{32}H_{46}O_{17}$ 702.703
Constit. of *Securidaca inappendiculata*. Yellow solid (MeOH). $[\alpha]_D^{25}$ -2.1 (c, 0.1 in MeOH). λ_{max} 225 (log ε 4.06); 276 (log ε 3.55) (MeOH)

3,3′-Di-Me ether, 7′-O-β-D-glucopyranoside
[1206731-41-0]
Constit. of the leaves of *Glochidion rubrum*. Amorph. powder. $[\alpha]_D^{23}$ -53.1 (c, 0.75 in MeOH). λ_{max} 207 (log ε 4.2); 226 (log ε 3.99); 279 (log ε 3.57) (MeOH)

3,3′-Di-Me ether, 9′-O-β-D-glucopyranoside
Camellignanoside A
$C_{26}H_{36}O_{12}$ 540.562
Constit. of Huangjinya green tea *Camellia sinensis*. Acetylcholinesterase inhibitor. Powder. $[\alpha]_D^{25}$ +4.5 (c, 0.4 in MeOH). λ_{max} 205; 228; 279 (MeOH)

3,3′-Di-Me ether, 9-O-(4-hydroxy-3-methoxy-E-cinnamoyl), 9′-O-(4-hydroxy-E-cinnamoyl)
erythro-**Carolignan X**
[1253289-76-7]
$C_{39}H_{40}O_{12}$ 700.737
Constit. of the bark of *Durio carinatus*. Amorph. solid. $[\alpha]_D^{22}$ -6.8 (c, 0.25 in MeOH). $[\alpha]_D^{22}$ +20.6 (c, 0.1 in $CHCl_3$)

3,3′-Di-Me ether, 9,9′-bis-O-(4-hydroxy-3-methoxy-E-cinnamoyl)
***(-)-erythro*-Carolignan E**
[344944-08-7]
$C_{40}H_{42}O_{13}$ 730.763
Constit. of the aerial parts of *Euphorbia sikkimensis*. Amorph. solid. $[\alpha]_D^{20}$ -16.4 (c, 0.1 in $CHCl_3$). Obt. from racemate by chiral hplc. λ_{max} 232 (log ε 4.36); 288 (log ε 3.92); 326 (log ε 4.64) (MeOH)

3,3′,4′-Tri-Me ether
3,3′,4′-Trimethoxy-4,8′-oxylignan-7′,9,9′-triol. 7′,9,9′-Trihydroxy-3,3′,4′-trimethoxy-4,8′-oxylignan. ***Schimalignan A***
[1309677-40-4]
$C_{21}H_{28}O_7$ 392.448
Constit. of the stems of *Schima superba*. Cytotoxic. Viscous oil. $[\alpha]_D^{18}$ +7.7 (c, 1.9 in MeOH). λ_{max} 230; 279 (MeOH)

3,3′,7′-Tri-Me ether
$[\alpha]_D^{20}$ -31 (c, 0.2 in MeOH)

3,3′,7′-Tri-Me ether, 9-O-β-D-glucopyranoside
Pinnatifidaninside B
[1820907-13-8]
$C_{27}H_{38}O_{12}$ 554.589
Constit. of the seeds of *Crataegus pinnatifida*. Exhibits strong free radical scavenging activity. Yellow oil. $[\alpha]_D^{20}$ -18.5 (c, 0.08 in MeOH). λ_{max} 231 (log ε 3.72); 279 (log ε 1.02) (MeOH)

3,3′,7′-Tri-Me ether, 9-O-(4-hydroxy-3-methoxy-E-cinnamoyl), 9′-O-(4-hydroxy-E-cinnamoyl)
erythro-**Carolignan Y**
[1253289-78-9]
$C_{40}H_{42}O_{12}$ 714.764
Constit. of the bark of *Durio oxleyanus*. Amorph. solid. $[\alpha]_D^{22}$ -2.6 (c, 0.2 in MeOH). $[\alpha]_D^{22}$ +48 (c, 0.5 in $CHCl_3$)

3,3′,7′-Tri-Me ether, 9,9′-bis-O-(4-hydroxy-3-methoxy-E-cinnamoyl)
7′-O-Methyl-(-)-*erythro*-carolignan E
[1851315-49-5]
$C_{41}H_{44}O_{13}$ 744.79
Constit. of the aerial parts of *Euphorbia sikkimensis*. Exhibits moderate anti-HIV activity. Amorph. solid. $[\alpha]_D^{20}$ -16.2 (c, 0.1 in $CHCl_3$). Obt. from racemate by chiral hplc. λ_{max} 231 (log ε 4.58); 288 (log ε 3.87); 327 (log ε 4.68) (MeOH)

4′-O-[2-Hydroxy-1-(hydroxymethyl)ethyl], 3,3′-di-Me ether
[1227162-60-8]
$C_{23}H_{32}O_9$ 452.5
Constit. of the stem bark of *Illicium difengpi*. Amorph. powder. $[\alpha]_D^{20}$ +4 (c, 0.08 in MeOH). λ_{max} 206 (log ε 4.3); 280 (log ε 3.7) (MeOH)

(7′S,8′R)-form

3-Me ether
3-Methoxy-4,8′-oxylignan-3′,4′,7′,9,9′-pentol. 3′,4′,7′,9,9′-Pentahydroxy-3-methoxy-4,8′-oxylignan
$C_{19}H_{24}O_7$ 364.394
Constit. of *Cephalotaxus koreana*. Amorph. powder. $[\alpha]_D^{24}$ +1.3 (c, 0.33 in MeOH). λ_{max} 227 (log ε 1.79); 278 (log ε 1.41) (MeOH)

3′-Me ether, 3-O-β-D-glucopyranoside
[1355467-98-9]
$C_{25}H_{34}O_{12}$ 526.535
Constit. of the roots of *Rhodiola crenulata*. Amorph. powder. $[\alpha]_D^{20}$ +19.3 (c, 0.03 in MeOH). λ_{max} 233; 278 (MeOH)

3′-Me ether, 4′-O-β-D-glucopyranoside
[1356023-60-3]
Constit. of the roots of *Rhodiola crenulata*. Amorph. powder. $[\alpha]_D^{20}$ +9.2 (c, 0.03 in MeOH). λ_{max} 235; 277 (MeOH)

3′-Me ether, 9-O-α-L-rhamnopyranoside
Pinusthunbergiside A
$C_{25}H_{34}O_{11}$ 510.536
Constit. of the needles of *Pinus thunbergii*. Moderately inhibits release of β-hexosaminidase induced by antigen stimulation in RBL-2H3 cells. Amorph. powder. $[\alpha]_D^{24}$ -31 (c, 0.02 in MeOH). λ_{max} 228 (log ε 4.15); 280 (log ε 3.78) (MeOH)

3,3′-Di-Me ether
[182265-18-5]
$C_{20}H_{26}O_7$ 378.421
Constit. of *Crataegus pinnatifida*

3,3′-Di-Me ether, 4′-O-β-D-glucopyranoside
[182057-06-3]
Constit. of *Arctium lappa*, *Lonicera gracilipes* var. *glandulosa* and *Symplocos caudata*. Amorph. powder. $[\alpha]_D$ -29.4 (c, 0.3 in MeOH). λ_{max} 224 (log ε 4.11); 276 (log ε 3.62) (MeOH)

3,3′-Di-Me ether, 7′-O-β-D-glucopyranoside
[182211-69-4]
Constit. of *Lonicera gracilipes* var. *glandulosa*. Amorph. powder. $[\alpha]_D$ +3.3 (c, 0.9 in MeOH). λ_{max} 227 (log ε 4.05); 279 (log ε 3.66) (MeOH)

3,3′-Di-Me ether, 9-O-α-L-rhamnopyranoside
[1380855-04-8]
$C_{26}H_{36}O_{11}$ 524.563
Constit. of *Epimedium pseudowushanese*. Exhibits moderate antiinflammatory activity. Amorph. powder. $[\alpha]^{20}_{D}$ +7.1 (c, 3.7 in MeOH). λ_{max} 230 (log ε 4.77); 280 (log ε 4.38) (MeOH)

3,3′-Di-Me ether, 9-O-β-D-glucopyranoside
[182211-71-8]
Constit. of *Lonicera gracilipes* var. *glandulosa*. Amorph. powder. $[\alpha]_{D}$ -2.9 (c, 0.3 in MeOH). λ_{max} 224 (log ε 4.1); 277 (log ε 3.65) (MeOH)

3,3′-Di-Me ether, 9′-O-β-D-glucopyranoside
Isodonoside II
[2095793-37-4]
$C_{26}H_{36}O_{12}$ 540.562
Constit. of the aerial parts of *Isodon japonicus*. Powder. $[\alpha]^{25}_{D}$ -35.3 (c, 0.3 in MeOH). λ_{max} 230 (log ε 3.6); 280 (log ε 3.2) (MeOH)

3,3′-Di-Me ether, 9-O-[β-D-apiofuranosyl-(1→6)-β-D-glucopyranoside]
$C_{31}H_{44}O_{16}$ 672.677
Constit. of the fruit of *Arctium lappa*. Amorph. powder. $[\alpha]^{25}_{D}$ -8.5 (c, 0.2 in MeOH). λ_{max} 230 (log ε 3.41); 280 (log ε 3.00) (MeOH)

3,3′-Di-Me ether, 9,9′-bis-O-(4-hydroxy-3-methoxy-E-cinnamoyl)
(+)-erythro-Carolignan E
[1851318-75-6]
$C_{40}H_{42}O_{13}$ 730.763
Constit. of the aerial parts of *Euphorbia sikkimensis*. Amorph. solid. $[\alpha]^{20}_{D}$ +16.6 (c, 0.1 in $CHCl_3$). Obt. from racemate by chiral hplc. λ_{max} 232 (log ε 4.36); 288 (log ε 2.92); 326 (log ε 4.64) (MeOH)

3,3′,4′-Tri-Me ether
[874448-64-3]
$C_{21}H_{28}O_7$ 392.448
Constit. of the flower buds of *Magnolia fargesii*. Oil. λ_{max} 279 (log ε 3.79) ($CHCl_3$)

3,3′,7′-Tri-Me ether
Pinnatifidanin BVI
[1462882-64-9]
$C_{21}H_{28}O_7$ 392.448
Constit. of *Clematis lasiandra* and *Crataegus pinnatifida*. Antibacterial agent. Yellow oil. $[\alpha]^{20}_{D}$ +29 (c, 0.2 in MeOH). λ_{max} 231 (log ε 2.75) (MeOH)

3,3′,7′-Tri-Me ether, 9,9′-bis-O-(4-hydroxy-3-methoxy-E-cinnamoyl)
7′-O-Methyl-(+)-carolignan E
[1851315-50-8]
$C_{41}H_{44}O_{13}$ 744.79
Constit. of the aerial parts of *Euphorbia sikkimensis*. Exhibits moderate anti-HIV activity. Amorph. solid. $[\alpha]^{20}_{D}$ +17.1 (c, 0.1 in $CHCl_3$). λ_{max} 231 (log ε 4.58); 288 (log ε 3.87); 327 (log ε 4.68) (MeOH)

9-Carboxylic acid, 3,3′-di-Me ether
Asprenol E
$C_{20}H_{24}O_8$ 392.404
Constit. of *Ilex asprella*. Amorph. light yellow powder. $[\alpha]^{25}_{D}$ -8.0 (c, 0.1 in MeOH). λ_{max} 207 (log ε 3.19); 226 (log ε 2.77); 280 (log ε 2.35) (MeOH)

(7′S,8′S)-form

3′-Me ether, 3-O-β-D-glucopyranoside
[1355468-00-6]
Constit. of the roots of *Rhodiola crenulata*. Amorph. powder. $[\alpha]^{20}_{D}$ +12.3 (c, 0.03 in MeOH). λ_{max} 225; 275 (MeOH)

3′-Me ether, 9,9′-bis-O-(4-hydroxy-3-methoxy-E-cinnamoyl)
3′-O-Demethyl-threo-carolignan E
[718606-94-1]
$C_{39}H_{40}O_{13}$ 716.736
Constit. of the stems of *Hibiscus taiwanensis*. Amorph. powder. $[\alpha]_{D}$ -4.5 (c, 0.12 in MeOH). Possesses (7*S*,8*S*)-config.. λ_{max} 204 (log ε 4.75); 232 (log ε 4.35); 289 (log ε 4.3); 326 (log ε 4.41) (MeOH)

3,3′-Di-Me ether, 4′-O-β-D-glucopyranoside
[1009838-87-2]
Constit. of *Arctium lappa* and *Symplocos caudata*. Amorph. powder. $[\alpha]^{25}_{D}$ +20 (c, 0.4 in MeOH). λ_{max} 225; 275 (MeOH)

3,3′-Di-Me ether, 7′-O-β-D-glucopyranoside
[182056-97-9]
Constit. of *Camellia amplexicaulis*, *Clethra barbinervis*, *Disporum viridescens* and *Lonicera gracilipes* var. *glandulosa*. Isol. from Huangjinya green tea *Camellia sinensis*. Osteoblast cell differentiation promotor. Acetylcholinesterase inhibitor. Amorph. powder. $[\alpha]_{D}$ -57.3 (c, 0.2 in MeOH). $[\alpha]^{25}_{D}$ +2.6 (c, 0.08 in MeOH). λ_{max} 226 (log ε 4.1); 278 (log ε 3.61) (MeOH)

3,3′-Di-Me ether, 9-O-β-D-glucopyranoside
[112295-60-0] (unspecified stereochem.)
Constit. of the roots of *Iodes cirrhosa*. Amorph. solid. $[\alpha]^{20}_{D}$ +2 (c, 1.2 in MeOH). λ_{max} 230 (log ε 4.1); 280 (log ε 3.4) (MeOH)

3,3′-Di-Me ether, 9-O-(4-hydroxy-3-methoxy-E-cinnamoyl), 9′-O-(4-hydroxy-E-cinnamoyl)
threo-Carolignan X
[1253289-75-6]
$C_{39}H_{40}O_{12}$ 700.737
Constit. of the bark of *Durio carinatus*. Amorph. solid. $[\alpha]^{22}_{D}$ -14.4 (c, 0.1 in MeOH). $[\alpha]^{22}_{D}$ -6.3 (c, 0.1 in $CHCl_3$)

3,3′-Di-Me ether, 9,9′-bis-O-(4-hydroxy-3-methoxy-E-cinnamoyl)
(-)-threo-Carolignan E
[344944-09-8]
$C_{40}H_{42}O_{13}$ 730.763
Constit. of the aerial parts of *Euphorbia sikkimensis*. Amorph. solid. $[\alpha]^{20}_{D}$ -29.8 (c, 0.1 in $CHCl_3$). λ_{max} 230 (log ε 4.51); 284 (log ε 2.75); 320 (log ε 4.83) (MeOH)

3,3′,7′-Tri-Me ether
$[\alpha]^{20}_{D}$ +34 (c, 0.1 in MeOH)

3,3′,7′-Tri-Me ether, 9-O-β-D-glucopyranoside
Pinnatifidaninside A
[1820907-12-7]
$C_{27}H_{38}O_{12}$ 554.589
Constit. of the seeds of *Crataegus pinnatifida*. Exhibits strong free radical scavenging activity. Yellow oil. $[\alpha]^{20}_{D}$ +8.2 (c, 0.10 in MeOH). λ_{max} 229 (log ε 3.72); 280 (log ε 1.39) (MeOH)

3,3′,7′-Tri-Me ether, 9-O-(4-hydroxy-3-methoxy-E-cinnamoyl), 9′-O-(4-hydroxy-E-cinnamoyl)
threo-Carolignan Y
[1253289-77-8]
$C_{40}H_{42}O_{12}$ 714.764
Constit. of the bark of *Durio oxleyanus*. Amorph. solid. $[\alpha]^{22}_{D}$ -16.1 (c, 0.2 in MeOH). $[\alpha]^{22}_{D}$ +32.6 (c, 0.2 in $CHCl_3$)

3,3′,7′-Tri-Me ether, 9,9′-bis-O-(4-hydroxy-3-methoxy-E-cinnamoyl)
7′-O-Methyl-(-)-threo-carolignan E
[1851315-51-9]
$C_{41}H_{44}O_{13}$ 744.79
Constit. of the aerial parts of *Euphorbia sikkimensis*. Amorph. solid. $[\alpha]_D^{20}$ -35.1 (c, 0.2 in $CHCl_3$). λ_{max} 232 (log ε 4.36); 288 (log ε 2.92); 326 (log ε 4.64) (MeOH)

(7′R*,8′R*)-form

3′-Me ether
[62137-27-3]
Constit. of *Pinus massoniana* and *Pinus sylvestris*. $[\alpha]_D^{20}$ +35.4 (c, 0.7 in EtOH)

3′-Me ether, 4′-O-β-D-xylopyranoside
[62223-56-7]
$C_{24}H_{32}O_{11}$ 496.509
Constit. of *Pinus massoniana*, *Pinus sylvestris* and *Picea abies*. Amorph.. Obt. as a mixt. of *erythro-* and *threo-*isomers

3′-Me ether, 7′-O-β-D-glucopyranoside
[62080-74-4]
$C_{25}H_{34}O_{12}$ 526.535
Constit. of *Pinus massoniana*, *Pinus sylvestris* and *Viburnum dilatum*. Amorph.. $[\alpha]_D^{20}$ -7.1 (c, 3 in EtOH)

3′-Me ether, 9′-O-β-D-glucopyranoside
[97055-95-3]
$C_{25}H_{34}O_{12}$ 526.535
Constit. of *Pinus massoniana* and *Epimedium grandiflorum*. Obt. as a mixt. of *erythro-* and *threo-*isomers

3′-Me ether, 9-O-α-L-rhamnopyranoside
[35128-76-8]
$C_{25}H_{34}O_{11}$ 510.536
Constit. of *Pinus massoniana*, *Pinus sylvestris*, *Epimedium grandiflorum* and *Thuja plicata*. Amorph.. $[\alpha]_D^{20}$ -58.1 (c, 2.8 in EtOH)

3′-Me ether, 9-O-(4-hydroxy-3-methoxy-E-cinnamoyl)
Sambucunol B
[1017252-24-2]
$C_{29}H_{32}O_{10}$ 540.565
Constit. of the stems of *Sambucus williamsii*. Yellowish oil. $[\alpha]_D^{26}$ +5.2 (c, 0.1 in MeOH). λ_{max} 231 (log ε 3.35); 284 (log ε 3.02) (MeOH)

3,3′-Di-Me ether, 4′-O-β-D-xylopyranoside
$C_{25}H_{34}O_{11}$ 510.536
Constit. of the root bark of *Picea abies*. $[\alpha]_D$ -46 (c, 0.3 in MeOH). A mixt. of *threo-* and *erythro-*isomers (1:7)

3,3′-Di-Me ether, 7′-O-α-L-rhamnopyranoside
[1309459-88-8]
$C_{26}H_{36}O_{11}$ 524.563
Constit. of *Illicium henryi*. Amorph. powder. $[\alpha]_D^{26}$ -36.4 (c, 0.05 in MeOH). λ_{max} 224 (log ε 4.16); 280 (log ε 3.71) ($CHCl_3$)

3,3′-Di-Me ether, 9-O-α-L-rhamnopyranoside
[97055-96-4]
$C_{26}H_{36}O_{11}$ 524.563
Constit. of *Pinus massoniana* and *Epimedium grandiflorum*. Obt. as a mixt. of *erythro-* and *threo-*isomers

3,3′-Di-Me ether, 9-O-β-D-glucopyranoside
[126253-43-8]
[97065-25-3] (unspecified stereochem.)
Constit. of *Pinus massoniana* and *Epimedium grandiflorum*. Amorph.. $[\alpha]_D^{22}$ -16.5 (c, 0.4 in MeOH)

3,3′-Di-Me ether, 9,9′-bis-O-(4-hydroxy-3-methoxycinnamoyl)(E,E-)
*threo-**Carolignan E***
$C_{40}H_{42}O_{13}$ 730.763
Constit. of *Durio carinatus*, *Ochroma lagopus* and *Hibiscus cannabinus*. Oil. $[\alpha]_D^{22}$ -4.2 (c, 0.38 in MeOH)

3,3′-Di-Me ether, 9,9′-bis-O-(4-hydroxy-3-methoxycinnamoyl)(E,Z-)
Carolignan D
[171866-18-5]
$C_{40}H_{42}O_{13}$ 730.763
Constit. of *Ochroma lagopus*

3,3′-Di-Me ether, 9,9′-bis-O-(4-hydroxy-3-methoxycinnamoyl)(Z,E-)
Carolignan C
[171868-03-4]
$C_{40}H_{42}O_{13}$ 730.763
Constit. of *Ochroma lagopus*

(7′R*,8′S*)-form

3′-Me ether, 9′-O-β-D-xylopyranoside
Taxuyunin A
[632336-68-6]
$C_{24}H_{32}O_{11}$ 496.509
Constit. of the bark of *Taxus yunnanensis*. Amorph. solid. $[\alpha]_D^{28}$ +58.3 (c, 0.15 in MeOH)

3′-Me ether, 9-O-(4-hydroxy-3-methoxy-E-cinnamoyl)
Sambucunol A
[1017252-23-1]
$C_{29}H_{32}O_{10}$ 540.565
Constit. of the stems of *Sambucus williamsii*. Yellowish oil. $[\alpha]_D^{26}$ +4.9 (c, 0.1 in MeOH). λ_{max} 230 (log ε 3.35); 284 (log ε 3.02) (MeOH)

3,3′-Di-Me ether, 4′-O-α-L-rhamnopyranoside
Rubrumoside A
[1357621-94-3]
$C_{26}H_{36}O_{11}$ 524.563
Constit. of the bark of *Acer rubrum*. Amorph. solid. $[\alpha]_D^{20}$ -10 (c, 0.1 in MeOH). λ_{max} 216 (log ε 3.93); 230 (sh); 277 (log ε 3.42) (MeOH)

3,3′-Di-Me ether, 4′-O-[β-D-glucopyranosyl-(1→2)-[β-D-glucopyranosyl-(1→6)]-β-D-glucopyranoside]
[1384878-53-8]
$C_{38}H_{56}O_{22}$ 864.844
Isol. from the exocarp of *Castanea henryi*. Powder. Mp 115 - 116°. $[\alpha]_D^{20}$ -33 (c, 0.001 in MeOH). λ_{max} 210 (log ε 4.58); 254 (log ε 3.92); 280 (log ε 4.23) (MeOH)

3,3′-Di-Me ether, 9′-O-(4-hydroxy-3-methoxy-E-cinnamoyl), 9-O-(4-hydroxy-E-cinnamoyl)
Turformosin A
[1415714-42-9]
$C_{39}H_{40}O_{12}$ 700.737
Constit. of the stems of *Turpinia formosana*. Amorph. solid. $[\alpha]_D^{25}$ +16.6 (c, 0.2 in MeOH). λ_{max} 229 (log ε 4.25); 289 (log ε 4.13); 317 (log ε 4.3) (MeOH)

3,3′-Di-Me ether, 9,9′-bis-O-(4-hydroxy-3-methoxy-E-cinnamoyl)
*erythro-**Carolignan E***
$C_{40}H_{42}O_{13}$ 730.763
Constit. of *Durio carinatus*, *Ochroma lagopus* and *Hibiscus cannabinus*. Oil

3,3′-Di-Me ether, 9-O-(4-hydroxy-3-methoxy-E-cinnamoyl)
*threo-**Carolignan H***
[344944-05-4]
$C_{30}H_{34}O_{10}$ 554.592
Constit. of *Clematis armandii* and *Hibiscus cannabinus*. Yellowish oil

(7′RS,8′RS)-form

3,3′,7′-Tri-Me ether
$C_{21}H_{28}O_7$ 392.448

Constit. of *Ailanthus altissima* and *Crataegus pinnatifida*. Light yellow gum. $[\alpha]^{20}_{D}$ +0.4 (c, 0.1 in MeOH). Partial racemate, resolved by chiral hplc

(7′RS,8′SR)-form

3,3′,7′-Tri-Me ether

Constit. of *Ailanthus altissima* and *Crataegus pinnatifida*. Light yellow gum. $[\alpha]^{20}_{D}$ -2.1 (c, 0.1 in MeOH). Partial racemate resolved by chiral hplc. λ_{max} 252 (log ε 3.64); 279 (log ε 3.90) (MeOH).

Popoff, T. *et al.*, *Acta Chem. Scand., Ser. B*, 1977, **31**, 329-337 (*Pinus sylvestris constits*).
Higuchi, R. *et al.*, *Phytochemistry*, 1977, **16**, 1587-1590 (*Pinus contorta glycosides*).
Lundgren, L.N. *et al.*, *Acta Chem. Scand., Ser. B*, 1985, **39**, 241-248 (*Pinus massoniana constit*).
Miyase, T. *et al.*, *Chem. Pharm. Bull.*, 1987, **35**, 3713-3719 (*Epimedium grandiflorum constits*).
Fang, J.-M. *et al.*, *Phytochemistry*, 1992, **31**, 3659-3661 (*Juniperus chinensis constit*).
Kuono, I. *et al.*, *Phytochemistry*, 1993, **32**, 1573-1577 (*Illicium difengpi constits*).
Pan, H. *et al.*, *Phytochemistry*, 1995, **39**, 1423-1428 (*Picea abies constits*).
Paula, V.F. *et al.*, *Tetrahedron*, 1995, **51**, 12453-12462 (*Carolignans*).
Matsuda, N. *et al.*, *Chem. Pharm. Bull.*, 1996, **44**, 1676-1679 (*Lonicera gracilipes constits, abs config*).
Seca, A.M.L. *et al.*, *Phytochemistry*, 2001, **56**, 759-767 (*Hibiscus cannabinus constits*).
Li, S.-H. *et al.*, *Chin. J. Chem.*, 2003, **21**, 926-930 (*Taxuyunin A*).
Wu, P.-L. *et al.*, *Bioorg. Med. Chem.*, 2004, **12**, 2193-2197 (*3-Demethylcarolignan E*).
Lee, J. *et al.*, *Bull. Korean Chem. Soc.*, 2005, **26**, 913-915 (*Magnolia fargesii tri-Me ether*).
Sinkkonen, J. *et al.*, *Magn. Reson. Chem.*, 2006, **44**, 633-636 (*3,3′-di-Me ether, pmr, cmr*).
Yang, X.-J. *et al.*, *J. Asian Nat. Prod. Res.*, 2007, **9**, 583-591 (*Sambucunols A,B*).
Gan, M. *et al.*, *J. Nat. Prod.*, 2008, **71**, 647-654 (*Iodes cirrhosa glucosides*).
Huo, C. *et al.*, *Phytochemistry*, 2008, **69**, 788-795 (*3,3′-di-Me 4-glucosides*).
Tung, N.H. *et al.*, *Chem. Pharm. Bull.*, 2009, **57**, 65-69 (*Camellioside A*).
Cai, W.-H. *et al.*, *J. Nat. Med.*, 2009, **63**, 408-414 (*Glochidion rubrum glucosides*).
Yoon, K.D. *et al.*, *Bull. Korean Chem. Soc.*, 2010, **31**, 495-496 (*Cephalotaxus koreana constit*).
Xu, W. *et al.*, *J. Asian Nat. Prod. Res.*, 2010, **12**, 874-878 (*Schimalignan A*).
Fang, L. *et al.*, *J. Nat. Prod.*, 2010, **73**, 818-824 (*Illicium difengpi constits*).
Rudiyansyah, *et al.*, *J. Nat. Prod.*, 2010, **73**, 1649-1654 (*Carolignans X,Y*).
Liu, J.-F. *et al.*, *Chem. Biodiversity*, 2011, **8**, 692-698 (*Illicium henryi rhamnoside*).
Nakamura, S. *et al.*, *Chem. Pharm. Bull.*, 2011, **59**, 1020-1028 (*Salacia chinensis constit, activity*).
Wu, B. *et al.*, *Carbohydr. Res.*, 2012, **355**, 45-49 (*Castanea henryi constit*).
Yang, Y.-N. *et al.*, *J. Agric. Food Chem.*, 2012, **60**, 964-972 (*Rhodiola crenulata glucosides*).
Huang, H.-C. *et al.*, *Molecules*, 2012, **17**, 1837-1851 (*Turformosin A*).
Yuan, T. *et al.*, *Tetrahedron*, 2012, **68**, 959-964 (*Rubrumoside A*).
Huang, X.-X. *et al.*, *Bioorg. Med. Chem. Lett.*, 2013, **23**, 5599-5604 (*Crataegus pinnatifida constits*).
Hong, S.S. *et al.*, *Fitoterapia*, 2014, **99**, 347-351 (*Pinusthunbergiside A, activity*).
Yang, Y.-N. *et al.*, *J. Agric. Food Chem.*, 2014, **62**, 9095-9102 (*Arctium lappa 4′-glucosides*).
Huang, X.-X. *et al.*, *Planta Med.*, 2014, **80**, 1732-1738 (*Pinnatifidaninsides A,B, activity*).
Huang, X.-Y. *et al.*, *J. Asian Nat. Prod. Res.*, 2015, **17**, 504-511 (*Arctium lappa glycosides*).
Jiang, C. *et al.*, *J. Nat. Prod.*, 2016, **79**, 578-583 (*Euphorbia sikkimensis constits, activity*).
Ti, H. *et al.*, *Nat. Prod. Res.*, 2017, **31**, 2621-2628 (*Epimedium pseudowushanese constit, activity*).
Matsumoto, T. *et al.*, *Phytochemistry*, 2017, **137**, 101-108 (*Isodonoside II*).
Yang, Y.-N. *et al.*, *Yaoxue Xuebao*, 2017, **52**, 779-784 (*3,3′-di-Me 9-glucoside*).
Du, B. *et al.*, *Fitoterapia*, 2018, **129**, 220-225 (*Asprenols E,F*).
Ji, J. *et al.*, *Fitoterapia*, 2019, **137**, 104271:1-7 (*Securidaca inappendiculata constit, cd*).
Du, Y.-Q. *et al.*, *Fitoterapia*, 2019, **139**, 104403:1-9 (*Ailanthus altissima constit, cd*).
Wu, H.-Y. *et al.*, *J. Agric. Food Chem.*, 2019, **67**, 11986-11993 (*Camellia sinensis constits, cd, activity*)

4,8′-Oxylignan-3,4′,5,7′,9,9′-hexol O52

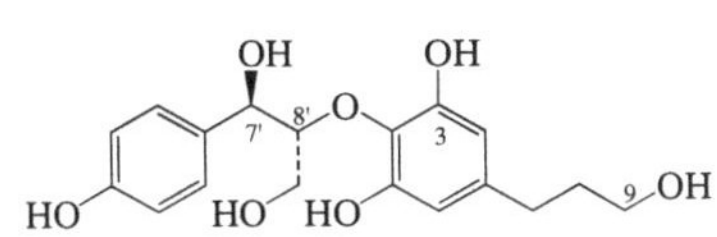

(7′R,8′R)-form

3,4′,5,7′,9,9′-Hexahydroxy-4,8′-oxyneolignan

$C_{18}H_{22}O_7$ 350.367

(7′R,8′R)-form

3,5,7′-Tri-Me ether

Amorph. powder. $[\alpha]^{20}_{D}$ -28 (c, 0.1 in MeOH)

(7′S,8′S)-form

3,5,7′-Tri-Me ether

Amorph. powder. $[\alpha]^{20}_{D}$ +32 (c, 0.1 in MeOH)

(7′RS,8′RS)-form

3,5,7′-Tri-Me ether

4′,9,9′-Trihydroxy-3,5,7′-trimethoxy-4,8′-oxylignan

$C_{21}H_{28}O_7$ 392.448

Constit. of the root bark of *Ailanthus altissima*. Amorph. powder. $[\alpha]^{20}_{D}$ +1.2 (c, 0.1 in MeOH). Resolved by chiral hplc. λ_{max} 254 (log ε 3.54); 274 (log ε 3.68) (MeOH).

Du, Y.-Q. *et al.*, *Fitoterapia*, 2019, **139**, 104403:1-9 (*Ailanthus altissima constit*)

4,8′-Oxylignan-3,3′,4′,5,5′,7,7′,9,9′-nonol O53

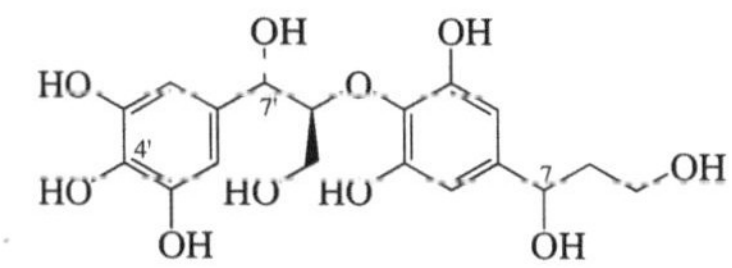

3,3′,4′,5,5′,7,7′,9,9′-Nonahydroxy-4,8′-oxyneolignan

$C_{18}H_{22}O_{10}$ 398.364

(7ξ,7′S,8′S)-form

7-Ketone, 3′,4′-di-Me ether

3,3′,5,7′,9,9′-Hexahydroxy-4′,5′-dimethoxy-4,8′-oxylignan-7-one. 3,3′,5,7′,9,9′-Hexahydroxy-4′,5′-dimethoxy-7-oxo-4,8′-oxylignan.

Caudatalignan A

$C_{20}H_{24}O_{10}$ 424.402

Constit. of the stem bark of *Lindera caudata*. $[\alpha]^{25}_{D}$ +35.6 (c, 0.25 in MeOH). λ_{max} 215 (log ε 4.08); 285 (log ε 3.68); 330 (log ε 3.26) (MeOH).

Lou, J. *et al.*, *Zhongcaoyao*, 2015, **46**, 955-957 (*Caudatalignan A*)

4,8′-Oxylignan-3,3′,4′,5,5′,7′,8,9,9′-nonol O54

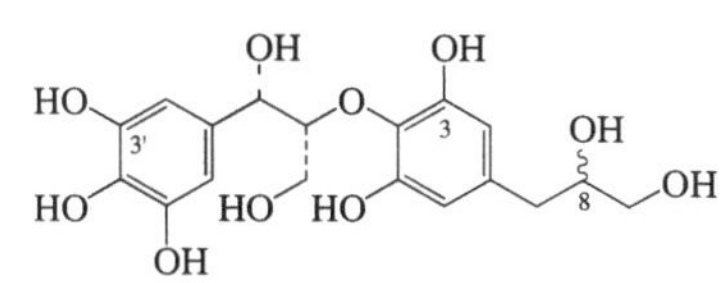

(7′S,8ξ,8′R)-form

3,3′,4′,5,5′,7′,8,9,9′-Nonahydroxy-4,8′-oxyneolignan

$C_{18}H_{22}O_{10}$ 398.364

(7′S,8ξ,8′R)-form

3,3′,5,5′-Tetra-Me ether

3,3′,5,5′-Tetramethoxy-4,8′-oxylignan-4′,7′,8,9,9′-pentol. 4′,7′,8,9,9′-Pentahydroxy-3,3′,5,5′-tetramethoxy-4,8′-oxylignan.

Casearialignan A

[1217193-06-0]

$C_{22}H_{30}O_{10}$ 454.472

Constit. of the leaves of *Casearia sylvestris*. Gum. $[\alpha]^{20}_{D}$ -15.1 (c, 0.09 in MeOH). λ_{max} 231 (log ε 3.39); 280 (log ε 2.89) (MeOH)

(7′S,8ξ,8′S)-form

3,3′,5,5′-Tetra-Me ether

Casearialignan B

[1217193-07-1]

$C_{22}H_{30}O_{10}$ 454.472
Constit. of the leaves of *Casearia sylvestris*. Amorph. powder. $[\alpha]_D^{20}$ +2.3 (c, 0.08 in MeOH). λ_{max} 229 (log ε 3.2); 279 (log ε 2.77) (MeOH).

Wang, W. *et al.*, *Helv. Chim. Acta*, 2010, **93**, 139-146 (*Casearialignans A,B*)

4,8′-Oxylignan-3,3′,4′,5,7,7′,8,9,9′-nonol O55

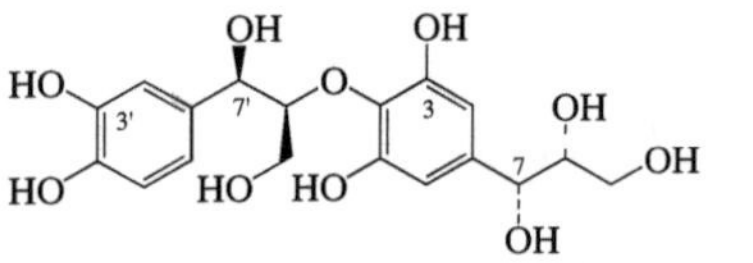

(7RS,7′RS,8RS,8′SR)-form

3,3′,4′,5,7,7′,8,9,9′-Nonahydroxy-4,8′-oxyneolignan
$C_{18}H_{22}O_{10}$ 398.364
(7RS,7′RS,8RS,8′SR)-form
3,3′,5-Tri-Me ether
3,3′,5-Trimethoxy-4,8′-oxylignan-4′,7,7′,8,9,9′-hexol. 4′,7,7′,8,9,9′-Hexahydroxy-3,3′,5-trimethoxy-4,8′-oxylignan. ***Oxyneolignan D***
[1313434-72-8] (stereoisomer)
$C_{21}H_{28}O_{10}$ 440.445
Constit. of the stems of *Bambusa tuldoides*. Also isol. from maple syrup. Resolved by chiral hplc
(7RS,7′RS,8SR,8′SR)-form
3,3′,5-Tri-Me ether
Oxyneolignan B
$C_{21}H_{28}O_{10}$ 440.445
Constit. of *Bambusa tuldoides*. Resolved by chiral hplc
(7RS,7′SR,8RS,8′RS)-form
3,3′,5-Tri-Me ether
Oxyneolignan C
$C_{21}H_{28}O_{10}$ 440.445
Constit. of *Bambusa tuldoides*. Resolved by chiral hplc
(7RS,7′SR,8SR,8′RS)-form
3,3′,5-Tri-Me ether
Oxyneolignan A
[1313434-74-0] (stereoisomer)
$C_{21}H_{28}O_{10}$ 440.445
Constit. of the stems of *Bambusa tuldoides*. Also isol. from maple syrup. Resolved by chiral hplc.

Li, L. *et al.*, *J. Agric. Food Chem.*, 2011, **59**, 7708-7716 (*maple constits*).
Sun, J. *et al.*, *Phytochem. Anal.*, 2015, **26**, 54-60 (*Oxyneolignans A-D*)

4,8′-Oxylignan-3,3′,5,5′,7,7′,8,9,9′-nonol O56

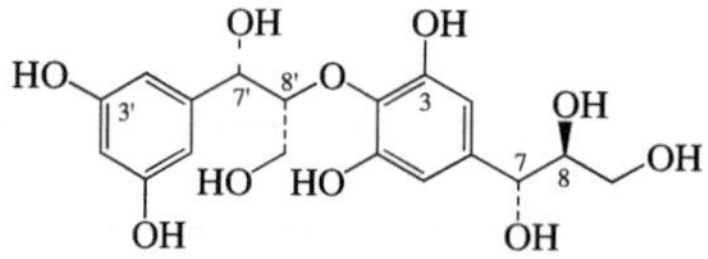

3,3′,5,5′,7,7′,8,9,9′-Nonahydroxy-4,8′-oxyneolignan
$C_{18}H_{22}O_{10}$ 398.364
Structures in this series may need revision. Phenyl 3,5-dioxygenated substn. pattern in doubt as a result of work done by Yang *et al* in 2015
(7R*,7′S,8S*,8′R)-form
3,3′,5,7′-Tetra-Me ether
3,5,5′,7′-Tetramethoxy-4,8′-oxylignan-3′,7,8,9,9′-pentol. 3′,7,8,9,9′-Pentahydroxy-3,5,5′,7′-tetramethoxy-4,8′-oxylignan.
Pinnatifidanin BII
[1462882-50-3]
$C_{22}H_{30}O_{10}$ 454.472
Constit. of *Crataegus pinnatifida*. Yellow oil. $[\alpha]_D^{20}$ +2.6 (c, 0.02 in MeOH). λ_{max} 233 (log ε 2.61) (MeOH).

Huang, X.-X. *et al.*, *Bioorg. Med. Chem. Lett.*, 2013, **23**, 5599-5604 (*Pinnatifidanin BII*).
Yang, Y.-N. *et al.*, *J. Nat. Prod.*, 2015, **78**, 705-711 (*pmr, struct*)

4,8′-Oxylignan-3,3′,4′,5,5′,7′,9,9′-octol O57

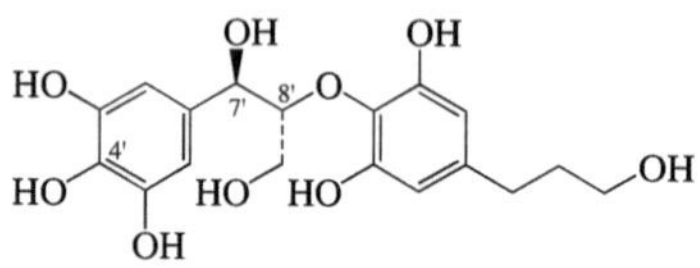

(7′R,8′R)-form

3,3′,4′,5,5′,7,9,9′-Octahydroxy-4,8′-oxyneolignan
$C_{18}H_{22}O_9$ 382.365
(7′R,8′R)-form
3,3′,5,5′-Tetra-Me ether, 4′-O-β-D-glucopyranoside
$C_{28}H_{40}O_{14}$ 600.614
Constit. of the rhizomes of *Smilax trinervula*. Powder (MeOH). $[\alpha]_D^{25}$ -31 (c, 0.002 in MeOH). λ_{max} 232 (log ε 0.25) (MeOH)
3,3′,5,5′-Tetra-Me ether, 9-O-α-L-rhamnopyranoside
Rhodomicranthoside D
$C_{28}H_{40}O_{13}$ 584.615
Constit. of *Rhododendron micranthum*. Powder. $[\alpha]_D^{20}$ -44.8 (c, 0.36 in MeOH)
(7′R,8′S)-form
3,3′,5,5′-Tetra-Me ether
Exhibits neuroprotective activity. $[\alpha]_D^{20}$ +32 (c, 0.1 in MeOH)
(7′S,8′R)-form
3,3′,5,5′-Tetra-Me ether, 4′-O-β-D-glucopyranoside
Symplocosneolignan A
[1421684-11-8]
$C_{28}H_{40}O_{14}$ 600.614
Constit. of *Symplocos cochinchinensis* var. *philippinensis*. Amorph. powder. $[\alpha]_D^{25}$ -13.5 (c, 0.79 in MeOH). λ_{max} 213 (log ε 4.31); 269 (log ε 3.59) (MeOH)
3,3′,5,5′-Tetra-Me ether
Exhibits neuroprotective activity. $[\alpha]_D^{20}$ -28.2 (c, 0.1 in MeOH)
(7′R*,8′R*)-form
3,3′,5,5′-Tetra-Me ether
3,3′,5,5′-Tetramethoxy-4,8′-oxylignan-4′,7′,9,9′-tetrol. 4′,7′,9,9′-Tetrahydroxy-3,3′,5,5′-tetramethoxy-4,8′-oxylignan.
Burseneolignan
[871711-76-1]
$C_{22}H_{30}O_9$ 438.473
Constit. of the roots of *Bursera tonkinensis*. Syrup. $[\alpha]_D^{20}$ +14 (c, 0.1 in MeOH). λ_{max} 217 (log ε 4) (MeOH)
(8′S)-form
7′-Ketone, 3,3′,5,5′-tetra-Me ether
4′,9,9′-Trihydroxy-3,3′,5,5′-tetramethoxy-4,8′-oxylignan-7′-one.
Asprenol B
$C_{22}H_{28}O_9$ 436.457
Constit. of *Ilex asprella*. Amorph. light yellow powder. $[\alpha]_D^{25}$ -4.0 (c, 0.1 in MeOH). λ_{max} 207 (log ε 3.15); 306 (log ε 2.39) (MeOH)
(7′RS,8′SR)-form
3,3′,5,5′-Tetra-Me ether
Constit. of the fruit of *Crataegus pinnatifida*. Gum. $[\alpha]_D^{20}$ +1.2 (c, 0.1 in MeOH). Partial racemate resolved by chiral hplc. λ_{max} 218 (log ε 2.58); 258 (log ε 0.23) (MeOH).

Jutiviboonsuk, A. *et al.*, *Phytochemistry*, 2005, **66**, 2745-2751 (*Burseneolignan*).
Cai, W.-H. *et al.*, *Am. J. Plant Sci.*, 2011, **2**, 609-618 (*Symplocosneolignan A*).
Shu, J. *et al.*, *Fitoterapia*, 2015, **104**, 64-68 (*Smilax trinervula constit*).

Du, B. *et al.*, *Fitoterapia*, 2018, **129**, 220-225 (*Asprenol B*).
Zhang, Z. *et al.*, *Fitoterapia*, 2019, **135**, 15-21 (*Rhodomicranthoside D*).
Zhao, P. *et al.*, *Fitoterapia*, 2019, **136**, 104164:1-7 (*Crataegus pinnatifida constit, cd, activity*)

4,8′-Oxylignan-3,3′,4′,5,7,7′,9,9′-octol O58

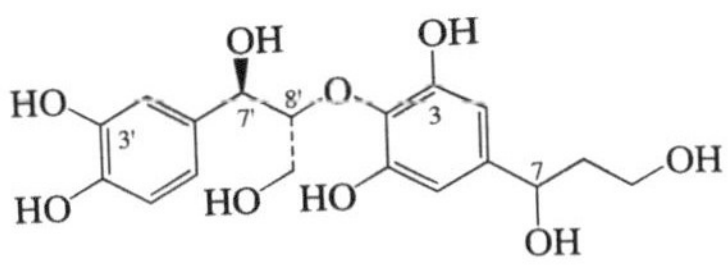

(7′R,8′R)-form

3,3′,4′,5,7,7′,9,9′-Octahydroxy-4,8′-oxyneolignan
$C_{18}H_{22}O_9$ 382.365
(7′R,8′R)-form
7-Ketone, 3,3′,5-tri-Me ether
4′,7′,9,9′-Tetrahydroxy-3,3′,5-trimethoxy-4,8′-oxylignan-7-one.
4′,7′,9,9′-Tetrahydroxy-3,3′,5-trimethoxy-7-oxo-4,8′-oxylignan.
Fordianole C
[2230065-40-2]
$C_{21}H_{26}O_9$ 422.430
Constit. of *Viburnum fordiae*. Amorph. powder. $[\alpha]_D^{23}$ +11.8 (c, 0.1 in MeOH). λ_{max} 220 (log ε 3.57); 282 (log ε 3.24) (MeOH)
(7′S,8′S)-form
7-Ketone
3,3′,4′,5,7′,9,9′-Heptahydroxy-4,8′-oxylignan-7-one.
3,3′,4′,5,7′,9,9′-Heptahydroxy-7-oxo-4,8′-oxylignan
7-Ketone, 3′-Me ether
3,4′,5,7′,9,9′-Hexahydroxy-3′-methoxy-4,8′-oxylignan-7-one.
Difenneolignan B
[1870003-60-3]
$C_{19}H_{22}O_9$ 394.376
Constit. of the stem bark of *Illicium difengpi*. Exhibits anti-HIV-1 activity. Amorph. solid. $[\alpha]_D^{24.9}$ +32.4 (c, 0.25 in MeOH). λ_{max} 210 (log ε 4.04); 282 (log ε 3.73); 328 (log ε 3.14) (MeOH)
7-Ketone, 3,3′,4′-tri-Me ether
3,7′,9,9′-Tetrahydroxy-3′,4′,5-trimethoxy-4,8′-oxylignan-7-one.
3,7′,9,9′-Tetrahydroxy-3′,4′,5-trimethoxy-7-oxo-4,8′-oxylignan.
Feddeiphenol C
[1383847-72-0]
$C_{21}H_{26}O_9$ 422.43
Constit. of the leaves and stems of *Daphne feddei*. Exhibits weak cytotoxic activity. Amorph. solid. $[\alpha]_D^{24}$ +28.5 (c, 0.2 in MeOH). λ_{max} 210 (log ε 4.87); 287 (log ε 4.16); 325 (log ε 2.86) (MeOH)
7-Ketone, 3,3′,5-tri-Me ether
Nicotphenol C
[1430237-67-4]
$C_{21}H_{26}O_9$ 422.43
Constit. of tobacco leaves, *Nicotiana tabacum*. Cytotoxic and shows modest anti-HIV activity. Amorph. solid. $[\alpha]_D^{24}$ +30.6 (c, 0.2 in MeOH). λ_{max} 210 (log ε 4.93); 286 (log ε 4.12); 325 (log ε 3.11) (MeOH).

Hu, Q.-F. *et al.*, *Chem. Pharm. Bull.*, 2011, **59**, 1421-1424 (*Feddeiphenol C*).
Chen, Y.-K. *et al.*, *J. Asian Nat. Prod. Res.*, 2012, **14**, 450-456 (*Nicotphenol C*).
Yang, Y. *et al.*, *Chem. Nat. Compd. (Engl. Transl.)*, 2016, **52**, 43-47 (*Difenneolignan B, activity*).
Chen, J. *et al.*, *Arch. Pharmacal Res.*, 2018, **41**, 625-632 (*Fordianole C*)

4,8′-Oxylignan-3,3′,4′,5,7′,8,9,9′-octol O59

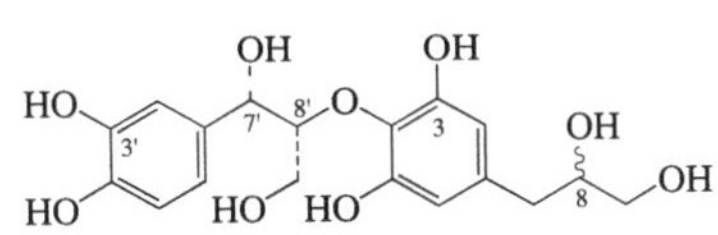

(7′S,8ξ,8′R)-form

3,3′,4′,5,7′,8,9,9′-Octahydroxy-4,8′-oxyneolignan
$C_{18}H_{22}O_9$ 382.365
(7′S,8ξ,8′R)-form
3,3′,5-Tri-Me ether
3,3′,5-Trimethoxy-4,8′-oxylignan-4′,7′,8,9,9′-pentol.
4′,7′,8,9,9′-Pentahydroxy-3,3′,5-trimethoxy-4,8′-oxylignan. ***Casearialignan C***
[1217193-08-2]
$C_{21}H_{28}O_9$ 424.446
Constit. of the leaves of *Casearia sylvestris*. Gum. $[\alpha]_D^{20}$ +46.6 (c, 0.13 in MeOH). λ_{max} 231 (log ε 3.42); 278 (log ε 3.04) (MeOH)
(7′S,8ξ,8′S)-form
3,3′,5-Tri-Me ether
Casearialignan D
[1217193-09-3]
$C_{21}H_{28}O_9$ 424.446
Constit. of the leaves of *Casearia sylvestris*. Gum. $[\alpha]_D^{20}$ -35.6 (c, 0.07 in MeOH). λ_{max} 231 (log ε 3.37); 279 (log ε 2.98) (MeOH).

Wang, W. *et al.*, *Helv. Chim. Acta*, 2010, **93**, 139-146 (*Casearialignans C,D*)

4,8′-Oxylignan-3,3′,4′,7,7′,8,9,9′-octol O60

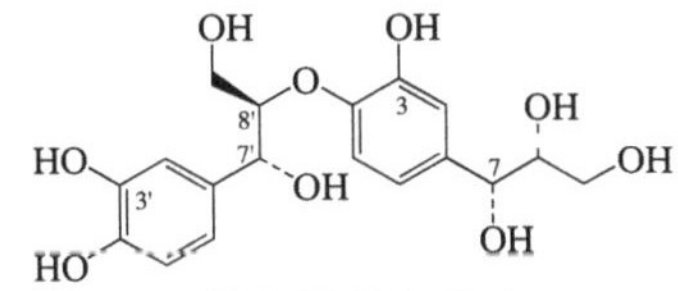

(7R*,7′R,8R*,8′R)-form

3,3′,4′,7,7′,8,9,9′-Octahydroxy-4,8′-oxyneolignan
$C_{18}H_{22}O_9$ 382.365
(7R*,7′R,8R*,8′R)-form
3,3′,7-Tri-Me ether
3,3′,7-Trimethoxy-4,8′-oxylignan-4′,7′,8,9,9′-pentol.
4′,7′,8,9,9′-Pentahydroxy-3,3′,7-trimethoxy-4,8′-oxylignan. ***Crataegusnin B***
$C_{21}H_{28}O_9$ 424.446
Constit. of the seeds of *Crataegus pinnatifida*. Exhibits a potent necrosis factor-α (TNF-α) inhibitory effect. Yellow oil. $[\alpha]_D^{20}$ -13.2 (c, 0.12 in MeOH). λ_{max} 229 (log ε 0.72); 286 (log ε 0.32) (MeOH)
(7R*,7′S,8R*,8′R)-form
3,3′,7′-Tri-Me ether
3,3′,7′-Trimethoxy-4,8′-oxylignan-4′,7,8,9,9′-pentol.
4′,7,8,9,9′-Pentahydroxy-3,3′,7′-trimethoxy-4,8′-oxylignan. ***Crataegusnin A***
$C_{21}H_{28}O_9$ 424.446
Constit. of the seeds of *Crataegus pinnatifida*. Exhibits a potent necrosis factor-α (TNF-α) inhibitory effect. Yellow oil. $[\alpha]_D^{20}$ -7.0 (c, 0.1 in MeOH). λ_{max} 228 (log ε 0.68); 287 (log ε 0.31) (MeOH)
(7R*,7′R*,8R*,8′R*)-form
3,3′-Di-Me ether
3,3′-Dimethoxy-4,8′-oxylignan-4′,7,7′,8,9,9′-hexol.
4′,7,7′,8,9,9′-Hexahydroxy-3,3′-dimethoxy-4,8′-oxylignan
[220014-17-5]
[14070-76-9]
$C_{20}H_{26}O_9$ 410.419
Constit. of *Zantedeschia aethiopica*. $[\alpha]_D$ +4.2 (c, 0.3 in EtOH).

Nimz, H. *Chem. Ber.*, 1967, **100**, 181-186 (*synth*).
Della Greca, M. *et al.*, *Phytochemistry*, 1998, **49**, 1299-1304 (*Zantedeschia aethiopica constit*).
Peng, Y. *et al.*, *Bioorg. Med. Chem. Lett.*, 2016, **26**, 5501-5506 (*Crataegusnins A,B, activity*)

4,8′-Oxylignan-3,3′,5,5′,7,7′,9,9′-octol O61

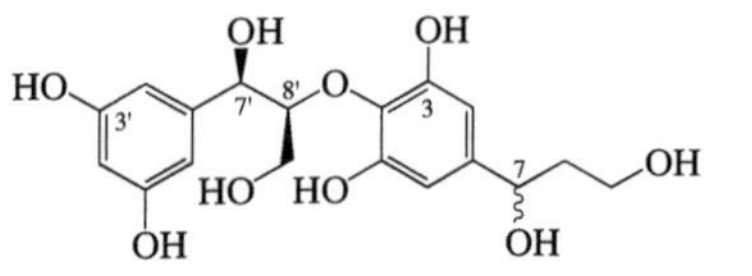

(7ξ,7′R,8′S)-form

3,3′,5,5′,7,7′,9,9′-Octahydroxy-4,8′-oxyneolignan

$C_{18}H_{22}O_9$ 382.365

Structures in this series may need revision. Phenyl 3,5-dioxygenated substn. pattern in doubt as a result of work done by Yang *et al* in 2015

(7ξ,7′R,8′S)-form

7′-Et, 3,3′,5-tri-Me ether

7′-Ethoxy-3,5,5′-trimethoxy-4,8′-oxylignan-3′,7,9,9′-tetrol. 7′-Ethoxy-3′,7,9,9′-tetrahydroxy-3,5,5′-trimethoxy-4,8′-oxylignan. ***Pinnatifidanin BIV***

[1462882-56-9]

$C_{23}H_{32}O_9$ 452.5

Constit. of the seeds of *Crataegus pinnatifida*. Yellow oil. $[\alpha]^{20}_{D}$ -2.1 (c, 0.02 in MeOH). λ_{max} 237 (log ε 2.92) (MeOH)

(7ξ,7′S,8′R)-form

7′-Et, 3,3′,5-tri-Me ether

Pinnatifidanin BIII

[1462882-54-7]

$C_{23}H_{32}O_9$ 452.5

Constit. of the seeds of *Crataegus pinnatifida*. Yellow oil. $[\alpha]^{20}_{D}$ +2.9 (c, 0.02 in MeOH). λ_{max} 231 (log ε 3.31) (MeOH).

Huang, X.-X. *et al.*, *Bioorg. Med. Chem. Lett.*, 2013, **23**, 5599-5604 (*Pinnatifidanins BIII,BIV*).

Yang, Y.-N. *et al.*, *J. Nat. Prod.*, 2015, **78**, 705-711 (*pmr, struct*)

4,8′-Oxylignan-3,3′,5,5′,7′,8,9,9′-octol O62

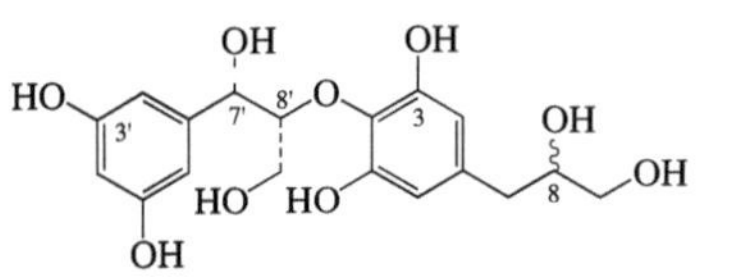

3,3′,5,5′,7′,8,9,9′-Octahydroxy-4,8′-oxyneolignan

$C_{18}H_{22}O_9$ 382.365

Structures in this series may need revision. Phenyl 3,5-dioxygenated substn. pattern in doubt as a result of work done by Yang *et al* in 2015

(7′S,8ξ,8′R)-form

3,3′,5,7′-Tetra-Me ether

3,5,5′,7′-Tetramethoxy-4,8′-oxylignan-3′,8,9,9′-tetrol. 3′,8,9,9′-Tetrahydroxy-3,5,5′,7′-tetramethoxy-4,8′-oxylignan. ***Pinnatifidanin BI***

[1462882-48-9]

$C_{22}H_{30}O_9$ 438.473

Constit. of the seeds of *Crataegus pinnatifida*. Yellow oil. $[\alpha]^{20}_{D}$ +3.6 (c, 0.03 in MeOH). λ_{max} 232 (log ε 2.20) (MeOH).

Huang, X.-X. *et al.*, *Bioorg. Med. Chem. Lett.*, 2013, **23**, 5599-5604 (*Pinnatifidanin BI*).

Yang, Y.-N. *et al.*, *J. Nat. Prod.*, 2015, **78**, 705-711 (*pmr, struct*)

7,7′-Oxylignan-2,2′,4,4′,5,5′,8,8′-octol O63

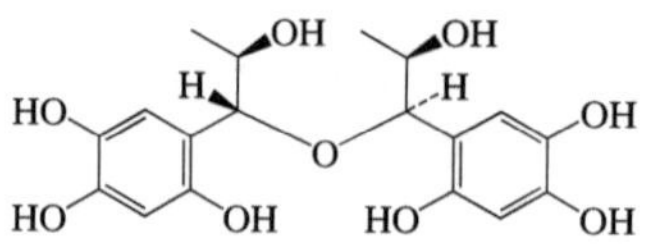

(7R,7′R,8R,8′R)-form

2,2′,4,4′,5,5′,8,8′-Octahydroxy-7,7′-oxyneolignan

$C_{18}H_{22}O_9$ 382.365

(7R,7′R,8R,8′R)-form

2,2′,4,4′,5,5′-Hexa-Me ether

2,2′,4,4′,5,5′-Hexamethoxy-7,7′-oxylignan-8,8′-diol. 8,8′-Dihydroxy-2,2′,4,4′,5,5′-hexamethoxy-7,7′-oxylignan. ***(+)-Asarolignan B***

$C_{24}H_{34}O_9$ 466.527

Constit. of the rhizomes of *Acorus tatarinowii*. Cryst.. $[\alpha]^{23}_{D}$ +173 (c, 0.5 in MeOH). λ_{max} 208 (log ε 4.46); 230 (log ε 4.14); 291 (log ε 3.92) (MeOH)

(7R,7′R,8S,8′R)-form

2,2′,4,4′,5,5′-Hexa-Me ether

(+)-Asarolignan C

$C_{24}H_{34}O_9$ 466.527

Constit. of the rhizomes of *Acorus tatarinowii*. Light yellow oil. $[\alpha]^{23}_{D}$ +167 (c, 0.5 in MeOH). λ_{max} 218 (log ε 3.99) (MeOH)

(7S,7′S,8R,8′S)-form

2,2′,4,4′,5,5′-Hexa-Me ether

(-)-Asarolignan C

$C_{24}H_{34}O_9$ 466.527

Constit. of the rhizomes of *Acorus tatarinowii*. Light yellow oil. $[\alpha]^{23}_{D}$ -161 (c, 0.5 in MeOH)

(7S,7′S,8S,8′S)-form

2,2′,4,4′,5,5′-Hexa-Me ether

(-)-Asarolignan B

$C_{24}H_{34}O_9$ 466.527

Constit. of the rhizomes of *Acorus tatarinowii*. Cryst.. $[\alpha]^{23}_{D}$ -174 (c, 0.5 in MeOH)

(7RS,7′SR,8RS,8′RS)-form

2,2′,4,4′,5,5′-Hexa-Me ether

meso-Asarolignan A

$C_{24}H_{34}O_9$ 466.527

Constit. of the rhizomes of *Acorus tatarinowii*. Cryst.. λ_{max} 205 (log ε 4.92); 231 (sh); 291 (MeOH).

Qin, D.-P. *et al.*, *RSC Adv.*, 2017, **7**, 8512-8520 (*Asarolignans A,B,C, cd, cryst struct*)

8,8′-Oxylignan-3,3′,4,4′,5,7′,9,9′-octol O64

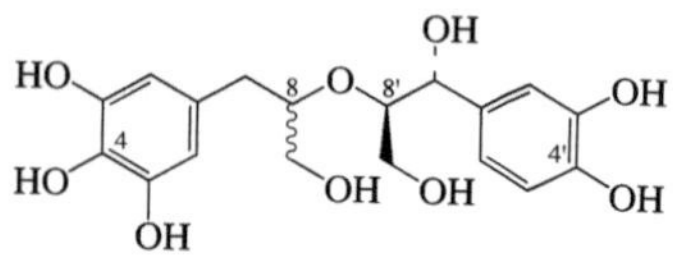

3,3′,4,4′,5,7′,9,9′-Octahydroxy-8,8′-oxyneolignan

$C_{18}H_{22}O_9$ 382.365

(7′R*,8ξ,8′R*)-form

3,3′,5,7′-Tetra-Me ether

3,3′,5,7′-Tetramethoxy-8,8′-oxylignan-4,4′,9,9′-tetrol. 4,4′,9,9′-Tetrahydroxy-3,3′,5,7′-tetramethoxy-8,8′-oxylignan

[1305346-21-7]

$C_{22}H_{30}O_9$ 438.473

Constit. of the leaves and twigs of *Euonymus alatus*. Syrup. $[\alpha]^{25}_{D}$ -5 (c, 0.5 in MeOH). λ_{max} 225 (log ε 3.4); 278 (log ε 2.4) (MeOH).

Jeong, E.J. *et al.*, *Bioorg. Med. Chem. Lett.*, 2011, **21**, 2283-2286 (*Euonymus alatus constit*)

8,8′-Oxylignan-3,3′,4,4′,5,7,9,9′-octol O65

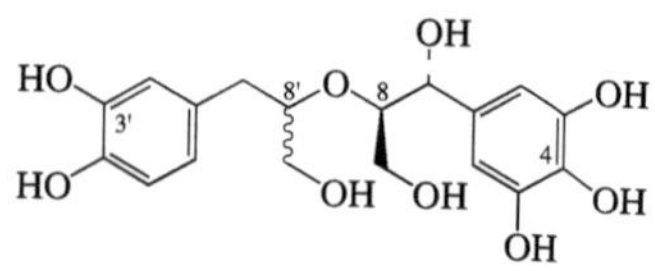

3,3′,4,4′,5,7,9,9′-Octahydroxy-8,8′-oxyneolignan

$C_{18}H_{22}O_9$ 382.365

(7R*,8R*,8′ξ)-form

3,3′,5,7-Tetra-Me ether

3,3′,5,7-Tetramethoxy-8,8′-oxylignan-4,4′,9,9′-tetrol. 4,4′,9,9′-Tetrahydroxy-3,3′,5,7-tetramethoxy-8,8′-oxylignan

[1305346-22-8]
$C_{22}H_{30}O_9$ 438.473
Constit. of the leaves and twigs of *Euonymus alatus*. Syrup. $[\alpha]_D^{25}$ -4.5 (c, 0.5 in MeOH). λ_{max} 224 (log ε 3.3); 279 (log ε 2.5) (MeOH).

Jeong, E.J. *et al.*, *Bioorg. Med. Chem. Lett.*, 2011, **21**, 2283-2286 (*Euonymus alatus constit*)

4,8′-Oxylignan-3,3′,4′,7′,9-pentol O66

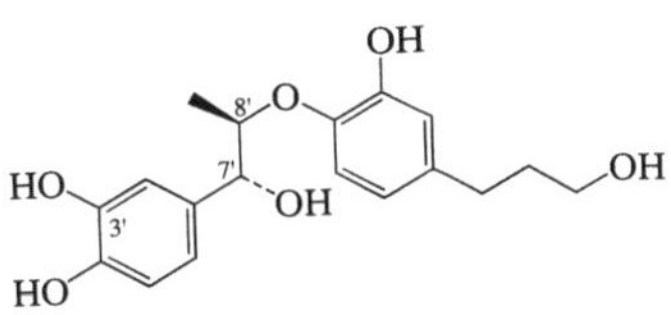

(7′R,8′R)-form

3,3′,4′,7′,9-Pentahydroxy-4,8′-oxyneolignan
$C_{18}H_{22}O_6$ 334.368

(7′R,8′R)-form

3′-Me ether

3′-Methoxy-4,8′-oxylignan-3,4′,7′,9-tetrol. 3,4′,7′,9-Tetrahydroxy-3′-methoxy-4,8′-oxylignan
[1890136-79-4]
$C_{19}H_{24}O_6$ 348.395
Constit. of the roots of *Paeonia lactiflora*. Exhibits moderate anti-β-amyloid aggregation activity. Pale yellow gum. $[\alpha]_D^{25}$ -21.7 (c, 0.06 in MeOH). Resolved by chiral hplc. λ_{max} 208 (log ε 4.46); 280 (log ε 3.81) (MeOH)

(7′R,8′S)-form

3′-Me ether
[1890136-00-1]
Constit. of the roots of *Paeonia lactiflora*. Exhibits moderate anti-β-amyloid aggregation activity. Gum. $[\alpha]_D^{25}$ +3.3 (c, 0.12 in MeOH). Resolved by chiral hplc. λ_{max} 208 (log ε 4.46); 280 (log ε 3.81) (MeOH)

(7′S,8′R)-form

3′-Me ether
[1890136-06-7]
Constit. of the roots of *Paeonia lactiflora*. Exhibits moderate anti-β-amyloid aggregation activity. Gum. $[\alpha]_D^{25}$ -4.7 (c, 0.14 in MeOH). Resolved by chiral hplc. λ_{max} 208 (log ε 4.46); 280 (log ε 3.81) (MeOH)

(7′S,8′S)-form

3′-Me ether
[1890136-30-7]
Constit. of the roots of *Paeonia lactiflora*. Exhibits strong anti-β-amyloid aggregation activity. Pale yellow gum. $[\alpha]_D^{25}$ +16 (c, 0.04 in MeOH). Resolved by chiral hplc. λ_{max} 208 (log ε 4.46); 280 (log ε 3.81) (MeOH)

(7′R*,8′R*)-form

3,3′,4′-Tri-Me ether

3,3′,4′-Trimethoxy-4,8′-oxylignan-7′,9-diol. 7′,9-Dihydroxy-3,3′,4′-trimethoxy-4,8′-oxylignan. ***Tatarinoid F***
[2222357-84-6]
$C_{21}H_{28}O_6$ 376.449
Constit. of the rhizomes of *Acorus tatarinowii*. $[\alpha]_D^{27}$ +12.8 (c, 0.5 in MeOH). Partial racemate. λ_{max} 205 (log ε 4.53); 230 (log ε 3.99); 279 (log ε 3.57) (MeOH).

Liu, X. *et al.*, *Fitoterapia*, 2015, **103**, 136-142 (*Paeonia lactiflora constits, activity*).
Zhang, W.-Y. *et al.*, *Bioorg. Med. Chem. Lett.*, 2018, **28**, 814-819 (*Tatarinoid F*)

4,8′-Oxylignan-3,3′,4′,9,9′-pentol O67

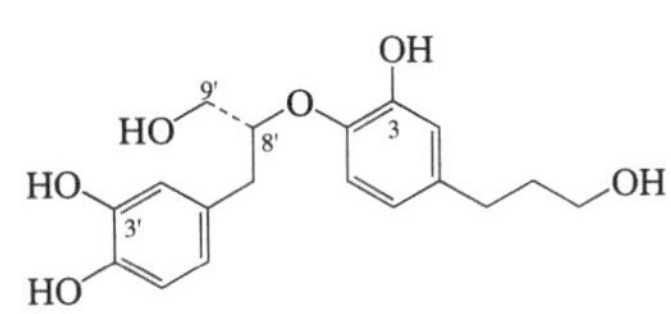

(R)-form

3,3′,4′,9,9′-Pentahydroxy-4,8′-oxyneolignan
$C_{18}H_{22}O_6$ 334.368

(R)-form

3,3′-Di-Me ether

3,3′-Dimethoxy-4,8′-oxylignan-4′,9,9′-triol. 4′,9,9′-Trihydroxy-3,3′-dimethoxy-4,8′-oxylignan. ***Ligraminol E***
[1335126-28-7]
$C_{20}H_{26}O_6$ 362.422
Isol. from the rhizomes of *Acorus gramineus*. Oil. $[\alpha]_D^{25}$ +18.3 (c, 0.1 in MeOH). λ_{max} 210 (log ε 3.6); 236 (log ε 2.2); 280 (log ε 1.8) (MeOH)

3,3′-Di-Me ether, 4′-O-β-D-xylopyranoside

Ligraminol E 4′-xyloside
[1493598-40-5]
$C_{25}H_{34}O_{10}$ 494.537
Constit. of the trunk of *Abies holophylla*. Gum. $[\alpha]_D^{25}$ +20.0 (c, 0.05 in MeOH). λ_{max} 231 (log ε 4.5); 279 (log ε 1.5) (MeOH)

(S)-form

3′-Me ether

3′-Methoxy-4,8′-oxylignan-3,4′,9,9′-tetrol. 3,4′,9,9′-Tetrahydroxy-3′-methoxy-4,8′-oxylignan
[144881-20-9]
$C_{19}H_{24}O_6$ 348.395
Constit. of the leaves of *Juniperus chinensis*. Oil. $[\alpha]_D^{25}$ +0.1 (c, 1.7 in MeOH). λ_{max} 217 (log ε 4.33); 278 (log ε 3.91) (MeOH)

3′-Me ether, 3-O-β-D-glucopyranoside
[1362738-64-4]
$C_{25}H_{34}O_{11}$ 510.536
Constit. of the stem bark of *Ginkgo biloba*. Amorph. powder. $[\alpha]_D^{25}$ -17.6 (c, 0.3 in MeOH)

3,3′,4′-Tri-Me ether

3,3′,4′-Trimethoxy-4,8′-oxylignan-9,9′-diol. 9,9′-Dihydroxy-3,3′,4-trimethoxy-4,8′-oxylignan. ***Ligraminol D***
[1335126-27-6]
$C_{21}H_{28}O_6$ 376.449
Isol. from the rhizomes of *Acorus gramineus*. Shows selective cytotoxicity and anti-neuroinflammatory activity. Oil. $[\alpha]_D^{25}$ +9.5 (c, 0.1 in MeOH). λ_{max} 206 (log ε 3.6); 232 (log ε 2.5); 280 (log ε 1.7) (MeOH)

4′-O-[2-hydroxy-1-(hydroxymethyl)ethyl], 3,3′-di-Me ether

Trapanatanol A
[1912455-24-3]
$C_{23}H_{32}O_8$ 436.501
Constit. of the pericarp of *Trapa natans*. Powder.

Fang, J.-M. *et al.*, *Phytochemistry*, 1992, **31**, 3659-3661 (*Juniperus chinensis constit*).
Kim, K.H. *et al.*, *J. Nat. Prod.*, 2011, **74**, 2187-2192 (*Ligraminols D,E*).
Ngan, N.T.T. *et al.*, *J. Agric. Food Chem.*, 2012, **60**, 2815-2824 (*Ginkgo biloba constit*).
Kim, C.S. *et al.*, *J. Nat. Prod.*, 2013, **76**, 2131-2135 (*Ligraminol E xyloside*).
Huang, H.-C. *et al.*, *J. Agric. Food Chem.*, 2016, **64**, 3794-3803 (*Trapanatanol A*).
Gangar, M. *et al.*, *ChemistrySelect*, 2017, **2**, 257-259 (*Ligraminol E, synth*)

4,8′-Oxylignan-3,4′,7′,9,9′-pentol O68

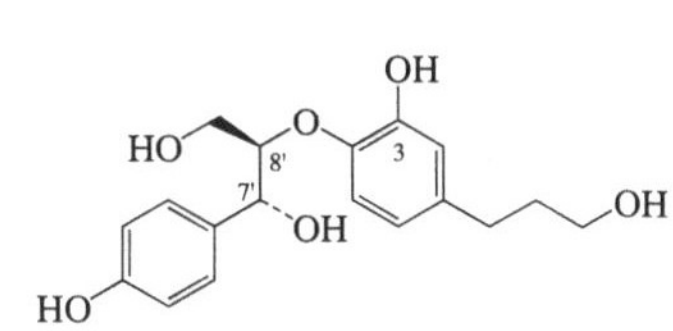

(7′R,8′R)-form

3,4′,7′,9,9′-Pentahydroxy-4,8′-oxyneolignan
$C_{18}H_{22}O_6$ 334.368
(7′R,8′R)-form
3-Me ether
[α]$^{20}_{D}$-28.5 (c, 0.1 in MeOH)
(7′R,8′S)-form
3-Me ether
[α]$^{20}_{D}$-30 (c, 0.1 in MeOH)
(7′S,8′R)-form
3-Me ether
[α]$^{20}_{D}$+29 (c, 0.1 in MeOH)
(7′S,8′S)-form
3-Me ether
[2088507-92-8]
[α]$^{20}_{D}$+26.9 (c, 0.11 in MeOH)
(7′RS,8′RS)-form
3-Me ether
3-Methoxy-4,8′-oxylignan-4′,7′,9,9′-tetrol. 4′,7′,9,9′-Tetrahydroxy-3-methoxy-4,8′-oxylignan. ***Chushizisin C***
[1142925-32-3]
$C_{19}H_{24}O_6$ 348.395
Constit. of the fruit of *Broussonetia papyrifera* and *Rubus idaeus*. Exhibits anti-β-amyloid aggregation and neuroprotective activities. Oil. [α]$^{20}_{D}$+4.3 (c, 0.2 in MeOH). The nat. prod. from *Rubus idaeus* was isol. as a partial racemate and resolved by chiral hplc. λ_{max} 226 (log ε 4.31); 277 (log ε 3.7) (MeOH)
(7′RS,8′SR)-form
3-Me ether
Chushizisin D
[1142925-33-4]
$C_{19}H_{24}O_6$ 348.395
Constit. of *Ailanthus altissima* and *Broussonetia papyrifera*. Oil or gum. [α]$^{20}_{D}$-1.3 (c, 0.1 in MeOH). λ_{max} 251 (log ε 3.76); 277 (log ε 4.49) (MeOH).

Mei, R.-Q. *et al.*, *J. Nat. Prod.*, 2009, **72**, 621-625 (*Chushizisins C,D*).
Zhou, L. *et al.*, *J. Funct. Foods*, 2017, **37**, 322-329 (*Rubus idaeus constit, activity*).
Du, Y.-Q. *et al.*, *Fitoterapia*, 2019, **139**, 104403:1-9 (*Ailanthus altissima constit, cd*)

4,8′-Oxylignan-4′,7′,9,9′-tetrol O69

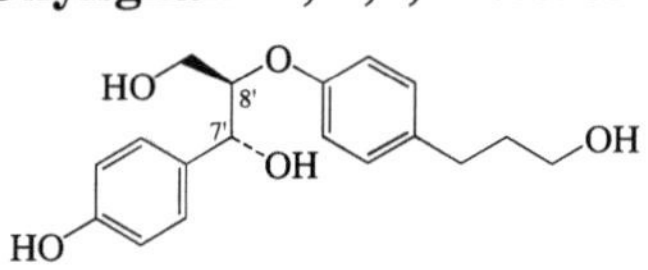

(7′R,8′R)-form

4′,7′,9,9′-Tetrahydroxy-4,8′-oxyneolignan
$C_{18}H_{22}O_5$ 318.369
The nat. prod. was isol. as a racemate and resolved by chiral hplc
(7′R,8′R)-form
Exhibits significant Aβ aggregation inhibition and weak neuroprotective effects. [α]$^{20}_{D}$-33.5 (c, 0.1 in MeOH)
(7′S,8′S)-form
Exhibits weak neuroprotective effects. [α]$^{20}_{D}$+36.7 (c, 0.1 in MeOH)
(7′RS,8′RS)-form
Constit. of the fruit of *Rubus idaeus*. Gum. [α]$^{20}_{D}$+0.3 (c, 0.18 in MeOH). λ_{max} 226 (log ε 3.62); 278 (log ε 0.51) (MeOH).

Zhou, L. *et al.*, *J. Funct. Foods*, 2017, **37**, 322-329 (*Rubus idaeus constit, activity*)

4,8′-Oxylign-7-ene-4′,7′-diol O70

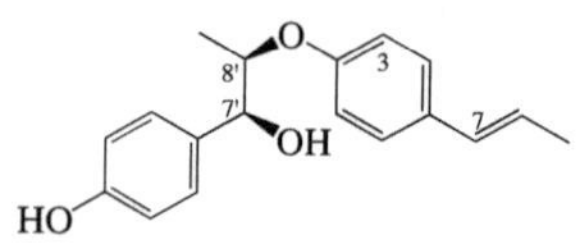

(7E,7′S,8′R)-form

1-(4-Hydroxyphenyl)-2-[4-(1-propenyl)phenoxy]-1-propanol. 4′,7′-Dihydroxy-4,8′-oxyneolign-7-ene
[110202-84-1]
$C_{18}H_{20}O_3$ 284.355
(7E,7′S,8′R)-form [851975-51-4]
Constit. of the stems of *Ribes fasciculatum* var. *chinense*. Inhibitor of NFAT transcription factor. Yellow oil. [α]$^{20}_{D}$+25 (c, 0.5 in $CHCl_3$). Authors' struct. and text do not agree. λ_{max} 258 (log ε 1.05) ($CHCl_3$)
(7E,7′R*,8′S*)-form [109194-73-2]
Constit. of *Krameria cytisoides* and *Krameria lanceolata*. Oil. [α]$^{21}_{D}$-6.4 (c, 0.44 in MeOH). Species given as Krameria cystisoides in ref.. λ_{max} 218 (sh); 227 (sh); 260 (log ε 4.38); 269 (sh); 285 (sh); 297 (sh); 309 (sh) (MeOH)
Ketone
1-(4-Hydroxyphenyl)-2-[4-(1-propenyl)phenoxy]-1-propanone. 4′-Hydroxy-4,8′-oxylign-7-en-7′-one. 4′-Hydroxy-7′-oxo-4,8′-oxylign-7-ene
[198206-52-9]
$C_{18}H_{18}O_3$ 282.339
Constit. of *Caryodaphnopsis baviensis*. [α]$^{22}_{D}$+26.7 (c, 1 in $CHCl_3$). Config. not determined.

Achenbach, H. *et al.*, *Phytochemistry*, 1987, **26**, 1159-1166 (*Krameria cytisoides constit*).
Anh, N.H. *et al.*, *Phytochemistry*, 1997, **46**, 569-571 (*ketone*).
Dat, N.T. *et al.*, *Chem. Pharm. Bull.*, 2005, **53**, 114-117 (*Ribes fasciculatum constit*)

4,8′-Oxylign-8-ene-4′,7′-diol O71

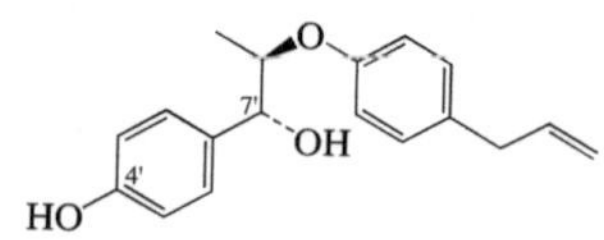

2-(4-Allylphenoxy)-1-(4-hydroxyphenyl)-1-propanol. 4′,7′-Dihydroxy-4,8′-oxyneolign-8-ene
$C_{18}H_{20}O_3$ 284.355
Lignan numbering shown
(7′R,8′R)-form
4′-Me ether
4′-Methoxy-4,8′-oxylign-8-en-7′-ol. 7′-Hydroxy-4′-methoxy-4,8′-oxylign-8-ene
4′-Me ether, 7′-Ac
Miliusamollin
[1529771-45-6]
$C_{21}H_{24}O_4$ 340.419
Constit. of the leaves of *Miliusa mollis*. Viscous liq.. [α]$^{20}_{D}$-55.4 (c, 0.20 in $CHCl_3$). λ_{max} 228 (log ε 4.31); 273 (log ε 3.84) (MeOH).

Sawasdee, K. *et al.*, *Fitoterapia*, 2013, **85**, 49-56 (*Miliusamollin*)

4,8′-Oxylign-7-ene-2,3′,4′,5,7′,9,9′-heptol O72

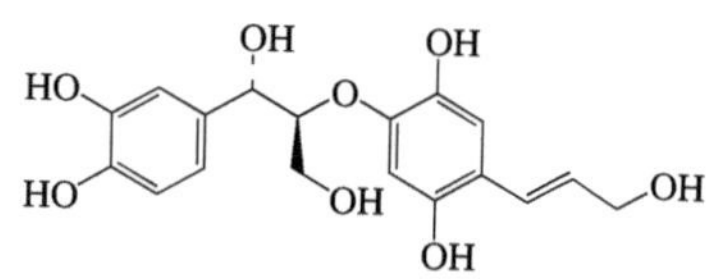

2,3′,4′,5,7′,9,9′-Heptahydroxy-4,8′-oxyneolign-7-ene
$C_{18}H_{20}O_8$ 364.350
(7E,7′S,8′S)-form
2,3′,5-Tri-Me ether
2,3′,5-Trimethoxy-4,8′-oxylign-7-ene-4′,7′,9,9′-tetrol
2,3′,5-Tri-Me ether, 4′-O-β-D-glucopyranoside
$C_{27}H_{36}O_{13}$ 568.572
Constit. of the branches of *Alangium barbatum*. Brown oil. [α]$^{20}_{D}$ -26.4 (c, 0.1 in MeOH).

Xie, Y.-G. *et al.*, *Phytochem. Lett.*, 2018, **28**, 64-68 (*Alangium barbatum constit*)

4,8′-Oxylign-7-ene-3,3′,4′,5,5′,7′,9-heptol O73

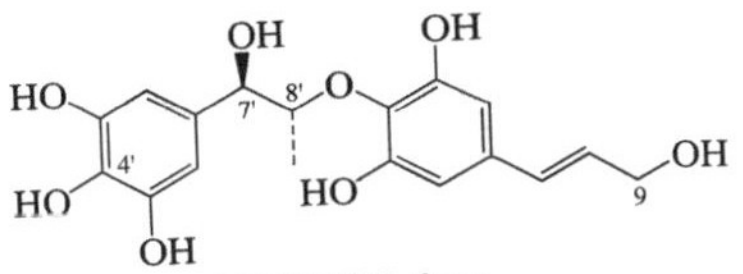

(7E,7′R,8′R)-form

3,3′,4′,5,5′,7′,9-Heptahydroxy-4,8′-oxyneolign-7-ene

$C_{18}H_{20}O_8$ 364.35

(7E,7′R,8′R)-form

3,3′,4′,5,5′-Penta-Me ether, 7′-Ac

Saurucinol E

[1521111-63-6]

$C_{25}H_{32}O_9$ 476.522

Constit. of *Saururus chinensis*. $[\alpha]^{20}_{D}$ -34 (c, 0.02 in $CHCl_3$). λ_{max} 228 (log ε 4.15); 268 (log ε 4.15) (MeOH)

7′-Ketone, 3,3′,4′,5,5′-penta-Me ether

9-Hydroxy-3,3′,4′,5,5′-pentamethoxy-4,8′-oxylign-7-en-7′-one. 9-Hydroxy-3,3′,4′,5,5′-pentamethoxy-7′-oxo-4,8′-oxylign-7-ene.

Saurucinol C

[1521111-57-8]

$C_{23}H_{28}O_8$ 432.469

Constit. of *Saururus chinensis*. Powder. $[\alpha]^{20}_{D}$ +44 (c, 0.01 in $CHCl_3$). λ_{max} 228 (log ε 4.36); 271 (log ε 4.34) (MeOH)

(7E,7′R,8′S)-form

9-Aldehyde, 3,3′,5,5′-tetra-Me ether

4′,7′-Dihydroxy-3,3′,5,5′-tetramethoxy-4,8′-oxylign-7-en-9-al. 4′,9′-Dihydroxy-3,3′,5,5′-tetramethoxy-9-oxo-4,8′-oxylign-7-ene.

Myrifralignan E

[1637669-74-9]

$C_{22}H_{26}O_8$ 418.442

Constit. of the seeds of *Myristica fragrans*. NO prodn. inhibitor. Yellow oil. $[\alpha]^{20}_{D}$ -6.0 (c, 1.0 in $CHCl_3$). λ_{max} 205 (log ε 4.42); 316 (log ε 3.72) (MeOH)

(7E,7′R*,8′S*)-form

3,3′,4′,5,5′-Penta-Me ether

3,3′,4′,5,5′-Pentamethoxy-4,8′-oxylign-7-ene-7′,9-diol. 7′,9-Dihydroxy-3,3′,4′,5,5′-pentamethoxy-4,8′-oxylign-7-ene.

Myristicanol A

[114892-44-3]

$C_{23}H_{30}O_8$ 434.485

Constit. of arils of *Myristica fragrans*. Oil.

Hattori, M. *et al.*, *Chem. Pharm. Bull.*, 1988, **36**, 648-653 (*Myristicanol A*).

Cui, H. *et al.*, *J. Nat. Prod.*, 2014, **77**, 100-110 (*Saurucinols C,E*).

Cao, G.-Y. *et al.*, *Food Chem.*, 2015, **173**, 231-237 (*Myrifralignan E, activity*)

4,8′-Oxylign-7-ene-3,3′,4′,5,5′,9,9′-heptol O74

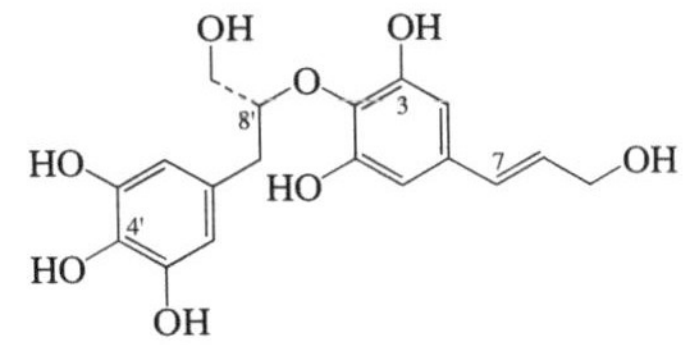

(R,E)-form

3,3′,4′,5,5′,9,9′-Heptahydroxy-4,8′-oxyneolign-7-ene

$C_{18}H_{20}O_8$ 364.35

(R,E)-form

3,3′,5,5′-Tetra-Me ether

3,3′,5,5′-Tetramethoxy-4,8′-oxylign-7-ene-4′,9,9′-triol. 4′,9,9′-Trihydroxy-3,3′,5,5′-tetramethoxy-4,8′-oxylign-7-ene

3,3′,5,5′-Tetra-Me ether, 4′,9′-di-O-β-D-glucopyranoside

[1647120-78-2]

$C_{34}H_{48}O_{18}$ 744.74

Constit. of the stems of *Dendrobium aurantiacum* var. *denneanum*. Gum. $[\alpha]^{20}_{D}$ -7.7 (c, 0.15 in MeOH). λ_{max} 226 (sh); 270 (log ε 3.66) (MeOH)

3,3′,5,5′,9-Penta-Me ether

3,3′,5,5′,9-Pentamethoxy-4,8′-oxylign-7-ene-4′,9′-diol. 4′,9′-Dihydroxy-3,3′,5,5′,9-pentamethoxy-4,8′-oxylign-7-ene

3,3′,5,5′,9-Penta-Me ether, 4′,9′-di-O-β-D-glucopyranoside

[1647120-80-6]

$C_{35}H_{50}O_{18}$ 758.767

Constit. of the stems of *Dendrobium aurantiacum* var. *denneanum*. Gum. $[\alpha]^{20}_{D}$ -12 (c, 0.1 in MeOH). λ_{max} 226 (sh); 273 (log ε 3.59) (MeOH)

(S,E)-form

3,3′,5,5′-Tetra-Me ether, 4′,9′-di-O-β-D-glucopyranoside

[1647120-79-3]

$C_{34}H_{48}O_{18}$ 744.74

Constit. of the stems of *Dendrobium aurantiacum* var. *denneanum*. Gum. $[\alpha]^{20}_{D}$ -34.4 (c, 0.1 in MeOH). λ_{max} 227 (sh); 270 (log ε 3.65) (MeOH)

(ξ,E)-form

3,3′,5,5′-Tetra-Me ether, 9′-O-β-D-glucopyranoside

Denchryside B

[729605-25-8]

$C_{28}H_{38}O_{13}$ 582.599

Constit. of *Dendrobium chrysanthum*. Yellowish oil. $[\alpha]^{20}_{D}$ -53 (c, 0.35 in MeOH).

Ye, Q.-H. *et al.*, *J. Asian Nat. Prod. Res.*, 2004, **6**, 39-43 (*Denchryside B*).

Li, X.-H. *et al.*, *Phytochem. Lett.*, 2014, **9**, 37-40 (*Dendrobium aurantiacum constits*)

4,8′-Oxylign-7-ene-3,3′,4′,5,7′,9,9′-heptol O75

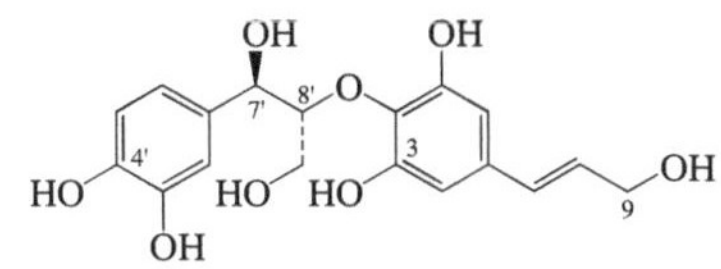

(7E,7′R,8′R)-form

3,3′,4′,5,7′,9,9′-Heptahydroxy-4,8′-oxyneolign-7-ene

$C_{18}H_{20}O_8$ 364.35

(7E,7′R,8′R)-form

3,3′,5-Tri-Me ether, 4′,9-di-O-β-D-glucopyranoside

5′-Methoxyligusinenoside B

[1415237-95-4]

$C_{33}H_{46}O_{18}$ 730.713

Constit. of *Senecio chrysanthemoides*

(7E,7′R,8′S)-form

3,3′,5-Tri-Me ether, 9′-O-β-D-glucopyranoside

[1082351-11-8]

$C_{27}H_{36}O_{13}$ 568.572

Constit. of the leaves of *Osmanthus ilicifolius*. Amorph. powder. $[\alpha]^{25}_{D}$ -36.1 (c, 0.17 in MeOH). λ_{max} 220 (log ε 4.5); 270 (log ε 4.2) (MeOH)

(7E,7′S,8′R)-form

3,3′-Di-Me ether

3′,5-Dimethoxy-4,8′-oxylign-7-ene-3,4′,7′,9,9′-pentol. 3,4′,7′,9,9′-Pentahydroxy-3′,5-dimethoxy-4,8′-oxylign-7-ene

3,3′-Di-Me ether, 5-O-β-D-glucopyranoside

[1018987-06-8]

$C_{26}H_{34}O_{13}$ 554.545

Constit. of the roots of *Iodes cirrhosa*. Amorph. powder. $[\alpha]^{20}_{D}$ -9 (c, 0.1 in MeOH). λ_{max} 221 (log ε 4.2); 269 (log ε 3.8) (MeOH)

3,3′,5-Tri-Me ether, 9-O-β-D-glucopyranoside
[1186295-11-3]
$C_{27}H_{36}O_{13}$ 568.572
Constit. of the leaves of *Syringa velutina*. Amorph. powder. $[\alpha]_D^{23}$ -9.1 (c, 0.2 in MeOH)

3,3′,5-Tri-Me ether, 9′-O-β-D-glucopyranoside
[1082351-10-7]
$C_{27}H_{36}O_{13}$ 568.572
Constit. of the leaves of *Osmanthus ilicifolius*. Amorph. powder. $[\alpha]_D^{25}$ -3.5 (c, 0.29 in MeOH). λ_{max} 220 (log ε 4.5); 271 (log ε 4.2) (MeOH)

9-Aldehyde, 3,3′,5-tri-Me ether, 4′-O-β-D-glucopyranoside
Armaoside
[872003-62-8]
$C_{27}H_{34}O_{13}$ 566.556
Constit. of *Pedicularis armata*. Amorph. powder. $[\alpha]_D^{20}$ +15 (c, 0.35 in MeOH)

(7E,7′S,8′S)-form

3,3′,5,7′,9-Penta-Me ether
3,3′,5,7′,9-Pentamethoxy-4,8′-oxylign-7-ene-4′,9′-diol
[1338232-86-2]
$C_{23}H_{30}O_8$ 434.485
Isol. from the leaves and stems of *Toona ciliata* var. *pubescens*. Shows mild antiproliferative activity against tumour cell lines. Yellow oil. $[\alpha]_D^{25}$ -10.5 (c, 1.0 in MeOH)

(7E,7′R*,8′R*)-form

3,3′,5-Tri-Me ether
3,3′,5-Trimethoxy-4,8′-oxylign-7-ene-4′,7′,9,9′-tetrol. Guaiacylglycerol β-sinapyl ether. ***Wikstroemol***
[613683-97-9]
$C_{21}H_{26}O_8$ 406.431
Constit. of *Brassica fruticulosa*, *Sinomenium acutum* and *Wikstroemia hainanensis*. Plant growth inhibitor. Powder. $[\alpha]_D^{21}$ -24 (c, 0.21 in $CHCl_3$). λ_{max} 220 (ε 19952); 270 (ε 9772) (MeOH)

3,3′,5-Tri-Me ether, 7′-O-β-D-glucopyranoside
[1186129-15-6]
$C_{27}H_{36}O_{13}$ 568.572
Constit. of *Eupatorium chinense* and *Osmanthus heterophyllus*. Amorph. powder. $[\alpha]_D^{27}$ -23.5 (c, 0.17 in MeOH). λ_{max} 219 (log ε 4.6); 270 (log ε 4.2) (MeOH)

3,3′,5-Tri-Me ether, 9′-Ac, 7′-O-β-D-glucopyranoside
Eupatorin A‡
$C_{29}H_{38}O_{14}$ 610.609
Constit. of the roots of *Eupatorium chinense*. Oil. $[\alpha]_D^{23}$ -20.8 (c, 0.24 in MeOH). λ_{max} 222 (log ε 4.53); 271 (log ε 4.02) (MeOH)

7′-Ketone, 3,3′,5-tri-Me ether
4′,9,9′-Trihydroxy-3,3′,5-trimethoxy-4,8′-oxylign-7-en-7′-one.
Wikstroemone
[878007-02-4]
$C_{21}H_{24}O_8$ 404.415
Constit. of *Wikstroemia hainanensis*. Powder. $[\alpha]_D^{21}$ +5 (c, 0.2 in $CHCl_3$). λ_{max} 205 (log ε 4.42); 223 (log ε 4.36); 272 (log ε 4.11); 306 (log ε 3.9) (MeOH)

(7E,7′R*,8′S*)-form

3,3′,5-Tri-Me ether
[905726-70-7]
$C_{21}H_{26}O_8$ 406.431
Constit. of *Brassica fruticulosa* and *Eucommia ulmoides*. $[\alpha]_D^{25}$ +9.2 (c, 2.5 in MeOH). Abs. config. of *B. fruticulosa* isolate revised in 2006

3,3′,5-Tri-Me ether, 4′-O-β-D-glucopyranoside
Citrusin B
[105279-10-5]
$C_{27}H_{36}O_{13}$ 568.572
Constit. of *Ailanthus altissima*, *Citrus limon*, *Eucommia ulmoides*, *Fortunella japonica* and *Ligularia veitchiana*. Exhibits moderate inhibitory activity against tobacco mosaic virus replication. Mp 105 - 106°. $[\alpha]_D^{20}$ -11.3 (c, 0.3 in MeOH)

3,3′,5-Tri-Me ether, 9,9′-bis-O-(4-hydroxybenzoyl)
Quiquelignan H
[1215818-73-7]
$C_{35}H_{34}O_{12}$ 646.645
Constit. of the stems of *Calamus quiquesetinervius*. NO prodn. inhibitor. Yellow powder. Mp 140°. $[\alpha]_D^{25}$ +10.4 (c, 0.5 in MeOH). λ_{max} 260; 276 (sh) (MeOH)

3,3′,5-Tri-Me ether, 9,9′-bis-O-(4-hydroxy-E-cinnamoyl)
Dadahol A
[405281-76-7]
$C_{39}H_{38}O_{12}$ 698.721
Constit. of *Artocarpus dadah*. Oil. $[\alpha]_D^{20}$ +20 (c, 0.07 in MeOH). λ_{max} 205 (log ε 4.02); 224 (log ε 3.98); 289 (log ε 3.78); 311 (log ε 3.81) (MeOH)

9-Aldehyde, 3,3′,5-tri-Me ether
4′,7′,9′-Trihydroxy-3,3′,5-trimethoxy-4,8′-oxylign-7-en-9-al.
Armaosigenin
[1471249-94-1]
$C_{21}H_{24}O_8$ 404.415
Constit. of the stem bark of *Daphne feddei* and *Chirita longgangensis* var. *hongyao*. Pale yellow oil. $[\alpha]_D^{25}$ +6.1 (c, 0.49 in MeOH) (0). λ_{max} 229 (log ε 3.56); 320 (log ε 3.37) (MeOH)

(7E,7′ξ,8′ξ)-form

3,3′,5-Tri-Me ether, 7′-O-β-D-glucopyranoside
[412029-03-9]
$C_{27}H_{36}O_{13}$ 568.572
Constit. of cane sugar, *Saccharum* sp.. Oil. λ_{max} 218 (sh); 271 (MeOH).

Sawabe, A. *et al., Nippon Kagaku Kaishi*, 1986, **60**, 593-599 (*Citrusins*).
Deyama, T. *et al., Chem. Pharm. Bull.*, 1987, **35**, 1803-1807 (*Citrusin B*).
Li, S. *et al., Phytochemistry*, 1998, **49**, 2125-2128 (*synth, pmr, cmr*).
Takara, K. *et al., Biosci., Biotechnol., Biochem.*, 2002, **66**, 29-35 (*cane sugar 3,3′-di-Me ether 7-glucosides*).
Su, B.-N. *et al., J. Nat. Prod.*, 2002, **65**, 163-169 (*Dadahols*).
Cutillo, F. *et al., J. Agric. Food Chem.*, 2003, **51**, 6165-6172 (*Brassica fruticulosa lignans, activity*).
Zhu, H. *et al., Z. Naturforsch., B*, 2004, **59**, 1063-1066 (*Ligularia veitchiana constit*).
Yuan, C.S. *et al., Chin. Chem. Lett.*, 2005, **16**, 781-782 (*Armaoside*).
Liao, S.-G. *et al., Helv. Chim. Acta*, 2006, **89**, 73-80 (*Wikstroemol, Wikstroemone*).
Yuan, C.-S. *et al., J. Asian Nat. Prod. Res.*, 2007, **9**, 673-677 (*Armaoside, config*).
Gan, M. *et al., J. Nat. Prod.*, 2008, **71**, 647-654 (*Iodes cirrhosa glucoside*).
Liang, S. *et al., J. Nat. Prod.*, 2008, **71**, 1902-1905 (*Armaosigenin*).
Machida, K. *et al., Magn. Reson. Chem.*, 2008, **46**, 990-994 (*3,3′,5-tri-Me ether 9′-glucosides*).
Machida, K. *et al., J. Nat. Med.*, 2009, **63**, 227-231 (*3,3′,5-tri-Me ether 7′-glucoside*).
Feng, X.-S. *et al., Molecules*, 2009, **14**, 953-958 (*Syringa velutina glucoside*).
Chang, C.-L. *et al., Bioorg. Med. Chem.*, 2010, **18**, 518-525 (*Quiquelignan H*).
Liao, P.-Y. *et al., Yunnan Zhiwu Yanjiu*, 2010, **32**, 183-188 (*Eupatorin A*).
Liu, Y.-B. *et al., Helv. Chim. Acta*, 2011, **94**, 1685-1691 (*penta-Me ether*).
Lin, S. *et al., Zhongguo Zhongyao Zazhi*, 2011, **36**, 1755-1762 (*5′-Methoxyligusinenoside B*).
Su, Y. *et al., J. Braz. Chem. Soc.*, 2012, **23**, 1925-1932 (*Chirita longgangensis aldehyde*)

4,8′-Oxylign-7-ene-3,3′,4′,5′,7′,9,9′-heptol O76

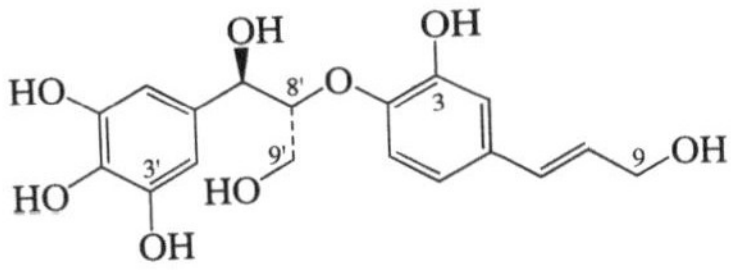

(7E,7′R,8′R)-form

3,3′,4′,5′,7′,9,9′-Heptahydroxy-4,8′-oxyneolign-7-ene

$C_{18}H_{20}O_8$ 364.35

(7E,7′R,8′R)-form

9-Aldehyde, 3,3′,5′-tri-Me ether

4′,7′,9′-Trihydroxy-3,3′,5′-trimethoxy-4,8′-oxylign-7-en-9-al. 4′,7′,9′-Trihydroxy-3,3′,5′-trimethoxy-9-oxo-4,8′-oxylign-7-ene. ***Cinncassin E***

$C_{21}H_{24}O_8$ 404.415

Constit. of the bark of *Cinnamomum cassia*. Potent inhibitor of NO prodn.. Amorph. powder. $[\alpha]^{20}_{D}$ -27 (c, 0.3 in MeOH). λ_{max} 205 (log ε 2.65); 303 (log ε 2.05); 337 (log ε 2.09) (MeOH)

(7E,7′R*,8′R*)-form

3,3′,5′-Tri-Me ether, 4′-O-β-D-glucopyranoside

Longifloroside B

[175556-09-9]

$C_{27}H_{36}O_{13}$ 568.572

Constit. of *Gnetum montanum* and *Pedicularis longiflora*. Amorph. powder. $[\alpha]^{14}_{D}$ +16.5 (c, 0.4 in MeOH)

(7E,7′R*,8′S*)-form

3,3′,5′-Tri-Me ether

3,3′,5′-Trimethoxy-4,8′-oxylign-7-ene-4′,7′,9,9′-tetrol. 4′,7′,9,9′-Tetrahydroxy-3,3′,5′-trimethoxy-4,8′-oxylign-7-ene

[905726-71-8]

$C_{21}H_{26}O_8$ 406.431

Constit. of *Eucommia ulmoides*

3,3′,5′-Tri-Me ether, 9,9′-bis-O-(4-hydroxybenzoyl)

Quiquelignan G

[1215818-72-6]

$C_{35}H_{34}O_{12}$ 646.645

Constit. of the stems of *Calamus quiquesetinervius*. NO prodn. inhibitor. Pale yellow powder. Mp 134°. $[\alpha]^{25}_{D}$ +1.9 (c, 0.5 in MeOH). λ_{max} 206; 261; 277 (sh) (MeOH).

Wang, C.Z. *et al.*, *Planta Med.*, 1997, **63**, 241-244 (*Longifloroside B*).
Lourith, N. *et al.*, *J. Wood Sci.*, 2005, **51**, 370-378 (*Eucommia ulmoides constit*).
Lourith, N. *et al.*, *J. Wood Sci.*, 2005, **51**, 379-386 (*Eucommia ulmoides constit, biosynth*).
Wang, L.Q. *et al.*, *Chem. Nat. Compd. (Engl. Transl.)*, 2009, **45**, 424-426 (*Longifloroside B*).
Chang, C.-L. *et al.*, *Bioorg. Med. Chem.*, 2010, **18**, 518-525 (*Quiquelignan G, activity*).
He, S. *et al.*, *Fitoterapia*, 2016, **112**, 153-160 (*Cinncassin E, cd, activity*)

4,8′-Oxylign-8-ene-3,3′,4′,5,5′,7′,9′-heptol O77

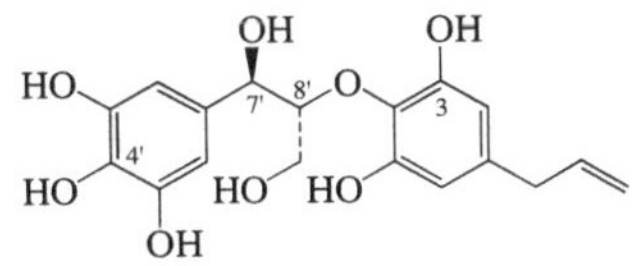

(7′R,8′R)-form

3,3′,4′,5,5′,7′,9′-Heptahydroxy-4,8′-oxyneolign-8-ene

$C_{18}H_{20}O_8$ 364.35

(7′R,8′R)-form

3,3′,4′,5,5′-Penta-Me ether

3,3′,4′,5,5′-Pentamethoxy-4,8′-oxylign-8-ene-7′,9′-diol. 7′,9′-Dihydroxy-3,3′,4′,5,5′-pentamethoxy-4,8′-oxylign-8-ene

3,3′,4′,5,5′-Penta-Me ether, 9′-Ac

Saurucinol D

[1521111-59-0]

$C_{25}H_{32}O_9$ 476.522

Constit. of *Saururus chinensis*. Powder. $[\alpha]^{20}_{D}$ -40 (c, 0.02 in $CHCl_3$). λ_{max} 228 (log ε 4.10); 270 (log ε 3.33) (MeOH)

(7′R,8′S)-form

3,3′,5,5′-Tetra-Me ether

3,3′,5,5′-Tetramethoxy-4,8′-oxylign-8-ene-4′,7′,9′-triol. 4′,7′,9′-Trihydroxy-3,3′,5,5′-tetramethoxy-4,8′-oxylign-8-ene

[1227063-98-0]

$C_{22}H_{28}O_8$ 420.458

Constit. of the stem bark of *Illicium difengpi*. Antioxidant. Exhibits moderate antiinflammatory activity. Amorph. powder. $[\alpha]^{20}_{D}$ +2.5 (c, 0.08 in MeOH). λ_{max} 209 (log ε 4.7); 278 (log ε 3.2) (MeOH).

Fang, L. *et al.*, *J. Nat. Prod.*, 2010, **73**, 818-824 (*Illicium difengpi constit, activity*).
Cui, H. *et al.*, *J. Nat. Prod.*, 2014, **77**, 100-110 (*Saurucinol D*)

3,4′-Oxylign-7-ene-3′,4,7′,8′,9,9′-hexol O78

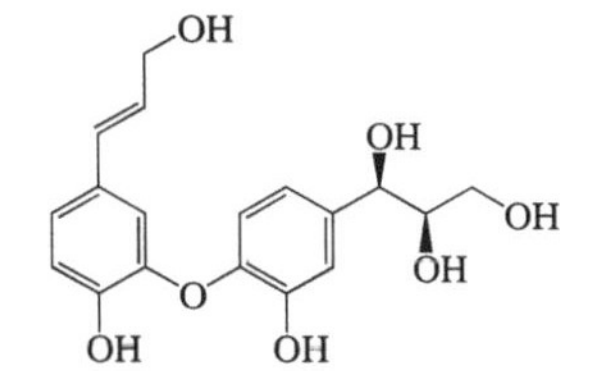

3′,4,7′,8′,9,9′-Hexahydroxy-3,4′-oxyneolign-7-ene

$C_{18}H_{20}O_7$ 348.351

(7E,7′R*,8′R*)-form

9,9′-Di-Me ether

9,9′-Dimethoxy-3,4′-oxylign-7-ene-3′,4,7′,8′-tetrol. 3′,4,7′,8′-Tetrahydroxy-9,9′-dimethoxy-3,4′-oxylign-7-ene. ***Taiwandimerol B***

[1449387-75-0]

[1449387-76-1] (3,4′-di-Me ether)

$C_{20}H_{24}O_7$ 376.405

Constit. of the roots of *Piper taiwanense*. Yellow oil. $[\alpha]^{25}_{D}$ -1.6 (c, 0.19 in $CHCl_3$). λ_{max} 216 (sh); 266 (log ε 4.09); 302 (sh); 314 (sh) (MeOH).

Chen, S. *et al.*, *Phytochemistry*, 2013, **93**, 203-209 (*Taiwandimerol B*)

3,8′-Oxylign-7-ene-3′,4,4′,7′,9,9′-hexol O79

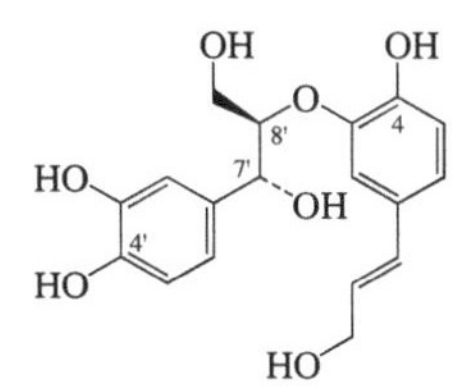

(7E,7′R,8′R)-form

3′,4,4′,7′,9,9′-Hexahydroxy-3,8′-oxyneolign-7-ene

$C_{18}H_{20}O_7$ 348.351

(7E,7′R,8′R)-form

3′,4-Di-Me ether

3′,4-Dimethoxy-3,8′-oxylign-7-ene-4′,7′,9,9′-tetrol. 4′,7′,9,9′-Tetrahydroxy-3′,4-dimethoxy-3,8′-oxylign-7-ene

3′,4-Di-Me ether, 4′-O-β-D-glucopyranoside

[1831849-99-0]

$C_{26}H_{34}O_{12}$ 538.546

Constit. of the stems and leaves of *Brandisia hancei*. Significantly decrease VEGF-induced human retinal endothelial cell migration. Zebra fish embryos show mild redn. of DLAV formn.. Powder.

$[\alpha]^{25}_{D}$ -23 (c, 0.1 in MeOH). λ_{max} 230; 272 (MeOH)

(7E,7′S,8′R)-form

3′-Me ether, 4-O-β-D-glucopyranoside

Morineoliganoside A

[2209122-26-7]

$C_{25}H_{32}O_{12}$ 524.519

Constit. of the stems of *Chrysanthemum morifolium*. Powder. $[\alpha]^{25}_{D}$ -57.9 (c, 0.57 in MeOH)

(7E,7′S,8′S)-form

3′,4-Di-Me ether

[2137027-33-7]

$C_{20}H_{24}O_7$ 376.405

Constit. of the grains of *Echinochloa utilis*

3′,4-Di-Me ether, 9′-Ac

Echinoutilin

[2135617-77-3]

$C_{22}H_{26}O_8$ 418.442

Constit. of the grains of *Echinochloa utilis*. α-Glucosidase inhibitor. Oil. $[\alpha]^{25}_{D}$ +15.1 (c, 0.1 in MeOH). λ_{max} 266 (log ε 3.66) (MeOH)

(7E,7′RS,8′RS)-form

3′,4-Di-Me ether, 9,9′-di-Ac

[1852570-00-3] (7′*S*,8′*S*-form)

[1852570-03-6] (7′*R*,8′*R*-form)

$C_{24}H_{28}O_9$ 460.479

Constit. of the grains of *Echinochloa utilis*. Oil. λ_{max} 203 (log ε 4.23); 259 (log ε 3.84) (MeOH)

(7E,7′R*,8′S*)-form

9-Aldehyde, 3′,4-di-Me ether

4′,7′,9′-Trihydroxy-3′,4-dimethoxy-3,8′-oxylign-7-en-9-al. 4′,7′,9′-Trihydroxy-3′,4-dimethoxy-9-oxo-3,8′-oxylign-7-ene

$C_{20}H_{22}O_7$ 374.389

Constit. of the stems of *Eleutherococcus senticosus*. Selectively inhibits DGAT1 but not DGAT2 (diacylglycerol acyltransferase). Amorph. brown powder. $[\alpha]^{25}_{D}$ -4.4 (c, 0.1 in MeOH). λ_{max} 201 (log ε 3.74); 231 (log ε 3.16); 280 (log ε 3.67) (MeOH).

Lee, I.S. *et al., Biosci., Biotechnol., Biochem.*, 2015, **79**, 581-586 (*Brandisia hancei constit, activity*).

Nguyen, D.H. *et al., J. Agric. Food Chem.*, 2016, **64**, 425-432 (*Echinochloa utilis constit*).

Li, J.-L. *et al., Phytochemistry*, 2016, **15**, 147-151 (*Eleutherococcus senticosus constit, activity*).

Nguyen, T.T. *et al., Bioorg. Chem.*, 2017, **74**, 221-227 (*Echinoutilin, cd, activity*).

Qu, L. *et al., Phytochem. Lett.*, 2017, **19**, 39-45 (*Morineoliganoside A*)

3,8′-Oxylign-7-ene-3′,4′,5,7′,9,9′-hexol **O80**

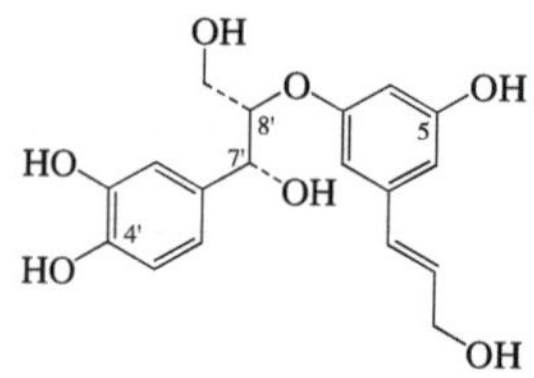

3′,4′,5,7′,9,9′-Hexahydroxy-3,8′-oxyneolign-7-ene

$C_{18}H_{20}O_7$ 348.351

Structures in this series may need revision. Phenyl 3,5-dioxygenated substn. pattern in doubt as a result of work done by Yang *et al* in 2015

(7E,7′R*,8′S*)-form

9-Aldehyde, 3′,5-di-Me ether

4′,7′,9′-Trihydroxy-3′,5-dimethoxy-3,8′-oxylign-7-en-9-al

[2417849-61-5]

$C_{20}H_{22}O_7$ 374.389

Constit. of the stems of *Eleutherococcus senticosus*. Exhibits selective inhibitory activity on protein tyrosine phosphatase 1B (PTP1B). Amorph. yellow powder. $[\alpha]^{25}_{D}$ -4.4 (c, 0.1 in MeOH). λ_{max} 225; 291 (MeOH).

Yang, Y.-N. *et al., J. Nat. Prod.*, 2015, **78**, 705-711 (*pmr, struct*).

Zhang, L. *et al., Fitoterapia*, 2017, **121**, 58-63 (*9-aldehyde 3′,5-di-Me ether, activity*)

3,8′-Oxylign-7-ene-3′,4′,6,7′,9,9′-hexol **O81**

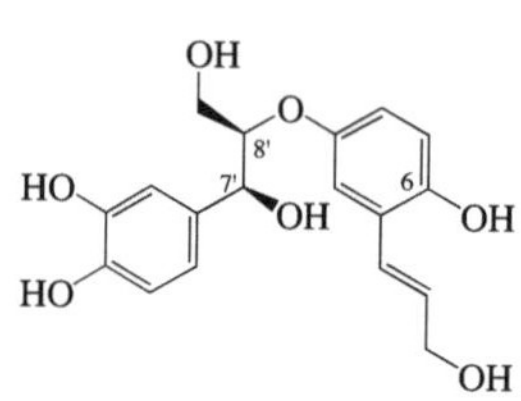

3′,4′,6,7′,9,9′-Hexahydroxy-3,8′-oxyneolign-7-ene

$C_{18}H_{20}O_7$ 348.351

(7E,7′S,8′R)-form

9-Carboxylic acid, 3′,6-di-Me ether

4′,7′,9′-Trihydroxy-3′,6-dimethoxy-3,8′-oxylign-7-en-9-oic acid

9-Carboxylic acid, 3′,6-di-Me ether, 9-β-D-glucopyranosyl ester

Neoasarininoside B

[2211144-27-1]

$C_{26}H_{32}O_{13}$ 552.529

Constit. of the roots and rhizomes of *Asarum heterotropoides*. Needles (MeOH). Mp 178 - 179°. $[\alpha]^{20}_{D}$ -23.3 (c, 0.6 in MeOH). λ_{max} 205 (log ε 4.1); 284 (log ε 3.89); 319 (log ε 3.91) (MeOH).

Jing, Y. *et al., Molecules*, 2017, **22**, 125:1-20 (*Neoasarininoside B*)

4,8′-Oxylign-7-ene-3,3′,4′,5,5′,7′-hexol **O82**

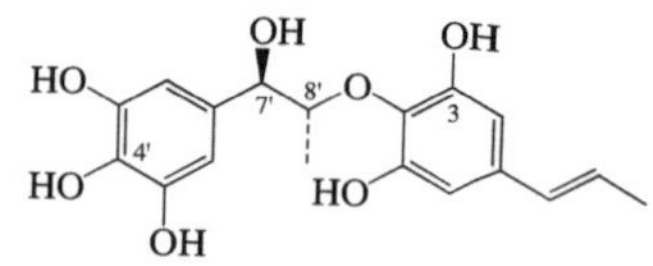

(7E,7′R,8′R)-form

3,3′,4′,5,5′,7′-Hexahydroxy-4,8′-oxyneolign-7-ene

$C_{18}H_{20}O_7$ 348.351

(7E,7′R,8′R)-form

3,3′,4′,5,5′-Penta-Me ether, 7′-Ac

[916264-20-5]

$C_{25}H_{32}O_8$ 460.523

Constit. of *Saururus chinensis*. Powder. $[\alpha]^{20}_{D}$ -24 (c, 0.02 in $CHCl_3$). λ_{max} 222 (log ε 4.15); 268 (log ε 4.15) (MeOH)

3′,4′-Methylene, 3,5,5′-tri-Me ether

3,3′,5-Trimethoxy-4′,5′-methylenedioxy-4,8′-oxylign-7-en-7′-ol. 7′-Hydroxy-3,3′,5-trimethoxy-4′,5′-methylenedioxy-4,8′-oxylign-7-ene. ***Saurucinol F***

[1521111-66-9]

$C_{22}H_{26}O_7$ 402.443

Constit. of *Saururus chinensis*. Oil. $[\alpha]^{20}_{D}$ -102 (c, 0.01 in $CHCl_3$). λ_{max} 225 (log ε 4.30); 266 (log ε 4.12) (MeOH)

(7E,7′S,8′R)-form

3,3′,5,5′-Tetra-Me ether, 7′-O-β-D-glucopyranoside

Pharbilignoside

[1393346-98-9]

$C_{28}H_{38}O_{12}$ 566.6

Constit. of the seeds of *Pharbitis nil*. Exhibits cytotoxic activity. Gum. $[\alpha]^{25}_{D}$ -13.2 (c, 0.1 in MeOH). λ_{max} 233 (log ε 3.98); 265 (log ε 3.25) (MeOH)

(7E,7′S,8′S)-form

3,3′,4′,5-Tetra-Me ether

3,4′,5,5′-Tetramethoxy-4,8′-oxylign-7-ene-3′,7′-diol. 3′,7′-Dihydroxy-3,4′,5,5′-tetramethoxy-4,8′-oxylign-7-ene. ***Fargesiphenol B***

[1436515-67-1]

$C_{22}H_{28}O_7$ 404.459

Constit. of *Magnolia fargesii*. Displays significant cytotoxic activity against tumour cells. Yellow gum. $[\alpha]_D^{24}$ +41.2 (c, 0.22 in MeOH). λ_{max} 210 (log ε 4.23); 282 (log ε 3.82); 320 (log ε 2.88) (MeOH)

3,3′,4′,5-Tetra-Me ether, 5′-Ac

Fargesiphenol C

[1435796-23-8]

$C_{24}H_{30}O_8$ 446.496

Constit. of *Magnolia fargesii*. Yellow gum. $[\alpha]_D^{24}$ +42.5 (c, 0.22 in MeOH). λ_{max} 210 (log ε 4.20); 280 (log ε 3.86); 318 (log ε 2.89) (MeOH)

(7E,7′RS,8′RS)-form

3,3′,4′,5,5′-Penta-Me ether

3,3′,4′,5,5′-Pentamethoxy-4,8′-oxylign-7-en-7′-ol. 7′-Hydroxy-3,3′,4′,5,5′-pentamethoxy-4,8′-oxylign-7-ene. ***Polysyphorin.*** *Polysphorin*

[137196-25-9]

[137196-27-1]

$C_{23}H_{30}O_7$ 418.486

Constit. of *Acorus gramineus, Magnolia fargesii, Piper polysyphorum* and *Rhaphidophora decursiva*. Antimalarial agent. Prisms. Mp 147 - 148°. Relative config. revised in 2003. λ_{max} 210 (log ε 4.66); 268 (log ε 3.83) (MeOH)

(7E,7′R*,8′R*)-form

3,3′,5′-Tri-Me ether

3,3′,5′-Trimethoxy-4,8′-oxylign-7-ene-4′,5,7′-triol. 4′,5,7′-Trihydroxy-3,3′,5′-trimethoxy-4,8′-oxylign-7-ene

$C_{21}H_{26}O_7$ 390.432

Constit. of the roots of *Adelostemma gracillimum*. Exhibits significant neuroprotection against NMDA induced cell death. Viscous oil. $[\alpha]_D^{20}$ -5.3 (c, 0.34 in MeOH). λ_{max} 260 (MeOH).

Ma, Y. *et al., Yaoxue Xuebao*, 1991, **26**, 345-350 (*Polysyphorin*).

Zhang, T. *et al., J. Nat. Prod.*, 2001, **64**, 772-777 (*Polysyphorin*).

Lee, A.-L. *et al., Org. Biomol. Chem.*, 2003, **1**, 3957-3966 (*Polysyphorin, synth, cryst struct*).

Kim, K.H. *et al., Chem. Pharm. Bull.*, 2011, **59**, 1425-1429 (*Pharbilignoside*).

Gao, X. *et al., J. Braz. Chem. Soc.*, 2012, **23**, 1274-1279 (*Fargesiphenols B,C*).

Cui, H. *et al., J. Nat. Prod.*, 2014, **77**, 100-110 (*Saururus chinensis constits*).

Zhang, J. *et al., Phytochem. Lett.*, 2016, **16**, 1-7 (*Adelostemma gracillimum constit, activity*)

4,8′-Oxylign-7-ene-3,3′,4′,5,5′,9-hexol O83

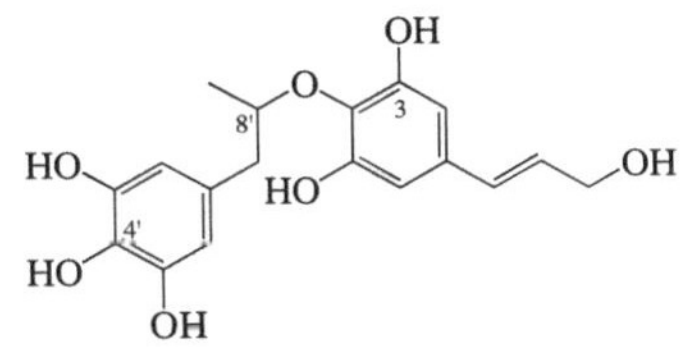

3,3′,4′,5,5′,9-Hexahydroxy-4,8′-oxyneolign-7-ene

$C_{18}H_{20}O_7$ 348.351

(±)-(E)-form

3,3′,4′,5,5′-Penta-Me ether

3,3′,4′,5,5′-Pentamethoxy-4,8′-oxylign-7-en-9-ol. 9-Hydroxy-3,3′,4′,5,5′-pentamethoxy-4,8′-oxylign-7-ene. ***Virolongin E***

[126176-80-5]

$C_{23}H_{30}O_7$ 418.486

Constit. of *Licaria chrysophylla* and *Liriodendron chinense*. Oil. λ_{max} 248 (ε 15750); 295 (ε 5100) (EtOH)

9-Aldehyde

3,3′,4′,5,5′-Pentahydroxy-4,8′-oxylign-7-en-9-al. 3,3′,4′,5,5′-Pentahydroxy-9-oxo-4,8′-oxylign-7-ene

9-Aldehyde, penta-Me ether

3,3′,4′,5,5′-Pentamethoxy-4,8′-oxylign-7-en-9-al. 3,3′,4′,5,5′-Pentamethoxy-9-oxo-4,8′-oxylign-7-ene.

Virolongin F

[126176-84-9]

$C_{23}H_{28}O_7$ 416.47

Constit. of *Licaria chrysophylla* and *Liriodendron chinense*. Oil. λ_{max} 231 (ε 21350); 280 (ε 4750); 326 (ε 2100) (EtOH).

Da Silva, M.S. *et al., Phytochemistry*, 1989, **28**, 3477-3482 (*Virolongins E,F*).

Su, Y. *et al., J. Chin. Chem. Soc. (Taipei)*, 2004, **51**, 1333-1337 (*Virolongins E,F, synth*)

4,8′-Oxylign-7-ene-3,3′,4′,5,7′,9-hexol O84

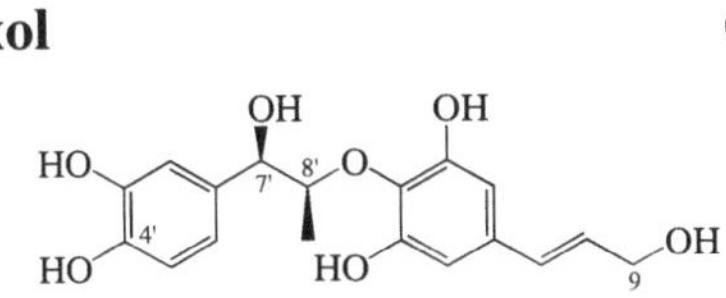

(7E,7′R,8′S)-form

3,3′,4′,5,7′,9-Hexahydroxy-4,8′-oxyneolign-7-ene

$C_{18}H_{20}O_7$ 348.351

(7E,7′R,8′S)-form

3,3′,4′,5-Tetra-Me ether

3,3′,4′,5-Tetramethoxy-4,8′-oxylign-7-ene-7′,9-diol. 7′,9-Dihydroxy-3,3′,4′,5-tetramethoxy-4,8′-oxylign-7-ene. ***Myristicanol B***

[114912-33-3]

$C_{22}H_{28}O_7$ 404.459

Constit. of the arils of *Myristica fragrans*. Lipid peroxidation inhibitor. Oil

9-Aldehyde, 3,3′,4′,5-tetra-Me ether

7′-Hydroxy-3,3′,4′,5-tetramethoxy-4,8′-oxylign-7-en-9-al. 7′-Hydroxy-3,3′,4′,5-tetramethoxy-9-oxo-4,8′-oxylign-7-ene

9-Aldehyde, 3,3′,4′,5-tetra-Me ether, 7′-Ac

Myrifralignan B

[1637669-68-1]

$C_{24}H_{28}O_8$ 444.48

Constit. of the seeds of *Myristica fragrans*. Yellow oil. $[\alpha]_D^{20}$ -8.4 (c, 1.0 in $CHCl_3$). λ_{max} 201 (log ε 4.45); 318 (log ε 3.65) (MeOH)

(7E,7′S,8′R)-form

9-Aldehyde, 3,3′,5-tri-Me ether

4′,7′-Dihydroxy-3,3′,5-trimethoxy-4,8′-oxylign-7-en-9-al. 4′,7′-Dihydroxy-3,3′,5-trimethoxy-9-oxo-4,8′-oxylign-7-ene. ***Maceneolignan G***

[1967042-73-4]

$C_{21}H_{24}O_7$ 388.416

Constit. of the arils of *Myristica fragrans*. Oil. $[\alpha]_D^{25}$ +22 (c, 0.6 in $CHCl_3$). λ_{max} 230 (log ε 4.05); 281 (log ε 3.68); 318 (log ε 3.75) (MeOH).

Hattori, M. *et al., Chem. Pharm. Bull.*, 1988, **36**, 648-653 (*Myristicanol B*).

Cao, G.-Y. *et al., Food Chem.*, 2015, **173**, 231-237 (*Myrifralignan B*).

Morikawa, T. *et al., J. Nat. Prod.*, 2016, **79**, 2005-2013 (*Maceneolignan H*)

4,8′-Oxylign-7-ene-3,3′,4′,5,9,9′-hexol O85

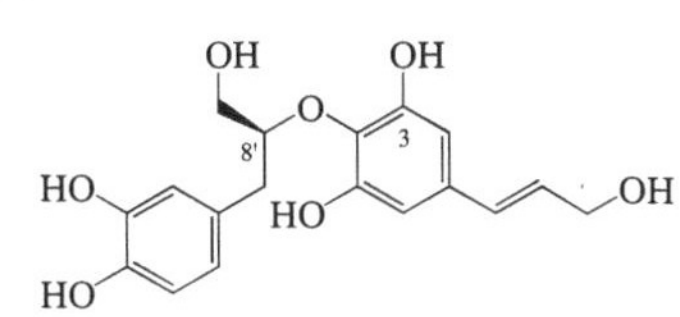

3,3′,4′,5,9,9′-Hexahydroxy-4,8′-oxyneolign-7-ene

$C_{18}H_{20}O_7$ 348.351

(S,E)-form

3,3′,5-Tri-Me ether

3,3′,5-Trimethoxy-4,8′-oxylign-7-ene-4′,9,9′-triol

3,3′,5-Tri-Me ether, 4′-O-β-D-glucopyranoside

[2329719-41-5]

$C_{27}H_{36}O_{12}$ 552.573
Constit. of *Murraya koenigii*. Inhibits the activation of IL-6 induced STAT3 in HepG2 cells. Powder. Mp 214.3 - 215.5°. $[\alpha]_D^{20}$ -11.2 (c, 0.3 in MeOH). λ_{max} 274 (MeOH).

Ma, Q. *et al.*, *Bioorg. Chem.*, 2019, **86**, 159-165 *(Murraya koenigii constit, activity)*

4,8′-Oxylign-7-ene-3,3′,4′,7′,9,9′-hexol O86

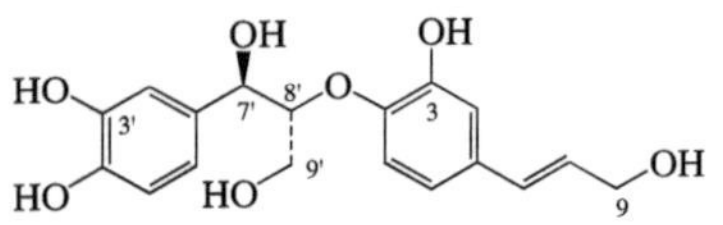

(7*E*,7′*R*,8′*R*)-form

3,3′,4′,7′,9,9′-Hexahydroxy-4,8′-oxyneolign-7-ene
$C_{18}H_{20}O_7$ 348.351
(7E,7′R,8′R)-form
3,3′-Di-Me ether
3,3′-Dimethoxy-4,8′-oxylign-7-ene-4′,7′,9,9′-tetrol. 4′,7′,9,9′-Tetrahydroxy-3,3′-dimethoxy-4,8′-oxylign-7-ene. Guaiacylglycerol β-coniferyl ether
[869799-76-8]
$C_{20}H_{24}O_7$ 376.405
Constit. of *Crataegus pinnatifida*
3,3′-Di-Me ether, 7′-O-β-D-glucopyranoside
Carpeside A
[1110658-94-0]
$C_{26}H_{34}O_{12}$ 538.546
Constit. of *Carpesium cernuum*. Amorph. powder. $[\alpha]_D^{23}$ -79 (c, 0.24 in MeOH)
3,3′-Di-Me ether, 4′,9-di-O-β-D-glucopyranoside
Ligusinenoside B
[927815-68-7]
$C_{32}H_{44}O_{17}$ 700.687
Constit. of the rhizomes of *Ligusticum sinensis*. Amorph. powder. $[\alpha]_D^{21}$ -27 (c, 0.32 in MeOH)
3,3′-Di-Me ether, 9-O-[β-D-apiofuranosyl-(1→6)-β-D-glucopyranoside]
Ligusinenoside C
[927815-69-8]
$C_{31}H_{42}O_{16}$ 670.661
Constit. of the rhizomes of *Ligusticum sinensis*. Amorph. powder. $[\alpha]_D^{21}$ -53 (c, 0.33 in MeOH)
3,3′-Di-Me ether, 7′,9-bis-O-[β-D-apiofuranosyl-(1→6)-β-D-glucopyranoside]
Lyciumlignan B
$C_{42}H_{60}O_{25}$ 964.917
Constit. of the root bark of *Lycium chinense*. $[\alpha]_D^{20}$ -66.7 (c, 0.1 in MeOH). λ_{max} 205 (log ε 4.54); 268 (log ε 3.95); 304 (log ε 3.32) (MeOH)
3,3′-Di-Me ether, 9,9′-bis-O-(3-methylbutanoyl)
(+)-Brachangobinan C
[1332863-89-4]
$C_{30}H_{40}O_9$ 544.641
Constit. of *Brachanthemum gobicum* and *Nannoglottis carpesioides*. Amorph. solid. $[\alpha]_D^{24}$ +7 (c, 0.03 in MeOH). λ_{max} 216 (log ε 4.5); 229 (log ε 4.5); 268 (log ε 4.3) (MeOH)
3,3′,7′-Tri-Me ether
3,3′,7′-Trimethoxy-4,8′-oxylign-7-ene-4′,9,9′-triol. 4′,9,9′-Trihydroxy-3,3′,7′-trimethoxy-4,8′-oxylign-7-ene
3,3′,7′-Tri-Me ether, 9,9′-bis-O-(3-methylbutanoyl)
[1332863-90-7]
$C_{31}H_{42}O_9$ 558.668
Constit. of the roots of *Nannoglottis carpesioides*. Oil. $[\alpha]_D^{20}$ +20 (c, 0.5 in MeOH). λ_{max} 217 (log ε 4.5); 268 (log ε 4.4) (MeOH)
3,3′,9-Tri-Me ether
3,3′,9-Trimethoxy-4,8′-oxylign-7-ene-4′,7′,9′-triol. 4′,7′,9′-Trihydroxy-3,3′,9-trimethoxy-4,8′-oxylign-7-ene
$[\alpha]_D^{20}$ -22 (c, 0.1 in MeOH)
3,3′,9-Tri-Me ether, 4′-O-β-D-glucopyranoside
Viburfordoside H
[2257425-93-5]
$C_{27}H_{36}O_{12}$ 552.573
Constit. of the fruit of *Viburnum fordiae*. Powder. $[\alpha]_D^{23}$ -21.2 (c, 0.1 in MeOH). λ_{max} 266 (log ε 4.38) (MeOH)
3,3′,7′,9-Tetra-Me ether
3,3′,7′,9-Tetramethoxy-4,8′-oxylign-7-ene-4′,9′-diol. 4′,9′-Dihydroxy-3,3′,7′,9-tetramethoxy-4,8′-oxylign-7-ene. ***Pinnatifidanin BIX***
[1462882-71-8]
$C_{22}H_{28}O_7$ 404.459
Constit. of *Crataegus pinnatifida*. Yellow oil. $[\alpha]_D^{20}$ +1.6 (c, 0.02 in MeOH). λ_{max} 233 (log ε 2.46) (MeOH)
9-Et, 3,3′-di-Me ether
9-Ethoxy-3,3′-dimethoxy-4,8′-oxylign-7-ene-4′,7′,9′-triol. 9-Ethoxy-4′,7′,9′-trihydroxy-3,3′-dimethoxy-4,8′-oxylign-7-ene. ***Huangnin A***
$C_{22}H_{28}O_7$ 404.459
Constit. of the seeds of *Crataegus pinnatifida*. Yellow oil. $[\alpha]_D^{20}$ +12.2 (c, 0.02 in MeOH). λ_{max} 231 (log ε 3.44); 280 (log ε 1.30) (MeOH)
9-Aldehyde
3,3′,4′,7′,9′-Pentahydroxy-4,8′-oxylign-7-en-9-al. 3,3′,4′,7′,9′-Pentahydroxy-9-oxo-4,8′-oxylign-7-ene
9-Aldehyde, 3,3′,7′-tri-Me ether
4′,9′-Dihydroxy-3,3′,7′-trimethoxy-4,8′-oxylign-7-en-9-al. 4′,9′-Dihydroxy-3,3′,7′-trimethoxy-9-oxo-4,8′-oxylign-7-ene. ***Pinnatifidanin BVIII***
[1462882-68-3]
$C_{21}H_{24}O_7$ 388.416
Constit. of *Crataegus pinnatifida*. Displays potent cytotoxic activity against human cancer cell lines. Yellow oil. $[\alpha]_D^{20}$ +3.3 (c, 0.02 in MeOH). λ_{max} 233 (log ε 2.16) (MeOH)
9-Carboxylic acid
3,3′,4′,7′,9′-Pentahydroxy-4,8′-oxylign-7-en-9-oic acid
9-Carboxylic acid, 3′-Me ether, Me ester, 3-O-β-D-glucopyranoside
Matteustruthioside A
[1826900-71-3]
$C_{26}H_{32}O_{13}$ 552.529
Constit. of the aerial parts of *Matteuccia struthiopteris*. Yellowish solid. $[\alpha]_D^{25}$ -1.62 (c, 0.5 in MeOH). λ_{max} 235 (log ε 4.29); 295 (log ε 4.25); 317 (log ε 4.25) (MeOH)
9-Carboxylic acid, 3,3′-di-Me ether
4′,7′,9′-Trihydroxy-3,3′-dimethoxy-4,8′-oxylign-7-en-9-oic acid
[1262045-76-0]
$C_{20}H_{22}O_8$ 390.388
Constit. of *Avicennia marina*
(7E,7′R,8′S)-form
3,3′-Di-Me ether, 7′-O-β-D-glucopyranoside
Carpeside B
[1110658-95-1]
$C_{26}H_{34}O_{12}$ 538.546

Constit. of *Carpesium cernuum*. Amorph. powder. $[\alpha]_D^{20}$ -26 (c, 0.8 in MeOH)

3,3′-Di-Me ether, 9′-O-β-D-glucopyranoside

Debilignanoside

[1353637-68-9]

$C_{26}H_{34}O_{12}$ 538.546

Constit. of *Equisetum debile*. Amorph. powder. $[\alpha]_D^{20}$ +5.5 (c, 0.58 in MeOH)

3,3′-Di-Me ether, 7′,9-di-O-β-D-glucopyranoside

Lyciumlignan C

$C_{32}H_{44}O_{17}$ 700.687

Constit. of the root bark of *Lycium chinense*. $[\alpha]_D^{20}$ +16.9 (c, 0.07 in MeOH). λ_{max} 205 (log ε 4.35); 271 (log ε 3.71); 301 (log ε 3.61) (MeOH)

3,3′-Di-Me ether, 9,9′-bis-O-(4-hydroxy-E-cinnamoyl)

(-)-erythro-Dadahol B

[2250284-21-8]

$C_{38}H_{36}O_{11}$ 668.695

Constit. of the twigs of *Sageretia theezans*. Gum. $[\alpha]_D^{20}$ -20 (c, 0.1 in MeOH). λ_{max} 220 (log ε 3.72); 312 (log ε 3.78) (MeOH)

3,3′-Di-Me ether, 9-O-(4-hydroxy-E-cinnamoyl), 9′-O-(4-hydroxy-Z-cinnamoyl)

[2250284-23-0]

$C_{38}H_{36}O_{11}$ 668.695

Constit. of the twigs of *Sageretia theezans*. Gum. $[\alpha]_D^{20}$ +59.3 (c, 0.1 in MeOH). λ_{max} 220 (log ε 3.59); 312 (log ε 3.57) (MeOH)

3,3′-Di-Me ether, 9-O-(4-hydroxy-Z-cinnamoyl), 9′-O-(4-hydroxy-E-cinnamoyl)

[2250284-25-2]

$C_{38}H_{36}O_{11}$ 668.695

Constit. of the twigs of *Sageretia theezans*. Gum. $[\alpha]_D^{20}$ -2.5 (c, 0.1 in MeOH). λ_{max} 215 (log ε 3.58); 310 (log ε 3.48) (MeOH)

3,3′,4′-Tri-Me ether

3,3′,4′-Trimethoxy-4,8′-oxylign-7-ene-7′,9,9′-triol. 7′,9,9′-Trihydroxy-3,3′,4′-trimethoxy-4,8′-oxylign-7-ene

[203726-22-1]

$C_{21}H_{26}O_7$ 390.432

Constit. of the buds of *Magnolia fargesii*. Oil. $[\alpha]_D^{20}$ +5 (c, 0.1 in MeOH). λ_{max} 206 (log ε 4.55); 267 (log ε 4.07) (MeOH)

3,3′,9-Tri-Me ether

$[\alpha]_D^{20}$ -18 (c, 0.2 in MeOH)

3,3′,9-Tri-Me ether, 4′-O-β-D-glucopyranoside

Glehlinoside H. ***Janceolaroside A***

[1335048-32-2]

$C_{27}H_{36}O_{12}$ 552.573

Constit. of the roots of *Glehnia littoralis* and *Jasminum lanceolarium*. Amorph. solid or oil. $[\alpha]_D^{20}$ -21.9 (c, 0.15 in MeOH). $[\alpha]_D^{20}$ -18.9 (c, 0.08 in MeOH). λ_{max} 236 (log ε 4.5); 258 (log ε 4.6) (MeOH)

3,3′,7′,9-Tetra-Me ether

[1338232-85-1]

Isol. from *Toona ciliata* var. *pubescens*. Weak antitumour agent. Yellow oil. $[\alpha]_D^{25}$ -4 (c, 1.3 in MeOH)

9-Aldehyde, 3,3′-di-Me ether

Fordiane A

[2171499-29-7]

$C_{20}H_{22}O_7$ 374.389

Constit. of the fruit of *Viburnum fordiae*. α-Glucosidase and NO prodn. inhibitor. Antioxidant. Yellow oil. $[\alpha]_D^{23}$ +38.5 (c, 0.1 in MeOH). λ_{max} 226 (log ε 4.20); 337 (log ε 4.12) (MeOH)

9-Carboxylic acid, 3′-Me ether, Me ester, 3-O-β-D-glucopyranoside

Matteustruthioside B

[1826900-72-4]

$C_{26}H_{32}O_{13}$ 552.529

Constit. of the aerial parts of *Matteuccia struthiopteris*. Yellowish solid. $[\alpha]_D^{25}$ -4.48 (c, 0.5 in MeOH). λ_{max} 235 (log ε 4.20); 292 (log ε 4.16); 318 (log ε 4.17) (MeOH)

(7E,7′S,8′R)-form

3′-Me ether

3′-Methoxy-4,8′-oxylign-7-ene-3,4′,7′,9,9′-pentol. 3,4′,7′,9,9′-Pentahydroxy-3′-methoxy-4,8′-oxylign-7-ene

3′-Me ether, 3-O-β-D-glucopyranoside

[1018987-04-6]

$C_{25}H_{32}O_{12}$ 524.519

Constit. of the roots of *Iodes cirrhosa*. Amorph. solid. $[\alpha]_D^{20}$ -6.2 (c, 0.08 in MeOH). λ_{max} 205 (log ε 4.3); 261 (log ε 3.9) (MeOH)

3,3′-Di-Me ether

[890317-92-7]

$C_{20}H_{24}O_7$ 376.405

Constit. of *Crataegus pinnatifida*

9-Carboxylic acid, 3,3′-di-Me ether, 7′-O-β-D-glucopyranoside

[1979182-11-0]

$C_{26}H_{32}O_{13}$ 552.529

Constit. of *Zanthoxylum armatum*. Oil. $[\alpha]_D^{20}$ -69.3 (c, 0.1 in H_2O)

3,3′-Di-Me ether, 7′-O-[β-D-apiofuranosyl-(1→6)-β-D-glucopyranoside]

Lyciumlignan A

$C_{31}H_{42}O_{16}$ 670.661

Constit. of the root bark of *Lycium chinense*. Powder. $[\alpha]_D^{20}$ -45.5 (c, 0.1 in MeOH). λ_{max} 206 (log ε 4.57); 268 (log ε 3.89); 301 (log ε 3.49) (MeOH)

3,3′-Di-Me ether, 9-O-[β-D-apiofuranosyl-(1→6)-β-D-glucopyranoside]

Ligusinenoside D

[1350889-91-6]

$C_{31}H_{42}O_{16}$ 670.661

Constit. of the rhizomes of *Ligusticum sinensis*. Amorph. powder. $[\alpha]_D^{21}$ -55 (c, 0.26 in MeOH)

3,3′,9-Tri-Me ether

$[\alpha]_D^{20}$ +20 (c, 0.2 in MeOH)

3,3′,9-Tri-Me ether, 4′-O-β-D-glucopyranoside

Viburfordoside G

[2257425-92-4]

$C_{27}H_{36}O_{12}$ 552.573

Constit. of the fruit of *Viburnum fordiae*. Powder. $[\alpha]_D^{23}$ -18.8 (c, 0.1 in MeOH). λ_{max} 267 (log ε 4.11) (MeOH)

3,3′,9-Tri-Me ether, 9′-O-β-D-glucopyranoside

Saikolignanoside A‡

$C_{27}H_{36}O_{12}$ 552.573

Constit. of the roots of *Bupleurum chinense*. Exhibits potent ABTS free radical scavenging activity. Yellow oil. $[\alpha]_D^{20}$ -17.3 (c, 0.1 in MeOH). λ_{max} 281 (log ε 2.96) (MeOH)

7′-Et, 3,3′-di-Me ether

Xanthiumnolic A

[2226004-44-8]

$C_{22}H_{28}O_7$ 404.459

Constit. of the fruit of *Xanthium sibiricum*. Exhibits moderate inhibitory activity against NO prodn.. Pale yellow viscous oil. $[\alpha]_D^{20}$ +12.1 (c, 1.0 in MeOH). λ_{max} 234 (log ε 4.22) (MeOH)

9-Et, 3,3′-di-Me ether
Huangnin B
$C_{22}H_{28}O_7$ 404.459
Constit. of the seeds of *Crataegus pinnatifida*. Yellow oil. $[\alpha]_D^{20}$ +7.2 (c, 0.02 in MeOH). λ_{max} 231 (log ε 4.53); 279 (log ε 1.24) (MeOH)

9-Carboxylic acid, 3,3′-di-Me ether
[1262045-73-7]
Constit. of *Avicennia marina*

(7E,7′S,8′S)-form

3,3′-Di-Me ether, 9,9′-di-O-(3-methylbutanoyl)
(-)-Brachangobinan C
[2357185-52-3]
Amorph. solid. $[\alpha]_D^{24}$ -7 (c, 0.03 in MeOH)

3,3′-Di-Me ether, 9,9′-bis-O-(4-hydroxy-E-cinnamoyl)
*(-)-threo-**Dadahol B***
[2250284-22-9]
$C_{38}H_{36}O_{11}$ 668.695
Constit. of the twigs of *Sageretia theezans*. Gum. $[\alpha]_D^{20}$ -291.2 (c, 0.1 in MeOH). λ_{max} 220 (log ε 3.66); 313 (log ε 3.71) (MeOH)

3,3′-Di-Me ether, 9-O-(4-hydroxy-E-cinnamoyl), 9′-O-(4-hydroxy-Z-cinnamoyl)
[2250284-24-1]
$C_{38}H_{36}O_{11}$ 668.695
Constit. of the twigs of *Sageretia theezans*. Gum. $[\alpha]_D^{20}$ -0.8 (c, 0.1 in MeOH). λ_{max} 224 (log ε 3.42); 311 (log ε 3.41) (MeOH)

3,3′-Di-Me ether, 9-O-(4-hydroxy-Z-cinnamoyl), 9′-O-(4-hydroxy-E-cinnamoyl)
[2250284-26-3]
$C_{38}H_{36}O_{11}$ 668.695
Constit. of the twigs of *Sageretia theezans*. Gum. $[\alpha]_D^{20}$ -0.8 (c, 0.1 in MeOH). λ_{max} 224 (log ε 3.42); 311 (log ε 3.41) (MeOH)

3,3′,9-Tri-Me ether
$[\alpha]_D^{20}$ +20 (c, 0.1 in MeOH)

3,3′,9-Tri-Me ether, 4′-O-β-D-glucopyranoside
Viburfordoside I
[2257425-94-6]
$C_{27}H_{36}O_{12}$ 552.573
Constit. of the fruit of *Viburnum fordiae*. Powder. $[\alpha]_D^{23}$ -17.8 (c, 0.1 in MeOH). λ_{max} 266 (log ε 4.11) (MeOH)

3,3′,7′,9-Tetra-Me ether
[1338232-84-0]
$C_{22}H_{28}O_7$ 404.459
Isol. from leaves and stems of *Toona ciliata* var. *pubescens*. Shows mild antiproliferative activity against tumour cell lines. Yellow oil. $[\alpha]_D^{25}$ -16.5 (c, 1.0 in MeOH)

9-Aldehyde, 3,3′-di-Me ether
Fordiane B
[2209905-55-3]
$C_{20}H_{22}O_7$ 374.389
Constit. of the fruit of *Viburnum fordiae*. α-Glucosidase and NO prodn. inhibitor. Antioxidant. Yellow oil. $[\alpha]_D^{23}$ -16.9 (c, 0.1 in MeOH)

(7E,7′R*,8′R*)-form

3,3′-Di-Me ether
[126061-41-4]
[1103-58-8] (unspecified stereochem.) [117703-32-9]
$C_{20}H_{24}O_7$ 376.405
Constit. of *Picea excelsa*, *Larix leptolepis*, and from *Brucea javanica*. Syrup. $[\alpha]_D^{22}$ -6.7 (c, 0.53 in MeOH). The *B. javanica* isolate was assigned an isomeric struct. in 1996 and then in 1998 reassigned as this compd.

3,3′-Di-Me ether, 4′-O-β-D-glucopyranoside
Citrusin A
[105279-09-2]
$C_{26}H_{34}O_{12}$ 538.546
Constit. of *Citrus limon*, *Fortunella japonica* and *Pedicularis alaschanica*. Mp 108 - 109°. $[\alpha]_D^{20}$ -17.2 (c, 0.3 in MeOH)

3,3′-Di-Me ether, 9-O-β-D-glucopyranoside
Hyuganoside IIIb
[838845-08-2]
$C_{26}H_{34}O_{12}$ 538.546
Constit. of *Angelica furcijuga* and *Breynia rostrata*. Powder. $[\alpha]_D^{23}$ -16.9 (c, 0.51 in MeOH). λ_{max} 270 (log ε 3.96) (MeOH)

3,3′-Di-Me ether, 9,9′-di-Ac
[210161-56-1]
$C_{24}H_{28}O_9$ 460.479
Constit. of Chilean propolis. Oil. $[\alpha]_D^{25}$ -15.6 (c, 0.85 in CH_2Cl_2). λ_{max} 268 (log ε 4.16); 300 (log ε 3.86) (MeOH)

3,3′-Di-Me ether, 9,9′-bis-O-(4-hydroxy-3-methoxy-E-cinnamoyl)
*threo-**Demethoxycarolignan Z***
$C_{40}H_{40}O_{13}$ 728.747
Constit. of the aerial parts of *Eryngium triquetrum*. Powder. Characterised as an inseparable *trans-erythro/threo* pair to which the data refers. λ_{max} 219 (log ε 4.12); 233 (log ε 4.02); 288 (log ε 3.93); 324 (log ε 4.02) (MeOH)

3,3′,9-Tri-Me ether
3,3′,9-Trimethoxy-4,8′-oxylign-7-ene-4′,7′,9′-triol. 4′,7′,9′-Trihydroxy-3,3′,9-trimethoxy-4,8′-oxylign-7-ene. ***Sabotenin B***
$C_{21}H_{26}O_7$ 390.432
Constit. of the seeds of *Opuntia ficus-indica* var. *saboten*. Unconfirmed struct. from 2008 congress report

7′-Et, 3,3′-di-Me ether
7′-Ethoxy-3,3′-dimethoxy-4,8′-oxylign-7-ene-4′,9,9′-triol. 7′-Ethoxy-4′,9,9′-trihydroxy-3,3′-dimethoxy-4,8′-oxylign-7-ene.
Hainanervalignan A
$C_{22}H_{28}O_7$ 404.459
Constit. of twigs and leaves of *Ervatamia hainanensis*. $[\alpha]_D^{25}$ +25.8 (c, 0.61 in MeOH). λ_{max} 273 (log ε 3.82) (MeOH)

7′-Et, 3,3′-di-Me ether, 9,9′-di-Ac
Jatrolignan A
$C_{26}H_{32}O_9$ 488.533
Constit. of *Jatropha curcas*. Yellow gum. $[\alpha]_D^{24}$ -5.6 (c, 0.3 in MeOH). λ_{max} 269 (log ε 4.04) (MeOH)

3′,4′-Methylene, 3-Me ether
3-Methoxy-3′,4′-methylenedioxy-4,8′-oxylign-7-ene-7′,9,9′-triol. 7′,9,9′-Trihydroxy-3-methoxy-3′,4′-methylenedioxy-4,8′-oxylign-7-ene

3′,4′-Methylene, 3-Me ether, 7′,9′-diangeloyl
Narynenol
[1028946-73-7]
$C_{30}H_{34}O_9$ 538.593
Constit. of the roots of *Ligularia narynensis*. Gum. $[\alpha]_D^{20}$ -8 (c, 0.7 in $CHCl_3$)

9-Aldehyde, 3,3′-di-Me ether
4′,7′,9′-Trihydroxy-3,3′-dimethoxy-4,8′-oxylign-7-en-9-al. 4′,7′,9′-Trihydroxy-3,3′-dimethoxy-9-oxo-4,8′-oxylign-7-ene. Guaiacylglycerol β-coniferyl aldehyde ether
[74474-54-7]
[17614-13-0] (unspecified stereochem.) [79389-55-2] (7ξ,7′R*,8′R*-form)

$C_{20}H_{22}O_7$ 374.389
Constit. of *Cornus kousa*, *Crataegus pinnatifida*, *Hevea brasiliensis*, *Catunaregam spinosa*, *Larix leptolepis*, *Torreya yunnanensis* and other plant spp.. Pale yellow oil. Abs. config. not determined for most isolates. λ_{max} 229; 287; 332 (MeOH)

9-Aldehyde, 3,3′,7′-tri-Me ether
Schimalignan B
[1309677-41-5]
$C_{21}H_{24}O_7$ 388.416
Constit. of the stems of *Schima superba*. Pale yellow oil. $[\alpha]^{22}_{D}$ -10 (c, 1.1 in MeOH). λ_{max} 230; 288 (MeOH)

9-Aldehyde, 7′-Et ether
7′-Ethoxy-3,3′,4′,9′-tetrahydroxy-4,8′-oxylign-7-en-9-al. 7′-Ethoxy-3,3′,4′,9′-tetrahydroxy-9-oxo-4,8′-oxylign-7-ene

9-Aldehyde, 7′-Et, 3,3′-di-Me ether
7′-Ethoxy-4′,9′-dihydroxy-3,3′-dimethoxy-4,8′-oxylign-7-en-9-al. 7′-Ethoxy-4′,9′-dihydroxy-3,3′-dimethoxy-9-oxo-4,8′-oxylign-7-ene
[1954754-09-6]
$C_{22}H_{26}O_7$ 402.443
Constit. of the fruit of *Melia toosendan*. Exhibits weak agonistic activity on human melatonin receptor MT_1. Amorph. pale yellow powder. Mp 59 - 61°. $[\alpha]^{20}_{D}$ -3.9 (c, 0.1 in MeOH). λ_{max} 337 (log ε 4.21) (MeOH)

9-Carboxylic acid, 3,3′-di-Me ether
[1372162-60-1]
$C_{20}H_{22}O_8$ 390.388
Constit. of *Catunaregam spinosa*

9-Carboxylic acid, 3,3′-di-Me ether, 4′-O-β-D-glucopyranoside
Glehlinoside C
[402565-04-2]
$C_{26}H_{32}O_{13}$ 552.529
Constit. of *Glehnia littoralis*. Off-white amorph. solid. $[\alpha]^{25}_{D}$ -34.3 (c, 0.09 in MeOH). λ_{max} 230; 275 (MeOH)

9-Carboxylic acid, 3,3′-di-Me ether, 2-(4-hydroxyphenyl)ethylamide
Cannabisin H
[403647-08-5]
$C_{28}H_{31}NO_8$ 509.555
Alkaloid from the bark of *Hibiscus cannabinus*. Yellowish oil

(7E,7′R*,8′S*)-form

9′-O-(4-Hydroxy-3-methoxy-E-cinnamoyl), 9-O-(4-hydroxy-E-cinnamoyl)
Carolignan K
[344944-06-5]
$C_{39}H_{38}O_{12}$ 698.721
Constit. of *Hibiscus cannabinus*. Yellowish oil. Incorr. indexed by CA and rel. config. misassigned by authors

3,3′-Di-Me ether
[126107-59-3]
[117703-33-0]
$C_{20}H_{24}O_7$ 376.405
Isol. from *Eucommia ulmoides* and *Pluchea indica*

3,3′-Di-Me ether, 4′-O-β-D-glucopyranoside
Alaschanioside A
[157446-63-4]
$C_{26}H_{34}O_{12}$ 538.546
Constit. of *Pedicularis alaschanica*. Amorph. powder. $[\alpha]^{23}_{D}$ -10.8 (c, 0.4 in MeOH)

3,3′-Di-Me ether, 4′-O-[4-hydroxy-3-methoxybenzoyl-(→6)-β-D-glucopyranoside]
Glehlinoside D
[1335048-36-6]
$C_{34}H_{40}O_{15}$ 688.679
Constit. of *Glehnia littoralis*. $[\alpha]^{20}_{D}$ -23.3 (c, 0.45 in MeOH)

3,3′-Di-Me ether, 9-O-β-D-glucopyranoside
Hyuganoside IIIa. *Terebinthacoside III*
[838845-07-1]
[214778-46-8] [249292-62-4]
$C_{26}H_{34}O_{12}$ 538.546
Constit. of the roots of *Angelica furcijuga*. Antitumour agent. Powder. $[\alpha]^{25}_{D}$ -6.1 (c, 0.22 in MeOH). Terebinthacoside III was named with unspecified stereochem.. λ_{max} 267 (log ε 4.16) (MeOH)

3,3′-Di-Me ether, 9,9′-di-Ac
[210161-55-0]
$C_{24}H_{28}O_9$ 460.479
Constit. of Chilean propolis. Oil. $[\alpha]^{25}_{D}$ +8.8 (c, 0.63 in CH_2Cl_2). λ_{max} 268 (log ε 4.49); 300 (log ε 4.2) (no solvent reported)

3,3′-Di-Me ether, 9,9′-bis-O-(4-hydroxybenzoyl)
[1456786-60-9]
$C_{34}H_{32}O_{11}$ 616.619
Constit. of *Posidonia oceanica*. Pale yellow oil. $[\alpha]_{D}$ +15.7 (c, 0.05 in MeOH)

3,3′-Di-Me ether, 9,9′-bis-O-(4-hydroxy-3-methoxy-E-cinnamoyl)
erythro-**Demethoxycarolignan Z**
$C_{40}H_{40}O_{13}$ 728.747
Constit. of the aerial parts of *Eryngium triquetrum*. Characterised as an inseparable *trans-erythro/threo* pair

3,3′,9-Tri-Me ether
Sabotenin A
$C_{21}H_{26}O_7$ 390.432
Constit. of the seeds of *Opuntia ficus-indica* var. *saboten*. Unconfirmed struct. from 2008 congress report

7′-Et, 3,3′-di-Me ether
7′-Ethoxy-3,3′-dimethoxy-4,8′-oxylign-7-ene-4′,9,9′-triol. 7′-Ethoxy-4′,9,9′-trihydroxy-3,3′-dimethoxy-4,8′-oxylign-7-ene

7′-Et, 3,3′-di-Me ether, 9,9′-di-Ac
Jatrolignan B
$C_{26}H_{32}O_9$ 488.533
Constit. of *Jatropha curcas*. Yellow gum. $[\alpha]^{24}_{D}$ -5.1 (c, 0.47 in MeOH). λ_{max} 270 (log ε 4.09) (MeOH)

7′-Et, 3,3′-di-Me ether, 9,9′-bis-O-(4-hydroxybenzoyl)
Quiquesetinerviusin C
[1250376-64-7]
$C_{36}H_{36}O_{11}$ 644.673
Constit. of the stems of *Calamus quiquesetinervius*. Pale yellow powder. Mp 92°. $[\alpha]^{25}_{D}$ -3.4 (c, 0.58 in MeOH). λ_{max} 203 (log ε 4.75); 261 (log ε 4.71); 275 (sh) (log ε 4.63) (MeOH)

9-Aldehyde, 3,3′-di-Me ether
[74474-55-8]
[79389-54-1] (7ξ,7′*R**,8′*S**-form)
$C_{20}H_{22}O_7$ 374.389
Constit. of *Cornus kousa*, *Crataegus pinnatifida*, *Hevea brasiliensis*, *Catunaregam spinosa*, *Larix leptolepis*, *Torreya yunnanensis* and other plant spp.. Pale yellow oil. Abs. config. not determined for most isolates

9-Aldehyde, 7′-Et, 3,3′-di-Me ether
[1954754-12-1]
$C_{22}H_{26}O_7$ 402.443
Constit. of the fruit of *Melia toosendan*. Exhibits moderate agonistic activity on human melatonin receptor MT_1. Amorph.

pale yellow powder. Mp 43 - 45°. $[\alpha]^{20}_{D}$ -6.1 (c, 0.1 in MeOH). λ_{max} 337 (log ε 3.89) (MeOH)

9-Carboxylic acid, 3,3′-di-Me ether
[1372162-61-2]
Constit. of *Catunaregam spinosa*

(7E,7′RS,8′RS)-form

3,3′,9-Tri-Me ether
$C_{21}H_{26}O_7$ 390.432
Constit. of the root bark of *Ailanthus altissima*. Light yellow gum. $[\alpha]^{20}_{D}$ -0.6 (c, 0.1 in MeOH). Partial racemate, resolved by chiral hplc. λ_{max} 250 (log ε 4.47); 275 (log ε 4.55) (MeOH)

3,3′-Di-Me ether, 9,9′-di-O-(3-methylbutanoyl)
Brachangobinan C
[2359598-23-3]
$C_{30}H_{40}O_9$ 544.641
Constit. of *Brachanthemum gobicum*. Exhibits moderate antitrypanosomal activity. Yellowish amorph. solid. The nat. prod. Brachangobinan C was isol. as a racemate and resolved by chiral hplc. λ_{max} 268 (log ε 3.67) (MeOH)

(7E,7′RS,8′SR)-form

3,3′,9-Tri-Me ether
Constit. of the root bark of *Ailanthus altissima*. Light yellow gum. $[\alpha]^{20}_{D}$ +2.3 (c, 0.1 in MeOH). Partial racemate resolved by chiral hplc

(7E,7′ξ,8′ξ)-form

3,3′-Di-Me ether, 7′-O-β-D-glucopyranoside
[415713-76-7]
$C_{26}H_{34}O_{12}$ 538.546
Constit. of *Urtica dioica* and cane sugar. Oil. λ_{max} 220 (sh); 266 (MeOH)

3,3′-Di-Me ether, 9,9′-bis-O-(4-hydroxy-E-cinnamoyl)
Dadahol B
[405281-77-8]
$C_{38}H_{36}O_{11}$ 668.695
Constit. of *Artocarpus dadah*. Oil. Isol. as a mixt. of *erythro-* and *threo-*stereoisomers, to which uv spectrum refers. λ_{max} 210 (log ε 4.11); 224 (log ε 3.96); 311 (log ε 3.98) (MeOH)

3,3′,7′-Tri-Me ether, 9,9′-di-Ac
Marylaurencinol D
[1792173-23-9]
$C_{25}H_{30}O_9$ 474.506
Constit. of the roots of *Cymbidium* Great Flower 'Mary Laurencin'. Amorph. powder. $[\alpha]^{25}_{D}$ -2.9 (c, 0.7 in $CHCl_3$). λ_{max} 214 (log ε 3.96); 266 (log ε 4.17) (MeOH)

(7Z,7′R*,8′R*)-form

3′-Me ether, 9-O-β-D-glucopyranoside
Semitortoside B
[190520-34-4]
$C_{25}H_{32}O_{12}$ 524.519
Constit. of *Pedicularis semitorta*

(7Z,7′R*,8′S*)-form

3′-Me ether, 9-O-β-D-glucopyranoside
Semitortoside A
[190520-35-5]
$C_{25}H_{32}O_{12}$ 524.519
Constit. of *Pedicularis semitorta*

(±)-(E)-form

7′-Ketone, 3,3′-di-Me ether, 9,9′-di-O-(3-methylbutanoyl)
Brachangobinan D
[2359598-24-4] (*R*-form)
$C_{30}H_{38}O_9$ 542.625
Constit. of *Brachanthemum gobicum*. Exhibits moderate antitrypanosomal activity. Amorph. yellowish solid. $[\alpha]^{22}_{D}$ +1 (c, 0.4 in MeOH). Partial racemate. λ_{max} 206 (log ε 4.15); 270 (log ε 3.84) (MeOH).

Freudenberg, K. *et al.*, *Chem. Ber.*, 1955, **88**, 626-633 (*3,3′-di-Me ether, synth, uv, ir*).
Nimz, H. *Chem. Ber.*, 1965, **98**, 533-537 (*di-Me ether*).
Miki, K. *et al.*, *Phytochemistry*, 1980, **19**, 449-453 (*di-Me ether*).
Sawabe, A. *et al.*, *Nippon Kagaku Kaishi*, 1986, **60**, 593-599 (*Citrusin A*).
Deyama, T. *et al.*, *Chem. Pharm. Bull.*, 1987, **35**, 1785-1789 (*9-aldehyde 3,3′-di-Me ethers*).
Uchiyama, T. *et al.*, *Phytochemistry*, 1989, **28**, 3369-3372 (*Pluchea indica 3,3′-di-Me ether*).
Matsubara, Y. *et al.*, *Agric. Biol. Chem.*, 1991, **55**, 647-650 (*Citrusin A*).
Della Greca, M. *et al.*, *Phytochemistry*, 1994, **35**, 777-779 (*di-Me ether*).
Gao, J. *et al.*, *Indian J. Chem., Sect. B*, 1995, **34**, 466-468 (*Alaschanioside A*).
Yoshikawa, K. *et al.*, *Phytochemistry*, 1995, **39**, 659-664 (*3,3′-di-Me ether*).
Wang, C.Z. *et al.*, *Indian J. Chem., Sect. B*, 1997, **36**, 150-153 (*Semitortosides*).
Valcic, S. *et al.*, *J. Nat. Prod.*, 1998, **61**, 771-775 (*di-Me ether di-Ac*).
Li, S. *et al.*, *Phytochemistry*, 1998, **49**, 2125-2128 (*synth, pmr, cmr*).
Japan. Pat., 1999, 99 310 589 (*Terebinthacoside III*).
Seca, A.M.L. *et al.*, *Phytochemistry*, 2001, **56**, 759-767 (*Carolignan K*).
Seca, A.M.L. *et al.*, *Phytochemistry*, 2001, **58**, 1219-1223 (*Cannabisin H*).
Takara, K. *et al.*, *Biosci., Biotechnol., Biochem.*, 2002, **66**, 29-35 (*cane sugar 3,3′-di-Me ether 7-glucoside*).
Yuan, Z. *et al.*, *Chem. Pharm. Bull.*, 2002, **50**, 73-77 (*Glehlinoside C*).
Su, B.-N. *et al.*, *J. Nat. Prod.*, 2002, **65**, 163-169 (*Dadahols*).
Li, S.-H. *et al.*, *J. Nat. Prod.*, 2003, **66**, 1002-1005 (*Torreya yunnanensis aldehydes*).
Morikawa, T. *et al.*, *Chem. Pharm. Bull.*, 2004, **52**, 1387-1390 (*Hyuganosides III*).
Lourith, N. *et al.*, *J. Wood Sci.*, 2005, **51**, 370-378 (*Eucommia ulmoides lignan*).
Lee, D.-Y. *et al.*, *Arch. Pharmacal Res.*, 2007, **30**, 402-407 (*Cornus kousa aldehydes*).
Ma, J.-P. *et al.*, *Helv. Chim. Acta*, 2007, **90**, 158-163 (*Ligusinenosides B,C*).
Gao, X. *et al.*, *Chin. Chem. Lett.*, 2008, **19**, 71-72 (*Narynenol*).
Ma, J.-P. *et al.*, *J. Asian Nat. Prod. Res.*, 2008, **10**, 565-569 (*Carpesides A,B*).
Gan, M. *et al.*, *J. Nat. Prod.*, 2008, **71**, 647-654 (*Iodes cirrhosa glucoside*).
Lee, Y.S. *et al.*, *Planta Med.*, (congress report), 2008, **74**, PB10 (*Sabotenins A,B*).
Wang, L. *et al.*, *Yaoxue Xuebao*, 2008, **43**, 1036-1039 (*Glehlinoside D*).
Sun, Y. *et al.*, *Beijing Daxue Xuebao, Yixueban*, 2009, **41**, 221-225 (*Avicennia marina acids*).
Lee, J. *et al.*, *Chem. Pharm. Bull.*, 2009, **57**, 298-301 (*3,3′,4′-tri-Me ether*).
Xu, W. *et al.*, *J. Asian Nat. Prod. Res.*, 2010, **12**, 874-878 (*Schimalignan B*).
Chang, C.-L. *et al.*, *J. Nat. Prod.*, 2010, **73**, 1482-1488 (*Quiquesetinerviusin C*).
Xu, Y. *et al.*, *Planta Med.*, 2010, **76**, 1706-1709 (*Glehlinosides C,D,H*).
Gao, G. *et al.*, *Zhongguo Yaoxue Zazhi*, 2010, **45**, 650-653 (*Catunaregum spinosa constits*).
Ma, J.-P. *et al.*, *Chin. Chem. Lett.*, 2011, **22**, 1454-1456 (*Ligusinenosides C,D*).
Liu, Y.-B. *et al.*, *Helv. Chim. Acta*, 2011, **94**, 1685-1691 (*tetra-Me ethers*).
Tan, J.-M. *et al.*, *J. Asian Nat. Prod. Res.*, 2011, **13**, 811-816 (*Debilignanoside*).
Lou, L. *et al.*, *Nat. Prod. Commun.*, 2011, **6**, 749-752 (*Janceolaroside A*).
Xue, C.-B. *et al.*, *Phytochemistry*, 2011, **72**, 1804-1813 (*Nannoglottis carpesioides constits*).
Bitam, F. *et al.*, *Phytochem. Lett.*, 2012, **5**, 696-699 (*Posidonia oceanica constit*).
Huang, X.-X. *et al.*, *Bioorg. Med. Chem. Lett.*, 2013, **23**, 5599-5604 (*Crataegus pinnatifida constits*).
Yoshikawa, K. *et al.*, *J. Nat. Med.*, 2014, **68**, 743-747 (*Marylaurencinol D*).
Wang, H. *et al.*, *Planta Med.*, 2015, **81**, 847-854 (*Melia toosendan 7′-Et ethers*).
Huang, X.-X. *et al.*, *Rec. Nat. Prod.*, 2015, **9**, 305-311 (*Huangnins A,B*).
Zhu, L.-J. *et al.*, *Chin. Chem. Lett.*, 2016, **27**, 63-65 (*Matteustruthiosides A,B*).
Bouzergoune, F. *et al.*, *Planta Med.*, 2016, **82**, 1438-1445 (*threo-Demethoxycarolignan Z, erythro-Demethoxycarolignan Z*).
Jiang, H. *et al.*, *Fitoterapia*, 2017, **117**, 11-15 (*Xanthiumnolic A, activity*).
Li, D.-Q. *et al.*, *J. Asian Nat. Prod. Res.*, 2017, **19**, 519-527 (*Saikolignanoside A, activity*).
Guo, T. *et al.*, *Nat. Prod. Res.*, 2017, **31**, 16-21 (*Zanthoxylum armatum constits*).
Yang, Y.-N. *et al.*, *RSC Adv.*, 2017, **7**, 805-812 (*Lyciumlignans A,B,C*).
Tang, B.-Q. *et al.*, *Zhongcaoyao*, 2017, **48**, 2371-2374 (*Hainanervalignan A*).

Zhao, C. *et al.*, *J. Agric. Food Chem.*, 2018, **66**, 10421-10430 (*Viburfordosides G,H,I, Fordianes A,B, activity*).
Kang, K.B. *et al.*, *J. Nat. Prod.*, 2018, **81**, 1819-1828 (*Sageretia theezans constits*).
Li, Y. *et al.*, *Chem. Nat. Compd. (Engl. Transl.)*, 2019, **55**, 614-617 (*Jatrolignans A,B*).
Du, Y.-Q. *et al.*, *Fitoterapia*, 2019, **139**, 104403:1-9 (*Ailanthus altissima constits, cd*).
Odonbayar, B. *et al.*, *J. Nat. Prod.*, 2019, **82**, 774-784 (*Brachangobinans C,D, activity*)

4,8′-Oxylign-8-ene-3,3′,4′,5,5′,7′-hexol O87

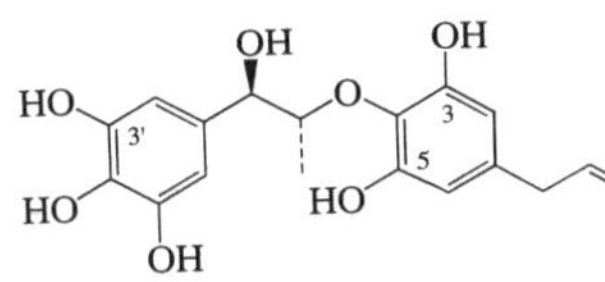

(7′R,8′R)-form

2-(4-Allyl-2,6-dihydroxyphenoxy)-1-(3,4,5-trihydroxyphenyl)-1-propanol. 3,3′,4′,5,5′,7′-Hexahydroxy-4,8′-oxyneolign-8-ene
$C_{18}H_{20}O_7$ 348.351
Structs. in this series may need revision. Phenyl 3,5-dioxygenated substn. pattern in doubt as a result of work done by Yang *et al* in 2015

3,3′,4′,5,5′-Penta-Me ether
3,3′,4′,5,5′-Pentamethoxy-4,8′-oxylign-8-en-7′-ol. 7′-Hydroxy-3,3′,4′,5,5′-pentamethoxy-4,8′-oxylign-8-ene. ***Rhaphidecursinol B****. Raphidecursinol B*
[52190-20-2]
$C_{23}H_{30}O_7$ 418.486
Constit. of *Rhaphidophora decursiva*. Racemic. Confusion in lit. over config. of stereomeric components. λ_{max} 210 (log ε 4.01) (MeOH)

(7′R,8′R)-form
3,3′,5,5′-Tetra-Me ether
3,3′,5,5′-Tetramethoxy-4,8′-oxylign-8-ene-4′,7′-diol. 4′,7′-Dihydroxy-3,3′,5,5′-tetramethoxy-4,8′-oxylign-8-ene
[1352451-13-8]
$C_{22}H_{28}O_7$ 404.459
$[\alpha]^{20}_{D}$ -12.5 (c, 0.1 in $CHCl_3$)

3,3′,4′,5,5′-Penta-Me ether
[658706-38-8]
[122211-56-7] (7′R,8′S-form)
[188584-91-0] (7′R*,8′R*-form)
[847476-41-9] (7′S,8′S-form)
$C_{23}H_{30}O_7$ 418.486
Constit. of *Myristica fragrans*. Cryst.. Mp 81 - 82°. $[\alpha]^{25}_{D}$ -58.2 (c, 5.26 in $CHCl_3$)

3,3′,4′,5,5′-Penta-Me ether, 7′-Ac
[916264-21-6]
$C_{25}H_{32}O_8$ 460.523
Constit. of *Saururus chinensis*. Amorph. solid. $[\alpha]^{20}_{D}$ -18 (c, 0.04 in $CHCl_3$). λ_{max} 217 (log ε 4.37); 270 (log ε 3.40) (MeOH)

3′,4′-Methylene, 3,5,5′-tri-Me ether
3,3′,5-Trimethoxy-4′,5′-methylenedioxy-4,8′-oxylign-8-en-7′-ol. ***Saurucinol G***
[1521111-70-5]
$C_{22}H_{26}O_7$ 402.443
Constit. of *Saururus chinensis*. Oil. $[\alpha]^{20}_{D}$ -111 (c, 0.02 in $CHCl_3$). λ_{max} 214 (log ε 4.54); 271 (log ε 2.97) (MeOH)

7′-Ketone, 3,3′,4′,5,5′-penta-Me ether
3,3′,4′,5,5′-Pentamethoxy-4,8′-oxylign-8-en-7′-one. 3,3′,4′,5,5′-Pentamethoxy-7′-oxo-4,8′-oxylign-8-ene
[657393-30-1]
[41535-98-2] (unspecified stereochem.)
$C_{23}H_{28}O_7$ 416.47
Constit. of *Saururus chinensis*. Powder. $[\alpha]^{20}_{D}$ +48 (c, 0.03 in $CHCl_3$). λ_{max} 222 (log ε 4.11); 281 (log ε 4.01) (MeOH)

(7′S,8′R)-form
3,3′,5,5′-Tetra-Me ether
Maceneolignan F
[1637648-08-8]
$C_{22}H_{28}O_7$ 404.459
Constit. of the seeds of *Myristica fragrans*. Inhibits CCR3-mediated chemotaxis. Potential candidate as CCR3 antagonist. Oil. $[\alpha]^{20}_{D}$ +5.6 (c, 1 in $CHCl_3$). $[\alpha]^{27}_{D}$ +14 (c, 1.1 in $CHCl_3$). λ_{max} 236 (sh); 273 (log ε 3.49); 317 (log ε 3.50) (MeOH)

3,3′,4′,5,5′-Penta-Me ether
[122211-57-8]
[41551-58-0]
$C_{23}H_{30}O_7$ 418.486
Constit. of *Myristica fragrans*. $[\alpha]_{D}$ +7.2 (c, 1 in $CHCl_3$)

(7′S,8′S)-form
3,3′,5,5′-Tetra-Me ether
Myrifralignan D
[1637669-72-7]
$C_{22}H_{28}O_7$ 404.459
Constit. of the seeds of *Myristica fragrans*. NO prodn. inhibitor. Yellow oil. $[\alpha]_{D}$ -14.4 (c, 1 in $CHCl_3$). λ_{max} 206 (log ε 4.51); 305 (log ε 3.61) (MeOH)

3′,4′-Methylene, 3,5,5′-tri-Me ether, 7′-Ac
Sauneolignan A
$C_{24}H_{28}O_8$ 444.480
Constit. of *Saururus chinensis*. Amorph. powder. λ_{max} 210 (log ε 4.70); 254 (log ε 4.01); 271 (sh) (MeOH)

7′-Ketone, 3,3′,4′,5,5′-penta-Me ether
Sauneolignan B
$C_{23}H_{28}O_7$ 416.470
Constit. of *Saururus chinensis*. Amorph. powder. $[\alpha]^{25}_{D}$ -40.2 (c, 0.1 in $CHCl_3$). λ_{max} 218 (log ε 4.10); 274 (log ε 3.96); 350 (log ε 3.10) (MeOH)

(7′R*,8′R*)-form
4′-Deoxy, 3,3′,5-tri-Me ether
2-(4-Allyl-2,6-dimethoxyphenoxy)-1-(3-hydroxy-5-methoxyphenyl)-1-propanol. 3,5,5′-Trimethoxy-4,8′-oxylign-8-ene-3′,7′-diol. 3′,7′-Dihydroxy-3,5,5′-trimethoxy-4,8′-oxylign-8-ene. ***Myrisisolignan***
[1246343-73-6]
$C_{21}H_{26}O_6$ 374.433
Constit. of the seeds of *Myristica fragrans*

(7′R*,8′S*)-form
3,3′,4′,5-Tetra-Me ether
3,4′,5,5′-Tetramethoxy-4,8′-oxylign-8-ene-3′,7′-diol
[108907-57-9]
$C_{22}H_{28}O_7$ 404.459
Constit. of oil of *Myristica fragrans*. Oil

3,3′,5,5′-Tetra-Me ether
[108907-53-5]
$C_{22}H_{28}O_7$ 404.459
Constit. of oil of *Myristica fragrans*

(7′ξ,8′ξ)-form
3,3′,4′,5,5′-Penta-Me ether
[41551-58-0]
$C_{23}H_{30}O_7$ 418.486
Constit. of *Myristica fragrans*. Inhibits growth of silkworm larvae. Mp 70 - 72° (65.5-66.5°). λ_{max} 269 (log ε 3.29) (MeOH).

Isogai, A. *et al.*, *Agric. Biol. Chem.*, 1973, **37**, 193-194 (*7′ξ,8′ξ-penta-Me ether*).
Isogai, A. *et al.*, *Agric. Biol. Chem.*, 1973, **37**, 889-895 (*7′ξ,8′ξ-penta-Me ether, activity*).
Isogai, A. *et al.*, *Agric. Biol. Chem.*, 1973, **37**, 1479-1486 (*7′ξ,8′ξ-penta-Me ether*).
Forrest, J.E. *et al.*, *J. Chem. Soc., Perkin Trans. 1*, 1974, 205-209 (*isol, struct*).
Hattori, H. *et al.*, *Chem. Pharm. Bull.*, 1987, **35**, 668-674 (*tetra-Me ethers*).

Zacchino, S.A. *et al.*, *J. Nat. Prod.*, 1988, **51**, 1261-1265 (*synth, ir, pmr, cmr, ms, abs config*).
Zhang, H.-J. *et al.*, *J. Nat. Prod.*, 2001, **64**, 772-777 (*Rhaphidecursinol B*).
Lee, A.-L. *et al.*, *Org. Biomol. Chem.*, 2003, **1**, 3957-3966 (*7′R,8′R-penta-Me ether, synth, cryst struct*).
Curti, C. *et al.*, *J. Org. Chem.*, 2006, **71**, 8552-8558 (*7′R,8′R-penta-Me ether, 7′S,8′R-penta-Me ether, synth*).
Yang, X. *et al.*, *Zhongguo Zhongyao Zazhi*, 2008, **33**, 397-402 (*Myrisisolignan*).
Yuan, H.J. *et al.*, *Chin. Chem. Lett.*, 2010, **21**, 127-130 (*Rhaphidecursinol B, synth*).
Das, S.K. *et al.*, *Eur. J. Org. Chem.*, 2010, 5100-5107 (*7′R,8′R-penta-Me ether, synth*).
Ding, J. *et al.*, *J. Chem. Res.*, 2011, **35**, 519-521 (*7′R,8′S-tetra-Me ether, synth*).
Nagaraju, M. *et al.*, *Synlett*, 2012, **23**, 1485-1488 (*7′R,8′R-penta-Me ether, synth*).
Cui, H. *et al.*, *J. Nat. Prod.*, 2014, **77**, 100-110 (*Saururus chinensis constits*).
Cao, G.-Y. *et al.*, *Food Chem.*, 2015, **173**, 231-237 (*Myrifralignan D*).
Yang, Y.-N. *et al.*, *J. Nat. Prod.*, 2015, **78**, 705-711 (*pmr, struct*).
Morikawa, T. *et al.*, *J. Nat. Prod.*, 2016, **79**, 2005-2013 (*Maceneolignan F, activity*).
Jin, Q. *et al.*, *J. Nat. Prod.*, 2019, **82**, 3002-3009 (*Sauneolignans A,B, cd*)

4,8′-Oxylign-8-ene-3,3′,4′,5,5′,9′-hexol O88

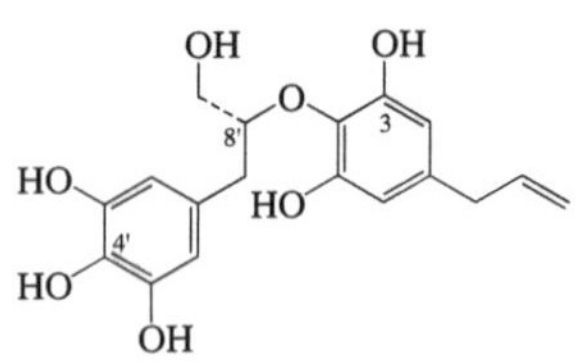

3,3′,4′,5,5′,9′-Hexahydroxy-4,8′-oxyneolign-8-ene
$C_{18}H_{20}O_7$ 348.351
(R)-form
3,3′,5,5′-Tetra-Me ether
3,3′,5,5′-Tetramethoxy-4,8′-oxylign-8-ene-4′,9′-diol. 4′,9′-Dihydroxy-3,3′,5,5′-tetramethoxy-4,8′-oxylign-8-ene.
4′-O-Demethylrhaphidecursinol A
$C_{22}H_{28}O_7$ 404.459
Constit. of the bark of *Cinnamomum cambodianum*. Oil. $[\alpha]^{20}_{D}$ +15.6 (c, 0.11 in $CHCl_3$). Unconfirmed stereochem.. λ_{max} 271 (log ε 0.61) (MeOH)
3,3′,4′,5,5′-Penta-Me ether
3,3′,4′,5,5′-Pentamethoxy-4,8′-oxylign-8-en-9′-ol. 9′-Hydroxy-3,3′,4′,5,5′-pentamethoxy-4,8′-oxylign-8-ene. ***Rhaphidecursinol A.*** *Raphidecursinol A*
[353760-04-0]
$C_{23}H_{30}O_7$ 418.486
Constit. of *Rhaphidophora decursiva*. $[\alpha]^{20}_{D}$ +2.6 (c, 0.12 in MeOH). λ_{max} 210 (log ε 3.99) (MeOH).

Zhang, H.-J. *et al.*, *J. Nat. Prod.*, 2001, **64**, 772-777 (*Rhaphidecursinol A*).
Ren, X. *et al.*, *J. Chin. Chem. Soc. (Taipei)*, 2004, **51**, 969-974 (*Rhaphidecursinol A, synth, abs config*).
Pel, P. *et al.*, *Bull. Korean Chem. Soc.*, 2015, **36**, 384-387 (*4′-O-Demethylrhaphidecursinol A*).
Gangar, M. *et al.*, *Tetrahedron Lett.*, 2016, **57**, 5931-5934 (*Rhaphidecursinol A, synth*)

4,8′-Oxylign-8-ene-3,3′,4′,5,7′,9′-hexol O89

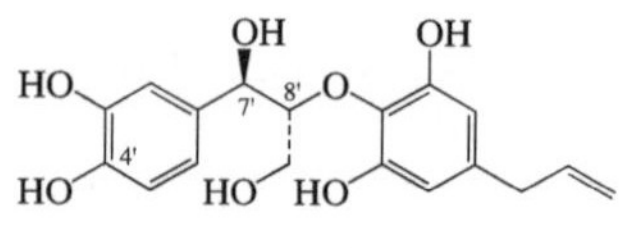

(7′R,8′R)-form

3,3′,4′,5,7′,9′-Hexahydroxy-4,8′-oxyneolign-8-ene
$C_{18}H_{20}O_7$ 348.351
(7′R,8′R)-form
3,3′,4′,5-Tetra-Me ether
3,3′,4′,5-Tetramethoxy-4,8′-oxylign-8-ene-7′,9′-diol. 7′,9′-Dihydroxy-3,3′,4′,5-tetramethoxy-4,8′-oxylign-8-ene.
(-)-Acortatarinowin D
[1807543-33-4]
$C_{22}H_{28}O_7$ 404.459
Constit. of *Acorus tatarinowii*. Light yellow gum. $[\alpha]^{20}_{D}$ -25 (c, 0.02 in MeOH). Obt. by resoln. of the partial racemate by chiral hplc
(7′S,8′S)-form
3,3′,4′,5-Tetra-Me ether
(+)-Acortatarinowin D
[1807543-32-3]
$C_{22}H_{28}O_7$ 404.459
Constit. of *Acorus tatarinowii*. Light yellow gum. $[\alpha]^{20}_{D}$ +23 (c, 0.01 in MeOH). Obt. by resoln. of the partial racemate by chiral hplc
(7′RS,8′RS)-form
3,3′,4′,5-Tetra-Me ether
Acortatarinowin D
[1807543-41-4]
$C_{22}H_{28}O_7$ 404.459
Constit. of *Acorus tatarinowii*. Light yellow gum. $[\alpha]^{20}_{D}$ -18 (c, 0.01 in MeOH). Partial racemate. Resolved by chiral hplc. λ_{max} 206 (log ε 4.70); 235 (log ε 4.02); 276 (log ε 3.40) (MeOH).

Lu, Y. *et al.*, *J. Nat. Prod.*, 2015, **78**, 2205-2214 (*Acortatarinowin D*)

2,8′-Oxylign-7-ene-3,3′,4′,5,5′,7′,9,9′-octol O90

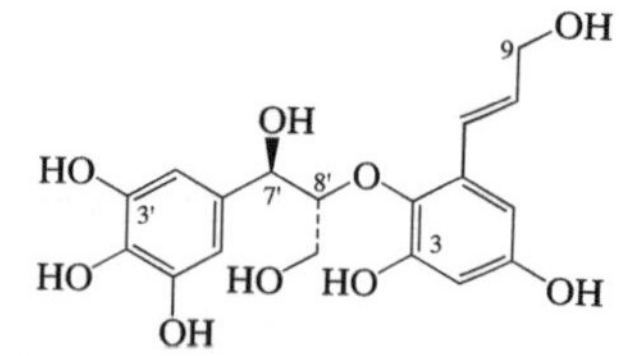

3,3′,4′,5,5′,7′,9,9′-Octahydroxy-2,8′-oxyneolign-7-ene
$C_{18}H_{20}O_9$ 380.349
(7E,7′R*,8′R*)-form
3,3′,4′,5,5′-Penta-Me ether
3,3′,4′,5,5′-Pentamethoxy-2,8′-oxylign-7-ene-7′,9,9′-triol. 7′,9,9′-Trihydroxy-3,3′,4′,5,5′-pentamethoxy-2,8′-oxylign-7-ene
3,3′,4′,5,5′-Penta-Me ether, 9-O-β-D-glucopyranoside
Woorenoside X
[195873-38-2]
$C_{29}H_{40}O_{14}$ 612.625
Constit. of the rhizomes of *Coptis japonica* var. *dissecta*. Amorph. solid. $[\alpha]_D$ -34.4 (c, 2.6 in MeOH). λ_{max} 210 (log ε 4.55); 274 (log ε 3.98) (EtOH).

Yoshikawa, K. *et al.*, *Nat. Med. (Tokyo)*, 1997, **51**, 244-248 (*Woorenoside X*)

4,8′-Oxylign-7-ene-3,3′,4′,5,5′,7′,9,9′-octol O91

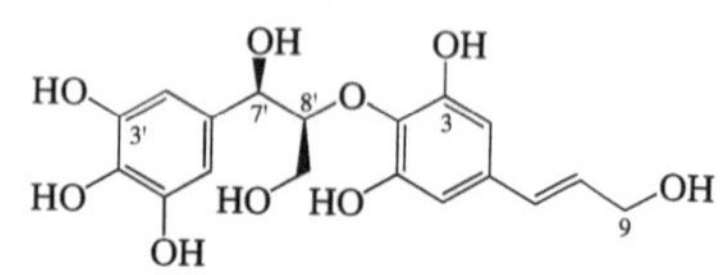

(7E,7′R,8′S)-form

3,3′,4′,5,5′,7′,9,9′-Octahydroxy-4,8′-oxyneolign-7-ene
$C_{18}H_{20}O_9$ 380.349
(7E,7′R,8′S)-form
3,3′,5,5′-Tetra-Me ether, 7′-O-β-D-glucopyranoside
[1919054-34-4]
$C_{28}H_{38}O_{14}$ 598.598
Constit. of the leaves of *Indocalamus latifolius*. Yellow powder. $[\alpha]_D$ +4.5 (c, 1 in MeOH). λ_{max} 230; 271 (MeOH)
3,3′,5,5′-Tetra-Me ether, 9′-O-β-D-glucopyranoside
[1082466-17-8]
$C_{28}H_{38}O_{14}$ 598.598
Constit. of the leaves of *Osmanthus ilicifolius*. Amorph. powder. $[\alpha]^{25}_{D}$

-20.6 (c, 0.15 in MeOH). λ_{max} 220 (sh); 269 (log ε 4.1) (MeOH)

(7E,7′S,8′R)-form

3,3′,5,5′-Tetra-Me ether

3,3′,5,5′-Tetramethoxy-4,8′-oxylign-7-ene-4′,7′,9,9′-tetrol. 4′,7′,9,9′-Tetrahydroxy-3,3′,5,5′-tetramethoxy-4,8′-oxylign-7-ene. **Syringylglycerol β-sinapyl ether.** *Syringylglycerol-8-O-4′-sinapyl alcohol*

[1304010-43-2]

$C_{22}H_{28}O_9$ 436.457

Constit. of *Euonymus alatus* and *Euonymus acanthocarpus*. $[\alpha]_D^{25}$ -1.9 (c, 0.1 in MeOH)

3,3′,5,5′-Tetra-Me ether, 4′-O-β-D-glucopyranoside

[1919054-32-2]

$C_{28}H_{38}O_{14}$ 598.598

Constit. of the leaves of *Indocalamus latifolius*. Yellow powder. $[\alpha]_D$ -15.9 (c, 1 in MeOH). λ_{max} 230; 270 (MeOH)

3,3′,5,5′-Tetra-Me ether, 9-O-β-D-glucopyranoside

[1186295-12-4]

$C_{28}H_{38}O_{14}$ 598.598

Constit. of the leaves of *Indocalamus latifolius* and *Syringa velutina*. Amorph. powder (MeOH). $[\alpha]_D^{23}$ -9.4 (c, 0.2 in MeOH)

3,3′,5,5′-Tetra-Me ether, 9′-O-β-D-glucopyranoside

[1082466-15-6]

Constit. of the leaves of *Osmanthus ilicifolius*. Amorph. powder. $[\alpha]_D^{25}$ -12.3 (c, 0.19 in MeOH). λ_{max} 220 (sh); 269 (log ε 4.1) (MeOH)

3,3′,5,5′-Tetra-Me ether, 4′,9-di-O-β-D-glucopyranoside

[2099167-55-0]

$C_{34}H_{48}O_{19}$ 760.739

Constit. of twigs of *Litsea cubeba*. Amorph. solid. $[\alpha]_D^{20}$ -11 (c, 0.3 in MeOH). λ_{max} 239 (log ε 1.02); 273 (log ε 3.36) (MeOH)

3,3′,5,5′-Tetra-Me ether, 7′,9-di-O-β-D-glucopyranoside

[1600527-51-2]

$C_{34}H_{48}O_{19}$ 760.739

Constit. of the stems of *Dendrobium aurantiacum* var. *denneanum*. Exhibits neuroprotective activity against glutamate-induced toxicity in PC12 cells. Gum. $[\alpha]_D^{20}$ -12.2 (c, 0.20 in MeOH). λ_{max} 226 (sh); 271 (log ε 3.67) (MeOH)

(7E,7′S,8′S)-form

3,3′,5,5′-Tetra-Me ether, 7′-O-β-D-glucopyranoside

[1919054-33-3]

Constit. of the leaves of *Indocalamus latifolius*. Yellow powder. $[\alpha]_D$ -3.1 (c, 1 in MeOH). λ_{max} 230; 269 (MeOH)

(7E,7′R*,8′R*)-form

3,3′,5′-Tri-Me ether

3′,5,5′-Trimethoxy-4,8′-oxylign-7-ene-3,4′,7′,9,9′-pentol. 3,4′,7′,9,9′-Pentahydroxy-3′,5,5′-trimethoxy-4,8′-oxylign-7-ene

3,3′,5′-Tri-Me ether, 9,9′-bis-O-(3-methylbutanoyl) **Patrineolignan B**

[1443211-45-7]

$C_{31}H_{42}O_{11}$ 590.666

Constit. of *Patrinia scabra*. Exhibits potent cytotoxic activity against tumour cell lines. Oil. $[\alpha]_D^{25}$ +103.5 (c, 0.13 in $CHCl_3$). λ_{max} 228 (log ε 4.10); 271 (log ε 4.22) (MeOH)

3,3′,5,5′-Tetra-Me ether, 4′-O-β-D-glucopyranoside **Picraquassioside C**

[169312-06-5]

$C_{28}H_{38}O_{14}$ 598.598

Constit. of *Dendrobium aurantiacum*, *Hemerocallis fulva*, *Picrasma quassioides* and *Viola tianshanica*. Amorph. powder. $[\alpha]_D^{20}$ -9.7 (c, 1 in MeOH)

3,3′,5,5′-Tetra-Me ether, 9,9′-bis-O-(3-methylbutanoyl) **Patrineolignan A**

[1443211-44-6]

$C_{32}H_{44}O_{11}$ 604.693

Constit. of *Patrinia scabra*. Exhibits potent cytotoxic activity against tumour cell lines. Oil. $[\alpha]_D^{15}$ +63.7 (c, 0.19 in $CHCl_3$). λ_{max} 238 (log ε 4.08); 274 (log ε 4.18) (MeOH)

(7E,7′R*,8′S*)-form

3,3′,5,5′-Tetra-Me ether

[613683-98-0]

$C_{22}H_{28}O_9$ 436.457

Constit. of *Brassica fruticulosa* and *Eucommia ulmoides*

3,3′,5,5′-Tetra-Me ether, 9,9′-bis-O-(4-hydroxybenzoyl) ***Quiquelignan F***

[1215818-70-4]

$C_{36}H_{36}O_{13}$ 676.671

Constit. of the stems of *Calamus quiquesetinervius*. NO prodn. inhibitor. Yellow powder. Mp 128°. $[\alpha]_D^{25}$ +2.1 (c, 0.9 in MeOH). λ_{max} 206; 260; 275 (sh) (MeOH)

(7E,7′ξ,8′ξ)-form

3,3′,5,5′-Tetra-Me ether, 7′-O-β-D-glucopyranoside

[415724-47-9]

Constit. of *Saccharum officinarum*. Oil. λ_{max} 221 (sh); 269 (MeOH)

3,3′,5,5′-Tetra-Me ether, 9-O-β-D-glucopyranoside

[545436-57-5]

$C_{28}H_{38}O_{14}$ 598.598

Constit. of non-centrifuged *Saccharum officinarum*. Oil

7′-Ketone, 3,3′,5,5′-tetra-Me ether

4′,9,9′-Trihydroxy-3,3′,5,5′-tetramethoxy-4,8′-oxylign-7-en-7′-one

7′-Ketone, 3,3′,5,5′-tetra-Me ether, 9,9′-bis-O-(4-hydroxybenzoyl) ***Quiquelignan D***

[1215818-67-9]

$C_{36}H_{34}O_{13}$ 674.655

Constit. of the stems of *Calamus quiquesetinervius*. NO prodn. inhibitor. Yellow powder. Mp 109°. $[\alpha]_D^{25}$ +2.7 (c, 0.7 in MeOH). λ_{max} 209; 261; 275 (sh); 312 (MeOH).

Yoshikawa, K. *et al.*, *Phytochemistry*, 1995, **40**, 253-256 (*Picraquassioside C*).

Takara, K. *et al.*, *Biosci., Biotechnol., Biochem.*, 2002, **66**, 29-35 ; 2003, **67**, 376-379 (*cane sugar glucosides*).

Cutillo, F. *et al.*, *J. Agric. Food Chem.*, 2003, **51**, 6165-6172 (*tetra-Me ether*).

Lourith, N. *et al.*, *J. Wood Sci.*, 2005, **51**, 370-378 (*Syringylglycerol β-sinapyl ether, biosynth*).

Lourith, N. *et al.*, *J. Wood Sci.*, 2005, **51**, 379-386 (*Syringylglycerol β-sinapyl ether, biosynth*).

Machida, K. *et al.*, *Magn. Reson. Chem.*, 2008, **46**, 990-994 (*3,3′,5,5′-tetra-Me ether 9′-glucosides*).

Feng, X.-S. *et al.*, *Molecules*, 2009, **14**, 953-958 (*Syringa velutina glucoside*).

Chang, C.-L. *et al.*, *Bioorg. Med. Chem.*, 2010, **18**, 518-525 (*Quiquelignans D,F, cd, activity*).

Jeong, E.J. *et al.*, *Bioorg. Med. Chem. Lett.*, 2011, **21**, 2283-2286 (*Euonymus alatus tetra-Me ether*).

Zhu, J.X. *et al.*, *Arch. Pharmacal Res.*, 2012, **35**, 1739-1747 (*Euonymus acanthocarpus tetra-Me-ether*).

Di, L. *et al.*, *Arch. Pharmacal Res.*, 2013, **36**, 1198-1203 (*Patrineolignans A,B*).

Xiong, L. *et al.*, *Molecules*, 2013, **18**, 6153-6160 (*Dendrobium 7′,9-diglucoside*).

Qin, Y. *et al.*, *Molecules*, 2013, **18**, 13636-13644 (*Picraquassioside C*).

Sun, J. *et al.*, *Molecules*, 2015, **20**, 15686-15700 (*Indocalamus latifolius constits*).

Wang, L.-Y. *et al.*, *J. Nat. Prod.*, 2017, **80**, 1808-1818 (*Litsea cubeba constit*)

2,8′-Oxylign-8-ene-3′,4,4′,5,5′-pentol O92

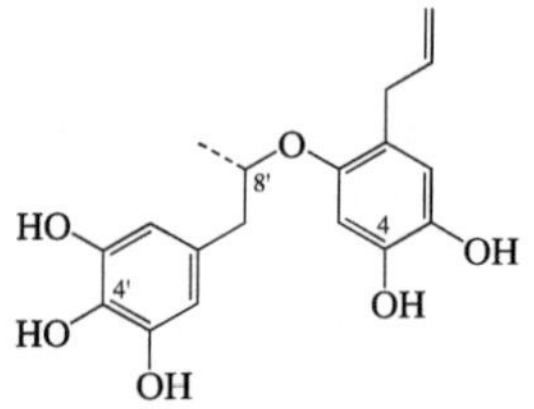

3′,4,4′,5,5′-Pentahydroxy-2,8′-oxyneolign-8-ene
$C_{18}H_{20}O_6$ 332.352
(S)-form
4,5-Methylene, 3′,4′,5′-tri-Me ether
3′,4′,5′-Trimethoxy-4,5-methylenedioxy-2,8′-oxylign-8-ene
[1651214-77-5]
$C_{22}H_{26}O_6$ 386.444
Constit. of the stems of *Ocotea cymosa*. Exhibits *in vitro* activity against *Aedes aegypti*. Oil. $[\alpha]_D^{25}$ +23 (c, 0.1 in MeOH). Tentative config. assigned. λ_{max} 234; 256; 300 (MeOH aq.).

Rakotondraibe, L.H. *et al.*, *J. Nat. Prod.*, 2015, **78**, 431-440 (*Ocotea cymosa constit*)

2,8′-Oxylign-8-ene-3′,4,4′,5,7′-pentol O93

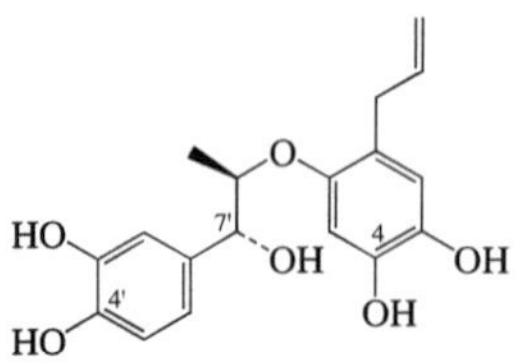

3′,4,4′,5,7′-Pentahydroxy-2,8′-oxyneolign-8-ene
$C_{18}H_{20}O_6$ 332.352
(7′R,8′R)-form
3′,4,4′,5-Tetra-Me ether
3′,4,4′,5-Tetramethoxy-2,8′-oxylign-8-en-7′-ol. 7′-Hydroxy-3′,4,4′,5-tetramethoxy-2,8′-oxylign-8-ene
3′,4,4′,5-Tetra-Me ether, 7′-Ac
Theiferin
[1335032-19-3]
$C_{24}H_{30}O_7$ 430.497
Constit. of the leaves and stems of *Taxillus theifer*. Oil. $[\alpha]_D^{25}$+28.4 (c, 1.48 in MeOH). λ_{max} 232 (log ε 4.28); 284 (log ε 3.83) (MeOH).

Lin, L.-C. *et al.*, *Nat. Prod. Res.*, 2011, **25**, 1319-1323 (*Theiferin*)

3,4′-Oxylign-8′-ene-3′,4,5′,7,8-pentol O94

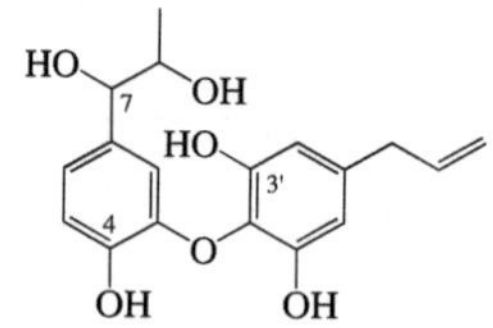

3′,4,5′,7,8-Pentahydroxy-3,4′-oxyneolign-8′-ene
$C_{18}H_{20}O_6$ 332.352
(7ξ,8ξ)-form
3′,4,5′-Tri-Me ether
1-[3-(4-Allyl-2,6-dimethoxyphenoxy)-4-methoxyphenyl]-1,2-propanediol. 3′,4,5′-Trimethoxy-3,4′-oxylign-8′-ene-7,8-diol. 7,8-Dihydroxy-3′,4,5′-trimethoxy-3,4′-oxylign-8′-ene
[1418540-47-2]
$C_{21}H_{26}O_6$ 374.433
Constit. of the stems of *Allamanda cathartica*.

Yamauchi, K. *et al.*, *Nat. Resour.*, 2011, **2**, 167-172 (*Allamanda cathartica constit*)

3,4′-Oxylign-8-ene-3′,4,7′,8′,9′-pentol O95

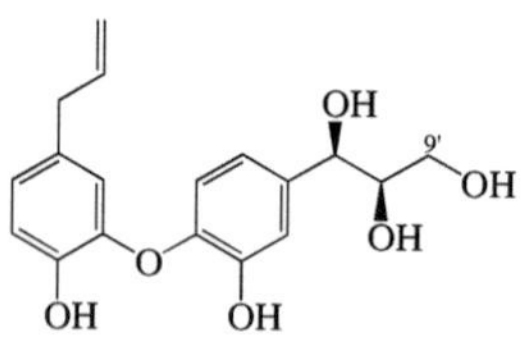

3′,4,7′,8′,9′-Pentahydroxy-3,4′-oxyneolign-8-ene
$C_{18}H_{20}O_6$ 332.352
(7′R*,8′R*)-form
9′-Me ether
9′-Methoxy-3,4′-oxylign-8-ene-3′,4,7′,8′-tetrol. 3′,4,7′,8′-Tetrahydroxy-9′-methoxy-3,4′-oxylign-8-ene. ***Taiwandimerol A***
[1449387-73-8]
$C_{19}H_{22}O_6$ 346.379
Constit. of the roots of *Piper taiwanense*. Yellow oil. $[\alpha]_D^{25}$-0.9 (c, 0.3 in $CHCl_3$). λ_{max} 222 (log ε 3.86); 281 (log ε 3.55) (MeOH).

Chen, S. *et al.*, *Phytochemistry*, 2013, **93**, 203-209 (*Taiwandimerol A*)

4,8′-Oxylign-7-ene-3,3′,4′,5,5′-pentol O96

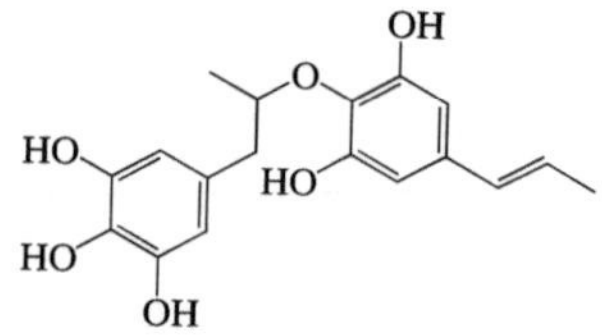

3,3′,4′,5,5′-Pentahydroxy-4,8′-oxyneolign-7-ene
$C_{18}H_{20}O_6$ 332.352
(+)-(E)-form
3,3′,5,5′-Tetra-Me ether
3,3′,5,5′-Tetramethoxy-4,8′-oxylign-7-en-4′-ol. 4′-Hydroxy-3,3′,5,5′-tetramethoxy-4,8′-oxylign-7-ene. ***Virolongin D***
[144049-94-5]
$C_{22}H_{28}O_6$ 388.46
Constit. of *Virola carinata*. Oil. $[\alpha]_D^{25}$+7.5 (c, 1.1 in $CHCl_3$). λ_{max} 233 (ε 16500) (MeOH)
Penta-Me ether
3,3′,4′,5,5′-Pentamethoxy-4,8′-oxylign-7-ene. ***(+)-Virolongin A***
[94608-22-7]
$C_{23}H_{30}O_6$ 402.487
Constit. of *Piper polysyphorum*
(-)-(E)-form
Penta-Me ether
(-)-Virolongin A
[169275-03-0]
$C_{23}H_{30}O_6$ 402.487
Constit. of *Bonamia spectabilis* and *Virola elongata*. Oil. $[\alpha]_D^{25}$-12.4 ($CHCl_3$). λ_{max} 227 (log ε 4.04); 268 (log ε 3.35) (MeOH).

MacRae, W.D. *et al.*, *Phytochemistry*, 1985, **24**, 561-566 (*Virolongin A*).
Cavalcante, S.H. *et al.*, *Phytochemistry*, 1985, **24**, 1051-1055 (*Virolongin D*).
Ma, Y. *et al.*, *Yaoxue Xuebao*, 1991, **26**, 345-350 (*(+)-Virolongin A*).
Ferri, P.H. *et al.*, *Phytochemistry*, 1992, **31**, 1375-1377 (*Virolongin D, synth*).
Kraft, C. *et al.*, *Phytochemistry*, 2002, **60**, 167-173 (*Bonamia spectabilis constit*)

4,8′-Oxylign-7-ene-3,3′,4′,5′,7′-pentol O97

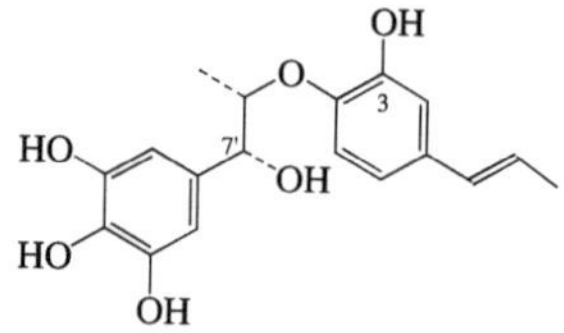

(7E,7′R,8′S)-form

3,3′,4′,5′,7′-Pentahydroxy-4,8′-oxyneolign-7-ene

$C_{18}H_{20}O_6$ 332.352

(7E,7′R,8′S)-form

3,3′,5′-Tri-Me ether

3,3′,5′-Trimethoxy-4,8′-oxylign-7-ene-4′,7′-diol. 4′,7′-Dihydroxy-3,3′,5′-trimethoxy-4,8′-oxylign-7-ene. ***Myrifralignan C***

[1637669-70-5]

$C_{21}H_{26}O_6$ 374.433

Constit. of the seeds of *Myristica fragrans*. NO prodn. inhibitor. Yellow oil. $[\alpha]^{20}_{D}$ -24.3 (c, 1.0 in $CHCl_3$). λ_{max} 207 (log ε 4.97); 260 (log ε 3.96) (MeOH)

(7E,7′S,8′R)-form

3,3′,4′-Tri-Me ether

3,4′,5′-Trimethoxy-4,8′-oxylign-7-ene-3′,7′-diol. 3′,7′-Dihydroxy-3,4′,5′-trimethoxy-4,8′-oxylign-7-ene. ***Longipedlignan R***

[2377872-22-3]

$C_{21}H_{26}O_6$ 374.433

Constit. of the roots of *Kadsura longipedunculata*. NO prodn. inhibitor. Amorph. powder. $[\alpha]^{20}_{D}$ -7 (c, 0.1 in MeOH). λ_{max} 206 (log ε 2.18); 268 (log ε 1.48) (MeOH)

(7E,7′S,8′S)-form

3,3′,4′-Tri-Me ether

Fargesiphenol A

[1394214-05-1]

$C_{21}H_{26}O_6$ 374.433

Constit. of the flower buds of *Magnolia fargesii*. Exhibits moderate cytotoxic activity. Pale yellow gum. $[\alpha]^{24}_{D}$ +38.5 (c, 0.2 in MeOH). λ_{max} 210 (log ε 4.27); 285 (log ε 3.78); 320 (log ε 2.82) (MeOH)

3,3′,4′,5′-Tetra-Me ether

3,3′,4′,5′-Tetramethoxy-4,8′-oxylign-7-en-7′-ol. 7′-Hydroxy-3,3′,4′,5′-tetramethoxy-4,8′-oxylign-7-ene. ***Surinamensin‡***

[68143-82-8]

$C_{22}H_{28}O_6$ 388.46

Constit. of *Daphne feddei* and *Virola surinamensis*. Exhibits antileishmanial activity. Oil. $[\alpha]_D$ +52 (c, 1 in $CHCl_3$)

(7E,7′R*,8′S*)-form

3,3′,4′,5′-Tetra-Me ether

erythro-***Surinamensin***

[845276-34-8]

$C_{22}H_{28}O_6$ 388.46

Constit. of the fruit of *Myristica fragrans*. Viscous liq.

(7E,7′ξ,8′ξ)-form

3,3′,5′-Tri-Me ether

[108907-56-8]

$C_{21}H_{26}O_6$ 374.433

Constit. of the oil of *Myristica fragrans*. Oil.

Hattori, M. *et al., Chem. Pharm. Bull.*, 1987, **35**, 668-674 (*Myristica fragrans constit*).
Santos, L.S. *et al., J. Braz. Chem. Soc.*, 1990, **1**, 47-49 (*Surinamensin, abs config*).
Barata, L.E.S. *et al., Phytochemistry*, 2000, **55**, 589-595 (*Surinamensin, activity*).
Hu, Q.-F. *et al., Chem. Pharm. Bull.*, 2011, **59**, 1421-1424 (*Daphne feddei constit, activity*).
Gao, X. *et al., J. Braz. Chem. Soc.*, 2012, **23**, 1274-1279 (*Fargesiphenol A, activity*).
Francis, K.S. *et al., Nat. Prod. Res.*, 2014, **28**, 1664-1668 (*erythro-Surinamensin*).
Cao, G.-Y. *et al., Food Chem.*, 2015, **173**, 231-237 (*Myrifralignan C, activity*).
Liu, J. *et al., J. Nat. Prod.*, 2019, **82**, 2842-2851 (*Longipedlignan R, cd, activity*)

4,8′-Oxylign-7-ene-3,3′,4′,5,7′-pentol O98

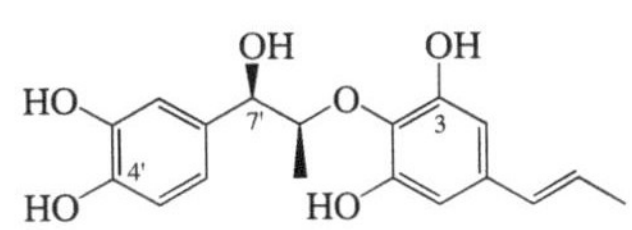

3,3′,4′,5,7′-Pentahydroxy-4,8′-oxyneolign-7-ene

$C_{18}H_{20}O_6$ 332.352

(7E,7′R,8′S)-form

3′,4′-Methylene, 3,5-di-Me ether

3,5-Dimethoxy-3′,4′-methylenedioxy-4,8′-oxylign-7-en-7′-ol. 7′-Hydroxy-3,5-dimethoxy-3′,4′-methylenedioxy-4,8′-oxylign-7-ene.

Myrifralignan A

[1637669-66-9]

$C_{21}H_{24}O_6$ 372.417

Constit. of seeds of *Myristica fragrans*. Yellow oil. $[\alpha]^{20}_{D}$ -59.3 (c, 1.0 in $CHCl_3$). λ_{max} 203 (log ε 4.59); 277 (log ε 3.87) (MeOH).

Cao, G.-Y. *et al., Food Chem.*, 2015, **173**, 231-237 (*Myrifralignan A*)

4,8′-Oxylign-7-ene-3,3′,4′,5,9-pentol O99

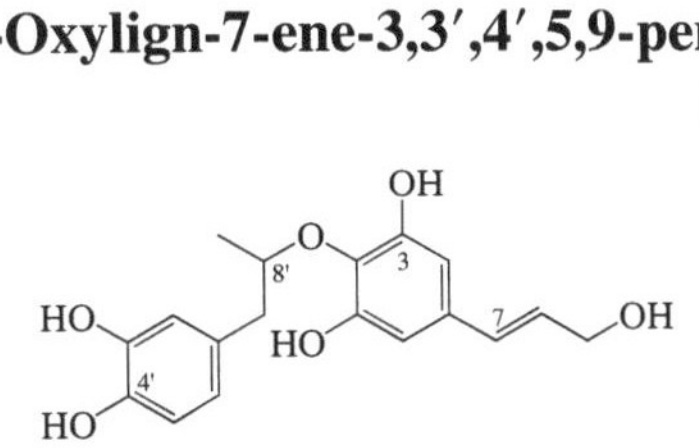

3,3′,4′,5,9-Pentahydroxy-4,8′-oxyneolign-7-ene

$C_{18}H_{20}O_6$ 332.352

(7E,8′ξ)-form

3,3′,4′,5-Tetra-Me ether

3,3′,4′,5-Tetramethoxy-4,8′-oxylign-7-en-9-ol. 9-Hydroxy-3,3′,4′,5-tetramethoxy-4,8′-oxylign-7-ene. ***Miliusfragranol A***

[1446741-18-9]

$C_{22}H_{28}O_6$ 388.46

Constit. of *Miliusa fragrans*. Viscous liq.. $[\alpha]^{20}_{D}$ -18.7 (c, 0.17 in $CHCl_3$). λ_{max} 215 (log ε 3.67); 277 (log ε 3.16) (MeOH).

Sawasdee, K. *et al., Tetrahedron Lett.*, 2013, **54**, 4259-4263 (*Miliusfragranol A*)

4,8′-Oxylign-7-ene-3,3′,4′,7′,9-pentol O100

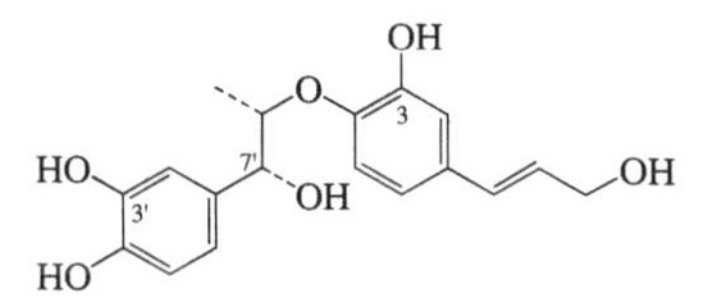

3,3′,4′,7′,9-Pentahydroxy-4,8′-oxyneolign-7-ene

$C_{18}H_{20}O_6$ 332.352

(7E,7′R*,8′S*)-form

3′,4′-Methylene, 3-Me ether

3-Methoxy-3′,4′-methylenedioxy-4,8′-oxylign-7-ene-7′,9-diol. 7′,9-Dihydroxy-3-methoxy-3′,4′-methylenedioxy-4,8′-oxylign-7-ene

3′,4′-Methylene, 3-Me ether, 7′-Ac

Machilin E

[110209-92-2]

$C_{22}H_{24}O_7$ 400.427

Constit. of the bark of *Machilus thunbergii*. Yellowish oil. $[\alpha]^{25}_{D}$ +29.2 (c, 0.11 in $CHCl_3$).

Shimomura, H. *et al., Phytochemistry*, 1987, **26**, 1513-1515 (*Machilin E*)

4,8′-Oxylign-7-ene-3,4′,7′,9,9′-pentol O101

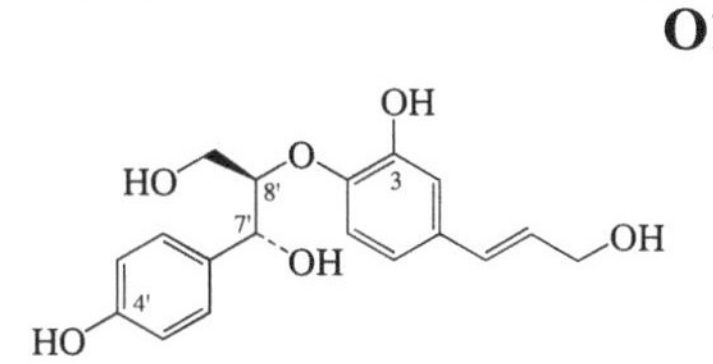

(7E,7′RS,8′RS)-form

3,4′,7′,9,9′-Pentahydroxy-4,8′-oxyneolign-7-ene

$C_{18}H_{20}O_6$ 332.352

(7E,7′RS,8′RS)-form

3-Me ether

3-Methoxy-4,8′-oxylign-7-ene-4′,7′,9,9′-tetrol.

4′,7′,9,9′-Tetrahydroxy-3-methoxy-4,8′-oxylign-7-ene. ***Chushizisin A***
[1142925-29-8]
$C_{19}H_{22}O_6$ 346.379
Constit. of the fruit of *Broussonetia papyrifera*. Exhibits antioxidant activity. Oil. λ_{max} 204 (log ε 4.63); 263 (log ε 4.05) (MeOH)
(7E,7′RS,8′SR)-form
3-Me ether
Chushizisin B
[1142925-31-2]
$C_{19}H_{22}O_6$ 346.379
Constit. of the fruit of *Broussonetia papyrifera*. Oil. λ_{max} 204 (log ε 4.5); 266 (log ε 4.07) (MeOH).

Mei, R.-Q. *et al., J. Nat. Prod.*, 2009, **72**, 621-625 (*Chushizisins A,B*)

4,8′-Oxylign-8-ene-3,3′,4′,5,5′-pentol O102

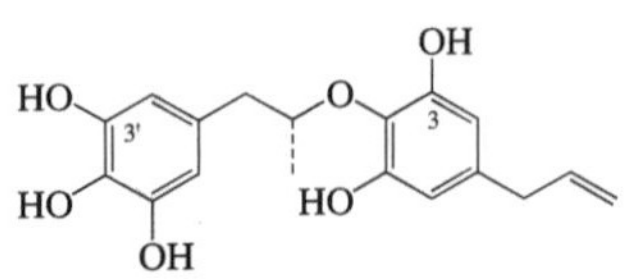

2-(4-Allyl-2,6-dihydroxyphenoxy)-1-(3,4,5-trihydroxyphenyl)propane. 3,3′,4′,5,5′-Pentahydroxy-4,8′-oxyneolign-8-ene
$C_{18}H_{20}O_6$ 332.352
(R)-form
Penta-Me ether
3,3′,4′,5,5′-Pentamethoxy-4,8′-oxylign-8-ene. β-O-Dilignol. ***Virolongin B***
[124151-41-3]
[41535-94-8] (unspecified stereochem.) [126456-04-0] (unspecified stereochem.)
$C_{23}H_{30}O_6$ 402.487
Constit. of *Aristolochia birostris, Licaria aurea, Miliusa fragrans, Ocotea cymosa, Virola carinata* and *Myristica fragrans*. Shows antiplasmodial activity. Inhibits growth of silkworm larvae. Oil. Identity of isolated samples not establ., may not be stereochemically identical. λ_{max} 225 (ε 13400); 280 (ε 5800) (EtOH)
3′,4′-Methylene, 3,5,5′-tri-Me ether
3,3′,5-Trimethoxy-4′,5′-methylenedioxy-4,8′-oxylign-8-ene. ***Virolongin C***
[126223-32-3]
$C_{22}H_{26}O_6$ 386.444
Constit. of *Aristolochia birostris, Licaria aurea* and *Virola carinata*. Oil. λ_{max} 240 (ε 16850); 267 (ε 12700) (MeOH).

Isogai, A. *et al., Agric. Biol. Chem.*, 1973, **37**, 193-194 (*Myristica fragrans constit*).
Isogai, A. *et al., Agric. Biol. Chem.*, 1973, **37**, 1479-1486 (*Myristica fragrans constit*).
Forrest, J.E. *et al., J. Chem. Soc., Perkin Trans. 1*, 1974, 205-209 (*Myristica fragrans constit*).
Calvacante, S.H. *et al., Phytochemistry*, 1985, **24**, 1051-1055 (*Virola carinata constits*).
Muhammed, I. *et al., J. Nat. Prod.*, 1989, **52**, 1177-1179 (*β-O-Dilignol*).
Barbosa-Filho, J.M. *et al., Phytochemistry*, 1989, **28**, 2209-2211 (*Virolongin B*).
Ren, X. *et al., J. Chin. Chem. Soc. (Taipei)*, 2004, **51**, 969-974 (*Virolongin B, synth*).
Patil, S.N. *et al., Tetrahedron Lett.*, 2016, **57**, 3371-3375 (*Virolongin B, synth*).
Gangar, M. *et al., Tetrahedron Lett.*, 2016, **57**, 5931-5934 (*Virolongin B, synth*)

4,8′-Oxylign-8-ene-3,3′,4′,5,7′-pentol O103

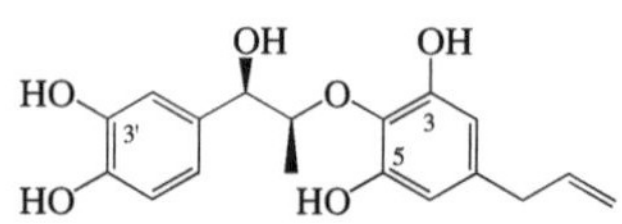

(7′R,8′S)-form

2-(4-Allyl-2,6-dihydroxyphenoxy)-1-(3,4-dihydroxyphenyl)-1-propanol. 3,3′,4′,5,7′-Pentahydroxy-4,8′-oxyneolign-8-ene
$C_{18}H_{20}O_6$ 332.352
(7′R,8′S)-form
3,3′,4′,5-Tetra-Me ether, 7′-Ac
Maceneolignan H
[1314042-85-7]
$C_{24}H_{30}O_7$ 430.497
Constit. of the arils of *Myristica fragrans*. Inhibits CCR3-mediated chemotaxis. Potential candidate as CCR3 antagonist. Oil. $[\alpha]_D^{24}$ -34 (c, 1.5 in $CHCl_3$). λ_{max} 230 (sh); 287 (log ε 3.50) (MeOH)
(7′S,8′R)-form
3,3′,4′,5-Tetra-Me ether, 7′-Ac
[1257094-34-0]
[41535-93-7] (7′R*,8′S*-form)
$C_{24}H_{30}O_7$ 430.497
Constit. of the fruit of *Myristica fragrans*. NO prodn. inhibitor. Oil. $[\alpha]_D^{20}$ +15.4 (c, 0.3 in $CHCl_3$)
(7′R*,8′S*)-form
3,3′,5-Tri-Me ether
3,3′,5-Trimethoxy-4,8′-oxylign-8-ene-4′,7′-diol. ***Myrislignan***
[52190-21-3]
[41535-95-9] (unspecified stereochem.) [162409-91-8] (7′R,8′S-form) [171485-39-5] (7′S,8′R-form)
$C_{21}H_{26}O_6$ 374.433
Constit. of *Cymbopogon nardus* and *Myristica fragrans*. NO prodn. inhibitor. Exhibits antiinflammatory and phytotoxic activities. Mp 96.5 - 97.5°. $[\alpha]_D^{26}$ +7.3 (c, 0.03 in MeOH)
3,3′,4′,5-Tetra-Me ether
3,3′,4′,5-Tetramethoxy-4,8′-oxylign-8-en-7′-ol
[93289-62-4]
[41535-92-6] (unspecified stereochem.) [170902-27-9] [170902-28-0]
$C_{22}H_{28}O_6$ 388.46
Constit. of *Myristica fragrans*. Cryst.. Mp 65 - 66°
3′,4′-Methylene, 3,5-di-Me ether
3,5-Dimethoxy-3′,4′-methylenedioxy-4,8′-oxylign-8-en-7′-ol
[52190-19-9]
[50354-29-5] (unspecified stereochem.)
$C_{21}H_{24}O_6$ 372.417
Constit. of *Myristica fragrans*. Oil
3′,4′-Methylene, 3,5-di-Me ether, 7′-Ac
[52190-22-4]
[85852-26-2] (unspecified stereochem.)
$C_{23}H_{26}O_7$ 414.454
Consit. of *Myristica fragrans*. Oil
3′,4′-Methylene, 3,5-di-Me ether, 7′-benzoyl
[52190-24-6]
$C_{28}H_{28}O_7$ 476.525
Constit. of *Myristica fragrans*. Oil.

Isogai, A. *et al., Agric. Biol. Chem.*, 1973, **37**, 1479-1486 (*Myristica fragrans constits*).
Forrest, J.E. *J. Chem. Soc., Perkin Trans. 1*, 1974, 205-209 (*Myristica fragrans constits*).
Hattori, M. *et al., Chem. Pharm. Bull.*, 1987, **35**, 668-674 (*Myristica fragrans constits*).
Kasahara, H. *et al., Phytochemistry*, 1995, **40**, 1515-1517 (*abs config*).
Duan, L. *et al., Planta Med.*, 2009, **75**, 1241-1245 (*tetra-Me ether 7′-Ac, cd*).
Xia, Y. *et al., Monatsh. Chem.*, 2011, **142**, 93-96 (*Myrislignan, synth*).
Jin, H. *et al., Phytother. Res.*, 2012, **26**, 1320-1326 (*Myrislignan, antiinflammatory activity*).
Cao, G.-Y. *et al., Food Chem.*, 2015, **173**, 231-237 (*Myrislignan, NO prodn inhibitory activity*).
Acuña, U.M. *et al., Food Chem.*, 2016, **202**, 269-275 (*tetra-Me ether 7′-Ac, activity*).
Morikawa, T. *et al., J. Nat. Prod.*, 2016, **79**, 2005-2013 (*Maceneolignan H, activity*).
Suwitchayanon, P. *et al., Nat. Prod. Commun.*, 2017, **12**, 1077-1078 (*Myrilislignan, activity*)

4,8′-Oxylign-8-ene-3,3′,4′,7′,9′-pentol O104

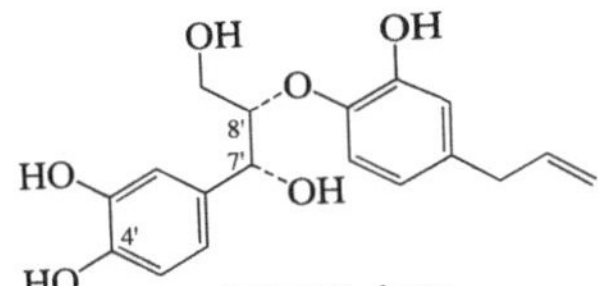

(7′R,8′S)-form

3,3′,4′,7′,9′-Pentahydroxy-4,8′-oxyneolign-8-ene
$C_{18}H_{20}O_6$ 332.352
(7′R,8′S)-form
3,3′-Di-Me ether
1-(4-Hydroxy-3-methoxyphenyl)-2-[2-methoxy-4-(2-propenyl) phenoxy]-1,3-propanediol. 3,3′-Dimethoxy-4,8′-oxylign-8-ene-4′,7′,9′-triol. 4′,7′,9′-Trihydroxy-3,3′-dimethoxy-4,8′-oxylign-8-ene. ***Neoasarinin B***
[2211144-24-8]
$C_{20}H_{24}O_6$ 360.406
Constit. of the roots and rhizomes of *Asarum heterotropoides*. Amorph. powder. $[\alpha]_D^{20}$ -3.6 (c, 0.55 in MeOH). λ_{max} 211 (log ε 4.03); 229 (log ε 3.95); 280 (log ε 3.55) (MeOH)
(7′S,8′S)-form
3,3′-Di-Me ether
Neoasarinin C
[2211144-25-9]
$C_{20}H_{24}O_6$ 360.406
Constit. of the roots and rhizomes of *Asarum heterotropoides*. Amorph. powder. $[\alpha]_D^{20}$ -5.3 (c, 0.38 in MeOH). λ_{max} 205 (log ε 3.33); 230 (log ε 2.77); 279 (log ε 2.38) (MeOH)
(7′ξ,8′ξ)-form
3,3′,4′-Tri-Me ether
1-(3,4-Dimethoxyphenyl)-2-[2-methoxy-4-(2-propenyl)phenoxy]-1,3-propanediol, CAS. *3,3′,4′-Trimethoxy-4,8′-oxylign-8-ene-7′,9′-diol. 7′,9′-Dihydroxy-3,3′,4′-trimethoxy-4,8′-oxylign-8-ene.* ***Carinatidiol***
[88640-86-2]
$C_{21}H_{26}O_6$ 374.433
Constit. of the bark of *Virola carinata*. Resinous mass. $[\alpha]_D$ +97.2 ($CHCl_3$). λ_{max} 248 (log ε 4.09); 285 (log ε 4.04) ($CHCl_3$).

Kawanishi, K. *et al.*, *Phytochemistry*, 1983, **22**, 2277-2280 (*Carinatidiol*).
Jing, Y. *et al.*, *Molecules*, 2017, **22**, 125:1-20 (*Neoasarinins B,C*)

3,4′-Oxylign-8′-ene-4,5,7,9-tetrol O105

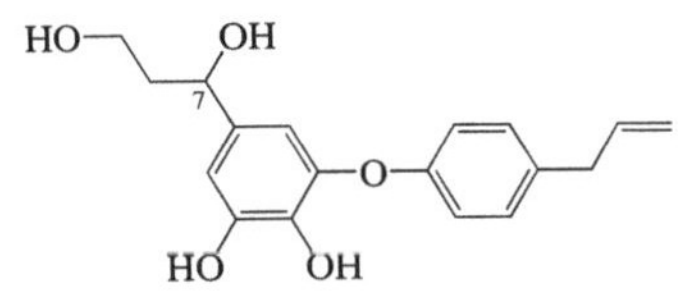

5-(1,3-Dihydroxypropyl)-3-[4-(2-propenyl)phenoxy]-1,2-benzenediol. 4,5,7,9-Tetrahydroxy-3,4′-oxyneolign-8′-ene
$C_{18}H_{20}O_5$ 316.353
(ξ)-form
7,9-Di-Me ether
7,9-Dimethoxy-3,4′-oxylign-8′-ene-4,5-diol. 4,5-Dihydroxy-7,9-dimethoxy-3,4′-oxylign-8′-ene. ***Obovatalignan B. Obovatoside B***
[1942113-40-7]
$C_{20}H_{24}O_5$ 344.407
Constit. of the fruit of *Magnolia obovata*. Exhibits neuroprotective activity. Inhibitor of NO prodn.. Amorph. powder. $[\alpha]_D^{18}$ +10.6 (c, 0.26 in MeOH). Assigned 2 synonyms. λ_{max} 226 (log ε 2.37); 273 (log ε 1.97) (MeOH).

Seo, K.-H. *et al.*, *Helv. Chim. Acta*, 2016, **99**, 411-415 (*Obovatalignan B, activity*)

3,8′-Oxylign-7-ene-3′,4′,5,7′-tetrol O106

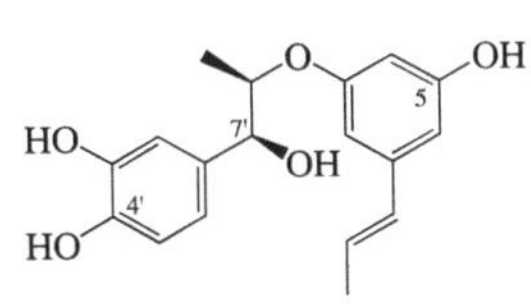

1-(3,4-Dihydroxyphenyl)-2-[3-hydroxy-5-(1-propenyl)phenoxy]-1-propanol. 3′,4′,5,7′-Tetrahydroxy-3,8′-oxyneolign-7-ene
$C_{18}H_{20}O_5$ 316.353
Structures in this series may need revision. Phenyl 3,5-dioxygenated substn. pattern in doubt as a result of work done by Yang *et al* in 2015
(7E,7′S,8′R)-form
3′,4′,5-Tri-Me ether
3′,4′,5-Trimethoxy-3,8′-oxylign-7-en-7′-ol. 7′-Hydroxy-3′,4′,5-trimethoxy-3,8′-oxylign-7-ene
[217082-56-9]
$C_{21}H_{26}O_5$ 358.434
Constit. of *Marrubium alysson*.

Hosny, M. *et al.*, *Al-Azhar J. Pharm. Sci.*, 1997, **19**, 165-178 (*Marrubium alysson constit*).
Yang, Y.-N. *et al.*, *J. Nat. Prod.*, 2015, **78**, 705-711 (*pmr, struct*)

4,4′-Oxylign-7′-ene-3,3′,7,8-tetrol O107

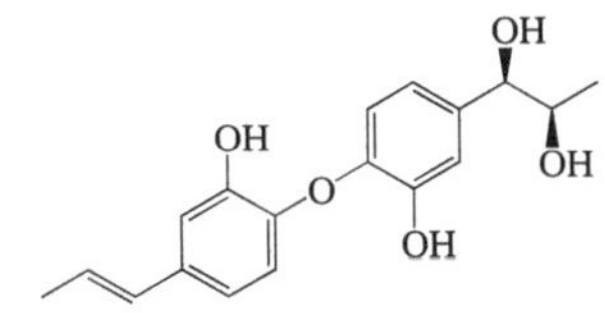

3,3′,7,8-Tetrahydroxy-4,4′-oxyneolign-7′-ene
$C_{18}H_{20}O_5$ 316.353
(7R*,7′E,8R*)-form
3,3′-Di-Me ether
3,3′-Dimethoxy-4,4′-oxylign-7′-ene-7,8-diol. 7,8-Dihydroxy-3,3′-dimethoxy-4,4′-oxylign-7′-ene
[2328018-93-3]
$C_{20}H_{24}O_5$ 344.407
Constit. of the fruit peel of *Elaeagnus rhamnoides*. Inhibits *Herpes simplex* type 2 virus replication. Amorph. solid. $[\alpha]_D$ +19 (c, 0.1 in MeOH).

Redei, D. *et al.*, *Tetrahedron*, 2019, **75**, 1364-1370 (*Elaeagnus rhamnoides constit, activity*)

4,8′-Oxylign-7-ene-3,3′,4′,5′-tetrol O108

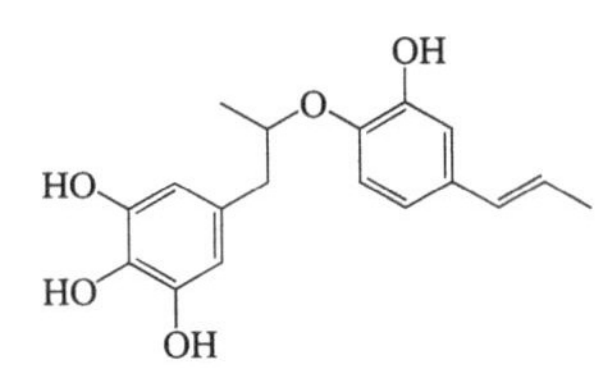

3,3′,4′,5′-Tetrahydroxy-4,8′-oxyneolign-7-ene
$C_{18}H_{20}O_5$ 316.353
(-)-(E)-form
Tetra-Me ether
3,3′,4′,5′-Tetramethoxy-4,8′-oxylign-7-ene. ***Tatarinoid G***
[2222357-85-7]
$C_{22}H_{28}O_5$ 372.461
Constit. of the rhizomes of *Acorus tatarinowii*. Gum. $[\alpha]_D^{27}$ -8.6 (c, 0.5 in MeOH). Partial racemate.

Zhang, W.-Y. *et al.*, *Bioorg. Med. Chem. Lett.*, 2018, **28**, 814-819 (*Tatarinoid G*)

4,8′-Oxylign-7-ene-3,3′,4′,7′-tetrol O109

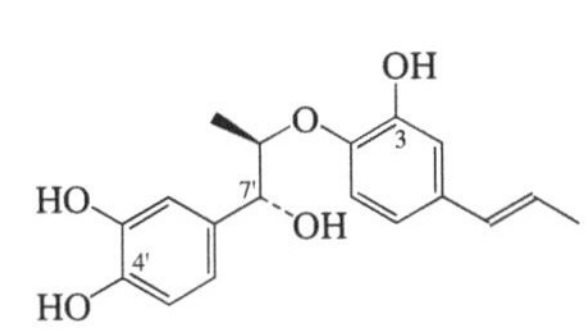

(7E,7′R,8′R)-form

3,3′,4′,7′-Tetrahydroxy-4,8′-oxyneolign-7-ene
$C_{18}H_{20}O_5$ 316.353
(7E,7′R,8′R)-form
3,3′-Di-Me ether
3,3′-Dimethoxy-4,8′-oxylign-7-ene-4′,7′-diol. 4′,7′-Dihydroxy-3,3′-dimethoxy-4,8′-oxylign-7-ene.
(-)-Machilin D
[50393-87-8] (rel. config.)
[138332-35-1] (rel. config.)
$C_{20}H_{24}O_5$ 344.407
Constit. of the roots of *Saururus chinensis*. Antioxidant. Oil. $[\alpha]_D$ -160 (c, 0.7 in $CHCl_3$). λ_{max} 211 (log ε 4.41); 267 (log ε 4.07) (MeOH)
3,3′,4′-Tri-Me ether, 7′-Ac
[916264-22-7]
$C_{23}H_{28}O_6$ 400.471
Constit. of *Taxillus theifer*. Oil. $[\alpha]_D^{25}$ -35.7 (c, 0.14 in MeOH). λ_{max} 259 (log ε 3.91); 269 (log ε 3.9); 286 (sh) (MeOH)
(7E,7′R,8′S)-form
3,3′-Di-Me ether
(-)-Machilin C
[110269-51-7]
[138332-33-9] (rel. config.)
$C_{20}H_{24}O_5$ 344.407
Constit. of the bark of *Machilus thunbergii*. Oil. $[\alpha]_D^{25}$ -16.5 (c, 0.27 in $CHCl_3$)
(7E,7′S,8′R)-form
3,3′-Di-Me ether
(+)-Machilin C. Odoratisol B
[891182-94-8]
[50393-88-9] (rel. config.)
$C_{20}H_{24}O_5$ 344.407
Constit. of the bark of *Machilus odoratissima*. Oil. $[\alpha]_D^{25}$ +18.6 (c, 0.7 in $CHCl_3$). λ_{max} 218 (log ε 3.87); 258 (log ε 3.68) (MeOH)
(7E,7′S,8′S)-form
3,3′-Di-Me ether
Machilin D
[110269-53-9]
$C_{20}H_{24}O_5$ 344.407
Constit. of *Machilus thunbergii*. Oil. $[\alpha]_D^{25}$ +38.1 (c, 0.07 in $CHCl_3$)
3,3′,4′-Tri-Me ether
3,3′,4′-Trimethoxy-4,8′-oxylign-7-en-7′-ol. 7′-Hydroxy-3,3′,4′-trimethoxy-4,8′-oxylign-7-ene. ***Virolin***
[68143-83-9]
$C_{21}H_{26}O_5$ 358.434
Constit. of *Virola surinamensis*. Glucan synthase inhibitor. Oil
3,3′,4′-Tri-Me ether, 7′-Ac
Piperbonin A
[2001069-74-3]
$C_{23}H_{28}O_6$ 400.471
Constit. of the aerial parts of *Piper bonii*. Exhibits weak activity against platelet aggregation induced by thrombin. Pale yellow oil. $[\alpha]_D^{23}$ +56 (c, 0.7 in MeOH). λ_{max} 230 (log ε 4.32); 278 (log ε 4.13); 309 (log ε 3.95) (MeOH)
(7E,7′R*,8′S*)-form
3-Me ether
3-Methoxy-4,8′-oxylign-7-ene-3′,4′,7′-triol. 3′,4′,7′-Trihydroxy-3-methoxy-4,8′-oxylign-7-ene
$C_{19}H_{22}O_5$ 330.38
Constit. of the roots of *Adelostemma gracillimum*. Light yellow oil. $[\alpha]_D^{20}$ +16.6 (c, 0.82 in MeOH). λ_{max} 259 (MeOH)
(7E,7′ξ,8′ξ)-form
3′-Me ether
3′-Methoxy-4,8′-oxylign-7-ene-3,4′,7′-triol. 3,4′,7′-Trihydroxy-3′-methoxy-4,8′-oxylign-7-ene.
Callislignan B
[1416369-72-6]
$C_{19}H_{22}O_5$ 330.38
Constit. of the leaves of *Callistemon lanceolatus*. Antibacterial agent. Yellowish gum. $[\alpha]_D^{29}$ -15.8 (c, 4.5 in MeOH). λ_{max} 205 (log ε 4.0); 227 (log ε 3.65); 269 (log ε 3.3) (MeOH)
7′-Ketone, 3,3′,4′-tri-Me ether
3,3′,4′-Trimethoxy-4,8′-oxylign-7-en-7′-one. 3,3′,4′-Trimethoxy-7′-oxo-4,8′-oxylign-7-ene.
Bennettinone
[139343-57-0]
$C_{21}H_{24}O_5$ 356.418
Constit. of *Eupomatia bennettii*. Needles (petrol). Mp 85 - 87°. Racemic. λ_{max} 231 (ε 23440); 279 (ε 12300); 309 (ε 8510) (EtOH)
(7Z,7′R,8′R)-form
3,3′,4′-Tri-Me ether, 7′-Ac
Ligraminol C
[1335126-26-5]
$C_{23}H_{28}O_6$ 400.471
Constit. of the rhizomes of *Acorus gramineus*. Shows selective cytotoxicity among tumour and normal cells and anti-neuroinflammatory activity. Oil. $[\alpha]_D^{25}$ -20.2 (c, 0.05 in MeOH). λ_{max} 204 (log ε 4.5); 233 (log ε 2.8); 257 (log ε 1.7); 287 (log ε 1.2) (MeOH).

Barata, L.E.S. *et al.*, *Phytochemistry*, 1978, **17**, 783-786 (*Virolin*).
Zacchino, S.A. *et al.*, *J. Nat. Prod.*, 1985, **48**, 830-832 (*Virolin, synth*).
Shimomura, H. *et al.*, *Phytochemistry*, 1987, **26**, 1513-1515 (*Machilins C,D*).
Carroll, A.R. *et al.*, *Aust. J. Chem.*, 1991, **44**, 1627-1633 (*Bennettinone*).
Zacchino, S.A. *J. Nat. Prod.*, 1994, **57**, 446-451 (*Virolin, synth, abs config*).
Sung, S.H. *et al.*, *Chem. Pharm. Bull.*, 2001, **49**, 1192-1194 (*(-)-Machilin D*).
Lee, W.S. *et al.*, *Bioorg. Med. Chem. Lett.*, 2004, **14**, 5623-5628 (*(-)-Machilin D, activity*).
Phan, M.G. *et al.*, *Chem. Pharm. Bull.*, 2006, **54**, 380-383 (*Odoratisol B*).
Xia, Y. *et al.*, *Mendeleev Commun.*, 2010, **20**, 151-152 (*Machilin C, synth*).
Kim, K.H. *et al.*, *J. Nat. Prod.*, 2011, **74**, 2187-2192 (*Ligraminol C, activity*).
Lin, L.-C. *et al.*, *Nat. Prod. Res.*, 2011, **25**, 1319-1323 (*Taxillus theifer constit*).
Rattanaburi, S. *et al.*, *Phytochem. Lett.*, 2012, **5**, 18-21 (*Callislignan B, activity*).
Cui, H. *et al.*, *J. Nat. Prod.*, 2014, **77**, 100-110 (*(-)-Machilin D*).
Zhang, J. *et al.*, *Phytochem. Lett.*, 2016, **16**, 1-7 (*Adelostemma gracillimum constit*).
Ding, D.-D. *et al.*, *Phytochemistry*, 2016, **129**, 36-44 (*Piperbonin A, activity*)

4,8′-Oxylign-7-ene-4′,7′,9,9′-tetrol O110

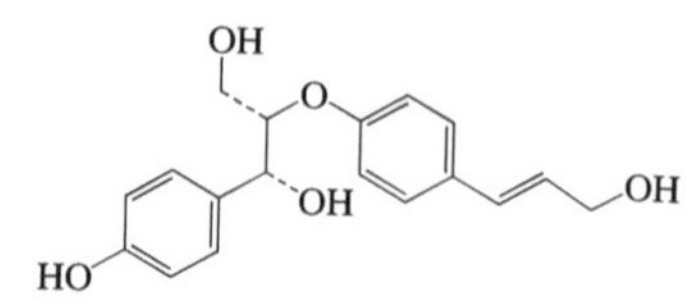

(7E,7′R,8′S)-form

4′,7′,9,9′-Tetrahydroxy-4,8′-oxyneolign-7-ene
$C_{18}H_{20}O_5$ 316.353
(7E,7′R,8′S)-form
9-Aldehyde
4′,7′,9′-Trihydroxy-4,8′-oxylign-7-en-9-al. 4′,7′,9′-Trihydroxy-9-oxo-4,8′-oxylign-7-ene
$C_{18}H_{18}O_5$ 314.337
Constit. of the fruit of *Juglans mandshurica*. Oil. $[\alpha]_D^{20}$ -31.1 (c, 0.2 in MeOH)
(7E,7′S,8′S)-form
9-Aldehyde
$C_{18}H_{18}O_5$ 314.337
Constit. of the fruit of *Juglans mandshurica*. Oil. $[\alpha]_D^{20}$ -11.2 (c, 0.2 in MeOH).

Park, S. *et al.*, *Phytochemistry*, 2017, **137**, 101-108 (*Juglans mandshurica constits, cd*)

4,8′-Oxylign-8-ene-3,3′,4′,5-tetrol **O111**

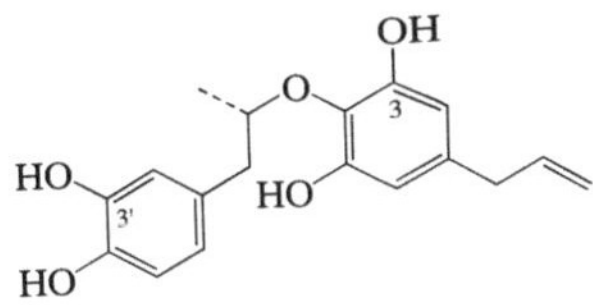

2-(4-Allyl-2,6-dihydroxyphenoxy)-1-(3,4-dihydroxyphenyl)propane. 3,3′,4′,5-Tetrahydroxy-4,8′-oxyneolign-8-ene

$C_8H_{20}O_5$ 196.243

(S)-form

Tetra-Me ether

3,3′,4′,5-Tetramethoxy-4,8′-oxylign-8-ene

[1246343-70-3]

$C_{22}H_{28}O_5$ 372.461

Constit. of the seeds of *Myristica fragrans*.

Yang, X. *et al.*, *Zhongguo Zhongyao Zazhi*, 2008, **33**, 397-402 (*Myristica fragrans constit*)

4,8′-Oxylign-8-ene-3,3′,4′,7′-tetrol **O112**

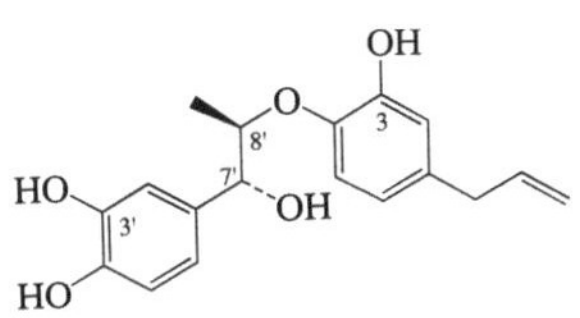

(7′R*,8′R*)-form

2-(4-Allyl-2-hydroxyphenoxy)-1-(3,4-dihydroxyphenyl)-1-propanol. 3,3′,4′,7′-Tetrahydroxy-4,8′-oxyneolign-8-ene

[83414-57-7]

$C_{18}H_{20}O_5$ 316.353

(7′R*,8′R*)-form

3,3′-Di-Me ether

2-(4-Allyl-2-methoxyphenoxy)-1-(4-hydroxy-3-methoxyphenyl)-1-propanol. 3,3′-Dimethoxy-4,8′-oxylign-8-ene-4′,7′-diol. 4′,7′-Dihydroxy-3,3′-dimethoxy-4,8′-oxylign-8-ene

[114394-18-2]

$C_{20}H_{24}O_5$ 344.407

Constit. of the aril of *Myristica fragrans*. Oil

(7′R*,8′S*)-form

3,3′-Di-Me ether

[108907-55-7]

Constit. of the aril of *Myristica fragrans*. Oil.

Hattori, M. *et al.*, *Chem. Pharm. Bull.*, 1987, **35**, 668-674 (*Myristica fragrans constit*).
Hada, S. *et al.*, *Phytochemistry*, 1988, **27**, 563-568 (*Myristica fragrans constit*)

4,8′-Oxylign-7-ene-3′,4′,7′-triol **O113**

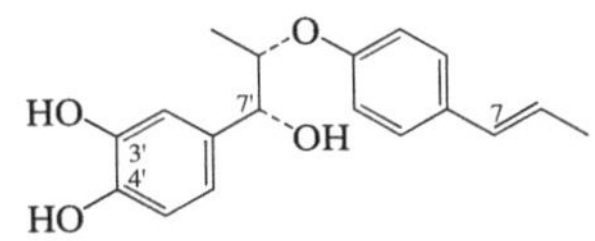

1-(3,4-Dihydroxyphenyl)-2-[4-(1-propenyl)phenoxy]-1-propanol. 3′,4′,7′-Trihydroxy-4,8′-oxyneolign-7-ene

$C_{18}H_{20}O_4$ 300.354

(7E,7′R*,8′S*)-form

3′,4′-Methylene ether

1-(3,4-Methylenedioxyphenyl)-2-[4-(1-propenyl)phenoxy]-1-propanol. 3′,4′-Methylenedioxy-4,8′-oxylign-7-en-7′-ol. 7′-Hydroxy-3′,4′-methylenedioxy-4,8′-oxylign-7-ene

[178560-42-4]

$C_{19}H_{20}O_4$ 312.365

Constit. of *Caryodaphnopsis tonkinensis*. Oil. $[\alpha]_D^{20}$ -36.1 (c, 0.7 in $CHCl_3$). Genus name given as Caryodaphnosis.

Anh, N.H. *et al.*, *Phytochemistry*, 1996, **42**, 1167-1169 (*Caryodaphnopsis tonkinensis constit*)

4,8′-Oxylign-7-ene-3,4′,7′-triol **O114**

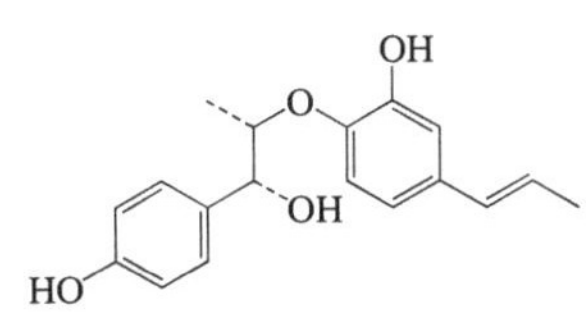

3,4′,7′-Trihydroxy-4,8′-oxyneolign-7-ene

$C_{18}H_{20}O_4$ 300.354

(7E,7′R*,8′S*)-form

3-Me ether

1-(4-Hydroxyphenyl)-2-[2-methoxy-4-(1-propenyl)phenoxy]-1-propanol. 3-Methoxy-4,8′-oxylign-7-ene-4′,7′-diol. 4′,7′-Dihydroxy-3-methoxy-4,8′-oxylign-7-ene

[109225-43-6]

$C_{19}H_{22}O_4$ 314.381

Constit. of *Krameria cytisoides*. Oil. $[\alpha]_D^{21}$ -11.6 (c, 0.79 in MeOH). Species given as Krameria cystisoides in ref.. λ_{max} 221 (sh); 259 (log ε 4.2); 266 (sh); 284 (sh); 301 (sh) (MeOH).

Achenbach, H. *et al.*, *Phytochemistry*, 1987, **26**, 1159-1166 (*Krameria cytisoides constit*)

4,8′-Oxylign-8-ene-3,3′,4′-triol **O115**

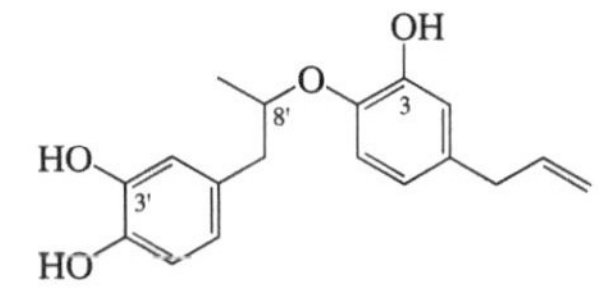

3,3′,4′-Trihydroxy-4,8′-oxyneolign-8-ene

$C_{18}H_{20}O_4$ 300.354

(8′ξ)-form

3,3′-Di-Me ether

3,3′-Dimethoxy-4,8′-oxylign-8-en-4′-ol. 4′-Hydroxy-3,3′-dimethoxy-4,8′-oxylign-8-ene. ***Miliusfragranol B***

[1446741-19-0]

$C_{20}H_{24}O_4$ 328.408

Constit. of *Miliusa fragrans*. Viscous liq.. $[\alpha]_D^{20}$ -20.0 (c, 0.11 in $CHCl_3$). λ_{max} 210 (log ε 4.20); 229 (log ε 4.07); 281 (log ε 3.70) (MeOH).

Sawasdee, K. *et al.*, *Tetrahedron Lett.*, 2013, **54**, 4259-4263 (*Miliusfragranol B*)

4,8′-Oxylign-8-en-4′-ol **O116**

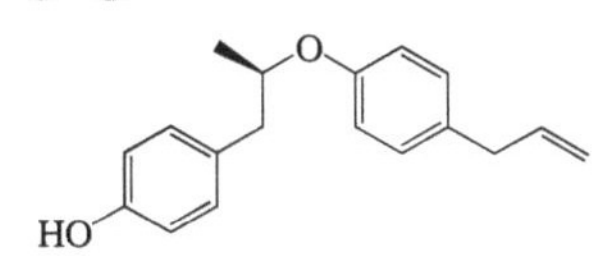

4′-Hydroxy-4,8′-oxyneolign-8-ene

$C_{18}H_{20}O_2$ 268.356

(R)-form

Sassarandainol [1643111-76-5]

Constit. of the stem of *Sassafras randaiense*. Yellowish oil. $[\alpha]_D^{23}$ -12.8 (c, 0.37 in $CHCl_3$). λ_{max} 206 (log ε 4.16); 228 (log ε 4.25); 278 (log ε 3.50) (MeOH).

Hou, Y.-L. *et al.*, *Nat. Prod. Res.*, 2015, **29**, 827-832 (*Sassarandainol*)

4′,8″-Oxy-8,8′-sesquilignan-3,3′,3″,4,4″,7″,9,9′,9″-nonol

O117

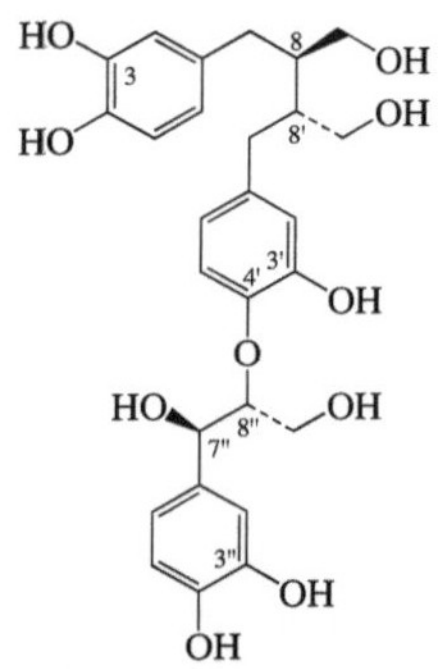

(7″R,8R,8′R,8″R)-form

3,3′,3″,4,4″,7″,9,9′,9″-Nonahydroxy-4′,8″-oxysesquineolignan

$C_{27}H_{32}O_{10}$ 516.543

(7″R,8R,8′R,8″R)-form

3,3′,3″-Tri-Me ether

3,3′,3″-Trimethoxy-4′,8″-oxy-8,8′-sesquilignan-4,4″,7″,9,9′,9″-hexol. 4,4″,7″,9,9′,9″-Hexahydroxy-3,3′,3″-trimethoxy-4′,8″-oxysesquilignan.

Sesquimarocanol B

[173356-78-0]

[173268-36-5] (hexa-Ac)

$C_{30}H_{38}O_{10}$ 558.624

Constit. of *Abies marocana*, *Campylotropis hirtella* and *Litchi chinensis*. Oil (as hexa-Ac). $[\alpha]_D^{25}$ +13 (c, 0.5 in $CHCl_3$) (hexa-Ac)

(7″S,8S,8′S,8″R)-form

3,3′,3″-Tri-Me ether

$C_{30}H_{38}O_{10}$ 558.624

Enzymic hydrol. prod. of Urariasinoside A. Pale yellow oil. $[\alpha]_D^{20}$ -7.1 (c, 0.03 in MeOH)

3,3′,3″-Tri-Me ether, 7″-O-β-D-glucopyranoside

Urariasinoside A

[1609488-99-4]

$C_{36}H_{48}O_{15}$ 720.765

Constit. of the aerial parts of *Uraria sinensis*. Pale yellow oil. $[\alpha]_D^{20}$ -29 (c, 0.79 in MeOH). λ_{max} 207 (log ε 4.03); 228 (log ε 3.68); 280 (log ε 3.32) (MeOH).

Barrero, A.F. *et al.*, *Phytochemistry*, 1996, **41**, 605-609 (*Sesquimarocanol B*).
Wen, L. *et al.*, *J. Funct. Foods*, 2014, **8**, 26-34 (*Litchi chinensis constit*).
Yang, Y. *et al.*, *Molecules*, 2014, **19**, 1178-1188 (*Urariasinoside A*)

Pachypostaudin B P1

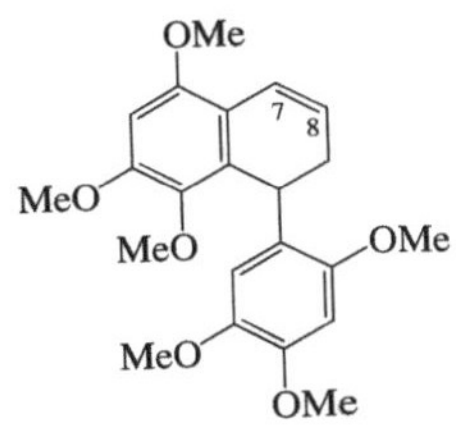

1,2-Dihydro-5,7,8-trimethoxy-1-(2,4,5-trimethoxyphenyl)naphthalene, CAS. *2,7′-Cyclo-2′,3,4,4′,5′,6-hexamethoxy-9,9′-dinor-8,8′-lign-7-ene. 2′,3,4,4′,5′,6-Hexamethoxy-9,9′-dinor-2,7′-cyclolign-7-ene*
$C_{22}H_{26}O_6$ 386.444
Lignan numbering shown
(ξ)-form [114542-59-5]
Constit. of *Pachypodanthium staudtii*. Shows activity against polio virus. Needles (EtOAc/hexane). Mp 120°. λ_{max} 235 (log ε 4.13); 295 (log ε 4.02) (MeOH)
7,8-Dihydro
2,7′-Cyclo-2′,3,4,4′,5′,6-hexamethoxy-9,9′-dinorlignan. 2′,3,4,4′,5′,6-Hexamethoxy-9,9′-dinor-2,7′-cyclolignan. ***Pachypostaudin A***
[114542-58-4]
$C_{22}H_{28}O_6$ 388.46
Constit. of *Pachypodanthium staudtii*. Shows activity against polio virus. Granules (EtOAc/hexane). Mp 130°. λ_{max} 225; 285 (MeOH)
7,8-Dihydro, 7-oxo
3,4-Dihydro-5,6,8-trimethoxy-4-(2,4,5-trimethoxyphenyl)-1(2H)-naphthalenone, CAS. *5,6,8-Trimethoxy-4-(2,4,5-trimethoxyphenyl)-1-tetralone. 2,7′-Cyclo-2′,3,4,4′,5′,6-hexamethoxy-9,9′-dinor-8,8′-lignan-7-one. 2′,3,4,4′,5′,6-Hexamethoxy-9,9′-dinor-7-oxo-2,7′-cyclolignan*
[878383-50-7]
Constit. of *Peperomia pellucida*. Shows moderate activity against human breast cancer MCF-7 cells. Yellowish gum. $[\alpha]_D^{20}$ -3 (c, 0.18 in Me_2CO). λ_{max} 234; 280; 324 (MeOH aq.).

Ngadjui, B.T. *et al., Phytochemistry*, 1989, **28**, 231-234 (*Pachypostaudins A,B*).
Xu, S. *et al., J. Nat. Prod.*, 2006, **69**, 247-250 (*Peperomia pellucida constit*)

Padocin P2

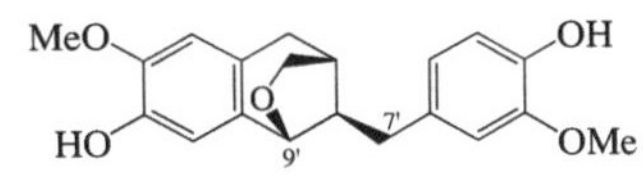

Relative Configuration

2,9′-Cyclo-9,9′-epoxy-3′,5-dimethoxy-8,8′-lignan-4,4′-diol
[160896-56-0]
$C_{20}H_{22}O_5$ 342.391
Constit. of *Haplophyllum cappadocicum, Syringa pinnatifolia* and *Viburnum foetidum* var. *foetidum*. Amorph. solid. $[\alpha]_D^{25}$ -11.3 (c, 0.27 in MeOH). λ_{max} 206 (ε 47860); 229 (ε 10960); 284 (ε 6025) (MeOH).

Gözler, B. *et al., Heterocycles*, 1994, **39**, 243-249 (*Padocin*).
Chen, X.-Q. *et al., Chem. Pharm. Bull.*, 2009, **57**, 1129-1131 (*Viburnum foetidum constit*)

Pahangin B P3

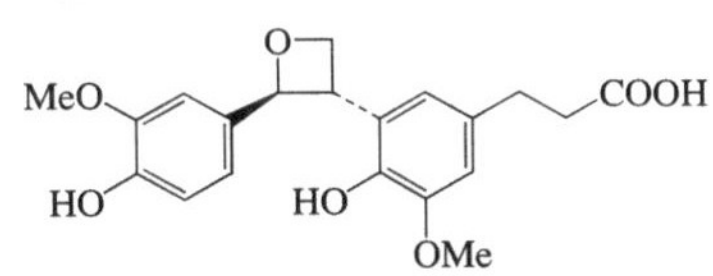

7′,9′-Epoxy-4,4′-dihydroxy-3′,5-dimethoxy-3,8′-lignan-9-oic acid
$C_{20}H_{22}O_7$ 374.389
Constit. of the bark of *Beilschmiedia glabra*. Amorph. yellow solid. $[\alpha]_D^{25}$ +12.5 (c, 0.1 in MeOH). λ_{max} 234 (log ε 4.7) (MeOH)
7,8E-Didehydro, amide
7′,9′-Epoxy-4,4′-dihydroxy-3′,5-dimethoxy-3,8′-lign-7-en-9-amide. ***Pahangin A***
$C_{20}H_{21}NO_6$ 371.389
Constit. of the bark of *Beilschmiedia glabra*. Amorph. yellow solid. $[\alpha]_D^{25}$ +37.5 (c, 0.1 in MeOH).
Related to Lyciumamide C, L119.

Sulaiman, S.N. *et al., Phytochem. Lett.*, 2018, **25**, 22-26 (*Pahangins A,B*)

Parabaroside A P4

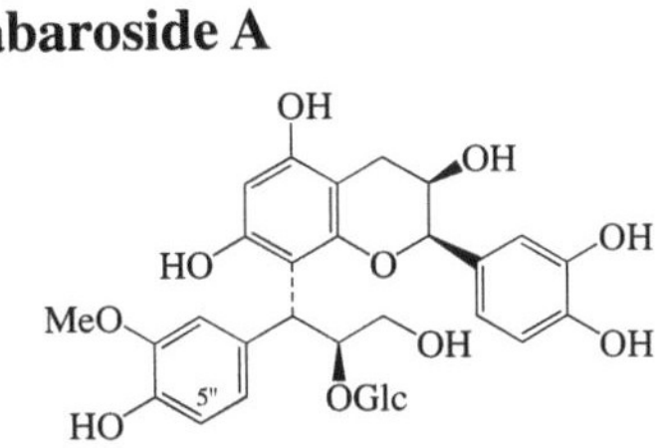

Absolute Configuration

[1262631-41-3]
$C_{31}H_{36}O_{15}$ 648.614
Constit. of the stems of *Parabarium huaitingii*. Amorph. brown powder. $[\alpha]_D^{27}$ -44.7 (c, 0.3 in MeOH)
5″-Methoxy
Parabaroside B
[1262631-43-5]
$C_{32}H_{38}O_{16}$ 678.64
Constit. of the stems of *Parabarium huaitingii*. Amorph. brown powder. $[\alpha]_D^{27}$ -11.8 (c, 0.2 in MeOH).

Tang, J.-S. *et al., Planta Med.*, 2009, **75**, 1586-1590 (*Parabarosides A,B*)

Pavonisol P5

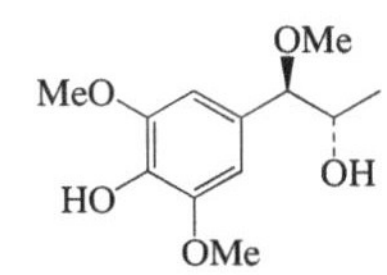

Relative Configuration

4-Hydroxy-β,3,5-trimethoxy-α-methylbenzeneethanol, CAS. *4-(2-Hydroxy-1-methoxypropyl)-2,6-dimethoxyphenol*
[144049-93-4]
$C_{12}H_{18}O_5$ 242.271
Constit. of the leaves of *Virola pavonis*. Cryst. (Et_2O). Mp 138 - 140°
Di-Ac
[144049-97-8]
$C_{16}H_{22}O_7$ 326.345
Cryst. (Et_2O). Mp 115 - 116°. $[\alpha]_D^{25}$ +86.7 (c, 0.85 in $CHCl_3$).

Ferri, P.H. *et al., Phytochemistry*, 1992, **31**, 1375-1377 (*Pavonisol*)

Pelliatin P6

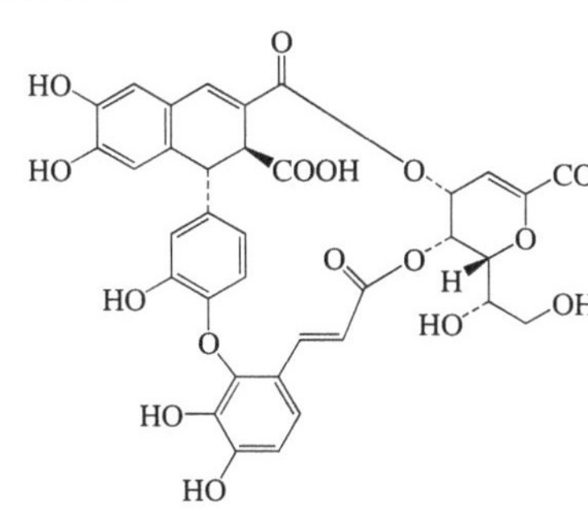

[173395-16-9]
$C_{35}H_{28}O_{17}$ 720.592
Constit. of the liverwort *Pellia epiphylla*. Yellow oil. $[\alpha]_D^{20}$ -106.7 (MeOH). λ_{max} 239; 294 (sh); 318 (br.) (MeOH).

Cullmann, F. *et al., Phytochemistry*, 1996, **41**, 611-615 (*Pelliatin*)

Penangianol A P7

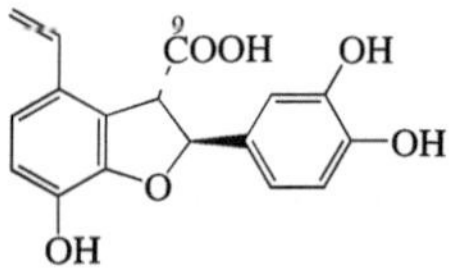

Relative Configuration

2-(3,4-Dihydroxyphenyl)-7-hydroxy-4-vinyl-2,3-dihydrobenzofuran-3-carboxylic acid. 3,7′-Epoxy-3′,4,4′-trihydroxy-9-nor-2,8′-lign-7-en-9′-oic acid

[1702426-07-0]

$C_{17}H_{14}O_6$ 314.293

Constit. of the rhizomes of *Abacopteris penangiana*. Pale yellow powder. $[\alpha]_D^{25}$ +118.2 (c, 0.5 in MeOH).

Jiao, Y. *et al.*, *Chem. Nat. Compd. (Engl. Transl.)*, 2015, **51**, 232-235 (*Penangianol A*)

Penangianol B P8

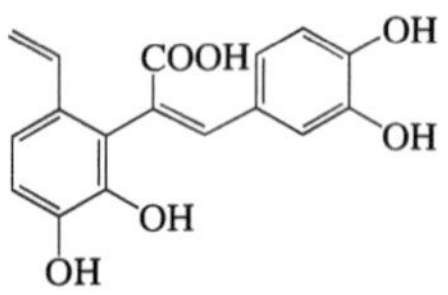

α-[(3,4-Dihydroxyphenyl)methylene]-6-ethenyl-2,3-dihydroxybenzeneacetic acid. 3,3′,4,4′-Tetrahydroxy-9-nor-2,8′-ligna-7,7′-dien-9′-oic acid

[1702426-11-6]

$C_{17}H_{14}O_6$ 314.293

Constit. of the rhizomes of *Abacopteris penangiana*. Pale yellow powder.

Jiao, Y. *et al.*, *Chem. Nat. Compd. (Engl. Transl.)*, 2015, **51**, 232-235 (*Penangianol B*)

1,2,3,4,5-Pentahydroxycyclohexanecarboxylic acid P9

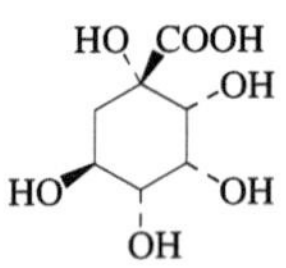

2-Hydroxyquinic acid

$C_7H_{12}O_7$ 208.166

(1αOH,2α,3α,4α,5β)-form

3,5-Bis-O-(3,4-dihydroxy-E-cinnamoyl), 4-O-(1-hydroxy-4-oxo-2,5-cyclohexadien-1-acetyl)

[953422-41-8]

$C_{33}H_{30}O_{16}$ 682.587

Constit. of *Hubertia ambavilla*.

Aldersberg, M. *et al.*, *J. Am. Chem. Soc.*, 1973, **95**, 887-896 (*synth*).

Sprogøe, K. *et al.*, *J. Nat. Prod.*, 2007, **70**, 1472-1477 (*Hubertia ambavilla constit*)

3′,4,4′,5,9′-Pentahydroxy-8,9-dinor-3,8′-lignan-7-oic acid P10

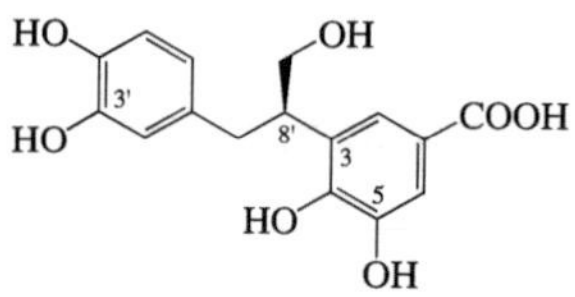

3-[2-(3,4-Dihydroxyphenyl)-1-(hydroxymethyl)ethyl]-4,5-dihydroxybenzoic acid

$C_{16}H_{16}O_7$ 320.297

(R)-form

3′,5-Di-Me ether

4,4′,9′-Trihydroxy-3′,5-dimethoxy-8,9-dinor-3,8′-lignan-7-oic acid

[1805782-08-4]

$C_{18}H_{20}O_7$ 348.351

Constit. of the rhizomes of *Acorus tatarinowii*. Amorph. powder. $[\alpha]_D^{20}$ -25.8 (c, 0.23 in MeOH). λ_{max} 268 (MeOH).

Liang, S. *et al.*, *Bioorg. Med. Chem. Lett.*, 2015, **25**, 4214-4218 (*Acorus tatarinowii constit*)

3,3′,4′,5′,7′-Pentahydroxy-8,9-dinor-4,8′-oxylignan-7-oic acid P11

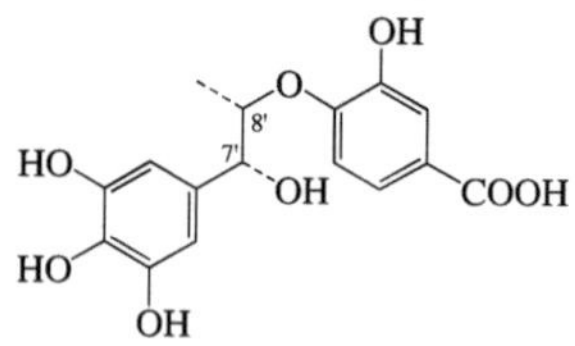

$C_{16}H_{16}O_8$ 336.296

(7′R,8′S)-form

3,3′,4′,5′-Tetra-Me ether

7′-Hydroxy-3,3′,4′,5′-tetramethoxy-8,9-dinor-4,8′-oxylignan-7-oic acid

3,3′,4′,5′-Tetra-Me ether, Me ester

Acortatarinowin N

$C_{21}H_{26}O_8$ 406.431

Constit. of the rhizomes of *Acorus tatarinowii*. Light yellow gum. $[\alpha]_D^{20}$ -23 (c, 0.1 in MeOH). λ_{max} 207 (log ε 4.77); 263 (log ε 4.22); 293 (log ε 3.93) (MeOH).

Lu, Y. *et al.*, *Sci. Rep.*, 2016, **6**, 22909 (*Acortatarinowin N*)

3,3′,4,4′,5-Pentahydroxy-8,8′-lignan-9,9′-dioic acid P12

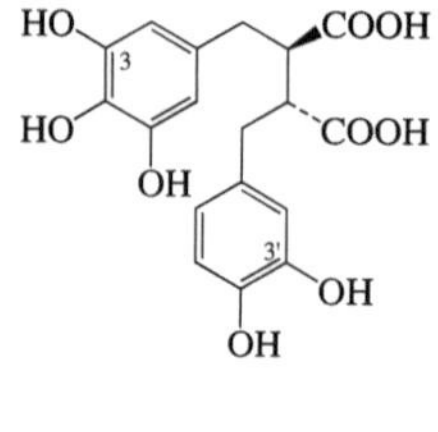

2-(3,4-Dihydroxybenzyl)-3-(3,4,5-trihydroxybenzyl)butanedioic acid

$C_{18}H_{18}O_9$ 378.333

(8R*,8′R*)-form

3,4:3′,4′-Bis(methylene), 5-Me ether, di-Me ester

Rhinacanthin F

[190850-42-1]

$C_{23}H_{24}O_9$ 444.436

Constit. of *Rhinacanthus nasutus*. Shows significant activity against influenza virus type A. Amorph. powder. $[\alpha]_D$ -5.2 (c, 0.06 in MeOH). λ_{max} 219 (log ε 3.02); 280 (log ε 3.42); 335 (log ε 3.15) (MeOH).

Kernan, M.R. *et al.*, *J. Nat. Prod.*, 1997, **60**, 635-637 (*Rhinacanthin F*)

3,3′,4,4′,8-Pentahydroxy-8,8′-lignan-9,9′-dioic acid P13

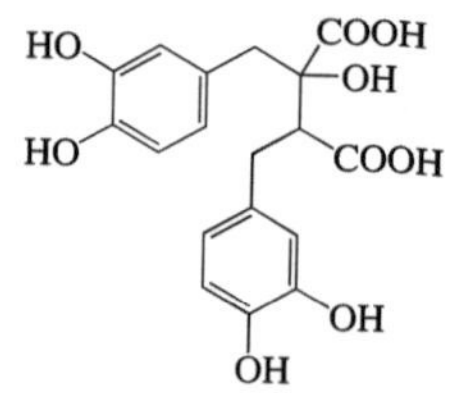

$C_{18}H_{18}O_9$ 378.333

(8ξ,8′ξ)-form

3,4:3′,4′-Bis(methylene) ether, di-Me ester

8-Hydroxyheliobuphthalmin

[102903-99-1]

$C_{22}H_{22}O_9$ 430.409

Constit. of *Heliopsis buphthalmoides*. Cryst.. Mp 129°.

Jakupovic, J. *et al.*, *Planta Med.*, 1986, **52**, 18-20 (*8-Hydroxyheliobuphthalmin*)

3,3′,4,4′,9′-Pentahydroxy-8,8′-lignan-9-oic acid P14

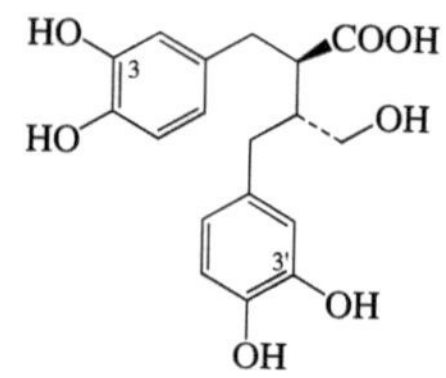

2,3-Bis(3,4-dihydroxybenzyl)-4-hydroxybutanoic acid

$C_{18}H_{20}O_7$ 348.351

(8R*,8′R*)-form

3,3′,4′-Tri-Me ether

4,9′-Dihydroxy-3,3′,4′-trimethoxy-8,8′-lignan-9-oic acid. ***Arctigenic acid***

[42320-79-6]

$C_{21}H_{26}O_7$ 390.432

Constit. of the seeds of *Arctium lappa*. Allelopathic agent. Hypoglycemic, antidiabetic agent. Mp 131°. $[\alpha]^{25}_{D}$ -3 (c, 0.2 in MeOH)

3,3′,4′-Tri-Me ether, 4-O-β-D-galactopyranoside

Aconiside

[1149341-21-8]

$C_{27}H_{36}O_{12}$ 552.573

Constit. of the seeds of *Onopordum acanthium*. Mp 80 - 81°. $[\alpha]^{25}_{D}$ -23.2 (c, 4.1 in $CHCl_3$). λ_{max} 281 ($CHCl_3$).

Higashinakasu, K. *et al.*, *Heterocycles*, 2005, **65**, 1431-1437 (*Arctigenic acid*).
Tyumkina, T.V. *et al.*, *Chem. Nat. Compd. (Engl. Transl.)*, 2009, **45**, 61-65 (*Aconiside*).
Xu, Z. *et al.*, *Phytomedicine*, 2015, **22**, 128-137 (*Arctigenic acid, activity*)

2,3,3′,4,4′-Pentahydroxy-8,8′-lignan-9,9′-olide P15

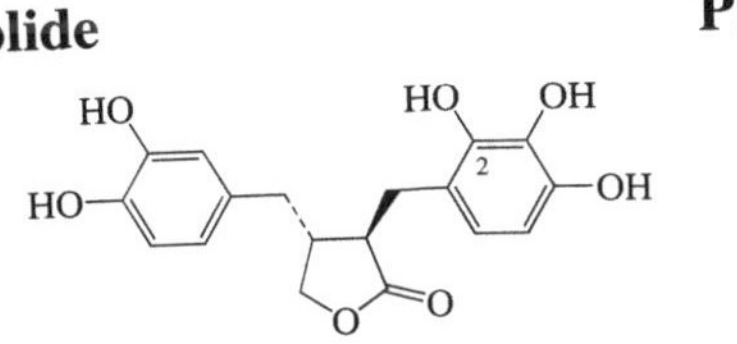

4-(3,4-Dihydroxybenzyl)-4,5-dihydro-3-(2,3,4-trihydroxybenzyl)-2(3H)-furanone

$C_{18}H_{18}O_7$ 346.335

Lignan numbering shown

(8R*,8′R*)-form

3,3′-Di-Me ether

2,4,4′-Trihydroxy-3,3′-dimethoxy-8,8′-lignan-9,9′-olide. ***Gnidifolin***

[58268-84-1]

$C_{20}H_{22}O_7$ 374.389

Constit. of *Gnidia latifolia*.

Bryan, R.F. *et al.*, *Acta Crystallogr., Sect. B*, 1978, **34**, 327-329 (*Gnidifolin, cryst struct*)

2,3′,4,4′,5-Pentahydroxy-8,8′-lignan-9,9′-olide P16

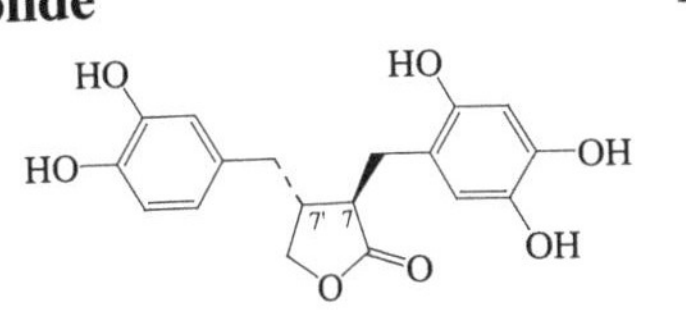

4-(3,4-Dihydroxybenzyl)-dihydro-3-(2,4,5-trihydroxybenzyl)-2(3H)-furanone

$C_{18}H_{18}O_7$ 346.335

(7R,7′R)-form

3′,4′:4,5-Bis(methylene) ether

2-Hydroxy-3′,4′:4,5-bis(methylenedioxy)-8,8′-lignan-9,9′-olide. ***6-Hydroxyhinokinin***

[935748-33-7]

$C_{20}H_{18}O_7$ 370.357

Constit. of the stems of *Kadsura heteroclita*. Yellow powder. Mp 68 - 70°. $[\alpha]^{20}_{D}$ +27 (c, 0.2 in MeOH). λ_{max} 235 (log ε 2.21); 294 (log ε 3.01) (MeOH)

3′,4′:4,5-Bis(methylene) ether, 2-O-β-D-glucopyranoside

Rubrifloside A

[724707-43-1]

$C_{26}H_{28}O_{12}$ 532.498

Constit. of *Kadsura heteroclita* and *Schisandra rubriflora*. Amorph. solid. Mp 101 - 102°. $[\alpha]^{20}_{D}$ +19 (c, 0.2 in MeOH). λ_{max} 234 (log ε 4.07); 290 (log ε 4) (MeOH)

3′,4′:4,5-Bis(methylene) ether, 2-O-[α-L-arabinofuranosyl-(1→6)-β-D-glucopyranoside]

Rubrifloside B

[724707-44-2]

$C_{31}H_{36}O_{16}$ 664.613

Constit. of the aerial parts of *Schisandra rubriflora*. Amorph. solid. $[\alpha]^{24}_{D}$ -22 (c, 0.45 in MeOH). λ_{max} 202 (log ε 4.33); 230 (log ε 3.66); 288 (log ε 3.51) (MeOH).

Li, G. *et al.*, *Heterocycles*, 2004, **63**, 1437-1441 (*Rubriflosides*).
Wang, W. *et al.*, *Nat. Prod. Commun.*, 2006, **1**, 109-112 (*Kadsura heteroclita constits*)

2′,3,4,4′,5′-Pentahydroxy-8,8′-lignan-9,9′-olide P17

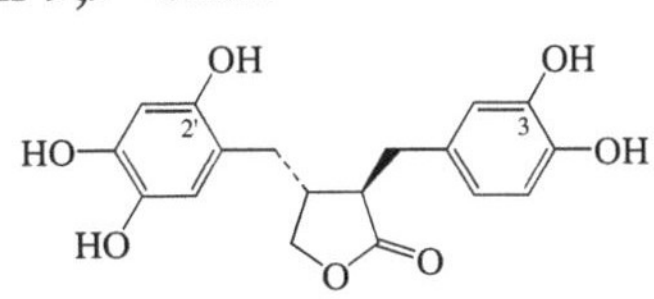

$C_{18}H_{18}O_7$ 346.335

(8R,8′R)-form

3,4:4′,5′-Bis(methylene) ether

2′-Hydroxy-3,4:4′,5′-bis(methylenedioxy)-8,8′-lignan-9,9′-olide. ***2′-Hydroxyhinokinin***

[222849-87-8]

$C_{20}H_{18}O_7$ 370.357

Constit. of *Taiwania cryptomerioides*. Mp 110 - 112°. $[\alpha]^{25}_{D}$ -20.2 (c, 1.6 in $CHCl_3$). λ_{max} 234 (ε 11300); 293 (ε 10100) (MeOH)

3,4:4′,5′-Bis(methylene) ether, 2′-O-β-D-glucopyranoside

[222849-89-0] (tetra-Ac)

$C_{26}H_{28}O_{12}$ 532.498

Constit. of *Taiwania cryptomerioides*. Mp 172 - 173° (as tetra-Ac). $[\alpha]^{23}_{D}$ -28.5 (c, 7.5 in $CHCl_3$) (tetra-Ac). λ_{max} 234 (ε 26600); 291 (ε 14900) (MeOH) (tetra-Ac).

Lin, W. *et al.*, *Phytochemistry*, 1999, **50**, 653-658 (*Taiwnia cryptomeroides constits*)

3,3′,4,4′,5-Pentahydroxy-8,8′-lignan-9,9′-olide P18

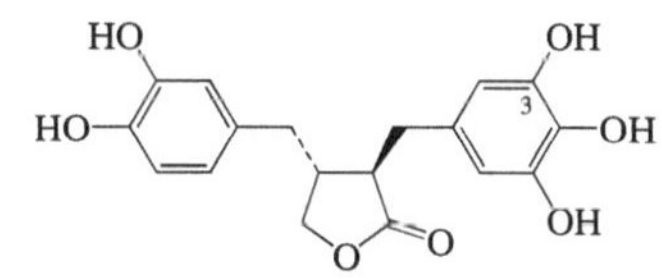

(8R,8′R)-form

4-(3,4-Dihydroxybenzyl)-dihydro-3-(3,4,5-trihydroxybenzyl)-2(3H)-furanone

$C_{18}H_{18}O_7$ 346.335

(8R,8′R)-form

3-Me ether

3,3′,4,4′-Tetrahydroxy-5-methoxy-8,8′-lignan-9,9′-olide

[807329-50-6]

$C_{19}H_{20}O_7$ 360.362

Constit. of the stems of *Macrococculus pomiferus*. Oil. $[\alpha]^{23}_{D}$ -27.5 (c, 0.1 in MeOH). λ_{max} 209 (ε 7244); 281 (ε 2630) (MeOH)

3,3′-Di-Me ether

3,4,4′-Trihydroxy-3′,5-dimethoxy-8,8′-lignan-9,9′-olide. ***Thujaplicatin***

[6512-66-9]

$C_{20}H_{22}O_7$ 374.389

Constit. of *Thuja plicata*. Amorph.

3,5-Di-Me ether

3′,4,4′-Trihydroxy-3,5-dimethoxy-8,8′-lignan-9,9′-olide

[807329-51-7]

$C_{20}H_{22}O_7$ 374.389

Constit. of the stems of *Macrococculus pomiferus*. Oil. $[\alpha]^{23}_{D}$ -6 (c, 0.12 in MeOH). λ_{max} 205 (ε 6606); 281 (ε 2691) (MeOH)

3,3′,5-Tri-Me ether

4,4′-Dihydroxy-3,3′,5-trimethoxy-8,8′-lignan-9,9′-olide. ***Thujaplicatin methyl ether***

[6512-67-0]

$C_{21}H_{24}O_7$ 388.416

From *Thuja plicata* and *Bupleurum salicifolium*. Toxic to brine shrimp. Cryst. (Me_2CO aq.). Mp 167 - 167.5°. $[\alpha]^{25}_{D}$ -48.7 (c, 3.94 in Me_2CO)

3,4,5-Tri-Me ether

3′,4′-Dihydroxy-3,4,5-trimethoxy-8,8′-lignan-9,9′-olide.

5-Methoxyguayarol

[846605-18-3]
$C_{21}H_{24}O_7$ 388.416
Constit. of the seeds of *Hernandia ovigera*. Pale yellow solid. $[\alpha]_D^{20}$ -63.8 (c, 0.5 in $CHCl_3$). λ_{max} 281 (log ε 3.71) (EtOH)

3,3′,4,4′-Tetra-Me ether
3-Hydroxy-3′,4,4′,5-tetramethoxy-8,8′-lignan-9,9′-olide
[129684-03-3]
$C_{22}H_{26}O_7$ 402.443
Constit. of *Virola elongata*. Oil. $[\alpha]_D^{25}$ -24.6 (c, 1.2 in $CHCl_3$)

3,3′,4,5-Tetra-Me ether
4′-Hydroxy-3,3′,4,5-tetramethoxy-8,8′-lignan-9,9′-olide. ***Hernanol***
[651340-40-8]
$C_{22}H_{26}O_7$ 402.443
Constit. of *Hernandia nymphaeifolia*. Semisolid. $[\alpha]_D^{24}$ -36.2 (c, 0.26 in MeOH). λ_{max} 203 (log ε 5); 225 (log ε 4.5); 278 (log ε 3.8) (MeOH)

3,3′,4′,5-Tetra-Me ether
4-Hydroxy-3,3′,4′,5-tetramethoxy-8,8′-lignan-9,9′-olide
[129684-02-2]
$C_{22}H_{26}O_7$ 402.443
Constit. of *Virola elongata*. Resinous solid. $[\alpha]_D^{25}$ -20.7 (c, 1.5 in $CHCl_3$)

Penta-Me ether
3,3′,4,4′,5-Pentamethoxy-8,8′-lignan-9,9′-olide. ***Thujaplicatin trimethyl ether***
[10516-73-1]
$C_{23}H_{28}O_7$ 416.47
Constit. of fruit of *Piper cubeba*. Cryst. (Et_2O). Mp 102 - 103°. $[\alpha]_D^{24}$ -37.4 (c, 0.64 in $CHCl_3$)

3′,4′-Methylene, 3,4-di-Me ether
3-Hydroxy-4,5-dimethoxy-3′,4′-methylenedioxy-8,8′-lignan-9,9′-olide. ***3-O-Demethylyatein***
[139014-46-3]
$C_{21}H_{22}O_7$ 386.4
Isol. from *Juniperus sabina*. $[\alpha]_D^{22}$ -10.3 (c, 0.9 in $CHCl_3$). λ_{max} 210 (ε 26600); 230 (ε 10400); 285 (ε 4200) (EtOH)

3′,4′-Methylene, 3,4-di-Me ether, 5-O-β-D-glucopyranoside
Wenchuanensin
[155961-85-6]
$C_{27}H_{32}O_{12}$ 548.541
Constit. of *Bupleurum wenchuanense*. Powder. Mp 90 - 92°. $[\alpha]_D$ -45 (c, 0.1 in MeOH)

3′,4′-Methylene, 3,5-di-Me ether
4-Hydroxy-3,5-dimethoxy-3′,4′-methylenedioxy-8,8′-lignan-9,9′-olide. ***4-O-Demethylyatein***
[162600-61-5]
$C_{21}H_{22}O_7$ 386.4
Constit. of *Thuja occidentalis*. Oil. $[\alpha]_D^{25}$ -25 (c, 0.32 in CH_2Cl_2). λ_{max} 213; 284 (EtOH)

3′,4′-Methylene, 3,4,5-tri-Me ether
3,4,5-Trimethoxy-3′,4′-methylenedioxy-8,8′-lignan-9,9′-olide. ***Yatein***. *Deoxypodorhizol. 7-Deoxypodorhizone*
[40456-50-6]
$C_{22}H_{24}O_7$ 400.427
Lignan from *Libocedrus yateensis*, fruit of *Piper cubeba* and seeds of *Hernandia ovigera*. Shows antiplatelet aggregation activity. Glass. $[\alpha]_D^{20}$ -28.4 (c, 0.32 in $CHCl_3$)

3,4:3′,4′-Bis(methylene), 5-Me ether
3-Methoxy-3′,4′:4,5-bis(methylenedioxy)-8,8′-lignan-9,9′-olide. ***5-Methoxyhinokinin***. *Dehydrotrichostin*
[112448-63-2]
$C_{21}H_{20}O_7$ 384.384
Isol. from fruits of *Piper cubeba* and from *Piper trichostachyon*. Viscous mass. $[\alpha]_D^{25}$ -37 (c, 0.15 in $CHCl_3$). Abs. config. of Dehydrotrichostin revised in 2006

(8S,8′R)-form

3,3′,5-Tri-Me ether
[1178561-80-2]
$C_{21}H_{24}O_7$ 388.416
Constit. of *Ipomoea cairica*. Pale yellow oil. $[\alpha]_D^{24}$ +10 (c, 0.15 in MeOH). λ_{max} 281 (log ε 3.71) (EtOH)

(8S,8′S)-form

3,3′,4-Tri-Me ether
3,4′-Dihydroxy-3′,4,5-trimethoxy-8,8′-lignan-9,9′-olide
[1574555-98-8]
$C_{21}H_{24}O_7$ 388.416
Constit. of the wood of *Cunninghamia konishii*. Yellow oil. $[\alpha]_D^{26}$ +33.5 (c, 0.31 in $CHCl_3$)

3′,4′-Methylene, 3,5-di-Me ether
[635752-19-1]
$C_{21}H_{22}O_7$ 386.4
Constit. of *Peperomia duclouxii*. Pale yellow gum. $[\alpha]_D^{25}$ +34.8 (c, 0.25 in $CHCl_3$). λ_{max} 241; 285 ($CHCl_3$)

3,4:3′,4′-Bis(methylene), 5-Me ether
[115794-38-2]
$C_{21}H_{20}O_7$ 384.384
Constit. of *Peperomia duclouxii*. Gum. $[\alpha]_D^{25}$ +24.5 (c, 0.65 in $CHCl_3$). λ_{max} 242; 286 ($CHCl_3$).

Maclean, H. *et al.*, *Can. J. Chem.*, 1966, **44**, 1541-1545 (*Thujaplicatin, Thujaplicatin methyl ether*).
Harmatha, J. *et al.*, *Coll. Czech. Chem. Comm.*, 1982, **47**, 644-633 (*Yatein*).
Badheka, L.P. *et al.*, *Phytochemistry*, 1986, **25**, 487-489 (*Piper cubeba constits*).
Tanoguchi, M. *et al.*, *Chem. Pharm. Bull.*, 1987, **35**, 4162-4168 (*Hernandia ovigera constits*).
Badheka, L.P. *et al.*, *Phytochemistry*, 1987, **26**, 2033-2036 (*5-Methoxyhinokinin*).
Kato, M.J. *et al.*, *Phytochemistry*, 1990, **29**, 1799-1810 (*Virola elongata constits*).
San Feliciano, A. *et al.*, *Phytochemistry*, 1991, **30**, 3483-3485 (*3-O-Demethylyatein*).
Luo, S.-Q. *et al.*, *Phytochemistry*, 1993, **33**, 193-196 (*Wenchuanensin*).
Kawai, S. *et al.*, *Phytochemistry*, 1994, **37**, 1699-1702 (*4-O-Demethylyatein*).
Chen, J.-J. *et al.*, *Planta Med.*, 2000, **66**, 251-256 (*Yatein, activity*).
Li, N. *et al.*, *J. Nat. Prod.*, 2003, **66**, 1421-1426 (*Peperomia duclouxii lignans, Dehydrotrichostin*).
Sakakibara, N. *et al.*, *Org. Biomol. Chem.*, 2003, 2474-2485 (*Yatein, biosynth*).
Pettit, G.R. *et al.*, *J. Nat. Prod.*, 2004, **67**, 214-220 (*Hernanol*).
Su, B.-N. *et al.*, *Phytochemistry*, 2004, **65**, 2861-2865 (*Macrococculus pomiferus lignans*).
Jang, D.S. *et al.*, *Planta Med.*, 2004, **70**, 893-896 (*5-Methoxyguayarol*).
Li, N. *et al.*, *J. Nat. Prod.*, 2006, **69**, 234-239 (*Peperomia duclouxii lignans*).
Lin, R.-J. *et al.*, *Nat. Prod. Res.*, 2008, **22**, 747-753 (*Ipomoea cairica tri-Me ether*).
Velázquez-Jiménez, R. *et al.*, *Phytochemistry*, 2011, **72**, 2237-2243 (*Yatein, abs config*).
Chang, C.-I. *et al.*, *Nat. Prod. Commun.*, 2013, **8**, 805-806 (*Cunninghamia konishii tri-Me ether*).
Hartmann, P. *et al.*, *J. Org. Chem.*, 2019, **84**, 5831-5837 (*synth*)

3,3′,4,4′,5′-Pentahydroxy-8,8′-lignan-9,9′-olide P19

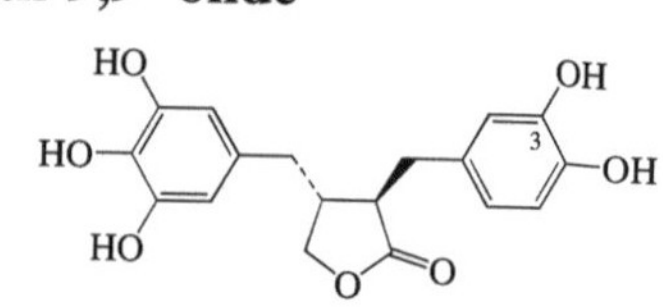

(8*R*,8′*R*-form)

3-(3,4-Dihydroxybenzyl)dihydro-4-(3,4,5-trihydroxybenzyl)-2(3H)-furanone
$C_{18}H_{18}O_7$ 346.335
Isomeric with 3,3′,4,4′,5-Pentahydroxy-8,8′-lignan-9,9′-olide, P18

(8R,8′R)-form

3-Me ether, 4-O-(3,4-dihydroxy-E-cinnamoyl)
[935509-40-3]

$C_{28}H_{26}O_{10}$ 522.506
Constit. of *Forsythia suspensa*

3,3′-Di-Me ether, 4-O-(3,4-dihydroxy-E-cinnamoyl)
[935509-43-6]
$C_{29}H_{28}O_{10}$ 536.533
Constit. of *Forsythia suspensa*

3,3′,5′-Tri-Me ether
4,4′-Dihydroxy-3,3′,5′-trimethoxy-8,8′-lignan-9,9′-olide. ***4′-De-O-methyltraxillagenin***
[362662-48-4]
[29643-16-1] (unspecified stereochem.)
$C_{21}H_{24}O_7$ 388.416
Constit. of the bark of *Torreya nucifera*. Neuroprotective agent. Powder. $[\alpha]_D^{18}$ -38.2 (c, 1 in $CHCl_3$). λ_{max} 231; 287 (MeOH)

3,3′,4′,5′-Tetra-Me ether
4-Hydroxy-3,3′,4′,5′-tetramethoxy-8,8′-lignan-9,9′-olide. ***Traxillagenin***
[79288-73-6]
$C_{22}H_{26}O_7$ 402.443
Constit. of *Trachelospermum axillare*, *Torreya nucifera*, *Trachelospermum asiatica* and *Saussurea conica*. Syrup. $[\alpha]_D$ -33 (c, 0.5 in EtOH) (synthetic). $[\alpha]_D$ -25.1 (c, 0.55 in EtOH) (natural). Struct. prob. in need of revision (2004). λ_{max} 281 (ε 4677) (EtOH)

3,3′,4′,5′-Tetra-Me ether, 4-O-β-D-glucopyranoside
Traxillaside
[149415-62-3]
$C_{28}H_{36}O_{12}$ 564.584
Constit. of the bark of *Trachelospermum axillare* and from *Torreya nucifera*. Neuroprotectant. Amorph. powder. Mp 88 - 92°. $[\alpha]_D^{20}$ -71.3 (c, 0.13 in MeOH). λ_{max} 225 (ε 16982); 277 (ε 2630) (EtOH)

Penta-Me ether
3,3′,4,4′,5′-Pentamethoxy-8,8′-lignan-9,9′-olide
[64767-71-1]
$C_{23}H_{28}O_7$ 416.47
Minor component of the heartwood of *Cinnamomum camphora*. Mp 120 - 121.5°. $[\alpha]_D$ -34 (c, 0.5 in $CHCl_3$) (-17.8). λ_{max} 226 (ε 9772); 279 (ε 2042) (EtOH)

3,4-Methylene, 3′,4′,5′-tri-Me ether
3′,4′,5′-Trimethoxy-3,4-methylenedioxy-8,8′-lignan-9,9′-olide. ***Isoyatein***
[101751-72-8]
$C_{22}H_{24}O_7$ 400.427
Constit. of *Piper cubeba*. Solid (EtOH) or viscous mass. Mp 113 - 116°. $[\alpha]_D^{25}$ -49.6 (c, 0.035 in $CHCl_3$) (-33). λ_{max} 286 (ε 2089) (MeOH)

(8S,8′S)-form

3′,4′-Methylene, 3,5′-di-Me ether
4-Hydroxy-3,3′-dimethoxy-4′,5′-methylenedioxy-8,8′-lignan-9,9′-olide
[635752-17-9]
$C_{21}H_{22}O_7$ 386.4
Constit. of *Peperomia duclouxii*. Gum. $[\alpha]_D^{25}$ +26.4 (c, 0.29 in $CHCl_3$). λ_{max} 241; 281 ($CHCl_3$)

3′,4′-Methylene, 3,4,5′-tri-Me ether
3,3′,4-Trimethoxy-4′,5′-methylenedioxy-8,8′-lignan-9,9′-olide. ***5′-Methoxybursehernin***
[1204520-90-0]
$C_{22}H_{24}O_7$ 400.427
Constit. of the leaves of *Phyllanthus amarus*. $[\alpha]_D^{28}$ +13.3 (c, 0.1 in $CHCl_3$). λ_{max} 232 (log ε 4.08); 279 (log ε 3.79) (MeOH)

3,4:3′,4′-Bis(methylene), 5′-Me ether
3′-Methoxy-3,4:4′,5′-bis(methylenedioxy)-8,8′-lignan-9,9′-olide
[878142-49-5]
$C_{21}H_{20}O_7$ 384.384
Constit. of *Peperomia duclouxii*. Gum. $[\alpha]_D^{25}$ +29.9 (c, 0.73 in $CHCl_3$). λ_{max} 242; 286 ($CHCl_3$).

Takaoka, D. *et al.*, *Bull. Chem. Soc. Jpn.*, 1977, **50**, 2821-2822 (*penta-Me ether, abs config*).
Duffley, R.P. *et al.*, *Org. Prep. Proced. Int.*, 1979, **11**, 281 (*penta-Me ether, synth*).
Nishibe, S. *et al.*, *Chem. Pharm. Bull.*, 1981, **29**, 2078-2082 (*Traxillagenin, penta-Me ether, uv*).
Badheka, L.P. *et al.*, *Phytochemistry*, 1986, **25**, 487-489 (*Isoyatein*).
Lalami, K. *et al.*, *J. Nat. Prod.*, 1991, **54**, 119-126 (*synth*).
Nishibe, S. *et al.*, *Phytochemistry*, 1993, **32**, 1579-1581 (*Traxillaside*).
Jang, Y.P. *et al.*, *Planta Med.*, 2001, **67**, 470-472 (*4′-Demethyltraxillagenin*).
Kim, S.H. *et al.*, *Biol. Pharm. Bull.*, 2003, **26**, 1202-1205 (*Torreya nucifera constits, activity*).
Li, N. *et al.*, *J. Nat. Prod.*, 2003, **66**, 1421-1426 (*Peperomia duclouxii lignans*).
Pettit, G.R. *et al.*, *J. Nat. Prod.*, 2004, **67**, 214-220 (*Traxillagenin, struct*).
Li, N. *et al.*, *J. Nat. Prod.*, 2006, **69**, 234-239 (*Peperomina duclouxii lignans*).
Fan, C.-Q. *et al.*, *Planta Med.*, 2006, **72**, 590-595 (*Saussurea conica constit*).
Guo, H. *et al.*, *Rapid Commun. Mass Spectrom.*, 2007, **21**, 715-729 (*Forsythia suspensa constits*).
Singh, M. *et al.*, *J. Asian Nat. Prod. Res.*, 2009, **11**, 562-568 (*5′-Methoxybursehernin*)

3,3′,4,4′,7′-Pentahydroxy-8,8′-lignan-9,9′-olide P20

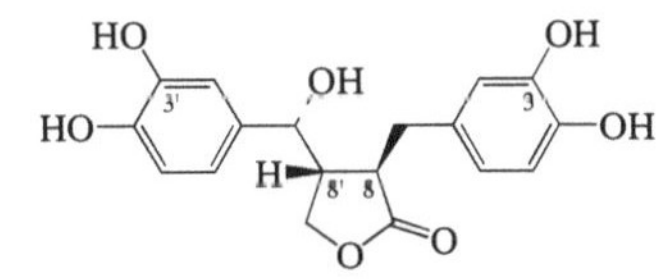

(7′R,8R,8′R)-form

$C_{18}H_{18}O_7$ 346.335

(7′R,8R,8′R)-form

3,3′-Di-Me ether
4,4′,7′-Trihydroxy-3,3′-dimethoxy-8,8′-lignan-9,9′-olide. Allo-7′-hydroxymatairesinol. ***Hydroxymatairesinol I***
[81623-30-5]
$C_{20}H_{22}O_7$ 374.389
Isol. from *Ptelea excelsa*. Cryst. (butanol). Mp 147 - 148°. $[\alpha]_D^{25}$ +4.9 (EtOH)

3,3′-Di-Me ether, 4-O-β-D-glucopyranoside
[1627676-88-3]
$C_{26}H_{32}O_{12}$ 536.53
Constit. of fruits of *Arctium lappa*. Amorph. powder. $[\alpha]_D^{20}$ -26.7 (c, 0.07 in MeOH)

3,3′-Di-Me ether, 4′-O-β-D-glucopyranoside
[1227736-03-9]
$C_{26}H_{32}O_{12}$ 536.53
Isol. from *Thellungiella salsuginea* elicited with uv radiation. Amorph. powder. $[\alpha]_D$ -108 (c, 0.04 in MeOH). λ_{max} 278 (log ε 4.14) (MeOH)

3,3′,4′-Tri-Me ether, 4-O-β-D-glucopyranoside
Rafanotrachelogenin 4-glucoside
[1069136-60-2]
$C_{27}H_{34}O_{12}$ 550.557
Constit. of the leaves of *Trachelospermum lucidum* and fruits of *Arctium lappa*. Yellow oil or amorph. powder. $[\alpha]_D^{25}$ -68.5 (c, 0.001 in MeOH). $[\alpha]_D^{25}$ -13.6 (c, 0.12 in MeOH). λ_{max} 208; 220 (MeOH). [neutral] λ_{max} 230 (log ε 3.41); 277 (log ε 3.08) (MeOH)

3,3′,4,4′-Tetra-Me ether
7′-Hydroxy-3,3′,4,4′-tetramethoxy-8,8′-lignan-9,9′-olide. ***Tupichilignan A***
[69586-96-5]
$C_{22}H_{26}O_7$ 402.443

Constit. of the rhizomes of *Tupistra chinensis*. Oil (natural); cryst. (semisynthetic). Mp 96 - 97°. $[\alpha]_D^{25}$ +59.8 (c, 2.0 in THF) (semisynthetic). $[\alpha]_D^{24}$ -2.8 (c, 1.45 in Me_2CO) (natural). The props. of Tupichilignan A differ considerably from those reported earlier for the semisynthetic di-Me ether of Hydroxymatairesinol. λ_{max} 233 (log ε 4.24); 280 (log ε 3.8) (EtOH) (natural)

7′-Ketone, 3,3′-di-Me ether, 4′-O-β-D-glucopyranoside

Wasabiside E

[2022212-56-0]

$C_{26}H_{30}O_{12}$ 534.514

Constit. of the roots of *Wasabia japonica*. Gum. $[\alpha]_D^{25}$+70 (c, 0.003 in MeOH). λ_{max} 230 (log ε 2.73); 274 (log ε 1.16) (MeOH)

7′-Ketone, bis(methylene) ether

3,4:3′,4′-Bis(methylenedioxy)-7′-oxo-8,8′-lignan-9,9′-olide.

7′-Oxohinokinin

[130837-92-2]

$C_{20}H_{16}O_7$ 368.341

Isol. from wood of *Calocedrus formosana*. Needles ($CHCl_3$/MeOH). Mp 109.5 - 111°. $[\alpha]_D^{25}$ +10 (c, 0.82 in $CHCl_3$). Has 8′*R*-config. (change of *RS*-priorities)

(7′R,8S,8′S)-form

3,3′-Di-Me ether, 4-O-β-D-glucopyranoside

Wasabiside D

[2022212-55-9]

$C_{26}H_{32}O_{12}$ 536.53

Constit. of the roots of *Wasabia japonica*. Gum. $[\alpha]_D^{25}$+57.7 (c, 0.5 in MeOH). λ_{max} 232 (log ε 2.20); 280 (log ε 1.12) (MeOH)

3,3′-Di-Me ether, 4′-O-β-D-glucopyranoside

Wasabiside B

[2022212-53-7]

$C_{26}H_{32}O_{12}$ 536.53

Constit. of the roots of *Wasabia japonica*. Gum. $[\alpha]_D^{25}$+30.4 (c, 0.01 in MeOH). λ_{max} 231 (log ε 2.14); 284 (log ε 1.17) (MeOH)

3,3′-Di-Me ether, 7′-O-β-D-glucopyranoside

Patrinian A. Wasabiside A

[2022212-52-6]

$C_{26}H_{32}O_{12}$ 536.53

Constit. of *Patrinia villosa* and *Wasabia japonica*. Exhibits neuroprotective activity. Antioxidant. Yellow oil. $[\alpha]_D^{25}$+9.6 (c, 0.5 in MeOH) (Wasabiside A). $[\alpha]_D^{20}$ -46.7 (c, 1.5 in MeOH) (Patrinian A). λ_{max} 232 (log ε 2.17); 280 (log ε 1.11) (MeOH)

(7′S,8R,8′R)-form

3,3′-Di-Me ether

7′-Hydroxymatairesinol

[20268-71-7]

$C_{20}H_{22}O_7$ 374.389

Obt. from *Podocarpus spicatus* infected with *Fomes annosus*. Antifungal agent. Amorph.. Sol. EtOH, $CHCl_3$; fairly sol. C_6H_6, Et_2O, toluene; poorly sol. hexane, H_2O. $[\alpha]_D^{25}$-6.3 (EtOH). $[\alpha]_D^{25}$-11 (c, 4 in THF)

3,3′-Di-Me ether, 4-O-β-D-glucopyranoside

[1627676-89-4]

Constit. of fruits of *Arctium lappa*. Amorph. powder. $[\alpha]_D^{20}$-33.9 (c, 0.07 in MeOH)

3,3′,4′-Tri-Me ether

4,7′-Dihydroxy-3,3′,4′-trimethoxy-8,8′-lignan-9,9′-olide.

7′-Hydroxyarctigenin

[74861-40-8]

$C_{21}H_{24}O_7$ 388.416

Constit. of *Centaurea calcitrapa*. Oil. $[\alpha]_D$ +11 (c, 3 in EtOH). C-7′ config. revised in 2004

3,3′,4′-Tri-Me ether, 4-O-β-D-glucopyranoside

[1807808-87-2]

$C_{27}H_{34}O_{12}$ 550.557

Constit. of the fruits of *Arctium lappa*. Exhibits strong antiinflammatory effect. Amorph. powder. $[\alpha]_D^{25}$-19.2 (c, 0.23 in MeOH). λ_{max} 228 (log ε 3.04); 278 (log ε 2.68) (MeOH)

7′-Et, 3,3′-di-Me ether

7′-Ethoxy-4,4′-dihydroxy-3,3′-dimethoxy-8,8′-lignan-9,9′-olide.

Daphnetone. *7-Ethoxymatairesinol*

[1395922-32-3]

$C_{22}H_{26}O_7$ 402.443

Constit. of the stems of *Daphne acutiloba*. Amorph. powder. $[\alpha]_D^{25}$ +14.5 (c, 1.53 in MeOH). λ_{max} 207 (log ε 4.39); 229 (log ε 4.07); 282 (log ε 3.75) (MeOH)

3,4:3′,4′-Bis(methylene) ether

7′-Hydroxy-3,4:3′,4′-bis(methylenedioxy)-8,8′-lignan-9,9′-olide. ***Parabenzlactone***

[27675-77-0]

Constit. of *Parabenzoin trilobum* and *Piper sanguineispicum*. Powder. Mp 159 - 161°. $[\alpha]_D^{20}$-16.5 (c, 2.3 in CH_2Cl_2). Stereochem. revised in 2006. λ_{max} 290 (CH_2Cl_2)

3,4:3′,4′-Bis(methylene) ether, Ac

Mp 142 - 145°

3,4:3′,4′-Bis(methylene) ether, 7′-O-(3-methylbutanoyl)

Haplotonin

[268742-13-8]

$C_{25}H_{26}O_8$ 454.475

Constit. of *Haplophyllum suaveolens*. Amorph. powder. λ_{max} 207; 234; 288 (MeOH)

(8R,8′S)-form

7′-Ketone, 3,3′-di-Me ether

4,4′-Dihydroxy-3,3′-dimethoxy-7′-oxo-8,8′-lignan-9,9′-olide.

Oxomatairesinol. *Ketomatairesinol*

[53250-61-6]

$C_{20}H_{20}O_7$ 372.373

Constit. of *Ptelea excelsa* and *Taxus cuspidata*. Cryst. (H_2O). Mp 74 - 76° (70-72°). $[\alpha]_D^{25}$+42 (c, 4 in THF)

7′-Ketone, 3,3′,4,4′-tetra-Me ether

3,3′,4,4′-Tetramethoxy-7′-oxo-8,8′-lignan-9,9′-olide.

4,4′-Di-O-methyloxomatairesinol

[1391860-62-0]

$C_{22}H_{24}O_7$ 400.427

Isol. from *Pseudolarix kaempferi*. Amorph. powder. $[\alpha]_D^{14}$+45.6 (c, 0.15 in $CHCl_3$). λ_{max} 229 (log ε 4.02); 279 (log ε 3.74); 307 (log ε 3.55) (MeOH).

Freudenberg, K. *et al.*, *Chem. Ber.*, 1957, **90**, 2857-2869 (*7′-Hydroxymatairesinol*).

Freudenberg, K. *et al.*, *Tetrahedron*, 1961, **15**, 115-128 (*nomencl*).

Rudman, P. *Holzforschung*, 1965, **19**, 57-58 (*activity*).

Duffield, A.M. *et al.*, *J. Heterocycl. Chem.*, 1967, **4**, 16-22 (*ms*).

Wada, K. *et al.*, *Tetrahedron Lett.*, 1970, **11**, 2017-2019 (*Parabenzlactone*).

Fang, J.M. *et al.*, *Phytochemistry*, 1990, **29**, 3048-3049 (*7′-Oxohinokinin*).

Marco, J.A. *et al.*, *Phytochemistry*, 1992, **31**, 3527-3530 (*7′-Hydroxyarctigenin*).

Miyata, O. *et al.*, *J. Org. Chem.*, 2000, **65**, 6922-6931 (*7′-Oxohinokinin, synth*).

Kostova, I. *et al.*, *Monatsh. Chem.*, 2000, **131**, 191-194 (*Haplotonin*).

Kawamura, F. *et al.*, *J. Wood Sci.*, 2004, **50**, 548-551 (*Oxomatairesinol*).

Fischer, J. *et al.*, *Org. Lett.*, 2004, **6**, 1345-1348 (*synth*).
Pan, W.B. *et al.*, *Chem. Pharm. Bull.*, 2006, **54**, 954-958 (*Tupichilignan A*).
Kousar, F. *et al.*, *J. Asian Nat. Prod. Res.*, 2008, **10**, 285-288 (*3,3′,4′-tri-Me ether 4-glucoside*).
Raffaelli, B. *et al.*, *Org. Biomol. Chem.*, 2008, **6**, 2619-2627 (*Parabenzlactone, synth, config*).
Cabanillas, B.J. *et al.*, *J. Nat. Prod.*, 2010, **73**, 1884-1890 (*Piper sanguineispicum constit*).
Pedras, M.S.C. *et al.*, *Phytochemistry*, 2010, **71**, 581-589 (*3,3′-di-Me ether 4′-glucoside*).
Hodgson, D.M. *et al.*, *Org. Lett.*, 2011, **13**, 2594-2597 (*7′-Hydroxymatairesinol, synth*).
He, W.-J. *et al.*, *Planta Med.*, 2011, **77**, 1924-1931 (*4,4′-Di-O-methyloxomatairesinol*).
Huang, S.Z. *et al.*, *Planta Med.*, 2012, **78**, 182-185 (*Daphnetone*).
Hajra, S. *et al.*, *Tetrahedron Lett.*, 2013, **54**, 2171-2173 (*7′-Hydroxymatairesinol, 7′-Hydroxyarctigenin, synth*).
Yang, Y.-N. *et al.*, *J. Agric. Food Chem.*, 2014, **62**, 9095-9102 (*Arctium lappa constits, cd*).
Yang, Y.-N. *et al.*, *J. Agric. Food Chem.*, 2015, **63**, 7958-7966 (*Arctium lappa glucosides, activity, cd*).
Kim, C.S. *et al.*, *J. Nat. Prod.*, 2016, **79**, 2652-2657 (*Wasabisides A,B,D,E, cd, activity*).
Bai, M. *et al.*, *J. Funct. Foods*, 2017, **28**, 106-113 (*Patrinian A, activity*)

3,3′,4,4′,7-Pentahydroxy-8,8′-lignan-9,9′-olide P21

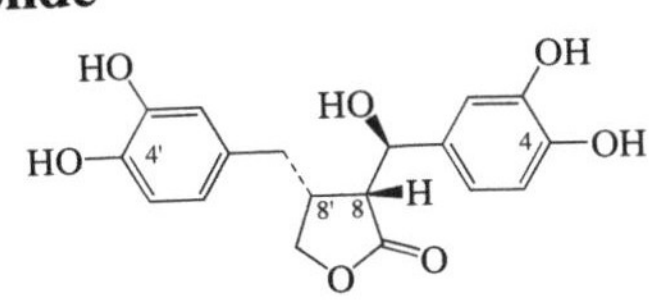

(7R,8R,8′R)-form

$C_{18}H_{18}O_7$ 346.335

(7R,8R,8′R)-form

3,3′,7-Tri-Me ether
4,4′-Dihydroxy-3,3′,7-trimethoxy-8,8′-lignan-9,9′-olide. ***7R-Methoxy-8-epimatairesinol***
[198827-23-5]
[198827-25-7]
$C_{21}H_{24}O_7$ 388.416
Constit. of *Himatanthus fallax*. Cryst. (C_6H_6). Mp 160 - 162°. $[\alpha]^{26}_{D}$+29.6 (c, 0.7 in MeOH). λ_{max} 214 (ε 11600); 225 (ε 4900); 280 (ε 4400) (MeOH)

7-Ketone, bis(methylene) ether
3,4:3′,4′-Bis(methylenedioxy)-7-oxo-8,8′-lignan-9,9′-olide. ***7-Oxo-(+)-hinokinin***
[165172-74-7]
$C_{20}H_{16}O_7$ 368.341
Constit. of the wood of *Protium tenuifolium*. Needles ($CHCl_3$). Mp 105 - 106°. $[\alpha]^{25}_{D}$+36.7 (c, 0.5 in $CHCl_3$)

(7R,8R,8′S)-form

3,3′-Di-Me ether, 4-O-α-L-rhamnopyranoside
Majusalignan B
$C_{26}H_{32}O_{11}$ 520.531
Constit. of *Illicium majus*. Exhibits inhibitory activity against the release of β-glucuronidase in rat polymorphonuclear leukocytes induced by platelet activating factor. Amorph. powder. $[\alpha]^{20}_{D}$-31.5 (c, 0.2 in MeOH). λ_{max} 229 (log ε 2.76); 278 (log ε 2.60) (MeOH)

(7R,8S,8′R)-form

3,3′-Di-Me ether, 4-O-β-D-glucopyranoside
[1807808-91-8]
$C_{26}H_{32}O_{12}$ 536.53
Constit. of fruits of *Arctium lappa*. Amorph. powder. $[\alpha]^{25}_{D}$-37.5 (c, 0.35 in MeOH). λ_{max} 230 (log ε 3.27); 280 (log ε 2.90) (MeOH)

3,3′,4′-Tri-Me ether, 4-O-β-D-glucopyranoside
[1807808-90-7]
$C_{27}H_{34}O_{12}$ 550.557
Constit. of fruits of *Arctium lappa*. Amorph. powder. $[\alpha]^{25}_{D}$-30.4 (c, 0.13 in MeOH). λ_{max} 229 (log ε 3.51); 279 (log ε 3.07) (MeOH)

(7S,8S,8′R)-form

3,3′-Di-Me ether
4,4′,7-Trihydroxy-3,3′-dimethoxy-8,8′-lignan-9,9′-olide. ***7-Hydroxymatairesinol***
[252333-69-0]
$C_{20}H_{22}O_7$ 374.389
Constit. of *Taxus mairei*. Oil. $[\alpha]_D$ +11 (c, 1.6 in $CHCl_3$)

3,3′,4′-Tri-Me ether
4,7-Dihydroxy-3,3′,4′-trimethoxy-8,8′-lignan-9,9′-olide
[1807808-89-4]
$C_{21}H_{24}O_7$ 388.416
Constit. of fruits of *Arctium lappa*. Exhibits strong antiinflammatory effect. Amorph. powder. $[\alpha]^{25}_{D}$-44.3 (c, 0.11 in MeOH). λ_{max} 231 (log ε 2.98); 279 (log ε 2.97) (MeOH)

3,3′,4′-Tri-Me ether, 4-O-β-D-glucopyranoside
[1807808-88-3]
Constit. of fruits of *Arctium lappa*. Amorph. powder. $[\alpha]^{25}_{D}$-48.5 (c, 0.21 in MeOH). λ_{max} 230 (log ε 3.48); 278 (log ε 3.04) (MeOH)

Bis(methylene) ether
7-Hydroxy-3,4:3′,4′-bis(methylenedioxy)-8,8′-lignan-9,9′-olide. ***Sventenin.*** *7-Hydroxyhinokinin*
[28168-95-8]
$C_{20}H_{18}O_7$ 370.357
Isol. from heartwood of *Taiwania cryptomeroides*. Also from *Juniperus chinensis*. Mp 129.5 - 130.5° (124.5-126°). $[\alpha]_D$ -63 (c, 3.46 in $CHCl_3$) (-43)

Bis(methylene) ether, 7-Ac
7-Acetyloxyhinokinin. *Sventenin acetate*
[34399-45-6]
$C_{22}H_{20}O_8$ 412.394
Constit. of *Juniperus chinensis* and *Ruta pinnata*. Mp 122 - 123.5°. $[\alpha]^{25}_{D}$-29 (c, 0.1 in $CHCl_3$)

(7R*,8R*,8′S*)-form

3,4-Methylene, 3′,4′-di-Me ether
7-Hydroxy-3′,4′-dimethoxy-3,4-methylenedioxy-8,8′-lignan-9,9′-olide

3,4-Methylene 3′,4′-di-Me ether,, 7-O-β-D-glucopyranoside
Plasiaticine F
$C_{27}H_{32}O_{12}$ 548.541
Constit. of the seeds of *Plantago asiatica*. Oil. $[\alpha]^{22}_{D}$-7.7 (c, 0.23 in MeOH). λ_{max} 230 (log ε 3.29); 283 (log ε 3.03) (MeOH)

(7R*,8S*,8′R*)-form

Bis(methylene) ether, 7-O-β-D-glucopyranoside
[1186119-09-4]
$C_{26}H_{28}O_{12}$ 532.498
Constit. of the leaves of *Koelreuteria henryi*. Amorph. powder. $[\alpha]^{21}_{D}$ +16 (c, 0.88 in MeOH). Incorrect synonym, Hinokinin 7-glucoside assigned in ref.. λ_{max} 236 (log ε 3.52); 286 (log ε 3.5) (MeOH)

(8R*,8′R*)-form

7-Ketone, 3,3′,4′-tri-Me ether
4-Hydroxy-3,3′,4′-trimethoxy-7-oxo-8,8′-lignan-9,9′-olide. ***Conicaol B.*** *7-Oxoarctigenin*
[916888-04-5]
$C_{21}H_{22}O_7$ 386.4
Constit. of *Aquilaria sinensis* and *Saussurea conica*. Gum. $[\alpha]^{20}_{D}$ +22.9 (c, 0.38 in MeOH). λ_{max} 231 (log ε 4.25); 281 (log ε 4.01); 311 (log ε 3.89) (MeOH).

Lee, C.L. *et al.*, *Mokuzai Gakkaishi*, 1974, **20**, 558-563 (*Sventenin, Sventenin acetate*).

Fang, J-M. *et al.*, *Phytochemistry*, 1992, **31**, 3659-3661 (*Sventenin acetate*).
Siqueira, J.B.G. *et al.*, *J. Nat. Prod.*, 1995, **58**, 730-732 (*7-Oxo-(+)-hinokinin*).
Abdel-Kader, M.S. *et al.*, *J. Nat. Prod.*, 1997, **60**, 1294-1297 (*7R-Methoxy-8-epimatairesinol*).
Yang, S.-J. *et al.*, *J. Chin. Chem. Soc. (Taipei)*, 1999, **46**, 811-818 (*7-Hydroxymatairesinol*).
Yamauchi, S. *et al.*, *J. Chem. Soc., Perkin Trans. 1*, 2001, 2158-2160 (*bismethylene ether, synth*).
Fan, C.-Q. *et al.*, *Planta Med.*, 2006, **72**, 590-595 (*Conicaol B*).
Lee, T.-H. *et al.*, *J. Nat. Med.*, 2009, **63**, 209-214 (*bismethylene 7-glucoside*).
Yang, Y.-N. *et al.*, *J. Agric. Food Chem.*, 2015, **63**, 7958-7966 (*Arctium lappa glucosides, activity, cd*).
Gao, Z. *et al.*, *Nat. Prod. Commun.*, 2016, **11**, 667-670 (*Plasiaticine F*).
Fang, Z.-F. *et al.*, *Fitoterapia*, 2018, **129**, 42-46 (*Majusalignan B, activity*)

3,3′,4,4′,8-Pentahydroxy-8,8′-lignan-9,9′-olide P22

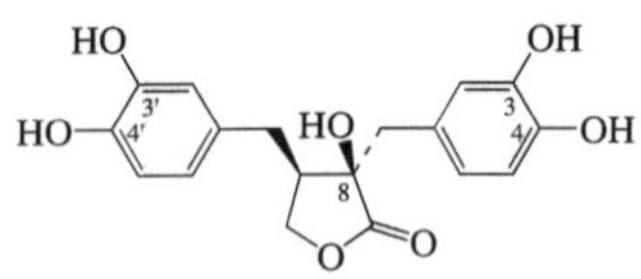

(8*R*,8′*R*)-*form*

$C_{18}H_{18}O_7$ 346.335
Lignan numbering shown
(8R,8′R)-form
3,3′-Di-Me ether
4,4′,8-Trihydroxy-3,3′-dimethoxy-8,8′-lignan-9,9′-olide. ***(+)-Nortrachelogenin****. Wikstromol. Pinopalustrin*
[61521-74-2]
$C_{20}H_{22}O_7$ 374.389
Isol. from *Wikstroemia* spp., *Passerina vulgaris* and *Pinus palustris*. Shows antifungal, moderate anti-HIV, moderate antineoplastic and inhibiting activity *in vivo* against Ehrlich ascites carcinoma and P-388 lymphocytic leukaemia. Yellow powder. $[\alpha]_D$ +15.4 ($CHCl_3$). Inactive as antileukaemic agent. λ_{max} 230; 282 (no solvent reported)
3,3′-Di-Me ether, 8-O-β-D-glucopyranoside
[1357560-23-6]
$C_{26}H_{32}O_{12}$ 536.53
Constit. of the roots of *Pulsatilla koreana*. Yellow gum. $[\alpha]_D^{25}$ -43.4 (c, 1.4 in MeOH). λ_{max} 278 (log ε 3.78) (EtOH)
3,3′,4′-Tri-Me ether, 4-O-β-D-glucopyranoside
Saussurenoside
[177930-79-9]
$C_{27}H_{34}O_{12}$ 550.557
Constit. of *Saussurea japonica*. Amorph. solid (as tetra-Ac). $[\alpha]_D$ -35.2 (c, 0.9 in $CHCl_3$) (tetra-Ac). λ_{max} 226 (log ε 4.37); 279 (log ε 3.93) (MeOH)
(8R,8′S)-form
3,3′-Di-Me ether
Epinortrachelogenin
[125072-69-7]
$C_{20}H_{22}O_7$ 374.389
Constit. of *Bupleurum salicifolium* and *Taxus* x *media* cv. Hicksii. Powder. Mp 171 - 172° (166-168°). $[\alpha]_D^{25}$ -21.1 (c, 0.4 in MeOH). $[\alpha]_D^{20}$ -5 (c, 0.5 in $CHCl_3$)
3,3′,4,4′-Tetra-Me ether, 8-Ac
[235743-17-6]
$C_{24}H_{28}O_8$ 444.48
Constit. of *Bupleurum acutifolium*. Amorph. powder. $[\alpha]_D^{25}$ -19.3 (c, 1 in $CHCl_3$)
3′,4′-Methylene, 3,4-di-Me ether
8-Hydroxy-3,4-dimethoxy-3′,4′-methylenedioxy-8,8′-lignan-9,9′-olide. ***Guayadequiol***
[129502-59-6]
$C_{21}H_{22}O_7$ 386.4
Constit. of *Bupleurum salicifolium*. λ_{max} 232; 283 (EtOH)
3′,4′-Methylene, 3,4-di-Me ether, 8-Ac
[129502-60-9]
$C_{23}H_{24}O_8$ 428.437
Constit. of *Bupleurum salicifolium*. Cryst.. Mp 128 - 129°. λ_{max} 234; 286 (EtOH)
(8S,8′R)-form
3,3′-Di-Me ether, 8-O-β-D-glucopyranoside
$C_{26}H_{32}O_{12}$ 536.53
Constit. of *Pulsatilla koreana*. Inhibitor of *Streptococcus mutans* sortase A. Amorph. solid. $[\alpha]_D^{25}$ -17.5 (c, 0.45 in MeOH). λ_{max} 230 (log ε 2.79); 282 (log ε 2.58) (MeOH)
(8S,8′S)-form
3,3′-Di-Me ether
(-)-Nortrachelogenin
[34444-37-6]
$C_{20}H_{22}O_7$ 374.389
Obt. from *Pinus palustris*, *Pinus sylvestris*, *Trachelospermum asiaticum*, *Carissa edulis*, *Cnicus benedictus* and *Ipomoea cairica*. *In vitro* anti-HIV and antileukaemic agent. Amorph.. $[\alpha]_D^{17}$ -16.8 (c, 0.178 in EtOH)LU3560500
3,3′-Di-Me ether, 4-O-β-D-glucopyranoside
Nortracheloside
[33464-78-7]
$C_{26}H_{32}O_{12}$ 536.53
Isol. from *Trachelospermum* spp. and *Cnicus benedictus*. Cryst. + H_2O. $[\alpha]_D^{19}$ -47.9 (c, 1.02 in EtOH)
3,3′-Di-Me ether, 8-O-β-D-glucopyranoside
[858127-38-5]
$C_{26}H_{32}O_{12}$ 536.53
Constit. of *Trachelospermum jasminoides*. Amorph. solid. $[\alpha]_D^{25}$ -58 (c, 0.57 in MeOH)
3,3′-Di-Me ether, 4,4′-di-O-β-D-glucopyranoside
[38976-09-9]
$C_{32}H_{42}O_{17}$ 698.671
Constit. of *Trachelospermum asiaticum*. Powder. Mp 137 - 140°. $[\alpha]_D^{14}$ -15.9 (c, 0.8 in EtOH)
3,3′-Di-Me ether, 9ξ-alcohol (lactol)
9′,9′-Epoxy-4,4′,8,9-tetrahydroxy-3,3′-dimethoxy-8,8′-lignan. 4,4′,8,9-Tetrahydroxy-3,3′-dimethoxy-9,9′-epoxylignan. Tetrahydro-3,4-bis[(4-hydroxy-3-methoxyphenyl)methyl]-2,3-furandiol, CAS. ***Carissanol***
[87402-76-4]
$C_{20}H_{24}O_7$ 376.405
Constit. of *Carissa edulis*, *Carissa spinarum* and *Carissa carandas*. Cytotoxic against breast MCF7 and lung A 549 cancer cell lines. Amorph. powder. $[\alpha]_D^{20}$ -14 (c, 0.7 in EtOH)
3′,4′-Di-Me ether
3,4,8-Trihydroxy-3′,4′-dimethoxy-8,8′-lignan-9,9′-olide
[147022-96-6]
$C_{20}H_{22}O_7$ 374.389
Isol. from *Phenax angustifolius*. $[\alpha]_D^{25}$ -39.1 (c, 0.5 in MeOH). λ_{max} 233 (log ε 4.15); 282 (log ε 3.75) (MeOH)
3′,4′-Di-Me ether, 4-O-β-D-glucopyranoside
[323586-29-4]
$C_{26}H_{32}O_{12}$ 536.53

Isol. from *Phenax angustifolius*. $[\alpha]_D^{25}$ -57.5 (c, 0.5 in MeOH)

3,3′,4′-Tri-Me ether

4,8-Dihydroxy-3,3′,4′-trimethoxy-8,8′-lignan-9,9′-olide. ***Trachelogenin***

[34209-69-3]

$C_{21}H_{24}O_7$ 388.416

Constit. of *Amberboa* sp., *Ipomoea cairica*, *Arctium* sp., *Cnicus* sp., *Saussurea* sp. and *Trachelospermum* sp.. Shows antitumour promoting effect, anti-HIV and antihypertensive activity. Antiinflammatory agent. Needles. Mp 139 - 141°. $[\alpha]_D^{23}$ -43.5 (c, 0.25 in EtOH). Log P 1.12 (calc). Reported in various papers in 1971-2 with uncertainties over struct. assignment. λ_{max} 229 (ε 8300); 280 (ε 4700) (MeOH) (Berdy)

3,3′,4′-Tri-Me ether, 4-O-β-D-glucopyranoside

Tracheloside

[33464-71-0]

$C_{27}H_{34}O_{12}$ 550.557

Constit. of *Trachelospermum* spp., *Carthamus tinctorius* and *Cirsium brachycephalum*. Cryst. (EtOAc). Mp 168 - 170°. $[\alpha]_D^{20}$ -60 (EtOH). Struct. revised in 1972-3

3,3′,4′-Tri-Me ether, 4-O-[β-D-glucopyranosyl-(1→6)-β-D-glucopyranoside]

Trachelogenin 4-gentiobioside

[106647-13-6]

$C_{33}H_{44}O_{17}$ 712.698

Constit. of *Trachelospermum asiaticum* and *Trachelospermum jasminoides*. Amorph. solid. $[\alpha]_D^{28}$ -70 (c, 2 in MeOH)

3,3′,4′-Tri-Me ether, 8-Ac

Daphnelignan A

[1240405-75-7]

$C_{23}H_{26}O_8$ 430.453

Constit. of *Daphne acutiloba*. Antioxidant. Abs. config. not confirmed

3,3′,4,4′-Tetra-Me ether

8-Hydroxy-3,3′,4,4′-tetramethoxy-8,8′-lignan-9,9′-olide.

Methyltrachelogenin

[33464-73-2]

Mp 97 - 98.5°. $[\alpha]_D^{28}$ -46.9 (c, 1.4 in EtOH)

3,4-Methylene, 3′,4′-di-Me ether, 8-angeloyl

Helianthoidin

[18884-21-4]

$C_{26}H_{28}O_8$ 468.502

Lignan from roots of *Heliopsis scabia*. Mp 134 - 135°. $[\alpha]_D^{25}$ -145.4 (c, 1.52 in $CHCl_3$)

3′,4′-Methylene, 3-Me ether

4,8-Dihydroxy-3-methoxy-3′,4′-methylenedioxy-8,8′-lignan-9,9′-olide. ***8-Hydroxypluviatolide***

[222849-85-6]

$C_{20}H_{20}O_7$ 372.373

Constit. of *Stellera chamaejasme* and *Taiwania cryptomerioides*. $[\alpha]_D^{25}$ -32.1 (c, 0.6 in $CHCl_3$). λ_{max} 232 (ε 17500); 285 (ε 9240) (MeOH)

3′,4′-Methylene, 3,4-di-Me ether

Benchequiol

[150448-65-0]

$C_{21}H_{22}O_7$ 386.4

Constit. of the roots of *Bupleurum salicifolium*. $[\alpha]_D^{20}$ -8 (c, 1.5 in $CHCl_3$)

Bis(methylene) ether

8-Hydroxy-3,4:3′,4′-bis(methylenedioxy)-8,8′-lignan-9,9′-olide. ***Meridinol***

[120051-54-9]

Constit. of *Zanthoxylum fagara*. Needles (EtOAc/petrol). Mp 122 - 123°. Mp 126.5° (synthetic). $[\alpha]_D^{25}$ -30 (c, 0.1 in $CHCl_3$). $[\alpha]_D^{20}$ -38.6 (c, 0.09 in $CHCl_3$) (synthetic)

(8R*,8′S*)-form

3,4-Methylene, 3′-Me ether

Stellerachamin A

$C_{20}H_{20}O_7$ 372.373

Constit. of the rhizomes of *Stellera chamaejasme*. Light yellow powder. $[\alpha]_D^{20}$ +253.8 (c, 0.1 in MeOH).

Burden, R.S. *et al.*, *J. Chem. Soc. D: Chem. Commun.*, 1969, 693-701 (*Helianthoidin*).

Nishibe, S. *et al.*, *Phytochemistry*, 1971, **10**, 2231-2232 (*(-)-Nortrachelogenin*).

Nishibe, S. *et al.*, *Chem. Pharm. Bull.*, 1973, **21**, 1108-1113 (*Tracheloside, Trachelogenin, Nortrachelogenin*).

Nishibe, S. *et al.*, *Chem. Pharm. Bull.*, 1973, **21**, 1114-1117 (*Nortrachelogenin 4,4′-diglucoside*).

Tandon, S. *et al.*, *Phytochemistry*, 1976, **15**, 1789-1791 (*(+)-Wikstromol, synth*).

Carnmalm, B. *et al.*, *Acta Chem. Scand., Ser. B*, 1977, **31**, 433 (*(-)-Pinopalustrin, struct*).

Kato, A. *et al.*, *J. Nat. Prod.*, 1979, **42**, 159-162 (*(+)-Nortrachelogenin*).

Torrance, S.J. *et al.*, *J. Pharm. Sci.*, 1979, **68**, 664-666 (*Wikstromol*).

Lee, K.-H. *et al.*, *J. Nat. Prod.*, 1981, **44**, 530-535 (*(+)-Nortrachelogenin, activity*).

Achenbach, H. *et al.*, *Phytochemistry*, 1983, **22**, 749-753 (*(-)-Nortrachelogenin, Carissanol*).

Abe, F. *et al.*, *Chem. Pharm. Bull.*, 1986, **34**, 4340-4345 (*Trachelogenin 4-gentiobioside*).

Belletire, J.L. *et al.*, *J. Org. Chem.*, 1988, **53**, 4724-4729 (*(+)-Wikstromol, synth*).

Amaro-Luis, J.M. *et al.*, *Phytochemistry*, 1988, **27**, 3933-3935 (*Meridinol*).

Khamlach, K. *et al.*, *Heterocycles*, 1990, **31**, 2195-2199 (*Carissanol, synth*).

Belletire, J.L. *J. Nat. Prod.*, 1990, **53**, 1587-1592 (*(+)-Wikstromol, synth, cryst struct, bibl*).

González, A.G. *et al.*, *Phytochemistry*, 1990, **29**, 1981-1983 (*Bupleurum salicifolium constits*).

Schröder, H.C. *Z. Naturforsch., C*, 1990, **45**, 1215-1221 (*HIV activity*).

John, L.M.D. *et al.*, *J. Nat. Prod.*, 1992, **55**, 1313-1314 (*Trachelogenin, cmr*).

Khamlach, K. *et al.*, *Tetrahedron*, 1992, **48**, 10115-10126 (*Trachelogenin, (-)-Nortrachelogenin, (+)-Wikstromol, synth*).

Estevez-Reyes, R. *et al.*, *J. Nat. Prod.*, 1993, **56**, 1177-1181 (*Benchequiol*).

Appendino, G. *et al.*, *J. Nat. Prod.*, 1994, **57**, 607-613 (*Epinortrachelogenin*).

Esterez-Braun, A. *et al.*, *Tetrahedron*, 1994, **50**, 5203-5210 (*Epinortrachelogenin*).

Kuo, Y.H. *et al.*, *J. Nat. Prod.*, 1996, **59**, 622-624 (*Saussurenoside*).

Moritani, Y. *et al.*, *J. Org. Chem.*, 1996, **61**, 6922-6930 (*Trachelogenin, Guayadequiol, synth*).

Takano, D. *et al.*, *J. Heterocycl. Chem.*, 1999, **36**, 221-224 (*Meridinol, synth, abs config*).

Barrero, A.F. *et al.*, *J. Nat. Prod.*, 1999, **62**, 946-948 (*tetra-Me 8-Ac*).

Hu, K. *et al.*, *Planta Med.*, 2000, **66**, 564-567 (*(+)-Nortrachelogenin, isol, activity*).

Rastrelli, L. *et al.*, *J. Nat. Prod.*, 2001, **64**, 79-81 (*Phenax angustifolius constits*).

Sefkow, M. *J. Org. Chem.*, 2001, **66**, 2343-2349 (*synth*).

Tan, X.-Q. *et al.*, *Planta Med.*, 2005, **71**, 93-95 (*Trachelospermum 8-glucoside*).

Sinkkonen, J. *et al.*, *Magn. Reson. Chem.*, 2006, **44**, 633-636 (*Pinus sylvestris constit, pmr, cmr*).

Schmidt, T.J. *et al.*, *Nat. Prod. Res.*, 2006, **20**, 443-453 (*Nortracheloside*).

Wangteeraprasert, R. *et al.*, *Helv. Chim. Acta*, 2009, **92**, 1217-1223 (*Carissa carandas constit*).

Chinese Pat., 2010, 101 787 004 (*Daphnelignan A*).

Cuong, T.D. *et al.*, *Planta Med.*, 2011, **77**, 66-69 (*Pulsatilla 8-glucoside*).

Wangteeraprasert, R. *et al.*, *Phytother. Res.*, 2012, **26**, 1496-1499 (*Carissanol, activity*).

Lee, S. *et al.*, *Bioorg. Med. Chem. Lett.*, 2014, **24**, 44-48 (*Pulsatilla koreana constit*).

Liu, S.-N. *et al.*, *Zhongguo Zhongyao Zazhi*, 2018, **43**, 1649-1653 (*Stellerachamin A, 8-Hydroxypluviatolide*)

3,3′,4,4′,8′-Pentahydroxy-8,8′-lignan-9,9′-olide P23

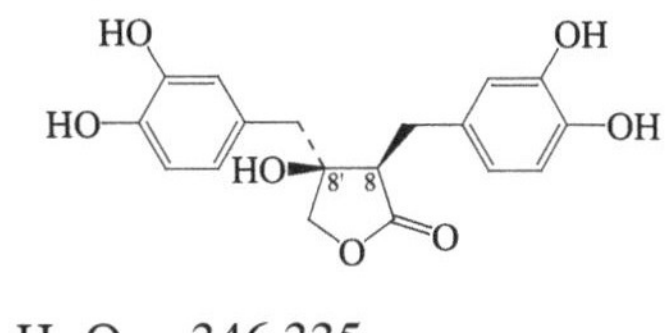

$C_{18}H_{18}O_7$ 346.335

(8R,8′S)-form

3,3′-Di-Me ether

4,4′,8′-Trihydroxy-3,3′-dimethoxy-8,8′-lignan-9,9′-olide

[1610736-05-4]

$C_{20}H_{22}O_7$ 374.389

Constit. of *Selaginella involven*. Exhibits potent protective effect against the injury of PC-12 cells induced by hypoxia/reoxygenation. Solid. λ_{max} 228; 282 (no solvent reported)

3,4:3′,4′-Bis(methylene) ether

8′-Hydroxy-3,4:3′,4′-bis(methylenedioxy)-8,8′-lignan-9,9′-olide. ***8′-Hydroxyhinokinin***

$C_{20}H_{18}O_7$ 370.357

Constit. of the heartwood of *Chamaecyparis obtusa* var. *formosana*. Pale yellow liq.. $[\alpha]_D^{19}$ -89.3 (c, 0.85 in $CHCl_3$). λ_{max} 234 (log ε 3.7); 286 (log ε 3.58) (MeOH).

Kuo, Y.-H. *et al.*, *Chem. Pharm. Bull.*, 2002, **50**, 978-980 (*8′-Hydroxyhinokinin*).

Long, H.-P. *et al.*, *Chin. Chem. Lett.*, 2014, **25**, 805-808 (*3,3′-di-Me ether, activity*)

3,3′,4,4′,9′-Pentahydroxy-8,8′-lignan-9,9′-olide P24

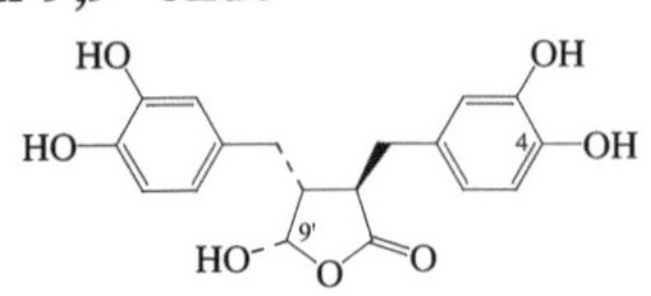

$C_{18}H_{18}O_7$ 346.335

Lignan numbering shown

(8R,8′R,9′S)-form

3,4:3′,4′-Bis(methylene), 9′-Me ether

9′-Methoxy-3,4:3′,4′-bis(methylenedioxy)-8,8′-lignan-9,9′-olide. ***Aristelegin A***

[477199-87-4]

$C_{21}H_{20}O_7$ 384.384

Constit. of *Aristolochia elegans*. Needles ($CHCl_3$). Mp 86 - 87°. $[\alpha]_D^{25}$ -51.5 (c, 0.04 in $CHCl_3$). λ_{max} 234 (log ε 3.66); 287 (log ε 3.78) (MeOH).

Wu, T.-S. *et al.*, *J. Nat. Prod.*, 2002, **65**, 1522-1525 (*Aristelegin A*)

3,3′,4,5′,8-Pentahydroxy-8,8′-lignan-9,9′-olide P25

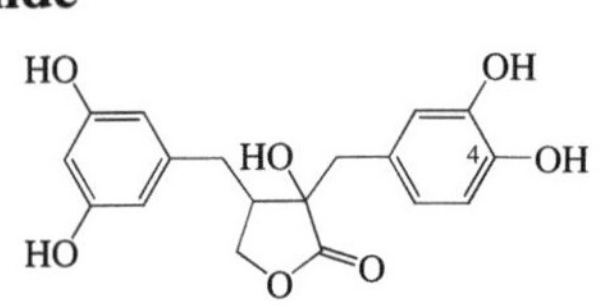

$C_{18}H_{18}O_7$ 346.335

Structures in this series may need revision. Phenyl 3,5-dioxygenated substn. pattern in doubt as a result of work done by Yang *et al* in 2015

(8ξ,8′ξ)-form

3,3′,5′-Tri-Me ether

4,8-Dihydroxy-3,3′,5′-trimethoxy-8,8′-lignan-9,9′-olide

$C_{21}H_{24}O_7$ 388.416

3,3′,5′-Tri-Me ether, 4-O-(2-O-methyl-β-D-glucopyranoside)

[1232491-45-0]

$C_{28}H_{36}O_{12}$ 564.584

Constit. of *Carthamus oxyacantha*. Yellowish gum. λ_{max} 217 (log ε 4.27); 281 (log ε 4.1) (MeOH).

Hassan, Z. *et al.*, *Nat. Prod. Commun.*, 2010, **5**, 419-422 (*Carthamus oxyacantha constit*).

Yang, Y.-N. *et al.*, *J. Nat. Prod.*, 2015, **78**, 705-711 (*pmr, struct*)

3′,4,4′,7,7′-Pentahydroxy-8,8′-lignan-9,9′-olide P26

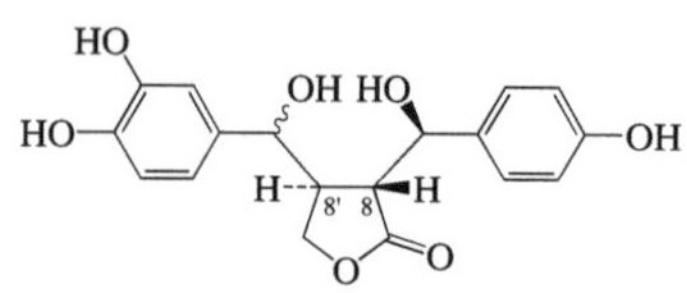

$C_{18}H_{18}O_7$ 346.335

(7R*,7′ξ,8R*,8′S*)-form

3′-Me ether

4,4′,7,7′-Tetrahydroxy-3′-methoxy-8,8′-lignan-9,9′-olide

$C_{19}H_{20}O_7$ 360.362

3′-Me ether, 7′-O-β-D-glucopyranoside

Bambulignan B

$C_{25}H_{30}O_{12}$ 522.503

Constit. of the dry leaves of *Pleioblastus amarus*. Amorph. powder. $[\alpha]_D^{20}$-14.2 (c, 0.15 in MeOH). λ_{max} 226; 277 (MeOH).

Sun, J. *et al.*, *J. Asian Nat. Prod. Res.*, 2014, **16**, 930-935 (*Bambulignan B*)

3,3′,4,4′,8-Pentahydroxy-8,8′-lignan-7-one P27

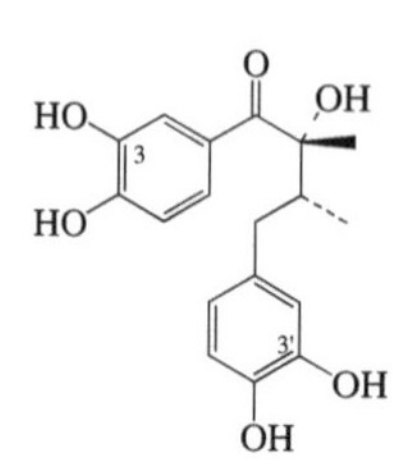

$C_{18}H_{20}O_6$ 332.352

(8R,8′R)-form

3′,4′-Methylene, 3,4-di-Me ether

8-Hydroxy-3,4-dimethoxy-3′,4′-methylenedioxy-8,8′-lignan-7-one

[1259294-28-4]

$C_{21}H_{24}O_6$ 372.417

Constit. of the fruit of *Schisandra sphenanthera*. Cryst.. $[\alpha]_D^{24}$+56.7 (c, 0.46 in MeOH). λ_{max} 232 (log ε 4.48); 278 (log ε 3.5); 290 (log ε 3.6); 325 (log ε 3.98) (MeOH).

Wang, C.R. *et al.*, *Nat. Prod. Commun.*, 2009, **4**, 1571-1574 (*Schisandra sphenanthera constit*)

3,3′,4,4′,5-Pentahydroxy-8,8′-lign-7-ene-9,9′-dioic acid P28

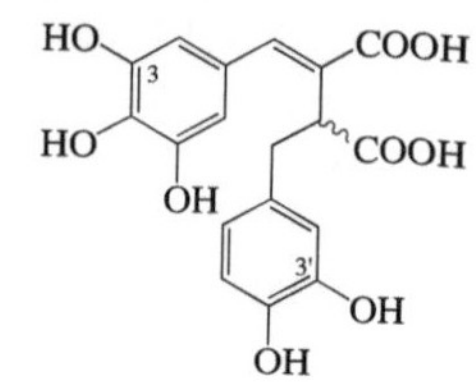

$C_{18}H_{16}O_9$ 376.317

(7E,8′ξ)-form

3,4:3′,4′-Bis(methylene), 5-Me ether, di-Me ester

Rhinacanthin E

[190850-41-0]

$C_{23}H_{22}O_9$ 442.42

Constit. of *Rhinacanthus nasutus*. Amorph. powder. $[\alpha]_D$ -30.2 (c, 0.08 in MeOH). λ_{max} 217 (log ε 3.17); 252 (log ε 3.71); 278 (log ε 3.67); 328 (log ε 3.66) (MeOH).

Kernan, M.R. *et al.*, *J. Nat. Prod.*, 1997, **60**, 635-637 (*Rhinacanthin E*)

3′,4,4′,5,5′-Pentahydroxy-2,8′-lign-7′-ene-9,9′-dioic acid P29

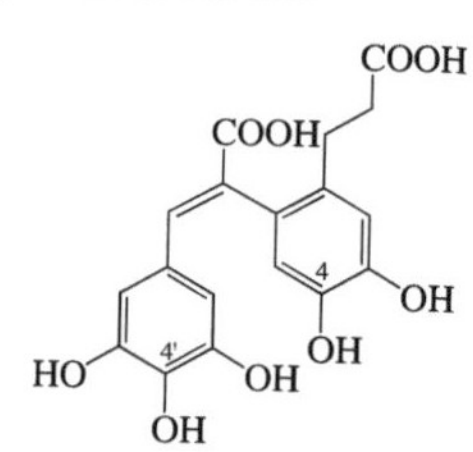

$C_{18}H_{16}O_9$ 376.317

3′,5′-Di-Me ether, 9-[2-(4-hydroxyphenyl)ethylamide], 9′-(4-acetamidobutyl)amide

[1416474-63-9]

$C_{34}H_{41}N_3O_9$ 635.714

Isol. from the root bark of *Lycium chinense*. Exhibits moderate radical scavenging activity against DPPH and superoxide radicals. Yellow gum. λ_{max} 318 (log ε 4.18) (MeOH)

3′,5′-Di-Me ether, 9,9′-bis[2-(4-hydroxyphenyl)ethylamide]
[1416474-62-8]
$C_{36}H_{38}N_2O_9$ 642.705
Isol. from the root bark of *Lycium chinense*. Exhibits moderate radical scavenging activity against DPPH and superoxide radicals. Yellow gum. λ_{max} 320 (log ε 4.22) (MeOH)

3′-Deoxy
3′,4,4′,5-Tetrahydroxy-2,8′-lign-7′-ene-9,9′-dioic acid

3′-Deoxy, 9,9′-bis[2-(4-hydroxyphenyl)ethylamide]
Lyciumamide G
$C_{34}H_{34}N_2O_8$ 598.652
Constit. of root bark of *Lycium chinense*. Amorph. yellow powder. λ_{max} 224 (log ε 4.23); 286 (log ε 3.90); 325 (log ε 3.69) (MeOH).

Zhang, J.-X. *et al.*, *J. Nat. Prod.*, 2013, **76**, 51-58 (*Lycium chinense amides, activity*).
Chen, H. *et al.*, *Fitoterapia*, 2017, **122**, 119-125 (*Lyciumamide G*)

2,3,3′,4,4′-Pentahydroxy-8,8′-lign-7-en-9,9′-olide P30

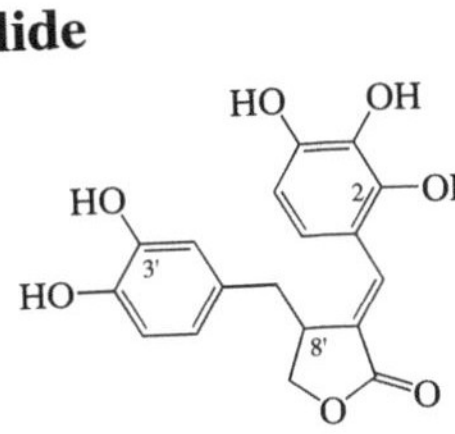

4-(3,4-Dihydroxyphenyl)dihydro-3-(3,4,5-trihydroxyphenylmethylene)-2(3H)-furanone
$C_{18}H_{16}O_7$ 344.319

(7E,8′R)-form

3,4-Methylene, 3′-Me ether
2,4′-Dihydroxy-3′-methoxy-3,4-methylenedioxy-8,8′-lign-7-en-9,9′-olide
[161961-48-4]
$C_{20}H_{18}O_7$ 370.357
Amorph. solid. $[\alpha]^{20}_{D}$ -39.4 (c, 0.16 in MeOH)

3,4-Methylene, 3′-Me ether, 2-O-β-D-glucopyranoside
Haplodoside
[161996-40-3]
$C_{26}H_{28}O_{12}$ 532.498
Constit. of *Haplophyllum cappadocicum*. Amorph. solid. $[\alpha]^{20}_{D}$ -69.5 (c, 0.1 in MeOH). λ_{max} 203 (ε 4900); 222 (sh); 252 (ε 1500); 292 (sh); 321 (ε 1600) (MeOH)

3,4-Methylene, 3′,4′-di-Me ether
2-Hydroxy-3′,4′-dimethoxy-3,4-methylenedioxy-8,8′-lign-7-en-9,9′-olide. ***Cappadocin***
[161996-41-4]
$C_{21}H_{20}O_7$ 384.384
Constit. of *Haplophyllum cappadocicum*. Amorph. solid. $[\alpha]^{20}_{D}$ -22.9 (c, 0.1 in MeOH). λ_{max} 204 (ε 55000); 225 (sh); 253 (ε 10000); 290 (sh); 323 (ε 12000) (MeOH)

3,4-Methylene, 3′,4′-di-Me ether, 2-O-β-D-glucopyranoside
Cappadoside
[161996-39-0]
$C_{27}H_{30}O_{12}$ 546.525
Constit. of *Haplophyllum cappadocicum* (Rutaceae). Cryst. (EtOAc/MeOH). Mp 139°. $[\alpha]^{20}_{D}$ -80 (c, 0.1 in MeOH). λ_{max} 204 (ε 38000); 225 (ε 15100); 252 (ε 11500); 292 (sh); 321 (ε 12900) (MeOH).

Goezler, B. *et al.*, *Phytochemistry*, 1994, **37**, 1693-1698 (*Haplodoside, Cappadocin, Cappadoside*)

2′,3,4,4′,5′-Pentahydroxy-8,8′-lign-7-en-9,9′-olide P31

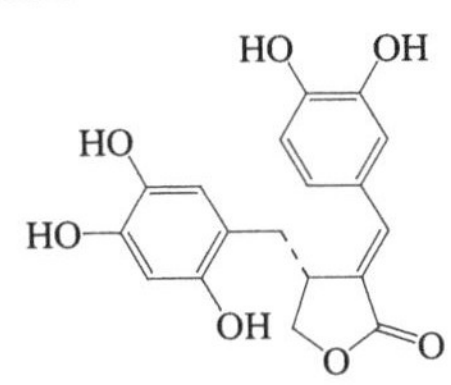

$C_{18}H_{16}O_7$ 344.319

(7E,8′R)-form

3,4:4′,5′-Bis(methylene) ether,
2′-Hydroxy-3,4:4′,5′-bis(methylenedioxy)-8,8′-lign-7-en-9,9′-olide

3,4:4′,5′-Bis(methylene) ether, 2′-O-β-D-glucopyranoside
2′-β-D-Glucopyranosyloxysavinin
$C_{26}H_{26}O_{12}$ 530.482
Constit. of *Taiwania cryptomerioides*. Mp 99 - 101° (as tetra-Ac). $[\alpha]^{23}_{D}$ +7.2 (c, 5.8 in $CHCl_3$) (tetra-Ac). λ_{max} 236 (ε 19800); 296 (ε 17400); 334 (ε 18300) (MeOH) (tetra-Ac).

Lin, W. *et al.*, *Phytochemistry*, 1999, **50**, 653-658 (*2′-Glucopyranosyloxysavinin*)

3,3′,4,4′,5-Pentahydroxy-8,8′-lign-7-en-9,9′-olide P32

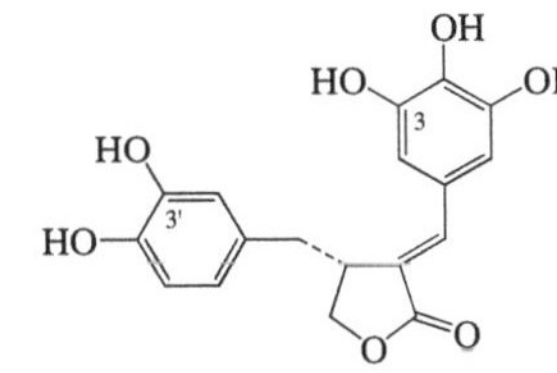

(7E,8′R)-form

$C_{18}H_{16}O_7$ 344.319

(7E,8′R)-form

3,3′,4-Tri-Me ether
3,4′-Dihydroxy-3′,4,5-trimethoxy-8,8′-lign-7-en-9,9′-olide
[130837-93-3]
$C_{21}H_{22}O_7$ 386.4
Isol. from wood of *Calocedrus formosana*. Needles. Mp 130 - 131°. $[\alpha]^{25}_{D}$ -94 (c, 0.15 in $CHCl_3$). λ_{max} 238 (ε 27410); 280 (ε 21700); 299 (ε 17300); 322 (ε 21620) (no solvent reported)

3,3′,5-Tri-Me ether
4,4′-Dihydroxy-3,3′,5-trimethoxy-8,8′-lign-7-en-9,9′-olide
$C_{21}H_{22}O_7$ 386.4
Isol. from wood of *Calocedrus formosana*. Needles. Mp 112 - 113°. $[\alpha]^{25}_{D}$ -16.3 (c, 0.3 in $CHCl_3$)

3′,4′-Methylene, 3,4,5-tri-Me ether
3,4,5-Trimethoxy-3′,4′-methylenedioxy-8,8′-lign-7-en-9,9′-olide. ***Nemerosin****. Anhydropodorhizol*
[17187-79-0]
$C_{22}H_{22}O_7$ 398.411
Constit. of *Anthriscus nemerosa, Juniperus thurifera* and *Podophyllum hexandrum*. Needles (EtOAc/petrol). Mp 113 - 115°. Mp 128 - 132°. $[\alpha]^{25}_{D}$ +123 (c, 0.7 in $CHCl_3$)

(7E,8′ξ)-form

3′,4′-Methylene, 3,4-di-Me ether
3-Hydroxy-4,5-dimethoxy-3′,4′-methylenedioxy-8,8′-lign-7-en-9,9′-olide
[224451-09-6]
$C_{21}H_{20}O_7$ 384.384
Constit. of *Bupleurum lanceolatum*. Amorph.. λ_{max} 240; 278; 298; 320 (MeOH)

(7Z,8′R)-form

3′,4′-Methylene, 3,4,5-tri-Me ether
Isochaihulactone*. Z-Nemerosin*
[652143-70-9]
$C_{22}H_{22}O_7$ 398.411

Constit. of the roots of *Bupleurum scorzonerifolium*. Needles (MeOH). Mp 137 - 138°. $[\alpha]^{25}_{D}$ -29 (c, 0.5 in $CHCl_3$). λ_{max} 247 (log ε 4.08); 298 (log ε 4.17); 327 (log ε 4.08) ($CHCl_3$)

(7Z,8′ξ)-form

3,3′-Di-Me ether

3,4,4′-Trihydroxy-3′,5-dimethoxy-8,8′-lign-7-en-9,9′-olide. ***γ-Thujaplicatene***

[29725-59-5]

$C_{20}H_{20}O_7$ 372.373

Constit. of the heartwood of the red cedar *Thuja plicata*. Cryst. (EtOH) (as tri-Ac). Mp 148 - 149° (tri-Ac). *Z*-Config. does not appear well established. It is the config. given in the diag. of the paper. λ_{max} 228 (log ε 3.39); 338 (log ε 4.02) (no solvent reported). [base] λ_{max} 248 (log ε 3.89); 292 (log ε 4.19) (alkali soln.).

MacDonald, B.F. *et al.*, *Can. J. Chem.*, 1970, **48**, 3144-3146 (*γ-Thujaplicatene*).

Kamil, W.M. *et al.*, *Phytochemistry*, 1986, **25**, 2093-2102 (*Nemerosin, biosynth*).

Fang, J.M. *et al.*, *Phytochemistry*, 1990, **29**, 3048-3049 (*Calocedrus formosana constits*).

Khetwal, K.S. *et al.*, *Indian J. Chem., Sect. B*, 1998, **37**, 1319-1320 (*Bupleurum lanceolatum constit*).

Chang, W.-L. *et al.*, *Phytochemistry*, 2003, **64**, 1375-1379 (*Nemerosin, Isochaihulactone*).

Chen, Y-L. *et al.*, *Biochem. Pharmacol.*, 2006, **72**, 308-319 (*Isochaihulactone*)

3,3′,4,4′,7′-Pentahydroxy-8,8′-lign-8(8′)-en-9,9′-olide P33

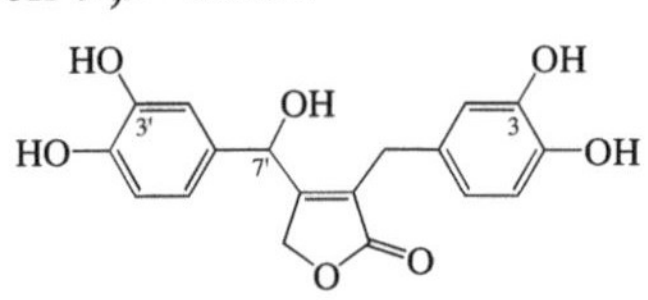

$C_{18}H_{16}O_7$ 344.319

(ξ)-form

3′,4′-Methylene, 4-Me ether

3,7′-Dihydroxy-4-methoxy-3′,4′-methylenedioxy-8,8′-lign-8(8′)-en-9,9′-olide. ***Chasnarolide***

[150417-73-5]

$C_{20}H_{18}O_7$ 370.357

Constit. of the roots of *Bupleurum salicifolium*. Orange oil. $[\alpha]^{20}_{D}$ -2.5 (c, 0.5 in $CHCl_3$). λ_{max} 286; 324 (EtOH)

3′,4′-Methylene, 3,4-di-Me ether

7′-Hydroxy-3,4-dimethoxy-3′,4′-methylenedioxy-8,8′-lign-8(8′)-en-9,9′-olide. ***Methylchasnarolide***

[150196-57-9]

$C_{21}H_{20}O_7$ 384.384

Constit. of the roots of *Bupleurum salicifolium*. Amorph.. $[\alpha]^{20}_{D}$ -6 (c, 0.5 in $CHCl_3$). λ_{max} 282; 324 (EtOH)

3,4:3′,4′-Bis(methylene) ether

7′-Hydroxy-3,4:3′,4′-bis(methylenedioxy)-8,8′-lign-8(8′)-en-9,9′-olide. ***Procumbiene***

[1624617-94-2]

$C_{20}H_{16}O_7$ 368.341

Constit. of *Justicia procumbens*. Cytotoxic agent. Cryst.. $[\alpha]^{20}_{D}$ +20.0 (c, 0.03 in MeOH). λ_{max} 256 (log ε 3.88); 275 (log ε 3.68); 288 (log ε 3.74) (MeOH).

Estevez-Reyes, R. *et al.*, *J. Nat. Prod.*, 1993, **56**, 1177-1181 (*Chasnarolide, Methylchasnarolide*).

Jin, H. *et al.*, *Fitoterapia*, 2014, **94**, 70-76 (*Procumbiene*)

3,3′,4,4′,8′-Pentahydroxy-8,8′-lign-7-en-9,9′-olide P34

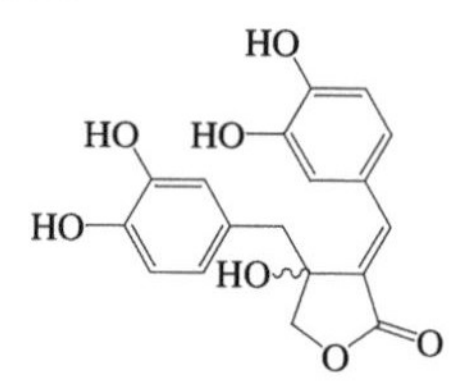

$C_{18}H_{16}O_7$ 344.319

(7E,8′ξ)-form

3,4:3′,4′-Bis(methylene) ether

8′-Hydroxy-3,4:3′,4′-bis(methylenedioxy)-8,8′-lign-7-en-9,9′-olide. ***8′-Hydroxysavinin***

[942505-97-7]

$C_{20}H_{16}O_7$ 368.341

Constit. of the roots of *Taiwania cryptomerioides*. Yellowish gum. $[\alpha]^{27}_{D}$ -34.9 (c, 0.12 in $CHCl_3$). λ_{max} 226 (log ε 4.01); 284 (log ε 3.65) (MeOH).

Chyu, C.-F. *et al.*, *Helv. Chim. Acta*, 2007, **90**, 738-747 (*8′-Hydroxysavinin*)

3,3′,4,4′,9-Pentahydroxy-8,8′-lign-7′-en-9′,7-olide P35

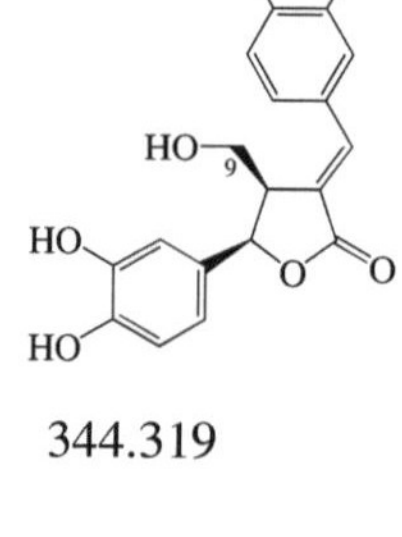

$C_{18}H_{16}O_7$ 344.319

(7R*,7′E,8R*)-form

9-Carboxylic acid

3,3′,4,4′-Tetrahydroxy-8,8′-lign-7′-en-9′,7-olid-9-oic acid

[1060663-38-8]

$C_{18}H_{14}O_8$ 358.302

Obt. from caffeic acid by biooxidn. with *Momordica charantia* peroxidase. Potent antioxidant. Amorph. powder. $[\alpha]_{D}$ +1.6 (c, 0.2 in MeOH). Has 8*S**-config. due to change in Cahn-Ingold-Prelog priorities. λ_{max} 336 (MeOH)

9-Carboxylic acid, Et ester

Pharbilignan D. *Pharbilignanol D*

[1620558-57-7]

$C_{20}H_{18}O_8$ 386.356

Constit. of the seeds of *Pharbitis nil*. Exhibits moderate cytotoxicity against SK-OV-3, SK-MEL-2 and HCT-15 human tumour cell lines. $[\alpha]^{25}_{D}$ +27.7 (c, 0.4 in EtOH). λ_{max} 258 (log ε 4.03); 310 (log ε 3.74) (MeOH)

(7R*,7′E,8S*)-form [1489173-20-7]

Isol. from *Morinda citrifolia*. Amorph. yellow powder. $[\alpha]^{25}_{D}$ -5.6 (c, 0.1 in MeOH). λ_{max} 233 (log ε 4.10); 284 (log ε 3.60) (MeOH)

3,3′-Di-Me ether

4,4′-Dihydroxy-3,3′-dimethoxy-8,8′-lign-7′-en-9′,7-olide. ***Impecylenolide***

[1622990-05-9]

$C_{20}H_{20}O_7$ 372.373

Constit. of *Imperata cylindrica*. Syrup. $[\alpha]^{20}_{D}$ +5.2 (c, 0.134 in MeOH). λ_{max} 233; 335 (MeOH)

(7RS,7′E,8SR)-form

9-Carboxylic acid

Phellinsin A

[307494-43-5]

[858659-90-2]

(7*R**,7′*E*,8*R**-form)

$C_{18}H_{14}O_8$ 358.302

Prod. by *Phellinus* sp. PL3 isol. from *Morus alba*. Chitin synthase inhibitor and antifungal agent. Inhibitor of low-density lipoprotein oxidation. Yellow powder. Racemic mix of (7*R*,7′*E*,8*R*)- and (7*S*,7′*E*,8*S*)-diastereomers (change in Cahn-Ingold-Prelog priorities). λ_{max} 205 (log ε 4.2); 336 (log ε 3.79) (MeOH).

Hwang, E.-I. *et al.*, *J. Antibiot.*, 2000, **53**, 903-911 (*Phellinsin A*).

Kim, E. *et al.*, *Synth. Commun.*, 2005, **35**, 1231-1238 (*Phellinsin A, rel config, synth*).
Wan, X. *et al.*, *Can. J. Chem.*, 2008, **86**, 821-830 (*Momordica charantia oxidn prod*).
Nguyen, P-H. *et al.*, *J. Nat. Prod.*, 2013, **76**, 2080-2087 (*Morinda citrifolia constit*).
Kim, K.H. *et al.*, *J. Agric. Food Chem.*, 2014, **62**, 7746-7752 (*Pharbillgnan D, activity*).
Liu, X. *et al.*, *J. Nat. Med.*, 2014, **68**, 295-301 (*Impecylenolide*)

3,3′,4,4′,7′-Pentahydroxy-3,8′-lign-8-en-6(3*H*)-one P36

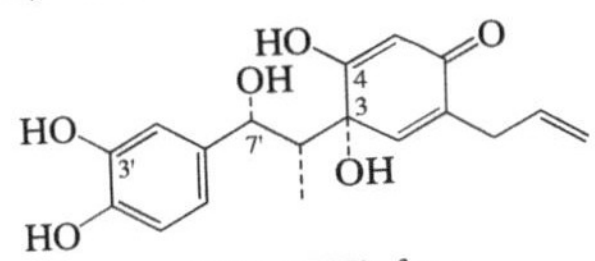

(3*R*,7′*R*,8′*R*)-*form*

$C_{18}H_{20}O_6$ 332.352
(3R,7′R,8′R)-form
3,3′,4,4′-Tetra-Me ether
7′-Hydroxy-3,3′,4,4′-tetramethoxy-3,8′-lign-8-en-6(3H)-one
3,3′,4,4′-Tetra-Me ether, 7′-Ac
[173241-75-3]
$C_{24}H_{30}O_7$ 430.497
Constit. of *Piper wightii*. Viscous oil. $[\alpha]^{24}_{D}$+50.5 (c, 0.05 in $CHCl_3$). λ_{max} 230; 275 (MeOH)
3′,4′-Methylene, 3,4-di-Me ether
7′-Hydroxy-3,4-dimethoxy-3′,4′-methylenedioxy-3,8′-lign-8-en-6(3H)-one
[155301-80-7]
$C_{21}H_{24}O_6$ 372.417
Constit. of the stems and fruit of *Piper wightii*. Pale yellow viscous oil. $[\alpha]^{24}_{D}$+89 (c, 0.37 in $CHCl_3$). λ_{max} 246; 286 ($CHCl_3$)
3′,4′-Methylene, 3,4-di-Me ether, 7′-Ac
[155301-79-4]
$C_{23}H_{26}O_7$ 414.454
Constit. of the stems of *Piper wightii*. Viscous oil. $[\alpha]^{24}_{D}$+43.3 (c, 0.2 in $CHCl_3$). λ_{max} 246; 286 ($CHCl_3$)
(3ξ,8′ξ)-form
7′-Ketone
3,3′,4,4′-Tetrahydroxy-3,8′-lign-8-ene-6,7′(3H)-dione
7′-Ketone, 3′,4′-methylene, 3,4-di-Me ether
3,4-Dimethoxy-3′,4′-methylenedioxy-3,8′-lign-8-ene-6,7′(3H)-dione. ***Fargesone C***
[116424-71-6]
$C_{21}H_{22}O_6$ 370.401
Constit. of *Magnolia fargesii*, *Magnolia obovata* and *Magnolia sprengeri*. Prisms. Mp 126 - 127°. $[\alpha]_{D}$ -206.5 (c, 1 in $CHCl_3$).

Chen, C.-C. *et al.*, *Chem. Pharm. Bull.*, 1988, **36**, 1791-1795 (*Fargesone C*).
Prasad, A.K. *et al.*, *Tetrahedron*, 1994, **50**, 2231-2240 (*Piper wightii constits*).
Prasad, A.K. *et al.*, *Tetrahedron*, 1994, **50**, 6721 (*corrigendum*).
Boll, P.M. *et al.*, *Recl. Trav. Chim. Pays-Bas*, 1996, **115**, 9-12 (*3,3′,4,4′-tetra-Me 7′-Ac*).
Sun, Y. *et al.*, *J. Chromatogr., B*, 2011, **879**, 3775-3779 (*Fargesone C*)

3,3′,4,4′,5′-Pentahydroxy-9-nor-8,8′-lign-7′-en-9′,7-olide P37

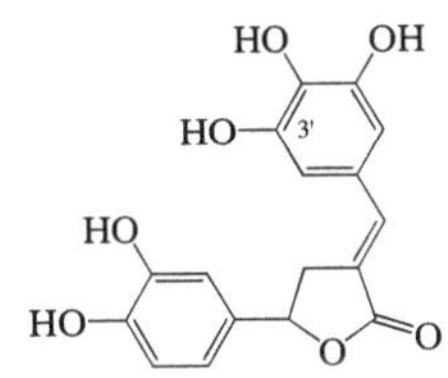

(7ξ,7′*E*)-*form*

5-(3,4-Dihydroxyphenyl)dihydro-3-(3,4,5-trihydroxybenzylidene)-2(3H)-furanone
$C_{17}H_{14}O_7$ 330.292
(7ξ,7′E)-form
3′,5′-Di-Me ether
3,4,4′-Trihydroxy-3′,5′-dimethoxy-9-nor-8,8′-lign-7′-en-9′,7-olide
[888011-15-2]
$C_{19}H_{18}O_7$ 358.346
Constit. of the leaves of *Cestrum parqui*. Oil. $[\alpha]^{25}_{D}$+3.5 (c, 0.23 in MeOH)
(7ξ,7′Z)-form
3′,5′-Di-Me ether
Tectonoelin B
[1422520-13-5]
$C_{19}H_{18}O_7$ 358.346
Constit. of the leaves of *Tectona grandis*. Brown oil. $[\alpha]^{25}_{D}$-10 (c, 0.1 in MeOH).

D'Abrosca, B. *et al.*, *Nat. Prod. Res.*, 2006, **20**, 293-298 (*Cestrum parqui constit*).
Lacret, R. *et al.*, *Phytochem. Lett.*, 2012, **5**, 382-386 (*Tectonoelin B*)

3,3′,4,4′,8-Pentahydroxy-9-nor-8,8′-lign-7′-en-9′,7-olide P38

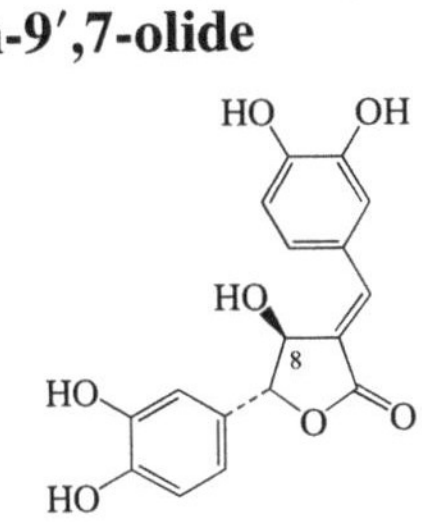

3-(3,4-Dihydroxybenzylidene)-5-(3,4-dihydroxyphenyl)dihydro-4-hydroxy-2(3H)-furanone
$C_{17}H_{14}O_7$ 330.292
(7R*,7′E,8S*)-form
8-Et ether
8-Ethoxy-3,3′,4,4′-tetrahydroxy-9-nor-8,8′-lign-7′-en-9′,7-olide.
Amauroderin A
$C_{19}H_{18}O_7$ 358.346
Constit. of the fruiting bodies of *Amauroderma rude*. Yellow oil. $[\alpha]^{25}_{D}$-2.4 (c, 0.001 in MeOH).

Ma, Z.-H. *et al.*, *Zhongcaoyao*, 2019, **50**, 315-317 (*Amauroderin A*)

3,3′,4′,7′,9′-Pentahydroxy-9-nor-4,8′-oxylignan-7-one P39

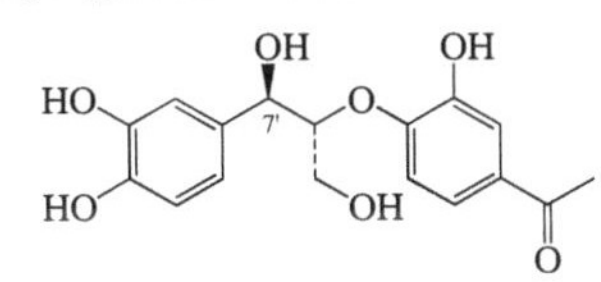

(7′*R*,8′*R*)-*form*

3,3′,4′,7′,9′-Pentahydroxy-9-nor-7-oxo-4,8′-oxyneolignan
$C_{17}H_{18}O_7$ 334.324
(7′R,8′R)-form
3,3′-Di-Me ether
$[\alpha]^{20}_{D}$-19 (c, 0.1 in MeOH)
(7′S,8′S)-form
3,3′-Di-Me ether
$[\alpha]^{20}_{D}$+17.2 (c, 0.1 in MeOH)
(7′RS,8′RS)-form
3,3′-Di-Me ether
4′,7′,9′-Trihydroxy-3,3′-dimethoxy-9-nor-4,8′-oxylignan-7-one.
4′,7′,9′-Trihydroxy-3,3′-dimethoxy-9-nor-7-oxo-4,8′-oxylignan
$C_{19}H_{22}O_7$ 362.378
Constit. of the fruit of *Crataegus pinnatifida*. Gum. $[\alpha]^{20}_{D}$+1.2 (c, 0.1 in MeOH). Partial racemate resolved by chiral hplc. λ_{max} 218 (log ε 2.58); 258 (log ε 0.23) (MeOH).

Zhao, P. *et al.*, *Fitoterapia*, 2019, **136**, 104164:1-7 (*Crataegus pinnatifida constit*)

3,3′,4,4′,8′-Pentahydroxy-7′-oxo-8,8′-lign-7-en-9,9′-olide P40

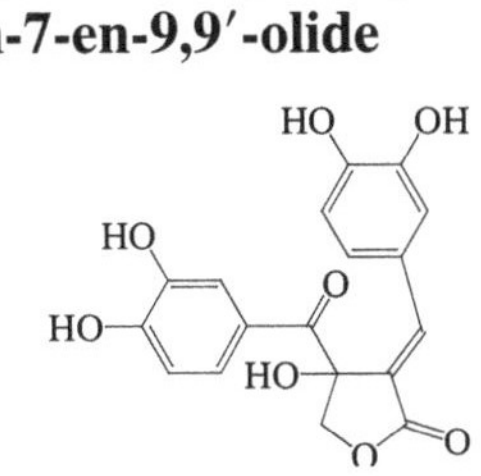

$C_{18}H_{14}O_8$ 358.302

(7E,8′ξ)-form

Actaealactone [874359-26-9]

Constit. of *Actaea racemosa* (*Cimicifuga racemosa*). Antioxidant. Weaks stimulant of MCF-7 breast cancer cell proliferation. Amorph. yellow powder. Mp 149 - 150°. $[\alpha]_D^{25}$ -23.6 (c, 0.0005 in MeOH). λ_{max} 237 (log ε 3.31); 295 (log ε 3.28); 339 (log ε 3.34) (MeOH).

Nuntanakorn, P. *et al.*, *J. Nat. Prod.*, 2006, **69**, 314-318 (*Actaealactone*)

3,3′,7′,8′,9′-Pentahydroxy-4,4′-oxylign-7-en-9-oic acid P41

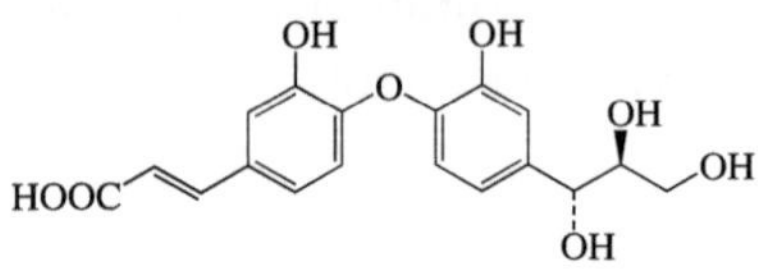

(7E,7′R,8′S)-form

$C_{18}H_{18}O_8$ 362.334

(7E,7′R,8′S)-form

3,3′-Di-Me ether

7′,8′,9′-Trihydroxy-3,3′-dimethoxy-4,4′-oxylign-7-en-9-oic acid

3,3′-Di-Me ether, Me ester

Rehmaneolignan A. Rhemaneolignan A

[2001061-00-1]

$C_{21}H_{24}O_8$ 404.415

Constit. of the roots of *Rehmannia glutinosa*. Amorph. powder. $[\alpha]_D^{20}$ -4.2 (c, 0.2 in MeOH). Synonym assigned 2 different spellings. λ_{max} 325 (MeOH)

(7E,7′S,8′S)-form

3,3′-Di-Me ether, Me ester

Rehmaneolignan B. Rhemaneolignan B

[2001061-01-2]

$C_{21}H_{24}O_8$ 404.415

Constit. of the roots of *Rehmannia glutinosa*. Amorph. powder. $[\alpha]_D^{20}$ +1.8 (c, 0.2 in MeOH). Synonyms assigned 2 different spellings. λ_{max} 322 (MeOH)

(7E,7′ξ,8ξ′)-form

5′-Methoxy, 3,3′-di-Me ether, Me ester

Lepidiumlignan A

$C_{22}H_{26}O_9$ 434.441

Constit. of the seeds of *Lepidium apetalum*. $[\alpha]_D^{20}$ -25.5 (c, 0.02 in MeOD). λ_{max} 206 (log ε 2.38); 238 (log ε 0.81); 288 (log ε 0.59); 328 (log ε 0.69) (MeOH).

Li, M. *et al.*, *Molecules*, 2015, **20**, 15192-15201 (*Rhemaneolignans A,B*).

Meng, L. *et al.*, *Zhongcaoyao*, 2018, **49**, 3981-3986 (*Lepidiumlignan A*)

Peperomin G P42

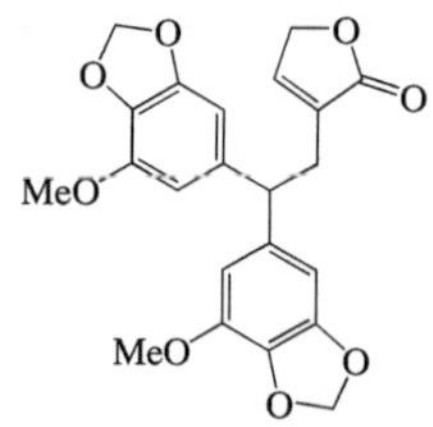

3-[2,2-Bis(7-methoxy-1,3-benzodioxol-5-yl)ethyl]-2(5H)-furanone

[1311291-65-2]

$C_{22}H_{20}O_8$ 412.394

Appears to be an abeo-seco-2,7-cyclolignan. Constit. of *Peperomia dindygulensis*. Powder. λ_{max} 252 (log ε 4.61); 280 (log ε 2.45) (MeOH).

Lin, M.-G. *et al.*, *Chem. Biodiversity*, 2011, **8**, 862-871 (*Peperomin G*)

Peperotetraphin P43

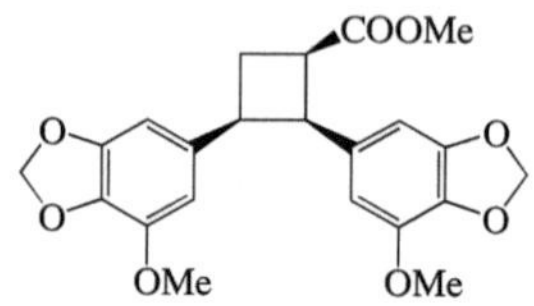

[1000017-87-7]

$C_{22}H_{22}O_8$ 414.41

Constit. of *Peperomia tetraphylla*. Gum. Racemic. λ_{max} 252; 275 ($CHCl_3$).

Li, Y.-Z. *et al.*, *Helv. Chim. Acta*, 2007, **90**, 2222-2226 (*Peperotetraphin*)

Peperotetraphin B P44

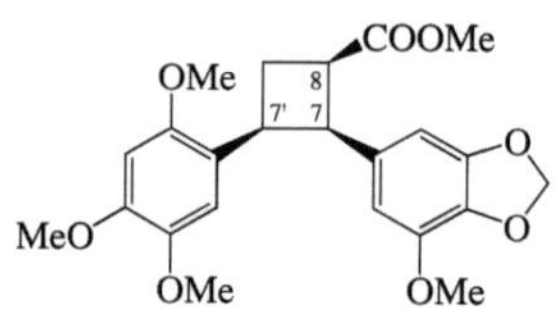

Relative Configuration

[1374216-65-5]

$C_{23}H_{26}O_8$ 430.453

Lignan numbering shown. Similar to Peperotetraphin, P43. Constit. of *Peperomia tetraphylla*. Gum. Racemic. λ_{max} 253 (log ε 4.2); 280 (log ε 4.05) ($CHCl_3$)

7-Epimer

Peperotetraphin C

[1374216-66-6]

$C_{23}H_{26}O_8$ 430.453

Constit. of *Peperomia tetraphylla*. Gum. $[\alpha]_D^{20}$ -10.5 (c, 0.01 in $CHCl_3$). λ_{max} 254 (log ε 4.2); 280 (log ε 4.05) ($CHCl_3$).

Li, Y.-Z. *et al.*, *Chem. Biodiversity*, 2012, **9**, 769-776 (*Peperotetraphins B,C*)

1-Phenyl-1,2,3-propanetriol, CAS P45

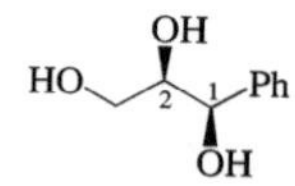

(1R,2R)-form

1-Phenylglycerol. Stycerine

[63157-81-3]

[116127-76-5] (1*S*,2*S*-form)

[116127-78-7] (1*R*,2*S*-form)

$C_9H_{12}O_3$ 168.192

(1R,2R)-form

[16354-93-1]

Syrup. $[\alpha]_D^{23}$ -38.6 (c, 3.5 in MeOH)

3-(4-Methylbenzenesulfonyl)

[1922156-03-3]

$C_{16}H_{18}O_5S$ 322.375

Pale yellow oil. $[\alpha]_D$ -18 (c, 2.3 in $CHCl_3$)

(1R,2S)-form

1-Benzoyl

$C_{16}H_{16}O_4$ 272.300

Constit. of *Desmos pedunculosus*. Oil. $[\alpha]_D^{21}$ +10.4 (c, 0.9 in $CHCl_3$)

3-Benzoyl

$C_{16}H_{16}O_4$ 272.300

Constit. of *Desmos pedunculosus*. Oil. $[\alpha]_D^{21}$ +9.6 (c, 0.8 in $CHCl_3$)

1-Me ether, 3-benzoyl

$C_{17}H_{18}O_4$ 286.327

Constit. of *Desmos pedunculosus*. Oil. $[\alpha]_D^{21}$ -39.2 (c, 0.5 in $CHCl_3$)

(1S,2R)-form

[16354-95-3]

Cryst. (Me_2CO). Mp 106 - 107°. $[\alpha]_D^{23}$ +19.6 (c, 6.34 in H_2O)

2-O-β-D-Glucopyranoside

Cinnacassoside D

$C_{15}H_{22}O_8$ 330.333

Constit. of the bark of *Cinnamomum cassica*. Amorph. powder. $[\alpha]_D^{20}$ +11.2 (c, 0.2 in MeOH). λ_{max} 207 (log ε 3.52); 255 (log ε 2.78) (MeOH)

(1RS,2RS)-form

[65914-68-3]

Tri-Ac

[65870-47-5]

Liq.. Bp 149 - 150° at 0.1 mm

(1RS,2SR)-form

[65914-69-4]

Solid. Mp 98.5°

Tri-Ac

[65870-50-0]

$C_{15}H_{18}O_6$ 294.303

Liq.. Bp 149 - 150° at 0.1 mm

Tribenzoyl

[110454-18-7]

$C_{30}H_{24}O_6$ 480.516

Solid. Mp 152°.

Bradley, W. *et al.*, *J. Chem. Soc.*, 1938, 1578-1582 (*synth*).

Krüger, K.H. *Chem. Ber.*, 1956, **89**, 1016-1019 (*synth*).

Delton, M.H. *et al.*, *J. Org. Chem.*, 1968, **33**, 2473-2477 (*synth, pmr, abs config*).

Citterio, A. *et al.*, *J. Chem. Soc., Perkin Trans. 1*, 1983, 891-896 (*synth, pmr, ms, tri-Ac*).

McCombie, S.W. *et al.*, *Tetrahedron Lett.*, 1987, **28**, 383-386 (*synth*).

He, S. *et al.*, *Fitoterapia*, 2016, **112**, 153-160 (*Cinnacassoside D*).

Patel, A.R. *Tetrahedron*, 2016, **72**, 3305-3317 (*1R,2R-form, synth, 1R,2R-3-tosyl*).

Clement, J.A. *et al.*, *Phytochem. Lett.*, 2017, **22**, 117-121 (*Desmos pedunculosus constits*)

1-Phenyl-2-(tetrahydro-4-hydroxy-6-phenyl-2*H*-pyran-2-yl)ethanone, CAS P46

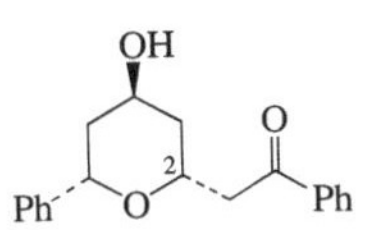

(2R,4S,6S)-form

$C_{19}H_{20}O_3$ 296.366

(2R,4S,6S)-form

Diospongin A [791104-63-7]

Constit. of the rhizomes of *Dioscorea spongiosa*. Amorph. solid. $[\alpha]^{25}_{D}$ -21.2 (c, 0.8 in $CHCl_3$)

(2S,4R,6R)-form

ent-Diospongin A [1192712-77-8]

Constit. of the rhizomes of *Dioscorea nipponica* ssp. *rosthornii*. Needles

(2S,4S,6S)-form

Diospongin B [791104-64-8]

Constit. of the rhizomes of *Dioscorea spongiosa*. Exhibits antiosteoporotic activity. Amorph. solid. $[\alpha]_D$ -23.4 (c, 0.6 in $CHCl_3$).

Yin, J. *et al.*, *Planta Med.*, 2004, **70**, 54-58 (*Diospongins A,B*).

Sawant, K.B. *et al.*, *J. Org. Chem.*, 2006, **71**, 7911-7914 (*synth, abs config*).

Chandrasekhar, S. *et al.*, *Tetrahedron Lett.*, 2006, **47**, 47-49 (*Diospongin B, synth*).

Bates, R.W. *et al.*, *Tetrahedron*, 2007, **63**, 4497-4499 (*synth*).

Hiebel, M.-A. *et al.*, *Tetrahedron*, 2007, **63**, 7874-7878 (*synth*).

Kawai, N. *et al.*, *Tetrahedron*, 2007, **63**, 9049-9056 (*synth*).

Kumaraswamy, G. *et al.*, *J. Org. Chem.*, 2009, **74**, 8468-8471 (*synth*).

Anada, M. *et al.*, *Eur. J. Org. Chem.*, 2010, 6850-6854 (*Diospongin B, synth*).

More, J.D. *Synthesis*, 2010, 2419-2423 (*synth*).

Kumar, R.N. *et al.*, *Tetrahedron Lett.*, 2011, **52**, 1003-1007 (*synth*).

Ho, T.-L. *et al.*, *J. Chin. Chem. Soc. (Taipei)*, 2012, **58**, 455-458 (*synth*).

Stefan, E. *et al.*, *Tetrahedron*, 2013, **69**, 7706-7712 (*synth*).

Zúñiga, A. *et al.*, *ARKIVOC*, 2015, **vii**, 195-215 (*synth*).

Rybak, T. *et al.*, *Org. Lett.*, 2015, **17**, 4156-4159 (*synth*).

Gharpure, S.J. *et al.*, *Isr. J. Chem.*, 2016, **56**, 553-557 (*synth*).

Li, Z. *et al.*, *Synthesis*, 2016, **48**, 1630-1636 (*Diospongin B, synth*).

Li, X. *et al.*, *Bioorg. Med. Chem. Lett.*, 2017, **27**, 3595-3601 (*ent-Diospongin A*).

Bharath, Y. *et al.*, *Org. Biomol. Chem.*, 2019, **17**, 9169-9181 (*synth*).

Vaithegi, K. *et al.*, *Tetrahedron*, 2020, **76**, 131625:1-6 (*synth*)

1-Phenyl-7-(3,4,5-trihydroxyphenyl)-4-hepten-3-one P47

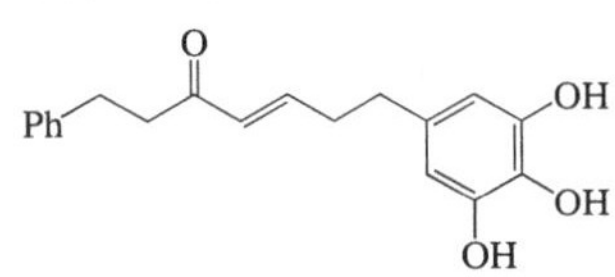

$C_{19}H_{20}O_4$ 312.365

(E)-form

3′-Me ether

7-(3,4-Dihydroxy-5-methoxyphenyl)-1-phenyl-4-hepten-3-one

[901118-74-9]

$C_{20}H_{22}O_4$ 326.392

Constit. of the rhizomes of *Alpinia officinarum*. Antibacterial agent. Pale yellow oil. $[\alpha]^{20}_{D}$ +7.5 (c, 0.5 in MeOH). λ_{max} 256; 280; 330; 380 (MeOH).

Zhang, B.-B. *et al.*, *Fitoterapia*, 2010, **81**, 948-952 (*Alpinia officinarum constit*)

Phlain IV P48

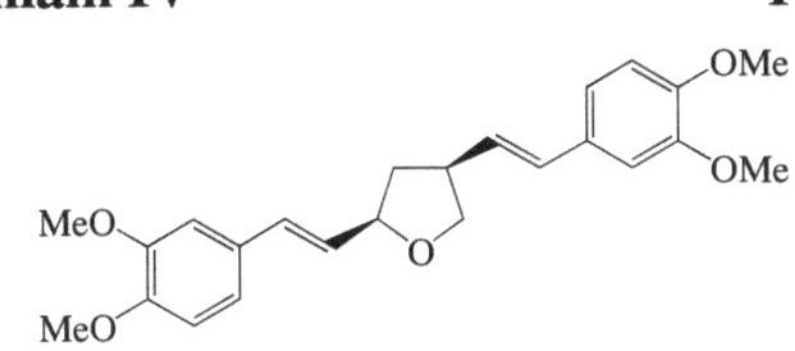

Relative Configuration

2,4-Bis[2-(3,4-dimethoxyphenyl) ethenyl]tetrahydrofuran

[1206701-59-8]

$C_{24}H_{28}O_5$ 396.483

Constit. of the rhizomes of *Zingiber cassumunar*. Pale yellow oil. $[\alpha]^{24}_{D}$ +8.3 (c, 0.21 in $CHCl_3$). λ_{max} 271 (log ε 4.2) ($CHCl_3$).

Nakamura, S. *et al.*, *Chem. Pharm. Bull.*, 2009, **57**, 1267-1272 (*Phlain IV*)

Phrymarolin V P49

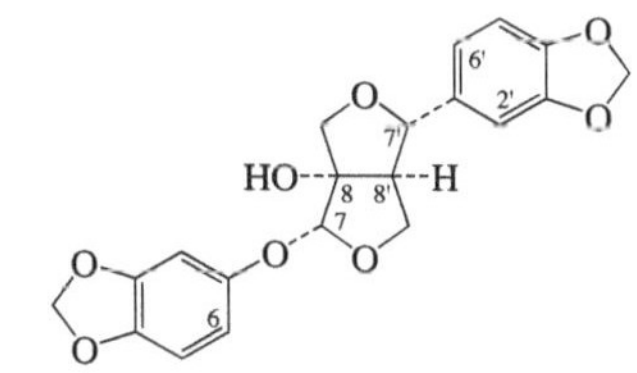

Absolute Configuration

1-Hydroxy-2-(3,4-methylenedioxyphenoxy)-6-(3,4-methylenedioxyphenyl)-3,7-dioxabicyclo[3.3.0]octane

[1449376-84-4]

$C_{20}H_{18}O_8$ 386.356

Possesses oxygen insertion in biosynth. pathway. Constit. of *Phryma leptostachya*. Powder. Mp 159 - 160°. $[\alpha]^{20}_{D}$ +155 (c, 3.6 in dioxan). λ_{max} 227 (log ε 3.73); 297 (log ε 4.06) (MeOH)

2′-Hydroxy

Phrymarolin III. *Phrymarolin B*

[1363160-29-5]

$C_{20}H_{18}O_9$ 402.355

Constit. of *Phryma leptostachya*. Exhibits insecticidal activity. Cryst. (EtOAc). Mp 143 - 144° (Phrymarolin B). Mp 160 - 161° (Phrymarolin III). $[\alpha]^{25}_{D}$ +155 (c, 3.6 in hexane) (Phrymarolin III). Stereochem. of Phrymarolin B not determined, therefore identity with Phrymarolin III tentative. λ_{max} 236 (log ε 3.97); 297 (log ε 4.06) (MeOH)

2′-Methoxy

Phrymarolin IV

[1363160-32-0]

$C_{21}H_{20}O_9$ 416.382

Constit. of *Phryma leptostachya*. Cryst. (EtOAc). Mp 160 - 161°. $[\alpha]^{25}_{D}$ +154 (c, 2.5 in dioxan). λ_{max} 234 (log ε 3.96); 297 (log ε 4.01) (MeOH)

6′-Methoxy, 8-Ac

1-Acetoxy-6-(2-methoxy-4,5-methylenedioxyphenyl)-2-(3,4-methylenedioxyphenoxy)-3,7-dioxabicyclo[3.3.0]octane.

Phrymarolin II

[23720-86-7]

$C_{23}H_{22}O_{10}$ 458.419

Constit. of the roots of *Phryma leptostachya*. Cryst. (Et_2O or EtOAc). Mp 161 - 162°. $[\alpha]_D$ +117.6 (c, 2.72 in dioxan). λ_{max} 236 (ε 8300); 296 (ε 10000) (EtOH)

6,6'-Dimethoxy, 8-Ac

Phrymarolin I

[38303-95-6]

$C_{24}H_{24}O_{11}$ 488.445

Constit. of the roots of *Phryma leptostachya*. Shows insecticidal activity. Cryst.. Mp 155 - 157°. $[\alpha]_D$ +131.3 (dioxan). λ_{max} 236 (ε 9300); 300 (ε 11600) (EtOH).

Taniguchi, E. *et al.*, *Agric. Biol. Chem.*, 1972, **36**, 1013-1025 (*Phrymarolin I*).
Taniguchi, E. *et al.*, *Agric. Biol. Chem.*, 1972, **36**, 1489-1496 (*Phrymarolin II*).
Ishibashi, F. *et al.*, *Bull. Chem. Soc. Jpn.*, 1988, **61**, 4361-4366 (*Phrymarolin I, synth, abs config*).
Taniguchi, E. *et al.*, *Chem. Lett.*, 1989, 313-316 (*Phrymarolin I, struct*).
Okazaki, M. *et al.*, *Biosci., Biotechnol., Biochem.*, 1997, **61**, 660-663 (*Phrymarolin I, synth*).
Xiao, X.-M. *et al.*, *Chem. Nat. Compd. (Engl. Transl.)*, 2012, **49**, 21-23 (*Phrymarolin B, activity*).
Chen, C. *et al.*, *Helv. Chim. Acta*, 2012, **95**, 333-338 (*Phrymarolins III,IV*).
Chen, C. *et al.*, *Helv. Chim. Acta*, 2013, **96**, 1392-1396 (*Phrymarolin V*)

Phyllanglaucin B P50

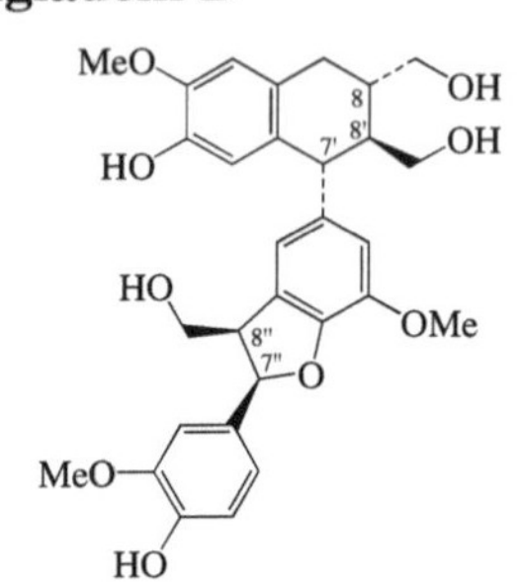

(7'R,7"R,8S,8'S,8"R)-form

$C_{30}H_{34}O_9$ 538.593

(7'R,7"R,8S,8'S,8"R)-form

Constit. of *Phyllanthus glaucus*. $[\alpha]_D^{20}$ -64.8 (c, 0.25 in MeOH). λ_{max} 206 (log ε 4.76); 231 (log ε 4.30); 283 (log ε 3.92) (MeOH)

(7'S,7"S,8R,8'R,8"S)-form

Constit. of *Phyllanthus glaucus*. $[\alpha]_D^{20}$ +62.6 (c, 0.06 in MeOH).

Wu, Z. *et al.*, *Sci. Rep.*, 2016, **6**, 24809 (*Phyllanglaucin B, abs config*)

Phyllostadimer A P51

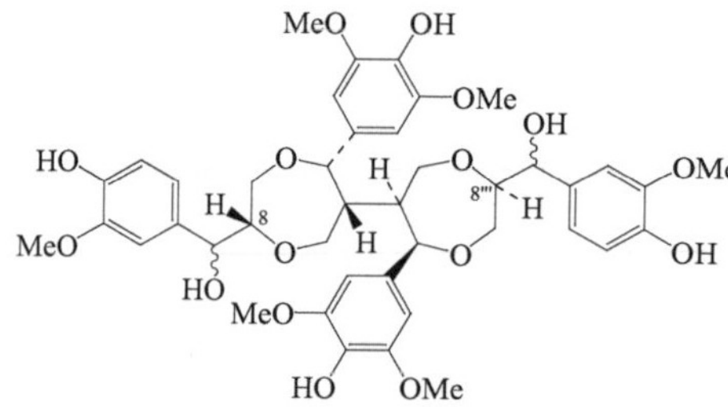

Relative Configuration

[638203-32-4]

$C_{42}H_{50}O_{16}$ 810.846

Constit. of the stems of bamboo, *Phyllostachys edulis*. Inhibits liposomal lipid peroxidation. Amorph. powder. Mp 117 - 119°. $[\alpha]_D$ -4 (c, 1 in $CHCl_3$)

8,8'''-Diepimer

Phyllostadimer B

[638203-38-0]

$C_{42}H_{50}O_{16}$ 810.846

Constit. of the stems of *Phyllostachys edulis*. Amorph. powder. Mp 118 - 120°. $[\alpha]_D$ +19 (c, 1 in $CHCl_3$).

Suga, A. *et al.*, *Phytochemistry*, 2003, **64**, 991-996 (*Phyllostadimers A,B*)

Pictalignan A P52

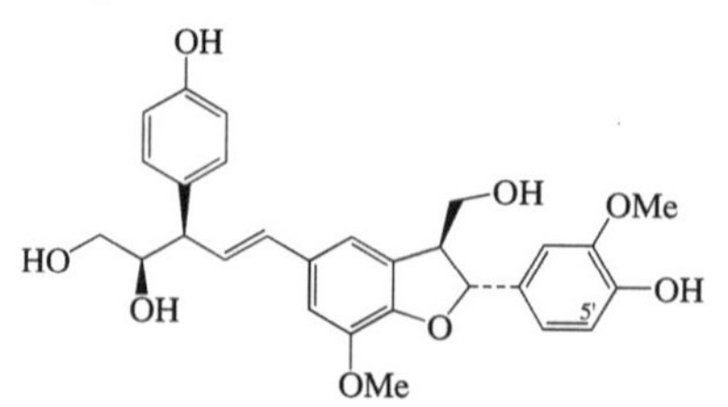

Absolute Configuration

$C_{28}H_{30}O_8$ 494.540

Constit. of *Selaginella picta*. Exhibits a moderate protective effect against the injury of HT-22 cells. Amorph. powder. $[\alpha]_D^{25}$ +2.5 (c, 0.12 in MeOH). λ_{max} 225 (log ε 2.63); 280 (log ε 2.42) (MeOH)

5'-Methoxy, 4'-O-β-D-glucopyranoside

Pictalignan B

$C_{35}H_{42}O_{14}$ 686.707

Constit. of *Selaginella picta*. Exhibits a moderate protective effect against the injury of HT-22 cells. Amorph. powder. $[\alpha]_D^{25}$ -28.4 (c, 0.11 in MeOH). λ_{max} 225 (log ε 2.48); 280 (log ε 2.27) (MeOH).

Cheng, F. *et al.*, *Fitoterapia*, 2018, **127**, 69-73 (*Pictalignans A,B, activity*)

Pictalignan C P53

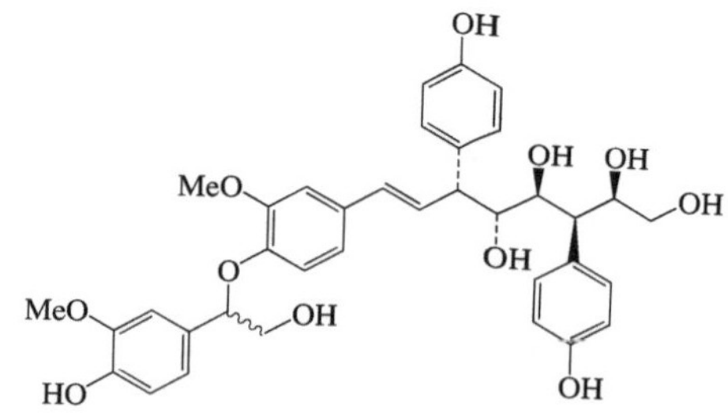

Absolute Configuration

$C_{36}H_{40}O_{11}$ 648.705

Constit. of *Selaginella picta*. Exhibits a moderate protective effect against the injury of HT-22 cells. Amorph. powder. $[\alpha]_D^{25}$ +22.7 (c, 0.11 in MeOH). λ_{max} 225 (log ε 2.69); 280 (log ε 2.44) (MeOH).

Cheng, F. *et al.*, *Fitoterapia*, 2018, **127**, 69-73 (*Pictalignan C, activity*)

Pinobatol P54

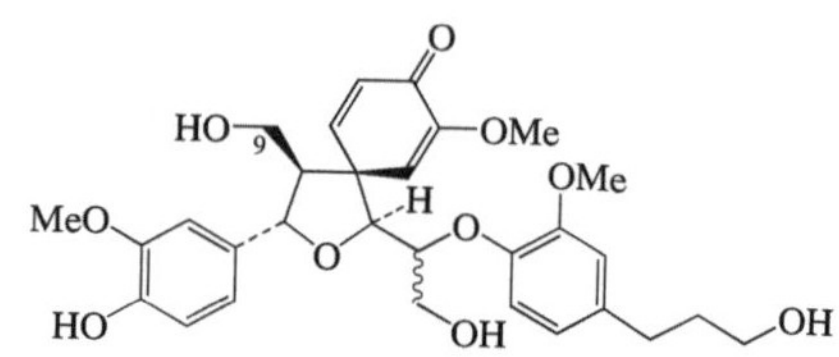

Relative Configuration

[945468-41-7]

$C_{30}H_{36}O_{10}$ 556.608

Spirosesquilignan. Related to Woorenol, W1. Constit. of *Abies nephrolepis*, *Kadsura longipeduculata* and *Pinus sylvestris*. $[\alpha]_D^{20}$ +6 (c, 0.8 in EtOH). λ_{max} 238; 259; 278 (EtOH)

9-O-β-D-Glucopyranoside

[2013596-07-9]

$C_{36}H_{46}O_{15}$ 718.749

Constit. of the stems of *Schisandra chinensis*. Amorph. powder. $[\alpha]_D^{20}$ +6 (c, 0.3 in MeOH). λ_{max} 258 (log ε 3.50) (MeOH).

Sinkkonen, J. *et al.*, *Angew. Chem., Int. Ed.*, 2007, **46**, 4148-4150 (*Pinobatol*).
Chen, J.-B. *et al.*, *Zhongcaoyao*, 2015, **46**, 178-184 (*Kadsura longipeduculata constit*).
Yang, B.-Y. *et al.*, *Chem. Biodiversity*, 2016, **13**, 1118-1125 (*9-glucoside*)

Piperarborenine C P55

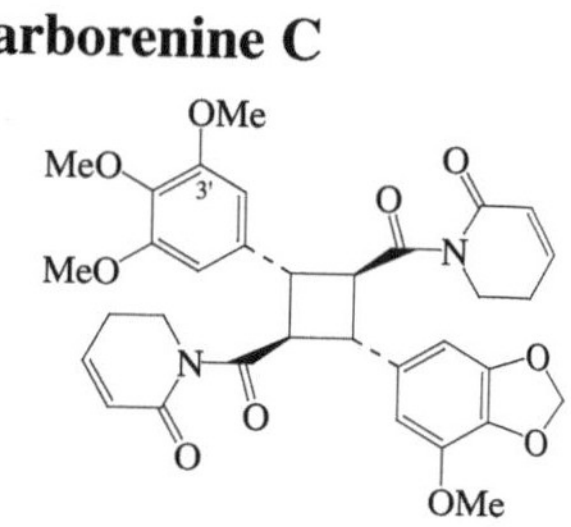

[870990-52-6]

$C_{33}H_{34}N_2O_{10}$ 618.639
Alkaloid from the stems of *Piper arborescens*. Brownish oil. Racemic. λ_{max} 219 (log ε 4.58); 260 (sh); 280 (sh) (MeOH)

3′-Demethoxy
Piperarborenine E
[870990-54-8]
$C_{32}H_{32}N_2O_9$ 588.613
Alkaloid from the stems of *Piper arborescens*. Brownish oil. Racemic. λ_{max} 210 (log ε 4.5); 280 (sh) (MeOH).

Tsai, I.-L. *et al.*, *Planta Med.*, 2005, **71**, 535-542 (*Piperarborenines C,E*)

Piperarborenine D P56

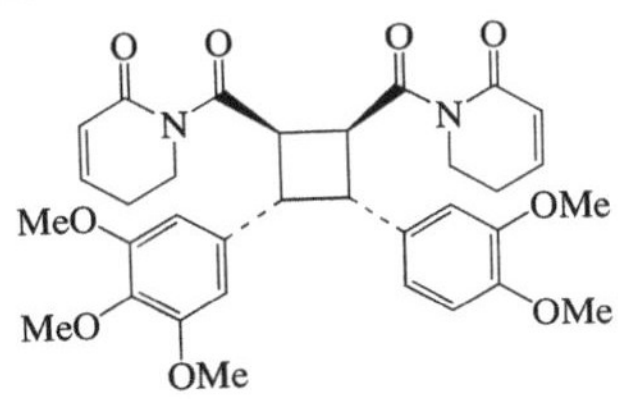

[870990-53-7]
$C_{33}H_{36}N_2O_9$ 604.656
Struct. revised in 2011. Alkaloid from the stems of *Piper arborescens*. Foam. λ_{max} 217 (log ε 4.55); 240 (log ε 4.33); 280 (log ε 3.73) (MeOH).

Tsai, I.L. *et al.*, *Planta Med.*, 2005, **71**, 535-542 (*Piperarborenine D*).
Gutekunst, W.R. *et al.*, *J. Am. Chem. Soc.*, 2011, **133**, 19076-19079 (*synth, struct*).
Frébault, F. *et al.*, *Angew. Chem., Int. Ed.*, 2012, **51**, 2815-2817 (*synth, struct*)

Piperarboresine P57

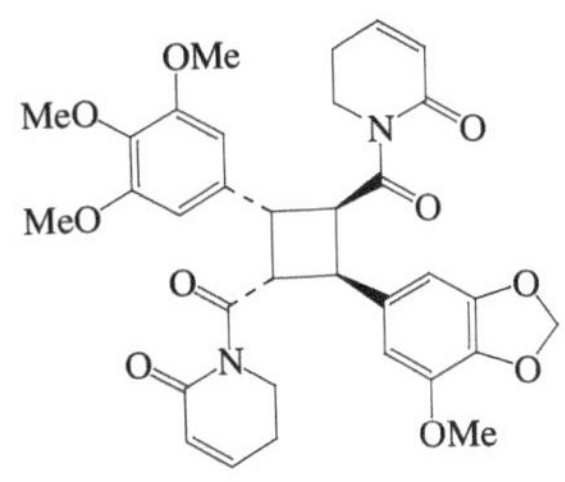

[741265-08-7]
$C_{33}H_{34}N_2O_{10}$ 618.639
Quasi-*meso*. Alkaloid from the stems and leaves of *Piper arborescens*. Needles (MeOH). Mp 182 - 184°. λ_{max} 213 (log ε 4.55); 240 (sh) (log ε 4.11) (MeOH).

Lee, F.-P. *et al.*, *Helv. Chim. Acta*, 2004, **87**, 463-468 (*Piperarboresine*)

Piperbonin C P58

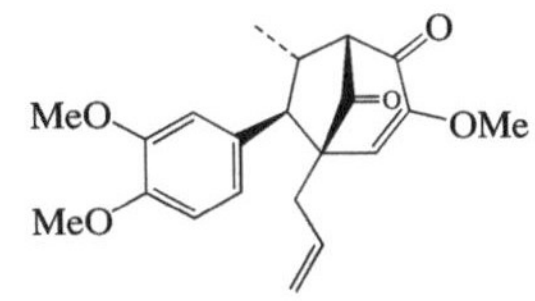

Relative Configuration

6-(3,4-Dimethoxyphenyl)-3-methoxy-7-methyl-5-(2-propenyl)bicyclo[3.2.1]oct-3-ene-2,8-dione, CAS. *3,8′-Cyclo-3′,4′,5-trimethoxy-1,7′-lign-8-ene-2,4(1H,3H)-dione*
[2001069-76-5]
$C_{21}H_{24}O_5$ 356.418
Constit. of the aerial parts of *Piper bonii*. Pale yellow oil. $[\alpha]^{20}_{D}$ -70 (c, 0.3 in MeOH). λ_{max} 228 (log ε 3.95); 280 (log ε 3.68) (MeOH).

Ding, D.-D. *et al.*, *Phytochemistry*, 2016, **129**, 36-44 (*Piperbonin C*)

Pipercyclobutanamide B P59

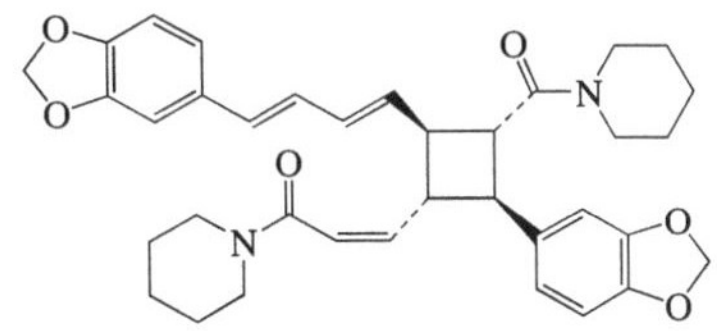

Relative Configuration

[342371-98-6]
$C_{36}H_{40}N_2O_6$ 596.724
Alkaloid from the fruit of *Piper nigrum*. Amorph. powder. Structure probably in need of revision. λ_{max} 285; 296; 309 (hexane).

Fujiwara, Y. *et al.*, *Tetrahedron Lett.*, 2001, **42**, 2497-2499 (*Pipercyclobutanamide B*)

Piperitylmagnolol P60

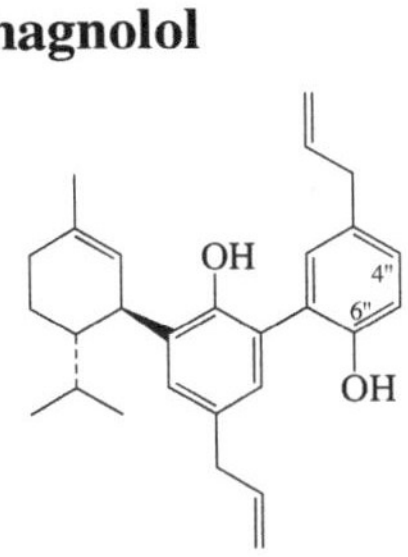

Absolute Configuration

Monoterpenylmagnolol
[135566-92-6]
$C_{28}H_{34}O_2$ 402.578
Constit. of *Magnolia officinalis*. Exhibits cytotoxic activity. Antitumour promotor. Antibacterial agent. Active against MRSA. Oil. $[\alpha]_D$ -138 (c, 1.03 in $CHCl_3$). λ_{max} 208 (log ε 3.85); 258 (log ε 3.80) (MeOH)

4″-Hydroxy, 6″-deoxy
Monoterpenylhonokiol.
Piperitylhonokiol
[92891-88-8]
$C_{28}H_{34}O_2$ 402.578
Constit. of *Magnolia officinalis*. Viscous oil. $[\alpha]^{24}_{D}$ -97 (c, 0.69 in $CHCl_3$).

Yahara, S. *et al.*, *Chem. Pharm. Bull.*, 1991, **39**, 2024-2036 (*Piperitylmagnolol: Piperitylhinokiol*).
Konoshima, T. *et al.*, *J. Nat. Prod.*, 1991, **54**, 816-822 (*Monoterpenylmagnolol, activity*).
Agharahimi, M.R. *et al.*, *J. Org. Chem.*, 1995, **60**, 1857-1863 (*Monoterpenylmagnolol, synth*).
Syu, W.-J. *et al.*, *Chem. Biodiversity*, 2004, **1**, 530-537 (*Piperitylmagnolol, activity*).
Youn, U.J. *et al.*, *Nat. Prod. Sci.*, 2011, **17**, 95-99 (*Piperitylmagnolol, activity*)

Piperkadsin B P61

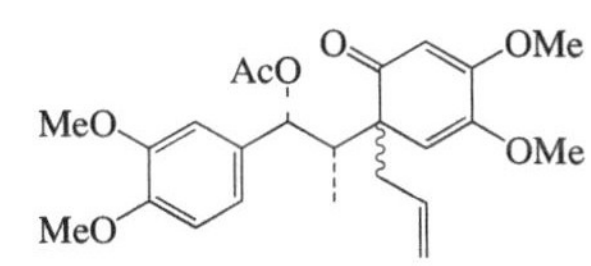

Absolute Configuration

7′-Acetoxy-3′,4,4′,5-tetramethoxy-1,8′-lign-8-en-2(1H)-one
[895543-37-0]
$C_{24}H_{30}O_7$ 430.497
Constit. of the stems of *Piper kadsura*. Exhibits antiinflammatory activity. Viscous oil. $[\alpha]^{24}_{D}$ +34.7 (c, 0.98 in $CHCl_3$). λ_{max} 235 (log ε 4.22); 280 (log ε 3.75); 318 (log ε 3.57) (MeOH).

Lin, L.-C. *et al.*, *J. Nat. Prod.*, 2006, **69**, 842-844 (*Piperkadsin B, cd, activity*)

Piperkadsin C P62

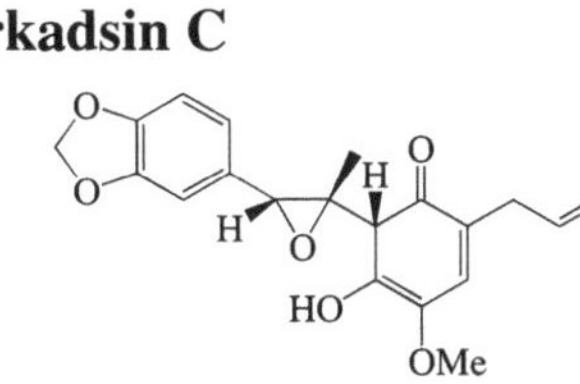

Absolute Configuration

7′,8′-Epoxy-4-hydroxy-5-methoxy-3′,4′-methylenedioxy-3,8′-lign-8-en-2(3H)-one
[1207645-08-6]
$C_{20}H_{20}O_6$ 356.374
Constit. of aerial parts of *Piper kadsura*. Potent inhibitor of NO production. Gum. $[\alpha]^{25}_{D}$ -26.2 (c, 0.1 in $CHCl_3$). λ_{max} 235 (log ε 4.2); 280 (log ε 3.5) (MeOH).

Kim, K.H. *et al.*, *Bioorg. Med. Chem. Lett.*, 2010, **20**, 409-412 (*Piperkadsin C*).
Kim, K.H. *et al.*, *Bioorg. Med. Chem. Lett.*, 2010, **20**, 3186-3187 (*corringendum*)

Piplartine dimer A P63

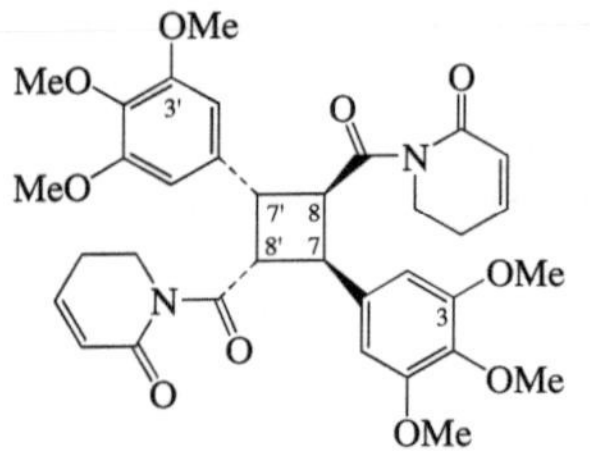

1,1′-[[2,4-Bis(3,4,5-trimethoxyphenyl)-1,3-cyclobutanediyl]dicarbonyl]bis[5,6-dihydro-2(1H)-pyridone]
[80248-71-1]
$C_{34}H_{38}N_2O_{10}$ 634.682
Lignan numbering shown. Alkaloid from *Piper arborescens*, *Piper rugosum* and *Piper tuberculatum*. Cytotoxic. Cryst. (MeOH). Mp 269 - 270° (262-264°). λ_{max} 214 (log ε 4.53); 240 (sh) (MeOH)

3′-Demethoxy
Piperarborenine B
[741265-07-6]
$C_{33}H_{36}N_2O_9$ 604.656
Alkaloid from the stems of *Piper arborescens*. Needles (MeOH). Mp 158 - 162°. λ_{max} 214 (log ε 4.52); 240 (sh); 280 (sh) (MeOH)

3,3′-Bis(demethoxy)
Piperarborenine A
[741265-05-4]
$C_{32}H_{34}N_2O_8$ 574.63
Alkaloid from the stems of *Piper arborescens*. Needles (MeOH). Mp 220 - 222°. $[\alpha]_D^{25}$ +5.5 (c, 0.05 in $CHCl_3$). λ_{max} 212 (log ε 4.53); 240 (sh); 280 (sh) (MeOH)

7,8′-Diepimer, 3,3′-bis(demethoxy)
[186391-17-3]
$C_{32}H_{34}N_2O_8$ 574.63
Alkaloid from *Piper longum*. Cryst. ($CHCl_3$). Mp 135 - 136°. λ_{max} 240; 271 (sh) ($CHCl_3$).

Filho, R.B. *et al.*, *Phytochemistry*, 1981, **20**, 345-346 (*Piplartine dimer A*).
Maxwell, A. *et al.*, *J. Nat. Prod.*, 1991, **54**, 1150-1152 (*Piper rugosum constit*).
Zhang, K. *et al.*, *Yunnan Zhiwu Yanjiu*, 1996, **18**, 353-355 (*Piper longum dimer*).
Lee, F.-P. *et al.*, *Helv. Chim. Acta*, 2004, **87**, 463-468 (*Piperarborenines A,B*).
Gutekunst, W.R. *et al.*, *J. Am. Chem. Soc.*, 2011, **133**, 19076-19070 (*Piperarborenine B, synth*).
Panish, R.A. *et al.*, *Angew. Chem., Int. Ed.*, 2016, **55**, 4983-4987 (*Piperarborenine B, synth*).
Hu, J.-L. *et al.*, *J. Am. Chem. Soc.*, 2016, **138**, 13151-13154 (*Piperarborenine B, synth*)

Plasiatine P64

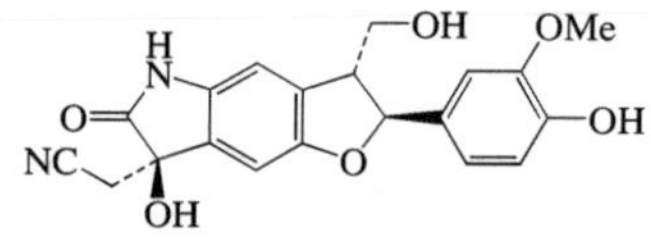

Absolute Configuration

[2172906-57-7]
$C_{20}H_{18}N_2O_6$ 382.372
Alkaloid from the seeds of *Plantago asiatica*. Potent activator of non-receptor protein tyrosine phosphatase Shp2. Amorph. powder. $[\alpha]_D^{23}$ -37.1 (c, 0.1 in MeOH). λ_{max} 277 (log ε 3.33); 320 (log ε 2.63) (MeOH).

Gao, Z.-H. *et al.*, *Sci. Rep.*, 2016, **6**, 24945:1-9 (*Plasiatine, cd, activity*)

Pleionesin C P65

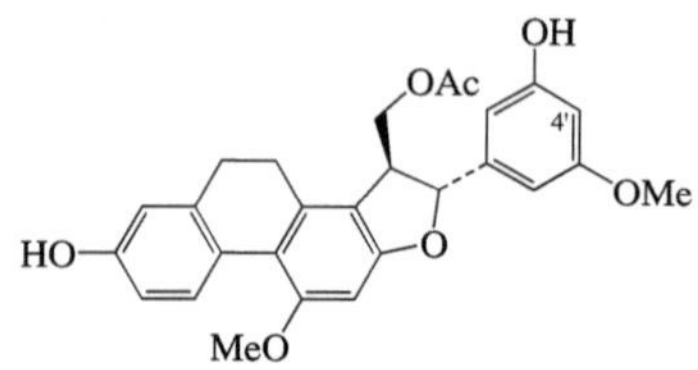

Relative Configuration

2-(3-Hydroxy-5-methoxyphenyl)-3-acetoxymethyl-10-methoxy-2,3,4,5-tetrahydrophenanthro[2,1-b]furan-7-ol
[1222077-25-9]
$C_{27}H_{26}O_7$ 462.498
Phenanthrenolignan. Structs. in this series may need revision. Phenyl 3,5-dioxygenated substn. pattern in doubt as a result of work done by Yang *et al* in 2015. Constit. of *Bletilla striata* and *Pleione yunnanensis*. Syrup. $[\alpha]_D^{25}$ -1.5 (c, 0.1 in MeOH). Partial racemate. Absolute configuration determined in 2019. λ_{max} 205 (log ε 4.74); 282 (log ε 4.35) (MeOH)

4′-Hydroxy, 3′-Me ether
Pleionesin B
[1222077-24-8]
$C_{28}H_{28}O_8$ 492.524
Constit. of the tubers of *Pleione yunnanensis*. Amorph. powder. $[\alpha]_D^{25}$ -2 (c, 0.1 in MeOH). λ_{max} 221 (log ε 4.48); 282 (log ε 4.2) (MeOH)

4′-Hydroxy, 3′-Me ether, O-de-Ac
Cyrtonesin B
[2241044-04-0]
$C_{26}H_{26}O_7$ 450.487
Constit. of *Bletilla striata* and *Cyrtopodium paniculatum*. Amorph. powder. $[\alpha]_D^{25}$ +26 (c, 0.02 in MeOH). Partial racemate. λ_{max} 207 (log ε 4.29); 292 (log ε 3.73); 301 (log ε 3.77) (MeOH)

3′-Deoxy, 4′-hydroxy
Shanciol G
[1114905-53-1]
$C_{27}H_{26}O_7$ 462.498
Constit. of *Eria bambusifolia* and *Pleione bulbocodioides*. Amorph. powder. $[\alpha]_D^{25}$ -5.3 (c, 0.1 in MeOH). Config. not determined. λ_{max} 216 (log ε 4.57); 283 (log ε 4.41); 306 (log ε 4.19) (MeOH)

3′-Deoxy, 4′-hydroxy, O-de-Ac
Erathrin B
[1454585-39-7]
$C_{25}H_{24}O_6$ 420.461
Constit. of *Bletilla straiata*, *Cremastra appendiculata* and *Eria bambusifolia*. Amorph. brown powder. $[\alpha]_D^{20}$ +5 (c, 0.133 in MeOH) (Cremastra constit). $[\alpha]_D^{19}$ -7 (c, 0.18 in MeOH) (Erathrin B). Partial racemate. Absolute configuration determined in 2019. λ_{max} 205 (log ε 4.39); 283 (log ε 4.00); 306 (log ε 3.81) (MeOH).

Liu, X.-Q. *et al.*, *J. Asian Nat. Prod. Res.*, 2008, **10**, 453-457 (*Shanciol G*).
Dong, H. *et al.*, *Magn. Reson. Chem.*, 2010, **48**, 256-260 (*Pleionesins B,C*).
Wang, Y. *et al.*, *Phytochemistry*, 2013, **94**, 268-276 (*Cremastra appendiculata constit*).
Yang, Y.-N. *et al.*, *J. Nat. Prod.*, 2015, **78**, 705-711 (*pmr, struct*).
Zhan, R. *et al.*, *Chin. J. Nat. Med.*, 2016, **14**, 621-625 (*Erathrin B*).
Auberon, F. *et al.*, *Molecules*, 2016, **21**, 1418:1-18 (*Cyrtonesin B*).
Zhou, D. *et al.*, *J. Nat. Prod.*, 2019, **82**, 2238-2245 (*Pleionesin C, Eranthrin B, abs config*)

Podotoxin P66

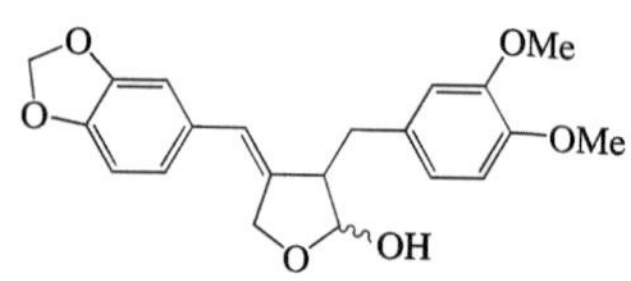

9,9′-Epoxy-3,4-dimethoxy-3′,4′-methylenedioxy-8,8-lign-7′-en-9-ol.
9-Hydroxy-3,4-dimethoxy-3′,4′-methylenedioxy-9,9′-epoxylign-7′-ene
[73191-25-0]
$C_{21}H_{22}O_6$ 370.401
Isol. from stem bark of *Zanthoxylum acanthopodium*. Germination inhibitor. Mp 140 - 142°. λ_{max} 232 (ε 39000); 282 (ε 6000) (EtOH) (Berdy).

Chakraborty, D.P. *et al.*, *Chem. Ind. (London)*, 1979, 667-668 (*Podotoxin*)

Pouzolignan A P67

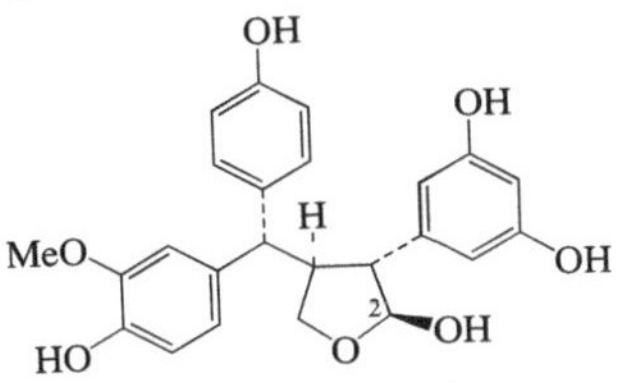

Relative Configuration

[1299486-05-7]
$C_{24}H_{24}O_7$ 424.449
Structures in this series may need revision. Phenyl 3,5-dioxygenated substn. pattern in doubt as a result of work done by Yang *et al* in 2015.
Constit. of *Pouzolzia occidentalis*. Brown gum. $[\alpha]^{25}_{D}$ -32.4 (c, 0.1 in MeOH). λ_{max} 270 (log ε 3.48) (MeOH)

2-Me ether

2-O-Methylpouzolignan A
[1299486-07-9]
$C_{25}H_{26}O_7$ 438.476
Constit. of *Pouzolzia occidentalis*. Brown gum. $[\alpha]^{25}_{D}$ -44.4 (c, 0.1 in MeOH). Probable artifact. λ_{max} 270 (log ε 3.89) (MeOH).

Mohammed, M. *et al.*, *Phytochem. Lett.*, 2010, **3**, 29-32 (*Pouzolignan A, 2-O-Methylpouzolignan A*).
Mohammed, M. *et al.*, *Phytochem. Lett.*, 2011, **4**, 56 (*erratum*).
Yang, Y.-N. *et al.*, *J. Nat. Prod.*, 2015, **78**, 705-711 (*pmr, struct*)

Pouzolignan D P68

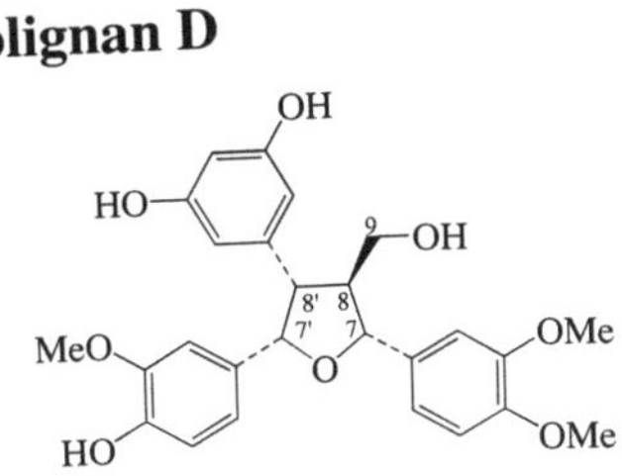

Relative Configuration

4-(3,5-Dihydroxyphenyl)-2-(3,4-dimethoxyphenyl)-5-(4-hydroxy-3-methoxyphenyl)-3-(hydroxymethyl) tetrahydrofuran
[1947492-06-9]
$C_{26}H_{28}O_8$ 468.502
Constit. of aerial parts of *Pouzolzia zeylanica*. Antiinflammatory agent. Amorph. brown powder. $[\alpha]^{20}_{D}$ +34 (c, 0.1 in MeOH). λ_{max} 274 (log ε 4.02) (MeOH).

Li, K.-Y. *et al.*, *Zhongguo Yaolixue Yu Dulixue Zazhi*, 2014, **28**, 731-736 (*activity*).
Zhong, C.-Q. *et al.*, *Heterocycles*, 2015, **91**, 1926-1936 (*Pouzolignan D*)

Precolpuchol P69

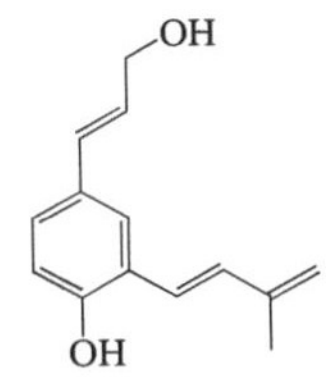

4-(3-Hydroxy-1-propenyl)-2-(3-methyl-1,3-butadienyl)phenol, CAS. *3-[4-Hydroxy-3-(3-methyl-1,3-butadienyl)phenyl]-2-propen-1-ol*
$C_{14}H_{16}O_2$ 216.28

(E,E)-form [194150-51-1]
Constit. of *Coleonema pulchellum*. Oil. λ_{max} 220; 269; 327 (Et_2O).

Brader, G. *et al.*, *Phytochemistry*, 1997, **45**, 1207-1212 (*Precolpuchol*)

Prepicrasmalignan P70

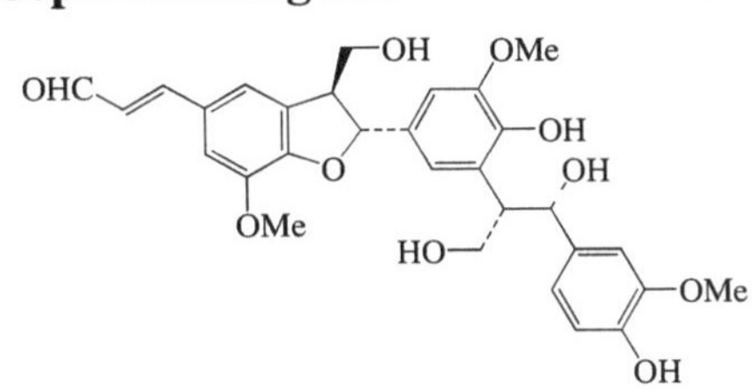

Absolute Configuration

[1887236-27-2]
$C_{30}H_{32}O_{10}$ 552.576
Sesquineolignan. Constit. of the fruit of *Cirsium eriophorum*. λ_{max} 236; 289; 347 (no solvent reported).

Sólyomváry, A. *et al.*, *Process Biochem.*, 2015, **50**, 853-858 (*Prepicrasmalignan, cd*).
Könye, R. *et al.*, *Fitoterapia*, 2018, **127**, 413-419 (*isol*)

Princepin P71

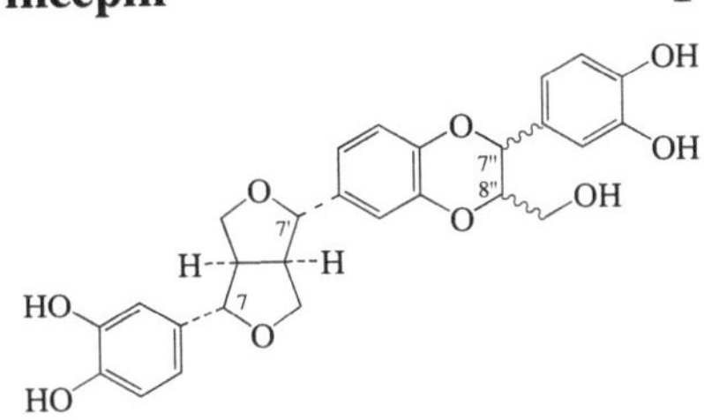

[571166-69-3]
$C_{27}H_{26}O_9$ 494.496
Sesquilignan. Related to Haedoxan B, H2. Constit. of *Capparis flavicans* and *Joannesia princeps*. Exhibits inhibitory activity on estrogen-responsive human breast cancer cell lines. Amorph. solid. Isol. as a mixt. of C-7″/8″-*trans*-isomers. λ_{max} 207 (log ε 4.97); 225 (sh) (log ε 4.36); 282 (log ε 4.07) (MeOH).

Waibel, R. *et al.*, *Phytochemistry*, 2003, **62**, 805-811 (*Princepin*).
Luecha, P. *et al.*, *J. Nat. Prod.*, 2009, **72**, 1954-1959 (*Capparis flavicola constit, activity*).
Kobayashi, M. *et al.*, *J. Org. Chem.*, 2019, **84**, 14227-14240 (*synth*)

2-(1-Propenyl)-1,4-benzenediol P72

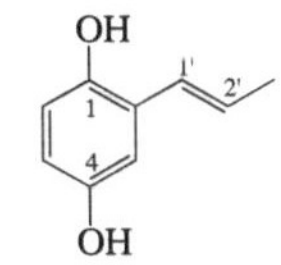

$C_9H_{10}O_2$ 150.177
Structs. of some nat. prods. revised in 1985. Formerly assigned as derivs. of 2-Methoxy-4-(1-propenyl)phenol, M37

(E)-form

4-Me ether

4-Methoxy-2-(1-propenyl)phenol. Pseudoisoeugenol
[98755-22-7]
[98755-19-2] (Z-form)
$C_{10}H_{12}O_2$ 164.204
Parent compd. of nat. prods. but does not appear to have been well characterised. Has been synthesised.

Dev, V. *et al.*, *J. Nat. Prod.*, 1987, **50**, 968-971 (*synth*).
Jimenez-Gonzalez, L. *et al.*, *Chem. - Eur. J.*, 2006, **12**, 8762-8769 (*synth*).
Daniels, R.N. *et al.*, *Org. Lett.*, 2008, **10**, 4097-4100 (*synth*)

2-(2-Propenyl)-1,4-benzenediol, CAS P73

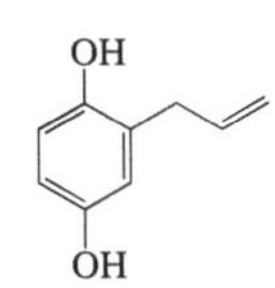

Allylhydroquinone, CAS
[5721-21-1]
$C_9H_{10}O_2$ 150.177
Cryst. ($CHCl_3$/petrol). Mp 92 - 93°

4-Me ether

4-Methoxy-2-(2-propenyl)phenol, CAS. *2-Allyl-4-methoxyphenol. ψ-Eugenol. Pseudoeugenol*
[584-82-7]
$C_{10}H_{12}O_2$ 164.204
Monomer in the biosynthesis of lignans. Bp 105 - 107° at 0.3 mm. Lignan monomer.

Smith, L.I. *et al.*, *J. Am. Chem. Soc.*, 1940, **62**, 1863-1869 (*synth*).
Darling, S.D. *et al.*, *J. Org. Chem.*, 1967, **32**, 2794-2796 (*4-Me ether, synth, pmr*).
Hegedus, L.S. *et al.*, *J. Am. Chem. Soc.*, 1976, **98**, 3901-3909 (*synth, ir, pmr*).
Pan, D. *et al.*, *Tetrahedron*, 2002, **58**, 2847-2852 (*synth, ir, pmr*).

Manetsch, R. *et al.*, *Chem. - Eur. J.*, 2004, **10**, 2487-2506 (*4-Me ether, synth, ir, uv, pmr, cmr*)

4-(1-Propenyl)-1,3-benzenediol P74

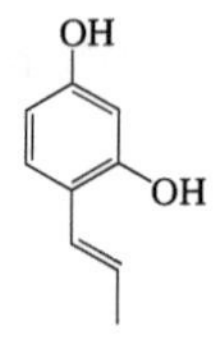

(*E*)-*form*

4-(1-Propenyl)resorcinol.
1-(2,4-Dihydroxyphenyl)-1-propene
$C_9H_{10}O_2$ 150.177

(E)-form

1-Me ether, 3-O-[α-L-rhamnopyranosyl-(1→6)-β-D-glucopyranoside]
[634920-40-4]
$C_{22}H_{32}O_{11}$ 472.487
Constit. of the fruit of *Illicium verum*. Amorph. yellow powder. Mp 109 - 111°. $[\alpha]_D^{25}$ +52.9 (c, 0.22 in MeOH). λ_{max} 224 (log ε 3.99); 281 (log ε 3.48) (MeOH)

Di-Me ether
2,4-Dimethoxy-1-(1-propenyl)benzene, CAS. ***Isoosmorhizol***
[50900-55-5]
[829-39-0]
$C_{11}H_{14}O_2$ 178.231
Constit. of *Myrrhis odorata* and *Scandix australis*. Strictly speaking the synonym Isoosmorhizol refers to the compd. with unspecified stereochem.

(Z)-form

Di-Me ether
(Z)-Isoosmorhizol. **Nothosmyrnol**
[17627-75-7]
$C_{11}H_{14}O_2$ 178.231
Constit. of *Nothosmyrnium japonicum*, *Osmorhiza aristata* and *Piper marginatum*. Bp 75 - 77° at 1 mm.

Huisgen, R. *et al.*, *Justus Liebigs Ann. Chem.*, 1964, **677**, 21-33 (*synth*).
Konoshima, M. *et al.*, *Yakugaku Zasshi*, 1967, **87**, 1138-1141 (*Nothosmyrnol*).
Saiki, Y. *et al.*, *Yakugaku Zasshi*, 1970, **90**, 344-351 (*Isoosmorhizol, synth*).
Fujita, H. *et al.*, *Yuki Gosei Kagaku Kyokaishi*, 1973, **31**, 770-771 (*synth*).
Lee, S.-W. *et al.*, *Arch. Pharmacal Res.*, 2003, **26**, 591-593 (*Illicium verum glycoside*)

4-(2-Propenyl)-1,2-benzenediol, CAS P75

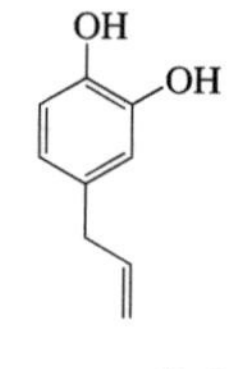

4-Allyl-1,2-benzenediol.
4-Allylpyrocatechol,
CAS. *4-Allylcatechol.*
3-(3,4-Dihydroxyphenyl)propene.
3-Hydroxychavicol
[1126-61-0]
$C_9H_{10}O_2$ 150.177
Constit. of *Piper betle*, *Piper taiwanense* and *Syzygium polyanthum*. Prod. by *Acrocarpospora punica* 04107M-2. Antioxidant, antimutagenic and antiinflammatory. Cytotoxic. Inhibitor of COX-1, COX-2, pancreatic lipase and xanthine oxidase. Platelet aggregation inhibitor. Cryst. (C_6H_6/petrol). Mp 50 - 52° (48-48.5°). Bp 120 - 122° at 2 mm

1-O-β-D-Glucopyranoside
[146087-04-9]
$C_{15}H_{20}O_7$ 312.318
Constit. of *Mentha spicata* var. *crispa*, *Alpinia officinarum*, *Magnolia denudata*, *Piper retrofractum* and *Piper taiwanense*. Antimicrobial agent for oral care formulations. $[\alpha]_D^{25}$ -108 (c, 0.18 in MeOH) (-52.6). λ_{max} 276 (log ε 2.89) (MeOH)

1-O-(6-O-Acetyl-β-D-glucopyranoside)
[1631960-12-7]
$C_{17}H_{22}O_8$ 354.355
Isol. from *Schizonepeta annua*. Amorph. powder

1-O-[2E-Butenoyl-(→6)-β-D-glucopyranoside]
[1631960-13-8]
$C_{19}H_{24}O_8$ 380.393
Isol. from *Schizonepeta annua*. Amorph. powder

2-O-β-D-Glucopyranoside
[454473-94-0]
$C_{15}H_{20}O_7$ 312.318
Constit. of the rhizomes of *Alpinia officinarum*. Brown wax. $[\alpha]_D^{25}$ -56.8 (c, 0.44 in MeOH). λ_{max} 256 (log ε 2.59); 278 (log ε 2.41) (MeOH)

2-O-(6-O-Acetyl-β-D-glucopyranoside)
[1278488-71-3]
$C_{17}H_{22}O_8$ 354.355
Constit. of *Illicium simonsii*. Amorph. solid. $[\alpha]_D^{28}$ -29.7 (c, 0.29 in MeOH). λ_{max} 280 (log ε 3.42) ($CHCl_3$)

2-O-β-D-Allopyranoside
Manglieside A. *Michehedyoside A*
[1093091-40-7]
$C_{15}H_{20}O_7$ 312.318
Constit. of *Manglietia phuthoensis* and *Michelia hedyosperma*. Amorph. powder. $[\alpha]_D^{25}$ -63 (c, 1 in MeOH). λ_{max} 204 (log ε 4.41); 284 (log ε 3.46) (MeOH)

1-O-[β-D-Apiofuranosyl-(1→6)-β-D-glucopyranoside]
[942602-10-0]
$C_{20}H_{28}O_{11}$ 444.433
Constit. of *Glechoma hederacea*. Amorph. powder. $[\alpha]_D^{26}$ -43.4 (c, 0.03 in MeOH). λ_{max} 220 (sh); 274 (log ε 3.6) (MeOH)

1-O-[β-D-Xylopyranosyl-(1→6)-β-D-glucopyranoside]
Forsythenside L
[1450732-43-0]
$C_{20}H_{28}O_{11}$ 444.433
Constit. of *Forsythia suspensa* and *Scilla scilloides*. Amorph. powder. $[\alpha]_D^{16}$ -64.8 (c, 4.2 in MeOH). λ_{max} 276 (log ε 3.31) (MeOH) (Scilla constit). [neutral] λ_{max} 205 (log ε 4.52); 254 (log ε 4.56) (MeOH) (Forsythenside L)

1-O-[α-L-Rhamnopyranosyl-(1→6)-β-D-glucopyranoside]
[267668-14-4]
$C_{21}H_{30}O_{11}$ 458.46
Constit. of *Dracaena cochinchinensis* and rhizomes of *Alpinia officinarum*. $[\alpha]_D$ -106 (c, 0.18 in MeOH). λ_{max} 275 (log ε 2.96) (MeOH)

1-O-[α-L-Rhamnopyranosyl-(1→6)-β-D-glucopyranosyl-(1→3)-[α-L-rhamnopyranosyl-(1→6)]-β-D-glucopyranoside]
Cambodianin F
[1800405-22-4]
$C_{33}H_{50}O_{20}$ 766.743
Constit. of the resin of *Dracaena cambodiana*. Brown wax. $[\alpha]_D^{20}$ -180 (c, 0.02 in MeOH). λ_{max} 208 (log ε 2.54); 215 (log ε 1.65); 278 (log ε 0.56) (MeOH)

1-O-[β-D-Glucopyranosyl-(1→4)-α-L-rhamnopyranosyl-(1→6)-β-D-allopyranoside]
Magnoloside R
[1946004-91-6]
$C_{27}H_{40}O_{16}$ 620.601
Constit. of stem bark of *Magnolia officinalis*. Amorph. solid. $[\alpha]_D^{20}$ -82.6 (c, 0.23 in MeOH). λ_{max} 278 (log ε 3.56) (H_2O)

1-O-[β-D-Glucopyranosyl-(1→4)-α-L-rhamnopyranosyl-(1→6)-β-D-glucopyranoside]
Magnoloside Q
[1946004-90-5]
$C_{27}H_{40}O_{16}$ 620.601
Constit. of stem bark of *Magnolia officinalis*. Amorph. solid. $[\alpha]_D^{20}$ -85.7 (c, 0.11 in MeOH). λ_{max} 278 (log ε 3.51) (H_2O)

2-O-[β-D-Apiofuranosyl-(1→6)-β-D-glucopyranoside]
Manglieside B
[1093091-41-8]
$C_{20}H_{28}O_{11}$ 444.433
Constit. of the leaves of *Manglietia phuthoensis*. Amorph. powder. $[\alpha]_D^{25}$ -85 (c, 1 in MeOH)

2-O-[β-D-Xylopyranosyl-(1→6)-β-D-glucopyranoside]
[1831849-00-3]
$C_{20}H_{28}O_{11}$ 444.433
Constit. of the roots of *Heracleum dissectum*. Exhibits moderate cytotoxicity against HCT-8 cells. Brown wax. $[\alpha]_D^{20}$ -92.5 (c, 0.1 in MeOH). λ_{max} 278 (log ε 2.84) (MeOH)

2-O-[α-L-Rhamnopyranosyl-(1→2)-β-D-glucopyranoside]
[319492-40-5]
$C_{21}H_{30}O_{11}$ 458.46
Constit. of the bark of *Magnolia rostrata*. Amorph. powder

2-O-[α-L-Rhamnopyranosyl-(1→6)-β-D-glucopyranoside]
[319492-39-2]
$C_{21}H_{30}O_{11}$ 458.46
Constit. of the bark of *Magnolia rostrata*. Amorph. powder. Mp 180 - 182°

1-Ac
Taiwanensol C
[91143-31-6]
$C_{11}H_{12}O_3$ 192.214
Constit. of roots of *Piper taiwanense*. Shows potent inhibitory activity against antiplatelet aggregation induced by collagen. Exhibits activity against *Mycobacterium tuberculosis* H37Rv. Oil. Obt. as a mixt. with 2-Ac to which biol. props. and uv refer. λ_{max} 234 (log ε 3.90); 276 (log ε 3.49) (MeOH)

2-Ac
[556814-10-9]
$C_{11}H_{12}O_3$ 192.214
Constit. of roots of *Piper taiwanense*. Obt. as component of Taiwanensol C

Di-Ac
1,2-Diacetoxy-4-allylbenzene
[13620-82-1]
$C_{13}H_{14}O_4$ 234.251
Constit. of *Piper betle* and *Piper taiwanense*. Prod. by *Acrocarpospora punica* 04107M-2. Platelet aggregation inhibitor. Sol. Me_2CO, C_6H_6; fairly sol. hexane; poorly sol. H_2O. Bp 150 - 152° at 3 mm. *Not* a lignan monomer

1-Me ether
see 2-Methoxy-5-(2-propenyl) phenol, M40

2-Me ether
see 2-Methoxy-4-(2-propenyl) phenol, M38

Di-Me ether
1,2-Dimethoxy-4-(2-propenyl)benzene. 4-Allylveratrole. ***Methyleugenol****. FEMA 2475*
[93-15-2]
$C_{11}H_{14}O_2$ 178.231
Present in many essential oils, e.g. nutmeg, mace and also many fruits, e.g. apple, banana, orange juice or peel, grapefruit, bilberry. First isol. in 1888 from *Asarum europaeum*. Perfumery and flavouring ingredient. Sedative. Insect pheromone. Synthetic attractant for the Oriental fruit fly. Shows antibacterial, myorelaxant and antispasmodic activities. $GABA_A$ receptor agonist. Liq. with clove/carnation odour and bitter, burning taste. Fp -4°. Mp 16 - 17°. Bp 255°. Bp 125 - 127° at 12 mm. Log P 2.67 (calc). As normally isol. contains the *Z*-isomer. *Not* a lignan monomer

▶ CY2450000
▶ Fl. p. 99°. Skin irritant. LD_{50} (rat, orl) 1179 mg/kg

2-Et ether
2-Ethoxy-4-(2-propenyl)phenol. 4-Allyl-2-ethoxyphenol
[1755-54-0]
$C_{11}H_{14}O_2$ 178.231
Constit. of *Cyperus conglomeratus*.

Aldrich Library of FT-IR Spectra: Vapor Phase, 1989, **3**, 986A (*ir*).
Aldrich Library of [13]*C and* [1]*H FT-NMR Spectra*, 1992, **2**, 209B (*nmr*).
Perkin, W.H. *et al.*, *J. Chem. Soc.*, 1927, 1663-1666 (*synth*).
Schöpf, C. *et al.*, *Justus Liebigs Ann. Chem.*, 1940, **544**, 30-62 (*synth*).
Sethi, S.C. *et al.*, *Indian J. Chem.*, 1964, **2**, 323-326 (*synth*).
Karrer, W. *Konstitution und Vorkommen der Organischen Pflanzenstoffe, 2nd edn., Birkhäuser Verlag*, 1976-1985, no. 190 (*di-Me ether, occur*).
Evans, P.H. *et al.*, *J. Agric. Food Chem.*, 1984, **32**, 1254-1256 (*di-Ac, Piper betle constit*).
Encyclopedia of Food and Color Additives, (ed. Burdock, G.A.), CRC Press , 1997, 1056-1058 (*occur, props*).
Islam, S.K.N. *et al.*, *Phytother. Res.*, 1997, **11**, 64-66 (*Methyleugenol, activity*).
Yamamura, S. *et al.*, *Phytochemistry*, 1998, **48**, 131-136 (*1-glucoside*).
Zhou, Z. *et al.*, *Zhongcaoyao*, 1999, **30**, 801-804 (*Dracaena cochinchinensis constit*).
Deng, S.M. *et al.*, *Chin. Chem. Lett.*, 2000, **11**, 1001-1002 (*2-rhamnosylglucosides*).
De Vincenzi, M. *et al.*, *Fitoterapia*, 2000, **71**, 216-221 (*Methyleugenol, rev*).
Pesticide Manual, 12th edn., 2000, 522 (*Methyleugenol*).
Abdel-Mogib, M. *et al.*, *Pharmazie*, 2000, **55**, 693-695 (*2-Et ether*).
Lima, C.C. *et al.*, *Planta Med.*, 2000, **66**, 408-411 (*Methyleugenol, activity*).
Ly, T.N. *et al.*, *J. Agric. Food Chem.*, 2002, **50**, 4919-4924 (*Alpinia officinarum glycosides*).
Miyazawa, M. *et al.*, *Nat. Prod. Res.*, 2005, **19**, 29-36 (*Methyleugenol, pmr, cmr*).
Chang, M.C. *et al.*, *Br. J. Pharmacol.*, 2007, **152**, 73-82 (*Hydroxychavicol, activity*).
Yamauchi, H. *et al.*, *Chem. Pharm. Bull.*, 2007, **55**, 346-347 (*1-apiosylglucoside*).
Kiem, P.V. *et al.*, *Chem. Pharm. Bull.*, 2008, **56**, 1270-1275 (*Mangliesides A,B*).
Murata, K. *et al.*, *J. Nat. Med.*, 2009, **63**, 355-359 (*Hydroxychavicol, activity*).
Ali, I. *et al.*, *Ann. Clin. Microbiol. Antimicrob.*, 2010, **9**, 7 (*Hydroxychavicol, activity*).
Liu, J.-F. *et al.*, *Planta Med.*, 2010, **76**, 1464-1467 (*2-6-acetylglucoside*).
Wang, X.-Y. *et al.*, *Food Chem.*, 2011, **126**, 1039-1043 (*Michehedyosides A,B*).
Ono, M. *et al.*, *Chem. Pharm. Bull.*, 2012, **60**, 1314-1319 (*Scilla scilloides xylosylglucoside*).
Tan, K.H. *et al.*, *J. Insect. Sci.*, 2012, **12**, Article 56 (*Methyleugenol, rev, occur*).
Shao, J.-H. *et al.*, *Biochem. Syst. Ecol.*, 2013, **51**, 83-85 (*Schizonepeta annua constits*).
Shen, H.-Y. *et al.*, *Chem. Res. Chin. Univ.*, 2013, **29**, 483-486 (*Cambodianin F*).
Kato, E. *et al.*, *Food Chem.*, 2013, **136**, 1239-1242 (*Hydroxychavicol, activity*).
Lin, C.-F. *et al.*, *Molecules*, 2013, **18**, 2563-2570 (*Hydroxychavicol*).
Chen, S. *et al.*, *Phytochemistry*, 2013, **93**, 203-209 (*Piper taiwanense constits, activity*).
Ding, J. *et al.*, *ACS Chem. Neurosci.*, 2014, **5**, 803-811 (*Methyleugenol, activity*).

Luyen, B.T.T. *et al.*, *Bioorg. Med. Chem. Lett.*, 2014, **24**, 4120-4124 (*Piper retrofractum constit*).
Chen, S. *et al.*, *Chem. Biodiversity*, 2014, **11**, 792-799 (*Taiwanensol C, 2-Ac, activity*).
Cheng, M.-J. *et al.*, *Chem. Nat. Compd. (Engl. Transl.)*, 2014, **50**, 606-610 (*Acrocarpospora constits*).
Gao, Y. *et al.*, *Phytochem. Lett.*, 2014, **10**, 276-280 (*2-6-xylosylglucoside, activity*).
Li, J. *et al.*, *Asian J. Chem.*, 2015, **27**, 463-466 (*Magnolia denudata constit*).
Kumar, S. *et al.*, *Eur. J. Med. Chem.*, 2015, **92**, 236-245 (*Hydroxychavicol*).
Chowdhry, B.Z. *et al.*, *J. Phys. Chem. A*, 2015, **119**, 11280-11292 (*Methyleugenol, ir, Raman, conformn*).
Xue, Z. *et al.*, *Phytochemistry*, 2016, **127**, 50-62 (*Magnolosides Q,R*).
Atkinson, R.G. *J. Agric. Food Chem.*, 2018, (*Methyleugenol, rev, biosynth, occur*).
Lewis, R.J. *Sax's Dangerous Properties of Industrial Materials, 12th edn., J. Wiley*, 2012, AGE250

4-(2-Propenyl)-1,3-benzenediol, CAS P76

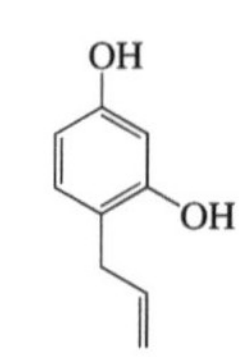

4-Allylresorcinol, CAS. *4-Allyl-1,3-dihydroxybenzene.*
3-(2,4-Dihydroxyphenyl)propene.
2-Hydroxychavicol
[1616-52-0]
$C_9H_{10}O_2$ 150.177
Constit. of the roots of *Piper betle*. Cryst. (C_6H_6/petrol). Mp 68 - 69°. λ_{max} 241 (log ε 3.47); 282 (log ε 3.66) (MeOH)

3-O-β-D-Glucopyranoside
[1416564-61-8]
$C_{15}H_{20}O_7$ 312.318
Constit. of the fruit of *Illicium simonsii*. Brown gum. $[\alpha]_D^{15}$ -56.9 (c, 0.16 in MeOH). λ_{max} 206 (log ε 5.28); 279 (log ε 4.54) (MeOH)

Di-Me ether
2,4-Dimethoxy-1-(2-propenyl)benzene, CAS. *1-Allyl-2,4-dimethoxybenzene*, CAS.
Osmorhizole
[3698-23-5]
$C_{11}H_{14}O_2$ 178.231
Constit. of *Anthriscus* sp., *Myrrhis* sp., *Nothosmyrnium* sp., *Osmorhiza* sp. and *Pinus* spp.. Bp 71 - 72° at 2 mm. *Not* a lignan monomer.

Kaufman, K.D. *J. Org. Chem.*, 1961, **26**, 117-121 (*synth*).
Kaufman, K.D. *et al.*, *J. Org. Chem.*, 1965, **30**, 1320-1321 (*synth*).
Nikiforova, I.S. *et al.*, *J. Org. Chem. USSR (Engl. Transl.)*, 1966, **2**, 1234-1237 (*synth*).
Wenkert, E. *et al.*, *Synthesis*, 1983, 701-703 (*di-Me ether, synth*).
Khan, V.A. *et al.*, *Chem. Nat. Compd. (Engl. Transl.)*, 1984, **20**, 420-423 (*Osmorhizole*).
Devakumar, C. *et al.*, *Indian J. Chem., Sect. B*, 1986, **25**, 368-372 (*Osmorhizole*).
Dieter, J.W. *et al.*, *Tetrahedron Lett.*, 1987, **28**, 5415-5418 (*di-Me ether, synth*).
Shih, T.L. *et al.*, *Tetrahedron Lett.*, 1987, **28**, 6015-6018 (*synth*).
Littke, A.F. *et al.*, *J. Am. Chem. Soc.*, 2002, **124**, 6343-6348 (*di-Me ether, synth*).
Ghosh, K. *et al.*, *Molecules*, 2005, **10**, 798-802 (*Piper betle constit*).
Coy, B. *et al.*, *Org. Biomol. Chem.*, 2010, **8**, 2003-2005 (*synth*).
Wang, J. *et al.*, *Org. Biomol. Chem.*, 2010, **9**, 663-666 (*di-Me ether, synth, pmr, cmr, ms*).
Wei, D.-D. *et al.*, *Chin. J. Nat. Med.*, 2012, **10**, 20-23 (*3-glucoside*)

5-(1-Propenyl)-1,2,3,4-benzenetetrol P77

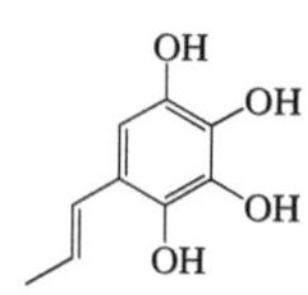

1-(2,3,4,5-Tetrahydroxyphenyl)-1-propene
$C_9H_{10}O_4$ 182.175
(E)-form

1,2,4-Tri-Me ether
2,3,6-Trimethoxy-5-(1-propenyl) phenol
[477800-93-4]
$C_{12}H_{16}O_4$ 224.256
Constit. of *Ceriops tagal* and *Smirnowia iranica*. Yellowish resin

2,3,4-Tri-Me ether
2,3,4-Trimethoxy-5-(1-propenyl) phenol
[1656285-68-5]
$C_{12}H_{16}O_4$ 224.256
Constit. of the roots of *Astragalus englerianus*. Exhibits antioxidant and antischistosomicidal activities. Pale brown-red oil. λ_{max} 218 (log ε 4.45); 258 (log ε 4.08); 304 (log ε 3.57) (MeOH).

Sairafianpour, M. *et al.*, *J. Nat. Prod.*, 2002, **65**, 1754-1758 (*1,2,4-tri-Me ether*).
Xiao, C.-J. *et al.*, *J. Asian Nat. Prod. Res.*, 2015, **17**, 772-777 (*2,3,4-tri-Me ether, activity*)

5-(2-Propenyl)-1,2,3,4-benzenetetrol P78

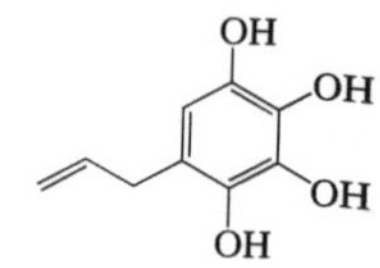

5-Allyl-1,2,3,4-benzenetetrol
$C_9H_{10}O_4$ 182.175

1,2-Methylene, 4-Me ether, 3-O-β-D-glucopyranoside
Perilloside E
[149380-62-1]
$C_{17}H_{22}O_9$ 370.354
Constit. of the leaves of *Perilla frutescens*. Antifungal agent. Needles. Mp 164 - 165°. $[\alpha]_D$ -22.8 (c, 0.3 in MeOH). λ_{max} 286 (ε 2750) (MeOH)

3,4-Methylene 1,2-di-Me ether
4,5-Dimethoxy-7-(2-propenyl)-1,3-benzodioxole, CAS.
1-Allyl-4,5-dimethoxy-2,3-methylenedioxybenzene, CAS.
Pseudodillapiole
[23724-27-8]
$C_{12}H_{14}O_4$ 222.240
Constit. of *Piper aduncum* and *Piper hispidum*. Antimicrobial agent. Oil. Bp 124 - 126° at 2 mm. λ_{max} 208 (ε 3340); 285 (ε 290) (EtOH)

Dallacker, F. *Chem. Ber.*, 1969, **102**, 2663-2676 (*synth*).
Burke, B. *et al.*, *Phytochemistry*, 1986, **25**, 1427-1430 (*Pseudodillapiole*).
Nair, M.G. *et al.*, *J. Agric. Food Chem.*, 1990, **38**, 1093-1096 (*Pseudodillapiole, synth, activity*).
Fujita, T. *et al.*, *Phytochemistry*, 1994, **37**, 543-546 (*Perilla frutescens constit, activity*)

5-(1-Propenyl)-1,2,3-benzenetriol, CAS P79

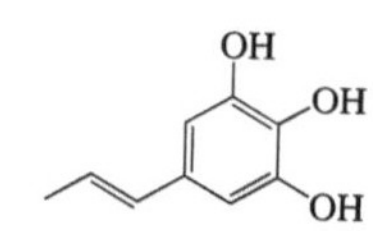

(*E*)-*form*

5-(1-Propenyl)pyrogallol.
1-(3,4,5-Trihydroxyphenyl)-2-propene
$C_9H_{10}O_3$ 166.176
(E)-form
Bp 153 - 156° at 10 mm

2-Me ether, 3-O-β-D-glucopyranoside
[1334396-88-1]
$C_{16}H_{22}O_8$ 342.344
Constit. of *Mediasia macrophylla*. Off-white powder. $[\alpha]_D$ -60.4 (c, 1.3 in MeOH)

1,2-Di-Me ether
2,3-Dimethoxy-5-(1-propenyl) phenol
[477800-92-3]
$C_{11}H_{14}O_3$ 194.23
Constit. of the roots of *Smirnowia iranica*. Resin

1,3-Di-Me ether
2,6-Dimethoxy-4-(1-propenyl) phenol. FEMA 3728
[6635-22-9]
$C_{11}H_{14}O_3$ 194.23
Flavouring ingredient. Viscous liq.. Bp 135° at 0.13 mm.

Fenaroli's Handbook of Flavor Ingredients, 3rd edn., (ed. Burdock, G.A.), CRC Press, 1995, **1**, 686 (*1,3-di-Me ether*).
Encyclopedia of Food and Color Additives, (ed. Burdock, G.A.), CRC Press , 1997, 2329 (*1,3-di-Me ether*).
Sairafianpour, M. *et al.*, *J. Nat. Prod.*, 2002, **65**, 1754-1758 (*1,2-di-Me ether*).
Kurimoto, S. *et al.*, *J. Nat. Med.*, 2011, **65**, 180-185 (*2-Me ether 3-glucoside*).
Patil, S.N. *et al.*, *Tetrahedron Lett.*, 2016, **57**, 3371-3375 (*1,3-di-Me, tri-Me ethers, synth*)

5-(2-Propenyl)-1,2,3-benzenetriol

P80

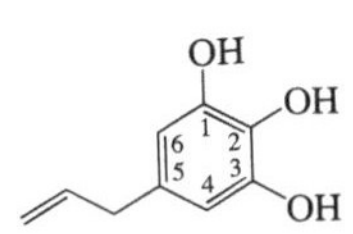

5-Allyl-1,2,3-benzenetriol. 5-Allylpyrogallol
$C_9H_{10}O_3$ 166.176
▶ CY2801000

1,2-Di-O-β-D-glucopyranoside
[2172431-63-7]
$C_{21}H_{30}O_{13}$ 490.458
Constit. of the tubers of *Ophiopogon japonicus*. Amorph. powder. $[\alpha]^{25}_{D}$ -96 (c, 0.07 in MeOH). λ_{max} 271 (MeOH)

1-Me ether
3-Methoxy-5-(2-propenyl)-1,2-benzenediol, CAS. *5-Allyl-3-methoxy-1,2-benzenediol. 5-Allyl-3-methoxypyrocatechol.* ***5-Hydroxyeugenol***
[4055-72-5]
$C_{10}H_{12}O_3$ 180.203
Constit. of *Bulbine capitata* and *Myristica fragrans*. Light yellow oil. Mp 75 - 76° (as di-Ac). Bp 206 - 210°. Bp 112.5 - 115° at 0.26 mm. λ_{max} 237 (log ε 1.21); 240 (log ε 1.5) ($CHCl_3$)

1-Me ether, 3-O-β-D-glucopyranoside
Michehedyoside D
[1283160-69-9]
$C_{16}H_{22}O_8$ 342.344
Constit. of the seeds of *Michelia hedyosperma*. Amorph. powder. $[\alpha]^{25}_{D}$ -53.6 (c, 0.43 in MeOH). λ_{max} 206 (log ε 4.57); 277 (log ε 3.36) (MeOH)

1-Me ether, 3-O-[4-hydroxy-3,5-dimethoxy-E-cinnamoyl (→6)-β-D-glucopyranoside]
[1007356-67-3]
$C_{27}H_{32}O_{12}$ 548.541
Constit. of the fruit of *Pimenta dioica*. Viscous liq.. $[\alpha]^{25}_{D}$ -21.2 (c, 0.82 in MeOH). λ_{max} 283 (log ε 3.75); 328 (log ε 3.94) (MeOH)

1-Me ether, 3-O-[3,4,5-trihydroxybenzoyl-(→6)-β-D-glucopyranoside]
Ericifolin. Pimentol
[141913-95-3]
$C_{23}H_{26}O_{12}$ 494.449
Constit. of *Pimenta officinalis* and *Melaleuca ericifolia*. Carcinogen-promotor inhibitor. Antioxidant. $[\alpha]^{24}_{D}$ -22.8 (c, 0.08 in MeOH) (Pimentol). $[\alpha]^{25}_{D}$ +77.8 (c, 0.14 in MeOH) (Ericifolin)

1-Me ether, 3-O-[β-D-apiofuranosyl-(1→6)-β-D-glucopyranoside]
Hovetrichoside G
[210050-42-3]
$C_{21}H_{30}O_{12}$ 474.459
Constit. of *Hovenia trichocarea*. Amorph. powder. $[\alpha]^{25}_{D}$ -75.7 (c, 1 in MeOH). λ_{max} 228 (log ε 4.1); 280 (log ε 3.5) (MeOH)

1-Me ether, 3-O-[α-L-arabinopyranosyl-(1→6)-β-D-glucopyranoside]
Shashenoside II
[138614-63-8]
$C_{21}H_{30}O_{12}$ 474.459
Constit. of the roots of *Adenophora tetraphylla*. Yellow powder. $[\alpha]^{18}_{D}$ -51.4 (c, 0.7 in MeOH)

1-Me ether, 3-O-[α-L-rhamnopyranosyl-(1→6)-β-D-glucopyranoside]
Illoliganoside C. Michehedyoside E
[1283160-70-2]
$C_{22}H_{32}O_{12}$ 488.486
Constit. of *Illicium oligandrum* and *Michelia hedyosperma*. Amorph. powder. $[\alpha]^{20}_{D}$ -73.9 (c, 0.66 in MeOH). λ_{max} 206 (log ε 4.53); 279 (log ε 3.38) (MeOH)

2-Me ether, 1-O-β-D-glucopyranoside
Sphalleroside A
[192324-34-8]
$C_{16}H_{22}O_8$ 342.344
Constit. of *Foeniculum vulgare* and *Sphallerocarpus gracilis*. Cryst.. Mp 185 - 186° (145-147°). $[\alpha]^{21}_{D}$ -48 (c, 0.8 in MeOH). Data of two isolates is at variance

1,2-Di-Me ether
2,3-Dimethoxy-5-(2-propenyl)phenol. 5-Allyl-2,3-dimethoxyphenol
[76773-99-4]
$C_{11}H_{14}O_3$ 194.23
Constit. of *Seseli vayredanum* and *Agastache rugosa*

1,3-Di-Me ether
2,6-Dimethoxy-4-(2-propenyl)phenol, CAS. *4-Allyl-2,6-dimethoxyphenol*, CAS. ***Methoxyeugenol.*** *4-Allylsyringol*
[6627-88-9]
$C_{11}H_{14}O_3$ 194.23
Found in oil of *Sassafras albidum* root, *Illicium anisatum*, *Myristica fragrans* and *Cinnamomum glanduiferum*. Oil. Bp 166 - 168° at 10 mm. Bp 123 - 125° at 2 mm.

Coscia, C.J. *et al.*, *J. Org. Chem.*, 1961, **26**, 5085-5091 (*1,3-di-Me, synth*).
Yamamura, S. *et al.*, *Tetrahedron Lett.*, 1973, **14**, 4877 (*1,3-di-Me, ir, pmr*).
Shibuya, M. *et al.*, *Chem. Pharm. Bull.*, 1978, **26**, 2671-2673 (*1,3-di-Me, isol, ir, ms, pmr*).
Kuang, H.-X. *et al.*, *Chem. Pharm. Bull.*, 1991, **39**, 2440-2442 (*Shashenosides*).
Oya, T. *et al.*, *Biosci., Biotechnol., Biochem.*, 1997, **61**, 263-266 (*Pimentol*).
Chen, N.Y. *et al.*, *Indian J. Chem., Sect. B*, 1997, **36**, 107-109 (*Sphalleroside A*).
Sy, L.K. *et al.*, *Phytochemistry*, 1997, **44**, 1099-1108 (*Methoxyeugenol, pmr, cmr*).
Bezabih, M. *et al.*, *Phytochemistry*, 1997, **46**, 1063-1067 (*1-Me ether*).
Kitajima, J. *et al.*, *Chem. Pharm. Bull.*, 1998, **46**, 1587-1590 (*2-Me 1-glucoside*).
Yoshikawa, K. *et al.*, *J. Nat. Prod.*, 1998, **61**, 786-790 (*Hovetrichoside G*).
Hussein, S.A.M. *et al.*, *Phytochemistry*, 2007, **68**, 1464-1470 (*Ericifolin, 5-Hydroxyeugenol*).
Kikuzaki, H. *et al.*, *J. Nat. Prod.*, 2008, **71**, 861-865 (*Pimenta sinapoylglucoside*).
Ma, S.-G. *et al.*, *Carbohydr. Res.*, 2011, **346**, 1165-1168 (*Illoliganoside C*).
Wang, X.-Y. *et al.*, *Food Chem.*, 2011, **126**, 1039-1043 (*Michehedyosides C,D,E*).
Liu, S.-Q. *et al.*, *J. Asian Nat. Prod. Res.*, 2016, **18**, 520-527 (*Ophiopogon japonicus 1,2-diglucoside*).
Patil, S.N. *et al.*, *Tetrahedron Lett.*, 2016, **57**, 3371-3375 (*Methoxyeugenol, synth*)

5-(2-Propenyl)-1,2,4-benzenetriol

P81

5-Allyl-1,2,4-benzenetriol. 3-(2,4,5-Trihydroxyphenyl)-1-propene

$C_9H_{10}O_3$ 166.176

1,2-Di-Me ether

4,5-Dimethoxy-2-(2-propenyl)phenol. 2-Allyl-4,5-dimethoxyphenol. 4-Hydroxy-5-allylveratrole

[59893-87-7]

$C_{11}H_{14}O_3$ 194.23

Obt. from branches of *Aniba lancifolia*. Mp 38 - 41°. Bp 143 - 145° at 2 mm

1,4-Di-Me ether

2,5-Dimethoxy-4-(2-propenyl)phenol, CAS. *4-Allyl-2,5-dimethoxyphenol*

[90377-06-3]

$C_{11}H_{14}O_3$ 194.23

Constit. of *Pentacalia andicola, Piper divaricatum, Santolina chamaecyparissus* and *Sassafras albidum*.

Diaz, P.P. *et al., Phytochemistry*, 1980, **19**, 285-288 (*Aniba lancifolia constit*).

van Otterlo, W.A.L. *et al., Tetrahedron Lett.*, 2003, **44**, 311-314 (*synth*).

Van, T.N. *et al., Tetrahedron Lett.*, 2003, **44**, 4199-4202 (*1,4-di-Me ether, synth*)

5′-(2-Propenyl)-2,2′,5-biphenyltriol P82

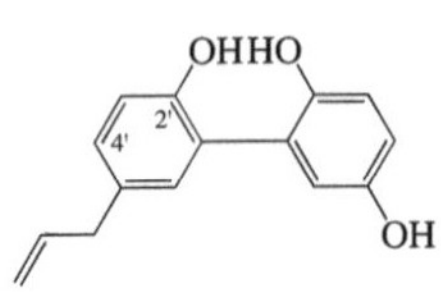

5′-Allyl-2,2′,5-trihydroxybiphenyl. ***Randaiol. Magnotriol A***

[87562-14-9]

$C_{15}H_{14}O_3$ 242.274

Constit. of *Sassafras randaiense* and *Magnolia officinalis*. Lipase inhibitor. Pale yellow powder

5-O-(1ξ-Hydroxy-2-propenyl) ether

5′-Allyl-5-(1-hydroxyallyloxy) biphenyl-2,2′-diol

[1416567-11-7]

$C_{18}H_{18}O_4$ 298.338

Isol. from *Magnolia officinalis*. Exhibits potent antiinflammatory activity. Oil. λ_{max} 209 (log ε 4.55); 255 (log ε 3.82); 295 (log ε 3.90) (MeOH)

2′-Deoxy, 4′-hydroxy

3′-(2-Propenyl)-2,4′,5-biphenyltriol. 3′-Allyl-2,4′,5-trihydroxybiphenyl. ***Magnotriol B***

[92829-71-5]

$C_{15}H_{14}O_3$ 242.274

Constit. of *Magnolia officinalis*. Pale yellow needles ($CHCl_3/C_6H_6$). Mp 99 - 100°.

Chen, F.-C. *et al., Phytochemistry*, 1983, **22**, 616-617 (*Randaiol*).

Yahara, S. *et al., Chem. Pharm. Bull.*, 1991, **39**, 2024 (*Magnotriol B*).

Kuo, W.-L. *et al., Phytochemistry*, 2013, **85**, 153-160 (*1-hydroxy-2-propenyl ether*)

4-(1-Propenyl)phenol, CAS P83

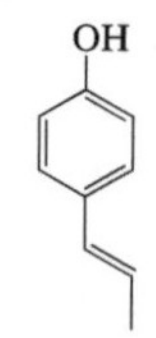

(*E*)-form

1-(4-Hydroxyphenyl)-1-propene. p-Hydroxy-β-methylstyrene. ***Anol***

[539-12-8]

[104-46-1]

$C_9H_{10}O$ 134.178

(E)-form [20649-39-2]

Constit. of roots of *Solidago odora*. Leaflets (H_2O). Mp 93°. Bp 140° at 14 mm. pK_{a1} 9.82 (25°). Dec. on standing in air

O-β-D-Glucopyranoside

[120396-85-2]

$C_{15}H_{20}O_6$ 296.319

Constit. of *Cleidion spiciflorum* and *Lilium cordatum*. Powder. $[\alpha]_D^{25}$ -55 (c, 1 in Py)

Me ether

1-Methoxy-4-(1-propenyl)benzene, CAS. *p-Propenylanisole,* CAS. ***Anethole,*** *USAN. Anistearoptene. Anise camphor. FEMA 2086*

[4180-23-8]

$C_{10}H_{12}O$ 148.205

Isol. from *Foeniculum vulgare* and *Illicium verum* and other plant oils. Semiochemical attractant of Southern and Western corn rootworms (*Diabrotica undecimpunctata* and *Diabrotica virgifera*). Extensively used in flavour industry. Possesses carminative, expectorant, insecticide props. Gastric stimulant. Pharmaceutic aid (flavour). Leaflets (EtOH) with anise odour and sweet taste. Sol. Me_2CO, C_6H_6. Mp 22.5°. Bp 235°. Log P 3.31 (calc). Undergoes cationic polym. *Not* a lignan monomer. λ_{max} 259 (ε 22300) (MeOH) (Berdy)BZ9275000

▶ LD_{50} (rat, orl) 2090 mg/kg. LD_{50} (mus, ipr) 1450 mg/kg. Exp. neoplastic agent. Fl. p. 90° :

(Z)-form [85960-81-2]

Me ether

(Z)-Anethole

[25679-28-1]

Constit. of numerous plant spp.. Liq.. d_4^{20} 0.99. Bp 63 - 65° at 2.5 mm. n_D^{20} +1.5550 (93% pure). Undergoes cationic polym.

▶ BZ9100000

▶ Skin irritant. LD_{50} (rat, orl) 150 mg/kg.

Aldrich Library of FT-IR Spectra, 1st edn., 1985, **1**, 1052D (*ir*).

Aldrich Library of FT-IR Spectra: Vapor Phase, 1989, **3**, 985C (*ir*).

Aldrich Library of ¹³C and ¹H FT-NMR Spectra, 1992, **2**, 208A (*nmr*).

Naves, Y.R. *Helv. Chim. Acta*, 1960, **43**, 230-232 (*cis-Anethole, synth*).

Briner, E. *et al., Helv. Chim. Acta*, 1963, **46**, 2249-2257 (*ir*).

Sørensen, J.S. *et al., Acta Chem. Scand.*, 1966, **20**, 992-1002 (*ir*).

Higashimura, T. *et al., J. Polym. Sci., Part A-1: Polym. Chem.*, 1972, **10**, 85-93 (*Me ether, polym*).

Morris, W.W. *J. Assoc. Off. Anal. Chem.*, 1973, **56**, 1037-1064 (*ir*).

Hamacher, H. *Arch. Pharm. (Weinheim, Ger.)*, 1974, **307**, 309-315 (*synth*).

Manitto, P. *et al., Tetrahedron Lett.*, 1974, **15**, 1567-1568 (*biosynth*).

Higashimura, T. *et al., Polym. J. (Tokyo)*, 1979, **11**, 855-862 (*polym, Me ether*).

Bohlmann, F. *et al., Phytochemistry*, 1980, **19**, 2655-2661 (*Solidago odora constit*).

Wenkert, E. *et al., Synthesis*, 1983, 701-703 (*synth*).

Lampman, R.L. *et al., J. Chem. Ecol.*, 1987, **13**, 959-975 (*Diabrotica constit*).

Sangster, S.A. *et al., Xenobiotica*, 1987, **17**, 1223-1232 (*metab*).

Cerrai, P. *et al., Ann. Chim. (Rome)*, 1989, **79**, 33-43 (*Me ether, polym*).

Lewis, R.J. *Food Additives Handbook*, Van Nostrand Reinhold International, New York , 1989, PMQ750.

Nakano, K. *et al., Phytochemistry*, 1989, **28**, 301-303 (*glucoside*).

Howes, A.J. *et al., Food Chem. Toxicol.*, 1990, **28**, 537-542 (*tox*).

Angle, S.R. *et al., J. Org. Chem.*, 1992, **57**, 5937-5947 (*Z-form Me ether*).

Mohan, R. *et al., J. Org. Chem.*, 1993, **58**, 2663-2669 (*Z-form, Z-Anethole, synth, pmr*).

Martindale, The Extra Pharmacopoeia, 30th edn., Pharmaceutical Press, 1993, 1337.

Naengchomnong, W. *et al., Phytochemistry*, 2006, **67**, 1029-1033 (*glucoside*).

Lewis, R.J. *Sax's Dangerous Properties of Industrial Materials, 8th edn., Van Nostrand Reinhold*, 1992, PMQ750 , PMR000, PMR250 (*Y*)

4-(2-Propenyl)phenol, CAS P84

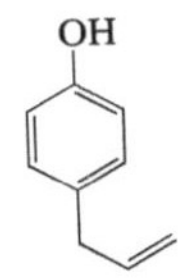

3-(p-Hydroxyphenyl)-1-propene. 4-Allylphenol. ***Chavicol.*** *FEMA 4075*
[501-92-8]
$C_9H_{10}O$ 134.178
Found in many essential oils, e.g. anise, bay leaf, clove and *Gardenia*. Found in *Alnus pendula*, *Malus domestica* and *Cucumis melo*. Pheromone of Southern and Western corn rootworms *Diabrotica undecimpunctata* and *Diabrotica virgifera*. Used in perfumery and flavours. Shows nematocidal props.. Light yellow oil. Sol. EtOH, hexane; poorly sol. H_2O. Mp 16°. Bp 237°. Bp 120° at 17 mm. pK_{a1} 10.23 (25°).

Aldrich Library of FT-IR Spectra, 1st edn., 1985, **1**, 1053B (*ir*).
Aldrich Library of FT-IR Spectra: Vapor Phase, 1989, **3**, 985D (*ir*).
Aldrich Library of ^{13}C and ^1H FT-NMR Spectra, 1992, **2**, 208B (*nmr*).
Palkin, S. *et al.*, *J. Am. Chem. Soc.*, 1933, **55**, 1556-1558 (*Chavicol, bay oil constit*).
Van der Zanden, J.M. *Recl. Trav. Chim. Pays-Bas*, 1938, **57**, 233-247 (*synth*).
Suga, T. *et al.*, *Bull. Chem. Soc. Jpn.*, 1972, **45**, 2058-2060 (*Alnus pendula constit*).
Fueno, T. *et al.*, *Bull. Chem. Soc. Jpn.*, 1973, **46**, 1418-1421 (*pmr*).
Karrer, W. *Konstitution und Vorkommen der Organischen Pflanzenstoffe, 2nd edn., Birkhäuser Verlag*, 1976-1985, nos. 165 , 166 (*occur*).
Waespe, H.R. *et al.*, *Helv. Chim. Acta*, 1978, **61**, 401-429 (*synth*).
Wenkert, E. *et al.*, *Synthesis*, 1983, 701-703 (*synth*).
Lampman, R.L. *et al.*, *J. Chem. Ecol.*, 1987, **13**, 959-975 (*activity*).
Sangster, S.A. *et al.*, *Xenobiotica*, 1987, **17**, 1223-1232 (*metab*).
Lewis, R.J. *Food Additives Handbook*, Van Nostrand Reinhold International, New York , 1989, AFW750 (*use*).
Howes, A.J. *et al.*, *Food Chem. Toxicol.*, 1990, **28**, 537-542 (*tox*).
Agharahimi, M.R. *et al.*, *J. Org. Chem.*, 1995, **60**, 1856-1863 (*synth*).
Vassäo, D.G. *et al.*, *Org. Biomol. Chem.*, 2006, **4**, 2733-2744 (*biosynth*).
Fenaroli's Handbook of Flavor Ingredients, 6th edn., (ed. Burdock, G.A.), CRC Press, 2010, 63-64 (*use, occur*).
da Silva, J.F.M. *et al.*, *Synth. Commun.*, 2015, **45**, 1995-2004 (*synth, pmr, cmr*).
Lewis, R.J. *Sax's Dangerous Properties of Industrial Materials, 12th edn., J. Wiley*, 2012, AFW750

5-Propyl-1,3-benzenediol, CAS P85

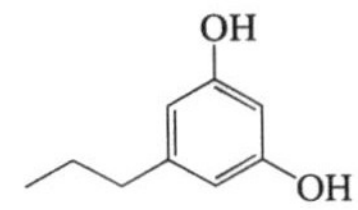

5-Propylresorcinol, CAS. ***Divarinol.*** *Divarin*
[500-49-2]
$C_9H_{12}O_2$ 152.193
Structures in this series may need revision. Phenyl 3,5-dioxygenated substn. pattern in doubt as a result of work done by Yang *et al* in 2015. Isol. from *Evernia divaricata* and *Protousnea poeppigii*. Hydrated leaflets (H_2O), anhyd. cryst. (C_6H_6). Mp 57° (hydrate). Mp 83 - 84° (anhyd.). Bp 127 - 129° at 3 mm. Hydrol. prod. of Imbricaric acid

O-β-D-Glucopyranoside
[1013648-94-6]
$C_{15}H_{22}O_7$ 314.334
Constit. of *Dryopteris fragrans* and *Humulus japonicus*. Powder. Mp 128°. $[\alpha]^{25}_{D}$ -32.3 (c, 0.06 in MeOH). λ_{max} 208 (log ε 5.37); 280 (log ε 4.57) (MeOH)

Di-Ac
[2096371-93-4]
$C_{13}H_{16}O_4$ 236.267
Mp 12 - 15°

Mono-Me ether
3-Methoxy-5-propylphenol. ***O-Methyldivarinol***
[55136-70-4]
$C_{10}H_{14}O_2$ 166.22
Isol. from *Evernia prunastri*

Di-Me ether
1,3-Dimethoxy-5-propylbenzene
[41395-10-2]
$C_{11}H_{16}O_2$ 180.247
Bp 147° at 29 mm. Bp 126 - 127° at 10 mm.

Sonn, A. *et al.*, *Ber. Dtsch. Chem. Ges.*, 1924, **57**, 959-961 (*synth*).
Kato, T. *et al.*, *Chem. Pharm. Bull.*, 1972, **20**, 1574-1578 (*synth*).
Chandrasekharan, V. *et al.*, *Indian J. Chem., Sect. B*, 1978, **16**, 970-972 (*synth*).
Azzena, U. *et al.*, *Synthesis*, 1989, 28-30 (*di-Me ether, synth*).
Elix, J.A. *et al.*, *Aust. J. Chem.*, 1997, **50**, 1145-1150 (*mono-Me ether, synth*).
Yu, B.C. *et al.*, *Arch. Pharmacal Res.*, 2008, **30**, 1471-1475 (*Humulus japonicus glucoside*).
Schmeda-Hirschmann, G. *et al.*, *Phytother. Res.*, 2008, **22**, 349-355 (*Protousnea poeppigii constit*).
Poldy, J. *et al.*, *Org. Biomol. Chem.*, 2009, **7**, 4296-4300 (*di-Me ether*).
Peng, B. *et al.*, *Zhongcaoyao*, 2013, **44**, 2347-2349 (*Dryopteris fragrans glucoside*).
Yang, Y.-N. *et al.*, *J. Nat. Prod.*, 2015, **78**, 705-711 (*pmr, struct*)

5-Propyl-1,2,3,4-benzenetetrol P86

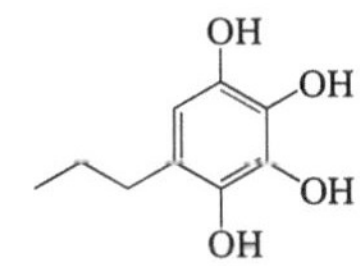

$C_9H_{12}O_4$ 184.191

2,3-Di-Me ether
2,3-Dimethoxy-5-propyl-1,4-benzenediol. 2,3-Dimethoxy-5-propylhydroquinone
[65162-41-6]
$C_{11}H_{16}O_4$ 212.245
Prod. by *Camarops microspora*. Needles (petrol/Et_2O). Mp 88 - 89°. λ_{max} 290 (log ε 3.58) (EtOH)

2,3,4-Tri-Me ether
2,3,4-Trimethoxy-5-propylphenol. ***Bruguierol D***
[1000779-07-6]
$C_{12}H_{18}O_4$ 226.272
Constit. of the mangrove *Bruguiera gymnorrhiza*. Oil. λ_{max} 282 (log ε 2.97) (MeOH). [acid] λ_{max} 277 (log ε 2.87) (MeOH/HCl). [base] λ_{max} 291 (log ε 3.06) (MeOH/NaOH).

Volc, J. *et al.*, *Coll. Czech. Chem. Comm.*, 1977, **42**, 2957-2961 (*Camarops microspora metab*).
Han, L. *et al.*, *J. Asian Nat. Prod. Res.*, 2007, **9**, 327-331 (*Bruguierol D*)

6-Propyl-1,2,4-benzenetriol P87

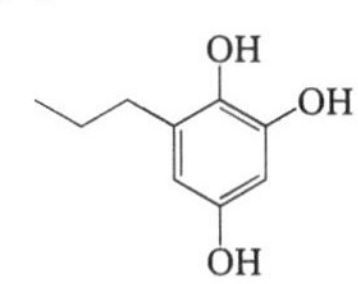

$C_9H_{12}O_3$ 168.192

1-O-β-D-Glucopyranoside
[1818337-03-9]
$C_{15}H_{22}O_8$ 330.333
Constit. of *Eugenia catharinae*

2-Me ether
2-Methoxy-6-propyl-1,4-benzenediol. 2-Methoxy-6-propylhydroquinone
[65162-40-5]
$C_{10}H_{14}O_3$ 182.219
Prod. by *Camarops microspora*. Needles (petrol). Mp 106°. λ_{max} 290 (log ε 3.56) (EtOH).

Dean, F.M. *et al.*, *J. Chem. Soc.*, 1955, 11-17 (*synth*).
Volc, J. *et al.*, *Coll. Czech. Chem. Comm.*, 1977, **42**, 2957-2961 (*Camarops microspora constit*).
Sharma, P.K. *et al.*, *Acta Chim. Hung.*, 1985, **120**, 163-166 (*synth*).

Wubshet, S.G. *et al.*, *J. Nat. Prod.*, 2015, **78**, 2657-2665 (*Eugenia glucoside*)

Pseudotsuganol P88

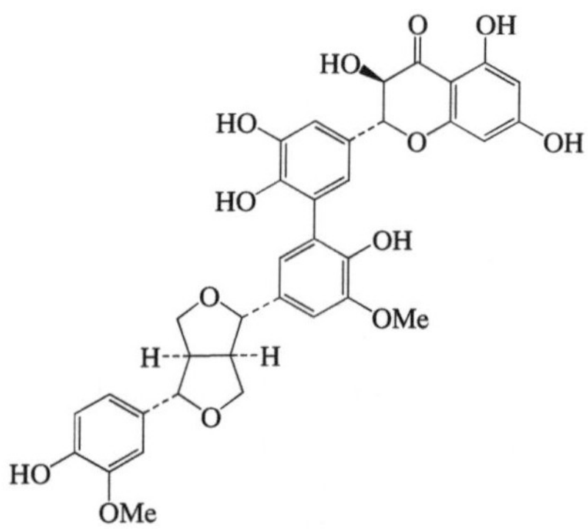

Relative Configuration

[120015-18-1]

$C_{35}H_{32}O_{13}$ 660.628

Constit. of the bark of *Pseudotsuga menziesii*. Exhibits marked antiparasitic activity towards *Leishmania donovani*. Shows immunomodulatory activity.

$[\alpha]_D$ +20.2 (c, 0.25 in MeOH).

Foo, L.Y. *et al.*, *J. Chem. Soc., Chem. Commun.*, 1989, 217-219 (*Pseudotsuganol*).
Kolodziej, H. *et al.*, *Biol. Pharm. Bull.*, 2001, **24**, 1016-1021 (*antiparasitic activity*).
Kolodziej, H. *et al.*, *Phytochemistry*, 2005, **66**, 2056-2071 (*immunomodulatory activity*)

Rabdosiin **R1**

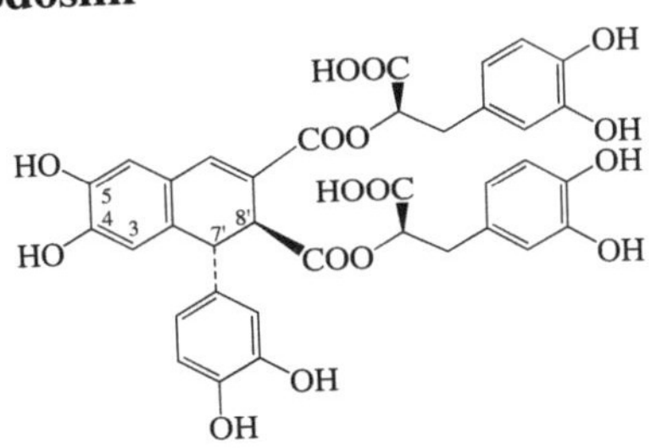

Absolute Configuration

[119152-54-4]
[164411-64-7] (Na salt) [164411-65-8] (K salt) [164411-70-5] (di-Na salt)
$C_{36}H_{30}O_{16}$ 718.62
Diester of Epiphyllic acid in C339. Constit. of *Arnebia euchroma, Eritrichium sericeum, Macrotomia euchroma, Mertensia maritima, Ocimum sanctum, Rabdosia japonica* and *Salvia miltiorrhiza*. Used for treatment of gastrointestinal disorders. Tanning agent. Exhibits anti-HIV activity. Antioxidant. Antiproliferative agent. Light brown amorph. powder. Mp 175 - 177°. $[\alpha]^{25}_{D}$ -78 (c, 3.5 in MeOH). λ_{max} 255 (log ε 4.23); 284 (log ε 4.05); 317 (sh); 346 (log ε 4.03) (MeOH)

4-O-β-D-Glucopyranoside
[164411-69-2]
[164577-22-4]
$C_{42}H_{40}O_{21}$ 880.761
Constit. of *Arnebia euchroma*. Off-white powder + $6H_2O$ (as di-K salt). $[\alpha]^{24}_{D}$ -125.4 (c, 0.5 in Me_2CO aq.) (di-K salt)

7′,8′-Diepimer
(+)-Rabdosiin
[130286-75-8]
$C_{36}H_{30}O_{16}$ 718.62
Constit. of *Arnebia euchroma, Macrotomia euchroma* and *Lithospermum erythrorhizon*. Off-white powder + $4H_2O$ (as K salt). $[\alpha]^{24}_{D}$ +132.8 (c, 1 in Me_2CO) (K salt). λ_{max} 255 (log ε 4.29); 284 (log ε 4.08); 317 (sh); 346 (log ε 4.04) (MeOH)

5-Deoxy, 3-hydroxy
Yunnaneic acid G
[199433-62-0]
$C_{36}H_{30}O_{16}$ 718.62
Constit. of *Salvia yunnanensis*. Amorph. powder + $2H_2O$. $[\alpha]^{23}_{D}$ -50.4 (c, 0.5 in MeOH).

Agata, I. *et al.*, *Chem. Pharm. Bull.*, 1988, **36**, 3223-3225 (*Rabdosiin*).
Agata, I. *et al.*, *Phytochemistry*, 1989, **28**, 2447-2450 (*Rabdosiin, pmr, cmr*).
Nishizawa, M. *et al.*, *Phytochemistry*, 1990, **29**, 2645-2649 (*Macrotomia euchroma constits*).
Kashiwada, Y. *et al.*, *J. Nat. Prod.*, 1995, **58**, 392-400 (*Arnebia euchroma constits*).
Bogucki, D.E. *et al.*, *Can. J. Chem.*, 1997, **75**, 1783-1794 (*Rabdosiin, 7′,8′-diepimer, synth*).
Tanaka, T. *et al.*, *Chem. Pharm. Bull.*, 1997, **45**, 1596-1600 (*Yunnaneic acid G*).
Yamamoto, H. *et al.*, *Phytochemistry*, 2000, **53**, 651-657 (*Lithospermum erythrorhizon constit*).
Fedoreyev, S.A. *et al.*, *Planta Med.*, 2005, **71**, 446-451 (*cd, struct*).
Yamada, Y. *et al.*, *Photochem. Photobiol.*, 2006, **82**, 1668-1676 (*Rabdosiin, activity*).
Flegkas, A. *et al.*, *Medicines*, 2019, **6**, 37:1-10 (*Ocimum sanctum constit, activity*)

Radiatinol **R2**

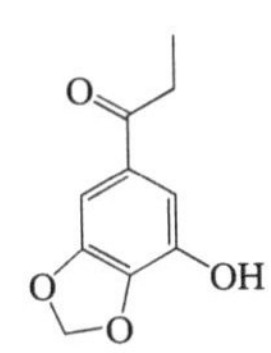

1-(7-Hydroxy-1,3-benzodioxol-5-yl)-1-propanone, CAS. *3-Hydroxy-4,5-methylenedioxypropiophenone. 2,3-Methylenedioxy-5-propanoylphenol*
[83016-68-6]
$C_{10}H_{10}O_4$ 194.186
Constit. of *Ferula* spp., *Laser trilobum* and others. Possesses insect antifeedant activity. Also possesses anticoagulant props.. Cryst. (EtOAc). Mp 161 - 162°

Ralfuranone I **R3**

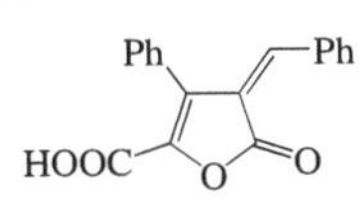

[1638617-55-6]
$C_{18}H_{12}O_4$ 292.29
Provisional lignan. Biosynth. pathway unclear. Prod. by *Ralstonia solanacearum* GMI1000. Yellow oil.

Pauly, J. *et al.*, *ChemBioChem*, 2013, **14**, 2169-2178 (*Ralfuranone I*).
Kai, K. *et al.*, *Biosci., Biotechnol., Biochem.*, 2016, **80**, 440-444 (*biosynth*)

Ramonanin A **R4**

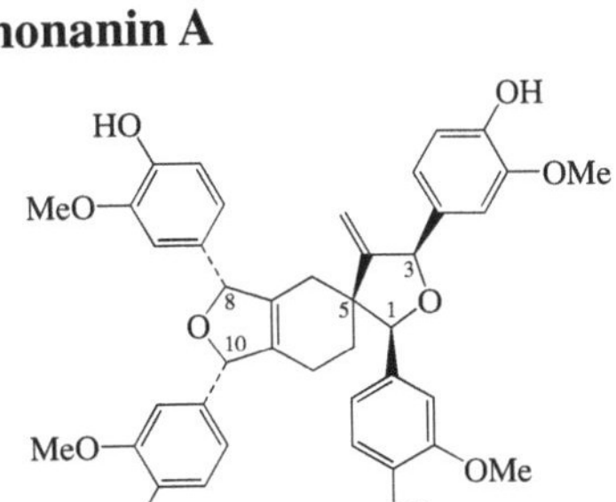

Relative Configuration

[1223582-19-1]
$C_{40}H_{40}O_{10}$ 680.75
Spirocyclic lignan. Constit. of the heartwood of *Guaiacum officinale*. Exhibits cytotoxic props.. Wax. $[\alpha]^{22}_{D}$ -2.2 (c, 0.6 in MeOH). λ_{max} 230 (log ε 4.1); 279 (log ε 3.6) (MeOH)

10-Epimer
Ramonanin D
[1283088-46-9]
$C_{40}H_{40}O_{10}$ 680.75
Constit. of the heartwood of *Guaiacum officinale*. Wax. $[\alpha]^{22}_{D}$ -5.1 (c, 0.3 in MeOH). λ_{max} 230 (log ε 4.1); 279 (log ε 3.6) (MeOH)

1,3-Diepimer
Ramonanin B
[1283088-44-7]
$C_{40}H_{40}O_{10}$ 680.75
Constit. of the heartwood of *Guaiacum officinale*. Exhibits cytotoxic props.. Oil. $[\alpha]^{22}_{D}$ -4 (c, 0.2 in MeOH). λ_{max} 230 (log ε 4.1); 279 (log ε 3.6) (MeOH)

8,10-Diepimer
Ramonanin C
[1283088-45-8]
$C_{40}H_{40}O_{10}$ 680.75
Constit. of the heartwood of *Guaiacum officinale*. Oil. $[\alpha]^{22}_{D}$ -5.2 (c, 0.2 in MeOH). λ_{max} 230 (log ε 4.1); 279 (log ε 3.6) (MeOH).

Chavez, K.J. *et al.*, *J. Nat. Prod.*, 2011, **74**, 1293-1297 (*Ramonanins A-D, activity*).
Harvey, R.S. *et al.*, *Angew. Chem., Int. Ed.*, 2015, **54**, 1795-1798 (*synth*)

Raphanuside C **R5**

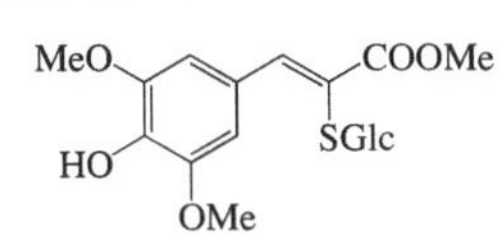

[1870838-50-8]
$C_{18}H_{24}O_{10}S$ 432.440
Possesses *Z*-config.. Constit. of the seeds of *Descurainia sophia*. Amorph. yellow powder. $[\alpha]^{20}_{D}$ +66.9 (c, 0.3 in MeOH). λ_{max} 244; 339 (MeOH).

Feng, W.-S. *et al.*, *Nat. Prod. Res.*, 2016, **30**, 1675-1681 (*Raphanuside C*)

Rashomonic acid A **R6**

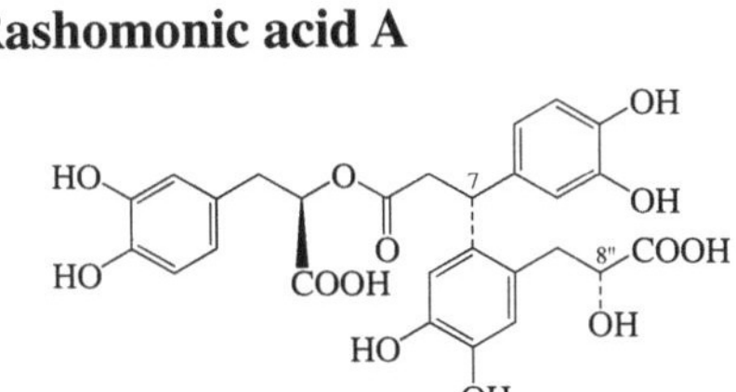

Absolute Configuration

[1387575-13-4]
$C_{27}H_{26}O_{13}$ 558.492
Constit. of *Meehania urticifolia*. Amorph. powder. $[\alpha]_D^{22}$ -17.6 (c, 0.74 in MeOH). λ_{max} 285 (log ε 4.58) (MeOH)

8″-O-(3,4-Dihydroxy-E-cinnamoyl)

Rashomonic acid C
[1387575-16-7]
$C_{36}H_{32}O_{16}$ 720.636
Constit. of *Meehania urticifolia*. Amorph. powder. $[\alpha]_D^{19}$ +27.7 (c, 1.1 in MeOH). λ_{max} 209 (log ε 5.67); 288 (log ε 5.19); 331 (log ε 5.05) (MeOH)

7-Epimer

Rashomonic acid B
[1387575-14-5]
$C_{27}H_{26}O_{13}$ 558.492
Constit. of *Meehania urticifolia*. Amorph. powder. $[\alpha]_D^{23}$ +5 (c, 0.44 in MeOH). λ_{max} 285 (log ε 4.63) (MeOH)

7-Epimer, 8″-O-(3,4-dihydroxy-E-cinnamoyl)

Rashomonic acid D
[1387575-17-8]
$C_{36}H_{32}O_{16}$ 720.636
Constit. of *Meehania urticifolia*. Amorph. powder. $[\alpha]_D^{21}$ +43.1 (c, 0.58 in MeOH). λ_{max} 203 (log ε 5.92); 288 (log ε 5.22); 333 (log ε 5.08) (MeOH).

Murata, T. *et al.*, *Chem. Pharm. Bull.*, 2011, **59**, 88-95 (*Rashomonic acids A-D*)

Rayalinol R7

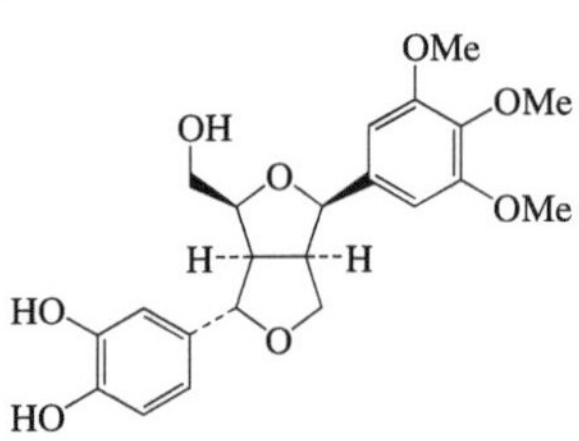

Relative Configuration

[1220189-37-6]
$C_{22}H_{26}O_8$ 418.442
Homolignan. Constit. of the aerial parts of *Eurotia ceratoides* (preferred genus name *Axyris*). Amorph. solid. $[\alpha]_D^{25}$ +0.05 (c, 0.04 in unstated solvent). λ_{max} 236 (log ε 7.34); 274 (log ε 6.94) (MeOH).

Siddiqui, B.S. *et al.*, *Tetrahedron*, 2010, **66**, 1716-1720 (*Rayalinol*)

Reevesiacoumarin R8

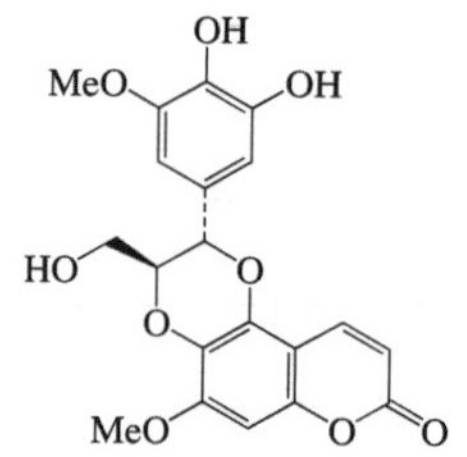

Absolute Configuration

$C_{20}H_{18}O_9$ 402.355
Constit. of the roots and stems of *Reevesia formosana*. Yellowish powder. $[\alpha]_D^{25}$ -16.8 (c, 0.24 in MeOH). λ_{max} 237 (sh); 320 (log ε 4.10) (MeOH).

Chang, H.-S. *et al.*, *RSC Adv.*, 2017, **7**, 27040-27047 (*Reevesiacoumarin*)

Rehmaglutoside E R9

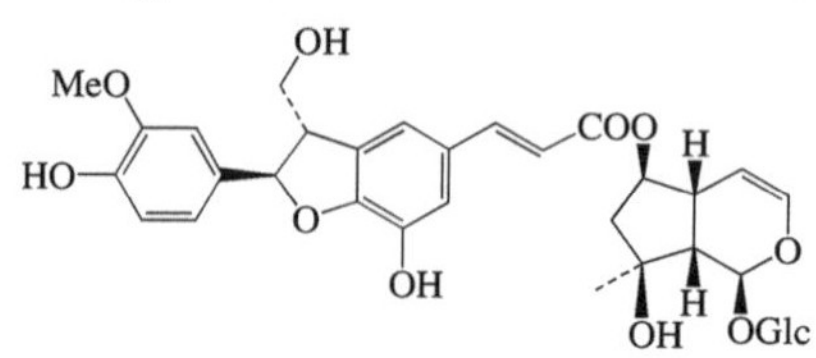

[1397694-39-1]
$C_{34}H_{40}O_{15}$ 688.679
Constit. of *Rehmannia glutinosa*. Amorph. powder. $[\alpha]_D^{20}$ -94 (c, 0.14 in MeOH). λ_{max} 224 (log ε 4.74); 330 (log ε 4.58) (MeOH).

Liu, Y.-F. *et al.*, *J. Nat. Prod.*, 2012, **75**, 1625-1631 (*Rehmaglutoside E*)

Renealtin A R10

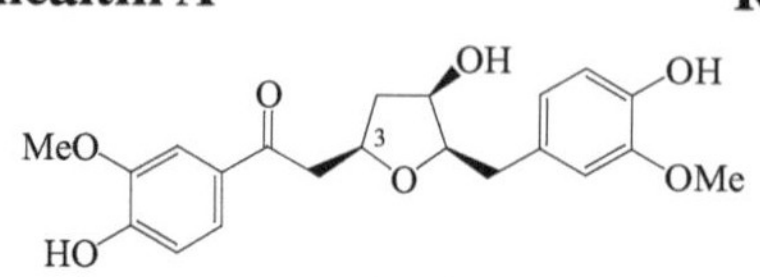

Absolute Configuration

[407602-77-1]
$C_{21}H_{24}O_7$ 388.416
Constit. of *Amomum kravanh* and *Renealmia exaltata*. Amorph. yellow solid. $[\alpha]_D^{23}$ +22.4 (c, 0.4 in MeOH). λ_{max} 228 (log ε 4.25); 279 (log ε 3.99); 305 (log ε 3.82) (MeOH)

3-Epimer

Renealtin B
[407602-78-2]
$C_{21}H_{24}O_7$ 388.416
Constit. of the seeds of *Renealmia exaltata*. Amorph. yellow solid. $[\alpha]_D^{23}$ +73.5 (c, 0.15 in MeOH). λ_{max} 228 (log ε 4.29); 279 (log ε 4.04); 304 (log ε 3.88) (MeOH).

Sekiguchi, M. *et al.*, *J. Nat. Prod.*, 2002, **65**, 375-376 (*Renealtins A,B*).
Katoh, T. *et al.*, *Synlett*, 2006, 2031-2034 (*synth*).
Sabitha, G. *et al.*, *Tetrahedron Lett.*, 2007, **48**, 8065-8086 (*synth*).
Yin, H. *et al.*, *Phytochem. Lett.*, 2013, **6**, 403-406 (*Amomum kravanh constit*)

Retipolide E R11

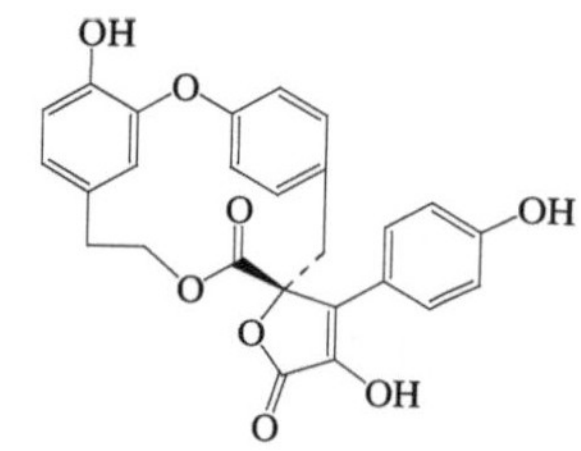

[945531-41-9]
$C_{26}H_{20}O_8$ 460.438
Provisional lignan. Biosynth. pathway unclear. Related to Butyrolactone I, B89. Isol. from *Retiboletus retipes*. Amorph. solid. Mp 195 - 197°. $[\alpha]_D^{20}$ +83 (c, 0.17 in MeCN). λ_{max} 223 (sh); 288 (log ε 3.97); 305 (log ε 3.98); 350 (sh) (MeCN).

Justus, K. *et al.*, *Eur. J. Org. Chem.*, 2007, 5560-5572 (*Retipolide E*).
Ingerl, A. *et al.*, *Tetrahedron*, 2007, **63**, 6548-6557 (*synth*)

Rhodiolin R12

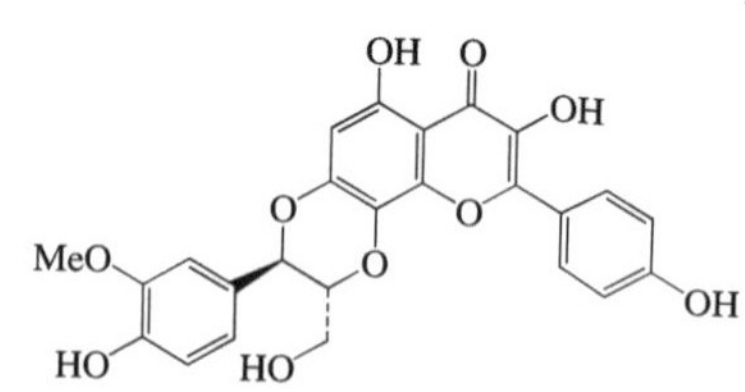

Relative Configuration

Rhodiolinin
[86831-53-0]
$C_{25}H_{20}O_{10}$ 480.425
Flavonolignan. Related to Scutellaprostin A, S38. Constit. of the rhizomes of *Rhodiola rosea* and *Rhodiola quadrifida*. Exhibits anti-influenza virus activity. Yellow cryst. (MeOH). Mp 235 - 237°. λ_{max} 281; 333; 384 (EtOH).

Zapesochnaya, G.G. *et al.*, *Chem. Nat. Compd. (Engl. Transl.)*, 1983, **19**, 21-29 (*Rhodiolin*).
Kurkin, V.A. *et al.*, *Chem. Nat. Compd. (Engl. Transl.)*, 1990, **26**, 710-711 (*synth*).
Wiedenfeld, H. *et al.*, *Pharmazie*, 2007, **62**, 308-311 (*Rhodiola quadrifida constit*).
Jeong, H.J. *et al.*, *Bioorg. Med. Chem.*, 2009, **17**, 6816-6823 (*Rhodiolinin, activity*)

Rhoiptelol B R13

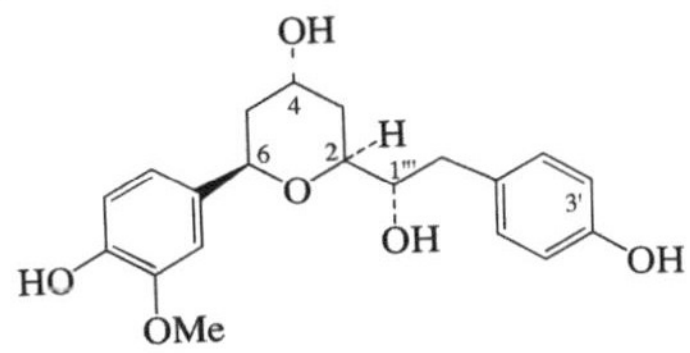

Absolute Configuration

[184485-69-6]
$C_{20}H_{24}O_6$ 360.406
Constit. of *Alnus hirsuta*, *Juglans mandshurica* and *Rhoiptelea chiliantha*. Amorph. powder + ¾H_2O. Mp 65 - 67°. Mp 68 - 70°. $[\alpha]^{12}_{D}$+97 (c, 0.3 in MeOH)

3′-Methoxy
Kravanhol A
[1607441-60-0]
$C_{21}H_{26}O_7$ 390.432
Constit. of *Amomum kravanh*. Yellow oil. $[\alpha]^{20}_{D}$-61.4 (c, 0.3 in MeOH). λ_{max} 227 (log ε 3.95); 280 (log ε 3.58) (MeOH)

3′-Methoxy, 4-deoxy
Kravanhol B
[1607441-61-1]
$C_{21}H_{26}O_6$ 374.433
Constit. of *Amomum kravanh*. NO prodn. inhibitor. Light yellow oil. $[\alpha]^{20}_{D}$-42.7 (c, 0.2 in MeOH). λ_{max} 227 (log ε 3.95); 280 (log ε 3.60) (MeOH)

1‴-Epimer, 3′-methoxy
Epikravanhol A
[1663475-00-0]
$C_{21}H_{26}O_7$ 390.432
Constit. of the roots of *Juglans mandshurica*. Light yellow oil. $[\alpha]^{20}_{D}$+89.4 (c, 0.2 in MeOH). λ_{max} 227 (log ε 0.4); 281 (log ε 0.15) (MeOH)

2,6-Diepimer
Engelheptanoxide D
[1397702-25-8]
$C_{20}H_{24}O_6$ 360.406
Constit. of the stems of *Engelhardtia roxburghiana*. Amorph. yellow solid. $[\alpha]^{25}_{D}$+58.7 (c, 0.12 in MeOH). λ_{max} 206 (log ε 4.2); 225 (log ε 4.03); 279 (log ε 3.51) (MeOH). [base] λ_{max} 209 (log ε 4.54); 243 (log ε 4.02); 287 (log ε 3.55) (MeOH/KOH)

2,6-Diepimer, 4-ketone
Engelheptanoxide B
[1397702-21-4]
$C_{20}H_{22}O_6$ 358.39
Constit. of the stems of *Engelhardtia roxburghiana*. Yellowish oil. $[\alpha]^{24}_{D}$+82.7 (c, 0.2 in MeOH). λ_{max} 207 (log ε 4.27); 225 (log ε 4.1); 279 (log ε 3.62) (MeOH). [base] λ_{max} 209 (log ε 4.53); 243 (log ε 4.03); 287 (log ε 3.67) (MeOH/KOH).

Jiang, Z.-H. *et al.*, *Phytochemistry*, 1996, **43**, 1049-1054 (*Rhoiptelol B*).
Jin, W.Y. *et al.*, *Arch. Pharmacal Res.*, 2007, **30**, 412-418 (*Alnus hirsuta constit*).
Reddy, C.R. *et al.*, *Eur. J. Org. Chem.*, 2010, 345-351 (*synth*).
Wu, H.-C. *et al.*, *Phytochemistry*, 2012, **82**, 118-127 (*Engelheptanoxides B,D*).
Yin, H. *et al.*, *Phytochem. Lett.*, 2013, **6**, 403-406 (*Kravanhols A,B*).
Yadav, J.S. *et al.*, *Synlett*, 2014, **25**, 661-664 (*synth*).
Jin, M. *et al.*, *Nat. Prod. Res.*, 2015, **29**, 1839-1844 (*Epikravanhol A*).
Tanaka, N. *et al.*, *Heterocycles*, 2018, **97**, 1157-1164 (*synth*).
Diao, S. *et al.*, *Nat. Prod. Res.*, 2019, **33**, 701-707 (*Juglans manshurica constit*)

Rhyncoside E R14

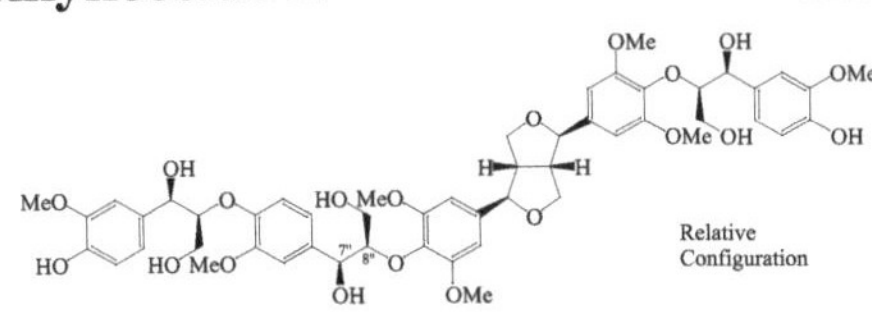

Relative Configuration

[1023336-07-3]
$C_{52}H_{62}O_{20}$ 1007.048
Constit. of the mangrove *Bruguiera sexangula* var. *rhynchopetala*. Pale yellow gum

7″- or 8″-Epimer
Rhyncoside F
[1023336-17-5]
$C_{52}H_{62}O_{20}$ 1007.048
Constit. of the mangrove *Bruguiera sexangula* var. *rhynchopetala*. Pale yellow gum.

Bao, S. *et al.*, *Chem. Pharm. Bull.*, 2007, **55**, 1175-1180 (*Rhyncosides E,F*)

Ribesin H R15

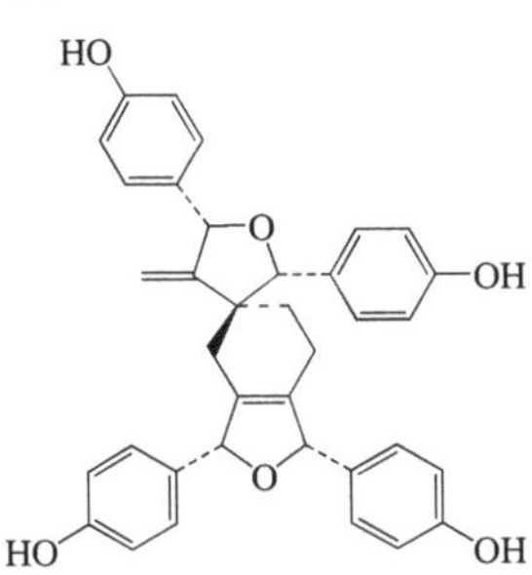

Relative Configuration

[1469818-69-6]
$C_{36}H_{32}O_6$ 560.646
Isol. from the leaves of *Ribes nigrum*. Solid. $[\alpha]^{25}_{D}$-20.6 (c, 0.52 in MeOH). λ_{max} 230 (log ε 4.12); 277 (log ε 3.62) (MeOH).

Sasaki, T. *et al.*, *Phytochemistry*, 2013, **95**, 333-340 (*Ribesin H*)

Rodgersinol R16

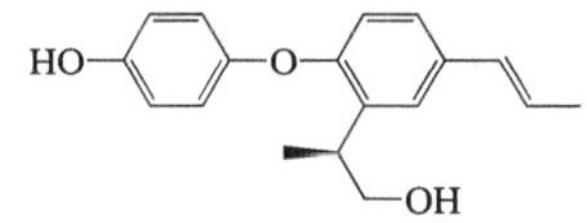

Absolute Configuration

2-(4-Hydroxyphenoxy)-β-methyl-5-(1-propenyl)benzeneethanol
[906670-79-9]
$C_{18}H_{20}O_3$ 284.355
Constit. of the aerial parts of *Rodgersia podophylla*. Inhibitor of iNOS and COX-2. Powder (MeOH). Mp 178 - 179°. $[\alpha]^{20}_{D}$-14.6 (c, 0.04 in MeOH). λ_{max} 261 (ε 8990) (MeOH).

Chin, Y.-W. *et al.*, *Bioorg. Med. Chem. Lett.*, 2006, **16**, 4600-4602 (*Rodgersinol, activity*).
Seo, S.-Y. *et al.*, *J. Org. Chem.*, 2007, **72**, 666-668 (*synth, abs config*).
Jung, J.-W. *et al.*, *Tetrahedron*, 2010, **66**, 6826-6831 (*synth*)

Celastrus hindsii Rosmarinic acid trimer(1) R17

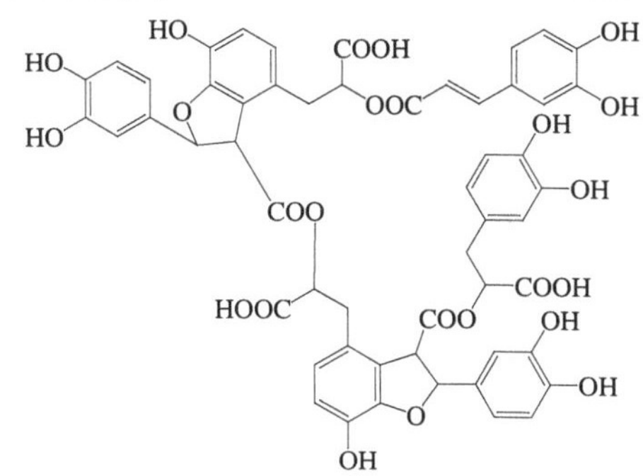

[894784-32-8]
$C_{54}H_{44}O_{24}$ 1076.922
Constit. of the leaves of *Celastrus hindsii*. Pale yellow-green wax +6H_2O. $[\alpha]^{25}_{D}$+57.7 (c, 0.27 in MeOH). λ_{max} 286 (log ε 4.52); 329 (log ε 4.35) (MeOH).

Ly, T.N. *et al.*, *J. Agric. Food Chem.*, 2006, **54**, 3786-3793 (*isol, pmr, cmr, ms*)

Celastrus hindsii Rosmarinic acid trimer(2) R18

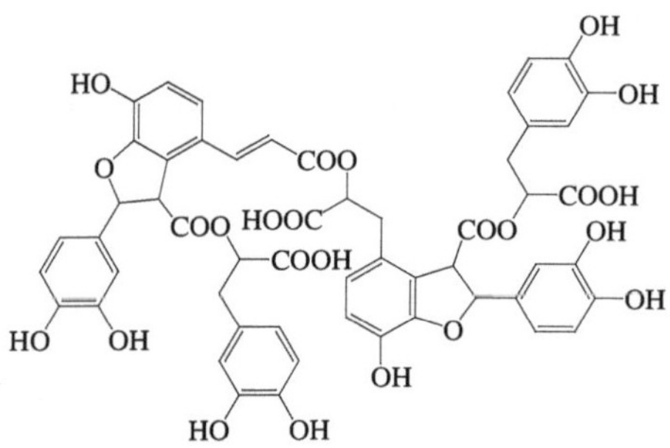

[894784-33-9]

$C_{54}H_{44}O_{24}$ 1076.922

Constit. of the leaves of *Celastrus hindsii*. Yellow wax. $[\alpha]_D^{25}$ +99 (c, 0.35 in MeOH). λ_{max} 287 (log ε 4.4) (MeOH).

Ly, T.N. *et al.*, *J. Agric. Food Chem.*, 2006, **54**, 3786-3793 (*isol, pmr, cmr, ms*)

Rubanol R19

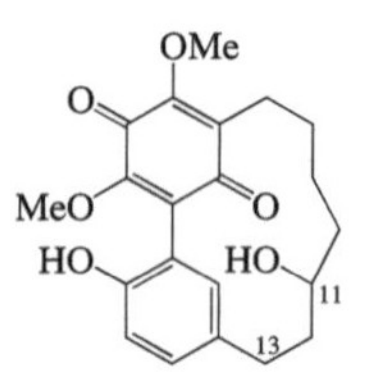

[955127-07-8]

$C_{21}H_{24}O_6$ 372.417

Related to Myricanol, M57. Constit. of the bark of *Myrica rubra*. Cytotoxic. Red oil (EtOAc). $[\alpha]_D^{23}$ +14 (c, 0.25 in MeOH). λ_{max} 228 (log ε 3.5); 280 (log ε 4.1); 385 (log ε 1.45) (MeOH)

11-Ketone

Rubanone

[1034410-14-4]

$C_{21}H_{22}O_6$ 370.401

Constit. of the bark of *Myrica rubra*. Red oil. λ_{max} 281 (log ε 4.5) (no solvent reported)

11-Deoxy, 12,13-didehydro(E-)

Nanaone

[1190618-72-4]

$C_{21}H_{22}O_5$ 354.402

Constit. of the bark of *Myrica nana*. Red cryst.. $[\alpha]_D^{23}$ +10.5 (c, 0.2 in MeOH). λ_{max} 225 (log ε 3.45); 284 (log ε 4.45); 380 (log ε 1.35) (MeOH).

Wang, D.Y. *et al.*, *Chin. Chem. Lett.*, 2008, **19**, 547-549 (*Rubanone*).
Wang, D.-Y. *et al.*, *Nat. Prod. Res.*, 2008, **22**, 292-295 (*Rubanol*).
Zeng, L. *et al.*, *J. Chem. Res.*, 2009, 337-338 (*Nanaone*)

Rubescenamine R20

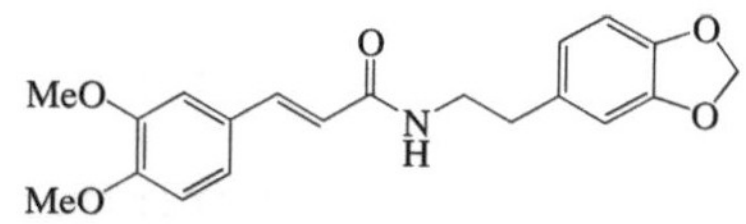

FEMA 4773

[125187-30-6]

$C_{20}H_{21}NO_5$ 355.39

Alkaloid from the stem bark of *Zanthoxylum rubescens*. Possesses Umami-taste modulating properties. Mp 145 - 147°

N-Me

Rubescenamide

[125187-29-3]

$C_{21}H_{23}NO_5$ 369.417

Alkaloid from the stem bark of *Zanthoxylum rubescens*. Amorph. solid.

Adesina, S.K. *Planta Med.*, 1989, **55**, 324-326 (*Rubescenamine*).
Backes, M. *et al.*, *J. Agric. Food Chem.*, 2015, **63**, 8694-8704 (*isol, activity*).
Amino, Y. *et al.*, *Biosci., Biotechnol., Biochem.*, 2017, **81**, 1690-1698 (*sar*)

Rupestrin A R21

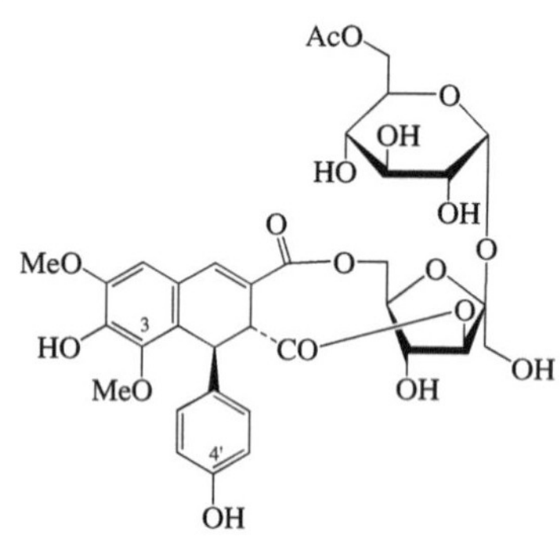

[880360-46-3]

$C_{34}H_{38}O_{18}$ 734.66

Constit. of *Eritrichium rupestre*. Powder (MeOH). $[\alpha]_D^{20}$ +79.4 (c, 0.61 in MeOH). λ_{max} 225 (log ε 4.24); 254 (log ε 4.29); 355 (log ε 4.1) (MeOH)

O-De-Ac

Rupestrin B

[880360-48-5]

$C_{32}H_{36}O_{17}$ 692.623

Constit. of *Eritrichium rupestre*. Yellow powder (MeOH). $[\alpha]_D^{20}$ +101.7 (c, 0.6 in MeOH). λ_{max} 224 (log ε 4.24); 253 (log ε 4.28); 353 (log ε 4.08) (MeOH)

4′-Deoxy, 2′,5′-dimethoxy, O³-de-Me

Rupestrin C

[880360-49-6]

$C_{35}H_{40}O_{19}$ 764.686

Constit. of *Eritrichium rupestre*. Light green powder (MeOH). $[\alpha]_D^{20}$ +78.5 (c, 0.39 in MeOH). λ_{max} 255 (log ε 4.34); 276 (log ε 4); 352 (log ε 3.99) (MeOH).

Suo, M.-R. *et al.*, *J. Nat. Prod.*, 2006, **69**, 682-684 (*Rupestrins A,B,C*)

Sacidumlignan D S1

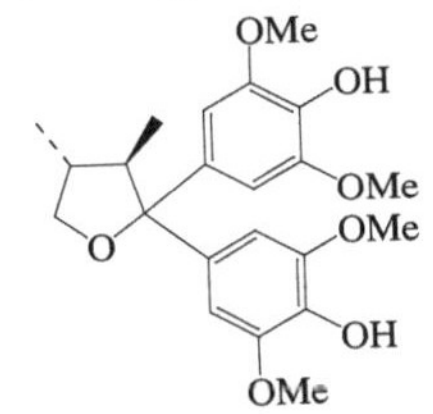

Relative Configuration

[848986-18-5]

$C_{22}H_{28}O_7$ 404.459

Rearranged epoxytetrahydrofuran lignan. Constit. of the whole plant of *Sarcostemma acidum*. Amorph. powder (EtOAc/pentane). Mp 112 - 114° (synthetic). Mp 149 - 153° (synthetic). $[\alpha]^{20}_{D}$-115 (c, 1.14 in Me_2CO). λ_{max} 204 (log ε 4.85); 273 (log ε 3.5) (MeOH).

Gan, L.-S. *et al.*, *J. Nat. Prod.*, 2005, **68**, 221-225 (*Sacidumlignan D*).

Pandey, S.K. *et al.*, *J. Org. Chem.*, 2011, **76**, 2315-2318 (*synth*).

Rout, J.K. *et al.*, *J. Org. Chem.*, 2012, **77**, 1566-1571 (*synth*).

Zhang, J.-J. *et al.*, *Org. Biomol. Chem.*, 2013, **11**, 2498-2513 (*synth*).

Xie, C. *et al.*, *Asian J. Org. Chem.*, 2014, **3**, 277-280 (*synth*).

Lei, X. *et al.*, *Angew. Chem., Int. Ed.*, 2021, **60**, 4221-4230 (*synth*)

Sacidumol A S2

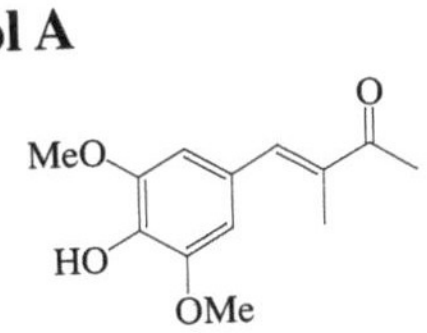

4-(4-Hydroxy-3,5-dimethoxyphenyl)-3-methyl-3-buten-2-one

[848986-19-6]

$C_{13}H_{16}O_4$ 236.267

Degraded lignan. Constit. of *Sarcostemma acidum*. Amorph. powder. λ_{max} 236 (log ε 4.05); 325 (log ε 4.08) (MeOH).

Gan, L.-S. *et al.*, *J. Nat. Prod.*, 2005, **68**, 221-225 (*Sacidumol A*)

Sacric acid B S3

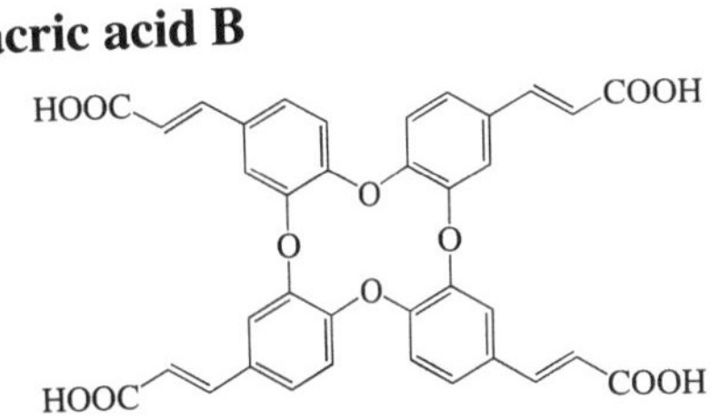

[2485841-19-6]

$C_{36}H_{24}O_{12}$ 648.576

Constit. of *Artemisia sacrorum*. Needles. Mp 245 - 248°. λ_{max} 259 (log ε 4.92) (MeOH).

Wang, Q. *et al.*, *Nat. Prod. Commun.*, 2016, **11**, 489-490 (*Sacric acid B*)

Sagecoumarin S4

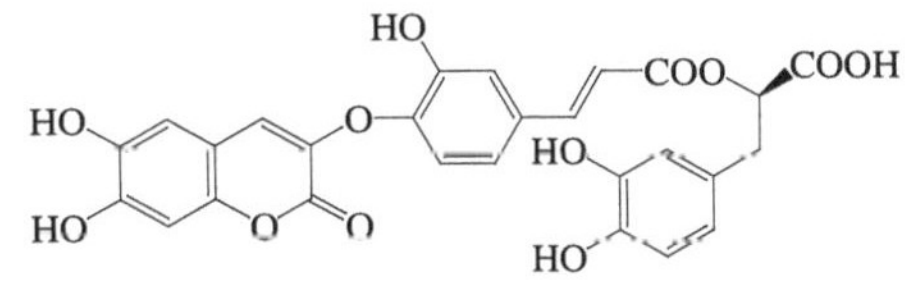

[260549-30-2]

$C_{27}H_{20}O_{12}$ 536.445

Caffeic acid trimer. Related to Salvianolic acid I, S6. Constit. of *Salvia officinalis*. Antioxidant. Light brown powder. $[\alpha]^{20}_{D}$+52 (c, 0.2 in MeOH). λ_{max} 288 (log ε 4.3); 332 (log ε 4.27) (MeOH).

Lu, Y. *et al.*, *Phytochemistry*, 1999, **52**, 1149-1152 (*Sagecoumarin*).

Lu, Y. *et al.*, *Food Chem.*, 2001, **75**, 197-202 (*activity*)

Sagerinic acid S5

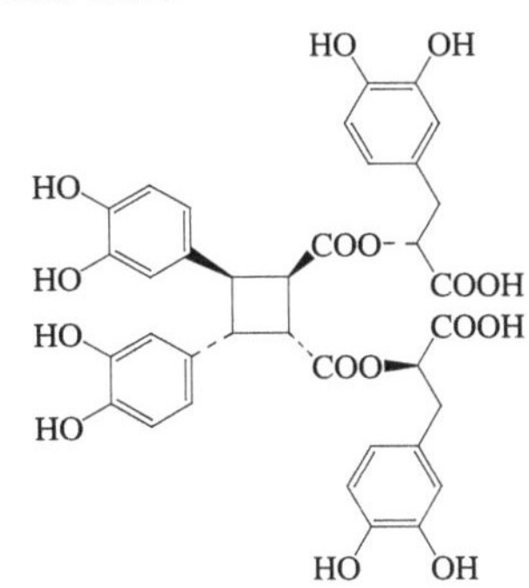

[238752-75-5]

$C_{36}H_{32}O_{16}$ 720.636

Derived from Rosmarinic acid in D209 and related to Shimobashiric acid C, S57. Constit. of *Melissa officinalis*, *Orthosiphon grandiflorus* and *Salvia officinalis*. Antioxidant. $[\alpha]^{20}_{D}$+4 (c, 0.2 in MeOH). λ_{max} 232 (log ε 4.08); 286 (log ε 4.09) (MeOH).

Lu, Y. *et al.*, *Phytochemistry*, 1999, **51**, 91-94 (*Sagerinic acid*).

Barros, L. *et al.*, *Food Chem.*, 2013, **136**, 1-8 (*Melissa officinalis constit*)

Salvianolic acid I S6

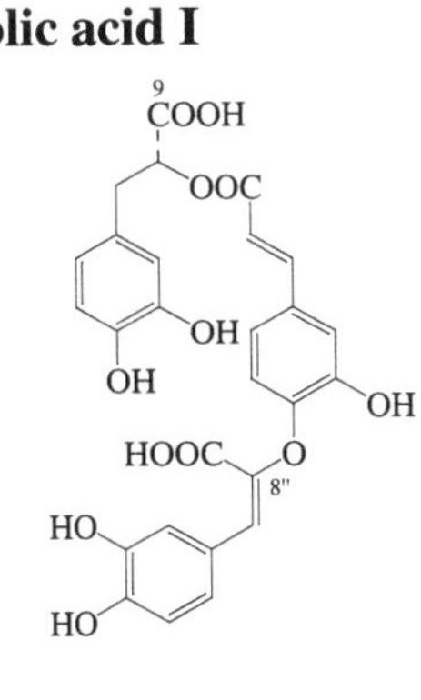

[150072-80-3]

$C_{27}H_{22}O_{12}$ 538.461

Regioisomer of Salvianolic acid H, S15. Constit. of *Salvia cavaleriei* and *Salvia officinalis*. Amorph. powder. $[\alpha]^{14}_{D}$+71 (c, 0.1 in EtOH)

8″Z-Isomer

Melitric acid A. *Schizotenuin E*

[153765-45-8]

$C_{27}H_{22}O_{12}$ 538.461

Constit. of *Melissa officinalis*, *Salvia officinalis*, *Satureja biflora* and *Schizonepeta tenuifolia*. Light brown powder + $1/2H_2O$. Mp 135 - 138°. $[\alpha]^{23}_{D}$+45 (c, 0.2 in MeOH). Related to Rosmarinic acid, in O209. λ_{max} 290 (log ε 4.32); 328 (log ε 4.34) (MeOH)

8″Z-Isomer, 8″-Me ester

Schizotenuin F. *Methyl melitrate A*

[127498-36-6]

$C_{28}H_{24}O_{12}$ 552.488

Constit. of *Salvia officinalis*, *Satureja biflora* and *Schizonepeta tenuifolia*. Antiinflammatory agent. Light brown powder. $[\alpha]^{20}_{D}$+40.5 (c, 0.2 in MeOH). λ_{max} 292 (log ε 4.33); 330 (log ε 4.36) (MeOH).

Agata, I. *et al.*, *Chem. Pharm. Bull.*, 1993, **41**, 1608-1611 (*Melitric acid A*).

Li, L. *Youji Huaxue*, 1993, **13**, 303-304 (*Salvianolic acid I, abs config*).

Zhang, H.-J. *et al.*, *Planta Med.*, 1994, **60**, 70-72 (*Salvianolic acid I*).

Matsuta, M. *et al.*, *Nat. Med. (Tokyo)*, 1996, **50**, 204-211 (*Schizotenuins*).

Lu, Y. *et al.*, *Phytochemistry*, 1999, **52**, 1149-1152 (*Melitric acid A*)

Salvianolic acid V S7

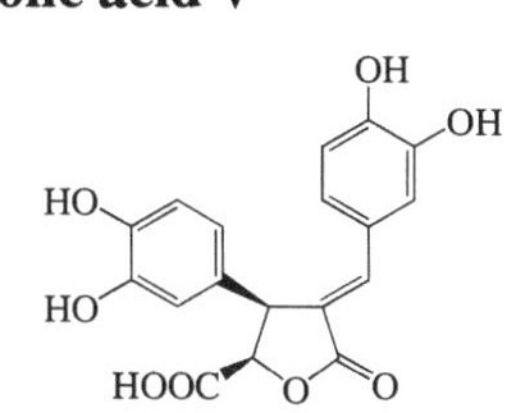

Relative Configuration

3,3′,4,4′-Tetrahydroxy-7,8′-lign-7′-en-9′,8-olid-9-oic acid

[2108687-52-9]

$C_{18}H_{14}O_8$ 358.302

Constit. of *Salvia miltiorrhiza*. Antibiotic potentiator. Amorph. yellow powder. $[\alpha]^{20}_{D}$+73.8 (c, 0.4 in MeOH).

Zhang, J. *et al.*, *Nat. Prod. Res.*, 2018, **32**, 1198-1202 (*Salvianolic acid V, activity*)

Salvianolic acid A S8

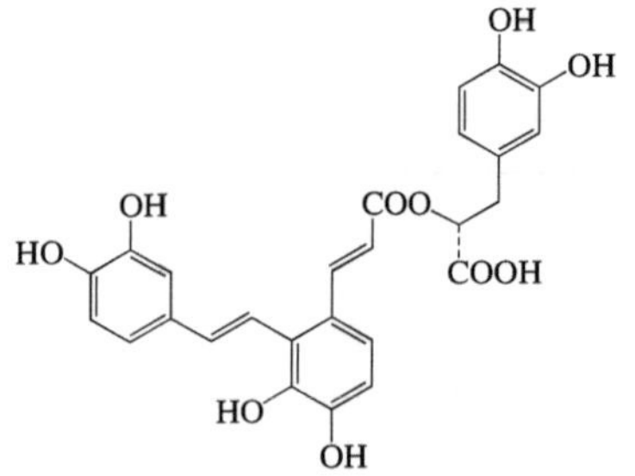

Absolute Configuration

[96574-01-5]

$C_{26}H_{22}O_{10}$ 494.452

Constit. of *Melissa officinalis*, *Origanum vulgare*, *Salvia miltiorrhiza*, *Salvia yunnanensis* and *Tournefortia sarmentosa*. Shows antiulcer activity. Inhibitor of protein–protein interactions mediated by the SH2 domains of Src-family kinases. Shows antiproliferative and hepatoprotectant activities. Displays cardioprotective props. Shows anti-LDL peroxidative activity. Matrix metalloproteinase-9 inhibitor. Antioxidant. Amorph. yellow powder

Me ester

[1015171-69-3]

$C_{27}H_{24}O_{10}$ 508.479

Constit. of the roots of *Salvia yunnanensis*. Yellow powder. $[\alpha]^{20}_{D}$ +35 (c, 0.11 in EtOH). λ_{max} 204 (log ε 4.72); 289 (log ε 4.44); 336 (log ε 4.36) (MeOH)

Et ester

[1015171-68-2]

$C_{28}H_{26}O_{10}$ 522.506

Constit. of the roots of *Salvia yunnanensis*. Yellow powder. $[\alpha]^{20}_{D}$ +35 (c, 0.1 in EtOH). λ_{max} 204 (log ε 4.7); 289 (log ε 4.45); 335 (log ε 4.33) (MeOH).

Li, L.-M. *et al., Planta Med.*, 1984, **50**, 227-228 (*Salvianolic acid A*).
Murakami, S. *et al., Planta Med.*, 1990, **56**, 360-363 (*activity*).
Lin, Y.-L. *et al., J. Nat. Prod.*, 2002, **65**, 745-747 (*Tournefortia sarmentosa constit, activity*).
Lin, Y.-L. *et al., J. Chin. Chem. Soc. (Taipei)*, 2003, **50**, 1079-1083 (*Origanum vulgare constit*).
Lin, Y.-L. *et al., J. Pharm. Pharmacol.*, 2006, **58**, 933-939 (*activity*).
Zhou, B. *et al., Anal. Chim. Acta*, 2008, **628**, 111-116 (*anal*).
Zhang, Z.-F. *et al., J. Asian Nat. Prod. Res.*, 2008, **10**, 391-396 (*Salvia yunnanensis constits*).
Sperl, B. *et al., Bioorg. Med. Chem. Lett.*, 2009, **19**, 3305-3309 (*activity*).
Zhang, Y.-Q. *et al., J. Asian Nat. Prod. Res.*, 2010, **12**, 899-915 (*activity*).
Barros, L. *et al., Food Chem.*, 2013, **136**, 1-8 (*Melissa officinalis constit*).
Zhang, T. *et al., Phytomedicine*, 2014, **21**, 1137-1145 (*activity*).
Zheng, Y. *et al., Tetrahedron*, 2016, **72**, 5047-5050 (*synth*).
Xu, K. *et al., Tetrahedron*, 2018, **74**, 5996-6002 (*synth*)

Salvianolic acid C S9

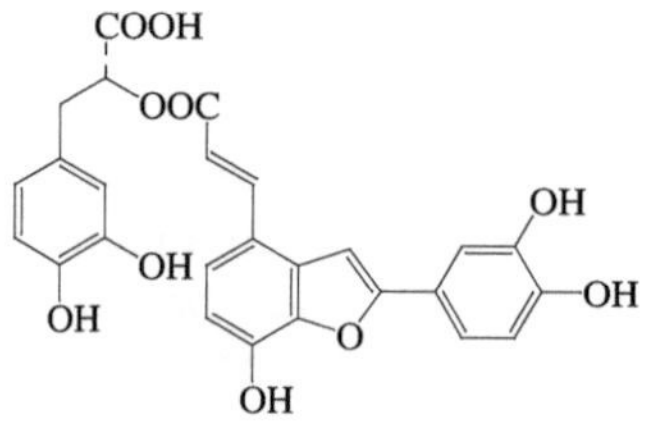

Absolute Configuration

[115841-09-3]

$C_{26}H_{20}O_{10}$ 492.436

Constit. of *Clerodendranthus spicatus*, *Origanum vulgare* and *Salvia miltiorrhiza*. Also found in *Melissa officinalis* prob. as a glycoside. Also prod. by *Phoma glomerata* D14 found on *Salvia miltiorrhiza*. Amorph. yellow powder. $[\alpha]^{14}_{D}$ +70 (c, 0.1 in EtOH).

Ai, C.-B. *et al., J. Nat. Prod.*, 1988, **51**, 145-149 (*Salvianolic acid C*).
Lin, Y.-L. *et al., J. Chin. Chem. Soc. (Taipei)*, 2003, **50**, 1079-1083 (*Origanum vulgare constit*).
Shen, S.-D. *et al., ARKIVOC*, 2012, **vi**, 204-213 (*synth*).
Barros, L. *et al., Food Chem.*, 2013, **136**, 1-8 (*Melissa officinalis constits*).
Alford, B.L. *et al., Org. Biomol. Chem.*, 2013, **11**, 2724-2727 (*synth*).
Sun, Z. *et al., Phytochem. Lett.*, 2014, **8**, 16-21 (*Clerodendranthus spicatus constit*).
López-Frías, G. *et al., Molecules*, 2015, **20**, 8654-8665 (*synth*).
Li, X. *et al., Curr. Microbiol.*, 2016, **73**, 31-37 (*Phoma glomerata constit*)

Salvianolic acid C_1 S10

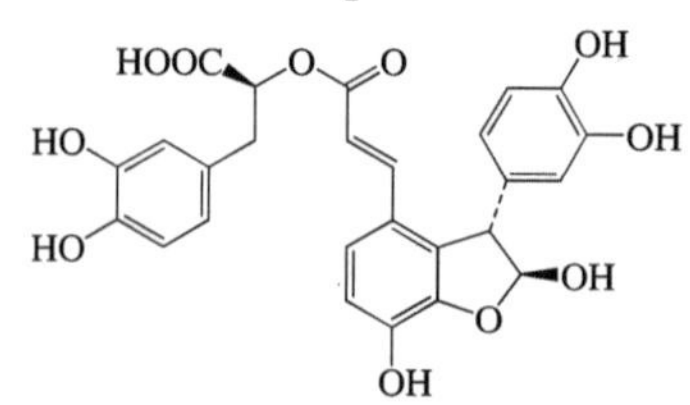

Absolute Configuration

$C_{26}H_{22}O_{11}$ 510.451

Closely related to Lithospermic acid, L114. Constit. of the roots of *Salvia miltiorrhiza*. Antioxidant. Amorph. powder.

Zhang, J. *et al., Fitoterapia*, 2017, **121**, 46-52 (*Salvianolic acid C_1, cd, activity*)

Salvianolic acid D S11

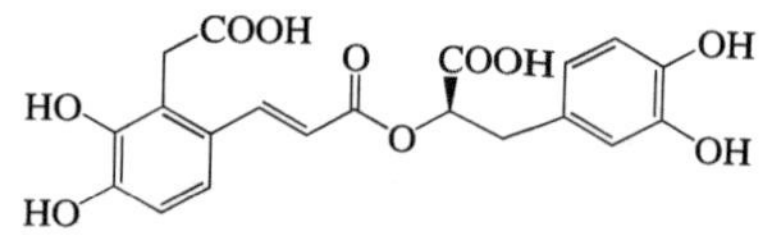

Absolute Configuration

[142998-47-8]

$C_{20}H_{18}O_{10}$ 418.354

Constit. of *Salvia miltiorrhiza*. Amorph. yellow powder. $[\alpha]_{D}$ +68.2 (EtOH)

γ-Lactone

Salvianolic acid D lactone

$C_{20}H_{16}O_{9}$ 400.339

Constit. of *Salvia miltiorrhiza*.

Ai, C.-B. *et al., Planta Med.*, 1992, **58**, 197-199 (*Salvianolic acid D*).
Tian, J.-F. *et al., Zhongcaoyao*, 2018, **49**, 5024-5028 (*Salvianolic acid D, Salvianolic acid D lactone*)

Salvianolic acid E S12

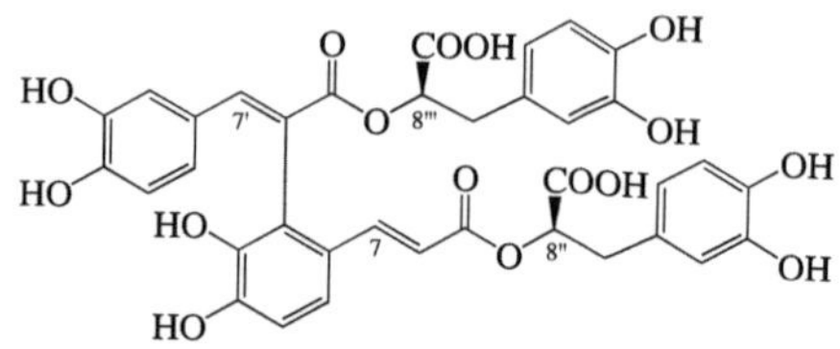

Absolute Configuration

[142998-46-7]

$C_{36}H_{30}O_{16}$ 718.62

Constit. of *Salvia miltiorrhiza*. Amorph. yellow powder. $[\alpha]_{D}$ +59.4 (EtOH)

8″,8‴-Diepimer, (7′Z)-isomer

[1835674-60-6]

$C_{36}H_{30}O_{16}$ 718.62

Minor constit. of *Salvia miltiorrhiza*. Brown powder. $[\alpha]^{20}_{D}$ +1.3 (c, 0.03 in MeOH aq.).

Ai, C.-B. *et al., Planta Med.*, 1992, **58**, 197-199 (*Salvianolic acid E*).
Quéffelec, C. *et al., Bioorg. Med. Chem. Lett.*, 2008, **18**, 4736-4740 (*synth*).
Gong, J. *et al., Molecules*, 2015, **20**, 683-692 (*biosynth*).
Si, Y. *et al., Bioorg. Med. Chem. Lett.*, 2016, **26**, 82-86 (*Salvianolic acid E isomer*)

Salvianolic acid F S13

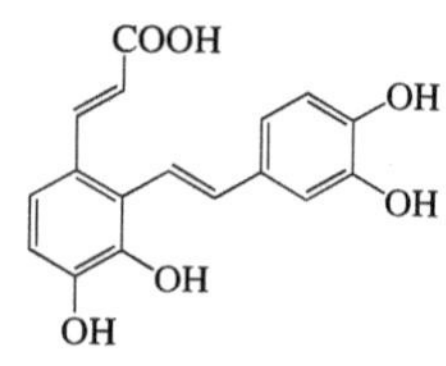

3,3′,4,4′-Tetrahydroxy-9′-nor-2,8′-ligna-7,7′-dien-9-oic acid

[158732-59-3]

$C_{17}H_{14}O_{6}$ 314.293

Constit. of *Melissa officinalis*, *Salvia miltiorrhiza* and *Tournefortia sarmentosa*.

Ai, C.B. *et al.*, *Chin. Chem. Lett.*, 1996, **7**, 427-430 (*synth*).
Dalla, V. *et al.*, *Tetrahedron*, 1999, **55**, 6923-6930 (*synth*).
Barros, L. *et al.*, *Food Chem.*, 2013, **136**, 1-8 (*Melissa officinalis constit*).
Sharol, A. *et al.*, *Eur. J. Org. Chem.*, 2016, 5941-594 (*synth*)

Salvianolic acid G S14

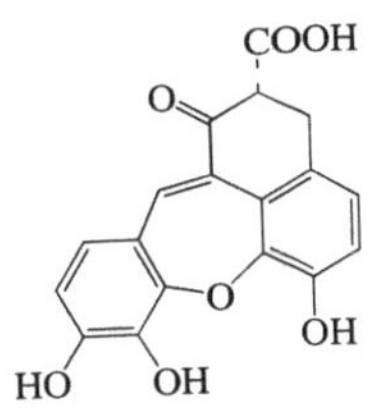

(*R*)-*form*

2,3-Dihydro-6,8,9-trihydroxy-1-oxo-1H-benzo[f]naphth[1,8-bc]oxepin-2-carboxylic acid, CAS
[136112-79-3]
$C_{18}H_{12}O_7$ 340.287
(R)-form [1602919-93-6]
Constit. of the roots of *Salvia miltiorrhiza*. Antioxidant. Yellow powder. $[\alpha]^{20}_{D}$ -138 (c, 0.02 in MeOH). λ_{max} 201; 395 (MeOH)
Me ester
[1602826-56-1]
$C_{19}H_{14}O_7$ 354.314
Constit. of *Salvia miltiorrhiza*. Yellow powder. $[\alpha]^{20}_{D}$ -163 (c, 0.02 in MeOH). λ_{max} 201; 218; 231; 395 (MeOH).

Ai, C. *et al.*, *Chin. Chem. Lett.*, 1991, **2**, 17-18 (*Salvianolic acid G*).
Liu, A.-H. *et al.*, *J. Chromatogr., A*, 2007, **1161**, 170-182 (*Salvianolic acid G*).
Liu, J. *et al.*, *Chem. Nat. Compd. (Engl. Transl.)*, 2014, **50**, 83-87 (*R-form, Me ester*)

Salvianolic acid H S15

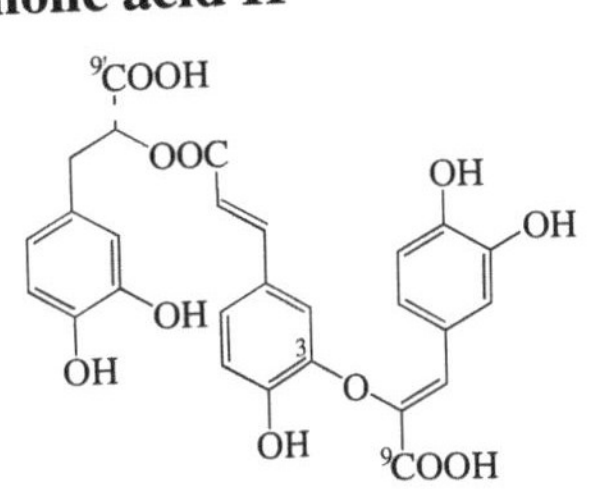

3′-O-(8″-Z-Caffeoyl)rosmarinic acid. *Isoschizotenuin E*. ***Isomelitric acid A***
[444179-57-1]
[155576-07-1] (deleted)
$C_{27}H_{22}O_{12}$ 538.461
Regioisomer of Salvianolic acid I, S6.
Constit. of *Lycopus lucidus*, *Melissa officinalis*, *Salvia cavaleriei* var. *simplicifolia* and *Thymus vulgaris*. $[\alpha]^{22}_{D}$ +40.3 (c, 1.2 in MeOH). λ_{max} 217 (sh); 231 (sh); 291; 322 (MeOH)
9-Me ester
[1331731-22-6]
$C_{28}H_{24}O_{12}$ 552.488
Constit. of *Lycopus lucidus* and *Pulmonaria officinalis*. Amorph. solid. $[\alpha]^{21}_{D}$ +65.5 (c, 0.55 in MeOH). λ_{max} 275 (sh); 324 (log ε 4.77) (MeOH)
9′-Me ester
Orthosiphoic acid D
[1638526-29-0]
$C_{28}H_{24}O_{12}$ 552.488
Constit. of the aerial parts of *Clerodendranthus spicatus*. Exhibits weak anti-HIV-1 protease activity. Amorph. brown powder. $[\alpha]^{20}_{D}$ +57.1 (c, 0.02 in MeOH). λ_{max} 288 (log ε 0.42); 320 (log ε 0.42) (MeOH)
Di-Me ester
Clerodendranoic acid
[1454839-07-6]
$C_{29}H_{26}O_{12}$ 566.515
Constit. of *Clerodendranthus spicatus*. Brown powder. $[\alpha]^{25}_{D}$ +100 (c, 1 in MeOH). λ_{max} 233 (log ε 0.41); 327 (log ε 0.57) (MeOH).

Zhang, H.-J. *et al.*, *Chin. Chem. Lett.*, 1993, **4**, 501-504 (*Salvianolic acid H*).
Dapkevicius, A. *et al.*, *J. Nat. Prod.*, 2002, **65**, 892-896 (*Isomelitric acid A*).
Murata, T. *et al.*, *Chem. Pharm. Bull.*, 2010, **58**, 394-397 (*9′-Me ester*).
Zheng, Q. *et al.*, *Molecules*, 2012, **17**, 13656-13661 (*Clerodendranoic acid*).
Sun, Z. *et al.*, *Phytochem. Lett.*, 2014, **8**, 16-21 (*Orthosiphoic acid D, activity*).
Krzyzanowska-Kowalczyk, J. *et al.*, *Molecules*, 2018, **23**, 2277:1-32 (*9-Me ester*)

Salvianolic acid J S16

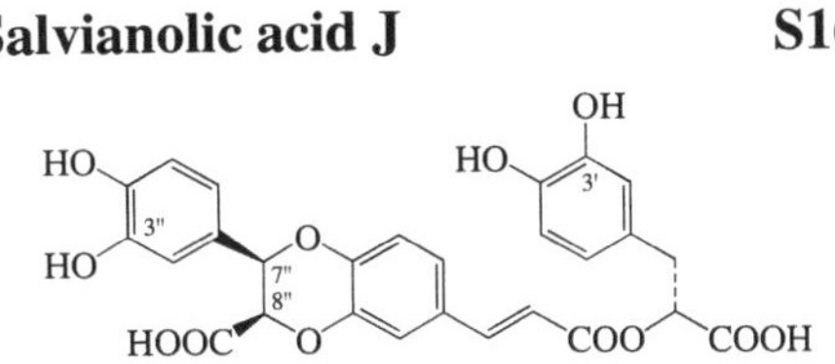

Absolute Configuration

Clinopodic acid E
[159736-38-6]
$C_{27}H_{22}O_{12}$ 538.461
Constit. of *Clinopodium chinense* var. *parviflorum*, *Lycopus lucidus*, *Salvia flava* and *Salvia miltiorrhiza*. Hyaluronidase inhibitor. Amorph. yellow powder. $[\alpha]^{27}_{D}$ +26 (c, 0.04 in EtOH). λ_{max} 218 (sh); 288 (log ε 4.24); 323 (log ε 4.21) (EtOH)
9″-[1R-Carboxy-2-(3,4-dihydroxyphenyl)ethyl] ester
Lycopic acid B
[1331784-39-4]
$C_{36}H_{30}O_{16}$ 718.62
Constit. of *Lycopus lucidus*. Hyaluronidase inhibitor. Amorph. solid. $[\alpha]^{21}_{D}$ +73.6 (c, 0.19 in MeOH). λ_{max} 287 (log ε 4.84); 320 (log ε 4.78) (MeOH)
9″-[2-Methoxy-6-(3-methyl-2-butenyl)-4-(2-propenyl)phenyl] ester
Citriodolic acid C
$C_{42}H_{40}O_{13}$ 752.769
Constit. of the leaves of *Eucalyptus citriodora*. Amorph. brown powder. $[\alpha]^{25}_{D}$ +31.8 (c, 0.05 in EtOH). λ_{max} 222 (log ε 4.54); 288 (log ε 4.26); 322 (log ε 4.22) (MeOH)
9″-[2,6-Dimethoxy-4-(2-propenyl) phenyl] ester
Citriodolic acid B
$C_{38}H_{34}O_{14}$ 714.676
Constit. of the leaves of *Eucalyptus citriodora*. Amorph. brown powder. $[\alpha]^{25}_{D}$ +25.1 (c, 0.05 in EtOH). λ_{max} 222 (log ε 4.54); 288 (log ε 4.25); 322 (log ε 4.21) (MeOH)
9″-[2,6-Dimethoxy-3-(3-methyl-2-butenyl)-4-(2-propenyl)phenyl] ester
Citriodolic acid A
$C_{43}H_{42}O_{14}$ 782.795
Constit. of the leaves of *Eucalyptus citriodora*. Amorph. brown powder. $[\alpha]^{25}_{D}$ +28.6 (c, 0.05 in EtOH). λ_{max} 222 (log ε 4.55); 288 (log ε 4.26); 322 (log ε 4.23) (MeOH)
3′-Me ether
Clinopodic acid F
[1174920-58-1]
$C_{28}H_{24}O_{12}$ 552.488
Constit. of *Clinopodium chinense* var. *parviflorum*. Amorph. solid. $[\alpha]^{23}_{D}$ +38.4 (c, 0.35 in MeOH). λ_{max} 222 (sh); 233 (sh); 288 (log ε 4.22); 323 (log ε 4.21) (MeOH)
3′,3″-Di-Me ether
Clinopodic acid G
[1174920-59-2]
$C_{29}H_{26}O_{12}$ 566.515
Constit. of *Clinopodium chinense* var. *parviflorum*. Amorph. solid.

$[\alpha]^{23}_{D}$ +59.8 (c, 0.31 in MeOH). λ_{max} 224 (sh); 233 (sh); 288 (log ε 4.28); 325 (log ε 4.28) (MeOH)

7″,8″-Diepimer

Clinopodic acid C. Salvianolic acid P

[1174920-56-9]

[1228442-68-9] (Salvianolic acid P)

$C_{27}H_{22}O_{12}$ 538.461

Constit. of *Clinopodium chinense* var. *parviflorum*, *Lycopus lucidus* and *Origanum dictamnus*. Matrix metalloproteinase-2 inhibitor. Hyaluronidase inhibitor. Shows antimicrobial activity. Amorph. yellow powder. $[\alpha]^{23}_{D}$ -14 (c, 1.1 in MeOH) (Clinopodic acid C). $[\alpha]^{23}_{D}$ +14.7 (c, 0.2 in MeOH) (Salvianolic acid P). Abs. config. claimed to be the same in both isolates. λ_{max} 221 (sh); 233 (sh); 289 (log ε 3.96); 326 (log ε 3.95) (MeOH) (Clinopodic acid C). [neutral] λ_{max} 236; 244 (sh); 286; 324 (MeOH) (Salvianolic acid P)

7″,8″-Diepimer, 3′-Me ether

Clinopodic acid D

[1174920-57-0]

$C_{28}H_{24}O_{12}$ 552.488

Constit. of *Clinopodium chinense* var. *parviflorum*. Amorph. solid. $[\alpha]^{23}_{D}$ -56.5 (c, 0.46 in MeOH). λ_{max} 223 (sh); 233 (sh); 288 (log ε 4.29); 324 (log ε 4.31) (MeOH)

7″,8″-Diepimer, 9″-[1R-carboxy-2-(3,4-dihydroxyphenyl)ethyl] ester

Lycopic acid A

[1331784-38-3]

$C_{36}H_{30}O_{16}$ 718.62

Constit. of *Lycopus lucidus*. Hyaluronidase inhibitor. Amorph. solid. $[\alpha]^{21}_{D}$ -6.6 (c, 0.16 in MeOH). λ_{max} 287 (log ε 4.72); 320 (log ε 4.68) (MeOH).

Ai, C.-B. *et al.*, *Phytochemistry*, 1994, **37**, 907-908 (*Salvianolic acid J*).

Hu, P. *et al.*, *Chem. Pharm. Bull.*, 2005, **53**, 677-683 (*Salvianolic acid J*).

Murata, T. *et al.*, *J. Nat. Prod.*, 2009, **72**, 1379-1384 (*Clinopodic acids C-G, activity*).

Murata, T. *et al.*, *Chem. Pharm. Bull.*, 2010, **58**, 394-397 (*Lycopic acids A,B, activity*).

Chatzopoulou, A. *et al.*, *J. Agric. Food Chem.*, 2010, **58**, 6064-6068 (*Salvianolic acid P, activity*).

Lin, S.-Q. *et al.*, *Chem. Pharm. Bull.*, 2016, **64**, 1641-1646 (*Citriodolic acids A,B,C*)

Salvianolic acid K S17

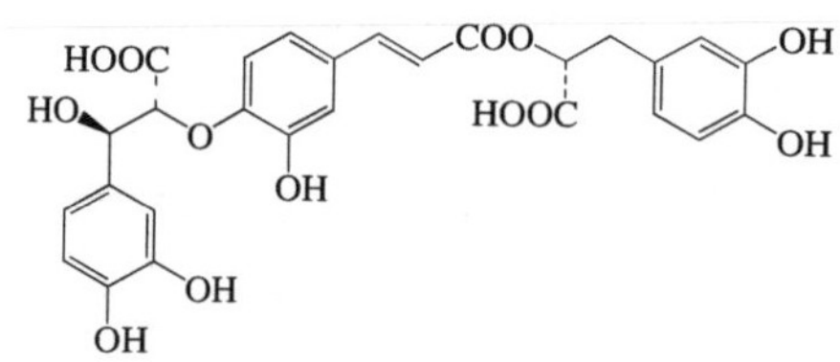

Absolute Configuration

[203733-40-8]

$C_{27}H_{24}O_{13}$ 556.476

Constit. of *Salvia deserta*, *Salvia miltiorrhiza* and *Salvia officinalis*. Moderate aldose reductase inhibitor. Amorph. solid. $[\alpha]^{26}_{D}$ +31.2 (c, 0.05 in H_2O). λ_{max} 210 (log ε 3.2); 220 (log ε 3.4); 230 (sh); 286 (log ε 2.9); 323 (log ε 2.4) (MeOH).

Tezuka, Y. *et al.*, *Chem. Pharm. Bull.*, 1998, **46**, 107-112 (*Salvia deserta constit*).

Kasimu, R. *et al.*, *Chem. Pharm. Bull.*, 1998, **46**, 500-504 (*activity*).

Lu, Y. *et al.*, *Phytochemistry*, 1999, **51**, 91-94 (*Salvia officinalis constit*)

Salvianolic acid L S18

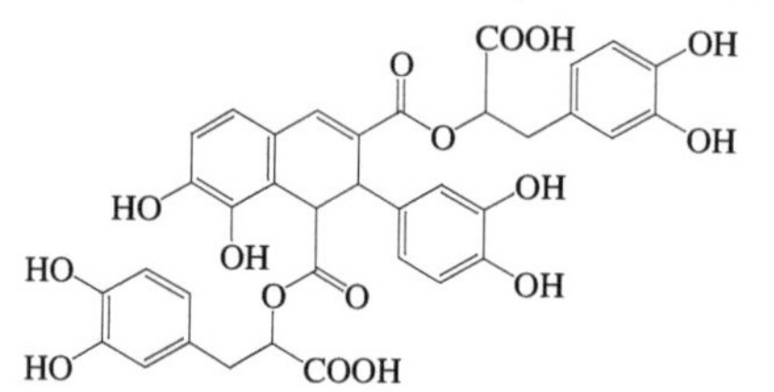

[389065-74-1]

$C_{36}H_{30}O_{16}$ 718.62

Constit. of *Mentha longifolia*, *Plectranthus amboinicus* and *Salvia officinalis*. Antioxidant. Brownish powder. $[\alpha]_{D}$ -177.5 (c, 0.2 in MeOH). Related to Rabdosiin, R1. λ_{max} 212 (log ε 4.48); 288 (log ε 4.05); 334 (log ε 4.12) (MeOH).

Lu, Y. *et al.*, *Tetrahedron Lett.*, 2001, **42**, 8223-8225 (*Salvianolic acid L*).

Chen, Y.-S. *et al.*, *Bioorg. Med. Chem.*, 2014, **22**, 1766-1772 (*Salvianolic acid L*)

Salvianolic acid N S19

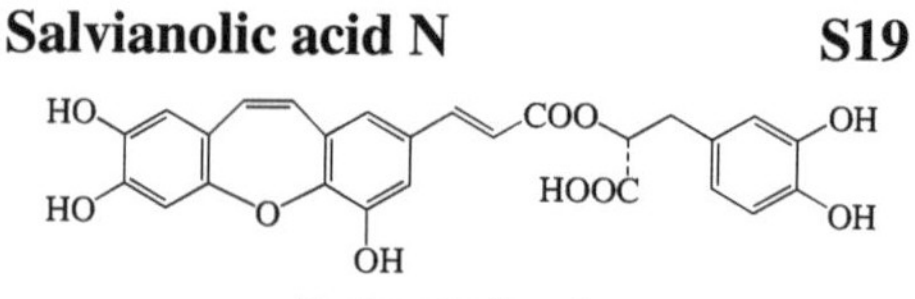

Absolute Configuration

[933776-37-5]

$C_{26}H_{20}O_{10}$ 492.436

Constit. of the roots of *Salvia yunnanensis*. Anti-HIV agent. Amorph. yellow powder. $[\alpha]^{20}_{D}$ +41 (c, 0.08 in MeOH). λ_{max} 290 (log ε 4.32); 328 (log ε 4.34) (MeOH).

Zhang, Z.-F. *et al.*, *J. Asian Nat. Prod. Res.*, 2008, **10**, 252-255 (*Salvianolic acid N*).

Wu, K. *et al.*, *Org. Biomol. Chem.*, 2018, **16**, 832-837 (*synth*)

Salvianolic acid T S20

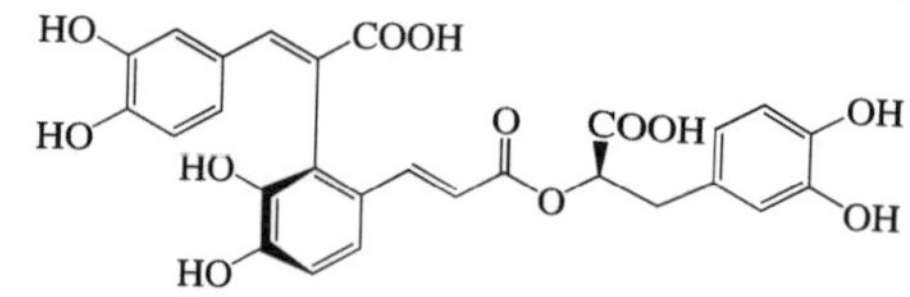

Absolute Configuration

α-[[3-[2-[1-Carboxy-2-(3,4-dihydroxyphenyl)ethenyl]-3,4-dihydroxyphenyl]-1-oxo-2-propenyl]oxy]-3,4-dihydroxybenzenepropanoic acid

[1568106-86-4]

$C_{27}H_{22}O_{12}$ 538.461

Constit. of the roots of *Salvia miltiorrhiza*. Amorph. yellow powder. $[\alpha]^{25}_{D}$ +196.6 (c, 0.25 in MeOH). λ_{max} 220; 326 (MeOH)

Atropisomer

Salvianolic acid U

$C_{27}H_{22}O_{12}$ 538.461

Constit. of the roots of *Salvia miltiorrhiza*. Amorph. yellow powder. $[\alpha]^{25}_{D}$ -157.6 (c, 0.16 in MeOH). λ_{max} 225; 325 (MeOH).

Li, W. *et al.*, *Fitoterapia*, 2014, **98**, 248-253 (*Salvianolic acids T,U, cd, abs config*)

Salvianolic acid Z S21

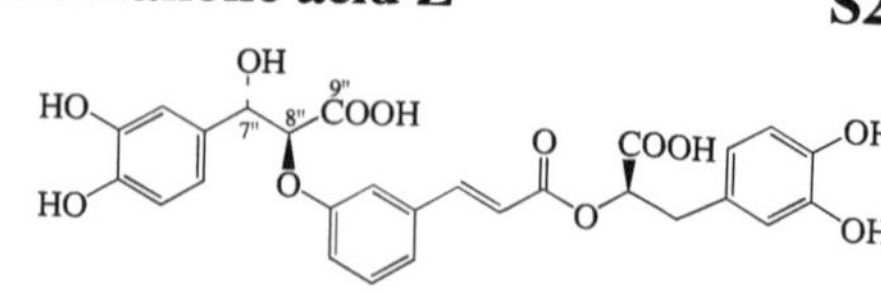

[1497207-93-8]

$C_{27}H_{24}O_{13}$ 556.476

Lignan numbering shown. Constit. of *Origanum dictamnus*

9″-Me ester

[1494662-61-1]

$C_{28}H_{26}O_{13}$ 570.503

Constit. of *Origanum dictamnus*

7″,8″-Diepimer, 9″-[2-(3,4-dihydroxyphenyl)-1S-hydroxyethyl] ester

Salvianolic acid S

[1497207-94-9]

$C_{35}H_{32}O_{16}$ 708.625

Constit. of *Origanum dictamnus*.

Exarchou, V. *et al.*, *Phytochem. Lett.*, 2013, **6**, 46-52 (*Salvianolic acids S,Z*)

Sambacolignoside S22

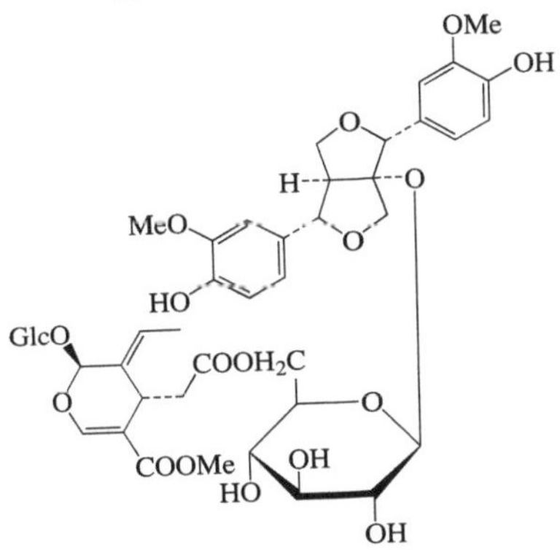

Absolute Configuration

[114449-12-6]

$C_{43}H_{54}O_{22}$ 922.883

Complex glycoside of 8-Hydroxypinoresinol

8-Hydroxypinoresinol in D87 and Oleoside. Isol. from leaves of *Jasminum sambac*. Powder. [α] D -97.1 (MeOH). λ_{max} 233 (ε 35481); 280 (ε 5129) (MeOH).

Tanahashi, T. *et al.*, *Chem. Pharm. Bull.*, 1987, **35**, 5032-5035 (*Sambacolignoside*)

Samin S23

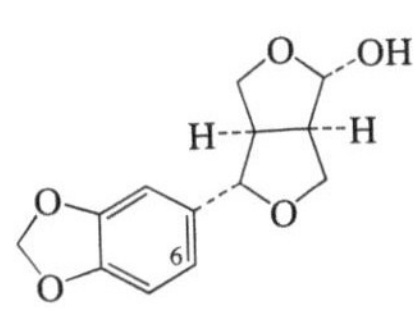

Absolute Configuration

[166239-82-3]

[138812-87-0] (unspecified stereochem.) [166239-83-4] (epimer)

$C_{13}H_{14}O_5$ 250.25

Constit. of the seeds of *Sesamum indicum*. Needles. Mp 183 - 185°. $[\alpha]_D^{25}$ +200 (c, 0.1 in $CHCl_3$). λ_{max} 203 (log ε 4.21); 238 (log ε 2.83); 287 (log ε 2.69) (MeOH)

O-(4-Hydroxy-3-methoxyphenyl)
see Sesamolinol, S51

O-(3,4-Methylenedioxyphenyl)

Sesamolin

[526-07-8]

$C_{20}H_{18}O_7$ 370.357

Constit. of *Justicia orbiculata*, *Sesamum indicum* and *Strychnos nux-blanda*. Plates (EtOH). Mp 94°. $[\alpha]_D^{20}$+212 (c, 1.13 in $CHCl_3$)

6-Hydroxy

Saminol

[914253-78-4]

$C_{13}H_{14}O_6$ 266.249

Constit. of the perisperm of *Sesamum indicum*. Amorph. solid. [α] D +30 (c, 0.2 in $CHCl_3$).

Carnmalm, B. *et al.*, *Acta Chem. Scand.*, 1955, **9**, 1111-1118 (*Sesamolin*).
Beroza, M. *J. Am. Chem. Soc.*, 1955, **77**, 3332-3334 (*Sesamolin, struct*).
Haslam, E. *J. Chem. Soc. C*, 1970, 2332-2334 (*abs config, biosynth*).
Samizu, K. *et al.*, *Chem. Lett.*, 1995, **24**, 543-544 (*Samin, synth*).
Maiti, G. *et al.*, *Tetrahedron*, 1995, **51**, 8389-8396 (*synth*).
Wirth, T. *et al.*, *J. Org. Chem.*, 1996, **61**, 2686-2689 (*Samin, synth*).
Marchand, P.A. *et al.*, *Can. J. Chem.*, 1997, **75**, 840-849 (*Samin, synth*).
Lutz, G. *et al.*, *Heterocycles*, 1997, **45**, 287-298 (*Sesamolin, conformn, cryst struct*).
Hull, H.M. *J. Chem. Soc., Perkin Trans. 1*, 1998, 1779-1787 (*Samin, synth*).
Kato, M.J. *et al.*, *Phytochemistry*, 1998, **47**, 583-591 (*Sesamolin, biosynth*).
Babu, G.V.R. *et al.*, *Indian J. Chem., Sect. B*, 2001, **40**, 864-866 (*Sesamolin, isol*).
Grougnet, R. *et al.*, *J. Agric. Food Chem.*, 2006, **54**, 7570-7574 (*Saminol*).
Kuo, P.-C. *et al.*, *J. Agric. Food Chem.*, 2011, **59**, 3214-3219 (*Samin*).
Dar, A.A. *et al.*, *Bioorg. Chem.*, 2013, **50**, 1-10 (*Sesamolin, rev, activity, biosynth*)

Santiaguine S24

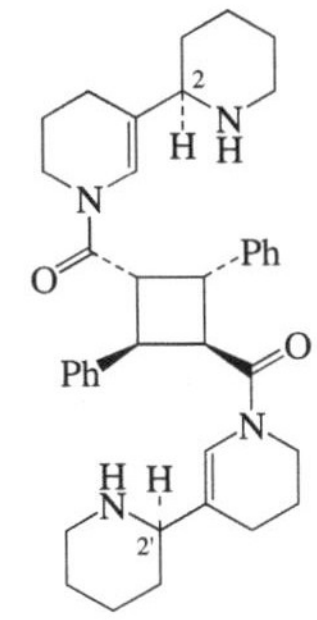

(*2R,2'R*)*-form*

[528-31-4]

$C_{38}H_{48}N_4O_2$ 592.828

α-Truxillate dimer of Adenocarpine. The truxillate residue is *meso*-, so there are four possible isomers; 2 enantiomers, one (±)-form and one *meso*-. All have been found in nature

(2R,2'R)-form [11016-82-3]

[11016-86-7] (hydrobromide (1:2))
[11016-96-9] (hydroiodide (1:2))

Alkaloid from *Adenocarpus commutatus* and *Adenocarpus grandiflorus*. Mp 240 - 245°. $[\alpha]_D^{20}$-18.42 (c, 9.6 in 85% HCOOH aq.). $[\alpha]_D^{17}$-21.81 (c, 3.24 in AcOH)

Hydrochloride (1:2)
[11016-90-3]
$[\alpha]_D^{18}$-16.66 (c, 3.96 in 85% HCOOH)

(2S,2'S)-form [11016-83-4]

[11016-89-0] (hydrobromide (1:2))
[11016-95-8] (hydroiodide (1:2))

Alkaloid from *Adenocarpus complicatus*, *Adenocarpus intermedius*, *Adenocarpus foliosus* and *Adenocarpus viscosus*. Cryst. (EtOH). Mp 235 - 236°. $[\alpha]_D^{17}$+3.3 (c, 3.68 in EtOH). $[\alpha]_D^{17}$+23.8 (c, 3.9 in AcOH)

Hydrochloride (1:2)
[11016-92-5]
Cryst. + $1H_2O$ (H_2O); cryst. (EtOH). Mp 238° (anhyd. 241°)

Diperchlorate
[11017-00-8]
Mp 244 - 245° dec.

Picrate
Mp 285° (explodes, prior blackening)

(2RS,2'RS)-form

[11016-84-5]
[11016-88-9] (hydrobromide (1:2))
[11016-94-7] (hydroiodide (1:2))

Alkaloid from *Adenocarpus commutatus*, *Adenocarpus grandiflorus*, *Adenocarpus foliosus* and *Adenocarpus viscosus*. Cryst. + $1H_2O$. Mp 237 - 239° dec.

Hydrochloride (1:2)
[11016-93-6]
Cryst. + $2H_2O$. Mp 245 - 250° dec.

Diperchlorate
Cryst. + $1½H_2O$. Mp 252 - 254° dec.

(2RS,2'SR)-form

[11016-85-6]
[11016-87-8] (hydrobromide (1:2))
[11016-97-0] (hydroiodide (1:2))

Alkaloid from *Adenocarpus commutatus*. Cryst. + $3H_2O$ (MeOH aq.). Mp 238 - 240° dec.

Hydrochloride (1:2)
[11016-91-4]
Cryst. + $3½H_2O$ (EtOH aq.). Mp 245 - 247° dec.

Diperchlorate
[11017-01-9]
Cryst. + $1½H_2O$. Mp 245 - 248° dec..

Ribas, I. *et al.*, *An. R. Soc. Esp. Fis. Quim., Ser. B*, 1950, **46**, 489-500 (*Adenocarpus complicatus constit*).
González, A.G. *et al.*, *An. R. Soc. Esp. Fis. Quim., Ser. B*, 1951, **47**, 730-733 (*Adenocarpus viscosus constit*).
Ribas, I. *et al.*, *Ann. Pharm. Fr.*, 1952, **10**, 54-61 (*Adenocarpus commutatus constit*).

González, A.G. *et al.*, *An. R. Soc. Esp. Fis. Quim., Ser. B*, 1953, **49**, 783-788 (*Adenocarpus foliosus constit*).
Domiguez, J. *et al.*, *An. R. Soc. Esp. Fis. Quim., Ser. B*, 1956, **52**, 133-138 (*synth*).
Méndez, M.R. *et al.*, *An. R. Soc. Esp. Fis. Quim., Ser. B*, 1958, **54**, 157-160 (*Adnocarpus grandiflorus constit*).
Ribas, I. *et al.*, *An. R. Soc. Esp. Fis. Quim., Ser. B*, 1966, **62**, 845-851 (*occur, abs config*).
Bernasconi, R. *et al.*, *Pharm. Acta Helv.*, 1970, **45**, 42-51 (*Adenocarpus mannii constit*).
O'Donovan, D.G. *et al.*, *J. Chem. Soc., Perkin Trans. 1*, 1974, 2524-2528 (*Adenocarpus foliosus constit, biosynth*).
Nehme, M. *et al.*, *An. Quim.*, 1977, **73**, 307-308 (*Adenocarpus complicatus constit*)

Sarusubine A S25

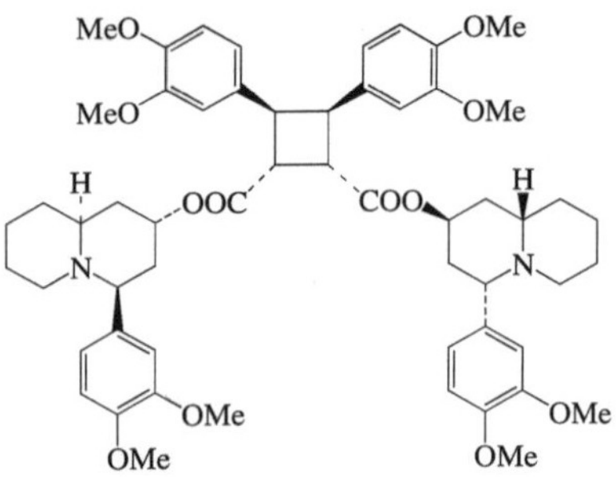

Relative Configuration

[956298-05-8]
$C_{56}H_{70}N_2O_{12}$ 963.178
Alkaloid from the leaves of *Lagerstroemia subcostata*. Pale yellow solid. $[\alpha]_D^{23}$ +40.1 (c, 1 in MeOH).

Watanabe, K. *et al.*, *Tetrahedron Lett.*, 2007, **48**, 7502-7504 (*Sarusubine A*)

Sativamide A S26

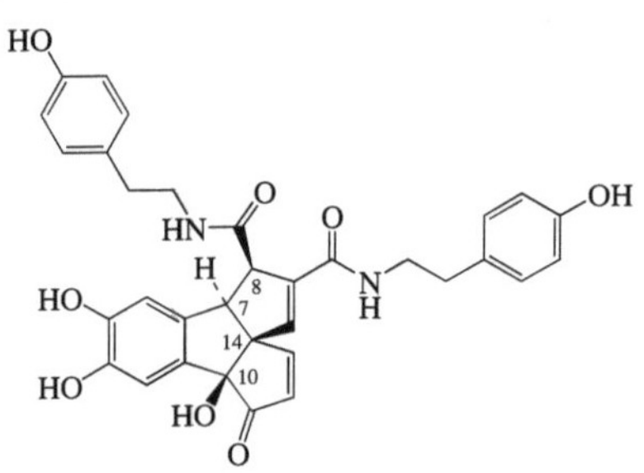

(7*R*,8*R*,10*R*,14*R*)-*form*

1,6,6a,10b-Tetrahydro-6a,8,9-trihydroxy-N[1],N[2]-bis[2-(4-hydroxyphenyl) ethyl]-6-oxo-dicyclopent[a,b]indene-1,2-dicarboxamide
$C_{33}H_{30}N_2O_8$ 582.609
The nat. prods. Sativamides A and B were isol. as partial racemates and resolved by chiral hplc
(7R,8R,10R,14R)-form [2211940-88-2]
Light yellow powder. $[\alpha]_D^{25}$ -25.1 (c, 0.5 in EtOH)
10-Me ether
[2211940-89-3]
Light yellow powder. $[\alpha]_D^{25}$ -24.1 (c, 0.5 in EtOH)
(7S,8S,10S,14S)-form [2211940-86-0]
Light yellow powder. $[\alpha]_D^{25}$ +23.5 (c, 0.5 in EtOH)
10-Me ether
[2211940-87-1]
Light yellow powder. $[\alpha]_D^{25}$ +22 (c, 0.5 in EtOH)
(7RS,8RS,10RS,14RS)-form
Constit. of the fruit of *Cannabis sativa*. Pre-treatment of neuroblastoma cells reduces the endoplasmic reticulum stress induced cytotoxic activity. Light yellow powder. $[\alpha]_D^{25}$ -1.9 (c, 1.0 in EtOH). λ_{max} 223 (log ε 4.48); 283 (log ε 3.70) (MeOH)
10-Me ether
Sativamide B
$C_{34}H_{32}N_2O_8$ 596.636
Constit. of the fruit of *Cannabis sativa*. Pre-treatment of neuroblastoma cells reduces the endoplasmic reticulum stress induced cytotoxic activity. $[\alpha]_D^{25}$ -2.8 (c, 1.0 in EtOH). λ_{max} 224 (log ε 4.56); 283 (log ε 3.77) (MeOH).

Zhu, G.-Y. *et al.*, *J. Org. Chem.*, 2018, **83**, 2376-2381 (*Sativamides A,B, activity*)

Saulangianin S27

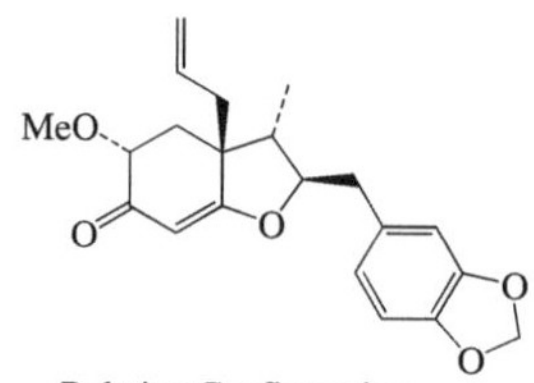

Relative Configuration

2,8'-Epoxy-9'a-homo-5,6-dihydro-5-methoxy-3',4'-methylenedioxy-1,9'-lign-8-en-4(1H)-one
[152753-14-5]
$C_{21}H_{24}O_5$ 356.418
Constit. of the flower buds of *Magnolia saulangiana*. Oil. $[\alpha]_D^{20}$ +2.5 (c, 0.5 in $CHCl_3$). λ_{max} 207; 235; 294 (MeOH).

Abdallah, O.M. *Phytochemistry*, 1993, **34**, 1185-1187 (*Saulangianin*)

Saurucinol A S28

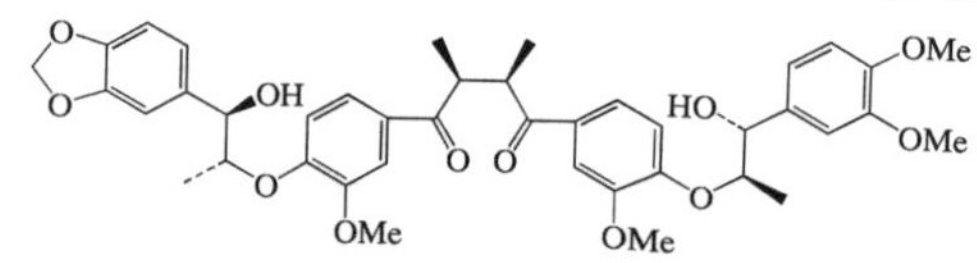

Absolute Configuration

7,7'-Dioxosecomanussantin B
[1521111-50-1]
$C_{41}H_{46}O_{12}$ 730.807
Constit. of *Saururus chinensis*. Powder. $[\alpha]_D^{20}$ -288 (c, 0.01 in $CHCl_3$). λ_{max} 212 (log ε 4.37); 230 (log ε 4.44); 280 (log ε 4.35) (MeOH).

Cui, H. *et al.*, *J. Nat. Prod.*, 2014, **77**, 100-110 (*Saurucinol A*)

Saurulignan E S29

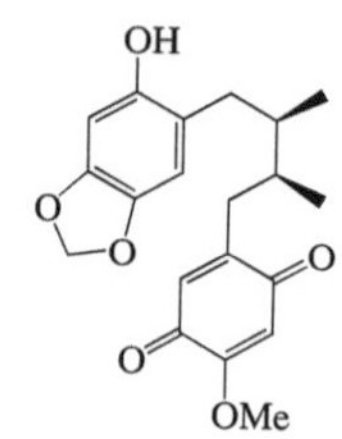

Relative Configuration

[1532522-54-5]
$C_{20}H_{22}O_6$ 358.39
Constit. of *Saururus chinensis*. Shows significant inhibition of platelet aggregation. Brown oil. $[\alpha]_D^{26}$ -24 (c, 0.5 in MeOH). λ_{max} 270 (log ε 3.91); 300 (log ε 3.58) (MeOH).

Tsai, W.-J. *et al.*, *J. Nat. Prod.*, 2014, **77**, 125-131 (*Saurulignan E*)

Scapaniapyrone A S30

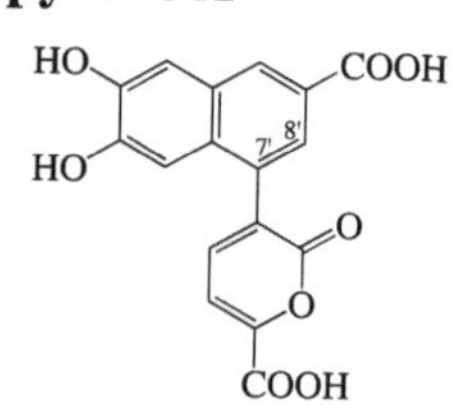

[123364-09-0]
$C_{17}H_{10}O_8$ 342.259
Seco-norlignan. Lignan numbering shown. Constit. of liverwort *Scapania undulata*. Yellow solid. λ_{max} 265 (ε 14600); 309 (ε 12800) (EtOH)
7'ξ8'-Dihydro
[171901-66-9]
$C_{17}H_{12}O_8$ 344.275
Isol. from the liverwort *Jamesoniella autumnalis*. Light yellow powder. $[\alpha]_D^{20}$ +135 (c, 0.1 in MeOH).

Mues, R. *et al.*, *Tetrahedron Lett.*, 1988, **29**, 6793-6796 (*Scapaniapyrone A*).

Tazaki, H. *et al.*, *Phytochemistry*, 1995, **40**, 1671-1675 (*Jamesoniella autumnalis constit*).
Tazaki, H. *et al.*, *Tetrahedron Lett.*, 1999, **40**, 101-104 (*biosynth*)

Schinifolisatin A S31

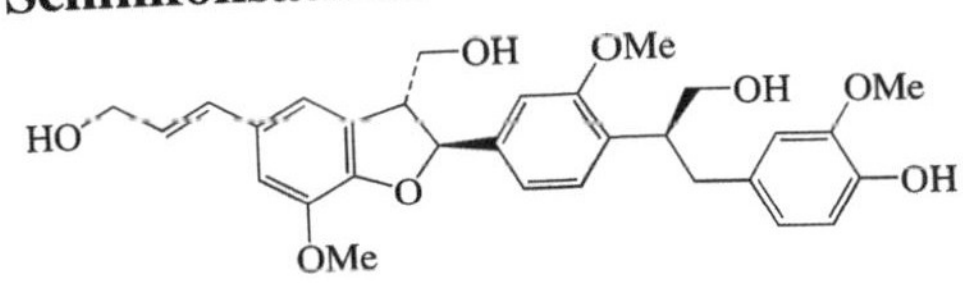

Absolute Configuration

[1478613-16-9]
$C_{30}H_{34}O_8$ 522.594
Constit. of *Zanthoxylum schinifolium*. Displays potent anticancer activity. Yellowish oil. $[\alpha]^{25}_{D}$ +42.6 (c, 0.1 in MeOH). λ_{max} 239; 282 (MeOH).
Li, W. *et al.*, *J. Agric. Food Chem.*, 2013, **61**, 10730-10740 (*Schinifolisatin A*)

Schinlignan A S32

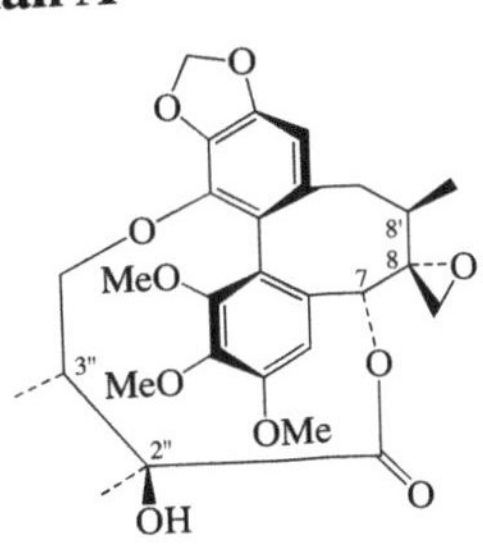

Absolute Configuration

[1714123-61-1]
$C_{28}H_{32}O_{10}$ 528.554
Constit. of the fruit of *Schisandra chinensis*. Amorph. powder. $[\alpha]^{29}_{D}$ -8.7 (c, 0.13 in MeOH). λ_{max} 213 (log ε 4.22); 292 (log ε 3.28) (MeOH).
Xue, Y. *et al.*, *Phytochemistry*, 2015, **116**, 253-261 (*Schinlignan A*)

Schischinone S33

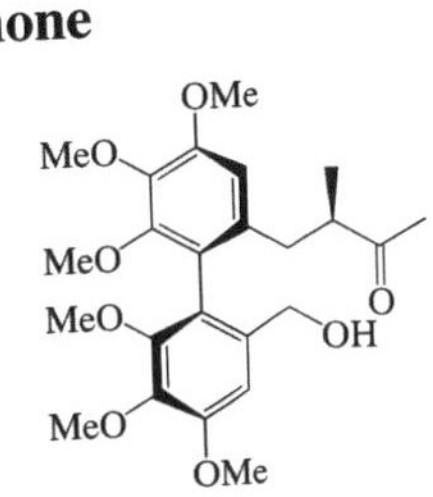

Absolute Configuration

[1714123-67-7]
$C_{24}H_{32}O_8$ 448.512
Constit. of the fruit of *Schisandra chinensis*. Yellowish gum. $[\alpha]^{29.2}_{D}$ -1.7 (c, 0.17 in MeOH). λ_{max} 210 (log ε 4.16); 283 (log ε 2.92) (MeOH).
Xue, Y. *et al.*, *Phytochemistry*, 2015, **116**, 253-261 (*Schischinone*)

Schiscupatetralin A S34

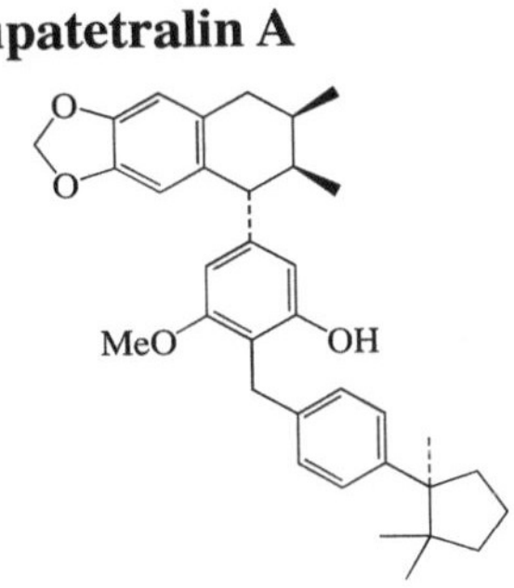

[1621683-65-5]
$C_{35}H_{42}O_4$ 526.717
Constit. of *Schisandra sphenanthera*. Amorph. powder. $[\alpha]^{24}_{D}$ -21 (c, 1.22 in MeOH). λ_{max} 216 (log ε 4.57); 241 (log ε 4.21); 290 (log ε 3.67) (MeOH).
Tsai, Y.-C. *et al.*, *Chem. Biodiversity*, 2014, **11**, 1053-1068 (*Schiscupatetralin A*)

Schisphenlignan E S35

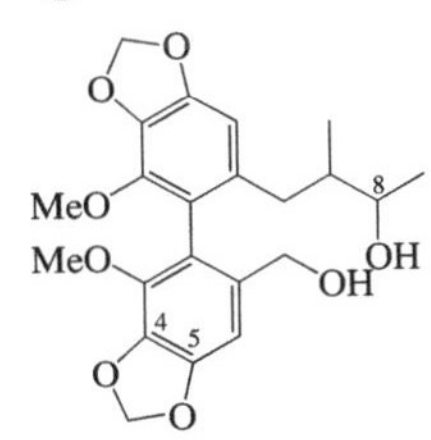

[1477666-11-7]
$C_{22}H_{26}O_8$ 418.442
Constit. of *Schisandra sphenanthera*. Amorph. powder. $[\alpha]^{26}_{D}$ +2.5 (c, 0.22 in $CHCl_3$). λ_{max} 240 (log ε 3.32) ($CHCl_3$)

8-Ketone, 4,5-demethylene, 4,5-di-Me ether

Schisphenone
[1311376-50-7]
$C_{23}H_{28}O_8$ 432.469
Constit. of the fruits of *Schisandra sphenanthera*. Exhibits antiinflammatory activity. $[\alpha]^{25}_{D}$ -82 (c, 0.1 in CH_2Cl_2). Has 8′*S*,S_{axial}-config.. λ_{max} 213 (log ε 3.72) (MeOH).

Huang, H.-C. *et al.*, *Food Chem.*, 2011, **128**, 348-357 (*Schisphenone, activity*).
Liang, C.Q. *et al.*, *Chem. Pharm. Bull.*, 2013, **61**, 96-100 (*Schisphenlignan E*)

Schitenoside C S36

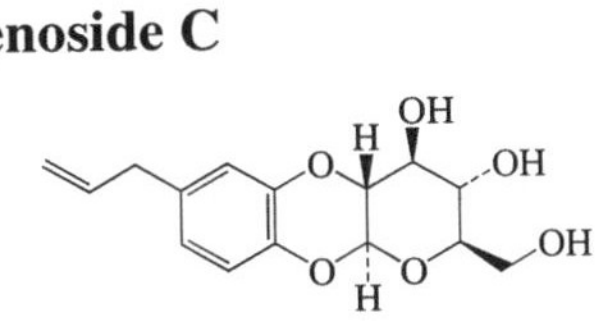

Probable Absolute Configuration

[2055172-91-1]
$C_{15}H_{18}O_6$ 294.303
Constit. of the whole plant of *Schizonepeta tenuifolia*. Yellow powder (MeOH). Tentative struct. assigned.
Huang, X.-H. *et al.*, *Chem. Nat. Compd. (Engl. Transl.)*, 2016, **52**, 1005-1007 (*Schitenoside C*)

Scopariusicide A S37

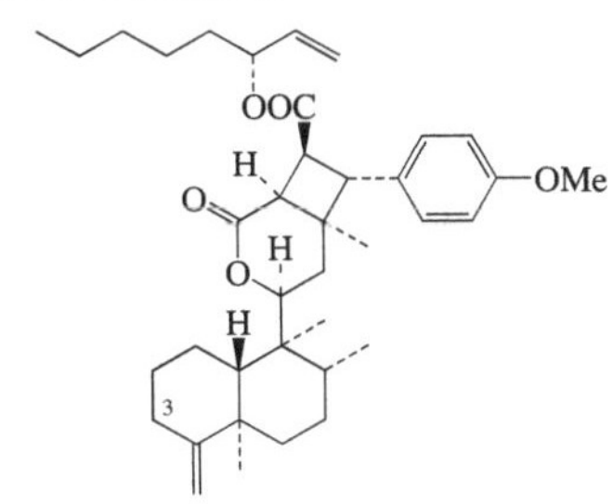

[1829579-13-6]
$C_{38}H_{54}O_5$ 590.845
Constit. of *Isodon scoparius*. Powder. $[\alpha]^{21}_{D}$ -35.5 (c, 0.12 in MeOH)

Δ³-Isomer

Scopariusicide B
[1829579-14-7]
$C_{38}H_{54}O_5$ 590.845
Constit. of *Isodon scoparius*. Powder. $[\alpha]^{22}_{D}$ -84.2 (c, 0.07 in MeOH).

Zhou, M. *et al.*, *Org. Lett.*, 2015, **17**, 6062-6065 (*Scopariusicides A,B, cryst struct*)

Scutellaprostin A S38

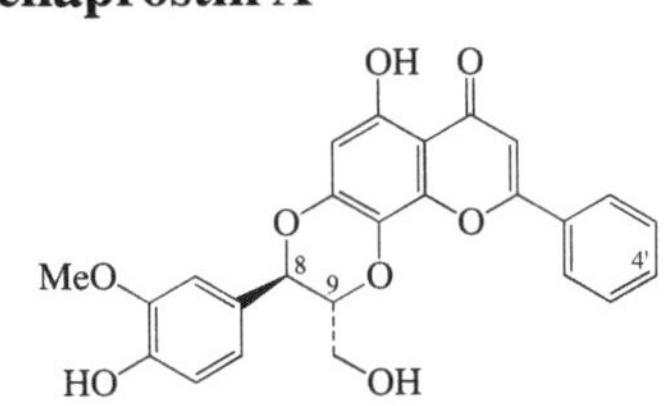

[137166-36-0]
[142674-24-6] (unspecified stereochem.)
$C_{25}H_{20}O_8$ 448.427
Similar to Rhodiolin, R12. Constit. of the roots of *Scutellaria prostrata*. Bacterial multidrug resistance (MDR) efflux pump inhibitor. Synergistic with Berberine on *Staphylococcus aureus*. Yellow needles (MeOH). Mp 221 - 222°. Racemic. λ_{max} 234 (log ε 4.24); 286 (log ε 4.45); 370 (log ε 3.40) (MeOH)

4′-Hydroxy

Scutellaprostin B
[137231-84-6]
[142674-26-8] (unspecified stereochem.)
$C_{25}H_{20}O_9$ 464.426
Constit. of the roots of *Scutellaria prostrata*. Bacterial multidrug

resistance (MDR) efflux pump inhibitor. Synergistic with Berberine on *Staphylococcus aureus*. Yellow needles (MeOH). Mp 255 - 256° dec.. λ_{max} 229 (log ε 4.45); 285 (log ε 4.41); 303 (sh); 310 (sh); 332 (sh) (MeOH)

3',4'-Dihydroxy

Scutellaprostin C

[137231-85-7]

[142674-28-0] (unspecified stereochem.)

$C_{25}H_{20}O_{10}$ 480.425

Constit. of the roots of *Scutellaria prostrata*. Yellow needles (MeOH). Mp 250 - 251° dec.. λ_{max} 230 (sh); 283 (log ε 4.74); 304 (sh); 348 (log ε 4.74) (MeOH).

Kikuchi, Y. *et al.*, *Chem. Pharm. Bull.*, 1991, **39**, 1466-1472 (*Scutellaprostins A,B,C*).
Kikuchi, Y. *et al.*, *Yakugaku Zasshi*, 1991, **111**, 424-435 (*synth*).
Guz, N.R. *et al.*, *J. Med. Chem.*, 2001, **44**, 261-268 (*Scutellaprostins A,B,C, activity*)

Scutellaprostin D S39

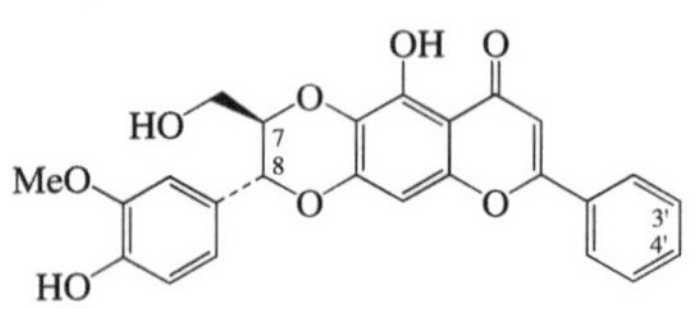

Relative Configuration

[137231-86-8]

[142674-92-8] (unspecified stereochem.) [149116-22-3] (*2S-trans*-form)

$C_{25}H_{20}O_8$ 448.427

Constit. of the roots of *Scutellaria prostrata*. Yellow needles (MeOH). Mp 234 - 235° dec.. Racemic. λ_{max} 219 (log ε 4.59); 239 (sh); 280 (log ε 4.49); 320 (log ε 4.27) (MeOH)

4'-Hydroxy

Scutellaprostin E

[137231-87-9]

$C_{25}H_{20}O_9$ 464.426

Constit. of the roots of *Scutellaria prostrata*. Yellow needles (MeOH). Mp 240 - 241° dec.. λ_{max} 220 (log ε 4.52); 236 (sh); 340 (log ε 4.40) (MeOH)

3',4'-Dihydroxy

Scutellaprostin F

[137231-88-0]

$C_{25}H_{20}O_{10}$ 480.425

Constit. of the roots of *Scutellaria prostrata*. Yellow needles (MeOH). Mp 204 - 205° dec.. λ_{max} 220 (sh); 237 (sh); 261 (log ε 4.12); 284 (log ε 4.28); 352 (log ε 4.39) (MeOH).

Kikuchi, Y. *et al.*, *Chem. Pharm. Bull.*, 1991, **39**, 1466-1472 (*Scutellaprostins D,E,F*).
Kikuchi, Y. *et al.*, *Yakugaku Zasshi*, 1991, **111**, 424-435 (*synth*)

Scutellaprostin G S40

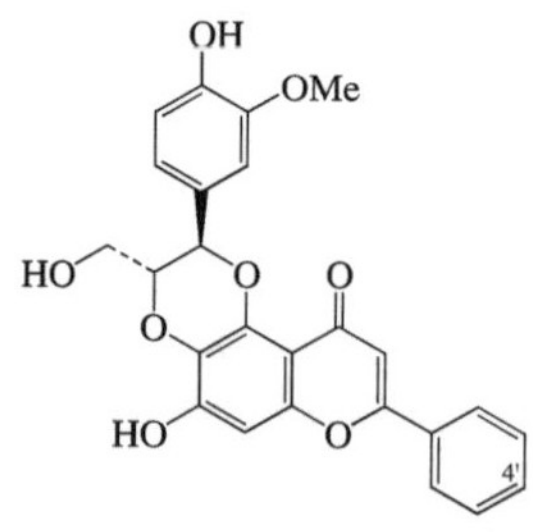

Relative Configuration

NSC 648335

[137231-89-1]

$C_{25}H_{20}O_8$ 448.427

Related to Scutellaprostin M, S41. Constit. of *Scutellaria prostrata*. Topoisomerase I inhibitor. Needles (MeOH). Mp 206 - 207° dec.. Racemate. λ_{max} 215 (log ε 4.37); 237 (log ε 4.07); 274 (log ε 4.26); 314 (log ε 3.98) (MeOH)

4'-Hydroxy

Scutellaprostin H

[137231-90-4]

$C_{25}H_{20}O_9$ 464.426

Constit. of *Scutellaria prostrata*. Needles (MeOH). Mp 235 - 236° dec.. Racemate. λ_{max} 219 (log ε 4.56); 278 (log ε 4.35); 328 (log ε 4.39) (MeOH)

3',4'-Dihydroxy

Scutellaprostin I

[137231-91-5]

$C_{25}H_{20}O_{10}$ 480.425

Constit. of *Scutellaria prostrata*. Needles (MeOH). Mp 213 - 214° dec.. Racemate. λ_{max} 222 (sh); 238 (log ε 4.35); 281 (log ε 4.24); 343 (log ε 4.35) (MeOH).

Kikuchi, Y. *et al.*, *Chem. Pharm. Bull.*, 1991, **39**, 1466-1472 (*synth*).
Drwal, M.N. *et al.*, *PLoS One*, 2011, **6**, e25150 (*activity*)

Scutellaprostin M S41

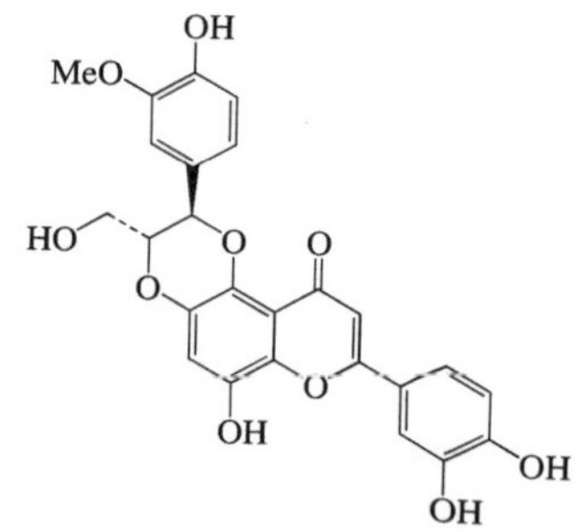

Relative Configuration

8-(3,4-Dihydroxyphenyl)-2,3-dihydro-6-hydroxy-2-(4-hydroxy-3-methoxyphenyl)-3-(hydroxymethyl)-10H-pyrano[3,2-f]-1,4-benzodioxin-10-one, CAS

[1620901-07-6]

$C_{25}H_{20}O_{10}$ 480.425

Constit. of roots of *Scutellaria prostrata*. Amorph. yellow powder.

Bhat, G. *et al.*, *Nat. Prod. Res.*, 2014, **28**, 1685-1690 (*Scutellaprostin M*)

Scyphiphorin D S42

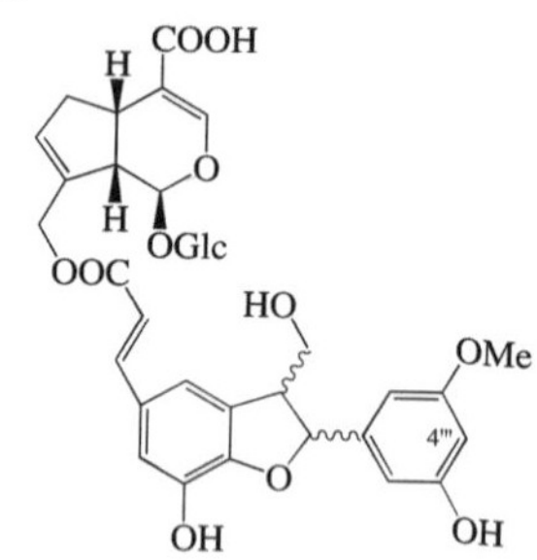

[1173922-01-4]

$C_{35}H_{38}O_{16}$ 714.673

Structures in this series may need revision. Phenyl 3,5-dioxygenated substn. pattern in doubt as a result of work done by Yang *et al* in 2015. Constit. of the stem bark of *Scyphiphora hydrophyllacea*. Amorph. powder. $[\alpha]^{20}_{D}$ +27 (c, 10 in Py). λ_{max} 228 (log ε 4.2); 331 (log ε 3.96) (MeOH)

4'''-Hydroxy, 3'''-Me ether

Scyphiphorin C

[1173921-99-7]

$C_{36}H_{40}O_{17}$ 744.699

Constit. of the stem bark of *Scyphiphora hydrophyllacea*. Amorph. powder. $[\alpha]^{20}_{D}$ +28 (c, 8 in MeOH). λ_{max} 230 (log ε 4.29); 329 (log ε 4.1) (MeOH).

Tao, S.-H. *et al.*, *Heterocycles*, 2009, **78**, 1557-1562 (*Scyphiphorins C,D*).
Yang, Y.-N. *et al.*, *J. Nat. Prod.*, 2015, **78**, 705-711 (*pmr, struct*)

Sebestenoid C S43

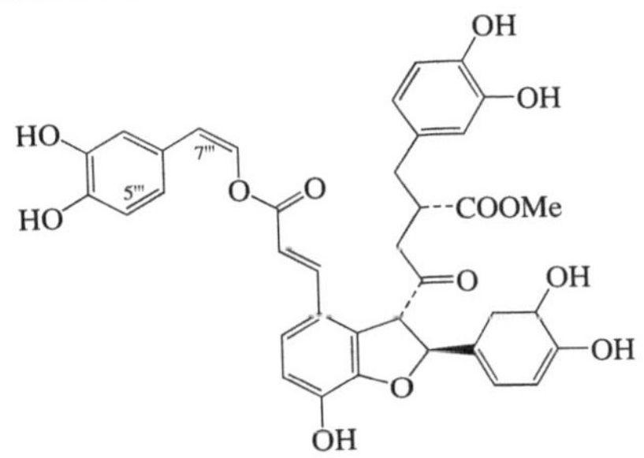

Absolute Configuration

[1261245-71-9]

$C_{36}H_{30}O_{14}$ 686.622

Constit. of the fruit of *Cordia sebestena*. Pale yellow powder. $[\alpha]_D^{22}$ +32 (c, 0.2 in MeOH). λ_{max} 257 (log ε 4.3); 290 (log ε 4.2); 335 (log ε 4.3) (MeOH)

7‴E-Isomer, 4‴-deoxy, 5‴-hydroxy

Sebestenoid D

[1261245-72-0]

$C_{36}H_{30}O_{14}$ 686.622

Constit. of the fruit of *Cordia sebestena*. Pale yellow powder. $[\alpha]_D^{22}$ +77 (c, 0.2 in MeOH). λ_{max} 223 (log ε 4.7); 253 (log ε 4.4); 292 (log ε 4.4); 337 (log ε 4.4) (MeOH).

Dai, J.-Q. *et al.*, *Phytochemistry*, 2010, **71**, 2168-2173 (*Sebestenoids C,D*)

7,8-Secoholostylone A S44

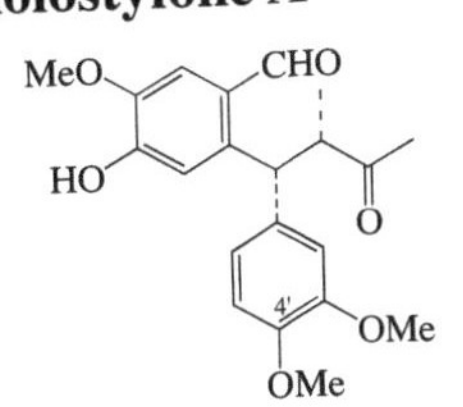

Relative Configuration

[694493-85-1]

$C_{21}H_{24}O_6$ 372.417

Secolignan. Constit. of the roots of *Holostylis reniformis*. Amorph. yellow solid. $[\alpha]_D^{25}$ -181.8 (c, 0.44 in $CHCl_3$). λ_{max} 211 (log ε 4.4); 235 (log ε 4.4); 279 (log ε 4.1) (MeOH)

4′-O-De-Me

Machilupin B

$C_{20}H_{22}O_6$ 358.390

Constit. of *Machilus philippinensis*. Syrup. $[\alpha]_D^{25}$ +68 (c, 0.02 in MeOH). Config. not determined. λ_{max} 237 (log ε 3.5); 282 (log ε 3.2); 317 (log ε 2.8) (MeOH).

Da Silva, T. *et al.*, *Phytochemistry*, 2004, **65**, 751-759 (*7,8-Secoholostylone A*).
Tai, S.-H. *et al.*, *RSC Adv.*, 2019, **9**, 21616-21625 (*Machilupin B*)

Secoretipolide E S45

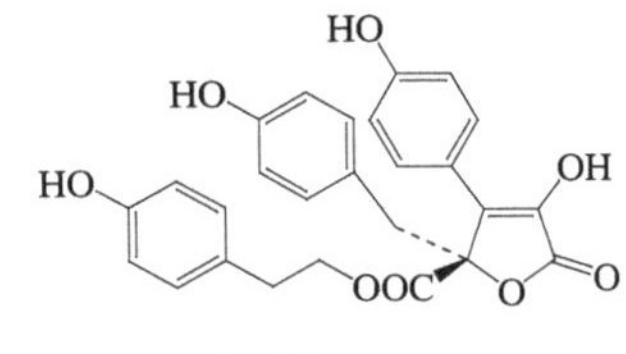

$C_{26}H_{22}O_8$ 462.454

Provisional lignan. Biosynth. pathway unclear. Related to Retipolide E, R11

(R)-form [1003192-83-3]

Isol. from *Retiboletus retipes*.

Justus, K. *et al.*, *Eur. J. Org. Chem.*, 2007, 5560-5572 (*Secoretipolide E, synth*)

Sequirin A S46

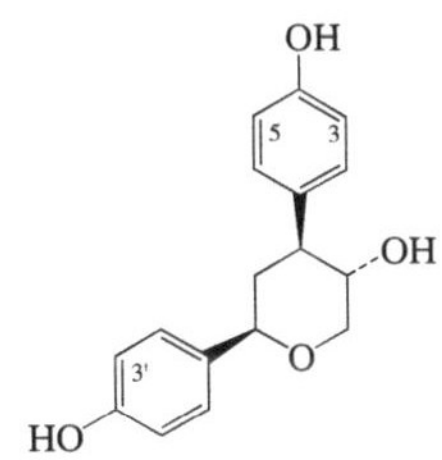

Tetrahydro-4,6-bis(4-hydroxyphenyl)-2H-pyran-3-ol, CAS. *Tetrahydro-5-hydroxy-2,4-bis(4-hydroxyphenyl)pyran. 7′,9-Epoxy-9′-nor-7,8′-lignan-4,4′,8-triol. 7′,9-Epoxy-4,4′,8-trihydroxy-9′-nor-7,8′-neolignan.* ***Sugiresinol***

[2141-07-3]

$C_{17}H_{18}O_4$ 286.327

Provisional norlignan. Biosynth. pathway unclear. Constit. of *Agathis australis*, *Cryptomeria japonica* and *Sequoia sempervirens*. Needles. Mp 250 - 251° (245°). $[\alpha]_D^{20}$ -18.7 (c, 0.25 in MeOH). λ_{max} 225; 277 (EtOH)

3-Hydroxy

7′,9-Epoxy-9′-nor-7,8′-lignan-3,4,4′,8-tetrol. 7′,9-Epoxy-3,4,4′,8-tetrahydroxy-9′-nor-7,8′-lignan. ***Hydroxysugiresinol. Sequirin B***

[15215-12-0]

[15215-14-2] (tri-Me ether)

$C_{17}H_{18}O_5$ 302.326

Constit. of *Cryptomeria japonica* and *Sequoia sempervirens*. Cryst. (H_2O). Mp 218° (212-213°). $[\alpha]_D^{25}$ -30 (c, 1 in MeOH)

3′-Hydroxy

7′,9-Epoxy-9′-nor-7,8′-lignan-3′,4,4′,8-tetrol. 7′,9-Epoxy-3′,4,4′,8-tetrahydroxy-9′-nor-7,8′-lignan. ***Sequirin E***

[68974-26-5]

[61243-87-6] (tri-Me ether)

$C_{17}H_{18}O_5$ 302.326

Constit. of the heartwood of *Sequoiadendron gigantea*. Cryst. (MeOH) (as tri-Me ether). Mp 125 - 126° (as tri-Me ether). Racemic

3′-Methoxy

7′,9-Epoxy-3′-methoxy-9′-nor-7,8′-lignan-4,4′,8-triol. 7′,9-Epoxy-4,4′,8-trihydroxy-3′-methoxy-9′-nor-7,8′-lignan. ***Sequosempervirin F***

[864719-21-1]

$C_{18}H_{20}O_5$ 316.353

Constit. of *Sequoia sempervirens*. Cryst. (MeOH). Mp 93 - 95°. $[\alpha]_D^{29}$ +2.2 (c, 0.15 in MeOH). λ_{max} 205 (log ε 4.77); 226 (log ε 4.53); 278 (log ε 3.98) (MeOH)

3,3′-Dihydroxy

7′,9-Epoxy-9′-nor-7,8′-lignan-3,3′,4,4′,8-pentol. 7′,9-Epoxy-3,3′,4,4′,8-pentahydroxy-9′-nor-7,8′-lignan. ***Sequirin F***

[61243-88-7] (tetra-Me ether)

$C_{17}H_{18}O_6$ 318.325

Constit. of the heartwood of *Sequoiadendron gigantea*. Gum (as tetra-Me ether)

3′,5′-Dimethoxy

7′,9-Epoxy-3′,5′-dimethoxy-9′-nor-7,8′-lignan-4,4′,8-triol. 7′,9-Epoxy-4,4′,8-trihydroxy-3′,5-dimethoxy-9′-nor-7,8′-lignan.

Sequosempervirin G

[864719-22-2]

$C_{19}H_{22}O_6$ 346.379

Constit. of *Sequoia sempervirens*. Oil. λ_{max} 208 (log ε 4.63); 276 (log ε 3.58); 359 (log ε 2.48) (MeOH)

3,3′,5-Trihydroxy

7′,9-Epoxy-9′-nor-7,8′-lignan-3,3′,4,4′,5,8-hexol. 7′,9-Epoxy-3,3′,4,4′,5,8-hexahydroxy-9′-nor-7,8′-lignan. ***Sequirin G***

[61243-89-8] (penta-Me ether)

$C_{17}H_{18}O_7$ 334.324

Constit. of *Sequoiadendron gigantea*. Amorph. solid (as penta-Me ether).

Balogh, B. *et al.*, *Phytochemistry*, 1965, **4**, 569-575 (*Sequirins A,B*).
Riffer, R. *et al.*, *Phytochemistry*, 1967, **6**, 1557-1562 (*pmr*).
Enzell, C.R. *et al.*, *Tetrahedron Lett.*, 1967, **8**, 793-798 (*Sugiresinol, abs config*).
Hatam, N.A.R. *et al.*, *J. Chem. Soc. C*, 1969, 1921-1932 (*Sequirin B*).
Horii, Z. *et al.*, *Chem. Pharm. Bull.*, 1971, **19**, 2212-2213 (*synth*).
Davies, R.V. *et al.*, *J. Chem. Soc., Perkin Trans. 1*, 1973, 2359-2365 (*synth*).

Henley-Smith, P. *et al., Phytochemistry*, 1976, **15**, 1285-1287 (*Sequirins E,F,G*).
Beracierta, A.P. *et al., J. Chem. Soc., Perkin Trans. 1*, 1978, 1257-1263 (*synth*).
Muraoka, O. *et al., J. Chem. Soc., Perkin Trans. 1*, 1996, 405-411 (*synth*).
Brown, E. *et al., Tetrahedron*, 1997, **53**, 9679-9694 (*synth*).
Matsuo, K. *et al., Heterocycles*, 2000, **53**, 1505-1513 (*synth*).
Evans, P.A. *et al., J. Am. Chem. Soc.*, 2003, **125**, 8974-8975 (*synth*).
Zhang, Y.-M. *et al., Chem. Biodiversity*, 2005, **2**, 497-505 (*Sequosempervirins F,G*).
Quan, W.-G. *et al., Chin. J. Chem.*, 2007, **25**, 688-693 (*synth*)

Sequirin C S47

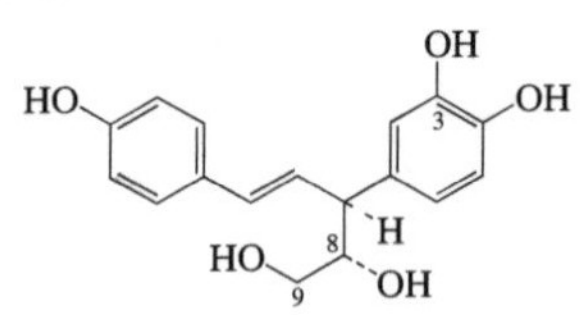

Absolute Configuration

4-[1-(1,2-Dihydroxyethyl)-3-(4-hydroxyphenyl)-2-propenyl]-1,2-benzenediol, CAS. *3-(3,4-Dihydroxyphenyl)-5-(4-hydroxyphenyl)-4-pentene-1,2-diol. 9′-Nor-7,8′-lign-7′-ene-3,4,4′,8,9-pentol. 3,4,4′,8,9-Pentahydroxy-9′-nor-7,8′-neolign-7′-ene*
[18194-29-1]
$C_{17}H_{18}O_5$ 302.326
Constit. of *Cryptomeria japonica, Sequoia sempervirens* and *Nicotiana tabacum*. Shows modest anti-HIV-1 activity. Long needles (H_2O). Mp 193°. $[\alpha]_D^{25}$ -90 (c, 1 in MeOH)

3-Me ether
3-Methoxy-9′-nor-7,8′-lign-7′-ene-4,4′,8,9-tetrol. 4,4′,8,9-Tetrahydroxy-3-methoxy-9′-nor-7,8′-lign-7′-ene. ***Sequosempervirin B***
[864719-17-5]
$C_{18}H_{20}O_5$ 316.353
Constit. of *Sequoia sempervirens*. Cryst. (MeOH). Mp 150 - 154°. $[\alpha]_D^{25}$ -25.3 (c, 1.75 in MeOH). λ_{max} 204 (log ε 4.45); 265 (log ε 4.2) (MeOH)

8,9-Isopropylidene, 3-Me ether
Sequosempervirin D
[864719-19-7]
$C_{21}H_{24}O_5$ 356.418
Constit. of *Sequoia sempervirens*. Oil. $[\alpha]_D^{26}$ -19.1 (c, 0.9 in MeOH). Possible artifact. λ_{max} 204 (log ε 4.7); 265 (log ε 4.42) (MeOH)

3-Deoxy
9′-Nor-7,8′-lign-7′-ene-4,4′,8,9-tetrol. 4,4′,8,9-Tetrahydroxy-9′-nor-7,8′-lign-7′-ene. ***Agatharesinol***
[7288-11-1]
$C_{17}H_{18}O_4$ 286.327
Constit. of *Agathis australis, Sequoia* spp., *Cryptomeria japonica* and *Athrotaxis selaginoides*. Mollusc antifeedant. Cryst.. Mp 105 - 109°. $[\alpha]_D^{25}$ -37.6 (c, 1.7 in Me_2CO)

3-Deoxy, tetra-Ac
[7287-90-3]
Resin. $[\alpha]_D$ -19 (c, 1 in $CHCl_3$)

3-Deoxy, 8,9-isopropylidene
Agatharesinol acetonide
[800389-33-7]
$C_{20}H_{22}O_4$ 326.392
Constit. of *Metasequoia glyptostroboides* and *Sequoia sempervirens*. Exhibits anticancer activity. $[\alpha]_D^{25}$ -13.3 (c, 1.2 in MeOH). Possible artifact

3-Deoxy, 3′-hydroxy
5-(3,4-Dihydroxyphenyl)-3-(4-hydroxyphenyl)-4-pentene-1,2-diol. 9′-Nor-7,8′-lign-7′-ene-3′,4,4′,8,9-pentol. 3′,4,4′,8,9-Pentahydroxy-9′-nor-7,8′-lign-7′-ene. ***Metasequirin C***
[800391-59-7]
$C_{17}H_{18}O_5$ 302.326
Constit. of the heartwood of *Metasequoia glyptostroboides*

3′-Methoxy, 3-Me ether
3,5-Bis(4-hydroxy-3-methoxyphenyl)-4-pentene-1,2-diol. 3,3′-Dimethoxy-9′-nor-7,8′-lign-7′-ene-4,4′,8,9-tetrol. 4,4′,8,9-Tetrahydroxy-3,3′-dimethoxy-9′-nor-7,8′-lign-7′-ene. ***Metasequirin D***
[1264694-96-3]
$C_{19}H_{22}O_6$ 346.379
Constit. of *Metasequoia glyptostroboides*. Gum. $[\alpha]_D^{25}$ -40.5 (c, 0.17 in MeOH). λ_{max} 194 (log ε 4.11); 204 (log ε 4.41); 269 (log ε 4.03) (MeOH)

5-Methoxy, 8,9-isopropylidene, 3-Me ether
Sequosempervirin E
[864719-20-0]
$C_{22}H_{26}O_6$ 386.444
Constit. of *Sequoia sempervirens*. Oil. $[\alpha]_D^{24}$ -19.1 (c, 1.45 in MeOH). Possible artifact. λ_{max} 206 (log ε 4.69); 265 (log ε 4.33) (MeOH)

5-Methoxy, 3-Me ether
3,5-Dimethoxy-9′-nor-7,8′-lign-7′-ene-4,4′,8,9-tetrol. 4,4′,8,9-Tetrahydroxy-3,5-dimethoxy-9′-nor-7,8′-lign-7′-ene. ***Sequosempervirin C***
[864719-18-6]
$C_{19}H_{22}O_6$ 346.379
Constit. of *Sequoia sempervirens*. Cryst. (MeOH). Mp 145 - 147°. $[\alpha]_D^{26}$ -63.3 (c, 0.1 in MeOH). λ_{max} 205 (log ε 4.81); 265 (log ε 4.46) (MeOH)

Z-Isomer, 3-deoxy
Isoagatharesinol
[835613-06-4]
$C_{17}H_{18}O_4$ 286.327
Constit. of the roots of *Asparagus gobicus* and *Asparagus cochinchinensis*. Gum. $[\alpha]_D^{21}$ +49.7 (c, 5.4 in Me_2CO)

Z-Isomer, 3-deoxy, 4′-O-β-D-glucopyranoside
Isoagatharesinoside
[1448462-78-9]
$C_{23}H_{28}O_9$ 448.468
Constit. of *Asparagus cochinchinensis*.

Enzell, C.R. *et al., Tetrahedron Lett.*, 1966, **7**, 2395-2402 (*Agatharesinol*).
Enzell, C.R. *et al., Tetrahedron Lett.*, 1967, **8**, 793-798 (*Agatharesinol, abs config*).
Hatam, N.A.R. *et al., J. Chem. Soc. C*, 1969, 1921-1932 (*Sequirin C, abs config*).
Beraciertia, A.P. *et al., J. Chem. Soc., Perkin Trans. 1*, 1978, 1257-1263 (*synth*).
Muraoka, O. *et al., J. Chem. Soc., Perkin Trans. 1*, 1996, 405-411 (*synth*).
Chen, X.H. *et al., Z. Naturforsch., C*, 2001, **56**, 249-252 (*Sequirin C, Agatharesinol, activity*).
Yang, C.-X. *et al., Planta Med.*, 2004, **70**, 446-451 (*Isoagatharesinol*).
Zhang, Y.-M. *et al., Chem. Biodiversity*, 2005, **2**, 497-505 (*Sequosempervirins, Agatharesinol acetonide, activity*).
Fang, G.Y. *et al., Angew. Chem., Int. Ed.*, 2007, **46**, 359-362 (*Isoagatharesinol, synth*).
Quan, W.-G. *et al., Chin. J. Chem.*, 2007, **25**, 688-693 (*Agatharesinol, synth*).
Imai, T. *et al., Phytochem. Lett.*, 2009, **2**, 196-200 (*Metasequirin C, Sequirin C, biosynth*).
Dong, L.-B. *et al., J. Nat. Prod.*, 2011, **74**, 234-239 (*Metasequirin D*).
Liao, Z. *et al., Asian J. Chem.*, 2012, **24**, 4895-4897 (*Nicotiana tabacum constit*).
Gao, X. *et al., Heterocycles*, 2012, **85**, 147-153 (*Sequirin C, activity*).
Li, X.N. *et al., Nat. Prod. Commun.*, 2012, **7**, 1357-1358 (*Isoagatharesinoside, Isoagatharesinol*)

Sequirin D S48

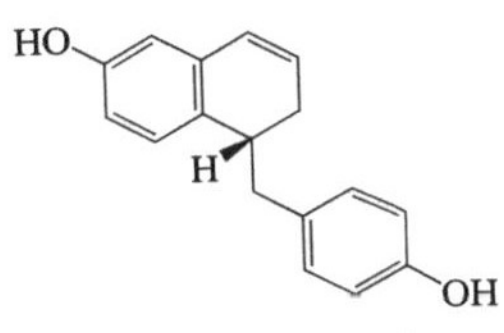

Absolute Configuration

5,6-Dihydro-5-[(4-hydroxyphenyl) methyl]-2-naphthalenol, CAS. *1,2-Dihydro-6-hydroxy-1-(4-hydroxybenzyl)naphthalene*

$C_{17}H_{16}O_2$ 252.313

(R)-form [50938-82-4]

Constit. of *Sequoia sempervirens*

Di-Me ether

[50469-99-3]

Needles (hexane). Mp 107.5 - 109°.

Begley, M.J. *et al.*, *J. Chem. Soc., Perkin Trans. 1*, 1978, 750-754 (*Sequirin D, di-Me ether, cryst struct*).

Hatam, N.A. *et al.*, *Tetrahedron Lett.*, 1978, **19**, 5145-5146 (*abs config, synth*).

Reddy, M.P. *et al.*, *J. Chem. Soc., Perkin Trans. 1*, 1981, 2662-2665 (*synth*).

Hatam, N.A.R. *et al.*, *J. Chem. Soc., Perkin Trans. 1*, 1982, 461-465 (*synth*)

Sequosempervirin A S49

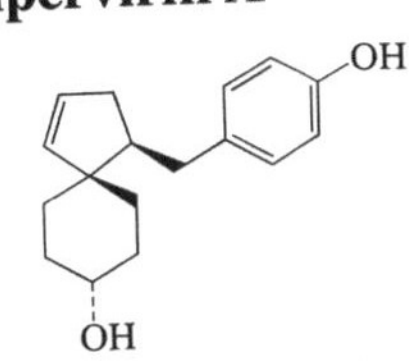

Absolute Configuration

8,9′-Cyclo-1,2,3,4,5,6-hexahydro-9-nor-1,8′-lign-7-ene-4,4′-diol. 8,9′-Cyclo-4,4′-dihydroxy-1,2,3,4,5,6-hexahydro-9-nor-1,8′-neolign-7-ene

[720685-82-5]

$C_{17}H_{22}O_2$ 258.361

Abs. config. revised in 2011. Constit. of *Sequoia sempervirens*. Cryst. (MeOH). Mp 172 - 174°. $[\alpha]^{26}_{D}$+24.7 (c, 0.6 in MeOH). λ_{max} 203 (log ε 3.91); 225 (log ε 3.76); 280 (log ε 3.1) (MeOH).

Zhang, Y.-M. *et al.*, *Tetrahedron Lett.*, 2004, **45**, 4319-4321 (*Sequosempervirin A*).

Maity, S. *et al.*, *Tetrahedron Lett.*, 2007, **48**, 3355-3358 (*synth*).

Ito, Y. *et al.*, *Org. Lett.*, 2011, **13**, 4640-4643 (*synth, abs config*)

Seric acid F S50

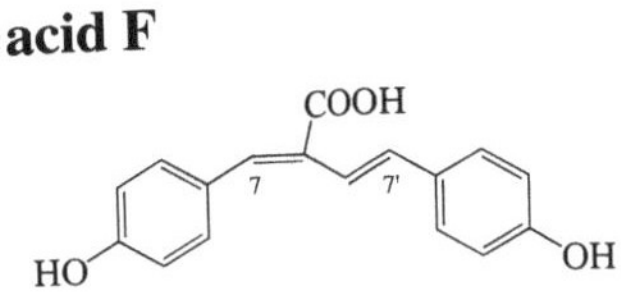

(7E,7′E)-form

2-(4-Hydroxybenzylidene)-4-(4-hydroxyphenyl)-3-butenoic acid. 4,4′-Dihydroxy-9′-nor-8,8′-ligna-7,7′-dien-9-oic acid

[2361326-67-0] (7*Z*,7′*E*-form)

$C_{17}H_{14}O_4$ 282.295

Lignan numbering shown. Easily interconverts between (7*E*,7′*E*)- and (7*Z*,7′*E*)-isomers. Predominant form shown. Constit. of *Oenanthe javanica*. Degranulation inhibitor. Amorph. pale yellow solid. λ_{max} 265 (log ε 4.01); 342 (log ε 4.35) (MeOH).

Murata, T. *et al.*, *J. Nat. Prod.*, 2019, **82**, 1518-1526 (*Seric acid F, activity*)

Sesamolinol S51

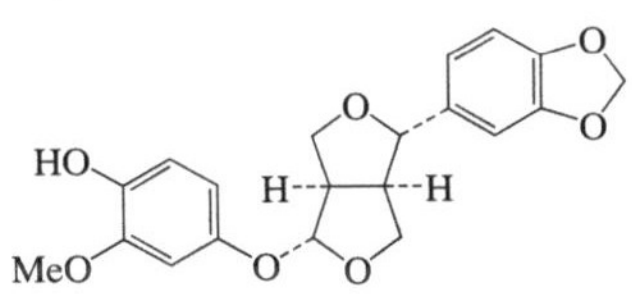

Relative Configuration

[100016-94-2]

$C_{20}H_{20}O_7$ 372.373

Derived from Pinoresinol and has unusual oxygen insertion in its biosynth. pathway. Constit. of the seeds of *Sesamum indicum*. Antioxidant. Exhibits antimutagenic activity. Oil. [α] D +184 (c, 1 in $CHCl_3$). λ_{max} 231 (ε 8912); 287 (ε 6310) (MeOH)

O-β-D-Glucopyranoside

Sesamolinol glucoside

[1353018-00-4]

$C_{26}H_{30}O_{12}$ 534.514

Constit. of the seeds of *Sesamum indicum*. Amorph. solid. [α] D +89 (c, 0.2 in MeOH)

O-[β-D-Glucopyranosyl-(1→6)-β-D-glucopyranoside]

Sesamolinol gentiobioside

[749874-95-1]

$C_{32}H_{40}O_{17}$ 696.655

Constit. of the seeds of *Sesamum indicum*.

Osawa, T. *et al.*, *Agric. Biol. Chem.*, 1985, **49**, 3351-3352 (*Sesamolinol*).

Marchand, P.A. *et al.*, *Can. J. Chem.*, 1997, **75**, 840-849 (*synth*).

Kang, M.-H. *et al.*, *Life Sci.*, 2000, **66**, 161-172 (*activity*).

Moazzami, A.A. *et al.*, *Biosci., Biotechnol., Biochem.*, 2006, **70**, 1478-1481 (*gentiobioside*).

Grougnet, R. *et al.*, *J. Agric. Food Chem.*, 2012, **60**, 108-111 (*Sesamolinol, activity, glucoside*).

Dar, A.A. *et al.*, *Bioorg. Chem.*, 2013, **50**, 1-10 (*rev*)

Seslignanoccidentaliol A S52

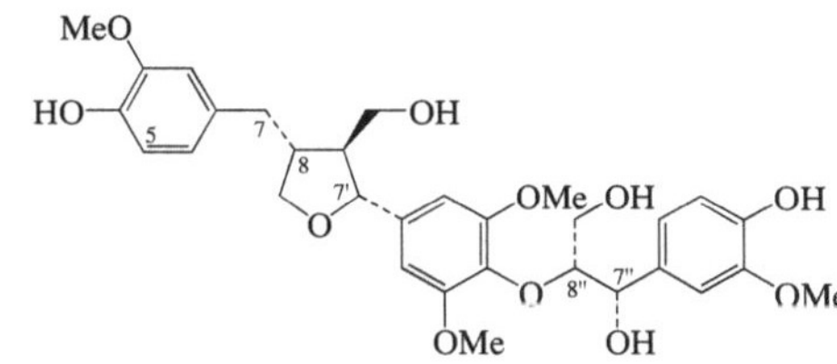

(7′R,7″S*,8R*,8′S*,8″R*)-form*

$C_{31}H_{38}O_{11}$ 586.634

Similar to Cerberalignan J, C51

(7′R*,7″S*,8R*,8′S*,8″R*)-form

[1134116-04-3]

Constit. of *Cassia occidentalis* and *Campylotropis hirtella*. Powder. Isolate from *C. hirtella* originally assigned different stereochem. in furan ring and revised in 2010. λ_{max} 204; 280 (MeOH)

7-Oxo, 5-methoxy

Seslignanoccidentaliol B

[1255922-29-2]

$C_{32}H_{38}O_{13}$ 630.643

Constit. of *Cassia occidentalis*. Powder. $[\alpha]^{24}_{D}$+33.3 (c, 0.09 in MeOH). λ_{max} 205; 283; 301 (MeOH)

(7′R*,7″S*,8S*,8′S*,8″R*)-form

5-Methoxy

[1954754-20-1]

$C_{32}H_{40}O_{12}$ 616.66

Constit. of the fruits of *Melia toosendan*. Amorph. pale yellow powder. Mp 92 - 94°. $[\alpha]^{20}_{D}$-9.0 (c, 0.2 in MeOH). λ_{max} 279 (log ε 3.73); 378 (log ε 2.74) (MeOH)

(7′R*,7″S*,8S*,8′S,8″S*)-form

Rosalaevin A [1340483-48-8]

Constit. of the fruit of *Rosa laevigata*. Antioxidant. Amorph. solid. $[\alpha]^{27}_{D}$ -10.9 (c, 0.2 in $CHCl_3$). λ_{max} 210 (log ε 3.48); 240 (log ε 3.90); 280 (log ε 3.67) (MeOH)

(7′R*,7″ξ,8S*,8′S*,8″ξ)-form

7″-O-β-D-Glucopyranoside

Glycopentoside F

[1805000-75-2]

$C_{37}H_{48}O_{16}$ 748.775

Constit. of the stems of *Glycosmis pentaphylla*. Oil. $[\alpha]^{20}_{D}$-13.9 (c, 0.22 in MeOH). C-7″/C-8″ config. not determined. λ_{max} 233 (log ε 3.83); 281 (log ε 3.68) (MeOH).

Han, H.-Y. *et al.*, *Nat. Prod. Res.*, 2008, **22**, 984-989 (*Campylotropis hirtella constit*).

Li, S.-F. *et al.*, *Helv. Chim. Acta*, 2010, **93**, 1795-1802 (*Seslignanoccidentaliols A,B*).

Li, X. *et al.*, *Food Chem.*, 2012, **130**, 575-580 (*Rosalaevin A, activity*).

Chen, Y. *et al.*, *Helv. Chim. Acta*, 2015, **98**, 1160-1166 (*Glycopentoside F*).
Wang, H. *et al.*, *Planta Med.*, 2015, **81**, 847-854 (*Melia toosendan constit*)

3,3′:8′,8″-Sesquilign-7-ene-3″,4,4′,4″,5,5′,9,9′,9″-nonol S53

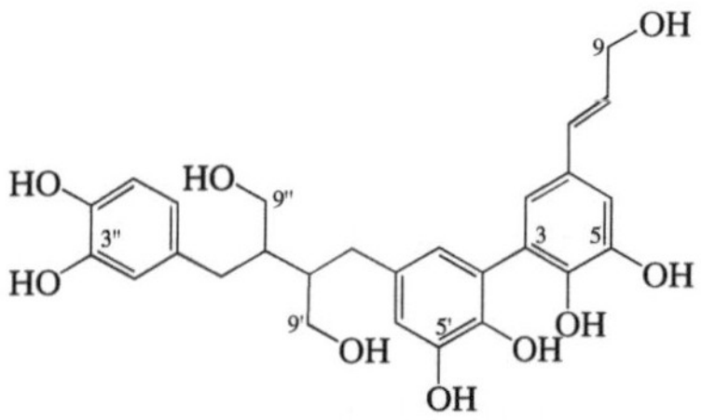

5′-[4-(3,4-Dihydroxyphenyl)-2,3-bis(hydroxymethyl)butyl]-5-(3-hydroxy-2-propenyl)-[1,1′-biphenyl]-2,2′,3,3′-tetrol
$C_{27}H_{30}O_9$ 498.528
Lignan numbering shown
3″,5,5′-Tri-Me ether, hexa-Ac
[221347-34-8]
$C_{42}H_{48}O_{15}$ 792.831
Constit. of degraded wood of *Pinus taeda*.

Peng, J. *et al.*, *Phytochemistry*, 1999, **50**, 659-666 (*Pinus taeda constit*)

Sesquipinsapol B S54

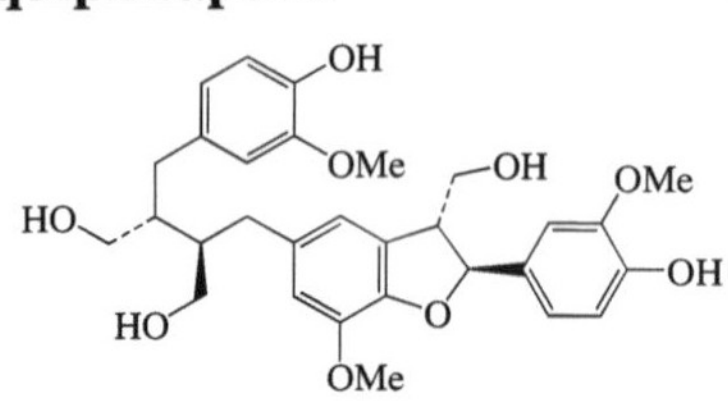

Relative Configuration

[155969-63-4]
$C_{30}H_{36}O_9$ 540.609
Constit. of *Abies pinsapo*, *Larix olgensis* var. *koreana* and *Litchi chinensis*.

Barrero, A.F. *et al.*, *Nat. Prod. Lett.*, 1993, **2**, 255-262 (*Abies pinsapo constit*).
Wen, L. *et al.*, *J. Funct. Foods*, 2014, **8**, 26-34 (*Litchi chinensis constit*)

Shanciol H S55

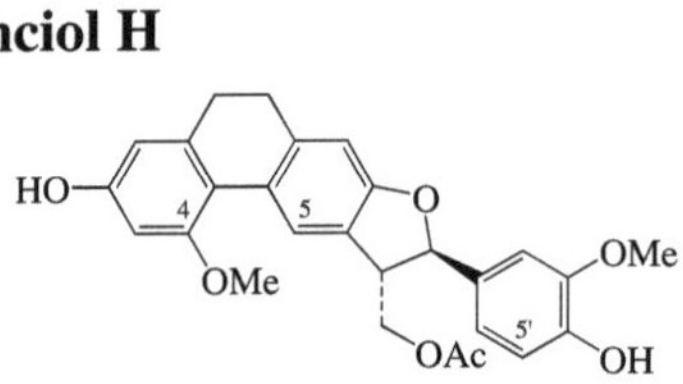

Relative Configuration

[1114905-55-3]
$C_{27}H_{26}O_7$ 462.498
Constit. of *Bletilla striata* and *Pleione bulbocodioides*. Amorph. powder. $[\alpha]_D^{25}$ -6.8 (c, 0.1 in MeOH). $[\alpha]_D^{20}$ +20 (c, 0.1 in MeOH) ((7′R,8′S)-form). Partial racemate. Absolute configuration determined in 2019. λ_{max} 223 (log ε 4.63); 282 (log ε 4.51); 307 (log ε 4.32) (MeOH)
5′-Methoxy
Pleionesin A
[1222077-23-7]
$C_{28}H_{28}O_8$ 492.524
Constit. of the tubers of *Pleione yunnanensis*. Amorph. powder (MeOH). $[\alpha]_D^{25}$ -3 (c, 0.1 in MeOH). λ_{max} 221 (log ε 4.51); 282 (log ε 4.22); 298 (log ε 4.16) (MeOH)
4-Demethoxy, 5-methoxy
Isoshanciol H
[1009566-73-7]
$C_{27}H_{26}O_7$ 462.498
Constit. of the tubers of *Pleione bulbocodioides*. Amorph. powder. $[\alpha]_D^{25}$ +25.8 (c, 0.5 in MeOH).

Liu, X.-Q. *et al.*, *Chin. Chem. Lett.*, 2007, **18**, 1089-1091 (*Isoshanciol H*).
Liu, X.-Q. *et al.*, *J. Asian Nat. Prod. Res.*, 2008, **10**, 453-457 (*Shanciol H*).
Dong, H. *et al.*, *Magn. Reson. Chem.*, 2010, **48**, 256-260 (*Pleionesin A*).
Zhou, D. *et al.*, *J. Nat. Prod.*, 2019, **82**, 2238-2245 (*Bletilla striata constit, abs config*)

Shegansu C S56

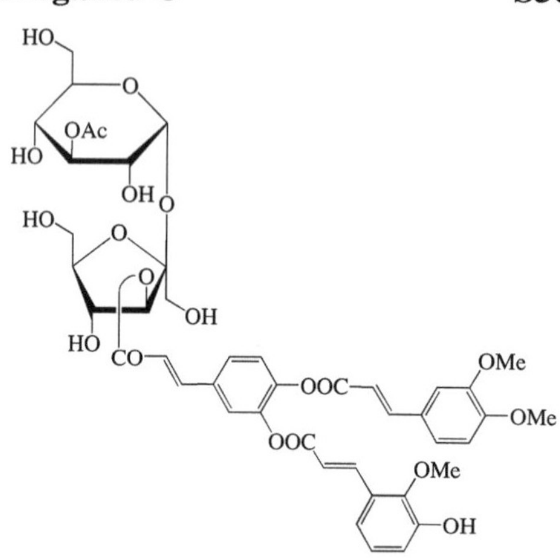

[221232-19-5]
$C_{44}H_{48}O_{21}$ 912.847
Glycoside of Sucrose. Constit. of the rhizomes of *Belamcanda chinensis*. Amorph. yellow powder. $[\alpha]_D^{31}$ +51.3 (c, 0.07 in EtOH). λ_{max} 218; 236; 300; 326 (EtOH).

Lin, M. *et al.*, *J. Asian Nat. Prod. Res.*, 1998, **1**, 67-75 (*Shegansu C*)

Shimobashiric acid C S57

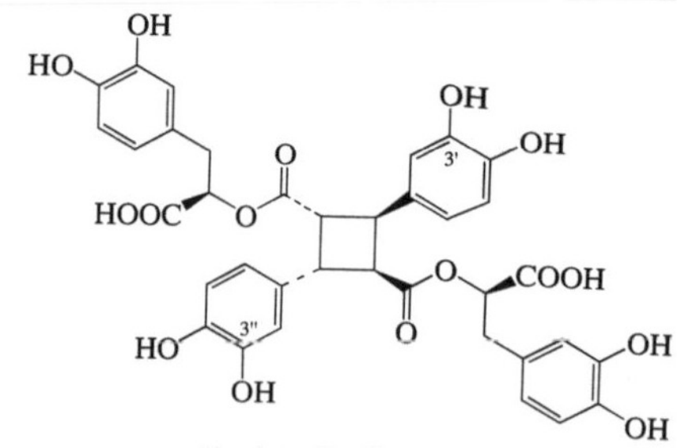

Absolute Configuration

[1393828-22-2]
$C_{36}H_{32}O_{16}$ 720.636
Related to Sagerinic acid, S5. Constit. of *Keiskea japonica*, *Plectranthus amboinicus* and *Pulmonaria officinalis*. Shows hyaluronidase inhibitory activity. Amorph. solid. $[\alpha]_D^{22}$ -4.1 (c, 1.21 in MeOH). λ_{max} 204 (log ε 4.92); 284 (log ε 4.05) (MeOH)
3′,3″-Di-Me ether
Shimobashiric acid D
[1393828-23-3]
$C_{38}H_{36}O_{16}$ 748.69
Constit. of *Keiskea japonica*. Amorph. solid. $[\alpha]_D^{21}$ -7.7 (c, 0.13 in MeOH). λ_{max} 204 (log ε 4.95); 284 (log ε 4.19) (MeOH).

Murata, T. *et al.*, *Chem. Pharm. Bull.*, 2012, **60**, 121-128 (*Shimobashiric acids C,D*).
Chen, Y.-S. *et al.*, *Bioorg. Med. Chem.*, 2014, **22**, 1766-1772 (*Shimobashiric acid C*)

Sibiricin A S58

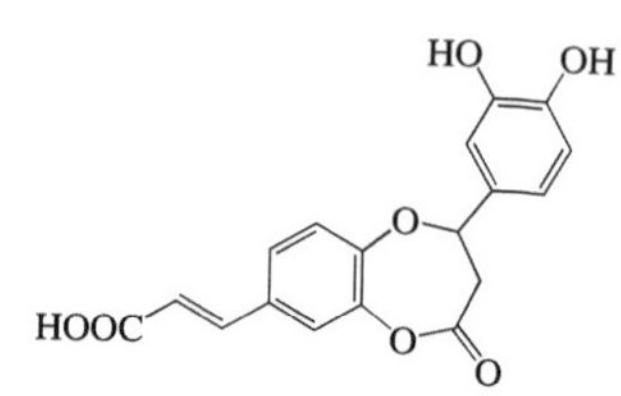

3-[2-(3,4-Dihydroxyphenyl)-3,4-dihydro-4-oxo-2H-1,5-benzodioxepin]-2-propenoic acid, CAS. *3′,4′-Dihydroxy-4,7′-oxylign-7-en-9′,3-olid-9-oic acid*.
Messerschmidin
[147362-49-0]
$C_{18}H_{14}O_7$ 342.303
Constit. of *Messerschmidia sibirica*. Amorph. powder
Et ester
Sibiricin B
[147362-50-3]
$C_{20}H_{18}O_7$ 370.357
Constit. of *Messerschmidia sibirica*. Pale yellow oil.

Song, Z. *et al.*, *Chin. Chem. Lett.*, 1992, **3**, 975-976 (*Messerschmidin*).

Song, Z. *et al.*, *Indian J. Chem., Sect. B*, 1996, **35**, 955-959 (*Sibiricins A,B*)

Sibiricumin A S59

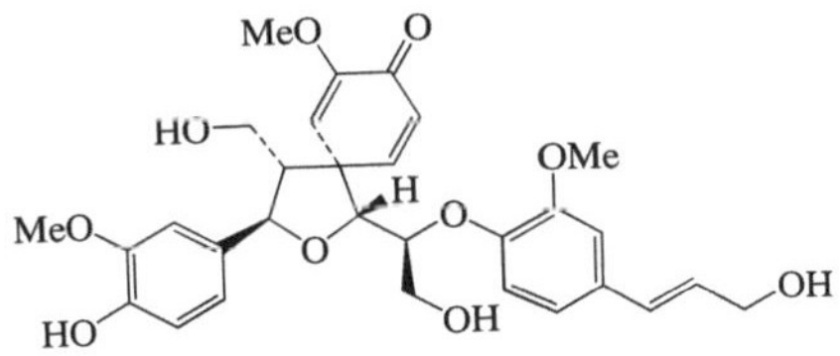

(+)-*form*

[1630002-77-5]

$C_{30}H_{34}O_{10}$ 554.592

Spirosesquineolignan. Similar to Pinobatol, P54. Abs. config. revised in 2016. Constit. of the fruit of *Xanthium sibiricum*. Cryst. (MeOH aq.). Isol. as racemate. Has been sepd. into one pair of enantiomers. λ_{max} 203 (log ε 2.59) (MeOH)

(+)-form [1629668-97-8]

Amorph. powder. [α] D +36 (c, 0.15 in MeOH)

(-)-form [1629668-98-9]

Amorph. powder. [α] D -36 (c, 0.15 in MeOH).

Shi, Y. *et al.*, *Org. Lett.*, 2014, **16**, 5406-5409 (*Sibiricumin A, cryst struct*).
Zhao, D. *et al.*, *Chirality*, 2016, **28**, 612-617 (*revised abs config*)

Silybin S60

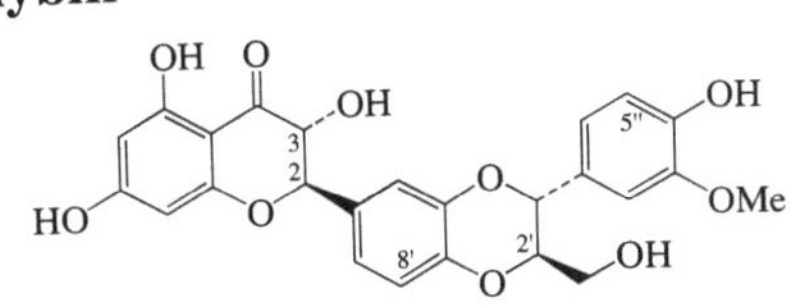

(2*R*,2'*R*,3*R*,3'*R*)-*form*

***Silibinin**, INN, USAN. Silybum substance E6. Silymarin.*

[87725-91-5]

$C_{25}H_{22}O_{10}$ 482.441

Flavonolignan. Obt. as a mixt. of two diastereoisomeric forms first characterised in 2003. Strictly Silymarin is a mixt. of various flavonolignan components and their diastereoisomers. Isol. from *Silybum marianum* and one of its fungal endophytes *Aspergillus iizukae*. Drug used for treatment of brain oedema and liver disorders. Hepatoprotective agent. Antioxidant. Stimulates hair regrowth. Dec. at 180°

(2R,2'R,3R,3'R)-form

Silybin A. Silymarin I [22888-70-6]

Constit. of *Silybum marianum, Anastatica hierochuntica, Euonymus acanthocarpus* and *Inula wissmanniana*. Prod. by *Aspergillus iizukae*. Yellow cryst. (MeOH aq. or EtOH). Mp 158 - 160°. Mp 162 - 163°. [α] D +6.1 (c, 0.3 in MeOH). [α] D +20 (c, 0.21 in Me_2CO). λ_{max} 217 (log ε 5.2); 230 (log ε 5.2); 288 (log ε 5.1) (MeOH)

2,3-Didehydro

2,3-Dehydrosilybin

[25166-14-7]

$C_{25}H_{20}O_{10}$ 480.425

Constit. of *Silybum marianum*. Mp 254 - 255°

3-Deoxy

Isosilandrin A

[848420-95-1]

$C_{25}H_{22}O_9$ 466.442

Constit. of the fruit of *Silybum marianum*. Antioxidant. Cryst.. Mp 169 - 172°. $[\alpha]^{25}_{D}$ -7.5 (c, 0.5 in MeOH). Possesses (2*S*,2'*R*,3'*R*)-config.. λ_{max} 284; 330 (no solvent reported)

5''-Methoxy, 8'-hydroxy

Hovenin A

[1423247-43-1]

$C_{26}H_{24}O_{12}$ 528.466

Constit. of the seeds of *Hovenia acerba*. Amorph. powder (MeOH). Mp 171 - 173°. $[\alpha]^{22}_{D}$ +22.8 (c, 0.04 in MeOH). λ_{max} 210 (log ε 4.72); 247 (log ε 4.07); 370 (log ε 3.81) (MeOH)

(2R,2'R,3R,3'S)-form

3-Deoxy

Isocisilandrin. *cis-Isosilandrin*

[1021297-35-7]

$C_{25}H_{22}O_9$ 466.442

Constit. of *Silybum marianum*. Possesses (2*S*,2'*R*,3'*S*)-config.

(2R,2'S,3R,3'S)-form

Silybin B [142797-34-0]

Constit of *Silybum marianum, Euonymus acanthocarpus* and *Inula wismanniana*. Prod. by *Aspergillus iizukae*. Yellow cryst. (MeOH aq.). Mp 158 - 160°. [α] D +0.6 (c, 0.3 in MeOH). [α] D -1.1 (c, 0.28 in Me_2CO). λ_{max} 217 (log ε 5.2); 230 (log ε 5.3); 288 (log ε 5.2) (MeOH)

3-Deoxy

Isosilandrin B

[1021297-34-6]

Constit. of *Silybum marianum*. Possesses (2*S*,2'*S*,3'*S*)-config.

5''-Methoxy, 8'-hydroxy

Hovenin B

[1423247-45-3]

$C_{26}H_{24}O_{12}$ 528.466

Constit. of the seeds of *Hovenia acerba*. Amorph. powder (MeOH). Mp 170 - 172°. $[\alpha]^{25}_{D}$ -3.4 (c, 0.04 in MeOH). λ_{max} 210 (log ε 4.73); 247 (log ε 4.08); 371 (log ε 3.79) (McOH).

Aldrich Library of FT-IR Spectra, 1st edn., 1985, **2**, 97C (*ir*).
Zoltan, O.T. *et al.*, *Arzneim.-Forsch.*, 1970, **20**, 1248-1249 (*pharmacol*).
Pelter, A. *et al.*, *Chem. Ber.*, 1975, **108**, 790-802 (*Silybin, struct*).
Hänsel, R. *et al.*, *Chem. Ber.*, 1975, **108**, 1482-1501 (*Silybin, 2,3-Dehydrosilybin*).
Merlini, L. *et al.*, *J. Chem. Soc., Perkin Trans. 1*, 1980, 775-778 (*synth*).
Lotter, H. *et al.*, *Z. Naturforsch., C*, 1983, **38**, 339-341 (*cryst struct*).
Tanaka, H. *et al.*, *Chem. Pharm. Bull.*, 1984, **33**, 1419-1423 (*Silybin, synth*).
Negwer, M. *Organic-Chemical Drugs and their Synonyms, 6th edn., Akademie-Verlag*, 1987, 7244 (*synonyms*).
Mericli, A.H. *Planta Med.*, 1988, **54**, 44-45 (*Dehydrosilybin*).
Delfini, M. *et al.*, *Spectrosc. Lett.*, 1990, **23**, 1007-1025 (*pmr, conformn*).
Martindale, The Extra Pharmacopoeia, 30th edn., Pharmaceutical Press, 1993, 693.
Morazzoni, P. *et al.*, *Fitoterapia*, 1995, **64**, 3-42 (*rev*).
Saller, R. *et al.*, *Drugs*, 2001, **61**, 2035-2063 (*rev*).
Kvasnička, F. *et al.*, *J. Chromatogr., A*, 2003, **990**, 239-245 (*chromatog, anal*).
Lee, D.Y-W. *et al.*, *J. Nat. Prod.*, 2003, **66**, 1171-1174 (*Silybins A,B, abs config*).
Kim, N.-C. *et al.*, *Org. Biomol. Chem.*, 2003, **1**, 1684-1689 (*Silybins A,B*).
Samu, Z. *et al.*, *Chem. Biodiversity*, 2004, **1**, 1668-1677 (*Isosilandrin*).
Victorrajmohan, C. *et al.*, *Drugs RD*, 2005, **6**, 395-400 (*pharmacol*).
Nyiredy, S. *et al.*, *J. Chromatogr. Sci.*, 2008, **46**, 93-96 (*Isosilandrin B, Isocisilandrin, biosynth*).
US Pat., 2008, 2008 033 037 (GreenPharma; Bioalternatives) (*Silymarin, activity*).
Nakashima , S. *et al.*, *Bioorg. Med. Chem.*, 2010, **18**, 2337-2345 (*Anastatica hierochuntica constit*).
Zhu, J.X. *et al.*, *Arch. Pharmacal Res.*, 2012, **35**, 1739-1747 (*Silybins A,B*).
Zhang, X.-Q. *et al.*, *Phytochem. Lett.*, 2012, **5**, 292-296 (*Hovenins A,B*).
Wang, C. *et al.*, *Arch. Pharmacal Res.*, 2013, **36**, 1516-1524 (*Silybins A,B*).
El-Elimat, T. *et al.*, *J. Nat. Prod.*, 2014, **77**, 193-199 (*Silybins A,B, Aspergillus iizukae*).
Biedermann, D. *et al.*, *Nat. Prod. Rep.*, 2014, **31**, 1138-1157 (*rev, chem*).
Marhol, P. *et al.*, *J. Funct. Foods*, 2015, **14**, 570-580 (*pharmacokinet*).
Federico, A. *et al.*, *Molecules*, 2017, **22**, 191:1-16 (*rev, activity*).
Bijak, M. *Molecules*, 2017, **22**, 1942:1-11 (*rev, activity*).
Pilkington, L.I. *et al.*, *J. Nat. Prod.*, 2018, **81**, 2630-2637 (*synth*).
Soleimani, V. *et al.*, *Phytother. Res.*, 2019, **33**, 1627-1638 (*rev, tox, safety*).

Gillessen, A. *et al.*, *Adv. Ther.*, 2020, **37**, 1279-1301 (*rev, activity*).
Petrášková, L *et al.*, *Foods*, 2020, **9**, 116-14 (*2,3-Dehydrosilybin, Silybins A,B*)

Silychristin S61

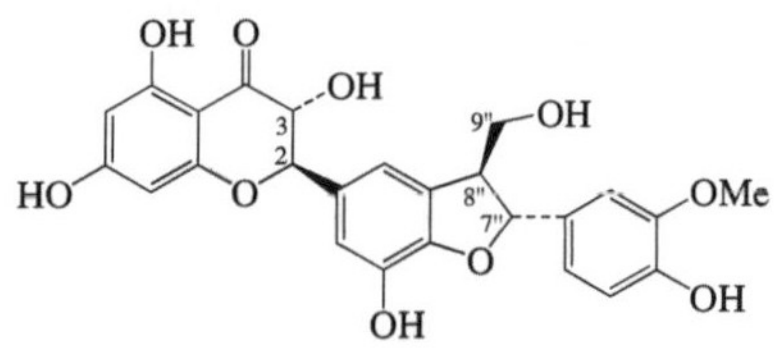

Absolute Configuration

***Silicristin**, INN, USAN. Silychristin A. Silymarin II*
[33889-69-9]
[119435-01-7] [125762-44-9]
$C_{25}H_{22}O_{10}$ 482.441
Flavonolignan. Two incorrect structs. prev. proposed. Related to Anthelminthicol A, A33 and Isohydnocarpin, I38. Constit. of fruit of *Silybum marianum*. Also from *Anastatica hierochuntica*, *Euonymus acanthocarpus* and *Inula wissmanniana*. Antioxidant. Inhibitor of horseradish peroxidases and lipoxygenase. Plant growth regulator. Inhibitor of murine B16 melanoma 4A5 cells and NS5B polymerase of hepatitis C virus JFH-1. Cryst. (H_2O). Mp 174 - 176°. $[\alpha]_D^{23}$ +81.4 (Py). $[\alpha]_D^{18}$ +66 (c, 0.2 in MeOH)

9″-Ac

9″-Acetylsilychristin
[1439933-97-7]
$C_{27}H_{24}O_{11}$ 524.478
Constit. of aerial parts of *Inula wissmanniana*. Exhibits weak antiinflammatory activity. Amorph. yellow powder. Mp 145 - 147°. $[\alpha]_D^{25}$ +34.6 (c, 0.05 in MeOH)

2,3-Didehydro

2,3-Dehydrosilychristin
[57499-41-9]
$C_{25}H_{20}O_{10}$ 480.425
Constit. of *Silybum marianum* and *Inula wissmanniana*. Mp 254 - 256°

3-Deoxy

3-Deoxysilychristin. ***Silyhermin***
[96238-88-9]
[96347-26-1]
$C_{25}H_{22}O_9$ 466.442
Constit. of the fruit of *Silybum marianum*. Cryst. (as penta-Ac). Mp 93 - 95° (penta-Ac). $[\alpha]_D$ +29.9 (c, 0.28 in $CHCl_3$) (penta-Ac). λ_{max} 262; 280 (sh); 305 (MeOH) (penta-Ac)

9″-Deoxy, 7″,8″-didehydro

Anhydrosilychristin
[2074606-99-6]
$C_{25}H_{20}O_9$ 464.426
Constit. of *Silybum marianum*. Pale yellow solid. $[\alpha]_D$ -4.3 (c, 0.007 in MeOH). May be an artifact. λ_{max} 290; 317 (MeOH)

9″-Deoxy, 2,3,7″,8″-tetradehydro

2,3-Dehydroanhydrosilychristin
[2087896-79-3]
$C_{25}H_{18}O_9$ 462.410
Red-brown solid. λ_{max} 267; 303; 425 (MeOH)

7″,8″-Diepimer

Silychristin B
[879325-58-3]
$C_{25}H_{22}O_{10}$ 482.441
Constit. of *Silybum marianum* and *Inula wissmanniana*. Exhibits antitumour props. on induced Epstein-Barr virus early antigen activation assay in Raji cells. Pale yellow gum. $[\alpha]_D^{19}$ +47 (c, 0.3 in MeOH). λ_{max} 289 (ε 10050) (MeOH).

Wagner, H. *et al.*, *Tetrahedron Lett.*, 1971, **12**, 1895-1899 (*Silychristin, struct*).
Bandopadhyay, M. *et al.*, *Indian J. Chem.*, 1972, **10**, 808-809 (*Silybum marianum constits*).
Greiml, A. *et al.*, *Experientia*, 1977, **33**, 1417-1418 (*activity*).
Wagner, H. *et al.*, *Tetrahedron Lett.*, 1978, **19**, 381-384 (*Silychristin, struct*).
Fiebrich, F. *et al.*, *Experientia*, 1979, **35**, 1548-1550 (*props*).
Zanarotti, A. *Heterocycles*, 1982, **19**, 1585-1586 (*stereochem, cd*).
Fiebig, M. *et al.*, *Planta Med.*, 1984, **51**, 310-313 (*Silyhermin*).
Mericli, A.H. *Planta Med.*, 1988, **54**, 44-45 (*Dehydrosilychristin*).
Tanaka, H. *et al.*, *Chem. Pharm. Bull.*, 1989, **37**, 1441-1445 (*synth*).
Delfini, M. *et al.*, *Spectrosc. Lett.*, 1990, **23**, 1007-1025 (*pmr, conformn*).
Kvasnicka, F. *et al.*, *J. Chromatogr., A*, 2003, **990**, 239-245 (*chromatog*).
Kim, N.-C. *et al.*, *Org. Biomol. Chem.*, 2003, **1**, 1684-1689 (*Silychristin, struct, cd*).
Smith, W.A. *et al.*, *Planta Med.*, 2005, **71**, 871-876 (*Silychristin B*).
Lee, J.I. *et al.*, *J. Chromatogr., A*, 2006, **1116**, 57-68 (*anal, detn*).
Lin, A.-S. *et al.*, *Pharm. Biol.*, 2007, **45**, 735-738 (*Silychristin B, activity*).
Nakashima, S. *et al.*, *Bioorg. Med. Chem.*, 2010, **18**, 2337-2345 (*Anastatica hierochuntica constit, activity*).
Polyak, S. *et al.*, *Proc. Natl. Acad. Sci. U.S.A.*, 2010, **107**, 5995-5999 (*activity*).
Zhu, J.X. *et al.*, *Arch. Pharmacal Res.*, 2012, **35**, 1739-1747 (*Euonymus acanthocarpus constit*).
Wang, C. *et al.*, *Arch. Pharmacal Res.*, 2013, **36**, 1516-1524 (*Silychristins A,B, 9″-Acetylsilychristin, 2,3-Dehydrosilychristin*).
Krenek, K. *et al.*, *Food Res. Int.*, 2014, **65**, 115-120 (*chromatog, sepn*).
Biedermann, D. *et al.*, *J. Nat. Prod.*, 2016, **79**, 3086-3092 (*Silychristins A,B, Silyhermin, Anhydrosilychristin, Dehydroanhydrosilychristin, activity*).
Viktorová, J. *et al.*, *Antioxidants*, 2019, **8**, 303:1-19 (*activity*).
Petrášková, l *et al.*, *Foods*, 2020, **9**, 11-14 (*2,3-Dehydrosilychristin, Silychristins A,B*)

Silydianin S62

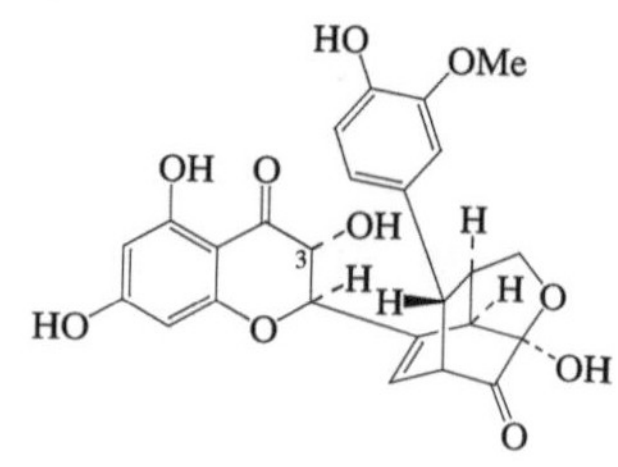

Absolute Configuration

***Silidianin,** INN. Silybum substance E5*
[29782-68-1]
$C_{25}H_{22}O_{10}$ 482.441
Flavonolignan. Constit. of *Silybum marianum*, *Cirsium japonicum* and *Inula wissmanniana*. Hepatoprotective agent. Antioxidant. Stimulates hair pigmentation and retards ageing of skin. Adipocyte differentiation inhibitor. Plant growth regulator. Mp 190 - 192°. $[\alpha]_D$ +256.4 (c, 0.004 in MeOH). $[\alpha]_D$ +218. λ_{max} 216 (log ε 5.4); 231 (sh); 289 (log ε 5.3) (MeOH)

3-Deoxy

Silymonin
[70815-31-5]
$C_{25}H_{22}O_9$ 466.442
Constit. of *Silybum marianum*. Stimulates hair pigmentation and retards ageing of skin. Cryst.. Mp 258 - 260°. $[\alpha]_D$ +127.

Janiak, B. *et al.*, *Planta Med.*, 1960, **8**, 71-84 (*Silybum substance E5*).
Abraham, D.J. *et al.*, *Tetrahedron Lett.*, 1970, **11**, 2675-2678 (*Silydianin*).
Hoelzl, J. *Z. Naturforsch., C*, 1974, **29**, 82-83 (*biosynth*).
Wagner, H. *et al.*, *Z. Naturforsch., B*, 1976, **31**, 876-884 (*Silydianin, cryst struct, abs config*).
Szilagi, I. *et al.*, *Planta Med.*, 1981, **43**, 121-127 (*Silymonin*).
Kvasnička, F. *et al.*, *J. Chromatogr., A*, 2003, **990**, 239-245 (*chromatog, anal*).
Kim, N.-C. *et al.*, *Org. Biomol. Chem.*, 2003, **1**, 1684-1689 (*Silydianin, struct, cd*).
US Pat., 2008, 2008 033 037 (*Silydianin, Silymonin, activity*).
Wang, C. *et al.*, *Arch. Pharmacal Res.*, 2013, **36**, 1516-1524 (*Inula wissmanniana constit*).
Park, H.-S. *et al.*, *J. Korean Soc. Appl. Biol. Chem.*, 2013, **56**, 709-713 (*Cirsium japonicum constit, activity*).

Biedermann, D. *et al.*, *Phytochem. Lett.*, 2019, **30**, 14-20 (*Silydianin, cryst struct*).
Petrásková, L. *et al.*, *Foods*, 2020, **9**, 116:1-14 (*anal, sepn*)

Simonin A S63

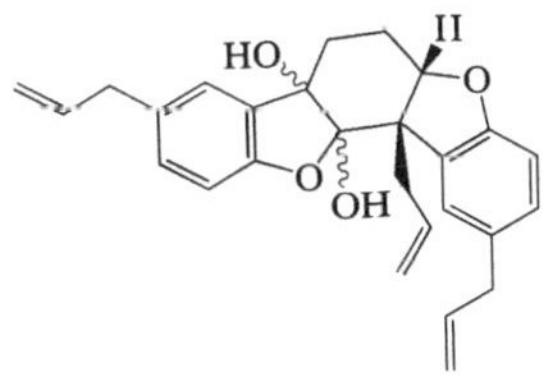

Relative Configuration

[1279018-93-7]
$C_{27}H_{28}O_4$ 416.517
Related to Fargenone A, F3. Constit. of *Illicium simonsii*. Shows antibacterial activity against oral microbial organisms. Gum. $[\alpha]^{26}_{D}$ +4.4 (c, 0.34 in MeOH). λ_{max} 223 (log ε 3.55); 241 (log ε 3.42); 288 (log ε 3.83) ($CHCl_3$).
Liu, J.-F. *et al.*, *Planta Med.*, 2010, **76**, 1464-1467 (*Simonin A, activity*)

Simonol A S64

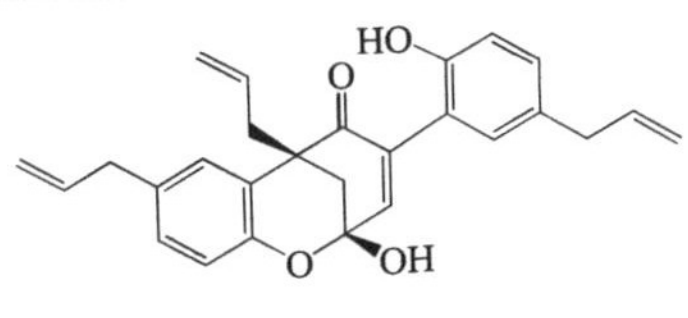

Absolute Configuration

[1535220-96-2]
$C_{27}H_{26}O_4$ 414.501
Sesquineolignan. Constit. of the fruit of *Illicium simonsii*. Oil. $[\alpha]^{20}_{D}$ +5.8 (c, 0.1 in MeOH). λ_{max} 199 (log ε 4.6); 287 (log ε 3.0) (MeOH).
Yin, P.-J. *et al.*, *Fitoterapia*, 2013, **88**, 31-37 (*Simonol A*)

Simonol B S65

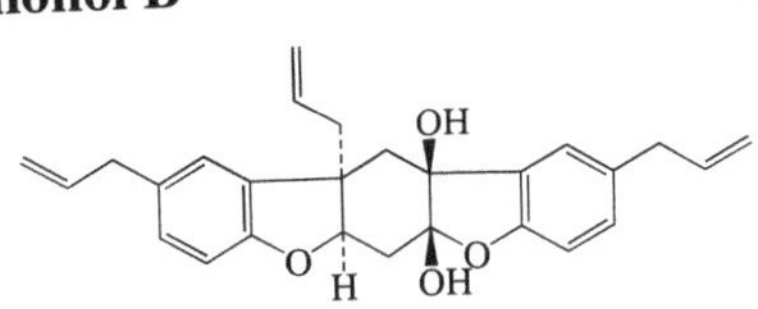

Absolute Configuration

[1537170-59-4]
$C_{27}H_{28}O_4$ 416.517
Sesquineolignan. Constit. of the fruit of *Illicium simonsii*. Oil. $[\alpha]^{23}_{D}$ +31.1 (c, 0.13 in MeOH). λ_{max} 207 (log ε 4.2); 225 (log ε 3.8); 285 (log ε 3.1) (MeOH).
Yin, P.-J. *et al.*, *Fitoterapia*, 2013, **88**, 31-37 (*Simonol B*)

Simonsienol A S66

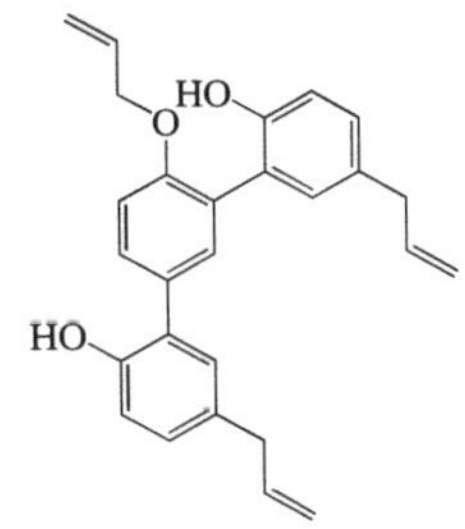

[1426534-52-2]
$C_{27}H_{26}O_3$ 398.502
Constit. of the stem bark of *Illicium simonsii*. Gum. $[\alpha]^{18}_{D}$ -10.3 (c, 0.001 in $CHCl_3$). λ_{max} 241 (log ε 3.84); 293 (log ε 3.52) ($CHCl_3$).
Dong, C.-F. *et al.*, *Nat. Prod. Bioprospect.*, 2012, **2**, 133-137 (*Simonsienol A*)

Simonsienol C S67

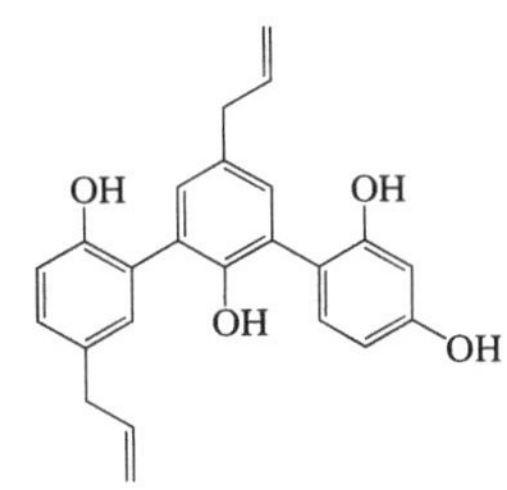

[1426534-54-4]
$C_{24}H_{22}O_4$ 374.436
Constit. of the stem bark of *Illicium simonsii*. Amorph. solid. $[\alpha]^{26}_{D}$ -2.9 (c, 0.003 in MeOH). λ_{max} 218 (log ε 3.95); 305 (log ε 3.26) (MeOH).
Dong, C.-F. *et al.*, *Nat. Prod. Bioprospect.*, 2012, **2**, 133-137 (*Simonsienol C*)

Simonsin A S68

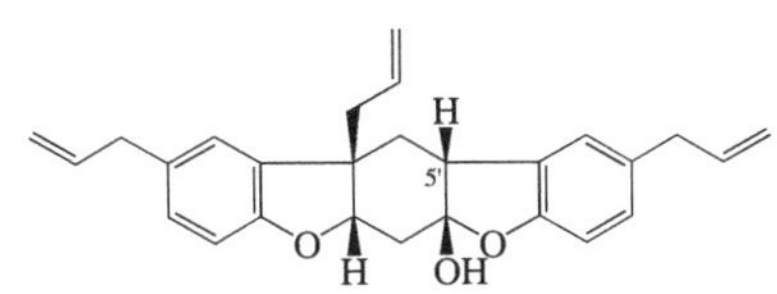

Relative Configuration

[1582261-56-0]
$C_{27}H_{28}O_3$ 400.518
Constit. of *Illicium simonsii*. Gum. $[\alpha]^{11}_{D}$ -15.1 (c, 0.4 in $CHCl_3$). λ_{max} 198 (log ε 2.56); 214 (log ε 2.56); 239 (log ε 2.99); 287 (log ε 3.11) ($CHCl_3$)

5′-Hydroxy

Simonsin B

[1582261-57-1]
$C_{27}H_{28}O_4$ 416.517
Constit. of *Illicium simonsii*. Gum. $[\alpha]^{12}_{D}$ +27 (c, 0.1 in $CHCl_3$). λ_{max} 198 (log ε 2.47); 210 (log ε 2.47); 215 (log ε 2.47); 239 (log ε 2.95); 288 (log ε 3.0) ($CHCl_3$).
Dong, C. *et al.*, *Planta Med.*, 2013, **79**, 338-347 (*Simonsins A,B*)

Simonsol A S69

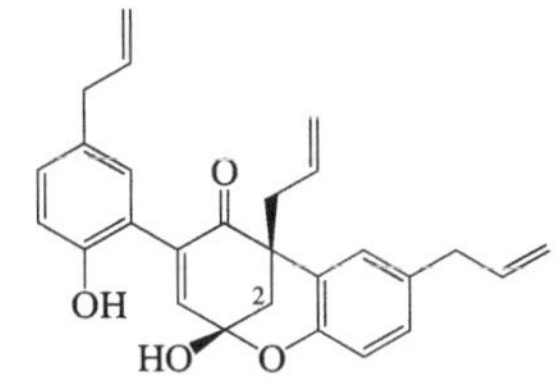

Relative Configuration

[1582261-51-5]
$C_{27}H_{26}O_4$ 414.501
Sesquineolignan. Constit. of *Illicium simonsii*. Yellow gum. $[\alpha]^{26}_{D}$ -15.5 (c, 0.2 in $CHCl_3$). λ_{max} 240 (log ε 3.13); 261 (log ε 3.01) ($CHCl_3$)

2β-Hydroxy

Simonsol B

[1582261-52-6]
$C_{27}H_{26}O_5$ 430.5
Constit. of *Illicium simonsii*. Gum. $[\alpha]^{26}_{D}$ -4.4 (c, 0.3 in $CHCl_3$). λ_{max} 240 (log ε 3.04); 268 (log ε 2.97) ($CHCl_3$).
Dong, C. *et al.*, *Planta Med.*, 2013, **79**, 338-347 (*Simonsols A,B*)

Simonsol C S70

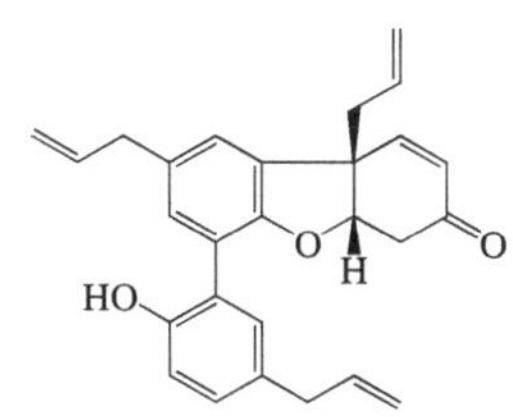

Relative Configuration

[1582261-53-7]
$C_{27}H_{26}O_3$ 398.502
Sesquineolignan. Constit. of *Illicium simonsii*. Amorph. solid. $[\alpha]^{18}_{D}$ -10.5 (c, 0.3 in $CHCl_3$). λ_{max} 240 (log ε 3.52); 302 (log ε 3.19) ($CHCl_3$).
Dong, C. *et al.*, *Planta Med.*, 2013, **79**, 338-347 (*Simonsol C*).
Nugent, J. *et al.*, *Org. Lett.*, 2016, **18**, 3798-3801 (*synth*)

Simonsol D S71

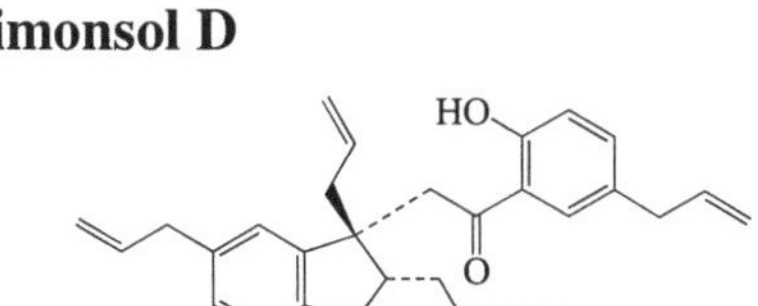

Relative Configuration

[1582261-54-8]

$C_{28}H_{30}O_5$ 446.543
Sesquineolignan. Constit. of *Illicium simonsii*. Gum. λ_{max} 199 (log ε 3.14); 238 (log ε 3.34); 261 (log ε 3.41); 284 (log ε 3.06); 345 (log ε 2.97) ($CHCl_3$).
Dong, C. *et al.*, *Planta Med.*, 2013, **79**, 338-347 (*Simonsol D*)

Simplexolin S72

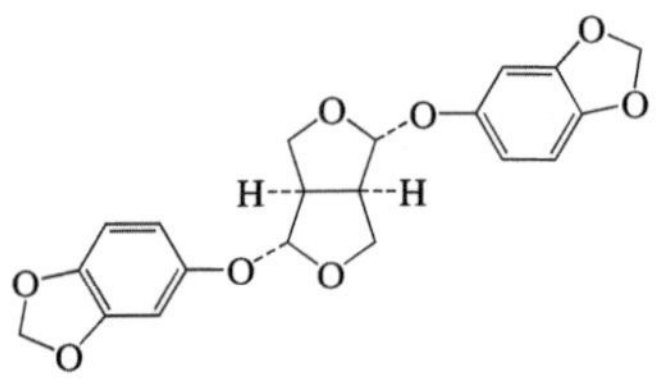

Relative Configuration

[71328-57-9]
$C_{20}H_{18}O_8$ 386.356
Isol. from whole plants of *Justicia orbiculata* and *Justicia simplex*. Mp 167 - 169°. $[\alpha]_D^{28}$ +254.8 ($CHCl_3$). λ_{max} 235 (log ε 4.12); 292 (log ε 4.04) (MeOH).
Ghosal, S. *et al.*, *Phytochemistry*, 1979, **18**, 503-505 (*Simplexolin*).
Babu, G.V.R. *et al.*, *Indian J. Chem., Sect. B*, 2001, **40**, 864-866 (*Justicia orbiculata constit*).
Jasemine, S. *et al.*, *Pharm. Biol.*, 2007, **45**, 274-277 (*isol, activity*)

Sinapine S73

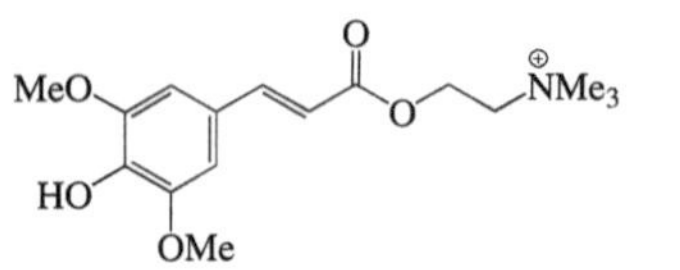

2-[[3-(4-Hydroxy-3,5-dimethoxyphenyl)-1-oxo-2-propenyl]oxy]-N,N,N-trimethylethanaminium(1+), CAS. *Choline 4-hydroxy-3,5-dimethoxycinnamate,* CAS. *O-Sinapoylcholine*
[18696-26-9]
[84123-21-7] (*Z*-form) [84123-22-8] (*E*-form)
$C_{16}H_{24}NO_5^{\oplus}$ 310.37 (ion)
Alkaloid from seeds of *Brassica nigra*, *Brassica alba* (*Sinapis alba*) and many other spp. in the Cruciferae. Also from the Chinese drug Ting Li (*Draba nemorosa*). Astringent antinutrient of practical importance, esp. in cattle feed, e.g. rape meal. Reported to have radioprotective effects in some cruciferous plants and in *Drosophila*

Iodide
[5655-06-1]
$C_{16}H_{24}INO_5$ 437.275
Mp 185 - 186° (178-179°)

Sulfate
[6509-38-2]
$C_{16}H_{25}NO_8S$ 391.435
Mp 186 - 188° (anhyd.). Mp 126.5 - 127.5° dec. (dihydrate)

O-β-D-Glucopyranoside
[21848-40-8]
[87206-57-3] [87206-59-5]
$C_{22}H_{34}NO_{10}^{\oplus}$ 472.511 (ion)
Alkaloid from the seeds of *Alliaria officinalis*. Semicryst. (as acetate salt). Obt. as a mix. of *E*- and *Z*-isomers. λ_{max} 315 (H_2O).

Gadamer, J. *Arch. Pharm. (Weinheim, Ger.)*, 1897, **235**, 44-80 (*Sinapis alba constit*).
Späth, E. *Monatsh. Chem.*, 1920, **41**, 271-285 (*synth*).
Kung, H.P. *et al.*, *J. Am. Chem. Soc.*, 1949, **71**, 1836-1837 (*Draba nemorosa constit*).
Schwartze, P. *Naturwissenschaften*, 1949, **36**, 88-89 (*Brassica constit*).
Tzagoloff, A. *Plant Physiol.*, 1963, **38**, 202-206 (*metab*).
Durkee, A.B. *et al.*, *J. Food Sci.*, 1975, **40**, 820-822 (*occur*).
Sosulski, F.W. *Dev. Food Proteins*, 1983, **2**, 109-132 (*rev*).
Larsen, L.M. *et al.*, *Phytochemistry*, 1983, **22**, 219-222 (*glucoside*).
Regenbrecht, J. *et al.*, *Phytochemistry*, 1985, **24**, 407-410 (*biochem*).
Mouterde, L.M.M. *et al.*, *J. Agric. Food Chem.*, 2020, **68**, 6998-7004 (*synth*)

Sinenside A S74

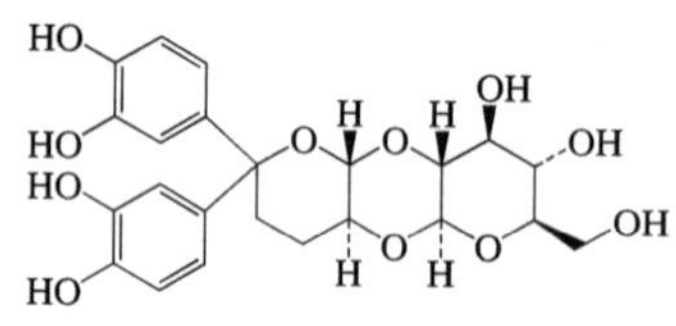

Absolute Configuration

[1363339-37-0]
$C_{23}H_{26}O_{11}$ 478.45
Constit. of the rhizomes of *Curculigo sinensis*. Exhibits strong radical scavenging props.. Amorph. powder. $[\alpha]_D^{21}$ -15.6 (c, 0.11 in MeOH). λ_{max} 205 (log ε 4.42); 258 (log ε 3.5); 285 (log ε 3.67) (MeOH).
Li, N. *et al.*, *Carbohydr. Res.*, 2012, **351**, 64-67 (*Sinenside A, activity*).
Vadhadiya, P.M. *et al.*, *Org. Lett.*, 2015, **17**, 1724-1727 (*synth*)

Sinenside B S75

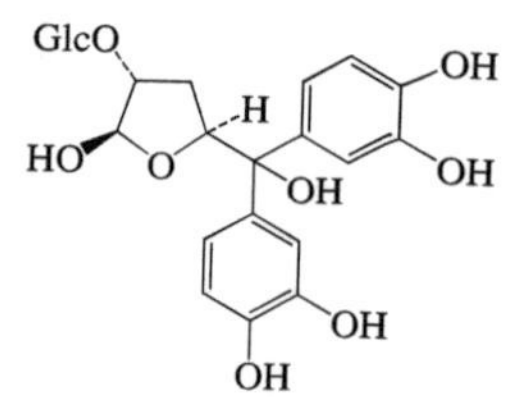

Absolute Configuration

[1363339-38-1]
$C_{23}H_{28}O_{13}$ 512.464
Rearranged norlignan. Constit. of the rhizomes of *Curculigo sinensis*. Strong radical scavenger. Amorph. powder. $[\alpha]_D^{24}$ -6.8 (c, 0.23 in MeOH). λ_{max} 206 (log ε 4.65); 283 (log ε 3.84) (MeOH).
Li, N. *et al.*, *Carbohydr. Res.*, 2012, **351**, 64-67 (*Sinenside B*)

Sinensigenin A S76

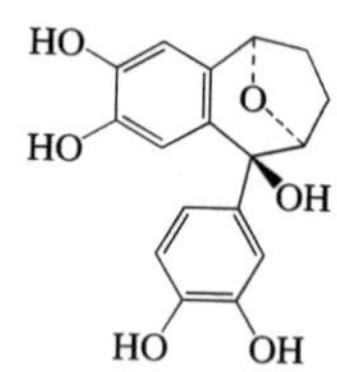

Relative Configuration

8,9′-Cyclo-7,8′-epoxy-9-nor-2,7′-lignan-3′,4,4′,5,7′-pentol. 8,9′-Cyclo-7,8′-epoxy-3′,4,4′,5,7′-pentahydroxy-9-nor-2,7′-neolignan
[1232141-03-5]
$C_{17}H_{16}O_6$ 316.309
Norlignan. Constit. of the rhizomes of *Curculigo sinensis*. Powder. $[\alpha]_D^{25}$ -75.7 (c, 0.12 in MeOH). λ_{max} 205 (log ε 4.22); 285 (log ε 3.43); 376 (log ε 2.84) (MeOH).
Li, N. *et al.*, *Helv. Chim. Acta*, 2010, **93**, 724-728 (*Sinensigenin A*).
Albury, A.M.M. *et al.*, *Tetrahedron Lett.*, 2015, **56**, 3057-3059 (*tetra-Me, synth*)

Sinensigenin C S77

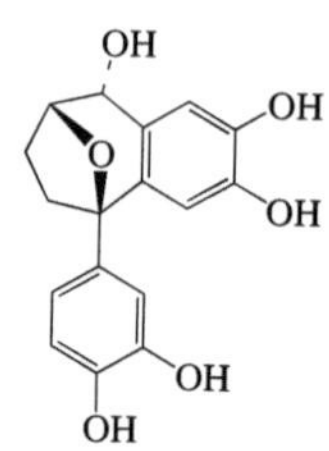

Relative Configuration

8,9′-Cyclo-7′,8-epoxy-9-nor-2,7′-lignan-3′,4,4′,5,7-pentol. 8,9′-Cyclo-7′,8-epoxy-3′,4,4′,5,7-pentahydroxy-9-nor-2,7′-neolignan
[1309355-43-8]
$C_{17}H_{16}O_6$ 316.309

Stereochem. unclear from the paper. Constit. of the rhizomes of *Curculigo capitulata*. Powder. $[\alpha]_D^{14}$ -12 (c, 0.2 in MeOH). λ_{max} 205 (log ε 4.55); 285 (log ε 3.66) (MeOH).

Wang, K.-J. *et al.*, *Fitoterapia*, 2010, **81**, 869-872 (*Sinensigenin C*)

Sophoratonin H S78

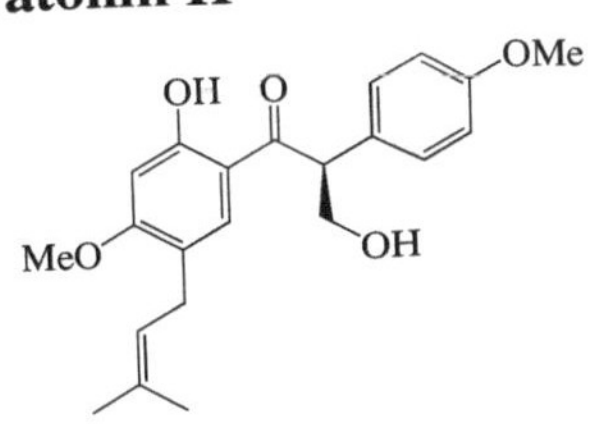

Absolute Configuration

3-Hydroxy-1-[2-hydroxy-4-methoxy-5-(3-methyl-2-butenyl)phenyl]-2-(4-methoxyphenyl)-1-propanone
[2303538-04-5]
$C_{22}H_{26}O_5$ 370.445
Constit. of the roots and rhizomes of *Sophora tonkinensis*. Exhibits antiinflammatory activity. Amorph. yellow solid. $[\alpha]_D^{20}$ +53.1 (c, 0.1 in MeOH). λ_{max} 234 (log ε 4.03); 277 (log ε 4.0); 330 (log ε 3.83) (MeOH).

Ahn, J. *et al.*, *J. Nat. Prod.*, 2019, **82**, 309-317 (*Sophoratonin H, activity*)

Sorbanin S79

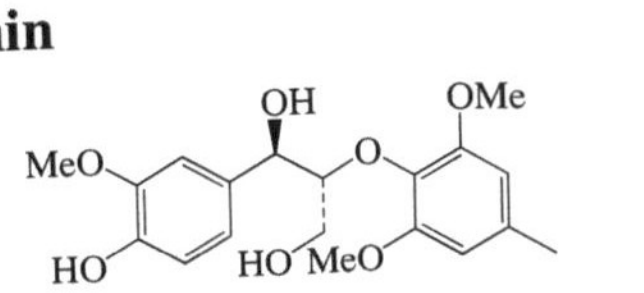

Relative Configuration

[1510846-63-5]
$C_{24}H_{26}O_7$ 426.465
Constit. of the stem wood of *Sorbus lanata*. Amorph. solid. λ_{max} 267 (MeOH).

Uddin, G. *et al.*, *Phytochem. Lett.*, 2013, **6**, 84-89 (*Sorbanin*)

Spathulatol S80

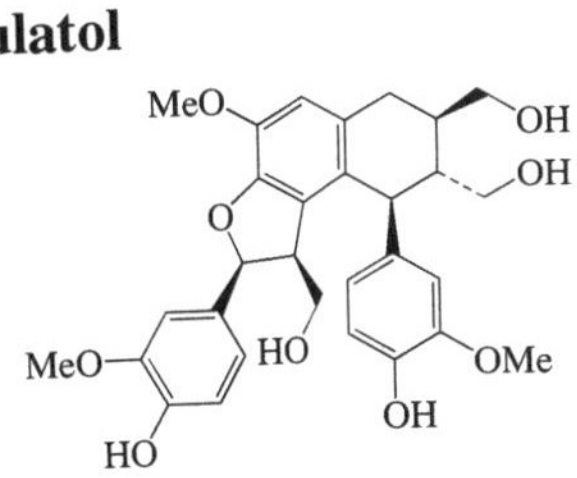

Relative Configuration

[1373888-27-7]
$C_{30}H_{34}O_9$ 538.593
Constit. of the fruit of *Illicium spathulatum*. Powder. Mp 114 - 116°. $[\alpha]_D$ -103 (c, 0.7 in MeOH). λ_{max} 283 (MeOH).

Lai, G. *et al.*, *Yunnan Zhiwu Yanjiu*, 2010, **32**, 281-284 (*Spathulatol*)

Stachysetin S81

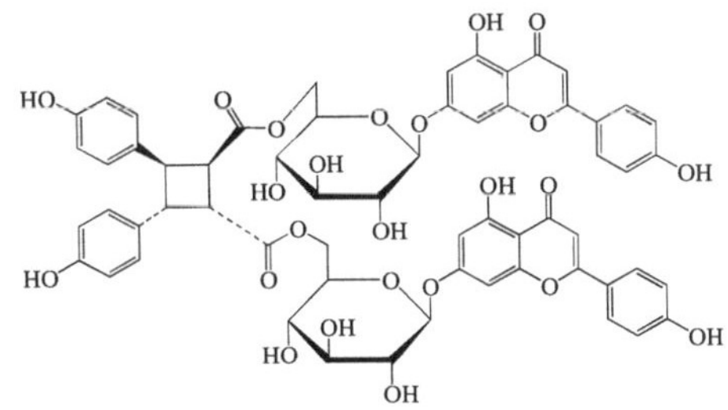

[171674-78-5]
$C_{60}H_{52}O_{24}$ 1157.052
Constit. of *Stachys aegyptiaca* and *Stachys lanata*. Off-white powder. λ_{max} 268; 328 (MeOH).

El-Ansari, M.A. *et al.*, *Phytochemistry*, 1995, **40**, 1543-1548 (*Stachysetin, struct*).
Murata, T. *et al.*, *J. Nat. Prod.*, 2008, **71**, 1768-1770 (*Stachys lanata constit*)

Steganoate A S82

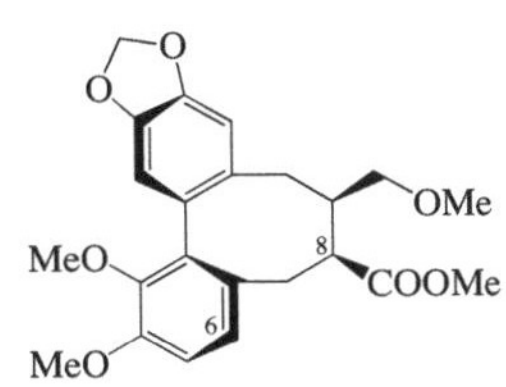

Absolute Configuration

[152645-86-8]
$C_{23}H_{26}O_7$ 414.454
Lignan numbering shown. Constit. of the stem bark of *Steganotaenia araliacea*. Cryst. (Et_2O). Mp 157 - 158°. $[\alpha]_D$ -63.5 (c, 0.9 in $CHCl_3$). λ_{max} 218 (log ε 4.37); 256 (log ε 3.86); 286 (log ε 3.56) (MeOH)

8-Epimer, 6-methoxy
Steganoate B
[152645-87-9]
$C_{24}H_{28}O_8$ 444.48
Constit. of the stem bark of *Steganotaenia araliacea*. Cytotoxic agent with activity against 11 different cell lines. Needles (Et_2O/petrol). Mp 122 - 123°. $[\alpha]_D$ -41.6 (c, 0.16 in $CHCl_3$). λ_{max} 222 (log ε 4.41); 259 (log ε 3.76); 290 (log ε 3.91) (MeOH).

Wickramaratne, D.B.M. *et al.*, *J. Nat. Prod.*, 1993, **56**, 2083-2090 (*Steganoates A,B*)

Steganol S83

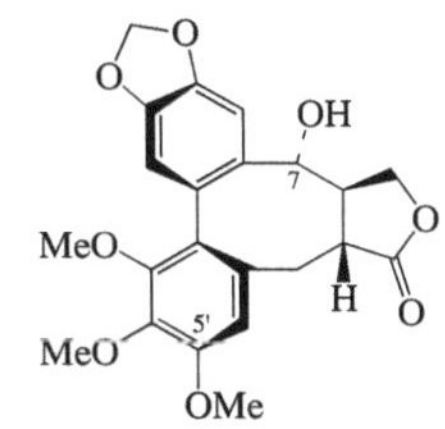

Absolute Configuration

[41451-71-2]
[58800-46-7] ((±)-form)
$C_{22}H_{22}O_8$ 414.41
Abs. config. revised in 1980 but the wrong assignment has been persistent in the lit.. Isol. from *Steganotaenia araliacea*. Antineoplastic agent. Cryst.. $[\alpha]_D^{23}$ -163 (c, 0.87 in $CHCl_3$). Log P -0.51 (calc). λ_{max} 255 (ε 11200); 287 (ε 5600) (EtOH) (Derep)
▸ DF8079800

Ac
Steganacin
[41451-68-7]
[58894-52-3] ((±)-form)
Isol. from *Steganotaenia araliacea*. Antineoplastic agent, disruptor of microtubule assembly. Cryst.. $[\alpha]_D^{23}$ -114 (c, 0.74 in $CHCl_3$). Log P 0.35 (calc). λ_{max} 255 (ε 10700); 287 (ε 5450) (EtOH) (Derep)
▸ DF8079500

Angeloyl
Steganangin
[41451-69-8]
$C_{27}H_{28}O_9$ 496.512
Isol. from *Steganotaenia araliacea*. Antineoplastic agent. Cryst.. Mp 142.5 - 143°. $[\alpha]_D^{23}$ -113 (c, 0.72 in $CHCl_3$). Log P 1.79 (uncertain value) (calc). λ_{max} 256 (ε 10600); 285 (ε 5180) (EtOH) (Derep)
▸ EM9253400

Ketone
Steganone
[41451-70-1]
[58800-45-6] ((±)-form)
Isol. from *Steganotaenia araliacea*. Antineoplastic agent. Mp 155 - 156°. $[\alpha]_D^{23}$ -202 (c, 0.76 in $CHCl_3$). Log P 0.81 (calc). λ_{max} 238 (ε 27600); 276 (ε 9200); 317 (ε 5710) (EtOH) (Derep)
▸ DF8079000

5′-Demethoxy, 7-deoxy
10-Demethoxystegane
[385764-61-4]
$C_{21}H_{20}O_6$ 368.385

Constit. of *Steganotaenia araliacea*. Solid. $[\alpha]_D^{25}$ -67.5 (c, 0.04 in MeOH)

7-Epimer, angeloyl

Episteganangin

[152698-42-5]

$C_{27}H_{28}O_9$ 496.512

Constit. of *Steganotaenia araliacea*. Amorph. solid. $[\alpha]_D$ -143.6 (c, 0.66 in $CHCl_3$). λ_{max} 215 (log ε 4.23); 255 (log ε 3.51); 291 (log ε 3.28) (MeOH).

Kupchan, S.M. *et al.*, *J. Am. Chem. Soc.*, 1973, **95**, 1335-1336 (*Steganacin, Steganangin*).
Kende, A.S. *et al.*, *J. Am. Chem. Soc.*, 1976, **98**, 267-268 (*Steganacin, synth*).
Wang, R.W. *et al.*, *Cancer Res.*, 1977, **37**, 3071-3079 (*Steganacin, activity*).
Becker, D. *et al.*, *J. Chem. Soc., Perkin Trans. 1*, 1977, 1674-1671 (*Steganacin, synth*).
Krow, G.R. *et al.*, *J. Org. Chem.*, 1978, **43**, 3950-3953 (*Steganone, synth*).
Brown, E. *et al.*, *Tetrahedron Lett.*, 1979, **20**, 733-736 (*Steganone, synth*).
Ziegler, F. *et al.*, *J. Am. Chem. Soc.*, 1980, **102**, 790-798 (*Steganacin, synth*).
Robin, J.P. *et al.*, *Tetrahedron Lett.*, 1980, **21**, 2709-2712 (*abs config*).
Larson, E.R. *et al.*, *J. Chem. Soc., Perkin Trans. 1*, 1982, 521-525 (*Steganone, synth*).
Magnus, P. *et al.*, *J. Chem. Soc., Chem. Commun.*, 1984, 1179-1180 (*Steganacin, Steganone, synth*).
Tomioka, K. *et al.*, *Tetrahedron*, 1984, **40**, 1303-1312 (*synth, abs config*).
Ishiguro, T. *et al.*, *Chem. Pharm. Bull.*, 1985, **33**, 609-617 (*Steganacin, synth*).
Magnus, P. *et al.*, *J. Am. Chem. Soc.*, 1985, **107**, 4984-4988 (*Steganone, synth*).
Aidhen, I.S. *et al.*, *Indian J. Chem., Sect. B*, 1993, **32**, 211 (*Stegnanone, synth*).
Wickramaratne, D.B.M. *et al.*, *J. Nat. Prod.*, 1993, **56**, 2083-2090 (*Episteganangin*).
Sackett, D.L. *Pharmacol. Ther.*, 1993, **59**, 163-228 (*Steganacin, rev, activity*).
Monovich, L.G. *et al.*, *J. Am. Chem. Soc.*, 2000, **122**, 52-57 (*Steganone, synth*).
Kamikawa, K. *et al.*, *Tetrahedron*, 2000, **56**, 2325-2337 (*Steganone, synth*).
Meragelman, K.M. *et al.*, *J. Nat. Prod.*, 2001, **64**, 1480-1482 (*Steganone, Demethoxystegane*).
Takeda, S. *et al.*, *Tetrahedron*, 2007, **63**, 396-408 (*Steganone, synth*).
Yalcouye, B. *et al.*, *Eur. J. Org. Chem.*, 2014, 6285-6294 (*Steganone, synth*).
Dherbassy, Q. *et al.*, *Tetrahedron*, 2016, **72**, 5238-5245 (*Steganone, synth*)

Steganolide A S84

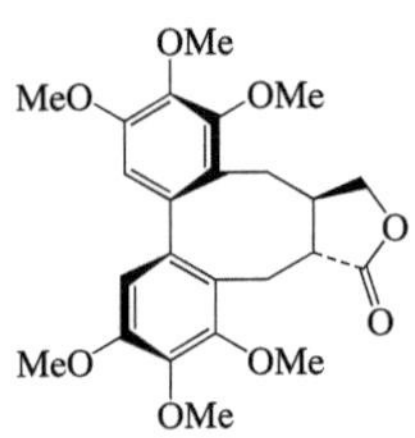

2,2′-Cyclo-4,4′,5,5′,6,6′-hexamethoxy-8,8′-lignan-9′,9-olide. 4,4′,5,5′,6,6′-Hexamethoxy-2,2′-cyclolignan-9′,9-olide

[105544-82-9]

$C_{24}H_{28}O_8$ 444.48

Constit. of *Steganotaenia araliacea*. Amorph.. Sol. MeOH, EtOH. $[\alpha]_D^{20}$ +67.9 (c, 1.65 in $CHCl_3$).

Taafrout, M. *et al.*, *Tetrahedron Lett.*, 1986, **27**, 1781 (*Steganolide A, isol, synth*)

Steganolide B S85

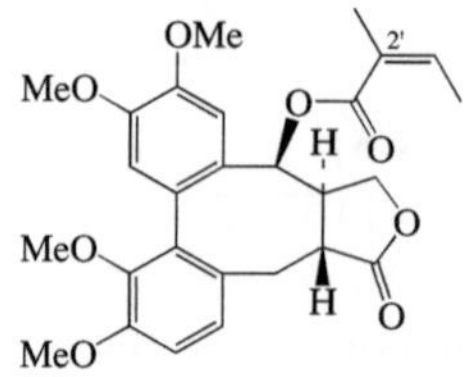

[103963-37-7]

$C_{27}H_{30}O_8$ 482.529

Constit. of *Steganotaenia araliacea*. Amorph.. $[\alpha]_D^{20}$ -92 (c, 2.26 in $CHCl_3$)

2′E-Isomer

Steganolide C

[104013-16-3]

$C_{27}H_{30}O_8$ 482.529

Constit. of *Steganotaenia araliacea*. Amorph.. $[\alpha]_D^{20}$ -150 (c, 0.07 in $CHCl_3$).

Robin, J.P. *et al.*, *Tetrahedron Lett.*, 1986, **27**, 2871-2874 (*Steganolides B,C*)

Stellerachama A S86

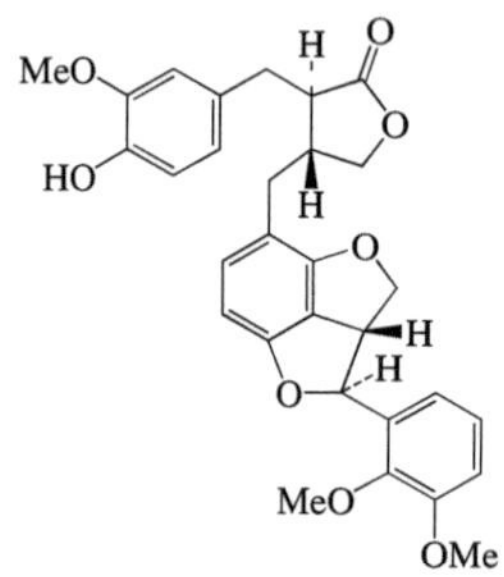

Relative Configuration

[1402543-75-2]

$C_{30}H_{30}O_8$ 518.562

Constit. of the roots of *Stellera chamaejasme*. Gum. λ_{max} 230 (log ε 4.53); 282 (log ε 4.1) (MeOH).

Liu, L.-P. *et al.*, *Chem. Nat. Compd. (Engl. Transl.)*, 2012, **48**, 559-561 (*Stellerachama A*)

Stelleralignan S87

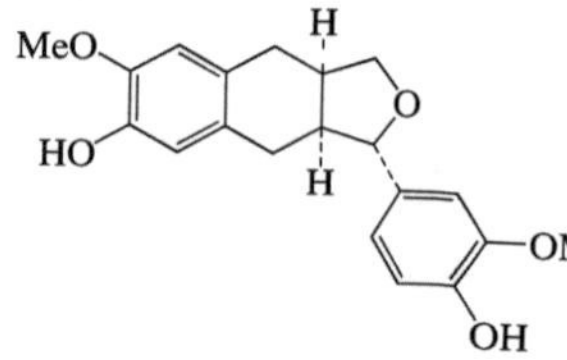

Relative Configuration

1,3,3a,4,9,9a-Hexahydro-3-(4-hydroxy-3-methoxyphenyl)-7-methoxynaphtho[2,3-c]furan-6-ol. 2,9′-Cyclo-7′,9-epoxy-3′,5-dimethoxy-8,8′-lignan-4,4′-diol. 7′,9-Epoxy-4,4′-dihydroxy-3′,5-dimethoxy-2,9′-cyclolignan

[1621377-29-4]

$C_{20}H_{22}O_5$ 342.391

Unusual type of lignan, apparently a carbocyclised 7′,9-epoxyfuranoid. Constit. of roots of *Stellera chamaejasme*. Exhibits inhibitory activity towards Topoisomerase II. Oil.

Liu, L.-P. *et al.*, *Bioorg. Med. Chem.*, 2014, **22**, 4198-4203 (*Stelleralignan, activity*)

Stenopalustroside E S88

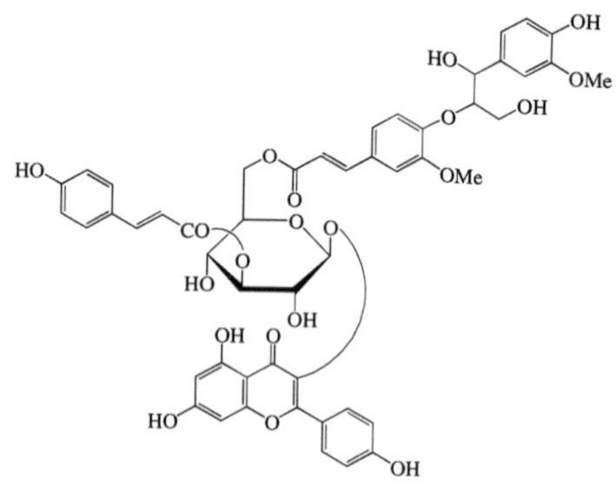

[218614-66-5]

$C_{50}H_{46}O_{20}$ 966.898

Flavonolignan. Constit. of the leaves of *Stenochlaena palustris*. Amorph. yellow powder. $[\alpha]_D^{20}$ -69 (c, 0.47 in MeOH).

Liu, H. *et al.*, *J. Nat. Prod.*, 1999, **62**, 70-75 (*Stenopalustroside E*)

Strangusin B S89

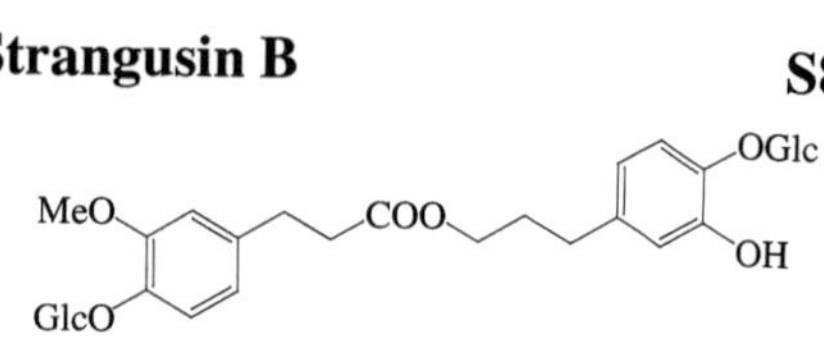

[432551-46-7]

$C_{31}H_{42}O_{16}$ 670.661

Constit. of *Serratula strangulata*. Gum.

Dai, J.Q. *et al.*, *Chin. Chem. Lett.*, 2002, **13**, 143-146 (*Strangusin B*)

Strebluslignanol F S90

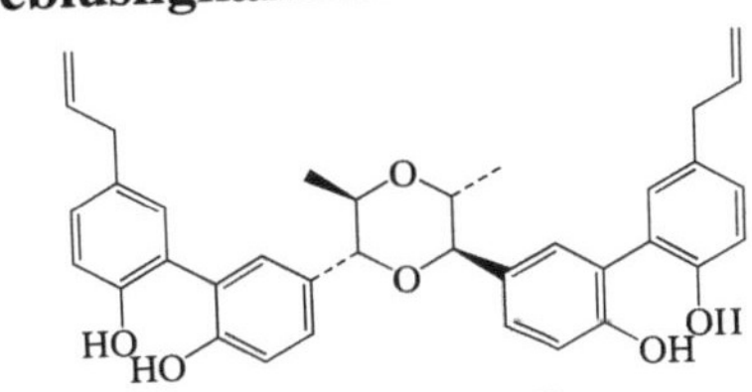

Relative Configuration

[1427279-39-7]

$C_{36}H_{36}O_6$ 564.678

Isol. from the roots of *Streblus asper*. Shows antiviral activity against hepatitis B. Powder (Me_2CO). Mp 248 - 249°. $[\alpha]_D^{25}$+6.7 (c, 0.1 in MeOH). λ_{max} 210 (log ε 4.68); 259 (log ε 2.78); 292 (log ε 2.89) (MeOH).

Li, J. *et al.*, *Bioorg. Med. Chem. Lett.*, 2013, **23**, 2238-2244 (*Strebluslignanol F*)

Strebluslignanol G S91

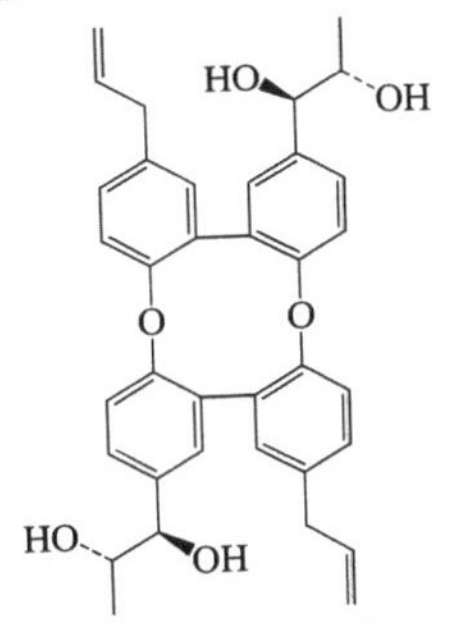

Absolute Configuration

[1427279-40-0]

$C_{36}H_{36}O_6$ 564.678

Isol. from the roots of *Streblus asper*. Shows antiviral activity against hepatitis B. Yellow solid (Me_2CO). Mp 215.5 - 216.5°. $[\alpha]_D^{25}$+22.2 (c, 0.1 in MeOH). λ_{max} 210 (log ε 4.75); 260 (log ε 2.82); 296 (log ε 2.94) (MeOH).

Li, J. *et al.*, *Bioorg. Med. Chem. Lett.*, 2013, **23**, 2238-2244 (*Strebluslignanol G*).

Li, C. *et al.*, *Rapid Commun. Mass Spectrom.*, 2014, **28**, 2363-2370 (*ms*)

Strebluslignanol H S92

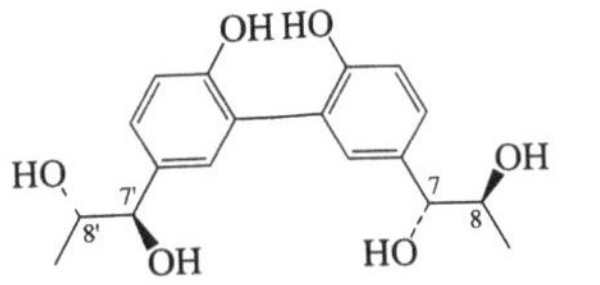

Absolute Configuration

3,3'-Lignan-4,4',7,7',8,8'-hexol. 4,4',7,7',8,8'-Hexahydroxy-3,3'-neolignan. ***Streblusol H***

[1426814-62-1]

$C_{18}H_{22}O_6$ 334.368

Lignan numbering shown. Constit. of the roots of *Streblus asper*. Powder. $[\alpha]_D^{25}$+25.1 (c, 1.5 in MeOH). λ_{max} 210 (log ε 4.72); 252 (log ε 3.36); 289 (log ε 1.39) (MeOH).

Chen, H. *et al.*, *Fitoterapia*, 2012, **83**, 643-649 (*Strebluslignanol H*)

Streblusol A S93

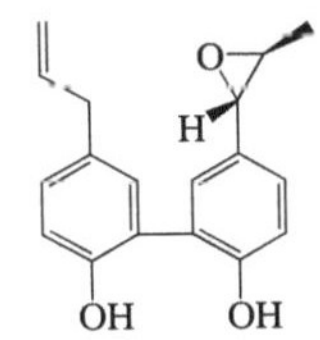

Absolute Configuration

5-(3-Methyl-2-oxiranyl)-5'-(2-propen-1-yl)-[1,1'-biphenyl]-2,2'-diol, CAS. *7,8-Epoxy-3,3'-lign-8'-ene-4,4'-diol. 7,8-Epoxy-4,4'-dihydroxy-3,3'-neolign-8'-ene*

[1399052-31-3]

$C_{18}H_{18}O_3$ 282.339

Constit. of the stem bark of *Streblus asper*. Displays anti-hepatitis B virus activity. Yellow oil. $[\alpha]_D^{25}$+45.2 (c, 2.2 in MeOH). λ_{max} 208 (log ε 4.54); 254 (log ε 2.65); 289 (log ε 2.33) (MeOH).

Li, J. *et al.*, *Phytochemistry*, 2012, **82**, 100-109 (*Streblusol A, activity*)

Streblusol D S94

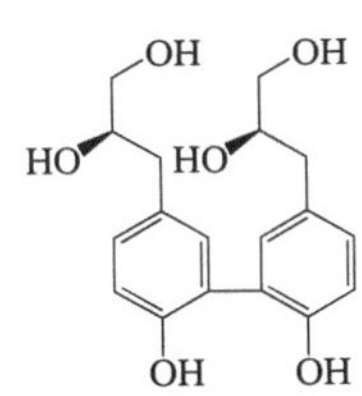

Absolute Configuration

5,5'-Bis(2,3-dihydroxypropyl)-[1,1'-biphenyl]-2,2'-diol, CAS. *3,3'-Lignan-4,4',8,8',9,9'-hexol. 4,4',8,8',9,9'-Hexahydroxy-3,3'-neolignan*

[1399052-36-8]

$C_{18}H_{22}O_6$ 334.368

Constit. of the stem bark of *Streblus asper*. Powder (MeOH). Mp 135 - 137°. $[\alpha]_D^{25}$+1.3 (c, 2.0 in MeOH). λ_{max} 209 (log ε 4.65); 254 (log ε 2.60); 290 (log ε 2.45) (MeOH).

Li, J. *et al.*, *Phytochemistry*, 2012, **82**, 100-109 (*Streblusol D*).

Li, C. *et al.*, *Rapid Commun. Mass Spectrom.*, 2014, **28**, 2363-2370 (*ms*)

Streblusol E S95

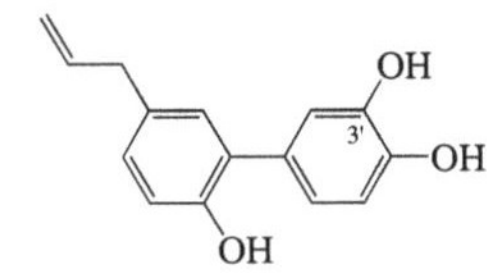

4-[5-(2-Propenyl)-2-hydroxyphenyl]-1,2-benzenediol. 5-Allyl-2,3',4'-trihydroxybiphenyl. 2-(3,4-Dihydroxyphenyl)-4-(2-propenyl)phenol. ***Strebluslignanphenol***

[1399052-37-9]

$C_{15}H_{14}O_3$ 242.274

Constit. of the stem bark of *Streblus asper*. Displays activity against hepatitis B virus. Purple oil. $[\alpha]_D^{25}$+3 (c, 1.5 in MeOH). λ_{max} 208 (log ε 4.54); 252 (log ε 2.05); 298 (log ε 1.83) (MeOH)

3'-Me ether

2-(4-Hydroxy-3-methoxyphenyl)-4-(2-propenyl)phenol. ***Houpulin M***

[1884599-75-0]

$C_{16}H_{16}O_3$ 256.301

Constit. of the roots of *Magnolia officinalis*. Syrup. λ_{max} 256 (log ε 3.85); 295 (log ε 3.79) (MeOH).

Li, J. *et al.*, *Phytochemistry*, 2012, **82**, 100-109 (*Streblusol E, activity*).

Li, C. *et al.*, *Rapid Commun. Mass Spectrom.*, 2014, **28**, 2363-2370 (*ms*).

Shih, H.-C. *et al.*, *Bioorg. Med. Chem.*, 2016, **24**, 1439-1445 (*Houpulin M*)

Striatoside A S96

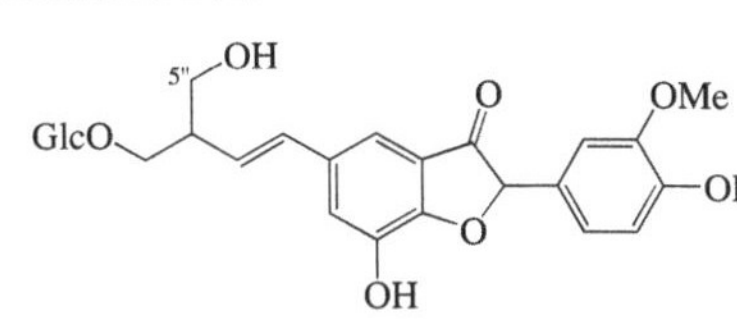

[149182-76-3]

$C_{26}H_{30}O_{12}$ 534.514

Constit. of *Pedicularis striata*. $[\alpha]_D^{15}$ +78.5 (c, 0.82 in MeOH)

5"-Me ether

Striatoside B

[149203-84-9]

$C_{27}H_{32}O_{12}$ 548.541

Constit. of *Pedicularis striata*.

Liu, Z. *et al.*, *Gaodeng Xuexiao Huaxue Xuebao*, 1993, **14**, 56-57 (*Striatoside A*)

Subaveniumin A S97

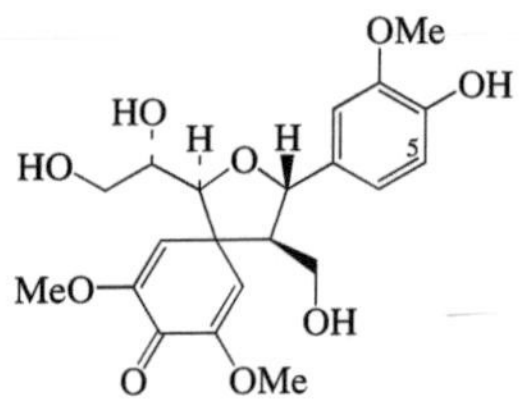

(+)-*form*

$C_{21}H_{26}O_9$ 422.43

(+)-form [1798799-55-9]

Constit. of the bark of *Cinnamomum subavenium*. Exhibits moderate inhibition against NO prodn.. $[\alpha]_D^{20}$ +62 (c, 0.12 in MeOH). Isol. as a racemate. Resolved by chiral hplc

5-Methoxy

(+)-Subaveniumin B

[1798799-57-1]

$C_{22}H_{28}O_{10}$ 452.456

Constit. of the bark of *Cinnamomum subavenium*. Exhibits moderate inhibition against NO prodn.. Powder. $[\alpha]_D^{20}$ +63 (c, 0.1 in MeOH). Isol. as a racemate. Resolved by chiral hplc

(-)-form [1798799-56-0]

Constit. of the bark of *Cinnamomum subavenium*. Exhibits moderate inhibition against NO prodn.. $[\alpha]_D^{20}$ -62 (c, 0.12 in MeOH). Isol. as a racemate. Resolved by chiral hplc

5-Methoxy

(-)-Subaveniumin B

[1798804-00-8]

$C_{22}H_{28}O_{10}$ 452.456

Constit. of the bark of *Cinnamomum subavenium*. Exhibits moderate inhibition against NO prodn.. Powder. $[\alpha]_D^{20}$ -63 (c, 0.1 in MeOH). Isol. as a racemate. Resolved by chiral hplc

(±)-form [1798799-53-7]

Constit. of the bark of *Cinnamomum subavenium*. Needles (MeOH). Mp 215.2 - 216.8°. λ_{max} 205 (log ε 4.42); 283 (log ε 4.0) (MeOH)

5-Methoxy

(±)-Subaveniumin B

[1798799-54-8]

$C_{22}H_{28}O_{10}$ 452.456

Constit. of the bark of *Cinnamomum subavenium*. Powder. Partial racemate; (+)-form predominates.

Lai, Y. *et al.*, *J. Nat. Prod.*, 2015, **78**, 1740-1744 (*Subaveniumins A,B, activity, cryst struct*)

Sucrose 1′,4′-(4,4′-dihydroxy-3,3′-dimethoxy-β-truxinate) S98

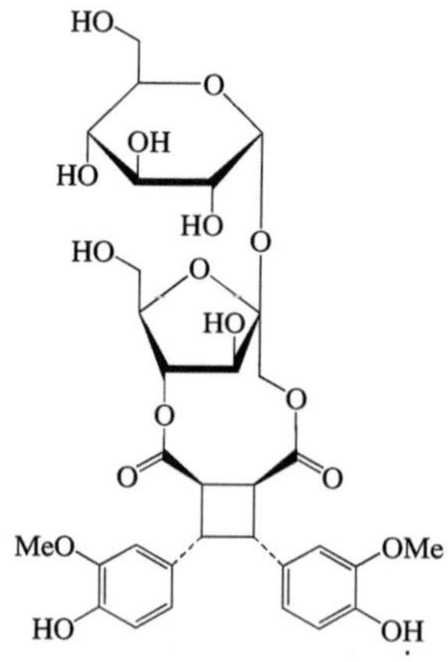

[349548-61-4]

$C_{32}H_{38}O_{17}$ 694.639

Constit. of oat grains, *Avena sativa*. Off-white solid. λ_{max} 208; 283 (MeOH).

Dimberg, L.H. *et al.*, *Phytochemistry*, 2001, **56**, 843-847 (*Avena sativa constit*)

Syripinnalignin A S99

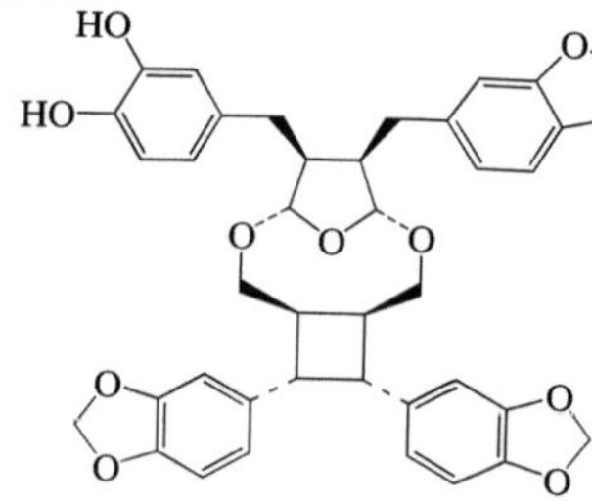

Relative Configuration

[1401688-14-9]

$C_{39}H_{36}O_{11}$ 680.706

Constit. of the stems of *Syringa pinnatifolia* var. *alashanensis*. Needles. Mp 256 - 257°. $[\alpha]_D^{25}$ +14.5 (c, 0.1 in MeOH). λ_{max} 219 (log ε 4.64); 267 (log ε 3.98); 314 (log ε 3.1) (MeOH)

Methylene ether

Syripinnalignin B

[1401688-15-0]

$C_{40}H_{36}O_{11}$ 692.717

Constit. of the stems of *Syringa pinnatifolia* var. *alashanensis*. Needles. Mp 256 - 257°. $[\alpha]_D^{25}$ +13.2 (c, 0.1 in MeOH). λ_{max} 219 (log ε 4.33); 267 (log ε 4.01); 314 (log ε 3.32) (MeOH).

Wang, Q.-H. *et al.*, *Helv. Chim. Acta*, 2012, **95**, 1680-1685 (*Syripinnalignins A,B*)

Tadehaginoside A T1

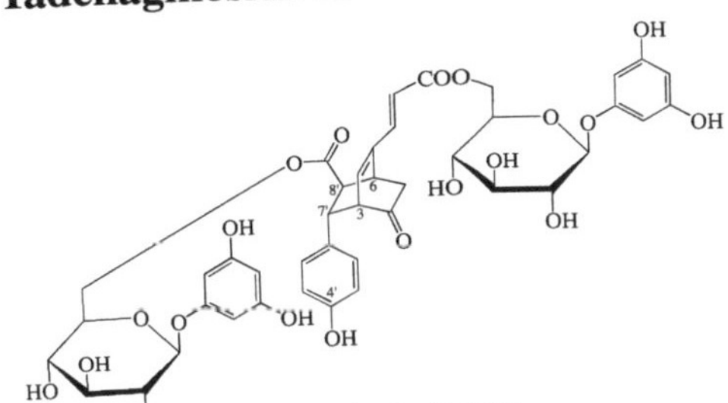

Absolute Configuration

[1912407-76-1]

$C_{42}H_{44}O_{20}$ 868.794

Constit. of the aerial parts of *Tadehagi triquetrum*. Amorph. powder (MeOH). $[\alpha]_D^{25}$ -28 (c, 0.1 in MeOH). λ_{max} 270 (log ε 2.81); 307 (log ε 0.15) (MeOH)

3,6,7′,8′-Tetraepimer

Tadehaginoside B

[1912407-77-2]

$C_{42}H_{44}O_{20}$ 868.794

Constit. of the aerial parts of *Tadehagi triquetrum*. Amorph. powder (MeOH). $[\alpha]_D^{25}$ -30 (c, 0.1 in MeOH). λ_{max} 269 (log ε 2.81); 308 (log ε 0.15) (MeOH).

Zhang, X. *et al.*, *J. Nat. Prod.*, 2016, **79**, 1249-1258 (*Tadehaginosides A,B*)

Tarennone T2

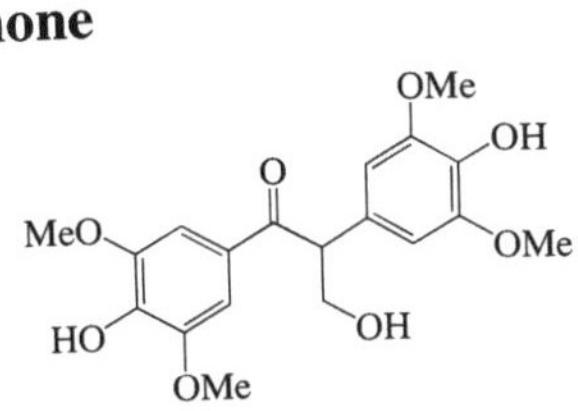

3-Hydroxy-1,2-bis(4-hydroxy-3,5-dimethoxyphenyl)-1-propanone. 4,4′,8′-Trihydroxy-3,3′,5,5′-tetramethoxy-8,9,9′-trinor-7,7′-lignan-7-one

[1006028-11-0]

$C_{19}H_{22}O_8$ 378.377

Similar to Evofolin B, E289. Constit. of *Tarenna attenuata*. Amorph. powder. $[\alpha]_D^{27}$ +4.7 (c, 0.43 in MeOH). λ_{max} 207 (log ε 4.55); 306 (log ε 3.87) (MeOH).

Yang, X.-W. *et al.*, *Planta Med.*, 2007, **73**, 496-498 (*Tarennone*)

Tatanan A T3

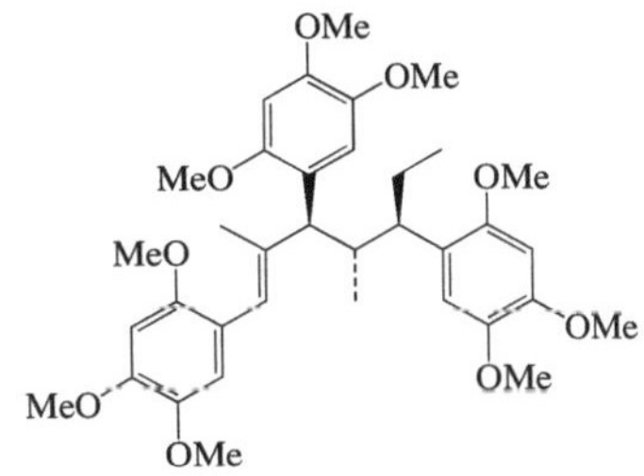

Absolute Configuration

2,2′,2″,4,4′,4″,5,5′,5″-Nonamethoxy-7,8′:7′,8″-sesquilign-7″-ene

[1280155-75-0]

$C_{36}H_{48}O_9$ 624.771

Constit. of the rhizomes of *Acorus tatarinowii*. Increases glucokinase enzymic activity. Cryst. (MeOH). Mp 125 - 127°. $[\alpha]_D^{20}$ +10 (c, 0.1 in MeOH). λ_{max} 204 (log ε 4.94); 232 (sh); 259 (log ε 4.21); 296 (log ε 4.32) (MeOH).

Ni, G. *et al.*, *J. Org. Chem.*, 2011, **76**, 2056-2061 (*Tatanan A, cryst struct*).

Xiao, Q. *et al.*, *Nat. Chem.*, 2013, **5**, 410-416 (*synth*).

Noble, A. *et al.*, *Angew. Chem., Int. Ed.*, 2016, **55**, 15920-15924 (*synth*)

Tatanan B T4

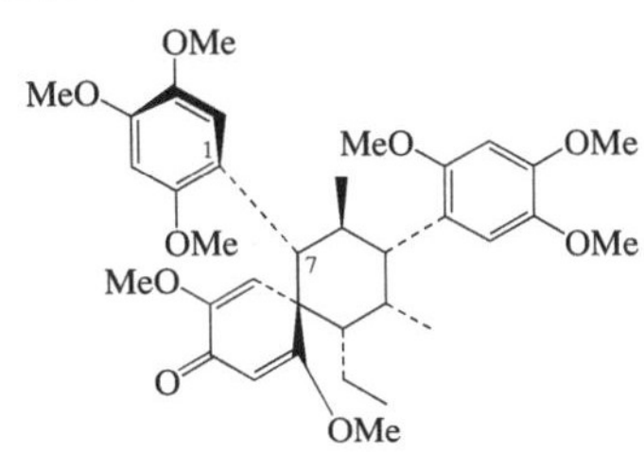

Relative Configuration

[1280155-77-2]

$C_{35}H_{46}O_9$ 610.744

Constit. of the rhizomes of *Acorus tatarinowii*. Increases glucokinase enzymic activity. Needles. Mp 216 - 218°. Slowly converts to Tatanan C on standing in soln.. λ_{max} 203 (log ε 5.08); 232 (log ε 4.44); 258 (log ε 4.39); 289 (log ε 4.26) (MeOH)

Atropisomer

Tatanan C

[1280155-79-4]

$C_{35}H_{46}O_9$ 610.744

Constit. of the rhizomes of *Acorus tatarinowii*. Increases glucokinase enzymic activity. Needles. Mp 166 - 168°. Isomer of restricted rotation about the C-1/C-7 bond. λ_{max} 205 (log ε 5.01); 230 (log ε 4.39); 260 (log ε 4.32); 289 (log ε 4.18) (MeOH).

Ni, G. *et al.*, *J. Org. Chem.*, 2011, **76**, 2056-2061 (*Tatanans B,C, cryst struct, activity*).

Xiao, Q. *et al.*, *Nat. Chem.*, 2013, **5**, 410-416 (*synth*)

Tatarinan T T5

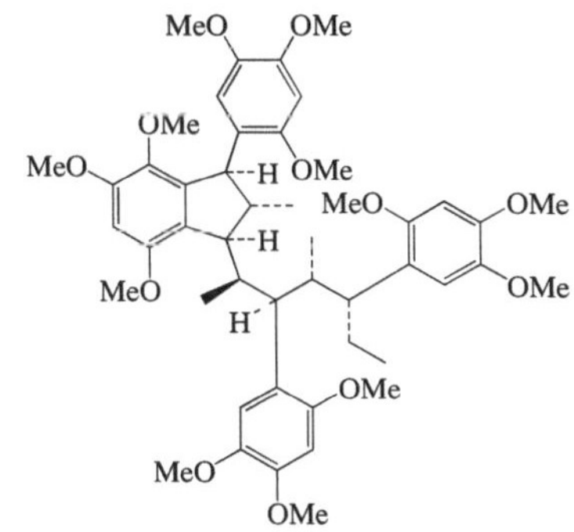

Relative Configuration

[2059918-83-9]

$C_{48}H_{64}O_{12}$ 833.028

Constit. of the roots of *Acorus tatarinowii*. Exhibits protective effect against β-amyloid toxicity. Needles (Me_2CO/petrol). $[\alpha]_D^{25}$ -19 (c, 0.05 in $CHCl_3$).

Luo, X.-H. *et al.*, *Fitoterapia*, 2016, **108**, 5-8 (*Tatarinan T, cryst struct, activity*)

Tatarinoid C T6

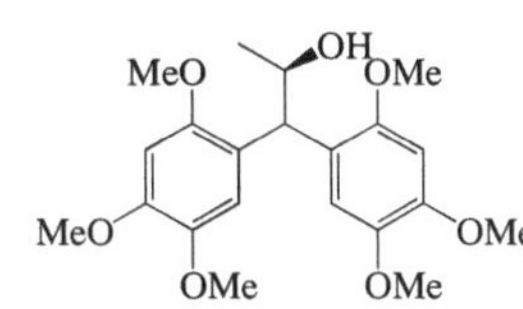

Absolute Configuration

1,1-Bis(2,4,5-trimethoxyphenyl)-2-propanol

[1229005-37-1]

$C_{21}H_{28}O_7$ 392.448

Constit. of the rhizomes of *Acorus tatarinowii*. Gum. $[\alpha]_D^{27}$ -10.2 (c, 0.17 in $CHCl_3$). λ_{max} 292 (log ε 3.99) (MeOH).

Tong, X.-G. *et al.*, *J. Nat. Prod.*, 2010, **73**, 1160-1163 (*Tatarinoid C*).

Slutskyy, Y. *et al.*, *Tetrahedron Lett.*, 2013, **54**, 210-212 (*synth*).

Babu, B.M. *et al.*, *Tetrahedron Lett.*, 2014, **55**, 1868-1872 (*synth*)

Taxodascendin T7

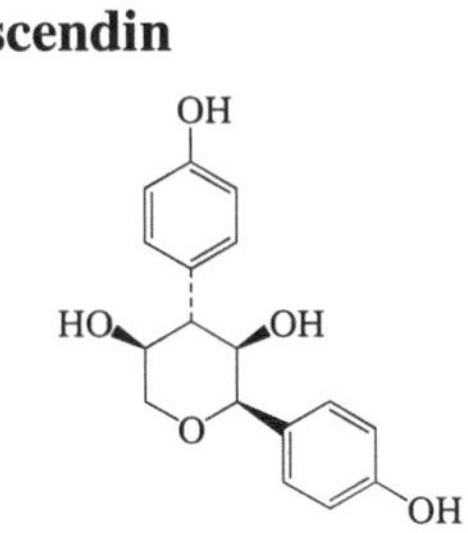

Absolute Configuration

Tetrahydro-2,4-bis(4-hydroxyphenyl)-2H-pyran-3,5-diol. 3,5-Dihydroxy-2,4-bis(4-hydroxyphenyl) tetrahydro-2H-pyran. 7′,9-Epoxy-9′-nor-7,8′-lignan-4,4′,8,8′-tetrol. 7′,9-Epoxy-4,4′,8,8′-tetrahydroxy-9′-nor-7,8′-neolignan
[1209486-77-0]
$C_{17}H_{18}O_5$ 302.326
Provisional norlignan. Biosynth. pathway unclear. Related to Sequirin A, S46. Constit. of *Taxodium ascendens*. Liq.. $[\alpha]_D^{17}$ +94.2 (c, 0.17 in MeOH). λ_{max} 202 (log ε 4.54); 227 (log ε 4.31); 277 (log ε 3.65) (MeOH).
Zhang, Y.M. *et al.*, *Fitoterapia*, 2009, **80**, 361-363 (*Taxodascendin*)

Taxuyunin B T8

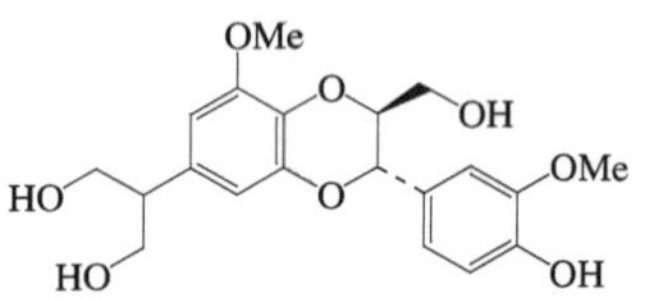

Relative Configuration

[632336-69-7]
$C_{20}H_{24}O_8$ 392.404
Constit. of the bark of *Taxus yunnanensis*. Needles. Mp 169°. $[\alpha]_D^{16}$ +5.6 (c, 0.4 in MeOH). λ_{max} 208 (log ε 5.2); 231 (log ε 4.7); 280 (log ε 4.1) (MeOH).
Li, S.-H. *et al.*, *Chin. J. Chem.*, 2003, **21**, 926-930 (*Taxuyunin B*)

Tedarene B T9

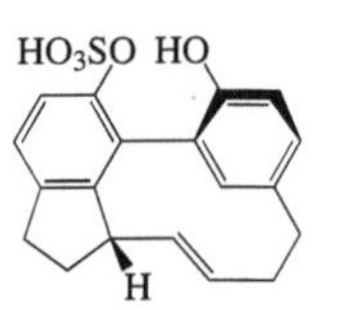

Absolute Configuration

[1372812-20-8]
$C_{19}H_{18}O_5S$ 358.408
Isol. from *Tedania ignis*. Amorph. solid. $[\alpha]_D^{20}$ +33 (c, 0.1 in MeOH). λ_{max} 281 (ε 5500) (MeOH).
Costantino, V. *et al.*, *J. Org. Chem.*, 2012, **77**, 6377-6383 (*Tedarene B*)

Terpecurcumin I T10

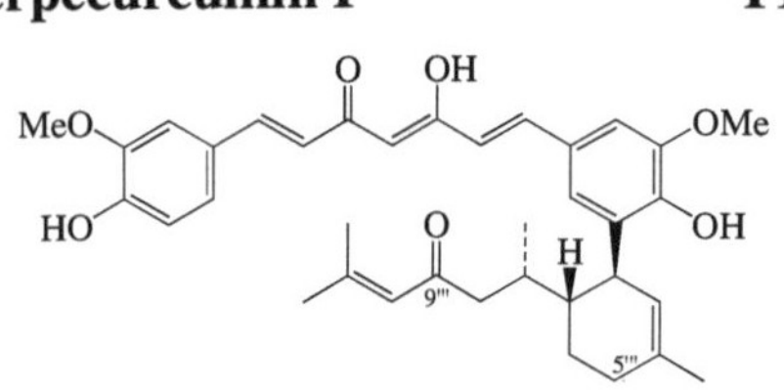

Absolute Configuration

[1410038-34-4]
$C_{36}H_{42}O_7$ 586.725
Enolised β-diketone. Isol. from the rhizomes of *Curcuma longa*. Amorph. orange powder. $[\alpha]_D^{25}$ +2.4 (c, 0.01 in MeCN). λ_{max} 412 (log ε 4.24) (MeCN)
9‴-Deoxo, 5‴S-hydroxy
Terpecurcumin H
[1410038-31-1]
$C_{36}H_{44}O_7$ 588.741
Isol. from the rhizomes of *Curcuma longa*. Amorph. orange powder. $[\alpha]_D^{25}$ +1.4 (c, 0.01 in MeCN). λ_{max} 420 (log ε 4.69) (MeCN).
Lin, X. *et al.*, *J. Nat. Prod.*, 2012, **75**, 2121-2131 (*Terpecurcumins H,I*)

Terpecurcumin B T11

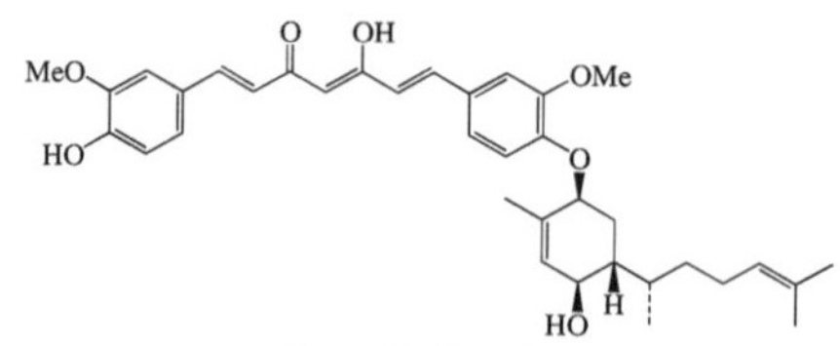

Absolute Configuration

[1413592-72-9]
$C_{36}H_{44}O_7$ 588.741
Hybrid of curcumins and bisabolones. Enolised β-diketone. Isol. from the rhizomes of *Curcuma longa*. Amorph. orange powder. $[\alpha]_D^{25}$ +2.6 (c, 0.01 in MeCN). λ_{max} 418 (log ε 4.77) (MeCN).
Lin, X. *et al.*, *J. Nat. Prod.*, 2012, **75**, 2121-2131 (*Terpecurcumin B*)

Terpecurcumin D T12

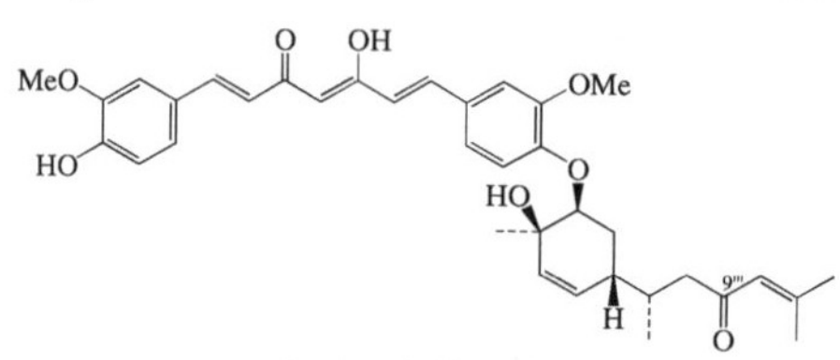

Absolute Configuration

[1413592-78-5]
$C_{36}H_{42}O_8$ 602.724
Hybrid of curcumins and bisabolanes. Enolised β-diketone. Isol. from the rhizomes of *Curcuma longa*. Shows cytotoxic activity against human cancer cell lines. Amorph. orange powder. $[\alpha]_D^{25}$ +2.5 (c, 0.01 in MeCN). λ_{max} 418 (log ε 4.65) (MeCN)
9‴-Deoxo
Terpecurcumin C
[1413592-75-2]
$C_{36}H_{44}O_7$ 588.741
Isol. from the rhizomes of *Curcuma longa*. Amorph. orange powder. $[\alpha]_D^{25}$ +3.5 (c, 0.01 in MeCN). λ_{max} 419 (log ε 4.75) (MeCN).
Lin, X. *et al.*, *J. Nat. Prod.*, 2012, **75**, 2121-2131 (*Terpecurcumins C,D*)

Terpecurcumin F T13

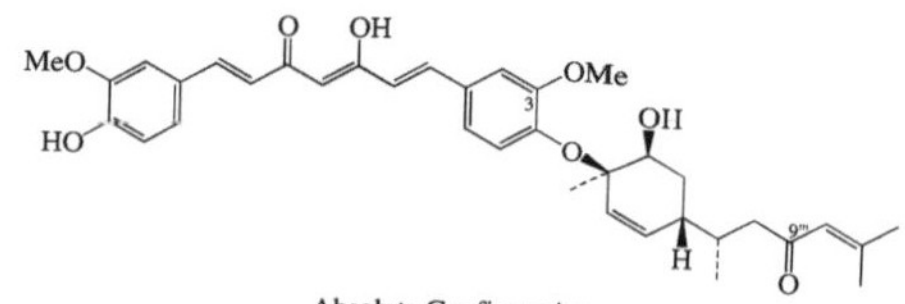

Absolute Configuration

[1410038-25-3]
$C_{36}H_{42}O_8$ 602.724
Enolised β-diketone. Isol. from the rhizomes of *Curcuma longa*. Shows cytotoxic activity against human cancer cell lines. Amorph. orange powder. $[\alpha]_D^{25}$ +3.8 (c, 0.01 in MeCN). λ_{max} 414 (log ε 4.61) (MeCN)
3-Demethoxy
Terpecurcumin G
[1410038-29-7]
$C_{35}H_{40}O_7$ 572.698
Isol. from the rhizomes of *Curcuma longa*. Shows cytotoxic activity against human cancer cell lines. Amorph. orange powder. $[\alpha]_D^{25}$ +1.9 (c, 0.01 in MeCN). λ_{max} 414 (log ε 4.71) (MeCN)
9‴-Deoxo
Terpecurcumin E
[1410038-23-1]
$C_{36}H_{44}O_7$ 588.741
Isol. from the rhizomes of *Curcuma longa*. Amorph. orange powder. $[\alpha]_D^{25}$ +4.1 (c, 0.01 in MeCN). λ_{max} 418 (log ε 4.76) (MeCN).
Lin, X. *et al.*, *J. Nat. Prod.*, 2012, **75**, 2121-2131 (*Terpecurcumins E,F,G*)

Terrelactone B T14

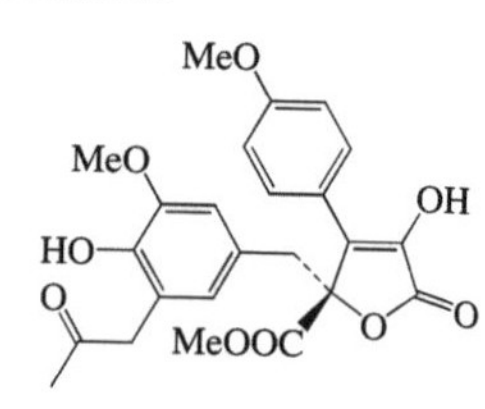

Absolute Configuration

[1675220-35-5]
$C_{24}H_{24}O_9$ 456.447
Provisional lignan. Biosynth. pathway is unclear. Related to Butyrolactone I, B89. Prod. by *Penicillium oxalicum* 0403. Active against tobacco mosaic virus. Amorph. powder. $[\alpha]_D^{25}$ +82.5 (c, 0.2 in MeOH). λ_{max} 210 (log ε 4.42);

255 (log ε 3.65); 316 (log ε 3.76) (MeOH).

Yuan, L. *et al.*, *Nat. Prod. Res.*, 2015, **29**, 1914-1919 (*Terrelactone B*)

Tetra-*O*-caffeoylquinic acid T15

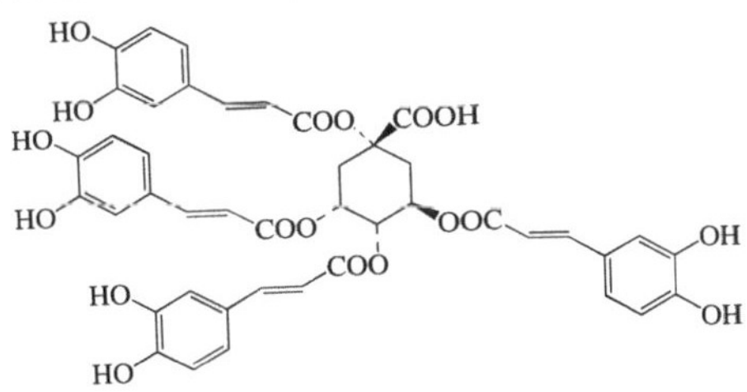

1,3,4,5-Tetrakis(3,4-dihydroxycinnamoyl)quinic acid
[158364-86-4]
$C_{43}H_{36}O_{18}$ 840.743
Constit. of *Bridelia ferruginea* and *Pluchea symphytifolia*. Shows antibacterial and anthelmintic activity. Yellowish-white amorph. powder. $[\alpha]^{20}_{D}$ -217 (c, 2.7 in MeOH aq.). λ_{max} 245 (ε 30903); 293 (ε 40738); 326 (ε 48978) (MeOH) (Berdy).

Scholz, E. *et al.*, *Planta Med.*, 1994, **60**, 360-364 (*Pluchea symphytifolia constit*).
Cimanga, K. *et al.*, *Planta Med.*, 1999, **65**, 213-217 (*Bridelia ferruginea constit*)

Tetracentronsin T16

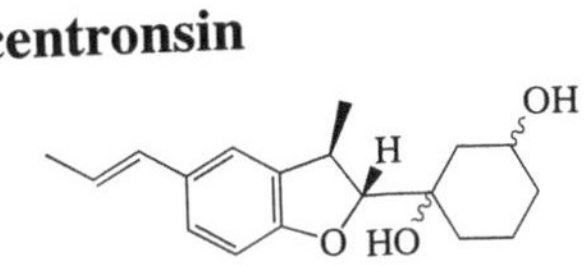

Relative Configuration

1-[2,3-Dihydro-3-methyl-5-(1-propenyl)-2-benzofuranyl]-1,3-cyclohexanediol. 4,7'-Epoxy-1',2',3',4',5',6'-hexahydro-3,8'-lign-7-ene-1',3'-diol. 4,7'-Epoxy-1',2',3',4',5',6'-hexahydro-1',3'-dihydroxy-3,8'-neolign-7-ene
[928039-38-7]
$C_{18}H_{24}O_3$ 288.387
Constit. of the stem bark of *Tetracentron sinense*. Powder. $[\alpha]^{27}_{D}$ +96 (c, 0.25 in $CHCl_3$). λ_{max} 264; 292 ($CHCl_3$).

Wang, Y.-F. *et al.*, *Chem. Biodiversity*, 2006, **3**, 1023-1030 (*Tetracentronsin*)

Tetra-*O*-demethylsaucerneol T17

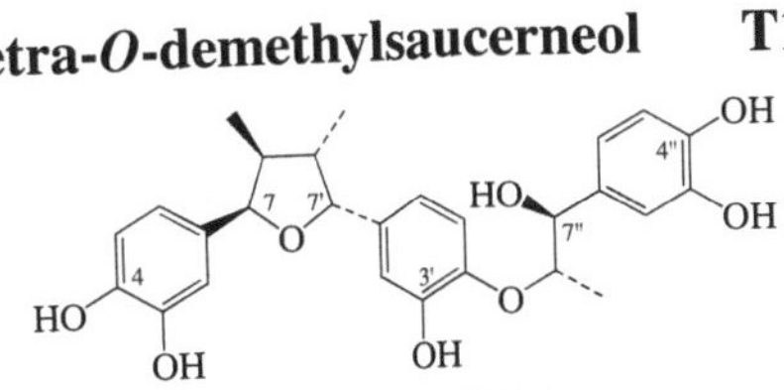

(7R,7'R,7''S,8S,8'S,8''S)-form

$C_{27}H_{30}O_8$ 482.529

Sesquineolignan. Related to Bonaspectin D, B78. Parent not isol.

(7R,7'R,7''S,8S,8'S,8''S)-form

9,9',9''-Trihydroxy, 3,3',3''-tri-Me ether
[2170866-08-5]
$C_{30}H_{36}O_{11}$ 572.607
Constit. of the rhizomes of *Iris tectorum*. Pale yellow powder. $[\alpha]^{20}_{D}$ +18.1 (c, 0.15 in MeOH)

(7S,7'S,7''R,8R,8'R,8''R)-form

3,3',3'',4''-Tetra-Me ether
Saucerneol
[88497-86-3]
$C_{31}H_{38}O_8$ 538.637
Constit. of the roots of *Saururus cernuus*. Exhibits vasorelaxant props.. $[\alpha]_{D}$ -52.3 (no solvent reported) (natural). $[\alpha]_{D}$ -91.8 (c, 1 in $CHCl_3$) (synthetic)

3,3',3'',4,4''-Penta-Me ether
4-O-Methylsaucerneol*. Saucerneol methyl ether*
[198836-15-6]
$C_{32}H_{40}O_8$ 552.664
Constit. of *Saururus chinensis*. Powder. Mp 72 - 74°. $[\alpha]^{25}_{D}$ -63 (c, 0.5 in $CHCl_3$). λ_{max} 207 (log ε 4.8); 235 (log ε 4.35); 285 (log ε 3.95) (MeOH)

3,4-Methylene, 3',4''-di-Me ether
Saucerneol E
[630096-18-3]
$C_{30}H_{34}O_8$ 522.594
Constit. of the roots of *Saururus chinensis*. Powder. Mp 76 - 78°. $[\alpha]^{25}_{D}$ -83 (c, 0.3 in $CHCl_3$). λ_{max} 210 (log ε 4.9); 232 (log ε 4.35); 282 (log ε 3.87) (MeOH)

3,4-Methylene, 3',3'',4''-tri-Me ether
Saucerneol D
[630096-17-2]
$C_{31}H_{36}O_8$ 536.621
Constit. of the roots of *Saururus chinensis*. Inhibits NO prodn. Exhibits vasorelaxant props. Active against Epstein-Barr virus lytic replication. Powder. Mp 75 - 76°. $[\alpha]^{25}_{D}$ -88.1 (c, 1.2 in $CHCl_3$). λ_{max} 208 (log ε 4.9); 236 (log ε 4.37); 286 (log ε 3.98) (MeOH)

3,4:3'',4''-Bis(methylene), 3'-Me ether
Saucerneol F
[1061273-57-1]
$C_{30}H_{32}O_8$ 520.578
Constit. of the roots of *Saururus chinensis*. Inhibits NO prodn. Shows cytotoxic activity. Brown powder (EtOAc/MeOH). Mp 59 - 61°. $[\alpha]^{25}_{D}$ -60.6 (c, 0.2 in $CHCl_3$). λ_{max} 234 (log ε 4.36); 284 (log ε 4.12) (MeOH)

(7R*,7'S*,7''R,8R*,8'R*,8''R)-form

3,3',3''-Tri-Me ether
Saucerneol C. Zijusesquilignan A
[373360-11-3]
$C_{30}H_{36}O_8$ 524.61
Constit. of *Saururus chinensis* and *Ziziphus jujuba*. Inhibitor of NO prodn. Exhibits cytotoxic activity. Amorph. powder. $[\alpha]_{D}$ -66 (c, 0.7 in $CHCl_3$). $[\alpha]^{24}_{D}$ -167.6 (c, 0.1 in Me_2CO) (Zijusesquilignan A). λ_{max} 232; 281 (MeOH)

(7R*,7'S*,7''R,8S*,8'R*,8''R)-form

3,3',3'',4,4''-Penta-Me ether
Saucerneol A
[373360-09-9]
$C_{32}H_{40}O_8$ 552.664
Constit. of *Saururus chinensis*. Amorph. powder. $[\alpha]_{D}$ -83 (c, 0.7 in $CHCl_3$). λ_{max} 232; 281 (MeOH)

3'',4''-Methylene, 3,3',4-tri-Me ether
Saucerneol B
[373360-10-2]
$C_{31}H_{36}O_8$ 536.621
Constit. of the roots of *Saururus chinensis*. ACAT inhibitor. Amorph. powder. $[\alpha]_{D}$ -58 (c, 0.6 in $CHCl_3$). λ_{max} 260 (MeOH)

(7R*,7'S*,7''S,8S*,8'S*,8''S)-form

3',4,4'',7''-Tetra-Me ether
Zijusesquilignan C
$C_{31}H_{38}O_8$ 538.637
Constit. of the fruit of *Ziziphus jujuba*. Inhibitor of NO prodn. Exhibits cytotoxic activity. Amorph. powder. $[\alpha]^{24}_{D}$ +120 (c, 0.1 in Me_2CO). λ_{max} 208 (log ε 2.17) (Me_2CO).

Rao, K.V. *et al.*, *Tetrahedron Lett.*, 1983, **24**, 4947-4950 (*Saucerneol*).
Sung, S.H. *et al.*, *Chem. Pharm. Bull.*, 2001, **49**, 1192-1194 (*Saucerneols A,B,C*).
Hwang, B.Y. *et al.*, *Phytochemistry*, 2003, **64**, 765-771 (*Saucerneols D,E, 4-O-Methylsaucerneol*).
Lee, W.S. *et al.*, *Bioorg. Med. Chem. Lett.*, 2004, **14**, 3109-3112 (*Saurcerneol B, activity*).
Hanessian, S. *et al.*, *Org. Lett.*, 2006, **8**, 5477-5480 (*synth, abs config*).
Seo, C.-S. *et al.*, *J. Nat. Prod.*, 2008, **71**, 1771-1774 (*Saucerneol F*).
Oh, K.S. *et al.*, *Planta Med.*, 2008, **74**, 233-238 (*Saucerneol, Saucerneol D, activity*).
Lu, Y. *et al.*, *Int. Immunopharmacol.*, 2012, **12**, 175-181 (*Saucerneol F, activity*).
Lu, Y. *et al.*, *Food Chem. Toxicol.*, 2013, **59**, 696-702 (*Saucerneol F, activity*).

Cui, H. *et al.*, *J. Nat. Prod.*, 2014, **77**, 100-110 (*Saucerneol D, activity*).
Zhang, C.-L. *et al.*, *Fitoterapia*, 2016, **108**, 93-97 (*Iris tectorum constit*).
Tran, H.N.K. *et al.*, *Fitoterapia*, 2019, **137**, 104261:1-7 (*Zijusesquilignans A,C, cd, activity*)

Tetrahydro-2,4-bis(4-hydroxyphenyl)-3-furanol T18

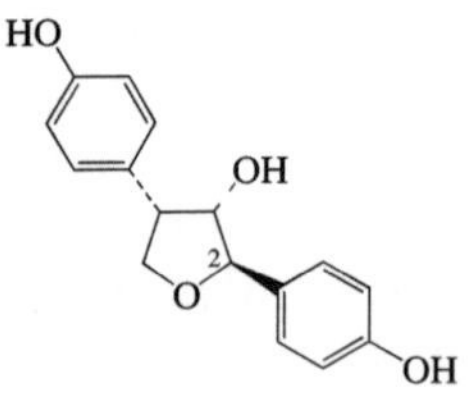

Tetrahydro-3-hydroxy-2,4-bis(4-hydroxyphenyl)furan
$C_{16}H_{16}O_4$ 272.300
(2R,3S,4S)-form
Mubiesin E
Constit. of seeds of *Momordica cochinchinensis*. Amorph. powder. $[\alpha]^{25}_{D}$ -7.9 (c, 0.32 in MeOH). Error in struct. diag. in ref.. λ_{max} 225 (log ε 3.41); 276 (log ε 2.29) (MeOH).

Wang, M. *et al.*, *Fitoterapia*, 2019, **139**, 104360:1-6 (*Mubiesin E*)

Tetrahydro-2,6-bis(4-hydroxyphenyl)-5-[3-(4-hydroxyphenyl)-2-propenyl]-2*H*-pyran-3-methanol T19

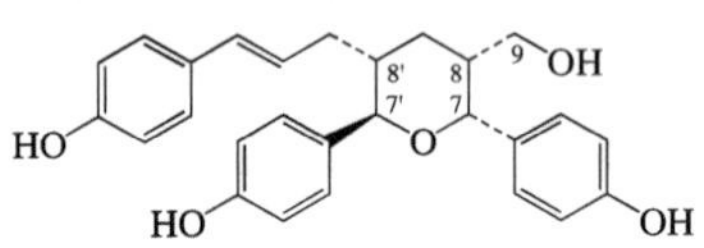

(7R,7'R*,8R*,8'R*)-form*

$C_{27}H_{28}O_5$ 432.516
Lignan numbering shown
(7R*,7'R*,8R*,8'R*)-form
9-Me ether
Conchignan B
[1588410-75-6]
$C_{28}H_{30}O_5$ 446.543
Constit. of *Alpinia conchigera*. Gum. $[\alpha]^{21}_{D}$ -4.9 (c, 0.51 in MeOH). λ_{max} 201 (log ε 4.55); 226 (log ε 4.31); 263 (log ε 4.18) (MeOH)
(7R*,7'S*,8R*,8'S*)-form
9-Me ether
Conchignan A
[1588410-74-5]
$C_{28}H_{30}O_5$ 446.543
Constit. of *Alpinia conchigera*. Gum. $[\alpha]^{21}_{D}$ -13.7 (c, 0.18 in MeOH). λ_{max} 201 (log ε 4.41); 262 (log ε 4.13) (MeOH)
9-Et ether
Conchignan C
[1588410-76-7]
$C_{29}H_{32}O_5$ 460.57
Constit. of *Alpinia conchigera*. Gum. $[\alpha]^{17}_{D}$ -12.4 (c, 0.15 in MeOH). λ_{max} 201 (log ε 4.43); 262 (log ε 4.11) (MeOH).

Xu, J.-J. *et al.*, *J. Asian Nat. Prod. Res.*, 2013, **15**, 833-839 (*Conchignans A,B,C*)

6,7,9,10-Tetrahydro-5-(4-hydroxy-3,5-dimethoxyphenyl)-5*H*-benzo[*g*][1,3]dioxolo[4',5':4,5]benzo[*de*]pyrido[3,2,1-*ij*]quinolin-7-one T20

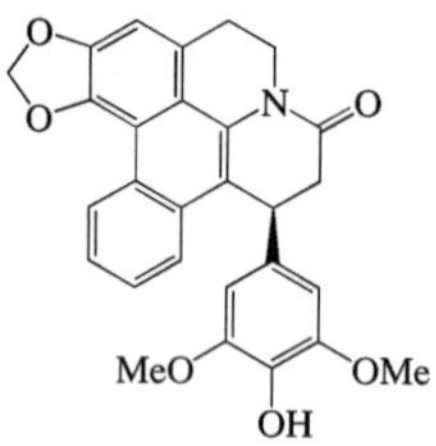

Absolute Configuration

[779355-68-9]
$C_{28}H_{23}NO_6$ 469.493
Lignoaporphine. Alkaloid from the wood of *Annona dioica*. Amorph. yellow solid. Mp 211 - 213°. $[\alpha]_D$ +326 (c, 2.5 in $CHCl_3$).

Dos Santos, P.R.D. *et al.*, *J. Braz. Chem. Soc.*, 2003, **14**, 396-400 (*Annona dioica constit*)

Tetrahydro-4-[4-[2-hydroxy-2-(4-hydroxy-3-methoxyphenyl)-1-(hydroxymethyl)ethoxy]-3,5-dimethoxyphenyl]-1*H*,3*H*-furo[3,4-*c*]furan-1-one T21

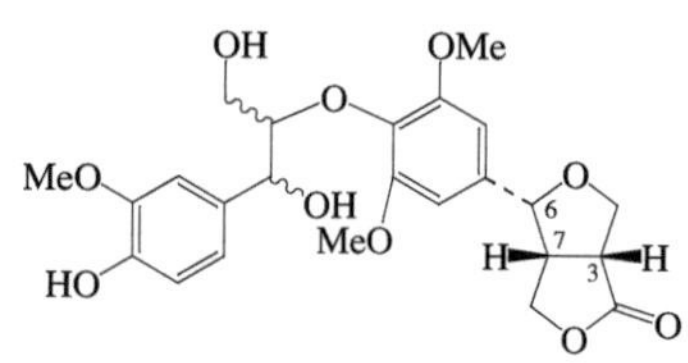

$C_{24}H_{28}O_{10}$ 476.478
(3R,6R,7R,7''ξ,8''ξ)-form [929688-50-6]
Constit. of *Tarenna attenuata*. Antioxidant. Amorph. powder. $[\alpha]^{20}_{D}$ -9.4 (c, 0.6 in MeOH). λ_{max} 280 (log ε 3.81); 344 (log ε 3.17) (MeOH).

Yang, X.-W. *et al.*, *J. Nat. Prod.*, 2007, **70**, 521-525 (*Tarenna attenuata constit, activity*)

2,3,4,5-Tetrahydro-7-hydroxy-2-(4-hydroxy-3-methoxyphenyl)-10-methoxyphenanthro[2,1-*b*]furan-3-carboxylic acid T22

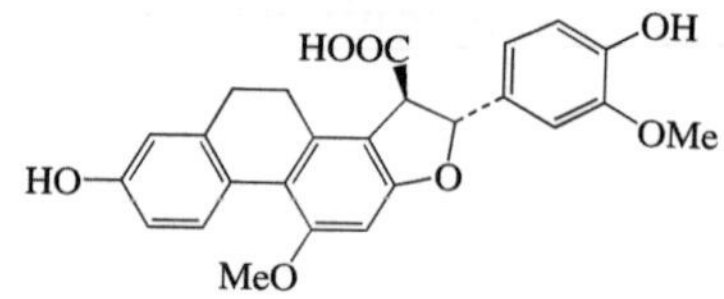

$C_{25}H_{22}O_7$ 434.444
(2R,3R)-form
Docosyl ester
Phaithrene
[2242422-68-8]
$C_{47}H_{66}O_7$ 743.038
Constit. of *Phaius mishmensis*. Amorph. powder. $[\alpha]_D$ -18.5 (c, 0.08 in $CHCl_3$). λ_{max} 240 (log ε 3.94); 282 (log ε 4.01) ($CHCl_3$).

Jao, C.-W. *et al.*, *Molecules*, 2016, **21**, 1605:1-10 (*Phaithrene, cd*)

Tetrahydro-1-hydroxy-4-(4-hydroxyphenyl)-1*H*,3*H*-furo[3,4-*c*]furan T23

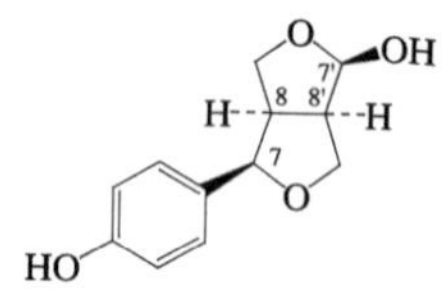

Tetrahydro-4-(4-hydroxyphenyl)-1H,3H-furo[3,4-c]furan-1-ol
$C_{12}H_{14}O_4$ 222.24
Norlignan. Lignan numbering shown. Cf. Samin, S23
(7R*,7'R*,8R*,8'R*)-form
7'-Me ether
Tetrahydro-1-(4-hydroxyphenyl)-4-methoxy-1H,3H-furo[3,4-c]furan
[870093-76-8]
$C_{13}H_{16}O_4$ 236.267
Constit. of the seeds of *Myrica rubra*. $[\alpha]_D$ -18.2 (c, 0.11 in $CHCl_3$).

Suga, A. *et al.*, *Nat. Med. (Tokyo)*, 2005, **59**, 70-75 (*Myrica rubra constit*)

5,6,7,8-Tetrahydro-5-hydroxy-2-methoxy-6,7-dimethyl-8-(2,4,5-trimethoxyphenyl)-1,4-naphthalenedione T24

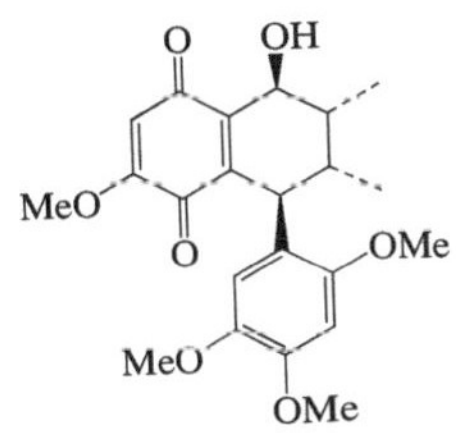

Relative Configuration

[259792-74-0]
$C_{22}H_{26}O_7$ 402.443
Constit. of *Sphallerocarpus gracilis*. Brown-yellow needles. Mp 202 - 204°.
Chen, N.Y. *et al.*, *Chin. Chem. Lett.*, 1999, **10**, 919-920 (*Sphallerocarpus gracilis constit*)

Tetrahydro-2-[2-(4-hydroxyphenyl) ethyl]-6-(4-hydroxyphenyl)-2*H*-pyran-4-ol T25

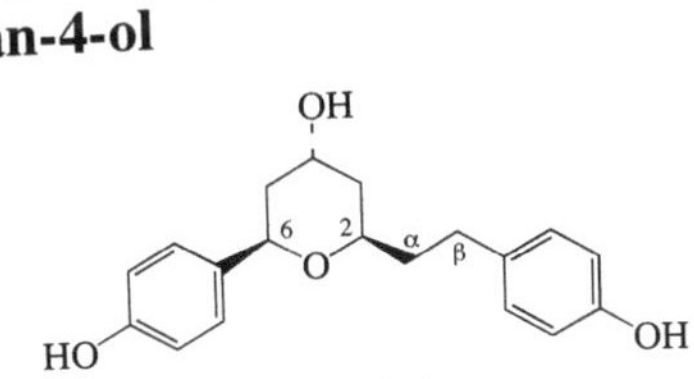

(2*R*,4*R*,6*R*)-form

1,5-Epoxy-1,7-bis(4-hydroxyphenyl)-3-heptanol
$C_{19}H_{22}O_4$ 314.381
(2R,4R,6R)-form [1415609-85-6]
Constit. of *Dioscorea villosa*. Amorph. powder. $[\alpha]_D^{22}$ +190.2 (c, 0.02 in MeOH). λ_{max} 224 (log ε 4.65); 276 (log ε 3.90) (MeOH)
α,β-*Didehydro(E-)*
1,5-Epoxy-3-hydroxy-1,7-bis(4-hydroxyphenyl)-6-heptene
[1415609-83-4]
$C_{19}H_{20}O_4$ 312.365
Constit. of *Dioscorea nipponica* and *Dioscorea villosa*. Amorph. powder. $[\alpha]_D^{22}$ +14.6 (c, 0.12 in MeOH). (2*R*,4*R*,6*S*)-config. due to change in Cahn-Ingold-Prelog priorities. λ_{max} 263 (log ε 4.49) (MeOH)
(2R,4S,6R)-form [1415609-87-8]
Constit. of *Dioscorea villosa*. Amorph. powder. $[\alpha]_D^{22}$ +32.9 (c, 0.22 in MeOH). λ_{max} 216 (log ε 4.39); 276 (log ε 3.54) (MeOH)
(2S,4R,6S)-form [1415609-86-7]
Constit. of *Dioscorea nipponica* and *Dioscorea villosa*. Amorph. powder. $[\alpha]_D^{22}$ -30.6 (c, 0.09 in MeOH). λ_{max} 216 (log ε 4.39); 276 (log ε 3.54) (MeOH)
(2S,4S,6S)-form [1415609-84-5]
Constit. of *Dioscorea nipponica* and *Dioscorea villosa*. Amorph. powder. $[\alpha]_D^{22}$ -166.7 (c, 0.02 in MeOH). λ_{max} 224 (log ε 4.65); 276 (log ε 3.90) (MeOH).

Dong, S.-H. *et al.*, *J. Nat. Prod.*, 2012, **75**, 2168-2177 (*Dioscorea villosa constits*).
Dong, S.-H. *et al.*, *J. Nat. Prod.*, 2013, **76**, 2005-2007 (*Dioscorea villosa constits, corrigendum*).
Yadav, J.S. *et al.*, *Tetrahedron Lett.*, 2015, **56**, 1360-1362 (*2R,4S,6R-form, synth*).
Kantee, K. *et al.*, *Tetrahedron Lett.*, 2016, **57**, 3505-3509 (*2S,4R,6S-form, 2S,4S,6S-form, synth*)

Tetrahydro-2-[2-(4-hydroxyphenyl) ethyl]-6-(3,4,5-trihydroxyphenyl)-2*H*-pyran-4-ol T26

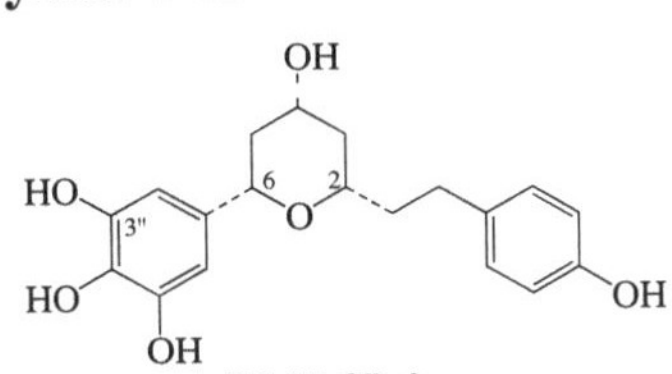

(2*S*,4*R*,6*S*)-form

1,5-Epoxy-7-(4-hydroxyphenyl)-1-(3,4,5-trihydroxyphenyl)-3-heptanol
$C_{19}H_{22}O_6$ 346.379
(2S,4R,6S)-form
3″-Me ether
[1007599-19-0]
$C_{20}H_{24}O_6$ 360.406
Constit. of the rhizomes of *Zingiber officinale*. Yellowish oil. $[\alpha]_D^{25}$ -32.5 (c, 0.75 in MeOH)
(2S,4S,6S)-form
3″,5″-Di-Me ether
[1869012-32-7]
$C_{21}H_{26}O_6$ 374.433
Constit. of the rhizomes of *Curcuma kwangsiensis*. Yellow oil. $[\alpha]_D^{25}$ +8.3 (c, 0.2 in MeOH). λ_{max} 224; 282 (MeOH).

Tao, Q.-F. *et al.*, *J. Nat. Prod.*, 2008, **71**, 12-17 (*Zingiber officinale constit*).
Chen, S.-D. *et al.*, *Fitoterapia*, 2015, **102**, 67-73 (*Curcuma kwangsiensis constit*)

4*a*,5,6,10*b*-Tetrahydro-3-(4-hydroxyphenyl)-9-methoxy-3*H*-naphtho[2,1-*b*]pyran-8-ol T27

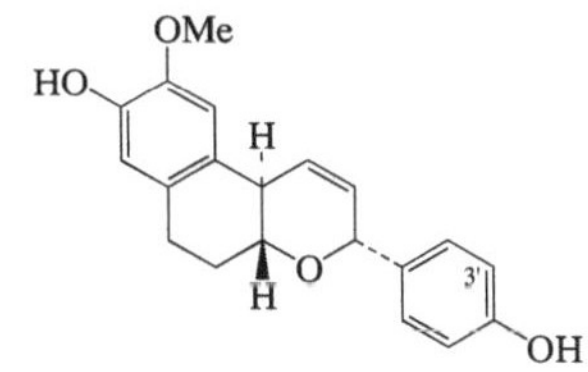

(3*R**,4*aR**,10*bR**)-form

$C_{20}H_{20}O_4$ 324.376
Bicyclic diarylheptanoid
(3R*,4aR*,10bR*)-form
Musellarin E [1392476-35-5]
Constit. of *Musella lasiocarpa*. Amorph. powder. $[\alpha]_D^{20}$ -111.5 (c, 0.19 in MeOH). λ_{max} 205 (log ε 4.23); 225 (log ε 3.84); 284 (log ε 3.31) (MeOH)
3′-Methoxy
Musellarin D
[1392476-34-4]
$C_{21}H_{22}O_5$ 354.402
Constit. of *Musella lasiocarpa*. Amorph. powder. $[\alpha]_D^{20}$ -41.9 (c, 0.29 in MeOH). λ_{max} 205 (log ε 4.21); 284 (log ε 3.26) (MeOH)
(3R*,4aS*,10bR*)-form
Musellarin A [477565-36-9]
Constit. of the fruit of *Musa* x *paradisiaca* cultivar and from *Musella lasiocarpa*. Cryst.. Mp 195 - 197°. $[\alpha]_D^{20}$ -292.8 (c, 0.77 in MeOH). λ_{max} 240 (log ε 3.28); 269 (log ε 3.23); 294 (log ε 3.4) (EtOH)
3′-Hydroxy, 4′-Me ether
Musellarin C
[1392476-33-3]
$C_{21}H_{22}O_5$ 354.402
Constit. of *Musella lasiocarpa*. Amorph. powder. $[\alpha]_D^{20}$ -176.8 (c, 0.3 in MeOH). λ_{max} 206 (log ε 4.42); 284 (log ε 3.53) (MeOH)
3′-Methoxy
Musellarin B
[1392476-32-2]
$C_{21}H_{22}O_5$ 354.402
Constit. of *Musella lasiocarpa*. Amorph. powder. $[\alpha]_D^{20}$ -223.3 (c, 0.17 in MeOH). λ_{max} 205 (log ε 4.51); 284 (log ε 3.63) (MeOH).

Jang, D.S. *et al.*, *J. Agric. Food Chem.*, 2002, **50**, 6330-6334 (*Musa paradisiaca constit, cryst struct*).
Dong, L.-B. *et al.*, *Nat. Prod. Bioprospect.*, 2011, **1**, 41-47 (*Musellarins A-E*).
Li, Z. *et al.*, *Chem. - Eur. J.*, 2015, **21**, 11152-11157 (*synth*)

Tetrahydro-1-hydroxy-4-(3,4,5-trihydroxyphenyl)-1*H*,3*H*-furo[3,4-*c*]furan **T28**

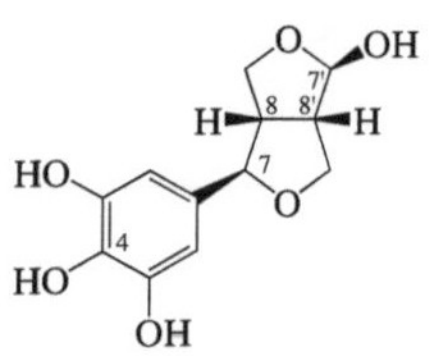

$C_{12}H_{14}O_6$ 254.238
Related to Samin, S23
(7R*,7′R*,8S*,8′S*)-form
3,5-Di-Me ether
Tetrahydro-4-(4-hydroxy-3,5-dimethoxyphenyl)-1H,3H-furo[3,4-c]furan-1-ol, CAS. *2-Hydroxy-6-(4-hydroxy-3,5-dimethoxyphenyl)-3,7-dioxabicyclo[3.3.0]octane*
[121008-45-5]
$C_{14}H_{18}O_6$ 282.292
Constit. of *Cinnamomum cassia, Incarvillea arguta* and *Tsoongiodendron odorum*.

Huang, Y. *et al., Biochem. Syst. Ecol.*, 2011, **39**, 209-212 (*Tsoongiodendron odorum constit*).
Luo, Q. *et al., Molecules*, 2013, **18**, 10930-10943 (*Cinnamomum cassia constit*)

1,4,5,7-Tetrahydroxy-1,7-bis(2-hydroxyphenyl)-4-hepten-3-one **T29**

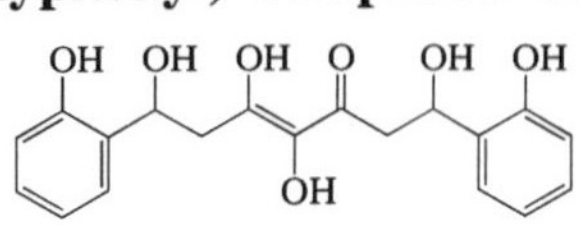

$C_{19}H_{20}O_7$ 360.362
(E)-form [2227433-86-3]
Constit. of the kernels of *Corylus avellana* infected with 'Cimiciato'.

Singldinger, B. *et al., J. Agric. Food Chem.*, 2018, **66**, 4660-4673 (*Corylus avellana constit*)

3,4′,7′,9′-Tetrahydroxy-8,9-dinor-4,8′-lignan-7-al **T30**

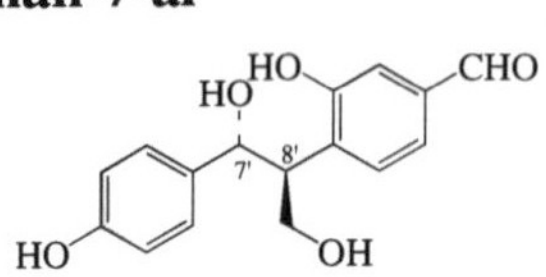

3-Hydroxy-4-[2-hydroxy-1-(hydroxymethyl)-2-(4-hydroxyphenyl)ethyl]benzaldehyde
$C_{16}H_{16}O_5$ 288.299
(7′R,8′S)-form
3,9′-Di-Me ether
4′,7′-Dihydroxy-3,9′-dimethoxy-8,9-dinor-4,8′-lignan-7-al
[2410993-12-1]
$C_{18}H_{20}O_5$ 316.353
Constit. of the flowers of *Urtica fissa*. Antioxidant. Amorph. powder. $[\alpha]_D^{25}$ +0.7 (c, 0.4 in MeOH). λ_{max} 230; 282 (MeOH).

Wang, M. *et al., Chem. Nat. Compd. (Engl. Transl.)*, 2019, **55**, 839-842 (*Urtica fissa constit, activity*)

3,3′,4′,7′-Tetrahydroxy-8,9-dinor-4,8′-oxylignan-7-al **T31**

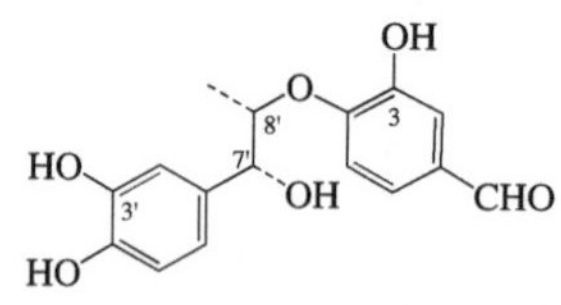

(7′R,8′S)-form

4-[2-(3,4-Dihydroxyphenyl)-2-hydroxy-1-methylethoxy]-3-hydroxybenzaldehyde. 3,3′,4′,7′-Tetrahydroxy-8,9-dinor-7-oxo-4,8′-oxyneolignan
$C_{16}H_{16}O_6$ 304.298
(7′R,8′S)-form
3,3′-Di-Me ether
4′,7′-Dihydroxy-3,3′-dimethoxy-8,9-dinor-4,8′-oxylignan-7-al. 4′,7′-Dihydroxy-3,3′-dimethoxy-8,9-dinor-7-oxo-4,8′-oxylignan.
Perseal A
[186033-11-4]
$C_{18}H_{20}O_6$ 332.352
Constit. of *Persea obovatifolia*. Cytotoxic agent. Oil. $[\alpha]_D$ +26 (c, 0.05 in $CHCl_3$). λ_{max} 230 (ε 30900); 280 (ε 19500); 310 (ε 15130) (MeOH) (Berdy). [base] λ_{max} 281 (ε 19050); 300 (ε 17780) (MeOH/NaOH) (Berdy)
(7′S,8′S)-form
3,3′-Di-Me ether
Perseal B
[186033-12-5]
$C_{18}H_{20}O_6$ 332.352
Constit. of *Persea obovatifolia*. Cytotoxic agent. Active against leishmaniasis. Amorph. powder. $[\alpha]_D$ +54 (c, 0.1 in $CHCl_3$). λ_{max} 230 (log ε 4.51); 280 (log ε 4.32); 309 (log ε 4.19) (EtOH). [base] λ_{max} 253 (ε 13200); 282 (ε 32600); 301 (ε 19500) (MeOH/NaOH) (Berdy)
(7′R*,8′R*)-form
3,3′,4′-Tri-Me ether, 7′-Ac
Piperbonin B
[2001069-75-4]
$C_{21}H_{24}O_7$ 388.416
Constit. of the aerial parts of *Piper bonii*. Pale yellow oil. $[\alpha]_D^{23}$ +72 (c, 0.3 in MeOH). λ_{max} 230 (log ε 4.32); 278 (log ε 4.13); 309 (log ε 3.95) (MeOH).

Tsai, I.-L. *et al., Phytochemistry*, 1996, **43**, 1261-1263 (*Perseals A,B*).
Xia, Y.-M. *et al., Chem. Pap.*, 2010, **64**, 630-636 (*Perseal A, synth*).
Xia, Y.-M. *et al., Turk. J. Chem.*, 2010, **34**, 375-380 (*Perseal B, synth*).
Ding, D.-D. *et al., Phytochemistry*, 2016, **129**, 36-44 (*Piperbonin B*)

3,3′,4,4′-Tetrahydroxy-8,8′-ligna-7,7′-diene-9,9′-dioic acid **T32**

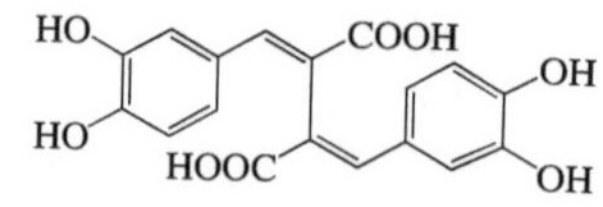

(7E,7′E)-form

2,3-Bis(3,4-dihydroxybenzylidene)butanedioic acid. 3,3′,4,4′-Tetrahydroxy-8,8′-dicinnamic acid
$C_{18}H_{14}O_8$ 358.302
(7E,7′E)-form
Mono-Et ester
Leontolignan A
$C_{20}H_{18}O_8$ 386.356
Constit. of *Leontopodium leontopodioides*. Amorph. powder. Mp 131 - 133°. λ_{max} 269; 341 (MeOH)
3,3′-Di-Me ether
Bis(4-hydroxy-3-methoxybenzylidene)butanedioic acid. 4,4′-Dihydroxy-3,3′-dimethoxy-8,8′-dicinnamic acid. 4,4′-Dihydroxy-3,3′-dimethoxy-8,8′-ligna-7,7′-diene-9,9′-dioic acid. ***8,8′-Diferulic acid****. 8,8′-Dehydrodiferulic acid*
[102767-61-3]
[102767-62-4] (Mono-Me ester)
[103780-25-2] (Di-Me ester)
[109648-34-2] (unspecified stereochem.)
$C_{20}H_{18}O_8$ 386.356
Constit. of various Gramineae cell walls and from *Beta vulgaris*. Yellow solid (MeOH aq. or EtOH aq.). Mp 171 - 172°. Mp 230 - 236° dec.
3,3′-Di-Me ether, bis[2-(4-hydroxyphenyl)ethylamide]
Cannabisin G
[163136-20-7]
$C_{36}H_{36}N_2O_8$ 624.69
Constit. of *Berberis vulgaris, Cannabis sativa, Chloranthus angustifolius, Hyoscyamus niger* and *Solanum melongena*. Antioxidant.

Exhibits cytoprotective activity. Amorph. powder

3,4-Methylene, 3′,4′-di-Me ether

3′,4′-Dimethoxy-3,4-methylenedioxy-8,8′-ligna-7,7′-diene-9,9′-dioic acid

[100169-45-7]

3,4-Methylene, 3′,4′-di-Me ether, di-Me ester

Jatrodiene

[173792-48-8]

$C_{23}H_{22}O_8$ 426.421

Constit. of the stems of *Jatropha gossypifolia*. Viscous yellow solid

3,4:3′,4′-Bis(methylene) ether

2,3-Bis(3,4-methylenedioxybenzylidene)butanedioic acid. 3,4:3′,4′-Bis(methylenedioxy)-8,8′-ligna-7,7′-diene-9,9′-dioic acid

[125130-64-5]

[24290-01-5] (unspecified stereochem.)

3,4:3′,4′-Bis(methylene) ether, di-Me ester

Dimethyl dipiperonylidenesuccinate. ***Gossypidiene***

[14833-90-0]

$C_{22}H_{18}O_8$ 410.378

Constit. of the stems of *Jatropha gossypifolia*. Pale yellow cryst. (EtOH or C_6H_6). Mp 184 - 185° (180-181°)

(7Z,7′Z)-form

3,3′-Di-Me ether, bis[2-(4-hydroxyphenyl)ethylamide]

Hyoscyamide

[404892-30-4]

$C_{36}H_{36}N_2O_8$ 624.69

Alkaloid from the seeds of *Hyoscyamus niger*. Amorph. yellow powder.

Al-Assal, L.S. *et al.*, *J. Chem. Soc.*, 1960, 849-852 (*Gossypidiene*).
Boshoff, P.R. *et al.*, *J. Chem. Res., Synop.*, 1986, 80 (*8,8′-Diferulic acid, synth*).
Ralph, J. *et al.*, *J. Chem. Soc., Perkin Trans. 1*, 1994, 3485-3498 (*Gramineae constit, cmr, synth*).
Sakakibara, I. *et al.*, *Phytochemistry*, 1995, **38**, 1003-1007 (*Cannabisin G*).
Das, B. *et al.*, *Phytochemistry*, 1996, **41**, 985-987 (*Jatrodiene*).
Micard, V. *et al.*, *Phytochemistry*, 1997, **44**, 1365-1368 (*sugar beet constit*).
Das, B. *et al.*, *Phytochemistry*, 1999, **51**, 115-117 (*Gossypidiene*).
Ma, C.-Y. *et al.*, *J. Nat. Prod.*, 2002, **65**, 206-209 (*Hyoscyamide*).
Rouau, X. *et al.*, *Phytochemistry*, 2003, **63**, 899-903 (*maize bran constit*).
Russell, W.R. *et al.*, *Bioorg. Med. Chem.*, 2005, **13**, 2537-2546 (*8,8′-Diferulic acid, synth, props*).
Tomosaka, H. *et al.*, *Phytother. Res.*, 2008, **22**, 979-981 (*Berberis vulgaris constit, activity*).
Li, D. *et al.*, *Bioorg. Med. Chem. Lett.*, 2010, **20**, 5095-5098 (*8,8′-Diferulic acid, struct, Cannabisin G, synth*).
Callipo, L. *et al.*, *J. Mass Spectrom.*, 2010, **45**, 1026-1040 (*8,8′-Diferulic acid, ms*).
Xia, Y. *et al.*, *J. Serb. Chem. Soc.*, 2010, **75**, 1617-1623 (*Cannabisin G, synth*).
Qi, C.-L. *et al.*, *Phytochem. Lett.*, 2017, **21**, 94-97 (*Leontolignan A*)

3,3′,4,4′-Tetrahydroxy-8,8′-ligna-7,7′-dien-9,9′-olide T33

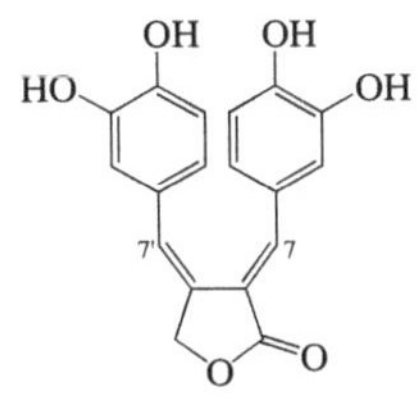

(7E,7′E)-form

$C_{18}H_{14}O_6$ 326.304

(7E,7′E)-form

3,4:3′,4′-Bis(methylene) ether

3,4:3′,4′-Bis(methylenedioxy)-8,8′-ligna-7,7′-dien-9,9′-olide. ***Taiwanin A****. Cytonin*

[4650-68-4]

$C_{20}H_{14}O_6$ 350.326

Constit. of the heartwood of *Taiwania cryptomeroides*. Interacts with microtubulin polymerisation. Antitumour agent. Shows cytotoxic and herbicidal activity. Clin. trial in China as antitumour drug under the name Cytonin (2001). No plans to invest in further development (2002). Yellow-orange cryst. (C_6H_6). Mp 202 - 203°

(7E,7′Z)-form

3,4:3′,4′-Bis(methylene) ether

Synthetic. Mp 175 - 176°.

Lin, Y.-T. *et al.*, *J. Chin. Chem. Soc. (Taipei)*, 1955, **2**, 87-90 (*isol, struct*).
Lin, Y.-T. *et al.*, *Chem. Commun. (London)*, 1965, 592-593 (*Taiwanin A*).
Swoboda, G.A. *et al.*, *J. Chem. Soc. C*, 1967, 161-162 (*Taiwanin A, synth*).
Matsubara, H. *et al.*, *Bull. Inst. Chem. Res., Kyoto Univ.*, 1972, **50**, 197-205 (*Taiwanin A, insecticide synergist activity*).
Sai, H. *et al.*, *Synthesis*, 1995, 582-586 (*Taiwanin A, synth, cryst struct*).
Jiaang, W.-T. *et al.*, *Heterocycles*, 2000, **53**, 1569-1572 (*synth, config*).
Chang, S.-T. *et al.*, *Phytochemistry*, 2000, **55**, 227-232 (*Taiwanin A, activity*).
Ho, P.-J. *et al.*, *Life Sci.*, 2007, **80**, 493-503 (*synth, activity*)

3,3′,4,4′-Tetrahydroxy-8,8′-lignan-9,9′-dioic acid T34

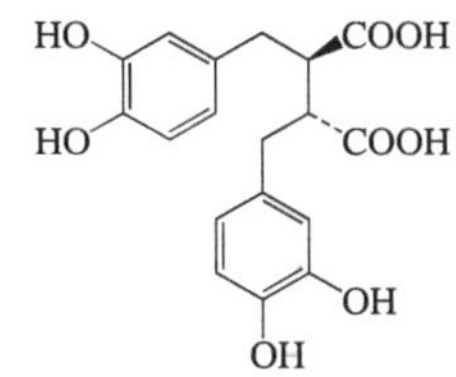

2,3-Bis(3,4-dihydroxybenzyl)butanedioic acid

$C_{18}H_{18}O_8$ 362.334

(8R*,8′R*)-form

3,4-Methylene ether, 3′,4′-di-Me ether, di-Me ester

[86926-69-4]

$C_{23}H_{26}O_8$ 430.453

Constit. of *Heliopsis buphthalmoides*. Oil

3,4:3′,4′-Bis(methylene) ether, di-Me ester

Dimethyl 2,3-bis(1,3-benzodioxol-5-ylmethyl)butanedioate. Dimethyl 2,3-bis(3,4-methylenedioxybenzyl)succinate.

Heliobuphthalmin

[66547-91-9]

$C_{22}H_{22}O_8$ 414.41

Constit. of *Heliopsis buphthalmoides*, *Pycnanthus angolensis* and *Justicia ciliata*. Antiproliferative agent. Cryst. (hexane). Mp 71 - 74°. $[\alpha]_D^{20}$ -5.4 (c, 0.25 in $CHCl_3$).

Bohlmann, F. *et al.*, *Phytochemistry*, 1978, **17**, 330-331 (*Heliopsis buphthalmoides constit*).
Bohlmann, F. *et al.*, *Liebigs Ann. Chem.*, 1983, 1202-1206 (*Heliopsis buphthalmoides constit*).
Day, S.H. *et al.*, *J. Nat. Prod.*, 1999, **62**, 1056-1058 (*Justicia ciliata constit*).
Abrantes, M. *et al.*, *Planta Med.*, 2008, **74**, 1408-1412 (*Pycnanthus angolensis constit*).
Duarte, N. *et al.*, *Planta Med.*, 2010, **76**, 975-980 (*activity*)

3,3′,5,5′-Tetrahydroxy-4,4′-lignan-9,9′-dioic acid T35

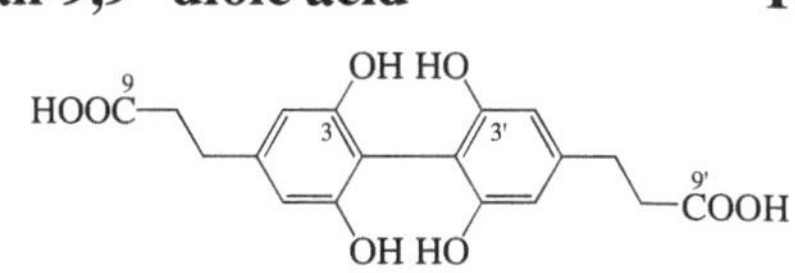

2,2′,6,6′-Tetrahydroxy-4,4′-biphenyldipropanoic acid

$C_{18}H_{18}O_8$ 362.334

Lignan numbering shown

9-Heptyl 9′-hexyl diester

[1195967-83-9]

$C_{31}H_{44}O_8$ 544.685

Constit. of the fruit of *Sorbus domestica*. Powder (MeOH). λ_{max} 274 (MeOH)

3,3′-Dideoxy

2,2′-Dihydroxy-4,4′-biphenyldipropanoic acid. 3,3′-Dihydroxy-4,4′-lignan-9,9′-dioic acid

3,3′-Dideoxy, 9-heptyl 9′-hexyl diester

[1195967-82-8]

$C_{31}H_{44}O_6$ 512.687

Constit. of the fruit of *Sorbus domestica*. Powder (MeOH). λ_{max} 278; 282 (sh) (MeOH).

Termentzi, A. *et al.*, *Food Chem.*, 2009, **116**, 371-381 (*Sorbus domestica constits*)

4,4′,5,5′-Tetrahydroxy-3,3′-lignan-9,9′-dioic acid T36

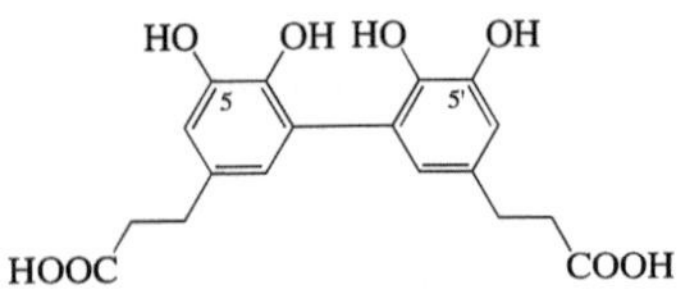

5,5′,6,6′-Tetrahydroxy[1,1′-biphenyl]-3,3′-dipropanoic acid

$C_{18}H_{18}O_8$ 362.334

5,5′-Di-Me ether

4,4′-Dihydroxy-5,5′-dimethoxy-3,3′-lignan-9,9′-dioic acid

5,5′-Di-Me ether, 4-O-β-D-glucopyranoside

Dichotomoside A

[845673-17-8]

$C_{26}H_{32}O_{13}$ 552.529

Constit. of the roots of *Stellaria dichotoma* var. *lanceolata*. Powder. $[\alpha]^{27}_{D}$ -2.3 (c, 0.3 in MeOH). λ_{max} 284 (log ε 3.78) (MeOH)

5,5′-Di-Me ether, 9-butyl ester, 4-O-β-D-glucopyranoside

Dichotomoside B

[845673-18-9]

$C_{30}H_{40}O_{13}$ 608.637

Constit. of the roots of *Stellaria dichotoma* var. *lanceolata*. Powder. $[\alpha]^{27}_{D}$ +8.4 (c, 0.5 in MeOH). λ_{max} 275 (log ε 4.09) (MeOH)

5,5′-Di-Me ether, 9′-butyl ester, 4-O-β-D-glucopyranoside

Dichotomoside C

[845673-19-0]

$C_{30}H_{40}O_{13}$ 608.637

Constit. of the roots of *Stellaria dichotoma* var. *lanceolata*. Powder. $[\alpha]^{27}_{D}$ +5.5 (c, 0.2 in MeOH). λ_{max} 283 (log ε 3.89) (MeOH)

5,5′-Di-Me ether, 9,9′-dibutyl ester, 4-O-β-D-glucopyranoside

Dichotomoside D

[845673-20-3]

$C_{34}H_{48}O_{13}$ 664.745

Constit. of the roots of *Stellaria dichotoma* var. *lanceolata*. Antiallergic agent. Inhibitor of β-hexosaminidase, tumour necrosis factor-α and interleukin-4. Powder. $[\alpha]^{27}_{D}$ +7.4 (c, 0.3 in MeOH). λ_{max} 283 (log ε 3.71) (MeOH).

Morikawa, T. *et al.*, *Chem. Pharm. Bull.*, 2004, **52**, 1194-1199 (*Dichotomosides A-D, activity*)

3,3′,4,4′-Tetrahydroxy-8,8′-lignan-9,7′:9′,7-diolide T37

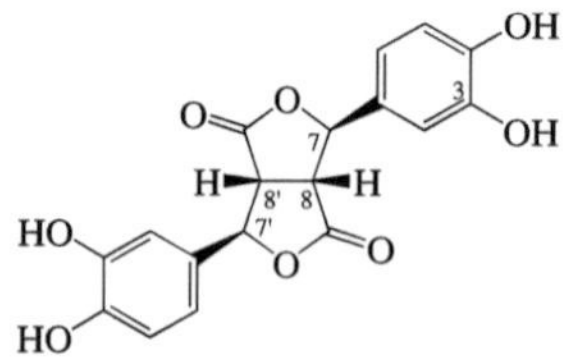

(7R,7′R,8R,8′R)-form

3,6-Bis(3,4-dihydroxyphenyl)tetrahydro-1H,4H-furo[3,4-c]furan-1,4-dione, CAS.

Dehydrodicaffeic acid dilactone

$C_{18}H_{14}O_8$ 358.302

(7R,7′R,8R,8′R)-form [60873-31-6]

Isol. from the mushroom *Inonotus* sp. K-1410. Inhibitor of catechol *O*-methyltransferase. Shows hypotensive props.. Yellow cryst. (C_6H_6/Me_2CO). Sol. MeOH, $CHCl_3$; poorly sol. H_2O. Mp 199 - 201° dec.. $[\alpha]^{20}_{D}$ -91 (c, 1 in MeOH)

(7S,7′S,8R,8′R)-form

3,4:3′,4′-Bis(methylene) ether

3,3′:4,4′-Bis(methylenedioxy)-8,8′-lignan-9,7′:9′,7-diolide. ***7,7′-Dioxodiasesamin***. *Diasesamin dilactone*

[816422-84-1]

$C_{20}H_{14}O_8$ 382.324

Constit. of *Wikstroemia lanceolata*. Needles (Me_2CO). Mp 188 - 189°. $[\alpha]^{26}_{D}$ -47.4 (c, 0.1 in $CHCl_3$). λ_{max} 241 (log ε 3.7); 286 (log ε 3.68) (MeOH)

(7S,7′S,8S,8′S)-form [60873-30-5]

[60804-37-7] (rel. config.)

Isol. from the mushroom *Inonotus* sp. K-1410. Inhibitor of catechol *O*-methyltransferase. Shows hypotensive props.. Yellow cryst. (C_6H_6/Me_2CO). Sol. MeOH, $CHCl_3$; poorly sol. H_2O. Mp 199 - 201° dec.. $[\alpha]^{20}_{D}$ +91 (c, 1 in MeOH)

(7ξ,7′ξ,8ξ,8′ξ)-form [164201-19-8]

3,3′-Di-Me ether

4,4′-Dihydroxy-3,3′-dimethoxy-8,8′-lignan-9,7′:9′,7-diolide. ***Dehydrodiferulic dilactone***

[13962-76-0]

$C_{20}H_{18}O_8$ 386.356

Constit. of the cell walls of wheat and barley. Potent trypanoside. Stereochem. not determined.

Kumada, Y. *et al.*, *J. Antibiot.*, 1976, **29**, 882-889 (*Inonotus constits*).
Nakamura, H. *et al.*, *Acta Crystallogr., Sect. B*, 1977, **33**, 1260-1263 (*abs config*).
Kumada, Y. *et al.*, *Agric. Biol. Chem.*, 1977, **41**, 877-885 (*synth*).
Quideau, S. *et al.*, *J. Chem. Soc., Perkin Trans. 1*, 1993, 653-659 (*synth*).
Turner, L.B. *et al.*, *Phytochemistry*, 1993, **33**, 791-796 (*cereal cell wall constit*).
Lin, R.-W. *et al.*, *Planta Med.*, 2004, **70**, 234-238 (*7,7′-Dioxodiasesamin*).
Andrade e Silva, M.L. *et al.*, *Phytochemistry*, 2014, **107**, 119-125 (*Dehydrodiferulic dilactone, activity*)

3,3′,4,4′-Tetrahydroxy-8,8′-lignan-9′,7-olide T38

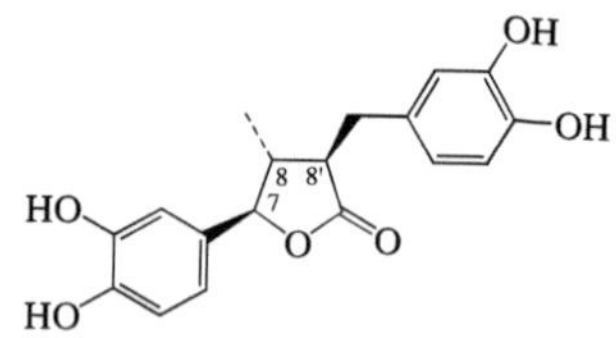

(7R,8R*,8′R*)-form*

$C_{18}H_{18}O_6$ 330.336

(7R*,8R*,8′R*)-form

Bis(methylene) ether

3,4:3′,4′-Bis(methylenedioxy)-8,8′-lignan-9′,7-olide. ***Pycnanolide A***

[1384932-18-6]

$C_{20}H_{18}O_6$ 354.358

Constit. of the leaves of *Pycnanthus angolensis*. Gum. λ_{max} 191 (log ε 6); 202 (log ε 0.87); 236 (log ε 0.15); 286 (log ε 0.11); 389 (log ε 0.01) ($CHCl_3$)

(7R*,8S*,8′R*)-form

Bis(methylene) ether

Pycnanolide B

[1384932-19-7]

Constit. of the leaves of *Pycnanthus angolensis*. λ_{max} 196 (log ε 6); 203 (log ε 1.43); 235 (log ε 0.21); 286 (log ε 0.19); 353 (log ε 0.02); 390 (log ε 0.02) (MeOH).

Onocha, P.A. *et al.*, *Res. J. Phytochem.*, 2011, **5**, 136-145 (*Pycnanolides A,B*)

3,3′,4,4′-Tetrahydroxy-8,8′-lignan-9,9′-olide **T39**

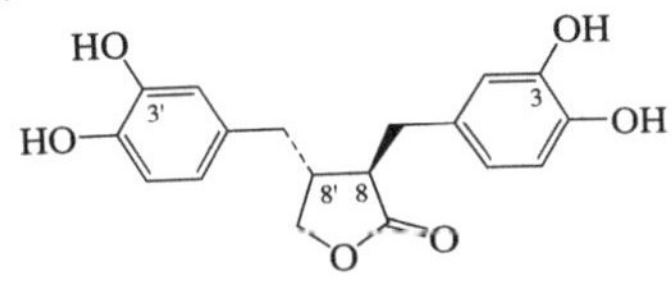

(8R,8′R)-form

3,4-Bis(3,4-dihydroxybenzyl)-4,5-dihydro-2(3H)-furanone
$C_{18}H_{18}O_6$ 330.336
(8R,8′R)-form [119098-95-2]
$[\alpha]^{25}_{D}$ -14 (c, 0.1 in MeOH). λ_{max} 220 (ε 11020); 283 (ε 5300) (MeOH)

3′-Me ether
3,4,4′-Trihydroxy-3′-methoxy-8,8′-lignan-9,9′-olide. ***Piperphilippinin I***
[479030-51-8]
$C_{19}H_{20}O_6$ 344.363
Constit. of the stems of *Piper philippinum*. Oil. $[\alpha]^{24}_{D}$ -45.3 (c, 5 in MeOH). λ_{max} 224 (log ε 4.51); 283 (log ε 4.23) (MeOH). [base] λ_{max} 244 (sh); 290 (log ε 4.25); 406 (log ε 3.45) (MeOH/KOH)

3,3′-Di-Me ether
see 4,4′-Dihydroxy-3,3′-dimethoxy-8,8′-lignan-9,9′-olide, D141

3,4-Di-Me ether
3′,4′-Dihydroxy-3,4-dimethoxy-8,8′-lignan-9,9′-olide. ***Guayarol***
[130288-64-1]
$C_{20}H_{22}O_6$ 358.39
Constit. of seeds of *Bupleurum salicifolium*. Brown oil. $[\alpha]^{25}_{D}$ -9.6 (c, 0.05 in $CHCl_3$)

4,4′-Di-Me ether
3,3′-Dihydroxy-4,4′-dimethoxy-8,8′-lignan-9,9′-olide. ***Prestegane B***
[93376-04-6]
$C_{20}H_{22}O_6$ 358.39
Constit. of *Steganotaenia araliacea* and *Thymelaea microphylla*. Antioxidant. Amorph. solid. $[\alpha]^{22}_{D}$ -25 (c, 0.18 in $CHCl_3$)

3,3′,4-Tri-Me ether
4′-Hydroxy-3,3′,4-trimethoxy-8,8′-lignan-9,9′-olide. ***Buplerol***
[119030-66-9]
$C_{21}H_{24}O_6$ 372.417
Constit. of *Aframomum melegueta* and *Bupleurum salicifolium*. Exhibits cytotoxic activity. Reddish oil. $[\alpha]^{25}_{D}$ -22.7 (c, 0.02 in $CHCl_3$)

3,3′,4-Tri-Me ether, 4′-O-β-D-glucopyranoside
Styraxjaponoside B
[848757-29-9]
$C_{27}H_{34}O_{11}$ 534.558
Constit. of the stem bark of *Styrax japonica*. Matrix metalloproteinase-1 inhibitor. Cryst.. Mp 84 - 86° dec.. $[\alpha]^{25}_{D}$ -5.1 (c, 0.3 in MeOH). λ_{max} 232 (log ε 2.35); 279 (log ε 1.1) (MeOH)

3,3′,4′-Tri-Me ether
see 4-Hydroxy-3,3′,4′-trimethoxy-8,8′-lignan-9,9′-olide, H141

3′,4,4′-Tri-Me ether
3-Hydroxy-3′,4,4′-trimethoxy-8,8′-lignan-9,9′-olide. ***Prestegane A***
[87605-56-9]
$C_{21}H_{24}O_6$ 372.417
Isol. from *Steganotaenia araliacea*. Amorph.. $[\alpha]^{21}_{D}$ -31 (c, 0.39 in $CHCl_3$)

Tetra-Me ether
3,3′,4,4′-Tetramethoxy-8,8′-lignan-9,9′-olide. ***Maculatin‡***. *Dimethylmatairesinol. Arctigenin methyl ether*
[25488-59-9]
$C_{22}H_{26}O_6$ 386.444
Constit. of *Ptelea trifoliata, Bupleurum salicifolium, Cinnamomum camphora, Steganotaenia araliacea* and others. Shows cytotoxic activity. Prisms (MeOH). Mp 129 - 131° (126-127°). $[\alpha]^{25}_{D}$ -30 (c, 1.3 in $CHCl_3$)

3,4-Methylene ether
3′,4′-Dihydroxy-3,4-methylenedioxy-8,8′-lignan-9,9′-olide. ***3′,4′-Demethylenehinokinin***
[475598-85-7]
$C_{19}H_{18}O_6$ 342.347
Constit. of the roots of *Taiwania cryptomerioides*. Pale yellow gum. $[\alpha]^{25}_{D}$ -20.2 (c, 0.24 in $CHCl_3$). This struct. was originally assigned to the isolate from *Chamaecyparis obtusa* but was revised in 2007. λ_{max} 230 (log ε 3.8); 285 (log ε 3.7) (MeOH)

3,4-Methylene, 3′-Me ether
4′-Hydroxy-3′-methoxy-3,4-methylenedioxy-8,8′-lignan-9,9′-olide. ***Haplomyrfolin***
[85404-48-4]
[84755-37-3] ((±)-form)
$C_{20}H_{20}O_6$ 356.374
Isol. from *Haplophyllum myrtifolium*. Oil. $[\alpha]^{25}_{D}$ -31.6 (c, 0.11 in $CHCl_3$). λ_{max} 230 (log ε 4.18); 284 (log ε 4) (MeOH)

3,4-Methylene, 3′,4′-di-Me ether
3′,4′-Dimethoxy-3,4-methylenedioxy-8,8′-lignan 9,9′ olide. ***Kusunokinin***
[58311-20-9]
$C_{21}H_{22}O_6$ 370.401
Isol. from *Aristolochia triangularis, Aristolochia malmeana, Cinnamomum camphora, Haplophyllum tuberculatum, Haplophyllum vulcanicum, Piper phillipinum, Smallanthus sonchifolius, Virola sebifera* and others. Anticoagulant. Inhibitor of human liver enzyme CYP3A4. Cryst. (MeOH). Mp 125 - 126°. $[\alpha]^{25}_{D}$ -34.5 (c, 0.52 in $CHCl_3$) (-8.8, -26.3). λ_{max} 242 (ε 5623); 282 (ε 5130); 286 (ε 5130) (EtOH) (Berdy)

3,4-Methylene, 4′-O-(3-methyl-2-butenyl), 3′-Me ether
3′-Methoxy-3,4-methylenedioxy-4′-prenyloxy-8,8′-lignan-9,9′-olide
[166600-26-6]
$C_{25}H_{28}O_6$ 424.493
Constit. of *Haplophyllum ptilostylum*

3′,4′-Methylene ether
3,4-Dihydroxy-3′,4′-methylenedioxy-8,8′-lignan-9,9′-olide. ***Piperphilippinin II***. *3,4-Demethylenehinokinin*
[942293-43-8]
$C_{19}H_{18}O_6$ 342.347
Constit. of the stems of *Piper philippinum* and the heartwood of *Chamaecyparis obtusa* var. *formosana*. Oil or pale yellow solid. $[\alpha]^{21}_{D}$ -24.1 (c, 0.11 in $CHCl_3$). $[\alpha]^{26}_{D}$ -21.8 (c, 0.61 in $CHCl_3$). Struct. revised in 2007. Originally assigned as 3′,4′-Demethylenehinokinin (see above). λ_{max} 234 (sh); 286 (log ε 3.94) (MeOH). [base] λ_{max} 240 (sh); 291 (log ε 3.97) (MeOH/KOH). [neutral] λ_{max} 233 (log ε 3.75); 286 (log ε 3.66) (MeOH)

3′,4′-Methylene, 3-Me ether
4-Hydroxy-3-methoxy-3′,4′-methylenedioxy-8,8′-lignan-9,9′-olide. ***Pluviatolide***
[28115-68-6]
$C_{20}H_{20}O_6$ 356.374

Isol. from bark of *Zanthoxylum pluviatile*, from *Aristolochia constricta* and *Taiwania cryptomeroides*. Cryst. (MeOH). Mp 160°. [α] D -35.5 ($CHCl_3$)

3′,4′-Methylene, 4-Me ether

3-Hydroxy-4-methoxy-3′,4′-methylenedioxy-8,8′-lignan-9,9′-olide.

Guamarolin

[145374-38-5]

[145274-64-2] (8*R*,8′*R*-form)

$C_{20}H_{20}O_6$ 356.374

Isol. from the roots of *Bupleurum salicifolium*. Yellow oil. $[\alpha]^{25}_{D}$ -27.4 (c, 0.06 in $CHCl_3$). λ_{max} 237; 290 (EtOH)

3′,4′-Methylene, 3,4-di-Me ether

see 3,4-Dimethoxy-3′,4′-methylenedioxy-8,8′-lignan-9,9′-olide, D265

3′,4′-Methylene, 4-O-(3-methyl-2-butenyl), 3-Me ether

3-Methoxy-3′,4′-methylenedioxy-4-prenyloxy-8,8′-lignan-9,9′-olide

[166600-25-5]

$C_{25}H_{28}O_6$ 424.493

Constit. of *Haplophyllum ptilostylum*

3,4:3′,4′-Bis(methylene) ether

see 3,4:3′,4′-Bis(methylenedioxy)-8,8′-lignan-9,9′-olide, B62

(8R,8′S)-form

3,4-Methylene, 3′,4′-di-Me ether

Isokusunokinin

[88198-99-6]

$C_{21}H_{22}O_6$ 370.401

Isol. from *Virola sebifera*. Cryst. (MeOH). Mp 125 - 126°. $[\alpha]^{25}_{D}$ -8.8 (c, 0.114 in $CHCl_3$). Config. uncertain. λ_{max} 230 (ε 22000); 279 (ε 10900) (MeOH)

(8S,8′S)-form

3′,4-Di-Me ether

3,4′-Dihydroxy-3′,4-dimethoxy-8,8′-lignan-9,9′-olide

3′,4-Di-Me ether, 4′-O-β-D-glucopyranoside

Styraxlignolide C

[823214-03-5]

$C_{26}H_{32}O_{11}$ 520.531

Constit. of *Juniperus chinensis* and *Styrax japonica*. Plates (MeOH). Mp 106 - 108°. $[\alpha]^{23}_{D}$ -8.7 (c, 0.23 in MeOH). λ_{max} 204 (log ε 4.69); 225 (log ε 4.11); 279 (log ε 3.7) (MeOH)

3,3′,4-Tri-Me ether, 4′-O-β-D-glucopyranoside

Styraxlignolide F

[823214-06-8]

$C_{27}H_{34}O_{11}$ 534.558

Constit. of the stem bark of *Styrax japonica*. Plates (MeOH). Mp 128 - 130°. $[\alpha]^{23}_{D}$ -24.8 (c, 0.25 in MeOH). λ_{max} 204 (log ε 4.68); 228 (log ε 4.15); 279 (log ε 3.68) (MeOH).

Reisch, J. *et al.*, *Tetrahedron Lett.*, 1969, **10**, 3803-3806 (*Arctigenin methyl ether*).

Corrie, J.E.T. *et al.*, *Aust. J. Chem.*, 1970, **23**, 133-145 (*Pluviatolide*).

Takaoka, D. *et al.*, *Nippon Kagaku Kaishi*, 1975, 2192-2196 (*Kusunokinin*).

Wenkert, E. *et al.*, *Phytochemistry*, 1976, **15**, 1547-1551 (*cmr*).

Ganeshpure, P.A. *et al.*, *Aust. J. Chem.*, 1982, **35**, 2175 (*synth*).

Lopes, L.M.X. *et al.*, *Phytochemistry*, 1983, **22**, 1516-1518 (*Kusunokinin, Isokusunokinin*).

Taafrout, M. *et al.*, *Tetrahedron Lett.*, 1983, **24**, 3237-3238 (*Prestegane A*).

Taafrout, M. *et al.*, *Tetrahedron Lett.*, 1984, **25**, 4127-4128 (*Prestegane B, Maculatin*).

Gözler, T. *et al.*, *Tetrahedron*, 1984, **40**, 1145-1150 (*Kusunokinin*).

Evcim, U. *et al.*, *Phytochemistry*, 1986, **25**, 1949-1951 (*Haplomyrfolin*).

Sheriha, G.M. *et al.*, *Phytochemistry*, 1987, **26**, 3339-3342 (*Haplophyllum tuberculatum constit*).

Brown, E. *et al.*, *Tetrahedron*, 1989, **45**, 141-154 (*Kusunokinin, synth*).

Gonzalez, A.G. *et al.*, *J. Chem. Res., Synop.*, A.G., 1990, 220-221 (*Bupleurum salicifolium constits*).

Lopes, L.M.X. *et al.*, *Phytochemistry*, 1990, **29**, 660-662 (*Aristolochia triangularis constit*).

Estévez-Reyes, R. *et al.*, *Phytochemistry*, 1992, **31**, 2841-2845 (*Guamarolin, synth*).

Pelter, A. *et al.*, *J. Chem. Soc., Perkin Trans. 1*, 1993, 2631-2638 (*Isokusunokinin, synth*).

Chimichi, S. *et al.*, *Magn. Reson. Chem.*, 1993, **31**, 1044-1047 (*Maculatin, pmr, cmr*).

Van Oeveren, A. *et al.*, *J. Org. Chem.*, 1994, **59**, 5999-6007 (*Pluviatolide, synth*).

Ulubelen, A. *et al.*, *Phytochemistry*, 1995, **39**, 417-422 (*Haplophyllum ptilostylum constits*).

Estévez-Braun, A. *et al.*, *Phytochemistry*, 1996, **43**, 885-886 (*Buplerol, Guayarol, cmr*).

Lin, W-H. *et al.*, *Phytochemistry*, 1999, **50**, 653-658 (*Taiwania cryptomeroides constit*).

Chang, S.-T. *et al.*, *Phytochemistry*, 2000, **55**, 227-232 (*Maculatin, activity*).

Meragelman, K.M. *et al.*, *J. Nat. Prod.*, 2001, **64**, 1480-1482 (*Prestegane B, cmr*).

Kuo, Y.-H. *et al.*, *Chem. Pharm. Bull.*, 2002, **50**, 978-980 (*3′,4′-Demethylenehinokinin*).

Xie, L-H. *et al.*, *Chem. Pharm. Bull.*, 2003, **51**, 508-515 (*8R,8′R-form, synth*).

Kim, M.-R. *et al.*, *Chem. Pharm. Bull.*, 2004, **52**, 1466-1469 (*Styraxjaponoside B*).

Min, B.-S. *et al.*, *J. Nat. Prod.*, 2004, **67**, 1980-1984 (*Styraxlignolides C,F*).

Chyu, C.-F. *et al.*, *Helv. Chim. Acta*, 2007, **90**, 738-747 (*Demethylenehinokinins*).

Chen, Y.-C. *et al.*, *Phytochemistry*, 2007, **68**, 2101-2111 (*Piperphilippinins I,II*).

Messiano, G.B. *et al.*, *J. Agric. Food Chem.*, 2008, **56**, 2655-2659 (*Aristolochia malmeana constit*).

Zhang, G. *et al.*, *J. Nat. Prod.*, 2008, **71**, 1167-1172 (*Aristolochia constricta constit*).

Yoo, H.H. *et al.*, *Planta Med.*, 2008, **74**, 822-827 (*Kusunokinin, activity*).

Hung, T.M. *et al.*, *J. Korean Soc. Appl. Biol. Chem.*, 2009, **52**, 560-563 (*Styraxlignolide F, cryst struct*).

Lu, J. *et al.*, *Asian J. Chem.*, 2012, **24**, 312-314 (*Smallanthus sonchifolius constit*).

Marques, J.V. *et al.*, *J. Biol. Chem.*, 2013, **288**, 466-479 (*Pluviatolide*)

3,3′,4,5′-Tetrahydroxy-8,8′-lignan-9,9′-olide T40

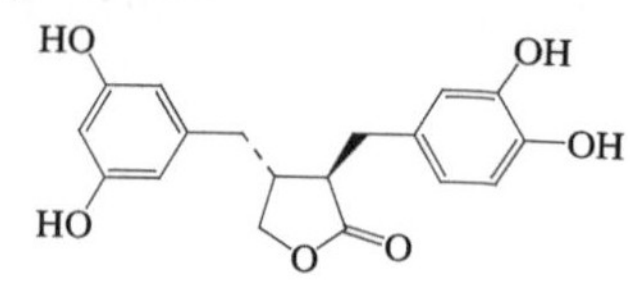

3-(3,4-Dihydroxybenzyl)-4-(3,5-dihydroxybenzyl)dihydro-2(3H)-furanone

$C_{18}H_{18}O_6$ 330.336

Structures in this series may need revision. Phenyl 3,5-dioxygenated substn. pattern in doubt as a result of work done by Yang *et al* in 2015

(8R,8′R)-form

3′,4-Di-Me ether

3,3′-Dihydroxy-4,5′-dimethoxy-8,8′-lignan-9,9′-olide

$C_{20}H_{22}O_6$ 358.39

3′,4-Di-Me ether, 5′-O-β-D-glucopyranoside

Apocynotrachelogenin 5′-glucoside

[1069136-59-9]

$C_{26}H_{32}O_{11}$ 520.531

Constit. of the leaves of *Trachelospermum lucidum*. Light yellow oil. $[\alpha]^{25}_{D}$ -70 (c, 0.001 in MeOH). λ_{max} 189; 207; 226 (MeOH).

Kousar, F. *et al.*, *J. Asian Nat. Prod. Res.*, 2008, **10**, 285-288 (*Apocynotrachelogenin glucoside*).

Yang, Y.-N. *et al.*, *J. Nat. Prod.*, 2015, **78**, 705-711 (*pmr, struct*)

3,4,4′,7-Tetrahydroxy-8,8′-lignan-9,9′-olide T41

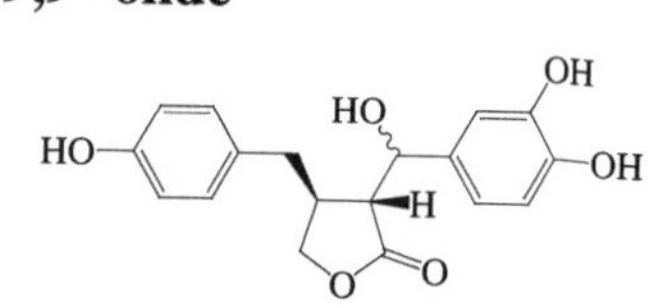

$C_{18}H_{18}O_6$ 330.336

(7ξ,8R,8′S*)-form*
3,4-Methylene ether
4′,7-Dihydroxy-3,4-methylene-dioxy-8,8′-lignan-9,9′-olide.
Taiwaninolide
[942505-96-6]
$C_{19}H_{18}O_6$ 342.347
Constit. of the roots of *Taiwania cryptomerioides*. Yellow gum. $[\alpha]_D^{25}$ -57.2 (c, 0.42 in $CHCl_3$). λ_{max} 226 (log ε 4.01); 284 (log ε 3.65) (MeOH).

Chyu, C.-F. *et al.*, *Helv. Chim. Acta*, 2007, **90**, 738-747 (*Taiwaninolide*)

3,3′,4,4′-Tetrahydroxy-8,8′-lign-7-ene-9,9′-dioic acid T42

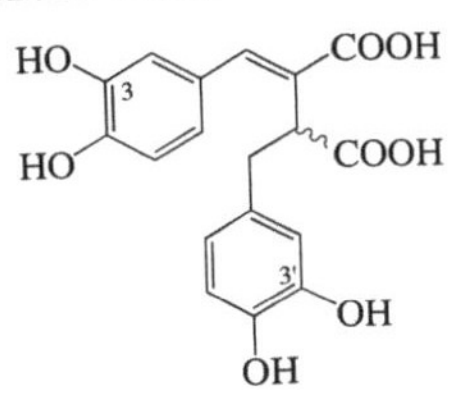

$C_{18}H_{16}O_8$ 360.318
(7E,8′ξ)-form
3,4:3′,4′-Bis(methylene) ether, di-Me ester
Dehydroheliobuphthalmin
[66547-92-0]
$C_{22}H_{20}O_8$ 412.394
Isol. from *Heliopsis buphthalmoides*. Needles (EtOAc/petrol). Mp 84 - 85°.

Yamashita, K. *et al.*, *Bull. Agric. Chem. Soc. Jpn.*, 1960, **24**, 160-162 (*synth*).
Bohlmann, F. *et al.*, *Phytochemistry*, 1978, **17**, 330-331 (*Heliopsis buphthalmoides constit*).
Bohlmann, F. *et al.*, *Liebigs Ann. Chem.*, 1983, 1202-1206 (*Heliopsis buphthalmoides constit*).
Jakupovic, J. *et al.*, *Planta Med.*, 1986, **52**, 18-20 (*Heliopsis buphthalmoides constit*)

3′,4,4′,5-Tetrahydroxy-2,8′-lign-7′-ene-9,9′-dioic acid T43

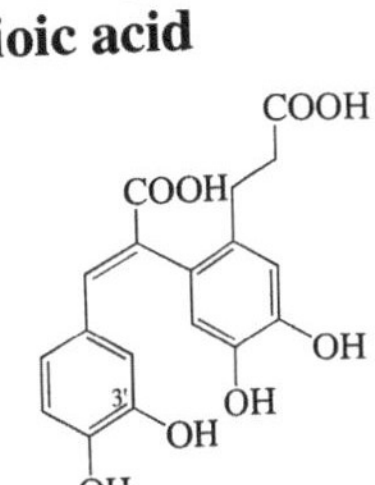

$C_{18}H_{16}O_8$ 360.318
3′-Me ether, 9,9′-bis[2-(4-hydroxy-phenyl)ethylamide]
$C_{35}H_{36}N_2O_8$ 612.679
Constit. of the stems of *Lycium barbarum*. Yellow powder. $[\alpha]_D^{19}$ -2.3 (c, 0.47 in MeOH). λ_{max} 217 (log ε 4.6); 289 (log ε 4.1); 324 (log ε 4.1) (MeOH)
3′-Me ether, 9-[2-(4-hydroxyphenyl) ethylamide], 9′-(4-acetamidobutyl) amide
[1416474-64-0]
$C_{33}H_{39}N_3O_8$ 605.688
Isol. from the root bark of *Lycium chinense*. Exhibits moderate radical scavenging activity against DPPH and superoxide radicals. Yellow gum. λ_{max} 291 (log ε 4.04); 321 (log ε 4.10) (MeOH).

Zhang, J.-X. *et al.*, *J. Nat. Prod.*, 2013, **76**, 51-58 (*Lycium chinense amide, activity*).
Zhu, P.-F. *et al.*, *Nat. Prod. Bioprospect.*, 2017, **7**, 421-431 (*Lycium barbarum constit*)

3,3′,4,4′-Tetrahydroxy-8,8′-lign-7-en-9,9′-olide T44

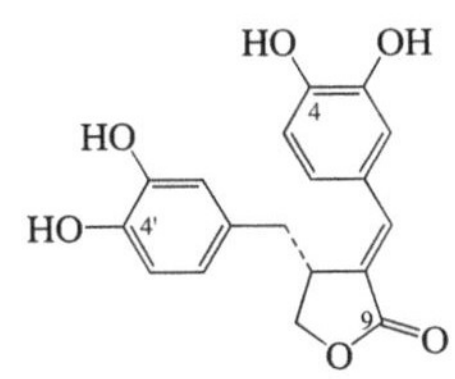

(7E,8′R)-form

$C_{18}H_{16}O_6$ 328.32
(7E,8′R)-form
3,3′-Di-Me ether
4,4′-Dihydroxy-3,3′-dime-thoxy-8,8′-lign-7-en-9,9′-olide.
Isosalicifolin
[156974-99-1]
$C_{20}H_{20}O_6$ 356.374
Constit. of the leaves of *Bupleurum salicifolium*. Yellow oil. $[\alpha]_D^{20}$ -26.5 (c, 0.07 in $CHCl_3$). λ_{max} 234; 290; 336 (EtOH)
3,3′,4′-Tri-Me ether
4-Hydroxy-3,3′,4′-trimethoxy-8,8′-lign-7-en-9,9′-olide. ***7,8-Didehydroarctigenin***. *Carthamogenin. Hemislin B*
[76015-62-8]
[160661-33-6]
$C_{21}H_{22}O_6$ 370.401
Constit. of *Cheirolophus* x *hortigenus*, *Ptilostemon gnaphaloides*, *Ptilostemon afer*, *Ptilostemon diacanthum* and fruit of *Leuzea carthamoides*. Gum or oil. $[\alpha]_D$ -25 (c, 1.7 in $CHCl_3$). $[\alpha]_D^{24}$ -58.3 (c. 0.7 in $CHCl_3$). Prob. config.. λ_{max} 288; 300 (sh); 332 (MeOH)
3,3′,4′-Tri-Me ether, 4-O-β-D-gluco-pyranoside
Hemislienoside
[420119-90-0]
$C_{27}H_{32}O_{11}$ 532.542
Constit. of *Hemisteptia lyrata*. Cryst. (MeOH). Mp 112 - 113°. $[\alpha]_D^{20}$ -9 (c, 0.18 in MeOH). Tentative config. assigned. λ_{max} 224 (log ε 4.24); 290 (log ε 4.1); 316 (log ε 4.16) (MeOH)
Tetra-Me ether
3,3′,4,4′-Tetramethoxy-8,8′-lign-7-en-9,9′-olide
[112820-15-2]
$C_{22}H_{24}O_6$ 384.428
Constit. of *Virola elongata*. Oil
3,4-Methylene, 3′,4′-di-Me ether
3′,4′-Dimethoxy-3,4-methylene-dioxy-8,8′-lign-7-en-9,9′-olide.
(+)-Suchilactone
[18884-23-6]
$C_{21}H_{20}O_6$ 368.385
Constit. of stem, roots and seeds of *Jatropha gossypifolia*. Cryst. (C_6H_6). Mp 129°. $[\alpha]_D^{25}$ +87 ($CHCl_3$). λ_{max} 223 (log ε 4.19); 288 (log ε 3.99); 297 (log ε 3.98); 333 (log ε 4.21) (EtOH)
3′,4′-Methylene ether
3,4-Dihydroxy-3′,4′-methylene-dioxy-8,8′-lign-7-en-9,9′-olide.
Piperphilippinin IV
[949102-29-8]
$C_{19}H_{16}O_6$ 340.331
Constit. of the stems of *Piper philippinum*. Shows anticoagulant props.. Yellowish oil. $[\alpha]_D^{23}$ -7 (c, 0.19 in MeOH). λ_{max} 237 (log ε 4.15); 295 (log ε 4.11); 338 (log ε 4.24) (MeOH). [base] λ_{max} 279 (log ε 4.05); 392 (log ε 4.32) (MeOH/KOH)
3′,4′-Methylene, 3-Me ether
4-Hydroxy-3-methoxy-3′,4′-methy-lenedioxy-8,8′-lign-7-en-9,9′-olide.
Isoguamarol‡
$C_{20}H_{18}O_6$ 354.358
Constit. of the roots of *Taiwania cryptomerioides*. Yellow gum. $[\alpha]_D^{25}$ -67.8 (c, 0.4 in $CHCl_3$). λ_{max} 237 (log ε 4.09); 295 (log ε 4.01); 331 (log ε 4.16) (MeOH)
3′,4′-Methylene, 4-Me ether
3-Hydroxy-4-methoxy-3′,4′-methy-lenedioxy-8,8′-lign-7-en-9,9′-olide.
Guamarol
[145274-61-9]
$C_{20}H_{18}O_6$ 354.358
Constit. of the roots of *Bupleurum salicifolium*. Yellow oil. $[\alpha]_D^{25}$ -44

(c, 0.04 in $CHCl_3$). λ_{max} 232; 291; 331 (EtOH)

3′,4′-Methylene, 3,4-di-Me ether

3,4-Dimethoxy-3′,4′-methylenedioxy-8,8′-lign-7-en-9,9′-olide. ***Kaerophyllin***. *Chaerophyllin‡*

[75590-33-9]

Isol. from *Chaerophyllum maculatum, Bupleurum lanceolatum, Bupleurum scorzonerifolium* and *Montrouzlera spheroidea*. Shows moderate antiparasitic activity against *Trypanosoma cruzi*. Needles (hexane or EtOAc/hexane). Mp 150°. $[\alpha]_D$ -91.8 (c, 0.39 in $CHCl_3$). λ_{max} 237; 293; 330 (EtOH)

Bis(methylene) ether

see 3,4:3′,4′-Bis(methylenedioxy)-8,8′-lign-7-en-9,9′-olide, B63

(7E,8′S)-form

3,3′-Di-Me ether

Acutissimalignan B

[1005390-47-5]

$C_{20}H_{20}O_6$ 356.374

Constit. of aerial parts of *Phyllanthus acutissima*. Shows moderate cytotoxicity against various human cancer cell lines and moderate anti-HIV activity. Powder (EtOAc/hexane). Mp 128 - 129°. $[\alpha]_D^{25}$ +29.8 (c, 0.63 in $CHCl_3$). λ_{max} 229 (log ε 3.46); 286 (log ε 3.28); 329 (log ε 3.51) (EtOH)

3,3′,4′-Tri-Me ether

(+)-Didehydroarctigenin

[887281-31-4]

$C_{21}H_{22}O_6$ 370.401

Constit. of the fruit of *Arctium lappa*. Resin. $[\alpha]_D^{20}$+16.8 (c, 0.1 in MeOH)

3,4-Methylene, 3′-Me ether

4′-Hydroxy-3′-methoxy-3,4-methylenedioxy-8,8′-lign-7-en-9,9′-olide. ***4′-O-Demethyl-(-)-suchilactone***

[162614-24-6]

$C_{20}H_{18}O_6$ 354.358

Constit. of *Taxus baccata*. Yellow oil. $[\alpha]_D^{25}$-38.3 (c, 0.24 in $CHCl_3$)

3,4-Methylene, 3′,4′-di-Me ether

(-)-Suchilactone. *Jatrophan*

[50816-74-5]

Isol. from whole plant of *Polygala chinensis* and by degradation of 3,3′,4,4′,8-Pentahydroxy-8,8′-lignan-9,9′-olide, P22. Constit. of *Haplophyllum popovii* and *Jatropha gossypifolia*. Mp 135°. $[\alpha]_D^{25}$-85.5 (c, 0.35 in $CHCl_3$)

3′,4′-Methylene ether, 3-Me ether

Songbosin

[1647073-96-8]

$C_{20}H_{18}O_6$ 354.358

Constit. of the aerial parts of *Phyllanthus songboiensis*. Amorph. powder. $[\alpha]_D^{20}$+99 (c, 0.1 in $CHCl_3$). Config. of double bond is uncertain from ref.. λ_{max} 236 (log ε 4.19); 295 (log ε 4.11); 332 (log ε 4.26) (MeOH)

3′,4′-Methylene, 3,4-di-Me ether

Gossypifan

[169873-74-9]

Constit. of *Jatropha gossypifolia* and *Zanthoxylum naranjillo*. Needles (C_6H_6). Mp 145 - 146°. $[\alpha]_D^{25}$+82.4 (c, 0.9 in $CHCl_3$) (+79.1). λ_{max} 235 (ε 15136); 288 (ε 10233); 296 (ε 9333); 334 (ε 16596) (EtOH)

(7Z,8′R)-form

3,3′-Di-Me ether

Salicifolin‡

[156957-19-6]

$C_{20}H_{20}O_6$ 356.374

Constit. of the leaves of *Bupleurum salicifolium*. Yellow oil. $[\alpha]_D^{20}$-61.3 (c, 0.08 in $CHCl_3$). λ_{max} 228; 288; 336 (EtOH)

3,3′,4′-Tri-Me ether

4′-O-Methylsalicifolin

[942293-42-7]

$C_{21}H_{22}O_6$ 370.401

Constit. of the roots of *Taiwania cryptomerioides*. Yellow gum. $[\alpha]_D^{25}$-26.4 (c, 1.1 in $CHCl_3$). λ_{max} 286 (log ε 2.92); 336 (log ε 3.12) (MeOH)

Tetra-Me ether

[235743-16-5]

Constit. of *Bupleurum acutifolium*. Cryst. (2-methoxy-2-methylpropane). Mp 120°. $[\alpha]_D^{25}$-227 (c, 1 in $CHCl_3$)

3′,4′-Methylene, 4-Me ether

Isoguamarol

[145274-62-0]

$C_{20}H_{18}O_6$ 354.358

Constit. of the roots of *Bupleurum salicifolium*. Yellow oil. λ_{max} 236; 288; 334 (EtOH)

3′,4′-Methylene, 3,4-di-Me ether

Isokaerophyllin. *Isochaerophyllin*

[126574-52-5]

Isol. from *Bupleurum salicifolium* and *Bupleurum scorzonerifolium*. Cryst. (EtOAc/hexane). Mp 130 - 131°. $[\alpha]_D^{25}$-76.4 ($CHCl_3$). λ_{max} 237; 293; 337 (EtOH)

(7Z,8′S)-form

3,4-Methylene, 3′,4′-di-Me ether

Isosuchilactone

[72598-33-5]

Formed by ready isom. of Suchilactone. Mp 133°

(7Z,8′ξ)-form

3′,4′-Methylene, 3,4-di-Me ether

Chisulactone

[50924-59-9]

Isol. from entire plant of *Polygala chinensis*. Abs. config. not detd.

(7ξ,8′ξ)-form

3,3′,4′-Tri-Me ether, 4-O-β-D-glucopyranoside

Carthamoside‡

[543725-49-1]

Constit. of the seeds of *Leuzea carthamoides*.

Ghosal, S. *et al., Phytochemistry*, 1974, **13**, 1933-1936 (*Chisulactone, isol, biosynth*).

Ghosal, S. *et al., J. Indian Chem. Soc.*, 1978, **55**, 1201-1203 (*synth, chromatog*).

Bohlmann, F. *et al., Phytochemistry*, 1980, **19**, 692-696 (*Didehydroarctigenin*).

Razakova, D.M. *et al., Khim. Prir. Soedin.*, 1981, **17**, 516-517 (*(-)-Suchilactone*).

Chatterjee, A. *et al., Phytochemistry*, 1981, **20**, 2047-2048 (*(+)-Suchilactone, cryst struct*).

Mikaya, G.A. *et al., Planta Med.*, 1981, **43**, 378-380 (*Kaerophyllin*).

González, A.G. *et al., Phytochemistry*, 1990, **29**, 675-678 (*Kaerophyllin, Isokaerophyllin*).

Kato, M.J. *et al., Phytochemistry*, 1990, **29**, 1799-1810 (*R,E-tetra-Me ether*).

Esteves-Reyes, R. *et al., Phytochemistry*, 1992, **31**, 2841-2845 (*Guamarol, Isoguamarol*).

Banerji, J. *et al., Indian J. Chem., Sect. B*, 1993, **32**, 709-712 (*synth*).

Khetwal, K.S. *et al., J. Nat. Prod.*, 1993, **56**, 318-328 (*abs config, cryst struct*).

Marco, J.A. *et al., Phytochemistry*, 1994, **37**, 1101-1107 (*Didehydroarctigenin*).

Estevez-Braun, A. *et al., Tetrahedron*, 1994, **50**, 5203-5210 (*Bupleurum constits*).

Das, B. *et al., Phytochemistry*, 1995, **38**, 715-717 (*4′-Demethylsuchilactone*).

Das, B. *et al., Phytochemistry*, 1995, **40**, 931-932 (*Gossypifan*).

Charlton, J.L. *et al., Can. J. Chem.*, 1997, **75**, 1076-1083 (*Gossipifan, synth*).

Barrero, A.F. *et al., J. Nat. Prod.*, 1999, **62**, 946-948 (*R,Z-tetra-Me ether*).

Bastos, J.K. *et al., Planta Med.*, 1999, **65**, 541-544 (*Kaerophyllin, activity*).

Bastos, J.K. *et al., Phytochemistry*, 2000, **53**, 1043-1046 (*Montrouzlera sphaeroidea constit*).

Ren, Y.-L. *et al., Yaoxue Xuebao*, 2001, **36**, 746-749 (*Hemislienoside*).

Ren, Y.L. *et al., Chin. Chem. Lett.*, 2002, **13**, 859-861 (*Hemislin B glucoside*).

Xia, Y.-M. *et al.*, *Chin. J. Chem.*, 2003, **21**, 1540-1542 (*synth*).
Chang, W.L. *et al.*, *Phytochemistry*, 2003, **64**, 1375-1379 (*Bupleurum scorzonerifolium constits*).
Da Silva, R. *et al.*, *Magn. Reson. Chem.*, 2005, **43**, 966-969 (*Gossypifan, Carthamogenin, pmr, cmr*).
Matsumoto, T. *et al.*, *Planta Med.*, 2006, **72**, 276-278 (*(+)-Didehydroarctigenin*).
Harmatha, J. *et al.*, *Coll. Czech. Chem. Comm.*, 2007, **72**, 334-336 (*Carthamoside*).
Chyu, C.-F. *et al.*, *Helv. Chim. Acta*, 2007, **90**, 738-747 (*Isoguamarol, 4′-O-Methylsalicifolin*).
Chen, Y.-C. *et al.*, *Phytochemistry*, 2007, **68**, 2101-2111 (*Piperphilippinin IV*).
Tuchinda, P. *et al.*, *J. Nat. Prod.*, 2008, **71**, 655-663 (*Acutissimalignan B*).
Solyomvary, A. *et al.*, *Nat. Prod. Res.*, 2014, **28**, 732-739 (*Leuzea carthamoides constits*).
Ren, Y. *et al.*, *Phytochemistry*, 2015, **111**, 132-140 (*Songbosin*)

3,3′,4,4′-Tetrahydroxy-8,8′-lign-7′-en-9,9′-olide T45

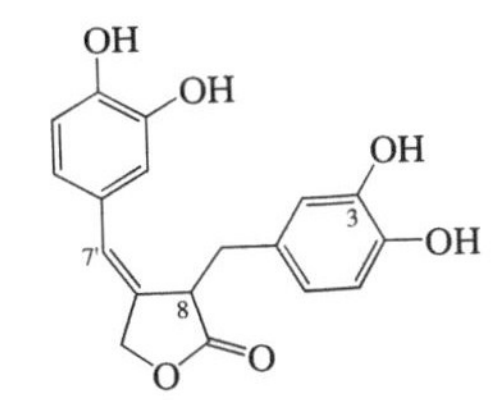

$C_{18}H_{16}O_6$ 328.32
(7′E,8ξ)-form
3′,4′-Methylene, 3,4-di-Me ether
3,4-Dimethoxy-3′,4′-methylenedioxy-8,8′-lign-7′-en-9,9′-olide
[313540-11-3]
$C_{21}H_{20}O_6$ 368.385
Constit. of aerial parts of *Bupleurum tenue*. Cryst.. Mp 136°. λ_{max} 233; 286; 296; 330 (MeOH).

Khetwal, K.S. *et al.*, *Indian J. Chem., Sect. B*, 2000, **39**, 448-450 (*Bupleurum tenue constit*)

3,3′,4,4′-Tetrahydroxy-8,8′-lign-8(8′)-en-9,9′-olide T46

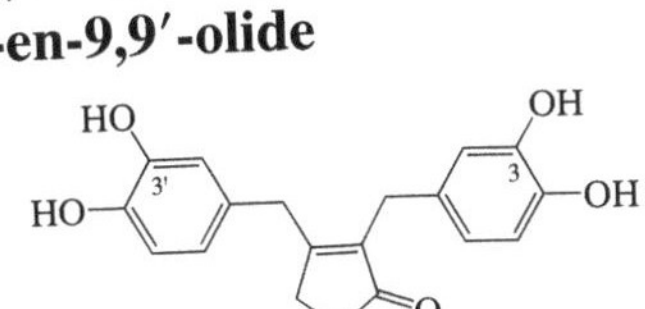

$C_{18}H_{16}O_6$ 328.32
3,4-Methylene, 3′,4′-di-Me ether
3′,4′-Dimethoxy-3,4-methylenedioxy-8,8′-lign-8(8′)-en-9,9′-olide. ***Guayadequiene***
[129502-61-0]
$C_{21}H_{20}O_6$ 368.385
Isol. from leaves of *Bupleurum salicifolium* and stems of *Zanthoxylum lemairei*, also by elimination from Guayadequiol in P22. Uncrystallisable. λ_{max} 256; 274 (EtOH)
3,4:3′,4′-Bis(methylene) ether
3,4:3′,4′-Bis(methylenedioxy)-8,8′-lign-8(8′)-en-9,9′-olide. ***Dihydrotaiwanin A***
[137809-97-3]
$C_{20}H_{16}O_6$ 352.342
Isol. from *Hyptis capitata, Zanthoxylum lemairei, Zanthoxylum nitidum* and from *Annulohypoxylon ilanense*, a fungus derived from *Cinnamomum* sp.. λ_{max} 231; 289 (no solvent reported).

Adesina, S.K. *et al.*, *Pharmazie*, 1977, **52**, 720-724 (*Zanthoxylum lemairei constits*).
Gonzalez, A.G. *et al.*, *Phytochemistry*, 1990, **29**, 1981-1983 (*Guaydequiene*).
Almtorp, G.T. *et al.*, *Phytochemistry*, 1991, **30**, 2753-2756 (*Hyptis capitata constit*).
Yang, C-H. *et al.*, *Chem. Biodiversity*, 2009, **6**, 846-857 (*Zanthoxylum nitidum constit*).
Wu, M.-N. *et al.*, *Chem. Biodiversity*, 2013, **10**, 493-505 (*Annulohypoxylon ilanense metab*)

3,4,4′,9-Tetrahydroxy-8,8′-lign-7′-en-9′,7-olide T47

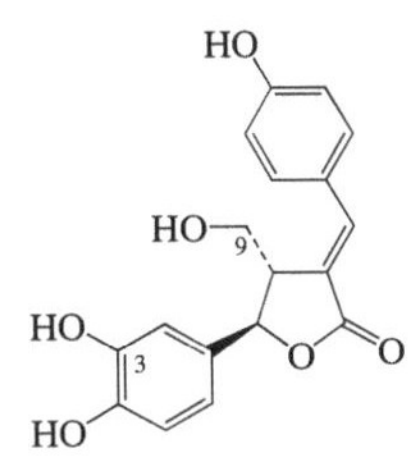

$C_{18}H_{16}O_6$ 328.320
(7R,7′E,8S)-form
9-Carboxylic acid, 3-Me ether
4,4′-Dihydroxy-3-methoxy-8,8′-lign-7′-en-9′,7-olid-9-oic acid
9-Carboxylic acid, 3-Me ether, 4-O-β-D-glucopyranoside
Ligusticumacid E. *Chuanxioside C*
[2573174-21-5]
$C_{25}H_{26}O_{12}$ 518.471
Constit. of the rhizomes of *Ligusticum chuanxiong*. Amorph. yellow powder. $[\alpha]_D^{20}$ +119 (c, 0.1 in MeOH). Possesses (7*R*,8*R*)-config. due to change in Cahn-Ingold-Prelog priorities. Ligusticumacid E and Chuanxioside C have the same optical rotation but are cliamed to be anantiomers. λ_{max} 230 (log ε 4.19); 321 (log ε 4.30) (MeOH)
9-Carboxylic acid, 3-Me ether, 4-O-[4-hydroxy-3-methoxybenzoyl-(1→6)-β-D-glucopyranoside]
Ligusticumacid F
[2573174-22-6]
$C_{33}H_{32}O_{15}$ 668.604
Constit. of the rhizomes of *Ligusticum chuanxiong*. Exhibits weak neuroprotective activity. Amorph. yellow powder. Possesses (7*R*,8*R*)-config. due to change in Cahn-Ingold-Prelog priorities. λ_{max} 221 (log ε 4.56); 266 (log ε 4.18); 321 (log ε 4.38) (MeOH).

Zhang, X. *et al.*, *Fitoterapia*, 2018, **125**, 147-154 (*Ligusticumacids E,F, activity*).
Yuan, X. *et al.*, *Bioorg. Chem.*, 2020, **101**, 104016:1-8 (*Chuanxioside C*)

2,3′,4′,5-Tetrahydroxy-1,8′-lign-8-en-4(1H)-one T48

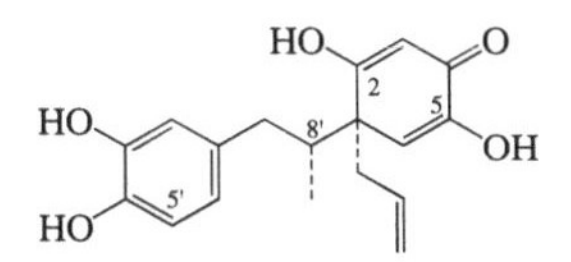

(1*R**,8′*R**)-*form*

$C_{18}H_{20}O_5$ 316.353
(1R*,8′R*)-form
3′,4′-Methylene, 2,5-di-Me ether
2,5-Dimethoxy-3′,4′-methylenedioxy-1,8′-lign-8-en-4(1H)-one. ***Hookerinone A***
[157110-22-0]
$C_{21}H_{24}O_5$ 356.418
Constit. of *Piper hookeri*. Cryst.. Mp 123 - 124°. $[\alpha]_D^{28}$ -20 (c, 0.5 in MeOH). λ_{max} 240 (log ε 4.68); 290 (log ε 4.39); 317 (log ε 4.14) (MeOH)
5′-Methoxy, tetra-Me ether
2,3′,4′,5,5′-Pentamethoxy-1,8′-lign-8-en-4(1H)-one. ***Tatarinone***
[1012781-38-2]
$C_{23}H_{30}O_6$ 402.487
Constit. of the rhizomes of *Acorus tatarinowii*. Needles (Me_2CO). Mp 178 - 179°. $[\alpha]_D^{20}$ +17 (c, 0.5 in $CHCl_3$)
(1R*,8′S*)-form
3′,4′-Methylene, 2,5-di-Me ether
Hookerinone B
[207916-52-7]
$C_{21}H_{24}O_5$ 356.418
Constit. of *Piper hookeri*. Liq.. $[\alpha]_D^{26}$ +27.3 (c, 0.59 in MeOH). λ_{max} 240 (log ε 4.68); 290 (log ε 4.39); 317 (log ε 4.14) (MeOH).

Pradhan, P. *et al.*, *Nat. Prod. Lett.*, 1994, **4**, 35-42 (*Hookerinone A*).
Pradhan, P. *et al.*, *Phytochem. Anal.*, 1998, **9**, 71-74 (*Hookerinones A,B*).
Nguyen, T.D. *et al.*, *Tap Chi Hoa Hoc*, 2007, **45**, 356-362 (*Tatarinone*)

3,3′,4,4′-Tetrahydroxy-3,8′-lign-8-en-6(3*H*)-one T49

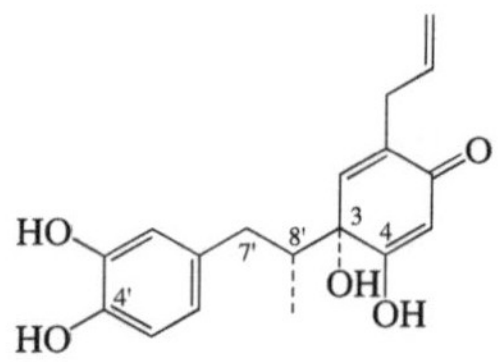

(3*R*,8′*R*)-form

$C_{18}H_{20}O_5$ 316.353

(3R,8′R)-form

3,3′,4-Tri-Me ether

4′-Hydroxy-3,3′,4-trimethoxy-3,8′-lign-8-en-6(3H)-one. ***Lancifolin B***

[74048-70-7]

$C_{21}H_{26}O_5$ 358.434

Constit. of *Aniba lancifolia*. Oil

Tetra-Me ether

3,3′,4,4′-Tetramethoxy-3,8′-lign-8-en-6(3H)-one. ***Lancifolin D***

[137196-26-0]

$C_{22}H_{28}O_5$ 372.461

Constit. of *Aniba lancifolia*, *Piper hookeri* and *Piper polysyphorum*. $[\alpha]^{18}_{D}$ +28.6 (c, 0.21 in Me_2CO). Abs. config. revised in 1998. λ_{max} 230 (log ε 4.18); 281 (log ε 3.87); 331 (log ε 4.24) (MeOH)

3′,4′-Methylene, 3,4-di-Me ether

3,4-Dimethoxy-3′,4′-methylenedioxy-3,8′-lign-8-en-6(3H)-one. ***Isodihydrofutoquinol A***. ***Lancifolin F***

[74163-94-3]

[62560-95-6] (3*R**,8′*R**-form)

$C_{21}H_{24}O_5$ 356.418

Constit. of *Aniba lancifolia*, *Magnolia denudata*, *Piper futokadsura*, *Piper schmidtii* and *Piper wightii*. Oil. $[\alpha]_D$ -19.6 (c, 1.3 in MeOH). λ_{max} 236 (ε 15600); 287 (ε 8070) (MeOH)

7′,8′-Didehydro(E-), 3,3′,4-tri-Me ether

4′-Hydroxy-3,3′,4-trimethoxy-3,8′-ligna-7′,8-dien-6(3H)-one. ***Hancinone B***. ***Piperkadsin A***

[111843-09-5]

[895543-36-9] (Piperkadsin A)

$C_{21}H_{24}O_5$ 356.418

Constit. of *Piper attenuatum*, *Piper kadsura* and *Piper hancei*. Oil. $[\alpha]^{24}_{D}$ -130.3 (c, 0.43 in $CHCl_3$). Stereochem. of Hancinone B was not determined. Data given is for Piperkadsin A. λ_{max} 238 (log ε 4.22); 287 (log ε 4.13) (MeOH)

7′,8′-Didehydro(E-), tetra-Me ether

3,3′,4,4′-Tetramethoxy-3,8′-ligna-7′,8-dien-6(3H)-one. ***Wallichinin***

[125292-97-9]

$C_{22}H_{26}O_5$ 370.445

Constit. of *Piper attenuatum*, *Piper polysyphorum*, *Piper hancei*, *Piper wightii* and *Piper wallichii*. Inhibits platelet-activating factor. Mild NO inhibitor. Oil. $[\alpha]^{24}_{D}$ -4.7 (c, 0.4 in $CHCl_3$). Reported to be racemic from *Piper wallichii*

(3S,8′R)-form

3,3′,4-Tri-Me ether

Lancifolin A

[74048-69-4]

$C_{21}H_{26}O_5$ 358.434

Constit. of *Aniba lancifolia*. Oil

Tetra-Me ether

Lancifolin C

[74048-71-8]

$C_{22}H_{28}O_5$ 372.461

Constit. of *Aniba lancifolia*, *Piper argyrophylum*, *Piper hookeri* and *Piper wightii*. Antioxidant. Oil. $[\alpha]^{26}_{D}$ +83.3 (c, 0.3 in $CHCl_3$). λ_{max} 230 (log ε 4.18); 281 (log ε 3.72) (MeOH)

3′,4′-Methylene, 3,4-di-Me ether

Isodihydrofutoquinol B. ***Lancifolin E***

[62499-71-2]

[74163-93-2] (deleted)

Constit. of *Aniba lancifolia*, *Magnolia denudata*, *Piper futokadsura*, *Piper schmidtii* and *Piper wightii*. Oil. $[\alpha]^{22}_{D}$ +108.5 (c, 0.15 in MeOH). λ_{max} 236 (ε 16800); 287 (ε 8660) (EtOH)

(3S,8′S)-form

Tetra-Me ether

(-)-Lancifolin D

[74048-72-9]

$C_{22}H_{28}O_5$ 372.461

Constit. of *Piper argyrophylum*, *Piper schmidtii* and *Piper wightii*. Pale yellow oil. $[\alpha]^{24}_{D}$ -26.8 (c, 1.5 in Me_2CO). Abs. config. revised in 1998

(8′ξ)-form

5′-Methoxy, 7′,8′-didehydro(E-), tetra-Me ether

3,3′,4,4′,5′-Pentamethoxy-3,8′-ligna-7′,8-dien-6(3H)-one. ***Hancinone C***

[111843-10-8]

$C_{23}H_{28}O_6$ 400.471

Constit. of *Piper hancei*. Stereochem. not determined.

Matsui, K. *et al.*, *Tetrahedron Lett.*, 1976, **17**, 4371-4374 (*Isodihydrofutoquinols A,B*).

Diaz, P.P. *et al.*, *Phytochemistry*, 1980, **19**, 285-288 (*Lancifolins A-F*).

Shizuri, Y. *et al.*, *Tetrahedron Lett.*, 1986, **27**, 727-730 (*Isodihydrofutoquinol A, synth*).

Li, S. *et al.*, *Zhiwu Xuebao*, 1987, **29**, 293-296 (*Hancinones*).

Han, G.Q. *et al.*, *Yaoxue Xuebao*, 1989, **24**, 438-443 (*Wallichinin*).

Tyagi, O.D. *et al.*, *Phytochemistry*, 1993, **32**, 445-448 (*Isodihydrofutoquinols A,B*).

Prasad, A.K. *et al.*, *Tetrahedron*, 1994, **50**, 2231-2240 (*Piper wightii constits*).

Prasad, A.K. *et al.*, *Tetrahedron*, 1994, **50**, 6721 (*corrigendum*).

Singh, S.K. *et al.*, *Phytochemistry*, 1996, **43**, 1355-1360 (*Piper argyrophylum constits*).

Pradhan, P. *et al.*, *Phytochem. Anal.*, 1998, **9**, 71-74 (*Lancifolins C,D*).

Lin, L.-C. *et al.*, *J. Nat. Prod.*, 2006, **69**, 842-844 (*Piperkadsins A,B*).

Reddy, S.D. *et al.*, *Pharmacogn. Mag.*, 2015, **11**, 235-241 (*Lancifolins C,D, Piperkadsin A, Wallichinin*)

3′,4,4′,5-Tetrahydroxy-1,8′-lign-8-en-2(1*H*)-one T50

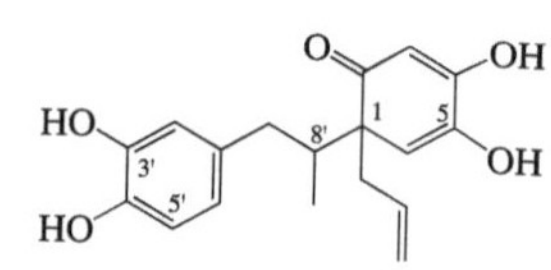

6-[2-(3,4-Dihydroxyphenyl)-1-methylethyl]-3,4-dihydroxy-6-(2-propenyl)-2,4-cyclohexadien-1-one

$C_{18}H_{20}O_5$ 316.353

(1ξ,8′ξ)-form

3′,4′-Methylene, 4,5-di-Me ether

4,5-Dimethoxy-3′,4′-methylenedioxy-1,8′-lign-8-en-2(1H)-one

[158506-58-2]

$C_{21}H_{24}O_5$ 356.418

Constit. of *Piper argyrophyllum* and *Piper wightii*. Gummy solid. $[\alpha]^{24}_{D}$ -2.3 (c, 1 in $CHCl_3$). λ_{max} 248; 289; 312 (sh) ($CHCl_3$)

3′,4′-Methylene, 4,5-di-Me ether, diastereoisomer

$C_{21}H_{24}O_5$ 356.418

Constit. of the stems of *Piper wightii*. Pale yellow oil. $[\alpha]^{24}_{D}$ -14.8 (c, 0.3 in $CHCl_3$). λ_{max} 245; 288 ($CHCl_3$)

4,5-Methylene, 3′-Me ether

4′-Hydroxy-3′-methoxy-4,5-methylenedioxy-1,8′-lign-8-en-2(1H)-one

[138965-85-2]

$C_{20}H_{22}O_5$ 342.391

Constit. of the roots of *Piper capense*. $[\alpha]_D^{23}$ -21.8 (c, 0.02 in $CDCl_3$)

4,5-Methylene, 3′,4′-di-Me ether

3′,4′-Dimethoxy-4,5-methylenedioxy-1,8′-lign-8-en-2(1H)-one. ***Cymosalignan B***

[1651840-78-6]

$C_{21}H_{24}O_5$ 356.418

Constit. of the stems of *Ocotea cymosa*. Oil. $[\alpha]_D^{25}$ -21 (c, 0.1 in MeOH). λ_{max} 258; 320 (MeOH aq.)

Bis(methylene) ether

3′,4′:4,5-Bis(methylenedioxy)-1,8′-lign-8-en-2(1H)-one

[138965-87-4]

$C_{20}H_{20}O_5$ 340.375

Constit. of the roots of *Piper capense*. $[\alpha]_D^{21}$ +3.3 (c, 0.007 in $CDCl_3$)

Bis(methylene) ether, diastereoisomer

$C_{20}H_{20}O_5$ 340.375

Constit. of the roots of *Piper capense*. $[\alpha]_D^{22}$ +11.2 (c, 0.007 in $CDCl_3$)

5′-Methoxy, 4,5-methylene, 3′-Me ether

4′-Hydroxy-3′,5′-dimethoxy-4,5-methylenedioxy-1,8′-lign-8-en-2(1H)-one. ***Cymosalignan C***

[1651840-79-7]

$C_{21}H_{24}O_6$ 372.417

Constit. of the stems of *Ocotea cymosa*. Oil. $[\alpha]_D^{25}$ -16 (c, 0.1 in MeOH). λ_{max} 217; 250; 318 (MeOH aq.)

5′-Methoxy, 4,5-methylene, 3′,4′-di-Me ether

3′,4′,5′-Trimethoxy-4,5-methylenedioxy-1,8′-lign-8-en-2(1H)-one. ***Cymosalignan A***

[1651840-77-5]

$C_{22}H_{26}O_6$ 386.444

Constit. of the stems of *Ocotea cymosa*. Oil. $[\alpha]_D^{25}$ -22 (c, 0.2 in MeOH). λ_{max} 256; 318 (MeOH aq.).

Green, T.P. *et al., Phytochemistry*, 1991, **30**, 3759-3762 (*Piper capense constits*).

Prasad, A.K. *et al., Tetrahedron*, 1994, **50**, 10579-10586 (*Piper wightii constits*).

Rakotondraibe, L.H. *et al., J. Nat. Prod.*, 2015, **78**, 431-434 (*Cymosalignans A,B,C*)

2,3′,4,4′-Tetrahydroxy-9′-nor-7,8′-lign-7-en-9,8′-olide T51

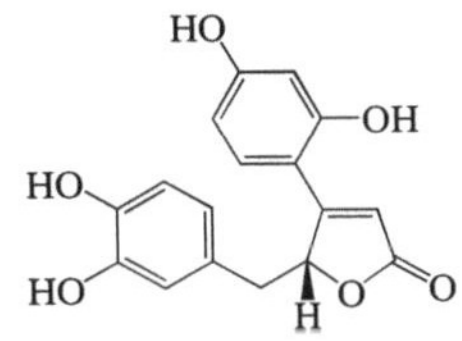

5-(3,4-Dihydroxybenzyl)-4-(2,4-dihydroxyphenyl)-2(5H)-furanone

$C_{17}H_{14}O_6$ 314.293

(R)-form

3′-Me ether

2,4,4′-Trihydroxy-3′-methoxy-9′-nor-7,8′-lign-7-en-9,8′-olide. ***Clitorienolactone C***

[1501651-76-8]

$C_{18}H_{16}O_6$ 328.32

Constit. of the roots of *Clitoria ternatea*. Yellow cryst. ($CHCl_3$/MeOH). Mp 222 - 224°. $[\alpha]_D^{25}$ +112.3 (c, 0.25 in MeOH). λ_{max} 327 (log ε 1.76) (MeOH)

2,3′-Di-Me ether

4,4′-Dihydroxy-2,3′-dimethoxy-9′-nor-7,8′-lign-7-en-9,8′-olide. ***Clitorienolactone A***

[1501651-70-2]

$C_{19}H_{18}O_6$ 342.347

Constit. of the roots of *Clitoria ternatea*. Possesses memory enhancing activity. Acetylcholinesterase inhibitor. Pale yellow cryst. ($CHCl_3$/MeOH). Mp 213 - 215°. $[\alpha]_D^{25}$ +42.4 (c, 0.37 in MeOH). λ_{max} 324 (log ε 1.77) (MeOH).

Vasisht, K. *et al., Tetrahedron Lett.*, 2016, **57**, 1758-1762 (*Clitorienolactones A,C, activity*)

3,3′,4,4′-Tetrahydroxy-9-nor-8,8′-lign-7′-en-9′,7-olide T52

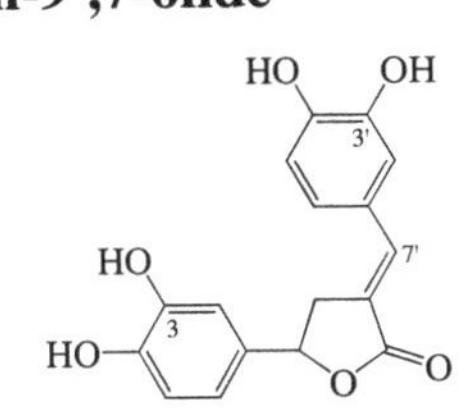

(7ξ,7′*E*)-form

3-(3,4-Dihydroxybenzylidene)-5-(3,4-dihydroxyphenyl)dihydro-2(3H)-furanone

$C_{17}H_{14}O_6$ 314.293

(7ξ,7′E)-form [1311319-77-3]

Constit. of the roots of *Scorzonera judaica*. Amorph. powder. $[\alpha]_D^{25}$ +18 (c, 0.1 in MeOH). λ_{max} 301 (sh); 330 (log ε 3.87) (MeOH)

3′-Me ether

3,4,4′-Trihydroxy-3′-methoxy-9-nor-8,8′-lign-7′-en-9′,7-olide. ***Tectonoelin A***

[1422520-12-4]

$C_{18}H_{16}O_6$ 328.32

Constit. of the leaves of *Tectona grandis*. Plant growth inhibitor. Brown oil. $[\alpha]_D$ 0

(7ξ,7′Z)-form [1311319-78-4]

Constit. of the roots of *Scorzonera judaica*. Amorph. powder. λ_{max} 266 (log ε 3.94); 425 (log ε 4.3) (MeOH).

Bader, A. *et al., J. Nat. Prod.*, 2011, **74**, 1421-1426 (*Scorzonera judaica constits*).

Lacret, R. *et al., Phytochem. Lett.*, 2012, **5**, 382-386 (*Tectonoelin A*)

3-(2,3,4,5-Tetrahydroxyphenyl)propanoic acid T53

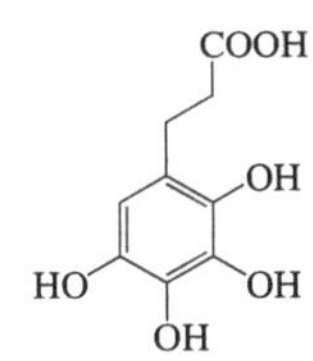

2,3,4,5-Tetrahydroxybenzenepropanoic acid.

2,3,4,5-Tetrahydroxydihydrocinnamic acid

$C_9H_{10}O_6$ 214.173

3,4,5-Tri-Me ether

3-(2-Hydroxy-3,4,5-trimethoxyphenyl)propanoic acid

3,4,5-Tri-Me ether, Et ester

$C_{14}H_{20}O_6$ 284.308

Constit. of the bark of *Cryptocarya bracteolata*. Brownish oily liq.. λ_{max} 301 (MeOH)

3,4,5-Tri-Me ether, Et ester, 2-O-β-D-glucopyranoside

$C_{20}H_{30}O_{11}$ 446.449

Constit. of the bark of *Cryptocarya bracteolata*. Amorph. solid. λ_{max} 294 (MeOH).

Saidi, N. *et al., Nat. Prod. Commun.*, 2016, **11**, 815-816 (*Cryptocarya bracteolata constits*)

1-(2,3,4,5-Tetrahydroxyphenyl)-1-propanone T54

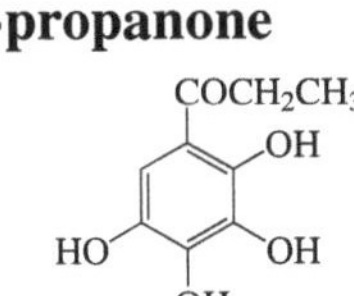

2′,3′,4′,5′-Tetrahydroxypropiophenone

$C_9H_{10}O_5$ 198.174

3,5-Di-Me ether

1-(2,4-Dihydroxy-3,5-dimethoxyphenyl)-1-propanone, CAS. *2',4'-Dihydroxy-3',5'-dimethoxypropiophenone*

[99964-98-4]

$C_{11}H_{14}O_5$ 226.228

Constit. of *Leucanthemopsis pallida* ssp. *flaveola*. Cryst.. Mp 122 - 123°. λ_{max} 217 (ε 5200); 284 (ε 3480) (MeOH).

Bellido, I.S. *et al.*, *Phytochemistry*, 1985, **24**, 2758-2760 (*Leucanthemopsis constits*)

3-(2,3,4,5-Tetrahydroxyphenyl)-2-propenoic acid T55

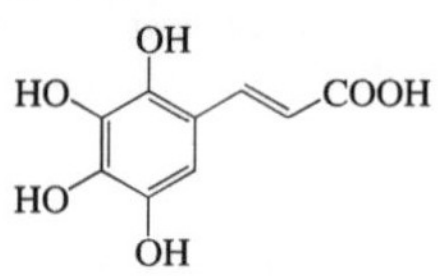

3-(2,3,4,5-Tetrahydroxyphenyl)acrylic acid. 2,3,4,5-Tetrahydroxycinnamic acid

$C_9H_8O_6$ 212.157

(E)-form

3'-Me ether

3-(2,4,5-Trihydroxy-3-methoxyphenyl)-2-propenoic acid. 2,4,5-Trihydroxy-3-methoxycinnamic acid

3'-Me ether, 5'-O-β-D-glucopyranoside

[1975992-81-4]

$C_{16}H_{20}O_{11}$ 388.325

Constit. of the seeds of *Dimocarpus longan*. Potent antioxidant. Yellow powder. λ_{max} 284; 320 (MeOH).

Chen, J.-Y. *et al.*, *J. Funct. Foods*, 2015, **17**, 872-880 (*3'-Me ether 5'-glucoside, activity*)

3,3',4,4'-Tetramethoxy-9,9'-dinor-7,7'-cyclolignan T56

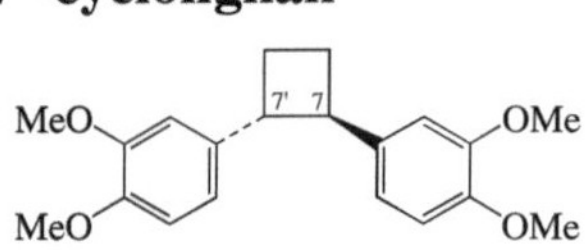

(7R*,7'R*)-form

1,1'-(1,2-Cyclobutanediyl)bis[3,4-dimethoxybenzene], CAS. *1,2-Bis(3,4-dimethoxyphenyl)cyclobutane*

$C_{20}H_{24}O_4$ 328.408

(7R*,7'R*)-form

[1337961-52-0]

Isol. from *Conocephalum japonicum*. Oil. $[\alpha]_D^{20}$ -35.2 (c, 0.1 in MeOH). λ_{max} 224 (log ε 4.4); 278 (log ε 3.09) (MeOH)

(7R*,7'S*)-form

[1337961-44-0]

Isol. from *Conocephalum japonicum*. Oil. Possible racemate. λ_{max} 231 (log ε 4.36); 282 (log ε 3.94) (MeOH).

Liu, N. *et al.*, *Nat. Prod. Commun.*, 2011, **6**, 49-52 (*Conocephalum japonicum constits*)

3,3',3",5'-Tetramethoxy-4',8"-oxy-8,8'-sesquilignan-4,4",7"-triol T57

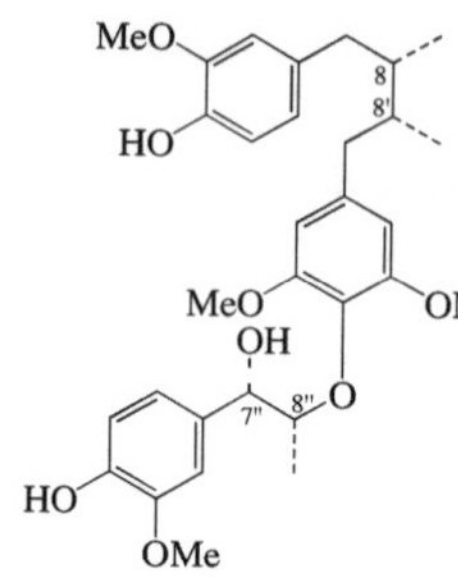

(7"S,8S,8'R,8"R)-form

4,4",7"-Trihydroxy-3,3',3",5'-tetramethoxy-4',8"-oxy-8,8'-sesquineolignan

$C_{31}H_{40}O_8$ 540.653

(7"S,8S,8'R,8"R)-form [1310453-55-4]

Isol. from the bark of *Machilus robusta*. Shows activity against HIV-1 replication *in vitro*. Amorph. solid. $[\alpha]_D^{20}$ +3.1 (c, 0.08 in $CHCl_3$). λ_{max} 203 (log ε 5.05); 232 (log ε 4.37); 280 (log ε 3.75) (MeOH)

(7"S,8S,8'R,8"S)-form [1310453-56-5]

Isol. from the bark of *Machilus robusta*. Shows activity against HIV-1 replication *in vitro*. Amorph. solid. $[\alpha]_D^{20}$ +44.1 (c, 0.13 in $CHCl_3$). λ_{max} 203 (log ε 4.84); 232 (log ε 4.08); 280 (log ε 3.4) (MeOH)

(7"S,8ξ,8'ξ,8"R)-form

5-Methoxy, 3,5'-di-O-de-Me, 4-Me ether

[1791438-96-4]

$C_{31}H_{40}O_9$ 556.652

Constit. of *Schisandra chinensis*. Yellow gum. $[\alpha]_D^{27}$ -21 (c, 0.05 in MeOH). λ_{max} 206 (log ε 4.28); 278 (log ε 2.97) (MeOH).

Li, Y. *et al.*, *J. Nat. Prod.*, 2011, **74**, 1444-1452 (*Machilus robusta constits*).

Shi, Y. *et al.*, *Chin. J. Chem.*, 2014, **32**, 734-740 (*Schisandra chinensis constit*)

Thesine T58

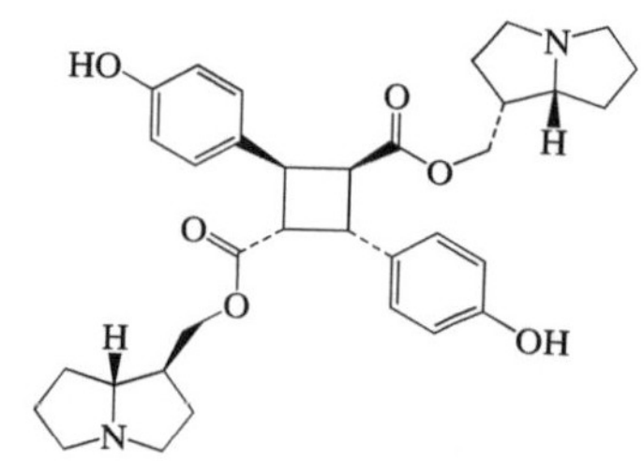

Absolute Configuration

Diisoretronecanyl p,p'-dihydroxy-α-truxillate

[528-37-0]

$C_{34}H_{42}N_2O_6$ 574.718

Alkaloid from *Thesium minkwitzianum* whole plant. Cryst. (EtOH). Mp 254 - 256°. Hydrol. gives (+)-isoretronecanol and thesinic acid (*p,p'*-dihydroxy-α-truxillic acid)

Sulfate

Cryst.. Mp 244 - 246°. $[\alpha]_D^{20}$ +33.4 (c, 1.6 in H_2O). Forms a tetrahydrate

Picrate

Fine yellow cryst. (EtOH). Mp 224 - 226° dec.

Methiodide (1:2)

Cryst.. Mp 140 - 150°. $[\alpha]_D^{20}$ +33.24 (c, 2.2 in EtOH).

Arendaruk, A.P. *et al.*, *J. Gen. Chem. USSR (Engl. Transl.)*, 1960, **30**, 507-510 (*Thesine*).

Arendaruk, A.P. *et al.*, *J. Gen. Chem. USSR (Engl. Transl.)*, 1960, **30**, 511-515 (*Thesine*).

Arendaruk, A.P. *et al.*, *J. Gen. Chem. USSR (Engl. Transl.)*, 1960, **30**, 693-698 (*Thesine, occur*)

Thoreliamide C T59

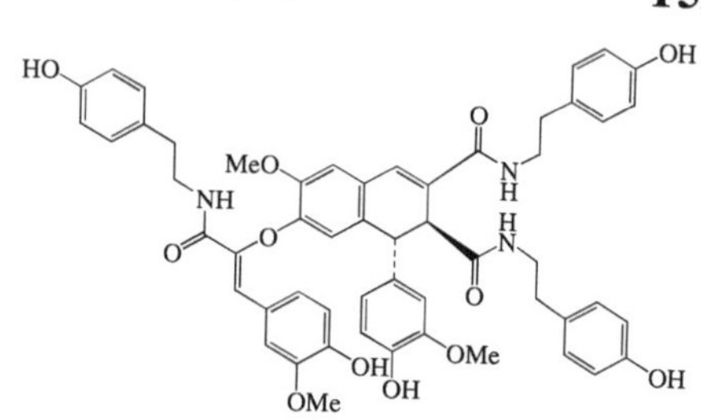

Relative Configuration

Melongenamide D

[1052115-07-7]

[1676050-29-5] (Melongenamide D)

$C_{54}H_{53}N_3O_{12}$ 936.027

Thoreliamide C and Melongenamide D assigned same gross struct. but have opposite reported opt. rotns., so may be enantiomers. Opt. rotn. solvs. were different however. Constit of *Mitrephora thorelii* and *Solanum melongena*. Inhibitor of NO prodn. in LPS-induced macrophages. Amorph. powder. $[\alpha]_D^{25}$ +24 (c, 0.1 in Me_2CO) (Thoreliamide C). $[\alpha]_D^{21}$ -8 (c, 0.1 in MeOH) (Melongenamide D). λ_{max} 223 (log ε 4.69); 287 (log ε 4.35); 321 (log ε 4.37) (MeOH) (Thoreliamide C). [neutral] λ_{max} 231 (log ε 4.84); 286 (log ε 4.54); 320 (log ε 4.71) (MeOH) (Melongenamide D).

Ge, F. *et al.*, *Helv. Chim. Acta*, 2008, **91**, 1023-1030 (*Thoreliamide C*).
Sun, J. *et al.*, *Fitoterapia*, 2014, **98**, 110-116 (*Melongenamide D*)

Thoreliin A T60

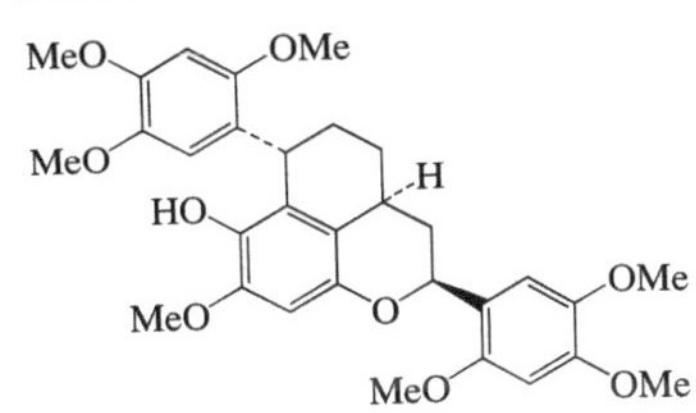

Relative Configuration

$C_{31}H_{36}O_9$ 552.620
Constit. of the rhizomes of *Boesenbergia thorelii*. Exhibits anti-HIV-1 activity. Plates (MeOH). Mp 187.7 - 189.8°. $[\alpha]_D^{25}$ -15.6 (c, 0.25 in MeOH). λ_{max} 232 (log ε 4.60); 291 (log ε 4.34) (MeOH).

Thongphichai, W. *et al.*, *Fitoterapia*, 2019, **139**, 104388:1-8 (*Thoreliin A, cryst struct, activity*)

Thoreliin B T61

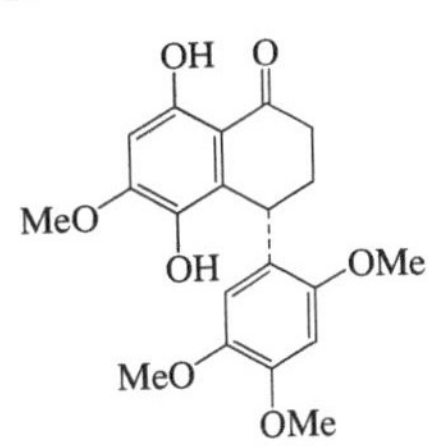

Absolute Configuration

3,4-Dihydro-5,8-dihydroxy-6-methoxy-4-(2,4,5-trimethoxyphenyl)-1(2H)-naphthalenone.
2,7'-Cyclo-3,6-dihydroxy-2',4,4',5'-tetramethoxy-9,9'-dinor-8,8'-lignan-7-one
$C_{20}H_{22}O_7$ 374.389
Constit. of the rhizomes of *Boesenbergia thorelii*. Needles (MeOH). Mp 179.2 - 179.9°. $[\alpha]_D^{25}$ -8.2 (c, 1.1 in $CHCl_3$). λ_{max} 236 (log ε 4.51); 283 (log ε 4.38); 356 (log ε 4.05) (MeOH).

Thongphichai, W. *et al.*, *Fitoterapia*, 2019, **139**, 104388:1-8 (*Thoreliin B*)

Tomentoside I‡ T62

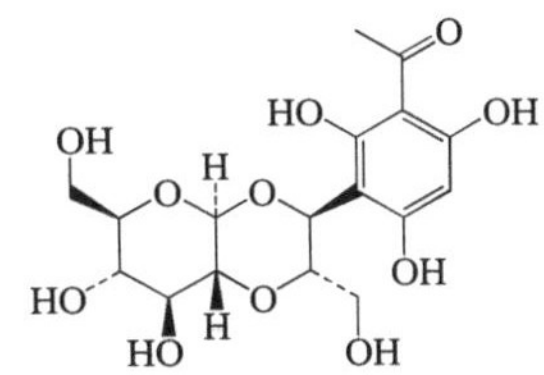

$C_{17}H_{22}O_{11}$ 402.352
Constit. of the roots of *Anemone tomentosa*. Cryst.. Mp 185 - 187°. $[\alpha]_D^{21}$ +21.2 (c, 0.54 in MeOH). λ_{max} 210 (log ε 2.12); 268 (log ε 3.57) (MeOH).

Hu, H.-B. *et al.*, *Helv. Chim. Acta*, 2011, **94**, 711-718 (*Tomentoside I*)

Torreyunlignan A T63

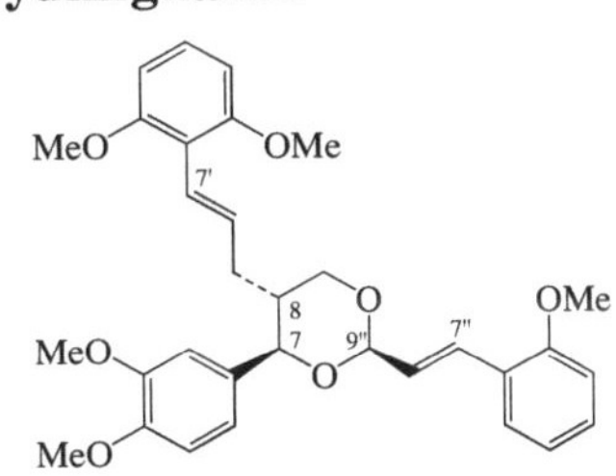

(7R,7'E,7''E,8R,9''R)-form

[1643111-46-9]
$C_{32}H_{36}O_7$ 532.633
(7R,7'E,7''E,8R,9''R)-form
[1643111-39-0]
Constit. of the trunk of *Torreya yunnanensis*. Phosphodiesterase-9A (PDE9A) inhibitor. Light yellow oil. $[\alpha]_D^{20}$ -4.2 (c, 0.3 in CH_2Cl_2). Resolved by chiral hplc
(7S,7'E,7''E,8S,9''S)-form
[1643111-38-9]
Constit. of the trunk of *Torreya yunnanensis*. Phosphodiesterase-9A (PDE9A) inhibitor. Light yellow oil. $[\alpha]_D^{20}$ +4.2 (c, 0.3 in CH_2Cl_2). Resolved by chiral hplc. λ_{max} 215 (log ε 4.72); 257 (log ε 4.55); 299 (log ε 3.98) (MeOH).

Cheng, Z.-B. *et al.*, *J. Nat. Prod.*, 2014, **77**, 2651-2657 (*Torreyunlignans A, activity*)

Torreyunlignan C T64

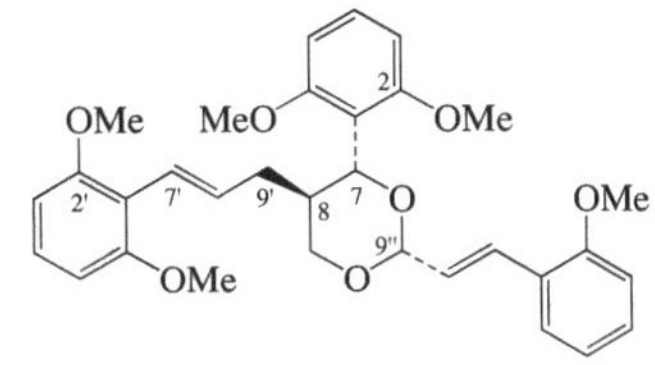

(7R,7'E,7''E,8R,9''R)-form

$C_{32}H_{36}O_7$ 532.633
Sesquineolignan

(7R,7'E,7''E,8R,9''R)-form
[1643111-43-6]
Constit. of the trunk of *Torreya yunnanensis*. Phosphodiesterase-9A (PDE9A) inhibitor. Light yellow oil. $[\alpha]_D^{20}$ +20.3 (c, 0.6 in CH_2Cl_2). Racemate. Resolved by chiral hplc

2-Demethoxy
(+)-Torreyunlignan B
[1643111-41-4]
$C_{31}H_{34}O_6$ 502.607
Constit. of the trunk of *Torreya yunnanensis*. Phosphodiesterase-9A (PDE9A) inhibitor. Light yellow oil. $[\alpha]_D^{20}$ +13.7 (c, 0.7 in CH_2Cl_2). Racemate. Resolved by chiral hplc

2,2'-Dimethoxy
(+)-Torreyunlignan D
[1643111-45-8]
$C_{30}H_{32}O_5$ 472.581
Constit. of the trunk of *Torreya yunnanensis*. Phosphodiesterase-9A (PDE9A) inhibitor. Light yellow oil. $[\alpha]_D^{20}$ +12.4 (c, 0.7 in CH_2Cl_2). Racemate. Resolved by chiral hplc

(7S,7'E,7''E,8S,8''S)-form
[1643111-42-5]
Constit. of the trunk of *Torreya yunnanensis*. Phosphodiesterase-9A (PDE9A) inhibitor. Light yellow oil. $[\alpha]_D^{20}$ -19.4 (c, 0.5 in CH_2Cl_2). Racemate. Resolved by chiral hplc. λ_{max} 211 (log ε 4.50); 256 (log ε 4.22); 299 (log ε 3.66) (MeOH)

2-Demethoxy
(-)-Torreyunlignan B
[1643111-40-3]
$C_{31}H_{34}O_6$ 502.607
Constit. of the trunk of *Torreya yunnanensis*. Phosphodiesterase-9A (PDE9A) inhibitor. Light yellow oil. $[\alpha]_D^{20}$ -13.2 (c, 0.7 in CH_2Cl_2). Racemate. Resolved by chiral hplc. λ_{max} 216 (log ε 4.59); 258 (log ε 4.40); 299 (log ε 3.85) (MeOH)

2,2'-Dimethoxy
(-)-Torreyunlignan D
[1643111-44-7]
$C_{30}H_{32}O_5$ 472.581
Constit. of the trunk of *Torreya yunnanensis*. Phosphodiesterase-9A (PDE9A) inhibitor. Light yellow oil. $[\alpha]_D^{20}$ -10 (c, 0.7 in CH_2Cl_2). Racemate. Resolved by chiral hplc. λ_{max} 214 (log ε 4.54);

254 (log ε 4.43); 299 (log ε 3.99) (MeOH).

Cheng, Z.-B. *et al.*, *J. Nat. Prod.*, 2014, **77**, 2651-2657 (*Torreyunlignans B,C,D, activity*)

Tournefolic acid A T65

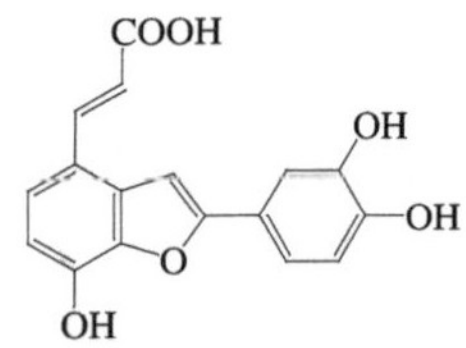

3-[2-(3,4-Dihydroxyphenyl)-7-hydroxy-4-benzofuranyl]-2-propenoic acid, CAS. *3,7′-Epoxy-3′,4,4′-trihydroxy-9′-nor-2,8′-ligna-7,7′-dien-9-oic acid*

$C_{17}H_{12}O_6$ 312.277

(E)-form [434337-16-3]

Constit. of the stems of *Tournefortia sarmentosa*. Anti-lipid-peroxidation agent. Yellow cryst. (EtOH). Mp 197 - 199°. λ_{max} 285 (log ε 4.08); 316 (log ε 4.01) (MeOH)

Et ester

[1675220-28-6]

Solid. Mp 156 - 160°.

Lin, Y.-L. *et al.*, *J. Nat. Prod.*, 2002, **65**, 745-747 (*Tournefolic acid A, activity*).

Shen, S.-D. *et al.*, *ARKIVOC*, 2012, **vi**, 204-213 (*synth*).

Hsieh, J.-F. *et al.*, *Eur. J. Med. Chem.*, 2015, **93**, 443-451 (*Et ester, synth*).

López-Frías, G. *et al.*, *Molecules*, 2015, **20**, 8654-8665 (*synth*)

Tournefolic acid B T66

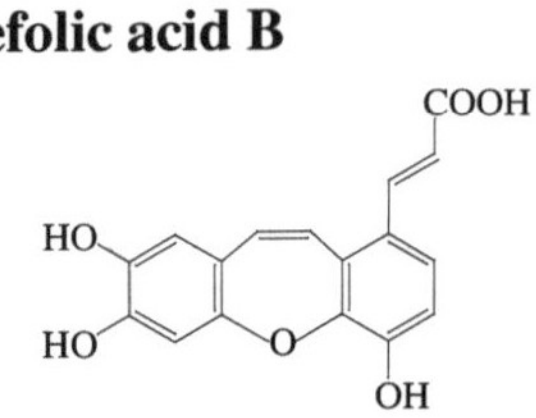

3-(4,7,8-Trihydroxydibenz[b,f]oxepin-1-yl)-2-propenoic acid. 2′,3-Epoxy-4,4′,5′-trihydroxy-9′-nor-2,8′-ligna-7,7′-dien-9-oic acid

[434337-17-4]

$C_{17}H_{12}O_6$ 312.277

Constit. of *Salvia miltiorrhiza* and *Tournefortia sarmentosa*. Anti-lipid-peroxidation agent. Yellow cryst. (EtOH). Mp 242 - 245°. λ_{max} 262 (sh); 285 (sh); 329 (log ε 3.95) (MeOH)

Et ester

[434333-50-3]

$C_{19}H_{16}O_6$ 340.331

Constit. of the stems of *Tournefortia sarmentosa*. Yellow cryst. (EtOH). Mp 225 - 227°. λ_{max} 260 (sh); 287 (sh); 329 (log ε 4.1) (MeOH).

Lin, Y.-L. *et al.*, *J. Nat. Prod.*, 2002, **65**, 745-747 (*Tournefolic acid B*).

Liu, J. *et al.*, *Chem. Nat. Compd. (Engl. Transl.)*, 2014, **50**, 83-87 (*Salvia miltiorrhiza constit*)

Trachelosiaside T67

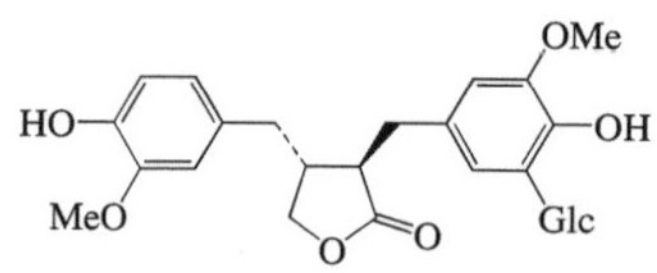

3-C-β-D-Glucopyranosyl-4,4′-dihydroxy-3′,5-dimethoxy-8,8′-lignan-9,9′-olide. Matairesinol 5-C-glucoside. 5-C-Glucosylmatairesinol

[106647-12-5]

$C_{26}H_{32}O_{11}$ 520.531

Constit. of *Gardneria multiflora*, *Trachelospermum asiaticum* and *Trachelospermum lucidum*. Amorph.. $[\alpha]^{27}_{D}$ -12.1 (c, 1.1 in MeOH).

Abe, F. *et al.*, *Chem. Pharm. Bull.*, 1986, **34**, 4340-4345 (*Trachelosiaside*).

Ahmad, V.U. *et al.*, *Z. Naturforsch., B*, 2005, **60**, 1287-1290 (*Trachelospermum lucidum constit*)

Triandrin B T68

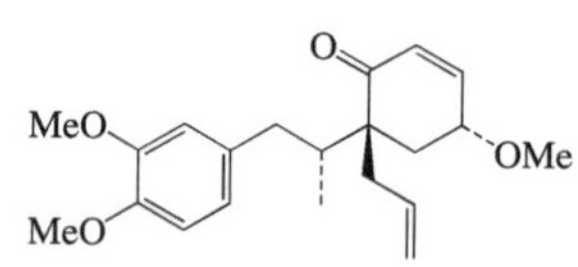

Absolute Configuration

5,6-Dihydro-3′,4′,5-trimethoxy-1,8′-lign-8-en-2(1H)-one

[117505-92-7]

$C_{21}H_{28}O_4$ 344.451

Constit. of the seeds of *Licaria triandra*. Viscous oil. $[\alpha]_D$ -4.7 (c, 6.4 in $CHCl_3$).

Castillo-Castro, O. *et al.*, *Rev. Latinoam. Quim.*, 1988, **19**, 60-62 (*Triandrin B*)

1,2,6-Tricaffeoylglucose T69

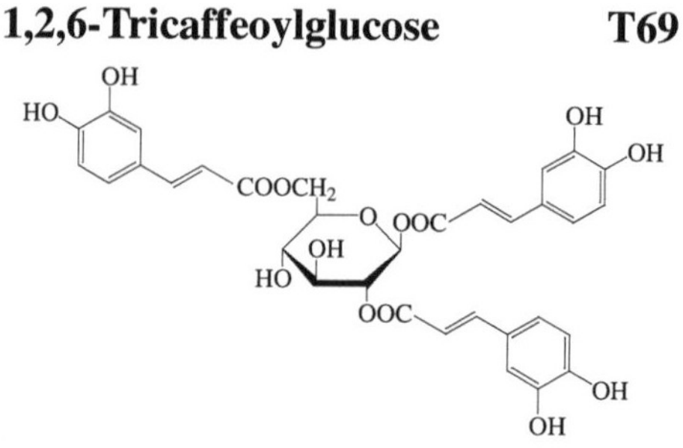

$C_{33}H_{30}O_{15}$ 666.588

β-D-***Pyranose-(E,E,E)-form***

[359819-40-2]

Constit. of *Balanophora japonica*. Amorph. yellow powder. $[\alpha]^{15}_{D}$ -218.6 (c, 0.4 in MeOH).

Jiang, Z.-H. *et al.*, *Chem. Pharm. Bull.*, 2001, **49**, 887-892 (*Balanophora japonica constit*)

1,3,4-Tricaffeoylglucose T70

$C_{33}H_{30}O_{15}$ 666.588

β-D-***Pyranose-(E,E,E)-form***

[1615694-33-1]

Constit. of the roots of *Aruncus sylvester*. Amorph. yellow solid (MeOH). $[\alpha]^{20}_{D}$ -208 (c, 0.48 in MeOH). λ_{max} 247 (log ε 4.88); 302 (sh); 334 (log ε 4.98) (MeOH).

Que, M. *et al.*, *J. Asian Nat. Prod. Res.*, 2014, **16**, 158-162 (*Aruncus sylvester constit*)

1,3,4-Tri-*O*-caffeoylquinic acid T71

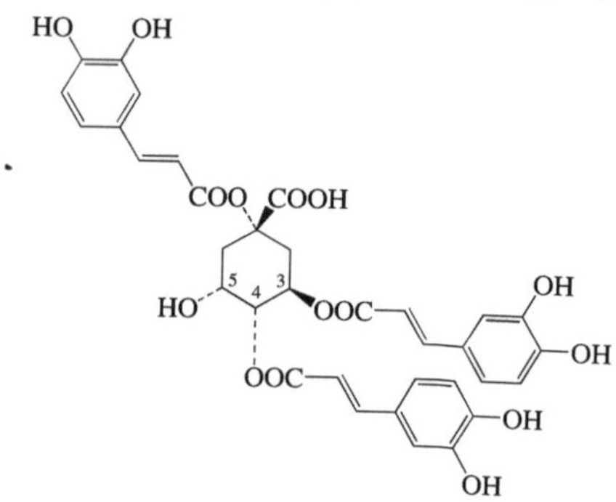

[145108-19-6]

$C_{34}H_{30}O_{15}$ 678.599

Descr. in the ref. as 1,4,5-tricaffeoyl, but is 1,3,4 according to the usual numbering scheme. Constit. of *Arnica chamissonis* and *Arnica montana*

Me ester

[1979138-93-6]

$C_{35}H_{32}O_{15}$ 692.626

Constit. of *Corchorus olitorius*. Amorph. brown powder.

Merfort, I. *Phytochemistry*, 1992, **31**, 2111-2113 (*Arnica constits*).

Taiwo, B.J. *et al.*, *Bioorg. Med. Chem. Lett.*, 2016, **26**, 3404-3410 (*Me ester*)

1,3,5-Tri-*O*-caffeoylquinic acid T72

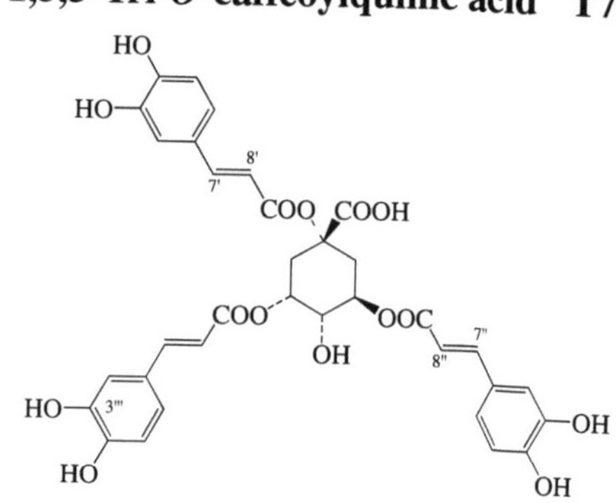

[150035-89-5]

$C_{34}H_{30}O_{15}$ 678.599
Isol. from fruit of *Xanthium strumarium* and florets of *Calendula officinalis*. Light brown amorph. powder + $2H_2O$. Mp 178 - 181°. $[\alpha]_D^{22}$ -27.5 (c, 0.8 in MeOH). λ_{max} 244 (log ε 4.49); 300 (sh); 327 (log ε 4.69) (MeOH)

4-O-(3-Carboxypropanoyl)
1,3,5-Tri-O-caffeoyl-4-O-succinoylquinic acid
[168002-99-1]
$C_{38}H_{34}O_{18}$ 778.672
Constit. of *Arctium lappa*

7′,7″,8′,8″-Tetrahydro, 3‴-Me ether
1,3-Bis-O-(dihydrocaffeoyl)-5-O-feruloylquinic acid. ***Feruloylpodospermic acid A***
[944746-95-6]
$C_{35}H_{36}O_{15}$ 696.658
Constit. of *Scorzonera divaricata*. Amorph. solid. $[\alpha]_D^{20}$ -22 (c, 0.11 in EtOH). λ_{max} 220; 286; 329 (MeOH aq.)

7′,7″,7‴,8′,8″,8‴-Hexahydro
1,3,5-Tris-O-(dihydrocaffeoyl) quinic acid. ***Podospermic acid***
[848191-63-9]
$C_{34}H_{36}O_{15}$ 684.647
Constit. of *Podospermum laciniatum* (preferred genus name *Scorzonera*). Amorph. solid. $[\alpha]_D^{20}$ -18 (c, 0.006 in MeOH). Glass transition at 123-129°. λ_{max} 283 (log ε 3.79) (MeOH).

Agata, I. *et al.*, *Phytochemistry*, 1993, **33**, 508-509 (*Xanthium strumarium constit*).
Maruta, Y. *et al.*, *J. Agric. Food Chem.*, 1995, **43**, 2592-2595 (*4-succinoyl*).
Zidorn, C. *et al.*, *Tetrahedron Lett.*, 2005, **46**, 1291-1294 (*Podospermic acid*).
Tsevegsuren, N. *et al.*, *J. Nat. Prod.*, 2007, **70**, 962-967 (*Feruloylpodospermic acid A*).
Olennikov, D. *et al.*, *Chem. Nat. Compd. (Engl. Transl.)*, 2013, **49**, 833-840 (*Calendula officinalis constit*)

1,4,5-Tri-*O*-caffeoylquinic acid T73

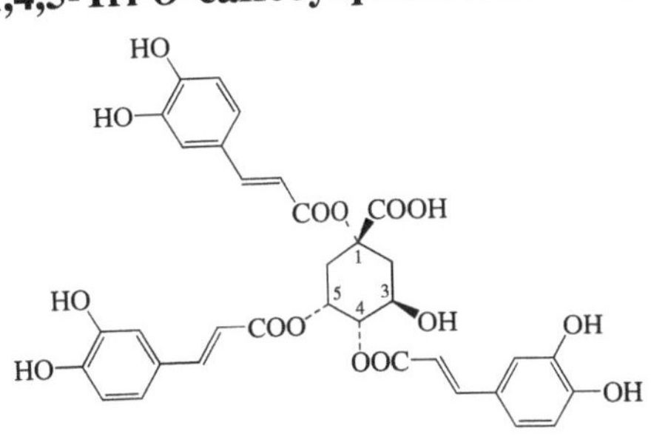

$C_{34}H_{30}O_{15}$ 678.599

Numbering systems vary. Constit. of the leaves of *Ligularia fischeri*. Amorph. brown powder

3-O-(Methoxycarbonylcarbonyl)
1,4,5-Tri-O-trans-caffeoyl-3-methoxalylquinic acid
[1290614-37-7]
$C_{37}H_{32}O_{18}$ 764.645
Constit. of the flowers of *Arnica montana*.

Jaiswal, R. *et al.*, *J. Agric. Food Chem.*, 2011, **59**, 4033-4039 (*Arnica montana constit*).
Kim, S.M. *et al.*, *J. Agric. Food Chem.*, 2012, **60**, 5597-5603 (*Ligularia constit*)

3,4,5-Tri-*O*-caffeoylquinic acid T74

[86632-03-3]
$C_{34}H_{30}O_{15}$ 678.599
Isol. from *Arnica chamissonis, Artemisia herba-alba, Ipomoea batatas, Chrysothamnus paniculatus, Ligularia fischeri, Calendula officinalis* and Brazilian propolis. Antihyperglycaemic agent. λ_{max} 218 (ε 36800); 236 (ε 27500); 244 (ε 28600); 298 (ε 34600); 323 (ε 36900) (MeOH)

1-O-[(Methoxycarbonyl)carbonyl]
3,4,5-Tri-O-trans-caffeoyl-1-methoxalylquinic acid
[1290614-36-6]
$C_{37}H_{32}O_{18}$ 764.645
Constit. of the flowers of *Arnica montana*.

Timmermann, B.N. *et al.*, *J. Nat. Prod.*, 1983, **46**, 365-368 (*Chrysothamnus paniculatus constit*).
Merfort, I. *et al.*, *Phytochemistry*, 1992, **31**, 2111-2113 (*Arnica chamissonis constit*).
Islam, M.S. *et al.*, *J. Agric. Food Chem.*, 2002, **50**, 3718-3722 (*Ipomoea batatas constit*).
Matsui, T. *et al.*, *Biol. Pharm. Bull.*, 2004, **27**, 1797-1803 (*propolis constit, activity*).
Miyamae, Y. *et al.*, *Chem. Pharm. Bull.*, 2011, **59**, 502-507 (*uv*).
Jaiswal, R. *et al.*, *J. Agric. Food Chem.*, 2011, **59**, 4033-4039 (*1-methoxalyl*).
Dahmani-Hamzaoui, N. *et al.*, *Helv. Chim. Acta*, 2012, **95**, 564-576 (*Artemisia herba-alba constit, pmr, cmr*).
Kim, S.M. *et al.*, *J. Agric. Food Chem.*, 2012, **60**, 5597-5603 (*Ligularia fischeri constit*).
Olennikov, D. *et al.*, *Chem. Nat. Compd. (Engl. Transl.)*, 2013, **49**, 833-840 (*Calendula officinalis constit*)

Tri-*O*-demethylzijusesquilignan B T75

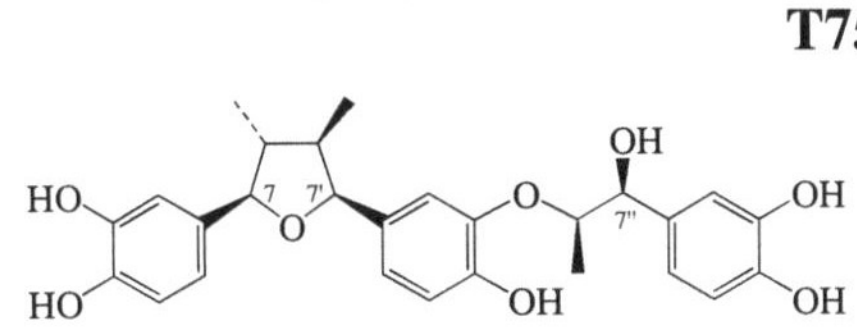

4-[5-[3-[[1-(3,4-Dihydroxyphenyl)-1-hydroxypropan-2-yl]oxy]-4-hydroxyphenyl]tetrahydro-3,4-dimethylfuran-2-yl]-1,2-benzenediol
$C_{27}H_{30}O_8$ 482.529
(7R*,7′S*,7″S,8R*,8′R*,8″R)-form

3″,4,4′-Tri-Me ether
Zijusesquilignan B
[2500744-64-7]
$C_{30}H_{36}O_8$ 524.610
Constit. of the fruit of *Ziziphus jujuba*. Inhibitor of NO prodn. Exhibits cytotoxic activity. Amorph. powder. $[\alpha]_D^{24}$ +18.9 (c, 0.1 in MeOH).

Tran, H.N.K. *et al.*, *Fitoterapia*, 2019, **137**, 104261:1-7 (*Zijusesquilignan B, activity*)

Trideoxy-8-asadaninene T76

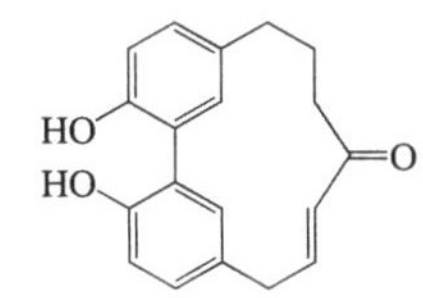

3,17-Dihydroxytricyclo[12.3.1.1^{2,6}] nonadeca-1(18),2,4,6(19),8,14,16-heptaen-10-one, CAS
[22761-12-2]
$C_{19}H_{18}O_3$ 294.35
Isol. from wood of *Ostrya japonica*. Needles (C_6H_6/Me_2CO). Mp 238 - 241°. λ_{max} 216; 230; 240 (sh); 297 (EtOH). [base] λ_{max} 327 (EtOH/NaOH aq.).

Sasaya, T. *et al.*, *Enshurin Kenkyu Hokoku (Hokkaido Daigaku Nogakubu)*, 1974, **31**, 23-50 (*Trideoxy-8-asadaninene*)

4-*O*-8′:5′,5″-Triferulic acid T77

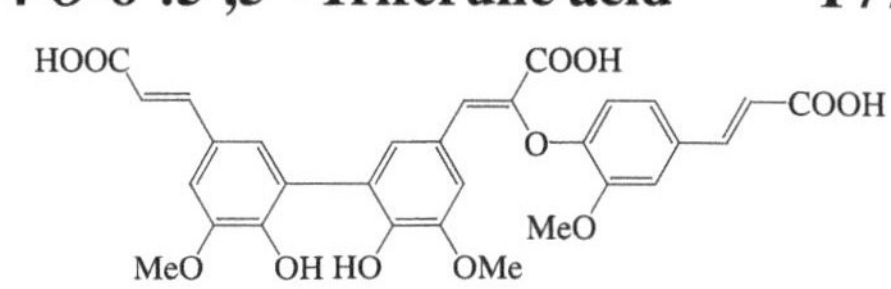

$C_{30}H_{26}O_{12}$ 578.526
(7E,7′Z,7″E)-form [610788-07-3]
Isol. from maize bran (*Zea mays*)

7′,8′-Dihydro, 7′ξ-hydroxy
7′,8′-Dihydro-7′-hydroxy-4-O-8′:5′:5″-triferulic acid
[865199-30-0]
$C_{30}H_{28}O_{13}$ 596.541
Isol. from maize bran (*Zea mays*). [acid] λ_{max} 317 (MeOH/TFA aq.).

Rouau, X. *et al.*, *Phytochemistry*, 2003, **63**, 899-903 (*Triferulic acid*).
Bunzel, M. *et al.*, *Tetrahedron Lett.*, 2005, **46**, 5845-5850 (*Dihydrohydroxytriferulic acid*)

Trigonotin C T78

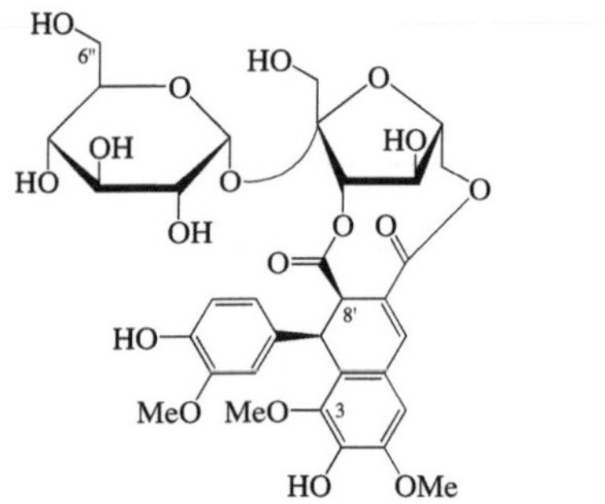

Absolute Configuration

[1036375-80-0]
$C_{33}H_{38}O_{18}$ 722.649
Constit. of the leaves of *Trigonotis peduncularis*. Amorph. yellow powder. $[\alpha]^{25}_{D}$+394 (c, 1.65 in MeOH). λ_{max} 252 (log ε 4.17); 350 (log ε 4.02) (MeOH)

6''-Ac

Trigonotin A
[1036375-78-6]
$C_{35}H_{40}O_{19}$ 764.686
Constit. of the leaves of *Trigonotis peduncularis*. Amorph. yellow powder. $[\alpha]^{25}_{D}$+491 (c, 1.03 in MeOH). λ_{max} 253 (log ε 4.25); 355 (log ε 4.09) (MeOH)

6''-O-[α-L-Rhamnopyranosyl-(1→4)-4-hydroxy-E-cinnamoyl]

Echiumin A
[2227328-37-0]
$C_{48}H_{54}O_{24}$ 1014.936
Constit. of *Echium angustifolium*. Exhibits cytotoxic activity. Amorph. yellowish-green powder. $[\alpha]^{20}_{D}$+88 (c, 0.04 in MeOH). λ_{max} 254 (ε 195); 282 (ε 167); 306 (ε 158); 358 (ε 61) (MeOH)

3'-Demethoxy, 6''-O-[α-L-rhamnopyranosyl-(1→4)-4-hydroxy-E-cinnamoyl]

Echiumin B
[2227328-41-6]
$C_{47}H_{52}O_{23}$ 984.910
Constit. of *Echium angustifolium*. Exhibits cytotoxic activity. Amorph. yellowish-green powder. $[\alpha]^{20}_{D}$+74 (c, 0.08 in MeOH). λ_{max} 254 (ε 120); 306 (ε 129); 358 (ε 63) (MeOH)

8'-Epimer, 3-demethoxy, 6''-Ac

Trigonotin B
[1036375-79-7]
$C_{34}H_{38}O_{18}$ 734.66
Constit. of the leaves of *Trigonotis peduncularis*. Amorph. yellow powder. $[\alpha]^{25}_{D}$+135 (c, 1.2 in MeOH). λ_{max} 249 (log ε 4.23); 340 (log ε 4.03) (MeOH).

Otsuka, H. *et al.*, *J. Nat. Prod.*, 2008, **71**, 1178-1181 (*Trigonotins A,B,C*).
El-Rokh, A.R. *et al.*, *Phytochemistry*, 2018, **149**, 155-160 (*Echiumins A,B, activity*)

2,3,9-Trihydroxy-7*H*-dibenzo[*a*,*c*]cycloheptene-5-methanol T79

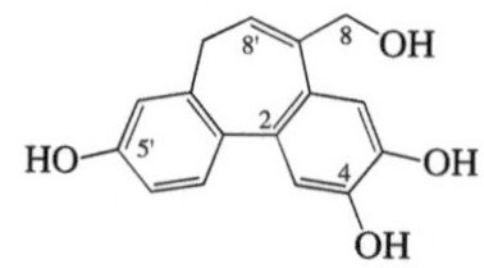

2,3,9-Trihydroxy-5-(hydroxymethyl)-7H-dibenzo[a,c]cycloheptene. 7,8'-Cyclo-9,9'-dinor-2,2'-lign-7(8')-ene-4,5,5',8-tetrol. 7,8'-Cyclo-4,5,5',8-tetrahydroxy-9,9'-dinor-2,2'-neolign-7(8')-ene
$C_{16}H_{14}O_4$ 270.284
Lignan numbering shown

4-Me ether

3,9-Dihydroxy-2-methoxy-7H-dibenzo[a,c]cycloheptene-5-methanol. 7,8'-Cyclo-4-methoxy-9,9'-dinor-2,2'-lign-7(8')-ene-5,5',8-triol. 7,8'-Cyclo-5,5',8-trihydroxy-4-methoxy-9,9'-dinor-2,2'-lign-7(8')-ene. ***Burmanol***
[1392224-29-1]
$C_{17}H_{16}O_4$ 284.311
Constit. of *Cinnamomum burmanii*. Amorph. powder. λ_{max} 235 (log ε 3.25); 255 (log ε 2.68); 290 (log ε 2.11) (MeCN)

5-Me ether, 4-O-α-L-rhamnopyranoside

Subavenoside F
[1402068-85-2] (2*R*-atropisomer)
[1402132-52-8] (2*S*-atropisomer)
$C_{23}H_{26}O_8$ 430.453
Constit. of the leaves of *Cinnamomum subavenium*. Mixt. of two inseparable diastereomers in ratio 1.1:1.0. λ_{max} 231; 265 (MeOH)

5-Me ether, 8-O-β-D-glucopyranoside

Subavenoside C
[1402068-82-9] (2*R*-atropisomer)
[1402132-49-3] (2*S*-atropisomer)
$C_{23}H_{26}O_9$ 446.452
Constit. of the leaves of *Cinnamomum subavenium*. Amorph. powder. Mixt. of two inseparable diastereoisomers in ratio 1.1:1.0. λ_{max} 232; 266 (MeOH)

5'-Me ether

7-(Hydroxymethyl)-3-methoxy-5H-dibenzo[a,c]cycloheptene-9,10-diol. 7,8'-Cyclo-5'-methoxy-9,9'-dinor-2,2'-lign-7(8')-ene-4,5,8-triol. 7,8'-Cyclo-4,5,8-trihydroxy-5'-methoxy-9,9'-dinor-2,2'-lign-7(8')-ene. ***Subamol***
[1235116-72-9]
$C_{17}H_{16}O_4$ 284.311
Constit. of the roots of *Cinnamomum subavenium*. Amorph. powder ($CHCl_3$). Incorrectly stated to be a sesquiterpenoid in paper. λ_{max} 235 (log ε 3.23); 255 (log ε 2.65); 290 (log ε 2.1) (MeCN)

5,5'-Di-Me ether, 8-O-β-D-glucopyranoside

Subavenoside B
[1402068-81-8] (2*R*-atropisomer)
[1402132-48-2] (2*S*-atropisomer)
$C_{24}H_{28}O_9$ 460.479
Constit. of the leaves of *Cinnamomum subavenium*. Amorph. powder. Mixt. of two inseparable diastereoisomers in ratio 1.0:0.9. λ_{max} 246 (log ε 4.75); 273 (log ε 4.36) (MeOH)

5,5',8-Tri-Me ether

7,8'-Cyclo-5,5',8-trimethoxy-9,9'-dinor-2,2'-lign-7(8')-en-4-ol. 7,8'-Cyclo-4-hydroxy-5,5',8-trimethoxy-9,9'-dinor-2,2'-lign-7(8')-ene. ***9,12-Di-O-methylsubamol***
[1401008-82-9]
$C_{19}H_{20}O_4$ 312.365
Constit. of the leaves of *Cinnamomum subavenium*. Amorph. powder. λ_{max} 246 (log ε 4.47); 285 (log ε 3.96) (MeOH)

5,5',8-Tri-Me ether, 4-O-β-D-glucopyranoside

Subavenoside E
[1402068-84-1] (2*R*-atropisomer)
[1402132-51-7] (2*S*-atropisomer)
$C_{25}H_{30}O_9$ 474.506
Constit. of the leaves of *Cinnamomum subavenium*. Amorph. powder. Mixt. of two inseparable diastereoisomers in ratio 2:1. λ_{max} 242 (log ε 4.46); 273 (log ε 4.02) (MeOH)

4,5-Methylene ether

9-Hydroxy-5-(hydroxymethyl)-2,3-methylenedioxy-7H-dibenzo[a,c]cycloheptene. 7,8'-Cyclo-4,5-methylenedioxy-

9,9′-dinor-2,2′-lign-7(8′)-ene-5′,8-diol. 7,8′-Cyclo-5′,8-dihydroxy-4,5-methylenedioxy-9,9′-dinor-2,2′-lign-7(8′)-ene. ***Reticuol***
[1103512-92-0]
$C_{17}H_{14}O_4$ 282.295
Constit. of the bark of *Cinnamomum burmanii* and *Cinnamomum reticulatum*. Inactivator of Cytochrome P450 3A4. Yellow powder

4,5-Methylene, 5′-Me ether
5-(Hydroxymethyl)-9-methoxy-2,3-methylenedioxy-7H-dibenzo[a,c]cycloheptene. 7,8′-Cyclo-5′-methoxy-4,5-methylenedioxy-9,9′-dinor-2,2′-lign-7(8′)-en-8-ol. 7,8′-Cyclo-8-hydroxy-5′-methoxy-4,5-methylenedioxy-9,9′-dinor-2,2′-lign-7(8′)-ene. ***Tenuifolin‡***
[1187955-78-7]
$C_{18}H_{16}O_4$ 296.322
Constit. of *Cinnamomum tenuifolium* and *Cinnamomum reticulatum*. Amorph. powder. λ_{max} 235 (log ε 3.23); 255 (log ε 2.65); 290 (log ε 2.11) (MeCN)

7R,8′-Dihydro, 5-Me ether
7,8′-Cyclo-5-methoxy-9,9′-dinor-2,2′-lignan-4,5′,8-triol. 7,8′-Cyclo-4,5′,8-trihydroxy-5-methoxy-9,9′-dinor-2,2′-lignan. ***Dihydroisosubamol***
[1402068-86-3]
[2058060-77-6] (unspecified stereochem.)
$C_{17}H_{18}O_4$ 286.327
Constit. of the leaves of *Cinnamomum subavenium*. Amorph. powder. Has R_{axial}-config.. λ_{max} 262 (log ε 4.30); 286 (log ε 4.06) (MeOH)

5′-Deoxy, 5-Me ether, 4-O-β-D-glucopyranoside
Subavenoside D
[1402068-83-0] (2*R*-atropisomer)
[1402132-50-6] (2*S*-atropisomer)
$C_{23}H_{26}O_8$ 430.453
Constit. of the leaves of *Cinnamomum subavenium*. Amorph. powder. $[\alpha]^{27}_{D}$ -4 (c, 0.1 in MeOH). Mixt. of two inseparable diastereoisomers in ratio 1.0:0.7. λ_{max} 240 (log ε 4.62); 270 (log ε 4.06) (MeOH)

5′-Deoxy, 5-Me ether, 8-O-β-D-glucopyranoside
Subavenoside A
[1402068-80-7] (2*R*-atropisomer)
[1402132-47-1] (2*S*-atropisomer)
$C_{23}H_{26}O_8$ 430.453
Constit. of the leaves of *Cinnamomum subavenium*. Amorph. powder. $[\alpha]^{27}_{D}$ -12 (c, 0.1 in MeOH). Mixt. of two inseparable diastereoisomers in ratio 1.1:1.0. λ_{max} 242 (log ε 4.47); 274 (log ε 3.95) (MeOH).

Subehan, *et al.*, *Planta Med.*, 2008, **74**, 1474-1480 (*Cinnamomum burmanii constit*).
Lin, R.-J. *et al.*, *J. Nat. Prod.*, 2009, **72**, 1816-1824 (*Tenuifolin*).
Chen, C.Y. *et al.*, *Nat. Prod. Res.*, 2010, **24**, 423-427 (*Subamol*).
Lin, I.J. *et al.*, *Chem. Nat. Compd. (Engl. Transl.)*, 2011, **47**, 43-45 (*Cinnamomum reticulatum constit*).
Chen, C.-Y. *et al.*, *Chem. Nat. Compd. (Engl. Transl.)*, 2011, **47**, 215-217 (*Subamol*).
Chia, Y.C. *et al.*, *Chem. Nat. Compd. (Engl. Transl.)*, 2011, **47**, 220-222 (*Reticuol*).
Tang, C. *et al.*, *Tetrahedron Lett.*, 2011, **52**, 3275-3278 (*Tenuifolin, synth*).
Lin, H.-C. *et al.*, *J. Nat. Prod.*, 2012, **75**, 1735-1743 (*Dimethylsubamol, Dihydroisosubamol, Subavenosides A-F*).
Chen, C.-Y. *et al.*, *Nat. Prod. Res.*, 2012, **26**, 1218-1223 (*Burmanol*).
Pflaesterer, D. *et al.*, *Adv. Synth. Catal.*, 2017, **359**, 866-874 ((±)-*Dihydroisosubamol, synth*)

4,4′,9′-Trihydroxy-3′,5-dimethoxy-8,9-dinor-7′-oxo-3,8′-lignan-7-al T80

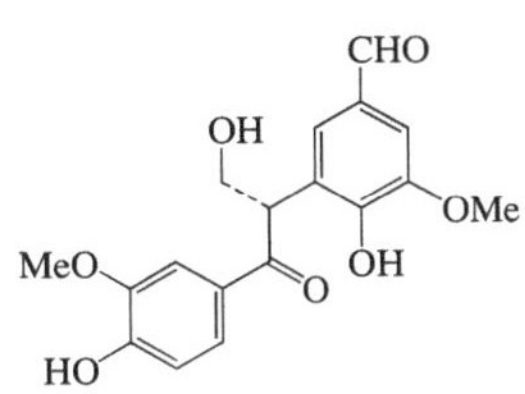

4-Hydroxy-3-[2-(4-hydroxy-3-methoxyphenyl)-1-(hydroxymethyl)-2-oxoethyl]-5-methoxybenzaldehyde, CAS. 4,4′,9′-Trihydroxy-3′,5-dimethoxy-8,9-dinor-7,7′-dioxo-3,8′-neolignan
$C_{18}H_{18}O_7$ 346.335
(S)-form [1829588-36-4]
Constit. of seeds of *Herpetospermum caudigerum*. Exhibits inhibitory activity on HBsAg and HBeAg secretion and cytotoxicity against HepG2 cells. Amorph. brown powder. $[\alpha]^{20}_{D}$ -40 (c, 0.25 in MeOH).

Yu, J.-Q. *et al.*, *Phytochem. Lett.*, 2014, **10**, 230-234 (*Herpetospermum caudigerum constit*)

3,5,7-Trihydroxy-8-[1-(4-hydroxy-3-methoxyphenyl)-2-propenyl]flavone T81

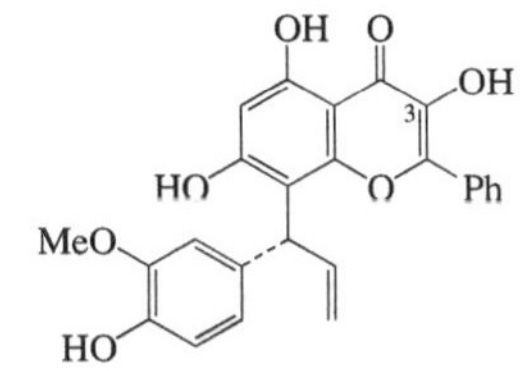

8-[1-(4-Hydroxy-3-methoxyphenyl)-2-propenyl]galangin
$C_{25}H_{20}O_7$ 432.428
Flavonolignan. Similar to 5,7-Dihydroxy-8-[1-(4-hydroxy-3-methoxyphenyl)-2-propenyl]flavanone, D46

(R)-form [1616279-88-9]
Constit. of Mexican propolis from hive of *Apis mellifera*. Exhibits cytotoxic activity against PANC-1 human pancreatic cancer cells. Amorph. yellow solid. $[\alpha]^{22}_{D}$ -14.3 (c, 0.3 in MeOH)

3-Deoxy
5,7-Dihydroxy-8-[1-(4-hydroxy-3-methoxyphenyl)-2-propenyl]flavone. 8-[1-(4-Hydroxy-3-methoxyphenyl)-2-propenyl]chrysin
[1616279-90-3]
$C_{25}H_{20}O_6$ 416.429
Constit. of Mexican propolis from hive of *Apis mellifera*. Yellow-brown solid. $[\alpha]^{22}_{D}$ -17.4 (c, 0.3 in $CHCl_3$).

Li, F. *et al.*, *J. Nat. Prod.*, 2010, **73**, 623-627 (*propolis constits, activity*).
Kim, H. *et al.*, *Tetrahedron*, 2014, **70**, 4738-4744 (*synth*)

3,3′,7′-Trihydroxy-8,8′-lignan-9,9′-olide T82

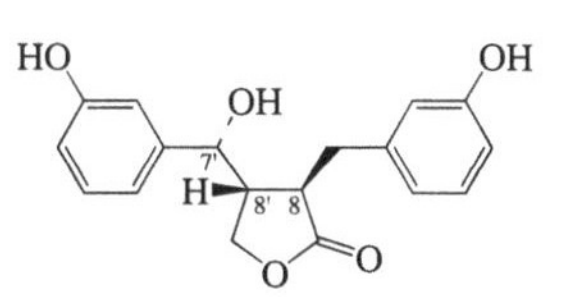

(7′R,8R,8′R)-form

Dihydro-4-[hydroxy(3-hydroxyphenyl)methyl]-3-[(3-hydroxyphenyl)methyl]-2(3H)-furanone, CAS. 7′-Hydroxyenterolactone
$C_{18}H_{18}O_5$ 314.337
(7′R,8R,8′R)-form [1191397-24-6]
[640280-00-8] (7′ξ,8*R*,8′*R*-form)
Found in human plasma. Semi-solid. $[\alpha]^{25}_{D}$ -13.04 (c, 1.50 in Me_2CO))
(7′S,8R,8′R)-form [353740-02-0]

Found in human plasma. Amorph. solid. $[\alpha]_D^{24}$ -1.3 (c, 0.31 in Me_2CO).

Heinonen, S. *et al.*, *J. Agric. Food Chem.*, 2001, **49**, 3178-3186 (*isol*).
Makela, T.H. *et al.*, *Steroids*, 2001, **66**, 777-784 (*7′R*,8R*,8′R*-form, 7′S*,8R*,8′R*-form, synth, sepn, pmr, cmr*).
Smeds, A. *et al.*, *J. Chromatogr., B*, 2003, **793**, 297-308 (*isol, chromatog, ms*).
Raffaelli, B. *et al.*, *Org. Biomol. Chem.*, 2006, **4**, 331-341 (*7′S,8R,8′R-form, synth*).
Hajra, S. *et al.*, *J. Org. Chem.*, 2009, **74**, 7971-7981 (*7′R,8R,8′R-form, synth, pmr, cmr*)

3,4,4′-Trihydroxy-8,8′-lignan-9,9′-olide T83

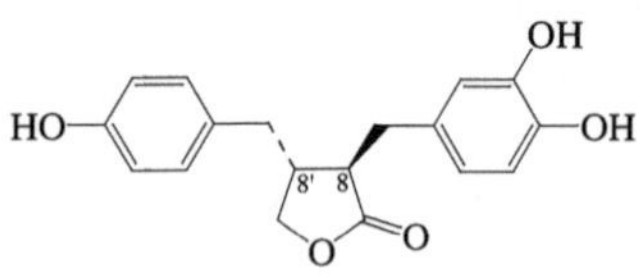

3-(3,4-Dihydroxybenzyl)dihydro-4-(4-hydroxybenzyl)-2(3H)-furanone
$C_{18}H_{18}O_5$ 314.337
(8R,8′R)-form
3,4-Methylene ether
4′-Hydroxy-3,4-methylenedioxy-8,8′-lignan-9,9′-olide. ***Chamalignolide***
[475635-02-0]
$C_{19}H_{18}O_5$ 326.348
Constit. of the heartwood of *Chamaecyparis obtusa* var. *formosana*. Pale yellow liq.. $[\alpha]_D^{22}$ -23.2 (c, 0.41 in $CHCl_3$). λ_{max} 225 (log ε 3.91); 285 (log ε 3.58) (MeOH).

Kuo, Y.-H. *et al.*, *Chem. Pharm. Bull.*, 2002, **50**, 978-980 (*Chamalignolide*)

3′,4,4′-Trihydroxy-8,8′-lignan-9,9′-olide T84

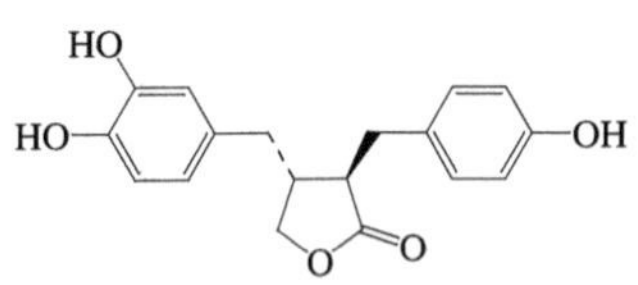

4-(3,4-Dihydroxybenzyl)dihydro-3-(4-hydroxybenzyl)-2(3H)-furanone
$C_{18}H_{18}O_5$ 314.337
(8R,8′R)-form
3′-Me ether
4,4′-Dihydroxy-3′-methoxy-8,8′-lignan-9,9′-olide. ***Piperphilippinin V***
[147960-94-9]
$C_{19}H_{20}O_5$ 328.364
Constit. of the stems of *Piper philippinum*. Oil. $[\alpha]_D^{20}$ -34.9 (c, 0.2 in $CHCl_3$). λ_{max} 226 (log ε 4.07); 280 (log ε 3.59) (MeOH). [base] λ_{max} 245 (log ε 4.09); 290 (log ε 3.62) (MeOH/KOH)
3′,4′-Di-Me ether
4-Hydroxy-3′,4′-dimethoxy-8,8′-lignan-9,9′-olide
[157238-68-1]
$C_{20}H_{22}O_5$ 342.391
Constit. of the roots of *Taiwania cryptomerioides*. Yellow gum. $[\alpha]_D^{25}$ -25.9 (c, 0.2 in $CHCl_3$). λ_{max} 226 (log ε 4.09); 280 (log ε 3.64) (MeOH).

Chyu, C.-F. *et al.*, *Helv. Chim. Acta*, 2007, **90**, 738-747 (*Taiwania cryptomerioides constit*).
Chen, Y.-C. *et al.*, *Phytochemistry*, 2007, **68**, 2101-2111 (*Piperphilippinin V*)

3′,4,4′-Trihydroxy-8,8′-lignan-7-one T85

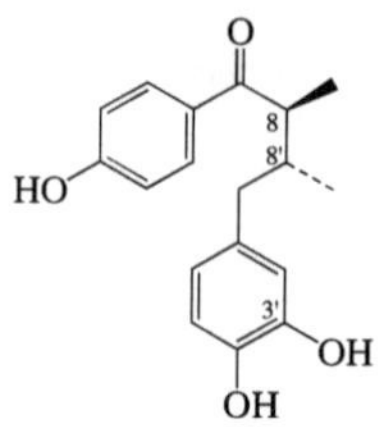

(8*S*,8′*R*)-*form*

$C_{18}H_{20}O_4$ 300.354
(8S,8′R)-form
3′,4′-Methylene ether
4-Hydroxy-3′,4′-methylenedioxy-8,8′-lignan-7-one. ***Demethoxyoleiferin D***
[1028810-09-4]
$C_{19}H_{20}O_4$ 312.365
Constit. of the stem bark of *Iryanthera lancifolia*. Shows weak estrogenic activity. Amorph. solid. $[\alpha]_D^{20}$ -2.5 (c, 0.2 in $CHCl_3$). Abs. config. probable, assigned by biogenetic analogy. λ_{max} 216 (log ε 3.5); 283 (log ε 3.5) (EtOH)
(8ξ,8′ξ)-form
3′-Me ether
4-(4-Hydroxy-3-methoxyphenyl)-1-(4-hydroxyphenyl)-2,3-dimethyl-1-butanone, CAS. *4,4′-Dihydroxy-3′-methoxy-8,8′-lignan-7-one*
[1319204-32-4]
$C_{19}H_{22}O_4$ 314.381
Constit. of the leaves of *Loropetalum chinensis*. Gum. $[\alpha]_D^{21}$ -6.4 (c, 0.51 in $CHCl_3$). λ_{max} 272 (log ε 4.3) ($CHCl_3$).

Mesa-Siverio, D. *et al.*, *Bioorg. Med. Chem.*, 2008, **16**, 3387-3394 (*Demethoxyoleiferin D*).
Zhou, X.-J. *et al.*, *Chem. Nat. Compd. (Engl. Transl.)*, 2011, **47**, 690-692 (*Loropetalum chinensis constit*)

3,7,11-Trihydroxy-6-methoxy-11-methyl-2-(1-propenyl)dibenz[*b,f*]oxepin-10(11*H*)-one T86

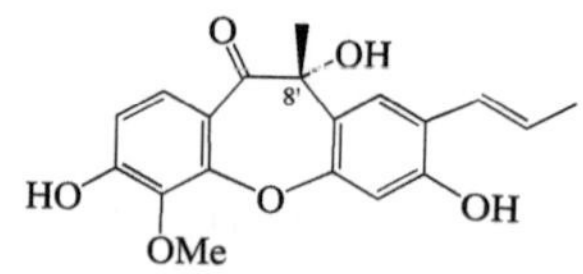

(*R,E*)-*form*

2′,4-Epoxy-4′,6,8′-trihydroxy-3′-methoxy-3,8′-lign-7-en-7′-one.
2′,4-Epoxy-4′,6,8′-trihydroxy-3′-methoxy-7′-oxo-3,8′-neolign-7-ene
$C_{19}H_{18}O_6$ 342.347
Lignan numbering shown
(R,E)-form
8′-Deoxy
2′,4-Epoxy-4′,6-dihydroxy-3′-methoxy-3,8′-lign-7-en-7′-one.
2′,4-Epoxy-4′,6-dihydroxy-3′-methoxy-7′-oxo-3,8′-lign-7-ene.
Penthorin A
[1832604-11-1]
$C_{19}H_{18}O_5$ 326.348
Constit. of *Penthorum chinense*. Gum. $[\alpha]_D^{20}$ -11.6 (c, 0.1 in MeOH). Possesses (*S*)-config. due to change in Cahn-Ingold-Prelog priorities. λ_{max} 214 (log ε 4.43); 254 (log ε 4.29); 297 (log ε 4.03) (MeOH)
(R,Z)-form
Constit. of *Penthorum chinense*. Amorph. powder
8′-Deoxy
Penthorin B
[1829587-76-9]
$C_{19}H_{18}O_5$ 326.348
Constit. of *Penthorum chinense*. Exhibits *in vitro* protective activities against acetaminophen-induced hepatocyte injury. Gum. $[\alpha]_D^{20}$ -7.7 (c, 0.1 in MeOH). Possesses (*S*)-config. due to change in Cahn-Ingold-Prelog priorities. λ_{max} 210 (log ε 4.41); 253 (log ε 3.95); 296 (log ε 3.72) (MeOH)
(ξ,E)-form [1017591-73-9]
Constit. of *Penthorum chinense*. Gum. $[\alpha]_D^{20}$ -56.8 (c, 0.34 in MeOH). λ_{max} 254 (log ε 4.66); 292 (log ε 4.39) (MeOH).

Zhang, T. *et al.*, *J. Integ. Plant Biol.*, 2007, **49**, 1611-1614 (*Penthorum chinense constit*).
He, Y. *et al.*, *Fitoterapia*, 2015, **100**, 7-10 (*Penthorins A,B, activity*)

4,4′,5-Trihydroxy-9′-nor-3,8′-ligna-7,7′-dien-9-oic acid T87

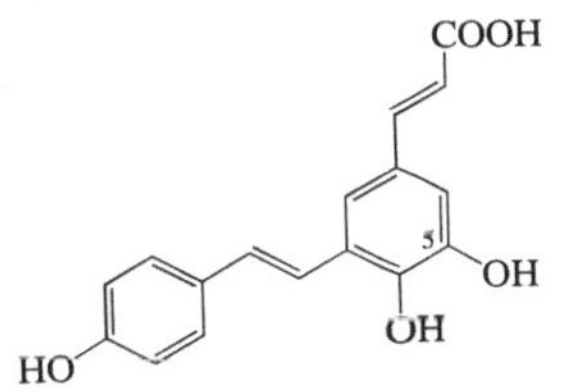

3-[3,4-Dihydroxy-5-[2-(4-hydroxyphenyl)ethenyl]phenyl]-2-propenoic acid
$C_{17}H_{14}O_5$ 298.294
(7E,7′E)-form
5-Me ether
4,4′-Dihydroxy-5-methoxy-9′-nor-3,8′-ligna-7,7′-dien-9-oic acid
5-Me ether, 4-O-β-D-glucopyranoside
Ligusticumacid B
[2573174-08-8]
$C_{24}H_{26}O_{10}$ 474.462
Constit. of the rhizomes of *Ligusticum chuanxiong*. Exhibits moderate neuroprotective activity. Amorph. yellow powder. $[\alpha]_D^{20}$ +70 (c, 0.1 in MeOH). λ_{max} 209 (log ε 4.46); 232 (log ε 4.31); 306 (log ε 4.63) (MeOH).

Zhang, X. *et al.*, *Fitoterapia*, 2018, **125**, 147-154 (*Ligusticumacid B, activity*)

2,4,4′-Trihydroxy-9′-nor-7,8′-lign-7-en-9,8′-olide T88

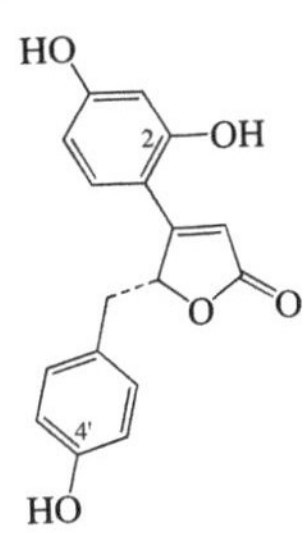

4-(2,4-Dihydroxyphenyl)-5-(4-hydroxybenzyl)-2(5H)-furanone. ***Puerol A.*** *Specionin‡*
[152784-32-2]
[126643-17-2] (Specionin)
$C_{17}H_{14}O_5$ 298.294
Provisional norlignan. Biosynth. pathway unclear. Struct. revised in 1993. Stereochem. assignments are confused in this series. Constit. of *Ononis speciosa*, *Pueraria lobata* and *Sophora japonica*. Needles. Mp 226 - 228°. $[\alpha]_D$ -75.8 (c, 1.1 in MeOH). λ_{max} 221 (ε 21700); 286 (ε 14100); 325 (ε 17500) (MeOH)

2-O-β-D-Glucopyranoside
Kuzubutenolide A. *Specioside A*
[200135-01-9]
[126617-61-6] (Specioside A)
$C_{23}H_{24}O_{10}$ 460.435
Constit. of *Pueraria lobata* and *Ononis speciosa*. Pale yellow prisms (MeOH aq.). Mp 169 - 171°. $[\alpha]_D$ +66.2 (c, 0.1 in DMSO). Struct. in ref. drawn as *S*-config. but stated to be *R*-config. Identity with Specioside A not certain. λ_{max} 200 (log ε 4.39); 219 (log ε 4.29); 227 (sh); 243 (sh); 289 (log ε 4.11); 316 (log ε 4.22) (EtOH)

O-β-D-Glucopyranoside
Specioside B
[126589-95-5]
Isol. from *Ononis speciosa*. Revised struct. Posn. of glycosidation not clear following struct. revision

2-O-[α-L-Rhamnopyranosyl-(1→6)-β-D-glucopyranoside]
Pueroside A
[100692-52-2]
$C_{29}H_{34}O_{14}$ 606.577
Constit. of *Pueraria lobata*. Needles. Mp 183 - 185°. $[\alpha]_D$ -107.5 (MeOH). Struct. of glycoside fragment revised in 1993

2-Me ether
4,4′-Dihydroxy-2-methoxy-9′-nor-7,8′-lign-7-en-9,8′-olide. ***Clitorienolactone B***
[1501651-73-5]
$C_{18}H_{16}O_5$ 312.321
Constit. of the roots of *Clitoria ternatea*. Possesses memory enhancing activity. Acetylcholinesterase inhibitor. Pale yellow cryst. ($CHCl_3$/MeOH). Mp 242 - 244°. λ_{max} 323 (log ε 1.73) (MeOH)

4-Me ether
Epipuerol B. (-)-Puerol B
[1431468-33-5]
$C_{18}H_{16}O_5$ 312.321
Constit. of the roots of *Dalea formosa*. Yellow solid. Mp > 180° dec.. $[\alpha]_D^{20}$ -43 (c, 0.1 in MeOH). *R*-Config. assigned in ref.. λ_{max} 218 (log ε 4.68); 243 (sh); 284 (log ε 4.38); 325 (log ε 4.42) (MeOH)

4-Me ether, 2-O-β-D-glucopyranoside
Pueroside C. *Sophoraside A*
[112343-16-5]
$C_{24}H_{26}O_{10}$ 474.462
Constit. of *Pueraria lobata* and *Sophora japonica*. Needles. Mp 234 - 236°. $[\alpha]_D^{25}$ +32 (c, 1.0 in DMSO)

4-Me ether, 4′-O-β-D-glucopyranoside
Pueroside D
[1145667-33-9]
$C_{24}H_{26}O_{10}$ 474.462
Constit. of the roots of *Pueraria lobata*. Possesses *R*-config.

4-Me ether, 2,4′-di-O-β-D-glucopyranoside
Pueroside B
[100692-54-4]
$C_{30}H_{36}O_{15}$ 636.603
Constit. of *Pueraria lobata*. Needles. Mp 227 - 229°. $[\alpha]_D$ -37.6 (MeOH)

Epimer
(+)-Puerol A
[1803270-52-1]
$C_{17}H_{14}O_5$ 298.294
Constit. of *Amorpha fruticosa* and *Ononis angustissima*. $[\alpha]_D$ +32 (c, 0.18 in MeOH). λ_{max} 287 (log ε 2.74); 326 (log ε 2.84) (MeOH)

Epimer, 4-Me ether
(+)-Puerol B
[112343-17-6]
$C_{18}H_{16}O_5$ 312.321
Constit. of the roots of *Pueraria lobata*. Plates. Mp 238 - 240°. $[\alpha]_D$ +68.2 (MeOH).

Kinjo, J. *et al.*, *Tetrahedron Lett.*, 1985, **26**, 6101-6102 (*Puerosides A,B*).
Shirataki, Y. *et al.*, *Chem. Pharm. Bull.*, 1987, **35**, 1637-1640 (*Sophoraside A*).
Barrero, A.E. *et al.*, *J. Nat. Prod.*, 1989, **52**, 1334-1337 (*Specionin, Speciosides A,B*).
Nohara, T. *et al.*, *Phytochemistry*, 1993, **33**, 1207 (*Puerosides, cryst struct*).
Hirakura, K. *et al.*, *Phytochemistry*, 1997, **46**, 921-928 (*Kuzubutenolide A*).
Wang, Y. *et al.*, *Yaoxue Xuebao*, 2007, **42**, 964-967 (*Pueroside D*).
Belofsky, G. *et al.*, *J. Nat. Prod.*, 2013, **76**, 915-925 (*Epipuerol B, cd*).
Ghribi, L. *et al.*, *Bioorg. Med. Chem. Lett.*, 2015, **25**, 3825-3830 (*(+)-Puerol A*).
Vasisht, K. *et al.*, *Tetrahedron Lett.*, 2016, **57**, 1758-1762 (*Clitorienolactone B*).
Muharini, R. *et al.*, *J. Nat. Prod.*, 2017, **80**, 169-180 (*Puerol A, cd, abs config*)

2-(2,4,6-Trihydroxyphenyl)-5-benzofurancarboxylic acid T89

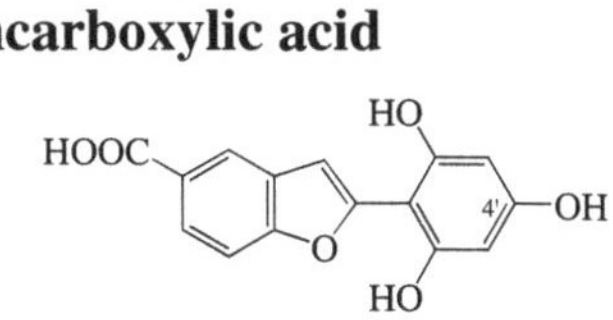

4,7′-Epoxy-2′,4′,6′-trihydroxy-8,9,9′-trinor-3,8′-lign-7′-en-7-oic acid
$C_{15}H_{10}O_6$ 286.239
2′,4′-Di-Me ether
4,7′-Epoxy-2′-hydroxy-4′,6′-dimethoxy-8,9,9′-trinor-3,8′-lign-7′-en-7-oic acid. ***Sobraline***
[1469981-58-5]
$C_{17}H_{14}O_6$ 314.293
Isol. from *Krameria tomentosa*. Powder. Mp 262 - 264°.

Madeiro, S.A.L. *et al.*, *J. Braz. Chem. Soc.*, 2012, **23**, 2021-2026 (*Sobraline*)

2-(2,4,6-Trihydroxyphenyl)-5-benzofuranpropanol T90

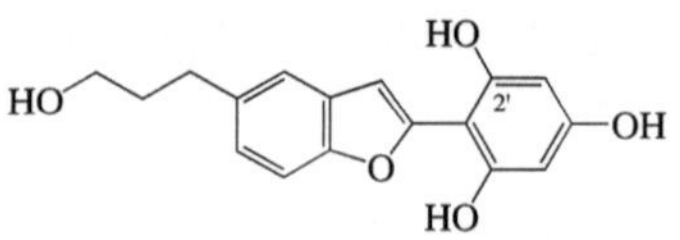

4,7′-Epoxy-9′-nor-3,8′-lign-7′-ene-2′,4′,6′,9-tetrol. 4,7′-Epoxy-2′,4′,6′,9-tetrahydroxy-9′-nor-3,8′-neolign-7′-ene
$C_{17}H_{16}O_5$ 300.31
2′-Me ether
2-(2,4-Dihydroxy-6-methoxyphenyl)-5-benzofuranpropanol. 4,7′-Epoxy-6′-methoxy-9′-nor-3,8′-lign-7′-ene-2′,4′,9-triol. 4,7′-Epoxy-2′,4′,9-trihydroxy-6′-methoxy-9′-nor-3,8′-lign-7′-ene
[184587-71-1]
$C_{18}H_{18}O_5$ 314.337
Constit. of *Krameria parvifolia*. Amorph.. λ_{max} 280 (sh); 292 (log ε 4.12); 306 (log ε 4.15); 328 (sh) (MeOH). [base] λ_{max} 240 (sh); 285 (sh); 327 (log ε 4.21) (MeOH/NaOH).

Achenbach, H. *et al.*, *Phytochemistry*, 1996, **43**, 1093-1095 (*Krameria parvifolia constit*)

1-(2,3,4-Trihydroxyphenyl)-1,2-propanediol T91

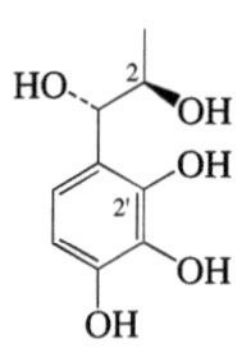

(1*S*,2*R*)-*form*

4-(1,2-Dihydroxypropyl)-1,2,3-benzenetriol
$C_9H_{12}O_5$ 200.190
(1S,2R)-form
3′,4′-Methylene, 1,2′-di-Me ether
1-Methoxy-1-(2-methoxy-3,4-methylenedioxyphenyl)-2-propanol
[2268715-08-6]
$C_{12}H_{16}O_5$ 240.255
Constit. of *Chloranthus anhuiensis*. Yellow oil. λ_{max} 210; 254 (MeOH)
3′,4′-Methylene, 1-Et, 2′-Me ether
1-Ethoxy-1-(2-methoxy-3,4-methylenedioxyphenyl)-2-propanol
[2268715-06-4]
$C_{13}H_{18}O_5$ 254.282
Constit. of *Chloranthus anhuiensis*. Oil. λ_{max} 210; 254 (MeOH)
(1S,2S)-form
3′,4′-Methylene, 1-Et, 2′-Me ether
[2268715-07-5]
$C_{13}H_{18}O_5$ 254.282
Constit. of *Chloranthus anhuiensis*. Oil. λ_{max} 210; 254 (MeOH).

Zhu, H.-L. *et al.*, *Chem. Biodiversity*, 2018, **15**, e1800249:1-6 (*Chloranthus anhuiensis constits*)

1-(3,4,5-Trihydroxyphenyl)-1,2-propanediol T92

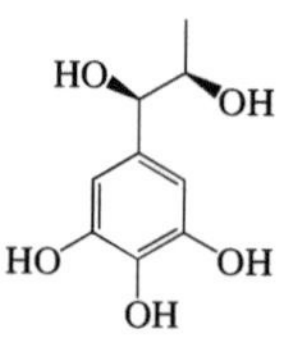

(1*R**,2*R**)-*form*

5-(1,2-Dihydroxypropyl)-1,2,3-benzenetriol
$C_9H_{12}O_5$ 200.19
(1R*,2R*)-form
3′,5′-Di-Me ether
1-(4-Hydroxy-3,5-dimethoxyphenyl)-1,2-propanediol. ***Pisoninol II***
[1346751-44-7]
$C_{11}H_{16}O_5$ 228.244
Constit. of the stems of *Pisonia umbellifera*. Needles. Mp 73 - 75°. $[\alpha]_D^{25}$ +75.2 (c, 0.03 in MeOH). λ_{max} 236 (log ε 3.75); 277 (log ε 2.54) (MeOH). [base] λ_{max} 242 (log ε 3.66); 351 (log ε 2.86) (MeOH/KOH aq.).

Kuo, H.-T. *et al.*, *Planta Med.*, 2011, **77**, 736-741 (*Pisoninol II*)

1-(3,4,5-Trihydroxyphenyl)-1,3-propanediol T93

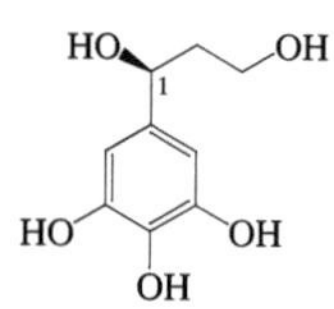

(*S*)-*form*

5-(1,3-Dihydroxypropyl)-1,2,3-benzenetriol
$C_9H_{12}O_5$ 200.19
(S)-form
1,3′,5′-Tri-Me ether
4-Hydroxy-γ,3,5-trimethoxybenzenepropanol. 4-(3-Hydroxy-1-methoxypropyl)-2,6-dimethoxyphenol. 3-(4-Hydroxy-3,5-dimethoxyphenyl)-3-methoxy-1-propanol
[923562-02-1]
[923579-54-8] (unspecified stereochem.)
$C_{12}H_{18}O_5$ 242.271
Constit. of the leaves of *Acer truncatum* and of *Sonchus uliginosus*. Gum. $[\alpha]_D^{25}$ -6.8 (c, 0.05 in MeOH). $[\alpha]_D^{20}$ -8 (c, 0.4 in $CHCl_3$). Abs. config. not demonstrated for the *S. uliginosus* isolate (negative opt. rotn. in $CHCl_3$) but appears probable. λ_{max} 220 (log ε 3.5); 241 (log ε 4); 271 (log ε 3.5) ($CHCl_3$)
(ξ)-form
3′,5′-Di-Me ether, 3-O-(4-hydroxy-E-cinnamoyl), 1-O-β-D-glucopyranoside
[1202751-77-6]
$C_{26}H_{32}O_{12}$ 536.53
Constit. of the leaves of *Sasa quelpaertensis*. Yellow solid. Mp 180 - 182°. $[\alpha]_D^{25}$ -30.7 (c, 0.16 in MeOH). λ_{max} 312 (MeOH).

Dong, L.-P. *et al.*, *Chem. Biodiversity*, 2006, **3**, 791-798 (*Acer truncatum constit*).
Zhang, Z.-X. *et al.*, *Helv. Chim. Acta*, 2006, **89**, 2927-2933 (*Sonchus uliginosus constits*).
Sultana, N. *et al.*, *Bull. Korean Chem. Soc.*, 2009, **30**, 1729-1732 (*Sasa quelpaertensis constit*)

3-(3,4,5-Trihydroxyphenyl)-1,2-propanediol T94

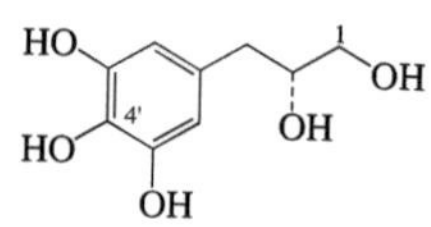

(*R*)-*form*

$C_9H_{12}O_5$ 200.19
(R)-form
3′,5′-Di-Me ether, 1-O-β-D-glucopyranoside
[2408995-29-7]
$C_{17}H_{26}O_{10}$ 390.385
Constit. of *Scutellaria barbata*. Amorph. powder. $[\alpha]_D^{23}$ +11.6 (c, 0.1 in MeOH)

(ξ)-form

3′,4′-Di-Me ether

3-(3-Hydroxy-4,5-dimethoxyphenyl)-1,2-propanediol, CAS. ***Todatriol***

[120021-92-3]

$C_{11}H_{16}O_5$ 228.244

Constit. of *Todaroa aurea*

3′,4′-Di-Me ether, 1-O-β-D-glucopyranoside

[360042-67-7]

$C_{17}H_{26}O_{10}$ 390.385

Constit. of the fruit of *Carum ajowan*. Amorph. powder. $[\alpha]_D^{23}$ -6 (c, 0.1 in MeOH)

3′,5′-Di-Me ether

3-(4-Hydroxy-3,5-dimethoxyphenyl)-1,2-propanediol. ***4′-Hydroxytodadiol***

[22805-15-8]

$C_{11}H_{16}O_5$ 228.244

Constit. of *Annona cherimola* and *Xylosma controversum*. Brown powder ($CHCl_3$). Mp 105 - 107°. $[\alpha]_D^{25}$ -61 (c, 0.15 in $CHCl_3$)

3′,5′-Di-Me ether, 1-O-β-D-glucopyranoside

Xylocoside A

[1042444-67-6]

$C_{17}H_{26}O_{10}$ 390.385

Constit. of the stems of *Xylosma controversum*. Amorph. solid. $[\alpha]_D^{25}$ -5.6 (c, 0.71 in MeOH). λ_{max} 208 (log ε 4.85); 273 (log ε 3.39) (MeOH)

3′,5′-Di-Me ether, 2-O-β-D-glucopyranoside

Xylocoside B

[1042444-78-9]

$C_{17}H_{26}O_{10}$ 390.385

Constit. of *Piper aduncum* and *Xylosma controversum*. Strong tyrosinase inhibitor. Amorph. solid. $[\alpha]_D^{25}$ -6.2 (c, 0.36 in MeOH). λ_{max} 211 (log ε 4.77); 272 (log ε 3.46) (MeOH)

3′,4′,5′-Tri-Me ether

3-(3,4,5-Trimethoxyphenyl)-1,2-propanediol. ***p-Methoxytodadiol. Kmeriol***

[140385-37-1]

[54306-10-4] (Kmeriol)

$C_{12}H_{18}O_5$ 242.271

Constit. of *Kmeria duperreana* and *Todaroa aurea*. Oil

4′,5′-Methylene, 3′-Me ether

3-(3-Methoxy-4,5-methylenedioxyphenyl)-1,2-propanediol. ***Suavediol***

[140385-38-2]

$C_{11}H_{14}O_5$ 226.228

Isol. from *Todaroa aurea*. Oil.

Gonzalez, A.G. *et al.*, *Biochem. Syst. Ecol.*, 1988, **16**, 641-645 (*Todatriol*).

Dong, X. *et al.*, *Pharm. Res.*, 1989, **6**, 637-640 (*Kmeriol*).

González, A.G. *et al.*, *Phytochemistry*, 1991, **30**, 4189-4190 (*p-Methoxytodadiol, Suavediol*).

Chen, C.-Y. *et al.*, *J. Chin. Chem. Soc. (Taipei)*, 1999, **46**, 77-86 (*4′-Hydroxytodadiol*).

Ishikawa, T. *et al.*, *Chem. Pharm. Bull.*, 2001, **49**, 840-844 (*3′,4′-di-Me 1-glucoside*).

Xu, Z.-R. *et al.*, *Helv. Chim. Acta*, 2008, **91**, 1346-1354 (*Xylocosides A,B*).

Luyen, B.T.T. *et al.*, *Med. Chem. Res.*, 2016, **26**, 220-226 (*Xylocoside B, activity*).

Gu, W.-Y. *et al.*, *Chem. Nat. Compd. (Engl. Transl.)*, 2019, **55**, 469-470 (*Scutellaria barbata constit, cd*)

1-(2,3,5-Trihydroxyphenyl)-1,2-propanedione T95

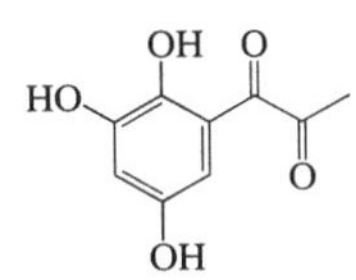

$C_9H_8O_5$ 196.158

2′-Me ether

1-(3,5-Dihydroxy-2-methoxyphenyl)-1,2-propanedione. ***Talaflavuol A***

[1823085-39-7]

$C_{10}H_{10}O_5$ 210.185

Prod. by *Talaromyces flavus* BYD07-13. Brown oil. λ_{max} 209 (log ε 4.41); 276 (log ε 3.93); 356 (log ε 3.53) (MeOH).

He, J.-W. *et al.*, *Nat. Prod. Commun.*, 2016, **11**, 805-808 (*Talaflavuol A*)

1-(3,4,5-Trihydroxyphenyl)-1,2-propanedione T96

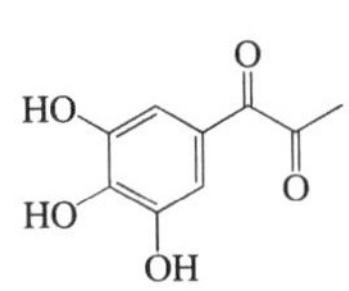

$C_9H_8O_5$ 196.158

3′,5′-Di-Me ether

1-(4-Hydroxy-3,5-dimethoxyphenyl)-1,2-propanedione. Methyl syringoyl ketone

[6925-65-1]

$C_{11}H_{12}O_5$ 224.212

Constit. of the rhizomes of *Daphne giraldii*. Pale orange needles (petrol). Mp 80 - 81°.

Kulka, M. *et al.*, *J. Am. Chem. Soc.*, 1941, **63**, 2371-2374 (*3′,5′-di-Me ether, synth*).

Pepper, J.M. *et al.*, *Can. J. Chem.*, 1964, **42**, 113-121 (*3′,5′-di-Me ether, synth*).

He, W. *et al.*, *Asian J. Tradit. Med.*, 2008, **3**, 117-119 (*Daphne giraldii constit*).

Li, P. *et al.*, *Zhongguo Yaowu Huaxue Zazhi*, 2010, **20**, 50-52 (*Daphne giraldii constit*)

3-(2,3,4-Trihydroxyphenyl)-propanoic acid T97

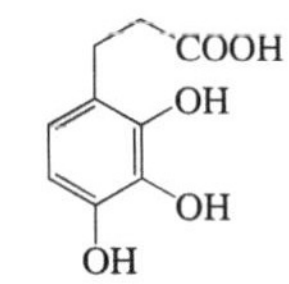

2,3,4-Trihydroxybenzenepropanoic acid, CAS. *2,3,4-Trihydroxydihydrocinnamic acid*

$C_9H_{10}O_5$ 198.174

3,4-Di-Me ether

3-(2-Hydroxy-3,4-dimethoxyphenyl)propanoic acid

[76474-94-7]

$C_{11}H_{14}O_5$ 226.228

Constit. of *Artemisia caruifolia*. Cryst. ($CHCl_3$/petrol). Mp 134°

Tri-Me ether

[33130-04-0]

$C_{12}H_{16}O_5$ 240.255

Cryst. (Et_2O/petrol). Mp 78.1 - 79.2°. Mp 76°

Tri-Me ether, Me ester

[76474-95-8]

$C_{13}H_{18}O_5$ 254.282

Amber oil

3,4-Methylene, 2-Me ether, amide

3-(2-Methoxy-3,4-methylenedioxyphenyl)propanamide. ***Boehmerine‡***

[1639836-61-5]

$C_{11}H_{13}NO_4$ 223.228

Alkaloid from *Piper boehmeriaefolium* var. *tonkinense*. Exhibits antidepressant activity. Cryst. (Me_2CO)

3-O-(4-Hydroxy-3ξ-methylbutyl), 4-Me ether

3-[2-Hydroxy-3-(4-hydroxy-3-methylbutoxy)-4-methoxyphenyl]propanoic acid

[1260087-67-9]

$C_{15}H_{22}O_6$ 298.335

Constit. of *Artemisia armeniaca*. Pale yellow solid.

Barger, G. *et al.*, *J. Chem. Soc., Trans.*, 1910, **97**, 2253-2261 (*tri-Me ether, synth*).

Barua, N.C. *et al.*, *Phytochemistry*, 1980, **19**, 2217-2218 (*Artemisia caruifolia constit*).

Reeder, M.D. *et al.*, *Org. Lett.*, 2005, **7**, 1089-1092 (*tri-Me ether Me ester, synth*).

Kumar, V. *et al.*, *Helv. Chim. Acta*, 2006, **89**, 483-495 (*tri-Me ether Me ester, synth*).

Mojarrab, M. *et al.*, *Nat. Prod. Commun.*, 2010, **5**, 1619-1622 (*Artemisia armeniaca constit*).

Ge, L. *et al., Chem. Nat. Compd. (Engl. Transl.)*, 2014, **50**, 1056-1059 (*Boehmerine*).
Chen, J. *et al., RSC Adv.*, 2015, **5**, 68128-68135 (*tri-Me ether, synth*)

3-(2,3,5-Trihydroxyphenyl)-propanoic acid T98

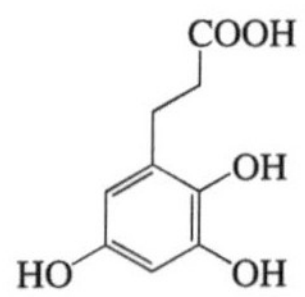

2,3,5-Trihydroxybenzenepropanoic acid, CAS. *2,3,5-Trihydroxydihydrocinnamic acid*
$C_9H_{10}O_5$ 198.174
5′-Me ether
3-(2,3-Dihydroxy-5-methoxyphenyl)propanoic acid
5′-Me ether, Me ester, 2′-O-β-D-glucopyranoside
[2185902-10-5]
$C_{17}H_{24}O_{10}$ 388.369
Constit. of *Lysimachia patungensis*. Antioxidant. Amorph. powder. $[\alpha]^{20}_{D}$+11 (c, 0.15 in MeOH). λ_{max} 281 (MeOH).

Gao, Y.-Y. *et al., J. Asian Nat. Prod. Res.*, 2017, **19**, 28-34 (*Lysimachia patungensis constit, activity*)

3-(2,4,5-Trihydroxyphenyl)propanoic acid T99

2,4,5-Trihydroxybenzenepropanoic acid, CAS. *2,4,5-Trihydroxydihydrocinnamic acid*
[70303-80-9]
$C_9H_{10}O_5$ 198.174
No phys. props. reported. Forms the lactone 3,4-Dihydro-6,7-dihydroxy-2*H*-1-benzopyran-2-one on heating
5-Me ether
2,4-Dihydroxy-5-methoxybenzenepropanoic acid. 3-(2,4-Dihydroxy-5-methoxyphenyl)propanoic acid
[824950-88-1]
$C_{10}H_{12}O_5$ 212.201
5-Me ether, Me ester
[1340493-23-3]
$C_{11}H_{14}O_5$ 226.228
Constit. of the fruit of *Morinda citrifolia*. Yellow gum. λ_{max} 204 (log ε 3.76); 290 (log ε 3.02) (MeOH)
5-Me ether, butyl ester
[1340493-88-0]
$C_{14}H_{20}O_5$ 268.309
Constit. of the fruit of *Morinda citrifolia*. Yellow gum. λ_{max} 204 (log ε 3.78); 290 (log ε 3.16) (MeOH).

Moore, C.W. *J. Chem. Soc., Trans.*, 1911, **99**, 1043-1048 (*5-Me ether, tri-Me ether*).
Wang, G. *et al., J. Asian Nat. Prod. Res.*, 2011, **13**, 238-241 (*Morinda citrifolia esters*)

3-(3,4,5-Trihydroxyphenyl)-1-propanol T100

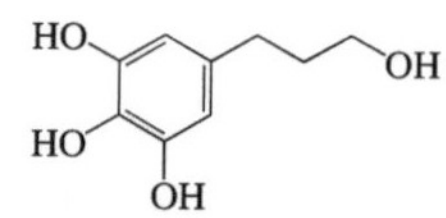

5-(3-Hydroxypropyl)-1,2,3-benzenetriol, CAS. *5-(3-Hydroxypropyl)pyrogallol. 5-Hydroxydihydroconiferyl alcohol*
$C_9H_{12}O_4$ 184.191
3′-Me ether
3-(3,4-Dihydroxy-5-methoxyphenyl)-1-propanol
[132473-95-1]
$C_{10}H_{14}O_4$ 198.218
Constit. of *Relhania fruticosa* and *Rhododendron primulaeflorum*
3′-Me ether, 5′-O-β-D-glucopyranoside
[1427310-41-5]
$C_{16}H_{24}O_9$ 360.359
Constit. of *Rhododendron primulaeflorum*. Powder. $[\alpha]^{20}_{D}$-20.8 (c, 0.12 in MeOH)
3′-Me ether, 1-octadecanoyl
5-Hydroxydihydroconiferyl stearate
[75838-63-0]
$C_{28}H_{48}O_5$ 464.687
Isol. from *Aristeguietia buddleaefolia*, as mixt. with 5-hydroxydihdroconiferylarachidate. Also from *Stomatanthes corumbensis*
3′-Me ether, 1-eicosanoyl
5-Hydroxydihydroconiferyl arachidate
[75838-64-1]
$C_{30}H_{52}O_5$ 492.741
Isol. from *Aristeguietia buddleaefolia*, as mixt. with 5-hydroxydihdroconiferyl stearate. Obt. only as a mixt. with the stearate (as acetates)
3′-Me ether, 1-O-(4-hydroxy-3-methoxy-E-cinnamoyl)
[132473-97-3]
$C_{20}H_{22}O_7$ 374.389
Constit. of *Relhania fruticosa*
3′-Me ether, 4′-O-(2-methylbutanoyl)
[271246-95-8]
$C_{15}H_{22}O_5$ 282.336
Constit. of *Schkuhria schkurioides*
3′-Me ether, 4′-O-(2-methylbutanoyl), 1-Ac
$C_{17}H_{24}O_6$ 324.373
Constit. of *Schkuhria pinnata* var. *wislizeni*
3′-Me ether, 5′-O-(2-methylbutanoyl)
[271246-96-9]
$C_{15}H_{22}O_5$ 282.336
Constit. of *Schkuhria schkuhrioides*
3′-Me ether, 5′-O-(2-methylbutanoyl), 1-Ac
$C_{17}H_{24}O_6$ 324.373
Constit. of *Schkuhria pinnata* var. *wislizeni*
3′,5′-Di-Me ether
3-(4-Hydroxy-3,5-dimethoxyphenyl)-1-propanol. 4-(3-Hydroxypropyl)-2,6-dimethoxyphenol. Dihydrosinapyl alcohol. ***Dihydrosyringenin***
[20736-25-8]
$C_{11}H_{16}O_4$ 212.245
Constit. of *Jurinea leptoloba*
3′,5′-Di-Me ether, 1-O-β-D-glucopyranoside
Dihydrosyringin*. Dihydrosinapyl β-D-glucoside*
[17609-07-3]
$C_{17}H_{26}O_9$ 374.386
Constit. of *Dendrobium devonianum, Ligustrum lucidum, Taraxacum formosanum* and *Vitis thunbergii*. Needles. Mp 160°
3′,5′-Di-Me ether, 4′-O-[3,4,5-trihydroxybenzoyl-(→6)-β-D-glucopyranoside]
[2101478-52-6]
$C_{24}H_{30}O_{13}$ 526.491
Constit. of the stem bark of *Schinopsis brasiliensis*. Amorph. solid
3′,5′-Di-Me ether, 4′-O-[3,4-dihydroxy-E-cinnamoyl-(→6)-β-D-glucopyranoside]
6′-trans-Caffeoyldihydrosyringin
[1827325-75-6]
$C_{26}H_{32}O_{12}$ 536.530
Constit. of *Aster glehni*. Shows antiinflammatory effects
3′,5′-Di-Me ether, 1-formyl
3-(4-Hydroxy-3,5-dimethoxyphenyl)-1-propyl formate
[923562-01-0]
$C_{12}H_{16}O_5$ 240.255
Constit. of the leaves of *Acer truncatum*. Gum. λ_{max} 240 (log ε 4.1); 272 (log ε 3.5); 355 (log ε 2.59) ($CHCl_3$)

3′,5′-Di-Me ether, 1-benzoyl
Dihydrosinapyl benzoate
$C_{18}H_{20}O_5$ 316.353
Constit. of *Croton hutchinsonianus*. Pale yellow solid. λ_{max} 228 (log ε 3.85); 272 (log ε 2.89) (MeOH)

3′,5′-Di-Me ether, 1-O-(4-hydroxy-E-cinnamoyl)
Dihydrosinapyl trans-p-coumarate
[80377-61-3]
$C_{20}H_{22}O_6$ 358.39
Constit. of *Eremanthus mollis*

3′,5′-Di-Me ether, 1-O-(4-hydroxy-Z-cinnamoyl)
Dihydrosinapyl cis-p-coumarate
[80377-62-4]
$C_{20}H_{22}O_6$ 358.39
Constit. of *Eremanthus mollis*

3′,5′-Di-Me ether, 1-O-(4-hydroxy-3-methoxy-E-cinnamoyl)
Dihydrosinapyl ferulate
$C_{21}H_{24}O_7$ 388.416
Constit. of *Peganum nigellastrum*. Pale yellow semi-solid. λ_{max} 206 (log ε 4.87); 230 (sh); 298 (sh); 325 (log ε 4.45) (MeOH)

3′,4′,5′-Tri-Me ether
3,4,5-Trimethoxybenzenepropanol, CAS
[53560-26-2]
$C_{12}H_{18}O_4$ 226.272
Oil. Bp 136 - 139° at 0.3 mm. n_D^{25} +1.5299

4′-O-[R-[2-(β-D-Glucopyranosyloxy)-1-(hydroxymethyl)ethyl]], 3′,5′-di-Me ether
[2264030-46-6]
$C_{20}H_{32}O_{11}$ 448.465
Constit. of *Pilea cavaleriei*. Amorph. powder. $[\alpha]_D^{20}$ +10 (c, 0.1 in MeOH). λ_{max} 208; 267 (MeOH).

Rapoport, H. *et al.*, *J. Am. Chem. Soc.*, 1951, **73**, 2239-2241 (*tri-Me ether*).
Bohlmann, F. *et al.*, *Phytochemistry*, 1980, **19**, 977-979 (*Aristeguietia buddlaefolia constits*).
Bohlmann, F. *et al.*, *Phytochemistry*, 1981, **20**, 1141-1143 (*Stomatanthes corumbensis constit*).
Bohlmann, F. *et al.*, *Phytochemistry*, 1981, **20**, 1609-1612 (*Eremanthus mollis constit*).
Tsichritzis, F. *et al.*, *Phytochemistry*, 1990, **29**, 3173-3187 (*Relhania fruticosa constit*).
Delgado, G. *et al.*, *Nat. Prod. Lett.*, 1998, **12**, 17-22 (*Schkuhria schkuhrioides constits*).
Kakuda, R. *et al.*, *CA*, 2000, **132**, 276640k (*Dihydrosyringin*).
Ma, Z.-Z. *et al.*, *Phytochemistry*, 2000, **53**, 1075-1078 (*Dihydrosinapyl ferulate*).
Dong, L.-P. *et al.*, *Chem. Biodiversity*, 2006, **3**, 791-798 (*3′,5′-di-Me ether 1-formyl*).
Athikomkulchai, S. *et al.*, *Chem. Pharm. Bull.*, 2006, **54**, 262-264 (*Dihydrosinapyl benzoate*).
Besong, G. *et al.*, *Org. Biomol. Chem.*, 2006, **4**, 2193-2207 (*tri-Me ether*).
León, A. *et al.*, *J. Mex. Chem. Soc.*, 2009, **53**, 193-200 (*Schkuhria pinnata constits*).
Zhou, X.-L. *et al.*, *Chin. Herb. Med.*, 2012, **4**, 81-83 (*Rhododendron primulaeflorum constits*).
Son, M.J. *et al.*, *Bull. Korean Chem. Soc.*, 2015, **36**, 503-512 (*6′-Caffeoyldihydrosyringin*).
Seo, S. *et al.*, *Bioorg. Med. Chem. Lett.*, 2016, **26**, 4592-4598 (*6′-Caffeoyldihydrosyringin, activity*).
Zhou, Y. *et al.*, *Fitoterapia*, 2017, **119**, 100-107 (*Pilea cavaleriei constit*).
Santos, C.C.S. *et al.*, *Phytochemistry*, 2017, **140**, 45-51 (*3′,5′-di-Me ether 4′-6-galloylglucoside*)

1-(2,4,5-Trihydroxyphenyl)-1-propanone, CAS T101

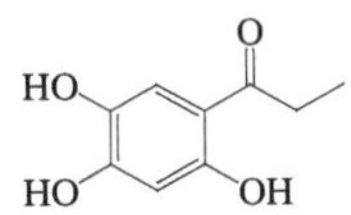

2′,4′,5′-Trihydroxypropiophenone
[79744-62-0]
$C_9H_{10}O_4$ 182.175
Cryst.. Mp 151 - 153°

4-Me ether
2′,5′-Dihydroxy-4′-methoxypropiophenone
[3839-58-5]
$C_{10}H_{12}O_4$ 196.202
Cryst. (EtOH). Mp 112 - 114°

4,5-Di-Me ether
1-(2-Hydroxy-4,5-dimethoxyphenyl)-1-propanone
[857226-16-5]
$C_{11}H_{14}O_4$ 210.229
Constit. of *Trigonostemon xyphophylloides*. Mp 118 - 119°

4,5-Methylene ether
2′-Hydroxy-4′,5′-methylenedioxypropiophenone, CAS. ***Kakuol***
[18607-90-4]
$C_{10}H_{10}O_4$ 194.186
Constit. of *Asarum heterotropoides* and *Asarum sieboldii*. Antifungal agent. Plates (petrol). Mp 110°.

Hardegger, E. *et al.*, *Helv. Chim. Acta*, 1964, **47**, 1996-2017 (*4-Me ether, tri-Me ether, synth*).
Saiki, Y. *et al.*, *Yakugaku Zasshi*, 1967, **87**, 1524-1528 (*Kakuol*).
Jha, H.C. *et al.*, *Can. J. Chem.*, 1981, **59**, 2266-2282 (*cmr*).
Lee, J.Y. *et al.*, *Pest Manag. Sci.*, 2005, **61**, 821-825 (*Kakuol, activity*).
Cen, C. *et al.*, *Acta Crystallogr., Sect. E*, 2009, **65**, o2815 (*Trigonostemon xyphophylloides constit, cryst struct*).
Musso, L. *et al.*, *Chem. Biodiversity*, 2010, **7**, 887-897 (*Kakuol, synth, activity*)

1-(3,4,5-Trihydroxyphenyl)-1-propanone, CAS T102

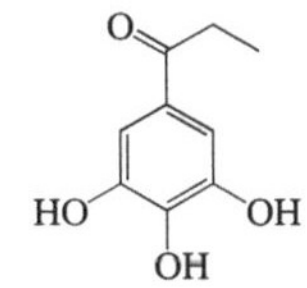

3′,4′,5′-Trihydroxypropiophenone. *5-Propionyl-1,2,3-benzenetriol*
$C_9H_{10}O_4$ 182.175

3′,4′-Di-Me ether
1-(3-Hydroxy-4,5-dimethoxyphenyl)-1-propanone. *3′-Hydroxy-4′,5′-dimethoxypropiophenone*
$C_{11}H_{14}O_4$ 210.229
Constit. of *Phaius mishmensis* and *Trigonostemon lutescens*. Needles. λ_{max} 238 (log ε 3.47); 274 (log ε 3.82) ($CHCl_3$)

3′,5′-Di-Me ether
1-(4-Hydroxy-3,5-dimethoxyphenyl)-1-propanone. *4′-Hydroxy-3′,5′-dimethoxypropiophenone*. ***Propiosyringone***
[5650-43-1]
$C_{11}H_{14}O_4$ 210.229
Constit. of *Baccharis* spp., *Eucalyptus globulus* and *Monstera deliciosa*. Lignin degradn. prod.. Cryst. (cyclohexane or EtOH aq.). Mp 109 - 110°. λ_{max} 299 (ε 11200) (EtOH)

3′,5′-Di-Me ether, 4′-O-[α-L-rhamnopyranosyl-(1→6)-β-D-glucopyranoside]
[2099167-58-3]
$C_{23}H_{34}O_{13}$ 518.512
Constit. of twigs of *Litsea cubeba*. Exhibits a reduction of acetaminophen-induced HepG2 cell injury. Amorph. powder. $[\alpha]_D^{20}$ -19 (c, 0.4 in MeOH). λ_{max} 279 (log ε 2.92) (MeOH)

3′,4′-Methylene ether
see Radiatinol, R2.

Biftu, T. *et al.*, *J. Chem. Soc., Perkin Trans. 1*, 1979, 2276-2281 (*synth, deriv*).
Becker, H.D. *et al.*, *J. Org. Chem.*, 1980, **45**, 1596-1600 (*synth*).
Zdero, C. *et al.*, *Phytochemistry*, 1986, **25**, 2841-2855 (*Baccharis constit*).
Joshi, B.P. *et al.*, *Can. J. Chem.*, 2005, **83**, 1826-1832 (*tri-Me ether, synth*).
Wang, L.-Y. *et al.*, *J. Nat. Prod.*, 2017, **80**, 1808-1818 (*Litsea cubeba constit, activity*)

3-(3,4,5-Trihydroxyphenyl)-2-propenal, CAS T103

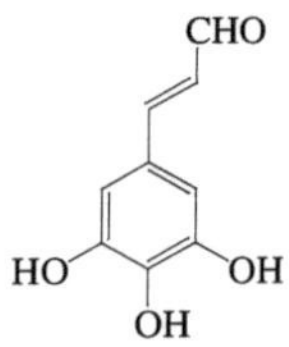

3,4,5-Trihydroxycinnamaldehyde
[951309-35-6]
$C_9H_8O_4$ 180.159
▶ GD6850000
(E)-form
3′-Me ether
3-(3,4-Dihydroxy-5-methoxyphenyl)-2-propenal. 2,4-Dihydroxy-5-methoxycinnamaldehyde. 5-Hydroxyconiferaldehyde
[266999-89-7]
[249647-14-1] (unspecified stereochem.)
$C_{10}H_{10}O_4$ 194.186
Intermed. in syringyl lignan biosynth. in angiosperms. Mp 112 - 116° (synthetic). Apparently not isol. as a nat. prod.
3′,5′-Di-Me ether
see 4-Hydroxy-3,5-dimethoxycinnamaldehyde, H76.

Olstein, R. *et al.*, *Aust. J. Chem.*, 1979, **32**, 681-686 (*synth*).
Liptaj, T. *et al.*, *Coll. Czech. Chem. Comm.*, 1980, **45**, 330-334 (*pmr, conformn*).
Bestmann, H.J. *et al.*, *Chem. Ber.*, 1982, **115**, 161-171 (*synth, ir, uv*).
Taylor, R.T. *et al.*, *J. Org. Chem.*, 1983, **48**, 5160-5164 (*synth*).
Tachibana, S. *et al.*, *Mokuzai Gakkaishi*, 1989, **35**, 1154-1158 (*synth*).
Jing, X-B. *et al.*, *J. Biol. Chem.*, 2000, **275**, 6537-6545 (*3′-Me ether, biosynth*).
Jing, X-B. *et al.*, *J. Chin. Chem. Soc. (Taipei)*, 2001, **48**, 59-63 (*3′-Me ether, synth*).
Jing, X-B. *et al.*, *J. Chin. Chem. Soc. (Taipei)*, 2004, **51**, 1001-1004 (*3′-Me ether, synth*).
Sakakibara, N. *et al.*, *Org. Biomol. Chem.*, 2007, **5**, 802-815 (*3′-Me ether, biosynth*).
Lewis, R.J. *Sax's Dangerous Properties of Industrial Materials, 12th edn., J. Wiley*, 2012, TLA250

3-(2,3,4-Trihydroxyphenyl)-2-propenoic acid, CAS T104

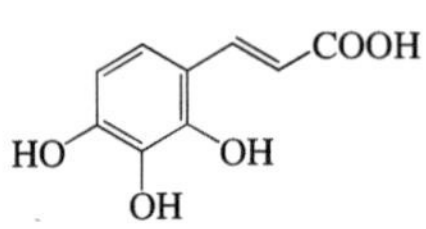

(E)-form

2,3,4-Trihydroxycinnamic acid, CAS. Daphnetic acid
[13058-13-4]
$C_9H_8O_5$ 196.158
The name Daphnetic acid found only in the old lit. as a degradation prod. (in the form of its derivs.) of 7,8-Dihydroxy-2*H*-1-benzopyran-2-one
(E)-form
3,4-Di-Me ether
2-Hydroxy-3,4-dimethoxycinnamic acid
[1195509-76-2]
[682805-22-7] (unspecified stereochem.)
$C_{11}H_{12}O_5$ 224.212
Solid (H_2O). Mp 209 - 210°. Probably *E*-form
Tri-Me ether
3-(2,3,4-Trimethoxyphenyl)-2-propenoic acid. 2,3,4-Trimethoxycinnamic acid
[116406-19-0]
[33130-03-9]
$C_{12}H_{14}O_5$ 238.239
Needles (EtOH). Mp 172°
Tri-Me ether, Me ester
[116406-21-4]
$C_{13}H_{16}O_5$ 252.266
Constit. of *Zanthoxylum nitidum*. Pale yellow needles. Mp 52°
Tri-Et ether
2,3,4-Triethoxycinnamic acid
$C_{15}H_{20}O_5$ 280.32
Cryst.. Mp 193°. Probably *E*-form
(Z)-form
Tri-Me ether
[882653-77-2]
$C_{12}H_{14}O_5$ 238.239
Constit. of the roots of *Zanthoxylum nitidum*.

Will, W. *et al.*, *Ber. Dtsch. Chem. Ges.*, 1884, **17**, 1081-1091 (*tri-Et ether*).
Slotta, K.H. *et al.*, *Ber. Dtsch. Chem. Ges.*, 1930, **63**, 3029-3044 (*E-form, tri-Me ether*).
Anet, F.A.L. *Aust. J. Sci. Res., Ser. A*, 1949, **2**, 127-131 (*3,4-di-Me ether, E-form, tri-Me ether*).
Ishikawa, T. *et al.*, *Chem. Pharm. Bull.*, 1995, **43**, 2014-2018 (*tri-Me ether Me ester, isol*).
Hu, J. *et al.*, *Zhongguo Zhongyao Zazhi*, 2006, **31**, 1689-1691 (*Z-form, tri-Me ether, isol*).
Srivastava, V. *et al.*, *Bioorg. Med. Chem.*, 2007, **15**, 508-525 (*tri-Me ether*)

3-(2,4,5-Trihydroxyphenyl)-2-propenoic acid T105

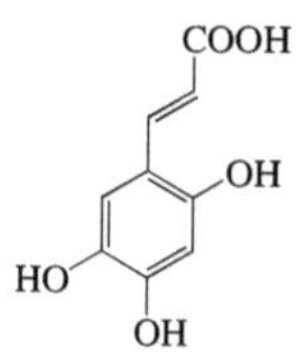

3-(2,4,5-Trihydroxyphenyl)acrylic acid. 2,4,5-Trihydroxycinnamic acid. Aesculetic acid. Esculetic acid
[56437-15-1]
$C_9H_8O_5$ 196.158
(E)-form
Me ester
[1824857-08-0]
$C_{10}H_{10}O_5$ 210.185
Mp 109°
5-Me ether
3-(2,4-Dihydroxy-5-methoxyphenyl)-2-propenoic acid. 2,4-Dihydroxy-5-methoxycinnamic acid
[1456512-03-0]
$C_{10}H_{10}O_5$ 210.185
Constit. of *Viola betonicifolia*. Needles ($CHCl_3$/EtOAc). λ_{max} 229; 345 (MeOH)
4,5-Di-Me ether, 2-O-β-D-glucopyranoside
Densifloroside
[57092-16-7]
$C_{17}H_{22}O_{10}$ 386.353
Isol. from *Dendrobium densiflorum*. Cryst. (H_2O). Mp 225 - 227°. $[\alpha]_D^{22}$ -119 (MeOH). The (*Z*)-form may also be present in the plant
Tri-Me ether
2,4,5-Trimethoxycinnamic acid. 3-(2,4,5-Trimethoxyphenyl)-2-propenoic acid
[73490-49-0]
[24160-53-0] (unspecified stereochem.)
$C_{12}H_{14}O_5$ 238.239
Constit. of *Piper brachystachyum*. Needles (EtOH aq.). Mp 169°
Tri-Me ether, Me ester
[80749-75-3]
$C_{13}H_{16}O_5$ 252.266
Constit. of *Piper brachystachyum*. Prisms (EtOH aq.). Mp 109°
Tri-Me ether, pyrrolidide
N-(2,4,5-Trimethoxycinnamoyl) pyrrolidine. ***Sarmentamide C***
[885677-76-9]
$C_{16}H_{21}NO_4$ 291.347
Constit. of the roots of *Piper sarmentosum*. Amorph. solid. Mp 159 - 162°. λ_{max} 203 (log ε 4.25); 230 (log ε 4.29); 305 (log ε 4.26) (MeOH)
Tri-Et ether
2,4,5-Triethoxycinnamic acid
$C_{15}H_{20}O_5$ 280.32
Cryst. (EtOH aq.). Mp 144°. Probably (*E*)-form

(Z)-form [1427534-96-0]
Constit. of the seeds of *Alisma orientale*
Tri-Me ether
[73490-48-9]
Constit. of *Piper brachystachyum*. Cryst.. Mp 124 - 125°.

Aldrich Library of Infrared Spectra, 3rd edn., 1981, 955H (*ir*).
Aldrich Library of ^{13}*C and* ^{1}*H FT-NMR Spectra*, 1992, **2**, 1060A (*nmr*).
Will, W. *Ber. Dtsch. Chem. Ges.*, 1883, **16**, 2106-2119 (*tri-Et ether*).
Dahmen, J. *et al.*, *Acta Chem. Scand., Ser. B*, 1975, **29**, 627-628 (*Densifloroside*).
Scharf, H.-D. *et al.*, *Justus Liebigs Ann. Chem.*, 1978, 573-579 (*synth, amides*).
Koul, S.K. *et al.*, *Phytochemistry*, 1988, **27**, 3523-3527 (*Piper brachystachyum constit*).
Sinha, A.K. *et al.*, *Indian J. Chem., Sect. B*, 2002, **41**, 635-638 (*tri-Me ether Me ester*).
Tuntiwachwuttikul, P. *et al.*, *Chem. Pharm. Bull.*, 2006, **54**, 149-151 (*Sarmentamide C*).
Zhao, M. *et al.*, *Nat. Prod. Commun.*, 2012, **7**, 785-787 (*Alisma orientale constit*).
Muhammad, N. *et al.*, *J. Enzyme Inhib. Med. Chem.*, 2013, **28**, 997-1001 (*5-Me ether*)

3-(3,4,5-Trihydroxyphenyl)-2-propenoic acid, CAS T106

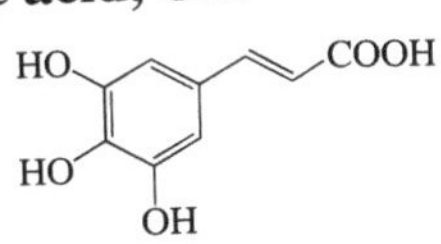

(*E*)-form

3-(3,4,5-Trihydroxyphenyl)acrylic acid. 3,4,5-Trihydroxycinnamic acid
[6093-59-0]
$C_9H_8O_5$ 196.158
Does not appear to have been found in nature in the free state despite extensive searches

(E)-form [709007-50-1]
Needles + $1H_2O$ (H_2O) becoming anhyd. at 120°; light yellow solid. Mp 186 - 188°. Mp 207 - 208°

4-O-β-D-Glucopyranoside
$C_{15}H_{18}O_{10}$ 358.299
Constit. of the fruit of *Psidium acutangulum*

Et ester
[709007-51-2]
$C_{11}H_{12}O_5$ 224.212
Exhibits antiproliferative activity towards human adenocarcinoma cells. Light yellow solid (Et_2O). Mp 176 - 179° (170°)

3-Me ether
3-(3,4-Dihydroxy-5-methoxyphenyl)-2-propenoic acid, CAS. 3,4-Dihydroxy-5-methoxycinnamic acid. 5-Hydroxyferulic acid. 3-Methoxycaffeic acid
[1782-55-4]
$C_{10}H_{10}O_5$ 210.185
Isol. from *Phyllostachys edulis* and *Sasa kirilensis*. Biosynthetic intermed. of lignin. Cryst. (H_2O). Mp 182° (168-170° dec.)

3-Me ether, Me ester
[123692-72-8]
[2041-35-2] (unspecified stereochem.)
$C_{11}H_{12}O_5$ 224.212
Cryst. (MeOH). Mp 122 - 123°

3-Me ether, docosyl ester
Docosyl 5-hydroxyferulate
[252366-72-6]
$C_{32}H_{54}O_5$ 518.779
Constit. of the pericarp of *Platycladus orientalis*. Mp 60 - 61°. λ_{max} 240 (log ε 4.22); 320 (log ε 3.91); 348 (log ε 3.74) (MeOH)

3,5-Di-Me ether
4-Hydroxy-3,5-dimethoxycinnamic acid. ***Sinapic acid****. Sinapinic acid*
[530-59-6]
$C_{11}H_{12}O_5$ 224.212
A common constit. of plants and fruits; isol. from *Sinapis alba*. Found by Bate-Smith in 26% of investigated dicotyledonous and 57% of monocotyledonous spp.. Exhibits antimicrobial, antiinflammatory, anticancer and antioxidant activities. Yellow needles (EtOH). Mp 198 - 200° (192°)

3,5-Di-Me ether, Me ester
Methyl sinapate
[42041-51-0]
$C_{12}H_{14}O_5$ 238.239
Constit. of *Helietta apiculata*, *Polygala vulgaris* and *Securidaca longipedunculata fres*. Needles (EtOH aq.). Mp 90 - 92° (88°)

3,5-Di-Me ether, Et ester
Ethyl sinapate
[41628-47-1]
[69113-98-0] (unspecified stereochem.)
$C_{13}H_{16}O_5$ 252.266
Constit. of *Descurainia sophia*, *Helietta apiculata* and *Thalictrum minus*. Needles (EtOH). Mp 81 - 82°. λ_{max} 240 (ε 18000); 330 (ε 19800) (EtOH)

3,5-Di-Me ether, isopropyl ester
Isopropyl sinapate
[389141-49-5]
$C_{14}H_{18}O_5$ 266.293
Constit. of the marine brown alga *Spatoglossum variabile*. Oil. λ_{max} 282 (log ε 4.6) ($CHCl_3$)

3,5-Di-Me ether, butyl ester
Butyl sinapate
[475211-36-0]
$C_{15}H_{20}O_5$ 280.32
Isol. from the marine brown alga *Spatoglossum variabile*. Yellow oil. λ_{max} 313 (log ε 5.07) ($CHCl_3$)

3,5-Di-Me ether, hexadecyl ester
Hexadecyl sinapate
[286938-83-8]
$C_{27}H_{44}O_5$ 448.644
Constit. of *Sophora flavescens*

3,5-Di-Me ether, eicosyl ester
Eicosyl sinapate. ***Ehretiate***
[1390640-63-7]
$C_{31}H_{52}O_5$ 504.752
Constit. of the roots of *Ehretia longiflora*. Oil. λ_{max} 240 (log ε 3.94); 329 (log ε 3.95) (EtOH). [base] λ_{max} 209 (log ε 4.91); 264 (log ε 3.82); 399 (log ε 4.13) (EtOH/KOH)

3,5-Di-Me ether, pyrrolide
4-Hydroxy-3,5-dimethoxy-trans-cinnamoylpyrrole. ***1-trans-Sinapoylpyrrole***
[1613629-67-6]
$C_{15}H_{15}NO_4$ 273.288
Alkaloid from *Piper hymenophyllum*. Amorph. yellow solid. λ_{max} 252; 353 (MeOH)

3,5-Di-Me ether, Ac
[90985-68-5]
$C_{13}H_{14}O_6$ 266.249
Pink cryst.. Mp 198 - 200°

4-Et ether, Me ester
Methyl 4-ethoxy-3,5-dihydroxycinnamate
[944162-76-9]
$C_{12}H_{14}O_5$ 238.239
Constit. of the stem bark of *Duguetia gardneriana*. Amorph. solid

3,4-Methylene ether, 2-methylpropylamide
3-Hydroxy-N-isobutyl-4,5-methylenedioxycinnamide. ***Heitziamide B***
[1193250-23-5]
$C_{14}H_{17}NO_4$ 263.293
Alkaloid from the stem bark of *Zanthoxylum heitzii*. Amorph. yellow powder. λ_{max} 215 (log ε 2.25); 235 (log ε 2.46); 290 (log ε 3.04); 325 (log ε 3.41) (MeOH)

(Z)-form

3,5-Di-Me ether

[7361-90-2]

Constit. of shoots of *Medicago sativa, Givotia rottleriformis, Artemisia* spp. and many other plants. Obt. as a mixt. with *E*-form by uv irradiation. Occurs only in a bound alkali-labile state.

Späth, E. *Monatsh. Chem.*, 1920, **41**, 271-285 (*Sinapinic acid, synth*).
Shinoda, J. *et al., Yakugaku Zasshi*, 1931, **51**, 249-254 (*synth*).
Kung, H.P. *et al., J. Am. Chem. Soc.*, 1949, **71**, 1836-1837 (*Sinapic acid, synth, uv*).
Freudenberg, *et al., Chem. Ber.*, 1951, **84**, 472-476 (*E-3,5-di-Me Et ester, synth*).
Bate-Smith, E.C. *Sci. Proc. R. Dublin Soc.*, 1956, **27**, 165-176 (*Sinapic acid, occur*).
Nakamura, T. *et al., Chem. Pharm. Bull.*, 1962, **10**, 281-290 (*3-Me ether, synth*).
Horner, L. *et al., Chem. Ber.*, 1965, **98**, 1233-1245 (*E-3-Me ether, synth*).
Karrer, W. *Konstitution und Vorkommen der Organischen Pflanzenstoffe, 2nd edn., Birkhäuser Verlag*, 1976-1985, no. 962 (*Sinapic acid, occur*).
Hartley, R.D. *et al., J. Chromatogr.*, 1975, **107**, 213-218 (*glc, ms*).
v. Massow, F. *et al., J. Chem. Soc., Perkin Trans. 2*, 1976, 977-980 (*cmr*).
Herrmann, K. *Fortschr. Chem. Org. Naturst.*, 1978, **35**, 73-132 (*rev*).
Newby, V.K. *et al., Phytochemistry*, 1980, **19**, 651-657 (*Ferulic acid, Sinapic acid, occur*).
Fujita, M. *et al., Chem. Pharm. Bull.*, 1984, **32**, 2622-2627 (*cmr*).
Mahmoud, Z.F. *Acta Pharm. Jugosl.*, 1985, **35**, 113-119 (*Thalictrum minus constit*).
Tachibana, S. *et al., Phytochemistry*, 1992, **31**, 3207-3209 (*3-Me ether, Phyllostachys edulis constit*).
Rabe, C. *et al., Phytochemistry*, 1994, **35**, 1559-1565 (*rooibos tea constit*).
Hashimoto, K. *et al., Planta Med.*, 1994, **60**, 124-127 (*Sinapic acid 4-glucoside*).
Tesaki, S. *et al., Biosci., Biotechnol., Biochem.*, 1998, **62**, 998-1000 (*Sinapic acid, activity*).
Kuo, Y.-H. *et al., J. Chin. Chem. Soc. (Taipei)*, 1999, **46**, 819-825 (*Docosyl 5-hydroxyferulate*).
Zhang, J. *et al., CA*, 2000, **133**, 132435v (*Hexadecyl sinapate*).
Hayat, S. *et al., Chem. Pharm. Bull.*, 2002, **50**, 1297-1299 (*Butyl sinapate, Isopropyl sinapate*).
Dall'Acqua, S. *et al., Chem. Pharm. Bull.*, 2002, **50**, 1499-1501 (*Methyl sinapate*).
Russell, W.R. *et al., Bioorg. Chem.*, 2003, **31**, 206-215 (*synth, pmr, cmr, E-3-Me ether*).
Sun, K. *et al., Chem. Pharm. Bull.*, 2004, **52**, 1483-1486 (*Descurainia sophia constit*).
Fiuza, S.M. *et al., J. Mol. Struct.*, 2004, **693**, 103-118 (*synth, struct, Raman, ir, pmr, cmr, ms*).
Srivastava, V. *et al., Bioorg. Med. Chem.*, 2007, **15**, 518-525 (*synth, ir, pmr*).
Mbaze, L.M. *et al., Phytochemistry*, 2009, **70**, 1442-1447 (*Heitziamide B*).
Aung, H.T. *et al., Bioorg. Med. Chem.*, 2011, **19**, 2392-2396 (*E-form, Z-form, synth, activity*).
Chien, Y.-C. *et al., Phytochemistry*, 2012, **80**, 50-57 (*Ehretiate*).
Dung, H.V. *et al., Bull. Korean Chem. Soc.*, 2014, **35**, 655-658 (*1-Sinapoylpyrrole*).
Niciforovic, N. *et al., Compr. Rev. Food Sci. Food Saf.*, 2014, **13**, 34-51 (*Sinapic acid, rev, occur, activity*).
Martins de Sá, L.Z.C. *J. Funct. Foods*, 2014, **8**, 169-179 (*jabuticaba berry juice constit*).
Ramos, A.S. *et al., Food Res. Int.*, 2015, **75**, 315-327 (*4-glucoside*)

3-(3,4,5-Trihydroxyphenyl)-2-propen-1-ol T107

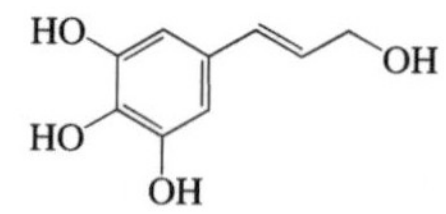

(*E*)-*form*

3′,4′,5′-Trihydroxycinnamyl alcohol

$C_9H_{10}O_4$ 182.175

(E)-form [1081808-21-0]

[876499-25-1] (unspecified stereochem.)

Mp 167°

3′-Me ether

3-(3,4-Dihydroxy-5-methoxyphenyl)-2-propen-1-ol. 5-Hydroxyconiferyl alcohol

[1782-47-4]

$C_{10}H_{12}O_4$ 196.202

Intermed. in lignin biosynth.. Mp 102°

3′-Me ether, 4′-O-β-D-glucopyranoside

De-O-methylsyringin

[183859-32-7]

$C_{16}H_{22}O_9$ 358.343

Constit. of *Wahlenbergia marginata*. Gum. λ_{max} 222 (log ε 3.31); 264 (log ε 3.81) (MeOH)

4′-Me ether, 1-O-(2E-decene-4,6-diynoyl)

4′-O-Methylsinapyl lachnophyllate

[79695-03-7]

$C_{20}H_{20}O_5$ 340.375

Constit. of *Dimerostemma asperatum*. Gum

3′,4′-Di-Me ether

3-(3-Hydroxy-4,5-dimethoxyphenyl)-2-propen-1-ol. 3′-Hydroxy-4′,5′-dimethoxycinnamyl alcohol

[135378-03-9]

$C_{11}H_{14}O_4$ 210.229

Constit. of *Croton xalapensis, Glycosmis pentaphylla* and *Ferula sinaica*. Amorph. yellow solid. λ_{max} 254; 280 (MeOH)

3′,5′-Di-Me ether

see 3-(4-Hydroxy-3,5-dimethoxyphenyl)-2-propen-1-ol, H81

3′,5′-Di-Me ether, 1-O-[α-L-rhamnopyranosyl-(1→2)-β-D-glucopyranosyl-(→4)-4-hydroxy-3-methoxy-E-cinnamoyl], 4′-O-[α-L-rhamnopyranosyl-(1→2)-β-D-glucopyranoside]

$C_{45}H_{62}O_{25}$ 1002.966

Constit. of *Solenostemma argel*. Moderate antioxidant. Amorph. yellow powder. $[\alpha]^{23}_{D}$ -11.2 (c, 0.49 in MeOH).

Freudenberg, K. *et al., Chem. Ber.*, 1953, **86**, 190-196 (*synth*).
Horner, L. *et al., Chem. Ber.*, 1965, **98**, 1233-1245 (*3′-Me ether*).
Bohlmann, F. *et al., Phytochemistry*, 1981, **20**, 1335-1338 (*lachnophyllate esters*).
Ahmed, A. *Phytochemistry*, 1991, **30**, 1207-1210 (*3′-Hydroxy-4′,5′-dimethoxycinnamyl alcohol*).
Tan, R.X. *et al., Phytochemistry*, 1998, **48**, 1245-1250 (*Demethylsyringin*).
Sakakibara, N. *et al., Org. Biomol. Chem.*, 2007, **5**, 802-815 (*3′-Me ether*).
Arevalo, C. *et al., Nat. Prod. Commun.*, 2009, **4**, 1697-1700 (*3′-Hydroxy-4′,5′-dimethoxycinnamyl alcohol*).
Wu, Y. *et al., J. Asian Nat. Prod. Res.*, 2012, **14**, 738-742 (*Glycosmis pentaphylla constit*)

4′,7′,9′-Trihydroxy-3,3′,5-trimethoxy-9-nor-4,8′-oxylignan-7-one T108

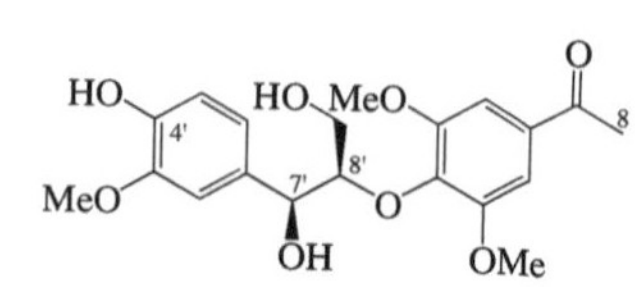

(7′*S*,8′*R*)-*form*

4′,7′,9′-Trihydroxy-3,3′,5-trimethoxy-9-nor-7-oxo-4,8′-oxyneolignan

$C_{20}H_{24}O_8$ 392.404

(7′S,8′R)-form

5-Demethoxy

Lycocernuaside D

[2301994-34-1]

$C_{19}H_{22}O_7$ 362.378

Constit. of *Palhinhaea cernua*. Moderate xanthine oxidase inhibitor. Amorph. yellow powder. $[\alpha]^{25}_{D}$ +21.3 (c, 0.2 in MeOH). λ_{max} 228 (log ε 4.66); 282 (log ε 4.15); 308 (log ε 3.79) (MeOH)

(7′ξ,8′ξ)-form [852621-25-1]

Prod. by elicitation of cell cultures of *Echinacea purpurea*. λ_{max} 220; 284 (no solvent reported)

4′-O-β-D-Glucopyranoside

[852621-26-2]

$C_{26}H_{34}O_{13}$ 554.545
Prod. by elicitation of cell cultures of *Echinacea purpurea*

8-Hydroxy
4′,7′,8,9′-Tetrahydroxy-3,3′,5-trimethoxy-9-nor-4,8′-oxylignan-7-one.
4′,7′,8,9′-Tetrahydroxy-3,3′,5-trimethoxy-9-nor-7-oxo-4,8′-oxylignan
[852621-24-0]
$C_{20}H_{24}O_9$ 408.403
Prod. by elicitation of cell cultures of *Echinacea purpurea*. λ_{max} 220; 284 (no solvent reported).

Li, W.-W. *et al.*, *Tetrahedron Lett.*, 2005, **46**, 2973-2977 (*Echinacea purpurea constits*).
Li, J. *et al.*, *Fitoterapia*, 2017, **119**, 45-50 (*Lycocernuaside D, activity*)

Trilobatin I — T109

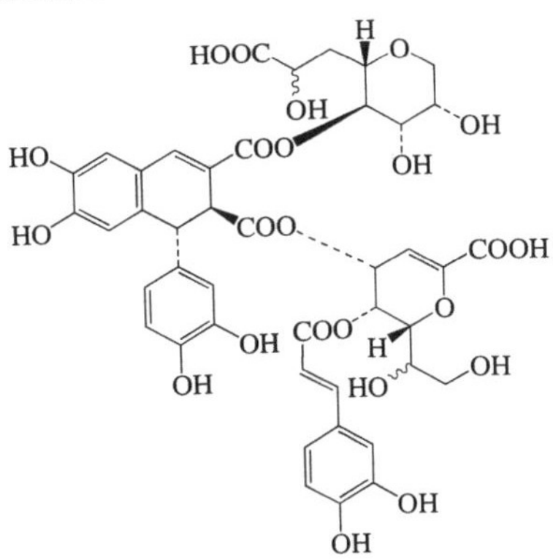

[543724-90-9]
$C_{43}H_{42}O_{23}$ 926.786
Ester of Epiphyllic acid in C339. Constit. of the liverwort *Bazzania trilobata*. $[\alpha]_D^{20}$ -84 (c, 0.07 in MeOH). λ_{max} 251 (log ε 3.25); 332 (log ε 3.25) (MeOH).

Scher, J.M. *et al.*, *Phytochemistry*, 2003, **62**, 769-777 (*Trilobatin I*)

Trilobatin A — T110

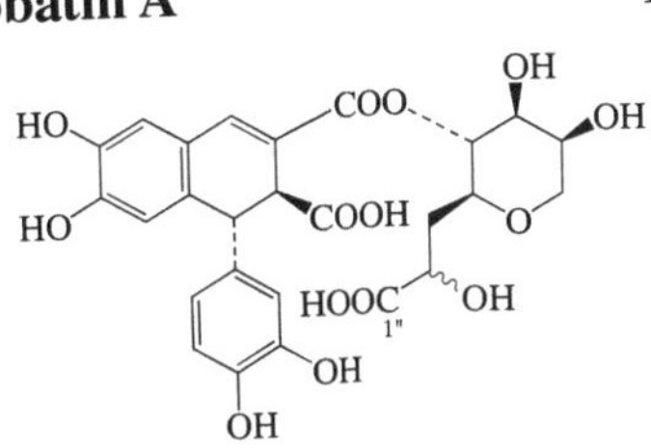

Relative Configuration

[217176-37-9]
$C_{26}H_{26}O_{14}$ 562.48
Ester of Epiphyllic acid, in C339. Isol. from the liverwort *Bazzania trilobata*. Oil. $[\alpha]_D^{20}$ -85.3 (c, 0.8 in MeOH). λ_{max} 252; 338 (MeOH)

1″-Me ester
[217176-39-1]
$C_{27}H_{28}O_{14}$ 576.507
Isol. from *Bazzania trilobata*. $[\alpha]_D^{20}$ -75.3 (c, 0.27 in MeOH). λ_{max} 253; 338 (MeOH).

Martini, U. *et al.*, *Phytochemistry*, 1998, **49**, 1139-1146 (*Trilobatin A, 1″-Me ester*)

Trilobatin B — T111

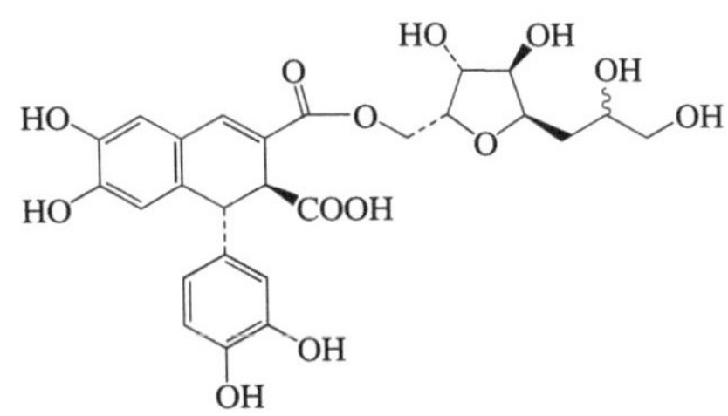

Relative Configuration

[217176-40-4]
$C_{26}H_{28}O_{13}$ 548.497
Ester of Epiphyllic acid, in C339. Isol. from the liverwort *Bazzania trilobata*. $[\alpha]_D^{20}$ -35.9 (c, 0.1 in MeOH). λ_{max} 253; 334 (MeOH).

Martini, U. *et al.*, *Phytochemistry*, 1998, **49**, 1139-1146 (*Trilobatin B*)

Trilobatin C — T112

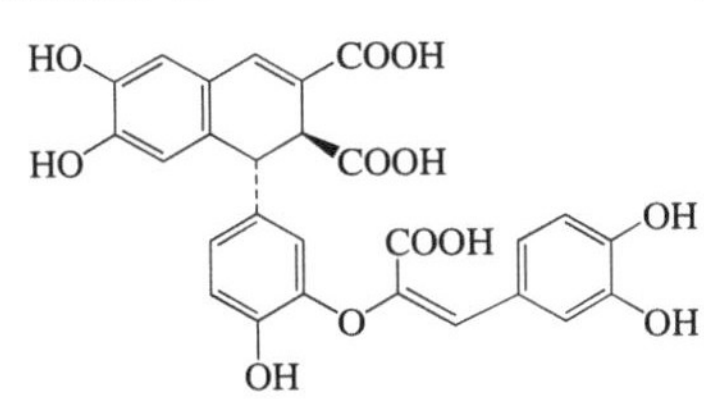

Relative Configuration

[217176-42-6]
$C_{27}H_{20}O_{12}$ 536.445
Deriv. of Epiphyllic acid, in C339. Isol. from the liverwort *Bazzania trilobata*. $[\alpha]_D^{20}$ -80.1 (c, 0.67 in MeOH). λ_{max} 247; 318 (MeOH).

Martini, U. *et al.*, *Phytochemistry*, 1998, **49**, 1139-1146 (*Trilobatin C*)

Trilobatin D — T113

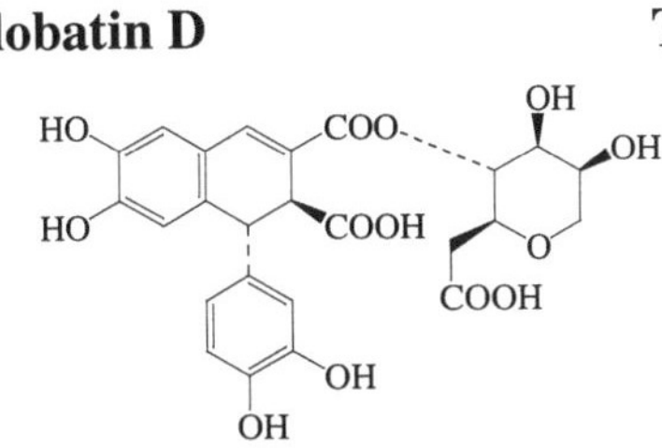

[543724-05-6]
$C_{25}H_{24}O_{13}$ 532.454
Ester of Epiphyllic acid, in C339. Constit. of the liverwort *Bazzania trilobata*. $[\alpha]_D^{20}$ -146.7 (c, 0.23 in MeOH). λ_{max} 253; 343 (MeOH).

Scher, J.M. *et al.*, *Phytochemistry*, 2003, **62**, 769-777 (*Trilobatin D*)

Trilobatin E — T114

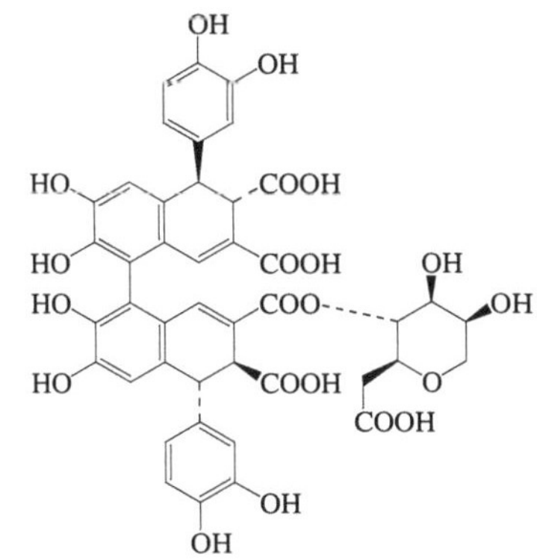

[543724-08-9]
$C_{43}H_{36}O_{21}$ 888.74
Ester of 5,5′-Bis[1-(3,4-dihydroxyphenyl)-1,2-dihydro-6,7-dihydroxy-2,3-naphthalenedicarboxylic acid], B28 or a dimer of Epiphyllic acid, in C339. Constit. of the liverwort *Bazzania trilobata*. λ_{max} 253; 348 (MeOH).

Scher, J.M. *et al.*, *Phytochemistry*, 2003, **62**, 769-777 (*Trilobatin E*)

Trilobatin F — T115

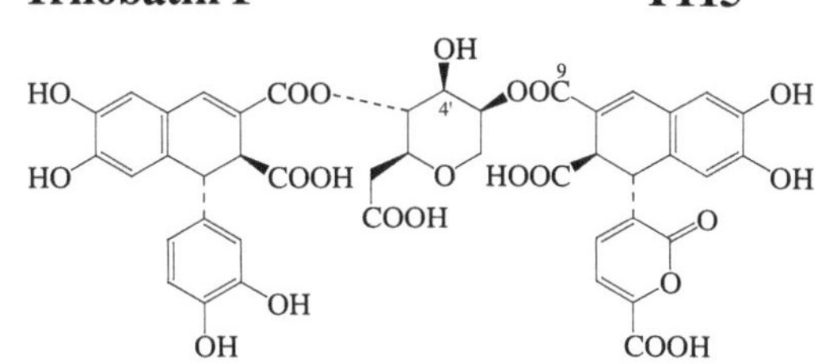

$C_{43}H_{34}O_{22}$ 902.723
Ester of Trilobatin D, T113 and Jamesopyrone, J2. Constit. of the liverwort *Bazzania trilobata*. $[\alpha]_D^{20}$ -75.3 (c, 0.27 in MeOH). λ_{max} 253 (log ε 4.14); 309 (log ε 4.32) (MeOH)

C-4′/C-9 Ester analogue
Trilobatin G
[543724-13-6]
$C_{43}H_{34}O_{22}$ 902.723
Constit. of *Bazzania trilobata*. $[\alpha]_D^{20}$ -20 (c, 0.11 in MeOH). λ_{max} 253; 312 (MeOH).

Scher, J.M. *et al.*, *Phytochemistry*, 2003, **62**, 769-777 (*Trilobatins F,G*)

Trilobatin H T116

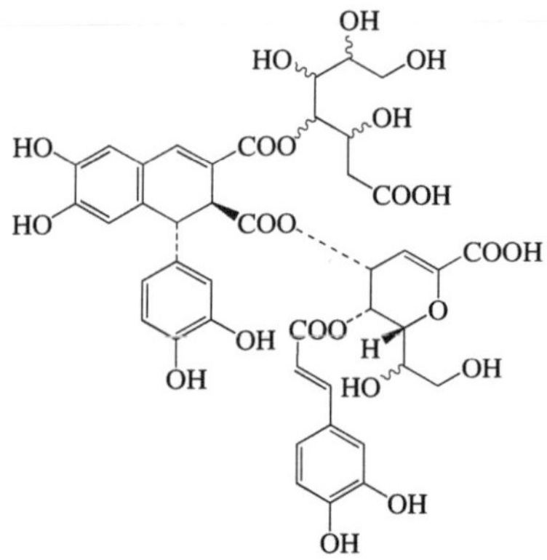

[543724-15-8]

$C_{42}H_{42}O_{23}$ 914.775

Diester of Epiphyllic acid, in C339.

Constit. of the liverwort *Bazzania trilobata*. $[\alpha]_D^{20}$ -164.6 (c, 0.33 in MeOH). λ_{max} 250 (log ε 3.39); 334 (log ε 3.28) (MeOH).

Scher, J.M. *et al.*, *Phytochemistry*, 2003, **62**, 769-777 (*Trilobatin H*)

Trilobatin J T117

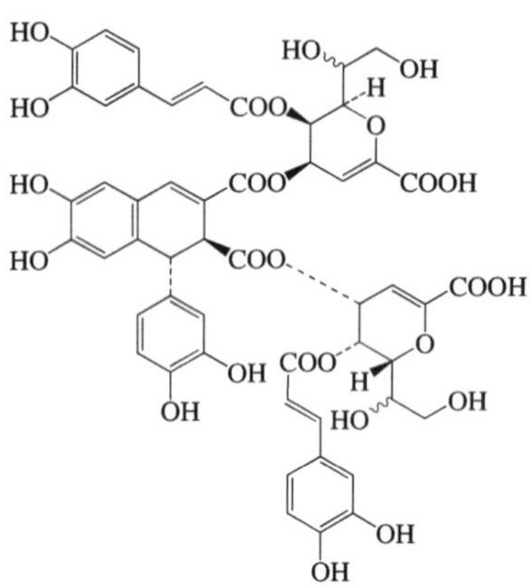

[543725-41-3]

$C_{52}H_{46}O_{26}$ 1086.914

Diester of Epiphyllic acid, in C339.

Constit. of the liverwort *Bazzania trilobata*. $[\alpha]_D^{20}$ -164.6 (c, 0.33 in MeOH).

Scher, J.M. *et al.*, *Phytochemistry*, 2003, **62**, 769-777 (*Trilobatin J*)

Trilobatin K T118

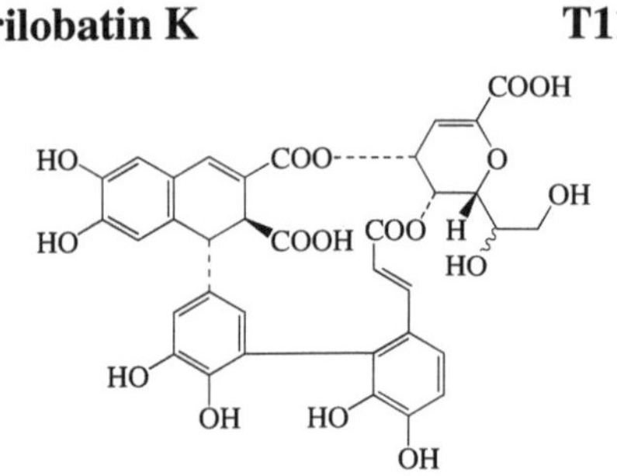

[548725-42-4]

$C_{35}H_{28}O_{17}$ 720.592

Ester of Epiphyllic acid, in C339.

Constit. of the liverwort *Bazzania trilobata*. $[\alpha]_D^{20}$ -12.3 (c, 0.33 in MeOH). λ_{max} 254 (log ε 4.56); 313 (log ε 4.37) (MeOH).

Scher, J.M. *et al.*, *Phytochemistry*, 2003, **62**, 769-777 (*Trilobatin K*)

7,8,9-Trinor-4,8′-oxylignan-1,3,3′,4′,5,7′,9′-heptol T119

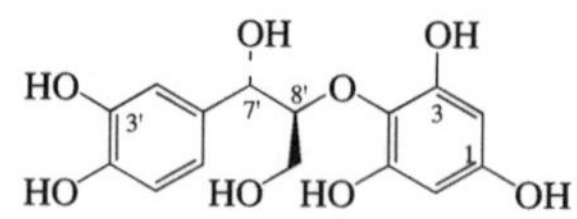

(7′S,8′S)-form

1,3,3′,4′,5,7′,9′-Heptahydroxy-7,8,9-trinor-4,8′-oxyneolignan

$C_{15}H_{16}O_8$ 324.285

(7′S,8′S)-form

3,3′,5-Tri-Me ether

3,3′,5-Trimethoxy-7,8,9-trinor-4,8′-oxylignan-1,4′,7′,9′-tetrol

[1290633-23-6]

$C_{18}H_{22}O_8$ 366.366

Constit. of the stems of *Sinocalamus affinis*. Amorph. powder. $[\alpha]_D^{20}$ +3.5 (c, 0.1 in MeOH). λ_{max} 204 (log ε 4.3); 234 (log ε 3.64); 280 (log ε 3.1) (MeOH)

(7′R*,8′S*)-form

3,3′,5-Tri-Me ether

Rosalaevin B

[1340483-49-9]

$C_{18}H_{22}O_8$ 366.366

Constit. of the fruit of *Rosa laevigata*. Amorph. solid. $[\alpha]_D^{27}$ -9.3 (c, 0.2 in MeOH). λ_{max} 204 (log ε 4.84); 226 (log ε 4.28); 280 (log ε 3.88) (MeOH).

Xiong, L. *et al.*, *J. Nat. Prod.*, 2011, **74**, 1188-1200 (*Sinocalamus affinis constit*).

Li, X. *et al.*, *Food Chem.*, 2012, **130**, 575-580 (*Rosalaevin B*)

7,8,9-Trinor-4,8′-oxylignan-1,3,3′,4′,7′,9′-hexol T120

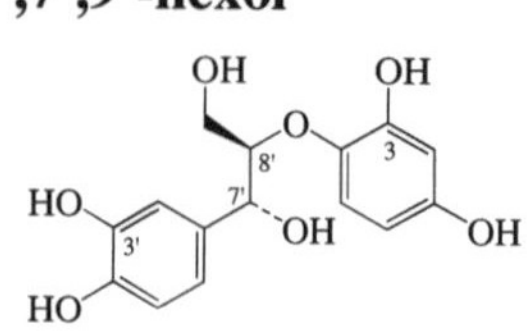

(7′R,8′R)-form

1,3,3′,4′,7′,9′-Hexahydroxy-7,8,9-trinor-4,8′-oxyneolignan

$C_{15}H_{16}O_7$ 308.286

(7′R,8′R)-form

3,3′-Di-Me ether

3,3′-Dimethoxy-7,8,9-trinor-4,8′-oxylignan-1,4′,7′,9′-tetrol.

1,4′,7′,9′-Tetrahydroxy-3,3′-dimethoxy-7,8,9-trinor-4,8′-oxyneolignan

3,3′-Di-Me ether, 1-O-β-D-glucopyranoside

Pinnatifidaninside C

[1820907-14-9]

$C_{23}H_{30}O_{12}$ 498.481

Constit. of the seeds of *Crataegus pinnatifida*. Yellow oil. $[\alpha]_D^{20}$ -16.2 (c, 0.1 in MeOH). λ_{max} 231 (log ε 3.62); 280 (log ε 1.31) (MeOH)

(7′S,8′R)-form

3,3′-Di-Me ether, 1-O-β-D-glucopyranoside

Pinnatifidaninside D

[1820907-15-0]

$C_{23}H_{30}O_{12}$ 498.481

Constit. of the seeds of *Crataegus pinnatifida*. Yellow oil. $[\alpha]_D^{20}$ -4.5 (c, 0.08 in MeOH). λ_{max} 230 (log ε 3.77); 282 (log ε 1.41) (MeOH).

Huang, X.-X. *et al.*, *Planta Med.*, 2014, **80**, 1732-1738 (*Pinnatifidaninsides C,D*)

Trogopterin C T121

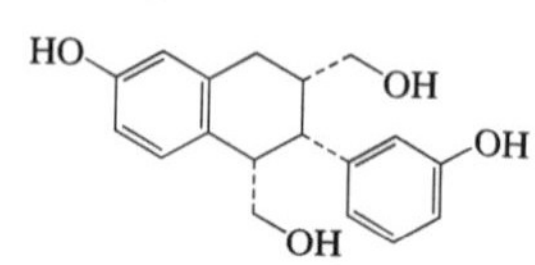

Relative Configuration

1,2,3,4-Tetrahydro-6-hydroxy-2-(3-hydroxyphenyl)-1,3-naphthalenedimethanol. 7′,8-Cyclo-2,8′-lignan-3′,5,9,9′-tetrol. 7′,8-Cyclo-3′,5,9,9′-tetrahydroxy-2,8′-neolignan

[1656992-35-6]

$C_{18}H_{20}O_4$ 300.354

Lignan numbering shown. Obt. from faeces of *Trogopterus xanthipes*. Exhibits moderate cytotoxicity against HL-60 cells. Light brown oil. $[\alpha]_D^{25}$ -14.8 (c, 1.0 in EtOH).

Baek, S. *et al.*, *Beilstein J. Org. Chem.*, 2014, **10**, 2955-2962 (*Trogopterin C, activity*)

Truncatulin A T122

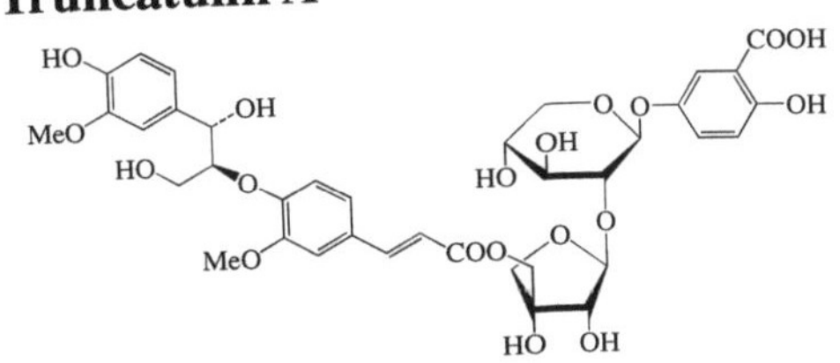

[1188414-27-8]

$C_{37}H_{42}O_{19}$ 790.724

Derivative of 4,8′-Oxylign-7-ene-3,3′,4′,7′,9,9′-hexol, O86. Constit. of the roots of *Medicago truncatula*. Amorph. brown powder. $[\alpha]^{20}_{D}$ -33.2 (c, 0.1 in MeOH). λ_{max} 326 (MeOH).

Stochmal, A. *et al.*, *Phytochemistry*, 2009, **70**, 1272-1276 (*Truncatulin A*)

Truncatulin C T123

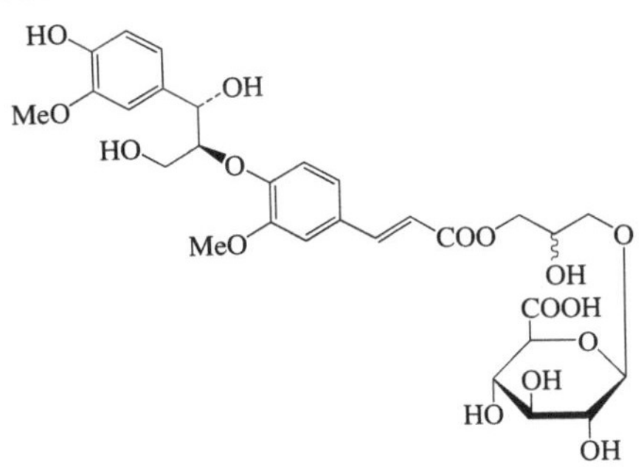

[1190119-51-7]

$C_{29}H_{36}O_{16}$ 640.591

Derivative of 4,8′-Oxylign-7-ene-3,3′,4′,7′,9,9′-hexol, O86. Constit. of the roots of *Medicago truncatula*. Amorph. brown powder. $[\alpha]^{20}_{D}$ -54.1 (c, 0.1 in MeOH). λ_{max} 326 (MeOH).

Stochmal, A. *et al.*, *Phytochemistry*, 2009, **70**, 1272-1276 (*Truncatulin C*)

α-Truxilline T124

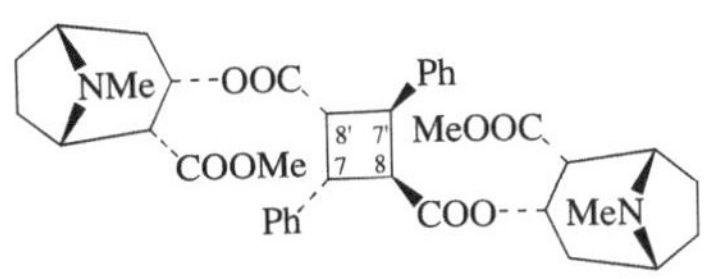

γ-Isotropylcocaine. Cocamine

[490-17-5]

$C_{38}H_{46}N_2O_8$ 658.792

Ester of α-truxillic acid with ecgonine methyl ester. Lignan numbering shown. Alkaloid from the leaves of *Erythroxylum coca*. Obt. as a mixt. with β-Truxilline. Obt. pure by synth.. Mp 63° (sinters). The mixt. with β-Truxilline was also known as Cocamine

Hydrochloride

Mp 274°

7-Epimer

*epi-**Truxilline**. Epitruxilline*

[113351-64-7]

$C_{38}H_{46}N_2O_8$ 658.792

Constit. of *Erythroxylum coca*

8′-Epimer

*γ-**Truxilline***

[113350-56-4]

$C_{38}H_{46}N_2O_8$ 658.792

Constit. of *Erythroxylum coca*

7,8-Diepimer

*ε-**Truxilline***

[113350-55-3]

$C_{38}H_{46}N_2O_8$ 658.792

Constit. of *Erythroxylum coca*

7,8′-Diepimer

*peri-**Truxilline***

[113350-53-1]

$C_{38}H_{46}N_2O_8$ 658.792

Constit. of *Erythroxylum coca*.

Liebermann, C. *Ber. Dtsch. Chem. Ges.*, 1889, **22**, 130-133 (*synth*).

Liebermann, C. *et al.*, *Ber. Dtsch. Chem. Ges.*, 1889, **22**, 680-683 (*synth*).

Hesse, O. *J. Prakt. Chem.*, 1902, **66**, 401-422 (*bibl*).

Novak, M. *et al.*, *J. Ethnopharmacol.*, 1984, **10**, 261-274 (*rev*).

Moore, J.M. *et al.*, *J. Chromatogr.*, 1987, **410**, 297-318 (*gc, detn*).

Lurie, I.S. *et al.*, *J. Chromatogr.*, 1990, **504**, 391-401 (*chromatog*)

Tulsinol C T125

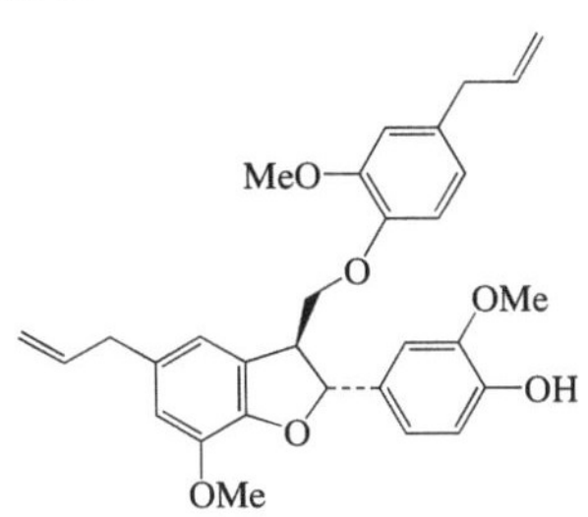

Relative Configuration

[1158305-95-3]

$C_{30}H_{32}O_6$ 488.58

Trimer of 2-Methoxy-4-(2-propenyl) phenol, M38. Constit. of the leaves of *Ocimum sanctum*. Exhibits strong antileishmanicidal activity. Pale yellow oil. $[\alpha]_D$ +4.3 (c, 0.03 in MeOH). λ_{max} 215 (ε 10300); 229 (ε 9100); 281 (ε 3500) (MeOH).

Suzuki, A. *et al.*, *Chem. Pharm. Bull.*, 2009, **57**, 245-251 (*Tulsinol C, activity*)

Tulsinol E T126

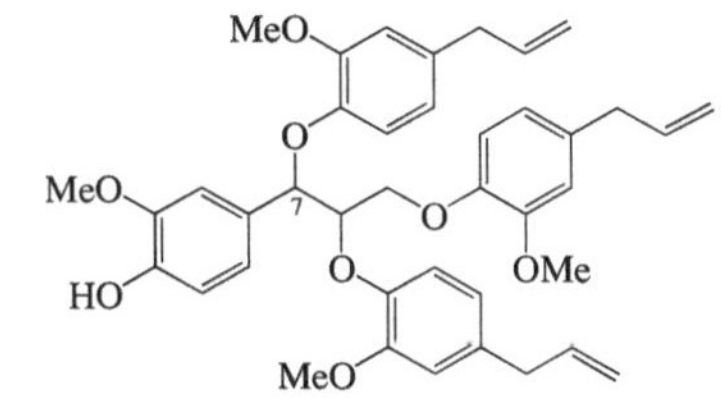

[1158305-98-6]

$C_{40}H_{44}O_8$ 652.784

Constit. of *Ocimum sanctum*. Pale yellow oil. $[\alpha]_D$ +8 (c, 0.01 in MeOH). λ_{max} 222 (sh); 234 (ε 13300); 279 (ε 6100) (MeOH)

7-O-Dearyl, O⁷-Me

Tulsinol D

[1158305-97-5]

$C_{31}H_{36}O_7$ 520.622

Constit. of *Ocimum sanctum*. Pale yellow oil.

Suzuki, A. *et al.*, *Chem. Pharm. Bull.*, 2009, **57**, 245-251 (*Tulsinols D,E*)

Tulsinol F T127

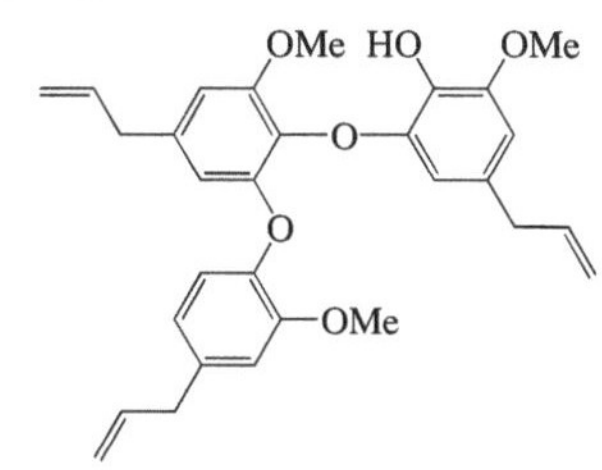

[1158306-01-4]

$C_{30}H_{32}O_6$ 488.58

Constit. of *Ocimum sanctum*. Exhibits moderate leishmanicidal activity. Pale yellow oil. λ_{max} 222 (ε 11300); 275 (ε 2400) (MeOH).

Suzuki, A. *et al.*, *Chem. Pharm. Bull.*, 2009, **57**, 245-251 (*Tulsinol F, activity*)

Tulsinol G T128

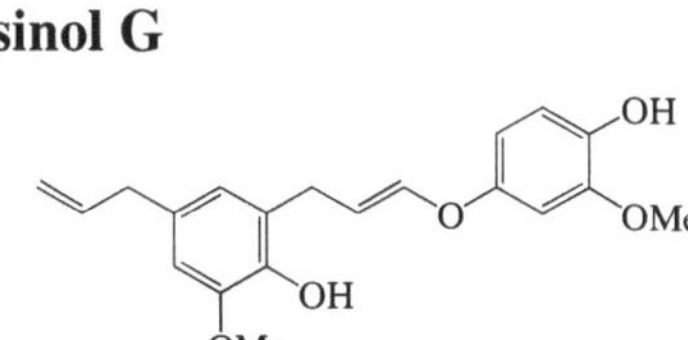

[1158306-03-6]

$C_{20}H_{22}O_5$ 342.391

Constit. of *Ocimum sanctum*. Viscous oil. λ_{max} 217 (ε 7200); 233 (ε 6800); 284 (ε 2300) (MeOH).

Suzuki, A. *et al.*, *Chem. Pharm. Bull.*, 2009, **57**, 245-251 (*Tulsinol G*)

Unciflavone C U1

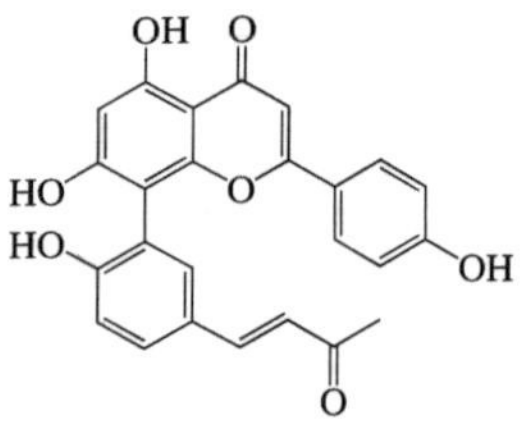

5,7-Dihydroxy-8-[2-hydroxy-5-(3-oxo-1-butenyl)phenyl]-2-(4-hydroxyphenyl)-4H-1-benzopyran-4-one, CAS
[1830290-25-9]
$C_{25}H_{18}O_7$ 430.412
Constit. of *Selaginella uncinata*.
Amorph. yellow powder. λ_{max} 230; 274; 326 (MeOH).

Zou, H. *et al.*, *Fitoterapia*, 2014, **99**, 328-333 (*Unciflavone C*)

Ussurienoside I U2

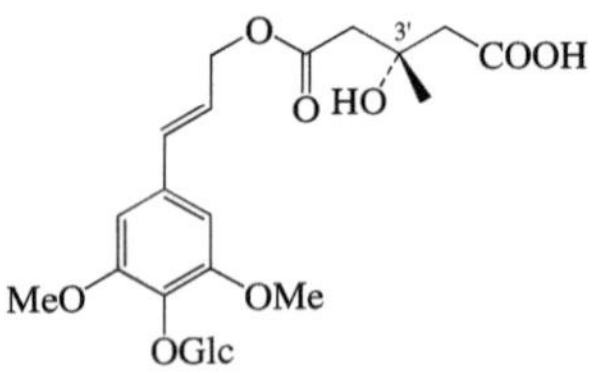

[150072-94-9]
$C_{23}H_{32}O_{13}$ 516.496
Ester of 3-(4-Hydroxy-3,5-dimethoxyphenyl)-2-propen-1-ol, H81 and 3-Hydroxy-3-methylpentanedioic acid. Isol. from *Codonopsis ussuriensis*. Mp 102 - 104°. *Not* a lignan monomer

O-De-Me
Wahlenoside A
[213008-37-8]
$C_{22}H_{30}O_{13}$ 502.469
Constit. of *Wahlenbergia marginata*. Gum. $[\alpha]_D$ -8.4 (c, 0.03 in MeOH). λ_{max} 223 (log ε 3.41); 266 (log ε 3.92) (MeOH)

O-De-Me, Me ester
Wahlenbergioside
[191089-97-1]
$C_{23}H_{32}O_{13}$ 516.496
Constit. of *Wahlenbergia marginata*. Powder. Mp 42 - 45°. $[\alpha]_D^{25}$ -9.6 (c, 0.01 in MeOH). λ_{max} 222 (log ε 3.34); 265 (log ε 3.82) (MeOH)

O-De-Me, 3′-O-β-D-glucopyranoside
Barbatoside A
[342641-86-5]
$C_{28}H_{40}O_{18}$ 664.61
Constit. of *Campanula barbata*. Amorph. powder. Mp 115 - 118°. $[\alpha]_D^{20}$ -24 (c, 0.3 in MeOH). λ_{max} 221 (log ε 4.38); 266 (log ε 4.06) (MeOH)

O-De-Me, 3′-O-[4-hydroxy-E-cinnamoyl-(→4)-β-D-glucopyranoside]
Barbatoside C
[342643-50-9]
$C_{37}H_{46}O_{20}$ 810.755
Constit. of *Campanula barbata*. Amorph. powder. Mp 121 - 122°. λ_{max} 222 (log ε 5.71); 266 (log ε 4.48); 314 (log ε 4.32) (MeOH)

O-De-Me, 3′-O-[4-hydroxy-Z-cinnamoyl-(→4)-β-D-glucopyranoside]
Barbatoside D
[342643-52-1]
$C_{37}H_{46}O_{20}$ 810.755
Constit. of *Campanula barbata*. Amorph. powder

O-De-Me, 3′-O-[4-methoxy-E-cinnamoyl-(→2)-β-D-glucopyranoside]
Barbatoside B
[342643-27-0]
$C_{38}H_{48}O_{20}$ 824.782
Constit. of *Campanula barbata*. Amorph. powder. Mp 124 - 125°. $[\alpha]_D^{20}$-40 (c, 0.1 in MeOH). λ_{max} 222 (log ε 4.69); 271 (log ε 4.43); 309 (log ε 4.33) (MeOH).

Lee, I.R. *et al.*, *Arch. Pharmacal Res.*, 1992, **15**, 289-291 (*Ussurienoside I*).
Ma, W.G. *et al.*, *Phytochemistry*, 1997, **45**, 411-415 (*Wahlenbergioside*).
Tan, R.X. *et al.*, *Phytochemistry*, 1998, **48**, 1245-1250 (*Wahlenoside A*).
Cuendet, M. *et al.*, *Phytochemistry*, 2001, **56**, 631-636 (*Barbatosides A-D*)

Vaccinin A V1

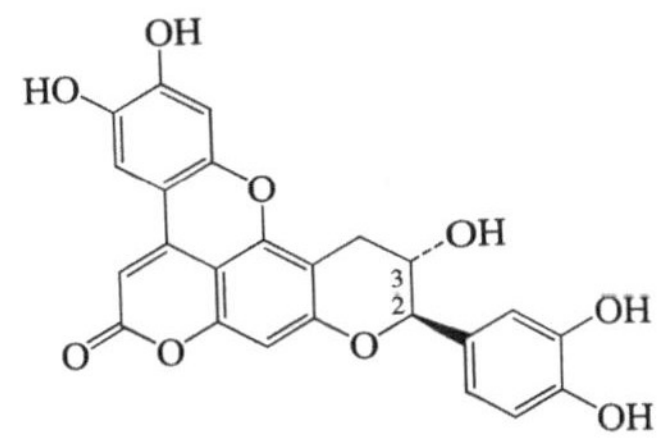

Absolute Configuration

$C_{24}H_{16}O_9$ 448.383
Constit. of *Vaccinium ashei*. Amorph. yellow powder. $[\alpha]_D^{20}$ -15.5 (c, 0.07 in Me_2CO aq. 9:1). λ_{max} 217 (log ε 4.58); 271 (log ε 4.02); 314 (log ε 4.08); 347 (sh); 374 (sh); 390 (log ε 4.07) (MeOH)

2,3-Diepimer

ent-Vaccinin A

$C_{24}H_{16}O_9$ 448.383
Constit. of *Cecropia obtusifolia*. Exhibits moderate antiplasmodial activity. Amorph. yellow powder. $[\alpha]_D^{20}$ +5.7 (c, 0.001 in Me_2CO).

Matsuo, Y. *et al.*, *Food Chem.*, 2010, **121**, 1073-1079 (*Vaccinin A*).
Rivera-Mondragon, A. *et al.*, *Phytochem. Lett.*, 2019, **31**, 118-120 (*ent-Vaccinin A, activity*)

Verimol D V2

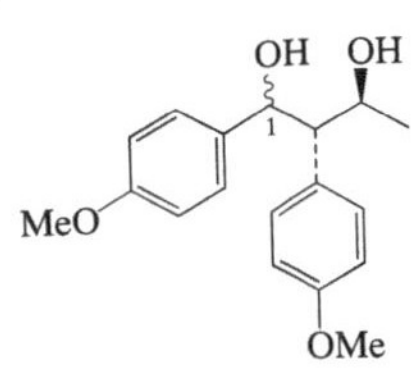

Relative Configuration

1,2-Bis(4-methoxyphenyl)-1,3-butanediol, CAS. *4,4'-Dimethoxy-8,9-dinor-7,7'-lignan-7,8'-diol. 7,8'-Dihydroxy-4,4'-dimethoxy-8,9-dinor-7,7'-neolignan*
[212516-37-5]
$C_{18}H_{22}O_4$ 302.37
Constit. of *Illicium verum*. Oil. $[\alpha]_D$ -1.5 (c, 0.58 in $CHCl_3$)

1-Epimer

Verimol E

[212516-38-6]
$C_{18}H_{22}O_4$ 302.37
Constit. of *Illicium verum*. Oil. $[\alpha]_D$ +7 (c, 0.43 in $CHCl_3$).

Sy, L.K. *et al.*, *J. Nat. Prod.*, 1998, **61**, 987-992 (*Verimols D,E*)

Verimol G V3

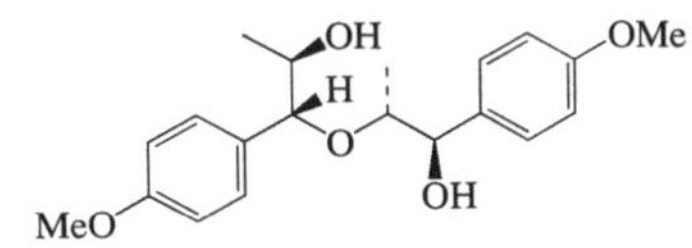

Relative Configuration

4,4'-Dimethoxy-7,8'-oxylignan-7',8-diol. 7',8-Dihydroxy-4,4'-dimethoxy-7,8'-oxyneolignan
[212516-40-0]
$C_{20}H_{26}O_5$ 346.423
Constit. of *Illicium verum*. Oil. $[\alpha]_D$ +1.6 (c, 1.67 in $CHCl_3$).

Sy, L.-K. *et al.*, *J. Nat. Prod.*, 1998, **61**, 987-992 (*Verimol G*)

Verimol H V4

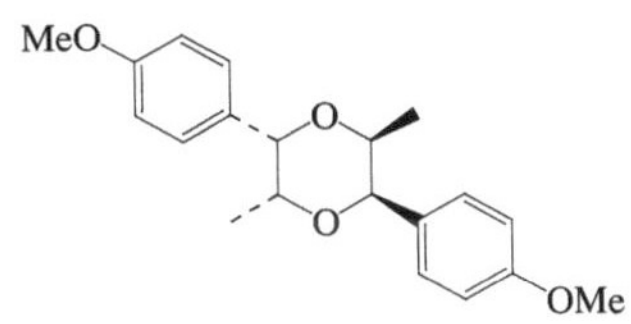

2,5-Bis(4-methoxyphenyl)-3,6-dimethyl-1,4-dioxane, CAS. *7',8-Epoxy-4,4'-dimethoxy-7,8'-oxylignan*
[212516-42-2]
$C_{20}H_{24}O_4$ 328.408
Constit. of *Illicium verum*. Oil. Confusion regarding stereochem. Text and diagrams in ref. do not agree.

Sy, L.-K. *et al.*, *J. Nat. Prod.*, 1998, **61**, 987-992 (*Verimol H*).
Liu, Y.N. *et al.*, *Chem. Biodiversity*, 2009, **6**, 963-989 (*rev*)

Vernicilignan A V5

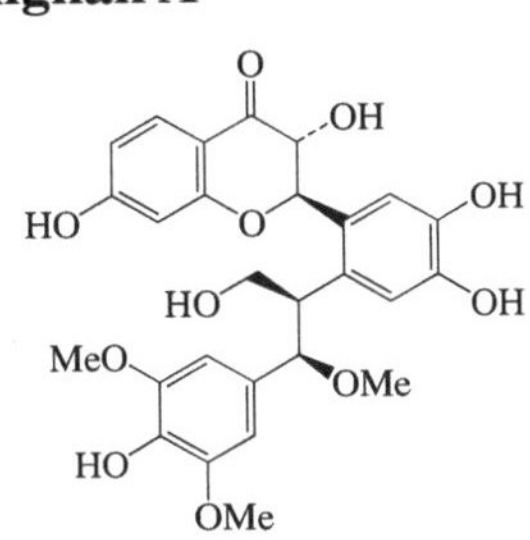

Absolute Configuration

$C_{27}H_{28}O_{11}$ 528.510
Flavonolignan. Constit. of the heartwood of *Toxicodendron vernicifluum*. Exhibits neuroprotective activity. Amorph. powder. $[\alpha]_D^{20}$ -27 (c, 0.1 in MeOH).

Li, M. *et al.*, *Fitoterapia*, 2019, **134**, 81-87 (*Vernicilignan A, activity*)

Versicolactone A‡ V6

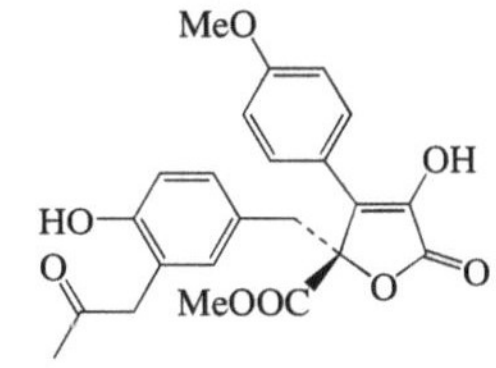

Absolute Configuration

[1803200-15-8]
$C_{23}H_{22}O_8$ 426.421
Provisional lignan. Biosynth. pathway unclear. Related to Butyrolactone I, B89. Prod. by *Aspergillus versicolor*. Active against tobacco mosaic virus. Amorph. powder. $[\alpha]_D^{25}$ +86.4 (c, 0.2 in MeOH). λ_{max} 210 (log ε 4.48); 252 (log ε 3.69); 314 (log ε 3.82) (MeOH).

Zhou, M. *et al.*, *Planta Med.*, 2015, **81**, 235-240 (*Versicolactone A*)

Virgatyne V7

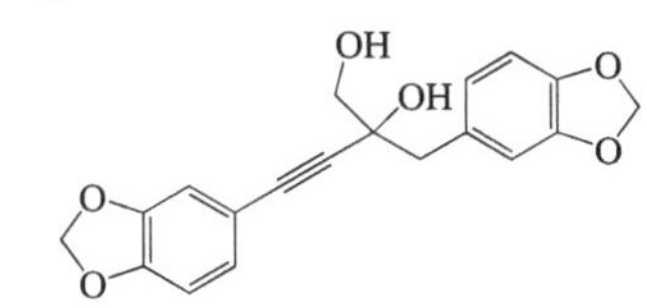

3,4:3',4'-Bis(methylenedioxy)-9'-nor-8,8'-lign-7'-yne-8,9-diol. 8,9-Dihydroxy-3,4:3',4'-bis(methylenedioxy)-9'-nor-7'-lignyne
[214554-70-8]
$C_{19}H_{16}O_6$ 340.331
Norlignan. Constit. of *Phyllanthus virgatus*. Needles (hexane). Mp 123 - 125°. $[\alpha]_D^{25}$ +30 (c, 1 in CH_2Cl_2). λ_{max} 228 (log ε 3.98); 265 (log ε 3.86); 294 (log ε 3.67) (CH_2Cl_2).

Huang, Y.-L. *et al.*, *J. Nat. Prod.*, 1998, **61**, 1194-1197 (*Virgatyne*)

Vitedoamine A V8

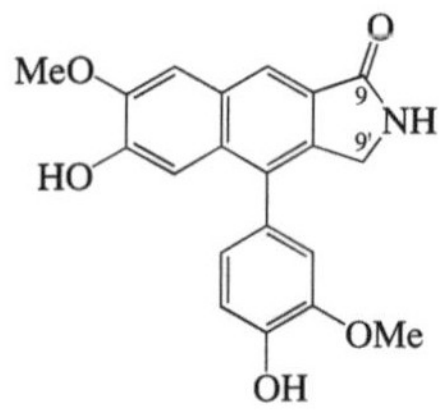

[819861-41-1]
$C_{20}H_{17}NO_5$ 351.358
Constit. of the seeds of *Vitex negundo*. Antioxidant. Powder. λ_{max} 202; 256; 314 (MeOH)

9-Deoxo, 9′-oxo

Vitedoamine B
[1186021-78-2]
$C_{20}H_{17}NO_5$ 351.358
Constit. of the seeds of *Vitex negundo*. Amorph. yellow powder. λ_{max} 201; 256 (MeOH).

Ono, M. *et al.*, *J. Nat. Prod.*, 2004, **67**, 2073-2075 (*Vitedoamine A*).
Zheng, C.-J. *et al.*, *J. Nat. Prod.*, 2009, **72**, 1627-1630 (*Vitedoamine B*)

Vitrofolal C V9

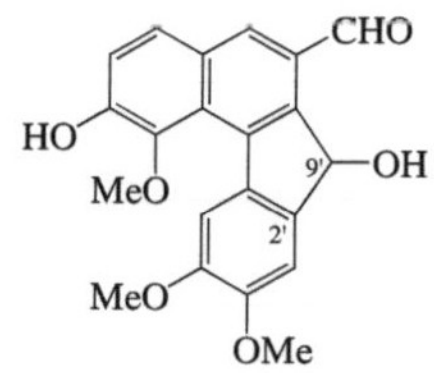

2,7′:2′,9′-Bicyclo-4,9′-dihydroxy-3,4′,5′-trimethoxy-8,8′-ligna-7,7′-dien-9-al
[261903-77-9]
$C_{21}H_{18}O_6$ 366.369
2′,9′-Cyclised naphthalenoid lignan. Lignan numbering shown. Constit. of the roots of *Vitex rotundifolia*. Amorph. red powder. $[\alpha]_D$ +3.7 (c, 0.53 in $CHCl_3$). λ_{max} 237 (log ε 4.5); 365 (log ε 4) (MeOH)

9′-Ketone

2,7′:2′,9′-Bicyclo-4-hydroxy-3,4′,5′-trimethoxy-9′-oxo-8,8′-ligna-7,7′-dien-9-al. ***Vitrofolal D***
[345902-00-3]
$C_{21}H_{16}O_6$ 364.353
Constit. of the roots of *Vitex rotundifolia*. Amorph. red solid. λ_{max} 259 (log ε 4.2) (dioxan).

Kawazoe, K. *et al.*, *Phytochemistry*, 1999, **52**, 1657-1659 (*Vitrofolal C*).
Kawazoe, K. *et al.*, *J. Nat. Prod.*, 2001, **64**, 588-591 (*Vitrofolal D*)

Woorenol W1

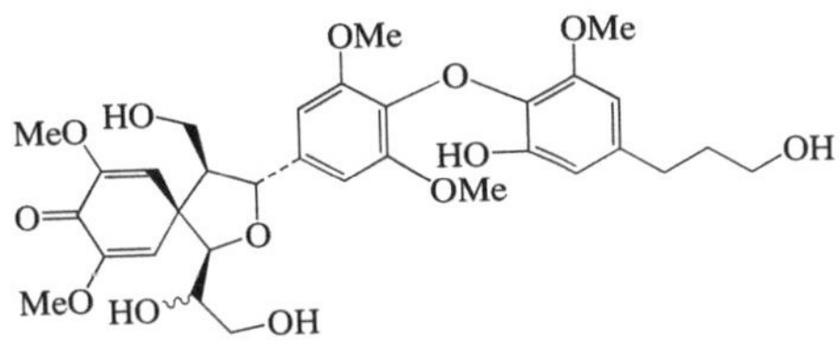

Relative Configuration

[189997-62-4]

$C_{32}H_{40}O_{13}$ 632.659

Spirosesquilignan. Related to Pinobatol, P54. Constit. of the rhizomes of *Coptis japonica* var. *dissecta*. Pale yellow solid. $[\alpha]_D^{25}$ +93.6 (c, 1.6 in MeOH). λ_{max} 207 (log ε 5.01); 230 (log ε 4.52); 278 (log ε 4.14) (MeOH).

Yoshikawa, K. *et al., J. Nat. Prod.*, 1997, **60**, 511-513 (*Woorenol*)

Wutaienin W2

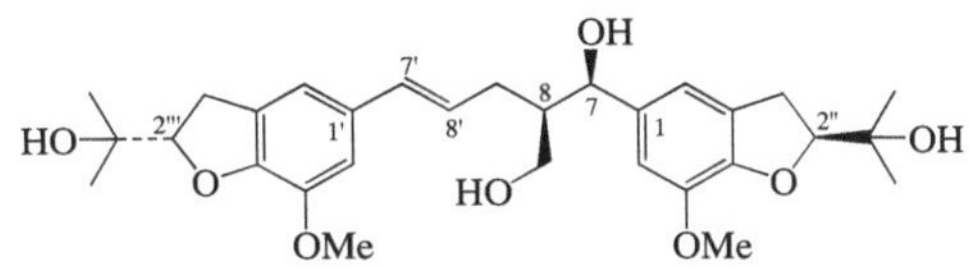

$C_{30}H_{40}O_8$ 528.642

Prenylated 8,9′-lignan

(+)-form [1638687-57-6]

Constit. of the roots of *Zanthoxylum wutaiense*. Pale yellow viscous oil. Isol. as a 1:1 mixt. of the diastereoisomers (2″*S*,2‴*S*,7*R*,8*R*)- and (2″*S*,2‴*S*,7*S*,8*S*)-forms

7-Me ether

7-O-Methylwutaienin

$C_{31}H_{42}O_8$ 542.669

Constit. of the roots of *Zanthoxylum wutaiense*. Isol. as a 1:1 mixt. of the diastereoisomers (2″*S*,2‴*S*,7*R*,8*R*) and (2″*S*,2‴*S*,7*S*,8*S*)-forms.

Takahashi, M. *et al., J. Nat. Prod.*, 2014, **77**, 2585-2589 (*Wutaienin, 7-O-Methylwutaienin, synth*)

Xanthocercin B X1

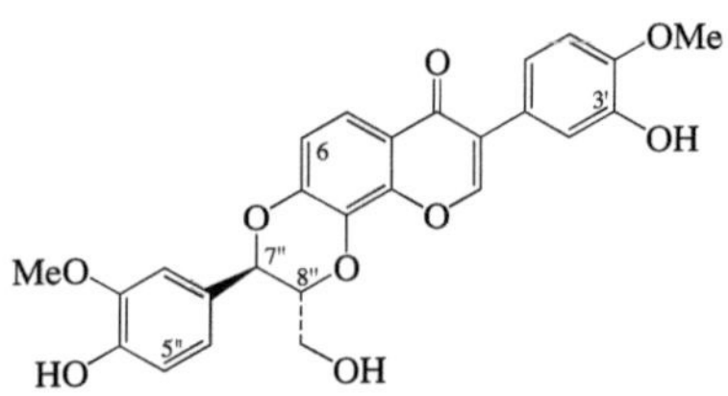

Relative Configuration

2,3-Dihydro-8-(3-hydroxy-4-methoxyphenyl)-3-(4-hydroxy-3-methoxyphenyl)-2-(hydroxymethyl)-7H-pyrano[2,3-f]-1,4-benzodioxin-7-one, CAS
[117678-13-4]
$C_{26}H_{22}O_9$ 478.453
Isoflavonolignoid. Constit. of the heartwood of *Xanthocercis zambesiaca*. Racemic

3′-Deoxy
Butesuperin A
[854008-79-0]
$C_{26}H_{22}O_8$ 462.454
Constit. of the roots of *Butea superba* and *Sophora tonkinensis*. Powder. Mp 135 - 136°. $[\alpha]_D^{21}$ +13.2 (c, 0.02 in Me_2CO). λ_{max} 259 (log ε 4.23) (MeOH)

5″-Methoxy
Xanthocercin A
[117678-12-3]
$C_{27}H_{24}O_{10}$ 508.479
Constit. of the heartwood of *Xanthocercis zambesiaca*. Amorph.. Racemic

5″-Methoxy, 3′-deoxy
Butesuperin B
[854008-80-3]
$C_{27}H_{24}O_9$ 492.48
Constit. of *Butea superba* and *Crotalaria sessiliflora*. Powder. Mp 116 - 118°. $[\alpha]_D^{21}$ +3.8 (c, 0.02 in Me_2CO). λ_{max} 259 (log ε 4.25) (MeOH)

5″-Methoxy, 3′-deoxy, 4″-O-β-D-glucopyranoside
Butesuperin B 4″-glucoside
$C_{33}H_{34}O_{14}$ 654.621
Constit. of the roots of *Sophora tonkinensis*. Pale yellow needles (MeOH). Mp 179 - 180°. $[\alpha]_D^{25}$ -15 (c, 0.04 in MeOH)

5,5″-Dimethoxy
5-Methoxyxanthocercin A
[537698-57-0]
$C_{28}H_{26}O_{11}$ 538.505
Constit. of the seeds of *Dipteryx odorata*. Cryst.. Mp 173 - 174°. $[\alpha]_D^{20}$ +11 (c, 0.1 in $CHCl_3$/MeOH). λ_{max} 256 (log ε 4.16); 295 (log ε 3.96); 313 (sh) (EtOH)

5″,6-Dimethoxy
6-Methoxyxanthocercin A. Dilmin
$C_{28}H_{26}O_{11}$ 538.505
Constit. of the resin of *Amburana cearensis*. Induces a significant prodn. of linear DNA acting as topo II poisons, which target the DNA-protein complex. Amorph. solid. Mp 245° (as tri-Ac). $[\alpha]_D^{20}$ -47 (c, 0.02 in MeOH/$CHCl_3$). λ_{max} 264 (MeOH).

Bezuidenhout, S.C. *et al.*, *J. Chem. Soc., Perkin Trans. 1*, 1988, 1237-1241 (*Xanthocercins A,B, synth*).
Jang, D.S. *et al.*, *J. Nat. Prod.*, 2003, **66**, 583-587 (*5-Methoxyxanthocercin A*).
Ma, K. *et al.*, *Heterocycles*, 2005, **65**, 893-900 (*Butesuperins A,B*).
Yang, R. *et al.*, *Rec. Nat. Prod.*, 2012, **6**, 212-217 (*Butesuperin B 4″-glucoside*).
Fan, C.-M. *et al.*, *Yaoxue Xuebao*, 2016, **51**, 775-779 (*Butesuperin B*).
Ganganna, B. *et al.*, *Bull. Korean Chem. Soc.*, 2017, **38**, 944-947 (*synth*).
De Oliveira, G.P. *et al.*, *Phytochem. Lett.*, 2017, **22**, 61-70 (*Dilmin, activity*)

Xylocoside C X2

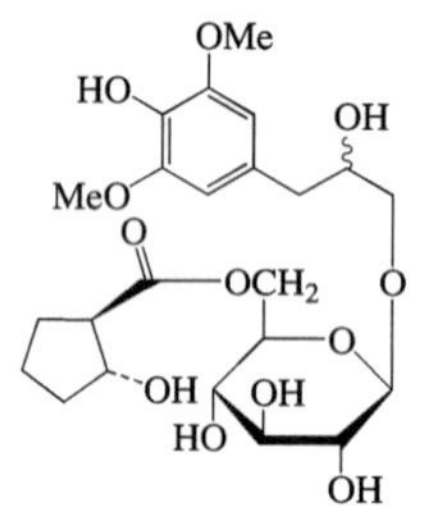

[1042444-92-7]
$C_{23}H_{34}O_{12}$ 502.513
Constit. of the stems of *Xylosma controversum*. Amorph. solid. $[\alpha]_D^{25}$ +2.2 (c, 0.09 in MeOH). λ_{max} 207 (log ε 5.36); 275 (log ε 2.46) (MeOH).

Xu, Z.-R. *et al.*, *Helv. Chim. Acta*, 2008, **91**, 1346-1354 (*Xylocoside C*)

Yunnaneic acid B Y1

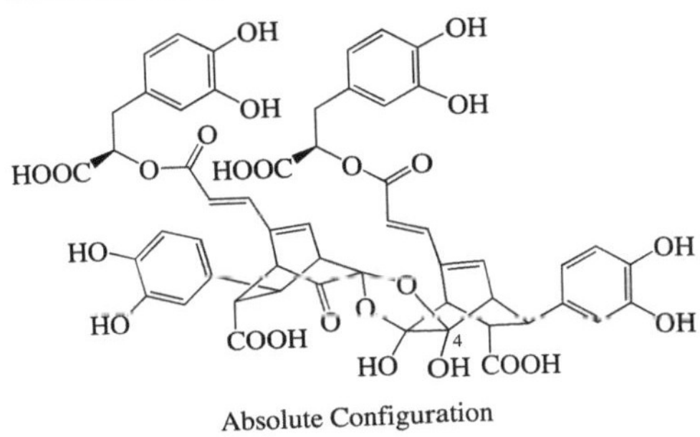

[180314-56-1]

$C_{54}H_{46}O_{25}$ 1094.937

Caffeic acid polymer. Constit. of the roots of *Salvia yunnanensis*. Tan powder. $[\alpha]_D^{25}$ +80.5 (c, 0.3 in MeOH). λ_{max} 225 (sh); 277 (log ε 4.6) (EtOH)

4-Deoxy

Yunnaneic acid A

[180314-55-0]

$C_{54}H_{46}O_{24}$ 1078.938

Constit. of the roots of *Salvia yunnanensis*. Tan powder. $[\alpha]_D^{25}$ +86.1 (c, 0.5 in MeOH). λ_{max} 225 (sh); 276 (log ε 4.6) (EtOH).

Tanaka, T. *et al.*, *J. Nat. Prod.*, 1996, **59**, 843-849 (*Yunnaneic acids A,B*)

Yunnaneic acid D Y2

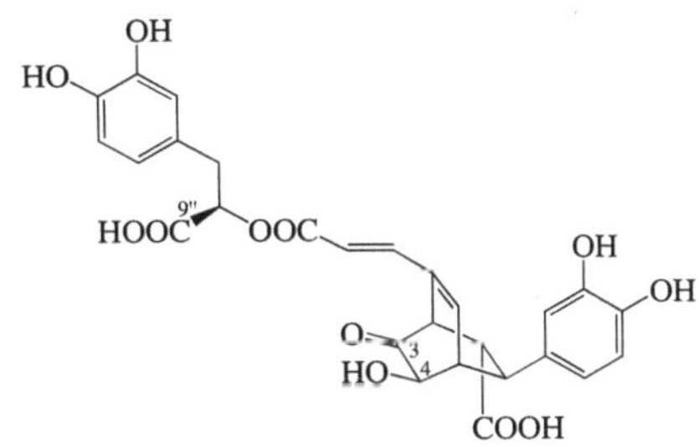

[180314-57-2]

$C_{27}H_{24}O_{12}$ 540.477

Caffeic acid trimer. Related to Helisterculin B in C87. Constit. of the roots of *Salvia yunnanensis*. Amorph. powder. $[\alpha]_D^{25}$ +243.4 (c, 0.7 in MeOH). λ_{max} 223 (log ε 4.3); 276 (log ε 4.3) (EtOH)

3α-Alcohol

Yunnaneic acid I

[1609652-14-3]

$C_{27}H_{26}O_{12}$ 542.493

Constit. of *Cordia americana*. Antioxidant. Brown oil. $[\alpha]_D^{25}$ +48.4 (c, 0.048 in MeOH). λ_{max} 218 (ε 7809); 275 (ε 9417) (MeOH)

3α-Alcohol, 9″-Me ester

Yunnaneic acid J

[1609652-15-4]

$C_{28}H_{28}O_{12}$ 556.52

Constit. of *Cordia americana*. Antioxidant. Brown oil. $[\alpha]_D^{25}$ +137.9 (c, 0.007 in MeOH). λ_{max} 220 (ε 44235); 268 (ε 60073) (MeOH)

4-Ketone

Yunnaneic acid C

[180314-58-3]

$C_{27}H_{22}O_{12}$ 538.461

Constit. of the roots of *Salvia yunnanensis*. Amorph. yellow powder. $[\alpha]_D^{25}$ +106.9 (c, 0.5 in MeOH). λ_{max} 230 (log ε 4.2); 277 (log ε 4.3) (EtOH).

Tanaka, Y. *et al.*, *J. Nat. Prod.*, 1996, **59**, 843-849 (*Yunnaneic acids C,D*).

Fernandez, L.R. *et al.*, *Planta Med.*, 2013, **79**, 1724-1729 (*Yunnaneic acids I,J*).

Griffith, D.R. *et al.*, *J. Org. Chem.*, 2014, **49**, 88-105 (*synth*)

Yunnaneic acid E Y3

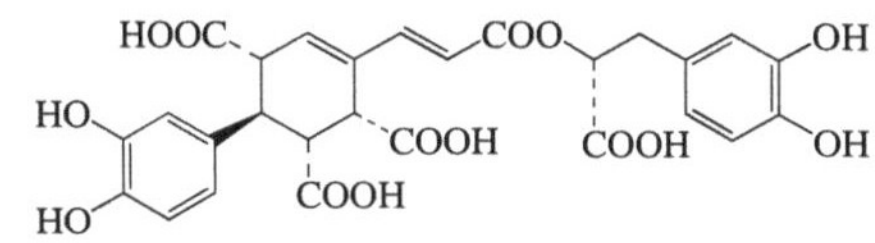

[199433-60-8]

$C_{27}H_{24}O_{14}$ 572.475

Constit. of *Salvia yunnanensis*. Amorph. powder + 1½ H_2O. $[\alpha]_D^{26}$ +81.3 (c, 0.5 in MeOH). λ_{max} 266 (ε 24450) (EtOH).

Tanaka, T. *et al.*, *Chem. Pharm. Bull.*, 1997, **45**, 1596-1600 (*Yunnaneic acid E*)

Yunnaneic acid F Y4

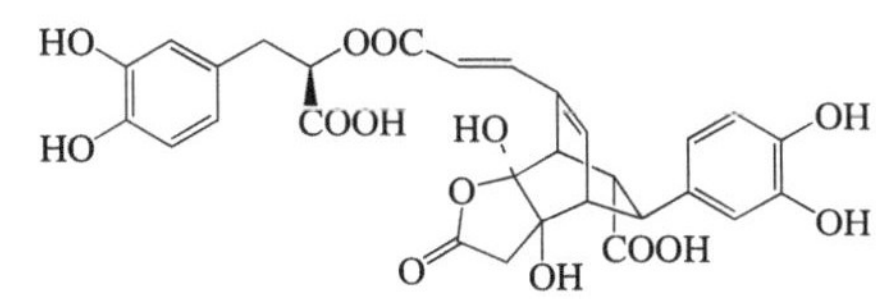

[199433-61-9]

$C_{29}H_{26}O_{14}$ 598.513

Constit. of *Melissa officinalis* and *Salvia yunnanensis*. Amorph. powder + 1H_2O. $[\alpha]_D^{26}$ +126.6 (c, 0.5 in MeOH). λ_{max} 273 (ε 22380) (EtOH).

Tanaka, T. *et al.*, *Chem. Pharm. Bull.*, 1997, **45**, 1596-1600 (*Yunnaneic acid F*).

Barros, L. *et al.*, *Food Chem.*, 2013, **136**, 1-8 (*Melissa officinalis constit*)

Zanthpodocarpin A Z1

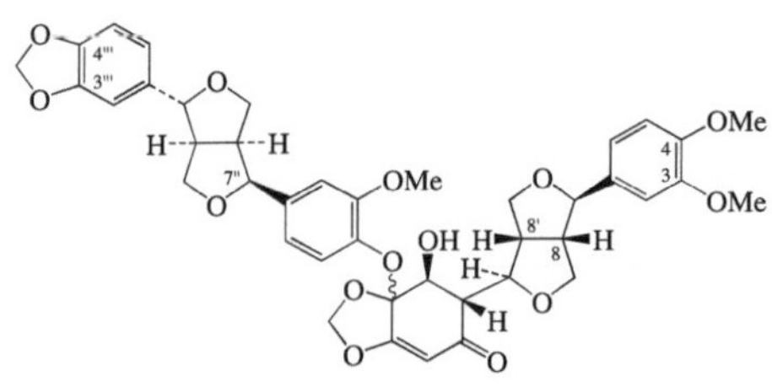

Relative Configuration

[1268468-79-6]
$C_{41}H_{42}O_{14}$ 758.773
Constit. of *Zanthoxylum planispinum* and *Zanthoxylum podocarpum*. NO prodn. inhibitor. Significantly reduces proliferation of HeLa cells. Powder. $[\alpha]_D^{24}$ -287 (c, 0.1 in $CHCl_3$). λ_{max} 242 (log ε 4.41); 283 (log ε 4.03) ($CHCl_3$)

4-O-De-Me
Zanthpodocarpin C
[1629056-37-6]
$C_{40}H_{40}O_{14}$ 744.746
Constit. of the bark of *Zanthoxylum simulans*. Exhibits antiinflammatory activity. Powder. $[\alpha]_D^{25}$ -245 (c, 0.1 in $CHCl_3$). λ_{max} 242 (log ε 4.33); 282 (log ε 4.01) (MeOH)

3,4-Di-O-de-Me, 3,4-methylene ether
Biplanispin A
[1189130-85-5]
$C_{40}H_{38}O_{14}$ 742.73
Constit. of the bark of *Zanthoxylum planispinum*. Amorph. powder. $[\alpha]_D$ -9.3 (c, 1 in DMSO)

3‴,4‴-Demethylene, 4‴-O-(3-methyl-2-butenyl), 3‴-Me ether
Bizanthplanispin A. Zanthpodocarpin G
[1629056-41-2]
$C_{46}H_{52}O_{14}$ 828.908
Constit. of *Zanthoxylum planispinum* and *Zanthoxylum simulans*. Exhibits antiinflammatory activity. Significantly reduces HeLa cell proliferation. Amorph. powder. $[\alpha]_D^{25}$ -290 (c, 0.1 in $CHCl_3$). λ_{max} 230 (log ε 4.27); 276 (log ε 4.03) (MeOH)

3‴,4‴-Demethylene, 4‴-O-(3-methyl-2-oxo-3-butenyl), 3‴-Me ether
Zanthpodocarpin H
[1629056-42-3]
$C_{46}H_{50}O_{15}$ 842.891
Constit. of the bark of *Zanthoxylum simulans*. Exhibits antiinflammatory activity. Powder. $[\alpha]_D^{25}$ -113 (c, 0.1 in $CHCl_3$). λ_{max} 241 (log ε 3.98); 279 (log ε 3.63) (MeOH)

7″ξ-Hydroxy
Zanthpodocarpin D
[1629056-38-7]
$C_{41}H_{42}O_{15}$ 774.772
Constit. of the bark of *Zanthoxylum simulans*. Exhibits antiinflammatory activity. Powder. $[\alpha]_D^{25}$ -255 (c, 0.1 in $CHCl_3$). λ_{max} 241 (log ε 4.42); 281 (log ε 4.06) (MeOH)

7″-Epimer
Biplanispin B
$C_{41}H_{42}O_{14}$ 758.773
Constit. of the bark of *Zanthoxylum planispinum*. $[\alpha]_D^{23}$ -173 (c, 0.74 in $CHCl_3$)

7″-Epimer, 3‴,4‴-demethylene, 4‴-O-(3-methyl-2-butenyl), 3‴-Me ether
Zanthpodocarpin F
[1629056-40-1]
$C_{46}H_{52}O_{14}$ 828.908
Constit. of the bark of *Zanthoxylum simulans*. Exhibits antiinflammatory activity. Powder. $[\alpha]_D^{25}$ -295 (c, 0.1 in $CHCl_3$). λ_{max} 230 (log ε 4.52); 276 (log ε 4.08) (MeOH)

8,8′-Diepimer, 3‴,4‴-demethylene, 3‴-Me ether
Isozanthpodocarpin B
[2209888-94-6]
$C_{41}H_{44}O_{14}$ 760.789
Constit. of *Zanthoxylum simulans*. NO prodn. inhibitor. Powder. $[\alpha]_D^{21}$ -139.9 (c, 0.04 in $CHCl_3$). λ_{max} 241 (log ε 3.26); 279 (log ε 2.93) ($CHCl_3$)

7″,8,8′-Triepimer, 3‴,4‴-demethylene, 3‴-Me ether
Zanthpodocarpin B
[1268468-80-9]
$C_{41}H_{44}O_{14}$ 760.789
Constit. of *Zanthoxylum planispinum* and *Zanthoxylum podocarpum*. NO prodn. inhibitor. Significantly reduces HeLa cell proliferation. Powder. $[\alpha]_D^{23}$ -107.2 (c, 0.11 in $CHCl_3$). λ_{max} 240 (log ε 4.14); 280 (log ε 3.76) ($CHCl_3$)

7″,8,8′-Triepimer, 3‴,4‴-demethylene, 4‴-O-(3,7-dimethyl-2E,6-octadienyl), 3‴-Me ether
Bizanthplanispin B
[1917346-95-2]
$C_{51}H_{60}O_{14}$ 897.027
Constit. of the roots of *Zanthoxylum planispinum*. Significantly reduces HeLa cell proliferation. Amorph. powder. $[\alpha]_D^{25}$ -115 (c, 0.1 in $CHCl_3$). λ_{max} 230; 287 (MeOH).

Yang, G.-Z. *et al.*, *Helv. Chim. Acta*, 2009, **92**, 1657-1664 (*Biplanispin A*).
Zhou, X.-J. *et al.*, *Bioorg. Med. Chem. Lett.*, 2011, **21**, 373-376 (*Zanthpodocarpins A,B, activity*).
Zhou, X.-J. *et al.*, *Bioorg. Med. Chem. Lett.*, 2013, **23**, 5452 (*corrigendum*).
Chen, Y. *et al.*, *Zhongcaoyao*, 2013, **44**, 3429-3434 (*Biplanispin B*).
Wang, J.-F. *et al.*, *Bioorg. Med. Chem. Lett.*, 2014, **24**, 4667-4671 (*Zanthpodocarpins C-H, activity*).
Su, G.-Y. *et al.*, *Phytochem. Lett.*, 2015, **11**, 120-126 (*Bizanthplanispins A,B, activity*).
Li, D.-X. *et al.*, *Zhongguo Zhongyao Zazhi*, 2015, **40**, 2843-2848 (*Isozanthpodocarpin B, activity*)

Zanthpodocarpin E Z2

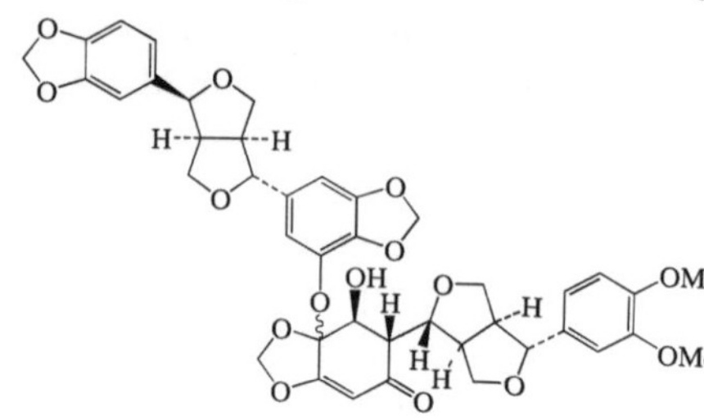

Relative Configuration

[1629056-39-8]
$C_{41}H_{40}O_{15}$ 772.756
Constit. of the bark of *Zanthoxylum simulans*. Exhibits antiinflammatory activity. Powder. $[\alpha]_D^{25}$ -273 (c, 0.1 in $CHCl_3$). λ_{max} 241 (log ε 4.42); 281 (log ε 4.06) (MeOH).

Wang, J.-F. *et al.*, *Bioorg. Med. Chem. Lett.*, 2014, **24**, 4667-4671 (*Zanthpodocarpin E, activity*)

Zeyherol Z3

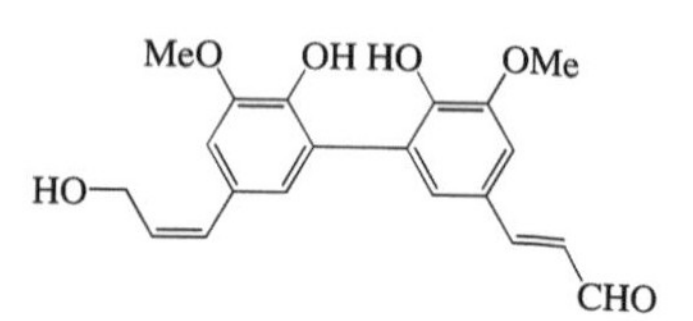

3-[2′,6-Dihydroxy-5′-(3-hydroxy-1-propenyl)-3′,5-dimethoxy[1,1′-biphenyl]-3-yl]-2-propenal, CAS. *4,4′,9′-Trihydroxy-5,5′-dimethoxy-3,3′-ligna-7,7′-dien-9-al. 4,4′,9′-Trihydroxy-5,5′-dimethoxy-9-oxo-3,3′-neoligna-7,7′-diene*
[57429-88-6]
$C_{20}H_{20}O_6$ 356.374
Related to 4,4′-Dihydroxy-5,5′-dimethoxy-3,3′-ligna-7,7′-diene-9,9′-dioic acid, D140 and Magnaldehyde C in M4. Constit. of *Zeyheria digitalis* and *Zeyheria tuberculosa*. Yellow oil. λ_{max} 230 (log ε 4.49); 286 (log ε 4.07); 344 (log ε 4.43) (EtOH).

Da Silveira, J.C. *et al.*, *Phytochemistry*, 1975, **14**, 1829-1830 (*Zeyherol*)

Zuihonin D

Z4

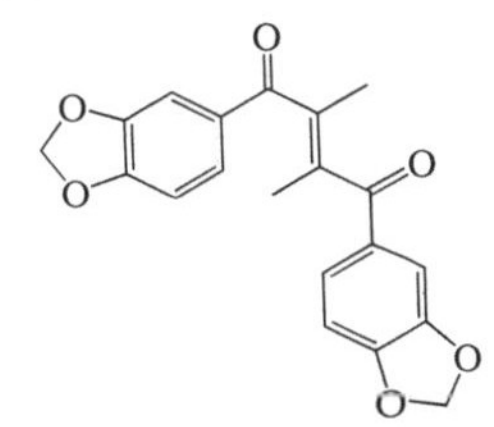

3,4:3′,4′-Bis(methylene)-8,8′-lign-8(8′)-ene-7,7′-dione. 2,3-Bis(3,4-methylenedioxybenzoyl)-2-butene.
Zuonin D
[84607-65-8]
$C_{20}H_{16}O_6$ 352.342
Isol. from bark of *Machilus zuihoensis*.
Lee, C.L. *CA*, 1983, **98**, 74043 (*Zuihonin D*)

Chemical Name Index

Type of Compound Index

Lignan monomers

Simple dibenzylbutane 8,8′-lignans

Side-chain oxygenated dibenzylbutane 8,8′-lignans

Saturated dibenzylbutyrolactone and -lactol 8,8′-lignans

Unsaturated dibenzylbutyrolactone and -lactol 8,8′-lignans

9,9′-Epoxyfuranoid 8,8′-lignans

7,9′-Epoxyfuranoid 8,8′-lignans

7,7′-Epoxyfuranoid 8,8′-lignans

Simple furofuranoid 8,8′-lignans

Side-chain oxygenated furofuranoid 8,8′-lignans

Miscellaneous furofuranoid 8,8′-lignans

1,2′-Cyclo-8,8′-lignans (Eupodienones)

Simple 2,7′-cyclo-8,8′-lignans (arylnaphthalenes)

2,7'-Cyclo-8,8'-lignans (side-chain oxygenated)

2,7'-Cyclo-9,9'-epoxy-8,8'-lignans

2,7'-Cyclo-8,8'-lignan-9,9'-olides

2,7′-Cyclo-8,8′-lignan-9′,9-olides

Miscellaneous naphthalenoid lignans

7,7′/7,8′-Cyclolignans (cyclobutanes)

2,2′-Cyclo-8,8′-lignans (dibenzocyclooctadienes)

Norlignans

Homolignans

Secolignans

Abeolignans

8,8′-Sesquilignans

Other sesquilignans

Sesterlignans

8,8′-Bilignans

Other bilignans

Flavonolignans

Stilbenolignans

Coumarinolignans

Xanthonolignans

Terpenoid lignans

Benzodioxin-containing lignans

Alkaloidal lignans

1,8′-Lignans

3,3′-Lignans

3,8′-Lignans

7,8′-Lignans

8,9′-Lignans

Other simple lignans

3,7′-Cyclo-1,8′-lignans

5,8′-Cyclo-3,7′-lignans

Other cyclolignans

2,7′-Epoxy-1,8′-lignans

2,7′-Epoxy-3,8′-lignans

2,8′-Epoxy-1,7′-lignans

3,7’-Epoxy-2,8’-lignans

3,7’-Epoxy-4,8′-lignans

4,7′-Epoxy-3,8′-lignans

Other simple epoxylignans

3,7′-Epoxy-4,8′-oxylignans

4,7′-Epoxy-3,8′-oxylignans

3,3′-Oxylignans

3,4′-Oxylignans

4,4′-Oxylignans

4,8′-Oxylignans

Other simple oxylignans

Rosmarinic/Salvianolic acid lignans

Linear diarylheptanoids

Diphenyl ether diarylheptanoids

Biphenyl diarylheptanoids

Tetrahydropyranoid diarylheptanoids

Miscellaneous diarylheptanoids

Lignans of unknown structure